Y0-BRC-044

One- and Three-Letter Symbols for the Amino Acids[a]

A	Ala	Alanine
B	Asx	Asparagine or aspartic acid
C	Cys	Cysteine
D	Asp	Aspartic acid
E	Glu	Glutamic acid
F	Phe	Phenylalanine
G	Gly	Glycine
H	His	Histidine
I	Ile	Isoleucine
K	Lys	Lysine
L	Leu	Leucine
M	Met	Methionine
N	Asn	Asparagine
P	Pro	Proline
Q	Gln	Glutamine
R	Arg	Arginine
S	Ser	Serine
T	Thr	Threonine
V	Val	Valine
W	Trp	Tryptophan
Y	Tyr	Tyrosine
Z	Glx	Glutamine or glutamic acid

[a]The one-letter symbol for an undetermined or nonstandard amino acid is X.

Thermodynamic Constants and Conversion Factors

Joule (J)
$$1\ J = 1\ kg \cdot m^2 \cdot s^{-2} \qquad 1\ J = 1\ C \cdot V\ (coulomb\ volt)$$
$$1\ J = 1\ N \cdot m\ (newton\ meter)$$

Calorie (cal)
1 cal heats 1 g of H_2O from 14.5 to 15.5°C
1 cal = 4.184 J

Large calorie (Cal)
1 Cal = 1 kcal $\qquad$ 1 Cal = 4184 J

Avogadro's number (N)
$N = 6.0221 \times 10^{23}$ molecules $\cdot$ mol^{-1}

Coulomb (C)
$1\ C = 6.241 \times 10^{18}$ electron charges

Faraday ($\mathscr{F}$)
$1\ \mathscr{F} = N$ electron charges
$1\ \mathscr{F} = 96{,}485\ C \cdot mol^{-1} = 96{,}485\ J \cdot V^{-1} \cdot mol^{-1}$

Kelvin temperature scale (K)
0 K = absolute zero $\qquad$ 273.15 K = 0°C

Boltzmann constant (k_B)
$k_B = 1.3807 \times 10^{-23}\ J \cdot K^{-1}$

Gas constant (R)
$R = Nk_B$ $\qquad$ $R = 1.9872\ cal \cdot K^{-1} \cdot mol^{-1}$
$R = 8.3145\ J \cdot K^{-1} \cdot mol^{-1}$ $\qquad$ $R = 0.08206\ L \cdot atm \cdot K^{-1} \cdot mol^{-1}$

The Standard Genetic Code

First Position (5′ end)	Second Position				Third Position (3′ end)
	U	C	A	G	
U	UUU Phe	UCU Ser	UAU Tyr	UGU Cys	U
	UUC Phe	UCC Ser	UAC Tyr	UGC Cys	C
	UUA Leu	UCA Ser	UAA Stop	UGA Stop	A
	UUG Leu	UCG Ser	UAG Stop	UGG Trp	G
C	CUU Leu	CCU Pro	CAU His	CGU Arg	U
	CUC Leu	CCC Pro	CAC His	CGC Arg	C
	CUA Leu	CCA Pro	CAA Gln	CGA Arg	A
	CUG Leu	CCG Pro	CAG Gln	CGG Arg	G
A	AUU Ile	ACU Thr	AAU Asn	AGU Ser	U
	AUC Ile	ACC Thr	AAC Asn	AGC Ser	C
	AUA Ile	ACA Thr	AAA Lys	AGA Arg	A
	AUG Met[a]	ACG Thr	AAG Lys	AGG Arg	G
G	GUU Val	GCU Ala	GAU Asp	GGU Gly	U
	GUC Val	GCC Ala	GAC Asp	GGC Gly	C
	GUA Val	GCA Ala	GAA Glu	GGA Gly	A
	GUG Val	GCG Ala	GAG Glu	GGG Gly	G

[a]AUG forms part of the initiation signal as well as coding for internal Met residues.

◼ BIOCHEMICAL INTERACTIONS

Biochemistry is bundled with *Biochemical Interactions,* a CD-ROM that expands upon the information presented in the textbook through the use of a variety of interactive three-dimensional molecular graphics displays and animations. These take the following forms, and they are all keyed to the text by a mouse icon (🖱).

Interactive Exercises, 56 Chime™-based molecular graphics displays of proteins and nucleic acids that can be interactively rotated and otherwise manipulated.

Kinemages, alternative types of molecular graphics displays. These are presented in the form of 21 Exercises comprising 54 kinemages that amplify specific aspects of protein and nucleic acid structures.

Guided Explorations, 31 more complex interactive computer graphics displays and computerized animations, dealing with specific subjects in the textbook.

Animated Figures, text figures that have been animated for better understanding.

The Interactive Exercises and Guided Explorations were produced by ScienceMedia Inc in collaboration with Donald Voet and Judith G. Voet. The Kinemages were produced by Donald Voet and Judith G. Voet. The Animated Figures were created by Super Nova.

For the student, the CD-ROM extends the learning process from the textbook to the multimedia environment by drawing upon motion, color, and three-dimensionality to illustrate aspects of molecular form and function that would otherwise be difficult to envision.

For the instructor, the CD-ROM is designed to be used as a teaching tool in computer presentation-equipped classrooms.

The Tables of Contents for these exercises (with text references in parentheses) *may be found on the last few pages of the text.*

BIOCHEMISTRY

3rd Edition

BIOCHEMISTRY

DONALD VOET
University of Pennsylvania

JUDITH G. VOET
Swarthmore College

WILEY JOHN WILEY & SONS, INC.

About the Cover: The cover is an illustration of horse heart cytochrome *c* designed to show the influence of amino acid side chains on the protein's three-dimensional folding pattern. It was drawn by Irving Geis, in collaboration with Richard Dickerson, in 1972, the same year that the Protein Data Bank was established and had one structure deposited. As of the beginning of 2003, there were 20,000 structures available for download and visualization on desktop and laptop computers that did not even exist when this drawing was created. It reminds us that biochemistry is a process that is driven by the creativity of the human mind. Our visualization tools have developed from pen, ink, and colored pencils to sophisticated computer software available to all. Without creativity, however, these tools have little use.

Executive Editor   *David Harris/Patrick Fitzgerald*

Senior Marketing Manager   *Robert Smith*

Developmental Editor   *Barbara Heaney*

Production Editor   *Sandra Dumas*

Photo Editor   *Hilary Newman*

Photo Researcher   *Elyse Rieder*

Cover/Text Designer   *Madelyn Lesure*

Production Management Services   *Suzanne Ingrao*

Illustration Editor   *Sigmund Malinowski*

Cover illustration and Part Openers, Irving Geis. Image from the Irving Geis Collection/ Howard Hughes Medical Institute. Rights owned by HHMI. Reproduction by permission only.

This book was typeset in 9.5/11.5 Times Ten Roman by TechBooks and printed and bound by Von Hoffmann Corporation. The cover was printed by Von Hoffmann Corporation.

The paper in this book was manufactured by a mill whose forest management programs include sustained yield harvesting of its timberlands. Sustained yield harvesting principles ensure that the number of trees cut each year does not exceed the amount of new growth.

This book is printed on acid-free paper. ∞

Copyright © 2004 by Donald Voet and Judith G. Voet. All rights reserved.

No part of this publication may be reproduced, stored in a retrieval system or transmitted in any form or by any means, electronic, mechanical, photocopying, recording, scanning or otherwise, except as permitted under Sections 107 or 108 of the 1976 United States Copyright Act, without either the prior written permission of the Publisher, or authorization through payment of the appropriate per-copy fee to the Copyright Clearance Center, 222 Rosewood Drive, Danvers, MA 01923, (978) 750-8400, fax (978) 646-8600. Requests to the Publisher for permission should be addressed to the Permissions Department, John Wiley & Sons, Inc., 111 River Street, Hoboken, NJ 07030, (201) 748-6011, fax (201) 748-6008.

To order books or for customer service please, call 1(800)-CALL-WILEY (225-5945).

Voet, Donald
Biochemistry, Third Edition Donald Voet, Judith G. Voet

ISBN 0-471-19350-x (cloth)

0-471-39223-5 (Wiley International Edition)

Printed in the United States of America.

10 9 8 7 6 5 4 3

To
Our parents, who encouraged us,
Our teachers, who enabled us, and
Our children, who put up with us.

PREFACE

Biochemistry is a field of enormous fascination and utility, arising, no doubt, from our own self-interest. Human welfare, particularly its medical and nutritional aspects, has been vastly improved by our rapidly growing understanding of biochemistry. Indeed, scarcely a day passes without the report of a biomedical discovery that benefits a significant portion of humanity. Further advances in this rapidly expanding field of knowledge will no doubt lead to even more spectacular gains in our ability to understand nature and to control our destinies. It is therefore essential that individuals embarking on a career in biomedical sciences be well versed in biochemistry.

This textbook is a distillation of our experiences in teaching undergraduate and graduate students at the University of Pennsylvania and Swarthmore College and is intended to provide such students with a thorough grounding in biochemistry. We assume that students who use this textbook have had the equivalent of one year of college chemistry and at least one semester of organic chemistry so that they are familiar with both general chemistry and the basic principles and nomenclature of organic chemistry. We also assume that students have taken a one-year college course in general biology in which elementary biochemical concepts were discussed. Students who lack these prerequisites are advised to consult the appropriate introductory textbooks in these subjects.

In the eight years since the second edition of *Biochemistry* was published, the field of biochemistry has continued its phenomenal and rapidly accelerating growth. This remarkable expansion of our knowledge, the work of thousands of talented and dedicated scientists, has been characterized by numerous new paradigms, as well as an enormous enrichment of almost every aspect of the field. For example, the number of known protein and nucleic acid structures as determined by X-ray and NMR techniques has increased by over fourfold. Moreover, the quality and complexity of these structures have significantly improved, thereby providing enormous advances in our understanding of structural biochemistry. Bioinformatics, an only recently coined word, has come to dominate the way that many aspects of biochemistry are conceived and practiced. When the second edition of *Biochemistry* was published, no genome had yet been sequenced. Now over 100 genome sequences, including that from humans, have been determined with a new one being reported almost weekly. Likewise, the state of knowledge has exploded in such subdisciplines as eukaryotic and prokaryotic molecular biology, metabolic control, protein folding, electron transport, membrane transport, immunology, signal transduction, etc. Indeed, these advances have affected our everyday lives in that they have changed the way that medicine is practiced, the way that we protect our own health, and the way in which food is produced.

■ THEMES

In writing this textbook we have emphasized several themes. First, biochemistry is a body of knowledge compiled by people through experimentation. In presenting what is known, we therefore stress how we have come to know it. The extra effort the student must make in following such a treatment, we believe, is handsomely repaid since it engenders the critical attitudes required for success in any scientific endeavor. Although science is widely portrayed as an impersonal subject, it is, in fact, a discipline shaped through the often idiosyncratic efforts of individual scientists. We therefore identify some of the major contributors to biochemistry (many of whom are still professionally active) and, in many cases, consider the approaches they have taken to solve particular biochemical puzzles. The student should realize, however, that most of the work described could not have been done without the dedicated and often indispensable efforts of numerous co-workers.

The unity of life and its variation through evolution is a second dominant theme running through the text. Certainly one of the most striking characteristics of life on earth is its enormous variety and adaptability. Yet, biochemical research has amply demonstrated that all living things are closely related at the molecular level. As a consequence, the molecular differences among the various species have provided intriguing insights into how organisms have evolved from one another and have helped delineate the functionally significant portions of their molecular machinery.

A third major theme is that biological processes are organized into elaborate and interdependent control networks. Such systems permit organisms to maintain relatively constant internal environments, to respond rapidly to external stimuli, and to grow and differentiate.

A fourth theme is that biochemistry has important medical consequences. We therefore frequently illustrate biochemical principles by examples of normal and abnormal human physiology and discuss the mechanisms of action of a variety drugs.

■ ORGANIZATION AND COVERAGE

As the information explosion in biochemistry has been occurring, teachers have been exploring more active learning methods such as problem-based learning, discovery-based learning, and cooperative learning. These new teaching and learning techniques involve more interaction among students and teachers and, most importantly, require more in-class time. In writing the third edition of this textbook, we have therefore been faced with the dual pressures of increased content and pedagogical innovation.

We have responded to this challenge by presenting the subject matter of biochemistry as thoroughly and accurately as we can so as to provide students and instructors alike with this information as they explore various innovative learning strategies. In this way we deal with the widespread concern that these novel methods of stimulating student learning tend to significantly diminish course content. We have thus written a textbook that permits teachers to direct their students to areas of content that can be explored outside of class as well as providing material for in-class discussion.

We have reported many of the advances that have occurred in the last eight years in the third edition of *Biochemistry* and have thereby substantially enriched nearly all of its sections. Nevertheless, with the several exceptions noted below, the basic organization of the third edition remains the same as those of the first and second editions.

The text is organized into five parts:

I. Introduction and Background: An introductory chapter followed by chapters that review the properties of aqueous solutions and the elements of thermodynamics.

II. Biomolecules: A description of the structures and functions of proteins, nucleic acids, carbohydrates, and lipids.

III. Mechanisms of Enzyme Action: An introduction to the properties, reaction kinetics, and catalytic mechanisms of enzymes.

IV. Metabolism: A discussion of how living things synthesize and degrade carbohydrates, lipids, amino acids, and nucleotides with emphasis on energy generation and consumption.

V. Expression and Transmission of Genetic Information: An expansion of the discussion of nucleic acid structure that is given in Part II followed by an exposition of both prokaryotic and eukaryotic molecular biology.

This organization permits us to cover the major areas of biochemistry in a logical and coherent fashion. Yet, modern biochemistry is a subject of such enormous scope that to maintain a relatively even depth of coverage throughout the text, we include more material than most one-year biochemistry courses will cover in detail. This depth of coverage, we feel, is one of the strengths of this book; it permits the instructor to teach a course of his/her own design and yet provide the student with a resource on biochemical subjects not emphasized in the course.

The order in which the subject matter of the text is presented more or less parallels that of most biochemistry courses. However, several aspects of the textbook's organization deserve comment:

1. Chapter 5 (Nucleic Acids, Gene Expression, and Recombinant DNA Technology) now introduces molecular biology early in the narrative in response to the central role that recombinant DNA technology has come to play in modern biochemistry. For the same reason, the chapter that contained the review of genetics and the discussion of how we came to know the role of DNA has been subsumed into Chapters 1 (Life) and 5 and the sections on nucleic acid sequencing and the synthesis of oligonucleotides now appear in Chapter 7 (Covalent Structures of Proteins and Nucleic Acids). Likewise, the burgeoning field of bioinformatics is discussed in a separate section of Chapter 7.

2. We have split our presentation of thermodynamics between two chapters. Basic thermodynamic principles—enthalpy, entropy, free energy, and equilibrium—are discussed in Chapter 3 because these subjects are prerequisites for understanding structural biochemistry, enzyme mechanisms, and kinetics. Metabolic aspects of thermodynamics—the thermodynamics of phosphate compounds and oxidation–reduction reactions—are presented in Chapter 16 since knowledge of these subjects is not required until the chapters that follow.

3. Techniques of protein purification are described in a separate chapter (Chapter 6) that precedes the discussion of protein structure and function. We have chosen this order so that students will not feel that proteins are somehow "pulled out of a hat." Nevertheless, Chapter 6 has been written as a resource chapter to be consulted repeatedly as the need arises. Techniques of nucleic acid purification are now also discussed in this chapter for the above-described reasons.

4. Chapter 10 describes the properties of hemoglobin in detail so as to illustrate concretely the preceding discussions of protein structure and function. This chapter introduces allosteric theory to explain the cooperative nature of hemoglobin oxygen binding. The subsequent extension of allosteric theory to enzymology in Chapter 13 is a relatively simple matter.

5. Concepts of metabolic control are presented in the chapters on glycolysis (Chapter 17) and glycogen metabolism (Chapter 18) through the consideration of flux generation, allosteric regulation, substrate cycles, covalent enzyme modification, cyclic cascades, and a newly added discussion of metabolic control analysis. We feel that these concepts are best understood when studied in metabolic context rather than as independent topics.

6. The rapid growth in our knowledge of biological signal transduction necessitates that this important subject now have its own chapter, Chapter 19.

7. There is no separate chapter on coenzymes. These substances, we feel, are more logically studied in the context of the enzymatic reactions in which they participate.

8. Glycolysis (Chapter 17), glycogen metabolism (Chapter 18), the citric acid cycle (Chapter 21), and electron transport and oxidative phosphorylation (Chapter 22) are detailed as models of general metabolic pathways with emphasis placed on many of the catalytic and control mechanisms of the enzymes involved. The principles illustrated in these chapters are reiterated in somewhat less detail in the other chapters of Part IV.

9. Consideration of membrane transport (Chapter 20) precedes that of mitochondrially based metabolic pathways such as the citric acid cycle, electron transport, and oxidative phosphorylation. In this manner, the idea of the compartmentalization of biological processes can be easily assimilated. We have moved the discussion of neurotransmission to this chapter because it is intimately involved with membrane transport.

10. Discussions of both the synthesis and the degradation of lipids have been placed in a single chapter (Chapter 25), as have the analogous discussions of amino acids (Chapter 26) and nucleotides (Chapter 28).

11. Energy metabolism is summarized and integrated in terms of organ specialization in Chapter 27, following the descriptions of carbohydrate, lipid, and amino acid metabolism.

12. The principles of both prokaryotic and eukaryotic molecular biology are expanded from their introduction in Chapter 5 in sequential chapters on DNA replication, repair and recombination (Chapter 30), transcription (Chapter 31), and translation (Chapter 32). Viruses (Chapter 33) are then considered as paradigms of more complex cellular functions, followed by discussions of eukaryotic gene expression (Chapter 34).

13. Chapter 35, the final chapter, is a series of minichapters that describe the biochemistry of a variety of well-characterized human physiological processes: blood clotting, the immune response, and muscle contraction.

The old adage that you learn a subject best by teaching it simply indicates that learning is an active rather than a passive process. The problems we provide at the end of each chapter are therefore designed to make students think rather than to merely regurgitate poorly assimilated and rapidly forgotten information. Few of the problems are trivial and some of them (particularly those marked with an asterisk) are quite difficult. Yet, successfully working out such problems can be one of the most rewarding aspects of the learning process. Only by thinking long and hard for themselves can students make a body of knowledge truly their own. The answers to the problems are worked out in detail in the solutions manual that accompanies this text. However, this manual can only be an effective learning tool if the student makes a serious effort to solve a problem before looking up its answer.

We have included lists of references at the end of every chapter to provide students with starting points for independent biochemical explorations. The enormity of the biochemical research literature prevents us from giving all but a few of the most seminal research reports. Rather, we list what we have found to be the most useful reviews and monographs on the various subjects covered in each chapter.

Finally, although we have made every effort to make this text error free, we are under no illusions that we have done so. Thus, we are particularly grateful to the many readers of the first and second editions, students and faculty alike, who have taken the trouble to write us with suggestions on how to improve the textbook and to point out errors they have found. We earnestly hope that the readers of the third edition will continue this practice.

Donald Voet
Judith G. Voet

■ ANCILLARY MATERIALS

The third edition of *Biochemistry* is accompanied by the following ancillary materials:

■ FOR THE STUDENT

• A CD-ROM that accompanies this textbook, which was produced by ScienceMedia, Inc. in collaboration with the authors. It contains an extensive series of computer-animated Interactive Exercises and Guided Explorations. The CD also contains a series of Kinemages by Donald Voet and Judith G. Voet. These are computer-animated color images of selected proteins and nucleic acids that are discussed in the text and which students can manipulate. Finally, the CD contains a series of animations of figures in

the textbook. All of these items are keyed to the textbook as indicated by a mouse icon (🐭).

• A *Solutions Manual* containing detailed solutions for all of the textbook's end-of-chapter problems.

■ FOR THE INSTRUCTOR

• A CD-ROM containing nearly all of the illustrations in the textbook. With computerized projection equipment, these full-color images can be shown in any prearranged order to provide "slide shows" to accompany lectures.

• A set of transparencies for overhead projection containing a selection of illustrations from the textbook.

ACKNOWLEDGMENTS

This textbook is the result of the dedicated effort of many individuals, several of whom deserve special mention:

David Harris, our Executive Editor, adroitly directed the entire project. Patrick Fitzgerald, our new Editor, helped us bring this edition successfully to market.

Barbara Heaney, our Developmental Editor, deftly coordinated both the art and the writing programs and kept our noses to the grindstone.

Suzanne Ingrao, our Production Editor, skillfully and patiently managed the production of the textbook.

Connie Parks, our Copy Editor, put the final polish on the manuscript and eliminated an enormous number of stylistic and typographical errors.

Laura Ierardi combined text, figures, and tables in designing each of the textbook's pages.

Madelyn Lesure designed the textbook's typography and its covers.

Hilary Newman and Elyse Reider acquired many of the photographs in this textbook and kept track of all of them.

Edward Starr and Sigmund Malinowski coordinated the illustration program, with the able help of Ken Liao.

Much of the art in this third edition of *Biochemistry* is the creative legacy of the drawings made for its first and second editions by John and Bette Woolsey and Patrick Lane of J/B Woolsey Associates.

Linda Muriello oversaw the development of the CD-ROM that accompanies this textbook.

The late Irving Geis provided us with his extraordinary molecular art and gave freely of his wise counsel.

The atomic coordinates we have used to draw many of the proteins and nucleic acids that appear in this textbook were obtained from the Protein Data Bank, which is managed by the Research Collaboratory for Structural Bioinformatics (RCSB). The drawings were created using the molecular graphics programs RIBBONS by Mike Carson; GRASP by Anthony Nicholls, Kim Sharp, and Barry Honig; INSIGHT II from BIOSYM Technologies; and RasMol by Roger Sayle. Many of the drawings generously contributed by others were made using these programs or MOLSCRIPT by Per Kraulis.

The interactive computer graphics diagrams that are presented in the CD-ROM that accompanies this textbook are either Chime images or Kinemages. Chime, which is based on the program RosMol, was developed and generously made publicly available by MDL Information Systems, Inc. Kinemages are displayed by the program MAGE, which was written and generously provided by David C. Richardson, who also wrote and provided the program PREKIN, which we used to help generate the Kinemages.

We wish especially to thank those colleagues who reviewed this textbook, in both its current and earlier editions, and provided us with their prudent advice:

Joseph Babitch, *Texas Christian University*

E.J. Berhman, *Ohio State University*

Karl D. Bishop, *Bucknell University*

Robert Blankenshop, *Arizona State University*

Charles L. Borders, Jr., *The College of Wooster*

Kenneth Brown, *University of Texas at Arlington*

Larry G. Butler, *Purdue University*

Carol Caparelli, *Fox Chase Cancer Center*

W. Scott Champney, *East Tennessee Stage University*

Paul F. Cook, *The University of Oklahoma*

Glenn Cunningham, *University of Central Florida*

Eugene Davidson, *Georgetown University*

Don Dennis, *University of Delaware*

Walter A. Deutsch, *Louisiana State University*

Kelsey R. Downum, *Florida International University*

William A. Eaton, *National Institutes of Health*

David Eisenberg, *University of California at Los Angeles*

Jeffrey Evans, *University of Southern Mississippi*

David Fahrney, *Colorado State University*

Paul Fitzpatrick, *Texas A&M University*

Robert Fletterick, *University of California at San Francisco*

Norbert C. Furumo, *Eastern Illinois University*

Scott Gilbert, *Swarthmore College*

Guido Guidotti, *Harvard University*

James H. Hageman, *New Mexico State University*

Lowell Hager, *University of Illinois at Urbana–Champaign*

James H. Hammons, *Swarthmore College*

Edward Harris, *Texas A&M University*

Angela Hoffman, *University of Portland*

Ralph A. Jacobson, *California Polytechnic State University*

Eileen Jaffe, *Fox Chase Cancer Center*

Jan G. Jaworski, *Miami University*

William P. Jencks, *Brandeis University*

Mary Ellen Jones, *University of North Carolina*

Jason D. Kahn, *University of Maryland*

Tokuji Kimura, *Wayne State University*

Barrie Kitto, *University of Texas at Austin*

Daniel J. Kosman, *State University of New York at Buffalo*

Robert D. Kuchta, *University of Colorado, Boulder*

Thomas Laue, *University of New Hampshire*

Albert Light, *Purdue University*

Dennis Lohr, *Arizona State University*

Larry Louters, *Calvin College*

Robert D. Lynch, *University of Lowell*

Harold G. Martinson, *University of California at Los Angeles*

Michael Mendenhall, *University of Kentucky*

Sabeeha Merchant, *University of California at Los Angeles*

Christopher R. Meyer, *California State University at Fullerton*

Ronald Montelaro, *Louisiana State University*

Scott Moore, *Boston University*

Harry F. Noller, *University of California at Santa Cruz*

John Ohlsson, *University of Colorado*

Gary L. Powell, *Clemson University*

Alan R. Price, *University of Michigan*

Paul Price, *University of California at San Diego*

Thomas I. Pynadath, *Kent State University*

Frank M. Raushel, *Texas A&M University*

Ivan Rayment, *University of Wisconsin*

Frederick Rudolph, *Rice University*

Raghupathy Sarma, *State University of New York at Stony Brook*

Paul R. Schimmel, *The Scripps Research Institute*

Thomas Schleich, *University of California at Santa Cruz*

Allen Scism, *Central Missouri State University*

Charles Shopsis, *Adelphi University*

Marvin A. Smith, *Brigham Young University*

Thomas Sneider, *Colorado State University*

Jochanan Stenish, *Western Michigan University*

Phyllis Strauss, *Northeastern University*

JoAnne Stubbe, *Massachusetts Institute of Technology*

William Sweeney, *Hunter College*

John Tooze, *European Molecular Biology Organization*

Mary Lynn Trawick, *Baylor University*

Francis Vella, *University of Saskatchewan*

Harold White, *University of Delaware*

William Widger, *University of Houston*

Ken Willeford, *Mississippi State University*

Lauren Williams, *Georgia Institute of Technology*

Jeffery T. Wong, *University of Toronto*

Beulah M. Woodfin, *The University of New Mexico*

James Zimmerman, *Clemson University*

D.V.
J.G.V.

BRIEF CONTENTS

(*Chapter 35 will be made available as a web-based chapter. Go to* www.wiley.com/college/voet *for more information.*)

CONTENTS

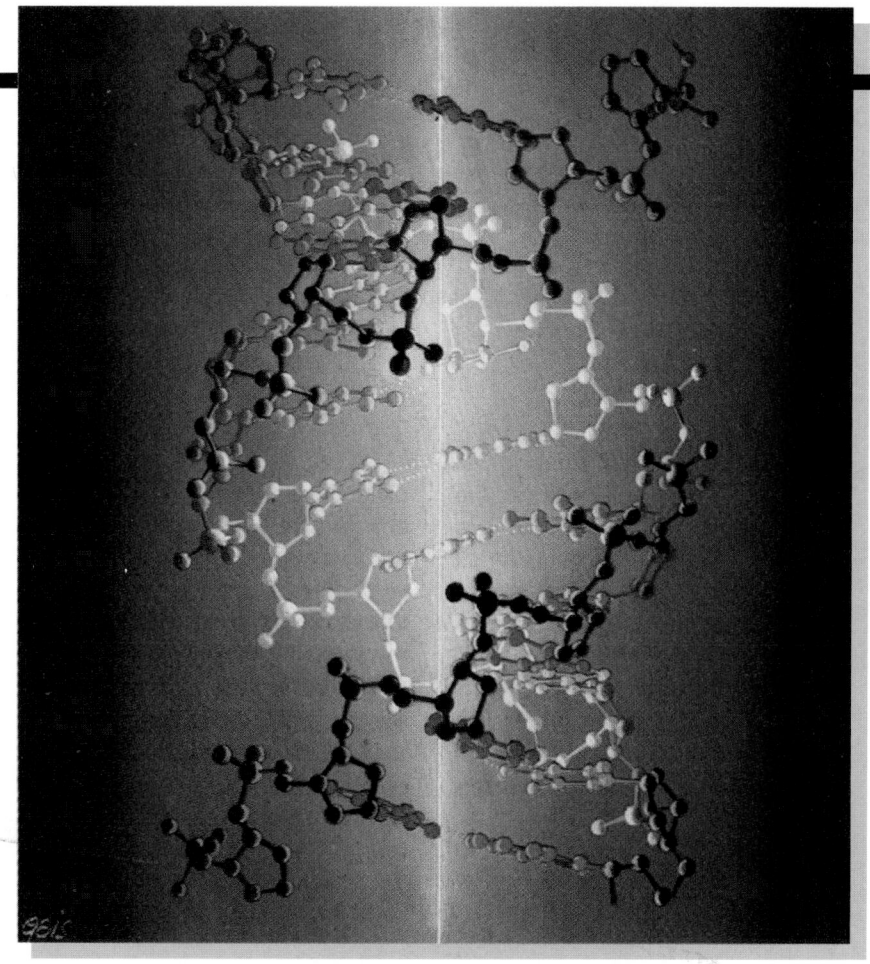
"Hot wire" DNA illuminated by its helix axis.

INTRODUCTION AND BACKGROUND

Chapter 1

Life

It is usually easy to decide whether or not something is alive. This is because living things share many common attributes, such as the capacity to extract energy from nutrients to drive their various functions, the power to actively respond to changes in their environment, and the ability to grow, to differentiate, and—perhaps most telling of all—to reproduce. Of course, a given organism may not have all of these traits. For example, mules, which are obviously alive, rarely reproduce. Conversely, inanimate matter may exhibit some lifelike properties. For instance, crystals may grow larger when immersed in a supersaturated solution of the crystalline material. Therefore, life, as are many other complex phenomena, is perhaps impossible to define in a precise fashion. Norman Horowitz, however, has proposed a useful set of criteria for living systems: *Life possesses the properties of replication, catalysis, and mutability.* Much of this text is concerned with the manner in which living organisms exhibit these properties.

Biochemistry is the study of life on the molecular level. The significance of such studies is greatly enhanced if they are related to the biology of the corresponding organisms or even communities of such organisms. This introductory chapter therefore begins with a synopsis of the biological realm. This is followed by an outline of biochemistry, a review of genetics, a discussion of the origin of life, and finally, an introduction to the biochemical literature.

1 ■ PROKARYOTES

It has long been recognized that life is based on morphological units known as **cells.** The formulation of this concept is generally attributed to an 1838 paper by Matthias Schleiden and Theodor Schwann, but its origins may be traced to the seventeenth century observations of early microscopists such as Robert Hooke. There are two major classifications of cells: the **eukaryotes** (Greek: *eu,* good or true + *karyon,* kernel or nut), which have a membrane-enclosed **nucleus** encapsulating their **DNA (deoxyribonucleic acid);** and the **prokaryotes** (Greek: *pro,* before), which lack this organelle. Prokaryotes, which comprise the various types of bacteria, have relatively simple structures and are invariably unicellular (although they may form filaments or colonies of independent cells). They are estimated to represent about half of Earth's biomass. Eukaryotes, which may be multicellular as well as unicellular, are vastly more complex than prokaryotes. (**Viruses,** which are much simpler entities than cells, are not classified as living because they lack the metabolic apparatus to reproduce outside their host cells. They are essentially large molecular aggregates.) This section is a discussion of prokaryotes. Eukaryotes are considered in the following section.

A. *Form and Function*

Prokaryotes are the most numerous and widespread organisms on Earth. This is because their varied and often highly adaptable metabolisms suit them to an enormous variety of habitats. Besides inhabiting our familiar temperate and aerobic environment, certain types of bacteria may thrive in or even require conditions that are hostile to eukaryotes such as unusual chemical environments, high temperatures (as high as 113°C), and lack of oxygen. Moreover, the rapid reproductive rate of prokaryotes (optimally <20 min per cell division for many species) permits them to take advantage of transiently favorable conditions, and conversely, the ability of many bacteria to form resistant **spores** allows them to survive adverse conditions.

a. Prokaryotes Have Relatively Simple Anatomies

Prokaryotes, which were first observed in 1683 by the inventor of the microscope, Antonie van Leeuwenhoek, have sizes that are mostly in the range 1 to 10 μm. They have one of three basic shapes (Fig. 1-1): spheroidal (**cocci**), rodlike (**bacilli**), and helically coiled (**spirilla**), but all have the same general design (Fig. 1-2). They are bounded, as are all cells, by an ~70-Å-thick **cell membrane (plasma membrane),** which consists of a lipid bilayer containing embedded proteins that control the passage of molecules in and out of the cell and catalyze a variety of reactions. The cells of most prokaryotic species are surrounded by a rigid, 30- to 250-Å-thick polysaccharide **cell wall** that mainly functions to protect the cell from mechanical injury and to prevent it from bursting in media more osmotically dilute than its contents. Some bacteria further encase themselves in a gelatinous polysaccharide **capsule** that protects them from the defenses of higher organisms. Although prokaryotes lack the membranous subcellular organelles characteristic of eukaryotes

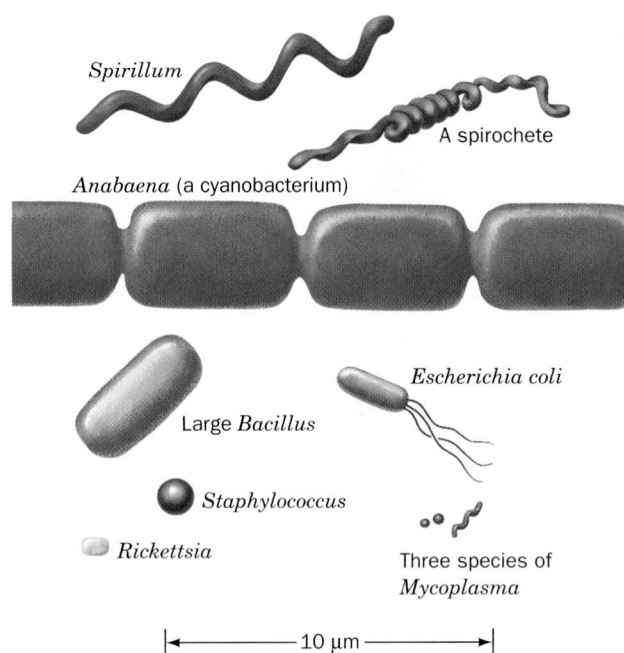

FIGURE 1-1   Scale drawings of some prokaryotic cells.

(Section 1-2), their plasma membranes may be infolded to form multilayered structures known as **mesosomes.** The mesosomes are thought to serve as the site of DNA replication and other specialized enzymatic reactions.

The prokaryotic **cytoplasm** (cell contents) is by no means a homogeneous soup. Its single **chromosome** (DNA molecule, several copies of which may be present in a rapidly growing cell) is condensed to form a body known as a **nucleoid.** The cytoplasm also contains numerous species of **RNA (ribonucleic acid),** a variety of soluble **enzymes** (proteins that catalyze specific reactions), and

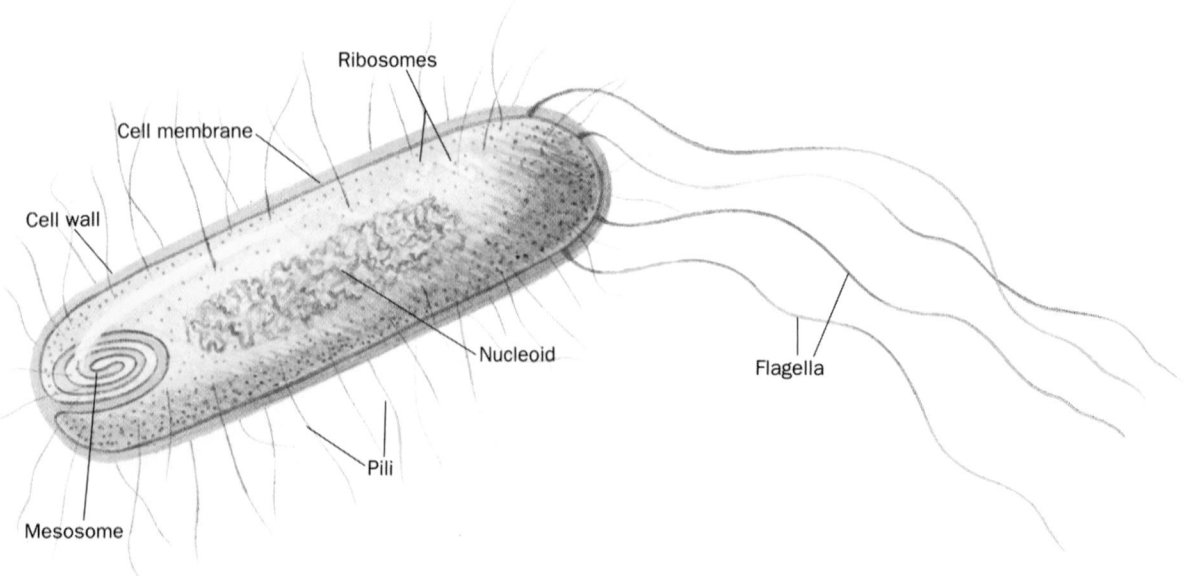

FIGURE 1-2   Schematic diagram of a prokaryotic cell.

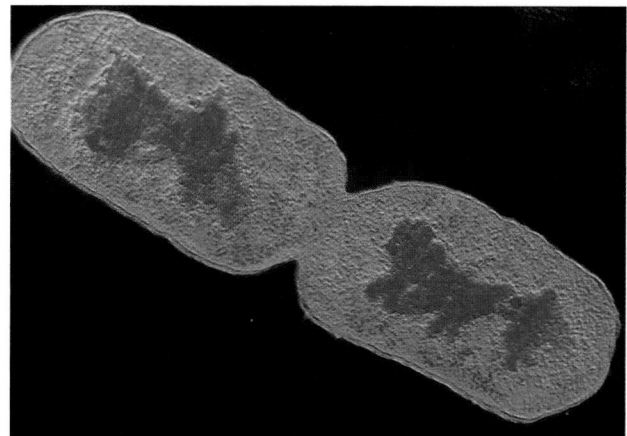

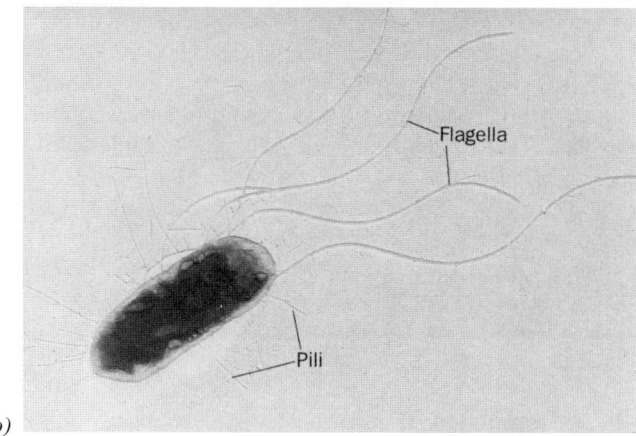

FIGURE 1-3   Electron micrographs of *E. coli* cells. (*a*) Stained to show internal structure. [CNRI.] (*b*) Stained to reveal flagella and pili. [Courtesy of Howard Berg, Harvard University.]

many thousands of 250-Å-diameter particles known as **ribosomes,** which are the sites of protein synthesis.

Many bacterial cells bear one or more whiplike appendages known as **flagella,** which are used for locomotion. Certain bacteria also have filamentous projections named **pili,** some types of which function as conduits for

TABLE 1-1   Molecular Composition of *E. Coli*

Component	Percentage by Weight
H_2O	70
Protein	15
Nucleic acids:	
DNA	1
RNA	6
Polysaccharides and precursors	3
Lipids and precursors	2
Other small organic molecules	1
Inorganic ions	1

Source: Watson, J.D., *Molecular Biology of the Gene* (3rd ed.), p. 69, Benjamin (1976).

DNA during sexual conjugation (a process in which DNA is transferred from one cell to another; prokaryotes usually reproduce by binary fission) or aid in the attachment of the bacterium to a host organism's cells.

The bacterium ***Escherichia coli*** (abbreviated ***E. coli*** and named after its discoverer, Theodor Escherich) is the biologically most well-characterized organism as a result of its intensive biochemical and genetic study over the past 60 years. Indeed, much of the subject matter of this text deals with the biochemistry of *E. coli.* Cells of this normal inhabitant of the higher mammalian colon (Fig. 1-3) are typically 2-μm-long rods that are 1 μm in diameter and weigh $\sim2 \times 10^{-12}$ g. Its DNA, which has a molecular mass of 2.5×10^9 **daltons (D),*** encodes ~4300 proteins (of which only ~60 to 70% have been characterized), although, typically, only ~2600 different proteins are present in a cell at any given time. Altogether an *E. coli* cell contains 3 to 6 thousand different types of molecules, including proteins, nucleic acids, polysaccharides, lipids, and various small molecules and ions (Table 1-1).

b.   Prokaryotes Employ a Wide Variety of Metabolic Energy Sources

The nutritional requirements of the prokaryotes are enormously varied. **Autotrophs** (Greek: *autos,* self + *trophikos,* to feed) can synthesize all their cellular constituents from simple molecules such as H_2O, CO_2, NH_3, and H_2S. Of course they need an energy source to do so as well as to power their other functions. **Chemolithotrophs** (Greek: *lithos,* stone) obtain their energy through the oxidation of inorganic compounds such as NH_3, H_2S, or even Fe^{2+}:

$$2\,NH_3 + 4\,O_2 \longrightarrow 2\,HNO_3 + 2\,H_2O$$
$$H_2S + 2\,O_2 \longrightarrow H_2SO_4$$
$$4\,FeCO_3 + O_2 + 6\,H_2O \longrightarrow 4\,Fe(OH)_3 + 4\,CO_2$$

Indeed, recent studies have revealed the existence of extensive albeit extremely slow-growing colonies of chemolithotrophs that live as far as 5 kilometers underground and whose aggregate biomass may rival that of surface-dwelling organisms.

Photoautotrophs are autotrophs that obtain energy via **photosynthesis** (Chapter 24), a process in which light energy powers the transfer of electrons from inorganic donors to CO_2 yielding **carbohydrates** $[(CH_2O)_n]$. In the most widespread form of photosynthesis, the electron donor in the light-driven reaction sequence is H_2O.

$$nCO_2 + nH_2O \longrightarrow (CH_2O)_n + nO_2$$

This process is carried out by **cyanobacteria** (e.g., the green slimy organisms that grow on the walls of aquariums;

---

*The **molecular mass** of a particle may be expressed in units of daltons, which are defined as 1/12th the mass of a ^{12}C atom [atomic mass units (amu)]. Alternatively, this quantity may be expressed in terms of **molecular weight,** a dimensionless quantity defined as the ratio of the particle mass to 1/12th the mass of a ^{12}C atom and symbolized M_r (for relative molecular mass). In this text, we shall refer to the molecular mass of a particle rather than to its molecular weight.

cyanobacteria were formerly known as **blue-green algae**), as well as by plants. This form of photosynthesis is thought to have generated the O_2 in Earth's atmosphere. Some species of cyanobacteria have the ability to convert N_2 from the atmosphere to organic nitrogen compounds. This **nitrogen fixation** capacity gives them the simplest nutritional requirements of all organisms: With the exception of their need for small amounts of minerals, they can literally live on sunlight and air.

In a more primitive form of photosynthesis, substances such as H_2, H_2S, thiosulfate, or organic compounds are the electron donors in light-driven reactions such as

$$nCO_2 + 2nH_2S \longrightarrow (CH_2O)_n + nH_2O + 2nS$$

The **purple** and the **green photosynthetic bacteria** that carry out these processes occupy such oxygen-free habitats as shallow muddy ponds in which H_2S is generated by rotting organic matter.

Heterotrophs (Greek: *hetero,* other) obtain energy through the oxidation of organic compounds and hence are ultimately dependent on autotrophs for these substances. **Obligate aerobes** (which include animals) must utilize O_2, whereas **anaerobes** employ oxidizing agents such as sulfate **(sulfate-reducing bacteria)** or nitrate **(denitrifying bacteria).** Many organisms can partially metabolize various organic compounds in intramolecular oxidation–reduction processes known as **fermentation. Facultative anaerobes** such as *E. coli* can grow in either the presence or the absence of O_2. **Obligate anaerobes,** in contrast, are poisoned by the presence of O_2. Their metabolisms are thought to resemble those of the earliest life-forms (which arose over 3.8 billion years ago when Earth's atmosphere lacked O_2; see Section 1-5B). At any rate, there are few organic compounds that cannot be metabolized by some prokaryotic organism.

B. *Prokaryotic Classification*

The traditional methods of **taxonomy** (the science of biological classification), which are based largely on the anatomical comparisons of both contemporary and fossil organisms, are essentially inapplicable to prokaryotes. This is because the relatively simple cell structures of prokaryotes, including those of ancient bacteria as revealed by their microfossil remnants, provide little indication of their phylogenetic relationships (**phylogenesis:** evolutionary development). Compounding this problem is the observation that prokaryotes exhibit little correlation between form and metabolic function. Moreover, the eukaryotic definition of a species as a population that can interbreed is meaningless for the asexually reproducing prokaryotes. Consequently, the conventional prokaryotic classification schemes are rather arbitrary and lack the implied evolutionary relationships of the eukaryotic classification scheme (Section 1-2B).

In the most widely used prokaryotic classification scheme, the **prokaryotae** (also known as **monera**) have two divisions: the cyanobacteria and the **bacteria.** The latter are further subdivided into 19 parts based on their various distinguishing characteristics, most notably cell structure, metabolic behavior, and staining properties.

A simpler classification scheme, which is based on cell wall properties, distinguishes three major types of prokaryotes: the **mycoplasmas,** the **gram-positive bacteria,** and the **gram-negative bacteria.** Mycoplasmas lack the rigid cell wall of other prokaryotes. They are the smallest of all living cells (as small as 0.12 μm in diameter, Fig. 1-1) and possess ~20% of the DNA of an *E. coli*. Presumably this quantity of genetic information approaches the minimum amount necessary to specify the essential metabolic machinery required for cellular life. Gram-positive and gram-negative bacteria are distinguished according to whether or not they take up **gram stain** (a procedure developed in 1884 by Christian Gram in which heat-fixed cells are successively treated with the dye crystal violet and iodine and then destained with either ethanol or acetone). Gram-negative bacteria possess a complex **outer membrane** that surrounds their cell wall and excludes gram stain, whereas gram-positive bacteria lack such a membrane (Section 11-3B).

The development, in recent decades, of techniques for determining amino acid sequences in proteins (Section 7-1) and base sequences in nucleic acids (Section 7-2A) has provided abundant indications as to the genealogical relationships between organisms. Indeed, these techniques make it possible to place these relationships on a quantitative basis, and thus to construct a phylogenetically based classification system for prokaryotes.

By the analysis of ribosomal RNA sequences, Carl Woese showed that a group of prokaryotes he named the **Archaea** (also known as the **archaebacteria**) appeared to be as distantly related to the other prokaryotes, the **Bacteria** (also called the **eubacteria**), as both of these groups are to the **Eukarya** (the eukaryotes). The Archaea initially appeared to constitute three different kinds of unusual organisms: the **methanogens,** obligate anaerobes that produce methane (marsh gas) by the reduction of CO_2 with H_2; the **halobacteria,** which can live only in concentrated brine solutions (>2 M NaCl); and certain **thermoacidophiles,** organisms that inhabit acidic hot springs (~90°C and pH < 2). However, recent evidence indicates that ~40% of the microorganisms in the oceans are Archaea, and hence, they may be the most common form of life on Earth.

On the basis of a number of fundamental biochemical traits that differ among the Archaea, the Bacteria, and the Eukarya, but that are common within each group, Woese proposed that these groups of organisms constitute the three primary **urkingdoms** or **domains** of evolutionary descent (rather than the traditional division into prokaryotes and eukaryotes). However, further sequence determinations have revealed that the Eukarya share sequence similarities with the Archaea that they do not share with the Bacteria. Evidently, the Archaea and the Bacteria diverged from some simple primordial life-form following which the Eukarya diverged from the Archaea, as the **phylogenetic tree** in Fig. 1-4 indicates.

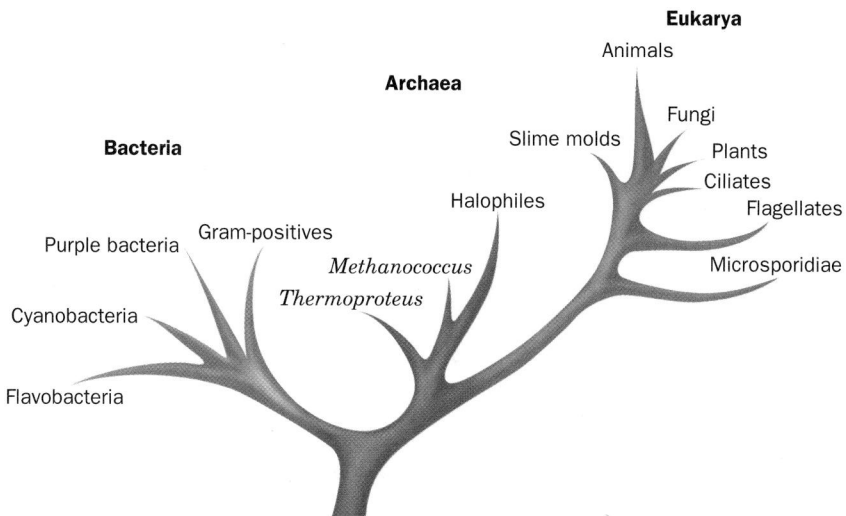

Archaea

Bacteria

Eukarya
Animals

Slime molds

Fungi
Plants
Ciliates
Flagellates

Halophiles

Microsporidiae

Purple bacteria   Gram-positives

Methanococcus
Thermoproteus

Cyanobacteria

Flavobacteria

FIGURE 1-4 Phylogenetic tree. This "family tree" indicates the evolutionary relationships among the three domains of life. The root of the tree represents the last common ancestor of all life on Earth. [After Wheelis, M.L., Kandler, O., and Woese, C.R., *Proc. Natl. Acad. Sci.* **89,** 2931 (1992).]

2 ■ EUKARYOTES

Eukaryotic cells are generally 10 to 100 μm in diameter and thus have a thousand to a million times the volume of typical prokaryotes. It is not size, however, but a profusion of membrane-enclosed organelles, each with a specialized function, that best characterizes eukaryotic cells (Fig. 1-5). In fact, *eukaryotic structure and function are more complex than those of prokaryotes at all levels of organization,* from the molecular level on up.

Eukaryotes and prokaryotes have developed according to fundamentally different evolutionary strategies. Prokaryotes have exploited the advantages of simplicity and miniaturization: Their rapid growth rates permit them

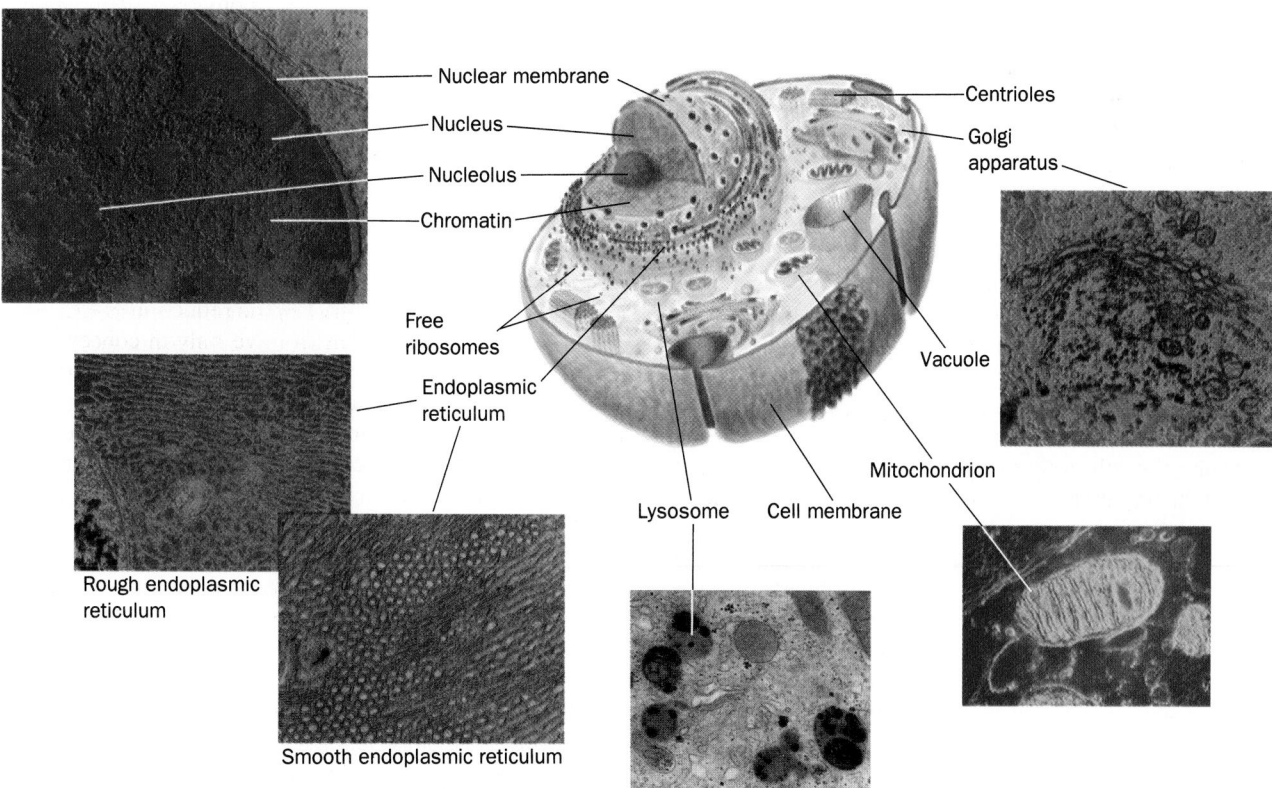

Nuclear membrane
Nucleus
Nucleolus
Chromatin

Centrioles
Golgi apparatus

Free ribosomes
Endoplasmic reticulum

Vacuole

Rough endoplasmic reticulum

Mitochondrion

Smooth endoplasmic reticulum

Lysosome   Cell membrane

FIGURE 1-5   Schematic diagram of an animal cell accompanied by electron micrographs of its organelles. [Nucleus: Tektoff-RM, CNRI/Photo Researchers; rough endoplasmic reticulum and Golgi apparatus: Secchi-Lecaque/Roussel-UCLAF/CNRI/Photo Researchers; smooth endoplasmic reticulum: David M. Phillips/Visuals Unlimited; mitochondrion: CNRI/Photo Researchers; lysosome: Biophoto Associates/Photo Researchers.]

FIGURE 1-6 [Drawing by T.A. Bramley, in Carlile, M., *Trends Biochem. Sci.* **7,** 128 (1982). Copyright © Elsevier Biomedical Press, 1982. Used by permission.]

to occupy ecological niches in which there may be drastic fluctuations of the available nutrients. In contrast, the complexity of eukaryotes, which renders them larger and more slowly growing than prokaryotes, gives them the competitive advantage in stable environments with limited resources (Fig. 1-6). It is therefore erroneous to consider prokaryotes as evolutionarily primitive with respect to eukaryotes. Both types of organisms are well adapted to their respective lifestyles.

The earliest known microfossils of eukaryotes date from ~1.4 billion years ago, some 2.4 billion years after life arose. This observation supports the classical notion that eukaryotes are descended from a highly developed prokaryote, possibly a mycoplasma. The differences between eukaryotes and modern prokaryotes, however, are so profound as to render this hypothesis improbable. Perhaps the early eukaryotes, which according to Woese's evidence evolved from a primordial life-form, were relatively unsuccessful and hence rare. Only after they had developed some of the complex organelles described in the following section did they become common enough to generate significant fossil remains.

A. Cellular Architecture

Eukaryotic cells, like prokaryotes, are bounded by a plasma membrane. The large size of eukaryotic cells results in their surface-to-volume ratios being much smaller than those of prokaryotes (the surface area of an object increases as the square of its radius, whereas volume does so as the cube). This geometrical constraint, coupled with the fact that many essential enzymes are membrane-associated, partially rationalizes the large amounts of intracellular membranes in eukaryotes (the plasma membrane typically constitutes <10% of the membrane in a eukaryotic cell). Since all the matter that enters or leaves a cell must somehow pass through its plasma membrane, the surface areas of many eukaryotic cells are increased by numerous projections and/or invaginations (Fig. 1-7). Moreover, portions of the plasma membrane often bud inward, in a process known as **endocytosis,** so that the cell surrounds portions of the external medium. Thus eukaryotic cells can engulf and digest food particles such as bacteria, whereas prokaryotes are limited to the absorption of individual nutrient molecules. The reverse of endocytosis, a process termed **exocytosis,** is a common eukaryotic secretory mechanism.

a. The Nucleus Contains the Cell's DNA

The nucleus, the eukaryotic cell's most conspicuous organelle, is the repository of its genetic information. This information is encoded in the base sequences of DNA molecules that form the discrete number of chromosomes characteristic of each species. The chromosomes consist of **chromatin,** a complex of DNA and protein. The amount of genetic information carried by eukaryotes is enormous;

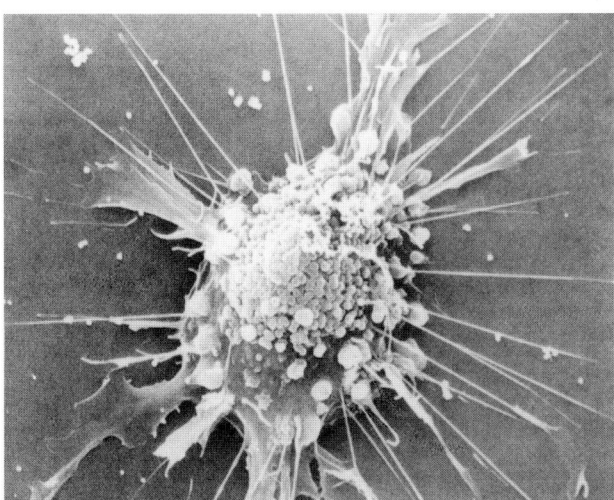

FIGURE 1-7   Scanning electron micrograph of a fibroblast.
[Courtesy of Guenther Albrecht-Buehler, Northwestern University.]

for example, a human cell has over 700 times the DNA of *E. coli* [in the terms commonly associated with computer memories, the **genome** (genetic complement) in each human cell specifies around 800 megabytes of information—about 200 times the information content of this text]. Within the nucleus, the genetic information encoded by the DNA is transcribed into molecules of RNA (Chapter 31), which, after extensive processing, are transported to the cytoplasm (in eukaroytes, the cell contents exclusive of the nucleus), where they direct the ribosomal synthesis of proteins (Chapter 32). The nuclear envelope consists of a double membrane that is perforated by numerous ~90-Å-wide pores that regulate the flow of proteins and RNA between the nucleus and the cytoplasm.

The nucleus of most eukaryotic cells contains at least one dark-staining body known as the **nucleolus,** which is the site of ribosomal assembly. It contains chromosomal segments bearing multiple copies of genes specifying ribosomal RNA. These genes are transcribed in the nucleolus, and the resulting RNA is combined with ribosomal proteins that have been imported from their site of synthesis in the **cytosol** (the cytoplasm exclusive of its membrane-bound organelles). The resulting immature ribosomes are then exported to the cytosol, where their assembly is completed. Thus protein synthesis can occur only in the cytosol.

b. The Endoplasmic Reticulum and the Golgi Apparatus Function to Modify Membrane-Bound and Secretory Proteins

The most extensive membrane in the cell, which was discovered in 1945 by Keith Porter, forms a labyrinthine compartment named the **endoplasmic reticulum.** A large portion of this organelle, which is called the **rough endoplasmic reticulum,** is studded with ribosomes that are engaged in the synthesis of proteins that are either membrane-bound or destined for secretion. The **smooth endoplasmic reticulum,** which is devoid of ribosomes, is the site of lipid synthesis. Many of the products synthesized in the endoplasmic reticulum are eventually transported to the **Golgi apparatus** (named after Camillo Golgi, who first described it in 1898), a stack of flattened membranous sacs in which these products are further processed (Section 23-3B).

c. Mitochondria Are the Site of Oxidative Metabolism

The **mitochondria** (Greek: *mitos,* thread + *chondros,* granule) are the site of cellular **respiration** (aerobic metabolism) in almost all eukaryotes. These cytoplasmic organelles, which are large enough to have been discovered by nineteenth century cytologists, vary in their size and shape but are often ellipsoidal with dimensions of around $1.0 \times 2.0 \ \mu m$—much like a bacterium. A eukaryotic cell typically contains on the order of 2000 mitochondria, which occupy roughly one-fifth of its total cell volume.

The mitochondrion, as the electron microscopic studies of George Palade and Fritjof Sjöstrand first revealed, has two membranes: a smooth outer membrane and a highly folded inner membrane whose invaginations are termed **cristae** (Latin: crests). Thus the mitochondrion contains two compartments, the **intermembrane space** and the internal **matrix space.** The enzymes that catalyze the reactions of respiration are components of either the gel-like **matrix** or the inner mitochondrial membrane. *These enzymes couple the energy-producing oxidation of nutrients to the energy-requiring synthesis of **adenosine triphosphate*** (**ATP;** Section 1-3B and Chapter 22). Adenosine triphosphate, after export to the rest of the cell, fuels its various energy-consuming processes.

Mitochondria are bacteria-like in more than size and shape. Their matrix space contains mitochondrion-specific DNA, RNA, and ribosomes that participate in the synthesis of several mitochondrial components. Moreover, they reproduce by binary fission, and the respiratory processes that they mediate bear a remarkable resemblance to those of modern aerobic bacteria. These observations led to the now widely accepted hypothesis championed by Lynn Margulis that mitochondria evolved from originally free-living gram-negative aerobic bacteria, which formed a symbiotic relationship with a primordial anaerobic eukaryote. The eukaryote-supplied nutrients consumed by the bacteria were presumably repaid severalfold by the highly efficient oxidative metabolism that the bacteria conferred on the eukaryote. This hypothesis is corroborated by the observation that the amoeba *Pelomyxa palustris,* one of the few eukaryotes that lack mitochondria, permanently harbors aerobic bacteria in such a symbiotic relationship.

d. Lysosomes and Peroxisomes Are Containers of Degradative Enzymes

Lysosomes, which were discovered in 1949 by Christian de Duve, are organelles bounded by a single membrane that are of variable size and morphology, although most have diameters in the range 0.1 to 0.8 μm. Lysosomes, which are essentially membranous bags containing a large variety of hydrolytic enzymes, function to digest materials ingested by endocytosis and to recycle cellular components (Section 32-6). Cytological investigations have revealed that lysosomes form by budding from the Golgi apparatus.

Peroxisomes (also known as **microbodies**) are membrane-enclosed organelles, typically 0.5 μm in diameter, that contain oxidative enzymes. They are so named because some peroxisomal reactions generate **hydrogen peroxide** (H_2O_2), a reactive substance that is either utilized in the enzymatic oxidation of other substances or degraded through a disproportionation reaction catalyzed by the enzyme **catalase:**

$$2 H_2O_2 \longrightarrow 2 H_2O + O_2$$

It is thought that peroxisomes function to protect sensitive cell components from oxidative attack by H_2O_2. Peroxisomes, like mitochondria, reproduce by fission and are therefore also thought to have descended from bacteria. Certain plants contain a specialized type of peroxisome, the **glyoxysome,** so named because it is the site of a series of reactions that are collectively termed the **glyoxylate pathway** (Section 23-2).

e. The Cytoskeleton Organizes the Cytosol

The cytosol, far from being a homogeneous solution, is a highly organized gel that can vary significantly in its composition throughout the cell. Much of its internal variability arises from the action of the **cytoskeleton,** an extensive array of filaments that gives the cell its shape and the ability to move and is responsible for the arrangement and internal motions of its organelles (Fig. 1-8).

The most conspicuous cytoskeletal components, the **microtubules,** are ~250-Å-diameter tubes that are composed of the protein **tubulin** (Section 35-3F). They form the supportive framework that guides the movements of organelles within a cell. For example, the **mitotic spindle** is an assembly of microtubules and associated proteins that participates in the separation of replicated chromosomes during cell division. Microtubules are also major constituents of **cilia,** the hairlike appendages extending from many cells, whose whiplike motions move the surrounding fluid past the cell or propel single cells through solution. Very long cilia, such as sperm tails, are termed **flagella** (prokaryotic flagella, which are composed of the protein **flagellin,** are quite different from and unrelated to those of eukaryotes). Mounting evidence suggests that cilia are also descended from free-living bacteria—perhaps spirochetes.

The **microfilaments** are ~90 Å in diameter fibers that consist of the protein **actin.** Microfilaments, as do microtubules, have a mechanically supportive function. Furthermore, through their interactions with the protein **myosin,** microfilaments form contractile assemblies that are responsible for many types of intracellular movements such as cytoplasmic streaming and the formation of cellular protuberances or invaginations. More conspicuously, however, actin and myosin are the major protein components of muscle (Section 35-3A).

The third major cytoskeletal component, the **intermediate filaments,** are protein fibers that are 100 to 150 Å in diameter. Their prominence in parts of the cell that are subject to mechanical stress suggests that they have a load-bearing function. For example, skin in higher animals con-

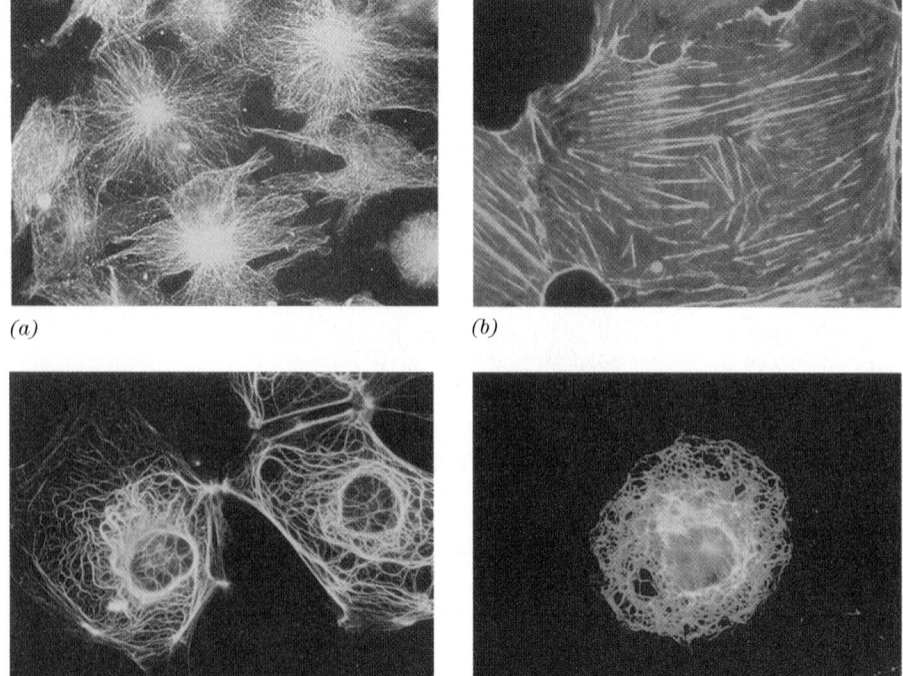

(a)

(b)

(c)

(d)

FIGURE 1-8 Immunofluorescence micrographs showing cytoskeletal components. Cells were stained with fluorescently labeled antibodies raised against (*a*) tubulin, (*b*) actin, (*c*) keratin, and (*d*) **vimentin** (a protein constituent of a type of intermediate filament). [*a* and *d*: K.G. Murti/Visuals Unlimited; *b*: M. Schliwa/Visuals Unlimited; *c*: courtesy of Mary Osborn, Max-Planck Institut für Molecular Biologie, Germany.]

tains an extensive network of intermediate filaments made of the protein **keratin** (Section 8-2A), which is largely responsible for the toughness of this protective outer covering. In contrast to the case with microtubules and microfilaments, the proteins forming intermediate filaments vary greatly in size and composition, both among the different cell types within a given organism and among the corresponding cell types in different organisms.

f. Plant Cells Are Enclosed by Rigid Cell Walls

Plant cells (Fig. 1-9) contain all of the previously described organelles. They also have several additional features, the most conspicuous of which is a rigid cell wall that lies outside the plasma membrane. These cell walls, whose major component is the fibrous polysaccharide **cellulose** (Section 11-2C), account for the structural strength of plants.

A **vacuole** is a membrane-enclosed space filled with fluid. Although vacuoles occur in animal cells, they are most prominent in plant cells, where they typically occupy 90%

of the volume of a mature cell. Vacuoles function as storage depots for nutrients, wastes, and specialized materials such as pigments. The relatively high concentration of solutes inside a plant vacuole causes it to take up water osmotically, thereby raising its internal pressure. This effect, combined with its cell walls' resistance to bursting, is largely responsible for the turgid rigidity of nonwoody plants.

g. Chloroplasts Are the Site of Photosynthesis in Plants

One of the definitive characteristics of plants is their ability to carry out photosynthesis. The site of photosynthesis is an organelle known as the **chloroplast,** which, although generally several times larger than a mitochondrion, resembles it in that both organelles have an inner and an outer membrane. Furthermore, the chloroplast's inner membrane space, the **stroma,** is similar to the mitochondrial matrix in that it contains many soluble enzymes. However, the inner chloroplast membrane is not folded into cristae. Rather, the stroma encloses a third membrane

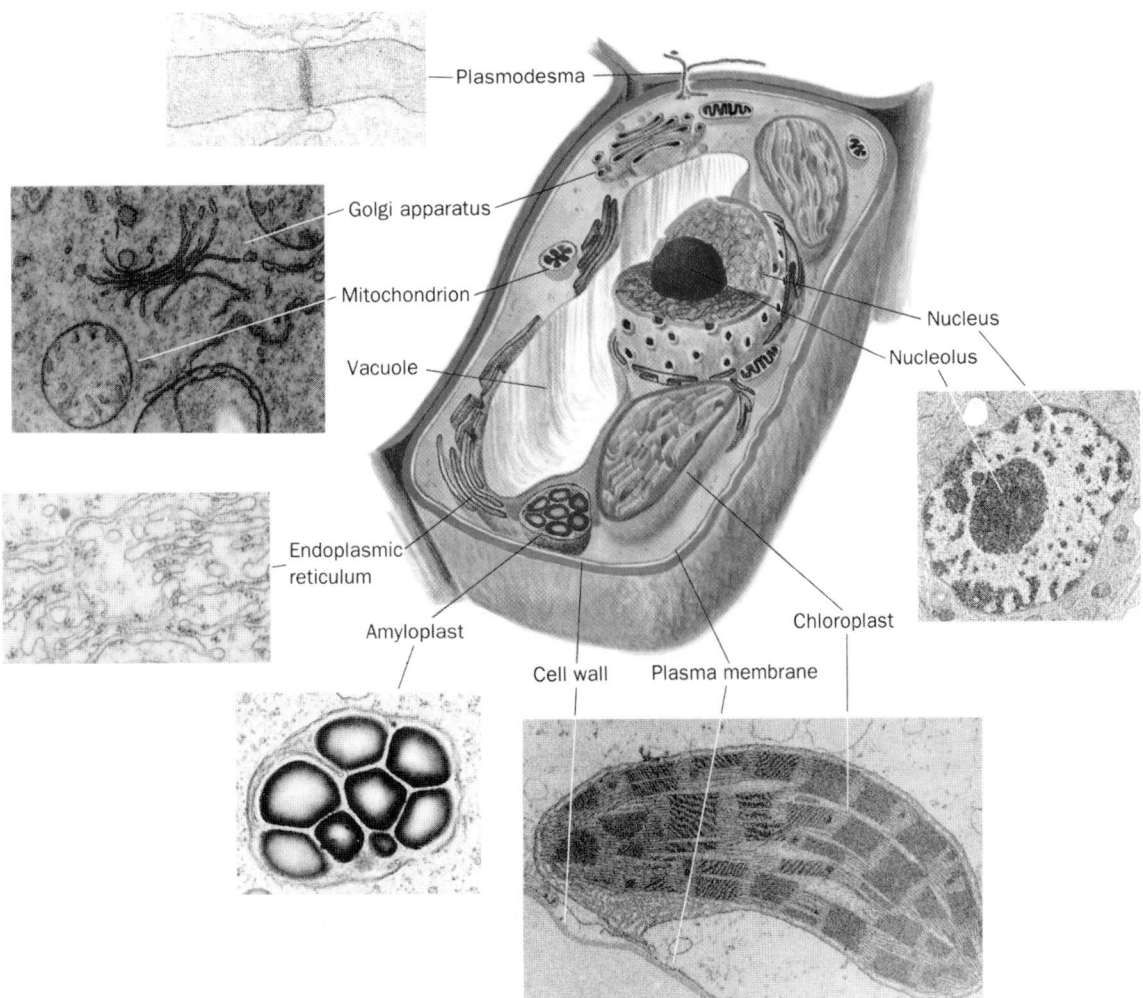

FIGURE 1-9   Drawing of a plant cell accompanied by electron micrographs of its organelles. [Plasmodesma: courtesy of Hilton Mollenhauer, USDA; nucleus: courtesy of Myron Ledbetter, Brookhaven National Laboratory; Golgi apparatus: courtesy of W. Gordon Whaley, University of Texas; chloroplast: courtesy of Lewis Shumway, College of Eastern Utah; amyloplast: Biophoto Associates; endoplasmic reticulum: Biophoto Associates/Photo Researchers.]

system that forms interconnected stacks of disklike sacs called **thylakoids,** which contain the photosynthetic pigment **chlorophyll.** The thylakoid uses chlorophyll-trapped light energy to generate ATP, which is used in the stroma to drive biosynthetic reactions forming carbohydrates and other products (Chapter 24).

Chloroplasts, as do mitochondria, contain their own DNA, RNA, and ribosomes, and they reproduce by fission. Apparently chloroplasts, much like mitochondria, evolved from an ancient cyanobacterium that took up symbiotic residence in an ancestral nonphotosynthetic eukaryote. In fact, several modern nonphotosynthetic eukaryotes have just such a symbiotic relationship with authentic cyanobacteria. Hence *most modern eukaryotes are genetic "mongrels" in that they simultaneously have nuclear, mitochondrial, peroxisomal, possibly ciliar, and—in the case of plants—chloroplast lines of descent.*

B. *Phylogeny and Differentiation*

One of the most remarkable characteristics of eukaryotes is their enormous morphological diversity, on both the cellular and organismal levels. Compare, for example, the architectures of the various human cells drawn in Fig. 1-10. Similarly, recall the great anatomical differences among, say, an amoeba, an oak tree, and a human being.

Taxonometric schemes based on gross morphology as well as on protein and nucleic acid sequences (Sections 7-1 and 7-2) indicate that eukaryotes may be classified into three kingdoms: **Fungi, Plantae** (plants), and **Animalia** (animals). The relative structural simplicity of many unicellular eukaryotes, however, makes their classification under this scheme rather arbitrary. Consequently, these organisms are usually assigned a fourth eukaryotic kingdom, the **Protista.** (Note that biological classification schemes are for the convenience of biologists; nature is rarely neatly categorized.) Figure 1-11 is a phylogenetic tree for eukaryotes.

Anatomical comparisons among living and fossil organisms indicate that the various kingdoms of multicellular organisms independently evolved from Protista (Fig. 1-11). The programs of growth, differentiation, and development followed by multicellular animals (the **metazoa)** in their transformation from fertilized ova to adult organisms provide a remarkable indication of this evolutionary history. For example, all vertebrates exhibit gill-like pouches in their early embryonic stages, which presumably reflect their common fish ancestry (Fig. 1-12). Indeed, these early embryos are similar in size and anatomy even though their respective adult forms are vastly different in these characteristics. Such observations led Ernst Haeckel to formulate his famous (although overstated) dictum: *Ontogeny recapitulates phylogeny* (on-

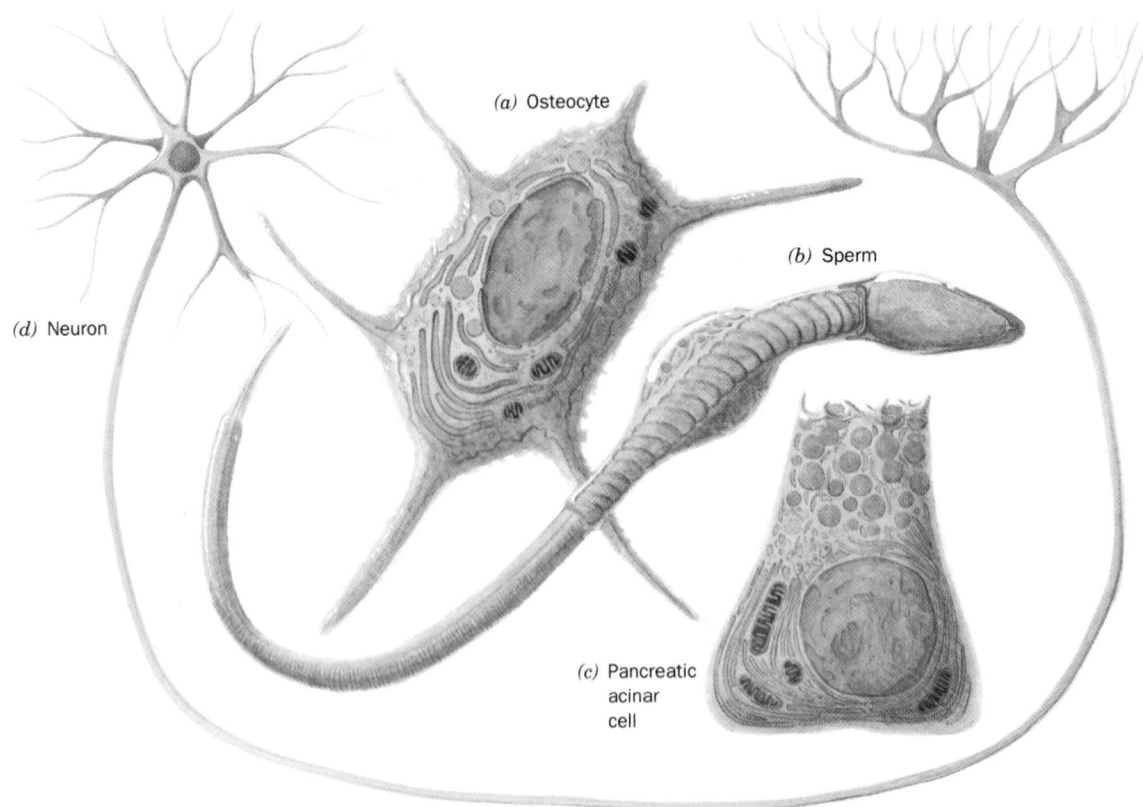

FIGURE 1-10   Drawings of some human cells. (*a*) An osteocyte (bone cell), (*b*) a sperm, (*c*) a pancreatic acinar cell (which secretes digestive enzymes), and (*d*) a neuron (nerve cell).

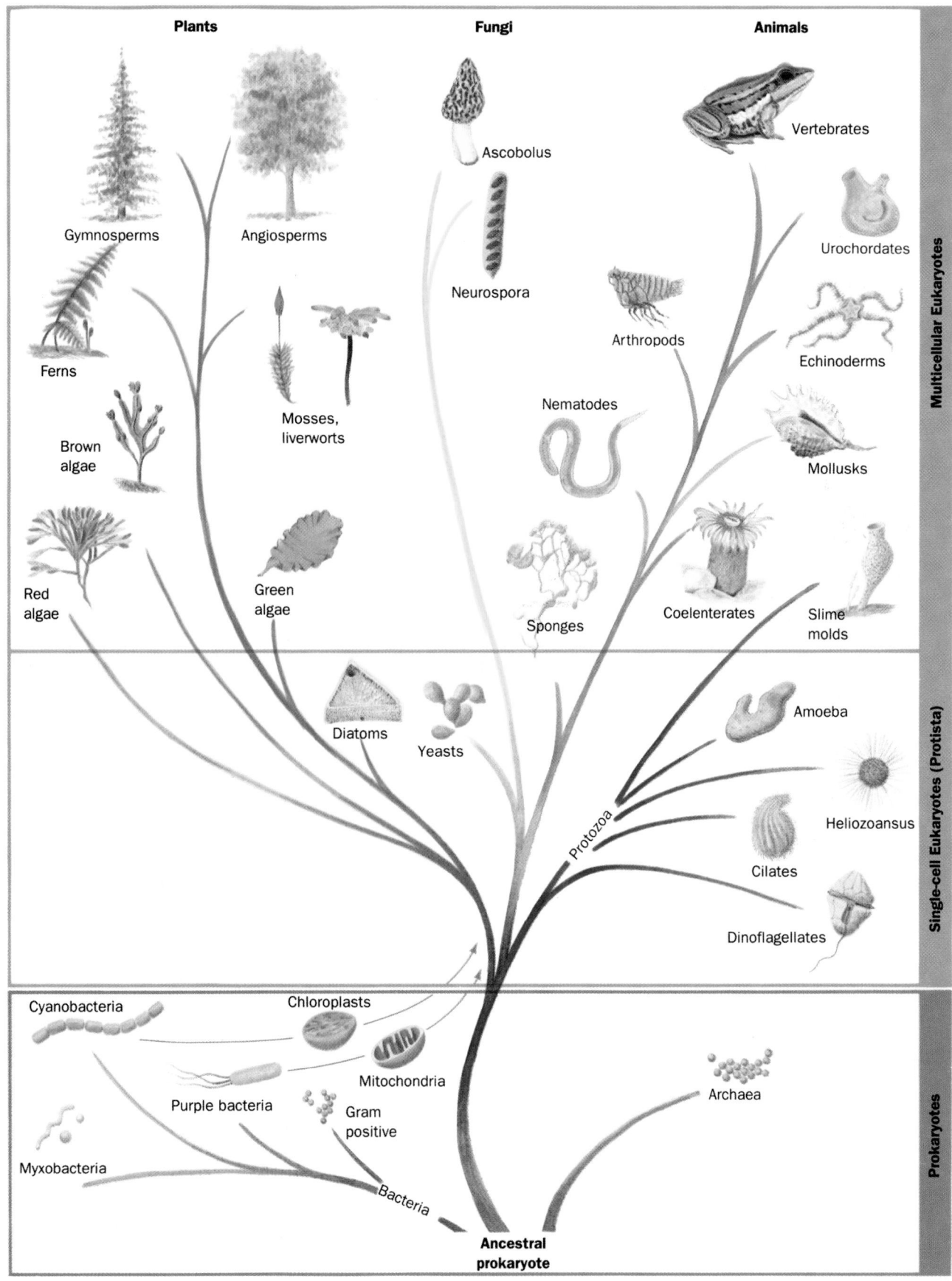

FIGURE 1-11 Evolutionary tree indicating the lines of descent of cellular life on Earth.

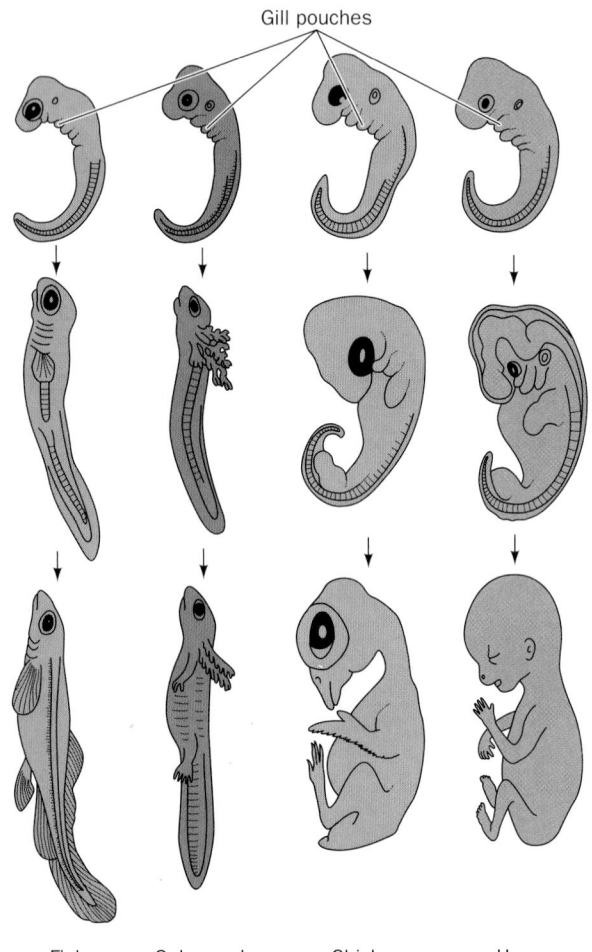

Gill pouches

Fish   Salamander   Chick   Human

FIGURE 1-12 Embryonic development of a fish, an amphibian (salamander), a bird (chick), and a mammal (human). At early stages they are similar in both size and anatomy (the top drawings have around the same scale), although it is now known that their similarites are not as great as these classic drawings indicate. Later they diverge in both these properties. [After Haeckel, E., *Anthropogenie oder Entwickelungsgeschichte des Menschen*, Engelmann (1874).]

togeny: biological development). The elucidation of the mechanism of cellular differentiation in eukaryotes is one of the major long-range goals of modern biochemistry.

3 ■ BIOCHEMISTRY: A PROLOGUE

Biochemistry, as the name implies, is the chemistry of life. It therefore bridges the gap between chemistry, the study of the structures and interactions of atoms and molecules, and biology, the study of the structures and interactions of cells and organisms. Since living things are composed of inanimate molecules, *life, at its most basic level, is a biochemical phenomenon.*

Although living organisms, as we have seen, are enormously diverse in their macroscopic properties, there is a remarkable similarity in their biochemistry that provides a unifying theme with which to study them. For example, hereditary information is encoded and expressed in an almost identical manner in all cellular life. Moreover, the series of biochemical reactions, which are termed **metabolic pathways,** as well as the structures of the enzymes that catalyze them are, for many basic processes, nearly identical from organism to organism. This strongly suggests that all known life-forms are descended from a single primordial ancestor in which these biochemical features first developed.

Although biochemistry is a highly diverse field, it is largely concerned with a limited number of interrelated issues. These are

1. What are the chemical and three-dimensional structures of biological molecules and assemblies, how do they form these structures, and how do their properties vary with them?

2. How do proteins work; that is, what are the molecular mechanisms of enzymatic catalysis, how do receptors recognize and bind specific molecules, and what are the intramolecular and intermolecular mechanisms by which receptors transmit information concerning their binding states?

3. How is genetic information expressed and how is it transmitted to future cell generations?

4. How are biological molecules and assemblies synthesized?

5. What are the control mechanisms that coordinate the myriad biochemical reactions that take place in cells and in organisms?

6. How do cells and organisms grow, differentiate, and reproduce?

These issues are previewed in this section and further illuminated in later chapters. However, as will become obvious as you read further, in all cases, our knowledge, extensive as it is, is dwarfed by our ignorance.

A. *Biological Structures*

Living things are enormously complex. As indicated in Section 1-1A, even the relatively simple *E. coli* cell contains some 3 to 6 thousand different compounds, most of which are unique to *E. coli* (Fig. 1-13). Higher organisms have a correspondingly greater complexity. *Homo sapiens* (human beings), for example, may contain 100,000 different types of molecules, although only a minor fraction of them have been characterized. One might therefore suppose that to obtain a coherent biochemical understanding of any organism would be a hopelessly difficult task. This, however, is not the case. *Living things have an underlying regularity that derives from their being constructed in a hierarchical manner.* Anatomical and cytological studies have

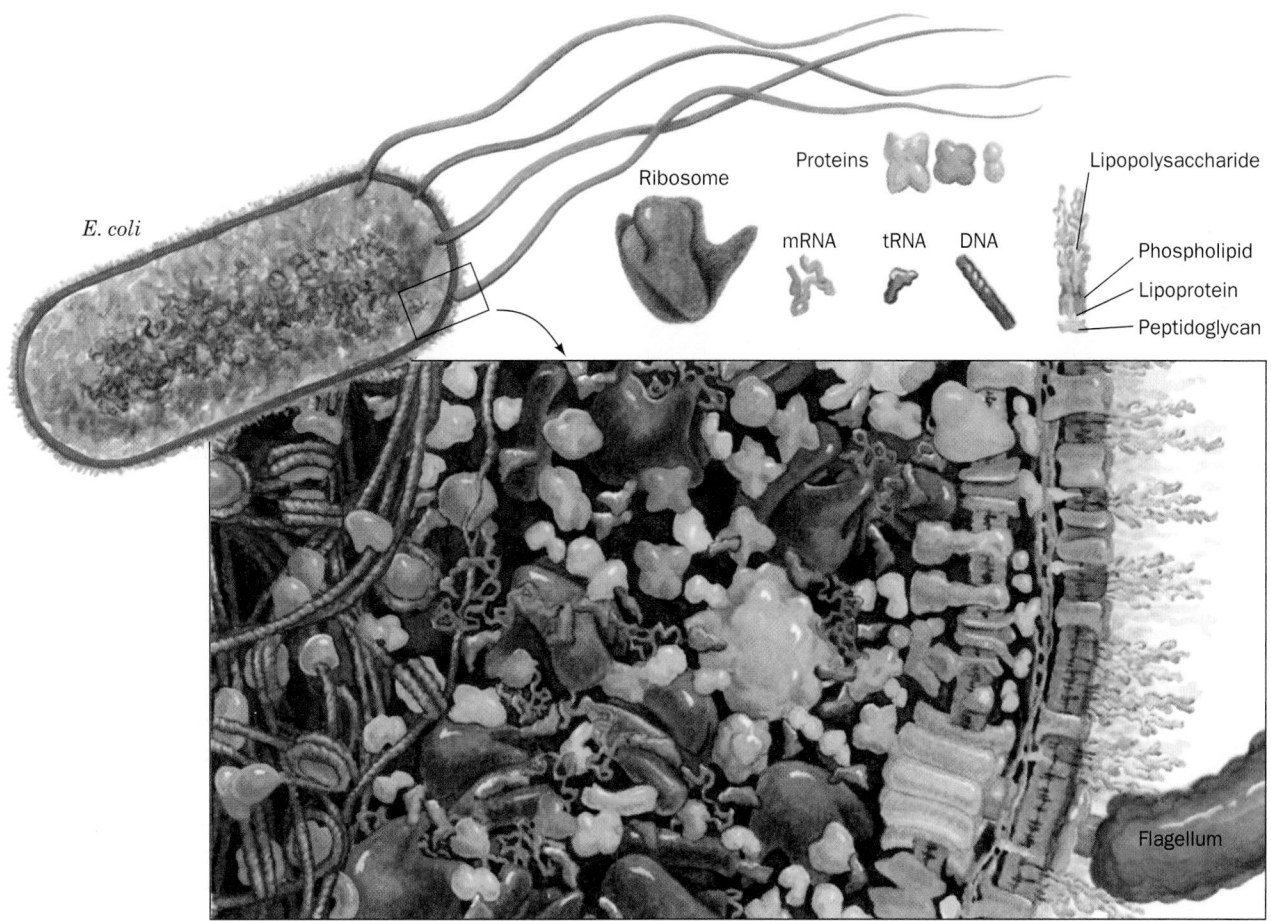

E. coli

Ribosome

Proteins

mRNA   tRNA   DNA

Lipopolysaccharide

Phospholipid

Lipoprotein

Peptidoglycan

Flagellum

FIGURE 1-13   Simulated cross section of an *E. coli* cell magnified around one millionfold. The right side of the drawing shows the multilayered cell wall and membrane, decorated on its exterior surface with lipopolysaccharides (Section 11-3B). A flagellum (*lower right*) is driven by a motor anchored in the inner membrane (Section 35-3G). The cytoplasm, which occupies the middle region of the drawing, is predominantly filled with ribosomes engaged in protein synthesis (Section 32-3). The left side of the drawing contains a dense tangle of DNA in complex with specific proteins. Only the largest macromolecules and molecular assemblies are shown. In a living cell, the remaining space in the cytoplasm would be crowded with smaller molecules and water (a water molecule would be about the size of the period at the end of this sentence). [After a drawing by David Goodsell, UCLA.]

shown that multicellular organisms are organizations of organs, which are made of tissues consisting of cells, composed of subcellular organelles (e.g., Fig. 1-14). At this point in our hierarchical descent, we enter the biochemical realm since organelles consist of **supramolecular assemblies,** such as membranes or fibers, that are organized clusters of **macromolecules** (polymeric molecules with molecular masses from thousands of daltons on up).

As Table 1-1 indicates, *E. coli,* and living things in general, contain only a few different types of macromolecules: **proteins** (Greek: *proteios,* of first importance), **nucleic acids,** and **polysaccharides** (Greek: *sakcharon,* sugar). *All of these substances have a modular construction; they consist of linked monomeric units that occupy the lowest level of our structural hierarchy.* Thus, as Fig. 1-15 indicates, proteins are polymers of amino acids (Section 4-1B), nucleic acids are polymers of nucleotides (Section 5-1), and poly-

saccharides are polymers of sugars (Section 11-2). **Lipids** (Greek: *lipos,* fat), the fourth major class of biological molecules, are too small to be classified as macromolecules but also have a modular construction (Section 12-1).

The task of the biochemist has been vastly simplified by the finding that *there are relatively few species of monomeric units that occur in each class of biological macromolecule.* Proteins are all synthesized from the same 20 species of **amino acids,** nucleic acids are made from 8 types of **nucleotides** (4 each in DNA and RNA), and there are ~8 commonly occurring types of **sugars** in polysaccharides. The great variation in properties observed among macromolecules of each type largely arises from the enormous number of ways its monomeric units can be arranged and, in many cases, derivatized.

One of the central questions in biochemistry is how biological structures are formed. As is explained in later

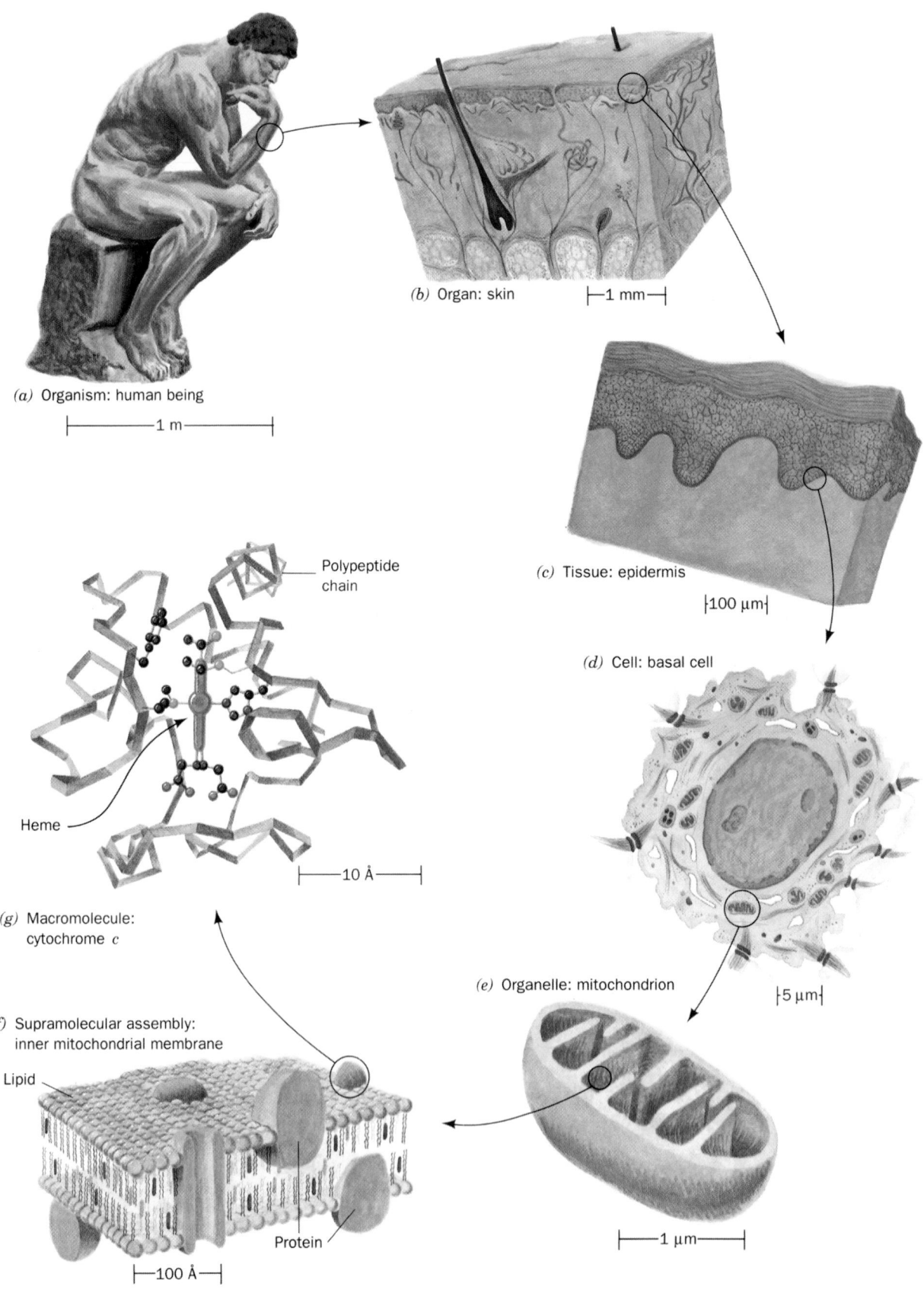

(a) Organism: human being

├────── 1 m ──────┤

(b) Organ: skin ├─1 mm─┤

(c) Tissue: epidermis

├100 μm┤

(d) Cell: basal cell

├5 μm┤

(e) Organelle: mitochondrion

├───── 1 μm─────┤

(g) Macromolecule: cytochrome *c*

Polypeptide chain

Heme

├───── 10 Å ─────┤

(f) Supramolecular assembly: inner mitochondrial membrane

Lipid

Protein

├──100 Å──┤

FIGURE 1-14   Example of the hierarchical organization of biological structures.

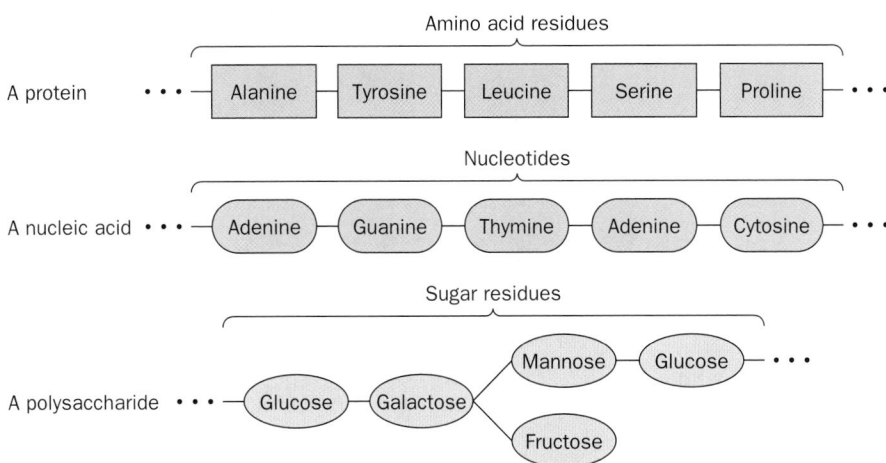

FIGURE 1-15 **Polymeric organization of proteins, nucleic acids, and polysaccharides.**

chapters, the monomeric units of macromolecules are either directly acquired by the cell as nutrients or enzymatically synthesized from simpler substances. Macromolecules are synthesized from their monomeric precursors in complex enzymatically mediated processes.

Newly synthesized proteins spontaneously fold to assume their native conformations (Section 9-1A); that is, they undergo **self-assembly.** Apparently their amino acid sequences specify their three-dimensional structures. Likewise, the structures of other types of macromolecules are specified by the sequences of their monomeric units. The principle of self-assembly extends at least to the level of supramolecular assemblies. However, the way in which higher levels of biological structures are generated is largely unknown. The elucidation of the mechanisms of cellular and organismal growth and differentiation is a major area of biological research.

B. *Metabolic Processes*

There is a bewildering array of chemical reactions that simultaneously occur in any living cell. Yet these reactions follow a pattern that organizes them into the coherent process we refer to as life. For instance, most biological reactions are members of a metabolic pathway; that is, they function as one of a sequence of reactions that produce one or more specific products. Moreover, one of the hallmarks of life is that the rates of its reactions are so tightly regulated that there is rarely an unsatisfied need for a reactant in a metabolic pathway or an unnecessary buildup of some product.

Metabolism has been traditionally (although not necessarily logically) divided into two major categories:

1. Catabolism or degradation, in which nutrients and cell constituents are broken down so as to salvage their components and/or to generate energy.

2. Anabolism or biosynthesis, in which biomolecules are synthesized from simpler components.

The energy required by anabolic processes is provided by catabolic processes largely in the form of **adenosine**

triphosphate (ATP). For instance, such energy-generating processes as photosynthesis and the biological oxidation of nutrients produce ATP from **adenosine diphosphate (ADP)** and a phosphate ion.

Adenosine diphosphate (ADP)

Adenosine triphosphate (ATP)

Conversely, such energy-consuming processes as biosynthesis, the transport of molecules against a concentration gradient, and muscle contraction are driven by the reverse of this reaction, the hydrolysis of ATP:

$$\text{ATP} + \text{H}_2\text{O} \rightleftharpoons \text{ADP} + \text{HPO}_4^{2-}$$

Thus, *anabolic and catabolic processes are coupled together through the mediation of the universal biological energy "currency," ATP.*

C. *Expression and Transmission of Genetic Information*

Deoxyribonucleic acid (DNA) is the cell's master repository of genetic information. This macromolecule, as is diagrammed in Fig. 1-16, consists of two strands of linked **nucleotides,** each of which is composed of a **deoxyribose** sugar residue, a phosphoryl group, and one of four bases: **adenine (A), thymine (T), guanine (G),** or **cytosine (C).** Genetic information is encoded in the sequence of these bases. Each DNA base is hydrogen bonded to a base on the opposite strand to form an entity known as a **base pair.** However, A can only hydrogen bond with T, and G with

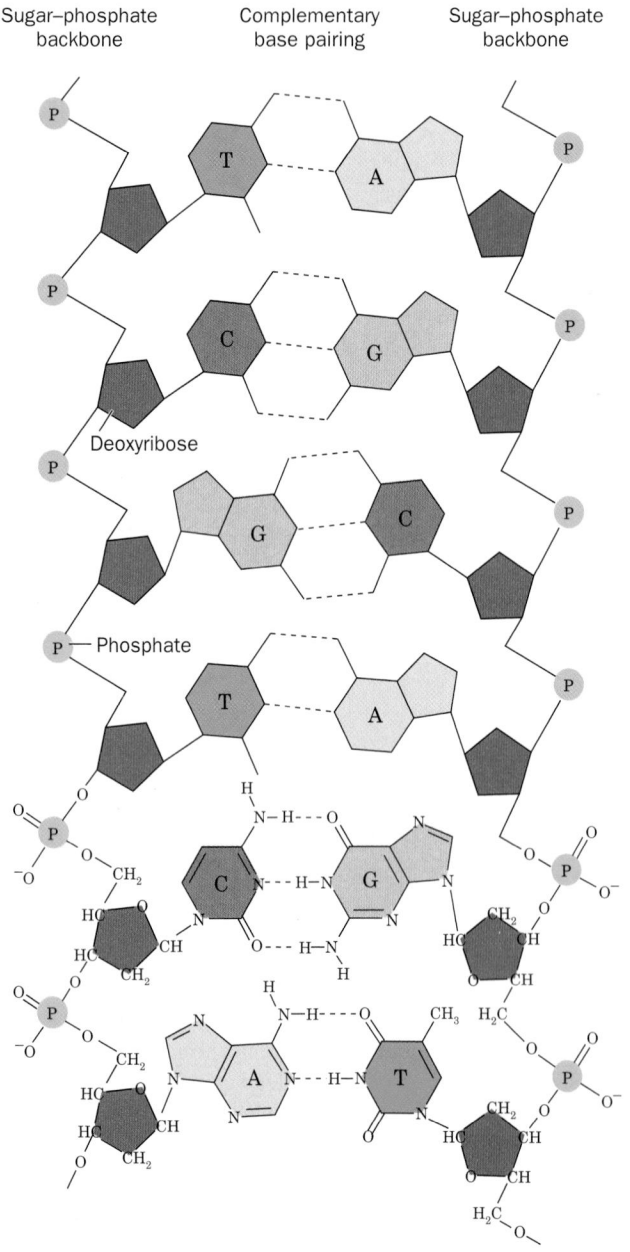

FIGURE 1-16   Double-stranded DNA. The two polynucleotide chains associate through complementary base pairing. A pairs with T, and G pairs with C by forming specific hydrogen bonds.

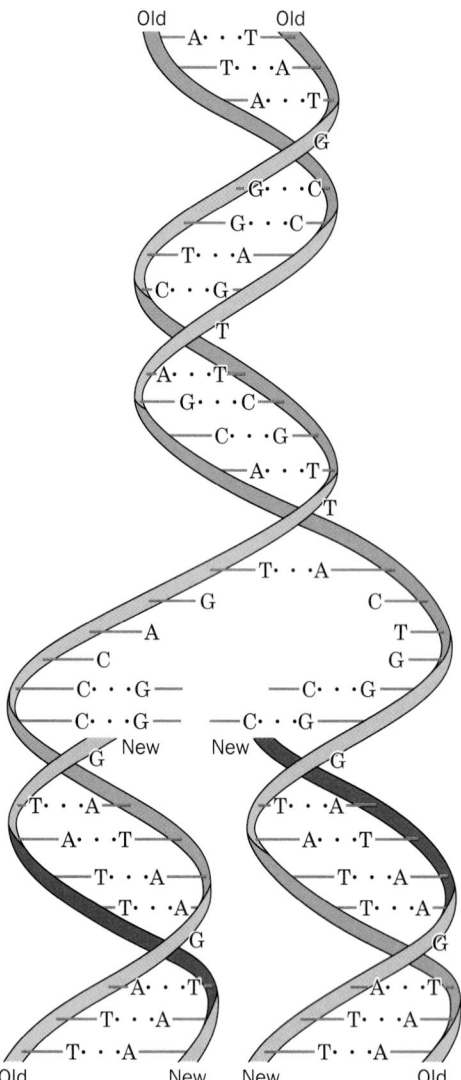

FIGURE 1-17   Schematic diagram of DNA replication. Each strand of parental DNA (*red*) acts as a template for the synthesis of a complementary daughter strand (*green*). Consequently, the resulting double-stranded molecules are identical.

C, so that the two strands are **complementary;** that is, the sequence of one strand implies the sequence of the other.

The division of a cell must be accompanied by the replication of its DNA. In this enzymatically mediated process, each DNA strand acts as a template for the formation of its complementary strand (Fig. 1-17; Section 5-4C). Consequently, every progeny cell contains a complete DNA molecule (or set of DNA molecules), each of which consists of one parental strand and one daughter strand. Mutations arise when, through rare copying errors or damage to a parental strand, one or more wrong bases are incorporated into a daughter strand. Most mutations are either innocuous or deleterious. Occasionally, however, one results in a new characteristic that confers some sort of selective advantage on its recipient. Individuals with such mutations, according to the tenets of the Darwinian theory of evolution, have an increased probability of reproducing. New species arise through a progression of such mutations.

The expression of genetic information is a two-stage process. In the first stage, which is termed **transcription,** a DNA strand serves as a template for the synthesis of a complementary strand of ribonucleic acid (RNA; Section 31-2). This nucleic acid, which is generally single stranded, differs chemically from DNA (Fig. 1-16) only in that it has **ribose** sugar residues in place of DNA's deoxyribose and **uracil (U)** replacing DNA's thymine base.

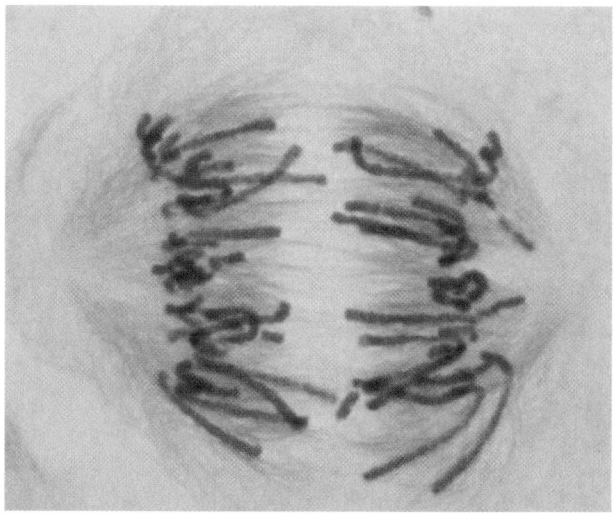

Ribose   **Uracil**

In the second stage of genetic expression, which is known as **translation,** ribosomes enzymatically link together amino acids to form proteins (Section 32-3). The order in which the amino acids are linked together is prescribed by the RNA's sequence of bases. Consequently, since proteins are self-assembling, the genetic information encoded by DNA serves, through the intermediacy of RNA, to specify protein structure and function. Just which genes are expressed in a given cell under a particular set of circumstances is controlled by complex regulatory systems whose workings are understood only in outline.

4 ■ GENETICS: A REVIEW

One has only to note the resemblance between parent and child to realize that physical traits are inherited. Yet the

mechanism of inheritance has, until recent decades, been unknown. The theory of **pangenesis,** which originated with the ancient Greeks, held that semen, which clearly has something to do with procreation, consists of representative particles from all over the body **(pangenes).** This idea was extended in the late eighteenth century by Jean Baptiste de Lamarck, who, in a theory known as **Lamarckism,** hypothesized that an individual's acquired characteristics, such as large muscles resulting from exercise, would be transmitted to his/her offspring. Pangenesis and at least some aspects of Lamarckism were accepted by most nineteenth century biologists, including Charles Darwin.

The realization, in the mid-nineteenth century, that all organisms are derived from single cells set the stage for the development of modern biology. In his **germ plasm theory,** August Weismann pointed out that sperm and ova, the **germ cells** (whose primordia are set aside early in embryonic development), are directly descended from the germ cells of the previous generation and that other cells of the body, the **somatic cells,** although derived from germ cells, do not give rise to them. He refuted the ideas of pangenesis and Lamarckism by demonstrating that the progeny of many successive generations of mice whose tails had been cut off had tails of normal length.

A. *Chromosomes*

In the 1860s, eukaryotic cell nuclei were observed to contain linear bodies that were named chromosomes (Greek: *chromos,* color + *soma,* body) because they are strongly stained by certain basic dyes (Fig. 1-18). There are normally two copies of each chromosome **(homologous pairs)** present in every somatic cell. The number of unique chromosomes (N) in such a cell is known as its **haploid number,** and the total number of chromosomes ($2N$) is its **diploid number.** Different species differ in their haploid number of chromosomes (Table 1-2).

FIGURE 1-18 Chromosomes. A photomicrograph of a plant cell (*Scadoxus katherinae* Bak.) during anaphase of mitosis showing its chromosomes being pulled to opposite poles of the cell by the mitotic spindle. The microtubules forming the mitotic spindle are stained red and the chromosomes are blue. [Courtesy of Andrew S. Bajer, University of Oregon.]

TABLE 1-2   Number of Chromosomes (2*N*) in Some Eukaryotes

Organism	Chromosomes
Human	46
Dog	78
Rat	42
Turkey	82
Frog	26
Fruit fly	8
Hermit crab	~254
Garden pea	14
Potato	48
Yeast	34
Green alga	~20

Source: Ayala, F.J. and Kiger, J.A., Jr., *Modern Genetics* (2nd ed.), *p. 9,* Benjamin/Cummings (1984).

a. Somatic Cells Divide by Mitosis

The division of somatic cells, a process known as **mitosis** (Fig. 1-19), is preceded by the duplication of each chromosome to form a cell with $4N$ chromosomes. During cell division, each chromosome attaches by its **centromere** to the **mitotic spindle** such that the members of each duplicate pair line up across the equatorial plane of the cell. The members of each duplicate pair are then pulled to opposite poles of the dividing cell by the action of the spindle to yield diploid daughter cells that each have the same $2N$ chromosomes as the parent cell.

b. Germ Cells Are Formed by Meiosis

The formation of germ cells, a process known as **meiosis** (Fig. 1-20), requires two consecutive cell divisions. Before the first meiotic division each chromosome replicates, but the resulting sister **chromatids** remain attached at their centromere. The homologous pairs of the doubled chromosomes then line up across the equatorial plane of the cell in zipperlike fashion, which permits an exchange of the corresponding sections of homologous chromosomes in a process known as **crossing-over.** The spindle then moves the members of each homologous pair to opposite poles of the cell so that, after the first meiotic division, each daughter cell contains N doubled chromosomes. In the second meiotic division, the sister chromatids separate to form chromosomes and move to opposite poles of the dividing cell to yield a total of four haploid cells that are known as **gametes.** Fertilization consists of the fusion of a male gamete (sperm) with a female gamete (ovum) to yield a diploid cell known as a **zygote** that has received N chromosomes from each of its parents.

B. *Mendelian Inheritance*

The basic laws of inheritance were reported in 1866 by Gregor Mendel. They were elucidated by the analysis of a series of **genetic crosses** between true-breeding strains (producing progeny that have the same characteristics as the parents) of garden peas, *Pisum sativum,* that differ in certain well-defined traits such as seed shape (round vs wrinkled), seed color (yellow vs green), or flower color (purple vs white). Mendel found that in crossing parents (*P*) that differ in a single trait, say seed shape, the progeny (F_1; first filial generation) all have the trait of only one of the parents, in this case round seeds (Fig. 1-21). The trait appearing in F_1 is said to be **dominant,** whereas the alternative trait is called **recessive.** In F_2, the progeny of F_1, three-quarters have the dominant trait and one-quarter have the recessive trait. Those peas with the recessive trait breed true; that is, self-crossing recessive F_2's results in progeny (F_3) that also have the recessive trait. The F_2's exhibiting the dominant trait, however, fall into two categories: One-third of them breed true, whereas the remainder have progeny with the same 3:1 ratio of dominant to recessive traits as do the members of F_2.

Mendel accounted for his observations by hypothesizing that *the various pairs of contrasting traits each result*

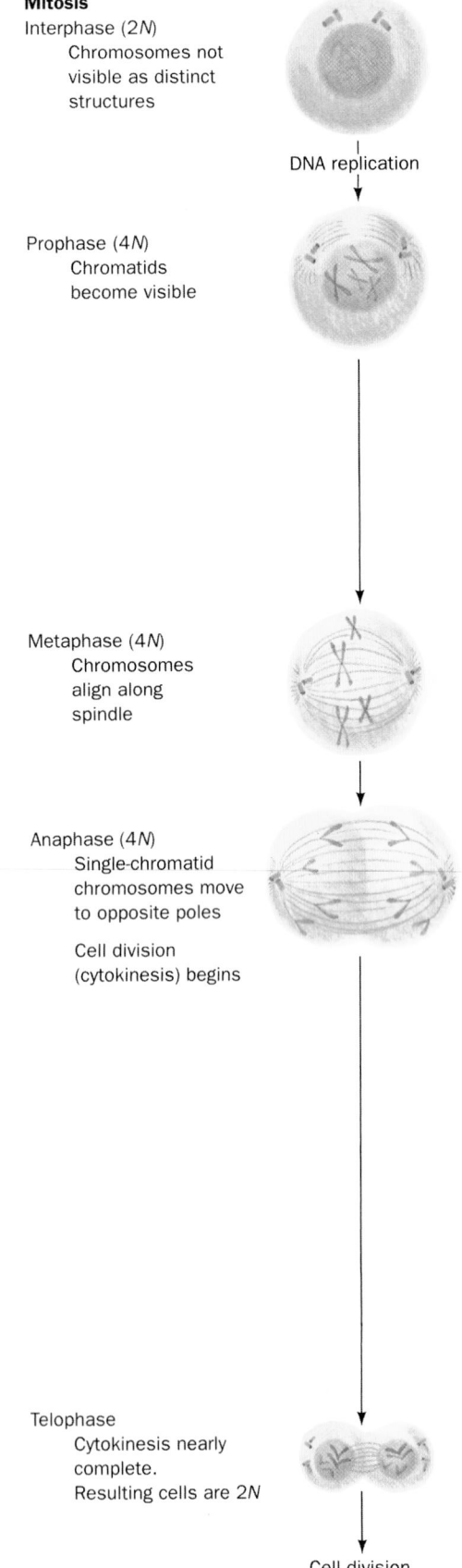

Mitosis

Interphase (2*N*)
   Chromosomes not
   visible as distinct
   structures

DNA replication

Prophase (4*N*)
   Chromatids
   become visible

Metaphase (4*N*)
   Chromosomes
   align along
   spindle

Anaphase (4*N*)
   Single-chromatid
   chromosomes move
   to opposite poles

   Cell division
   (cytokinesis) begins

Telophase
   Cytokinesis nearly
   complete.
   Resulting cells are 2*N*

Cell division

FIGURE 1-19   Mitosis, the usual form of cell division in eukaryotes. Mitosis yields two daughter cells, each with the same chromosomal complement as the parental cell.

Meiosis

Interphase (2*N*)

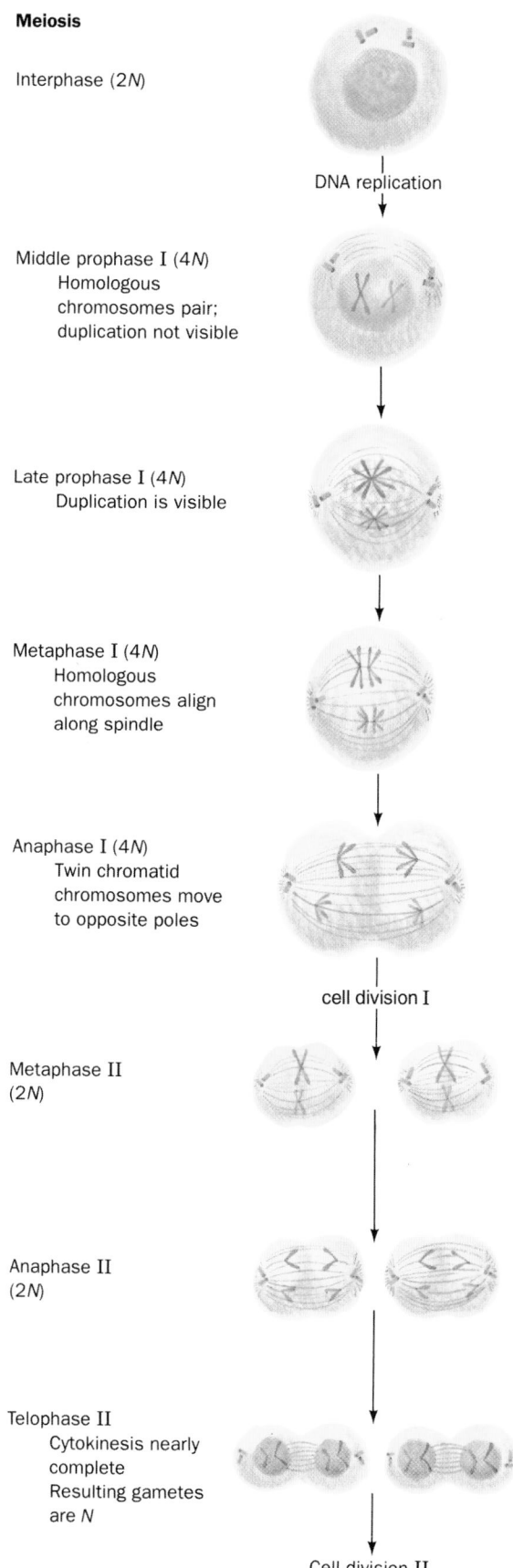

DNA replication

Middle prophase I (4*N*)
Homologous
chromosomes pair;
duplication not visible

Late prophase I (4*N*)
Duplication is visible

Metaphase I (4*N*)
Homologous
chromosomes align
along spindle

Anaphase I (4*N*)
Twin chromatid
chromosomes move
to opposite poles

cell division I

Metaphase II
(2*N*)

Anaphase II
(2*N*)

Telophase II
Cytokinesis nearly
complete
Resulting gametes
are *N*

Cell division II

**FIGURE 1-20   Meiosis, which leads to the formation of
gametes (sex cells).** Meiosis comprises two consecutive cell
divisions to yield four daughter cells, each with half of the
chromosomal complement of its parental cell.

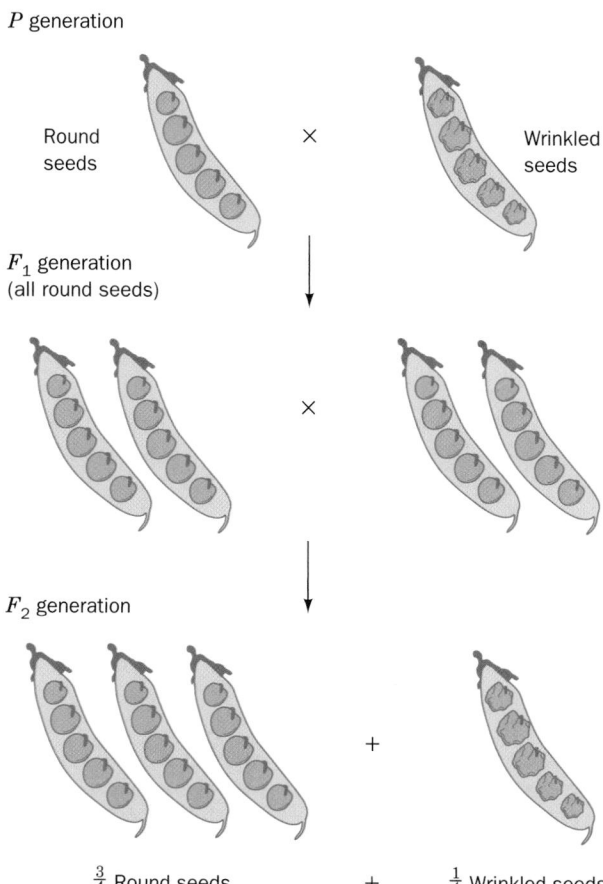

P generation

Round
seeds   ×   Wrinkled
seeds

*F*₁ generation
(all round seeds)

×

*F*₂ generation

+

$\frac{3}{4}$ Round seeds   + $\frac{1}{4}$ Wrinkled seeds

FIGURE 1-21   Genetic crosses, Crossing a pea plant that has
round seeds with one that has wrinkled seeds yields *F*₁ progeny
that all have round seeds. Crossing these *F*₁ peas yields an *F*₂
generation, of which three-quarters have round seeds and one-
quarter have wrinkled seeds.

from a factor (now called a **gene**) that has alternative forms
(alleles). *Every plant contains a pair of genes governing a
particular trait, one inherited from each of its parents.* The
alleles for seed shape are symbolized *R* for round seeds
and *r* for wrinkled seeds (gene symbols are generally given
in italics). The pure-breeding plants with round and wrinkled seeds, respectively, have *RR* and *rr* **genotypes** (genetic
composition) and are both said to be **homozygous** in seed
shape. Plants with the *Rr* genotype are **heterozygous** in
seed shape and have the round seed **phenotype** (appearance or character) because *R* is dominant over *r*. The
two alleles do not blend or mix in any way in the plant and
are independently transmitted through gametes to progeny
(Fig. 1-22).

Mendel also found that *different traits are independently
inherited.* For example, crossing peas that have round yellow seeds (*RRYY*) with peas that have wrinkled green
seeds (*rryy*) results in *F*₁ progeny (*RrYy*) that have round
yellow seeds (yellow seeds are dominant over green seeds).
The *F*₂ phenotypes appear in the ratio 9 round yellow : 3
round green : 3 wrinkled yellow : 1 wrinkled green. This
result indicates that there is no tendency for the genes from

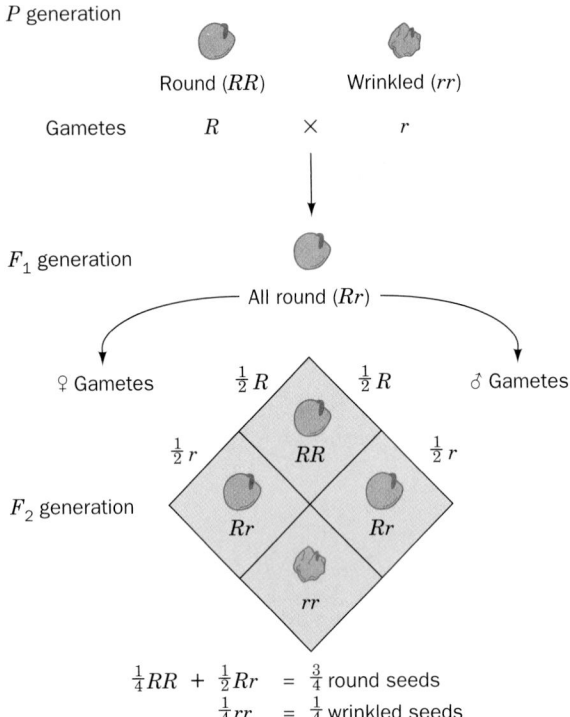

P generation

Round (RR)   Wrinkled (rr)

Gametes R × r

F_1 generation

All round (Rr)

♀ Gametes $\frac{1}{2}R$ $\frac{1}{2}R$ ♂ Gametes

$\frac{1}{2}r$ RR $\frac{1}{2}r$

F_2 generation Rr Rr

rr

$$\frac{1}{4}RR + \frac{1}{2}Rr = \frac{3}{4} \text{ round seeds}$$
$$\frac{1}{4}rr = \frac{1}{4} \text{ wrinkled seeds}$$

FIGURE 1-22   Genotypes and phenotypes. In a genetic cross between peas with round seeds and peas with wrinkled seeds, the F_1 generation has the round seed phenotype because the round seed genotype is dominant over the wrinkled seed genotype. Three-fourths of the F_2 generation's seeds are round and one-fourth are wrinkled because the genes for these alleles are independently transmitted by haploid gametes.

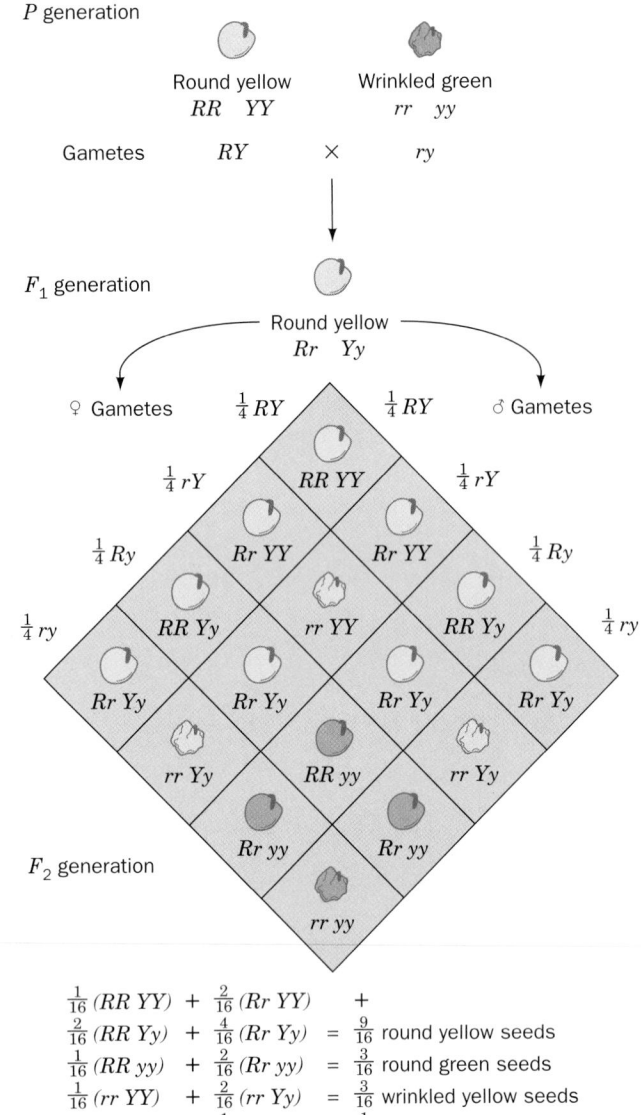

P generation

Round yellow   Wrinkled green
RR YY rr yy

Gametes RY × ry

F_1 generation

Round yellow
Rr Yy

♀ Gametes $\frac{1}{4}RY$ $\frac{1}{4}RY$ ♂ Gametes

$\frac{1}{4}rY$ $RR\ YY$ $\frac{1}{4}rY$

$\frac{1}{4}Ry$ $Rr\ YY$ $Rr\ YY$ $\frac{1}{4}Ry$

$\frac{1}{4}ry$ $RR\ Yy$ $rr\ YY$ $RR\ Yy$ $\frac{1}{4}ry$

$Rr\ Yy$ $Rr\ Yy$ $Rr\ Yy$ $Rr\ Yy$

$rr\ Yy$ $RR\ yy$ $rr\ Yy$

F_2 generation $Rr\ yy$ $Rr\ yy$

$rr\ yy$

$$\frac{1}{16}(RR\ YY) + \frac{2}{16}(Rr\ YY) + $$
$$\frac{2}{16}(RR\ Yy) + \frac{4}{16}(Rr\ Yy) = \frac{9}{16} \text{ round yellow seeds}$$
$$\frac{1}{16}(RR\ yy) + \frac{2}{16}(Rr\ yy) = \frac{3}{16} \text{ round green seeds}$$
$$\frac{1}{16}(rr\ YY) + \frac{2}{16}(rr\ Yy) = \frac{3}{16} \text{ wrinkled yellow seeds}$$
$$\frac{1}{16}(rr\ yy) = \frac{1}{16} \text{ wrinkled green seeds}$$

FIGURE 1-23   Independent assortment. The genes for round (R) versus wrinkled (r) and yellow (Y) versus green (y) pea seeds assort independently. The F_2 progeny consist of nine genotypes comprising the four possible phenotypes.

any parent to assort together (Fig. 1-23). It was later shown, however, that *only genes that occur on different chromosomes exhibit such independence.*

The dominance of one trait over another is a common but not universal phenomenon. For example, crossing a pure-breeding red variety of the snapdragon *Antirrhinum* with a pure-breeding white variety results in pink-colored F_1 progeny. The F_2 progeny have red, pink, and white flowers in 1:2:1 ratio because the flowers of homozygotes for the red color (AA) contain more red pigment than do the heterozygotes (Aa; Fig. 1-24). The red and white traits are therefore said to be **codominant.** In the case of codominance, the phenotype reveals the genotype.

A given gene may have multiple alleles. A familiar example is the human **ABO blood group system** (Section 12-3E). A person may have type A, type B, type AB, or type O blood depending on whether his/her red blood cells bear A antigens, B antigens, both, or neither. The A and B antigens are specified by the codominant I^A and I^B alleles, respectively, and the O phenotype is homozygous for the recessive i allele.

C. *Chromosomal Theory of Inheritance*

Mendel's theory of inheritance was almost universally ignored by his contemporaries. This was partially because in

analyzing his data he used probability theory, an alien subject to most biologists of the time. The major reason his theory was ignored, however, is that it was ahead of its time: Contemporary knowledge of anatomy and physiology provided no basis for its understanding. For instance, mitosis and meiosis had yet to be discovered. Yet, after Mendel's work was rediscovered in 1900, it was shown that his principles explained inheritance in animals as well as in plants. In 1903, as a result of the realization that chromosomes and genes behave in a parallel fashion, Walter Sutton formulated the **chromosomal theory of inheritance** in which he hypothesized that genes are parts of chromosomes.

The first trait to be assigned a chromosomal location was that of sex. *In most eukaryotes, the cells of females each*

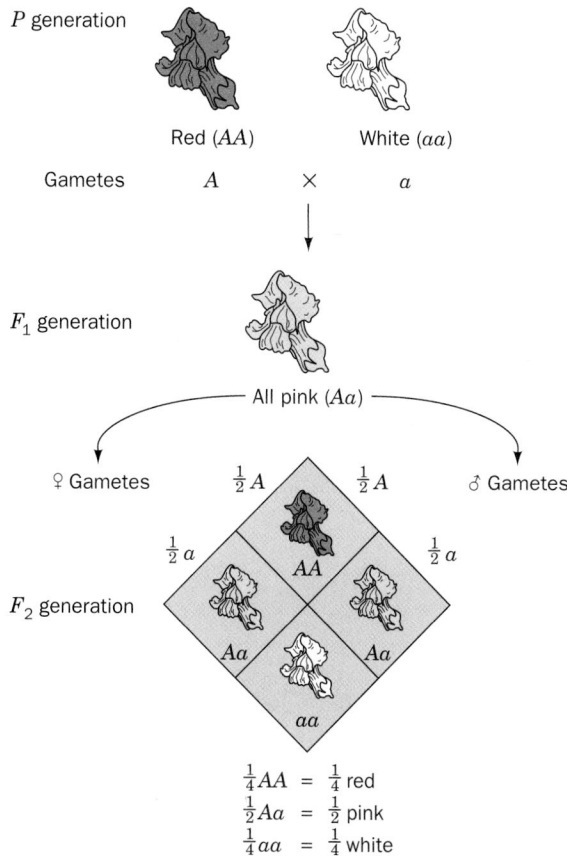

FIGURE 1-24 **Codominance.** In a cross between snapdragons with red (AA) and white (aa) flowers, the F_1 generation is pink (Aa), which demonstrates that the two alleles, A and a, are codominant. The F_2 flowers are red, pink, and white in a 1:2:1 ratio.

*contain two copies of the **X chromosome** (XX), whereas male cells contain one copy of X and a morphologically distinct **Y chromosome** (XY; Fig. 1-25).* Ova must therefore contain a single X chromosome, and sperm contain either an X or a Y chromosome (Fig. 1-25). Fertilization by an X-bearing sperm therefore results in a female zygote and by a Y-bearing sperm yields a male zygote. This explains the observed 1:1 ratio of males to females in most species. The X and Y chromosomes are referred to as **sex chromosomes;** the others are known as **autosomes.**

a. Fruit Flys Are Favorite Genetic Subjects

The pace of genetic research greatly accelerated after Thomas Hunt Morgan began using the fruit fly *Drosophila melanogaster* as an experimental subject. This small prolific insect (Fig. 1-26), which is often seen hovering around ripe fruit in summer and fall, is easily maintained in the laboratory, where it produces a new generation every 14 days. With *Drosophila,* the results of genetic crosses can be determined some 25 times faster than they can with peas. *Drosophila* is presently the genetically best characterized higher organism.

The first known mutant strain of *Drosophila* had white eyes rather than the red eyes of the **wild type** (occurring

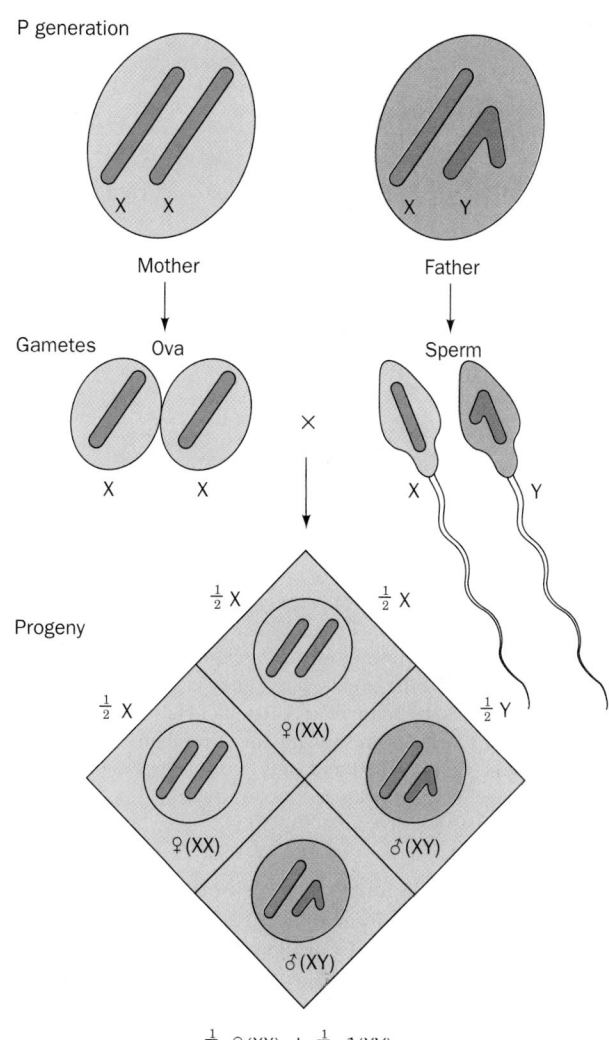

FIGURE 1-25 **Independent segregation.** The independent segregation of the sex chromosomes, X and Y, results in a 1:1 ratio of males to females.

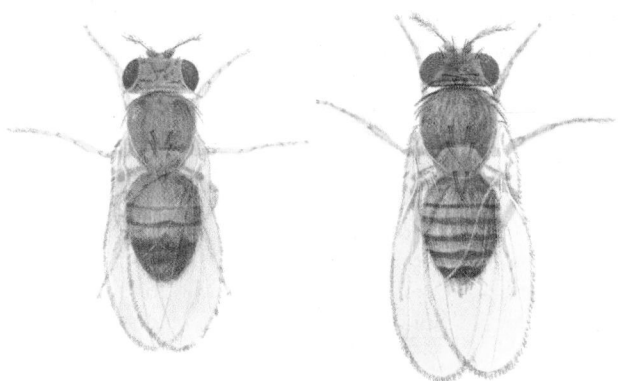

FIGURE 1-26 **The fruit fly *Drosophila melanogaster.*** The male (*left*) and the female (*right*) are shown in their relative sizes; they are actually ~2 mm long and weigh ~1 mg.

(a)

FIGURE 1-27   Crossing-over. *(a)* An electron micrograph, together with an interpretive drawing, of two homologous pairs of chromatids during meiosis in the grasshopper *Chorthippus parallelus*. Nonsister chromatids *(different colors)* may recombine at any of the points where they cross over. [Courtesy of Bernard John, The Australian National University.] *(b)* A diagram showing the recombination of pairs of allelic genes (*A*, *B*) and (*a*, *b*) during crossover.

(b)

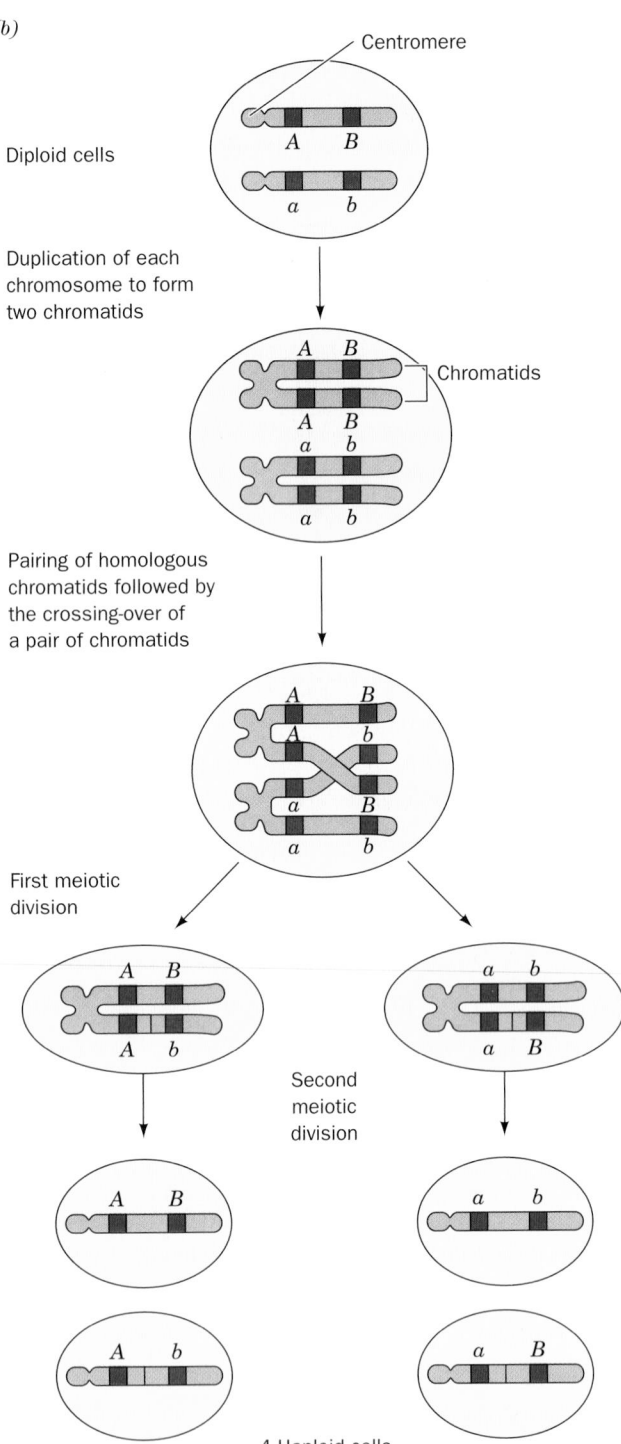

in nature). Through genetic crosses of the white eye strain with the wild type, Morgan showed that the distribution of the white eye gene (*wh*) parallels that of the X chromosome. This indicates that the *wh* gene is located on the X chromosome and that the Y chromosome does not contain it. The *wh* gene is therefore said to be **sex linked.**

b. Genetic Maps Can Be Constructed from an Analysis of Crossover Rates

In succeeding years, the chromosomal locations of many *Drosophila* genes were determined. Those genes that reside on the same chromosome do not assort independently. However, any pair of such **linked** genes **recombine** (exchange relative positions with their allelic counterparts on the homologous chromosome) with a characteristic frequency. The cytological basis of this phenomenon was found to occur at the start of meiosis when the homologous doubled chromosomes line up in parallel (metaphase I; Fig. 1-20). Homologous chromatids then exchange equivalent sections by crossing-over (Fig. 1-27). The chromosomal location of the crossover point varies nearly randomly from event to event. Consequently, *the crossover frequency of a pair of linked genes varies directly with their physical separation along the chromosome.* Morgan and Alfred Sturtevant made use of this phenomenon to **map** (locate) the relative positions of genes on *Drosophila*'s four unique chromosomes. Such studies have demonstrated that *chromosomes are linear unbranched structures.* We now know that such **genetic maps** (Fig. 1-28) parallel the corresponding base sequences of the DNA within the chromosomes.

c. Nonallelic Genes Complement One Another

Whether or not two recessive traits that affect similar functions are allelic (different forms of the same gene) can be determined by a **complementation test.** In this test, a homozygote for one of the traits is crossed with a homozygote for the other. If the two traits are nonallelic, the progeny will have the wild-type phenotype because each of the homologous chromosomes supplies the wild-type function that the other lacks; that is, they complement each

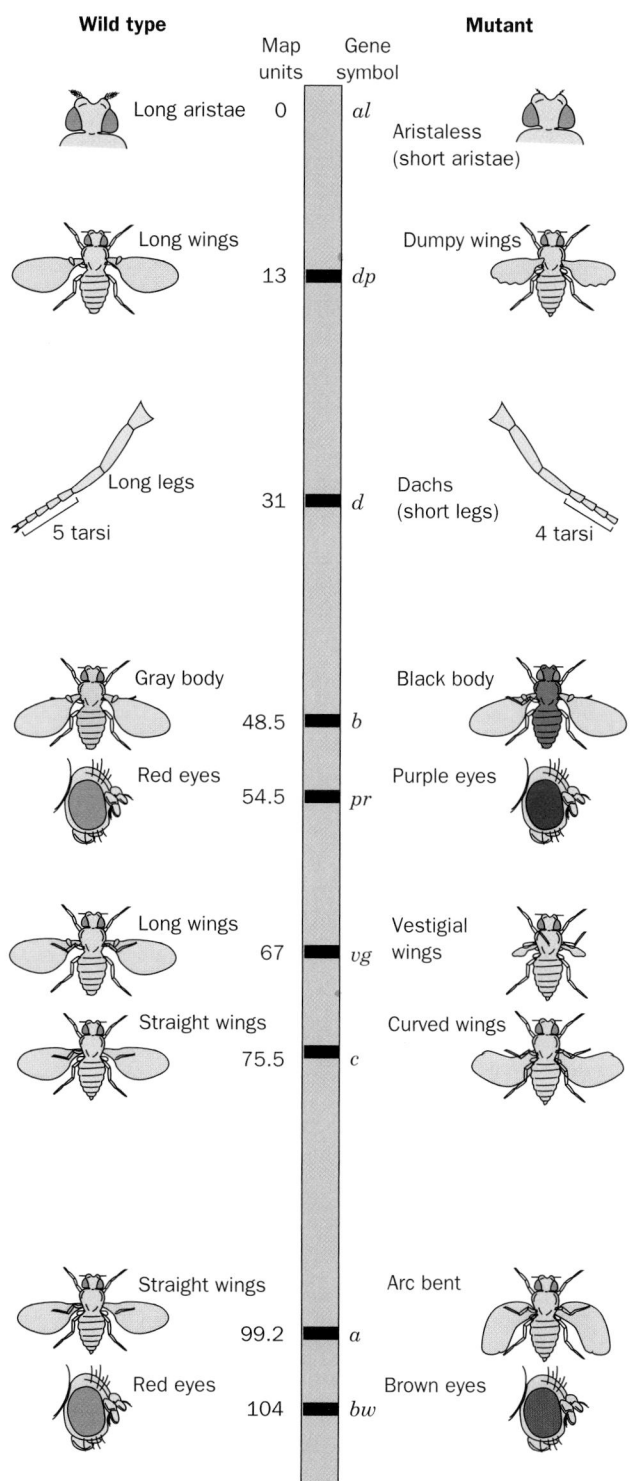

Wild type — Map units — Gene symbol — **Mutant**

Wild type	Map units	Gene symbol	Mutant
Long aristae	0	*al*	Aristaless (short aristae)
Long wings	13	*dp*	Dumpy wings
Long legs 5 tarsi	31	*d*	Dachs (short legs) 4 tarsi
Gray body	48.5	*b*	Black body
Red eyes	54.5	*pr*	Purple eyes
Long wings	67	*vg*	Vestigial wings
Straight wings	75.5	*c*	Curved wings
Straight wings	99.2	*a*	Arc bent
Red eyes	104	*bw*	Brown eyes

FIGURE 1-28 Portion of the genetic map of chromosome 2 of *Drosophila*. The positions of the genes are given in map units. Two genes separated by *m* map units recombine with a frequency of *m*%.

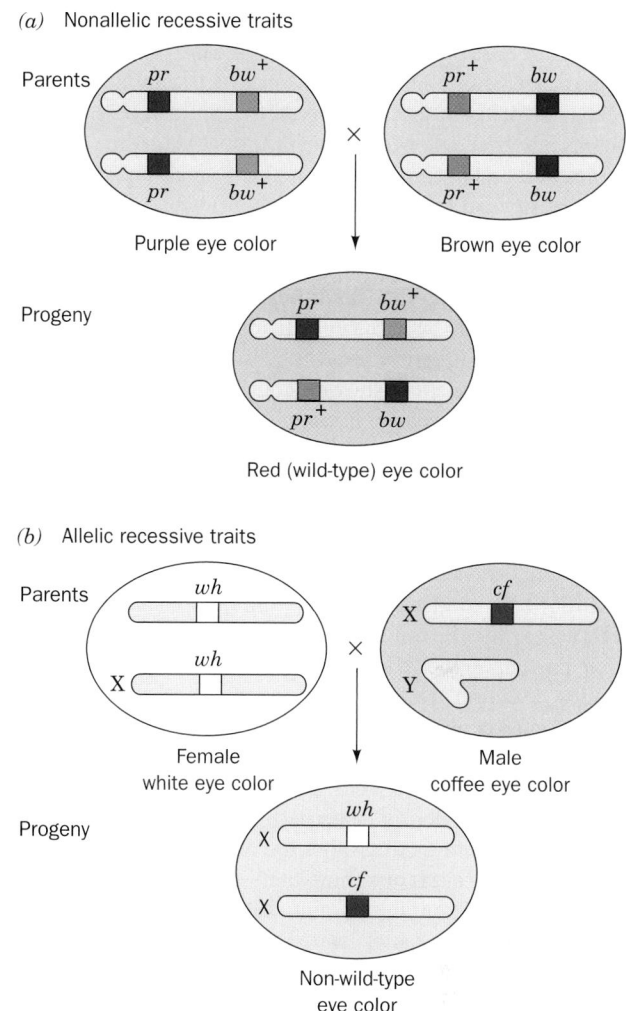

(a) Nonallelic recessive traits

Parents — Purple eye color × Brown eye color

Progeny — Red (wild-type) eye color

(b) Allelic recessive traits

Parents — Female white eye color × Male coffee eye color

Progeny — Non-wild-type eye color

FIGURE 1-29 The complementation test indicates whether two recessive traits are allelic. Two examples in *Drosophila* are shown. (*a*) Crossing a homozygote for purple eye color (*pr*) with a homozygote for brown eye color (*bw*) yields progeny with wild-type eye color. This indicates that *pr* and *bw* are nonallelic. Here the superscript "+" indicates the wild-type allele. (*b*) In crossing a female that is homozygous for the sex-linked white eye color gene *wh* with a male bearing the sex-linked coffee eye color gene *cf*, the female progeny do not have the wild-type eye color. The *wh* and *cf* genes must therefore be allelic.

other. For example, crossing a *Drosophila* that is homozygous for an eye color mutation known as purple (*pr*) with a homozygote for another eye color mutation known as brown (*bw*) yields progeny with wild-type eye color,

thereby demonstrating that these two genes are not allelic (Fig. 1-29*a*). In contrast, in crossing a female *Drosophila* that is homozygous for the sex-linked white eye color allele (*wh*) with a male carrying the sex-linked coffee eye color allele (*cf*), the female progeny do not have wild-type eye color (Fig. 1-29*b*). The *wh* and *cf* genes must therefore be allelic.

d. Genes Direct Protein Expression

The question of how genes control the characteristics of organisms took some time to be answered. Archibald Garrod was the first to suggest a specific connection between genes and enzymes. Individuals with **alkaptonuria** produce urine that darkens alarmingly on exposure to air, a consequence of the oxidation of the **homogentisic acid** they excrete (Section 16-3A). In 1902, Garrod showed that this rather benign metabolic disorder (its only adverse effect is arthritis in later life) results from a recessive trait that is inherited in a Mendelian fashion. He further demonstrated that alkaptonurics are unable to metabolize the homogentisic acid fed to them and therefore concluded that *they lack an enzyme that metabolizes this substance.* Garrod described alkaptonuria and several other inherited human diseases he had studied as **inborn errors of metabolism.**

Beginning in 1940, George Beadle and Edward Tatum, in a series of investigations that mark the beginning of biochemical genetics, showed that *there is a one-to-one correspondence between a mutation and the lack of a specific enzyme.* The wild-type mold *Neurospora* grows on a "minimal medium" in which the only sources of carbon and nitrogen are glucose and NH_3. Certain mutant varieties of *Neurospora* that were generated by means of irradiation with X-rays, however, require an additional substance in order to grow. Beadle and Tatum demonstrated, in several cases, that the mutants lack a normally present enzyme that participates in the biosynthesis of the required substance (Section 16-3A). This resulted in their famous maxim **one gene–one enzyme.** Today we know this principle to be only partially true since many genes specify proteins that are not enzymes and many proteins consist of several independently specified subunits (Section 8-5). A more accurate dictum might be **one gene–one polypeptide.** Yet even that is not completely correct because RNAs with structural and functional roles are also genetically specified.

D. *Bacterial Genetics*

Bacteria offer several advantages for genetic study. Foremost of these is that *under favorable conditions, many have generation times of under 20 min. Consequently, the results of a genetic experiment with bacteria can be ascertained in a matter of hours rather than the weeks or years required for an analogous study with higher organisms. The tremendous number of bacteria that can be quickly grown* ($\sim 10^{10}\ mL^{-1}$) *permits the observation of extremely rare biological events.* For example, an event that occurs with a frequency of 1 per million can be readily detected in bacteria with only a few minutes' work. To do so in *Drosophila* would be an enormous and probably futile effort. Moreover, bacteria are usually haploid, so their phenotype indicates their genotype. Nevertheless, the basic principles of genetics were elucidated from the study of higher plants and animals. This is because bacteria do not reproduce sexually in the manner of higher organisms, so the basic technique of classical genetics, the genetic cross, is not normally applicable to bacteria. In fact, before it was shown that DNA is the carrier of hereditary information, it was not altogether clear that bacteria had chromosomes.

The study of bacterial genetics effectively began in the 1940s when procedures were developed for isolating bacterial mutants. Since bacteria have few easily recognized morphological features, *their mutants are usually detected*

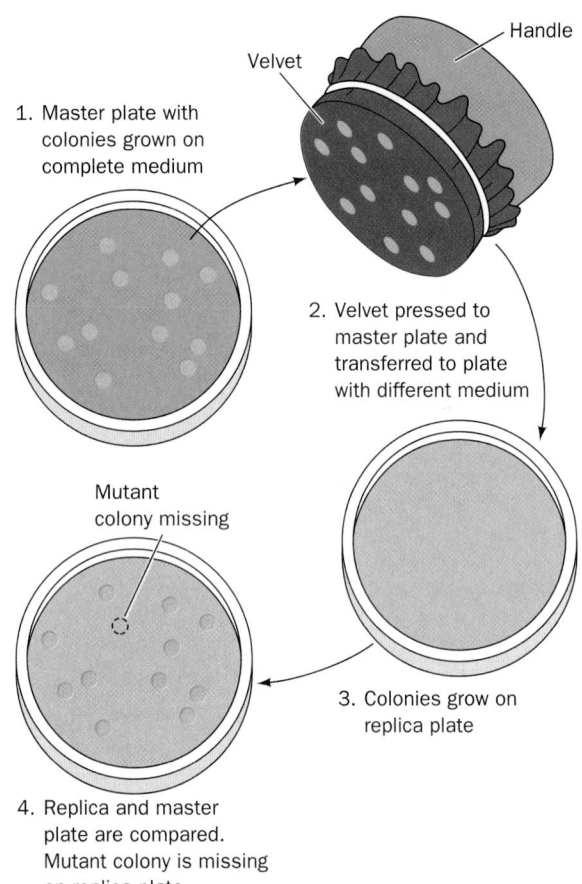

1. Master plate with colonies grown on complete medium

Velvet

Handle

2. Velvet pressed to master plate and transferred to plate with different medium

Mutant colony missing

3. Colonies grow on replica plate

4. Replica and master plate are compared. Mutant colony is missing on replica plate

FIGURE 1-30 Replica plating. A technique for rapidly and conveniently transferring colonies from a "master" culture plate (Petri dish) to a different medium on another culture plate. Since the colonies on the master plate and on the replicas should have the same spatial distribution, it is easy to identify the desired mutants.

(selected for) by their ability or inability to grow under certain conditions. For example, wild-type *E. coli* can grow on a medium in which glucose is the only carbon source. Mutants that are unable to synthesize the amino acid **leucine,** for instance, require the presence of leucine in their growth media. Mutants that are resistant to an antibiotic, say **ampicillin,** can grow in the presence of that antibiotic, whereas the wild type cannot. Mutants in which an essential protein has become temperature sensitive grow at 30 but not at 42°C, whereas the wild type grows at either temperature. By using a suitable screening protocol, a bacterial colony containing a particular mutation or combination of mutations can be selected. This is conveniently done by the method of **replica plating** (Fig. 1-30).

E. *Viral Genetics*

Viruses are infectious particles consisting of a nucleic acid molecule enclosed by a protective **capsid** *(coat) that consists largely or entirely of protein.* A virus specifically adsorbs to a susceptible cell into which it insinuates its nucleic acid. Over the course of the infection (Fig. 1-31), the

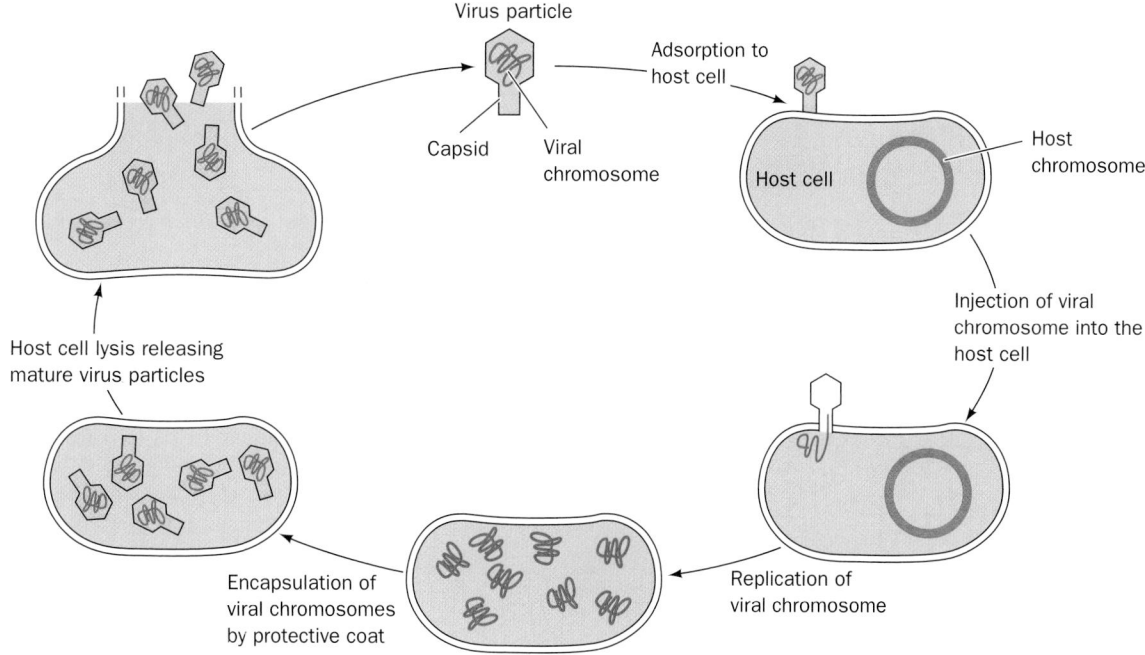

FIGURE 1-31 The life cycle of a virus.

viral chromosome redirects the cell's metabolism so as to produce new viruses. A viral infection usually culminates in the **lysis** (breaking open) of the host cell, thereby releasing large numbers (tens to thousands) of mature virus particles that can each initiate a new round of infection. Viruses, having no metabolism of their own, are the ultimate parasites. They are not living organisms since, in the absence of their host, they are as biologically inert as any other large molecule.

a. Viruses Are Subject to Complementation and Recombination

The genetics of viruses can be studied in much the same way as that of cellular organisms. Since viruses have no metabolism, however, their presence is usually detected by their ability to kill their host. The presence of viable **bacteriophages** (viruses infecting bacteria, **phages** for short; Greek: *phagein,* to eat) is conveniently indicated by **plaques** (clear spots) on a "lawn" of bacteria on a culture plate (Fig. 1-32). Plaques mark the spots where single phage particles had multiplied with the resulting lysis of the bacteria in the area. A mutant phage, which can produce progeny under certain **permissive conditions,** is detected by its inability to do so under other **restrictive conditions** in which the wild-type phage is viable. These conditions usually involve differences in the strain of the bacterial host employed or in the temperature.

Viruses are subject to complementation. Simultaneous infection of a bacterium by two different mutant varieties of a phage may yield progeny under conditions in which neither variety by itself can reproduce. If this occurs, then each mutant phage must have supplied a function that could not be supplied by the other. Each such mutation is

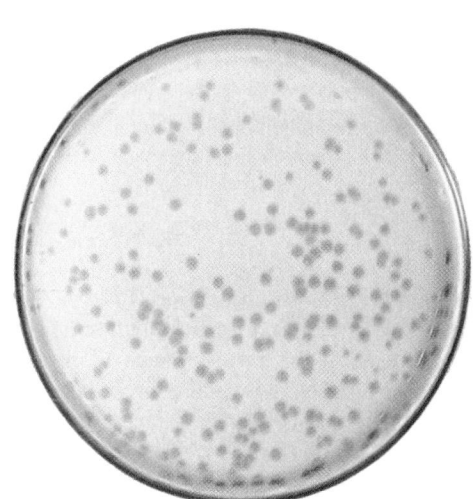

FIGURE 1-32 Screening for viral mutants. A culture plate covered with a lawn of *E. coli* on which bacteriophage T4 has formed plaques. [Bruce Iverson.]

said to belong to a different **complementation group,** a term synonymous for gene.

Viral chromosomes are also subject to recombination. This occurs when a single cell is simultaneously infected by two mutant strains of a virus (Fig. 1-33). The dynamics of viral recombination differ from those in eukaryotes or bacteria because the viral chromosome undergoes recombination throughout the several rounds of DNA replication that occur during the viral life cycle. Recombinant viral progeny therefore consist of many if not all of the possible recombinant types.

b. The Recombinational Unit Is a Base Pair

The enormous rate at which bacteriophages reproduce permits the detection of recombinational events that occur

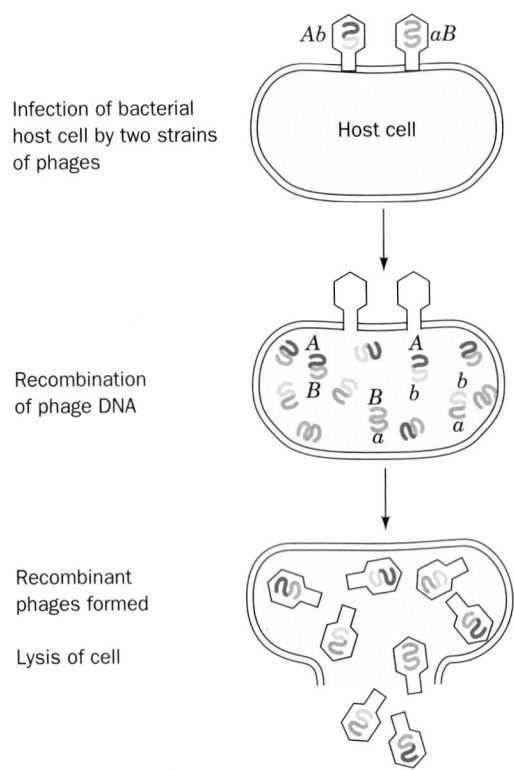

FIGURE 1-33   Viral recombination. Recombination of bacteriophage chromosomes occurs on simultaneous infection of a bacterial host by two phage strains carrying the genes *Ab* and *aB*.

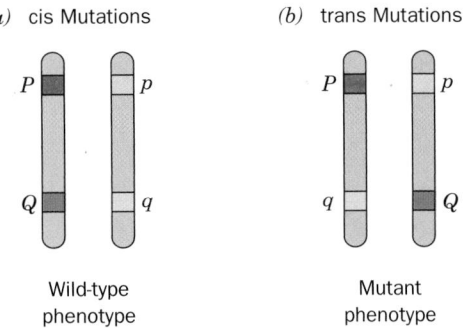

FIGURE 1-34   The cis–trans test. Consider a chromosome that is present in two copies in which two positions on the same gene, *P* and *Q*, have defective (recessive) mutants, *p* and *q*, respectively. (*a*) If the two mutations are cis (physically on the same chromosome), one gene will be wild type, so the organism will have a wild-type phenotype. (*b*) If the mutations are trans (on physically different chromosomes), both genes will be defective and the organism will have a mutant phenotype.

with a frequency of as little as 1 in 10^8. In the 1950s, Seymour Benzer carried out high-resolution genetic studies of the *rII* region of the **bacteriophage T4** chromosome. This ~4000 base pair region, which represents ~2% of the T4 chromosome, consists of two adjacent complementation groups designated *rIIA* and *rIIB*. In a permissive host, *E. coli* B, a mutation that inactivates the product of either gene, causes the formation of plaques that are easily identified because they are much larger than those of the wild-type phage (the designation *r* stands for rapid lysis). However, only the wild-type phage will lyse the restrictive host, *E. coli* K12(λ). The presence of plaques in an *E. coli* K12(λ) culture plate that had been simultaneously infected with two different *rII* mutants in the same complementation group demonstrated that *recombination can take place within a gene*. This refuted a then widely held model of the chromosome in which genes were thought to be discrete entities, rather like beads on a string, such that recombination could take place only between intact genes. The genetic mapping of mutations at over 300 distinguishable sites in the *rIIA* and *rIIB* regions indicated that *genes, as are chromosomes, are linear unbranched structures*.

Benzer also demonstrated that a complementation test between two mutations in the same complementation group yields progeny in the restrictive host when the two mutations are in the **cis** configuration (on the same chromosome; Fig. 1-34*a*), but fails to do so when they are in the **trans** configuration (on physically different chromo-

somes; Fig. 1-34*b*). This is because only when both mutations physically occur in the same gene will the other gene be functionally intact. The term **cistron** was coined to mean a functional genetic unit defined according to this **cis–trans test.** This word has since become synonymous with gene or complementation group.

The recombination of pairs of *rII* mutants was observed to occur at frequencies as low as 0.01% (although frequencies as low as 0.0001% could, in principle, have been detected). Since a recombination frequency in T4 of 1% corresponds to a 240-bp separation of mutation sites, the unit of recombination can be no larger than $0.01 \times 240 = 2.4$ bp. For reasons having to do with the mechanism of recombination, this is an upper-limit estimate. On the basis of high-resolution genetic mapping, it was therefore concluded that *the unit of recombination is about the size of a single base pair*.

5 ■ THE ORIGIN OF LIFE

People have always pondered the riddle of their existence. Indeed, all known cultures, past and present, primitive and sophisticated, have some sort of a creation myth that rationalizes how life arose. Only in the modern era, however, has it been possible to consider the origin of life in terms of a scientific framework, that is, in a manner subject to experimental verification. One of the first to do so was Charles Darwin, the originator of the theory of evolution. In 1871, he wrote in a letter to a colleague:

> *It is often said that all the conditions for the first production of a living organism are now present, which could ever have been present. But if (and oh what a big if) we could conceive in some warm little pond, with all sorts of ammonia and phosphoric salts, light, heat, electricity, etc., present, that a protein compound was chemically formed ready to undergo still more complex changes, at the present day such matter*

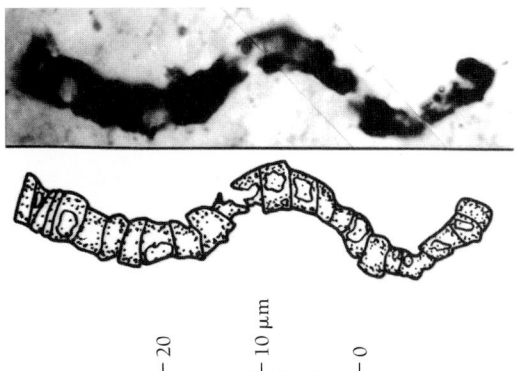

FIGURE 1-35 Microfossil of filamentous bacterial cells.
This fossil (shown with its interpretive drawing) is from ~3400 million-year-old rock from western Australia. [Courtesy of J. William Schopf, UCLA.]

would be instantly devoured, or absorbed, which would not have been the case before living creatures were formed.

Radioactive dating studies indicate that Earth formed ~4.6 billion years ago but, due to the impacts of numerous large objects, its surface remained too hot to support life for several hundred million years thereafter. The earliest known fossil evidence of life, which was generated by organisms resembling modern bacteria (Fig. 1-35), is ~3.5 billion years old. However, the oldest known sedimentary rocks on Earth, which are ~3.8 billion years old, have been subject to such extensive metamorphic forces (500°C and 5000 atm) that any microfossils they contained would have been obliterated. Nevertheless, geochemical analysis indicates (although not without dispute) that these rocks contain carbonaceous inclusions that are likely to be of biological origin and hence that life must have existed at the time these sedimentary rocks were laid down. If so, life on Earth must have arisen within a window of as little as a hundred million years that opened up ~4 billion years ago.

Since the prebiotic era left no direct record, *we cannot hope to determine exactly how life arose. Through laboratory experimentation, however, we can at least demonstrate what sorts of abiotic chemical reactions may have led to the formation of a living system.* Moreover, we are not entirely without traces of prebiotic development. The underlying biochemical and genetic unity of modern organisms suggests that life as we know it arose but once (if life arose more than once, the other forms must have rapidly died out, possibly because they were "eaten" by the present form). Thus, by comparing the corresponding genetic messages of a wide variety of modern organisms it may be possible to derive reasonable models of the primordial messages from which they have descended.

It is generally accepted that the development of life occupied three stages (Fig. 1-36):

1. Chemical evolution, in which simple geologically occurring molecules reacted to form complex organic polymers.

2. The self-organization of collections of these polymers to form replicating entities. At some point in this process, the transition from a lifeless collection of reacting molecules to a living system occurred.

3. Biological evolution to ultimately form the complex web of modern life.

In this section, we outline what has been surmised about these processes. We precede this discussion by a consideration of why only carbon, of all the elements, is suitable as the basis of the complex chemistry required for life.

A. *The Unique Properties of Carbon*

Living matter, as Table 1-3 indicates, consists of a relatively small number of elements. C, H, O, N, P, and S, all of which readily form covalent bonds, comprise 92% of the dry weight of living things (most organisms are ~70% water). The balance consists of elements that are mainly present as ions and for the most part occur only in trace quantities (they usually carry out their functions at the active sites of enzymes). Note, however, that there is no known biological requirement for 65 of the 90 naturally occurring elements. Conversely, with the exceptions of oxygen and calcium, the biologically most abundant elements are but minor constituents of Earth's crust (the most abundant components of which are O, 47%; Si, 28%; Al, 7.9%; Fe, 4.5%; and Ca, 3.5%).

The predominance of carbon in living matter is no doubt a result of its tremendous chemical versatility compared with all the other elements. Carbon has the unique ability to form a virtually infinite number of compounds as a result of its capacity to make as many as four highly stable covalent bonds (including single, double, and triple bonds) combined with its ability to form covalently linked C—C chains of unlimited extent. Thus, of the over 17 million chemical

TABLE 1-3 Elemental Composition of the Human Body

Element	Dry Weight (%)[a]	Elements Present in Trace Amounts
C	61.7	B
N	11.0	F
O	9.3	Si
H	5.7	V
Ca	5.0	Cr
P	3.3	Mn
K	1.3	Fe
S	1.0	Co
Cl	0.7	Cu
Na	0.7	Zn
Mg	0.3	Se
		Mo
		Sn
		I

[a] Calculated from Frieden, E., *Sci. Am.* **227**(1), 54–55 (1972).

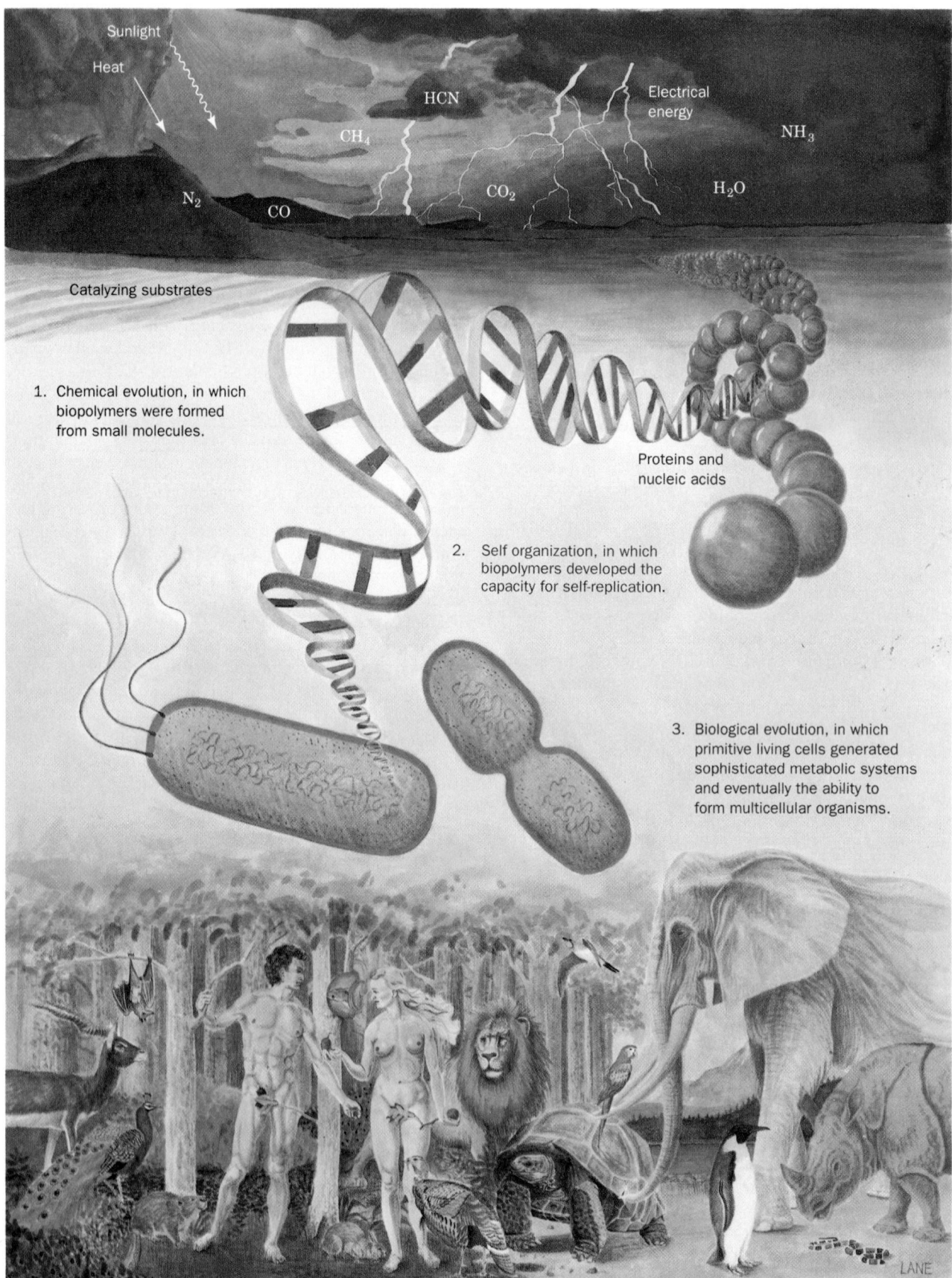

FIGURE 1-36 **The three stages in the evolution of life.**

compounds that are presently known, nearly 90% are organic (carbon-containing) substances. Let us examine the other elements in the periodic table to ascertain why they lack these combined properties.

Only five elements, B, C, N, Si, and P, have the capacity to make three or more bonds each and thus to form chains of covalently linked atoms that can also have pendant side chains. The other elements are either metals, which tend to form ions rather than covalent bonds; noble gases, which are essentially chemically inert; or atoms such as H or O that can each make only one or two covalent bonds. However, although B, N, Si, and P can each participate in at least three covalent bonds, they are, for reasons indicated below, unsuitable as the basis of a complex chemistry.

Boron, having fewer valence electrons (3) than valence orbitals (4), is electron deficient. This severely limits the types and stabilities of compounds that B can form. Nitrogen has the opposite problem; its 5 valence electrons make it electron rich. The repulsions between the lone pairs of electrons on covalently bonded N atoms serve to greatly reduce the bond energy of an N—N bond ($171 \, kJ \cdot mol^{-1}$ vs $348 \, kJ \cdot mol^{-1}$ for a C—C single bond) relative to the unusually stable triple bond of the N_2 molecule ($946 \, kJ \cdot mol^{-1}$). Even short chains of covalently bonded N atoms therefore tend to decompose, usually violently, to N_2. Silicon and carbon, being in the same column of the periodic table, might be expected to have similar chemistries. Silicon's large atomic radius, however, prevents two Si atoms from approaching each other closely enough to gain much orbital overlap. Consequently Si—Si bonds are weak ($177 \, kJ \cdot mol^{-1}$) and the corresponding multiple bonds are rarely stable. Si—O bonds, in contrast, are so stable ($369 \, kJ \cdot mol^{-1}$) that chains of alternating Si and O atoms are essentially inert (silicate minerals, whose frameworks consist of such bonds, form Earth's crust). Science fiction writers have speculated that **silicones,** which are oily or rubbery organosilicon compounds with backbones of linked Si—O units, for example, **methyl silicones,**

$$\cdots - \underset{\underset{CH_3}{|}}{\overset{\overset{CH_3}{|}}{Si}} - O - \underset{\underset{CH_3}{|}}{\overset{\overset{CH_3}{|}}{Si}} - O - \underset{\underset{CH_3}{|}}{\overset{\overset{CH_3}{|}}{Si}} - O - \underset{\underset{CH_3}{|}}{\overset{\overset{CH_3}{|}}{Si}} - O - \cdots$$

could form the chemical basis of extraterrestrial life-forms. Yet the very inertness of the Si—O bond makes this seem unlikely. Phosphorus, being below N in the periodic table, forms even less stable chains of covalently bonded atoms.

The foregoing does not imply that heteronuclear bonds are unstable. On the contrary, proteins contain C—N—C linkages, carbohydrates have C—O—C linkages, and nucleic acids possess C—O—P—O—C linkages. However, *these heteronuclear linkages are less stable than are C—C bonds. Indeed, they usually form the sites of chemical cleavage in the degradation of macromolecules and, conversely, are the bonds formed when monomer units are linked together to form macromolecules.* In the same vein, homonuclear linkages other than C—C bonds are so reactive that

they are, with the exception of S—S bonds in proteins, extremely rare in biological systems.

B. *Chemical Evolution*

In the remainder of this section, we describe the most widely favored scenario for the origin of life. *Keep in mind, however, that there are valid scientific objections to this scenario as well as to the several others that have been seriously entertained, so that we are far from certain as to how life arose.*

The solar system is thought to have formed by the gravitationally induced collapse of a large interstellar cloud of dust and gas. The major portion of this cloud, which was composed mostly of hydrogen and helium, condensed to form the sun. The rising temperature and pressure at the center of the protosun eventually ignited the self-sustaining thermonuclear reaction that has since served as the sun's energy source. The planets, which formed from smaller clumps of dust, were not massive enough to support such a process. In fact the smaller planets, including Earth, consist of mostly heavier elements because their masses are too small to gravitationally retain much H_2 and He.

The primordial Earth's atmosphere was quite different from what it is today. It could not have contained significant quantities of O_2, a highly reactive substance. Rather, in addition to the H_2O, N_2, and CO_2 that it presently has, the atmosphere probably contained smaller amounts of CO, CH_4, NH_3, SO_2, and possibly H_2, all molecules that have been spectroscopically detected in interstellar space. The chemical properties of such a gas mixture make it a **reducing atmosphere** in contrast to Earth's present atmosphere, which is an **oxidizing atmosphere** (although recent contradictory evidence suggests that the primordial Earth had an oxidizing atmosphere).

In the 1920s, Alexander Oparin and J. B. S. Haldane independently suggested that *ultraviolet (UV) radiation from the sun [which is presently largely absorbed by an ozone (O_3) layer high in the atmosphere] or lightning discharges caused the molecules of the primordial reducing atmosphere to react to form simple organic compounds such as amino acids, nucleic acid bases, and sugars.* That this process is possible was first experimentally demonstrated in 1953 by Stanley Miller and Harold Urey, who, in the apparatus diagrammed in Fig. 1-37, simulated effects of lightning storms in the primordial atmosphere by subjecting a refluxing mixture of H_2O, CH_4, NH_3, and H_2 to an electric discharge for about a week. The resulting solution contained significant amounts of water-soluble organic compounds, the most abundant of which are listed in Table 1-4, together with a substantial quantity of insoluble tar (polymerized material). Several of the soluble compounds are amino acid components of proteins, and many of the others, as we shall see, are also of biochemical significance. Similar experiments in which the reaction conditions, the gas mixture, and/or the energy source were varied have resulted in the synthesis of many other amino acids. This, together with the observation that carbonaceous meteorites

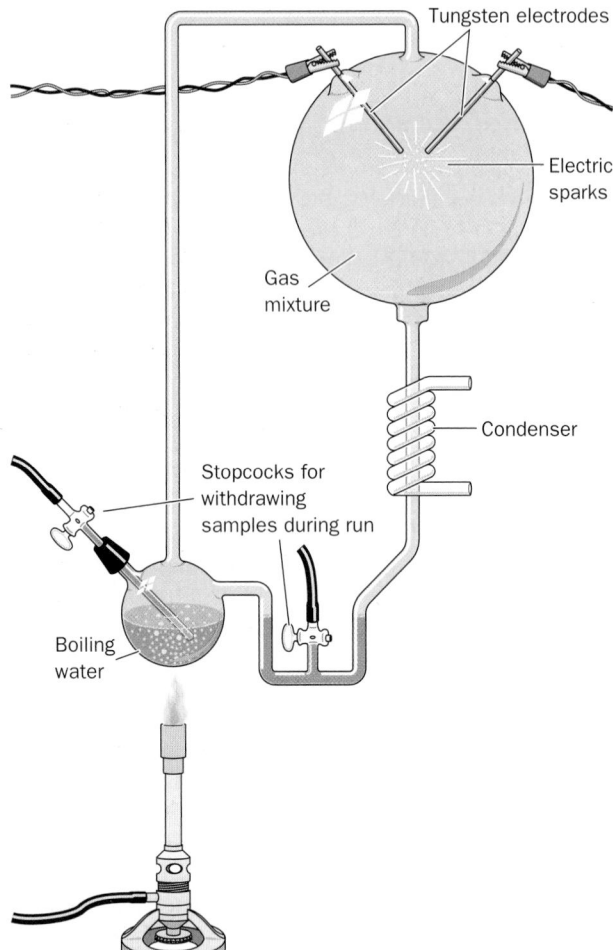

FIGURE 1-37 Apparatus for emulating the synthesis of organic compounds on the prebiotic Earth. A mixture of gases thought to resemble the primitive Earth's reducing atmosphere is subjected to an electric discharge, to simulate the effects of lightning, while the water in the flask is refluxed so that the newly formed compounds dissolve in the water and accumulate in the flask. [After Miller, S.L., and Orgel, L.E., *The Origins of Life on Earth*, p. 84, Prentice-Hall (1974).]

TABLE 1-4 Yields from Sparking a Mixture of CH$_4$, NH$_3$, H$_2$O, and H$_2$

Compound	Yield (%)
Glycine[a]	2.1
Glycolic acid	1.9
Sarcosine	0.25
Alanine[a]	1.7
Lactic acid	1.6
N-Methylalanine	0.07
α-Amino-n-butyric acid	0.34
α-Aminoisobutyric acid	0.007
α-Hydroxybutyric acid	0.34
β-Alanine	0.76
Succinic acid	0.27
Aspartic acid[a]	0.024
Glutamic acid[a]	0.051
Iminodiacetic acid	0.37
Iminoaceticpropionic acid	0.13
Formic acid	4.0
Acetic acid	0.51
Propionic acid	0.66
Urea	0.034
N-Methylurea	0.051

[a] Amino acid constituent of proteins.

Source: Miller, S.J. and Orgel, L.E., *The Origins of Life on Earth*, p. 85, Prentice-Hall (1974).

contain many of the same amino acids, strongly suggests that these substances were present in significant quantities on the primordial Earth. Indeed, it seems likely that large quantities of organic molecules were delivered to the primordial Earth by the meteorites and dust that so heavily bombarded it.

Nucleic acid bases can also be synthesized under supposed prebiotic conditions. In particular, adenine is formed by the condensation of HCN, a plentiful component of the prebiotic atmosphere, in a reaction catalyzed by NH$_3$ [note that the chemical formula of adenine is (HCN)$_5$]. The other bases have been synthesized in similar reactions involving HCN and H$_2$O. Sugars have been synthesized by the polymerization of formaldehyde (CH$_2$O) in reactions catalyzed by divalent cations, alumina, or clays. It is probably no accident that these compounds are the basic components of biological molecules. *They were apparently the most common organic substances in prebiotic times.*

The above described prebiotic reactions probably occurred over a period of hundreds of millions of years. Ultimately, it has been estimated, the oceans attained the organic consistency of a thin bouillon soup. Of course there must have been numerous places, such as tidal pools and shallow lakes, where the prebiotic soup became much more concentrated. In such environments its component organic molecules could have condensed to form, for example, polypeptides and polynucleotides (nucleic acids). Quite possibly these reactions were catalyzed by the adsorption of the reactants on minerals such as clays. If, however, life were to have formed, the rates of synthesis of these complex polymers would have had to be greater than their rates of hydrolysis. Therefore, the "pond" in which life arose may have been cold rather than warm, possibly even below 0°C (seawater freezes solidly only below −21°C), since hydrolysis reactions are greatly retarded at such low temperatures.

C. *The Rise of Living Systems*

Living systems have the ability to replicate themselves. The inherent complexity of such a process is such that no man-made device has even approached having this capacity.

Clearly there is but an infinitesimal probability that a collection of molecules can simply gather at random to form a living entity (the likelihood of a living cell forming spontaneously from simple organic molecules has been said to be comparable to that of a modern jet aircraft being assembled by a tornado passing through a junkyard). How then did life arise? The answer, most probably, is that it was guided according to the Darwinian principle of the survival of the fittest as it applies at the molecular level.

a. Life Probably Arose through the Development of Self-Replicating RNA Molecules

The primordial self-replicating system is widely believed to have been a collection of nucleic acid molecules because such molecules, as we have seen in Section 1-3C, can direct the synthesis of molecules complementary to themselves. RNA, as does DNA, can direct the synthesis of a complementary strand. In fact, RNA serves as the hereditary material of many viruses (Chapter 33). The polymerization of the progeny molecules would, at first, have been a simple chemical process and hence could hardly be expected to be accurate. The early progeny molecules would therefore have been only approximately complementary to their parents. Nevertheless, repeated cycles of nucleic acid synthesis would eventually exhaust the supply of free nucleotides so that the synthesis rate of new nucleic acid molecules would be ultimately limited by the hydrolytic degradation rate of old ones. Suppose, in this process, a nucleic acid molecule randomly arose that, through folding, was more resistant to degradation than its cousins. The progeny of this molecule, or at least its more faithful copies, could then propagate at the expense of the nonresistant molecules; that is, the resistant molecules would have a Darwinian advantage over their fellows. Theoretical studies suggest that such a system of molecules would evolve so as to optimize its replication efficiency under its inherent physical and chemical limitations.

In the next stage of the evolution of life, it is thought the dominant nucleic acids evolved the capacity to influence the efficiency and accuracy of their own replication. This process occurs in living systems through the nucleic acid–directed ribosomal synthesis of enzymes that catalyze nucleic acid synthesis. How nucleic acid–directed protein synthesis could have occurred before ribosomes arose is unknown because nucleic acids are not known to interact selectively with particular amino acids. This difficulty exemplifies the major problem in tracing the pathway of prebiotic evolution. Suppose some sort of rudimentary nucleic acid–influenced system arose that increased the efficiency of nucleic acid replication. This system must have eventually been replaced, presumably with almost no trace of its existence, by the much more efficient ribosomal system. Our hypothetical nucleic acid synthesis system is therefore analogous to the scaffolding used in the construction of a building. After the building has been erected the scaffolding is removed, leaving no physical evidence that it ever was there. *Most of the statements in this section must therefore be taken as educated guesses.* Without our having witnessed the event, it seems unlikely that we shall ever be certain of how life arose.

A plausible hypothesis for the evolution of self-replicating systems is that they initially consisted entirely of RNA, a scenario known as the "RNA world." This idea is based, in part, on the observation that certain species of RNA exhibit enzymelike catalytic properties (Section 31-4A). Moreover, since ribosomes are approximately two-thirds RNA and only one-third protein, it is plausible that the primordial ribosomes were entirely RNA. A cooperative relationship between RNA and protein might have arisen when these self-replicating proto-ribosomes evolved the ability to influence the synthesis of proteins that increased the efficiency and/or the accuracy of RNA synthesis. *From this point of view, RNA is the primary substance of life; the participation of DNA and proteins were later refinements that increased the Darwinian fitness of an already established self-replicating system.*

The types of systems that we have so far described were bounded only by the primordial "pond." A self-replicating system that developed a more efficient component would therefore have to share its benefits with all the "inhabitants" of the "pond," a situation that minimizes the improvement's selective advantage. Only through compartmentalization, that is, the generation of cells, could developing biological systems reap the benefits of any improvements that they might have acquired. Of course, cell formation would also hold together and protect any self-replicating system and therefore help it spread beyond its "pond" of origin. Indeed, the importance of compartmentalization is such that it may have preceded the development of self-replicating systems. The erection of cell boundaries is not without its price, however. Cells, as we shall see in later chapters, must expend much of their metabolic effort in selectively transporting substances across their cell membranes. How cell boundaries first arose, or even what they were made from, is presently unknown. However, one plausible theory holds that membranes first arose as empty vesicles whose exteriors could serve as attachment sites for such entities as enzymes and chromosomes in ways that facilitated their function. Evolution then flattened and folded these vesicles so that they enclosed their associated molecular assemblies, thereby defining the primordial cells.

b. Competition for Energy Resources Led to the Development of Metabolic Pathways, Photosynthesis, and Respiration

At this stage in their development, the entities we have been describing already fit Horowitz's criteria for life (exhibiting replication, catalysis, and mutability). The polymerization reactions through which these primitive organisms replicated were entirely dependent on the environment to supply the necessary monomeric units and the energy-rich compounds such as ATP or, more likely, just polyphosphates, that powered these reactions. As some of the essential components in the prebiotic soup became scarce, organisms developed the enzymatic systems that could synthesize these substances from simpler but more abundant precursors. As a consequence, energy-producing metabolic pathways arose. This latter development only postponed an "energy crisis," however, because these pathways consumed other preexisting energy-rich substances. The increasing scarcity of all such substances ultimately stimulated the development of photosynthesis to take advantage of a practically inexhaustible energy supply, the sun. Yet this process, as we saw in Section 1-1A, consumes reducing agents such as H_2S. The eventual exhaustion of these substances led to the refinement of the photosynthetic process so that it used the ubiquitous H_2O as its reducing agent, thereby yielding O_2 as a by-product. The recent discovery, in ~3.5-billion-year-old rocks, of fossilized cyanobacteria-like microorganisms strongly suggests that oxygen-producing photosynthesis developed very early in the history of life.

The development of oxygen-producing photosynthesis led to yet another problem. The accumulation of the highly reactive O_2, which over the eons converted the reducing atmosphere of the prebiotic Earth to the modern oxidizing atmosphere (21% O_2), eventually interfered with the existing metabolic apparatus, which had evolved to operate under reducing conditions. The O_2 accumulation therefore stimulated the development of metabolic refinements that protected organisms from oxidative damage. More importantly, it led to the evolution of a much more efficient form of energy metabolism than had previously been possible, **respiration** (oxidative metabolism), which used the newly available O_2 as an oxidizing agent.

As previously outlined, the basic replicative and metabolic apparatus of modern organisms evolved quite early in the history of life on Earth. Indeed, many modern prokaryotes appear to resemble their very ancient ancestors. The rise of eukaryotes, as Section 1-2 indicates, occurred perhaps 2 billion years after prokaryotes had become firmly established. Multicellular organisms are a relatively recent evolutionary innovation, having not appeared, according to the fossil record, until ~700 million years ago.

6 ■ THE BIOCHEMICAL LITERATURE

The biochemical literature contains the results of the work of tens of thousands of scientists extending well over a century. Consequently a biochemistry textbook can report only selected highlights of this vast amount of information. Moreover, the tremendous rate at which biochemical knowledge is presently being acquired, which is perhaps greater than that of any other intellectual endeavor, guarantees that there will have been significant biochemical advances even in the year or so that it took to produce this text from its final draft. A serious student of biochemistry must therefore regularly read the biochemical literature to flesh out the details of subjects covered in (or omitted from) this text, as well as to keep abreast of new developments. This section provides a few suggestions on how to do so.

A. *Conducting a Literature Search*

The primary literature of biochemistry, those publications that report the results of biochemical research, is presently being generated at a rate of tens of thousands of papers per year appearing in over 200 periodicals. An individual can therefore only read this voluminous literature in a highly selective fashion. Indeed, most biochemists tend to "read" only those publications that are likely to contain reports pertaining to their interests. By "read" it is meant that they scan the tables of contents of these journals for the titles of articles that seem of sufficient interest to warrant further perusal.

It is difficult to learn about a new subject by beginning with its primary literature. Instead, to obtain a general overview of a particular biochemical subject it is best to first peruse appropriate reviews and monographs. These usually present a synopsis of recent (at the time of their writing) developments in the area, often from the authors' particular point of view. There are more or less two types of reviews: those that are essentially a compilation of facts and those that critically evaluate the data and attempt to place them in some larger context. The latter type of review is of course more valuable, particularly for a novice in the field. Most reviews are published in specialized books and journals, although many journals that publish research reports also occasionally print reviews. Table 1-5 provides a list of many of the important biochemical review publications.

Monographs and reviews relevant to a subject of interest are usually easy to find through the use of a library catalog and the subject indexes of the major review publications (the chapter-end references of this text may also be helpful in this respect). An important part of any review is its reference list. It usually has previous reviews in the same or allied fields as well as indicating the most significant research reports in the area. Note the authors of these articles and the journals in which they tend to publish. When the most current reviews and research articles you have found tend to refer to the same group of earlier articles, you can be reasonably confident that your search for these earlier articles is largely complete. Finally, to familiarize yourself with the latest developments in the field, search the recent primary literature for the work of its most active research groups.

Biological Abstracts, *Chemical Abstracts*, and *Science Citation Index* are useful aids for locating references. These compendia list the articles in the many journals they cover by both author and subject (permuted title index). *Biological Abstracts* and *Chemical Abstracts* contain short English-language abstracts of the articles listed (including many of foreign-language articles). *Science Citation Index* lists all articles in a given year that cite a particular earlier paper, so it can be used to follow the developments in a field that build on a particular body of work.

Most academic libraries subscribe to World Wide Web–based reference search services such as those of

TABLE 1-5   Some Biochemical Review Publications

Accounts of Chemical Research

Advances in Enzymology and Related Areas of Molecular Biology

Advances in Protein Chemistry

Angewandte Chemie, International Edition in English[a]

Annual Review of Biochemistry

Annual Review of Biophysics and Biomolecular Structure

Annual Review of Cell and Developmental Biology

Annual Review of Genetics

Annual Review of Immunology

Annual Review of Medicine

Annual Review of Microbiology

Annual Review of Physiology

Annual Review of Plant Physiology and Plant Molecular Biology

Biochemical Journal[a]

BioEssays

Cell[a]

Chemistry and Biology

Critical Reviews in Biochemistry and Molecular Biology

Critical Reviews in Eukaryotic Gene Expression

Current Biology

Current Opinion in Biotechnology[a]

Current Opinion in Cell Biology

Current Opinion in Genetics and Development

Current Opinion in Structural Biology

Essays in Biochemistry

FASEB Journal[a]

Harvey Lectures

Journal of Biological Chemistry[a]

Methods in Enzymology

Nature[a]

Nature Reviews Molecular Cell Biology

Nature Structural Biology[a]

Proceedings of the National Academy of Sciences USA[a]

Progress in Biophysics and Molecular Biology

Progress in Nucleic Acid Research and Molecular Biology

Protein Science[a]

Quarterly Reviews of Biophysics

Science[a]

Scientific American

Structure[a]

Trends in Biochemical Sciences

[a] Periodicals that mainly publish research reports.

Chemical Abstracts Services, Current Contents, MedLine, and Science Citation Index. MedLine can also be accessed free of charge through the National Center for Biotechnology Information (NCBI) (http://www.ncbi.nlm.nih.gov/PubMed) and BioMedNet (http://www.bmn.com).

If used properly, these bibliographic search services can be highly efficient tools for locating specific information.

B. *Reading a Research Article*

Research reports more or less all have the same five-part format. They usually have a short abstract or summary located before (or, in some journals, after) the main body of the paper. The paper then continues (or begins) with an introduction, which often contains a short synopsis of the field, the motivation for the research reported, and a preview of its conclusions. The next section contains a description of the methods used to obtain the experimental data. This is followed by a presentation of the results of the investigation. Finally, there is a discussion section wherein the conclusions of the investigation are set forth and placed in the context of other work in the field. Most articles are "full papers," which may be tens of pages long. However, many journals also contain "communications," which are usually only a page or two in length and are often published more quickly than are full papers.

It is by no means obvious how to read a scientific paper. Perhaps the worst way to do so is to read it from beginning to end as if it were some kind of a short story. In fact, most practicing scientists only occasionally read a research article in its entirety. It simply takes too long and is rarely productive. Rather, they scan selected parts of a paper and only dig deeper if it appears that to do so will be profitable. The following paragraph describes a reasonably efficient scheme for reading scientific papers. *This should be an active process in which the reader is constantly evaluating what is being read and relating it to his/her previous knowledge.* Moreover, the reader should maintain a healthy skepticism since there is a reasonable probability that any paper, particularly in its interpretation of experimental data and in its speculations, may be erroneous.

If the title of a paper indicates that it may be of interest, then this should be confirmed by a reading of its abstract. For many papers, even those containing useful information, it is unnecessary to read further. If you choose to continue, it is probably best to do so by scanning the introduction so as to obtain an overview of the work reported. At this point most experienced scientists scan the conclusions section of the paper to gain a better understanding of what was found. If further effort seems warranted, they scan the results section to ascertain whether the experimental data support the conclusions. The methods section is usually not read in detail because it is often written in a condensed form that is only fully interpretable by an expert in the field. However, for such experts, the methods section may be the most valuable part of the paper. At this point, what to read next, if anything, is largely dictated by the remaining points of confusion. In many cases this confusion can only be eliminated by reading some of the references given in the paper. At any rate, unless you plan to repeat or extend some of the work described, it is rarely necessary to read an article in detail. To do so in a critical manner, you will find, takes several hours for a paper of even moderate size.

CHAPTER SUMMARY

1 ■ Prokaryotes Prokaryotes are single-celled organisms that lack a membrane-enclosed nucleus. Most prokaryotes have similar anatomies: a rigid cell wall surrounding a cell membrane that encloses the cytoplasm. The cell's single chromosome is condensed to form a nucleoid. *Escherichia coli,* the biochemically most well-characterized organism, is a typical prokaryote. Prokaryotes have quite varied nutritional requirements. The chemolithotrophs metabolize inorganic substances. Photolithotrophs, such as cyanobacteria, carry out photosynthesis. Heterotrophs, which live by oxidizing organic substances, are classified as aerobes if they use oxygen in this process and as anaerobes if some other oxidizing agent serves as their terminal electron acceptor. Traditional prokaryotic classification schemes are rather arbitrary because of poor correlation between bacterial form and metabolism. Sequence comparisons of nucleic acids and proteins, however, have established that all life-forms can be classified into three domains of evolutionary descent: the Archaea (archaebacteria), the Bacteria (eubacteria), and the Eukarya (eukaryotes).

2 ■ Eukaryotes Eukaryotic cells, which are far more complex than those of prokaryotes, are characterized by having numerous membrane-enclosed organelles. The most conspicuous of these is the nucleus, which contains the cell's chromosomes, and the nucleolus, where ribosomes are assembled. The endoplasmic reticulum is the site of synthesis of lipids and of proteins that are destined for secretion. Further processing of these products occurs in the Golgi apparatus. The mitochondria, wherein oxidative metabolism occurs, are thought to have evolved from a symbiotic relationship between an aerobic bacterium and a primitive eukaryote. The chloroplast, the site of photosynthesis in plants, similarly evolved from a cyanobacterium. Other eukaryotic organelles include the lysosome, which functions as an intracellular digestive chamber, and the peroxisome, which contains a variety of oxidative enzymes including some that generate H_2O_2. The eukaryotic cytoplasm is pervaded by a cytoskeleton whose components include microtubules, which consist of tubulin; microfilaments, which are composed of actin; and intermediate filaments, which are made of different proteins in different types of cells. Eukaryotes have enormous morphological diversity on the cellular as well as on the organismal level. They have been classified into four kingdoms: Protista, Plantae, Fungi, and Animalia. The pattern of embryonic development in multicellular organisms partially mirrors their evolutionary history.

3 ■ Biochemistry: A Prologue Organisms have a hierarchical structure that extends down to the submolecular level. They contain but three basic types of macromolecules: proteins, nucleic acids, and polysaccharides, as well as lipids, each of which is constructed from only a few different species of monomeric units. Macromolecules and supramolecular assemblies form their native biological structures through a process of self-assembly. The assembly mechanisms of higher biological structures are largely unknown. Metabolic processes are organized into a series of tightly regulated pathways. These are classified as catabolic or anabolic depending on whether they participate in degradative or biosynthetic processes. The common energy "currency" in all these processes is ATP, whose synthesis is the product of many catabolic pathways and whose hydrolysis drives most anabolic pathways. DNA, the cell's hereditary molecule, encodes genetic information in its sequence of bases. The complementary base sequences of its two strands permit them to act as templates for their own replication and for the synthesis of complementary strands of RNA. Ribosomes synthesize proteins by linking amino acids together in the order specified by the base sequences of RNAs.

4 ■ Genetics: A Review Eukaryotic cells contain a characteristic number of homologous pairs of chromosomes. In mitosis each daughter cell receives a copy of each of these chromosomes, but in meiosis each resulting gamete receives only one member of each homologous pair. Fertilization is the fusion of two haploid gametes to form a diploid zygote. The Mendelian laws of inheritance state that alternative forms of true-breeding traits are specified by different alleles of the same gene. Alleles may be dominant, codominant, or recessive depending on the phenotype of the heterozygote. Different genes assort independently unless they are on the same chromosome. The linkage between genes on the same chromosome, however, is never complete because of crossing-over among homologous chromosomes during meiosis. The rate at which genes recombine varies with their physical separation because crossing-over occurs essentially at random. This permits the construction of genetic maps. Whether two recessive traits are allelic may be determined by the complementation test. The nature of genes is largely defined by the dictum "one gene–one polypeptide." Mutant varieties of bacteriophages are detected by their ability to kill their host under various restrictive conditions. The fine structure analysis of the *rII* region of the bacteriophage T4 chromosome has revealed that recombination may take place within a gene, that genes are linear unbranched structures, and that the unit of mutation is ~1 bp.

5 ■ The Origin of Life Life is carbon based because only carbon, among all the elements in the periodic table, has a sufficiently complex chemistry together with the ability to form virtually infinite stable chains of covalently bonded atoms. Reactions among the molecules in the reducing atmosphere of the prebiotic Earth are thought to have formed the simple organic precursors from which biological molecules developed. Eventually, in reactions that may have been catalyzed by minerals such as clays, polypeptides and polynucleotides formed. These evolved under the pressure of competition for the available monomeric units. Ultimately, a nucleic acid, most probably RNA, developed the capability of influencing its own replication by directing the synthesis of proteins that catalyze polynucleotide synthesis. This was followed by the development of cell membranes so as to form living entities. Subsequently, metabolic processes evolved to synthesize necessary intermediates from available precursors as well as the high-energy compounds required to power these reactions. Likewise, photosynthesis and respiration arose in response to environmental pressures brought about by the action of living organisms.

6 ■ The Biochemical Literature The sheer size and rate of increase of the biochemical literature requires that it be read to attain a thorough understanding of any aspect of biochemistry. The review literature provides an *entrée* into a given subspeciality. To remain current in any field, however, requires a regular perusal of its primary literature. This should be read in a critical but highly selective fashion.

REFERENCES

PROKARYOTES AND EUKARYOTES

Becker, W.M., Kleinsmith, L.J., and Hardin, J., *The World of the Cell* (4th ed.), Benjamin/Cummings (2000). [A highly readable cell biology text.]

Campbell, N.A., Reece, J.B., and Mitchell, L.G., *Biology* (5th ed.) Benjamin/Cummings (1999). [A comprehensive general biology text. There are several others available of similar content.]

Dulbecco, R., *The Design of Life,* Yale University Press (1987). [An incisive introduction to modern concepts of biology and biochemistry.]

Fredrickson, J.K. and Onstott, T.C., Microbes deep inside the earth, *Sci. Am.* **274**(4), 68–73 (1996).

Frieden, E., The chemical elements of life, *Sci. Am.* **227**(1), 52–60 (1972).

Goodsell, D.S., *The Machinery of Life,* Springer-Verlag (1993); A look inside the living cell; *and* Inside a living cell, *Trends Biochem. Sci.* **16,** 203–206 (1991).

Holt, J.G., Krieg, N.R., Sneath, H.A., Staley, J.T., and Williams, S.T., *Bergey's Manual of Determinative Bacteriology* (9th ed.), Williams and Wilkins (1994).

Holtzman, E. and Novikoff, A.B., *Cells and Organelles* (3rd ed.), Holt, Rinehart, & Winston (1984).

Madigan, M.T., Martinko, J.M., and Parker, J., *Brock Biology of Microorganisms* (9th ed.), Prentice-Hall (2000).

Margulis, L. and Schwartz, K.V., *Five Kingdoms. An Illustrated Guide to the Phyla of Life on Earth* (3rd ed.), Freeman (1998).

Pace, N.R., A molecular view of microbial diversity and the biosphere, *Science* **276**, 734–740 (1997).

Stanier, R.Y., Ingrahan, J.L., Wheelis, M.L., and Painter, P.R., *The Microbial World* (5th ed.), Prentice-Hall (1986).

Whitman, W.B., Coleman, D.C., and Wiebe, W.J., Prokaryotes: The unseen majority, *Proc. Natl. Acad. Sci.* **95**, 6578–6583 (1998). [Estimates the number of prokaryotes on Earth $(4-6 \times 10^{30}$ cells) and the aggregate mass of their cellular carbon $(3.5-5.5 \times 10^{14}$ kg, which therefore comprises 66–100% of the carbon in plants).]

GENETICS

Benzer, S., The fine structure of the gene, *Sci Am.* **206**(1), 70–84 (1962).

Cairns, J., Stent, G.S., and Watson, J. (Eds.), *Phage and the Origins of Molecular Biology* (expanded ed.), Cold Spring Harbor Laboratory (1992). [A series of scientific memoirs by many of the pioneers of molecular biology.]

Russell, P.J., *Genetics* (5th ed.), Addison Wesley Longman (1998).

Snustad, D.P. and Simmons, M.J., *Principles of Genetics* (3rd ed.), Wiley (2001).

ORIGIN OF LIFE

Berstein, M.P., Sandford, S.A., and Allamandola, S.A., Life's far-flung raw materials, *Sci. Am.* **281**(1), 42–49 (1999). [A discussion of the possibility that the complex organic molecules which provided the starting materials for life were delivered to the primordial Earth by meteorites and dust.]

Brack, A. (Ed.), *The Molecular Origins of Life,* Cambridge University Press (1998).

de Duve, C., *Vital Dust. Life as a Cosmic Imperative,* Basic Books (1995).

Doolittle, F.W., Phylogenetic classification and the universal tree, *Science* **284**, 2124–2128 (1999). [A discussion of how lateral gene transfer among the various forms of life may have confounded the ability to elucidate the "universal tree of life" if, in fact, such a tree is a reasonable model of the history of life.]

Dyson, F., *Origins of Life,* Cambridge University Press (1985). [A fascinating philosophical discourse on theories of life's origins by a respected theoretical physicist.]

Fraústo da Silva, J.R. and Williams, R.J.P., *The Biological Chemistry of the Elements,* Oxford (1991).

Knoll, A.H., The early evolution of eukaryotes: A geological perspective, *Science* **256**, 622–627 (1992).

Lahav, N., *Biogenesis. Theories of Life's Origins,* Oxford University Press (1999).

Lazcano, A. and Miller, S.L., The origin and early evolution of life: Prebiotic chemistry, the pre-RNA world, and time, *Cell* **85,** 793–798 (1996).

Lifson, S., On the crucial stages in the origin of animate matter, *J. Mol. Evol.* **44,** 1–8 (1997).

Mojzsis, S.J., Arrhenius, G., McKeegan, K.D., Harrison, T.M., Nutman, A.P., and Friend, C.R.L., Evidence for life on Earth before 3,800 million years ago, *Nature* **384**, 55–57 (1996).

Orgel, L.E., The origin of life—a review of facts and speculations, *Trends Biochem. Sci.* **23**, 491–495 (1998). [Reviews the most widely accepted hypotheses on the origin of life and discusses the evidence supporting them and their difficulties.]

Schopf, J.W., *Cradle of Life. The Discovery of the World's Oldest Fossils,* Princeton University Press (1999). [But also see Brasier, M.D., Green, O.R., Jephcoat, A.P., Kleppe, A.K., Van Kranendonk, M.J., Linday, J.F., Steel, A., and Grassineau, N.V., Questioning the evidence of the Earth's oldest fossils, *Nature* **416**, 76–81 (2002).]

Sci. Am. **271**(4) (1994). [A special issue on "Life in the Universe."]

Shapiro, R., *Origins. A Skeptic's Guide to the Creation of Life on Earth,* Summit Books (1986). [An incisive and entertaining critique of the reigning theories of the origin of life.]

Woese, C.R. and Pace, N.R., Probing RNA structure, function, and history by comparative analysis, *in* Gesteland, R.F. and Atkins, J.F. (Eds.), *The RNA World, pp.* 91–117, Cold Spring Harbor Laboratory (1993).

PROBLEMS

It is very difficult to learn something well without somehow participating in it. The chapter-end problems are therefore an important part of this book. They contain few problems of the regurgitation type. Rather they are designed to make you think and to offer insights not discussed in the text. Their difficulties range from those that require only a few moments' reflection to those that might take an hour or more of concentrated effort to work out. The more difficult problems are indicated by a leading asterisk (*). The answers to the problems are worked out in detail in the *Solutions Manual to Accompany Biochemistry* (3rd ed.) by Donald Voet and Judith G. Voet. You should, of course, make every effort to work out a problem before consulting the *Solutions Manual.*

1. Under optimal conditions for growth, an *E. coli* cell will divide around every 20 min. If no cells died, how long would it take a single *E. coli* cell, under optimal conditions in a 10-L culture flask, to reach its maximum cell density of 10^{10} cells $\cdot$ mL^{-1} (a "saturated" culture)? Assuming that optimum conditions could be maintained, how long would it take for the total volume of the cells alone to reach 1 km^3? (Assume an *E. coli* cell to be a cylinder 2 μm long and 1 μm in diameter.)

2. Without looking them up, draw schematic diagrams of a bacterial cell and an animal cell. What are the functions of their various organelles? How many lines of descent might a typical animal cell have?

3. Compare the surface-to-volume ratios of a typical *E. coli* cell (its dimensions are given in Problem 1) and a spherical eukaryotic cell that is 20 μm in diameter. How does this difference affect the lifestyles of these two cell types? In order to improve their ability to absorb nutrients, the **brush border cells** of the intestinal epithelium have velvetlike patches of **microvilli** facing into the intestine. How does the surface-to-volume ratio of this eukaryotic cell change if 20% of its surface area is covered with cylindrical microvilli that are 0.1 μm in diameter, 1 μm in length, and occur on a square grid with 0.2-μm center-to-center spacing?

4. Many proteins in *E. coli* are normally present at concentrations of two molecules per cell. What is the molar concentration of such a protein (the dimensions of *E. coli* are given in Problem 1)? Conversely, how many glucose molecules does an *E. coli* cell contain if it has an internal glucose concentration of 1.0 mM?

5. The DNA of an *E. coli* chromosome measures 1.6 mm in length, when extended, and 20 Å in diameter. What fraction of an *E. coli* cell is occupied by its DNA (the dimensions of *E. coli* are given in Problem 1)? A human cell has some 700 times the DNA of an *E. coli* cell and is typically spherical with a diameter of 20 μm. What fraction of such a human cell is occupied by its DNA?

*6. A new planet has been discovered that has approximately the same orbit about the sun as Earth but is invisible from Earth because it is always on the opposite side of the sun. Interplanetary probes have already established that this planet has a significant atmosphere. The National Aeronautics and Space Administration is preparing to launch a new unmanned probe that will land on the surface of the planet. Outline a simple experiment for this lander that will test for the presence of life on the surface of this planet (assume that the life-forms, if any, on the planet are likely to be microorganisms and therefore unable to walk up to the lander's video cameras and say "Hello").

7. It has been suggested that an all-out nuclear war would so enshroud Earth with clouds of dust and smoke that the entire surface of the planet would be quite dark and therefore intensely cold (well below 0°C) for several years (the so-called nuclear winter). In that case, it is thought, eukaryotic life would die out and bacteria would inherit Earth. Why?

8. One method that Mendel used to test his laws is known as a **testcross.** In it, F_1 hybrids are crossed with their recessive par-

ent. What is the expected distribution of progeny and what are their phenotypes in a testcross involving peas with different-colored seeds? What is it for snapdragons with different flower colors (use the white parent in this testcross)?

9. The disputed paternity of a child can often be decided on the basis of blood tests. The M, N, and MN blood groups (Section 12-3E) result from two alleles, L^M and L^N; the Rh^+ blood group arises from a dominant allele, *R*. Both sets of alleles occur on a different chromosome from each other and from the alleles responsible for the ABO blood groups. The following table gives the blood types of three children, their mother, and the two possible fathers. Indicate, where possible, each child's paternity and justify your answer.

Child 1	B	M	Rh^-
Child 2	B	MN	Rh^+
Child 3	AB	MN	Rh^+
Mother	B	M	Rh^+
Male 1	B	MN	Rh^+
Male 2	AB	N	Rh^+

10. The most common form of color blindness, red–green color blindness, afflicts almost only males. What are the genotypes and phenotypes of the children and grandchildren of a red–green color-blind man and a woman with no genetic history of color blindness? Assume the children mate with individuals who also have no history of color blindness.

11. Green and purple photosynthetic bacteria are thought to resemble the first organisms that could carry out photosynthesis. Speculate on the composition of Earth's atmosphere when these organisms first arose.

12. Explore your local biochemistry library (it may be disguised as a biology, chemistry, or medical library). Locate where the current periodicals, the bound periodicals, and the books are kept. Browse through the contents of a current major biochemistry journal, such as *Biochemistry, Cell,* or *Proceedings of the National Academy of Sciences,* and pick a title that interests you. Scan the corresponding paper and note its organization. Likewise, peruse one of the articles in the latest volume of *Annual Review of Biochemistry.*

13. Using MedLine, look up the publications over the past five years of your favorite biomedical scientist. This person might be a recent Nobel Prize winner or someone at your college/university. If you use the BioMedNet website, you will have to join the organization (without charge) to obtain a user name and password. Note that even if the person you choose has an unusual name, it is likely that publications by other individuals with the same name will be included in your initial list.

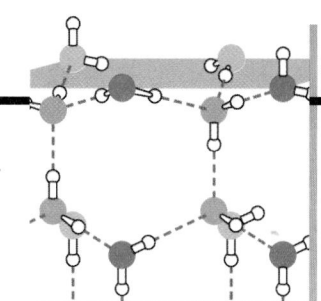

Chapter

2

Aqueous Solutions

Life, as we know it, occurs in aqueous solution. Indeed, terrestrial life apparently arose in some primordial sea (Section 1-5B) and, as the fossil record indicates, did not venture onto dry land until comparatively recent times. Yet even those organisms that did develop the capacity to live out of water still carry the ocean with them: The compositions of their intracellular and extracellular fluids are remarkably similar to that of seawater. This is true even of organisms that live in such unusual environments as saturated brine, acidic hot sulfur springs, and petroleum.

Water is so familiar, we generally consider it to be a rather bland fluid of simple character. It is, however, a chemically reactive liquid with such extraordinary physical properties that, if chemists had discovered it in recent times, it would undoubtedly have been classified as an exotic substance.

The properties of water are of profound biological significance. *The structures of the molecules on which life is based—proteins, nucleic acids, lipids, and complex carbohydrates—result directly from their interactions with their aqueous environment. The combination of solvent properties responsible for the intramolecular and intermolecular associations of these substances is peculiar to water; no other solvent even resembles water in this respect.* Although the hypothesis that life could be based on organic polymers other than proteins and nucleic acids seems plausible, it is all but inconceivable that the complex structural organization and chemistry of living systems could exist in other than an aqueous medium. Indeed, direct observations on the surface of Mars, the only other planet in the solar system with temperatures compatible with life, indicate that it is devoid both of water and of life.

Biological structures and processes can only be understood in terms of the physical and chemical properties of water. We therefore begin this chapter with a discussion of the molecular and solvent properties of water. In the following section we review its chemical behavior, that is, the nature of aqueous acids and bases.

1 ■ PROPERTIES OF WATER

Water's peculiar physical and solvent properties stem largely from its extraordinary internal cohesiveness compared to that of almost any other liquid. In this section, we explore the physical basis of this phenomenon.

A. *Structure and Interactions*

The H_2O molecule has a bent geometry with an O—H bond distance of 0.958 Å and an H—O—H bond angle of 104.5° (Fig. 2-1). The large electronegativity difference between H and O confers a 33% ionic character on the O—H bond as is indicated by water's dipole moment of 1.85 debye units. Water is clearly a highly polar molecule, a phenomenon with enormous implications for living systems.

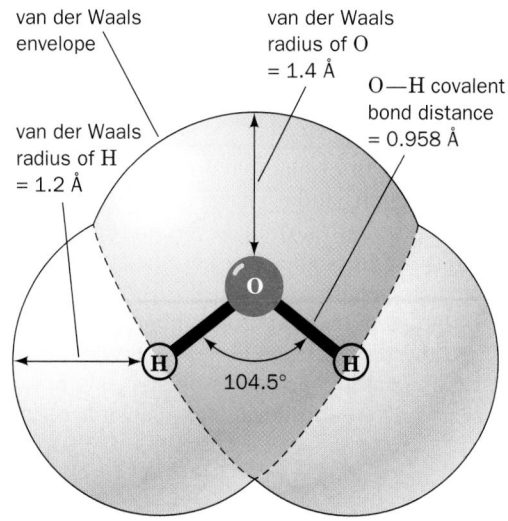

FIGURE 2-1 Structure of the water molecule. The outline represents the van der Waals envelope of the molecule (where the attractive components of the van der Waals interactions balance the repulsive components). The skeletal model of the molecule indicates its covalent bonds.

a. Water Molecules Associate through Hydrogen Bonds

The electrostatic attractions between the dipoles of two water molecules tend to orient them such that the O—H bond on one water molecule points toward a lone-pair electron cloud on the oxygen atom of the other water molecule. This results in a directional intermolecular association known as a **hydrogen bond** (Fig. 2-2), an interaction that is crucial both to the properties of water itself and to its role as a biochemical solvent. In general, *a hydrogen bond may be represented as D—H ··· A, where D—H is a weakly acidic "donor group" such as N—H or O—H, and A is a lone-pair-bearing and thus weakly basic "acceptor atom" such as N or O.* Hence, a hydrogen bond is better represented as $^{\delta-}D—H^{\delta+} \cdots {^{\delta-}}A$, where the charge separation in the D—H bond arises from the greater electronegativity of D relative to H. The peculiar requirement of a central hydrogen atom rather than some other atom in a hydrogen bond stems from the hydrogen atom's small size: Only a hydrogen nucleus can approach the lone-pair electron cloud of an acceptor atom closely enough to permit an electrostatic association of significant magnitude. Moreover, as recent X-ray scattering measurements have revealed, hydrogen bonds are partially (~10%) covalent in character.

Hydrogen bonds are structurally characterized by an H ··· A distance that is at least 0.5 Å shorter than the calculated van der Waals distance (distance of closest approach between two nonbonded atoms) between the atoms. In water, for example, the O ··· H hydrogen bond distance is ~1.8 Å versus 2.6 Å for the corresponding van der Waals distance. The energy of a hydrogen bond (~20 kJ · mol^{-1} in H_2O) is small compared to covalent bond energies (for instance, 460 kJ · mol^{-1} for an O—H covalent bond). Nevertheless, most biological molecules have so many hydrogen bonding groups that hydrogen bonding is of paramount importance in determining their three-dimensional structures and their intermolecular associations. Hydrogen bonding is further discussed in Section 8-4B.

b. The Physical Properties of Ice and Liquid Water Largely Result from Intermolecular Hydrogen Bonding

The structure of ice provides a striking example of the cumulative strength of many hydrogen bonds. X-ray and neutron diffraction studies have established that water molecules in ice are arranged in an unusually open structure. Each water molecule is tetrahedrally surrounded by four nearest neighbors to which it is hydrogen bonded (Fig. 2-3). In two of these hydrogen bonds the central H_2O molecule is the "donor," and in the other two it is the "acceptor." As a consequence of its open structure, water is one of the very few substances that expands on freezing (at 0°C, liquid water has a density of 1.00 g · mL^{-1}, whereas ice has a density of 0.92 g · mL^{-1}).

The expansion of water on freezing has overwhelming consequences for life on Earth. Suppose that water contracted on freezing, that is, became more dense rather than less dense. Ice would then sink to the bottoms of lakes and oceans rather than float. This ice would be insulated from the sun so that oceans, with the exception of a thin surface

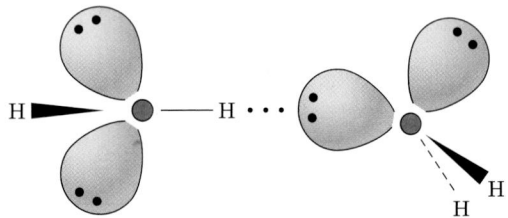

FIGURE 2-2 Hydrogen bond between two water molecules. The strength of this interaction is maximal when the O—H covalent bond points directly along a lone-pair electron cloud of the oxygen atom to which it is hydrogen bonded.

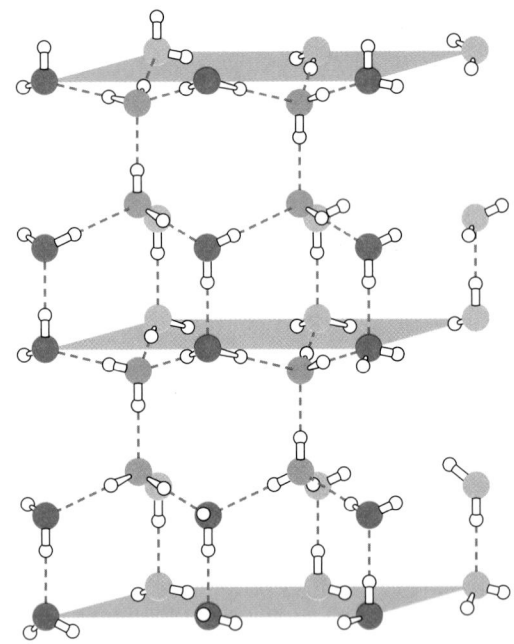

FIGURE 2-3 Structure of ice. The tetrahedral arrangement of the water molecules is a consequence of the roughly tetrahedral disposition of each oxygen atom's sp^3-hybridized bonding and lone-pair orbitals (Fig. 2-2). Oxygen and hydrogen atoms are represented, respectively, by red and white spheres, and hydrogen bonds are indicated by dashed lines. Note the open structure that gives ice its low density relative to liquid water. [After Pauling, L., *The Nature of the Chemical Bond* (3rd ed.), p. 465, Cornell University Press (1960).]

layer of liquid in warm weather, would be permanently frozen solid (the water at great depths even in tropical oceans is close to 4°C, its temperature of maximum density). The reflection of sunlight by these frozen oceans and their cooling effect on the atmosphere would ensure that land temperatures would also be much colder than at present; that is, Earth would have a permanent ice age. Furthermore, since life apparently evolved in the ocean, it seems unlikely that life could have developed at all if ice contracted on freezing.

Although the melting of ice is indicative of the cooperative collapse of its hydrogen bonded structure, hydrogen bonds between water molecules persist in the liquid state. The heat of sublimation of ice at 0°C is 46.9 kJ · mol^{-1}. Yet only ~6 kJ · mol^{-1} of this quantity can be attributed to the kinetic energy of gaseous water molecules. The remaining 41 kJ · mol^{-1} must therefore represent the energy required to disrupt the hydrogen bonding interactions holding an ice crystal together. The heat of fusion of ice (6.0 kJ · mol^{-1}) is ~15% of the energy required to disrupt the ice structure. *Liquid water is therefore only ~15% less hydrogen bonded than ice at 0°C.* Indeed, the boiling point of water is 264°C higher than that of methane (CH$_4$), a substance with nearly the same molecular mass as H$_2$O but which is incapable of hydrogen bonding (in the absence of intermolecular associations, substances with equal molecular masses should have similar boiling points). This reflects the extraordinary internal cohesiveness of liquid water resulting from its intermolecular hydrogen bonding.

c. Liquid Water Has a Rapidly Fluctuating Structure

X-ray and neutron scattering measurements of liquid water reveal a complex structure. Near 0°C, water exhibits an average nearest-neighbor O · · · O distance of 2.82 Å, which is slightly greater than the corresponding 2.76-Å distance in ice despite the greater density of the liquid. The X-ray data further indicate that each water molecule is surrounded by an average of about 4.4 nearest neighbors, which strongly suggests that the short-range structure of liquid water is predominantly tetrahedral in character. This picture is corroborated by the additional intermolecular distances in liquid water of around 4.5 and 7.0 Å, which are near the expected second and third nearest-neighbor distances in an icelike tetrahedral structure. Liquid water, however, also exhibits a 3.5-Å intermolecular distance, which cannot be rationalized in terms of an icelike structure. These average distances, moreover, become less sharply defined as the temperature increases into the physiologically significant range, thereby signaling the thermal breakdown of the short-range water structure.

The structure of liquid water is not simply described. This is because each water molecule reorients about once every 10^{-12} s, which makes the determination of water's instantaneous structure an experimentally and theoretically difficult problem (very few experimental techniques can make measurements over such short time spans). Indeed, only with the advent of modern computational methods have theoreticians felt that they are beginning to have a reasonable understanding of liquid water at the molecular level.

For the most part, molecules in liquid water are each hydrogen bonded to four nearest neighbors as they are in ice. These hydrogen bonds are distorted, however, so that the networks of linked molecules are irregular and varied, with the number of hydrogen bonds formed by each water molecule ranging from 3 to 6. Thus, for example, 3- to 7-membered rings of hydrogen bonded molecules commonly occur in liquid water (Fig. 2-4), in contrast to the cyclohexane-like 6-membered rings characteristic of ice (Fig. 2-3). Moreover, these networks are continually breaking up and re-forming over time periods on the order of 2 × 10^{-11} s. *Liquid water therefore consists of a rapidly fluctuating, space-filling network of hydrogen bonded H$_2$O molecules that, over short distances, resembles that of ice.*

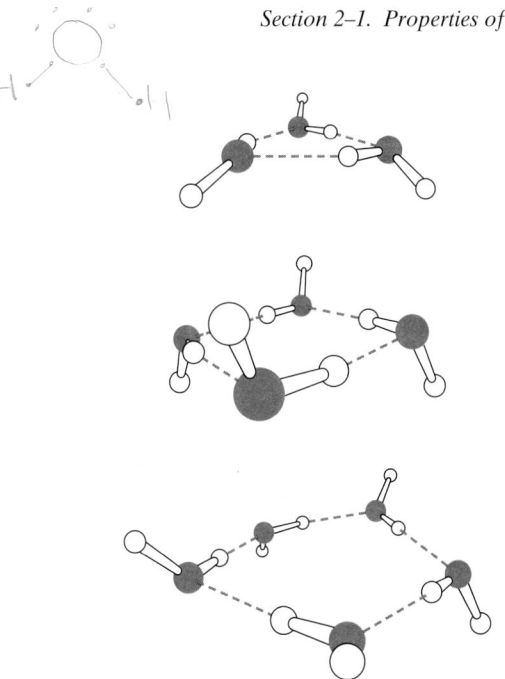

FIGURE 2-4 Theoretically predicted and spectroscopically confirmed structures of the water trimer, tetramer, and pentamer. Note that these rings are all essentially planar, with each water molecule acting as both a hydrogen bonding donor and acceptor and with the free hydrogens located above and below the planes of the rings. [After Liu, K., Cruzan, J.D., and Saykelly, R.J., *Science* **271**, 930 (1996).]

B. *Water as a Solvent*

Solubility depends on the ability of a solvent to interact with a solute more strongly than solute particles interact with each other. Water is said to be the "universal solvent." Although this statement cannot literally be true, water certainly dissolves more types of substances and in greater amounts than any other solvent. In particular, the polar character of water makes it an excellent solvent for polar and ionic materials, which are therefore said to be **hydrophilic** (Greek: *hydor*, water + *philos*, loving). On the other hand, nonpolar substances are virtually insoluble in water ("oil and water don't mix") and are consequently described as being **hydrophobic** (Greek: *phobos*, fear). Nonpolar substances, however, are soluble in nonpolar solvents such as CCl$_4$ or hexane. This information is summarized by another maxim, "like dissolves like."

Why do salts dissolve in water? Salts, such as NaCl or K$_2$HPO$_4$, are held together by ionic forces. The ions of a salt, as do any electrical charges, interact according to **Coulomb's law:**

$$F = \frac{kq_1q_2}{Dr^2} \qquad [2.1]$$

where F is the force between two electrical charges, q_1 and q_2, that are separated by the distance r, D is the **dielectric constant** of the medium between them, and k is a proportionality constant (8.99 × 10^9 J · m · C^{-2}). Thus, as the dielectric constant of a medium increases, the force between its embedded charges decreases; that is, the dielectric constant of a solvent is a measure of its ability to keep opposite charges apart. In a vacuum, D is unity and in air, it is only negligibly larger. The dielectric constants of several common solvents, together with their permanent molecular dipole moments, are listed in Table 2-1. Note that these quantities tend to increase together, although not in any regular way.

The dielectric constant of water is among the highest of any pure liquid, whereas those of nonpolar substances, such as hydrocarbons, are relatively small. The force between two ions separated by a given distance in nonpolar liquids such as hexane or benzene is therefore 30 to 40 times greater than that in water. Consequently, in nonpolar solvents (low D), ions of opposite charge attract each other so strongly that they coalesce to form a salt, whereas the much weaker forces between ions in water solution (high D) permit significant quantities of the ions to remain separated.

An ion immersed in a polar solvent attracts the oppositely charged ends of the solvent dipoles, as is diagrammed in Fig. 2-5 for water. The ion is thereby surrounded by several concentric shells of oriented solvent molecules. Such ions are said to be **solvated** or, if water is the solvent, to be **hydrated.** The electric field produced by the solvent dipoles opposes that of the ion so that, in effect, the ionic charge is spread over the volume of the solvated complex. This arrangement greatly attenuates the coulombic forces between ions, which is why polar solvents have such high dielectric constants.

The orienting effect of ionic charges on dipolar molecules is opposed by thermal motions, which continually tend to randomly reorient all molecules. The dipoles in a solvated complex are therefore only partially oriented. The reason why the dielectric constant of water is so much greater than that of other liquids with comparable dipole moments is that liquid water's hydrogen bonded structure permits it to form oriented structures that resist thermal randomization, thereby more effectively distributing ionic charges.

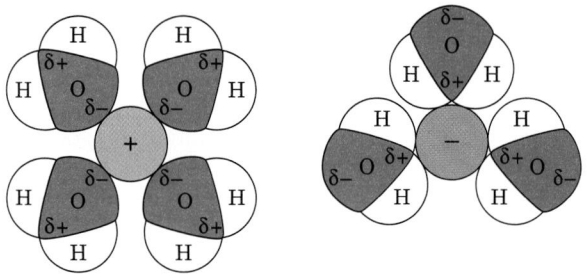

FIGURE 2-5   Solvation of ions by oriented water molecules.

The bond dipoles of uncharged polar molecules make them soluble in aqueous solutions for the same reasons that ionic substances are water soluble. The solubilities of polar and ionic substances are enhanced if they carry functional groups, such as hydroxyl (—OH), keto (C=O), carboxyl (—CO_2H or COOH), or amino (—NH_2) groups, that can form hydrogen bonds with water, as is illustrated in Fig. 2-6. Indeed, water-soluble biomolecules such as proteins, nucleic acids, and carbohydrates bristle with just such groups. Nonpolar substances, in contrast, lack both hydrogen bonding donor and acceptor groups.

a. Amphiphiles Form Micelles and Bilayers

Most biological molecules have both polar (or ionically charged) and nonpolar segments and are therefore simultaneously hydrophilic and hydrophobic. Such molecules, for example, **fatty acid** ions (soap ions; Fig. 2-7), are said to be **amphiphilic** or, synonymously, **amphipathic** (Greek: *amphi*, both + *pathos*, passion). How do amphiphiles interact with an aqueous solvent? Water, of course, tends to hydrate the

TABLE 2-1   Dielectric Constants and Permanent Molecular Dipole Moments of Some Common Solvents

Substance	Dielectric Constant	Dipole Moment (debye)
Formamide	110.0	3.37
Water	78.5	1.85
Dimethyl sulfoxide	48.9	3.96
Methanol	32.6	1.66
Ethanol	24.3	1.68
Acetone	20.7	2.72
Ammonia	16.9	1.47
Chloroform	4.8	1.15
Diethyl ether	4.3	1.15
Benzene	2.3	0.00
Carbon tetrachloride	2.2	0.00
Hexane	1.9	0.00

Source: Brey, W.S., *Physical Chemistry and Its Biological Applications,* p. 26, Academic Press (1978).

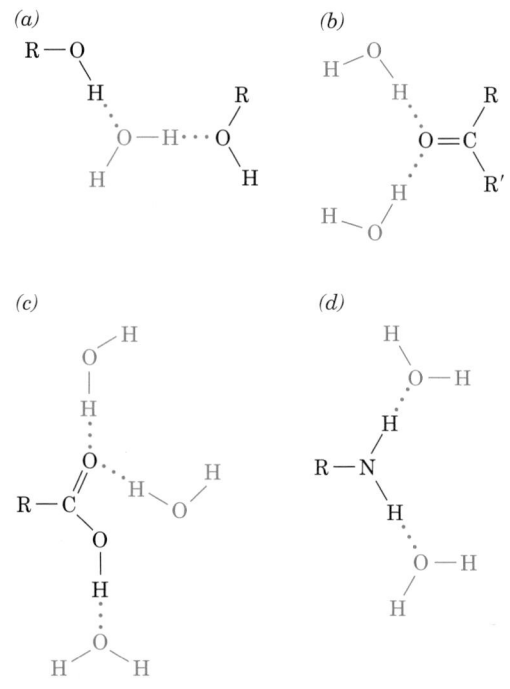

FIGURE 2-6   Hydrogen bonding by functional groups.
Hydrogen bonds form between water and (*a*) hydroxyl groups, (*b*) keto groups, (*c*) carboxyl groups, and (*d*) amino groups.

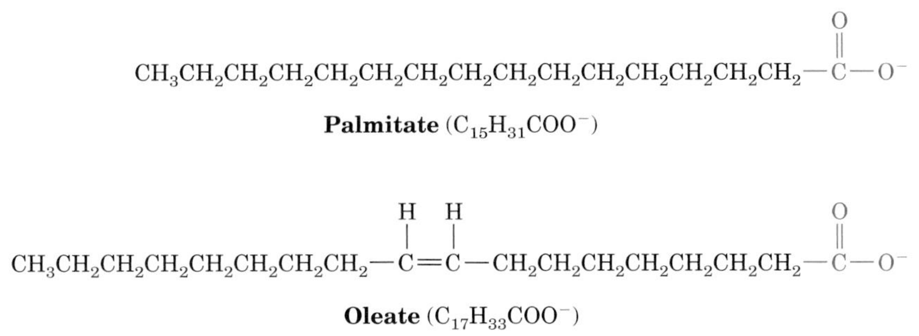

$$CH_3CH_2CH_2CH_2CH_2CH_2CH_2CH_2CH_2CH_2CH_2CH_2CH_2CH_2CH_2 \overset{\overset{\displaystyle O}{\|}}{-C}-O^-$$

Palmitate ($C_{15}H_{31}COO^-$)

$$CH_3CH_2CH_2CH_2CH_2CH_2CH_2CH_2 \overset{\overset{\displaystyle H}{|}}{-C}=\overset{\overset{\displaystyle H}{|}}{C}-CH_2CH_2CH_2CH_2CH_2CH_2CH_2 \overset{\overset{\displaystyle O}{\|}}{-C}-O^-$$

Oleate ($C_{17}H_{33}COO^-$)

FIGURE 2-7   Examples of fatty acid anions. They consist of a polar carboxylate group coupled to a long nonpolar hydrocarbon chain.

hydrophilic portion of an amphiphile, but it also tends to exclude its hydrophobic portion. Amphiphiles consequently tend to form water-dispersed structurally ordered aggregates. Such aggregates may take the form of **micelles,** which are globules of up to several thousand amphiphiles arranged with their hydrophilic groups at the globule surface so that they can interact with the aqueous solvent while the hydrophobic groups associate at the center so as to exclude solvent (Fig. 2-8*a*). Alternatively, the amphiphiles may arrange themselves to form bilayered sheets or vesicles (Fig. 2-8*b*) in which the polar groups face the aqueous phase.

The interactions stabilizing a micelle or bilayer are collectively described as **hydrophobic forces** or **hydrophobic interactions** to indicate that they result from the tendency of water to exclude hydrophobic groups. Hydrophobic interactions are relatively weak compared to hydrogen bonds and lack directionality. Nevertheless, hydrophobic interactions are of pivotal biological importance because, as we shall see in later chapters, they are largely responsible for the structural integrity of biological macromolecules (Sections 8-4C and 29-2C), as well as that of supramolecular aggregates such as membranes. Note that hydrophobic interactions are peculiar to an aqueous environment. Other polar solvents do not promote such associations.

C. *Proton Mobility*

When an electrical current is passed through an ionic solution, the ions migrate toward the electrode of opposite polarity at a rate proportional to the electrical field and in-

TABLE 2-2   Ionic Mobilities[a] in H₂O at 25°C

Ion	Mobility $\times 10^{-5}$ ($cm^2 \cdot V^{-1} \cdot s^{-1}$)
H_3O^+	362.4
Li^+	40.1
Na^+	51.9
K^+	76.1
NH_4^+	76.0
Mg^{2+}	55.0
Ca^{2+}	61.6
OH^-	197.6
Cl^-	76.3
Br^-	78.3
CH_3COO^-	40.9
SO_4^{2-}	79.8

[a] Ionic mobility is the distance an ion moves in 1 s under the influence of an electric field of $1\ V \cdot cm^{-1}$.

Source: Brey, W.S., *Physical Chemistry and Its Biological Applications,* p. 172, Academic Press (1978).

versely proportional to the frictional drag experienced by the ion as it moves through the solution. This latter quantity, as Table 2-2 indicates, varies with the size of the ion. Note, however, that the ionic mobilities of both H_3O^+ and OH^- are anomalously large compared to those of other

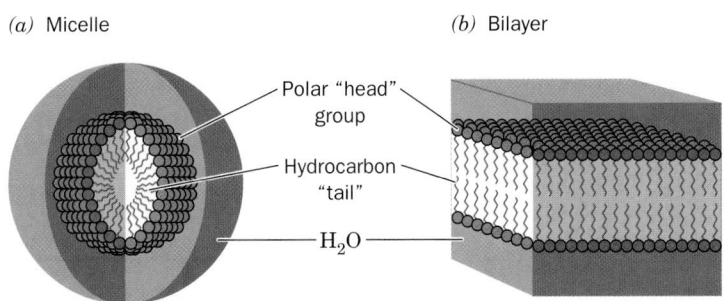

(a) Micelle   *(b)* Bilayer

Polar "head" group

Hydrocarbon "tail"

H_2O

FIGURE 2-8   Associations of amphipathic molecules in aqueous solutions. The polar "head" groups are hydrated, whereas the nonpolar "tails" aggregate so as to exclude the aqueous solution. (*a*) A spheroidal aggregate of amphipathic

molecules known as a micelle. (*b*) An extended planar aggregate of amphipathic molecules called a **bilayer.** The bilayer may form a closed spheroidal shell, known as a vesicle, that encloses a small amount of aqueous solution.

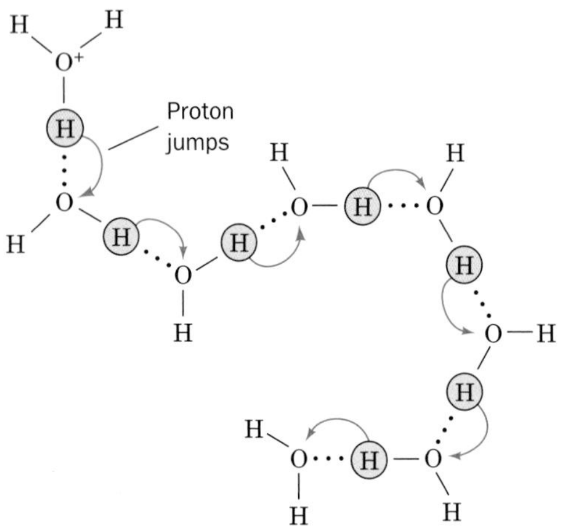

FIGURE 2-9   Mechanism of hydronium ion migration in aqueous solution via proton jumps. Proton jumps, which mostly occur at random, take place rapidly compared with direct molecular migration, thereby accounting for the observed high ionic mobilities of hydronium and hydroxyl ions in aqueous solutions.

ions. For H_3O^+ (the **hydronium ion,** which is abbreviated H^+; a bare proton has no stable existence in aqueous solution), this high migration rate results from the ability of protons to jump rapidly from one water molecule to another, as is diagrammed in Fig. 2-9. Although a given hydronium ion can physically migrate through solution in the manner of, say, an Na^+ ion, the rapidity of the proton-jump mechanism makes the H_3O^+ ion's effective ionic mobility much greater than it otherwise would be (the mean lifetime of a given H_3O^+ ion is 10^{-12} s at 25°C). The anomalously high ionic mobility of the OH^- ion is likewise accounted for by the proton-jump mechanism but, in this case, the apparent direction of ionic migration is opposite to the direction of proton jumping. Proton jumping is also responsible for the observation that *acid–base reactions are among the fastest reactions that take place in aqueous solutions,* and it is probably of importance in biological proton-transfer reactions.

2 ■ ACIDS, BASES, AND BUFFERS

Biological molecules, such as proteins and nucleic acids, bear numerous functional groups, such as carboxyl and amino groups, that can undergo acid–base reactions. Many properties of these molecules therefore vary with the acidities of the solutions in which they are immersed. In this section we discuss the nature of acid–base reactions and how acidities are controlled, both physiologically and in the laboratory.

A. *Acid–Base Reactions*

Acids and **bases,** in a definition coined in the 1880s by Svante Arrhenius, are, respectively, substances capable of donating protons and hydroxide ions. This definition is rather limited, because, for example, it does not account

for the observation that NH_3, which lacks an OH^- group, exhibits basic properties. In a more general definition, which was formulated in 1923 by Johannes Brønsted and Thomas Lowry, *an acid is a substance that can donate protons (as in the Arrhenius definition) and a base is a substance that can accept protons.* Under this definition, in every acid–base reaction,

$$HA + H_2O \rightleftharpoons H_3O^+ + A^-$$

a **Brønsted acid** (here HA) reacts with a **Brønsted base** (here H_2O) to form the **conjugate base** of the acid (A^-) and the **conjugate acid** of the base (H_3O^+) (this reaction is usually abbreviated $HA \rightleftharpoons H^+ + A^-$ with the participation of H_2O implied). Accordingly, the acetate ion (CH_3COO^-) is the conjugate base of acetic acid (CH_3COOH) and the ammonium ion (NH_4^+) is the conjugate acid of ammonia (NH_3). (In a yet more general definition of acids and bases, Gilbert Lewis described a **Lewis acid** as a substance that can accept an electron pair and a **Lewis base** as a substance that can donate an electron pair. This definition, which is applicable to both aqueous and nonaqueous systems, is unnecessarily broad for describing most biochemical phenomena.)

a. The Strength of an Acid Is Specified by Its Dissociation Constant

The above acid dissociation reaction is characterized by its **equilibrium constant,** which, for acid–base reactions, is known as a **dissociation constant,**

$$K = \frac{[H_3O^+][A^-]}{[HA][H_2O]} \qquad [2.2]$$

a quantity that is a measure of the relative proton affinities of the HA/A^- and H_3O^+/H_2O conjugate acid–base pairs. Here, as throughout the text, quantities in square brackets symbolize the molar concentrations of the indicated substances. Since in dilute aqueous solutions the water concentration is essentially constant with $[H_2O] = 1000$ g · $L^{-1}/18.015$ g · $mol^{-1} = 55.5M$, this term is customarily combined with the dissociation constant, which then takes the form

$$K_a = K[H_2O] = \frac{[H^+][A^-]}{[HA]} \qquad [2.3]$$

For brevity, however, we shall henceforth omit the subscript "*a.*" The dissociation constants for acids useful in preparing biochemical solutions are listed in Table 2-3.

Acids may be classified according to their relative strengths, that is, according to their abilities to transfer a proton to water. Acids with dissociation constants smaller than that of H_3O^+ (which, by definition, is unity in aqueous solutions) are only partially ionized in aqueous solutions and are known as **weak acids** ($K < 1$). Conversely, **strong acids** have dissociation constants larger than that of H_3O^+ so that they are almost completely ionized in aqueous solutions ($K > 1$). The acids listed in Table 2-3 are all weak acids. However, many of the so-called mineral acids, such as $HClO_4$, HNO_3, HCl, and H_2SO_4 (for the first ion-

TABLE 2-3 Dissociation Constants and pK's at 25°C of Some Acids in Common Laboratory Use as Biochemical Buffers

Acid	K (M)	pK
Oxalic acid	5.37×10^{-2}	1.27 (pK_1)
H_3PO_4	7.08×10^{-3}	2.15 (pK_1)
Citric acid	7.41×10^{-4}	3.13 (pK_1)
Formic acid	1.78×10^{-4}	3.75
Succinic acid	6.17×10^{-5}	4.21 (pK_1)
Oxalate$^-$	5.37×10^{-5}	4.27 (pK_1)
Acetic acid	1.74×10^{-5}	4.76
Citrate$^-$	1.74×10^{-5}	4.76 (pK_2)
Citrate^{2-}	3.98×10^{-6}	5.40 (pK_3)
Succinate$^-$	2.29×10^{-6}	5.64 (pK_2)
2-(N-Morpholino)ethanesulfonic acid (MES)	8.13×10^{-7}	6.09
Cacodylic acid	5.37×10^{-7}	6.27
H_2CO_3	4.47×10^{-7}	6.35 (pK_1)
N-(2-Acetamido)iminodiacetic acid (ADA)	2.69×10^{-7}	6.57
Piperazine-N,N'-bis(2-ethanesulfonic acid) (PIPES)	1.74×10^{-7}	6.76
N-(2-Acetamido)-2-aminoethanesulfonic acid (ACES)	1.58×10^{-7}	6.80
$H_2PO_4^-$	1.51×10^{-7}	6.82 (pK_2)
3-(N-Morpholino)propanesulfonic acid (MOPS)	7.08×10^{-8}	7.15
N-2-Hydroxyethylpiperazine-N'-2-ethanesulfonic acid (HEPES)	3.39×10^{-8}	7.47
N-2-Hydroxyethylpiperazine-N'-3-propanesulfonic acid (HEPPS)	1.10×10^{-8}	7.96
N-[Tris(hydroxymethyl)methyl]glycine (Tricine)	8.91×10^{-9}	8.05
Tris(hydroxymethyl)aminomethane (TRIS)	8.32×10^{-9}	8.08
Glycylglycine	5.62×10^{-9}	8.25
N,N-Bis(2-hydroxyethyl)glycine (Bicine)	5.50×10^{-9}	8.26
Boric acid	5.75×10^{-10}	9.24
NH_4^+	5.62×10^{-10}	9.25
Glycine	1.66×10^{-10}	9.78
HCO_3^-	4.68×10^{-11}	10.33 (pK_2)
Piperidine	7.58×10^{-12}	11.12
HPO_4^{2-}	4.17×10^{-13}	12.38 (pK_3)

Source: Mostly Dawson, R.M.C., Elliott, D.C., Elliott, W.H., and Jones, K.M., *Data for Biochemical Research* (3rd ed.), *pp.* 424–425, Oxford Science Publications (1986); *and* Good, N.E., Winget, G.D., Winter, W., Connolly, T.N., Izawa, S., and Singh, R.M.M., *Biochemistry* **5,** 467 (1966).

ization), are strong acids. Since strong acids rapidly transfer all their protons to H_2O, the strongest acid that can stably exist in aqueous solutions is H_3O^+. Likewise, there can be no stronger base in aqueous solutions than OH^-.

Water, being an acid, has a dissociation constant:

$$K = \frac{[H^+][OH^-]}{[H_2O]}$$

As above, the constant $[H_2O] = 55.5M$ can be incorporated into the dissociation constant to yield the expression for the ionization constant of water,

$$K_w = [H^+][OH^-] \qquad [2.4]$$

The value of K_w at 25°C is $10^{-14}M^2$. Pure water must contain equimolar amounts of H^+ and OH^- so that $[H^+] = [OH^-] = (K_w)^{1/2} = 10^{-7}M$. Since $[H^+]$ and $[OH^-]$ are reciprocally related by Eq. [2.4], if $[H^+]$ is greater than this value, $[OH^-]$ must be correspondingly less and vice versa.

Solutions with $[H^+] = 10^{-7}M$ are said to be **neutral,** those with $[H^+] > 10^{-7}M$ are said to be **acidic,** and those with $[H^+] < 10^{-7}M$ are said to be **basic.** Most physiological solutions have hydrogen ion concentrations near neutrality. For example, human blood is normally slightly basic, with $[H^+] = 4.0 \times 10^{-8}M$.

The values of $[H^+]$ for most solutions are inconveniently small and difficult to compare. A more practical quantity, which was devised in 1909 by Søren Sørensen, is known as the **pH:**

$$pH = -\log[H^+] \qquad [2.5]$$

The pH of pure water is 7.0, whereas acidic solutions have pH < 7.0 and basic solutions have pH > 7.0. For a $1M$ solution of a strong acid, pH = 0 and for a $1M$ solution of a strong base, pH = 14. Note that if two solutions differ in pH by one unit, they differ in $[H^+]$ by a factor of 10. The pH of a solution may be accurately and easily determined through electrochemical measurements with a device known as a **pH meter.**

b. The pH of a Solution Is Determined by the Relative Concentrations of Acids and Bases

The relationship between the pH of a solution and the concentrations of an acid and its conjugate base can be easily derived by rearranging Eq. [2.3]

$$[H^+] = K\left(\frac{[HA]}{[A^-]}\right)$$

and substituting it into Eq. [2.5]

$$pH = -\log K + \log\left(\frac{[A^-]}{[HA]}\right)$$

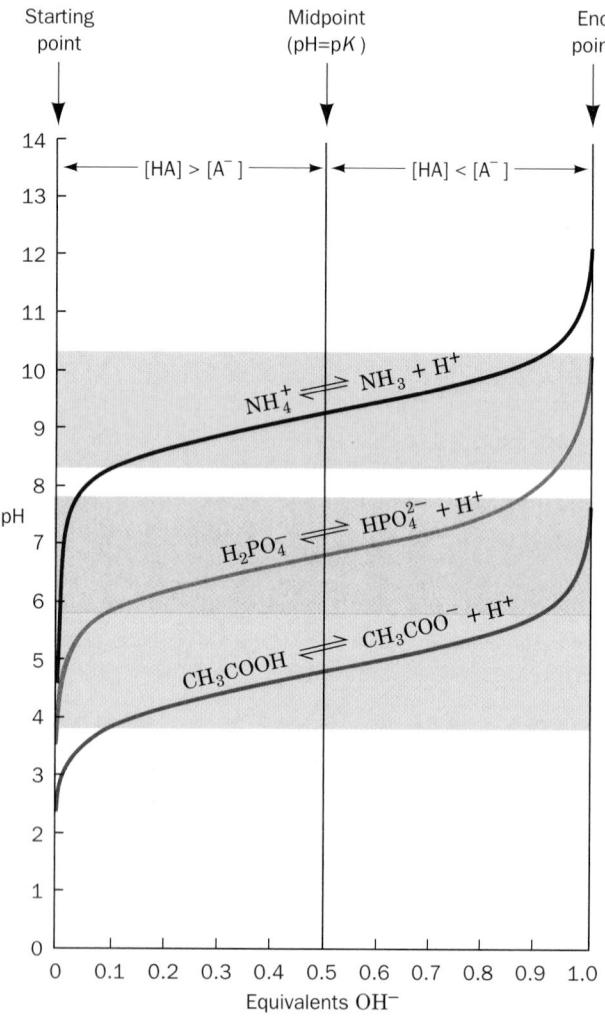

Defining $pK = -\log K$ in analogy with Eq. [2.5], we obtain the **Henderson–Hasselbalch equation:**

$$pH = pK + \log\left(\frac{[A^-]}{[HA]}\right) \qquad [2.6]$$

This equation indicates that *the pK of an acid is numerically equal to the pH of the solution when the molar concentrations of the acid and its conjugate base are equal.* Table 2-3 lists the pK values of several acids.

B. Buffers

A 0.01-mL droplet of $1M$ HCl added to 1 L of pure water changes the water's pH from 7 to 5, which represents a 100-fold increase in $[H^+]$. Yet, since the properties of biological substances vary significantly with small changes in pH, they require environments in which the pH is insensitive to additions of acids or bases. To understand how this is possible, let us consider the titration of a weak acid with a strong base.

Figure 2-10 shows how the pH values of 1-L solutions of $1M$ acetic acid, $(H_2PO_4^-)$, and ammonium ion (NH_4^+), vary with the quantity of OH^- added. Titration curves such as those in Fig. 2-10, as well as distribution curves such as those in Fig. 2-11, may be calculated using the Henderson–Hasselbalch equation. Near the beginning of the titration, a significant fraction of the A^- present arises from the dissociation of HA. Similarly, near the end point, much of the HA derives from the reaction of A^- with H_2O. Throughout most of the titration, however, the OH^- added reacts essentially completely with the HA to form A^- so that

$$[A^-] = \frac{x}{V} \qquad [2.7]$$

where x represents the equivalents of OH^- added and V is the volume of the solution. Then, letting c_0 represent the equivalents of HA initially present,

$$[HA] = \frac{c_0 - x}{V} \qquad [2.8]$$

Incorporating these relationships into Eq. [2.6] yields

$$pH = pK + \log\left(\frac{x}{c_0 - x}\right) \qquad [2.9]$$

which accurately describes a titration curve except near its wings (these regions require more exact treatments that take into account the ionizations of water).

Several details about the titration curves in Fig. 2-10 should be noted:

1. The curves have similar shapes but are shifted vertically along the pH axis.

2. The pH at the **equivalence point** of each titration (where the equivalents of OH^- added equal the equivalents of HA initially present) is >7 because of the reaction of A^- with H_2O to form $HA + OH^-$; similarly, each initial pH is <7.

FIGURE 2-10   Acid–base titration curves of 1-L solutions of $1M$ acetic acid, $H_2PO_4^-$, and NH_4^+ by a strong base. At the starting point of each titration, the acid form of the conjugate acid–base pair overwhelmingly predominates. At the midpoint of the titration, where $pH = pK$, the concentration of the acid is equal to that of its conjugate base. Finally, at the end point of the titration, where the equivalents of strong base added equal the equivalents of acid at the starting point, the conjugate base is in great excess over acid. The shaded bands indicate the pH ranges over which the corresponding solution can function effectively as a buffer.
🎞 See the Animated Figures.

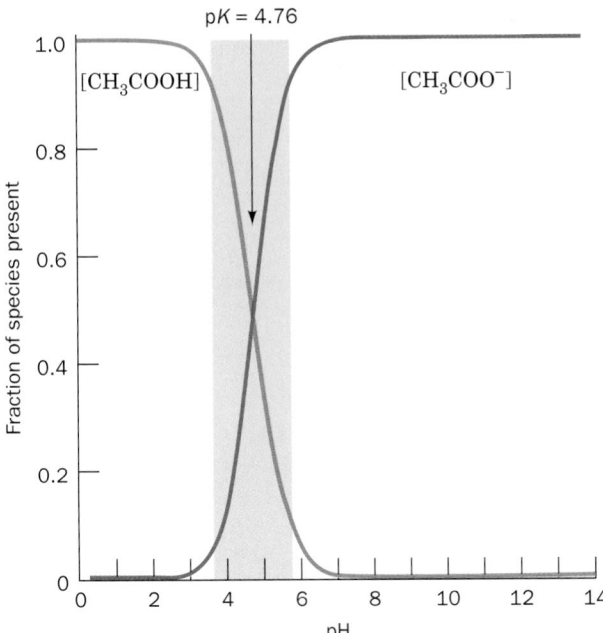

FIGURE 2-11 Distribution curves for acetic acid and acetate ion. The fraction of species present is given as the ratio of the concentration of CH_3COOH or CH_3COO^- to the total concentrations of these two species. The customarily accepted useful buffer range of $pK \pm 1$ is indicated by the shaded region.

3. The pH at the midpoint of each titration is numerically equal to the pK of its corresponding acid; here, according to the Henderson–Hasselbalch equation, $[HA] = [A^-]$.

4. The slope of each titration curve is much less near its midpoint than it is near its wings. This indicates that *when $[HA] \approx [A^-]$, the pH of the solution is relatively insensitive to the addition of strong base or strong acid. Such a solution, which is known as an* **acid–base buffer**, *is resistant to pH changes because small amounts of added H^+ or OH^-, respectively, react with the A^- or HA present without greatly changing the value of $\log([A^-]/[HA])$.*

a. Buffers Stabilize a Solution's pH

The ability of a buffer to resist pH changes with added acid or base is directly proportional to the total concentration of the conjugate acid–base pair, $[HA] + [A^-]$. It is maximal when $pH = pK$ and decreases rapidly with a change in pH from that point. A good rule of thumb is that *a weak acid is in its useful buffer range within 1 pH unit of its pK* (the shaded regions of Figs. 2-10 and 2-11). Above this range, where the ratio $[A^-]/[HA] > 10$, the pH of the solution changes rapidly with added strong base. A buffer is similarly impotent with addition of strong acid when its pK exceeds the pH by more than a unit.

Biological fluids, both those found intracellularly and extracellularly, are heavily buffered. For example, the pH of the blood in healthy individuals is closely controlled at pH 7.4. The phosphate and carbonate ions that are components of most biological fluids are important in this respect because they have pK's in this range (Table 2-3). Moreover, many biological molecules, such as proteins, nucleic acids, and lipids, as well as numerous small organic molecules, bear multiple acid–base groups that are effective as buffer components in the physiological pH range.

The concept that the properties of biological molecules vary with the acidity of the solution in which they are dissolved was not fully appreciated before the beginning of the twentieth century so that the acidities of biochemical preparations made before that time were rarely controlled. Consequently these early biochemical experiments yielded poorly reproducible results. More recently, biochemical preparations have been routinely buffered to simulate the properties of naturally occurring biological fluids. Many of the weak acids listed in Table 2-3 are commonly used as buffers in biochemical preparations. In practice, the chosen weak acid and one of its soluble salts are dissolved in the (nearly equal) mole ratio necessary to provide the desired pH and, with the aid of a pH meter, the resulting solution is fine-tuned by titration with strong acid or base.

C. *Polyprotic Acids*

Substances that bear more than one acid–base group, such as H_3PO_4 or H_2CO_3, as well as most biomolecules, are known as **polyprotic acids.** The titration curves of such substances, as is illustrated in Fig. 2-12 for H_3PO_4, are characterized by multiple pK's, one for each ionization step. Exact calculations of the concentrations of the various ionic species present at a given pH is clearly a more complex task than for a **monoprotic acid.**

The pK's of two closely associated acid–base groups are not independent. The ionic charge resulting from a proton dissociation electrostatically inhibits further proton dissociation from the same molecule, thereby increasing the values of the corresponding pK's. This effect, according to Coulomb's law, decreases as the distance between the ionizing groups increases. For example, the pK's of **oxalic acid's** two adjacent carboxyl groups differ by 3 pH units (Table 2-3), whereas those of **succinic acid,** in which the carboxyl groups are separated by two methylene groups, differ by 1.4 units.

$$\underset{\textbf{Oxalic acid}}{\text{H}-\text{O}-\overset{\text{O}}{\overset{\|}{\text{C}}}-\overset{\text{O}}{\overset{\|}{\text{C}}}-\text{O}-\text{H}} \qquad \underset{\textbf{Succinic acid}}{\text{H}-\text{O}-\overset{\text{O}}{\overset{\|}{\text{C}}}-\text{CH}_2\text{CH}_2-\overset{\text{O}}{\overset{\|}{\text{C}}}-\text{O}-\text{H}}$$

Likewise, successive ionizations from the same center, such as in H_3PO_4 or H_2CO_3, have pK's that differ by 4 to 5 pH units. If the pK's for successive ionizations of a polyprotic acid differ by at least 3 pH units, it can be accurately assumed that, at a given pH, only the members of the conjugate acid–base pair characterized by the nearest pK are present in significant concentrations. This, of course, greatly simplifies the calculations for determining the concentrations of the various ionic species present.

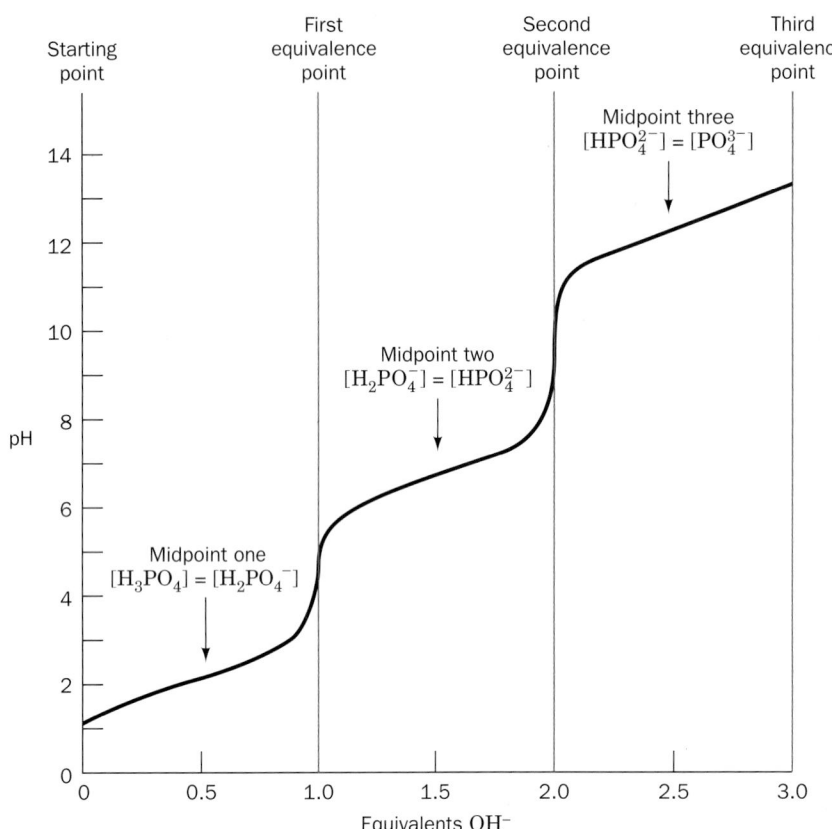

FIGURE 2-12   Titration curve of a 1-L solution of 1*M* H₃PO₄.
The two intermediate equivalence points occur at the steepest
parts of the curve. Note the flatness of the curve near its starting points and end points in comparison with the curved ends
of the titration curves in Fig. 2-10. This indicates that H_3PO_4
(pK_1 = 2.15) is verging on being a strong acid and PO_4^{3-} (pK_3
= 12.38) is verging on being a strong base.
See the Animated Figures.

a. Polyprotic Acids with Closely Spaced pK's Have Molecular Ionization Constants

If the p*K*'s of a polyprotic acid differ by less than ~2 pH
units, as is true in perhaps the majority of biomolecules,
the ionization constants measured by titration are not true
group ionization constants but, rather, reflect the average
ionization of the groups involved. The resulting ionization
constants are therefore known as **molecular ionization
constants.**

Consider the acid–base equilibria shown in Fig. 2-13 in
which there are two nonequivalent protonation sites. Here,
the quantities K_A, K_B, K_C, and K_D, the ionization constants for each group, are alternatively called **microscopic
ionization constants.** The molecular ionization constant for
the removal of the first proton from HAH is

$$K_1 = \frac{[H^+]([AH^-] + [HA^-])}{[HAH]} = K_A + K_B \quad [2.10]$$

Similarly, the molecular ionization constant K_2 for the removal of the second proton is

$$K_2 = \frac{[H^+][A^{2-}]}{[AH^-] + [HA^-]} = \frac{1}{(1/K_C) + (1/K_D)}$$

$$= \frac{K_C K_D}{K_C + K_D} \quad [2.11]$$

If $K_A \gg K_B$, then $K_1 \approx K_A$; that is, the first molecular
ionization constant is equal to the microscopic ionization
constant of the more acidic group. Likewise, if $K_D \gg K_C$,
then $K_2 \approx K_C$, so that the second molecular ionization constant is the microscopic ionization constant of the less
acidic group. If the ionization steps differ sufficiently in
their p*K*'s, the molecular ionization constants, as expected,
become identical to the microscopic ionization constants.

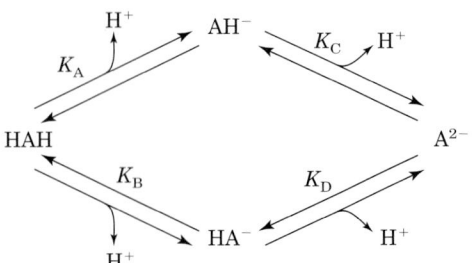

**FIGURE 2-13   Ionization of an acid that has two nonequivalent
protonation sites.**

CHAPTER SUMMARY

1 ■ Properties of Water Water is an extraordinary substance, the properties of which are of great biological importance. A water molecule can simultaneously participate in as many as four hydrogen bonds: two as a donor and two as an acceptor. These hydrogen bonds are responsible for the open, low-density structure of ice. Much of this hydrogen bonded structure exists in the liquid phase, as is evidenced by the high boiling point of water compared to those of other substances of similar molecular masses. Physical and theoretical evidence indicates that liquid water maintains a rapidly fluctuating, hydrogen bonded molecular structure that, over short ranges, resembles that of ice. The unique solvent properties of water derive from its polarity as well as its hydrogen bonding properties. In aqueous solutions, ionic and polar substances are surrounded by multiple concentric hydration shells of oriented water dipoles that act to attenuate the electrostatic interactions between the charges in the solution. The thermal randomization of the oriented water molecules is resisted by their hydrogen bonding associations, thereby accounting for the high dielectric constant of water. Nonpolar substances are essentially insoluble in water. However, amphipathic substances aggregate in aqueous solutions to form micelles and bilayers due to the combination of hydrophobic interactions among the nonpolar portions of these molecules and the hydrophilic interactions of their polar groups with the aqueous solvent. The H_3O^+ and OH^- ions have anomalously large ionic mobilities in aqueous solutions because the migration of these ions through solution occurs largely via proton jumping from one H_2O molecule to another.

2 ■ Acids, Bases, and Buffers A Brønsted acid is a substance that can donate protons, whereas a Brønsted base can accept protons. On losing a proton, a Brønsted acid becomes its conjugate base. In an acid–base reaction, an acid donates its proton to a base. Water can react as an acid to form hydroxide ion, OH^-, or as a base to form hydronium ion, H_3O^+. The strength of an acid is indicated by the magnitude of its dissociation constant, K. Weak acids, which have a dissociation constant less than that of H_3O^+, are only partially dissociated in aqueous solution. Water has the dissociation constant $10^{-14} M$ at 25°C. A practical quantity for expressing the acidity of a solution is pH = $-\log[H^+]$. The relationship between pH, pK, and the concentrations of the members of its conjugate acid–base pair is expressed by the Henderson–Hasselbalch equation. An acid–base buffer is a mixture of a weak acid with its conjugate base in a solution that has a pH near the pK of the acid. The ratio $[A^-]/[HA]$ in a buffer is not very sensitive to the addition of strong acids or bases, so that the pH of a buffer is not greatly affected by these substances. Buffers are operationally effective only in the pH range of p$K \pm 1$. Outside of this range, the pH of the solution changes rapidly with the addition of strong acid or base. Buffer capacity also depends on the total concentration of the conjugate acid–base pair. Biological fluids are generally buffered near neutrality. Many acids are polyprotic. However, unless the pK's of their various ionizations differ by less than 2 or 3 pH units, pH calculations can effectively treat them as if they were a mixture of separate weak acids. For polyprotic acids with pK's that differ by less than this amount, the observed molecular ionization constants are simply related to the microscopic ionization constants of the individual dissociating groups.

REFERENCES

Cooke, R. and Kuntz, I.D., The properties of water in biological systems, *Ann. Rev. Biophys. Bioeng.* **3,** 95–126 (1974).

Edsall, J.T. and Wyman, J., *Biophysical Chemistry,* Vol. 1, Chapters 2, 8, and 9, Academic Press (1958). [Contains detailed treatments on the structure of water and on acid–base equilibria.]

Eisenberg, D. and Kauzman, W., *The Structure and Properties of Water,* Oxford University Press (1969). [A comprehensive monograph with a wealth of information.]

Franks, F., *Water,* The Royal Society of Chemistry (1993).

Gestein, M. and Levitt, M., Simulating water and the molecules of life, *Sci. Am.* **279**(5), 100–105 (1998).

Martin, T.W. and Derewenda, Z.S., The name is bond—H bond, *Nature Struct. Biol.* **6,** 403–406 (1999). [Reviews the history and nature of the hydrogen bond and describes the X-ray scattering experiments that demonstrated that hydrogen bonds have a partially covalent character.]

Segel, I.H., *Biochemical Calculations* (2nd ed.), Chapter 1, Wiley (1976). [An intermediate-level discussion of acid–base chemistry with worked-out problems.]

Stillinger, F.H., Water revisited, *Science* **209,** 451–457 (1980). [An outline of water structure on an elementary level.]

Tanford, C., *The Hydrophobic Effect: Formation of Micelles and Biological Membranes* (2nd ed.), Chapters 5 and 6, Wiley–Interscience (1980). [Discussion of the structure of water and of micelles.]

Westhof, E., *Water and Biological Macromolecules,* CRC Press (1993).

Zumdahl, S.S., *Chemical Principles* (4th ed.), Chapters 7 and 8, Houghton Mifflin (2002). [Discusses acid–base chemistry. Most other general chemistry textbooks contain similar information.]

PROBLEMS

1. Draw the hydrogen bonding pattern that water forms with acetamide (CH_3CONH_2) and with pyridine (benzene with a CH group replaced by N).

2. Explain why the dielectric constants of the following pairs of liquids have the order given in Table 2-1: (a) carbon tetrachloride and chloroform; (b) ethanol and methanol; and (c) acetone and formamide.

3. "Inverted" micelles are made by dispersing amphipathic molecules in a nonpolar solvent, such as benzene, together with a small amount of water (counterions are also provided if the head groups are ionic). Draw the structure of an inverted micelle and describe the forces that stabilize it.

*4. Amphipathic molecules in aqueous solutions tend to concentrate at surfaces such as liquid–solid or liquid–gas interfaces. They are therefore said to be **surface-active molecules** or **surfactants.** Rationalize this behavior in terms of the properties of the amphiphiles and indicate the effect that surface-active molecules have on the surface tension of water (surface tension is a measure of the internal cohesion of a liquid as manifested by the force necessary to increase its surface area). Explain why surfactants such as soaps and detergents are effective in dispersing oily substances and oily dirt in aqueous solutions. Why do aqueous solutions of surfactants foam and why does the presence of oily substances reduce this foaming?

5. Indicate how hydrogen bonding forces and hydrophobic forces vary with the dielectric constant of the medium.

6. Using the data in Table 2-2, indicate the times it would take a K^+ and an H^+ ion to each move 1 cm in an electric field of $100 \text{ V} \cdot \text{cm}^{-1}$.

7. Explain why the mobility of H^+ in ice is only about an order of magnitude less than that in liquid water, whereas the mobility of Na^+ in solid NaCl is zero.

8. Calculate the pH of: (a) $0.1M$ HCl; (b) $0.1M$ NaOH; (c) $3 \times 10^{-5}M$ HNO_3; (d) $5 \times 10^{-10}M$ $HClO_4$; and (e) $2 \times 10^{-8}M$ KOH.

9. The volume of a typical bacterial cell is on the order of 1.0 μm^3. At pH 7, how many hydrogen ions are contained inside a bacterial cell? A bacterial cell contains thousands of macromolecules, such as proteins and nucleic acids, that each bear multiple ionizable groups. What does your result indicate about the common notion that ionizable groups are continuously bathed with H^+ and OH^- ions?

10. Using the data in Table 2-3, calculate the concentrations of all molecular and ionic species and the pH in aqueous solutions that have the following formal compositions: (a) $0.01M$ acetic acid; (b) $0.25M$ ammonium chloride; (c) $0.05M$ acetic acid $+0.10$M sodium acetate; and (d) $0.20M$ boric acid $[B(OH)_3]$ + $0.05M$ sodium borate $[NaB(OH)_4]$.

11. Acid–base indicators are weak acids that change color on changing ionization states. When a small amount of an appropriately chosen indicator is added to a solution of an acid or base being titrated, the color change "indicates" the **end point** of the

titration. **Phenolphthalein** is a commonly used acid–base indicator that, in aqueous solutions, changes from colorless to red-violet in the pH range between 8.2 and 10.0. Referring to Figs. 2-10 and 2-12, describe the effectiveness of phenolphthalein for accurately detecting the end point of a titration with strong base of: (a) acetic acid; (b) NH_4Cl; and (c) H_3PO_4 (at each of its three equivalence points).

*12. The formal composition of an aqueous solution is $0.12M$ K_2HPO_4 + $0.08M$ KH_2PO_4. Using the data in Table 2-3, calculate the concentrations of all ionic and molecular species in the solution and the pH of the solution.

13. Distilled water in equilibrium with air contains dissolved carbon dioxide at a concentration of $1.0 \times 10^{-5}M$. Using the data in Table 2-3, calculate the pH of such a solution.

14. Calculate the formal concentrations of acetic acid and sodium acetate necessary to prepare a buffer solution of pH 5 that is $0.20M$ in total acetate. The pK of acetic acid is given in Table 2-3.

15. In order to purify a certain protein, you require $0.1M$ glycine buffer at pH 9.4. Unfortunately, your stockroom has run out of glycine. However, you manage to find two $0.1M$ glycine buffer solutions, one at pH 9.0 and the other at pH 10.0. What volumes of these two solutions must you mix in order to obtain 200 mL of your required buffer?

16. An enzymatic reaction takes place in 10 mL of a solution that has a total citrate concentration of 120 mM and an initial pH of 7.00. During the reaction (which does not involve citrate), 0.2 milliequivalents of acid are produced. Using the data in Table 2-3, calculate the final pH of the solution. What would the final pH of the solution be in the absence of the citrate buffer assuming that the other components of the solution have no significant buffering capacity and that the solution is initially at pH 7?

*17. A solution's **buffer capacity,** β, is defined as the ratio of an incremental amount of base added, in equivalents, to the corresponding pH change. This is the reciprocal of the slope of the titration curve, Eq. [2.9]. Derive the equation for β and show that it is maximal at pH = pK.

18. Using the data in Table 2-3, calculate the microscopic ionization constants for oxalic acid and for succinic acid. How do these values compare with their corresponding molecular ionization constants?

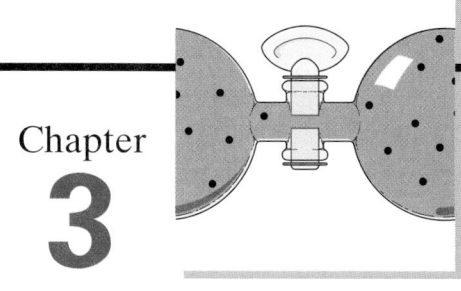

Chapter 3

Thermodynamic Principles: A Review

You can't win.
First law of thermodynamics
You can't even break even.
Second law of thermodynamics
You can't stay out of the game.
Third law of thermodynamics

Living things require a continuous throughput of energy. For example, through photosynthesis, plants convert radiant energy from the sun, the primary energy source for life on Earth, to the chemical energy of carbohydrates and other organic substances. The plants, or the animals that eat them, then metabolize these substances to power such functions as the synthesis of biomolecules, the maintenance of concentration gradients, and the movement of muscles. These processes ultimately transform the energy to heat, which is dissipated to the environment. A considerable portion of the cellular biochemical apparatus must therefore be devoted to the acquisition and utilization of energy.

Thermodynamics (Greek: *therme,* heat + *dynamis,* power) is a marvelously elegant description of the relationships among the various forms of energy and how energy affects matter on the macroscopic as opposed to the molecular level; that is, it deals with amounts of matter large enough for their average properties, such as temperature and pressure, to be well defined. Indeed, the basic principles of thermodynamics were developed in the nineteenth century before the atomic theory of matter had been generally accepted.

With a knowledge of thermodynamics we can determine whether a physical process is possible. Thermodynamics is therefore essential for understanding why macromolecules fold to their native conformations, how metabolic pathways are designed, why molecules cross biological membranes, how muscles generate mechanical force, and so on. The list is endless. Yet the reader should be cautioned that thermodynamics does not indicate the rates at which possible processes actually occur. For instance, although thermodynamics tells us that glucose and oxygen react with the release of copious amounts of energy, it does not indicate that this mixture is indefinitely stable at room temperature in the absence of the appropriate enzymes. The prediction of reaction rates requires, as we shall see in Section 14-1C, a mechanistic description of molecular processes. Yet thermodynamics is also an indispensable guide in formulating such mechanistic models because such models must conform to thermodynamic principles.

Thermodynamics, as it applies to biochemistry, is most frequently concerned with describing the conditions under which processes occur *spontaneously* (by themselves). We shall consequently review the elements of thermodynamics that enable us to predict chemical and biochemical spontaneity: the first and second laws of thermodynamics, the concept of free energy, and the nature of processes at equilibrium. Familiarity with these principles is indispensable for understanding many of the succeeding discussions in this text. We shall, however, postpone consideration of the thermodynamic aspects of metabolism until Sections 16-4 through 16-6.

1 ■ FIRST LAW OF THERMODYNAMICS: ENERGY IS CONSERVED

In thermodynamics, a **system** is defined as that part of the universe that is of interest, such as a reaction vessel or an organism; the rest of the universe is known as the **surroundings.** A system is said to be **open, closed,** or **isolated** according to whether or not it can exchange matter and energy with its surroundings, only energy, or neither matter nor energy. Living organisms, which take up nutrients, release waste products, and generate work and heat, are examples of open systems; if an organism were sealed inside an uninsulated box, it would, together with the box, constitute a closed system, whereas if the box were perfectly insulated, the system would be isolated.

A. *Energy*

The **first law of thermodynamics** is a mathematical statement of the law of conservation of energy: *Energy can be neither created nor destroyed.*

$$\Delta U = U_{final} - U_{initial} = q - w \qquad [3.1]$$

Here U is energy, q represents the **heat** absorbed *by* the system *from* the surroundings, and w is the **work** done *by* the system *on* the surroundings. Heat is a reflection of random molecular motion, whereas work, which is defined as force times the distance moved under its influence, is associated with organized motion. Force may assume many different forms, including the gravitational force exerted by one mass on another, the expansional force exerted by a gas, the tensional force exerted by a spring or muscle fiber, the electrical force of one charge on another, or the dissipative forces of friction and viscosity. Processes in which the system releases heat, which by convention are assigned a negative q, are known as **exothermic processes** (Greek: *exo,* out of); those in which the system gains heat (positive q) are known as **endothermic processes** (Greek: *endon,* within). Under this convention, work done by the system against an external force is defined as a positive quantity.

The SI unit of energy, the **joule (J),** is steadily replacing the **calorie (cal)** in modern scientific usage. The **large calorie (Cal,** with a capital C) is a unit favored by nutritionists. The relationships among these quantities and other units, as well as the values of constants that will be useful throughout this chapter, are collected in Table 3-1.

a. State Functions Are Independent of the Path a System Follows

Experiments have invariably demonstrated that the energy of a system depends only on its current properties or state, not on how it reached that state. For example, the state of a system composed of a particular gas sample is completely described by its pressure and temperature. The energy of this gas sample is a function only of these so-called **state functions** (quantities that depend only on the state of the system) and is therefore a state function itself. Consequently, there is no net change in energy ($\Delta U = 0$)

TABLE 3-1 **Thermodynamic Units and Constants**

Joule (J)
   $1 \text{ J} = 1 \text{ kg} \cdot \text{m}^2 \cdot \text{s}^{-2}$ $1 \text{ J} = 1 \text{ C} \cdot \text{V (coulomb volt)}$
   $1 \text{ J} = 1 \text{ N} \cdot \text{m (newton meter)}$

Calorie (cal)
   1 cal heats 1 g of H_2O from 14.5 to 15.5°C
   $1 \text{ cal} = 4.184 \text{ J}$

Large calorie (Cal)
   $1 \text{ Cal} = 1 \text{ kcal}$ $1 \text{ Cal} = 4184 \text{ J}$

Avogadro's number (N)
   $N = 6.0221 \times 10^{23}$ molecules $\cdot$ mol^{-1}

Coulomb (C)
   $1 \text{ C} = 6.241 \times 10^{18}$ electron charges

Faraday ($\mathcal{F}$)
   $1 \mathcal{F} = N$ electron charges
   $1 \mathcal{F} = 96{,}485 \text{ C} \cdot \text{mol}^{-1} = 96{,}485 \text{ J} \cdot \text{V}^{-1} \cdot \text{mol}^{-1}$

Kelvin temperature scale (K)
   $0 \text{ K} = $ absolute zero $273.15 \text{ K} = 0°C$

Boltzmann constant (k_B)
   $k_B = 1.3807 \times 10^{-23} \text{ J} \cdot \text{K}^{-1}$

Gas constant (R)
   $R = Nk_B$ $R = 1.9872 \text{ cal} \cdot \text{K}^{-1} \cdot \text{mol}^{-1}$
   $R = 8.3145 \text{ J} \cdot \text{K}^{-1} \cdot \text{mol}^{-1}$ $R = 0.08206 \text{ L} \cdot \text{atm} \cdot \text{K}^{-1} \cdot \text{mol}^{-1}$

for any process in which the system returns to its initial state (a **cyclic process**).

Neither heat nor work is separately a state function because each is dependent on the **path** followed by a system in changing from one state to another. For example, in the process of changing from an initial to a final state, a gas may do work by expanding against an external force, or do no work by following a path in which it encounters no external resistance. If Eq. [3.1] is to be obeyed, heat must also be path dependent. It is therefore meaningless to refer to the heat or work content of a system (in the same way that it is meaningless to refer to the number of one dollar bills and ten dollar bills in a bank account containing $85.00). To indicate this property, the heat or work produced during a change of state is never referred to as Δq or Δw but rather as just q or w.

B. *Enthalpy*

Any combination of only state functions must also be a state function. One such combination, which is known as **enthalpy** (Greek: *enthalpein,* to warm in), is defined

$$H = U + PV \qquad [3.2]$$

where V is the volume of the system and P is its pressure. Enthalpy is a particularly convenient quantity with which to describe biological systems because *under constant pressure, a condition typical of most biochemical processes, the enthalpy change between the initial and final states of a process, ΔH, is the easily measured heat that it generates or absorbs.* To show this, let us divide work into two categories: pressure–volume (P–V) work, which is work performed by expansion against an external pressure ($P\Delta V$), and all other work (w'):

$$w = P\Delta V + w' \qquad [3.3]$$

Then, by combining Eqs. [3.1], [3.2], and [3.3], we see that

$$\Delta H = \Delta U + P\Delta V = q_P - w + P\Delta V = q_P - w' \qquad [3.4]$$

where q_P is the heat transferred at constant pressure. Thus if $w' = 0$, as is often true of chemical reactions, $\Delta H = q_P$. Moreover, the volume changes in most biochemical processes are negligible, so that the differences between their ΔU and ΔH values are usually insignificant.

We are now in a position to understand the utility of state functions. For instance, suppose we wished to determine the enthalpy change resulting from the complete oxidation of 1 g of glucose to CO_2 and H_2O by muscle tissue. To make such a measurement directly would present enormous experimental difficulties. For one thing, the enthalpy changes resulting from the numerous metabolic reactions not involving glucose oxidation that normally occur in living muscle tissue would greatly interfere with our enthalpy measurement. Since enthalpy is a state function, however, we can measure glucose's enthalpy of combustion in any apparatus of our choosing, say, a constant pressure calorimeter rather than a muscle, and still obtain the same value. This, of course, is true whether or not we know the mechanism through which muscle converts glucose to CO_2 and H_2O, as long as we can establish that these substances actually are the final metabolic products. *In general, the change of enthalpy in any hypothetical reaction pathway can be determined from the enthalpy change in any other reaction pathway between the same reactants and products.*

We stated earlier in the chapter that thermodynamics serves to indicate whether a particular process occurs spontaneously. Yet the first law of thermodynamics cannot, by itself, provide the basis for such an indication, as the following example demonstrates. If two objects at different temperatures are brought into contact, we know that heat spontaneously flows from the hotter object to the colder one, never vice versa. Yet either process is consistent with the first law of thermodynamics since the aggregate energy of the two objects is independent of their temperature distribution. Consequently, we must seek a criterion of spontaneity other than only conformity to the first law of thermodynamics.

2 ■ SECOND LAW OF THERMODYNAMICS: THE UNIVERSE TENDS TOWARD MAXIMUM DISORDER

When a swimmer falls into the water (a spontaneous process), the energy of the coherent motion of his body is converted to that of the chaotic thermal motion of the surrounding water molecules. The reverse process, the swimmer being ejected from still water by the sudden coherent motion of the surrounding water molecules, has never been witnessed even though such a phenomenon violates neither the first law of thermodynamics nor Newton's laws of motion. This is because *spontaneous processes are characterized by the conversion of order (in this case the coherent motion of the swimmer's body) to chaos (here the random thermal motion of the water molecules).* The **second law of thermodynamics,** which expresses this phenomenon, therefore provides a criterion for determining whether a process is spontaneous. Note that thermodynamics says nothing about the rate of a process; that is the purview of **chemical kinetics** (Chapter 14). Thus a spontaneous process might proceed at only an infinitesimal rate.

A. *Spontaneity and Disorder*

The second law of thermodynamics states, in accordance with all experience, that *spontaneous processes occur in directions that increase the overall **disorder** of the universe,* that is, of the system and its surroundings. Disorder, in this context, is defined as the number of equivalent ways, W, of arranging the components of the universe. To illustrate this point, let us consider an isolated system consisting of two bulbs of equal volume containing a total of N identical molecules of ideal gas (Fig. 3-1). When the stopcock connecting the bulbs is open, there is an equal probability that a given molecule will occupy either bulb, so there are a total of 2^N equally probable ways that the N molecules may be distributed among the two bulbs. Since the gas molecules are indistinguishable from one another, there are only ($N + 1$) different states of the system: those with 0, 1, 2, . . . , ($N-1$), or N molecules in the left bulb. Probability theory indicates that the number of (indistinguishable) ways, W_L, of placing L of the N molecules in the left bulb is

$$W_L = \frac{N!}{L!(N - L)!}$$

The probability of such a state occurring is its fraction of the total number of possible states: $W_L/2^N$.

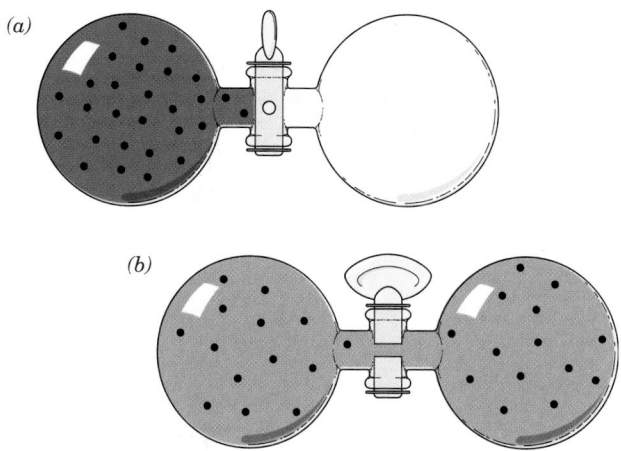

(a)

(b)

FIGURE 3-1 Two bulbs of equal volumes connected by a stopcock. In (*a*), a gas occupies the left bulb, the right bulb is evacuated, and the stopcock is closed. When the stopcock is opened (*b*), the gas molecules diffuse back and forth between the bulbs and eventually become distributed, so that half of them occupy each bulb.

For any value of N, the state that is most probable, that is, the one with the highest value of W_L, is the one with half of the molecules in one bulb ($L = N/2$ for N even). As N becomes large, the probability that L is nearly equal to $N/2$ approaches unity: For instance, when $N = 10$ the probability that L is within 20% of $N/2$ (that is, 4, 5, or 6) is 0.66, whereas for $N = 50$ this probability (that L is in the range 20–30) is 0.88. For a chemically significant number of molecules, say $N = 10^{23}$, the probability that the number of molecules in the left bulb differs from those in the right by as insignificant a ratio as 1 molecule in every 10 billion is 10^{-434}, which, for all intents and purposes, is zero. Therefore, the reason the number of molecules in each bulb of the system in Fig. 3-1*b* is always observed to be equal is not because of any law of motion; the energy of the system is the same for any arrangement of the molecules. *It is because the aggregate probability of all other states is so utterly insignificant* (Fig. 3-2). By the same token, the reason that our swimmer is never thrown out of the water or even noticeably disturbed by the chance coherent motion of the surrounding water molecules is that the probability of such an event is nil.

B. *Entropy*

In chemical systems, W, the number of equivalent ways of arranging a system in a particular state, is usually inconveniently immense. For example, when the above twin-bulb system contains N gas molecules, $W_{N/2} \approx 10^{N \ln 2}$ so that for $N = 10^{23}$, $W_{5 \times 10^{22}} \approx 10^{7 \times 10^{22}}$. In order to be able to deal with W more easily, we define, as did Ludwig Boltzmann in 1877, a quantity known as **entropy** (Greek: *en*, in + *trope*, turning):

$$S = k_B \ln W \qquad [3.5]$$

that increases with W but in a more manageable way. Here k_B is the **Boltzmann constant** (Table 3-1). For our twin-bulb system, $S = k_B N \ln 2$, so the entropy of the system in its most probable state is proportional to the number of gas

molecules it contains. Note that *entropy is a state function because it depends only on the parameters that describe a state.*

The laws of random chance cause any system of reasonable size to spontaneously adopt its most probable arrangement, the one in which entropy is a maximum, simply because this state is so overwhelmingly probable. For example, assume that all N molecules of our twin-bulb system are initially placed in the left bulb (Fig. 3-1*a*; $W_N = 1$ and $S = 0$ since there is only one way of doing this). After the stopcock is opened, the molecules will randomly diffuse in and out of the right bulb until eventually they

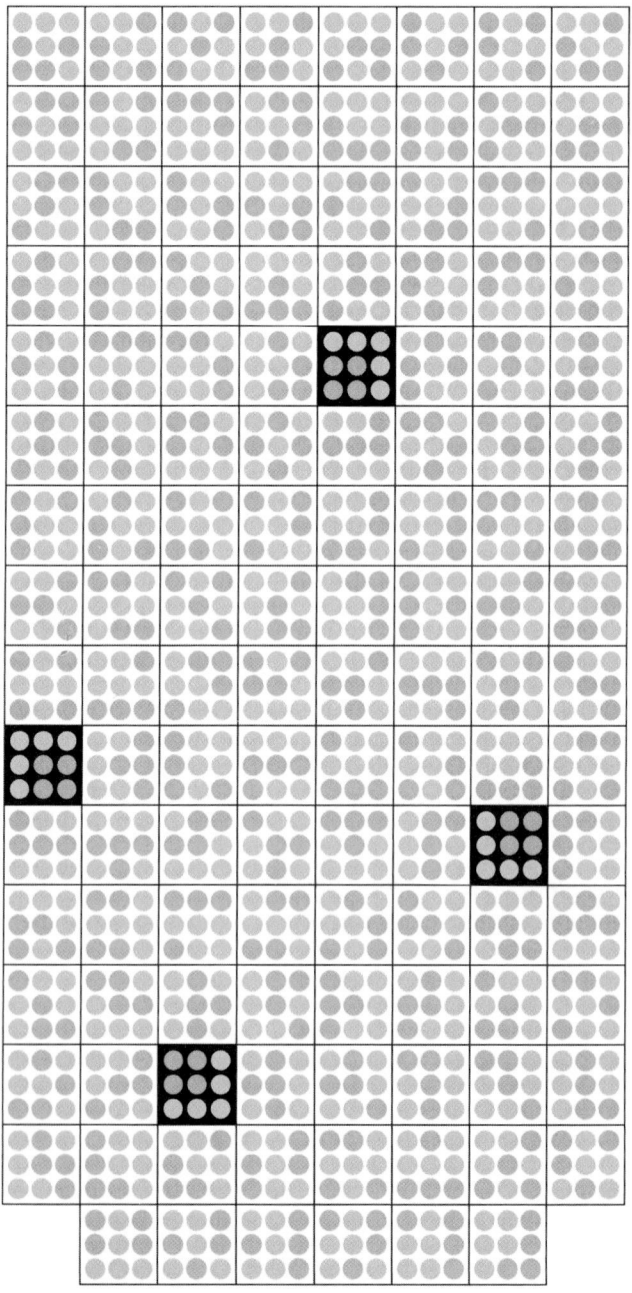

FIGURE 3-2   The improbability of even a small amount of order. Consider a simple "universe" consisting of a square array of 9 positions that collectively contain 4 identical "molecules" (*red dots*). If the 4 molecules are arranged in a square, we shall call the arrangement a "crystal"; otherwise we shall call it a "gas." The total number of distinguishable arrangements of our 4 molecules in 9 positions is given by

$$W = \frac{9 \cdot 8 \cdot 7 \cdot 6}{4 \cdot 3 \cdot 2 \cdot 1} = 126$$

Here, the numerator indicates that the first molecule may occupy any of the universe's 9 positions, the second molecule may occupy any of the 8 remaining unoccupied positions, and so on, whereas the denominator corrects for the number of indistinguishable arrangements of the 4 identical molecules. Of the 126 arrangements this universe can have, only 4 are crystals (*black squares*). Thus, even in this simple universe, there is a more than 30-fold greater probability that it will contain a disordered gas, when arranged at random, than an ordered crystal. [Illustration, Irving Geis/Geis Archives Trust. Copyright Howard Hughes Medical Institute. Reproduced with permission.]

A crystal   A gas

achieve their most probable (maximum entropy) state, that with half of the molecules in each bulb. The gas molecules will subsequently continue to diffuse back and forth between the bulbs, but there will be no further macroscopic (net) change in the system. The system is therefore said to have reached **equilibrium.**

According to Eq. [3.5], the foregoing spontaneous expansion process causes the system's entropy to increase. In general, *for any constant energy process ($\Delta U = 0$), a spontaneous process is characterized by $\Delta S > 0$.* Since the energy of the universe is constant (energy can assume different forms but can be neither created nor destroyed), *any spontaneous process must cause the entropy of the universe to increase:*

$$\Delta S_{system} + \Delta S_{surroundings} = \Delta S_{universe} > 0 \qquad [3.6]$$

Equation [3.6] is the usual expression for the second law of thermodynamics. It is a statement of the general tendency of all spontaneous processes to disorder the universe; that is, *the entropy of the universe tends toward a maximum.*

The conclusions based on our twin-bulb apparatus may be applied to explain, for instance, why blood transports O_2 and CO_2 between the lungs and the tissues. Solutes in solution behave analogously to gases in that they tend to maintain a uniform concentration throughout their occupied volume because this is their most probable arrangement. In the lungs, where the concentration of O_2 is higher than that in the venous blood passing through them, more O_2 enters the blood than leaves it. On the other hand, in the tissues, where the O_2 concentration is lower than that in the arterial blood, there is net diffusion of O_2 from the blood to the tissues. The reverse situation holds for CO_2 transport since the CO_2 concentration is low in the lungs but high in the tissues. Keep in mind, however, that thermodynamics says nothing about the rates at which O_2 and CO_2 are transported to and from the tissues. The rates of these processes depend on the physicochemical properties of the blood, the lungs, and the cardiovascular system.

Equation [3.6] does not imply that a particular system cannot increase its degree of order. As is explained in Section 3-3, however, *a system can only be ordered at the expense of disordering its surroundings to an even greater extent by the application of energy to the system.* For example, living organisms, which are organized from the molecular level upward and are therefore particularly well ordered, achieve this order at the expense of disordering the nutrients they consume. Thus, *eating is as much a way of acquiring order as it is of gaining energy.*

A state of a system may constitute a distribution of more complicated quantities than those of gas molecules in a bulb or simple solute molecules in a solvent. For example, if our system consists of a protein molecule in aqueous solution, its various states differ, as we shall see, in the conformations of the protein's amino acid residues and in the distributions and orientations of its associated water molecules. The second law of thermodynamics applies here because a protein molecule in aqueous solution assumes its native conformation largely in response to the tendency of its surrounding water structure to be maximally disordered (Section 8-4C).

C. *Measurement of Entropy*

In chemical and biological systems, it is impractical, if not impossible, to determine the entropy of a system by counting the number of ways, *W,* it can assume its most probable state. An equivalent and more practical definition of entropy was proposed in 1864 by Rudolf Clausius: For spontaneous processes

$$\Delta S \geq \int_{initial}^{final} \frac{dq}{T} \qquad [3.7]$$

where T is the absolute temperature at which the change in heat occurs. The proof of the equivalence of our two definitions of entropy, which requires an elementary knowledge of statistical mechanics, can be found in many physical chemistry textbooks. It is evident, however, that any system becomes progressively disordered (its entropy increases) as its temperature rises (e.g., Fig. 3-3). The equality in Eq. [3.7] holds only for processes in which the

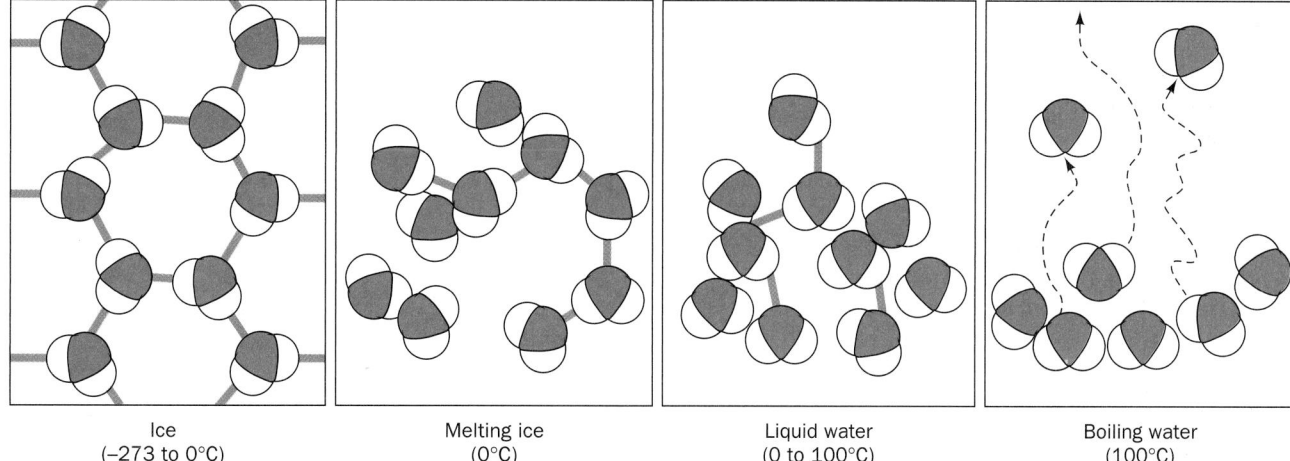

Ice	Melting ice	Liquid water	Boiling water
(−273 to 0°C)	(0°C)	(0 to 100°C)	(100°C)

FIGURE 3-3 Relationship of entropy and temperature. The structure of water, or any other substance, becomes increasingly disordered, that is, its entropy increases, as its temperature rises.

system remains in equilibrium throughout the change; these are known as **reversible processes.**

For the constant temperature conditions typical of biological processes, Eq. [3.7] reduces to

$$\Delta S \geq \frac{q}{T} \qquad [3.8]$$

Thus the entropy change of a reversible process at constant temperature can be determined straightforwardly from measurements of the heat transferred and the temperature at which this occurs. However, since a process at equilibrium can only change at an infinitesimal rate (equilibrium processes are, by definition, unchanging), real processes can approach, but can never quite attain, reversibility. Consequently, *the universe's entropy change in any real process is always greater than its ideal (reversible) value.* This means that when a system departs from and then returns to its initial state via a real process, the entropy of the universe must increase even though the entropy of the system (a state function) does not change.

3 ■ FREE ENERGY: THE INDICATOR OF SPONTANEITY

The disordering of the universe by spontaneous processes is an impractical criterion for spontaneity because it is rarely possible to monitor the entropy of the entire universe. Yet the spontaneity of a process cannot be predicted from a knowledge of the system's entropy change alone. This is because exothermic processes ($\Delta H_{system} < 0$) may be spontaneous even though they are characterized by $\Delta S_{system} < 0$. For example, 2 mol of H_2 and 1 mol of O_2, when sparked, react in a decidedly exothermic reaction to form 2 mol of H_2O. Yet two water molecules, each of whose three atoms are constrained to stay together, are more ordered than are the three diatomic molecules from which they formed. Similarly, under appropriate conditions, many **denatured** (unfolded) proteins will spontaneously fold to assume their highly ordered **native** (normally folded) conformations (Section 9-1A). What we really want, therefore, is a state function that predicts whether or not a given process is spontaneous. In this section, we consider such a function.

A. *Gibbs Free Energy*

The **Gibbs free energy,**

$$G = H - TS \qquad [3.9]$$

which was formulated by J. Willard Gibbs in 1878, is the required indicator of spontaneity for constant temperature and pressure processes. For systems that can only do pressure–volume work ($w' = 0$), combining Eqs. [3.4] and [3.9] while holding T and P constant yields

$$\Delta G = \Delta H - T\Delta S = q_P - T\Delta S \qquad [3.10]$$

But Eq. [3.8] indicates that $T\Delta S \geq q$ for spontaneous processes at constant T. Consequently, $\Delta G \leq 0$ *is the cri-*

TABLE 3-2 **Variation of Reaction Spontaneity (Sign of ΔG) with the Signs of ΔH and ΔS**

ΔH	ΔS	$\Delta G = \Delta H - T\Delta S$
−	+	The reaction is both enthalpically favored (exothermic) and entropically favored. It is spontaneous (exergonic) at all temperatures.
−	−	The reaction is enthalpically favored but entropically opposed. It is spontaneous only at temperatures *below $T = \Delta H/\Delta S$.*
+	+	The reaction is enthalpically opposed (endothermic) but entropically favored. It is spontaneous only at temperatures *above $T = \Delta H/\Delta S$.*
+	−	The reaction is both enthalpically and entropically opposed. It is *un*spontaneous (endergonic) at all temperatures.

terion of spontaneity we seek for the constant T and P conditions that are typical of biochemical processes.

Spontaneous processes, that is, those with negative ΔG values, are said to be **exergonic** (Greek: *ergon*, work); they can be utilized to do work. Processes that are not spontaneous, those with positive ΔG values, are termed **endergonic;** they must be driven by the input of free energy (through mechanisms discussed in Section 3-4C). Processes at equilibrium, those in which the forward and backward reactions are exactly balanced, are characterized by $\Delta G = 0$. Note that the value of ΔG varies directly with temperature. This is why, for instance, the native structure of a protein, whose formation from its denatured form has both $\Delta H < 0$ and $\Delta S < 0$, predominates below the temperature at which $\Delta H = T\Delta S$ (the **denaturation temperature**), whereas the denatured protein predominates above this temperature. The variation of the spontaneity of a process with the signs of ΔH and ΔS is summarized in Table 3-2.

B. *Free Energy and Work*

When a system at constant temperature and pressure does non-P–V work, Eq. [3.10] must be expanded to

$$\Delta G = q_P - T\Delta S - w' \qquad [3.11]$$

or, because $T\Delta S \geq q_P$ (Eq. [3.8]),

$$\Delta G \leq -w'$$

so that

$$\Delta G \geq w' \qquad [3.12]$$

Since P–V work is unimportant in biological systems, ΔG *for a biological process represents its maximum recoverable work.* The ΔG of a process is therefore indicative of the maximum charge separation it can establish, the maximum concentration gradient it can generate (Section 3-4A), the maximum muscular activity it can produce, and so on. In fact, for real processes, which can only approach reversibility, the inequality in Eq. [3.12] holds, so that *the work put into any system can never be fully recovered.* This is indicative of the inherent dissipative character of nature. Indeed, as we have seen, it is precisely this dissipative character that provides the overall driving force for any change.

It is important to reiterate that a large negative value of ΔG does not ensure a chemical reaction will proceed at a measurable rate. This depends on the detailed mechanism of the reaction, which is independent of ΔG. For instance, most biological molecules, including proteins, nucleic acids, carbohydrates, and lipids, are thermodynamically unstable to hydrolysis but, nevertheless, spontaneously hydrolyze at biologically insignificant rates. Only with the introduction of the proper enzymes will the hydrolysis of these molecules proceed at a reasonable pace. Yet a catalyst, which by definition is unchanged by a reaction, cannot affect the ΔG of a reaction. Consequently, *an enzyme can only accelerate the attainment of thermodynamic equilibrium; it cannot, for example, promote a reaction that has a positive ΔG.*

4 ■ CHEMICAL EQUILIBRIA

The entropy (disorder) of a substance increases with its volume. For example, as we have seen for our twin-bulb apparatus (Fig. 3-1), a collection of gas molecules, in occupying all of the volume available to it, maximizes its entropy. Similarly, dissolved molecules become uniformly distributed throughout their solution volume. Entropy is therefore a function of concentration.

If entropy varies with concentration, so must free energy. Thus, as is shown in this section, the free energy change of a chemical reaction depends on the concentrations of both its reactants and its products. This phenomenon is of great biochemical significance because enzymatic reactions can proceed in either direction depending on the relative concentrations of their reactants and products. Indeed, the directions of many enzymatically catalyzed reactions depend on the availability of their **substrates** (reactants) and on the metabolic demand for their products (although most metabolic pathways operate unidirectionally; Section 16-6C).

A. Equilibrium Constants

The relationship between the concentration and the free energy of a substance A, which is derived in the appendix to this chapter, is approximately

$$\overline{G}_A - \overline{G}_A^\circ = RT \ln[A] \qquad [3.13]$$

where $\overline{G}_A$ is known equivalently as the **partial molar free energy** or the **chemical potential** of A (the bar indicates the quantity per mole), $\overline{G}_A^\circ$ is the partial molar free energy of A in its **standard state** (see Section 3-4B), R is the gas constant (Table 3-1), and [A] is the molar concentration of A. Thus for the general reaction,

$$a\text{A} + b\text{B} \rightleftharpoons c\text{C} + d\text{D}$$

since free energies are additive and the free energy change of a reaction is the sum of the free energies of the products less those of the reactants, the free energy change for this reaction is

$$\Delta G = c\overline{G}_C + d\overline{G}_D - a\overline{G}_A - b\overline{G}_B \qquad [3.14]$$

Substituting this relationship into Eq. [3.13] yields

$$\Delta G = \Delta G^\circ + RT \ln\left(\frac{[\text{C}]^c [\text{D}]^d}{[\text{A}]^a [\text{B}]^b}\right) \qquad [3.15]$$

where ΔG° is the free energy change of the reaction when all of its reactants and products are in their standard states. Thus the expression for the free energy change of a reaction consists of two parts: (1) a constant term whose value depends only on the reaction taking place, and (2) a variable term that depends on the concentrations of the reactants and the products, the stoichiometry of the reaction, and the temperature.

For a reaction at equilibrium, there is no *net* change because the free energy of the forward reaction exactly balances that of the backward reaction. Consequently, $\Delta G = 0$, so that Eq. [3.15] becomes

$$\Delta G^\circ = -RT \ln K_{eq} \qquad [3.16]$$

where K_{eq} is the familiar **equilibrium constant** of the reaction:

$$K_{eq} = \frac{[\text{C}]_{eq}^c [\text{D}]_{eq}^d}{[\text{A}]_{eq}^a [\text{B}]_{eq}^b} = e^{-\Delta G^\circ/RT} \qquad [3.17]$$

and the subscript "eq" in the concentration terms indicates their equilibrium values. (The equilibrium condition is usually clear from the context of the situation, so that equilibrium concentrations are often expressed without this subscript.) *The equilibrium constant of a reaction may therefore be calculated from standard free energy data and vice versa.* Table 3-3 indicates the numerical relationship between ΔG° and K_{eq}. Note that a 10-fold variation of K_{eq} at 25°C corresponds to a $5.7 \text{ kJ} \cdot \text{mol}^{-1}$ change in ΔG°, which is less than half of the free energy of even a weak hydrogen bond.

Equations [3.15] through [3.17] indicate that when the reactants in a process are in excess of their equilibrium

TABLE 3-3   Variation of K_{eq} with ΔG° at 25°C

K_{eq}	$\Delta G^\circ (\text{kJ} \cdot \text{mol}^{-1})$
10^6	-34.3
10^4	-22.8
10^2	-11.4
10^1	-5.7
10^0	0.0
10^{-1}	5.7
10^{-2}	11.4
10^{-4}	22.8
10^{-6}	34.3

concentrations, the net reaction will proceed in the forward direction until the excess reactants have been converted to products and equilibrium is attained. Conversely, when products are in excess, the net reaction proceeds in the reverse reaction so as to convert products to reactants until the equilibrium concentration ratio is likewise achieved. Thus, as **Le Châtelier's principle** states, *any deviation from equilibrium stimulates a process that tends to restore the system to equilibrium. All isolated systems must therefore inevitably reach equilibrium.* Living systems escape this thermodynamic cul-de-sac by being open systems (Section 16-6A).

The manner in which the equilibrium constant varies with temperature is seen by substituting Eq. [3.10] into Eq. [3.16] and rearranging:

$$\ln K_{eq} = \frac{-\Delta H^\circ}{R}\left(\frac{1}{T}\right) + \frac{\Delta S^\circ}{R} \qquad [3.18]$$

where H° and S° represent enthalpy and entropy in the standard state. If ΔH° and ΔS° are independent of temperature, as they often are to a reasonable approximation, a plot of $\ln K_{eq}$ versus $1/T$, known as a **van't Hoff plot,** yields a straight line of slope $-\Delta H^\circ/R$ and intercept $\Delta S^\circ/R$. This relationship permits the values of ΔH° and ΔS° to be determined from measurements of K_{eq} at two (or more) different temperatures. Calorimetric data, which until recently have been quite difficult to measure for biochemical processes, are therefore not required to obtain the values of ΔH° and ΔS°. Consequently, most biochemical thermodynamic data have been obtained through the application of Eq. [3.18]. However, the recent development of the **scanning microcalorimeter** has made the direct measurement of $\Delta H(q_P)$ for biochemical processes a practical alternative. Indeed, a discrepancy between the values of ΔH° for a reaction as determined calorimetrically and from a van't Hoff plot suggests that the reaction occurs via one or more intermediate states in addition to the initial and final states implicit in the formulation of Eq. [3.18].

B. *Standard Free Energy Changes*

Since only free energy differences, ΔG, can be measured, not free energies themselves, it is necessary to refer these differences to some standard state in order to compare the free energies of different substances (likewise, we refer the elevations of geographic locations to sea level, which is arbitrarily assigned the height of zero). By convention, the free energy of all pure elements in their standard state of 25°C, 1 atm, and in their most stable form (e.g., O_2 not O_3), is defined to be zero. The **free energy of formation** of any nonelemental substance, ΔG_f°, is then defined as the change in free energy accompanying the formation of 1 mol of that substance, in its standard state, from its component elements in their standard states. The standard free energy change for any reaction can be calculated according to

$$\Delta G^\circ = \sum \Delta G_f^\circ(\text{products}) - \sum \Delta G_f^\circ(\text{reactants}) \qquad [3.19]$$

TABLE 3-4 **Free Energies of Formation of Some Compounds of Biochemical Interest**

Compound	$-\Delta G_f^\circ$ (kJ $\cdot$ mol^{-1})
Acetaldehyde	139.7
Acetate$^-$	369.2
Acetyl-CoA	374.1[a]
cis-Aconitate^{3-}	920.9
$CO_2(g)$	394.4
$CO_2(aq)$	386.2
HCO_3^-	587.1
Citrate^{3-}	1166.6
Dihydroxyacetone phosphate^{2-}	1293.2
Ethanol	181.5
Fructose	915.4
Fructose-6-phosphate^{2-}	1758.3
Fructose-1, 6-bisphosphate^{4-}	2600.8
Fumarate^{2-}	604.2
α-D-Glucose	917.2
Glucose-6-phosphate^{2-}	1760.2
Glyceraldehyde-3-phosphate^{2-}	1285.6
H^+	0.0
$H_2(g)$	0.0
$H_2O(\ell)$	237.2
Isocitrate^{3-}	1160.0
α-Ketoglutarate^{2-}	798.0
Lactate$^-$	516.6
L-Malate^{2-}	845.1
OH^-	157.3
Oxaloacetate^{2-}	797.2
Phosphoenolpyruvate^{3-}	1269.5
2-Phosphoglycerate^{3-}	1285.6
3-Phosphoglycerate^{3-}	1515.7
Pyruvate$^-$	474.5
Succinate^{2-}	690.2
Succinyl-CoA	686.7[a]

[a] For formation from free elements + free CoA (coenzyme A).

Source: Metzler, D.E., *Biochemistry, The Chemical Reactions of Living Cells, pp.* 162–164, Academic Press (1977).

Table 3-4 provides a list of standard free energies of formation, ΔG_f°, for a selection of substances of biochemical significance.

a. Standard State Conventions in Biochemistry

The standard state convention commonly used in physical chemistry defines the standard state of a solute as that with unit **activity** at 25°C and 1 atm (activity is concentration corrected for nonideal behavior, as is explained in the appendix to this chapter; for the dilute solutions typical of biochemical reactions in the laboratory, such corrections are small, so activities can be replaced by concentrations). However, because biochemical reactions usually occur in dilute aqueous solutions near neutral pH, a somewhat different standard state convention for biological systems has been adopted:

■ Water's standard state is defined as that of the pure liquid, so that the activity of pure water is taken to be unity despite the fact that its concentration is $55.5M$. In essence, the $[H_2O]$ term is incorporated into the value of the equilibrium constant. This procedure simplifies the free energy expressions for reactions in dilute aqueous solutions involving water as a reactant or product because the $[H_2O]$ term can then be ignored.

■ The hydrogen ion activity is defined as unity at the physiologically relevant pH of 7 rather than at the physical chemical standard state of pH 0, where many biological substances are unstable.

■ The standard state of a substance that can undergo an acid–base reaction is defined in terms of the total concentration of its naturally occurring ion mixture at pH 7. In contrast, the physical chemistry convention refers to a pure species whether or not it actually exists at pH 0. The advantage of the biochemistry convention is that the total concentration of a substance with multiple ionization states, such as most biological molecules, is usually easier to measure than the concentration of one of its ionic species. Since the ionic composition of an acid or base varies with pH, however, the standard free energies calculated according to the biochemistry convention are valid only at pH 7.

Under the biochemistry convention, the standard free energy changes of substances are customarily symbolized by $\Delta G^{\circ\prime}$ in order to distinguish them from physical chemistry standard free energy changes, ΔG° (note that the value of ΔG for any process, being experimentally measurable, is independent of the chosen standard state; i.e., $\Delta G = \Delta G'$). Likewise, the biochemical equilibrium constant, which is defined by using $\Delta G^{\circ\prime}$ in place of ΔG° in Eq. [3.17], is represented by K'_{eq}.

The relationship between $\Delta G^{\circ\prime}$ and ΔG° is often a simple one. There are three general situations:

1. If the reacting species include neither H_2O nor H^+, the expressions for $\Delta G^{\circ\prime}$ and ΔG° coincide.

2. For a reaction in dilute aqueous solution that yields $n H_2O$ molecules:

$$A + B \rightleftharpoons C + D + n H_2O$$

Eqs. [3.16] and [3.17] indicate that

$$\Delta G^\circ = -RT \ln K_{eq} = -RT \ln\left(\frac{[C][D][H_2O]^n}{[A][B]}\right)$$

Under the biochemistry convention, which defines the activity of pure water as unity,

$$\Delta G^{\circ\prime} = -RT \ln K'_{eq} = -RT \ln\left(\frac{[C][D]}{[A][B]}\right)$$

Therefore

$$\Delta G^{\circ\prime} = \Delta G^\circ + nRT \ln[H_2O] \qquad [3.20]$$

where $[H_2O] = 55.5M$ (the concentration of water in aqueous solution), so that for a reaction at 25°C which yields 1 mol of H_2O, $\Delta G^{\circ\prime} = \Delta G^\circ + 9.96\,kJ \cdot mol^{-1}$.

3. For a reaction involving hydrogen ions, such as

$$A + B \rightleftharpoons C + HD$$
$$\Big\Updownarrow K$$
$$D^- + H^+$$

where

$$K = \frac{[H^+][D^-]}{[HD]}$$

manipulations similar to those above lead to the relationship

$$\Delta G^{\circ\prime} = \Delta G^\circ - RT \ln(1 + K/[H^+]_0) + RT \ln[H^+]_0 \qquad [3.21]$$

where $[H^+]_0 = 10^{-7}M$, the only value of $[H^+]$ for which this equation is valid. Of course, if more than one ionizable species participates in the reaction and/or if any of them are polyprotic, Eq. [3.21] is correspondingly more complicated.

C. Coupled Reactions

The additivity of free energy changes allows an endergonic reaction to be driven by an exergonic reaction under the proper conditions. This phenomenon is the thermodynamic basis for the operation of metabolic pathways, since most of these reaction sequences comprise endergonic as well as exergonic reactions. Consider the following two-step reaction process:

(1) $A + B \rightleftharpoons C + D \qquad \Delta G_1$
(2) $D + E \rightleftharpoons F + G \qquad \Delta G_2$

If $\Delta G_1 \geq 0$, Reaction (1) will not occur spontaneously. However, if ΔG_2 is sufficiently exergonic so that $\Delta G_1 + \Delta G_2 < 0$, then although the equilibrium concentration of D in Reaction (1) will be relatively small, it will be larger than that in Reaction (2). As Reaction (2) converts D to products, Reaction (1) will operate in the forward direction to replenish the equilibrium concentration of D. The highly exergonic Reaction (2) therefore *drives* the endergonic Reaction (1), and the two reactions are said to be **coupled** through their common intermediate, D. That these coupled reactions proceed spontaneously (although not necessarily at a finite rate) can also be seen by summing Reactions (1) and (2) to yield the overall reaction

(1 + 2) $A + B + E \rightleftharpoons C + F + G \qquad \Delta G_3$

where $\Delta G_3 = \Delta G_1 + \Delta G_2 < 0$. *As long as the overall pathway (reaction sequence) is exergonic, it will operate in the forward direction.* Thus, the free energy of ATP hydrolysis, a highly exergonic process, is harnessed to drive many otherwise endergonic biological processes to completion (Section 16-4C).

■ APPENDIX: CONCENTRATION DEPENDENCE OF FREE ENERGY

To establish that the free energy of a substance is a function of its concentration, consider the free energy change of an ideal gas during a reversible pressure change at constant temperature ($w' = 0$, since an ideal gas is incapable of doing non-P–V work). Substituting Eqs. [3.1] and [3.2] into Eq. [3.9] and differentiating the result yields

$$dG = dq - dw + P\,dV + V\,dP - T\,dS \qquad [3.A1]$$

On substitution of the differentiated forms of Eqs. [3.3] and [3.8] into this expression, it reduces to

$$dG = V\,dP \qquad [3.A2]$$

The ideal gas equation is $PV = nRT$, where n is the number of moles of gas. Therefore

$$dG = nRT\frac{dP}{P} = nRT\,d\ln P \qquad [3.A3]$$

This gas phase result can be extended to the more biochemically relevant area of solution chemistry by application of **Henry's law** for a solution containing the volatile solute A in equilibrium with the gas phase:

$$P_A = K_A X_A \qquad [3.A4]$$

Here P_A is the partial pressure of A when its mole fraction in the solution is X_A, and K_A is the **Henry's law constant** of A in the solvent being used. It is generally more convenient, however, to express the concentrations of the relatively dilute solutions of chemical and biological systems in terms of molarity rather than mole fractions. For a dilute solution

$$X_A \approx \frac{n_A}{n_{solvent}} = \frac{[A]}{[solvent]} \qquad [3.A5]$$

where the solvent concentration, [solvent], is approximately constant. Thus

$$P_A \approx K'_A [A] \qquad [3.A6]$$

where $K'_A = K_A/[solvent]$. Substituting this expression into Eq. [3.A3] yields

$$dG_A = n_A RT\,d(\ln K'_A + \ln[A]) = n_A RT\,d\ln[A] \quad [3.A7]$$

Free energy, as are energy and enthalpy, is a relative quantity that can only be defined with respect to some arbitrary standard state. The standard state is customarily taken to be 25°C, 1 atm pressure, and, for the sake of mathematical simplicity, [A] = 1. The integration of Eq. [3.A7] from the standard state, [A] = 1, to the final state, [A] = [A], results in

$$G_A - G_A^\circ = n_A RT \ln[A] \qquad [3.A8]$$

where G_A° is the free energy of A in the standard state and

[A] really represents the concentration ratio [A]/1. Since Henry's law is valid for real solutions only in the limit of infinite dilution, however, the standard state is defined as the entirely hypothetical state of $1M$ solute with the properties that it has at infinite dilution.

The free energy terms in Eq. [3.A8] may be converted from **extensive quantities** (those dependent on the amount of material) to **intensive quantities** (those independent of the amount of material) by dividing both sides of the equation by n_A. This yields

$$\overline{G}_A - \overline{G}_A^\circ = RT \ln[A] \qquad [3.A9]$$

Equation [3.A9] has the limitation that it refers to solutions that exactly follow Henry's law, although real solutions only do so in the limit of infinite dilution if the solute is, in fact, volatile. These difficulties can all be eliminated by replacing [A] in Eq. [3.A9] by a quantity, a_A, known as the **activity** of A. This is defined

$$a_A = \gamma_A [A] \qquad [3.A10]$$

where γ_A is the **activity coefficient** of A. Equation [3.A9] thereby takes the form

$$\overline{G}_A - \overline{G}_A^\circ = RT \ln a_A \qquad [3.A11]$$

in which all departures from ideal behavior, including the provision that the system may perform non-P–V work, are incorporated into the activity coefficient, which is an experimentally measurable quantity. Ideal behavior is only approached at infinite dilution; that is, $\gamma_A \to 1$ as $[A] \to 0$. The standard state in Eq. [3.A11] is redefined as that of unit activity.

The concentrations of reactants and products in most laboratory biochemical reactions are usually so low (on the order of millimolar or less) that the activity coefficients of these various species are nearly unity. Consequently, the activities of most biochemical species under laboratory conditions can be satisfactorily approximated by their molar concentrations:

$$\overline{G}_A - \overline{G}_A^\circ = RT \ln[A] \qquad [3.13]$$

However, the activity coefficient of a particular species varies with the total concentration of all other species present as well as with its own concentration. Thus, despite the low concentrations of most biochemical species in the cell, their extraordinarily high combined concentrations (e.g., see Fig. 1-13) make the activity coefficients of the individual species deviate significantly from unity. Unfortunately, it is difficult to determine the values of these quantities in a cellular compartment (where it is likewise difficult to determine the concentration of any given species).

CHAPTER SUMMARY

1 ■ First Law of Thermodynamics: Energy Is Conserved
The first law of thermodynamics,

$$U = q - w \qquad [3.1]$$

where q is heat and w is work, is a statement of the law of conservation of energy. Energy is a state function because the energy of a system depends only on the state of the system. Enthalpy,

$$H = U + PV \qquad [3.2]$$

where P is pressure and V is volume, is a closely related state function that represents the heat at constant pressure under conditions where only pressure–volume work is possible.

2 ■ Second Law of Thermodynamics: The Universe Tends toward Maximum Disorder Entropy, which is also a state function, is defined

$$S = k_B \ln W \qquad [3.5]$$

where W, the disorder, is the number of equivalent ways the system can be arranged under the conditions governing it and k_B is the Boltzmann constant. The second law of thermodynamics states that the universe tends toward maximum disorder and hence $\Delta S_{universe} > 0$ for any real process.

3 ■ Free Energy: The Indicator of Spontaneity The Gibbs free energy of a system

$$G = H - TS \qquad [3.9]$$

decreases in a spontaneous, constant pressure process. In a process at equilibrium, the system suffers no net change, so that $\Delta G = 0$. An ideal process, in which the system is always at equilibrium, is said to be reversible. All real processes are irreversible since processes at equilibrium can only occur at an infinitesimal rate.

4 ■ Chemical Equilibria For a chemical reaction

$$aA + bB \rightleftharpoons cC + dD$$

the change in the Gibbs free energy is expressed

$$\Delta G = \Delta G^\circ + RT \ln\left(\frac{[C]^c[D]^d}{[A]^a[B]^b}\right) \qquad [3.15]$$

where ΔG°, the standard free energy change, is the free energy change at 25°C, 1 atm pressure, and unit activities of reactants and products. The biochemical standard state, $\Delta G^{\circ\prime}$, is similarly defined but in dilute aqueous solution at pH 7 in which the activities of water and H^+ are both defined as unity. At equilibrium

$$\Delta G^{\circ\prime} = -RT \ln K'_{eq} = -RT \ln\left(\frac{[C]^c_{eq}[D]^d_{eq}}{[A]^a_{eq}[B]^b_{eq}}\right) \qquad [3.17a]$$

where K'_{eq} is the equilibrium constant under the biochemical convention. An endergonic reaction ($\Delta G > 0$) may be driven by an exergonic reaction ($\Delta G < 0$) if they are coupled and if the overall reaction is exergonic.

REFERENCES

Atkins, P.W., *The Second Law,* Scientific American Books (1984). [An insightful but nonmathematical exposition of the second law of thermodynamics.]

Atkins, P.W. and de Paula, J., *Physical Chemistry* (7th ed.), Chapters 1–10, Freeman (2002). [Most physical chemistry texts treat thermodynamics in some detail.]

Dickerson, R.E., *Molecular Thermodynamics,* Benjamin (1969).

Edsall, J.T. and Gutfreund, H., *Biothermodynamics,* Wiley (1983).

Eisenberg, D. and Crothers, D., *Physical Chemistry with Applications to Life Sciences,* Chapters 1–5, Benjamin (1979).

Hammes, G.G., *Thermodynamics and Kinetics for Biological Sciences,* Wiley (2000).

Nash, L.K., *CHEMTHERMO: A Statistical Approach to Classical Chemical Thermodynamics,* Addison–Wesley (1971). [A delightfully written text on an elementary level.]

Segel, I.H., *Biochemical Calculations* (2nd ed.), Chapter 3, Wiley (1976). [Contains instructive problems accompanied by detailed solutions.]

Tinoco, I., Jr., Sauer, K., Wang, J.C., and Puglisi, J.C., *Physical Chemistry. Principles and Applications in Biological Sciences* (4th ed.), Chapters 2–5, Prentice-Hall (2002).

van Holde, K.E., Johnson, W.C., and Ho, P.S., *Principles of Physical Biochemistry,* Chapters 1–3, Prentice-Hall (1998). [The equivalence of the Boltzmann and Clausius formulations of the second law of thermodynamics is demonstrated in Section 2.3.]

Wood, W.B., Wilson, J.H., Benbow, R.M., and Hood, L.E., *Biochemistry, A Problems Approach* (2nd ed.), Chapter 9, Benjamin/Cummings (1981). [A question-and-answer book.]

PROBLEMS

1. A common funeral litany is the Biblical verse: "Ashes to ashes, dust to dust." Why might a bereaved family of thermodynamicists be equally comforted by a recitation of the second law of thermodynamics?

2. How many flights of 4-m high stairs must an overweight person weighing 75 kg climb to atone for the indiscretion of eating a 500-Cal hamburger? Assume that there is a 20% efficiency in converting nutritional energy to mechanical energy. The gravitational force of an object of mass m kg is $F = mg$, where the gravitational constant g is $9.8 \text{ m} \cdot \text{s}^{-2}$.

3. In terms of thermodynamic concepts, why is it more difficult to park a car in a small space than it is to drive it out from such a space?

4. It has been said that an army of dedicated monkeys, typing at random, would eventually produce all of Shakespeare's works. How long, on average, would it take 1 million monkeys, each typing on a 46-key typewriter (space included but no shift key) at the rate of 1 keystroke per second, to type the phrase "to be or not to be"? How long, on average, would it take one monkey to do so at a computer if the computer would only accept the correct letter in the phrase and then would shift to its next letter (i.e., the computer knew what it wanted)? What do these results indicate about the probability of order randomly arising from disorder versus order arising through a process of evolution?

5. Show that the transfer of heat from an object of higher temperature to one of lower temperature, but not the reverse process, obeys the second law of thermodynamics.

6. Carbon monoxide crystallizes with its CO molecules arranged in parallel rows. Since CO is a very nearly ellipsoidal molecule, in the absence of polarity effects, adjacent CO molecules could equally well line up in a head-to-tail or a head-to-head fashion. In a crystal consisting of 10^{23} CO molecules, what is the entropy of all the CO molecules being aligned head to tail?

7. The U.S. Patent Office has received, and continues to receive, numerous applications for perpetual motion machines. Perpetual motion machines have been classified as those of the first kind, which violate the first law of thermodynamics, and those of the second kind, which violate the second law of thermodynamics. The fallacy in a perpetual motion machine of the first kind is generally easy to detect. An example would be a motor-driven electrical generator that produces energy in excess of that input by the motor. The fallacy in a perpetual motion machine of the second type, however, is usually more subtle. Take, for example, a ship that uses heat energy extracted from the sea by a heat pump to boil water so as to power a steam engine that drives the ship as well as the heat pump. Show, in general terms, that such a propulsion system would violate the second law of thermodynamics.

8. Using the data in Table 3-4, calculate the values of $\Delta G°$ at 25°C for the following metabolic reactions:

(a) $C_6H_{12}O_6 + 6\,O_2 \rightleftharpoons 6\,CO_2(aq) + 6\,H_2O(\ell)$
   Glucose

(b) $C_6H_{12}O_6 \rightleftharpoons 2(CH_3CH_2OH) + 2\,CO_2(aq)$
   Glucose   **Ethanol**

(c) $C_6H_{12}O_6 \rightleftharpoons 2(CH_3CHOHCOO^-) + 2\,H^+$
   Glucose   **Lactate**

[These reactions, respectively, constitute oxidative metabolism, alcoholic fermentation in yeast deprived of oxygen, and homolactic fermentation in skeletal muscle requiring energy faster than oxidative metabolism can supply it (Section 17-3B).]

9. The native and denatured forms of a protein are generally in equilibrium as follows:

$$\text{Protein } (denatured) \rightleftharpoons \text{protein } (native)$$

For a certain solution of the protein **ribonuclease A,** in which the total protein concentration is $2.0 \times 10^{-3}M$, the concentrations of the denatured and native proteins at both 50 and 100°C are given in the following table:

Temperature (°C)	[Ribonuclease A (*denatured*)] (*M*)	[Ribonuclease A (*native*)] (*M*)
50	5.1×10^{-6}	2.0×10^{-3}
100	2.8×10^{-4}	1.7×10^{-3}

(a) Determine $\Delta H°$ and $\Delta S°$ for the folding reaction assuming that these quantities are independent of temperature. (b) Calculate $\Delta G°$ for ribonuclease A folding at 25°C. Is this process spontaneous under standard state conditions at this temperature? (c) What is the denaturation temperature of ribonuclease A under standard state conditions?

10. Using the data in Table 3–4, calculate $\Delta G_f°{}'$ for the following compounds at 25°C: (a) $H_2O(\ell)$; (b) sucrose (sucrose + $H_2O \rightleftharpoons$ glucose + fructose: $\Delta G°{}' = -29.3$ kJ · mol^{-1}); and (c) ethyl acetate (ethyl acetate + $H_2O \rightleftharpoons$ ethanol + acetate$^-$ + H^+: $\Delta G°{}' = -19.7$ kJ · mol^{-1}; the pK of acetic acid is 4.76).

11. Calculate the equilibrium constants for the hydrolysis of the following compounds at pH 7 and 25°C: (a) phosphoenolpyruvate ($\Delta G°{}' = -61.9$ kJ · mol^{-1}); (b) pyrophosphate ($\Delta G°{}' = -33.5$ kJ · mol^{-1}); and (c) glucose-1-phosphate ($\Delta G°{}' = -20.9$ kJ · mol^{-1}).

The digestive enzyme bovine carboxypeptidase A showing its central β sheet.

PART

II

BIOMOLECULES

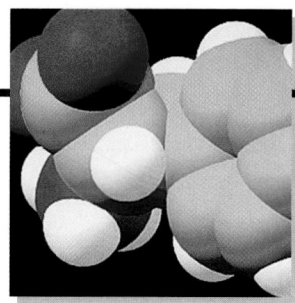

Chapter

4

Amino Acids

It is hardly surprising that much of the early biochemical research was concerned with the study of proteins. Proteins form the class of biological macromolecules that have the most well-defined physicochemical properties, and consequently they were generally easier to isolate and characterize than nucleic acids, polysaccharides, or lipids. Furthermore, proteins, particularly in the form of enzymes, have obvious biochemical functions. The central role that proteins play in biological processes has therefore been recognized since the earliest days of biochemistry. In contrast, the task of nucleic acids in the transmission and expression of genetic information was not realized until the late 1940s and their catalytic function only began to come to light in the 1980s, the role of lipids in biological membranes was not appreciated until the 1960s, and the biological functions of polysaccharides are still somewhat mysterious.

In this chapter we study the structures and properties of the monomeric units of proteins, the **amino acids.** It is from these substances that proteins are synthesized through processes that we discuss in Chapter 32. Amino acids are also energy metabolites and, in animals, many of them are essential nutrients (Chapter 26). In addition, as we shall see, many amino acids and their derivatives are of biochemical importance in their own right (Section 4-3B).

1 ■ THE AMINO ACIDS OF PROTEINS

The analyses of a vast number of proteins from almost every conceivable source have shown that *all proteins are composed of the 20 "standard" amino acids listed in Table 4-1.* These substances are known as **α-amino acids** because, with the exception of **proline,** they have a primary amino group and a carboxylic acid group substituent on the same carbon atom (Fig. 4-1; proline has a secondary amino group).

A. *General Properties*

The pK values of the 20 "standard" α-amino acids of proteins are tabulated in Table 4-1. Here pK_1 and pK_2, respectively, refer to the α-carboxylic acid and α-amino groups, and pK_R refers to the side groups with acid–base properties. Table 4-1 indicates that the pK values of the α-carboxylic acid groups lie in a small range around 2.2 so that above pH 3.5 these groups are almost entirely in their carboxylate forms. The α-amino groups all have pK values near 9.4 and are therefore almost entirely in their ammonium ion forms below pH 8.0. This leads to an important structural point: *In the physiological pH range, both the carboxylic acid and the amino groups of α-amino acids are completely ionized (Fig. 4-2).* An amino acid can therefore act as either an acid or a base. Substances with this property are said to be **amphoteric** and are referred to as **ampholytes** (*ampho*teric electro*lytes*). In Section 4-1D, we

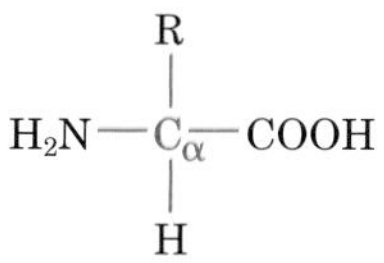

$$H_2N - \underset{\underset{H}{|}}{\overset{\overset{R}{|}}{C_\alpha}} - COOH$$

FIGURE 4-1   General structural formula for α-amino acids. There are 20 different R groups in the commonly occurring amino acids (Table 4-1).

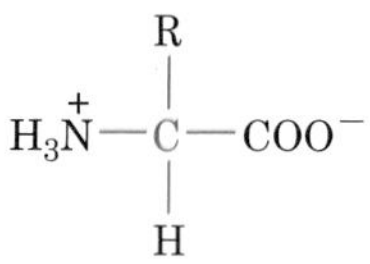

$$H_3\overset{+}{N} - \underset{\underset{H}{|}}{\overset{\overset{R}{|}}{C}} - COO^-$$

FIGURE 4-2   Zwitterionic form of the α-amino acids that occur at physiological pH values.

TABLE 4-1 Covalent Structures and Abbreviations of the "Standard" Amino Acids of Proteins, Their Occurrence, and the p*K* Values of Their Ionizable Groups

Name, Three-letter Symbol, and One-letter Symbol	Structural Formula[a]	Residue Mass (D)[b]	Average Occurrence in Proteins (%)[c]	pK$_1$ α-COOH[d]	pK$_2$ α-NH$_3^{+}$[d]	pK$_R$ Side Chain[d]
Amino acids with nonpolar side chains						
Glycine Gly G	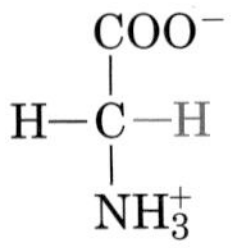	57.0	6.8	2.35	9.78	
Alanine Ala A	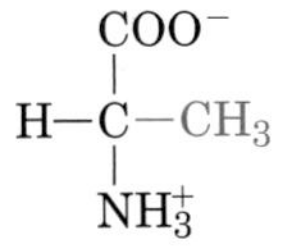	71.1	7.6	2.35	9.87	
Valine Val V	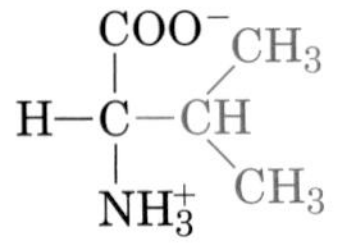	99.1	6.6	2.29	9.74	
Leucine Leu L	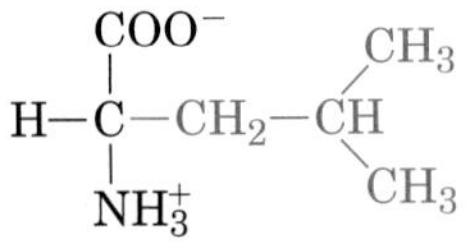	113.2	9.5	2.33	9.74	
Isoleucine Ile I	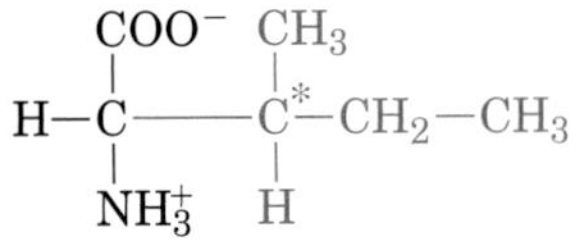	113.2	5.8	2.32	9.76	
Methionine Met M	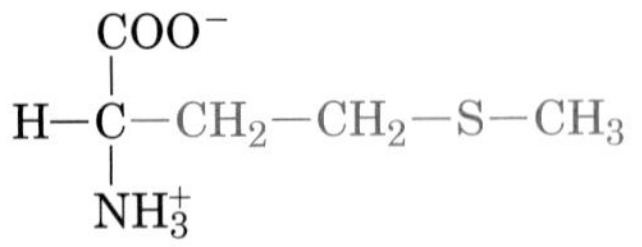	131.2	2.4	2.13	9.28	
Proline Pro P	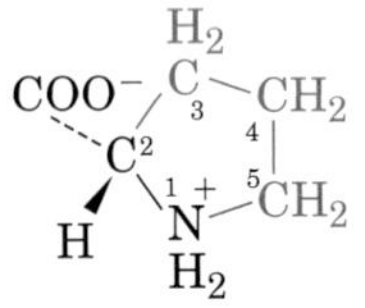	97.1	5.0	1.95	10.64	
Phenylalanine Phe F	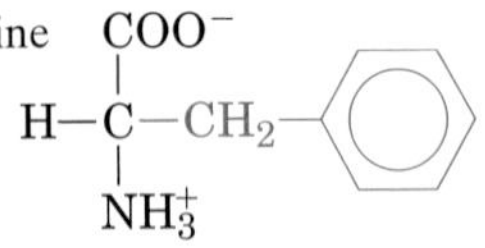	147.2	4.1	2.20	9.31	
Tryptophan Trp W	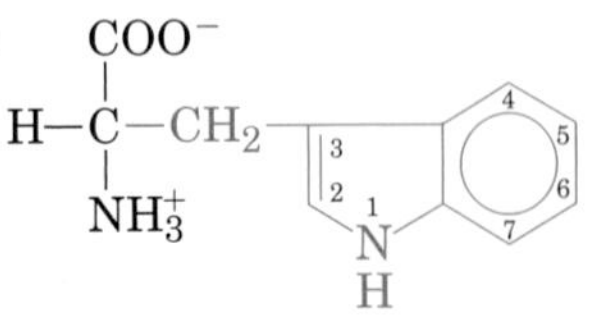	186.2	1.2	2.46	9.41	

(continued)

[a]The ionic forms shown are those predominating at pH 7.0 (except for that of histidine[e]), although residue mass is given for the neutral compound. The C$_\alpha$ atoms, as well as those atoms marked with an asterisk, are chiral centers with configurations as indicated according to Fischer projection formulas. The standard organic numbering system is provided for heterocycles.

[b]The residue masses are given for the neutral residues. For molecular masses of the parent amino acids, add 18.0 D, the molecular mass of H$_2$O, to the residue masses. For side chain masses, subtract 56.0 D, the formula mass of a peptide group, from the residue masses.

[c]The average amino acid composition in the complete SWISS-PROT database (http://www.expasy.ch/sprot), Release 40.7.

[d]From Dawson, R.M.C., Elliott, D.C., Elliott, W.H., and Jones, K.M., *Data for Biochemical Research* (3rd ed.), *pp.* 1–31, Oxford Science Publications (1986).

[e]Both the neutral and protonated forms of histidine are present at pH 7.0 because its pK$_R$ is close to 7.0. The imidazole ring of histidine is numbered here according to the biochemistry convention. In the IUPAC convention, N3 of the biochemistry convention is designated N1 and the numbering increases clockwise around the ring.

[f]The three- and one-letter symbols for asparagine *or* aspartic acid are Asx and B, whereas for glutamine *or* glutamic acid they are Glx and Z. The one-letter symbol for an undetermined or "nonstandard" amino acid is X.

TABLE 4-1 *(continued)*

Name Three-letter Symbol, and One-letter Symbol	Structural Formula[a]	Residue Mass (D)[b]	Average Occurrence in Proteins (%)[c]	pK_1 α-COOH[d]	pK_2 α-NH$_3^+$[d]	pK_R Side Chain[d]
Amino acids with uncharged polar side chains						
Serine Ser S	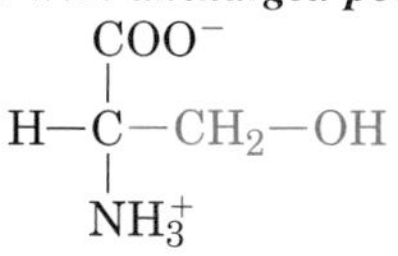	87.1	7.1	2.19	9.21	
Threonine Thr T	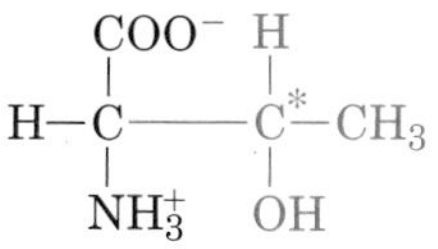	101.1	5.6	2.09	9.10	
Asparagine[f] Asn N	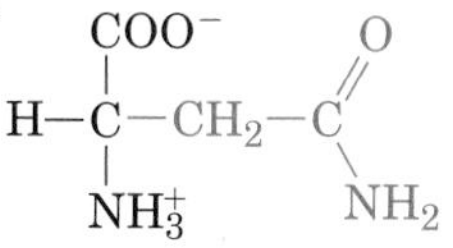	114.1	4.3	2.14	8.72	
Glutamine[f] Gln Q	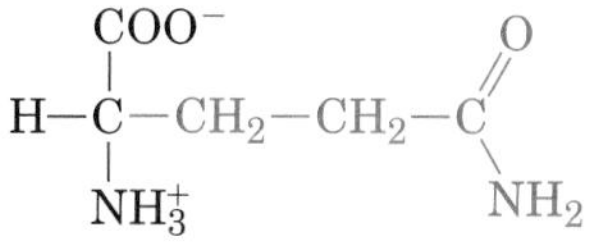	128.1	3.9	2.17	9.13	
Tyrosine Tyr Y	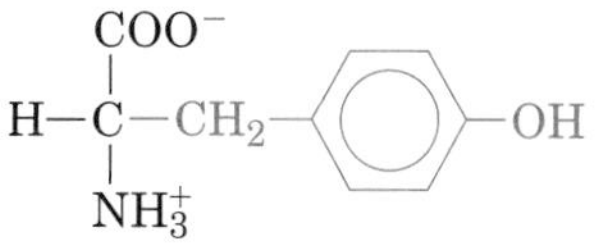	163.2	3.2	2.20	9.21	10.46 (phenol)
Cysteine Cys C	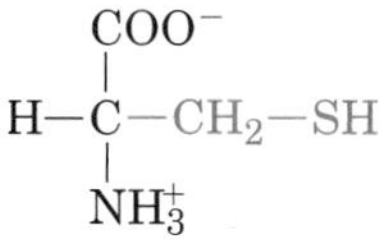	103.1	1.6	1.92	10.70	8.37 (sulfhydryl)
Amino acids with charged polar side chains						
Lysine Lys K	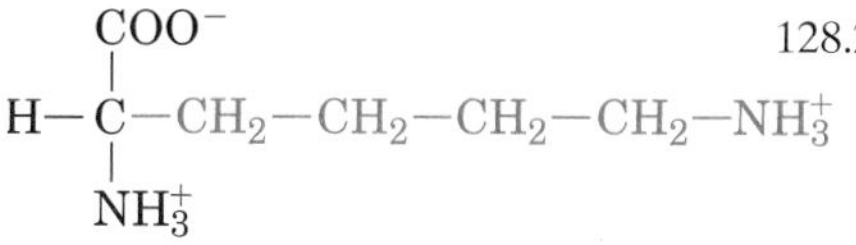	128.2	6.0	2.16	9.06	10.54 (ε-NH$_3^+$)
Arginine Arg R	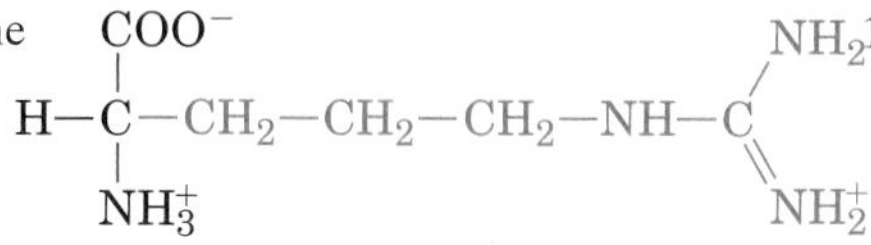	156.2	5.2	1.82	8.99	12.48 (guanidino)
Histidine[e] His H	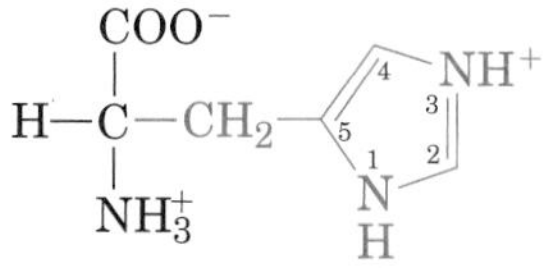	137.1	2.2	1.80	9.33	6.04 (imidazole)
Aspartic acid[f] Asp D	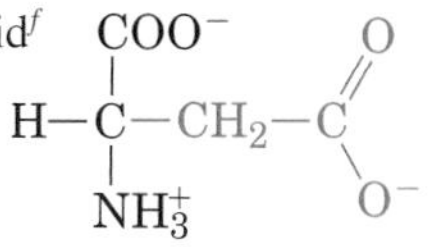	115.1	5.2	1.99	9.90	3.90 (β-COOH)
Glutamic acid[f] Glu E	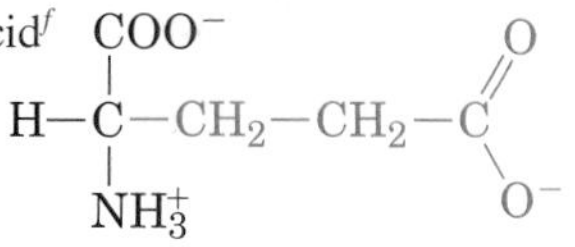	129.1	6.5	2.10	9.47	4.07 (γ-COOH)

shall delve a bit deeper into the acid–base properties of the amino acids.

Molecules that bear charged groups of opposite polarity are known as **zwitterions** (German: *zwitter*, hybrid) or **dipolar ions.** The zwitterionic character of the α-amino acids has been established by several methods including spectroscopic measurements and X-ray crystal structure determinations (in the solid state the α-amino acids are zwitterionic because the basic amine group abstracts a proton from the nearby acidic carboxylic acid group). Because amino acids are zwitterions, their physical properties are characteristic of ionic compounds. For instance, most α-amino acids have melting points near 300°C, whereas their nonionic derivatives usually melt around 100°C. Furthermore, amino acids, like other ionic compounds, are more soluble in polar solvents than in nonpolar solvents. Indeed, most α-amino acids are very soluble in water but are largely insoluble in most organic solvents.

B. *Peptide Bonds*

The α-amino acids polymerize, at least conceptually, through the elimination of a water molecule as is indicated in Fig. 4-3. The resulting CO—NH linkage, which was independently characterized in 1902 by Emil Fisher and Franz Hofmeister, is known as a **peptide bond.** Polymers composed of two, three, a few (3–10), and many **amino acid residues** (alternatively called **peptide units**) are known, respectively, as **dipeptides, tripeptides, oligopeptides,** and **polypeptides.** These substances, however, are often referred to simply as "peptides." *Proteins are molecules that consist of one or more polypeptide chains.* These polypeptides range in length from ~40 to ~33,000 amino acid residues (although few have more than 1500 residues) and, since the average mass of an amino acid residue is ~110 D, have molecular masses that range from ~4 to over ~3600 kD.

*Polypeptides are **linear polymers;*** that is, each amino acid residue is linked to its neighbors in a head-to-tail fashion rather than forming branched chains. This observation reflects the underlying elegant simplicity of the way living systems construct these macromolecules for, as we shall see, the nucleic acids that encode the amino acid sequences of polypeptides are also linear polymers. This permits the direct correspondence between the monomer (nucleotide) sequence of a nucleic acid and the monomer (amino acid) sequence of the corresponding polypeptide without the added complication of specifying the positions and sequences of any branching chains.

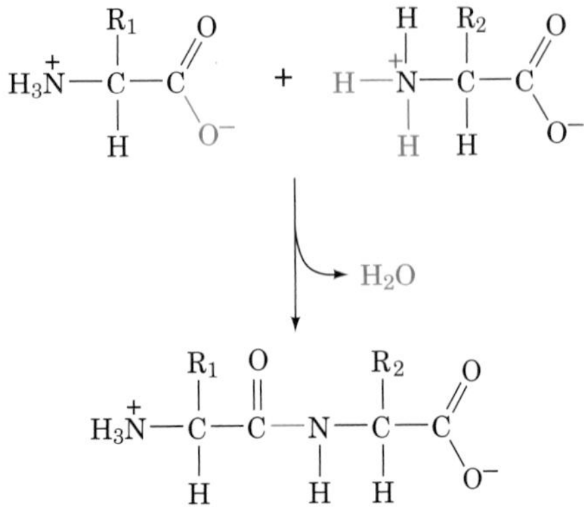

With 20 different choices available for each amino acid residue in a polypeptide chain, it is easy to see that a huge number of different protein molecules can exist. For example, for dipeptides, each of the 20 different choices for the first amino acid residue can have 20 different choices for the second amino acid residue, for a total of $20^2 = 400$ distinct dipeptides. Similarly, for tripeptides, there are 20 possibilities for each of the 400 choices of dipeptides to yield a total of $20^3 = 8000$ different tripeptides. A relatively small protein molecule consists of a single polypeptide chain of 100 residues. There are $20^{100} = 1.27 \times 10^{130}$ possible unique polypeptide chains of this length, a quantity vastly greater than the estimated number of atoms in the universe (9×10^{78}). Clearly, nature can have made only a tiny fraction of the possible different protein molecules. Nevertheless, *the various organisms on Earth collectively synthesize an enormous number of different protein molecules whose great range of physicochemical characteristics stem largely from the varied properties of the 20 "standard" amino acids.*

C. *Classification and Characteristics*

The most common and perhaps the most useful way of classifying the 20 "standard" amino acids is according to the polarities of their side chains **(R groups).** This is because proteins fold to their native conformations largely in response to the tendency to remove their hydrophobic side chains from contact with water and to solvate their hydrophilic side chains (Chapters 8 and 9). According to this classification scheme, there are three major types of amino acids: (1) those with nonpolar R groups, (2) those with uncharged polar R groups, and (3) those with charged polar R groups.

a. The Nonpolar Amino Acid Side Chains Have a Variety of Shapes and Sizes

Nine amino acids are classified as having nonpolar side chains. **Glycine** (which, when it was found to be a component of gelatin in 1820, was the first amino acid to be identified in protein hydrolyzates) has the smallest possible side chain, an H atom. **Alanine, valine, leucine,** and **isoleucine** have aliphatic hydrocarbon side chains ranging in size from a methyl group for alanine to isomeric butyl groups for leucine and isoleucine. **Methionine** has a thiol ether side chain that resembles an *n*-butyl group in many of its physical properties (C and S have nearly equal electronegativities and S is about the size of a methylene group). **Proline,** a cyclic secondary amino acid, has conformational constraints imposed by the cyclic nature of its pyrrolidine side group, which is unique among the "standard" 20 amino acids. **Phenylalanine,** with its phenyl moiety (Fig. 4-4), and **tryptophan,** with its indole group, contain aromatic

FIGURE 4-3 **Condensation of two α-amino acids to form a dipeptide.** The peptide bond is shown in red.

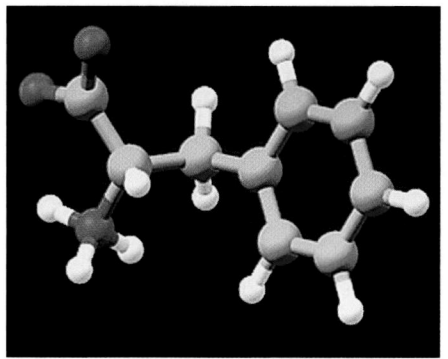 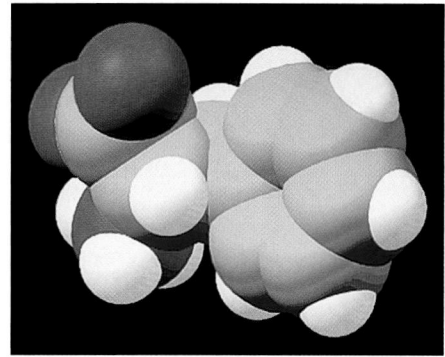

(a) (b)

FIGURE 4-4 Structure of phenylala-nine. Computer graphics drawings showing the α-amino acid phenylala-nine represented in (*a*) ball-and-stick form and (*b*) space-filling form. The molecule, which is colored according to atom type (C green, H white, N blue, and O red), has the same conformation and orientation in both drawings.

side groups, which are characterized by bulk as well as nonpolarity.

b. Uncharged Polar Side Chains Have Hydroxyl, Amide, or Thiol Groups

Six amino acids are commonly classified as having uncharged polar side chains. **Serine** and **threonine** bear hydroxylic R groups of different sizes. **Asparagine** and **glut-amine** have amide-bearing side chains of different sizes. **Tyrosine** has a phenolic group, which, together with the aromatic groups of phenylalanine and tryptophan, accounts for most of the UV absorbance and fluorescence exhibited by proteins. **Cysteine** has a thiol group that is unique among the 20 amino acids in that it often forms a disulfide bond to another cysteine residue (Fig. 4-5) through the oxidation of their thiol groups. This dimeric compound is referred to in the older biochemical literature as the amino acid **cystine.** The disulfide bond has great importance in protein structure: *It can join separate polypeptide chains or cross-link two cysteines in the same chain.* The confusing similarity between the names cysteine and cystine has led to the former occasionally being referred to as a **half-cystine** residue. However, the realization that cystine arises through the cross-linking of two

cysteine residues after polypeptide biosynthesis has occurred has caused the name cystine to become less commonly used.

c. Charged Polar Side Chains May Be Positively or Negatively Charged

Five amino acids have charged side chains. The basic amino acids are positively charged at physiological pH values; they are **lysine,** which has a butylammonium side chain, **arginine,** which bears a guanidino group, and **histidine,** which carries an imidazolium moiety. Of the 20 α-amino acids, only histidine, with $pK_R = 6.0$, ionizes within the physiological pH range. At pH 6.0, its imidazole side group is only 50% charged so that histidine is neutral at the basic end of the physiological pH range. As a consequence, histidine side chains often participate in the catalytic reactions of enzymes. The acidic amino acids, **aspartic acid** and **glutamic acid,** are negatively charged above pH 3; in their ionized state, they are often referred to as **aspartate** and **glutamate.** Asparagine and glutamine are, respectively, the amides of aspartic acid and glutamic acid.

The allocation of the 20 amino acids among the three different groups is, of course, somewhat arbitrary. For example, glycine and alanine, the smallest of the amino acids, and tryptophan, with its heterocyclic ring, might just as well be classified as uncharged polar amino acids. Similarly, tyrosine and cysteine, with their ionizable side chains, might also be thought of as charged polar amino acids, particularly at higher pH's, whereas asparagine and glutamine are nearly as polar as their corresponding carboxylates, aspartate and glutamate.

The 20 amino acids vary considerably in their physicochemical properties such as polarity, acidity, basicity, aromaticity, bulk, conformational flexibility, ability to cross-link, ability to hydrogen bond, and chemical reactivity. These several characteristics, many of which are interrelated, are largely responsible for proteins' great range of properties.

D. *Acid–Base Properties*

Amino acids and proteins have conspicuous acid–base properties. The α-amino acids have two or, for those with ionizable side groups, three acid–base groups. The titra-

FIGURE 4-5 Structure of cystine. The cystine residue consists of two disulfide-linked cysteine residues.

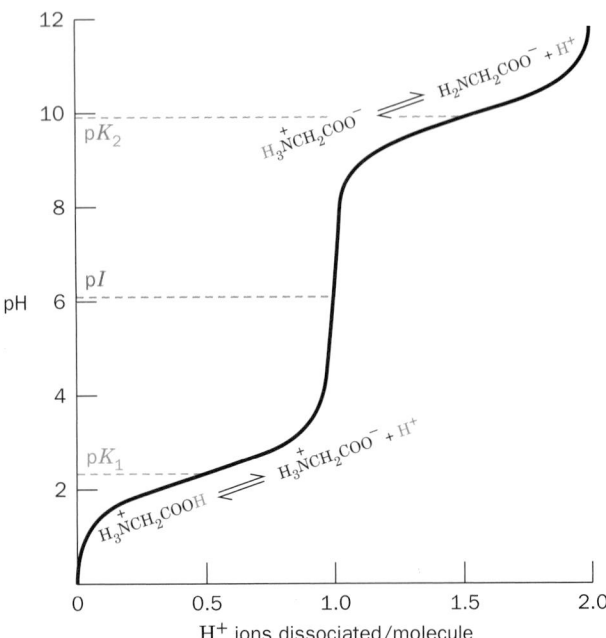

FIGURE 4-6   Titration curve of glycine. Other monoamino, monocarboxylic acids ionize in a similar fashion. [After Meister, A., *Biochemistry of the Amino Acids* (2nd ed.), Vol. 1, *p.* 30, Academic Press (1965).] ✏ **See the Animated Figures**

tion curve of glycine, the simplest amino acid, is shown in Fig. 4-6. At low pH values, both acid–base groups of glycine are fully protonated so that it assumes the cationic form $^+H_3NCH_2COOH$. In the course of the titration with a strong base, such as NaOH, glycine loses two protons in the stepwise fashion characteristic of a polyprotic acid.

The pK values of glycine's two ionizable groups are sufficiently different so that the Henderson–Hasselbalch equation:

$$pH = pK + \log\left(\frac{[A^-]}{[HA]}\right) \qquad [2.6]$$

closely approximates each leg of its titration curve. Consequently, the pK for each ionization step is that of the midpoint of its corresponding leg of the titration curve (Sections 2-2A & 2-2C): At pH 2.35 the concentrations of the cationic form, $^+H_3NCH_2COOH$, and the zwitterionic form, $^+H_3NCH_2COO^-$, are equal; similarly, at pH 9.78 the concentrations of the zwitterionic form and the anionic form, $H_2NCH_2COO^-$, are equal. Note that *amino acids never assume the neutral form in aqueous solution.*

The pH at which a molecule carries no net electric charge is known as its **isoelectric point, pI.** For the α-amino acids, the application of the Henderson–Hasselbalch equation indicates that, to a high degree of precision,

$$pI = \frac{1}{2}(pK_i + pK_j) \qquad [4.1]$$

where K_i and K_j are the dissociation constants of the two ionizations involving the neutral species. For monoamino, monocarboxylic acids such as glycine, K_i and K_j represent

K_1 and K_2. However, for aspartic and glutamic acids, K_i and K_j are K_1 and K_R, whereas for arginine, histidine, and lysine, these quantities are K_R and K_2.

Acetic acid's pK (4.76), which is typical of aliphatic monocarboxylic acids, is ~2.4 pH units higher than the pK_1 of its α-amino derivative glycine. This large difference in pK values of the same functional group is caused, as is discussed in Section 2-2C, by the electrostatic influence of glycine's positively charged ammonium group; that is, its $—NH_3^+$ group helps repel the proton from its COOH group. Conversely, glycine's carboxylate group increases the basicity of its amino group ($pK_2 = 9.78$) with respect to that of glycine methyl ester (pK = 7.75). However, the $—NH_3^+$ groups of glycine and its esters are significantly more acidic than are aliphatic amines (pK ≈ 10.7) because of the electron-withdrawing character of the carboxyl group.

The electronic influence of one functional group on another is rapidly attenuated as the distance between the groups increases. Hence, the pK values of the α-carboxylate groups of amino acids and the side chain carboxylates of aspartic and glutamic acids form a series that is progressively closer in value to the pK of an aliphatic monocarboxylic acid. Likewise, the ionization constant of lysine's side chain amino group is indistinguishable from that of an aliphatic amine.

a.   Proteins Have Complex Titration Curves

The titration curves of the α-amino acids with ionizable side chains, such as that of glutamic acid, exhibit the expected three pK values. However, the titration curves of polypeptides and proteins, an example of which is shown in Fig. 4-7, rarely provide any indication of individual pK values because of the large numbers of ionizable groups they represent (typically 30% of a protein's amino acid side chains are ionizable; Table 4-1). Furthermore, the covalent and three-dimensional structure of a protein may cause the pK of each ionizable group to shift by as much as several pH units from its value in the free α-amino acid as a result of the electrostatic influence of nearby charged groups, medium effects arising from the proximity of groups of low dielectric constant, and the effects of hydrogen bonding associations. The titration curve of a protein is also a function of the salt concentration, as is shown in Fig. 4-7, because the salt ions act electrostatically to shield the side chain charges from one another, thereby attenuating these charge–charge interactions.

E. *A Few Words on Nomenclature*

The three-letter abbreviations for the 20 amino acid residues are given in Table 4-1. It is worthwhile memorizing these symbols because they are widely used throughout the biochemical literature, including this text. These abbreviations are, in most cases, taken from the first three letters of the corresponding amino acid's name; they are conversationally pronounced as read.

The symbol **Glx** means Glu or Gln and, similarly, **Asx** means Asp or Asn. These ambiguous symbols stem from

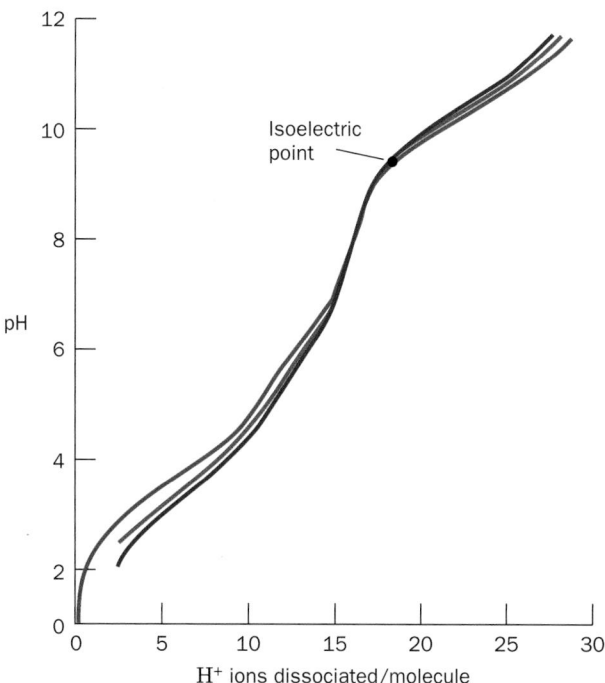

FIGURE 4-7   Titration curves of the enzyme ribonuclease A at 25°C. The concentration of KCl is 0.01*M* for the blue curve, 0.03*M* for the red curve, and 0.15*M* for the green curve. [After Tanford, C. and Hauenstein, J.D., *J. Am. Chem. Soc.* **78**, 5287 (1956).]

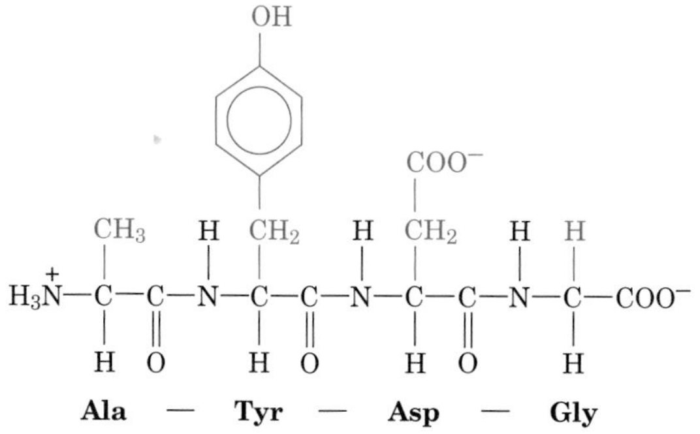

FIGURE 4-8   The tetrapeptide Ala-Tyr-Asp-Gly.

laboratory experience: Asn and Gln are easily hydrolyzed to aspartic acid and glutamic acid, respectively, under the acidic or basic conditions that are usually used to excise them from proteins (Section 7-1D). Therefore, without special precautions, we cannot determine whether a detected Glu was originally Glu or Gln, and likewise for Asp and Asn.

The one-letter symbols for the amino acids are also given in Table 4-1. This more compact code is often used when comparing the amino acid sequences of several similar proteins and hence should also be memorized by the serious student. Note that the one-letter symbols are usually the first letter of the amino acid residue's name. However, for those sets of residues that have the same first letter, this is only true of the most abundant residue of the set.

Amino acid residues in polypeptides are named by dropping the suffix **-ine** in the name of the amino acid and replacing it by **-yl.** Polypeptide chains are described by starting at the amino terminus (known as the **N-terminus**) and sequentially naming each residue until the carboxyl terminus (the **C-terminus**) is reached. The amino acid at the C-terminus is given the name of its parent amino acid. Thus the compound shown in Fig. 4-8 is alanyltyrosylaspartylglycine. Of course such names for polypeptide chains of more than a few residues are extremely cumbersome. The use of abbreviations for amino acid residues partially relieves this problem. Thus the foregoing tetrapeptide is Ala-Tyr-Asp-Gly using the three-letter abbreviations and AYDG using the one-letter symbols. Note that these ab-

breviations are always written so that the N-terminus of the polypeptide chain is to the left and the C-terminus is to the right.

The various nonhydrogen atoms of the amino acid side chains are often named in sequence with the Greek alphabet (α, β, γ, δ, ε, ζ, η, …) starting at the carbon atom adjacent to the peptide carbonyl group (the C_α atom). Therefore, as Fig. 4-9 indicates, Glu has a γ-carboxyl group and Lys has a ζ-amino group (alternatively known as an ε-amino group because the N atom is substituent to C_ε). Unfortunately, this labeling system is ambiguous for several amino acids. Consequently, standard numbering schemes for organic molecules are also employed. These are indicated in Table 4-1 for the heterocyclic side chains.

2 ■ OPTICAL ACTIVITY

The amino acids as isolated by the mild hydrolysis of proteins are, with the exception of glycine, all **optically active;** that is, they rotate the plane of plane-polarized light (see below).

Optically active molecules have an asymmetry such that they are not superimposable on their mirror image in the same way that a left hand is not superimposable on its mirror image, a right hand. This situation is characteristic of substances that contain tetrahedral carbon atoms that have four different substituents. The two such molecules depicted in

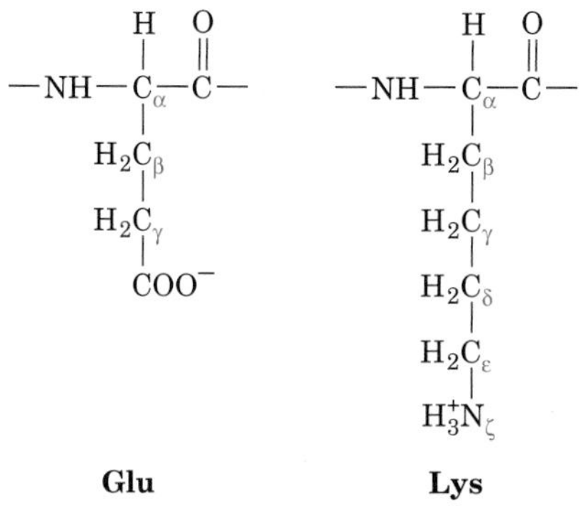

FIGURE 4-9   Greek lettering scheme used to identify the atoms in the glutamyl and lysyl R groups.

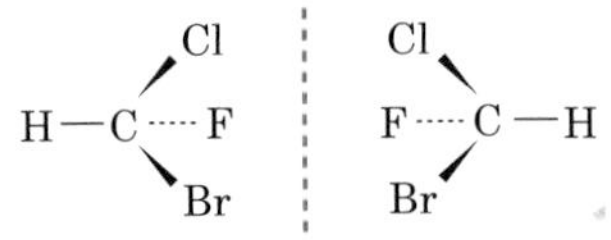

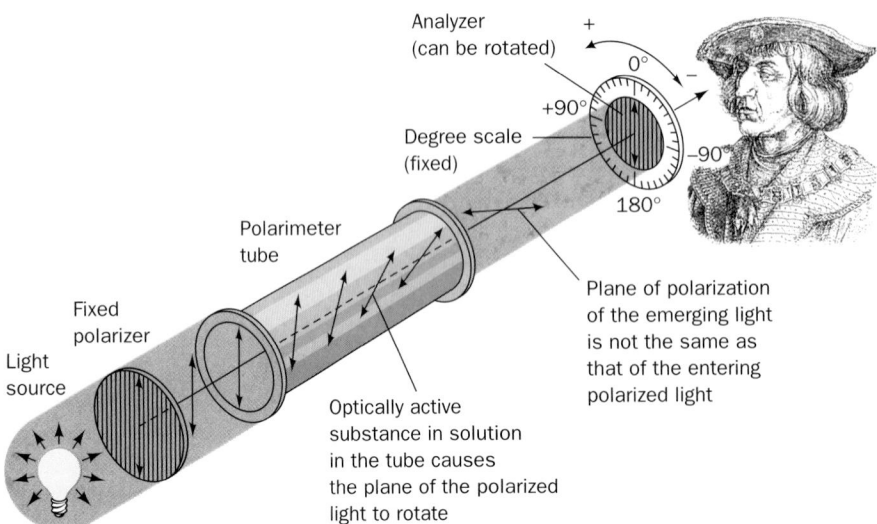

FIGURE 4-10   The two enantiomers of fluorochlorobromo-methane. The four substituents are tetrahedrally arranged about the central atom with the dotted lines indicating that a substituent lies behind the plane of the paper, a triangular line indicating that it lies above the plane of the paper, and a thin line indicating that it lies in the plane of the paper. The mirror plane relating the enantiomers is represented by a vertical dashed line.

Fig. 4-10 are not superimposable since they are mirror images. The central atoms in such atomic constellations are known as **asymmetric centers** or **chiral centers** and are said to have the property of **chirality** (Greek: *cheir,* hand). The C_α atoms of all the amino acids, with the exception of glycine, are asymmetric centers. Glycine, which has two H atoms substituent to its C_α atom, is superimposable on its mirror image and is therefore not optically active.

Molecules that are nonsuperimposable mirror images are known as **enantiomers** of one another. Enantiomeric molecules are physically and chemically indistinguishable by most techniques. *Only when probed asymmetrically, for example, by plane-polarized light or by reactants that also contain chiral centers, can they be distinguished and/or differentially manipulated.*

There are three commonly used systems of nomenclature whereby a particular stereoisomer of an optically active molecule can be classified. These are explained in the following sections.

A. *An Operational Classification*

Molecules are classified as **dextrorotatory** (Greek: *dexter,* right) or **levorotatory** (Greek: *laevus,* left) depending on whether they rotate the plane of plane-polarized light clockwise or counterclockwise from the point of view of the observer. This can be determined by an instrument known as a **polarimeter** (Fig. 4-11). A quantitative measure of the optical activity of the molecule is known as its **specific rotation:**

$$[\alpha]_D^{25} = \frac{\text{observed rotation (degrees)}}{\frac{\text{optical path}}{\text{length (dm)}} \times \frac{\text{concentration}}{(g \cdot cm^{-3})}} \qquad [4.2]$$

where the superscript 25 refers to the temperature at which polarimeter measurements are customarily made (25°C) and the subscript D indicates the monochromatic light that is traditionally employed in polarimetry, the so-called D-line in the spectrum of sodium (589.3 nm). Dextrorotatory and levorotatory molecules are assigned positive and negative values of $[\alpha]_D^{25}$. Dextrorotatory molecules are therefore designated by the prefix (+) and their levorotatory enantiomers have the prefix (−). In an equivalent but archaic nomenclature, the lowercase letters *d (dextro)* and *l (levo)* are used.

The sign and magnitude of a molecule's specific rotation depend on the structure of the molecule in a complicated and poorly understood manner. It is not yet possible to predict reliably the magnitude or even the sign of a given molecule's specific rotation. For example, proline, leucine, and arginine, which are isolated from proteins, have specific rotations in pure aqueous solutions of −86.2°, −10.4°, and +12.5°, respectively. Their enantiomers exhibit values of $[\alpha]_D^{25}$ of the same magnitude but of opposite signs. As might be expected from the acid–base nature of the amino acids, these quantities vary with the solution pH.

A problem with this operational classification system for optical isomers is that it provides no presently interpretable indication of the **absolute configuration** (spatial arrangement) of the chemical groups about a chiral center. Furthermore, a molecule with more than one asymmetric center may have an optical rotation that is not obviously related to the rotatory powers of the individual asymmet-

FIGURE 4-11   Schematic diagram of a polarimeter. This device is used to measure optical rotation.

ric centers. For this reason, the following relative classification scheme is more useful.

B. *The Fischer Convention*

In this system, the configuration of the groups about an asymmetric center is related to that of **glyceraldehyde,** a molecule with one asymmetric center. By a convention introduced by Fischer in 1891, the (+) and (−) stereoisomers of glyceraldehyde are designated **D-glyceraldehyde** and **L-glyceraldehyde,** respectively (note the use of small uppercase letters). With the realization that there was only a 50% chance that he was correct, Fischer assumed that the configurations of these molecules were those shown in Fig. 4-12. Fischer also proposed a convenient shorthand notation for these molecules, known as **Fischer projections,** which are also given in Fig. 4-12. In the Fischer convention, horizontal bonds extend above the plane of the paper and vertical bonds extend below the plane of the paper as is explicitly indicated by the accompanying geometrical formulas.

The configuration of groups about a chiral center can be related to that of glyceraldehyde by chemically converting these groups to those of glyceraldehyde using reactions of known stereochemistry. For α-amino acids, the arrangement of the amino, carboxyl, R, and H groups about the C_α atom is related to that of the hydroxyl, aldehyde, CH_2OH, and H groups, respectively, of glyceraldehyde. In this way, L-glyceraldehyde and L-α-amino acids are said to have the same relative configurations

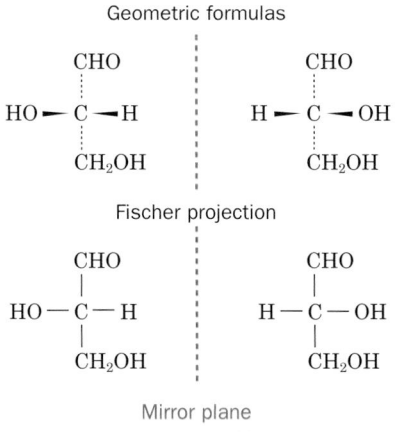

Geometric formulas

CHO   CHO

HO — C — H   H — C — OH

CH₂OH   CH₂OH

Fischer projection

CHO   CHO

HO — C — H   H — C — OH

CH₂OH   CH₂OH

Mirror plane

L-Glyceraldehyde   D-Glyceraldehyde

FIGURE 4-12   Fischer convention configurations for naming the enantiomers of glyceraldehyde. Glyceraldehyde enantiomers are represented by geometric formulas (*top*) and their corresponding Fischer projection formulas (*bottom*). Note that in Fischer projections, all horizontal bonds point above the page and all vertical bonds point below the page. The mirror planes relating the enantiomers are represented by a vertical dashed line. (Fischer projection formulas, as traditionally presented, omit the central C symbolizing the chiral carbon atom. The Fischer projection formulas in this text, however, will generally have a central C.)

CHO   COO⁻

HO — C — H $H_3\overset{+}{N}$ — C — H

CH₂OH   R

L-Glyceraldehyde   L-α-Amino acid

FIGURE 4-13   Configurations of L-glyceraldehyde and L-α-amino acids.

(Fig. 4-13). Through the use of this method, the configurations of the α-amino acids can be described without reference to their specific rotations.

All α-amino acids derived from proteins have the L *stereochemical configuration;* that is, they all have the same relative configuration about their C_α atoms. In 1949, it was demonstrated by a then new technique in X-ray crystallography that Fischer's arbitrary choice was correct: The designation of the relative configuration of chiral centers is the same as their absolute configuration. The absolute configuration of L-α-amino acid residues may be easily remembered through the use of the "CORN crib" mnemonic that is diagrammed in Fig. 4-14.

a. Diastereomers Are Chemically and Physically Distinguishable

A molecule may have multiple asymmetric centers. For such molecules, the terms **stereoisomers** and **optical isomers** refer to molecules with different configurations about at least one of their chiral centers, but that are otherwise identical. The term enantiomer still refers to a molecule that is the mirror image of the one under consideration, that is, different in all its chiral centers. Since each asymmetric center in a chiral molecule can have two possible configurations, a molecule with n chiral centers has 2^n different possible stereoisomers and 2^{n-1} enantiomeric pairs. Threonine and isoleucine each have two chiral centers and hence $2^2 = 4$ possible stereoisomers. The forms of threonine and isoleucine that are isolated from proteins, which are by convention called the L forms, are indicated in Table 4-1. The mirror images of the L forms are the D forms. Their other two optical isomers are said to be **diastereomers** (or **allo** forms) of the enantiomeric D and L forms.

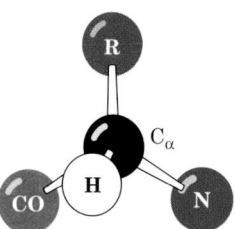

FIGURE 4-14   "CORN crib" mnemonic for the hand of L-amino acids. Looking at the C_α atom from its H atom substituent, its other substituents should read CO—R—N in the clockwise direction as shown. Here CO, R, and N, respectively, represent the carbonyl group, side chain, and main chain nitrogen atom. [After Richardson, J.S., *Adv. Protein Chem.* **34,** 171 (1981).]

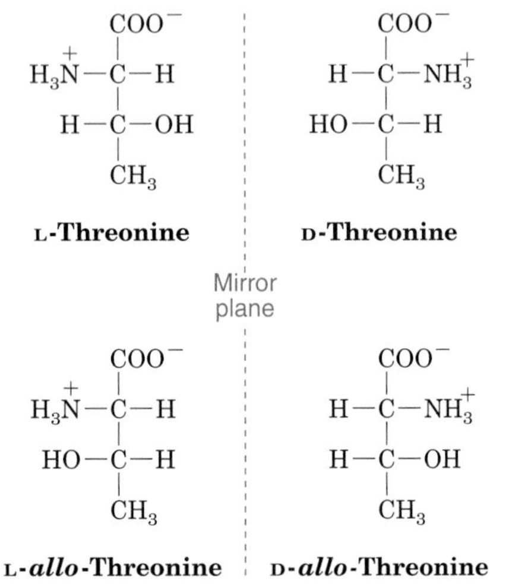

FIGURE 4-15 **Fischer projections of threonine's four stereoisomers.** The D and L forms are mirror images as are the D-*allo* and L-*allo* forms. D- and L-threonine are each diastereomers of both D-*allo*- and L-*allo*-threonine.

The relative configurations of all four stereoisomers of threonine are given in Fig. 4-15. Note the following points:

1. The D-*allo* and L-*allo* forms are mirror images of each other, as are the D and L forms. Neither allo form is symmetrically related to either of the D or L forms.

2. In contrast to the case for enantiomeric pairs, diastereomers are physically and chemically distinguishable from one another by ordinary means such as melting points, spectra, and chemical reactivity; that is, they are really different compounds in the usual sense.

A special case of diastereoisomerism occurs when the two asymmetric centers are chemically identical. Two of the four Fischer projections of the sort shown in Fig. 4-15 then represent the same molecule. This is because the two asymmetric centers in this molecule are mirror images of each other. Such a molecule is superimposible on its mirror image and is therefore optically inactive. This so-called **meso** form is said to be **internally compensated.** The three optical isomers of cystine are shown in Fig. 4-16, where it can be seen that the D and L isomers are mirror images of each other as before. Only L-cystine occurs in proteins.

C. *The Cahn–Ingold–Prelog System*

Despite its usefulness, the Fischer scheme is awkward and often ambiguous for molecules with more than one asymmetric center. For this reason, the following absolute nomenclature scheme was formulated in 1956 by Robert Cahn, Christopher Ingold, and Vladimir Prelog. In this system, the four groups surrounding a chiral center are ranked according to a specific although arbitrary priority scheme: *Atoms of higher atomic number bonded to a chiral center are ranked above those of lower atomic number.* For example, the oxygen atom of an OH group takes precedence over the carbon atom of a CH_3 group that is bonded to the same chiral C atom. If any of the first substituent atoms are of the same element, the priority of these groups is established from the atomic numbers of the second, third, etc., atoms outward from the asymmetric center. Hence a CH_2OH group takes precedence over a CH_3 group. There are other rules (given in the references and in many organic chemistry textbooks) for assigning priority ratings to substituents with multiple bonds or differing isotopes. The order of priority of some common functional groups is

$$SH > OH > NH_2 > COOH > CHO$$
$$> CH_2OH > C_6H_5 > CH_3 > {}^2H > {}^1H$$

Note that each of the groups substituent to a chiral center must have a different priority rating; otherwise the center could not be asymmetric.

The prioritized groups are assigned the letters W, X, Y, Z such that their order of priority rating is $W > X > Y > Z$. To establish the configuration of the chiral center, it is viewed from the asymmetric center toward the Z group (lowest priority). *If the order of the groups* $W \rightarrow X \rightarrow Y$ *as seen from this direction is clockwise, then the configuration of the asymmetric center is designated (R)* (Latin: *rectus,* right). *If the order of* $W \rightarrow X \rightarrow Y$ *is counterclockwise, the asymmetric center is designated (S)* (Latin: *sinister,* left). L-Glyceraldehyde is therefore designated (*S*)-glyceraldehyde (Fig. 4-17) and, similarly, L-alanine is (*S*)-alanine (Fig. 4-18). In fact, all the L-amino acids from proteins are (*S*)-amino acids, with the exception of L-cysteine, which is (*R*)-cysteine.

A major advantage of this so-called **Cahn–Ingold–Prelog** or **(RS) system** is that the chiralities of compounds with multiple asymmetric centers can be unambiguously described. Thus, in the (*RS*) system, L-threonine is (2*S*,3*R*)-threonine, whereas L-isoleucine is (2*S*,3*S*)-isoleucine (Fig. 4-19).

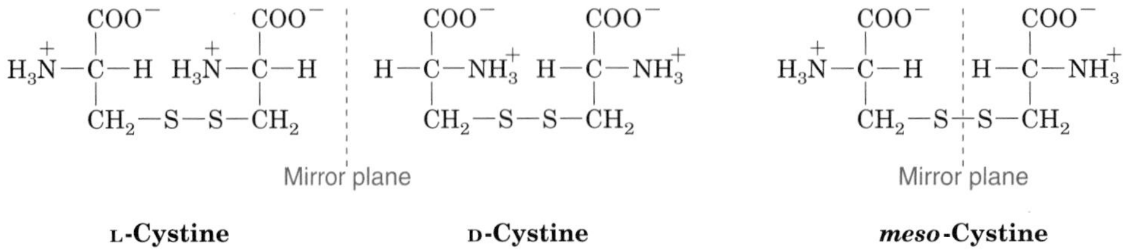

FIGURE 4-16 **The three stereoisomers of cystine.** The D and L forms are related by mirror symmetry, whereas the meso form has internal mirror symmetry and therefore lacks optical activity.

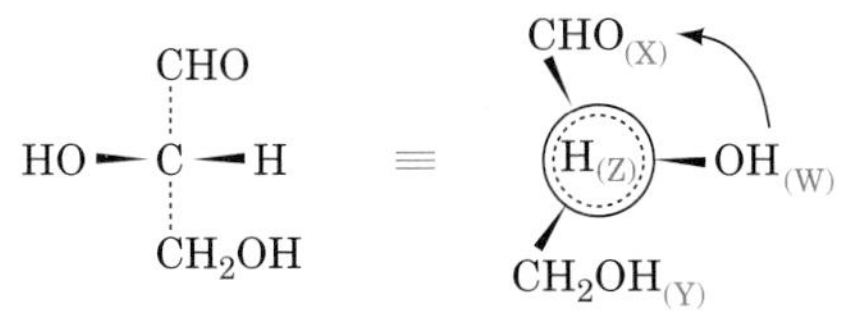

L-Glyceraldehyde   **(S)-Glyceraldehyde**

FIGURE 4-17 The structural formula of L-glyceraldehyde. Its equivalent (*RS*) system representation indicates that it is (*S*)-glyceraldehyde. In the latter drawing, the chiral C atom is represented by the large circle, and the H atom, which is located behind the plane of the paper, is represented by the smaller concentric dashed circle.

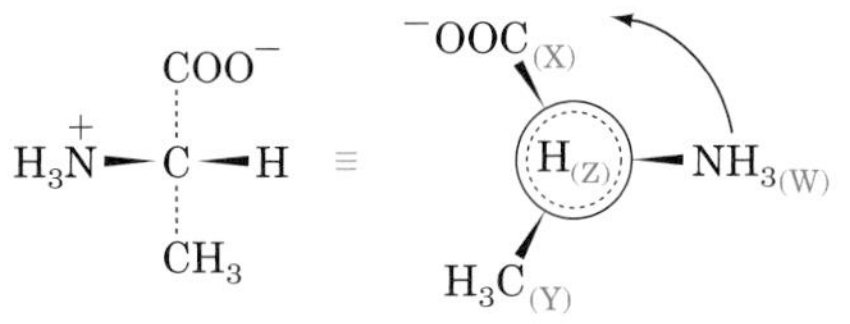

L-Alanine   **(S)-Alanine**

FIGURE 4-18 The structural formula of L-alanine. Its equivalent (*RS*) system representation indicates that it is (*S*)-alanine.

a. Prochiral Centers Have Distinguishable Substituents

Two chemically identical substituents to an otherwise chiral tetrahedral center are geometrically distinct; that is, the center has no rotational symmetry so that it can be unambiguously assigned left and right sides. Consider, for example, the substituents to the C1 atom of ethanol (the CH_2 group; Fig. 4-20a). If one of the H atoms were converted to another group (not CH_3 or OH), C1 would be a chiral center. The two H atoms are therefore said to be **prochiral.** If we arbitrarily assign the H atoms the subscripts *a* and *b* (Fig. 4-20), then H_b is said to be ***pro-R*** because in sighting from C1 toward H_a (as if it were the Z group of a chiral center), the order of priority of the other substituents decreases in a clockwise direction (Fig. 4-20b). Similarly, H_a is said to be ***pro-S*** (Fig. 4-20c).

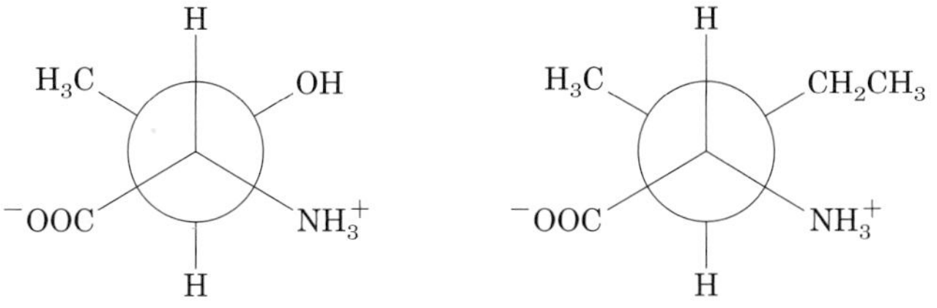

(2S, 3R)-Threonine   **(2S, 3S)-Isoleucine**

FIGURE 4-19 Newman projection diagrams of the stereoisomers of threonine and isoleucine derived from proteins. Here the C_α—C_β bond is viewed end on. The nearer atom, C_α, is represented by the confluence of the three bonds to its substituents, whereas the more distant atom, C_β, is represented by a circle from which its three substituents project.

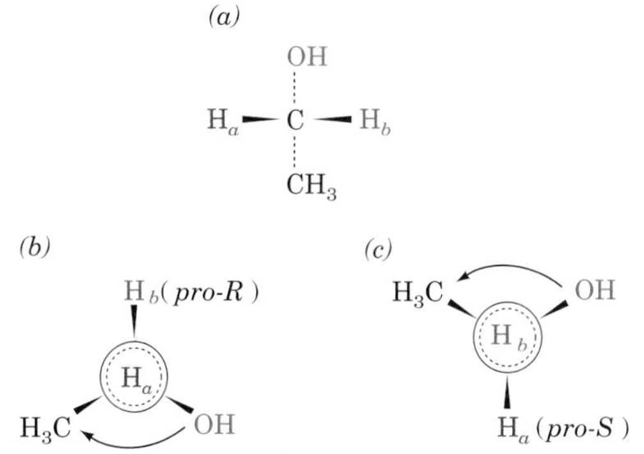

FIGURE 4-20 Views of ethanol. (*a*) Note that H_a and H_b, although chemically identical, are distinguishable: Rotating the molecule by 180° about the vertical axis so as to interchange these two hydrogen atoms does not yield an indistinguishable view of the molecule because the rotation also interchanges the chemically different OH and CH_3 groups. (*b*) Looking from C1 to H_a, the *pro-S* hydrogen atom (the dotted circle). (*c*) Looking from C1 to H_b, the *pro-R* hydrogen atom.

Planar objects with no rotational symmetry also have the property of prochirality. For example, in many enzymatic reactions, stereospecific addition to a trigonal carbon atom occurs from a particular side of that carbon atom to yield a chiral center (Section 13-2A). If a trigonal carbon is facing the viewer such that the order of priority of its substituents decreases in a clockwise manner (Fig. 4-21a), that face is designated as the ***re* face** (after *rectus*). The opposite face is designated as the ***si* face** (after *sinister*) since the priorities of its substituents decrease in the counterclockwise direction (Fig. 4-21b). Comparison of Figs. 4-20b and 4-21a indicates that an H atom adding to the *re* side of acetaldehyde atom C1 occupies the *pro-R* position of the resulting tetrahedral center. Conversely, a *pro-S* H atom is generated by *si* side addition to this trigonal center (Figs. 4-20c and 4-21b).

Closely related compounds that have the same configurational representation under the Fischer DL convention may have different representations under the (*RS*) system. Consequently, we shall use the Fischer convention in most cases. The (*RS*) system, however, is indispensable for describing prochirality and stereospecific reactions, so we shall find it invaluable for describing enzymatic reactions.

D. *Chirality and Biochemistry*

The ordinary chemical synthesis of chiral molecules produces **racemic** mixtures of these molecules (equal amounts of each member of an enantiomeric pair) because ordinary

FIGURE 4-21 Views of acetaldehyde. (*a*) Its *re* face and (*b*) its *si* face.

chemical and physical processes have no stereochemical bias. Consequently, there are equal probabilities for an asymmetric center of either hand to be produced in any such process. In order to obtain a product with net optical activity, a chiral process must be employed. This usually takes the form of using chiral reagents, although, at least in principle, the use of any asymmetric influence such as light that is plane polarized in one direction can produce a net asymmetry in a reaction product.

One of the most striking characteristics of life is its production of optically active molecules. *The biosynthesis of a substance possessing asymmetric centers almost invariably produces a pure stereoisomer.* The fact that the amino acid residues of proteins all have the L configuration is just one example of this phenomenon. This observation has prompted the suggestion that a simple diagnostic test for the past or present existence of extraterrestrial life, be it on moon rocks or in meteorites that have fallen to Earth, would be the detection of net optical activity in these materials. Any such finding would suggest that the asymmetric molecules thereby detected had been biosynthetically produced. Thus, even though α-amino acids have been extracted from carbonaceous meteorites, the observation that they come in racemic mixtures suggests that they are of chemical rather than biological origin.

One of the enigmas of the origin of life is why terrestrial life is based on certain chiral molecules rather than their enantiomers, that is, on L-amino acids, for example, rather than D-amino acids. Arguments that physical effects such as polarized light might have promoted significant net asymmetry in prebiotically synthesized molecules (Section 1-5B) have not been convincing. Perhaps L-amino acid–based life-forms arose at random and simply "ate" any D-amino acid–based life-forms.

3 ■ "NONSTANDARD" AMINO ACIDS

The 20 common amino acids are by no means the only amino acids that occur in biological systems. "Nonstandard" amino acid residues are often important constituents of proteins and biologically active polypeptides. Many amino acids, however, are not constituents of proteins. Together with their derivatives, they play a variety of biologically important roles.

A. *Amino Acid Derivatives in Proteins*

The "universal" genetic code, which is nearly identical in all known life-forms (Section 5-4B), specifies only the 20 "standard" amino acids of Table 4-1. Nevertheless, many other amino acids, a selection of which is given in Fig. 4-22, are components of certain proteins. *In all known cases but one (Section 32-2D), however, these unusual amino acids result from the specific modification of an amino acid residue after the polypeptide chain has been synthesized.* Among the most prominent of these modified amino acid residues are **4-hydroxyproline** and **5-hydroxylysine.** Both of these amino acid residues are im-

portant structural constituents of the fibrous protein **collagen,** the most abundant protein in mammals (Section 8-2B). Amino acids of proteins that form complexes with nucleic acids are often modified. For example, ribosomal proteins (Section 32-3A) and the chromosomal proteins known as **histones** (Section 34-1A) may be specifically methylated, acetylated, and/or phosphorylated. Several of these derivatized amino acid residues are presented in Fig. 4-22. **N-Formylmethionine** is initially the N-terminal residue of all prokaryotic proteins, but is usually removed as part of the protein maturation process (Section 32-3C). **γ-Carboxyglutamic acid** is a constituent of several proteins involved in blood clotting (Section 35-1B). Note that in most cases, these modifications are important, if not essential, for the function of the protein.

D-Amino acid residues are components of many of the relatively short (< 20 residues) bacterial polypeptides that are enzymatically rather than ribosomally synthesized. These polypeptides are perhaps most widely distributed as constituents of bacterial cell walls (Section 11-3B), which D-amino acids render less susceptible to attack by the **peptidases** (enzymes that hydrolyze peptide bonds) that many organisms employ to digest bacterial cell walls. Likewise, D-amino acids are components of many bacterially produced peptide antibiotics including **valinomycin, gramicidin A** (Section 20-2C), and **actinomycin D** (Section 31-2D). D-Amino acid residues are also functionally essential components of several ribosomally synthesized polypeptides of eukaryotic as well as prokaryotic origin. These D-amino acid residues are posttranslationally formed, most probably through the enzymatically mediated inversion of the preexisting L-amino acid residues.

B. *Specialized Roles of Amino Acids*

Besides their role in proteins, amino acids and their derivatives have many biologically important functions. A few examples of these substances are shown in Fig. 4-23. This alternative use of amino acids is an example of the biological opportunism that we shall repeatedly encounter: *Nature tends to adapt materials and processes that are already present to new functions.*

Amino acids and their derivatives often function as chemical messengers in the communications between cells. For example, glycine, **γ-aminobutyric acid (GABA;** a glutamate decarboxylation product), and **dopamine** (a tyrosine product) are neurotransmitters (substances released by nerve cells to alter the behavior of their neighbors; Section 20-5C); **histamine** (the decarboxylation product of histidine) is a potent local mediator of allergic reactions; and **thyroxine** (a tyrosine product) is an iodine-containing thyroid hormone that generally stimulates vertebrate metabolism (Section 19-1D).

Certain amino acids are important intermediates in various metabolic processes. Among them are **citrulline** and **ornithine,** intermediates in urea biosynthesis (Section 26-2B); **homocysteine,** an intermediate in amino acid metabolism (Section 26-3E); and **S-adenosylmethionine,** a biological methylating reagent (Section 26-3E).

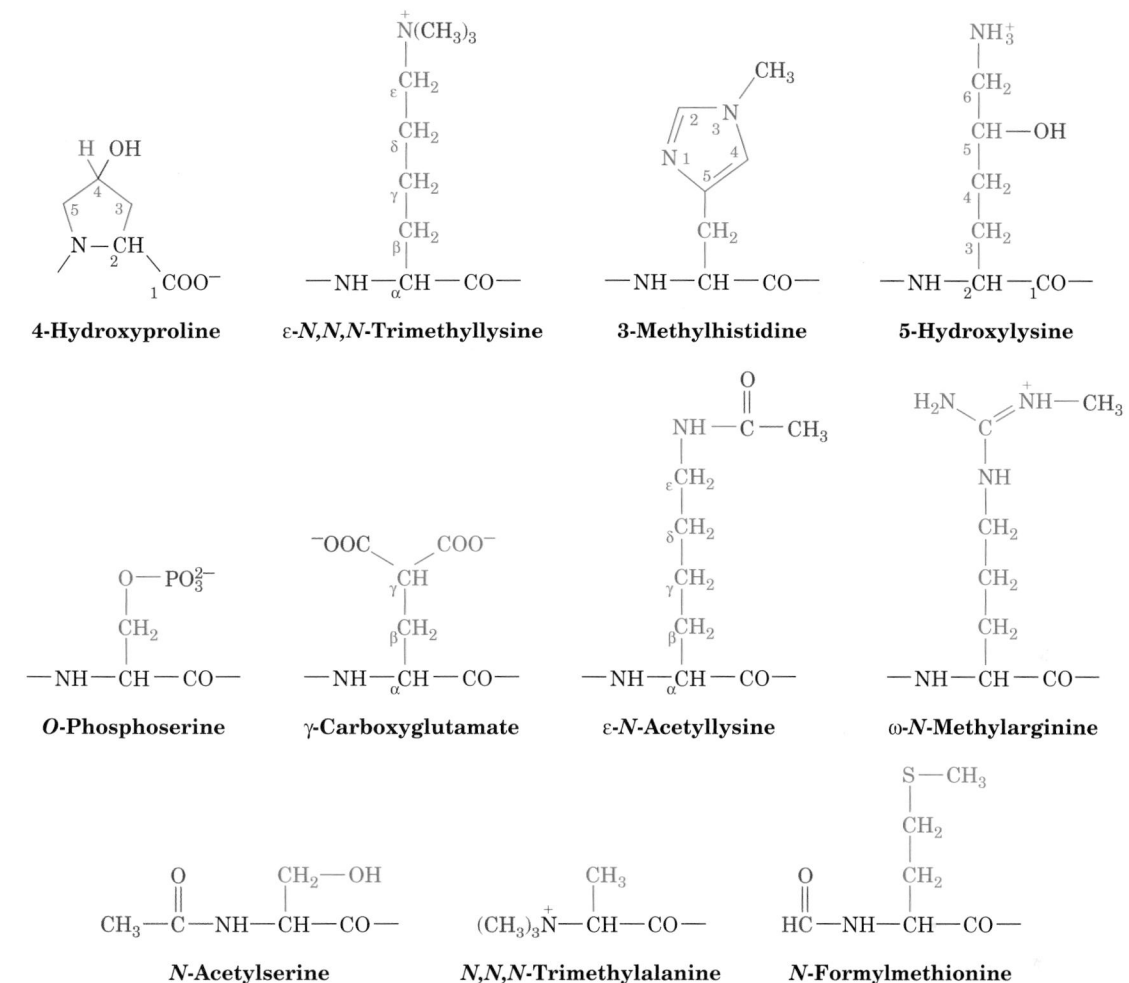

FIGURE 4-22 Some uncommon amino acid residues that are components of certain proteins. All of these residues are modified from one of the 20 "standard" amino acids after polypeptide chain biosynthesis. Those amino acid residues that are derivatized at their N_α position occur at the N-termini of proteins.

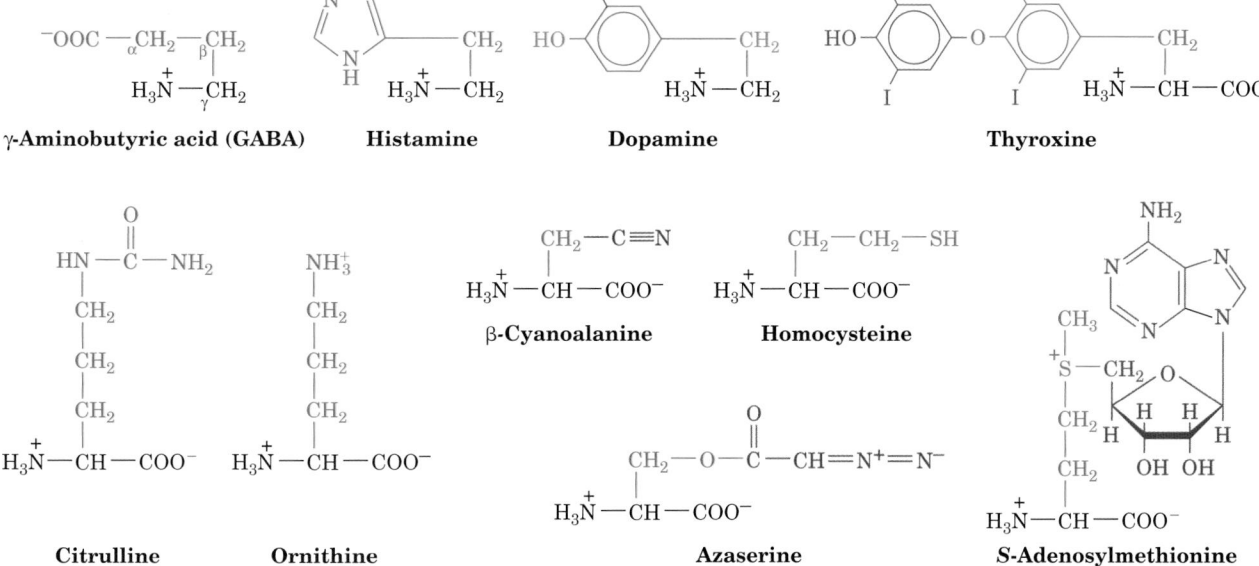

FIGURE 4-23 Some biologically produced derivatives of "standard" amino acids and amino acids that are not components of proteins.

Nature's diversity is remarkable. Over 700 different amino acids have been found in various plants, fungi, and bacteria, most of which are α-amino acids. For the most part, their biological roles are obscure although the fact that many are toxic suggests that they have a protective function. Indeed, some of them, such as **azaserine,** are medically useful antibiotics. Many of these amino acids are simple derivatives of the 20 "standard" amino acids although some of them, including azaserine and **β-cyanoalanine** (Fig. 4-23), have unusual structures.

CHAPTER SUMMARY

1 ■ The Amino Acids of Proteins Proteins are linear polymers that are synthesized from the same 20 "standard" α-amino acids through their condensation to form peptide bonds. These amino acids all have a carboxyl group with a pK near 2.2 and an amino substituent with a pK near 9.4 attached to the same carbon atom, the C_α atom. The α-amino acids are zwitterionic compounds, ^+H_3N—CHR—COO^-, in the physiological pH range. The various amino acids are usually classified according to the polarities of their side chains, R, which are also substituent to the C_α atom. Glycine, alanine, valine, leucine, isoleucine, methionine, proline (which is really a secondary amino acid), phenylalanine, and tryptophan are nonpolar amino acids; serine, threonine, asparagine, glutamine, tyrosine, and cysteine are uncharged and polar; and lysine, arginine, histidine, aspartic acid, and glutamic acid are charged and polar. The side chains of many of these amino acids bear acid–base groups, and hence the properties of the proteins containing them are pH dependent.

2 ■ Optical Activity The C_α atoms of all α-amino acids except glycine each bear four different substituents and are therefore chiral centers. According to the Fischer convention, which relates the configuration of D- or L-glyceraldehyde to that of the asymmetric center of interest, all the amino acids of proteins have the L configuration; that is, they all have the same absolute configuration about their C_α atom. According to the Cahn–Ingold–Prelog (*RS*) system of chirality nomenclature, they are, with the exception of cysteine, all (*S*)-amino acids. The side chains of threonine and isoleucine also contain chiral centers. A prochiral center has no rotational symmetry, and hence its substituents, in the case of a central atom, or its faces, in the case of a planar molecule, are distinguishable.

3 ■ "Nonstandard" Amino Acids Amino acid residues other than the 20 from which proteins are synthesized also have important biological functions. These "nonstandard" residues result from the specific chemical modifications of amino acid residues in preexisting proteins. Amino acids and their derivatives also have independent biological roles such as neurotransmitters, metabolic intermediates, and poisons.

REFERENCES

HISTORY

Vickery, H.B. and Schmidt, C.L.A., The history of the discovery of amino acids, *Chem. Rev.* **9,** 169–318 (1931).

Vickery, H.B., The history of the discovery of the amino acids. A review of amino acids discovered since 1931 as components of native proteins, *Adv. Protein Chem.* **26,** 81–171 (1972).

PROPERTIES OF AMINO ACIDS

Barrett, G.C. and Elmore, D.T., *Amino Acids and Peptides,* Chapters 1–4, Cambridge University Press (1998).

Cohn, E.J. and Edsall, J.T., *Proteins, Amino Acids and Peptides as Ions and Dipolar Ions,* Academic Press (1943). [A classic work in its field.]

Davies, J.S. (Ed.), *Amino Acids and Peptides,* Chapman & Hall (1985). [A sourcebook on amino acids.]

Edsall, J.T. and Wyman, J., *Biophysical Chemistry,* Vol. 1, Academic Press (1958). [A detailed treatment of the physical chemistry of amino acids.]

Jakubke, H.-D. and Jeschkeit, H., *Amino Acids, Peptides and Proteins,* translated into English by Cotterrell, G.P., Wiley (1977).

Meister, A., *Biochemistry of the Amino Acids* (2nd ed.), Vol. 1, Academic Press (1965). [A compendium of information on amino acid properties.]

OPTICAL ACTIVITY

Cahn, R.S., An introduction to the sequence rule, *J. Chem. Ed.* **41,** 116–125 (1964). [A presentation of the Cahn–Ingold–Prelog system of nomenclature.]

Huheey, J.E., A novel method for assigning *R,S* labels to enantiomers, *J. Chem. Ed.* **63,** 598–600 (1986).

Lamzin, V.S., Dauter, Z., and Wilson, K.S., How nature deals with stereoisomers, *Curr. Opin. Struct. Biol.* **5,** 830–836 (1995). [Discusses proteins synthesized from D-amino acids.]

Mislow, K., *Introduction to Stereochemistry,* Benjamin (1966).

Solomons, T.W.G., *Organic Chemistry* (7th ed. upgrade), Chapter 5, Wiley (2001). [A discussion of chirality. Most other organic chemistry textbooks contain similar material.]

"NONSTANDARD" AMINO ACIDS

Amino Acids, Peptides, and Proteins, The Royal Society of Chemistry. [An annual series containing literature reviews on amino acids.]

Fowden, L., Lea, P.J., and Bell, E.A., The non-protein amino acids of plants, *Adv. Enzymol.* **50,** 117–175 (1979).

Fowden, L., Lewis, D., and Tristram, H., Toxic amino acids: their action as antimetabolites, *Adv. Enzymol.* **29,** 89–163 (1968).

Kleinkauf, H. and Döhren, H., Nonribosomal polypeptide formation on multifunctional proteins, *Trends Biochem. Sci.* **8,** 281–283 (1993).

Mor, A., Amiche, M., and Nicholas, P., Enter a new posttranscriptional modification: D-amino acids in gene-encoded peptides, *Trends Biochem. Sci.* **17,** 481–485 (1992).

Thompson, J. and Donkersloot, J.A., N-(Carboxyalkyl)amino acids: Occurrence, synthesis, and functions, *Annu. Rev. Biochem.* **61,** 517–557 (1992).

PROBLEMS

1. Name the 20 standard amino acids without looking them up. Give their three-letter and one-letter symbols. Identify the two standard amino acids that are isomers and the two others that, although not isomeric, have essentially the same molecular mass for the neutral molecules.

2. Draw the following oligopeptides in their predominant ionic forms at pH 7: (a) Phe-Met-Arg, (b) tryptophanyllysyl-aspartic acid, and (c) Gln-Ile-His-Thr.

3. How many different pentapeptides are there that contain one residue each of Gly, Asp, Tyr, Cys, and Leu? $5! = 120$

4. Draw the structures of the following two oligopeptides with their cysteine residues cross-linked by a disulfide bond: Val-Cys, Ser-Cys-Pro.

*5. What are the concentrations of the various ionic species in a 0.1M solution of lysine at pH 4, 7, and 10?

6. Derive Eq. [4.1] for a monoamino, monocarboxylic acid (use the Henderson–Hasselbalch equation).

*7. The **isoionic point** of a compound is defined as the pH of a pure water solution of the compound. What is the isoionic point of a 0.1M solution of glycine?

8. Normal human hemoglobin has an isoelectric point of 6.87. A mutant variety of hemoglobin, known as **sickle-cell hemoglobin,** has an isoelectric point of 7.09. The titration curve of hemoglobin indicates that, in this pH range, 13 groups change ionization states per unit change in pH. Calculate the difference in ionic charge between molecules of normal and sickle-cell hemoglobin.

$$\Delta z = 13\,\Delta pH = 13(7.09-6.87) \approx 3$$

9. Indicate whether the following familiar objects are chiral, prochiral, or nonchiral.

(a) A glove
(b) A tennis ball
(c) A good pair of scissors
(d) A screw
(e) This page
(f) A toilet paper roll
(g) A snowflake
(h) A spiral staircase
(i) A flight of normal stairs
(j) A paper clip
(k) A shoe
(l) A pair of glasses

10. Draw four equivalent Fischer projection formulas for L-alanine (see Figs. 4-12 and 4-13).

*11. (a) Draw the structural formula and the Fischer projection formula of (S)-3-methylhexane. (b) Draw all the stereoisomers of 2,3-dichlorobutane. Name them according to the (RS) system and indicate which of them has the meso form.

12. Identify and name the prochiral centers or faces of the following molecules:

(a) Acetone
(b) Propene
(c) Glycine
(d) Alanine
(e) Lysine
(f) 3-Methylpyridine

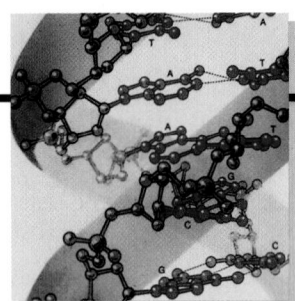

Chapter
5

Nucleic Acids, Gene Expression, and Recombinant DNA Technology

Knowledge of how genes are expressed and how they can be manipulated is becoming increasingly important for understanding nearly every aspect of biochemistry. Consequently, although we do not undertake a detailed discussion of these processes until Part V of this textbook, we outline their general principles in this chapter. We do so by describing the chemical structures of nucleic acids, how we have come to know that DNA is the carrier of genetic information, the structure of the major form of DNA, and the general principles of how the information in genes

directs the synthesis of RNA and proteins (how genes are expressed) and how DNA is replicated. The chapter ends with a discussion of how DNA is experimentally manipulated and expressed, processes that are collectively referred to as genetic engineering. These processes have revolutionized the practice of biochemistry.

1 ■ NUCLEOTIDES AND NUCLEIC ACIDS

Nucleotides and their derivatives are biologically ubiquitous substances that participate in nearly all biochemical processes:

   1. They form the monomeric units of nucleic acids and thereby play central roles in both the storage and the expression of genetic information.

   2. Nucleoside triphosphates, most conspicuously ATP (Section 1-3B), are the "energy-rich" end products of the majority of energy-releasing pathways and the substances whose utilization drives most energy-requiring processes.

   3. Most metabolic pathways are regulated, at least in part, by the levels of nucleotides such as ATP and ADP. Moreover, certain nucleotides, as we shall see, function as intracellular signals that regulate the activities of numerous metabolic processes.

   4. Nucleotide derivatives, such as **nicotinamide adenine dinucleotide** (Section 13-2A), **flavin adenine dinucleotide** (Section 16-2C), and **coenzyme A** (Section 21-2), are required participants in many enzymatic reactions.

   5. As components of the enzymelike nucleic acids known as **ribozymes,** nucleotides have important catalytic activities themselves.

A. *Nucleotides, Nucleosides, and Bases*

*Nucleotides are phosphate esters of a five-carbon sugar (which is therefore known as a **pentose;** Section 11-1A) in*

(a) *(b)*

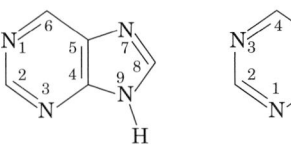

Ribonucleotides   **Deoxyribonucleotides**

FIGURE 5-1 Chemical structures of (*a*) ribonucleotides and (*b*) deoxyribonucleotides.

which a nitrogenous base is covalently linked to C1′ of the sugar residue. In **ribonucleotides** (Fig. 5-1*a*), the monomeric units of RNA, the pentose is **D-ribose,** whereas in **deoxyribonucleotides** (or just **deoxynucleotides;** Fig. 5-1*b*), the monomeric units of DNA, the pentose is **2′-deoxy-D-ribose** (note that the "primed" numbers refer to the atoms of the ribose residue; "unprimed" numbers refer to the nitrogenous base). The phosphate group may be bonded to C5′ of the pentose to form a **5′-nucleotide** (Fig. 5-1) or to its C3′ to form a **3′-nucleotide.** If the phosphate group is absent, the compound is known as a **nucleoside.** A 5′-nucleotide, for example, may therefore be referred to as a **nucleoside-5′-phosphate.** In all naturally occurring nucleotides and nucleosides, the bond linking the nitrogenous base to the pentose C1′ atom (which is called a glycosidic bond; Section 11-1C) extends from the same side of the ribose ring as does the C4′—C5′ bond (the so-called β configuration; Section 11-1B) rather than from the opposite side (the α configuration). Note that nucleotide phosphate groups are doubly ionized at physiological pH's; that is, *nucleotides are moderately strong acids.*

*The nitrogenous bases are planar, aromatic, heterocyclic molecules which, for the most part, are derivatives of either **purine** or **pyrimidine.***

Purine   **Pyrimidine**

The structures, names, and abbreviations of the common bases, nucleosides, and nucleotides are given in Table 5-1. The major purine components of nucleic acids are **adenine** and **guanine** residues; the major pyrimidine residues are those of **cytosine, uracil** (which occurs mainly in RNA), and **thymine** (5-methyluracil, which occurs mainly in DNA). The purines form glycosidic bonds to ribose via their N9 atoms, whereas pyrimidines do so through their N1 atoms (note that purines and pyrimidines have dissimilar atom numbering schemes).

B. *The Chemical Structures of DNA and RNA*

The chemical structures of the nucleic acids were elucidated by the early 1950s largely through the efforts of

Phoebus Levine, followed by the work of Alexander Todd. *Nucleic acids are, with few exceptions, linear polymers of nucleotides whose phosphate groups bridge the 3′ and 5′ positions of successive sugar residues* (e.g., Fig. 5-2). The phosphates of these **polynucleotides,** the **phosphodiester** groups, are acidic, so that, *at physiological pH's, nucleic acids are polyanions.* Polynucleotides have directionality, that is, each has a **3′ end** (the end whose C3′ atom is not linked to a neighboring nucleotide) and a **5′ end** (the end whose C5′ atom is not linked to a neighboring nucleotide).

a. DNA's Base Composition Is Governed by Chargaff's Rules

DNA has equal numbers of adenine and thymine residues (A = T) and equal numbers of guanine and cytosine residues (G = C). These relationships, known as **Chargaff's rules,** were discovered in the late 1940s by Erwin Chargaff, who first devised reliable quantitative methods for the separation and analysis of DNA hydrolysates. Chargaff also found that the base composition of DNA from a given organism is characteristic of that organism; that is, it is independent of the tissue from which the DNA is taken as well as the age of the organism, its nutritional state, or any other environmental factor. The structural basis for Chargaff's rules is that in double-stranded DNA, G is always hydrogen bonded (forms a **base pair**) with C, whereas A always forms a base pair with T (Fig. 1-16).

DNA's base composition varies widely among different organisms. It ranges from ~25% to 75% G + C in different species of bacteria. It is, however, more or less constant among related species; for example, in mammals G + C ranges from 39% to 46%.

RNA, which usually occurs as single-stranded molecules, has no apparent constraints on its base composition. However, double-stranded RNA, which comprises the genetic material of several viruses, also obeys Chargaff's rules (here A base pairs with U in the same way it does with T in DNA; Fig. 1-16). Conversely, single-stranded DNA, which occurs in certain viruses, does not obey Chargaff's rules. On entering its host organism, however, such DNA is replicated to form a double-stranded molecule, which then obeys Chargaff's rules.

b. Nucleic Acid Bases May Be Modified

Some DNAs contain bases that are chemical derivatives of the standard set. For example, dA and dC in the DNAs of many organisms are partially replaced by *N*[6]-**methyl-dA** and **5-methyl-dC,** respectively.

N [6]-**Methyl-dA**   **5-Methyl-dC**

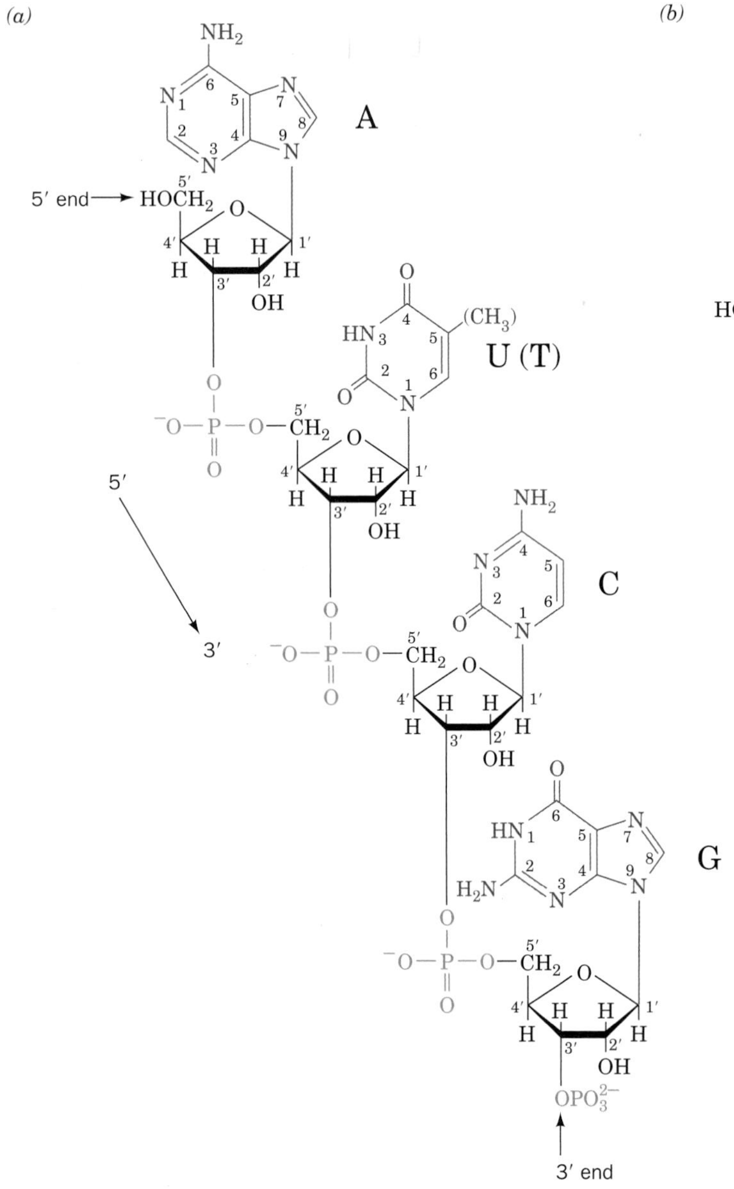

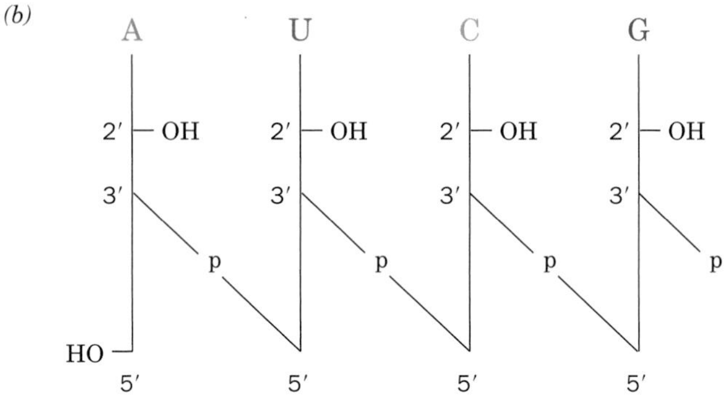

(a)

(b)

FIGURE 5-2   Chemical structure of a nucleic acid.
(a) The tetranucleotide adenyl-3′,5′-uridyl-3′,5′-cytidyl-3′,5′-guanylyl-3′-phosphate. The sugar atom numbers are primed to distinguish them from the atomic positions of the bases. By convention, a polynucleotide sequence is written with its 5′ end at the left and its 3′ end to the right. Thus, reading left to right, the phosphodiester bond links neighboring ribose residues in the 5′ → 3′ direction. The above sequence may be abbreviated ApUpCpGp or just AUCGp (where a "p" to the left and/or right of a nucleoside symbol indicates a 5′ and/or a 3′ phosphate group, respectively; see Table 5-1 for other symbol definitions). The corresponding deoxy-tetranucleotide, in which the 2′-OH groups are all replaced by H and the base uracil (U) is replaced by thymine (5-methyluracil; T), is abbreviated d(ApTpCpGp) or d(ATCGp). *(b)* A schematic representation of AUCGp. Here a vertical line denotes a ribose residue, its attached base is indicated by the corresponding one-letter abbreviation, and a diagonal line flanking an optional "p" represents a phosphodiester bond. The atom numbering of the ribose residues, which is indicated here, is usually omitted. The equivalent representation of deoxypolynucleotides differs only by the absence of the 2′-OH groups and the replacement of U by T.

The altered bases are generated by the sequence-specific enzymatic modification of normal DNA (Sections 5-5A and 30-7). The modified DNAs obey Chargaff's rules if the derivatized bases are taken as equivalent to their parent bases. Likewise, many bases in RNAs and, in particular, those in **transfer RNAs (tRNAs;** Section 32-2A), are derivatized.

c. RNA but Not DNA Is Susceptible to Base-Catalyzed Hydrolysis

RNA is highly susceptible to base-catalyzed hydrolysis by the reaction mechanism diagrammed in Fig. 5-3 so as to yield a mixture of 2′ and 3′ nucleotides. In contrast, DNA, which lacks 2′-OH groups, is resistant to base-catalyzed hydrolysis and is therefore much more chemically stable than RNA. This is probably why DNA rather than RNA evolved to be the cellular genetic archive.

2 ■ DNA IS THE CARRIER OF GENETIC INFORMATION

Nucleic acids were first isolated in 1869 by Friedrich Miescher and so named because he found them in the nuclei of **leukocytes** (pus cells) from discarded surgical bandages. The presence of nucleic acids in other cells was demonstrated within a few years, but it was not until some 75 years after their discovery that their biological function was elucidated. Indeed, in the 1930s and 1940s it was widely held, in what was termed the **tetranucleotide hypothesis,** that nucleic acids have a monotonously repeating sequence of all four bases, so that they were not suspected of having a genetic function. Rather, it was generally assumed that genes were proteins since proteins were the only biochemical entities that, at that time, seemed capable of the

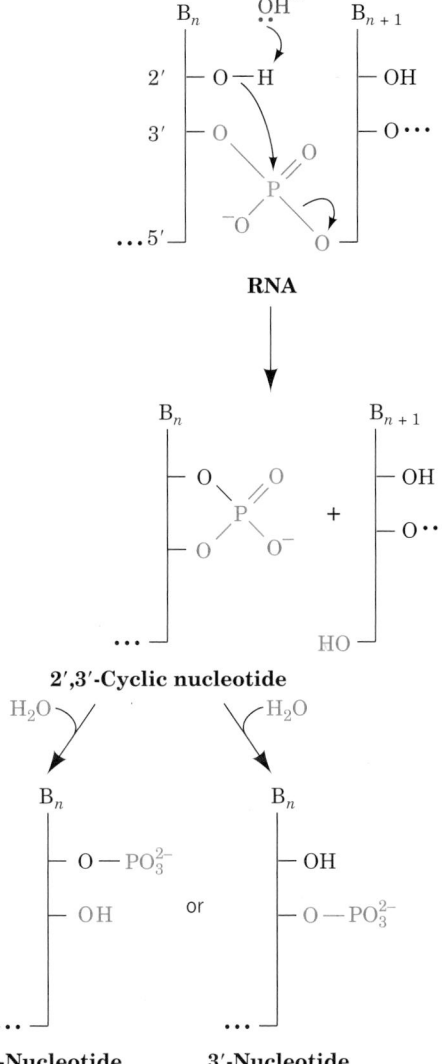

FIGURE 5-3 Mechanism of base-catalyzed RNA hydrolysis. The base-induced deprotonation of the 2′-OH group facilitates its nucleophilic attack on the adjacent phosphorus atom, thereby cleaving the RNA backbone. The resultant 2′,3′-cyclic phosphate group subsequently hydrolyzes to either the 2′ or the 3′ phosphate.

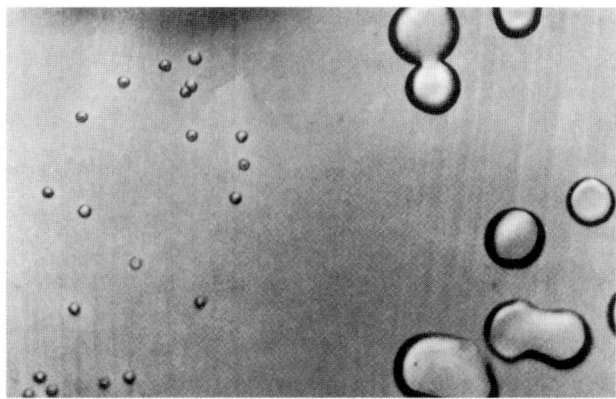

FIGURE 5-4 Pneumococci. The large glistening colonies are virulent S-type pneumococci that resulted from the transformation of nonpathogenic R-type pneumococci (smaller colonies) by DNA from heat-killed S pneumococci. [From Avery, O.T., MacLeod, C.M., and McCarty, M., *J. Exp. Med.* **79**, 153 (1944). Copyright © 1944 by Rockefeller University Press.]

required specificity. In this section, we outline the experiments that established DNA's genetic role.

A. *Transforming Principle Is DNA*

The virulent (capable of causing disease) form of pneumococcus (*Diplococcus pneumoniae*), a bacterium that causes pneumonia, is encapsulated by a gelatinous polysaccharide coating that contains the binding sites (known as **O-antigens;** Section 11-3B) through which it recognizes the cells it infects. Mutant pneumococci that lack this coating, because of a defect in an enzyme involved in its formation, are not pathogenic (capable of causing disease). The virulent and nonpathogenic pneumococci are known as the S and R forms, respectively, because of the smooth

and rough appearances of their colonies in culture (Fig. 5-4).

In 1928, Frederick Griffith made a startling discovery. He injected mice with a mixture of live R and heat-killed S pneumococci. This experiment resulted in the death of most of the mice. More surprising yet was that the blood of the dead mice contained live S pneumococci. The dead S pneumococci initially injected into the mice had somehow **transformed** the otherwise innocuous R pneumococci to the virulent S form. Furthermore, the progeny of the transformed pneumococci were also S; the transformation was permanent. Eventually, it was shown that the transformation could also be made *in vitro* (outside a living organism; literally "in glass") by mixing R cells with a cell-free extract of S cells. The question remained: What is the nature of the **transforming principle?**

In 1944, Oswald Avery, Colin MacLeod, and Maclyn McCarty, after a 10-year investigation, reported that *transforming principle is DNA.* The conclusion was based on the observations that the laboriously purified (few modern fractionation techniques were then available) transforming principle had all the physical and chemical properties of DNA, contained no detectable protein, was unaffected by enzymes that catalyze the hydrolysis of proteins and RNA, and was totally inactivated by treatment with an enzyme that catalyzes the hydrolysis of DNA. *DNA must therefore be the carrier of genetic information.*

Avery's discovery was another idea whose time had not yet come. This seminal advance was initially greeted with skepticism and then largely ignored. Indeed, even Avery did not directly state that DNA is the hereditary material but merely that it has "biological specificity." His work, however, influenced several biochemists, including Erwin Chargaff, whose subsequent accurate determination of DNA base ratios using the then newly developed technique of **paper chromatography** (Section 6-3D) refuted the tetranucleotide hypothesis and thereby indicated that DNA could be a complex molecule.

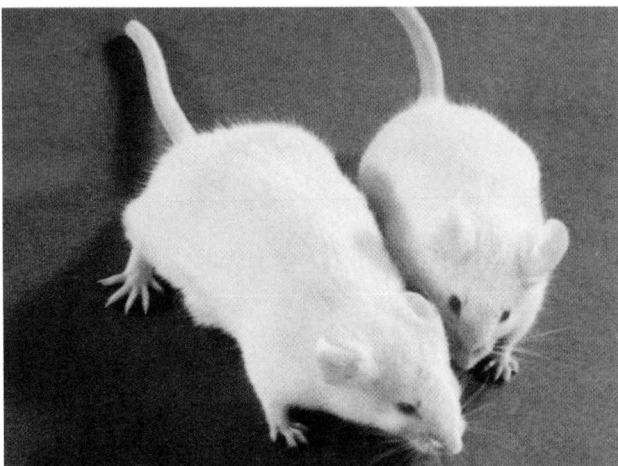

FIGURE 5-5   Transgenic mouse. The gigantic mouse (*left*) grew from a fertilized ovum that had been microinjected with DNA bearing the rat growth hormone gene. His normal-sized littermate (*right*) is shown for comparison. [Courtesy of Ralph Brinster, University of Pennsylvania.]

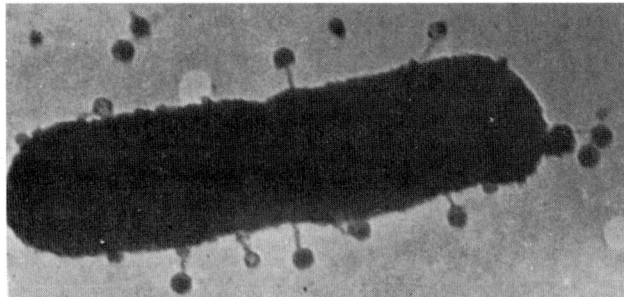

FIGURE 5-6   Bacteriophages attached to the surface of a bacterium. An early electron micrograph of an *E. coli* cell to which **bacteriophage T5** are adsorbed by their tails. [Courtesy of Thomas F. Anderson, Fox Chase Cancer Center.]

It was eventually demonstrated that eukaryotes are also subject to transformation by DNA. Thus DNA, which cytological studies had shown resides in the chromosomes, must also be the hereditary material of eukaryotes. In a spectacular demonstration of eukaryotic transformation, Ralph Brinster, in 1982, microinjected DNA bearing the gene for rat **growth hormone** (a polypeptide) into the nuclei of fertilized mouse eggs (a technique discussed in Section 5-5H) and implanted these eggs into the uteri of foster mothers. The resulting "supermice" (Fig. 5-5), which

had high levels of rat growth hormone in their serum, grew to nearly twice the weight of their normal littermates. Such genetically altered animals are said to be **transgenic.**

B. *The Hereditary Molecule of Many Bacteriophages Is DNA*

Electron micrographs of bacteria infected with bacteriophages show empty-headed phage "ghosts" attached to the bacterial surface (Fig. 5-6). This observation led Roger Herriott to suggest that "the virus may act like a little hypodermic needle full of transforming principle," which it injects into the bacterial host (Fig. 5-7). This proposal was tested in 1952 by Alfred Hershey and Martha Chase as is diagrammed in Fig. 5-8. **Bacteriophage T2** was grown on *E. coli* in a medium containing the radioactive isotopes ^{32}P and ^{35}S. This labeled the phage capsid, which contains no

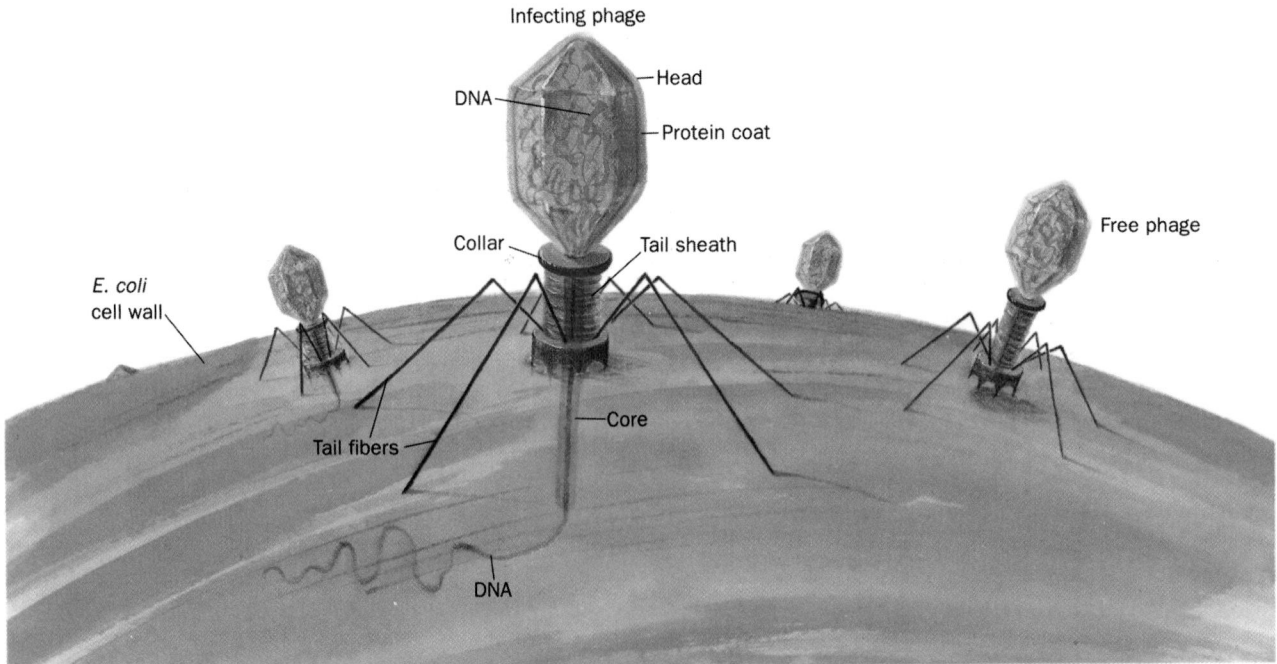

FIGURE 5-7   Diagram of T2 bacteriophage injecting its DNA into an *E. coli* cell.

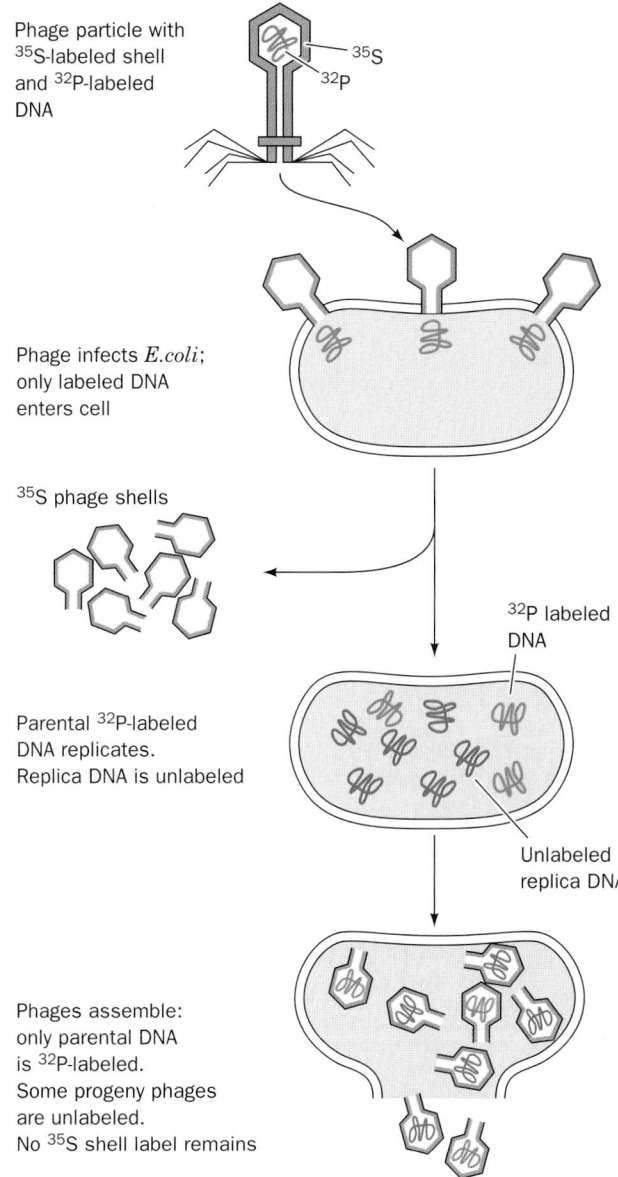

Phage particle with
^{35}S-labeled shell
and ^{32}P-labeled
DNA

^{35}S

^{32}P

Phage infects *E.coli*;
only labeled DNA
enters cell

^{35}S phage shells

^{32}P labeled
DNA

Parental ^{32}P-labeled
DNA replicates.
Replica DNA is unlabeled

Unlabeled
replica DNA

Phages assemble:
only parental DNA
is ^{32}P-labeled.
Some progeny phages
are unlabeled.
No ^{35}S shell label remains

FIGURE 5-8   The Hershey–Chase experiment. This experiment
demonstrates that only the nucleic acid component of
bacteriophages enters the bacterial host during phage
infection.

P, with ^{35}S, and its DNA, which contains no S, with ^{32}P.
These phages were added to an unlabeled culture of *E. coli*
and, after sufficient time was allowed for the phages to infect the bacterial cells, the culture was agitated in a kitchen
blender so as to shear the phage ghosts from the bacterial
cells. This rough treatment neither injured the bacteria nor
altered the course of the phage infection. When the phage
ghosts were separated from the bacteria (by centrifugation;
Section 6-5), the ghosts were found to contain most of the
^{35}S, whereas the bacteria contained most of the ^{32}P.
Furthermore, 30% of the ^{32}P appeared in the progeny
phages but only 1% of the ^{35}S did so. Hershey and Chase
therefore concluded that only the phage DNA was essen-

tial for the production of progeny. *DNA therefore must be
the hereditary material.* In later years it was shown that, in
a process known as **transfection,** purified phage DNA can,
by itself, induce a normal phage infection of a properly
treated bacterial host (transfection differs from transformation in that the latter results from the recombination of
the bacterial chromosome with a fragment of homologous
DNA).

In 1952, the state of knowledge of biochemistry was such
that Hershey's discovery was much more readily accepted
than Avery's identification of the transforming principle
had been some 8 years earlier. Within a few months, the
first speculations arose as to the nature of the **genetic code**
(the correspondence between the base sequence of a gene
and the amino acid sequence of a protein, Section 5-4B),
and James Watson and Francis Crick were inspired to investigate the structure of DNA. In 1955, it was shown that
the somatic cells of eukaryotes have twice the DNA of the
corresponding germ cells. When this observation was proposed to be a further indicator of DNA's genetic role, there
was little comment even though the same could be said of
any other chromosomal component.

3 ■ DOUBLE HELICAL DNA

The determination of the structure of DNA by Watson and
Crick in 1953 is often said to mark the birth of modern
molecular biology. The **Watson–Crick structure** of DNA is
of such importance because, in addition to providing the
structure of what is arguably the central molecule of life,
it suggested the molecular mechanism of heredity. Watson
and Crick's accomplishment, which is ranked as one of science's major intellectual achievements, tied together the
less than universally accepted results of several diverse
studies:

1. Chargaff's rules. At the time, the relationships
A = T and G = C were quite obscure because their significance was not apparent. In fact, even Chargaff did not
emphasize them.

2. Correct tautomeric forms of the bases. X-Ray, nuclear magnetic resonance (NMR), and spectroscopic investigations have firmly established that the nucleic acid
bases are overwhelmingly in the keto tautomeric forms
shown in Table 5-1. In 1953, however, this was not generally appreciated. Indeed, guanine and thymine were widely
believed to be in their enol forms (Fig. 5-9) because it was
thought that the resonance stability of these aromatic molecules would thereby be maximized. Knowledge of the
dominant tautomeric forms, which was prerequisite for the
prediction of the correct hydrogen bonding associations of
the bases, was provided by Jerry Donohue, an office mate
of Watson and Crick and an expert on the X-ray structures
of small organic molecules.

3. Information that DNA is a helical molecule. This was
provided by an X-ray diffraction photograph of a DNA

TABLE 5-1   Names and Abbreviations of Nucleic Acid Bases, Nucleosides, and Nucleotides

Base Formula	Base (X = H)	Nucleoside (X = ribose[a])	Nucleotide[b] (X = ribose phosphate[a])
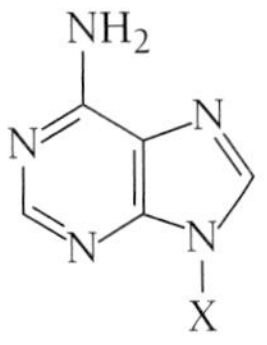 NH$_2$	Adenine Ade A	Adenosine Ado A	Adenylic acid Adenosine monophosphate AMP
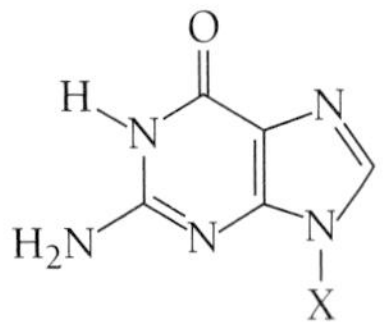 (Guanine structure)	Guanine Gua G	Guanosine Guo G	Guanylic acid Guanosine monophosphate GMP
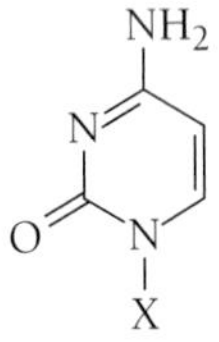 (Cytosine structure)	Cytosine Cyt C	Cytidine Cyd C	Cytidylic acid Cytidine monophosphate CMP
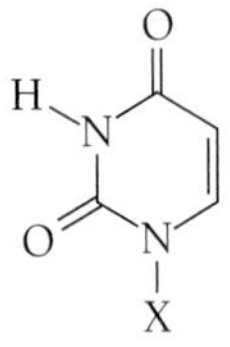 (Uracil structure)	Uracil Ura U	Uridine Urd U	Uridylic acid Uridine monophosphate UMP
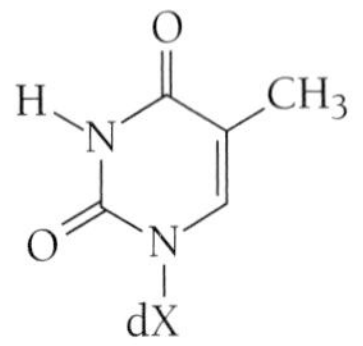 (Thymine structure)	Thymine Thy T	Deoxythymidine dThd dT	Deoxythymidylic acid Deoxythymidine monophosphate dTMP

[a] The presence of a 2'-deoxyribose unit in place of ribose, as occurs in DNA, is implied by the prefixes "deoxy" or "d." For example, the deoxynucleoside of adenine is deoxyadenosine or dA. However, for thymine-containing residues, which rarely occur in RNA, the prefix is redundant and may be dropped. The presence of a ribose unit may be explicitly implied by the prefixes "ribo" or "r." Thus the ribonucleotide of thymine is ribothymidine or rT.

[b] The position of the phosphate group in a nucleotide may be explicitly specified as in, for example, 3'-AMP and 5'-GMP.

fiber taken by Rosalind Franklin (Fig. 5-10; DNA, being a threadlike molecule, does not crystallize but, rather, can be drawn out in fibers consisting of parallel bundles of molecules). A description of the photograph enabled Crick, an X-ray crystallographer by training who had earlier derived the equations describing diffraction by helical molecules, to deduce (a) that DNA is a helical molecule and (b) that its planar aromatic bases form a stack of parallel rings which is parallel to the fiber axis.

This information only provided a few crude landmarks that guided the elucidation of the DNA structure. It mostly sprang from Watson and Crick's imaginations through model building studies. Once the Watson–Crick model had been published, however, its basic simplicity combined with its obvious biological relevance led to its rapid acceptance. Later investigations have confirmed the essential correctness of the Watson–Crick model, although its details have been modified.

A. *The Watson–Crick Structure: B-DNA*

Fibers of DNA assume the so-called B conformation, as indicated by their X-ray diffraction patterns, when the counterion is an alkali metal such as Na$^+$ and the relative humidity is >92%. ***B-DNA** is regarded as the **native** (biologically functional) form of DNA because, for example, its X-ray pattern resembles that of the DNA in intact sperm heads.*

(a)

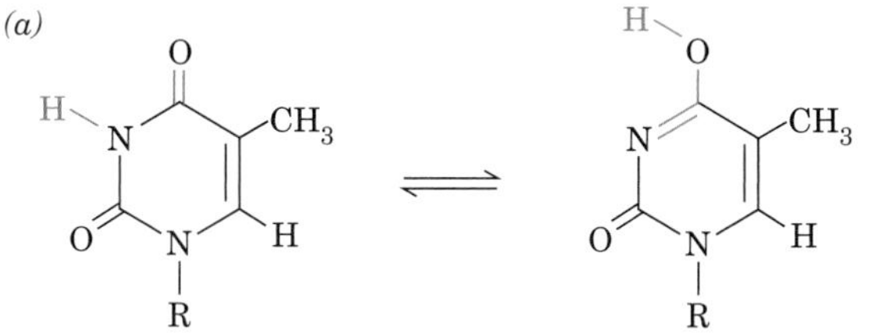

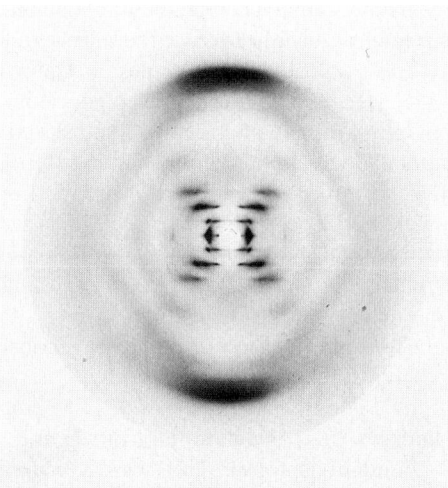

Thymine
(keto *or* lactam form)   Thymine
(enol *or* lactim form)

(b)

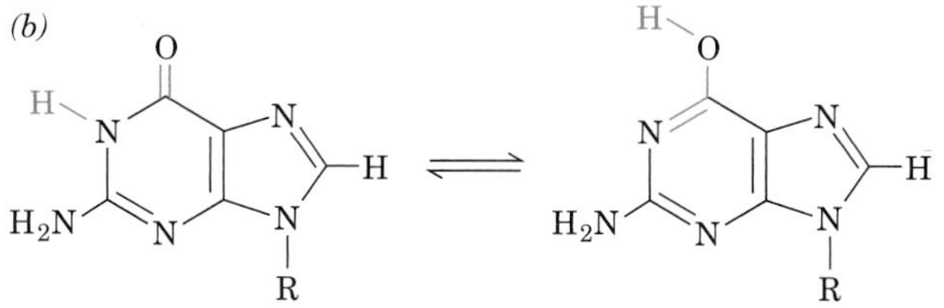

Guanine
(keto *or* lactam form)   Guanine
(enol *or* lactim form)

FIGURE 5-9   Some possible tautomeric conversions for bases. (*a*) Thymine and (*b*) guanine residues. Cytosine and adenine residues can undergo similar proton shifts.

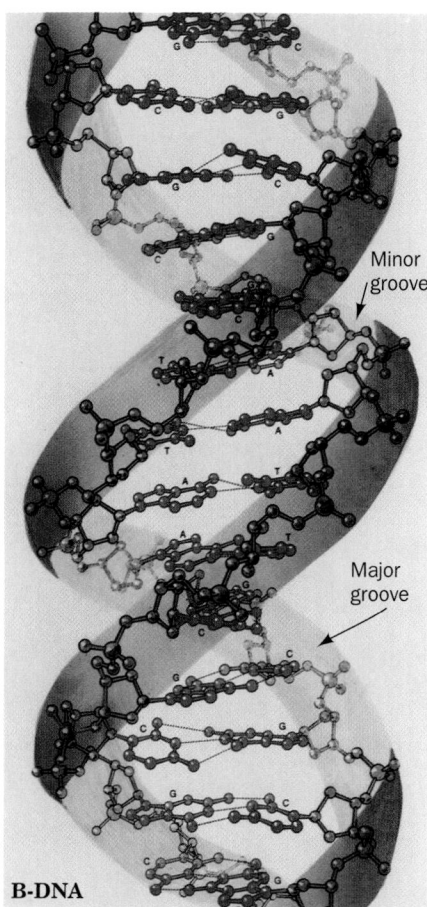

B-DNA

FIGURE 5-11   Three-dimensional structure of B-DNA. The repeating helix in this ball-and-stick drawing is based on the X-ray structure of the self-complementary dodecamer d(CGCGAATTCGCG) determined by Richard Dickerson and Horace Drew. The view is perpendicular to the helix axis. The sugar–phosphate backbones (*blue with blue-green ribbon outlines*) wind about the periphery of the molecule in opposite directions. The bases (*red*), which occupy its core, form hydrogen bonded base pairs. H atoms have been omitted for clarity. [Illustration, Irving Geis/Geis Archives Trust. Copyright Howard Hughes Medical Institute. Reproduced with permission.] ♦♦ See the Interactive exercises ♦♦ See Kinemage 2-1

FIGURE 5-10   X-ray diffraction photograph of a vertically oriented Na$^+$ DNA fiber in the B conformation taken by Rosalind Franklin. This is the photograph that provided key information for the elucidation of the Watson–Crick structure. The central X-shaped pattern of spots is indicative of a helix, whereas the heavy black arcs on the top and bottom of the diffraction pattern correspond to a distance of 3.4 Å and indicate that the DNA structure largely repeats every 3.4 Å along the fiber axis. [Courtesy of Maurice Wilkins, King's College, London.]

The Watson–Crick structure of B-DNA has the following major features:

1. *It consists of two polynucleotide strands that wind about a common axis with a right-handed twist to form an ~20-Å-diameter double helix (Fig. 5-11). The two strands are antiparallel (run in opposite directions) and* wrap around each other such that they cannot be separated without unwinding the helix. The bases occupy the core of the helix and the sugar–phosphate chains are coiled about its periphery, thereby minimizing the repulsions between charged phosphate groups.

2. The planes of the bases are nearly perpendicular to the helix axis. Each base is hydrogen bonded to a base on the opposite strand to form a planar base pair (Fig. 5-11). It is these hydrogen bonding interactions, a phenomenon known as **complementary base pairing,** that result in the specific association of the two chains of the double helix.

3. The "ideal" B-DNA helix has 10 base pairs (**bp**) per turn (a helical twist of 36° per bp) and, since the aromatic bases have van der Waals thicknesses of 3.4 Å and are partially stacked on each other (**base stacking,** Fig. 5-11), the helix has a **pitch** (rise per turn) of 34 Å.

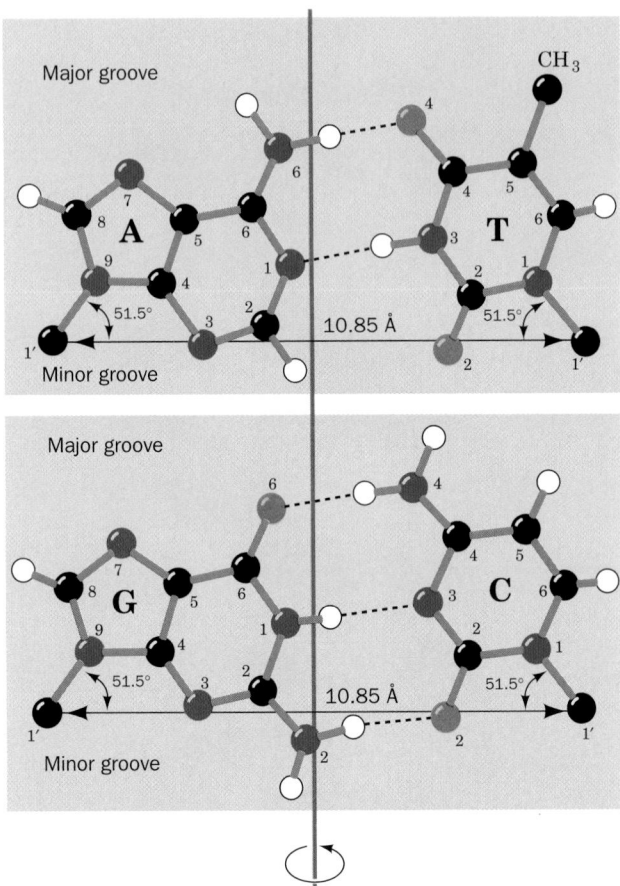

FIGURE 5-12 Watson–Crick base pairs. The line joining the C1′ atoms is the same length in both base pairs and makes equal angles with the glycosidic bonds to the bases. This gives DNA a series of pseudo-twofold symmetry axes (often referred to as **dyad axes**) that pass through the center of each base pair (*red line*) and are perpendicular to the helix axis. Note that A · T base pairs associate via two hydrogen bonds, whereas C · G base pairs are joined by three hydrogen bonds. [After Arnott, S., Dover, S.D., and Wonacott, A.J., *Acta Cryst.* **B25**, 2196 (1969).] 🖢 See Kinemages 2-2 and 17-2

The most remarkable feature of the Watson–Crick structure is that *it can accommodate only two types of base pairs: Each adenine residue must pair with a thymine residue and vice versa, and each guanine residue must pair with a cytosine residue and vice versa.* The geometries of these A · T and G · C base pairs, the so-called **Watson–Crick base pairs,** are shown in Fig. 5-12. It can be seen that *both of these base pairs are interchangeable in that they can replace each other in the double helix without altering the positions of the sugar–phosphate backbone's C1′ atoms. Likewise, the double helix is undisturbed by exchanging the partners of a Watson–Crick base pair, that is, by changing a G · C to a C · G or an A · T to a T · A.* In contrast, any other combination of bases (e.g., A · G or A · C) would significantly distort the double helix since the formation of a non-Watson–Crick base pair would require considerable reorientation of the sugar–phosphate chain.

B-DNA has two deep exterior grooves that wind between its sugar–phosphate chains as a consequence of the

helix axis passing through the approximate center of each base pair. However, the grooves are of unequal size (Fig. 5-11) because (1) the top edge of each base pair, as drawn in Fig. 5-12, is structurally distinct from the bottom edge; and (2) the deoxyribose residues are asymmetric. The **minor groove** exposes that edge of a base pair from which its C1′ atoms extend (opening toward the bottom in Fig. 5-12), whereas the **major groove** exposes the opposite edge of each base pair (the top of Fig. 5-12).

Although B-DNA is, by far, the most prevalent form of DNA in the cell, double helical DNAs and RNAs can assume several distinct structures. The structures of these other double helical nucleic acids are discussed in Section 29-1B.

B. *DNA Is Semiconservatively Replicated*

The Watson–Crick structure can accommodate any sequence of bases on one polynucleotide strand if the opposite strand has the complementary base sequence. This immediately accounts for Chargaff's rules. More importantly, *it suggests that hereditary information is encoded in the sequence of bases on either strand.* Consequently, each polynucleotide strand can act as a template for the formation of its complementary strand through base pairing interactions (Fig. 1-17). The two strands of the parent molecule must therefore separate so that a complementary daughter strand may be enzymatically synthesized on the surface of each parent strand. This results in two molecules of **duplex** (double-stranded) DNA, each consisting of one polynucleotide strand from the parent molecule and a newly synthesized complementary strand. Such a mode of replication is termed **semiconservative** in contrast with **conservative** replication, which, if it occurred, would result in a newly synthesized duplex copy of the original DNA molecule with the parent DNA molecule remaining intact. The mechanism of DNA replication is the main subject of Chapter 30.

The semiconservative nature of DNA replication was elegantly demonstrated in 1958 by Matthew Meselson and Franklin Stahl. The density of DNA was increased by labeling it with ^{15}N, a heavy isotope of nitrogen (^{14}N is the naturally abundant isotope). This was accomplished by growing *E. coli* for 14 generations in a medium that contained $^{15}NH_4Cl$ as the only nitrogen source. The labeled bacteria were then abruptly transferred to an ^{14}N-containing medium, and the density of their DNA was monitored as a function of bacterial growth by **equilibrium density gradient ultracentrifugation** (a technique for separating macromolecules according to their densities, which Meselson, Stahl, and Jerome Vinograd had developed for the purpose of distinguishing ^{15}N-labeled DNA from unlabeled DNA; Section 6-5B).

The results of the Meselson–Stahl experiment are displayed in Fig. 5-13. After one generation (doubling of the cell population), all of the DNA had a density exactly halfway between the densities of fully ^{15}N-labeled DNA and unlabeled DNA. This DNA must therefore contain equal amounts of ^{14}N and ^{15}N, as is expected after one generation of semiconservative replication. Conservative DNA replication, in contrast, would result in the preservation of the parental DNA, so that it maintained its original density, and

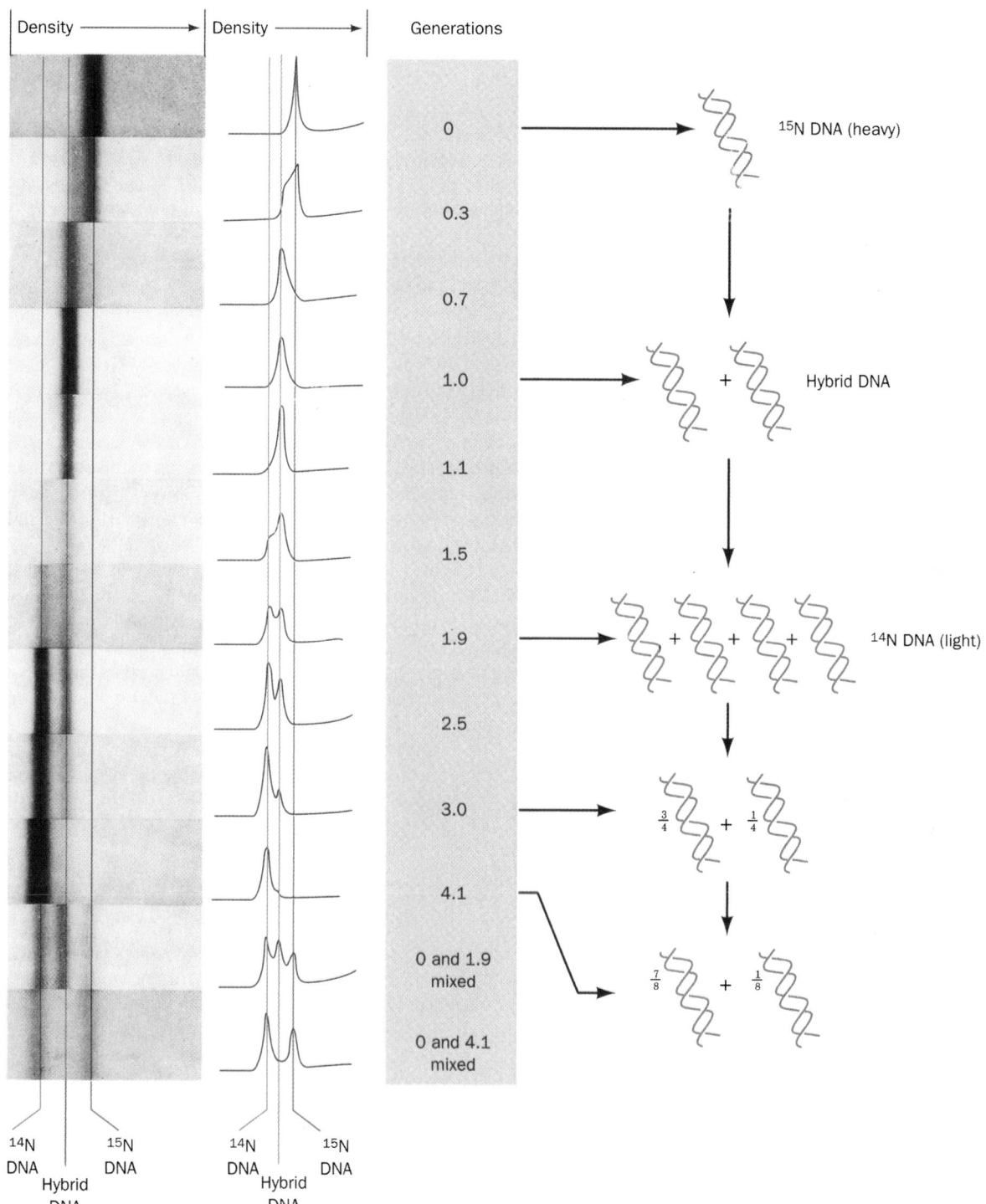

FIGURE 5-13 Demonstration of the semiconservative nature of DNA replication in *E. coli* by density gradient ultracentrifugation. The DNA was dissolved in an aqueous CsCl solution of density 1.71 g · cm^{-3} and was subjected to an acceleration of 140,000 times that of gravity in an analytical ultracentrifuge (a device in which the rapidly spinning sample can be optically observed). This enormous acceleration induced the CsCl to redistribute in the solution such that its concentration increased with its radius in the ultracentrifuge. Consequently, the DNA migrated within the resulting density gradient to its position of buoyant density. The left panels are ultraviolet absorption photographs of ultracentrifuge cells (DNA strongly absorbs ultraviolet light) and are arranged such that regions of equal density have the same horizontal positions. The middle panels are microdensitometer traces of the corresponding photographs in which the vertical displacement is proportional to the DNA concentration. The buoyant density of DNA increases with its ^{15}N content. The bands farthest to the right (of greatest radius and density) arise from DNA that is fully ^{15}N labeled, whereas unlabeled DNA, which is 0.014 g · cm^{-3} less dense, forms the leftmost bands. The bands in the intermediate position result from duplex DNA in which one strand is ^{15}N labeled and the other strand is unlabeled. The accompanying interpretive drawings (*right*) indicate the relative numbers of DNA strands at each generation donated by the original parents (*blue*, ^{15}N labeled) and synthesized by succeeding generations (*red*, unlabeled). [From Meselson, M. and Stahl, F.W., *Proc. Natl. Acad. Sci.* **44,** 674 (1958).] ⚙ **See the Animated Figures**

the generation of an equal amount of unlabeled DNA. After two generations, half of the DNA molecules were unlabeled and the remainder were ^{14}N–^{15}N hybrids. This is also in accord with the predictions of the semiconservative replication model and in disagreement with the conservative replication model. In succeeding generations, the amount of unlabeled DNA increased relative to the amount of hybrid DNA, although the hybrid never totally disappeared. This is again in harmony with semiconservative replication but at odds with conservative replication, which predicts that the fully labeled parental DNA will always be present and that hybrid DNA never forms.

C. *Denaturation and Renaturation*

When a solution of duplex DNA is heated above a characteristic temperature, its native structure collapses and its two complementary strands separate and assume a flexible and rapidly fluctuating conformational state known as a **random coil** *(Fig. 5-14)*. This **denaturation** process is accompanied by a qualitative change in the DNA's physical properties. For instance, the characteristic high viscosity of native DNA solutions, which arises from the resistance to deformation of its rigid and rodlike duplex molecules, drastically decreases when the duplex DNA decomposes (denatures) to two relatively freely jointed single strands.

a. DNA Denaturation Is a Cooperative Process

The most convenient way of monitoring the native state of DNA is by its ultraviolet (UV) absorbance spectrum. When DNA denatures, its UV absorbance, which is almost entirely due to its aromatic bases, increases by ~40% at all wavelengths (Fig. 5-15). This phenomenon, which is known as the **hyperchromic effect** (Greek: *hyper,* above + *chroma,* color), results from the disruption of the electronic interactions among nearby bases. DNA's hyperchromic shift, as monitored at a particular wavelength (usually 260 nm), occurs over a narrow temperature range (Fig. 5-16). This indicates that the collapse of one part of the duplex DNA's structure destabilizes the remainder, a phenomenon known as a **cooperative process.** The denaturation of DNA may be described as the melting of a one-dimensional solid, so Fig. 5-16 is referred to as a **melting curve** and the temperature at its midpoint is known as its **melting temperature, T_m.**

The stability of the DNA double helix, and hence its T_m, depends on several factors, including the nature of the

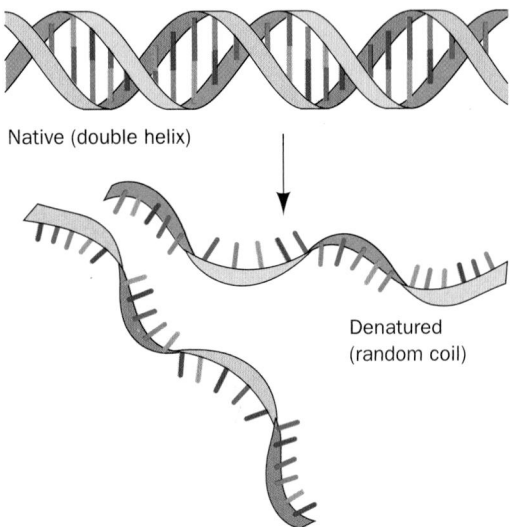

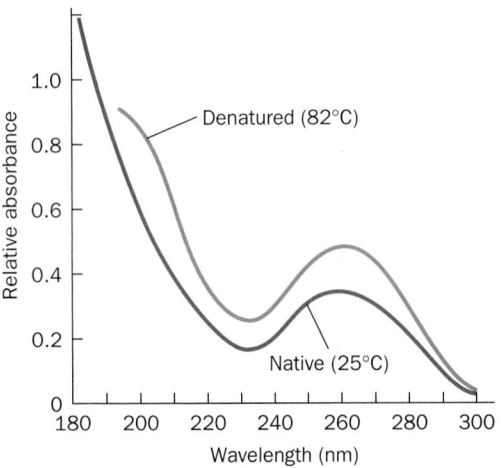

FIGURE 5-14   Schematic representation of the strand separation in duplex DNA resulting from its heat denaturation.

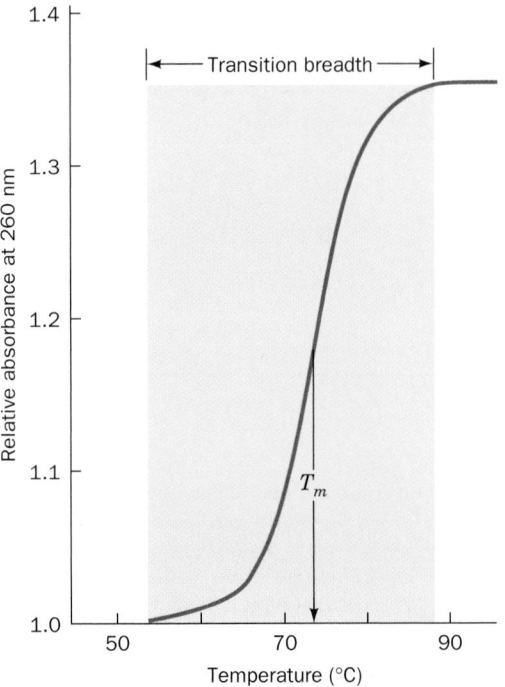

FIGURE 5-15   UV absorbance spectra of native and heat-denatured *E. coli* DNA. Note that denaturation does not change the general shape of the absorbance curve but only increases its intensity. [After Voet, D., Gratzer, W.B., Cox, R.A., and Doty, P., *Biopolymers* **1,** 205 (1963).] *⬥ See the* **Animated Figures**

FIGURE 5-16   Example of a DNA melting curve. The relative absorbance is the ratio of the absorbance (customarily measured at 260 nm) at the indicated temperature to that at 25°C. The melting temperature, T_m, is defined as the temperature at which half of the maximum absorbance increase is attained. *⬥ See the Animated Figures*

solvent, the identities and concentrations of the ions in solution, and the pH. For example, duplex DNA denatures (its T_m decreases) under alkaline conditions that cause some of the bases to ionize and thereby disrupt their base pairing interactions. The T_m increases linearly with the mole fraction of G · C base pairs (Fig. 5-17), which indicates that triply hydrogen bonded G · C base pairs are more stable than doubly hydrogen bonded A · T base pairs.

b. Denatured DNA Can Be Renatured

If a solution of denatured DNA is rapidly cooled to well below its T_m, the resulting DNA will be only partially base paired (Fig. 5-18) because its complementary strands will not have had sufficient time to find each other before the partially base paired structures become effectively "frozen in." If, however, the temperature is maintained ~25°C below the T_m, enough thermal energy is available for short base paired regions to rearrange by melting and reforming but not enough to melt out long complementary stretches. Under such **annealing conditions,** as Julius Marmur discovered in 1960, denatured DNA eventually completely renatures. Likewise, complementary strands of RNA and DNA, in a process known as **hybridization,** form RNA–DNA hybrid double helices that are only slightly less stable than the corresponding DNA double helices.

D. *The Size of DNA*

DNA molecules are generally enormous (Fig. 5-19). The molecular mass of DNA has been determined by a variety of techniques including ultracentrifugation (Section 6-5) and through length measurements by electron microscopy [a base pair of Na$^+$ B-DNA has an average molecular mass of 660 D and a length (thickness) of 3.4 Å] and **autoradiography** (a technique in which the position of a radioactive substance in a sample is recorded by the blackening

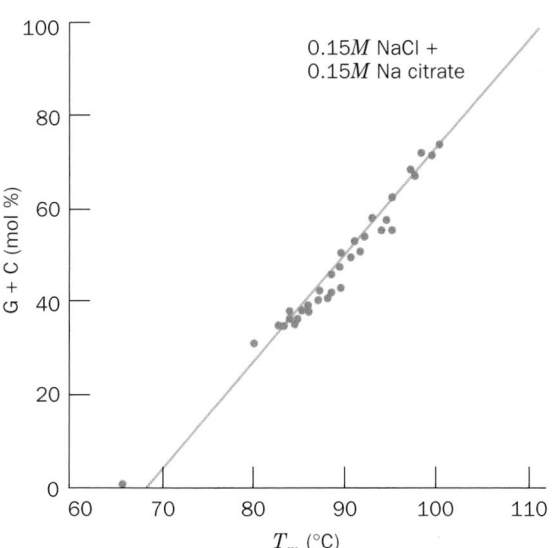

FIGURE 5-17 Variation of the melting temperatures, T_m, of various DNAs with their G + C content. The DNAs were dissolved in a solution containing 0.15*M* NaCl and 0.015*M* sodium citrate. [After Marmur, J. and Doty, P., *J. Mol. Biol.* **5,** 113 (1962).]

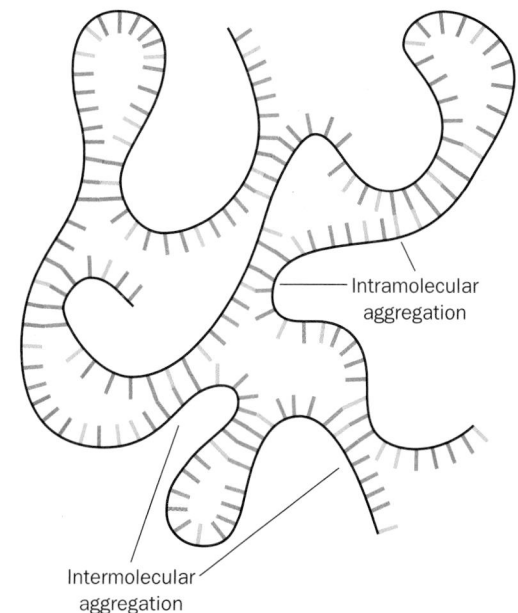

FIGURE 5-18 Partially renatured DNA. This schematic representation shows the imperfectly base paired structures assumed by DNA that has been heat denatured and then rapidly cooled to well below its T_m. Note that both intramolecular and intermolecular aggregation may occur.

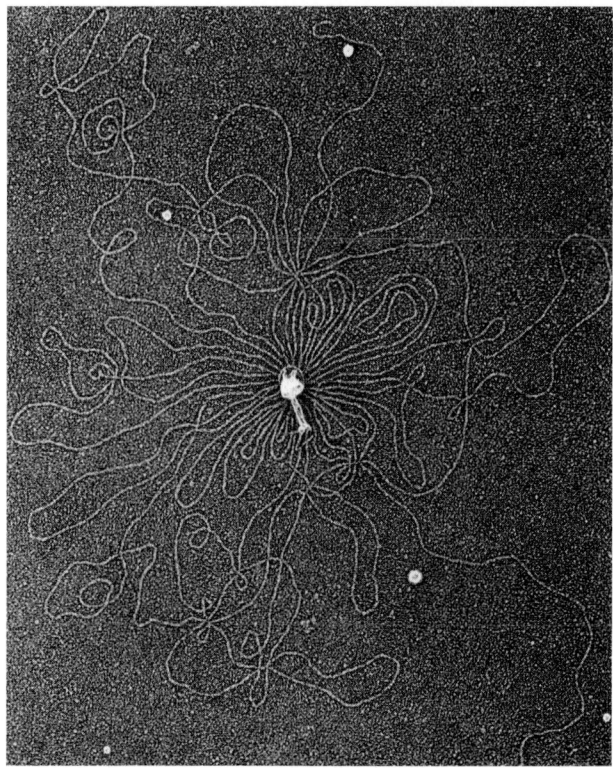

FIGURE 5-19 Electron micrograph of a T2 bacteriophage and its DNA. The phage has been osmotically lysed (broken open) in distilled water so that its DNA spilled out. Without special treatment, duplex DNA, which is only 20 Å in diameter, is difficult to visualize in the electron microscope. In the **Kleinschmidt procedure** used here, DNA is fattened to ~200 Å in diameter by coating it with a denatured basic protein. [From Kleinschmidt, A.K., Lang, D., Jacherts, D., and Zahn, R.K., *Biochim. Biophys. Acta* **61,** 861 (1962).]

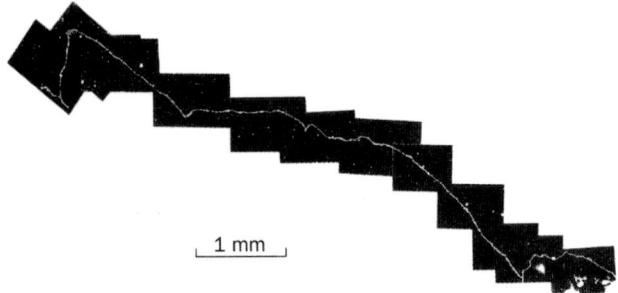

⌞ 1 mm ⌟

FIGURE 5-20 Autoradiograph of *Drosophila melanogaster* DNA. Lysates of *D. melanogaster* cells that had been cultured with ^{3}H-labeled thymidine were spread on a glass slide and covered with a photographic emulsion that was developed after a 5-month exposure. The white curve, which resulted from the radioactive decay of the ^{3}H, traces the path of the DNA in this photographic positive. The DNA's measured contour length is 1.2 cm. [From Kavenoff, R., Klotz, L.C., and Zimm, B.H., *Cold Spring Harbor Symp. Quant. Biol.* **38,** 4 (1973). Copyright © 1974 by Cold Spring Harbor Laboratory Press.]

of a photographic emulsion that the sample is laid over or embedded in; Fig. 5-20). The number of base pairs and the **contour lengths** (the end-to-end lengths of the stretched-out native molecules) of the DNAs from a selection of organisms of increasing complexity are listed in Table 5-2. Not surprisingly, an organism's haploid quantity (unique amount) of DNA varies more or less with its complexity (although there are notable exceptions to this generalization, such as the last entry in Table 5-2).

The visualization of DNAs from prokaryotes has demonstrated that their entire **genome** (complement of genetic information) is contained on a single, usually circular, length of DNA. Similarly, Bruno Zimm demonstrated that the *largest chromosome of the fruit fly Drosophila melanogaster contains a single molecule of DNA* by com-

paring the molecular mass of this DNA with the cytologically measured length of DNA contained in the chromosome. Presumably other eukaryotic chromosomes also contain only single molecules of DNA.

The highly elongated shape of duplex DNA (recall B-DNA is only 20 Å in diameter), together with its stiffness, make it extremely susceptible to mechanical damage outside the cell's protective environment (for instance, if the *Drosophila* DNA of Fig. 5-20 were enlarged by a factor of 500,000, it would have the shape and some of the mechanical properties of a 6-km-long strand of uncooked spaghetti). The hydrodynamic shearing forces generated by such ordinary laboratory manipulations as stirring, shaking, and pipetting break DNA into relatively small pieces so that the isolation of an intact molecule of DNA requires extremely gentle handling. Before 1960, when this was first realized, the measured molecular masses of DNA were no higher than ~10 million D (~15 **kb,** where 1 kb = 1 kilobase pair = 1000 bp). DNA fragments of uniform molecular mass and as small as a few hundred base pairs may be generated by **shear degrading** DNA in a controlled manner; for instance, by pipetting, through the use of a high-speed blender, or by **sonication** (exposure to intense high-frequency sound waves).

4 ■ GENE EXPRESSION AND REPLICATION: AN OVERVIEW

⚭ See Guided Exploration 1: Overview of Transcription and Translation How do genes function, that is, how do they direct the synthesis of RNA and proteins, and how are they replicated? The answers to these questions form the multifaceted subdiscipline known as **molecular biology.** In 1958, Crick neatly encapsulated the broad outlines of this process in a flow scheme he called the **central dogma of molecular biology:** *DNA directs its own replication and its* ***transcription*** *to yield RNA which, in turn, directs its*

TABLE 5-2 Sizes of Some DNA Molecules

Organism	Number of base pairs (kb)[a]	Contour length (μm)
Viruses		
Polyoma, SV40	5.2	1.7
λ Bacteriophage	48.6	17
T2, T4, T6 bacteriophage	166	55
Fowlpox	280	193
Bacteria		
Mycoplasma hominis	580	260
Escherichia coli	4,639	1,600
Eukaryotes		
Yeast (in 17 haploid chromosomes)	11,700	4,100
Drosophila (in 4 haploid chromosomes)	137,000	61,000
Human (in 23 haploid chromosomes)	3,200,000	1,100,000
Lungfish (in 19 haploid chromosomes)	102,000,000	35,000,000

[a]kb = kilobase pair = 1000 base pairs (bp).

Source: Mainly Kornberg, A. and Baker, T.A., *DNA Replication* (2nd ed.), p. 20, Freeman (1992).

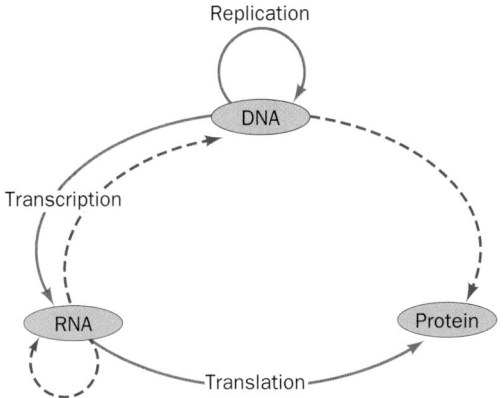

Replication

DNA

Transcription

RNA

Protein

Translation

FIGURE 5-21 The central dogma of molecular biology. Solid arrows indicate the types of genetic information transfers that occur in all cells. Special transfers are indicated by the dashed arrows: **RNA-directed RNA polymerase** occurs both in certain RNA viruses and in some plants (where it is of unknown function); **RNA-directed DNA polymerase (reverse transcriptase)** occurs in other RNA viruses; and DNA directly specifying a protein is unknown but does not seem beyond the realm of possibility. However, the missing arrows are information transfers the central dogma postulates never occur: protein specifying either DNA, RNA, or protein. In other words, *proteins can only be the recipients of genetic information.* [After Crick, F., *Nature* **227**, 562 (1970).]

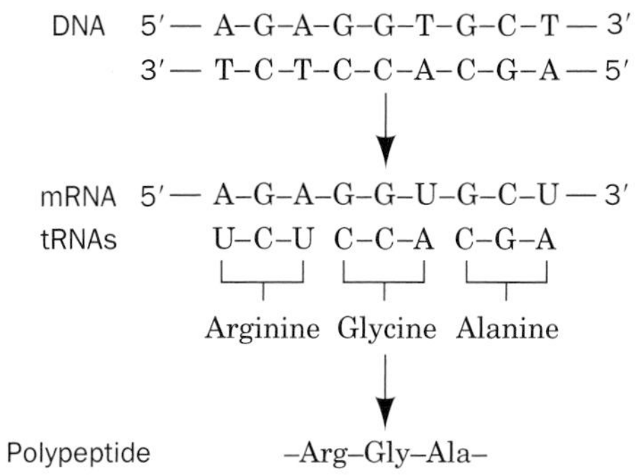

DNA 5′— A–G–A–G–G–T–G–C–T — 3′
      3′— T–C–T–C–C–A–C–G–A — 5′

mRNA 5′— A–G–A–G–G–U–G–C–U — 3′
tRNAs   U–C–U   C–C–A   C–G–A

          Arginine Glycine Alanine

Polypeptide   –Arg–Gly–Ala–

FIGURE 5-22 Gene expression. One strand of DNA directs the synthesis of RNA, a process known as transcription. The base sequence of the transcribed RNA is complementary to that of the DNA strand. The RNAs known as **messenger RNAs (mRNAs)** are translated when molecules of **transfer RNA (tRNA)** align with the mRNA via complementary base pairing between 3-nucleotide segments known as codons. Each type of tRNA carries a specific amino acid. These amino acids are covalently joined by the ribosome to form a polypeptide. Thus, the sequence of bases in DNA specifies the sequence of amino acids in a protein.

translation to form proteins (Fig. 5-21). Here the term "transcription" indicates that in transferring information from DNA to RNA, the "language" encoding the information remains the same, that of base sequences, whereas the term "translation" indicates that in transferring information from RNA to proteins, the "language" changes from that of base sequences to that of amino acid sequences (Fig. 5-22). The machinery required to carry out the complex tasks of gene expression and DNA replication in an organized manner and with high fidelity occupies a major portion of every cell. In this section we summarize how gene expression and replication occur to provide the background for understanding the techniques of recombinant DNA

technology (Section 5-5). This subject matter is explored in considerably greater detail in Chapters 29 to 34.

A. *RNA Synthesis: Transcription*

The enzyme that synthesizes RNA is named **RNA polymerase.** It catalyzes the DNA-directed coupling of the **nucleoside triphosphates (NTPs) adenosine triphosphate (ATP), cytidine triphosphate (CTP), guanosine triphosphate (GTP),** and **uridine triphosphate (UTP)** in a reaction that releases pyrophosphate ion ($P_2O_7^{4-}$):

$$(RNA)_{n \text{ residues}} + NTP \rightarrow (RNA)_{n+1 \text{ residues}} + P_2O_7^{4-}$$

RNA synthesis proceeds in a stepwise manner in the $5' \rightarrow 3'$ direction, that is, the incoming nucleotide is appended to the free 3′-OH group of the growing RNA chain (Fig. 5-23). RNA polymerase selects the nucleotide it in-

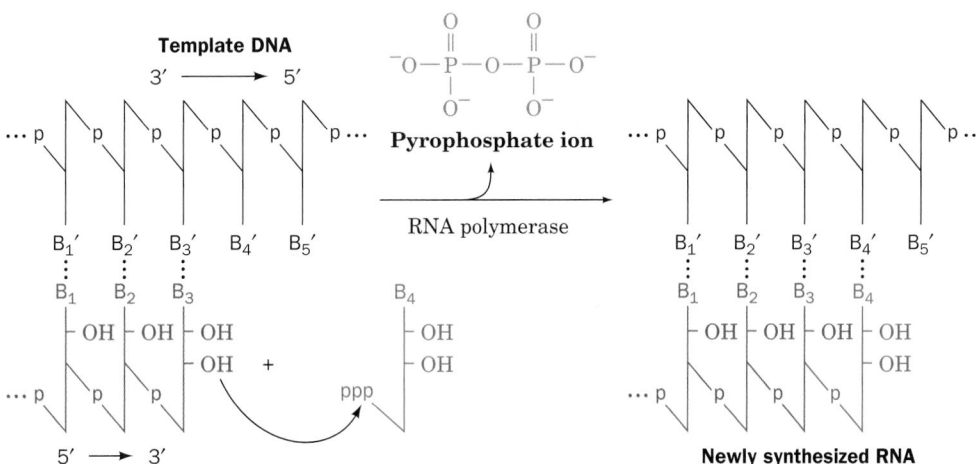

FIGURE 5-23 Action of RNA polymerases. These enzymes assemble incoming ribonucleoside triphosphates on templates

consisting of single-stranded segments of DNA such that the growing strand is elongated in the 5′ to 3′ direction.

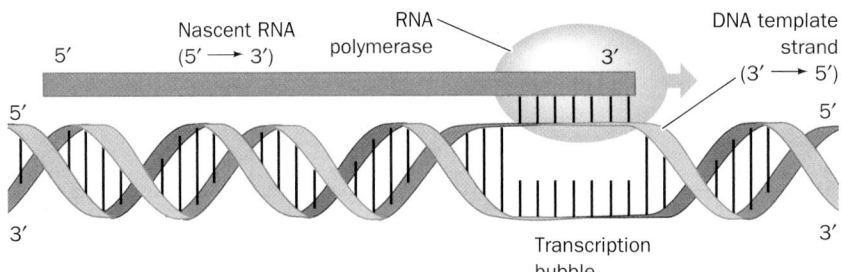

FIGURE 5-24 Function of the transcription bubble. RNA polymerase unwinds the DNA double helix by about a turn in the region being transcribed, thereby permitting the DNA's template strand to form a short segment of DNA–RNA hybrid double helix with the RNA's newly synthesized 3′ end. As the RNA polymerase advances along the DNA template (here to the right), the DNA unwinds ahead of the RNA's growing 3′ end and rewinds behind it, thereby stripping the newly synthesized RNA from the template strand.

corporates into the nascent (growing) RNA chain through the requirement that it form a Watson–Crick base pair with the DNA strand that is being transcribed, the **template strand** (only one of duplex DNA's two strands is transcribed at a time). This is possible because, as the RNA polymerase moves along the duplex DNA it is transcribing, it separates a short (~10 bp) segment of its two strands to form a so-called **transcription bubble,** thereby permitting this portion of the template strand to transiently form a short DNA–RNA hybrid helix with the newly synthesized RNA (Fig. 5-24). Like duplex DNA, a DNA–RNA hybrid helix consists of antiparallel strands, and hence the DNA's template strand is read in the 3′ → 5′ direction.

All cells contain RNA polymerase. In bacteria, one species of this enzyme synthesizes nearly all of the cell's RNA. Certain viruses generate RNA polymerases that synthesize only virus-specific RNAs. Eukaryotic cells contain four or five different types of RNA polymerases that each synthesize a different class of RNA.

a. Transcriptional Initiation Is a Precisely Controlled Process

The DNA template strand contains control sites consisting of specific base sequences that specify both the site at which RNA polymerase initiates transcription (the site on the DNA at which the RNA's first two nucleotides are joined) and the rate at which RNA polymerase initiates transcription at this site. Specific proteins known in prokaryotes as **activators** and **repressors** and in eukaryotes as **transcription factors** bind to these control sites or to other such proteins that do so and thereby stimulate or inhibit transcriptional initiation by RNA polymerase. For the RNAs that encode proteins, which are named **messenger RNAs (mRNAs),** these control sites precede the initiation site (that is, they are "upstream" of the initiation site relative to the RNA polymerase's direction of travel).

The rate at which a cell synthesizes a given protein, or even whether the protein is synthesized at all, is mainly governed by the rate at which the synthesis of the corresponding mRNA is initiated. The way that prokaryotes regulate the rate at which many genes undergo transcriptional initiation can be relatively simple. For example, the transcriptional initiation of numerous prokaryotic genes re-

quires only that RNA polymerase bind to a control sequence, known as a **promoter,** that precedes the transcriptional initiation site. However, not all promoters are created equal: RNA polymerase initiates transcription more often at so-called efficient promoters than at those with even slightly different sequences. Thus the rate at which a gene is transcribed is governed by the sequence of its associated promoter.

A more complex way in which prokaryotes control the rate of transcriptional initiation is exemplified by the *E. coli lac* **operon,** a cluster of three consecutive genes *(Z, Y,* and *A)* encoding proteins that the bacterium requires to metabolize the sugar **lactose** (Section 11-2B). In the absence of lactose, a protein named the *lac* **repressor** specifically binds to a control site in the *lac* operon known as an **operator** (Section 31-3B). This prevents RNA polymerase from initiating the transcription of *lac* operon genes (Fig. 5-25a), thereby halting the synthesis of unneeded proteins. However, when lactose is available, the bacterium metabolically modifies a small amount of it to form the related sugar **allolactose.** This so-called **inducer** specifically binds to the *lac* repressor, thereby causing it to dissociate from the operator DNA so that RNA polymerase can initiate the transcription of the *lac* operon genes (Fig. 5-25b).

In eukaryotes, the control sites regulating transcriptional initiation can be quite extensive and surprisingly distant from the transcriptional initiation site (by as much as several tens of thousands of base pairs; Section 34-3). Moreover, the eukaryotic transcriptional machinery that binds to these sites and thereby induces RNA polymerase to commence transcription can be enormously complex (consisting of up to 50 or more proteins; Section 34-3).

b. Transcriptional Termination Is a Relatively Simple Process

The site on the template strand at which RNA polymerase terminates transcription and releases the completed RNA is governed by the base sequence in this region. However, the control of transcriptional termination is rarely involved in the regulation of gene expression. In keeping with this, the cellular machinery that mediates transcriptional termination is relatively simple compared with that involved in transcriptional initiation (Section 31-2D).

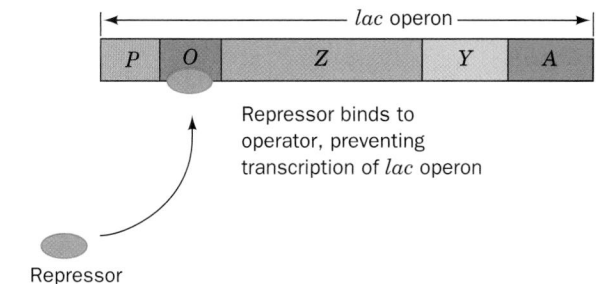

(a) **Absence of inducer**

Repressor binds to
operator, preventing
transcription of *lac* operon

Repressor

(b) **Presence of inducer**

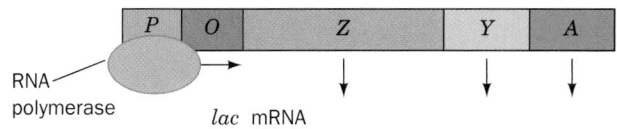

RNA
polymerase

lac mRNA

Inducer

Transcription of
lac structural genes

Inducer–repressor
complex does not
bind to operator

FIGURE 5-25 Control of transcription of the *lac* operon. (*a*) In
the absence of an inducer such as allolactose, the *lac* repressor
binds to the operator (*O*), thereby preventing RNA polymerase
from transcribing the *Z, Y,* and *A* genes of the *lac* operon.
(*b*) On binding inducer, the *lac* repressor dissociates from the
operator, which permits RNA polymerase to bind to the
promoter (*P*) and transcribe the *Z, Y,* and *A* genes. **See
Guided Exploration 2: Regulation of gene expression by the *lac*
repressor system**

c. Eukaryotic RNA Undergoes Post-Transcriptional Modifications

Most prokaryotic mRNA transcripts participate in
translation without further alteration. However, most primary transcripts in eukaryotes require extensive **posttranscriptional modifications** to become functional. For
mRNAs, these modifications include the addition of a 7-
methylguanosine-containing "cap" that is enzymatically

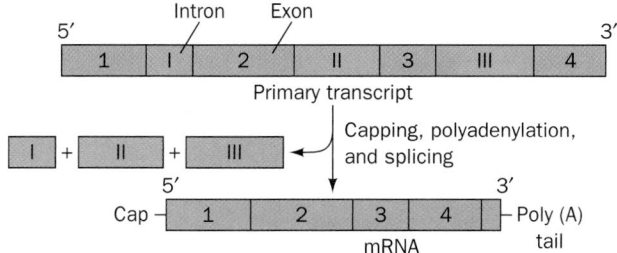

**FIGURE 5-26 Post-transcriptional processing of eukaryotic
mRNAs.** Most primary transcripts require further covalent
modification to become functional, including the addition of a
5′ cap and a 3′ poly(A) tail, and splicing to excise its introns
from between its exons.

appended to the transcript's 5′ end and ~250-nucleotide
polyadenylic acid [poly(A)] "tail" that is enzymatically appended to its 3′ end. However, the most striking modification that most eukaryotic transcripts undergo is a process
called **gene splicing** in which one or more often lengthy
RNA segments known as **introns** (for "intervening sequences") are precisely excised from the RNA and the remaining **exons** (for "expressed sequences") are rejoined in
their original order to form the mature mRNA (Fig. 5-26;
Section 31-4A). Different mRNAs can be generated from
the same gene through the selection of alternate transcriptional initiation sites and/or alternative splice sites,
leading to the production of somewhat different proteins,
usually in a tissue-specific manner (Section 34-3C).

B. *Protein Synthesis: Translation*

Polypeptides are synthesized under the direction of the corresponding mRNA by **ribosomes,** numerous cytosolic organelles that consist of about two-thirds RNA and one-third
protein and have molecular masses of ~2500 kD in prokaryotes and ~4200 kD in eukaryotes. Ribosomal RNAs
(**rRNAs**), of which there are several kinds, are transcribed
from DNA templates, as are all other kinds of RNA.

a. Transfer RNAs Deliver Amino Acids to the Ribosome
mRNAs are essentially a series of consecutive 3-
nucleotide segments known as **codons,** each of which specifies a particular amino acid. However, codons do not bind
amino acids. Rather, on the ribosome, *they specifically bind
molecules of **transfer RNA (tRNA)** that are each covalently
linked to the corresponding amino acid* (Fig. 5-27). A tRNA

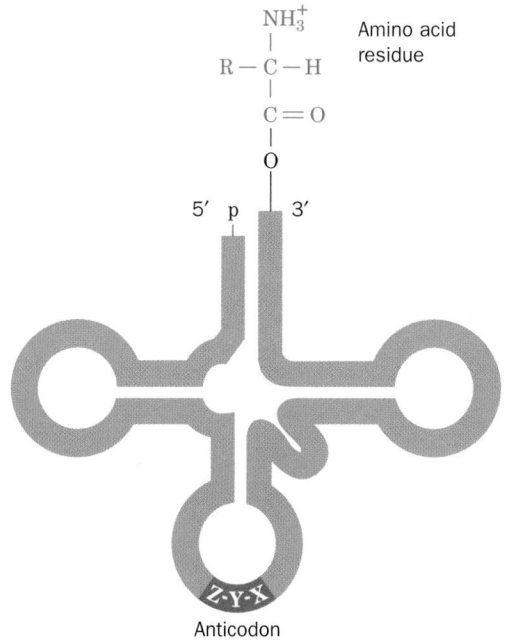

**FIGURE 5-27 Transfer RNA (tRNA) drawn in its "cloverleaf"
form.** Its covalently linked amino acid residue forms an
aminoacyl-tRNA (*top*), and its anticodon (*bottom*), a
trinucleotide segment, base pairs with the complementary codon
on mRNA during translation.

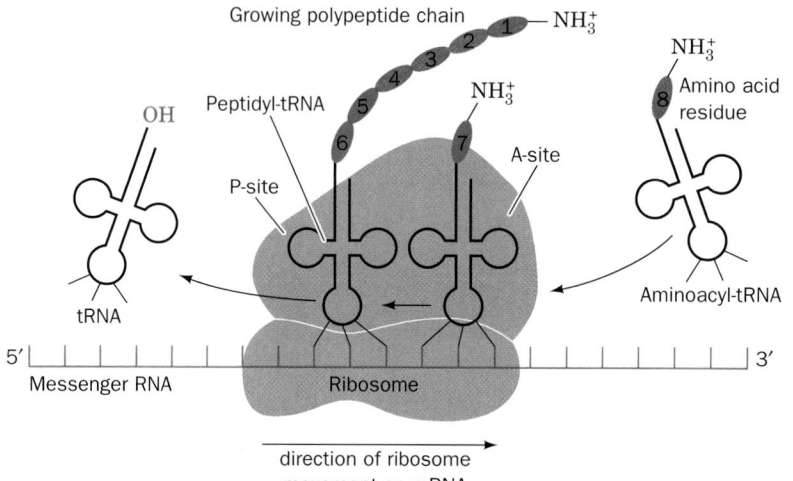

FIGURE 5-28   Schematic diagram of translation. The ribosome binds an mRNA and two tRNAs and facilitates their specific association through consecutive codon–anticodon interactions. The ribosomal binding site closer to the 5′ end of the mRNA binds a **peptidyl-tRNA** (*left,* a tRNA to which the growing polypeptide chain is covalently linked) and is therefore known as the **P-site,** whereas the ribosomal site closer to the 3′ end of the mRNA binds an aminoacyl-tRNA (*right*) and is hence called the **A-site.** The ribosome catalyzes the transfer of the polypeptide from the peptidyl-tRNA to the aminoacyl-tRNA, thereby forming a new peptidyl-tRNA whose polypeptide chain has increased in length by one residue at its C-terminus. The discharged tRNA in the P-site is then ejected, and the peptidyl-tRNA, together with its bound mRNA, is shifted from the A-site to the P-site, thereby permitting the next codon to bind its corresponding aminoacyl-tRNA in the ribosomal A-site.

typically consists of ~76 nucleotides (which makes it comparable in mass and structural complexity to a medium-sized protein) and contains a trinucleotide sequence, its **anticodon,** which is complementary to the codon(s) specifying its appended amino acid (see below). An amino acid is covalently linked to the 3′ end of its corresponding tRNA to form an **aminoacyl-tRNA** (a process called "charging") through the action of an enzyme that specifically recognizes both the tRNA and the amino acid (see below). During translation, the mRNA is passed through the ribosome such that each codon, in turn, binds its corresponding aminoacyl-tRNA (Fig. 5-28). As this occurs, the ribosome transfers the amino acid residue on the tRNA to the C-terminal end of the growing polypeptide chain (Fig. 5-29). Hence, *the polypeptide grows from its N-terminus to its C-terminus.*

b. The Genetic Code

The correspondence between the sequence of bases in a codon and the amino acid residue it specifies is known

FIGURE 5-29   The ribosomal reaction forming a peptide bond. The amino group of the aminoacyl-tRNA in the ribosomal A-site nucleophilically displaces the tRNA of the peptidyl-tRNA ester in the ribosomal P-site, thereby forming a new peptide bond and transferring the growing polypeptide to the A-site tRNA.

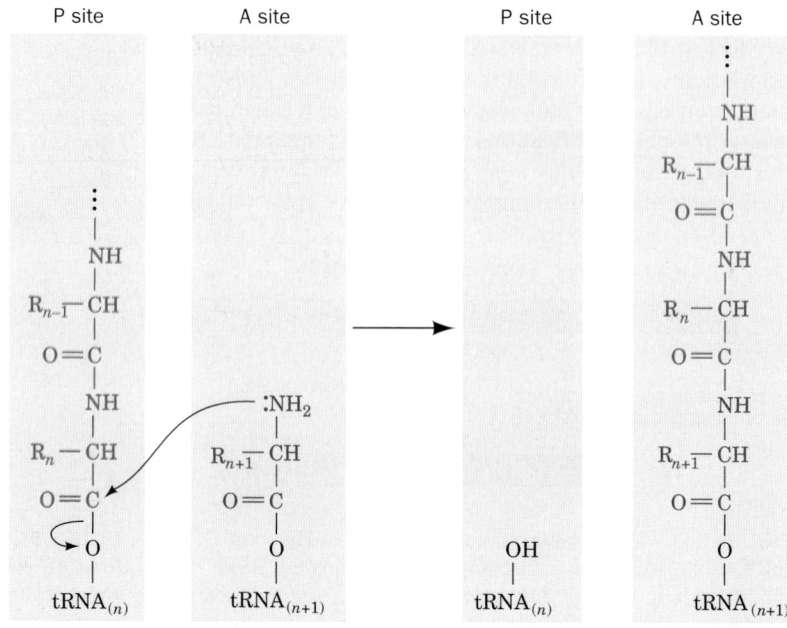

as the **genetic code** (Table 5-3). Its near universality among all forms of life is compelling evidence that life on Earth arose from a common ancestor and makes it possible, for example, to express human genes in *E. coli* (Section 5-5G). There are four possible bases (U, C, A, and G) that can occupy each of the three positions in a codon, and hence there are $4^3 = 64$ possible codons. Of these codons, 61 specify amino acids (of which there are only 20) and the remaining three, UAA, UAG, and UGA, are **Stop codons** that instruct the ribosome to cease polypeptide synthesis

TABLE 5-3.   The "Standard" Genetic Code[a]

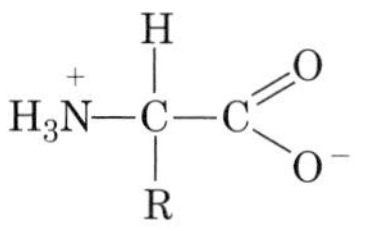

First position (5' end)	Second position				Third position (3' end)
	U	**C**	**A**	**G**	
U	UUU Phe / UUC	UCU Ser / UCC / UCA / UCG	UAU Tyr / UAC	UGU Cys / UGC	U / C
	UUA Leu / UUG		UAA STOP / UAG STOP	UGA STOP	A
				UGG Trp	G
C	CUU / CUC Leu / CUA / CUG	CCU / CCC Pro / CCA / CCG	CAU His / CAC	CGU / CGC Arg / CGA / CGG	U / C
			CAA Gln / CAG		A / G
A	AUU Ile / AUC / AUA	ACU / ACC Thr / ACA / ACG	AAU Asn / AAC	AGU Ser / AGC	U / C
	AUG Met[b]		AAA Lys / AAG	AGA Arg / AGG	A / G
G	GUU / GUC Val / GUA / GUG	GCU / GCC Ala / GCA / GCG	GAU Asp / GAC	GGU / GGC Gly / GGA / GGG	U / C
			GAA Glu / GAG		A / G

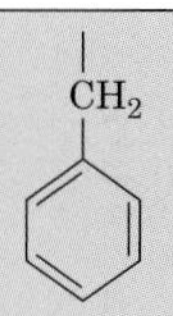

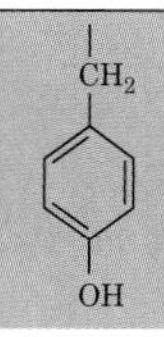

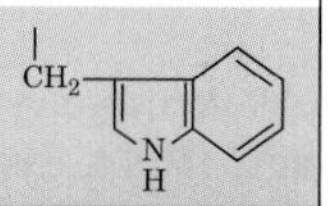

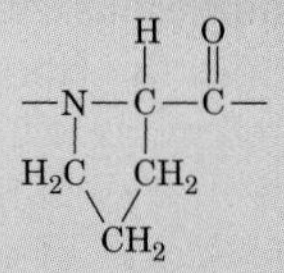

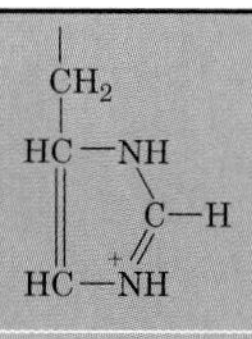

[a]Nonpolar residues are tan, basic residues are blue, acidic residues are red, and polar uncharged residues are purple.

[b]AUG forms part of the initiation signal as well as coding for internal Met residues.

and release the resulting transcript. All but two amino acids (Met and Trp) are specified by more than one codon and three (Leu, Ser, and Arg) are specified by six codons. Consequently, in a term borrowed from mathematics, the genetic code is said to be **degenerate** (taking on several discrete values).

Note that the arrangement of the genetic code is non-random: Most codons that specify a given amino acid, which are known as **synonyms,** occupy the same box in Table 5-3, that is, they differ in sequence only in their third (3′) nucleotide. Moreover, most codons specifying nonpolar amino acid residues have a G in their first position and/or a U in their second position (Table 5-3).

A tRNA may recognize as many as three synonymous codons because the 5′ base of a codon and the 3′ base of a corresponding anticodon do not necessarily interact via a Watson–Crick base pair (Section 32-2D; keep in mind that the codon and the anticodon associate in an antiparallel fashion to form a short segment of an RNA double helix). Thus, cells can have far fewer than the 61 tRNAs that would be required for a 1:1 match with the 61 amino acid–specifying codons, although, in fact, some eukaryotic cells contain as many as 150 different tRNAs.

c. tRNAs Acquire Amino Acids through the Actions of Aminoacyl-tRNA Synthetases

In synthesizing a polypeptide, a ribosome does not recognize the amino acid appended to a tRNA but only whether its anticodon binds to the mRNA's codon (the anticodon and the amino acid on a charged tRNA are actually quite distant from one another, as Fig. 5-27 suggests). Thus, *the charging of a tRNA with the proper amino acid is as critical a step for accurate translation as is the proper recognition of a codon by its corresponding anticodon.* The enzymes that catalyze these additions are known as **aminoacyl-tRNA synthetases (aaRSs).** Cells typically contain 20 aaRSs, one for each amino acid, and therefore a given aaRS will charge all the tRNAs that bear codons specifying its corresponding amino acid. Consequently, each aaRS must somehow differentiate its cognate (corresponding) tRNAs from among the many other types of structurally and physically quite similar tRNAs that each cell contains. Although many aaRSs recognize the anticodons of their cognate tRNAs, not all of them do so. Rather, they recognize other sites on their cognate tRNAs.

d. Translation Is Initiated at Specific AUG Codons

Ribosomes read mRNAs in their 5′ to 3′ direction (from "upstream" to "downstream"). The initiating codon is AUG, which specifies a Met residue. However, the tRNA that recognizes this initiation codon differs from the tRNA that delivers a polypeptide's internal Met residues to the ribosome, although both types of tRNA are charged by the same **methionyl-tRNA synthetase (MetRS).**

If a polypeptide is to be synthesized with the correct amino acid sequence, it is essential that the ribosome maintain the proper register between the mRNA and the incoming tRNAs, that is, that the ribosome maintain the correct **reading frame.** As is illustrated in Fig. 5-30, a shift

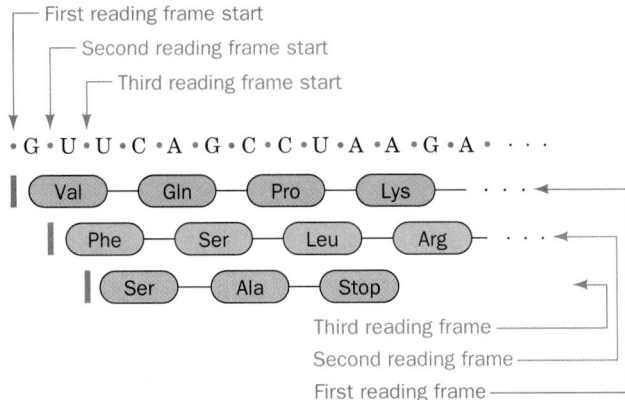

FIGURE 5-30   Nucleotide reading frames. An mRNA might be read in any of three different reading frames, each of which yields a different polypeptide.

of even one nucleotide along an mRNA will lead to the synthesis of an entirely different polypeptide from the point of the shift onward. Thus, the AUG codon that initiates polypeptide synthesis also sets the polypeptide's reading frame. Yet AUG also specifies a polypeptide's internal Met residues, and an mRNA is likely to contain numerous AUGs in different reading frames. How then does the ribosome select the initiation codon from among the many AUGs in an mRNA? In prokaryotes, the answer is that each mRNA contains a sequence on the upstream (5′) side of the initiating codon (a region that does not encode polypeptide chain) through which the ribosome identifies this codon. In eukaryotes, the answer is simpler; the initiating codon is usually the first AUG that is downstream of the mRNA's 5′ cap.

e. Prokaryotic mRNAs Have Short Lifetimes

In prokaryotes, transcription and translation both take place in the same cellular compartment, the cytosol (Figs. 1-2 and 1-13). Consequently ribosomes often attach to the 5′ end of an mRNA before its synthesis is complete and commence synthesizing the corresponding polypeptide. This is essential because, since the mRNAs in prokaryotes have average lifetimes of only 1 to 3 minutes before being hydrolytically degraded by enzymes known as **nucleases,** the 5′ end of an mRNA may be degraded before its 3′ end is synthesized. This rapid turnover of its mRNAs permits a prokaryote to respond quickly to changes in its environment by synthesizing the proteins appropriate for its new situation within minutes of the change (recall that prokaryotes are adapted to live in environments in which there are rapid fluctuations in the available nutrients; Section 1-2).

Eukaryotic cells, in contrast, mostly lead a more sedentary existence. Their RNAs are transcribed and post-transcriptionally modified in the nucleus, whereas ribosomes occupy the cytosol where translation takes place (Fig. 1-5). Hence, mature mRNAs must be transported from the nucleus to the cytosol in order to participate in translation. Eukaryotic mRNAs therefore tend to have lifetimes on the order of several days.

f. Proteins Are Subject to Post-Translational Modifications and Degradation

Newly synthesized polypeptides often require post-translational modifications to become functional. In many proteins, the leading (N-terminal) Met residue that was specified by its mRNA's initiating codon is excised by a specific **protease** (an enzyme that hydrolytically cleaves peptide bonds). Proteins are then subject to numerous other chemical modifications at specific residues, including specific proteolytic cleavages, acylation, hydroxylation, methylation, and phosphorylation (Section 4-3A). In addition, eukaryotic proteins, but not prokaryotic proteins, are subject to **glycosylation** (the addition of polysaccharides) at specific sites (Sections 11-3C and 23-3B). Indeed, **glycoproteins** (proteins that have been glycosylated) are the most common type of eukaryotic protein and can consist of up to 90% or more by mass of polysaccharide groups.

All cells have several mechanisms for degrading proteins to their component amino acids. This enables cells to eliminate damaged or abnormal proteins, destroy proteins that are no longer needed, and utilize proteins as nutrients. The lifetime of a protein in a cell can be surprisingly short, as little as a fraction of a minute, although many proteins in eukaryotes have lifetimes of days or weeks. Thus cells are dynamic entities that are constantly turning over most of their components, in particular their RNA and proteins.

C. DNA Replication

The chemical reaction by which DNA is replicated (Fig. 5-31) is nearly identical to that synthesizing RNA (Fig. 5-23), but with two major differences: (1) deoxynucleoside triphosphates (**dNTPs**) rather than nucleoside triphosphates are the reactants and (2) the enzyme that catalyzes the reaction is **DNA polymerase** rather than RNA polymerase. The properties of DNA polymerase result in a third major difference between RNA and DNA

synthesis: Whereas RNA polymerase can link together two nucleotides on a DNA template, *DNA polymerase can only extend (in the 5' to 3' direction) an existing polynucleotide that is base paired to the DNA's template strand.* Thus, whereas RNA polymerase can initiate RNA synthesis *de novo* (from the beginning), *DNA polymerase requires an oligonucleotide* **primer**, *which it lengthens.*

a. Primers Are RNA

If DNA polymerase cannot synthesize DNA *de novo*, where do primers come from? It turns out that they are not DNA, as might be expected, but rather RNA. In *E. coli*, these RNA primers are synthesized by both RNA polymerase (the same enzyme that synthesizes all other RNAs) and by a special RNA polymerase known as **primase.** DNA polymerase then extends this RNA primer, which is eventually excised and replaced by DNA, as is explained below. This extra complexity in DNA synthesis increases the fidelity of DNA replication. Whereas a cell makes many copies of an RNA and hence can tolerate an occasional mistake in its synthesis, a mistake (mutation) in the synthesis of DNA, the archive of genetic information, may be passed on to all of the cell's descendants. Since a Watson–Crick base pair is partially stabilized by its neighboring base pairs (a cooperative interaction), the first few base pairs that are formed in a newly synthesized polynucleotide will initially be less stable than the base pairs that are formed later. Consequently, these first few bases are more likely to be erroneously incorporated due to mispairing than those at the end of a longer chain. If a primer were DNA, there would be no way to differentiate it from other DNA so as to selectively replace it with more accurately synthesized DNA. Since the primer is RNA, however, it is readily identified and replaced.

b. DNA's Two Strands Are Replicated in Different Ways

A fourth major difference between RNA and DNA synthesis is that, whereas only one DNA strand at a time is transcribed, in most cases both of its strands are simultaneously replicated. This takes place at a **replication fork,**

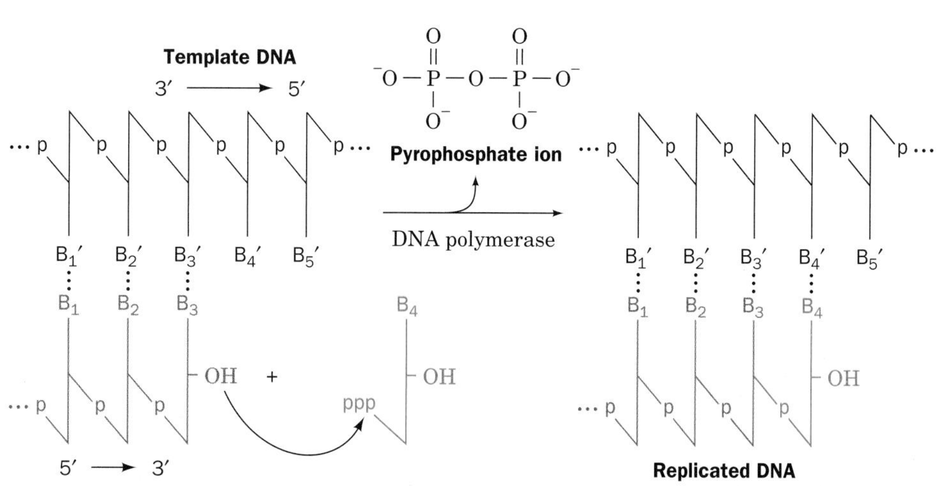

FIGURE 5-31 Action of DNA polymerases. DNA polymerases assemble incoming deoxynucleoside triphosphates on single-stranded DNA templates such that the growing strand is elongated in the 5' to 3' direction.

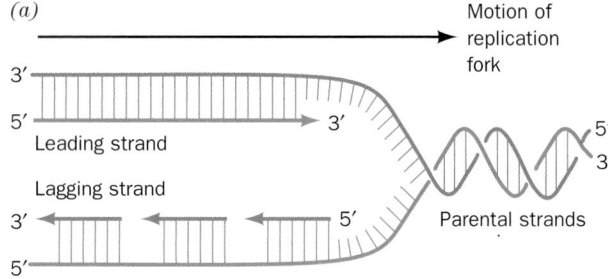

(a)

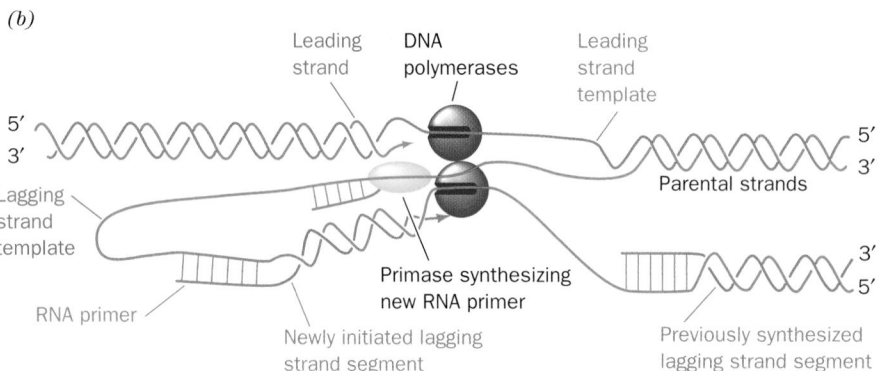

(b)

FIGURE 5-32   Replication of duplex DNA in *E. coli*. (*a*) Since the two DNA polymerase molecules at the replication fork are linked together and DNA polymerase can only synthesize DNA in its 5′ to 3′ direction, the leading strand can be synthesized continuously but the lagging strand must be synthesized discontinuously, that is, in segments. (*b*) This is because the lagging strand template can only be copied if it loops around so as to feed through the DNA polymerase in its 3′ to 5′ direction. Consequently, when the DNA polymerase that is synthesizing the lagging strand encounters the previously synthesized lagging strand segment, it releases the lagging strand template and rebinds to it farther upstream so as to extend the next RNA primer to be synthesized.

the junction where the two strands of the parental DNA are pried apart and where the two daughter strands are synthesized (Fig. 1-17), each by a different molecule of DNA polymerase. One of these DNA polymerase molecules continuously copies the parental strand that extends in its 3′ to 5′ direction from the replication fork, thereby synthesizing the resulting daughter strand, which is known as the **leading strand,** in its 5′ to 3′ direction. However, since the second DNA polymerase at the replication fork also synthesizes DNA in the 5′ to 3′ direction and yet must travel with the replication fork, how does it copy the parental strand that extends from the replication fork in its 5′ to 3′ direction? The answer is that *it synthesizes the so-called **lagging strand** discontinuously, that is, in pieces (Fig. 5-32a).* It does so by binding the looped-around lagging strand template so as to extend its newly synthesized RNA primer in its 5′ to 3′ direction (Fig. 5-32b; in effect, reversing its direction of travel) until it encounters the previously synthesized primer. The DNA polymerase then disengages from the lagging strand template and rebinds to it upstream of its previous position, where it then extends the next RNA primer to be synthesized. Thus the lagging strand is synthesized discontinuously, whereas the leading strand is synthesized continuously. The synthesis of lagging strand primers in *E. coli* is catalyzed by primase, which accompanies the replication fork (Fig. 5-32b), whereas the synthesis of leading strand primers, a much rarer event, occurs most efficiently when both primase and RNA polymerase are present.

c. Lagging Strand Synthesis Requires Several Enzymes

Escherichia coli contains two species of DNA polymerase that are essential for its survival. Of these, **DNA**

polymerase III (Pol III) is the DNA replicase, that is, it synthesizes the leading strand and most of the lagging strand. **DNA polymerase I (Pol I)** has a different function, that of removing the RNA primers and replacing them with DNA. Pol I can do so because it has a second enzymatic activity besides that of a DNA polymerase; it is also a **5′→3′ exonuclease** (an exonuclease hydrolytically removes one or more nucleotides from the end of a polynucleotide rather than cleaving it at an internal position). The 5′→3′ exonuclease function binds to single-strand nicks (places where successive nucleotides are not covalently linked such as on the 5′ side of an RNA primer after the succeeding lagging strand segment has been synthesized). It then excises a 1- to 10-nucleotide segment of the nicked strand in the 5′ to 3′ (5′→3′) direction past the nick (Fig. 5-33). Pol I's 5′→3′ exonuclease and DNA polymerase activities work in concert, so as *Pol I's 5′→3′ exonuclease*

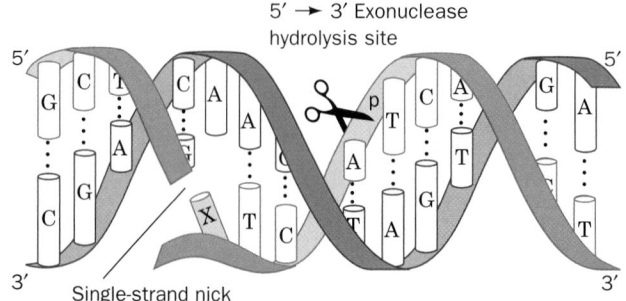

FIGURE 5-33   The 5′→3′ exonuclease function of DNA polymerase I. This enzymatic activity excises up to 10 nucleotides from the 5′ end of a single-strand nick. The nucleotide immediately past the nick (X) may or may not be base paired.

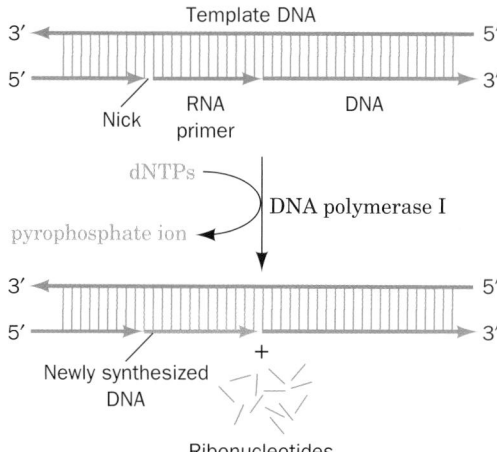

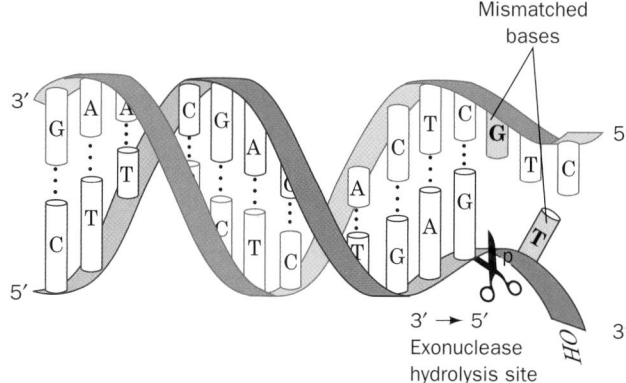

FIGURE 5-36 **The 3′→5′ exonuclease function of DNA polymerase I and DNA polymerase III.** In *E. coli*, this enzymatic activity excises mispaired nucleotides from the 3′ end of a growing DNA strand.

FIGURE 5-34 **Replacement of RNA primers by DNA in lagging strand synthesis.** In *E. coli*, the RNA primer on the 5′ end of a newly synthesized DNA segment is excised through the action of DNA polymerase I's 5′→3′ exonuclease activity and is simultaneously replaced by DNA as catalyzed by the enzyme's DNA polymerase activity.

removes the primer, its DNA polymerase activity replaces this RNA with DNA (Fig. 5-34).

The synthesis of the leading strand is completed by the replacement of its single RNA primer with DNA. However, the completion of lagging strand synthesis requires that the nicks between its multiple discontinuously synthesized segments be sealed. This is the job of an independent enzyme named **DNA ligase** that covalently links adjacent 3′-OH and 5′-phosphate groups (Fig. 5-35).

d. Errors in DNA Sequences Are Subject to Correction

In *E. coli*, RNA polymerase has an error rate of ~1 wrong base for every 10^4 nucleotides it transcribes. In contrast, newly replicated DNA contains only ~1 error per 10^8 to 10^{10} base pairs. We have already seen that the use of RNA primers increases the fidelity of lagging strand synthesis. However, the main reason for the enormous fidelity

of DNA replication is that both Pol I and Pol III have **3′→5′ exonuclease** activities. The 3′→5′ exonuclease degrades the newly synthesized 3′ end of a daughter strand one nucleotide at a time (Fig. 5-36), thereby annulling the polymerase reaction. This enzymatic function is activated by non-Watson–Crick base pairing and consequently acts to edit out the occasional mistakes made by the polymerase function, thereby greatly increasing the fidelity of replication. However, in addition to this proofreading function on both Pol I and Pol III, all cells contain a battery of enzymes that detect and correct residual errors in replication as well as damage which DNA incurs through the action of such agents as UV radiation and **mutagens** (substances that damage DNA by chemically reacting with it) as well as by spontaneous hydrolysis. In *E. coli*, Pol I also functions to replace the damaged DNA segments that these enzymes have excised.

5 ■ MOLECULAR CLONING

A major problem in almost every area of biochemical research is obtaining sufficient quantities of the substance of interest. For example, a 10-L culture of *E. coli* grown to its maximum titer of ~10^{10} cells · mL^{-1} contains, at most, 7 mg of DNA polymerase I, and many of its proteins are present in far lesser amounts. Yet it is rare that even as much as half of any protein originally present in an organism can be recovered in pure form (Chapter 6). Eukaryotic proteins may be even more difficult to obtain because many eukaryotic tissues, whether acquired from an intact organism or grown in tissue culture, are available only in small quantities. As far as the amount of DNA is concerned, our 10-L *E. coli* culture would contain ~0.1 mg of any 1000-bp length of chromosomal DNA (a length sufficient to contain most prokaryotic genes), but its purification in the presence of the rest of the chromosomal DNA (which consists of 4.6 million bp) would be an all but impossible task. These difficulties have been largely eliminated through the development of **molecular cloning** techniques (a **clone** is a col-

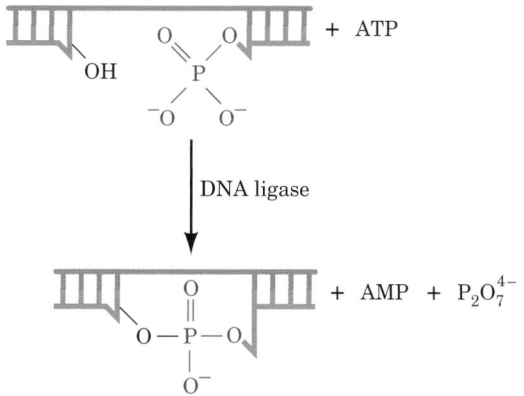

FIGURE 5-35 **Function of DNA ligase.** DNA ligase seals single-strand nicks in duplex DNA. It does so in a reaction that is powered by the hydrolysis of ATP or a similar compound.

lection of identical organisms that are derived from a single ancestor). These methods, which are also referred to as **genetic engineering** and **recombinant DNA** technology, deserve much of the credit for the enormous progress in biochemistry and the dramatic rise of the biotechnology industry since the late 1970s.

The main idea of molecular cloning is to insert a DNA segment of interest into an autonomously replicating DNA molecule, a so-called **cloning vector** *or* **vehicle,** *so that the DNA segment is replicated with the vector.* Cloning such a **chimeric vector** (*chimera:* a monster in Greek mythology that has a lion's head, a goat's body, and a serpent's tail) in a suitable **host organism** such as *E. coli* or yeast results in the production of large amounts of the inserted DNA segment. If a cloned gene is flanked by the properly positioned control sequences for transcription and translation, the host may also produce large quantities of the RNA and protein specified by that gene. The techniques of genetic engineering, whose understanding is prerequisite to understanding many of the experiments discussed in this textbook, are outlined in this section.

A. *Restriction Endonucleases*

In order to effectively carry out molecular cloning, it is necessary to be able to manipulate precisely sequence-defined DNA fragments. This is done through the use of enzymes known as **restriction endonucleases.**

Bacteriophages that propagate efficiently on one bacterial strain, such as *E. coli* K12, have a very low rate of infection (~0.001%) in a related bacterial strain such as *E. coli* B. However, the few viral progeny of this latter infection propagate efficiently in the new host but only poorly in the original host. Evidently, the new host modifies these bacteriophages in some way. What is the molecular basis of this **host-specific modification?** Werner Arber showed that it results from a **restriction–modification system** in the bacterial host which consists of a restriction endonuclease (alternatively, **restriction enzyme;** endonucleases are enzymes that hydrolytically cleave polynucleotides at internal sites) and a matched **modification methylase.** *Restriction endonucleases recognize a specific base sequence of four to eight bases in double-stranded DNA and cleave both strands of the duplex.* Modification methylases methylate a specific base (at the amino group of an adenine or either the 5 position or the amino group of a cytosine) in the same base sequence recognized by the matched restriction enzyme.

A restriction enzyme does not cleave its corresponding methylated DNA. A newly replicated strand of bacterial DNA, which is protected from degradation by the methylated parent strand with which it forms a duplex, is modified before the next cycle of replication. A restriction–modification system therefore protects the bacterium against invasion by foreign (usually viral) DNAs which, once they have been cleaved by a restriction endonuclease, are further degraded by bacterial exonucleases. Invading DNAs are only rarely modified before being attacked by restriction enzymes. Yet if a viral DNA does become modified, it is able to reproduce in its new host. Its progeny, however, are no longer modified in the way that permits them to propagate in the original host (which has different restriction–modification systems).

There are three known types of restriction endonucleases. **Type I** and **Type III** restriction enzymes each carry both the endonuclease and the methylase activity on a single protein molecule. Type I restriction enzymes cleave the DNA at a possibly random site located at least 1000 bp from the recognition sequence, whereas Type III enzymes do so 24 to 26 bp distant from the recognition sequence. However, **Type II** restriction enzymes, which were discovered and characterized by Hamilton Smith and Daniel Nathans in the late 1960s, are separate entities from their corresponding modification methylases. *They cleave DNAs at specific sites within the recognition sequence, a property that makes Type II restriction enzymes indispensable biochemical tools for DNA manipulation.* In what follows, we discuss only Type II restriction enzymes.

Over 3000 species of Type II restriction enzymes with over 200 differing sequence specificities and from a variety of bacteria have been characterized. Several of the more widely used species are listed in Table 5-4. A restriction endonuclease is named by the first letter of the genus of the bacterium that produced it and the first two letters of its species, followed by its serotype or strain designation, if any, and a roman numeral if the bacterium expresses more than one type of restriction enzyme. For example, *Eco***RI** is produced by *E. coli* strain RY13.

a. Most Restriction Endonucleases Recognize Palindromic DNA Sequences

Most restriction enzyme recognition sites possess exact twofold rotational symmetry, as is diagrammed in Fig. 5-37. Such sequences are known as **palindromes.**

A palindrome is a word, verse, or sentence that reads the same backward and forward. Two examples are "Madam, I'm Adam" and "Sex at noon taxes."

Many restriction enzymes, such as *Eco*RI (Fig. 5-37*a*), catalyze cleavage of the two DNA strands at positions that are symmetrically staggered about the center of the palindromic recognition sequence. This yields restriction fragments with complementary single-stranded ends that are from one to four nucleotides in length. Restriction fragments with such **cohesive** or **sticky ends** can associate by complementary base pairing with other restriction fragments generated by the same restriction enzyme. Some restriction cuts, such as that of *Eco*RV (Fig. 5-37*b*), pass through the twofold axis of the palindrome to yield restriction fragments with fully base paired **blunt ends.** Since a given base has a one-fourth probability of occurring at any nucleotide position (assuming the DNA has equal proportions of all bases), a restriction enzyme with an n-base pair recognition site produces restriction fragments that are, on average, 4^n base pairs long. Thus *Alu*I (4-bp recognition sequence) and *Eco*RI (6-bp recognition sequence) restriction fragments should average $4^4 = 256$ and $4^6 = 4096$ bp in length, respectively.

TABLE 5-4   Recognition and Cleavage Sites of Some Type II Restriction Enzymes

Enzyme	Recognition Sequence[a]	Microorganism
*Alu*I	AG↓C*T	*Arthrobacter luteus*
*Bam*HI	G↓GATC*C	*Bacillus amyloliquefaciens* H
*Bgl*I	GCCNNNN↓NGCC	*Bacillus globigii*
*Bgl*II	A↓GATCT	*Bacillus globigii*
*Eco*RI	G↓AA*TTC	*Escherichia coli* RY13
*Eco*RII	↓CC*(^A_T)GG	*Escherichia coli* R245
*Eco*RV	GA*T↓ATC	*Escherichia coli* J62 pLG74
*Hae*II	RGCGC↓Y	*Haemophilus aegyptius*
*Hae*III	GG↓C*C	*Haemophilus aegyptius*
*Hin*dIII	A*↓AGCTT	*Haemophilus influenzae* R$_d$
*Hpa*II	C↓C*GG	*Haemophilus parainfluenzae*
*Msp*I	C*↓CGG	*Moraxella* species
*Pst*I	CTGCA*↓G	*Providencia stuartii* 164
*Pvu*II	CAG↓C*TG	*Proteus vulgaris*
*Sal*I	G↓TCGAC	*Streptomyces albus* G
*Taq*I	T↓CGA*	*Thermus aquaticus*
*Xho*I	C↓TCGAG	*Xanthomonas holcicola*

[a]The recognition sequence is abbreviated so that only one strand, reading 5′ to 3′, is given. The cleavage site is represented by an arrow (↓) and the modified base, where it is known, is indicated by an asterisk (A* is N^6-methyladenine and C* is 5-methylcytosine). R, Y, and N represent purine nucleotide, pyrimidine nucleotide, and any nucleotide, respectively.

Source: Roberts, R.J. and Macelis, D., REBASE. The restriction enzyme database. http://rebase.neb.com.

b. Restriction Maps Provide a Means of Characterizing a DNA Molecule

The treatment of a DNA molecule with a restriction endonuclease produces a series of precisely defined fragments that can be separated according to size by **gel electrophoresis** (Fig. 5-38). (In gel electrophoresis, charged molecules are applied to one end of a thin slab of polyacrylamide or agarose gel and are separated through the application of an electric field. Under the conditions used to separate DNA fragments, the molecules move accord-

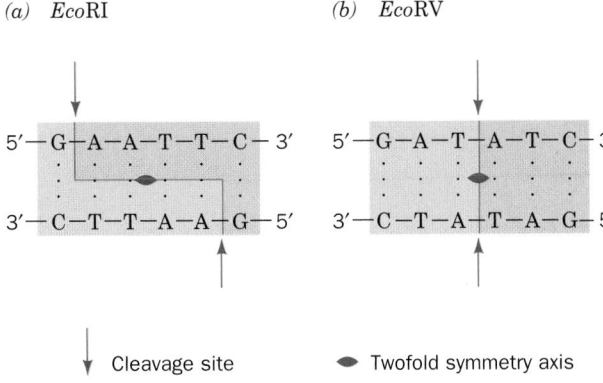

FIGURE 5-37   Restriction sites. The recognition sequences of the restriction endonucleases (*a*) *Eco*RI and (*b*) *Eco*RV have twofold (palindromic) symmetry (*red symbol*). The cleavage sites are indicated (*arrows*). Note that *Eco*RI generates DNA fragments with sticky ends, whereas *Eco*RV generates blunt-ended fragments.

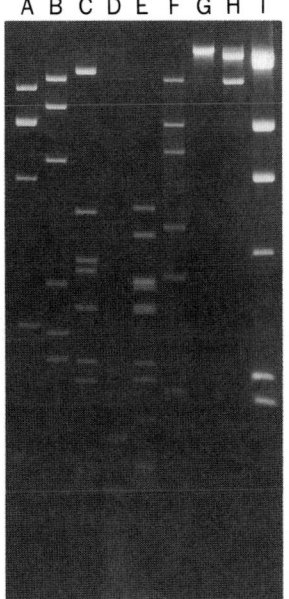

FIGURE 5-38   Agarose gel electrophoretogram of restriction digests. The *Agrobacterium radiobacter* plasmid pAgK84 was digested with (A) *Bam*HI, (B) *Pst*I, (C) *Bgl*II, (D) *Hae*III, (E) *Hin*cII, (F) *Sac*I, (G) *Xba*I, and (H) *Hpa*I. Lane I contains λ phage DNA digested with *Hin*dIII as standards since these fragments have known sizes. The DNA fragments in the electrophoretogram are made visible by fluorescence against a black background. [From Slota, J.E. and Farrand, S.F., *Plasmid* **8**, 180 (1982). Copyright © 1982 by Academic Press.]

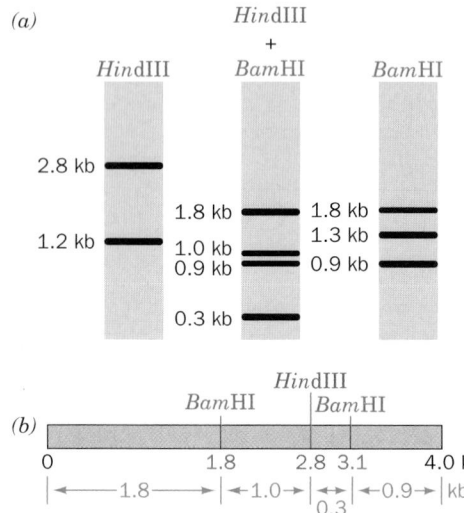

(a)

HindIII

HindIII
+
BamHI

BamHI

2.8 kb

1.2 kb

1.8 kb
1.0 kb
0.9 kb

0.3 kb

1.8 kb
1.3 kb

0.9 kb

(b)

HindIII
BamHI   BamHI

0   1.8   2.8 3.1   4.0 kb

|◄——1.8——►|◄—1.0—►|◄►|◄—0.9—►| kb
                         0.3

FIGURE 5-39   Construction of a restriction map. (*a*) The gel electrophoretic patterns of digests of a hypothetical DNA molecule with *Hind*III, *Bam*HI, and their mixture. The lengths of the various fragments are indicated. (*b*) The restriction map of the DNA resulting from the information in *a*. This map is equivalent to one that has been reversed, right to left. The green numbers indicate the sizes, in kilobase pairs, of the corresponding restriction fragments.

ing to size, with the smallest fragments moving fastest. Gel electrophoresis is further discussed in Section 6-4B.) Complementary single strands can be separated either by melting the DNA and subjecting it to gel electrophoresis, or by using density gradient ultracentrifugation in alkaline CsCl solution (recall that DNA is denatured under alkaline conditions).

A diagram of a DNA molecule showing the relative positions of the cleavage sites of various restriction enzymes is known as its **restriction map.** Such a map is generated by subjecting the DNA to digestion with two or more restriction enzymes, both individually and in mixtures. By comparing the lengths of the fragments in the various digests, as determined, for instance, by their electrophoretic mobilities relative to standards of known molecular masses, a restriction map can be constructed. For example, consider the 4-kb linear DNA molecule that *Bam*HI, *Hind*III, and their mixture cut to fragments of the lengths indicated in Fig. 5-39*a*. This information is sufficient to deduce the positions of the restriction sites in the intact DNA and hence to construct the restriction map diagrammed in Fig. 5-39*b*. The restriction map of the 5243-bp **simian virus 40 (SV40)** chromosome is shown in Fig. 5-40. The restriction sites are physical reference points on a DNA molecule that are easily located. *Restriction maps therefore constitute a convenient framework for locating particular*

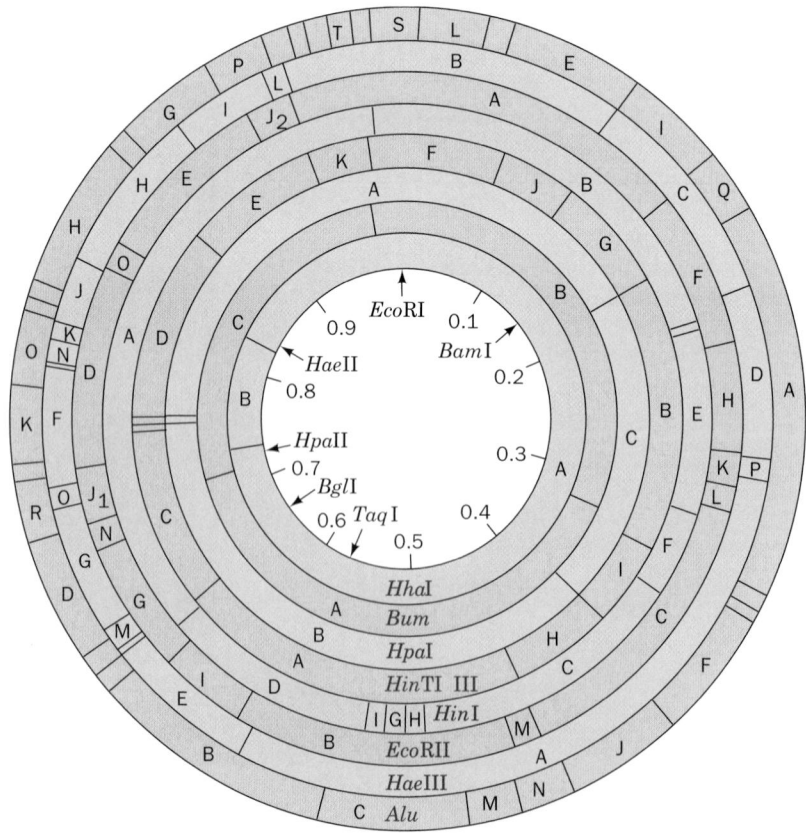

FIGURE 5-40   Restriction map for the 5243-bp circular DNA of SV40. The central circle indicates the fractional map coordinates with respect to the single *Eco*RI restriction site. The letters A, B, C, . . . in each ring represent the various restriction fragments of the corresponding enzyme in order of decreasing length. [After Nathans, D., *Science* **206,** 905 (1979).]

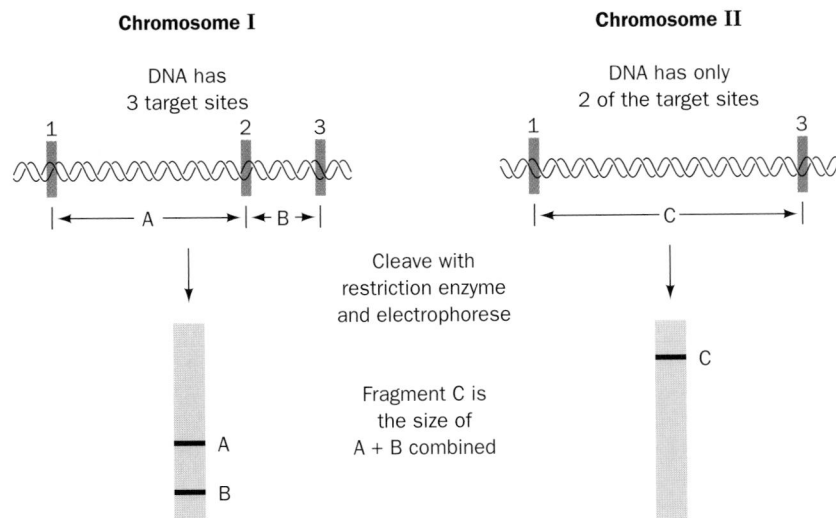

Chromosome I

DNA has
3 target sites

1   2   3

|◄──── A ────►|◄ B ►|

Chromosome II

DNA has only
2 of the target sites

1   3

|◄──────── C ────────►|

Cleave with
restriction enzyme
and electrophorese

Fragment C is
the size of
A + B combined

A

B

C

FIGURE 5-41   Restriction-fragment length polymorphisms. A mutational change that affects a restriction site in a DNA segment alters the number and sizes of its restriction fragments.

base sequences on a chromosome and for estimating the degree of difference between related chromosomes.

c. Restriction-Fragment Length Polymorphisms Provide Markers for Characterizing Genes

Individuality in humans and other species derives from their genetic polymorphism; homologous human chromosomes differ in sequence, on average, every 1250 bp. These genetic differences create and eliminate restriction sites (Fig. 5-41). Restriction enzyme digests of the corresponding segments from homologous chromosomes therefore contain fragments with different lengths; that is, these DNAs have **restriction-fragment length polymorphisms** (**RFLPs;** Fig. 5-42). Since, with the exception of identical twins, each individual has a unique set of RFLPs (its **haplotype**), RFLPs have been used for purposes of identification.

RFLPs are medically valuable for diagnosing inherited diseases for which the molecular defect is unknown. If a particular RFLP is so closely linked to a defective gene that there is little chance the two will recombine from generation to generation (recall that the probability of recombination between two genes increases with their physical separation on a chromosome; Section 1-4C), then the detection of that RFLP in an individual indicates that the individual has also inherited the defective gene. For example, **Huntington's disease,** a progressive and invariably fatal neurological deterioration, whose symptoms first appear around age 40, is caused by a dominant but until recently unknown genetic defect (Section 30-7). The identification of an RFLP that is closely linked to the defective Huntington's gene has permitted the children of Huntington's disease victims (50% of whom inherit this devastating condition) to make informed decisions in ordering their lives. Such genetic testing of fetal cells has similarly permitted the prenatal detection of serious genetic defects. (Note that the availability of fetal genetic testing has actually increased the number of births because many couples who knew they had a high risk of conceiving a genetically defective child previously chose not to have children.)

B. *Cloning Vectors*

Plasmids, viruses, and artificial chromosomes are used as cloning vectors in genetic engineering.

a. Plasmid-Based Cloning Vectors

Plasmids are circular DNA duplexes of 1 to 200 kb that contain the requisite genetic machinery, such as a **replication origin** (a site at which DNA replication is initiated; Section 30-3C), to permit their autonomous propagation in a bacterial host or in yeast. Plasmids may be considered

Pedigree and genotypes

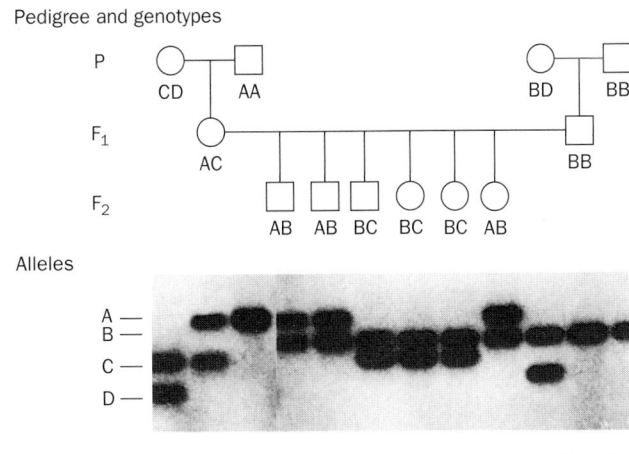

FIGURE 5-42   Inheritance of RFLPs according to the rules of Mendelian genetics. Four alleles of a particular gene, each characterized by different restriction markers, can occur in all possible pairwise combinations and segregate independently in each generation (circles represent females and squares represent males). In the *P* (parental) generation, two individuals have heterozygous haplotypes (CD and BD) and the other two have homozygous haplotypes (AA and BB) for the gene in question. Their children, the *F₁* generation, have the haplotypes AC or BB. Consequently, every individual in the *F₂* generation (grandchildren) inherited either an A or a C from their mother and a B from their father. [Courtesy of Ray White, University of Utah Medical School.]

molecular parasites but in many instances they benefit their host by providing functions, such as resistance to antibiotics, that the host lacks. Indeed, the widespread and alarming appearance, since antibiotics came into use, of antibiotic-resistant pathogens is partially the result of the rapid proliferation among these organisms of plasmids containing genes that confer resistance to antibiotics.

Some types of plasmids, which are present in one or a few copies per cell, replicate once per cell division as does the bacterial chromosome; their replication is said to be under **stringent control.** Most plasmids used in molecular cloning, however, are under **relaxed control;** they are normally present in 10 to as many as 700 copies per cell. Moreover, if protein synthesis in the bacterial host is inhibited, for example, by the antibiotic **chloramphenicol** (Section 32-3G), thereby preventing cell division, these plasmids continue to replicate until 2 or 3 thousand copies have accumulated per cell (which represents about half of the cell's total DNA). The plasmids that have been constructed (by genetic engineering techniques; Section 5-5C) for use in molecular cloning are relatively small, replicate under relaxed control, carry genes specifying resistance to one or more antibiotics, and contain a number of conveniently located restriction endonuclease sites into which the DNA to be cloned may be inserted. Indeed, many plasmid vectors contain a strategically located short (<100 bp)

segment of DNA known as a **polylinker** that has been synthesized to contain a variety of restriction sites that are not present elsewhere in the plasmid. The *E. coli* plasmid designated **pUC18** (Fig. 5-43) is representative of the cloning vectors presently in use ("pUC" stands for "plasmid-Universal Cloning").

The expression of a chimeric plasmid in a bacterial host was first demonstrated in 1973 by Herbert Boyer and Stanley Cohen. The host bacterium takes up a plasmid when the two are mixed together in a process that is greatly enhanced by the presence of divalent cations such as Ca^{2+} and brief heating to ~42°C (which increases cell membrane permeability to DNA; such cells are said to be **transformation competent**). Nevertheless, an absorbed plasmid vector becomes permanently established in its bacterial host (transformation) with an efficiency of only ~0.1%.

Plasmid vectors cannot be used to clone DNAs of more than ~10 kb. This is because the time required for plasmid replication increases with plasmid size. Hence intact plasmids with large unessential (to them) inserts are lost through the faster proliferation of plasmids that have eliminated these inserts by random deletions.

b. Virus-Based Cloning Vectors

Bacteriophage λ (Fig. 5-44) is an alternative cloning vehicle that can be used to clone DNAs of up to 16 kb.

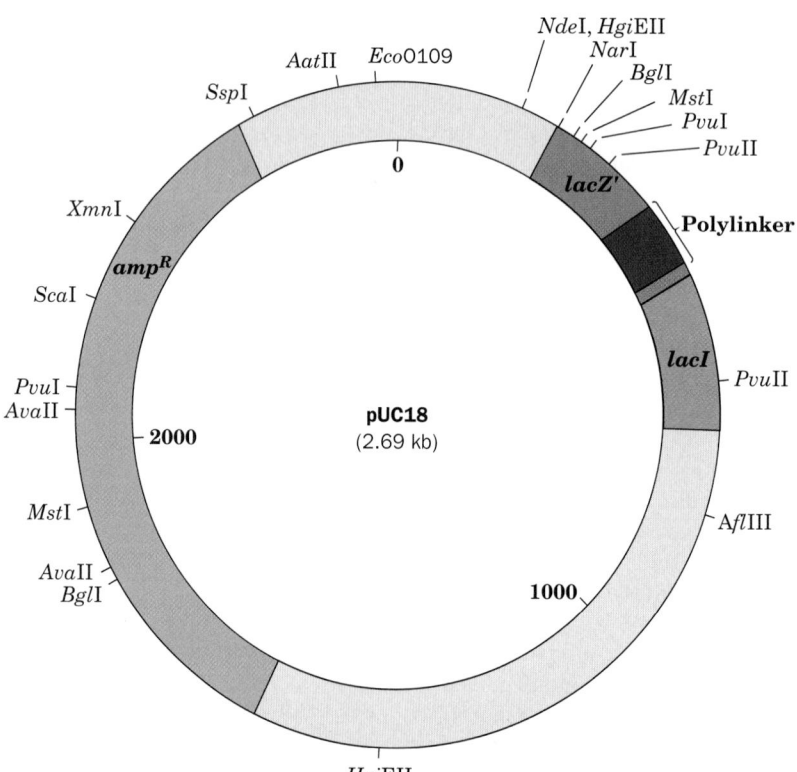

FIGURE 5-43 The pUC18 cloning vector. A restriction map of the plasmid pUC18 indicates the positions of its *amp*^{R}, *lacZ'*, and *lacI* genes. The *amp*^{R} gene confers resistance to the antibiotic **ampicillin** (a penicillin derivative; Section 11-3B); *lacZ'* is a modified form of the *lacZ* gene, which encodes the enzyme **β-galactosidase** (Section 11-2B); and *lacI* encodes the

lac **repressor,** a protein that controls the transcription of *lacZ*, as is discussed in Sections 31-1A and 31-3B. The polylinker, which encodes an 18-residue polypeptide segment inserted near the N-terminus of β-galactosidase, incorporates 13 different restriction sites that do not occur elsewhere in the plasmid.

The central third of this virus's 48.5-kb genome is not required for phage infection (Section 33-3A) and can therefore be replaced by foreign DNAs of up to slightly greater size using techniques discussed in Section 5-5C. The chimeric phage DNA can then be introduced into the host cells by infecting them with phages formed from the DNA by an *in vitro* packaging system (Section 33-3B). The use of phages as cloning vectors has the additional advantage that the chimeric DNA is produced in large amounts and in easily purified form.

λ Phages can be used to clone even longer DNA inserts. The viral apparatus that packages DNA into phage heads requires only that the DNA have a specific 16-bp sequence known as a ***cos* site** located at each end and that these ends be 36 to 51 kb apart (Section 33-3B). Placing two *cos* sites the proper distance apart on a plasmid vector yields, via an *in vitro* packaging system, a so-called **cosmid** vector, which can contain foreign DNA of up to ~49 kb. Cosmids have no phage genes and hence, on introduction into a host cell via phage infection, reproduce as plasmids.

The **filamentous bacteriophage M13** (Fig. 5-45) is also a useful cloning vector. It has a single-stranded circular

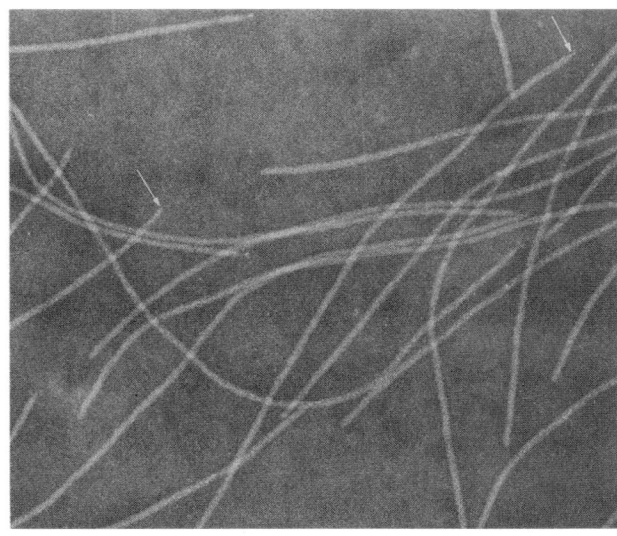

FIGURE 5-45   Electron micrograph of the filamentous bacteriophage M13. Note that some filaments appear to be pointed at one end (*arrows*). [Courtesy of Robley Williams, Stanford University, and Harold Fisher, University of Rhode Island.]

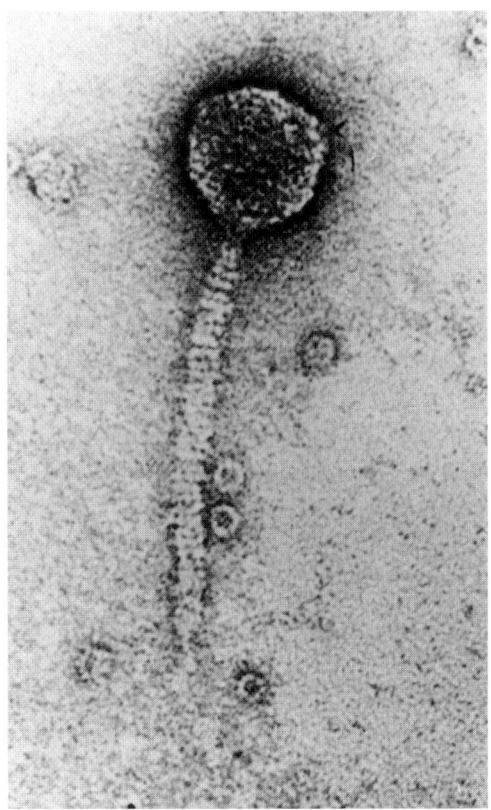

FIGURE 5-44   Electron micrograph of bacteriophage λ. Bacteriophage λ reproduces in certain strains of *E. coli*. On binding to a susceptible *E. coli*, the DNA contained in the "head" of the phage particle is injected, through its "tail," into the bacterial cell, where it is replicated ~100 times and packaged to form progeny phage (Section 33-3). [Courtesy of A.F. Howatson. From Lewin, B., *Gene Expression*, Vol. 3, Fig. 5.23, Wiley (1977).]

DNA that is contained in a protein tube composed of ~2700 helically arranged identical protein subunits. This number is controlled, however, by the length of the phage DNA being coated; insertion of foreign DNA in a nonessential region of the M13 chromosome results in the production of longer phage particles. Although M13 cloning vectors cannot stably maintain DNA inserts of >1 kb, they are widely used in the production of DNA for sequence analysis (Section 7-2A) because these phages directly produce the single-stranded DNA that the technique requires.

Baculoviruses are a large and diverse group of pathogenic viruses that infect mainly insects (but not vertebrates, so that they are safe for laboratory use) and hence can be grown in cultures of insect cells. A segment of the double-stranded DNA that forms the genome of some of these viruses is unnecessary for viral replication in tissue cultures of insect cells and hence can be replaced by a foreign DNA of up to 15 kb.

c. YAC and BAC Vectors

DNA segments larger than those that can be carried by cosmids may be cloned in **yeast artificial chromosomes (YACs)** and in **bacterial artificial chromosomes (BACs)**. YACs are linear DNA segments that contain all the molecular paraphernalia required for replication in yeast: a replication origin [known as an **autonomously replicating sequence (ARS)**], a **centromere** (the chromosomal segment attached to the spindle during mitosis and meiosis), and **telomeres** (the ends of linear chromosomes that permit their replication; Section 30-4D). BACs, which replicate in *E. coli*, are derived from circular plasmids that normally replicate long regions of DNA and are maintained at a level of approximately one copy per cell (properties similar to those of actual chromosomes). These vectors

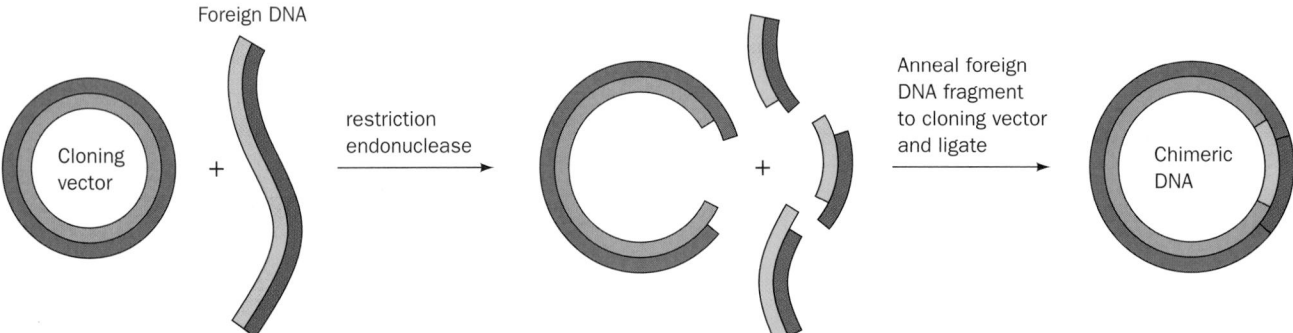

FIGURE 5-46   Construction of a recombinant DNA molecule.
A restriction fragment is inserted in a cloning vector's
corresponding restriction cut. The sticky ends of the vector
and the foreign DNA anneal and are subsequently covalently
joined by DNA ligase to yield a chimeric DNA.   🕭 **See the
Animated Figures**

contain the minimal sequences required for autonomous
replication, copy-number control, and proper partitioning
of the plasmid during cell division. YACs and BACs containing inserts of several hundred kb have been successfully cloned.

C. *Gene Manipulation*

*A DNA to be cloned is, in many cases, obtained as a
sequence-defined fragment through the application of
restriction endonucleases* (for M13 vectors, the restriction
enzymes' requirement of duplex DNA necessitates converting this phage DNA to its double-stranded form
through the use of DNA polymerase I). Recall that most
restriction endonucleases cleave duplex DNA at specific
palindromic sites so as to yield single-stranded ends that
are complementary to each other (cohesive or sticky ends;
Section 5-5A). Therefore, as Janet Mertz and Ron Davis
first demonstrated in 1972, *a restriction fragment may be
inserted into a cut made in a cloning vector by the same restriction enzyme (Fig. 5-46). The complementary (cohesive)
ends of the two DNAs specifically associate under annealing
conditions and are covalently joined (spliced) through the
action of DNA ligase* (Fig. 5-35; the DNA ligase produced
by **bacteriophage T4** must be used for blunt-ended restriction cuts such as those generated by *Alu*I, *Eco*RV, or
*Hae*III; Table 5-4). *A great advantage of using a restriction
enzyme to construct a chimeric vector is that the DNA insert can be precisely excised from the cloned vector by cleaving it with the same restriction enzyme.*

If the foreign DNA and cloning vector have no common restriction sites at innocuous positions, they may still
be spliced, using a procedure pioneered by Dale Kaiser
and Paul Berg, through the use of **terminal deoxynu-
cleotidyl transferase (terminal transferase).** This mammalian enzyme adds nucleotides to the 3'-terminal OH
group of a DNA chain; it is the only known DNA polymerase that does not require a template. Terminal transferase and dTTP, for example, can build up poly(dT) tails
of ~100 residues on the 3' ends of the DNA segment to
be cloned (Fig. 5-47). The cloning vector is enzymatically
cleaved at a specific site and the 3' ends of the cleavage

site are similarly extended with poly(dA) tails. The complementary homopolymer tails are annealed, any gaps resulting from differences in their lengths filled in by DNA
polymerase I, and the strands joined by DNA ligase.

A disadvantage of the above technique is that it eliminates the restriction sites that were used to generate the
foreign DNA insert and to cleave the vector. It may therefore be difficult to recover the insert from the cloned vector. This difficulty is circumvented by a technique in which
a chemically synthesized palindromic "linker" having a restriction site matching that of the cloning vector is appended to both ends of the foreign DNA (the chemical
synthesis of oligonucleotides is discussed in Section 7-6).
The linker is attached to the foreign DNA by blunt end
ligation with T4 DNA ligase and then cleaved with the appropriate restriction enzyme to yield the correct cohesive
ends for ligation to the vector (Fig. 5-48).

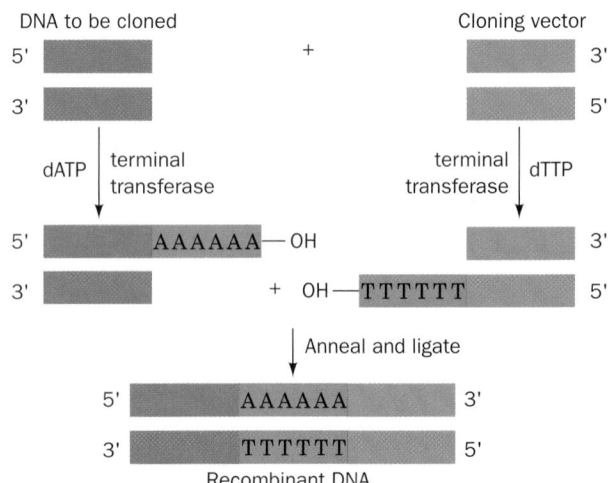

FIGURE 5-47   Splicing DNA using terminal transferase. Two
DNA fragments may be joined through the generation of
complementary homopolymer tails via the action of the enzyme
terminal transferase. The poly(dA) and poly(dT) tails shown in
this example may be replaced by poly(dC) and poly(dG) tails.

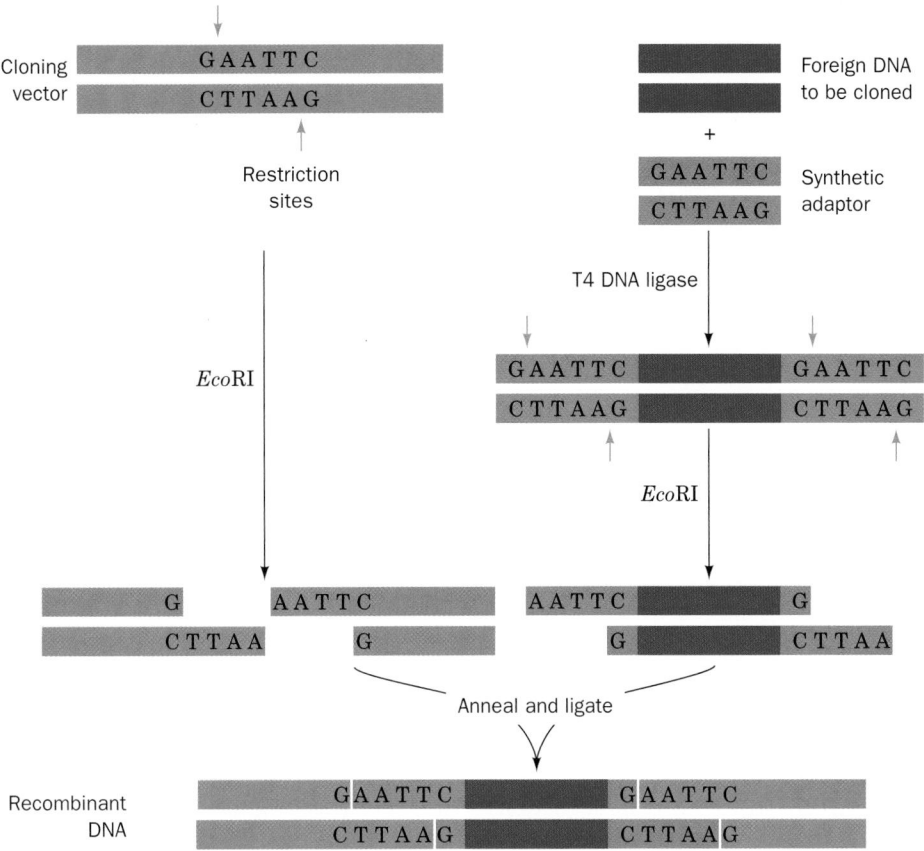

FIGURE 5-48 Construction of a recombinant DNA molecule through the use of synthetic oligonucleotide adaptors. In this example, the adaptor and the cloning vector have *Eco*RI restriction sites (*red arrows*).

a. Properly Transformed Cells Must Be Selected

Both transformation and the proper construction of chimeric vectors occur with low efficiency. How can one select only those host organisms that have been transformed by the properly constructed vector? In the case of plasmid transformation, this is usually done through a double screen using antibiotics and/or chromogenic (color-producing) substrates. For example, the pUC18 plasmid contains the ***lacZ′*** gene (Fig. 5-43; a modified form of the *lac* operon's Z gene; Fig. 5-25). The *lacZ′* gene encodes the enzyme **β-galactosidase,** which catalyzes the hydrolysis of the bond from O1 of the sugar β-D-galactose to a substituent.

HOCH₂ ... O—R → β-galactosidase → HOCH₂ ... OH + ROH

β-D-Galactose

Thus, when grown in the presence of **5-bromo-4-chloro-3-indolyl-β-D-galactoside** (commonly known as **X-gal**), a colorless substance which when hydrolyzed by β-galactosidase yields a blue product,

5-Bromo-4-chloro-3-indolyl-β-D-galactoside (X-gal)
(*colorless*)

H_2O → β-galactosidase

β-D-Galactose   **5-Bromo-4-chloro-3-hydroxyindole**
(*blue*)

E. coli transformed by an unmodified pUC18 plasmid form blue colonies. However, *E. coli* transformed by a pUC18 plasmid containing a foreign DNA insert in its polylinker

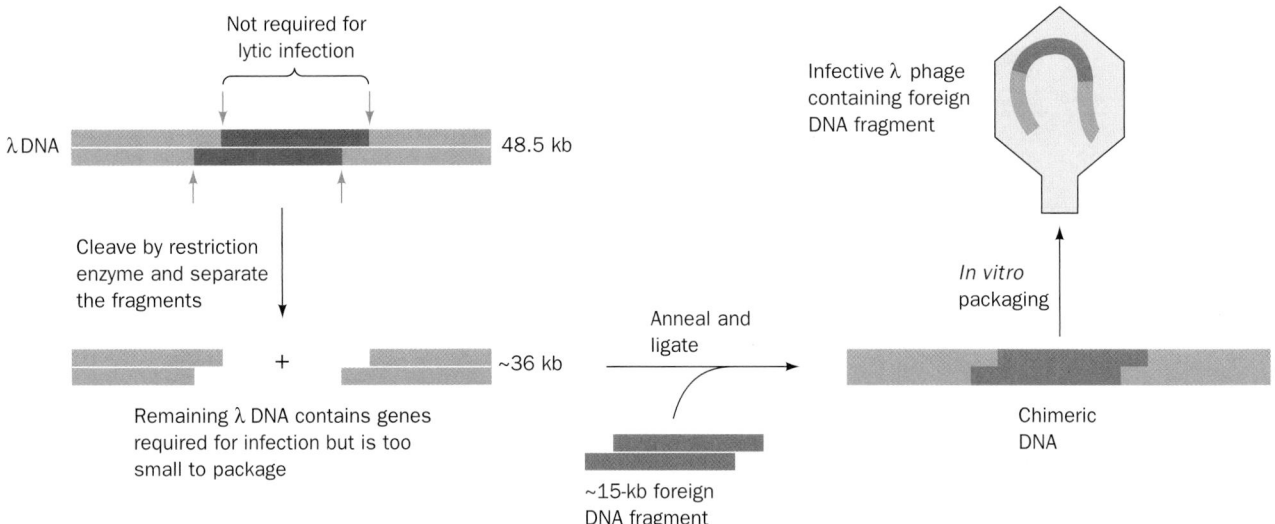

FIGURE 5-49 Cloning of foreign DNA in λ phages. A nonessential portion of the phage genome can be replaced by a foreign DNA and packaged to form an infectious phage particle only if the foreign DNA is approximately the same size as the DNA segment it replaced. ⚡ **See the Animated Figures**

region form colorless colonies because the insert interrupts the protein-encoding sequence of the *lacZ'* gene and hence they lack β-galactosidase activity. Bacteria that have failed to take up any plasmid, which would otherwise also form colorless colonies in the presence of X-gal, are excluded by adding the antibiotic **ampicillin** (Fig. 11-25) to the growth medium. Bacteria that do not contain the plasmid are sensitive to ampicillin, whereas bacteria containing the plasmid will grow, because the plasmid's intact *amp*^{R} gene confers ampicillin resistance. Genes such as *amp*^{R} are therefore known as **selectable markers.**

Genetically engineered λ phage variants contain restriction sites that flank the dispensable central third of the phage genome (Section 5-5B). This segment may therefore be replaced, as is described above, by a foreign DNA insert (Fig. 5-49). DNA is only packaged in λ phage heads if its length is from 75 to 105% of the 48.5-kb wild-type λ genome. Consequently, λ phage vectors that have failed to acquire a foreign DNA insert are unable to propagate because they are too short to form infectious phage particles. Cosmid vectors are subject to the same limitation. Moreover, cloned cosmids are harvested by repackaging them into phage particles. Hence, any cosmids that have lost sufficient DNA through random deletion to make them shorter than the above limit are not recovered. This is why cosmids can support the proliferation of large DNA inserts, whereas most other types of plasmids cannot.

D. *The Identification of Specific DNA Sequences: Southern Blotting*

DNA with a specific base sequence may be identified through a procedure developed by Edwin Southern known as the **Southern transfer technique** or, more colloquially, as **Southern blotting** (Fig. 5-50). This procedure takes advantage of the valuable property of nitrocellulose that it tenaciously binds single-stranded (but not duplex) DNA [**nylon**

and **polyvinylidine difluoride (PVDF)** membranes also have this property]. Following the gel electrophoresis of double-stranded DNA, the gel is soaked in 0.5*M* NaOH solution, which converts the DNA to the single-stranded form. The gel is then overlaid by a sheet of nitrocellulose paper which, in turn, is covered by a thick layer of paper towels, and the entire assembly is compressed by a heavy plate. The liquid in the gel is thereby forced (blotted) through the nitrocellulose so that the single-stranded DNA binds to it at the same position it had in the gel (the transfer to nitrocellulose can alternatively be accomplished by an electrophoretic process called **electroblotting**). After vacuum drying the nitrocellulose at 80°C, which permanently fixes the DNA in place, the nitrocellulose sheet is moistened with a minimal quantity of solution containing ^{32}P-labeled single-stranded DNA or RNA (the "probe") that is complementary in sequence to the DNA of interest. The moistened sheet is held at a suitable renaturation temperature for several hours to permit the probe to anneal to its target sequence(s), washed to remove the unbound radioactive probe, dried, and then autoradiographed by placing it for a time over a sheet of X-ray film. The positions of the molecules that are complementary to the radioactive sequences are indicated by a blackening of the developed film.

A DNA segment containing a particular base sequence (e.g., an RFLP) may, in this manner, be detected and isolated. The radioactive probe used in this procedure can be the corresponding mRNA if it is produced in sufficient quantity to be isolated [e.g., **reticulocytes** (immature red blood cells), which produce little protein besides **hemoglobin** (the red protein that transports O$_2$ in the blood) and are rich in the mRNAs that specify hemoglobin]. Alternatively, the gene specifying a protein of known amino acid sequence may be found by synthesizing a radiolabeled probe that is a mixture of all oligonucleotides, which, according to the genetic code (Table 5-3), can specify a segment of the gene that has low degeneracy (Fig. 5-51).

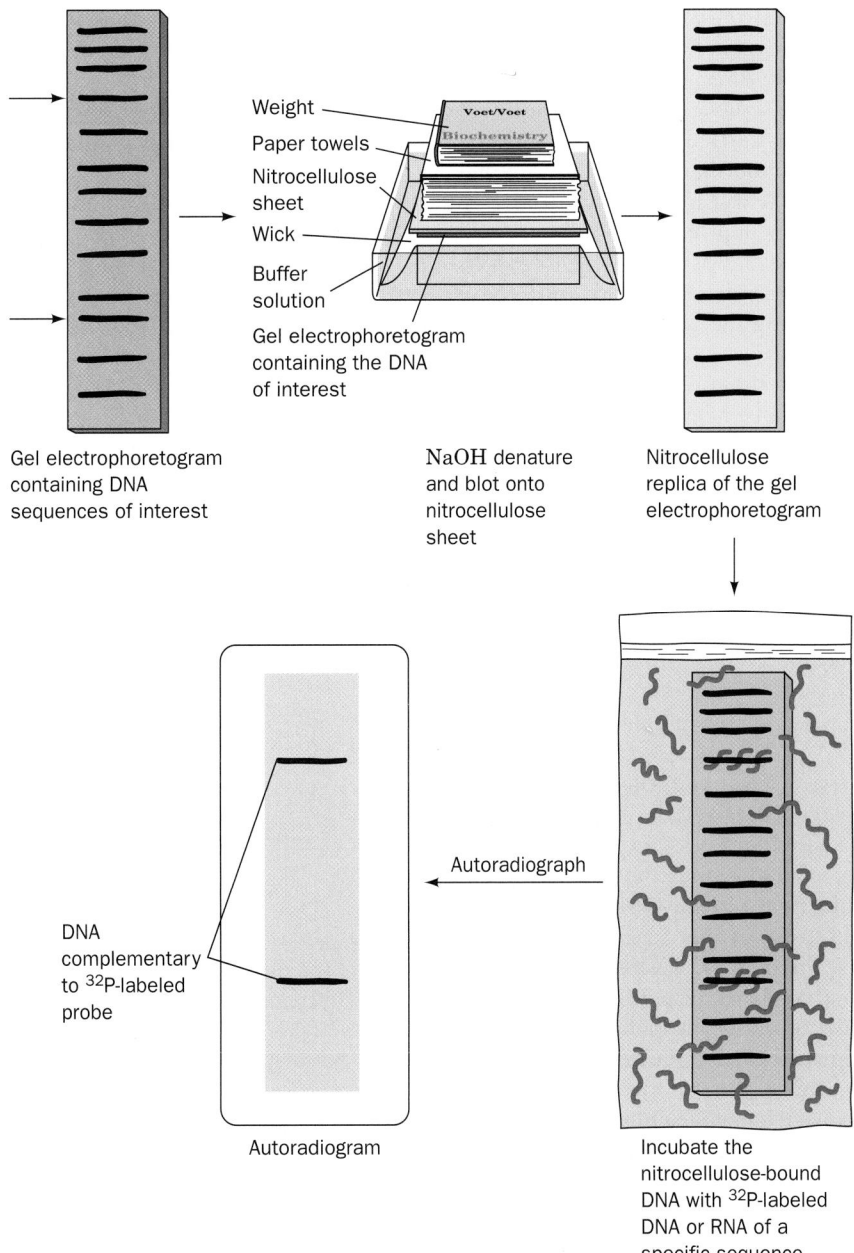

FIGURE 5-50 Detection of DNAs containing specific base sequences by the Southern transfer technique.

— Trp – Lys – Gln – Cys – Met —Peptide segment

^{32}P–UGG–AAA–CAA–UGU–AUG

^{32}P–UGG–AAG–CAA–UGU–AUG

^{32}P–UGG–AAA–CAG–UGU–AUG

^{32}P–UGG–AAG–CAG–UGU–AUG

^{32}P–UGG–AAA–CAA–UGC–AUG

^{32}P–UGG–AAG–CAA–UGC–AUG

^{32}P–UGG–AAA–CAG–UGC–AUG

^{32}P–UGG–AAG–CAG–UGC–AUG

Mixture of all oligonucleotides that can encode the peptide

FIGURE 5-51 A degenerate oligonucleotide probe. Such a probe is a mixture of all oligonucleotides that can encode a polypeptide segment of known sequence. In practice, such a segment is chosen to contain a high proportion of residues specified by low-degeneracy codons. In the pentapeptide segment shown here, Trp and Met are each specified by only one codon and Lys, Gln, and Cys are each specified by two codons that differ in only their third positions (*blue and red;* Table 5-3) for a total of $1 \times 2 \times 2 \times 2 \times 1 = 8$ oligonucleotides. The oligonucleotides are ^{32}P-labeled for use in Southern blotting.

Southern blotting may be used for the diagnosis and prenatal detection of genetic diseases. These diseases often result from a specific change in a single gene such as a base substitution, deletion, or insertion. The temperature at which probe hybridization is carried out may be adjusted so that only an oligonucleotide that is perfectly complementary to a length of DNA will hybridize to it. Even a single base mismatch, under appropriate conditions, will result in a failure to hybridize. For example, the genetic disease **sickle-cell anemia** arises from a single A → T base change in the gene specifying the β subunit of hemoglobin, which causes the amino acid substitution Glu β6 → Val (Section 7-3A). A 19-residue oligonucleotide that is complementary to the sickle-cell gene's mutated segment hybridizes, at the proper temperature, to DNA from homozygotes for the sickle-cell gene but not to DNA from normal individuals. An oligonucleotide that is complementary to the gene encoding the β subunit for normal hemoglobin yields opposite results. DNA from sickle-cell heterozygotes (who have one hemoglobin β gene bearing the sickle-cell anemia mutation and one that is normal) hybridizes to both probes but in reduced amounts relative to the DNAs from homozygotes. The oligonucleotide probes may consequently be used in the prenatal diagnosis of sickle-cell anemia. DNA probes are also rapidly replacing the much slower and less accurate culturing techniques for the identification of pathogenic bacteria.

In a variation of the Southern blotting procedure, specific DNAs may be detected by linking the probe to an enzyme that generates a colored or fluorescent deposit on the blot when exposed to the proper reagents. Such nonradioactive detection techniques are desirable in a clinical setting because of health hazards, disposal problems, and the more cumbersome nature of autoradiographic methods. Specific RNA sequences may be detected through a different variation of the Southern transfer, punningly named a **Northern transfer (Northern blot),** in which the RNA is immobilized on nitrocellulose paper and detected through the use of complementary radiolabeled RNA or DNA probes.

E. *Genomic Libraries*

In order to clone a particular DNA fragment, it must first be obtained in relatively pure form. The magnitude of this task may be appreciated when it is realized that, for example, a 1-kb fragment of human DNA represents only 0.000029% of the 3.2 billion-bp human genome. A DNA fragment may be identified by Southern blotting of a restriction digest of the genomic DNA under investigation (Section 5-5F). In practice, however, it is usually more difficult to identify a particular gene from an organism and then clone it than it is to clone the organism's entire genome as DNA fragments and then identify the clone(s) containing the sequence(s) of interest. Such a set of cloned fragments is known as a **genomic library.** A genomic library of a particular organism need only be made once since it can be perpetuated for use whenever a new probe becomes available.

Genomic libraries are generated according to a procedure known as **shotgun cloning.** The chromosomal DNA of the organism of interest is isolated, cleaved to fragments of clonable size, and inserted in a cloning vector by the methods described in Section 5-5B. The DNA is fragmented by partial rather than exhaustive restriction digestion (permitting the restriction enzyme to act for only a short time) so that the genomic library contains intact representatives of all the organism's genes, including those whose sequences contain restriction sites. Shear fragmentation by rapid stirring of a DNA solution or by sonication is also used but requires further treatment of the fragments to insert them into cloning vectors. Genomic libraries have been established for numerous organisms including yeast, *Drosophila,* mice, and humans.

a. Many Clones Must Be Screened to Obtain a Gene of Interest

The number of random cleavage fragments that must be cloned to ensure a high probability that a given sequence is represented at least once in the genomic library is calculated as follows: The probability P that a set of N clones contains a fragment that constitutes a fraction f, in bp, of the organism's genome is

$$P = 1 - (1 - f)^N \qquad [5.1]$$

Consequently,

$$N = \log(1 - P)/\log(1 - f) \qquad [5.2]$$

Thus, to have $P = 0.99$ with fragments averaging 10 kb in length, for the 4639-kb *E. coli* chromosome ($f = 0.00216$), $N = 2134$ clones, whereas for the 180,000-kb *Drosophila* genome ($f = 0.0000606$), $N = 83,000$. The use of YAC- or BAC-based genomic libraries therefore greatly reduces the effort necessary to obtain a given gene segment from a large genome.

Since a genomic library lacks an index, it must be screened for the presence of a particular gene. This is done by a process known as **colony** or ***in situ* hybridization** (Fig. 5-52; Latin: *in situ,* in position). The cloned yeast colonies, bacterial colonies, or phage plaques to be tested are transferred, by replica plating (Fig. 1-30), from a master plate to a nitrocellulose filter. The filter is treated with NaOH, which lyses the cells or phages and denatures the DNA so that it binds to the nitrocellulose (recall that single-stranded DNA is preferentially bound to nitrocellulose). The filter is then dried to fix the DNA in place, treated under annealing conditions with a radioactive probe for the gene of interest, washed, and autoradiographed. *Only those colonies or plaques containing the sought-after gene will bind the probe and thereby blacken the film.* The corresponding clones can then be retrieved from the master plate. Using this technique, even an ~1 million clone human genomic library can be readily screened for the presence of one particular DNA segment.

Many eukaryotic genes and gene clusters span enormous tracts of DNA (Section 34-2); some consist of >1000 kb. With the use of plasmid-, phage-, or cosmid-

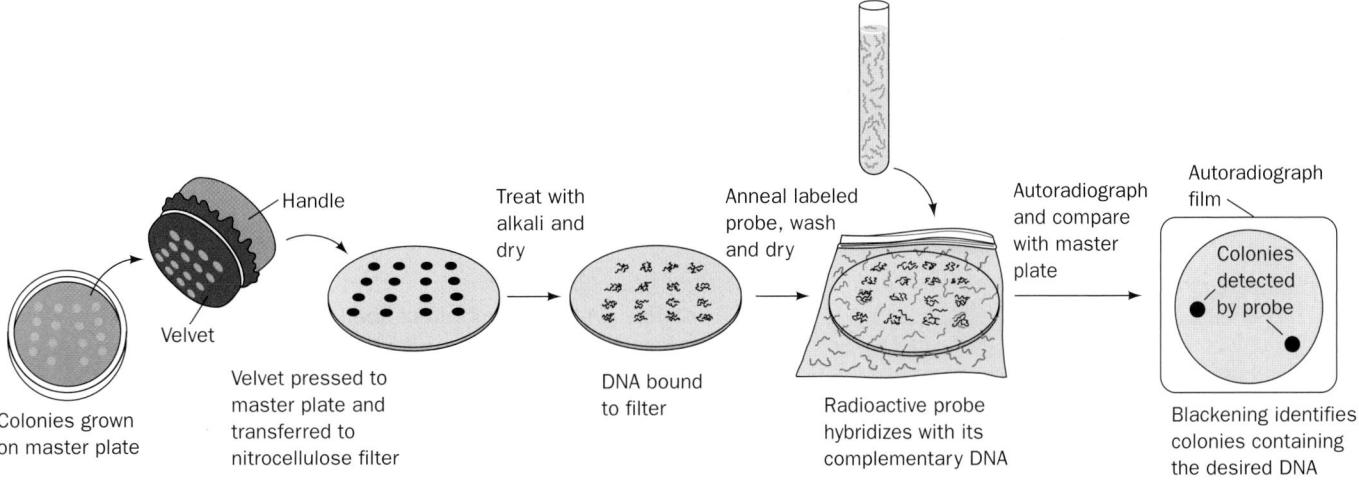

FIGURE 5-52   Colony (*in situ*) hybridization. This technique identifies the clones containing a DNA of interest.

based genomic libraries, such long DNAs can only be obtained as a series of overlapping fragments (Fig. 5-53). Each gene fragment that has been isolated is, in turn, used as a probe to identify a successive but partially overlapping fragment of that gene, a process called **chromosome walking.** The use of YACs and BACs, however, greatly reduces the need for this laborious and error-prone process.

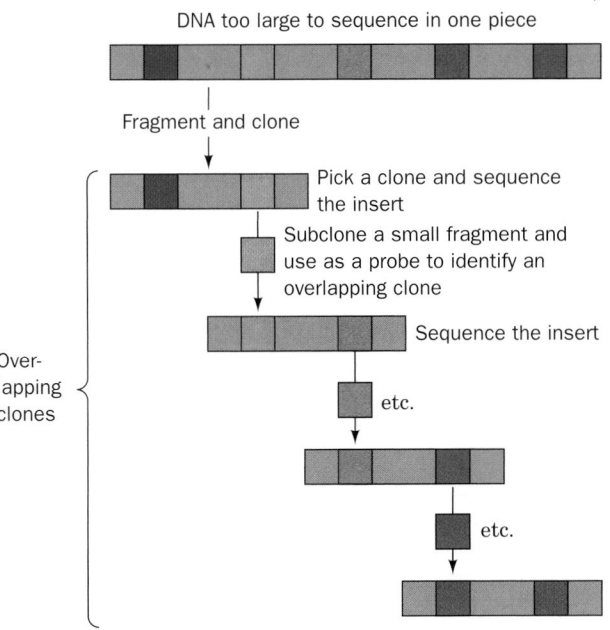

FIGURE 5-53   Chromosome walking. A DNA segment too large to sequence in one piece is fragmented and cloned. A clone is picked and the DNA insert it contains is sequenced. A small fragment of the insert near one end is subcloned (cloned from a clone) and used as a probe to select a clone containing an overlapping insert, which, in turn, is sequenced. The process is repeated so as to "walk" down the chromosome. Chromosome walking can, of course, extend in both directions.

F. *The Polymerase Chain Reaction*

📖 **See Guided Exploration 3: PCR and site-directed mutagenesis** Although molecular cloning techniques are indispensable to modern biochemical research, the use of the **polymerase chain reaction (PCR)** offers a faster and more convenient method of amplifying a specific DNA segment of up to 6 kb. In this technique (Fig. 5-54), which was formulated in 1985 by Kerry Mullis, a heat denatured (strand-separated) DNA sample is incubated with DNA polymerase, dNTPs, and two oligonucleotide primers whose sequences flank the DNA segment of interest so that they direct the DNA polymerase to synthesize new complementary strands. Multiple cycles of this process, each doubling the amount of DNA present, geometrically amplify the DNA starting from as little as a single gene copy. In each cycle, the two strands of the duplex DNA are separated by heat denaturation at 95°C, the temperature is then lowered to permit the primers to anneal to their complementary segments on the DNA, and the DNA polymerase directs the synthesis of the complementary strands (Section 5-4C). The use of a heat-stable DNA polymerase, such as those from the thermophilic bacteria *Thermus aquaticus* (**Taq DNA polymerase**) or *Pyroccocus furiosus* (**Pfu DNA polymerase**), both of which are stable at 95°C, eliminates the need to add fresh enzyme after each heat denaturation step. Hence, in the presence of sufficient quantities of primers and dNTPs, the PCR is carried out simply by cyclically varying the temperature.

Twenty cycles of PCR amplification theoretically increase the amount of the target sequence around one millionfold with high specificity (in practice, the number of copies of target sequence doubles with each PCR cycle until more primer–template complex accumulates than the DNA polymerase can extend during a cycle, whereon the rate of increase of target DNA becomes linear rather than geometric). Indeed, the method has been shown to amplify a target DNA present only once in a sample of 10^5 cells, thereby demonstrating that the method can be used without prior DNA purification (although, as a consequence

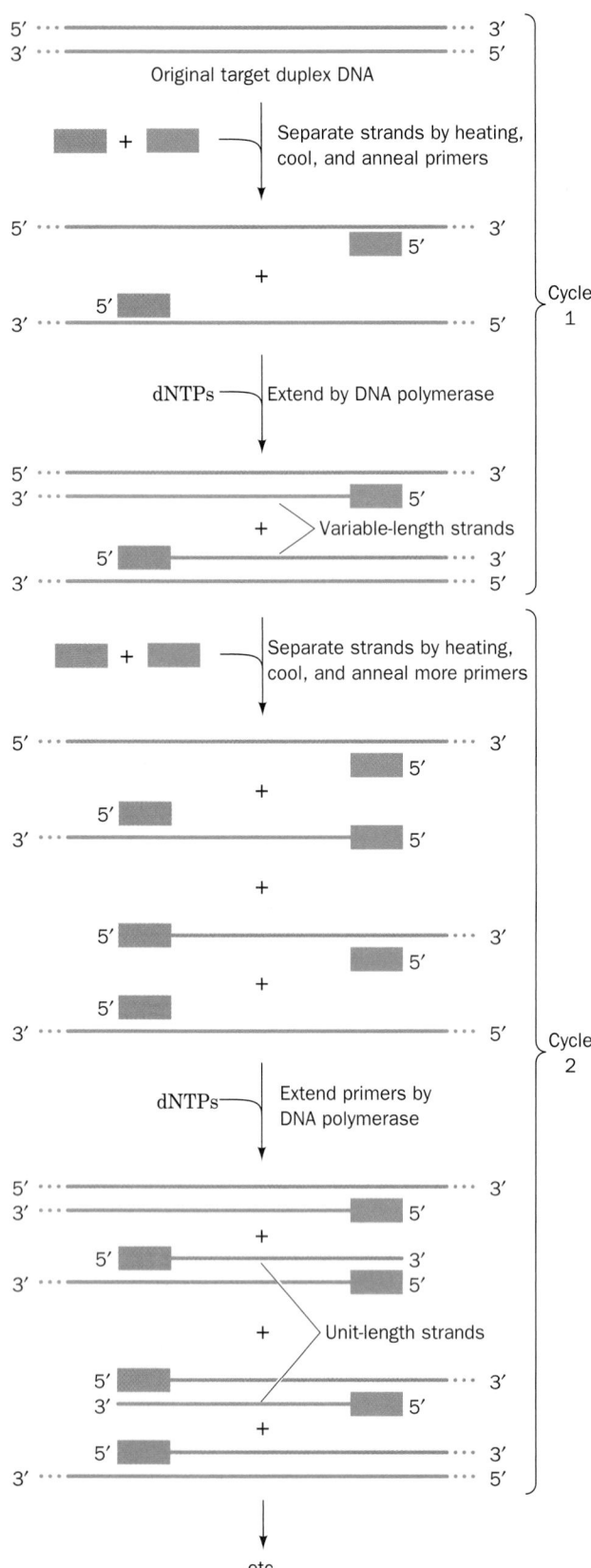

FIGURE 5-54   The polymerase chain reaction (PCR). In each cycle of the reaction, the strands of the duplex DNA are separated by heat denaturation, the preparation is cooled such that synthetic DNA primers anneal to a complementary segment on each strand, and the primers are extended by DNA polymerase. The process is then repeated for numerous cycles. The number of "unit-length" strands doubles with every cycle after the second cycle.

of this enormous amplification, particular care must be taken that the DNA sample of interest is not contaminated by extraneous DNA that is similar in sequence to that under investigation). The amplified DNA can be characterized by a variety of techniques including RFLP analysis, Southern blotting, and direct sequencing (Section 7-2A). PCR amplification is therefore a form of "cell-free molecular cloning" that can accomplish in an automated *in vitro* procedure requiring as little as 30 minutes what would otherwise take days or weeks via the cloning techniques discussed above.

a. PCR Has Many Uses

PCR amplification has become an indispensable tool in a great variety of applications. Clinically, it is used for the rapid diagnosis of infectious diseases and the detection of rare pathological events such as mutations leading to cancer (Section 19-3B). Forensically, the DNA from a single hair, sperm, or drop of blood can be used to identify the donor. This is most commonly done through the analysis of **short tandem repeats (STRs),** segments of DNA that contain repeating sequences of 2 to 7 bp such as $(CA)_n$ and $(ATGC)_n$ and that are scattered throughout the genome [e.g., the human genome contains $\sim$100,000$(CA)_n$ STRs]. The number of tandem repeats, *n*, in many STRs is genetically variable [*n* varies from 1 to 40 for particular $(CA)_n$ STRs] and hence such repeats are markers of individuality (much as are RFLPs). The DNA of a particular STR can be PCR-amplified through the use of primers that are complementary to the unique (nonrepeating) sequences flanking the STR. The number of tandem repeats in that STR from a particular individual can then be determined, either by the measurement of its molecular mass through polyacrylamide gel electrophoresis (Section 6-6C) or by direct sequencing (Section 7-2A). The determination of this number for several well-characterized STRs (those whose numbers of repeats have been determined in numerous individuals of multiple ethnicities), that is, the DNA's haplotype, can unambiguously identify the DNA's donor.

STRs are also widely used to prove or disprove familial relationships. For example, oral tradition suggests that Thomas Jefferson, the third American president, fathered a son, Eston Hemmings (born in 1808), with his slave Sally Hemmings (Eston Hemmings was said to bear a striking physical resemblance to Jefferson). Only the tips of the Y chromosome undergo recombination (with the X chromosome) and the rest of it is passed unchanged from father to son (except for occasional mutations). The finding that the Y chromosomes of male-line descendants of both Eston Hemmings and Jefferson's father's brother (Jefferson had no surviving legitimate sons) had identical STR-based haplotypes indicates that Thomas Jefferson was probably Eston Hemmings' father (although this could also be true of any of Jefferson's contemporary male-line relatives).

RNA may also be amplified via PCR by first reverse-transcribing it into a complementary strand of DNA **(cDNA)** through the action of an enzyme named **RNA-directed DNA polymerase** (commonly known as **reverse transcriptase**). This enzyme, which is produced by certain RNA-containing viruses known as **retroviruses** (Section 30-4C), uses an RNA template but is otherwise similar in the reaction it catalyzes to DNA polymerase I.

Variations on the theme of PCR have found numerous applications. For instance, single-stranded DNA (which is required for DNA sequencing; Section 7-2A) can be rapidly generated via **asymmetric PCR,** in which such a small amount of one primer is used that it is exhausted after several PCR cycles. In subsequent cycles, only the strand extended from the other primer, which is present in excess, is synthesized (note that PCR amplification becomes linear rather than geometric after one primer is used up). In cases that primers may anneal to more than one site in the target DNA, **nested primers** can be used to ensure that only the target sequence is amplified. In this technique, PCR amplification is normally carried out using one pair of primers. The products of this amplification are then further amplified through the use of a second pair of primers that anneal to the target DNA within its amplified region. It is highly unlikely that both pairs of primers will incorrectly anneal in a nested fashion to a nontarget DNA, and hence only the target DNA will be amplified.

b. Neandertals Are Not Ancestors of Modern Humans

PCR is also largely responsible for the budding science of **molecular archeology.** For example, PCR-based techniques have been used by Svante Pääbo to determine whether or not Neandertals form a different species from modern human beings. Neandertals (also called Neanderthals) are extinct hominids that were about 30% larger than are modern humans, apparently had great muscular strength, and had low foreheads and protruding brows. According to the radiodated fossil record, they became extinct ~28,000 years ago after having inhabited Europe and Western Asia for over 100,000 years. During the latter part of this period they coexisted with our direct ancestors (who might well have been responsible for their demise). Thus, an important anthropological issue is whether *Homo Neandertalensis* constituted an ancient race of *Homo sapiens* ancestral to modern humans or were a separate species. The morphological evidence has been cited as supporting both possibilities. A convincing way to settle this dispute would be by the comparison of the DNA sequences of modern humans with those of Neandertals.

The DNA was extracted from a 0.4-g sample of a Neandertal bone, and its mitochondrial DNA **(mtDNA)** was amplified by PCR (mtDNA rather than nuclear DNA was amplified because cells contain numerous mitochon-

dria and hence an mtDNA sequence is 100- to 1000-fold more abundant than is any particular sequence of nuclear DNA). The sequence of the Neandertal mtDNA was compared to those of 986 modern human lineages of a wide variety of ethnicities and 16 common lineages of chimpanzees (the closet living relatives of modern humans). A phylogenetic tree based on their sequence differences indicates that humans and chimpanzees diverged (had their last common ancestor) about 4 million years ago, humans and Neandertals diverged around 600,000 years ago, and modern humans diverged from one another about 150,000 years ago. These sequence comparisons indicate that Neandertals did not contribute significant genetic information to modern humans during their many thousand–year coexistence and hence that *Homo Neandertalensis* and *Homo sapiens* are separate species. This conclusion was confirmed by a similar analysis of a second Neandertal sample from a geographically distant location.

c. DNA Decays Quickly on the Geological Time Scale

There have been reports in the literature of DNAs that were PCR-amplified from fossils that were several million years old and from amber-entombed insects (amber is fossilized tree resin) that were as old as 135 million years (a phenomenon that formed the "basis" for the novel and movie *Jurassic Park*). Yet, over geological time spans, DNA decomposes, mostly through hydrolysis of the sugar–phosphate backbone and oxidative damage to the bases. How old can a fossil become before its DNA has decayed beyond recognition?

The amino acid residues in hydrated proteins racemize at a rate similar to the rate at which DNA decomposes. Since proteins in organisms are far more abundant than are specific DNA sequences, the enantiomeric (D/L) ratios of an amino acid residue can be determined directly (rather than requiring some sort of amplification, as in the case of DNA). The determination, in a variety of archeological specimens whose age could be authenticated, of the enantiomeric ratio of Asp (the fastest racemizing amino acid residue) revealed that DNA sequences can only be retrieved from samples in which the Asp D/L ratio is less than 0.08. These studies indicate that the survival of recognizable DNA sequences is limited to a few thousand years in warm regions such as Egypt and to as much as 100,000 years in cold regions such as Siberia. It therefore appears that the putatively very ancient DNAs, in reality, resulted from the artifactual amplification of contaminating modern DNAs (particularly those from the human operators carrying out the PCR amplifications). Indeed, the DNA in the above Neandertal fossil had decomposed to the point that it was unlikely that its nuclear DNA could have been successfully PCR-amplified, which is why its mtDNA was amplified instead.

Despite the foregoing, there is credible evidence that certain bacterial spores can remain viable almost indefinitely. Bacterial spores, which several bacterial groups including bacilli form under adverse conditions, function to permit the bacterium's survival until conditions become favorable for growth. Bacterial spores have thick protective protein coats, their cytoplasm is partially dehydrated and mineralized, and their DNA is specifically stabilized by specialized proteins (Section 29-1B). Thus, a bacillus was cultured from a 25- to 40-million-year-old (Myr) amber-entombed bee after the surface of the amber had been chemically sterilized. Similarly, a halophilic (salt-loving) bacillus was cultured from a tiny (~9 μL) brine-filled inclusion in a surface-sterilized salt crystal from a 250-Myr salt deposit.

G. *Production of Proteins*

One of the most important uses of recombinant DNA technology is the production of large quantities of scarce and/or novel proteins. This is a relatively straightforward procedure for bacterial proteins: A cloned **structural gene** (a gene that encodes a protein) is inserted into an **expression vector,** a plasmid or virus that contains the properly positioned transcriptional and translational control sequences for the protein's expression. With the use of a relaxed control plasmid and an efficient promoter, the production of a protein of interest may reach 30% of the host bacterium's total cellular protein. Such genetically engineered organisms are called **overproducers.**

Bacterial cells often sequester such large amounts of useless (to the bacterium) protein as insoluble and denatured **inclusion bodies** (Fig. 5-55). A protein extracted from such inclusion bodies must therefore be renatured, usually by dissolving it in a solution of **urea** or **guanidinium ion** (substances that induce proteins to denature)

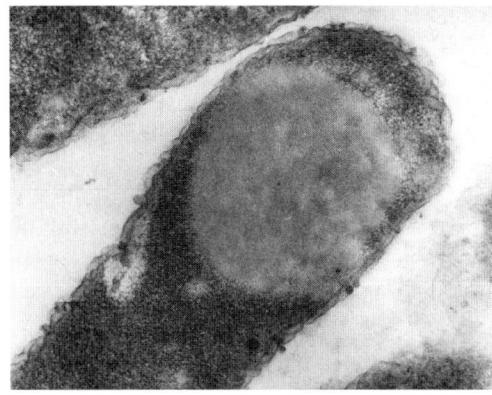

FIGURE 5-55   Electron micrograph of an inclusion body of the protein prochymosin in an *E. coli* cell. [Courtesy of Teruhiko Beppu, Nikon University, Japan.]

$$
\underset{\textbf{Urea}}{N_2H-\overset{\overset{\displaystyle O}{\|}}{C}-NH_2}
\qquad
\underset{\textbf{Guanidinium ion}}{N_2H-\overset{\overset{\displaystyle NH_2^+}{\|}}{C}-NH_2}
$$

and then slowly removing the denaturant via a membrane through which the denaturant but not the protein can pass [**dialysis** or **ultrafiltration** (Section 6-3B); protein denaturation and renaturation are discussed in Section 9-1A].

A strategy for avoiding the foregoing difficulty is to engineer the gene for the protein of interest so that is preceded with a bacterial **signal sequence** that directs the protein synthesizing machinery of gram-negative bacteria such as *E. coli* to secrete the protein to their **periplasmic space** (the compartment between their plasma membrane and cell wall; signal sequences are discussed in Section 12-4B). The signal sequence is then removed by a specific bacterial protease. Secreted proteins, which are relatively few in number, can be released into the medium by the osmotic disruption (Section 6-1B) of the bacterial outer membrane (Section 1-1B; the bacterial cell wall is porous), so their purification is greatly simplified relative to that of intracellular proteins.

Another problem encountered when producing a foreign protein is that the protein may be toxic to the host cell (e.g., producing a protease may destroy the cell's proteins), thus killing the bacterial culture before sufficient amounts of the protein can be generated. One way to cir-

cumvent this problem is to place the gene encoding the toxic protein under the control of an inducible promoter, for example, the *lac* promoter in a plasmid that also includes the gene for the *lac* repressor (Section 5-4A). Then, the binding of the *lac* repressor to the *lac* promoter will prevent the expression of the foreign protein in the same way that it prevents the expression of the *lac* operon genes (Fig. 5-25*a*). However, after the cells have grown to a high concentration, an inducer is added that releases the repressor from the promoter and permits the expression of the foreign protein (Fig. 5-25*b*). The cells are thereby killed but not before they have produced large amounts of the foreign protein. For the *lac* repressor, the inducer of choice is **isopropylthiogalactoside (IPTG;** Section 31-1A), a synthetic, nonmetabolizable analog of the *lac* repressor's natural inducer, allolactose.

A problem associated with inserting a DNA segment into a vector, as is indicated in Fig. 5-46, is that any pair of sticky ends that have been made by a given restriction enzyme can be ligated together. Consequently, the products of a ligation reaction will include tandemly (one-after-the-other) linked vectors, inserts, and their various combinations in both linear and circular arrangements. Moreover, in the case of expression systems, 50% of the structural genes that are inserted into circular expression vectors will be installed backward with respect to the expression vector's transcriptional and translational control sequences and hence will not be properly expressed. The efficiency of the ligation process can be greatly enhanced through the use of **directional cloning** (Fig. 5-56). In this process, two different restriction enzymes are employed to generate two different types of sticky ends on both the insert and the vector. In expression systems, these are arranged such that the structural gene can only be inserted into the expression vector in the correct orientation for expression.

a. Eukaryotic Proteins Can Be Produced in Bacteria and in Eukaryotic Cells

The synthesis of a eukaryotic protein in a prokaryotic host presents several problems not encountered with prokaryotic proteins:

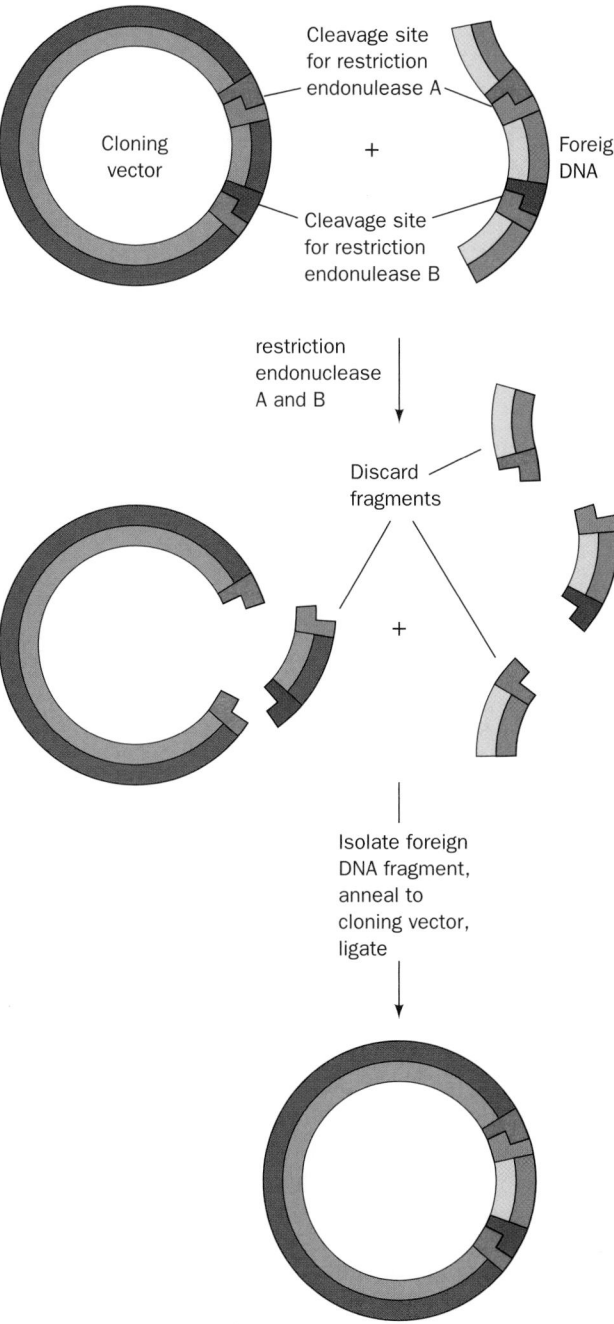

FIGURE 5-56 Construction of a recombinant DNA molecule by directional cloning. Two restriction enzymes, which yield different sticky ends, are used so that the foreign DNA fragment can only be inserted into the cloning vector in one orientation.

1. The eukaryotic control elements for RNA and protein synthesis are not recognized by bacterial hosts.

2. Bacteria lack the cellular machinery to excise the introns that are present in most eukaryotic transcripts, that is, bacteria cannot carry out gene splicing (Section 5-4A).

3. Bacteria lack the enzyme systems to carry out the specific post-translational processing that many eukaryotic proteins require for biological activity (Section 32-5). Most conspicuously, bacteria do not glycosylate proteins (al-

though, in many cases, glycosylation does not seem to affect protein function).

4. Eukaryotic proteins may be preferentially degraded by bacterial proteases (Section 32-6A).

The problem of nonrecognition of eukaryotic control elements can be eliminated by inserting the protein-encoding portion of a eukaryotic gene into a vector containing correctly placed bacterial control elements. The need to excise introns can be circumvented by cloning the cDNA of the protein's mature mRNA. Alternatively, genes encoding small proteins of known sequence can be chemically synthesized (Section 7-6A). Neither of these strategies is universally applicable, however, because few mRNAs are sufficiently abundant to be isolated and the genes encoding many eukaryotic proteins are much larger than can presently be reliably synthesized. Likewise, no general approach has been developed for the post-translational modification of eukaryotic proteins.

The preferential bacterial proteolysis of certain eukaryotic proteins may be prevented by inserting the eukaryotic gene after a bacterial gene such that both have the same reading frame. The resulting **hybrid** or **fusion protein** has an N-terminal polypeptide of bacterial origin that, in some cases, prevents bacterial proteases from recognizing the eukaryotic segment as being foreign. The purification of a fusion protein may be greatly facilitated by the specific binding properties of its N-terminal portion via a process known as **affinity chromatography** (Section 6-3C). Moreover, the formation of a fusion protein may render soluble its otherwise insoluble C-terminal portion. The two polypeptide segments can later be separated by treatment with a protease that specifically cleaves a susceptible site that had been engineered into the boundary between the segments (see below).

The development of cloning vectors that propagate in eukaryotic hosts, such as yeast or cultured animal cells, has led to the elimination of many of the above problems (although post-translational processing, and in particular glycosylation, may vary among different eukaryotes). Baculovirus-based vectors, which replicate in cultured insect cells, have been particularly successful in this regard. Moreover, **shuttle vectors** are available that can propagate in both yeast and *E. coli* and thus transfer (shuttle) genes between these two types of cells.

b. Recombinant Protein Production Has Important Practical Consequences

The ability to synthesize a given protein in large quantities has had an enormous medical, agricultural, and industrial impact. Those that are in routine clinical use include **human insulin** (a polypeptide hormone that regulates fuel metabolism and whose administration is required for survival by certain types of diabetics; Section 27-3B), human growth hormone (**somatotropin,** which induces the proliferation of muscle, bone, and cartilage and is used to stimulate growth in children of abnormally short stature; Section 19-1J), **erythropoietin** (a protein growth factor secreted by the kidney that stimulates the production of red blood cells and is used in the treatment of anemia arising from kidney disease), several types of **colony-stimulating factors** (which stimulate the production and activation of white blood cells and are used clinically to counter the white cell–killing effects of chemotherapy and to facilitate bone marrow transplantation), and **tissue-type plasminogen activator** (**t-PA,** which is used to promote the dissolution of the blood clots responsible for heart attacks and stroke; Section 35-1F). Synthetic vaccines consisting of harmless but immunogenic components of pathogens are eliminating the risks attendant in using killed or attenuated viruses or bacteria in vaccines as well as making possible new strategies of vaccine development. The use of recombinant **blood clotting factors** in treating individuals with the inherited disease **hemophilia** (in which these factors are defective; Section 35-1C) has replaced the need to extract these scarce proteins from large quantities of human blood and has thereby eliminated the high risk that hemophiliacs previously faced of contracting such blood-borne diseases as hepatitis and AIDS. Bovine somatotropin (**bST**) has long been known to stimulate milk production in dairy cows by ~15%. Its use has been made cost-effective, however, by the advent of recombinant DNA technology since bST could previously only be obtained in small quantities from cow pituitaries. Recombinant porcine somatotropin (**pST**), which is administered to growing pigs, induces ~15% greater growth on ~20% less feed while producing leaner meat.

c. Site-Directed and Cassette Mutagenesis Generate Proteins with Specific Sequence Changes

Of equal importance to protein production is the ability to tailor proteins to specific applications by altering their amino acid sequences at specific sites. This is frequently done via a method pioneered by Michael Smith known as **site-directed mutagenesis.** In this technique, an oligonucleotide containing a short gene segment with the desired altered base sequence corresponding to the new amino acid sequence (and synthesized by techniques discussed in Section 7-5A) is used as a primer in the DNA polymerase I–mediated replication of the gene of interest. Such a primer can be made to hybridize to the corresponding wild-type sequence if there are only a few mismatched base pairs, and its extension, by DNA polymerase I, yields the desired altered gene (Fig. 5-57). The altered gene can then be inserted in a suitable organism via techniques discussed in Section 5-5C and grown (cloned) in quantity. Similarly, PCR may be used as a vehicle for site-directed mutagenesis simply by using a mutagenized primer in amplifying a gene of interest so that the resulting DNA contains the altered sequence.

Using site-directed mutagenesis, the development of a variant form of the bacterial protease **subtilisin** (Section 15-3B) in which Met 222 has been changed to Ala (Met 222 → Ala) has permitted its use in laundry detergent that contains bleach (which largely inactivates wild-type subtilisin by oxidizing Met 222). **Monoclonal antibodies** (a single species of antibody produced by a clone of an antibody-producing cell; Sections 6-1D and 35-2B) can be targeted

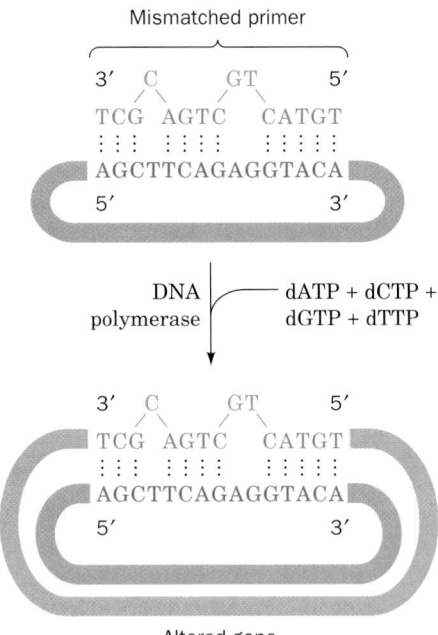

FIGURE 5-57   Site-directed mutagenesis. A chemically synthesized oligonucleotide incorporating the desired base changes is hybridized to the DNA encoding the gene to be altered (*green strand*). The mismatched primer is then extended by DNA polymerase I, thereby generating the mutated gene (*blue strand*). The mutated gene can subsequently be inserted into a suitable host organism so as to yield the mutant DNA, or its corresponding RNA, in quantity, produce a specifically altered protein, and/or generate a mutant organism. 🔎 **See the Animated Figures**

against specific proteins and hence, it is hoped, be used as antitumor agents. However, since monoclonal antibodies, as presently made, are mouse proteins, they are ineffective as therapeutic agents in humans because humans mount an immune response against mouse proteins. These difficulties may be rectified by "humanizing" monoclonal antibodies by replacing their mouse-specific sequences with those of humans (which the human immune system ignores) through site-directed mutagenesis.

In an alternative mutagenesis technique called **cassette mutagenesis,** complementary oligonucleotides containing the mutation(s) of interest are chemically synthesized (Section 7-6A) and annealed to create a duplex "cassette." The cassette is then ligated into the target gene, which must therefore contain an appropriately placed unique restriction site (which can be introduced through site-directed mutagenesis; the cassette must, of course, have the corresponding sticky ends) or, if the cassette is to replace an existing segment of the target gene, two possibly different restriction sites flanking the replaceable segment. Cassette mutagenesis is particularly useful for the insertion of short peptide sequences into a protein of interest (e.g., for the introduction of a proteolytic target site for the cleavage of a fusion protein), when a specific region of the protein is to be subjected to extensive and/or repeated mutagenesis, and for the generation of proteins containing all possible sequences in a short segment (by synthesizing a mixture of cassettes containing all possible variants of the corresponding codons; Section 7-6C).

We will see numerous instances throughout this textbook of protein function being mutagenically characterized through the replacement of a specific residue(s) or a polypeptide segment suspected of having an important mechanistic or structural role. Indeed, mutagenesis has become an indispensable tool in the practice of enzymology.

d. Reporter Genes Can Be Used to Monitor Transcriptional Activity

The rate at which a structural gene is expressed depends on its upstream control sequences. Consequently, the rate of expression of a gene can be monitored by replacing its protein-encoding portion with or fusing it in frame to a gene expressing a protein whose presence can be easily determined. An already familiar example of such a **reporter gene** is the *lacZ* gene in the presence of X-gal (Section 5-5C) because its level of expression is readily quantitated by the intensity of the blue color that is generated. Although numerous reporter genes have been developed, that which has recently gained the greatest use encodes **green fluorescent protein (GFP).** GFP, a product of the bioluminescent jellyfish *Aequorea victoria,* fluoresces with a peak wavelength of 508 nm (green light) when irradiated by UV or blue light (optimally 400 nm). This nontoxic protein is intrinsically fluorescent; it requires no substrate or small molecule cofactor to fluoresce as do other highly fluorescent proteins. Its presence can therefore be monitored through the use of UV light or a fluorometer, and its cellular location can be determined through fluorescence microscopy (Fig. 5-58). Consequently, when the GFP gene is placed under control of the gene expressing a particular protein (GFP's fluorescence is unaffected by the formation of a fusion protein), the protein's expressional activity can be readily determined. In fact, a number of GFP variants with distinct sets of excitation and emission wavelengths have been developed through genetic engineering, and hence the expressional activities of several different genes can be monitored simultaneously. Moreover, the development of pH-sensitive GFP variants permits the monitoring of the pH in subcellular compartments.

H. *Transgenic Organisms and Gene Therapy*

For many purposes it is preferable to tailor an intact organism rather than just a protein—true genetic engineering. Multicellular organisms expressing a foreign (from another organism) gene are said to be **transgenic** and their transplanted foreign genes are often referred to as **transgenes.** For the change to be permanent, that is, heritable, a transgene must be stably integrated into the organism's germ cells. For mice, this is accomplished by microinjecting (transfecting) DNA encoding the desired altered characteristics into a **pronucleus** of a fertilized ovum (Fig. 5-59; a fertilized ovum contains two pronuclei, one from the sperm and the other from the ovum, which eventually fuse to form the nucleus of the one-celled em-

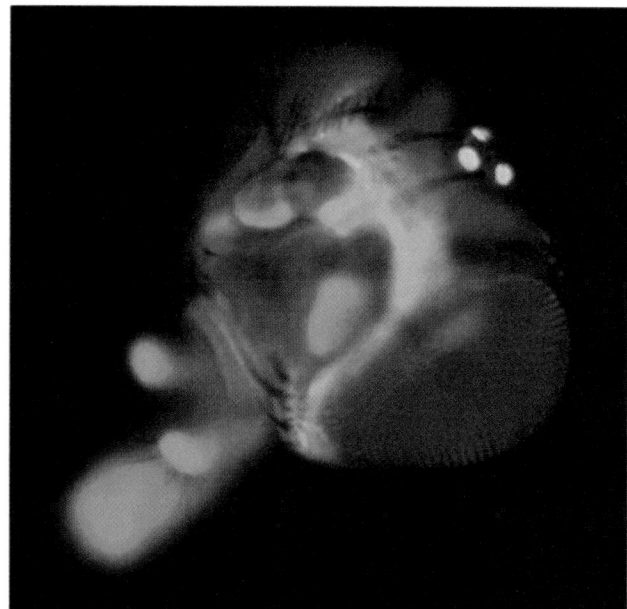

FIGURE 5-58   Use of green fluorescent protein (GFP) as a reporter gene. The gene for GFP was placed under the control of the *Drosophila per* gene promoter and transformed into *Drosophila*. The *per* gene encodes a so-called clock protein that is involved in controlling the fruit fly's circadian (daily) rhythm. The intensity of the green fluorescence of the isolated fly head seen here, which also occurs in other body parts, follows a daily oscillation that can be reset by light. These observations indicate that individual cells in *Drosophila* have photoreceptors and suggest that each of these cells possesses an independent clock. Evidently the head, which was previously thought to be the fly's master oscillator, does not coordinate all of its rhythms. [Courtesy of Steve A. Kay, The Scripps Research Institute, La Jolla, California.]

bryo) and implanting it into the uterus of a foster mother. The DNA integrates at a random position in the genome of the pronucleus through a poorly understood mechanism. Alternatively, an **embryonic stem cell** (an undifferentiated

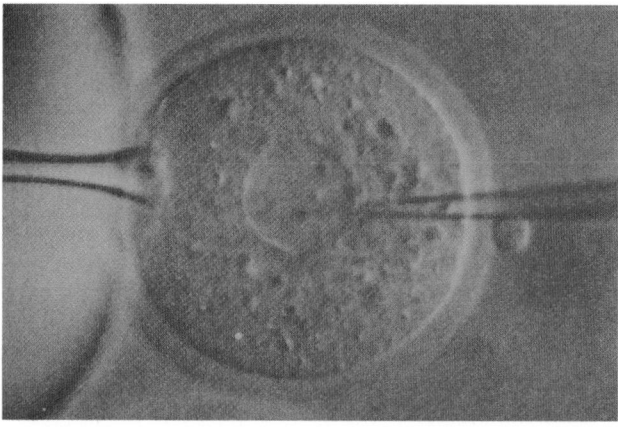

FIGURE 5-59   Microinjection of DNA into the pronucleus of a fertilized mouse ovum. The ovum is being held in place by gentle suction from the pipette on the left. [Science Vu/Visuals Unlimited.]

embryonic cell that can give rise, *in utero*, to an entire organism) may be transfected with an altered gene, which occasionally replaces its normal counterpart via recombination. A normal gene may, in this manner, be "knocked out" (permanently inactivated) by transfection with a defective version of the gene. With either method, mating heterozygotes for the altered gene yields progeny that are homozygotes for the altered gene. The use of transgenic mice, and in particular **knockout mice,** has greatly enhanced our understanding of vertebrate gene expression (Section 34-4C).

a. Transgenic Organisms Have Many Uses

Procedures are being developed to generate transgenic farm animals such as cows, goats, pigs, and sheep. Animals may thus be induced to grow larger on lesser amounts of feed and/or to be resistant to particular diseases, although this will require a greater understanding of the genes involved than is presently available. An intriguing application of transgenic farm animals is for them to secrete pharmaceutically useful proteins, such as human growth hormone and blood clotting factors, into their milk. Such a transgenic cow, it is expected, will yield several grams of a foreign protein per liter of milk (tens of kilograms per year), which can thereby be produced far more economically than it can by bacteria. A small herd of such "pharm animals" could satisfy the world's need for a particular medicinally useful protein.

The transplantation between humans of organs such as hearts, lungs, livers, and kidneys (a process known as **allotransplantation;** Greek: *allos,* other) has saved tens of thousands of lives over the past three decades. However, the demand for transplantable organs has so outstripped the supply that as little as 5% of the organs needed in the United States become available. This organ shortage could be entirely eliminated if organs from human-sized animals such as pigs could be transplanted into humans (a process known as **xenotransplantation;** Greek: *xenos,* strange or foreign). However, the xenotransplantation of a pig organ into a human results in the destruction of that organ in as little as a few minutes through a series of **complement system**–mediated reactions that are triggered by the foreign antigens lining the blood vessels of the xenograft (the complement system constitutes the body's first line of immunological defenses; Section 35-2F). This **hyperacute rejection** occurs because the porcine tissue lacks the human proteins that inhibit the human complement system. However, when the organs from pigs that were made transgenic in these human inhibitory proteins were transplanted into primates, the hyperacute rejection of these organs did not occur. Thus, although not all the problems of xenotransplantation have been eliminated, it now seems likely that genetic engineering techniques will eventually make xenotransplantation a practical alternative to allotransplantation.

Transgenic plants are becoming increasingly available, promising a significant extension of the "green revolution" that has changed the face of agriculture throughout the world over the past three decades. For example, during sporulation, various strains of the soil microbe *Bacillus thuringiensis* **(Bt)** express proteins that specifically bind to the intestinal cells of certain insects in a manner that lyses these cells, thereby killing the insect through starvation and infection. These so-called **δ-endotoxins** (also known as **crystal proteins** because Bt spores contain them in microcrystalline form) are innocuous to vertebrates and, hence, Bt spores have been used to control such pests as the **gypsy moth.** Unfortunately, Bt decays after a short time. However, the gene for a δ-endotoxin has been cloned into corn, where, for example, it confers protection against the **European corn borer** (a commercially significant pest that, for much of its life cycle, lives inside the corn plant, where it is largely inaccessible to chemical insecticides). The use of such **Bt corn,** which is now widely planted in the United States, has greatly reduced the need for chemical insecticides. δ-Endotoxin genes have likewise been successfully cloned into a variety of agriculturally significant plants including potatoes, soybeans, and cotton. Among the properties of crop plants that have been generated through genetic engineering are increased herbicide resistance (which permits the more selective use of herbicides to control weeds); resistance to viruses, bacteria, and fungi; control of ripening (to permit a crop to be brought to market at the optimum time); altered plant architecture such as height (which can improve crop productivity); increased tolerance to environmental stresses such as cold, heat, water, and salinity; and modified or additional vitamins, starch, proteins, and oils (for improved nutritional properties and the production of sustainable supplies of raw materials).

b. Gene Therapy Has Enormous Medical Potential

Gene therapy, the transfer of new genetic material to the cells of an individual resulting in therapeutic benefit to that individual, has been under clinical investigation since 1990, when W. French Anderson and Michael Blaese employed this technology with two children (Section 28-4A) in an effort to alleviate their **severe combined immunodeficiency disease (SCID;** any of several genetic diseases that so impair the immune system that a victim must be kept in a sterile environment in order to survive). Around 4000 genetic diseases are presently known and are thereby potential targets of gene therapy.

Although several types of gene-transfer protocols are currently under development for use in gene therapy, those which have been most widely employed utilize retroviral vectors. Retroviruses are RNA-containing viruses that, on entering the host cell, use virally encoded reverse transcriptase to transcribe the viral RNA to its complementary DNA, thereby forming an RNA–DNA hybrid helix. The enzyme then uses the newly synthesized DNA as a template to synthesize the complementary DNA while degrading the original RNA (Section 30-4C). The resulting duplex DNA is then integrated into a host chromosome (becomes part of that chromosome), a characteristic that makes retroviral vectors of great value in gene transfer.

The retroviral RNAs that are used in these procedures have been engineered so as to replace the genes encoding essential viral proteins with therapeutic genes. Hence, cells that have been infected by these "viruses" contain the therapeutic genes in their chromosomes but they lack the genetic information to replicate the virus.

There are three categories of gene therapy:

1. In the *ex vivo* (out of the body) approach, cells are removed from the body, incubated with a vector, and then returned to the body. This procedure is usually done with bone marrow cells, which are blood cell precursors.

2. In the *in situ* approach, the vector is applied directly to affected tissues. Such methods are being developed, for example, to treat tumors by injecting into the tumor a vector bearing the gene for a toxin or a gene that would make the tumor susceptible to a chemotherapeutic agent or to attack by the immune system; and to treat **cystic fibrosis,** by inhaling an aerosol containing a vector encoding the normal protein. (Cystic fibrosis, one of the most common genetic diseases, is caused by a defect in a protein involved in the secretion of chloride ion in the lungs and other tissues. This causes the secretion of abnormally thick mucus, which results in recurrent and often damaging lung infections leading to an early death.)

3. In the *in vivo* (in the body) approach, the vector would be injected directly into the bloodstream. There are, as yet, no clinical examples of this approach, although vectors to do so must ultimately be developed if gene therapy is to fulfill its promise.

Recently, in the first well-documented clinical success of gene therapy, Alain Fischer reported that two infants appear to have been cured, via the *ex vivo* treatment of their bone marrow cells, of a form of SCID called **SCID-X1** (which is caused by a mutation in the gene encoding the **γc cytokine receptor,** a receptor for certain protein growth factors, whose proper function is essential for the differentiation, growth, and survival of the white blood cells known as *T* **cells;** Section 35-2D). Although no other gene therapy protocol has, as yet, been fully successful in curing disease, steady progress has been made in our understanding of the requirements for the construction of effective vectors. It has therefore been predicted that, over the next 25 years, gene therapy will revolutionize the way that medicine is practiced.

I. *Social, Ethical, and Legal Considerations*

In the early 1970s, when strategies for genetic engineering were first being discussed, it was realized that little was known about the safety of the proposed experiments. Certainly it would be foolhardy to attempt experiments such as introducing the gene for **diphtheria toxin** (Section 32-3G) into *E. coli* so as to convert this human symbiont into a deadly pathogen. But what biological hazards would result, for example, from cloning tumor virus genes in *E. coli* (a useful technique for analyzing these viruses)?

Consequently, in 1975, molecular biologists declared a voluntary moratorium on molecular cloning experiments until these risks could be assessed. There ensued a spirited debate, at first among molecular biologists and later in the public arena, between two camps: those who thought that the enormous potential benefits of recombinant DNA research warranted its continuation once adequate safety precautions had been instituted, and those who felt that its potential dangers were so great that it should not be pursued under any circumstances.

The former viewpoint eventually prevailed with the promulgation, in 1976, of a set of U.S. government regulations for recombinant DNA research. Experiments that are obviously dangerous were forbidden. In other experiments, the escape of laboratory organisms was to be prevented by both physical and biological containment. By biological containment it is meant that vectors will only be cloned in host organisms with biological defects that prevent their survival outside the laboratory. For example, χ1776, the first approved "safe" strain of *E. coli,* has among its several defects the requirement for diaminopimelic acid, an intermediate in lysine biosynthesis (Section 26-5B), which is neither present in human intestines nor commonly available in the environment.

As experience with recombinant DNA research accumulated, it became evident that the foregoing reservations were largely groundless. No genetically altered organism yet reported has caused an unexpected health hazard. Indeed, recombinant DNA techniques have, in many cases, eliminated the health hazards of studying dangerous pathogens such as the virus causing **acquired immune deficiency syndrome (AIDS).** Consequently, since 1979, the regulations governing recombinant DNA research have been gradually relaxed.

There are other social, ethical, and legal considerations that will have to be faced as new genetic engineering techniques become available (Fig. 5-60). Recombinant erythropoietin is now routinely prescribed to treat the effects of certain types of kidney disease. However, should athletes be permitted to use this protein, as many reportedly have, to increase the number of red cells in their blood and hence its oxygen-carrying capacity (a dangerous procedure if uncontrolled since the increased number of cells in the blood can put a great strain on the heart)? Few would dispute the use of gene therapy, if it can be developed, to cure such devastating genetic defects as **sickle-cell anemia** (a painful and debilitating condition caused by deformed red blood cells that often results in early death; Section 10-3B) and **Tay−Sachs disease** (which is caused by the absence of the lysosomal enzyme **hexoseaminidase A** and results in progressive neuronal dysfunction that is invariably fatal by around age 3; Section 25-8C). If, however, it becomes possible to alter complex (i.e., multigene) traits such as athletic ability or intelligence, which changes would be considered desirable, under what circumstances would they be made, and who would decide whether to make them? Should gene therapy be used on individuals with inherited diseases only to correct defects in their somatic cells or

FIGURE 5-60   [Drawing by T.A. Bramley, *in* Andersen, K., Shanmugam, K.T., Lim, S.T., Csonka, L.N., Tait, R., Hennecke, H., Scott, D.B., Hom, S.S.M., Haury, J.F., Valentine, A., and Valentine, R.C., *Trends Biochem. Sci.* **5,** 35 (1980). Copyright © Elsevier Biomedical Press, 1980. Used by permission.]

should it also be used to alter genes in their germ cells, which could then be transmitted to succeeding generations? Animals such as sheep, cows, and mice have already been cloned. Should humans with particularly desirable traits, either naturally occurring or generated through genetic engineering, be cloned? When it becomes easy to determine an individual's genetic makeup, should this information be used, for example, in evaluating applications for educational and employment opportunities, or in assessing a person's eligibility for health insurance? Under present U.S. laws, newly sequenced human genes of unknown function may be patented. But to what extent will such proprietary rights impede the development of therapeutic techniques based on those genes? These conundrums have led to the advent of a branch of philosophy named **bioethics** designed to deal with them.

CHAPTER SUMMARY

1 ■ Nucleotides and Nucleic Acids   A nucleotide consists of either a ribose or a 2′-deoxyribose residue whose C1′ atom forms a glycosidic bond with a nitrogenous base and whose 3′ or 5′ position is esterified to a phosphate group. Nucleosides lack the phosphate groups of nucleotides. The nitrogenous bases in the great majority of nucleotides are the purines adenine and guanine and the pyrimidines cytosine and either thymine (in DNA) or uracil (in RNA). Nucleic acids are linear polymers of nucleotides containing either ribose residues in RNA or deoxyribose residues in DNA and whose

3′ and 5′ positions are linked by phosphate groups. In double-helical DNAs and RNAs, the base compositions obey Chargaff's rules: A = T(U) and G = C. RNA, but not DNA, is susceptible to base-catalyzed hydrolysis.

2 ■ DNA Is the Carrier of Genetic Information Extracts of virulent S-type pneumococci transform nonpathogenic R-type pneumococci to the S form. The transforming principle is DNA. Similarly, radioactive labeling has demonstrated that the genetically active substance of bacteriophage T2 is its DNA. The viral capsid serves only to protect its enclosed DNA and to inject it into the bacterial host. This establishes that DNA is the hereditary molecule.

3 ■ Double Helical DNA B-DNA consists of a right-handed double helix of antiparallel sugar–phosphate chains with ~10 bp per turn of 34 Å and with its bases nearly perpendicular to the helix axis. Bases on opposite strands hydrogen bond in a geometrically complementary manner to form A · T and G · C Watson–Crick base pairs. DNA replicates in a semiconservative manner, as has been demonstrated by the Meselson–Stahl experiment. When heated past its melting temperature, T_m, DNA denatures and undergoes strand separation. This process may be monitored by the hyperchromism of the DNA's UV spectrum. Denatured DNA can be renatured by maintaining it at ~25°C below its T_m. DNA occurs in nature as molecules of enormous lengths which, because they are also quite stiff, are easily mechanically cleaved by laboratory manipulations.

4 ■ Gene Expression and Replication: An Overview Genes are expressed according to the central dogma of molecular biology: DNA directs its own replication and its transcription to yield RNA, which, in turn, directs its translation to form proteins. RNA is synthesized from ribonucleoside triphosphates on DNA templates by RNA polymerase, a process in which the DNA template strand is read in its 3′ to 5′ direction and the RNA is synthesized in its 5′ to 3′ direction. The rate at which a particular gene is transcribed is governed by control sites, which, for mRNAs, are located upstream of the transcriptional initiation site and can be quite extensive, particularly in eukaryotes. Eukaryotic mRNAs often require substantial post-transcriptional modifications, including gene splicing (the excision of introns and the rejoining of their flanking exons), to become functional.

mRNAs direct the ribosomal synthesis of polypeptides. In this process, ribosomes facilitate the binding of the mRNA's codons to the anticodons of tRNAs bearing their cognate amino acids, and the ribosomes then catalyze the formation of peptide bonds between successive amino acids. The correspondence between codons and the amino acid carried by the tRNAs that bind to them is called the genetic code. Enzymes known as aminoacyl-tRNA synthetases covalently link their corresponding tRNAs to their cognate amino acids. The selection of the correct initiation site on the mRNA also sets the reading frame for the polypeptide being synthesized. Newly synthesized proteins often require post-translational modifications to be functional, including specific proteolytic cleavages and, in eukaryotes only, glycosylation. The lifetime of a protein in a cell varies from fractions of a minute to days or weeks.

DNA is synthesized from deoxynucleoside triphosphates by DNA polymerase, an enzyme that can only extend existing polynucleotides bound to the template DNA and hence requires a primer. In cells, primers are RNA, which are synthesized on DNA templates by an RNA polymerase. The replication of both strands of duplex DNA takes place at a replication fork. In *E. coli*, duplex DNA replication is carried out by two molecules of DNA polymerase III, one of which synthesizes the leading strand and the other of which synthesizes the lagging strand. The leading strand is synthesized continuously. However, since all DNA polymerases can only extend DNA in its 5′ to 3′ direction, the lagging strand template must loop around to be read in its 3′ to 5′ direction, which requires that the lagging strand be synthesized discontinuously. The RNA primers for the lagging strand are synthesized by primase, and once a lagging strand segment has been synthesized, its primer is replaced through the combined actions of DNA polymerase I's 5′→3′ exonuclease and DNA polymerase activities. The single-strand nicks between successive lagging strand segments are then sealed by DNA ligase. Both DNA polymerase I and DNA polymerase III also have 3′→5′ exonuclease activities that function to proofread the newly synthesized DNA for mispairing errors and excise the mispaired nucleotides.

5 ■ Molecular Cloning Molecular cloning techniques have revolutionized the practice of biochemistry. Defined DNA fragments are generated through the use of Type II restriction endonucleases (restriction enzymes), which cleave DNA at specific and usually palindromic sequences of four to six bases. Restriction maps provide easily located physical reference points on a DNA molecule. Restriction-fragment length polymorphisms (RFLPs) provide markers for identifying chromosomal differences and hence have become invaluable for such purposes as identity tests, establishing familial relationships, and diagnosing inherited diseases. A DNA fragment may be produced in large quantities by inserting it, using recombinant DNA techniques, into a suitable cloning vector. These may be genetically engineered plasmids, viruses, cosmids, yeast artificial chromosomes (YACs), or bacterial artificial chromosomes (BACs). The DNA to be cloned is usually obtained as a restriction fragment so that it can be specifically ligated into a corresponding restriction cut in the cloning vector. Gene splicing may also occur through the generation of complementary homopolymer tails on the DNA fragment and the cloning vector or through the use of synthetic palindromic linkers containing restriction sequences. The introduction of a recombinant cloning vector into a suitable host organism permits the foreign DNA segment to be produced in nearly unlimited quantities. Those cells that have been properly transformed by a vector can be chosen through the use of selectable markers and chromogenic substrates. Specific base sequences may be detected in DNA by Southern blotting and in RNA by the similar Northern blotting. A particular gene may be isolated through the screening of a genomic library of the organism producing the gene. The polymerase chain reaction (PCR) is a particularly fast and convenient method of identifying and obtaining specific sequences of DNA. Genetic engineering techniques may be used to produce otherwise scarce or specifically altered proteins in large quantities and to monitor gene expression through the use of reporter genes such as green fluorescent protein. They are also used to produce transgenic plants and animals and in gene therapy. The development of recombinant DNA techniques has generated numerous social, ethical, and legal issues whose resolution will govern how biotechnology is to be used.

REFERENCES

USEFUL WEBSITES

REBASE. The restriction enzyme database. http://rebase.neb.com

THE ROLE OF DNA

Avery, O.T., MacLeod, C.M., and McCarty, M., Studies on the chemical nature of the substance inducing transformation of pneumococcal types, *J. Exp. Med.* **79,** 137–158 (1944). [The milestone report identifying transforming principle as DNA.]

Hershey, A.D. and Chase, M., Independent functions of viral proteins and nucleic acid in growth of bacteriophage, *J. Gen. Physiol.* **36,** 39–56 (1952).

McCarty, M., *The Transforming Principle,* Norton (1985). [A chronicle of the discovery that genes are DNA.]

Palmiter, R.D., Brinster, R.L., Hammer, R.E., Trumbauer, M.E., Rosenfeld, M.G., Birmberg, N.C., and Evans, R.M., Dramatic growth of mice that develop from eggs microinjected with metallothionein–growth hormone fusion genes, *Nature* **300,** 611–615 (1982).

Stent, G.S., Prematurity and uniqueness in scientific discovery, *Sci. Am.* **227**(6): 84–93 (1972). [A fascinating philosophical discourse on what it means for discoveries such as Avery's to be "ahead of their time" and on the nature of creativity in science.]

STRUCTURE AND PROPERTIES OF B-DNA

Bloomfield, V.A., Crothers, D.M., and Tinoco, I., Jr., *Nucleic Acids. Structures, Properties, and Functions,* University Science Books (2000).

Crick, F., *What Mad Pursuit,* Basic Books (1988). [A scientific autobiography.]

Judson, H.F., *The Eighth Day of Creation,* Expanded edition, Part I, Cold Spring Harbor Laboratory Press (1996). [A fascinating narrative on the discovery of the DNA double helix.]

Meselson, M. and Stahl, F.W., The replication of DNA in Escherichia coli, *Proc. Natl. Acad. Sci.* **44,** 671–682 (1958). [The classic paper establishing the semiconservative nature of DNA replication.]

Saenger, W., *Principles of Nucleic Acid Structure,* Springer-Verlag (1984).

Sayre, A., *Rosalind Franklin and DNA,* Norton (1975) [A biography which argues that Rosalind Franklin, who died in 1958, deserves far more credit than is usually accorded her for the discovery of the structure of DNA.]; *and* Piper, A., Light on a dark lady, *Trends Biochem. Sci.* **23,** 151–154 (1998). [A biographical memoir on Rosalind Franklin.]

Schlenk, F., Early nucleic acid chemistry, *Trends Biochem. Sci.* **13,** 67–68 (1988).

Voet, D. and Rich, A., The crystal structures of purines, pyrimidines and their intermolecular structures, *Prog. Nucleic Acid Res. Mol. Biol.* **10,** 183–265 (1970).

Watson, J.D., *The Double Helix,* Atheneum (1968). [A provocative autobiographical account of the discovery of the DNA structure.]

Watson, J.D. and Crick, F.H.C., Molecular structure of nucleic acids, *Nature* **171,** 737–738 (1953); *and* Genetical implications of the structure of deoxyribonucleic acid, *Nature* **171,** 964–967 (1953). [The seminal papers that are widely held to mark the origin of modern molecular biology.]

Wing, R., Drew, H., Takano, T., Broka, C., Tanaka, S., Itakura, K., and Dickerson, R.E., Crystal structure analysis of a complete turn of B-DNA, *Nature* **287,** 755–758 (1980). [The first X-ray crystal structure of a segment of B-DNA, which largely confirmed the less reliable fiber diffraction-based structure of B-DNA proposed by Watson and Crick.]

Zimm, B.H., One chromosome: one DNA molecule, *Trends Biochem. Sci.* **24,** 121–123 (1999). [A scientific reminiscence on how it was established that chromosomes each contain only one piece of DNA.]

MOLECULAR CLONING

Anderson, W.F., Human gene therapy, *Nature* **392** April 30 Supp., 25–30 (1998).

Ausbel, F.M., Brent, R., Kingston, R.E., Moore, D.D., Seidman, J.G., Smith, J.A., and Struhl, K., *Short Protocols in Molecular Biology* (4th ed.), Wiley (1999).

Birren, B., Green, E.D., Klapholz, S., Myers, R.M., and Roskams, J. (Eds.), *Genome Analysis. A Laboratory Manual,* Cold Spring Harbor Laboratory Press (1997). [A four-volume set.]

Burden, D.W. and Whitney, D.B., *Biotechnology: Proteins to PCR,* Birkhäuser (1995).

Carey, P.R. (Ed.), *Protein Engineering and Design,* Part II: Production, Academic Press (1996). [Contains chapters on *E. coli* and yeast expression systems and on mutagenesis techniques.]

Cavazzana-Calvo, M., et al., Gene therapy of human severe combined immunodeficiency (SCID)-X1 disease, *Science* **288,** 669–672 (2000).

Coombs, G.S. and Corey, D.R., Site-directed mutagenesis and protein engineering, *in* Angeletti, R.H. (Ed.), *Proteins. Analysis and Design,* Chapter 4, Academic Press (1998).

Cooper, A. and Wayne, R., New uses for old DNA, *Curr. Opin. Biotech.* **9,** 49–53 (1998). [Reviews the successes and pitfalls of studying ancient DNAs.]

Cooper, D.K.C., Gollackner, B., and Sachs, D.H., Will the pig solve the transplantation backlog? *Annu. Rev. Med.* **53,** 133–147 (2002).

Dieffenbach, C.W. and Dveksler, G.S. (Eds.), *PCR Primer: A Laboratory Manual,* Cold Spring Harbor Laboratory Press (1995).

Erlich, H.A. and Arnheim, N., Genetic analysis using the polymerase chain reaction, *Annu. Rev. Genet.* **26,** 479–506 (1992).

Fersht, A. and Winter, G., Protein engineering, *Trends Biochem. Sci.* **17,** 292–294 (1992).

Foster, E.A., Jobling, M.A., Taylor, P.G., Donnelly, P., de Knijff, P., Mieremet, R., Zerjal, T., and Tyler-Smith, C., Jefferson fathered slave's last child, *Nature* **396,** 27–28 (1998).

Gadowski, P.J. and Henner, D. (Eds.), *Protein Overproduction in Heterologous Systems, Methods.* **4**(2) (1992).

Glick, B.R. and Pasternak, J.J., *Molecular Biotechnology. Principles and Applications of Recombinant DNA* (2nd ed.), ASM Press (1998).

Glover, D.M. and Hames, B.D. (Eds.), *DNA Cloning. A Practical Approach,* IRL Press (1995). [A four-volume set.]

Goeddel, D.V. (Ed.), *Gene Expression Technology, Methods Enzymol.* **185** (1990).

Howe, C., *Gene Cloning and Manipulation*, Cambridge University Press (1995).

Krings, M., Stone, A., Schmitz, R.W., Krainitzki, H., Stoneking, M., and Pääbo, S., Neandertal DNA sequences and the origin of modern humans; *and* Lindahl, T., Facts *and* artifacts of ancient DNA, *Cell* **90,** 19–30; *and* 1–3 (1997); *and* Ovchinnikov, I.V., Götherström, A., Romanova, G.P., Kharitonov, V.M., Lidén, K., and Goodwin, W., Molecular analysis of

Neandertal DNA from the northern Caucasus, *Nature* **404,** 490–493 (2000).

Mullis, K.B., The unusual origin of the polymerase chain reaction. *Sci. Am.* **262**(4): 56–65 (1990).

Nicholl, D.S.T., *An Introduction to Genetic Engineering* (2nd ed.) Cambridge University Press (2003).

Platt, J.L., New directions for organ transplantation, *Nature* **392** April 30 Supp., 11–17 (1998).

Rees, A.R., Sternberg, M.J.E., and Wetzel, R. (Eds.), *Protein Engineering. A Practical Approach*, IRL Press (1992).

Sambrook, J. and Russel, D.W., *Molecular Cloning* (3rd ed.), Cold Spring Harbor Laboratory (2001). [A three-volume "bible" of laboratory protocols with accompanying background explanations.]

Tsien, R.Y., The green fluorescent protein, *Annu. Rev. Biochem.* **67,** 509–544 (1998).

Verma, I.M. and Somia, N., Gene therapy—promises, problems, prospects, *Nature* **289,** 239–242 (1997).

Vreeland, R.H., Rosenzweig, W.D., and Powers, D.W., Isolation of a 250 million-year-old halotolerant bacterium from a primary salt crystal, *Nature* **407,** 897–900 (2000).

Watson, J.D., Gilman, M., Witkowski, J., and Zoller, M., *Recombinant DNA* (2nd ed.), Freeman (1992). [A detailed exposition of the methods, findings, and results of recombinant DNA technology and research.]

Wells, R.D., Klein, R.D., and Singleton, C.K., Type II restriction enzymes, *in* Boyer, P.D. (Ed.), *The Enzymes* (3rd ed.), Vol. 14, *pp.* 137–156, Academic Press (1981).

Wu, R., Grossman, L., and Moldave, K. (Eds.), *Recombinant DNA,* Parts A–I, *Methods Enzymol.* **68, 100, 101, 153–155,** and **216–218** (1979, 1983, 1987, 1992, and 1993).

■ PROBLEMS

1. The base sequence of one of the strands of a 20-bp duplex DNA is:

5′-GTACCGTTCGACGGTACATC-3′

What is the base sequence of its complementary strand?

2. Non-Watson–Crick base pairs are of biological importance. For example: (a) **Hypoxanthine** (6-oxopurine) is often one of the bases of the anticodon of tRNA. With what base on mRNA is hypoxanthine likely to pair? Draw the structure of this base pair. (b) The third position of the codon–anticodon interaction between tRNA and mRNA is often a G · U base pair. Draw a plausible structure for such a base pair. (c) Many species of tRNA contain a hydrogen bonded U · A · U assembly. Draw two plausible structures for this assembly in which each U forms at least two hydrogen bonds with the A. (d) Mutations may arise during DNA replication when mispairing occurs as a result of the transient formation of a rare tautomeric form of a base. Draw the structure of a base pair with proper Watson–Crick geometry that contains a rare tautomeric form of adenine. What base sequence change would be caused by such mispairing?

3. (a) What is the molecular mass and contour length of a segment of B-DNA that specifies a 40-kD protein? (b) How many helical turns does this DNA have and what is its axial ratio (length to width ratio)?

*4. The antiparallel orientation of complementary strands in duplex DNA was elegantly demonstrated in 1960 by Arthur Kornberg by **nearest-neighbor analysis.** In this technique, DNA is synthesized by DNA polymerase I from one deoxynucleoside triphosphate whose α-phosphate group is radioactively labeled with ^{32}P and three unlabeled deoxynucleoside triphosphates. The resulting product is treated with an enzyme, DNase I, that catalyzes the hydrolysis of the phosphodiester bonds on the 3′ sides of all deoxynucleotides.

$$ppp^*A + pppC + pppG + pppT$$

$$\downarrow \text{DNA polymerase} \quad PP_i \nwarrow$$

$$\cdots pCpTp^*ApCpCp^*ApGp^*Ap^*ApTp\cdots$$

$$\downarrow \text{DNase I} \quad H_2O \nwarrow$$

$$\cdots + Cp + Tp^* + Ap + Cp + Cp^* + Ap + Gp^* + Ap^* + Ap + Tp + \cdots$$

In this example, the relative frequencies of occurrence of ApA, CpA, GpA, and TpA in the DNA can be determined by measuring the relative amounts of Ap*, Cp*, Gp*, and Tp*, respectively, in the product (where p* represents a ^{32}P-labeled phosphate group). The relative frequencies with which the other 12 dinucleotides occur may likewise be determined by labeling, in turn, the other 3 nucleoside triphosphates in the above reactions. There are equivalencies between the amounts of certain pairs of dinucleotides. However, the identities of these equivalencies depend on whether the DNA consists of parallel or antiparallel strands. What are these equivalences in both cases?

5. What would be the effect of the following treatments on the melting curve of B-DNA dissolved in 0.5*M* NaCl solution? Explain. (a) Decreasing the NaCl concentration. (b) Squirting the DNA solution, at high pressure, through a very narrow orifice. (c) Bringing the solution to 0.1*M* adenine. (d) Heating the solution to 25°C above the DNA's melting point and then rapidly cooling it to 35°C below the DNA's melting point.

6. What is the mechanism of alkaline denaturation of duplex DNA? [Hint: Some of the bases are relatively acidic.]

7. The following duplex DNA is transcribed from right to left as is printed here.

5′-TCTGACTATTCAGCTCTCTGGCACATAGCA-3′
3′-AGACTGATAAGTCGAGAGACCGTGTATCGT-5′

(a) Identify the template strand. (b) What is the amino acid sequence of the polypeptide that this DNA sequence encodes? Assume that translation starts at the first initiation codon. (c) Why doesn't the UGA sequence in the mRNA transcript cause transcription to terminate?

8. After undergoing splicing, a mature mRNA has the following sequence, where the vertical line indicates the position of the splice junction (the nucleotides from between which an intron had been removed).

5′-CUAGAUGGUAG|

GUACGGUUAUGGGAUAACUCUG-3′

(a) What is the sequence of the polypeptide specified by this mRNA? Assume that translation starts at the first initiation codon. (b) What would the polypeptide sequence be if the splicing system had erroneously deleted the GU on the 3′ side of the splice junction? (c) What would the polypeptide sequence be if the splicing system had erroneously failed to excise a G at the splice junction? (d) Is there any relationship between the polypeptides specified in b and c and, if so, why?

9. Explain why the charging of a tRNA with the correct amino acid is equally as important for accurate translation as is the correct recognition of a codon by its corresponding aminoacyl-tRNA.

10. In DNA replication, the leading strand and lagging strand are so named because any particular portion of the lagging strand is always synthesized after the corresponding portion of the leading strand has been synthesized. Explain why this must be the case.

11. SV40 DNA is a circular molecule of 5243 bp that is 40% G + C. In the absence of sequence information, how many restriction cuts would *Taq*I, *Eco*RII, *Pst*I, and *Hae*II be expected to make, on average, in SV40 DNA? (Figure 5-40 indicates the number of restriction cuts that three of these enzymes actually make.)

12. Which of the restriction endonucleases listed in Table 5-4 produce blunt ends? Which sets of them are **isoschizomers** (enzymes that have the same recognition sequence but do not necessarily cleave at the same sites; Greek: *isos*, equal + *schizein*, to cut); which of them are **isocaudamers** (enzymes that produce identical sticky ends; Latin: *cauda*, tail)?

13. In investigating a newly discovered bacterial species that inhabits the sewers of Berkeley, you isolate a plasmid which you suspect carries genes that confer resistance to several antibiotics. To characterize this plasmid you decide to make its restriction map. The sizes of the plasmid's restriction fragments, as determined from their electrophoretic mobilities on agarose gels, are given in the following table. From the data, construct the restriction map of the plasmid.

Sizes of Restriction Fragments from a Plasmid DNA

Restriction Enzyme	Fragment Sizes (kb)
*Eco*RI	5.4
*Hind*III	2.1, 1.9, 1.4
*Sal*I	5.4
*Eco*RI + *Hind*III	2.1, 1.4, 1.3, 0.6
*Eco*RI + *Sal*I	3.2, 2.2
*Hind*III + *Sal*I	1.9, 1.4, 1.2, 0.9

14. The plasmid pBR322 contains the *amp*[R] and *tet*[R] genes, which respectively confer resistance to the antibiotics ampicillin and **tetracycline** (Section 32-3G). The *tet*[R] gene contains a cleavage site for the restriction enzyme *Sal*I, the only such site in the entire plasmid. Describe how one can select for *E. coli* that had been transformed by pBR322 that contains a foreign DNA insert in its *Sal*I site.

15. How many yeast DNA fragments of average length 5 kb must be cloned in order to be 90, 99, and 99.9% certain that a genomic library contains a particular segment? The yeast genome consists of 12,100 kb.

16. Many of the routine operations in genetic engineering are carried out using commercially available "kits." Genbux Inc., a prospective manufacturer of such kits, has asked your advice on the feasibility of supplying a kit of intact λ phage cloning vectors with the nonessential central section of their DNA already removed. Presumably a "gene jockey" could then grow the required amount of phage, isolate its DNA, and restriction cleave it without having to go to the effort of separating out the central section. What advice would you give the company?

17. Indicate the sequences of the two 10-residue primers that could be used to amplify the central 40-nucleotide region of the following 50-nucleotide single-stranded DNA by PCR.

5'-AGCTGGACCACTGATCATTGACTGCTAGCGTCA
GTCCTAGTAGACTGACG-3'

18. A protein segment of sequence -Phe-Cys-Gly-Val-Leu-His-Lys-Met-Glu-Thr- is encoded by the following DNA segment:

5'-TTGTGCGGAGTCCTACACAAGATGGAGACA-3'

Design an 18-base oligonucleotide that could be used to change the protein's Leu-His segment to Ile-Pro via site-directed mutagenesis.

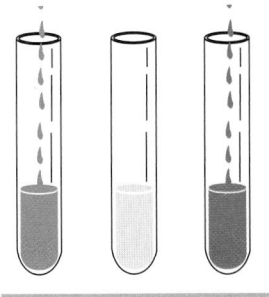

Chapter 6

Techniques of Protein and Nucleic Acid Purification

A major portion of most biochemical investigations involves the purification of the materials under consideration because these substances must be relatively free of contaminants if they are to be properly characterized. This is often a formidable task because a typical cell contains thousands of different substances, many of which closely resemble other cellular constituents in their physical and chemical properties. Furthermore, the material of interest may be unstable and exist in vanishingly small amounts. Typically, a substance that comprises $<0.1\%$ of a tissue's dry weight must be brought to $\sim 98\%$ purity. Purification problems of this magnitude would be considered unreasonably difficult by most synthetic chemists. It is therefore hardly surprising that our understanding of biochemical processes has by and large paralleled our ability to purify biological materials.

This chapter presents an overview of the most commonly used techniques for the isolation, the purification, and, to some extent, the characterization of proteins and nucleic acids, as well as other types of biological molecules. These methods are the basic tools of biochemistry whose operation dominates the day-to-day efforts of the practicing biochemist. Furthermore, many of these techniques are routinely used in clinical applications. Indeed, *a basic comprehension of the methods described here is necessary for an appreciation of the significance and the limitations of much of the information presented in this text*. This chapter should therefore be taken as reference material to be consulted repeatedly as the need arises while reading other chapters. Many of the techniques used for protein and nucleic acid fractionation are similar. Consequently, we shall first focus on how proteins are purified and only then concentrate on how these techniques are used in nucleic acid fractionation.

1 ■ PROTEIN ISOLATION

Proteins constitute a major fraction of the mass of all organisms. A particular protein, such as **hemoglobin** in red blood cells, may be the dominant substance present in a tissue. Alternatively, a protein such as the ***lac* repressor** of *E. coli* (Section 31-3B) may normally have a population of only a few molecules per cell. Similar techniques are used for the isolation and purification of both proteins, although, in general, the lower the initial concentration of a substance, the more effort is required to isolate it in pure form.

In this section we discuss the care and handling of proteins and outline the general strategy for their purification. For many proteins, the isolation and purification procedure is an exacting task requiring days of effort to obtain only a few milligrams or less of the desired product. However, as we shall see, modern analytical techniques have achieved such a high degree of sensitivity that this small amount of material is often sufficient to characterize a protein extensively. You should note that the techniques described in this chapter are applicable to the separations of most types of biological molecules.

A. *Selection of a Protein Source*

Proteins with identical functions generally occur in a variety of organisms. For example, most of the enzymes that mediate basic metabolic processes or that are involved in the expression and transmission of genetic information are common to all cellular life. Of course, there is usually considerable variation in the properties of a particular protein from various sources. In fact, different variants of a given protein may occur in different tissues from the same organism or even in different compartments in the same cell. Therefore, if flexibility of choice is possible, the isolation of a protein may be greatly simplified by a judicious choice of the protein source. This choice should be based on such criteria as the ease of obtaining sufficient quantities of the tissue from which the protein is to be isolated, the amount of the chosen protein in that tissue, and any properties peculiar to the specific protein chosen that would aid in its stabilization and isolation. Tissues from domesticated animals such as chickens, cows, pigs, or rats are often chosen. Alternative sources might be easily obtainable microorganisms such as *E. coli* or **baker's yeast (*Saccharomyces cerevisiae*).** We shall see, however, that proteins from a vast variety of organisms have been studied.

Molecular cloning methods (Section 5-5) have rapidly become equally if not more important protein production techniques. Almost any protein-encoding gene can be isolated from its parent organism, specifically altered (genetically engineered) if desired, and expressed at high levels (overproduced) in a conveniently grown organism such as *E. coli* or yeast, where it may constitute up to 30% of the overproducer's total cell protein. This high level of protein production generally renders the cloned protein far easier to isolate than it would be from its parent organism (in which it may normally occur in vanishingly small amounts).

B. *Methods of Solubilization*

The first step in the isolation of a protein, or any other biological molecule, is to get it into solution. In some cases, such as with blood serum proteins, nature has already done so. However, a protein must usually be liberated from the cells that contain it. The method of choice for this procedure depends on the mechanical characteristics of the source tissue as well as on the location of the required protein in the cell.

If the protein of interest is located in the cytosol of the cell, its liberation requires only the breaking open **(lysis)** of the cell. In the simplest and gentlest method of doing so, which is known as **osmotic lysis,** the cells are suspended in a **hypotonic solution;** that is, a solution in which the total molar concentration of solutes is less than that inside the cell in its normal physiological state. Under the influence of osmotic forces, water diffuses into the more concentrated intracellular solution, thereby causing the cells to swell and burst. This method works well with animal cells, but with cells that have a cell wall, such as bacteria or plant cells, it is usually ineffective. The use of an enzyme, such as **lysozyme,** which chemically degrades bacterial cell walls (Section 15-2), is sometimes effective with such cells. Detergents or organic solvents such as acetone or toluene are also useful in lysing cells, but care must be exercised in their use as they may denature the protein of interest (Section 8-4E).

Many cells require some sort of mechanical disruption process to break them open. This may include grinding with sand or alumina, or the use of a high-speed blender (similar to the familiar kitchen appliance), a **homogenizer** (an implement for crushing tissue between a closely fitting piston and sleeve), a **French press** (a device that shears open cells by squirting them at high pressure through a small orifice), or a **sonicator** (which breaks open cells through ultrasonic vibrations). Once the cells have been broken open, the crude **lysate** may be filtered or centrifuged to remove the particulate cell debris, thereby leaving the protein of interest in the supernatant solution.

If the required protein is a component of subcellular assemblies such as membranes or mitochondria, a considerable purification of the protein can be effected by first separating the subcellular assembly from the rest of the cellular material. This is usually accomplished by **differential centrifugation,** a process in which the cell lysate is centrifuged at a speed that removes only the cell components denser than the desired organelle followed by centrifugation at a speed that spins down the component of interest. The required protein is then usually separated from the purified subcellular component by extraction with concentrated salt solutions or, in the case of proteins tightly bound to membranes, with the use of detergent solutions or organic solvents, such as butanol, that solubilize lipids.

C. *Stabilization of Proteins*

Once a protein has been removed from its natural environment, it becomes exposed to many agents that can irreversibly damage it. These influences must be carefully controlled at all stages of a purification process or the yield of the desired protein may be greatly reduced or even eliminated.

The structural integrity of many proteins is sensitive to pH as a consequence of their numerous acid–base groups. To prevent damage to biological materials due to variations in pH, they are routinely dissolved in buffer solutions effective in the pH range over which the material is stable.

Proteins are easily denatured by high temperatures. Although the thermal stabilities of proteins vary widely, many of them slowly denature above 25°C. Therefore, the purification of proteins is normally carried out at temperatures near 0°C. However, there are numerous proteins that require lower temperatures, some even lower than −100°C, for stability. Conversely, some **cold-labile** proteins become unstable below characteristic temperatures.

The thermal stability characteristics of a protein can sometimes be used to advantage in its purification. A heat-stable protein in a crude mixture can be greatly purified by briefly heating the mixture so as to denature and precipitate most of the contaminating proteins without affecting the desired protein.

Cells contain **proteases** (enzymes that catalyze the hydrolytic cleavage of peptide bonds) and other degradative enzymes that, on cell lysis, are liberated into solution along with the protein of interest. Care must be taken that the protein is not damaged by these enzymes. Degradative enzymes may often be rendered inactive at pH's and temperatures that are not harmful to the protein of interest. Alternatively, these enzymes can often be specifically inhibited by chemical agents without affecting the desired protein. Of course, as the purification of a protein progresses, more and more of these degradative enzymes are eliminated.

Some proteins are more resistant than others to proteolytic degradation. The purification of a protein that is particularly resistant to proteases may be effected by maintaining conditions in a crude protein mixture under which the proteolytic enzymes present are active. This so-called **autolysis** technique simplifies the purification of the resistant protein because it is generally far easier to remove selectively the degradation products of contaminating proteins than it is the intact proteins.

Many proteins are denatured by contact with the air–water interface, and, at low concentrations, a significant fraction of the protein present may be lost by adsorption to surfaces. Hence, a protein solution should be handled so as to minimize frothing and should be kept relatively concentrated. There are, of course, other factors to which a protein may be sensitive, including the oxidation of cysteine residues to form disulfide bonds; heavy metal contaminants, which may irreversibly bind to the protein; and the salt concentration and polarity of the solution, which must be kept within the stability range of the protein. Finally, many microorganisms consider proteins to be delicious, so protein solutions should be stored under conditions that inhibit the growth of microorganisms [e.g., in a refrigerator and/or with small amounts of a toxic substance that does not react with proteins, such as **sodium azide** (NaN_3)].

D. *Assay of Proteins*

To purify any substance, some means must be found for quantitatively detecting its presence. A protein rarely comprises more than a few percent by weight of its tissue of origin and is usually present in much smaller amounts. Yet much of the material from which it is being extricated closely resembles the protein of interest. Accordingly, an assay must be specific for the protein being purified and highly sensitive to its presence. Furthermore, the assay must be convenient to use because it may be done repeatedly, often at every stage of the purification process.

Among the most straightforward of protein assays are those for enzymes that catalyze reactions with readily detectable products. Perhaps such a product has a characteristic spectroscopic absorption or fluorescence that can be monitored. Alternatively, the enzymatic reaction may consume or generate acid so that the enzyme can be assayed by acid–base titrations. If an enzymatic reaction product is not easily quantitated, its presence may still be revealed by further chemical treatment to yield a more readily observable product. Often, this takes the form of a **coupled enzymatic reaction,** in which the product of the enzyme being assayed is converted, by an added enzyme, to an observable substance.

Proteins that are not enzymes may be assayed through their ability to bind specific substances or through observation of their biological effects. For example, receptor proteins are often assayed by incubating them with a radioactive molecule that they specifically bind, passing the mixture through a protein-retaining filter, and then measuring the amount of radioactivity bound to the filter. The presence of a hormone may be revealed by its effect on some standard tissue sample or on a whole organism. The latter type of assays are usually rather lengthy procedures because the response elicited by the assay may take days to develop. In addition, their reproducibility is often less than satisfactory because of the complex behavior of living systems. Such assays are therefore used only when no alternative procedure is available.

a. Immunochemical Techniques Can Readily Detect Small Quantities of Specific Proteins

Immunochemical procedures provide protein assay techniques of high sensitivity and discrimination. These methods employ **antibodies,** proteins that are produced by an animal's immune system in response to the introduction of a foreign protein and that specifically bind to the foreign protein (antibodies and the immune system are discussed in Section 35-2).

Antibodies extracted from the blood serum of an animal that has been immunized against a particular protein are the products of many different antibody-producing cells. They therefore form a heterogeneous mixture of molecules, which vary in their exact specificities and binding affinities for their target protein. Antibody-producing cells normally die after a few cell divisions, so one of them cannot be cloned to produce a single species of antibody in useful quantities. Such **monoclonal antibodies** may be obtained, however, by fusing a cell producing the desired antibody with a cell of an immune system cancer known as a **myeloma** (Section 35-2B). The resulting **hybridoma** cell has an unlimited capacity to divide and, when raised in cell culture, produces large quantities of the monoclonal antibody.

A protein can be directly detected, or even isolated, through its precipitation by its corresponding antibodies. Alternatively, in a so-called **radioimmunoassay,** a protein can be indirectly detected by determining the degree with which it competes with a radioactively labeled standard for binding to the antibody (Section 19-1A). In an **enzyme-linked immunosorbent assay (ELISA;** Fig. 6-1):

1. An antibody against the protein of interest is immobilized on an inert solid such as polystyrene.

2. The solution being assayed for the protein is applied to the antibody-coated surface under conditions in which the antibody binds the protein. The unbound protein is then washed away.

3. The resulting protein–antibody complex is further reacted with a second protein-specific antibody to which an easily assayed enzyme has been covalently linked.

4. After washing away any unbound antibody-linked enzyme, the enzyme in the immobilized antibody–protein–antibody–enzyme complex is assayed, thereby indicating the amount of the protein present.

Both radioimmunoassays and ELISAs are widely used to detect small amounts of specific proteins and other biological substances in both laboratory and clinical applications. For example, a commonly available pregnancy test, which is reliably positive within a few days postconception, uses an ELISA to detect the placental hormone **chorionic gonadotropin** (Section 19-1I) in the mother's urine.

E. *General Strategy of Protein Purification*

The fact that proteins are well-defined substances was not widely accepted until after 1926, when James Sumner first crystallized an enzyme, jack bean **urease.** Before that, it was thought that the high molecular masses of proteins resulted from a colloidal aggregation of rather ill-defined and mysterious substances of lower molecular mass. Once it was realized that it was possible, in principle, to purify proteins, work to do so began in earnest.

In the first half of the twentieth century, the protein purification methods available were extremely crude by today's standards. Protein purification was an arduous task that was as much an art as a science. Usually, the development of a satisfactory purification procedure for a given protein was a matter of years of labor ultimately involving huge quantities of starting material. Nevertheless, by 1940, ~20 enzymes had been obtained in pure form.

Since then, tens of thousands of proteins have been purified and characterized to varying extents. Modern techniques of separation have such a high degree of discrimination that one can now obtain, in quantity, a series of proteins with such similar properties that only a few years ago their mixture was thought to be a pure substance. Nevertheless, the development of an efficient procedure for the purification of a given protein may still be an intellectually challenging and time-consuming task.

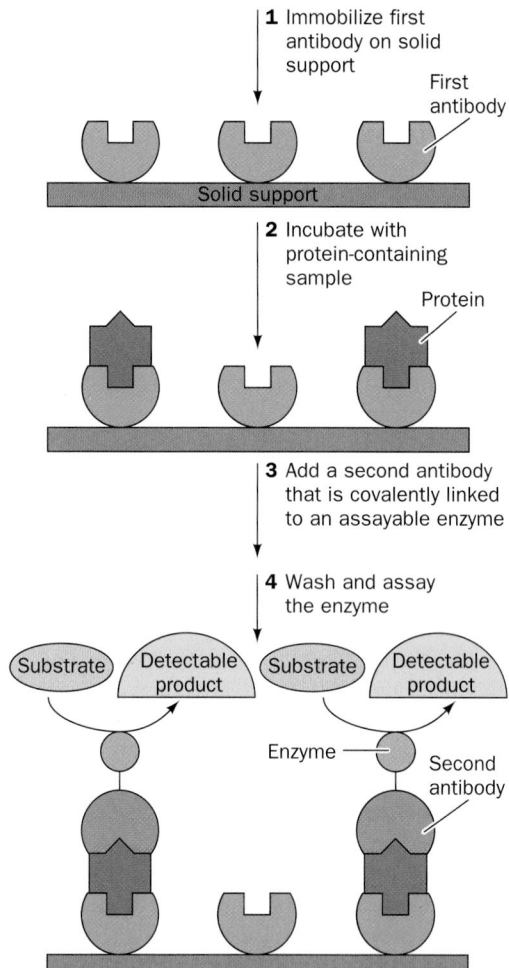

FIGURE 6-1   An enzyme-linked immunosorbent assay (ELISA). ⌖ See the Animated Figures

Proteins are purified by fractionation procedures. In a series of independent steps, the various physicochemical properties of the protein of interest are utilized to separate it progressively from other substances. The idea here is not necessarily to minimize the loss of the desired protein, but to eliminate selectively the other components of the mixture so that only the required substance remains.

It may not be philosophically possible to prove that a substance is pure. However, *the operational criterion for establishing purity takes the form of the method of exhaustion: the demonstration, by all available methods, that the sample of interest consists of only one component.* Therefore, as new separation techniques are devised, standards of purity may have to be revised. Experience has shown that when a sample of material previously thought to be a pure substance is subjected to a new separation technique, it occasionally proves to be a mixture of several components.

The characteristics of proteins and other biomolecules that are utilized in the various separation procedures are solubility, ionic charge, polarity, molecular size, and binding specificity for other biological molecules. Some of the procedures we shall discuss and the protein characteristics they depend on are as follows:

Characteristic	Procedure
Solubility:	1. Salting in
	2. Salting out
Ionic Charge:	1. Ion exchange chromatography
	2. Electrophoresis
	3. Isoelectric focusing
Polarity:	1. Adsorption chromatography
	2. Paper chromatography
	3. Reverse-phase chromatography
	4. Hydrophobic interaction chromatography
Molecular Size:	1. Dialysis and ultrafiltration
	2. Gel electrophoresis
	3. Gel filtration chromatography
	4. Ultracentrifugation
Binding Specificity:	1. Affinity chromatography

In the remainder of this chapter, we discuss these separation procedures.

2 ■ SOLUBILITIES OF PROTEINS

A protein's multiple acid–base groups make its solubility properties dependent on the concentrations of dissolved salts, the polarity of the solvent, the pH, and the temperature. Different proteins vary greatly in their solubilities under a given set of conditions: Certain proteins precipitate from solution under conditions in which others remain quite soluble. This effect is routinely used as a basis for protein purification.

A. *Effects of Salt Concentrations*

The solubility of a protein in aqueous solution is a sensitive function of the concentrations of dissolved salts (Figs. 6-2 through 6-4). The salt concentration in Figs. 6-2 and 6-3 is expressed in terms of the **ionic strength,** *I,* which is defined

$$I = \frac{1}{2} \sum c_i Z_i^2 \qquad [6.1]$$

where c_i is the molar concentration of the *i*th ionic species and Z_i is its ionic charge. The use of this parameter to account for the effects of ionic charges results from theoretical considerations of ionic solutions. However, as Fig. 6-3 indicates, a protein's solubility at a given ionic strength varies with the types of ions in solution. The order of effectiveness of these various ions in influencing protein solubility is quite similar for different proteins and is apparently mainly due to the ions' size and hydration.

The solubility of a protein at low ionic strength generally increases with the salt concentration (left side of Fig. 6-3 and the different curves of Fig. 6-4). The explanation of this **salting in** phenomenon is that as the salt concentration of the protein solution increases, the additional counter-

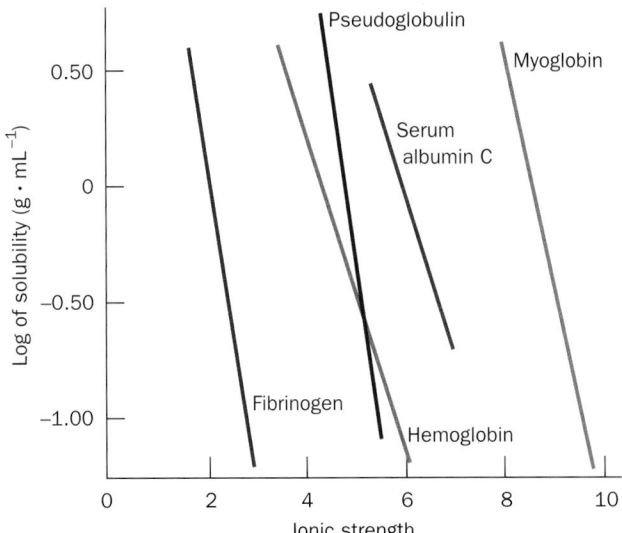

FIGURE 6-2   Solubilities of several proteins in ammonium sulfate solutions. [After Cohn, E.J. and Edsall, J.T., *Proteins, Amino Acids and Peptides*, p. 602, Academic Press (1943).]

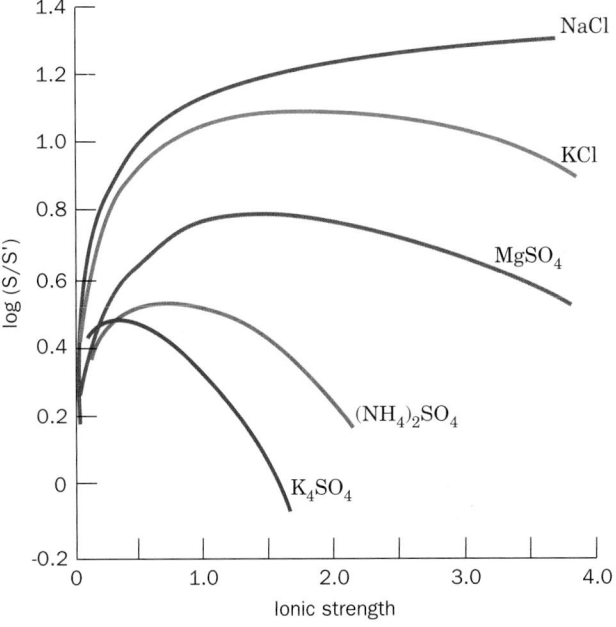

FIGURE 6-3   Solubility of carboxy-hemoglobin at its isoelectric point as a function of ionic strength and ion type. Here *S* and *S′* are, respectively, the solubilities of the protein in the salt solution and in pure water. The logarithm of their ratios is plotted so that the solubility curves can be placed on a common scale. [After Green, A.A., *J. Biol. Chem.* **95,** 47 (1932).]

ions more effectively shield the protein molecules' multiple ionic charges and thereby increase the protein's solubility.

At high ionic strengths, the solubilities of proteins, as well as those of most other substances, decrease. This effect, known as **salting out,** is primarily a result of the

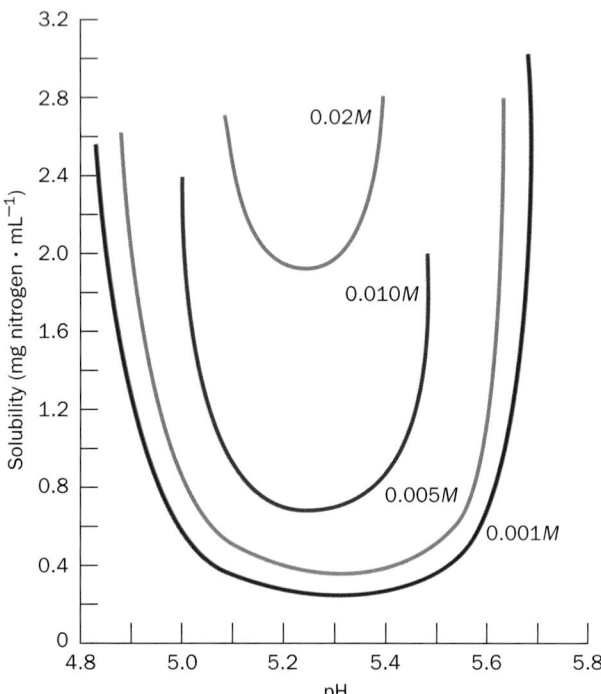

FIGURE 6-4   Solubility of β-lactoglobulin as a function of pH at several NaCl concentrations. [After Fox, S. and Foster, J.S., *Introduction to Protein Chemistry, p.* 242, Wiley (1975).]

competition between the added salt ions and the other dissolved solutes for molecules of solvation. At high salt concentrations, so many of the added ions are solvated that the amount of bulk solvent available becomes insufficient to dissolve other solutes. In thermodynamic terms, the solvent's activity (effective concentration; appendix to Chapter 3) is decreased. Hence, solute–solute interactions become stronger than solute–solvent interactions and the solute precipitates.

Salting out is the basis of one of the most commonly used protein purification procedures. Figure 6-2 shows that the solubilities of different proteins vary widely as a function of salt concentration. For example, at an ionic strength of 3, fibrinogen is much less soluble than the other proteins in Fig. 6-2. *By adjusting the salt concentration in a solution containing a mixture of proteins to just below the precipitation point of the protein to be purified, many unwanted proteins can be eliminated from the solution. Then, after the precipitate is removed by filtration or centrifugation, the salt concentration of the remaining solution is increased so as to precipitate the desired protein.* In this manner, a significant purification and concentration of large quantities of protein can be conveniently effected. Consequently, salting out is often the initial step in protein purification procedures. Ammonium sulfate is the most commonly used reagent for salting out proteins because its high solubility 3.9*M* in water at 0°C) permits the achievement of solutions with high ionic strengths (up to 23.4 in water at 0°C). Certain ions, notably I⁻, ClO₄⁻, SCN⁻, Li⁺, Mg²⁺, Ca²⁺, Ba²⁺, increase the solubilities of proteins rather than

salting them out. These ions also tend to denature proteins (Section 8-4E). Conversely, ions that decrease the solubilities of proteins stabilize their native structures, so that proteins which have been salted out are not denatured.

B. *Effects of Organic Solvents*

Water-miscible organic solvents, such as acetone and ethanol, are generally good protein precipitants because their low dielectric constants lower the solvating power of their aqueous solutions for dissolved ions such as proteins. The different solubilities of proteins in these mixed solvents form the basis of a useful fractionation technique. This procedure is normally used near 0°C or less because, at higher temperatures, organic solvents tend to denature proteins. The lowering of the dielectric constant by organic solvents also magnifies the differences in the salting out behavior of proteins, so that these two techniques can be effectively combined. Some water-miscible organic solvents, however, such as dimethyl sulfoxide (DMSO) or *N,N*-dimethylformamide (DMF), are rather good protein solvents because of their relatively high dielectric constants.

C. *Effects of pH*

Proteins generally bear numerous ionizable groups that have a variety of p*K*'s. At a pH characteristic for each protein, the positive charges on the molecule exactly balance its negative charges. At this pH, the protein's isoelectric point, p*I* (Section 4-1D), the protein molecule carries no net charge and is therefore immobile in an electric field.

Figure 6-4 indicates that the solubility of the protein β-lactoglobulin is a minimum near its p*I* of 5.2 in dilute NaCl solutions and increases more or less symmetrically about the p*I* with changes in pH. This solubility behavior, which is shared by most proteins, is easily explained. Physicochemical considerations suggest that the solubility properties of uncharged molecules are insensitive to the salt concentration. To a first approximation, therefore, a protein at its isoelectric point should not be subject to salting in. Conversely, as the pH is varied from a protein's p*I*, that is, as the protein's net charge increases, it should be increasingly subject to salting in because the electrostatic interactions between neighboring molecules that promote aggregation and precipitation should likewise increase. Hence, *in solutions of moderate salt concentrations, the solubility of a protein as a function of pH is expected to be at a minimum at the protein's pI and to increase about this point with respect to pH.*

Proteins vary in their amino acid compositions and therefore, as Table 6-1 indicates, in their p*I*'s. This phenomenon is the basis of a protein purification procedure known as **isoelectric precipitation** in which the pH of a protein mixture is adjusted to the p*I* of the protein to be isolated so as to selectively minimize its solubility. In practice, this technique is combined with salting out so that the protein being purified is usually salted out near its p*I*.

TABLE 6-1 Isoelectric Points of Several Common Proteins

Protein	Isoelectric pH
Pepsin	<1.0
Ovalbumin (hen)	4.6
Serum albumin (human)	4.9
Tropomyosin	5.1
Insulin (bovine)	5.4
Fibrinogen (human)	5.8
γ-Globulin (human)	6.6
Collagen	6.6
Myoglobin (horse)	7.0
Hemoglobin (human)	7.1
Ribonuclease A (bovine)	7.8
Cytochrome *c* (horse)	10.6
Histone (bovine)	10.8
Lysozyme (hen)	11.0
Salmine (salmon)	12.1

D. *Crystallization*

Once a protein has been brought to a reasonable state of purity, it may be possible to crystallize it. This is usually done by bringing the protein solution just past its saturation point with the types of precipitating agents discussed above. On standing for a time (as little as a few minutes, as much as several months), often while the concentration of the precipitating agent is being slowly increased, the pro-

(a)   (b)   (c)

(d)   (e)   (f)

FIGURE 6-5 Protein crystals. (*a*) Azurin from *Pseudomonas aeruginosa,* (*b*) flavodoxin from *Desulfovibrio vulgaris,* (*c*) rubredoxin from *Clostridium pasteurianum,* (*d*) azidomet myohemerythrin from the marine worm *Siphonosoma funafuti,* (*e*) lamprey hemoglobin, and (*f*) bacteriochlorophyll *a* protein from *Prosthecochloris aestuarii.* These proteins are colored because of their associated chromophores (light-absorbing groups); proteins are colorless in the absence of such bound groups. [Parts *a–c* courtesy of Larry Sieker, University of Washington; Parts *d* and *e* courtesy of Wayne Hendrickson, Columbia University; and Part *f* courtesy of John Olsen, Brookhaven National Laboratories, and Brian Matthews, University of Oregon.]

tein may precipitate from the solution in crystalline form. It may be necessary to attempt the crystallization under different solution conditions and with various precipitating agents before crystals are obtained. The crystals may range in size from microscopic to 1 mm or more across. Crystals of the latter size, which generally require great care to grow, may be suitable for X-ray crystallographic analysis (Section 8-3A). Several such crystals are shown in Fig. 6-5.

3 ■ CHROMATOGRAPHIC SEPARATIONS

In 1903, the Russian botanist Mikhail Tswett described the separations of plant leaf pigments in solution through the use of solid adsorbents. He named this process **chromatography** (Greek: *chroma,* color + *graphein,* to write), presumably because of the colored bands that formed in the adsorbents as the components of the pigment mixtures separated from one another (and possibly because Tswett means "color" in Russian).

Modern separation methods rely heavily on chromatographic procedures. In all of them, a mixture of substances to be fractionated is dissolved in a liquid or gaseous fluid known as the **mobile phase.** The resultant solution is percolated through a column consisting of a porous solid matrix known as the **stationary phase,** which in certain types of chromatography may be associated with a bound liquid. The interactions of the individual solutes with the stationary phase act to retard their progress through the matrix in a manner that varies with the properties of each solute. If the mixture being fractionated starts its journey through the column in a narrow band, the different retarding forces on each component that cause them to migrate at different rates will eventually cause the mixture to separate into bands of pure substances.

The power of chromatography derives from the continuous nature of the separation processes. A single purification step (or "theoretical plate" as it is often termed in analogy with distillation processes) may have very little tendency to separate a mixture into its components. However, since this process is applied in a continuous fashion so that it is, in effect, repeated hundreds or even hundreds of thousands of times, the segregation of the mixture into its components ultimately occurs. The separated components can then be collected into separate fractions for analysis and/or further fractionation.

The various chromatographic methods are classified according to their mobile and stationary phases. For example, in gas–liquid chromatography the mobile and stationary phases are gaseous and liquid, respectively, whereas in liquid–liquid chromatography they are immiscible liquids, one of which is bound to an inert solid support. Chromatographic methods may be further classified according to the nature of the dominant interaction between the stationary phase and the substances being separated. For example, if the retarding force is ionic in character, the separation technique is referred to as **ion exchange chromatography,** whereas if it is a result of the adsorption of the solutes onto a solid stationary phase, it is known as **adsorption chromatography.**

As has been previously mentioned, a cell contains huge numbers of different components, many of which closely resemble one another in their various properties. Therefore, the isolation procedures for most biological substances incorporate a number of independent chromatographic steps in order to purify the substance of interest according to several criteria. In this section, the most commonly used of these chromatographic procedures are described.

A. Ion Exchange Chromatography

*In the process of **ion exchange,** ions that are electrostatically bound to an insoluble and chemically inert matrix are reversibly replaced by ions in solution.*

$$R^+A^- + B^- \rightleftharpoons R^+B^- + A^-$$

Here, R^+A^- is an **anion exchanger** in the A^- form and B^- represents anions in solution. **Cation exchangers** similarly bear negatively charged groups that reversibly bind cations.

Polyanions and polycations therefore bind to anion and cation exchangers, respectively. However, proteins and other **polyelectrolytes** (polyionic polymers) that bear both positive and negative charges can bind to both cation and anion exchangers depending on their net charge. *The affinity with which a particular polyelectrolyte binds to a given ion exchanger depends on the identities and concentrations of the other ions in solution because of the competition among these various ions for the binding sites on the ion exchanger. The binding affinities of polyelectrolytes bearing acid–base groups are also highly pH dependent because of the variation of their net charges with pH.* These principles are used to great advantage in isolating biological molecules by **ion exchange chromatography** (Fig. 6-6), as described below.

In purifying a given protein (or some other polyelectrolyte), the pH and the salt concentration of the buffer solution in which the protein is dissolved are chosen so that the desired protein is strongly bound to the selected ion exchanger. A small volume of the impure protein solution is applied to the top of a column in which the ion exchanger has been packed, and the column is washed with this buffer solution.

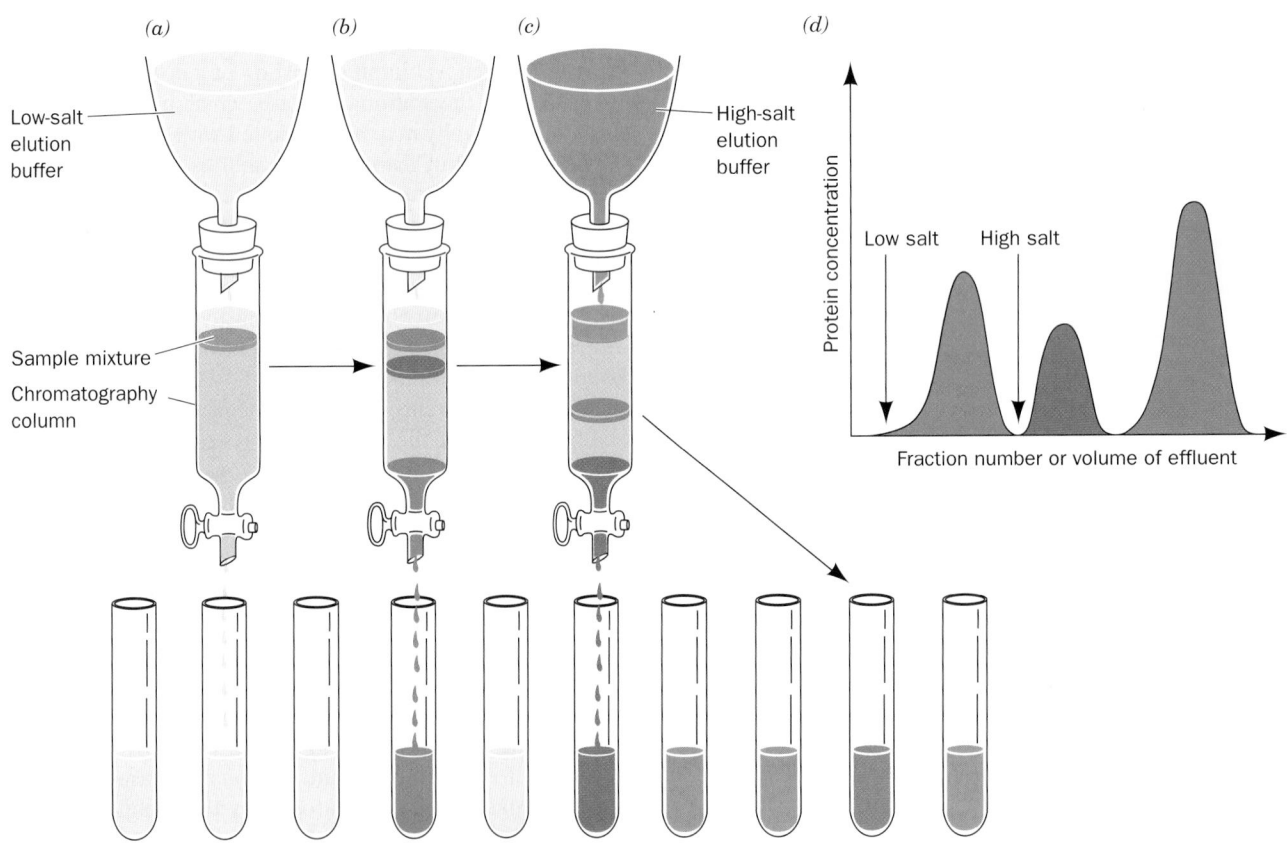

Fractions sequentially collected

FIGURE 6-6 Ion exchange chromatography using stepwise elution. Here the tan region of the column represents the ion exchanger and the colored bands represent the various proteins. (*a*) The protein mixture is bound to the topmost portion of the ion exchanger in the chromatography column. (*b*) As the elution progresses, the various proteins separate into discrete bands as a consequence of their different affinities for the ion exchanger under the prevailing solution conditions.

Here the first band of protein (*red*) has passed through the column and is being isolated as a separate fraction, whereas the other, less mobile, bands remain near the top of the column. (*c*) The salt concentration in the elution buffer is increased to increase the mobility of and thus elute the remaining bands. (*d*) The elution diagram of the protein mixture from the column. ✎ See the Animated Figures

Various proteins bind to the ion exchanger with different affinities. As the column is washed with the buffer, a process known as **elution,** *those proteins with relatively low affinities for the ion exchanger move through the column faster than the proteins that bind to the ion exchanger with higher affinities.* This occurs because the progress of a given protein through the column is retarded relative to that of the solvent due to interactions between the protein molecules and the ion exchanger.

The greater the binding affinity of a protein for the ion exchanger, the more it will be retarded. Thus, proteins that bind tightly to the ion exchanger can be eluted by changing the elution buffer to one with a higher salt concentration (and/or a different pH), a process called **stepwise elution.**

With the use of a fraction collector, purification of a substance can be effected by selecting only those fractions of the column effluent that contain it. Chromatographically separated materials may be detected in a variety of ways. The contents of the column effluent may be directly monitored through column-mounted detectors according to its UV absorbance at a specific wavelength [often 280 nm for proteins (because the aromatic side chains of His, Phe, Trp, and Tyr have strong absorbances at this wavelength) and

260 nm for nucleic acids (their absorption maximum; Fig. 5-15)], its fluorescence, its radioactivity, its refractive index, its pH, or its electrical conductivity. These properties may also be measured for the individual column fractions after the chromatographic run has been completed. In addition, biomolecules may be detected through their enzymatic and biological activities, as is discussed in Section 6-1D.

a. Gradient Elution Improves Chromatographic Separations

The purification process can be further improved by washing the protein-loaded column using the method of **gradient elution.** Here the salt concentration and/or pH is continuously varied as the column is eluted so as to release sequentially the various proteins that are bound to the ion exchanger. This procedure generally leads to a better separation of proteins than does elution of the column by a single solution or stepwise elution.

Many different types of elution gradients have been successfully employed in purifying biological molecules. The most widely used of these is the **linear gradient,** in which the concentration of the eluant solution varies linearly with the volume of solution passed. A simple device for generating such a gradient is illustrated in Fig. 6-7. Here the solute concentration, c, in the solution being withdrawn from the mixing chamber, is expressed by

$$c = c_2 - (c_2 - c_1)f \qquad [6.2]$$

where c_1 is the solution's initial concentration in the mixing chamber, c_2 is its concentration in the reservoir chamber, and f is the remaining fraction of the combined volumes of the solutions initially present in both reservoirs. Linear gradients of increasing salt concentration are probably more commonly used than all other means of column elution. However, gradients of different shapes can be generated by using two or more chambers of different cross-sectional areas or programmed mixing devices.

b. Several Types of Ion Exchangers Are Available

Ion exchangers consist of charged groups covalently attached to a support matrix. The chemical nature of the charged groups determines the types of ions that bind to the ion exchanger and the strength with which they bind. The chemical and mechanical properties of the support matrix govern the flow characteristics, ion accessibility, and stability of the ion exchanger.

Several classes of materials, colloquially referred to as **resins,** are in general use as support matrices for ion exchangers in protein purification, including cellulose (Fig. 6-8), polystyrene, agarose gels, and cross-linked dextran

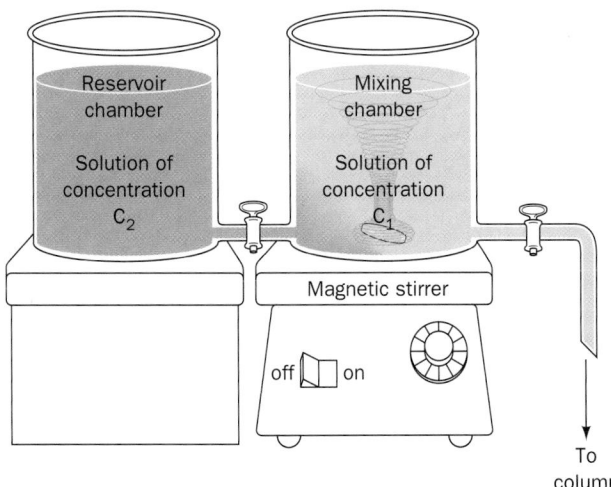

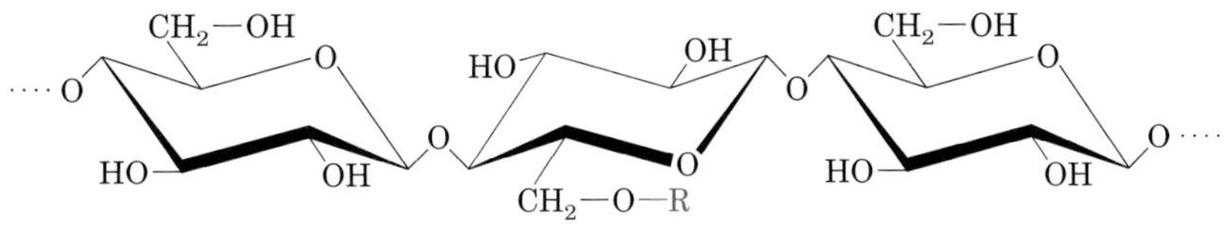

FIGURE 6-7 Device for generating a linear concentration gradient. Two connected open chambers, which have identical cross-sectional areas, are initially filled with equal volumes of solutions of different concentrations. As the solution of concentration c_1 drains out of the mixing chamber, it is partially replaced by a solution of concentration c_2 from the reservoir chamber. The concentration of the solution in the mixing chamber varies linearly from its initial concentration, c_1, to the final concentration, c_2, as is expressed by Eq. [6.2]

FIGURE 6-8 Molecular formulas of cellulose-based ion exchangers.

DEAE: $R = -CH_2-CH_2-\overset{+}{N}H(CH_2CH_3)_2$
CM: $R = -CH_2-COO^-$

TABLE 6-2   Some Biochemically Useful Ion Exchangers

Name[a]	Type	Ionizable group	Remarks
DEAE-cellulose	Weakly basic	Diethylaminoethyl —$CH_2CH_2N(C_2H_5)_2$	Used to separate acidic and neutral proteins
CM-cellulose	Weakly acidic	Carboxymethyl —CH_2COOH	Used to separate basic and neutral proteins
P-cellulose	Strongly and weakly acidic	Phosphate —OPO_3H_2	Dibasic; binds basic proteins strongly
Bio-Rex 70	Weakly acidic, polystyrene-based	Carboxylic acid —$COOH$	Used to separate basic proteins and amines
DEAE-Sephadex	Weakly basic cross-linked dextran gel	Diethylaminoethyl —$CH_2CH_2N(C_2H_5)_2$	Combined chromatography and gel filtration of acidic and neutral proteins
SP-Sepharose	Strongly acidic cross-linked agarose gel	Methyl sulfonate —CH_2SO_3H	Combined chromatography and gel filtration of basic proteins
CM Bio-Gel A	Weakly acidic cross-linked agarose gel	Carboxymethyl —CH_2COOH	Combined chromatography and gel filtration of basic and neutral proteins

[a]Sephadex and Sepharose gels are manufactured by Amersham Biosciences, Piscataway, New Jersey; Bio-Rex resins and Bio-Gels are manufactured by BioRad Laboratories, Hercules, California.

gels (see Section 6-3B). Table 6-2 contains descriptions of some commercially available ion exchangers in common use.

Cellulosic ion exchangers are among the materials most commonly employed to separate biological molecules. The cellulose, which is derived from wood or cotton, is lightly derivatized with ionic groups to form the ion exchanger. The most often used cellulosic anion exchanger is **diethylaminoethyl (DEAE)-cellulose,** whereas **carboxymethyl (CM)-cellulose** is the most popular cellulosic cation exchanger (Fig. 6-8).

Gel-type ion exchangers can have the same sorts of charged groups as do cellulosic ion exchangers. The advantage of gel-type ion exchangers is that they combine the separation properties of gel filtration (Section 6-3B) with those of ion exchange. Because of their high degree of substitution of charged groups, which results from their porous structures, these gels have a higher loading capacity than do cellulosic ion exchangers.

One disadvantage of cellulosic and gel-type matrices is that they are easily compressed (usually by the high pressures resulting from attempts to increase the eluant flow rate), thereby greatly reducing eluant flow. This problem has been alleviated by the fabrication of noncompressible matrices such as derivatized silica or coated glass beads. Such materials allow very high flow rates and pressures, even when they are very finely powdered, and hence permit more effective chromatographic separations (see HPLC in Section 6-3D).

B. *Gel Filtration Chromatography*

In gel filtration chromatography, which is also called size exclusion and molecular sieve chromatography, molecules are separated according to their size and shape. The stationary phase in this technique consists of beads of a hy-

drated, spongelike material containing pores that span a relatively narrow size range of molecular dimensions. If an aqueous solution containing molecules of various sizes is passed through a column containing such "molecular sieves," the molecules that are too large to pass through the pores are excluded from the solvent volume inside the gel beads. These larger molecules therefore traverse the column more rapidly, that is, in a smaller eluant volume, than the molecules that pass through the pores (Fig. 6-9).

The molecular mass of the smallest molecule unable to penetrate the pores of a given gel is said to be the gel's **exclusion limit.** This quantity is to some extent a function of molecular shape because elongated molecules, as a consequence of their higher radius of hydration, are less likely to penetrate a given gel pore than spherical molecules of the same molecular volume.

The behavior of a molecule on a particular gel column can be quantitatively characterized. If V_x is the volume occupied by the gel beads and V_0, the **void volume,** is the volume of the solvent space surrounding the beads, then V_t, the total **bed volume** of the column, is simply their sum:

$$V_t = V_x + V_0 \qquad [6.3]$$

V_0 is typically ~35% of V_t.

The **elution volume** of a given solute, V_e, is the volume of solvent required to elute the solute from the column after it has first contacted the gel. The void volume of a column is easily measured as the elution volume of a solute whose molecular mass is larger than the exclusion limit of the gel. The behavior of a particular solute on a given gel is therefore characterized by the ratio V_e/V_0, the **relative elution volume,** a quantity that is independent of the size of the particular column used.

Molecules with molecular masses ranging below the exclusion limit of a gel will elute from the gel in the order of

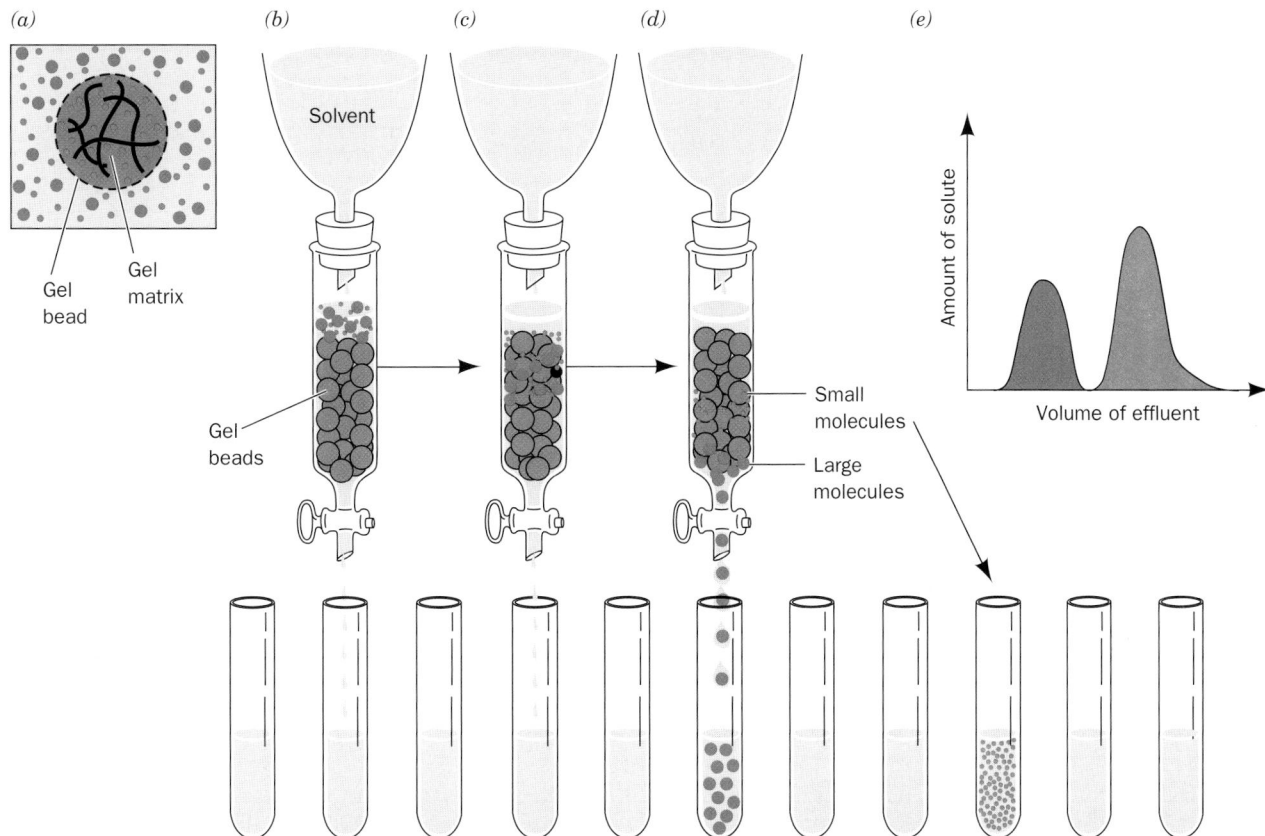

FIGURE 6-9   Gel filtration chromatography. (*a*) A gel bead, whose periphery is represented by a dashed line, consists of a gel matrix (*wavy solid lines*) that encloses an internal solvent space. Smaller molecules (*red dots*) can freely enter the internal solvent space of the gel bead from the external solvent space. However, larger molecules (*blue dots*) are too large to penetrate the gel pores. (*b*) The sample solution begins to enter the gel column (in which the gel beads are now represented by brown spheres). (*c*) The smaller molecules can penetrate the gel and consequently migrate through the column more slowly than the larger molecules that are excluded from the gel. (*d*) The larger molecules emerge from the column to be collected separately from the smaller molecules, which require additional solvent for elution from the column. (*e*) The elution diagram of the chromatogram indicating the complete separation of the two components, with the larger component eluting first.
🐁 See the Animated Figures

their molecular masses, with the largest eluting first. This is because the pore sizes in any gel vary over a limited range, so that larger molecules have less of the gel's interior volume available to them than do smaller molecules. This effect is the basis of gel filtration chromatography.

a. Gel Filtration Chromatography Can Be Used to Estimate Molecular Masses

There is a linear relationship between the relative elution volume of a substance and the logarithm of its molecular mass over a considerable molecular mass range (Fig. 6-10). If a plot such as Fig. 6-10 is made for a particular gel filtration column using macromolecules of known molecular masses, *the molecular mass of an unknown substance can be estimated from its position on the plot. The precision of this technique is limited by the accuracy of the underlying assumption that the known and unknown macromolecules have identical shapes.* Nevertheless, gel filtration chromatography is often used to estimate molecular masses because it can be applied to quite impure samples (providing that the molecule of interest can be identified) and because it can be rapidly carried out using simple equipment.

b. Most Gels Are Made from Dextran, Agarose, or Polyacrylamide

The most commonly used materials for making chromatographic gels are **dextran** (a high molecular mass polymer of glucose produced by the bacterium *Leuconostoc mesenteroides*), **agarose** (a linear polymer of alternating D-galactose and 3,6-anhydro-L-galactose from red algae), and **polyacrylamide** (see Section 6-4B). The properties of several gels that are commonly employed in separating biological molecules are listed in Table 6-3. The porosity of dextran-based gels, sold under the trade name Sephadex, is controlled by the molecular mass of the dextran used and the introduction of glyceryl ether units that cross-link the hydroxyl groups of the polyglucose chains. The several classes of Sephadex that are available have exclusion limits between 0.7 and 600 kD. The pore size in polyacrylamide

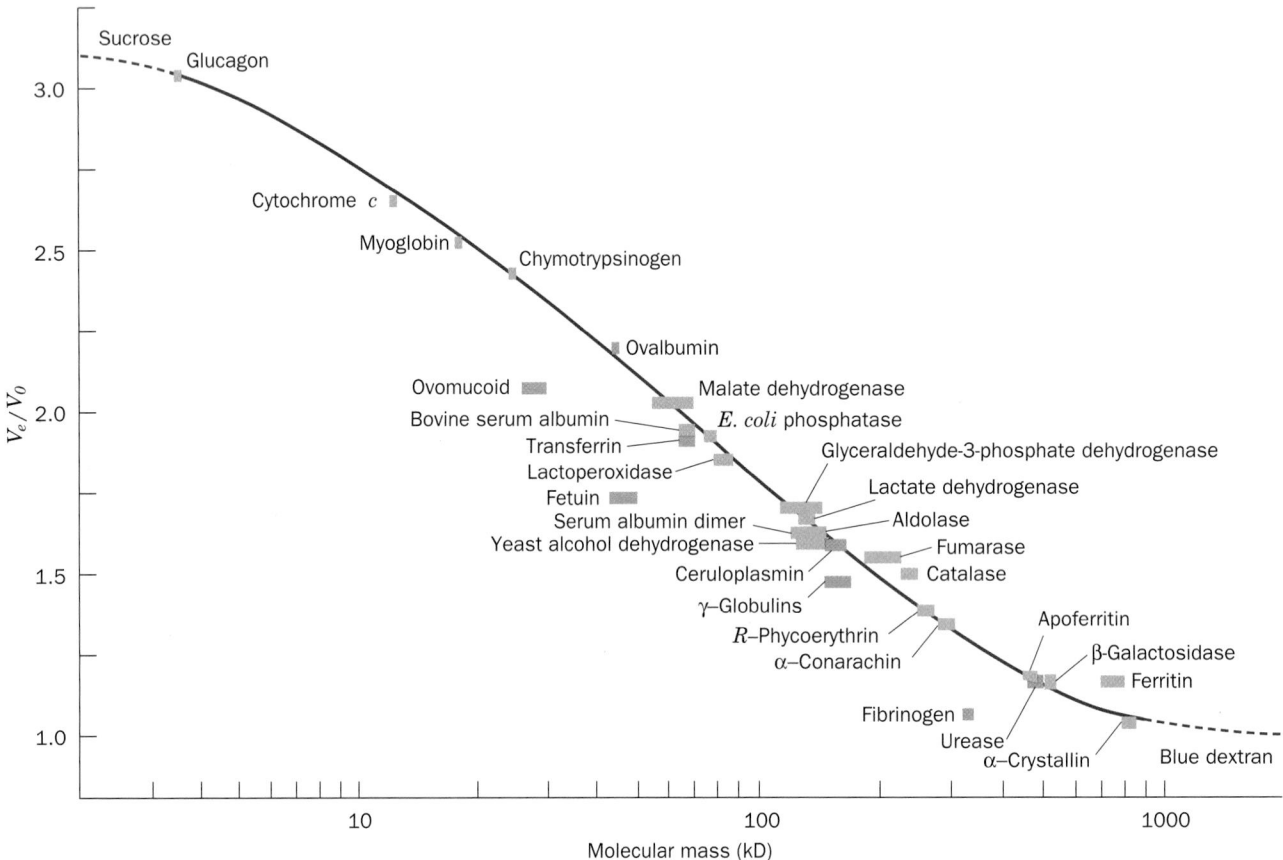

FIGURE 6-10 Molecular mass determination by gel filtration chromatography. The graph shows the relative elution volume versus the logarithm of molecular mass for a variety of proteins on a cross-linked dextran column (Sephadex G-200) at pH 7.5. Orange bars represent glycoproteins (proteins with attached carbohydrate groups). [After Andrews, P., *Biochem. J.* **96,** 597 (1965).]

TABLE 6-3   Some Commonly Used Gel Filtration Materials

Name[a]	Type	Fractionation Range (kD)
Sephadex G-10	Dextran	0.05–0.7
Sephadex G-25	Dextran	1–5
Sephadex G-50	Dextran	1–30
Sephadex G-100	Dextran	4–150
Sephadex G-200	Dextran	5–600
Bio-Gel P-2	Polyacrylamide	0.1–1.8
Bio-Gel P-6	Polyacrylamide	1–6
Bio-Gel P-10	Polyacrylamide	1.5–20
Bio-Gel P-30	Polyacrylamide	2.4–40
Bio-Gel P-100	Polyacrylamide	5–100
Bio-Gel P-300	Polyacrylamide	60–400
Sepharose 6B	Agarose	10–4,000
Sepharose 4B	Agarose	60–20,000
Sepharose 2B	Agarose	70–40,000
Bio-Gel A-5	Agarose	10–5000
Bio-Gel A-50	Agarose	100–50,000
Bio-Gel A-150	Agarose	1000–150,000

[a]Sephadex and Sepharose gels are products of Amersham Biosciences; Bio-Gel gels are manufactured by BioRad Laboratories.

gels is similarly controlled by the extent of cross-linking of neighboring polyacrylamide molecules (Section 6-4B). They are commercially available under the trade name of Bio-Gel P and have exclusion limits between 0.2 and 400 kD. Very large molecules and supramolecular assemblies can be separated using agarose gels, sold under the trade names Sepharose and Bio-Gel A, which have exclusion limits ranging up to 150,000 kD.

Gel filtration is often used to "desalt" a protein solution. For example, an ammonium sulfate–precipitated protein can be easily freed of ammonium sulfate by dissolving the protein precipitate in a minimum volume of suitable buffer and applying this solution to a column of gel with an exclusion limit less than the molecular mass of the protein. On elution of the column with buffer, the protein will precede the ammonium sulfate through the column.

Dextran and agarose gels can be derivatized with ionizable groups such as DEAE and CM to form ion exchange gels (Section 6-3A). Substances that are chromatographed on these gels are therefore subject to separation according to their ionic charges as well as their sizes and shapes.

c. Dialysis Is a Form of Molecular Filtration

Dialysis *is a process that separates molecules according to size through the use of semipermeable membranes containing pores of less than macromolecular dimensions.* These pores allow small molecules, such as those of solvents, salts, and small metabolites, to diffuse across the membrane but block the passage of larger molecules. **Cellophane** (cellulose acetate) is the most commonly used dialysis material, although several other substances such as cellulose and **collodion** are similarly employed. These are available in a wide variety of **molecular weight cutoff** values (the size of the smallest particle that cannot penetrate the membrane) that range from 0.5 to 500 kD.

Dialysis (which is not considered to be a form of chromatography) is routinely used to change the solvent in which macromolecules are dissolved. A macromolecular solution is sealed inside a dialysis bag (usually made by knotting dialysis membrane tubing at both ends), which is immersed in a relatively large volume of the new solvent (Fig. 6-11a). After several hours of stirring, the solutions will have equilibrated, but with the macromolecules remaining inside the dialysis bag (Fig. 6-11b). The process can be repeated several times to replace one solvent system completely by another.

Dialysis can be used to concentrate a macromolecular solution by packing a filled dialysis bag in a polymeric desiccant, such as **polyethylene glycol** [$HOCH_2$ $(CH_2—O—CH_2)_nCH_2OH$], which cannot penetrate the membrane. Concentration is effected as water diffuses across the membrane to be absorbed by the polymer. A related technique that is used to concentrate macromolecular solutions is known as **ultrafiltration.** Here a macromolecular solution is forced, under pressure or by centrifugation, through a semipermeable membranous disk. Solvent and small solutes pass through the membrane, leaving behind a more concentrated macromolecular solution. Since ultrafiltration membranes with different pore sizes are available, ultrafiltration can also be used to separate different-sized macromolecules. Solvent may also be removed from a sample solution through **lyophilization** (freeze-drying), a process in which the solution is frozen and the solvent sublimed away under vacuum.

C. *Affinity Chromatography*

A striking characteristic of many proteins is their ability to bind specific molecules tightly but noncovalently. This property can be used to purify such proteins by **affinity chromatography** (Fig. 6-12). In this technique, a molecule, known as a **ligand** (in analogy with the ligands of coordination compounds), which specifically binds to the protein of interest, is covalently attached to an inert and porous matrix. *When an impure protein solution is passed through this chromatographic material, the desired protein binds to the immobilized ligand, whereas other substances are washed through the column with the buffer. The desired protein can then be recovered in highly purified form by changing the elution conditions such that the protein is released from the chromatographic matrix.* The great advantage of

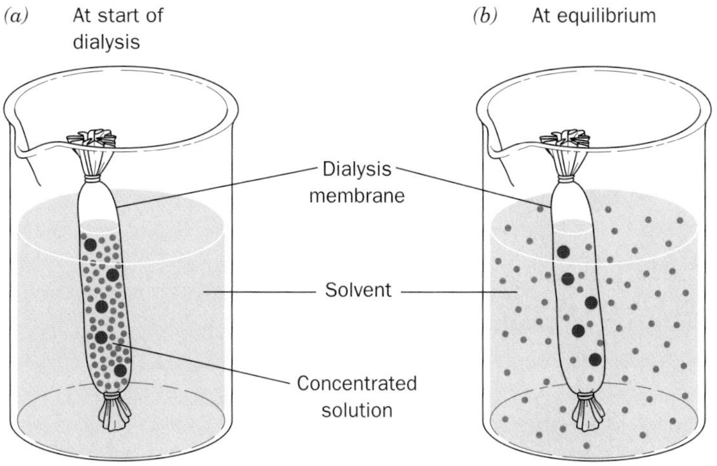

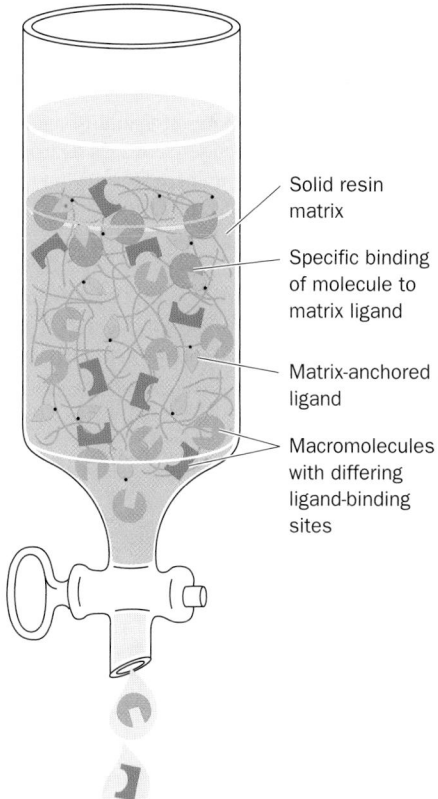

Solid resin matrix

Specific binding of molecule to matrix ligand

Matrix-anchored ligand

Macromolecules with differing ligand-binding sites

FIGURE 6-11 Use of dialysis to separate small and large molecules. (*a*) Only small molecules can diffuse through the pores in the bag, which is shown here as a tube knotted at both ends. (*b*) At equilibrium, the concentrations of small molecules are nearly the same inside and outside the bag, whereas the macromolecules remain in the bag.

(*a*) At start of dialysis
(*b*) At equilibrium
Dialysis membrane
Solvent
Concentrated solution

FIGURE 6-12 Affinity chromatography. A ligand (*yellow*) is covalently anchored to a porous matrix. The sample mixture (whose ligand-binding sites are represented by the cutout squares, semicircles, and triangles) is passed through the column. Only certain molecules (represented by orange circles) specifically bind to the ligand; the others are washed through the column.

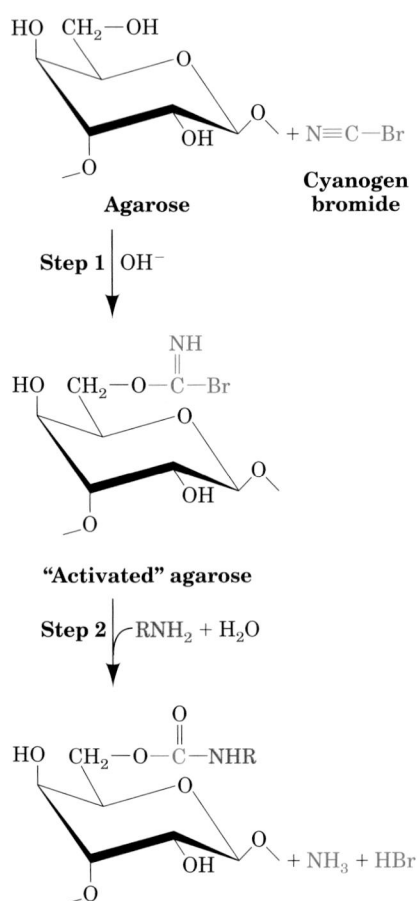

FIGURE 6-13   Covalent linking of ligand to agarose. The formation of cyanogen bromide–activated agarose (*top*) and its reaction with a primary amine to form a covalently attached ligand for affinity chromatography (*bottom*).

The chromatographic matrix in affinity chromatography must be chemically inert, have high porosity, and have large numbers of functional groups capable of forming covalent linkages to ligands. Of the few materials available that meet these criteria, agarose, which has numerous free hydroxyl groups, is by far the most widely used. If the ligand has a primary amino group that is not essential for its binding to the protein of interest, the ligand can be covalently linked to the agarose in a two-step process (Fig. 6-13):

1. Agarose is reacted with **cyanogen bromide** to form an "activated" but stable intermediate (which is commercially available).

2. Ligand reacts with the activated agarose to form covalently bound product.

Many proteins are unable to bind their cyanogen bromide–coupled ligands due to steric interference with the agarose matrix. This problem is alleviated by attaching the ligand to the agarose by a flexible "spacer" group. This is conveniently done through the use of commercially available activated resins. One such resin is "epoxy-activated" agarose, in which a spacer group (containing, e.g., a chain of 12 atoms) links the resin to a reactive epoxy group. The epoxy group can react with many of the nucleophilic groups on ligands, thereby permitting the ligand of choice to be covalently linked to the agarose via a tether of defined length (Fig. 6-14).

The ligand used in the affinity chromatography isolation of a particular protein must have an affinity for the protein high enough to immobilize it on the agarose gel but not so high as to prevent its subsequent release. If the ligand is a substrate for an enzyme being isolated, the chromatography conditions must be such that the enzyme does not function catalytically or the ligand will be destroyed.

After a protein has been bound to an affinity chromatography column and washed free of impurities, it must be released from the column. One method of doing so is to elute the column with a solution of a compound that has

affinity chromatography is its ability to exploit the desired protein's unique biochemical properties rather than the small differences in physicochemical properties between proteins that other chromatographic methods must utilize.

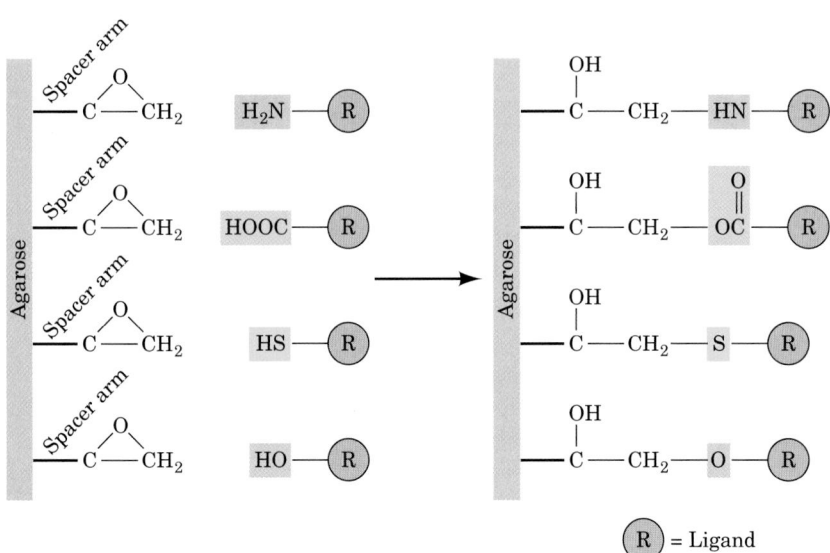

FIGURE 6-14   Derivatization of epoxy-activated agarose. Various types of nucleophilic groups can be covalently attached to epoxy-activated agarose via reaction with its epoxide groups.

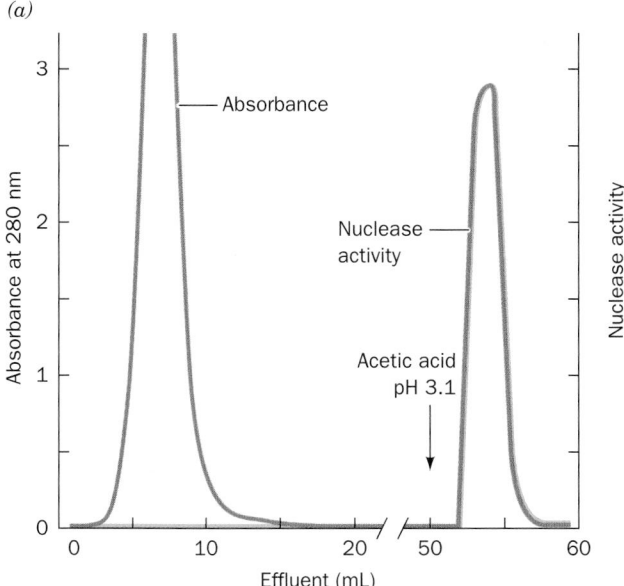

FIGURE 6-15 (*a*) The purification of **staphylococcal nuclease** (a DNA-hydrolyzing enzyme) by affinity chromatography. The compound shown in Part *b*, whose bisphosphothymidine moiety specifically binds to the enzyme, was covalently linked to cyanogen bromide–activated agarose. The column was equilibrated with 0.05*M* borate buffer, pH 8.0, containing 0.01*M* CaCl$_2$, and approximately 40 mg of partially purified material was applied to the column. After 50 mL of buffer had been passed through the column to wash away the unbound material, 0.1*M* acetic acid was added to elute the enzyme. All of the original enzymatic activity, comprising 8.2 mg of pure nuclease, was recovered. [After Cuatrecasas, P., Wilchek, M., and Anfinsen, C.B., *Proc. Natl. Acad. Sci.* **61,** 636 (1968).]

higher affinity for the protein-binding site than the bound ligand. Another is to alter the solution conditions such that the protein–ligand complex is no longer stable, for example, by changes in pH, ionic strength, and/or temperature. However, care must be taken that the solution conditions are not so inhospitable to the protein being isolated that it is irreversibly damaged. An example of protein purification by affinity chromatography is shown in Fig. 6-15.

Affinity chromatography has been used to isolate such substances as enzymes, antibodies, transport proteins, hormone receptors, membranes, and even whole cells. For instance, the protein hormone **insulin** (Section 7-1) has been covalently attached to agarose and used to isolate **insulin receptor** (Section 19-3A), a cell-surface protein whose other properties were previously unknown and which is present in tissues in only very small amounts. Genetic engineering techniques (Section 5-5G) have permitted the affinity purification of proteins for which there is no useful ligand by forming a fusion protein with (linking them to) a protein for which a useful ligand is available. For example, fusion proteins whose N-terminal portions consist of the enzyme **glutathione-*S*-transferase (GST;** Section 25-7C) tightly bind the tripeptide **glutathione** (Section 21-2B) and hence are readily purified by affinity chromatography on glutathione–agarose.

The separation power of affinity chromatography for a specific protein is often far greater than that of other chromatographic techniques (e.g., Table 6-4). Indeed, the replacement of many chromatographic steps in a tried-and-true protein isolation protocol by a single affinity chromatographic step often results in purer protein in higher yield.

a. Immunoaffinity Chromatography Employs the Binding Specificity of Monoclonal Antibodies

A melding of immunochemistry with affinity chromatography has generated a powerful method for purifying biological molecules. Cross-linking monoclonal antibodies (Section 6-1D) to a suitable column material yields a substance that will bind only the protein against which the antibody has been raised. Such **immunoaffinity chromatography** can achieve a 10,000-fold purification in a single step. Disadvantages of immunoaffinity chromatography include the technical difficulty of producing monoclonal antibodies and the harsh conditions that are often required to elute the bound protein.

D. *Other Chromatographic Techniques*

A number of other chromatographic techniques are of biochemical value. These are briefly discussed below.

a. Adsorption Chromatography Separates Nonpolar Substances

In **adsorption chromatography** (the original chromatographic method), molecules are physically adsorbed on the surface of an insoluble substance such as **alumina** (Al$_2$O$_3$), charcoal, **diatomaceous earth** (also called **kieselguhr,** the

TABLE 6-4 **Purification of Rat Liver Glucokinase**

Stage	Specific Activity (nkat · g^{-1})[a]	Yield (%)	Fold[b] Purification
Scheme A: A "traditional" chromatographic procedure			
1. Liver supernatant	0.17	100	1
2. (NH$_4$)$_2$SO$_4$ precipitate	c	c	c
3. DEAE-Sephadex chromatography by stepwise elution with KCl	4.9	52	29
4. DEAE-Sephadex chromatography by linear gradient elution with KCl	23	45	140
5. DEAE-cellulose chromatography by linear gradient elution with KCl	44	33	260
6. Concentration by stepwise KCl elution from DEAE-Sephadex	80	15	480
7. Bio-Gel P-225 chromatography	130	15	780
Scheme B: An affinity chromatography procedure			
1. Liver supernatant	0.092	100	1
2. DEAE-cellulose chromatography by stepwise elution with KCl	20.1	104	220
3. Affinity chromatography[d]	**420**	**83**	**4500**

[a] A **katal** (abbreviation **kat**) is the amount of enzyme that catalyzes the transformation of 1 mol of substrate per second under standard conditions. One nanokatal (nkat) is 10^{-9} kat.

[b] Calculated from specific activity; the first step is arbitrarily assigned unity.

[c] The activity could not be accurately measured at this stage because of uncertainty in correcting for contamination by other enzymes.

[d] The affinity chromatography material was made by linking glucosamine (an inhibitor of glucokinase) through a 6-aminohexanoyl spacer arm to NCBr-activated agarose.

Source: Cornish-Bowden, A., *Fundamentals of Enzyme Kinetics*, p. 48, Butterworth (1979), as adapted from Parry, M.J. and Walker, D.G., *Biochem. J.* **99**, 266 (1966) for Scheme A and from Holroyde, M.J., Allen, B.M., Storer, A.C., Warsey, A.S., Chesher, J.M.E., Trayer, I.P., Cornish-Bowden, A., and Walker, D.G., *Biochem. J.* **153**, 363 (1976) for Scheme B.

siliceous fossils of unicellular organisms known as diatoms), finely powdered sucrose, or **silica gel** (silicic acid), through van der Waals and hydrogen bonding associations. The molecules are then eluted from the column by a pure solvent such as chloroform, hexane, or ethyl ether or by a mixture of such solvents. The separation process is based on the partition of the various substances between the polar column material and the nonpolar solvent. This procedure is most often used to separate nonpolar molecules rather than proteins.

b. Hydroxyapatite Chromatography Separates Proteins

Proteins are adsorbed by gels of crystalline **hydroxyapatite,** an insoluble form of calcium phosphate with empirical formula Ca$_5$(PO$_4$)$_3$OH. The separation of the proteins occurs on gradient elution of the column with phosphate buffer (the presence of other anions is unimportant). The physicochemical basis of this fractionation procedure is not fully understood but apparently involves the adsorption of anions to the Ca^{2+} sites and cations to the PO$_4^{3-}$ sites of the hydroxyapatite crystalline lattice.

c. Paper Chromatography Separates Small Polar Molecules

Paper chromatography, developed in 1941 by Archer Martin and Richard Synge, has played an indispensable role in biochemical analysis due to its ability to efficiently separate small molecules such as amino acids, oligopeptides, nucleotides, and oligonucleotides and its requirement for only the simplest of equipment. Although paper chromatography has been largely supplanted by the more modern techniques discussed in this chapter, we briefly describe it here because of its historical importance and because many of its principles and ancillary techniques are directly applicable to these more modern techniques.

In paper chromatography (Fig. 6-16), a few drops of solution containing a mixture of the components to be separated are applied (spotted) ∼2 cm above one end of a strip of filter paper. After drying, that end of the paper is dipped into a solvent mixture consisting of aqueous and organic components; for example, water/butanol/acetic acid in a 4:5:1 ratio, 77% aqueous ethanol, or 6:7:7 water/*t*-amyl alcohol/pyridine. The paper should also be in contact with the equilibrium vapors of the solvent. The solvent soaks into the paper by capillary action because of the fibrous nature of the paper. The aqueous component of the solvent binds to the cellulose of the paper and thereby forms a stationary gel-like phase with it. The organic component of the solvent continues migrating, thus forming the mobile phase.

The rates of migration of the various substances being separated are governed by their relative solubilities in the polar stationary phase and the nonpolar mobile phase. In a single step of the separation process, a given solute is distributed between the mobile and stationary phases ac-

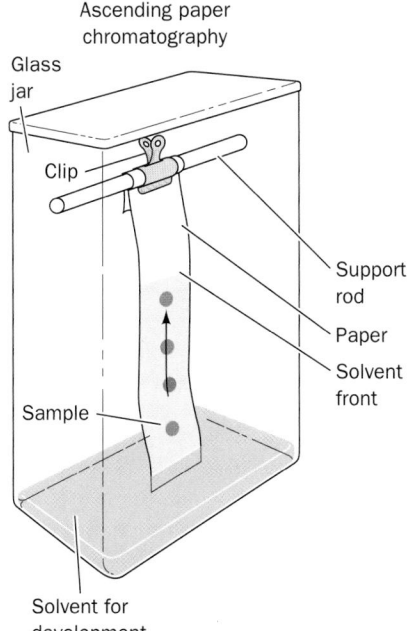

FIGURE 6-16 **Experimental arrangement for paper chromatography.**

cording to its **partition coefficient,** an equilibrium constant defined as

$$K_p = \frac{\text{concentration in stationary phase}}{\text{concentration in mobile phase}} \quad [6.4]$$

The molecules are therefore separated according to their polarities, with nonpolar molecules moving faster than polar ones.

After the solvent front has migrated an appropriate distance, the **chromatogram** is removed from the solvent and dried. The separated materials, if not colored, may be detected by such means as their radioactivity, their fluorescence or ability to quench the normal fluorescence of paper under UV light, or by spraying the chromatogram with a solution of a substance that forms a colored product on reaction with the substance(s) under investigation.

The migration rate of a substance may be expressed according to the ratio

$$R_f = \frac{\text{distance traveled by substance}}{\text{distance traveled by solvent front}} \quad [6.5]$$

For a given solvent system and paper type, each substance has a characteristic R_f value.

A complex mixture that is incompletely separated in a single paper chromatogram can often be fully resolved by **two-dimensional paper chromatography** (Fig. 6-17). In this technique, a chromatogram is made as previously described except that the sample is spotted onto one corner of a sheet of filter paper and the chromatogram is run parallel to an edge of the paper. After the chromatography has been completed and the paper dried, the chromatogram is rotated 90° and is chromatographed parallel to the second edge using another solvent system. Since each compound

migrates at a characteristic rate in a given solvent system, the second chromatographic step should greatly enhance the separation of the mixture into its components.

d. Thin Layer Chromatography Is Used to Separate Organic Molecules

In **thin layer chromatography (TLC),** a thin (~ 0.25 mm) coating of a solid material spread on a glass or plastic plate is utilized in a manner similar to that of the paper in paper chromatography. In the case of TLC, however, the chromatographic material can be a variety of substances such as ion exchangers, gel filtration agents, and physical adsorbents. According to the choice of solvent for the mobile phase, the separation may be based on adsorption, partition, gel filtration, ion exchange processes, or some combination of these. The advantages of thin layer chromatography in convenience, rapidity, and high resolution have led to its routine use in the analysis of organic molecules.

e. Reverse-Phase Chromatography Separates Nonpolar Substances Including Denatured Proteins

Reverse-phase chromatography (RPC) is a form of liquid–liquid partition chromatography in which the polar character of the phases is reversed relative to that of paper chromatography: The stationary phase consists of a nonpolar liquid immobilized on a relatively inert solid, and the mobile phase is a more polar liquid. Reverse-phase chromatography was first developed to separate mixtures of nonpolar substances such as lipids but has also been found to be effective in separating polar substances such as oligonucleotides and proteins, provided that they have exposed nonpolar areas. Although nonpolar side chains tend to inhabit the water-free interiors of native proteins (Section 8-3B), denaturation results in the exposure of these side chains to the solvent. Even when the protein is still in the native state, a significant

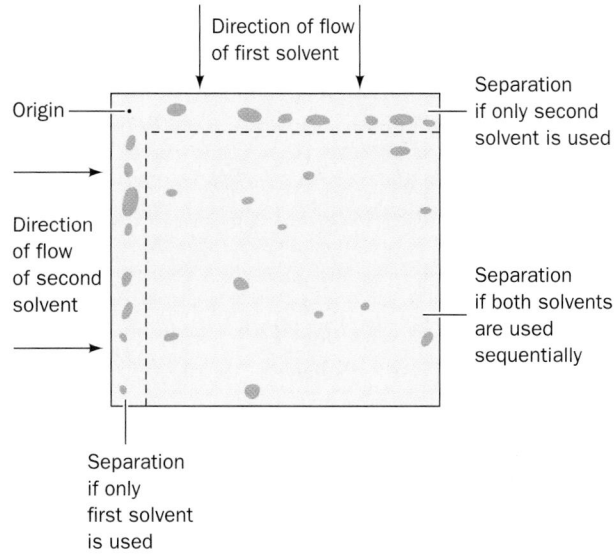

FIGURE 6-17 **Two-dimensional paper chromatography.**

fraction of these hydrophobic groups are at least partially exposed to the solvent at the protein surface. Consequently, under suitable conditions, proteins hydrophobically interact with the nonpolar groups on an immobilized matrix. The hydrophobic interactions in RPC are strong, so the eluting mobile phase must be highly nonpolar (containing high concentrations of organic solvents such as acetonitrile) to dislodge adsorbed substances from the stationary phase. RPC therefore usually denatures proteins.

f. Hydrophobic Interaction Chromatography Separates Native Proteins on the Basis of Surface Hydrophobicity

Hydrophobic interactions form the basis not only of RPC but of **hydrophobic interaction chromatography (HIC).** However, whereas the stationary phase in RPC is strongly hydrophobic in character, often resulting in protein denaturation, in HIC it is a hydrophilic substance, such as an agarose gel, that is only lightly substituted with hydrophobic groups, usually octyl or phenyl residues. The resulting hydrophobic interactions in HIC are therefore relatively weak, so proteins maintain their native structures. The eluants in HIC, whose gradients must progressively reduce these weak hydrophobic interactions, are aqueous buffers with, for example, decreasing salt concentrations (hydrophobic interactions are strengthened by increased ionic strength; Section 6-2A), increasing concentrations of detergents, or increasing pH. Thus, HIC separates native proteins according to their degree of surface hydrophobicity, a criterion that differs from those on which other types of chromatography are based.

g. HPLC Has Permitted Greatly Improved Separations

In **high-performance liquid chromatography (HPLC),** a separation may be based on adsorption, ion exchange, size exclusion, HIC, or RPC as previously described. The separations are greatly improved, however, through the use of high-resolution columns, and the column retention times are much reduced. The narrow and relatively long columns are packed with a noncompressible matrix of fine glass or plastic beads coated with a thin layer of the stationary phase. Alternatively, the matrix may consist of **silica,**

```
 OH      OH      OH
  |       |       |
Si — O — Si — O — Si
  |   |   |   |   |
  O — Si — O — Si — O
  |   |   |   |   |
Si — O — Si — O — Si
```

Silica

whose available hydroxyl groups can be derivatized with many of the commonly used functional groups of ion exchange chromatography, RPC, HIC, or affinity chromatography. The mobile phase is one of the solvent systems previously discussed, including gradient elutions with

binary or even ternary mixtures. In the case of HPLC, however, the mobile phase is forced through the tightly packed column at pressures of up to 5000 psi (pounds per square inch), leading to greatly reduced analysis times. The elutants are detected as they leave the column according to their UV absorption, refractive index, or fluorescence. The advantages of HPLC are

1. Its high resolution, which permits the routine purification of mixtures that have defied separation by other techniques.

2. Its speed, which permits most separations to be accomplished in significantly less than 1 h.

3. Its high sensitivity, which, in favorable cases, permits the quantitative estimation of less than picomole quantities of materials.

4. Its capacity for automation.

Thus, few biochemistry laboratories now function without access to an HPLC system. HPLC is also often utilized in the clinical analyses of body fluids because it can rapidly, routinely, and automatically yield reliable quantitative estimates of nanogram quantities of biological materials such as vitamins, steroids, lipids, and drug metabolites.

4 ■ ELECTROPHORESIS

Electrophoresis, the migration of ions in an electric field, is widely used for the analytical separation of biological molecules. The laws of electrostatics state that the electrical force, $F_{electric}$, on an ion with charge q in an electric field of strength E is expressed by

$$F_{electric} = qE \qquad [6.6]$$

The resulting electrophoretic migration of the ion through the solution is opposed by a frictional force

$$F_{friction} = vf \qquad [6.7]$$

where v is the rate of migration (velocity) of the ion and f is its **frictional coefficient.** *The frictional coefficient is a measure of the drag that the solution exerts on the moving ion and is dependent on the size, shape, and state of solvation of the ion as well as on the viscosity of the solution (Section 6-5A).* In a constant electric field, the forces on the ion balance each other:

$$qE = vf \qquad [6.8]$$

so that each ion moves with a constant characteristic velocity. An ion's **electrophoretic mobility,** μ, is defined

$$\mu = \frac{v}{E} = \frac{q}{f} \qquad [6.9]$$

The electrophoretic (ionic) mobilities of several common small ions in H_2O at 25°C are listed in Table 2-2.

Equation [6.9] really applies only to ions at infinite dilution in a nonconducting solvent. In aqueous solutions, polyelectrolytes such as proteins are surrounded by a cloud of counterions, which impose an additional electric field of such magnitude that Eq. [6.9] is, at best, a poor approximation of reality. Unfortunately, the complexities of ionic solutions have, so far, precluded the development of a theory that can accurately predict the mobilities of polyelectrolytes. Equation [6.9], however, correctly indicates that molecules at their isoelectric points, pI, have zero electrophoretic mobility. Furthermore, for proteins and other polyelectrolytes that have acid–base properties, the ionic charge, and hence the electrophoretic mobility, is a function of pH.

The use of electrophoresis to separate proteins was first reported in 1937 by the Swedish biochemist Arne Tiselius. The technique he introduced, **moving boundary electrophoresis,** was one of the few powerful analytical techniques available in the early years of protein chemistry. However, since this method takes place entirely in solution, preventing the convective mixing of the migrating proteins necessitates a cumbersome apparatus that requires very large samples. Moving boundary electrophoresis has therefore been supplanted by **zone electrophoresis,** a technique in which the sample is constrained to move in a solid support such as filter paper, cellulose acetate, or, most commonly, a gel. This largely eliminates the convective mixing of the sample that limits the resolution achievable by moving boundary electrophoresis. Moreover, in zone electrophoresis, the various sample components migrate as discrete bands (zones) and hence only small quantities of materials are required.

A. *Paper Electrophoresis*

In **paper electrophoresis,** the sample is applied to a point on a strip of filter paper or cellulose acetate moistened with buffer solution. The ends of the strip are immersed in separate reservoirs of buffer in which the electrodes are placed (Fig. 6-18). On application of a direct current (often of ~20 V · cm^{-1}), the ions of the sample migrate toward the electrodes of opposite polarity at characteristic rates to eventually form discrete bands. An ion's migration rate is influenced, to some extent, by its interaction with the support matrix but is largely a function of its charge. On completion of the electrophoretogram (which usually takes several hours), the strip is dried and the sample components are located using the same detection methods employed in paper chromatography (Section 6-3D).

Paper electrophoresis and paper chromatography are superficially similar. However, *paper electrophoresis separates ions largely on the basis of their ionic charges, whereas paper chromatography separates molecules on the basis of their polarities.* The two methods can be combined in a two-dimensional technique called **fingerprinting** in which a sample is first treated as in two-dimensional paper chromatography (Section 6-3D) but is subjected to electrophoresis in place of the second chromatographic step.

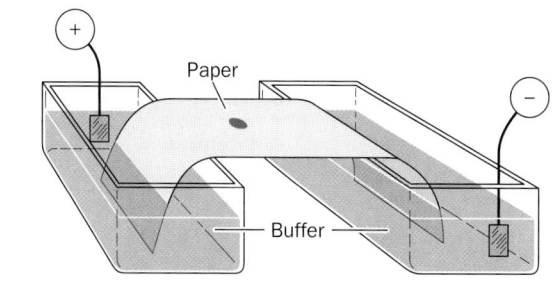

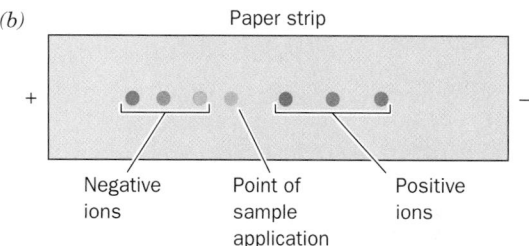

FIGURE 6-18   Paper electrophoresis. (*a*) A diagram of the apparatus used. The sample is applied to a point on the buffer-moistened paper. The ends of the paper are dipped into reservoirs of buffer in which the electrodes are immersed, and an electric field is applied. (*b*) The completed paper electrophoretogram. Note that positive ions (cations) have migrated toward the cathode and negative ions (anions) have migrated toward the anode. Uncharged molecules remain at the point of sample application.

Molecules are thereby separated according to both their charge and their polarity.

B. *Gel Electrophoresis*

Gel electrophoresis, which is among the most powerful and yet conveniently used methods of macromolecular separation, has largely supplanted paper electrophoresis. The gels in common use, polyacrylamide and agarose, have pores of molecular dimensions whose sizes can be specified. *The molecular separations are therefore based on gel filtration as well as the electrophoretic mobilities of the molecules being separated.* The gels in gel electrophoresis, however, retard large molecules relative to smaller ones, the reverse of what occurs in gel filtration chromatography, because there is no solvent space in gel electrophoresis analogous to that between the gel beads in gel filtration chromatography (electrophoretic gels are often directly cast in the electrophoresis device). Since the molecules in a sample cannot leave the gel, the electrophoretic movement of larger molecules is impeded relative to that of smaller molecules.

In **polyacrylamide gel electrophoresis (PAGE),** gels are made by the free radical–induced polymerization of **acrylamide** and *N,N*′-**methylenebisacrylamide** in the buffer of

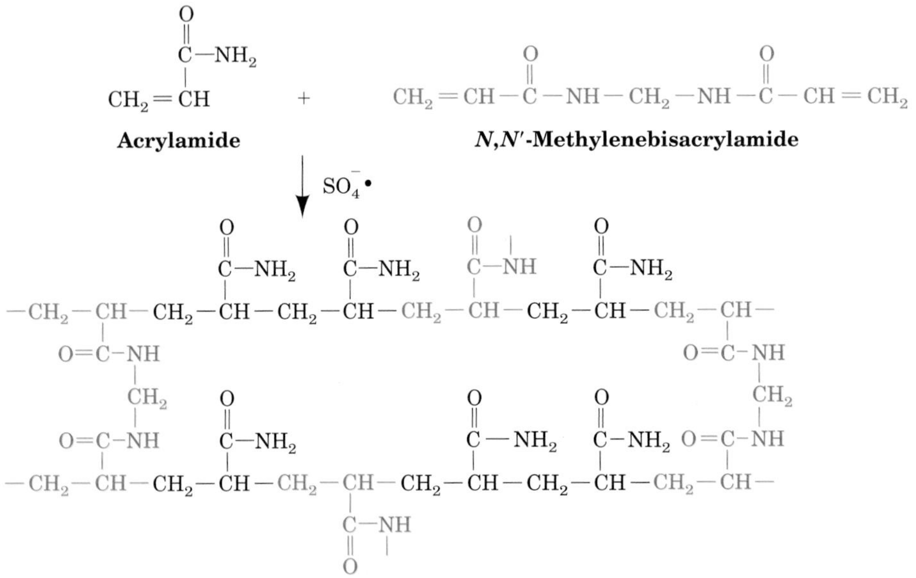

FIGURE 6-19   Polymerization of acrylamide and *N,N'*-methylenebisacrylamide to form a cross-linked polyacrylamide gel. The polymerization is induced by free radicals resulting from the chemical decomposition of **ammonium persulfate** $(S_2O_8^{2-} \rightarrow 2SO_4^{-} \cdot)$ or the photodecomposition of riboflavin in the presence of traces of O_2. In either case, ***N,N,N',N'*-tetramethylethylenediamine (TEMED),** a free radical stabilizer, is usually added to the gel mixture. The physical properties of the gel and its pore size are controlled by the proportion of polyacrylamide in the gel and its degree of cross-linking. The most commonly used polyacrylamide concentrations are in the range 3 to 15%, with the amount of TEMED usually fixed at 5% of the total acrylamide present.

choice (Fig. 6-19). The gel is usually cast as a thin rectangular slab in which several samples can be simultaneously analyzed in parallel lanes (Fig. 6-20), a good way of comparing similar samples. The buffer, which is the same in both reservoirs and the gel, has a pH (usually ~9 for proteins) such that the macromolecules have net negative charges and hence migrate to the anode in the lower reservoir. Each sample, which can contain as little as 10 μg of macromolecular material, is dissolved in a minimal amount

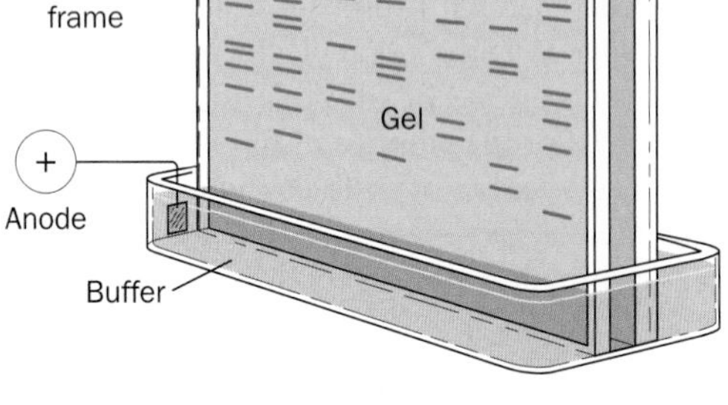

FIGURE 6-20   Apparatus for slab gel electrophoresis. Samples, applied in slots that have been cast in the top of the gel, are electrophoresed in parallel lanes.

of a relatively dense glycerol or sucrose solution to prevent it from mixing with the buffer in the upper reservoir and is applied in preformed slots at the top of the gel in Fig. 6-20. Alternatively, the sample may be contained in a short length of "sample gel," whose pores are too large to impede macromolecular migration. A direct current of ~300 V is passed through the gel for a time sufficient to separate the macromolecular components into a series of discrete bands (30–90 min), the gel is removed from its holder, and the bands are visualized by an appropriate method (see below). Using this technique, a protein mixture of 0.1 to 0.2 mg can be resolved into as many as 20 discrete bands.

a. Disc Electrophoresis Has Improved Resolution

The narrowness of the bands in the foregoing method, and therefore the resolution of the separations, is limited by the length of the sample column as it enters the gel. The bands are greatly sharpened by an ingenious technique known as **discontinuous pH** or **disk electrophoresis,** which requires a two-gel system and several different buffers (Fig. 6-21). The "running gel," in which the separation takes place, is prepared as described previously and then overlayered by a short (1 cm), large-pored "stacking" or "spacer gel." The buffer in the lower reservoir and in the running gel is as described before, while that in the sample solution (or gel) and in the stacking gel has a pH about two units less than that of the lower reservoir. The pH of the buffer in the upper reservoir, which must contain a weak acid (usually glycine, $pK_2 = 9.78$), is adjusted to a pH near that of the lower reservoir.

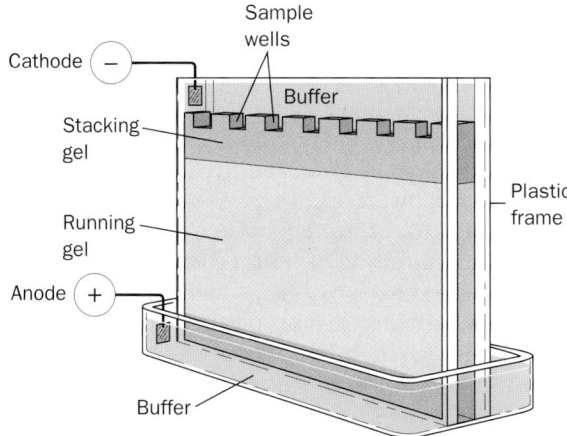

FIGURE 6-21 Diagram of a disc electrophoresis apparatus.

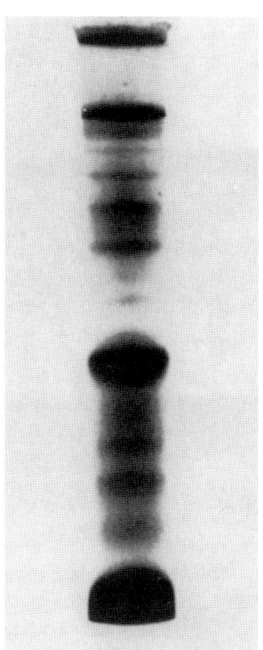

FIGURE 6-22 **Disc electrophoresis of human serum in a 0.5 × 4.0-cm polyacrylamide gel column.** The proteins were visualized by staining them with **amido black.** [Courtesy of Robert W. Hartley, NIH.]

When the current is switched on, the buffer ions from the upper reservoir migrate into the stacking gel as the stacking gel buffer ions migrate ahead of them. As this occurs, the upper reservoir buffer ions encounter a pH that is much lower than their pK. They therefore assume their uncharged (or, in the case of glycine, zwitterionic) form and become electrophoretically immobile. This causes a deficiency of charge carriers, that is, a high electrical resistance R, in this region which, because of the requirement of a constant current I throughout the electrical circuit, results, according to Ohm's law ($E = IR$), in a highly localized increase in the electric field, E. In response to this increased field, the macromolecular anions migrate rapidly until they reach the region containing the stacking gel buffer ions, where they slow down because at that point there is no ion deficiency. *This effect causes the macromolecular ions to approach the running gel as stacks of very narrow (~0.01 mm thick) bands or disks that are ordered according to their mobilities* and lie between the migrating ions of the upper reservoir and those of the stacking gel. As the macromolecular ions enter the running gel, they slow down as a result of gel filtration effects. This permits the upper reservoir buffer ions to overtake the macromolecular bands and, because of the running gel's higher pH, assume their fully charged form as they too enter the gel. The charge carrier deficiency therefore disappears and from this point on the electrophoretic separation proceeds normally. However, *the compactness of the macromolecular bands entering the running gel greatly increases the resolution of the macromolecular separations* (e.g., Fig. 6-22).

b. Agarose Gels Are Used to Separate Large Molecules Electrophoretically

The very large pores needed for the PAGE of large molecular mass compounds (>200 kD) requires gels with such low polyacrylamide concentrations (<2.5%) that they are too soft to be usable. This difficulty is circumvented by using agarose (Fig. 6-13). For example, a 0.8% agarose gel is used for the electrophoretic separation of nucleic acids with molecular masses of up to 50,000 kD.

c. Gel Bands May Be Detected by Staining, Radioactive Counting, or Immunoblotting

Bands resulting from a gel electrophoretic separation can be located by a variety of techniques. Proteins are often visualized by staining. **Coomassie brilliant blue,**

$$^-O_3S \qquad\qquad SO_3^-$$

R250: R = H
G250: R = CH$_3$

OC$_2$H$_5$

Coomassie brilliant blue

which is the most widely used dye for this purpose, is applied by soaking the gel in an acidic, alcoholic solution containing the dye. This fixes the protein in the gel by denaturing it and complexes the dye to the protein. Excess dye is removed by extensively washing the gel with an acidic solution or by electrophoretic destaining. Protein bands containing as little as 0.1 μg can thereby be detected. Gel bands containing less than this amount of protein may be visualized with **silver stain,** which is ~50 times more sensitive but more difficult to apply. **Fluorescamine,** an alternative type of protein stain, is a nonfluorescent molecule that reacts with primary amines, such as lysine, to yield an addition product that is highly fluorescent under UV irradiation.

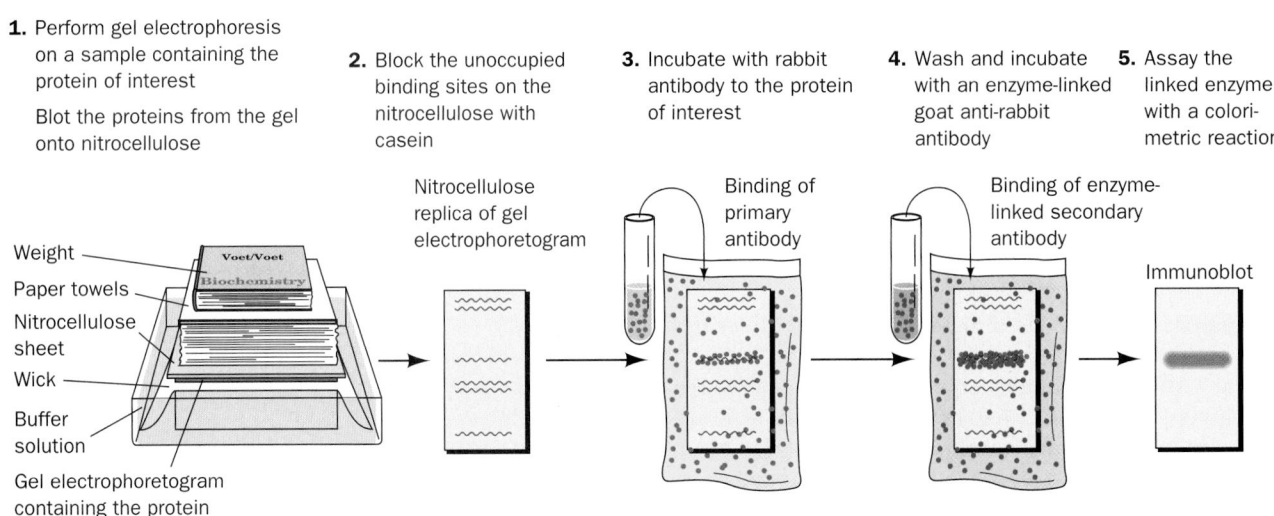

Fluorescamine (nonfluorescent)   **Fluorescamine adduct** (highly fluorescent)

Proteins, as well as other substances, can be detected through the UV absorption of a gel along its length. If the sample is radioactive, the gel may be dried under vacuum to form a cellophane-like material or, alternatively, covered with plastic wrap, and then clamped over a sheet of X-ray film. After a time (from a few minutes to many weeks depending on the radiation intensity) the film is developed, and the resulting **autoradiograph** shows the positions of the radioactive components by a blackening of the film [alternatively, a position-sensitive radiation detector (electronic film) can be used to reveal the locations of the radioactive components within even a few seconds]. A gel may also be sectioned widthwise into many slices and the level of radioactivity in each slice determined using a **scintillation counter.** The latter method yields quantitatively more accurate results than autoradiography. Sample materials can also be eluted from gel slices for identification and/or further treatment.

If an antibody to the protein of interest is available, it is possible to specifically detect this protein on a gel in the presence of many other proteins by an **immunoblot** (also known as a **Western blot**). This procedure is a variation of Southern blotting (Section 5-5D) that uses a technique similar to ELISA (Section 6-1D) to detect the protein(s) of interest (Fig. 6-23):

1. A completed gel electrophoretogram is blotted onto a sheet of nitrocellulose (much like in Fig. 5-50), which strongly and nonspecifically binds proteins [nylon or polyvinylidine difluoride (PVDF) membranes may also be used].

2. The excess adsorption sites on the nitrocellulose are blocked with a nonspecific protein such as **casein** (milk protein; nonfat milk itself is often used) to prevent the nonspecific adsorption of the antibodies (which are also proteins) used in Steps 3 and 4.

3. The blot is treated with antibody to the protein of interest (the primary antibody). This is usually a rabbit antibody.

4. After washing away the unbound primary antibody, the blot is incubated with a goat antibody, directed against all rabbit antibodies, to which an easily assayed enzyme has been covalently linked (the secondary antibody).

5. After washing away the unbound secondary antibody, the enzyme in the bound secondary antibody is assayed with a color-producing reaction, causing colored bands to appear on the nitrocellulose where the protein of interest is bound.

1. Perform gel electrophoresis on a sample containing the protein of interest

Blot the proteins from the gel onto nitrocellulose

2. Block the unoccupied binding sites on the nitrocellulose with casein

3. Incubate with rabbit antibody to the protein of interest

4. Wash and incubate with an enzyme-linked goat anti-rabbit antibody

5. Assay the linked enzyme with a colorimetric reaction

FIGURE 6-23 Detection of proteins by immunoblotting.

Alternatively, the primary antibody used in Step 3 may be labeled with the radioactive isotope ^{125}I, the unbound antibody washed away, and the position of the bound protein on the blot revealed by autoradiography.

C. *SDS–PAGE*

Soaps and detergents are amphipathic molecules (Section 2-1B) that are strong protein denaturing agents for reasons explained in Section 8-4E. **Sodium dodecyl sulfate (SDS),**

$$[CH_3-(CH_2)_{10}-CH_2-O-SO_3^-]Na^+$$
Sodium dodecyl sulfate (SDS)

a detergent that is often used in biochemical preparations, binds quite tenaciously to proteins, causing them to assume a rodlike shape. Most proteins bind SDS in the same ratio, 1.4 g of SDS per gram of protein (about one SDS molecule for every two amino acid residues). The large negative charge that the SDS imparts masks the protein's intrinsic charge so that SDS-treated proteins tend to have

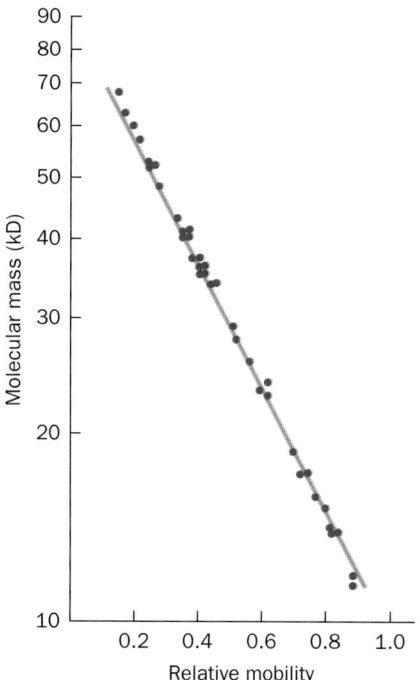

FIGURE 6-25   Logarithmic relationship between the molecular mass of a protein and its relative electrophoretic mobility in SDS–PAGE. This relationship is plotted for 37 polypeptides ranging from 11 to 70 kD. [After Weber, K. and Osborn, M., *J. Biol. Chem.* **244,** 4406 (1969).]

identical charge-to-mass ratios and similar shapes. Consequently, *the electrophoresis of proteins in an SDS-containing polyacrylamide gel separates them in order of their molecular masses because of gel filtration effects.* Figure 6-24 provides an example of the resolving power and the reproducibility of **SDS–PAGE.**

The molecular masses of "normal" proteins are routinely determined to an accuracy of 5 to 10% through SDS–PAGE. The relative mobilities of proteins on such gels vary linearly with the logarithm of their molecular masses (Fig. 6-25). In practice, a protein's molecular mass is determined by electrophoresing it together with several "marker" proteins of known molecular masses which bracket that of the protein of interest.

Many proteins consist of more than one polypeptide chain (Section 8-5A). SDS treatment disrupts the noncovalent interactions between these subunits. Therefore, SDS–PAGE yields the molecular masses of the protein's subunits rather than that of the intact protein unless the subunits are disulfide linked. However, mercaptoethanol is usually added to SDS–PAGE gels so as to reductively cleave these disulfide bonds (Section 7-1B).

D. *Isoelectric Focusing*

A protein has charged groups of both polarities and therefore has an isoelectric point, p*I*, the pH at which it is immobile in an electric field (Section 4-1D). *If a mixture of proteins is electrophoresed through a solution having a stable pH gradient in which the pH smoothly increases from*

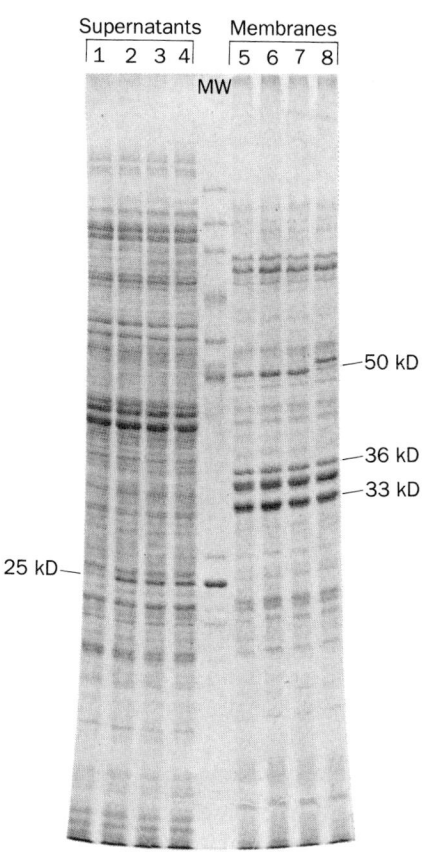

FIGURE 6-24   SDS–PAGE. The SDS–polyacrylamide disc electrophoretogram shows separation of proteins from the supernatant (*left*) and membrane fractions (*right*) of various strains of the bacterium *Salmonella typhimurium*. Samples of 200 μg of protein each were run in parallel lanes on a 35-cm-long, 0.8-mm-thick slab gel containing 10% polyacrylamide. The lane marked MW contains molecular weight standards. [Courtesy of Giovanna F. Ames, University of California at Berkeley.]

anode to cathode, each protein will migrate to the position in the pH gradient corresponding to its isoelectric point. If a protein molecule diffuses away from this position, its net charge will change as it moves into a region of different pH, and the resulting electrophoretic forces will move it back to its isoelectric position. Each species of protein is thereby "focused" into a narrow band about its isoelectric point that may be as thin as 0.01 pH unit. Hence, this process is called **isoelectric focusing (IEF).**

A pH gradient produced by mixing two different buffers together in continuously varying ratios is unstable in an electric field because the buffer ions migrate to the electrode of opposite polarity. Rather, the pH gradient in IEF is formed by a mixture of low molecular mass (600–900 D) oligomers bearing aliphatic amino and carboxylic acid groups (Fig. 6-26) that have a range of isoelectric points. Under the influence of an electric field in solution, these **ampholytes** (amphoteric electrolytes) will each migrate to their isoelectric points. Consequently, the most acidic ampholytes gather at the anode and the progressively more basic ones position themselves ever closer to the cathode. The pH gradient, which is maintained by the ~1000 V electric field, arises from the buffering action of these ampholytes. The convection of the pH gradient is prevented by carrying out IEF in a lightly cross-linked polyacrylamide gel in the form of a rod or slab. IEF gels often contain ~$6M$ urea, a powerful protein denaturing agent that, unlike SDS, is uncharged and hence cannot directly affect the charge of a protein.

An alternative form of IEF utilizes gels containing **immobilized pH gradients.** These are made from acrylamide derivatives that are covalently linked to ampholytes. Through the use of a gradient maker (e.g., Fig. 6-7), a gel containing an immobilized pH gradient is polymerized from a continuously varying mixture of acrylamide derivatives with different pK's so that the gel's pH varies smoothly from one end to the other.

The fact that IEF separates proteins into sharp bands makes it a useful analytical and preparative tool. In fact, many protein preparations once thought to be homogeneous have been resolved into several components by IEF. IEF can be combined with electrophoresis in an extremely powerful two-dimensional separation technique named **two-dimensional (2D) gel electrophoresis;** Fig. 6-27). Up to 5000 proteins can be observed on a single two-dimensional electrophoretogram. Hence two-dimensional gel electrophoresis is a valuable tool for **proteomics** (the study of

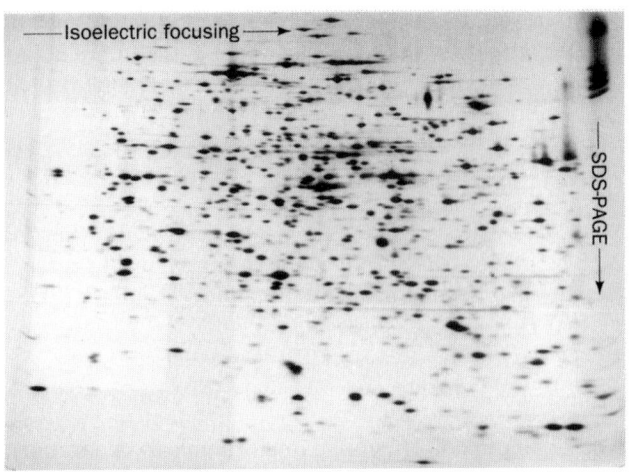

FIGURE 6-27   Two-dimensional (2D) gel electrophoresis. This autoradiogram shows the separation of *E. coli* proteins by 2D gel electrophoresis (isoelectric focusing horizontally and SDS–PAGE vertically). A 10-μg sample of proteins from *E. coli* that had been labeled with ^{14}C-amino acids was subjected to isoelectric focusing in a 2.5 × 130-mm tube of a urea-containing polyacrylamide gel. The gel was then extruded from its tube, placed in contact with one edge of an SDS–polyacrylamide slab gel, and subjected to electrophoresis. Over 1000 spots were counted on the original autoradiogram, which resulted from an 825-h exposure. [Courtesy of Patrick O'Farrell, University of California at San Francisco.]

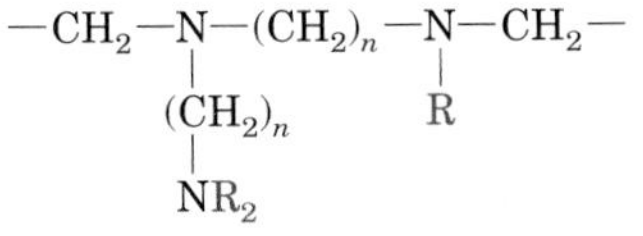

the **proteome,** which, in analogy with the term "genome," is defined as the aggregate of all the proteins expressed by a cell or organism, but with emphasis on their quantitation, localization, modifications, interactions, and activities, as well as their identification). Individual protein bands in a stained gel can be cut out from the gel (with a scalpel or by a robot guided by a digitized image of the gel acquired by an optical scanner or a digital camera), destained, and the protein eluted from the gel fragment for identification and/or characterization, often by mass spectrometry (Section 7-1J). Variant proteins can be found by comparing the positions and intensities of the bands in 2D gels of similar preparations. This can be done with the aid of a computer after acquiring digitized images of the stained gels. Numerous reference 2D gels are publicly available for this purpose in the Web-accessible databases listed at http://www.expasy.ch/ch2d/2d-index.html. These databases contain images of 2D gels of a variety of organisms and tissues and identify many of their component proteins.

E. *Capillary Electrophoresis*

Although gel electrophoresis in its various forms is a common and highly effective method for separating charged molecules, it typically requires an hour or more for a run and is difficult to quantitate and automate. These disadvantages are largely overcome through the use of **capillary electrophoresis (CE),** a technique in which electrophore-

$$-CH_2-N-(CH_2)_n-N-CH_2-$$
$$| \qquad\qquad |$$
$$(CH_2)_n \qquad R$$
$$|$$
$$NR_2$$

$n = 2$ or 3
$R = H$ or $-(CH_2)_n-COOH$

FIGURE 6-26   General formula of the ampholytes used in isoelectric focusing.

sis is carried out in very thin (20 to 100 μm inner diameter) capillary tubes made of quartz, glass, or plastic. Such narrow capillaries rapidly dissipate heat and hence permit the use of high electric fields (typically 100 to 300 V · cm^{-1}, about 10 times that of most other electrophoretic techniques), which reduces separation times to a few minutes. These rapid separations, in turn, minimize band broadening caused by diffusion, thereby yielding extremely sharp separations. Capillaries can be filled with buffer (as in moving boundary electrophoresis, but here the capillary's narrow bore all but eliminates convective mixing), SDS–polyacrylamide gel (separation according to molecular mass; Section 6-4C), or ampholytes (isoelectric focusing; Section 6-4D). These CE techniques have extremely high resolution and can be automated in much the same way as is HPLC, that is, with automatic sample loading and on-line sample detection. Since CE can only separate small amounts of material, it is largely limited to use as an analytical technique.

5 ■ ULTRACENTRIFUGATION

If a container of sand and water is shaken and then allowed to stand quietly, the sand will rapidly sediment to the bottom of the container due to the influence of Earth's gravity (an acceleration g equal to 9.81 m · s^{-2}). Yet macromolecules in solution, which experience the same gravitational field, do not exhibit any perceptible sedimentation because their random thermal (Brownian) motion keeps them uniformly distributed throughout the solution. *Only when they are subjected to enormous accelerations will the sedimentation behavior of macromolecules begin to resemble that of sand grains.*

The ultracentrifuge, which was developed around 1923 by the Swedish biochemist The Svedberg, can attain rotational speeds as high as 80,000 rpm (revolutions per minute) so as to generate centrifugal fields in excess of 600,000g. Using this instrument, Svedberg first demonstrated that proteins are macromolecules with homogeneous compositions and that many proteins are composed of subunits. More recently, ultracentrifugation has become an indispensable tool for the isolation of proteins, nucleic acids, and subcellular particles. In this section we outline the theory and practice of ultracentrifugation.

A. *Sedimentation*

The rate at which a particle sediments in the ultracentrifuge is related to its mass. The force, $F_{sedimentation}$, acting to sediment a particle of mass m that is located a distance r from a point about which it is revolving with angular velocity ω (in radians · s^{-1}) is the centrifugal force ($m\omega^2 r$) on the particle less the buoyant force ($V_p\rho\omega^2 r$) exerted by the solution:

$$F_{sedimentation} = m\omega^2 r - V_p\rho\omega^2 r \qquad [6.10]$$

Here V_p is the particle volume and ρ is the density of the solution. However, the motion of the particle through the solution, as we have seen in our study of electrophoresis, is opposed by the frictional force:

$$F_{friction} = vf \qquad [6.7]$$

where $v = dr/dt$ is the rate of migration of the sedimenting particle and f is its frictional coefficient. The particle's frictional coefficient can be determined from measurements of its rate of diffusion.

Under the influence of gravitational (centrifugal) force, the particle accelerates until the forces on it exactly balance:

$$m\omega^2 r - V_p\rho\omega^2 r = vf \qquad [6.11]$$

The mass of 1 mol of particles, M, is

$$M = mN \qquad [6.12]$$

where N is Avogadro's number (6.022 × 10^{23}). Thus, a particle's volume, V_p, may be expressed in terms of its molar mass:

$$V_p = \bar{V}m = \frac{\bar{V}M}{N} \qquad [6.13]$$

where $\bar{V}$, the particle's **partial specific volume,** is the volume change when 1 g (dry weight) of particles is dissolved in an infinite volume of the solute. For most proteins dissolved in pure water at 20°C, $\bar{V}$ is near 0.73 cm^3 · g^{-1} (Table 6-5). Indeed, for proteins of known amino acid composition, $\bar{V}$ is closely approximated by the sum of the partial specific volumes of its component amino acid residues, thereby indicating that the atoms in proteins are closely packed (Section 8-3B).

a. A Particle May Be Characterized by Its Sedimentation Rate

Substituting Eqs. [6.12] and [6.13] into Eq. [6.11] yields

$$vf = \frac{M(1 - \bar{V}\rho)\omega^2 r}{N} \qquad [6.14]$$

Now define the **sedimentation coefficient,** s, as

$$s = \frac{v}{\omega^2 r} = \frac{1}{\omega^2}\left(\frac{d\ln r}{dt}\right) = \frac{M(1 - \bar{V}\rho)}{Nf} \qquad [6.15]$$

The sedimentation coefficient, a quantity that is analogous to the electrophoretic mobility (Eq. [6.9]) in that it is a velocity per unit force, is usually expressed in units of 10^{-13} s, which are known as **Svedbergs (S).** For the sake of uniformity, the sedimentation coefficient is customarily corrected to the value that would be obtained at 20°C in a solvent with the density and viscosity of pure water. This is symbolized $s_{20,w}$. Table 6-5 and Fig. 6-28 indicate the values of $s_{20,w}$ in Svedbergs for a variety of biological materials.

Equation [6.15] indicates that a particle's mass, $m = M/N$, can be determined from the measurement of

TABLE 6-5   **Physical Constants of Some Proteins**

Protein	Molecular Mass (kD)	Partial Specific Volume, $\bar{V}_{20,w}$ ($cm^3 \cdot g^{-1}$)	Sedimentation Coefficient, $s_{20,w}$ (S)	Frictional Ratio, f/f_0
Lipase (milk)	6.7	0.714	1.14	1.190
Ribonuclease A (bovine pancreas)	12.6	0.707	2.00	1.066
Cytochrome *c* (bovine heart)	13.4	0.728	1.71	1.190
Myoglobin (horse heart)	16.9	0.741	2.04	1.105
α-Chymotrypsin (bovine pancreas)	21.6	0.736	2.40	1.130
Crotoxin (rattlesnake)	29.9	0.704	3.14	1.221
Concanavalin B (jack bean)	42.5	0.730	3.50	1.247
Diphtheria toxin	70.4	0.736	4.60	1.296
Cytochrome oxidase (*P. aeruginosa*)	89.8	0.730	5.80	1.240
Lactate dehydrogenase H (chicken)	150	0.740	7.31	1.330
Catalase (horse liver)	222	0.715	11.20	1.246
Fibrinogen (human)	340	0.725	7.63	2.336
Hemocyanin (squid)	612	0.724	19.50	1.358
Glutamate dehydrogenase (bovine liver)	1015	0.750	26.60	1.250
Turnip yellow mosaic virus protein	3013	0.740	48.80	1.470

Source: Smith, M.H., *in* Sober, H.A. (Ed.), *Handbook of Biochemistry and Molecular Biology* (2nd ed.), p. C-10, CRC Press (1970).

its sedimentation coefficient, *s,* and the solution density, ρ, if its frictional coefficient, *f,* and its partial specific volume, $\bar{V}$, are known. Indeed, before about 1970, most macromolecular mass determinations were made using the **analytical ultracentrifuge,** a device in which the sedimentation rates of molecules under centrifugation can be optically measured (the masses of macromolecules are too high to be accurately determined by such classic physical techniques as melting point depression or osmotic pressure measurements). Although the advent of much simpler molecular mass determination methods, such as gel filtration chromatography (Section 6-3B) and SDS–PAGE (Section 6-4C), had caused analytical ultracentrifugation to largely fade from use, recently developed instrumentation has led to a resurgence in the use of analytical ultracentrifugational measurements. They are particularly useful in characterizing systems of associating macromolecules.

b. The Frictional Ratio Is Indicative of Molecular Solvation and Shape

For an unsolvated spherical particle of radius r_p, the frictional coefficient is determined according to the **Stokes equation:**

$$f = 6\pi\eta r_p \qquad [6.16]$$

where η is the **viscosity** of the solution. Solvation increases the frictional coefficient of a particle by increasing its effective or **hydrodynamic volume.** Furthermore, *f* is minimal when the particle is a sphere. This is because a nonspherical particle has a larger surface area than a sphere of equal volume and therefore must, on the average, present a greater surface area toward the direction of movement than a sphere.

The frictional coefficient, *f,* of a particle of known mass and partial specific volume can be ultracentrifugationally determined using Eq. [6.15]. The effective or **Stokes radius**

of a particle in solution can be calculated by solving Eq. [6.16] for r_p, given the experimentally determined values of *f* and η. Conversely, the minimal frictional coefficient of a particle, f_0, can be calculated from the mass and the partial specific volume of the particle by assuming it to be spherical ($V_p = \frac{4}{3}\pi r_p^3$) and unsolvated:

$$f_0 = 6\pi\eta\left(\frac{3M\bar{V}}{4\pi N}\right)^{1/3} \qquad [6.17]$$

If the **frictional ratio,** f/f_0, of a particle is much greater than unity, it must be concluded that the particle is highly solvated and/or significantly elongated. The frictional ratios of a selection of proteins are presented in Table 6-5. The globular proteins, which are known from structural studies to be relatively compact and spheroidal (Section 8-3B), have frictional ratios ranging up to ~1.5. Fibrous molecules such as DNA and the blood clotting protein **fibrinogen** (Section 35-1A) have larger frictional ratios. On denaturation, the frictional coefficients of globular proteins increase by as much as twofold because denatured proteins assume flexible and fluctuating **random coil** conformations in which all parts of the molecule are in contact with solvent (Section 8-1D).

B. *Preparative Ultracentrifugation*

Preparative ultracentrifuges, which as their name implies are designed for sample preparation, differ from analytical ultracentrifuges in that they lack sample observation facilities. Preparative rotors contain cylindrical sample tubes whose axes may be parallel, at an angle, or perpendicular to the rotor's axis of rotation, depending on the particular application (Fig. 6-29).

In the derivation of Eq. [6.15], it was assumed that sedimentation occurred through a homogeneous medium. Sedimentation may be carried out in a solution of an in-

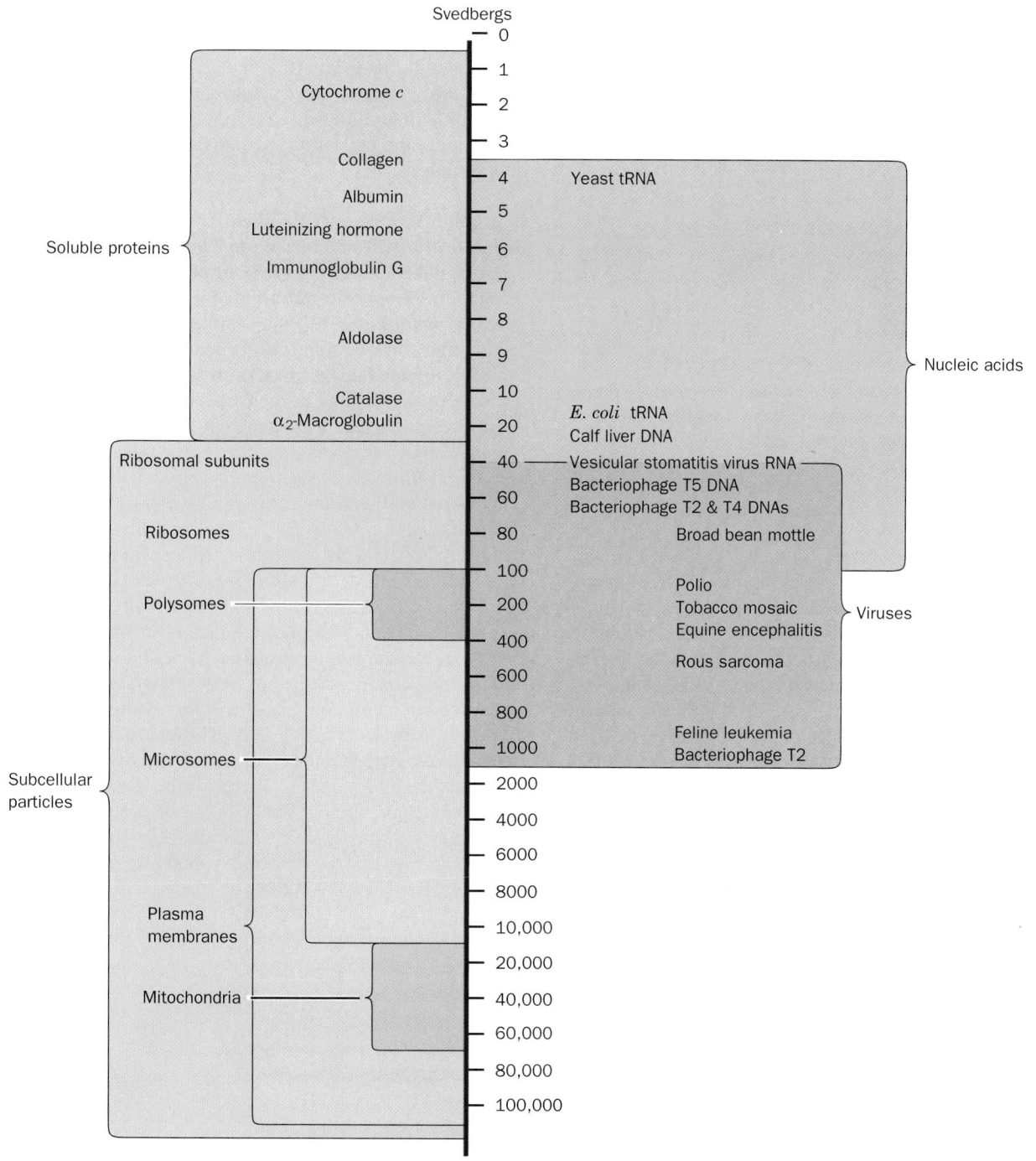

FIGURE 6-28 **Sedimentation coefficients in Svedbergs (S) for some biological materials.**
[After a diagram supplied by Beckman Instruments, Inc.]

ert substance, however, such as sucrose or CsCl, in which the concentration, and therefore the density, of the solution increases from the top to the bottom of the centrifuge tube. The use of such **density gradients** greatly enhances the resolving power of the ultracentrifuge. Two applications of density gradients are widely employed: (1) **zonal ultracentrifugation** and (2) **equilibrium density gradient ultracentrifugation.**

a. Zonal Ultracentrifugation Separates Particles According to Their Sedimentation Coefficients

In zonal ultracentrifugation, a macromolecular solution is carefully layered on top of a density gradient prepared by use of a device resembling that diagrammed in Fig. 6-7. The purpose of the density gradient is to allow smooth passage of the various macromolecular zones by damping out convective mixing of the solution. Sucrose, which forms a

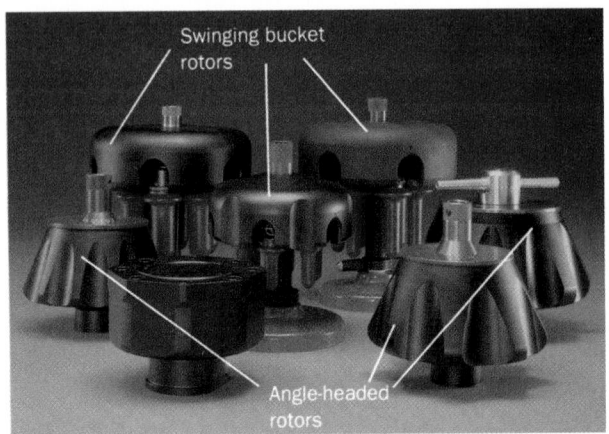

FIGURE 6-29   A selection of preparative ultracentrifuge rotors. The sample tubes of the swinging bucket rotors (*rear*) are hinged so that they swing from the vertical to the horizontal position as the rotor starts spinning, whereas the sample tubes of the other rotors have a fixed angle relative to the rotation axis. [Courtesy of Beckman Instruments, Inc.]

syrupy and biochemically benign solution, is commonly used to form a density gradient for zonal ultracentrifugation. The density gradient is normally rather shallow because the maximum density of the solution must be less than that of the least dense macromolecule of interest. Nevertheless, consideration of Eq. [6.15] indicates that the sedimentation rate of a macromolecule is a more sensitive function of molecular mass than density. Consequently, *zonal ultracentrifugation separates similarly shaped macromolecules largely on the basis of their molecular masses.*

During centrifugation, each species of macromolecule moves through the gradient at a rate largely determined

by its sedimentation coefficient and therefore travels as a zone that can be separated from other such zones as is diagrammed in Fig. 6-30. After centrifugation, fractionation is commonly effected by puncturing the bottom of the celluloid centrifuge tube with a needle, allowing its contents to drip out, and collecting the individual zones for subsequent analysis.

b. Equilibrium Density Gradient Ultracentrifugation Separates Particles According to Their Densities

In **equilibrium density gradient ultracentrifugation** [alternatively, **isopycnic ultracentrifugation** (Greek: *isos,* equal + *pyknos,* dense)], the sample is dissolved in a relatively concentrated solution of a dense, fast-diffusing (and therefore low molecular mass) substance, such as CsCl or Cs_2SO_4, and is spun at high speed until the solution achieves equilibrium. *The high centrifugal field causes the low molecular mass solute to form a steep density gradient (Fig. 6-31) in which the sample components band at positions where their densities are equal to that of the solution;* that is, where $(1 - \overline{V}\rho)$ in Eq. [6.15] is zero (Fig. 6-32). These bands are collected as separate fractions when the sample tube is drained as described above. The salt concentration in the fractions and hence the solution density is easily determined with an **Abbé refractometer,** an optical instrument that measures the refractive index of a solution. *The equilibrium density gradient technique is often the method of choice for separating mixtures whose components have a range of densities (e.g., Fig. 5-13).* These substances include nucleic acids, viruses, and certain subcellular organelles such as ribosomes. However, isopycnic ultracentrifugation is rather ineffective for the fractionation of protein mixtures because most proteins have similar densities (high salt concentrations also salt out or possibly denature proteins).

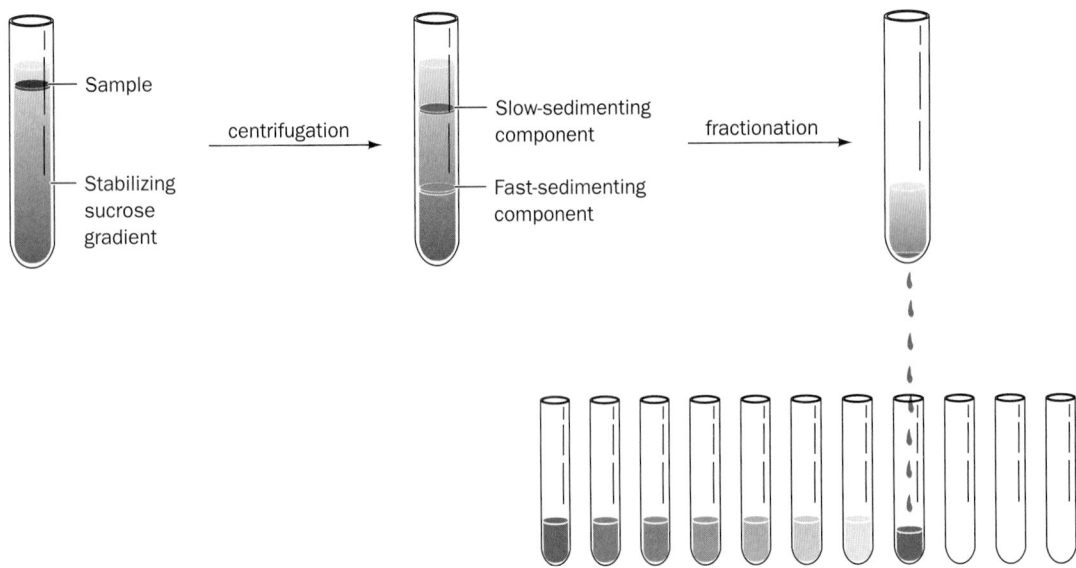

FIGURE 6-30   Zonal ultracentrifugation. The sample is layered onto a sucrose gradient (*left*). During centrifugation (*middle*), each particle sediments at a rate that depends largely on its mass. After the end of the run, the centrifugation tube is punctured and the separated particles (zones) are collected (*right*).

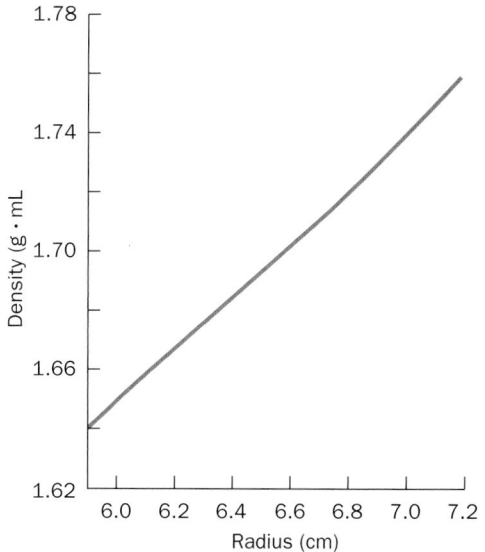

FIGURE 6-31 Equilibrium density distribution of a CsCl solution in an ultracentrifuge spinning at 39,460 rpm. The initial density of the solution was 1.7 g · mL^{-1}. [After Ifft, J.B., Voet, D.H., and Vinograd, J., *J. Phys. Chem.* **65,** 1138 (1961).]

6 ■ NUCLEIC ACID FRACTIONATION

In the preceding parts of this chapter we considered the most commonly used procedures for isolating and, to some extent, characterizing proteins. Most of these methods, often with some modification, are also regularly used to fractionate nucleic acids according to size, composition, and se-

quence. There are also many techniques that are applicable only to nucleic acids. In this section we outline some of the most useful of the procedures employed in the separation of nucleic acids.

A. *Solution Methods*

Nucleic acids in cells are invariably associated with proteins. Once cells have been broken open (Section 6-1B), their nucleic acids must be deproteinized. This can be accomplished by shaking (very gently if high molecular mass DNA is being isolated; Section 5-3D) the aqueous solution containing the protein–nucleic acid complex with a 25:24:1 mixture of phenol, chloroform, and isoamyl alcohol. The protein is thereby denatured and extracted into the water-immiscible organic phase, which is separated from the nucleic acid–containing aqueous phase by centrifugation (when a large amount of protein is present, it forms a white precipitate between the organic and aqueous phases). Alternatively (or in addition), protein can be dissociated from the nucleic acids by such denaturing agents as detergents, guanidinium chloride, or high salt concentrations, and/or it can be enzymatically degraded by proteases. In all cases, the nucleic acids, a mixture of RNA and DNA, can then be isolated by precipitation with ethanol. The RNA can be recovered from such a precipitate by treating it with **pancreatic DNase** to eliminate the DNA. Conversely, the DNA can be freed of RNA by treatment with **RNase.** Alternatively, RNA and DNA may be separated by ultracentrifugation (Section 6-5B).

In all these and subsequent manipulations, the nucleic acids must be protected from degradation by nucleases that occur both in the experimental materials and on human

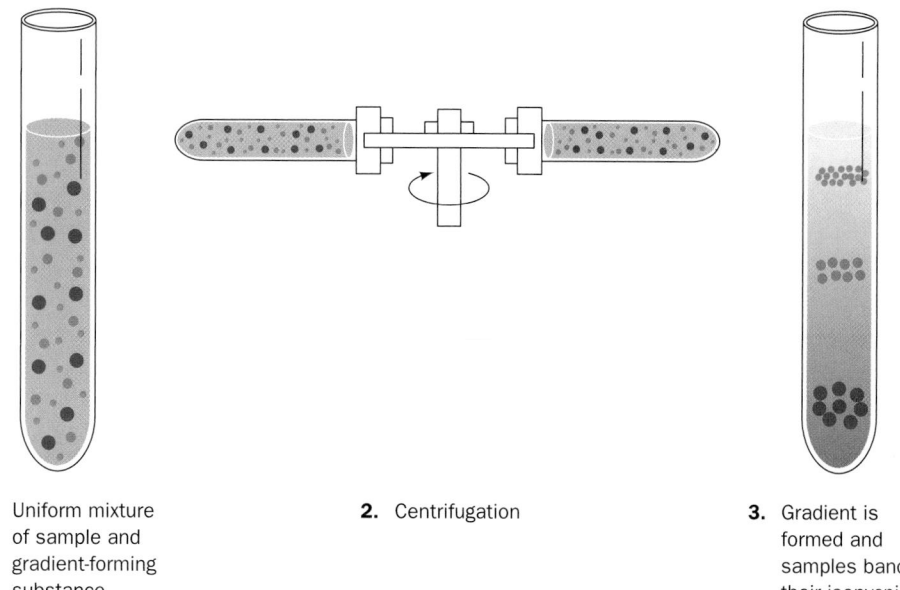

1. Uniform mixture of sample and gradient-forming substance

2. Centrifugation

3. Gradient is formed and samples band at their isopycnic positions

FIGURE 6-32 Isopycnic ultracentrifugation. The centrifugation starts with a uniform mixture of a macromolecular sample dissolved in a solution of a dense, fast-diffusing solute such as CsCl (*left*). At equilibrium in a centrifugal field, the solute forms a density gradient in which the macromolecules migrate to their positions of buoyant density (*right*).

hands. Nucleases may be inhibited by the presence of chelating agents such as **ethylenediaminetetraacetic acid (EDTA),**

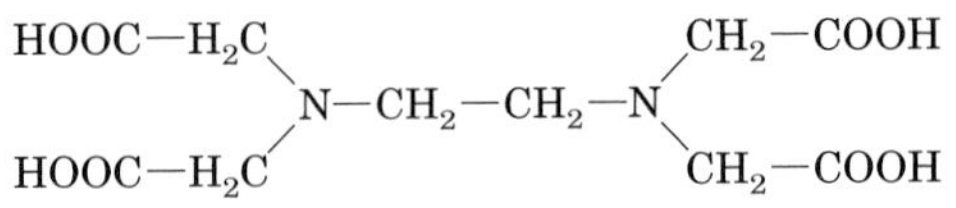

$$HOOC-H_2C \underset{HOOC-H_2C}{\diagdown} N-CH_2-CH_2-N \underset{CH_2-COOH}{\diagup} CH_2-COOH$$

Ethylene diamine tetraacetic acid (EDTA)

which sequester the divalent metal ions that nucleases require for activity. In cases where no nuclease activity can be tolerated, all glassware must be autoclaved to heat denature the nucleases and the experimenter should wear plastic gloves. Nevertheless, nucleic acids are generally easier to handle than proteins because their lack, in most cases, of a complex tertiary structure makes them relatively tolerant of extreme conditions.

B. *Chromatography*

Many of the chromatographic techniques that are used to separate proteins (Section 6-3) are also applicable to nucleic acids. Paper chromatography and thin layer chromatography are useful in fractionating oligonucleotides. They have been largely replaced, however, by the more powerful techniques of HPLC, particularly those using reverse-phase chromatography. Larger nucleic acids are often separated by procedures that include ion exchange chromatography and gel filtration chromatography.

a. Hydroxyapatite Can Be Used to Isolate and Fractionate DNA

Hydroxyapatite (a form of calcium phosphate; Section 6-3D) is particularly useful in the chromatographic purification and fractionation of DNA. Double-stranded DNA binds to hydroxyapatite more tightly than do most other molecules. Consequently, DNA can be rapidly isolated by passing a cell lysate through a hydroxyapatite column, washing the column with a phosphate buffer of concentration low enough to release only the RNA and proteins, and then eluting the DNA with a more concentrated phosphate solution. In addition, single-stranded DNA elutes from hydroxyapatite at a lower phosphate concentration than does double-stranded, DNA.

b. Messenger RNAs Can Be Isolated by Affinity Chromatography

Affinity chromatography (Section 6-3C) is useful in isolating specific nucleic acids. For example, most eukaryotic messenger RNAs (mRNAs) have a poly(A) sequence at their 3' ends (Section 5-4A). They can be isolated on an agarose or cellulose matrix to which poly(U) has been covalently attached. The poly(A) sequences specifically bind to the complementary poly(U) in high salt and at low temperatures and can later be released by altering these conditions. Moreover, if the (partial) sequence of an mRNA is known (e.g., as inferred from the corresponding protein's amino acid sequence), the complementary DNA strand may be synthesized (via methods discussed in Section 7-6A) and used to isolate that particular mRNA.

C. *Electrophoresis*

Nucleic acids of a given type may be separated by polyacrylamide gel electrophoresis (Sections 6-4B and 6-4C) because their electrophoretic mobilities in such gels vary inversely with their molecular masses. However, DNAs of more than a few thousand base pairs are too large to penetrate even a weakly cross-linked polyacrylamide gel. This difficulty is partially overcome through the use of agarose gels. By using gels with an appropriately low agarose content, relatively large DNAs in various size ranges may be fractionated. In this manner, plasmids, for example, may be separated from the larger chromosomal DNA of bacteria.

a. Duplex DNA Is Detected by Selectively Staining It with Intercalation Agents

The various DNA bands in a gel must be detected if they are to be isolated. Double-stranded DNA is readily stained by planar aromatic cations such as **ethidium ion, acridine orange,** or **proflavin.**

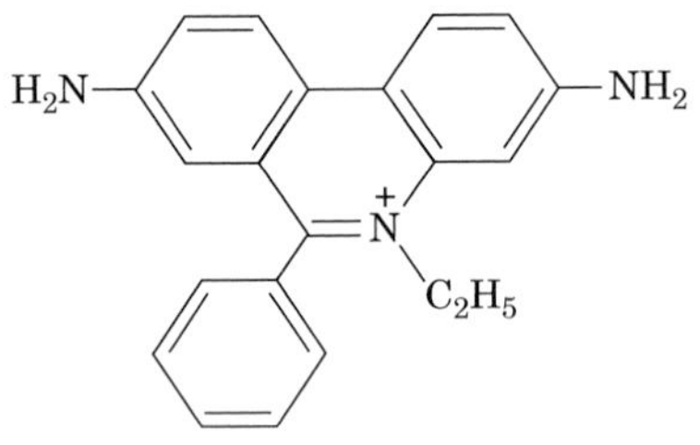

Ethidium

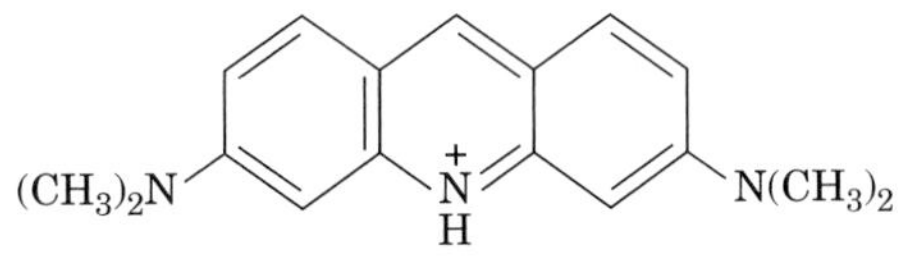

Acridine orange

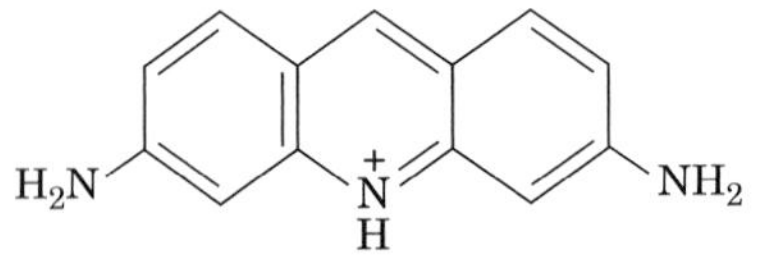

Proflavin

These dyes bind to duplex DNA by **intercalation** (slipping in between the stacked base pairs), where they exhibit a fluorescence under UV light that is far more intense than that of the free dye. As little as 50 ng of DNA may be detected in a gel by staining it with ethidium bromide (Fig. 6-33). Single-stranded DNA and RNA also stimulate the fluorescence of ethidium ion but to a lesser extent than does duplex DNA.

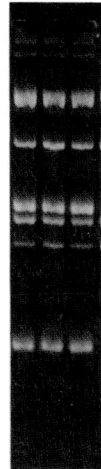

FIGURE 6-33 Agarose gel electrophoretogram of double heli-cal DNA. After electrophoresis, the gel was soaked in a solution of ethidium bromide, washed, and photographed under UV light. The fluorescence of the ethidium cation is strongly enhanced by binding to DNA, so each fluorescent band marks a different sized DNA fragment. The three parallel lanes contain identical DNA samples so as to demonstrate the technique's reproducibility. [Photo by Elizabeth Levine. From Freifelder, D., *Biophysical Chemistry. Applications to Biochemistry and Molecular Biology* (2nd ed.), p. 292, Freeman (1982). Used by permission.]

b. Very Large DNAs Are Separated by Pulsed-Field Gel Electrophoresis

The sizes of the DNAs that can be separated by conventional gel electrophoresis are limited to ~100,000 bp, even when gels containing as little as 0.1% agarose (which makes an extremely fragile gel) are used. However, the development of **pulsed-field gel electrophoresis (PFGE)** by Charles Cantor and Cassandra Smith extended this limit to DNAs with up to 10 million bp (6.6 million kD). The electrophoresis apparatus used in PFGE has two or more pairs of electrodes arrayed around the periphery of an agarose slab gel. The different electrode pairs are sequentially pulsed for times varying from 0.1 to 1000 s depending on the sizes of the DNAs being separated. Gel electrophoresis of DNA requires that these elongated molecules worm their way through the gel's labyrinthine channels more or less in the direction from the cathode to the anode. If the direction of the electric field abruptly changes, these DNAs must reorient their long axes along the new direction of the field before they can continue their passage through the gel. The time required to reorient very long gel-embedded DNA molecules evidently increases with their size. Consequently, a judicious choice of electrode distribution and pulse lengths causes shorter DNAs to migrate through the gel faster than longer DNAs, thereby effecting their separation (Fig. 6-34).

D. *Ultracentrifugation*

Equilibrium density gradient ultracentrifugation (Section 6-5B) in CsCl constitutes one of the most commonly used

DNA separation procedures. The buoyant density, ρ, of double-stranded Cs⁺ DNA depends on its base composition:

$$\rho = 1.660 + 0.098X_{G+C} \qquad [6.18]$$

so that a CsCl density gradient fractionates DNA according to its base composition. For example, eukaryotic DNAs often contain minor fractions that band separately from the major species. Some of these **satellite bands** consist of mitochondrial and chloroplast DNAs. Another important class of satellite DNA is composed of **repetitive sequences** that are short segments of DNA tandemly repeated hundreds, thousands, and in some cases millions of times in a genome (Section 34-2B). Likewise, plasmids may be separated from bacterial chromosomal DNA by equilibrium density gradient ultracentrifugation.

Single-stranded DNA is ~0.015 g · cm⁻³ denser than the corresponding double-stranded DNA so that the two may be separated by equilibrium density gradient ultracentrifugation. RNA is too dense to band in CsCl but does so in Cs₂SO₄ solutions. RNA–DNA hybrids will band in

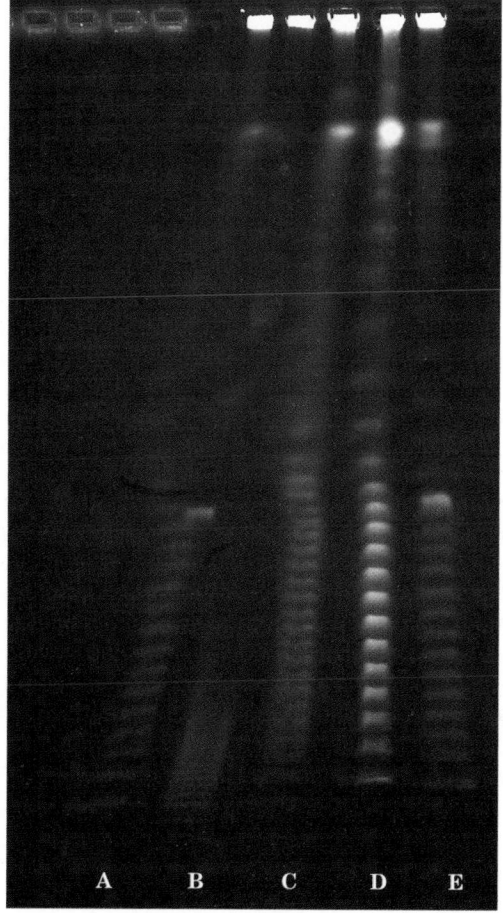

FIGURE 6-34 Pulsed-field gel electrophoresis (PFGE) of a set of different sized bacteriophage DNAs. Lanes A and E contain the same sample. The sizes of the DNAs decrease (their mobilities increase) toward the bottom of the electrophoretogram. [Courtesy of Charles Cantor, Boston University.]

CsCl but at a higher density than the corresponding duplex DNA.

RNA may be fractionated by zonal ultracentrifugation through a sucrose gradient (Section 6-5B). RNAs are separated by this technique largely on the basis of their size.

In fact, ribosomal RNA, which constitutes the major portion of cellular RNA, is classified according to its sedimentation rate; for example, the RNA of the *E. coli* small ribosomal subunit is known as **16S RNA** (Section 32-3A).

CHAPTER SUMMARY

1 ■ Protein Isolation Macromolecules in cells are solubilized by disrupting the cells by various chemical or mechanical means such as detergents or blenders. Partial purification by differential centrifugation is used after cell lysis to remove cell debris or to isolate a desired subcellular component. When out of the protective environment of the cell, proteins and other macromolecules must be treated so as to prevent their destruction by such influences as extremes of pH and temperature, enzymatic and chemical degradation, and rough mechanical handling. The state of purity of a substance being isolated must be monitored throughout the purification procedure by a specific assay.

2 ■ Solubilities of Proteins Proteins are conveniently purified on a large scale by a fractional precipitation process called salting out, in which protein solubilities are varied by changing the salt concentration or pH.

3 ■ Chromatographic Separations Ion exchange chromatography employs support materials such as cellulose or cross-linked dextran gels. Separations are based on differential electrostatic interactions between charged groups on the ion exchange materials and those on the substances being separated. Molecules may be located through their UV absorbance, fluorescence, radioactivity, or enzymatic activity. In gel filtration chromatography, molecules are separated according to their size and shape through the use of cross-linked dextran, polyacrylamide, or agarose beads that have pores of molecular dimensions. A calibrated gel filtration column can be used to estimate the molecular masses of macromolecules. Affinity chromatography separates biomolecules according to their unique biochemical abilities to bind other molecules specifically. High-performance liquid chromatography (HPLC) utilizes any of the foregoing separation techniques but uses high-resolution chromatographic materials, high solvent pressures, and automatic solvent mixing and monitoring systems so as to obtain much greater degrees of separation than are achieved with the more conventional chromatographic procedures. Adsorption chromatography, paper chromatography, thin layer chromatography (TLC), reverse-phase chromatography (RPC), and hydrophobic interaction chromatography (HIC) also have valuable biochemical applications.

4 ■ Electrophoresis In electrophoresis, charged molecules are separated according to their rates of migration in an electric field on a solid support such as paper, cellulose acetate, cross-linked polyacrylamide, or agarose. Gel electrophoresis employs a cross-linked polyacrylamide or agarose gel support, so that molecules are separated according to size by gel filtration as well as according to charge. The separated molecules may be visualized by means of stains, autoradiography, or immunoblotting. The anionic detergent sodium dodecyl sulfate (SDS) denatures proteins and uniformly coats them so as to give most proteins a similar charge density and shape. SDS–PAGE may be used to estimate macromolecular masses. In isoelectric focusing (IEF), macromolecules are immersed in a stable pH gradient and subjected to an electric field that causes them to migrate to their isoelectric positions. In capillary electrophoresis, the use of thin capillary tubes and high electric fields permits rapid and highly resolved separations of small amounts of material.

5 ■ Ultracentrifugation In ultracentrifugation, molecules are separated by subjecting them to gravitational fields large enough to counteract diffusional forces. Molecules may be separated and their molecular masses estimated from their rates of sedimentation through a solvent or a preformed gradient of an inert low molecular mass material such as sucrose. Alternately, molecules may be separated according to their buoyant densities in a solution with a density gradient of a dense, fast-diffusing substance such as CsCl. The deviation of a molecule's frictional ratio from unity is indicative of its degrees of solvation and elongation.

6 ■ Nucleic Acid Fractionation Nucleic acids can be fractionated by many of the techniques that are used to separate proteins. Hydroxyapatite chromatography separates single-stranded DNA from double-stranded DNA. Polyacrylamide or agarose gel electrophoresis separates DNA largely on the basis of size. Very large DNAs can be separated by pulsed-field gel electrophoresis (PFGE) on agarose gels. DNAs may be fractionated according to their base composition by CsCl density gradient ultracentrifugation. Different species of RNA can be separated by zonal ultracentrifugation through a sucrose gradient.

REFERENCES

GENERAL

Bollag, D.M., Rozycki, D., and Edelstein, S.J., *Protein Methods* (2nd ed.), Wiley-Liss (1996).

Boyer, R.F., *Modern Experimental Biochemistry* (3rd ed.), Benjamin Cummings (2001).

Burden, D.W. and Whitney, D.B., *Biotechnology: Proteins to PCR,* Chapters 2–6, and 8, Birkhäuser (1995).

Freifelder, D., *Physical Biochemistry* (2nd ed.), Freeman (1983). [A textbook on the techniques used in biophysical analysis.]

Harding, S.E. and Chowdhry, B.Z. (Eds.)., *Protein Ligand Interactions: Structure and Spectroscopy. A Practical Approach,* Oxford University Press (2001). [Contains descriptions of a variety of physical techniques for studying proteins and their interactions with other molecules.]

Janson, J.C. and Rydén, L. (Eds.), *Protein Purification: Principles, High Resolution Methods, and Applications,* Wiley (1998). [Contains detailed discussions of a variety of chromatographic and electrophoretic separation techniques.]

Karger, B.L. and Hancock, W.S. (Eds.), *High Resolution Separation and Analysis of Biological Macromolecules, Part A. Fundamentals; Part B. Applications, Methods Enzymol.* **270** and **271** (1996).

Marshak, D.R., Kadonaga, J.T., Burgess, R.R., Knuth, M.W., Brennan, W.A., Jr., and Lin, S.-H., *Strategies for Protein Purification and Characterization,* Cold Spring Harbor Laboratory Press (1996).

Pingoud, A., Urbanke, C., Hoggett, J., and Jeltsch, A., *Biochemical Methods. A Concise Guide for Students and Researchers,* Wiley–VCH (2002).

Roe, S. (Ed.), *Protein Purification Techniques. A Practical Approach* (2nd ed.); and *Protein Purification Applications. A Practical Approach* (2nd ed.), Oxford University Press (2001).

Rosenberg, I.M., *Protein Analysis and Purification: Benchtop Techniques,* Birkhäuser (1996).

Scopes, R., *Protein Purification: Principles and Practice* (3rd ed.), Springer-Verlag (1994).

Switzer, R. and Garrity, L., *Experimental Biochemistry. Theory and Exercises in Fundamental Methods* (3rd ed.), Freeman (1999).

Tinoco, I., Sauer, K., Wang, J.C., and Puglisi, J.C., *Physical Chemistry. Principles and Applications in Biological Sciences* (4th ed.), Chapter 6, Prentice-Hall (2002).

Walker, J.M. (Ed.), *The Protein Protocols Handbook* (2nd ed.), Humana Press (2002).

SOLUBILITY AND CRYSTALLIZATION

Arakawa, T. and Timasheff, S.N., Theory of protein solubility, *Methods Enzymol.* **114,** 49–77 (1985).

Ducruix, A. and Giegé, R.(Eds.), *Crystallization of Nucleic Acids and Proteins. A Practical Approach,* (2nd ed.), Oxford University Press (1999).

Edsall, J.T. and Wyman, J., *Biophysical Chemistry,* Vol. 1, Academic Press (1958). [A classic, detailed treatise on the acid–base and electrostatic properties of amino acids and proteins.]

McPherson, A., *Crystallization of Biological Macromolecules,* Cold Spring Harbor Laboratory Press (1999).

CHROMATOGRAPHY

Dean, P.D.G., Johnson, W.S., and Middle, F.A. (Eds.), *Affinity Chromatography. A Practical Approach,* IRL Press (1985).

Fischer, L., Gel filtration chromatography (2nd ed.), *in* Work, T.S. and Burdon, R.H. (Eds.), *Laboratory Techniques in Biochemistry and Molecular Biology,* Vol. 1, Part II, North-Holland Biomedical Press (1980).

Meyer, V.R., *Practical High-Performance Liquid Chromatography* (2nd ed.), Wiley (1994).

Oliver, R.W.A. (Ed.), *HPLC of Macromolecules. A Practical Approach* (2nd ed.), IRL Press (1998).

Rossomando, E.F., *HPLC in Enzymatic Analysis* (2nd ed.), Wiley (1998).

Schott, H., *Affinity Chromatography,* Dekker (1984).

Weston, A. and Brown, P.R., *HPLC and CE. Principles and Practice,* Academic Press (1997).

ELECTROPHORESIS

Altria, K.D., *Capillary Electrophoresis Guidebook,* Humana Press (1996).

Baker, D.R., *Capillary Electrophoresis,* Wiley (1995).

Burmeister, M. and Ulanovsky, L., *Pulsed-Field Gel Electrophoresis,* Humana Press (1992).

Cantor, C.R. and Schimmel, P.R., *Biophysical Chemistry,* Chapter 12, Freeman (1980).

Gersten, D.M., *Gel Electrophoresis: Proteins,* Wiley (1996).

Griffin, T.J. and Aebersold, R., Advances in proteome analysis by mass spectrometry, *J. Biol. Chem.* **276,** 45497–45500 (2001).

Hames, B.D. (Ed.), *Gel Electrophoresis of Proteins. A Practical Approach* (3rd ed.), IRL Press (1998).

Jones, P., *Gel Electrophoresis: Essential Techniques,* Wiley (1999).

Karger, B.L., Chu, Y.-H., and Foret, F., Capillary electrophoresis of proteins and nucleic acids, *Annu. Rev. Biophys. Biomol. Struct.* **24,** 579–610 (1995).

Monaco, A.P. (Ed.), *Pulsed Field Gel Electrophoresis. A Practical Approach,* IRL Press (1995).

Righetti, P.G., Immobilized pH gradients: Theory and methodology, *in* Burdon, R.H. and van Knippenberg, P.H. (Eds.), *Laboratory Techniques in Biochemistry and Molecular Biology,* Vol. 20, Elsevier (1990). [Discusses isoelectric focusing.]

Strahler, J.R. and Hanash, S.M., Immobilized pH gradients: Analytical and preparative use, *Methods* **3,** 109–114 (1991).

Wehr, T., Rodríeguez-Diaz, R., and Zhu, M., *Capillary Electrophoresis of Proteins,* Marcel Dekker (1999).

Westermeier, R., *Electrophoresis in Practice* (2nd ed.), VCH (1997).

ULTRACENTRIFUGATION

Cantor, C.R. and Schimmel, P.R., *Biophysical Chemistry,* Chapters 10 and 11, Freeman (1980).

Harding, S.E., Rowe, A.J., and Horton, J.C. (Eds.), *Analytical Ultracentrifugation in Biochemistry and Polymer Science,* Royal Society of Chemistry (1992).

Hesley, P., Defining the structure and stability of macromolecular assemblages in solution: The re-emergence of analytical ultracentrifugation as a practical tool, *Structure* **4,** 367–373 (1996).

Hinton, R. and Dobrata, M., Density gradient ultracentrifugation, *in* Work, T.S. and Work, E. (Eds.), *Laboratory Techniques in Biochemistry and Molecular Biology,* Vol. 6, Part I, North-Holland (1978).

Laue, T., Biophysical studies by ultracentrifugation, *Curr. Opin. Struct. Biol.* **11,** 579–583 (2001); *and* Laue, T.M. and Stafford, W.F., III, Modern applications of analytical ultracentrifugation, *Annu. Rev. Biophys. Biomol. Struct.* **28,** 75–100 (1999).

Schachman, H.K., *Ultracentrifugation in Biochemistry,* Academic Press (1959). [A classic treatise on ultracentrifugation.]

Schuster, T.M. and Toedt, J.M., New revolutions in the evolution of analytical ultracentrifugation, *Curr. Opin. Struct. Biol.* **6,** 650–658 (1996).

Stafford, W.F., III, Sedimentation velocity spins a new weave for an old fabric, *Curr. Opin. Biotech.* **8,** 14–24 (1997).

van Holde, K.E., Sedimentation analyses of proteins, *in* Neurath, H. and Hill, R.L. (Eds.), *The Proteins* (3rd ed.), Vol. 1, pp. 225–291, Academic Press (1975).

PROBLEMS

1. What are the ionic strengths of $1.0M$ solutions of NaCl, $(NH_4)_2SO_4$, and K_3PO_4? In which of these solutions would a protein be expected to be most soluble; least soluble?

2. An **isotonic saline solution** (one that has the same salt concentration as blood) is 0.9% NaCl. What is its ionic strength?

3. In what order will the following amino acids be eluted from a column of P-cellulose ion exchange resin by a buffer at pH 6: arginine, aspartic acid, histidine, and leucine?

4. In what order will the following proteins be eluted from a CM-cellulose ion exchange column by an increasing salt gradient at pH 7: fibrinogen, hemoglobin, lysozyme, pepsin, and ribonuclease A (see Table 6-1)?

5. What is the order of elution of the following proteins from a Sephadex G-50 column: catalase, α-chymotrypsin, concanavalin B, lipase, and myoglobin (see Table 6-5)?

6. Estimate the molecular mass of an unknown protein that elutes from a Sephadex G-50 column between cytochrome c and ribonuclease A (see Table 6-5).

7. A gel chromatography column of Bio-Gel P-30 with a bed volume of 100 mL is poured. The elution volume of the protein hexokinase (96 kD) on this column is 34 mL. That of an unknown protein is 50 mL. What are the void volume of the column, the volume occupied by the gel, and the relative elution volume of the unknown protein?

8. What chromatographic method would be suitable for separating the following pairs of substances? (a) Ala-Phe-Lys, Ala-Ala-Lys; (b) lysozyme, ribonuclease A (see Table 6-1); and (c) hemoglobin, myoglobin (see Table 6-1).

9. What is the order of the R_f values of the following amino acids in their paper chromatography with a water/butanol/acetic acid solvent system in which the pH of the aqueous phase is 4.5: alanine, aspartic acid, lysine, glutamic acid, phenylalanine, and valine?

10. The neurotransmitter γ-aminobutyric acid is thought to bind to specific receptor proteins in nerve tissue. Design a procedure for the partial purification of such a receptor protein.

11. A mixture of amino acids consisting of arginine, cysteine, glutamic acid, histidine, leucine, and serine is applied to a strip of paper and subjected to electrophoresis using a buffer at pH 7.5. What are the directions of migration of these amino acids and what are their relative mobilities?

*12. Sketch the appearance of a fingerprint of the following tripeptides: Asn-Arg-Lys, Asn-Leu-Phe, Asn-His-Phe, Asp-Leu-Phe, and Val-Leu-Phe. Assume the paper chromatographic step is carried out using a water/butanol/acetic acid solvent system (pH 4.5) and the electrophoretic step takes place in a buffer at pH 6.5.

13. What is the molecular mass of a protein that has a relative electrophoretic mobility of 0.5 in an SDS–polyacrylamide gel such as that of Fig. 6-25.

14. Explain why the molecular mass of fibrinogen is significantly overestimated when measured using a calibrated gel filtration column (Fig. 6-10) but can be determined with reasonable accuracy from its electrophoretic mobility on an SDS–polyacrylamide gel (see Table 6-5).

15. (a) What would be the relative arrangement of the following proteins after they had been subjected to isoelectric focusing: insulin, cytochrome c, histone, myoglobin, and ribonuclease A? (b) Sketch the appearance of a two-dimensional gel electrophoretogram of cytochrome c, myoglobin, and ribonuclease A (see Tables 6-1 and 6-5).

16. Calculate the centrifugal acceleration, in gravities (g's), on a particle located 6.5 cm from the axis of rotation of an ultracentrifuge rotating at 60,000 rpm ($1\ g = 9.81\ m \cdot s^{-2}$).

17. In a dilute buffer solution at 20°C, rabbit muscle aldolase has a frictional coefficient of $8.74 \times 10^{-8}\ g \cdot s^{-1}$, a sedimentation coefficient of 7.35 S, and a partial specific volume of $0.742\ cm^3 \cdot g^{-1}$. Calculate the molecular mass of aldolase assuming the density of the solution to be $0.998\ g \cdot cm^{-3}$.

*18. The sedimentation coefficient of a protein was measured by observing its sedimentation at 20°C in an ultracentrifuge spinning at 35,000 rpm.

Time, t (min)	Distance of Boundary from Center of Rotation, r (cm)
4	5.944
6	5.966
8	5.987
10	6.009
12	6.032

The density of the solution is $1.030\ g \cdot cm^{-3}$, the partial specific volume of the protein is $0.725\ cm^3 \cdot g^{-1}$, and its frictional coefficient is $3.72 \times 10^{-8}\ g \cdot s^{-1}$. Calculate the protein's sedimentation coefficient, in Svedbergs, and its molecular mass.

Chapter 7

Covalent Structures of Proteins and Nucleic Acids

Proteins are at the center of the action in biological processes. They function as enzymes, which catalyze the complex set of chemical reactions that are collectively referred to as life. Proteins serve as regulators of these reactions, both directly as components of enzymes and indirectly in the form of chemical messengers, known as hormones, as well as the receptors for those hormones. They act to transport and store biologically important substances such as metal ions, O_2, glucose, lipids, and many other molecules. In the form of muscle fibers and other contractile assemblies, proteins generate the coordinated mechanical motion of numerous biological processes, such as the separation of chromosomes during cell division and the movement of your eyes as you read this page. Proteins, such as **rhodopsin** in the retina of your eye, acquire sensory information that is processed through the action of nerve cell proteins. The proteins of the immune system, such as the **immunoglobulins,** form an essential biological defense system in higher animals. Proteins are major active elements in, as well as products of, the expression of genetic information. However, proteins also have important passive roles, such as that of **collagen,** which provides bones, tendons, and ligaments with their characteristic tensile strength. Clearly, there is considerable validity to the old cliché that proteins are the "building blocks" of life.

The function of DNA as the genetic archive and the association of RNA with protein synthesis have been known since the mid-twentieth century. However, it was not until the 1970s that it became clear that RNA can form structures whose complexities rival those of proteins, and it was not until the mid-1980s that it was shown that RNA has biologically important catalytic functions.

Protein and nucleic acid function can best be understood in terms of their structures, that is, the three-dimensional relationships between their component atoms. The structural descriptions of proteins and nucleic acids, as well as

(a)   – Lys – Ala – His – Gly – Lys – Lys – Val – Leu – Gly - Ala –
Primary structure (amino acid sequence in a polypeptide chain)

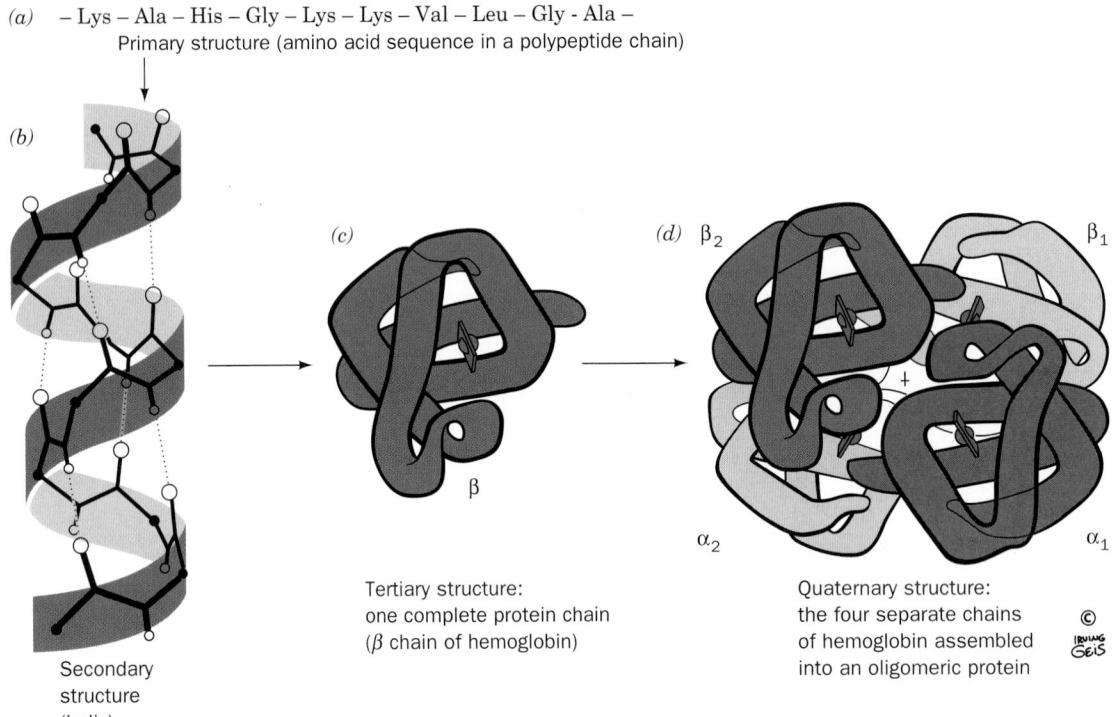

(b)

Secondary
structure
(helix)

(c)

Tertiary structure:
one complete protein chain
(β chain of hemoglobin)

(d)

Quaternary structure:
the four separate chains
of hemoglobin assembled
into an oligomeric protein

FIGURE 7-1   The structural hierarchy in proteins. (*a*) Primary structure, (*b*) secondary structure, (*c*) tertiary structure, and (*d*) quaternary structure. [Illustration, Irving Geis/Geis Archives Trust. Copyright Howard Hughes Medical Institute. Reproduced with permission.]

those of other polymeric materials, have been traditionally described in terms of four levels of organization (Fig. 7-1):

1. Primary structure (1° structure), which for a protein is the amino acid sequence of its polypeptide chain(s) and for a nucleic acid is its base sequence.

2. Secondary (2°) structure, which is the local spatial arrangement of a polypeptide's or a nucleic acid's backbone atoms without regard to the conformations of their side chains.

3. Tertiary (3°) structure, which refers to the three-dimensional structure of an entire polypeptide or polynucleotide chain. The distinction between secondary and tertiary structures is, of necessity, somewhat vague; in practice, the term "secondary structure" alludes to easily characterized structural entities such as helices.

4. Most proteins are composed of two or more polypeptide chains, loosely referred to as **subunits,** which associate through noncovalent interactions and, in some cases, disulfide bonds. A protein's **quaternary (4°) structure** refers to the spatial arrangement of its subunits. A nucleic acid's quaternary structure is similarly defined.

In this chapter, we discuss the 1° structures of proteins and nucleic acids: how they are elucidated and their biological and evolutionary significance. We also survey the field of bioinformatics as well as methods of chemically synthesizing polypeptide and oligonucleotide chains. The 2°, 3° and 4° structures of proteins and nucleic acids, as we shall

see, are a consequence of their 1° structures. For proteins, these topics are treated in Chapters 8 and 9, whereas for nucleic acids, they discussed mainly in Chapters 29, 31, and 32.

1 ■ PRIMARY STRUCTURE DETERMINATION OF PROTEINS

⬙ See Guided Exploration 4: Protein Sequence Determination
The first determination of the complete amino acid sequence of a protein, that of the bovine polypeptide hormone **insulin** by Frederick Sanger in 1953, was of enormous biochemical significance in that it definitively established that proteins have unique covalent structures. Since that time, the amino acid sequences of tens of thousands of proteins have been elucidated. This extensive information has been of central importance in the formulation of modern concepts of biochemistry for several reasons:

1. The knowledge of a protein's amino acid sequence is essential for an understanding of its molecular mechanism of action as well as being prerequisite for the elucidation of its X-ray and nuclear magnetic resonance (NMR) structures (Section 8-3A).

2. Sequence comparisons among analogous proteins from the same individual, from members of the same species, and from members of related species have yielded important insights into how proteins function and have in-

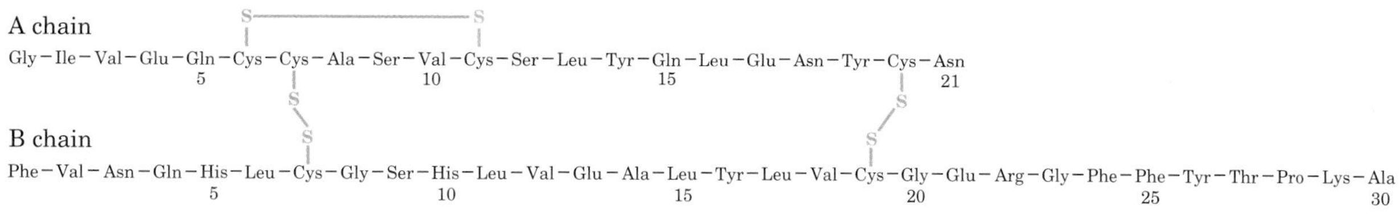

A chain

Gly−Ile−Val−Glu−Gln−Cys−Cys−Ala−Ser−Val−Cys−Ser−Leu−Tyr−Gln−Leu−Glu−Asn−Tyr−Cys−Asn

B chain

Phe−Val−Asn−Gln−His−Leu−Cys−Gly−Ser−His−Leu−Val−Glu−Ala−Leu−Tyr−Leu−Val−Cys−Gly−Glu−Arg−Gly−Phe−Phe−Tyr−Thr−Pro−Lys−Ala

FIGURE 7-2 Primary structure of bovine insulin. Note the intrachain and interchain disulfide bond linkages.

dicated the evolutionary relationships among the proteins and the organisms that produce them. These analyses, as we shall see in Section 7-3, complement and extend corresponding taxonometric studies based on anatomical comparisons.

3. Amino acid sequence analyses have important clinical applications because many inherited diseases are caused by mutations leading to an amino acid change in a protein. Recognition of this fact has led to the development of valuable diagnostic tests for many such diseases and, in many cases, to symptom-relieving therapy.

The elucidation of the 51-residue primary structure of insulin (Fig. 7-2) was the labor of many scientists over a period of a decade that altogether utilized ~100 g of protein. Procedures for primary structure determination have since been so refined and automated that proteins of similar size can be sequenced by an experienced technician in a few days using only a few micrograms of protein. The sequencing of the 1021-residue enzyme **β-galactosidase** in 1978 signaled that the sequence analysis of almost any protein could be reasonably attempted. Despite these technical advances, the basic procedure for primary structure determination using the techniques of protein chemistry is that developed by Sanger. The procedure consists of three conceptual parts, each of which requires several laboratory steps:

1. **Prepare the protein for sequencing:**
   a. Determine the number of chemically different polypeptide chains (subunits) in the protein.
   b. Cleave the protein's disulfide bonds.
   c. Separate and purify the unique subunits.
   d. Determine the subunits' amino acid compositions.

2. **Sequence the polypeptide chains:**
   a. Fragment the individual subunits at specific points to yield peptides small enough to be sequenced directly.
   b. Separate and purify the fragments.
   c. Determine the amino acid sequence of each peptide fragment.
   d. Repeat Step 2a with a fragmentation process of different specificity so that the subunit is cleaved at peptide bonds different from before. Separate these peptide fragments as in Step 2b and determine their amino acid sequences as in Step 2c.

3. **Organize the completed structure:**
   a. Span the cleavage points between one set of peptide fragments by the other. By comparison, the sequences of these sets of polypeptides can be arranged in the order that they occur in the subunit, thereby establishing its amino acid sequence.
   b. Elucidate the positions of the disulfide bonds, if any, between and within the subunits.

We discuss these various steps in the following sections.

A. *End Group Analysis: How Many Different Types of Subunits?*

Each polypeptide chain (if it is not chemically blocked or circular) has an N-terminal residue and a C-terminal residue. By identifying these **end groups,** we can establish the number of chemically distinct polypeptides in a protein. For example, insulin has equal amounts of the N-terminal residues Phe and Gly, which indicates that it has equal numbers of two nonidentical polypeptide chains.

a. N-Terminus Identification

There are several effective methods by which a polypeptide's N-terminal residue may be identified. **1-Dimethyl-amino-naphthalene-5-sulfonyl chloride (dansyl chloride)** reacts with primary amines (including the ε-amino group of Lys) to yield dansylated polypeptides (Fig. 7-3). Acid hydrolysis (Section 6-1D) liberates the N-terminal residue as a **dansylamino acid,** which exhibits such intense yellow fluorescence that it can be chromatographically identified from as little as 100 picomoles of material [1 picomole (pmol) = 10^{-12} mol].

In the most useful method of N-terminal residue identification, the **Edman degradation** (named after its inventor, Pehr Edman), **phenylisothiocyanate (PITC, Edman's reagent)** reacts with the N-terminal amino groups of proteins under mildly alkaline conditions to form their **phenylthiocarbamyl (PTC)** adduct (Fig. 7-4). This product is treated with an anhydrous strong acid such as trifluoroacetic acid, which cleaves the N-terminal residue as its **thiazolinone** derivative but does not hydrolyze other peptide bonds. *The Edman degradation therefore releases the N-terminal amino acid residue but leaves intact the rest of the polypeptide chain.* The thiazolinone-amino acid is selectively extracted into an organic solvent and is converted

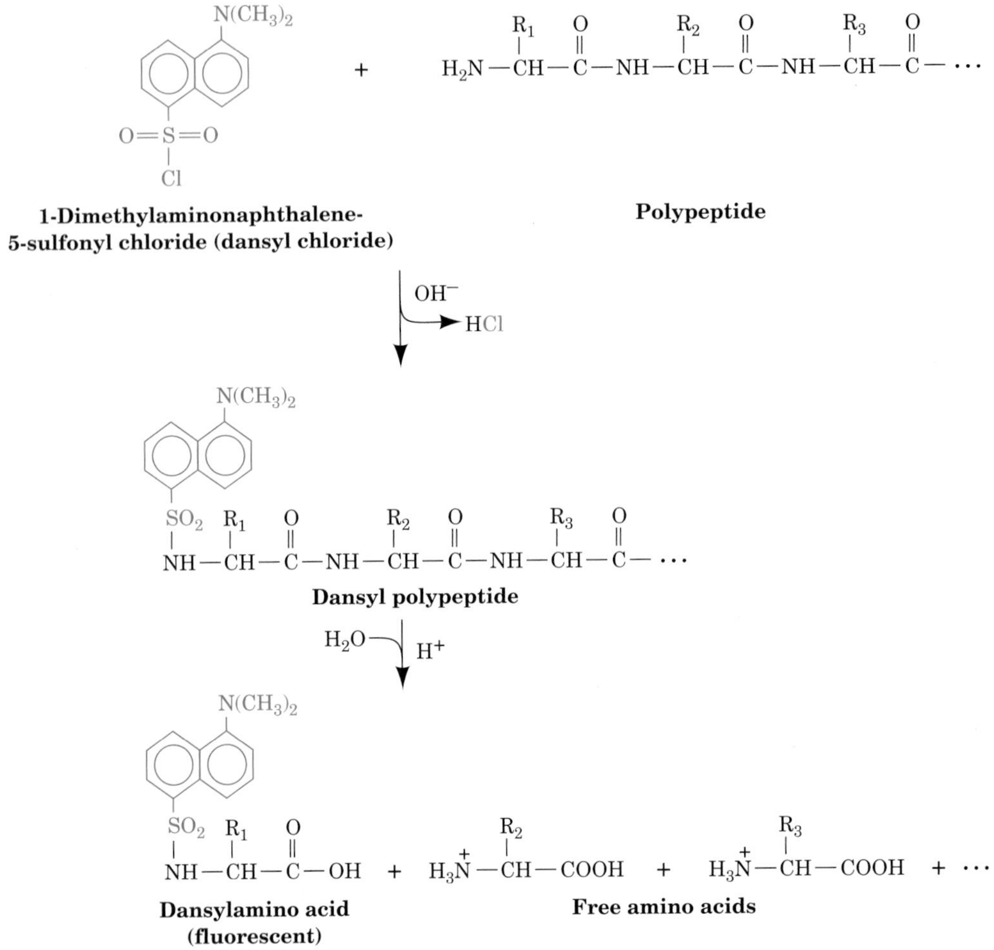

FIGURE 7-3   The reaction of dansyl chloride in end group analysis.

to the more stable **phenylthiohydantoin (PTH)** derivative by treatment with aqueous acid. This PTH-amino acid is most commonly identified by comparing its retention time on HPLC with those of known PTH-amino acids.

The most important difference between the Edman degradation and other methods of N-terminal residue identification is that *we can determine the amino acid sequence of a polypeptide chain from the N-terminus inward by subjecting the polypeptide to repeated cycles of the Edman degradation and, after every cycle, identifying the newly liberated PTH-amino acid.* This technique has been automated, resulting in great savings of time and materials (Section 7-1G).

b. C-Terminus Identification

There is no reliable chemical procedure comparable to the Edman degradation for the sequential end group analysis from the C-terminus of a polypeptide. This can be done enzymatically, however, using **exopeptidases** (enzymes that cleave a terminal residue from a polypeptide). One class of exopeptidases, the **carboxypeptidases,** catalyzes the hydrolysis of the C-terminal residues of polypeptides:

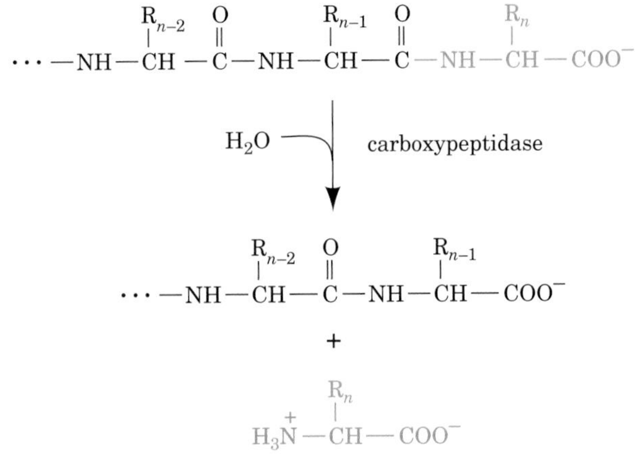

Carboxypeptidases, like all enzymes, are highly specific (selective) for the chemical identities of the substances whose reactions they catalyze (Section 13-2). The side chain specificities of the various carboxypeptidases in common use are listed in Table 7-1. The second type of

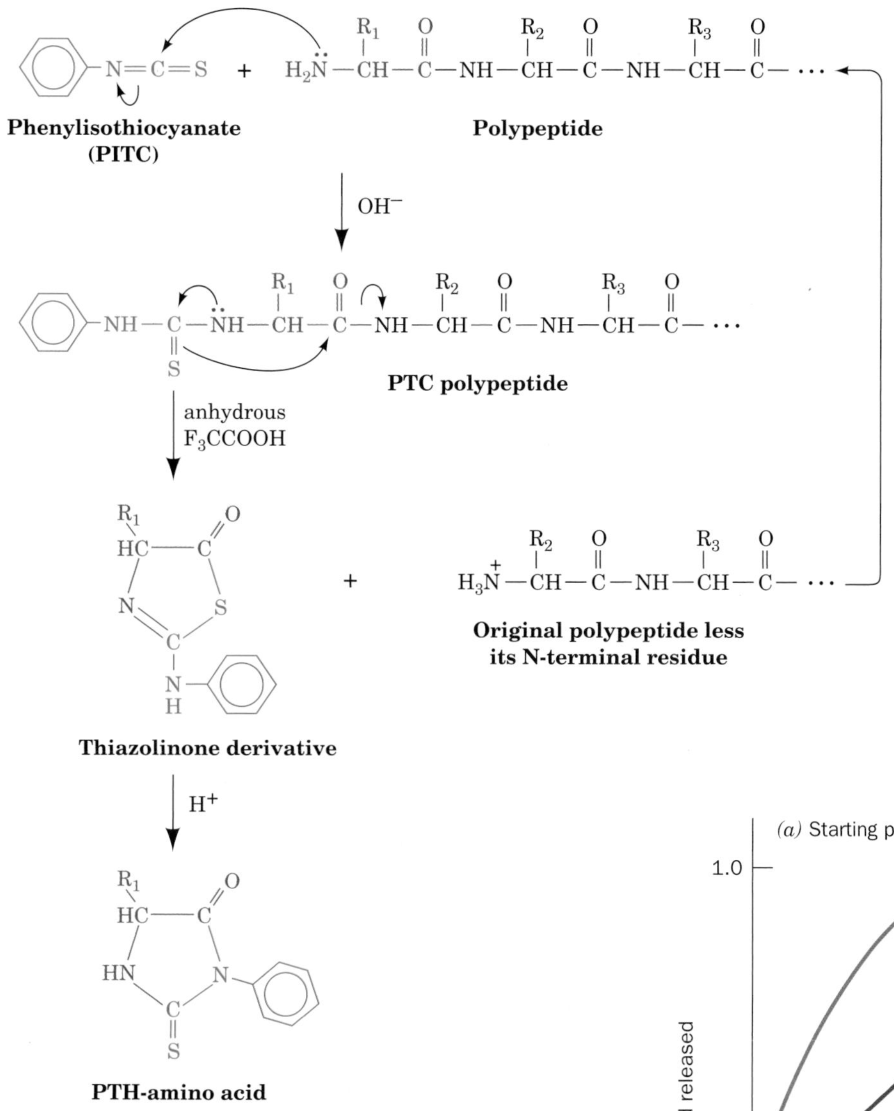

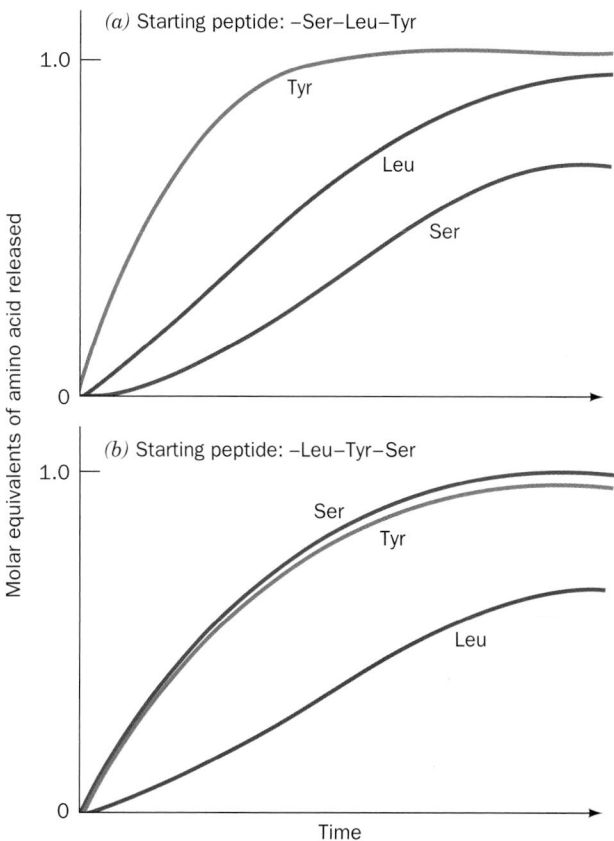

**Phenylisothiocyanate
(PITC)**

Polypeptide

OH⁻

PTC polypeptide

anhydrous
F₃CCOOH

Thiazolinone derivative

+

**Original polypeptide less
its N-terminal residue**

H⁺

PTH-amino acid

FIGURE 7-4   The Edman degradation. Note that the reaction occurs in three separate stages that each require quite different conditions. Amino acid residues can therefore be sequentially removed from the N-terminus of a polypeptide in a controlled stepwise fashion. *See the Animated Figures*

exopeptidase listed in Table 7-1, the **aminopeptidases,** sequentially cleave amino acids from the N-terminus of a polypeptide and have been similarly used to determine N-terminal sequences.

Why can't carboxypeptidases be used to determine amino acid sequences? If a carboxypeptidase cleaved all C-terminal residues at the same rate, irrespective of their identities, then by following the course of appearance of the various free amino acids in the reaction mixture (Fig. 7-5a), the sequence of several amino acids at the C-terminus could be determined. If, however, the second amino acid residue, for example, were cleaved at a much faster rate than the first, both amino acids would appear to be released simultaneously (Fig. 7-5b). Carboxypeptidases, in fact, exhibit selectivity toward side chains, so their use, either singly or in

(a) Starting peptide: –Ser–Leu–Tyr

(b) Starting peptide: –Leu–Tyr–Ser

Molar equivalents of amino acid released

Time

FIGURE 7-5   The hypothetical rate of the carboxypeptidase-catalyzed release of amino acids from peptides having the indicated C-terminal sequences. (*a*) All bonds cleaved at the same rate. (*b*) Ser removed slowly, Tyr cleaved rapidly, and Leu cleaved at an intermediate rate.

TABLE 7-1   Specificities of Various Exopeptidases

Enzyme	Source	Specificity[a]
Carboxypeptidase A	Bovine pancreas	$R_n \neq$ Arg, Lys, Pro; $R_{n-1} \neq$ Pro
Carboxypeptidase B	Bovine pancreas	$R_n =$ Arg, Lys; $R_{n-1} \neq$ Pro
Carboxypeptidase C	Citrus leaves	All free C-terminal residues; pH optimum = 3.5
Carboxypeptidase Y	Yeast	All free C-terminal residues, but slowly with $R_n =$ Gly
Leucine aminopeptidase	Porcine kidney	$R_1 \neq$ Pro
Aminopeptidase M	Porcine kidney	All free N-terminal residues

[a] R_1 = the N-terminal residue; R_n = the C-terminal residue.

mixtures, rarely reveals the order of more than the first few C-terminal residues of a polypeptide.

C-Terminal residues with a preceding Pro residue are not subject to cleavage by carboxypeptidases A and B (Table 7-1). Chemical methods are therefore usually employed to identify their C-terminal residue. In the most reliable such chemical method, **hydrazinolysis,** a polypeptide is treated with anhydrous **hydrazine** at 90°C for 20 to 100 h in the presence of a mildly acidic ion exchange resin (which acts as a catalyst):

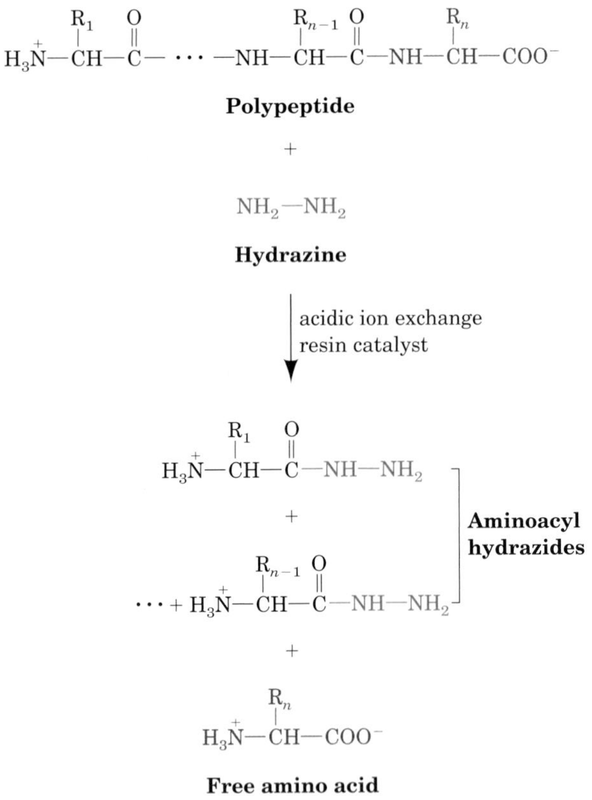

$$
\begin{array}{ccccc}
& R_1 & O & & R_{n-1} & O & & R_n \\
& | & || & & | & || & & | \\
H_3\overset{+}{N}-CH-C- & \cdots & -NH-CH-C-NH-CH-COO^-
\end{array}
$$

Polypeptide

+

$$NH_2-NH_2$$

Hydrazine

↓ acidic ion exchange resin catalyst

$$
\begin{array}{ccc}
& R_1 & O \\
& | & || \\
H_3\overset{+}{N}-CH-C-NH-NH_2 \\
\end{array}
$$

+

$$
\begin{array}{ccc}
& R_{n-1} & O \\
& | & || \\
\cdots + H_3\overset{+}{N}-CH-C-NH-NH_2 \\
\end{array}
$$
⎤ **Aminoacyl hydrazides**

+

$$
\begin{array}{cc}
& R_n \\
& | \\
H_3\overset{+}{N}-CH-COO^-
\end{array}
$$

Free amino acid

All the peptide bonds are thereby cleaved, yielding the aminoacyl hydrazides of all the amino acid residues except that of the C-terminal residue, which is released as the free amino acid and therefore can be identified chromatographically. Unfortunately, hydrazinolysis is subject to a great many side reactions that have largely limited its application to carboxypeptidase-resistant polypeptides.

B. *Cleavage of the Disulfide Bonds*

The next step in the sequence analysis is to cleave the disulfide bonds between Cys residues. This is done for two reasons:

1. To permit the separation of polypeptide chains (if they are disulfide linked).

2. To prevent the native protein conformation, which is stabilized by disulfide bonds, from obstructing the action of the proteolytic (protein-cleaving) agents used in primary structure determinations (Section 7-1E).

Disulfide bond locations are established in the final step of the sequence analysis (Section 7-1I).

Disulfide bonds are most often cleaved reductively by treatment with **2-mercaptoethanol**

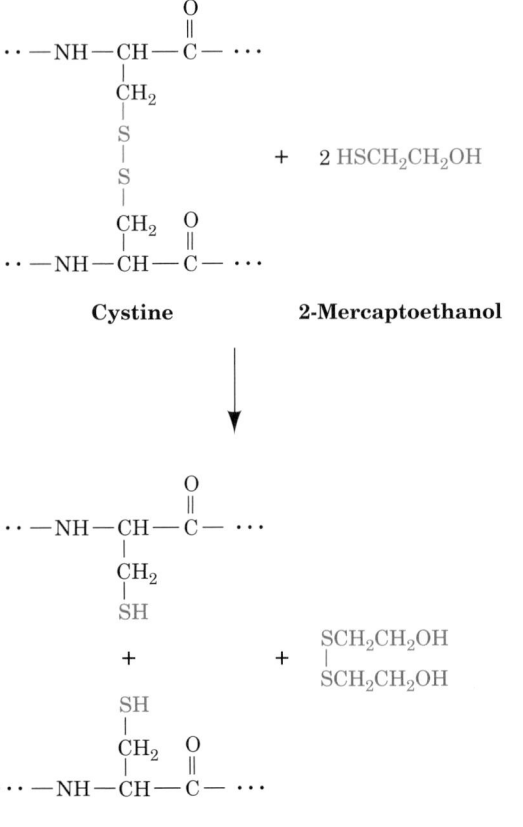

Cystine   **2-Mercaptoethanol**

Cysteine

or either of the diastereomers **dithiothreitol** or **dithioery-thritol (Cleland's reagent)**:

$$
\begin{array}{c}
\text{O} \\
\parallel \\
\cdots -\text{NH}-\text{CH}-\text{C}- \cdots \\
\mid \\
\text{CH}_2 \\
\mid \\
\text{S} \\
\mid \\
\text{S} \\
\mid \\
\text{CH}_2 \quad \text{O} \\
\mid \quad \parallel \\
\cdots -\text{NH}-\text{CH}-\text{C}- \cdots
\end{array}
\qquad + \qquad
\begin{array}{c}
\text{CH}_2\text{SH} \\
\mid \\
\text{CHOH} \\
\mid \\
\text{CHOH} \\
\mid \\
\text{CH}_2\text{SH}
\end{array}
$$

Cystine   **Dithiothreitol** or **Dithioerythritol**

$$
\downarrow
$$

$$
\begin{array}{c}
\text{O} \\
\parallel \\
\cdots -\text{NH}-\text{CH}-\text{C}- \cdots \\
\mid \\
\text{CH}_2 \\
\mid \\
\text{SH} \\
+ \\
\text{SH} \\
\mid \\
\text{CH}_2 \quad \text{O} \\
\mid \quad \parallel \\
\cdots -\text{NH}-\text{CH}-\text{C}- \cdots
\end{array}
\qquad + \qquad
\begin{array}{c}
\text{HO} \\
\text{HO}
\end{array}
$$

In order to expose all disulfide groups to the reducing agent, the reaction is usually carried out under conditions that denature the protein. The resulting free sulfhydryl groups are alkylated, usually by treatment with **iodoacetic acid,**

$$
\text{Cys}-\text{CH}_2-\text{SH} \quad + \quad \text{ICH}_2\text{COO}^-
$$

Cysteine   **Iodoacetate**

$$
\longrightarrow
$$

$$
\text{Cys}-\text{CH}_2-\text{S}-\text{CH}_2\text{COO}^- \quad + \quad \text{HI}
$$

S-Carboxymethylcysteine

to prevent the reformation of disulfide bonds through oxidation by O_2. *S*-Alkyl derivatives are stable in air and under the conditions used for the subsequent cleavage of peptide bonds.

C. *Separation, Purification, and Characterization of the Polypeptide Chains*

A protein's nonidentical polypeptides must be separated and purified in preparation for their amino acid sequence

determination. Subunit dissociation, as well as denaturation, occurs under acidic or basic conditions, at low salt concentrations, at elevated temperatures, or through the use of denaturing agents such as urea, guanidinium ion (Section 5-5G), or detergents such as sodium dodecyl sulfate (SDS; Section 6-4C). The dissociated subunits can then be separated by methods described in Chapter 6 that capitalize on small differences in polypeptide size and polarity. Ion exchange and gel filtration chromatography are most often used.

It is, of course, desirable to know the number of residues in the polypeptide to be sequenced, which can be estimated from its molecular mass ($\sim$110 D/residue). Molecular mass can be measured with an accuracy of no better than 5 to 10% by the usual laboratory techniques of gel filtration chromatography and SDS–PAGE (Sections 6-3B and 6-4C). In recent years, however, mass spectrometry (Section 7-1J) has provided a faster and far more accurate means to determine the molecular masses of macromolecules. Mass spectrometry can determine the molecular masses of picomolar amounts of >100 kD polypeptides with accuracies of $\sim$0.01%.

D. *Amino Acid Composition*

Before we begin the actual sequencing of a polypeptide chain, it is useful to know its amino acid composition, that is, the number of each type of amino acid residue present. *The amino acid composition of a subunit is determined by its complete hydrolysis followed by the quantitative analysis of the liberated amino acids.* Polypeptide hydrolysis can be accomplished by either chemical (acid or base) or enzymatic means, although none of these methods alone is fully satisfactory. For acid-catalyzed hydrolysis, the polypeptide is dissolved in 6*M* HCl, sealed in an evacuated tube to prevent the air oxidation of the sulfur-containing amino acids, and heated at 100 to 120°C for 10 to 100 h. The long hydrolysis times are required for the complete liberation of the aliphatic amino acids Val, Leu, and Ile. Unfortunately, not all side chains are impervious to these harsh conditions. Ser, Thr, and Tyr are partially degraded, although, by following their disappearance as a function of hydrolysis time, correction factors for these losses can be established. A more serious problem is that acid hydrolysis largely destroys Trp residues. Moreover, Gln and Asn are converted to Glu and Asp plus NH_4^+, so that only the amounts of Asx (= Asp + Asn), Glx (= Glu + Gln), and NH_4^+ (= Asn + Gln) can be independently measured after acid hydrolysis.

Base-catalyzed hydrolysis of polypeptides is carried out in 2 to 4*M* NaOH at 100°C for 4 to 8 h. This treatment is even more problematic because it causes the decomposition of Cys, Ser, Thr, and Arg and partially deaminates and racemizes the other amino acids. Hence alkaline hydrolysis is principally used to determine Trp content.

The complete enzymatic digestion of a polypeptide requires mixtures of peptidases because individual pepti-

TABLE 7-2   Specificities of Various Endopeptidases

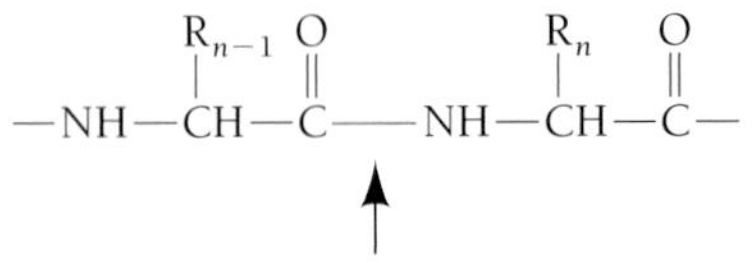

Enzyme	Source	Specificity	Comments

$$-NH-\underset{\underset{R_{n-1}}{|}}{CH}-\underset{\underset{O}{\|}}{C}-NH-\underset{\underset{R_n}{|}}{CH}-\underset{\underset{O}{\|}}{C}-$$

**Scissile
peptide bond**

Enzyme	Source	Specificity	Comments
Trypsin	Bovine pancreas	R_{n-1} = positively charged residues: Arg, Lys; $R_n \neq$ Pro	Highly specific
Chymotrypsin	Bovine pancreas	R_{n-1} = bulky hydrophobic residues: Phe, Trp, Tyr; $R_n \neq$ Pro	Cleaves more slowly for R_{n-1} = Asn, His, Met, Leu
Elastase	Bovine pancreas	R_{n-1} = small neutral residues: Ala, Gly, Ser, Val; $R_n \neq$ Pro	
Thermolysin	*Bacillus thermoproteolyticus*	R_n = Ile, Met, Phe, Trp, Tyr, Val; $R_{n-1} \neq$ Pro	Occasionally cleaves at R_n = Ala, Asp, His, Thr; heat stable
Pepsin	Bovine gastric mucosa	R_n = Leu, Phe, Trp, Tyr; $R_{n-1} \neq$ Pro	Also others; quite nonspecific; pH optimum 2
Endopeptidase Arg-C	Mouse submaxillary gland	R_{n-1} = Arg	May cleave at R_{n-1} = Lys
Endopeptidase Asp-N	*Pseudomonas fragi*	R_n = Asp	May cleave at R_n = Glu
Endopeptidase Glu-C	*Staphylococcus aureus*	R_{n-1} = Glu	May cleave at R_{n-1} = Gly
Endopeptidase Lys-C	*Lysobacter enzymogenes*	R_{n-1} = Lys	May cleave at R_{n-1} = Asn

dases do not cleave all peptide bonds. Tables 7-1 and 7-2 indicate the specificities of the exopeptidases and **endopeptidases** (enzymes that catalyze the hydrolysis of internal peptide bonds) commonly used for this purpose. **Pronase,** a mixture of relatively nonspecific proteases from *Streptomyces griseus,* is also often used to effect complete proteolysis. The amount of enzyme used is limited to ~1% by weight of the polypeptide to be hydrolyzed because proteolytic enzymes, being proteins themselves, are self-degrading, so that they will, if used too generously, significantly contaminate the final digest. Enzymatic digestion is most often used for determining the amounts of Trp, Asn, and Gln in a polypeptide, which are destroyed by the harsher chemical methods.

a. Amino Acid Analysis Has Been Automated

The amino acid content of a polypeptide hydrolysate can be quantitatively determined through the use of an automated **amino acid analyzer.** Such an instrument separates amino acids by ion exchange chromatography, a technique pioneered by William Stein and Stanford Moore, or by reverse-phase chromatography using HPLC (Section 6-3D). The amino acids are pre- or postcolumn derivatized by treatment with either dansyl chloride, Edman's reagent, or *o*-phthalaldehyde (OPA) + 2-mercaptoethanol. The latter reagents react with amino acids to form highly fluorescent adducts:

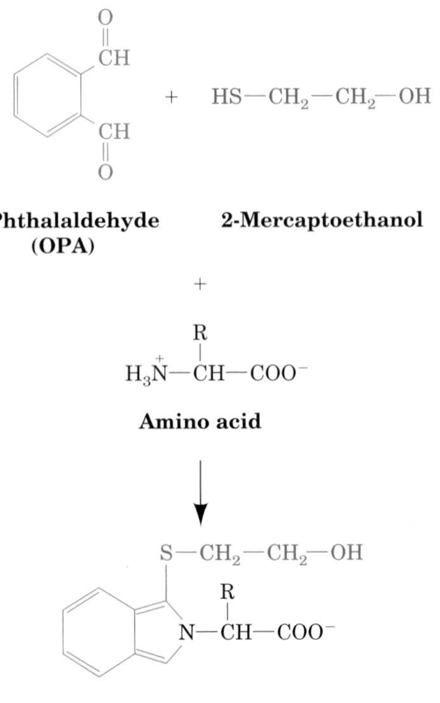

o-Phthalaldehyde
(OPA)   2-Mercaptoethanol

+

$$H_3\overset{+}{N}-\underset{\underset{R}{|}}{CH}-COO^-$$

Amino acid

Fluorescent adduct

The amino acids are then identified according to their characteristic elution volumes (retention times on HPLC;

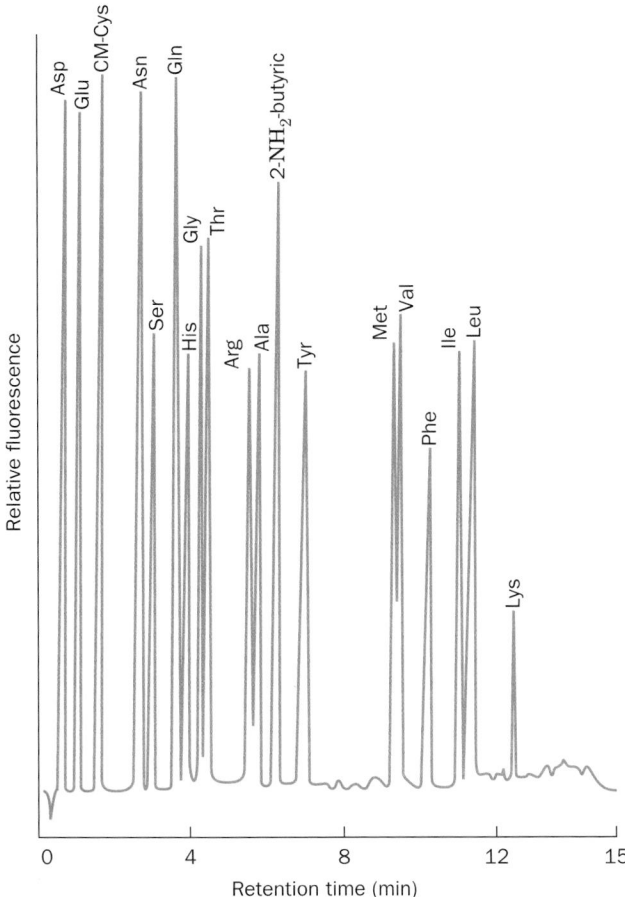

FIGURE 7-6 Amino acid analysis. The reverse-phase HPLC separation of precolumn OPA-derivatized amino acids. [After Hunkapiller, M.W., Strickler, J.E., and Wilson, K.J., *Science* **226**, 309 (1984).]

Fig. 7-6) and the amounts present determined by their fluorescence intensities (UV absorbances for PTC-amino acids). With modern amino acid analyzers, the complete analysis of a protein digest can be performed in <1 h with a sensitivity that can detect <1 pmol of each amino acid.

b. The Amino Acid Compositions of Proteins Are Indicative of Their Structures

The amino acid analysis of a vast number of proteins indicates that they have considerable variation with respect to their amino acid compositions. Leu, Ala, Gly, Ser, Val, and Glu are the most common amino acid residues (>6% abundance; Table 4-1), whereas His, Met, Cys, and Trp occur least frequently (<3% abundance). Indeed, many proteins lack one or more of the "standard" amino acid residues. The ratio of polar to nonpolar residues is generally >1 for globular proteins and tends to decrease with increasing protein size. This is because, as we shall see in Chapter 8, globular proteins have a hydrophobic core and a hydrophilic exterior; that is, they have a micellelike structure. Nonpolar residues predominate in membrane-bound proteins, however, because these proteins, being immersed in a nonpolar environment (Section 12-3A), must also have a hydrophobic exterior.

E. *Specific Peptide Cleavage Reactions*

Polypeptides that are longer than 40 to 100 residues cannot be directly sequenced (Section 7-1G). Polypeptides of greater length must therefore be cleaved, either enzymatically or chemically, to fragments small enough to be sequenced. In either case, the cleavage process must be complete and highly specific so that the aggregate sequence of a subunit's peptide fragments, when correctly ordered, is that of the intact subunit.

a. Trypsin Specifically Cleaves Peptide Bonds after Positively Charged Residues

Endopeptidases, like exopeptidases, have side chain requirements for the residues flanking the **scissile** (to be cleaved) peptide bond. The side chain specificities of the endopeptidases most commonly used to fragment polypeptides are listed in Table 7-2. The digestive enzyme **trypsin** has the greatest specificity and is therefore the most valuable member of the arsenal of peptidases used to fragment polypeptides. It cleaves peptide bonds on the C-side (toward the carboxyl terminus) of the positively charged residues Arg and Lys if the next residue is not Pro:

Since trypsin cleaves peptide bonds that follow positively charged residues, trypsin cleavage sites may be added to or deleted from a polypeptide by chemically adding or deleting positive charges to or from its side chains. For example, the positive charge on Lys is eliminated by treatment with a dicarboxylic anhydride such as **citraconic anhydride.** The reagent forms a negatively charged derivative of the Lys ε-amino group that trypsin does not recognize:

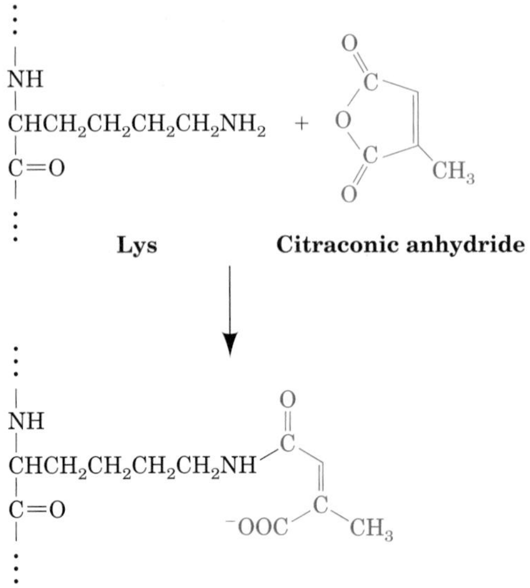

Lys   **Citraconic anhydride**

After trypsin hydrolysis of the polypeptide, the Lys residue can be deblocked for identification by mild acid (pH 2–3) hydrolysis. Conversely, Cys may be **aminoalkylated** by a **β-haloamine** to yield a positively charged residue that is subject to tryptic cleavage:

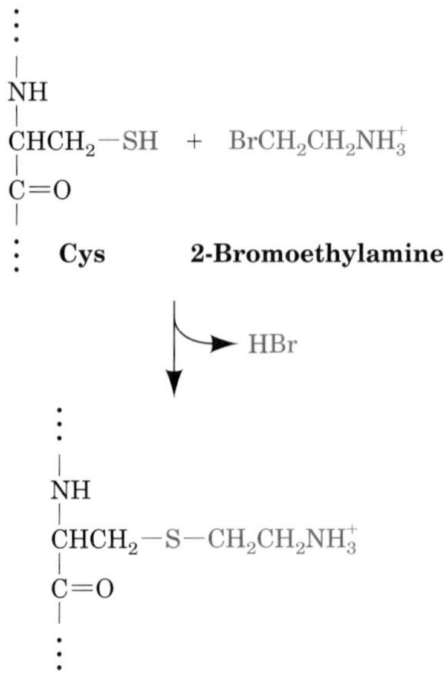

Cys   **2-Bromoethylamine**

Such reactions extend the use of trypsin to take further advantage of its great specificity.

The other endopeptidases listed in Table 7-2 exhibit broader side chain specificities than trypsin and often yield a series of peptide fragments with overlapping sequences. However, through **limited proteolysis,** that is, by adjusting reaction conditions and limiting reaction times, these less specific endopeptidases can yield useful peptide fragments. This is because the complex native structure of a protein (subunit) buries many otherwise enzymatically susceptible peptide bonds beneath the surface of the protein molecule. With proper conditions and reaction times, only those peptide bonds in the native protein that are initially accessible to the peptidase will be hydrolyzed. Limited proteolysis is often employed to generate peptide fragments of useful size from subunits that have too many or too few

Arg and Lys residues to do so with trypsin (although if too many are present, limited proteolysis with trypsin may also yield useful fragments).

b. Cyanogen Bromide Specifically Cleaves Peptide Bonds after Met Residues

Several chemical reagents promote peptide bond cleavage at specific residues. The most useful of these, **cyanogen bromide** (CNBr), causes specific and quantitative cleavage on the C-side of Met residues to form a **peptidyl homoserine lactone:**

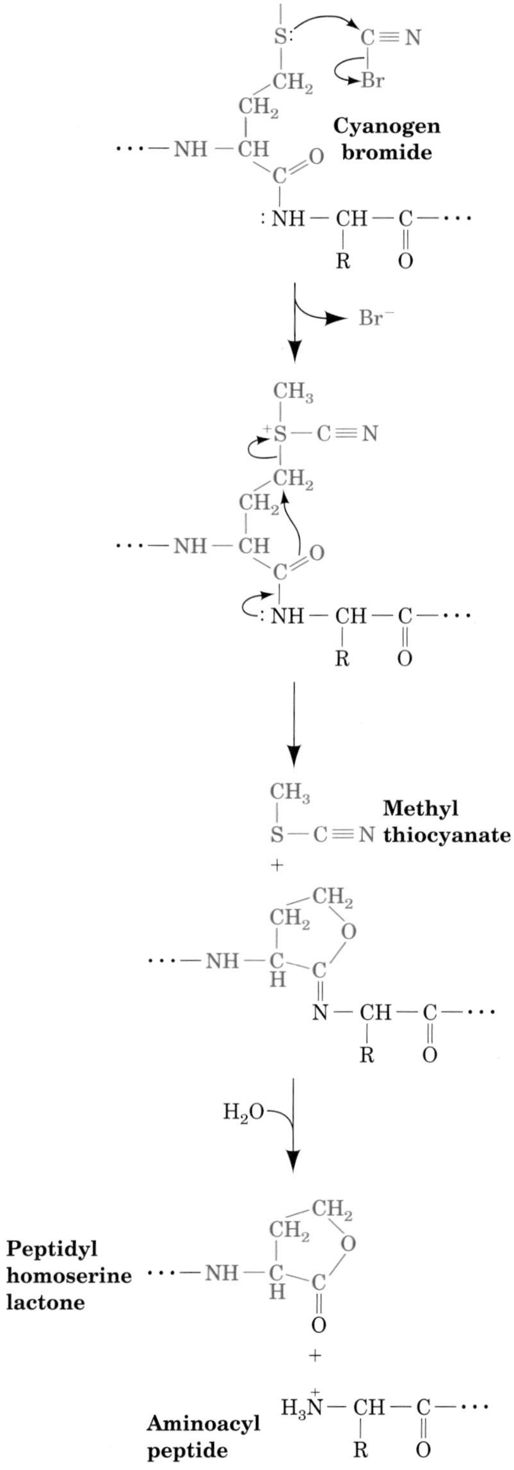

The reaction is performed in an acidic solvent ($0.1M$ HCl or 70% formic acid) that denatures most proteins so that cleavage normally occurs at all Met residues.

A peptide fragment generated by a specific cleavage process may still be too large to sequence. In that case, after its purification, it can be subjected to a second round of fragmentation using a different cleavage process.

F. *Separation and Purification of the Peptide Fragments*

Once again we must employ separation techniques, this time to isolate the peptide fragments of specific cleavage operations for subsequent sequence determinations. The nonpolar residues of peptide fragments are not excluded from the aqueous environment as they are in native proteins (Chapter 8). Consequently, many peptide fragments aggregate, precipitate, and/or strongly adsorb to chromatographic materials, which can result in unacceptable peptide losses. Until around 1980, the trial-and-error development of methods that could satisfactorily separate a mixture of peptide fragments constituted the major technical challenge of a protein sequence determination, as well as its most time-consuming step. Such methods involved the use of denaturants, such as urea and SDS, to solubilize the peptide fragments, and the selection of chromatographic materials and conditions that would reduce their adsorptive losses. The advent of reverse-phase chromatography by HPLC (Section 6-3D), however, has largely reduced the separation of peptide fragments to a routine procedure.

G. *Sequence Determination*

Once the manageably sized peptide fragments that were formed through specific cleavage reactions have been isolated, their amino acid sequences can be determined. *This is done through repeated cycles of the Edman degradation* (Section 6-1A). An automated device for doing so was first developed by Edman and Geoffrey Begg. In modern sequencers, the peptide sample is adsorbed onto a polyvinyl-

idine difluoride (PVDF) membrane or dried onto glass fiber paper which is impregnated with **polybrene** (a polymeric quaternary ammonium salt). In either case, the peptide is immobilized but is readily accessible to Edman reagents. Accurately measured quantities of reagents, either in solution or as vapors in a stream of argon (which minimizes peptide loss), are then delivered to the reaction cell at programmed intervals. The thiazolinone-amino acids are automatically removed, converted to the corresponding PTH-amino acids (Fig. 7-4), and identified via HPLC. Such instruments are capable of processing up to one residue per hour.

A peptide's 40 to 60 N-terminal residues can usually be identified (100 or more with the most advanced systems) before the cumulative effects of incomplete reactions, side reactions, and peptide loss make further amino acid identification unreliable. As little as 0.1 pmol of a PTH-amino acid can be reliably identified by the UV detector–equipped reverse-phase HPLC systems used in advanced sequencers. Consequently, a peptide's N-terminal 5 to 25 residues can, respectively, be determined with as little as 1 to 10 pmol of the peptide—invisibly small amounts.

H. *Ordering the Peptide Fragments*

With the peptide fragments individually sequenced, what remains is to elucidate the order in which they are connected in the original polypeptide. *We do so by comparing the amino acid sequences of one set of peptide fragments with those of a second set whose specific cleavage sites overlap those of the first set (Fig. 7-7).* The overlapping peptide segments must be of sufficient length to identify each cleavage site uniquely, but as there are 20 possibilities for each amino acid residue, an overlap of only a few residues is usually enough.

I. *Assignment of Disulfide Bond Positions*

The final step in an amino acid sequence analysis is to determine the positions (if any) of the disulfide bonds. This is done by cleaving a sample of the native protein under con-

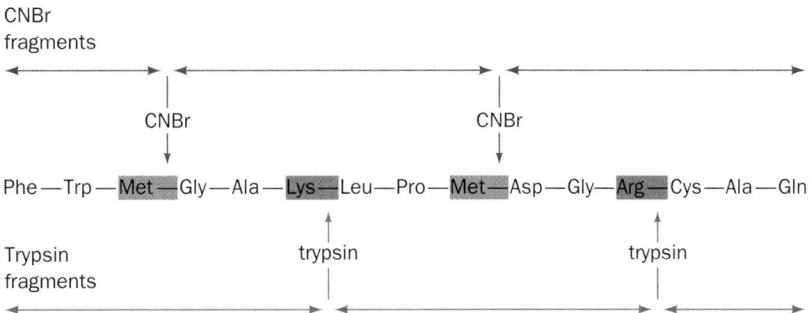

FIGURE 7-7 The amino acid sequence of a polypeptide chain is determined by comparing the sequences of two sets of mutually overlapping peptide fragments. In this example, the two sets of peptide fragments are generated by cleaving the polypeptide after all its Arg and Lys residues with trypsin and, in a separate reaction, after all its Met residues by treatment with cyanogen bromide. The order of the first two tryptic peptides is established, for example, by the observation that the Gly-Ala-Lys-Leu-Pro-Met cyanogen bromide peptide has its N- and C-terminal sequences in common with the C- and N-termini, respectively, of the two tryptic peptides. In this manner the order of the peptide fragments in their parent polypeptide chain is established. *꧂* **See the Animated Figures**

ditions that leave its disulfide bonds intact. The resulting peptide fragments are then separated by reverse-phase HPLC, and those that contain Cys residues are identified by determining their amino acid compositions (Section 7-1D). The Cys residues that bear a free sulfhydryl group can be identified by labeling them with radioactive iodoacetate (Section 7-1B). The Cys-containing fragments are then subjected to Edman degradation. Although disulfide-linked peptides yield two PTH-amino acids in each step of this process (at least initially), their locations within the predetermined amino acid sequence of the protein are readily surmised, thereby establishing the positions of the disulfide bonds.

J. *Peptide Characterization and Sequencing by Mass Spectrometry*

Mass spectrometry (MS) has emerged as an important technique for characterizing and sequencing polypeptides. *MS accurately measures the mass-to-charge (m/z) ratio for ions in the gas phase* (where m is the ion's mass and z is its charge). Yet, until about 1985, macromolecules such as proteins and nucleic acids could not be analyzed by MS. This was because the method by which mass spectrometers produced gas phase ions destroyed macromolecules: vaporization by heating followed by ionization via bombardment

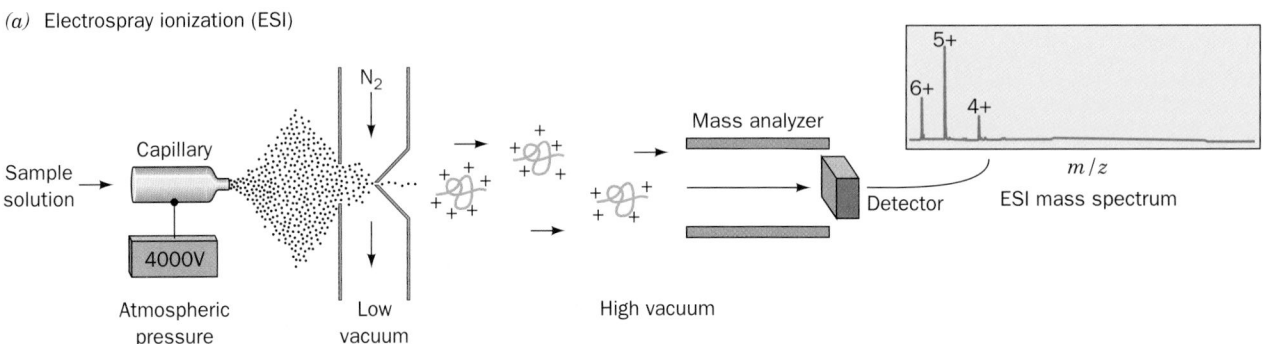

(a) Electrospray ionization (ESI)

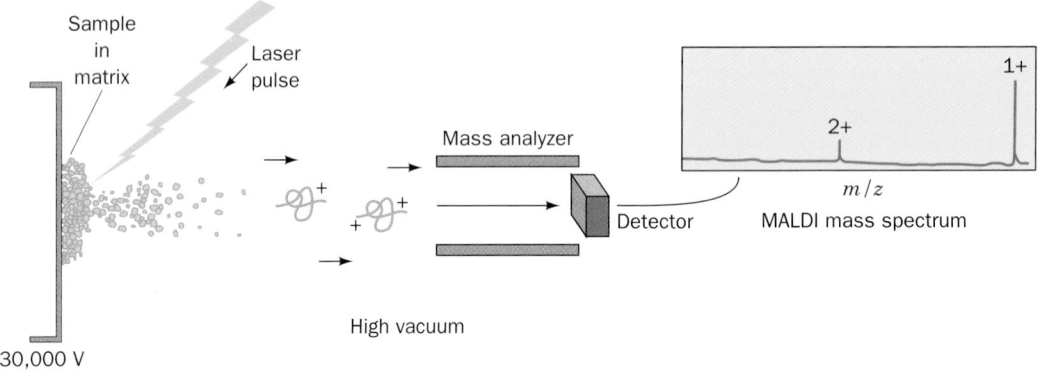

(b) Matrix-assisted laser desorption/ionization (MALDI)

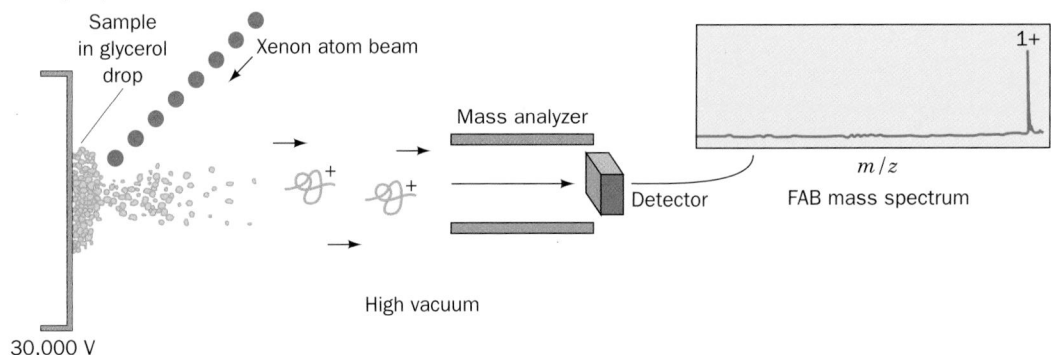

(c) Fast atom bombardment (FAB)

FIGURE 7-8   The generation of the gas phase ions required for the mass spectrometric analysis of proteins. (*a*) By electrospray ionization (ESI), (*b*) by matrix-assisted laser desorption/ionization (MALDI), and (*c*) by fast atom bombardment (FAB). In ESI, a stream of dry N_2 or some other gas is used to promote the evaporation of the solvent from the droplets. [After Fitzgerald, M.C. and Siuzdak, G., *Chem. Biol.* **3**, 708 (1996).]

with electrons. However, the development of three techniques has eliminated this roadblock:

1. Electrospray ionization (**ESI;** Fig 7-8*a, opposite*), a technique pioneered by John Fenn in which a solution of a macromolecule such as a peptide is sprayed from a narrow capillary tube maintained at high voltage (~4000 V), forming fine highly charged droplets from which the solvent rapidly evaporates. This yields a series of gas phase macromolecular ions that typically have ionic charges in the range +0.5 to +2 per kD. For polypeptides, the ionic charges result from the protonation of basic side chains such as Lys and Arg $[(M + nH)^{n+}$ ions].

2. Matrix-assisted laser desorption/ionization (**MALDI;** Fig 7-8*b, opposite*), in which the macromolecule is embedded in a crystalline matrix of a low molecular mass organic molecule (usually prepared by drying a droplet of solution containing the macromolecule and a large excess of the organic molecule) and irradiated with intense short (ns) pulses of laser light at a wavelength absorbed by the matrix material but not the macromolecule. The energy absorbed by the matrix ejects the intact macromolecules from its surface into the gas phase, usually with a charge of +1 but, with larger molecules, occasionally with charges of +2, +3, etc. For polypeptides, **gentisic acid** (2,5-dihydroxybenzoic acid) is one of the few substances found to have satisfactory properties as a matrix. Through the use of MALDI, polypeptides of >300 kD have been characterized.

3. Fast atom bombardment (**FAB;** Fig 7-8*c, opposite*), in which the polypeptide is dissolved in a low-volatility solvent such as **glycerol** and then bombarded with a low energy beam of Ar or Xe atoms or Cs^+ ions. This ejects the macromolecule into the gas phase with a charge of +1. FAB has a practical mass limit of ~7 kD and hence its use is limited to polypeptides with <70 residues.

In each of these techniques, the gas phase macromolecular ions are directed into the mass spectrometer, which measures their m/z values with an accuracy of >0.01%. Consequently, if an ion's z value can be determined, its molecular mass can be determined with far greater accuracy than by any other method. For example, Fig. 7-9 shows the ESI-based mass spectrum (**ESI-MS**) of the 16,951-D protein **myoglobin.** Note that successive peaks in this spectrum differ by a single ionic charge with the rightmost peak corresponding to an $(M + 9H)^{9+}$ ion. Consequently, for the mass spectrum of a macromolecule of molecular mass M containing two adjacent peaks with m/z values of p_1 and p_2 arising from ions with charges of z_1 and $z_1 - 1$,

$$p_1 = \frac{M + z_1}{z_1} \qquad [7.1]$$

and

$$p_2 = \frac{M + z_1 - 1}{z_1 - 1} \qquad [7.2]$$

These two linear equations can therefore be readily solved for their two unknowns, M and z_1.

Since most mass spectrometers are limited to detecting ions with m/z values less than several thousand, the use of ESI-MS has the advantage that the high ionic charges of the ions it produces has permitted the analysis of compounds with molecular masses >100 kD. Another advantage of ESI-MS is that it can be configured to operate in a continuous flow mode with the effluent of an HPLC or CE system. ESI-MS is used in this way, for example, to characterize the tryptic digest of a protein by determining the molecular masses of its component peptides (Section 7-1K).

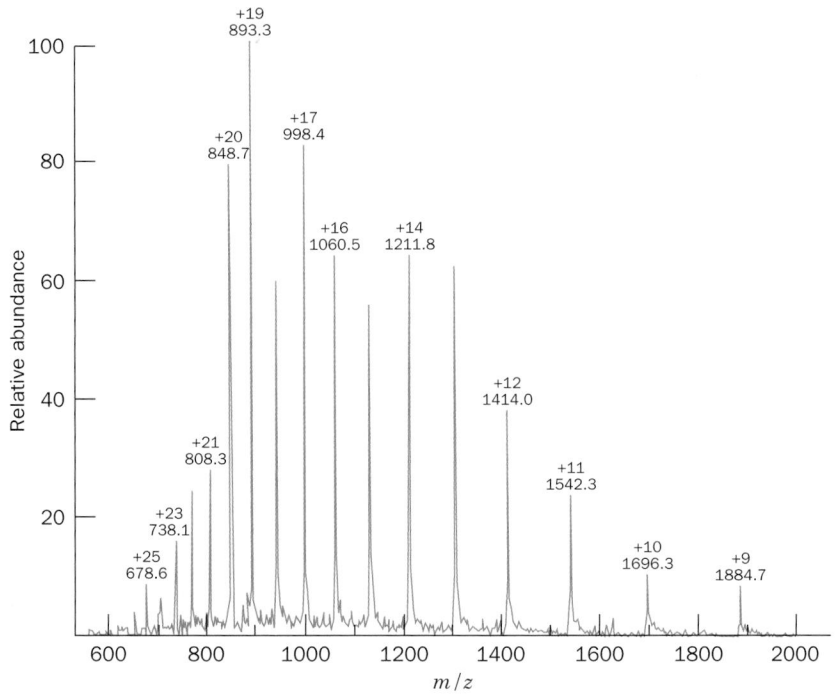

FIGURE 7-9 The ESI-MS spectrum of the 16,951-D horse heart protein apomyoglobin. The measured m/z ratios and the inferred charges for most of the peaks are indicated. Note the bell-shaped distribution of the peaks, which is typical of ESI-MS spectra. The peaks all have shoulders because the polypeptide's component elements contain small admixtures of heavier isotopes (e.g., naturally abundant carbon consists of 98.9% ^{12}C and 1.1% ^{13}C and naturally abundant sulfur consists of 0.8% ^{33}S, 4.2% ^{34}S, and 95.0% ^{35}S). [After Yates, J.R., *Methods Enzymol.* **271,** 353 (1996).]

(a)

Electrospray ionization tandem mass spectrometer

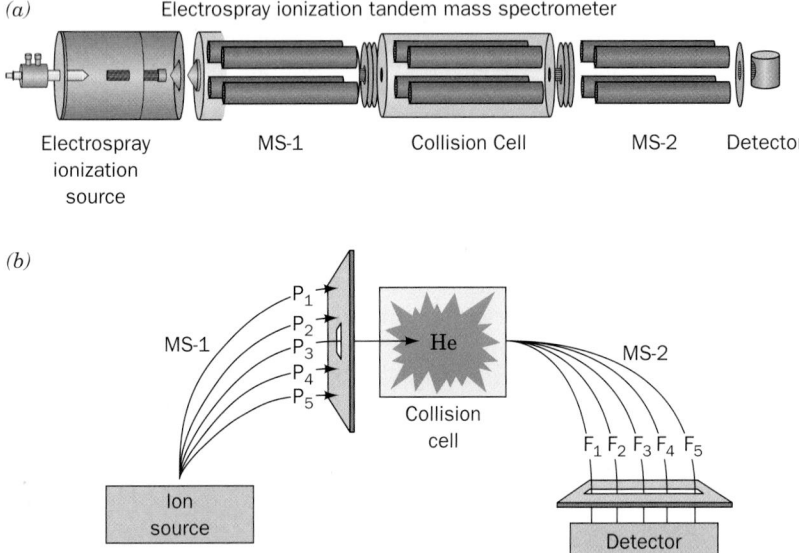

Electrospray   MS-1   Collision Cell   MS-2   Detector
ionization
source

(b)

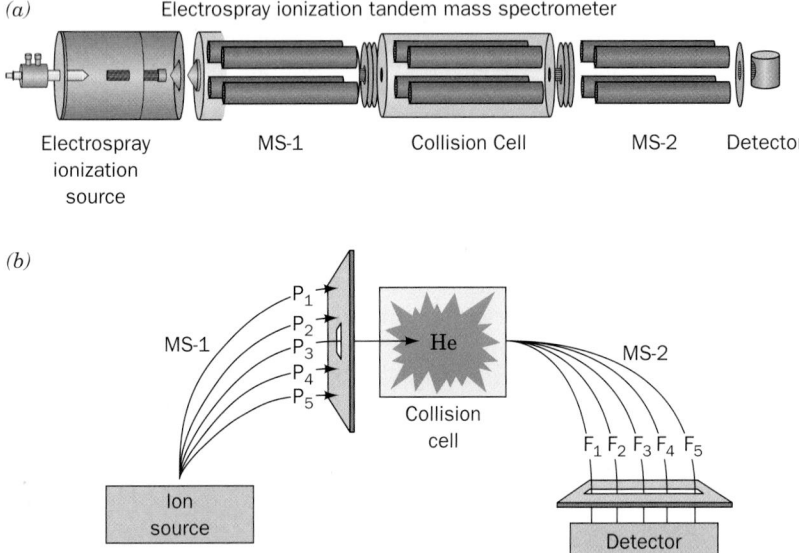

FIGURE 7-10 The use of a tandem mass spectrometer (MS/MS) in amino acid sequencing. (*a*) An MS/MS consists of an ion source (here shown as an ESI system), a first mass spectrometer (MS-1), a collision cell, a second mass spectrometer (MS-2), and a detector. (*b*) The ion source generates gas phase peptide ions, P_1, P_2, etc., from a digest of the protein being analyzed. These peptides are separated by the MS-1 according to their m/z values and one of them, here P_3, is directed into the collision cell, where it collides with helium atoms. This treatment induces the breakdown of the polypeptide ion to yield the fragments F_1, F_2, etc., which are directed into the MS-2, where their m/z values are determined. [Part *a* after Yates, J.R., *Methods Enzymol.* **271**, 358 (1996) and Part *b* after Biemann, K. and Scoble, H.A., *Science* **237**, 992 (1987).]

a. Peptide Sequencing by Mass Spectrometry

Short polypeptides (<25 residues) can be directly sequenced though the use of a **tandem mass spectrometer (MS/MS;** two mass spectrometers coupled in series; Fig. 7-10). The first mass spectrometer functions to select the peptide ion of interest from other peptide ions as well as any contaminants that may be present. The selected peptide ion (P_3 in Fig. 7-10*b*) is then passed into a collision cell, where it collides with chemically inert atoms such as helium. The energy thereby imparted to a peptide ion causes it to fragment predominantly at only one of its several peptide bonds, yielding one or two charged fragments (Fig. 7-11). The molecular masses of the charged fragments are then determined by the second mass spectrometer.

By comparing the molecular masses of successively larger members of a family of fragments, the molecular masses and therefore the identities of the corresponding amino acid residues can be determined. The sequence of an entire polypeptide can thus be elucidated (although MS cannot distinguish the isomeric residues Ile and Leu because they have exactly the same mass and cannot always reliably distinguish Gln and Lys residues because their molecular masses differ by only 0.036 D). Computerization of this comparison process has reduced the time required

(a)

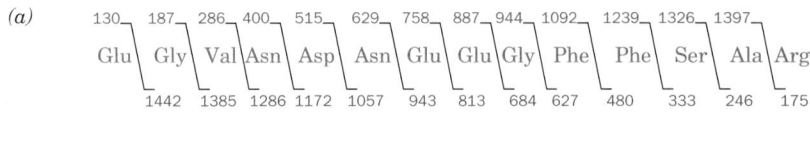

130	187	286	400	515	629	758	887	944	1092	1239	1326	1397
Glu	Gly	Val	Asn	Asp	Asn	Glu	Glu	Gly	Phe	Phe	Ser	Ala	Arg
1442	1385	1286	1172	1057	943	813	684	627	480	333	246	175

(b)

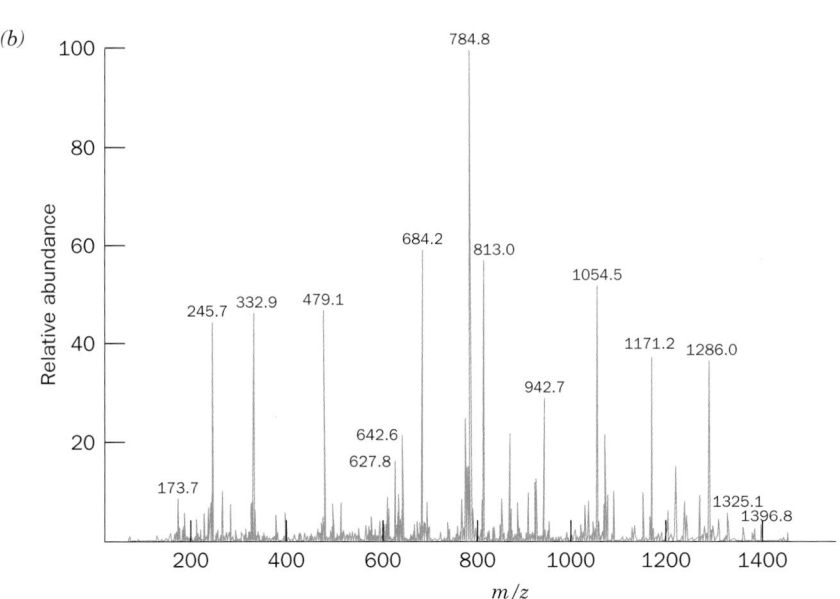

FIGURE 7-11 The tandem mass spectrum of the doubly charged ion of the 14-residue human [Glu¹]fibrinopeptide B (*m/z* = 786). (*a*) The peptide's sequence. The upper and lower rows of numbers are the molecular masses of the charged N-terminal and C-terminal fragments, respectively, that are formed by the cleavage indicated by their connecting diagonal line. (*b*) The mass spectrum of the fragmented peptide with the m/z values of the most abundant fragments indicated above the corresponding peaks. The energy of the collisions in the collision cell has been adjusted so that each peptide ion fragments an average of only once. Note that the predominant fragments under these conditions have $z = 1$ and contain the intact peptide's C-terminus. [After Yates, J.R., *Methods Enzymol.* **271**, 354 (1996).]

to sequence a (short) polypeptide to only a few minutes as compared to the 30 to 50 min required per cycle of Edman degradation. The reliability of this process has been increased through the computerized matching of a measured mass spectrum with those of peptides of known sequence as maintained in databases.

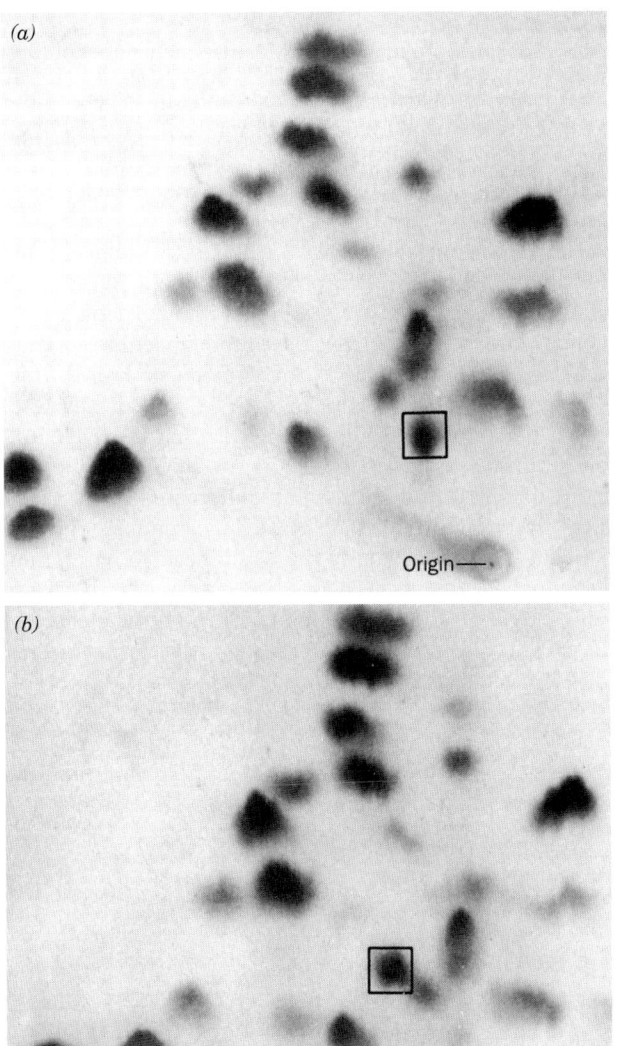

FIGURE 7-12 **Peptide mapping.** A comparison of the fingerprints of trypsin-digested (*a*) hemoglobin A (HbA) and (*b*) hemoglobin S (HbS) shows two peptides that differ in these two forms of hemoglobin (*boxes*). These peptides constitute the eight N-terminal residues of the β subunit of hemoglobin. Their amino acid sequences are

Hemoglobin A Val− His− Leu− Thr− Pro− **Glu**− Glu− Lys
Hemoglobin S Val− His− Leu− Thr− Pro− **Val**− Glu− Lys
      β1   2   3   4   5   **6**   7   8

[Courtesy of Corrado Baglioni, State University of New York at Albany.]

The sequences of several polypeptides in a mixture can be determined, even in the presence of contaminants, by sequentially selecting the corresponding polypeptide ions in the first mass spectrometer of the tandem instrument. Hence, in separating and purifying the polypeptide fragments of a protein digest in preparation for their sequencing, less effort need be expended for MS-based as compared to Edman techniques. Finally, MS can be used to sequence peptides with chemically blocked N-termini (a common eukaryotic post-translational modification that prevents Edman degradation) and to characterize other post-translational modifications such as phosphorylations (Section 4-3A) and glycosylations (Section 11-3C). Thus, although Edman techniques remain the mainstay of amino acid sequence determination (due to their high sensitivity, lack of ambiguity, and ability to sequence peptides of up to 100 residues), MS has become an important tool in the characterization of polypeptides.

K. *Peptide Mapping*

The sequence determination of a protein can be an exacting and time-consuming process. Once the primary structure of a protein has been elucidated, however, that of a nearly identical protein, such as one arising from a closely related species, a mutation, or a chemical modification, can be more easily determined. This was originally done through the combined paper chromatography and paper electrophoresis (Section 6-4A) of partial protein digests, a technique synonymously known as **fingerprinting** or **peptide mapping.** The peptide fragments incorporating the amino acid variations migrate to different positions on their fingerprint (peptide map) than do the corresponding peptides of the original protein (Fig. 7-12). The variant peptides could then be eluted and sequenced, thereby identifying the changes in the protein without the need to sequence it in its entirety.

In more recent times, peptide mapping has come to mean any method that fragments a protein in a reproducible manner and then separates the resulting peptides to yield a pattern that can be used to distinguish differences between related proteins. Thus, peptide mapping can be carried out by two-dimensional gel electrophoresis or by such high resolution one-dimensional techniques as HPLC, SDS–PAGE, IEF, or CE (Sections 6-3 and 6-4). With any of these methods, variant peptides can be isolated and sequenced to establish the sequence differences between the related proteins.

2 ■ NUCLEIC ACID SEQUENCING

The basic strategy of nucleic acid sequencing is identical to that of protein sequencing (Section 7-1). It involves

1. The specific degradation and fractionation of the polynucleotide of interest to fragments small enough to be fully sequenced.

2. The sequencing of the individual fragments.

3. The ordering of the fragments by repeating the preceding steps using a degradation procedure that yields a set of polynucleotide fragments that overlap the cleavage points in the first such set.

Before about 1975, however, nucleic acid sequencing techniques lagged far behind those of protein sequencing, largely because there were no available endonucleases that were specific for sequences greater than a nucleotide. Rather, RNAs were cleaved into relatively short fragments by partial digestion with enzymes such as **ribonuclease T1** (from *Aspergillus oryzae*), which cleaves RNA after guanine residues, or pancreatic **ribonuclease A,** which does so after pyrimidine residues. Moreover, there is no reliable polynucleotide reaction analogous to the Edman degradation for proteins (Section 7-1A). Consequently, the polynucleotide fragments were sequenced by their partial digestion with either of two exonucleases: **snake venom phosphodiesterase,** which removes residues from the 3′ end of polynucleotides (Fig. 7-13), or **spleen phosphodiesterase,** which does so from the 5′ end. The resulting oligonucleotide fragments were identified from their chromatographic and electrophoretic mobilities. Sequencing RNA in this manner is a lengthy and painstaking procedure.

The first biologically significant nucleic acid to be sequenced was that of yeast **alanine tRNA** (Section 32-2A). The sequencing of this 76-nucleotide molecule by Robert Holley, a labor of 7 years, was completed in 1965, some 12 years after Frederick Sanger had determined the amino acid sequence of insulin. This was followed, at an accelerating pace, by the sequencing of numerous species of tRNAs and the **5S ribosomal RNAs** (Section 32-3A) from several organisms. The art of RNA sequencing by these techniques reached its zenith in 1976 with the sequencing, by Walter Fiers, of the entire 3569-nucleotide genome of the **bacteriophage MS2.** In contrast, DNA sequencing was in a far more primitive state because of the lack of avail-

able DNA endonucleases with any sequence specificity.

After 1975, dramatic progress was made in nucleic acid sequencing technology. Three advances made this possible:

1. The discovery of restriction endonucleases to enable the cleavage of DNA at specific sequences (Section 5-5A).

2. The development of molecular cloning techniques to permit the acquisition of almost any identifiable DNA segment in the amounts required for sequencing (Section 5-5).

3. The development of DNA sequencing techniques.

These procedures are largely responsible for the enormous advances in our understanding of molecular biology that have been made over the past two decades and which we discuss in succeeding chapters. DNA sequencing techniques are the subject of this section.

The pace of nucleic acid sequencing has become so rapid that directly determining a protein's amino acid sequence is far more time-consuming than determining the base sequence of its corresponding gene (although amino acid and base sequences provide complementary information; Section 7-2C). There has been such a flood of DNA sequence data—over 35 billion nucleotides in nearly 22 million sequences as of the beginning of 2003, and approximately doubling every year—that only computers can keep track of them. The first complete cellular genome sequence to be determined, that of the gram-negative bacterium *Haemophilus influenzae,* was reported in 1995 by Craig Venter. By early 2003, the complete genome sequences of 110 prokaryotes had been reported (with many more being determined) as well as those of 11 eukaryotes including *Saccharomyces cerevisiae* (baker's yeast), *Caenorhabditis elegans* (a nematode worm), *Drosophila melanogaster* (a fruit fly), *Arabidopsis thaliana* (a flowering plant), mouse, and human (Table 7-3).

A. *Chain-Terminator Method*

🖑 See Guided Exploration 5: DNA Sequence Determination by the Chain Terminator Method After 1975, several methods were developed for the rapid sequencing of long stretches of DNA. Here we discuss the **chain-terminator** procedure of Frederick Sanger (the same individual who pioneered the amino acid sequencing of proteins), which is largely responsible for the vast number of DNA sequences that have been elucidated.

*The chain-terminator method (alternatively called the **dideoxy method**) utilizes the E. coli enzyme DNA polymerase I (Section 5-4C) to synthesize complementary copies of the single-stranded DNA being sequenced.* As we have previously seen, under the direction of the strand being replicated (the template strand), DNA polymerase I assembles the four deoxynucleoside triphosphates (dNTPs), dATP, dCTP, dGTP, and dTTP, into a complementary polynucleotide chain that it elongates in the 5′ to 3′ direction (Fig. 5-30). To initiate DNA synthesis, DNA polymerase requires a primer in a stable base paired complex with the template DNA. If the DNA being sequenced is a

```
G C A C U U G A
        │  snake venom
        │  phosphodiesterase
        ▼
G C A C U U G A
G C A C U U G
G C A C U U
G C A C U
G C A C
G C A
G C   + Mononucleotides
```

FIGURE 7-13   Sequence determination of an oligonucleotide by partial digestion with snake venom phosphodiesterase. This enzyme sequentially cleaves the nucleotides from the 3′ end of a polynucleotide that has a free 3′-OH group. Partial digestion of an oligonucleotide with snake venom phosphodiesterase yields a mixture of fragments of all lengths, as indicated, that may be chromatographically separated. Comparison of the base compositions of pairs of fragments that differ in length by one nucleotide establishes the identity of the 3′-terminal nucleotide of the larger fragment. In this way the base sequence of the oligonucleotide may be elucidated.

TABLE 7-3 Some Sequenced Genomes

Organism	Haploid Genome Size (kb)	Number of Chromosomes
Mycoplasma genitalium (human parasite)	580	1
Rickettsia prowazeki (bacterium, cause of typhus, putative mitochondial relative)	1,112	1
Borrelia burgdorferi (Lyme disease spirochaete)	1,444	1
Methanococcus jannaschii (thermophilic methanogenic archaeon)	1,665	1
Haemophilus influenzae (bacterium, human pathogen)	1,830	1
Archaeoglobus fulgidus (hyperthermophilic, sulfate-reducing archaeon)	2,178	1
Synechocystis sp. (cyanobacterium)	3,573	1
Mycobacterium tuberculosis (cause of tuberculosis)	4,412	1
Escherichia coli (bacterium, human symbiont)	4,639	1
Schizosaccharomyces pombe (fission yeast)	13,800	3
Saccharomyces cerevisiae (baker's or budding yeast)	11,700	16
Plasmodium falciparum (protozoan, cause of malaria)	30,000	14
Caenorhabditis elegans (nematode worm)	97,000	6
Drosophila melanogaster (fruit fly)	137,000	4
Arabidopsis thaliana (flowering plant)	117,000	5
Oryza sativa (rice)	430,000	12
Mus musculus (mouse)	2,500,000	20
Ratus norvegicus (rat)	2,600,000	21
Homo sapiens (human)	3,200,000	23

Source: Mainly http://www.ncbi.nlm.nih.gov:80/PMGifs/Genomes/org.html

restriction fragment, as it usually is, it begins and ends with a restriction site. The primer can therefore be a short DNA segment containing this restriction site annealed to the strand being replicated. The template DNAs are obtained in sufficient quantity to sequence by cloning them in M13-based vectors (Section 5-5B) or by PCR (Section 5-5F), both of which yield the required single-stranded DNAs.

DNA polymerase I's $5' \rightarrow 3'$ exonuclease activity (Fig. 5-33) is catalyzed by a separate active site from those which mediate its polymerase and $3' \rightarrow 5'$ exonuclease (Fig. 5-36) functions. This is demonstrated by the observation that on proteolytic cleavage of the enzyme into two fragments, the larger C-terminal fragment, which is known as the **Klenow fragment,** possesses the full polymerase and $3' \rightarrow 5'$ exonuclease activities of the intact enzyme, whereas the smaller N-terminal fragment has its $5' \rightarrow 3'$ exonuclease activity. Only the Klenow fragment is used in DNA sequencing to ensure that all replicated chains have the same $5'$ terminus.

a. The Synthesis of Labeled DNA by DNA Polymerase Is Terminated after Specific Bases

In the chain-terminator technique (Fig. 7-14), *the DNA to be sequenced is incubated with the Klenow fragment of DNA polymerase I, a suitable primer, and the four deoxynucleoside triphosphates (dNTPs). Either at least one dNTP (usually dATP) or the primer is α-^{32}P-labeled. In addition, a small amount of the* **2′, 3′-dideoxynucleoside triphosphate (ddNTP)**

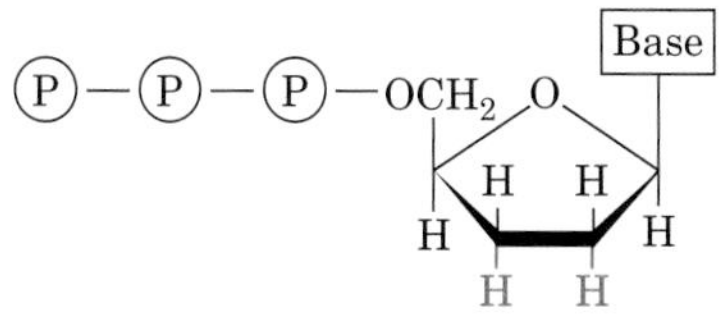

2′,3′-Dideoxynucleoside triphosphate

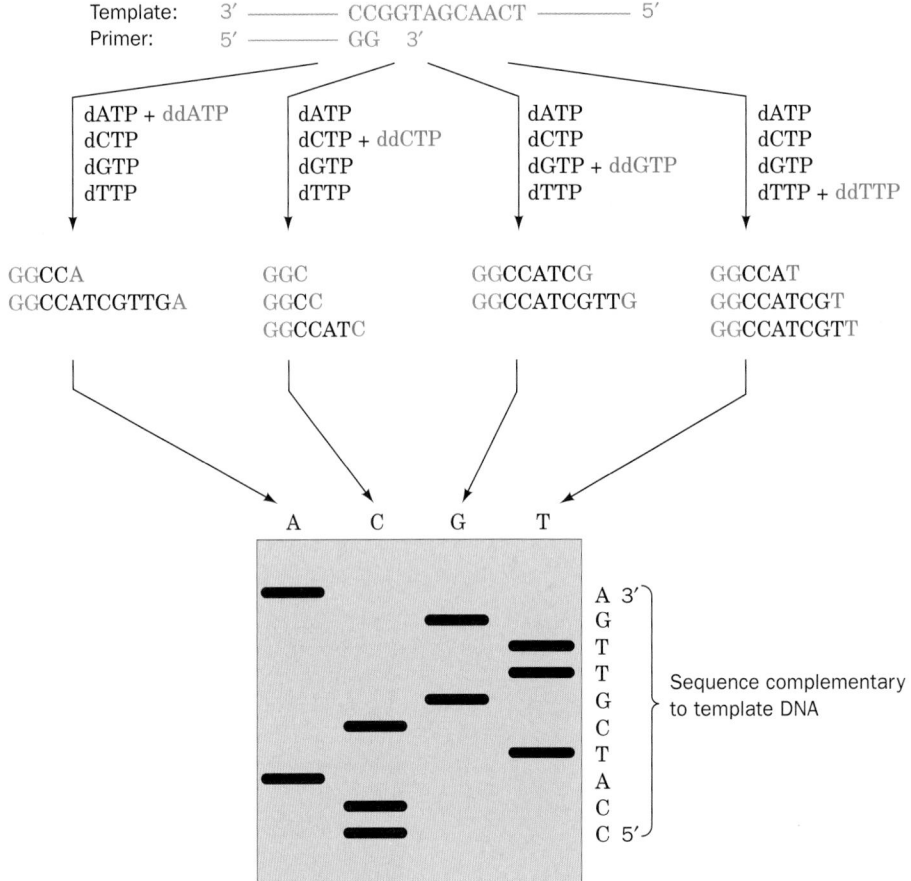

Template: 3' ———— CCGGTAGCAACT ———— 5'
Primer: 5' ———— GG 3'

dATP + ddATP	dATP	dATP	dATP
dCTP	dCTP + ddCTP	dCTP	dCTP
dGTP	dGTP	dGTP + ddGTP	dGTP
dTTP	dTTP	dTTP	dTTP + ddTTP

GGCCA	GGC	GGCCATCG	GGCCAT
GGCCATCGTTGA	GGCC	GGCCATCGTTG	GGCCATCGT
	GGCCATC		GGCCATCGTT

A   C   G   T

A 3'
G
T
T
G
C
T
A
C
C 5'

Sequence complementary to template DNA

FIGURE 7-14 Flow diagram of the chain-terminator (dideoxy) method of DNA sequencing. The symbol ddATP represents dideoxyadenosine triphosphate, etc. The sequence that is determined by reading the gel from bottom to top (from the smallest to the largest fragment) is complementary to the sequence of the template DNA.

of one of the bases is added to the reaction mixture. When the dideoxy analog is incorporated in the growing polynucleotide in place of the corresponding normal nucleotide, chain growth is terminated because of the absence of a 3'-OH group. By using only a small amount of the ddNTP, *a series of truncated chains is generated, each of which is terminated by the dideoxy analog at one of the positions occupied by the corresponding base.* Each of the four ddNTPs is reacted in a separate vessel.

The four reaction mixtures are simultaneously electrophoresed in parallel lanes on a **sequencing gel.** This is a long, thin (as little as 0.1 mm by up to 100 cm) polyacrylamide slab. It contains ~7M urea and is run at ~70°C so as to eliminate all hydrogen bonding associations. *These conditions ensure that the DNA fragments separate only according to their size. The sequence of the DNA that is complementary to the template DNA can then be directly read off an autoradiogram of the sequencing gel, from bottom to top,* as is indicated in Fig. 7-15. Indeed, computerized devices are available to aid in doing so. However, a single gel is incapable of resolving much more than 300 to 400 consecutive fragments. This limitation is circumvented by generating two sets of gels, one run for a longer time and per-

haps at a higher voltage than the other, to obtain the sequence of an up to 800-bp DNA fragment.

Improved gels can be obtained through the use of dNTPs whose α-phosphate groups are radioactively labeled with ^{35}S rather than ^{32}P:

$$^-O-\overset{\displaystyle O}{\underset{\displaystyle O^-}{\overset{\|}{P}}}-O-\overset{\displaystyle O}{\underset{\displaystyle O^-}{\overset{\|}{P}}}-O-\overset{\displaystyle ^{35}S}{\underset{\displaystyle O^-}{\overset{\|}{P}}}-O-CH_2$$

Base

OH H

α-Thio-[^{35}S]-dNTP

This is because the β particles emitted by ^{35}S nuclei have less energy and hence shorter path lengths than those of ^{32}P, thereby yielding sharper gel bands. More readily interpretable gels may also be obtained by replacement of the Klenow fragment with DNA polymerases either from **bacteriophage T7 (T7 DNA polymerase,** which is less sensitive to the presence of ddNTPs than is Klenow fragment and hence yields gel bands of more even intensities) or

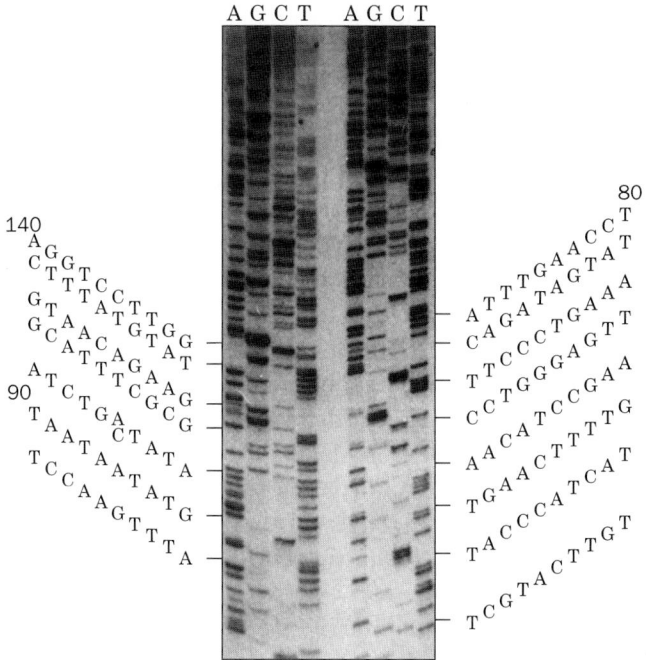

FIGURE 7-15 Autoradiograph of a sequencing gel. DNA fragments were produced by the chain-terminator method of DNA sequencing. A second loading of the gel (*right*) was made 90 min after the initial loading (*left*). The deduced sequence of 140 nucleotides is written alongside. [From Hindley, J., DNA sequencing, *In* Work, T.S. and Burdon, R.H. (Eds.), *Laboratory Techniques in Biochemistry and Molecular Biology,* Vol. 10, *p.* 82, Elsevier (1983). Used by permission.]

from thermophilic bacteria such as *Thermus aquaticus* (*Taq* polymerase; Section 5-5F) that are stable above 90°C and hence can be used at the temperatures required to denature particularly stable segments of DNA.

With a few hours' effort by a skilled operator, the chain-terminator method can sequence a DNA segment of up to

800 nucleotides. Indeed, the major obstacle to sequencing a very long DNA molecule is ensuring that all of its fragments have been obtained (Section 5-5E) rather than sequencing them once they are in hand.

b. RNA May Be Sequenced through Its Transcription to cDNA

RNA can be readily sequenced by only a slight modification of the above DNA sequencing procedures. The RNA to be sequenced is transcribed into a complementary strand of DNA (cDNA) through the action of reverse transcriptase (Section 5-5F). The resulting cDNA may then be sequenced normally.

c. The Chain-Terminator Method Has Been Automated

In order to sequence large tracts of DNA such as entire chromosomes, the chain-terminator method has been greatly accelerated through automation. This required that the above-described radiolabeling techniques, which are not readily automated, be replaced by fluorescence-labeling techniques (with the added benefit of eliminating the health hazards and storage problems of using radiolabeled nucleotides). Two types of fluorescence-labeling systems are in use in automated DNA sequencers:

1. Four reaction/one gel systems The primers used in each of the four chain extension reactions are 5′-linked to a differently fluorescing dye. The separately reacted mixtures are combined and subjected to sequencing gel electrophoresis in a single lane. As each fragment exits the gel, its terminal base is identified according to its characteristic fluorescence spectrum by a laser-activated fluorescence detection system (Fig. 7-16).

2. One reaction/one gel systems Each of the four ddNTPs used to terminate chain extension is covalently linked to a differently fluorescing dye, the chain-extension reaction is carried out in a single vessel, the resulting fragment mixture is subjected to sequencing gel electrophore-

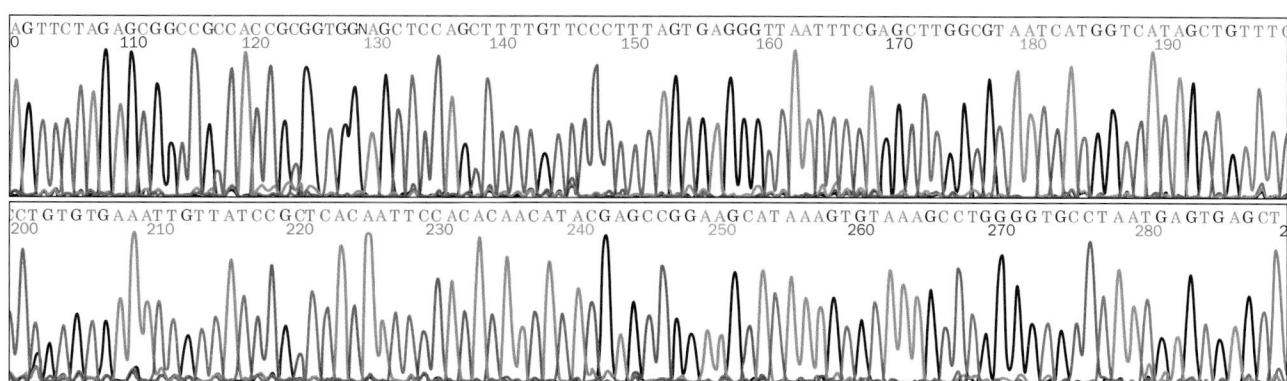

FIGURE 7-16 A portion of the output of a four reaction/one gel sequencing system. Each of the four differently colored curves indicates the fluorescence intensity of a particular dye that is linked to the primer used with a specific ddNTP in terminating the primer extension reaction (green, red, black, and blue with ddATP, ddTTP, ddGTP, and ddCTP, respectively).

The 3′-terminal base of each terminated oligonucleotide, which the gel separates according to size, is identified by the fluorescence of its gel band (letters above the bands; the numbers indicate the positions of the bases in the DNA segment being sequenced). [Courtesy of Mark Adams, The Institute for Genomic Research, Rockville, Maryland.]

sis in a single lane, and the terminal base on each fragment is identified according to its characteristic fluorescence spectrum.

The fluorescence detectors used in these devices, which have error rates of ~1%, are computer-controlled and hence data acquisition is automated. In the most advanced systems, the sequencing gel is contained in an array of up to 96 capillary tubes rather than in a slab-shaped apparatus, sample preparation and sample loading are performed by robotic systems, and electrophoresis and data analysis are fully automated. These systems can simultaneously sequence 96 DNA samples averaging ~600 bases each with a turnaround time of ~2.5 h and hence can identify up to 550,000 bases per day—all with only ~15 min of human attention (vs the ~25,000 bases per year that a skilled operator can identify using the above-described manual methods). Nevertheless, one such system would require ~30 years of uninterrupted operation to sequence the 3.2 billion-bp human genome with only two sets of overlapping fragments. However, to ensure the complete coverage of a large tract of DNA (Section 5-5E) and to reduce its error rate to <0.01%, at least 10 sets of overlapping segments must be sequenced (Section 7-2B). Hence, the major se-

quencing centers, where most genome sequencing is carried out, each have over 100 advanced sequencing systems.

B. *Genome Sequencing*

The major technical hurdle in sequencing a genome is not the DNA sequencing itself but, rather, assembling the tens of thousands to tens of millions of sequenced segments (depending on the size of the genome) into contiguous blocks (called **contigs**) and assigning them to their proper positions in the genome. One way that contigs might be ordered is through chromosome walking (Section 5-5E). However, to do so for a eukaryotic genome would be prohibitively time-consuming and expensive (e.g., to "walk" the 125 million-bp length of an average length human chromosome using ~10-kb inserts from a plasmid library would require a minimum of $1.25 \times 10^8/10,000 = 12,500$ labor-intensive "steps").

a. Conventional Genome Sequencing

A more efficient strategy for genome sequencing (Fig. 7-17a) was developed in the late 1980s. In this approach, low resolution physical maps of each chromosome are pre-

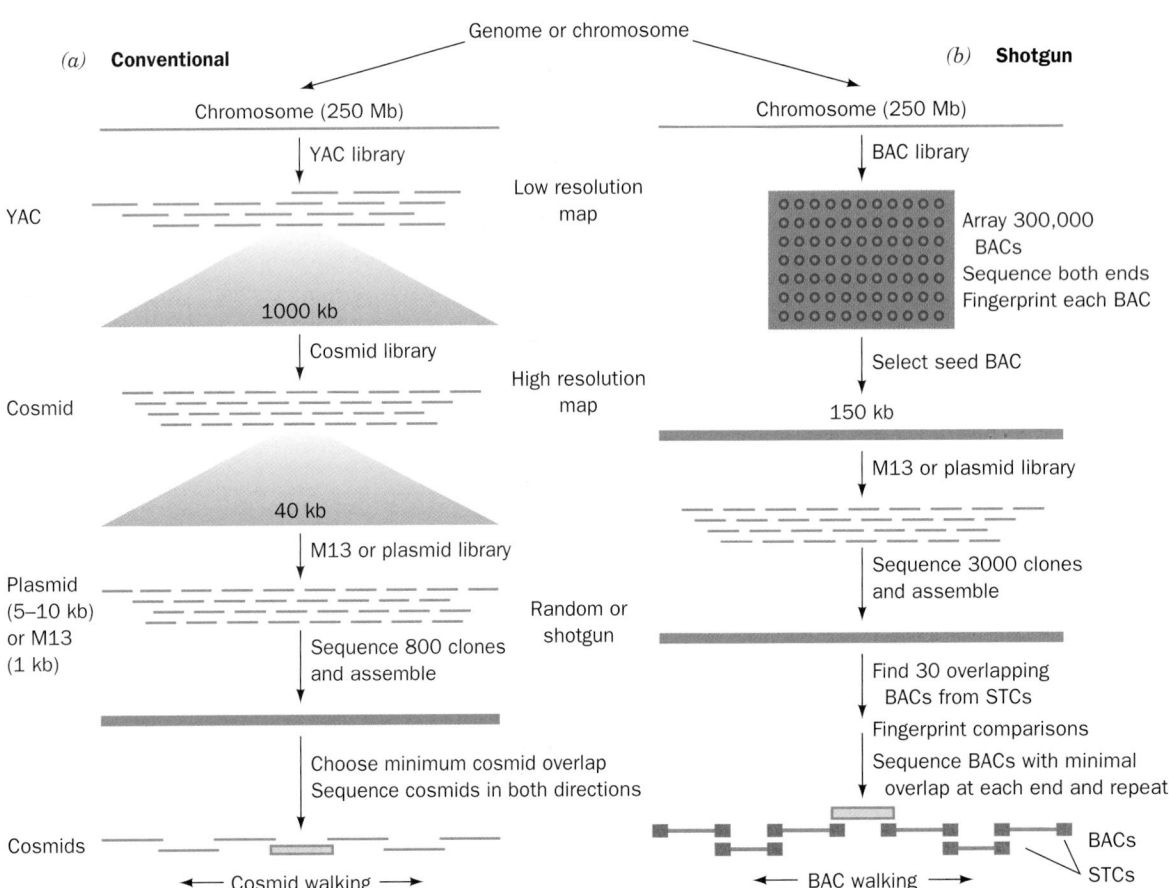

FIGURE 7-17   Genome sequencing strategies. (*a*) The conventional strategy, which uses three sets of progressively smaller cloned inserts and assembles the sequenced inserts through cosmid walking and the use of landmarks such as STSs and ESTs (see below). (*b*) The shotgun strategy, which uses only two levels of cloning and employs sophisticated computer algorithms as well as STCs to assemble the sequenced inserts into finished chromosomes. [After Venter, J.C., Smith, H.O., and Hood, L., *Science* **381**, 365 (1996).]

pared by identifying shared landmarks on overlapping ~250-kb inserts that are cloned in yeast artificial chromosomes (YACs). These landmarks often take the form of 200- to 300-bp segments known as **sequence-tagged sites (STSs),** whose exact sequence occurs nowhere else in the genome. Hence, two clones that contain the same STS must overlap. The STS-containing inserts are then randomly fragmented (usually by sonication; Section 5-3D) into ~40-kb segments that are subcloned into cosmid vectors so that a high resolution map can be constructed by identifying their landmark overlaps. The cosmid inserts are then randomly fragmented into overlapping 5- to 10-kb or 1-kb segments for insertion into plasmid or M13 vectors (shotgun cloning; Section 5-5E). These inserts (~800 M13 clones per cosmid) are then sequenced (~400 bp per clone) and assembled computationally into contigs to yield the sequence of their parent cosmid insert (with a redundancy of 400 bp per clone × 800 clones per cosmid/40,000 bp per cosmid = 8). Finally, the cosmid inserts are assembled, through **cosmid walking** (the computational analog of chromosome walking; Fig. 5-53), using their landmark overlaps, to yield the sequences of the YAC inserts which are then assembled, using their STSs, to yield the chromosome's sequence.

According to this conventional strategy, the first stage in sequencing a genome is to identify the requisite number of evenly spaced STSs. However, the genomes of most complex eukaryotes contain numerous tracts of **repetitive sequences,** that is, short segments of DNA that are tandemly repeated hundreds, thousands, and in some cases millions of times (Section 34-2B). The presence in a genome of extensive amounts of repetitive DNA greatly exacerbates the difficulty of finding STSs (the human genome consists of ~50% repetitive sequences; their function, if any, is unknown). To circumvent this difficulty, STS-like sequences of cDNAs, known as **expressed sequence tags (ESTs),** are used in place of STSs. Since the mRNAs from which cDNAs are reverse transcribed encode proteins, they are unlikely to contain repetitive sequences.

b. The Shotgun Strategy

Although the initial goal of the **human genome project** of identifying STSs and ESTs every ~100 kb in the human genome was achieved, advances in computational and cloning technology permitted a more straightforward sequencing approach that eliminates the need for both the low resolution (YAC) and high resolution (cosmid) mapping steps. In this so-called shotgun strategy, which was formulated by Venter, Hamilton Smith, and Leroy Hood, a genome is randomly fragmented, a large number of cloned fragments are sequenced, and the genome is assembled by identifying overlaps between pairs of fragments. Statistical considerations indicate that, using this strategy, the probability that a given base is not sequenced is e^{-c}, where c is the redundancy of coverage ($c = LN/G$, where L is the average length of the cloned inserts in base pairs, N is the number of inserts sequenced, and G is the length of the genome in base pairs), the aggregate length of the gaps between contigs is $G e^{-c}$, and the average gap size is G/N.

For bacterial genomes, the shotgun strategy is carried out straightforwardly by sequencing tens of thousands of fragments and assembling them (a task that required the development of computer algorithms capable of assembling contigs from very large numbers of sequenced fragments). The gaps between contigs are then filled in by synthesizing PCR primers complementary to the ends of the contigs and using them to isolate the missing segments (chromosome walking), a task known as "finishing."

For eukaryotic genomes, their much greater sizes require that the shotgun strategy be carried out in stages as follows (Fig. 7-17b). A bacterial artificial chromosome (BAC) library of ~150-kb inserts is generated (for the human genome, an ~15-fold redundancy, which would still leave ~900 bases unsequenced, would require ~300,000 such clones; BACs are used because they are subject to less technical difficulties than are YACs). The insert in each of these BAC clones is identified by sequencing ~500 bp in from each end to yield segments known as **sequence-tagged connectors** (**STCs** or **BAC-ends;** which for the above 300,000 clones would collectively comprise ~300,000 kb, that is, 10% of the entire human genome). One BAC insert is then fragmented, shotgun cloned into plasmid or M13 vectors (so as to yield ~3000 overlapping clones), and the fragments are sequenced and assembled into contigs. The sequence of this "seed" BAC is then compared with the database of STCs to identify the ~30 overlapping BAC clones. The two with minimal overlap at either end are then selected, sequenced, and the operation repeated until the entire chromosome is sequenced (BAC walking), a process that for the human genome required the sequencing of ~20,000 BAC inserts. The databases of STSs and ESTs can then be used to verify the final assemblies.

The shotgun strategy is readily automated through robotics and hence is faster and less expensive than the conventional strategy. Indeed, most known genome sequences were determined using the shotgun strategy, many in a matter of a few months, and its advent reduced the time to sequence the human genome by several years.

c. The Human Genome Has Been Sequenced

The "rough draft" of the human genome was reported in early 2001 by two independent groups: the International Human Genome Sequencing Consortium, which was led by Francis Collins, Eric Lander, and John Sulston; and a group, mainly from Celera Genomics, which was led by Venter. This stunning achievement, the culmination of over a decade of intense effort by hundreds of scientists, promises to revolutionize the way both biochemistry and medicine are viewed and practiced. However, a considerable amount of work is still required to "finish" and annotate this ~3.2-billion nucleotide sequence, which is only ~90% complete (many highly repetitive and hence difficult to sequence regions have not been sequenced). Nevertheless, numerous important conclusions have already be drawn, including:

1. About half the human genome consists of repeating sequences of various types.

2. Only ~28% of the genome is transcribed to RNA.

3. Only 1.1% to 1.4% of the genome (~5% of the transcribed RNA) encodes protein.

4. The human genome appears to contain only ~30,000 protein-encoding genes [also known as **open reading frames (ORFs)**] rather than the 50,000 to 140,000 ORFs that had previously been predicted based mainly on extrapolations. This compares with the ~6000 ORFs in yeast, ~13,000 in *Drosophila*, ~18,000 in *C. elegans*, and ~26,000 in *Arabadopsis*. Note that these numbers will almost certainly change as our presently imperfect ability to recognize ORFs improves.

5. Only a small fraction of human protein families is unique to vertebrates; most occur in other if not all life forms.

6. Two randomly selected human genomes differ, on average, by only 1 nucleotide per 1250; that is, any two people are likely to be >99.9% genetically identical.

The obviously greater complexity of humans (vertebrates) relative to "lower" (nonvertebrate) forms of life is unlikely to be due to the not much larger numbers of ORFs that vertebrates encode. Rather, it appears, that vertebrate proteins themselves are more complex than those of nonvertebrates; that is, vertebrate proteins tend to have more domains (modules) than invertebrate proteins and these modules are more often selectively expressed through differential gene splicing (Section 5-4A). Thus, many vertebrate genes encode several different although similar proteins.

The human genome can be explored at http://www.ncbi.nlm.nih.gov/Genomes/index.html (the International Human Genome Sequencing Consortium sequences) and http://publication.celera.com/ (the Celera sequences).

C. *Nucleic Acid Sequencing versus Amino Acid Sequencing*

The amino acid sequences of proteins are specified by the base sequences of nucleic acids (Section 5-4B). Consequently, with a knowledge of the genetic code (Table 5-3) and the nature of transcriptional and translational initiation sequences (Sections 31-3 and 32-3C), a protein's primary structure can be inferred from that of a corresponding nucleic acid. Techniques for sequencing nucleic acids initially lagged far behind those for proteins, but by the late 1970s, DNA sequencing methods had advanced to the point that it became easier to sequence a DNA segment than the protein it specified. Although protein primary structures are now routinely inferred from DNA sequences, direct protein sequencing remains an indispensable biochemical tool for several important reasons:

1. Disulfide bonds can be located only by protein sequencing.

2. Many proteins are modified after their biosynthesis by the excision of certain residues and by the specific derivatization of others (Section 32-5). The identities of these modifications, which are often essential for the protein's biological function, can be determined only by directly sequencing the protein.

3. It is often difficult to identify and isolate a nucleic acid that encodes the protein of interest. Indeed, one of the most effective ways of doing so is to determine the amino acid sequence of at least a portion of the protein, infer the base sequence of the DNA segment that encodes this polypeptide segment, chemically synthesize this DNA, and use it to identify and isolate the gene(s) containing its base sequence through Southern blotting or PCR (Sections 5-5D and 5-5F). This process is known as **reverse genetics** because, in prokaryotes, genetics has been traditionally used to characterize proteins rather than vice versa. Of course, for organisms whose genomes have been sequenced, this process can be carried out *in silico* (by computer).

4. A common error in DNA sequencing is the inadvertent insertion or deletion of a single nucleotide. This changes the gene's apparent reading frame (Section 5-4A) and thus changes the predictions for all the amino acid residues past the point of error. Double checking the predicted amino acid sequence by directly sequencing a series of oligopeptides scattered throughout the protein readily detects such errors.

5. The "standard" genetic code is not universal: Those of mitochondria and certain protozoa are slightly different (Section 32-1D). In addition, in certain species of protozoa, the RNA transcripts are "edited"; that is, their sequences are altered before they are translated (Section 31-4A). These genetic code anomalies were discovered by comparing the amino acid sequences of proteins and the base sequences of their corresponding genes. If there are other genetic code anomalies, they will no doubt be discovered in a like manner.

3 ■ CHEMICAL EVOLUTION

Individuals, as well as whole species, are characterized by their inherited genetic compositions. An organism's genetic complement specifies the amino acid sequences of all of its proteins together with their quantity and schedule of appearance in each cell. An organism's protein composition is therefore the direct expression of its genetic composition.

In this section, we concentrate on the evolutionary aspects of amino acid sequences, the study of the **chemical evolution** of proteins. Evolutionary changes, which stem from random mutational events, often alter a protein's primary structure. A mutational change in a protein, if it is to be propagated, must somehow increase, or at least not decrease, the probability that its owner will survive to reproduce. Many mutations are deleterious and often lethal in their effects and therefore rapidly die out. On rare occasions, however, a mutation arises that, as we shall see below, improves the fitness of its host in its natural environment.

A. *Sickle-Cell Anemia: The Influence of Natural Selection*

Hemoglobin, the red blood pigment, is a protein whose major function is to transport oxygen throughout the body. A molecule of hemoglobin is an $\alpha_2\beta_2$ tetramer; that is, it consists of two identical α chains and two identical β chains (Fig. 7-1*d*). Hemoglobin is contained in the **erythrocytes** (red blood cells; Greek: *erythrose,* red + *kytos,* a hollow vessel) of which it forms ~33% by weight in normal individuals, a concentration that is nearly the same as it has in the crystalline state. In every cycle of their voyage through the circulatory system, the erythrocytes, which are normally flexible biconcave disks (Fig. 7-18*a*), must squeeze through capillary blood vessels smaller in diameter than they are.

In individuals with the inherited disease **sickle-cell anemia,** many erythrocytes assume an irregular crescent-like shape under conditions of low oxygen concentration typical of the capillaries (Fig. 7-18*b*). This "sickling" increases the erythrocytes' rigidity, which hinders their free passage through the capillaries. The sickled cells therefore impede the flow of blood in the capillaries such that, in a sickle-cell "crisis," the blood flow in some areas may be completely blocked, thereby giving rise to extensive tissue damage and excruciating pain. Moreover, individuals with sickle-cell anemia suffer from severe **hemolytic anemia** (a condition characterized by red cell destruction) because the increased mechanical fragility of their erythrocytes halves the normal 120-day lifetime of these cells. The debilitating effects of this disease are such that, before the latter half of the twentieth century, individuals with sickle-cell anemia rarely survived to maturity (although modern treatments by no means constitute a cure).

a. Sickle-Cell Anemia Is a Molecular Disease

In 1945, Linus Pauling correctly hypothesized that *sickle-cell anemia, which he termed a **molecular disease,** is a result of the presence of a mutant hemoglobin.* Pauling and his co-workers subsequently demonstrated, through electrophoretic studies, that normal human hemoglobin **(HbA)** has an anionic charge that is around two units more negative than that of sickle-cell hemoglobin **(HbS;** Fig. 7-19).

In 1956, Vernon Ingram developed the technique of fingerprinting peptides (Section 7-1K) in order to pinpoint the difference between HbA and HbS. Ingram's fingerprints of tryptic digests of HbA and HbS revealed that their α subunits are identical but that their β subunits differ by a variation in one tryptic peptide (Fig. 7-12). Sequencing studies eventually indicated that this difference arises from the replacement of the Glu $\beta6$ of HbA (the Glu in the sixth position of each β chain) with Val in HbS (Glu $\beta6 \rightarrow$ Val), thus accounting for the charge difference observed by Pauling. This was the first time an inherited disease was shown to arise from a specific amino acid change in a protein. *This mutation causes deoxygenated HbS to aggregate into filaments of sufficient size and stiffness to deform erythrocytes*—a remarkable example of the influence of primary structure on quaternary structure. The structure of these filaments is further discussed in Section 10-3B.

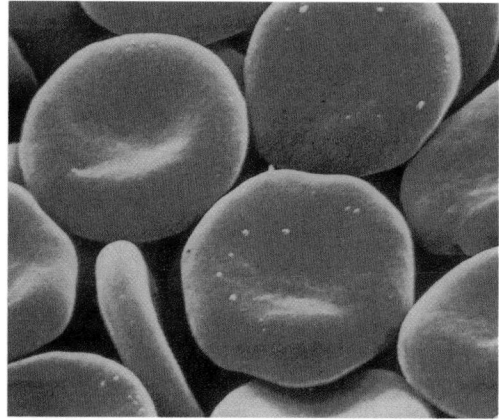

(*a*)

(*b*)

FIGURE 7-18 **Scanning electron micrographs of human erythrocytes.** (*a*) Normal human erythrocytes revealing their biconcave disklike shape. [David M. Phillips/Visuals Unlimited.] (*b*) Sickled erythrocytes from an individual with sickle-cell anemia. [Bill Longcore/Photo Researchers, Inc.]

b. The Sickle-Cell Trait Confers Resistance to Malaria

Sickle-cell anemia is inherited according to the laws of Mendelian genetics (Section 1-4B). The hemoglobin of individuals who are homozygous for sickle-cell anemia is almost entirely HbS. In contrast, individuals heterozygous for sickle-cell anemia have hemoglobin that is ~40% HbS (Fig. 7-19). Such persons, who are said to have the **sickle-cell trait,** lead a normal life even though their erythrocytes have a shorter lifetime than those of normal individuals.

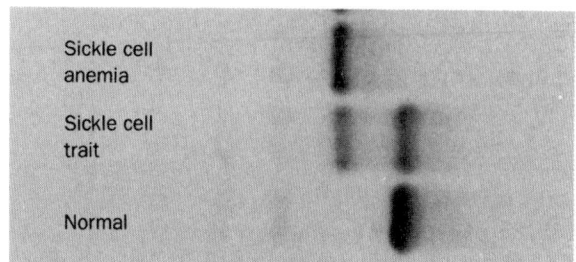

Sickle cell anemia

Sickle cell trait

Normal

FIGURE 7-19 The electrophoretic pattern of hemoglobins from normal individuals and from those with the sickle-cell trait and sickle-cell anemia. [From Montgomery, R., Dryer, R.L., Conway, T.W., and Spector, A.A., *Biochemistry, A Case Oriented Approach* (4th ed.), *p.* 87. Copyright © 1983 C.V. Mosby Company, Inc.]

TABLE 7-4   **Amino Acid Sequences of Cytochromes *c* from 38 Species**

Group	Species	Sequence (positions −9 … 40)
Mammals	Human, chimpanzee	a GDVEKGKKI FIMKCSQCHTVEKGGKHKTGPNLHGLFGRKTGQA
	Rhesus monkey	a GDVEKGKKI FIMKCSQCHTVEKGGKHKTGPNLHGLFGRKTGQA
	Horse	a GDVEKGKKI FVQKCAQCHTVEKGGKHKTGPNLHGLFGRKTGQA
	Donkey	a GDVEKGKKI FVQKCAQCHTVEKGGKHKTGPNLHGLFGRKTGQA
	Cow, pig, sheep	a GDVEKGKKI FVQKCAQCHTVEKGGKHKTGPNLHGLFGRKTGQA
	Dog	a GDVEKGKKI FVQKCAQCHTVEKGGKHKTGPNLHGLFGRKTGQA
	Rabbit	a GDVEKGKKI FVQKCAQCHTVEKGGKHKTGPNLHGLFGRKTGQA
	California gray whale	a GDVEKGKKI FVQKCAQCHTVEKGGKHKTGPNLHGLFGRKTGQA
	Great gray kangaroo	a GDVEKGKKI FVQKCAQCHTVEKGGKHKTGPNINGIFGRKTGQA
Other vertebrates	Chicken, turkey	a GDIEKGKKI FVQKCSQCHTVEKGGKHKTGPNLHGLFGRKTGQA
	Pigeon	a GDIEKGKKI FVQKCSQCHTVEKGGKHKTGPNLHGLFGRKTGQA
	Pekin duck	a GDVEKGKKI FVQKCSQCHTVEKGGKHKTGPNLHGLFGRKTGQA
	Snapping turtle	a GDVEKGKKI FVQKCAQCHTVEKGGKHKTGPNLNGLIGRKTGQA
	Rattlesnake	a GDVEKGKKI FTMKCSQCHTVEKGGKHKTGPNLHGLFGRKTGQA
	Bullfrog	a GDVEKGKKI FVQKCAQCHTCEKGGKHKVGPNLYGLIGRKTGQA
	Tuna	a GDVAKGKKT FVQKCAQCHTVENGGKHKVGPNLWGLFGRKTGQA
	Dogfish	a GDVEKGKKV FVQKCAQCHTVENGGKHKTGPNLSGLFGRKTGQA
Insects	*Samia cynthia* (a moth)	h GVPAGNAENGKKI FVQRCAQCHTVEAGGKHKVGPNLHGFYGRKTGQA
	Tobacco hornworm moth	h GVpAGNADNGKKI FVQRCAQCHTVEAGGKHKVGPNLHGFFGRKTGQA
	Screwworm fly	h GVPAGDVEKGKKI FVQRCAQCHTVEAGGKHKVGPNLHGLFGRKTGQA
	Drosophila (fruit fly)	h GVPAGDVEKGKKL FVQRCAQCHTVEAGGKHKVGPNLHGLIGRKTGQA
Fungi	Baker's yeast	h TEFKAGSAKKGATLFKTRCLQCHTVEKGGPHKVGPNLHGIFGRHSGQA
	Candida krusei (a yeast)	h PAPFEQGSAKKGATLFKTRCAQCHTIEAGGPHKVGPNLHGIFSRHSGQA
	Neurospora crassa (a mold)	h GFSAGDSKKGANLFKTRCAQCHTLEEGGGNKIGPALHGLFGRKTGSV
Higher plants	Wheat germ	a ASFSEAPPGNPDAGAKIFKTKCAQCHTVDAGAGHKQGPNLHGLFGRQSGTT
	Buckwheat seed	a ATFSEAPPGNIKSGEKIFKTKCAQCHTVEKGAGHKQGPNLNGLFGRQSGTT
	Sunflower seed	a ASFAEAPPGDPTTGAKIFKTKCAQCHTVEKGAGHKQGPNLNGLFGRQSGTT
	Mung bean	a ASFBEAPPGBSKSGEKIFKTKCAQCHTVDKGAGHKQGPNLNGLFGRQSGTT
	Cauliflower	a ASFBEAPPGBSKAGEKIFKTKCAQCHTVDKGAGHKQGPNLNGLFGRQSGTT
	Pumpkin	a ASFBEAPPGBSKAGEKIFKTKCAQCHTVDKGAGHKQGPNLNGLFGRQSGTT
	Sesame seed	a ASFBEAPPGBVKSGEKIFKTKCAQCHTVDKGAGHKQGPNLNGLFGRQSGTT
	Castor bean	a ASFBEAPPGBVKAGEKIFKTKCAQCHTVEKGAGHKQGPNLNGLFGRQSGTT
	Cottonseed	a ASFZEAPPGBAKAGEKIFKTKCAQCHTVDKGAGHKQGPNLNGLFGRQSGTT
	Abutilon seed	a ASFZEAPPGBAKAGEKIFKTKCAQCHTVEKGAGHKQGPNLNGLFGRQSGTT

Number of different amino acids: 1 3 5 5 5 1 3 3 4 1 4 3 2 1 3 1 1 1 1 4 2 4 1 2 3 2 1 4 1 1 2 1 5 1 3 3 2 1 3 2 1 3 3

[a]The amino acid side chains have been shaded according to their polarity characteristics so that an invariant or conservatively substituted residue is identified by a vertical band of a single color. The letter a at the beginning of the chain indicates that the N-terminal amino group is acetylated; an h indicates that the acetyl group is absent.

Source: Dickerson, R.E., *Sci. Am.* **226**(4), 58–72 (1972), with corrections from Dickerson, R.E. and Timkovich, R., *in* Boyer, P.D. (Ed.), *The Enzymes* (3rd ed.), Vol. 11, *pp.* 421–422, Academic Press (1975). Table copyrighted © by Irving Geis.

The sickle-cell trait and disease occur mainly in persons of equatorial African descent. The regions of equatorial Africa where **malaria** is a major cause of death (contributing to childhood mortality rates as high as 50%), as Fig. 7-20 indicates, coincide closely with those areas where the sickle-cell gene is prevalent (possessed by as much as 40% of the population in some places). This observation led Anthony Allison to the discovery that *individuals heterozygous for HbS are resistant to malaria, that is, they are less likely to die of a malarial infection.*

Malaria is one of the most lethal infectious diseases that presently afflict humanity: Of the 2.5 billion people living within malaria-endemic areas, 100 million are clinically ill with the disease at any given time and at least 1 million, mostly very young children, die from it each year. In Africa, malaria is caused by the mosquito-borne protozoan *Plasmodium falciparum,* which resides within an erythrocyte during much of its 48-h life cycle. Plasmodia increase the acidity of the erythrocytes they infect by ~0.4 pH units

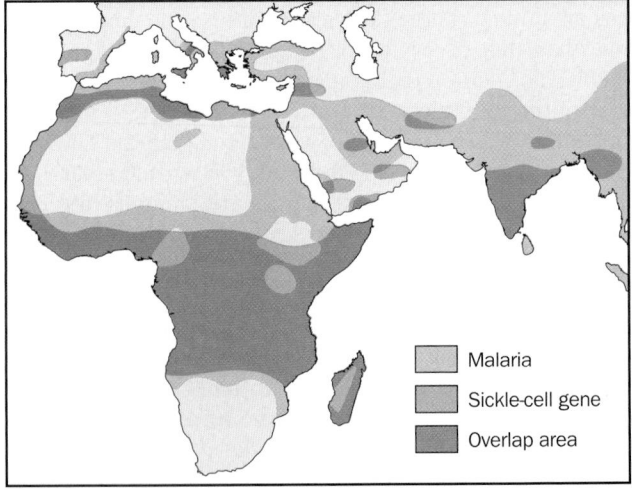

FIGURE 7-20   A map indicating the regions of the world where malaria caused by *P. falciparum* was prevalent before 1930, together with the distribution of the sickle-cell gene.

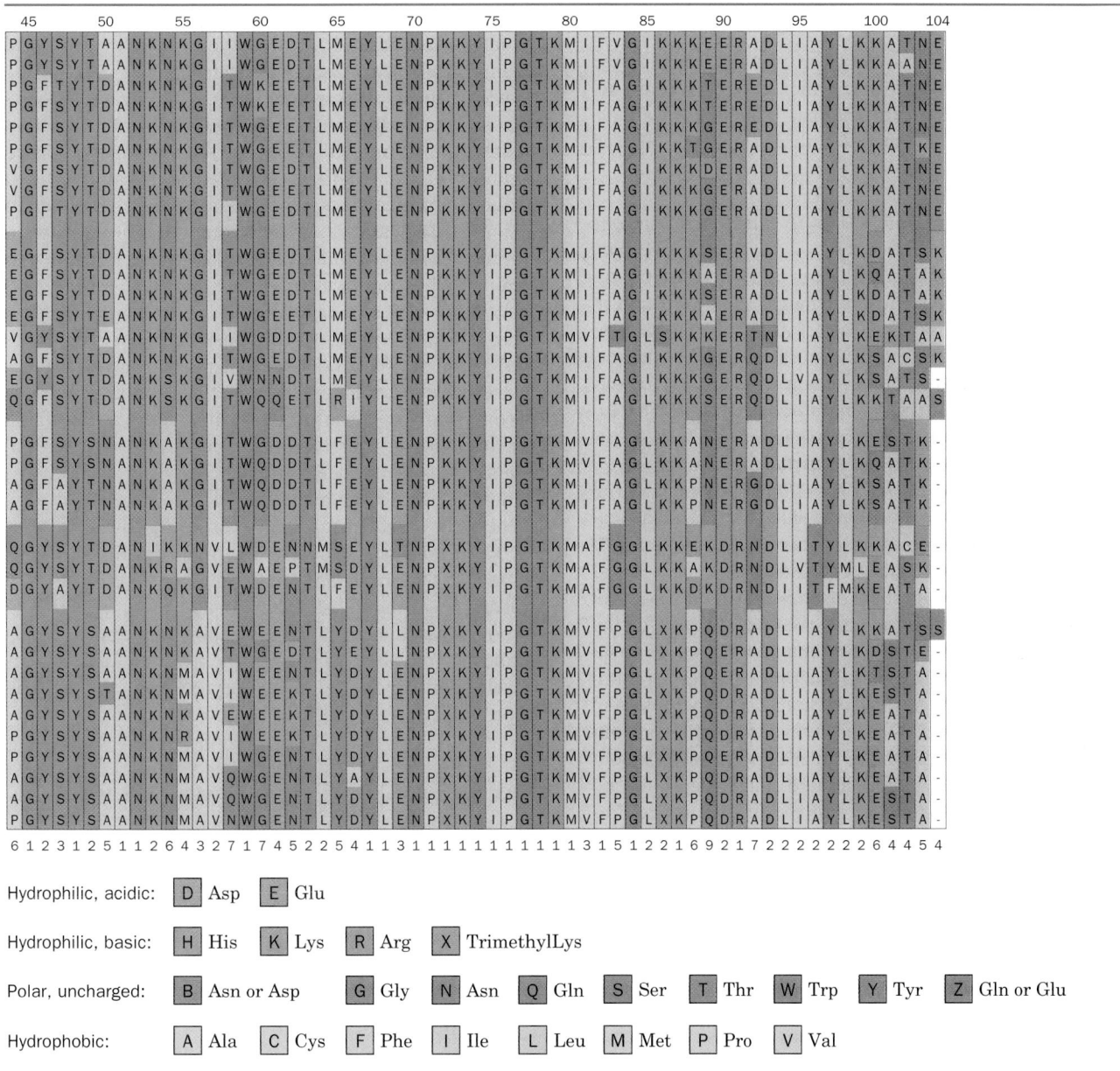

Hydrophilic, acidic:	D Asp	E Glu							
Hydrophilic, basic:	H His	K Lys	R Arg	X TrimethylLys					
Polar, uncharged:	B Asn or Asp	G Gly	N Asn	Q Gln	S Ser	T Thr	W Trp	Y Tyr	Z Gln or Glu
Hydrophobic:	A Ala	C Cys	F Phe	I Ile	L Leu	M Met	P Pro	V Val	

and cause them to adhere to a specific protein lining capillary walls by protein knobs that develop on the erythrocyte surfaces (the spleen would otherwise remove the infected erythrocytes from the circulation, thereby killing the parasites). Death often results when so many erythrocytes are lodged in a vital organ (such as the brain in cerebral malaria) that its blood flow is significantly impeded.

How does the sickle-cell trait confer malarial resistance? Normally, ~2% of the erythrocytes of individuals with the sickle-cell trait are observed to sickle under the low oxygen concentration conditions found in the capillaries. However, the lowered pH of infected erythrocytes increases their proportion of sickling in the capillaries to ~40%. Thus, during the early stages of a malarial infection, parasite-enhanced sickling probably causes the pref-

erential removal of infected erythrocytes from the circulation. In the later stages of infection, when the parasitized erythrocytes are attached to the capillary walls, the sickling induced by this low oxygen environment may mechanically and/or metabolically disrupt the parasite. Consequently, bearers of the sickle-cell trait in a malarial region have an adaptive advantage: The fractional population of heterozygotes (sickle-cell trait carriers) in such areas increases until their reproductive advantage becomes balanced by the inviability of the correspondingly increasing proportion of homozygotes (those with sickle-cell disease). Thus *sickle-cell anemia provides a classic Darwinian example of a single mutation's adaptive consequences in the ongoing biological competition among organisms for the same resources.*

B. *Species Variations in Homologous Proteins: The Effects of Neutral Drift*

The primary structures of a given protein from related species closely resemble one another. If one assumes, according to evolutionary theory, that related species have evolved from a common ancestor, then it follows that each of their proteins must have likewise evolved from the corresponding protein in that ancestor.

A protein that is well adapted to its function, that is, one that is not subject to significant physiological improvement, nevertheless continues evolving. The random nature of mutational processes will, in time, change such a protein in ways that do not significantly affect its function, a process called **neutral drift** (deleterious mutations are, of course, rapidly rejected through natural selection). *Comparisons of the primary structures of **homologous proteins** (evolutionarily related proteins) therefore indicate which of the proteins' residues are essential to its function, which are of lesser significance, and which have little specific function.* If, for example, we find the same side chain at a particular position in the amino acid sequence of a series of related proteins, we can reasonably conclude that the chemical and/or structural properties of that so-called **invariant residue** uniquely suit it to some essential function of the protein. Other amino acid positions may have less stringent side chain requirements so that only residues with similar characteristics (e.g., those with acidic properties: Asp and Glu) are required; such positions are said to be **conservatively substituted.** On the other hand, many different amino acid residues may be tolerated at a particular amino acid position, which indicates that the functional requirements of that position are rather nonspecific. Such a position is called **hypervariable.**

a. Cytochrome c Is a Well-Adapted Protein

To illustrate these points, let us consider the primary structure of a nearly universal eukaryotic protein, **cytochrome c.** Cytochrome c has a single polypeptide chain that, in vertebrates, consists of 103 or 104 residues, but in other phyla has up to 8 additional residues at its N-terminus. It occurs in the mitochondrion as part of the **electron-transport chain,** a complex metabolic system that functions in the terminal oxidation of nutrients to produce adenosine triphosphate (ATP) (Section 22-2). The role of cytochrome c is to transfer electrons from a large enzyme complex known as **cytochrome c reductase** to one called **cytochrome c oxidase.**

It is believed that the electron-transport chain took its present form between 1.5 and 2 billion years ago as organisms evolved the ability to respire (Section 1-5C). Since that time, the components of this multienzyme system have changed very little, as is evidenced by the observation that the cytochrome c from any eukaryotic organism, say a pigeon, will react *in vitro* with the cytochrome oxidase from any other eukaryote, for instance, wheat. Indeed, hybrid cytochromes c consisting of covalently linked fragments from such distantly related species as horse and yeast (prepared via techniques of genetic engineering) exhibit biological activity.

b. Protein Sequence Comparisons Yield Taxonometric Insights

Emanuel Margoliash, Emil Smith, and others elucidated the amino acid sequences of the cytochromes c from over 100 widely diverse eukaryotic species ranging in complexity from yeast to humans. The sequences from 38 of these organisms are arranged in Table 7-4 (*page 184*) so as to maximize the similarities between vertically aligned residues (methods of sequence alignment are discussed in Section 7-4B). The various residues in the table have been colored according to their physical properties in order to illuminate the conservative character of the amino acid substitutions. Inspection of Table 7-4 indicates that cytochrome c is an evolutionarily conservative protein. A total of 38 of its 105 residues (23 in all that have been sequenced) are invariant and most of the remaining residues are conservatively substituted (see the bottom row of Table 7-4). In contrast, there are eight positions that each accommodate six or more different residues and, accordingly, are described as being hypervariable.

The clear biochemical role of certain residues makes it easy to surmise why they are invariant. For instance, His 18 and Met 80 form ligands to the redox-active Fe atom of cytochrome c; the substitution of any other residues in these positions inactivates the protein. However, the biochemical significance of most of the invariant and conservatively substituted residues of cytochrome c can only be profitably assessed in terms of the protein's three-

TABLE 7-5 **Amino Acid Difference Matrix for 26 Species of Cytochrome c[a]**

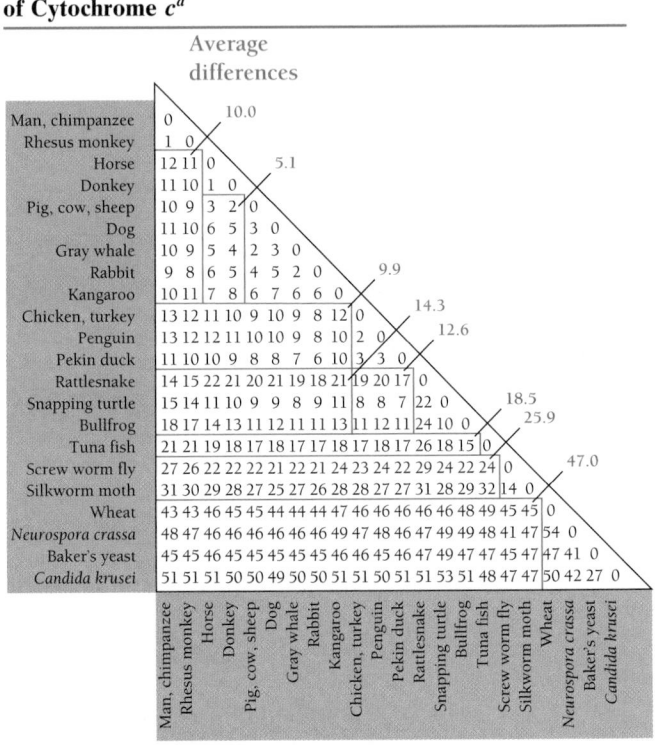

	Man, chimpanzee	Rhesus monkey	Horse	Donkey	Pig, cow, sheep	Dog	Gray whale	Rabbit	Kangaroo	Chicken, turkey	Penguin	Pekin duck	Rattlesnake	Snapping turtle	Bullfrog	Tuna fish	Screw worm fly	Silkworm moth	Wheat	Neurospora crassa	Baker's yeast	Candida krusei
Man, chimpanzee	0																					
Rhesus monkey	1	0																				
Horse	12	11	0																			
Donkey	11	10	1	0																		
Pig, cow, sheep	10	9	3	2	0																	
Dog	11	10	6	5	3	0																
Gray whale	10	9	5	4	2	3	0															
Rabbit	9	8	6	5	4	5	2	0														
Kangaroo	10	11	7	8	6	7	6	6	0													
Chicken, turkey	13	12	11	10	9	9	8	9	12	0												
Penguin	13	12	12	11	10	10	9	8	10	2	0											
Pekin duck	11	10	10	9	8	8	7	6	10	3	3	0										
Rattlesnake	14	15	22	21	20	21	19	18	21	19	20	17	0									
Snapping turtle	15	14	11	10	9	9	8	9	11	8	8	7	22	0								
Bullfrog	18	17	14	13	11	12	11	11	13	11	12	11	24	10	0							
Tuna fish	21	21	19	18	17	18	17	17	18	17	18	17	26	18	15	0						
Screw worm fly	27	26	22	22	22	21	22	21	24	23	24	22	29	24	22	24	0					
Silkworm moth	31	30	29	28	27	25	27	26	28	28	27	31	28	29	32	14	0					
Wheat	43	43	46	45	45	44	44	47	46	46	46	46	48	49	45	45	0					
Neurospora crassa	48	47	46	46	46	46	46	46	49	47	48	46	47	49	49	48	41	47	54	0		
Baker's yeast	45	45	46	45	45	45	45	46	46	45	46	47	49	47	45	45	47	41	0			
Candida krusei	51	51	51	50	50	49	50	51	51	51	50	51	51	53	51	48	47	47	50	42	27	0

[a]Each table entry indicates the number of amino acid differences between the cytochromes c of the species noted to the left of and below that entry.

[Table copyrighted © by Irving Geis.]

dimensional structure and is therefore deferred until Section 9-6A. In what follows, we consider what insights can be gleaned solely from the comparisons of the amino acid sequences of related proteins. The conclusions we draw are surprisingly far reaching.

The easiest way to compare the evolutionary differences between two homologous proteins is simply to count the amino acid differences between them (more realistically, we should infer the minimum number of DNA base changes to convert one protein to the other but, because of the infrequency with which mutations are accepted, counting amino acid differences yields similar information). Table 7-5 (*opposite*) is a tabulation of the amino acid sequence differences among 22 of the cytochromes *c* listed in Table 7-4. It has been boxed off to emphasize the relationships among groups of similar species. The order of these differences largely parallels that expected from classical taxon-

omy. Thus primate cytochromes *c* more nearly resemble those of other mammals than they do, for example, those of insects (8–12 differences for mammals vs 26–31 for insects). Similarly, the cytochromes *c* of fungi differ as much from those of mammals (45–51 differences) as they do from those of insects (41–47) or higher plants (47–54).

Through the analysis of data such as those in Table 7-5, *a phylogenetic tree (Section 1-1A) can be constructed that indicates the ancestral relationships among the organisms which produced the proteins* (the methods used to construct phylogenetic trees are discussed in Section 7-4C). That for cytochrome *c* is sketched in Fig. 7-21. Similar trees have been derived for other proteins. Each branch point of a tree indicates the probable existence of a common ancestor for all the organisms above it. The relative evolutionary distances between neighboring branch points are expressed as the number of amino acid differences per 100

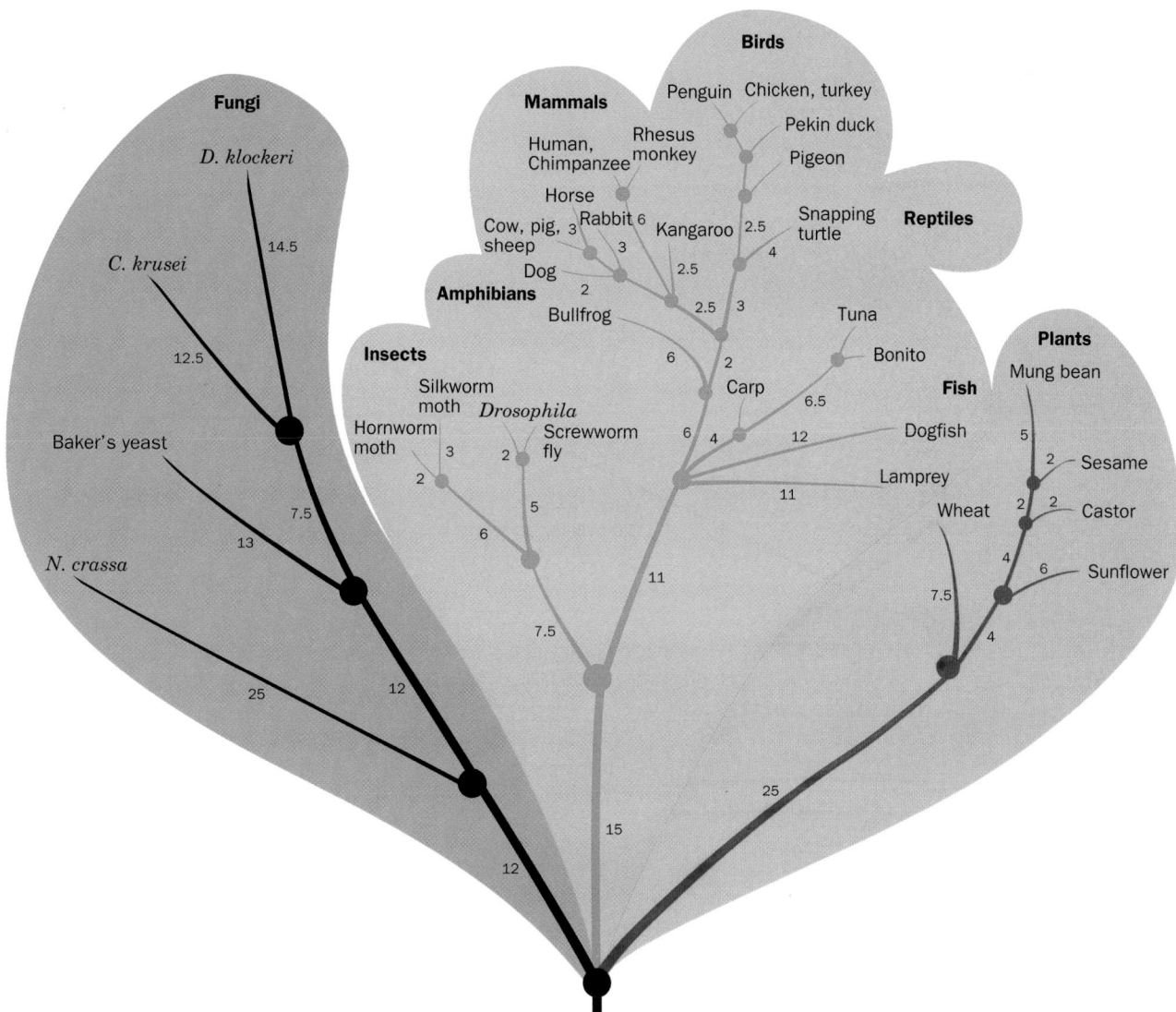

FIGURE 7-21 Phylogenetic tree of cytochrome *c*. The tree was generated by the computer-aided analysis of difference data such as that in Table 7-5 (see Section 7-4C). Each branch point indicates the existence of an organism deduced to be ancestral to the species connected above it. The numbers beside each branch indicate the inferred differences, in PAM units, between the cytochromes *c* of its flanking branch points or species. [After Dayhoff, M.O., Park, C.M., and McLaughlin, P.J., *in* Dayhoff, M.O. (Ed.), *Atlas of Protein Sequence and Structure*, p. 8, National Biomedical Research Foundation (1972).]

residues of the protein (*percentage* of *accepted* point *mutations*, or **PAM units**). This furnishes a quantitative measure of the degree of relatedness of the various species that macroscopic taxonomy cannot provide. Note that the evolutionary distances of modern cytochromes *c* from the lowest branch point on their tree are all approximately equal. Evidently, the cytochromes *c* of the so-called lower forms of life have evolved to the same extent as those of the higher forms.

c. Proteins Evolve at Characteristic Rates

The evolutionary distances between various species can be plotted against the time when, according to radiodated fossil records, the species diverged. For cytochrome *c*, this plot is essentially linear, thereby indicating that cytochrome *c* has accumulated mutations at a constant rate over the geological time scale (Fig. 7-22). This is also true for the other three proteins whose rates of evolution are plotted in Fig. 7-22. Each has its characteristic rate of

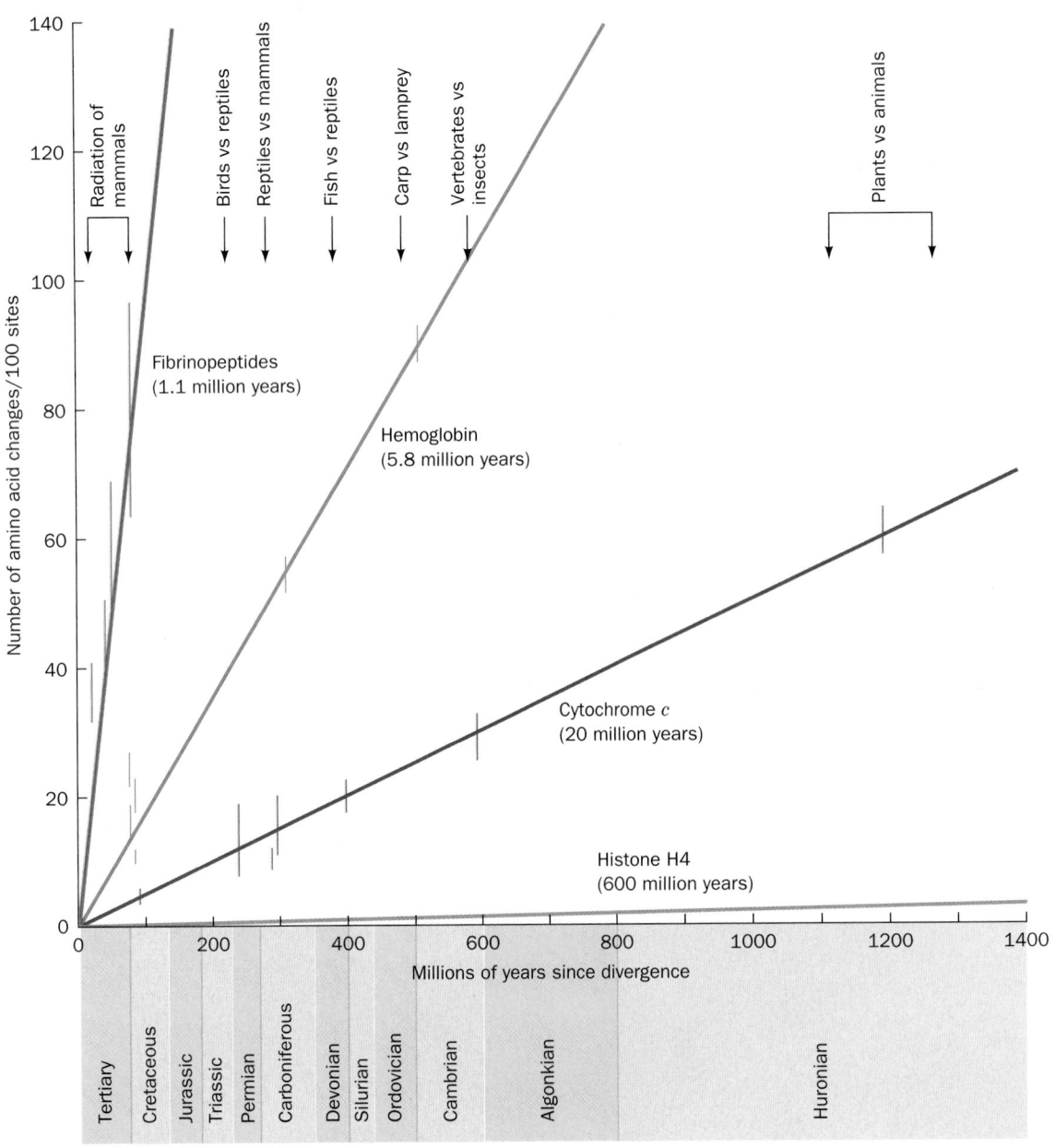

FIGURE 7-22   Rates of evolution of four unrelated proteins. The graph was constructed by plotting the average differences, in PAM units, of the amino acid sequences on two sides of a branch point of a phylogenetic tree (corrected to allow for more than one mutation at a given site) versus the time, according to the fossil record, since the corresponding species diverged from their common ancestor. The error bars indicate the experimental scatter of the sequence data. Each protein's rate of evolution, which is inversely proportional to the slope of its line, is indicated beside the line as its unit evolutionary period. [Figure copyrighted © by Irving Geis.]

change, known as a **unit evolutionary period,** which is defined as the time required for the amino acid sequence of a protein to change by 1% after two species have diverged. For cytochrome *c,* the unit evolutionary period is 20.0 million years. Compare this with the much less variant **histone H4** (600 million years) and the more variant hemoglobin (5.8 million years) and **fibrinopeptides** (1.1 million years).

The foregoing information does not imply that the rates of mutation of the DNAs specifying these proteins differ, but rather that *the rate that mutations are accepted into a protein depends on the extent that amino acid changes affect its function.* Cytochrome *c,* for example, is a rather small protein that, in carrying out its biological function, must interact with large protein complexes over much of its surface area. Any mutational change to cytochrome *c* will, most likely, affect these interactions unless, of course, the complexes simultaneously mutate to accommodate the change, a very unlikely occurrence. This accounts for the evolutionary stability of cytochrome *c.* Histone H4 is a protein that binds to DNA in eukaryotic chromosomes (Section 34-1A). Its central role in packaging the genetic archives evidently makes it extremely intolerant of any mutational changes. Indeed, histone H4 is so well adapted to its function that the histones H4 from peas and cows, species that diverged 1.2 billion years ago, differ by only two conservative changes in their 102 amino acids. Hemoglobin, like cytochrome *c,* is an intricate molecular machine (Section 10-2). It functions as a free floating molecule, however, so that its surface groups are usually more tolerant of change than are those of cytochrome *c* (although not in the case of HbS; Section 10-3B). This accounts for hemoglobin's greater rate of evolution. The fibrinopeptides are polypeptides of ~20 residues that are proteolytically cleaved from the vertebrate protein **fibrinogen** when it is converted to **fibrin** in the blood clotting process (Section 35-1A). Once they have been excised, the fibrinopeptides are discarded, so there is relatively little selective pressure on them to maintain their amino acid sequence and thus their rate of variation is high. If it is assumed that the fibrinopeptides are evolving at random, then the foregoing unit evolutionary periods indicate that in hemoglobin only 1.1/5.8 = 1/5 of the random amino acid changes are acceptable, that is, innocuous, whereas this quantity is 1/18 for cytochrome *c* and 1/550 for histone H4.

d. Mutational Rates Are Constant in Time

Amino acid substitutions in a protein mostly result from single base changes in the gene specifying the protein (Section 5-4B). If such **point mutations** mainly occur as a consequence of errors in the DNA replication process, then the rate at which a given protein accumulates mutations would be constant with respect to numbers of cell generations. If, however, the mutational process results from the random chemical degradation of DNA, then the mutation rate would be constant with absolute time. To choose between these alternative hypotheses, let us compare the rate of cytochrome *c* divergence in insects with that in mammals.

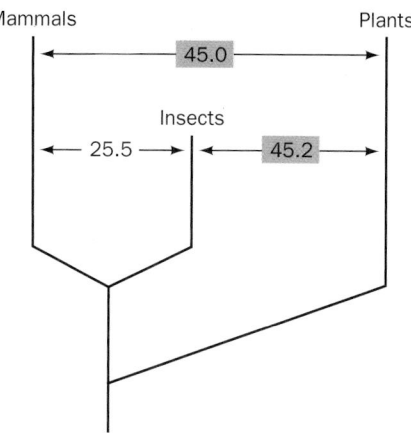

FIGURE 7-23 **A phylogenetic tree for cytochrome *c.*** The tree shows the average number of amino acid differences between cytochromes *c* from mammals, insects, and plants. Mammals and insects have diverged equally far from plants since their common branch point. [Adapted from Dickerson, R.E. and Timkovitch, R., *in* Boyer, P.D. (Ed.), *The Enzymes* (3rd ed.), Vol. 11, *p.* 447, Academic Press (1975).]

Insects have shorter generation times than mammals. Therefore, if DNA replication were the major source of mutational error, then from the time the insect and mammalian lines diverged, insects would have evolved further from plants than have mammals. However, a simple phylogenetic tree (Fig. 7-23) indicates that the average number of amino acid differences between the cytochromes *c* of insects and plants (45.2) is essentially the same as that between mammals and plants (45.0). We must therefore conclude that cytochrome *c* accumulates mutations at a uniform rate with respect to time rather than number of cell generations. This, in turn, implies that *point mutations in DNA accumulate at a constant rate with time, that is, through random chemical change, rather than resulting mainly from errors in the replication process.*

e. Sequence Comparisons Indicate when the Major Kingdoms of Life Diverged

Estimates of when two species diverged, that is, when they last had a common ancestor, are based largely on the radiodated fossil record. However, the macrofossil record only extends back ~600 million years (after multicellular organisms arose) and phylogenetic comparisons of microfossils (fossils of single-celled organisms) based on their morphology are unreliable. Thus, previous estimates of when the major groupings of organisms (animals, plants, fungi, protozoa, eubacteria, and archaea; Figs. 1-4 and 1-11) diverged from one another (e.g., the right side of Fig. 7-22) are only approximations based mainly on considerations of shared characteristics.

The burgeoning databases of amino acid sequences (Section 7-4A) permitted Russell Doolittle to compare the sequences of a large variety of enzymes that each have homologous representatives in many of the above major groupings (531 sequences in 57 different enzymes). This analysis is consistent with the existence of a **molecular clock** that can provide reliable estimates of when these groupings diverged. This molecular clock, which is based on the supposition that homologous sequences diverge at a uniform rate, was calibrated using sequences from vertebrates for which there is a reasonably reliable fossil record. This analysis indicates that animals, plants, and fungi last had a common ancestor ~1 billion years ago, with plants having diverged from animals slightly before fungi; that the major protozoan lineages separated from those of other eukarya ~1.2 billion years ago; that eukarya last shared a common ancestor with archaea ~1.8 billion years ago and with bacteria slightly more than 2 billion years ago; and that gram-positive and gram-negative bacteria diverged ~1.4 billion years ago.

f. Protein Evolution Is Not the Basis of Organismal Evolution

Despite the close agreement between phylogenetic trees derived from sequence similarities and classic taxonometric analyses, it appears that protein sequence evolution is not the only or even the most important basis of organismal evolution. There is, for example, more than a 99% sequence identity between the corresponding proteins of humans and our closest relative, the chimpanzee (e.g., their cytochromes *c* are identical). This is the level of homology observed among sibling species of fruit flies and mammals. Yet the anatomical and behavioral differences between human and chimpanzee are so great that these species have been classified in separate families. *This suggests that the rapid divergence of human and chimpanzee stems from relatively few mutational changes in the segments of DNA that control gene expression, that is, how much of each protein will be made, where, and when.* Such mutations do not change protein sequences but can result in major organismal alterations.

C. *Evolution through Gene Duplication*

Most proteins have extensive sequence similarities with other proteins from the same organism. Such proteins arose through **gene duplication,** a result, it is thought, of an aberrant genetic recombination event in which a single chromosome acquired both copies of the primordial gene in question (the mechanism of genetic recombination is discussed in Section 30-6A). *Gene duplication is a particularly efficient mode of evolution because one of the duplicated genes can evolve a new functionality through natural selection while its counterpart continues to direct the synthesis of the presumably essential ancestral protein.*

The **globin** family of proteins, which includes hemoglobin and **myoglobin,** provides an excellent example of evolution through gene duplication. Hemoglobin transports oxygen from the lungs (or gills or skin) to the tissues. Myoglobin, which occurs in muscles, facilitates rapid oxygen diffusion through these tissues and also functions as an oxygen storage protein. *The sequences of hemoglo-*

bin's α and β subunits (recall that hemoglobin is an $\alpha_2\beta_2$ tetramer) and myoglobin (a monomer) are quite similar.

The globin family's phylogenetic tree indicates that its members, in humans, arose through the following chain of events (Fig. 7-24):

1. The primordial globin probably functioned simply as an oxygen-storage protein. Indeed, the globins in certain modern invertebrates still have this function. For example, treating a *Planorbis* snail with CO (the binding of which prevents globins from binding O_2; Section 10-1A) does not affect its behavior in well-aerated water, but if the oxygen concentration is reduced, a poisoned *Planorbis* becomes even more sluggish than a normal one.

2. Duplication of a primordial globin gene, ~1.1 billion years ago, permitted the resulting two genes to evolve separately so that, largely by a series of point mutations, a monomeric hemoglobin arose that had the lower oxygen-binding affinity required for it to transfer oxygen to the developing myoglobin. Such a monomeric hemoglobin can still be found in the blood of the **lamprey,** a primitive vertebrate that, according to the fossil record, has maintained its eel-like morphology for over 425 million years.

3. Hemoglobin's tetrameric character is a structural feature that greatly increases its ability to transport oxygen efficiently (Section 10-2C). This provided the adaptive advantage that gave rise to the evolution of the β chain from a duplicated α chain.

4. In fetal mammals, oxygen is obtained from the maternal circulation. **Fetal hemoglobin,** an $\alpha_2\gamma_2$ tetramer in which the **γ chain** is a gene-duplicated β chain variant, evolved to have an oxygen-binding affinity between that of normal adult hemoglobin and myoglobin.

5. Human embryos, in their first 8 weeks postconception, make a $\zeta_2\varepsilon_2$ hemoglobin in which the ζ and ε **chains** are, respectively, gene-duplicated α and β variants.

6. In primates, the β chain has undergone a relatively recent duplication to form a **δ chain.** The $\alpha_2\delta_2$ hemoglobin, which occurs as a minor hemoglobin component in normal adults (~1%), has no known unique function. Perhaps it may eventually evolve one (although the human genome contains the relics of globin genes that are no longer expressed; Section 34-2F).

Homologous proteins in the same organism and the genes that encode them are said to be **paralogous** (Greek: *para,* alongside), whereas homologous proteins/genes in different organisms that arose through species divergence (e.g., the various cytochomes *c*) are said to be **orthologous** (Greek: *ortho,* straight). Hence, the α- and β-globins and myoglobin are paralogs, whereas the α-globins from different species are orthologs.

Our discussion of the globin family indicates that protein evolution through gene duplication leads to proteins of similar structural and functional properties. Another well-documented example of this phenomenon has resulted in the formation of a family of endopeptidases, which include trypsin, chymotrypsin, and elastase. These paralogous digestive enzymes, which are all secreted by the pancreas into the small intestine, are quite similar in their properties, differing mainly in their side chain specificities (Table 7-2). We examine how these functional variations are structurally rationalized in Section 15-3B. Individually, these three enzymes are limited in their abilities to degrade a protein, but in concert, they form a potent digestive system.

As we have stated previously and will explore in detail in Section 9-1, the *three-dimensional structure of a protein, and hence its function, is dictated by its amino acid sequence.* Most proteins that have been sequenced are more or less similar to several other known proteins. In fact, *many proteins are mosaics of sequence motifs that occur in a variety of other proteins.* It therefore seems likely that

most of the myriads of proteins in any given organism have arisen through gene duplications. This suggests that the appearance of a protein with a novel sequence and function is an extremely rare event in biology—one that may not have occurred since early in the history of life.

4 ■ BIOINFORMATICS: AN INTRODUCTION

☙ **See Guided Exploration 6: Bioinformatics** The enormous profusion of sequence and structural data that have been generated over the last decades has led to the creation of a new field of inquiry, **bioinformatics,** which is loosely de-

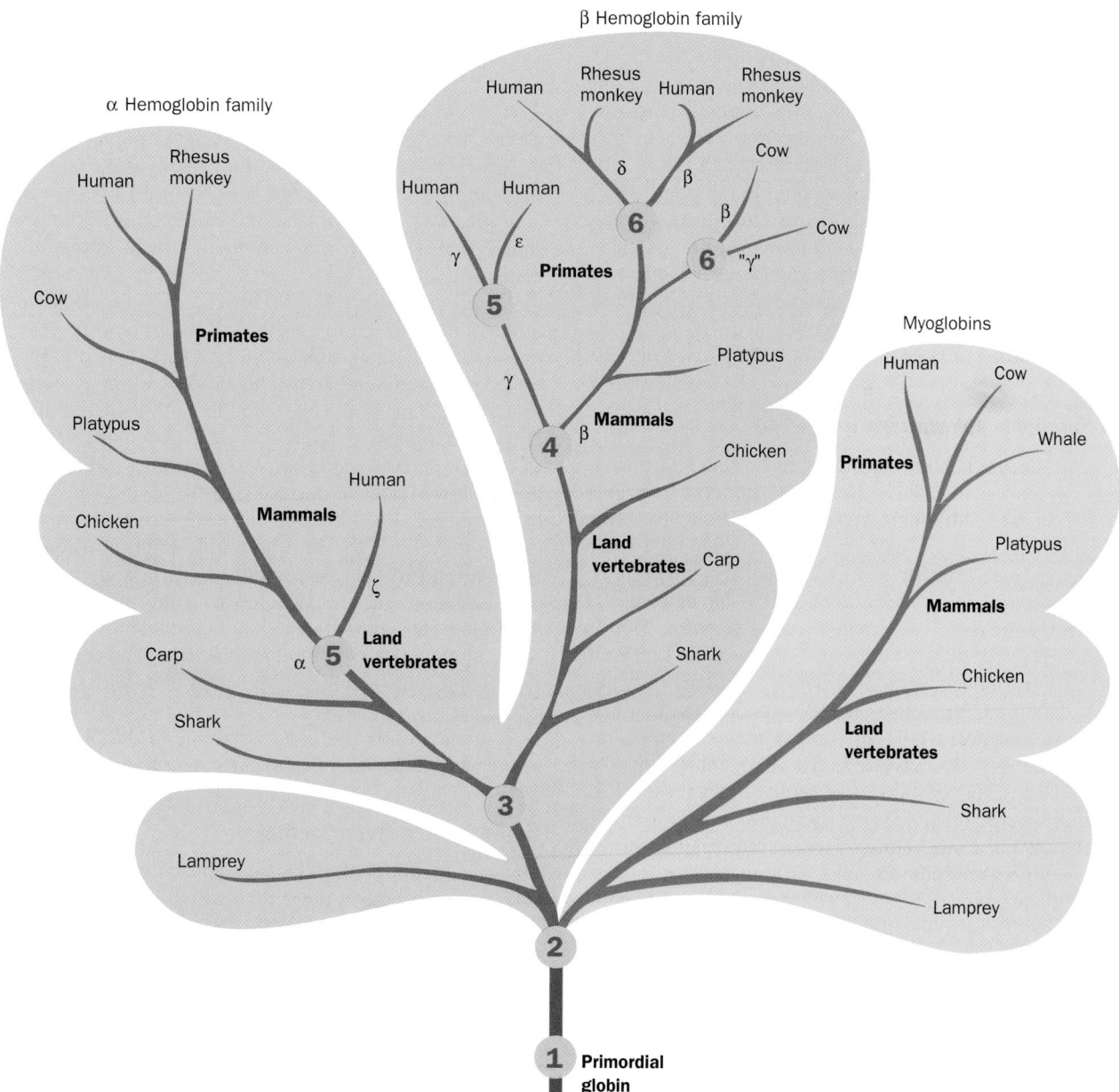

FIGURE 7-24 Phylogenetic tree of the globin family. The circled branch points represent gene duplications and unmarked branch points are species divergences. [After Dickerson, R.E. and Geis, I., *Hemoglobin, p.* 82, Benjamin/Cummings (1983).]

fined as being at the intersection of biotechnology and computer science. It is the computational tools that the practitioners of bioinformatics have produced that have permitted the "mining" of this treasure trove of biological data, thereby yielding surprisingly far reaching biomolecular insights.

As we have seen in the previous section, the alignment of the sequences of homologous proteins yields valuable clues as to which of the proteins' residues are essential for function and is indicative of the evolutionary relationships among these proteins. Since proteins are encoded by nucleic acids, the alignment of homologous DNA or RNA sequences provides similar information. Moreover, DNA sequence alignment is an essential task for assembling chromosomal sequences (contigs) from large numbers of sequenced segments (Section 7-2B).

If the sequences of two proteins or nucleic acids are closely similar, one can usually line them up by eye. In fact, this is the way that the cytochrome *c* sequences in Table 7-4 were aligned. But how can one correctly align sequences that are so distantly related that their sequence similarities are no longer readily apparent? In this section we discuss the computational techniques through which this is done, preceded by a short introduction to publicly accessible sequence databases. In doing so, we shall concentrate on techniques of peptide alignment. We end with a short discussion of how phylogenetic trees are generated. Those aspects of bioinformatics concerned with the analysis of structures are postponed until Chapters 9 and 10.

A. Sequence Databases

Since it became possible to elucidate protein and nucleic acid sequences, they have been determined at an ever increasing rate. Although, at first, these sequences were printed in research journals, their enormous numbers and lengths (particularly for genome sequences) make it no longer practical to do so. Moreover, it is far more useful to have sequences in computer-accessible form. Hence, researchers now directly deposit sequences, via the Web, into various publicly accessible databases, many of which share data on a daily basis. The Web addresses [Uniform Resource Locators (URLs)] for the major protein and DNA sequence databases are listed in Table 7-6. The URLs for a variety of specialized sequence databases (e.g., those of specific organisms or organelles) can be found at "Amos' WWW Links" (http://www.expasy.ch/alinks.html). This website also contains links to numerous other biochemically useful databases as well a great variety of computer tools for biomolecular analyses, bibliographic references, tutorials, and many other websites of biomedical interest. (Note that websites evolve far faster than organisms: Even well established websites change addresses or even disappear with little warning and useful new websites appear on an almost daily basis.)

As an example of a sequence database, let us describe (in cursory detail) the annotated protein sequence database named SWISS-PROT. A sequence record in SWISS-PROT begins with the proteins' ID code of the form X_Y

TABLE 7-6   World Wide Web Addresses for the Major Protein and DNA Sequence Data Banks

Data Banks Containing Protein Sequences
ExPASy Molecular Biology Server (SWISS-PROT):
http://expasy.ch/
Protein Information Resource (PIR):
http://pir.georgetown.edu/
Protein Research Foundation (PRF):
http://www.prf.or.jp/en/

Data Banks Containing Gene Sequences
GenBank:
http://www.ncbi.nlm.nih.gov/Genbank/GenbankSearch.html
European Bioinformatics Institute (EBI):
http://srs.ebi.ac.uk/
DBGET/LinkDB Integrated Database Retrieval System:
http://www.genome.ad.jp/dbget/

where X is an up-to-four-character mnemonic indicating the protein's name (e.g., CYC for cytochrome *c* and HBA for hemoglobin α chain) and Y is an up-to-five-character identification code indicating the protein's biological source that usually consists of the first three letters of the genus and the first two letters of the species [e.g., CANFA for *Canis familiaris* (*dog*)]. However, for the most commonly encountered organisms, Y is instead a self-explanatory code (e.g., BOVIN or ECOLI). This is followed by an accession number such as P04567, which is assigned by the database to ensure a stable way of identifying an entry from release to release, even if it becomes necessary to change its ID code. The entry continues with the date the entry was entered into SWISS-PROT and when it was last modified and annotated, a list of pertinent references (which are linked to MedLine), a description of the protein, and its links to other databases. A Feature Table describes regions or sites of interest in the protein such as disulfide bonds, post-translational modifications, elements of local secondary structure, binding sites, and conflicts between different references. The entry ends with the length of the peptide in residues, its molecular weight, and finally its sequence using the one-letter code (Table 4-1). Other sequence databases are similarly constructed.

B. Sequence Alignment

One can quantitate the sequence similarity of two polypeptides or two DNAs by determining their number of aligned residues that are identical. For example, human and dog cytochromes *c*, which differ in 11 of their 104 residues (Table 7-5) are $[(104 - 11)/104] \times 100 = 89\%$ identical, whereas human and baker's yeast cytochromes *c* are $[(104 - 45)/104] \times 100 = 57\%$ identical. Table 7-4 indicates that baker's yeast cytochrome *c* has 5 residues at its N-terminus that human cytochrome *c* lacks but lacks the human protein's C-terminal residue. When determining percent identity, the length of the shorter peptide/DNA is, by convention, used in the denominator. One can likewise calculate the percent similarity between two peptides, once it is decided which amino acid residues are to be considered similar (e.g., Asp and Glu).

a. The Homology of Distantly Related Proteins May Be Difficult to Recognize

Let us examine how proteins evolve by considering a simple model. Assume that we have a 100-residue protein in which all point mutations have an equal probability of being accepted and occur at a constant rate. Thus at an evolutionary distance of one PAM unit (Section 7-3B), the original and evolved proteins are 99% identical. At an evolutionary distance of two PAM units, they are $(0.99)^2 \times 100 = 98\%$ identical, whereas at 50 PAM units they are $(0.99)^{50} \times 100 = 61\%$ identical. Note that the latter quantity is not 50%, as one might naively expect. This is because *mutation is a stochastic (probablistic or random) process: At every stage of evolution, each residue has an equal chance of mutating.* Hence some residues may change twice or more before others change even once. Consequently, a plot of percent identity vs evolutionary distance (Fig. 7-25*a*) is an exponential curve that approaches but never equals zero. Even at quite large evolutionary distances, original and evolved proteins still have significant sequence identities.

Real proteins evolve in a more complex manner than our simple model predicts. This is in part because certain amino acid residues are more likely to form accepted mutations than others and in part because the distribution of amino acids in proteins is not uniform (e.g., 9.5% of the residues in proteins, on average, are Leu but only 1.2% are Trp; Table 4-1). Consequently real proteins evolve even more slowly than in our simple model (Fig. 7-25*b*).

At what point in the evolutionary process does homology become unrecognizable? If identical length polypeptides of random sequences were of uniform amino acid composition, that is, if they consisted of 5% of each of the 20 amino acids, then they would exhibit, on average, 5% identity. However, since mutations occur at random, there is considerable variation in such numbers. Thus, statistical considerations reveal that there is a 95% probability that such 100-residue peptides are between 0 and 10% identical. But, as we have seen for cytochrome *c*, homologous peptides may have different lengths because one may have more or fewer residues at its N- or C-termini. If we therefore permit our random 100-residue peptides to shift in their alignment by up to 5 residues, the average expected identity for the best alignment increases to 8%, with 95% of such comparisons falling between 4% and 12%. Consequently, one in 20 of such comparisons will be out of this range (>12% or <4%) and one in 40 will exhibit >12% sequence identity.

But this is not the whole story because mutational events may result in the insertion or deletion of one or more residues within a chain. Thus, one chain may have gaps relative to another. Yet, if we permit an unlimited number of gaps, we can always get a perfect match between any two chains. For example, two 15-residue peptides that have only one match (using the one-letter code; Table 4-1)

SQMC I LFKAQMN Y GH
MF Y ACRLPMGAH Y WL

would have a perfect match over their aligned portions if we allowed unlimited gapping:

SQMC I L F KAQMN Y GH
– – M – – – F – – – – – Y – – ACRLPMGAHYWL

Thus we cannot allow unlimited gapping to maximize a match between two peptides, but neither can we forbid all gapping because insertions and deletions (also called **indels**) really do occur. Consequently, for each allowed gap, we must impose some sort of penalty in our alignment algorithm that strikes a balance between finding the best alignment between distantly related peptides and rejecting improper alignments. But if we do so (using methods discussed below), *unrelated proteins will exhibit sequence identities in the range 15% to 25%. Yet distantly related proteins may have similar levels of sequence identity. This is*

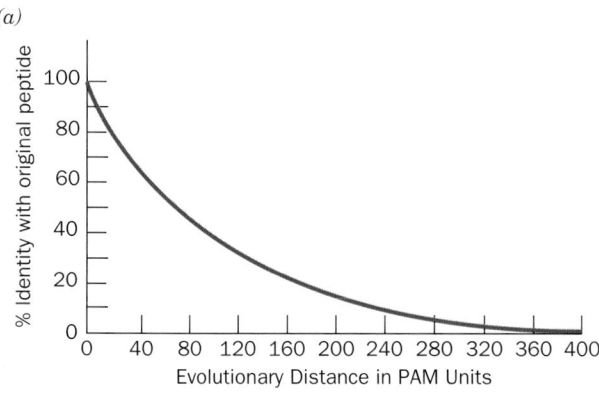

(a)

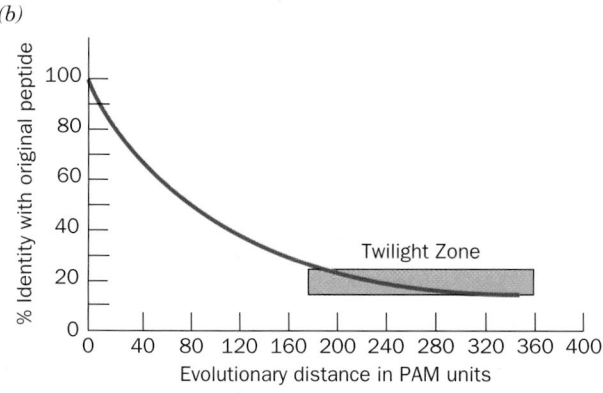

(b)

FIGURE 7-25   Rate of sequence change in evolving proteins. (*a*) For a protein that is evolving at random and that initially consists of 5% of each of the 20 "standard" amino acid residues. (*b*) For a protein of average amino acid composition evolving as is observed in nature, that is, with certain residue changes more likely to be accepted than others and with occasional insertions and deletions. [Part *b* after Doolittle, R.F., *Methods Enzymol.* **183**, 103 (1990).]

the origin of the **twilight zone** in Fig. 7-25*b*. It requires the sophisticated alignment algorithms we discuss below to differentiate homologous proteins in the twilight zone from those that are unrelated.

b. Sequence Alignment Using Dot Matrices

How does one perform a sequence alignment between two polypeptides (a **pairwise alignment**)? The simplest way is to construct a **dot matrix** (alternatively, a **dot plot** or **diagonal plot**): Lay the sequence of one polypeptide horizontally and that of the other vertically and place a dot in the resulting matrix wherever the residues are identical. A dot plot of a peptide against itself results in a square matrix with a row of dots along the diagonal and a scattering of dots at points where there are chance identities. If the peptides are closely similar, there are only a few absences along the diagonal (e.g., Fig. 7-26*a*), whereas distantly related peptides have a large number of absences along the diagonal and a shift in its position wherever one peptide has a gap relative to the other (e.g., Fig. 7-26*b*).

Once an alignment has been established, it should be scored in some way to determine if it has any relationship to reality. A simple but effective way to calculate an **alignment score (AS)** is to add 10 for every identity but those of Cys, which count 20 (because Cys residues often have indispensable functions), and then subtract 25 for every gap. Furthermore, we can calculate the **normalized alignment score (NAS)** by dividing the AS by the number of residues in the shortest of the two polypeptides and multiplying by 100. Thus, for the alignment of the human hemoglobin α chain (141 residues) and human myoglobin (153 residues; Fig. 7-27), AS = $37 \times 10 + 1 \times 20 - 1 \times 25 = 365$ and NAS = $(365/141) \times 100 = 259$. Statistical analysis (Fig. 7-28) indicates that this NAS is indicative of homology. Note that a perfect match would result in NAS = 1000 in the absence of Cys residues or gaps. An acceptable NAS decreases with peptide length because a high proportion of matches is more likely to occur between short peptides than between long ones (e.g., 2 matches in 10 residues is more likely to occur at random than 20 matches in 100 residues, although both have NAS = 200).

c. Alignments Should Be Weighted According to the Likelihood of Residue Substitutions

The foregoing techniques can easily (although tediously) be carried out by hand, particularly when there is an obvious alignment. But what if we have numerous polypeptides with which we want to align a new sequence (and, typically, one checks newly determined sequences against all other known sequences). Moreover, alignments in the twilight zone are difficult to discern. We must therefore turn to computerized statistical analyses so as to be able to distinguish distant evolutionary relationships from chance similarities with maximum sensitivity.

(a) Tuna fish cytochrome *c*

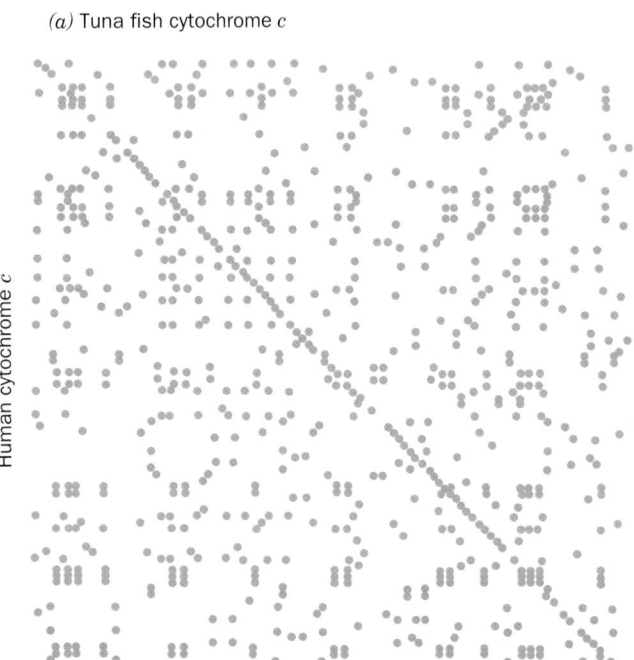

(b) Rhodospirillum rubrum cytochrome *c*₂

FIGURE 7-26   Sequence alignment with dot matrices. Dots plots show alignments of (*a*) human cytochrome *c* (104 residues) vs tuna fish cytochrome *c* (103 residues) and (*b*) human cytochrome *c* vs *Rhodospirillum rubrum* cytochrome *c*₂ (a bacterial *c*-type cytochrome consisting of 112 residues). The N-termini of these peptides are at the top and left of the diagrams. The two proteins in Part *a* have 82 identities, whereas those in Part *b* have 40 identities. The diagonal in Part *b* is more clearly seen if the diagram is viewed edgewise from its lower right corner. Note that there are two horizontal displacements of this diagonal, one near its center and other toward the C-terminus. This is indicative of inserts in the *Rhodospirillum* protein relative to the human protein. [After Gibbs, A.J. and McIntyre, G.A., *Eur. J. Biochem.* **16**, 2 (1970).]

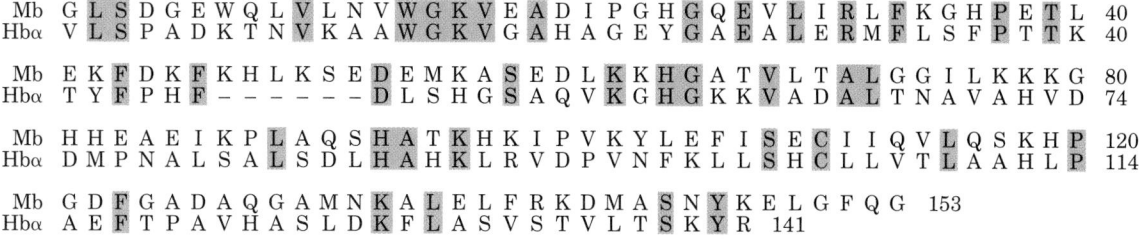

| Mb | G L S D G E W Q L V L N V W G K V E A D I P G H G Q E V L I R L F K G H P E T L | 40 |
| Hbα | V L S P A D K T N V K A A W G K V G A H A G E Y G A E A L E R M F L S F P T T K | 40 |

| Mb | E K F D K F K H L K S E D E M K A S E D L K K H G A T V L T A L G G I L K K K G | 80 |
| Hbα | T Y F P H F – – – – – – D L S H G S A Q V K G H G K K V A D A L T N A V A H V D | 74 |

| Mb | H H E A E I K P L A Q S H A T K H K I P V K Y L E F I S E C I I Q V L Q S K H P | 120 |
| Hbα | D M P N A L S A L S D L H A H K L R V D P V N F K L L S H C L L V T L A A H L P | 114 |

| Mb | G D F G A D A Q G A M N K A L E L F R K D M A S N Y K E L G F Q G | 153 |
| Hbα | A E F T P A V H A S L D K F L A S V S T V L T S K Y R | 141 |

AS = 365   NAS = 259   % ID = 27.0

FIGURE 7-27   The optimal alignments of human myoglobin (Mb, 153 residues) and the human hemoglobin α chain (Hbα, 141 residues). Identical residues are boxed and gaps are indicated by dashes. [After Doolittle, R.F., *Of URFs and ORFs*, University Science Books (1986).]

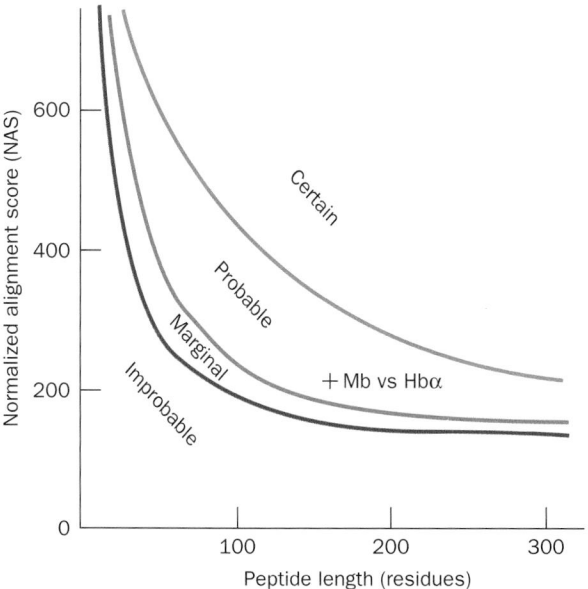

FIGURE 7-28   A guide to the significance of normalized alignment scores (NAS) in the comparison of peptide sequences. Note how the significance of an NAS varies with peptide length. The position of the Mb vs Hbα alignment (Fig. 7-27) is indicated. [After Doolittle, R.F., *Methods Enzymol.* **183**, 102 (1990).]

A dot matrix can be cast in a more easily mathematized form by replacing each dot (match) with a 1 and each nonmatch with a 0. Then a self-dot matrix would become a square diagonal matrix (having all 1's along its diagonal) with a few off-diagonal 1's, and two closely related peptides would have several diagonal positions with 0's. But this is a particularly rigid system: It does not differentiate between conservative substitutions and those that are likely to be hypervariable. Yet it is clear that certain substitutions are more readily accepted than others. What are these favored substitutions, how can we obtain a quantitative measure of them, and how can we use this information to increase the confidence with which we can align distantly related peptides?

One way we might assign a weight (a quantity that increases with the probability of occurrence) to a residue change is according to the genetic code (Table 5-3). Thus residue changes that require only a single base change [e.g., Leu (CUX) → Pro (CCX)] are likely to occur more often and hence would be assigned a greater weight than a residue change that requires two base changes [e.g., Leu (CUX) → Thr (ACX)], which likewise would be assigned a greater weight than a residue change that requires three base changes [e.g., His (CA^U_C) → Trp (UGG)]. Of course, no change (the most probable event) would be assigned the greatest weight of all. However, such a scheme only weights the probability of a mutation occurring, not that of a mutation being accepted, which depends on its Darwinian fitness. In fact, over half of the possible single-base residue changes are between physically dissimilar residues, which are therefore less likely to be accepted.

A more realistic weighting scheme would be to assign some sort of relative probability to two residues being exchanged according to their physical similarity. Thus it would seem that a Lys → Arg mutation is more likely to be accepted than, say, a Lys → Phe mutation. However, it is by no means obvious how to formulate such a weighting scheme based on theoretical considerations because it is unclear how to evaluate the various different sorts of properties that suit the different residues to the many functions they have in a large variety of proteins.

d. PAM Substitution Matrices Are Based on Observed Rates of Protein Evolution

An experimentally based method of determining the rates of acceptance of the various residue exchanges is to weight them according to the frequencies with which they are observed to occur. Margaret Dayhoff did so by comparing the sequences of a number of closely related proteins (>85% identical; similar enough so that we can be confident that their alignments are correct and that there are insignificant numbers of multiple residue changes at a single site) and determining the relative frequency of the $20 \times 19/2 = 190$ different possible residue changes (we divide by two to account for the fact that changes in either direction, A → B or B → A, are equally likely). From these data one can prepare a symmetric square matrix, 20 elements on a side, whose elements, M_{ij}, indicate the probability that, in a related sequence, amino acid *i* will replace amino acid *j* after some specified evolutionary interval—usually one PAM unit. Using this **PAM-1** matrix, one can generate a mutation probability matrix for other evolutionary distances, say N PAM units, by multiplying the matrix by itself N times ($[M]^N$), thereby generating a PAM-N matrix. Then an element of the **relatedness odds matrix,** R, is

$$R_{ij} = M_{ij}/f_i \qquad [7.3]$$

where M_{ij} is now an element of the PAM-N matrix and f_i is the probability that the amino acid i will occur in the second sequence by chance. R_{ij} is the probability that amino acid i will replace amino acid j (or vice versa) per occurrence of i per occurrence of j. When two polypeptides are compared with each other, residue by residue, the R_{ij}'s for each position are multiplied to obtain the **relatedness odds** for the entire polypeptide. For example, when the hexapeptide A-B-C-D-E-F evolves to the hexapeptide P-Q-R-S-T-U,

$$\text{Relatedness odds} = R_{AP} \times R_{BQ} \times R_{CR} \times R_{DS} \atop \times R_{ET} \times R_{FU} \qquad [7.4]$$

A more convenient way of making this calculation is to take the logarithm of each R_{ij}, thereby yielding the **log odds substitution matrix.** Then we add the resulting matrix elements rather than multiplying them to obtain the **log odds.** Thus for our hexapeptide pair:

$$\log \text{odds} = \log R_{AP} + \log R_{BQ} + \log R_{CR} \atop + \log R_{DS} + \log R_{ET} + \log R_{FU} \qquad [7.5]$$

It is a peptide pair's log odds that we wish to maximize in obtaining their best alignment, that is, we will use log odds values as our alignment scores.

Table 7-7 is the **PAM-250** log odds substitution matrix with all of its elements multiplied by 10 for readability (which only adds a scale factor). Each diagonal element in

the matrix indicates the mutability of the corresponding amino acid, whereas the off-diagonal elements indicate their exchange probabilities. A neutral (random) score is 0, whereas an amino acid pair with a score of -3 exchanges with only $10^{-3/10} = 0.50$ of the frequency expected at random. This substitution matrix has been arranged such that the amino acid residues most likely to replace each other in related proteins (the pairs that have the highest log R_{ij} values) are grouped together. Note that this grouping is more or less what is expected from their physical properties.

Identities (no replacement) tend to have the highest values in Table 7-7. Trp and Cys (diagonal values 17 and 12) are the residues least likely to be replaced, whereas Ser, Ala, and Asn (all 2) are the most readily mutated. The residue pair least likely to exchange is Cys and Trp (-8), whereas the pair most likely to exchange is Tyr and Phe (7), although these latter residues are among the least likely to exchange with other residues (mostly negative entries). Similarly, charged and polar residues are unlikely to exchange with nonpolar residues (entries nearly always negative).

The confidence that one can align sequences known to be distantly related has been investigated as a function of PAM values (N). The PAM-250 log odds substitution matrix tends to yield the best alignments, that is, the highest alignment scores relative to those derived using substitution matrices based on larger or smaller PAM values. Note that Fig. 7-25b indicates that at 250 PAMs, 80% of the residues in the original polypeptide have been replaced.

e. Sequence Alignment Using the Needleman–Wunsch Algorithm

The use of a log odds substitution matrix to find an alignment is straightforward (although tedious). When comparing two sequences, rather than just making a matrix with 1's at all matching positions, one enters the appropriate value in the log odds substitution matrix at every position. Such a matrix represents all possible pair combinations of the two sequences. In Fig. 7-29a, we use the PAM-250 log odds matrix with a 10-residue peptide horizontal and an 11-residue peptide vertical. Thus, the alignment of these two peptides must have at least one gap or overhang, assuming a significant alignment can be found at all.

An algorithm for finding the best alignment between two polypeptides (that with the highest log odds value) was formulated by Saul Needleman and Christian Wunsch. One starts at the lower right corner (C-termini) of the matrix, position (M, N) (where in Fig. 7-29a, $M = 11$ and $N = 10$), and adds its value (here 2) to the value at position $(M - 1, N - 1)$ [here 12, so that the value at position $(M - 1, N - 1)$, that is, $(10, 9)$, becomes 14 in the transformed matrix]. Continuing this process in an iterative manner, add to the value of the element at position (i, j) the maximum value of the elements $(p, j + 1)$, where $p = i + 1, i + 2, \ldots, M$, and those of $(i + 1, q)$, where $q = j + 1, j + 2, \ldots, N$. Figure 7-29b shows this process at an intermediate stage with the original value of the $(6, 5)$ position in a small box and the transformed values of the

TABLE 7-7 The PAM-250 Log Odds Substitution Matrix

	C	S	T	P	A	G	N	D	E	Q	H	R	K	M	I	L	V	F	Y	W
C Cys	12																			
S Ser	0	2																		
T Thr	-2	1	3																	
P Pro	-3	1	0	6																
A Ala	-2	1	1	1	2															
G Gly	-3	1	0	-1	1	5														
N Asn	-4	1	0	-1	0	0	2													
D Asp	-5	0	0	-1	0	1	2	4												
E Glu	-5	0	0	-1	0	0	1	3	4											
Q Gln	-5	-1	-1	0	0	-1	1	2	2	4										
H His	-3	-1	-1	0	-1	-2	2	1	1	3	6									
R Arg	-4	0	-1	0	-2	-3	0	-1	-1	1	2	6								
K Lys	-5	0	0	-1	-1	-2	1	0	0	1	0	3	5							
M Met	-5	-2	-1	-2	-1	-3	-2	-3	-2	-1	-2	0	0	6						
I Ile	-2	-1	0	-2	-1	-3	-2	-2	-2	-2	-2	-2	-2	2	5					
L Leu	-6	-3	-2	-3	-2	-4	-3	-4	-3	-2	-2	-3	-3	4	2	6				
V Val	-2	-1	0	-1	0	-1	-2	-2	-2	-2	-2	-2	-2	2	4	2	4			
F Phe	-4	-3	-3	-5	-4	-5	-4	-6	-5	-5	-2	-4	-5	0	1	2	-1	9		
Y Tyr	0	-3	-3	-5	-3	-5	-2	-4	-4	-4	0	-4	-4	-2	-1	-1	-2	7	10	
W Trp	-8	-2	-5	-6	-6	-7	-4	-7	-7	-5	-3	2	-3	-4	-5	-2	-6	0	0	17
	Cys	Ser	Thr	Pro	Ala	Gly	Asn	Asp	Glu	Gln	His	Arg	Lys	Met	Ile	Leu	Val	Phe	Tyr	Trp

Source: Dayhoff, M.O. (Ed.), *Atlas of Protein Sequence and Structure,* Vol. 5, Supplement 3, *p.* 352, National Biomedical Research Foundation (1978).

(a) Comparison matrix

	V	E	D	Q	K	L	S	K	C	N
V	4	−2	−2	−2	−2	2	−1	−2	−2	−2
E	−2	4	3	2	0	−3	0	0	−5	1
N	−2	1	2	1	1	−3	1	1	−4	2
K	−2	0	0	1	5	−3	0	5	−5	1
L	2	−3	−4	−2	−3	6	−3	−3	−6	−3
T	0	0	0	−1	0	−2	1	0	−2	0
R	−2	−1	−1	1	3	−3	0	3	−4	0
P	−1	−1	−1	0	−1	−3	1	0	−3	0
K	−2	0	0	1	5	−3	0	5	−5	0
C	−2	−5	−5	−5	−5	−6	0	−5	12	−4
D	−2	3	4	2	0	−4	0	0	−5	2

(c) Transformed matrix

	V	E	D	Q	K	L	S	K	C	N
V	(41)	33	31	29	24	22	18	12	0	−2
E	31	(37)	35	33	26	17	19	14	−3	1
N	29	32	(33)	32	27	17	20	15	−2	2
K	24	26	26	27	(31)	17	19	19	−3	1
L	25	20	18	21	17	(26)	16	11	−4	−3
T	23	23	23	22	19	18	(20)	14	0	0
R	18	19	19	21	23	17	(19)	17	−2	0
P	18	18	18	19	18	16	(20)	14	−1	0
K	12	14	14	15	19	11	14	(19)	−3	0
C	2	−1	−3	−3	−3	−4	2	−3	(14)	−4
D	−2	3	4	2	0	−4	0	0	−5	(2)

(b) Transforming the matrix by the
Needleman–Wunsch alignment scheme

	V	E	D	Q	K	L	S	K	C	N
V	4	−2	−2	−2	−2	2	−1	−2	−2	−2
E	−2	4	3	2	0	−3	0	0	−5	1
N	−2	1	2	1	1	−3	1	1	−4	2
K	−2	0	0	1	5	−3	0	5	−5	1
L	2	−3	−4	−2	3	6	−3	−3	−6	−3
T	0	0	0	−1	[0]	−2	1	0	−2	0
R	−2	−1	−1	1	−3	17	19	17	−2	0
P	−1	−1	−1	0	−1	16	20	14	−1	0
K	−2	0	0	1	5	11	14	19	−3	0
C	−2	−5	−5	−5	−5	−4	2	−3	14	−4
D	−2	3	4	2	0	−4	0	0	−5	2

(d) Alignment

VEDQKLS‑‑KCN
VEN‑KLTRPKCD

or

VEDQKL‑‑SKCN
VEN‑KLTRPKCD

FIGURE 7-29   Use of the Needleman–Wunsch alignment algorithm in the alignment of a 10-residue peptide (*horizontal*) with an 11-residue peptide (*vertical*). (*a*) The comparison matrix, whose elements are the corresponding entries in the PAM-250 log odds substitution matrix (Table 7-7). (*b*) The Needleman–Wunsch transformation after several steps starting from the lower right. The numbers in red have already been transformed. The Needleman–Wunsch score of the T–K alignment (*small box*) is the sum of its PAM-250 value (0) plus the maximum of the quantities in the L-shaped box (19). The text explains the mechanics of the transformation process. (*c*) The completed Needleman–Wunsch matrix. The best alignment follows the ridgeline of the matrix as is described in the text. The aligned residues are those whose corresponding elements are circled. Note the ambiguitity in this alignment. (*d*) The resulting two equivalent peptide alignments, with the aligned identical residues colored green.

positions (*p*, 6), where *p* = 7, 8, . . ., 11, together with positions (7, *q*), where *q* = 6, 7, . . ., 10, in the L-shaped box. The maximum value of the matrix elements in this L-shaped box is 19 and hence this is the value to add to the value (0) at position (6, 5) to yield the value 19 in the transformed matrix. This process is iterated, from the lower right toward the upper left of the matrix, until all its

elements have been so treated, yielding the fully transformed matrix shown in Fig. 7-29*c*. The **Needleman–Wunsch algorithm** thereby yields the log odds values for all possible alignments of the two sequences.

The best alignment (that with the highest log odds value) is found by tracing the ridgeline of the transformed matrix (Fig. 7-29*c*) from its maximum value at or near the upper left (N-terminus) to that at or near the lower right (C-terminus). This is because the alignment of a particular residue pair is independent of the alignment of any other residue pair, and hence the best score up to any point in an alignment is the best score through the previous step plus the incremental score of the new step. This additive scoring scheme is based on the assumption that mutations at different sites are independently accepted, which appears to be an adequate characterization of protein evolution even though specific interactions between residues are known to have critical structural and functional roles in proteins.

The line connecting the aligned residue pairs (those circled in Fig. 7-29c) must always extend down and to the right. This is because a move up or to the left, or even straight down or straight to the right, would imply that a residue in one peptide aligned with more than one residue in the other peptide. Any allowed deviation from a move of $(+1, +1)$ implies the presence of a gap. The best alignment of the two polypeptides, that connected by the lines in Fig. 7-29c, is indicated in Fig. 7-29d. Note that this alignment is not unambiguous; the alignment of S in the 10-mer with either T or P in the 11-mer yields the same log odds value, and hence we have insufficient information to choose between them. The overall alignment score is the maximum value of the transformed matrix, here 41, which occurs at the upper left of the alignment (Fig. 7-29c).

The Needleman–Wunsch algorithm optimizes the global alignment of two peptides, that is, it maximizes the alignment score over the whole of the two sequences (and does so even if it has no biological significance). However, since many proteins are modularly constructed from sequence motifs that occur in a variety of other proteins, a better approach would be to optimize the local alignment of two peptides, that is, maximize the alignment score only over their homologous regions. A variant of the Needleman–Wunsch algorithm, formulated by Temple Smith and Michael Waterman, is widely used to do so. This **Smith–Waterman algorithm** exploits the property of the substitution matrix-based scoring system that the cumulative score for an alignment path decreases in regions in which the sequences are poorly matched. Where the cumulative score drops to zero, the Smith–Waterman algorithm terminates the extension of an alignment path. Two peptides may have several such local alignments.

f. Gap Penalties

If there are gaps in the alignment, one should now subtract the gap penalty from the overall alignment score to obtain the final alignment score. Since a single mutational event can insert or delete more than one residue, a long gap should be penalized only slightly more than a short gap. Consequently, gap penalties have the form $a + bk$, where a is the penalty for opening the gap, k is the length of the gap in residues, and b is the penalty for extending the gap by one residue. Current statistical theory provides little guidance for optimizing a and b, but empirical studies suggest that $a = -8$ and $b = -2$ are appropriate val-

ues for use with the PAM-250 matrix. Thus the final alignment score for both alignments in Fig. 7-29d (which have both a 1-residue gap and a 2-residue gap) is $41 - (8 + 2 \times 1) - (8 + 2 \times 2) = 19$.

g. Pairwise Alignments Using BLAST and FASTA

The Needleman–Wunsch algorithm and later the Smith–Waterman algorithm (in their computerized forms) were widely used in the 1970s and 1980s to find relationships between proteins. However, the need to compare every newly determined sequence with the huge and rapidly growing number of sequences in publicly available databases requires that this process be greatly accelerated. Modern computers are up to this task but, equally important, the most widely used sequence alignment programs employ sophisticated **heuristic algorithms** (algorithms that make educated "guesses," thereby greatly increasing the speed of the program at the risk of obtaining suboptimal results; in the case of sequence alignments, the heuristic algorithms are based on knowledge of how sequences evolve). Consequently, in what follows, we shall describe how these programs are used rather than how they work.

The PAM-250 substitution matrix is based on an extrapolation: Its calculation assumes that the rate of mutation over one PAM unit of evolutionary distance is the same as that over 250 PAM units. This may not be the case. After all, homologous proteins that are separated by large evolutionary distances may diverge in their function and hence their rates of evolution may change (recall that different proteins have different rates of evolution; Fig. 7-22). To account for this possibility, and because of the huge amount of sequence data that had become available since the PAM matrices were calculated in the mid-1970s, a log odds substitution matrix based on ~2000 blocks of aligned sequences from ~500 groups of related proteins was calculated. The substitution matrix that gives the most sensitive performance for ungapped alignments is called **BLOSUM62** (for *blo*ck *su*bstitution *m*atrix; the 62 indicates that all blocks of aligned polypeptides in which there is ≥62% identity are weighted as a single sequence in order to reduce contributions from closely related sequences), whereas **BLOSUM50** appears to perform better for alignments with gaps. Sequence alignments based on the BLOSUM62 or BLOSUM50 matrices are more sensitive than are those based on the PAM-250 matrix.

There are two publicly available software packages that are most widely used for making pairwise sequence alignments—both for polypeptides and for polynucleotides. These heuristic programs, **BLAST** (for *b*asic *l*ocal *a*lignment *s*earch *t*ool) and **FASTA,** have different search philosophies. Consequently, although in a given database both will find the same peptides/DNAs that have high sequence similarity to a given query sequence, the homologous but low similarity sequences that these programs find may differ. Both programs attempt to obtain the optimum mix of sensitivity (the ability to identify distantly related sequences) and selectivity (the avoidance of unrelated sequences with spuriously high alignment scores).

Let us first discuss the BLAST program package as it compares peptides. This program, which was originated by Stephen Altschul, is publicly available for interactive use over the Web (http://www.ncbi.nlm.nih.gov/BLAST/) on a server at the National Center for Biotechnology Information (NCBI; the program can be manipulated, free of charge, over the Web using a desktop computer but it actually runs on NCBI computers). BLAST pairwise aligns up to a user-selected number of subject sequences in the chosen database(s) that are the most similar to an input query sequence.

Protein databases presently contain ~900,000 peptide sequences. BLAST therefore minimizes the time it spends on a sequence region whose similarity with the query sequence has little chance of exceeding some minimal alignment score. Pairwise alignments (e.g., Fig. 7-30*a*), which are found using BLOSUM62, are listed in order of decreasing statistical significance and are presented in a manner that indicates the positions of both the identical residues and similar residues in the query and subject sequences. The number of identical residues, positives (those residue pairs whose exchange has a positive value in the

(a) **BLAST pairwise alignment**

>sp|P38524|HPI2_ECTVA HIGH POTENTIAL IRON-SULFUR PROTEIN, ISOZYME 2 (HIPIP 2)
       Length = 71

Score = 50.4 bits (118), Expect = 6e-07
Identities = 27/69 (39%), Positives = 35/69 (50%), Gaps = 4/69 (5%)

```
Query: 1   EPRAEDGHAHDYVNEAADASG--HPRYQEGQLCENCAFWGEAVQDGWGRCTHPDFDEVLVKAEGWCSVY  67
           E  +ED  A    +   DAS   HP Y+EGQ C NC  +  +A   WG C+    F   LV A GWC+ +
Sbjct: 2   ERLSEDDPAAQALEYRHDASSVQHPAYEEGQTCLNCLLYTDASAQDWGPCS--VFPGKLVSANGWCTAW  68
```

(b) **FASTA pairwise alignment**

>>SWALL:HPI2_ECTVA P38524 HIGH POTENTIAL IRON-SULFUR PRO (71 aa)
initn: 102 init1: 77 opt: 116 Z-score: 278.0 expect() 4e-08
Smith-Waterman score: 116; 39.130% identity in 69 aa overlap (1-67:2-68)

```
              10        20        30        40        50        60        70
Sequen  EPRAEDGHAHDYVNEAADASG--HPRYQEGQLCENCAFWGEAVQDGWGRCTHPDFDEVLVKAEGWCSVYAPAS
        :  .:: .    .      :::.  :: :.::: ::  . .::    :: ..    ::  :. ..:: .
SWALL:  MERLSEDDPAAQALEYRHDASSVQHPAYEEGQTCLNCLLYTDASAQDWGPCSV--FPGKLVSANGWCTAWVAR
           10        20        30        40        50        60        70
```

(c) **CLUSTAL X multiple-sequence alignment**

```
                                   *  *  :..... :.      *. * ** ::      * *  . *   *  :***.: .
1  sp|P38524|HPI2  |----MERLSEDDPAAQALEYRHDASSVQ-HPAYE---EGQTCLNCLLYTDASAQDWGPC--SVFPGKLVSANGWCTAWVAR-
2  sp|P38941|HPI1  |----AERLDENSPEALALNYKHDGASVD-HPSHA---AGQKCINCLLYTDPSATEWGGC--AVFPNKLVNANGWCTAYVARG
3  sp|P00265|HPIS  |------APVDEKNPQAVALGYVSDAAKAD-KAKYKQFVAGSHCGNCALFQGKATDAVGGC--PLFAGKQVANKGWCSAWAKKA
4  sp|P04168|HPI1  |---------EPRAEDGHAHDYVNEAADASGHPRYQ---EGQLCENCAFWGEAVQDGWGRCTHPDFDEVLVKAEGWCSVYAPAS
5  sp|P04169|HPI2  |GLPDGVEDLPKAEDDHAHDYVNDAADTD-HARFQ---EGQLCENCQFWVDYVN-GWGYCQHPDFTDVLVRGEGWCSVYAPA-
```

FIGURE 7-30   Examples of peptide sequence alignments.
(*a*) BLAST, (*b*) FASTA, and (*c*) CLUSTAL. The proteins being aligned are **high potential iron–sulfur proteins (HIPIPs),** small bacterial proteins whose sequences are archived in the SWISS-PROT database. The amino acid residues are indicated by their one-letter codes (Table 4-1) and gaps are indicated by dashes. In Parts *a* and *b*, the query sequence (Sequen in *b*) is HIPIP Isozyme 1 from *Ectothiorhodospira halophila* (HPI1_ECTHA, SWISS-PROT accession number P04168; **isozymes** are catalytically and structurally similar but genetically distinct enzymes from the same organism) and the subject sequence (SWALL in *b*) is HIPIP Isozyme 2 from *Ectothiorhodospira vacuolata* (HPI2_ECTVA, SWISS-PROT accession number P38524). In *a* and *b*, the first line(s) (*green*) identifies the subject sequence and indicates its length in residues. This is followed by an assortment of alignment statistics (*black*). The query and subject sequences are then shown vertically aligned (*blue*) with the line between them (*black*) indicating residues that are identical [by their one-letter codes in BLAST and a colon (:) in FASTA] and similar [by a plus (+) in BLAST and a period (.) in FASTA]. BLAST and FASTA outputs consist of a series of such pairwise alignments. Part *c* shows the alignment of five HIPIP sequences, the two foregoing sequences, their corresponding Isozymes 1 and 2, and HIPIP from *Rhodocyclus gelatinosus* (P00265). The residues in this multiple sequence alignment are colored according to residue type and the confidence level of their alignment. The upper line (*black*) has an asterisk (*) for columns in which all residues are identical, a colon for columns in which all residues are strongly similar (e.g., MILV), and a period for columns in which they are all weakly similar (e.g., CSA). Note that in Part *c* the alignment of the first 21 residues of P04168 (the query sequence in *a* and *b*) with P38524 differs from that in *a* and *b*.

substitution matrix used), and gaps over the length of the alignment are indicated. BLAST assesses the statistical significance of an alignment in terms of its "E value" (E for expectation), which is the number of alignments with at least the same score that would have been expected to occur in the database by chance. For example, an alignment with an E value of 5 is statistically insignificant, whereas one with an E value of 0.01 is significant, and one with an E value of 1×10^{-20} offers extremely high confidence that the query and subject sequences are homologous. BLAST also reports a "bit score" for each alignment, which is a type of normalized alignment score.

FASTA, which was originated by William Pearson, is publicly available for interactive use over the Web (http://www.ebi.ac.uk/fasta33/) on a server at the European Bioinformatics Institute (EBI). FASTA's input is similar to that of BLAST but permits the user to choose the substitution matrix (including a selection of PAM and BLOSUM matrices; the default matrix is BLOSUM50) and the gap penalty parameters that the program uses (similar options are available with the "advanced" BLAST search). For peptide alignments, the user has the choice of 1 or 2 for the value of the parameter *ktup* (for *k*-tuple), the number of consecutive residues in the "words" that FASTA uses to search for identities (*ktup* may be as high as 6 for nucleic acid alignments). The smaller the value of *ktup*, the more sensitive the alignment search but the more computer time it requires. *ktup*'s default value of 2 is sufficiently sensitive for most peptide sequence alignment searches. The FASTA output (e.g., Fig. 7-30b) contains information similar to that in a BLAST output.

h. Multiple Sequence Alignments with CLUSTAL

BLAST and FASTA make only pairwise alignments. To simultaneously align more than two sequences, that is, to obtain a **multiple sequence alignment** such as Table 7-4, a different program must used. Perhaps the most widely used such program, CLUSTAL, is publicly available for interactive use over the Web at http://www2.ebi.ac.uk/clustalw/. The input for the program is a file containing all the sequences (either peptides or DNA) to be aligned. As with FASTA and advanced BLAST, the user can select the substitution matrix and the gap penalty parameters that CLUSTAL uses. CLUSTAL begins by finding all possible pairwise alignments of the input sequences. This permits the program to determine the relationships of the input sequences with one another based on their similarity scores and hence generate a crude phylogenetic tree called a dendrogram. Then, starting with the highest scoring pairwise alignment, it sequentially carries out realignments on the basis of the remaining sequences, which it adds in order of their decreasing relationships with the previously added sequences, while opening up gaps as necessary. The output of CLUSTAL is the aligned sequences (e.g., Fig. 7-30c). Since multiple sequence alignment programs are easily confounded by such anomalies as sequences that are not homologous or that contain homologous segments in different orders, multiple sequence alignments should be carefully inspected to determine if they are sensible and, if necessary, corrected and trimmed by hand. Indeed, in Figs. 7-30a and b, the alignment of the first 21 residues of the query sequence (P04168) with the subject sequence (P38524) differs from that in Fig. 7-30c.

i. The Use of Profiles Extends the Sensitivities of Sequence Alignments

Multiple sequence alignments can be used to improve the sensitivity of similarity searches, that is, to detect weak but significant sequence similarities. For example, in pairwise alignments, peptide A may appear to be similar to peptide B and peptide B to be similar to peptide C but peptides A and C may not appear to be similar. However, a multiple sequence alignment of peptides A, B, and C will reveal the similarities between peptides A and C. This idea has been extended through the construction of **profiles** (also called **position-specific score matrices**) in which, at each residue position in a multiple sequence alignment, highly conserved residues are assigned a large positive score, weakly conserved positions are assigned a score near zero, and unconserved residues are assigned a large negative score. Many profile-generating algorithms are based on statistical models called **hidden Markov models (HMMs).** The use of such conservation patterns has been successful in finding sequences that are so distantly related to a query sequence (so far into the twilight zone) that BLAST and FASTA would not consider them to have significant sequence similarity.

The program **PSI-BLAST** (for Position-Specific Iterated BLAST), which is publicly available for use at http://www.ncbi.nlm.nih.gov/BLAST/, uses the results of a BLAST search with a query sequence to generate a profile and then employs the profile to search for new alignments. This is an iterative process in which the profile generated after each alignment search is used to make a new alignment search, etc., until no further significant alignments are found. For instance, for the query sequence used in Figs. 7-30a and b (HPI1_ECTHA; SWISS-PROT accession number P04168), BLAST and FASTA both find only five sequences (called hits) that have E values less than 0.001 in the SWISS-PROT database (those in Fig. 7-30c, which includes a self-alignment). In contrast, PSI-BLAST finds 22 such hits after two iterations (and no additional hits in the third iteration, whereupon the search is said to have converged). Thus the use of profile analysis makes possible the detection of subtle but significant sequence relationships that, as we shall see in succeeding chapters, lead to considerable evolutionary and functional insights.

j. Structural Genes Should Be Aligned as Polypeptides

In many cases, only the base sequence of the DNA encoding a protein is known. Indeed, most of the known protein sequences have been inferred from DNA sequences. Although BLAST, FASTA, and CLUSTAL are all capable of aligning nucleic acid sequences (and are routinely used to do so), one should compare the inferred amino acid sequences of structural genes rather than just their base sequences. This is because amino acid sequence comparisons can routinely identify sequences that shared a common ancestor over 1 billion years ago (e.g., those of cytochrome *c* and histone H4; Fig. 7-22), whereas it is rare to detect homologies in noncoding DNA sequences that diverged >200 million years (Myr) ago and in coding sequences that diverged >600 Myr ago. The reasons for this are threefold:

1. DNA has only four different bases, whereas peptides consist of 20 different amino acid residues. Consequently, it is much easier to find spurious alignments with DNA, at least for short segments, than with peptides (a dot plot of two unrelated DNAs would have, on average, 25% of its spaces filled vs 5% for unrelated polypeptides).

2. DNA evolves much more quickly than proteins. In the coding regions of structural genes, 24% of single base changes encode the same amino acid. Hence there are few evolutionary constraints to maintain the sequence identity of these bases or of the gene's noncoding regions (e.g., those containing introns). Consequently, the evolutionary constraints on proteins are more stringent than those on DNA.

3. Direct DNA sequence alignments do not use amino acid substitution matrices such as PAM-250 and BLOSUM62 and, hence, are not constrained by the evolutionary information implicit in these matrices (although there are analogous 4 × 4 matrices for base substitutions).

If the base sequence of a structural gene is known, its putative control regions, particularly its start and stop codons, can usually be identified. This, in turn, reveals which of two complementary DNA strands is the so-called **sense strand** (which has the same sequence as the mRNA transcribed from the DNA) and indicates its correct reading frame. Even if it is unclear that a DNA segment that is flanked by what appear to be start and stop codons actually encodes a protein, one can compare the amino acid sequences from all six possible reading frames (three each from the DNA's two complementary strands). In fact, both BLAST and FASTA do so automatically when aligning peptide sequences based on DNA sequences.

C. *Construction of Phylogenetic Trees*

Phylogenetic trees were first made by Linnaeus, the eighteenth century biologist who originated the system of taxonomy (biological classification) still in use today. These trees (e.g., Fig. 1-4) were originally based on morphological characteristics, whose measurements were largely subjective. It was not until the advent of sequence analysis, however, that the generation of phylogenetic trees was put on a firm quantitative basis (e.g., Fig. 7-21). In the following paragraphs we discuss the characteristics of phylogenetic trees and outline how they are generated.

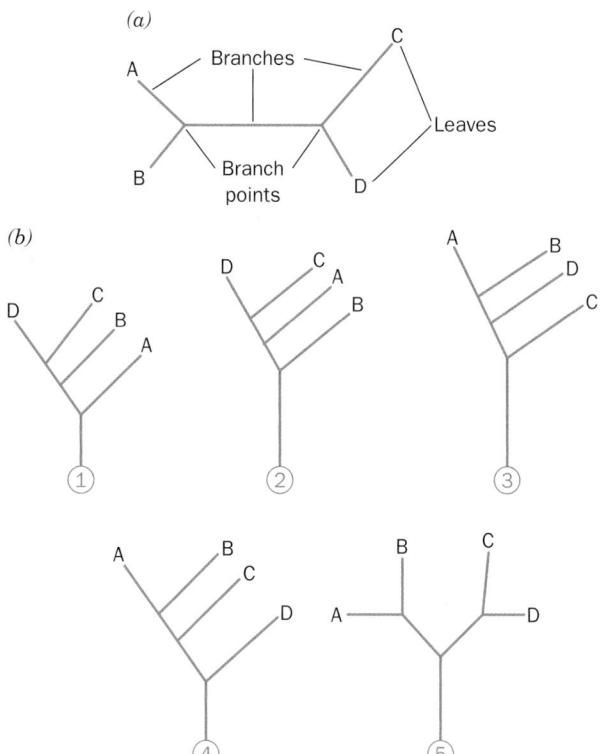

FIGURE 7-31 Phylogenetic trees. (*a*) An unrooted tree with four leaves (A, B, C, and, D) and two branch points. (*b*) The five rooted trees that can be generated from the unrooted tree in Part *a*. The roots are drawn and numbered in red.

Figure 7-31*a* is a phylogenetic tree diagramming the evolutionary relationships between four homologous genes, A, B, C, and D. The tree consists of four **leaves** or **external nodes,** each representing one of these genes, and two **branch points** or **internal nodes,** which represent ancestral genes, with the length of each **branch** indicating the degree of difference between the two nodes they connect. All branch points are binary, that is, one gene is considered to give rise to only two descendents at a time so that branches can only bifurcate (although branch points may turn out to be so close together that their order cannot be determined; e.g., the root of Fig. 7-21). Note that this is an **unrooted tree,** that is, it indicates the relationships between the four genes but provides no information as to the evolutionary events through which they arose. The five different evolutionary pathways that are possible for our unrooted tree are diagrammed in Fig. 7-31*b* as different **rooted trees,** where the node at which the root joins the tree represents the four genes' last common ancestor. With knowledge of only the A, B, C, and D genes, phylogenetic analysis cannot distinguish between these rooted trees. In order to find a tree's root, it is necessary to obtain the sequence of an **outgroup,** a homologous gene that is less closely related to the genes in the tree than they are to each other. This permits the root of the tree to be identified and hence the pathway through which the genes evolved to be elucidated.

The number of different bifurcating trees with the same n leaves increases extremely rapidly with n (e.g., for $n = 10$, it is over 2 million). Unfortunately, *there is no exact method for generating an optimal phylogenetic tree.* Indeed, there is no general agreement on what constitutes an optimal tree. Consequently, numerous methods have been formulated for constructing phylogenetic trees based on sequence alignments.

In one class of methods for constructing phylogenetic trees, the sequence data are converted into a **distance matrix,** which is a table showing the evolutionary distances between all pairs of genes in the data set (e.g., Table 7-5). Evolutionary distances are the number of sequence differences between two genes (ideally corrected for the possibility of multiple mutations at a given site). These quantities are used to calculate the lengths of the branches in a tree under the assumption that they are additive, that is, the distance between any pair of leaves is the sum of the lengths of the branches connecting them.

Perhaps the conceptually simplest (if it can be called that) way of generating a phylogenetic tree is the **neighbor-joining (N-J) method,** in which it is initially assumed that there is only one internal node, Y, and hence that all N leaves radiate from it in a starlike pattern (Fig. 7-32a). The branch lengths in the star are then calculated according to such relationships as $d_{AB} = d_{AY} + d_{BY}$ (where d_{AB} is the total length of the branches connecting leaves A and B, etc.), $d_{AC} = d_{AY} + d_{CY}$, and $d_{BC} = d_{BY} + d_{CY}$, so that, for instance, $d_{AY} = \frac{1}{2}(d_{AB} + d_{AC} - d_{BC})$. A pair of leaves is then transferred from the star to a new internal node, X, that is connected to the center of the star by a new branch, XY (Fig. 7-32b), and the sum of all the branch lengths, S_{AB}, in this revised tree is calculated:

$$S_{AB} = d_{AX} + d_{BX} + d_{XY} + \sum_{k \neq A,B}^{N} d_{kY}$$
$$= \frac{d_{AB}}{2} + \frac{[2Q - R_A - R_B]}{2(N-2)} \qquad [7.6]$$

where

$$Q = \sum_{i=1}^{N} \sum_{j=1}^{i-1} d_{ij} \qquad [7.7]$$

(that is, the sum of all the off-diagonal elements in the unique half of the distance matrix),

$$R_A = \sum_{i=1}^{N} d_{Ai} \qquad [7.8]$$

(that is, the sum of the elements in the Ath row of the distance matrix), and

$$R_B = \sum_{i=1}^{N} d_{Bi} \qquad [7.9]$$

The two leaves are then returned to their initial positions, replaced by a second pair of leaves, and the total branch length again calculated. The process is repeated until all possible $N(N-1)/2$ pairs of leaves have been so treated. The pair yielding the smallest value of S_{ij} (the shortest total branch length) in this process, which will be nearest neighbors in the final tree, are combined into a single unit of their average length, yielding a star with one less branch. If leaves A and B are chosen as neighbors, then the lengths of the branches connecting them are estimated to be

$$d_{AX} = \frac{d_{AB}}{2} + \frac{R_A - R_B}{2(N-2)} \qquad [7.10]$$

$$d_{BX} = d_{AB} - d_{AX} \qquad [7.11]$$

and

$$d_{XY} = \frac{(N-1)(R_i + R_j) - 2Q - (N^2 - 3N + 2)d_{AB}}{2(N-2)(N-3)} \qquad [7.12]$$

Assuming that S_{AB} has the lowest value of all the S_{ij}, a new distance matrix is calculated whose elements, d'_{ij}, are the same as the d_{ij} with the exception that $d'_{A-B,i} = d'_{i,A-B} = (d_{Ai} + d_{Bi})/2$, where $d'_{A-B,i}$ is distance between the averaged leaves A and B and leaf i. The entire process is then iterated so as to find all pairs of nearest-neighbor sequences, thereby generating a phylogenetic tree. Figure 7-33 is an unrooted phylogenetic tree that CLUSTAL generated from the multiple sequence alignment shown in Fig. 7-30c using the N-J method.

The N-J method is representative of **distance-based** tree-building procedures. There are two other types of tree-building criteria that are in widespread use:

1. Maximum parsimony (MP), which is based on the principle of "Occam's razor": The best explanation of the

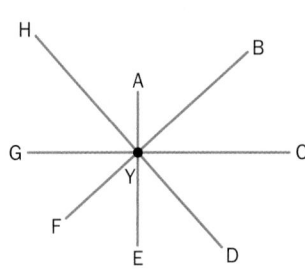

(a)

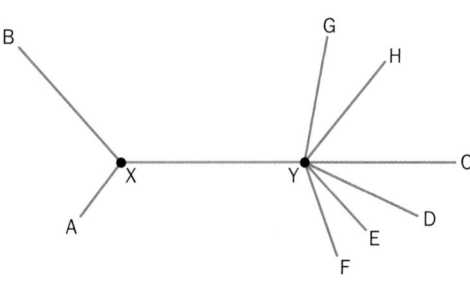

(b)

FIGURE 7-32 Manipulations employed in the neighbor-joining method for the construction of a phylogenetic tree. (*a*) The starting configuration. (*b*) The transfer of leaves A and B to a new branch point that is connected to the central star (*red*).

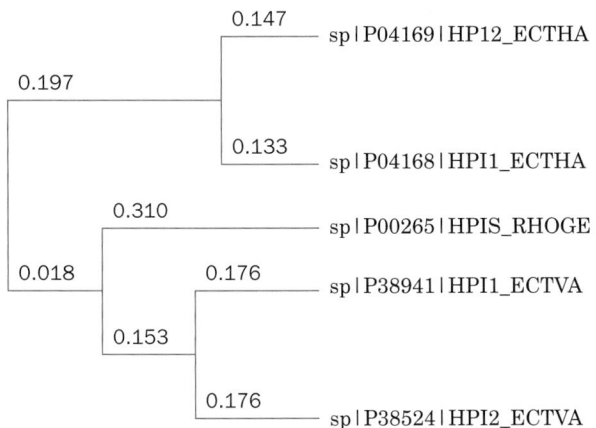

FIGURE 7-33 An unrooted phylogenetic tree of the five HIPIP sequences that are aligned in Fig. 7-30c. The tree was generated by CLUSTAL using the neighbor-joining method. The numbers indicate the relative lengths of the associated branches.

data is the simplest. Thus, MP-based methods assume (perhaps inaccurately) that evolution occurs via the fewest possible genetic changes and hence that the best phylogenetic tree requires the smallest number of sequence changes to account for a multiple sequence alignment.

2. Maximum likelihood (ML), which finds the tree and branch lengths that have the highest probability of yielding the observed multiple sequence alignment. This, in turn, requires an evolutionary model that indicates the probability of occurrence for each type of residue change (e.g., the PAM substitution matrices).

Since the number of possible trees increases very rapidly with the number of leaves, the construction of a phylogenetic tree is a computer-intensive task for even relatively small sets of aligned sequences (e.g., $N = 20$, although distance-based methods require far fewer computations than do MP- or ML-based methods). And, because of the ambiguities inherent in all known tree-building procedures, statistical tests have been developed to check the validity of any particular tree.

5 ■ CHEMICAL SYNTHESIS OF POLYPEPTIDES

In this section we describe methods for the chemical synthesis of polypeptides from amino acids. The ability to manufacture polypeptides not available in nature has considerable biomedical potential:

1. To investigate the properties of polypeptides by systematically varying their side chains.

2. To obtain polypeptides with unique properties, particularly those with nonstandard side chains or with isotopic labels incorporated in specific residues (neither of which is easily accomplished using biological methods).

3. To manufacture pharmacologically active polypeptides that are biologically scarce or nonexistent.

One of the most promising applications of polypeptide synthesis is the production of synthetic vaccines. Vaccines, which have consisted of viruses that have been "killed" (inactivated) or attenuated ("live" but mutated so as not to cause disease in humans), stimulate the immune system to synthesize antibodies specifically directed against these viruses, thereby conferring immunity to them (the immune response is discussed in Section 35-2A). The use of such vaccines, however, is not without risk; attenuated viruses, for example, may mutate to a virulent form and "killed" virus vaccines have, on several occasions, caused disease because they contained "live" viruses. Moreover, it is difficult to culture many viruses and therefore to obtain sufficient material for vaccine production. Such problems would be eliminated by preparing vaccines from synthetic polypeptides that have the amino acid sequences of viral antigenic determinants (molecular groupings that stimulate the immune system to manufacture antibodies against them). Indeed, several such synthetic vaccines are already in general use.

The first polypeptides to be chemically synthesized were composed of only one type of amino acid and are therefore known as **homopolypeptides.** Such compounds as **polyglycine, polyserine,** and **polylysine** are easily synthesized according to classic methods of polymer chemistry. They have served as valuable model compounds in studying the physicochemical properties of polypeptides, such as conformational behavior and interactions with the aqueous environment.

The first chemical synthesis of a biologically active polypeptide was that of the nonapeptide (9-amino acid residue) hormone **oxytocin** by Vincent du Vigneaud in 1953:

$$Gly-Leu-Pro-\underset{\underset{\displaystyle |}{\overset{\displaystyle \lceil}{Cys}}}{}-Asn-Gln-Ile-Tyr-\underset{\underset{\displaystyle \rfloor}{\overset{\displaystyle S-S}{Cys}}}{}$$

Oxytocin

Improvements in polypeptide synthesis methodology since then have led to the synthesis of numerous biologically active polypeptides and several proteins.

A. *Synthetic Procedures*

Polypeptides are chemically synthesized by covalently linking (coupling) amino acids, one at a time, to the terminus of a growing polypeptide chain. Imagine that a polypeptide is being synthesized from its C-terminus toward its N-terminus; that is, the growing chain ends with a free amino group. Then each amino acid being added to the chain must already have its own α-amino group chemically protected (blocked) or it would react with other like molecules as well as with the N-terminal amino group of the chain. Once the new amino acid is coupled, its now N-terminal amino group must be deprotected (deblocked) so that the next peptide bond can be formed. *Every cycle of amino acid ad-*

dition therefore requires a coupling step and a deblocking step. Furthermore, reactive side chains must be blocked to prevent their participation in the coupling reactions, and then deblocked in the final step of the synthesis.

The reactions that were originally developed for synthesizing polypeptides such as oxytocin take place entirely in solution. The losses that are incurred on isolation and purification of the reaction product in each of the many steps, however, contribute significantly to the low yields of final polypeptide. This difficulty was ingeniously circumvented in 1962 by Bruce Merrifield, through his development of **solid phase peptide synthesis (SPPS).** In SPPS, a growing polypeptide chain is covalently anchored, usually by its C-terminus, to an insoluble solid support such as beads of polystyrene resin, and the appropriately blocked amino acids and reagents are added in the proper sequence (Fig. 7-34). This permits the quantitative recovery and purification of intermediate products by simply filtering and washing the beads.

When polypeptide chains are synthesized by amino acid addition to their N-terminus (the opposite direction to that in protein biosynthesis; Section 5-4B), the α-amino group of each sequentially added amino acid must be chemically protected during the coupling reaction. The ***tert*-butyloxy-carbonyl (Boc)** group is frequently used for this purpose,

$$(CH_3)_3C-O-\overset{\overset{\displaystyle O}{\|}}{C}-Cl + H_2N-\overset{\overset{\displaystyle R}{|}}{C}H-C\overset{\nearrow O}{\underset{\searrow O^-}{}}$$

***t*-Butyloxycarbonyl chloride**   **α-Amino acid**

$\longrightarrow$ HCl

$$(CH_3)_3C-O-\overset{\overset{\displaystyle O}{\|}}{C}-NH-\overset{\overset{\displaystyle R}{|}}{C}H-C\overset{\nearrow O}{\underset{\searrow O^-}{}}$$

Boc-amino acid

as is the **9-fluorenylmethoxycarbonyl (Fmoc)** group:

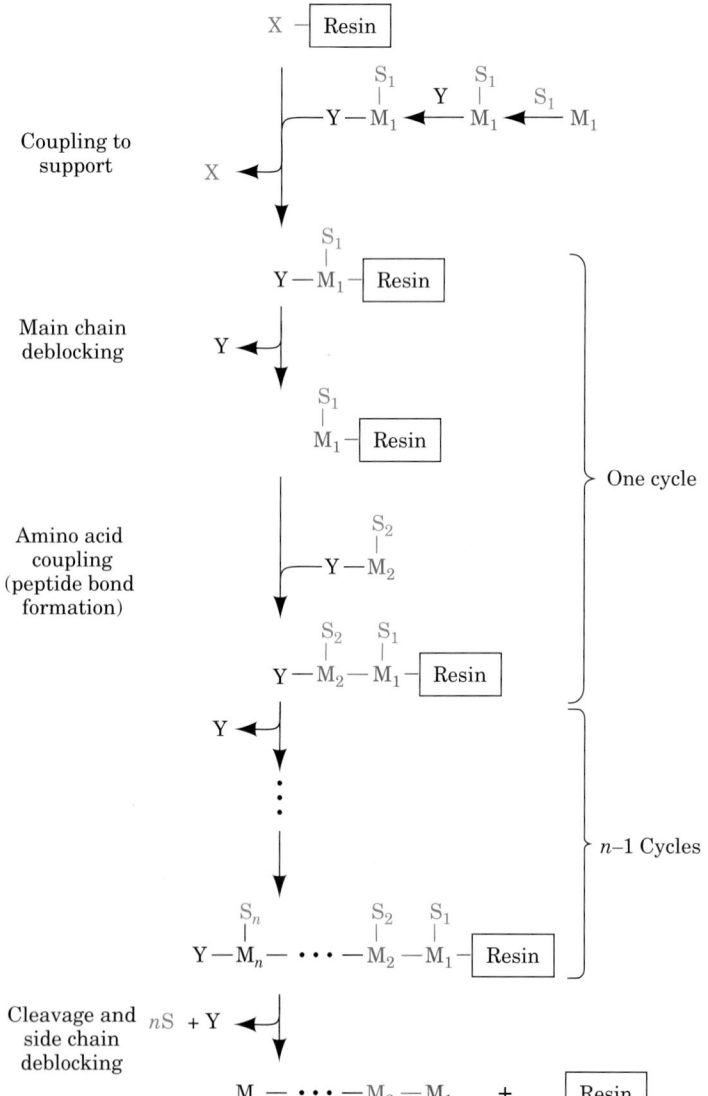

Coupling to support

Main chain deblocking

Amino acid coupling (peptide bond formation)

One cycle

n–1 Cycles

Cleavage and side chain deblocking

FIGURE 7-34 Flow diagram for polypeptide synthesis by the solid phase method. The symbol M_i represents the *i*th amino acid residue to be added to the polypeptide, S_i is its side chain protecting group, and Y represents the main chain protecting group. The specific reactions are discussed in the text. [After Erikson, B.W. and Merrifield, R.B., *in* Neurath, H. and Hill, R.L. (Eds.), *The Proteins* (3rd ed.), Vol. 2, p. 259, Academic Press (1979).]

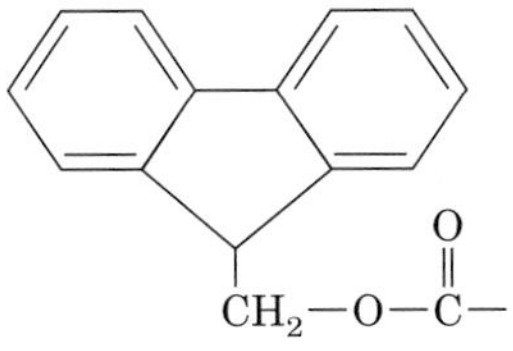

**9-Fluorenylmethoxycarbonyl
(Fmoc) group**

Both groups undergo analogous reactions, although, in what follows, we shall only discuss the Boc group.

a. Anchoring the Chain to the Inert Support

The first step in SPPS is the coupling of the C-terminal amino acid to a solid support. The most commonly used support is a cross-linked polystyrene resin with pendant chloromethyl groups. Resin coupling occurs through the following reaction:

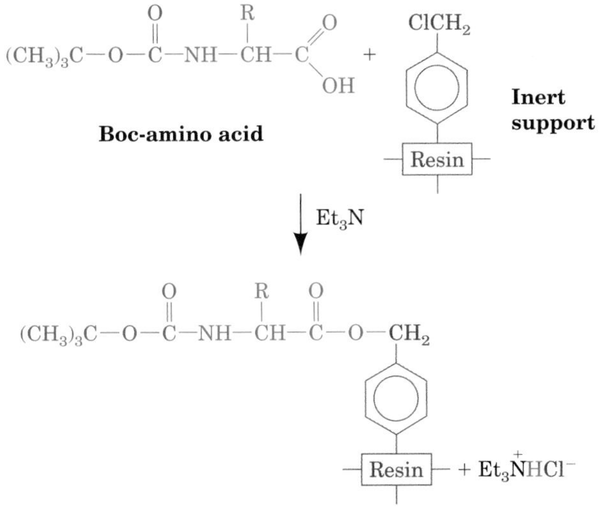

and the resulting α-amino acid-derivatized resin is filtered and washed. The amino group is then deblocked by treatment with an anhydrous acid such as trifluoroacetic acid, which leaves intact the alkylbenzyl ester bond to the support resin:

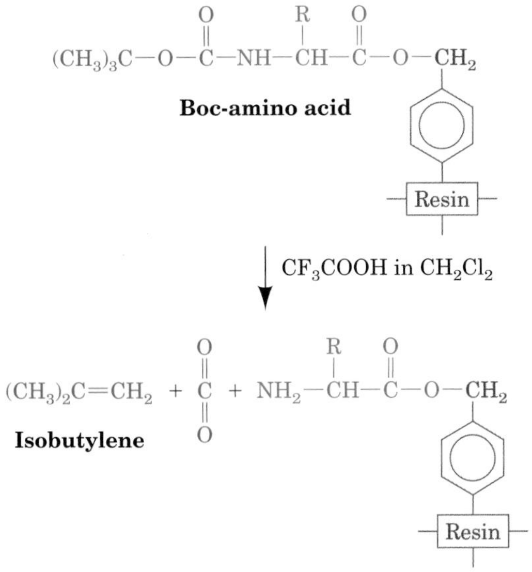

b. Coupling the Amino Acids

The reaction coupling two amino acids through a peptide bond is endergonic and therefore must be activated to obtain significant yields. **Carbodiimides** ($R-N=C=N-R'$) such as **dicyclohexylcarbodiimide (DCCD)** are commonly used coupling agents:

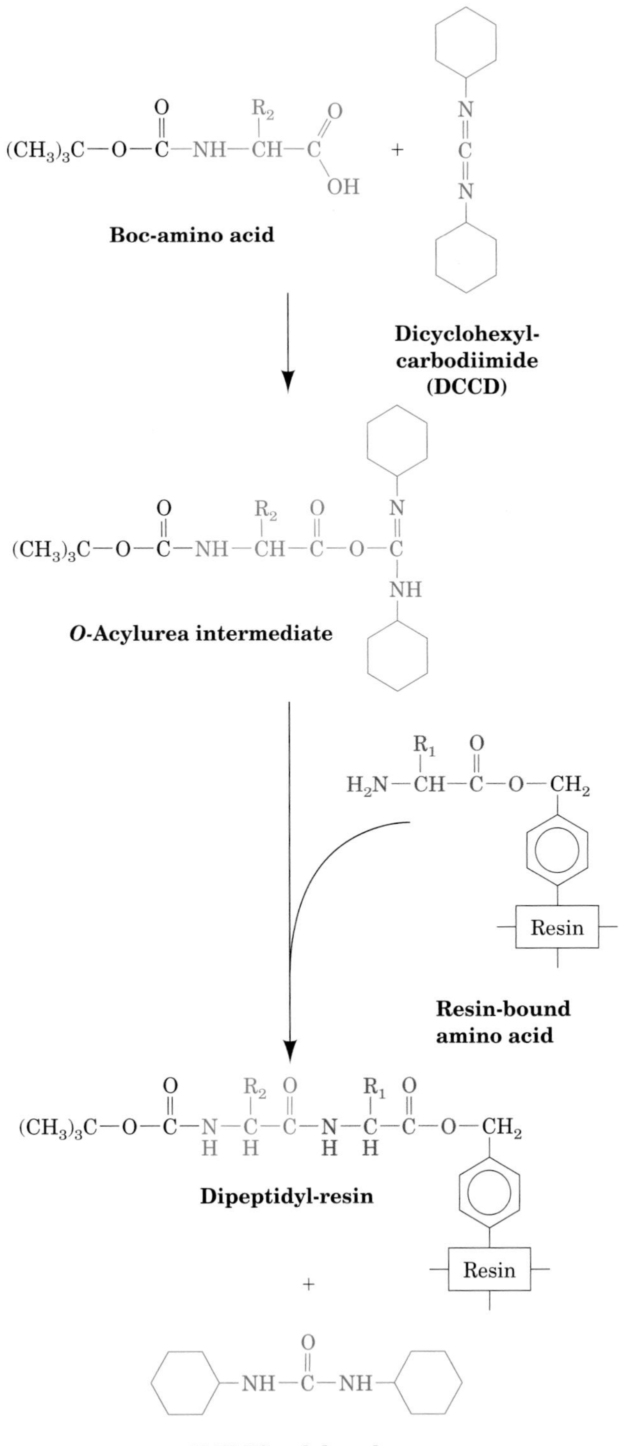

The O-acylurea intermediate that results from the reaction of DCCD with the carboxyl group of a Boc-protected α-amino acid readily reacts with the resin-bound α-amino acid to form the desired peptide bond in high yield. By subsequently alternating the deblocking and coupling reactions, a polypeptide with the desired amino acid sequence can be synthesized. *The repetitive nature of these operations allows the SPPS method to be easily automated.*

During the course of peptide synthesis, many of the side chains also require protection to prevent their reaction with the coupling agent. Although there are many different blocking groups, the benzyl group is the most widely used (Fig. 7-35).

c. Releasing the Polypeptide from the Resin

The final step in SPPS is the cleavage of the polypeptide from the solid support. The benzyl ester link from the polypeptide's C-terminus to the support resin may be cleaved by treatment with liquid HF:

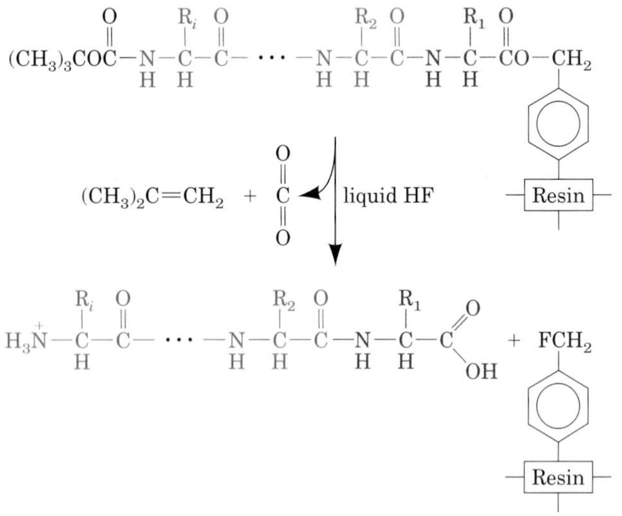

The Boc group linked to the polypeptide's N-terminus, as well as the benzyl groups protecting its side chains, are also removed by this treatment.

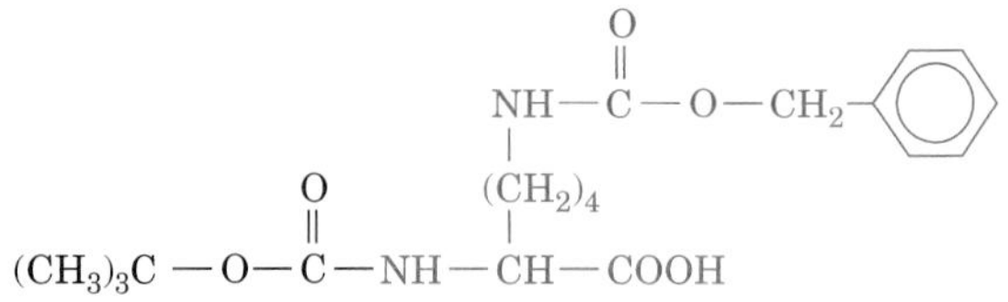

B. *Problems and Prospects*

The steps just outlined seem simple enough, but they are not as straightforward as we have implied. A major difficulty with the entire procedure is its low cumulative yield. Let us examine the reasons for this. To synthesize a polypeptide chain with n peptide bonds requires at least $2n$ reaction steps—one for coupling and one for deblocking each residue. If a protein-sized polypeptide is to be synthesized in reasonable yield, then each reaction step must be essentially quantitative; anything less greatly reduces the yield of final product. For example, in the synthesis of a 101-residue polypeptide chain, in which each reaction step occurs with an admirable 98% yield through 200 reaction steps, the overall yield is only $0.98^{200} \times 100 = 2\%$. Therefore, although oligopeptides can be routinely made, the synthesis of large polypeptides requires almost fanatical attention to chemical detail.

Boc, N^ε-benzyloxycarbonyl-Lys

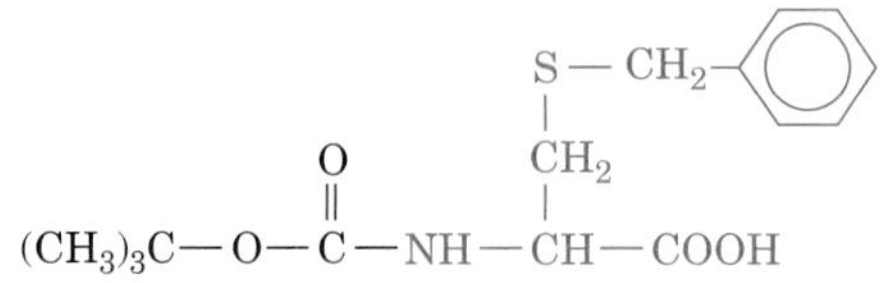

Boc, S-benzyl-Cys

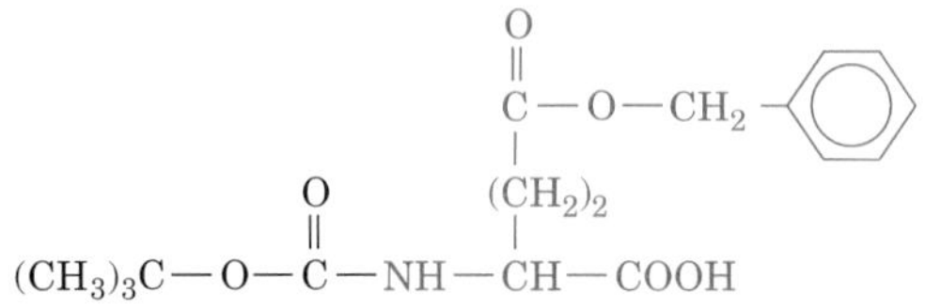

Boc-Glu, γ-Benzyl ester

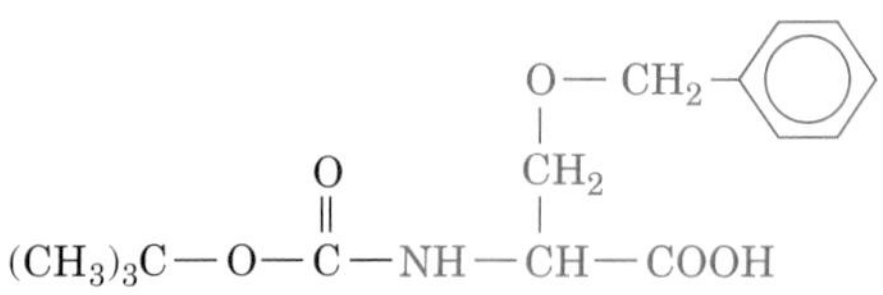

Boc, O-benzyl-Ser

FIGURE 7-35 **A selection of amino acids with benzyl-protected side chains and a Boc-protected α-amino group.** These substances can be used directly in the coupling reactions forming peptide bonds.

An ancillary problem is that the newly liberated synthetic polypeptide must be purified. This may be a difficult task because a significant level of incomplete reactions and/or side reactions at every stage of SPPS will result in almost a continuum of closely related products for large polypeptides. The use of reverse-phase HPLC techniques (Section 6-3D), however, greatly facilitates this purification process, and the quality of both intermediate and final products can be readily assessed through mass spectrometric techniques (Section 7-1J).

Using automated SPPS, Merrifield synthesized the nonapeptide hormone **bradykinin** in 85% yield:

Arg—Pro—Pro—Gly—Phe—Ser—Pro—Phe—Arg
Bradykinin

However, it was only in 1988, through steady progress in improving reaction yields (to >99.5% on average) and eliminating side reactions, that it became possible to synthesize ~100-residue polypeptides of reasonable quality. Thus, Stephen Kent synthesized the 99-residue **HIV protease** [an enzyme that is essential for the maturation of **human immunodeficiency virus (HIV,** the **AIDS virus;** Section 15-4C)] in such high yield and purity that, after being renatured (folded to its native conformation; Section 9-1A), it exhibited full biological activity. Indeed, this synthetic protein was crystallized and its X-ray structure was shown to be identical to that of biologically synthesized HIV protease. Kent also synthesized HIV protease from D-amino acids and experimentally verified, for the first time, that such a protein has the opposite chirality of its biologically produced counterpart. Moreover, this D-amino acid protease catalyzes the cleavage of its target polypeptide made from D-amino acids but not that made from L-amino acids as does naturally occurring HIV protease.

Despite the foregoing, the accumulation of resin-bound side products limits the sizes of the polypeptides that can routinely be synthesized by SPPS to ~60 residues. Kent has partially circumvented this limitation through the development of the so-called **native chemical ligation** reaction, which links together two polypeptides in a peptide bond to routinely yield polypeptides as large as ~120 residues (Fig. 7-36). Moreover, several peptide segments can be consecutively linked by native chemical ligation, so that the chemical synthesis of polypeptides consisting of several hundred residues can be anticipated. Of course,

each peptide bond formed in this way is limited to having a Cys residue on its C-terminal side.

6 ■ CHEMICAL SYNTHESIS OF OLIGONUCLEOTIDES

Molecular cloning techniques (Section 5-5) have permitted the genetic manipulation of organisms in order to investigate their cellular machinery, change their characteristics, and produce scarce or specifically altered proteins in large quantities. *The ability to chemically synthesize DNA oligonucleotides of specified base sequences is an indispensable part of this powerful technology.* Thus, as we have seen, specific oligonucleotides are required as probes in Southern blotting (Section 5-5D) and *in situ* hybridization (Section 5-5E), as primers in PCR (Section 5-5F), and to carry out site-directed mutagenesis (Section 5-5G).

A. *Synthetic Procedures*

The basic strategy of oligonucleotide synthesis is analogous to that of polypeptide synthesis (Section 7-5A): *A suitably protected nucleotide is coupled to the growing end of the oligonucleotide chain, the protecting group is removed, and the process is repeated until the desired oligonucleotide has been synthesized.* The first practical technique for DNA synthesis, the **phosphodiester method,** which was devel-

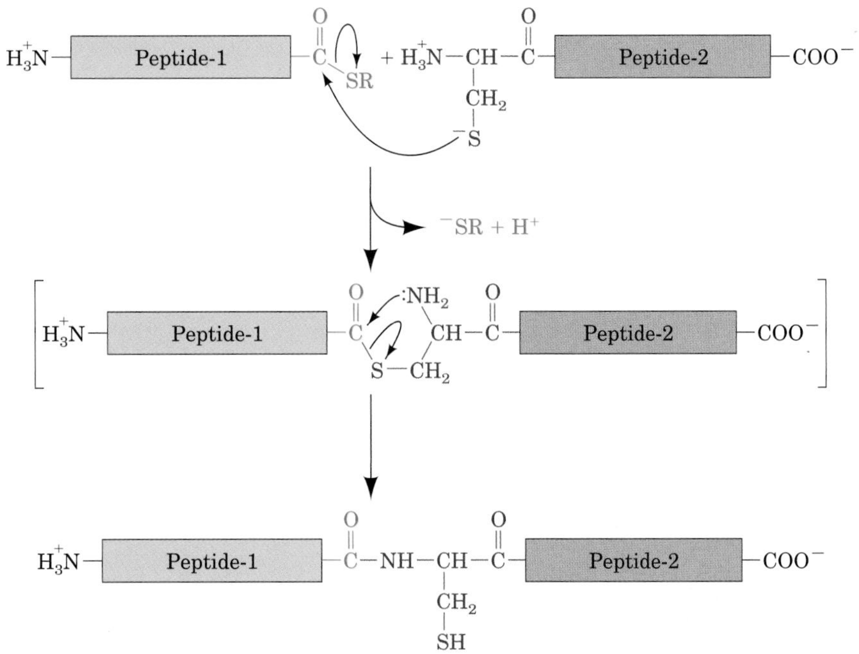

FIGURE 7-36 The native chemical ligation reaction. Peptide-1 has an C-terminal thioester group (R is an alkyl group) and Peptide-2 has an N-terminal Cys residue. The reaction, which occurs in aqueous solution at pH 7, is initiated by the nucleophilic attack of Peptide-2's Cys thiol group on Peptide-1's thioester group to yield, in a thiol exchange reaction, a new thioester group. This intermediate (as is indicated by the square brackets), undergoes a rapid intramolecular nucleophilic attack to yield a native peptide bond at the ligation site. [After Dawson, P.E., Muir, T.W., Clark-Lewis, I., and Kent, S.B.H., *Science* **266,** 777 (1994).]

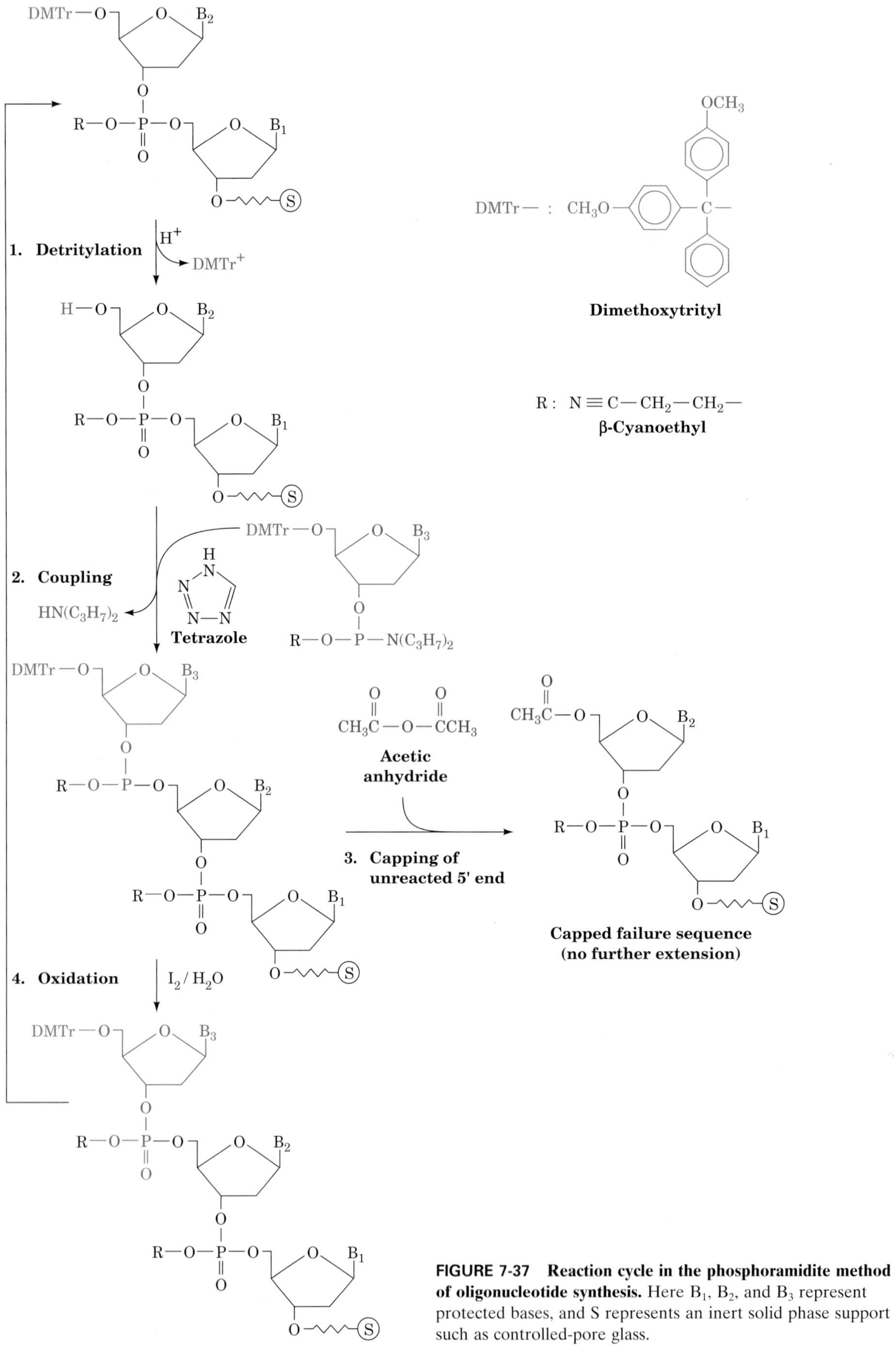

FIGURE 7-37   Reaction cycle in the phosphoramidite method of oligonucleotide synthesis. Here B_1, B_2, and B_3 represent protected bases, and S represents an inert solid phase support such as controlled-pore glass.

oped by H. Gobind Khorana in the 1960s, is a laborious process in which all reactions are carried out in solution and the products must be isolated at each stage of the multistep synthesis. Khorana, nevertheless, used this method, in combination with enzymatic techniques, to synthesize a 126-nucleotide tRNA gene, a project that required several years of intense effort by numerous skilled chemists.

a. The Phosphoramidite Method

By the early 1980s, these difficult and time-consuming processes had been supplanted by much faster solid phase methodologies that permitted oligonucleotide synthesis to be automated. The presently most widely used chemistry, which was formulated by Robert Letsinger and further developed by Marvin Caruthers, is known as the **phosphoramidite method.** This nonaqueous reaction sequence adds a single nucleotide to a growing oligonucleotide chain as follows (Fig. 7-37):

1. The **dimethoxytrityl (DMTr)** protecting group at the 5′ end of the growing oligonucleotide chain (which is anchored via a linking group at its 3′ end to a solid support,

S) is removed by treatment with an acid such as **trichloroacetic acid** (Cl_3CCOOH).

2. The newly liberated 5′ end of the oligonucleotide is coupled to the 3′-phosphoramidite derivative of the next deoxynucleoside to be added to the chain. The coupling agent in this reaction is **tetrazole,** which protonates the incoming nucleotide's diisopropylamine moiety so that it becomes a good leaving group. Modified nucleosides (e.g., containing a fluorescent label) can be incorporated into the growing oligonucleotide at this stage. Likewise, a mixture of oligonucleotides containing different bases at this position can be synthesized by adding the corresponding mixture of nucleosides.

3. Any unreacted 5′ end (the coupling reaction has a yield of over 99%) is capped by acetylation so as to block its extension in subsequent coupling reactions. This prevents the extension of erroneous oligonucleotides (failure sequences).

4. The phosphite triester group resulting from the coupling step is oxidized with I_2 to the more stable phosphotriester, thereby yielding a chain that has been lengthened by one nucleotide.

This reaction sequence, in commercially available automated synthesizers, can be repeated up to ~150 times with a cycle time of 40 min or less. Once an oligonucleotide of desired sequence has been synthesized, it is treated with NH_4OH to release it from its support and remove its various blocking groups, including those protecting the exocyclic amines on the bases. The product can then be separated from the failure sequences and protecting groups by reverse-phase HPLC and/or gel electrophoresis.

B. *DNA Chips*

The determination of the human genome sequence (Section 7-2B) was really only the means to a highly complex end. The questions of real biochemical significance are: What are the functions of these ~30,000 genes; in which cells, under what circumstances, and to what extent are each of them expressed; how do their gene products interact to yield a functional organism; and what are the medical consequences of variant genes? The traditional method of addressing such questions, the one-gene-at-a-time approach, is simply incapable of acquiring the vast amounts of data necessary to answer these questions. What will be required are methods that can globally analyze biological processes, that is, techniques that can simultaneously monitor all the components of a biological system.

A recently developed technology that shows great promise for making such global assessments involves the use of **DNA chips** (also called **DNA microarrays** and **gene chips;** Fig. 7-38). These are arrays of different DNA oligonucleotides anchored to a glass or nylon substrate (surface) in an ~1-cm-wide square grid. In one of several methodologies that are presently used to manufacture DNA chips, large numbers (up to ~1 million) of different

FIGURE 7-38   A DNA chip. This ~6000-gene array contains most of the genes from baker's yeast, one per spot. The chip had been hybridized to the cDNAs derived from mRNAs extracted from yeast. The cDNAs derived from cells that were grown in glucose were labeled with a red-fluorescing dye, whereas the cDNAs derived from cells grown in the absence of glucose were labeled with a green-fluorescing dye. Thus the red and green spots, respectively, reveal those genes that are transcriptionally activated by the presence or absence of glucose, whereas the yellow spots (*red plus green*) indicate genes whose expression is unaffected by the level of glucose. [Courtesy of Patrick Brown, Stanford University School of Medicine.]

oligonucleotides are simultaneously synthesized via a combination of photolithography (the process used to fabricate electronic chips) and solid phase DNA synthesis. In this process (Fig. 7-39), which was developed by Stephen Fodor, the nucleotides from which the oligonucleotides are synthesized each have a photochemically removable protective group at their 5' end that has the same function as the DMTr group in conventional solid phase DNA synthesis (Fig. 7-37). At a given stage in the synthetic procedure, oligonucleotides that, for example, require a T at their next position are deprotected by shining light on them through a mask that blocks the light from hitting those grid positions that require a different base at this next position. The chip is then incubated with a solution of activated thymidine nucleotide, which couples only to the deprotected oligonucleotides. After washing away the unreacted thymidine nucleotide, the process is repeated with different

masks for each of the remaining three bases. By repeating these 4 steps N times, an array of all 4^N possible N-residue sequences can be synthesized simultaneously in $4N$ coupling cycles, where $N \leq 30$. In an alternative DNA chip fabrication technique, which uses conventional phosphoramidite chemistry to synthesize oligonucleotides (Section 7-6A), nanoliter-sized droplets of reagents are delivered to the proper site on a chip using a device similar to an ink-jet printer. A series of DNA molecules such as cDNAs, PCR products, or synthetic oligonucleotides may also be robotically deposited (spotted) onto a chip.

In one application of DNA chips, L-residue oligonucleotides (the probes) are arranged in an array of L columns by 4 rows for a total of $4L$ sequences. The probe in the array's Mth column has the "standard" sequence with the exception of the probe's Mth position, where it has a different base, A, C, G, or T, in each row. Thus, one

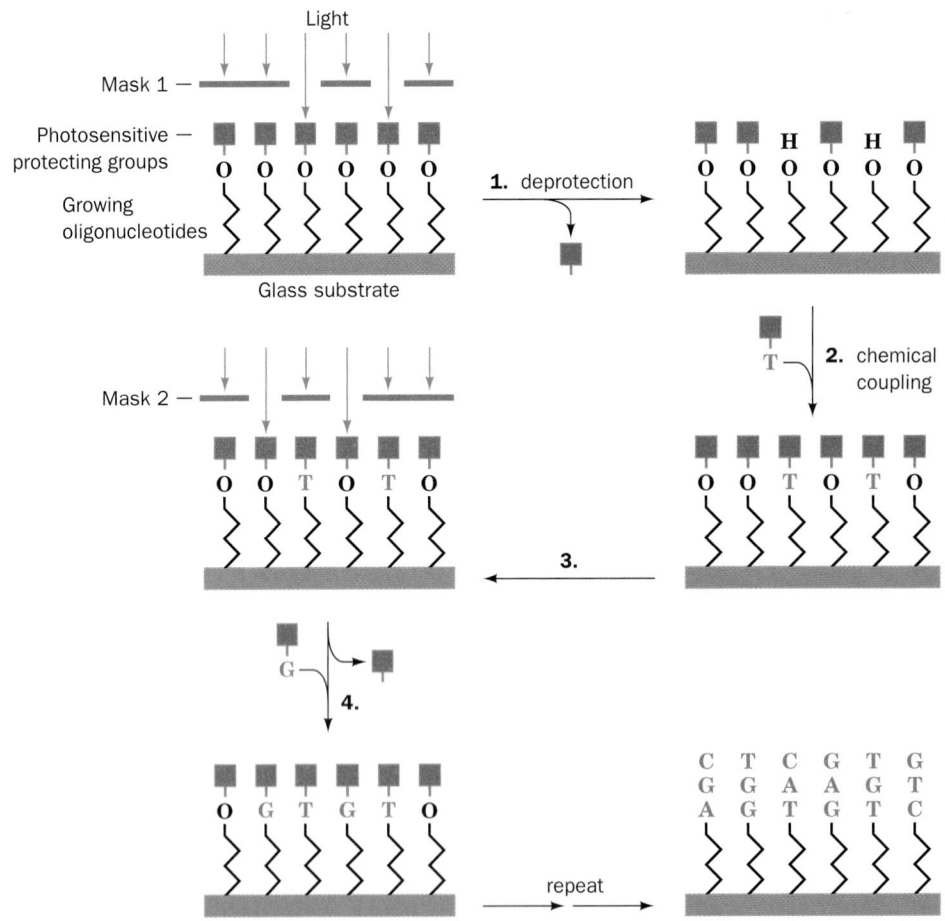

FIGURE 7-39   The photolithographic synthesis of a DNA chip.
In Step 1 of the process, oligonucleotides that are anchored to a glass substrate, and each having a photosensitive protecting group (*filled red square*) at their 5' ends, are exposed to light through a mask that only permits the illumination of the oligonucleotides destined to be coupled, for example, to a T residue. The light deprotects these oligonucleotides so that only they react with the activated T nucleotide that is incubated with

the chip in Step 2. The entire process is repeated in Steps 3 and 4 with a different mask for G residues, and in subsequent reaction cycles for A and C residues, thereby extending all of the oligonucleotides by one residue. This quadruple cycle is then repeated for as many nucleotides as are to be added to form the final set of oligonucleotides. [After Pease, A.C., Solas, D., Sullivan, E.J., Cronin, M.T., Holmes, C.P., and Fodor, S.P.A., *Proc. Natl. Acad. Sci.* **91,** 5023 (1994).]

of the four probe DNAs in every column will have the standard sequence, whereas the other three will differ from the standard DNA by only one base change. The probe array is then hybridized with the complementary DNA or RNA, whose variation relative to the standard DNA is to be determined, and the unhybridized DNA or RNA is washed away. This "target" DNA or RNA is fluorescently labeled so that, when interrogated by a laser, the positions on the probe array to which it binds are revealed by fluorescent spots. Since hybridization conditions can be adjusted so that a single base mismatch will significantly reduce the level of binding, a target DNA or RNA that varies from the complement of the standard DNA by a single base change at its Mth position, say C to A, would be readily detected by an increased fluorescence at the row corresponding to A in the Mth column relative to that at other positions [a target DNA or RNA that was exactly complementary to the standard DNA would exhibit high fluorescence in each of its columns at the base position (row) corresponding to the standard sequence]. The intensity of the fluorescence at every position in the array, and hence the sequence variation from the standard DNA, is rapidly determined with a computerized fluorescence scanning device. In this manner, sequence polymorphisms on the single-base level [**single nucleotide polymorphisms (SNPs;**

pronounced "snips")] can be automatically determined. It is becoming increasing apparent that genetic variations, and in particular SNPs, are largely responsible for an individual's susceptibility to many diseases as well as to adverse reactions to drugs (side effects; Section 15-4B).

In an alternative DNA chip methodology, up to 10,000 different DNAs are robotically deposited in an array on a coated glass surface. These probe DNAs most often consist of PCR-amplified inserts from cDNA clones or expressed sequence tags (ESTs), which are usually robotically synthesized. Such DNA microarrays can be used to monitor the level of expression of the corresponding genes in a tissue of interest by the degree of hybridization of its fluorescently labeled mRNA or cDNA population. Hence, they can be used to determine the pattern of gene expression (the **expression profile**) in different tissues of the same organism (e.g., Fig. 7-40) and how specific diseases and drugs (or drug candidates) affect gene expression. Moreover, DNA chips can be used to specifically identify infectious agents by detecting unique segments of their DNA. Consequently, DNA chips hold enormous promise for understanding the interplay of genes during cell growth and changes in the environment, for the characterization and diagnosis of both infectious and noninfectious diseases (e.g., cancers), for identifying their genetic risk factors and

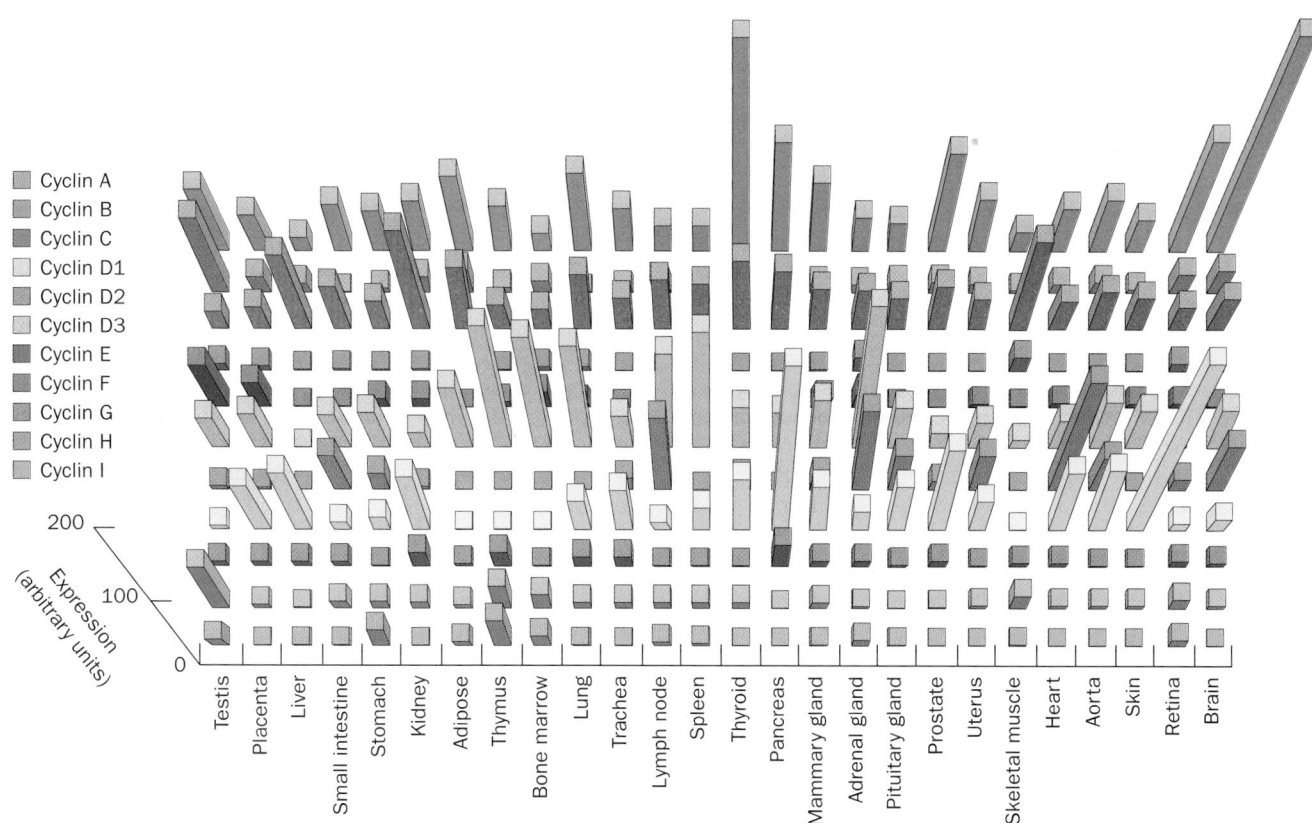

FIGURE 7-40 Variation in the expression of genes that encode proteins known as cyclins (Section 34-4C) in human tissues. The levels of hybridization of the various cyclin mRNAs produced in each tissue were measured through the use of DNA chips. [After Gerhold, D., Rushmore, T., and Caskey, C.T., *Trends Biochem. Sci.* **24,** 172 (1999).]

designing specific treatments, and as tools in drug development.

C. *SELEX*

Although nucleic acids such as DNA and mRNA are widely thought of as being passive molecular "tapes" whose structures have only incidental significance, it is now abundantly clear that certain nucleic acids, for example tRNAs and ribosomal RNAs (Sections 32-2B and 32-3A), have functions that require that these molecules maintain well-defined three-dimensional structures. This has led to the development of a process called **SELEX** (for *Systematic Evolution of Ligands by EXponential enrichment*) for generating single-stranded oligonucleotides that specifically bind target molecules with high affinity and specificity.

SELEX, a technique pioneered by Larry Gold, starts with a library of polynucleotides that have fixed known sequences at their 5' and 3' ends and a central region of randomized sequences. Such sequences are synthesized via solid phase methods by adding a mixture of all four nucleoside triphosphates rather than a single nucleoside triphosphate at the positions to be randomized. For a variable region that is 30 nucleotides long, this will yield a mixture of $4^{30} \approx 10^{18}$ different oligonucleotides. The next step in the SELEX process is to select those sequences in the mixture that selectively bind to a target molecule, X. Those oligonucleotides that bind to X with higher than average affinity can be separated from the other oligonucleotides in the mixture by a variety of techniques including affinity chromatography (Section 6-3C) with X linked to a suitable matrix and the precipitation of X–oligonucleotide complexes by anti-X antibodies. The oligonucleotides are then separated from X and amplified by cloning or PCR. This selection and amplification process is then repeated for several cycles (usually 10–15), thereby yielding oligonucleotides that tenaciously bind X. Several such enrichment cycles are necessary because those few oligonucleotides in a library with the highest affinity for X must compete for binding X with the far more abundant weak binders.

The results of SELEX are remarkable. Its oligonucleotide products, which are known as **aptamers,** bind to their target molecule, X, with dissociation constants that are typically around $10^{-9}\,M$ if X is a protein and between $10^{-3}\,M$ and $10^{-6}\,M$ if X is a small molecule (for the reaction $A + B \rightleftharpoons A \cdot B$, the dissociation constant, $K_D = [A][B]/[A \cdot B]$). Moreover, aptamers bind to their target molecules with great specificity. Thus, aptamers selected for their affinity toward a particular member of a protein family do not bind other members of that family. Aptamers bind a wide variety of target molecules including simple ions, small molecules, proteins, organelles, and even whole cells. These various properties give aptamers great promise both as diagnostic tools and as therapeutic agents.

■ CHAPTER SUMMARY

1 ■ Primary Structure Determination of Proteins   The initial step in the amino acid sequence determination of a protein is to ascertain its content of chemically different polypeptides by end group analysis. The protein's disulfide bonds are then chemically cleaved, the different polypeptides are separated and purified, and their amino acid compositions are determined. Next, the purified polypeptides are specifically cleaved, by enzymatic or chemical means, to smaller peptides that are separated, purified, and then sequenced by (automated) Edman degradation. Repetition of this process, using a cleavage method of different specificity, generates overlapping peptides whose amino acid sequences, when compared to those of the first group of peptide fragments, indicate their order in the parent polypeptide. The primary structure determination is completed by establishing the positions of the disulfide bonds. This requires the degradation of the protein with its disulfide bonds intact. Then, by sequencing the pairs of disulfide-linked peptide fragments, their positions in the intact protein can be deduced. Mass spectrometry using electrospray ionization or matrix-assisted laser desorption/ionization to vaporize and ionize peptides can be used to determine their molecular masses. By using these techniques in a tandem mass spectrometer equipped with a collision cell, the sequences of short peptides can be rapidly determined. Once the primary structure of a protein is known, its minor variants, which may arise from mutations or chemical modifications, can be readily identified by peptide mapping.

2 ■ Nucleic Acid Sequencing   Nucleic acids may be sequenced by the same basic strategy used to sequence proteins. In the chain-terminator method, the DNA fragment to be sequenced is replicated by a DNA polymerase in the presence of all four deoxynucleoside triphosphates, one of which is α-^{32}P-labeled, and a small amount of the dideoxy analog of one of the deoxynucleoside triphosphates. This results in a series of ^{32}P-labeled chains that are terminated after the various positions occupied by the corresponding base. The electrophoresis of the four differently terminated DNA samples in parallel lanes of a sequencing gel resolves fragments that differ in size by one nucleotide. An autoradiograph of the sequencing gel containing the four sets of fragments reveals the base sequence of the complementary DNA. RNA may be sequenced by determining the sequence of its corresponding cDNA. Automated methods, which use fluorescently labeled DNA fragments, are greatly speeding up DNA sequence determinations. They are being used to sequence very large tracts of DNA such as the human genome. In the conventional strategy for genome sequencing, the genome is mapped by sequencing large numbers of landmark sequences such as sequence-tagged sites (STSs) and expressed sequence tags (ESTs). Landmark-containing YAC clones are then fragmented, the fragments subcloned and sequenced, and then computationally assembled into contigs, which are further assembled into chromosomes based on the landmark sequences. Advances in cloning and computational techniques now permit genome sequencing via a shotgun strategy. For bacterial genomes this is carried out straightforwardly. For eukaryotic genomes, large segments are cloned in bacterial artificial chromosomes (BACs), whose ends are sequenced to yield sequence-tagged connectors (STCs). The BAC inserts are fragmented, shotgun cloned, and the clones sequenced and assembled into contigs. The STCs are used to further assemble the sequenced BAC inserts into chromosomes via BAC walking.

3 ■ Chemical Evolution Sickle-cell anemia is a molecular disease of individuals who are homozygous for a gene specifying an altered β chain of hemoglobin. Fingerprinting and sequencing studies have identified this alteration as arising from a point mutation that changes Glu β6 to Val. In the heterozygous state, the sickle-cell trait confers resistance to malaria without causing deleterious effects. This accounts for its high incidence in populations living in malarial regions. The cytochromes *c* from many eukaryotic species contain many amino acid residues that are invariant or conservatively substituted. Hence, this protein is well adapted to its function. The amino acid differences between the various cytochromes *c* have permitted the generation of their phylogenetic tree, which closely parallels that determined by classical taxonometry. The number of sequence differences between homologous proteins from related species plotted against the time when, according to the fossil record, these species diverged from a common ancestor reveals that point mutations are accepted into proteins at a constant rate. Proteins whose functions are relatively intolerant to sequence changes evolve more slowly than those that are more tolerant to such changes. Phylogenetic analysis of the globin family—myoglobin and the α and β chains of hemoglobin—reveals that these proteins arose through gene duplication. In this process, the original function of the protein is maintained, while the duplicated copy evolves a new function. Many, if not most, proteins have evolved through gene duplication.

4 ■ Bioinformatics: An Introduction The sequences of nucleic acids and proteins are archived in large and rapidly growing databases that are publicly available over the World Wide Web. The alignment of homologous sequences provides insights into their function and evolution. The alignment of distantly related sequences requires statistically and computationally sophisticated algorithms such as BLAST, FASTA, and CLUSTAL. The construction of an optimal phylogenetic tree based on aligned sequences is a computationally intensive process for which there is no exact solution.

5 ■ Chemical Synthesis of Polypeptides The strategy of polypeptide chemical synthesis involves coupling amino acids, one at a time, to the N-terminus of a growing polypeptide chain. The α-amino group of each amino acid must be chemically protected during the coupling reaction and then unblocked before the next coupling step. Reactive side chains must also be chemically protected but then unblocked at the conclusion of the synthesis. The difficulty in recovering the intermediate product of each of the many steps of such a synthesis has been eliminated by the development of solid phase synthesis techniques. These methods have led to the synthesis of numerous biologically active polypeptides and have recently become capable of synthesizing small proteins in useful amounts. The use of the chemical ligation technique significantly increases the sizes of the polypeptides that can be chemically synthesized.

6 ■ Chemical Synthesis of Oligonucleotides Oligonucleotides are indispensable to recombinant DNA technology. They are used to identify normal and mutated genes and to alter specific genes through site-directed mutagenesis. Oligonucleotides of defined sequence are efficiently synthesized by the phosphoramidite method, a cyclic, nonaqueous, solid phase process that has been automated. DNA chips, which can be manufactured through photolithographically based DNA synthesis or by the robotic deposition of DNA on a glass substrate, are finding increasing numbers of applications. These include the detection of single nucleotide polymorphisms, monitoring patterns of gene expression, and identifying infectious agents. SELEX is a technique for selecting and amplifying those members of a large polynucleotide library that specifically bind to a target molecule. The resulting aptamers, which have remarkably high affinity and specificity for their target molecules, may be useful diagnostic and therapeutic agents.

REFERENCES

PROTEIN SEQUENCING

Aebersold, R. and Patterson, S.D., Current problems and technical solutions in protein biochemistry, *in* Angeletti, R.H. (Ed.), *Proteins. Anaysis and Design,* Chapter 1, Academic Press (1998). [Describes advanced methods for protein characterization.]

Allen, G., Sequencing of proteins and peptides, in Burdon, R.M. and van Knippenberg, P.H. (Eds.), *Laboratory Techniques in Biochemistry and Molecular Biology,* Vol. 9 (2nd revised ed.), Elsevier (1989).

Barrett, G.C. and Elmore, D.T., *Amino Acids and Peptides,* Chapter 5, Cambridge University Press (1998).

Bhown, A.S. (Ed.), *Protein/Peptide Sequence Analysis: Current Methodologies,* CRC Press (1988).

Bogusky, M.S., Ostell, J., and States, D.J., Molecular sequence data-bases and their uses, *in* Rees, A.R., Sternberg, M.J.E., and Wetzel, R. (Eds.), *Protein Engineering. A Practical Approach, pp.* 57–88, IRL Press (1992).

Chapman, J.R. (Ed.), *Protein and Peptide Analysis by Mass Spectrometry,* Humana Press (1996).

Costello, C.E., Bioanalytic applications of mass spectrometry, *Curr. Opin. Biotech.* **10,** 22–28 (1999).

Creighton, T.E., *Proteins* (2nd ed.), Chapters 1 and 3, Freeman (1993).

Findlay, J.B.C. and Geisow, M.J. (Eds.), *Protein Sequencing, A Practical Approach,* IRL Press (1989).

James, P. (Ed.), *Proteome Research: Mass Spectrometry,* Springer-Verlag (2001).

Karger, B.L. and Hancock, W.S. (Eds.), *Methods Enzymol.* **270** *and* **271,** Section III. Mass spectrometry (1996).

Mann, M., Hendrickson, R.C., and Pandey, A., Analysis of proteins and proteomes by mass spectrometry, *Annu. Rev. Biochem.* **70,** 437–473 (2001).

Matsudaira, P.T. (Ed.), *A Practical Guide to Protein and Peptide Purification and Microsequencing,* Academic Press (1989).

Rappsilber, J. and Mann, M., What does it mean to identify a protein in proteomics? *Trends Biochem. Sci.* **27,** 74–78 (2002).

Sanger, F., Sequences, sequences, and sequences, *Annu. Rev. Biochem.* **57,** 1–28 (1988). [A scientific autobiography that provides a glimpse of the early difficulties in sequencing proteins.]

Simpson, R.J. and Reid, G.E., Sequence analysis of gel-resolved proteins, *in* Hames, B.D. (Ed.), *Gel Electrophoresis of Proteins. A Practical Approach, pp.* 237–267 (1998).

Siuzdak, G., *Mass Spectrometry for Biotechnology,* Academic Press (1996).

Wilm, M., Mass spectrometric analysis of proteins, *Adv. Prot. Chem.* **54,** 1–30 (2000).

NUCLEIC ACID SEQUENCING

Ansorge, W., Voss, H., and Zimmermann, J. (Eds.), *DNA Sequencing Strategies. Automated and Advanced Approaches,* Wiley (1997).

Brown, T.A., *Essentials of Medical Genomics,* Wiley-Liss (2003).

Brown, T.A., *Genomes,* (2nd ed.), Wiley-Liss (2002).

Cantor, C.R. and Smith, C.L., *Genomics. The Science and Technology Behind the Human Genome Project,* Wiley-Interscience (1999).

Collins, F.S., Patrinos, A., Jordan, E., Chakravarti, A., Gesteland, R., Walters, L., and the members of the DOE and NIH planning groups, New goals for the U.S. human genome project: 1998–2003, *Science* **282,** 682–689 (1998).

DeLoukis, P., et al., A physical map of 30,000 human genes, *Science* **282,** 744–746 (1998). [The gene maps for the individual chromosomes can be seen at http://www.ncbi.nlm.nih.gov/genemap99.]

Graham, C.A. and Hill, A.J.M. (Eds.), *DNA Sequencing Protocols* (2nd ed.), Humana Press (2001).

Howe, C.J. and Ward, E.S., *Nucleic Acid Sequencing. A Practical Approach,* IRL Press (1989).

Lipschutz, R.J. and Fodor, S.P.A., Advanced DNA sequencing technologies, *Curr. Opin. Struct. Biol.* **4,** 376–380 (1994).

Primrose, S.B., *Principles of Genome Analysis* (2nd ed.), Blackwell Science (1998).

Roe, B.A. (Ed.), *DNA Sequencing, Methods* **3**(1) (1991).

Venter, J.C., Smith, H.O., and Hood, L., A new strategy for genome sequencing, *Science* **381,** 364–366 (1996). [Describes the shotgun strategy for genome sequencing.]

SOME GENOME SEQUENCES

Adams, M.D., et al., The genome sequence of *Drosophila melanogaster, Science* **287,** 2185–2195 (2000).

Anderson, S.E.V., et al., The genome sequence of *Rickettsia prowazekii* and the origin of mitochondria, *Nature* **396,** 133–140 (1998).

Blattner, F.R., et al., The complete genome sequence of *Escherichia coli* K-12, *Science* **277,** 1453–1474 (1997).

Bult, C.J., et al., Complete genome sequence of the methanogen archeon, *Methanococcus jannaschii, Science* **273,** 1058–1073 (1996).

Cole, S.T., et al., Massive genetic decay in the leprosy bacillus, *Nature* **409,** 1007–1011 (2001).

Cole, S.T., et al., Deciphering the biology of *Mycobacterium tuberculosis* from the complete genome sequence, *Nature* **393,** 537–544 (1998).

Fleischman, R.D., et al., Whole-genome random sequencing and assembly of *Haemophilus influenzae* Rd., *Science* **269,** 496–512 (1995).

Fraser, C.M., et al., The minimal gene complement of *Mycoplasma genitalium, Science* **270,** 397–403 (1995).

Fraser, C.M., et al., Genomic sequence of a Lyme disease spirochaete, *Borrelia burgdorferi, Nature* **390,** 580–586 (1997).

Gardiner, M.J., et al., Genome sequence of the human malaria parasite *Plasmodium falciparum, Nature* **419,** 498–511 (2002).

Goffeau, A., et al., The yeast genome directory, *Nature* **387** (May 29, 1997 Suppl.), 5–105 (1997).

International Human Genome Sequencing Consortium, Initial sequencing and analysis of the human genome, *Nature* **409,** 860–921 (2001).

Kaneko, T., et al., Sequence analysis of the genome of the unicellular cyanobacterium *Synechocystis* sp. strain PCC6803. II. Sequence determination of the entire genome and assignment of potential protein-coding regions, *DNA Res.* **3,** 109–136 (1996).

Mouse Genome Sequencing Consortium, Initial sequencing and comparative analysis of the mouse genome, *Nature* **420,** 520–562 (2002).

The Arabidopsis Genome Initiative, The analysis of the genome sequence of the flowering plant *Arabidopsis thaliana, Nature* **408,** 796–815 (2000).

The *C. elegans* Sequencing Consortium, Genome sequence of the nematode *C. elegans.* A platform for investigating biology, *Science* **282,** 2012–2017 (1998).

Venter, J.C., et al., The sequence of the human genome, *Science* **291,** 1304–1351 (2001).

Wood, V., et al., The genome sequence of *Schizosaccharomyces pombe, Nature.* **415,** 871–880 (2002).

Yu, J., et al., A draft sequence of the rice genome (*Oryza sativa* L. spp. *indica*); *and* Goff, S.A., et al., A draft sequence of the rice genome (*Oryza sativa* L. spp. *japonica*), *Science* **296,** 79–92; *and* 92–113 (2002).

CHEMICAL EVOLUTION

Allison, A.C., The discovery of resistance to malaria of sickle-cell heterozygotes, *Biochem. Mol. Biol. Educ.* **30,** 279–287 (2002).

Dickerson, R.E., The structure and history of an ancient protein, *Sci. Am.* **226**(4), 58–72 (1972). [Discusses the evolution of cytochrome *c.*]

Dickerson, R.E. and Geis, I. *Hemoglobin,* Chapter 3, Benjamin/Cummings (1983). [A detailed discussion of globin evolution.]

Dickerson, R.E. and Timkovich, R., Cytochromes *c, in* Boyer, P.D. (Ed.), *The Enzymes* (3rd ed.), Vol. 11, *pp.* 397–547, Academic Press (1975). [Contains a detailed analysis of cytochrome *c* sequence studies.]

Doolittle, R.F., Feng, D.F., Tsang, S., Cho, G., and Little, E., Determining divergence times of the major kingdoms of living organisms with a protein clock, *Science* **271,** 470–477 (1996).

Ingram, V.M., A case of sickle-cell anaemia: A commentary, *Biochim. Biophys. Acta* **1000,** 147–150 (1989). [A scientific memoir on the development of fingerprinting to characterize sickle-cell hemoglobin.]

Kimura, M., The neutral theory of molecular evolution, *Sci. Am.* **241**(5), 98–126 (1979).

King, M.C. and Wilson, A.C., Evolution at two levels in humans and chimpanzees, *Science* **188**, 107–116 (1975).

Moore, G.R. and Pettigrew, G.W., *Cytochromes c. Evolutionary, Structural and Physicochemical Aspects,* Springer-Verlag (1990).

Nagel, R.L. and Roth, E.F., Jr., Malaria and red cell genetic defects, *Blood* **74**, 1213–1221 (1989). [An informative review of the various genetic "defects," including sickle-cell anemia, that inhibit malaria and how they might do so.]

Strasser, B.J., Sickle cell anemia, a molecular disease, *Science* **286**, 1488–1490 (1999). [A historical account of Pauling's work on sickle-cell anemia.]

Wilson, A.C., The molecular basis of evolution, *Sci Am.* **253**(4), 164–173 (1985).

BIOINFORMATICS

Altschul, S.F. and Koonin, E.V., Iterated profile searches with PSI-BLAST—a tool for discovery in protein data bases, *Trends Biochem. Sci.* **23**, 444–447 (1998).

Baxevanis, A.D. and Ouellette, B.F.F. (Eds.), *Bioinformatics* (2nd ed.), Wiley-Interscience (2001). [A collection of authoritative articles.]

Bork, P. (Ed.), *Analysis of Amino Acid Sequences, Adv. Prot. Chem.* **54** (2000). [Contains informative articles on sequence alignment and phylogenetic tree generation.]

Database Issue, *Nucleic Acids Res.* **31**(1) (2003). [Annually updated descriptions of numerous databases of biomolecular interest. For the past several years, the database issue has been this journal's first issue of the year.]

Doolittle, R.F. (Ed.), *Molecular Evolution: Computer Analysis of Proteins and Nucleic Acids;* and *Computing Methods for Macromolecular Sequence Analysis, Methods Enzymol.* **183** (1990); and **266** (1996).

Doolittle, R.F., *Of Urfs and Orfs. A Primer of How to Analyze Derived Amino Acid Sequences,* University Science Books (1986).

Durbin, R., Eddy, S., Krogh, A., and Mitchison, G., *Biological Sequence Analysis. Probabilistic Models of Proteins and Nucleic Acids,* Cambridge University Press (1998).

Gibson, G. and Muse, S.V., *A Primer of Genomic Science,* Sinauer Associates (2002).

Henikoff, S., Scores for sequence searches and alignments, *Curr. Opin. Struct. Biol.* **6**, 353–360 (1996).

Higgins, D. and Taylor, W. (Eds.), *Bioinformatics. Sequence, Structure and Databanks,* Oxford University Press (2000).

Jeanmougin, F. and Thompson, J.D., Multiple sequence alignment with Clustal X, *Trends Biochem. Sci.* **23**, 403–405 (1998).

Jones, D.T. and Swindells, M.B., Getting the most from PSI-BLAST, *Trends Biochem. Sci.* **27**, 161–164 (2002).

Koonin, E.V., Aravind, L., and Kondrashov, A.S., The impact of comparative genomics on our understanding of evolution, *Cell* **101**, 573–576 (2000).

Mann, M. and Pandey, A., Use of mass spectrometry–derived data to annotate nucleotide and protein sequence databases, *Trends Biochem. Sci.* **26**, 54–61 (2001).

Misener, S. and Krawetz, S.A. (Eds.), *Bioinformatics Methods and Protocols,* Humana Press (2000). [A sourcebook.]

Mount, D.W., *Bioinformatics. Sequence and Genome Analysis,* Cold Spring Harbor Laboratory Press (2001).

Needleman, S.B. and Wunsch, C.D., A general method applicable to the search for similarities in the amino acid sequence of two proteins, *J. Mol. Biol.* **48**, 443–453 (1970). [The formulation of the Needleman–Wunsch algorithm.]

Pagel, M., Inferring the historical patterns of biological evolution, *Nature* **401**, 877–884 (1999). [A review.]

Trends Guide to Bioinformatics, Trends Supplement 1998, Elsevier (1998). [A valuable collection of tutorials published as a supplement to the "Trends" journals such as *Trends Biochem. Sci.*]

POLYPEPTIDE SYNTHESIS

Atherton, E. and Sheppard, R.C., *Solid Phase Peptide Synthesis. A Practical Approach,* IRL Press (1989).

Barrett, G.C. and Elmore, D.T., *Amino Acids and Peptides,* Chapter 7, Cambridge University Press (1998).

Bodanszky, M., *Principles of Peptide Synthesis,* Springer-Verlag (1993).

Dawson, P.E. and Kent, S.B.H., Synthesis of native proteins by chemical ligation, *Annu. Rev. Biochem.* **69**, 923–960 (2000).

Fields, G.B. (Ed.), *Solid-Phase Peptide Synthesis, Meth. Enzymol.* **289** (1997).

Kent, S.B.H., Alewood, D., Alewood, P., Baca, M., Jones, A., and Schnölzer, M., Total chemical synthesis of proteins: Evolution of solid phase synthetic methods illustrated by total chemical synthesis of the HIV-1 protease, *in* Epton, R. (Ed.), *Innovation & Perspectives in Solid Phase Synthesis,* SPPC Ltd. (1992); *and* Milton, R.C. deL., Milton, S.C.F., and Kent, S.B.H., Total chemical synthesis of a D-enzyme: The enantiomers of HIV-1 protease show demonstration of reciprocal chiral substrate specificity, *Science* **256**, 1445–1448 (1992).

Merrifield, B., Solid phase synthesis, *Science* **232**, 342–347 (1986).

Wilken, J. and Kent, S.B.H., Chemical protein synthesis, *Curr. Opin. Biotech.* **9**, 412–426 (1998).

CHEMICAL SYNTHESIS OF OLIGONUCLEOTIDES

Caruthers, M.H., Beaton, G., Wu, J.V., and Wiesler, W., Chemical synthesis of deoxynucleotides and deoxynucleotide analogs, *Methods Enzymol.* **211**, 3–20 (1992); *and* Caruthers, M.H., Chemical synthesis of DNA and DNA analogues, *Acc. Chem. Res.* **24**, 278–284 (1991).

Gait, M.J. (Ed.), *Oligonucleotide Synthesis. A Practical Approach,* IRL Press (1984).

Gerhold, D., Rushmore, T., and Caskey, C.T., DNA chips: Promising toys have become powerful tools, *Trends Biochem. Sci.* **24**, 168–173 (1999).

Gold, L., The SELEX process: A surprising source of therapeutic and diagnostic compounds, *Harvey Lectures* **91**, 47–57 (1997).

Hermann, T. and Patel, D.J., Adaptive recognition by nucleic acid aptamers, *Science* **287**, 820–825 (2000). [A review.]

Nature Genetics Supplement **21**, 1–60 (January, 1999). [Contains a series of authoritative articles on DNA chips.]

Ramsey, G., DNA chips: State-of-the-art, *Nature Biotech.* **16**, 40–44 (1999).

Schena, M. (Ed.), *DNA Microarrays. A Practical Approach,* Oxford University Press (1999).

Wilson, D.S. and Szostak, J.W., In vitro selection of functional nucleic acids, *Annu. Rev. Biochem.* **68**, 611–647 (1999). [Discusses SELEX].

Young, R., Biomedical discovery with DNA arrays, *Cell* **102**, 9–15 (2000).

PROBLEMS

Note: Amino acid compositions of polypeptides with unknown sequences are written in parentheses with commas separating amino acid abbreviations such as in (Gly, Tyr, Val). Known amino acid sequences are written with residue names in order and separated by hyphens; for example, Tyr-Val-Gly.

1. State the cleavage pattern of the following polypeptides by the indicated agents.

a. Ser-Ala-Phe-Lys-Pro by chymotrypsin
b. Thr-Cys-Gly-Met-Asn by NCBr
c. Leu-Arg-Gly-Asp by carboxypeptidase A
d. Gly-Phe-Trp-Asp-Phe-Arg by endopeptidase Asp-N
e. Val-Trp-Lys-Pro-Arg-Glu by trypsin

2. A protein is subjected to end group analysis by dansyl chloride. The liberated dansylamino acids are found to be present with a molar ratio of two parts Ser to one part Ala. What conclusions can be drawn about the nature of the protein?

3. A protein is subjected to degradation by carboxypeptidase B. Within a short time, Arg and Lys are liberated, following which no further change is observed. What does this information indicate concerning the primary structure of the protein?

4. Consider the following polypeptide:

Asp-Trp-Val-Arg-Asn-Ser-Phe-Cys-Gln-Gly-Pro-Tyr-Met

(a) What amino acids would be liberated on its complete acid hydrolysis? (b) What amino acids would be liberated on its complete alkaline hydrolysis?

5. Before the advent of the Edman degradation, the primary structures of proteins were elucidated through the use of partial acid hydrolysis. The resulting oligopeptides were separated and their amino acid compositions were determined. Consider a polypeptide with amino acid composition (Ala$_2$, Asp, Cys, Leu, Lys, Phe, Pro, Ser$_2$, Trp$_2$). Treatment with carboxypeptidase A released only Leu. Oligopeptides with the following compositions were obtained by partial acid hydrolysis:

(Ala, Lys)	(Ala, Ser$_2$)	(Cys, Leu)
(Ala, Lys, Trp)	(Ala, Trp)	(Cys, Leu, Pro)
(Ala, Pro)	(Asp, Lys, Phe)	(Phe, Ser, Trp)
(Ala, Pro, Ser)	(Asp, Phe)	(Ser, Trp)
(Ser$_2$, Trp)		

Determine the amino acid sequence of the polypeptide.

*6. A polypeptide is subjected to the following degradative techniques resulting in polypeptide fragments with the indicated amino acid sequences. What is the amino acid sequence of the entire polypeptide?

I. Cyanogen bromide treatment:

1. Asp-Ile-Lys-Gln-Met
2. Lys
3. Lys-Phe-Ala-Met
4. Tyr-Arg-Gly-Met

II. Trypsin hydrolysis:

5. Gln-Met-Lys
6. Gly-Met-Asp-Ile-Lys
7. Phe-Ala-Met-Lys
8. Tyr-Arg

7. Treatment of a polypeptide by dithiothreitol yields two polypeptides that have the following amino acid sequences:

1. Ala-Phe-Cys-Met-Tyr-Cys-Leu-Trp-Cys-Asn
2. Val-Cys-Trp-Val-Ile-Phe-Gly-Cys-Lys

Chymotrypsin-catalyzed hydrolysis of the intact polypeptide yields polypeptide fragments with the following amino acid compositions:

3. (Ala, Phe)
4. (Asn, Cys$_2$, Met, Tyr)
5. (Cys, Gly, Lys)
6. (Cys$_2$, Leu, Trp$_2$, Val)
7. (Ile, Phe, Val)

Indicate the positions of the disulfide bonds in the original polypeptide.

*8. The following treatments of a polypeptide yielded the indicated results. What is the primary structure of the polypeptide?

I. Acid hydrolysis:

1. (Arg, Asx, Cys$_2$, Gly, Ile, Leu, Lys, Met, Phe, Pro, Ser)

II. Edman degradation (one cycle):

2. (Leu)
3. (Ser)

III. Carboxypeptidase A (sufficient time to remove one residue per chain):

4. (Asp)

IV. Dithioerythritol + iodoacetic acid followed by trypsin hydrolysis:

5. (Arg, Ser)
6. (Asp, Met)
7. (Cys, Gly, Ile, Leu, Phe, Pro)
8. (Cys, Lys)

V. Dithioerythritol + 2-bromoethylamine followed by trypsin hydrolysis:

9. (Arg, Ser)
10. (Asp, Met)
11. (Cys)
12. (Cys, Gly, Leu)
13. (Ile, Phe, Pro)
14. (Lys)

VI. Chymotrypsin:

15. No fragments

VII. Pepsin:

16. (Arg, Asp, Cys$_2$, Gly, Leu, Lys, Met, Ser)
17. (Ile, Phe, Pro)

9. A polypeptide was subjected to the following treatments with the indicated results. What is its primary structure?

I. Acid hydrolysis:

1. (Ala, Arg, Cys, Glx, Gly, Lys, Leu, Met, Phe, Thr)

II. Aminopeptidase M:

2. No fragments

III. Carboxypeptidase A + carboxypeptidase B:

3. No fragments

IV. Trypsin followed by Edman degradation of the separated products:

4. Cys-Gly-Leu-Phe-Arg
5. Thr-Ala-Met-Glu-Lys

*10. While on an expedition to the Amazon jungle, you isolate a polypeptide you suspect of being the growth hormone of a newly discovered species of giant spider. Unfortunately, your portable sequencer was so roughly treated by the airport baggage handlers that it refuses to provide the sequence of more than four consecutive amino acid residues. Nevertheless, you persevere and obtain the following data:

I. Hydrazinolysis:

1. (Val)

II. Dansyl chloride treatment followed by acid hydrolysis:

2. (Dansyl-Pro)

III. Trypsin followed by Edman degradation of the separated fragments:

3. Gly-Lys
4. Phe-Ile-Val
5. Pro-Gly-Ala-Arg
6. Ser-Arg

(a) Provide as much information as you can concerning the amino acid sequence of the polypeptide. (b) Considering the poor condition of your sequencer, what additional analytical technique would most conveniently permit you to complete the sequence determination of the polypeptide?

11. In taking the electrospray ionization mass spectrum of an unknown protein, you find that four successive peaks have *m/z* values of 953.9, 894.4, 841.8, and 795.1. What is the molecular mass of the protein and what are the ionic charges of the ions responsible for the four peaks?

12. Figure 7-41 pictures an autoradiograph of the sequencing gel of a DNA that was treated according to the chain-terminator method of DNA sequencing. What is the sequence of the template strand corresponding to bases 50 to 100? If there are any positions on the gel where a band seems to be absent, leave a question mark in the sequence for the indeterminate base.

13. Using Table 7-5, compare the relatedness of fungi to higher plants and to animals. Fungi are often said to be nongreen plants. In light of your analysis, is this a reasonable classification?

14. The inherited hemoglobin disease **β-thalassemia** is common to people from around the Mediterranean Sea and areas of Asia where malaria is prevalent (Fig. 7-20). The disease is characterized by a reduction in the amount of synthesis of the β chain of hemoglobin. Heterozygotes for the β-thalassemia gene, who are said to have **thalassemia minor,** are only mildly affected with adverse symptoms. Homozygotes for this gene, however, suffer from **Cooley's anemia** or **thalassemia major;** they are so severely afflicted that they do not survive their childhood. About 1% of the children born in the malarial regions around the Mediterranean Sea have Cooley's anemia. Why do you suppose the β-thalassemia gene is so prevalent in this area? Justify your answer.

15. Leguminous plants synthesize a monomeric oxygen-binding globin known as **leghemoglobin.** (Section 26-6). From your knowledge of biology, sketch the evolutionary tree of the globins (Fig. 7-24) with leghemoglobin included in its most likely position.

16. Since point mutations mostly arise through random chemical change, it would seem that the rate at which mutations appear in a gene expressing a protein should vary with the size of the gene (number of amino acid residues the gene expresses). Yet, even though the rates at which proteins evolve vary quite widely, these rates seem to be independent of protein size. Explain.

17. Sketch the self-dot matrices of the following: (a) A 100-residue peptide with nearly identical segments spanning its residues 20 to 40 and 60 to 80. (b) A 100-nucleotide DNA that is palindromic.

*18. (a) Using the PAM-250 log odds substitution matrix and the Needleman–Wunsch algorithm, find the best alignment of the two peptides PQRSTV and PDLRSCPSV. (b) What is its alignment score using a gap penalty of −8 for opening a gap and −2 for every "residue" in the gap? (c) What is its normalized alignment score (NAS) using the scoring system of 10 for each identity, 20 for each identity involving Cys, and −25 for every gap? (d) Is this NAS indicative of a homology? Explain.

19. You have been given an unknown protein to identify. You digest it with trypsin and, by using Edman degradation, find that one of the resulting peptide fragments has the sequence GIIWGEDTLMEYLENPK. Using BLAST, find the most probable identity (or identities) of the unknown protein. [To carry out a search on the BLAST server, go to http://www.ncbi.nlm.nih.gov/blast/ and, under "Protein BLAST," click on the link "standard protein-protein BLAST [blastp]" (which is BLAST for comparing an amino acid query sequence against a protein sequence database). In the window that comes up, enter the above sequence (without spaces or punctuation) into the sequence box, from the "Choose database" pulldown menu select "nr," and click on the "BLAST!" button.]

*20. A desiccated dodo bird in a reasonable state of preservation has been found in a cave on Mauritius Island. You have been given a tissue sample in order to perform biochemical analyses and have managed to sequence its cytochrome *c*. The amino

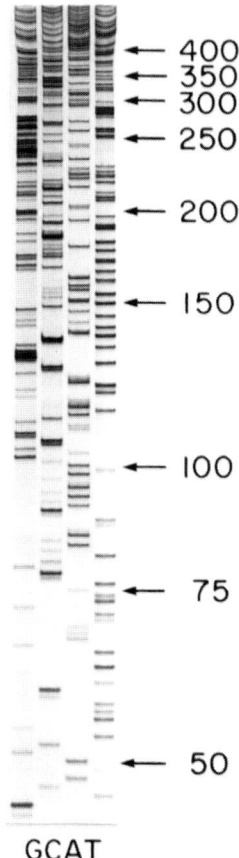

FIGURE 7-41 [Courtesy of Barton Slatko, New England Biolabs Inc., Beverly, Massachusetts.]

acid difference matrix for a number of birds including the dodo is shown here.

Chicken, turkey	0				
Penguin	2	0			
Pigeon	4	4	0		
Pekin duck	3	3	3	0	
Dodo	4	4	2	3	0

(a) Find the phylogenetic tree for these species using the neighbor-joining method. (b) To which of the other birds does the dodo appear most closely related? (c) What additional information do you require to find the root of this tree? Without making further computations, indicate the most likely possibilities.

21. In a tragic accident, the desiccated dodo discussed in Problem 20 has been eaten by a deranged cat. The sequence of the dodo cytochrome *c* that you had determined before the accident led you to suspect that this cytochrome *c* has some unique biochemical properties. To test this hypothesis, you are forced to synthesize dodo cytochrome *c* by chemical means. As with other known avian cytochromes *c,* that of the dodo consists of 104 amino acid residues. In planning a solid phase synthesis, you expect a 99.7% yield for each coupling step and a 99.3% yield for each deblocking step. The cleavage of the completed polypeptide from the resin and the side chain deblocking step should yield an 80% recovery of product. (a) What percentage of the original resin-bound C-terminal amino acid can you expect to form unmodified dodo cytochrome *c* if you synthesize the entire polypeptide in a single run? (b) You have found that dodo cytochrome *c* has a Cys residue at its position 50. What would be your overall yield if you used the native chemical ligation reaction (assume it has a 75% yield) to synthesize dodo cytochrome *c*? Compare this yield with that from Part a and discuss the implications of this comparison for synthesizing long polypeptides.

22. You have manufactured a DNA chip consisting of 4 rows and 10 columns in which the *M*th column contains DNAs of the sequence 5′-GACCTGACGT-3′ but with a different base in the *M*th position for each of the 4 rows (from top to bottom, G, A, T, and C). Draw the appearance of the chip after it is hybridized to fluorescently labeled RNA of sequence (a) 5′-ACGUCAGGUC-3′ and (b) 5′-ACGUCUGGUC-3′.

Three-Dimensional Structures of Proteins

The properties of a protein are largely determined by its three-dimensional structure. One might naively suppose that since proteins are all composed of the same 20 types of amino acid residues, they would be more or less alike in their properties. Indeed, **denatured** (unfolded) proteins have rather similar characteristics, a kind of homogeneous "average" of their randomly dangling side chains. However, the three-dimensional structure of a **native** (physiologically folded) protein is specified by its primary structure, so that it has a unique set of characteristics.

In this chapter, we shall discuss the structural features of proteins, the forces that hold them together, and their hierarchical organization to form complex structures. This will form the basis for understanding the structure–function relationships necessary to comprehend the biochemical roles of proteins. Detailed consideration of the dynamic behavior of proteins and how they fold to their native structures is deferred until Chapter 9.

1 ■ SECONDARY STRUCTURE

A polymer's **secondary structure (2° structure)** is defined as the local conformation of its backbone. For proteins, this has come to mean the specification of regular polypeptide backbone folding patterns: helices, pleated sheets, and turns. However, before we begin our discussion of these basic structural motifs, let us consider the geometrical properties of the peptide group because its understanding is prerequisite to that of any structure containing it.

A. *The Peptide Group*

In the 1930s and 1940s, Linus Pauling and Robert Corey determined the X-ray structures of several amino acids and dipeptides in an effort to elucidate the structural constraints on the conformations of a polypeptide chain. These studies indicated that *the peptide group has a rigid, planar* structure (Fig. 8-1), which, Pauling pointed out, is a consequence of resonance interactions that give the peptide bond an ~40% double-bond character:

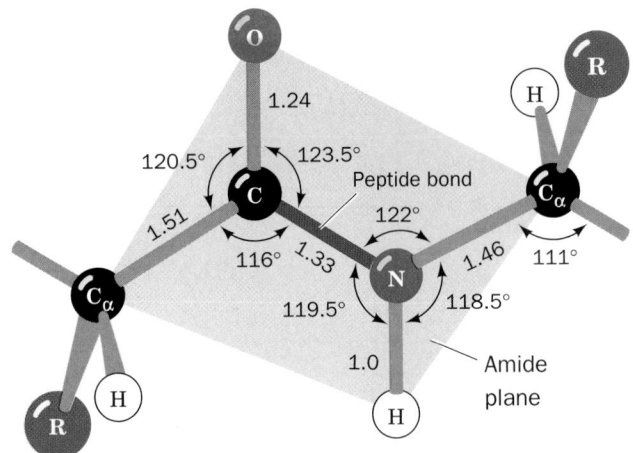

FIGURE 8-1   The trans-peptide group. The standard dimensions (in angstroms, Å, and degrees, °) of this planar group were derived by averaging the corresponding quantities in the X-ray crystal structures of amino acids and peptides. [After Marsh, R.E. and Donohue, J., *Adv. Protein Chem.* **22**, 249 (1967).] 🔊 **See Kinemage Exercise 3-1**

This explanation is supported by the observations that a peptide's C—N bond is 0.13 Å shorter than its N—C$_\alpha$ single bond and that its C=O bond is 0.02 Å longer than that of aldehydes and ketones. The peptide bond's resonance energy has its maximum value, ~85 kJ · mol^{-1}, when the peptide group is planar because its π-bonding overlap is maximized in this conformation. This overlap, and thus the resonance energy, falls to zero as the peptide bond is twisted to 90° out of planarity, thereby accounting for the planar peptide group's rigidity. (The positive charge on the above resonance structure should be taken as a formal charge; quantum mechanical calculations indicate that the peptide N atom, in fact, has a partial negative charge arising from the polarization of the C—N σ bond.)

Peptide groups, with few exceptions, assume the trans conformation: that in which successive C$_\alpha$ atoms are on opposite sides of the peptide bond joining them (Fig. 8-1). This is partly a result of steric interference, which causes the cis conformation (Fig. 8-2) to be ~8 kJ · mol^{-1} less stable than the trans conformation (this energy difference is somewhat less in peptide bonds followed by a Pro residue and, in fact, ~10% of the Pro residues in proteins follow a cis peptide bond, whereas cis peptides are otherwise extremely rare).

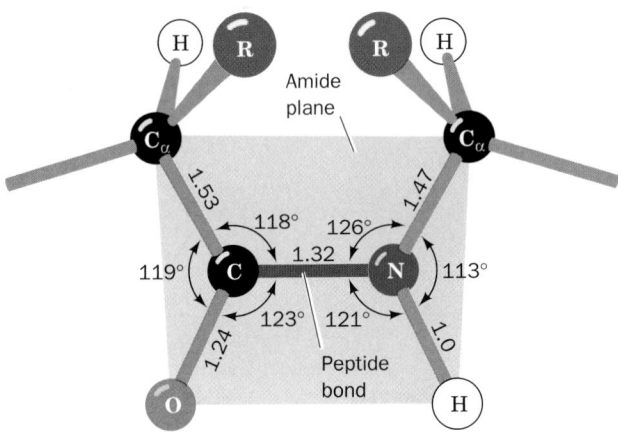

FIGURE 8-2   The cis-peptide group. 🔊 **See Kinemage Exercise 3-1**

fully extended (all-trans) conformation and increase for a clockwise rotation when viewed from C$_\alpha$ (Fig. 8-4).

There are several steric constraints on the torsion angles, φ and ψ, of a polypeptide backbone that limit its conformational range. The electronic structure of a single (σ) bond, such as a C—C bond, is cylindrically symmetrical

a. Polypeptide Backbone Conformations May Be Described by Their Torsion Angles

The above considerations are important because they indicate that *the backbone of a protein is a linked sequence of rigid planar peptide groups* (Fig. 8-3). We can therefore specify a polypeptide's backbone conformation by the **torsion angles** (rotation angles or **dihedral angles**) about the C$_\alpha$—N bond (φ) and the C$_\alpha$—C bond (ψ) of each of its amino acid residues. These angles, φ and ψ, are both defined as 180° when the polypeptide chain is in its planar,

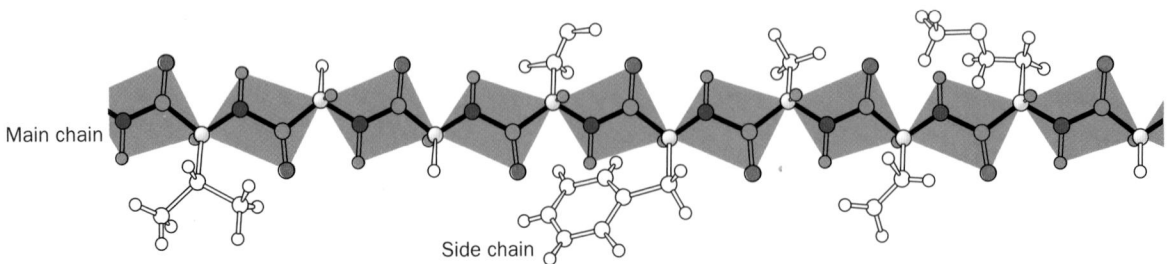

FIGURE 8-3   A polypeptide chain in its fully extended conformation showing the planarity of each of its peptide groups. [Illustration, Irving Geis/Geis Archives Trust. Copyright Howard Hughes Medical Institute. Reproduced with permission.]

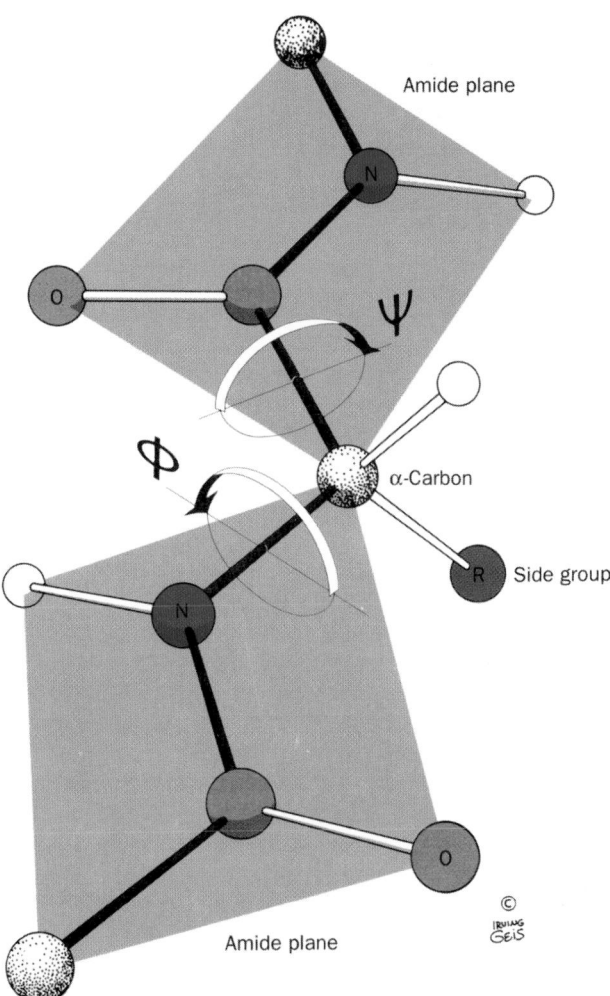

FIGURE 8-4 The torsional degrees of freedom in a peptide unit. The only reasonably free movements are rotations about the C_α—N bond (ϕ) and the C_α—C bond (ψ). The torsion angles are both 180° in the conformation shown and increase, as is indicated, in a clockwise manner when viewed from C_α. [Illustration, Irving Geis/Geis Archives Trust. Copyright Howard Hughes Medical Institute. Reproduced with permission.] ⚫ See Kinemage Exercise 3-1

about its bond axis, so that we might expect such a bond to exhibit free rotation. If this were the case, then in ethane, for example, all torsion angles about the C—C bond would be equally likely. Yet certain conformations in ethane are favored due to quantum mechanical effects arising from the interactions of its molecular orbitals. The **staggered conformation** (Fig. 8-5a; torsion angle = 180°) is ethane's most stable arrangement, whereas the **eclipsed conformation** (Fig. 8-5b; torsion angle = 0°) is least stable. The energy difference between the staggered and eclipsed conformations in ethane is ~12 kJ · mol^{-1}, a quantity that represents an **energy barrier** to free rotation about the C—C single bond. Substituents other than hydrogen exhibit greater steric interference; that is, they increase the size of this energy barrier due to their greater bulk. Indeed, with large substituents, some conformations may be sterically forbidden.

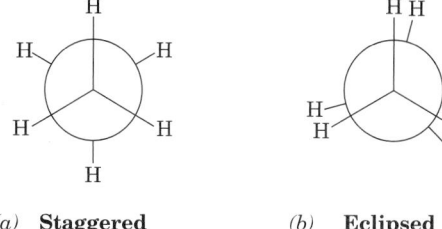

(a) **Staggered** *(b)* **Eclipsed**

FIGURE 8-5 Conformations of ethane. Newman projections indicating the (a) staggered conformation and (b) eclipsed conformation of ethane.

b. Allowed Conformations of Polypeptides Are Indicated by the Ramachandran Diagram

The sterically allowed values of ϕ and ψ can be determined by calculating the distances between the atoms of a tripeptide at all values of ϕ and ψ for the central peptide unit. Sterically forbidden conformations, such as that shown in Fig. 8-6, are those in which any nonbonding interatomic distance is less than its corresponding van der Waals distance. Such information is summarized in a **conformation map** or **Ramachandran diagram** (Fig. 8-7), which was invented by G. N. Ramachandran.

Figure 8-7 indicates that ~75% of the Ramachandran diagram (most combinations of ϕ and ψ) is conformationally inaccessible to a polypeptide chain. The particular regions of the Ramachandran diagram that represent allowed conformations depend on the van der Waals radii chosen to calculate it. But with any realistic set of values, such as that in Table 8-1, *only three small regions of the conformational map are physically accessible to a polypeptide chain.* Nevertheless, as we shall see, all of the common types of regular secondary structures found in proteins fall within allowed regions of the Ramachandran diagram.

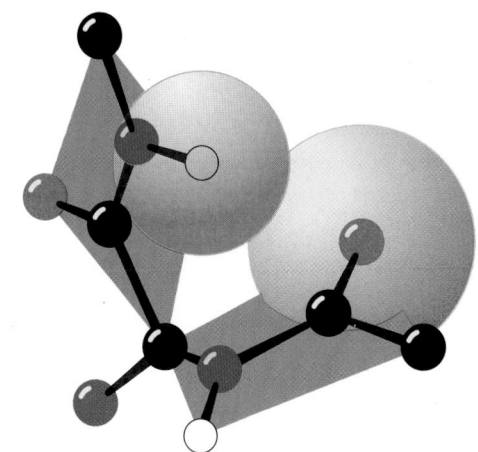

FIGURE 8-6 Steric interference between adjacent residues. The collision between a carbonyl oxygen and the following amide hydrogen prevents the conformation $\phi = -60°$, $\psi = 30°$. [Illustration, Irving Geis/Geis Archives Trust. Copyright Howard Hughes Medical Institute. Reproduced with permission.] ⚫ See Kinemage Exercise 3-1

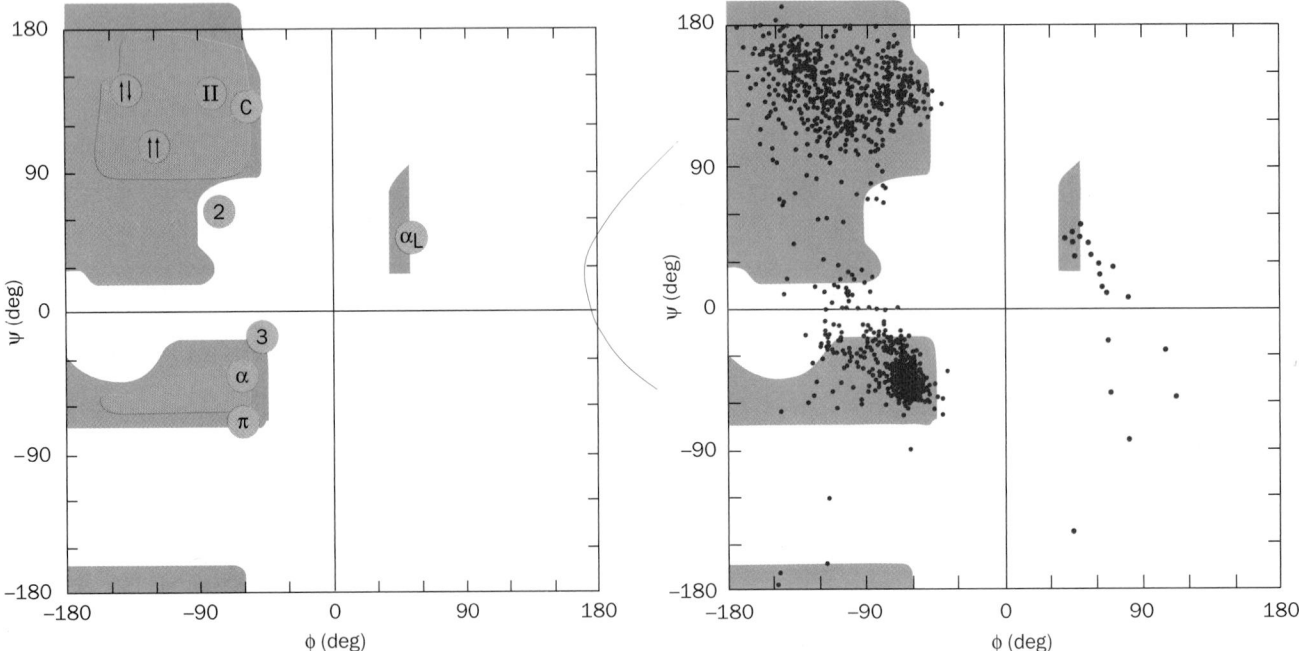

FIGURE 8-7   The Ramachandran diagram. It shows the sterically allowed φ and ψ angles for poly-L-alanine and was calculated using the van der Waals distances in Table 8-1. Regions of "normally allowed" φ and ψ angles are shaded in blue, whereas green-shaded regions correspond to conformations having "outer limit" van der Waals distances. The conformation angles, φ and ψ, of several secondary structures are indicated below:

Secondary Structure	φ (deg)	ψ (deg)
Right-handed α helix (α)	−57	−47
Parallel β pleated sheet (↑↑)	−119	113
Antiparallel β pleated sheet (↑↓)	−139	135
Right-handed 3_{10} helix (3)	−49	−26
Right-handed π helix (π)	−57	−70
2.2_7 ribbon (2)	−78	59
Left-handed polyglycine II and poly-L-proline II helices (II)	−79	150
Collagen (C)	−51	153
Left-handed α helix ($α_L$)	57	47

[After Flory, P.J., *Statistical Mechanics of Chain Molecules,* p. 253, Interscience (1969); *and* IUPAC-IUB Commission on Biochemical Nomenclature, *Biochemistry* **9**, 3475 (1970).]

FIGURE 8-8   Conformation angles in proteins. The conformation angle distribution of all residues but Gly and Pro in 12 precisely determined high-resolution X-ray structures is superimposed on the Ramachandran diagram. [After Richardson, J.S. and Richardson, D.C., *in* Fasman, G.D. (Ed.), *Prediction of Protein Structure and the Principles of Protein Conformation,* p. 6, Plenum Press (1989).]

TABLE 8-1   van der Waals Distances for Interatomic Contacts

Contact Type	Normally Allowed (Å)	Outer Limit (Å)
H ⋯ H	2.0	1.9
H ⋯ O	2.4	2.2
H ⋯ N	2.4	2.2
H ⋯ C	2.4	2.2
O ⋯ O	2.7	2.6
O ⋯ N	2.7	2.6
O ⋯ C	2.8	2.7
N ⋯ N	2.7	2.6
N ⋯ C	2.9	2.8
C ⋯ C	3.0	2.9
C ⋯ CH_2	3.2	3.0
CH_2 ⋯ CH_2	3.2	3.0

Source: Ramachandran, G.N. and Sasisekharan, V., *Adv. Protein Chem.* **23**, 326 (1968).

Indeed, the observed conformational angles of most non-Gly residues in proteins whose X-ray structures have been determined lie in these allowed regions (Fig. 8-8).

Most points that fall in forbidden regions of Fig. 8-8 lie between its two fully allowed areas near ψ = 0. However, these "forbidden" conformations, which arise from the collision of successive amide groups, are allowed if twists of only a few degrees about the peptide bond are permitted. This is not unreasonable since the peptide bond offers little resistance to small deformations from planarity.

Gly, the only residue without a $C_β$ atom, is much less sterically hindered than the other amino acid residues. This is clearly apparent in comparing the Ramachandran diagram for Gly in a polypeptide chain (Fig. 8-9) with that of

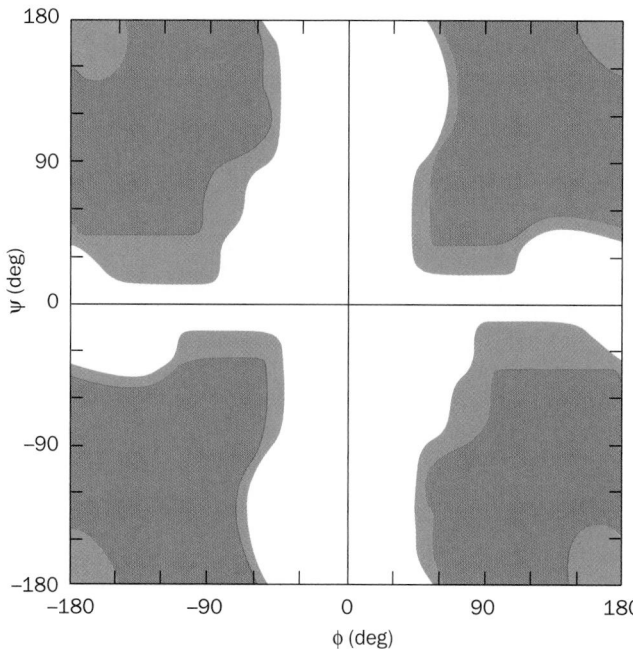

FIGURE 8-9   The Ramachandran diagram of Gly residues in a polypeptide chain. "Normally allowed" regions are shaded in blue, whereas green-shaded regions correspond to "outer limit" van der Waals distances. Gly residues have far greater conformational freedom than do other (bulkier) amino acid residues, as the comparison of this figure with Fig. 8-7 indicates. [After Ramachandran, G.N. and Sasisekharan, V., *Adv. Protein Chem.* **23,** 332 (1968).]

other residues (Fig. 8-7). In fact, Gly often occupies positions where a polypeptide backbone makes a sharp turn which, with any other residue, would be subject to steric interference.

Figure 8-7 was calculated for three consecutive Ala residues. Similar plots for larger residues that are unbranched at C_β, such as Phe, are nearly identical. In Ramachandran diagrams of residues that are branched at C_β, such as Thr, the allowed regions are somewhat smaller than for Ala. The cyclic side chain of Pro limits its ϕ to the range $-60° \pm 25°$, making it, not surprisingly, the most conformationally restricted amino acid residue. The conformations of residues in chains longer than tripeptides are even more restricted than the Ramachandran diagram indicates because a polypeptide chain with all its ϕ and ψ angles allowed nevertheless cannot assume a conformation in which it passes through itself. We shall see, however, that despite the great restrictions that peptide bond planarity and side chain bulk place on the conformations of a polypeptide chain, different unique primary structures have correspondingly unique three-dimensional structures.

B. *Helical Structures*

🔊 **See Guided Exploration 7: Stable Helices in Proteins: The α-Helix**
Helices are the most striking elements of protein 2° structure. If a polypeptide chain is twisted by the same amount about each of its C_α atoms, it assumes a helical conformation. As an alternative to specifying its ϕ and ψ angles, a helix may be characterized by the number, *n,* of peptide units per helical turn and by its **pitch,** *p,* the distance the helix rises along its axis per turn. Several examples of helices are diagrammed in Fig. 8-10. Note that a helix has chi-

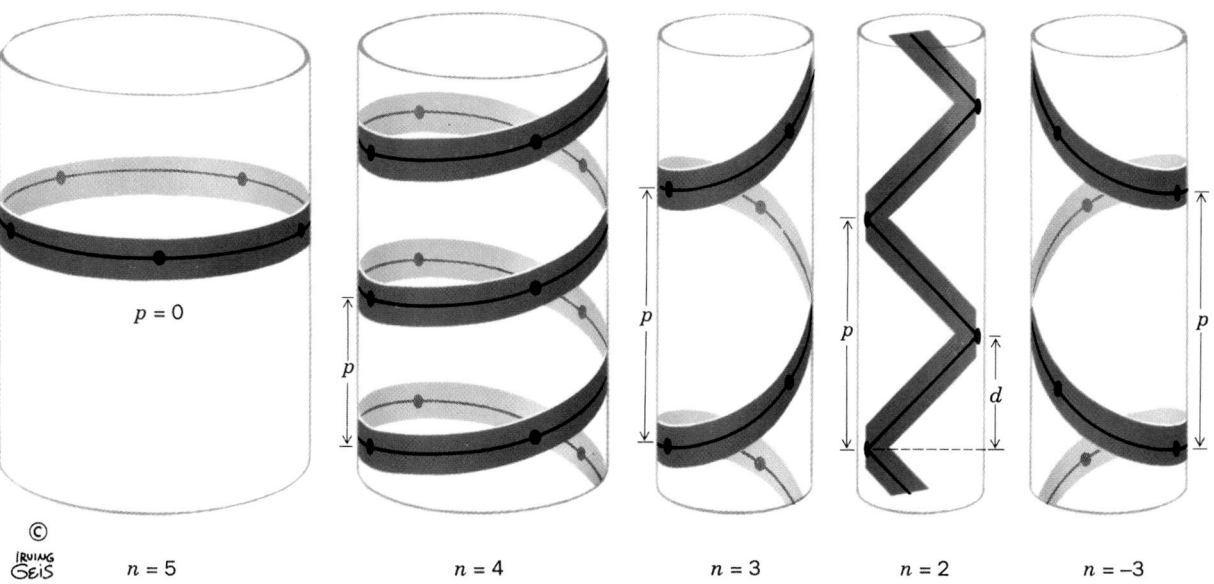

FIGURE 8-10   Examples of helices. These provide definitions of the helical pitch, *p,* the number of repeating units per turn, *n,* and the helical rise per repeating unit, $d = p/n$. Right- and left-handed helices are defined, respectively, as having positive and negative values of *n.* For $n = 2$, the helix degenerates to a nonchiral ribbon. For $p = 0$, the helix degenerates to a closed ring. [Illustration, Irving Geis/Geis Archives Trust. Copyright Howard Hughes Medical Institute. Reproduced with permission.]

A polypeptide helix must, of course, have conformation angles that fall within the allowed regions of the Ramachandran diagram. As we have seen, this greatly limits the possibilities. Furthermore, if a particular conformation is to have more than a transient existence, it must be more than just allowed, it must be stabilized. The "glue" that holds polypeptide helices and other 2° structures together is, in part, hydrogen bonds.

a. The α Helix

*Only one helical polypeptide conformation has simultaneously allowed conformation angles and a favorable hydrogen bonding pattern: the **α helix** (Fig. 8-11), a particularly rigid arrangement of the polypeptide chain.* Its discovery through model building, by Pauling in 1951, ranks as one of the landmarks of structural biochemistry.

For a polypeptide made from L-α-amino acid residues, the α helix is right handed with torsion angles $\phi = -57°$ and $\psi = -47°$, $n = 3.6$ residues per turn, and a pitch of 5.4 Å. (An α helix of D-α-amino acid residues is the mirror image of that made from L-amino acid residues: It is left handed with conformation angles $\phi = +57°$, $\psi = +47°$, and $n = -3.6$ but with the same value of p.)

Figure 8-11 indicates that the hydrogen bonds of an α helix are arranged such that the peptide N—H bond of the nth residue points along the helix toward the peptide C=O group of the $(n-4)$th residue. This results in a strong hydrogen bond that has the nearly optimum N ··· O distance of 2.8 Å. In addition, the core of the α helix is tightly packed; that is, its atoms are in van der Waals contact across the helix, thereby maximizing their association energies (Section 8-4A). The R groups, whose positions, as we saw, are not fully dealt with by the Ramachandran diagram, all project backward (downward in Fig. 8-11) and outward from the helix so as to avoid steric interference with the polypeptide backbone and with each other. Such an arrangement can also be seen in Fig. 8-12. Indeed, a major reason why the left-handed α helix has never been observed (its helical parameters are but mildly forbidden; Fig. 8-7) is that its side chains contact its polypeptide backbone too closely. Note, however, that 1 to 2% of the individual non-Gly residues in proteins assume this conformation (Fig. 8-8).

The α helix is a common secondary structural element of both fibrous and globular proteins. In globular proteins, α helices have an average span of ~12 residues, which corresponds to over three helical turns and a length of 18 Å. However, α helices with as many as 282 residues (78 turns) have been found.

b. Other Polypeptide Helices

Figure 8-13 indicates how hydrogen bonded polypeptide helices may be constructed. The first two, the **2.2₇ ribbon** and the **3₁₀ helix,** are described by the notation, n_m, where n, as before, is the number of residues per helical turn and m is the number of atoms, including H, in the ring that is closed by the hydrogen bond. With this notation, an α helix is a 3.6_{13} helix.

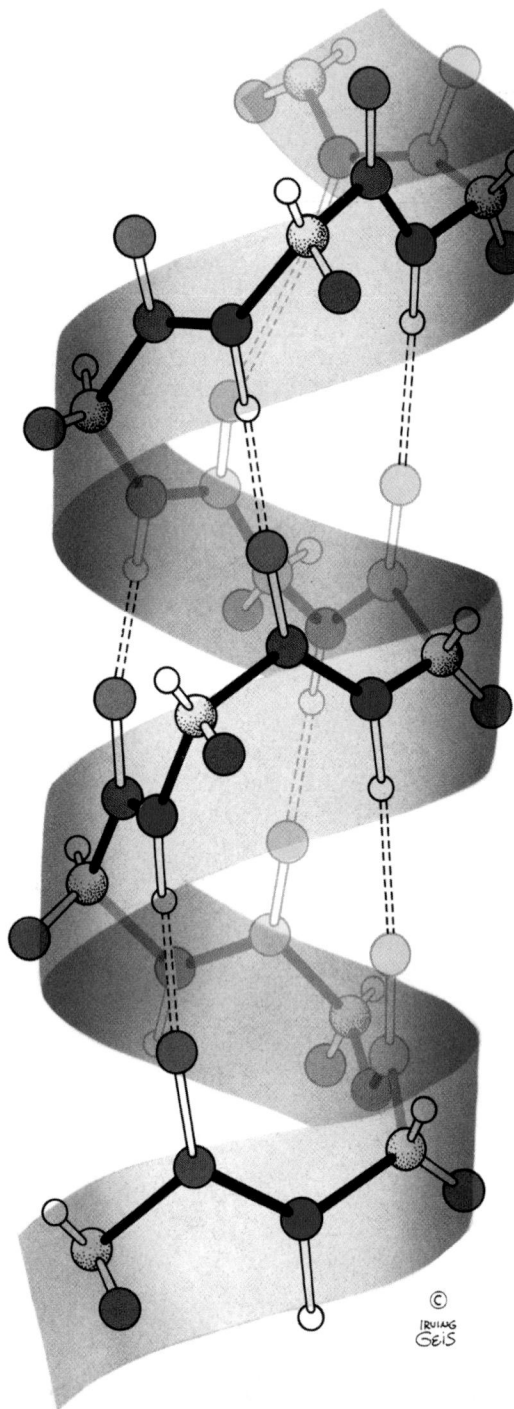

FIGURE 8-11 The right-handed α helix. Hydrogen bonds between the N—H groups and the C=O groups that are four residues back along the polypeptide chain are indicated by dashed lines. [Illustration, Irving Geis/Geis Archives Trust. Copyright Howard Hughes Medical Institute. Reproduced with permission.] 🖉 **See Kinemage Exercise 3-2 and the Animated Figures**

rality; that is, it may be either right handed or left handed (a right-handed helix turns in the direction that the fingers of a right hand curl when its thumb points along the helix axis in the direction that the helix rises). In proteins, moreover, n need not be an integer and, in fact, rarely is.

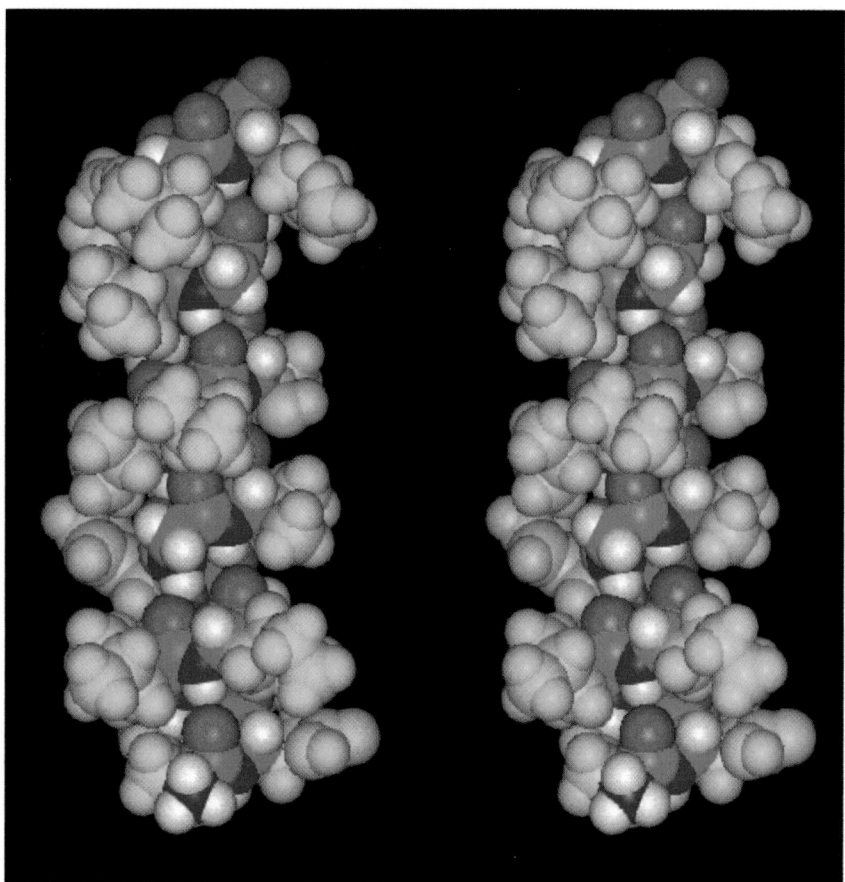

FIGURE 8-12 Stereo, space-filling representation of an α helical segment of sperm whale myoglobin (its E helix) as determined by X-ray crystal structure analysis. In the main chain, carbon atoms are green, nitrogen atoms are blue, oxygen atoms are red, and hydrogen atoms are white. The side chains are yellow. Instructions for viewing stereo diagrams are given in the appendix to this chapter.

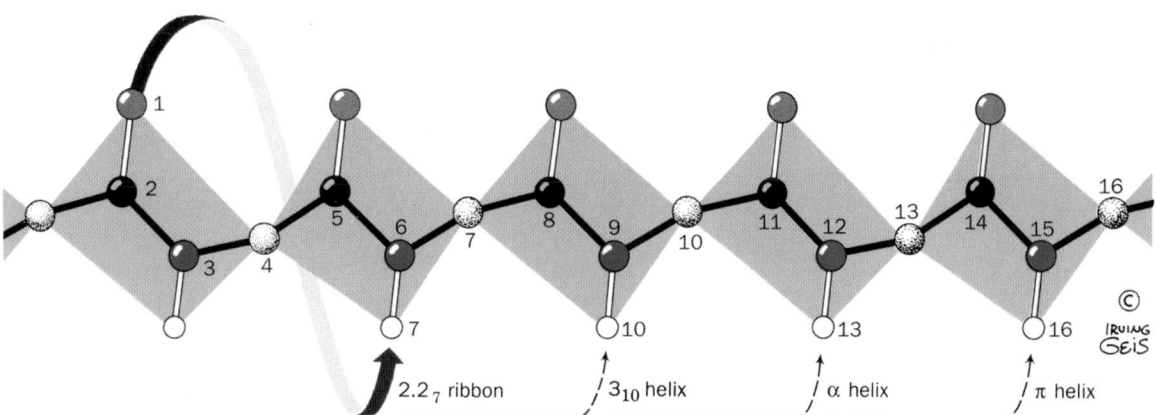

FIGURE 8-13 The hydrogen bonding pattern of several polypeptide helices. In the cases shown, the polypeptide chain is helically wound such that the N—H group on residue n forms a hydrogen bond with the C=O groups on residues $n - 2$, $n - 3$, $n - 4$, or $n - 5$. [Illustration, Irving Geis/Geis Archives Trust. Copyright Howard Hughes Medical Institute. Reproduced with permission.]

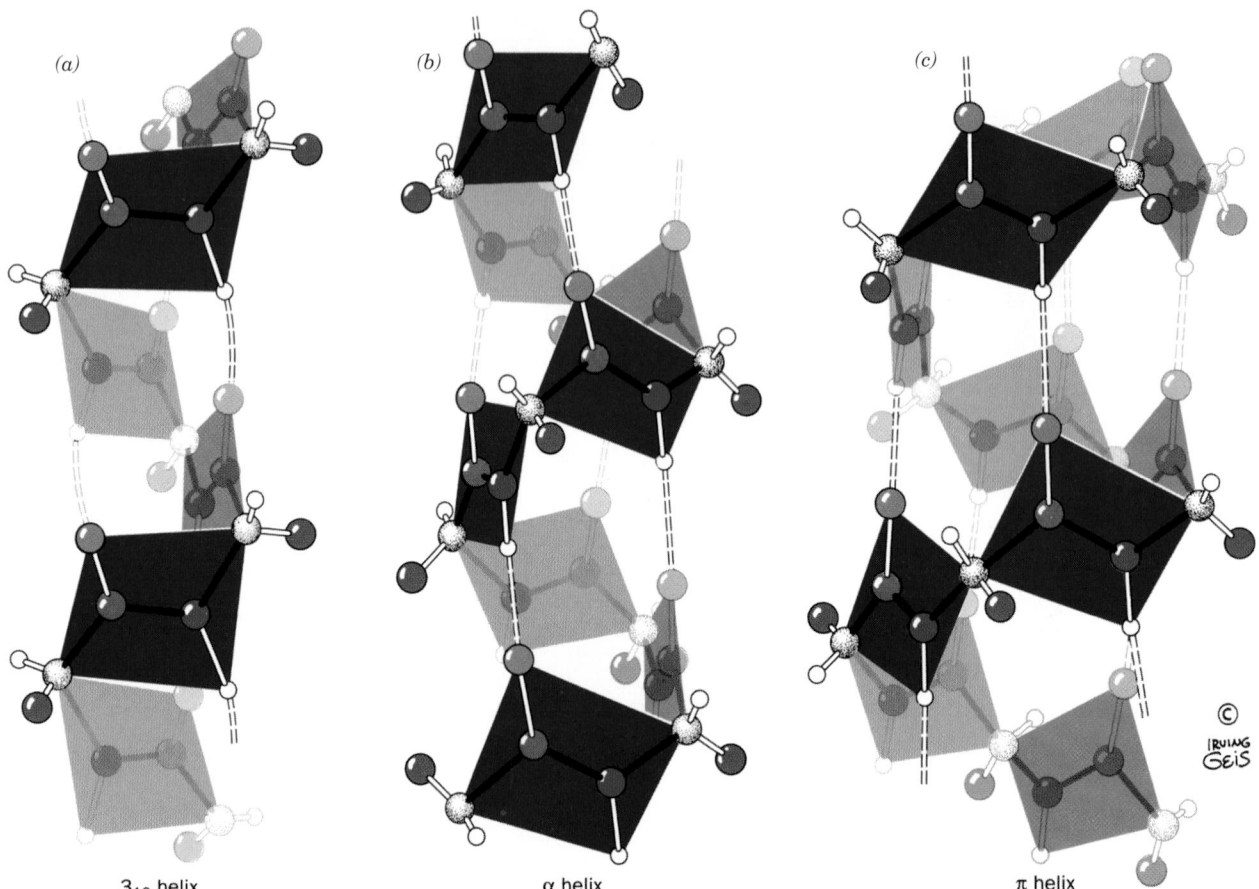

(a) *(b)* *(c)*

3₁₀ helix   α helix   π helix

**FIGURE 8-14 Comparison of the two polypeptide helices that
occasionally occur in proteins with the commonly occurring α
helix.** (*a*) The 3_{10} helix, which has 3.0 peptide units per turn and
a pitch of 6.0 Å, making it thinner and more elongated than the
α helix. (*b*) The α helix, which has 3.6 peptide units per turn
and a pitch of 5.4 Å (also see Fig. 8-11). (*c*) The π helix, which
has 4.4 peptide units per turn and a pitch of 5.2 Å, making it
wider and shorter than the α helix. The peptide planes are
indicated. [Illustration, Irving Geis/Geis Archives Trust. Copyright
Howard Hughes Medical Institute. Reproduced with permission.]

The right-handed 3_{10} helix (Fig. 8-14*a*), which has a pitch
of 6.0 Å, is thinner and rises more steeply than does the α
helix (Fig. 8-14*b*). Its torsion angles place it in a mildly forbidden zone of the Ramachandran diagram that is rather
near the position of the α helix (Fig. 8-7), and its R groups
experience some steric interference. This explains why the
3_{10} helix is only occasionally observed in proteins, and then
mostly in short segments that are frequently distorted from
the ideal 3_{10} conformation. The 3_{10} helix most often occurs
as a single-turn transition between one end of an α helix
and the adjoining portion of a polypeptide chain.

The **π helix** (4.4_{16} helix), which also has a mildly forbidden conformation (Fig. 8-7), has only rarely been observed and then only as segments of longer helices. This is
probably because its comparatively wide and flat conformation (Fig. 8-14*c*) results in an axial hole that is too small
to admit water molecules but too wide to allow van der
Waals associations across the helix axis; this greatly reduces
its stability relative to more closely packed conformations.

The 2.2_7 ribbon, which, as Fig. 8-7 indicates, has strongly
forbidden conformation angles, has never been observed.

Certain synthetic homopolypeptides assume conformations that are models for helices in particular proteins.
Polyproline is unable to assume any common secondary
structure due to the conformational constraints imposed
by its cyclic pyrrolidine side chains. Furthermore, the lack
of a hydrogen substituent on its backbone nitrogen precludes any polyproline conformation from being knit
together by hydrogen bonding. Nevertheless, under the
proper conditions, polyproline precipitates from solution
as a left-handed helix of all-trans peptides that has 3.0
residues per helical turn and a pitch of 9.4 Å (Fig. 8-15).
This rather extended conformation permits the Pro side
chains to avoid each other. Curiously, **polyglycine,** the least
conformationally constrained polypeptide, precipitates
from solution as a helix whose parameters are essentially
identical to those of polyproline, the most conformationally constrained polypeptide (although the polyglycine

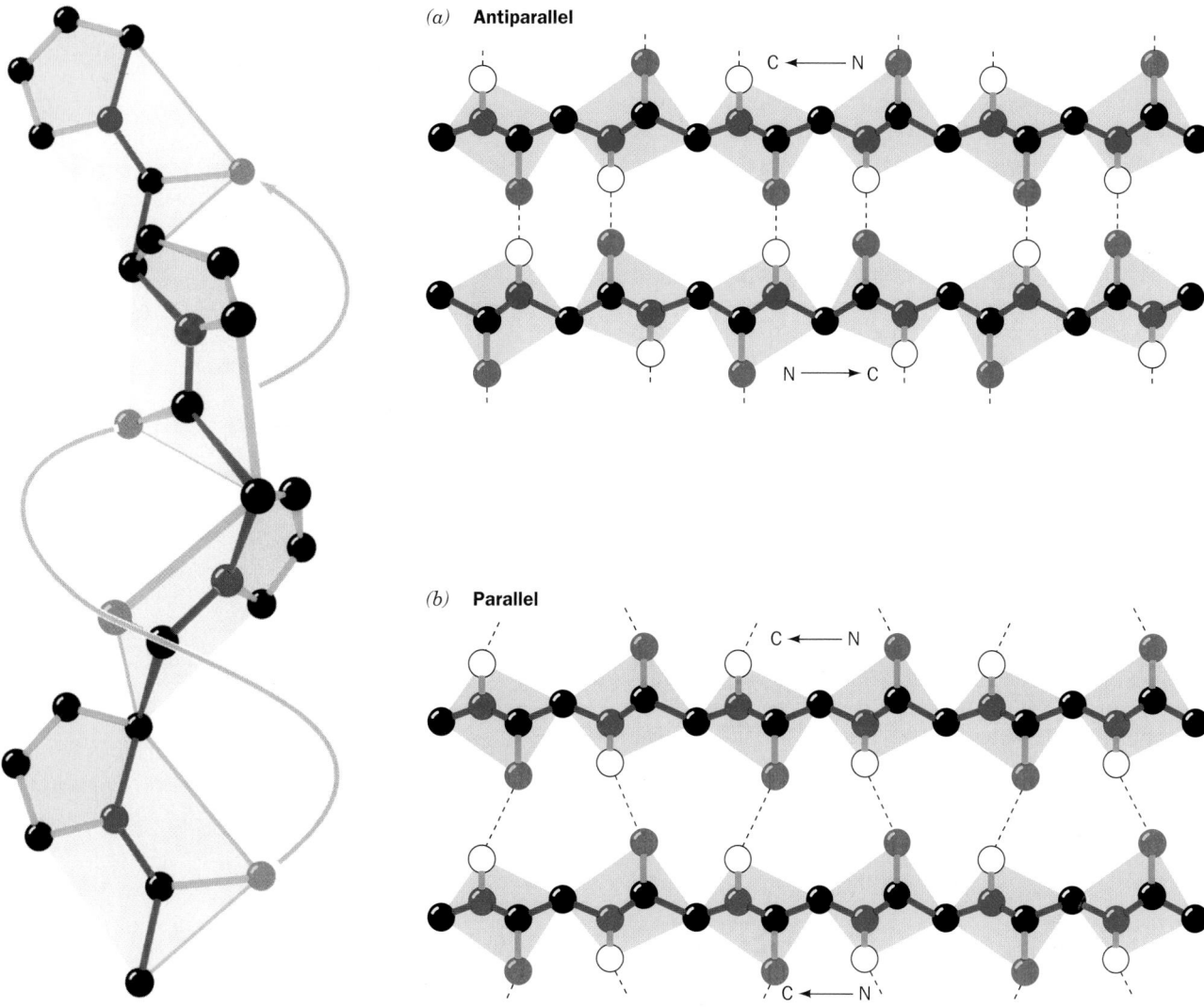

FIGURE 8-15 The polyproline II helix.
Polyglycine forms a nearly identical helix
(polyglycine II). [Illustration, Irving
Geis/Geis Archives Trust. Copyright
Howard Hughes Medical Institute.
Reproduced with permission.]

FIGURE 8-16 β pleated sheets. Hydrogen bonds are indicated by dashed lines and
side chains are omitted for clarity. (*a*) The antiparallel β pleated sheet. (*b*) The
parallel β pleated sheet. [Illustration, Irving Geis/Geis Archives Trust. Copyright
Howard Hughes Medical Institute. Reproduced with permission.] **See Kinemage
Exercise 3-3 and the Animated Figures**

helix may be either right or left handed because Gly is
nonchiral). The structures of the polyglycine and polyproline helices are of biological significance because they form
the basic structural motif of collagen, a structural protein
that contains a remarkably high proportion of both Gly
and Pro (Section 8-2B).

C. *Beta Structures*

**See Guided Exploration 8: Hydrogren Bonding in Sheets and
Guided Exploration 9: Secondary Structures in Proteins** In 1951,
the year that they proposed the α helix, Pauling and Corey
also postulated the existence of a different polypeptide secondary structure, the **β pleated sheet.** As with the α helix,
the β pleated sheet's conformation has repeating φ and ψ

angles that fall in the allowed region of the Ramachandran
diagram (Fig. 8-7) and utilizes the full hydrogen bonding
capacity of the polypeptide backbone. *In β pleated sheets,
however, hydrogen bonding occurs between neighboring
polypeptide chains* rather than within one as in α helices.

β Pleated sheets come in two varieties:

1. The antiparallel β pleated sheet, in which neighboring hydrogen bonded polypeptide chains run in opposite
directions (Fig. 8-16*a*).

2. The parallel β pleated sheet, in which the hydrogen
bonded chains extend in the same direction (Fig. 8-16*b*).

The conformations in which these β structures are optimally hydrogen bonded vary somewhat from that of a fully
extended polypeptide ($\phi = \psi = \pm180°$), as indicated in

Fig. 8-7. They therefore have a rippled or pleated edge-on appearance (Fig. 8-17), which accounts for the appellation "pleated sheet." In this conformation, successive side chains of a polypeptide chain extend to opposite sides of the pleated sheet with a two-residue repeat distance of 7.0 Å.

β Sheets are common structural motifs in proteins. They consist of from 2 to as many as 22 polypeptide strands, the average being 6 strands, which have an aggregate width of ~25 Å. The polypeptide chains in a β sheet are known to be up to 15 residues long, with the average being 6 residues that have a length of ~21 Å. A 6-stranded antiparallel β sheet, for example, occurs in the jack bean protein concanavalin A (Fig. 8-18).

Parallel β sheets of less than five strands are rare. This observation suggests that parallel β sheets are less stable than antiparallel β sheets, possibly because the hydrogen bonds of parallel sheets are distorted in comparison to those of the antiparallel sheets (Fig. 8-16). Mixed parallel–antiparallel β sheets are common but, nevertheless, only ~20% of the strands in β sheets have parallel bonding on one side and antiparallel bonding on the other (vs an expected 50% for the random mixing of strand directions).

The β pleated sheets in globular proteins invariably exhibit a pronounced right-handed twist when viewed along their polypeptide strands (e.g., Fig. 8-19). Such twisted β sheets are important architectural features of globular proteins since β sheets often form their central cores

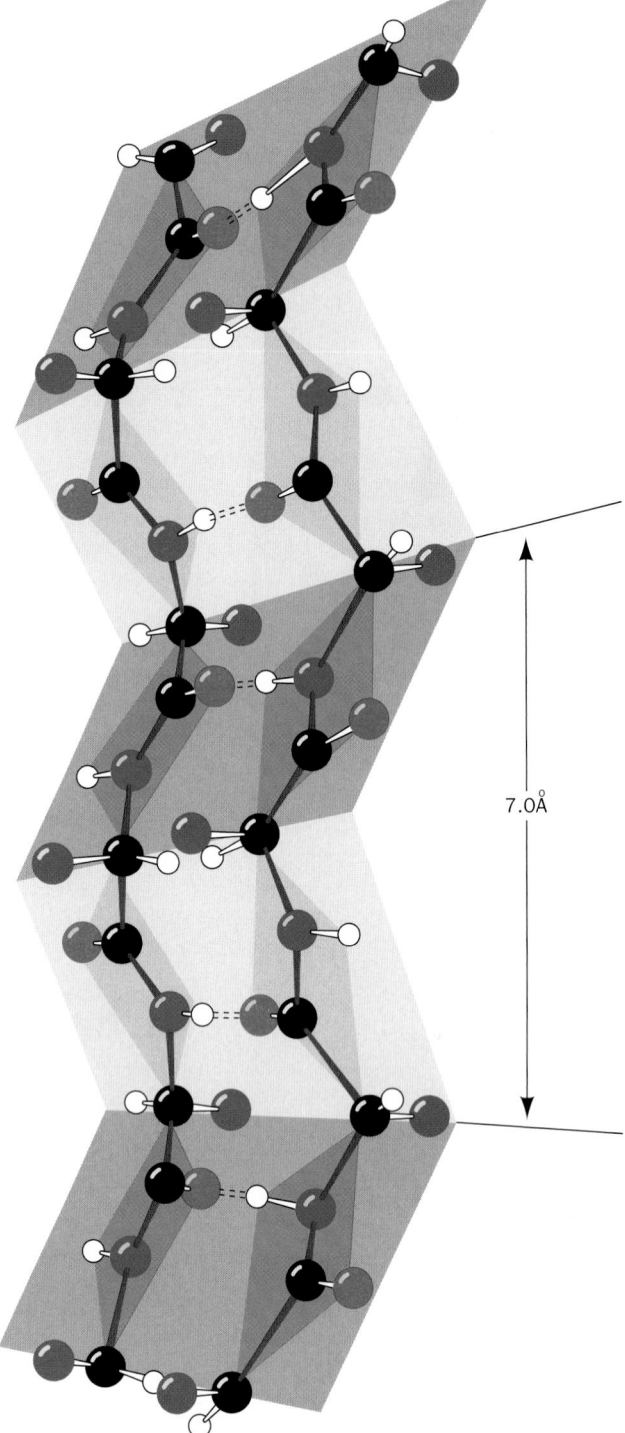

7.0 Å

FIGURE 8-17   A two-stranded β antiparallel pleated sheet drawn to emphasize its pleated appearance. Dashed lines indicate hydrogen bonds. Note that the R groups (*purple balls*) on each polypeptide chain alternately extend to opposite sides of the sheet and that they are in register on adjacent chains. [Illustration, Irving Geis/Geis Archives Trust. Copyright Howard Hughes Medical Institute. Reproduced with permission.] ✍ **See Kinemage Exercise 3-3**

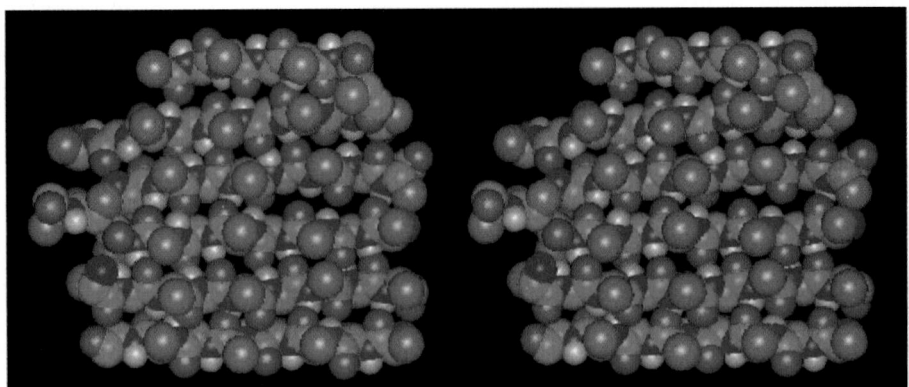

FIGURE 8-18   Stereo, space-filling representation of the 6-stranded antiparallel β pleated sheet in jack bean concanavalin A as determined by X-ray crystal structure analysis. In the main chain, carbon atoms are green, nitrogen atoms are blue, oxygen atoms are red, and hydrogen atoms are white. R groups are represented by large purple balls. Instructions for viewing stereo drawings are given in the appendix to this chapter. ✍ **See Kinemage Exercise 3-3**

(a)

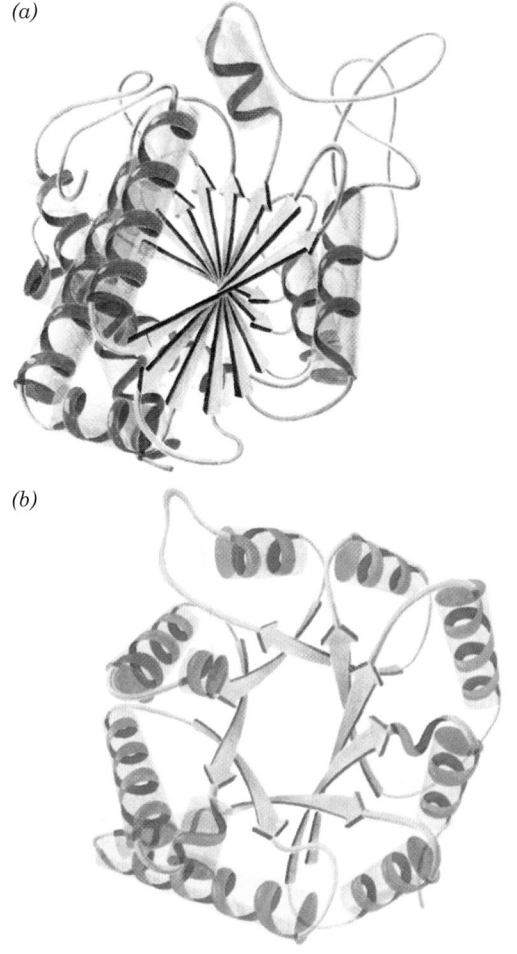

(b)

FIGURE 8-19 Polypeptide chain folding in proteins illustrating the right-handed twist of β sheets. Here the polypeptide backbones are represented by ribbons with α helices shown as coils and the strands of β sheets indicated by arrows pointing toward the C-terminus. Side chains are not shown. (*a*) Bovine carboxypeptidase A, a 307-residue protein, contains an 8-stranded mixed β sheet that forms a saddle-shaped curved surface with a right-handed twist. (*b*) Chicken muscle **triose phosphate isomerase,** a 247-residue enzyme, contains an 8-stranded parallel β sheet that forms a cylindrical structure known as a **β barrel** [here viewed from the top (*left*) and from the side (*right*)]. Note that the crossover connections between successive strands of the β barrel, which each consist predominantly of an α helix, are outside the β barrel and have a right-handed helical sense. [After drawings by Jane Richardson, Duke University. Part *a* based on an X-ray structure by William Lipscomb, Harvard University. PDBid 3CPA. Part *b* based on an X-ray structure by David Phillips, Oxford University, U.K., PDBid 1TIM (for the definition of "PDBid", see Section 8-3C).] **⏃ See the Interactive Exercises**

(Fig. 8-19). Conformational energy calculations indicate that a β sheet's right-handed twist is a consequence of nonbonded interactions between the chiral L-amino acid residues in the sheet's extended polypeptide chains. These interactions tend to give the polypeptide chains a slight right-handed helical twist (Fig. 8-19) which distorts and hence weakens the β sheet's interchain hydrogen bonds. A particular β sheet's geometry is thus the result of a compromise between optimizing the conformational energies of its polypeptide chains and preserving its hydrogen bonds.

The **topology** (connectivity) of the polypeptide strands in a β sheet can be quite complex; the connecting links of these assemblies often consist of long runs of polypeptide chain which frequently contain helices (e.g., Fig. 8-19). The link connecting two consecutive antiparallel strands is topologically equivalent to a simple hairpin turn (Fig. 8-20*a*). However, tandem parallel strands must be linked by a crossover connection that is out of the plane of the β sheet. Such crossover connections almost always have a right-handed helical sense (Fig. 8-20*b*), which is

(a)   *(b)*

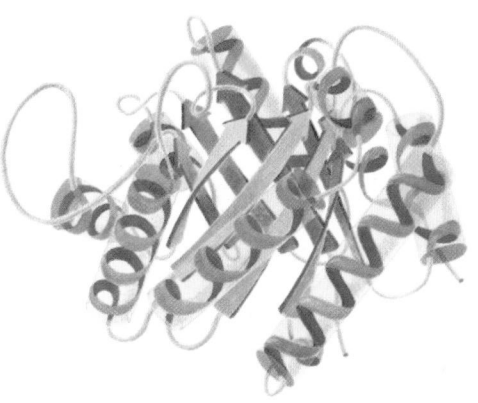

(c)

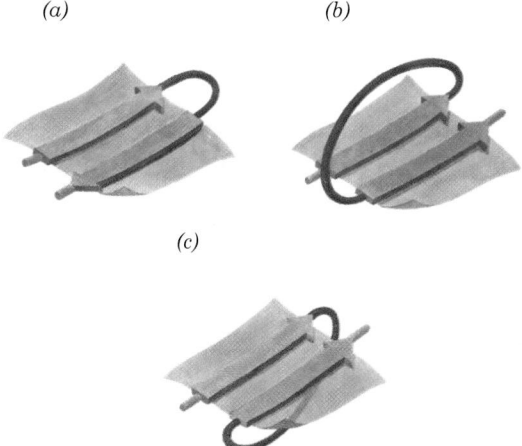

FIGURE 8-20 Connections between adjacent polypeptide strands in β pleated sheets. (*a*) The hairpin connection between antiparallel strands is topologically in the plane of the sheet. (*b*) A right-handed crossover connection between successive strands of a parallel β sheet. Nearly all such crossover connections in proteins have this chirality (see, e.g., Fig. 8-19*b*). (*c*) A left-handed crossover connection between parallel β sheet strands. Connections with this chirality are rare. [After Richardson, J.S., *Adv. Protein Chem.* **34,** 290, 295 (1981).]

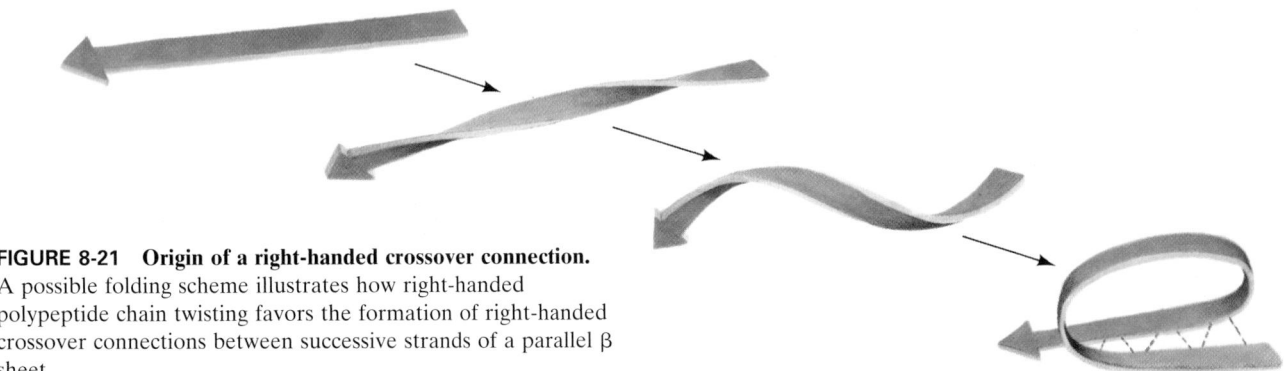

FIGURE 8-21 Origin of a right-handed crossover connection. A possible folding scheme illustrates how right-handed polypeptide chain twisting favors the formation of right-handed crossover connections between successive strands of a parallel β sheet.

thought to better fit the β sheets' inherent right-handed twist (Fig. 8-21).

D. *Nonrepetitive Structures*

Regular secondary structures—helices and β sheets—comprise around half of the average globular protein. The protein's remaining polypeptide segments are said to have a **coil** or **loop conformation.** That is not to say, however, that these nonrepetitive secondary structures are any less ordered than are helices or β sheets; they are simply irregular and hence more difficult to describe. You should therefore not confuse the term coil conformation with the term **random coil,** which refers to the totally disordered and rapidly fluctuating set of conformations assumed by denatured proteins and other polymers in solution.

Globular proteins consist largely of approximately straight runs of secondary structure joined by stretches of polypeptide that abruptly change direction. Such **reverse turns** or **β bends** (so named because they often connect successive strands of antiparallel β sheets) almost always occur at protein surfaces; indeed, they partially define these surfaces. Most reverse turns involve four successive amino acid residues more or less arranged in one of two ways, Type I and Type II, that differ by a 180° flip of the peptide unit linking residues 2 and 3 (Fig. 8-22). Both types of β bends contain a hydrogen bond, although deviations from these ideal conformations often disrupt this hydrogen bond. Type I β bends may be considered to be distorted sections of 3_{10} helix. In Type II β bends, the oxygen atom of residue 2 crowds the C_β atom of residue 3, which is therefore usually Gly. Residue 2 of either type of β bend is often Pro since it can facilely assume the required conformation.

Almost all proteins of >60 residues contain one or more loops of 6 to 16 residues that are not components of helices or β sheets and whose end-to-end distances are <10 Å. Such **Ω loops** (so named because they have the necked-in

(a) **Type I β bend**   *(b)* **Type II β bend**

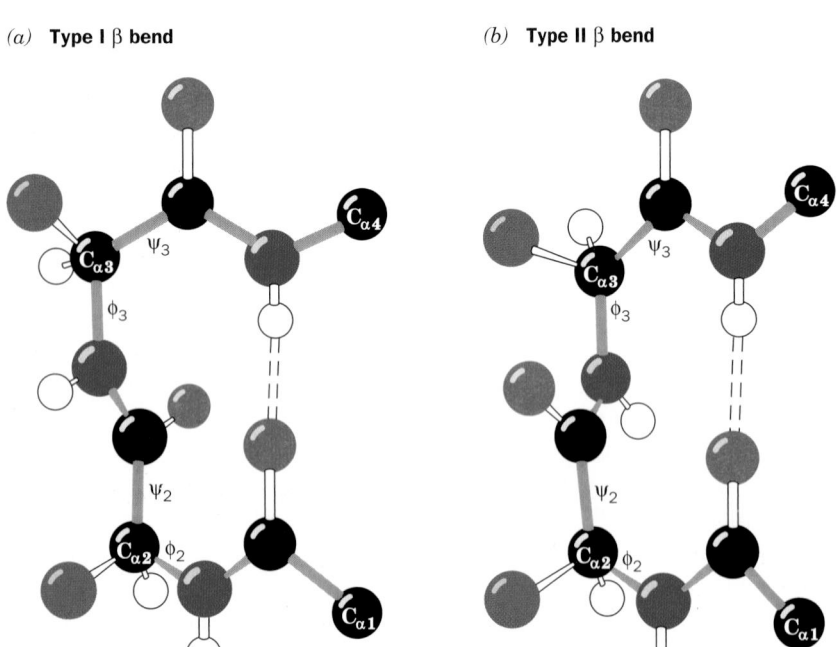

FIGURE 8-22 Reverse turns in polypeptide chains. (*a*) A Type I β bend, which has the following torsion angles:

$$\phi_2 = -60°, \qquad \psi_2 = -30°,$$
$$\phi_3 = -90°, \qquad \psi_2 = 0°.$$

(*b*) A Type II β bend, which has the following torsion angles:

$$\phi_2 = -60°, \qquad \psi_2 = 120°,$$
$$\phi_3 = 90°, \qquad \psi_3 = 0°.$$

Variations from these ideal conformation angles by as much as 30° are common. Hydrogen bonds are represented by dashed lines. [Illustration, Irving Geis/Geis Archives Trust. Copyright Howard Hughes Medical Institute. Reproduced with permission.] *See Kinemage Exercise 3-4*

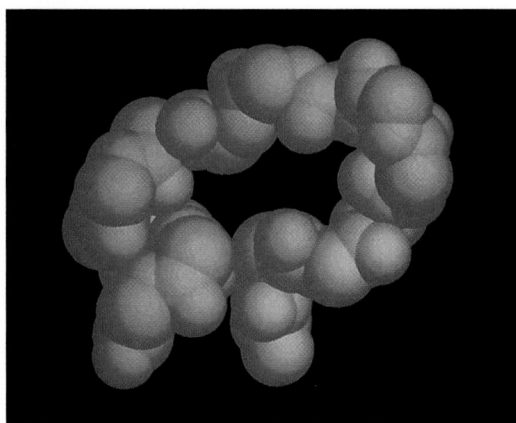

FIGURE 8-23 Space-filling representation of an Ω loop comprising residues 40 to 54 of cytochrome *c.* Only backbone atoms are shown; the addition of side chains would fill in the loop. [Courtesy of George Rose, Washington University School of Medicine.]

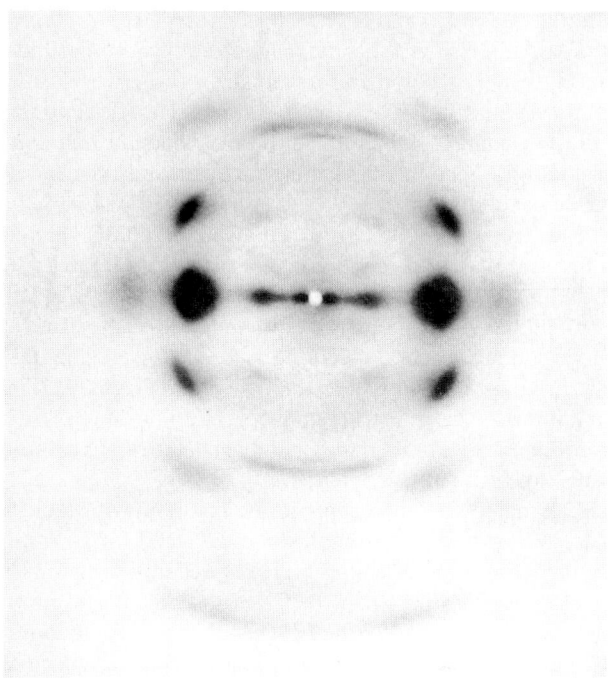

FIGURE 8-24 X-Ray diffraction photograph of a fiber of silkworm *(Bombyx mori)* silk. The photograph was obtained by shining a collimated beam of monochromatic X-rays through the silk fiber and recording the diffracted X-rays on a sheet of photographic film placed behind the fiber. The photograph has only a few spots and thus contains little structural information. [From March, R.E., Corey, R.B., and Pauling, L., *Biochim. Biophys. Acta* **16,** 5 (1955).]

shape of the Greek uppercase letter omega; Fig. 8-23), which may contain reverse turns, are compact globular entities because their side chains tend to fill in their internal cavities. Since Ω loops are almost invariably located on the protein surface, they may have important roles in biological recognition processes.

Many proteins have regions that are truly disordered. Extended, charged surface groups such as Lys side chains or the N- or C-termini of polypeptide chains are good examples: They often wave around in solution because there are few forces to hold them in place (Section 8-4). Sometimes entire peptide chain segments are disordered. Such segments may have functional roles, such as the binding of a specific molecule, so they may be disordered in one state of the protein (molecule absent) and ordered in another (molecule bound). This is one mechanism whereby a protein can interact flexibly with another molecule in the performance of its biological function.

2 ■ FIBROUS PROTEINS

Fibrous proteins are highly elongated molecules whose secondary structures are their dominant structural motifs. Many fibrous proteins, such as those of skin, tendon, and bone, function as structural materials that have a protective, connective, or supportive role in living organisms. Others, such as muscle and ciliary proteins, have motive functions. In this section, we shall discuss structure–function relationships in two common and well-characterized fibrous proteins: keratin and collagen (muscle and ciliary proteins are considered in Section 35-3). The structural simplicity of these proteins relative to those of globular proteins (Section 8-3) makes them particularly amenable to understanding how their structures suit them to their biological roles.

Fibrous molecules rarely crystallize and hence are usually not subject to structural determination by single-

crystal X-ray structure analysis (Section 8-3A). Rather than crystallizing, they associate as fibers in which their long molecular axes are more or less parallel to the fiber axis but in which they lack specific orientation in other directions. The X-ray diffraction pattern of such a fiber, Fig. 8-24, for example, contains little information, far less than would be obtained if the fibrous protein could be made to crystallize. Consequently, the structures of fibrous proteins are not known in great detail. Nevertheless, the original X-ray studies of proteins were carried out in the early 1930s by William Astbury on such easily available protein fibers as wool and tendon. Since the first X-ray crystal structure of a protein was not determined until the late 1950s, these fiber studies constituted the first tentative steps in the elucidation of the structural principles governing proteins and formed much of the experimental basis for Pauling's formulation of the α helix and β pleated sheet.

A. α Keratin—A Helix of Helices

Keratin is a mechanically durable and chemically unreactive protein that occurs in all higher vertebrates. It is the principal component of their horny outer epidermal layer, comprising up to 85% of the cellular protein, and its related appendages such as hair, horns, nails, and feathers. Keratins have been classified as either **α keratins,** which occur in mammals, or **β keratins,** which occur in birds and reptiles. Mammals have over 30 keratin genes, which are expressed in a tissue-specific manner and whose products are classified as belonging to families of relatively acidic (Type I) and relatively basic (Type II) polypeptides. Keratin filaments, which form the intermediate filaments of skin cells (Section 1-2A), must contain at least one member of each type.

Electron microscopic studies indicate that hair, which is composed mainly of α keratin, consists of a hierarchy of structures (Figs. 8-25 and 8-26). A typical hair is ~20 μm in diameter and is constructed from dead cells, each of which contains packed **macrofibrils** (~2000 Å in diameter) that are oriented parallel to the hair fiber (Fig. 8-25). The macrofibrils are constructed from **microfibrils** (~80 Å wide) that are cemented together by an amorphous protein matrix of high sulfur content.

Moving to the molecular level, the X-ray diffraction pattern of α keratin resembles that expected for an α helix (hence the name α keratin). Yet α keratin exhibits a 5.1-Å spacing rather than the 5.4-Å distance corresponding to the pitch of the α helix. This observation, together with a variety of physical and chemical evidence, suggests that *α keratin polypeptides form closely associated pairs of α helices in which each pair is composed of a Type I and a Type II keratin chain twisted in parallel into a left-handed coil (Fig. 8-26a).* The normal 5.4-Å repeat distance of each α helix in the pair is thereby tilted with respect to the axis of this assembly, yielding the observed 5.1-Å spacing. This assembly is said to have a **coiled coil** structure because each α helix axis itself follows a helical path.

The conformation of α keratin's coiled coil is a consequence of its primary structure: The central ~310-residue segment of each polypeptide chain has a heptad (7-residue)

pseudorepeat, *a-b-c-d-e-f-g*, with nonpolar residues predominating at positions *a* and *d*. Since an α helix has 3.6 residues per turn, α keratin's *a* and *d* residues line up on one side of the α helix to form a hydrophobic strip that promotes its lengthwise association with a similar strip on another such α helix (Fig. 8-27; hydrophobic residues, as we shall see in Section 8-4C, have a strong tendency to associate). Indeed, the slight discrepancy between the 3.6 residues per turn of a normal α helix and the ~3.5-residue repeat of a coiled coil's hydrophobic strip is responsible for the coiled coil's coil. The resulting 18° inclination of the α helices relative to one another permits the helical ridges formed by the side chains on one helix to fit into the grooves between these ridges on the other helix, thereby greatly increasing their favorable interactions. Coiled coils, as we shall see, are common components of globular proteins as well as of other fibrous proteins.

The higher order substructure of α keratin is poorly understood. The N- and C-terminal portions of each polypeptide probably have a flexible conformation and facilitate the assembly of the coiled coils into ~30-Å-wide protofilaments. These are thought to consist of two staggered antiparallel rows of head-to-tail aligned coiled coils (Fig. 8-26b). Two such protofilaments are thought to comprise a ~50-Å-wide protofibril, four of which, in turn, coil around each other to form a microfibril (Fig. 8-26c).

α Keratin is rich in Cys residues, which form disulfide bonds that cross-link adjacent polypeptide chains. This

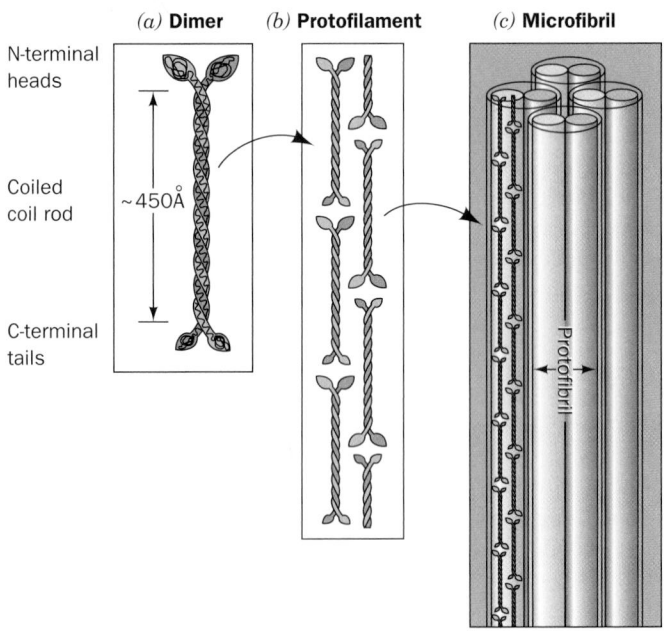

FIGURE 8-26   The structure of α keratin. (*a*) The central ~310 residues of one polypeptide chain each of Types I and II α keratins associate in a dimeric coiled coil. The conformations of the polypeptides' globular N- and C-terminal domains are unknown. (*b*) Protofilaments are formed from two staggered and antiparallel rows of associated head-to-tail coiled coils. (*c*) The protofilaments dimerize to form a protofibril, four of which form a microfibril. The structures of these latter assemblies are poorly characterized but are thought to form helical arrays.

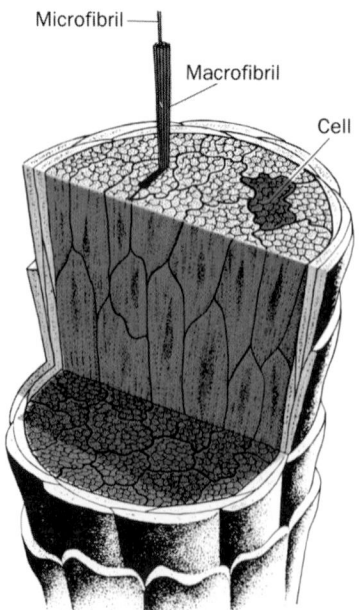

FIGURE 8-25   **The macroscopic organization of hair.**
[Illustration, Irving Geis/Geis Archives Trust. Copyright Howard Hughes Medical Institute. Reproduced with permission.]

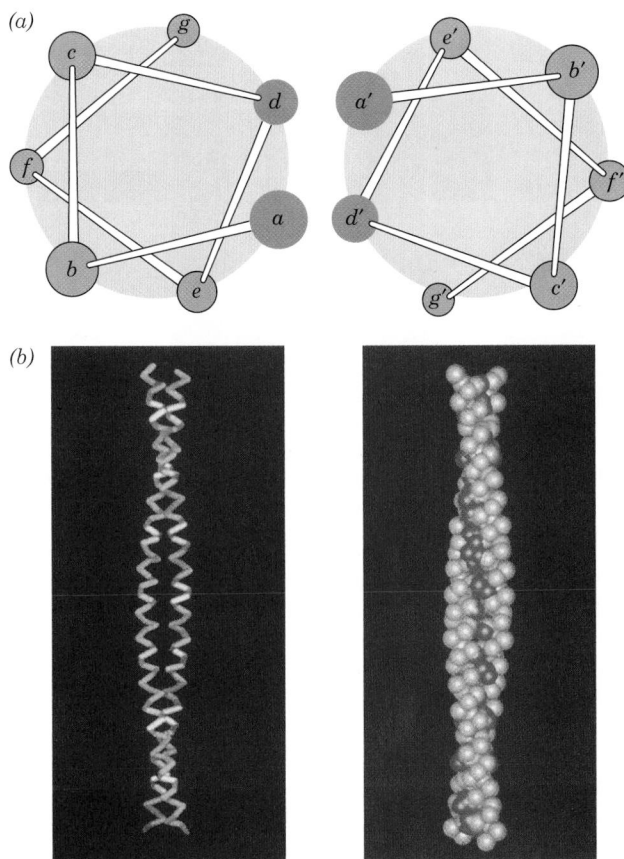

force is relaxed. After some of its disulfide bonds have been cleaved, however, an α keratin fiber can be stretched to over twice its original length by the application of moist heat. In this process, as X-ray analysis indicates, the α helical structure extends with concomitant rearrangement of its hydrogen bonds to form a β pleated sheet. β Keratin, such as that of feathers, exhibits a similar X-ray pattern in its native state (hence the name β sheet).

a. Keratin Defects Result in a Loss of Skin Integrity

The inherited skin diseases **epidermolysis bullosa simplex (EBS)** and **epidermolytic hyperkeratosis (EHK)** are characterized by skin blistering arising from the rupture of epidermal basal cells (Fig. 1-14*d*) and suprabasal cells, respectively, as caused by mechanical stresses that normally would be harmless. Symptomatic variations in these conditions range from severely incapacitating, particularly in early childhood, to barely noticeable. In families afflicted with EBS, sequence abnormalities may be present in either keratin 14 or keratin 5, the dominant Types I and II keratins in basal skin cells. EHK is similarly caused by defects in keratins 1 or 10, the dominant Types I and II keratins in suprabasal cells (which arise through the differentiation of basal cells, a process in which the synthesis of keratins 14 and 5 is switched off and that of keratins 1 and 10 is turned on). These defects evidently interfere with normal filament formation, thereby demonstrating the function of the keratin cytoskeleton in maintaining the mechanical integrity of the skin.

B. *Collagen—A Triple Helical Cable*

Collagen (Greek: *kolla,* glue) occurs in all multicellular animals and is the most abundant protein of vertebrates. It is an extracellular protein that is organized into insoluble fibers of great tensile strength. This suits collagen to its role as the major stress-bearing component of **connective tissues** such as bone, teeth, cartilage, tendon, ligament, and the fibrous matrices of skin and blood vessels. Collagen occurs in virtually every tissue.

Mammals have at least 33 genetically distinct polypeptide chains comprising at least 20 distinct collagen types that occur in different tissues of the same individual. The most prominent of these are listed in Table 8-2. A single molecule of Type I collagen is composed of three polypeptide chains with an aggregate molecular mass of ~285 kD. It has a rodlike shape with a length of ~3000 Å and a width of ~14 Å.

FIGURE 8-27 **The two-stranded coiled coil.** (*a*) View down the coil axis showing the interactions between the nonpolar edges of the α helices. The α helices have the pseudorepeating heptameric sequence *a-b-c-d-e-f-g* in which residues *a* and *d* are predominantly nonpolar. [After McLachlan, A.D. and Stewart, M., *J. Mol. Biol.* **98**, 295 (1975).] (*b*) Side view of a coiled coil in the muscle protein tropomyosin (Section 35-3A) in which the polypeptide backbone is represented in skeletal (*left*) and space-filling (*right*) forms. Note the interlocking of the contacting nonpolar side chains (shown as red spheres) in the space-filling model. [Courtesy of Carolyn Cohen, Brandeis University.] **See Kinemage Exercises 4-1 and 4-2**

accounts for α keratin's insolubility and resistance to stretching, two of its most important biological properties. The α keratins are classified as "hard" or "soft" according to whether they have a high or low sulfur content. Hard keratins, such as those of hair, horn, and nail, are less pliable than soft keratins, such as those of skin and callus, because the disulfide bonds resist any forces tending to deform them. The disulfide bonds can be reductively cleaved with mercaptans (Section 7-1B). Hair so treated can be curled and set in a "permanent wave" by application of an oxidizing agent which reestablishes the disulfide bonds in the new "curled" conformation. Although the insolubility of α keratin prevents most animals from digesting it, the clothes moth larva, which has a high concentration of mercaptans in its digestive tract, can do so to the chagrin of owners of woolen clothing.

The springiness of hair and wool fibers is a consequence of the coiled coil's tendency to untwist when stretched and to recover its original conformation when the external

TABLE 8-2 **The Most Abundant Types of Collagen**

Type	Chain Composition	Distribution
I	[α1(I)]₂α2(I)	Skin, bone, tendon, blood vessels, cornea
II	[α1(II)]₃	Cartilage, intervertebral disk
III	[α1(III)]₃	Blood vessels, fetal skin

Source: Eyre, D.R., *Science* **207**, 1316 (1980).

*Collagen has a distinctive amino acid composition: Nearly one-third of its residues are Gly; another 15 to 30% of them are Pro and **4-hydroxyprolyl (Hyp)** residues:*

**4-Hydroxyprolyl residue
(Hyp)**   **3-Hydroxyprolyl residue**

5-Hydroxylysyl residue (Hyl)

3-Hydroxyprolyl and **5-hydroxylysyl (Hyl)** residues also occur in collagen but in smaller amounts. Radioactive labeling experiments have established that these nonstandard hydroxylated amino acids are not incorporated into collagen during polypeptide synthesis: If ^{14}C-labeled 4-hydroxyproline is administered to a rat, the collagen synthesized is not radioactive, whereas radioactive collagen is produced if the rat is fed ^{14}C-labeled proline. The hydroxylated residues appear after the collagen polypeptides are synthesized, when certain Pro residues are converted to Hyp in a reaction catalyzed by the enzyme **prolyl hydroxylase.**

Hyp confers stability on collagen, probably through intramolecular hydrogen bonds that involve bridging water molecules. If, for example, collagen is synthesized under conditions that inactivate prolyl hydroxylase, it loses its native conformation (denatures) at 24°C, whereas normal collagen denatures at 39°C (heat-denatured collagen is known as **gelatin**). Prolyl hydroxylase requires **ascorbic acid (vitamin C)**

Ascorbic acid (vitamin C)

to maintain its enzymatic activity. In the vitamin C deficiency disease **scurvy,** the collagen synthesized cannot

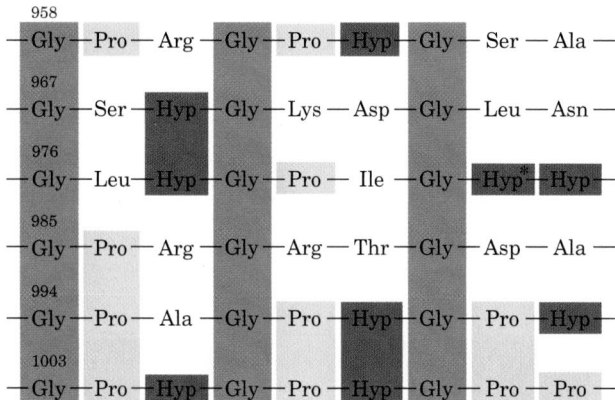

FIGURE 8-28   The amino acid sequence at the C-terminal end of the triple helical region of the bovine α1(I) collagen chain. Note the repeating triplets Gly-X-Y, where X is often Pro and Y is often Hyp. Here Gly is shaded in purple, Pro in tan, and Hyp and Hyp* (3-hydroxyPro) in brown. [From Bornstein, P. and Traub, W., *in* Neurath, H. and Hill, R.L. (Eds.), *The Proteins* (3rd ed.), Vol. 4, p. 483, Academic Press (1979).]

form fibers properly. This results in the skin lesions, blood vessel fragility, and poor wound healing that are symptomatic of this ultimately fatal vitamin deficiency disease.

a. Collagen Has a Triple Helical Structure

The amino acid sequence of bovine collagen α1(I), which is similar to that of other collagens, consists of monotonously repeating triplets of sequence Gly-X-Y over a continuous 1011-residue stretch of its 1042-residue polypeptide chain (Fig. 8-28). Here X is often Pro and Y is often Hyp. The restriction of Hyp to the Y position stems from the specificity of prolyl hydroxylase. Hyl is similarly restricted to the Y position.

The high Gly, Pro, and Hyp content of collagen suggests that its polypeptide backbone conformation resembles those of the polyglycine II and polyproline II helices (Fig. 8-15). X-Ray fiber diffraction and model building studies by Alexander Rich and Francis Crick led them to propose, in 1955, that *collagen's three polypeptide chains, which individually resemble polyproline II helices, are parallel and wind around each other with a gentle, right-handed, rope-like twist to form a triple helical structure (Fig. 8-29).* It was not until 1994, however, that Helen Berman and Barbara Brodsky confirmed this model through their X-ray crystal structure determination of the collagenlike polypeptide (Pro-Hyp-Gly)$_{10}$ in which the fifth Gly is replaced by Ala (Fig. 8-30a). In this structure, every third residue of each polypeptide chain passes through the center of the triple helix, which is so crowded that only a Gly side chain can fit there (Fig. 8-30b). This crowding explains the absolute requirement for a Gly at every third position of a collagen polypeptide chain (Fig. 8-28). It also requires that the three polypeptide chains be staggered so that the Gly, X, and Y residues from the three chains occur at similar levels (Fig.

FIGURE 8-29 The triple helix of collagen. This diagram indicates how the left-handed polypeptide helices are twisted together to form a right-handed superhelical structure. Ropes and cables are similarly constructed from hierarchies of fiber bundles that are alternately twisted in opposite directions. An individual collagen polypeptide helix has 3.3 residues per turn and a pitch of 10.0 Å (in contrast to polyproline II's 3.0 residues per turn and pitch of 9.4 Å; Fig. 8-15). The collagen triple helix has 10 Gly-X-Y units per turn and a pitch of 86.1 Å. [Illustration, Irving Geis/Geis Archives Trust. Copyright Howard Hughes Medical Institute. Reproduced with permission.]

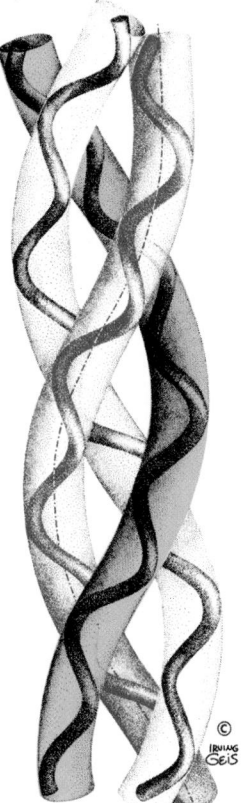

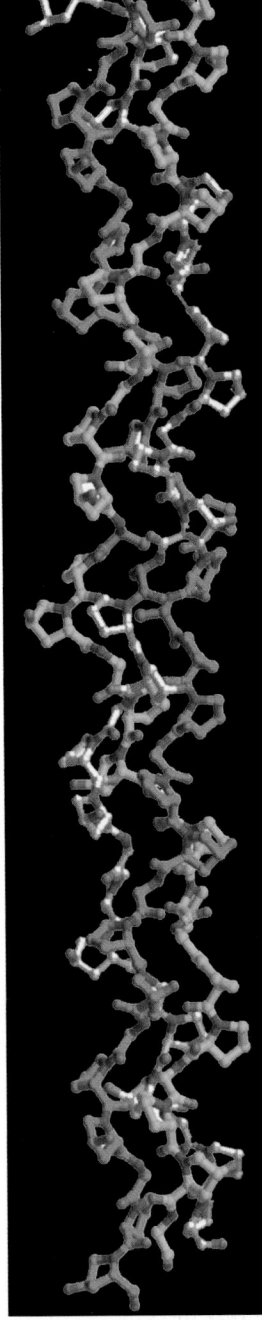

FIGURE 8-30 X-Ray structure of the triple helical collagen model peptide (Pro-Hyp-Gly)$_{10}$ in which the fifth Gly is replaced by Ala. (*a*) A ball-and-stick representation of the triple helix viewed with its helix axis vertical and its N-termini above. The C atoms of the three different polypeptide chains are colored gold, magenta, and white, except those of the Ala residues, which are green. The N and O atoms on all chains are blue and red. Compare this structure with Fig. 8-29. (*b*) A view approximately along the helix axis from the N-termini showing the interchain hydrogen bonding associations in a Gly-containing region of the triple helix. Three consecutive residues from each chain are shown in ball-and-stick form with C green, N blue, and O red. Hydrogen bonds are represented by dashed white lines from Gly N atoms to Pro O atoms in adjacent chains. The backbone atoms of the central residue in each chain are enclosed by dot surfaces drawn at their van der Waals radii. Note the close packing of the atoms along the helix axis. The substitution of a centrally located Gly C$_\alpha$ atom (CH$_2$ group) by any other residue would distort the triple helix. [Parts *a* and *b* are based on an X-ray structure by Helen Berman, Rutgers University, and Barbara Brodsky, UMDNJ–Robert Wood Johnson Medical School. PDBid 1CAG (for the definition of "PDBid", see Section 8-3C).] 🔗 **See Kinemage Exercises 4-3 and 4-4**

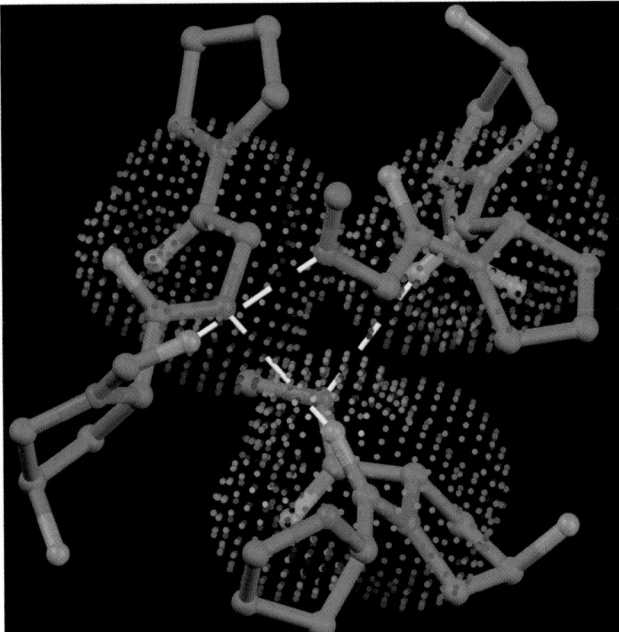

(a)   *(b)*

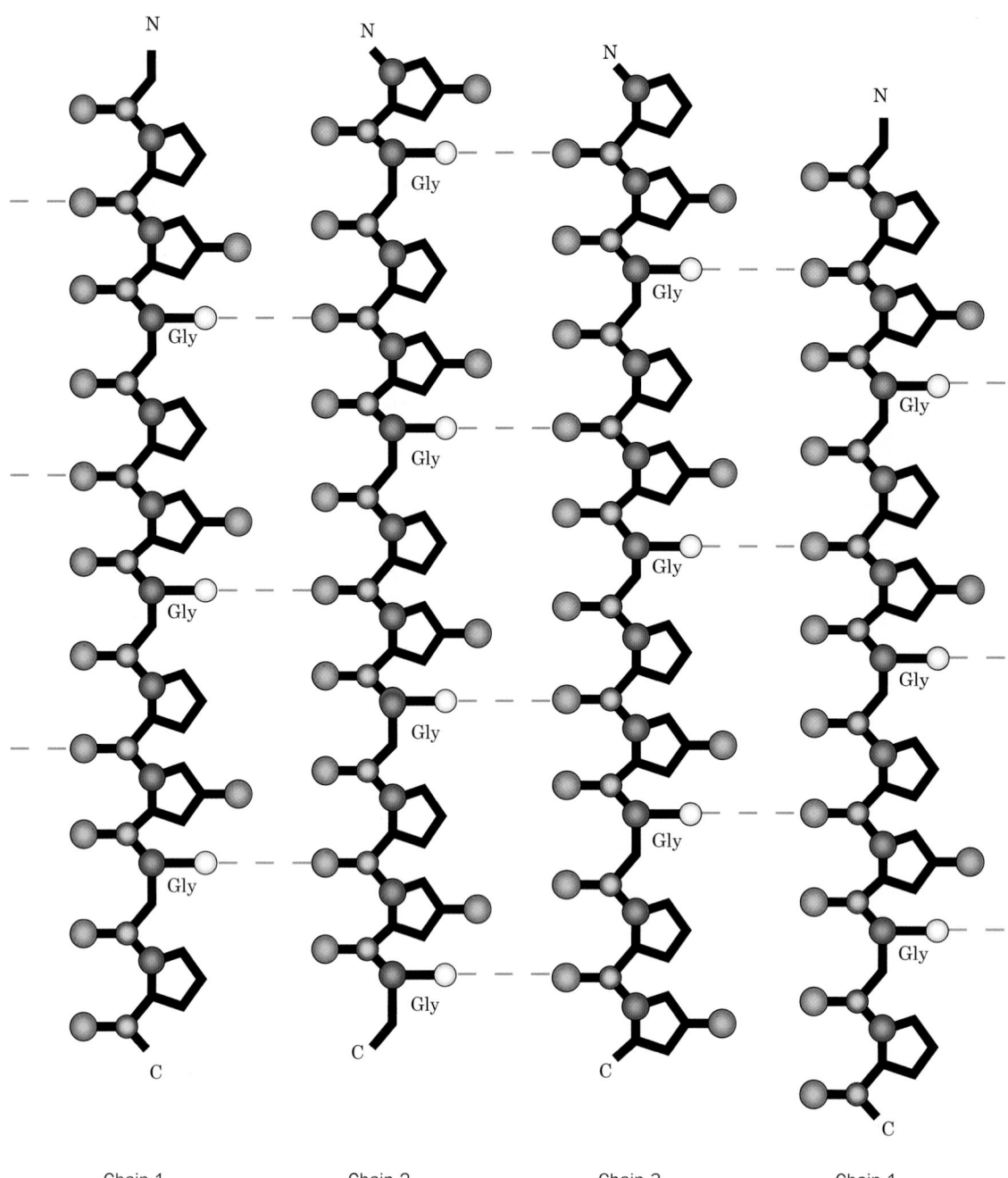

Chain 1   Chain 2   Chain 3   Chain 1

FIGURE 8-30   (*continued*)   (*c*) A schematic diagram showing the interchain hydrogen bonding (*dashed lines*) in the Gly-containing regions of the triple helix. This is a cylindrical projection with Chain 1 repeated on the right for clarity. Note that the three chains are each vertically staggered by one residue so that a Gly, a Pro, and a Hyp from the three different chains occur on the same level. [Part *c* is after Bella, J., Eaton, M., Brodsky, B., and Berman, H.M., *Science* **266,** 78 (1994).]

8-30*c*). The staggered peptide groups are oriented such that the N—H of each Gly makes a strong hydrogen bond with the carbonyl oxygen of an X (Pro) residue on a neighboring chain. The bulky and relatively inflexible Pro and Hyp residues confer rigidity on the entire assembly.

Collagen's well-packed, rigid, triple helical structure is responsible for its characteristic tensile strength. As with the twisted fibers of a rope, the extended and twisted polypeptide chains of collagen convert a longitudinal tensional force to a more easily supported lateral compressional force on the almost incompressible triple helix. This occurs because the oppositely twisted directions of collagen's polypeptide chains and triple helix (Fig. 8-29) prevent the twists from being pulled out under tension (note that successive levels of fiber bundles in ropes and cables are like-wise oppositely twisted). The successive helical hierarchies in other fibrous proteins exhibit similar alternations of twist directions, for example, keratin (Section 8-2A) and muscle (Section 35-3A).

b. Collagen Is Organized into Fibrils

Types I, II, III, V, and XI collagens form distinctive banded fibrils (Fig. 8-31) that are mostly, if not entirely, composed of several different types of collagens. These fib-rils have a periodicity of 680 Å and a diameter of 100 to 2000 Å depending on the types of collagen they contain and their tissue of origin (the other collagen types form different sorts of aggregates such as networks; we will not discuss them further). X-Ray fiber diffraction studies re-veal that the molecules in fibrils of Type I collagen are packed in a hexagonal array. Computerized model build-ing studies further indicate that these collagen molecules are precisely staggered parallel to the fibril axis (Fig. 8-32). The darker portions of the banded structures correspond to the 400-Å "holes" on the surface of the fibril between head-to-tail aligned collagen molecules. Structural and en-ergetic considerations suggest that the conformations of individual collagen molecules, much like those of individ-ual α helices and β sheets, are but marginally stable (Section 8-4). The driving force for the assembly of colla-gen molecules into a fibril is apparently provided by the added hydrophobic interactions within the fibrils in a man-ner analogous to the packing of secondary structural ele-ments to form a globular protein (Section 8-3B).

Collagen contains covalently attached carbohydrates in amounts that range from ~0.4 to 12% by weight, depend-ing on the collagen's tissue of origin. The carbohydrates, which consist mostly of glucose, galactose, and their dis-accharides, are covalently attached to collagen at its Hyl residues by specific enzymes:

Galactose

CH₂OH

HO

Hydroxylysine residue

Glucose

Although the function of carbohydrates in collagen is un-known, the observation that they are located in the "hole" regions of the collagen fibrils suggests that they are in-volved in directing fibril assembly.

FIGURE 8-31 **Electron micrograph of collagen fibrils from skin.** [Courtesy of Jerome Gross, Massachusetts General Hospital.]

Collagen molecule

Packing of molecules

Hole zone ——— 0.6D   Overlap zone 0.4D

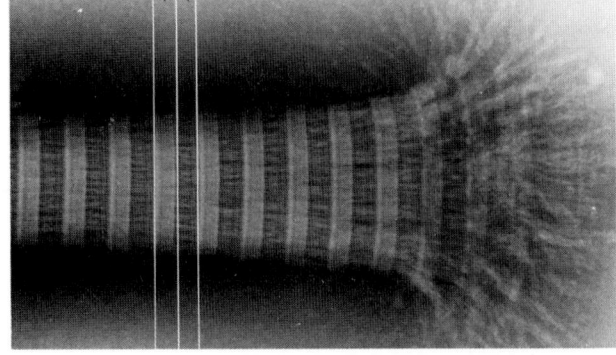

FIGURE 8-32 **Banded appearance of collagen fibrils.** The banded appearance in the electron microscope arises from the schematically represented staggered arrangement of collagen molecules (*top*) that results in a periodically indented surface. *D*, the distance between cross striations, is ~680 Å, so the length of a 3000-Å-long collagen molecule is 4.4*D*. [Courtesy of Karl A. Piez, Collagen Corporation.]

c. Collagen Fibrils Are Covalently Cross-Linked

Collagen's insolubility in solvents that disrupt hydrogen bonding and ionic interactions is explained by the observation that it is both intramolecularly and intermolecularly covalently cross-linked. The cross-links cannot be disulfide bonds, as in keratin, because collagen is almost devoid of Cys residues. Rather, they are derived from Lys and His side chains in reactions such as those in Fig. 8-33. **Lysyl oxidase,** a Cu-containing enzyme that converts Lys residues to those of the aldehyde **allysine,** is the only enzyme implicated in this cross-linking process. Up to four side chains can be covalently bonded to each other. The cross-links do not form at random but, instead, tend to occur near the N- and C-termini of the collagen molecules.

The importance of cross-linking to the normal functioning of collagen is demonstrated by the disease **lathyrism,** which occurs in humans and other animals as a result of the regular ingestion of seeds from the sweet pea *Lathyrus odoratus.* The symptoms of this condition are serious abnormalities of the bones, joints, and large blood vessels, which are caused by an increased fragility of the collagen fibers. The causative agent of lathyrism, **β-aminopropionitrile,**

$$N \equiv C - CH_2 - CH_2 - NH_3^+$$
β-Aminopropionitrile

inactivates lysyl oxidase by covalently binding to its active site. This results in markedly reduced cross-linking in the collagen of lathrytic animals.

The degree of cross-linking of the collagen from a particular tissue increases with the age of the animal. This is why meat from older animals is tougher than that from younger animals. In fact, individual molecules of collagen (called **tropocollagen**) can only be extracted from the tissues of very young animals. Collagen cross-linking is not the central cause of aging, however, as is demonstrated by the observation that lathyrogenic agents do not slow the aging process.

The collagen fibrils in various tissues are organized in ways that largely reflect the functions of the tissues (Table 8-3). Thus tendons (the "cables" connecting muscles to bones), skin (a tear-resistant outer fabric), and cartilage (which has a load-bearing function) must support stress in predominantly one, two, and three dimensions, respectively, and their component collagen fibrils are arrayed accordingly. How collagen fibrils are laid down in these arrangements is unknown. However, some of the factors

TABLE 8-3   The Arrangement of Collagen Fibrils in Various Tissues

Tissue	Arrangement
Tendon	Parallel bundles
Skin	Sheets of fibrils layered at many angles
Cartilage	No distinct arrangement
Cornea	Planar sheets stacked crossways so as to minimize light scattering

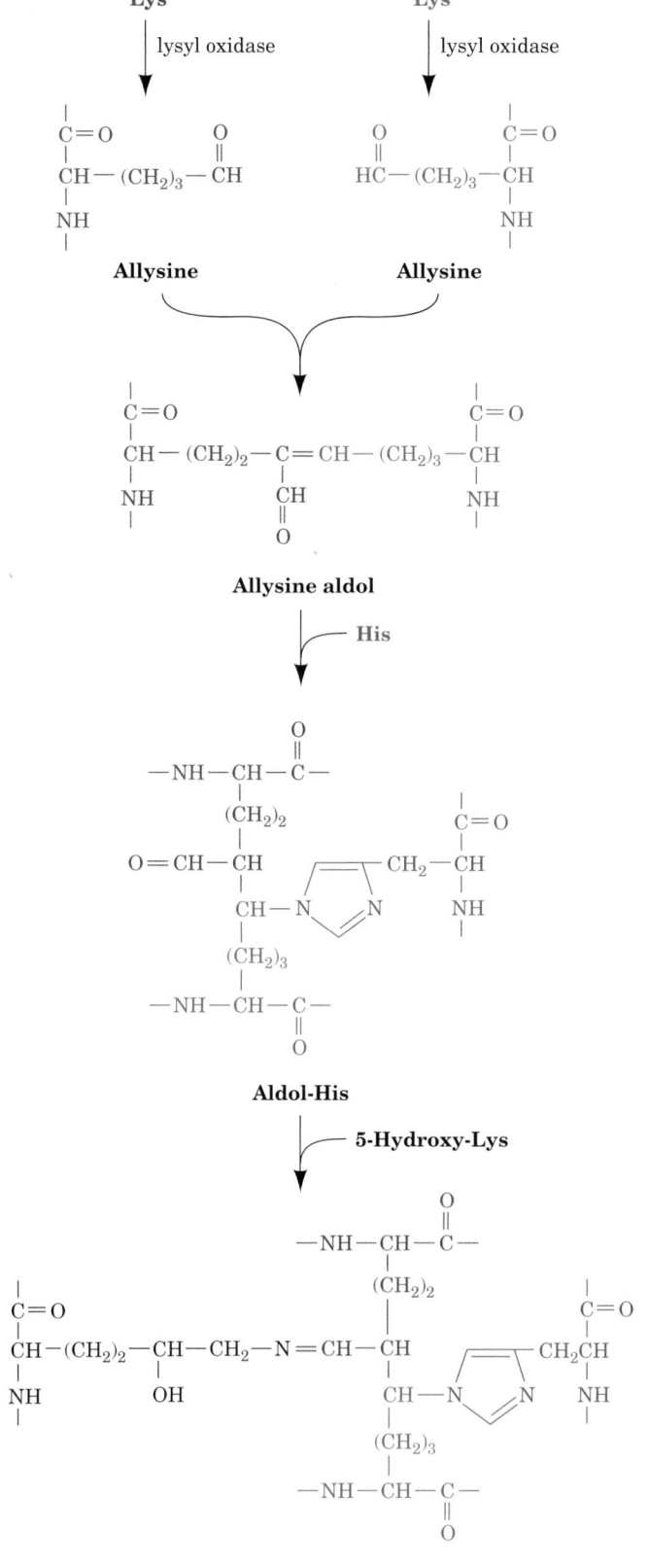

FIGURE 8-33   A biosynthetic pathway for cross-linking Lys, Hyl, and His side chains in collagen. The first step in the reaction is the lysyl oxidase-catalyzed oxidative deamination of Lys to form the aldehyde allysine. Two such aldehydes then undergo an aldol condensation to form **allysine aldol.** This product can react with His to form **aldol histidine.** This, in turn, can react with 5-hydroxylysine to form a Schiff base (an imine bond), thereby cross-linking four side chains.

guiding collagen molecule assembly are discussed in Sections 32-5A and 32-5B.

d. Collagen Defects Are Responsible for a Variety of Human Diseases

Several rare heritable disorders of collagen are known. Mutations of Type I collagen, which constitutes the major structural protein in most human tissues, usually result in **osteogenesis imperfecta (brittle bone disease).** The severity of this disease varies with the nature and position of the mutation: Even a single amino acid change can have lethal consequences. Mutations may affect the structure of the collagen molecule or how it forms fibrils. For example, the substitution of Ala for the central Gly in each polypeptide chain of the structure shown in Fig. 8-30, which re-

duces the denaturation temperature of this model compound from 62°C to 29°C, locally distorts the collagen triple helix. The need to accommodate the three additional methyl groups in the tightly packed interior of the triple helix pries apart the polypeptide chains in the region of the substitutions so as to rupture the hydrogen bonds that would otherwise link the main chain N—H group of each Ala (normally Gly) to the carbonyl oxygen of the adjacent Pro in a neighboring chain (Fig. 8-34). Rather, these hydrogen bonding groups are bridged by water molecules that insinuate themselves into the distorted part of the structure. Similar distortions almost certainly occur in the Gly → X mutated collagens responsible for such diseases as osteogenesis imperfecta. Such mutations tend to be dominant because they affect either the folding of the triple

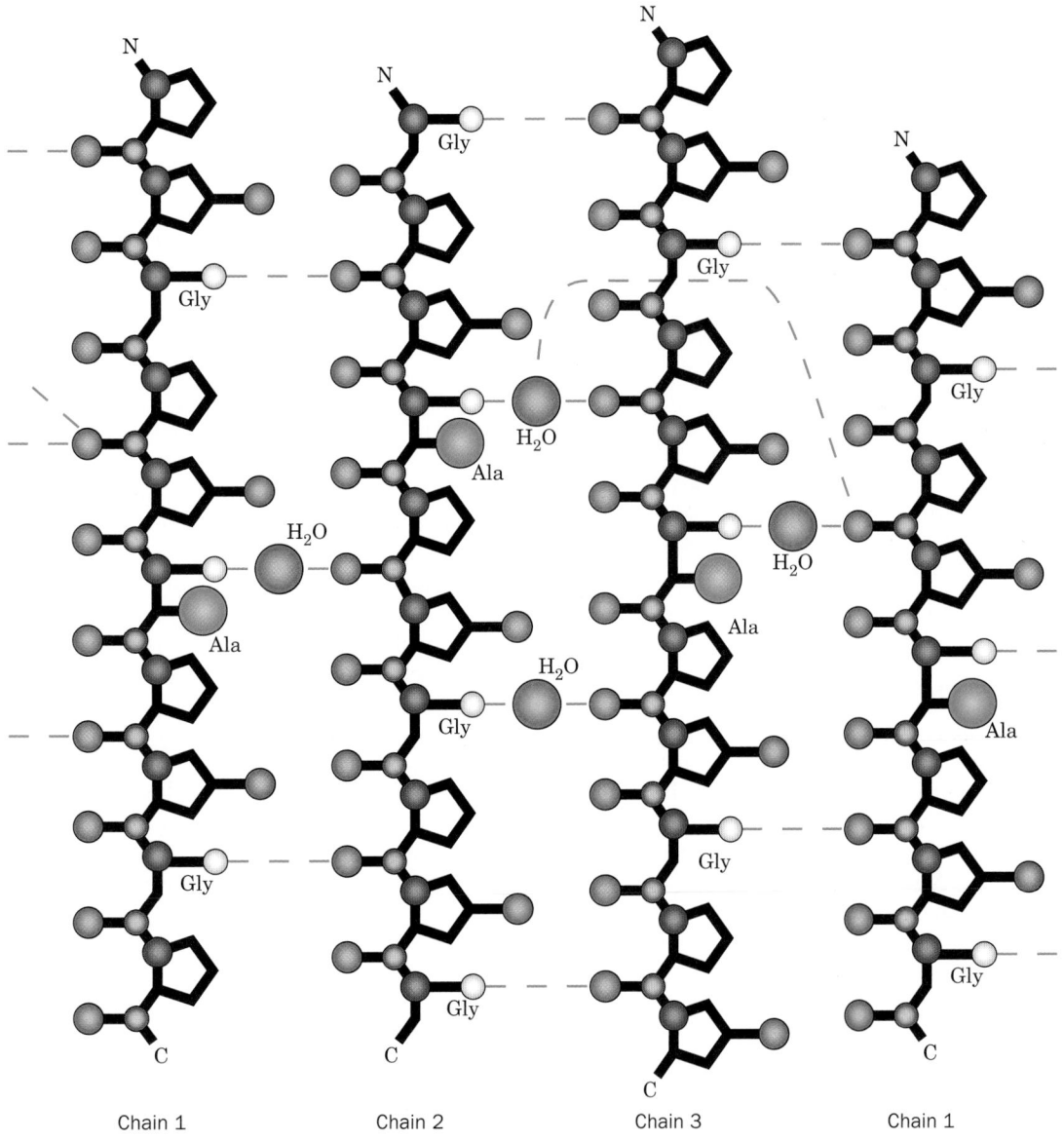

FIGURE 8-34 Distorted structure in abnormal collagen. This schematic diagram shows hydrogen bonding interactions in the Ala-containing portions of the X-ray structure of (Pro-Hyp-Gly)$_{10}$ in which the fifth Gly is replaced by Ala. This cylindrical projection is in the style of Fig. 8-30c. Note how the Ala side chains (*large green balls*) distort the triple helix so as to disrupt the normally occurring Gly NH ··· Pro O hydrogen bonds and replace them with water-bridged hydrogen bonds. [After Bella, J., Eaton, M., Brodsky, B., and Berman, H.M., *Science* **266,** 78 (1994).]

helix or fibril formation even when normal chains are also members of the triple helix. All known amino acid changes within Type I collagen's triple helical region result in abnormalities, indicating that the structural integrity of this region is essential for proper collagen function.

Many collagen disorders are characterized by deficiencies in the amount of a particular collagen type synthesized or by abnormal activities of collagen-processing enzymes such as lysyl hydroxylase or lysyl oxidase. One group of at least 10 different collagen deficiency diseases, the **Ehlers–Danlos syndromes,** are all characterized by hyperextensibility of the joints (really the ligaments holding them together) and skin. This is because these tissues also contain large amounts of **elastin,** a protein with rubberlike elastic properties. Consequently, the loss of the rigidity normally conferred by collagen coupled with the presence of elastin results in the hyperextensibility of the affected tissues. Several degenerative diseases exhibit collagen abnormalities in certain tissues, including cartilage in **osteoarthritis** and the fibrous **atherosclerotic plaques** in human arteries.

3 ■ GLOBULAR PROTEINS

Globular proteins comprise a highly diverse group of substances that, in their native states, exist as compact spheroidal molecules. Enzymes are globular proteins, as are transport and receptor proteins. In this section we consider the tertiary structures of globular proteins. However, since most of our detailed structural knowledge of proteins, and thus to a large extent their function, has resulted from X-ray crystal structure determinations of globular proteins and, more recently, from their nuclear magnetic resonance (NMR) structure determinations, we begin this section with a discussion of the capabilities and limitations of these powerful techniques.

A. *Interpretation of Protein X-Ray and NMR Structures*

X-Ray crystallography is a technique that directly images molecules. X-Rays must be used to do so because, according to optical principles, the uncertainty in locating an object is approximately equal to the wavelength of the radiation used to observe it (covalent bond distances and the wavelengths of the X-rays used in structural studies are both ~1.5 Å; individual molecules cannot be seen in a light microscope because visible light has a minimum wavelength of 4000 Å). There is, however, no such thing as an X-ray microscope because there are no X-ray lenses. Rather, a crystal of the molecule to be imaged is exposed to a collimated beam of X-rays and the consequent diffraction pattern is recorded by a radiation counter or, now infrequently, on photographic film (Fig. 8-35). The X-rays used in structural studies are produced by laboratory X-ray generators or, increasingly often, by **synchrotrons,** a type of particle accelerator that produces X-rays of far greater intensity. The intensities of the diffraction maxima (dark-

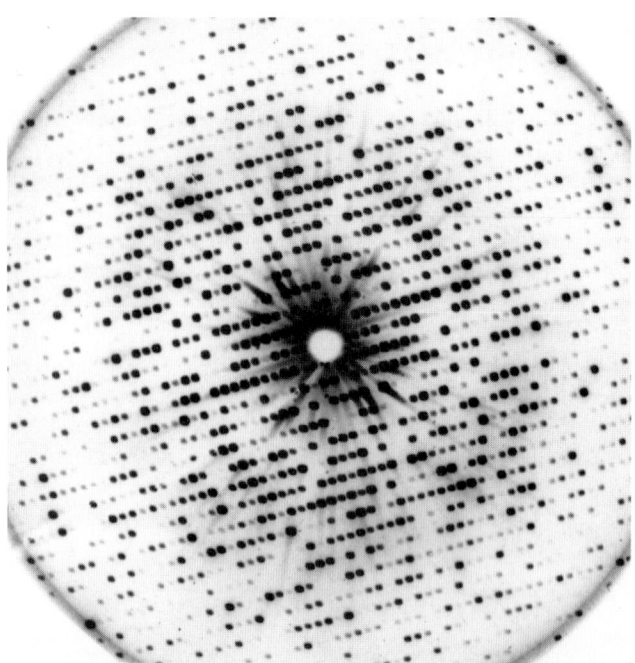

FIGURE 8-35   X-Ray diffraction photograph of a single crystal of sperm whale myoglobin. The intensity of each diffraction maximum (the darkness of each spot) is a function of the myoglobin crystal's electron density. The photograph contains only a small fraction of the total diffraction information available from a myoglobin crystal. [Courtesy of John Kendrew, Cambridge University, U.K.]

ness of the spots on a film) are then used to construct mathematically the three-dimensional image of the crystal structure through methods that are beyond the scope of this text. In what follows, we discuss some of the special problems associated with interpreting the X-ray crystal structures of proteins.

X-Rays interact almost exclusively with the electrons in matter, not the nuclei. An X-ray structure is therefore an image of the **electron density** of the object under study. Such **electron density maps** may be presented as a series of parallel sections through the object. On each section, the electron density is represented by contours (Fig. 8-36a) in the same way that altitude is represented by the contours on a topographic map. A stack of such sections, drawn on transparencies, yields a three-dimensional electron density map (Fig. 8-36b). Modern structural analysis, however, is carried out with the aid of computers, that graphically display electron density maps that are contoured in three dimensions (Fig. 8-36c).

a. Most Protein Crystal Structures Exhibit Less than Atomic Resolution

The molecules in protein crystals, as in other crystalline substances, are arranged in regularly repeating three-dimensional lattices. Protein crystals, however, differ from those of most small organic and inorganic molecules in being highly hydrated; they are typically 40 to 60% water by volume. The aqueous solvent of crystallization is necessary for the structural integrity of the protein crystals, as J. D.

(a)

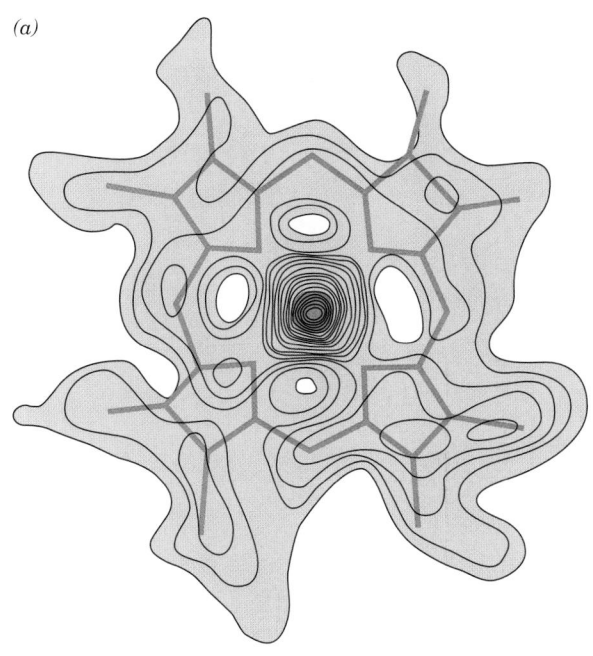

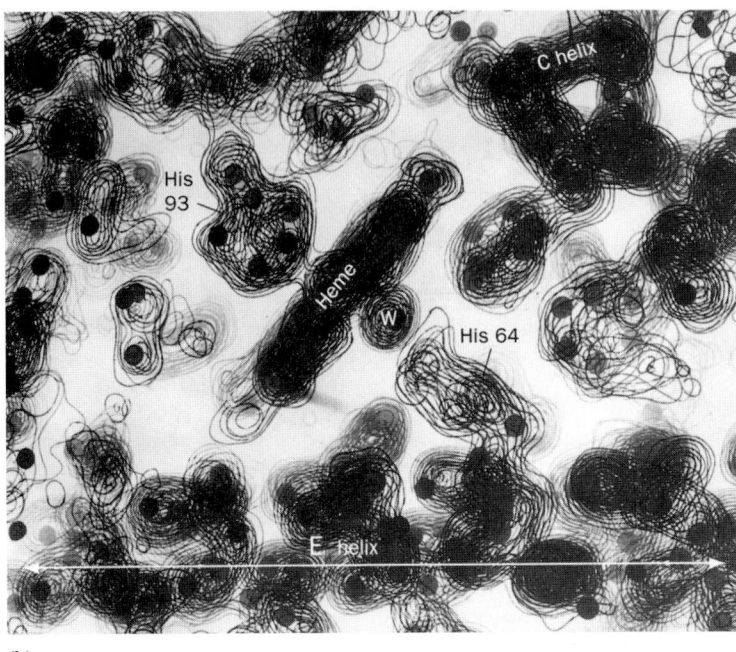

(b)

FIGURE 8-36   Electron density maps of proteins. (*a*) A section through the 2.0-Å-resolution electron density map of sperm whale myoglobin, which contains the heme group (*red*). The large peak at the center of the map represents the electron-dense Fe atom. [After Kendrew, J.C., Dickerson, R.E., Strandberg, B.E., Hart, R.G., Davies, D.R., Phillips, D.C., and Shore, V.C., *Nature* **185**, 434 (1960).] (*b*) A portion of the 2.4-Å-resolution electron density map of myoglobin constructed from a stack of contoured transparencies. Dots have been placed at the positions deduced for the nonhydrogen atoms. The heme group is seen edge-on together with its two associated His residues and a water molecule, W. An α helix, the so-called E helix (Fig. 8-12), extends across the bottom of the map. Another α helix, the C helix, extends into the plane of the paper on the upper right. Note the hole along its axis. [Courtesy of John Kendrew, Cambridge University.] (*c*) A thin section through the 1.5-Å-resolution electron density map of *E. coli* 6-hydroxymethyl-7,8-dihydropterin pyrophosphokinase (which catalyzes the first reaction in the biosynthesis of folic acid; Section 26-3H) contoured in three dimensions. Only a single contour level (*cyan*) is shown, together with an atomic model of the corresponding polypeptide segments colored according to atom type (C yellow, O red, and N blue, with a water molecule represented by a red sphere). [Courtesy of Xinhua Ji, NCI-Frederick Cancer Research and Development Center, Frederick, Maryland.]

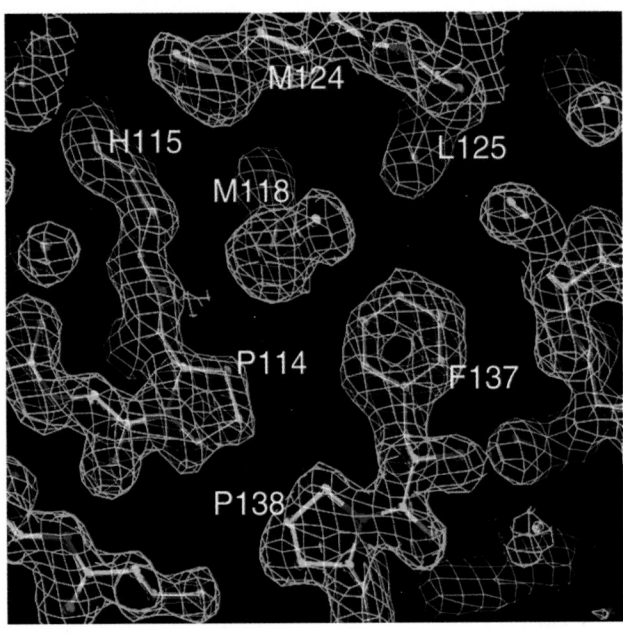

(c)

Bernal and Dorothy Crowfoot Hodgkin first noted in 1934 when they carried out the original X-ray studies of protein crystals. This is because water is required for the structural integrity of native proteins themselves (Section 8-4).

The large solvent content of protein crystals gives them a soft, jellylike consistency, so that their molecules usually lack the rigid order characteristic of crystals of small molecules such as NaCl or glycine. The molecules in a protein crystal are typically disordered by more than an angstrom, so that the corresponding electron density map lacks information concerning structural details of smaller size. The

crystal is therefore said to have a resolution limit of that size. Protein crystals typically have resolution limits in the range 1.5 to 3.0 Å, although some are better ordered (have higher resolution, that is, a lesser resolution limit) and many are less ordered (have lower resolution).

Since an electron density map of a protein must be interpreted in terms of its atomic positions, the accuracy and even the feasibility of a crystal structure analysis depends on the crystal's resolution limit. Indeed, the ability to obtain crystals of sufficiently high resolution is a major limiting factor in determining the X-ray crystal structure of a

(a)   **6.0-Å resolution**   *(b)*   **2.0-Å resolution**   *(c)*   **1.5-Å resolution**   *(d)*   **1.1-Å resolution**

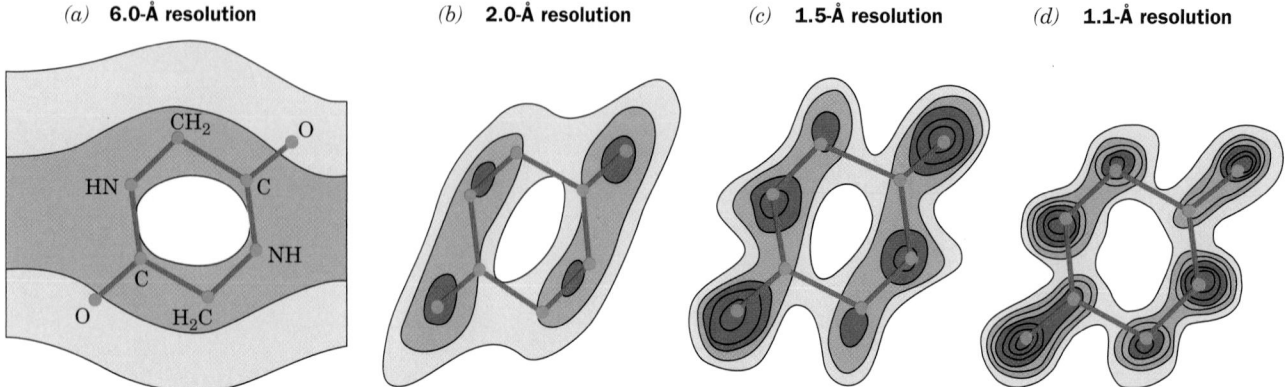

FIGURE 8-37   Sections through the electron density map of diketopiperazine calculated at the indicated resolution levels. Hydrogen atoms are not apparent in this map because of their low electron density. [After Hodgkin, D.C., *Nature* **188,** 445 (1960).]

protein or other macromolecule. Figure 8-37 indicates how the quality (degree of focus) of an electron density map varies with its resolution limit. At 6-Å resolution, the presence of a molecule the size of diketopiperazine is difficult to discern. At 2.0-Å resolution, its individual atoms cannot yet be distinguished, although its molecular shape has become reasonably evident. At 1.5-Å resolution, which roughly corresponds to a bond distance, individual atoms become partially resolved. At 1.1-Å resolution, atoms are clearly visible.

Most protein crystal structures are too poorly resolved for their electron density maps to reveal clearly the positions of individual atoms (e.g., Fig. 8-36). Nevertheless, the distinctive shape of the polypeptide backbone usually permits it to be traced, which, in turn, allows the positions and orientations of its side chains to be deduced (e.g., Fig. 8-37*c*). Yet side chains of comparable size and shape, such as those of Leu, Ile, and Thr, cannot always be differentiated with a reasonable degree of confidence (hydrogen atoms, having but one electron, are only visible in the few macromolecular X-ray structures with resolution limits less than ~1.2 Å), so that a protein structure cannot be elucidated from its electron density map alone. Rather, the primary structure of the protein must be known, thereby permitting the sequence of amino acid residues to be fitted to its electron density map. Mathematical refinement can then reduce the uncertainty in the crystal structure's atomic positions to as little as 0.1 Å (in contrast, positional errors in the most accurately determined small molecule X-ray structures are as little as 0.001 Å).

b. Most Crystalline Proteins Maintain Their Native Conformations

What is the relationship between the structure of a protein in a crystal and that in solution, where globular proteins normally function? Several lines of evidence indicate that *crystalline proteins assume very nearly the same structures that they have in solution:*

1. A protein molecule in a crystal is essentially in solution because it is bathed by the solvent of crystallization over all of its surface except for the few, generally small patches that contact neighboring protein molecules. In fact, the 40 to 60% water content of typical protein crystals is similar to that of many cells (e.g., see Fig. 1-13).

2. A protein may crystallize in one of several forms or "habits," depending on crystallization conditions, that differ in how the protein molecules are arranged in space relative to each other. In the numerous cases in which different crystal forms of the same protein have been independently analyzed, the molecules have virtually identical conformations. Similarly, in the several cases in which both the X-ray crystal structure and the solution NMR structure of the same protein have been determined, the two structures are, for the most part, identical to within experimental error (see below). Evidently, crystal packing forces do not greatly perturb the structures of protein molecules.

3. The most compelling evidence that crystalline proteins have biologically relevant structures, however, is the observation that many enzymes are catalytically active in the crystalline state. The catalytic activity of an enzyme is very sensitive to the relative orientations of the groups involved in binding and catalysis (Chapter 15). Active crystalline enzymes must therefore have conformations that closely resemble their solution conformations.

c. Protein Structure Determination by NMR

The determination of the three-dimensional structures of small globular proteins in aqueous solution has become possible, since the mid 1980s, through the development of **two-dimensional (2D) NMR spectroscopy** (and, more recently, of 3D and 4D techniques), in large part by Kurt Wüthrich. Such NMR measurements, whose description is beyond the scope of this text, yield the interatomic distances between specific protons that are <5 Å apart in a protein of known sequence. The interproton distances may be either through space, as determined by nuclear Overhauser effect spectroscopy (NOESY, Fig. 8-38*a*), or through bonds, as determined by correlated spectroscopy (COSY). These distances, together with known geometric constraints such as covalent bond distances and angles, group planarity, chirality, and van der Waals radii, are used to compute the protein's three-dimensional structure. However, since interproton distance measurements are imprecise, they are insufficient to imply a unique structure. Rather, they are consistent with an ensemble of closely related structures. Consequently, an NMR structure of a protein (or any other macromolecule with a well-defined structure) is often presented as a representative sample of structures that are consistent with the constraints (e.g., Figure 8-38*b*). The "tightness" of a bundle of such structures is indicative both of the accuracy with which the structure is known, which in the most favorable cases is roughly comparable to that of an X-ray crystal structure with a resolution of 2 to 2.5 Å, and of the conformational fluctuations that the protein under-

goes (Section 9-4). Although present NMR methods are limited to determining the structures of macromolecules with molecular masses no greater than ~40 kD, recent advances in NMR technology suggest that this limit may soon increase to ~100 kD or more.

In most of the several cases in which both the NMR and X-ray crystal structures of a particular protein have been determined, the two structures are in good agreement. There are, however, a few instances in which there are real differences between the corresponding X-ray and NMR structures. These, for the most part, involve surface residues that, in the crystal, participate in intermolecular contacts and are thereby perturbed from their solution conformations. NMR methods, besides providing mutual cross-checks with X-ray techniques, can determine the structures of proteins and other macromolecules that fail

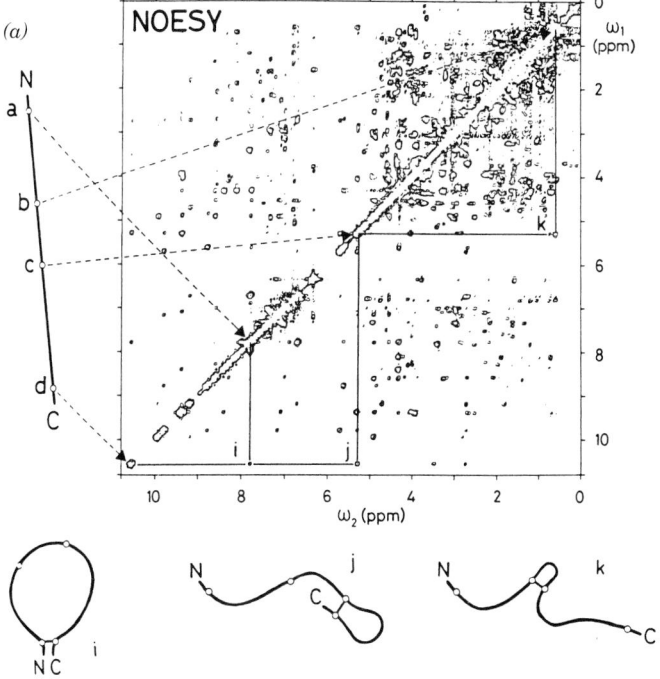

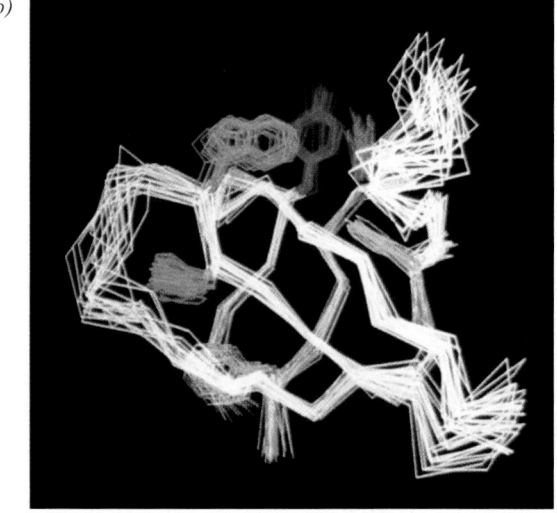

FIGURE 8-38   The 2D proton NMR structures of proteins.
(*a*) A NOESY spectrum of a protein presented as a contour plot with two frequency axes, ω_1 and ω_2. The conventional 1D-NMR spectrum of the protein, which occurs along the diagonal of the plot ($\omega_1 = \omega_2$), is too crowded with peaks to be directly interpretable (even a small protein has hundreds of protons). The off-diagonal peaks, the so-called cross peaks, each arise from the interaction of two protons that are <5 Å apart in space and whose 1D-NMR peaks are located where the horizontal and vertical lines through the cross peak intersect the diagonal [a **nuclear Overhauser effect (NOE)**]. For example, the line to the left of the spectrum represents the extended polypeptide chain with its N- and C-terminal ends identified by the letters N and C and with the positions of four protons, a to d, represented by small circles. The dashed arrows indicate the diagonal NMR peaks to which these protons give rise. Cross peaks, such as i, j, and k, which are each located at the intersections of the horizontal and vertical lines through two diagonal peaks, are indicative of an NOE between the

corresponding two protons, indicating that they are <5 Å apart. These distance relationships are schematically indicated by the three looped structures drawn below the spectrum. Note that the assignment of a distance relationship between two protons in a polypeptide requires that the NMR peaks to which they give rise and their positions in the polypeptide be known, which requires that the polypeptide's amino acid sequence has been previously determined. [After Wüthrich, K., *Science* **243,** 45 (1989).] (*b*) The NMR structure of a 64-residue polypeptide comprising the **Src protein SH3 domain** (Section 19-3C). The drawing represents 20 superimposed structures that are consistent with the 2D- and 3D-NMR spectra of the protein (each calculated from a different, randomly generated starting structure). The polypeptide backbone, as represented by its connected C_α atoms, is white and its Phe, Tyr, and Trp side chains are yellow, red, and blue, respectively. It can be seen that the polypeptide backbone folds into two three-stranded antiparallel β sheets that form a sandwich. [Courtesy of Stuart Schreiber, Harvard University.]

to crystallize. Moreover, since NMR can probe motions over time scales spanning 10 orders of magnitude, it can be used to study protein folding and dynamics (Chapter 9).

d. Protein Molecular Structures Are Most Effectively Illustrated in Simplified Form

The several hundred nonhydrogen atoms of even a small protein makes understanding a protein's detailed structure a considerable effort. This complexity makes building a skeletal (ball-and-stick) model of a protein such a time-consuming task that such models are rarely available. Moreover, a drawing of a protein showing all its non-hydrogen atoms (e.g., Fig. 8-39*a*) is too complicated to be

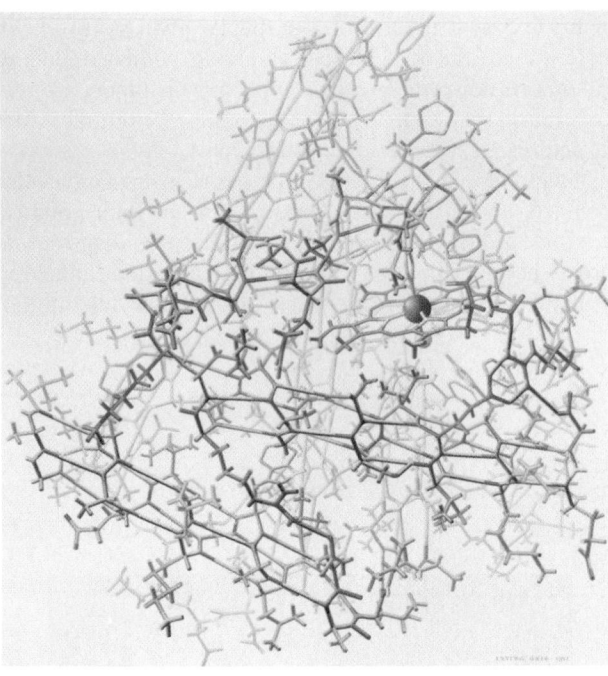

(a)

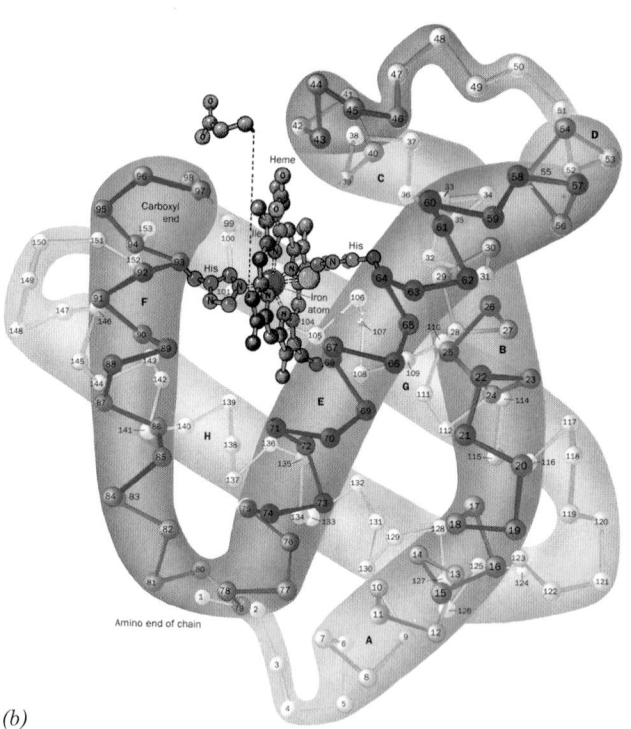

(b)

FIGURE 8-39   Representations of the X-ray structure of sperm whale myoglobin. (*a*) The protein and its bound heme are drawn in stick form, with protein C atoms green, heme C atoms red, N atoms blue, and O atoms red. The Fe and its bound water molecule are shown as orange and gray spheres and hydrogen bonds are gray. In this one-of-a-kind painting of the first known protein structure, the artist has employed "creative distortions" to emphasize the protein's structural features, particularly its α helices. (*b*) A diagram in which the protein is represented by its computer-generated C$_\alpha$ backbone, with its C$_\alpha$ atoms, shown as balls, consecutively numbered from the N-terminus. The 153-residue polypeptide chain is folded into eight α helices (highlighted here by hand-drawn envelopes), designated A through H, that are connected by short polypeptide links. The protein's bound heme group (*purple,* with its Fe atom represented by a red sphere), in complex with a water molecule (*orange sphere*), is shown together with its two closely associated His side chains (*blue*). One of the heme group's propionic acid side chains has been displaced for clarity. Hydrogen atoms are not visible in the X-ray structure. (*c*) A computer-generated cartoon drawing in an orientation similar to that of Part *b*, emphasizing the protein's secondary structure. Here helices are green and the intervening coil regions are yellow. The heme group with its bound O$_2$ molecule and its two associated His side chains are shown in ball-and-stick form with C magenta, N blue, O red, and Fe orange. [Parts *a* and *b* are based on an X-ray structure by John Kendrew, MRC Laboratory of Molecular Biology, Cambridge, U.K. PDBid 1MBN. Illustrations, Irving Geis/Geis Archives Trust. Copyright Howard Hughes Medical Institute. Reproduced with permission. Part *c* is based on an X-ray structure by Simon Phillips, University of Leeds, Leeds, U.K. PDBid 1MBO.]
🔖 **See Kinemage Exercise 6-1**

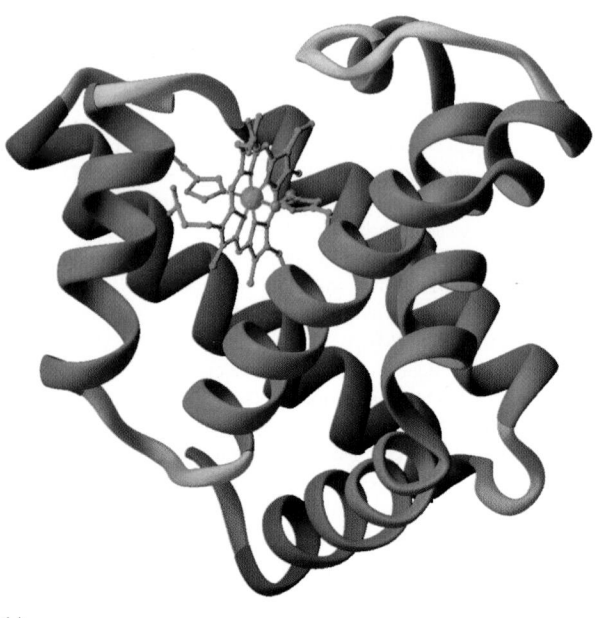

(c)

of much use. In order to be intelligible, drawings of proteins must be selectively simplified. One way of doing so is to represent the polypeptide backbone only by its C_α atoms (its **C_α backbone**) and to display only a few key side chains (e.g, Fig. 8-39*b*). A further level of abstraction may be obtained by representing the protein in a cartoon form that emphasizes its secondary structure (e.g., Figs. 8-39*c* and 8-19). Computer-generated drawings of space-filling models, such as Figs. 8-12 and 8-18, may also be employed to illustrate certain features of protein structures. However, the most instructive way to examine a macromolecular structure is through the use of interactive computer graphics programs. The use of such programs is discussed in Section 8-3C.

B. *Tertiary Structure*

The **tertiary structure (3° structure)** of a protein is its three-dimensional arrangement; that is, the folding of its 2° structural elements, together with the spatial dispositions of its side chains. The first protein X-ray structure, that of sperm whale **myoglobin,** was elucidated in the late 1950s by John

Kendrew and co-workers. Its polypeptide chain follows such a tortuous, wormlike path (Fig. 8-39) that these investigators were moved to indicate their disappointment at its lack of regularity. In the intervening years, over 18,000 protein structures have been reported. Each of them is a unique, highly complicated entity. Nevertheless, their tertiary structures have several outstanding features in common, as we shall see below.

a. Globular Proteins May Contain Both α Helices and β Sheets

The major types of secondary structural elements, α helices and β pleated sheets, commonly occur in globular proteins but in varying proportions and combinations. Some proteins, such as myoglobin, consist only of α helices spanned by short connecting links that have coil conformations (Fig. 8-39). Others, such as concanavalin A, have a large proportion of β sheets but are devoid of α helices (Fig. 8-40). Most proteins, however, have significant amounts of both types of secondary structure (on average, ~31% α helix and ~28% β sheet, with the total content of helices, sheets, turns, and Ω loops comprising ~90% of the average protein). Human **carbonic anhydrase** (Fig. 8-41) as well as carboxypeptidase and triose phosphate isomerase (Fig. 8-19) are examples of such proteins.

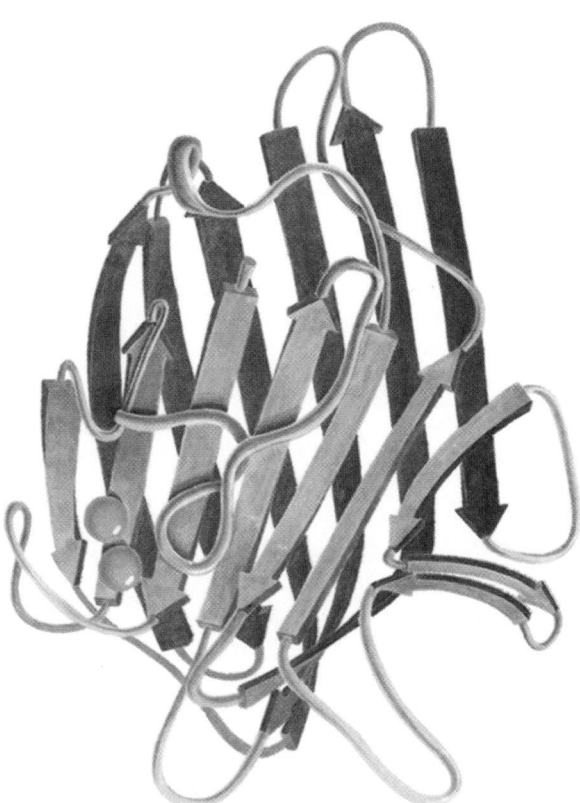

FIGURE 8-40 The X-ray structure of jack bean protein concanavalin A. This protein largely consists of extensive regions of antiparallel β pleated sheet, here represented by flat arrows pointing toward the polypeptide chain's C-terminus. The balls represent protein-bound metal ions. The back sheet is shown in a space-filling representation in Fig. 8-18. [After a drawing by Jane Richardson, Duke University. Based on an X-ray structure by George Reeke, Jr., Joseph Becker, and Gerald Edelman, The Rockefeller University. PDBid 2CNA.]

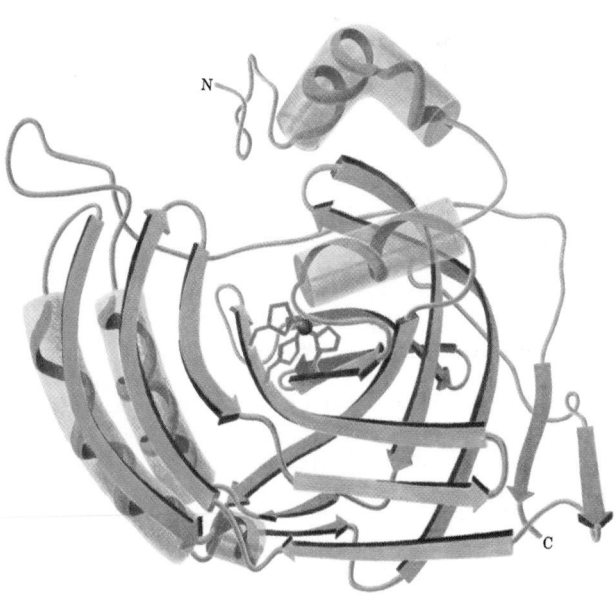

FIGURE 8-41 Human carbonic anhydrase. The α helices are represented as cylinders, and each strand of β sheet is drawn as an arrow pointing toward the polypeptide's C-terminus. The gray ball in the center represents a Zn^{2+} ion that is coordinated by three His side chains (*blue*). Note that the C-terminus is tucked through the plane of a surrounding loop of polypeptide chain, so that carbonic anhydrase is one of the rare native proteins in which a polypeptide chain forms a knot. [After Kannan, K.K., et al. *Cold Spring Harbor Symp. Quant. Biol.* **36,** 221 (1971). PDBid 2CAB.] ✑ **See the Interactive Exercises**

b. Side Chain Location Varies with Polarity

The primary structures of globular proteins generally lack the repeating or pseudorepeating sequences that are responsible for the regular conformations of fibrous proteins. The amino acid side chains in globular proteins are, nevertheless, spatially distributed according to their polarities:

1. *The nonpolar residues Val, Leu, Ile, Met, and Phe largely occur in the interior of a protein, out of contact with the aqueous solvent.* The hydrophobic interactions that promote this distribution, which are largely responsible for the three-dimensional structures of native proteins, are further discussed in Section 8-4C.

2. *The charged polar residues Arg, His, Lys, Asp, and Glu are largely located on the surface of a protein in contact with the aqueous solvent.* This is because the immersion of an ion in the virtually anhydrous interior of a protein results in the uncompensated loss of much of its hydration energy. In the instances that these groups occur in the interior of a protein, they often have a specific chemical function such as promoting catalysis or participating in metal ion binding (e.g., the metal ion–ligating His residues in Figs. 8-39 and 8-41).

3. The uncharged polar groups Ser, Thr, Asn, Gln, Tyr, and Trp are usually on the protein surface but frequently occur in the interior of the molecule. In the latter case, these residues are almost always hydrogen bonded to other groups in the protein. In fact, *nearly all buried hydrogen bond donors form hydrogen bonds with buried acceptor groups;* in a sense, the formation of a hydrogen bond "neutralizes" the polarity of a hydrogen bonding group.

Such a side chain distribution is clearly apparent in Fig. 8-42, which displays the X-ray structure of cytochrome *c.* This arrangement is also seen in Fig. 8-43, which shows the surface and interior exposures of the amino acid side chains of myoglobin's H helix, and in Fig. 8-44, which shows one of the antiparallel β pleated sheets of concanavalin A.

c. Globular Protein Cores Are Efficiently Arranged with Their Side Chains in Relaxed Conformations

Globular proteins are quite compact; there is very little space inside them, so that water is largely excluded from their interiors. The micellelike arrangement of their side chains (polar groups outside, nonpolar groups inside) has led to their description as "oil drops with polar coats." This

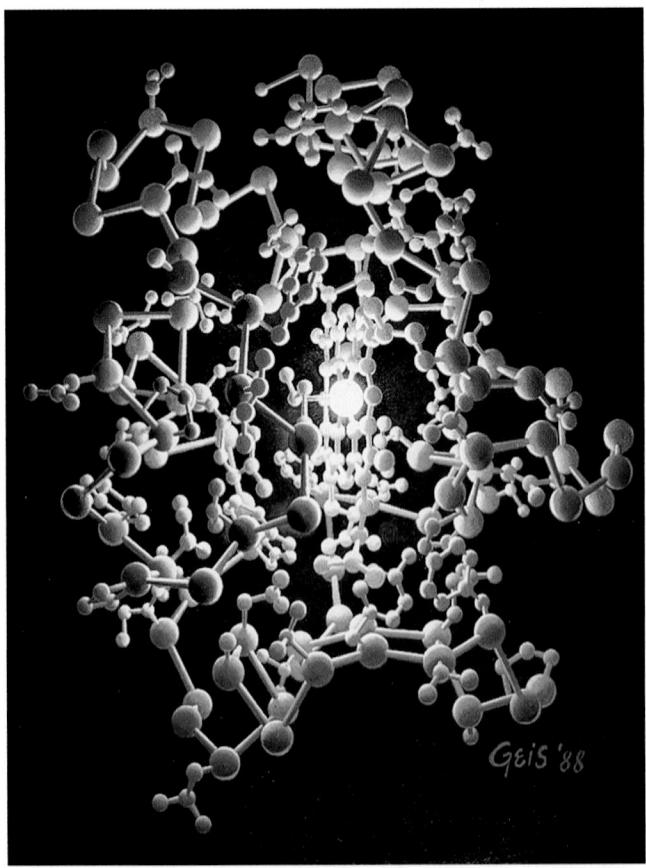

(a)

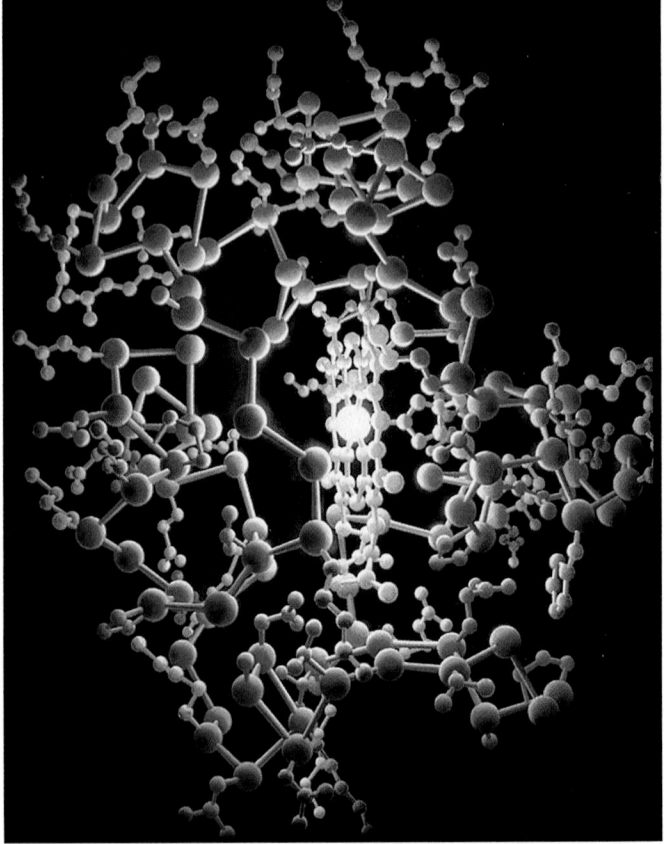

(b)

FIGURE 8-42   The X-ray structure of horse heart cytochrome *c.* The protein (*blue*) is illuminated by the Fe atom of its heme group (*orange*). In Part *a,* the hydrophobic side chains are red, and in Part *b,* the hydrophilic side chains are green. [Based on an X-ray structure by Richard Dickerson, UCLA; Illustration, Irving Geis/Geis Archives Trust. Copyright Howard Hughes Medical Institute. Reproduced with permission.] See **Kinemage Exercise 5**

(a)

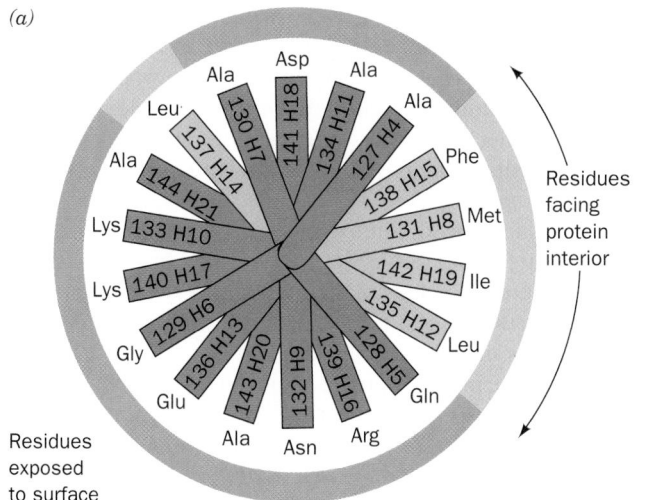

Residues
exposed
to surface

(b)

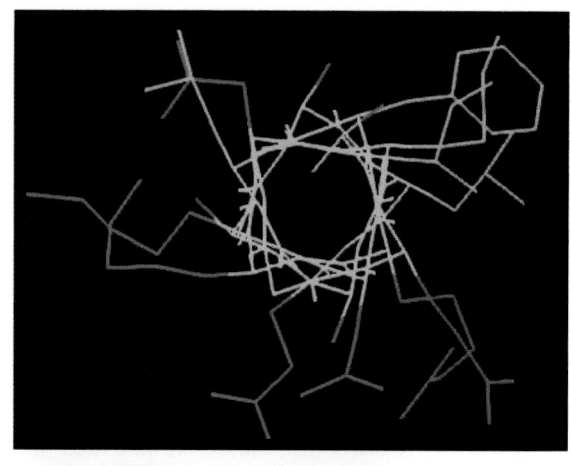

Residues
facing
protein
interior

(c)

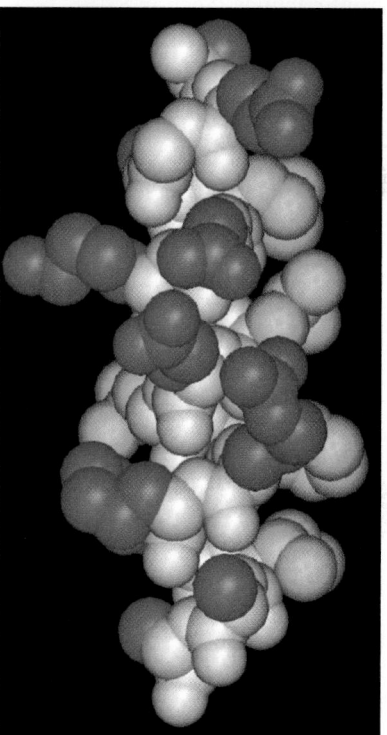

FIGURE 8-43 **The H helix of sperm whale myoglobin.** (*a*) A **helical wheel** representation in which side chain positions about the α helix are projected down the helix axis onto a plane. Here each residue is identified both according to its sequence in the polypeptide chain and according to its position in the H helix. The residues lining the side of the helix facing the protein's interior regions are all nonpolar (*yellow*). The other residues, except Leu 137, which contacts the protein segment linking helices E and F (Fig. 8-39*b*), are exposed to the solvent and are all more or less polar (*purple*). (*b*) A skeletal model, viewed as in Part *a*, in which the main chain is white, nonpolar side chains are yellow, and polar side chains are purple. (*c*) A space-filling model, viewed from the bottom of the page in Parts *a* and *b* and colored as in Part *b*. See Kinemage Exercise 3-2 for another example.

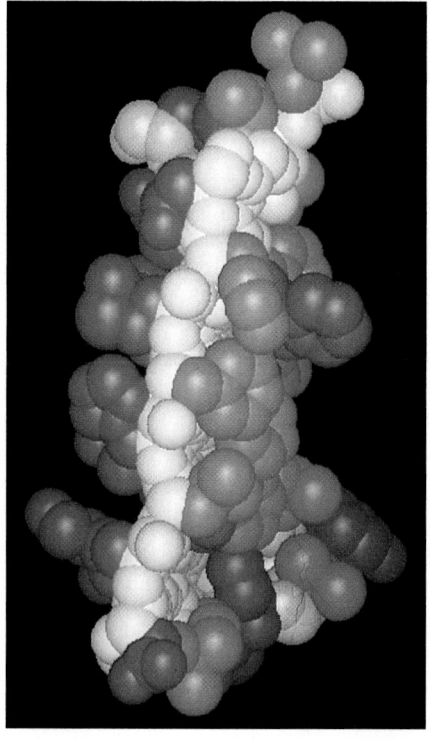

FIGURE 8-44 **A space-filling model of an antiparallel β sheet from concanavalin A.** The β sheet is shown in side view with the interior of the protein (the surface of a second antiparallel β sheet; Fig. 8-40) to the right and the exterior to the left. The main chain is white, nonpolar side chains are brown, and polar side chains are purple. See Kinemage Exercise 3-3

generalization, although picturesque, lacks precision. The **packing density** (ratio of the volume enclosed by the van der Waals envelopes of the atoms in a region to the total volume of the region) of the internal regions of globular proteins averages ~0.75, which is in the same range as that of molecular crystals of small organic molecules. In comparison, equal-sized close-packed spheres have a packing density of 0.74, whereas organic liquids (oil drops) have packing densities that are mostly between 0.60 and 0.70. *The interior of a protein is therefore more like a molecular crystal than an oil drop; that is, it is efficiently packed.*

The bonds of protein side chains, including those occupying protein cores, almost invariably have low-energy staggered torsion angles (Fig. 8-5*a*). Evidently, interior side chains adopt relaxed conformations despite their profusion of intramolecular interactions (Section 8-4).

d. Large Polypeptides Form Domains

Polypeptide chains that consist of more than ~200 residues usually fold into two or more globular clusters known as **domains,** which give these proteins a bi- or multilobal appearance. Most domains consist of 100 to 200 amino acid residues and have an average diameter of

~25 Å. Each subunit of the enzyme **glyceraldehyde-3-phosphate dehydrogenase,** for example, has two distinct domains (Fig. 8-45). A polypeptide chain wanders back and forth within a domain, but neighboring domains are usually connected by one, or less commonly two, polypeptide segments. *Domains are therefore structurally independent units that each have the characteristics of a small globular protein.* Indeed, limited proteolysis of a multidomain protein often liberates its domains without greatly altering their structures or enzymatic activities. Nevertheless, the domain structure of a protein is not always obvious since its domains may make such extensive contacts with each other that the protein appears to be a single globular entity.

An inspection of the various protein structures diagrammed in this chapter reveals that domains consist of two or more layers of secondary structural elements. The reason for this is clear: At least two such layers are required to seal off a domain's hydrophobic core from the aqueous environment.

Domains often have a specific function, such as the binding of a small molecule. In Fig. 8-45, for example, the dinucleotide **nicotinamide adenine dinucleotide (NAD⁺)**

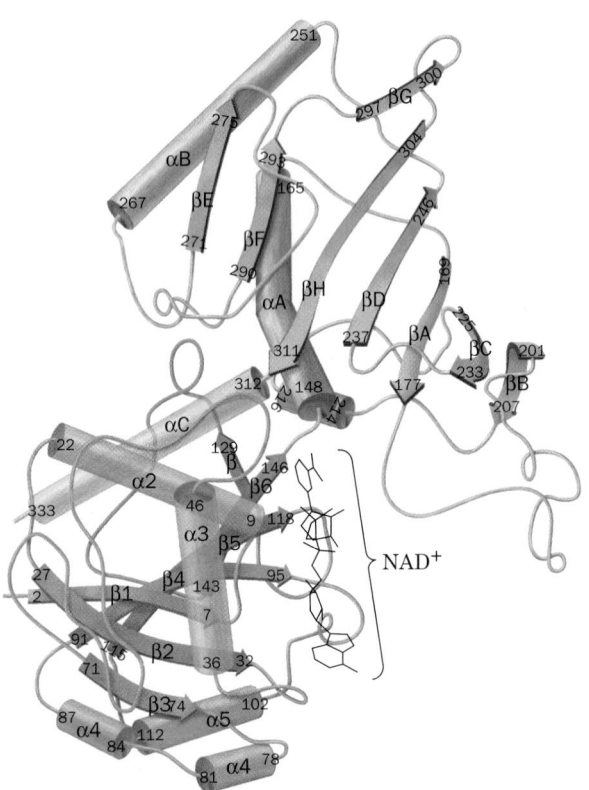

FIGURE 8-45   One subunit of the enzyme glyceraldehyde-3-phosphate dehydrogenase from *Bacillus stearothermophilus.* The polypeptide folds into two distinct domains. The first domain (*red,* residues 1–146) binds NAD⁺ (*black*) near the C-terminal ends of its parallel β strands, and the second domain (*green*) binds glyceraldehyde-3-phosphate (not shown). [After Biesecker, G., Harris, J.I., Thierry, J.C., Walker, J.E., and Wonacott, A., *Nature* **266,** 331 (1977).] ✒ See the Interactive Exercises

Nicotinamide adenine dinucleotide (NAD⁺)

binds to the first domain of the enzyme glyceraldehyde-3-phosphate dehydrogenase. Small molecule binding sites in multidomain proteins often occur in the clefts between domains; that is, the small molecules are bound by groups from two domains. This arrangement arises, in part, from the need for a flexible interaction between the protein and the small molecule that the relatively pliant covalent connection between the domains can provide.

e. Supersecondary Structures Are the Building Blocks of Proteins

Certain groupings of secondary structural elements, named **supersecondary structures** or **motifs,** occur in many unrelated globular proteins:

1. The most common form of supersecondary structure is the **βαβ motif** (Fig. 8-46*a*), in which the usually right-handed crossover connection between two consecutive parallel strands of a β sheet consists of an α helix.

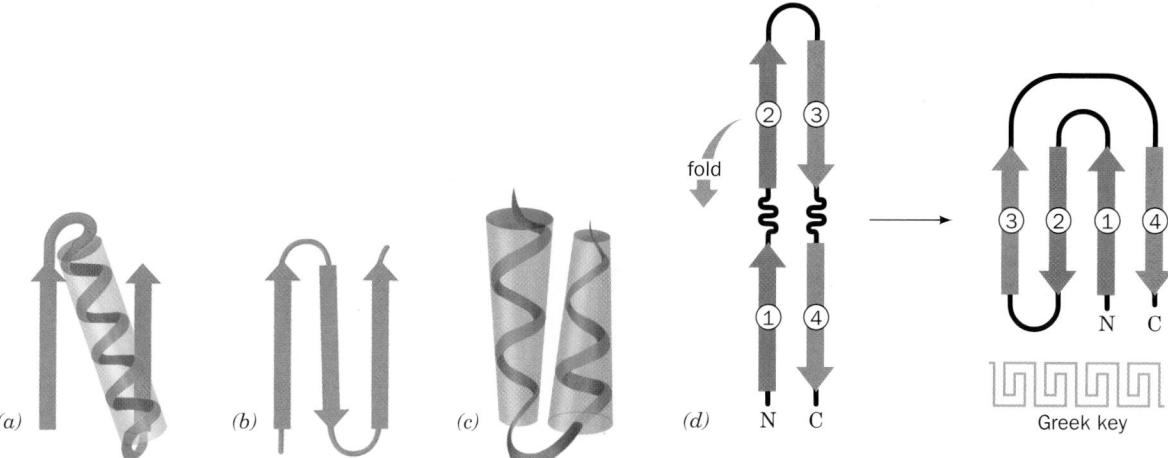

FIGURE 8-46 Schematic diagrams of supersecondary structures. (*a*) A βαβ motif, (*b*) a β hairpin motif, (*c*) an αα motif, and (*d*) a Greek key motif, showing how it is constructed from a folded-over β hairpin.

2. Another common supersecondary structure, the **β hairpin motif** (Fig. 8-46*b*), consists of an antiparallel β sheet formed by sequential segments of polypeptide chain that are connected by relatively tight reverse turns.

3. In an **αα motif** (Fig. 8-46*c*), two successive antiparallel α helices pack against each other with their axes inclined so as to permit their contacting side chains to interdigitate efficiently. Such energetically favorable associations stabilize the coiled coil conformation of α keratin (Section 8-2A).

4. In the **Greek key motif** (Fig. 8-46*d;* named after an ornamental design commonly used in ancient Greece; see inset), a β hairpin is folded over to form a four-stranded antiparallel β sheet. Of the 10 possible ways of connecting the strands of a four-stranded antiparallel β sheet, the two that form Greek key motifs are, by far, the most common in proteins of known structure.

Groups of motifs combine in overlapping and nonoverlapping ways to form the tertiary structure of a domain, which is called a **fold.**

The number of possible different folds would, of course, seem to be unlimited. However, comparisons of the now large number of known protein structures have revealed that *few protein folds are unique; that is, most proteins of known structure have folds that also occur in unrelated proteins.* Indeed, the rate at which new folds are being discovered relative to the rate at which protein structures are being determined suggests that there are only about ~1000 naturally occurring folds. Of these, ~600 have already been observed.

Although there are numerous ways in which domain structures might be categorized (see, e.g., Section 8-3C), perhaps the simplest way to do so is to classify them as **α domains** (containing secondary structural elements that are exclusively α helices), **β domains** (containing only β

sheets), and **α/β domains** (containing both α helices and β sheets). The α/β domain category may be further divided into two main groups: **α/β barrels** and **open β sheets.** In the following paragraphs we describe some of the most common folds in each of these domain categories.

f. α Domains

We are already familiar with a fold that contains only α helices: the **globin fold,** which contains 8 helices in two layers and occurs in myoglobin (Fig. 8-39) and in both the α and β chains of hemoglobin (Section 10-2B). In another common all-α fold, two αα motifs combine to form a **4-helix bundle** such as occurs in **cytochrome b_{562}** (Fig. 8-47*a*). The helices in this fold are inclined such that their contacting side chains intermesh and are therefore out of contact with the surrounding aqueous solution. They are consequently largely hydrophobic. The 4-helix bundle is a relatively common fold that occurs in a variety of proteins. However, not all of them have the up-down-up-down topology (connectivity) of cytochrome b_{562}. For example, human **growth hormone** is a 4-helix bundle with up-up-down-down topology (Fig. 8-47*b*). The successive parallel helices in this fold are, of necessity, joined by longer loops than those joining successive antiparallel helices.

Different types of α domains occur in **transmembrane proteins.** We study these proteins in Section 12-3A.

g. β Domains

β domains contain from 4 to >10 predominantly antiparallel β strands that are arranged into two sheets that pack against each other to form a **β sandwich.** For example, the **immunoglobulin fold** (Fig. 8-48), which forms the basic domain structure of most immune system proteins (Section 35-2B), consists of a 4-stranded antiparallel β sheet in face-to-face contact with a 3-stranded antipar-

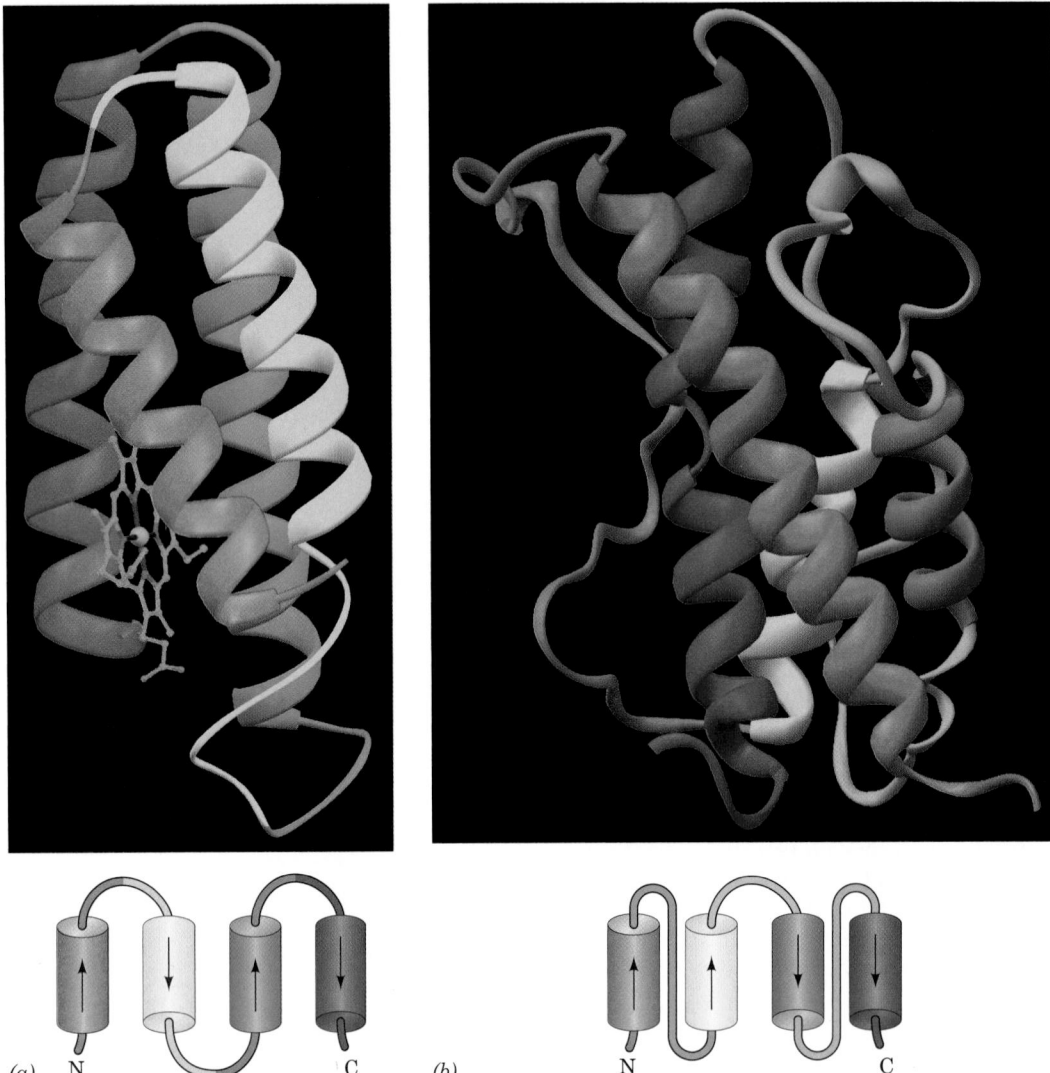

FIGURE 8-47   X-Ray structures of 4-helix bundle proteins.
(*a*) *E. coli* cytochrome b_{562} and (*b*) human growth hormone. The proteins are represented by their peptide backbones, drawn in ribbon form, in which the helices and their connecting links are colored, from N- to C-terminus, in rainbow order, from red to blue. Cytochrome b_{562}'s bound heme group is shown in ball-and-stick form with C magenta, N blue, O red, and Fe orange. The inset below each ribbon diagram is a topological diagram indicating the connectivity of the α helices in each 4-helix bundle. Cytochrome b_{562} (106 residues) has up-down-up-down topology, whereas human growth hormone (191 residues) has up-up-down-down topology. Note that the N- and C-terminal helices of human growth hormone are longer than its other two helices, so that, at one end, these longer helices associate as an αα motif. [Based on X-ray structures by (*a*) F. Scott Matthews, Washington University School of Medicine, and (*b*) Alexander Wlodawer, National Cancer Institute, Frederick, Maryland. PDBids (*a*) 256B and (*b*) 1HGU.]

allel β sheet. Note that the strands in the two sheets are not parallel to one another, a characteristic of stacked β sheets. The side chains between the two stacked β sheets are out of contact with the aqueous medium and thereby form the domain's hydrophobic core. Since successive residues in a β strand alternately extend to opposite sides of the β sheet (Fig. 8-17), these residues are alternately hydrophobic and hydrophilic.

The inherent curvature of β sheets (Section 8-1C) often causes sheets of more than 6 strands to roll up into **β barrels**. Indeed, β sandwiches may be considered to be flattened β barrels. Several different β barrel topologies have been observed, most commonly:

1. The **up-and-down β barrel,** which consists of 8 successive antiparallel β strands that are arranged like the staves of a barrel. An example of an up-and-down β barrel occurs in **retinol binding protein** (Fig 8-49), which functions to transport the nonpolar visual pigment precursor **retinol (vitamin A)** in the bloodstream:

Retinol

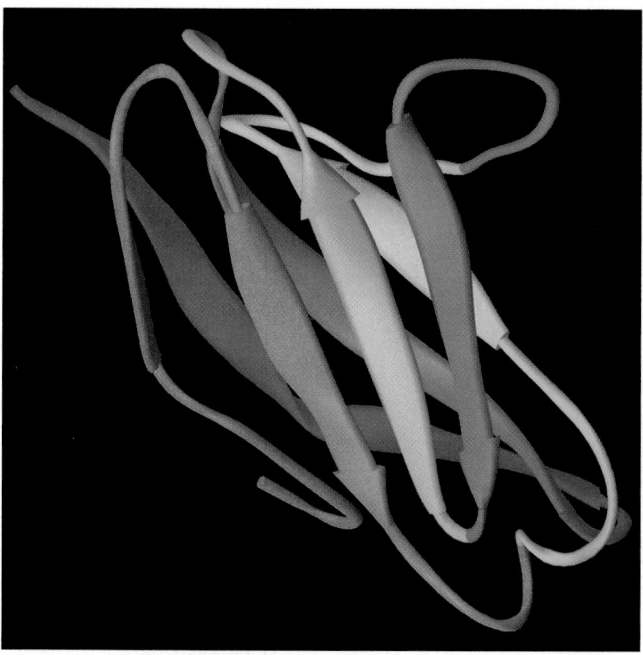

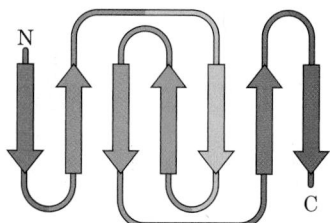

FIGURE 8-48   The immunoglobulin fold. The X-ray structure of the N-terminal domain of the human immunoglobulin fragment **Fab New** shows its immunoglobulin fold. The peptide backbone of this 103-residue domain is drawn in ribbon form with its β strands represented by flat arrows pointing toward the C-terminus and colored, from N- to C-terminus, in rainbow order, from red to blue. The inset is the topological diagram of the immunoglobulin fold showing the connectivity of its stacked 4-stranded and 3-stranded antiparallel β sheets. [Based on an X-ray structure by Roberto Poljak, The Johns Hopkins School of Medicine. PDBid 7FAB.]

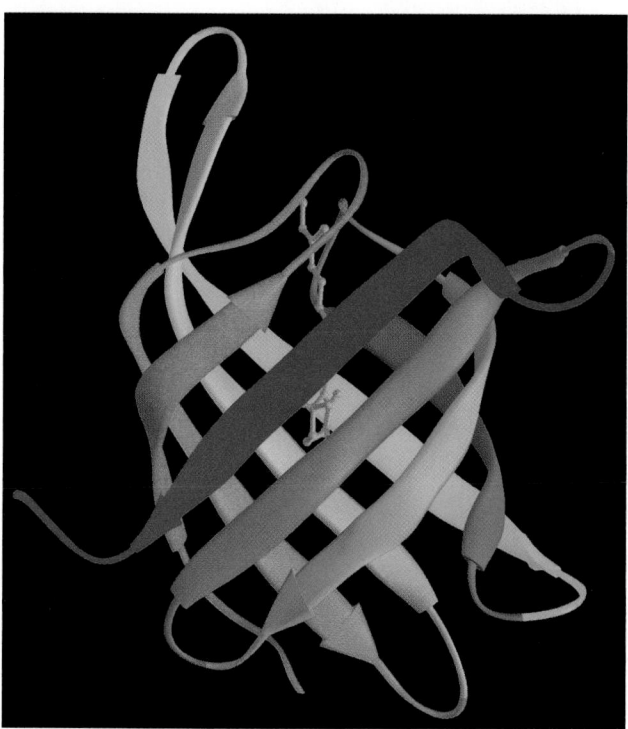

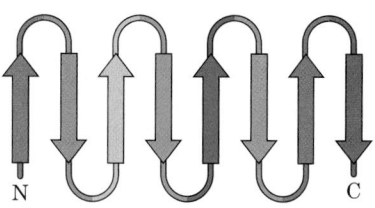

FIGURE 8-49   Retinol binding protein. Its X-ray structure shows its up-and-down β barrel (residues 1–142 of this 182-residue protein). Its peptide backbone is drawn in ribbon form, with its β strands colored, from N- to C-terminus, in rainbow order, red to blue. Note that each β strand is linked via a short loop to its clockwise-adjacent strand as seen from the top. The protein's bound retinol molecule is represented by a white ball-and-stick model. The inset is the protein's topological diagram. [Based on an X-ray structure by T. Alwyn Jones, Biomedical Center, Uppsala, Sweden. PDBid 1RBP.]

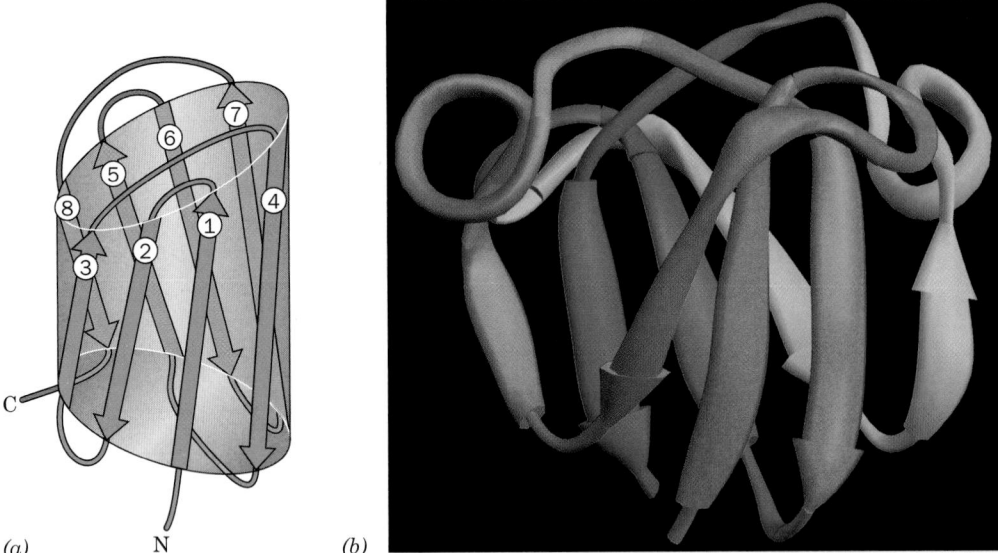

(a)   N   *(b)*

**FIGURE 8-50   X-Ray structure of the C-terminal domain of
bovine γ-B crystallin.** (*a*) A topological diagram showing how its
two Greek key motifs are arranged in a β barrel. One Greek key
motif (*red*) is formed by β strands 1 to 4 and the other (*blue*) is
formed by β strands 5 to 8. [After Branden, C. and Tooze, J.,
Introduction to Protein Structure (2nd ed.), *p.* 75, Garland (1999).]
(*b*) The 83-residue peptide backbone displayed in ribbon form.

Here the members of an antiparallel pair of β strands in a Greek
key motif are colored alike with the N-terminal Greek key colored
red (strands 1 & 4) and orange (2 & 3) and the C-terminal Greek
key colored blue (5 & 8) and cyan (6 & 7). The N-terminal
domain of this two-domain protein is nearly superimposable on
its C-terminal domain. [Based on an X-ray structure by Tom
Blundell, Birkbeck College, London, U. K. PDBid 4GCR.]

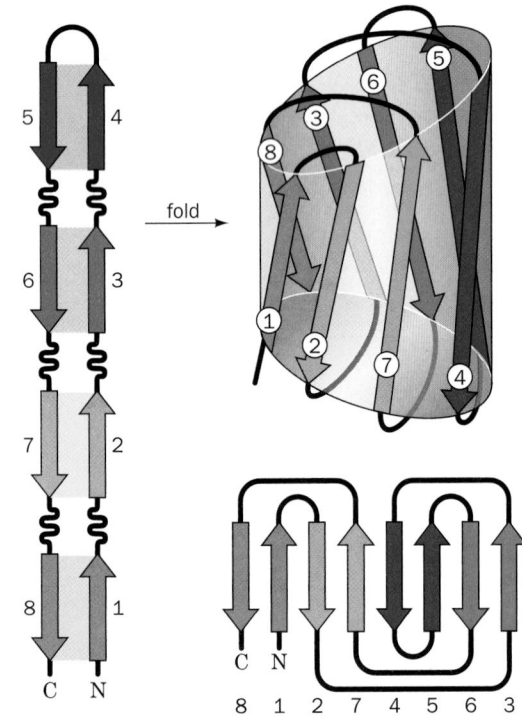

(a)

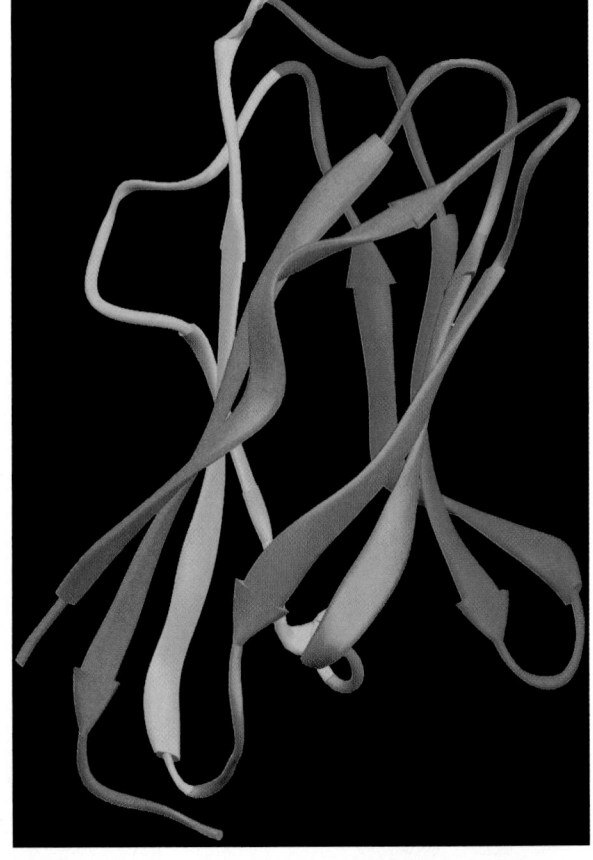

(b)

**FIGURE 8-51   X-Ray structure of the enzyme peptide-N^4-
(N-acetyl-β-D-glucosaminyl)asparagine amidase F from
Flavobacterium meningosepticum.** (*a*) A diagram indicating how
an 8-stranded β barrel is formed by rolling up a 4-segment β
hairpin. A topological diagram of the jelly roll barrel is also
shown. [After Branden, C. and Tooze, J., *Introduction to Protein
Structure* (2nd ed.), *pp.* 77–78, Garland (1999).] (*b*) A ribbon
diagram of the domain formed by residues 1 to 140 of this
314-residue enzyme. Here the two β strands in each segment of

the β hairpin are colored alike, with strands 1 & 8 (the N- and
C-terminal strands) red, strands 2 & 7 orange, strands 3 & 6 cyan,
and strands 4 & 5 blue. [Based on an X-ray structure by Patrick
Van Roey, New York State Department of Health, Albany, New
York. PDBid 1PNG.]

2. A fold consisting of two Greek key motifs, which thereby constitutes an alternative way of connecting the strands of an 8-stranded antiparallel β barrel. Figure 8-50 indicates how two Greek key motifs in the C-terminal domain of the eye lens protein **γ-B crystallin** are arranged to form a β barrel.

3. The **jelly roll** or **Swiss roll barrel** (which is named for its topological resemblance to these rolled up pastries), in which a 4-segment β hairpin is rolled up into an 8-stranded antiparallel β barrel of yet different topology, as is diagrammed in Fig. 8-51*a*. The X-ray structure of the enzyme **peptide-N^4-(N-acetyl-β-D-glucosaminyl)asparagine amidase F,** which was determined by Patrick Van Roey, contains a domain consisting of a jelly roll barrel (Fig. 8-51*b*).

h. α/β Barrels

In α/β domains, a central parallel or mixed β sheet is flanked by α helices. The α/β barrel, which is diagrammed in Fig. 8-19*b,* is a remarkably regular structure that consists of 8 tandem βα units (essentially 8 overlapping βαβ motifs) wound in a right-handed helical sense to form an inner 8-stranded parallel β barrel concentric with an outer barrel of 8 α helices. Each β strand is approximately antiparallel to the succeeding α helix and all are inclined at around the same angle to the barrel axis. Figure 8-52 shows the X-ray structure of chicken **triose phosphate isomerase (TIM),** determined by David Phillips, which consists of an α/β barrel. This is the first known structure of an α/β barrel, which is therefore also called a **TIM barrel.**

The side chains that point inward from the α helices interdigitate with the side chains that point outward from the β strands. A large fraction (~40%) of these side chains are those of the branched aliphatic residues Ile, Leu, and Val. The side chains that point inward from the β strands tend to be bulky and hence fill the core of the β barrel (contrary to the impression that Figs. 8-19*b* and 8-52 might provide, α/β barrels, with one known exception, do not have hollow cores). Those side chains that fill the ends of the barrel are in contact with solvent and hence tend to be polar, whereas those in its center are out of contact with solvent and are therefore nonpolar. Thus, α/β barrels have four layers of polypeptide backbone interleaved by regions of hydrophobic side chains. In contrast, both α domains and β domains consist of two layers of polypeptide backbone sandwiching a hydrophobic core.

Around 10% of known enzyme structures contain an α/β barrel, making it the most common fold assumed by enzymes. Moreover, nearly all known α/β barrel proteins are enzymes. Intriguingly, the active sites of α/β barrel enzymes are almost all located in funnel-shaped pockets formed by the loops that link the C-termini of the β strands to the succeeding α helices and hence surround the mouth of the β barrel, an arrangement that has no obvious structural rationale. Thus, despite the observation that few α/β barrel proteins exhibit significant sequence identity, it has been postulated that all of them descended from a common ancestor and are therefore (distantly) related by **divergent evolution.** On the other hand, it has been argued that the α/β barrel is structurally so well suited for its enzymatic roles that it independently arose several times and hence that α/β enzymes are related by **convergent evolution** (i.e., nature has discovered the same fold on several occasions). Convincing evidence supporting either view

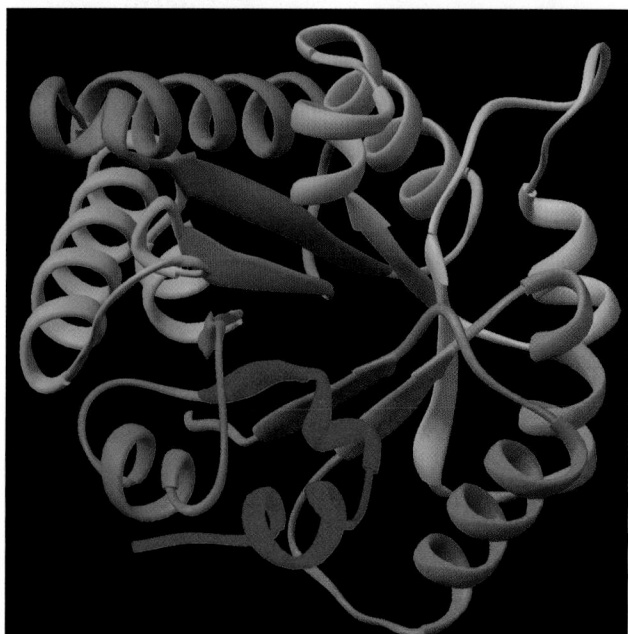

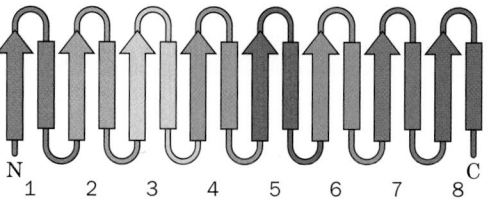

FIGURE 8-52 The X-ray structure of the 247-residue enzyme triose phosphate isomerase (TIM) from chicken muscle. The protein is viewed approximately along the axis of its α/β barrel. The peptide backbone is shown as a ribbon diagram with its successive βα units colored, from N- to C-terminus, in rainbow order, red to blue. The inset is its topological diagram (with α helices represented by rectangles). [Based on an X-ray structure by David Phillips, Oxford University, U.K. PDBid 1TIM.]

has not been forthcoming, so that the nature of the evolutionary relationships among the α/β barrel enzymes remains an open question.

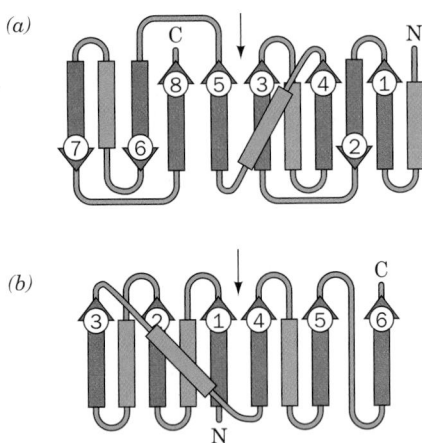

FIGURE 8-53   Topological diagrams of (*a*) carboxypeptidase A and (*b*) the N-terminal domain of glyceraldehyde-3-phosphate dehydrogenase. The X-ray structures of these proteins are diagrammed in Figs. 8-19*a* and 8-45. The thin black vertical arrows mark the proteins' topological switch points.

i. Open β Sheets

We have previously encountered examples of an open β sheet in the structures of carboxypeptidase A (Fig. 8-19*a*) and the N-terminal domain of glyceraldehyde-3-phosphate dehydrogenase (Fig. 8-45). Their topological diagrams are drawn in Fig. 8-53. The X-ray structures and topological diagrams of two other such proteins, those of the enzymes **lactate dehydrogenase** (N-terminal domain) and **adenylate kinase,** are shown in Fig. 8-54. Such folds consist of a central parallel or mixed β sheet flanked on both sides by α helices that form the right-handed crossover connections between successive parallel β strands (Fig. 8-20*b*). The strands of such a β sheet do not follow their order in the peptide sequence. Rather, the β sheet has a long crossover that reverses the direction of the succeeding section of sheet and turns it upside down, thereby putting its helical crossovers on the opposite side of the sheet from those of the preceding section (Fig. 8-55). Such assemblies are therefore also known as **doubly wound sheets** (in contrast to singly wound α/β barrels, whose helices are all on the

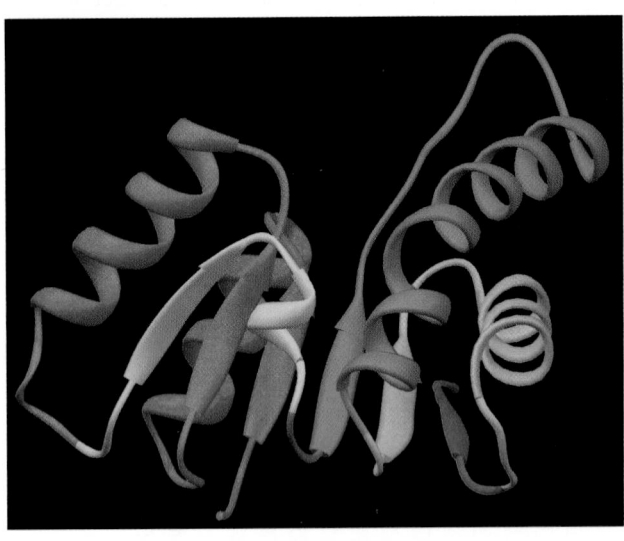

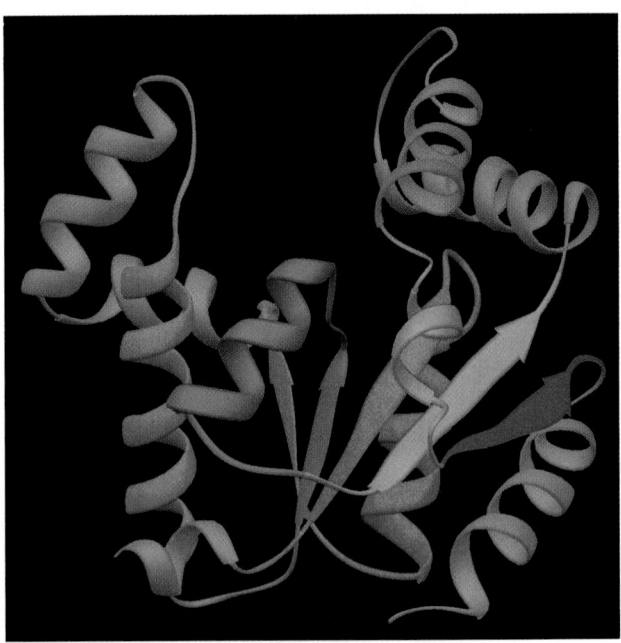

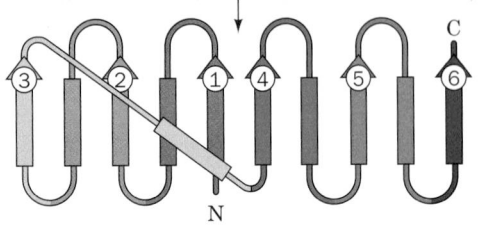

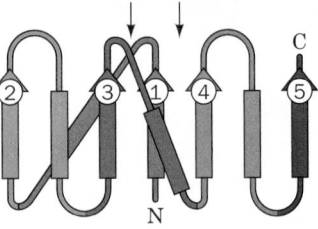

FIGURE 8-54   X-Ray structures of open β sheet–containing enzymes. (*a*) Dogfish lactate dehydrogenase, N-terminal domain (residues 20–163 of this 330-residue protein) and (*b*) porcine adenylate kinase (195 residues). The peptide backbones are shown as ribbon diagrams with successive βα units colored, from N- to C-terminus, in rainbow order, red to blue. In Part *b*, structural elements that are not components of

the open β sheet are gray. The insets are topological diagrams of these proteins, with the thin black vertical arrows marking their topological switch points. [Based on X-ray structures by (*a*) Michael Rossmann, Purdue University, and (*b*) Georg Schulz, Institut für Organische Chemie und Biochemie, Freiburg, Germany. PDBids (*a*) 6LDH & (*b*) 3ADK.]

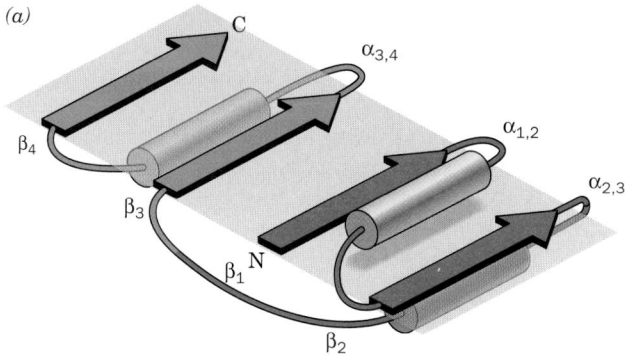

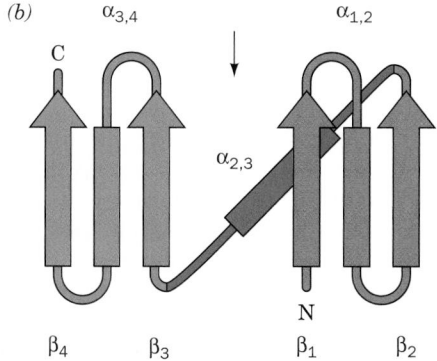

FIGURE 8-55 Doubly wound sheets. (*a*) A schematic diagram of a doubly wound sheet indicating how the long crossover (*green*) between its N- and C-terminal segments (*red and blue*) reverses the direction of the sheet's C-terminal segment and places its α helical crossover connections on the opposite side of the sheet. (*b*) The corresponding topological diagram, with the thin black vertical arrow marking the topological switch point. [After Branden, C. and Tooze, J., *Introduction to Protein Structure* (2nd ed.), *p.* 49, Garland (1999).]

same side of their β sheets). Doubly wound sheets consist of three layers of polypeptide backbone interspersed by regions of hydrophobic side chains (in contrast to four such layers for α/β barrels and two for α domains and β domains). Note that both types of domains containing parallel β sheets are hydrophobic on both sides of the sheet, whereas antiparallel sheets are hydrophobic on only one side. This additional stabilization of parallel β sheets probably compensates for the reduced strength of their nonlinear hydrogen bonds relative to the linear hydrogen bonds of antiparallel β sheets (Fig. 8-16).

There are few geometric constraints on the number of strands in open β sheets; they have been observed to contain from 4 to 10 β strands, with 6 β strands being the most common. Since the position where the chain reverses its winding direction, the so-called **topological switch point,** can occur between any consecutive α helix and β strand, doubly wound sheets can have many different folds. Moreover, some β strands may run in an antiparallel direction to yield mixed sheets (e.g., carboxypeptidase A; Figs. 8-19*a* and 8-53*a*), and in several instances, there is

more than one topological switch point (e.g., adenylate kinase; Fig. 8-54*b*).

The open β sheet is the most common domain structure that occurs in globular proteins. Moreover, nearly all such proteins are enzymes, many of which bind mono- or dinucleotides. In fact, the fold exemplified by lactate dehydrogenase (**LDH;** Fig. 8-54*a*) is known as the **dinucleotide-binding fold** or **Rossmann fold** (after Michael Rossmann, who first pointed out its significance). This is because mononucleotide units are commonly bound by βαβαβ units, of which LDH (which binds the dinucleotide NAD$^+$) has two. In some proteins, the second α helix in a βαβαβ unit is replaced by a length of nonhelical polypeptide. This occurs, for example, between the β5 and β6 strands of glyceraldehyde-3-phosphate dehydrogenase (Figs. 8-45 and 8-53*b*), which also binds NAD$^+$.

At the topological switch point of an open β sheet, the loops emerging from the C-terminal ends of the flanking β strands go to opposite sides of the sheet and thereby form a crevice between them. In nearly all of the >100 known structures of open β sheet–containing enzymes, as Carl-Ivar Brändén first pointed out, this crevice forms at least part of the enzyme's active site. Thus the active sites of both α/β barrels and the open β sheet enzymes are formed by loops emanating from the C-terminal ends of the β strands. In contrast, the active sites of enzymes that have other types of domain structures exhibit no apparent regularities in the positions of their active sites.

C. *Structural Bioinformatics*

In Section 7-4, we discussed the rapidly developing field of inquiry known as bioinformatics as it applies to the sequences of proteins and nucleic acids, that is, how sequence alignments are determined and how phylogenetic trees are generated. An equally important aspect of bioinformatics, which we discuss below, is how macromolecular structures are displayed and compared.

a. The Protein Data Bank

The atomic coordinates of most known macromolecular structures are archived in the **Protein Data Bank (PDB).** Indeed, most scientific journals that publish macromolecular structures now require that authors deposit their structure's coordinates in the PDB. The PDB contains the atomic coordinates of over 20,000 macromolecular structures (proteins, nucleic acids, and carbohydrates as determined by X-ray and other diffraction-based techniques, NMR, electron microscopy, and theoretical modeling) and is growing exponentially at a rate that is presently ~2500 structures per year. The PDB's website (URL), from which these coordinates are publicly available, is listed in Table 8-4.

Each independently determined structure in the PDB is assigned a unique four-character identifier (its **PDBid**) in which the first character must be a digit (1–9) and no distinction is made between uppercase and lowercase letters (e.g., 1MBO is the PDBid of the myoglobin structure illustrated in Fig. 8-39*c*, although PDBids do not necessarily have a relationship to the macromolecule's name).

TABLE 8-4   Structural Bioinformatics Websites (URLs)

Structural Databases

Protein Data Bank (PDB):
http://www.rcsb.org/pdb/

Nucleic Acid Databank:
http://ndbserver.rutgers.edu/NDB/ndb.html

Molecular Modeling Database (MMDB):
http://www.ncbi.nlm.nih.gov/Structure/index.shtml

PQS Protein Quaternary Structure Query Form at the EBI:
http://pqs.ebi.ac.uk/

Molecular Graphics Programs/Plug-ins

Chime:
http://www.mdli.com/cgi/dynamic/welcome.html

Cn3D:
http://www.ncbi.nlm.nih.gov/Structure/CN3D/cn3d.shtml

MAGE:
http://kinemage.biochem.duke.edu/

Protein Explorer:
http://www.umass.edu/microbio/chime/explorer/index.htm

RasMol:
http://www.bernstein-plus-sons.com/software/rasmol/ *and*
    http://www.umass.edu/microbio/rasmol/index.html

Virtual Reality Modeling Language (VRML):
Requires a VRML plug-in, available
    through http://www.web3d.org/vrml/vrml.htm

Structural Classification Algorithms

CATH (class, architecture, topology and homologous superfamily):
http://www.biochem.ucl.ac.uk/bsm/cath_new/index.html

CE (combinatorial extension of optimal pathway):
http://cl.sdsc.edu/

FSSP (fold classification based on structure–structure alignment of proteins):
http://www2.ebi.ac.uk/dali/fssp/

SCOP (structural classification of proteins):
http://scop.mrc-lmb.cam.ac.uk/scop/

VAST (vector alignment search tool):
http://www.ncbi.nlm.nih.gov/Structure/VAST/vast.shtml

A coordinate file begins with information that identifies/describes the macromolecule, the date the coordinate file was deposited, its source (the organism from which it was obtained), the author(s) who determined the structure, and key journal references. The file continues with a synopsis of how the structure was determined together with indicators of its accuracy and information that could be helpful in its interpretation, such as a description of its symmetry and which residues, if any, were not observed. The sequences of the structure's various chains are then listed together with the descriptions and formulas of its so-called HET (for heterogen) groups, which are molecular entities that are not among the "standard" amino acid or nucleotide residues (e.g., organic molecules such as the heme group, nonstandard residues such as Hyp, metal ions, and bound water molecules). The positions of the structure's

secondary structural elements and its disulfide bonds are then provided.

The bulk of a PDB file consists of a series of ATOM (for "standard" residues) and HETATM (for heterogens) records (lines), each of which provides the coordinates for one atom in the structure. An ATOM or HETATM record identifies its corresponding atom according to its serial number (usually just its sequence in the list), atom name (e.g., C and O for an amino acid residue's carbonyl C and O atoms, CA and CB for C_α and C_β atoms, N1 for atom N1 of a nucleic acid base, C4* for atom C4′ of a ribose or deoxyribose residue), residue name [e.g., PHE, G (for a guanosine residue), HEM (for a heme group), MG (for an Mg^{2+} ion), and HOH (for a water molecule)], chain identifier (e.g., A, B, C, etc., for structures consisting of more than one chain, whether or not the chains are chemically identical), and the residue sequence number in the chain. The record then continues with the atom's Cartesian (orthogonal) coordinates (X, Y, Z), in angstroms relative to an arbitrary origin, the atom's occupancy (which is the fraction of sites that actually contain the atom in question, a quantity that is usually 1.00 but, for groups that have multiple conformations or for molecules/ions that are only partially bound to a protein, may be a positive number less than 1.00), and its isotropic temperature factor (a quantity that is indicative of the atom's thermal motion, with larger numbers denoting a greater degree of motion). The ATOM records are listed in the order of the residues in a chain. For NMR-based structures, the PDB file contains a full set of ATOM and HETATM records for each member of the ensemble of structures that were calculated in solving the structure (Section 8-3A; the most representative member of such a coordinate set can be obtained from http://msd.ebi.ac.uk/pqs-nmr.html). PDB files usually end with CONECT (connectivity) records, which denote the nonstandard connectivities between atoms such as disulfide bonds and hydrogen bonds.

A particular PDB file may be located according to its PDBid or, if this is unknown, through a search that can identify files according to a variety of criteria including a protein's name, its author(s), its source, and/or the experimental technique used to determine its structure. Selecting a particular macromolecule in the PDB invokes the "Structure Explorer," which initially displays a summary page but in which the selected structure can be statically or interactively displayed (see below), its atomic coordinate file displayed and downloaded to the user's computer, and the structure classified and compared to other structures and/or analyzed in terms of its geometric properties and sequence (see below).

b. The Nucleic Acid Database

The **Nucleic Acid Database (NDB)** archives the atomic coordinates of structures that contain nucleic acids. Its coordinate files have substantially the same format as do those of the PDB, where this information is also kept. However, the NDB's organization and search algorithms are specialized for dealing with nucleic acids. This is useful, in part, because many nucleic acids of known structure are identified only by their sequences rather than by names, as are proteins (e.g., myoglobin), and consequently could easily be overlooked in a search of the PDB.

c. Viewing Macromolecular Structures in Three Dimensions

The most informative way to examine a macromolecular structure is through the use of molecular graphics programs. Such programs permit the user to interactively rotate a macromolecule and thereby perceive its three-dimensionality. This impression may be further enhanced by simultaneously viewing the macromolecule in stereo. Nearly all molecular graphics programs use PDB files as input. The programs described below can be downloaded from the websites listed in Table 8-4, some of which also provide instructions for the program's use.

RasMol, a widely used molecular graphics program written by Roger Sayle, is publicly available for use on a variety of computer platforms (Windows, MacOS, and UNIX). Its Web browser–based counterpart (plug-in) is named **Chime.** RasMol and Chime allow the user to simultaneously display different user-selected portions of a macromolecule in a variety of colors and formats (e.g., wireframe, ball and stick, backbone, space-filling, and cartoons). Many of the exercises on the CD-ROM that accompanies this textbook use Chime (which can also be downloaded from this CD-ROM). Moreover, the PDB provides the facility for viewing user-selected structures over the Web using the Chime-based program **Protein Explorer** by Eric Martz or, alternatively, **Virtual Reality Modeling Language (VRML).** Another molecular graphics progam, **MAGE,** which was written by David Richardson, displays the so-called **Kinemages** on the above CD-ROM (from which MAGE can also be downloaded). MAGE provides a generally more author-directed user environment than does RasMol or Chime.

d. Structural Classification and Comparison

Most proteins are structurally related to other proteins. Indeed, as we shall see in Section 9-6, evolution tends to conserve the structures of proteins rather than their sequences. The following paragraphs describe several publicly available websites that contain computational tools for the classification and comparison of protein structures. They can be accessed directly via their websites (Table 8-4) or linked to a selected protein in the PDB from the "Structural Neighbors" window of the Structure Explorer. Studies using these programs yield functional insights, reveal distant evolutionary relationships that are not apparent from sequence comparisons (Section 7-4B), generate libraries of unique folds for structure prediction, and provide indications as to why certain types of structures are preferred over others.

1. CATH (for *C*lass, *A*rchitecture, *T*opology and *H*omologous superfamily), as its name suggests, categorizes proteins in a four-level structural hierarchy. (1) "Class," the highest level, places the selected protein in one of four categories of gross secondary structure: Mainly α, Mainly β,

α/β (having both α helices and β sheets), and Few Secondary Structures. (2) "Architecture" is the description of the gross arrangement of secondary structure independent of topology. (3) "Topology" is indicative of both the overall shape and connectivity of the protein's secondary structures. (4) "Homologous superfamily" is those proteins of known structure that are homologous (share a common ancestor) to the selected protein. A static or an interactive (Chime/RasMol or VRML) drawing of each of the proteins can be displayed. For 1MBO (sperm whale myoglobin), the CATH classification is Class (C): Mainly alpha; Architecture (A): Orthogonal bundle; Topology (T): Globin-like; and Homologous superfamily (H): Globin. CATH permits the user to navigate up and down the various hierarchies and thereby structurally compare them.

2. CE (for *C*ombinatorial *E*xtension of the optimal path) finds all proteins in the PDB that can be structurally aligned with the query structure to within user-specified geometric criteria. The amino acid sequences of any or all of these proteins can be aligned on the basis of this structural alignment rather than sequence alignment (Section 7-4B). The structurally aligned proteins can be simultaneously displayed via RasMol, Protein Explorer, or a Java applet called Compare 3D that displays both the aligned C_α backbones and the aligned sequences. CE can likewise optimally align and display two user-selected structures.

3. FSSP (*F*old classification based on *S*tructure–*S*tructure alignment of *P*roteins) lists the protein structures in the PDB which, at least in part, structurally resemble that of the query protein based on continuously updated all-against-all comparisons of the protein structures in the PDB. These structural comparisons are made by a program called **Dali** based on the distances between the various atoms in each domain of a protein. One or more proteins, with their similar portions structurally aligned, can be simultaneously displayed using Chime. The structure-based sequence alignments of selected proteins or an entire family of proteins can also be displayed.

4. SCOP (*S*tructural *C*lassification *O*f *P*roteins) classifies protein structures based mainly on manually generated topological considerations according to a 6-level hierarchy: Class [All-α, All-β, α/β (having α helices and β strands that are largely interspersed), α+β (having α helices and β strands that are largely segregated), and Multi-domain (having domains of different classes)], Fold (groups that have similar arrangements of secondary structural elements), Superfamily (indicative of distant evolutionary relationships based on structural criteria and functional features), Family (indicative of near evolutionary relationships based on sequence as well as on structure), Protein, and Species. For 1MBO these are Class: All-α; Fold: Globin-like; Superfamily: Globin-like; Family: Globins; Protein: Myoglobin; and Species: Sperm whale (*Physeter catodon*). SCOP permits the user to navigate through its treelike hierarchical organization and lists the known members of any particular branch. Thus, with 1MBO, SCOP displays a list of all 137 structures in the PDB that contain sperm whale myoglobin (a protein that has received far more structural study than most). A selected protein or a member of a selected branch can be displayed using Chime.

5. VAST (*Vector Alignment Search Tool*), a component of the National Center for Biotechnology Information (NCBI) Entrez system, reports a precomputed list of proteins of known structure that structurally resemble the query protein ("structure neighbors"). The VAST system uses the **Molecular Modeling Database (MMDB),** an NCBI-generated database that is derived from PDB coordinates but in which molecules are represented by connectivity graphs rather than sets of atomic coordinates. VAST displays the superposition of the query protein in its structural alignment with up to five other proteins using **Cn3D** [a molecular graphics program that displays MMDB files and that is publicly available for a variety of computer platforms (Table 8-4)] or with only one other protein using MAGE. VAST also reports a precomputed list of proteins that are similar to the query protein in sequence ("sequence neighbors") and provides links from a selected protein to several bibliographic databases including MedLine.

A number of other structural analysis/classification/comparison tools can be invoked from the "Other Sources" window of the Structure Explorer. The "Sequence Details" window provides the sequence of each chain in the structure and, for polypeptides, indicates the secondary structure of each of its residues.

4 ■ PROTEIN STABILITY

Incredible as it may seem, thermodynamic measurements indicate that *native proteins are only marginally stable entities under physiological conditions*. The free energy required to denature them is ~0.4 kJ · mol^{-1} of amino acid residues, so that 100-residue proteins are typically stable by only around 40 kJ · mol^{-1}. In contrast, the energy required to break a typical hydrogen bond is ~20 kJ · mol^{-1}. The various noncovalent influences to which proteins are subject—electrostatic interactions (both attractive and repulsive), hydrogen bonding (both intramolecular and to water), and hydrophobic forces—each have energetic magnitudes that may total thousands of kilojoules per mole over an entire protein molecule. Consequently, *a protein structure arises from a delicate balance among powerful countervailing forces*. In this section we discuss the nature of these forces and end by considering protein denaturation, that is, how these forces can be disrupted.

A. Electrostatic Forces

Molecules are collections of electrically charged particles and hence, to a reasonable degree of approximation, their interactions are determined by the laws of classical electrostatics (more exact calculations require the application of quantum mechanics). The energy of association, U, of two electric charges, q_1 and q_2, that are separated by the distance r is found by integrating the expression for Coulomb's law, Eq. [2.1], to determine the work necessary to separate these charges by an infinite distance:

$$U = \frac{kq_1q_2}{Dr} \qquad [8.1]$$

Here $k = 9.0 \times 10^9$ J · m · C^{-2} and D is the dielectric constant of the medium in which the charges are immersed (recall that $D = 1$ for a vacuum and, for the most part, increases with the polarity of the medium; Table 2-1). The dielectric constant of a molecule-sized region is difficult to estimate. For the interior of a protein, it is usually taken to be in the range 3 to 5 in analogy with the measured dielectric constants of substances that have similar polarities, such as benzene and diethyl ether.

Coulomb's law is only valid for point or spherically symmetric charges that are immersed in a medium of constant D. However, proteins are by no means spherical and their internal D values vary with position. Moreover, a protein in solution associates with mobile ions such as Na$^+$ and Cl$^-$, which modulate the protein's electrostatic potential. Consequently, calculating the electrostatic potential of a protein requires mathematically sophisticated and computationally intensive algorithms that are beyond the scope of this text. These methods are widely used to calculate the surface electrostatic potentials of proteins using a program called **GRASP** (for Graphical Representation and Analysis of Surface Properties) written by Anthony Nicholls, Kim Sharp, and Barry Honig. Figure 8-56 shows

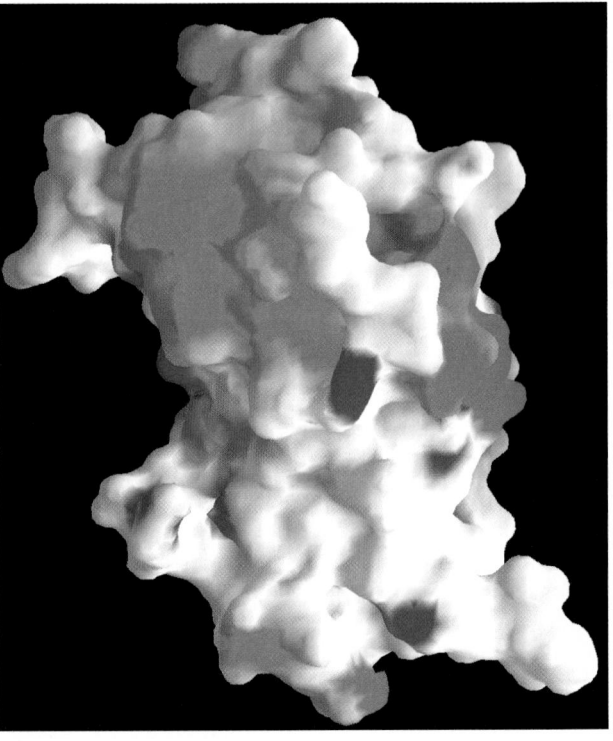

FIGURE 8-56   A GRASP diagram of human growth hormone. The diagram shows the protein's surface colored according to its electrostatic potential, with its most negative areas dark red, its most positive areas dark blue, and its neutral areas white. The protein's orientation is the same as that in Fig. 8-47b. [Based on an X-ray structure by Alexander Wlodawer, National Cancer Institute, Frederick, Maryland. PDBid 1HGU.]

a GRASP diagram of human growth hormone in which the protein's surface is colored according to its electrostatic potential. Such diagrams are useful for assessing how a protein might associate with charged molecules such as other proteins, nucleic acids, and substrates. Similar computations are used to predict the pK's of protein surface groups, which can have significant application in the elucidation of an enzyme's mechanism of action (Section 15-1).

a. Ionic Interactions Are Strong but Do Not Greatly Stabilize Proteins

The association of two ionic protein groups of opposite charge is known as an **ion pair** or **salt bridge.** According to Eq. [8.1], the energy of a typical ion pair, say the carboxyl group of Glu and the ammonium group of Lys, whose charge centers are separated by 4.0 Å in a medium of dielectric constant 4, is $-86 \text{ kJ} \cdot \text{mol}^{-1}$ (one electronic charge $= 1.60 \times 10^{-19}$ C). However, free ions in aqueous solution are highly solvated, and the formation of a salt bridge has the entropic penalty of localizing the salt bridge's charged side chains. Consequently, the free energy of solvation of two separated ions is about equal to the free energy of formation of their unsolvated ion pair. *Ion pairs therefore contribute little stability toward a protein's native structure.* This accounts for the observations that although ~75% of charged residues occur in ion pairs, very few ion pairs are buried (unsolvated), and ion pairs that are exposed to the aqueous solvent tend to be but poorly conserved among homologous proteins.

b. Dipole–Dipole Interactions Are Weak but Significantly Stabilize Protein Structures

The noncovalent associations between electrically neutral molecules, collectively known as **van der Waals forces,** arise from electrostatic interactions among permanent and/or induced dipoles. These forces are responsible for numerous interactions of varying strengths between nonbonded neighboring atoms. (The hydrogen bond, a special class of dipolar interaction, is considered separately in Section 8-4B.)

Interactions among permanent dipoles are important structural determinants in proteins because many of their groups, such as the carbonyl and amide groups of the peptide backbone, have permanent dipole moments. These interactions are generally much weaker than the charge–charge interactions of ion pairs. Two carbonyl groups, for example, each with dipoles of 4.2×10^{-30} C · m (1.3 debye units) that are oriented in an optimal head-to-tail arrangement (Fig. 8-57a) and separated by 5 Å in a medium of dielectric constant 4, have a calculated attractive energy of only $-9.3 \text{ kJ} \cdot \text{mol}^{-1}$. Furthermore, these energies vary with r^{-3}, so they rapidly attenuate with distance. In α helices, however, the negative ends of the dipolar amide and carbonyl groups of the polypeptide backbone all point in the same direction (Fig. 8-11), so that their interactions and bond dipoles are additive (these groups, of course, also form hydrogen bonds, but here we are concerned with their residual electric fields). The α helix therefore has a significant dipole moment that is positive toward the N-terminus

and negative toward the C-terminus. Consequently, *in the low dielectric constant core of a protein, dipole–dipole interactions significantly influence protein folding.*

A permanent dipole also induces a dipole moment on a neighboring group so as to form an attractive interaction (Fig. 8-57b). Such dipole–induced dipole interactions are generally much weaker than are dipole–dipole interactions.

Although nonpolar molecules are nearly electrically neutral, at any instant they have a small dipole moment resulting from the rapid fluctuating motions of their electrons. This transient dipole moment polarizes the electrons in a neighboring group, thereby giving rise to a dipole mo-

(a) Interactions between permanent dipoles

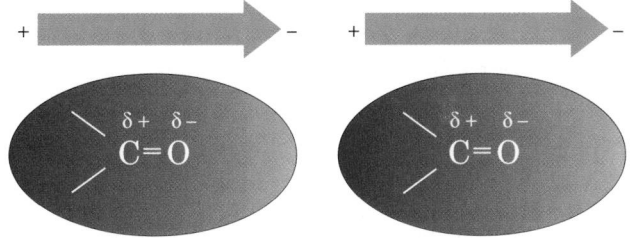

(b) Dipole–induced dipole interactions

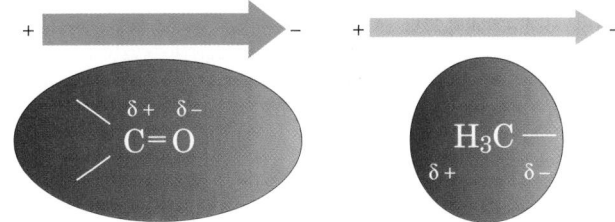

(c) London dispersion forces

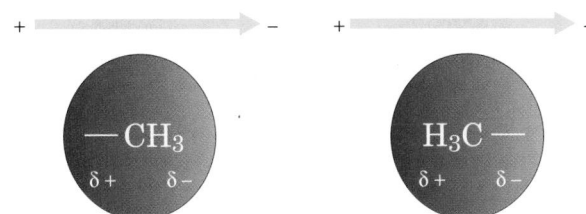

FIGURE 8-57 Dipole–dipole interactions. The strength of each dipole is represented by the thickness of the accompanying arrow. (*a*) Interactions between permanent dipoles. These interactions, here represented by carbonyl groups lined up head to tail, may be attractive, as shown, or repulsive, depending on the relative orientations of the dipoles. (*b*) Dipole–induced dipole interactions. A permanent dipole (here shown as a carbonyl group) induces a dipole in a nearby group (here represented by a methyl group) by electrostatically distorting its electron distribution (*shading*). This always results in an attractive interaction. (*c*) London dispersion forces. The instantaneous charge imbalance (*shading*) resulting from the motions of the electrons in a molecule (*left*) induces a dipole in a nearby group (*right*); that is, the motions of the electrons in neighboring groups are correlated. This always results in an attractive interaction.

ment (Fig. 8-57c) such that, near their van der Waals contact distances, the groups are attracted to one another (a quantum mechanical effect that cannot be explained in terms of only classical physics). These so-called **London dispersion forces** are extremely weak. The 8.2-kJ $\cdot$ mol^{-1} heat of vaporization of CH_4, for example, indicates that the attractive interaction of a nonbonded H $\cdots$ H contact between neighboring CH_4 molecules is roughly -0.3 kJ $\cdot$ mol^{-1} (in the liquid, a CH_4 molecule touches its 12 nearest neighbors with $\sim$2 H $\cdots$ H contacts each).

London forces are only significant for contacting groups because their association energy is proportional to r^{-6}. Nevertheless, *the great numbers of interatomic contacts in the closely packed interiors of proteins make London forces a major influence in determining their conformations.* London forces also provide much of the binding energy in the sterically complementary interactions between proteins and the molecules that they specifically bind.

B. *Hydrogen Bonding Forces*

Hydrogen bonds (D—H $\cdots$ A), as we discussed in Section 2-1A, are predominantly electrostatic interactions (but with $\sim$10% covalent character) between a weakly acidic donor group (D—H) and an acceptor (A) that bears a lone pair of electrons. In biological systems, D and A can both be the highly electronegative N and O atoms and occasionally S atoms. In addition, a relatively acidic C—H group (e.g., a C_α—H group) can act as a weak hydrogen bond donor, and the polarizable π electron system of an aromatic ring (e.g., that of Trp) can act as a weak acceptor.

Hydrogen bonds have association energies that are normally in the range -12 to -40 kJ $\cdot$ mol^{-1} (but only around -8 to -16 kJ $\cdot$ mol^{-1} for C—H $\cdots$ A and D—H $\cdots$ π hydrogen bonds and -2 to -4 kJ $\cdot$ mol^{-1} for C—H $\cdots$ π hydrogen bonds), values which are between those for covalent bonds and van der Waals forces. Hydrogen bonds **(H bonds)** are much more directional than are van der Waals forces but less so than are covalent bonds. The D $\cdots$ A distance is normally in the range 2.7 to 3.1 Å, although since H atoms are unseen in all but the very highest resolution macromolecular X-ray structures, a possible D—H $\cdots$ A interaction (where D and A are either N or O) is assumed to be a H bond if its D $\cdots$ A distance is significantly less than the 3.7 Å sum of a D—H bond length ($\sim$1.0 Å) and the van der Waals contact distance between H and A ($\sim$2.7 Å). Keep in mind, however, that there is no rigid cutoff distance beyond which H bonds cease to exist because the energy of an H bond, which is mainly electrostatic in character, varies inversely with the distance between the negative and positive centers (Eq. [8.1]).

H bonds tend to be linear, with the D—H bond pointing along the acceptor's lone pair orbital (or, in D—H $\cdots$ π hydrogen bonds, roughly perpendicular to the aromatic ring and pointing at its center with the distance from the D atom to the center of the aromatic ring normally in the range 3.2–3.8 Å). Large deviations from this ideal geometry are not unusual, however. For example, in the H bonds of both α helices (Fig. 8-11) and antiparallel β

pleated sheets (Fig. 8-16a), the N—H bonds point approximately along the C=O bonds rather than along an O lone pair orbital, and in parallel β pleated sheets (Fig. 8-16b), the H bonds depart significantly from linearity. Indeed, many of the H bonds in proteins are members of networks in which each donor is H bonded to two acceptors (a **bifurcated hydrogen bond**) and each acceptor is H bonded to two donors. For example, although the H bonds in ideal α helices form between the N—H group at residue n and the C=O group at residue $n - 4$ ($n \rightarrow n - 4$ H bonds), many of the N—H groups in real α helices associate via bifurcated H bonds with two adjacent C=O groups to form both $n \rightarrow n - 4$ and $n \rightarrow n - 3$ H bonds.

a. Hydrogen Bonds Only Weakly Stabilize Proteins

A protein's internal H bonding groups are arranged such that most possible H bonds are formed (Section 8-3B). Clearly, H bonding has a major influence on the structures of proteins. However, an unfolded protein makes most of its H bonds with the water molecules of the aqueous solvent (water, it will be recalled, is a strong H bonding donor and acceptor). The free energy of stabilization that internal H bonds confer on a native protein is therefore equal to the difference in the free energy of H bonding between the native protein and the unfolded protein. Consequently, it might be expected that H bonds do not stabilize (and perhaps even slightly destabilize) the structure of a native protein relative to its unfolded state. However, since H bonding interactions are largely electrostatic in nature, they are likely to be stronger in the low polarity interior of a protein than they are in the high polarity aqueous medium. Moreover, there may be an entropic effect that destabilizes the H bonds between water and an unfolded polypeptide relative to intraprotein H bonds: The water molecules that are H bonded to a polypeptide are likely to be more positionally and orientationally constrained (ordered) than those that are H bonded to only other water molecules, thus favoring the formation of intraprotein H bonds. These effects may very well account for the observation that the mutagenic removal of an H bond from a protein generally reduces the protein's stability by -2 to 8 kJ $\cdot$ mol^{-1}.

Despite their low stability, *a protein's hydrogen bonds provide a structural basis for its native folding pattern:* If a protein folded in a way that prevented some of its internal H bonds from forming, their free energy would be lost and such conformations would be less stable than those that are fully H bonded. In fact, the formation of α helices and β sheets efficiently satisfies the polypeptide backbone's H bonding requirements. This argument also applies to the van der Waals forces discussed in the previous section.

b. Most Hydrogen Bonds in Proteins Are Local

How can as complex a molecule as a protein fold so as to make nearly all of its potential H bonds? The answer to this question was revealed by a survey of the H bonds in high resolution protein X-ray structures by Ken Dill and George Rose: *Most of the H bonds in a protein are local, that is, they involve donors and acceptors that are close* together in sequence and hence can readily find their H bonding mates.

1. On average, 68% of the H bonds in proteins are between backbone atoms. Of these, ~1/3 form $n \rightarrow n - 4$ H bonds (as in ideal α helices), ~1/3 form $n \rightarrow n - 3$ H bonds (as in reverse turns and ideal 3_{10} helices), and ~1/3 are between paired strands in β sheets. In fact, only ~5% of the H bonds between backbone atoms are not wholly within a helix, sheet, or turn.

2. Hydrogen bonds between side chains and backbones are clustered at **helix-capping** positions. In an α helix, the first four N—H groups and the last four C=O groups cannot form H bonds within the helix (which accounts for half the potential H bonds involving backbone atoms in an α helix of 12 residues, the average length of α helices). These potential H bonds are often made with nearby side chains. In particular, ~1/2 of the N-terminal N—H groups of α helices form H bonds with polar side chains that are 1 to 3 residues distant, and ~1/3 of their C-terminal C=O groups form H bonds with polar side chains that are 2 to 5 residues distant.

3. Over half the H bonds between side chains are between charged residues (i.e., they form salt bridges) and are therefore located on protein surfaces between and within surface loops. However, ~85% of the remaining side chain–side chain H bonds are between side chains that are 1 to 5 residues apart. Hence with the exception of those in salt bridges, side chain–side chain H bonds also tend to be local.

C. *Hydrophobic Forces*

*The **hydrophobic effect** is the name given to those influences that cause nonpolar substances to minimize their contacts with water and amphipathic molecules, such as soaps and detergents, to form micelles in aqueous solutions (Section 2-1B).* Since native proteins form a sort of intramolecular micelle in which their nonpolar side chains are largely out of contact with the aqueous solvent, *hydrophobic interactions must be an important determinant of protein structures.*

The hydrophobic effect derives from the special properties of water as a solvent, only one of which is its high dielectric constant. In fact, other polar solvents, such as dimethyl sulfoxide (DMSO) and *N,N*-dimethylformamide (DMF), tend to denature proteins. The thermodynamic data of Table 8-5 provide considerable insight as to the origin of the hydrophobic effect because the transfer of a hydrocarbon from water to a nonpolar solvent resembles the transfer of a nonpolar side chain from the exterior of a protein in aqueous solution to its interior. The isothermal Gibbs free energy changes ($\Delta G = \Delta H - T\Delta S$) for the transfer of a hydrocarbon from an aqueous solution to a nonpolar solvent is negative in all cases, which indicates, as we know to be the case, that such transfers are spontaneous processes (oil and water don't mix). What is perhaps unexpected is that these transfer processes are endothermic (positive ΔH) for aliphatic compounds and athermic ($\Delta H = 0$) for aromatic compounds; that is, *it is enthalpically more or equally favorable for nonpolar molecules to dissolve in water than in nonpolar media.* In contrast, the

TABLE 8-5 **Thermodynamic Changes for Transferring Hydrocarbons from Water to Nonpolar Solvents at 25°C**[a]

Process	ΔH (kJ · mol^{-1})	$-T\Delta S_u$ (kJ · mol^{-1})	ΔG_u (kJ · mol^{-1})
CH_4 in H_2O $\rightleftharpoons$ CH_4 in C_6H_6	11.7	−22.6	−10.9
CH_4 in H_2O $\rightleftharpoons$ CH_4 in CCl_4	10.5	−22.6	−12.1
C_2H_6 in H_2O $\rightleftharpoons$ C_2H_6 in benzene	9.2	−25.1	−15.9
C_2H_4 in H_2O $\rightleftharpoons$ C_2H_4 in benzene	6.7	−18.8	−12.1
C_2H_2 in H_2O $\rightleftharpoons$ C_2H_2 in benzene	0.8	−8.8	−8.0
Benzene in H_2O $\rightleftharpoons$ liquid benzene[b]	0.0	−17.2	−17.2
Toluene in H_2O $\rightleftharpoons$ liquid toluene[b]	0.0	−20.0	−20.0

[a]ΔG_u, the **unitary Gibbs free energy change,** is the Gibbs free energy change, ΔG, corrected for its concentration dependence so that it reflects only the inherent properties of the substance in question and its interaction with solvent. This relationship, according to Equation [3.13], is

$$\Delta G_u = \Delta G - nRT \ln\frac{[A_f]}{[A_i]}$$

where $[A_i]$ and $[A_f]$ are the initial and final concentrations of the substance under consideration, respectively, and n is the number of moles of that substance. Since the second term in this equation is a purely entropic term (concentrating a substance increases its order), ΔS_u, the **unitary entropy change,** is expressed

$$\Delta S_u = \Delta S + nR \ln\frac{[A_f]}{[A_i]}$$

[b]Data measured at 18°C.

Source: Kauzmann, W., *Adv. Protein Chem.* **14,** 39 (1959).

entropy component of the unitary free energy change, $-T\Delta S_u$ (see footnote *a* to Table 8-5), is large and negative in all cases. Evidently, *the transfer of a hydrocarbon from an aqueous medium to a nonpolar medium is entropically driven. The same is true of the transfer of a nonpolar protein group from an aqueous environment to the protein's nonpolar interior.*

What is the physical mechanism whereby nonpolar entities are excluded from aqueous solution? Recall that entropy is a measure of the order of a system; it decreases with increasing order (Section 3-2). Thus the decrease in entropy when a nonpolar molecule or side chain is solvated by water (the reverse of the foregoing process) must be due to an ordering process. This is an experimental observation, not a theoretical conclusion. The magnitudes of the entropy changes are too large to be attributed only to changes in the conformations of the hydrocarbons; rather, as Henry Frank and Marjorie Evans pointed out in 1945, *these entropy changes mainly arise from some sort of ordering of the water structure.*

Liquid water has a highly ordered and extensively H bonded structure (Section 2-1A). The insinuation of a nonpolar group into this structure disrupts it: A nonpolar group can neither accept nor donate H bonds, so the water molecules at the surface of the cavity occupied by the nonpolar group cannot H bond to other molecules in their usual fashion. In order to recover the lost H bonding energy, these surface waters must orient themselves so as to form an H bonded network enclosing the cavity (Fig. 8-58). This orientation constitutes an ordering of the water structure, since the number of ways that water molecules can form H bonds about the surface of a nonpolar group is less than the number of ways that they can H bond in bulk water.

Unfortunately, the complexity of liquid water's basic structure (Section 2-1A) has not yet allowed a detailed structural description of this ordering process. One model that has been proposed is that water forms quasi-crystalline H bonded cages about the nonpolar groups similar to those of **clathrates** (Fig. 8-59). The magnitudes of the entropy changes that result when nonpolar substances are dissolved in water, however, indicate that the resulting water structures can only be slightly more ordered than bulk water. They also must be quite different from that of ordinary ice, because, for instance, the solvation of nonpolar groups by water causes a large decrease in water volume (e.g., the transfer of CH_4 from hexane to water shrinks the water solution by 22.7 mL · mol^{-1} of CH_4), whereas the freezing of water results in a 1.6-mL · mol^{-1} expansion.

The unfavorable free energy of hydration of a nonpolar substance caused by its ordering of the surrounding water molecules has the net result that *the nonpolar substance is excluded from the aqueous phase.* This is because the surface area of a cavity containing an aggregate of nonpolar molecules is less than the sum of the surface areas of the cavities that each of these molecules would individually occupy. The aggregation of the nonpolar groups thereby minimizes the surface area of the cavity and therefore the entropy loss of the entire system. In a sense, the nonpolar groups are squeezed out of the aqueous phase by the hy-

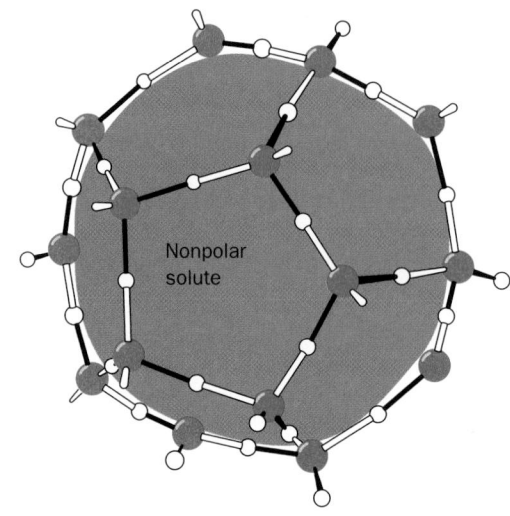

FIGURE 8-58   The orientational preference of water molecules next to a nonpolar solute. In order to maximize their H bonding energy, these water molecules tend to straddle the inert solute such that two or three of their tetrahedral directions are tangential to its surface. This permits them to form H bonds (*black*) with neighboring water molecules lining the nonpolar surface. This ordering of water molecules extends several layers of water molecules beyond the first hydration shell of the nonpolar solute. [Illustration, Irving Geis/Geis Archives Trust. Copyright Howard Hughes Medical Institute. Reproduced with permission.]

drophobic interactions. Thermodynamic measurements indicate that the free energy change of removing a —CH_2— group from an aqueous solution is about -3 kJ · mol^{-1}. Although this is a relatively small amount of free energy,

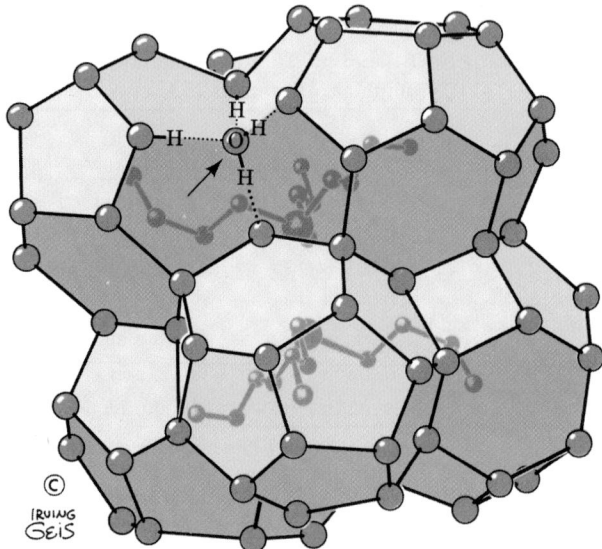

FIGURE 8-59   Structure of the clathrate (n-C_4H_9)$_3$S$^+$F$^-$ · 23H$_2$O. Clathrates are crystalline complexes of nonpolar compounds with water (usually formed at low temperatures and high pressures) in which the nonpolar molecules are enclosed, as shown, by a polyhedral cage of tetrahedrally H bonded water molecules (here represented by only their oxygen atoms). The H bonding interactions of one such water molecule (*arrow*) are shown in detail. [Illustration, Irving Geis/Geis Archives Trust. Copyright Howard Hughes Medical Institute. Reproduced with permission.]

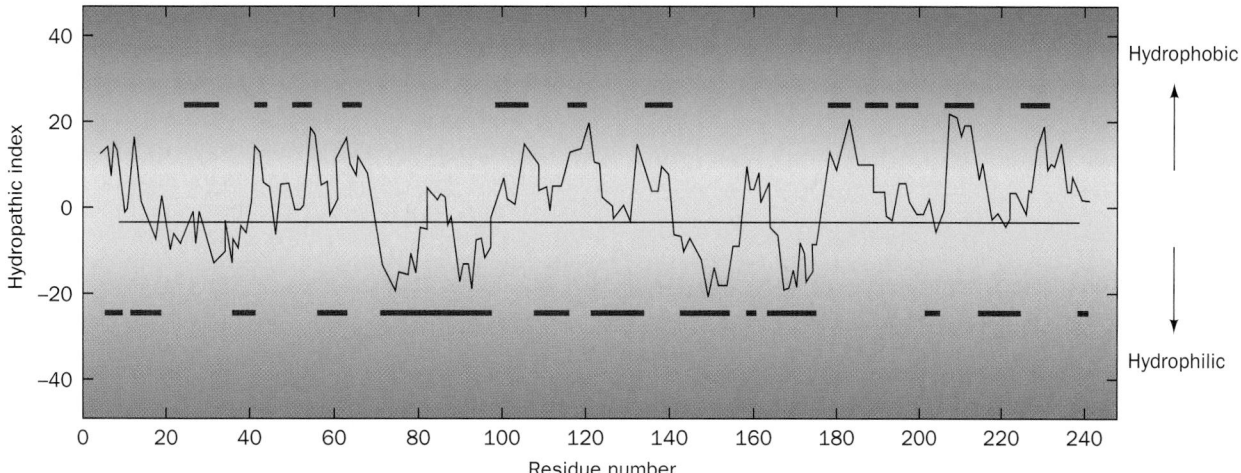

FIGURE 8-60   Hydropathic index plot for bovine chymotrypsinogen. The sum of the hydropathies of nine consecutive residues (see Table 8-6) are plotted versus the residue sequence number. A large positive hydropathic index is indicative of a hydrophobic region of the polypeptide chain, whereas a large negative value is indicative of a hydrophilic region. The bars above the midpoint line denote the protein's interior regions, as determined by X-ray crystallography, and the bars below the midpoint line indicate the protein's exterior regions. [After Kyte, J. and Doolittle, R.F., *J. Mol. Biol.* **157,** 111 (1982).]

in molecular assemblies involving large numbers of non-polar contacts, hydrophobic interactions are a potent force.

Walter Kauzmann pointed out in 1958 that *hydrophobic forces are a major influence in causing proteins to fold into their native conformations.* Figure 8-60 indicates that the amino acid side chain **hydropathies** (indexes of combined hydrophobic and hydrophilic tendencies; Table 8-6) are, in fact, good predictors of which portions of a polypeptide chain are inside a protein, out of contact with the aqueous solvent, and which portions are outside, in contact with the aqueous solvent. In proteins, the effects of hydrophobic forces are often termed **hydrophobic bonding,** presumably to indicate the specific nature of protein folding under the influence of the hydrophobic effect. You should keep in mind, however, that hydrophobic bonding does not generate the directionally specific interactions usually associated with the term "bond."

D. *Disulfide Bonds*

Since disulfide bonds form as a protein folds to its native conformation (Section 9-1A), they function to stabilize its three-dimensional structure. The relatively reducing chemical character of the cytoplasm, however, greatly diminishes the stability of intracellular disulfide bonds. In fact, almost all proteins with disulfide bonds are secreted to more oxidized extracellular destinations, where their disulfide bonds are effective in stabilizing protein structures [secreted proteins fold to their native conformations—and hence form their disulfide bonds—in the endoplasmic reticulum (Section 12-4B), which, unlike other cellular compartments, has an oxidizing environment]. Apparently, the relative "hostility" of extracellular environments toward proteins (e.g., uncontrolled temperatures and pH's) requires the additional structural stability conferred by disulfide bonds.

E. *Protein Denaturation*

The low conformational stabilities of native proteins make them easily susceptible to denaturation by altering the balance of the weak nonbonding forces that maintain the native conformation. When a protein in solution is heated,

TABLE 8-6   Hydropathy Scale for Amino Acid Side Chains

Side Chain	Hydropathy
Ile	4.5
Val	4.2
Leu	3.8
Phe	2.8
Cys	2.5
Met	1.9
Ala	1.8
Gly	−0.4
Thr	−0.7
Ser	−0.8
Trp	−0.9
Tyr	−1.3
Pro	−1.6
His	−3.2
Glu	−3.5
Gln	−3.5
Asp	−3.5
Asn	−3.5
Lys	−3.9
Arg	−4.5

Source: Kyte, J. and Doolitle, R.F., *J. Mol. Biol.* **157,** 110 (1982).

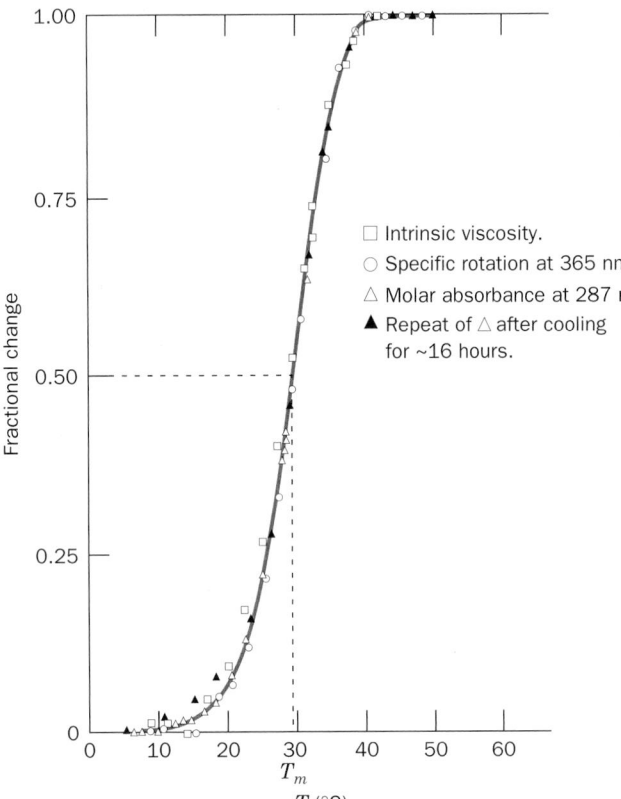

FIGURE 8-61   Protein denaturation. The heat-induced denaturation of bovine pancreatic ribonuclease A (RNase A) in an HCl–KCl solvent at pH 2.1 and 0.019 ionic strength was monitored by several conformationally sensitive techniques. The curve is drawn only through the points △. The melting temperature, T_m, is defined as the temperature at the midpoint of the transition. Compare the shape of this melting curve with that of duplex DNA (Fig. 5-16). [After Ginsburg, A. and Carroll, W.R., *Biochemistry* **4**, 2169 (1965).]

its conformationally sensitive properties, such as optical rotation (Section 4-2A), viscosity, and UV absorption, change abruptly over a narrow temperature range (e.g., Fig. 8-61). *Such a nearly discontinuous change indicates that the native protein structure unfolds in a cooperative manner: Any partial unfolding of the structure destabilizes the remaining structure, which must simultaneously collapse to the random coil.* The temperature at the midpoint of this process is known as the protein's **melting temperature,** T_m, in analogy with the melting of a solid. Most proteins have T_m values well below 100°C. Recall that nucleic acids likewise have characteristic T_m's (Section 5-3C).

In addition to high temperatures, proteins are denatured by a variety of other conditions and substances:

1. pH variations alter the ionization states of amino acid side chains (Table 4-1), which changes protein charge distributions and H bonding requirements.

2. Detergents, some of which significantly perturb protein structures at concentrations as low as $10^{-6}M$,

hydrophobically associate with the nonpolar residues of a protein, thereby interfering with the hydrophobic interactions responsible for the protein's native structure.

3. High concentrations of water-soluble organic substances, such as aliphatic alcohols, interfere with the hydrophobic forces stabilizing protein structures through their own hydrophobic interactions with water. Organic substances with several hydroxyl groups, such as ethylene glycol or sucrose,

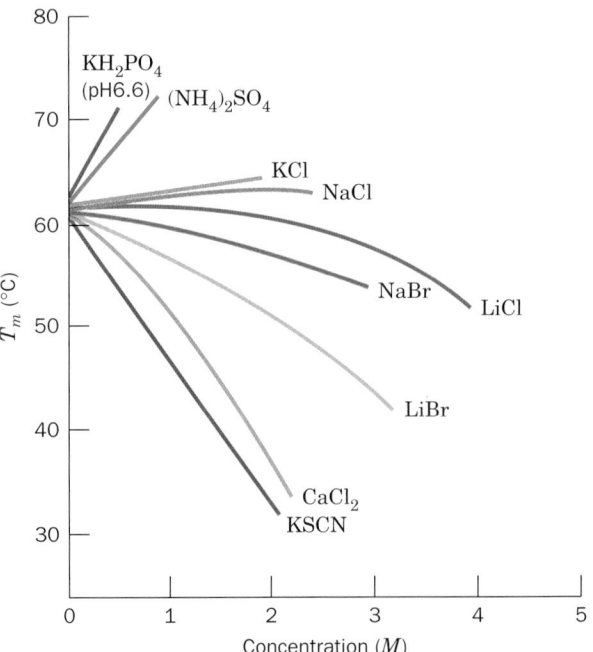

however, are relatively poor denaturants because their H bonding ability renders them less disruptive of water structure.

The influence of salts is more variable. Figure 8-62 shows the effects of a number of salts on the T_m of bovine pancreatic **ribonuclease A (RNase A).** Some salts, such as $(NH_4)_2SO_4$ and KH_2PO_4, stabilize the native protein structure (raise its T_m); others, such as KCl and NaCl, have little effect; and yet others, such as KSCN and LiBr, destabilize it. The order of effectiveness of the various ions in stabilizing a protein, which is largely independent of the identity

FIGURE 8-62   Melting temperature of RNase A as a function of the concentrations of various salts. All solutions also contained 0.15M KCl and 0.013M sodium cacodylate buffer, pH 7. [After von Hippel, P.J. and Wong, K.Y., *J. Biol. Chem.* **10**, 3913 (1965).]

of the protein, parallels their capacity to salt out proteins (Section 6-2A). This order is known as the **Hofmeister series:**

Anions: $SO_4^{2-} > H_2PO_4^- > CH_3COO^- > Cl^-$
$> Br^- > I^- > ClO_4^- > SCN^-$

Cations: $NH_4^+, Cs^+, K^+, Na^+ > Li^+ > Mg^{2+}$
$> Ca_2^+ > Ba^{2+}$

The ions in the Hofmeister series that tend to denature proteins, I^-, ClO_4^-, SCN^-, Li^+, Mg^{2+}, Ca^{2+}, and Ba^{2+}, are said to be **chaotropic.** This list should also include the guanidinium ion (Gu^+) and the nonionic urea, which, in concentrations in the range 5 to $10M$, are the most commonly used protein denaturants. The effect of the various ions on proteins is largely cumulative: GuSCN is a much more potent denaturant than the often used GuCl, whereas Gu_2SO_4 stabilizes protein structures.

Chaotropic agents increase the solubility of nonpolar substances in water. Consequently, their effectiveness as denaturing agents stems from their ability to disrupt hydrophobic interactions, although the manner in which they do so is not well understood. Conversely, those substances listed that stabilize proteins strengthen hydrophobic forces, thus increasing the tendency of water to expel proteins. This accounts for the correlation between the abilities of an ion to stabilize proteins and to salt them out.

F. *Explaining the Stability of Thermostable Proteins*

Certain species of bacteria known as **hyperthermophiles** grow at temperatures near 100°C (they live in such places as hot springs and submarine hydrothermal vents, with the most extreme, the archaeon *Pyrolobus fumarii*, able to grow at temperatures as high as 113°C). These organisms have many of the same metabolic pathways as do **mesophiles** (organisms that grow at "normal" temperatures). Yet, most mesophilic proteins denature at the temperatures at which hyperthermophiles thrive. What is the structural basis for the thermostability of hyperthermophilic proteins?

The difference in the thermal stabilities of the corresponding (hyper)thermophilic and mesophilic proteins does not exceed ~100 kJ $\cdot$ mol^{-1}, the equivalent of a few noncovalent interactions. This is probably why comparisons of the X-ray structures of hyperthermophilic enzymes with their mesophilic counterparts have failed to reveal any striking differences between them. These proteins exhibit some variations in secondary structure but no more so than is often the case for homologous proteins from distantly related mesophiles. However, several of these thermostable enzymes have a superabundance of salt bridges on their surfaces, many of which are arranged in extensive networks. Indeed, one such network from *Pyrococcus furiosis* **glutamate dehydrogenase** consists of 18 side chains.

The idea that salt bridges can stabilize a protein structure appears to contradict the conclusion of Section 8-4A that ion pairs are, at best, marginally stable. The key to this apparent paradox is that the salt bridges in thermostable proteins form networks. Thus, the gain in charge–charge free energy on associating a third charged group with an ion pair is comparable to that between the members of this ion pair, whereas the free energy lost on desolvating and immobilizing the third side chain is only about half that lost in bringing together the first two side chains. The same, of course, is true for the addition of a fourth, fifth, etc., side chain to a salt bridge network.

Not all thermostable proteins have such a high incidence of salt bridges. Structural comparisons suggest that these proteins are stabilized by a combination of small effects, the most important of which are an increased size in the protein's hydrophobic core, an increased size in the interface between its domains and/or subunits, and a more tightly packed core as evidenced by a reduced surface-to-volume ratio.

The fact that the proteins of hyperthermophiles and mesophiles are homologous and carry out much the same functions indicates that mesophilic proteins are by no means maximally stable. This, in turn, strongly suggests *that the marginal stability of most proteins under physiological conditions (averaging ~0.4 kJ/mol of amino acid residues) is an essential property that has arisen through evolutionary design.* Perhaps this marginal stability helps confer the structural flexibility that many proteins require to carry out their physiological functions (Section 9-4). Other possibilities are that it may facilitate the elimination of otherwise stable non-native conformations (Section 9-2C), it may promote the unfolding of proteins so as to permit their insertion into or transport through membranes (Section 12-4E), and/or it may expedite their programmed degradation (Section 32-6).

5 ■ QUATERNARY STRUCTURE

Proteins, because of their multiple polar and nonpolar groups, stick to almost anything; anything, that is, but other proteins. This is because the forces of evolution have arranged the surface groups of proteins so as to prevent their association under physiological conditions. If this were not the case, their resulting nonspecific aggregation would render proteins functionally useless (recall, e.g., the consequences of sickle-cell anemia; Section 7-3A). In his pioneering ultracentrifugational studies on proteins, however, The Svedberg discovered that some proteins are composed of more than one polypeptide chain. Subsequent studies established that this is, in fact, true of most proteins, including nearly all those with molecular masses >100 kD. Furthermore, these polypeptide **subunits** associate in a geometrically specific manner. The spatial arrangement of these subunits is known as a protein's **quaternary structure (4° structure).**

There are several reasons why multisubunit proteins are so common. In large assemblies of proteins, such as collagen fibrils, the advantages of subunit construction over the synthesis of one huge polypeptide chain are analogous to those of using prefabricated components in constructing a building: Defects can be repaired by simply replacing the flawed subunit rather than the entire protein, the site of subunit manufacture can be different from the site of assembly into the final product, and the only genetic information necessary to specify the entire edifice is that specifying its few different self-assembling subunits. In the case of enzymes, increasing a protein's size tends to better fix the three-dimensional positions of the groups forming the enzyme's active site. Increasing the size of an enzyme through the association of identical subunits is more efficient, in this regard, than increasing the length of its polypeptide chain since each subunit has an active site. Additionally, in some multimeric enzymes, the active site occurs at the interface between subunits where it is comprised of groups from two or more subunits. More importantly, however, the subunit construction of many enzymes provides the structural basis for the regulation of their activities. Mechanisms for this indispensable function are discussed in Sections 10-4 and 13-4.

In this section we discuss how the subunits of multisubunit proteins associate, what sorts of symmetries they have, and how their stoichiometries may be determined.

A. *Subunit Interactions*

A multisubunit protein may consist of identical or nonidentical polypeptide chains. Recall that hemoglobin, for example, has the subunit composition $\alpha_2\beta_2$. We shall refer to proteins with identical subunits as **oligomers** and to these identical subunits as **protomers.** A protomer may therefore consist of one polypeptide chain or several unlike polypeptide chains. In this sense, hemoglobin is a **dimer** (oligomer of two protomers) of $\alpha\beta$ protomers (Fig. 8-63).

The association of two subunits typically buries 1000 to 2000 Å^2 of surface area that would otherwise be exposed to solvent. The resulting contact regions resemble the interior of a single subunit protein: They contain closely packed nonpolar side chains, hydrogen bonds involving the polypeptide backbones and their side chains, and, in some cases, interchain disulfide bonds. However, interfaces between subunits tend to have hydrophobicities between those of protein interiors and exteriors. In particular, the

subunit interfaces of proteins that dissociate *in vivo* have lesser hydrophobicities than do permanent interfaces. Moreover, protein–protein interfaces often contain salt bridges that contribute to the specificity as well as to the stability of subunit associations.

B. *Symmetry in Proteins*

In the vast majority of oligomeric proteins, the protomers are symmetrically arranged; that is, the protomers occupy geometrically equivalent positions in the oligomer. This implies that each protomer has exhausted its capacity to bind to other protomers; otherwise, higher oligomers would form. As a result of this limited binding capacity, protomers pack about a single point to form a closed shell, a phenomenon known as **point symmetry.** Proteins cannot have inversion or mirror symmetry, however, because such symmetry operations convert chiral L-residues to D-residues. Thus, *proteins can only have rotational symmetry.*

Various types of rotational symmetry occur in proteins, as X-ray crystal structure determinations have shown:

1. Cyclic symmetry

In the simplest type of rotational symmetry, **cyclic symmetry,** subunits are related (brought to coincidence) by a single axis of rotation (Fig. 8-64*a*). Objects with 2, 3, . . ., or *n*-fold rotational axes are said to have C_2, C_3, . . ., or C_n symmetry, respectively. An oligomer with C_n symmetry consists of *n* protomers that are related by $(360/n)°$ rotations. C_2 symmetry is the most common symmetry in proteins; higher cyclic symmetries are relatively rare.

A common mode of association between protomers related by a twofold rotation axis is the continuation of a β sheet across subunit boundaries. In such cases, the twofold axis is perpendicular to the β sheet so that two symmetry equivalent β strands hydrogen bond in an antiparallel fashion. In this manner, the sandwich of two four-stranded β sheets in a protomer of **transthyretin** (also known as **prealbumin**) is extended across a twofold axis to form a sandwich of two eight-stranded β sheets (Fig. 8-65). Hemoglobin's two αβ protomers are also related by C_2 symmetry (Fig. 8-63).

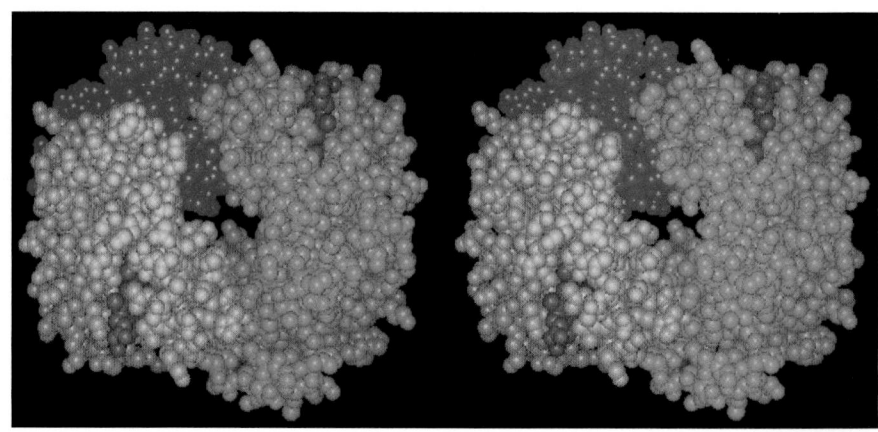

FIGURE 8-63   The quaternary structure of hemoglobin. The α_1, α_2, β_1, and β_2 subunits in this stereo, space-filling drawing are colored yellow, green, cyan, and magenta, respectively. Heme groups are red. The protein is viewed along its molecular 2-fold rotation axis, which relates the $\alpha_1\beta_1$ protomer to the $\alpha_2\beta_2$ protomer. Instructions for viewing stereo drawings are given in the appendix to this chapter. [Based on an X-ray structure by Max Perutz, MRC Laboratory of Molecular Biology, Cambridge, U.K. PDBid 2DHB.]

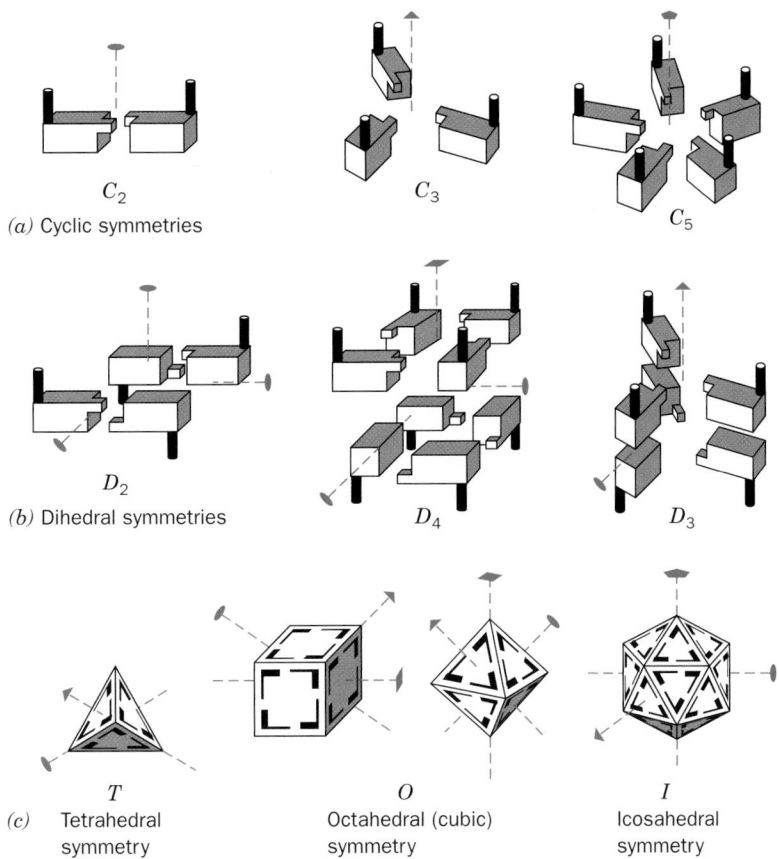

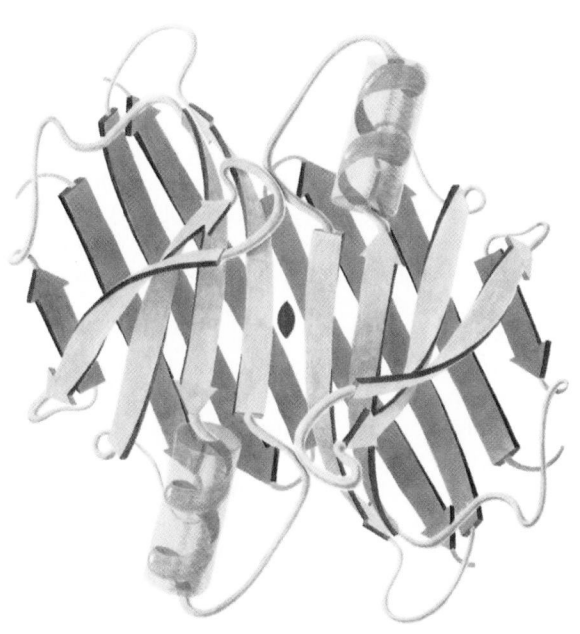

FIGURE 8-64 Some possible symmetries of proteins with identical protomers. The lenticular shape, the triangle, the square, and the pentagon at the ends of the dashed lines indicate, respectively, the unique 2-fold, 3-fold, 4-fold, and 5-fold rotational axes of the objects shown. (*a*) Assemblies with the cyclic symmetries C_2, C_3, and C_5. (*b*) Assemblies with the dihedral symmetries D_2, D_4, and D_3. In these objects, a twofold axis is perpendicular to the vertical 2-, 4-, and 3-fold axes. (c) Assemblies with *T, O,* and *I* symmetry. Note that the tetrahedron has some but not all of the symmetry elements of the cube, and that the cube and the octahedron have the same symmetry. [Illustration, Irving Geis/Geis Archives Trust. Copyright Howard Hughes Medical Institute. Reproduced with permission.] 🔊 **See the Animated Figures**

2. Dihedral symmetry

Dihedral symmetry (D_n), a more complicated type of rotational symmetry, is generated when an *n*-fold rotation axis and a two-fold rotation axis intersect at right angles (Fig. 8-64*b*). An oligomer with (D_n) symmetry consists of $2n$ protomers. D_2 symmetry is, by far, the most common type of dihedral symmetry in proteins.

Hemoglobin's α and β subunits have such similar structures that, in the hemoglobin $\alpha_2\beta_2$ tetramer, they are related by pseudo-two-fold rotational axes that are perpendicular to the tetramer's exact two-fold axis (lie in the

FIGURE 8-65 A dimer of transthyretin as viewed down its twofold axis (*red lenticular symbol*). Each protomer consists of a β barrel (really a β sandwich) containing two Greek keys (Fig. 8-50*a*). Note how both of its β sheets are continued in an antiparallel fashion in the symmetry-related protomer to form a sandwich of two eight-stranded β sheets. Two of these dimers associate back to back in the native protein to form a tetramer with D_2 symmetry. This protein was previously known as **prealbumin.** [After a drawing by Jane Richardson, Duke University. Based on an X-ray structure by Colin Blake, Oxford University, U.K. PDBid 2PAB.]

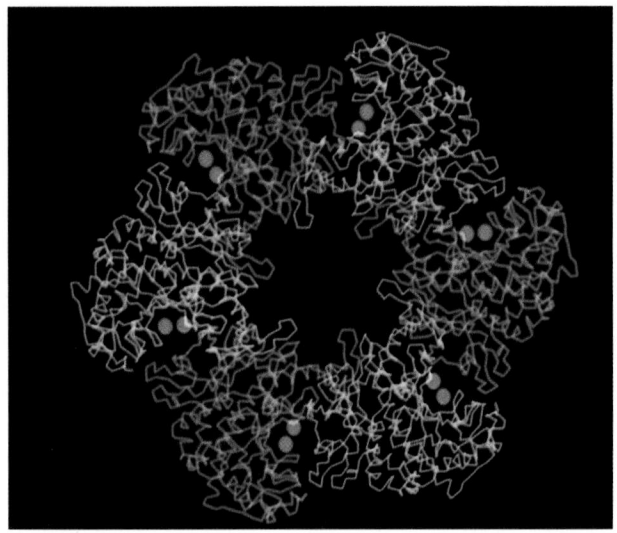

(a)

(b)

FIGURE 8-66 X-Ray structure of glutamine synthetase from *Salmonella typhimurium.* The enzyme consists of 12 identical subunits, here represented by their C_α backbones, arranged with D_6 symmetry. (*a*) View down the 6-fold axis of symmetry showing only the six subunits of the upper ring in alternating blue and green. The subunits of the lower ring are roughly directly below those of the upper ring. The protein, including its side chains (*not shown*), has a diameter of 143 Å. The six active sites shown are marked by pairs of bound Mn^{2+} ions (*red spheres*). (*b*) Side view along one of the protein's twofold axes showing only its six nearest subunits. The molecule extends 103 Å along the sixfold axis, which is vertical in this view. [Courtesy of David Eisenberg, UCLA. PDBid 2GLS.]

plane of Fig. 8-63; see Section 10-2B). Hence the tetramer is said to be **pseudosymmetric** and have pseudo-D_2 symmetry. The X-ray structure of **glutamine synthetase** reveals that this enzyme consists of 12 identical subunits that are related by D_6 symmetry (Fig. 8-66).

Under the proper conditions, many oligomers with D_n symmetry dissociate into two oligomers, each with C_n symmetry (and which are related by the twofold rotation axes in the D_n oligomer). These, in turn, dissociate to their component protomers under more stringent denaturing conditions.

3. Other rotational symmetries

The only other types of rotationally symmetric objects are those that have the rotational symmetries of a tetrahedron (*T*), a cube or octahedron (*O*), or an icosahedron (*I*), and have 12, 24, and 60 equivalent positions, respectively (Fig. 8-64*c*). Certain multienzyme complexes are based on octahedral symmetry (Section 21-2A), whereas the protein coats of the so-called spherical viruses are based on icosahedral symmetry (Section 33-2A).

a. Helical Symmetry

Some protein oligomers have **helical symmetry** (Fig. 8-67). The chemically identical subunits in a helix are not strictly equivalent because, for instance, those at the ends of the helix have a different environment than those in the middle. Nevertheless, the surroundings of all subunits in a long helix, except those near its ends, are sufficiently similar that the subunits are said to be **quasi-equivalent.** The subunits of many structural proteins, for example, those

of **actin** (Section 35-3A) and **tubulin** (Section 35-3F), assemble into fibers with helical symmetry.

b. Obtaining the Atomic Coordinates of a Biologically Functional Quaternary Structure

A Protein Data Bank (PDB) entry for an X-ray crystal structure contains the atomic coordinates of its unit cell's asymmetric unit (a **unit cell** is the smallest portion of a crystal lattice that is repeated by translation). The entire crystal structure can then be generated through the application of its **crystallographic symmetry** (the combination of the unit cell's point symmetry with its translational symmetry). Thus, for a protein that has quaternary structure, its PDB entry might contain the coordinates of only one of its several protomers or perhaps those of two or more protomers that are portions of different biologically func-

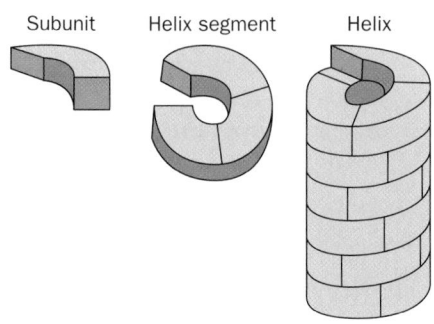

Subunit   Helix segment   Helix

FIGURE 8-67 A helical structure composed of a single kind of subunit.

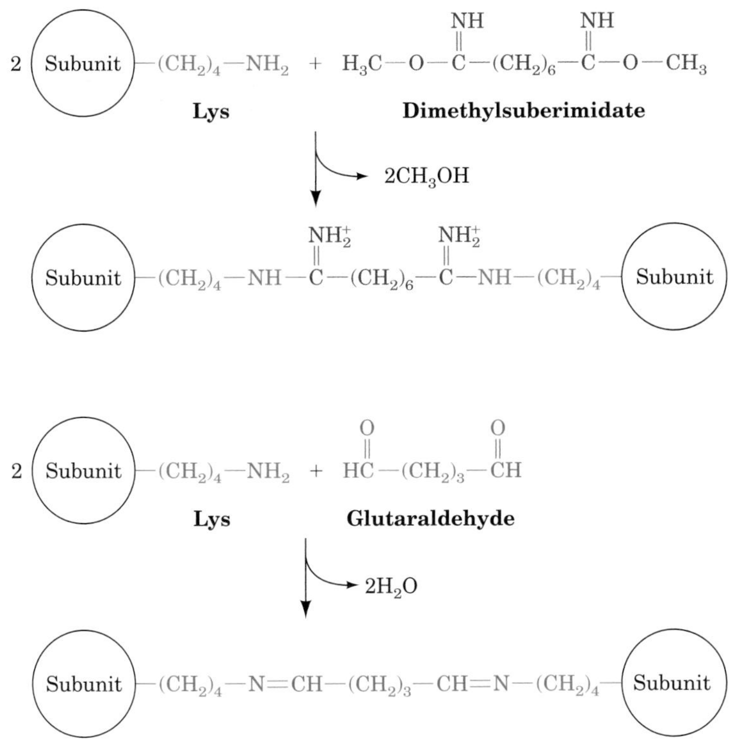

$$2 \; \text{(Subunit)} - (CH_2)_4 - NH_2 \; + \; H_3C - O - \overset{\overset{\displaystyle NH}{\|}}{C} - (CH_2)_6 - \overset{\overset{\displaystyle NH}{\|}}{C} - O - CH_3$$

Lys　　　　　　　　　**Dimethylsuberimidate**

$$\downarrow \; 2CH_3OH$$

$$\text{(Subunit)} - (CH_2)_4 - NH - \overset{\overset{\displaystyle NH_2^+}{\|}}{C} - (CH_2)_6 - \overset{\overset{\displaystyle NH_2^+}{\|}}{C} - NH - (CH_2)_4 - \text{(Subunit)}$$

$$2 \; \text{(Subunit)} - (CH_2)_4 - NH_2 \; + \; H\overset{\overset{\displaystyle O}{\|}}{C} - (CH_2)_3 - \overset{\overset{\displaystyle O}{\|}}{C}H$$

Lys　　　　　　　　　**Glutaraldehyde**

$$\downarrow \; 2H_2O$$

$$\text{(Subunit)} - (CH_2)_4 - N = CH - (CH_2)_3 - CH = N - (CH_2)_4 - \text{(Subunit)}$$

FIGURE 8-68　Cross-linking agents.
Dimethylsuberimidate and glutaraldehyde are bifunctional reagents that react to covalently cross-link two Lys residues.

tional molecules. Moreover, the application of the symmetry transformations that are often given in the PDB file's header may yield the coordinates of portions of different molecules. To alleviate these difficulties, a computerized procedure has been devised to generate the coordinates of the most probable biologically functional molecule based on several criteria, including maximizing the solvent-accessible surface area that is buried on forming the oligomer. An atlas of the most probable quaternary structures of macromolecules whose structures have been determined by X-ray crystallography is publicly available at http://pqs.ebi.ac.uk/.

C. *Determination of Subunit Composition*

In the absence of an X-ray or NMR structure, the number of different types of subunits in an oligomeric protein may be determined by end group analysis (Section 7-1A). In principle, the subunit composition of a protein may be determined by comparing its molecular mass with those of its component subunits. In practice, however, experimental difficulties, such as the partial dissociation of a supposedly intact protein and uncertainties in molecular mass determinations, often provide erroneous results.

a. Cross-Linking Agents Stabilize Oligomers

A method for 4° structure analysis, which is especially useful for oligomeric proteins that decompose easily, employs **cross-linking agents** such as **dimethylsuberimidate** or **glutaraldehyde** (Fig. 8-68). If carried out at sufficiently low protein concentrations to eliminate intermolecular reactions, cross-linking reactions will covalently join only

the subunits in a molecule that are no farther apart than the length of the cross-link (assuming, of course, that the proper amino acid residues are present). The molecular mass of a cross-linked protein therefore places a lower limit on its number of subunits. Such studies can also provide some indication of the distance between subunits, particularly if a series of cross-linking agents with different lengths is employed.

Appendix: ■ VIEWING STEREO PICTURES

Although we live in a three-dimensional world, the images that we see have been projected onto the two-dimensional plane of our retinas. Depth perception therefore involves binocular vision: The slightly different views perceived by each eye are synthesized by the brain into a single three-dimensional impression.

Two-dimensional pictures of complex three dimensional objects are difficult to interpret because most of the information concerning the third dimension is suppressed. This information can be recovered by presenting each eye with the image only it would see if the three-dimensional object were actually being viewed. A **stereo pair** therefore consists of two images, one for each eye. Corresponding points of stereo pairs are generally separated by ~6 cm, the average distance between human eyes. Stereo drawings are usually computer generated because of the required precision of the geometric relationship between the members of a stereo pair.

FIGURE 8-69   Stereo drawing of a tetrahedron inscribed in a cube. When properly viewed, the apex of the tetrahedron should appear to be pointing toward the viewer.

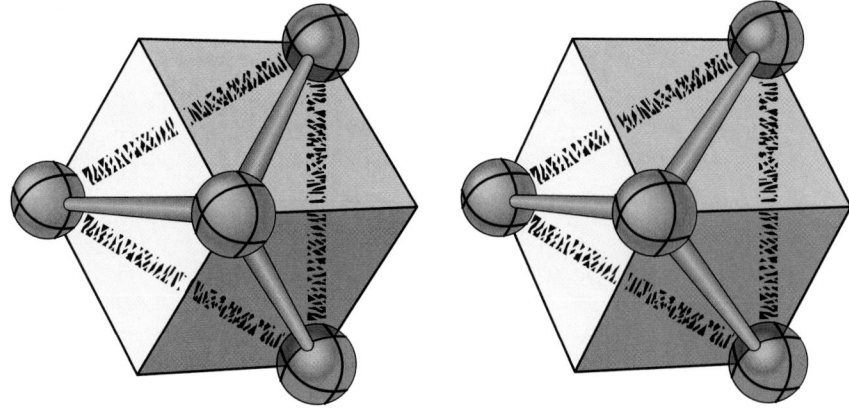

In viewing a stereo picture, one must overcome the visual habits of a lifetime because each eye must see its corresponding view independently. Viewers are commercially available to aid in this endeavor. However, with some training and practice, equivalent results can be obtained without their use.

To train yourself to view stereo pictures, you should become aware that each eye sees a separate image. Hold your finger up about a foot (30 cm) before your eyes while fixing your gaze on some object beyond it. You may realize that you are seeing two images of your finger. If, after some concentration, you are still aware of only one image, try blinking your eyes alternately to ascertain which of your eyes is seeing the image you perceive. Perhaps alternately covering and uncovering this dominant eye while staring past your finger will help you become aware of the independent workings of your eyes.

The principle involved in seeing a stereo picture is to visually fuse the left member of the stereo pair seen by the left eye with the right member seen by the right eye. To do this, sit comfortably at a desk, center your eyes about a foot over a stereo drawing such as Fig. 8-69 and stare through it at a point about a foot below the drawing. Try to visually fuse the central members of the four out-of-focus images you see. When you have succeeded, your vi-

sual system will "lock onto" it and this fused central image will appear three-dimensional. Ignore the outer images. You may have to slightly turn the book, which should be held perfectly flat, or your head in order to bring the two images to the same level. It may help to place the book near the edge of a desk, center your finger about a foot below the drawing, and fixate on your finger while concentrating on the stereo pair. Another trick is to hold your flattened hand or an index card between your eyes so that the left eye sees only the left half of the stereo pair and the right eye sees only the right half and then fuse the two images you see.

The final step in viewing a stereo picture is to focus on the image while maintaining fusion. This may not be easy because our ingrained tendency is to focus on the point at which our gaze converges. It may help to move your head closer to or farther from the picture. Most people (including the authors) require a fair amount of practice to become proficient at seeing stereo without a viewer. However, the three-dimensional information provided by stereo pictures, not to mention their aesthetic appeal, makes it worth the effort. In any case, the few stereo figures used in this textbook have been selected for their visual clarity without the use of stereo; stereo will simply enhance their impression of depth.

CHAPTER SUMMARY

1 ■ Secondary Structure   The peptide group is constrained by resonance effects to a planar, trans conformation. Steric interactions further limit the conformations of the polypeptide backbone by restricting the torsion angles, ϕ and ψ, of each peptide group to three small regions of the Ramachandran diagram. The α helix, whose conformation angles fall within the allowed regions of the Ramachandram diagram, is held together by hydrogen bonds. The 3_{10} helix, which is more tightly coiled than the α helix, lies in a mildly forbidden region of the Ramachandran diagram. Its infrequent occurrences are most often as single-turn terminations of α helices. In the parallel and antiparallel β pleated sheets,

two or more almost fully extended polypeptide chains associate such that neighboring chains are hydrogen bonded. These β sheets have a right-handed curl when viewed along their polypeptide chains. The polypeptide chain often reverses its direction through a β bend. Other arrangements of the polypeptide chain, which are collectively known as coil conformations, are more difficult to describe but are no less ordered than are α or β structures.

2 ■ Fibrous Proteins   The mechanical properties of fibrous proteins can often be correlated with their structures. Keratin, the principal component of hair, horn, and nails, forms protofibrils that consist of two pairs of α helices in which the members of each pair are twisted together into a left-handed coil. The pliability of keratin decreases as the content of disulfide cross-links between the protofibrils increases. Collagen is the major protein component of connective tissue. Its every third residue is Gly, and many of the others are Pro and Hyp. This permits collagen to form a ropelike triple helical structure that has great tensile strength. Collagen molecules aggregate in a staggered array to form fibrils that are covalently cross-linked by groups derived from their His and Lys side chains. Mutations in collagen or the inactivation of the enzymes that process it usually cause a loss in the structural integrity of the affected tissues.

3 ■ Globular Proteins The accuracies of protein X-ray structure determinations are limited by crystalline disorder to resolutions that are mostly in the range 1.5 to 3.0 Å. This requires that a protein's structure be determined by fitting its primary structure to its electron density map. Several lines of evidence indicate that protein crystal structures are nearly identical to their solution structures. The structures of small proteins may also be determined in solution by 2D- or multi-dimensional-NMR techniques which, for the most part, yield results similar to those of X-ray crystal structures. A globular protein's 3° structure is the arrangement of its various elements of 2° structure together with the spatial dispositions of its side chains. Its amino acid residues tend to segregate according to residue polarity. Nonpolar residues preferentially occupy the interior of a protein out of contact with the aqueous solvent, whereas charged polar residues are located on its surface. Uncharged polar residues may occur at either location but, if they are internal, they form hydrogen bonds with other protein groups. The interior of a protein molecule resembles a crystal of an organic molecule in its packing efficiency. Larger proteins often fold into two or more domains that may have functionally and structurally independent properties. Certain groupings of secondary structural elements, known as motifs or supersecondary structures, repeatedly occur as components of globular proteins. They combine in numerous ways to form a fold, that is, the tertiary structure of a domain. Among the common folds in globular proteins are the globin fold, 4-helix bundles, various β barrels (up-and-down, Greek key, and jelly roll), the α/β barrel, and the various open β sheets. The atomic coordinates of the >20,000 proteins and nucleic acids whose structures have been determined are archived in the Protein Data Bank (PDB). These macromolecules may be visually examined through the use of a variety of molecular graphics programs and may be analyzed by programs that compare and classify their structures in various ways.

4 ■ Proteins Stability Proteins have marginally stable native structures that form as a result of a fine balance among the various noncovalent forces to which they are subject: ionic and dipolar interactions, hydrogen bonding, and hydrophobic forces. Ionic interactions are relatively weak in aqueous solutions due to the solvating effects of water. The various interactions among permanent and induced dipoles, which are collectively referred to as van der Waals forces, are even weaker and are effective only at short range. Nevertheless, because of their large numbers, they cumulatively have an important influence on protein structures. H bonding forces are far more directional than are other noncovalent forces. They add little stability to a protein structure, however, because the H bonds that native proteins form internally are only marginally stronger than those that unfolded proteins form with water. Yet, because most H bonds are local, a protein can only fold stably in ways such that almost all of its possible internal H bonds are formed. Hence H bonding is important in specifying the native structure of a protein. Hydrophobic forces arise from the unfavorable ordering of water structure that results from the hydration of nonpolar groups. By folding such that its nonpolar groups are out of contact with the aqueous solvent, a protein minimizes these unfavorable interactions. The fact that most protein denaturants interfere with the hydrophobic effect demonstrates the importance of hydrophobic forces in stabilizing native protein structures. Disulfide bonds often stabilize the native structures of extracellular proteins. The proteins from (hyper)thermophilic organisms have higher T_m's than their mesophilic counterparts. Several hyperthermophilic proteins are stabilized by extensive surface networks of salt bridges.

5 ■ Quaternary Structure Many proteins consist of non-covalently linked aggregates of subunits in which the subunits may or may not be identical. Most oligomeric proteins are rotationally symmetric. The protomers in many fibrous proteins are related by helical symmetry. The quaternary structures of proteins are best elucidated using X-ray or NMR techniques. In their absence, cross-linking studies can indicate the subunit composition of an oligomeric protein.

■ REFERENCES

GENERAL

Branden, C. and Tooze, J., *Introduction to Protein Structure* (2nd ed.), Garland (1999).

Creighton, T.E., *Proteins* (2nd ed.), Chapters 4–6, Freeman (1993).

Dickerson, R.E. and Geis, I., *The Structure and Action of Proteins*, Benjamin/Cummings (1969). [A classic and marvelously illustrated exposition of the fundamentals of protein structure.]

Finkelstein, A.V. and Ptitsyn, O.B., *Protein Physics. A Course of Lectures*, Academic Press (2002).

Kyte, J., *Structure in Protein Chemistry*, Garland (1995).

Lesk, A.M., *Introduction to Protein Architecture*, Oxford University Press (2001).

Perutz, M., *Protein Structure. New Approaches to Disease and Therapy*, Freeman (1992). [A series of short articles on the structures of a variety of proteins and their biomedical implications.]

Shirley, B.A. (Ed.), *Protein Stability and Folding*, Humana Press (1995).

Tanford, C. and Reynolds, J., *Nature's Robots*. Oxford University Press (2001). [A history of proteins.]

SECONDARY STRUCTURE

Leszczynski, J.F. and Rose, G.D., Loops in globular proteins: A novel category of secondary structure, *Science* **234,** 849–855 (1986).

Milner-White, E.J., The partial charge of the nitrogen atom in peptide bonds, *Protein Science* **6,** 2477–2482 (1997). [Discusses the origin of the peptide N atom's partial negative charge.]

Toniolo, C. and Benedetti, E., The polypeptide 3₁₀-helix, *Trends Biochem. Sci.* **16,** 350–353 (1991).

FIBROUS PROTEINS

Baum, J. and Brodsky, B., Folding of peptide models of collagen and misfolding in disease, *Curr. Opin. Struct. Biol.* **9,** 122–128 (1999).

Bella, J., Eaton, M., Brodsky, B., and Berman, H.M., Crystal and molecular structure of a collagen-like peptide at 1.9-Å resolution, *Science* **266,** 75–81 (1994); *and* Bella, J., Brodsky, B., and Berman, H.M., Hydration structure of a collagen peptide, *Structure* **3,** 893–906 (1995).

Byers, P.H., Disorders of collagen synthesis and structure, *in* Scriver, C.R., Beaudet, A.L., Sly, W.S., and Valle, D. (Eds.), *The Metabolic and Molecular Bases of Inherited Disease* (8th ed.), *pp.* 5241–5286, McGraw-Hill (2001).

Engel, J. and Prokop, D.J., The zipper-like folding of collagen triple helices and the effects of mutations that disrupt the zipper, *Annu. Rev. Biophys. Biophys. Chem.* **20,** 137–152 (1991).

Fuchs, E. and Cleveland, D.W., A structural scaffolding of intermediate filaments in health and disease, *Science* **279,** 514–519 (1998); *and* Fuchs, E. and Weber, K., Intermediate filaments: Structure, dynamics, function, and disease, *Annu. Rev. Biochem.* **63,** 345–382 (1994).

Kadler, K.E., Holmes, D.F., Trotter, J.A., and Chapman, J.A., Collagen fibril formation, *Biochem. J.* **316,** 1–11 (1996).

Kramer, R.Z., Bella, J., Mayville, P., Brodsky, B., and Berman, H.M., Sequence dependent conformational variations of collagen triple-helical structure, *Nature Struct. Biol.* **6,** 454–457 (1999).

Orgel, J.P.R.O., Miller, A., Irving, T.C., Fischetti, R.F., Hammersley, A.P., and Wess, T.J., The in situ supermolecular structure of type I collagen, *Structure* **9,** 1061–1069 (2001).

Parry, D.A.D. and Steinert, P.M., Intermediate filaments: Molecular architecture, assembly, dynamics and polymorphism, *Quart. Rev. Biophys.* **32,** 99–187 (1999).

Prockop, D.J. and Kivirikko, K.I., Collagens: Molecular biology, diseases, and potentials for therapy, *Annu. Rev. Biochem.* **64,** 403–434 (1995).

van der Rest, M. and Bruckner, P., Collagens: Diversity at the molecular and supramolecular levels, *Curr. Opin. Struct. Biol.* **3,** 430–436 (1993).

MACROMOLECULAR STRUCTURE DETERMINATION

Blow, D., *Outline of Crystallography for Biologists,* Oxford University Press (2002).

Brünger, A.T., X-Ray crystallography and NMR reveal complementary views of structure and dynamics, *Nature Struct. Biol.* **4,** 862–865 (1997).

Carey, P.R. (Ed.), *Protein Engineering and Design,* Part III: Characterization, Academic Press (1996). [Contains chapters on 3- and 4-dimensional NMR, protein crystallography, and spectroscopic and calorimetric methods for characterizing proteins.]

Carter, C.W., Jr. and Sweet, R.M. (Eds.), *Macromolecular Crystallography,* Parts A and B, *Methods Enzymol.* **276** and **277** (1997).

Cavanaugh, J., Fairbrother, W.J., Palmer, A.G., III, and Skelton, N.J., *Protein NMR Spectroscopy. Principles and Practice,* Academic Press (1996).

Clore, G.M. and Gronenborn, A.M., NMR structures of proteins and protein complexes beyond 20,000 M_r, *Nature Struct. Biol.* **4,** 849–853 (1997); *and* Kay, L.E. and Gardiner, K.H., Solution

NMR spectroscopy beyond 25 kDa, *Curr. Opin. Struct. Biol.* **7,** 722–731 (1997).

Drenth, J., *Principles of Protein X-Ray Crystallography* (2nd. ed.), Springer-Verlag (1999).

Ferentz, A.E. and Wagner, G., NMR spectroscopy: A multifaceted approach to macromolecular structure, *Quart. Rev. Biophys.* **33,** 29–65 (2000).

Glusker, J.P., Lewis, M., and Rossi, M., *Crystal Structure Analysis for Chemists and Biologists,* VCH Publishers (1994).

McPherson, A., *Macromolecular Crystallography,* Wiley (2002).

McRee, D.E., *Practical Protein Crystallography* (2nd ed.), Elsevier Science (2002).

Mozzarelli, A. and Rossi, G.L., Protein function in the crystal, *Annu. Rev. Biophys. Biomol. Struct.* **25,** 343–365 (1996).

Reid, D.G. (Ed.), *Protein NMR Techniques,* Humana Press (1997).

Reik, R., Pervushkin, K., and Wüthrich, K., TROSY and CRINEPT: NMR with large molecular and supramolecular structures in solution, *Trends Biochem. Sci.* **25,** 462–468 (2000).

Rhodes, G., *Crystallography Made Crystal Clear: A Guide for Users of Macromolecular Models* (2nd ed.), Academic Press (2000).

Wider, G. and Wüthrich, K., NMR spectroscopy of large molecules and multimolecular assemblies in solution, *Curr. Opin. Struct. Biol.* **9,** 594–601 (1999).

Wüthrich, K., NMR—This other method for protein and nucleic structure determination, *Acta Cryst.* **D51,** 249–270 (1995); *and* Protein structure determination in solution by nuclear magnetic resonance spectroscopy, *Science* **243,** 45–50 (1989).

GLOBULAR PROTEINS

Bork, P., Gellerich, J., Groth, H., Hooft, R., and Martin, F., Divergent evolution of β/α-barrel subclass: Detection of numerous phosphate-binding sites by motif search, *Protein Science,* **4,** 268–274 (1995).

Bourne, P. and Wessig, H. (Eds.), *Structural Bioinformatics,* Wiley-Interscience (2003).

Chothia, C. and Finkelstein, A.V., The classification and origins of protein folding patterns, *Annu. Rev. Biochem.* **59,** 1007–1039 (1990).

Cohen, C. and Parry, D.A.D., α-Helical coiled coils and bundles: How to design an α-helical protein, *Proteins* **7,** 1–15 (1990).

Farber, G.K., An α/β-barrel full of evolutionary trouble, *Curr. Opin. Struct. Biol.* **3,** 409–412 (1993); *and* Farber, G.K. and Petsko, G.A., The evolution of α/β barrel enzymes, *Trends Biochem. Sci.* **15,** 228–234 (1990).

Hadley, C. and Jones, J.T., A systematic comparison of protein structure classifications: SCOP, CATH, and FSSP, *Structure* **7,** 1099–1112 (1999).

Hogue, C.W.V., Structure data bases, *in* Baxevanis, A.D. and Ouellette, B.F.F. (Eds.), *Bioinformatics* (2nd ed.), Chapter 5, Wiley-Interscience (2001).

Orengo, C.A., Todd, A.E., and Thornton, J.M., From protein structure to function, *Curr. Opin. Struct. Biol.* **9,** 374–382 (1999); *and* Swindells, M.B., Orengo, C.A., Jones, D.T., Hutchinson, E.G., and Thornton, J.M., Contemporary approaches to protein structure classification, *BioEssays* **20,** 884–891 (1998).

Richards, F.M., Areas, volumes, packing, and protein structure, *Annu. Rev. Biophys. Bioeng.* **6,** 151–176 (1977).

Richardson, J.S., The anatomy and taxonomy of protein structures, *Adv. Protein Chem.* **34,** 168–339 (1981). [A detailed discussion of the structural principles governing globular proteins accompanied by an extensive collection of their cartoon representations.]

Richardson, J.S. and Richardson, D.C., Principles and patterns of protein conformation, *in* Fasman, G.D. (Ed.), *Prediction of Protein Structure and the Principles of Protein Conformation*, pp. 1–98, Plenum Press (1989). [A comprehensive account of protein conformations based on X-ray structures.]

PROTEIN STABILITY

Alber, T., Stabilization energies of protein conformation, *in* Fasman, G.D. (Ed.), *Prediction of Protein Structure and the Principles of Protein Conformation*, pp. 161–192, Plenum Press (1989).

Burley, S.K. and Petsko, G.A., Weakly polar interactions in proteins, *Adv. Protein Chem.* **39,** 125–189 (1988).

Creighton, T.E., Stability of folded proteins, *Curr. Opin. Struct. Biol.* **1,** 5–16 (1991).

Derewenda, Z.S., Lee, L., and Derewenda, U., The occurrence of C—H $\cdots$ O hydrogen bonds in proteins, *J. Mol. Biol.* **252,** 248–262 (1995).

Edsall, J.T. and McKenzie, H.A., Water and proteins, *Adv. Biophys.* **16,** 51–183 (1983).

Eigenbrot, C. and Kossiakoff, A.A., Structural consequences of mutation, *Curr. Opin. Biotech.* **3,** 333–337 (1992).

Fersht, A., *Structure and Mechanism in Protein Science,* Chapter 11, Freeman (1999).

Fersht, A.R. and Serrano, L., Principles of protein stability derived from protein engineering experiments, *Curr. Opin. Struct. Biol.* **3,** 75–83 (1993). [Discusses how the roles of specific side chains in proteins can be quantitatively determined by mutationally changing them and calorimetrically measuring the stabilities of the resulting proteins.]

Goldman, A., How to make my blood boil, *Structure* **3,** 1277–1279 (1995). [Discusses the stabilities of hyperthermophilic proteins.]

Hendsch, Z. and Tidor, B., Do salt bridges stabilize proteins? A continuum electrostatic analysis, *Prot. Sci.* **3,** 211–226 (1994).

Honig, B. and Nichols, A., Classical electrostatics in biology and chemistry, *Science* **268,** 1144–1149 (1995).

Jaenicke, R. and Böhm, G., The stability of proteins in extreme environments, *Curr. Opin. Struct. Biol.* **8,** 738–748 (1998).

Jeffrey, G.A. and Saenger, W., *Hydrogen Bonding in Biological Structures,* Springer-Verlag (1991).

Jones, S. and Thornton, J.M., Principles of protein–protein interactions, *Proc. Natl. Acad. Sci.* **93,** 13–20 (1996).

Karshikoff, A. and Ladenstein, R., Ion pairs and the thermotolerance of proteins from hyperthermophiles: A 'traffic rule' for hot roads, *Trends Biochem. Sci.* **26,** 550–556 (2001).

Kauzmann, W., Some factors in the interpretation of protein denaturation, *Adv. Protein Chem.* **14,** 1–63 (1958). [A classic review that first pointed out the importance of hydrophobic bonding in stabilizing proteins.]

Martin, T.W. and Derewenda, Z.S., The name is bond—H bond, *Nature Struct. Biol.* **6,** 403–406 (1999). [Reviews the history of the concept of hydrogen bonding and discusses X-ray scattering experiments that demonstrated that hydrogen bonds have ~10% covalent character.]

Matthews, B.W., Studies on protein stability with T4 lysozyme, *Adv. Protein Chem.* **46,** 249–278 (1995). [A distillation of the results of stability studies on a large number of mutant varieties of lysozyme from bacteriophage T4, many of which have also been structurally determined by X-ray analysis.]

Mattos, C., Protein–water interactions in a dynamic world, *Trends Biochem. Sci.* **27,** 203–208 (2002).

Ramachandran, G.N. and Sasisekharan, V., Conformation of polypeptides and proteins, *Adv. Protein Chem.* **23,** 283–437 (1968). [A classic paper.]

Rees, D.C. and Adams, M.W.W., Hyperthermophiles: Taking the heat and loving it, *Structure* **3,** 251–254 (1995).

Richards, F.M., Folded and unfolded proteins: An introduction, *in* Creighton, T.E. (Ed.), *Protein Folding*, pp. 1–58, Freeman (1992).

Schellman, J.A., The thermodynamic stability of proteins, *Annu. Rev. Biophys. Biophys. Chem.* **16,** 115–137 (1987).

Steiner, T. and Koellner, G., Hydrogen bonds with π-acceptors in proteins: Frequencies and role in stabilizing local 3D structures, *J. Mol. Biol.* **305,** 535–557 (2001).

Stickle, D.F., Presta, L.G., Dill, K.A., and Rose, G.D., Hydrogen bonding in globular proteins, *J. Mol. Biol.* **226,** 1143–1159 (1992).

Tanford, C., How protein chemists learned about the hydrophobic factor, *Protein Science* **6,** 1358–1366 (1997). [A historical narrative.]

Teeter, M.M., Water–protein interactions: Theory and experiment, *Annu. Rev. Biophys. Biophys. Chem.* **20,** 577–600 (1991).

Weiss, M.S., Brandl, M., Sühnel, J., Pal, D., and Hilgenfeld, R., More hydrogen bonds for the (structural) biologist, *Trends Biochem. Sci.* **26,** 521–523 (2001). [Discusses C—H $\cdots$ O and C—H $\cdots$ π hydrogen bonds.]

Yang, A.-S. and Honig, B., Electrostatic effects on protein stability, *Curr. Opin. Struct. Biol.* **2,** 40–45 (1992).

QUATERNARY STRUCTURE

Eisenstein, E. and Schachman, H.K., Determining the roles of subunits in protein function, *in* Creighton, T.E. (Ed.), *Protein Function. A Practical Approach*, pp. 135–176, IRL Press (1989).

Goodsell, D.S. and Olson, J., Structural symmetry and protein function, *Annu. Rev. Biophys. Biomol. Struct.* **29,** 105–153 (2000).

Sheinerman, F.B., Norel, R., and Honig, B., Electrostatic aspects of protein–protein interactions, *Curr. Opin. Struct. Biol.* **10,** 153–159 (2000).

PROBLEMS

1. What is the length of an α helical section of a polypeptide chain of 20 residues? What is its length when it is fully extended (all trans)? $\left(\dfrac{n}{20} \times 5.4 \overset{\circ}{A}\right) = 3.6 = 30 \overset{\circ}{A}$

*2. From an examination of Figs. 8-7 and 8-8, it is apparent that the polypeptide conformation angle φ is more constrained than is ψ. By referring to Fig. 8-4, or better yet, by examining a molecular model, indicate the sources of the steric interference that limit the allowed values of φ when ψ = 180°.

3. For a polypeptide chain made of γ-amino acids, state the nomenclature of the helix analogous to the 3_{10} helix of α-amino acids. Assume the helix has a pitch of 9.9 Å and a rise per residue of 3.2 Å.

*4. Table 8-7 (page 274) gives the torsion angles, φ and ψ, of hen egg white lysozyme for residues 24–73 of this 129-residue protein. (a) What is the secondary structure of residues 26–35? (b) What is the secondary structure of residues 42–53? (c) What is the probable identity of residue 54? (d) What is the secondary structure of residues 56–68? (e) What is the secondary structure of residues 69–71? (f) What additional information besides the torsion angles, φ and ψ, of each of its residues is required to define the three-dimensional structure of a protein?

TABLE 8-7   Torsion Angles (ϕ, ψ) for Residues 24 to 73 of Hen Egg White Lysozyme

Residue Number	Amino Acid	ϕ (deg)	ψ (deg)	Residue Number	Amino Acid	ϕ (deg)	ψ (deg)
24	Ser	−60	147	49	Gly	95	−75
25	Leu	−49	−32	50	Ser	−18	138
26	Gly	−67	−34	51	Thr	−131	157
27	Asn	−58	−49	52	Asp	−115	130
28	Trp	−66	−32	53	Tyr	−126	146
29	Val	−82	−36	54	xxx	67	−179
30	Cys	−69	−44	55	Ile	−42	−37
31	Ala	−61	−44	56	Leu	−107	14
32	Ala	−72	−29	57	Gln	35	54
33	Lys	−66	−65	58	Ile	−72	133
34	Phe	−67	−23	59	Asn	−76	153
35	Glu	−81	−51	60	Ser	−93	−3
36	Ser	−126	−8	61	Arg	−83	−19
37	Asn	68	27	62	Trp	−133	−37
38	Phe	79	6	63	Trp	−91	−32
39	Asn	−100	109	64	Cys	−151	143
40	Thr	−70	−18	65	Asn	−85	140
41	Glu	−84	−36	66	Asp	133	8
42	Ala	−30	142	67	Gly	73	−8
43	Thr	−142	150	68	Arg	−135	17
44	Asn	−154	121	69	Thr	−122	83
45	Arg	−91	136	70	Pro	−39	−43
46	Asn	−110	174	71	Gly	−61	−11
47	Thr	−66	−20	72	Ser	−45	122
48	Asp	−96	36	73	Arg	−124	146

Source: Imoto, T., Johnson, L.N., North, A.C.T., Phillips, D.C., and Rupley, J.A., *in* Boyer, P.D. (Ed.), *The Enzymes* (3rd ed.), Vol. 7, *pp.* 693–695, Academic Press (1972).

5. Hair splits most easily along its fiber axis, whereas fingernails tend to split across the finger rather than along it. What are the directions of the keratin fibrils in hair and in fingernails? Explain your reasoning.

6. What is the growth rate, in turns per second, of the α helices in a hair that is growing 15 cm · year^{-1}? 10^8 Å/cm

7. Can polyproline form a collagenlike triple helix? Explain.

8. As Mother Nature's chief engineer, you have been asked to design a five-turn α helix that is destined to have half its circumference immersed in the interior of a protein. Indicate the helical wheel projection of your prototype α helix and its amino acid sequence (see Fig. 8-43).

9. β-Aminopropionitrile is effective in reducing excessive scar tissue formation after an injury (although its use is contraindicated by side effects). What is the mechanism of action of this lathyrogen?

*10. Using your network browser, visit the Protein Data Bank (PDB) at http://www.rcsb.org/pdb/. To explore the structure of γ-B crystalline, enter 4GCR in the "Enter a PDB id or keyword" field, check the "query by PDB id only" box, and click on "Find a structure". In the "Structure Explorer" window that comes up, click on "Download/Display File" and in the window that then comes up, click on "HTML" to the right of "complete with co-

ordinates". Inspect the file that appears. (a) How many residues does this protein have and how many water molecules have been found to be associated with it in this crystal structure? (b) Draw skeletal diagrams of an Arg, a Glu, and a Tyr residue and label their atoms using the nomenclature employed in the PDB file. (c) What are the atomic coordinates of the S atom of Cys 32? What is the identity of atom 1556, which is labeled "OXT"? (d) Return to the Structure Explorer and click on "View Structure" and examine the structure of the protein using the viewer of your choice (making sure that you have previously downloaded the necessary program/plug-in to do so onto your computer). Can you see that the protein is composed of two well-separated and apparently similar domains? (e) To structurally classify this protein, return to the Structure Explorer and click on "Structural Neighbors". State the different ways that CATH and SCOP classify this protein. (f) To structurally compare the protein's two domains, return to Structural Neighbors and click on CE and, in the subsequent window, click on "Two Chains" (in the box at the bottom). In the window that comes up, enter 4GCR for both Chain 1 and Chain 2. For both chains, click on the "PDB:" button and check the "Use Fragment From:" box. Then for Chain 1 enter the range 1 to 83 and for Chain 2 enter the range 84 to 174. Then click on the "Calculate Alignment" button (at the top). The window that appears shows the aligned sequences of the two peptides based on their structure. What is the percent identity of the two peptides? Describe the gaps, if any, in the alignment. Now click on the "Press to Start Compare3D" button. The window that comes up shows the superimposed C_α backbones of the two peptide segments in blue and magenta. Rotate this model by clicking on it and dragging your mouse. Describe what you see. What is the "RMSD (A)" (root mean square deviation in Å) of the superimposed backbone segments? What is the significance of the short white segment of chain? What are the two gray segments?

*11. Using a molecular graphics viewer, inspect the structures of the following proteins as indicated by their PDBids. Draw the corresponding topological diagrams and name the fold, if it is a standard one, for each domain in the protein. (a) 1RCP, (b) 1RCB, (c) 1TNF, (d) 2CMD, (e) 1RHD, and (f) 2TAA. [Note: Although all of these proteins consist of only one type of subunit, some of the PDB files contain the coordinates for more than one chain. Your task will be simplified if you display only one of these chains (their structures are essentially identical). If you use RasMol as your viewer, this is easily done by typing, for example, "restrict *a <return>" into the Command Line window to restrict the display to the A chain. For proteins consisting of more than one domain, inspect each domain individually using the same technique (for example, after typing "restrict *a <return>", typing "restrict 1-20 <return>" will display only residues 1-20 of the A chain). A polypeptide chain is most easily followed if it is displayed in Backbone, Ribbons, or Cartoons form and assigned Group colors (which colors the chain, from N- to C-terminus, in rainbow order, blue to red).]

12. It is often stated that proteins are quite large compared to the molecules they bind. However, what constitutes a large molecule depends on your point of view. Calculate the ratio of the volume of a hemoglobin molecule (65 kD) to that of the four O_2 molecules it binds. Also calculate the ratio of the volume of a typical office ($4 \times 4 \times 3$ m) to that of a typical office worker (70 kg) that occupies it. Assume that the molecular volumes of hemoglobin and O_2 are in equal proportion to their molecular masses and that the office worker has a density of 1.0 g $\cdot$ cm^{-3}. Compare these ratios. Is this the answer you expected?

13. Why are London dispersion forces always attractive?

14. Membrane-bound proteins are generally closely associated with the nonpolar groups of lipid molecules (Section 12-3A). Explain how detergents affect the structural integrity of membrane-bound proteins in comparison to their effects on normal globular proteins.

15. The coat protein of **tomato bushy stunt virus** consists of 180 chemically identical subunits, each of which is composed of ~386 amino acid residues. The probability that a wrong amino acid residue will be biosynthetically incorporated in a polypeptide chain is 1 part in 3000 per residue. Calculate the average number of coat protein subunits that would have to be synthesized in order to produce a perfect viral coat. What would this number be if the viral coat were a single polypeptide chain with the same number of residues that it actually has?

16. State the rotational symmetries of each of the following objects: (a) a starfish, (b) a square pyramid, (c) a rectangular box, and (d) a trigonal bipyramid.

*17. Through the use of your favorite molecular graphics viewer, state the rotational symmetries of the proteins with the following PDBids: (a) 1TIM, (b) 1TNF, (c) 6PFK, and (d) 1AIY. [Note: When using RasMol or Chime, individual polypeptide chains are most easily differentiated by displaying the protein in Backbone, Ribbons, or Cartoons form and assigning it Chain colors (which makes each polypeptide chain a different color).]

18. Myoglobin and the subunits of hemoglobin are polypeptides of similar size and structure. Compare the expected ratio of nonpolar to polar amino acid residues in myoglobin and in hemoglobin.

19. Sickle-cell hemoglobin (HbS) differs from normal human adult hemoglobin (HbA) by a single mutational change, Glu β6 → Val, which causes the HbS molecules to aggregate under proper conditions (Section 7-3A). Under certain conditions, the HbS filaments that form at body temperature disaggregate when the temperature is lowered to 0°C. Explain.

20. Indicate experimental evidence that is inconsistent with the hypothesis that urea and guanidinium ion act to denature proteins by competing for their internal hydrogen bonds.

21. Proteins in solution are often denatured if the solution is shaken violently enough to cause foaming. Indicate the mechanism of this process. (*Hint:* The nonpolar groups of detergents extend into the air at air–water interfaces.)

22. An oligomeric protein in a dilute buffer at pH 7 dissociates to its component subunits when exposed to the following agents. Which of these observations would not support the contention that the quaternary structure of the protein is stabilized exclusively by hydrophobic interactions? Explain. (a) $6M$ guanidinium chloride, (b) 20% ethanol, (c) $2M$ NaCl, (d) temperatures below 0°C, (e) 2-mercaptoethanol, (f) pH 3, and (g) $0.01M$ SDS.

*23. The SDS–polyacrylamide gel electrophoresis of a protein yields two bands corresponding to molecular masses of 10 and 17 kD. After cross-linking this protein with dimethylsuberimidate under sufficient dilution to eliminate intermolecular cross-linking, SDS–polyacrylamide gel electrophoresis of the product yields 12 bands with molecular masses 10, 17, 20, 27, 30, 37, 40, 47, 54, 57, 64, and 74 kD. Assuming that dimethylsuberimidate can cross-link only contacting subunits, diagram the quaternary structure of the protein.

*24. Mammals possess two genetically distinct but closely related forms of lactate dehydrogenase (LDH), the M type (which predominates in skeletal muscle) and the H type (which predominates in heart tissue). Before the X-ray structure of LDH was known, the oligomeric state of LDH was determined by dissociating M type and H type LDH to their component subunits and then reconstituting the mixture. This treatment yielded five electrophoretically distinct **isozymes** (catalytically and structurally similar but genetically distinct enzymes from the same organism), M_4, M_3H, M_2H_2, MH_3, and H_4, thereby demonstrating that LDH is a tetramer. What are the relative amounts of each isozyme formed when equimolar amounts of M_4 and H_4 are so hybridized?

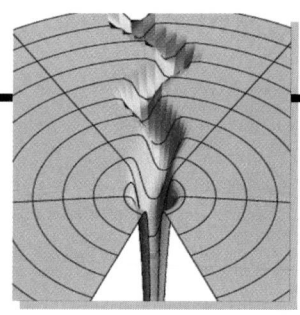

Chapter 9

Protein Folding, Dynamics, and Structural Evolution

In the preceding chapters, we saw how proteins are constructed from their component parts. This puts us in a similar position to a mechanic who has learned to take apart and put together an automobile engine without any inkling of how the engine works. What we need in order to understand the workings of a protein is knowledge of the types of internal motions it can and must undergo in order to carry out its biological function as well as how it achieves its native structure. Put in terms of our deprived auto mechanic, we need to understand the operations of the "gears" and "levers" with which proteins carry out their function. This is a problem of enormous complexity whose solution we have only glimpsed in outline. We shall see in later chapters, for example, that even though the catalytic mechanisms of many enzymes of known structure have been studied in great detail, it cannot be said that we fully understand any of these mechanisms. This is because our comprehension of the ways in which a protein's component groups interact is far from complete. As far as proteins are concerned, we have not greatly surpassed our hypothetical mechanic's level of understanding.

In this third of four chapters on protein structure, we consider the temporal behavior of proteins. Specifically, we first take up the problem of how random coil polypeptides fold to their native structures and how this process is facilitated by other proteins. This is followed by a consideration of the progress that has been made in predicting protein structures based on their sequences and in understanding their dynamic properties, that is, the nature and functional significance of their internal motions. Next we consider the diseases caused by proteins taking up alternate conformations. We end by extending the discussions we began in Section 7-3 on protein evolution but do so in terms of the three-dimensional structures of proteins.

1 ■ PROTEIN FOLDING: THEORY AND EXPERIMENT

Early notions of protein folding postulated the existence of "templates" that somehow caused proteins to assume their native conformations. Such an explanation begs the question of how proteins fold because, even if it were true, one would still have to explain how the template achieved its conformation. In fact, *proteins spontaneously fold into their native conformations under physiological conditions.* This implies that *a protein's primary structure dictates its three-dimensional structure.* In general, under the proper conditions, biological structures are **self-assembling** so that they have no need of external templates to guide their formation.

A. *Protein Renaturation*

Although evidence had been accumulating since the 1930s that proteins could be reversibly denatured, it was not until

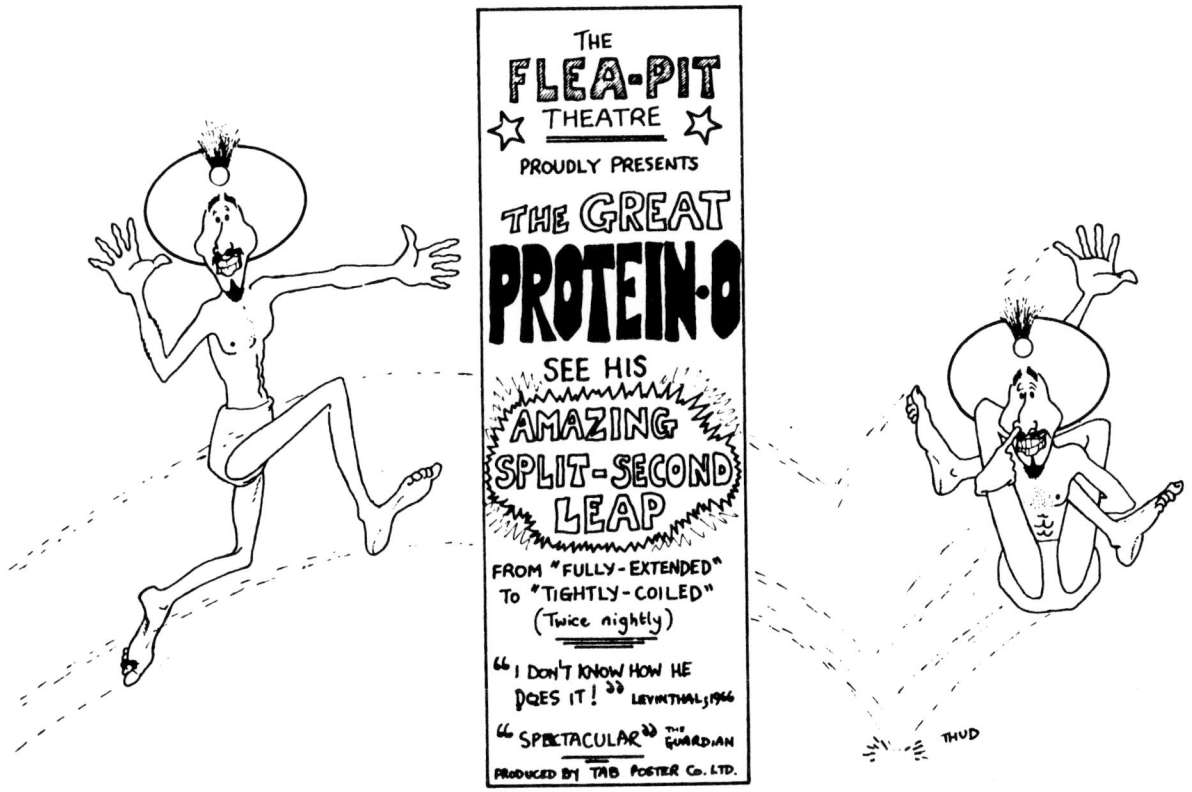

FIGURE 9-1 [Drawing by T.A. Bramley, *in* Robson, B., *Trends Biochem. Sci.* **1,** 50 (1976). Copyright © Elsevier Biomedical Press, 1976. Used by permission.]

1957 that the elegant experiments of Christian Anfinsen on bovine pancreatic **RNase A** put **protein renaturation** on a quantitative basis. RNase A, a 124-residue single-chain protein, is completely unfolded and its four disulfide bonds reductively cleaved in an $8M$ urea solution containing

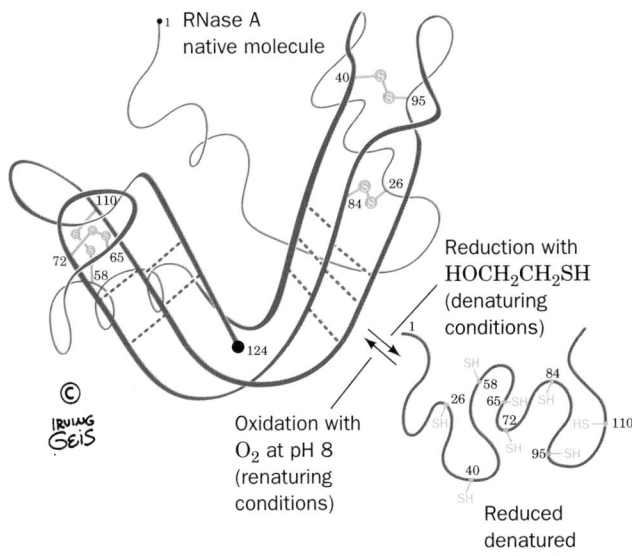

FIGURE 9-2 Reductive denaturation and oxidative renaturation of RNase A. [Illustration, Irving Geis/Geis Archives Trust. Copyright Howard Hughes Medical Institute. Reproduced with permission.]

2-mercaptoethanol (Fig. 9-2). Yet dialyzing away the urea and exposing the resulting solution to O_2 at pH 8 yields a protein that is virtually 100% enzymatically active and physically indistinguishable from native RNase A. The protein must therefore have spontaneously renatured. Any reservations that this occurs only because RNase A is really not totally denatured by $8M$ urea have been satisfied by the chemical synthesis of enzymatically active RNase A (Section 7-5).

The renaturation of RNase A demands that its four disulfide bonds reform. The probability of one of the eight Cys residues from RNase A randomly reforming a disulfide bond with its proper (native) mate among the other seven Cys residues is $\frac{1}{7}$; that of one of the remaining six Cys residues then randomly reforming its proper disulfide bond is $\frac{1}{5}$; etc. The overall probability of RNase A reforming its four native disulfide links at random is

$$\frac{1}{7} \times \frac{1}{5} \times \frac{1}{3} \times \frac{1}{1} = \frac{1}{105}$$

Clearly, the disulfide bonds from RNase A do not randomly reform under renaturing conditions.

If the RNase A is reoxidized in $8M$ urea so that its disulfide bonds reform while the polypeptide chain is a random coil, then after removal of the urea, the RNase A is, as expected, only ~1% enzymatically active. This "scrambled" RNase A can be made fully active by exposing it to a trace of 2-mercaptoethanol, which, over about a 10-h period, catalyzes disulfide bond interchange reactions until the na-

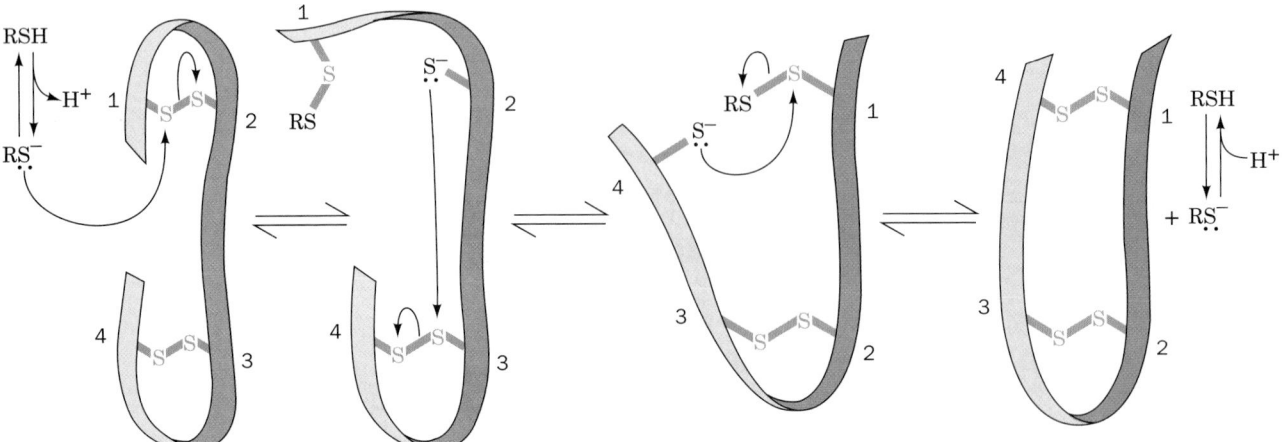

FIGURE 9-3   Plausible mechanism for the thiol- or enzyme-catalyzed disulfide interchange reaction in a protein. The purple ribbon represents the polypeptide backbone of the protein. The attacking thiol group must be in its ionized thiolate form.

tive structure is achieved (Fig. 9-3). The native state of RNase A under physiological conditions is therefore, most probably, its thermodynamically most stable conformation (if the protein has a conformation that is more stable than the native state, conversion to it must involve such a large activation barrier so as to make it kinetically inaccessible; rate processes are discussed in Section 14-1C).

The time for renaturation of "scrambled" RNase A is reduced to ~2 min through the use of an enzyme, **protein disulfide isomerase (PDI),** that catalyzes disulfide interchange reactions. (In fact, the supposition that *in vivo* folding to the native state requires no more than a few minutes prompted the search that led to this enzyme's discovery.) PDI itself contains two active site Cys residues, which must be in the —SH form for the isomerase to be active. The enzyme evidently catalyzes the random cleavage and reformation of a protein's disulfide bonds (Fig. 9-3), thereby interchanging them as the protein progressively attains thermodynamically more favorable conformations. PDI is further discussed in Section 9-2A.

a. Posttranslationally Modified Proteins May Not Readily Renature

Many "scrambled" proteins are renatured through the action of PDI and are unaffected by it in their native state (their PDI-cleaved disulfide bonds rapidly reform because these native proteins are in their most stable local conformations). In posttranslationally modified proteins, however, the disulfide bonds may serve to hold the protein in its otherwise unstable native state. For instance, the 51-residue polypeptide hormone **insulin,** which consists of two polypeptide chains joined by two disulfide bonds (Fig. 7-2), is inactivated by PDI. This observation led to the discovery that insulin is derived from a single-chain, 84-residue precursor named **proinsulin** (Fig. 9-4). Only after its disulfide bonds have formed is proinsulin converted to the two-chained active hormone by the specific proteolytic excision of an internal 33-residue segment known as its C chain. Nevertheless, two sets of observations suggest that the C

chain does not direct the folding of the A and B chains but, rather, simply holds them together while they form their native disulfide bonds: (1) Under proper renaturing conditions, native insulin is obtained from scrambled insulin in 25 to 30% yield, which increases to 75% when the A and

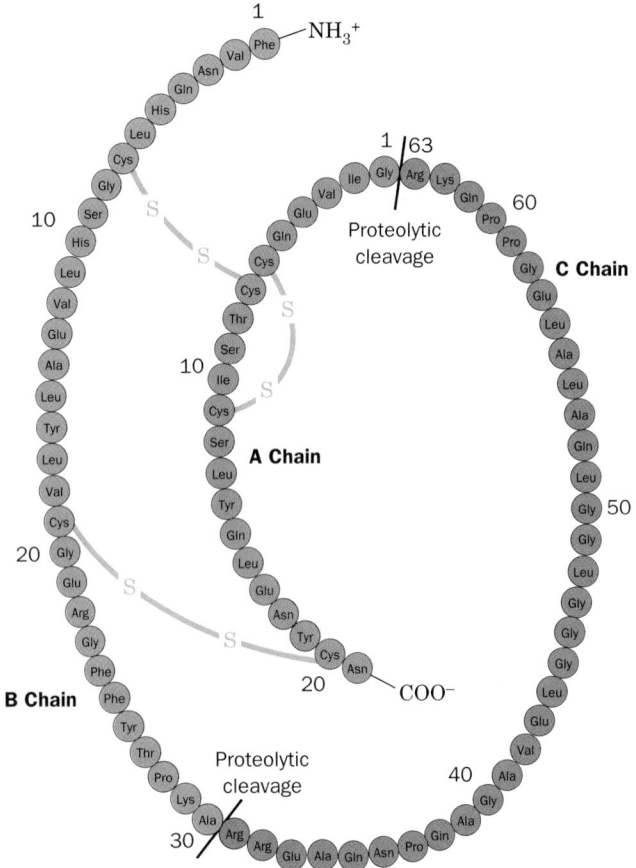

FIGURE 9-4   Primary structure of porcine proinsulin. Its C chain (*brown*) is proteolytically excised from between its A and B chains to form the mature hormone. [After Chance, R.E., Ellis, R.M., and Brommer, W.W., *Science* **161,** 165 (1968).]

B chains are chemically cross-linked; and (2) sequence comparisons of proinsulins from a variety of species indicate that mutations are accepted into the C chain at a rate which is eight times that for the A and B chains.

B. *Determinants of Protein Folding*

In Section 8-4, we discussed the various interactions that stabilize native protein structures. In this section we extend this discussion by considering how these interactions are organized in native proteins. Keep in mind that only a small fraction of the myriads of possible polypeptide sequences are likely to have unique stable conformations. Evolution has, of course, selected such sequences for use in biological systems.

a. Helices and Sheets May Predominate in Proteins Simply because They Fill Space Efficiently

Why do proteins contain such a high proportion (~60%, on average) of α helices and β pleated sheets? Hydrophobic interactions, although the dominant influence responsible for the compact nonpolar cores of proteins, lack the specificity to restrict polypeptides to particular conformations. Similarly, the observation that polypeptide segments in the coil conformation are no less hydrogen bonded than helices and sheets suggests that the conformations available to polypeptides are not greatly limited by their hydrogen bonding requirements. Rather, as Ken Dill has shown, it appears that helices and sheets form largely as a consequence of steric constraints in compact polymers. Exhaustive simulations of the conformations which simple flexible chains (such as a string of pearls) can assume indicate that the proportion of helices and sheets increases dramatically with a chain's level of compaction (number of intrachain contacts); that is, helices and sheets are particularly compact entities. Thus, most ways to compact a chain involve the formation of helices and sheets. In native proteins, such elements of secondary structure are fine tuned to form α helices and β sheets by short-range forces such as hydrogen bonding, ion pairing, and van der Waals interactions. It is probably these less dominant but more specific forces that "select" the unique native structure of a protein from among its relatively small number of hydrophobically generated compact conformations (recall that most hydrogen bonds in proteins link residues that are close together in sequence; Section 8-4B).

b. Protein Folding Is Directed Mainly by Internal Residues

Numerous protein modification studies have been aimed at determining the role of various classes of amino acid residues in protein folding. In one particularly revealing study, the free primary amino groups of RNase A (Lys residues and the N-terminus) were derivatized with 8-residue chains of poly-DL-alanine. Intriguingly, these large, water-soluble poly-Ala chains could be simultaneously coupled to RNase's 11 free amino groups without significantly altering the protein's native conformation or its ability to refold. Since these free amino groups are all located on the exterior of RNase A, this observation suggests that *it is largely a protein's internal residues that direct its folding to the native conformation*. Similar conclusions have been reached from studies of protein structure and evolution (Section 9-6): Mutations that change surface residues are accepted more frequently and are less likely to affect protein conformations than are changes of internal residues. It is therefore not surprising that the perturbation of protein folding by limited concentrations of denaturing agents indicates that *protein folding is driven by hydrophobic forces*.

c. Protein Structures Are Hierarchically Organized

Large protein subunits consist of domains, that is, of contiguous, compact, and physically separable segments of the polypeptide chain. Furthermore, as George Rose showed, domains consist of subdomains, which in turn consist of sub-subdomains, etc. Conceptually, this means that if a polypeptide segment of any length in a native protein is viewed as a tangle of string, a single plane can be found that divides the string into only two segments rather than many smaller segments (such as would happen if a ball of yarn were cut in this way). This is readily demonstrated by coloring the first $n/2$ residues of an n-residue domain red and the second $n/2$ residues blue. If this process is iterated, as is shown in Fig. 9-5 for high potential iron–sulfur protein (HiPIP), it is clear that at every stage of the process,

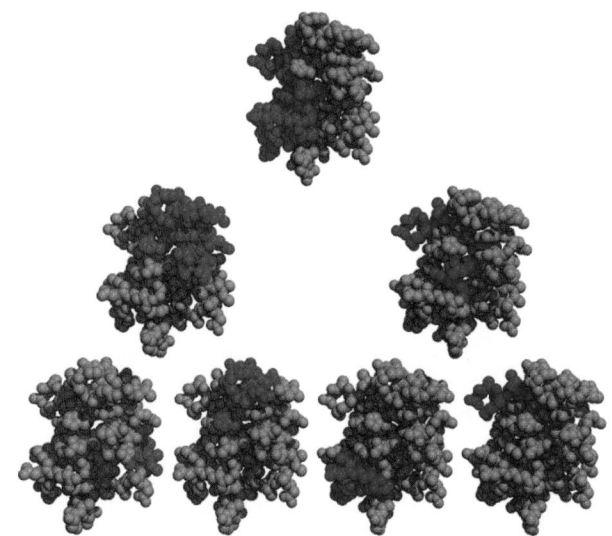

FIGURE 9-5 Hierarchical organization of globular proteins. Here the X-ray structure of high potential iron protein (HiPIP) is represented by its C_α atoms shown as spheres. In the top drawing, the first $n/2$ residues of this n-residue protein (where $n = 71$) are colored red and the remaining $n/2$ residues are colored blue. In the second row, the process is iterated such that, on the right, for example, the first and last halves of the second half of the protein are red and blue, with the remainder of the chain gray. In the third row, the process is again iterated. Note that at each stage of this hierarchy, the red and blue regions do not intermingle. [Courtesy of George Rose, The Johns Hopkins University School of Medicine, and Robert Baldwin, Stanford University School of Medicine.]

the red and blue regions do not interpenetrate. Evidently, *protein structures are organized hierarchically,* that is, polypeptide chains form locally compact structures that associate with similar adjacent (in sequence) structures to form larger compact structures, etc. This structural organization is, of course, consistent with the observation that hydrogen bonding interactions in proteins are mostly local (Section 8-4B). It also has important implications for how polypeptides fold to form native proteins (Section 9-1C).

d. Protein Structures Are Highly Adaptable

Globular proteins have packing densities comparable to those of organic crystals (Section 8-3B) because the side chains in a protein's interior fit together with exquisite complementarity. To ascertain whether this phenomenon is an important determinant of protein structure, Eaton Lattman and Rose analyzed 67 globular proteins of known structure for the existence of preferred interactions between side chains. They found none, thereby indicating that, at least in globular proteins, *the native fold determines the packing but packing does not determine the native fold.* This view is corroborated by the widespread occurrence of protein families whose members assume the same fold even though they may be so distantly related as to have no recognizable sequence similarity (e.g., the α/β barrel proteins; Section 8-3B).

The foregoing study indicates that *there are a large number of ways in which a protein's internal residues can pack together efficiently.* This was perhaps most clearly shown by Brian Matthews in an extensive series of studies on **T4 lysozyme** (a product of bacteriophage T4) in which the X-ray structures of over 300 mutant varieties of this 164-residue monomeric enzyme were compared. Replacements of one or a few residues in T4 lysozyme's hydrophobic core were accommodated mainly by local shifts in the protein backbone rather than by any global structural changes. In many cases, T4 lysozyme could accommodate the insertion of up to four residues without a major structural change or even a loss of enzymatic activity. Moreover, assays of the enzymatic activities of 2015 single-residue substitutions in T4 lysozyme indicated that only 173 of these mutants had significantly decreased enzymatic activity. Clearly protein structures are highly resilient.

e. Secondary Structure Can Be Context-Dependent

The structure of a native protein is determined by its amino acid sequence, but to what extent is the conformation of a given polypeptide segment influenced by the surrounding protein? The NMR structure of **protein GB1** (the B1 domain of streptococcal **protein G,** which helps the bacterium evade the host's immunological defenses by binding to the antibody protein **immunoglobulin G**) reveals that this 56-residue domain, which lacks disulfide bonds, consists of a long α helix lying across a 4-stranded mixed β sheet (Fig. 9-6). In mutagenesis experiments by Peter Kim, the 11-residue "chameleon" sequence AWTVEKAFKTF was made to replace either residues 23 to 33 of GB1's α helix (AATAEKFVFQY in GB1; a 7-residue change) to yield Chm-α, or residues 42 to 52 of

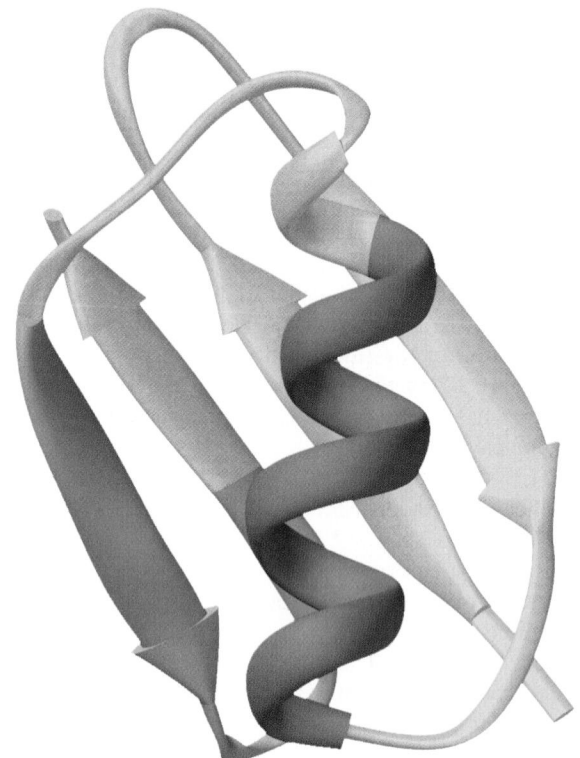

FIGURE 9-6   NMR structure of protein GB1. Residues 23 to 33 are green and residues 42 to 53 are cyan. The 11-residue chameleon sequence AWTVEKAFKTF can occupy either of these positions without significantly altering the native protein's backbone conformation. [NMR structure by Angela Gronenborn and Marius Clore, National Institutes of Health, Bethesda, Maryland. PDBid 1GB1.]

its C-terminal β hairpin (EWTYDDATKTF in GB1; a 5-residue change) to yield Chm-β. Both Chm-α and Chm-β display reversible thermal unfolding typical of compact single-domain globular proteins, and their 2D NMR spectra indicate that each assumes a structure similar to that of native GB1. Yet NMR measurements also demonstrate that the isolated chameleon peptide (Ac-AWTVEKAFKTF-NH$_2$, where Ac is acetyl) is unfolded in solution, which indicates that this sequence has no strong preference for either an α helix or a β sheet conformation. This suggests that the information specifying α helix or β sheet secondary structures can be nonlocal; that is, context-dependent effects may be important in protein folding (but see Section 9-1C).

f. Changing the Fold of a Protein

Proteins that share as little as ~20% sequence identity may be structurally similar. To what degree must a protein's sequence be changed in order to convert its fold to that of another protein? This question was answered, at least for the protein GB1, by the finding that changing 50% of its 56 residues converted its fold to that of **Rop protein** (Rop for *repressor of primer;* a transcriptional regulator).

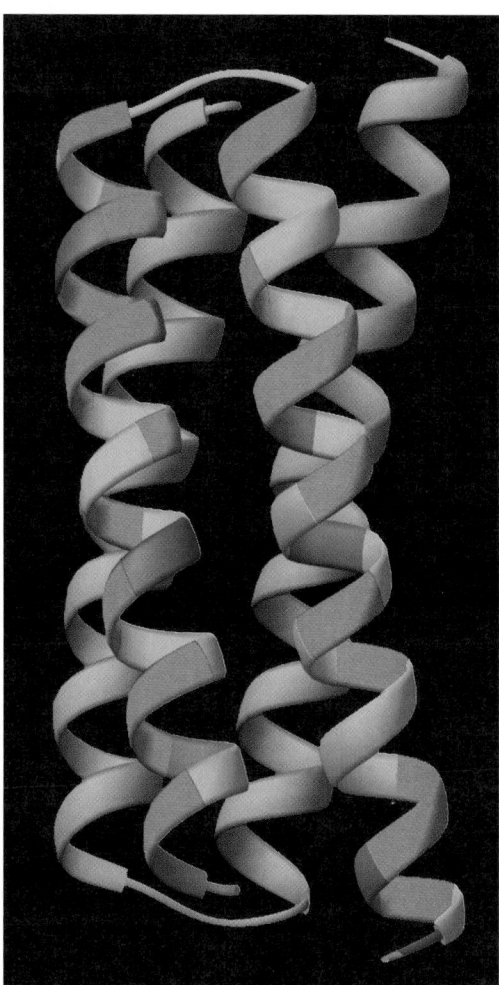

FIGURE 9-7 X-Ray structure of Rop protein, a homodimer of αα motifs that associate to form a 4-helix bundle. On a change of 50% of its residues, protein GB1, whose structure is shown in Fig. 9-6, assumes the structure of Rop protein. One of the subunits of the structure shown here is colored according to the sequence of the GB1-derived polypeptide with purple residues identical in both native proteins, magenta residues unchanged from native GB1, cyan residues identical to those in native Rop, and green residues different from those in either native protein. The N-terminus of this subunit is at the lower right. [X-Ray structure by Demetrius Tsernoglou, Università di Roma, Rome, Italy. PDBid 1ROP.]

Rop is a homodimer whose 63-residue subunits each form an αα motif (Fig. 8-46c) that dimerizes with its 2-fold axis perpendicular to the helix axes to form a 4-helix bundle (Fig. 9-7). 50% of the residues of GB1 were changed based largely on a secondary structure prediction algorithm (Section 9-3), energy minimization, and visual modeling to yield a new polypeptide named Janus (after the 2-faced Roman god of new beginnings) that is 41% identical to Rop. In this manner, GB1 residues with high helix-forming propensities were retained, whereas in regions required to be α helical, a number of residues with high β sheet–forming propensities were replaced (helix- and sheet-forming propensities are discussed in Section 9-3); hy-

drophobic residues were incorporated at the appropriate *a* and *d* positions of a heptad repeat (Fig. 8-27) to form the core of Rop's 4-helix bundle; and residue changes were made to mimic Rop's distribution of surface charges. Fluorescence and NMR measurements reveal that Janus assumes a stable Rop-like conformation. These studies indicate that not all residues have equally important roles in specifying a particular fold. Indeed, the Janus sequence is more closely related to that of GB1 (50% identity) than to that of Rop (41% identity), even though Janus structurally resembles Rop but not GB1.

C. Folding Pathways

How does a protein fold to its native conformation? We, of course, cannot hope to answer this question in detail until we better understand why native protein structures are stable. Moreover, as one might guess, the folding process itself is one of enormous complexity. Nevertheless, as we shall see below, the broad outlines of how proteins fold to their native conformations are beginning to come into focus.

The simplest folding mechanism one might envision is that a protein randomly explores all of the conformations available to it until it eventually "stumbles" onto its native conformation. A "back-of-the-envelope" calculation first made by Cyrus Levinthal, however, convincingly demonstrates that this cannot possibly be the case: Assume that the $2n$ backbone torsional angles, ϕ and ψ, of an n-residue protein each have three stable conformations. This yields $3^{2n} \approx 10^n$ possible conformations for the protein, which is a gross underestimate, if only because the side chains are ignored. If a protein can explore new conformations at the rate at which single bonds can reorient, it can find $\sim 10^{13}$ conformations per second, which is, no doubt, an overestimate. We can then calculate the time, t, in seconds, required for a protein to explore all the conformations available to it:

$$t = \frac{10^n}{10^{13} \text{ s}^{-1}} \qquad [9.1]$$

For a small protein of $n = 100$ residues, $t = 10^{87}$ s, which is immensely more than the apparent age of the universe ($\sim$20 billion years = 6×10^{17} s).

It would obviously take even the smallest protein an absurdly long time fold to its native conformation by randomly exploring all its possible conformations, an inference known as the **Levinthal paradox.** Yet many proteins fold to their native conformations in less than a few seconds. Therefore, as Levinthal suggested, *proteins must fold by some sort of ordered pathway or set of pathways in which the approach to the native state is accompanied by sharply increasing conformational stability (decreasing free energy).*

a. Rapid Measurements Are Required to Monitor Protein Folding

Folding studies on several small single-domain proteins, including RNase A, cytochrome *c*, and **apomyoglobin** (myoglobin that lacks its heme group), indicate that these

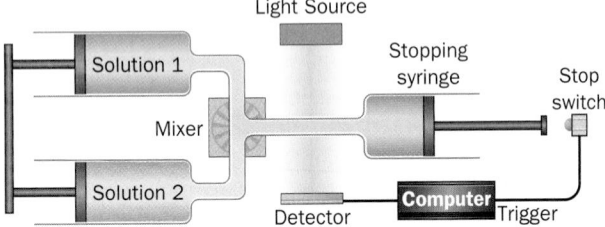

FIGURE 9-8   A stopped-flow device. The reaction is initiated by simultaneously and rapidly discharging the contents of both syringes through the mixer. On hitting the stop switch, the stopping syringe triggers the computer to commence optically monitoring the reaction (via its UV/visible, fluorescence, or CD spectrum).

proteins fold to a significant degree within one millisecond or less of being brought to native conditions. Hence, if the earliest phases of the folding process are to be observed, denatured proteins must be brought to native conditions in significantly less time. This is most often done using a rapid mixing device such as a **stopped-flow** apparatus (Fig. 9-8) in which a protein solution at a pH that denatures the protein or containing guanidinium chloride or urea at a concentration that does so is rapidly changed in pH or diluted to initiate folding. Such instruments have "dead times" (the interval between the times when mixing is initiated and meaningful measurements can first be made) of >0.5 ms. However, recently developed ultrarapid mixing devices have dead times of at little as 40 µs.

An alternative technique involves the refolding of **cold denatured proteins** [For proteins whose folding has both ΔH and ΔS positive, a decrease in temperature is destabilizing (Table 3-2). Since $\Delta G = \Delta H - T\Delta S$, these proteins are unstable, that is denature, when $T < \Delta H/\Delta S$. For many of these proteins, solution conditions can be found for which this temperature is >0°C]. The refolding of the cold-denatured protein is initiated by a so-called **temperature-jump** in which the solution is heated with an infrared laser pulse by 10 to 30°C in <100 ns.

With either of the above methods, the folding protein must be monitored by a technique that can report rapid structural changes in a protein. The two such techniques that have been most extensively used are (1) optical techniques, particularly **circular dichroism (CD)** spectroscopy, and (2) **pulsed H/D exchange** followed by 2D-NMR spectroscopy. We discuss these methods below.

b.   The Circular Dichroism Spectrum of a Protein Is Indicative of Its Conformation

A solution containing a solute that absorbs light does so according to the **Beer–Lambert law,**

$$A = \log\left(\frac{I_0}{I}\right) = \varepsilon c l \qquad [9.2]$$

where A is the solute's **absorbance** (alternatively, its **optical density**), I_0 is the intensity of the incident light at a given wavelength λ, I is its transmitted intensity at λ, ε is the **molar extinction coefficient** of the solute at λ, c is its

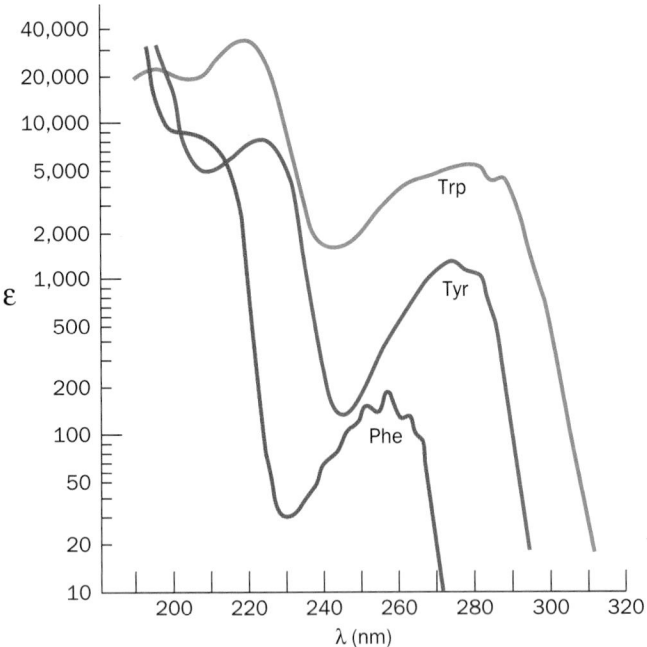

FIGURE 9-9   UV absorbance spectra of the three aromatic amino acids, phenylalanine, tryptophan, and tyrosine. Note that the molar absorbance, ε, is displayed on a log scale. [After Wetlaufer, D.B., *Adv. Prot. Chem.* **7,** 310 (1962).]

molar concentration, and l is the length of the light path in cm. The value of ε varies with λ; a plot of ε vs λ for the solute is called its **absorbance spectrum.**

Polypeptides absorb strongly in the ultraviolet (UV) region of the spectrum ($\lambda = 100$ to 400 nm) largely because their aromatic side chains (those of Phe, Trp, and Tyr) have particularly large molar extinction coefficients in this spectral region (ranging into the tens of thousands; Fig. 9-9). However, polypeptides do not absorb visible light ($\lambda = 400$ to 800 nm), so that they are colorless.

For chiral molecules such as proteins, ε has different values for left and right circularly polarized light, ε_L and ε_R. The variation with λ of the difference in these quantities, $\Delta\varepsilon = \varepsilon_L - \varepsilon_R$, constitutes the **CD spectrum** of the solute of interest (for nonchiral molecules $\varepsilon_L = \varepsilon_R$ and hence they have no CD spectrum). In proteins, α helices, β sheets, and random coils exhibit characteristic CD spectra (Fig. 9-10). Hence the CD spectrum of a polypeptide provides a rough estimate of its secondary structure.

c.   Pulsed H/D Exchange Provides Structural Details on How Proteins Fold

Pulsed H/D exchange, a method devised by Walter Englander and Robert Baldwin, is the only known technique that can follow the time course of individual residues in a folding protein. Weakly acidic protons (^{1}H), such as those of amine and hydroxyl groups (X—H), exchange with those of water, a process known as **hydrogen exchange** that can be demonstrated with the use of deuterated water [D_2O; deuterium (D or ^{2}H) is a stable isotope of ^{1}H]:

$$X—H + D_2O \rightleftharpoons X—D + HOD$$

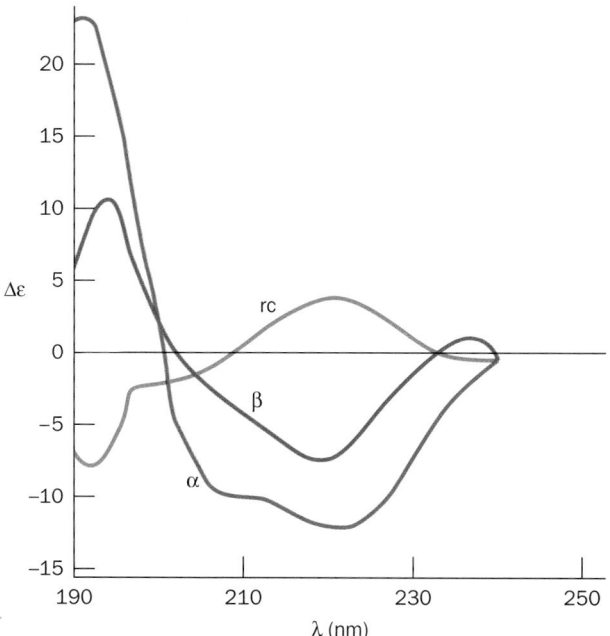

FIGURE 9-10   Circular dichroism (CD) spectra of polypeptides.
Polypeptides in the α helix, β sheet, and random coil (rc) conformations were determined from the CD spectra of proteins of known X-ray structures. By comparing these spectra with the absorption spectra in Fig. 9-9, it can be seen that $\Delta\varepsilon = \varepsilon_L - \varepsilon_R$ is a small difference of two large numbers. [After Saxena, V.P. and Wetlaufer, D.B., *Proc. Nat'l. Acad. Sci.* **66**, 971 (1971).]

Since ^{1}H has an NMR spectrum in a different frequency range from that of D, the exchange of ^{1}H for D can be readily followed by NMR spectroscopy. Under physiological conditions, small organic molecules, such as amino acids and dipeptides, completely exchange their weakly acidic protons for D in times ranging from milliseconds to seconds. Proteins bear numerous exchangeable protons such as those of its backbone amide groups. However, protons that are engaged in hydrogen bonding do not exchange with solvent and, moreover, groups in the interior of a native protein are not in contact with solvent.

Through the use of 2D-NMR (Section 8-3A), pulsed H/D exchange can be used to follow the time course of protein folding. The protein of interest, usually with its native disulfide bonds intact, is denatured by guanidinium chloride or urea in D_2O solution such that all of the protein's peptide nitrogen atoms become deuterated (N—D). Folding is then initiated in a stopped-flow apparatus by diluting the denaturant solution with ^1H_2O while the pH is simultaneously lowered so as to arrest hydrogen exchange (near neutrality, hydrogen exchange reactions are catalyzed by OH$^-$ and, therefore, their rates are highly pH dependent). After a preset folding time, t_f, the pH is rapidly increased (using a third independently triggered syringe; the so-called labeling pulse) to initiate hydrogen exchange. Peptide nitrogen atoms whose D atoms have not formed hydrogen bonds by time t_f exchange with ^{1}H, whereas those that are hydrogen bonded at t_f, and hence unavailable for

hydrogen exchange, remain deuterated. After a short time (10 to 40 ms), the labeling pulse is terminated by rapidly lowering the pH (with a fourth syringe). Folding is then allowed to go to completion and the H/D ratio at each exchangeable site is determined by 2D-NMR (the peaks in the 2D proton NMR spectrum must have been previously assigned). By repeating the analysis for several values of t_f, the time course of hydrogen bond formation at each residue can be determined.

Pulsed H/D exchange–NMR studies do not directly indicate the structures of the folding intermediates. However, if the native structure of the protein under investigation is known (as it almost always is for proteins whose folding is being investigated) and if it is assumed that the protein folds without forming secondary structures not present in the native protein, then the 2D-NMR spectra reveal the time course of the formation of the elements of the native structure together with how fast they are excluded from the bulk solvent.

d. The Earliest Protein Folding Events Are Initiated by a Hydrophobic Collapse

Stopped-flow–CD measurements indicate that *for many, if not all, small single-domain proteins, much of the secondary structure that is present in native proteins forms within a few milliseconds of when folding is initiated.* This is called the **burst phase** because subsequent folding events occur over much longer time intervals. Pulsed H/D exchange measurements of these small proteins show that some protection against hydrogen exchange in some secondary structural elements develops by ~5 ms after folding initiation.

Since globular proteins contain a compact hydrophobic core, it seems likely that the driving force in protein folding is a so-called **hydrophobic collapse,** in which the protein's hydrophobic groups coalesce so as to expel most of their surrounding water molecules. The polypeptide's radius of gyration is thereby dramatically reduced (from ~30 to ~15 Å for a 100-residue polypeptide), a phenomenon that is generally characteristic of polymers on being transferred from a good to a poor solvent.

This hydrophobic collapse mechanism is consistent with the observation that the hydrophobic dye **8-anilino-1-naphthalenesulfonate (ANS)**

**8-Anilino-1-naphthalene
sulfonate (ANS)**

binds to folding proteins. ANS undergoes a significant enhancement of its fluorescence when it occupies a nonpolar environment, an enhancement that is observed within the burst phase when ANS is present in a solution of a folding protein. Since ANS is expected to preferentially bind to hydrophobic groups, this indicates that the hydrophobic core of a protein rapidly forms once folding has been initiated.

The initial collapsed state of a folding protein is known as a **molten globule.** Such a species has a radius of gyration that is only 5 to 15% greater than that of the native protein and has significant amounts of the native secondary structure and overall fold. However, a molten globule's side chains are extensively disordered, its structure fluctuates far more than that of the native protein, and it has only marginal thermodynamic stability. Nevertheless, to continue folding toward its native state, the polypeptide chain need not undergo large rearrangements in the crowded core of the partially folded protein.

e. Nativelike Tertiary Structure Appears during Intermediate Folding Events

After the burst phase, small proteins exhibit increased ANS binding, further changes in their CD spectrum, and enhanced protection against H/D exchange. These intermediate folding events occur over a time interval of 5 to 1000 ms. This is the stage at which the protein's secondary structure becomes stabilized and its tertiary structure begins to form. These nativelike elements are thought to take the form of subdomains that are not yet properly docked to each other. Side chains are probably still mobile, so that, at this stage of folding, the protein can be described as an ensemble of closely related and rapidly interconverting structures.

f. Final Folding Events Require Several Seconds

In the final stage of folding, a protein achieves its native structure. To do so, the polypeptide must undergo a series of complex motions that permit the attainment of the relatively rigid native core packing and hydrogen bonding, while expelling the remaining water molecules from its hydrophobic core. For small single-domain proteins, this takes place over a time interval of several seconds or less.

g. Landscape Theory of Protein Folding

The classic view of protein folding was that proteins fold through a series of well-defined intermediates. The folding of a random coil polypeptide was thought to begin with the random formation of short stretches of $2°$ structure, such as α helices and β turns, that acted as **nuclei** (scaffolding) for the stabilization of additional ordered regions of the protein. Nuclei with the proper nativelike structure then grew by the diffusion, random collision, and adhesion of two or more such nuclei. The stabilities of these ordered regions were thought to increase with size, so, after having randomly reached a certain threshold size, they spontaneously grew in a cooperative fashion until they formed a nativelike domain. Finally, through a series of relatively small conformational adjustments, the domain re-

arranged to the more compact $3°$ structure of the native conformation.

The advent of experimental methods that could observe early events in protein folding led to a somewhat different view of how proteins fold. In this so-called **landscape theory,** which was formulated in large part by Peter Wolynes, Baldwin, and Dill, folding is envisioned to occur on an **energy surface** or landscape that represents the conformational energy states available to a polypeptide under the prevailing conditions. The horizontal coordinates of a point on this surface represent a particular conformation of the polypeptide, that is, the values of φ and ψ for each of its amino acid residues and the torsion angles for each of its side chains (but here projected onto two dimensions from its multidimensional space). The vertical coordinate of a point on the energy surface represents the polypeptide's internal free energy in this conformation. The above-described measurements indicate that the energy surface of a folding polypeptide is funnel-shaped, with the native state represented by the bottom of the funnel, the global (overall) free energy minimum (Fig. 9-11*a*). The width of the funnel at any particular height (free energy) above the native state is indicative of the number of conformational states with that free energy, that is, the entropy of the polypeptide.

Polypeptides fold via a series of conformational adjustments that reduce their free energy and entropy until the native state is reached. Since a collection of unfolded polypeptides all have different conformations (have different positions on the folding funnel), they cannot follow precisely the same pathway in folding to the native state. If the polypeptide actually folded to its native state via a random conformational search, as Levinthal conjectured, its energy surface would resemble a flat disk with a single small hole, much like the surface of a golf course (Fig. 9-11*b*). Thus, it would take an enormously long time for a polypeptide (a golf ball) to achieve the native state (to fall in the hole) via a random conformational search (by rolling about aimlessly on the surface of the golf course).

The energy surface of a protein that follows the classic view of protein folding would have a deep radial groove in its disklike surface that slopes toward the hole representing the native state (Fig. 9-11*c*). The extent of the conformational search to randomly find this groove would be much reduced relative to the Levinthal model, so that such a polypeptide would readily fold to its native state. However, the conformational search for the pathway (groove) leading to the native state would still take time, so that the polypeptide would require perhaps several seconds to start down the folding pathway.

The observation that many polypeptides acquire significant nativelike structure within fractions of a millisecond after folding commences indicates that their energy surfaces are, in fact, funnel-shaped; that is, they tend to slope toward the native conformation at all points. Thus, the various pathways followed by initially unfolded polypeptides in folding to their native state are analogous to the various trajectories that could be taken by skiers initially distributed around the top of a bowl-shaped valley to reach the valley's lowest point. Apparently, *there is no single pathway or closely related set of pathways that a polypeptide must follow in folding to its native conformation.*

The foregoing does not imply that the surface of the folding funnel is necessarily smooth, as is drawn in Fig. 9-11*a*. Indeed, landscape theory suggests that this energy surface has a relatively rugged topography, that is, has many local energy minima and maxima (Fig. 9-11*d*). Consequently, in following any particular folding pathway, a polypeptide is likely to become trapped in a local minimum until it randomly acquires sufficient thermal energy to surmount this kinetic barrier and continue the folding process. Thus, in landscape theory, the local energy max-

ima (transition states; Section 14-1C) that govern the rate of folding are not specific structures as the classic theory of protein folding suggests but, rather, are ensembles of structures.

h. Protein Folding Is Hierarchical

The observation that protein structures are hierarchically organized (Section 9-1B) suggests that they also fold in a hierarchic manner. By this it is meant that the folding process begins with the formation of marginally stable structures that are local in sequence and that these local structures locally interact to yield intermediates of increasing complexity that sequentially grow to form the native

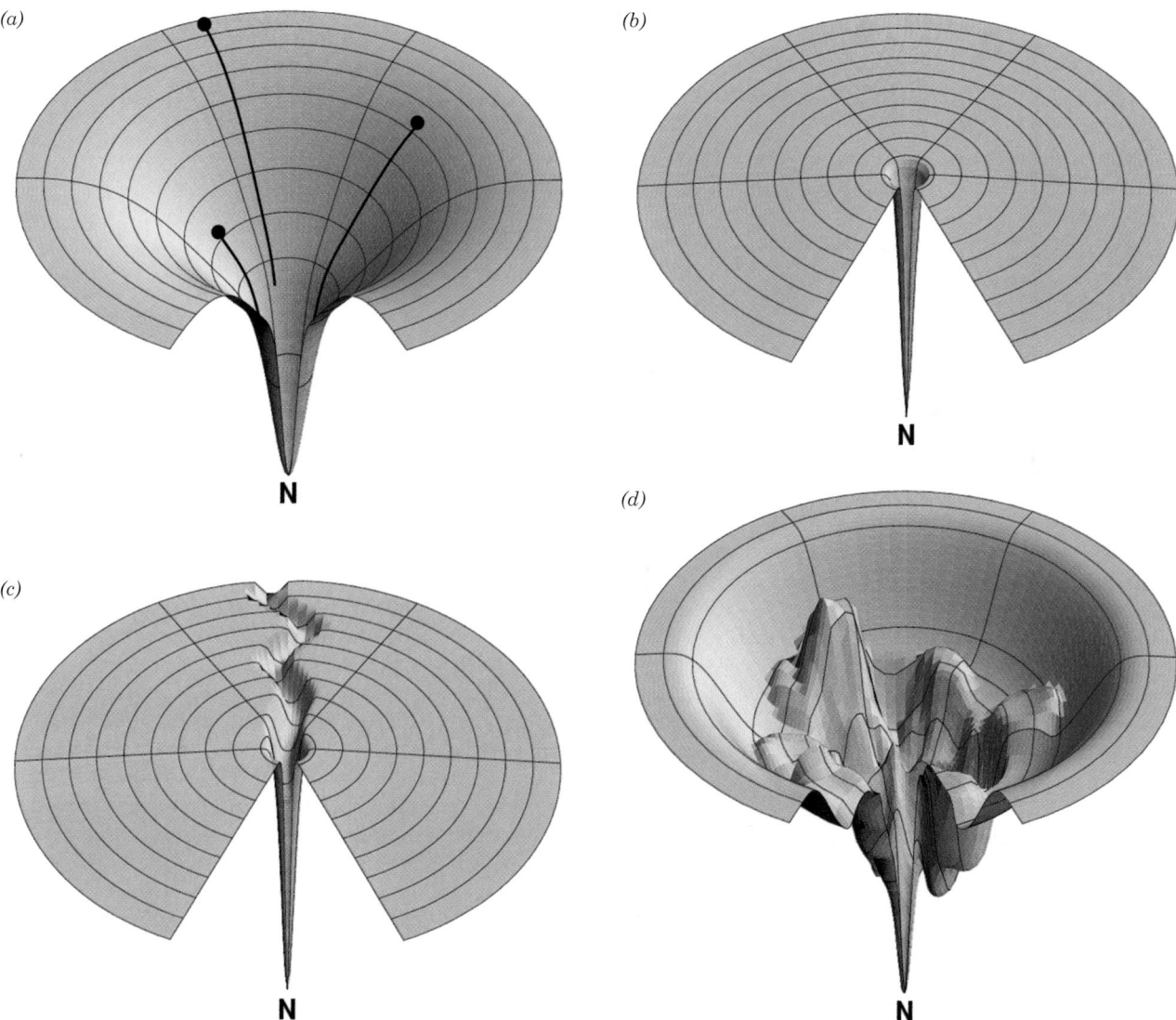

FIGURE 9-11   Folding funnels. (*a*) An idealized funnel landscape. As the chain forms increasing numbers of intrachain contacts, its internal free energy (its height above the native state, N) decreases together with its conformational freedom (the width of the funnel). Polypeptides with differing conformations (*black dots*) follow different pathways (*black lines*) in achieving the native fold. (*b*) The Levinthal "golf course" landscape in which the chain must search for the native fold (the hole) randomly, that is, on a level energy surface. (*c*) The classic folding landscape in which the chain must search at random on a level energy surface until it encounters the canyon that leads it to the native state. (*d*) A rugged energy surface containing local minima in which a folding polypeptide can become transiently trapped. The folding funnels of real proteins are thought to have such topographies. [Courtesy of Ken Dill, University of California at San Francisco.]

protein. In contrast, in nonhierarchical folding, a protein's tertiary structure would not only stabilize its local structures but also determine them. Landscape theory is consistent with hierarchical folding, whereas the classic theory of protein folding is more in accord with nonhierarchical folding. Moreover, since a polypeptide *in vivo* begins folding as it is being synthesized, that is, as it is extruded from the ribosome, it would seem that it would most readily achieve its native state if it folded in a hierarchical manner.

Several lines of evidence indicate that proteins, in fact, fold in a hierarchical manner.

1. Many peptide fragments excised from proteins either form or exhibit a tendency to form nativelike folds in the absence of long-range (3°) interactions. Moreover, when proteins such as cytochrome *c* and apomyoglobin are brought to a pH sufficiently low to destabilize their native structures, their nativelike 2° structural elements persist.

2. The intermediates observed in protein folding are consistent with a hierarchical process.

3. The boundaries of helices in native proteins are fixed by their flanking sequences (Section 9-3) rather than by 3° interactions.

4. LINUS (for Local Independently Nucleated Units of Structure), a program designed by Rose that simulates hierarchical folding, successfully approximates the 2° structures in several proteins, even when all long-range interactions are suppressed.

In Section 9-1B we saw that in protein GB1 (Fig. 9-6), the 11-residue "chameleon" sequence assumed either an α helix or a β hairpin, depending on its position in the protein. Thus, its conformation appears to be determined by its context rather than by local interactions. However, simulations by LINUS indicate that the conformation of the chameleon sequence is actually determined by local interactions beyond its boundaries.

The folds of native proteins, as we have seen, are highly resistant to sequence changes. Evidently, *the sequence information specifying a particular fold is both distributed throughout the polypeptide chain and highly overdetermined.* It is these characteristics that appear to be responsible for hierarchical folding.

i. BPTI Folds to Its Native Conformation via an Ordered Pathway

We have discussed the general principles of protein folding but have not described the folding of any particular protein. Of course, if landscape theory is correct, this would only be possible in a statistical sense, at least for the early stages of folding. Indeed, many small proteins appear to fold to their native conformations without any detectable (stable) intermediates. Such proteins (e.g., cytochrome *c*) are said to fold by a two-state mechanism (only their unfolded and native states are stable). However, many other proteins appear to fold via well-defined intermediates, at least in their later stages of folding. This has, perhaps, been most clearly shown by renaturation studies on **bovine pancreatic trypsin inhibitor (BPTI;** Fig. 9-12), a

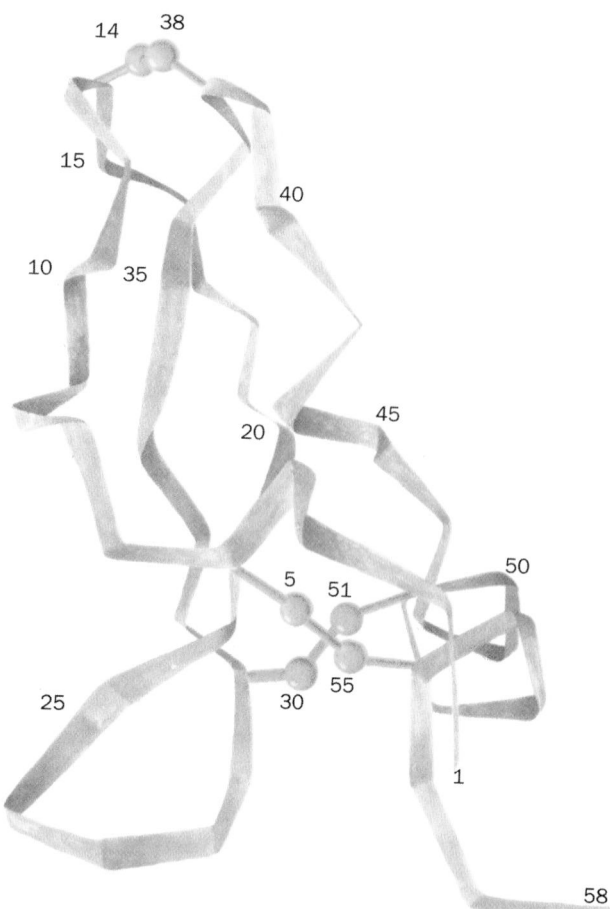

FIGURE 9-12   Polypeptide backbone and disulfide bonds of native BPTI. [After a drawing by Michael Levitt, *in* Creighton, T.E., *J. Mol. Biol.* **95,** 168 (1975).]

58-residue monomeric protein that has three disulfide bonds (BPTI functions to bind and thereby inactivate trypsin in the pancreas, thus protecting this secretory organ from self-digestion; Section 15-3E).

Fully reduced BPTI (which has no disulfide bonds) is completely unfolded (a random coil) under physiological conditions. This permitted Thomas Creighton and Kim to determine the order of disulfide bond formation in BPTI by initiating folding through the addition of a sulfhydryl reagent such as oxidized dithiothreitol (Section 7-1B), allowing the protein to refold for a time and then trapping the resulting intermediate(s) (e.g., by lowering the pH, which inhibits the formation of the thiolate ion required for the disulfide interchange reactions; Fig. 9-3). The intermediates formed at a given time were then separated chromatographically, the positions of their disulfide bonds determined (Section 7-1I), and several of their structures characterized by NMR. Each distinguishable disulfide-bonded species represents a subset of the conformations that the BPTI polypeptide can assume, so that by following the time course of the appearance of these various species, the approximate conformational path taken by the renaturing protein was deduced.

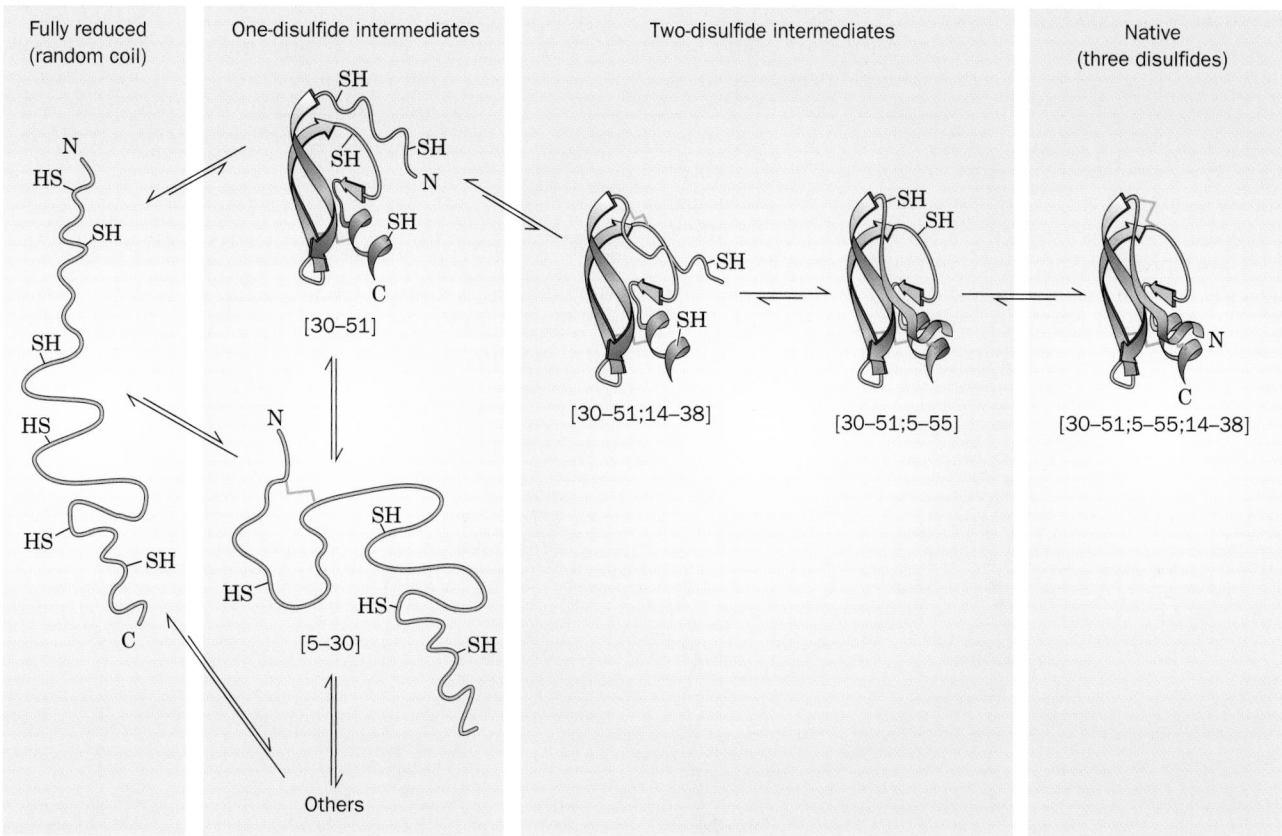

| Fully reduced (random coil) | One-disulfide intermediates | Two-disulfide intermediates | Native (three disulfides) |

FIGURE 9-13   Renaturation of BPTI. The renaturation pathway of BPTI shows the conformations of its polypeptide backbone as deduced from disulfide trapping experiments and NMR measurements (note that these views of the protein differ from that in Fig. 9-12 by a slight rotation about the vertical axis). The sequence numbers of the Cys residues involved in each disulfide bond are given in brackets below the diagram representing each folding intermediate. The two one-disulfide intermediates, [5–30] and [30–51], are in rapid equilibrium. [After Creighton, T.E., *Biochem. J.* **270,** 12 (1990).]

These studies indicated that *BPTI follows a limited set of pathways in folding to its native structure (Fig. 9-13):*

1. The six Cys residues of the fully reduced BPTI are equally likely to participate in forming the initial disulfide bond. Yet, after the molecule has equilibrated through a series of rapid internal disulfide interchange reactions, only 2 of the 15 possible one-disulfide intermediates, [5–30] and [30–51] (where the numbers indicate the Cys residues participating in a disulfide bond), exist in significant quantities. (An intermediate's relative abundance at equilibrium is indicative of its thermodynamic stability relative to other intermediates; Section 3-4A.) Of these, only [30–51] has a disulfide bond that occurs in the native protein and only this species reacts in significant amounts with sulfhydryl reagents to form a second disulfide bond. A variety of conformational studies indicate that [30–51] is a conformationally fluctuating molecule that, much of the time, forms the native protein's β sheet and C-terminal α helix, the two major elements of BPTI's 2° structure, which comprise much of its hydrophobic core. Note that the 30–51 disulfide bond links these two elements of 2° structure.

2. Of the 45 possible 2-disulfide intermediates, only [30–51;14–38] occurs in a significant quantity even though all four free SH groups of [30–51] are equally reactive. Note that this intermediate contains a new native disulfide bond (14–38). Nevertheless, in order to fold to the native conformation, [30–51;14–38] must first convert to [30–51;5–55], which also contains a second native disulfide bond (5–55). This conversion is relatively slow, which is indicative of a large conformational rearrangement. Evidently, the conformation necessary to form the 5–55 disulfide bond is difficult to achieve (requires a high energy of activation). NMR studies indicate that [30–51;5–55] has a nativelike conformation in that it exhibits all the 2° structural elements of the native protein.

3. This conclusion is corroborated by the rapid formation of BPTI's third disulfide bond (14–38) to yield [30–51; 5–55;14–38], native BPTI.

j. Primary Structures Determine Protein Folding Pathways as Well as Structures

The above discussions suggest that *protein primary structures evolved to specify efficient folding pathways as well as stable native conformations.* Evidence corroborating this hypothesis has been obtained by Jonathan King in his study of the renaturation of the **tail spike protein** of bacteriophage P22. The tail spike protein is a trimer of identical 76-kD polypeptides, whose $T_m = 88°C$. However, certain mutant varieties of the protein fail to renature at 39°C. Nevertheless, at 30°C, these mutant proteins fold to structures whose properties, including their T_m's, are indistinguishable from that of the wild-type tail spike protein. The amino acid changes causing these temperature-sensitive folding mutations apparently act to destabilize intermediate states in the folding process but do not affect the native protein's stability. This observation suggests that *a protein's amino acid sequence dictates its native structure by specifying how it folds to its native conformation.* This hypothesis is supported by the observation that, in native proteins, a greater number of polar residues than would be randomly expected occupy helix-capping positions (Section 8-4B) even though they do not make helix-capping hydrogen bonds. This suggests that they do so as the helix forms so as to facilitate the protein's proper folding.

2 ■ FOLDING ACCESSORY PROTEINS

Most unfolded proteins renature *in vitro* over periods ranging from minutes to days and, quite often, with low efficiency, that is, with a large fraction of the polypeptide

(a)

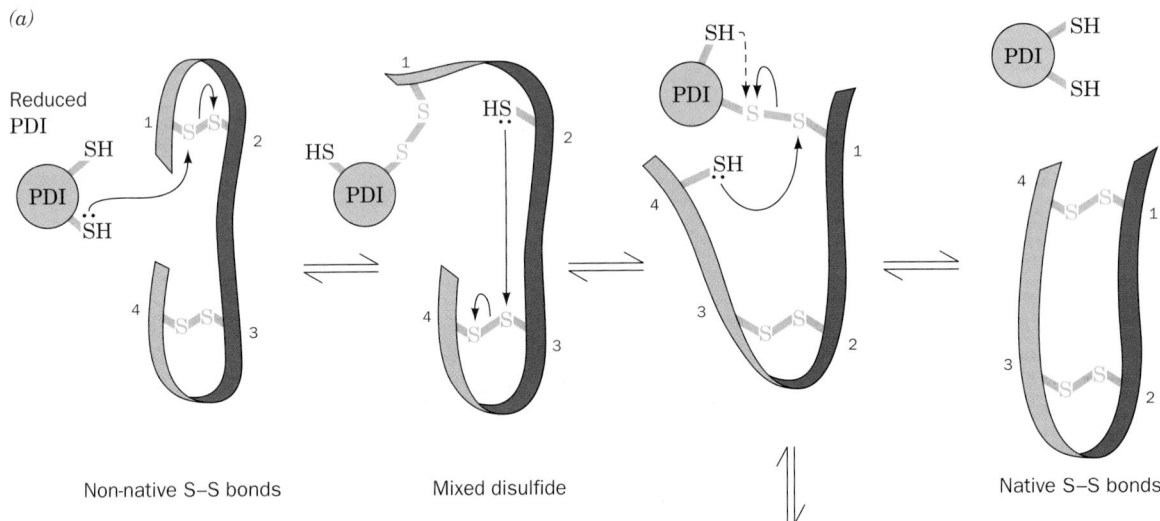

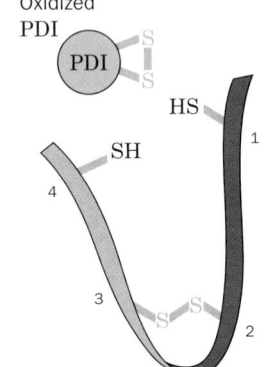

FIGURE 9-14   Reactions catalyzed by protein disulfide isomerase (PDI). (*a*) Reduced PDI catalyzes the rearrangement of the non-native disulfide bonds in a substrate protein (*purple ribbon*) via disulfide interchange to yield the native disulfide bonds (*horizontal reactions*). If a disulfide bond between PDI and the substrate protein is resistant to disulfide interchange, it is reduced by PDI's second SH group to yield reduced substrate protein and oxidized PDI (*vertical reaction and dashed curly arrow*). (*b*) The oxidized PDI-dependent synthesis of disulfide bonds in proteins. The reaction occurs with the intermediate formation of a mixed disulfide between PDI and the protein. The reduced PDI reaction product reacts with cellular oxidizing agents to regenerate oxidized PDI.   ✍ **See the Animated Figures**

(b)

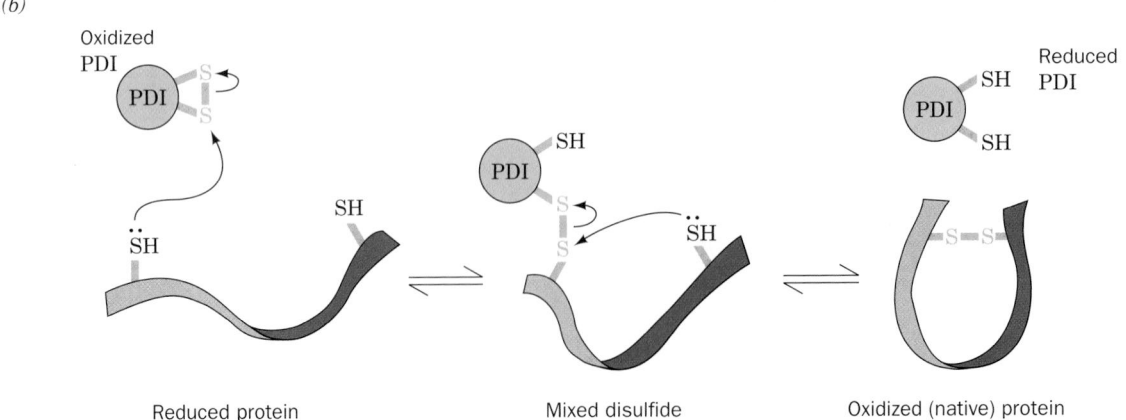

chains assuming quasi-stable non-native conformations and/or forming nonspecific aggregates. *In vivo*, however, polypeptides efficiently fold to their native conformations as they are being synthesized, a process that normally requires a few minutes or less. This is because all cells contain three types of accessory proteins that function to assist polypeptides in folding to their native conformations and in assembling to their 4° structures: protein disulfide isomerases, peptidyl prolyl cis–trans isomerases, and molecular chaperones. We discuss these essential proteins in this section.

A. *Protein Disulfide Isomerases*

Protein disulfide isomerase (PDI), which we encountered in Section 9-1A, is a homodimeric eukaryotic enzyme of 486-residue subunits (prokaryotes contain similar proteins). In its reduced form, PDI catalyzes disulfide interchange reactions, thereby facilitating the shuffling of the disulfide bonds in proteins (Fig. 9-14*a, horizontal reactions*) until they achieve their native pairings, which are resistant to further rearrangement. Moreover, PDI must facilitate the correct folding of those proteins that denature in the absence of their native disulfide bonds. Intriguingly, PDI is also the β subunit of the $\alpha_2\beta_2$ heterotetramer prolyl hydroxylase, the enzyme that hydroxylates the Pro residues of collagen (Section 8-2B). The significance of this latter finding is unknown.

Sequence comparisons indicate that a PDI subunit consists of four ~100-residue domains that are arranged, from N- to C-terminus, as *a–b–b′–a′*, in which domains *a* and *a′* are homologs, as are domains *b* and *b′*. The *a* and *a′* domains (PDI-*a* and PDI-*a′*) are homologous to the ubiquitous disulfide-containing redox protein **thioredoxin** (Section 28-3A), and hence these domains belong to the thioredoxin superfamily.

PDI-*a* and PDI-*a′* each contain the active site sequence motif -Cys-Gly-His-Cys-, in which the first Cys residue, in its —SH form, participates in the disulfide interchange reaction diagrammed in Fig. 9-14*a*. If the second Cys residue is mutated, PDI's isomerization activity drops to <1% of the wild type and it accumulates in disulfide-linkage to substrate proteins. This suggests that this second Cys residue functions, in its —SH form, to release PDI from the otherwise stable disulfide bonds that its first Cys residue occasionally forms with substrate proteins, thereby yielding reduced substrate proteins and PDI with a disulfide bond linking its two active site Cys residues (Fig. 9-14*a, vertical reaction*).

Although the structure of intact PDI is unknown, the NMR structures of PDI-*a* and PDI-*b* have been determined by Creighton and Johan Kemmink. PDI-*a* (120 residues) forms an open β sheet (Fig. 9-15*a*) which, as expected, closely resembles that of thioredoxin. Disulfide bonds in native proteins are usually buried and frequently occur in hydrophobic environments. However, the S atom of oxidized PDI-*a*'s Cys 36 (the first of its two active site Cys residues) is exposed on the protein surface, where it is centered in a hydrophobic (uncharged) patch (Fig.

9-15*b*). Thus, PDI-*a*'s active site occupies a region that appears capable of binding unfolded polypeptide segments. Moreover, even though disulfide bonds almost always sta-

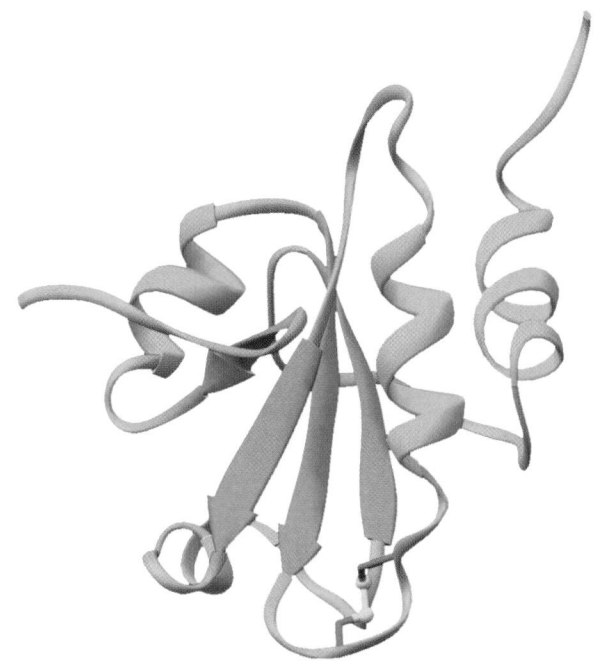

(a)

(b)

FIGURE 9-15 NMR structure of the *a* domain of human protein disulfide isomerase (PDI-*a*) in its oxidized form.
(*a*) The polypeptide backbone is shown in ribbon form with its helices cyan and its β strands magenta. The side chains of the active site Cys residues (Cys 36 and Cys 39), which form a disulfide bond, are shown in ball-and-stick form with C green and S yellow. (*b*) The molecular surface as viewed from the bottom of Part *a*. The surface is colored according to its electrostatic potential with dark blue most positive, dark red most negative, and white uncharged. The S atom of Cys 36 is yellow. [Based on an NMR structure by Johan Kemmink and Thomas Creighton, European Molecular Biology Laboratory, Heidelberg, Germany. PDBid 1MEK.]

bilize proteins (Section 8-4D) and are usually unreactive, oxidized PDI-*a* is less stable than reduced PDI-*a* and has a highly reactive, that is, strongly oxidizing, disulfide bond. This permits oxidized PDI-*a* to directly introduce disulfide bonds into newly synthesized and hence reduced polypeptides via a disulfide interchange mechanism (Fig. 9-14*b*). For this latter process to continue, reduced PDI must be reoxidized (its disulfide bond reformed) by cellular oxidizing agents.

PDI-*b* (110 residues) is only 11% identical to PDI-*a* and is devoid of Cys residues. It was therefore a surprise when the NMR structure of this catalytically inactive domain revealed that it too adopts the thioredoxin fold. Evidently, PDI consists of two catalytically active thioredoxin-like domains linked by two catalytically inactive such domains. Although oxidized PDI-*a* and PDI-*a'* are both good catalysts of disulfide bond formation, efficient catalysis of disulfide bond rearrangement requires that reduced PDI be intact. The reason for this must await the structural determination of intact PDI.

B. *Peptidyl Prolyl Cis–Trans Isomerases*

Although polypeptides are probably biosynthesized with almost all of their Xaa–Pro peptide bonds (where Xaa is any amino acid residue) in the trans conformation, ~10% of these bonds assume the cis conformation in globular proteins because, as we have seen in Section 8-1A, the energy difference between their cis and trans conformations is relatively small. **Peptidyl prolyl cis–trans isomerases (PPIs;** alternatively known as **rotamases)** catalyze the otherwise slow interconversion of Xaa–Pro peptide bonds between their cis and trans conformations, thereby accelerating the folding of Pro-containing polypeptides. Two structurally unrelated families of PPIs, collectively named the **immunophilins,** have been characterized: the **cyclophilins** (so named because they are inhibited by the immunosuppressive drug **cyclosporin A,**

Cyclosporin A

a fungally produced 11-residue cyclic peptide) and the family for which the 12-kD **FK506 binding protein (FKBP12)** is prototypic (**FK506**

FK506

is a fungally produced macrocyclic lactone that is also an immunosuppressive drug; medicinal chemists tend to identify the often huge numbers of related drug candidates they deal with by serial numbers rather than by trivial names).

The X-ray structure of human cyclophilin in complex with succinyl-Ala-Ala-Pro-Phe-*p*-nitroanilide reveals that this model substrate binds to the enzyme with its Ala–Pro peptide bond in the cis conformation and that it could not do so if it had the trans conformation. This suggests that the enzyme predominantly catalyzes the trans to cis isomerization of peptidyl-prolyl amide bonds. In addition, the Arg 55 → Ala mutation in cyclophilin reduces its enzymatic activity 100-fold. This, together with the observation that Arg 55 is positioned so that it could hydrogen bond to the N atom of the Ala–Pro peptide bond (although it does not do so in the crystal structure) suggests that the formation of a hydrogen bond from Arg 55 to this N atom facilitates the cis–trans isomerization by deconjugating and hence weakening the peptidyl-prolyl amide bond.

Cyclosporin A and FK506 are highly effective agents for the treatment of autoimmune disorders and for preventing organ-transplant rejection. Indeed, until the advent of cyclosporin A in the early 1980s, the long-term survival of a transplanted organ (and its recipient) was a rare occurrence. The more recently discovered FK506 is an even more potent immunosuppressant. The immunosuppressive properties of both cyclosporin A and FK506 stem from the abilities of their respective complexes with cyclophilin and FKBP12 to prevent the expression of genes involved in the activation of ***T* lymphocytes** (the immune system cells responsible for **cellular immunity;** the immune response is discussed in Section 35-2) by interfering with these cells' intracellular signaling pathways. Enigmatically, there is no obvious relationship between the immunophilins' immunosuppressive properties and rotamase activities: Both cyclosporin A and FK506 are effective immunosuppressants at concentrations far below those of the cyclophilin and FKBP12 in cells; and mutational changes that destroy cyclophilin's rotamase activity do not eliminate its ability to bind cyclosporin A or the ability of the resulting complex to interfere with *T* lymphocyte signaling. This conundrum is explained in Section 19-3F.

C. *Molecular Chaperones: The GroEL/ES System*

Newly synthesized and hence unfolded proteins contain numerous solvent-exposed hydrophobic groups. Moreover, proteins *in vivo* fold in the presence of extremely high concentrations of other macromolecules (~300 g/L, which occupy ~25% of the available volume). Consequently, unfolded proteins *in vivo* have a great tendency to form both intramolecular and intermolecular aggregates. **Molecular chaperones** are proteins that function to prevent or reverse such improper associations, particularly in multidomain and multisubunit proteins. They do so by binding to an unfolded or aggregated polypeptide's solvent-exposed hydrophobic surfaces and subsequently releasing them, often repeatedly, in a manner that facilitates their proper folding and/or 4° assembly. Many molecular chaperones are **ATPases** (enzymes that catalyze ATP hydrolysis), which bind to unfolded polypeptides and apply the free energy of ATP hydrolysis to effect their release in a favorable manner. Thus it appears, as John Ellis has pointed out, that molecular chaperones function analogously to their human counterparts: *They inhibit inappropriate interactions between potentially complementary surfaces and disrupt unsuitable liasons so as to facilitate more favorable associations.*

The molecular chaperones comprise several unrelated classes of proteins that have somewhat different functions including:

1. The **heat shock proteins 70 (Hsp70),** which are highly conserved in both prokaryotes and eukaryotes (where different species occur in the cytosol, the endoplasmic reticulum, mitochondria, and chloroplasts). They are so-named because the rate of synthesis of these ~70-kD monomeric proteins greatly increases at elevated temperatures (the *E. coli* Hsp70 is called **DnaK** because it was discovered through the isolation of mutants that do not support the growth of bacteriophage λ and hence was initially thought to participate in DNA replication). They function as monomers in an ATP-driven process to reverse the denaturation and aggregation of proteins (processes that are accelerated at elevated temperatures), to facilitate the proper folding of newly synthesized polypeptides as they emerge from the ribosome, to unfold proteins in preparation for their transport through membranes (Section 12-E), and to subsequently refold them. Hsp70 works in association with the **cochaperone** protein **Hsp40** (**DnaJ** in *E. coli*) to bind and release small hydrophobic regions of misfolded proteins.

2. The **chaperonins,** which form large, multisubunit, cagelike assemblies, and are universal components of bacteria and eukaryotes. They bind improperly folded globular proteins via their exposed hydrophobic surfaces and then, in an ATP-driven process, induce the protein to fold while enveloping it in an internal cavity, thereby protecting the folding protein from nonspecific aggregation with other unfolded proteins (see below).

3. The **Hsp90** proteins, which are mainly involved in the folding of proteins that are involved in signal transduction, including **steroid hormone receptors** (Section

34-3B). They are among the most abundant proteins in eukaryotes, constituting ~1% of their soluble proteins.

4. The **nucleoplasmins,** which are acidic nuclear proteins whose presence is required for the proper *in vivo* assembly of **nucleosomes** (particles in which eukaryotic DNA is packaged) from their component DNA and histones (Section 34-1B).

Certain classes of chaperone proteins may work in concert. For example, the chaperonins often finish the job that the Hsp70 proteins have begun. In the following paragraphs we concentrate on the structure and function of the chaperonins, as these are the best characterized molecular chaperones. This discussion also constitutes our introduction to the dynamic functions of proteins, that is, to proteins as molecular machines.

a. The GroEL/ES System Forms a Large Cavity in Which Substrate Protein Folds

The chaperonins consist of two families of proteins that work in concert: (1) The **Hsp60** proteins (**GroEL** in *E. coli* and **Cpn60** in chloroplasts), which, as electron microscopic images first revealed, consist of 14 identical ~60-kD subunits arranged in two apposed rings of 7 subunits each (Fig. 9-16); and (2) the **Hsp10** proteins (**GroES** in *E. coli* and **Cpn10** in chloroplasts), which form single heptameric

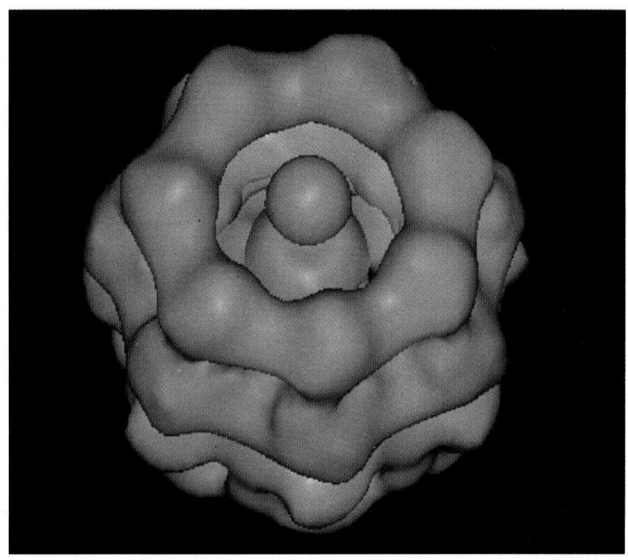

FIGURE 9-16   Electron micrograph–derived 3D image of the Hsp60 chaperonin from the photosynthetic bacterium *Rhodobacter sphaeroides*. Hsp60 consists of 14 identical ~60-kD subunits arranged to form two apposed rings of 7 subunits, each surrounding a central cavity. The image of Hsp60, which is viewed with its 7-fold axis tipped toward the viewer, indicates that each subunit consists of two major domains, one in contact with the opposing heptameric ring, and the other at the end of the cylindrical protein molecule. The spherical density occupying the protein's central cavity is thought to represent a bound polypeptide. The cavity provides a protected microenvironment in which a polypeptide can fold. [Courtesy of Helen Saibil and Steve Wood, Birkbeck College, London, U.K.]

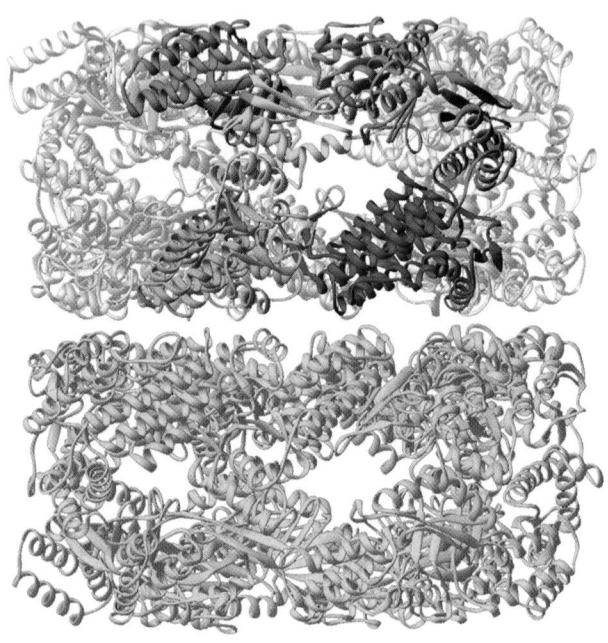

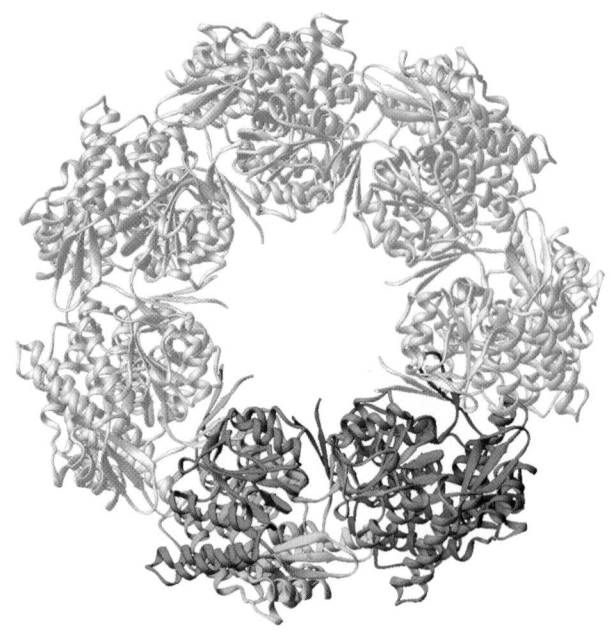

(a)

(b)

FIGURE 9-17   X-Ray structure of GroEL. (*a*) Side view perpendicular to the 7-fold axis in which the seven identical subunits of the lower ring are gold and those of the upper ring are silver, with the exception of the two subunits nearest the viewer, whose equatorial, intermediate, and apical domains are colored blue, green, and red on the right subunit and cyan, yellow, and magenta on the left subunit. The two rings of the complex are held together through side chain interactions that are not seen in this drawing. (*b*) Top view along the 7-fold axis in which only the upper ring is shown for the sake of clarity. Note the large central channel that appears to run the length of the protein. [Based on an X-ray structure by Axel Brünger, Arthur Horwich, and Paul Sigler, Yale University. PDBid 1OEL.]

rings of identical ~10-kD proteins. These proteins, which are essential to the survival of *E. coli* under all conditions tested, facilitate the folding of improperly folded proteins to their native conformations.

The X-ray structure of GroEL (Fig. 9-17), determined by Arthur Horwich and Paul Sigler, showed, as expected, that GroEL's 14 identical 547-residue subunits associate to form a porous thick-walled hollow cylinder that consists of two 7-fold symmetric rings of subunits stacked back to back with 2-fold symmetry to yield a complex with D_7 symmetry (Section 8-5B). Each GroEL subunit consists of three domains: a large equatorial domain (residues 1–135 and 410–547) that forms the waist of the protein and holds its subunits together through both intra- and inter-ring interactions, a loosely structured apical domain (residues 191–376) that forms the open ends of the GroEL cylinder, and a small intermediate domain (residues 136–190 and 377–409) that connects the equatorial and apical domains. The X-ray structure suggests that GroEL encloses an ~45-Å-diameter central channel that runs the length of the complex. We shall see below that this channel, in part, forms the chambers in which partially folded proteins fold to their native states. However, both electron microscopy–based images and neutron scattering studies indicate that the channel is obstructed in its equatorial region, so that proteins cannot pass between two GroEL rings. The obstruction is apparently caused by each subunit's N-terminal 5 residues and C-terminal 22 residues, which are not

seen in the X-ray structure and hence are almost certainly disordered.

The X-ray structure of GroEL with **ATPγS** bound to each subunit (ATPγS is a poorly hydrolyzable analog of ATP in which S replaces one of the O atoms substituent to P_γ)

$$^-O-\underset{\underset{O^-}{\|}}{\overset{\overset{S}{\|}}{P}}-O-\underset{\underset{O^-}{\|}}{\overset{\overset{O}{\|}}{P}}-O-\underset{\underset{O^-}{\|}}{\overset{\overset{O}{\|}}{P}}-O-CH_2$$

ATPγS

indicates that ATP binds to a pocket in the equatorial domain that opens onto the central channel. The residues forming this pocket are highly conserved among chaperonins. The only significant differences between the structures of the GroEL–ATPγS complex and that of GroEL alone are modest movements of the residues in the vicinity of the ATP pocket.

The X-ray structure of GroES (Fig. 9-18), determined by Lila Gierasch and Johann Deisenhofer, shows that this protein's 7 identical 97-residue subunits form a domelike structure with C_7 symmetry. Each GroES subunit consists of an irregular antiparallel β barrel from which two β hair-

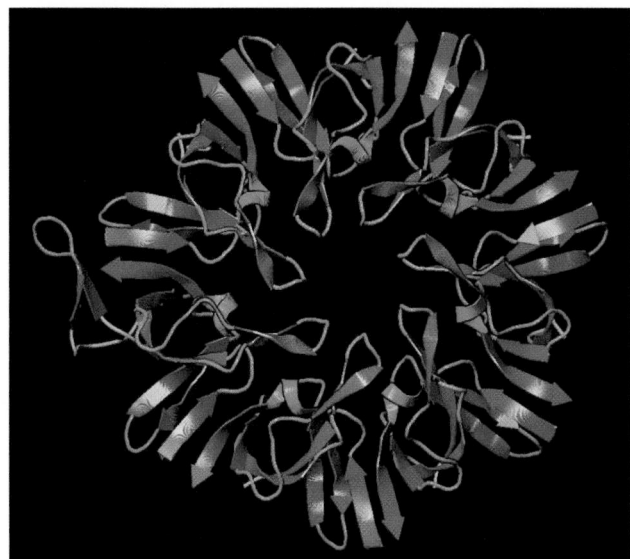

FIGURE 9-18 X-Ray structure of GroES as viewed along its 7-fold axis. The mobile loop of only one of the protein's 7 identical subunits (*left*) is visible in the structure. The polypeptide segments that flank the mobile loop are yellow. [Courtesy of Johann Deisenhofer, University of Texas Southwest Medical Center, Dallas.]

pins project. One of these β hairpins (residues 47–55) extends from the top of the β barrel toward the protein's 7-fold axis, where it interacts with the other such β hair-

pins to form the roof of the dome. The second β hairpin (residues 16–33) extends from the opposite side of the β barrel outward from the bottom outer rim of the dome. This so-called mobile loop is seen in only one of GroES's 7 subunits; it is apparently disordered in the other subunits in agreement with the results of NMR studies of uncomplexed GroES in solution. The inner surface of the GroES dome is lined with hydrophilic residues.

Both electron microscopic and neutron scattering studies reveal that partially unfolded proteins bind in the mouth of the GroEL barrel in a manner reminiscent of a cork in a champagne bottle (Fig. 9-16). Mutations that impair polypeptide binding to GroEL all map to a poorly resolved (and presumably flexible) segment at the top of the apical domain that, in the structure of GroEL alone, faces the central channel. In fact, changing any of eight highly conserved hydrophobic residues in this region to Glu or Ser abolishes polypeptide binding. It therefore seems likely that these residues provide the binding site(s) for non-native polypeptides. Interestingly, mutations of these same residues also abolish the binding of GroES.

The X-ray structure of the GroEL–GroES–(ADP)$_7$ complex (Fig. 9-19), also determined by Horwich and Sigler, provides considerable insight into how this chaperonin carries out its function. In this complex, a GroES heptamer and the 7 ADPs are bound to the same GroEL ring (the so-called cis ring; the opposing GroEL ring is known as the trans ring) such that the GroES cap closes over the GroEL cis ring barrel like a lid on a pot, thereby forming a bullet-shaped complex with C_7 symmetry. The trans ring

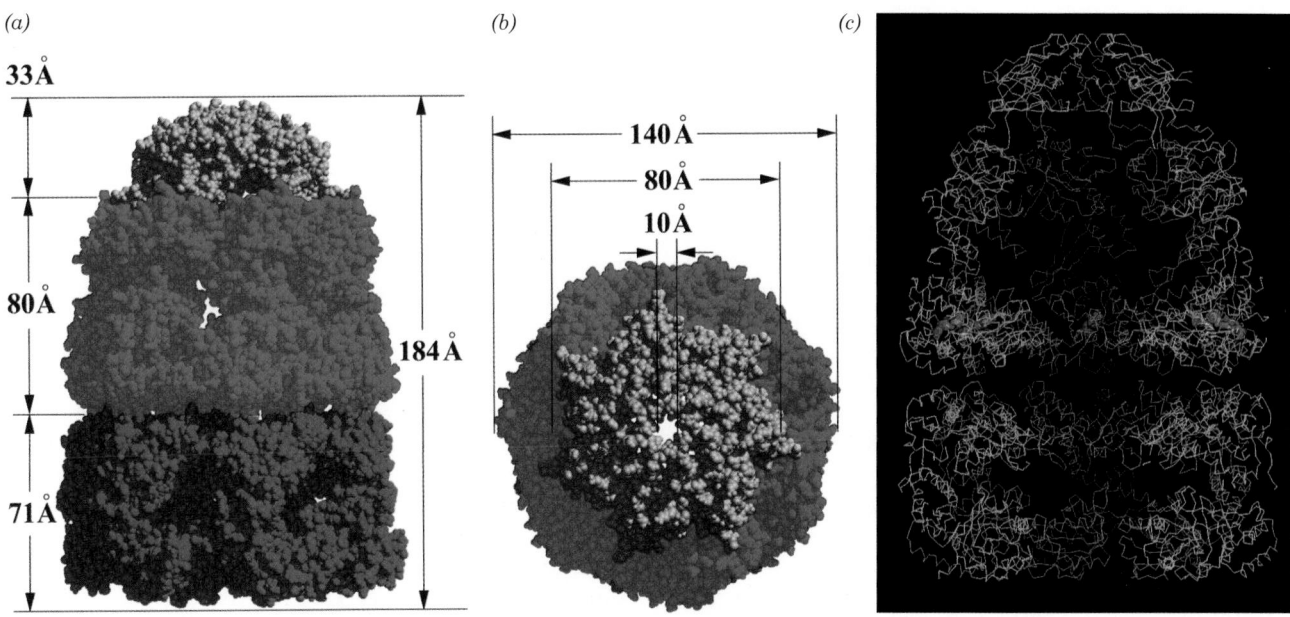

(a) *(b)* *(c)*

33 Å

140 Å
80 Å
10 Å

80 Å

184 Å

71 Å

FIGURE 9-19 X-Ray structure of the GroEL–GroES–(ADP)$_7$ complex. (*a*) A space-filling drawing as viewed perpendicularly to the complex's 7-fold axis, with the GroES ring gold, the cis ring of GroEL green, and the trans ring of GroEL red. The dimensions of the complex are indicated. Note the different conformations of the two GroEL rings. (*b*) As in Part *a* but viewed along the 7-fold axis. (*c*) The C$_α$ backbone of the complex viewed as in Part *a* but which is cut away along the plane containing the complex's 7-fold axis. The ADPs bound to the cis ring of GroEL are shown in space-filling form. Note the much larger size of the cavity formed by the cis ring and GroES in comparison to that of the trans ring. [Courtesy of Paul Sigler, Yale University. PDBid 1AON.]

(a)

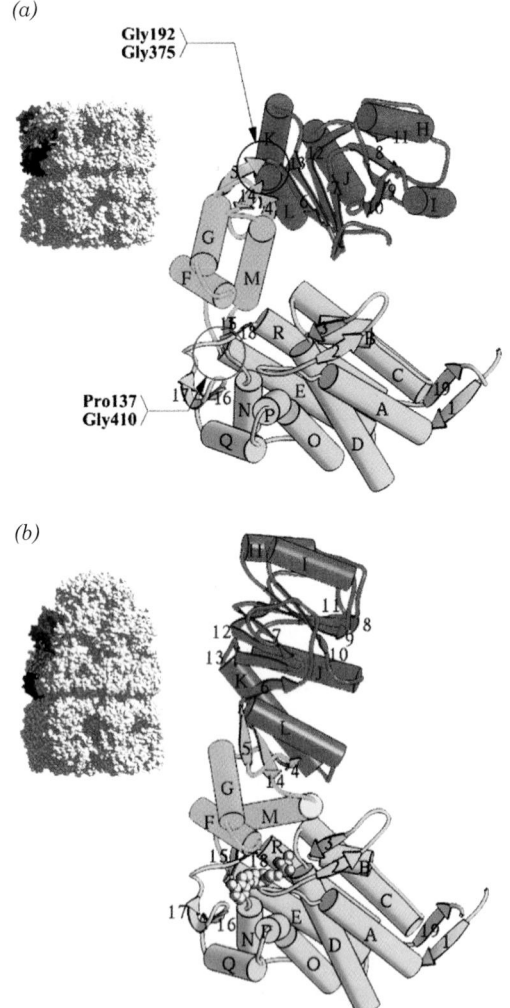

(b)

FIGURE 9-20   Domain movements in GroEL. (*a*) Ribbon diagram of a single subunit of GroEL in the X-ray structure of GroEL alone. Its equatorial, intermediate, and apical subunits are colored blue, green, and red. The inset shows a space-filling drawing of GroEL with the colored subunit oriented identically. Circles and arrows indicate the pivot points for domain movements. (*b*) A GroEL subunit in the X-ray structure of GroEL–GroES–(ADP)$_7$ displayed as in Part *a*. The ADP, which is bound in a pocket at the top of the equatorial domain, is shown in space-filling form in yellow. (*c*) Schematic diagram indicating the conformational changes in GroEL when it binds GroES. Its equatorial (E), intermediate (I), and apical (A) domains are colored as in Part *a* and GroES is yellow. The arrows indicate the extent of the domain movements in the cis ring of GroEL. [Parts *a* and *b* courtesy of Arthur Horwich, Yale University; Part *c* after Richardson, A., Landry, S.J., and Georgopoulos, C., *Trends Biochem. Sci.* **23**, 138 (1998).]

(c)

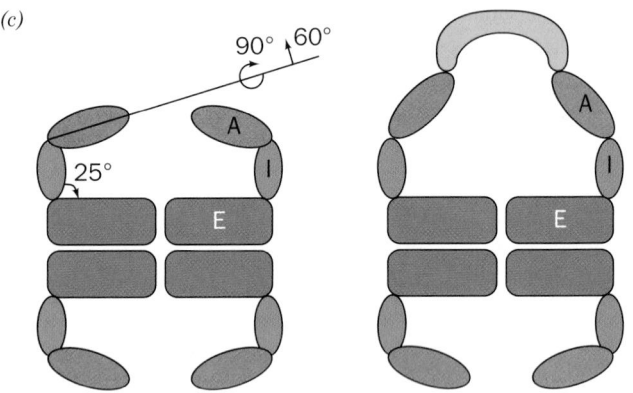

subunits have conformations that closely resemble those in the structure of GroEL alone. In contrast, the apical and intermediate domains of the cis ring have undergone large *en bloc* movements relative to their positions in GroEL alone (Fig. 9-20). This widens and elongates the cis cavity in a way that more than doubles its volume (from 85,000 to 175,000 Å^3; Fig. 9-19*c*), thereby permitting it to enclose a partially folded substrate protein of at least 70 kD. *These en bloc movements are concerted, that is, they occur simultaneously in all seven subunits of a GroEL ring, most probably because if one GroEL subunit did not undergo these conformational shifts, it would mechanically block its adjacent subunits from doing so.*

In forming the GroEL–GroES–(ADP)$_7$ complex, the ADP becomes completely enclosed by protein through the collapse of the intermediate domain onto the equatorial domain (Fig. 9-20*b*). This movement activates GroEL's ATPase function by shifting the side chain of its catalytically essential Asp 398, which extends from the L helix of the equatorial domain, into its catalytically active position near the ADP's β phosphate group. Electron microscopy studies at 10-Å resolution by Horwich and Helen Saibil reveal that similar movements occur when ATP binds to GroEL.

The hydrophobic groups that line the inner surface of the trans ring's apical domain presumably bind to the improperly exposed hydrophobic groups of substrate proteins. Indeed, an X-ray structure of the apical domain of GroEL in complex with a 12-residue peptide that binds strongly to GroEL reveals that this peptide binds to these exposed hydrophobic groups (Fig. 9-21). However, in the cis ring of the GroEL–GroES–(ADP)$_7$ complex, these hydrophobic groups participate either in binding GroES via its flexible loops or in stabilizing the newly formed interface between the rotated and elevated apical domains. Consequently, *these hydrophobic groups are no longer exposed on the inner surface of the cis cavity (Fig. 9-22), thereby depriving a substrate protein of its binding sites.*

b. GroEL/ES Undergoes Coordinated Conformational Changes That Are Paced by ATP Binding and Hydrolysis

The binding of ATP and GroES to the cis ring of GroEL strongly inhibits their binding to the trans ring. The X-ray structure of the GroEL–GroES–(ADP)$_7$ complex suggests that this occurs through concerted small conformational shifts in the GroEL equatorial domains that apparently prevent the trans ring from assuming the conformation of

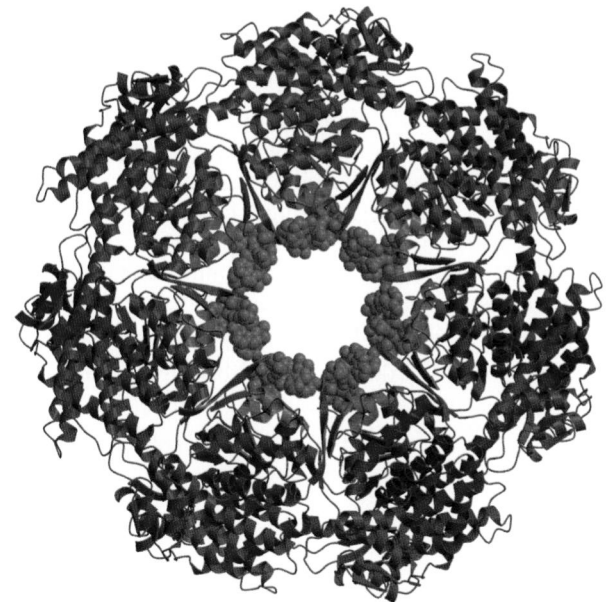

FIGURE 9-21 **Apical domain of GroEL in complex with a tight-binding 12-residue polypeptide (SWMTTPWGFLHP).** To generate this drawing, the C_α atoms of the apical domain in the X-ray structure of the complex were superimposed on those of the apical domains in the X-ray structure of GroEL alone (Fig. 9-17). Each apical subunit is represented by a ribbon diagram in which the two helices involved in binding the polypeptide (helices H and I in Fig. 9-20*a*) are red and the remainder of the subunit is green. The polypeptides are shown in space-filling form in red. [Courtesy of Lingling Chen, Yale University. PDBid 1DKD.]

the cis ring. However, once the cis ring has hydrolyzed its bound ATP (which it is committed to do once its nucleotide binding sites close off and its ATPase active sites form), the trans ring can bind ATP and the resulting conformational shifts release GroES from the cis ring. This explains why a mutant form of GroEL that has only one ring can bind substrate protein and GroES but does not release them after it hydrolyzes its bound ATP. *The proper functioning of GroEL requires two rings, even though their central cavities are unconnected.*

A mutant form of GroEL, D398A (in which Asp 398 has been changed to Ala), binds but cannot hydrolyze ATP. In the presence of ATP, D398A GroEL binds GroES together with substrate protein. However, it does not release GroES or the protein when the trans ring is exposed to ATP, as is the case when the cis ring can hydrolyze ATP. Evidently, *ATP's γ phosphate group provides strong contacts that stabilize the GroEL–GroES interaction. When the ATP in the cis ring is hydrolyzed, the resulting phosphate group is released and these interactions are lost.*

c. ATP Hydrolysis in the Cis Ring Must Occur before Substrate Protein and GroES Can Bind to the Trans Ring

The foregoing indicates that *events in the cis and trans rings of the GroEL–GroES complex are coordinated through concerted conformational changes in one ring that influence the conformation of the opposing ring.* What is the sequence of events in the trans ring relative to those in the cis ring, that is, at what stage of the folding cycle in

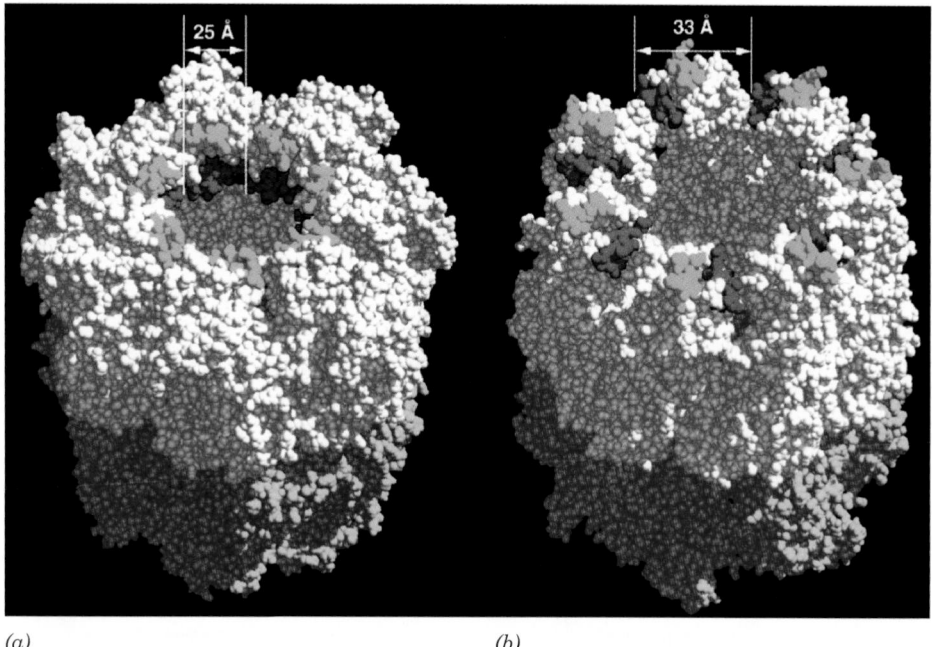

(a)   *(b)*

FIGURE 9-22 **Movements of the polypeptide-binding helices of GroEL.** (*a*) A space-filling drawing of GroEL in the structure of GroEL alone and (*b*) in the structure of GroEL–GroES–(ADP)$_7$. The GroEL cis and trans rings are white and yellow and the cis ring's H and I helices (Figs. 9-20*a, b*), which form the hydrophobic binding sites for improperly folded proteins, are green and red. On the addition of GroES and ATP to GroEL, neighboring binding sites separate by 8 Å and non-neighboring sites separate by up to 20 Å. A substrate protein initially bound to two of these sites will be forcibly stretched and hence partially unfolded before being released as the binding sites become occluded. [Courtesy of Walter Englander, University of Pennsylvania.]

the cis ring do substrate protein and GroES bind to the trans ring? Horwich answered this question using fluorescence labeling techniques. D398A GroEL that had been mixed with ADP and GroES so as to form a stable complex [D398A GroEL–GroES–(ADP)$_7$] was then mixed with a substrate protein to which a fluorescent group had been covalently linked. When this mixture was subjected to gel filtration chromatography (Section 6-3B), the label migrated with the GroEL, thereby indicating that the substrate protein had bound to the complex's trans ring. However, when the initial complex was instead made with ATP (recall that D398A GroEL cannot hydrolyze ATP), the substrate protein did not associate with the GroEL. In similar experiments, fluorescently labeled GroES associated with preformed D398A GroEL–GroES–(ADP)$_7$ in the presence of ATP but not with preformed D398A GroEL–GroES–(ATP)$_7$. Evidently, *the cis ring of the GroEL–GroES complex must hydrolyze its bound ATP before the trans ring can bind either substrate protein or GroES + ATP.*

d. The GroEL/ES System Functions as a Two-Stroke Engine

Taken together, all of the preceding observations indicate how the GroEL/ES system functions (Fig. 9-23):

1. A GroEL ring that is binding 7 ATP and an improperly folded substrate protein via the hydrophobic patches on its apical domains (Fig. 9-23, *upper left*) binds GroES. This induces a conformational change in the now cis GroEL ring, thereby releasing the substrate protein into the resulting enlarged and closed cavity, where the substrate protein commences folding. The cavity, which is now lined only with hydrophilic groups, provides the substrate protein with an isolated microenvironment that prevents it from nonspecifically aggregating with other unfolded proteins (a so-called **Anfinsen cage**).

2. Within ~13 s (the time the substrate protein has to fold), the cis ring catalyzes the hydrolysis of its 7 bound ATPs to ADP + P$_i$ (where P$_i$ is the symbol for inorganic phosphate) and the P$_i$ is released. The absence of ATP's γ-phosphate group weakens the interactions that bind GroES to GroEL.

3. A second molecule of substrate protein binds to the trans ring followed by 7 ATP.

4. The binding of substrate protein and ATP to the trans ring induces the cis ring to release its bound GroES, 7 ADP, and the presumably now better folded substrate protein. This leaves only ATP and substrate protein bound to the previous trans ring of GroEL, which becomes the

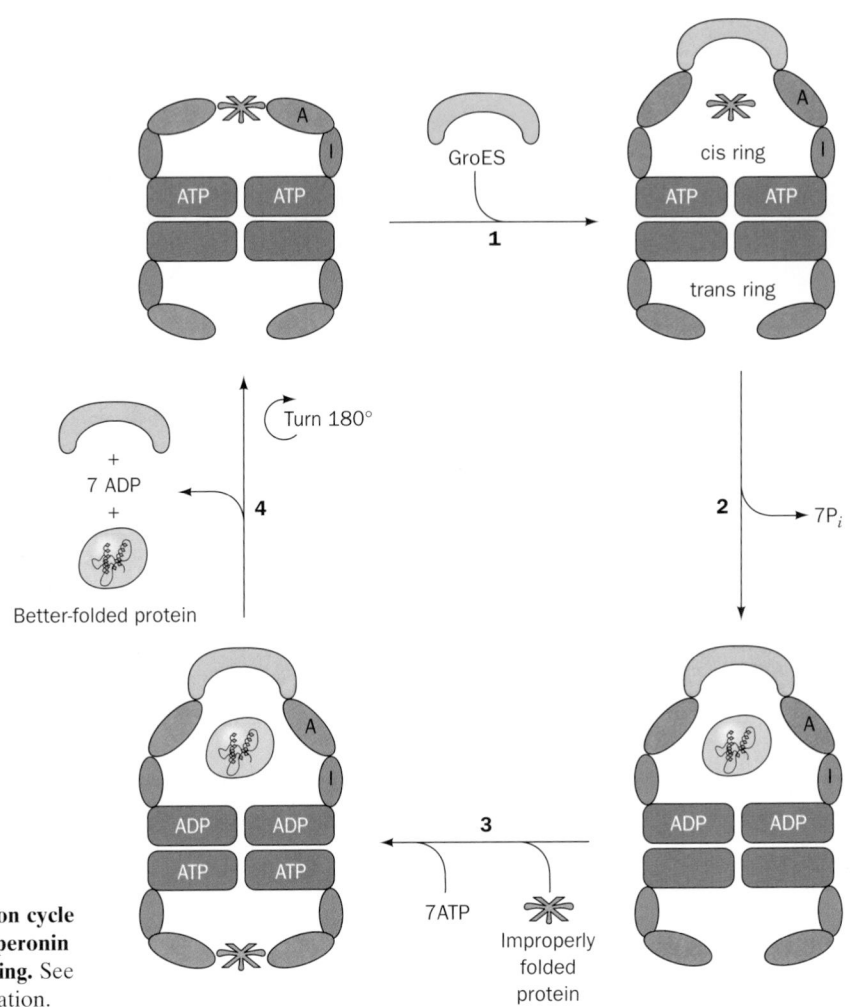

FIGURE 9-23   Reaction cycle of the GroEL/ES chaperonin system in protein folding. See the text for an explanation.

cis ring on binding GroES when the complex again cycles through Step 1.

Thus, the GroEL/ES system expends 7 ATPs per folding cycle. If the released substrate protein has not achieved its native state or is not committed to do so, it may subsequently rebind to GroEL (see below). A substrate protein that has achieved its native fold lacks exposed hydrophobic groups and hence cannot bind to GroEL.

e. GroEL/ES Partially Unfolds Its Substrate Protein and Releases It after Every Turnover of ATP

How does the foregoing cycle promote the proper folding of an improperly folded protein? Two models, not mutually exclusive, have received the most consideration:

1. The Anfinsen cage model, in which the GroEL/ES complex provides the substrate protein with a protected microenvironment in which it can fold to its native conformation without interference by nonspecific aggregation with other misfolded proteins.

2. The **iterative annealing** model, in which the ATP-driven unfolding of a misfolded and conformationally trapped substrate protein followed by its release permits it to resume folding to its native state. In the GroEL/ES system, this presumably occurs through the binding of a misfolded protein to the hydrophobic patches on two or more of the seven apical domains of GroEL, followed by the stretching and ultimate release of the protein as GroEL changes conformation on binding ATP and GroES [note that these patches are further apart in the GroEL–GroES–(ADP)$_7$ complex than they are in GroEL alone; Fig. 9-22]. In terms of landscape theory (Section 9-1C), this stretching expels the substrate protein (raises its free energy) from a local energy minimum in which it had become trapped (Fig. 9-11*d*) and thereby permits it to continue (although not necessarily complete) its conformational journey down the folding funnel toward its global energy minimum, that is, its native state.

George Lorimer and Englander have employed hydrogen exchange studies (Section 9-1C) to differentiate these two models. They studied the GroEL/ES-induced folding of the enzyme **ribulose-1,5-bisphosphate carboxylase–oxygenase (RuBisCO;** Sections 24-3A and 24-3C), whose exchangeable protons had been replaced with tritium (^{3}H; a radioactive isotope of ^{1}H). In the absence of the GroEL/ES system and under conditions such that unfolded RuBisCO aggregates rather than spontaneously folds to its native state, RuBisCO rapidly (within 2 min) exchanges all but 12 of its exchangeable ^{3}H atoms with unlabeled water protons as determined by radioactive counting techniques (black curve in Fig. 9-24). The 12 highly protected hydrogens exchange with half-lives of 30 min or more, which indicates that they are located on hydrogen bonded backbone amide groups rather than on side chains (recall that exchangeable hydrogen atoms participating in hydrogen bonds do not exchange with the solvent).

The rate of exchange of the protected hydrogens is the same in the presence of GroEL alone as it is in solution.

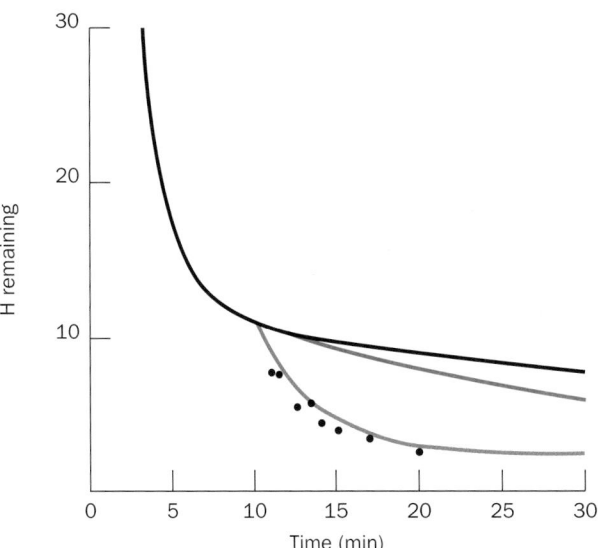

FIGURE 9-24   Rate of hydrogen–tritium exchange of tritiated RuBisCO. At time 0, a solution of RuBisCO containing a denaturing concentration of urea was diluted into conditions (pH 8, 22°C) under which the native state of RuBisCO is stable but RuBisCO is unable to fold without the assistance of a chaperonin. The black curve indicates the measured rate of hydrogen exchange in the absence of any or all of the components of the GroEL/GroES/ATP system. At 10 min (when all but 12 tritium atoms per RuBisCO had been exchanged out), GroES, ATP, and a limiting amount of GroEL (GroEL:GroES:RuBisCO = 0.05:1.2:1.0) were added to the solution. The red curve is the predicted rate of hydrogen exchange assuming that RuBisCO is released from the complex only when it reaches its native state after an average of 24 turnovers (cycles) of the GroEL/ES system (at 13 s per turnover), even though it loses all but 2.5 of its protected tritium atoms in the first turnover. The green curve indicates the predicted exchange rate if each RuBisCO molecule loses all but 2.5 of its protected tritium atoms in its first turnover and is then ejected into solution while still unfolded so that it must compete with other unfolded RuBisCO molecules for rebinding to GroEL. The green predicted curve is in agreement with the measured exchange data (*black dots*). [After Shtilerman, M., Lorimer, G.H., and Englander, S.W., *Science* **284**, 824 (1999).]

Likewise, it is unaffected by the addition of GroEL + GroES, GroEL + ADP, or GroEL + ATP. However, the incubation of RuBisCO with GroEL + GroES + ATP results in the rapid exchange of all but 2.5 of the protected hydrogens. Moreover, the replacement of ATP in the latter experiment by its nonhydrolyzable analog **adenosine-5′-(β,γ-imido)triphosphate (ADPNP)**

Adenosine-5′-(β, γ-imido)triphosphate (ADPNP)

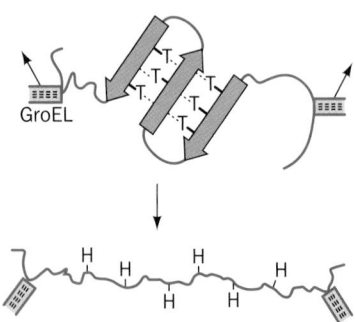

FIGURE 9-25   Schematic diagram of the mechanism of stretch-induced hydrogen exchange by the GroEL/ES system. In GroEL alone (*top*), a partially folded substrate protein is tethered between the binding sites on two of GroEL's apical domains. The substrate protein's hydrogen bonded secondary structural elements, here shown as a β sheet, protect the radiolabeled amide hydrogen atoms (T) from exchanging with solvent. As the apical domains of the cis ring of GroEL move apart during the formation of the GroEL–GroES complex (*bottom*), they stretch out and thereby unfold the substrate protein. This breaks the hydrogen bonds formed by the amide hydrogens, which then rapidly exchange with solvent hydrogens (H). [After Shtilerman, M., Lorimer, G.H., and Englander, S.W., *Science* **284,** 823 (1999).]

also induces the nearly instantaneous exchange of the protected hydrogens. Evidently, *the formation of the GroEL–GroES–ATP complex partially unfolds RuBisCO (Fig. 9-25).*

In order to differentiate whether RuBisCO is released from GroEL after every turnover, as Fig. 9-23 implies, or remains bound to GroEL until it attains its native fold (which requires an average of 24 turnovers), exchange experiments were performed with a limiting amount of GroEL (the molar ratio of GroEL:GroES:RuBisCO was 0.05:1.2:1.0). On the addition of ATP, tritium loss occurs over a 10-min period because multiple turnovers of each GroEL/ES complex are required to process the excess RuBisCO (13 s per turnover). However, the rate of tritium loss is in agreement with the model that GroEL releases RuBisCO at the end of each turnover (green curve in Fig. 9-24) but not with the model that RuBisCO remains bound to GroEL until it attains its native state (red curve in Fig. 9-24). Similar results were obtained when GroES rather than GroEL was limiting (GroEL:GroES:RuBisCO = 1.2:0.04:1.0). It therefore appears that *GroEL/ES releases its bound polypeptide at the end of each turnover, whether or not it has achieved its native fold.* Note that the 168 ATPs hydrolyzed during the 24 turnovers required, on average, to properly fold RuBisCO, while an apparently profligate use of ATP, is but a small fraction of the ~2000 ATPs expended in ribosomally synthesizing the ~500-residue RuBisCO from its component amino acids (4 ATPs per residue; Sections 32-2C and 3D), not to mention the numerous ATPs required to synthesize these amino acids (Section 26-5).

f. Folding Inside the GroEL/ES Cage Is Significantly Faster than Folding in Solution

Despite the foregoing, Ulrich Hartl and Manajit Hayer-Hartl have shown that *proteins fold much faster when inside the GroEL/ES cage than when free in solution.* They

did so by mutagenically replacing all three Cys residues of wild-type GroEL with Ala and replacing Asn 229, which is located in the opening of the GroEL cylinder, with Cys. This Cys side chain was then covalently linked to **biotin** (a substance that normally participates in enzymatically mediated carboxylation reactions; Section 21-1A). The resulting biotinylated GroEL was shown to be fully functional in mediating the folding of RuBisCO. A protein named **streptavidin,** which binds biotin rapidly and extremely tightly, was then added to this system at various times during the folding cycle. The streptavidin, as electron microscopy studies revealed, completely occludes the mouth of the GroEL cylinder and hence blocks the rebinding of RuBisCO that had been released in solution by the GroEL/ES system. Under conditions in which RuBisCO does not spontaneously fold to its native state, the addition of streptavidin to the RuBisCO–GroEL/ES system inhibits the further folding of RuBisCO, as indicated by the refolded RuBisCO's level of enzymatic activity. Moreover, under conditions in which RuBisCO folds spontaneously although more slowly than when assisted by the GroEL/ES system, the addition of streptavidin to the RuBisCO–GroEL/ES system immediately slows the folding of RuBisCO to the spontaneous rate. These results suggests that the confinement of at least some unfolded proteins to the narrow hydrophilic spaces of a GroEL/ES cage accelerates their folding, perhaps by smoothing the energy landscapes of their folding funnels (Section 9-1C).

g. GroEL/ES Interacts Strongly with ~300 *E. coli* Proteins *in Vivo*

The GroEL/ES system only interacts *in vivo* with a subset of *E. coli* proteins. What are their characteristics? Hartl identified these proteins by supplying *E. coli* with [^{35}S] methionine (^{35}S is a radioactive isotope of S) for the 15 s it takes to ribosomally synthesize an average length polypeptide, followed by the addition of an excess of unlabeled methionine (a **pulse–chase** experiment) for varying lengths of time. The cells were then lysed in the presence of EDTA, which sequesters the Mg^{2+} required by the ATPase function of GroEL, thereby preventing GroEL/ES from releasing its bound substrate protein. The GroEL–GroES–substrate complexes were immunoprecipitated with anti-GroEL antibodies and the GroEL-bound proteins were separated by 2D-gel electrophoresis (Section 6-4D). Only ~300 proteins were reproducibly observed (by autoradiography; Section 6-4B) to be associated with GroEL vs ~2500 cytoplasmic proteins that could be detected by 2D gel electrophoresis. The former proteins were isolated and 52 of them were unequivocally identified via the mass spectrometry of their tryptic fragments (Section 7-1J).

Nearly all of the GroEL substrate proteins that were identified are enzymes that participate in a wide variety of metabolic functions or in transcription or translation. Their molecular masses are mostly in the range 20 to 60 kD. The analysis, via SCOP and CATH (Section 8-3C), of those proteins of known structure or with homologs of known structure revealed that they preferentially contain two or more αβ domains that mainly consist of open β sheets (Section 8-3B). Such a protein is expected to only slowly fold to its native state because the formation of its hydrophobic β sheets requires the assembly of a large number of specific long-range interactions in their proper orientations. Moreover, such a protein may easily misfold or become kinetically trapped due to the improper packing of its helices and sheets in one domain or, more likely, between domains. Presumably the proteins that do not bind to GroEL fold to their native states so fast that they bury their hydrophobic groups before they can bind to GroEL.

The level of radioactive labeling in most of the various GroEL-associated proteins drops to zero after an ~5 min chase with unlabeled methionine, indicating that they permanently dissociate from GroEL after having achieved their native folds. However, ~100 of these proteins remain partially associated with GroEL, even after a 2-hr chase. Evidently, these proteins repeatedly return to GroEL for conformational maintenance, which strongly suggests that they are structurally labile and/or prone to aggregation.

h. The Concept of Self-Assembly Must Take Accessory Proteins into Account

Many proteins can fold/assemble to their native conformations in the absence of accessory proteins, albeit with low efficiency. Moreover, accessory proteins are not components of the native proteins whose folding/assembly they facilitate. Hence, accessory proteins must mediate the proper folding/assembly of a polypeptide to a conformation/complex governed solely by the polypeptide's amino acid sequence. Nevertheless, the concept that proteins are self-assembling entities must be modified to incorporate the effects of accessory proteins.

3 ■ PROTEIN STRUCTURE PREDICTION AND DESIGN

Since the primary structure of a protein specifies its three-dimensional structure, it should be possible, at least in principle, to predict the native structure of a protein from a knowledge of only its amino acid sequence. This might be done using theoretical methods based on physicochemical principles, or by empirical methods in which predictive schemes are distilled from the analyses of known protein structures. Theoretical methods, which usually attempt to determine the minimum energy conformation of a protein, are mathematically quite sophisticated and require extensive computations. The enormous difficulty in making such calculations sufficiently accurate and yet computationally tractable has, so far, limited their success. Nevertheless, an understanding of how and why proteins fold to their native structures must ultimately be based on such theoretical methods. In this section we outline various methods that have been used to predict the secondary and tertiary structures of proteins and end with a discussion of a related technique, that of designing proteins that will have a particular structure.

A. *Secondary Structure Prediction*

The most reliable way to determine the secondary structure taken up by a polypeptide is to map its amino acid

sequence onto that of a homolog of known structure. If, however, no such structure is available, the above-mentioned predictive methods must be employed. Here we discuss the use of empirical methods for secondary structure prediction. The theoretical methods discussed in the following section to predict a polypeptide's tertiary structure will, of necessity, also predict its secondary structure.

a. The Chou–Fasman Method

Empirical methods have had reasonable success in secondary structure prediction. Clearly, certain amino acid sequences limit the conformations available to a polypeptide chain in an easily understood manner. For example, a Pro residue cannot fit into the interior portions of a regular α helix or β sheet because its pyrrolidine ring would fill the space normally occupied by part of an abutting segment of chain and because it lacks the backbone N—H group with which to contribute a hydrogen bond. Likewise, steric interactions between several consecutive amino acid residues with side chains branched at C_β (e.g., Ile and Thr) will destabilize an α helix. Furthermore, there are more subtle effects that may not be apparent without a detailed analysis of known protein structures. Here we shall discuss simple empirical methods for predicting the positions of α helices, β sheets, and reverse turns in proteins of known sequence.

The empirical structure prediction scheme developed by Peter Chou and Gerald Fasman is widely used because it can be readily applied by hand and is reasonably reliable. Its use requires two definitions. The frequency, f_α, with which a given residue occurs in an α helix in a set of protein structures is defined as

$$f_\alpha = \frac{n_\alpha}{n} \qquad [9.3]$$

where n_α is the number of amino acid residues of the given type that occur in α helices and n is the total number of residues of this type in the set. The propensity of a particular amino acid residue to occur in an α helix is defined as

$$P_\alpha = \frac{f_\alpha}{\langle f_\alpha \rangle} \qquad [9.4]$$

where $\langle f_\alpha \rangle$ is the average value of f_α for all 20 residues. Accordingly, a value of $P_\alpha > 1$ indicates that a residue occurs with greater than average frequency in an α helix. The propensity, P_β, of a residue to occur in a β sheet is similarly defined.

Table 9-1 contains a list of α and β propensities based on the analysis of 29 X-ray structures. In accordance with its value of a given propensity, a residue is classified as a strong former (*H*), former (*h*), weak former (*I*), indifferent former (*i*), breaker (*b*), or strong breaker (*B*) of that secondary structure. Using these data, Chou and Fasman formulated the following empirical rules (the **Chou–Fasman method**) to predict the secondary structures of proteins:

1. A cluster of four helix-forming residues (H_α or h_α, with I_α counting as one-half h_α) out of six contiguous residues will nucleate a helix. The helix segment propagates in both directions until the average value of P_α for a

TABLE 9-1   Propensities and Classifications of Amino Acid Residues for α Helical and β Sheet Conformations

Residue	P_α	Helix Classification	P_β	Sheet Classification
Ala	1.42	H_α	0.83	i_β
Arg	0.98	i_α	0.93	i_β
Asn	0.67	b_α	0.89	i_β
Asp	1.01	I_α	0.54	B_β
Cys	0.70	i_α	1.19	h_β
Gln	1.11	h_α	1.10	h_β
Glu	1.51	H_α	0.37	B_β
Gly	0.57	B_α	0.75	b_β
His	1.00	I_α	0.87	h_β
Ile	1.08	h_α	1.60	H_β
Leu	1.21	H_α	1.30	h_β
Lys	1.16	h_α	0.74	b_β
Met	1.45	H_α	1.05	h_β
Phe	1.13	h_α	1.38	h_β
Pro	0.57	B_α	0.55	B_β
Ser	0.77	i_α	0.75	b_β
Thr	0.83	i_α	1.19	h_β
Trp	1.08	h_α	1.37	h_β
Tyr	0.69	b_α	1.47	H_β
Val	1.06	h_α	1.70	H_β

Source: Chou, P.Y. and Fasman, G.D., *Annu. Rev. Biochem.* **47**, 258 (1978).

tetrapeptide segment falls below 1.00. A Pro residue, however, can occur only at the N-terminus of an α helix.

2. A cluster of three β sheet formers (H_β or h_β) out of five contiguous residues nucleates a sheet. The sheet is propagated in both directions until the average value of P_β for a tetrapeptide segment falls below 1.00.

3. For regions containing both α- and β-forming sequences, the overlapping region is predicted to be helical if its average value of P_α is greater than its average value of P_β; otherwise a sheet conformation is assumed.

These easily applied empirical rules predict the α helix and β sheet strand positions in a protein with an average reliability of ~50% and, in the most favorable cases, ~80% (Fig. 9-26; note that since proteins consist, on average, of ~31% α helix and ~28% β sheet, random predictions of these secondary structures would average ~30% correct).

b.   Reverse Turns Are Characterized by a Minimum in Hydrophobicity along a Polypeptide Chain

The positions of reverse turns can also be predicted by the Chou–Fasman method. However, since a reverse turn usually consists of four consecutive residues, each with a different conformation (Section 8-1D), their prediction algorithm is necessarily more cumbersome than those for sheets and helices.

Rose has proposed a simpler empirical method for predicting the positions of reverse turns. Reverse turns nearly

always occur on the surface of a protein and, in part, define that surface. Since the core of a protein consists of hydrophobic groups and its surface is relatively hydrophilic, reverse turns occur at positions along a polypeptide chain where the hydropathy (Table 8-6) is a minimum. Using these criteria for partitioning a polypeptide chain, we can deduce the positions of most reverse turns by inspection (Fig. 9-26). Since this method often predicts reverse turns to occur in helical regions (helices are all turns), it should be applied only to regions that are not predicted to be helical.

c.   Physical Basis of α Helix Propensity

Why do amino acid residues have such different propensities for forming α helices? This question has been answered, in part, by Matthews through the structural and thermodynamic analysis of the T4 lysozyme (Section 9-1B) in which Ser 44, a solvent-exposed residue in the middle of a 12-residue (3.3-turn) α helix, was mutagenically replaced, in turn, by all 19 other amino acids. The X-ray structures of 13 of these variant proteins revealed that, with the exception of Pro, the substitutions caused no significant distortion to the α helix backbone and, hence, that differences in α helix propensities are unlikely to arise from strain. However, for 17 of the amino acids (all but Pro, Gly, and Ala), the stability of the α helix increases with the amount of side chain hydrophobic surface that is buried (brought out of contact with the solvent) when residue 44 is transferred from a fully extended state to an α helix. The low α helix propensity of Pro is due to the strain generated by its presence in an α helix, and that of Gly arises from the entropy cost associated with restricting this most conformationally flexible of residues to an α helical conformation (compare Figs. 8-7 and 8-9) and its lack of hydrophobic stabilization. The high α helix propensity of Ala, however, is caused by its lack of a γ substituent (possessed by all residues but Gly and Ala) and hence the absence of the entropy cost associated with conformationally restricting such a group within an α helix together with its small amount of hydrophobic stabilization.

d.   Computer-Based Secondary Structure Prediction Algorithms

A number of sophisticated computer-based secondary structure prediction algorithms have been developed. Most of them, like the Chou–Fasman method, employ a set of parameters whose values are estimated by the statistical analysis of (learning from) a set of nonhomologous proteins with known structures. By combining the predictions of four of the most reliable of these algorithms [all of which provide a 3-state secondary structure classification: helix (H), sheet (E), and coil (C)] via a simple majority wins method, the accuracy of secondary structure prediction has been improved to ~76%. The use of this combination algorithm, **Jpred,** is publicly available over the Web at http://www.compbio.dundee.ac.uk/~www-jpred/. It requires as input either the sequence of a single polypeptide or a multiple sequence alignment. However, since a single sequence-based secondary structure prediction is less reli-

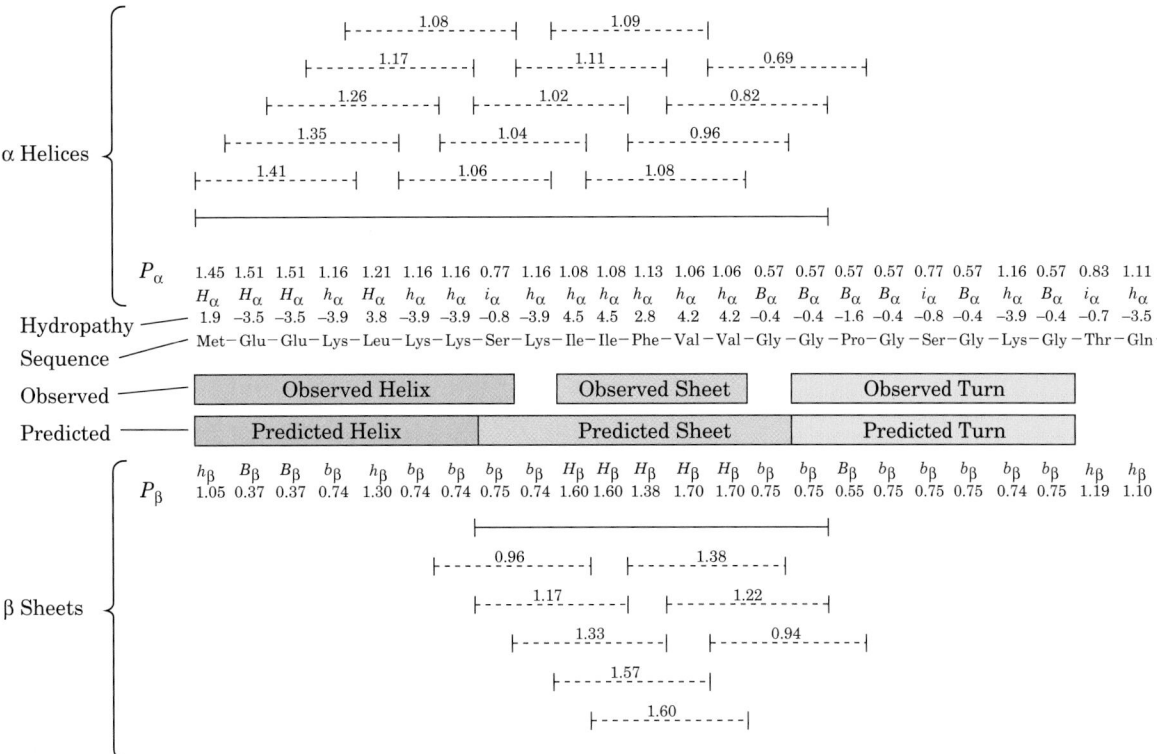

FIGURE 9-26 Secondary structure prediction. The prediction of α helices and β sheets was made by the Chou–Fasman method and the prediction of reverse turns by the method of Rose for the N-terminal 24 residues of adenylate kinase. The helix and sheet propensities and classifications are taken from Table 9-1. The solid lines indicate all hexapeptide sequences that can nucleate an α helix (*top*) and all pentapeptide sequences that can nucleate a β sheet (*bottom*), as is explained in the text. The average helix and sheet propensities for each tetrapeptide segment in the helix and sheet regions are given above the corresponding dashed lines. Twelve of the 15 residues are observed to have their predicted secondary structures (*middle*), so that the prediction accuracy, in this case, is 80%. Reverse turns are predicted to occur in sequences in which the hydropathy (Table 8-6) is a minimum and which do not occur in regions predicted to be helical. The region that matches this criterion is observed to have a reverse turn. [After Schultz, G.E. and Schirmer, R.H., *Principles of Protein Structure, p.* 121, Springer-Verlag (1979).]

able than a multiple sequence alignment-based prediction, if Jpred is supplied with only a single sequence, it will first extract the homologous sequences from a nonredundant sequence database using PSI-BLAST and then align them using CLUSTAL (Section 7-4B).

Although we have seen that secondary structure is mainly dictated by local sequences, we have also seen that tertiary structure can influence secondary structure (Section 9-1B). The inability of sophisticated secondary structure prediction schemes to surpass 76% average reliability is therefore partially explained by the failure of all of them to take tertiary interactions into account.

B. *Tertiary Structure Prediction*

The sequence databases (Section 7-4A) contain the sequences of ~900,000 polypeptides, and the rapid rate at which entire genomes are being sequenced (Section 7-2) promises that many more such sequences will soon be known. Yet the structures of only around 20,000 proteins are known (Section 8-3C). Moreover, around 40% of the **open reading frames** (**ORFs;** nucleic acid sequences that

appear to encode proteins) in known genome sequences specify proteins whose function is unknown. Thus, there is an increasingly pressing need to be able to predict the 3-dimensional structure of a protein based on its amino acid sequence. In the following paragraphs we discuss the progress that has been made in solving this difficult problem.

There are currently several major approaches to tertiary structure prediction. The simplest and most reliable approach, **comparative** or **homology modeling,** aligns the sequence of interest with the sequences of one or more homologous proteins of known structure, compensating for amino acid substitutions, insertions, and deletions through modeling and energy minimization calculations. For proteins with as little as 30% sequence identity, this method can yield a root-mean-square deviation (rmsd) between the predicted and observed positions of corresponding C_α atoms of the "unknown" protein (once its structure has been determined) of as little as ~2.0 Å. However, the accuracy of this method decreases precipitously (the rmsd's rapidly increase) as the degree of sequence identity drops below 30%. Conversely, for polypeptides that are >60% identical, a homology model may have rmsd's of ~1 Å (the accuracy of the atomic positions in an ~2.5-Å-resolution X-ray structure).

There are numerous instances of proteins that are structurally similar even though their sequences have diverged to such an extent that they have no apparent similarity. **Fold recognition** or **threading** is a computational technique that attempts to determine the unknown fold of a protein by ascertaining whether its sequence is compatible with any of the members of a library of known protein structures. It does so by placing the "unknown" protein's residues along the backbone of a known protein structure, determining the stability of the side chains of the unknown protein in that arrangement, and then sliding (threading) the sequence of the unknown protein along that of the known protein by one residue and repeating the calculation, etc. If the "correct" fold can be found (there is no guarantee that the fold of the unknown protein will resemble that of any member of the library), the resulting model can be improved via homology modeling. This method has yielded encouraging results, although it cannot yet be considered to be reliable. Of course, as sequence alignment algorithms (Section 7-4B) improve in their ability to recognize distant homologs, sequences that previously would have been candidates for fold recognition can instead be directly treated by comparative modeling.

Since the native structure of a protein depends only on its amino acid sequence, it should be possible, in principle, to predict the structure of a protein based only on its physicochemical properties (e.g., the hydrophobicity, size, hydrogen bonding propensity, and charge of each of its amino acid residues). A major problem faced by such ***ab initio*** (from the beginning) **methods** is that polypeptide chains have astronomical numbers of non-native low-energy conformations, so that it is presently quite difficult, even with the fastest available computers, to determine a polypeptide's lowest energy conformation. Nevertheless, these computer-intensive algorithms have been moderately successful in predicting simple structures such as a single α helix. However, they have only had sporadic success in predicting the folds of larger polypeptides whose structures have later been experimentally determined (although this is a major improvement over the *ab initio* predictions of only a few years earlier, which were no more reliable than random guesses). Thus, the ability to predict the native structure of a polypeptide from only its sequence remains one of the most important unrealized goals of biochemistry.

C. *Protein Design*

Although we cannot yet reliably predict the native fold of a protein that is unrelated to a protein of known structure, considerable progress has been made in solving the inverse problem: generating polypeptide sequences to assume specific 3-dimensional structures, that is, **protein design.** This is probably because a polypeptide can be "overengineered" to take up a desired conformation. Consequently, protein design has provided insights into protein folding and stability, and it promises to yield useful proteins that

are "made to order." Protein design begins with a target structure such as a 4-helix bundle and attempts to find an amino acid sequence that will form this structure. The designed polypeptide is then synthesized and its structure elucidated.

Successful design requires not only that the desired fold be stable but that other folds be significantly less stable (by ~15–40 kJ · mol^{-1}). Otherwise a sequence that has been found to be the most stable in the desired conformation may actually be more stable in other conformations. Before such **negative design** concepts were implemented, efforts to design proteins typically yielded an ensemble of molten globulelike states rather than the desired folds.

The first wholly successful *de novo* (from the beginning) protein design, accomplished by Stephen Mayo, was that of a 28-residue ββα motif which has a backbone conformation designed to resemble that of a **zinc finger** motif of the DNA-binding mouse protein **Zif268** (Fig. 9-27a), but which contains no stabilizing metal ions (zinc fingers are largely held together by their tetrahedral liganding of Zn^{2+} ions; Section 34-3B). The polypeptide was computationally designed. The side chain selection algorithm quantitatively considered the interactions between side chains and between side chains and backbone, as well as the solvation of the protein. It screened all possible amino acid sequences and, in order to take into account side chain flexibility, considered all sets of energetically allowed torsion angles for each side chain (each of which is known as a **rotamer**). By restricting core residues to be only hydrophobic and surface residues to be only hydrophilic, but permitting residues at the interface between the core and surface to be hydrophobic or hydrophilic, the number of amino acid sequences that was considered was limited to 1.9×10^{27}, which collectively had 1.1×10^{62} possible rotamers. The search therefore required a particularly computationally efficient algorithm and a fast computer.

The optimal sequence, called **FSD-1,** has only 6 of its 28 residues (21%) identical to those of Zif268 and an additional 5 residues (18%) similar. FSD-1's 8 core and boundary residues were predicted to form a well-packed cluster with Phe residues replacing the two zinc-liganding His residues in Zif268 and filling the void created by the loss of the Zn^{2+} ion. FSD-1 was chemically synthesized and its NMR structure (Fig. 9-27b) was found to closely resemble its predicted structure and to have a backbone conformation nearly superimposable on that of Zif268 (Fig. 9-27c). FSD-1's small size makes it but marginally stable. However, it is the smallest known polypeptide that is capable of folding into a unique structure without the aid of disulfide bonds, metal ions, or other subunits, thereby demonstrating the power of the protein design algorithm.

4 ■ PROTEIN DYNAMICS

The fact that X-ray studies yield time-averaged "snapshots" of proteins may leave the false impression that proteins have fixed and rigid structures. In fact, as is becoming increasing clear, *proteins are flexible and rapidly*

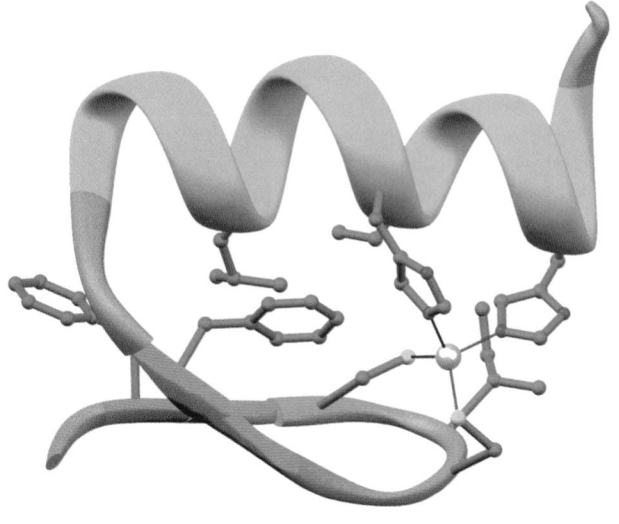

(a)

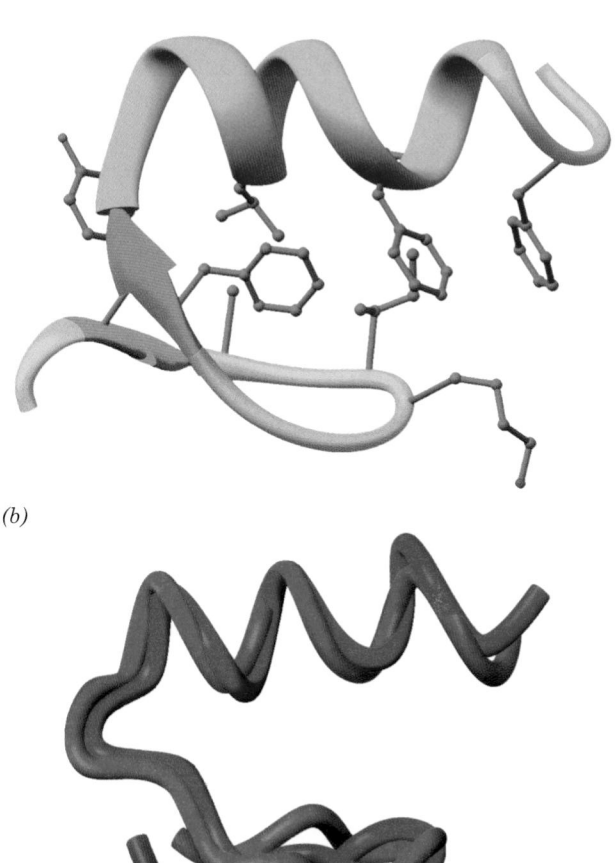

(b)

(c)

FIGURE 9-27   Comparison of the structures of the second zinc finger motif of Zif268 and FSD-1. (*a*) The X-ray structure of Zif268 showing the 9 side chains comprising its hydrophobic core and its zinc binding site in which a Zn^{2+} ion (*silver sphere*) is tetrahedrally liganded (*thin gray lines*) by the side chains of two His residues and two Cys residues. (*b*) An NMR structure of FSD-1 showing the 9 side chains of its hydrophobic core. (*c*) The best-fit superpositions of the polypeptide backbones of Zif268 (*red*) and FSD-1 (*blue*). [Part *a* based on an X-ray structure by Carl Pabo, MIT. Part *b* based on an NMR structure by and Part *c* courtesy of Stephen Mayo, California Institute of Technology. PDBids (*a*) 1ZAA and (*b*) 1FSD.]

fluctuating molecules whose structural mobilities have considerable functional significance. For example, X-ray studies indicate that the heme groups of myoglobin and

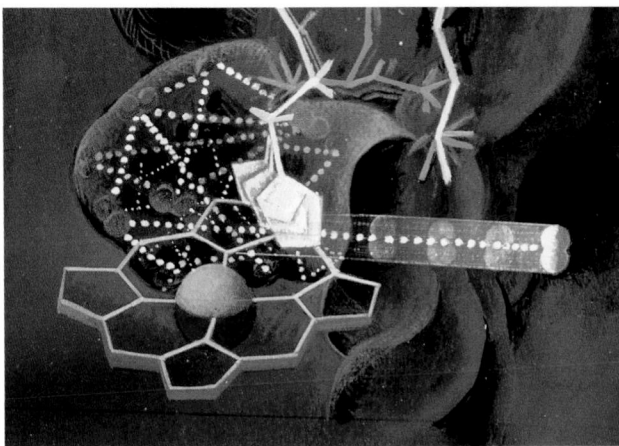

FIGURE 9-28   Conformational fluctuations in myoglobin. An artist's conception of the "breathing" motions in myoglobin that permit the escape of its bound O_2 molecule (*double red spheres*). The dotted lines trace a trajectory an O_2 molecule might take in worming its way through the rapidly fluctuating protein before finally escaping. O_2 binding presumably resembles the reverse of this process. [Illustration, Irving Geis/Geis Archives Trust. Copyright Howard Hughes Medical Institute. Reproduced with permission.]

hemoglobin are so surrounded by protein that there is no clear path for O_2 to approach or escape from its binding pocket. Yet we know that myoglobin and hemoglobin readily bind and release O_2. These proteins must therefore undergo conformational fluctuations, **breathing motions,** that permit O_2 reasonably free access to their heme groups (Fig. 9-28). The three-dimensional structures of myoglobin and hemoglobin undoubtedly evolved the flexibility to facilitate the diffusion of O_2 to its binding pocket.

The intramolecular motions of proteins have been classified into three broad categories according to their coherence:

1. Atomic fluctuations, such as the vibrations of individual bonds, which have time periods ranging from 10^{-15} to 10^{-11} s and spatial displacements between 0.01 and 1 Å.

2. Collective motions, in which groups of covalently linked atoms, which vary in size from amino acid side chains to entire domains, move as units with time periods ranging from 10^{-12} to 10^{-3} s and spatial displacements between 0.01 and >5 Å. Such motions may occur frequently or infrequently compared with their characteristic time period.

3. Triggered conformational changes, in which groups of atoms varying in size from individual side chains to complete subunits move in response to specific stimuli such as the binding of a small molecule, for example, the binding of ATP to GroEL (Section 9-2C). Triggered conformational changes occur over time spans ranging from 10^{-9} to 10^3 s and result in atomic displacements between 0.5 and >10 Å.

In this section, we discuss how these various motions are characterized and their structural and functional significance. We shall mainly be concerned with atomic fluctuations and collective motions; triggered conformational changes are considered in later chapters in connection with specific proteins.

a. Proteins Have Mobile Structures

X-Ray crystallographic analysis is a powerful technique for the analysis of motion in proteins; it reveals not only the average positions of the atoms in a crystal, but also their mean-square displacements from these positions. X-Ray analysis indicates, for example, that myoglobin has a rigid core surrounding its heme group and that the regions toward the periphery of the molecule have a more mobile character. Similarly, the apical domain of GroEL and the mobile loop of GroES are both highly flexible in the individual proteins, but when they interact in the

GroEL–GroES–(ADP)$_7$ complex, they become significantly more rigid (Fig. 9-29; Section 9-2C). Indeed, as we shall see, portions of the binding sites of many proteins rigidify on binding their target molecules. Evidently, *this initial flexibility facilitates the formation of a specific complex between a protein and its binding partner* (see below).

Molecular dynamics simulations, a theoretical technique pioneered by Martin Karplus, has revealed the nature of the atomic motions in proteins. In this technique, the atoms of a protein of known structure and its surrounding solvent are initially assigned random motions with velocities that are collectively characteristic of a chosen temperature. Then, after a time step of ~1 femtosecond (1 fs = 10^{-15} s), the aggregate effects of the various interatomic forces in the system (those due to departures from ideal covalent bond lengths, angles, and torsion angles as well as noncovalent interactions) on the velocities of each of its atoms are calculated according to Newton's equations of motion. Since all the atoms in the system will have

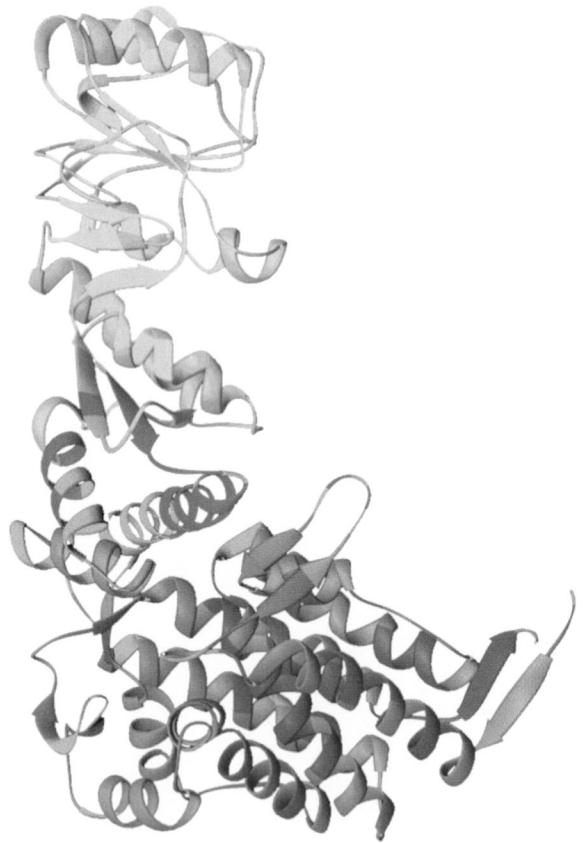

(a)   *(b)*

FIGURE 9-29   The mobility of the GroEL subunit in (*a*) the X-ray structure of GroEL alone and (*b*) the X-ray structure of GroEL–GroES–(ADP)$_7$. The polypeptide backbone is colored in rainbow order according to its degree of thermal motion, with blue being the least mobile (cool) and red being the most mobile (hot). The subunits are oriented as in Fig. 9-20*a*, *b*. Note that the outer end of the apical domain, which functions to bind both substrate protein and the mobile loop of GroES (Section 9-2C), is more mobile in GroEL alone (*red and red-orange*) than it is in GroEL–GroES–(ADP)$_7$ (*orange and yellow*). [Based on X-ray structures by Axel Brünger, Arthur Horwich, and Paul Sigler, Yale University. PDBids (*a*) 1OEL and (*b*) 1AON.]

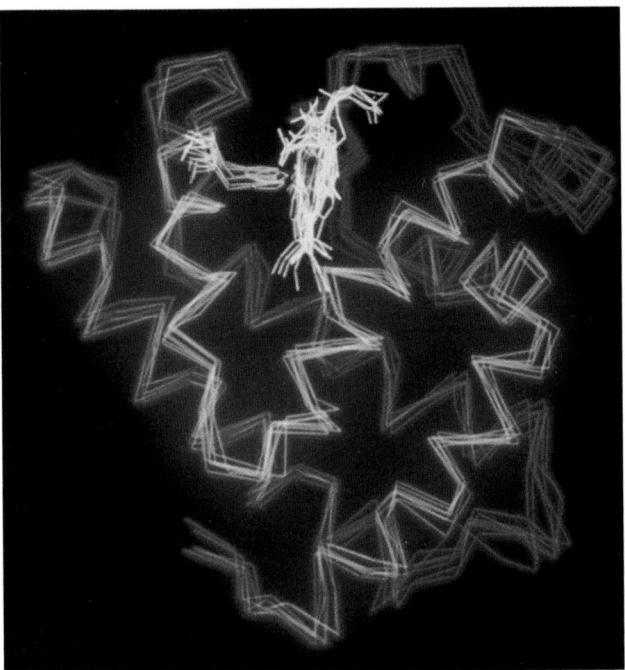

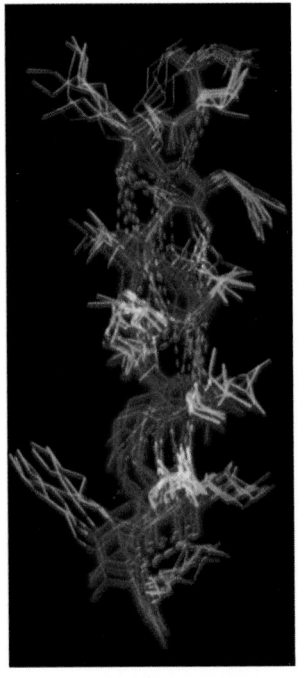

(a)   *(b)*

**FIGURE 9-30   The internal motions of myoglobin as
determined by a molecular dynamics simulation.** Several
"snapshots" of the molecule calculated at intervals of 5×10^{-12}
s are superimposed. (*a*) The C_α backbone and the heme group.
The backbone is shown in blue, the heme in yellow, and the

His residue liganding the Fe in orange. (*b*) An α helix. The
backbone is shown in blue, the side chains in green, and the
helix hydrogen bonds as dashed orange lines. Note that the
helices tend to move in a coherent fashion so as to retain their
shape. [Courtesy of Martin Karplus, Harvard University.]

moved after this time step (by a distance that is only a small
fraction of a bond length), the interatomic forces (their potential field) on any given atom will likewise have changed.
Then, using this altered potential field together with the
new positions and velocities of the atoms, the calculation
is repeated for an additional time step. This computationally intensive process has been iterated for up to ~100 ns
for ~100 residue proteins (a time that is increasing with
the available computational power), thereby yielding a
record of the positions and velocities of all the atoms in
the system over this time interval.

Molecular dynamics simulations (e.g., Fig. 9-30) have
revealed that *a protein's native structure really consists of
a large collection of conformational substates that have essentially equal stabilities.* These substates, which each have
slightly different atomic arrangements, randomly interconvert at rates that increase with temperature. Consequently, the interior of a protein typically has a fluidlike
character for structural displacements of up to ~2 Å, that
is, over excursions that are somewhat larger than a bond
length.

Gregory Petsko and Dagmar Ringe have demonstrated
the functional significance of the internal motions in proteins. Both experimental and theoretical evidence indicates
that below ~220 K ($-53°$C), collective motions in proteins
are arrested, leaving atomic fluctuations as the dominant

intramolecular motions. For example, X-ray studies have
shown that, at 228 K, the enzyme RNase A, in its crystalline
form, readily binds an unreactive substrate analog (protein
crystals generally contain large solvent-filled channels
through which small molecules rapidly diffuse; at low temperatures, the water is prevented from freezing by the addition of an antifreeze such as methanol). Yet, when the
same experiment is performed at 212 K, the substrate analog does not bind to the enzyme, even after 6 days of exposure. Likewise, at 228 K, substrate-free solvent washes
bound substrate analog out of the crystal within minutes
but, if the temperature is first lowered to 212 K, the substrate analog remains bound to the crystalline enzyme for
at least 2 days. Evidently, RNase A assumes a glasslike state
below 220 K that is too rigid to bind or release substrate.

b. Protein Core Mobility Is Revealed by Aromatic Ring Flipping

The rate at which an internal Phe or Tyr ring in a protein undergoes 180° "flips" about its C_β—C_γ bonds is
indicative of the surrounding protein's rigidity. This is because, in the close packed interior of a protein, these bulky
asymmetric groups can move only when the surrounding
groups move aside transiently (although note that these
rings have the shape of flattened ellipsoids rather than thin
disks).

NMR spectroscopy can determine the mobilities of protein groups over a wide range of time scales. Consequently, the rate at which a particular aromatic ring in a protein flips is best inferred from an analysis of its NMR spectrum (infrequent motions such as ring flipping are not detected by X-ray crystallography since this technique only reveals the average structure of a protein). NMR measurements indicate that the ring flipping rate varies from over 10^6 s^{-1} to one of immobility ($<1 \text{ s}^{-1}$) depending on both the protein and the location of the aromatic ring within the protein. Thus, at 4°C, four of BPTI's eight Phe and Tyr rings flip at rates $>5 \times 10^4 \text{ s}^{-1}$, whereas the remaining four rings flip at rates ranging between 30 and $<1 \text{ s}^{-1}$. These ring-flipping rates sharply increase with temperature, as expected.

c. Infrequent Motions Can Be Detected through Hydrogen Exchange

Conformational changes occurring over time spans of more than several seconds can be chemically characterized through hydrogen exchange studies (Sections 9-1C and 9-2C). These show that the exchangeable protons of native proteins exchange at rates that vary from milliseconds to many years (Fig. 9-31). Protein interiors, as we have seen (Section 8-3B), are largely excluded from contact with their surrounding aqueous solvent, and, moreover, protons cannot exchange with solvent while they are engaged in hydrogen bonding. The observation that the internal protons of proteins do, in fact, exchange with solvent must therefore be a consequence of transient local unfolding or "breathing" that physically and chemically exposes these exchangeable protons to the solvent. Hence, *the rate at which a particular proton undergoes hydrogen exchange is a reflection of the conformational mobility of its surroundings.* This hypothesis is corroborated by the observation that the hydrogen exchange rates of proteins decrease as their denaturation temperatures increase and that these exchange rates are sensitive to the proteins' conformational states (Fig. 9-31).

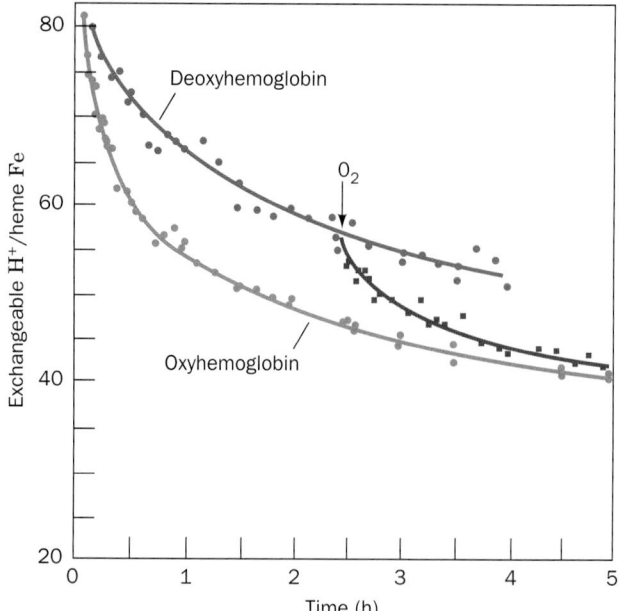

FIGURE 9-31   The hydrogen–tritium "exchange-out" curve for hemoglobin that has been preequilibrated with tritiated water. The vertical axis expresses the ratio of exchangeable protons to heme Fe atoms. Exchange-out was initiated by replacing the protein's tritiated water solvent with untritiated water through rapid gel filtration (Section 6-3B). As the exchange-out proceeded, additional gel filtration separations were performed and the amount of tritium remaining bound to the protein was measured. At the arrow, O_2 was added to exchanging deoxyhemoglobin (hemoglobin lacking bound O_2). The changing slopes of these curves indicate that the hydrogen exchange rates of the ~80 exchangeable protons of each hemoglobin subunit vary by factors of many decades and that O_2 binding increases the exchange rates for ~10 of these protons (the structural changes that O_2 binding induces in hemoglobin are discussed in Section 10-2). [After Englander, S.W. and Mauel, C., *J. Biol. Chem.* **247,** 2389 (1972).]

5 ■ CONFORMATIONAL DISEASES: AMYLOID AND PRIONS

Most proteins in the body maintain their native conformations or, if they become partially denatured, are either renatured through the auspices of molecular chaperones (Section 9-2C) or are proteolytically degraded (Section 32-6). However, at least 18 different, often fatal, human diseases are associated with the extracellular deposition of normally soluble proteins in certain tissues in the form of insoluble aggregates known as **amyloid** (starchlike; a misnomer because it was originally thought that this material resembled starch). These include **Alzheimer's disease,** a neurodegenerative disease that mainly strikes the elderly; the **transmissible spongiform encephalopathies (TSEs),** a family of infectious neurodegenerative diseases that are

propagated in a most unusual way; and the **amyloidoses,** a series of diseases caused by the deposition of an often mutant protein in organs such as the heart, liver, or kidney. The deposition of amyloid interferes with normal cellular function, resulting in cell death and eventual organ failure.

Although the various types of amyloidogenic proteins are unrelated and their native structures have widely different folds, their amyloid forms have remarkably similar core structures: Each consists of an array of ~10-nm-diameter **amyloid fibrils** (Fig. 9-32*a*) in which, as infrared and X-ray diffraction methods indicate, the proteins consist mainly of β sheets whose β strands are perpendicular to the fibril axis (Fig. 9-32*b, c*). Thus, *these proteins each have two radically different stable conformations, their native forms and their amyloid forms.*

We begin this section with a discussion of the amyloidoses as exemplified by certain mutants forms of lysozyme.

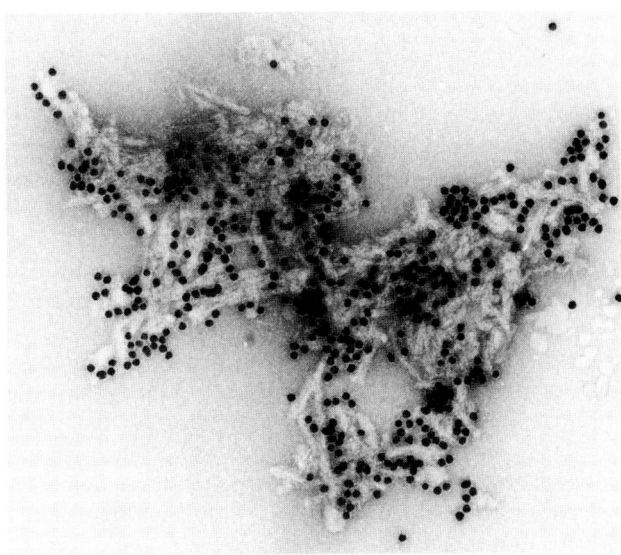

(a)

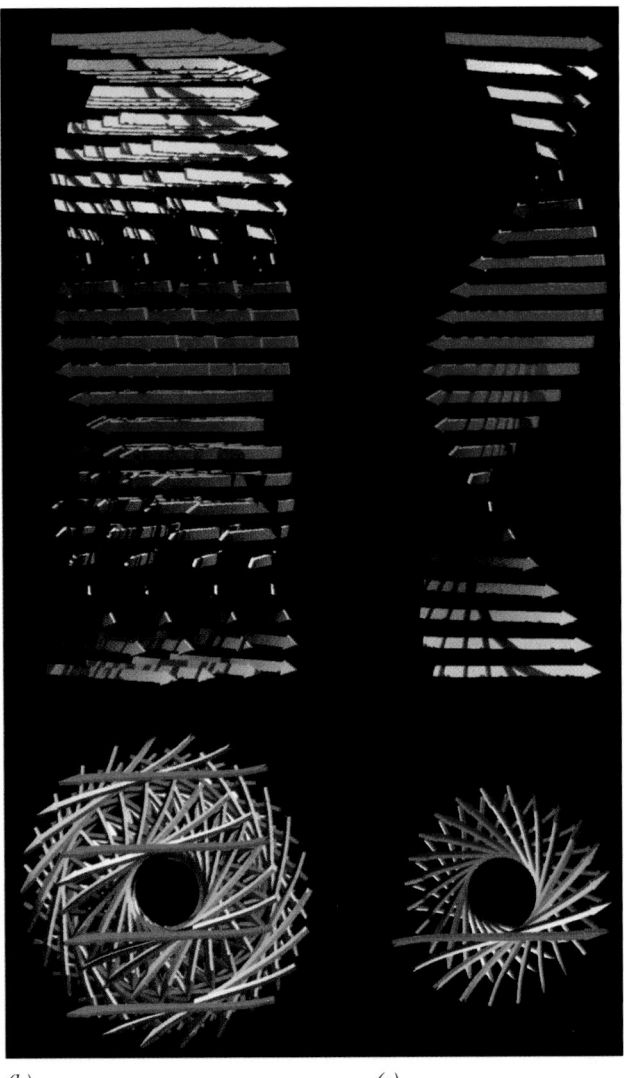

(b)   (c)

FIGURE 9-32   Amyloid fibrils. (*a*) An electron micrograph of amyloid fibrils of the protein PrP 27–30 (Section 9-5C). These amyloid fibrils are visually indistinguishable from those of other proteins. The black dots are colloidal gold beads that are coupled to anti-PrP antibodies that are adhering to the PrP 27–30. (*b*) A model, based on X-ray fiber diffraction measurements, of an amyloid fibril protofilament viewed normal to the filament axis (*above*) and along the filament axis (*below*). The arrows indicate the paths but not necessarily the directions of the β strands. The β strands form four β sheets that are parallel to the filament axis. In a given β sheet, adjacent β strands are related by a twist of 15° about the filament axis to form a continuous β sheet helix. (*c*) An isolated β sheet, which is shown for clarity. The loop regions connecting the β strands have unknown structures. Two amyloid protofilaments wrap around each other in a left-handed helix to form an amyloid fibril. [Part *a* courtesy of Stanley Prusiner, University of California at San Francisco Medical Center; Parts *b* and *c* courtesy of Colin Blake, Oxford University, U.K. and Louise Serpell, University of Cambridge, U.K.]

We then consider Alzheimer's disease and finally the TSEs and their bizarre mode of propagation.

A. *Amyloid Diseases*

Many amyloidogenic proteins are mutant forms of normally occurring proteins. These include lysozyme (an enzyme that hydrolyzes bacterial cell walls; Section 15-2) in the disease **familial visceral amyloidosis, transthyretin** [Fig. 8-65; a blood plasma protein that functions as a carrier for both the water-insoluble thyroid hormone **thyroxin** (Section 19-1D) and retinol binding protein (Section 8-3B)] in **familial amyloid polyneuropathy,** and **fibrinogen** (the precursor of **fibrin,** which forms blood clots; Section 35-1A) in **hereditary renal amyloidosis.** Most such diseases do not present (become symptomatic) until the 3rd to 7th decades

of life and typically progress over 5 to 15 years ending in death.

a. Amyloidogenic Lysozyme Variants Have Conformationally Flexible Native Structures

There are two known amyloidogenic variants of the 130-residue human lysozyme, I56T and D67H. These form amyloid fibrils that are deposited in the viscera (internal organs), usually resulting in death by the 5th decade. The amyloid fibrils consist exclusively of the variant lysozymes, thereby explaining why these mutations are dominant. Structural studies on these variant proteins have shed light on how they form amyloid fibrils.

The X-ray structures of both mutant lysozymes resemble that of the wild-type enzyme. However, the replacement of Asp 67 by His interrupts a network of hydrogen

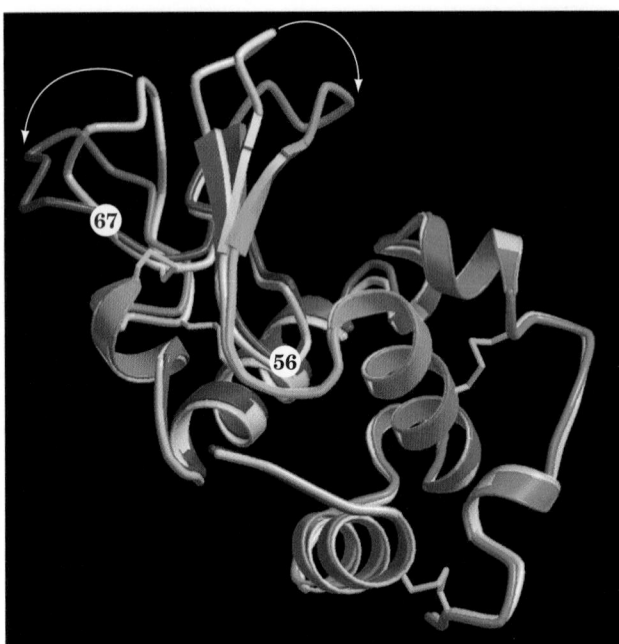

FIGURE 9-33   Superpositions of wild-type human lysozyme and its D67H mutant. Wild-type lysozyme is gray and its D76H mutant is color-ramped in rainbow order from blue at its N-terminus to red and back to blue at its C-terminus. The white arrows indicate the conformational shifts of residues 45 to 54 and 67 to 75 in the D67H mutant relative to those in the wild-type protein. The four disulfide bonds present in both the wild-type and mutant protein are shown in yellow. The positions of residues 56 and 67 are indicated. [Courtesy of Colin Blake, Oxford University, U.K., and Louise Serpell, University of Cambridge, U.K.]

bonds that stabilizes the domain containing the structure's only β sheet (its so-called β domain), resulting in the movements of the β sheet and an adjoining loop away from each other by displacements of up to 11 Å (Fig. 9-33). Although the replacement of Ile 56 by Thr causes only subtle changes in the protein structure, it insinuates a hydrophilic residue in a critical hydrophobic interface that links the protein's two domains.

The melting temperatures (T_m's) of both variants are at least 10°C less than those of the wild-type enzyme, and both variants eventually lose all enzymatic activity when incubated at physiological temperature and pH (37°C and 7.4), conditions under which wild-type lysozyme remains fully active. The variants also aggregate on heating *in vitro*, and a variety of physical measurements indicate that, in doing so, they form amyloidlike fibrils. In hydrogen exchange experiments (Section 9-1C), wild-type lysozyme strongly protects 55 protons from exchange with D_2O under conditions (37°C and pH 5) that these protons are essentially unprotected in the amyloidogenic variants, thereby confirming that the two mutations greatly loosen the native protein's tertiary structure. This suggests that the partially folded, aggregation-prone forms are in dynamic equilibrium with the native conformation, even under conditions in which the native state is thermodynami-

cally stable [keep in mind that the ratio of unfolded (U) to native (N) protein molecules in the reaction N $\rightleftharpoons$ U is governed by Eq. [3.17]: $[U]/[N] = e^{-\Delta G^{\circ\prime}/RT}$, where $\Delta G^{\circ\prime}$ is the free energy of unfolding, so that as $\Delta G^{\circ\prime}$ decreases, the proportion of U increases]. It has therefore been proposed that lysozyme fibrillogenesis is initiated by the association of the β domains of two partially unfolded lysozyme variants to form a more extensive β sheet. This would provide a template or nucleus for the recruitment of additional polypeptide chains to form the growing fibril in a process that may involve the conformational conversion of α helices to β strands. Such an autocatalytic refolding process may be a general mechanism for amyloid fibrillogenesis. However, the several decades that many hereditary amyloid diseases require to become symptomatic suggest that the spontaneous generation of an amyloid nucleus is a rare event, that is, has a high free energy of activation (activation barriers and their relationship to reaction rates are discussed in Section 14-1C).

B. *Alzheimer's Disease*

Alzheimer's disease (AD), a neurodegenerative condition that strikes mainly the elderly (it affects an estimated ~10% of those over the age of 65 and ~50% of those over 85), causes devastating mental deterioration and eventual death. It is characterized by brain tissue containing abundant amyloid plaques (deposits) surrounded by dead and dying neurons. In addition, many neuronal cell bodies contain abnormal ~20-nm-diameter fibers known as **neurofibrillary tangles.** The amyloid plaques consist mainly of amyloid fibrils of a 40- to 42-residue protein named **amyloid-β protein (Aβ)** [the neurofibrillary tangles, which we shall not further discuss, consist of a hyperphosphorylated form of a protein named **tau** that is normally associated with microtubules (Section 1-2A)].

The sequence of the gene encoding Aβ, which was identified via reverse genetics (Section 7-2C) based on the sequence of Aβ, reveals that Aβ is a segment of a 770-residue transmembrane protein named **Aβ precursor protein (βPP;** transmembrane proteins are discussed in Section 12-3A). βPP has a receptorlike sequence (Section 19-2B) although its normal function is unknown. Aβ is excised from βPP in a multistep process through the actions of two membrane-anchored proteolytic enzymes dubbed **β-** and **γ-secretases.**

It had been hotly debated whether Aβ causes AD or is merely a product of its neurodegenerative processes. This argument was largely put to rest by the observation that microinjecting 200 pg of fibrillar but not soluble Aβ (the approximate quantity in a single Aβ plaque) into the cerebral cortexes of aged but not young rhesus monkeys causes marked neuronal loss and other microscopic changes characteristic of AD as far as 1.5 mm from the injection site. Evidently, *the neurotoxic agents in AD are the Aβ-containing amyloid fibrils before their deposition in amyloid plaques.*

The age-dependence of AD suggests that **β-amyloid** deposition is an ongoing process, at least in the later decades of life. Indeed, there are several rare variants of the βPP gene with mutations in their Aβ regions that result in the onset of AD as early as the fourth decade of life. These mutatations have been shown to affect the proteolytic processing of βPP in a way that increases the rate of Aβ production. A similar phenomenon is seen in **Down's syndrome,** a condition characterized by mental retardation and a distinctive physical appearance caused by the trisomy (3 copies per cell) of chromosome 21 rather than the normal two copies. Individuals with Down's syndrome invariably develop AD by their fortieth year. This is because the gene encoding βPP is located on chromosome 21, and hence individuals with Down's syndrome produce βPP and presumably Aβ at an accelerated rate.

A second gene that has been implicated in the premature onset of AD encodes the cholesterol transport protein **apolipoprotein E** (**apoE;** Section 12-5B). The *apoE* gene has several normally occurring variants (alleles) in the population, one of which, *apoE4,* is a major risk factor for both the development of AD and its earlier onset. Moreover, AD victims with *apoE4* have significantly higher densities of β-amyloid plaques in their brain tissue than AD victims with other apoE variants. These observations motivated experiments showing that ApoE4 induces enhanced aggregation of synthetic Aβ *in vitro.* This suggests that ApoE4 facilitates Aβ aggregation *in vivo* (although another possibility is that ApoE4 inhibits the clearance of Aβ from the extracellular spaces).

There is, at present, no known treatment that arrests the progress of AD. However, the foregoing suggests several strategies for therapeutic intervention, including decreasing the rate of production of Aβ through the administration of substances that inhibit the action of β- or γ-secretase and through the administration of agents that would interfere with the formation of β-amyloid fibrils from soluble Aβ.

C. *Prion Diseases*

Certain infectious diseases that affect the mammalian central nervous system were originally classified as being caused by "slow viruses" because they take months, years, or even decades to develop. Among them are **scrapie,** a neurological disorder of sheep and goats, so named for the tendency of infected sheep to scrape off their wool [they rub against fences in an effort to stay upright due to ataxia (loss of muscle coordination)]; **bovine spongiform encephalopathy (BSE** or **mad cow disease),** which similarly afflicts cattle; and **kuru,** a degenerative brain disease that occurred among the Fore people of Papua New Guinea (kuru means trembling) and that was transmitted by ritual cannibalism. There is also a sporadic (apparently spontaneously arising) human disease with similar symptoms, **Creutzfeldt–Jakob disease (CJD),** a rare, progressive, cerebellar disorder, which resembles and may be identical to kuru. These diseases, all of which are ultimately fatal, have similar symptoms, which suggests that they are closely related. Since, in all of these diseases, neurons develop large vacuoles that gives brain tissue a spongelike micro-

scopic appearance, they are collectively known as **transmissible spongiform encephalopathies (TSEs).** None of the TSEs exhibit any sign of an inflammatory process or fever, which indicates that the immune system, which is not impaired by the disease, is not activated by it.

The classic technique for isolating the agent causing an infectious disease involves the fractionation of diseased tissue as monitored by assays for the disease. The long incubation time for scrapie, the most extensively studied "slow virus" disease, enormously hampered initial efforts to characterize its disease agent. Indeed, in the early work on scrapie in the 1930s, an entire herd of sheep and several years of observation were necessary to evaluate the results of a single fractionation. Assays for scrapie were greatly accelerated, however, by the discovery that, after intracerebral inoculation of the scrapie agent, Syrian hamsters develop the disease in a time, minimally 60 days, that decreases as the dose given is increased. Using a hamster assay, Stanley Prusiner has purified the scrapie agent to a high degree and has been instrumental in characterizing it.

a. Scrapie Is Caused by Prion Protein

The scrapie agent apparently is a single species of protein. This astonishing conclusion was established by the observations that the scrapie agent is inactivated by substances that modify proteins, such as proteases, detergents, phenol, urea, and reagents that react with specific amino acid side chains, whereas it is unaffected by agents that alter nucleic acids, such as nucleases, UV irradiation, and substances that specifically react with nucleic acids. For example, scrapie agent is inactivated by treatment with **diethylpyrocarbonate,** which carboxyethylates the His residues of proteins (Fig. 9-34*a*), but is unaltered by the cytosine-specific reagent **hydroxylamine** (Fig. 9-34*b*). In fact, the infectivity of diethylpyrocarbonate-inactivated scrapie agent is restored by treatment with hydroxylamine, presumably by the reaction shown in Fig. 9-34*c*.

The novel properties of scrapie agent, which distinguish it from viruses and plasmids, have resulted in its being termed a **prion** (for *pr*oteinaceous *in*fectious particle that lacks nucleic acid). The scrapie protein, which is named **PrP** (for *Pr*ion *P*rotein), consists of 280 mostly hydrophobic residues. This hydrophobicity, as we shall see below, causes partially proteolyzed PrP to aggregate as clusters of rodlike particles. There is a close resemblance between these clusters and the amyloid fibrils that are seen on electron microscopic examination of prion-infected brain tissue (Fig. 9-32*a*). In fact, brain tissue from CJD victims contains protease-resistant protein that cross-reacts with antibodies raised against scrapie PrP.

b. PrP Is a Widely Expressed Product of a Normal Cellular Gene That Has No Known Function

The bizarre composition of prions immediately raises the question: How are they synthesized? Three possibilities have been suggested:

1. Despite all evidence to the contrary, prions contain a nucleic acid genome that is somehow shielded from

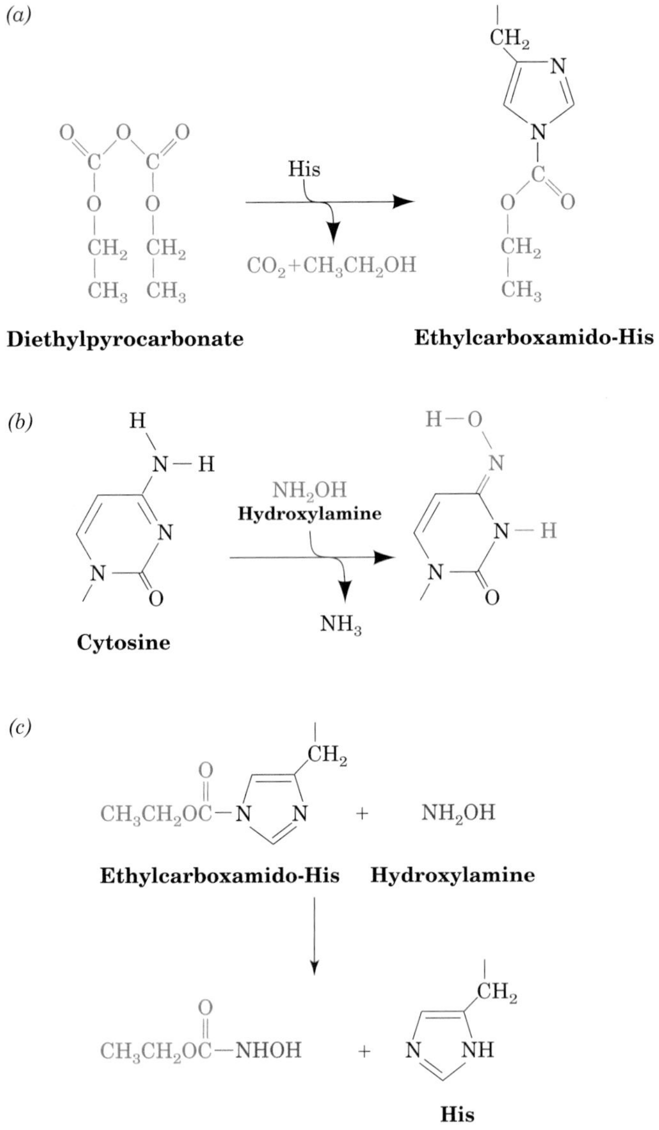

(a)

Diethylpyrocarbonate   **Ethylcarboxamido-His**

(b)

Cytosine NH_3

(c)

Ethylcarboxamido-His   **Hydroxylamine**

His

FIGURE 9-34   Evidence that the scrapie agent is a protein.
(a) Scrapie agent is inactivated by treatment with diethyl-pyrocarbonate, which specifically reacts with His side chains.
(b) Scrapie agent is unaffected by treatment by hydroxylamine, which reacts with cytosine residues. (c) However, hydroxylamine rescues diethylpyrocarbonate-inactivated scrapie reagent, presumably by the reaction shown.

detection; that is, prions are conventional viruses. The enormous and still rapidly growing body of information concerning the nature of prions, however, makes this notion increasingly untenable.

2. Prions might somehow specify their own amino acid sequence by "reverse translation" to yield a nucleic acid that is normally translated by the cellular system. Such a process, of course, would directly contravene the "central dogma" of molecular biology (Section 5-4), which states that genetic information flows unidirectionally from nu-

cleic acids to proteins. Alternatively, prions might directly catalyze their own synthesis. Such protein-directed protein synthesis is likewise unknown (although many small bacterial polypeptides are enzymatically rather than ribosomally synthesized).

3. Susceptible cells carry a gene that codes for the corresponding PrP. Infection of such cells by prions activates this gene and/or alters its protein product in some autocatalytic way.

The latter hypothesis seems to be the most plausible mechanism of prion replication. Indeed, the use of oligonucleotide probes complementary to the PrP gene (which is named ***Prn-p*** for *prion protein*), as inferred from the amino acid sequence of PrP's N-terminus (Section 7-2C), established that the brains of both scrapie-infected and normal mice contain *Prn-p*. The most surprising discovery, however, is that *Prn-p is transcribed at similar levels in both normal and scrapie-infected brain tissue.* Moreover, the use of the above probes has revealed that *Prn-p genes occur in all vertebrates so far tested, including humans, as well as in invertebrates such as Drosophila.* This evolutionary conservation suggests that PrP, a membrane-anchored protein (via glycosylphosphatidylinositol groups; Section 12-3B) that occurs mainly on neuron surfaces, has an important function. Thus it came as a further surprise that knockout mice (Section 5-5H) in which both *Prn-p* genes have been disrupted appear to be normal and that mating two such $Prn-p^{0/0}$ mice gives rise to normal $Prn-p^{0/0}$ progeny (although there is some evidence that $Prn-p^{0/0}$ mice develop neurological abnormalities late in life). Nevertheless, evidence is accumulating that PrP is normally a cell-surface signal receptor, although the identity of its corresponding signal and its consequences are as yet unknown.

c. Scrapie Disease Requires the Expression of the Corresponding PrP^C Protein

$Prn-p^{0/0}$ mice remain completely free of scrapie symptoms after inoculation with a dose of mouse scrapie PrP (PrP^{Sc}; Sc for scrapie) that causes wild-type ($Prn-p^{+/+}$) mice to die of scrapie within 6 months after inoculation. Evidently, PrP^{Sc} *induces the conversion of normal PrP (PrP^C; C for cellular) to PrP^{Sc}.* This unorthodox notion, the so-called **prion hypothesis,** is supported by the observation that when wild-type mice are inoculated with PrP^{Sc} that has been continuously passaged (incubated) in hamsters, the incubation time for developing disease symptoms is, at first, 500 days but then, in all further passages in mice, diminishes to 140 days. Conversely, when PrP^{Sc} that has been passaged in mice is inoculated into hamsters, the incubation time is first 400 days but subsequently shortens to 75 days. This suggests that the conversion of host PrP^C (whose sequence in mice differs from that in hamsters) to PrP^{Sc} by a foreign PrP^{Sc} is a rare event; once it has occurred, however, the newly formed host PrP^{Sc} catalyzes the conversion much more efficiently. Indeed, after inoculation with hamster PrP^{Sc}, transgenic mice expressing hamster PrP have incubation times that are reduced to between 48 and 250 days, depending on the transgenic line.

d. Mutant *Prn-p* Genes Give Rise to Prion Diseases

Three dominantly inherited neurodegenerative disorders in humans have been traced to mutations in the *Prn-p* gene. These are **familial CJD, Gerstmann–Sträussler–Scheinker syndrome (GSS),** and **fatal familial insomnia (FFI).** All of them are extremely rare. In fact, FFI has been found in only five families. The mutant PrP^{Sc}s causing these diseases are nevertheless infectious.

e. PrP^{Sc} Is a Stable Conformational Variant of PrP^{C}

The NMR structure of residues 23 to 230 of the 280-residue human PrP^{C} (Fig. 9-35a), determined by Kurt Wüthrich, consists of a flexibly disordered (and hence unobserved) 98-residue N-terminal "tail" and a 110-residue C-terminal globular domain containing three α helices and a short 2-stranded antiparallel β sheet. As expected, this structure closely resembles those of the homologous mouse and hamster PrP^{C}s.

How does PrP^{Sc} differ from PrP^{C}? The direct sequencing of PrP^{Sc} indicates that its amino acid sequence is identical to that deduced from the *Prn-p* gene sequence, thereby eliminating any post-transcriptional sequence variation as a possible cause for the pathogenic properties of PrP^{Sc}. Furthermore, mass spectrometric studies on PrP^{Sc} designed to reveal previously uncharacterized posttranslational modifications indicated that, in fact, PrP^{Sc} and PrP^{C} are chemically identical. Thus, although the possibility that only a small fraction of PrP^{Sc} is chemically modified has not been eliminated, it seems more likely that PrP^{Sc} and PrP^{C} differ in their secondary and/or tertiary structures. Unfortunately, the insolubility of PrP^{Sc} (see below) has precluded its structural determination. However, CD measurements show that, in fact, the conformations of PrP^{Sc} and PrP^{C} are quite different: PrP^{C} has a high (~40%) α helix content but little (~3%) β sheet content (in good agreement with the NMR structure of its globular domain), whereas PrP^{Sc} has a lesser (~30%) α helix content but a

high (~45%) β sheet content. In a plausible model of PrP^{Sc} (Fig. 9-35b), its N-terminal region has refolded to form a 4-stranded mixed β sheet; only helices 2 and 3 of PrP^{C}, which are joined by a disulfide bond, maintain their original conformation. The high β sheet content of PrP^{Sc} would, presumably, facilitate the aggregation of PrP^{Sc} as amyloid fibrils. *Evidently, the PrP^{C} → PrP^{Sc} conformational change is autocatalytic; that is, PrP^{Sc} induces PrP^{C} to convert to PrP^{Sc}.* In fact, PrP^{Sc} in a cell-free system has been shown to catalyze the conversion of PrP^{C} from an uninfected source to PrP^{Sc}. However, there is growing evidence that *in vivo* the PrP^{C} → PrP^{Sc} conversion is facilitated by an as yet unidentified molecular chaperone dubbed **protein X.**

In cells, PrP^{Sc} is deposited in cytosolic vesicles rather than being anchored to the cell-surface membrane as is PrP^{C}. Both PrP^{C} and PrP^{Sc} are subject to eventual proteolytic degradation in the cell (Section 32-6). However, although PrP^{C} is completely degraded, PrP^{Sc} only loses its N-terminal 67 residues to form a 27- to 30-kD protease-resistant core, known as **PrP 27–30,** which still exhibits a high β sheet content. *PrP 27–30 then aggregates to form the amyloid plaques that appear to be directly responsible for the neuronal degeneration characteristic of prion diseases.*

According to the prion hypothesis, sporadically occurring prion diseases such as CJD (which strikes one person per million per year) arise from the spontaneous although infrequent conversion of sufficient quantities of PrP^{C} to PrP^{Sc} to support the autocatalytic conformational isomerization reaction. This model is corroborated by the observation that transgenic mice that overexpress wild-type *Prn-p* invariably develop scrapie late in life. The prion hypothesis similarly explains inherited prion diseases such as FFI as arising from a lower free energy barrier and hence higher rate for the conversion of the mutant PrP^{C} to PrP^{Sc} relative to that of normal PrP^{C}.

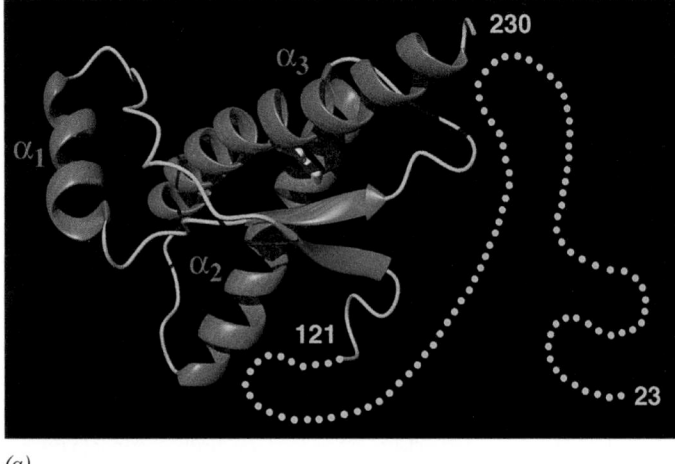

(a)

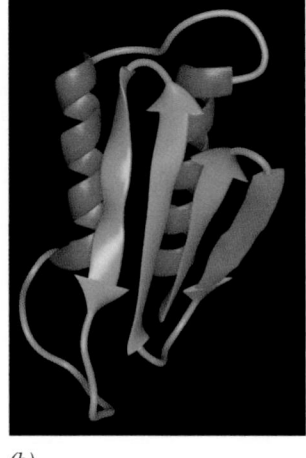

(b)

FIGURE 9-35 Prion protein conformations. (a) The NMR structure of human prion protein (PrP^{C}). Its flexibly disordered N-terminal "tail" (residues 23–121) is represented by yellow dots (the protein's N-terminal 23 residues have been post-trans-

lationally excised). (b) A plausible model for the structure of PrP^{Sc}. [Part *a* courtesy of Kurt Wüthrich, Eidgenössische Technische Hochschule, Zurich, Switzerland; Part *b* courtesy of Fred Cohen, University of California at San Francisco.]

f. Prions Have Different Strains

Prions from different sources, when passaged in mice or hamsters, reproducibly exhibit characteristic incubation times, neurological symptoms, and neuropathologies. *Evidently, there are different strains of prions, each of whose corresponding PrP^Sc^s must have a different stable conformation and induce PrP^C^ to take up this conformation.* The existence of different prion strains (as many as 30 for scrapie in sheep and at least 4 for CJD in humans) has been cited as evidence against the prion hypothesis. However, there are several known instances of proteins that can assume multiple stable conformations, including tubulin, the protein from which microtubules are composed (Section 35-3F).

> BSE or mad cow disease was first reported in the U.K. in late 1985. It soon became an epidemic and at its peak, in 1993, over 3000 cases of BSE were being reported each month, an annual incidence of 1% of the British cattle population. BSE is surmised to have arisen as a consequence of feeding cattle meat-and-bone meal made from scrapie-infected sheep (and eventually from BSE-infected cattle). BSE, which has an ~5-year incubation period, was unknown before 1985, most likely because the process for manufacturing meat-and-bone meal was changed in the late 1970s from a way that fully inactivates scrapie prions to one that fails to do so. In 1988, the U.K. banned the feeding of ruminants with ruminant-derived protein (other than milk), so that, following its peak in 1993, the BSE epidemic rapidly abated (a process accelerated by the slaughter of large numbers of cattle at risk of having BSE). However, since humans consumed meat from BSE-infected cattle for over a decade, the question remained, had BSE been transmitted to humans? It should be noted that scrapie-infected sheep have long been consumed worldwide and yet the incidence of CJD in mainly meat-eating countries such as the U.K. (in which sheep are particularly abundant) is no greater than that in largely vegetarian countries such as India. However, in 1994, several cases of CJD in teenagers and young adults were reported in the U.K., although heretofore CJD before the age of 40 was extremely rare (its average age of onset is ~64). Individuals with this **new variant CJD (vCJD** or **nvCJD),** of which there have been over 110 cases yet reported, almost entirely in the U.K., have neurological symptoms and neuropathology that are atypical for sporadic CJD. Moreover, when transmitted to mice expressing bovine PrP^C^, vCJD has an incubation time, neurological symptoms, and neuropathology indistinguishable from that caused by BSE. It therefore seems highly likely that vCJD is caused by a prion strain that humans acquired by eating meat products from BSE-infected cattle.

g. Prions Occur in Yeast

Although prions were originally defined to be scrapie-like infectious pathogens, it is now evident that this definition must be broadened to include all proteins with stable conformational variants that catalyze their own formation from "wild-type" protein. For example, *Saccharomyces cerevisiae* (baker's yeast) can harbor a genetic element designated **[URE3]** that, in sexual reproduction with cells that lack [URE3], is inherited by all progeny rather than according to the rules of Mendelian genetics (Section 1-4B). Yet [URE3] is a chromosomal gene rather than a plasmid-based or mitochondrial gene (which would account for its non-Mendelian inheritance).

[URE3] is identical to the chromosomal gene, **URE2,** which specifies a protein, **Ure2p,** that in the presence of yeast's preferred nitrogen sources (ammonia or glutamine) represses the expression of the proteins required to metabolize yeast's less preferred nitrogen sources (e.g., proline). Yeast that have the [URE3] phenotype (trait) lack this regulation of nitrogen metabolism (nitrogen metabolism is discussed in Chapter 26). However, [URE3] yeast can be "cured" of this defect by treatment with 5mM guanidinium chloride; that is, they and their progeny then exhibit normal regulation of nitrogen metabolism. Nevertheless, about one yeast cell per million spontaneously reverts to the [URE3] phenotype. Evidently, Ure2p has two stable conformational states, the "wild type," which regulates nitrogen metabolism, and the [URE3] form, which catalyzes its own formation from "wild-type" Ure2p to yield an amyloid that does not influence nitrogen metabolism. Thus *Ure2p is a type of prion.*

The yeast genetic element **[PSI]** encodes a protein, **Sup35p,** with similar prionlike properties that participates in transcriptional termination (Section 32-3E). Indeed, the introduction, via liposomes, of Sup35p in its [PSI] conformation into the cytoplasm of yeast containing "wild-type" Sup35p induced the formation of the [PSI] phenotype. This study constitutes the first direct evidence that supports the prion hypothesis.

6 ■ STRUCTURAL EVOLUTION

Proteins, as we discussed in Section 7-3, evolve through point mutations and gene duplications. Over eons, through processes of natural selection and/or neutral drift, homologous proteins thereby diverge in character and develop new functions. How these primary structure changes affect function, of course, depends on the protein's three-dimensional structure. In this section, we explore the effects of evolutionary change on protein structures.

A. Structures of Cytochromes c

🔖 **See Guided Exploration 10: Protein Evolution** The *c*-type cytochromes are small globular proteins that contain a covalently bound heme group (**iron–protoporphyrin IX;** Fig. 9-36). The X-ray structures of the cytochromes *c* from horse (Fig. 8-42), tuna, bonito, rice, and yeast are closely similar and thus permit the structural significance of cytochrome *c*'s amino acid sequences (Section 7-3B) to be assessed. The internal residues of cytochrome *c*, particularly those lining its heme pocket, tend to be invariant or conservatively substituted, whereas surface positions have greater variability. This observation is, in part, an indication of the more exacting packing requirements of a protein's internal regions compared to those of its surface (Section 8-3B).

Certain invariant or highly conserved residues (Table 7-4) have specific structural and/or functional roles in cytochrome *c*:

1. The invariant Cys 14, Cys 17, His 18, and Met 80 residues form covalent bonds with the heme group (Fig. 9-36).

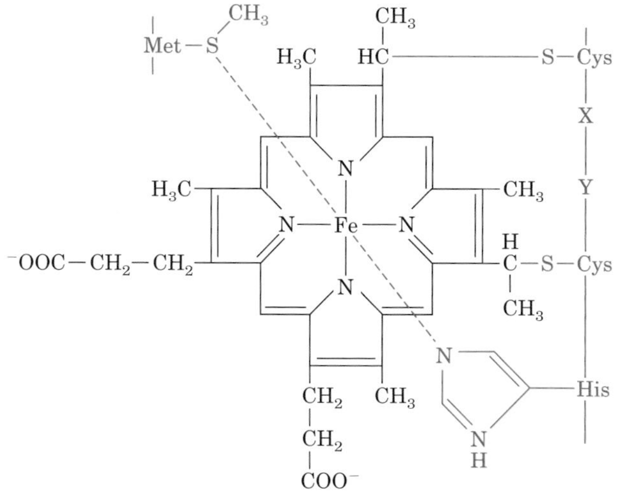

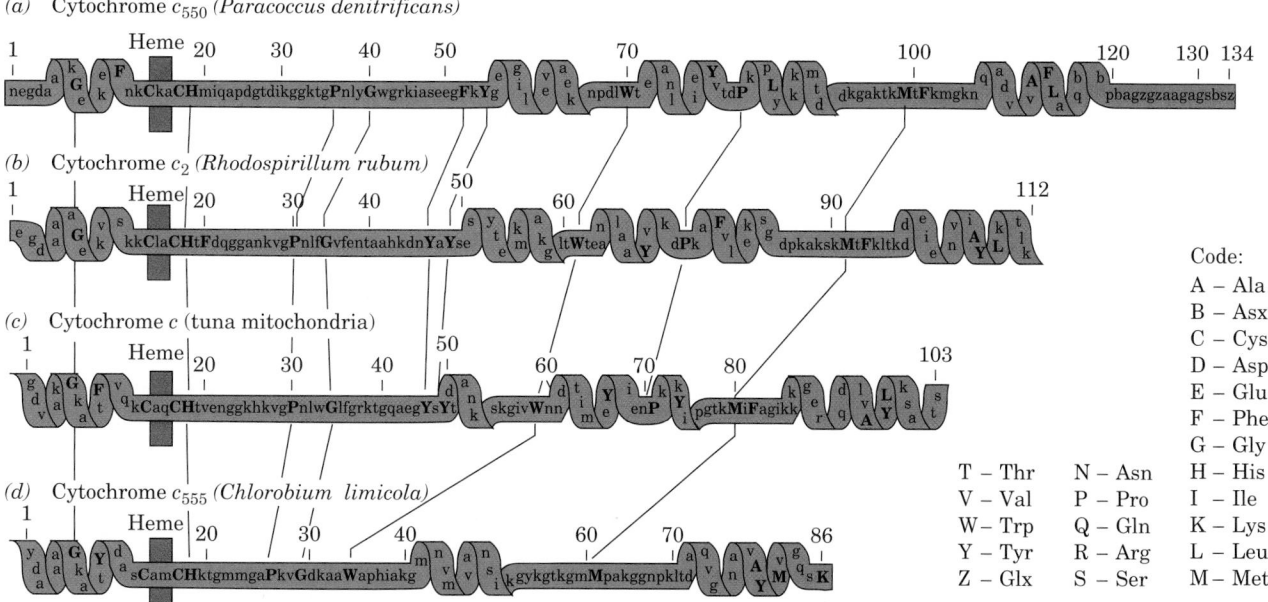

FIGURE 9-36 Molecular formula of iron–protoporphyrin IX (heme). In *c*-type cytochromes, the heme is covalently bound to the protein (*red*) by two thioether bonds linking what were the heme vinyl groups to two Cys residues that occur in the sequence Cys-X-Y-Cys-His (residues 14–18 in Table 7-4). Here X and Y symbolize any amino acid residues. A fifth and sixth ligand to the Fe atom, both normal to the heme plane, are formed by a side chain nitrogen of His 18 and the sulfur of Met 80. The iron atom, which is thereby octahedrally liganded, can stably assume either the Fe(II) or the Fe(III) oxidation state. Heme also occurs in myoglobin and hemoglobin but without the thioether bonds or the Met ligand.

2. The nine invariant or highly conserved Gly residues occupy close-fitting positions in which larger side chains would significantly alter the protein's three-dimensional structure.

3. The highly conserved Lys residues 8, 13, 25, 27, 72, 73, 79, 86, and 87 are distributed in a ring around the exposed edge of the otherwise buried heme group. There is considerable evidence that this unusual constellation of positive charges specifically associates with complementary sets of negative charges on the physiological reaction partners of cytochrome *c*, cytochrome *c* reductase, and cytochrome *c* oxidase (Section 22-2C).

a. Prokaryotic *c*-Type Cytochromes Are Structurally Related to Cytochrome *c*

Although cytochrome *c* occurs only in eukaryotes, similar proteins known as ***c*-type cytochromes** are common in prokaryotes, where they function to transfer electrons at analogous positions in a variety of respiratory and photosynthetic electron-transport chains. Unlike the eukaryotic proteins, however, the prokaryotic *c*-type cytochromes exhibit considerable sequence variability among species. For example, the >30 bacterial *c*-type cytochromes whose primary structures are known have from 82 to 134 amino acid residues, whereas eukaryotic cytochromes *c* have a narrower range, from 103 to 112 residues. The primary structures of several representative *c*-type cytochromes have few obvious similarities (Fig. 9-37). Yet their X-ray structures closely resemble each

FIGURE 9-37 Primary structures of some representative c-type cytochromes. (*a*) Cytochrome c_{550} (the subscript indicates the protein's peak absorption wavelength in visible light, in nm) from *Paracoccus denitrificans*, a respiring bacterium that can use nitrate as an oxidant. (*b*) Cytochrome c_2 (the subscript has only historical significance) from *Rhodospirillum rubrum*, a purple photosynthetic bacterium.

(*c*) Cytochrome *c* from tuna mitochondria. (*d*) Cytochrome c_{555} from *Chlorobium limicola*, a green photosynthetic bacterium that utilizes H_2S as a hydrogen source. Thin lines connect structurally significant or otherwise invariant residues (*uppercase*). Helical regions are indicated to facilitate structural comparisons with Fig. 9-38. [After Salemme, F.R., *Annu. Rev. Biochem.* **46**, 307 (1977).]

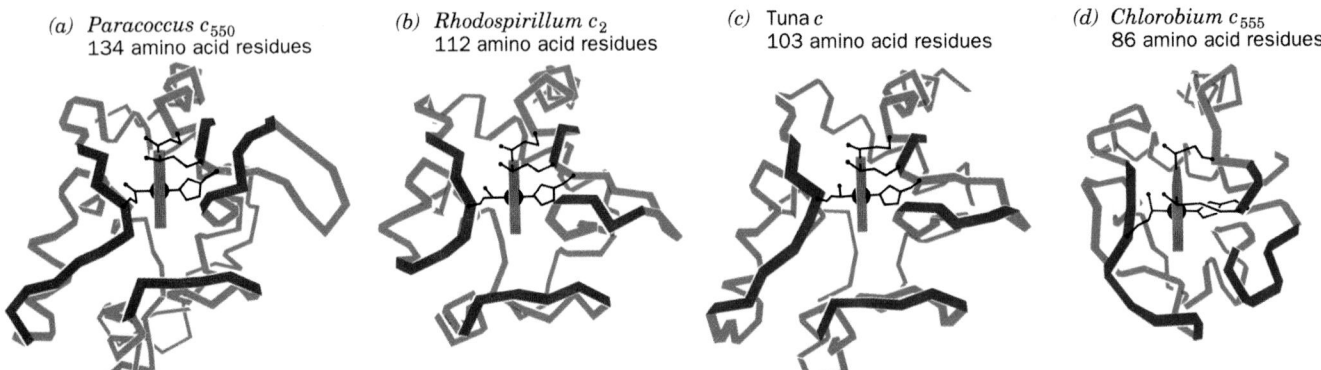

(a) *Paracoccus* c_{550}
134 amino acid residues

(b) *Rhodospirillum* c_2
112 amino acid residues

(c) Tuna *c*
103 amino acid residues

(d) *Chlorobium* c_{555}
86 amino acid residues

FIGURE 9-38   Three-dimensional structures of the *c*-type cytochromes whose primary structures are displayed in Fig. 9-37. The polypeptide backbones (*blue*) are shown in analogous orientations such that their heme groups (*red*) are viewed edge on. The Cys, Met, and His side chains that covalently link the heme to the protein are also shown. (*a*) Cytochrome c_{550} from *P. denitrificans*. (*b*) Cytochrome c_2 from *Rs. rubrum*. (*c*) Tuna cytochrome *c*. (*d*) Cytochrome c_{555} from *C. limicola*. [Illustration, Irving Geis/Geis Archives Trust. Copyright Howard Hughes Medical Institute. Reproduced with permission.]

other, particularly in their backbone conformations and side chain packing in the regions surrounding the heme group (Fig. 9-38). Furthermore, most of them have aromatic rings in analogous positions and orientations relative to their heme groups as well as similar distributions of positively charged Lys residues about the perimeters of their heme crevices. The major structural differences among these *c*-type cytochromes stem from various loops of polypeptide chain that are located on their surfaces.

Before the advent of sophisticated sequence alignment algorithms such as BLAST and FASTA (Section 7-4B), the correct alignments of analogous *c*-type cytochrome residues (thin lines in Fig. 9-37) could not have been made on the basis of only their primary structures: These proteins have diverged so far that their three-dimensional structures were essential guides for this task. Three-dimensional structures are evidently more indicative of the similarities

among these distantly related proteins than are primary structures. *It is the essential structural and functional elements of proteins, rather than their amino acid residues, that are conserved during evolutionary change.*

B. *Gene Duplication*

Gene duplication may promote the evolution of new functions through structural evolution (Section 7-3C). In over half of the multidomain proteins of known structure, two of the domains are structurally quite similar. Consider, for example, the two domains of the bovine liver enzyme **rhodanese** (Fig. 9-39). It seems highly improbable that its two complex but conformationally similar domains could have independently evolved their present structures (a process known as **convergent evolution**). More likely, they arose through the duplication of the gene specifying an ancestral

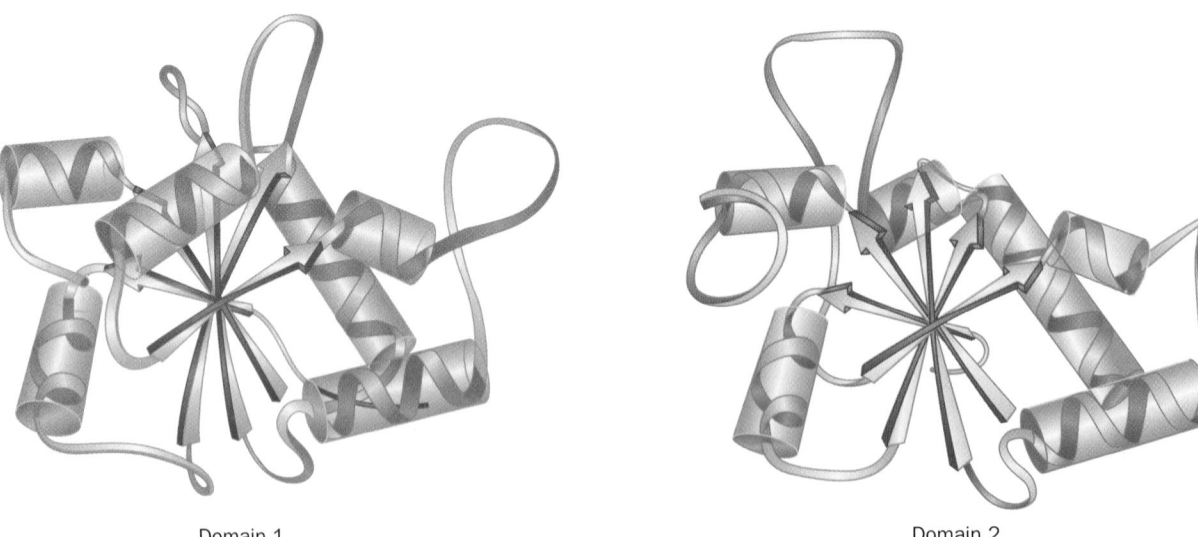

Domain 1   Domain 2

FIGURE 9-39   The two structurally similar domains of rhodanese. Note that both are open β sheets (Section 8-3B) with identical topologies. [After drawings provided by Jane Richardson, Duke University. Based on an X-ray structure by Wim Hol, University of Washington. PDBid 1RHD.]

domain followed by the fusion of the resulting two genes to yield a single gene specifying a polypeptide that folds into two similar domains. The differences between the two domains is therefore due to their **divergent evolution.**

Structurally similar domains often occur in proteins whose other domains bear no resemblance to one another. The redox enzymes known as **dehydrogenases,** for example, each consist of two domains: a domain that binds redox-active dinucleotides such as NAD^+ and which is structurally similar in all the dehydrogenases, and a dissimilar substrate-binding domain that determines the specificity and mode of action of each enzyme. Indeed, in some dehydrogenases, such as glyceraldehyde-3-phosphate dehydrogenase (Fig. 8-45), the dinucleotide-binding domain occurs at the N-terminal end of the polypeptide chain, whereas in others it occurs at the C-terminal end. Each of these dehydrogenases must have arisen by the fusion of the gene specifying an ancestral dinucleotide-binding domain with a gene encoding a proto–substrate-binding domain. This must have happened very early in evolutionary history, perhaps in the precellular stage (Section 1-5C), because there are no significant sequence similarities among these dinucleotide-binding domains. Evidently, *a domain is as much a unit of evolution as it is a unit of structure. By genetically combining these structural modules in various ways, nature can develop new functions far more rapidly than it can do so by the evolution of completely new structures through point mutations.*

CHAPTER SUMMARY

1 ■ Protein Folding: Theory and Experiment Under renaturing conditions, many proteins fold to their native structures in a matter of seconds. Helices and sheets, which together constitute ~60% of the average protein, are so common because they efficiently fill space. Proteins are hierarchically organized, that is, they consist of domains, which consist of subdomains, etc. They are highly tolerant of sequence changes, to which they adapt by local rather than global structural alterations. The rapidity with which proteins renature indicates that they fold in an ordered manner rather than via a random search of all their possible conformations. Thus the study of protein folding requires rapid mixing and observational techniques such as stopped-flow devices, circular dichroism, and pulsed H/D exchange followed by NMR. The folding of small single-domain proteins is initiated by a hydrophobic collapse to yield a molten globule, which appears within ~5 ms. This is followed by the stabilization of secondary structure and then the formation of tertiary structure to yield the native protein in a matter of several seconds. Folding is thought to follow landscape theory, which postulates that a polypeptide folds via a folding funnel and hence can take any of a great variety of pathways to reach its native state. This is consistent with the finding that proteins fold in a hierarchical manner. However, in the final stages of folding, bovine pancreatic trypsin inhibitor (BPTI) follows an ordered pathway, as has been found by determining the order in which its disulfide bonds form. The sequence of a protein appears to specify its folding pathway as well as its native structure.

2 ■ Folding Accessory Proteins Even though it is clear that a protein's primary structure dictates its three-dimensional structure, many proteins require the assistance of accessory proteins such as protein disulfide isomerase (PDI), peptidyl prolyl cis–trans isomerases, and molecular chaperones to fold/assemble to their native structures. PDI consists of four thioredoxinlike domains, two of which contain exposed Cys residues that form disulfide bonds, either internally or with another protein in a disulfide interchange reaction. Two families of peptidyl prolyl cis–trans isomerases have been characterized, the cyclophilins, which bind cyclosporin A, and FK506 binding protein, which binds FK506. The chaperonins, such as GroEL and GroES, stimulate the proper folding of misfolded proteins through a cyclic sequence of concerted conformational changes that is driven by the binding and hydrolysis of ATP. GroES is a cap-shaped heptamer and GroEL is a 14-mer arranged in two apposed heptameric rings that form two unconnected end-to-end hollow barrels. Together, GroEL and GroES form a bullet-shaped complex that contains a closed cavity in which misfolded proteins can fold without interference by aggregation with other misfolded proteins. In doing so, GroEL/ES partially unfolds a misfolded and conformationally trapped protein of up to 70 kD and releases it so as to enable it to continue its progress down its folding funnel toward its native fold. Such proteins are bound and released an average of 24 times before achieving their native folds. Many of the proteins whose folding is facilitated by GroEL/ES contain open β sheets, whose structural complexity is largely responsible for their misfolding. Many such proteins repeatedly return to GroEL/ES for conformational maintenance.

3 ■ Protein Structure Prediction and Design The prediction of protein secondary structures from only amino acid sequences has been reasonably successful using empirical techniques such as the Chou–Fasman method. However, sophisticated computational techniques yield somewhat more reliable predictions. Comparative (homology) modeling can provide accurate tertiary structures for polypeptides with >30% identity to a protein of known structure. Fold recognition (threading) techniques have only been marginally successful for determining the structures of proteins that have no apparent homology with proteins of known structure, whereas *ab initio* structure determination methods are only beginning to yield useful models. However, computationally based protein design has, of late, been successful, in part, because one can "overengineer" a protein to take up a desired conformation.

4 ■ Protein Dynamics Proteins are flexible and fluctuating molecules whose group motions have characteristic periods ranging from 10^{-15} to over 10^3 s. X-Ray analysis, which reveals the average atomic mobilities in a protein, indicates that proteins tend to be more mobile at their peripheries than in their interiors. Molecular dynamics simulations indicate that native protein structures each consist of a large number of closely related and rapidly interconverting conformational substates of nearly equal stabilities. Without this flexibility, enzymes would be nonfunctional. The rates of aromatic ring flipping, as revealed by NMR measurements, indicate that internal group mobilities within proteins vary both with the protein and with the position within the protein. The exchange of a protein's internal protons with solvent requires its transient local unfolding. Hydrogen exchange studies therefore demonstrate that proteins have a great variety of infrequently occurring internal motions.

5 ■ Conformational Diseases: Amyloids and Prions A number of often fatal human diseases are associated with the deposition of amyloid in the brain and other organs. Although the various amyloidogenic proteins are unrelated in both sequence and native structure, all form similar amyloid fibrils that consist mainly of β strands that extend perpendicularly to the fibril axis. The two known human lysozyme variants that have amyloidogenic properties are conformationally much looser than wild-type lysozyme. In Alzheimer's disease, a neurodegenerative disease of mainly the elderly, the proteolysis of Aβ-precursor protein (βPP) in brain tissue yields the 40- to 42-residue amyloid-β protein (Aβ), which forms the amyloid fibrils that kill neurons. Humans and other mammals are subject to infectious neurodegenerative diseases such as scrapie, which are caused by prions. Prions appear to consist of only a single species of protein named PrP. PrP exists in two forms: the normal cellular form, PrP^C, a conserved membrane-anchored cell-surface protein on neurons; and PrP^{Sc}, which although chemically identical to PrP^C, has a different conforma-

tion. PrP^{Sc} autocatalytically converts PrP^C to PrP^{Sc}, thereby accounting for the infectious properties of PrP^{Sc} and the observation that $Prn-p^{0/0}$ mice are resistant to scrapie. PrP^{Sc} is proteolytically degraded in the cell to form a protease-resistant core, PrP 27–30, that aggregates to form the neurotoxic amyloid fibrils thought to be responsible for the symptoms of prion diseases. Yeast also have proteins with prionlike properties.

6 ■ Structural Evolution The X-ray structures of eukaryotic cytochromes *c* demonstrate that internal residues and those having specific structural and functional roles tend to be conserved during evolution. Prokaryotic *c*-type cytochromes from a variety of organisms structurally resemble each other and those of eukaryotes even though they have little sequence similarity. This indicates that the three-dimensional structures of proteins rather than their amino acid sequences are conserved during evolutionary change. The structural similarities between the domains in many multidomain proteins indicate that these proteins arose through the duplication of a gene specifying an ancestral domain followed by their fusion. Similarly, the structural resemblance between the dinucleotide-binding domains of dehydrogenases suggests that these proteins arose by duplication of a primordial dinucleotide-binding domain followed by its fusion with a gene specifying a proto–substrate-binding domain. In this manner, proteins with new functions can evolve much faster than by a series of point mutations.

REFERENCES

PROTEIN FOLDING

Anfinsen, C.B., Principles that govern the folding of protein chains, *Science* **181**, 223–230 (1973). [A Nobel laureate explains how he got his prize.]

Aurora, R. and Rose, G.D., Helix capping, *Protein Sci.* **7**, 21–38 (1998). [Summarizes the evidence that helix capping interactions stabilize helices.]

Baldwin, R.L., Pulsed H/D-exchange studies of folding intermediates, *Curr. Opin. Struct. Biol.* **3**, 84–91 (1993).

Baldwin, R.L., Protein folding from 1961 to 1982, *Nature Struct. Biol.* **6**, 814–817 (1999). [An intellectual history.]

Baldwin, R.L. and Rose, G.D., Is protein folding hierarchic? I. Local structure and peptide folding; *and* II. Folding intermediates and transition states, *Trends Biochem. Sci.* **24**, 26–33; *and* 77–83 (1999).

Behe, M., Lattman, E.E., and Rose, G.D., The protein folding problem: The native fold determines the packing but does packing determine the native fold? *Proc. Natl. Acad. Sci.* **88**, 4195–4199 (1991).

Betts, S. and King, J., There's a right way and a wrong way: *in vivo* and *in vitro* folding, misfolding and subunit assembly of the P22 tailspike, *Structure* **7**, R131–R139 (1999).

Bukau, B. and Horwich, A.L., The Hsp70 and Hsp60 chaperone machines, *Cell* **92**, 351–366 (1998).

Creighton, T.E., Protein folding, *Biochem. J.* **270**, 1–16 (1990).

Creighton, T.E. (Ed.), *Protein Folding*, Freeman (1992). [A series of authoritative reviews.]

Creighton, T.E., *Proteins* (2nd ed.), Chapter 7, Freeman (1993).

Dalal, S., Balasubramanian, S., and Regan, L., Protein alchemy: Changing β-sheet into α-helix, *Nature Struct. Biol.* **4**, 548–552 (1997). [Reports the sequence changes in protein GB1 that cause it to assume the fold of Rop protein.]

Dill, K.A. and Chan, H.S., From Levinthal to pathways to funnels, *Nature Struct. Biol.* **4**, 10–19 (1997). [Reviews the landscape theory of protein folding.]

Dinner, A.R., Sali, A., Smith, L.J., Dobson, C.M., and Karplus, M., Understanding protein folding via free-energy surfaces from theory and experiment, *Trends Biochem. Sci.* **25**, 331–339 (2000).

Dobson, C.M. and Karplus, M., The fundamentals of protein folding: Bringing together theory and experiment, *Curr. Opin. Struct. Biol.* **9**, 92–101 (1999).

Eaton, W.A., Thompson, P.A, Chan, C.K., Hagen, S.J., and Hofrichter, J., Fast events in protein folding, *Structure* **4**, 1133–1139 (1996).

Englander, S.W., Protein folding intermediates and pathways studied by hydrogen exchange, *Annu. Rev. Biophys. Biomol. Struct.* **29**, 213–238 (2000); *and* Englander, S.W., Sosnick, T.R., Englander, J.J., and Mayne, L., Mechanisms and uses of hydrogen exchange, *Curr. Opin. Struct. Biol.* **6**, 18–23 (1996).

Fersht, A., *Structure and Mechanism in Protein Science*, Chapters 17–19, Freeman (1999).

Fink, A.L., Compact intermediate states in protein folding, *Annu. Rev. Biophys. Biomol. Struct.* **24**, 495–522 (1995).

Frydman, J., Folding of newly translated proteins in vivo: The role of molecular chaperones, *Annu. Rev. Biochem.* **70**, 603–649 (2001).

Lattman, E.E. and Rose, G.D., Protein folding—what's the question? *Proc. Natl. Acad. Sci.* **90**, 439–441 (1993).

Levitt, M., Gerstein M., Huang, E., Subbiah, S., and Tsai, J., Protein folding: The endgame, *Annu. Rev. Biochem.* **66**, 549–579 (1997).

Matthews, B.W., Studies on protein stability with T4 lysozyme, *Adv. Prot. Chem.* **46**, 249–278 (1995).

Matthews, C.R., Pathways of protein folding, *Annu. Rev. Biochem.* **62,** 653–684 (1993).

Minor, D.L., Jr. and Kim, P.S., Context-dependent secondary structure formation of a designed protein sequence, *Nature* **380,** 730–734 (1996). [Describes the position-dependent conformation of the chameleon sequence in protein GB1.]

Miranker, A.D. and Dobson, C.M., Collapse and cooperativity in protein folding, *Curr. Opin. Struct. Biol.* **6,** 31–42 (1996).

Pain, R.H. (Ed.), *Mechanisms of Protein Folding* (2nd ed.), Oxford University Press (2000).

Raschke, T.M. and Marqusee, S., Hydrogen exchange studies of protein structure, *Curr. Opin. Biotech.* **9,** 80–86 (1998).

Roder, H. and Shastry, M.C.R., Methods for exploring early events in protein folding, *Curr. Opin. Struct. Biol.* **9,** 620–626 (1999).

Rose, G.D. and Wolfenden, R., Hydrogen bonding, hydrophobicity, packing, and protein folding, *Annu. Rev. Biophys. Biomol. Struct.* **22,** 381–415 (1993).

Srinivasan, R. and Rose, G.D., LINUS: A hierarchic procedure to predict the fold of a protein, *Proteins* **22,** 81–99 (1995); *and* Ab initio prediction of protein structures using LINUS, *Proteins* **47,** 489–495 (2002).

Wang, C.C. and Tsou, C.L., The insulin A and B chains contain sufficient structural information to form the native molecule, *Trends Biochem. Sci.* **16,** 279–281 (1991).

Weissman, J.S. and Kim, P.S., Reexamination of the folding of BPTI: Predominance of native intermediates, *Science* **253,** 1386–1393 (1991); *and* Kinetic role of nonnative species in the folding of bovine pancreatic trypsin inhibitor, *Proc. Natl. Acad. Sci.* **89,** 9900–9904 (1992).

Wolynes, P.G., Luthey-Schulten, Z., and Onuchic, J.N., Fast-folding experiments and the topography of protein folding energy landscapes, *Chem. Biol.* **3,** 425–432 (1996).

FOLDING ACCESSORY PROTEINS

Accessory Folding Proteins, Adv. Protein Chem. **44** (1993). [Contains authoritative articles on protein disulfide isomerase, peptidyl prolyl cis–trans isomerase, and several types of molecular chaperones.]

Brinker, A., Pfeifer, G., Kerner, M.J., Naylor, D.J., Hartl, F.U., and Hayer-Hartl, M., Dual function of protein confinement in chaperonin-assisted protein folding, *Cell* **107,** 223–233 (2001).

Chen, L. and Sigler, P.B., The crystal structure of a GroEL/peptide complex: Plasticity as a basis for substrate diversity, *Cell* **99,** 757–768 (1999).

Ellis, R.J., Macromolecular crowding: Obvious but underappreciated. *Trends Biochem. Sci.* **26,** 597–604 (2001).

Ellis, R.J. and Hartl, F.U., Principles of protein folding in the cellular environment, *Curr. Opin. Struct. Biol.* **9,** 102–110 (1999).

Galat, A. and Metcalfe, S.M., Peptidylprolyl *cis/trans* isomerases, *Prog. Biophys. Mol. Biol.* **63,** 67–118 (1995). [A detailed review.]

Gilbert, H.F., Protein disulfide isomerase and assisted protein folding, *J. Biol. Chem.* **272,** 29399–29402 (1997).

Hartl, F.U., and Hayer-Hartl, M., Molecular chaperones in the cytosol: From nascent chain to unfolded protein, *Science* **295,** 1852–1858 (2002).

Hartl, F.-U., Hlodan, R., and Langer, T., Molecular chaperones in protein folding: The art of avoiding sticky situations, *Trends Biochem. Sci.* **19,** 20–25 (1994).

Horwich, A.R. (Ed.), *Protein Folding in the Cell, Adv. Prot. Chem.* **59,** (2002). [Contains authoritative articles on a variety of folding accessory proteins.]

Houry, W.A., Frishman, D., Ekersorn, C., Lottspeich, F., and Hartl, F.U., The identification of *in vivo* substrates of the chaperonin GroEL, *Nature* **402,** 147–154 (1999).

Kemmink, J., Darby, N.J., Dijkstra, K., Nilges, M., and Creighton, T.E., Structure determination of the N-terminal thioredoxin-like domain of protein disulfide isomerase using multidimensional heteronuclear ^{13}C/^{15}N NMR spectroscopy, *Biochemistry* **35,** 7684–7691 (1996); *and* Kemmink, J., Dijkstra, K., Mariano, M., Scheek, R.M., Penka, E., Nilges, M., and Darby, N.J., The structure in solution of the *b* domain of protein disulfide isomerase, *J. Biomol. NMR* **13,** 357–368 (1999).

Lund, P. (Ed.), *Molecular Chaperones in the Cell,* Oxford University Press (2001).

Pearl, L.H. and Prodromou, C., Structure and *in vivo* function of Hsp90, *Curr. Opin. Struct. Biol.* **10,** 46–51 (2000).

Raina, S. and Missiakas, D., Making and breaking disulfide bonds, *Annu. Rev. Microbiol.* **51,** 179–202 (1997).

Ransom, N.A., Farr, G.W., Roseman, A.M., Gowen, B., Fenton, W.A., Horwich, A.L., and Saibil, H.R., ATP-bound states of GroEL captured by cryo-electron microscopy, *Cell* **107,** 869–879 (2001).

Rye, H.S., Roseman, A.M., Chen, S., Furtak, K., Fenton, W.A., Saibil, H.R., and Horwich, A.L., GroEL-GroES cycling: ATP and nonnative polypeptide direct alternation of folding-active rings, *Cell* **97,** 325–338 (1999).

Saibil, H., Molecular chaperones: Containers and surfaces for folding, stabilising or unfolding proteins, *Curr. Opin. Struct. Biol.* **10,** 251–258 (2000).

Schiene, C. and Fischer, G., Enzymes that catalyse the restructuring of proteins, *Curr. Opin. Struct. Biol.* **10,** 40–45 (2000). [Discussess protein disulfide isomerases and peptidyl prolyl cis–trans isomerases.]

Schreiber, S.L., Chemistry and biology of immunophilins and their immunosuppressive ligands, *Science* **251,** 238–287 (1991).

Shtilerman, M., Lorimer, G.H., and Englander, S.W., Chaperonin function: Folding by forced unfolding, *Science* **284,** 822–825 (1999).

Sigler, P.B., Xu, Z., Rye, H.S., Burston, S.G., Fenton, W.A., and Horwich, A.L., Structure and function in GroEL-mediated protein folding, *Annu. Rev. Biochem.* **67,** 581–608 (1998).

Thirumalai, D. and Lorimer, G.H., Chaperone-mediated protein folding, *Annu. Rev. Biophys. Biomol. Struct.* **30,** 245–269 (2001).

Walsh, C.T., Zydowsky, L.D., and McKeon, F.D., Cyclosporin A, the cyclophilin class of peptidylprolyl isomerases, and blockade of T cell signal transduction, *J. Biol. Chem.* **267,** 13115–13118 (1992).

Xu, Z., Horwich, A.L., and Sigler, P.B., The crystal structure of the asymmetric GroEL–GroES–(ADP)$_7$ chaperonin complex, *Nature* **388,** 741–750 (1997).

Zhao, Y. and Ke, H., Crystal structure implies that cyclophilin predominantly catalyzes the *trans* to *cis* isomerization, *Biochemistry* **35,** 7356–7361 (1996).

PROTEIN STRUCTURE PREDICTION AND DESIGN

Blaber, M., Zhang, X., and Matthews, B.W., Structural basis of amino acid α helix propensity, *Science* **260,** 1637–1640 (1993).

Branden, C. and Tooze, J., *Introduction to Protein Structure* (2nd ed.), Chapter 17, Garland (1999).

Chou, P.Y. and Fasman, G.D., Empirical predictions of protein structure, *Annu. Rev. Biochem.* **47,** 251–276 (1978); *and* Prediction of the secondary structure of proteins from their amino acid sequence, *Adv. Enzymol.* **47,** 45–148 (1978). [Expositions of a widely used and particularly simple method of protein secondary structure prediction.]

Cuff, J.A. and Barton, G.J., Evaluation and improvement of multiple sequence methods for protein secondary structure prediction, *Proteins* **34**, 508–519 (1999).

Dahiyat, B.I. and Mayo, S.L., De novo protein design: Fully automated sequence selection, *Science* **278**, 82–87 (1997). [Describes the design of FSD-1.]

DeGrado W.F., Summa, S.M., Pavone, V., Nastri, F., and Lombardi, A., De novo design and structural characterization of proteins and metalloproteins, *Annu. Rev. Biochem.* **68**, 779–819 (1999).

Klemba, M.W., Munson, M., and Regan, L., *De novo* design of protein structure and function, *in* Angeletti, R.H. (Ed.), *Proteins. Analysis and Design, pp.* 313–353, Academic Press (1998).

Mirny, L. and Shakhnovitch, E., Protein folding theory: From lattice to all-atom models, *Annu. Rev. Biophys. Biomol. Struct.* **30**, 361–396 (2001).

Rose, G.D., Prediction of chain turns in globular proteins on a hydrophobic basis, *Nature* **272**, 586–590 (1978).

Street, A.G. and Mayo, S.L., Computational protein design, *Structure* **7**, R105–R109 (1999).

Webster, D.M. (Ed.), *Protein Structure Prediction,* Humana Press (2000).

PROTEIN DYNAMICS

Dagget, V., Long timescale simulations, *Curr. Opin. Struct. Biol.* **10**, 160–164 (2000).

Huber, R., Flexibility and rigidity of proteins and protein-pigment complexes, *Angew. Chem. Int. Ed. Engl.* **27**, 79–88 (1988).

Karplus, M. and McCammon, A., Molecular dynamics simulations of biomolecules, *Nature Struct. Biol.* **9**, 646–651 (2002).

Palmer, A.G., III, Probing molecular motion by NMR, *Curr. Opin. Struct. Biol.* **7**, 732–737 (1997).

Rasmussen, B.F., Stock, A.M., Ringe, D., and Petsko, G.A., Crystalline ribonuclease A loses function below the dynamical transition at 220 K, *Nature* **357**, 423–424 (1992).

Ringe, D. and Petsko, G.A., Mapping protein dynamics by X-ray diffraction, *Prog. Biophys. Mol. Biol.* **45**, 197–235 (1985).

Rogero, J.R., Englander, J.J., and Englander, S.W., Measurement and identification of breathing units in hemoglobin by hydrogen exchange, *in* Sarma R.H. (Ed.), *Biomolecular Stereodynamics,* Vol. 2, *pp.* 287–298, Adenine Press (1981).

CONFORMATIONAL DISEASES

Aguzzi, A., Montrasio, F., and Kaeser, P.S., Prions: health scare and biological challenge, *Nature Rev. Mol. Cell Biol.* **2**, 118–126 (2001).

Blake, C. and Serpell, L., Synchrotron X-ray studies suggest that the core of the transthyretin amyloid fibril is a continuous β-sheet helix, *Structure* **4**, 989–998 (1996).

Booth, D.R., et al., Instability, unfolding and aggregation of human lysozyme variants underlying amyloid fibrillogenesis, *Nature* **385**, 787–793 (1997); *and* Funahashi, J., Takano, K., Ogasahara, K., Yamagata, Y., and Yutani, K., The structure, stability, and folding process of amyloidogenic mutant lysozyme, *J. Biochem.* **120**, 1216–1223 (1996).

Büeler, H., Aguzzi, A., Sailer, A., Greiner, R.A., Autenreid, P., Aguet, M., and Weissmann, C., Mice devoid of PrP are resistant to scrapie, *Cell* **73**, 1339–1347 (1993); *and* Büeler, H., Fischer, M., Lang, Y., Bluethmann, H., Lipp, H.-P., DeArmond, S.J., Prusiner, S.B., Aguet, M., and Weissmann, C., Normal development and behaviour of mice lacking the neuronal cell-surface PrP protein, *Nature* **356**, 577–582 (1992).

Buxbaum, J.N. and Tagoe, C.E., The genetics of amyloidoses, *Annu. Rev. Med.* **51**, 543–569 (2000).

Carrell, R.W. and Lomas, D.A., Conformational diseases, *Lancet* **350**, 134–138 (1997); *and* Carrell, R.W. and Gooptu, B., Conformational changes and disease—serpins, prions and Alzheimer's, *Curr. Opin. Struct. Biol.* **8**, 799–809 (1998).

Caughey, B., Interactions between prion protein isoforms: The kiss of death? *Trends Biochem. Sci.* **26**, 235–242 (2001).

Daggett, V., Structure-function aspects of prion proteins, *Curr. Opin. Biotech.* **9**, 359–365 (1998).

Geula, C., Wu, C.-K., Saroff, D., Lorenzo, A., Yuan, M., and Yankner, B.A., Aging renders the brain vulnerable to amyloid β-protein neurotoxicity, *Nature Medicine* **4**, 827–831 (1998).

Hardy, J. and Selkoe, D.J., The amyloid hypothesis of Alzheimer's disease: Progress and problems on the road to therapeutics, *Science* **297**, 353–356 (2002).

Haywood, A.M., Transmissible spongiform encephalopathies, *New Engl. J. Med.* **337**, 1821–1828 (1997).

Horiuchi, M. and Caughey, B., Prion protein interconversions and the transmissible spongiform encephalopathies, *Structure* **7**, R231–R240 (1999).

Jackson, G.S. and Clarke, A.R., Mammalian prion proteins, *Curr. Opin. Struct. Biol.* **10**, 69–74 (2000).

Kaytor, M.D. and Warren, S.T., Aberrant protein deposition and neurological disease, *J. Biol. Chem.* **274**, 37507–37510 (1999).

Kelly, J.W., The alternative conformations of amyloidogenic proteins and their multi-step assembly pathways, *Curr. Opin. Struct. Biol.* **8**, 101–106 (1998).

Kiselevsky, R. and Fraser, P.E., Aβ amyloidogenesis: Unique, or variation on a systemic theme? *Crit. Rev. Biochem. Molec. Biol.* **32**, 361–404 (1997).

Kocisko, D.A., Come, J.H., Priola, S.A., Chesebro, B., Raymond, G.J., Lansbury, P.T., and Caughey, B., Cell-free formation of protease-resistant prion protein, *Nature* **370**, 471–474 (1994).

Mouillet-Richard, S., Ermonval, M., Chebassier, C., Laplanche, J.L., Lehmann, S., Launay, J.M., and Kellermann, O., Signal transduction through prion protein, *Science* **289**, 1925–1928 (2000).

Pan, K.M., Baldwin, M., Nguyen, J., Gasset, M., Serban, A., Groth, D., Mehlhorn, I., Huang, Z., Fletterick, R.J., Cohen, F.E., and Prusiner, S.B., Conversion of α-helices into β-sheets features in the formation of the scrapie prion proteins, *Proc. Natl. Acad. Sci.* **90**, 10962–10966 (1993).

Prusiner, S.B., Prions, *Proc. Natl. Acad. Sci.* **95**, 13363–13383 (1998); *and* Prusiner, S.B., Scott, M.R., DeArmond, S.J., and Cohen, F.E., Prion protein biology, *Cell* **93**, 337–348 (1998).

Prusiner, S.B. (Ed.), *Prion Biology and Diseases,* Cold Spring Harbor Laboratory Press (1999).

Rochet, J.C. and Lansbury, P.T., Jr., Amyloid fibrillogenesis: Themes and variations, *Curr. Opin. Struct. Biol.* **10**, 60–68 (2000).

Selkoe, D.J. Amyloid β-protein and the genetics of Alzheimer's disease, *J. Biol. Chem.* **271**, 18295–18298 (1996).

Sparrer, H.E., Santoso, A., Szoka, F.C., Jr., and Weissman, J.S., Evidence for the prion hypothesis: Induction of the yeast [*PSI*+] factor by in vitro-converted Sup35 protein, *Science* **289**, 595–599 (2000).

Sunde, M. and Blake, C.C.F., From the globular to the fibrous state: Protein structure and structural conversion in amyloid formation, *Quart. Rev. Biophys.* **31**, 1–39 (1998).

Terry, R.D., Katzman, R., Bick, K.L., and Sisodia, S.S. (Eds.), *Alzheimer Disease,* Lippincott, Williams, & Wilkins (1999).

Tuite, M.F., Yeast prions and their prion-forming domain, *Cell* **100**, 289–292 (2000).

Weissmann, C., Molecular genetics of transmissible spongiform encephalopathies, *J. Biol. Chem.* **274**, 3–6 (1999).

Wickner, R.B., *Prion Diseases of Mammals and Yeast: Molecular Mechanisms and Genetic Features,* Chapman & Hall (1997).

Zahn, R., Liu, A., Lührs, T., Riek, R., von Schroetter, C., Garcia, F.L., Billeter, M., Calzolai, L., Wider, G., and Wüthrich, K., NMR solution structure of the human prion protein, *Proc. Natl. Acad. Sci.* **97**, 145–150 (2000); *and* Liu, H., Farr-Jones, S., Ulyanov, N.B., Llinas, M., Marqusee, S., Groth, D., Cohen, F.E., Prusiner, S.B., and James, T.L., Solution structure of Syrian hamster prion protein rPrP(90–231), *Biochemistry* **38**, 5362–5377 (1999).

STRUCTURAL EVOLUTION

Bajaj, M. and Blundell, T., Evolution and the tertiary structure of proteins, *Annu. Rev. Biophys. Bioeng.* **13**, 453–492 (1983).

Dickerson, R.E., The structure and history of an ancient protein, *Sci. Am.* **226**(4), 58–72 (1972); *and* Cytochrome *c* and the evolution of energy metabolism, *Sci. Am.* **242**(3), 137–149 (1980).

Dickerson, R.E., Timkovitch, R., and Almassy, R.J., The cytochrome fold and the evolution of bacterial energy metabolism, *J. Mol. Biol.* **100**, 473–491 (1976).

Eventhoff, W. and Rossmann, M., The structures of dehydrogenases, *Trends Biochem. Sci.* **1**, 227–230 (1976).

Lesk, A.M., NAD-binding domains of dehydrogenases, *Curr. Opin. Struct. Biol.* **5**, 775–783 (1995).

Salemme, R., Structure and function of cytochromes *c*, *Annu. Rev. Biochem.* **46**, 299–329 (1977).

Scott, R.A. and Mauk, A.G. (Eds.), *Cytochrome c. A Multidiscipinary Approach,* University Science Books (1996).

PROBLEMS

1. How long will it take the polypeptide backbone of a 6-residue folding nucleus to explore all its possible conformations? Repeat the calculation for 10-, 15-, and 20-residue folding nuclei. Why, in the classic view of protein folding, are folding nuclei thought to be no larger than 15 residues?

*2. Consider a protein with 10 Cys residues. On air oxidation, what fraction of the denatured and reduced protein will randomly reform the native set of disulfide bonds if: (a) The native protein has five disulfide bonds? (b) The native protein has three disulfide bonds?

3. Why are β sheets more commonly found in the hydrophobic interiors of proteins than on their surfaces?

4. Under physiological conditions, polylysine assumes a random coil conformation. Under what conditions might it form an α helix?

5. Explain how landscape theory is consistent with the observation that many small proteins appear to fold to their native conformations without detectable intermediates, that is, via 2-state mechanisms.

6. Explain why Pro residues can occupy the N-terminal turn of an α helix.

7. Explain why β sheets are less likely to form than α helices during the earliest stages of protein folding.

8. Molten globules are thought to be predominantly stabilized by hydrophobic forces. Why aren't hydrogen bonding forces implicated in doing so?

*9. The GroEL/ES cycle diagrammed in Fig. 9-23 only circulates in the clockwise direction. Explain the basis for this irreversibility in terms of the sequence of structural and binding changes in the GroEL/ES system.

*10. Predict the secondary structure of the C peptide of proinsulin (Fig. 9-4) using the methods of Chou–Fasman and Rose. Is it likely to have a supersecondary structure?

11. As Mother Nature's chief engineer, now certified as a master helix builder, you are asked to repeat Problem 8-8 with the stipulation that the α helix really be helical. Use Table 9-1.

12. Indicate the probable effects of the following mutational changes on the structure of a protein. Explain your reasoning. (a) Changing a Leu to a Phe, (b) changing a Lys to a Glu, (c) changing a Val to a Thr, (d) changing a Gly to an Ala, and (e) changing a Met to a Pro.

13. Explain why Trp rings are usually completely immobile in proteins that have rapidly flipping Phe and Tyr rings.

14. Explain why *Prn-p*[0/0] mice are resistant to scrapie. What might be the susceptibility of heterozygous *Prn-p*[+/0] mice to scrapie?

*15. Discuss the merits of the hypothesis that the dinucleotide-binding domains of the dehydrogenases arose by convergent evolution.

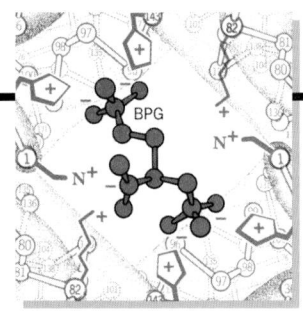

Chapter

10

Hemoglobin: Protein Function in Microcosm

associated with a specific physiological function (that of oxygen transport), and, in sickle-cell anemia, the first in which a point mutation was demonstrated to cause a single amino acid change (Section 7-3A). Theories formulated to account for the cooperative binding of oxygen to hemoglobin (Section 10-4) have also been successful in explaining the control of enzyme activity. The first protein X-ray structures to be elucidated were those of hemoglobin and myoglobin. This central role in the development of protein chemistry together with its enzymelike O_2-binding properties have caused hemoglobin to be dubbed an "honorary enzyme."

Hemoglobin is not just a simple oxygen tank. Rather, it is a sophisticated oxygen delivery system that provides the proper amount of oxygen to the tissues under a wide variety of circumstances. In this chapter, we discuss hemoglobin's properties, structure, and mechanism of action, both to understand the workings of this physiologically essential molecule and to illustrate the principles of protein structure that we have developed in the preceding chapters. We also consider the properties of abnormal hemoglobins and their relationship to human disease. Finally, we discuss theories of cooperative interactions among proteins, both to better understand the properties of hemoglobin and to set the stage for our later consideration of how enzyme action is regulated.

The existence of hemoglobin, the red blood pigment, is evident to every child who scrapes a knee. Its brilliant red color, widespread occurrence, and ease of isolation have made it an object of inquiry since ancient times. Indeed, the early history of protein chemistry is essentially that of hemoglobin. The observation of crystalline hemoglobin was first reported by Friedrich Hünefeld in 1840, and by 1909 Edward Reichert and Amos Brown had published a photographic atlas of hemoglobin crystals from several hundred species. In contrast, it was not until 1926 that crystals of an enzyme, those of jack bean **urease,** were first reported. Hemoglobin was one of the first proteins to have its molecular mass accurately determined, the first protein to be characterized by ultracentrifugation, the first to be

1 ■ HEMOGLOBIN FUNCTION

Hemoglobin **(Hb),** as we have seen in Chapters 7 and 8, is a heterotetramer, $\alpha_2\beta_2$ (alternatively, a dimer of $\alpha\beta$ protomers). The α and β subunits are structurally and evolutionarily related to each other and to myoglobin **(Mb),** the monomeric oxygen-binding protein of muscle (Section 7-3C).

Hemoglobin transports oxygen from the lungs, gills, or skin of an animal to its capillaries for use in respiration. Very small organisms do not require such a protein because their respiratory needs are satisfied by the simple passive diffusion of O_2 through their bodies. However, since the transport rate of a diffusing substance varies inversely with the square of the distance it must diffuse, the O_2 diffusion rate through tissue thicker than ~1 mm is too slow to support life. The evolution of organisms as large and complex as annelids (e.g., earthworms) therefore required the development of circulatory systems that actively transport O_2 and nutrients to the tissues. The blood of such organisms must contain an oxygen transporter such as Hb because the solubility of O_2 in **blood plasma** (the fluid component of blood) is too low (~$10^{-4}M$ under physiological conditions) to carry sufficient O_2 for metabolic needs. In contrast, whole blood, which normally contains ~150 g of Hb $\cdot$ L^{-1}, can carry O_2 at concentrations as high as $0.01M$, about the same as in air. [Although many invertebrate species have hemoglobin-based oxygen transport systems, others produce one of two alternative types of O_2-binding proteins: (1) **hemocyanin**, a Cu-containing protein that is blue in complex with oxygen and colorless otherwise; or (2) **hemerythrin**, a nonheme Fe-containing protein that is burgundy colored in complex with oxygen and colorless otherwise. Antarctic icefish, the only adult vertebrates that lack hemoglobin—their blood is colorless—are viable because of their reduced need for O_2 at low temperatures combined with the relatively high aqueous solubility of O_2 at the $-1.9°C$ temperature of their environment (recall that the solubilities of gases increase with decreasing temperature).]

Although Mb was originally assumed to function only to store oxygen, this function now appears to be significant only in aquatic mammals such as seals and whales, which have Mb concentrations in their muscles ~10-fold greater than those in terrestrial mammals. It seems more likely that Mb's major physiological role in terrestrial mammals is to facilitate oxygen transport in rapidly respiring muscle. The rate at which O_2 can diffuse from the capillaries to the tissues, and thus the level of respiration, is limited by the oxygen's low solubility in aqueous solution. Mb increases the effective solubility of O_2 in muscle, the most rapidly respiring tissue under conditions of high exertion. Hence, in rapidly respiring muscle, Mb could function as a kind of molecular bucket brigade to facilitate O_2 diffusion. It therefore came as a surprise when knockout mice lacking the gene encoding Mb exhibited no obvious abnormalities (except for the pale color of their muscles), reproduced normally, and exhibited normal exercise capacity and response to low oxygen levels. Recently, however, an alternative physiological function for Mb has come to light: the detoxification of the highly reactive biological signaling molecule **nitric oxide (NO)** through its conversion to NO_3^- (see below).

In this section, we begin our discussions of hemoglobin by considering its chemical and physical properties and how they relate to its physiological function. Hemoglobin structure and the mechanisms by which it carries out these physiological functions are discussed in Section 10-2.

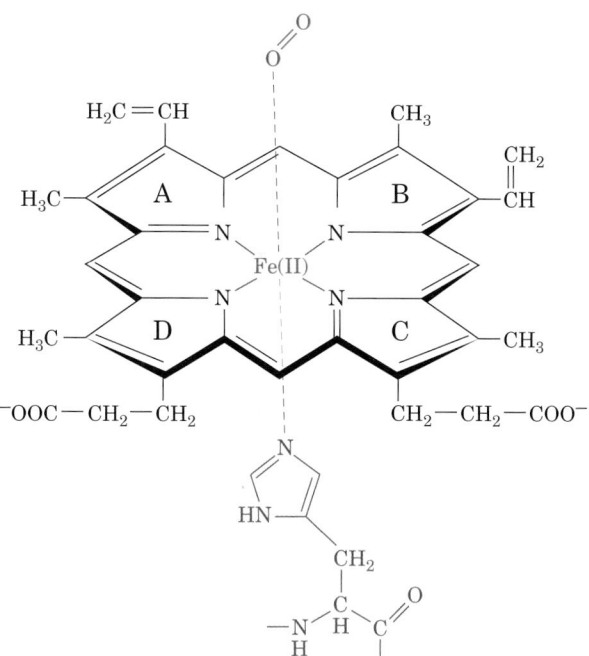

FIGURE 10-1   The heme group. Fe(II)–heme (ferroprotoporphyrin IX) is shown liganded to His and O_2 as it is in oxygenated myoglobin and oxygenated hemoglobin. Note that the heme is a conjugated system so that, although two of its Fe—N bonds are coordinate covalent bonds (bonds in which the bonding electron pair is formally contributed by only one of the atoms forming the bond), all of the Fe—N bonds are equivalent. The pyrrole ring lettering scheme is shown.

A. *Heme*

*Myoglobin and each of the four subunits of hemoglobin noncovalently bind a single **heme** group (Fig. 10-1; spelled "haem" in British English).* This is the same group that occurs in the cytochromes (Section 9-6A) and in certain redox enzymes such as **catalase.** Heme is responsible for the characteristic red color of blood and is the site at which each **globin** monomer binds one molecule of O_2 (globins are the heme-free proteins of Hb and Mb). The heterocyclic ring system of heme is a **porphyrin** derivative; it consists of four **pyrrole** rings (lettered A–D in Fig. 10-1) linked by methene bridges. The porphyrin in heme, with its particular arrangement of four methyl, two propionate, and two vinyl substituents, is known as **protoporphyrin IX.** Heme, then, is protoporphyrin IX with a centrally bound iron atom. *In Hb and Mb, the iron atom normally remains in the Fe(II) (ferrous) oxidation state whether or not the heme is oxygenated (binds O_2).*

The Fe atom in deoxygenated Hb and Mb is 5-coordinated by a square pyramid of N atoms: four from the porphyrin and one from a His side chain of the protein. On oxygenation, the O_2 binds to the Fe(II) on the opposite side of the porphyrin ring from the His ligand, so that the Fe(II) is octahedrally coordinated; that is, the ligands occupy the six corners of an octahedron centered on the Fe atom (Fig. 10-1). *Oxygenation changes the electronic state of the Fe(II)–heme, as is indicated by the color change of blood from the dark purplish hue characteristic of ve-*

nous blood to the brilliant scarlet color of arterial blood and blood from a cut finger (Fig. 10-2).

Certain small molecules, such as CO, NO, and H_2S, coordinate to the sixth liganding position of the Fe(II) in Hb and Mb with much greater affinity than does O_2. This, together with their similar binding to the hemes of cytochromes, accounts for the highly toxic properties of these substances.

The Fe(II) of Hb or Mb can be oxidized to Fe(III) to form **methemoglobin (metHb)** or **metmyoglobin (metMb).** MetHb does not bind O_2; its Fe(III) is already octahedrally coordinated with an H_2O molecule in the sixth liganding position. The brown color of dried blood and old meat is that of metHb and metMb. Erythrocytes (red blood cells) contain the enzyme **methemoglobin reductase,** which converts the small amount of metHb that spontaneously forms back to the Fe(II) form.

Nitric oxide (NO), which is synthesized in several tissues, functions as a locally active signaling molecule, most notably to induce vasodilation (Section 19-1L). Once NO has delivered its message, it is important that it be rapidly eliminated to prevent interference with subsequent NO signals (or lack of them). Moreover, NO is a highly reactive and hence toxic substance. In muscle, NO is detoxified through its reaction with oxygenated myoglobin **(oxyMb)** to yield nitrate ion and metmyoglobin:

$$NO + MbO_2 \rightarrow NO_3^- + metMb$$

Since the metMb is subsequently reconverted to Mb through the action of an intracellular **metmyoglobin reductase,** myoglobin functions as an enzyme in this process. The NO that is present in the blood is similarly detoxified through the mediation of oxygenated hemoglobin **(oxyHb).**

B. *Oxygen Binding*

The binding of O_2 to myoglobin is described by a simple equilibrium reaction

$$Mb + O_2 \rightleftharpoons MbO_2$$

with dissociation constant

$$K = \frac{[Mb][O_2]}{[MbO_2]} \qquad [10.1]$$

(biochemists usually express equilibria in terms of dissociation constants, the reciprocals of the more chemically traditional association constants). The O_2 dissociation of Mb may be characterized by its **fractional saturation,** Y_{O_2}, defined as the fraction of O_2-binding sites occupied by O_2.

$$Y_{O_2} = \frac{[MbO_2]}{[Mb] + [MbO_2]} = \frac{[O_2]}{K + [O_2]} \qquad [10.2]$$

Since O_2 is a gas, its concentration is conveniently expressed by its partial pressure, pO_2 (also called the **oxygen tension**). Equation [10.2] may therefore be expressed:

$$Y_{O_2} = \frac{pO_2}{K + pO_2} \qquad [10.3]$$

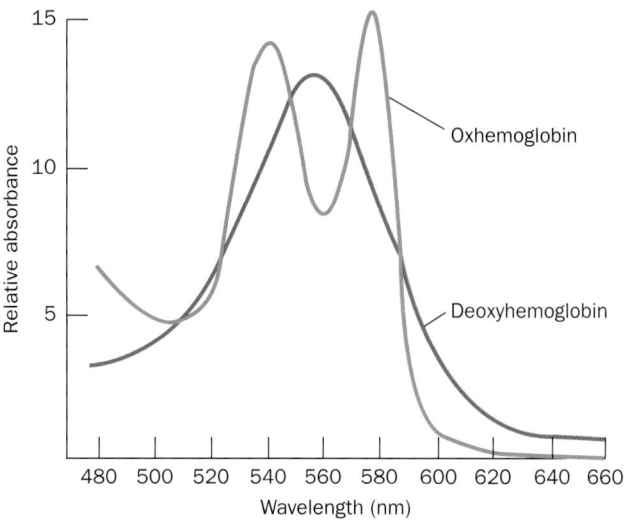

FIGURE 10-2   The visible absorption spectra of oxygenated and deoxygenated hemoglobins.

Now define p_{50} as the value of pO_2 when $Y_{O_2} = 0.50$, that is, when half of myoglobin's O_2-binding sites are occupied. Substituting this value into Eq. [10.3] and solving for K yields $K = p_{50}$. Hence our expression for the fractional saturation of Mb finally becomes:

$$Y_{O_2} = \frac{pO_2}{p_{50} + pO_2} \qquad [10.4]$$

a. Hemoglobin Cooperatively Binds O_2

Myoglobin's O_2-dissociation curve (Fig. 10-3) closely follows the hyperbolic curve described by Eq. [10.4]; its p_{50} is 2.8 torr (1 torr = 1 mm Hg at 0°C = 0.133 kPa; 760 torr = 1 atm). Mb therefore gives up little of its bound O_2 over the normal physiological range of pO_2 in blood (100 torr in arterial blood and 30 torr in venous blood); for example, $Y_{O_2} = 0.97$ at $pO_2 = 100$ torr and 0.91 at 30 torr. In contrast, hemoglobin's O_2-dissociation curve (Fig. 10-3), which has a **sigmoidal** shape (S shape) that Eq. [10.4] does not describe, indicates that the amount of O_2 bound by Hb changes significantly over the normal physiological range of pO_2 in blood. For example, $Y_{O_2} = 0.95$ at 100 torr and 0.55 at 30 torr in whole blood for a difference in Y_{O_2} of 0.40. Mb therefore binds O_2 under conditions in which Hb releases it. Thus, the two proteins form a sophisticated O_2 transport system that delivers O_2 from lung to muscle (where pO_2 may be <20 torr). Hemoglobin's sigmoidal O_2-dissociation curve is of great physiological importance; *it permits the blood to deliver much more O_2 to the tissues than it could if Hb had a hyperbolic O_2-dissociation curve with the same p_{50} (26 torr; dashed curve in Fig. 10-3).* Such a hyperbolic curve has $Y_{O_2} = 0.79$ at 100 torr and 0.54 at 30 torr for a difference in Y_{O_2} of only 0.25.

A sigmoidal dissociation curve is diagnostic of a **cooperative interaction** between a protein's small molecule binding sites; that is, the binding of one small molecule affects the binding of others. In this case, the binding of O_2

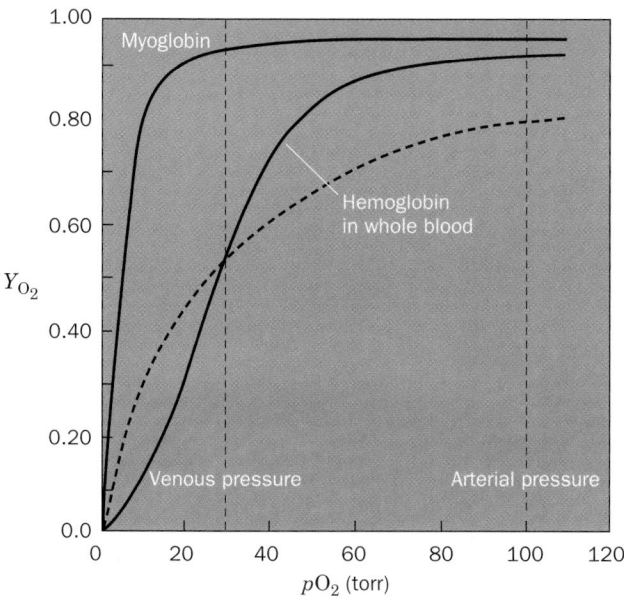

FIGURE 10-3   Oxygen dissociation curves of Mb and of Hb in whole blood. The normal sea level values of human arterial and venous pO_2 values are indicated. The dashed curve is a hyperbolic O_2-dissociation curve with the same p_{50} as Hb (26 torr).

increases the affinity of Hb for binding additional O_2. The structural mechanism of hemoglobin cooperativity is discussed in Section 10-2C.

b. The Hill Equation Phenomenologically Describes Hemoglobin's O_2-Binding Curve

The earliest attempt to analyze hemoglobin's sigmoidal O_2-dissociation curve was formulated by Archibald Hill in 1910. We shall follow his analysis in general form because it is useful for characterizing the cooperative behavior of oligomeric enzymes as well as that of hemoglobin.

Consider a protein E consisting of n subunits that can each bind a molecule S, which, in analogy with the substituents of metal ion complexes, is known as a **ligand.** Assume that the ligand binds with infinite cooperativity,

$$E + nS \rightleftharpoons ES_n$$

that is, the protein either has all or none of its ligand-binding sites occupied, so that there are no observable intermediates ES_1, ES_2, etc. The dissociation constant for this reaction is

$$K = \frac{[E][S]^n}{[ES_n]} \qquad [10.5]$$

and, as before, its fractional saturation is expressed:

$$Y_S = \frac{n[ES_n]}{n([E] + [ES_n])} \qquad [10.6]$$

Combining Eqs. [10.5] and [10.6] yields

$$Y_S = \frac{[E][S]^n/K}{[E](1 + [S]^n/K)}$$

which on algebraic rearrangement and cancellation of terms becomes the **Hill equation:**

$$Y_S = \frac{[S]^n}{K + [S]^n} \qquad [10.7]$$

which, in a manner analogous to Eq. [10.4], describes the degree of saturation of a multisubunit protein as a function of ligand concentration.

Infinite ligand-binding cooperativity (n equal to the number of protein subunits), as assumed in deriving the Hill equation, is a physical impossibility. Nevertheless, n may be taken to be a noninteger parameter related to the degree of cooperativity among interacting ligand-binding sites rather than the number of subunits per protein. The Hill equation then becomes a useful empirical curve-fitting relationship rather than an indicator of a particular model of ligand binding. *The quantity n, the **Hill constant,** increases with the degree of cooperativity of a reaction and thereby provides a convenient, although simplistic, characterization of a ligand-binding reaction.* If $n = 1$, Eq. [10.7] describes a hyperbola, as do Eqs. [10.3] and [10.4] for Mb, and the ligand-binding reaction is said to be **noncooperative.** A reaction with $n > 1$ is described as **positively cooperative:** Ligand binding increases the affinity of E for further ligand binding (cooperativity is infinite in the limit that n is equal to the number of ligand-binding sites in E). Conversely, if $n < 1$, the reaction is termed **negatively cooperative:** Ligand binding reduces the affinity of E for subsequent ligand binding.

c. Hill Equation Parameters May Be Graphically Evaluated

The Hill constant, n, and the dissociation constant, K, that best describe a saturation curve can be graphically determined by rearranging Eq. [10.7] as follows:

$$\frac{Y_S}{1 - Y_S} = \frac{\dfrac{[S]^n}{K + [S]^n}}{1 - \dfrac{[S]^n}{K + [S]^n}} = \frac{[S]^n}{K}$$

and then taking the log of both sides to yield a linear equation:

$$\log\left(\frac{Y_S}{1 - Y_S}\right) = n \log[S] - \log K \qquad [10.8]$$

The linear plot of $\log[Y_s/(1 - Y_s)]$ versus $\log[S]$, the **Hill plot,** has a slope of n and an intercept on the $\log[S]$ axis of $(\log K)/n$ (recall that the linear equation $y = mx + b$ describes a line with a slope of m and an x intercept of $-b/m$).

For Hb, if we substitute pO_2 for [S] as was done for Mb, the Hill equation becomes:

$$Y_{O_2} = \frac{(pO_2)^n}{K + (pO_2)^n} \qquad [10.9]$$

As in Eq. [10.4], let us define p_{50} as the value of pO_2 at $Y_{O_2} = 0.50$. Then, substituting this value into Eq. [10.9],

$$0.50 = \frac{(p_{50})^n}{K + (p_{50})^n}$$

so that

$$K = (p_{50})^n \qquad [10.10]$$

Substituting this result back into Eq. [10.9] yields

$$Y_{O_2} = \frac{(pO_2)^n}{(p_{50})^n + (pO_2)^n} \qquad [10.11]$$

(*Note:* Equation [10.4] is a special case of Eq. [10.11] with $n = 1$.) Equation [10.8] for the Hill plot of Hb therefore takes the form

$$\log\left(\frac{Y_{O_2}}{1 - Y_{O_2}}\right) = n \log pO_2 - n \log p_{50} \qquad [10.12]$$

so that *this plot has a slope of n and an intercept on the log pO₂ axis of log p₅₀.*

Figure 10-4 shows the Hill plots for Mb and Hb. For Mb it is linear with a slope of 1, as expected. Although Hb does not bind O_2 in a single step as is assumed in deriving the Hill equation, its Hill plot is essentially linear for values of Y_{O_2} between 0.1 and 0.9. Its maximum slope, which occurs near $pO_2 = p_{50}$ [$Y_{O_2} = 0.5$; $Y_{O_2}/(1 - Y_{O_2}) = 1$], is normally taken to be the Hill constant. For normal human Hb, the Hill constant is between 2.8 and 3.0; that is, hemoglobin oxygen binding is highly, but not infinitely, cooperative. Many abnormal hemoglobins exhibit smaller Hill constants (Section 10-3A), indicating that they have a less than normal degree of cooperativity. At Y_{O_2} values near 0, when few Hb molecules have bound even one O_2 molecule, the Hill plot of Hb assumes a slope of 1 (Fig. 10-4, lower asymptote) because the Hb subunits independently compete for O_2 as do molecules of Mb. At Y_{O_2} values near 1, when at least three of each of hemoglobin's four

O_2-binding sites are occupied, the Hill plot also assumes a slope of 1 (Fig. 10-4, upper asymptote) because the few remaining unoccupied sites are on different molecules and therefore bind O_2 independently.

Extrapolating the lower asymptote in Fig. 10-4 to the horizontal axis indicates, according to Eq. [10.11], that $p_{50} = 30$ torr for binding the first O_2 to Hb. Likewise, extrapolating the upper asymptote yields $p_{50} = 0.3$ torr for binding hemoglobin's fourth O_2. Thus *the fourth O_2 to bind to Hb does so with 100-fold greater affinity than the first.* This difference, as we shall see in Section 10-2C, is entirely due to the influence of the globin on the O_2 affinity of heme. It corresponds to a free energy difference of $11.4 \text{ kJ} \cdot \text{mol}^{-1}$ between binding the first and binding the last O_2 to Hb (Section 3-4A).

More sophisticated mathematical models than the Hill equation have been developed for analyzing the cooperative binding of ligands to proteins. We examine some of them in Section 10-4.

d. Globin Prevents Oxyheme from Autooxidizing

Globin not only modulates the O_2-binding affinity of heme, but makes reversible O_2 binding possible. Fe(II)–heme by itself is incapable of binding O_2 reversibly. Rather, in the presence of O_2, it autooxidizes irreversibly to the Fe(III) form through the intermediate formation of a complex consisting of an O_2 bridging the Fe atoms of two hemes. This reaction can be inhibited by derivatizing the heme with bulky groups that sterically prevent the close face-to-face approach of two hemes. Such **picket-fence** Fe(II)–porphyrin complexes (Fig. 10-5), which James Collman first synthesized, bind O_2 reversibly. The back side of this porphyrin is unhindered and is complexed with a substituted imidazole in a manner similar to that in Mb and Hb. In fact, the O_2 affinity of the picket-fence complex is similar to that of Mb. Thus, the globins of Mb and

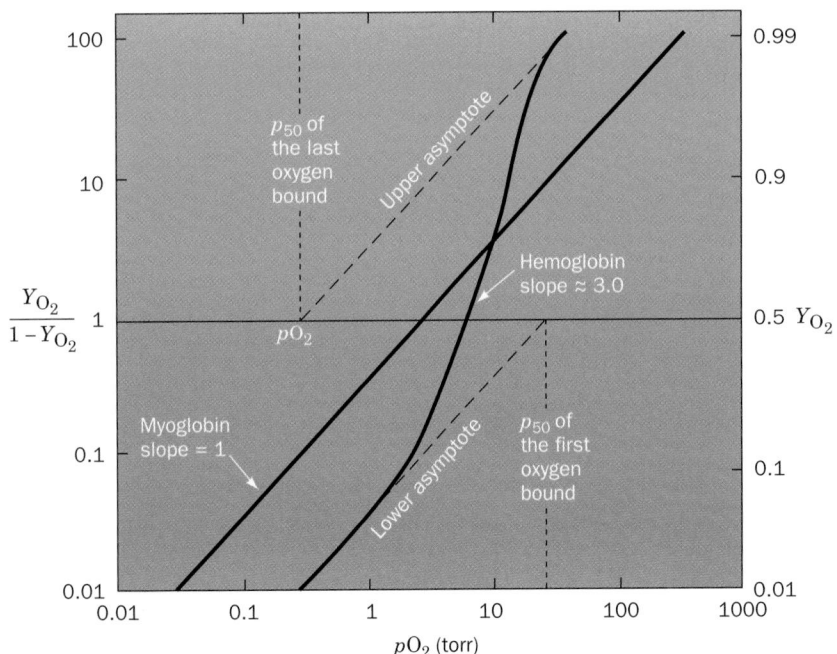

FIGURE 10-4 Hill plots for Mb and purified ("stripped") Hb. Note that this is a log–log plot. Hence the horizontal axis, $\log [Y_{O_2}/(1 - Y_{O_2})] = 0$, occurs where $Y_{O_2}/(1 - Y_{O_2}) = 1$ (and $pO_2 = p_{50}$).

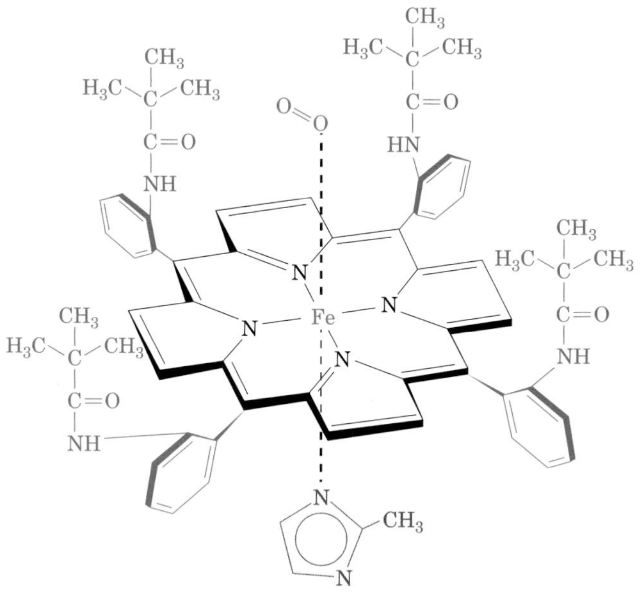

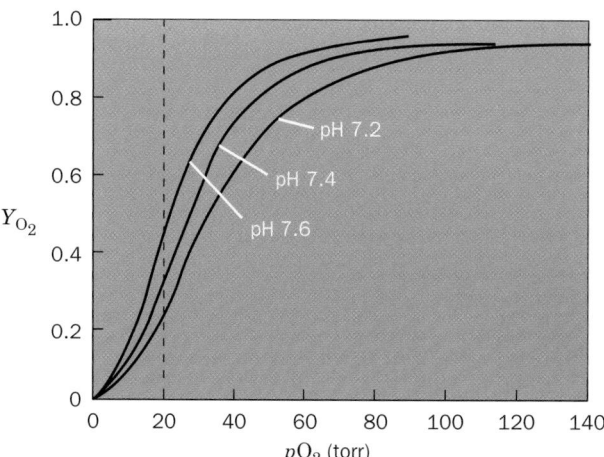

FIGURE 10-5 **A picket-fence Fe(II)–porphyrin complex with bound O_2.** [After Collman, J.P., Brauman, J.I., Rose, E., and Suslick, K.S., *Proc. Natl. Acad. Sci.* **75**, 1053 (1978).]

FIGURE 10-6 **Effect of pH on the O_2-dissociation curve of Hb: the Bohr effect.** The vertical dashed line indicates the pO_2 in actively respiring muscle tissue. [After Benesch, R.E. and Benesch, R., *Adv. Protein Chem.* **28**, 212 (1974).] See the Animated Figures.

Hb function to prevent the autooxidation of oxyheme by surrounding it, rather like a hamburger bun surrounds a hamburger, so that only its propionate side chains are exposed to the aqueous solvent (Section 10-2B).

C. *Carbon Dioxide Transport and the Bohr Effect*

In addition to being an O_2 carrier, *Hb plays an important role in the transport of CO_2 by the blood.* When Hb (but not Mb) binds O_2 at physiological pH's, it undergoes a conformational change (Section 10-2B) that makes it a slightly stronger acid. It therefore releases protons on binding O_2:

$$Hb(O_2)_nH_x + O_2 \rightleftharpoons Hb(O_2)_{n+1} + xH^+$$

where $n = 0, 1, 2,$ or 3 and $x \approx 0.6$ under physiological conditions. Conversely, *increasing the pH, that is, removing protons, stimulates Hb to bind O_2* (Fig. 10-6). This phenomenon, whose molecular basis is discussed in Section 10-2E, is known as the **Bohr effect** after Christian Bohr (the father of the pioneering atomic physicist Niels Bohr), who first reported it in 1904.

a. The Bohr Effect Facilitates O_2 Transport

The ~0.8 molecules of CO_2 formed per molecule of O_2 consumed by respiration diffuse from the tissues to the capillaries largely as dissolved CO_2 as a result of the slowness of the reaction forming bicarbonate:

$$CO_2 + H_2O \rightleftharpoons H^+ + HCO_3^-$$

This reaction, however, is catalyzed in the erythrocyte by carbonic anhydrase (Fig. 8-41). Accordingly, most of the CO_2 in the blood is carried in the form of bicarbonate (in the absence of carbonic anhydrase, the hydration of CO_2 would

equilibrate 100-fold more slowly, so bubbles of the only slightly soluble CO_2 would form in the blood and tissues).

In the capillaries, where pO_2 is low, the H^+ generated by bicarbonate formation is taken up by Hb, which is thereby induced to unload its bound O_2. This H^+ uptake, moreover, facilitates CO_2 transport by stimulating bicarbonate formation. Conversely, in the lungs, where pO_2 is high, O_2 binding by Hb releases the Bohr protons, which drive off the CO_2. These reactions are closely matched, so they cause very little change in blood pH.

The Bohr effect provides a mechanism whereby additional O_2 can be supplied to highly active muscles. Such muscles generate acid (Section 17-3A) so fast that they lower the pH of the blood passing through them from 7.4 to 7.2. At pH 7.2, Hb releases ~10% more O_2 at the <20 torr pO_2 in these muscles than it does at pH 7.4 (Fig. 10-6).

b. CO_2 and Cl^- Modulate Hemoglobin's O_2 Affinity

CO_2 modulates O_2 binding directly and by combining reversibly with the N-terminal amino groups of blood proteins to form **carbamates:**

$$R-NH_2 + CO_2 \rightleftharpoons R-NH-COO^- + H^+$$

The conformation of deoxygenated Hb **(deoxyHb),** as we shall see in Section 10-2B, is significantly different from that of oxygenated Hb **(oxyHb).** Consequently, deoxyHb binds more CO_2 as carbamate than does oxyHb. CO_2, like H^+, is therefore a modulator of hemoglobin's O_2 affinity: A high CO_2 concentration, as occurs in the capillaries, stimulates Hb to release its bound O_2. Note the complexity of this Hb–O_2–CO_2–H^+ equilibrium: The protons released by carbamate formation are, in part, taken up through the Bohr effect, thereby increasing the amount of O_2 that Hb would otherwise release. Although the difference in CO_2 binding between the oxy and deoxy states of hemoglobin accounts for only ~5% of the total blood CO_2, it is nevertheless responsible for around half of the CO_2 transported by blood. This is because only ~10% of the total blood CO_2 turns over in each circulatory cycle.

Cl^- is also bound more tightly to deoxyHb than to oxyHb (Section 10-2E). Accordingly, hemoglobin's O_2 affinity also varies with $[Cl^-]$. HCO_3^- freely permeates the erythrocyte membrane (Section 12-3D), so that once formed, it equilibrates with the surrounding plasma. The need for charge neutrality on both sides of the membrane, however, requires that Cl^-, which also freely permeates the membrane, replace the HCO_3^- that leaves the erythrocyte (the erythrocyte membrane is impermeable to cations). Consequently, $[Cl^-]$ in the erythrocyte is greater in the venous blood than it is in the arterial blood. *Cl^- is therefore also a modulator of hemoglobin's O_2 affinity.*

D. *Effect of BPG on O_2 Binding*

Purified (stripped) hemoglobin has a much greater O_2 affinity than does hemoglobin in whole blood (Fig. 10-7). This observation led Joseph Barcroft, in 1921, to speculate that blood contains some other substance that complexes with Hb so as to reduce its O_2 affinity. In 1967, Reinhold and Ruth Benesch demonstrated that this substance is **D-2,3-bisphosphoglycerate (BPG)**

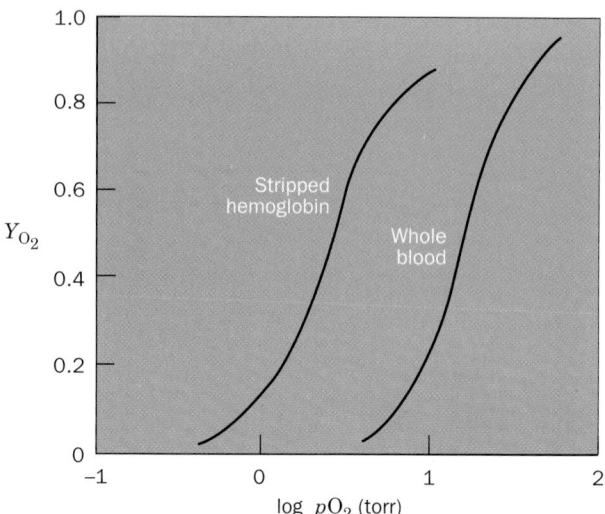

D-2,3-Bisphosphoglycerate (BPG)

[previously known as **2,3-diphosphoglycerate (DPG)**]. BPG binds tightly to deoxyHb in a 1:1 mole ratio $(K = 1.5 \times 10^{-5}M)$ but only weakly to oxyHb. The presence of BPG therefore decreases hemoglobin's oxygen affinity by keeping it in the deoxy conformation; for example, the p_{50} of stripped hemoglobin is increased from 12 to 22 torr by 4.7 mM BPG, its normal concentration in erythrocytes (similar to that of Hb). Organic polyphosphates, such as **inositol hexaphosphate (IHP)**

Inositol hexaphosphate (IHP)

and ATP, also have this effect on Hb. In fact, in birds, IHP functionally replaces BPG and ATP does so in fish and in most amphibians. The ~2 mM ATP normally present in mammalian erythrocytes is prevented from binding to Hb by its complexation with Mg^{2+}.

FIGURE 10-7   Comparison of the O_2-dissociation curves of "stripped" Hb and whole blood in 0.01M NaCl at pH 7.0. [After Benesch, R.E. and Benesch, R., *Adv. Protein Chem.* **28,** 217 (1974).]

BPG has an indispensable physiological function: In arterial blood, where pO_2 is ~100 torr, Hb is ~95% saturated with O_2, but in venous blood, where pO_2 is ~30 torr, it is only ~55% saturated (Fig. 10-3). Consequently, in passing through the capillaries, Hb unloads ~40% of its O_2. *In the absence of BPG, little of this O_2 is released since hemoglobin's O_2 affinity is increased, thus shifting the O_2-dissociation curve significantly toward lower pO_2 (Fig. 10-8, left).*

CO_2 and BPG independently modulate hemoglobin's O_2 affinity. Figure 10-8 indicates that stripped Hb can be made to have the same oxygen-dissociation curve as the Hb in whole blood by adding CO_2 and BPG in the concentrations found in erythrocytes (the pH and $[Cl^-]$ are also the same). Hence, *the presence of these four substances in whole blood—BPG, CO_2, H^+, and Cl^-—accounts for the O_2-binding properties of Hb.*

a. Increased BPG Levels Are Partially Responsible for High-Altitude Adaptation

High-altitude adaptation is a complex physiological process that involves an increase in the amount of hemoglobin per erythrocyte and in the number of erythrocytes. It normally requires several weeks to complete. Yet, as is clear to anyone who has quickly climbed to high altitude, even a 1-day stay there results in a noticeable degree of adaptation. This effect results from a rapid increase in the erythrocyte BPG concentration (Fig. 10-9; BPG, which cannot pass through the erythrocyte membrane, is synthesized in the erythrocyte; Section 17-2H). The consequent decrease in O_2-binding affinity, as indicated by its elevated p_{50}, increases the amount of O_2 that hemoglobin unloads in the capillaries (Fig. 10-10). Similar increases in BPG concentration occur in individuals suffering from disorders that limit the oxygenation of the blood **(hypoxia),** such as various anemias and cardiopulmonary insufficiency.

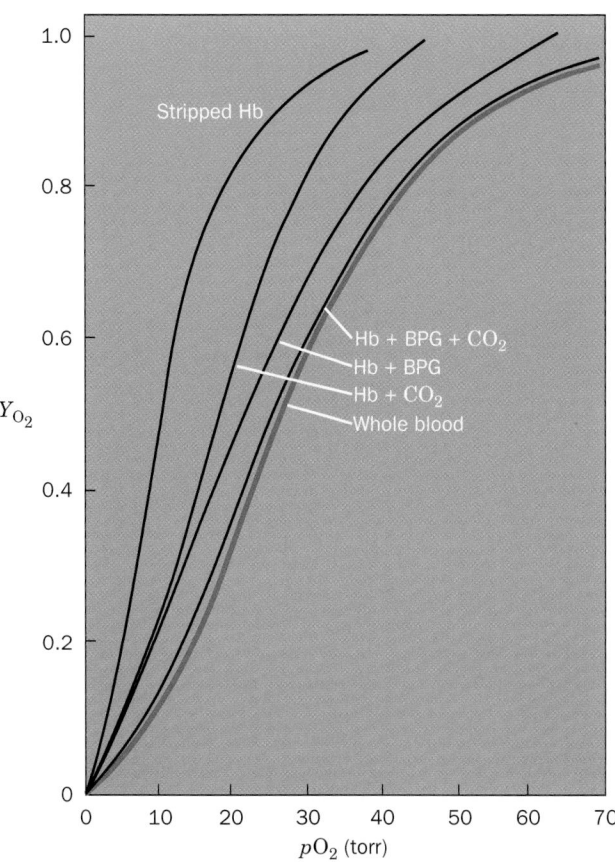

FIGURE 10-8   The effects of BPG and CO_2, both separately and combined, on hemoglobin's O_2-dissociation curve compared with that of whole blood (*red curve*). In the Hb solutions, which were $0.1 M$ KCl and pH 7.22, $pCO_2 = 40$ torr and the BPG concentration was 1.2 times that of Hb. The blood had $pCO_2 = 40$ torr and a plasma pH of 7.40, which corresponds to a pH of 7.22 inside the erythrocyte. [After Kilmartin, J.V. and Rossi-Bernardi, L., *Physiol. Rev.* **53**, 884 (1973).] 🔊 **See the Animated Figures.**

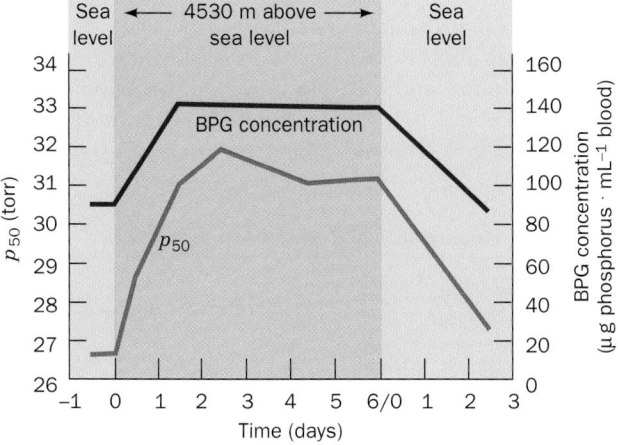

FIGURE 10-9   The effect of high-altitude exposure on the p_{50} and the BPG concentration of blood in sea level–adapted individuals. The region on the right marked "Sea level" indicates the effects of exposure to sea level on high altitude–adapted individuals. [After Lenfant, C., Torrance, J.D., English, E., Finch, C.A., Reynafarje, C., Ramos, J., and Faura, J., *J. Clin. Invest.* **47**, 2653 (1968).]

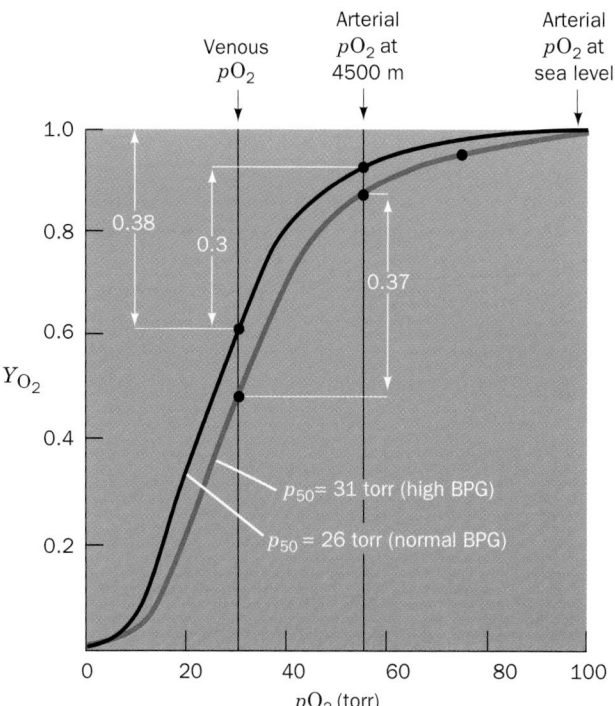

FIGURE 10-10   The O_2-dissociation curves of blood adapted to sea level (*black curve*) and to high altitude (*red curve*). Between the sea level arterial and venous pO_2 values of 100 and 30 torr, respectively, Hb normally unloads 38% of the O_2 it can maximally carry. However, when the arterial pO_2 drops to 55 torr, as it does at an altitude of 4500 m, this difference is reduced to 30% in nonadapted blood. High-altitude adaptation increases the BPG concentration in erythrocytes, which shifts the O_2-dissociation curve of Hb to the right. The amount of O_2 that Hb delivers to the tissues is thereby restored to 37% of its maximum load.

b. Fetal Hemoglobin Has a Low BPG Affinity

The effects of BPG also help supply the fetus with oxygen. A fetus obtains its O_2 from the maternal circulation via the placenta. This process is facilitated because fetal hemoglobin (HbF) has a higher O_2 affinity than does maternal hemoglobin (HbA; recall that HbF has the subunit composition $\alpha_2\gamma_2$, in which the γ subunit is a variant of HbA's β subunit; Section 7-3C). BPG occurs in about the same concentrations in adult and fetal erythrocytes but binds more tightly to deoxyHbA than to deoxyHbF; this accounts for HbF's greater O_2 affinity. In the next section we shall develop the structural rationale for the effect of BPG and for the other aspects of O_2 binding.

2 ■ STRUCTURE AND MECHANISM

The determination of the first protein X-ray structures, those of sperm whale myoglobin by John Kendrew in 1959 and of human deoxyhemoglobin and horse methemoglobin by Max Perutz shortly thereafter, ushered in a revolution in biochemical thinking that has reshaped our understanding of the chemistry of life. Before the advent of protein crystallography, macromolecular structures, if they were considered at all, were thought of as having a rather hazy existence of uncertain biological significance. However, as the elucidation of macromolecular structures has continued at an ever quickening pace, it has become clear that *life is based on the interactions of complex, structurally well-defined macromolecules.*

The story of hemoglobin's structural determination is a tale of enormous optimism and tenacity. Perutz began this study in 1937 at Cambridge University as a graduate student of J. D. Bernal (who, with Dorothy Crowfoot Hodgkin, had taken the first X-ray diffraction photographs of hydrated protein crystals in 1934). In 1937, the X-ray crystal structure determination of even the smallest molecule required many months of hand computation, and the largest structure yet determined was that of the dye phthalocyanin, which has 40 nonhydrogen atoms. Since hemoglobin has ~4500 nonhydrogen atoms, it must have seemed to Perutz's colleagues that he was pursuing an impossible goal. Nevertheless, the laboratory director, Lawrence Bragg (who in 1912, with his father William Bragg, had determined the first X-ray structure, that of NaCl), realized the tremendous biological significance of determining a protein structure and supported the project.

It was not until 1953 that Perutz finally hit on the method that would permit him to solve the X-ray structure of hemoglobin, that of isomorphous replacement. Kendrew, a colleague of Perutz, used this technique to solve the X-ray structure of sperm whale myoglobin, first at low resolution in 1957, and then at high resolution in 1959. Hemoglobin's greater complexity delayed its low-resolution structural determination until 1959, and it was not until 1968, over 30 years after he had begun the project, that Perutz and his associates obtained the high-resolution X-ray structure of horse methemoglobin. Those of human and horse deoxyhemoglobins followed shortly thereafter. Since then, the X-ray structures of hemoglobins from numerous different species, from mutational variants, and with different bound ligands have been elucidated. This, together with many often ingenious physicochemical investigations, has made hemoglobin the most intensively studied, and perhaps the best understood, of proteins.

In this section, we examine the molecular structures of myoglobin and hemoglobin and consider the structural ba-

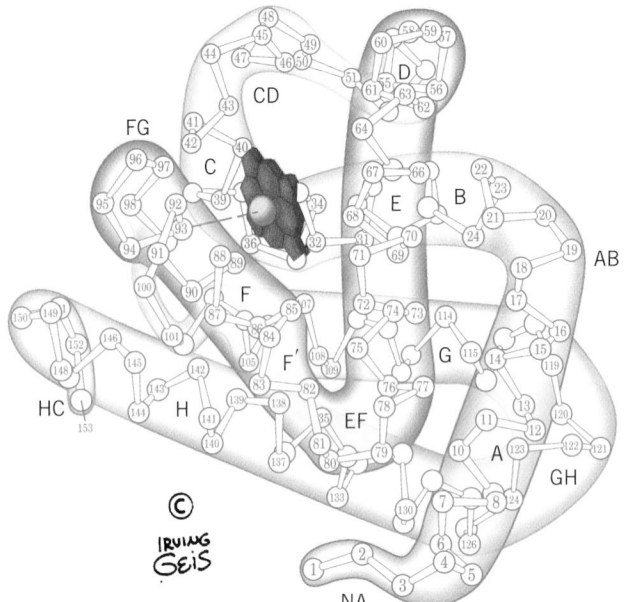

FIGURE 10-11   Structure of sperm whale myoglobin. Its 153 C_α positions are numbered from the N-terminus and its eight helices are sequentially labeled A through H. The last half of the EF corner is now regarded as a turn of helix and is therefore designated the F′ helix. The heme group is shown in red. Also see Fig. 8-39. [Illustration, Irving Geis/Geis Archives Trust. Copyright Howard Hughes Medical Institute. Reproduced with permission. Based on an X-ray structure by John Kendrew, MRC Laboratory of Molecular Biology, Cambridge, U.K. PDBid 1MBN.] *See Kinemage Exercise 6-1*

sis of hemoglobin's oxygen-binding cooperativity, the Bohr effect, and BPG binding.

A. *Structure of Myoglobin*

Myoglobin consists of eight helices (labeled A–H) that are linked by short polypeptide segments to form an ellipsoidal molecule of approximate dimensions 44 × 44 × 25 Å (Fig. 10-11; see also Fig. 8-39). The helices range in length from 7 to 26 residues and incorporate 121 of myoglobin's 153 residues (Table 10-1). They are largely α helical but with some distortions from this geometry such as a tightening

TABLE 10-1   The Amino Acid Sequences of the α and β Chains of Human Hemoglobin and of Human Myoglobin[a,b]

Helix Boundaries	A1		A16 B1		B16 C1	C7	D1	D7 E1	
Hb α	V-LSPADKTNVKAAWGKVG	AHAGEYGAEALERMFLS	FPTTKTYFPHF-DLSH-----G	SAQVKGHGKKVADALT					
Hb β	VHLTPEEKSAVTALWGKV--NVDEVGGEALGRLLVVYPWT	QRFFESFGDLSTPDAVMG	NPKVKAHGKKVLGAFS						
Mb	G-LSDGEWQLVLNVWGKVEADIPGHGQE	VLIRLFKGHPETLEKFDKFKHLKSEDEMKA	SEDLKKHGATVLTALG						

[a]The residues have been aligned in structurally analogous positions. The blue boxes shade the residues that are identical in both Hb chains, the purple boxes shade the residues that are identical in both Hb chains and in Mb, and the dark purple boxes shade residues that are invariant in all vertebrate Hb and Mb chains (Thr C4, Phe CD1, Leu F4, His F8, and Tyr HC2).

[b]The first and last residues in helices A–H are indicated, whereas the residues between helices constitute the intervening "segments." The refined Hb structure reveals that much of what is designated the EF segment is really helical in both chains: It encompasses residues EF4–F2 and is designated the F′ helix.

FIGURE 10-12 Stereo drawings of the heme complex in oxyMb. In the upper drawing, atoms are represented as spheres of van der Waals radii. The lower drawing shows the corresponding skeletal model with a dashed line representing the hydrogen bond between the distal His and the bound O_2. Instructions for viewing stereo drawings are given in the Appendix to Chapter 8. [After Phillips, S.E.V., *J. Mol. Biol.* **142**, 544 (1980). PDBid 1MBO.] ⚛ **See Kinemage Exercise 6-1**

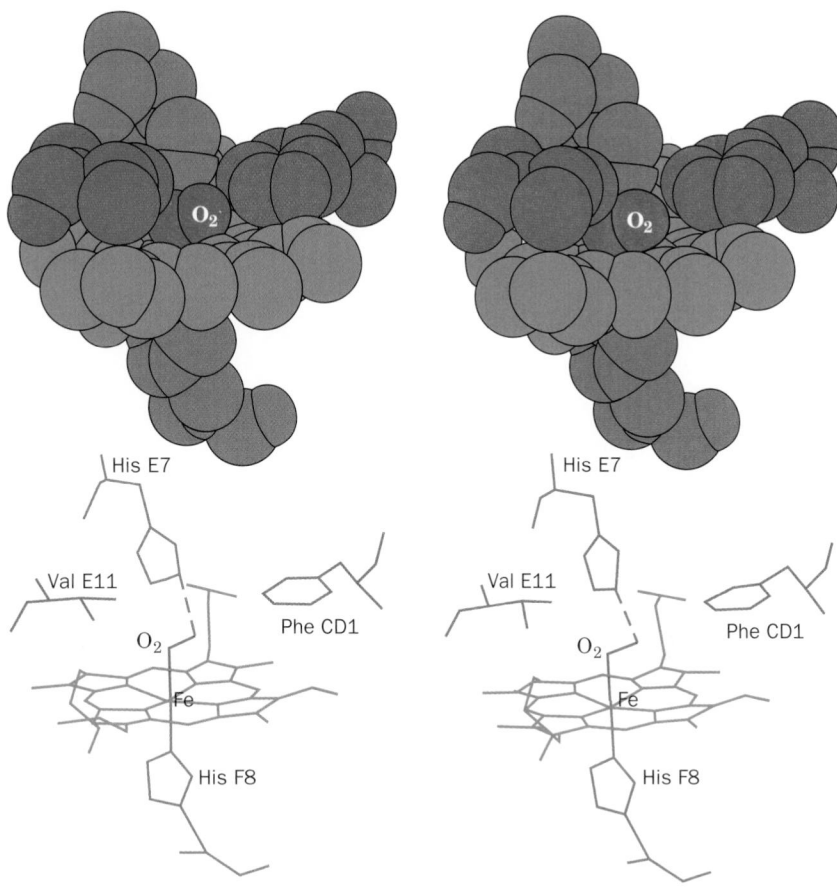

of the final turns of helices A, C, E, and G to form segments of 3_{10} helix.

In a helix numbering convention peculiar to Mb and Hb, residues are designated according to their position in a helix or interhelical segment. For example, residue B5 is the fifth residue from the N-terminus of the B helix and residue FG3 is the third residue from the N-terminus in the nonhelical segment connecting helices F and G. The nonhelical N- and C-terminal segments are designated NA and HC, respectively. The usual convention of sequentially numbering all amino acid residues from the N-terminal residue of the polypeptide is also used, and often both conventions

are used together. For example, Glu EF7(83) of human Mb is the 83rd residue from its N-terminus and the 7th residue in the nonhelical segment connecting its E and F helices.

The heme is tightly wedged in a hydrophobic pocket formed mainly by helices E and F but which includes contacts with helices B, C, G, and H as well as the CD and FG segments. The fifth ligand of the heme Fe(II) is His F8, the **proximal** (near) **histidine.** In oxyMb, the Fe(II) is positioned 0.22 Å out of the heme plane on the side of the proximal His and is coordinated by O_2 with the bent geometry shown in Fig. 10-12. His E7, the **distal** (distant) **histidine,** hydrogen bonds to the O_2. In deoxyMb, the sixth

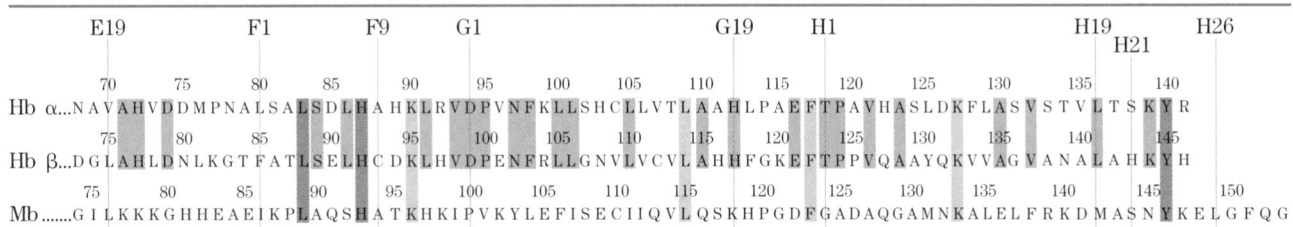

Source: Dickerson, R.E. and Geis, I., *Hemoglobin*, pp. 68–69, Benjamin/Cummings (1983).

(*a*)

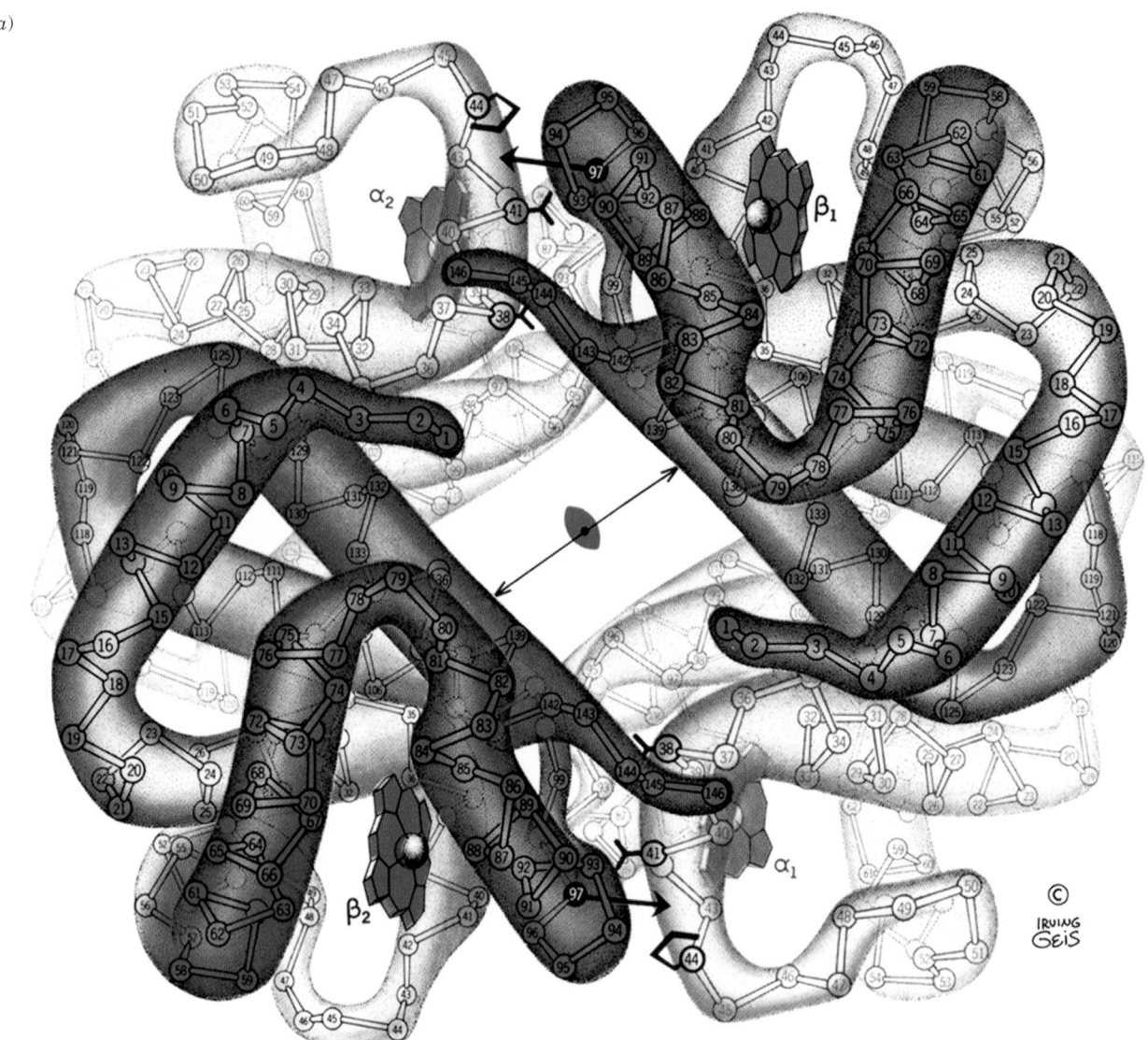

FIGURE 10-13   The X-ray structures of (*a*) deoxyHb and (*b*) oxyHb as viewed down their exact 2-fold axes. The C_α atoms, numbered from each N-terminus, and the heme groups are shown. The Hb tetramer contains a solvent-filled central channel paralleling its 2-fold axis, whose flanking β chains draw closer together on oxygenation (compare the lengths of the double-headed arrows). In the deoxy state, His FG4(97)β (*small single-headed arrow*) fits between Thr C6(41)α and Pro CD2(44)α (*lower right and upper left*). The relative movements of the two αβ protomers on oxygenation (*large gray arrows in Part b*) shift His FG4(97)β to a new position between Thr C3(38)α and Thr C6(41)α. See Fig. 8-63 for a similarly viewed space-filling model of deoxyHb. [Illustration, Irving Geis/Geis Archives Trust. Copyright Howard Hughes Medical Institute. Reproduced with permission. Based on X-ray structures by Max Perutz, MRC Laboratory of Molecular Biology, Cambridge, U.K. PDBids (*a*) 2DHB and (*b*) 2MHB.]
📁 **See Kinemage Exercises 6-2 and 6-3**

liganding position of the Fe(II) is unoccupied because the distal His is too far away from the Fe(II) to coordinate with it. Furthermore, the Fe(II) has moved to a point 0.55 Å out of the heme plane. Other structural changes in Mb on changing oxygenation states consist of small motions of various chain segments and slight readjustments of side chain conformations. *By and large, however, the structures of oxy- and deoxyMb are nearly superimposable.*

B. *Structure of Hemoglobin*

The hemoglobin tetramer is a spheroidal molecule of dimensions 64 × 55 × 50 Å. Its two αβ protomers are symmetrically related by a twofold rotation (Fig. 10-13; see also Fig. 8-63). *The tertiary structures of the α and β subunits are remarkably similar, both to each other and to that of Mb (Figs. 10-11 and 10-13), even though only 18% of the*

(b)

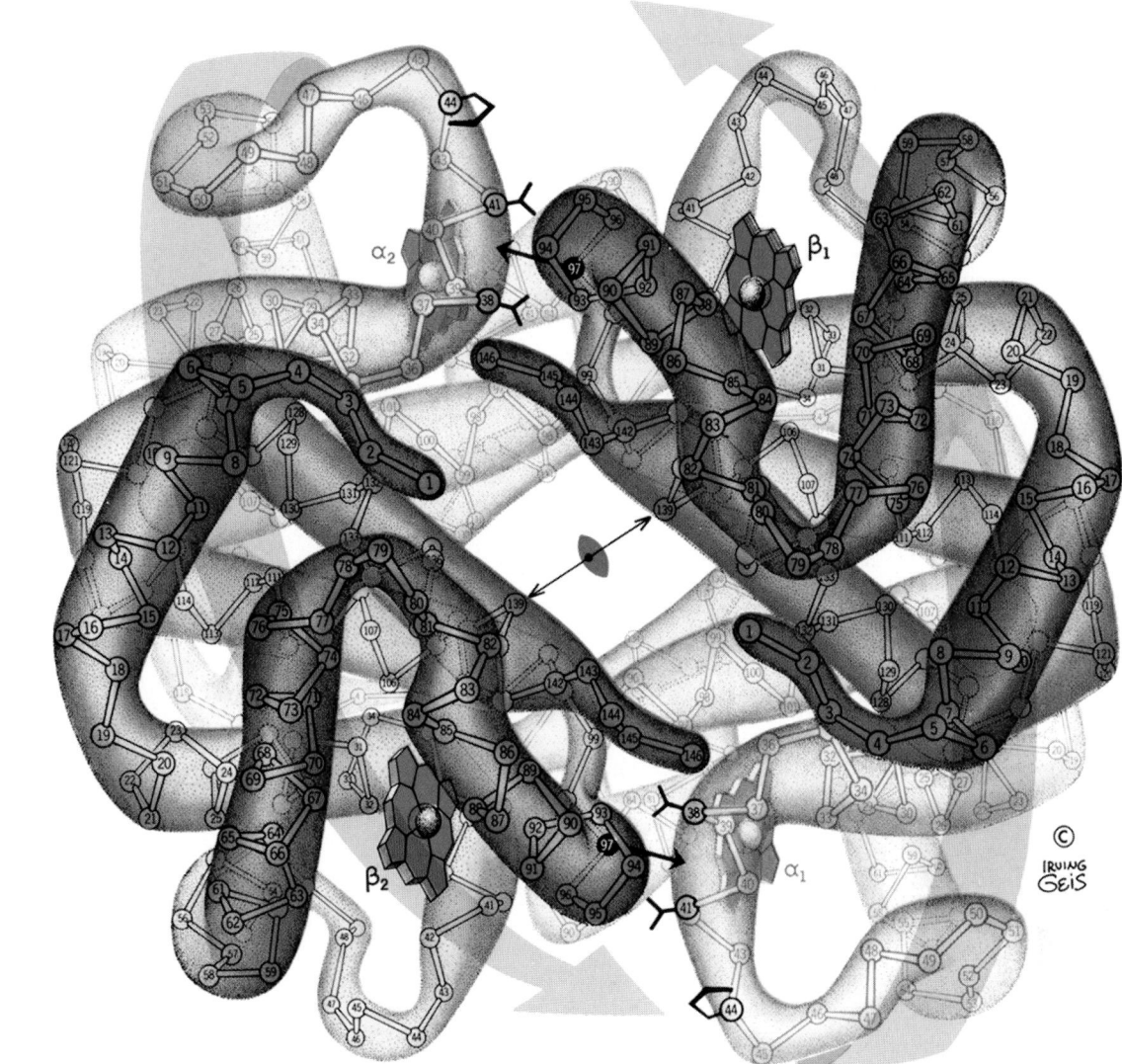

FIGURE 10-13 (*continued*)

corresponding residues are identical among these three polypeptides (Table 10-1) and there is no D helix in hemoglobin's α subunit. Indeed, *the α and β subunits in the tetramer are related by pseudo- (inexact) 2-fold rotations so that the subunits occupy the vertices of a tetrahedron (pseudo-D_2 symmetry; Section 8-5B).*

The polypeptide chains of Hb are arranged such that there are extensive interactions between unlike subunits. The α_1–β_1 interface (and its α_2–β_2 symmetry equivalent) involves 35 residues, whereas the α_1–β_2 (and α_2–β_1) interface involves 19 residues. These associations are predominantly hydrophobic in character, although numerous hydrogen bonds and several ion pairs are also involved (Section 10-2C). In contrast, contacts between like subunits, α_1–α_2 and β_1–β_2, are few and largely polar in character. This is because like subunits face each other across an ~20-Å-diameter solvent-filled channel that par-

allels the 50-Å length of the exact 2-fold axis (Figs. 8-63 and 10-13).

a. Oxy- and Deoxyhemoglobins Have Different Quaternary Structures

Oxygenation causes such extensive quaternary structural changes to Hb that oxy- and deoxyHb have different crystalline forms; indeed, crystals of deoxyHb shatter on exposure to O_2. The crystal structures of hemoglobin's oxy and deoxy forms therefore had to be determined independently. *The quaternary structural change preserves hemoglobin's exact 2-fold symmetry and takes place entirely across its α_1–β_2 (and α_2–β_1) interface. The α_1–β_1 (and α_2–β_2) contact is unchanged,* presumably as a result of its more extensive close associations. This contact provides a convenient frame of reference from which the oxy and deoxy conformations may be compared. Viewed in this way, oxygenation rotates the $\alpha_1\beta_1$ dimer ~15° with respect to the $\alpha_2\beta_2$ dimer (Fig. 10-14), so that some atoms at the α_1–β_2 interface shift by as much as 6 Å relative to each other (compare Fig. 10-13a and b).

The quaternary conformation of deoxyHb is named the **T state** (T for "tense"). That of oxyHb, which is essentially independent of the ligand used to induce it (e.g., O_2, met, CO, CN^-, and NO hemoglobins all have the same quaternary structure), is called the **R state** (R for "relaxed"). Similarly, the tertiary conformational states for the deoxy and liganded subunits are designated as the **t** and **r states,** respectively. The structural differences between the quaternary and tertiary conformations are described in the following subsection in terms of hemoglobin's O_2-binding mechanism.

C. *Mechanism of Oxygen-Binding Cooperativity*

The positive cooperativity of O_2 binding to Hb arises from the effect of the ligand-binding state of one heme on the ligand-binding affinity of another. Yet the distances of 25 to 37 Å between the hemes in an Hb molecule are too large for these heme–heme interactions to be electronic in character. Rather, *they are mechanically transmitted by the protein.* The elucidation of how this occurs has motivated much of the structural research on Hb for the past three decades.

X-Ray crystal structure analysis has provided "snapshots" of the R and T states of Hb in various states of ligation but does not indicate how the protein changes states. It is difficult to determine the sequence of events that result in such transformations because to do so requires an understanding of the inner workings of proteins that is presently lacking. It is as if you were asked to explain the mechanism of a complicated mechanical clock from its out-of-focus photographs when you had only a hazy notion of how gears, levers, and springs might function. Nevertheless, largely on the basis of the X-ray structures of Hb, Perutz formulated the following mechanism of Hb oxygenation, the **Perutz mechanism.**

a. The Movement of Fe(II) into the Heme Plane Triggers the T→R Conformational Shift

In the t state, the Fe(II) is situated ~0.6 Å out of the heme plane on the side of the proximal His because of a pyramidal doming of the porphyrin skeleton and because the Fe—$N_{porphyrin}$ bonds are too long to allow the Fe to lie in the porphyrin plane (Figs. 10-15 and 10-16). The change in the heme's electronic state on binding O_2, however, causes the doming to subside and the Fe—$N_{porphyrin}$ bonds to contract by ~0.1 Å. Consequently, on changing from the t to the r state, the Fe(II) moves to the center of the heme plane (Fig. 10-16) where O_2 can coordinate it without steric interference from the porphyrin. The Fe's movement drags the proximal His along with it, which tilts the attached F helix and translates it ~1 Å across the heme plane (Fig. 10-16). This lateral translation occurs because, in the t state, the imidazole ring of the proximal His is oriented such that its direct movement of ~0.6 Å toward the heme

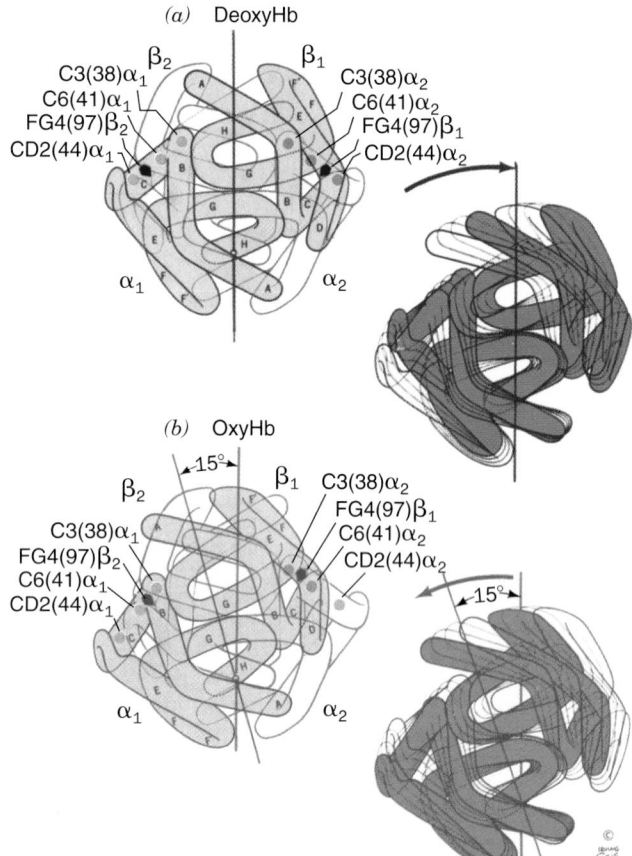

FIGURE 10-14   The major structural differences between the quaternary conformations of (a) deoxyHb and (b) oxyHb. On oxygenation, the $\alpha_1\beta_1$ (*shaded*) and $\alpha_2\beta_2$ (*outline*) dimers move, as indicated on the right, as rigid units such that there is an ~15° off-center rotation of one protomer relative to the other that preserves the molecule's exact 2-fold symmetry. Note how the position of His FG4β (*pentagons*) changes with respect to Thr C3α, Thr C6α, and Pro CD2α (*yellow dots*) at the α_1–β_2 and α_2–β_1 interfaces. The view is from the right side relative to that in Fig. 10-13. [Illustration, Irving Geis/Geis Archives Trust. Copyright Howard Hughes Medical Institute. Reproduced with permission.]

plane would cause it to collide with the heme (Figs. 10-15 and 10-16); however, the F helix shift reorients the imidazole ring, thereby permitting the Fe(II) to move into the heme plane. In addition, in the t state of the β but not the α subunits, Val E11 partially occludes the O_2-binding pocket so that it must be moved aside before O_2 binding can occur.

b. The α_1–β_2 and α_2–β_1 Contacts Have Two Stable Positions

As we saw above, the difference between hemoglobin's R and T conformations occurs mainly in the α_1–β_2 (and the symmetrically related α_2–β_1) interface, which consists of the C helix and FG segment of α_1, respectively, contacting the FG segment and C helix of β_2. The quaternary change results in a 6-Å relative shift at the α_1C–β_2FG interface (Fig. 10-14). In the T state, His FG4(97)β is in contact with Thr C6(41)α (Fig. 10-13a and 10-17a), whereas in the R

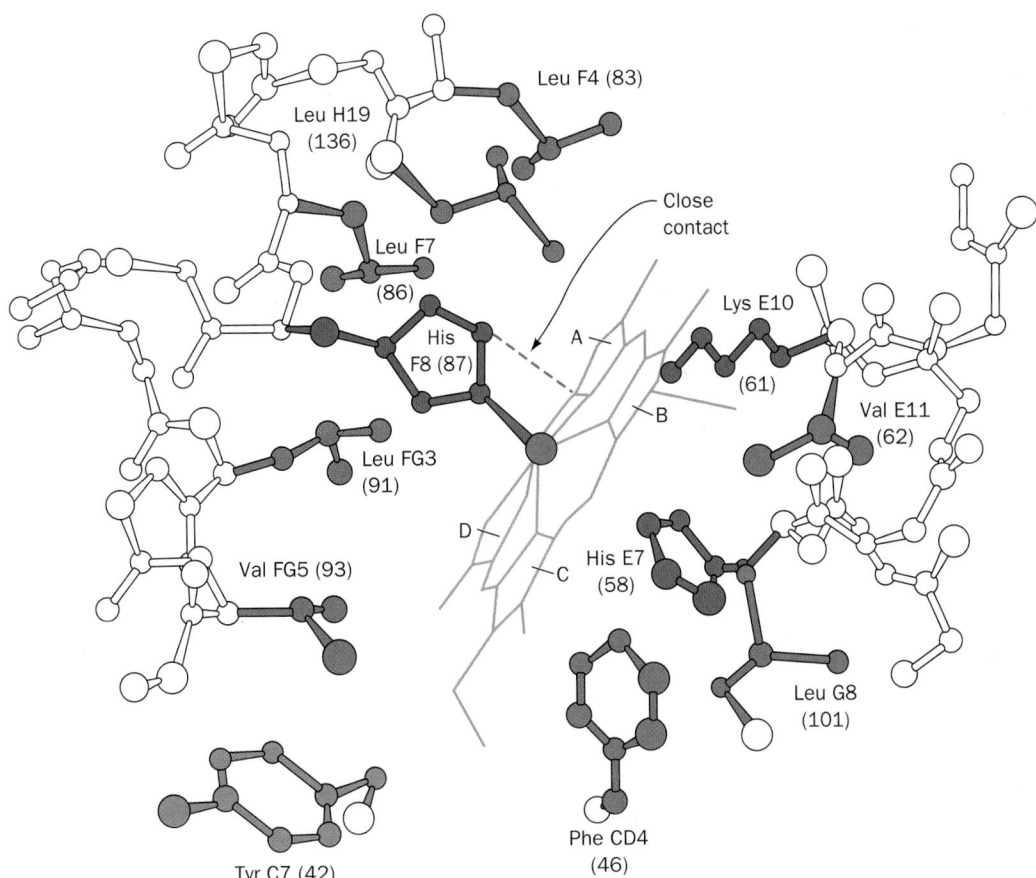

FIGURE 10-15 **The heme group and its environment in the unliganded α chain of human Hb.** Only selected side chains are shown and the heme D propionate group is omitted for clarity. The F helix runs along the left side of the drawing. The close contact between the proximal His and the heme group that inhibits oxygenation of t-state hemes is indicated by a dashed red line. [After Gelin, B.R., Lee, A.W.N., and Karplus, M., *J. Mol. Biol.* **171,** 542 (1983). PDBid 2HHB.] *See Kinemage* **Exercise 6-4**

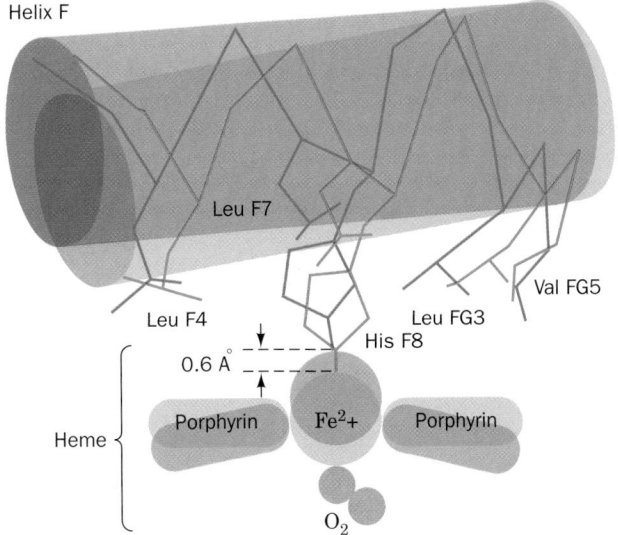

FIGURE 10-16 **Triggering mechanism for the T→R transition in Hb.** In the T form (*blue*), the Fe is ~0.6 Å above the mean plane of the domed porphyrin ring. On assuming the R form (*red*), the Fe moves into the plane of the now undomed porphyrin, where it can readily bind O_2 and, in doing so, pulls the proximal His F8 and its attached F helix with it. The Fe—O_2 bond is thereby strengthened because of the relaxation of the steric interference between the O_2 and the heme. *See* **Kinemage Exercise 6-4 and the animated figures**

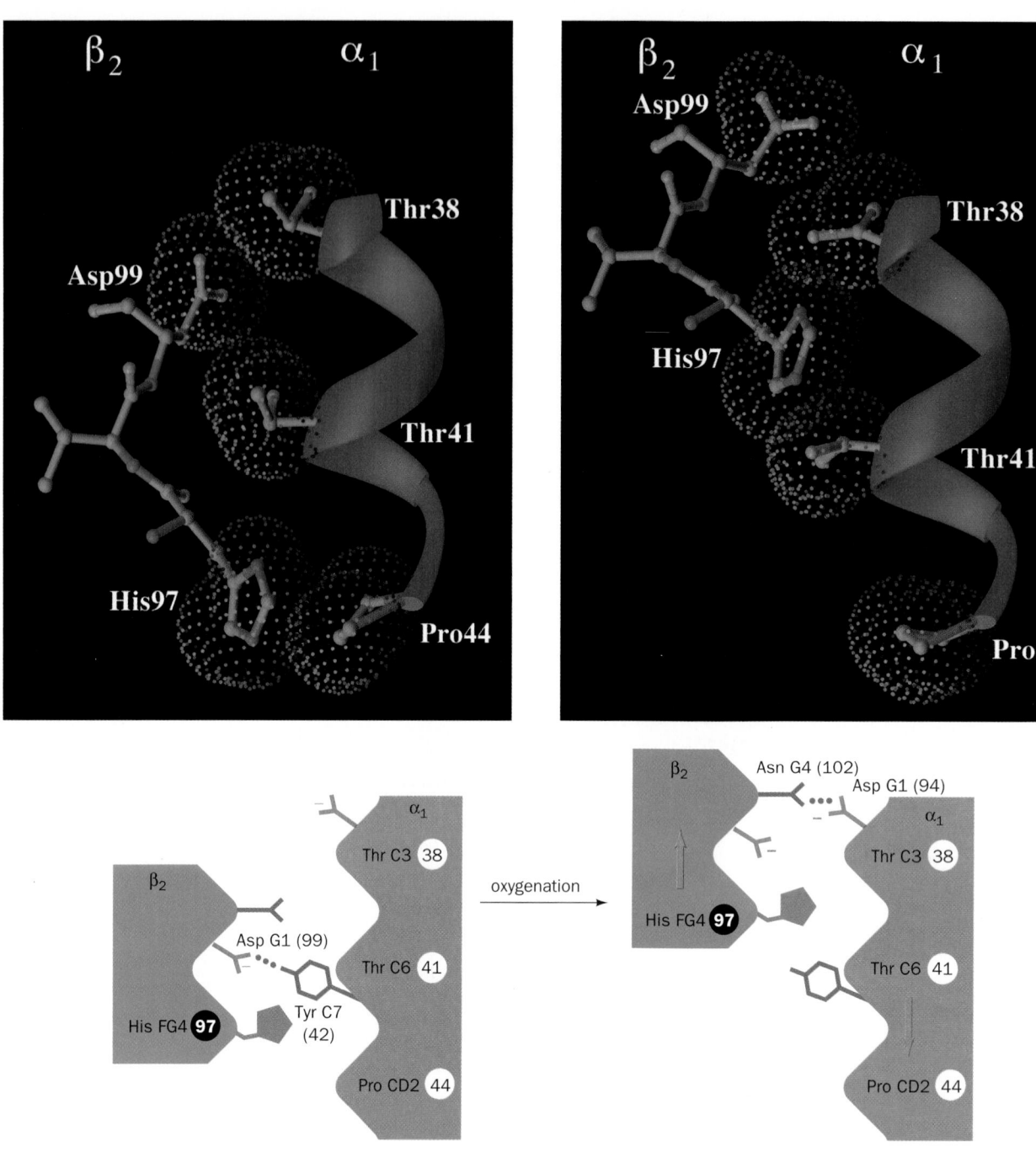

(a) **T State (deoxy)**

oxygenation

(b) **R State (oxy)**

FIGURE 10-17 The α₁C–β₂FG interface of Hb in (a) the T state and (b) the R state. The upper drawings show the C helix in ribbon form (*purple*) and its contacting portion of the FG region in ball-and-stick form colored according to atom type (C green, N blue, and O red). The dots outline the contacting van der Waals surfaces and are also colored according to atom type. The lower drawings are the corresponding schematic diagrams of the α₁C–β₂FG contact. On a T→R transformation, this contact snaps from one position to the other with no stable intermediate (note how, in both conformations, the knobs

formed by the side chains of His 97β and Asp 99β fit between the grooves on the C helix formed by the side chains of Thr 38α, Thr 41α, and Pro 44α). The subunits are joined by different hydrogen bonds in the two quarternary states. Figures 10-13, 10-14, and 10-18 provide additional structural views of these interactions. [Based on X-ray structures by Giulio Fermi, Max Perutz, and Boaz Shaanan, MRC Laboratory of Molecular Biology, Cambridge, U.K. PDBids (a) 2HHB and (b) 1HHO.] See Kinemage Exercise 6-5

state it is in contact with Thr C3(38) α, one turn back along the C helix (Figs. 10-13b and 10-17b). In both conformations, the "knobs" on one subunit mesh nicely with the

"grooves" on the other (Fig. 10-17). An intermediate position, however, would be severely strained because it would bring His FG4(97)β and Thr C6(41)α too close to-

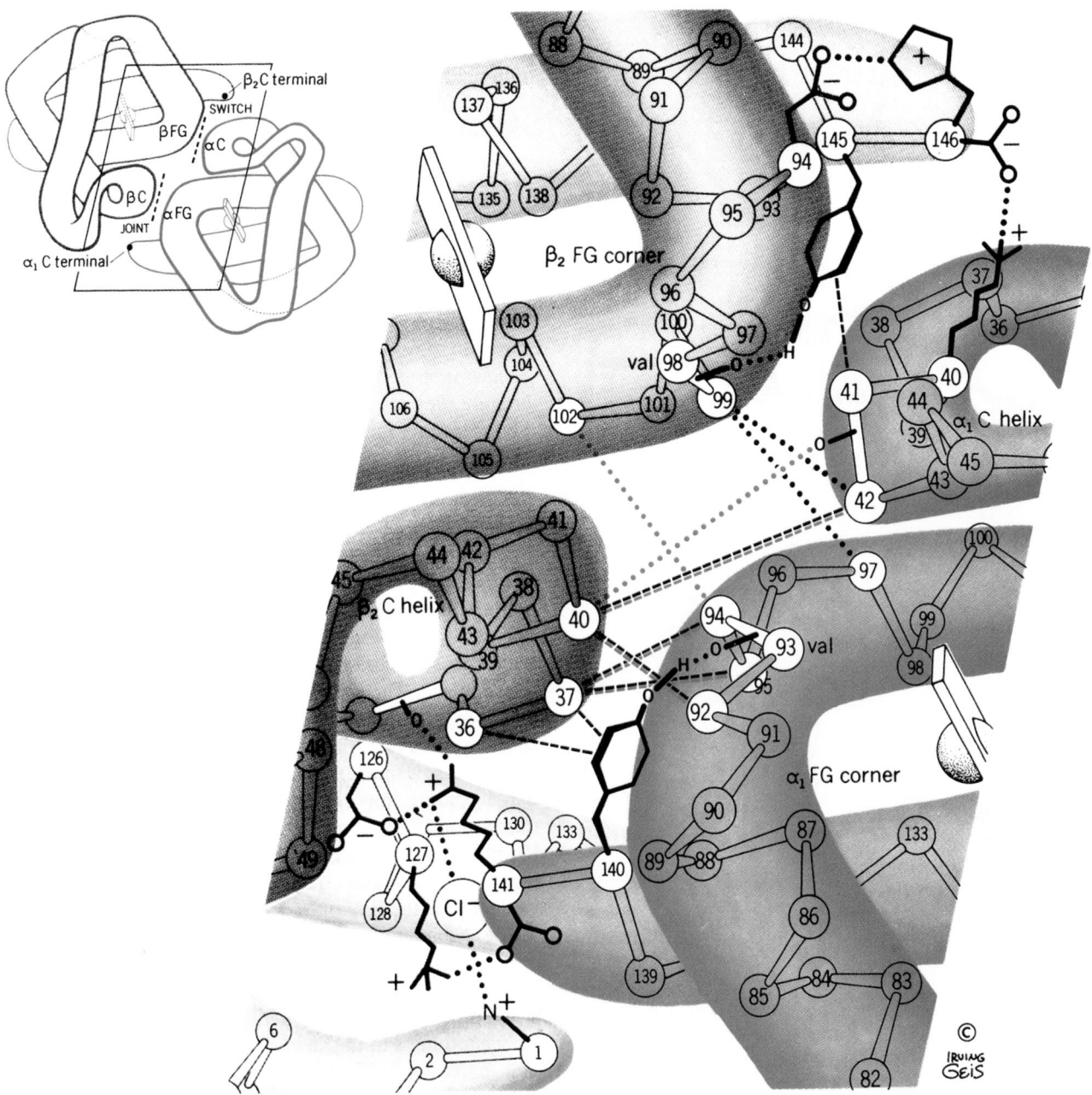

FIGURE 10-18 **The hemoglobin $\alpha_1\beta_2$ interface as viewed perpendicularly to Fig. 10-13.** The boxed area on the left is shown in greater detail on the right. Hydrogen bonds and salt bridges are represented by dotted lines, black for deoxyHb and blue for oxyHb, whereas van der Waals contacts are likewise indicated by dashed lines. Note that the α_1C–β_2FG interface (the "switch" region) undergoes significant readjustment in the T→R transition, whereas the pseudosymmetrically related α_1FG–β_2C interface (the "flexible joint") only undergoes small reorientations. Also note that the T-state salt bridges involving the C-terminal residues [Arg 141α (*below*) and His 146β (*above*)] are ruptured by the T→R transition. [Illustration, Irving Geis/Geis Archives Trust. Copyright Howard Hughes Medical Institute. Reproduced with permission.]

gether (knobs on knobs). Hence *these contacts, which are joined by different but equivalent sets of hydrogen bonds in the two states (Figs. 10-17 and 10-18), act as a binary switch that permits only two stable positions of the subunits relative to each other.* In contrast, the quaternary change causes only a 1-Å shift at the α_1FG–β_2C contact, so its side chains maintain the same associations throughout the change (Fig. 10-18). *These side chains therefore act as flexible joints or hinges about which the α_1 and β_2 subunits pivot during the quaternary change.*

c. The T State Is Stabilized by a Network of Salt Bridges That Must Break to Form the R State

The R state is stabilized by ligand binding. But in the absence of ligand, why is the T state more stable than the R state? In the electron density maps of R-state Hb, the C-terminal residues of each subunit (Arg 141α and His 146β) appear as a blur, which suggests that these residues are free to wave about in solution. Maps of the T form, however, show these residues firmly anchored in place via several intersubunit and intrasubunit salt bridges, which

(*a*) α Chains

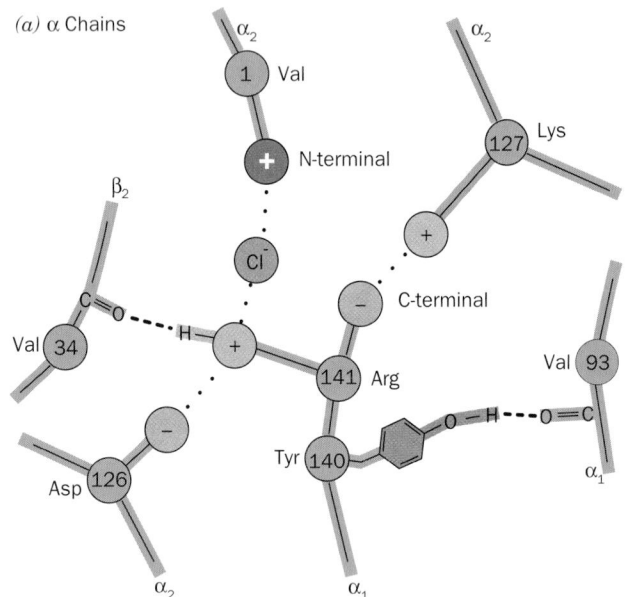

(*b*) β Chains

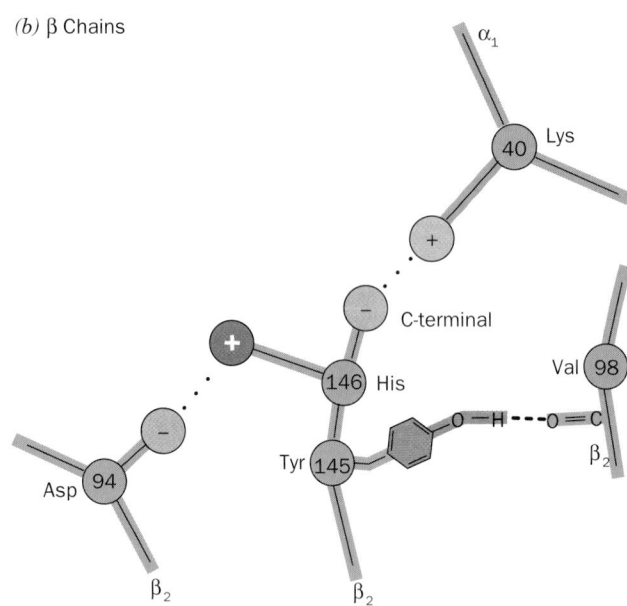

FIGURE 10-19   Networks of salt bridges and hydrogen bonds in deoxyHb. These bonds, which involve the last two residues of (*a*) the α chains and (*b*) the β chains, are all ruptured in the T→R transition. The two groups that participate in the Bohr effect by becoming partially deprotonated in the R state are indicated by white plus signs. [Illustration, Irving Geis/Geis Archives Trust. Copyright Howard Hughes Medical Institute. Reproduced with permission.]

evidently help stabilize the T state (Figs. 10-18 and 10-19). *The structural changes accompanying the T→R transition tear away these salt bridges in a process driven by the Fe—O$_2$ bonds' energy of formation.*

d. Hemoglobin's O$_2$-Binding Cooperativity Derives from the T→R Conformational Shift

The hemoglobin molecule resembles a finely tooled mechanism that has very little slop. The binding of O$_2$ requires a series of tightly coordinated movements:

1. The Fe(II) of any subunit cannot move into its heme plane without the reorientation of its proximal His so as to prevent this residue from bumping into the porphyrin ring.

2. The proximal His is so tightly packed by its surrounding groups that it cannot reorient unless this movement is accompanied by the previously described translation of the F helix across the heme plane.

3. The F helix translation is only possible in concert with the quaternary shift that steps the α$_1$C–β$_2$FG contact one turn along the α$_1$C helix.

4. The inflexibility of the α$_1$–β$_1$ and α$_2$–β$_2$ interfaces requires that this shift simultaneously occur at both the α$_1$–β$_2$ and the α$_2$–β$_1$ interfaces.

Consequently, *no one subunit or dimer can greatly change its conformation independently of the others. Indeed, the two stable positions of the α$_1$C–β$_2$FG contact limit the Hb molecule to only two quaternary forms, R and T.*

We are now in a position to structurally rationalize hemoglobin's O$_2$-binding cooperativity. Any deoxyHb sub-

unit binding O$_2$ is constrained to remain in the t state by the T conformation of the tetramer. However, *the t state has reduced O$_2$ affinity, most probably because its Fe—O$_2$ bond is stretched beyond its normal length by the steric repulsions between the heme and the O$_2$ and in the β subunits, by the need to move Val E11 out of the O$_2$-binding site.* As more O$_2$ is bound to the Hb tetramer, this strain, which derives from the Fe—O$_2$ bond energy, accumulates in the liganded subunits until it is of sufficient strength to snap the molecule into the R conformation. *All the subunits are thereby converted to the r state whether or not they are liganded. Unliganded subunits in the r state have an increased O$_2$ affinity because they are already in the O$_2$-binding conformation.* This accounts for the high O$_2$ affinity of nearly saturated Hb.

e. Hemoglobin's Sigmoidal O$_2$-Binding Curve Is a Composite of Its Hyperbolic R- and T-State Curves

The relative stabilities of the T and R states, as indicated by their free energies, vary with fractional saturation (Fig. 10-20*a*). In the absence of ligand, the T state is more stable than the R state, and vice versa when all ligand-binding sites are occupied. The formation of Fe—O$_2$ bonds causes the free energy of both the T and the R states to decrease (become more stable) with oxygenation, although the rate of this decrease is smaller for the T state as a result of the strain that liganding imposes on t-state subunits. The R ⇌ T transformation is, of course, an equilibrium process, so that Hb molecules, at intermediate levels of fractional saturation (1, 2, or 3 bound O$_2$ molecules), continually interconvert between the R and the T states.

The O$_2$-binding curve of Hb can be understood as a composite of those of its R and T states (Fig. 10-20*b*). For

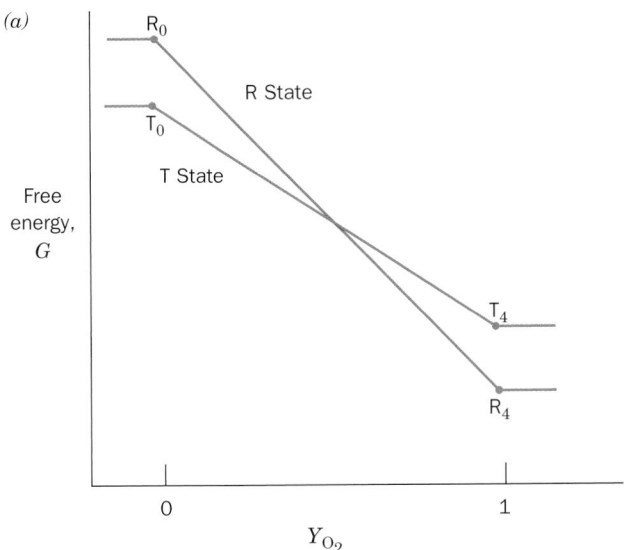

(a)

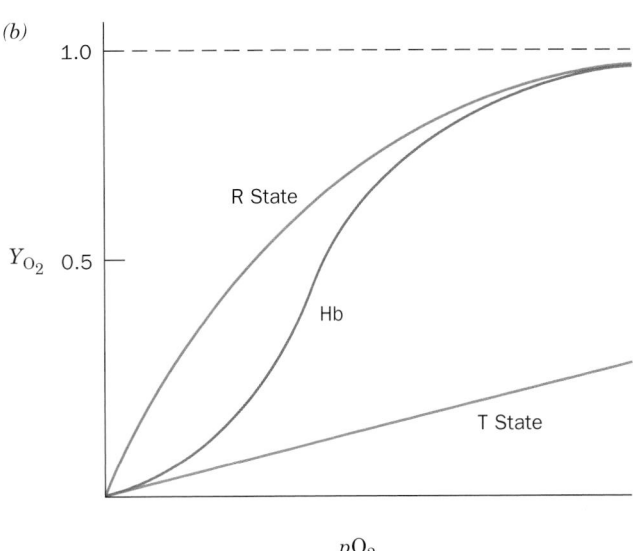

(b)

FIGURE 10-20 Free energy and saturation curves for O_2 binding to hemoglobin. (*a*) The variation of the free energies of hemoglobin's T and R states with their fractional saturation, Y_{O_2}. In the absence of O_2, the T state is more stable, and when saturated with O_2, the R state is more stable. The free energy of both states is reduced with increasing oxygenation as a consequence of O_2 liganding. The Fe(II)—O_2 bonding is more exergonic in the R state than it is in the T state, however, so that the relative stabilities of these two states reverse order at intermediate levels of oxygenation. (*b*) The sigmoid O_2-binding curve of Hb (*purple*) is a composite of its hyperbolic R-state (*red*) and T-state (*blue*) binding curves: It is more T-like at lower pO_2 values and more R-like at higher pO_2 values.

pure states, such as R or T, these curves are hyperbolic because ligand binding at one protomer is unaffected by the state of other protomers in the absence of a quaternary structural change. At low pO_2's, Hb follows the low-affinity T-state curve and at high pO_2's, it follows the high-affinity R-state curve. At intermediate pO_2's, Hb exhibits an O_2 affinity that changes from T-like to R-like as pO_2 increases. The switchover results in the sigmoidal shape of hemoglobin's O_2-binding curve.

D. *Testing the Perutz Mechanism*

The Perutz mechanism is a description of the dynamic behavior of Hb that is largely based on the static structures of its R and T end states. Accordingly, without the direct demonstration that Hb actually follows the postulated pathway in changing conformational states, the Perutz mechanism must be taken as being at least partially conjectural. Unfortunately, the physical methods that can follow dynamic changes in proteins are, as yet, incapable of providing detailed descriptions of these changes. Nevertheless, certain aspects of the Perutz mechanism are supported by static measurements, as is described below and in Section 10-3.

a. C-Terminal Salt Bridges Are Required to Maintain the T State

The proposed function of the C-terminal salt bridges in stabilizing the T state has been corroborated by chemically modifying human Hb. Removal of the C-terminal Arg 141α (by treating isolated α chains with carboxypeptidase B followed by reconstitution) drastically reduces the co-operativity of O_2 binding (Hill constant of 1.7, reduced from its normal value of 2.8). Cooperativity is abolished by the further removal of the other C-terminal residue, His 146β (Hill constant of ~1.0). Apparently, in the absence of its C-terminal salt bridges, the T form of Hb is unstable. Indeed, human deoxy-Hb, with its C-terminal residues removed, crystallizes in a form very similar to that of normal human oxyHb.

b. Fe—O_2 Bond Tension Has Been Spectroscopically Demonstrated

If movement of the Fe into the heme plane on oxygenation is mechanically coupled via the proximal His to the T→R transformation, then conversely, forcing oxyHb into the T form must exert a tension on the Fe, through the proximal His, that tends to pull the Fe out of the heme plane. Perutz demonstrated the existence of this tension as follows. IHP's six phosphate groups cause it to bind to deoxyHb with much greater affinity than does BPG (the structural basis of BPG binding to Hb is discussed in Section 10-2F); the presence of IHP therefore tends to force Hb into the T state. Conversely, nitric oxide (NO) binds to Hb far more strongly than does O_2 and thereby tends to force Hb into the R state. Spectroscopic analysis indicates the consequences of simultaneously binding both NO and IHP to Hb:

1. The NO, as expected, pulls the Fe into the plane of the heme.

2. The IHP forces the Hb molecule into the T state, which through the "gears and levers" coupling the 4° and 3° conformational changes, pulls the proximal His in the opposite direction, away from the Fe.

The bond between the proximal His and the Fe lacks the strength to withstand these two opposing "irresistible" forces; it simply breaks. The spectroscopic observation of this phenomenon therefore confirms the existence of the heme–protein tension predicted by the Perutz mechanism.

c. Detaching the Proximal His from the F Helix Eliminates Most Cooperativity

In a further experimental investigation of the origin of cooperativity in hemoglobin, Chien Ho mutagenically changed the proximal His residue to Gly on only the α subunits, on only the β subunits, and on both the α and β subunits. The missing imidazole ring of the proximal His was then replaced by imidazole (which a variety of evidence indicates ligands the heme Fe as does the proximal His). This, in effect, detaches the proximal His from the protein, thereby cutting the covalent bond that, according to the Perutz model, links the ligand-induced movement of the Fe into the heme plane to the accompanying movement of helix F. In all three cases, this proximal detachment, in agreement with the Perutz model, significantly increases hemoglobin's ligand-binding affinity, reduces its cooperativity, and prevents its T→R quaternary switch. However, these mutant hemoglobins exhibit a small amount of residual cooperativity, suggesting that the heme groups also communicate via pathways that do not require covalent coupling between the F helix and the proximal His. These pathways may involve movements of protein groups in contact with the heme (see Figs. 10-12 and 10-15) in response to the subsidence of heme doming on ligand binding. They may also involve movements of the distal His residues of the α and β subunits, and/or the movement of Val E11 of the β subunits, all of whose side chains must move aside when ligand binds to Hb.

E. *Origin of the Bohr Effect*

The Bohr effect, hemoglobin's release of H^+ on binding O_2, is also observed when Hb binds other ligands. *It arises from pK changes of several groups caused by changes in their local environments that accompany hemoglobin's T→R transition.* The groups involved include the N-terminal amino groups of the α subunits and the C-terminal His of the β subunits. These have been identified through chemical and structural studies, and their quantitative contributions to the Bohr effect have been estimated.

Reaction of the α subunits of Hb with **cyanate** results in the specific **carbamoylation** of the N-terminal amino groups (Fig. 10-21). When such carbamoylated α subunits are mixed with normal β subunits, the resulting reconstituted Hb lacks 20 to 30% of the normal Bohr effect. The reason for this is seen on comparing the X-ray structure of deoxyHb with that of carbamoylated deoxyHb. In deoxyHb, a Cl^- ion binds between the N-terminal amino group of Val $1\alpha_2$ and the guanidino group of Arg $141\alpha_1$ (the C-terminal residue; Fig. 10-19*a*). This Cl^- is absent in carbamoylated deoxyHb. It is also absent in normal R-state Hb because its C-terminal residues are not held in place by salt bridges (which partially accounts for the preferen-

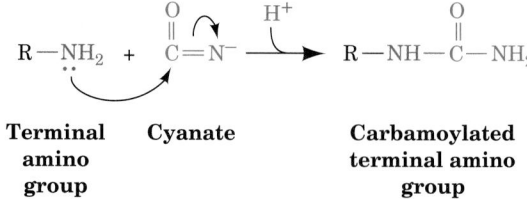

FIGURE 10-21 **Reaction of cyanate with the unprotonated (nucleophilic) forms of primary amino groups.** At physiological pH's, N-terminal amino groups, which have pK's near 8.0, readily react with cyanate. Lys ε-amino groups (pK≈10.8), however, are fully protonated under these conditions and are therefore unreactive.

tial binding of Cl^- to deoxyHb; Section 10-1C). N-Terminal amino groups of polypeptides normally have pK's near 8.0. On deoxyHb α subunits, however, the N-terminal amino group is electrostatically influenced by its closely associated Cl^- to increase its positive charge by binding protons more tightly, that is, to increase its pK. Since at the pH of blood (7.4) N-terminal amino groups are normally only partially charged, this pK shift causes them to bind significantly more protons in the T state than in the R state.

The Hb β chain also contributes to the Bohr effect. Removal of its C-terminal residue, His 146β, reduces the Bohr effect by 40%. In normal deoxyHb, the imidazole ring of His 146β associates with the carboxylate of Asp 94β on the same subunit (Figs. 10-18 and 10-19*b*) to form a salt bridge that is absent in the R state. Proton NMR measurements indicate that formation of this salt bridge increases the pK of the imidazole group from 7.1 to 8.0. This effect more than accounts for His 146β's share of the Bohr effect.

We have not yet accounted for about 30 to 40% of the Bohr effect. It is almost certainly due to small contributions from the numerous surface-exposed His residues whose environments are altered on hemoglobin's T→R transition [since His is the only residue with an intrinsic pK (6.04) in the physiological range, small changes in its pK will significantly alter the number of protons it binds]. Indeed, NMR measurements by Ho indicate that the T→R transition induces small shifts in the pK's of these various His residues, although, interestingly, some of these shifts are in the direction that diminishes the magnitude of the Bohr effect.

F. *Structural Basis of BPG Binding*

BPG decreases the oxygen-binding affinity of Hb by preferentially binding to its deoxy state (Section 10-1D). The binding of the physiologically quadruply charged BPG to deoxyHb is weakened by high salt concentrations, which suggests that this association is ionic in character. This explanation is corroborated by the X-ray structure of a BPG–deoxyHb complex, which indicates that BPG binds in the central cavity of deoxyHb on its 2-fold axis (Fig. 10-22). The anionic groups of BPG are within hydrogen bonding

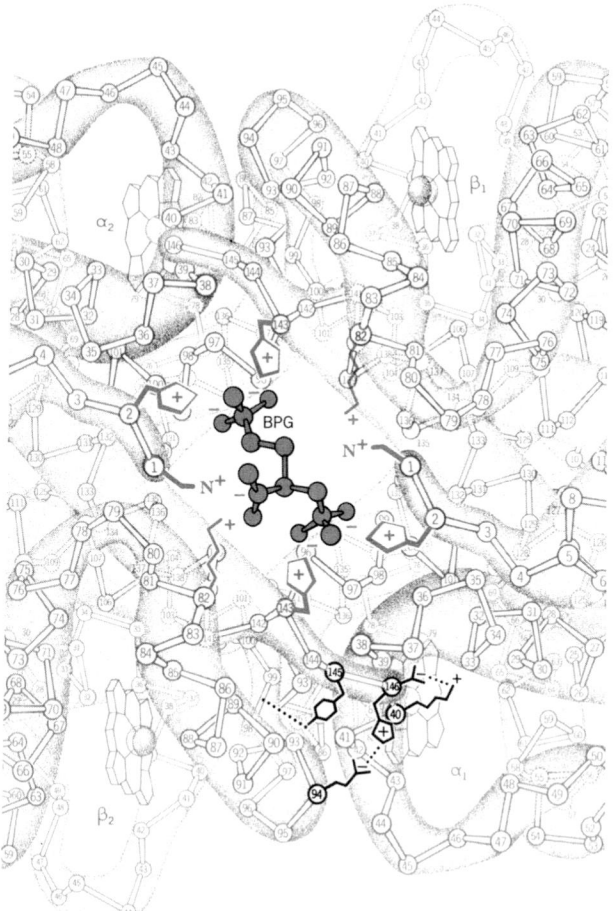

FIGURE 10-22   Binding of BPG to deoxyHb. The view is down the molecule's exact twofold axis (the same view as in Fig. 10-13*a*). BPG (*red*), with its five anionic groups, binds in the central cavity of deoxyHb, where it is surrounded by a ring of eight cationic side chains (*blue*) extending from the two β subunits. In the R state, the central cavity is too narrow to admit BPG (Fig. 10-13*b*). The arrangement of salt bridges and hydrogen bonds between the α₁ and β₂ subunits that partially stabilizes the T state (Figs. 10-18 and 10-19*b*) is indicated at the lower right. [Illustration, Irving Geis/Geis Archives Trust. Copyright Howard Hughes Medical Institute. Reproduced with permission.] *See Kinemage Exercise 6-3*

and salt bridging distances of the cationic Lys EF6(82), His H21(143), His NA2(2), and N-terminal amino groups of both β subunits (Fig. 10-22). The T→R transformation brings the two β H helices together, which narrows the central cavity (compare Fig. 10-13*a, b*) and expels the BPG. It also widens the distance between the β N-terminal amino groups from 16 to 20 Å, which prevents their simultaneous hydrogen bonding with BPG's phosphate groups. BPG therefore stabilizes the T conformation of Hb by cross-linking its β subunits. This shifts the T ⇌ R equilibrium toward the T state, which lowers hemoglobin's O₂ affinity.

The structure of the BPG–deoxyHb complex also indicates why fetal hemoglobin (HbF) has a reduced affinity for BPG relative to HbA (Section 10-1D). The cationic His H21(143)β of HbA is changed to an uncharged Ser residue

in HbF's β-like γ subunit, thereby eliminating a pair of ionic interactions stabilizing the BPG–deoxyHb complex (Fig. 10-22).

The excess positive charge lining Hb's central cavity is also partially responsible for the allosteric effect of Cl⁻ ions in stabilizing the T state relative to the R state (the remainder being due to the participation of Cl⁻ in the T-state salt bridge networks; Fig. 10-19*a*). The central cavity is larger in the T state than in the R state (Fig. 10-13), so that more Cl⁻ ions occupy this channel in the T state than in the R state. The additional Cl⁻ ions, through electrostatic shielding, reduce the mutual repulsions of the positive charges, thereby stabilizing the T state.

G. *Role of the Distal Histidine Residue*

O₂ binding paradoxically protects the heme iron from autooxidation: The rate of Mb oxidation decreases as the partial pressure of O₂ increases. This is because heme iron oxidation is catalyzed by protons that are reduced by the heme iron and that in turn reduce O₂ in the solvent to **superoxide ion** (O₂⁻). Bound O₂ evidently shields the Fe from the attacking protons.

The replacement, using genetic engineering techniques, of the distal His residue in Mb by any other residue reduces Mb's oxygen affinity and increases its rate of autooxidation. Asp, a proton source, at this position increases the rate of Mb autooxidation by 350-fold, the largest increase of all residue replacements, whereas Phe, Met, and Arg provide only 50-fold accelerations, the smallest observed increases. The imidazole ring of the distal His, which has a pK of 5.5 and is therefore neutral at neutral pH and whose unprotonated Nε atom faces the heme pocket (Fig. 10-12), acts as a proton trap, thereby protecting the Fe from protons. Thus, to quote Perutz, "Evolution is a brilliant chemist."

3 ■ ABNORMAL HEMOGLOBINS

Mutant hemoglobins provided the original opportunity to study structure–function relationships in proteins because Hb is a readily isolated protein of known structure that has a large number of well-characterized naturally occurring variants. The examination of individuals with physiological disabilities, together with the routine electrophoretic screening of human blood samples, has led to the discovery of over 860 variant hemoglobins, >90% of which result from single amino acid substitutions in a globin polypeptide chain (a compendium of variant human hemoglobins is located at http://globin.cse.psu.edu). In this section, we consider the nature of these **hemoglobinopathies.** Hemoglobin diseases characterized by defective globin synthesis, the **thalassemias,** are the subject of Section 34-2G. It should be noted that ~300,000 individuals with serious hemoglobin disorders are born every year and that ~5% of the world's population are carriers of an inherited variant hemoglobin.

A. *Molecular Pathology of Hemoglobin*

The physiological effect of an amino acid substitution on Hb can, in most cases, be understood in terms of its molecular location:

1. Changes in surface residues

Changes of surface residues are usually innocuous because most of these residues have no specific functional role [although sickle-cell Hb (HbS) is a glaring exception to this generalization; Section 10-3B]. For example, **HbE** [Glu B8(26)β → Lys], the most common human Hb mutant after HbS (possessed by up to 10% of the population in parts of Southeast Asia), has no clinical manifestations in either heterozygotes or homozygotes. About half of the known Hb mutations are of this type and have been discovered only accidentally or through surveys of large populations.

2. Changes in internally located residues

Changing an internal residue often destabilizes the Hb molecule. The degradation products of these hemoglobins, particularly those of heme, form granular precipitates (known as **Heinz bodies**) that are hydrophobically adsorbed to the erythrocyte cell membrane. The membrane's permeability is thereby increased, causing premature cell lysis. Carriers of unstable hemoglobins therefore suffer from **hemolytic anemia** of varying degrees of severity.

The structure of Hb is so delicately balanced that small structural changes may render it nonfunctional. This can occur through the weakening of the heme–globin association or as a consequence of other conformational changes. For instance, the heme group is easily dislodged from its closely fitting hydrophobic binding pocket. This occurs in **Hb Hammersmith** (Hb variants are often named after the locality of their discovery), in which Phe CD1(42)β, an invariant residue that wedges the heme into its pocket (see Fig. 10-12), is replaced by Ser. The resulting gap permits water to enter the heme pocket, which causes the hydrophobic heme to drop out easily (Phe CD1 and the proximal His F8 are the only invariant residues among all known hemoglobins). Similarly, in **Hb Bristol,** the substitution of Asp for Val E11(67)β, which partially occludes the O_2 pocket, places a polar group in contact with the heme. This weakens the binding of the heme to the protein, probably by facilitating the access of water to the subunit's otherwise hydrophobic interior.

Hb may also be destabilized by the disruption of elements of its 2°, 3°, and/or 4° structures. The instability of **Hb Bibba** results from the substitution of a helix-breaking Pro for Leu H19(136)α. Likewise, the instability of **Hb Savannah** is caused by the substitution of Val for the highly conserved Gly B6(24)β, which is located on the B helix where it crosses the E helix with insufficient clearance for side chains larger than an H atom (Fig. 10-13). The α_1–β_1 contact, which does not significantly dissociate under physiological conditions, may do so on structural alteration. This occurs in **Hb Philly,** in which Tyr C1(35)β, which participates in the hydrogen bonded network that helps knit together the α_1–β_1 interface, is replaced by Phe.

3. Changes stabilizing methemoglobin

Changes at the O_2-binding site that stabilize the heme in the Fe(III) oxidation state eliminate the binding of O_2 to the defective subunits. Such methemoglobins are designated **HbM** and individuals carrying them are said to have **methemoglobinemia.** These individuals usually have bluish skin, a condition known as **cyanosis,** which results from the presence of deoxyHb in their arterial blood.

All known methemoglobins arise from substitutions that provide the Fe atom with an anionic oxygen atom ligand. In **Hb Boston,** the substitution of Tyr for His E7(58)α (the distal His, which protects the heme from oxidation; Section 10-2G) results in the formation of a 5-coordinate Fe(III) complex, with the phenolate ion of the mutant Tyr E7 displacing the imidazole ring of His F8(87) as the apical ligand (Fig. 10-23a). In **Hb Milwaukee,** the γ-carboxyl group of the Glu that replaces Val E11(67)β forms an ion pair with a 5-coordinate Fe(III) complex (Fig. 10-23b). Both the phenolate and glutamate ions in these methemoglobins so stabilize the Fe(III) oxidation state that methemoglobin reductase is ineffective in converting them to the Fe(II) form.

Individuals with HbM are alarmingly cyanotic and have blood that is chocolate brown, even when their normal subunits are oxygenated. In northern Japan, this condition is named "black mouth" and has been known for centuries; it is caused by the presence of **HbM Iwate** [His F8(87)α→ Tyr]. Methemoglobins have Hill constants of ~1.2. This indicates a reduced cooperativity in comparison with HbA even though HbM, which can bind only two oxygen molecules, can have a maximum Hill constant of 2 (the unmutated α or β chains remain functional). Surprisingly, heterozygotes with HbM, which have an average of one nonfunctional α or β subunit per Hb molecule, have no apparent physical disabilities. Evidently, the amount of O_2 released in their capillaries is within normal limits. Homozygotes of HbM, however, are unknown; this condition is, no doubt, lethal.

4. Changes at the α_1–β_2 contact

Changes at the α_1–β_2 contact often interfere with hemoglobin's quaternary structural changes. Most such hemoglobins have an increased O_2 affinity so that they release less than normal amounts of O_2 in the tissues. Individuals with such defects compensate for it by increasing the concentration of erythrocytes in their blood. This condition, which is named **polycythemia,** often gives them a ruddy complexion. Some amino acid substitutions at the α_1–β_2 interface instead result in a reduced O_2 affinity. Individuals carrying such hemoglobins are cyanotic.

Amino acid substitutions at the α_1–β_2 contact may change the relative stabilities of hemoglobin's R and T forms, thereby altering its O_2 affinity. For example, the replacement of Asp G1(99)β by His in **Hb Yakima** eliminates the hydrogen bond at the α_1–β_2 contact that stabilizes the T form of Hb (Fig. 10-17a). The interloping

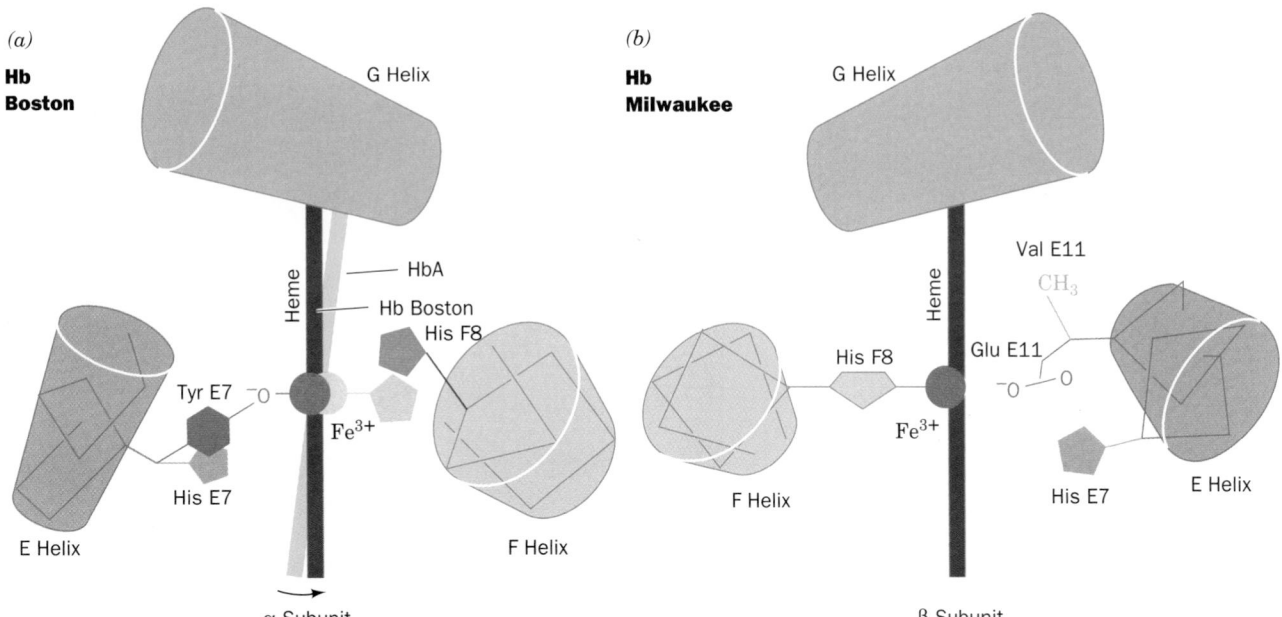

(a)

Hb Boston

G Helix

Heme

HbA

Hb Boston

His F8

Tyr E7

Fe^{3+}

His E7

E Helix

F Helix

α Subunit

(b)

Hb Milwaukee

G Helix

Heme

Val E11

CH$_3$

His F8

Glu E11

O

Fe^{3+}

O

His E7

E Helix

F Helix

β Subunit

FIGURE 10-23 Mutations stabilizing the Fe(III) oxidation state of heme. (*a*) Alterations in the heme pocket of the α subunit on changing from deoxyHbA to Hb Boston [His E7(58)α → Tyr]. The phenolate ion of the mutant Tyr becomes the fifth ligand of the Fe atom, thereby displacing the proximal His [F8(87)α]. [After Pulsinelli, P.D., Perutz, M.F., and Nagel, R.L., *Proc. Natl. Acad. Sci.* **70**, 3872 (1973).] (*b*) The structure of the heme pocket of the β subunit in Hb Milwaukee [Val E11(67)β → Glu]. Here the mutant Glu residue's carboxyl group forms an ion pair with the heme iron atom so as to stabilize its Fe(III) state. [From Perutz, M.F., Pulsinelli, P.D., and Ranney, H.M., *Nature* **237**, 260 (1972).]

imidazole ring also acts as a wedge that pushes the subunits apart and displaces them toward the R state. This change shifts the T→R equilibrium almost entirely to the R state, which results in Hb Yakima having an increased O$_2$ affinity (p_{50} = 12 torr under physiological conditions vs 26 torr for HbA) and a total lack of cooperativity (Hill constant = 1.0). In contrast, the replacement of Asn G4(102)β by Thr in **Hb Kansas** eliminates the hydrogen bond in the α$_1$–β$_2$ contact that stabilizes the R state (Fig. 10-17*b*), so that this Hb variant remains in the T state on binding O$_2$. Hb Kansas therefore has a low O$_2$ affinity (p_{50} = 70 torr) and a low cooperativity (Hill constant = 1.3).

B. *Molecular Basis of Sickle-Cell Anemia*

Most harmful Hb variants occur in only a few individuals, in many of whom the mutation apparently originated. However, ~10% of American blacks and as many as 25% of African blacks are heterozygotes for **sickle-cell hemoglobin (HbS)**. HbS arises, as we have seen (Section 7-3A), from the substitution of a hydrophobic Val residue for the hydrophilic surface residue Glu A3(6)β (Fig. 10-13). The prevalence of HbS results from the protection it affords heterozygotes against malaria. However, homozygotes for HbS, of which there are ~50,000 in the USA, are severely afflicted by hemolytic anemia together with painful, debilitating, and sometimes fatal blood flow blockages caused by the irregularly shaped and inflexible erythrocytes characteristic of the disease (Fig. 7-18*b*).

a. HbS Fibers Are Stabilized by Intermolecular Contacts Involving Val β6 and Other Residues

The sickling of HbS-containing erythrocytes results from the aggregation (polymerization) of deoxyHbS into rigid fibers that extend throughout the length of the cell (Fig. 10-24). Electron microscopy indicates that these fibers are ~220-Å-diameter elliptical rods consisting of 14 hexagonally packed and helically twisting strands of deoxyHbS

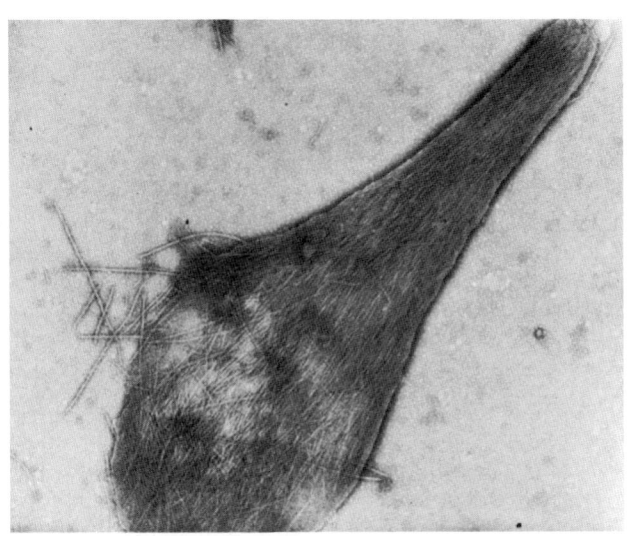

FIGURE 10-24 Electron micrograph of deoxyHbS fibers spilling out of a ruptured erythrocyte. [Courtesy of Robert Josephs, University of Chicago.]

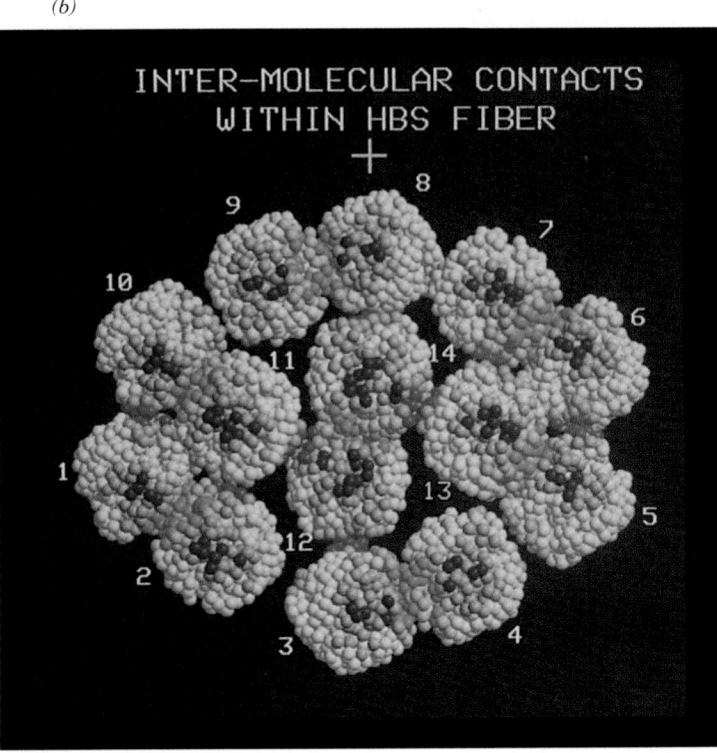

FIGURE 10-25   220-Å in diameter fibers of deoxyHbS.
(*a*) An electron micrograph of a negatively stained fiber. The accompanying cutaway interpretive drawing indicates the relationship between the inner and outer strands; each sphere represents an individual HbS molecule. The fiber has a layer repeat distance of 64 Å and a moderate twist such that it repeats every 350 Å along the fiber axis. [Courtesy of Stuart Edelstein, University of Geneva.] (*b*) A model, viewed in cross section, of the HbS fiber based on the crystal structure of HbS and three-dimensional reconstructions of electron micrographs of HbS fibers. The residues in the 14 HbS molecules are represented by spheres centered on their C_α positions. The residues making inter-double strand, intra-double strand lateral, and intra-double strand axial contacts are colored red, green, and blue, respectively, with lighter and darker toned residues making intermolecular contacts of <8 Å and <5 Å, respectively. The α and β chain residues outside the contact regions are colored white. [Courtesy of Stanley Watowich, Leon Gross, and Robert Josephs, University of Chicago.]

molecules that associate in parallel pairs (Figs. 10-25 and 10-26*a*).

The structural relationship among the HbS molecules in the pairs of parallel HbS strands has been established by the X-ray structure analysis of deoxyHbS crystals. When this crystal structure was first determined, it was unclear whether the intermolecular contacts in the crystal resembled those in the fiber. However, the subsequent observation that HbS fibers slowly convert to these crystals with little change in their overall X-ray diffraction pattern indicates that the fibers structurally resemble the crystals. The crystal structure of deoxyHbS consists of double filaments of HbS molecules whose several different intermolecular contacts are diagrammed in Fig. 10-26*b*. Only one of the two Val 6β's per Hb molecule contacts a neighboring molecule. In this contact, the mutant Val side chain occupies a hydrophobic surface pocket on the β subunit of an adjacent molecule whose Val 6β does not make an intermolecular contact (Fig. 10-26*c*). This pocket is absent in oxyHb. Other contacts involve residues that also occur in HbA, including Asp 73β and Glu 23α (Fig. 10-26*b*). The observation that deoxyHbA does not aggregate into fibers, however, even at very high concentrations, indicates that *the contact involving Val 6β is essential for fiber formation.* This conclusion is corroborated by the observation that a genetically engineed human Hb in which Glu 6β is replaced by Ile (which differs from Val by an additional CH_2 group and is therefore even more hydrophobic) has half the solubility of HbS in 1.8*M* phosphate.

The importance of the other intermolecular contacts to the structural integrity of HbS fibers has been demonstrated by studying the effects of other mutant hemoglobins on HbS gelation (polymerization). For example, the doubly mutated **Hb Harlem** (Glu 6β→Val + Asp 73β→ Asn) requires a higher concentration to gel than does HbS (Glu 6β→Val); similarly, mixtures of HbS and **Hb Korle- Bu** (Asp 73β→Asn) gel less readily than equivalent mixtures of HbS and HbA. These observations suggest that

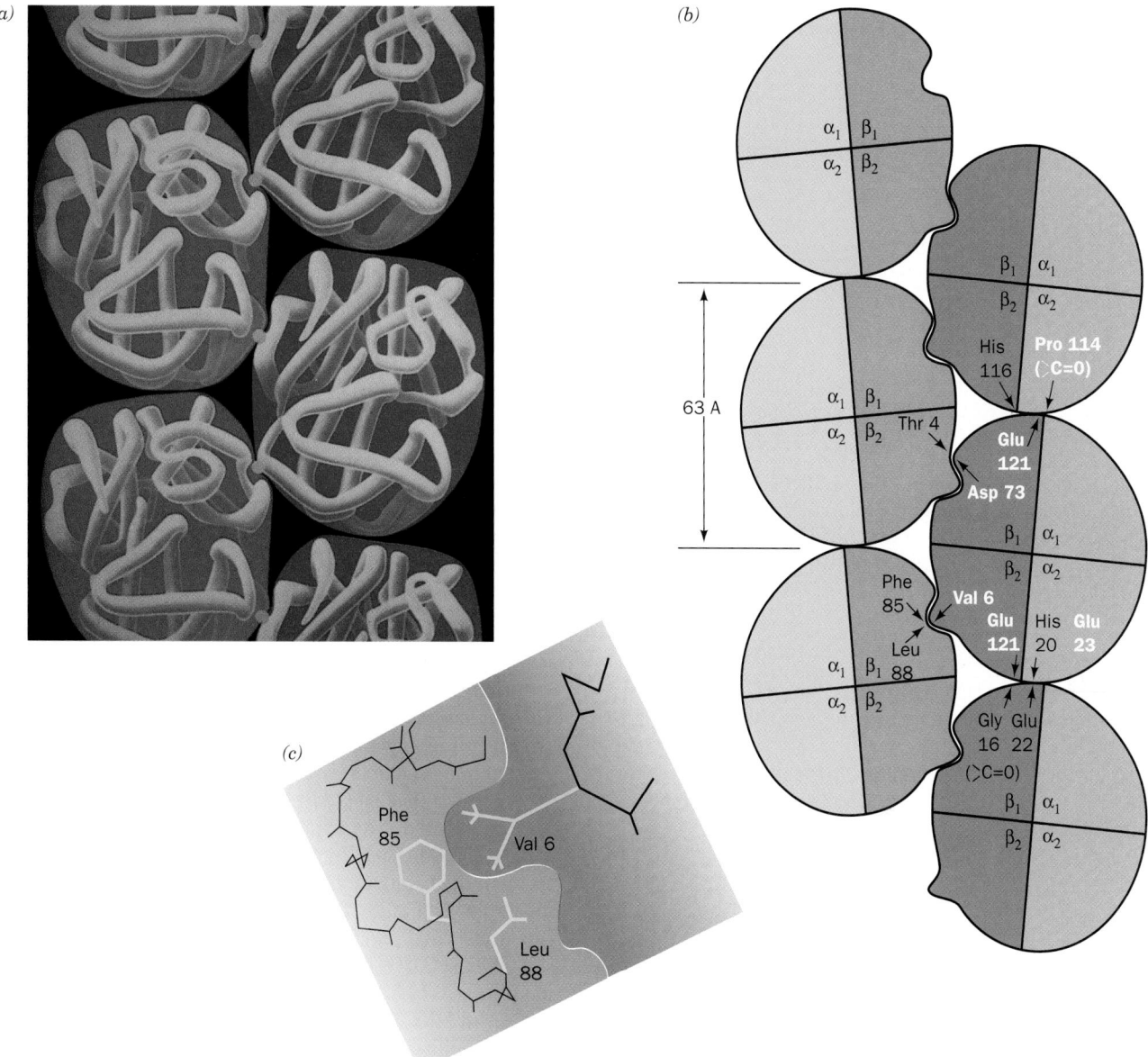

FIGURE 10-26 Structure of the deoxyHbS fiber. (*a*) The arrangement of the deoxyHbS molecules in the fiber. [Illustration, Irving Geis/Geis Archives Trust. Copyright Howard Hughes Medical Institute. Reproduced with permission.] (*b*) A schematic diagram indicating the intermolecular contacts in the crystal structure of deoxyHbS. The white-lettered residues are implicated in forming these contacts. Note that the only intermolecular association in which the mutant residue Val 6β participates involves subunit β₂; Val 6 of subunit β₁ is free. [After Wishner, B.C., Ward, K.B.,

Lattman, E.E., and Love, W.E., *J. Mol. Biol.* **98,** 192 (1975).] (*c*) The mutant Val 6β₂ fits neatly into a hydrophobic pocket formed mainly by Phe 85 and Leu 88 of an adjacent β₁ subunit. This pocket, which is located between helices E and F at the periphery of the heme pocket, is absent in oxyHb and is too hydrophobic to contain the normally occurring Glu 6β side chain. [Illustration, Irving Geis/Geis Archives Trust. Copyright Howard Hughes Medical Institute. Reproduced with permission.]

Asp 73β occupies an important intermolecular contact site in HbS fibers (Fig. 10-26*b*). Likewise, the observation that hybrid tetramers consisting of α subunits from **Hb Memphis** (Glu 23α→Gln) and β subunits from HbS gel less readily than does HbS indicates that Glu 23α also participates in the polymerization of HbS fibers (Fig. 10-26*b*). The other white-lettered residues in Fig. 10-26*b* have been similarly implicated in sickling interactions.

b. The Initiation of HbS Gelation Is a Complex Process

The gelation of HbS, both in solution and within the red cell, follows an unusual time course. A solution of HbS can be brought to conditions under which it will gel by lowering the pO_2, raising the HbS concentration, and/or raising the temperature. *On achieving gelation conditions, there is a reproducible delay that varies according to conditions from milliseconds to days: During this time, no HbS*

(a)

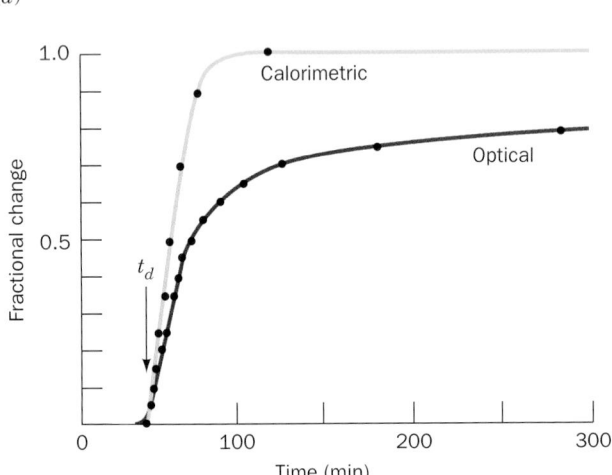

(b)

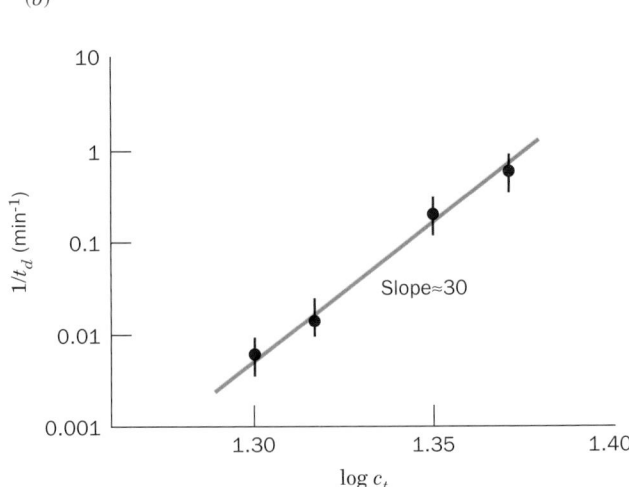

FIGURE 10-27   Time course of deoxyHbS gelation. (*a*) The extent of gelation as monitored calorimetrically (*yellow*) and optically (*purple*). Gelation of the 0.233 g · mL^{-1} deoxyHbS solution was initiated by rapidly increasing the temperature from 0°C, where HbS is soluble, to 20°C; t_d is the delay time.

(*b*) A log–log plot showing the concentration dependence of $1/t_d$ for the gelation of deoxyHbS at 30°C. The slope of this line is ~30. [After Hofrichter, J., Ross, P.D., and Eaton, W.A., *Proc. Natl. Acad. Sci.* **71**, 4865, 4867 (1974).]

fibers can be detected. Only after the delay do fibers first appear, and gelation is then completed in about half the delay time (Fig. 10-27*a*).

William Eaton and James Hofrichter discovered that the delay time, t_d, has a concentration dependence described by

$$\frac{1}{t_d} = k\left(\frac{c_t}{c_s}\right)^n \qquad [10.13]$$

where c_t is the total deoxyHbS concentration prior to gelation, c_s is the solubility of deoxyHbS measured after gelation is complete, and k and n are constants. Graphical analysis of the data indicates that $k \approx 10^{-7}\,\mathrm{s}^{-1}$ and that n is between 30 and 50 (Fig. 10-27*b*). This is a remarkable result: *No other known solution process even approaches a 30th power concentration dependence.*

A two-stage process accounts for Eq. [10.13]:

1. At first, HbS molecules sequentially aggregate to form a **nucleus** consisting of *m* HbS molecules (Fig. 10-28*a*):

$$\mathrm{HbS} \rightleftharpoons (\mathrm{HbS})_2 \rightleftharpoons (\mathrm{HbS})_3 \rightleftharpoons \cdots$$
$$\rightleftharpoons (\mathrm{HbS})_m \to \mathrm{Growth}$$

Prenuclear aggregates are unstable and easily decompose, but once a nucleus has formed it assumes a stable structure that rapidly elongates to form an HbS fiber.

2. Once a fiber has formed, it can nucleate the growth of other fibers (Fig. 10-28*b*). These newly formed fibers, in turn, nucleate the growth of yet other fibers, etc., so that this latter process is autocatalytic.

The initial **homogeneous nucleation** process (taking place in solution) accounts for the very high concentration dependence in Eq. [10.13], whereas the secondary **heterogeneous nucleation** process (taking place on a surface—

that of a fiber in this case) is responsible for the rapid onset of gelation (Fig. 10-27*a*).

The foregoing kinetic hypothesis suggests why sickle-cell anemia is characterized by episodic "crises" caused by blood flow blockages. HbS fibers dissolve essentially instantaneously on oxygenation, so that none are present in arterial blood. Erythrocytes take from 0.5 to 2 s to pass through the capillaries, where deoxygenation renders HbS insoluble. If the delay time, t_d, for sickling is greater than this transit time, no blood flow blockage occurs (although sickling that occurs in the veins damages the erythrocyte membrane). However, Eq. [10.13] indicates that small increases in HbS concentration, c_t, and/or small decreases in HbS solubility, c_s, caused by conditions known to trigger sickle-cell crises, such as dehydration, O$_2$ deprivation, and fever, result in significant decreases of t_d. Once a blockage occurs, the resulting lack of O$_2$ and slowdown of blood flow in the area compound the situation.

The kinetic hypothesis of sickling has profound clinical implications for the treatment of sickle-cell anemia. Heterozygotes of HbS, whose blood usually contains ~60% HbA and 40% HbS, rarely show any symptoms of sickling. The t_d for the gelation of their Hb is ~10^6-fold greater than that of homozygotes. Accordingly, a treatment of sickle-cell anemia that increases t_d by this amount, which corresponds to decreasing the ratio c_t/c_s by a factor of ~1.6, would relieve the symptoms of this disease. This has suggested three different therapeutic strategies to increase t_d, and thus inhibit HbS gelation:

1. The disruption of intermolecular interactions, thus increasing c_s. Of particular interest are compounds that have been designed with the aid of the X-ray structure of HbS to bind stereospecifically to its intermolecular contact regions. However, a large amount of any such compound would be necessary to bind to the ~400 g of hemoglobin

(a) Homogeneous nucleation

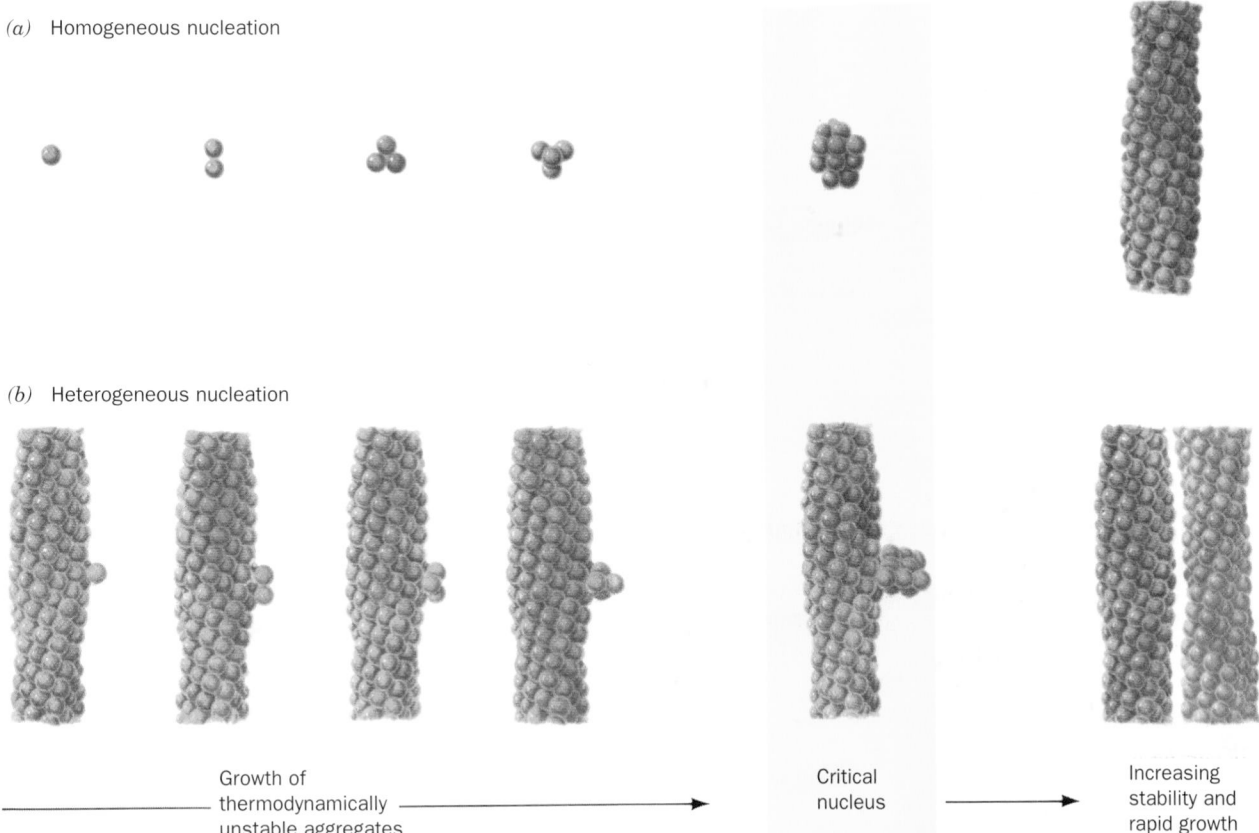

(b) Heterogeneous nucleation

Growth of thermodynamically unstable aggregates ⟶

Critical nucleus ⟶

Increasing stability and rapid growth

FIGURE 10-28 Double nucleation mechanism for deoxyHbS gelation. *(a)* The initial aggregation of HbS molecules *(circles)* occurs very slowly because this process is thermodynamically unfavorable and hence the intermediates tend to decompose rather than grow. However, once an aggregate reaches a certain size, the **critical nucleus,** its further growth becomes thermodynamically favorable, leading to rapid fiber formation. *(b)* Each fiber, in turn, can nucleate the growth of other fibers, leading to the explosive appearance of polymer. [After Ferrone, F.A., Hofrichter, J., and Eaton, W.A., *J. Mol. Biol.* **183,** 614 (1985).]

in the human body. Consequently, no antisickling drug yet tested has had a sufficiently high ratio of efficacy to toxicity to merit clinical use.

2. The use of agents that increase hemoglobin's O_2 affinity, thus decreasing c_t. For example, the administration of cyanate carbamoylates the N-terminal amino groups of Hb (Fig. 10-21). This treatment eliminates some of the salt bridges that stabilize the T state (Section 10-2E) and thereby increases the O_2 affinity of Hb. Although cyanate is an effective *in vitro* antisickling agent, its clinical use has been discontinued because of toxic side effects, cataract formation and peripheral nervous system damage, that probably result from the carbamoylation of proteins other than Hb.

3. Lowering the HbS concentration (c_t) in erythrocytes. Agents that alter erythrocyte membrane permeability so as to permit the influx of water have promise in this regard.

The first, and as yet the only, effective treatment for sickle-cell anemia is a variation of the latter strategy through the administration of **hydroxyurea.**

$$
\begin{array}{c}
\overset{\displaystyle O}{\overset{\displaystyle \|}{H_2N-C-NH-OH}}
\end{array}
$$

Hydroxyurea

Adults with sickle-cell anemia have two types of red blood cells: S cells, which contain only HbS; and F cells, which contain ~20% HbF and the remainder HbS. In most adults, the fraction of F cells is ~30%. However, in those treated with hydroxyurea, this fraction increases to ~50%. Although the mechanism by which hydroxyurea stimulates the production of F cells is unknown, the mechanism by which increased levels of F cells prevent sickling seems clear. F cells contain three species of hemoglobin: HbS ($\alpha_2\beta_2^S$), HbF ($\alpha_2\gamma_2$), and their hybrid ($\alpha_2\beta^S\gamma$), where β^S subunits are the sickle-cell variants of the normal β subunits. Since neither HbF nor the $\alpha_2\beta^S\gamma$ hybrid Hb can form sickle-cell fibers, they act to dilute the HbS in a cell. This, in turn, increases the time it takes the F cells to sickle by a factor of ~1000, so that F cells do not significantly sickle in the period (10–20 s) it takes them pass from the tissues to the lungs, where they are oxygenated. Thus, the greater the proportion of F cells in the blood, the smaller the proportion of S cells that can sickle.

4 ■ ALLOSTERIC REGULATION

One of the outstanding characteristics of life is the high degree of control exercised in almost all of its processes. Through a great variety of regulatory mechanisms, the exploration of which constitutes a significant portion of this textbook, an organism is able to respond to changes in its environment, maintain intra- and intercellular communications, and execute an orderly program of growth and development. Regulation is exerted at every organizational level in living systems, from the control of rates of reactions on the molecular level, through the control of expression of genetic information on the cellular level, to the control of behavior on the organismal level. It is therefore not surprising that many, if not most, diseases are caused by aberrations in biological control processes.

Our exploration of the structure and function of hemoglobin continues with a theoretical discussion of the regulation of ligand binding to proteins through **allosteric interactions** (Greek: *allos,* other + *stereos,* solid or space). These cooperative interactions occur when the binding of one ligand at a specific site is influenced by the binding of another ligand, known as an **effector** or **modulator,** at a different (allosteric) site on the protein. If the ligands are identical, this is known as a **homotropic effect,** whereas if they are different, it is described as a **heterotropic effect.** These effects are termed **positive** or **negative** depending on whether the effector increases or decreases the protein's ligand-binding affinity.

Hemoglobin, as we have seen, exhibits both homotropic and heterotropic effects. The binding of O_2 to Hb results in a positive homotropic effect since it increases hemoglobin's O_2 affinity. In contrast, BPG, CO_2, H^+, and Cl^- are negative heterotropic effectors of O_2 binding to Hb because they decrease its affinity for O_2 (negative) and are chemically different from O_2 (heterotropic). The O_2 affinity of Hb, as we have seen, depends on its quaternary structure. *In general, allosteric effects result from interactions among subunits of oligomeric proteins.*

Even though hemoglobin catalyzes no chemical reaction, it binds ligands in the same manner as do enzymes. Since an enzyme cannot catalyze a reaction until after it has bound its **substrate** (the molecule undergoing reaction), the enzyme's catalytic rate varies with its substrate-binding affinity. Consequently, the cooperative binding of O_2 to Hb is taken as a model for the allosteric regulation of enzyme activity. Indeed, in this section, we shall consider several models of allosteric regulation that, for the most part, were formulated to explain the O_2-binding properties of Hb. Following this, we shall compare these models with the realities of Hb behavior.

A. *The Adair Equation*

The derivation of the Hill equation (Section 10-1B) is predicated on the assumption of all-or-none O_2 binding. The observation of partially oxygenated Hb molecules, however, led Gilbert Adair, in 1924, to propose that the binding of ligands to proteins occurs sequentially with dissociation constants that are not necessarily equal. The expression for the saturation function under this model is straightforwardly derived.

For a protein such as Hb with four ligand-binding sites, the reaction sequence is

$$
\begin{aligned}
E + S &\rightleftharpoons ES & k_1 &= 4K_1 \\
ES + S &\rightleftharpoons ES_2 & k_2 &= \tfrac{3}{2}K_2 \\
ES_2 + S &\rightleftharpoons ES_3 & k_3 &= \tfrac{2}{3}K_3 \\
ES_3 + S &\rightleftharpoons ES_4 & k_4 &= \tfrac{1}{4}K_4
\end{aligned}
$$

where the K_i are the **macroscopic** or **apparent dissociation constants** for binding the ith ligand to the protein,

$$
K_i = \frac{[ES_{i-1}][S]}{[ES_i]} \qquad [10.14]
$$

and the k_i are the **microscopic** or **intrinsic dissociation constants,** that is, the individual dissociation constants for the ligand-binding sites. The intrinsic dissociation constants are equal to the apparent dissociation constants multiplied by **statistical factors,** $4, \tfrac{3}{2}, \tfrac{2}{3}$, and $\tfrac{1}{4}$, that account for the number of ligand-binding sites on the protein molecule. The statistical factor 4 derives from the fact that a tetrameric protein E bears four sites that can bind ligand to form ES (that is, the concentration of ligand-binding sites is 4[E]) but only one site from which ES can dissociate ligand to form E (that is, the concentration of bound ligand is 1[E]); the statistical factor $\tfrac{3}{2}$ is a result of there being three remaining sites on ES that can bind ligand to form ES_2 and two sites from which ES_2 can dissociate ligand to form ES; etc. In general, for a protein with n equivalent binding sites:

$$
k_i = \frac{(n - i + 1)[ES_{i-1}][S]}{i[ES_i]} = \left(\frac{n - i + 1}{i}\right)K_i \qquad [10.15]
$$

since $(n - i + 1)[ES_{i-1}]$ is the concentration of free ligand-binding sites in ES_{i-1} and $i[ES_i]$ is the concentration of bound ligand on ES_i. Therefore, solving sequentially for the concentration of each protein–ligand species in a tetrameric protein, we obtain:

$$
\begin{aligned}
[ES] &= [E][S]/K = 4[E][S]/k_1 \\
[ES_2] &= [ES][S]/K_2 = \tfrac{3}{2}[ES][S]/k_2 = 6[E][S]^2/k_1 k_2 \\
[ES_3] &= [ES_2][S]/K_3 = \tfrac{2}{3}[ES_2][S]/k_3 = 4[E][S]^3/k_1 k_2 k_3 \\
[ES_4] &= [ES_3][S]/K_4 = \tfrac{1}{4}[ES_3][S]/k_4 = [E][S]^4/k_1 k_2 k_3 k_4
\end{aligned}
$$

The fractional saturation of ligand binding, the fraction of occupied ligand-binding sites divided by the total concentration of ligand-binding sites, is expressed

$$
Y_S = \frac{[ES] + 2[ES_2] + 3[ES_3] + 4[ES_4]}{4([E] + [ES] + [ES_2] + [ES_3] + [ES_4])} \qquad [10.16]
$$

so that, substituting in the above relationships and canceling terms, we obtain

$$
Y_S = \frac{\dfrac{[S]}{k_1} + \dfrac{3[S]^2}{k_1 k_2} + \dfrac{3[S]^3}{k_1 k_2 k_3} + \dfrac{[S]^4}{k_1 k_2 k_3 k_4}}{1 + \dfrac{4[S]}{k_1} + \dfrac{6[S]^2}{k_1 k_2} + \dfrac{4[S]^3}{k_1 k_2 k_3} + \dfrac{[S]^4}{k_1 k_2 k_3 k_4}} \qquad [10.17]
$$

This is the **Adair equation** for four ligand-binding sites. Equations describing ligand binding to proteins with different numbers of binding sites are similarly derived.

If the microscopic dissociation constants of the Adair equation are not equal, the fractional saturation curve will

TABLE 10-2 Adair Constants for Hemoglobin A at pH 7.40

Solution	k_1 (torr)	k_2 (torr)	k_3 (torr)	k_4 (torr)
Stripped	8.8	6.1	0.85	0.25
0.1M NaCl	41.	13.	12.	0.14
2 mM BPG	74.	112.	23.	0.24
0.1M NaCl + 2mM BPG	97.	43.	119.	0.09

Source: Tyuma, I., Imai, K., and Shimizu, K., *Biochemistry* **12**, 1493, 1495 (1973).

describe cooperative ligand binding. Decreasing and increasing values of these constants lead to positive and negative cooperativity, respectively. Of course, the values of the microscopic dissociation constants may also alternate so that, for example, $k_1 < k_2 > k_3 < k_4$.

In our discussion of the O_2-dissociation curve of Hb (Section 10-1B), we have seen how its values of k_1 and k_4 may be obtained by extrapolating the lower and upper asymptotes of the Hill plot to the log pO_2 axis. The remaining microscopic dissociation constants can be evaluated by fitting Eq. [10.17] to the Hill plot. The values of these **Adair constants** for Hb are given in Table 10-2. Note that k_4 is relatively insensitive to the presence of BPG. Hb therefore binds and releases its last O_2 almost independently of the BPG concentration.

Although the Adair equation is the most general relationship describing ligand binding to a protein and is widely used to do so, it provides no physical insight as to why the various microscopic dissociation constants differ from each other. Yet, if the protein consists, as so many do, of identical subunits that are symmetrically related, it is desirable to understand how ligand binding at one site influences the ligand-binding affinity at a seemingly identical site. This need led to the development of models for ligand binding that rationalize how the binding sites of oligomeric proteins can exhibit different affinities. Two of these models are described in the following sections.

B. *The Symmetry Model*

Perhaps the most elegant model for describing cooperative ligand binding to a protein is the **symmetry model** of allosterism, which was formulated in 1965 by Jacques Monod, Jeffries Wyman, and Jean-Pierre Changeux. This model, alternatively termed the **MWC model,** is defined by the following rules:

1. An allosteric protein is an oligomer of protomers that are symmetrically related (for hemoglobin, we shall assume, for the sake of algebraic simplicity, that all four subunits are functionally identical).

2. Each protomer can exist in (at least) two conformational states, designated T and R; these states are in equilibrium whether or not ligand is bound to the oligomer.

3. The ligand can bind to a protomer in either confor-

mation. *Only the conformational change alters the affinity of a protomer for the ligand.*

4. *The molecular symmetry of the protein is conserved during conformational change.* Protomers must therefore change conformation in a concerted manner, which implies that the conformation of each protomer is constrained by its association with the other protomers; in other words, there are no oligomers that simultaneously contain R- and T-state protomers.

For a ligand S and an allosteric protein consisting of n protomers, these rules imply the following equilibria for conformational conversion and ligand-binding reactions (for the sake of brevity, $T_i \equiv TS_i$ and $R_i \equiv RS_i$).

$$T_0 \rightleftharpoons R_0$$
$$T_0 + S \rightleftharpoons T_1 \qquad R_0 + S \rightleftharpoons R_1$$
$$T_1 + S \rightleftharpoons T_2 \qquad R_1 + S \rightleftharpoons R_2 \qquad [10.18]$$
$$\vdots \qquad\qquad \vdots$$
$$T_{n-1} + S \rightleftharpoons T_n \qquad R_{n-1} + S \rightleftharpoons R_n$$

This is illustrated in Fig. 10-29 for a tetramer.

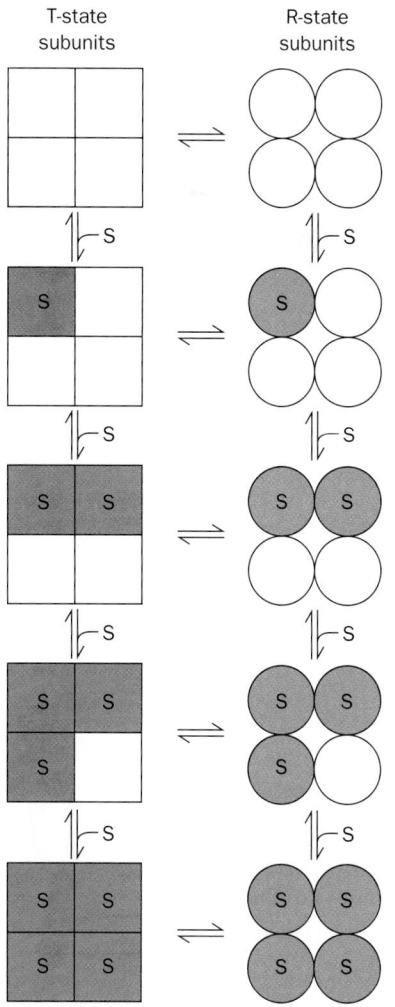

T-state subunits **R-state subunits**

FIGURE 10-29 The species and reactions permitted under the symmetry model of allosterism. Squares and circles represent T- and R-state protomers, respectively.

The equilibrium constant L for the conformational interconversion of the oligomeric protein in the absence of ligand is expressed

$$L = \frac{[T_0]}{[R_0]} \quad [10.19]$$

The microscopic dissociation constant for the R state, k_R, which according to Rule 3 is independent of the number of ligands bound to R, is expressed according to Eq. [10.15]:

$$k_R = \left(\frac{n - i + 1}{i}\right)\frac{[R_{i-1}][S]}{[R_i]} \quad (i = 1, 2, 3, \ldots, n) \quad [10.20]$$

The microscopic dissociation constant for ligand binding to the T state, k_T, is similarly expressed. The fractional saturation, Y_s, for ligand binding is

$$Y_s =$$
$$\frac{([R_1] + 2[R_2] + \cdots + n[R_n]) + ([T_1] + 2[T_2] + \cdots + n[T_n])}{n\{([R_0] + [R_1] + \cdots + [R_n]) + ([T_0] + [T_1] + \cdots [T_n])\}}$$
$$[10.21]$$

We shall make two definitions:

$$\alpha = [S]/k_R \qquad c = k_R/k_T$$

α may be considered a normalized ligand concentration. c is the ratio of the ligand-binding dissociation constants; c increases with the ligand-binding affinity of the T state relative to that of the R state. Then, combining the foregoing relationships as is shown in Section A of the Appendix to this chapter, we obtain the equation describing the symmetry model of allosterism for homotropic interactions:

$$Y_s = \frac{\alpha(1 + \alpha)^{n-1} + Lc\alpha(1 + c\alpha)^{n-1}}{(1 + \alpha)^n + L(1 + c\alpha)^n} \quad [10.22]$$

Note that this equation depends on three parameters, α, c, and L, which are, respectively, the normalized ligand concentration, the relative affinities of the T and R states for ligand, and the relative stabilities of the T and R states. In

contrast, the Hill equation (Section 10-1B) has but two parameters, K and n, whereas the number of parameters in the Adair equation is equal to the number of ligand-binding sites on the protein.

a. Homotropic Interactions

Let us examine the nature of the symmetry model by plotting Eq. [10.22] for a tetramer ($n = 4$) as a function of α for different values of the parameters L and c (Fig. 10-30). Three major points are evident from an inspection of these plots:

1. The degree of upward curvature exhibited by the initial sections of these sigmoid curves is indicative of their level of cooperativity.

2. When only the R state binds ligand ($c = 0$), the ligand-binding cooperativity increases as the oligomer's conformational preference for the non–ligand-binding T state increases (L increases; Fig. 10-30a). *For high L values, if a single ligand is to bind, it must "force" the protein into its less preferred R state. The requirement that all protomers change their conformational states in a concerted manner causes the remaining three ligand-binding sites to become available.* The binding of the first ligand therefore promotes the binding of subsequent ligands, which is the essence of a positive homotropic effect. Note that cooperativity and ligand-binding affinity are different quantities; in fact, for $c = 0$, curves indicative of high ligand-binding affinity (those with low L) exhibit low cooperativity and vice versa.

3. When the T state is highly preferred (L is large), ligand-binding cooperativity increases with the R state's ligand-binding affinity relative to that of the T state (decreasing c; Fig. 10-30b). At low ligand concentrations (low α) the amount of ligand bound (Y_S) increases with the ligand-binding affinity of the T state (increasing c) since the protein is largely in the T state. As α increases, however, the amount of ligand bound to the intrinsically less stable R state eventually surpasses that of the T state,

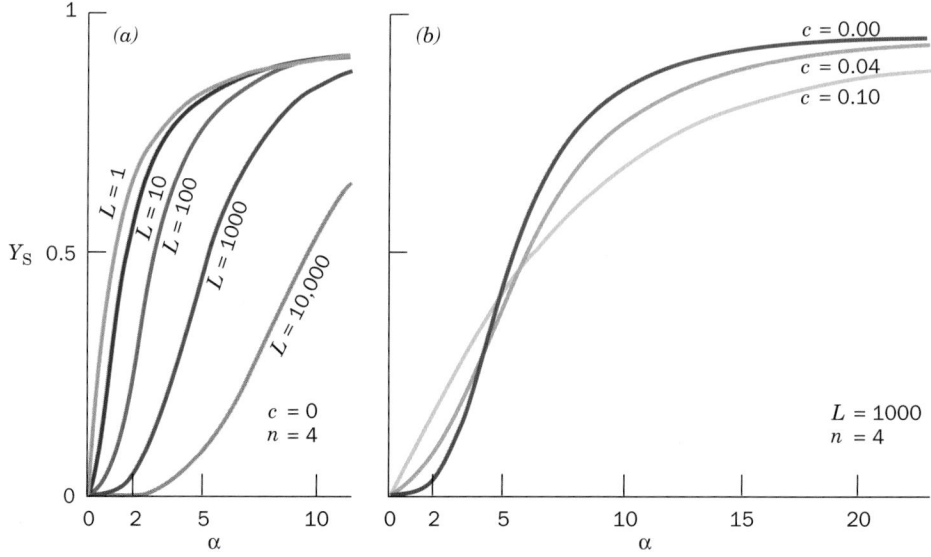

FIGURE 10-30   Symmetry model saturation function curves for tetramers according to Eq. [10.22]. Here $L = [T_0]/[R_0]$, $c = k_R/k_T$, and $\alpha = [S]/k_R$. (a) Their variation with L when $c = 0$. (b) Their variation with c when $L = 1000$. [After Monod, J., Wyman, J., and Changeux, J.P., *J. Mol. Biol.* **12**, 92 (1965).]

thereby resulting in a cooperative effect. This is because *the free energy of ligand binding stabilizes the R state with respect to the T state.*

b. Heterotropic Interactions

The symmetry model of allosterism is also capable of accounting for heterotropic effects. This comes about by assuming that each protomer has specific and independent binding sites for the three types of ligands: a substrate, S, that for simplicity let us assume binds only to the R state ($c = 0$); an **activator, A,** that also binds only to the R state; and an **inhibitor, I,** that binds only to the T state (Fig. 10-31). Then, through the derivation in Section B of the Appendix to this chapter, we obtain a more general equation for the symmetry model that describes heterotropic interactions as well as homotropic interactions:

$$Y_S = \frac{\alpha(1 + \alpha)^{n-1}}{(1 + \alpha)^n + \dfrac{L(1 + \beta)^n}{(1 + \gamma)^n}} \qquad [10.23]$$

where $\alpha = [S]/k_R$ as before and, analogously, $\beta = [I]/k_I$ and $\gamma = [A]/k_A$.

Note that this equation differs from Eq. [10.22] for $c = 0$ only in that the second term in the denominator is modulated by terms related to the amounts of activator and inhibitor bound to the oligomer.

Figure 10-32 indicates the consequences of effector binding to a tetramer that follows this model:

1. Activator binding ($\gamma > 0$) increases the concentration of the substrate-binding R state (the second term in the denominator of Eq. [10.23] decreases) because it is the only state capable of binding activator. *The presence of activator therefore increases the protein's substrate-binding affinity* (a positive heterotropic effect), although it decreases the protein's degree of substrate-binding cooperativity (compare Curves 1 and 2 in Fig. 10-32). (*Note:* There is nothing in the derivation of Eq. [10.23] that differentiates the roles of substrate and activator; consequently, the substrate and the activator each bind to the protein with a positive homotropic effect as well as being positive heterotropic effectors of each other.)

2. *The presence of inhibitor* ($\beta > 0$), which only binds to the T state, reduces the binding affinity for substrate (a negative heterotropic effect) by increasing the concentration of the T state (the second term in the denominator of Eq. [10.23] increases). Therefore, since substrate must "work harder" to convert the oligomer to the substrate-binding R state, inhibitor increases the cooperativity of substrate binding (compare Curves 2 and 3 of Fig. 10-32), as well as that for activator binding.

The model derived here is a rather simple one. In a more realistic but algebraically much more complicated symmetry model, all types of ligands would bind to both conformational states of the oligomer. Nevertheless, this model demonstrates that *both homotropic and heterotropic effects can be explained solely by the requirement that the molecular symmetry of the oligomer be conserved rather*

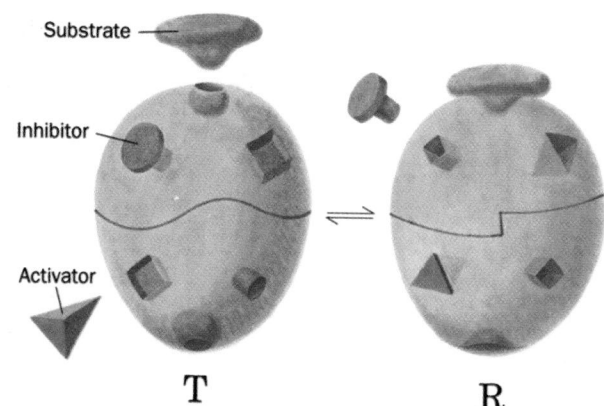

FIGURE 10-31   Heterotropic interactions in the symmetry model of allosterism. Heterotropic effects arise when substrates and activators bind exclusively (or at least preferentially) to the R state (*right*), and inhibitors bind exclusively (or at least preferentially) to the T state (*left*). The binding of substrate and/or activator to the oligomer therefore facilitates the further binding of substrate and activator. Conversely, the binding of inhibitor prevents (or at least inhibits) the oligomer from binding substrate or activator.

than by the existence of any direct interactions between ligands. In Section 10-4D, we compare the theoretical predictions of the symmetry model with our experimentally based model of hemoglobin oxygen binding.

C. *The Sequential Model*

The symmetry model provides a reasonable rationalization for the ligand-binding properties of many proteins. There are, however, several valid objections to it. Foremost of

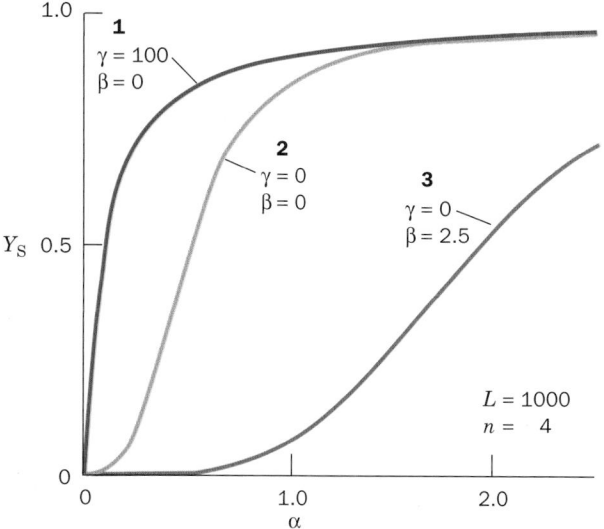

FIGURE 10-32   The effects of allosteric activator ($\gamma = [A]/k_A$) and inhibitor ($\beta = [I]/k_I$) on the shape of the fractional saturation curve for substrate ($\alpha = [S]/k_R$) according to Eq. [10.23] for tetramers. [After Monod, J., Wyman, J., and Changeux, J.-P., *J. Mol. Biol.* **12**, 94 (1965).]

these is that it is difficult to believe that oligomeric symmetry is invariably preserved in all proteins so that there are never any hybrid conformations such as $R_{n-2}T_2$. Furthermore, there are well-established instances of negative homotropic effects (e.g., in the GroEL–GroES complex, the binding of ATP to the cis ring of GroEL prevents ATP from binding to the trans ring; Section 9-2C), although the symmetry model, which permits only positive homotropic effects, is unable to account for them.

The symmetry model implicitly assumes Emil Fischer's "lock-and-key" model of ligand binding in which ligand-binding sites of proteins are rigid and complementary in shape to their ligand (Fig. 10-33, *left*). A more sophisticated extension of this "lock-and-key" model, known as the **induced-fit hypothesis,** postulates that *a flexible interaction between ligand and protein induces a conformational change in the protein, which results in its increased ligand-binding affinity* (Fig. 10-33, *right*). The observation, through X-ray crystal structure analysis, that such conformational changes occur in numerous proteins has established the validity of the induced-fit hypothesis.

Daniel Koshland, George Némethy, and David Filmer adapted the induced-fit hypothesis to explain allosteric effects. *In the resulting **sequential model** (alternatively, the **induced-fit** or **KNF model**), ligand binding induces a conformational change in a subunit; cooperative interactions arise through the influence that these conformational changes have on neighboring subunits (Fig. 10-34).* If, for example, they increase the neighbor's ligand-binding affinity, then ligand binding is positively cooperative. *The strengths of these interactions depend on the degree of mechanical coupling between subunits.* In the limit of very strong coupling, conformational changes become concerted, so that the oligomer maintains its symmetry (the symmetry model). With looser coupling, however, conformational changes occur sequentially as more and more ligand is bound (Fig. 10-35). Thus, *the essence of the sequential model is that a protein's ligand-binding affinity varies with its number of bound ligands, whereas in the symmetry model this affinity depends only on the protein's quaternary state.*

The degree of coupling between oligomer subunits depends on how these subunits are arranged, that is, on the protein's symmetry. Consequently, in the sequential model, the fractional saturation has a different algebraic form for each oligomeric symmetry. The form of the Adair equation (Eq. [10.17] for a tetramer) similarly depends on the number of subunits in the protein. In fact, the sequential model of allosterism may be considered an extension of the Adair model that provides a physical rationalization for the values of its microscopic dissociation constants, k_i.

D. *Hemoglobin Cooperativity*

Hemoglobin's fractional saturation curve is closely approximated by both the symmetry model and the sequential model (Fig. 10-36). Clearly such curves cannot by themselves be used to differentiate between these two models, if, in fact, either is correct. It is of interest, however, to compare these models with the mechanistic model of Hb we developed in Section 10-2C.

Hb, of course, is not composed of identical subunits, as the symmetry model demands. At least to a first approximation, however, the functional differences of Hb's closely related α and β subunits may be ignored (although their structural differences are essential to the molecular mechanism of Hb cooperativity). *To this approximation, Hb largely follows the symmetry model, although it also exhibits some features of the sequential model.* The quaternary T→R conformation change is concerted as the symmetry model requires. Yet ligand binding to the T state does cause small tertiary structural changes as the sequential model predicts. This phenomenon is evident in the X-ray struc-

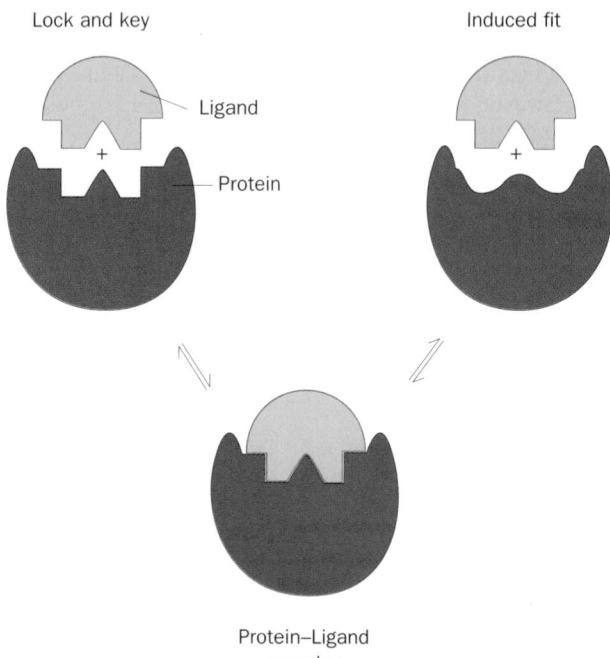

Lock and key

Induced fit

Ligand

Protein

Protein–Ligand complex

FIGURE 10-33 Models of ligand binding. In the lock-and-key mechanism of ligand binding (*left*), proteins are postulated to have preformed ligand-binding sites that are complementary in shape to their ligand. Under the induced-fit mechanism, a protein does not have this complementary binding site in the absence of ligand (*right*). Rather, the ligand induces a conformational change at the binding site that results in the complementary interaction.

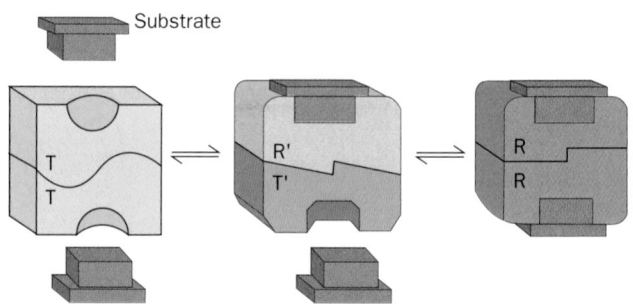

Substrate

FIGURE 10-34 The sequential model of allosterism. Substrate binding to the low-affinity T state induces conformational changes in unliganded subunits that give them ligand-binding affinities between those of the low-affinity T state and the high-affinity R state.

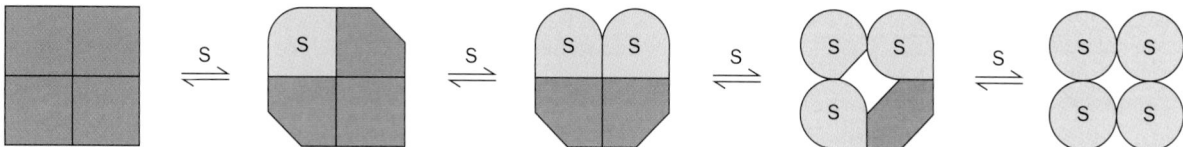

FIGURE 10-35 Sequential binding of ligand in the sequential model of allosterism. Ligand binding progressively induces conformational changes in the subunits, with the greatest changes occurring in those subunits that have bound ligand. The coupling between subunits is not necessarily of sufficient strength to maintain the symmetry of the oligomer as it is in the symmetry model.

ture of crystals of human Hb whose α subunits are fully oxygenated and whose β subunits are unliganded. This particular crystal form constrains the partially liganded Hb to remain in the T state. Nevertheless, its α subunit Fe's are 0.15 Å closer to the still domed porphyrins than they are in deoxyHb (25% of the total distance moved in the T→R transition). *Such tertiary structural changes are undoubtedly responsible for the buildup of strain that eventually triggers the T→R transition.*

A more telling but more difficult question to answer is does ligand-binding cooperativity in Hb arise solely from the T→R transition, in agreement with the symmetry model, or do the T and R states themselves exhibit at least some degree of cooperativity, in accord with the sequential model? Put another way, does the ligand-binding affinity of Hb subunits depend only on Hb's quaternary state (symmetry model) or does this affinity vary with the number of ligands bound to Hb (sequential model)? Although this matter has been under experimental scrutiny and a subject of spirited debate for over three decades, it is still not fully resolved. For example, the determination, by Andrea Mozzarelli and Eaton, of the Hill plot of the above-described single crystals of T-state Hb via sophisticated optical spectroscopic techniques is indicative of noncoop-

erative ligand binding, in accord with the symmetry model. Likewise, time-resolved spectroscopic measurements covering the picosecond to microsecond time scales after ligand binding was initiated indicate that the conformational changes that Hb initially undergoes on binding ligand are consistent with the symmetry model. However, detailed thermodynamic analysis, by Gary Ackers, of the interac-

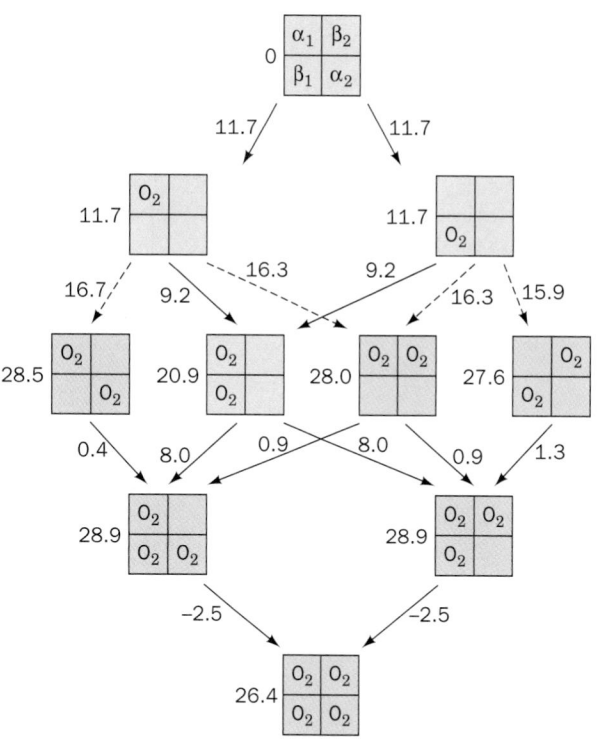

FIGURE 10-37 Free energy penalties for binding O_2 to various ligation states of Hb tetramers relative to O_2-binding to noncooperative Hb αβ dimers. Only the ten unique liganding states are diagrammed (an additional six states are related to those shown by Hb's 2-fold symmetry). Penalties, in kJ · mol^{-1}, for individual binding steps are shown beside the arrows. Cumulative free energy penalties are shown to the left of each Hb tetramer state. Ligation states that predominantly assume the T state are blue and those that predominantly assume the R state are red. The preferred pathways, those in which the free energy penalty of ligand binding progressively decreases with each successively bound ligand (those with all solid arrows), pass through the T state in which O_2 is bound to both sites on one αβ dimer before its conversion to the R state. Note that the T→R transition predominantly occurs via pathways in which at least one subunit in each αβ protomer is liganded. [Based on data from Ackers, G.K., *Adv. Prot. Chem.* **51,** 193 (1998).]

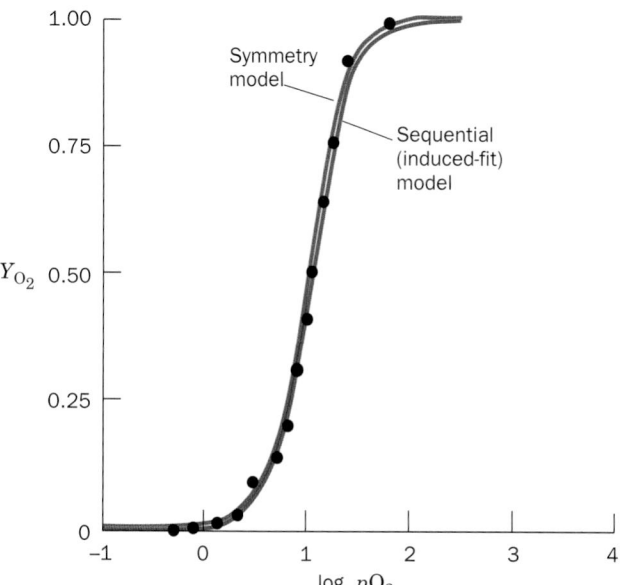

FIGURE 10-36 The sequential and the symmetry models of allosterism can provide equally good fits to the measured O_2-dissociation curve of Hb. [After Koshland, D.E., Jr., Némethy, G., and Filmer, D., *Biochemistry* **5,** 382 (1966).]

tions associated with the formation of each of Hb's 10 different ligation microstates (Fig. 10-37) indicates that the Hb tetramer undergoes a T→R transition only when at least one liganding site on each of its αβ protomers is occupied. This heretofore unrecognized symmetry is inconsistent with the symmetry model. Evidently, cooperativity arises both from concerted quaternary switching (as called for by the symmetry model) and from sequential modulation of ligand binding within each quaternary state through ligand-induced alterations in tertiary structure (in accord with the sequential model).

Appendix: ■ DERIVATIONS OF SYMMETRY MODEL EQUATIONS

A. *Homotropic Interactions — Equation [10.22]*

The fractional saturation Y_S for ligand binding is expressed:

$$Y_S = \frac{([R_1] + 2[R_2] + \cdots + n[R_n]) + ([T_1] + 2[T_2] + \cdots + n[T_n])}{n\{([R_0] + [R_1] + \cdots + [R_n]) + ([T_0] + [T_1] + \cdots + [T_n])\}} \quad [10.21]$$

Defining $\alpha = [S]/k_R$ and $c = k_R/k_T$, and using Eq. [10.20] to substitute $[R_{n-1}]$ for $[R_n]$, $[R_{n-2}]$ for $[R_{n-1}]$ etc., the terms enclosed by the first parentheses in the numerator of Eq. [10.21] are reduced to

$$[R_0]\left\{n\alpha + \frac{2n(n-1)\alpha^2}{2} + \cdots + \frac{n\,n!\alpha^n}{n!}\right\}$$

$$= [R_0]\alpha n\left\{1 + \frac{2(n-1)\alpha}{2} + \cdots + \frac{n(n-1)!\alpha^{n-1}}{n(n-1)!}\right\}$$

$$= [R_0]\alpha n(1 + \alpha)^{n-1}$$

and similarly, the terms in the first parentheses of the denominator of Eq. [10.21] become

$$[R_0]\left\{1 + n\alpha + \cdots + \frac{n!\alpha^n}{n!}\right\} = [R_0](1 + \alpha)^n$$

Likewise, the terms in the second parentheses of the numerator and the denominator of Eq. [10.21] assume the respective forms

$$[T_0]([S]/k_T)n(1 + [S]/k_T)^{n-1} = L[R_0]c\alpha n(1 + c\alpha)^{n-1}$$

and

$$[T_0](1 + [S]/k_T)^n = L[R](1 + c\alpha)^n$$

Accordingly,

$$Y_S = \frac{[R_0]\alpha n(1 + \alpha)^{n-1} + L[R_0]c\alpha n(1 + c\alpha)^{n-1}}{n\{[R_0](1 + \alpha)^n + L[R_0](1 + c\alpha)^n\}}$$

which, on cancellation of terms, yields the equation describing the symmetry model for homotropic interactions:

$$Y_S = \frac{\alpha(1 + \alpha)^{n-1} + Lc\alpha(1 + c\alpha)^{n-1}}{(1 + \alpha)^n + L(1 + c\alpha)^n} \quad [10.22]$$

B. *Heterotropic Interactions—Equation [10.23]*

For an oligomer that binds activator A and substrate S to only its R state, and inhibitor I to only its T state, the fractional saturation for substrate, Y_S, the fraction of substrate-binding sites occupied by substrate, is expressed:

$$Y_S = \frac{\sum_{i=1}^{n} \sum_{j=0}^{n} i[R_{i,j}]}{n\left(\sum_{i=0}^{n} \sum_{j=0}^{n} [R_{i,j}] + \sum_{k=0}^{n} [T_k]\right)}$$

Here the subscripts i, j, and k indicate the respective numbers of S, A, and I molecules that are bound to one oligomer; that is, $R_{i,j} \equiv RS_iA_j$ and $T_k \equiv TI_k$. Then defining $\alpha = [S]/k_R$ and following the foregoing derivation of Eq. [10.22]:

$$Y_S = \frac{\left(\sum_{j=0}^{n} [R_{0,j}]\right)\alpha n(1 + \alpha)^{n-1}}{n\left\{\left(\sum_{j=0}^{n} [R_{0,j}]\right)(1 + \alpha)^n + \sum_{k=0}^{n} [T_k]\right\}} = \frac{\alpha(1 + \alpha)^{n-1}}{(1 + \alpha)^n + L'}$$

where

$$L' = \sum_{k=0}^{n} [T_k] / \sum_{j=0}^{n} [R_{0,j}]$$

In analogy with the definition of α, we define $\beta = [I]/k_I$ and $\gamma = [A]/k_A$, and again follow the derivation of Eq. [10.22] to obtain:

$$\sum_{k=0}^{n} [T_k] = [T_0](1 + \beta)^n$$

and

$$\sum_{j=0}^{n} [R_{0,j}] = [R_{0,0}](1 + \gamma)^n$$

so that

$$L' = \frac{L(1 + \beta)^n}{(1 + \gamma)^n}$$

The symmetry model equation extended to include heterotropic effects is therefore expressed:

$$Y_S = \frac{\alpha(1 + \alpha)^{n-1}}{(1 + \alpha)^n + \dfrac{L(1 + \beta)^n}{(1 + \gamma)^n}} \quad [10.23]$$

■ CHAPTER SUMMARY

1 ■ Hemoglobin Function The heme group in myoglobin and in each subunit of hemoglobin reversibly binds O_2. In deoxyHb, the Fe(II) is five-coordinated to the four pyrrole nitrogen atoms of the protoporphyrin IX and to the protein's proximal His. On oxygenation, O_2 becomes the sixth ligand of Fe(II). Mb has a hyperbolic fractional saturation curve (Hill constant, $n = 1$). However, that of Hb is sigmoidal ($n \approx 2.8$) as a consequence of its cooperative O_2 binding: Hb binds its fourth O_2 with 100-fold greater affinity than its first O_2. The variation of O_2 affinity with pH, the Bohr effect, causes Hb to release O_2 in the tissues in response to the binding of protons liberated by the hydration of CO_2 to HCO_3^-. Hb facilitates the transport of CO_2, both directly, by binding CO_2 as N-terminal carbamate, and indirectly, by increasing the concentration of HCO_3^- through the Bohr effect. The presence of BPG in erythrocytes, which only binds to deoxyHb, further modulates the O_2 affinity of Hb. Short-term high-altitude adaptation results from an increase of BPG concentration in the erythrocytes, which increases the amount of O_2 delivered to the tissues by decreasing hemoglobin's O_2 affinity.

2 ■ Structure and Mechanism The α and β subunits of Hb consist mostly of seven or eight consecutive helices arranged to form a hydrophobic pocket that almost completely envelops the heme. Oxygen binding moves the Fe(II) from a position ~0.6 Å out of the heme plane on the side of the proximal His to the center of the heme, thereby relieving the steric interference that would otherwise occur between the bound O_2 and the porphyrin. The Fe(II) pulls the attached proximal His after it in a motion that can only occur if its imidazole ring reorients so as to avoid collision with the heme. In the T→R conformational transition, the symmetry equivalent $\alpha_1 C$–$\beta_2 FG$ and $\alpha_2 C$–$\beta_1 FG$ contacts simultaneously shift between two stable positions. Intermediate positions are sterically prevented, so that these contacts act as a two-position conformational switch. The Perutz mechanism of O_2 binding proposes that the low O_2 affinity of the T state arises from strain that prevents the Fe(II) from moving into the heme plane to form a strong Fe—O_2 bond. This strain is relieved by the concerted 4° shift of the Hb molecule to the high O_2 affinity R state. The quaternary shift is opposed by a network of salt bridges in the T state that involve the C-terminal carboxyl groups and that are ruptured in the R state. The stability of the R state relative to the T state increases with the degree of oxygenation as a result of the strain of binding O_2 in the T state. The existence of this strain has been demonstrated through the breakage of the Fe(II)–proximal His bond on hemoglobin's simultaneous binding of IHP, a tight-binding BPG analog that forces Hb into the T state, and NO, a strong ligand that forces Hb into the R state. Conversely, mutagenically detaching the proximal His from the protein eliminates most of hemoglobin's cooperativity. The Bohr effect results from increases in the pK's of the α N-terminal amino group and His 146β on forming the T-state salt bridges. Surface-exposed His residues also participate in the Bohr effect. BPG

binding occurs in the central cavity of T-state Hb through several salt bridges. The distal His residue protects deoxyHb from autooxidation by taking up the protons that would otherwise catalyze the oxidation of the heme Fe.

3 ■ Abnormal Hemoglobins Over 860 mutant varieties of Hb are known. About half of them are innocuous because they result in surface residue changes. However, alterations of internal residues often disrupt the structure of Hb, which causes hemolytic anemia. Changes at the O_2-binding site that stabilize the Fe(III) state eliminate O_2 binding to these subunits, which results in cyanosis. Mutations affecting subunit interfaces may stabilize either the R state or the T state, which, respectively, increase and decrease hemoglobin's O_2 affinity. Sickle-cell anemia is caused by the homozygous Hb mutant Glu 6β→Val, which promotes the gelation of the resulting deoxyHbS to form rigid 14-strand fibers that deform erythrocytes. Under gelation conditions, fiber growth occurs via a two-stage nucleation mechanism, resulting in a delay time that varies with the 30th to 50th power of the initial HbS concentration. Agents that increase this delay time to longer than the transit times of erythrocytes through the capillaries should therefore prevent sickle-cell crises and thus relieve the symptoms of sickle-cell anemia.

4 ■ Allosteric Regulation The Adair equation rationalizes the O_2-binding cooperativity of Hb by assigning a separate dissociation constant to each O_2 bound. Positive cooperativity results if these constants decrease sequentially. However, the Adair equation offers no physical insight as to why this occurs. The symmetry model proposes that symmetrical oligomers can exist in one of two conformational states, R and T, that differ in ligand-binding affinity. Ligand binding to the high-affinity state forces the oligomer to assume this conformation, which facilitates the binding of additional ligand. This homotropic model is extended to heterotropic effects by postulating that activator and substrate can bind only to the R state and inhibitor can bind only to the T state. The binding of activator forces the oligomer into the R state, which facilitates the binding of substrate and additional activator. The binding of inhibitor, however, forces the oligomer into the T state, which prevents substrate and activator binding. The sequential model postulates that an induced fit between ligand and substrate confers conformational strain on the protein that alters its affinity for binding other ligands without requiring the oligomer to maintain its symmetry. The Perutz mechanism for O_2 binding to Hb is structurally largely consistent with the symmetry model but exhibits some elements of the sequential model. However, Hb's ligand-binding cooperativity is in full accord with the symmetry model.

REFERENCES

GENERAL

Bunn, F.H. and Forget, B.G., *Hemoglobin: Molecular, Genetic and Clinical Aspects,* Saunders (1986). [A valuable compendium on normal and abnormal hemoglobins.]

Dickerson, R.E. and Geis, I., *Hemoglobin,* Benjamin/Cummings (1983). [A beautifully written and lavishly illustrated treatise on the structure, function, and evolution of hemoglobin.]

Everse, J., Vandegriff, K.K., and Winslow, R.M., *Hemoglobins,* Parts B and C, *Methods Enzymol.* **231** *and* **232** (1994).

Judson, H.F., *The Eighth Day of Creation* (expanded edition), Chapters 9 and 10, Cold Spring Harbor Laboratory Press

(1996). [Includes a fascinating historical account of how our present perception of hemoglobin structure and function came about.]

STRUCTURES OF MYOGLOBIN, HEMOGLOBIN, AND MODEL COMPOUNDS

Brunori, M., Nitric oxide moves myoglobin to center stage, *Trends Biochem. Sci.* **26,** 209–210 (2001).

Fermi, G., Perutz, M.F., Shaanan, B., and Fourme, R., The crystal structure of human deoxyhaemoglobin at 1.74 Å, *J. Mol. Biol.* **175,** 159–174 (1984).

Garry, D.J., Ordway, A.G., Lorenz, J.N., Radford, N., Chin, E.R., Grange, R.W., Bassel-Duby, R., and Williams, R.S., Mice without myoglobin, *Nature* **395**, 905–908 (1998).

Jameson, G.B., Molinaro, F.S., Ibers, J.A., Collman, J.P., Brauman, J.I., Rose, E., and Suslick, K.S., Models for the active site of oxygen-binding hemoproteins. Dioxygen binding properties and the structures of (2-methylimidazole)-*meso*-tetra(α,α,α,α,-*o*-pivalamidophenyl)porphinato iron(II)- ethanol and its dioxygen adduct, *J. Am. Chem. Soc.* **102**, 3224–3237 (1980). [The picket-fence complex.]

Liddington, R., Derewenda, Z., Dodson, G., and Harris, D., Structure of the liganded T state of haemoglobin identifies the origin of cooperative oxygen binding, *Nature* **331**, 725–728 (1988).

Phillips, S.E.V., Structure and refinement of oxymyoglobin at 1.6 Å resolution, *J. Mol. Biol.* **142**, 531–554 (1980).

Shaanan, B., Structure of human oxyhaemoglobin at 2.1 Å resolution, *J. Mol. Biol.* **171**, 31–59 (1983).

Takano, T., Structure of myoglobin refined at 2.0 Å resolution, *J. Mol. Biol.* **110**, 537–568, 569–584 (1977).

MECHANISM OF HEMOGLOBIN OXYGEN BINDING

Baldwin, J. and Chothia, C., Haemoglobin: The structural changes related to ligand binding and its allosteric mechanism, *J. Mol. Biol.* **129**, 175–220 (1979). [The exposition of a detailed mechanism of O₂ binding to Hb based on the structures of oxyHb and deoxyHb.]

Barrick, D., Ho, N.T., Simplaceanu, V., Dahlquist, F.W., and Ho, C., A test of the role of the proximal histidines in the Perutz model for cooperativity in haemoglobin, *Nature Struct. Biol.* **4**, 78–83 (1997). [Describes the experiments in which the proximal His is detached from the F helix.]

Gelin, B.R., Lee, A.W.-N., and Karplus, M., Haemoglobin tertiary structural change on ligand binding, *J. Mol. Biol.* **171**, 489–559 (1983). [A theoretical study of the dynamics of O₂ binding to Hb.]

Perutz, M.F., Stereochemistry of cooperative effects in haemoglobin, *Nature* **228**, 726–734 (1970). [The landmark paper in which the Perutz mechanism was first proposed. Although many of its details have since been modified, the basic model remains intact.]

Perutz, M.F., Regulation of oxygen affinity of hemoglobin, *Annu. Rev. Biochem.* **48**, 327–386 (1979). [An examination of the Perutz mechanism in light of structural and spectroscopic data.]

Perutz, M.F., Mechanisms of cooperativity and allosteric regulation in proteins, *Quart. Rev. Biophys.* **22**, 139–236 (1989). [Contains a detailed structural description of allosterism in hemoglobin.]

Perutz, M.F., Wilkinson, A.J., Paoli, M., and Dodson, G.G., The stereochemical mechanism of the cooperative effects in hemoglobin revisited, *Annu. Rev. Biophys. Biomol. Struct.* **27**, 1–34 (1998).

BOHR EFFECT AND BPG BINDING

Arnone, A., X-Ray studies of the interaction of CO₂ with human deoxyhaemoglobin, *Nature* **247**, 143–145 (1974).

Benesch, R.E. and Benesch, R., The mechanism of interaction of red cell organic phosphates with hemoglobin, *Adv. Protein Chem.* **28**, 211–237 (1974).

Kilmartin, J.V. and Rossi-Bernardi, L., Interactions of hemoglobin with hydrogen ion, carbon dioxide and organic phosphates, *Physiol. Rev.* **53**, 836–890 (1973).

Lenfant, C., Torrance, J., English, E., Finch, C.A., Reynafarje, C., Ramos, J., and Faura, J., Effect of altitude on oxygen binding by hemoglobin and on organic phosphate levels, *J. Clin. Invest.* **47**, 2652–2656 (1968).

Perutz, M.F., Kilmartin, J.V., Nishikura, K., Fogg, J.H., and Butler,

P.J.G., Identification of residues contributing to the Bohr effect of human haemoglobin, *J. Mol. Biol.* **138**, 649–670 (1980).

Richard, V., Dodson, G.G., and Mauguen, Y., Human deoxy-haemoglobin-2,3-diphosphoglycerate complex low-salt structure at 2.5 Å resolution, *J. Mol. Biol.* **233**, 270–274 (1993).

Sun, D.P., Zou, M., Ho, N.T., and Ho, C., Contribution of surface histidyl residues in the α-chain of the Bohr effect of human normal adult hemoglobin: Roles of global electrostatic effects, *Biochemistry* **36**, 6663–6673 (1997).

ABNORMAL HEMOGLOBINS

Allison, A.C., The discovery of resistance to malaria of sickle-cell heterozygotes, *Biochem. Molec. Biol. Educ.* **30**, 279–287 (2002).

Baudin-Chich, V., Pagnier, J., Marden, M., Bohn, B., Lacaze, N., Kister, J., Schaad, O., Edelstein, S.J., and Poyart, C., Enhanced polymerization of recombinant human deoxyhemoglobin β6 Glu→Ile, *Proc. Natl. Acad. Sci.* **87**, 1845–1849 (1990).

Bunn, F.H., Pathogenesis and treatment of sickle cell disease, *New Engl. J. Med.* **337**, 762–769 (1997).

Bunn, F.H., Human hemoglobins: sickle hemoglobin and other mutants, *in* Stamatoyannopoulos, G., Majerus, P.W., Perlmutter, R.M., and Varmus, H. (Eds.), *The Molecular Basis of Blood Diseases* (3rd ed.), Chapter 7, Elsevier (2001).

Eaton, W.A. and Hofrichter, J., Sickle cell hemoglobin polymerization, *Adv. Prot. Chem.* **40**, 63–279 (1990). [An authoritative and exhaustive review of HbS polymerization.]

Eaton, W.A. and Hofrichter, J., The biophysics of sickle cell hydroxyurea therapy, *Science* **268**, 1142–1143 (1995).

Harrington, D.J., Adachi, K., and Royer, W.E., Jr., The high resolution crystal structure of deoxyhemoglobin S, *J. Mol. Biol.* **272**, 398–407 (1997).

Nagel, R.L., Haemoglobinopathies due to structural mutations, *in* Provan, D. and Gribben, J. (Eds.), *Molecular Haematology*, pp. 121–133, Blackwell Science (2000).

Perutz, M., *Protein Structure. New Approaches to Disease and Therapy*, Chapter 6, Freeman (1992).

Perutz, M.F. and Lehmann, H., Molecular pathology of human haemoglobin, *Nature* **219**, 902–909 (1968). [A ground-breaking study correlating the clinical symptoms and inferred structural alterations of numerous mutant hemoglobins.]

Steinberg, M.H., Management of sickle cell disease, *New Engl. J. Med.* **340**, 1021–1030 (1999).

Strasser, B.J., Sickle-cell anemia, a molecular disease, *Science* **286**, 1488–1490 (1999). [A short history of Pauling's characterization of sickle-cell anemia.]

Watowich, S.J., Gross, L.J., and Josephs, R., Intermolecular contacts within sickle hemoglobin fibers, *J. Mol. Biol.* **209**, 821–828 (1989).

Weatherall, D.J., Clegg, J.B., Higgs, D.R., and Wood, W.G., The hemoglobinopathies, *in* Scriver, C.R., Beaudet, A.L., Sly, W.S., and Valle, D. (Eds.), *The Metabolic & Molecular Bases of Inherited Disease* (8th ed.), *pp.* 4571–4436, McGraw-Hill (2001). [A detailed review of abnormal hemoglobins.]

ALLOSTERIC REGULATION

Ackers, G.A., Deciphering the molecular code of hemoglobin allostery, *Adv. Protein Chem.* **51**, 185–253 (1998). [Presents thermodynamic arguments that both of Hb's αβ dimers must be ligated for quaternary switching to occur.]

Eaton, W.A., Henry, E.R., Hofrichter, J., and Mozzarelli, A., Is cooperative oxygen binding by hemoglobin really understood? *Nature Struct. Biol.* **6**, 351–359 (1999). [An incisive review.]

Fersht, A., *Structure and Mechanism in Protein Science*, Chapter 10, Freeman (1999).

Koshland, D.E., Jr., Némethy, G., and Filmer, D., Comparison of experimental binding data and theoretical models in proteins containing subunits, *Biochemistry* **5**, 365–385 (1966). [The formulation of the sequential model of allosteric regulation.]

Monod, J., Wyman, J., and Changeux, J.P., On the nature of allosteric transitions: A plausible model, *J. Mol. Biol.* **12**, 88–118 (1965). [The exposition of the symmetry model of allosteric regulation.]

PROBLEMS

1. The urge to breathe in humans results from a high blood CO_2 content; there are no direct physiological sensors of blood pO_2. Skindivers often **hyperventilate** (breathe rapidly and deeply for several minutes) just before making a protracted dive in the belief that they will thereby increase the O_2 content of their blood. This belief results from the fact that hyperventilation represses the breathing urge by expelling significant quantities of CO_2 from the blood. In light of what you know about the properties of hemoglobin, is hyperventilation a useful procedure? Is it safe? Explain.

2. Explain why the Hill constant, n, can never be larger than the number of ligand-binding sites on the protein.

*3. In the Bohr effect, protonation of the N-terminal amino groups of hemoglobin's α chains is responsible for ~30% of the 0.6 mol of H^+ that combine with Hb on the release of 1 mol of O_2 at pH 7.4. Assuming that this group has pK = 7.0 in oxyHb, what is its pK in deoxyHb?

4. As one of the favorites to win the La Paz, Bolivia, marathon, you have trained there for the several weeks it requires to become adapted to its 3700-m altitude. A manufacturer of running equipment who sponsors an opponent has invited you for the weekend to a prerace party at a beach house near Lima, Peru, with the assurance that you will be flown back to La Paz at least a day before the race. Is this a token of his respect for you or an underhanded attempt to handicap you in the race? Explain (see Fig. 10-9).

5. In active muscles, the pO_2 may be 10 torr at the cell surface and 1 torr at the mitochondria (the organelles where oxidative metabolism occurs). How might myoglobin (p_{50} = 2.8 torr) facilitate the diffusion of O_2 through these cells? Active muscles consume O_2 much faster than do other tissues. Could myoglobin also be an effective O_2 transport protein in other tissues? Explain.

6. Erythrocytes that have been stored for over a week in standard acid–citrate–dextrose medium become depleted in BPG. Discuss the merits of using fresh versus week-old blood in blood transfusions.

7. The following fractional saturation data have been measured for a certain blood sample:

pO_2	Y_{O_2}	pO_2	Y_{O_2}
20	0.14	60	0.59
30	0.26	70	0.66
40	0.39	80	0.72
50	0.50	90	0.76

What are the Hill constant and the p_{50} of this blood sample? Are they normal?

8. An anemic individual, whose blood has only half the normal Hb content, may appear to be in good health. Yet a normal individual is incapacitated by exposure to sufficient carbon monoxide to occupy half of his/her heme sites (pCO of 1 torr for ~1 h; CO binds to Hb with 200-fold greater affinity than does O_2). Explain.

*9. The X-ray structure of Hb Rainier (Tyr 145β→Cys) indicates that the mutant Cys residue forms a disulfide bond with Cys 93β of the same subunit. This holds the β subunit's C-terminal residue in a quite different orientation than it assumes in HbA. How would the following quantities for Hb Rainier compare with those of HbA? Explain. (a) The oxygen affinity, (b) the Bohr effect, (c) the Hill constant, and (d) the BPG affinity.

10. The crocodile, which can remain under water without breathing for up to 1 h, drowns its air-breathing prey and then dines at its leisure. An adaptation that aids the crocodile in doing so is that it can utilize virtually 100% of the O_2 in its blood, whereas humans, for example, can extract only ~65% of the O_2 in their blood. Crocodile Hb does not bind BPG. However, crocodile deoxyHb preferentially binds HCO_3^-. How does this help the crocodile obtain its dinner?

11. The gelation time of an equimolar mixture of HbA and HbS is less than that of a solution of only HbS in the same concentration that it has in the mixture. What does this observation imply about the participation of HbA in the gelation of HbS?

12. The severely anemic condition of homozygotes for HbS results in an elevated BPG content in their erythrocytes. Discuss whether or not this is a beneficial effect.

13. As organizer of an expedition that plans to climb several very high mountains, it is your responsibility to choose its members. Each of the applicants for one of the positions on the team is a heterozygote for one of the following variant hemoglobins: (1) HbS, (2) **Hb Hyde Park** [His F8(92)β→Tyr], (3) **Hb Riverdale–Bronx** [Gly B6(24)β→Arg], (4) **Hb Memphis** [Glu B4(23)α→Gln], and (5) **Hb Cowtown** [His HC3(146)β→Leu]. Assuming that all of these candidates are equal in ability at low altitudes, which one would you choose for the position? Explain your reasoning.

14. Show that the Adair equation for a tetramer reduces to the Hill equation for $k_1 \approx k_2 \approx k_3 \gg k_4$ and to a hyperbolic relationship for $k_1 = k_2 = k_3 = k_4$.

15. Derive the equilibrium constant for the reaction $R_2 \rightleftharpoons T_2$ for a symmetry model n-mer in terms of the parameters L, c, and α.

*16. Derive the equation for the fraction of protein molecules in the R state, $\overline{R}$, for the homotropic symmetry model in terms of the parameters n, L, c, and α. Plot this function versus α for $n = 4$, $L = 1000$, and $c = 0$ and discuss its physical significance.

17. In the symmetry model of allosterism, why must an inhibitor (which causes a negative heterotropic effect with the substrate) undergo a positive homotropic effect?

18. At low concentrations, the hemoglobin tetramer reversibly dissociates into two $\alpha_1\beta_1$ dimers. What is the Hill constant for O_2 binding to these dimers? Explain.

19. Describe the nature of the allosteric changes (homotropic or heterotropic, positive or negative) that take place in the GroEL/ES system during the various stages of its catalytic cycle (Fig. 9-23).

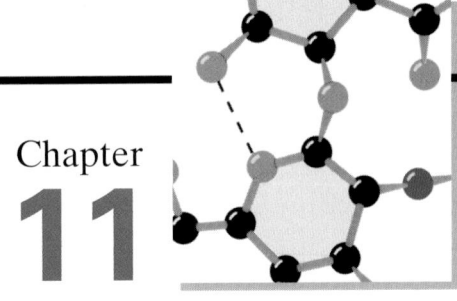

Chapter

11

Sugars and Polysaccharides

Carbohydrates or **saccharides** (Greek: *sakcharon,* sugar) are essential components of all living organisms and are, in fact, the most abundant class of biological molecules. The name carbohydrate, which literally means "carbon hydrate," stems from their chemical composition, which is roughly $(C \cdot H_2O)_n$, where $n \geq 3$. The basic units of carbohydrates are known as **monosaccharides.** Many of these compounds are synthesized from simpler substances in a process named **gluconeogenesis** (Section 23-1). Others (and ultimately nearly all biological molecules) are the products of **photosynthesis** (Section 24-3), the light-powered combination of CO_2 and H_2O through which plants and certain bacteria form "carbon hydrates." The metabolic breakdown of monosaccharides (Chapters 17 and 21) provides much of the energy used to power biological processes. Monosaccharides are also principal components of nucleic acids (Section 5-1A), as well as important elements of complex lipids (Section 12-1D).

Oligosaccharides consist of a few covalently linked monosaccharide units. They are often associated with proteins (**glycoproteins**) and lipids (**glycolipids**) in which they have both structural and regulatory functions (glycoproteins and glycolipids are collectively called **glycoconjugates**). **Polysaccharides** consist of many covalently linked monosaccharide units and have molecular masses ranging well into the millions of daltons. They have indispensable structural functions in all types of organisms but are most conspicuous in plants because **cellulose,** their principal

structural material, comprises up to 80% of their dry weight. Polysaccharides such as **starch** in plants and **glycogen** in animals serve as important nutritional reservoirs.

The elucidation of the structures and functions of carbohydrates has lagged well behind those of proteins and nucleic acids. This can be attributed to several factors. Carbohydrate compounds are often heterogeneous, both in size and in composition, which greatly complicates their physical and chemical characterization. They are not subject to the types of genetic analysis that have been invaluable in the study of proteins and nucleic acids because saccharide sequences are not genetically specified but are built up through the sequential actions of specific enzymes (Section 23-3B). Furthermore, it has been difficult to establish assays for the biological activities of polysaccharides because of their largely passive roles. Nevertheless, it is abundantly clear that carbohydrates are essential elements in many, if not most, biological processes.

In this chapter, we explore the structures, chemistry, and, to a limited extent, the functions of carbohydrates, alone and in association with proteins. Glycolipid structures are considered in Section 12-1D. The biosynthesis of complex carbohydrates is discussed in Section 23-3.

1 ■ MONOSACCHARIDES

Monosaccharides or **simple sugars** are aldehyde or ketone derivatives of straight-chain polyhydroxy alcohols containing at least three carbon atoms. Such substances, for example, **D-glucose** and **D-ribulose,** cannot be hydrolyzed to form simpler saccharides.

$$
\begin{array}{cc}
\overset{1}{C}\!\!\diagup^{O}_{\diagdown H} & \\
\overset{2}{H-C-OH} & {}^{1}CH_2OH \\
\overset{3}{HO-C-H} & \overset{2}{C=O} \\
\overset{4}{H-C-OH} & \overset{3}{H-C-OH} \\
\overset{5}{H-C-OH} & \overset{4}{H-C-OH} \\
{}^{6}CH_2OH & {}^{5}CH_2OH \\
\textbf{D-Glucose} & \textbf{D-Ribulose}
\end{array}
$$

In this section, the structures of the monosaccharides and some of their biologically important derivatives are discussed.

A. *Classification*

Monosaccharides are classified according to the chemical nature of their carbonyl group and the number of their C atoms. If the carbonyl group is an aldehyde, as in glucose, the sugar is an **aldose.** If the carbonyl group is a ketone, as in ribulose, the sugar is a **ketose.** The smallest monosaccharides, those with three carbon atoms, are **trioses.** Those with four, five, six, seven, etc., C atoms are, respectively, **tetroses, pentoses, hexoses, heptoses,** etc.

These terms may be combined so that, for example, glucose is an **aldohexose,** whereas ribulose is a **ketopentose.**

Examination of D-glucose's molecular formula indicates that all but two of its six C atoms—Cl and C6—are chiral centers, so that D-glucose is one of $2^4 = 16$ stereoisomers that comprise all possible aldohexoses. In general, n-carbon aldoses have 2^{n-2} stereoisomers. The stereochemistry and names of the D-aldoses are presented in Fig. 11-1. Emil Fischer elucidated these configurations for the aldohexoses in 1896. According to the Fischer convention (Section 4-2B), *D sugars have the same absolute configuration at the asymmetric center farthest removed from their carbonyl group as does D-glyceraldehyde.* The L sugars, in accordance with this convention, are mirror images of their D counter-

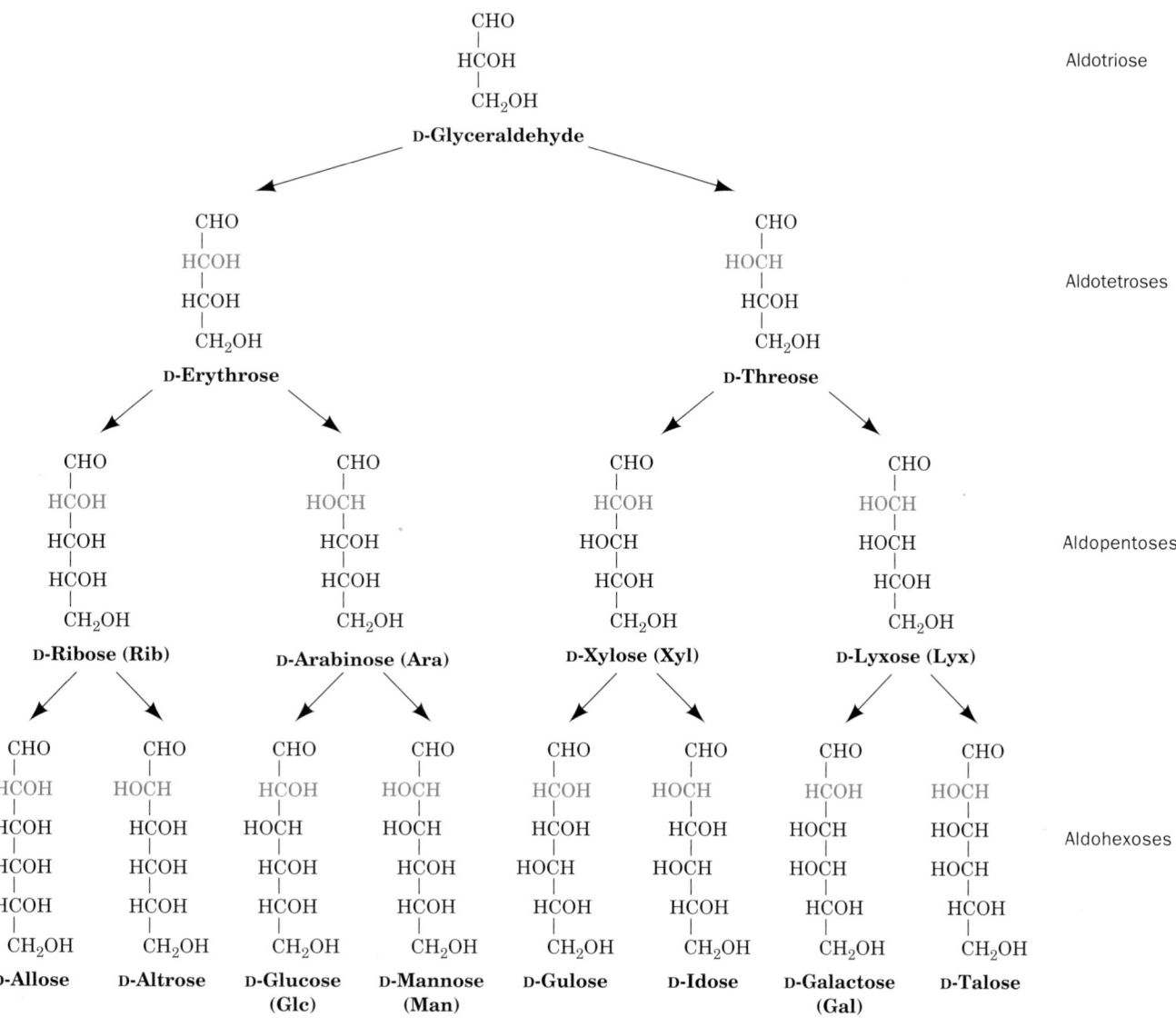

FIGURE 11-1   The stereochemical relationships, shown in Fischer projection, among the D-aldoses with three to six carbon atoms. The configuration about C2 (*red*) distinguishes the members of each pair.

parts, as is shown below in Fischer projection for glucose.

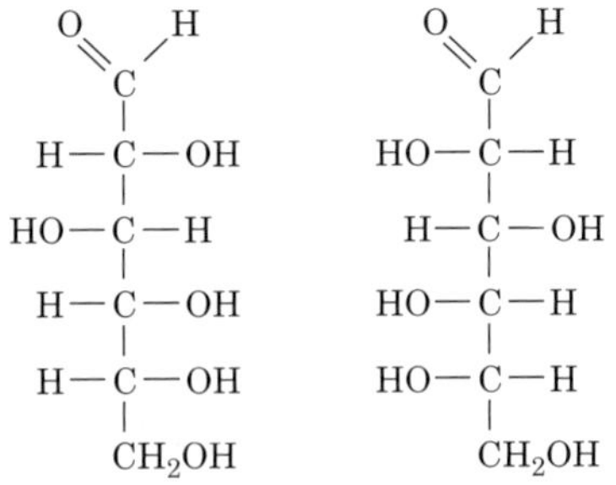

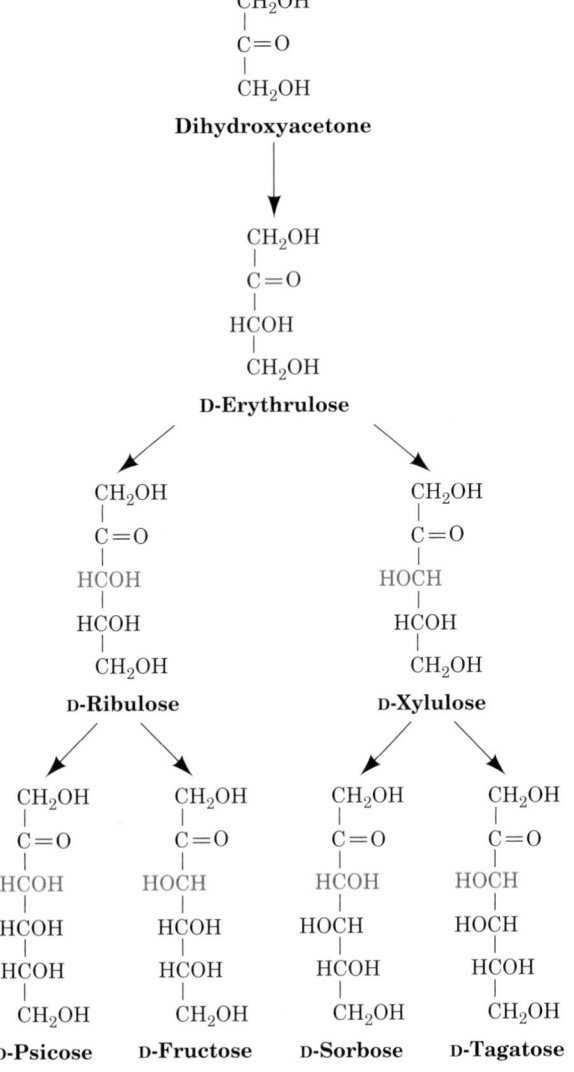

D-Glucose   **L-Glucose**

Sugars that differ only by the configuration about one C atom are known as **epimers** of one another. Thus D-glucose and **D-mannose** are epimers with respect to C2, whereas D-glucose and **D-galactose** are epimers with respect to C4 (Fig. 11-1). However, D-mannose and D-galactose are not epimers of each other because they differ in configuration about two of their C atoms.

$$
\begin{array}{c}
\text{CH}_2\text{OH} \\
| \\
\text{C}=\text{O} \\
| \\
\text{CH}_2\text{OH}
\end{array}
$$

Dihydroxyacetone

$$
\begin{array}{c}
\text{CH}_2\text{OH} \\
| \\
\text{C}=\text{O} \\
| \\
\text{HCOH} \\
| \\
\text{CH}_2\text{OH}
\end{array}
$$

D-Erythrulose

D-Ribulose

D-Xylulose

D-Psicose   D-Fructose   D-Sorbose   D-Tagatose

FIGURE 11-2   The stereochemical relationships among the D-ketoses with three to six carbon atoms. The configuration about C3 (*red*) distinguishes the members of each pair.

D-*Glucose is the only aldose that commonly occurs in nature as a monosaccharide.* However, it and several other monosaccharides including D-glyceraldehyde, D-ribose, D-mannose, and D-galactose are important components of larger biological molecules. L Sugars are biologically much less abundant than D sugars.

The position of their carbonyl group gives ketoses one less asymmetric center than their isomeric aldoses (e.g., compare D-fructose and D-glucose). *n*-Carbon ketoses therefore have 2^{n-3} stereoisomers. Those with their ketone function at C2 are the most common form (Fig. 11-2). Note that some of these ketoses are named by the insertion of *-ul-* before the suffix *ose* in the name of the corresponding aldose; thus D-xylulose is the ketose corresponding to the aldose D-xylose. Dihydroxyacetone, D-fructose, D-ribulose, and D-xylulose are the biologically most prominent ketoses.

B. Configurations and Conformations

Alcohols react with the carbonyl groups of aldehydes and ketones to form **hemiacetals** and **hemiketals,** respectively (Fig. 11-3). The hydroxyl and either the aldehyde or the ketone functions of monosaccharides can likewise react intramolecularly to form cyclic hemiacetals and hemiketals (Fig. 11-4). The configurations of the substituents to each carbon atom of these sugar rings are conveniently represented by their **Haworth projection formulas.**

A sugar with a 6-membered ring is known as a **pyranose** in analogy with **pyran,** the simplest compound containing such a ring. Similarly, sugars with 5-membered rings are designated **furanoses** in analogy with **furan.**

Pyran   **Furan**

The cyclic forms of glucose and fructose with 6- and 5-membered rings are therefore known as **glucopyranose** and **fructofuranose,** respectively.

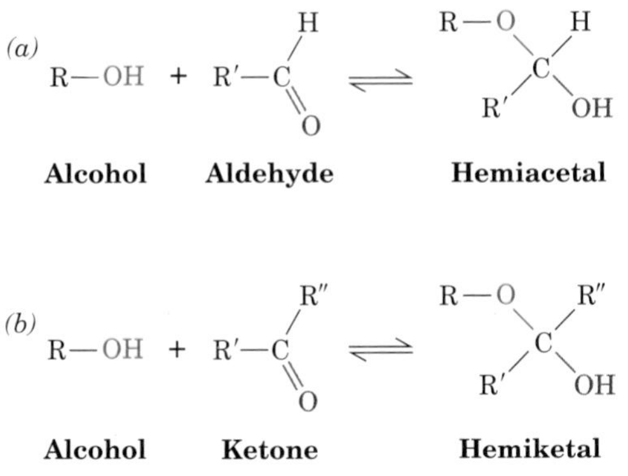

(a)

R—OH + R′—C   ⇌   Hemiacetal

Alcohol   Aldehyde   Hemiacetal

(b)

R—OH + R′—C   ⇌   Hemiketal

Alcohol   Ketone   Hemiketal

FIGURE 11-3   The reactions of alcohols with (*a***) aldehydes to form hemiacetals and (***b***) ketones to form hemiketals.**

(a)

D-Glucose
(linear form)

α-D-Glucopyranose
(Haworth projection)

(b)

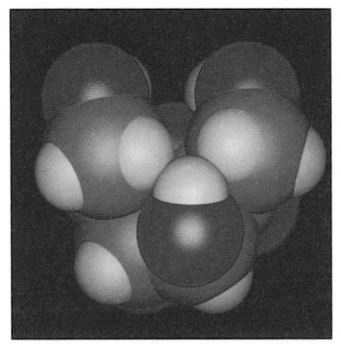

D-Fructose
(linear form)

α-D-Fructofuranose
(Haworth projection)

FIGURE 11-4   Cyclization reactions for hexoses. (*a*) D-Glucose in its linear form reacts to yield the cyclic hemiacetal α-D-glucopyranose and (*b*) D-fructose in its linear form reacts to yield the hemiketal α-D-fructofuranose. The cyclic sugars are shown as both Haworth projections and space-filling models. [Space-filling models courtesy of Robert Stodola, Fox Chase Cancer Center.]

a. Cyclic Sugars Have Two Anomeric Forms

The Greek letters preceding the names in Fig. 11-4 still need to be explained. The cyclization of a monosaccharide renders the former carbonyl carbon asymmetric. The resulting pair of diastereomers are known as **anomers** and the hemiacetal or hemiketal carbon is referred to as the **anomeric** carbon. In the α anomer, the OH substituent to the anomeric carbon is on the opposite side of the sugar ring from the CH$_2$OH group at the chiral center that des-

ignates the D or L configuration (C5 in hexoses). The other anomer is known as the β form (Fig. 11-5).

The two anomers of D-glucose, as any pair of diastereomers, have different physical and chemical properties. For example, the values of the specific optical rotation, $[\alpha]_D^{20}$, for α-D-glucose and β-D-glucose are, respectively, +112.2° and +18.7°. When either of these pure substances is dissolved in water, however, the specific optical rotation of the solution slowly changes until it reaches an equilibrium

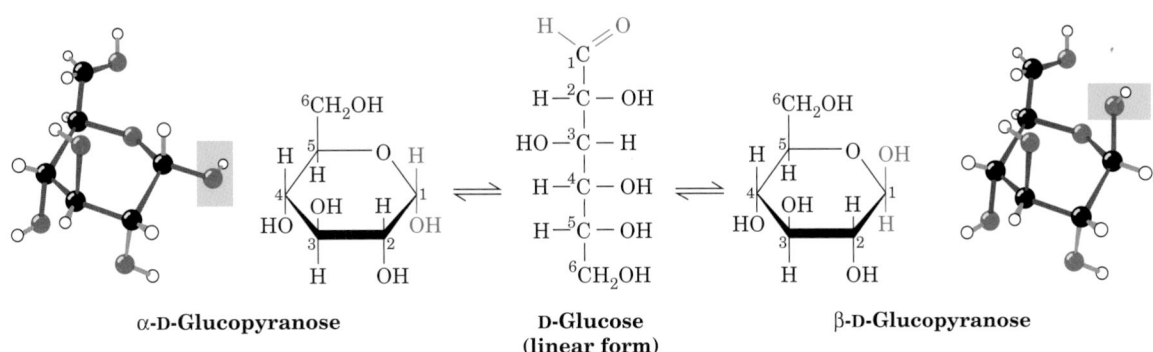

α-D-Glucopyranose

D-Glucose
(linear form)

β-D-Glucopyranose

FIGURE 11-5   The anomeric monosaccharides α-D-glucopyranose and β-D-glucopyranose, drawn as both Haworth projections and ball-and-stick models. These pyranose sugars interconvert through the linear form of D-glucose and differ only by the configurations about their anomeric carbon atoms.
C1. ✍ **See Kinemage Exercise 7-1**

value of $[\alpha]_D^{20} = +52.7°$. This phenomenon is known as **muta-rotation;** in glucose, it results from the formation of an equilibrium mixture consisting of 63.6% of the β anomer and 36.4% of the α anomer (the optical rotations of separate molecules in solution are independent of each other so that the optical rotation of a solution is the weighted average of the optical rotations of its components). The interconversion between these anomers occurs via the linear form of glucose (Fig. 11-5). Yet, since the linear forms of these monosaccharides are normally present in only minute amounts, these carbohydrates are accurately described as cyclic polyhydroxy hemiacetals or hemiketals.

b. Sugars Are Conformationally Variable

Hexoses and pentoses may each assume pyranose or furanose forms. The equilibrium composition of a particular monosaccharide depends somewhat on conditions but mostly on the identity of the monosaccharide. For instance, NMR measurements indicate that whereas glucose almost exclusively assumes its pyranose form in aqueous solutions, fructose is 67% pyranose and 33% furanose, and ribose is 75% pyranose and 25% furanose (although in polysaccharides, glucose, fructose, and ribose residues are exclusively in their respective pyranose, furanose, and furanose forms). Although, in principle, hexoses and larger sugars can form rings of seven or more atoms, such rings are rarely

observed because of the greater stabilities of the 5- and 6-membered rings that these sugars can also form. The internal strain of 3- and 4-membered sugar rings makes them unstable with respect to linear forms.

The use of Haworth formulas may lead to the erroneous impression that furanose and pyranose rings are planar. This cannot be the case, however, because all of the atoms in these rings are tetrahedrally (sp^3) hybridized. The pyranose ring, like the cyclohexane ring, may assume a **boat** or a **chair** conformation (Fig. 11-6). The relative stabilities of these various conformations depend on the stereochemical interactions between the substituents on the ring. The boat conformer crowds the substituents on its "bow" and "stern" and eclipses those along its sides, so that in cyclohexane it is ~25 kJ · mol^{-1} less stable than the chair conformer. The ring substituents on the chair conformer (Fig. 11-6b) fall into two geometrical classes: the rather close-fitting **axial** groups that extend parallel to the ring's threefold rotational axis and the staggered, and therefore minimally encumbered, **equatorial** groups. Since the axial and equatorial groups on a cyclohexane ring are conformationally interconvertible, a given ring has two alternative chair forms (Fig. 11-7); the one that predominates usually has the lesser crowding among its axial substituents. The conformational situation of a group directly affects its chemical reactivity. For example, equatorial OH groups on pyranoses esterify more readily than do axial OH groups. Note that β-D-glucose is the only D-aldohexose that can simultaneously have all five non-H substituents in the equatorial position (left side of Fig. 11-7). Perhaps this is why glucose is the most abundant naturally occurring monosaccharide. The conformational properties of furanose rings are discussed in Section 29-2B in relation to their effects on the conformations of nucleic acids.

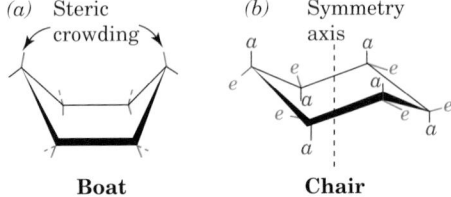

FIGURE 11-6   Conformations of the cyclohexane ring. (*a*) In the boat conformation, substituents at the "bow" and "stern" (*red*) are sterically crowded, whereas those along its sides (*green*) are eclipsed. (*b*) In the chair conformation, the substituents that extend parallel to the ring's threefold rotation axis are designated axial [*a*] and those that extend roughly outward from this symmetry axis are designated equatorial [*e*]. The equatorial substituents about the ring are staggered so that they alternately extend above and below the mean plane of the ring.

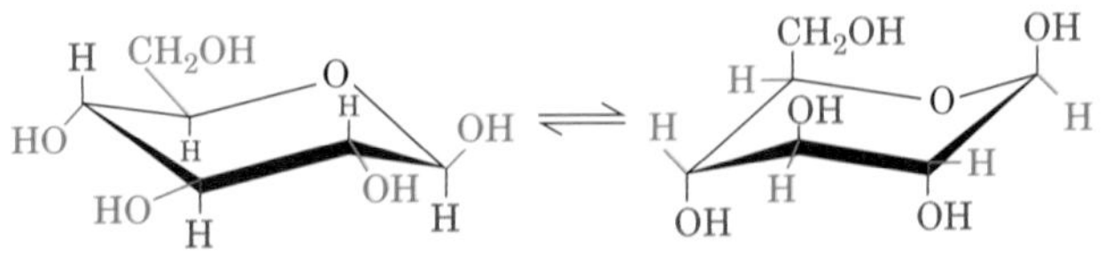

FIGURE 11-7   The two alternative chair conformations of β-D-glucopyranose. In the conformation on the left, which predominates, the relatively bulky OH and CH$_2$OH substituents all occupy equatorial positions, whereas in that on the right (drawn in ball-and-stick form in Fig. 11-5, *right*) they occupy the more crowded axial positions. See Kinemage Exercise 7-1

C. Sugar Derivatives

a. Polysaccharides Are Held Together by Glycosidic Bonds

The chemistry of monosaccharides is largely that of their hydroxy and carbonyl groups. For example, in an acid-catalyzed reaction, the anomeric hydroxyl of a sugar reversibly condenses with alcohols to form α- and β-**glycosides** (Greek: *glykys,* sweet) (Fig. 11-8). The bond connecting the anomeric carbon to the acetal oxygen is termed a **glycosidic bond.** *Polysaccharides are held together by glycosidic bonds between neighboring monosaccharide units.* The glycosidic bond is therefore the carbohydrate analog of the peptide bond in proteins. The bond in a nucleoside linking its ribose residue to its base is also a glycosidic bond (Section 5-1A).

The hydrolysis of glycosidic bonds is catalyzed by enzymes known as **glycosidases** that differ in specificity according to the identity and anomeric configuration of the glycoside but are often rather insensitive to the identity of the alcohol residue. Under basic or neutral conditions and in the absence of glycosidases, however, the glycosidic bond is stable, so glycosides do not undergo mutarotation as do

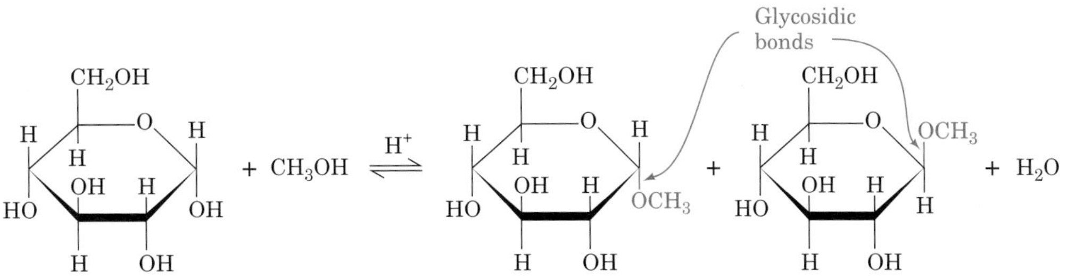

FIGURE 11-8 The acid-catalyzed condensation of α-D-glucose with methanol to form an anomeric pair of methyl-D-glucosides.

monosaccharides. The methylation of the non-anomeric OH groups of monosaccharides requires more drastic conditions than is required for the formation of methyl glycosides, such as treatment with dimethyl sulfate.

b. Oxidation–Reduction Reactions

Because the cyclic and linear forms of aldoses and ketoses interconvert so readily, these sugars undergo reactions typical of aldehydes and ketones. Mild oxidation of an aldose, either chemically or enzymatically, results in the conversion of its aldehyde group to a carboxylic acid function, thereby yielding an **aldonic acid** such as **gluconic acid.** Aldonic acids are named by appending the suffix *-onic acid* to the root name of the parent aldose.

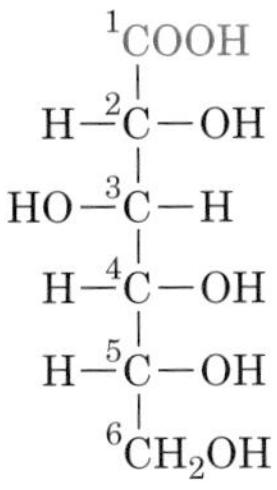

D-Gluconic acid

Saccharides bearing anomeric carbon atoms that have not formed glycosides are termed **reducing sugars** because of the facility with which the aldehyde group reduces mild oxidizing agents. A standard test for the presence of a reducing sugar is the reduction of Ag^+ in an ammonia solution (**Tollens' reagent**) to yield a metallic silver mirror lining on the inside of the reaction vessel.

The specific oxidation of the primary alcohol group of aldoses yields **uronic acids,** which are named by appending *-uronic acid* to the root name of the parent aldose. **D-Glucuronic acid, D-galacturonic acid,** and **D-mannuronic acid** are important components of many polysaccharides.

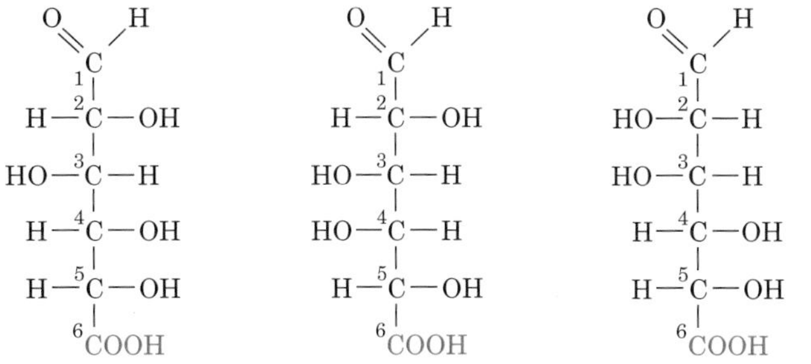

D-Glucuronic acid **D-Galacturonic acid** **D-Mannuronic acid**

Uronic acids can assume the pyranose, furanose, and linear forms.

Both aldonic and uronic acids have a strong tendency to internally esterify so as to form five- and six-membered lactones (Fig. 11-9). **Ascorbic acid (vitamin C,** Fig. 11-10)

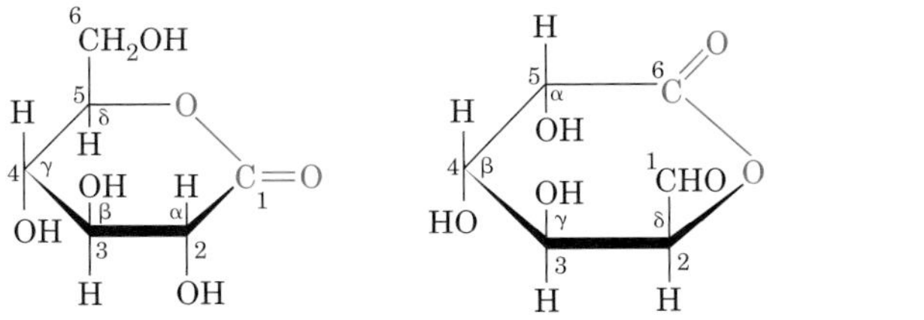

D-Glucono-δ-lactone **D-Glucurono-δ-lactone**

FIGURE 11-9 D-Glucono-δ-lactone and D-glucurono-δ-lactone are, respectively, the lactones of D-gluconic acid and D-glucuronic acid. The "δ" indicates that the O atom closing the lactone ring is also substituent to C_δ.

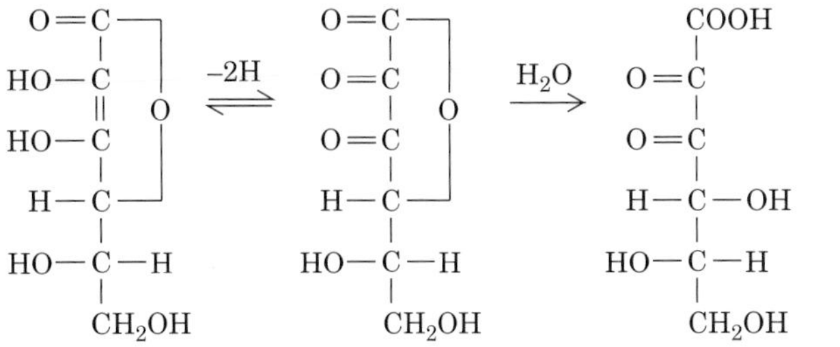

L-Ascorbic acid **L-Dehydroascorbic acid** **L-Diketogulonic acid**

FIGURE 11-10 The reversible oxidation of L-ascorbic acid to L-dehydroascorbic acid. This is followed by the physiologically irreversible hydrolysis of its lactone ring to form L-diketogulonic acid.

is a γ-lactone that is synthesized by plants and almost all animals except primates and guinea pigs. Its prolonged deficiency in the diet of humans results in the disease known as **scurvy,** which is caused by the impairment of collagen formation (Section 8-2B). Scurvy generally results from a lack of fresh food. This is because, under physiological conditions, ascorbic acid is reversibly oxidized to **dehydro-ascorbic acid,** which, in turn, is irreversibly hydrolyzed to the vitamin-inactive **diketogulonic acid** (Fig. 11-10).

Aldoses and ketoses may be reduced under mild conditions, for example, by treatment with $NaBH_4$, to yield acyclic polyhydroxy alcohols known as **alditols,** which are named by appending the suffix *-itol* to the root name of the parent aldose. **Ribitol** is a component of flavin coenzymes (Section 16-2C), and **glycerol** and the cyclic polyhydroxy alcohol *myo*-**inositol** are important lipid components (Section 12-1). **Xylitol** is a sweetener that is used in "sugarless" gum and candies.

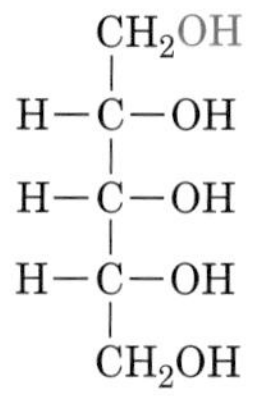

Ribitol

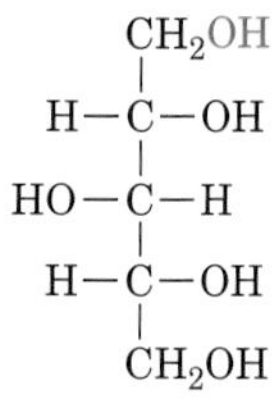

Xylitol

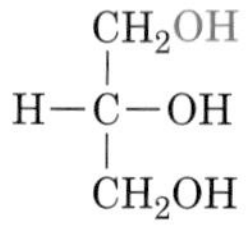

Glycerol

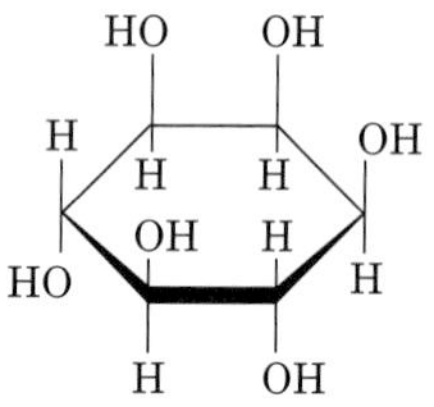

myo-**Inositol**

c. Other Biologically Important Sugar Derivatives

Monosaccharide units in which an OH group is replaced by H are known as **deoxy sugars.** The biologically most important of these is **β-D-2-deoxyribose,** the sugar component of DNA's sugar–phosphate backbone (Section 5-1A). **L-Rhamnose** and **L-fucose** are widely occurring polysaccharide components.

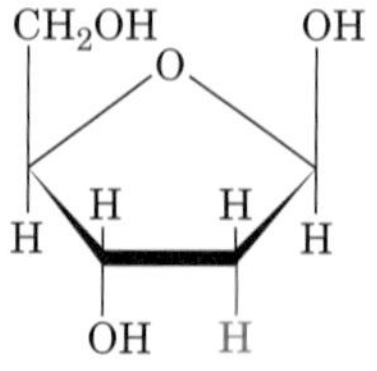

β-D-2-Deoxyribose

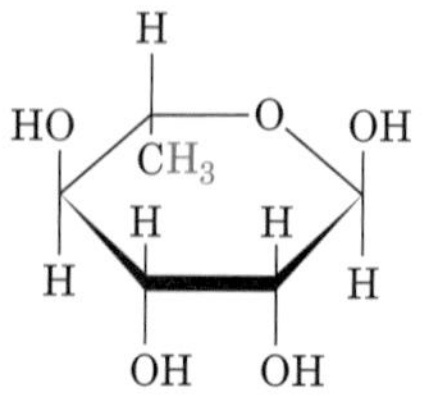

α-L-Rhamnose
(6-deoxy-L-mannose)

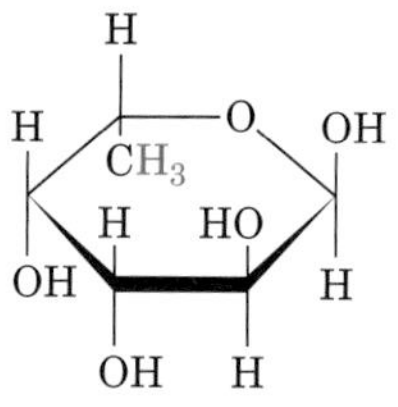

α-L-Fucose
(6-deoxy-L-galactose)

In **amino sugars,** one or more OH groups are replaced by an often acetylated amino group. **D-Glucosamine** and **D-galactosamine** are components of numerous biologically important polysaccharides.

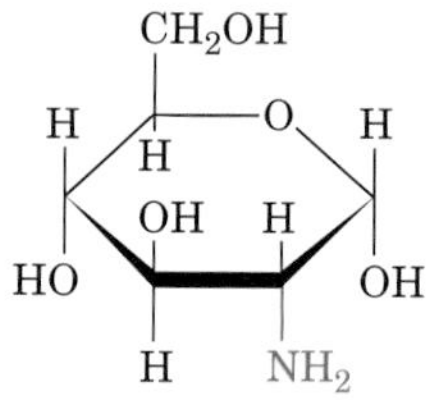

α-D-Glucosamine
(2-amino-2-deoxy-α-D-glucopyranose)

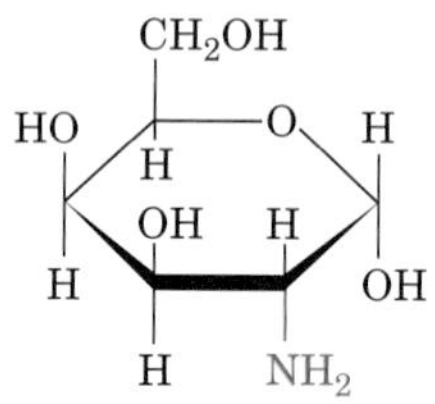

α-D-Galactosamine
(2-amino-2-deoxy-α-D-galactopyranose)

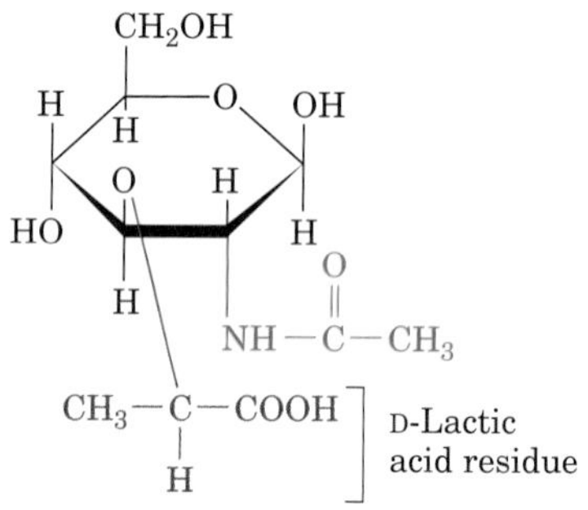

N-Acetylmuramic acid (NAM)

The amino sugar derivative **N-acetylmuramic acid,** which consists of **N-acetyl-D-glucosamine** in an ether linkage with **D-lactic acid,** is a prominent component of bacterial cell walls (Section 11-3B). *N*-Acetylneuraminic acid, which is derived from *N*-acetylmannosamine and pyruvic acid (Fig. 11-11), is an important constituent of glycoproteins (Section 11-3C) and glycolipids (Section 12-1D). *N*-Acetylneuraminic acid and its derivatives are often referred to as **sialic acids.**

2 ■ POLYSACCHARIDES

Polysaccharides, which are also known as **glycans,** consist of monosaccharides linked together by glycosidic bonds. They are classified as **homopolysaccharides** or **heteropolysaccharides** if they consist of one type or more than

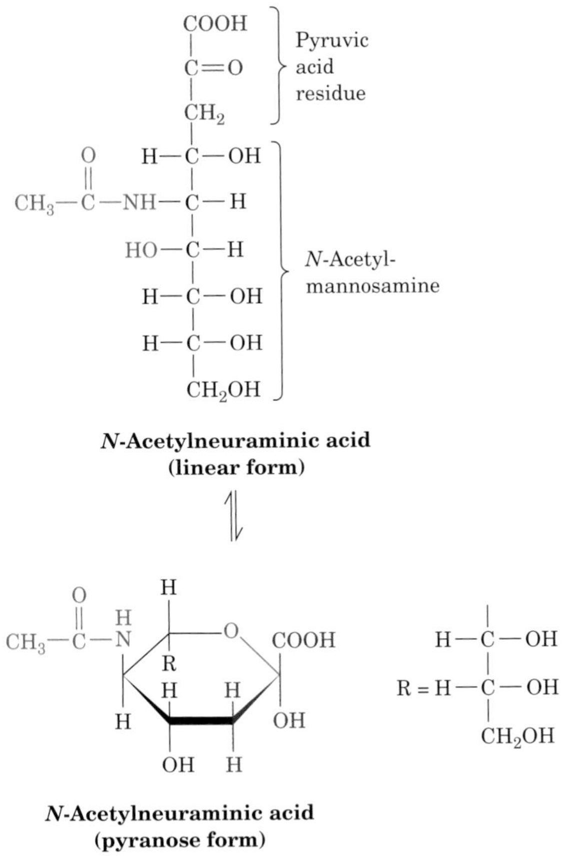

COOH
|
C=O   } Pyruvic
| } acid
CH₂   } residue

O   H—C—OH
‖
CH₃—C—NH—C—H
|
HO—C—H   } N-Acetyl-
| } mannosamine
H—C—OH
|
H—C—OH
|
CH₂OH

N-Acetylneuraminic acid
(linear form)

⇅

O   H
‖   H   O
CH₃—C—N   COOH   H—C—OH
R |
H   H   R = H—C—OH
H   OH |
OH   H   CH₂OH

N-Acetylneuraminic acid
(pyranose form)

FIGURE 11-11 *N*-**Acetylneuraminic acid in its linear and pyranose forms.** Note that its pyranose ring incorporates the pyruvic acid residue (*blue*) and part of the mannose moiety.

one type of monosaccharide residue. Homopolysaccharides may be further classified according to the identity of their monomeric unit. For example, **glucans** are polymers of glucose, whereas **galactans** are polymers of galactose. Although monosaccharide sequences of heteropolysaccharides can, in principle, be as varied as those of proteins, they are usually composed of only a few types of monosaccharides that alternate in a repetitive sequence.

Polysaccharides, in contrast to proteins and nucleic acids, form branched as well as linear polymers. This is because glycosidic linkages can be made to any of the hydroxyls of a monosaccharide. Fortunately for structural biochemists, most polysaccharides are linear and those that branch do so in only a few well-defined ways.

In this section, we discuss the structures of the simplest polysaccharides, the disaccharides, and then consider the structures and properties of the most abundant classes of polysaccharides. We begin by outlining how polysaccharide structures are elucidated.

A. Carbohydrate Analysis

The purification of carbohydrates can, by and large, be effected by chromatographic and electrophoretic procedures

similar to those used in protein purification (Sections 6-3 and 6-4). Affinity chromatography (Section 6-3C), using immobilized proteins known as **lectins,** is a particularly powerful technique in this regard. Lectins are sugar-binding proteins that are usually of plant origin but that also occur in animals and bacteria. Among the best known lectins are jack bean **concanavalin A** (Fig. 8-40), which specifically binds α-D-glucose and α-D-mannose residues, and wheat germ **agglutinin** (so named because it causes cells to agglutinate or clump together), which specifically binds β-*N*-acetylmuramic acid and α-*N*-acetylneuraminic acid.

Characterization of an oligosaccharide requires that the identities, anomers, linkages, and order of its component monosaccharides be elucidated. The linkages of the monosaccharides may be determined by **methylation analysis,** a technique pioneered by Norman Haworth in the 1930s: *Methyl ethers not at the anomeric C atom are resistant to acid hydrolysis but glycosidic bonds are not. Consequently, if an oligosaccharide is exhaustively methylated and then hydrolyzed, the free OH groups on the resulting methylated monosaccharides mark the former positions of the glycosidic bonds.* Methylated monosaccharides are often identified by **gas–liquid chromatography** (GLC; a technique in which the stationary phase is an inert solid, such as diatomaceous earth, impregnated with a low-volatility liquid, such as silicone oil, and the mobile phase is an inert gas, such as He, into which the sample has been flash evaporated) combined with mass spectrometry (GLC/MS). Other mass spectrometric techniques for analyzing nonvolatile substances are discussed in Section 7-1J (although note, e.g., that all the aldoses and ketoses in Figs. 11-1 and 11-2 are isomers and hence have identical molecular masses). HPLC techniques may similarly be used.

The sequence and anomeric configurations of the monosaccharides in an oligosaccharide can be determined through the use of specific **exoglycosidases.** These enzymes specifically hydrolyze their corresponding monosaccharides from the nonreducing ends of oligosaccharides (the ends lacking a free anomeric carbon atom) in a manner analogous to the actions of exopeptidases on proteins (Section 7-1A). For example, **β-galactosidase** excises the terminal β anomers of galactose, whereas **α-mannosidase** does so with the α anomers of mannose. Some of these exoglycosidases also exhibit specificity for the **aglycone,** the sugar chains to which the monosaccharide to be excised (the **glycone**) is linked. Through the use of mass spectrometry, the sequence of a polysaccharide may be deduced from the mass decrements generated by exoglycosidases. The use of **endoglycosidases** (hydrolases that cleave glycosidic bonds between nonterminal sugar residues) of varying specificities can also supply useful sequence information. The proton and ¹³C NMR of oligosaccharides can provide the complete sequence of an oligosaccharide if sufficient material is available. Moreover, two-dimensional NMR techniques can can reveal oligosaccharide conformations (Section 8-3A).

B. *Disaccharides*

We begin our studies of polysaccharides by considering disaccharides (Fig. 11-12). **Sucrose,** the most abundant disaccharide, occurs throughout the plant kingdom and is familiar to us as common table sugar. Its structure (Fig. 11-12)

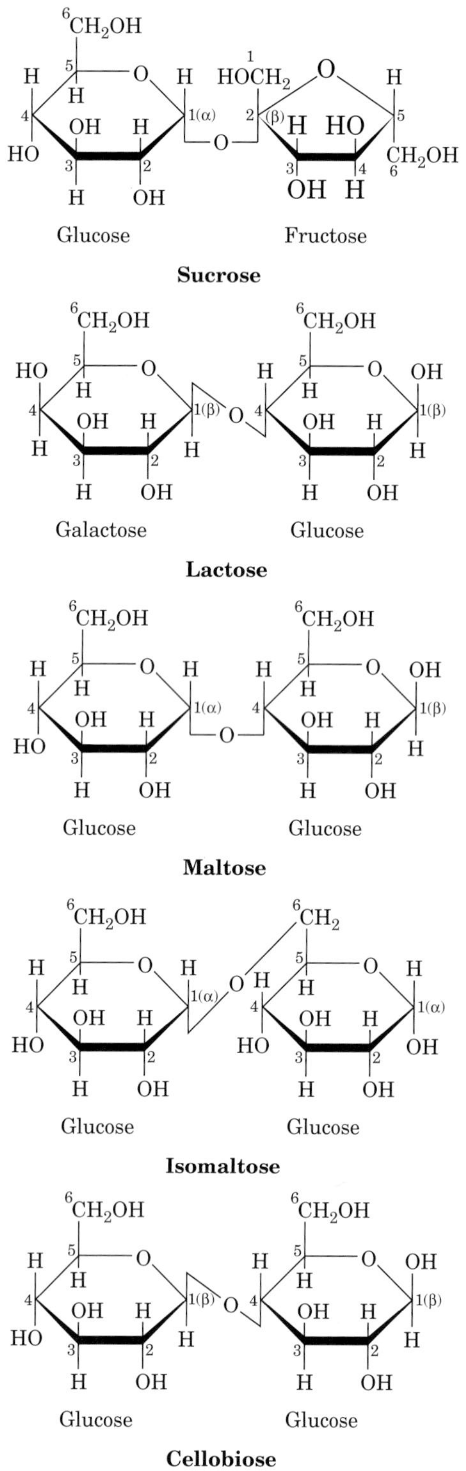

FIGURE 11-12   Several common disaccharides. 🙢 See
Kinemage Exercise 7-2

was established by methylation analysis as described above and was later confirmed by its X-ray structure. To name a polysaccharide systematically, one must specify its component monosaccharides, their ring types, their anomeric forms, and how they are linked together. Sucrose is therefore O-α-D-glucopyranosyl-$(1 \rightarrow 2)$-β-D-fructofuranoside, where the symbol $(1 \rightarrow 2)$ indicates that the glycosidic bond links C1 of the glucose residue to C2 of the fructose residue. Note that since these two positions are the anomeric carbon atoms of their respective monosaccharides, sucrose is not a reducing sugar (as the suffix *-ide* implies).

The hydrolysis of sucrose to D-glucose and D-fructose is accompanied by a change in optical rotation from *dextro* to *levo*. Consequently, hydrolyzed sucrose is sometimes called **invert sugar** and the enzyme that catalyzes this process, **α-D-glucosidase,** is archaically named **invertase.**

Lactose [O-β-D-galactopyranosyl-$(1 \rightarrow 4)$-D-glucopyranose] or milk sugar occurs naturally only in milk, where its concentration ranges from 0 to 7% depending on the species. The free anomeric carbon of its glucose residue makes lactose a reducing sugar.

> Infants normally have the intestinal enzyme **β-D-galactosidase** or **lactase** that catalyzes the hydrolysis of lactose to its component monosaccharides for absorption into the bloodstream. Many adults, however, including most Africans and almost all Asians, have a low level of this enzyme (as do most adult mammals). Consequently, much of the lactose in any milk they drink moves through their digestive tract to the colon, where its bacterial fermentation produces large quantities of CO_2, H_2, and irritating organic acids. This results in an embarrassing and often painful digestive upset termed **lactose intolerance.** Perhaps this is why Chinese cuisine, which is noted for the wide variety of foodstuffs it employs, is devoid of milk products. However, modern food technology has come to the aid of milk lovers who develop lactose intolerance: Milk products in which the lactose has been hydrolyzed enzymatically are now widely available.

There are several common glucosyl–glucose disaccharides. These include **maltose** [O-α-D-glucopyranosyl-$(1 \rightarrow 4)$-D-glucopyranose], an enzymatic hydrolysis product of starch; **isomaltose,** its $\alpha(1 \rightarrow 6)$ isomer; and **cellobiose,** its $\beta(1 \rightarrow 4)$ isomer, the repeating disaccharide of cellulose.

Only a few tri- or higher oligosaccharides occur in significant natural abundance. Not surprisingly, they all occur in plants.

C. *Structural Polysaccharides: Cellulose and Chitin*

Plants have rigid cell walls (Fig. 1-9) that, in order to maintain their shapes, must be able to withstand osmotic pressure differences between the extracellular and intracellular spaces of up to 20 atm. In large plants, such as trees, the cell walls also have a load-bearing function. Cellulose, the primary structural component of plant cell walls (Fig. 11-13), accounts for over half of the carbon in the biosphere: $\sim10^{15}$ kg of cellulose are estimated to be synthesized and degraded annually. Although cellulose is predominantly of vegetable origin, it also occurs in the stiff

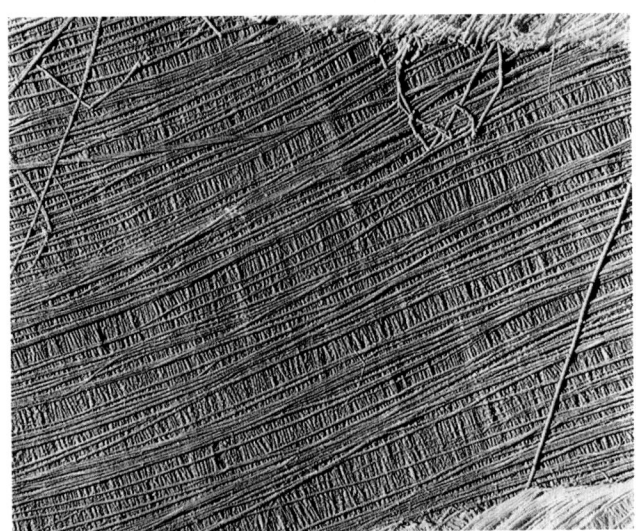

CH₂OH and glucose structure shown.

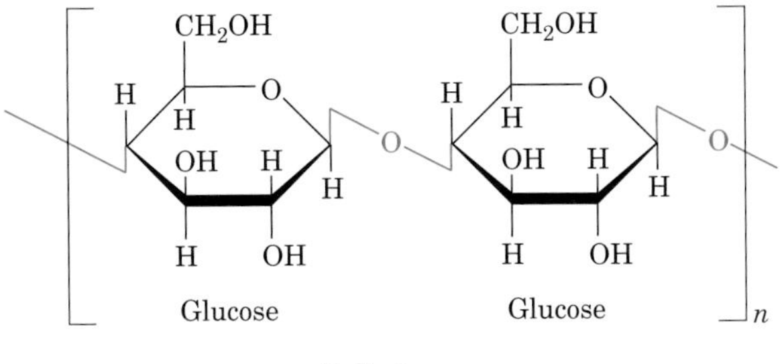

Cellulose

FIGURE 11-14 The primary structure of cellulose. Here *n* may be several thousand.

of up to 15,000 D-glucose residues (a glucan) linked by $\beta(1 \rightarrow 4)$ glycosidic bonds (Fig. 11-14). As is generally true of large polysaccharides, it has no defined size since, in contrast to proteins and nucleic acids, there is no genetically determined template that directs its synthesis.

X-Ray studies of cellulose fibers led Anatole Sarko to tentatively propose the structure diagrammed in Fig. 11-15. This highly cohesive, hydrogen bonded structure gives cellulose fibers exceptional strength and makes them water insoluble despite their hydrophilicity.

In plant cell walls, the cellulose fibers are embedded in and cross-linked by a matrix of several polysaccharides that are composed of glucose as well as other monosaccharides. In wood, this cementing matrix also contains a large pro-

FIGURE 11-13 Electron micrograph of the cellulose fibers in the cell wall of the alga *Chaetomorpha melagonium*. Note that the cell wall consists of layers of parallel fibers. [Biophoto Associates/Photo Researchers.]

outer mantles of marine invertebrates known as **tunicates** (urochordates; Fig. 1-11).

The primary structure of cellulose was determined through methylation analysis. Cellulose is a linear polymer

FIGURE 11-15 Proposed structural model of cellulose. Cellulose fibers consist of ~40 parallel glucan chains arranged in an extended fashion. Each of the $\beta(1 \rightarrow 4)$-linked glucose units in a chain is turned over with respect to its preceding residue and is held in this position by intrachain hydrogen bonds (*dashed lines*). The glucan chains line up laterally to form sheets, and these sheets stack vertically such that they are staggered by half the length of a glucose unit. The entire assembly is stabilized by intermolecular hydrogen bonds between glucose units of neighboring chains. Hydrogen atoms not participating in hydrogen bonds have been omitted for clarity.

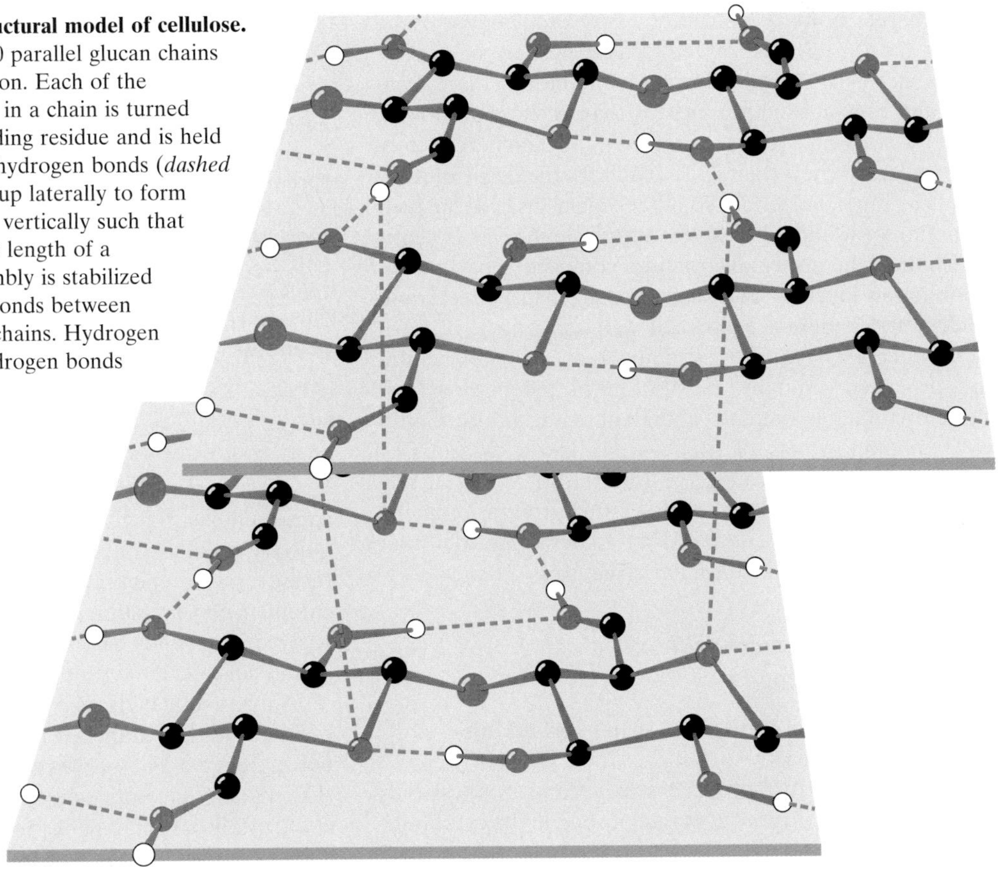

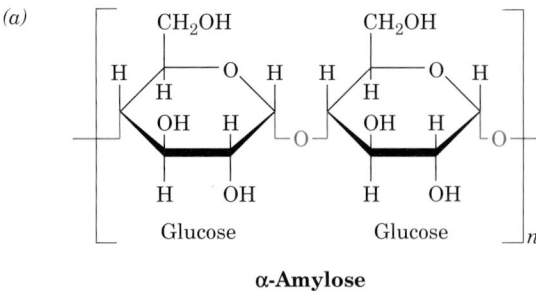

N-Acetylglucosamine   *N*-Acetylglucosamine

Chitin

FIGURE 11-16   Structure of chitin. Chitin is a $\beta(1 \rightarrow 4)$-linked homopolymer of *N*-acetyl-D-glucosamine.

portion of **lignin,** a plasticlike phenolic polymer. One has only to watch a tall tree in a high wind to realize the enormous strength of plant cell walls. In engineering terms, they are "composite materials," as is concrete reinforced by steel rods. Composite materials can withstand large stresses because the matrix evenly distributes the stresses among the reinforcing elements.

Although vertebrates themselves do not possess an enzyme capable of hydrolyzing the $\beta(1 \rightarrow 4)$ linkages of cellulose, the digestive tracts of herbivores contain symbiotic microorganisms that secrete a series of enzymes, collectively known as **cellulase,** that do so. The same is true of termites. Nevertheless, the degradation of cellulose is a slow process because its tightly packed and hydrogen bonded glucan chains are not easily accessible to cellulase and do not separate readily even after many of their glycosidic bonds have been hydrolyzed. The digestion of fibrous plants such as grass by herbivores is therefore a more complex and time-consuming process than is the digestion of meat by carnivores (cows, e.g., have multichambered stomachs and must chew their cud). Similarly, the decay of dead plants by fungi, bacteria, and other organisms, and the consumption of wooden houses by termites, often takes years.

Chitin is the principal structural component of the exoskeletons of invertebrates such as crustaceans, insects, and spiders and is also a major cell wall constituent of most fungi and many algae. It is estimated that $\sim 10^{14}$ kg of chitin are produced annually, most of it in the oceans, and therefore that it is almost as abundant as is cellulose. Chitin is a homopolymer of $\beta(1 \rightarrow 4)$-linked *N*-acetyl-D-glucosamine residues (Fig. 11-16). It differs chemically from cellulose only in that each C2-OH group is replaced by an acetamido function. X-Ray analysis indicates that chitin and cellulose have similar structures.

D. *Storage Polysaccharides: Starch and Glycogen*

a. Starch Is a Food Reserve in Plants and a Major Nutrient for Animals

Starch is a mixture of glucans that plants synthesize as their principal food reserve. It is deposited in the cytoplasm of plant cells as insoluble granules composed of **α-amylose** and **amylopectin.** α-Amylose is a linear polymer of several

(a)

Glucose   Glucose

α-Amylose

(b)

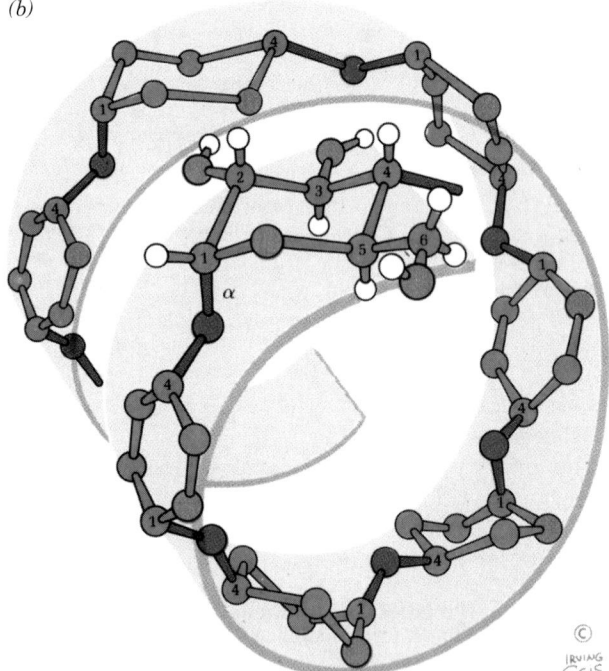

FIGURE 11-17   α-Amylose. (*a*) The D-glucose residues of α-amylose are linked by $\alpha(1 \rightarrow 4)$ bonds (*red*). Here *n* is several thousand. (*b*) This regularly repeating polymer forms a left-handed helix. Note the great differences in structure and properties that result from changing α-amylose's $\alpha(1 \rightarrow 4)$ linkages to the $\beta(1 \rightarrow 4)$ linkages of cellulose (Fig. 11-15). [Illustration, Irving Geis/Geis Archives Trust. Copyright Howard Hughes Medical Institute. Reproduced with permission.]

thousand glucose residues linked by $\alpha(1 \rightarrow 4)$ bonds (Fig. 11-17a). Note that although α-amylose is an isomer of cellulose, it has very different structural properties. This is because cellulose's β-glycosidic linkages cause each successive glucose residue to flip 180° with respect to the preceding residue, so that the polymer assumes an easily packed, fully extended conformation (Fig. 11-15). In contrast, α-amylose's α-glycosidic bonds cause it to adopt an irregularly aggregating helically coiled conformation (Fig. 11-17b).

Amylopectin consists mainly of $\alpha(1 \rightarrow 4)$-linked glucose residues but is a branched molecule with $\alpha(1 \rightarrow 6)$ branch points every 24 to 30 glucose residues on average (Fig. 11-18). Amylopectin molecules contain up to 10^6 glucose residues, which makes them among the largest molecules occurring in nature. The storage of glucose as starch greatly reduces the large intracellular osmotic pressures

(a)

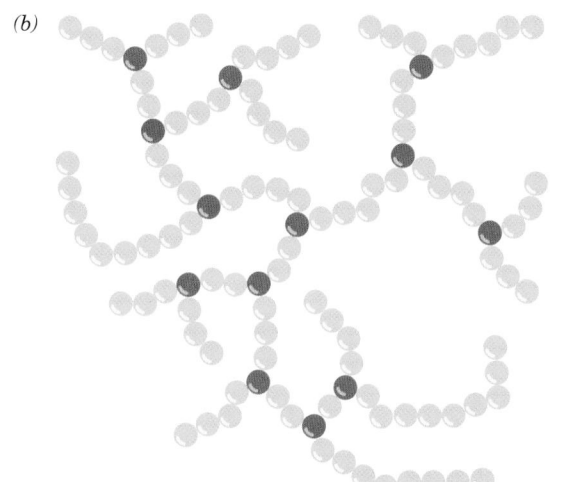

Amylopectin

(b)

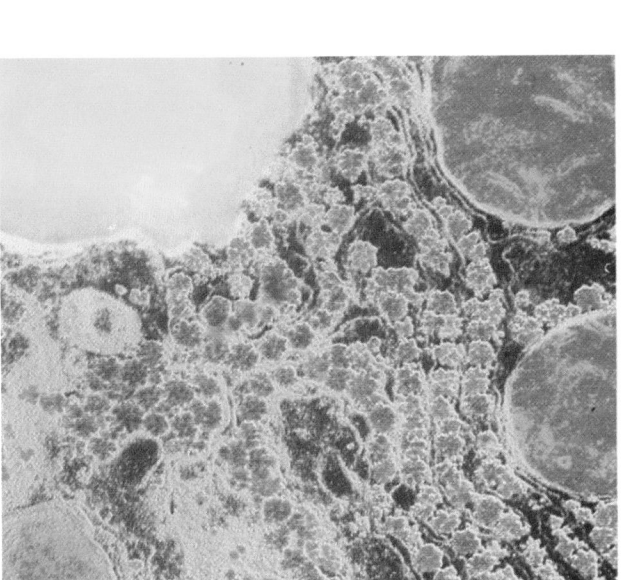

FIGURE 11-18   **Amylopectin.** (*a*) Its primary structure near one of its $\alpha(1 \to 6)$ branch points (*red*). (*b*) Its bushlike structure with glucose residues at branch points indicated in red. The actual distance between branch points averages 24 to 30 glucose residues. Glycogen has a similar structure but is branched every 8 to 14 residues.

FIGURE 11-19   **Photomicrograph showing the glycogen granules (*pink*) in the cytoplasm of a liver cell.** The greenish objects are mitochondria and the yellow object is a fat globule. Note that the glycogen granules tend to aggregate. The glycogen content of liver may reach as high as 10% of its net weight. [CNRI.]

that would result from its storage in monomeric form because osmotic pressure is proportional to the number of solute molecules in a given volume.

b.   Starch Digestion Occurs in Stages

The digestion of starch, the main carbohydrate source in the human diet, begins in the mouth. Saliva contains **α-amylase,** which randomly hydrolyzes all the $\alpha(1 \to 4)$ glucosidic bonds of starch except its outermost bonds and those next to branches. By the time thoroughly chewed food reaches the stomach, where the acidity inactivates α-amylase, the average chain length of starch has been reduced from several thousand to fewer than eight glucose units. Starch digestion continues in the small intestine under the influence of pancreatic α-amylase, which is similar to the salivary enzyme. This enzyme degrades starch to a mixture of the disaccharide maltose, the trisaccharide **maltotriose,** which contains three $\alpha(1 \to 4)$-linked glucose residues, and oligosaccharides known as **dextrins** that contain the $\alpha(1 \to 6)$ branches. These oligosaccharides are hydrolyzed to their component monosaccharides by specific enzymes contained in the brush border membranes of the intestinal mucosa: an **α-glucosidase,** which removes one glucose residue at a time from oligosaccharides, an **α-dextrinase** or **debranching enzyme,** which hydrolyzes $\alpha(1 \to 6)$ and $\alpha(1 \to 4)$ bonds, a **sucrase,** and, at least in infants, a lactase. The resulting monosaccharides are absorbed by the intestine and transported to the bloodstream (Section 20-4A).

c.   Glycogen Is "Animal Starch"

Glycogen, the storage polysaccharide of animals, is present in all cells but is most prevalent in skeletal muscle and liver, where it occurs as cytoplasmic granules (Fig. 11-19). The primary structure of glycogen resembles that of amylopectin, but glycogen is more highly branched, with branch points occurring every 8 to 14 glucose residues. Glycogen's degree of polymerization is nevertheless similar to that of amylopectin. In the cell, glycogen is degraded for metabolic use by **glycogen phosphorylase,** which phosphorolytically cleaves glycogen's $\alpha(1 \rightarrow 4)$ bonds sequentially inward from its nonreducing ends to yield **glucose-1-phosphate.** Glycogen's highly branched structure, which has many nonreducing ends, permits the rapid mobilization of glucose in times of metabolic need. The $\alpha(1 \rightarrow 6)$ branches of glycogen are cleaved by a debranching enzyme. These enzymes play an important role in glucose metabolism and are discussed further in Section 18-1.

E. *Glycosaminoglycans*

The extracellular spaces, particularly those of connective tissues such as cartilage, tendon, skin, and blood vessel walls, consist of collagen and elastin fibers (Section 8-2B) embedded in a gel-like matrix known as **ground substance.** Ground substance is composed largely of **glycosaminoglycans** (alternatively, **mucopolysaccharides**), unbranched polysaccharides of alternating uronic acid and hexosamine residues. Solutions of glycosaminoglycans have a slimy, mucuslike consistency that results from their high viscosity and

elasticity. In the following paragraphs, we discuss the structural origin of these important mechanical properties.

a. Hyaluronic Acid

Hyaluronic acid (also called **hyaluronan**) is an important glycosaminoglycan component of ground substance, synovial fluid (the fluid that lubricates the joints), and the vitreous humor of the eye. It also occurs in the capsules surrounding certain, usually pathogenic, bacteria. Hyaluronic acid molecules are composed of 250 to 25,000 $\beta(1 \rightarrow 4)$-linked disaccharide units that consist of D-glucuronic acid and N-acetyl-D-glucosamine linked by a $\beta(1 \rightarrow 3)$ bond (Fig. 11-20). The anionic character of its glucuronic acid residues causes hyaluronic acid to bind cations such as K^+, Na^+, and Ca^{2+} tightly. X-Ray fiber analysis indicates that Ca^{2+} hyaluronate forms an extended left-handed single-stranded helix with three disaccharide units per turn (Fig. 11-21).

Hyaluronate's structural features suit it to its biological function. Its high molecular mass and numerous mutually repelling anionic groups make hyaluronate a rigid and highly hydrated molecule which, in solution, occupies a volume ~1000 times that in its dry state. Hyaluronate solutions therefore have a viscosity that is shear dependent (an object under shear stress has equal and opposite forces

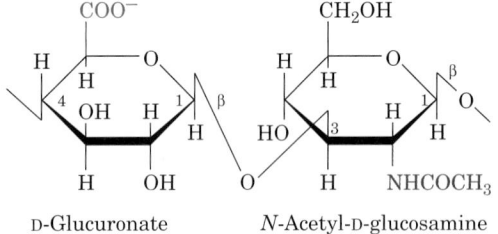

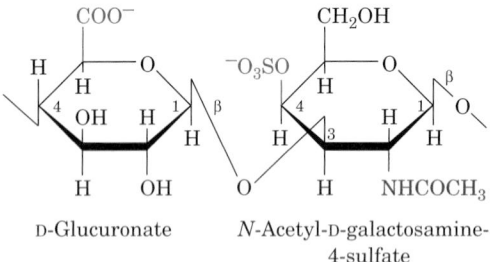

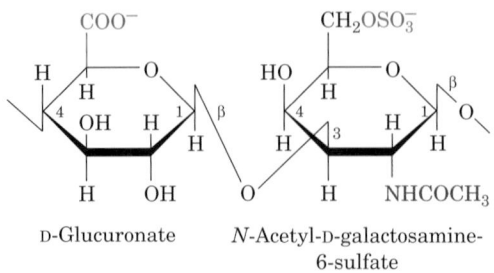

FIGURE 11-20   The disaccharide repeating units of the common glycosaminoglycans. The anionic groups are shown in red and the N-acetylamido groups are shown in blue. 🔹 See Kinemage Exercise 7-3

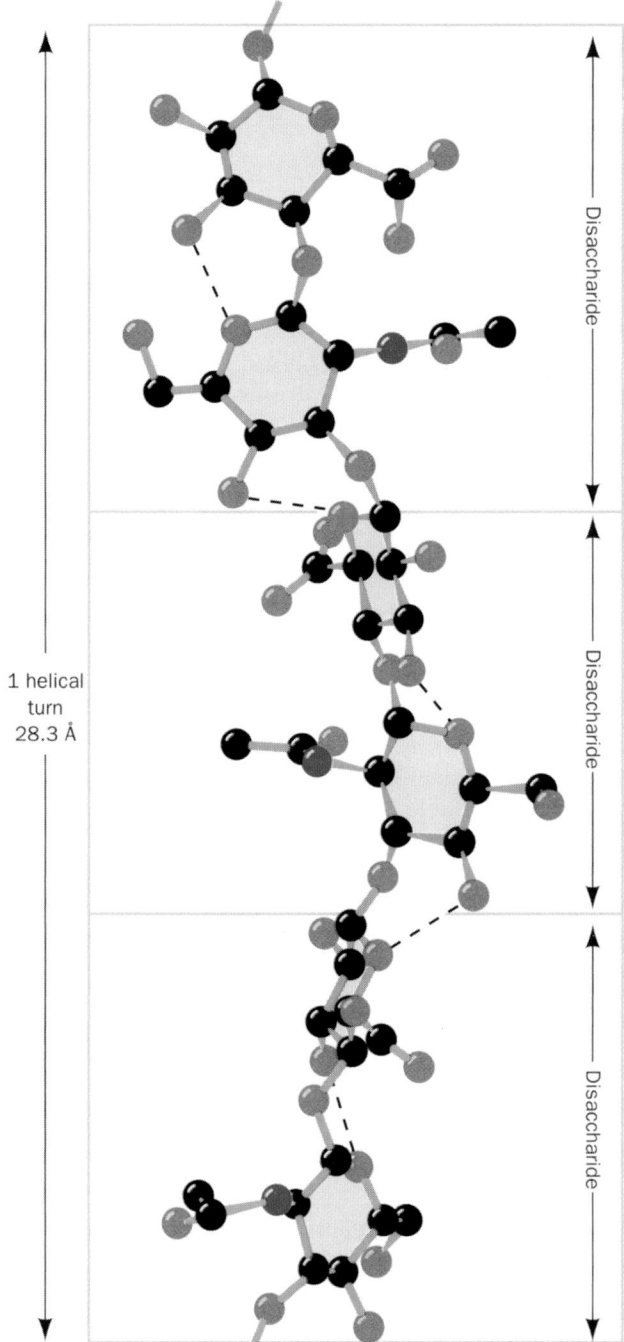

FIGURE 11-21   X-Ray fiber structure of Ca²⁺ hyaluronate. The hyaluronate polyanion forms an extended left-handed single-stranded helix with three disaccharide units per turn that is stabilized by intramolecular hydrogen bonds (*dashed lines*). H atoms and Ca²⁺ ions are omitted for clarity. [After Winter, W.T. and Arnott, S., *J. Mol. Biol.* **117,** 777 (1977) PDBid 4HYA.]

applied across its opposite faces). At low shear rates, hyaluronate molecules form tangled masses that greatly impede flow; that is, the solution is quite viscous. As the shear rate increases, the stiff hyaluronate molecules tend to line up with the flow and thus offer less resistance to it. This viscoelastic behavior makes hyaluronate solutions excellent biological shock absorbers and lubricants.

Hyaluronic acid and other glycosaminoglycans (see below) are degraded by **hyaluronidase,** which hydrolyzes their β(1 → 4) linkages. Hyaluronidase occurs in a variety of animal tissues, in bacteria (where it presumably expedites their invasion of animal tissue), and in snake and insect toxins.

b. Other Glycosaminoglycans

Other glycosaminoglycan components of ground substance consist of 50 to 1000 sulfated disaccharide units which occur in proportions that are both tissue and species dependent. The most prevalent structures of these generally heterogeneous substances are indicated below (Fig. 11-20):

1. Chondroitin-4-sulfate (Greek: *chondros,* cartilage), a major component of cartilage and other connective tissue, has *N*-acetyl-D-galactosamine-4-sulfate residues in place of hyaluronate's *N*-acetyl-D-glucosamine residues.

2. Chondroitin-6-sulfate is instead sulfated at the C6 position of its *N*-acetyl-D-galactosamine residues. The two chondroitin sulfates occur separately or in mixtures depending on the tissue.

3. Dermatan sulfate, which is so named because of its prevalence in skin, differs from chondroitin-4-sulfate only by an inversion of configuration about C5 of the β-D-glucuronate residues to form α-L-iduronate. This results from the enzymatic epimerization of these residues after the formation of chondroitin. The epimerization is usually incomplete, so dermatan sulfate also contains glucuronate residues.

4. Keratan sulfate (not to be confused with the protein keratin) contains alternating β(1 → 4)-linked D-galactose and *N*-acetyl-D-glucosamine-6-sulfate residues. It is the most heterogeneous of the major glycosaminoglycans in that its sulfate content is variable and it contains small amounts of fucose, mannose, *N*-acetylglucosamine, and sialic acid.

5. Heparin is a variably sulfated glycosaminoglycan that consists predominantly of alternating α(1 → 4)-linked residues of D-iduronate-2-sulfate and *N*-sulfo-D-glucosamine-6-sulfate. It has an average of 2.5 sulfate residues per disaccharide unit, which makes it the most negatively charged polyelectrolyte in mammalian tissues. Heparin, in contrast to the above glycosaminoglycans, is not a constituent of connective tissue, but occurs almost exclusively in the intracellular granules of the **mast cells** that line arterial walls, especially in the liver, lungs, and skin. It inhibits the clotting of blood, and its release, through injury, is thought to prevent runaway clot formation (Section 35-1E). Heparin is therefore in wide clinical use to inhibit blood clotting, for example, in postsurgical patients. **Heparan sulfate,** a ubiquitous cell-surface component as well as an extracellular substance in blood vessel walls and brain, resembles heparin but has a far more variable composition with fewer *N*- and *O*-sulfate groups and more *N*-acetyl groups.

3 ■ GLYCOPROTEINS

Until about 1960, carbohydrates were thought to be rather dull compounds that were probably some sort of inert filler. Protein chemists therefore considered them to be a nui-

TABLE 11-1   **Properties of Some Proteoglycans**

Proteoglycan	Approximate Core Protein Molecular Mass (kD)	Glycosaminoglycan Type (Number)[a]
Proteoglycans interacting with hyaluronic acid		
Aggrecan	220	CS (~100), KS (~30)
Versican	265–370	CS/DS (10–30)
Neurocan	136	CS (3–7)
Proteoglycans of the basal laminae		
Perlecan	400–467	Heparan sulfate/CS (3)
Agrin	250	Heparan sulfate (3)
Bamacan	138	CS (3)
Small leucine-rich proteoglycans		
Decorin	40	DS/CS (1)
Fibromodulin	42	KS (2–3)
Osteoglycin	35	KS (2–3)

[a]Abbreviations: CS, chondroitin sulfate; DS, dermatan sulfate; KS, keratan sulfate.

Source: Iozzo, R.V., *Annu. Rev. Biochem.* **67,** 611, 626, and 624 (1998).

sance that complicated protein "purification." In fact, most eukaryotic proteins are **glycoproteins,** that is, they are covalently associated with carbohydrates. Glycoproteins vary in carbohydrate content from <1% to >90% by weight. They occur in all forms of life and have functions that span the entire spectrum of protein activities, including those of enzymes, transport proteins, receptors, hormones, and structural proteins. Their carbohydrate moieties, as we shall see, have several important biological roles, but in many cases their functions remain enigmatic.

The polypeptide chains of glycoproteins, like those of all proteins, are synthesized under genetic control. Their carbohydrate chains, in contrast, are enzymatically generated and covalently linked to the polypeptide without the rigid guidance of nucleic acid templates. The processing enzymes are generally not available in sufficient quantities to ensure the synthesis of uniform products. Glycoproteins therefore have variable carbohydrate compositions, a phenomenon known as **microheterogeneity,** that compounds the difficulties in their purification and characterization.

In this section we consider the structures and properties of glycoproteins. In particular, we shall study the glycoproteins of connective tissues, those of bacterial cell walls, and several soluble glycoproteins. We end by discussing the general principles of glycoprotein structure and function.

A. *Proteoglycans*

Proteins and glycosaminoglycans in ground substance, in **basal laminae [basement membranes;** the thin matlike extracellular matrix separating **epithelial cells** (the cells lining body cavities and free surfaces) from underlying cells], and in cell-surface membranes aggregate covalently and noncovalently to form a diverse group of macromolecules known as **proteoglycans.** *Proteoglycans consist of a core protein to which at least one glycosaminoglycan chain, most*

often keratan sulfate and/or chondroitin sulfate, is covalently linked. Numerous types of core proteins have been characterized (Table 11-1). Proteoglycans appear to have multiple roles, most notably as organizers of tissue morphology via their interactions with molecules such as collagen; as selective filters that regulate the traffic of molecules according to their size and/or charge; and as regulators of the activities of other proteins, particularly those involved in signaling (see below).

Electron micrographs such as Fig. 11-22a together with reconstitution experiments indicate that proteoglycans can form huge complexes. For example, **aggrecan,** the main proteoglycan component of cartilage, has a bottlebrushlike

FIGURE 11-22 (*Opposite*) **Proteoglycans.** (*a*) An electron micrograph showing a central strand of hyaluronic acid, which runs down the field of view, supporting numerous projections, each of which consists of a core protein to which many bushy polysaccharide protrusions are linked. [From Caplan, A.I., *Sci. Am.* **251**(4); 87 (1984). Copyright © 1984 Scientific American, Inc. Used by permission.] (*b*) The bottlebrush model of the proteoglycan aggrecan. The core proteins, one of which is shown extending down through the middle of the diagram, project from the central hyaluronic acid strand. The core is noncovalently anchored to the hyaluronic acid via its globular N-terminal end in an association that is stabilized by link protein. The core has three saccharide-binding regions: (1) the inner region predominantly binds oligosaccharides via the side chain N atoms of Asn residues; (2) the central region binds oligosaccharides, many of which bear keratan sulfate chains, via the side chain O atoms of Ser and Thr residues; and (3) the outer region mainly binds chondroitin sulfate chains that are linked to the core protein via a galactose–galactose–xylose trisaccharide that is bonded to side chain O atoms of Ser residues in the sequence Ser-Gly. The C-terminal end of the aggrecan core protein consists of a lectinlike sequence.

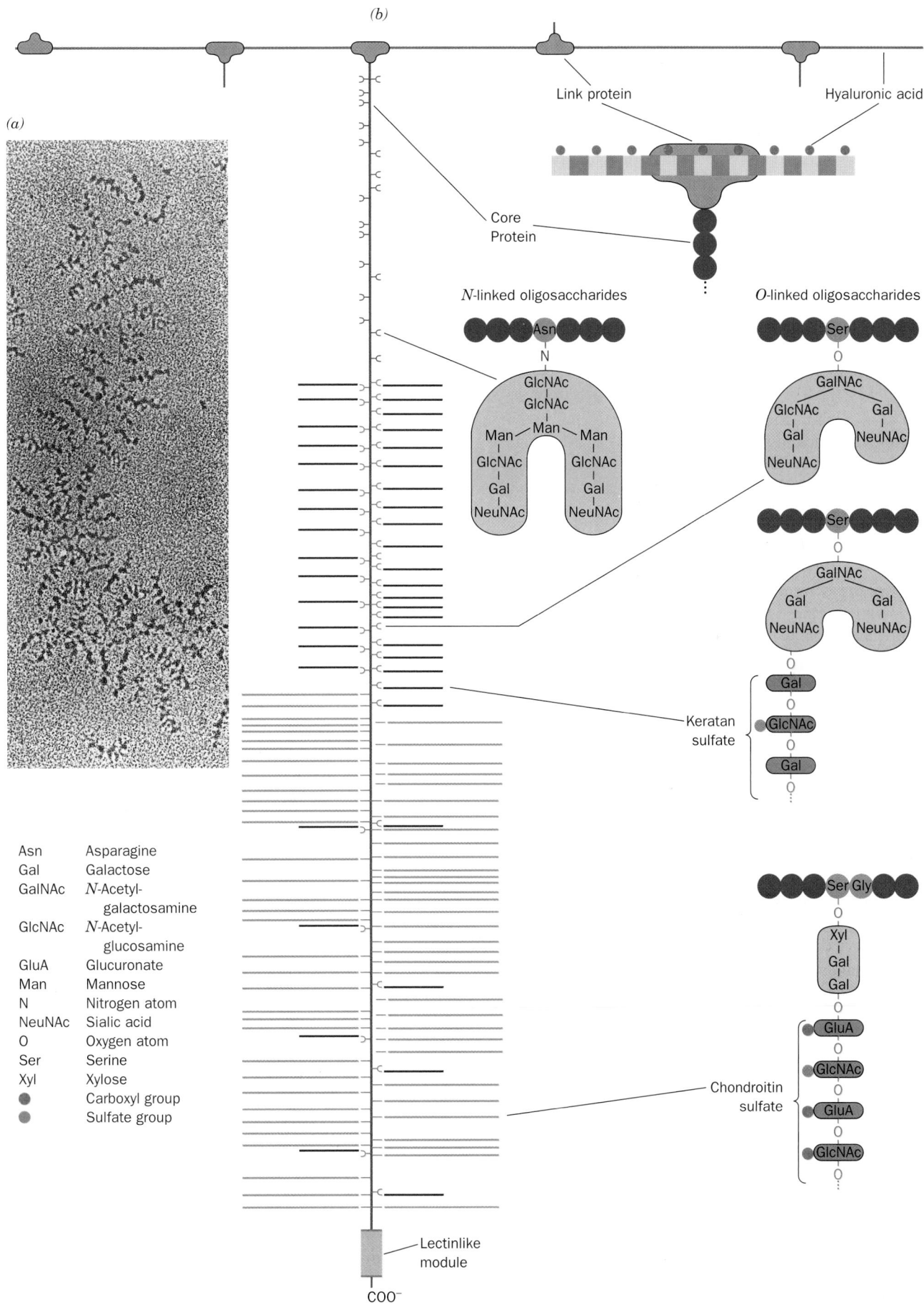

(b)

(a)

Link protein

Hyaluronic acid

Core Protein

N-linked oligosaccharides

O-linked oligosaccharides

Asn
N
GlcNAc
GlcNAc
Man — Man — Man
GlcNAc   GlcNAc
Gal   Gal
NeuNAc   NeuNAc

Ser
O
GalNAc
GlcNAc   Gal
Gal   NeuNAc
NeuNAc

Ser
O
GalNAc
Gal   Gal
NeuNAc   NeuNAc
O
Gal
O
GlcNAc
O
Gal

Keratan sulfate

Ser Gly
O
Xyl
Gal
Gal
GluA
GlcNAc
GluA
GlcNAc

Chondroitin sulfate

Asn	Asparagine
Gal	Galactose
GalNAc	*N*-Acetyl-galactosamine
GlcNAc	*N*-Acetyl-glucosamine
GluA	Glucuronate
Man	Mannose
N	Nitrogen atom
NeuNAc	Sialic acid
O	Oxygen atom
Ser	Serine
Xyl	Xylose
●	Carboxyl group
●	Sulfate group

Lectinlike module

COO⁻

molecular architecture (Fig. 11-22*b*), whose **proteoglycan subunit** "bristles" are noncovalently attached to a filamentous hyaluronic acid "backbone" at intervals of 200 to 300 Å. Aggrecan has three domains. Its N-terminal domain forms a globular region of 60 to 70 kD that binds noncovalently to hyaluronic acid. This attachment is stabilized by the 40- to 60-kD **link protein,** which is similar in sequence to aggrecan's N-terminal domain. Aggrecan's highly extended central domain is covalently linked to a series of polysaccharides, which comprise nearly 90% of this glycoprotein's mass. They divide the central domain into three regions:

1. An N-terminal region, which overlaps the globular hyaluronic acid–binding domain, binds a relatively few carbohydrate chains. These tend to be oligosaccharides that are covalently bonded to the protein via the amide N atoms of specific Asn residues (Section 11-3C).

2. A region rich in oligosaccharides, many of which serve as anchor points for keratan sulfate chains. These oligosaccharides are covalently bonded to side chain O atoms of Ser and Thr residues.

3. A C-terminal region rich in chondroitin sulfate chains, which are covalently linked to the side chain O atoms of Ser residues in Ser-Gly dipeptides via galactose–galactose–xylose trisaccharides.

Aggrecan's C-terminal domain contains a lectinlike module, which binds certain monosaccharide units. Thus, aggrecan probably functions to bind together various constituents of the cell surface and the extracellular matrix (see below).

Altogether, a central strand of hyaluronic acid, which varies in length from 4000 to 40,000 Å, noncovalently binds up to 100 associated aggrecan chains, each of which covalently binds ~30 keratan sulfate chains of up to 250 disaccharide units each and ~100 chondroitin sulfate chains of up to 1000 disaccharide units each. This accounts for the enormous molecular masses of the aggrecans, which range up to 200,000 kD, and for their high degree of polydispersity (range of molecular masses). Note, however, that many proteoglycans do not bind to hyaluronic acid (Table 11-1) and hence function as monomers.

a. Cartilage's Mechanical Properties Are Explained by Its Molecular Structure

Cartilage consists largely of a meshwork of collagen fibrils that is filled in by proteoglycans whose chondroitin sulfate and core protein components specifically interact with the collagen. The tensile strength of cartilage and other connective tissues is, as we have seen (Section 8-2B), a consequence of their collagen content. Cartilage's characteristic resilience, however, results from its high proteoglycan content. The extended brushlike structure of proteoglycans, together with the polyanionic character of keratan sulfate and chondroitin sulfate, cause this complex to be highly hydrated. The application of pressure on cartilage squeezes water away from these charged regions until charge–charge repulsions prevent further compression.

When the pressure is released, the water returns. Indeed, the cartilage in the joints, which lack blood vessels, is nourished by this flow of liquid brought about by body movements. This explains why long periods of inactivity cause joint cartilage to become thin and fragile.

b. Proteoglycans Modulate the Effects of Protein Growth Factors

Proteoglycans have been implicated in a variety of cellular processes. For example, **fibroblast growth factor (FGF;** growth factors are proteins that function to induce their specific target cells to grow and/or differentiate; Section 19-3A) binds to heparin or to the heparan sulfate chains of proteoglycans and is only bound to its cell-surface receptor in complex with these glycosaminoglycans. Since the binding of FGF to heparin or heparan sulfate protects FGF from degradation, the release of this growth factor from the extracellular matrix by the proteolysis of proteoglycan core proteins or by the partial degradation of heparan sulfate probably provides an important source of active FGF–glycosaminoglycan complexes. Several other growth factors interact similarly with proteoglycans. Apparently, the abundant and ubiquitous distribution of proteoglycans limits the action of these growth factors on their target cells to short distances from the cells secreting the growth factors, a phenomenon that probably greatly influences the formation and maintenance of tissue architecture.

B. *Bacterial Cell Walls*

Bacteria are surrounded by rigid cell walls (Fig. 1-13) that give them their characteristic shapes (Fig. 1-1) and permit them to live in hypotonic (less than intracellular salt concentration) environments that would otherwise cause them to swell osmotically until their plasma (cell) membranes lysed (burst). Bacterial cell walls are of considerable medical significance because they are responsible for bacterial **virulence** (disease-evoking power). In fact, the symptoms of many bacterial diseases can be elicited in animals merely by the injection of bacterial cell walls. Furthermore, the characteristic **antigens** (immunological markers; Section 35-2) of bacteria are components of their cell walls, so that injection of bacterial cell wall preparations into an animal often invokes its immunity against these bacteria.

Bacteria are classified as **gram-positive** or **gram-negative** depending on whether or not they take up gram stain (Section 1-1B). Gram-positive bacteria (Fig. 11-23*a*) have a thick (~250 Å) cell wall surrounding their plasma membrane, whereas gram-negative bacteria (Fig. 11-23*b*) have a thin (~30 Å) cell wall covered by a complex outer membrane.

a. Bacterial Cell Walls Have a Peptidoglycan Framework

The cell walls of both gram-positive and gram-negative bacteria consist of covalently linked polysaccharide and polypeptide chains which form a baglike molecule that completely encases the cell. This framework, whose structure

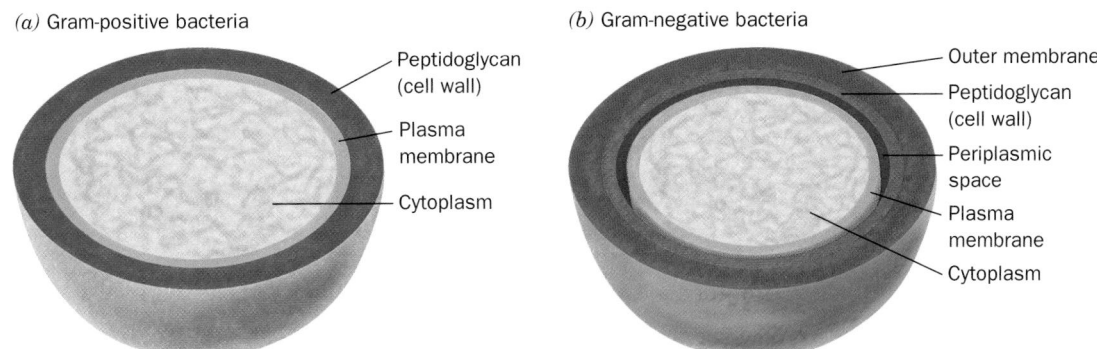

FIGURE 11-23 Schematic diagram comparing the cell envelopes of (*a*) gram-positive bacteria and (*b*) gram-negative bacteria.

was elucidated in large part by Jack Strominger, is known as a **peptidoglycan** or **murein** (Latin: *murus,* wall). Its polysaccharide component consists of linear chains of alternating $\beta(1 \rightarrow 4)$-linked *N*-acetylglucosamine **(NAG)** and *N*-acetylmuramic acid **(NAM).** The NAM's lactic acid

residue forms an amide bond with a D-amino acid–containing tetrapeptide to form the peptidoglycan repeating unit (Fig. 11-24). Neighboring parallel peptidoglycan chains are covalently cross-linked through their tetrapeptide side chains. In the gram-positive bacterium

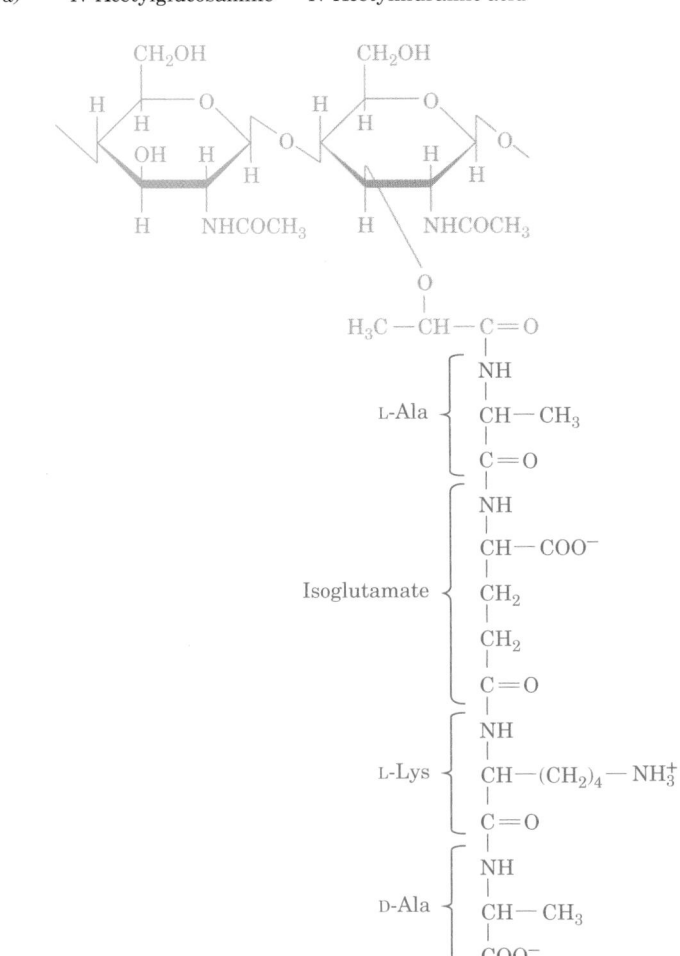

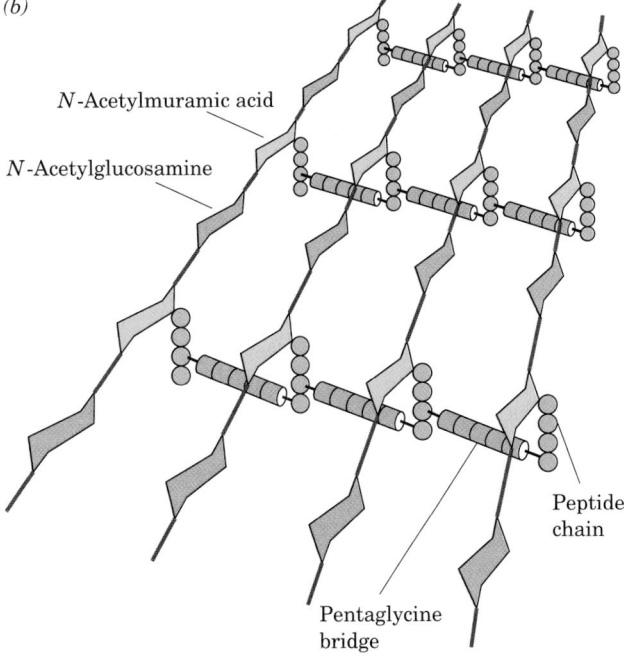

FIGURE 11-24 Chemical structure of peptidoglycan. (*a*) The repeating unit of peptidoglycan is an NAG–NAM disaccharide whose lactyl side chain forms an amide bond with a tetrapeptide. The tetrapeptide of *S. aureus* is shown. The isoglutamate is so designated because it forms an amide link via its γ-carboxyl group. In some species, its α-carboxylate group is replaced by an amide group to form D-isoglutamine and/or the L-Lys residue may have a carboxyl group appended to its C_ε to form **diaminopimelic acid.** (*b*) The *S. aureus* bacterial cell wall peptidoglycan. In other gram-positive bacteria, the Gly_5 connecting bridges shown here may contain different amino acid residues such as Ala or Ser. In gram-negative bacteria, the peptide chains are directly linked via peptide bonds.

Staphylococcus aureus, whose tetrapeptide has the sequence L-Ala-D-isoglutamyl-L-Lys-D-Ala, this cross-link consists of a pentaglycine chain that extends from the terminal carboxyl group of one tetrapeptide to the ε-amino group of the Lys in a neighboring tetrapeptide. The bacterial cell wall consists of several concentric layers of peptidoglycan that are probably cross-linked in the third dimension; gram-positive bacteria have up to 20 such layers.

The D-amino acids of peptidoglycans render them resistant to proteases. However, **lysozyme,** an enzyme which is present in tears, mucus, and other vertebrate body secretions, as well as in egg whites, catalyzes the hydrolysis of the β(1 → 4) glycosidic linkage between NAM and NAG. Consequently, treatment of gram-positive bacteria with lysozyme degrades their cell walls, which results in their lysis (gram-negative bacteria are resistant to lysozyme degradation). Lysozyme was discovered in 1922 by the British bacteriologist Alexander Fleming after he noticed that a bacterial culture had dissolved where mucus from a sneeze had landed. It was Fleming's hope that lysozyme would be a universal antibiotic but, unfortunately, it is clinically ineffective against pathogenic bacteria. The structure and mechanism of lysozyme are examined in detail in Section 15-2.

b. Penicillin Kills Bacteria by Inhibiting Cell Wall Biosynthesis

In 1928, Fleming noticed that the chance contamination of a bacterial culture plate with the mold *Penicillium notatum* lysed nearby bacteria (a clear demonstration of Pasteur's maxim that chance favors a prepared mind). This was caused by the presence of **penicillin** (Fig. 11-25), an antibiotic secreted by the mold. Yet the difficulties of isolating and characterizing penicillin, owing to its instability, led to the passage of over 15 years before penicillin was available for routine clinical use. Penicillin specifically binds to and inactivates enzymes that function to cross-link the peptidoglycan strands of bacterial cell walls. Since cell wall expansion also requires the action of enzymes that degrade cell walls, *exposure of growing bacteria to penicillin results in their lysis;* that is, penicillin disrupts the normal balance between cell wall biosynthesis and degradation. However, since no human enzyme binds penicillin, it is of low human toxicity, a therapeutic necessity.

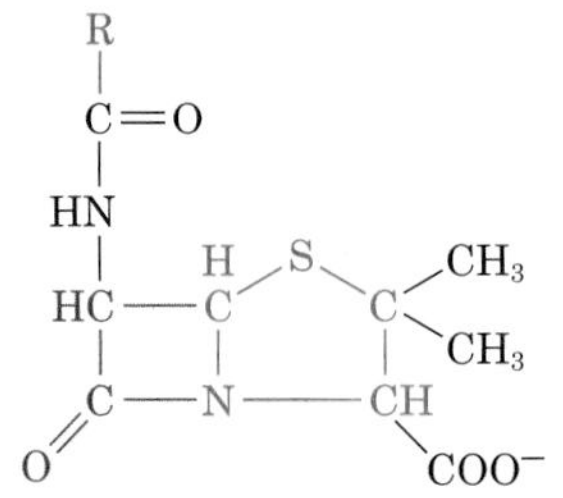

Penicillin

FIGURE 11-25 Structure of penicillin. Penicillin contains a thiazolidine ring (*red*) fused to a β-lactam ring (*blue*). A variable R group is bonded to the β-lactam ring via a peptide link. In benzyl penicillin (penicillin G), one of several naturally occurring derivatives that are clinically effective, R is the benzyl group (—CH$_2$φ). In **ampicillin,** a semisynthetic derivative, R is the aminobenzyl group [—CH(NH$_2$)φ].

Penicillin-treated bacteria that are kept in a hypertonic medium remain intact, even though they have no cell wall. Such bacteria, which are called **protoplasts** or **spheroplasts,** are spherical and extremely fragile because they are encased by only their plasma membranes. Protoplasts immediately lyse on transfer to a normal medium.

Most bacteria that are resistant to penicillin secrete **penicillinase,** which inactivates penicillin by cleaving the amide bond of its β-lactam ring (Fig. 11-26). However, the observation that penicillinase activity varies with the nature of penicillin's R group has prompted the semisynthesis of penicillins, such as **ampicillin** (Fig. 11-25), which are clinically effective against penicillin-resistant strains of bacteria.

c. Bacterial Cell Walls Are Studded with Antigenic Groups

The surfaces of gram-positive bacteria are covered by **teichoic acids** (Greek: *teichos,* city walls), which account for up to 50% of the dry weight of their cell walls. Teichoic acids are polymers of glycerol or ribitol linked by phosphodiester bridges (Fig. 11-27). The hydroxyl groups of this sugar–phosphate chain are substituted by D-Ala residues and saccharides such as glucose or NAG. Teichoic acids are anchored to the peptidoglycans via phosphodiester

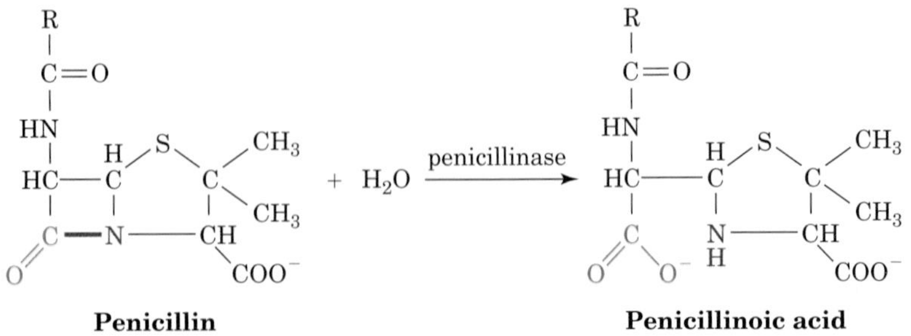

Penicillin   **Penicillinoic acid**

FIGURE 11-26 Enzymatic inactivation of penicillin. Penicillinase inactivates penicillin by catalyzing the hydrolysis of its β-lactam ring to form **penicillinoic acid.**

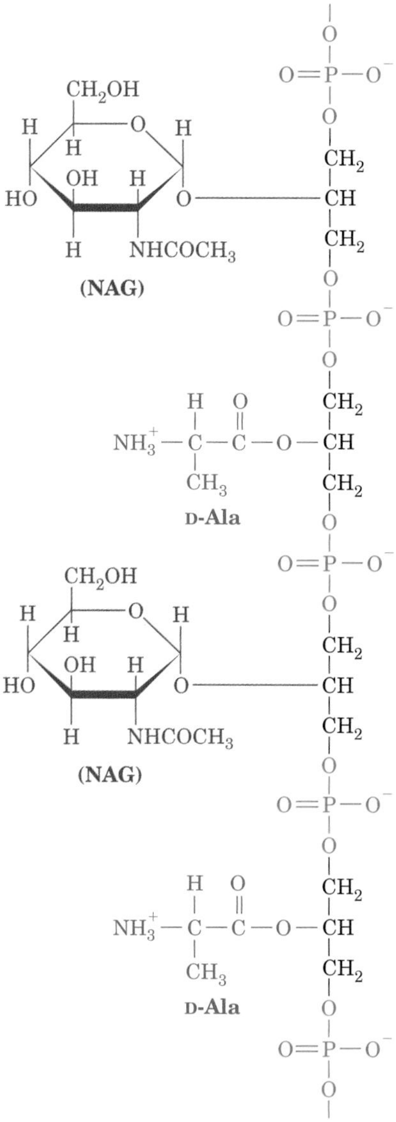

FIGURE 11-27   Structure of teichoic acid. A segment of a teichoic acid molecule with a glycerol phosphate backbone that bears alternating residues of D-Ala and NAG.

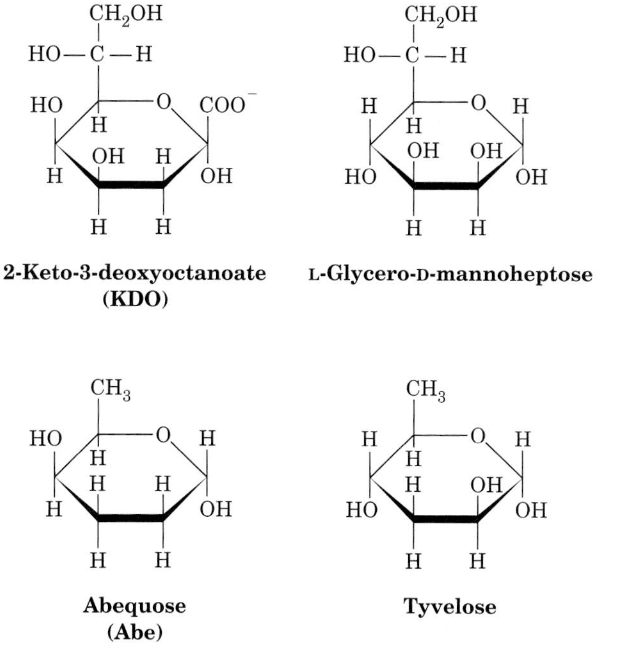

FIGURE 11-28   Some of the unusual monosaccharides that occur in the O-antigens of gram-negative bacteria. These sugars rarely occur in other organisms.

bonds to the C6-OH groups of their NAG residues. They often terminate in **lipopolysaccharides** (lipids that contain polysaccharides; Section 12-1).

The outer membranes of gram-negative bacteria (Fig. 11-23b) are composed of complex lipopolysaccharides, proteins, and phospholipids that are organized in a complicated manner. The **periplasmic space,** an aqueous compartment that lies between the plasma membrane and the peptidoglycan cell wall, contains proteins that transport sugars and other nutrients. The outer membrane functions as a barrier to exclude harmful substances (such as gram stain). This accounts for the observation that gram-negative bacteria are less affected by lysozyme and penicillin, as well as by other antibiotics, than are gram-positive bacteria.

The outer surfaces of gram-negative bacteria are coated with complex and often unusual polysaccharides known as **O-antigens** that uniquely mark each bacterial strain (Fig. 11-28). The observation that mutant strains of pathogenic bacteria lacking O-antigens are nonpathogenic suggests that O-antigens participate in the recognition of host cells. O-Antigens, as their name implies, are also the means by which a host's immunological defense system recognizes invading bacteria as foreign (Section 35-2A). As part of the ongoing biological warfare between pathogen and host, O-antigens are subject to rapid mutational alteration so as to generate new bacterial strains that the host does not initially recognize (the mutations are in the genes specifying the enzymes that synthesize the O-antigens).

C. *Glycoprotein Structure and Function*

a. **Glycoprotein Carbohydrate Chains Are Highly Diverse**

Almost all the secreted and membrane-associated proteins of eukaryotic cells are glycosylated. Indeed, protein glycosylation is more abundant than all other types of post-translational modifications combined. Oligosaccharides form two types of direct attachments to these proteins: **N-linked** and **O-linked.** Sequence analyses of glycoproteins have led to the following generalizations about these attachments.

1. *In N-glycosidic (N-linked) attachments, an NAG is invariably β-linked to the amide nitrogen of an Asn in the sequence Asn-X-Ser or Asn-X-Thr, where X is any amino*

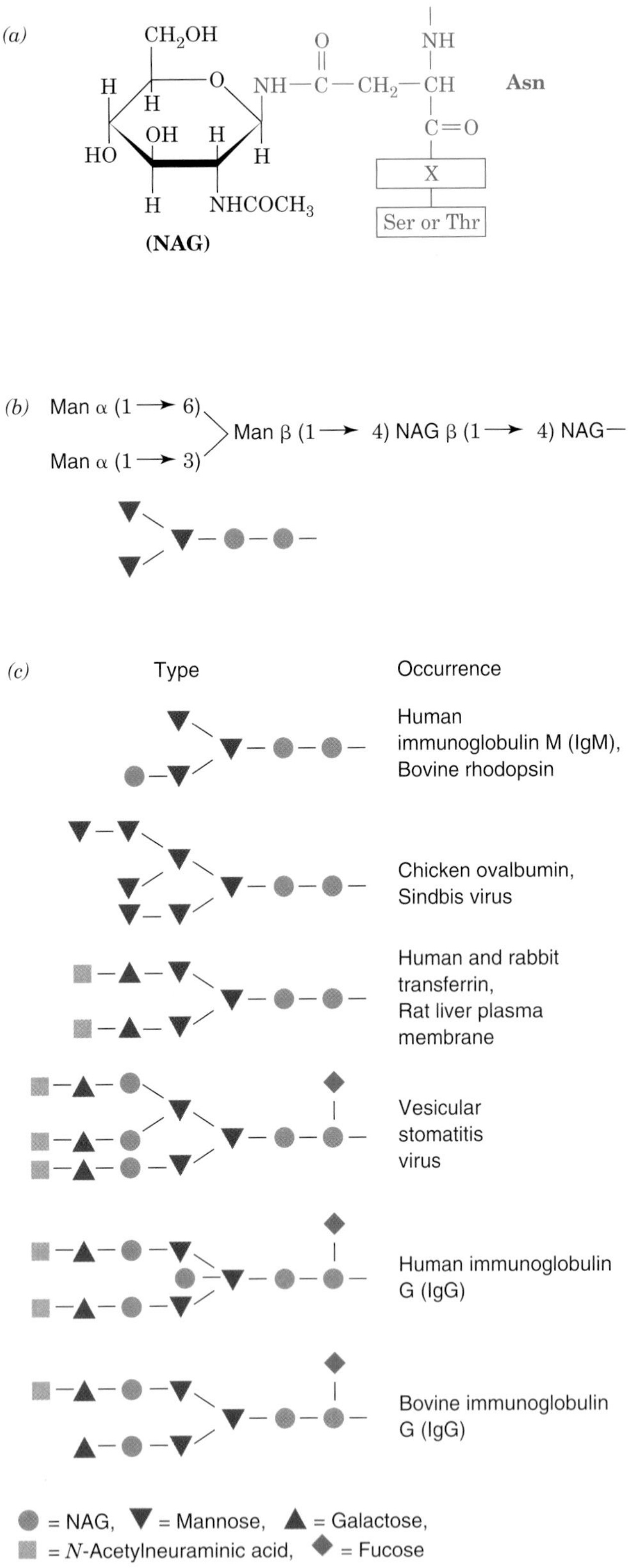

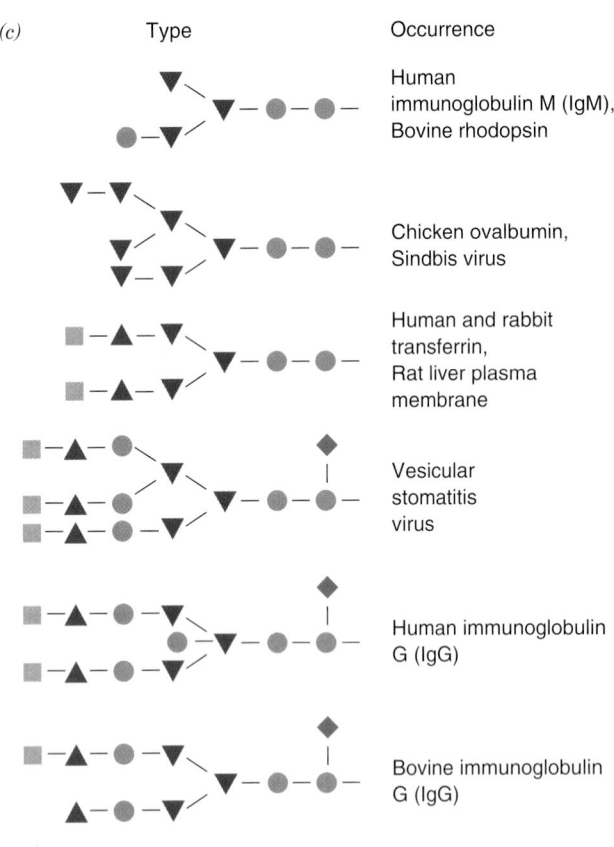

(a)

(NAG)

Asn

X

Ser or Thr

(b) Man α (1 ⟶ 6)

Man α (1 ⟶ 3) ⟩ Man β (1 ⟶ 4) NAG β (1 ⟶ 4) NAG—

(c)

Type	Occurrence
	Human immunoglobulin M (IgM), Bovine rhodopsin
	Chicken ovalbumin, Sindbis virus
	Human and rabbit transferrin, Rat liver plasma membrane
	Vesicular stomatitis virus
	Human immunoglobulin G (IgG)
	Bovine immunoglobulin G (IgG)

● = NAG, ▼ = Mannose, ▲ = Galactose,
■ = N-Acetylneuraminic acid, ◆ = Fucose

FIGURE 11-29 *N*-**Linked oligosaccharides.** (*a*) All *N*-glycosidic protein attachments occur through a β-*N*-acetylglucosamino–Asn bond in which the Asn occurs in the sequence Asn-X-Ser/Thr (*red*) where X is any amino acid. (*b*) *N*-Linked oligosaccharides usually have the branched (mannose)$_3$(NAG)$_2$ core shown. (*c*) Some examples of *N*-linked oligosaccharides. [After Sharon, N. and Lis, H., *Chem. Eng. News* **59**(13), 28 (1981).] 🔎 **See Kinemage Exercise 7-4**

acid residue except Pro and only rarely Asp, Glu, Leu, or Trp (Fig. 11-29a). The oligosaccharides in these linkages usually have a distinctive **core** *(innermost sequence; Fig. 11-29b) whose peripheral mannose residues are linked to either mannose or NAG residues. These latter residues may, in turn, be linked to yet other sugar residues, so that an enormous diversity of N-linked oligosaccharides is possible. Examples of such oligosaccharides are shown in Fig. 11-29c.*

2. *The most common O-glycosidic (O-linked) attachment involves the disaccharide core β-galactosyl-(1 → 3)-α-N-acetylgalactosamine α-linked to the OH group of either Ser or Thr (Fig. 11-30a). Less commonly, galactose, mannose, or xylose forms α-O-glycosides with Ser or Thr (Fig. 11-30b). Galactose also forms O-glycosidic bonds to the 5-hydroxylysyl residues of collagen (Section 8-2B). However, there seem to be few, if any, additional generalizations that can be made about O-glycosidically linked oligosaccharides. They vary in size from a single galactose residue in collagen to chains of up to 1000 disaccharide units in proteoglycans.*

Oligosaccharides tend to attach to proteins at sequences that form β bends. Taken with their hydrophilic character, this observation suggests that *oligosaccharides extend from the surfaces of proteins rather than participate in their internal structures.* Indeed, the relatively few glycoprotein X-ray structures that have yet been reported, for example, those of **immunoglobulin G** (Section 35-2B) and the influenza virus **hemagglutinin** (Section 33-4B), are consistent with this hypothesis. This accounts for the observation that the protein structures of glycoproteins are unaffected by the removal of their associated oligosaccharides.

Both experimental and theoretical studies indicate that oligosaccharides have mobile and rapidly fluctuating con-

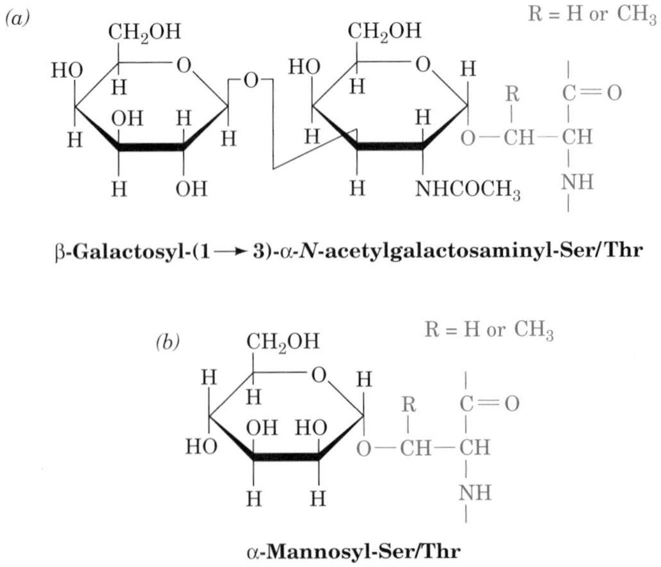

(a)

R = H or CH$_3$

β-Galactosyl-(1 ⟶ 3)-α-N-acetylgalactosaminyl-Ser/Thr

(b)

R = H or CH$_3$

α-Mannosyl-Ser/Thr

FIGURE 11-30 **Some common *O*-glycosidic attachments of oligosaccharides to glycoproteins (*red*).**

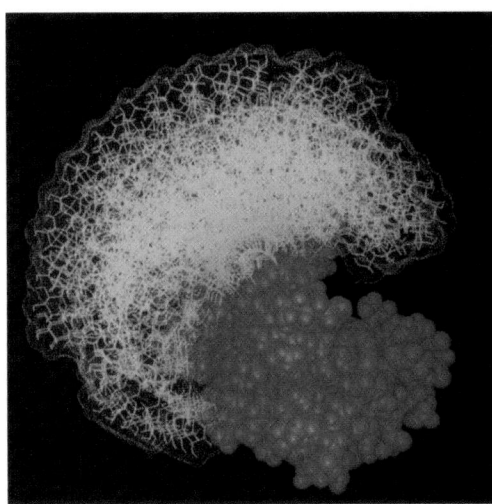

FIGURE 11-31   Model of oligosaccharide dynamics in bovine pancreatic ribonuclease B (RNase B). The allowed conformations of the (mannose)₅(NAG)₂ oligosaccharide (*yellow*) that is linked to a single site on the protein (*purple*) are shown in superimposed snapshots. [Courtesy of Raymond Dwek, Oxford University.]

formations (Fig. 11-31). Thus, representations in which oligosaccharides are shown as having fixed three-dimensional structures do not tell the whole story.

b. *N*-Linked Glycoproteins Exhibit Numerous Glycoforms

Cells tend to synthesize a large repertoire of a given N-linked glycoprotein, in which each variant species **(glycoform)** *differs somewhat in the sequences, locations, and numbers of its covalently attached oligosaccharides.* For example, one of the simplest glycoproteins, bovine pancreatic **ribonuclease B (RNase B),** differs from the well-characterized and carbohydrate-free enzyme RNase A (Section 9-1A) only by the attachment of a single *N*-glycosidically linked oligosaccharide chain. The oligosaccharide has the core sequence diagrammed in Fig. 11-32 with considerable microheterogeneity in the position of a sixth mannose residue. Nevertheless, the oligosaccharide does not appear to affect the conformation, substrate specificity, or catalytic properties of RNase A. In contrast, human **granulocyte–macrophage colony-stimulating factor (GM-CSF),** a 127-amino acid residue protein growth fac-

tor that promotes the development, activation, and survival of the white blood cells known as **granulocytes** and **macrophages,** is variably glycosylated at two *N*-linked sites and five *O*-linked sites. Through the generation of mutant varieties of GM-CSF that lack one or both of the *N*-glycosylation sites, it was found that the lifetime of GM-CSF in the bloodstream increases with its level of glycosylation. However, GM-CSF that is produced by *E. coli* and hence is unglycosylated (bacteria do not glycosylate the proteins they synthesize) has a 20-fold higher specific biological activity than does the naturally occurring glycoprotein.

As the foregoing examples suggest, no generalization can be made about the effects of glycosylation on protein properties; they must be experimentally determined on a case-by-case basis. Nevertheless, it is becoming increasingly evident that glycosylation can affect protein properties in many ways, including protein folding, oligomerization, physical stability, specific bioactivity, rate of clearance from the bloodstream, and protease resistance. Thus, *the species-specific and tissue-specific distribution of glycoforms that each cell synthesizes endows it with a characteristic spectrum of biological properties.*

c. *O*-Linked Glycoproteins Often Serve Protective Functions

O-Linked polysaccharides tend not to be uniformly distributed along polypeptide chains. Rather, they are clustered into heavily glycosylated (65–85% carbohydrate by weight) segments in which glycosylated Ser and Thr residues comprise 25 to 40% of the sequence. The carbohydrates' hydrophilic and steric interactions cause these heavily glycosylated regions, which are also rich in Pro and other helix-breaking residues, to assume extended conformations. For example, **mucins,** the protein components of **mucus,** are *O*-linked glycoproteins that can be exceedingly large (up to ~10⁷ D) and whose carbohydrate chains are often sulfated and hence mutually repelling. Mucins, which may be membrane-bound or secreted, therefore consist of stiff chains that are devoid of secondary structure and which occupy time-averaged volumes approximating those of small bacteria. Consequently, mucins, at their physiological concentrations, form intertangled networks that comprise the viscoelastic gels that protect and lubricate the mucous membranes that produced them.

Eukaryotic cells, as we shall see in Section 12-3D, have a thick and fuzzy coating of glycoproteins and **glycolipids** named the **glycocalyx** that prevents the close approach of macromolecules and other cells. How, then, can cells interact? Many cell-surface proteins, such as the receptors for various macromolecules, have relatively short and presumably stiff *O*-glycosylated regions that link these glycoproteins' membrane-bound domains to their functional domains. This arrangement is thought to extend the functional domains in a lollipoplike manner above the cell's densely packed glycocalyx, thereby permitting the functional domain to interact with extracellular macromolecules that cannot penetrate the glycocalyx.

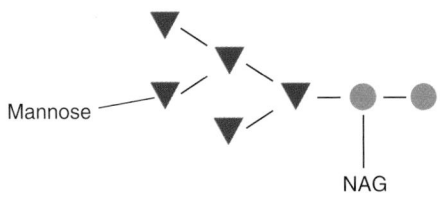

FIGURE 11-32   The microheterogeneous *N*-linked oligosaccharide of RNase B has the (mannose)₅(NAG)₂ core shown. A sixth mannose residue occurs at various positions on this core.

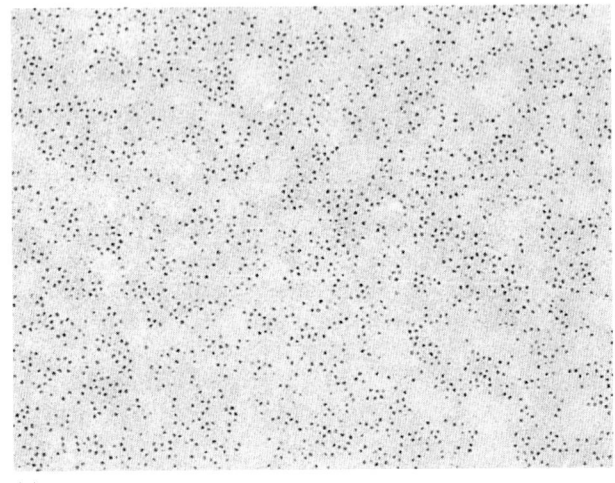

(a)

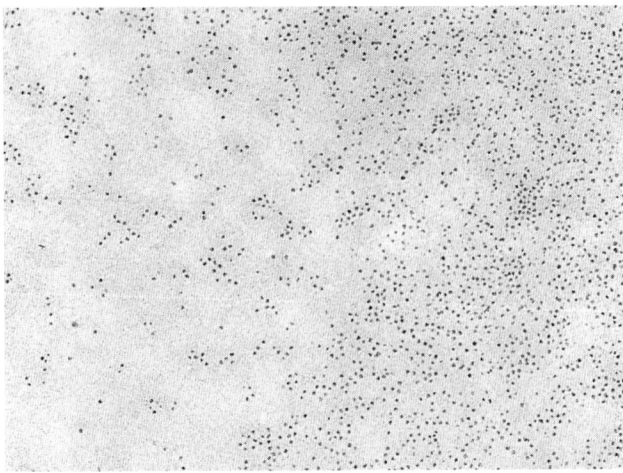

(b)

FIGURE 11-33   The surfaces of (*a*) a normal mouse cell and (*b*) a cancerous cell as seen in the electron microscope. Both cells were incubated with the ferritin-labeled lectin concanavalin A. The lectin is evenly dispersed on the normal cell but is aggregated into clusters on the cancerous cell. [Courtesy of Garth Nicolson, The Institute for Molecular Medicine, Huntington Beach, California.]

d. Oligosaccharide Markers Mediate a Variety of Intercellular Interactions

Glycoproteins are important constituents of plasma membranes (Section 12-3). The location of their carbohydrate moieties can be determined by electron microscopy. The glycoproteins are labeled with lectins that have been conjugated (covalently cross-linked) to **ferritin,** an iron-transporting protein that is readily visible in the electron microscope because of its electron-dense iron hydroxide core. Such experiments, with lectins of different specificities and with a variety of cell types, have demonstrated that *the carbohydrate groups of membrane-bound glycoproteins are, for the most part, located on the external surface of the cell membrane.* Thus, the viability of cultured cells from multicellular organisms that have any of a large number of glycosylation mutations and the infrequent viability of whole organisms that bear such mutations indicate that oligosaccharides are important for intercellular communications but not for intracellular housekeeping functions.

A further indication that oligosaccharides function as biological markers is the observation that the carbohydrate content of a glycoprotein often governs its metabolic fate. For example, the excision of sialic acid residues from certain radioactively labeled blood plasma glycoproteins by treatment with **sialidase** greatly increases the rate at which these glycoproteins are removed from the circulation. The glycoproteins are taken up and degraded by the liver in a process that depends on the recognition by liver cell receptors of sugar residues such as galactose and mannose which are exposed by the sialic acid excision. A diverse series of receptors, each specific for a particular type of sugar residue, participates in removing any particular glycoprotein from the blood. A variety of glycoforms for a given glycoprotein therefore probably ensures that it has a range of lifetimes in the blood. *Similar "ticketing" mechanisms probably govern the compartmentation and degradation of glycoproteins within cells.*

The observation that cancerous cells are more susceptible to agglutination by lectins than are normal cells led to the discovery that *there are significant differences between the cell-surface carbohydrate distributions of cancerous and noncancerous cells* (Fig. 11-33). Normal cells stop growing when they touch each other, a phenomenon

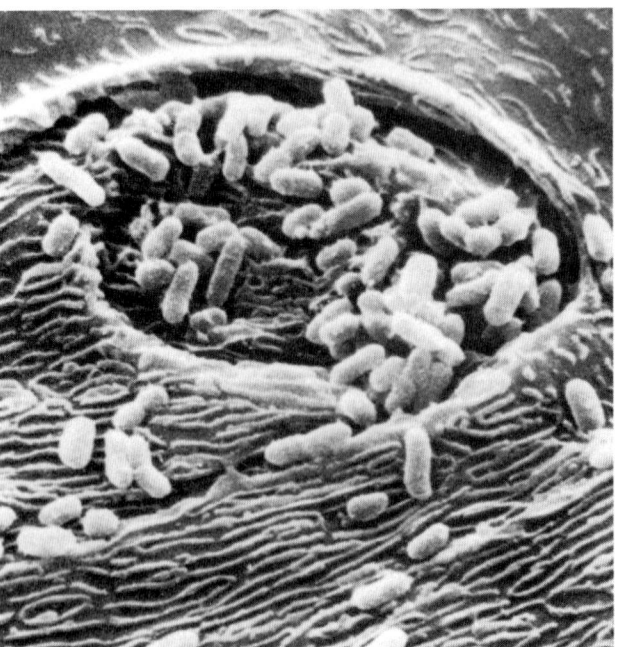

FIGURE 11-34   Scanning electron micrograph of tissue from the inside of a human cheek. The white cylindrical objects are *E. coli.* The bacteria adhere to mannose residues that are incorporated in the plasma membrane of cheek cells. This is the first step of a bacterial infection. [Courtesy of Fredric Silverblatt and Craig Kuehn, Veterans Administration Hospital, Sepulveda, California.]

known as **contact inhibition.** Cancer cells, however, are under no such control and therefore form **malignant tumors.**

Carbohydrates are important mediators of cell–cell recognition and have been implicated in related processes such as fertilization, cellular differentiation, the aggregation of cells to form organs, and the infection of cells by bacteria and viruses. For example, bacteria initiate infections by attaching to host cells (Fig. 11-34) via bacterial proteins known as **adhesins,** which each specifically bind certain host cell molecules (the adhesins' receptors). In gram-negative bacteria such as *E. coli,* adhesins are often

minor components of the heteropolymeric rodlike organelles called pili (Fig. 1-3*b*). The so-called P pili that mediate the attachment of the *E. coli* causing urinary tract infections in humans do so via an adhesin named **PapG** protein. This protein specifically binds to the α-D-galactopyranosyl-(1 → 4)-β-D-galactopyranose groups that are present on the surfaces of urinary tract epithelial cells. Electron microscopy studies revealed that the PapG adhesin is located at the end of the P pili's flexible tip, thereby providing this adhesin with considerable steric freedom in binding to its digalactoside receptor.

CHAPTER SUMMARY

Carbohydrates are polyhydroxy aldehydes or ketones of approximate composition $(C \cdot H_2O)_n$ that are important components of biological systems.

1 ■ Monosaccharides The various monosaccharides, such as ribose, fructose, glucose, and mannose, differ in their number of carbon atoms, the positions of their carbonyl groups, and their diastereomeric configurations. These sugars exist almost entirely as cyclic hemiacetals and hemiketals, which, for 5- and 6-membered rings, are respectively known as furanoses and pyranoses. The two anomeric forms of these cyclic sugars may interconvert by mutarotation. Pyranose sugars have nonplanar rings with boat and chair conformations similar to those of substituted cyclohexanes. Polysaccharides are held together by glycosidic bonds between neighboring monosaccharide units. Glycosidic bonds do not undergo mutarotation. Monosaccharides can be oxidized to aldonic and glycuronic acids or reduced to alditols. An OH group is replaced by H in deoxy sugars and by an amino group in amino sugars.

2 ■ Polysaccharides Carbohydrates can be purified by electrophoretic and chromatographic procedures. Affinity chromatography using lectins has been particularly useful in this regard. The sequences and linkages of polysaccharides may be determined by methylation analysis and by the use of specific exoglycosidases. Similar information may be obtained through NMR spectroscopy and/or mass spectrometric techniques. Cellulose, the structural polysaccharide of plant cell walls, is a linear polymer of β(1 → 4)-linked D-glucose residues. It forms a fibrous hydrogen bonded structure of exceptional strength that in plant cells is embedded in an amorphous matrix. Starch, the food storage polysaccharide of plants, consists of a mixture of the linear α(1 → 4)-linked glucan α-amylose and the α(1 → 6)-branched and α(1 → 4)-linked glucan amylopectin. Glycogen, the animal storage polysaccharide, resembles amylopectin but is more highly branched. Digestion of starch and glycogen is initiated by α-amylase and is completed by specific membrane-bound intestinal enzymes.

3 ■ Glycoproteins Proteoglycans of ground substance are mostly high molecular mass aggregates, many of which

structurally resemble a bottlebrush. Their proteoglycan subunits consist of a core protein to which glycosaminoglycans, usually chondroitin sulfate and keratan sulfate, are covalently linked. The rigid framework of a bacterial cell wall consists of chains of alternating β(1 → 4)-linked NAG and NAM that are cross-linked by short polypeptides to form a bag-shaped peptidoglycan molecule that encloses the bacterium. Lysozyme cleaves the glycosidic linkages between NAM and NAG of peptidoglycan. Penicillin specifically inactivates enzymes involved in the cross-linking of peptidoglycans. Both of these substances cause the lysis of susceptible bacteria. Gram-positive bacteria have teichoic acids that are linked covalently to their peptidoglycans. Gram-negative bacteria have outer membranes that bear complex and unusual polysaccharides known as O-antigens. These participate in the recognition of host cells and are important in the immunological recognition of bacteria by the host. Oligosaccharides attach to eukaryotic proteins in only a few ways. In *N*-glycosidic attachments, an NAG is invariably bound to the amide nitrogen of Asn in the sequence Asn-X-Ser(Thr). *O*-Glycosidic attachments are made to Ser or Thr in most proteins and to hydroxylysine in collagen.

Oligosaccharides are located on the surfaces of glycoproteins. Glycoproteins have functions that span the entire range of protein activities, although the roles of their carbohydrate moieties are just beginning to be understood. For example, ribonuclease B differs from the functionally indistinguishable and carbohydrate-free ribonuclease A only by the attachment of a single oligosaccharide of somewhat variable sequence, whereas the biological properties of granulocyte–macrophage colony-stimulating factor are significantly affected by its multiple oligosaccharide chains. The viscoelastic and hence protective properties of mucus largely result from the numerous negatively charged oligosaccharide groups of its component mucins. The carbohydrate moieties of glycoproteins in plasma membranes are invariably located on the external surfaces of the membranes. A glycoprotein's carbohydrate moieties may direct its metabolic fate by governing its uptake by certain cells or cell compartments. Glycoproteins are also important mediators of cell–cell recognition and, in many cases, are the receptors for bacterial attachment, via adhesins, in the initial stages of infection.

REFERENCES

GENERAL

Allen, H.J. and Kisailus, E.C. (Eds.), *Glycoconjugates. Composition, Structure, and Function,* Marcel Dekker (1992).

Aspinall, G.O. (Ed.), *The Polysaccharides,* Vols. 1–3, Academic Press (1982, 1983, and 1985).

Gabius, H.-J. and Gabius, S. (Eds.), *Glycosciences. Status and Perspectives,* Chapman & Hall (1997).

Solomons, T.W.G. and Fryhle, C., *Organic Chemistry* (7th ed.), Chapter 22, Wiley (2000). [A general discussion of carbohydrate nomenclature and chemistry. Other comprehensive organic chemistry texts have similar material.]

Varki, A., Cummings, R., Esko, J., Freeze, H., Hart, G., and Marth, J. (Eds.), *Essentials of Glycobiology,* Cold Spring Harbor Laboratory Press (1999).

OLIGOSACCHARIDES AND POLYSACCHARIDES

Bayer, E.A., Chanzy, H., Lamed, R., and Shoham, Y., Cellulose, cellulases, and cellulosomes, *Curr. Opin. Struct. Biol.* **8,** 548 (1998).

Carver, J.P., Experimental structure determination of oligosaccharides, *Curr. Opin. Struct. Biol.* **1,** 716–720 (1991).

Kretchmer, M., Lactose and lactase, *Sci. Am.* **227**(4), 74–78 (1972).

Weis, W.I. and Drickamer, K., Structural basis of lectin–carbohydrate recognition, *Annu. Rev. Biochem.* **65,** 441–473 (1996).

GLYCOPROTEINS

Bernfield, M., Götte, M., Park, P.W., Reizes, O., Fitxgerald, M.L., Linecum, J., and Zako, M., Functions of cell surface heparan sulfate proteoglycans, *Annu. Rev. Biochem.* **68,** 729–777 (1999).

Bush, C.A., Martin-Pastor, M., and Imberty, A., Structure and conformation of complex carbohydrates of glycoproteins, glycolipids, and bacterial polysaccharides, *Annu. Rev. Biophys. Biomol. Struct.* **28,** 269–293 (1999).

Caplan, A.I., Cartilage, *Sci. Am.* **251**(4), 84–94 (1984).

Chain, E., Fleming's contribution to the discovery of penicillin, *Trends Biochem. Sci.* **4,** 143–146 (1979). [A historical account by one of the biochemists who characterized penicillin.]

Devine, P.L. and McKenzie, F.C., Mucins: Structure, function, and associations with malignancy, *BioEssays* **14,** 619–624 (1992).

Drickamer, K., Clearing up glycoprotein hormones, *Cell* **67,** 1029–1032 (1991).

Drickamer, K. and Taylor, M.E., Evolving views of protein glycosylation, *Trends Biochem. Sci.* **23,** 321–324 (1998).

Hardingham, T.E. and Fosang, A.J., Proteoglycans: Many forms and many functions, *FASEB J.* **6,** 861–870 (1992).

Hart, G.W., Glycosylation, *Curr. Opin. Cell Biol.* **4,** 1017–1023 (1992).

Iozzo, R.V., Matrix proteoglycans: From molecular design to cellular function, *Annu. Rev. Biochem.* **67,** 609–652 (1998); *and* The biology of the small leucine-rich proteoglycans, *J. Biol. Chem.* **274,** 18843–18846 (1999).

Jentoft, N., Why are proteins O-glycosylated? *Trends Biochem. Sci.* **15,** 291–294 (1990).

Jollés, P. (Ed.), *Proteoglycans,* Birkhäuser (1994).

Kjellén, L. and Lindahl, U., Proteoglycans: Structure and interactions, *Annu. Rev. Biochem.* **60,** 443–475 (1991). [Indicates the scope of the extensive structural and functional heterogeneity in proteoglycans.]

Kuehn, M.J., Heuser, J., Normark, S., and Hultgren, S.J., P pili in uropathic *E. coli* are composite fibres with distinct fibrillar adhesive tips, *Nature* **356,** 252–255 (1992).

Parekh, R.B., Effects of glycosylation on protein function, *Curr. Opin. Struct. Biol.* **1,** 750–754 (1991).

Perez-Vilar, J. and Hill, R.L., The structure and assembly of secreted mucins, *J. Biol. Chem.* **274,** 31751–31754 (1999).

Rademacher, T.W., Parekh, R.B., and Dwek, R.A., Glycobiology, *Annu. Rev. Biochem.* **57,** 787–838 (1988). [The role of *N*-linked oligosaccharides in mediating protein-specific biological activity.]

Rasmussen, J.R., Effect of glycosylation on protein function, *Curr. Opin. Struct. Biol.* **2,** 682–686 (1992).

Rudd, P.M. and Dwek, R.A., Rapid, sensitive sequencing of oligosaccharides from glycoproteins, *Curr. Opin. Biotech.* **8,** 488–497 (1997); *and* Dwek, R.A., Edge, C.J., Harvey, D.J., and Parekh, R.B., Analysis of glycoprotein-associated oligosaccharides, *Annu. Rev. Biochem.* **62,** 65–100 (1993).

Ruoslahti, E. and Yamaguchi, Y., Proteoglycans as modulators of growth factor activities, *Cell* **64,** 867–869 (1991).

Schauer, R., Sialic acids and their role as biological masks, *Trends Biochem. Sci.* **10,** 357–360 (1985).

Sharon, N. and Lis, H., Carbohydrates in cell recognition, *Sci. Am.* **268**(1), 82–89 (1993).

Wormald, M.R. and Dwek, R.A., Glycoproteins: Glycan presentation and protein-fold stability, *Structure* **7,** R155–R160 (1999).

PROBLEMS

1. The trisaccharide shown is named **raffinose.** What is its systematic name? Is it a reducing sugar?

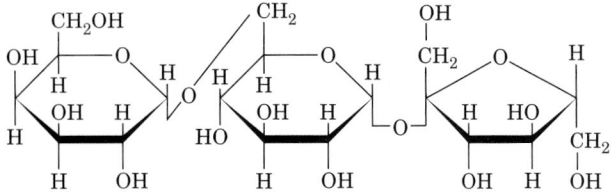

Raffinose

2. The systematic name of **melezitose** is *O*-α-D-glucopyranosyl-(1 → 3)-*O*-β-D-fructofuranosyl-(2 → 1)-α-D-glucopyranoside. Draw its molecular formula. Is it a reducing sugar?

3. Name the linear form of D-glucose using the (*RS*) chirality nomenclature system. [See Section 4-2C. *Hint:* The branch toward C1 has higher priority than the branch toward C6.]

***4.** Draw the α-furanose form of D-talose and the β-pyranose form of L-sorbose.

5. The NaBH₄ reduction product of D-glucose may be named L-sorbitol or D-glucitol. Explain.

6. How many different disaccharides of D-glucopyranose are possible? How many trisaccharides?

7. A molecule of amylopectin consists of 1000 glucose residues and is branched every 25 residues. How many reducing ends does it have?

8. Most paper is made by removing the lignin from wood pulp and forming the resulting mass of largely unoriented cellulose fibers into a sheet. Untreated paper loses most of its strength when wet with water but maintains its strength when wet with oil. Explain.

*9. Write a chemical mechanism for the acid-catalyzed mutarotation of glucose.

10. The values of the specific rotation, $[\alpha]_D^{20}$, for the α and β anomers of D-galactose are 150.7° and 52.8°, respectively. A mixture that is 20% α-D-galactose and 80% β-D-galactose is dissolved in water at 20°C. What is its initial specific rotation? After several hours, the specific rotation of this mixture reached an equilibrium value of 80.2°. What is its anomeric composition?

11. Name the epimers of D-gulose.

12. Exhaustive methylation of a trisaccharide followed by acid hydrolysis yields equimolar quantities of 2,3,4,6-tetra-*O*-methyl-D-galactose, 2,3,4-tri-*O*-methyl-D-mannose, and 2,4,6-tri-*O*-methyl-D-glucose. Treatment of the trisaccharide with β-galactosidase yields D-galactose and a disaccharide. Treatment of this disaccharide with α-mannosidase yields D-mannose and D-glucose. Draw the structure of the trisaccharide and state its systematic name.

13. The enzyme β-amylase cleaves successive maltose units from the nonreducing end of $\alpha(1 \rightarrow 4)$ glucans. It will not cleave at glucose residues that have an $\alpha(1 \rightarrow 6)$ bond. The end products of the exhaustive digestion of amylopectin by β-amylase are known as **limit dextrins.** Draw a schematic diagram of an amylopectin molecule and indicate what part(s) of it constitutes limit dextrins.

14. One demonstration of P.T. Barnum's maxim that there's a sucker born every minute is that new "reducing aids" regularly appear on the market. An eat-all-you-want nostrum, which was touted as a "starch blocker" [and which the Food and Drug Administration (FDA) eventually banned], contained an α-amylase-inhibiting protein extracted from beans. If this substance had really worked as advertised, which it did not, what unpleasant side effects would have resulted from its ingestion with a starch-containing meal? Discuss why this substance, which inhibited α-amylase *in vitro*, would not do so in the intestines after oral ingestion.

*15. Treatment of a 6.0-g sample of glycogen with Tollens' reagent followed by exhaustive methylation and then hydrolysis yields 3.1 mmol of 2,3-di-*O*-methylglucose and 0.0031 mmol of 1,2,3-tri-*O*-methylgluconic acid as well as other products. (a) What fraction of glucose residues occur at $(1 \rightarrow 6)$ branch points, and what is the average number of glucose residues per branch? (b) What are the other products of the methylation–hydrolysis treatment and in what amounts are they formed? (c) What is the average molecular mass of the glycogen?

16. The lysis of a culture of *E. coli* yields a solution with mucuslike viscosity. Adding DNase to the solution greatly reduces this viscosity. What is the physical basis of the viscosity?

17. Instilling methyl-α-D-mannoside into the bladder of a mouse prevents the colonization of its urinary tract by *E. coli.* What is the reason for this effect?

Chapter 12

Lipids and Membranes

Membranes function to organize biological processes by compartmentalizing them. Indeed, the cell, the basic unit of life, is essentially defined by its enveloping plasma membrane. Moreover, in eukaryotes, many subcellular organelles, such as nuclei, mitochondria, chloroplasts, the endoplasmic reticulum, and the Golgi apparatus (Fig. 1-5), are likewise membrane bounded.

Biological membranes are organized assemblies of lipids and proteins with small amounts of carbohydrate. Yet they are not impermeable barriers to the passage of materials. Rather, they regulate the composition of the intracellular medium by controlling the flow of nutrients, waste products, ions, etc., into and out of the cell. They do this through membrane-embedded "pumps" and "gates" that transport specific substances against an electrochemical gradient or permit their passage with such a gradient (Chapter 20).

Many fundamental biochemical processes occur on or in a membranous scaffolding. For example, electron transport and oxidative phosphorylation (Chapter 22), processes that oxidize nutrients with the concomitant generation of ATP, are mediated by an organized battery of enzymes that are components of the inner mitochondrial membrane. Likewise, photosynthesis, in which light energy powers the chemical combination of H_2O and CO_2 to form carbohydrates (Chapter 24), occurs in the inner membranes of chloroplasts. The processing of information, such as sensory stimuli or intercellular communications, is generally a membrane-based phenomenon. Thus nerve impulses are mediated by nerve cell membranes (Section 20-5) and the presence of certain substances such as hormones and nutrients is detected by specific membrane-bound receptors (Chapter 19).

In this chapter, we examine the compositions, structures, and formation of biological membranes and related substances. Specific membrane-based biochemical processes, such as those mentioned above, are dealt with in later chapters.

1 ■ LIPID CLASSIFICATION

Lipids (Greek: lipos, fat) are substances of biological origin that are soluble in organic solvents such as chloroform and methanol but are only sparingly soluble, if at all, in water. Hence, they are easily separated from other biological materials by extraction into organic solvents and may be further fractionated by such techniques as adsorption chromatography, thin layer chromatography, and reverse-phase chromatography (Section 6-3D). Fats, oils, certain vitamins and hormones, and most nonprotein membrane components are lipids. In this section, we discuss the structures and physical properties of the major classes of lipids.

A. Fatty Acids

Fatty acids are carboxylic acids with long-chain hydrocarbon side groups (Fig. 12-1). They are rarely free in nature but, rather, occur in esterified form as the major components of the various lipids described in this chapter. The more common biological fatty acids are listed in Table 12-1. In higher plants and animals, the predominant fatty acid residues are those of the C_{16} and C_{18} species **palmitic, oleic, linoleic,** and **stearic acids.** Fatty acids with <14 or >20 carbon atoms are uncommon. *Most fatty acids have an even number of carbon atoms because they are usually biosynthesized by the concatenation of C_2 units (Section 25-4C).* Over half of the fatty acid residues of plant and animal lipids are unsaturated (contain double bonds) and are often polyunsaturated (contain two or more double bonds). Bacterial fatty acids are rarely polyunsaturated but are commonly branched, hydroxylated, or contain cyclopropane rings. Unusual fatty acids also occur as components of the oils and **waxes** (esters of fatty acids and long-chain alcohols) produced by certain plants.

a. The Physical Properties of Fatty Acids Vary with Their Degree of Unsaturation

Table 12-1 indicates that the first double bond of an unsaturated fatty acid commonly occurs between its C9 and C10 atoms counting from the carboxyl C atom (a Δ^9- or 9-double bond). In polyunsaturated fatty acids, the double bonds tend to occur at every third carbon atom toward the methyl terminus of the molecule (such as $-CH=CH-CH_2-CH=CH-$). Double bonds in polyunsaturated fatty acids are almost never conjugated (as in $-CH=CH-CH=CH-$). Triple bonds rarely occur in fatty acids or any other compound of biological origin. Two important classes of polyunsaturated fatty acids are denoted $n-3$ (or $\omega-3$) and $n-6$ (or $\omega-6$) fatty acids. This nomenclature identifies the last double-bonded carbon atom as counted from the methyl terminal (ω) end of the chain.

Saturated fatty acids are highly flexible molecules that can assume a wide range of conformations because there is relatively free rotation about each of their C—C bonds. Nevertheless, their fully extended conformation is that of minimum energy because this conformation has the least amount of steric interference between neighboring methylene groups. The melting points (mp) of saturated fatty

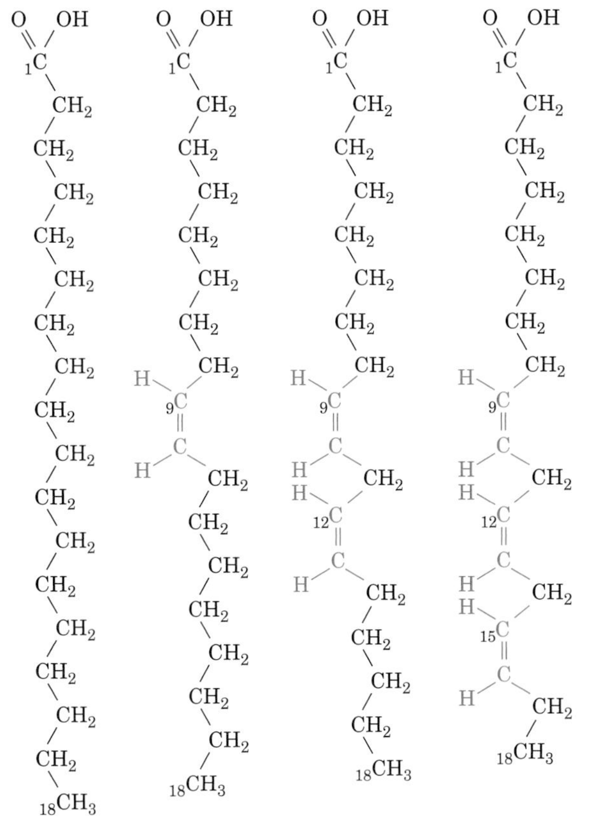

Stearic acid   Oleic acid   Linoleic acid   α-Linolenic acid

FIGURE 12-1   Structural formulas of some C_{18} fatty acids. The double bonds all have the cis configuration.

TABLE 12-1   The Common Biological Fatty Acids

Symbol[a]	Common Name	Systematic Name	Structure	mp (°C)
Saturated fatty acids				
12:0	Lauric acid	Dodecanoic acid	$CH_3(CH_2)_{10}COOH$	44.2
14:0	Myristic acid	Tetradecanoic acid	$CH_3(CH_2)_{12}COOH$	52
16:0	Palmitic acid	Hexadecanoic acid	$CH_3(CH_2)_{14}COOH$	63.1
18:0	Stearic acid	Octadecanoic acid	$CH_3(CH_2)_{16}COOH$	69.6
20:0	Arachidic acid	Eicosanoic acid	$CH_3(CH_2)_{18}COOH$	75.4
22:0	Behenic acid	Docosanoic acid	$CH_3(CH_2)_{20}COOH$	81
24:0	Lignoceric acid	Tetracosanoic acid	$CH_3(CH_2)_{22}COOH$	84.2
Unsaturated fatty acids (all double bonds are cis)				
$16:1n-7$	Palmitoleic acid	9-Hexadecenoic acid	$CH_3(CH_2)_5CH=CH(CH_2)_7COOH$	−0.5
$18:1n-9$	Oleic acid	9-Octadecenoic acid	$CH_3(CH_2)_7CH=CH(CH_2)_7COOH$	13.4
$18:2n-6$	Linoleic acid	9,12-Octadecadienoic acid	$CH_3(CH_2)_4(CH=CHCH_2)_2(CH_2)_6COOH$	−9
$18:3n-3$	α-Linolenic acid	9,12,15-Octadecatrienoic acid	$CH_3CH_2(CH=CHCH_2)_3(CH_2)_6COOH$	−17
$18:3n-6$	γ-Linolenic acid	6,9,12-Octadecatrienoic acid	$CH_3(CH_2)_4(CH=CHCH_2)_3(CH_2)_3COOH$	
$20:4n-6$	Arachidonic acid	5,8,11,14-Eicosatetraenoic acid	$CH_3(CH_2)_4(CH=CHCH_2)_4(CH_2)_2COOH$	−49.5
$20:5n-3$	EPA	5,8,11,14,17-Eicosapentaenoic acid	$CH_3CH_2(CH=CHCH_2)_5(CH_2)_2COOH$	−54
$22:6n-3$	DHA	4,7,10,13,16,19-Docosahexaenoic acid	$CH_3CH_2(CH=CHCH)_6CH_2COOH$	
$24:1n-9$	Nervonic acid	15-Tetracosenoic acid	$CH_3(CH_2)_7CH=CH(CH_2)_{13}COOH$	39

[a]Number of carbon atoms : number of double bonds. For unsaturated fatty acids, the quantity "$n-x$" indicates the position of the last double bond in the fatty acid, where n is its number of C atoms and x is the position of the last double-bonded C atom counting from the methyl terminal (ω) end of the chain.

Source: Dawson, R.M.C., Elliott, D.C., Elliott, W.H., and Jones, K.M., *Data for Biochemical Research* (3rd ed.), Chapter 8, Clarendon Press (1986).

acids, like those of most substances, increase with molecular mass (Table 12-1).

Fatty acid double bonds almost always have the cis configuration (Fig. 12-1). This puts a rigid 30° bend in the hydrocarbon chain of unsaturated fatty acids that interferes with their efficient packing to fill space. The consequent reduced van der Waals interactions cause fatty acid melting points to decrease with their degree of unsaturation (Table 12-1). Lipid fluidity likewise increases with the degree of unsaturation of their component fatty acid residues. This phenomenon, as we shall see in Section 12-3B, has important consequences for membrane properties.

B. *Triacylglycerols*

The fats and oils that occur in plants and animals consist largely of mixtures of **triacylglycerols** (also referred to as **triglycerides** or **neutral fats**). *These nonpolar, water-insoluble substances are fatty acid triesters of* **glycerol:**

$$
\begin{array}{ll}
^{1}CH_{2}-OH & \quad ^{1}CH_{2}-O-\overset{\overset{O}{\|}}{C}-R_{1} \\[1em]
^{2}CH-OH & \quad ^{2}CH-O-\overset{\overset{O}{\|}}{C}-R_{2} \\[1em]
^{3}CH_{2}-OH & \quad ^{3}CH_{2}-O-\overset{\overset{O}{\|}}{C}-R_{3}
\end{array}
$$

$\quad\quad$ **Glycerol** $\quad\quad\quad$ **Triacylglycerol**

Triacylglycerols function as energy reservoirs in animals and are therefore their most abundant class of lipids even though they are not components of biological membranes.

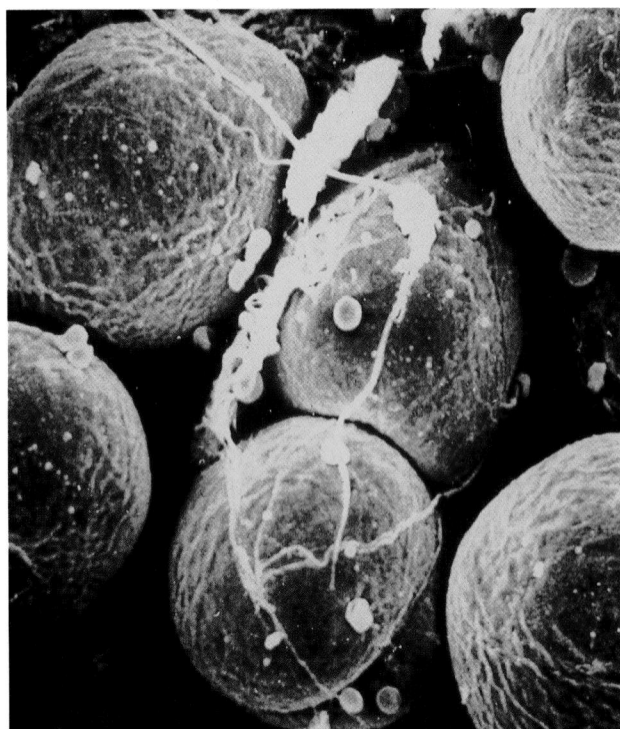

FIGURE 12-2 Scanning electron micrograph of adipocytes. Each contains a fat globule that occupies nearly the entire cell. [Fred E. Hossler/Visuals Unlimited.]

Triacylglycerols differ according to the identity and placement of their three fatty acid residues. The so-called **simple triacylglycerols** contain one type of fatty acid residue and are named accordingly. For example, **tristearoylglycerol** or **tristearin** contains three stearic acid residues, whereas **trioleoylglycerol** or **triolein** has three oleic acid residues. The more common **mixed triacylglycerols** contain two or three different types of fatty acid residues and are named according to their placement on the glycerol moiety.

**1-Palmitoleoyl-2-linoleoyl-
3-stearoyl-glycerol**

Fats and oils (which differ only in that fats are solid and oils are liquid at room temperature) are complex mixtures of simple and mixed triacylglycerols whose fatty acid compositions vary with the organism that produced them. Plant oils are usually richer in unsaturated fatty acid residues than are animal fats, as the lower melting points of oils imply.

a. Triacylglycerols Are Efficient Energy Reserves

Fats are a highly efficient form in which to store metabolic energy. This is because fats are less oxidized than are carbohydrates or proteins and hence yield significantly more energy on oxidation. Furthermore, fats, being nonpolar substances, are stored in anhydrous form, whereas glycogen, for example, binds about twice its weight of water under physiological conditions. Fats therefore provide about six times the metabolic energy of an equal weight of hydrated glycogen.

In animals, **adipocytes** (fat cells; Fig. 12-2) are specialized for the synthesis and storage of triacylglycerols.

Whereas other types of cells have only a few small droplets of fat dispersed in their cytosol, adipocytes may be almost entirely filled with fat globules. **Adipose tissue** is most abundant in a subcutaneous layer and in the abdominal cavity. The fat content of normal humans (21% for men, 26% for women) enables them to survive starvation for 2 to 3 months. In contrast, the body's glycogen supply, which functions as a short-term energy store, can provide for the body's metabolic needs for less than a day. The subcutaneous fat layer also provides thermal insulation, which is particularly important for warm-blooded aquatic animals, such as whales, seals, geese, and penguins, which are routinely exposed to low temperatures.

C. *Glycerophospholipids*

Glycerophospholipids (or *phosphoglycerides*) *are the major lipid components of biological membranes.* They consist of *sn-glycerol-3-phosphate* (Fig. 12-3*a*) esterified at its C1 and C2 positions to fatty acids and at its phosphoryl group to a group, X, to form the class of substances diagrammed in Fig. 12-3*b*. *Glycerophospholipids are therefore amphiphilic molecules with nonpolar aliphatic "tails" and polar phosphoryl-X "heads."* The simplest glycerophospholipids, in which X = H, are **phosphatidic acids;** they are present only in small amounts in biological membranes. *In the glycerophospholipids that commonly occur in biological membranes, the head groups are derived from polar alcohols (Table 12-2).* Saturated C_{16} and C_{18} fatty acids usually

(*a*)

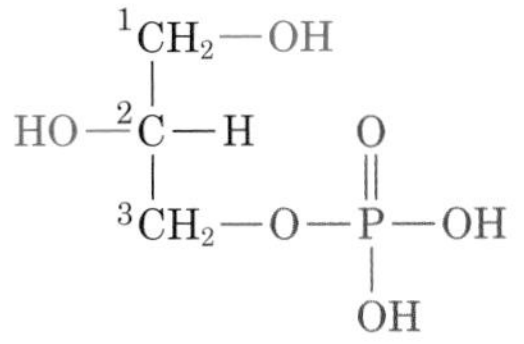

sn-**Glycerol-3-phosphate**

(*b*)

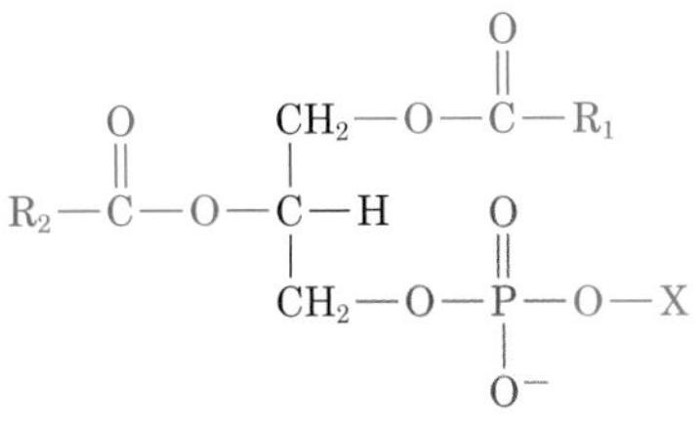

Glycerophospholipid

FIGURE 12-3 Molecular formula of glycerophospholipids.
(*a*) The compound shown in Fischer projection (Section 4-2B) can be equivalently referred to as L-glycerol-3-phosphate or D-glycerol-1-phosphate. However, using **stereospecific numbering** (*sn*), which assigns the 1-position to the group occupying the *pro-S* position of a prochiral center (see Section 4-2C for a discussion of prochirality), the compound is unambiguously named *sn*-glycerol-3-phosphate. (*b*) The general formula of the glycerophospholipids. R_1 and R_2 are long-chain hydrocarbon tails of fatty acids and X is derived from a polar alcohol (see Table 12-2).

TABLE 12-2   The Common Classes of Glycerophospholipids

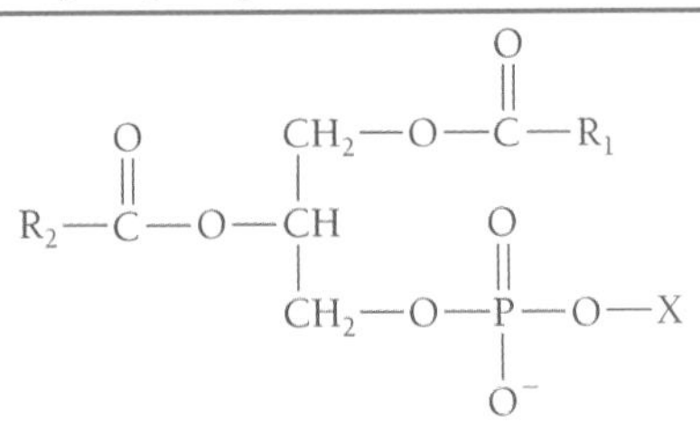

Name of X—OH	Formula of —X	Name of Phospholipid
Water	—H	Phosphatidic acid
Ethanolamine	$-CH_2CH_2NH_3^+$	Phosphatidylethanolamine
Choline	$-CH_2CH_2N(CH_3)_3^+$	Phosphatidylcholine (lecithin)
Serine	$-CH_2CH(NH_3^+)COO^-$	Phosphatidylserine
myo-Inositol	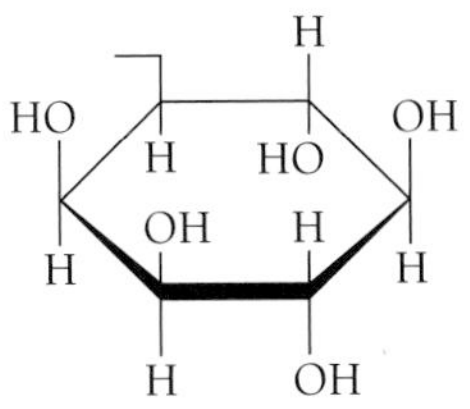	Phosphatidylinositol
Glycerol	$-CH_2CH(OH)CH_2OH$	Phosphatidylglycerol
Phosphatidylglycerol	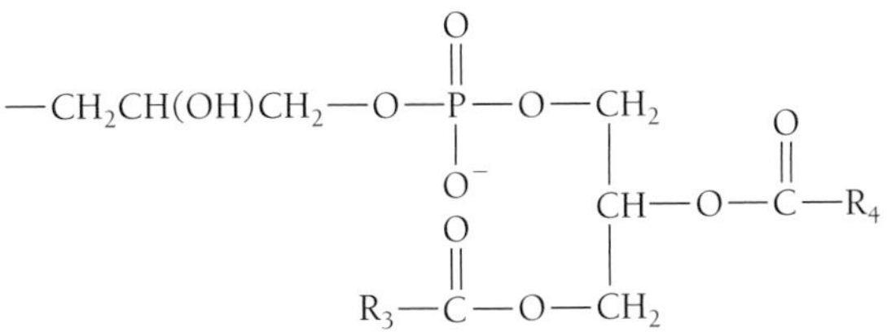	Diphosphatidylglycerol (cardiolipin)

occur at the C1 position of glycerophospholipids, and the C2 position is often occupied by an unsaturated C_{16} to C_{20} fatty acid. Glycerophospholipids are, of course, also named according to the identities of these fatty acid residues (Fig. 12-4). Some glycerophospholipids have common names. For example, phosphatidylcholines are known as **lecithins;** diphosphatidylglycerols, the "double" glycerophospholipids, are known as **cardiolipins** (because they were first isolated from heart muscle).

Plasmalogens

X
|
O
|
O=P—O⁻
|
O
|
CH₂—CH—CH₂
| |
O   O
| |
CH   C=O
‖ |
CH   R₂
|
R₁

A plasmalogen

are glycerophospholipids in which the C1 substituent to the glycerol moiety is bonded to it via an α,β-unsaturated ether linkage in the cis configuration rather than through an es-

ter linkage. **Ethanolamine, choline,** and serine form the most common plasmalogen head groups.

D. *Sphingolipids*

Sphingolipids, which are also major membrane components, are derivatives of the C_{18} amino alcohols **sphingosine, dihydrosphingosine** (Fig. 12-5), and their C_{16}, C_{17}, C_{19}, and C_{20} homologs. Their *N*-acyl fatty acid derivatives, **ceramides,**

OH H   OH
| | |
H₂C—C——C—H
| |
NH   CH
| ‖
O=C   HC
Fatty acid | |
residue   R   (CH₂)₁₂
|
CH₃

A ceramide

occur only in small amounts in plant and animal tissues but form the parent compounds of more abundant sphingolipids:

1. Sphingomyelins, the most common sphingolipids, are ceramides bearing either a phosphocholine (Fig. 12-6) or a phosphoethanolamine moiety, so that they can also be

(a)

CH₃
|
H₃C—N⁺—CH₃
|
CH₂
|
CH₂
|
O
|
⁻O—P=O
|
O   H
| |
³CH₂—²C———¹CH₂
| |
O   O
| |
C=O   C=O
| |
(CH₂)₇   (CH₂)₁₆
| |
C—H   CH₃
‖
C—H
|
(CH₂)₇
|
CH₃

1-Stearoyl-2-oleoyl-3-phosphatidylcholine

(b)

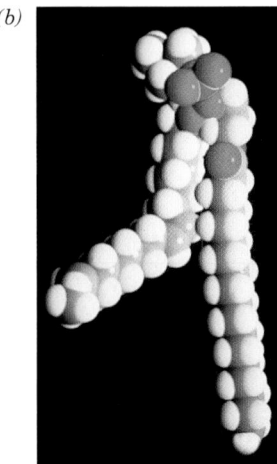

FIGURE 12-4   The glycerophospholipid 1-stearoyl-2-oleoyl-3-phosphatidylcholine. (*a*) Molecular formula in Fischer projection and (*b*) space-filling model with H white, C gray, O red, and P green. [Courtesy of Richard Pastor, FDA, Bethesda, Maryland.]

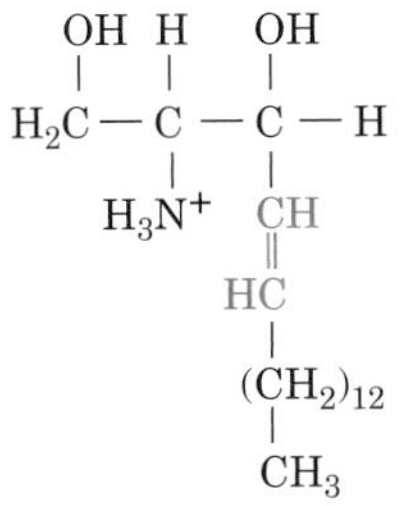

OH H   OH
| | |
H₂C — C — C — H
| |
H₃N⁺   CH
‖
HC
|
(CH₂)₁₂
|
CH₃

Sphingosine

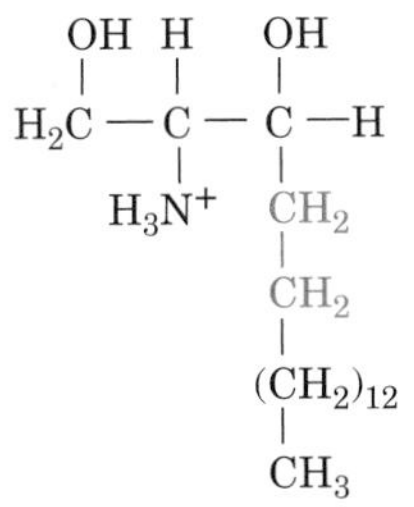

OH H   OH
| | |
H₂C — C — C —H
| |
H₃N⁺   CH₂
|
CH₂
|
(CH₂)₁₂
|
CH₃

Dihydrosphingosine

FIGURE 12-5   Molecular formulas of sphingosine and dihydrosphingosine. The chiral centers at C2 and C3 of sphingosine and dihydrosphingosine have the configurations shown in Fischer projection. The double bond in sphingosine has the trans configuration.

groups that consist of a single sugar residue. **Galactocerebrosides,** which are most prevalent in the neuronal cell membranes of the brain, have a β-D-galactose head group.

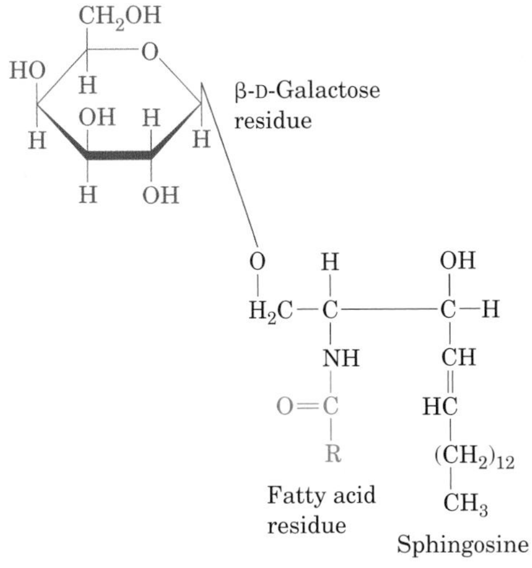

CH₂OH
O
HO
H   β-D-Galactose
OH H   residue
H   H
H   OH

O   H   OH
| | |
H₂C — C   C—H
| |
NH   CH
‖
O=C   HC
| |
R   (CH₂)₁₂
|
Fatty acid   CH₃
residue
Sphingosine

A galactocerebroside

Glucocerebrosides, which instead have a β-D-glucose residue, occur in the membranes of other tissues. *Cerebrosides, in contrast to phospholipids, lack phosphate groups and hence are most frequently nonionic compounds.* The galactose residues of some galactocerebrosides, however, are sulfated at their C3 positions to form ionic compounds known as **sulfatides.** More complex sphingoglycolipids have unbranched oligosaccharide head groups of up to four sugar residues.

3. Gangliosides form the most complex group of sphingoglycolipids. They are ceramide oligosaccharides that include among their sugar groups at least one sialic acid

classified as **sphingophospholipids.** *Although sphingomyelins differ chemically from phosphatidylcholine and phosphatidylethanolamine, their conformations and charge distributions are quite similar.* The membranous myelin sheath that surrounds and electrically insulates many nerve cell axons (Section 20-5B) is particularly rich in sphingomyelin.

2. **Cerebrosides,** the simplest **sphingoglycolipids** (alternatively **glycosphingolipids**), are ceramides with head

(a)

CH₃
|⁺
CH₃ —N—CH₃
|
CH₂
|
Phosphocholine   CH₂
head group |
O
|
O=P—O⁻
|
O   H   OH
| | |
CH₂—C————C—H
| |
NH   CH
‖
O=C   HC
| |
Palmitate   (CH₂)₁₄ (CH₂)₁₂
residue | |
CH₃   CH₃

A sphingomyelin

(b)

FIGURE 12-6   A sphingomyelin. *(a)* Molecular formula in Fischer projection and *(b)* space-filling model with H white, C gray, N blue, and O red. Note its conformational resemblance to glycerophospholipids (Fig. 12-4). [Courtesy of Richard Pastor, FDA, Bethesda, Maryland.]

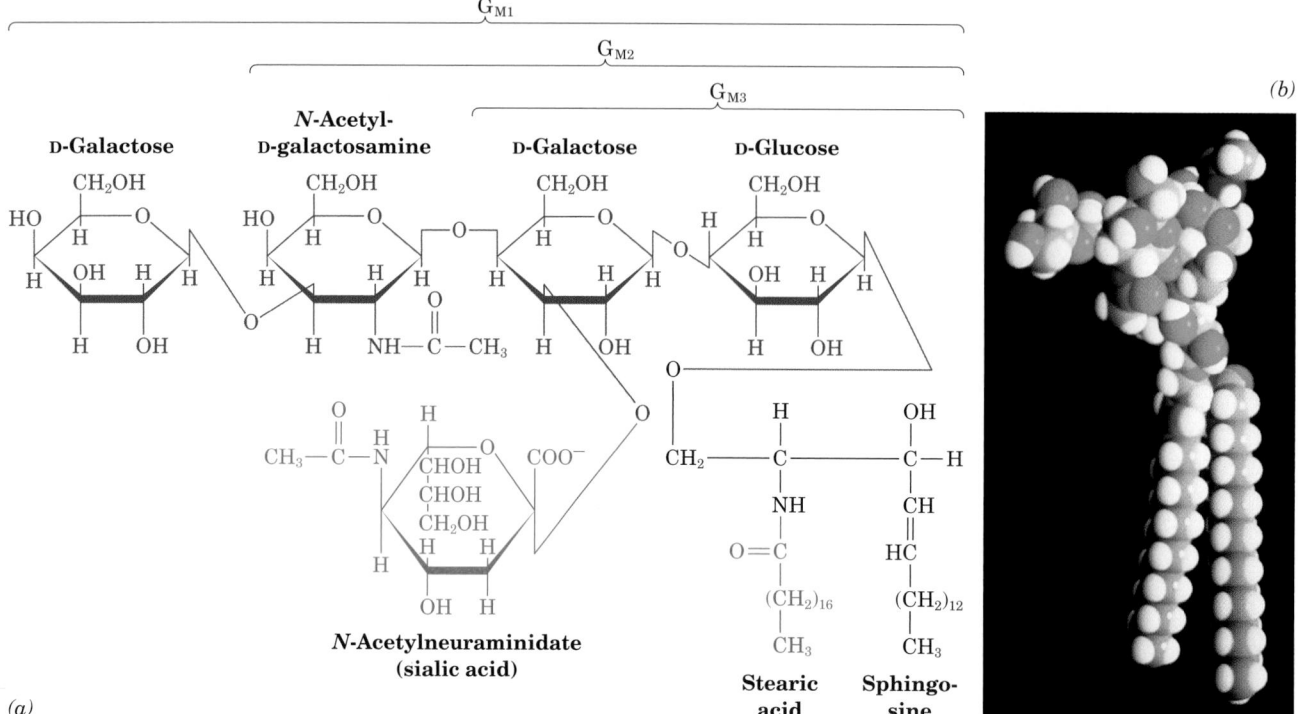

(a)

FIGURE 12-7   Ganglioside G$_{M1}$. (*a*) Structural formula with its sphingosine residue in Fischer projection and (*b*) space-filling model with H white, C gray, N blue, and O red. Gangliosides G$_{M2}$ and G$_{M3}$ differ from G$_{M1}$ only by the sequential absences of the terminal D-galactose and N-acetyl-D-galactosamine residues. Other gangliosides have different oligosaccharide head groups. [Courtesy of Richard Venable, FDA, Bethesda, Maryland.]

residue (N-acetylneuraminic acid and its derivatives; Section 11-1C). The structures of gangliosides G$_{M1}$, G$_{M2}$, and G$_{M3}$, three of the over 60 that are known, are shown in Fig. 12-7. Gangliosides are primarily components of cell-surface membranes and constitute a significant fraction (6%) of brain lipids. Other tissues also contain gangliosides but in lesser amounts.

Gangliosides have considerable physiological and medical significance. Their complex carbohydrate head groups, which extend beyond the surfaces of cell membranes, act as specific receptors for certain pituitary glycoprotein hormones that regulate a number of important physiological functions (Section 19-1). Gangliosides are also receptors for bacterial protein toxins such as cholera toxin (Section 19-2C). There is considerable evidence that gangliosides are specific determinants of cell–cell recognition, so they probably have an important role in the growth and differentiation of tissues as well as in carcinogenesis (cancer generation). Disorders of ganglioside breakdown are responsible for several hereditary **sphingolipid storage diseases,** such as **Tay-Sachs disease,** which are characterized by an invariably fatal neurological deterioration (Section 25-8C).

E. *Cholesterol*

Steroids, which are mostly of eukaryotic origin, are derivatives of **cyclopentanoperhydrophenanthrene** *(Fig. 12-8).* The much maligned **cholesterol** (Fig. 12-9), the most abundant steroid in animals, is further classified as a **sterol** because of its C3-OH group and its branched aliphatic side chain of 8 to 10 carbon atoms at C17.

Cholesterol is a major component of animal plasma membranes and occurs in lesser amounts in the membranes of their subcellular organelles. Its polar OH group gives it a weak amphiphilic character, whereas its fused ring system provides it with greater rigidity than other membrane lipids. Cholesterol is therefore an important determinant of membrane properties. It is also abundant in blood plasma lipoproteins (Section 12-5), where ~70% of it is

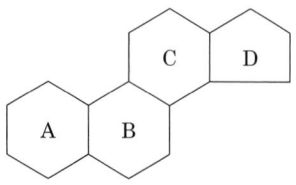

Cyclopentanoperhydrophenanthrene

FIGURE 12-8   Cyclopentanoperhydrophenanthrene, the parent compound of steroids. It consists of four fused saturated rings. The standard ring labeling system is indicated.

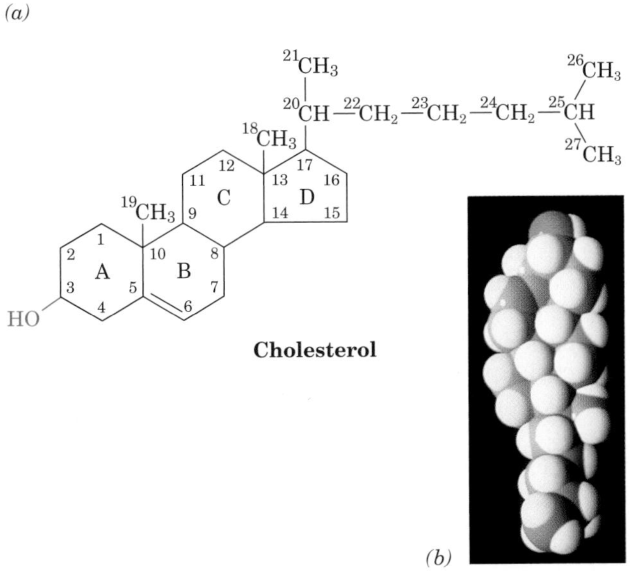

(a)

Cholesterol

(b)

FIGURE 12-9 Cholesterol. (*a*) Structural formula with the standard numbering system and (*b*) space-filling model with H white, C gray, and O red. Cholesterol's rigid ring system makes it far less conformationally flexible than are membrane lipids: Its cyclohexane rings can adopt either the boat or the chair conformations (Fig. 11-6) but the chair conformation is highly preferred. [Courtesy of Richard Pastor, FDA, Bethesda, Maryland.]

esterified to long-chain fatty acids to form **cholesteryl esters.**

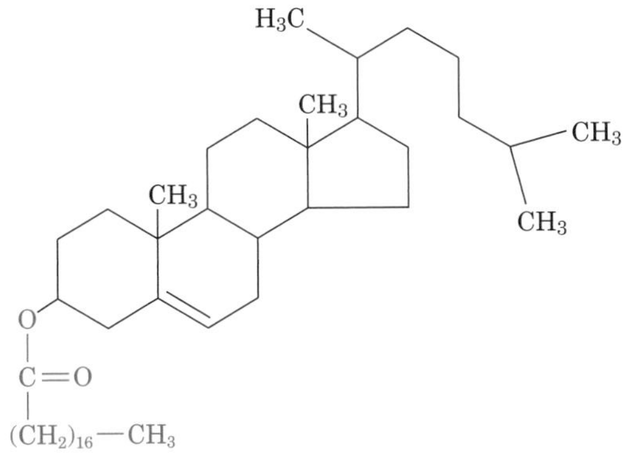

Cholesteryl stearate

Cholesterol is the metabolic precursor of **steroid hormones,** substances that regulate a great variety of physiological functions including sexual development and carbohydrate metabolism (Section 19-1G). The much debated role of cholesterol in heart disease is examined in Section 12-5C. Cholesterol metabolism and the biosynthesis of steroid hormones are discussed in Section 25-6.

Plants contain little cholesterol. Rather, the most common sterol components of their membranes are

stigmasterol and **β-sitosterol**

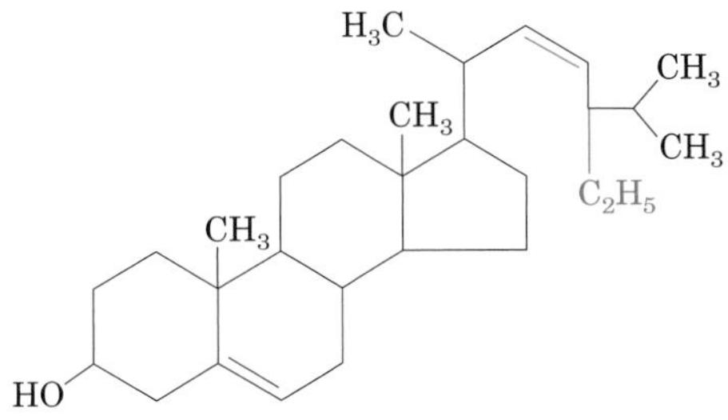

Stigmasterol

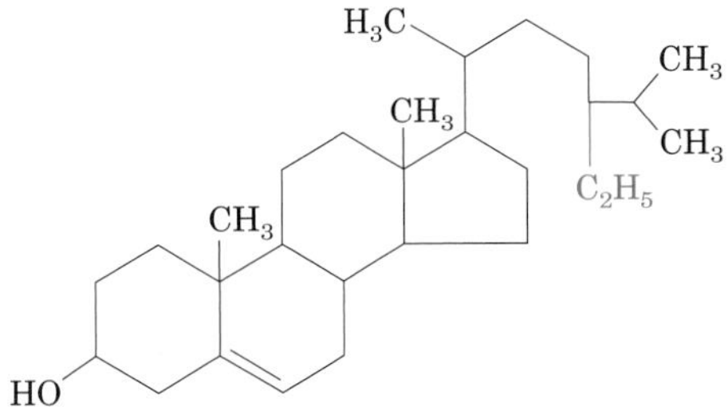

β-Sitosterol

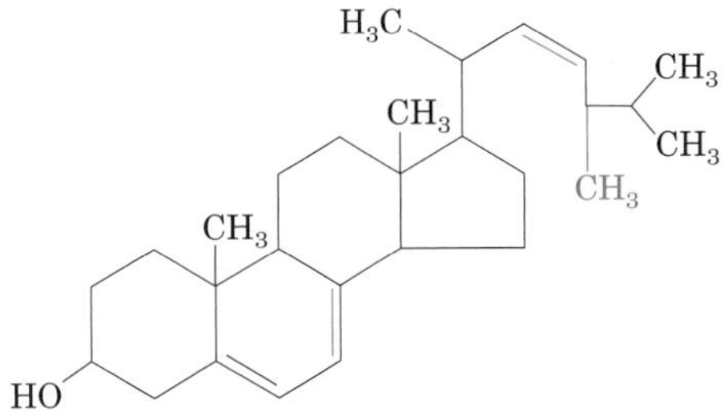

Ergosterol

which differ from cholesterol only in their aliphatic side chains. Yeast and fungi have yet other membrane sterols such as **ergosterol,** which has a C7 to C8 double bond. Prokaryotes, with the exception of mycoplasmas (Section 1-1B), contain little, if any, sterol.

2 ■ PROPERTIES OF LIPID AGGREGATES

The first recorded experiments on the physical properties of lipids were made in 1774 by the American statesman and scientist Benjamin Franklin. In investigating the well-known (at least among sailors) action of oil in calming waves, Franklin wrote:

At length being at Clapham [in London] where there is, on the common, a large pond, which I observed to be one day very rough with the wind, I fetched out a cruet of oil [probably olive oil] and dropt a little of it on the water. I saw it spread itself with surprising swiftness upon the surface. . . . I then went to the windward side, where [the waves] began to form; and there the oil, though not more than a teaspoonful, produced an instant calm over a space several yards square, which spread amazingly, and extended itself gradually till it reached the lee side, making all that quarter of the pond, perhaps half an acre, as smooth as a looking glass.

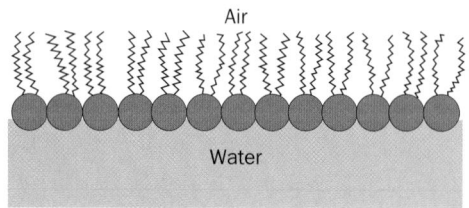

FIGURE 12-10   An oil monolayer at the air–water interface. The hydrophobic tails of the lipids avoid association with water by extending into the air.

This is sufficient information to permit the calculation of the oil layer's thickness (although there is no indication that Franklin made this calculation, we can; see Problem 4). We now know that oil forms a monomolecular layer on the surface of water in which the polar heads of the amphiphilic oil molecules are immersed in the water and their hydrophobic tails extend into the air (Fig. 12-10).

The calming effect of oil on rough water is a consequence of a large reduction in the water's surface tension. An oily surface film has the weak intermolecular cohesion characteristic of hydrocarbons rather than the strong intermolecular attractions of water responsible for its normally large surface tension. Oil, nevertheless, calms only smaller waves; it does not, as Franklin later observed, affect the larger swells.

In this section, we discuss how lipids aggregate to form micelles and bilayers. We shall also be concerned with the physical properties of lipids in bilayers because these aggregates form the structural basis for biological membranes.

A. *Micelles and Bilayers*

In aqueous solutions, amphiphilic molecules, such as soaps and detergents, form micelles (globular aggregates whose hydrocarbon groups are out of contact with water; Section 2-1B). This molecular arrangement eliminates unfavorable

contacts between water and the hydrophobic tails of the amphiphiles and yet permits the solvation of the polar head groups. Micelle formation is a cooperative process: An assembly of just a few amphiphiles cannot shield its tails from contact with water. Consequently, dilute aqueous solutions of amphiphiles do not form micelles until their concentration surpasses a certain **critical micelle concentration (cmc).** Above the cmc, almost all the added amphiphile aggregates to form micelles. The value of the cmc depends on the identity of the amphiphile and the solution conditions. For amphiphiles with relatively small single tails, such as dodecyl sulfate ion, $CH_3(CH_2)_{11}OSO_3^{2-}$, the cmc is ~1 m$M$. Those of biological lipids, most of which have two large hydrophobic tails, are generally <$10^{-6}M$.

a.   Single-Tailed Lipids Tend to Form Micelles

The approximate size and shape of a micelle can be predicted from geometrical considerations. Single-tailed amphiphiles, such as soap anions, form spheroidal or ellipsoidal micelles because of their conical shapes (their hydrated head groups are wider than their tails; Fig. 12-11a, b). The number of molecules in such micelles depends on the amphiphile, but for many substances, it is on the order of several hundred. For a given amphiphile, these numbers span a narrow range: Less would expose the hydrophobic core of the micelle to water, whereas more would give the micelle an energetically unfavorable hollow center (Fig. 12-11c). Of course, a large micelle could flatten out to eliminate this hollow center, but the resulting decrease of curvature at the flattened surfaces would also generate empty spaces (Fig. 12-11d).

b.   Glycerophospholipids and Sphingolipids Tend to Form Bilayers

The two hydrocarbon tails of glycerophospholipids and sphingolipids give these amphiphiles a more or less cylindrical shape (Fig. 12-12a). The steric requirements of packing such molecules together yields large disklike micelles (Fig. 12-12b) that are really extended bimolecular leaflets.

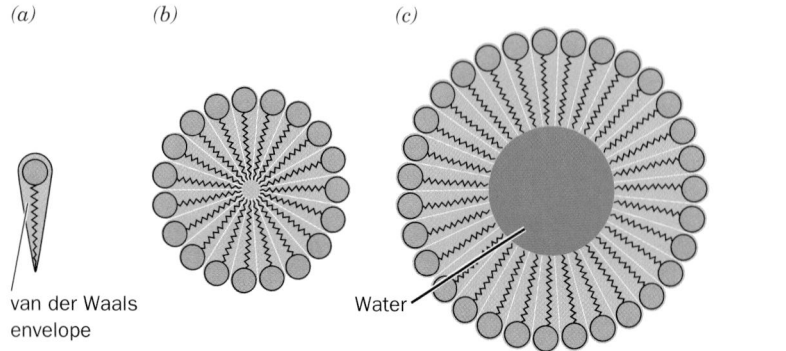
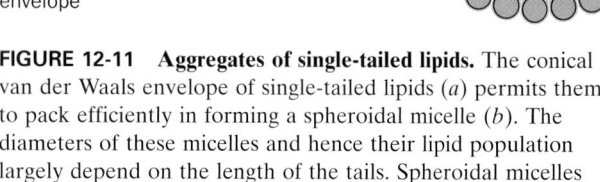
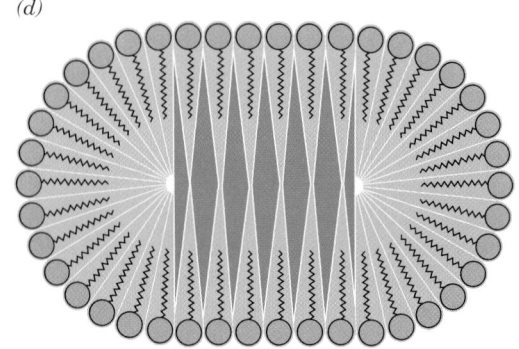

FIGURE 12-11   Aggregates of single-tailed lipids. The conical van der Waals envelope of single-tailed lipids (a) permits them to pack efficiently in forming a spheroidal micelle (b). The diameters of these micelles and hence their lipid population largely depend on the length of the tails. Spheroidal micelles composed of many more lipid molecules than the optimal number would have an unfavorable water-filled center (*blue*) (c). Such micelles could flatten out to collapse the hollow center, but as such ellipsoidal micelles become elongated they also develop water-filled spaces (d).

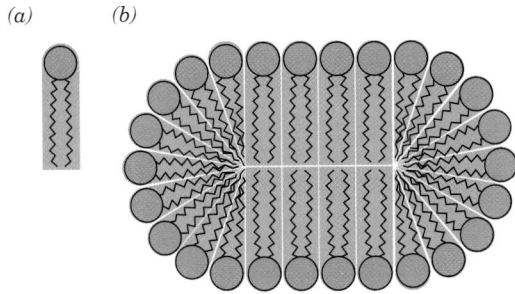

(a)   *(b)*

FIGURE 12-12   Bilayer formation by phospholipids. The cylindrical van der Waals envelope of phospholipids (*a*) causes them to form extended disklike micelles (*b*) that are better described as lipid bilayers.

The existence of such **lipid bilayers** was first proposed in 1925 by E. Gorter and F. Grendel, on the basis of their observation that lipids extracted from erythrocytes covered twice the area when spread as a monolayer at the air–water interface (Fig. 12-10) than in the erythrocyte plasma membrane (the erythrocyte's only membrane). Lipid bilayers typically have thicknesses of ~60 Å, as measured by electron microscopy and X-ray diffraction techniques. Since their two head group layers are each expected to be ~15 Å thick, their ~15-Å-long hydrocarbon tails must be nearly fully extended. We shall see below that *lipid bilayers form the structural basis of biological membranes.*

B. *Liposomes*

A suspension of phospholipids in water forms multilamellar vesicles that have an onionlike arrangement of lipid bilayers (Fig. 12-13*a*). On **sonication** (agitation by ultrasonic vibrations), these structures rearrange to form **liposomes**—closed, self-sealing, solvent-filled vesicles that are bounded by only a single bilayer (Fig. 12-13*b*). They usually have diameters of several hundred Å and, in a given preparation, are rather uniform in size. Liposomes with diameters of ~1000 Å can be made by injecting an ethanolic solution of phospholipid into water or by dissolving phospholipid in a detergent solution and then dialyzing out the detergent. Once formed, liposomes are quite stable and, in fact, may be separated from the solution in which they reside by dialysis, gel filtration chromatography, or centrifugation. Liposomes with differing internal and external environments can therefore be readily prepared. *Biological membranes consist of lipid bilayers with which proteins are associated (Section 12-3A).* Liposomes composed of synthetic lipids and/or lipids extracted from biological sources (e.g., lecithin from egg yolks) have therefore been extensively studied as models for biological membranes.

a. **Lipid Bilayers Are Impermeable to Most Polar Substances**

Since biological membranes form cell and organelle boundaries, it is important to determine their ability to partition two aqueous compartments. The permeability of a

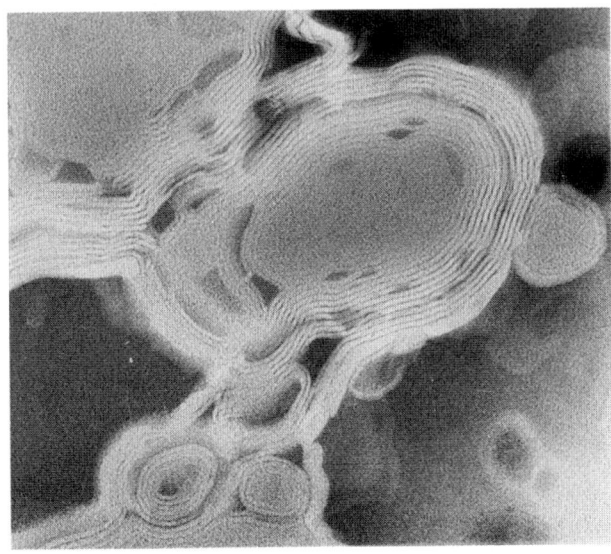

(a)

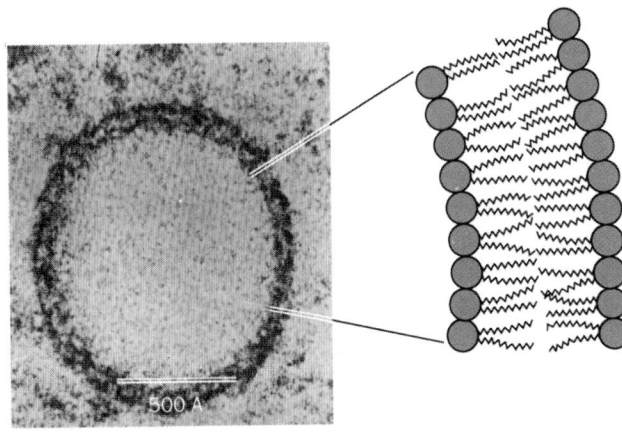

(b)

FIGURE 12-13   Lipid bilayers. (*a*) An electron micrograph of a multilamellar phospholipid vesicle in which each layer is a lipid bilayer. [Courtesy of Alec D. Bangham, Institute of Animal Physiology, Cambridge, U.K.] (*b*) An electron micrograph of a liposome. Its wall, as the accompanying diagram indicates, consists of a bilayer. [Courtesy of Walter Stoeckenius, University of California at San Francisco.]

lipid bilayer to a given substance may be determined by forming liposomes in a solution containing the substance of interest, changing the external aqueous solution, and then measuring the rate at which the substance appears in the new external solution. It has been found in this way that *lipid bilayers are extraordinarily impermeable to ionic and polar substances and that the permeabilities of such substances increase with their solubilities in nonpolar solvents.* This suggests that to penetrate a lipid bilayer, a solute molecule must shed its hydration shell and become solvated by the bilayer's hydrocarbon core. Such a process is highly unfavorable for polar molecules, so that even the ~30-Å thickness of a lipid bilayer's hydrocarbon core forms an effective barrier for polar substances. However, measurements using tritiated water indicate that lipid bilayers are

(a) Transverse diffusion (flip-flop)

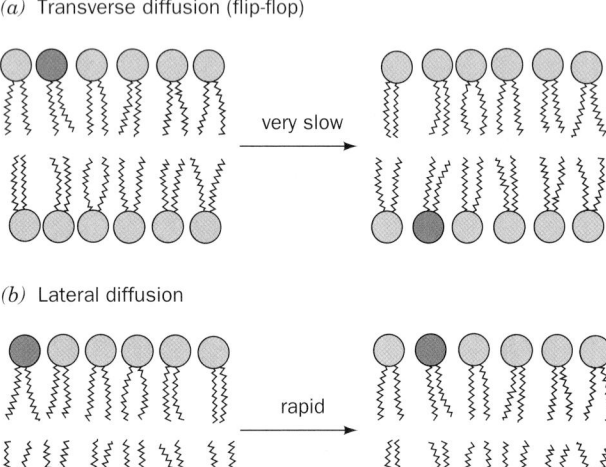

(b) Lateral diffusion

FIGURE 12-14 Phospholipid diffusion in a lipid bilayer.
(*a*) Transverse diffusion (flip-flop) is defined as the transfer of
a phospholipid molecule from one bilayer leaflet to the other.
(*b*) Lateral diffusion is defined as the pairwise exchange of
neighboring phospholipid molecules in the same bilayer leaflet.

appreciably permeable to water. Despite the polarity of
water, its small molecular size makes it significantly soluble in the hydrocarbon core of lipid bilayers and therefore
able to permeate them.

> The stability of liposomes and their impermeability to many
> substances makes them promising vehicles for the delivery
> of therapeutic agents, such as drugs, enzymes, and genes (for
> gene therapy), to particular tissues. Liposomes are absorbed
> by many cells through fusion with their plasma membranes.
> If methods can be developed for targeting liposomes to spe-
> cific cell populations, then the desired substances could be
> directed toward particular tissues through liposome mi-

croencapsulation. Indeed, a number of liposome-delivered
anticancer agents and antibiotics are already in use.

C. *Bilayer Dynamics*

a. **Lipid Bilayers Are Two-Dimensional Fluids**

The transfer of a lipid molecule across a bilayer
(Fig. 12-14*a*), a process termed **transverse diffusion** or a
flip-flop, is an extremely rare event. This is because a flip-
flop requires the polar head group of the lipid to pass
through the hydrocarbon core of the bilayer. The flip-flop
rates of phospholipids, as measured by several techniques,
are characterized by half-times that are minimally several
days.

In contrast to their low flip-flop rates, *lipids are highly
mobile in the plane of the bilayer* (**lateral diffusion**,
Fig. 12-14*b*). The X-ray diffraction patterns of bilayers at
physiological temperatures have a diffuse band, centered
at a spacing of 4.6 Å, whose width is a measure of the distribution of lateral spacings between the hydrocarbon
chains in the bilayer plane. This band, which resembles one
in the X-ray diffraction patterns of liquid paraffins, is indicative that *the bilayer is a two-dimensional fluid in which
the hydrocarbon chains undergo rapid fluxional (continuously changing) motions involving rotations about their
C—C bonds.*

The lateral diffusion rate of lipid molecules can be
quantitatively determined from the rate of **fluorescence
photobleaching recovery** (Fig. 12-15). A fluorescent group
(**fluorophore**) is specifically attached to a bilayer component, and an intense laser pulse focused on a very small
area ($\sim$3 μm^2) is used to destroy (bleach) the fluorophore
there. The rate at which the bleached area recovers its
fluorescence, as monitored by fluorescence microscopy,
indicates the rate at which unbleached and bleached
fluorescence-labeled molecules laterally diffuse into and
out of the bleached area, respectively. Such observations

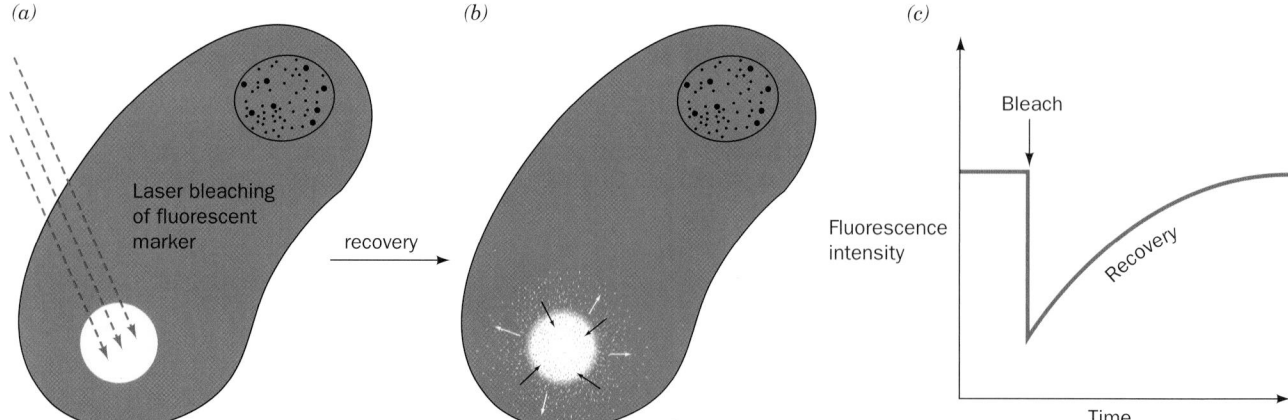

**FIGURE 12-15 The fluorescence photobleaching recovery
technique.** (*a*) An intense laser light pulse bleaches the
fluorescent markers (*green*) from a small region of an
immobilized cell that has a fluorescence-labeled membrane
component. (*b*) The fluorescence of the bleached area, as
monitored by fluorescence microscopy, recovers as the bleached

molecules laterally diffuse out of it and intact fluorescence-
labeled molecules diffuse into it. (*c*) The fluorescence recovery
rate depends on the diffusion rate of the labeled molecule.
 See Guided Exploration 11: Membrane Structure and the Fluid
Mosaic Model

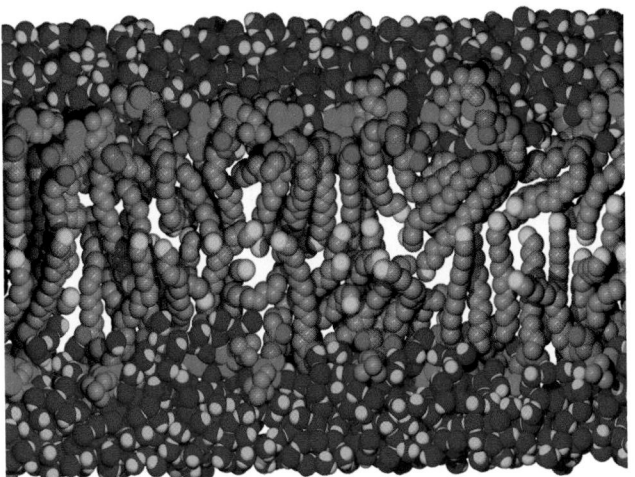

FIGURE 12-16 Snapshot of a molecular dynamics simulation of a lipid bilayer consisting of dipalmitoyl phosphatidylcholine surrounded by water. Atom colors are chain C gray (except for terminal methyl C, which is yellow), glycerol C brown, ester O red, phosphate P and O green, choline C and N pale violet, water O dark blue, and water H cyan. Lipid H atoms have been omitted for clarity. [Courtesy of Richard Pastor and Richard Venable, FDA, Bethesda, Maryland.] ♒ **See Guided Exploration 11: Membrane Structure and the Fluid Mosaic Model**

the more rigid head groups. Note that the methyl ends of the tails from opposite leaflets of the bilayer are frequently interdigitated rather than forming entirely separate layers, as Fig. 12-14 might be taken to suggest. This is particularly true in biological membranes because their various lipid molecules have tails of different lengths and/or are kinked due to the presence of double bonds. Molecular dynamics simulations also indicate that a lipid bilayer is flanked by several layers of ordered water molecules. Moreover, as Fig. 12-16 indicates, water molecules commonly penetrate well below the level of the head groups and glycerol residues. Hence, *a lipid bilayer typically consists of an ~30-Å-thick hydrocarbon core bounded on both sides by ~15-Å-thick interface regions containing rapidly fluctuating conglomerations of head groups, water, glycerol, carbonyl, and methylene groups.*

b. Bilayer Fluidity Varies with Temperature

*As a lipid bilayer cools below a characteristic **transition temperature,** it undergoes a sort of phase change, termed an **order–disorder transition,** in which it becomes a gel-like solid (Fig. 12-17);* that is, it loses its fluidity. Below the transition temperature, the 4.6-Å X-ray diffraction band characteristic of the lateral spacing between hydrocarbon chains in a liquid-crystalline bilayer is replaced by a sharp 4.2-Å band similar to that exhibited by crystalline paraffins. This indicates that the hydrocarbon chains in a bilayer become fully extended and packed in a hexagonal array as in crystalline paraffins.

The transition temperature of a bilayer increases with the chain length and the degree of saturation of its component fatty acid residues for the same reasons that the melting temperatures of fatty acids increase with these quantities. The transition temperatures of most biological membranes are in the range 10 to 40°C. *Cholesterol, which by itself does not form a bilayer, decreases membrane fluidity near the*

indicate, as do magnetic resonance measurements, that lipids in bilayers have lateral mobilities similar to those of the molecules in a light machine oil. Lipids in bilayers can therefore diffuse the 1-μm length of a bacterial cell in ~1 s.

Molecular dynamics simulations (Section 9-4) of lipid bilayers (Fig. 12-16) indicate that their lipid tails are highly conformationally mobile due to rotations about their C—C bonds. However, the viscosity of these tails sharply increases closer to the lipid head groups because their lateral mobilities are more constrained by interactions with

(a) Above transition temperature

(b) Below transition temperature

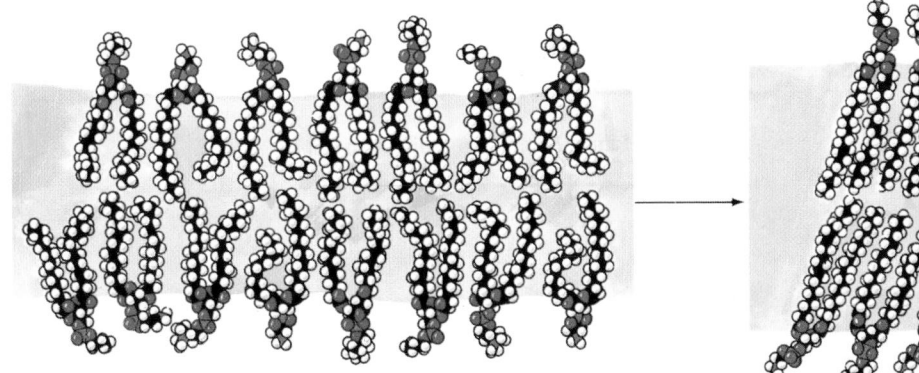

FIGURE 12-17 Structure of a lipid bilayer composed of phosphatidylcholine and phosphatidylethanolamine as the temperature is lowered below the bilayer's transition temperature. (*a*) Above the transition temperature, both the lipid molecules as a whole and their nonpolar tails are highly mobile in the plane of the bilayer. Such a state of matter, which

is ordered in some directions but not in others, is called a liquid crystal. (*b*) Below the transition temperature, the lipid molecules form a much more orderly array to yield a gel-like solid. [After Robertson, R.N., *The Lively Membranes,* pp. 69–70, Cambridge University Press (1983).]

membrane surface because cholesterol's rigid steroid ring system interferes with the motions of the fatty acid tails. However, because cholesterol does not extend into the membrane as far as most lipids, it also acts as a spacer that facilitates the increased mobility of the fatty acid tails near their methyl ends. Cholesterol also broadens the temperature range of the order–disorder transition and in high concentrations totally abolishes it. This behavior occurs because cholesterol inhibits the crystallization (cooperative aggregation into ordered arrays) of fatty acid tails by fitting in between them. Thus cholesterol functions as a kind of membrane plasticizer.

The fluidity of biological membranes is one of their important physiological attributes since it permits their embedded proteins to interact (Section 12-3B). The transition temperatures of mammalian membranes are well below body temperatures and hence these membranes all have a fluidlike character. Bacteria and poikilothermic (cold-blooded) animals such as fish modify (through lipid biosynthesis and degradation) the fatty acid compositions of their membrane lipids with the ambient temperature so as to maintain membrane fluidity. For example, the membrane viscosity of *E. coli* at its growth temperature remains constant as the growth temperature is varied from 15 to 43°C.

Gaseous anesthetics, such as diethyl ether, cyclopropane, **halothane** (2-bromo-2-chloro-1,1,1-trifluoroethane), and the noble gas Xe, act by interfering with the transmission of nerve impulses in the central nervous system. Since the body excretes these general anesthetics unchanged, it appears that they do not act by chemical means. Rather, experimental evidence, such as the linear correlation of their anesthetic effectiveness with their lipid solubilities, suggests that these nonpolar substances alter the structures of membranes by dissolving in their hydrocarbon cores. Nerve impulse transmission, which is a membrane-based phenomenon (Section 20-5), is disrupted by these structural changes to which neuronal membranes seem particularly sensitive.

3 ◼ BIOLOGICAL MEMBRANES

Biological membranes are composed of proteins associated with a lipid bilayer matrix. Their lipid fractions consist of complex mixtures that vary according to the membrane source (Table 12-3) and, to some extent, with the diet and environment of the organism that produced the membrane. *Membrane proteins carry out the dynamic processes associated with membranes, and therefore specific proteins occur only in particular membranes.* Protein-to-lipid ratios in membranes vary considerably with membrane function, as is indicated by Table 12-4, although most membranes are at least one-half protein. The myelin membrane, which functions passively as an insulator around certain nerve fibers (Section 20-5B), is a prominent exception to this generalization in that it contains only 18% protein.

In this section, we discuss the properties of membrane proteins and their behavior in biological membranes. Following this, we examine specific aspects of biological membranes, namely, the erythrocyte cytoskeleton, the nature of blood groups, gap junctions, and channel-forming proteins. We consider how membranes are assembled and how their component proteins are directed to them in Section 12-4.

A. *Membrane Proteins*

[handwritten note: Integral Proteins — (intrinsic) ~~xxxx~~ tightly bound to membrane by hydrophobic forces.

To separate them from the membrane requires the denaturing of membrane]

TABLE 12-3   Lipid Compositions of Some B

Lipid	Human Erythrocyte			
Phosphatidic acid	1.5	0.5	0	0
Phosphatidylcholine	19	10	39	0
Phosphatidylethanolamine	18	20	27	65
Phosphatidylglycerol	0	0	0	18
Phosphatidylinositol	1	1	7	0
Phosphatidylserine	8.5	8.5	0.5	0
Cardiolipin	0	0	22.5	12
Sphingomyelin	17.5	8.5	0	0
Glycolipids	10	26	0	0
Cholesterol	25	26	3	0

*The values given are weight percent of total lipid.

Source: Tanford, C., *The Hydrophobic Effect*, p. 109, Wiley (1980).

TABLE 12-4 **Compositions of Some Biological Membranes**

Membrane	Protein (%)	Lipid (%)	Carbohydrate (%)	Protein to Lipid Ratio
Plasma membranes:				
Mouse liver cells	46	54	2–4	0.85
Human erythrocyte	49	43	8	1.1
Amoeba	52	42	4	1.3
Rat liver nuclear membrane	59	35	2.0	1.6
Mitochondrial outer membrane	52	48	$(2–4)^a$	1.1
Mitochondrial inner membrane	76	24	$(1–2)^a$	3.2
Myelin	18	79	3	0.23
Gram-positive bacteria	75	25	$(10)^a$	3.0
Halobacterium purple membrane	75	25		3.0

^{a}Deduced from the analyses.

Source: Guidotti, G., *Annu. Rev. Biochem.* **41,** 732 (1972).

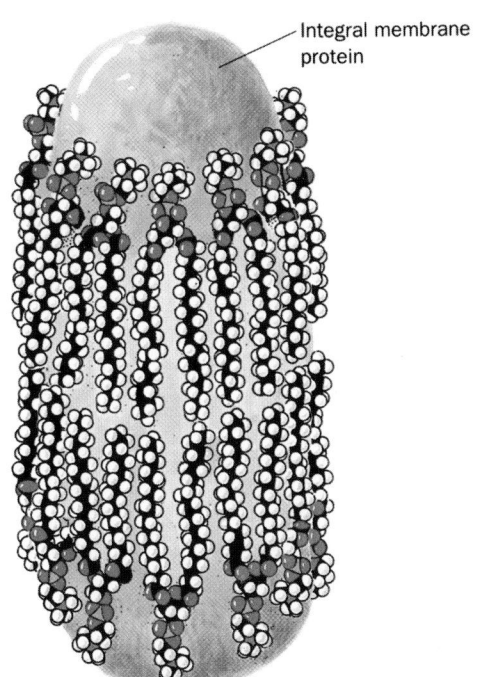

FIGURE 12-18 **Model of an integral membrane protein.**
Integral proteins in a lipid bilayer are "solvated" by lipids
through hydrophobic interactions between the protein and the
lipids' nonpolar tails. The polar head groups may also associate
with the protein through hydrogen bonding and salt bridges.
[After Robertson, R.N., *The Lively Membranes, p.* 56,
Cambridge University Press (1983).]

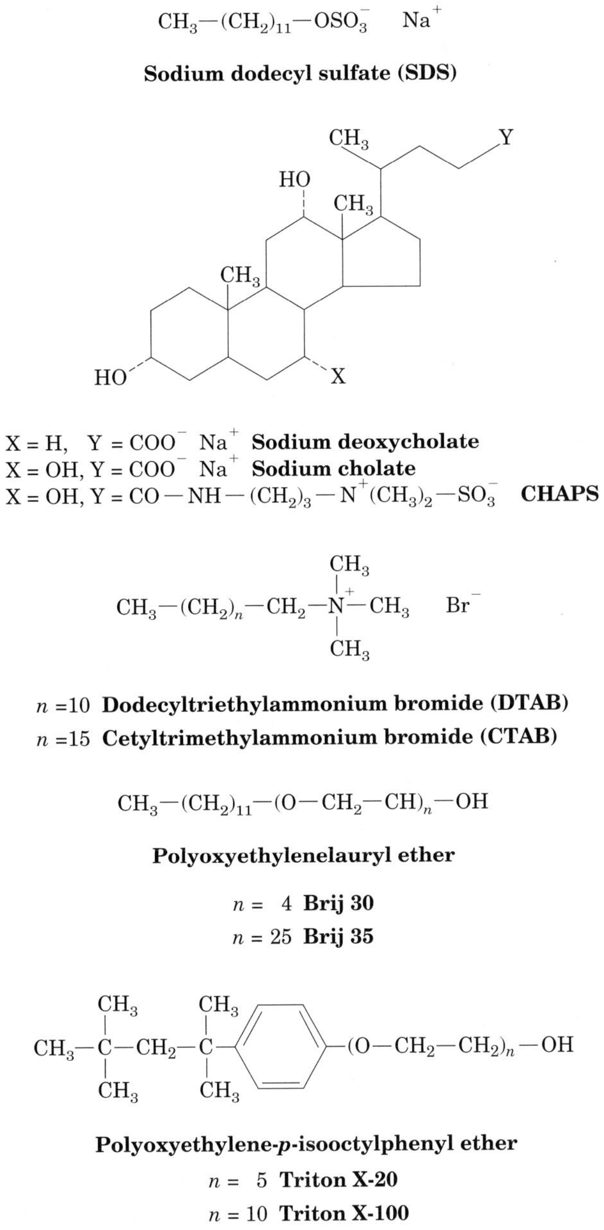

$$CH_3—(CH_2)_{11}—OSO_3^-\quad Na^+$$

Sodium dodecyl sulfate (SDS)

X = H,　Y = COO^- Na^+ **Sodium deoxycholate**
X = OH, Y = COO^- Na^+ **Sodium cholate**
X = OH, Y = $CO—NH—(CH_2)_3—N^+(CH_3)_2—SO_3^-$　**CHAPS**

$$CH_3—(CH_2)_n—CH_2—\overset{\overset{\textstyle CH_3}{|}}{\underset{\underset{\textstyle CH_3}{|}}{N^+}}—CH_3 \quad Br^-$$

n =10 **Dodecyltriethylammonium bromide (DTAB)**
n =15 **Cetyltrimethylammonium bromide (CTAB)**

$$CH_3—(CH_2)_{11}—(O—CH_2—CH)_n—OH$$

Polyoxyethylenelauryl ether

n = 4 **Brij 30**
n = 25 **Brij 35**

Polyoxyethylene-*p*-isooctylphenyl ether

n = 5 **Triton X-20**
n = 10 **Triton X-100**

FIGURE 12-19 **A selection of the detergents used in
biochemical manipulations.** Note that they may be anionic,
cationic, zwitterionic, or uncharged. Ionic detergents are
strongly amphiphilic and therefore tend to denature proteins,
whereas neutral detergents are unlikely to do so.

unless they are solubilized by detergents or water-miscible organic solvents such as butanol or glycerol. Some integral proteins bind lipids so tenaciously that they can be freed from them only under denaturing conditions. Solubilized integral proteins can be purified by many of the protein fractionation methods discussed in Chapter 6.

2. *Peripheral* or *extrinsic proteins* *are dissociated from membranes by relatively mild procedures that leave the membrane intact,* such as exposure to high ionic strength salt solutions (e.g., 1*M* NaCl), metal chelating agents, or pH changes. Peripheral proteins, for example, cytochrome *c,* are stable in aqueous solution and do not bind lipid. They associate with a membrane by binding at its surface to its lipid head groups and/or its integral proteins through electrostatic and hydrogen bonding interactions. Membrane-free peripheral proteins behave as water-soluble globular proteins and can be purified as such (Chapter 6).

In this subsection we concentrate on integral proteins.

a. Integral Proteins Are Asymmetrically Oriented Amphiphiles

All biological membranes contain integral proteins, which typically comprise ~25% of the proteins encoded by a genome. Their locations on a membrane may be determined through **surface labeling,** a technique employing agents that react with proteins but cannot penetrate membranes. For example, an integral protein on the outer surface of an intact cell membrane binds antibodies elicited against it, but a protein on the membrane's inner surface

can do so only if the membrane has been ruptured. Membrane-impermeable protein-specific reagents that are fluorescent or radioactively labeled may be similarly employed. Using such surface-labeling reagents, it has been shown that *some integral proteins are exposed only to a specific surface of a membrane, whereas others, known as* **transmembrane proteins,** *span the membrane.* However, no protein is known to be completely buried in a membrane; that is, all have some exposure to the aqueous environment. Such studies have also established that *biological membranes are asymmetric in that a particular membrane protein is invariably located on only one particular face of a membrane or, in the case of a transmembrane protein, oriented in only one direction with respect to the membrane* (Fig. 12-20).

Integral proteins are amphiphilic; the protein segments immersed in a membrane's nonpolar interior have predominantly hydrophobic surface residues, whereas those portions that extend into the aqueous environment are by and large sheathed with polar residues. For example, proteolytic digestion and chemical modification studies indicate that the erythrocyte transmembrane protein **glycophorin A** (Fig. 12-21) has three domains: (1) a 72-residue externally located N-terminal domain that bears 16 carbohydrate chains; (2) a 19-residue sequence, consisting almost entirely of hydrophobic residues, that spans the erythrocyte cell membrane; and (3) a 40-residue cytoplasmic C-terminal domain that has a high proportion of charged and polar residues. The transmembrane domain, as is common in many integral proteins, almost certainly

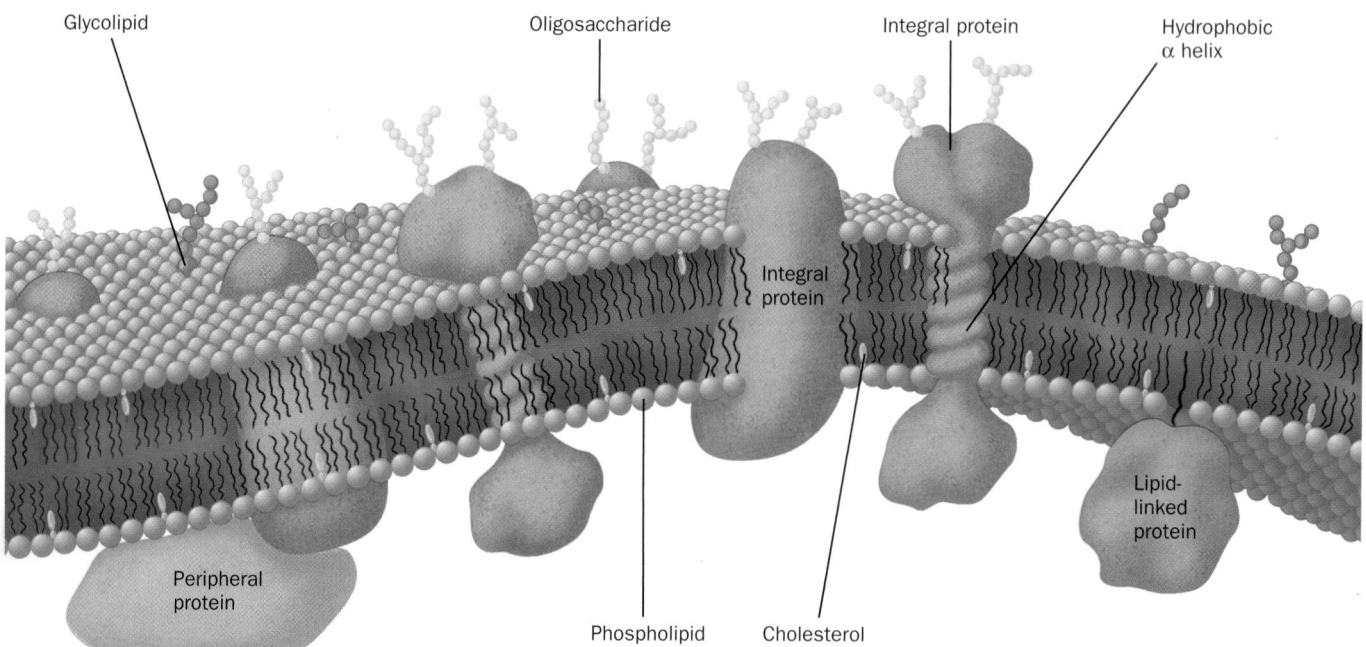

FIGURE 12-20 Schematic diagram of a plasma membrane. Integral proteins (*orange*) are embedded in a bilayer composed of phospholipids (*blue spheres with two wiggly tails;* shown, for clarity, in much greater proportion than they have in biological membranes) and cholesterol (*yellow*). The carbohydrate components of glycoproteins (*yellow beaded chains*) and glycolipids (*green beaded chains*) occur only on the external face of the membrane. **See Guided Exploration 11: Membrane Structure and the Fluid Mosaic Model**

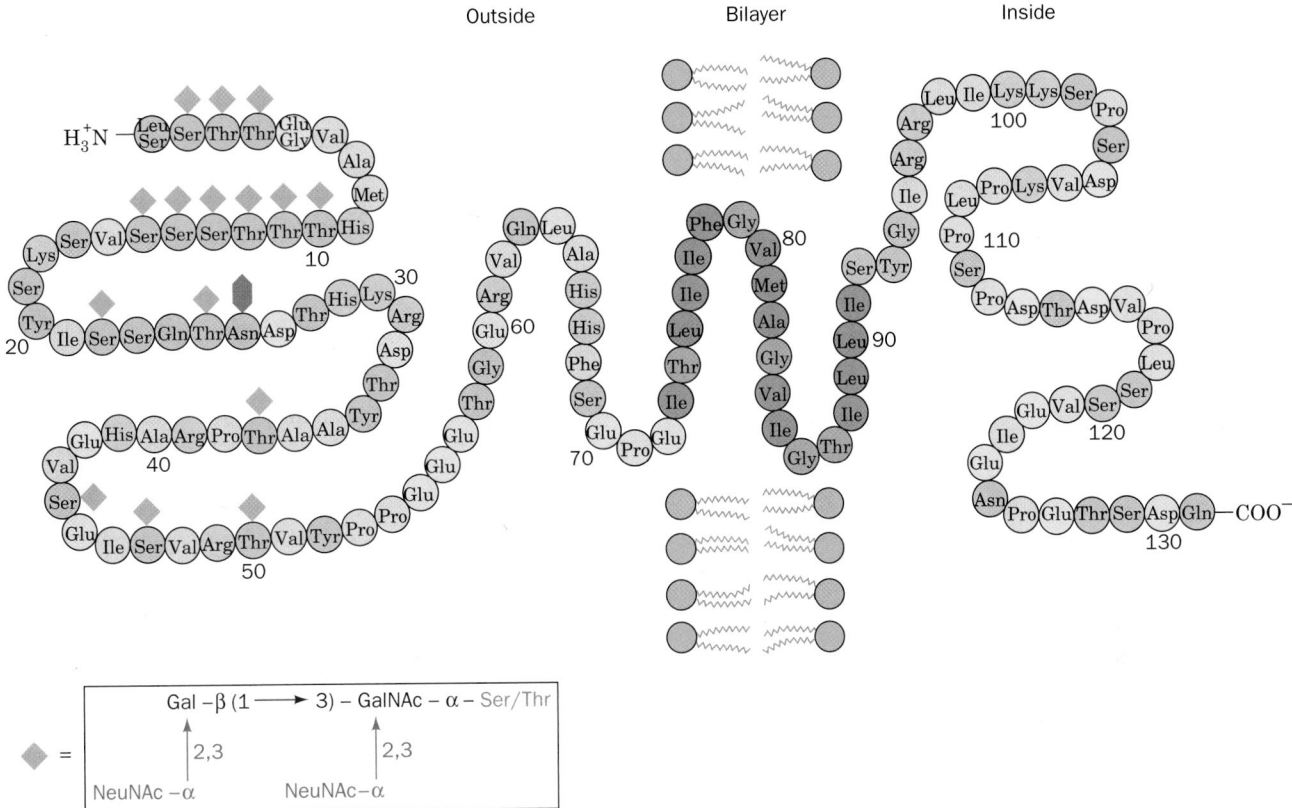

FIGURE 12-21 The amino acid sequence and membrane location of human erythrocyte glycophorin A. The protein, which is ~60% carbohydrate by weight, bears 15 *O*-linked oligosaccharides (*green diamonds*) and one that is *N*-linked (*dark green hexagon*). The predominant sequence of the *O*-linked oligosaccharides is given below. The protein's transmembrane portion (*brown and purple*) consists of 19 sequential predominantly hydrophobic residues. Its C-terminal portion, which is located on the membrane's cytoplasmic face, is rich in anionic (*pink*) and cationic (*blue*) amino acid residues. There are two common genetic variants of glycophorin A: glycophorin A^M has Ser and Gly at positions 1 and 5, respectively, whereas they are Leu and Glu in glycophorin A^N. [Abbreviations: Gal = galactose, GalNAc = *N*-acetylgalactosamine, NeuNAc = *N*-acetylneuraminic acid (sialic acid)]. [After Marchesi, V.T., *Semin. Hematol.* **16**, 8 (1979).]

forms an α helix, thereby satisfying the hydrogen bonding requirements of its polypeptide backbone. Indeed, the existence of glycophorin A's single transmembrane helix is predicted by computing the free energy change in transferring α helically folded polypeptide segments from the nonpolar interior of a membrane to water (Fig. 12-22). Similar computations on other integral proteins have also identified their transmembrane helices.

In many integral proteins, the hydrophobic segment(s) anchors the active region of the protein to the membrane. For instance, trypsin cleaves the membrane-bound enzyme **cytochrome b_5** into a polar, enzymatically active ~85-residue N-terminal fragment and an ~50-residue C-terminal fragment that remains embedded in the mem-

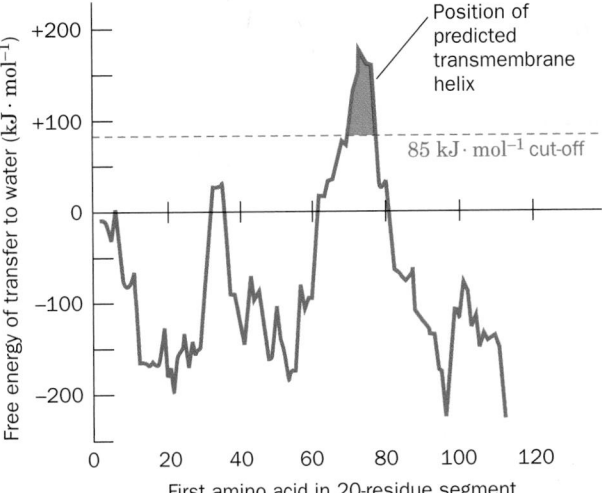

FIGURE 12-22 A plot, for glycophorin A, of the calculated free energy change in transferring 20-residue-long α helical segments from the interior of a membrane to water versus the position of the segment's first residue. Peaks higher than +85 kJ · mol⁻¹ are indicative of a transmembrane helix. [After Engleman, D.M., Steitz, T.A., and Goldman, A., *Annu. Rev. Biophys. Biophys. Chem.* **15**, 343 (1986).]

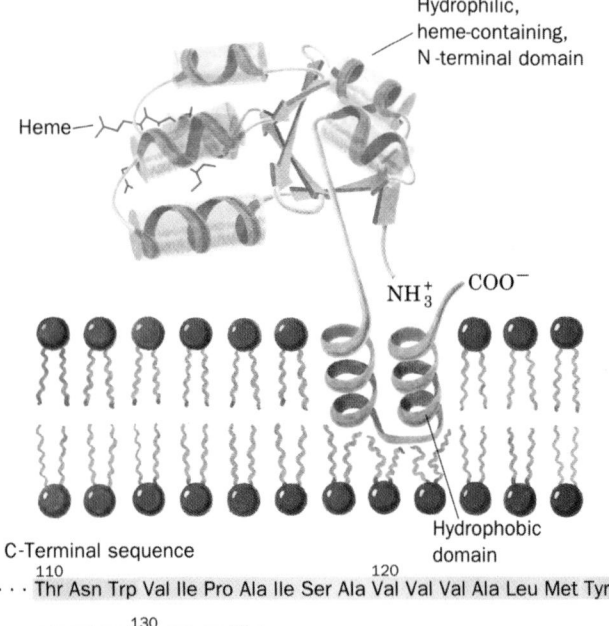

C-Terminal sequence

110   120
··· Thr Asn Trp Val Ile Pro Ala Ile Ser Ala Val Val Val Ala Leu Met Tyr

130
Arg Ile Tyr Thr Ala Glu Asp COO⁻

FIGURE 12-23 **Liver cytochrome b_5 in association with a membrane.** The protein's enzymatically active N-terminal domain (*purple*), whose X-ray structure has been determined, is anchored in the membrane by a hydrophobic and presumably α helical C-terminal segment (*brown*) that begins and ends with hydrophilic segments (*purple*). The amino acid sequence of the horse enzyme indicates that this hydrophobic anchor consists of a 13-residue segment ending 9 residues from the polypeptide's C-terminus (*below*). [Ribbon diagram of the N-terminal domain after a drawing by Jane Richardson, Duke University. Amino acid sequence from Ozols, J. and Gerard, C., *J. Biol. Chem.* **253,** 8549 (1977).]

brane (Fig. 12-23). *The asymmetric orientation of integral proteins in the membrane is maintained by their infinitesimal flip-flop rates (even slower than those of lipids), which result from the greater sizes of the membrane protein "head groups" in comparison to those of lipids.* The origin of this asymmetry is discussed in Section 12-4.

Relatively few integral proteins have yet been crystallized—and then only in the presence of detergents, which are but poor substitutes for lipid bilayers. Thus, despite their biological abundance, only ~0.7% of the proteins of known structure are integral proteins. In the remainder of

this subsection, we discuss the structures of four integral proteins, bacteriorhodopsin, the bacterial photosynthetic reaction center, porins, and cyclooxygenase.

b. Bacteriorhodopsin Contains a Bundle of Seven Hydrophobic Helical Rods

One of the structurally best characterized integral proteins is **bacteriorhodopsin (BR)** from the halophilic (salt loving) bacterium *Halobacterium salinarium* that inhabits such salty places as the Dead Sea (it grows best in 4.3*M* NaCl and is nonviable below 2.0*M* NaCl; seawater contains 0.6*M* NaCl). Under low O_2 conditions, its cell membrane develops ~0.5-μm-wide patches of· **purple membrane** whose only protein component is BR. This 247-residue protein is a light-driven proton pump; it generates a proton concentration gradient across the membrane that powers the synthesis of ATP (by a mechanism discussed in Section 22-3B). Bacteriorhodopsin's light-absorbing element, **retinal,** is covalently bound to its Lys 216 (Fig. 12-24). This **chromophore** (light-absorbing group), which is responsible for the membrane's purple color, is also the light-sensitive element in vision.

The purple membrane, which is 75% protein and 25% lipid, has an unusual structure compared to most other membranes (Section 12-3B): Its BR molecules are arranged in a highly ordered two-dimensional array (a two-dimensional crystal). This permitted Richard Henderson and Nigel Unwin, through **electron crystallography** (a technique they devised, resembling X-ray crystallography, in which the electron beam of an electron microscope is used to elicit diffraction from two-dimensional crystals), to determine the structure of BR to near-atomic resolution (3.0 Å). The more recently determined 1.9-Å-resolution X-ray structure of BR, based on single crystals of BR dissolved in lipidic cubic phases (mixtures of lipids and water that form a highly convoluted but continuous bilayer that is interpenetrated by aqueous channels), closely resembles that determined by electron crystallography.

Bacteriorhodopsin forms a homotrimer. Each of its subunits consists mainly of a bundle of seven ~25-residue α helical rods that each span the lipid bilayer in directions almost perpendicular to the bilayer plane (Fig. 12-25). BR is therefore said to be **polytopic** (multispanning; Greek: *topos,* place). The ~20-Å spaces between the protein molecules in the purple membrane are occupied by this bilayer (Fig. 12-25*b*). Adjacent α helices, which are largely hy-

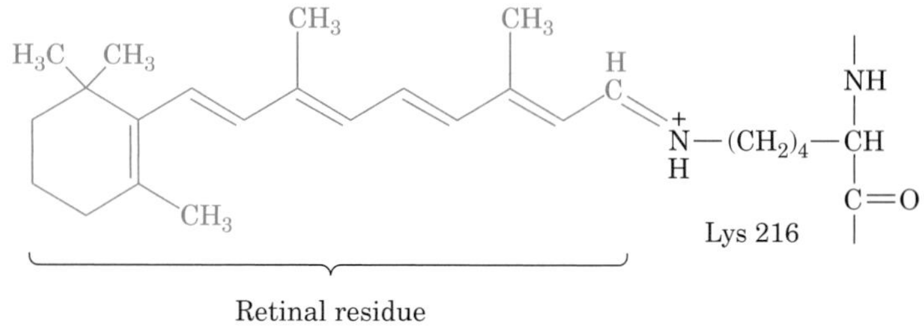

FIGURE 12-24 **Molecular formula of retinal.** Retinal, the prosthetic group of bacteriorhodopsin, forms a Schiff base with Lys 216 of the protein. A similar linkage occurs in **rhodopsin,** the photoreceptor of the eye.

Retinal residue

Lys 216

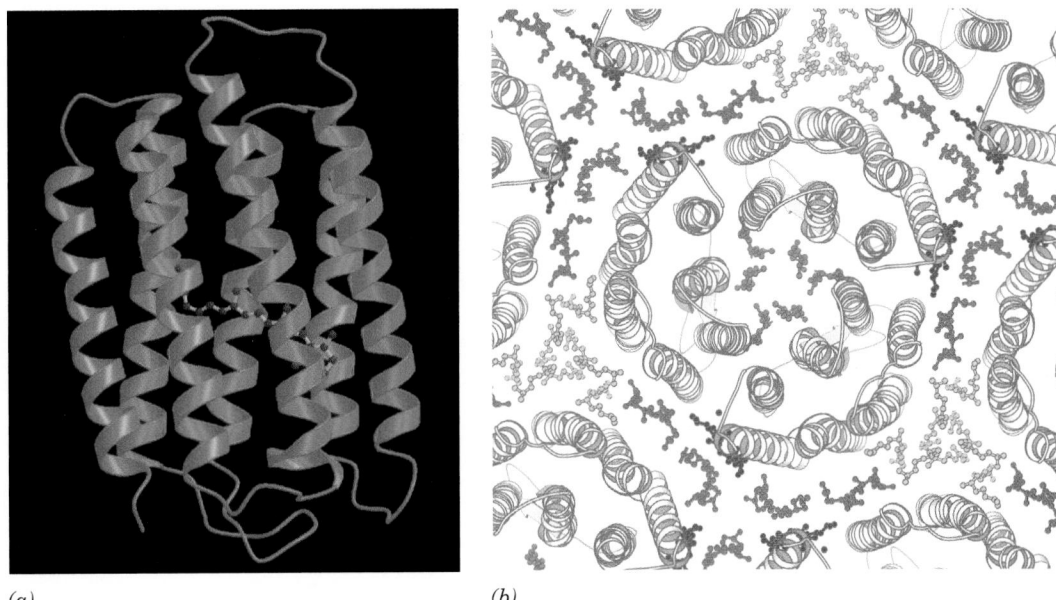

(a)   *(b)*

FIGURE 12-25   Structure of bacteriorhodopsin. (*a*) The
electron crystallography–based structure as viewed from within
the purple membrane with its extracellular side below. The
polypeptide backbone is cyan, and its covalently bound retinal
is shown in yellow and gray ball-and-stick form. The N-terminus
as seen here is on the lower left. [Courtesy of Nikolaus
Grigorieff and Richard Henderson, MRC Laboratory of
Molecular Biology, Cambridge, U.K.] (*b*) The X-ray structure
of a bacteriorhodopsin trimer with portions of its surrounding
trimers as viewed from the extracellular side of the membrane.
The protein molecules are shown in ribbon form (*gray*) and

their associated lipid tails are shown in ball-and-stick form in
different colors with symmetry-related lipid tails the same color
(the lipid head groups are disordered and hence not seen).
Only the lipids in the extracellular leaflet are shown; those in
the cytoplasmic leaflet have a similar distribution. Note how
the 7 antiparallel α helices in each BR monomer are cyclically
arranged in two layers of 4 and 3 helices with helices adjacent
in sequence also adjacent in space (the N to C direction
circulates clockwise in this view). [Courtesy of Eva Pebay-
Peyroula, Université Joseph Fourier, Grenoble, France. PDBid
1AP9.] *See Kinemage Exercise 8-1*

drophobic in character, are connected in a head-to-tail
fashion by short polypeptide loops. This arrangement
places the protein's charged residues near the surfaces of
the membrane in contact with the aqueous solvent. The internal charged residues line the center of the helix bundle
of each monomer so as to form a hydrophilic channel that
facilitates the passage of protons. Other membrane pumps
and channels (Chapter 20) probably have similar structures.

c. The Photosynthetic Reaction Center Contains Eleven Transmembrane Helices

The primary photochemical process of photosynthesis
in purple photosynthetic bacteria is mediated by the so-
called **photosynthetic reaction center** (**PRC;** Section
24-2B), a transmembrane (**TM**) protein consisting of at
least three nonidentical ~300-residue subunits that collectively bind four **chlorophyll** molecules, four other chromophores, and a nonheme Fe atom. The 1187-residue
photosynthetic reaction center of *Rhodopseudomonas
(Rps.) viridis,* whose X-ray structure was determined in
1984 by Hartmut Michel, Johann Deisenhofer, and Robert
Huber, was the first TM protein to be described in atomic

detail (Fig. 12-26). The polytopic protein's TM portion
consists of 11 α helices that form a 45-Å-long flattened
cylinder with the expected hydrophobic surface.

d. Porins Are Channel-Forming Proteins That Contain Transmembrane β Barrels

The outer membranes of gram-negative bacteria
(Section 11-3B) protect them from hostile environments
but must nevertheless be permeable to small polar solutes
such as nutrients and waste products. These outer membranes consequently contain embedded channel-forming
proteins called **porins,** which are usually trimers of identical 30- to 50-kD subunits that permit the passage of solutes
of less than ~600 D. Porins also occur in eukaryotes in the
outer membranes of mitochondria and chloroplasts
(thereby providing a further indication that these organelles are descended from bacteria; Section 1-2A).

The X-ray structures of several different porins have
been elucidated, among them the *Rhodobacter (Rb.) capsulatus* porin, determined by Georg Schulz, and the *E. coli*
OmpF and **PhoE** porins, determined by Johan Jansonius.
The 340- and 330-residue OmpF and PhoE porins share
63% sequence identity but have little sequence similarity

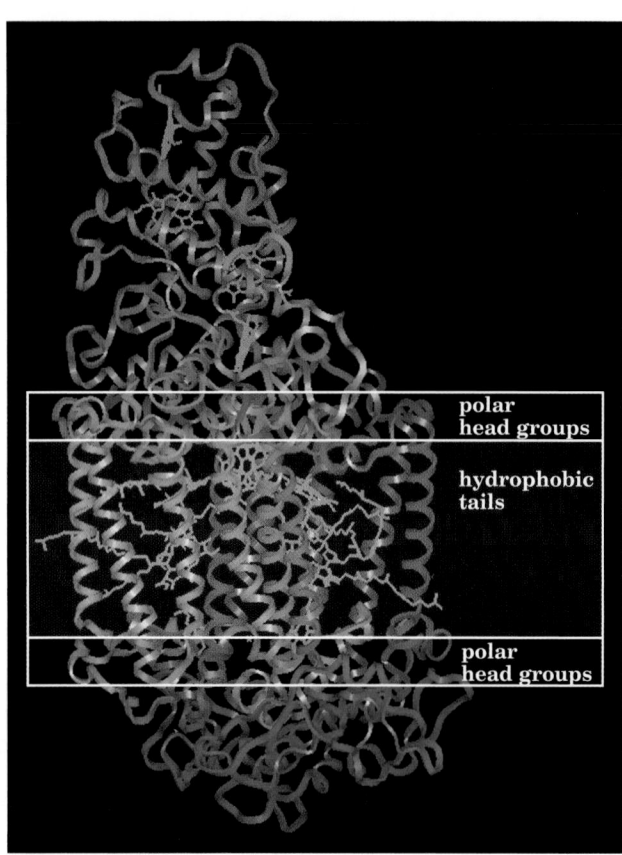

(a)

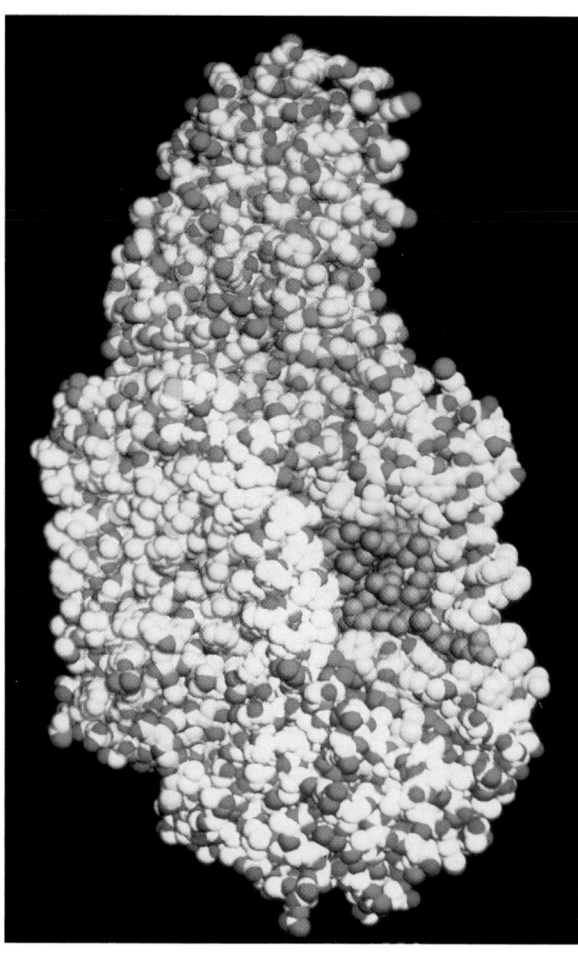

(b)

FIGURE 12-26   X-Ray structure of the photosynthetic reaction center of *Rps. viridis.* (*a*) A ribbon diagram in which only the C_α backbone and the prosthetic groups (*yellow*) are shown. The H, M, and L subunits (*pink, blue,* and *orange,* respectively) collectively have 11 transmembrane helices. The 4-heme *c*-type cytochrome (*green*), which does not occur in all species of photosynthetic bacteria, is bound to the external face of the complex. The position that the transmembrane protein is thought to occupy in the lipid bilayer is indicated schematically. [Based on an X-ray structure by Johann Deisenhofer, Robert Huber, and Hartmut Michel, Max-Planck-Institut für Biochemie,

Martinsried, Germany.] (*b*) A space-filling model, viewed as in Part a, in which nitrogen atoms are blue, oxygens are red, sulfurs are yellow, and the carbon atoms of the H, M, L, and cytochrome subunits are tinted pink, blue, orange, and green, respectively. Exposed portions of prosthetic groups are brown. Note how few polar groups (nitrogens and oxygens) are externally exposed in the portion of the protein that is immersed in the nonpolar region of the lipid bilayer. [From Deisenhofer, J. and Michel, H., *Les Prix Nobel* (1989). PDBid 1PRC.] **See Kinemage Exercise 8-2**

with the 301-residue *Rb. capsulatus* porin. Nevertheless, all three porins have closely similar structures. Each monomer of these homotrimeric proteins predominantly consists of a 16-stranded antiparallel β barrel which forms a solvent-accessible pore along the barrel axis that has a length of ~55 Å and a minimum diameter of ~7 Å (Fig. 12-27; although note that β barrel membrane proteins with 8, 12, 18, and 22 strands are also known). In the OmpF and PhoE porins, the N- and C-termini associate via a salt bridge in the 16th β strand, thereby forming a pseudocyclic structure (Fig. 12-27*a*). Note that a β barrel fully satisfies the polypeptide backbone's hydrogen bonding potential, as does an α helix. As expected, the side chains at the protein's membrane-exposed surface are nonpolar, thereby forming an ~27-Å-high hydrophobic band encircling the trimer (Fig 12-27*c*). In contrast, the side chains at the

solvent-exposed surface of the protein, including those lining the walls of the aqueous channel, are polar. Possible mechanisms for solute selectivity by these porins are discussed in Section 20-2D.

e. Cyclooxygenase Only Binds to One Bilayer Leaflet

Not all integral proteins are TM proteins. For example, the enzyme **cyclooxygenase (COX;** also known as **prostaglandin H_2 synthase),** which participates in the synthesis of hormonelike substances known as **prostaglandins** (Section 25-7B), is an integral protein that binds to the luminal leaflet of the endoplasmic reticulum. It is therefore said to be **monotopic** as is cytochrome b_5 (Fig. 12-23). COX's X-ray structure, determined by Michael Garavito, reveals that each 576-residue subunit of this homodimer consists of 3 domains (Fig. 12-28): a 48-residue N-terminal

(a)

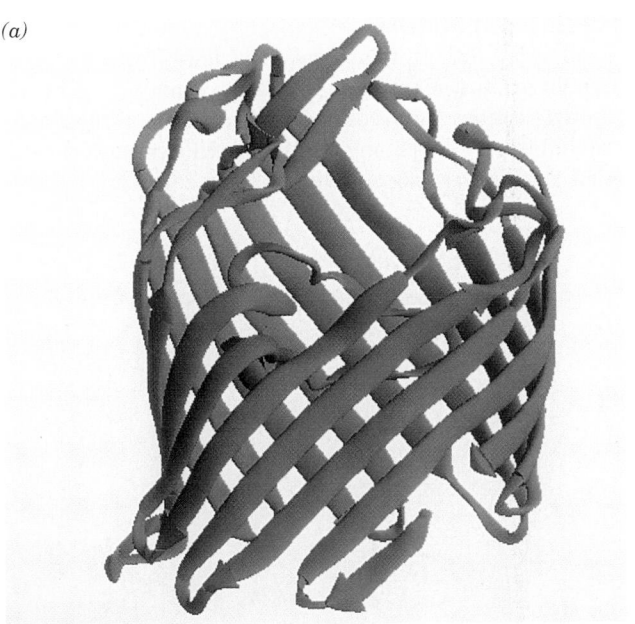

(b)

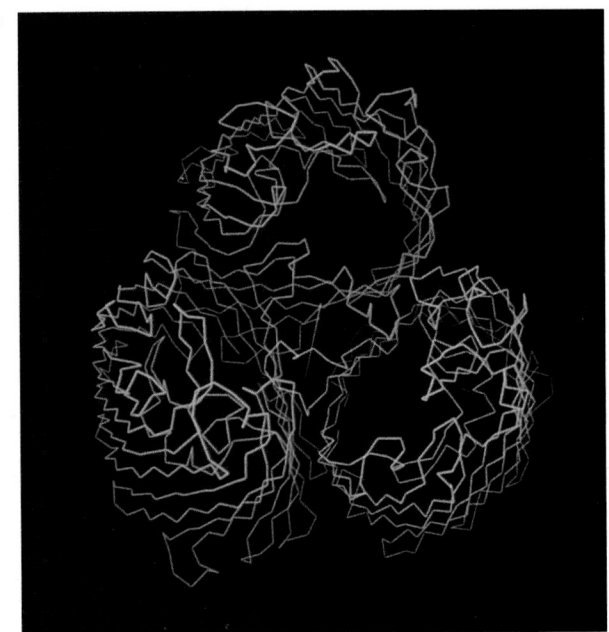

(c)

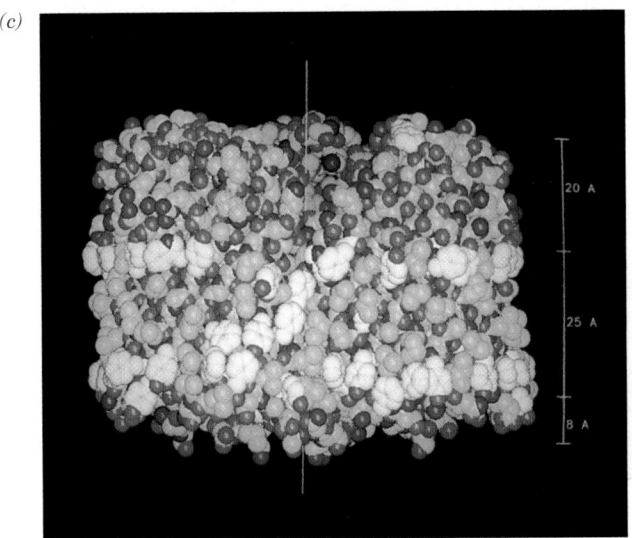

**FIGURE 12-27   X-Ray crystal structure of the *E. coli* OmpF
porin.** (*a*) A ribbon diagram of the monomer. Each strand of this
16-stranded antiparallel β barrel is inclined by ~45° to the barrel
axis. Its C-terminal strand is continued by the N-terminal
segment (*bottom right*), thereby forming a pseudocontinuous
strand. All porins of known structure have similar structural
properties. [Based on an X-ray structure by Johan Jansonius.
PDBid 1OPF.] (*b*) The C_α backbone of the trimer viewed ~30°
from its threefold axis of symmetry showing the pore through
each subunit. The subunits are differently colored. It can be seen
at the interface between the blue and green subunits that the
strands in adjacent β sheets extend essentially perpendicularly to
each other. (*c*) A space-filling model of the trimer viewed
perpendicular to its 3-fold axis (*vertical green line*). N atoms are
blue, O atoms are red, and C atoms are yellow, except those in
the side chains of aromatic residues, which are white. The
aromatic groups appear to delimit an ~25-Å-high hydrophobic
band (*scale at right*) that is immersed in the nonpolar portion of
the bacterial outer membrane (with the cell's exterior at the tops
of Parts *a* and *c*). Compare the hydrophobic band in this figure
with that in Fig. 12-26*b*. [Parts *b* and *c* courtesy of Tilman
Schirmer and Johan Jansonius, University of Basel, Switzerland.
PDBid 1OPF.] ♫ **See Kinemage Exercise 8-3**

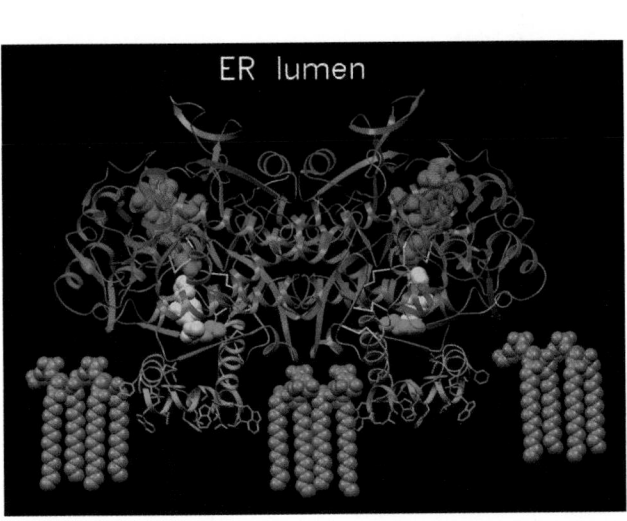

**FIGURE 12-28   X-Ray structure of sheep cyclooxygenase-1
showing its proposed disposition in the endoplasmic reticulum
membrane.** The homodimeric enzyme is viewed along the
plane of the membrane (*gray*) with its 2-fold axis of symmetry
vertical. The EGF-like motif of each subunit is green, the
membrane-binding motif together with several of its
hydrophobic side chains is orange, the enzymatic domain is
blue, and disulfide bonds are yellow. Several groups are
shown in space-filling form: The active site heme is red, the
mechanistically implicated residues Arg 120 and Tyr 385
(Section 25-7B) are green and magenta, and a bound inhibitor
of the enzyme, **flurbiprofen,** is yellow. [Courtesy of Michael
Garavito, Michigan State University. PDBid 1CQE.]

module that structurally resembles **epidermal growth factor (EGF;** a hormonally active polypeptide that stimulates cell proliferation; Section 19-3); a 44-residue central membrane-binding motif; and a 484-residue C-terminal heme-containing enzymatic domain. The membrane-binding motif consists of four amphipathic α helices arranged in a right-handed spiral, whose hydrophobic surfaces face outward from the body of the protein. These nonpolar residues, many of which are aromatic, are flanked by basic residues, which are postulated to interact electrostatically with the membrane's phospholipid head groups.

f. Integral Proteins Have Common Structural Features

Hydrophobic forces, as we have seen in Section 8-4, are the dominant interactions stabilizing the three-dimensional structures of water-soluble globular proteins. However, since the membrane-exposed regions of integral proteins are immersed in nonpolar environments, what stabilizes their structures? Analysis of the preceding integral proteins indicates that their membrane-exposed regions have a hydrophobic organization opposite to that of water-soluble proteins: Their membrane-exposed residues are more hydrophobic, on average, than their interior residues, even though these interior residues have average hydrophobicities and packing densities comparable to those of water-soluble proteins. Evidently, *the structures of integral and water-soluble proteins are both stabilized by the exclusion of their interior residues from the surrounding solvent, although in the case of integral proteins, the solvent is the lipid bilayer.*

In the foregoing TM proteins, those portions of the transmembrane secondary structural elements (helices in BR, the PRC, and COX, and β strands in the porins) that contact the bilayer's hydrocarbon core consist mainly of the hydrophobic residues Ile, Leu, Val, and Phe. The flanking residues, which penetrate the bilayer's interface region, are enriched with Phe, Trp, and Tyr. Hence, *TM proteins' hydrophobic transmembrane bands are bordered by rings of aromatic side chains (e.g., Fig. 12-27c) that delineate the water–bilayer interface.*

In each of the above TM proteins, the secondary structural elements that are adjacent in sequence are also adjacent in structure and hence tend to be antiparallel. This relatively simple up–down topology may result from the constraints associated with the insertion of a folding polypeptide chain into the lipid bilayer (see Section 12-4B).

B. *Lipid-Linked Proteins*

*Lipids and proteins associate covalently to form **lipid-linked proteins,** whose lipid portions anchor their attached proteins to membranes and mediate protein–protein interactions.* Proteins form covalent attachments with three classes of lipids: (1) isoprenoid groups such as farnesyl and geranylgeranyl residues, (2) fatty acyl groups such as myristoyl and palmitoyl residues, and (3) glycoinositol phospholipids (GPIs). In this subsection, we discuss the properties of these lipid-linked proteins.

a. Prenylated Proteins

A variety of proteins have covalently attached **isoprenoid groups,** mainly the C_{15} **farnesyl** and C_{20} **geranylgeranyl** residues (**isoprene,** a C_5 hydrocarbon, is the chemical unit from which many lipids, including cholesterol and other steroids, are constructed; Section 25-6A).

Isoprene

Farnesyl residue

Geranylgeranyl residue

The most common isoprenylation (or just **prenylation**) site in proteins is the C-terminal tetrapeptide CaaX, where "C" is Cys, "a" is often an aliphatic amino acid residue, and "X" is any amino acid. However, the identity of X is a major prenylation determinant: Proteins are farnesylated when X is Gln, Met, or Ser and geranylgeranylated when X is Leu. In both cases, the prenyl group is enzymatically linked to the Cys sulfur atom via a thioether linkage. The aaX tripeptide is then proteolytically excised and the newly exposed terminal carboxyl group is esterified with a methyl group (Fig. 12-29).

Two other types of prenylation sites have also been characterized: (1) the C-terminal sequence CXC, in which both Cys residues are geranylgeranylated and the terminal carboxyl group is methyl esterified; and (2) the C-terminal sequence CC in which one or both Cys residues are geranylgeranylated but the carboxyl group is not methylated. Proteins that are so prenylated are almost exclusively members of the **Rab** family of small GTP-binding proteins that participate in intracellular membrane trafficking (Section 12-4D).

What functions are served by protein prenylation? Many prenylated proteins are associated with intracellular membranes, and mutating their Cys prenylation sites blocks their membrane localization. Evidently, *the hydrophobic prenyl group can act to anchor its attached protein to a membrane.* However, this can only be part of the story since proteins with the same prenyl groups may be localized to different intracellular membranes. Moreover, fusing the CaaX motif from a normally prenylated protein to the C-terminus of a normally unprenylated protein yields a hybrid protein that is correctly prenylated and carboxyl methylated but which remains cytosolic. These observations suggest that prenylated proteins may interact with specific membrane-bound receptor proteins and hence that *prenylation also facilitates protein–protein in-*

(a)

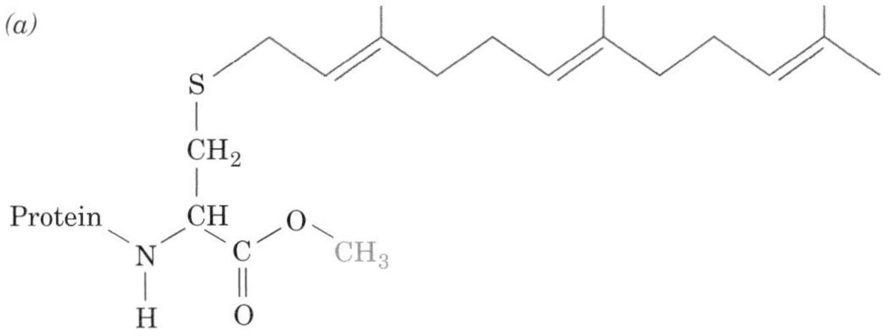

S–Farnesyl cysteine methyl ester

(b)

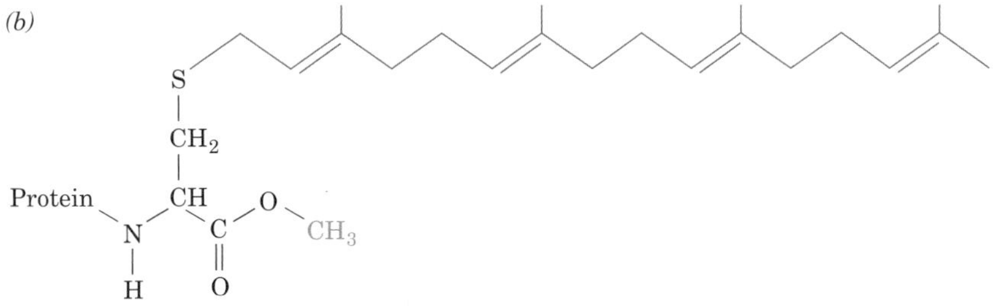

S–Geranylgeranyl cysteine methyl ester

FIGURE 12-29 Prenylated proteins. (*a*) A farnesylated protein and (*b*) a geranylgeranylated protein. In both cases, the protein is synthesized with the C-terminal sequence CaaX, where "C" is Cys, "a" is often an aliphatic amino acid, and "X" is any amino acid. After the prenyl group is appended to the protein in thioether linkage with the Cys residue, the aaX tripeptide is hydrolytically cleaved away and the new carboxyl terminus is methyl esterified. When X is Ala, Met, or Ser, the protein is farnesylated and when X is Leu, it is geranylgeranylated.

teractions. This idea is corroborated by the observation that, in certain proteins involved in intracellular signaling [for example, **Ras** (Section 19-3C) and the so-called **G proteins** (Section 19-2)], prenylation and carboxyl methylation enhance the intersubunit associations that mediate signal transmission.

b. Fatty Acylated Proteins

Two fatty acids are known to be covalently linked to eukaryotic proteins:

1. Myristic acid, a biologically rare saturated C_{14} fatty acid (Table 12-1), which is appended to a protein in amide linkage to the α-amino group of an N-terminal Gly residue. Myristoylation almost always occurs cotranslationally (as the protein is being synthesized), and this attachment is stable, that is, the myristoyl group has a half-life similar to that of the protein to which it is appended.

2. Palmitic acid, a biologically common saturated C_{16} fatty acid, which is joined to a protein in thioester linkage to a specific Cys residue. In some cases, the palmitoylated protein is also prenylated. For example, Ras must be farnesylated and carboxyl methylated as described above before it is palmitoylated at a Cys residue that precedes the protein's C-terminus by several residues. Palmitoylation occurs posttranslationally in the cytosol and is reversible.

Fatty acyl groups are thought to function as membrane anchors for proteins, much as do prenyl groups. However,

the requirement of many proteins for specific fatty acyl residues suggests that these groups also participate in targeting their attached proteins to specific cellular locations. Indeed, palmitoylated proteins occur almost exclusively on the cytoplasmic face of the plasma membrane, whereas myristoylated proteins are found in a number of subcellular compartments including the cytosol, endoplasmic reticulum, Golgi apparatus, plasma membrane, and nucleus. Many fatty acylated proteins participate in intracellular signaling processes through protein–protein interactions in a manner similar to prenylated proteins. Since the membrane affinities and biological activities of many proteins are enhanced by palmitoylation, the reversibility of the palmitoylation appears to be involved in controlling intracellular signaling processes.

c. GPI-Linked Proteins

Glycosylphosphatidylinositol (GPI) groups function to anchor a wide variety of proteins to the exterior surface of the eukaryotic plasma membrane. There is no obvious relationship among the numerous proteins that have GPI anchors, which include enzymes, receptors, immune system proteins, and recognition antigens. Evidently, *GPI groups simply provide an alternative to transmembrane polypeptide domains in binding proteins to the plasma membrane.*

The core structure of GPI anchors consists of phosphatidylinositol (Table 12-2) glycosidically linked to a linear tetrasaccharide composed of three mannose residues

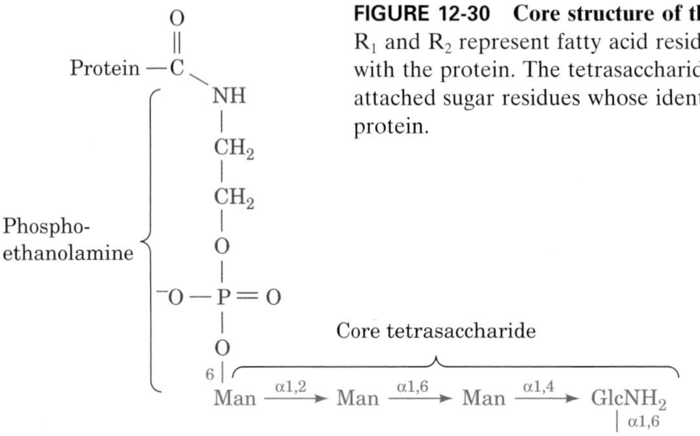

FIGURE 12-30 Core structure of the GPI anchors of proteins. R_1 and R_2 represent fatty acid residues whose identities vary with the protein. The tetrasaccharide may have a variety of attached sugar residues whose identities also vary with the protein.

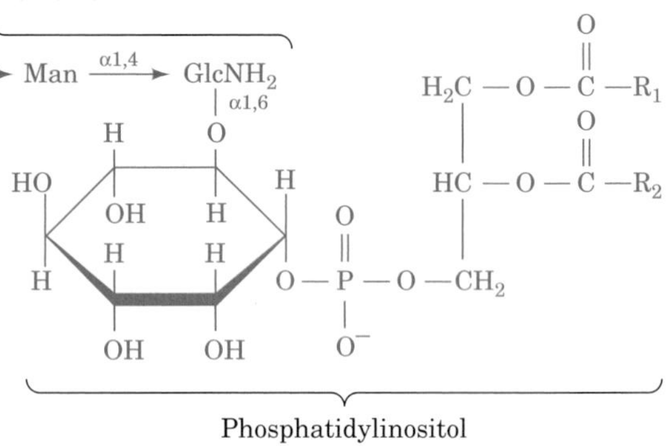

and one glucosaminyl residue (Fig. 12-30). The mannose at the nonreducing end of this assembly forms a phosphoester bond with a phosphoethanolamine residue which, in turn, is amide-linked to the protein's C-terminal carboxyl group. The core tetrasaccharide is generally substituted with a variety of sugar residues that vary with the identity of the protein. There is likewise considerable diversity in the fatty acid residues. The synthesis of GPI anchors is discussed in Section 23-3B.

GPI-anchored proteins occur on the exterior surface of the plasma membrane for the same reason as do the carbohydrate residues of glycoproteins (which we discuss in Section 12-4C). Proteins destined to be GPI-anchored are synthesized with membrane-spanning C-terminal sequences of 20 to 30 hydrophobic residues (as described in Section 12-4B) that are removed during GPI addition. This is corroborated by the observation that GPI-anchored proteins are released from the plasma membrane by treatment with phosphatidylinositol-specific **phospholipases** (Section 19-4B), thereby demonstrating that the mature polypeptides are not embedded in the lipid bilayer.

C. *Fluid Mosaic Model of Membrane Structure*

See Guided Exploration 11: Membrane Structure and the Fluid Mosaic Model The demonstrated fluidity of artificial lipid bilayers suggests that biological membranes have similar properties. This idea was proposed in 1972 by S. Jonathan Singer and Garth Nicholson in their unifying theory of membrane structure known as the **fluid mosaic model.** The theory postulates that integral proteins resemble "icebergs" floating in a two-dimensional lipid "sea" (Fig. 12-20) and that these proteins freely diffuse laterally in the lipid matrix unless their movements are restricted by associations with other cell components.

a. The Fluid Mosaic Model Has Been Verified Experimentally

The validity of the fluid mosaic model has been established in several ways. Perhaps the most vivid is an experiment by Michael Edidin (Fig. 12-31). Cultured mouse cells were fused with human cells by treatment with **Sendai virus** to yield a hybrid cell known as a **heterokaryon.** The mouse cells were labeled with mouse protein–specific antibodies

to which a green-fluorescing dye had been covalently linked **(immunofluorescence).** The proteins on the human cells were similarly labeled with a red-fluorescing marker. On cell fusion, the mouse and human proteins, as seen under the fluorescence microscope, were segregated on the two halves of the heterokaryon. After 40 min at 37°C, however, these proteins had thoroughly intermingled. The addition of substances that inhibit metabolism or protein synthesis did not slow this process, but lowering the temperature below 15°C did. These observations indicate that the mixing process is independent of both metabolic energy and the insertion into the membrane of newly synthesized proteins. Rather, it is a result of the diffusion of existing proteins throughout the fluid membrane, a process that slows as the temperature is lowered.

Fluorescence photobleaching recovery measurements (Fig. 12-15) indicate that membrane proteins vary in their lateral diffusion rates. Some 30 to 90% of these proteins are freely mobile; they diffuse at rates only an order of magnitude or so slower than those of the much smaller lipids, so that they typically take from 10 to 60 min to diffuse the 20-μm length of a eukaryotic cell. Other proteins diffuse more slowly, and some, because of submembrane attachments, are essentially immobile.

The distribution of proteins in membranes may be visualized through electron microscopy using the **freeze-fracture** and **freeze-etch** techniques. In the freeze-fracture procedure, which was devised by Daniel Branton, a membrane specimen is rapidly frozen to near liquid nitrogen temperatures (−196°C). This immobilizes the sample and thereby minimizes its disturbance by subsequent manipulations. The specimen is then fractured with a cold microtome knife, which often splits the bilayer into monolayers (Fig. 12-32). Since the exposed membrane itself would be

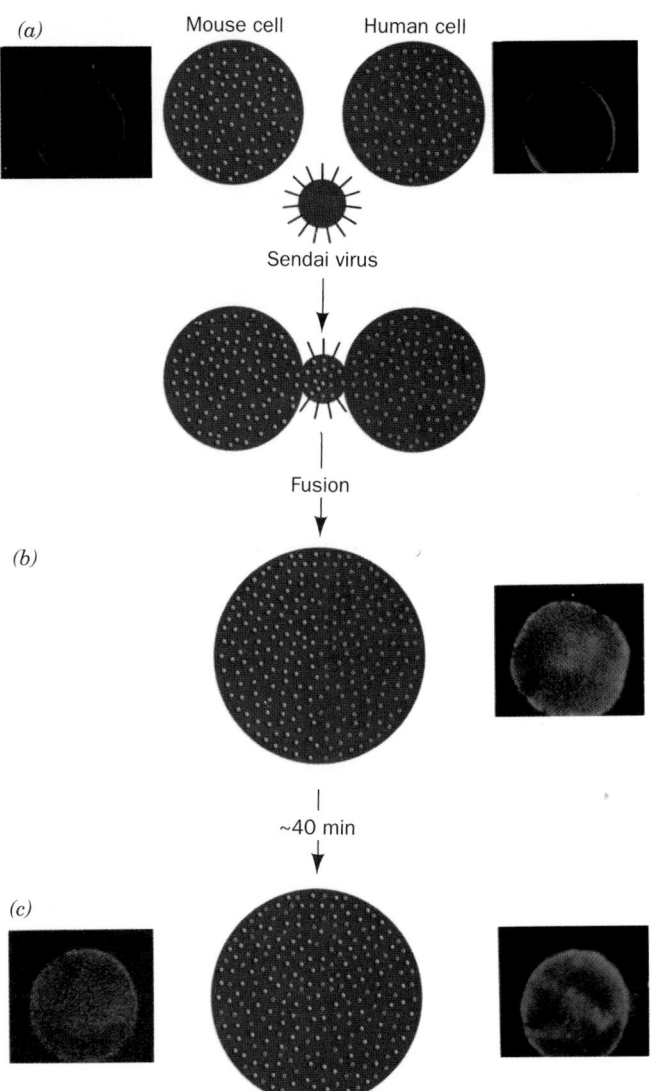

(a) Mouse cell   Human cell

Sendai virus

Fusion

(b)

~40 min

(c)

FIGURE 12-31 Sendai virus–induced fusion of a mouse cell with a human cell and the subsequent intermingling of their cell-surface components as visualized by immunofluorescence. Human and mouse antigens are labeled with red and green fluorescent markers, respectively. (*a*) The membrane-encapsulated Sendai virus specifically binds to cell-surface receptors on both types of cells and subsequently fuses to their cell membranes. (*b*) This results in the formation of a cytoplasmic bridge between the cells that expands so as to form the heterokaryon. (*c*) After 40 min, the red and green markers are fully intermingled. The photomicrographs were taken through filters that allowed only red or green light to reach the camera; that in Part *b* is a double exposure and those in Part *c* are of the same cell. [Immunofluorescence photomicrographs courtesy of Michael Edidin, The Johns Hopkins University.]

destroyed by an electron beam, its metallic replica is made by coating the membrane with a thin layer of carbon, shadowing it (covering it by evaporative deposition under high vacuum) with platinum, and removing the organic matter by treatment with acid. Such a metallic replica can be examined by electron microscopy. In the freeze-etch procedure, the external surface of the membrane adjacent to the

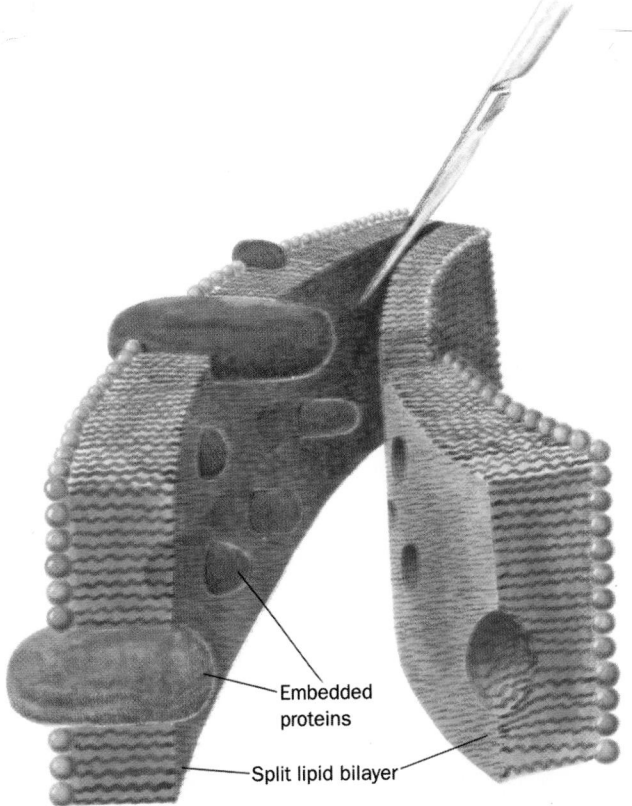

Embedded proteins

Split lipid bilayer

FIGURE 12-32 The freeze-fracture technique. A membrane that has been split by freeze-fracture, as is schematically diagrammed, exposes the interior of the lipid bilayer and its embedded proteins.

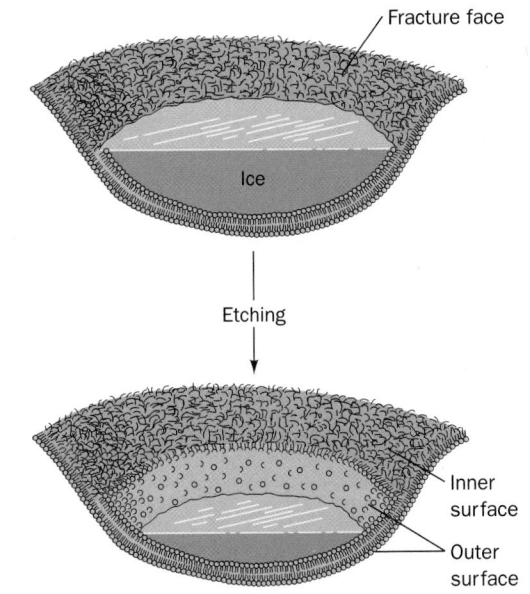

Fracture face

Ice

Etching

Inner surface

Outer surface

FIGURE 12-33 The freeze-etch procedure. The ice that encases a freeze-fractured membrane (*top*) is partially sublimed away so as to expose the outer membrane surface (*bottom*) for electron microscopy.

cleaved area revealed by freeze fracture may also be visualized by first subliming (etching) away, at −100°C, some of the ice in which it is encased (Fig. 12-33).

Freeze-etch electron micrographs of most biological membranes show an inner fracture face that is studded with embedded 50- to 85-Å-diameter globular particles (Fig. 12-34) that appear to be distributed randomly. These particles correspond to membrane proteins, as is demonstrated by their disappearance when the membrane is treated with proteases before its freeze fracture. This is further corroborated by the observation that the myelin membrane, which has a low protein content, as well as liposomes composed of only lipids, have smooth inner fracture faces. Outer membrane surfaces also have a relatively smooth appearance (Fig. 12-34) because integral proteins do not protrude very far beyond them. The distributions of individual external proteins may be visualized by staining procedures, such as the use of ferritin-labeled antibodies, to yield electron micrographs similar in appearance to Fig. 11-33.

b. Membrane Lipids and Proteins Are Asymmetrically Distributed

The distribution of lipids between the different sides of biological membranes has been established through the use of phospholipid-hydrolyzing enzymes known as **phospholipases.** Phospholipases cannot pass through membranes, so that only phospholipids on the external surfaces of intact cells are susceptible to their action. Such studies indicate that *the lipids in biological membranes, like the proteins, are asymmetrically distributed between the leaflets of a bilayer* (e.g., Fig. 12-35). Carbohydrates, as we have seen (Section 11-3C), are located almost exclusively on the external surfaces of plasma membranes.

Lipids and proteins in plasma membranes may also be laterally organized. Thus, the plasma membranes of most cells have two or more distinct domains that have different functions. For example, the plasma membranes of **epithelial cells** (the cells lining body cavities and free surfaces) have an **apical domain,** which faces the lumen of the cavity and often has a specialized function (e.g., the absorption of nutrients in intestinal brush border cells; Section 20-4A), and a **basolateral domain,** which covers the remainder of the cell. These two domains, which do not intermix, have different compositions of both lipids and proteins.

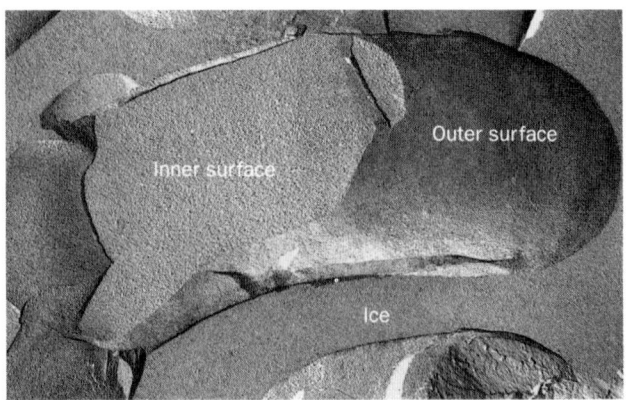

FIGURE 12-34   Freeze-etch electron micrograph of a human erythrocyte plasma membrane. The exposed interior face of the membrane is studded with numerous globular particles that are integral proteins (see Fig. 12-32). The outer surface of the membrane appears smoother than the inner surface because proteins do not project very far beyond the outer membrane surface. [Courtesy of Vincent Marchesi, Yale University.]

A variety of measurements indicate that the hundreds of different lipids and proteins within a given plasma membrane domain are not uniformly mixed but instead often segregate to form **microdomains** that contain only certain types of lipids and proteins. This may occur for several reasons:

1. Certain integral proteins associate to form aggregates or patches in the membrane (e.g., BR), which in turn may preferentially associate with specific lipids. Alternatively, some integral proteins are localized by attachments to elements of the cytoskeleton (which underlies the plasma membrane; Section 1-2A) or are trapped within the spaces enclosed by the resulting "fences."

2. Integral proteins may specifically interact with particular lipids. For example, mismatches between the length of an integral protein's hydrophobic TM band and the average thickness of a lipid bilayer may result in the selective accumulation of certain phospholipids around the protein in an annulus of 10 to 20 layers.

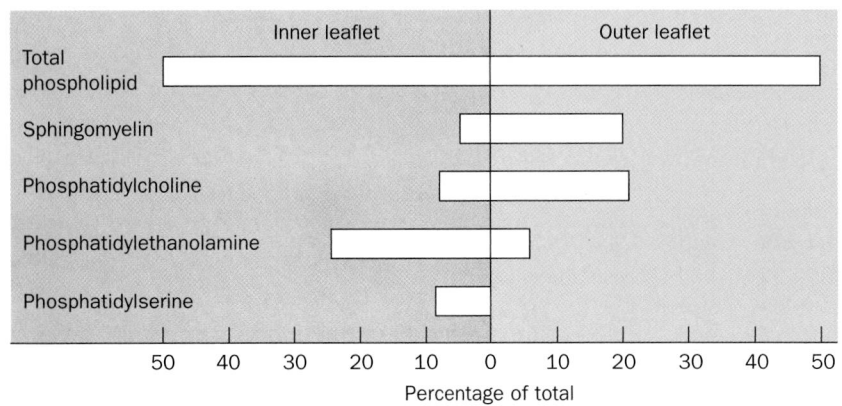

FIGURE 12-35   Asymmetric distribution of phospholipids in the human erythrocyte membrane. The phospholipid content is expressed as mol %. [After Rothman, J.E. and Lenard, J., *Science* **194,** 1744 (1977).]

3. Divalent metal ions, notably Ca^{2+}, selectively ligate negatively charged head groups such as those of phosphatidylserine, thereby causing these phospholipids to aggregate in the membrane. Such metal ion–induced phase separations are known to regulate the activities of certain membrane-bound enzymes.

4. Glycosphingolipids (which occur only in the outer leaflet of the plasma membrane) and cholesterol pack together to form mobile **rafts** and ~75-nm-diameter flask-shaped indentations named **caveolae** (Latin for small caves) with which specific proteins preferentially associate. Glycosphingolipids, by themselves, do not form bilayers because their large head groups prevent the requisite close packing of their predominantly saturated hydrophobic tails. Conversely, cholesterol by itself does not form a bilayer due to its small head group. It therefore appears that the glycosphingolipids in these microdomains associate laterally via weak interactions between their carbohydrate head groups with the voids between their tails filled in by cholesterol. The sphingolipid–cholesterol rafts and caveolae are not solublized at 4°C by uncharged detergents such as Triton X-100 (Fig. 12-19). The low density of the resulting **detergent-insoluble glycolipid-enriched complexes (DIGs)** allows their isolation by sucrose density gradient ultracentrifugation (Section 6-5B), thereby permitting their associated proteins to be identified. Many of the proteins that participate in transmembrane signaling processes (Chapter 19), including GPI-linked proteins, preferentially associate with DIGs. Caveolae, which appear to be rafts with which one or more homologous proteins named **caveolins** are associated, are likewise enriched with proteins that participate in signaling.

It should be noted that all of these aggregates are highly dynamic structures that rapidly exchange both proteins and lipids with their surrounding membrane as a consequence of the weak and transient interactions between membrane components.

D. *The Erythrocyte Membrane*

The erythrocyte membrane's relative simplicity, availability, and ease of isolation have made it the most extensively studied and best understood biological membrane. It is therefore a model for the more complex membranes of other cell types. A mature mammalian erythrocyte is devoid of organelles and carries out few metabolic processes; it is essentially a membranous bag of hemoglobin. Erythrocyte membranes can therefore be obtained by osmotic lysis, which causes the cell contents to leak out. The resultant membranous particles are known as erythrocyte **ghosts** because, on return to physiological conditions, they reseal to form colorless particles that retain their original shape. Indeed, by transferring sealed ghosts to another medium, their contents can be made to differ from the external solution.

a. Erythrocyte Membranes Contain a Variety of Proteins

The erythrocyte membrane has a more or less typical plasma membrane composition of about half protein, somewhat less lipid, and the remainder carbohydrate (Table 12-4). Its proteins may be separated by SDS–polyacrylamide gel electrophoresis (Section 6-4C) after first solubilizing the membrane in a 1% SDS solution. The resulting electrophoretogram for a human erythrocyte membrane exhibits seven major and many minor bands when stained with Coomassie brilliant blue (Fig. 12-36). If the electrophoretogram is instead treated with **periodic acid–Schiff's reagent (PAS),** which stains carbohydrates, four so-called PAS bands become evident. The polypeptides corresponding to bands 1, 2, 4.1, 4.2, 5, and 6 are readily extracted from the membrane by changes in ionic strength or pH and hence are peripheral proteins. These proteins are located on the inner side of the membrane, as is indicated by the observation that they are not altered by the incubation of intact erythrocytes or sealed ghosts with proteolytic enzymes or membrane-impermeable protein labeling reagents. These proteins are altered, however, if "leaky" ghosts are so treated.

In contrast, bands 3, 7, and all four PAS bands correspond to integral proteins; they can be released from the membrane only by extraction with detergents or organic solvents. Of these, band 3 and PAS bands 1 and 2 correspond to TM proteins, as indicated by their different labeling patterns when intact cells are treated with membrane-impermeable protein-labeling reagents and when these reagents are introduced inside sealed ghosts.

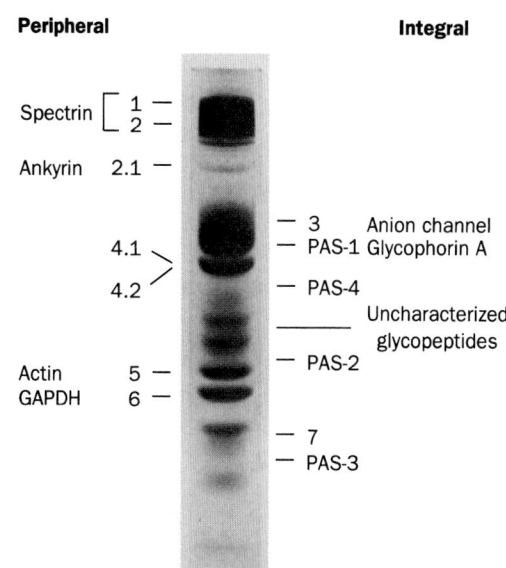

FIGURE 12-36 SDS–PAGE electrophoretogram of human erythrocyte membrane proteins as stained by Coomassie brilliant blue. The bands designated 4.1 and 4.2 are not separated with the 1% SDS concentration used. The minor bands are not labeled for the sake of clarity. The positions of the four sialoglycoproteins that would be revealed by PAS staining are indicated. [Courtesy of Vincent Marchesi, Yale University.]

The PAS band 1 is a dimer of glycophorin A, which is formed through an SDS-resistant association between the TM helices of the polypeptide chains (Fig. 12-21); this dimer is the protein's native form. The PAS band 2 protein is the monomeric form of glycophorin A.

The transport of CO_2 in blood (Section 10-1C) requires that the erythrocyte membrane be permeable to HCO_3^- and Cl^- (the maintenance of electroneutrality requires that for every HCO_3^- that enters a cell, a Cl^- or some other anion must leave the cell; Section 10-1C). The rapid transport of these and other anions across the erythrocyte membrane is mediated by a specific **anion channel** of which there are ~1 million/cell (comprising >30% of the membrane protein). Band 3 protein (929 residues and 5–8% carbohydrate) specifically reacts with anionic protein-labeling reagents that block the anion channel, thereby indicating that the anion channel is composed of band 3 protein. Furthermore, cross-linking studies with bifunctional reagents (Section 8-5C) demonstrate that the anion channel is at least a dimer. Hemoglobin and the glycolytic (glucose metabolizing) enzymes **aldolase, phosphofructokinase (PFK),** and the band 6 protein **glyceraldehyde-3-phosphate dehydrogenase (GAPDH;** Section 17-2F) all specifically and reversibly bind to band 3 protein on the cytoplasmic side of the membrane. The functional significance of this observation is unknown.

b. The Erythrocyte's Cytoskeleton Is Responsible for Its Shape and Flexibility

A normal erythrocyte's biconcave disklike shape (Fig. 7-18a) assures the rapid diffusion of O_2 to its hemoglobin molecules by placing them no farther than 1 μm from the cell surface. However, the rim and the dimple regions of an erythrocyte do not occupy fixed positions on the cell membrane. This can be demonstrated by anchoring an erythrocyte to a microscope slide by a small portion of its surface and inducing the cell to move laterally with a gentle flow of isotonic buffer. A point originally on the rim of the erythrocyte will move across the dimple to the rim on the opposite side of the cell from where it began. Evidently, the membrane rolls across the cell while maintaining its shape, much like the tread of a tractor. This remarkable mechanical property of the erythrocyte membrane results from the presence of a submembranous network of proteins that function as a membrane "skeleton"—the cell's cytoskeleton. Indeed, this property is partially duplicated by a mechanical model consisting of a geodesic sphere (a spheroidal cage) that is freely jointed at the intersections of its struts but constrained from collapsing much beyond a flat surface. When placed inside an evacuated plastic bag, this cage also assumes a biconcave disklike shape.

The fluidity and flexibility imparted to an erythrocyte by its cytoskeleton has important physiological consequences. A slurry of solid particles of a size and concentration equal to that of red cells in blood has the flow characteristics approximating that of sand. Consequently, in order for blood to flow at all, much less for its erythrocytes to squeeze through capillary blood vessels smaller in diameter than they are, erythrocyte membranes, together with their cytoskeletons, must be fluidlike and easily deformable.

The protein **spectrin,** so called because it was discovered in erythrocyte ghosts, accounts for ~75% of the erythrocyte cytoskeleton. It is composed of two similar polypeptide chains, band 1 (α subunit; 2418 residues) and band 2 (β subunit; 2137 residues), which sequence analysis indicates each consist of repeating 106-residue segments that are predicted to fold into triple-stranded α helical coiled coils (Fig. 12-37a, b). Electron microscopy indicates that these large polypeptides are loosely intertwined to form a flexible wormlike αβ dimer that is ~1000 Å long (Fig. 12-37c). Two such heterodimers further associate in a head-to-head manner to form an $(\alpha\beta)_2$ heterotetramer. These tetramers, of which there are ~100,000/cell, are cross-linked at both ends by attachments to bands 4.1 and 5 to form a dense and irregular protein meshwork that underlies the erythrocyte plasma membrane (Fig. 12-37c, d). Band 5, a globular protein that forms filamentous oligomers, has been identified as **actin,** a common cytoskeletal element in other cells (Section 1-2A) and a major component of muscle (Section 35-3A). Spectrin also as-

FIGURE 12-37 (*Opposite*) **The human erythrocyte cytoskeleton.** (*a*) Structure of an αβ dimer of spectrin. Both of these antiparallel polypeptides contain multiple 106-residue repeats, which are thought to form flexibly connected triple helical bundles. Two of these heterodimers join, head to head, to form an $(\alpha\beta)_2$ heterotetramer. [After Speicher, D.W. and Marchesi, V., *Nature* **311**, 177 (1984).] (*b*) X-Ray structure of two consecutive repeats of chicken brain α-spectrin. Each of these 106-residue repeats consists of a down–up–down triple helical bundle in which the C-terminal helix of first repeat (R16; *red*) is continuous, via a 5-residue helical linker (*green*), with the N-terminal helix of the second repeat (R17; *blue*). The helices within each triple helical bundle wrap around each other in a gentle left-handed supercoil that is hydrophobically stabilized by the presence of nonpolar residues at the *a* and *d* positions of heptad repeats on all three of its component α helices (Fig. 8-27). Despite the expected rigidity of α helices, there is considerable evidence that spectrin is a flexible wormlike molecule. [Courtesy of Alfonso Mondragón, Northwestern University. PDBid 1CUN.] (*c*) Electron micrograph of an erythrocyte cytoskeleton that has been stretched to an area 9 to 10 times greater than that of the native membrane. Stretching makes it possible to obtain clear images of the cytoskeleton, which in its native state is so densely packed and irregularly flexed that it is difficult to pick out individual molecules and to ascertain how they are interconnected. Note the predominantly hexagonal network composed of spectrin tetramers cross-linked by junctions containing actin and band 4.1 protein. [Courtesy of Daniel Branton, Harvard University.] (*d*) Model of the erythrocyte cytoskeleton. The so-called junctional complex, which is magnified in this drawing, contains actin, **tropomyosin** (which, in muscle, also associates with actin; Section 35-3A), and band 4.1 protein, as well as **adducin, dematin,** and **tropomodulin** (not shown). [After Goodman, S.R., Krebs, K.E., Whitfield, C.F., Riederer, B.M., and Zagen, I.S., *CRC Crit. Rev. Biochem.* 23, 196 (1988).]

(a)

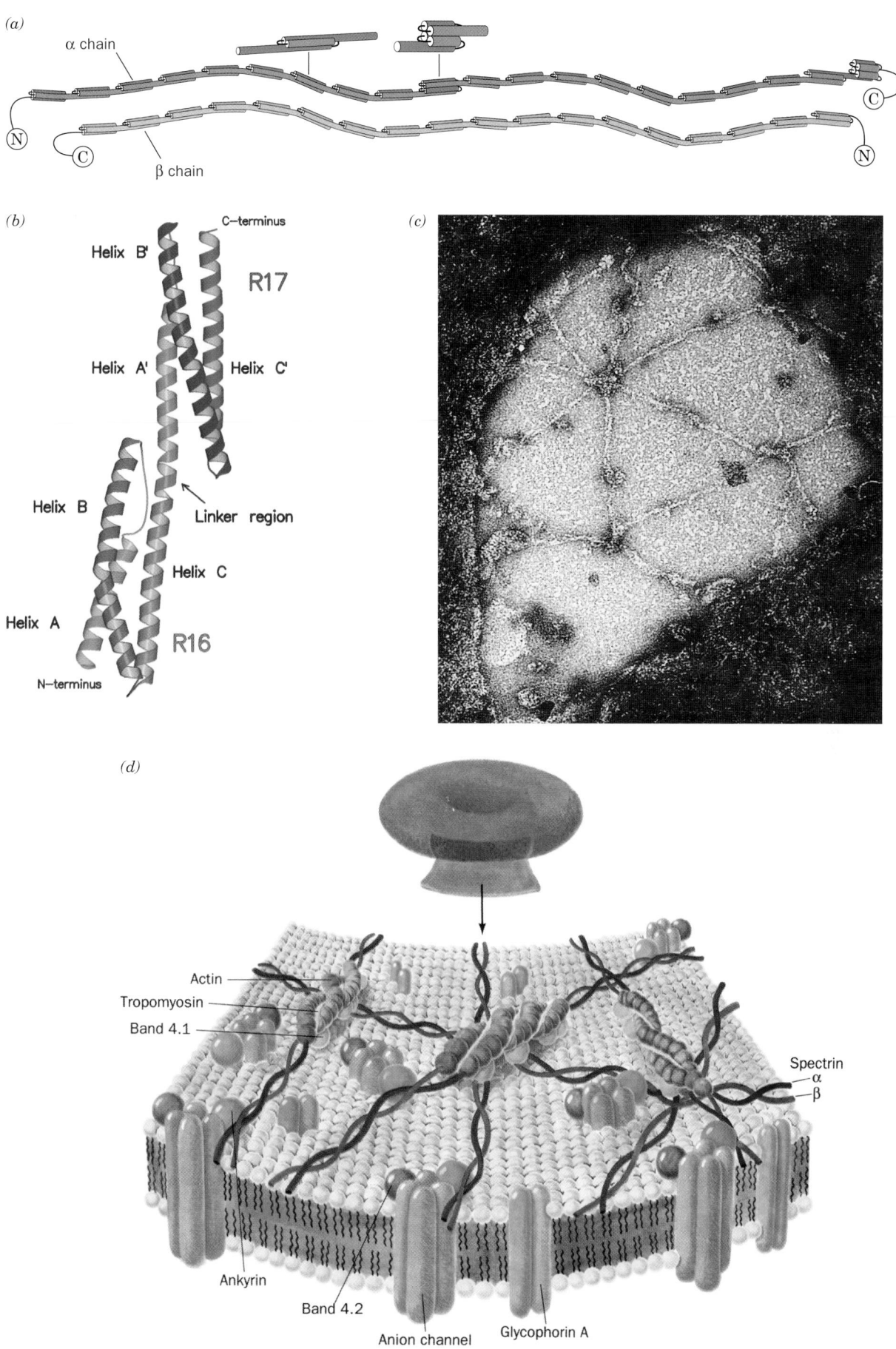

α chain

N

C

β chain

(b)

C−terminus

Helix B′

R17

Helix A′

Helix C′

Helix B

Linker region

Helix C

Helix A

R16

N−terminus

(c)

(d)

Actin

Tropomyosin

Band 4.1

Spectrin

α

β

Ankyrin

Band 4.2

Anion channel

Glycophorin A

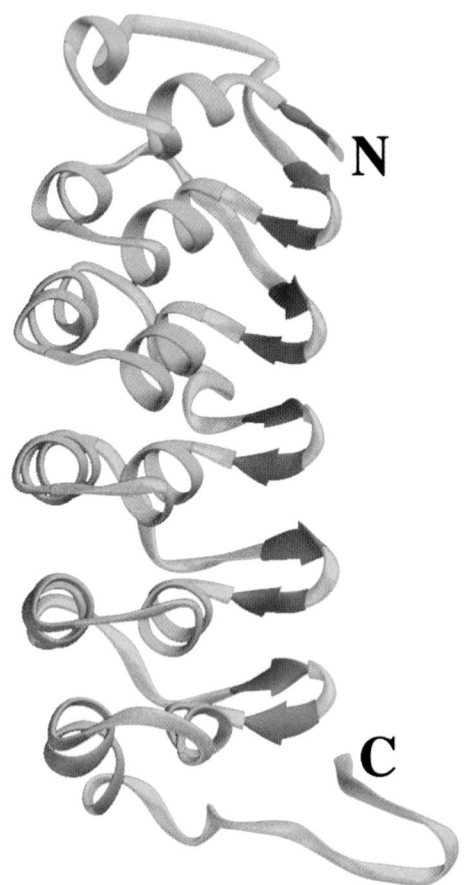

N

C

FIGURE 12-38 X-Ray structure of the transcriptional regulator IκBα. This 236-residue protein consists almost entirely of six ~33-residue ankyrin repeats, each of which contains a β strand, followed by two α helices, and then a β strand so as to form a single turn of a right-handed superhelix. Successive ankyrin repeats are connected by a tight hairpin turn between the C-terminal β strand of one repeat and the N-terminal β strand of the following repeat. The structure of the N-terminal segment of ankyrin, which consists of 24 tandem ankyrin repeats, presumably resembles a quadruply extended version of IκBα. [X-Ray structure by Gourisankar Ghosh, University of California at San Diego. PDBid 1IKN.]

sociates with band 2.1, an 1880-residue monomer known as **ankyrin,** which, in turn, binds to band 3, the anion channel protein. This attachment anchors the cytoskeleton to the membrane. Indeed, on solubilization of spectrin and actin by low ionic strength solutions, erythrocyte ghosts lose their biconcave shape and their integral proteins, which normally occupy fixed positions in the membrane plane, and become laterally mobile.

Ankyrin's N-terminal segment consists almost entirely of 24 tandem ~33-residue repeats known as **ankyrin repeats,** which also occur in a variety of other proteins (Fig. 12-38). Each ankyrin repeat consists of a β strand, two α helices, and a second β strand, all separated by short loops so as to form a single turn of a right-handed superhelix. The β strands in this assembly are arranged so as to yield an extended and nearly flat antiparallel β sheet which, together with the adjoining platform of parallel α helices,

forms an elongated concave surface that is postulated to form the binding sites for various integral proteins. Immunochemical studies have revealed spectrinlike, ankyrinlike, and band 4.1–like proteins in the cytoskeletons of a variety of tissues.

c. Hereditary Spherocytosis and Elliptocytosis Arise from Defects in the Erythrocyte Cytoskeleton

Individuals with **hereditary spherocytosis** have spheroidal erythrocytes that are relatively fragile and inflexible. These individuals suffer from hemolytic anemia because the spleen, a labyrinthine organ with narrow passages that normally filters out aged erythrocytes (which lose flexibility toward the end of their ~120-day lifetime), prematurely removes spherocytotic erythrocytes. The hemolytic anemia may be alleviated by the spleen's surgical removal. However, the primary defects in spherocytotic cells are reduced synthesis of spectrin, the production of an abnormal spectrin that binds band 4.1 protein with reduced affinity, or the absence of band 4.1 protein.

Hereditary elliptocytosis (having elongated or elliptical red cells; also known as **hereditary ovalcytosis**), a condition that is common in certain areas of Southeast Asia and Melanesia, confers resistance to malaria in heterozygotes (but apparently is lethal in homozygotes). This condition arises from defects in the erythrocyte anion channel. A common such defect consists of a 9-residue deletion that inactivates this TM protein. The consequent reduced capacity of red cells to import phosphate or sulfate ions may inhibit the intraerythrocytotic growth of rapidly developing malarial parasites.

> The camel, the renowned "ship of the desert," provides a striking example of adaptation involving the erythrocyte membrane. This remarkable animal is still active after a loss of water constituting 30% of its body weight and, when thus dehydrated, can drink sufficient water in a few minutes to become fully rehydrated. The rapid uptake of such a large amount of water by the blood, which must deliver it to the cells, would lyse the erythrocytes of most animals. Yet camel erythrocytes, which have the shape of flattened ellipsoids rather than biconcave disks, are resistant to osmotic lysis. Camel spectrin binds to its membrane with particular tenacity, but on spectrin removal, which requires a strong denaturing agent such as guanidinium chloride, camel erythrocytes assume a spherical shape.

E. Blood Groups

The outer surfaces of erythrocytes and other eukaryotic cells are covered with complex carbohydrates that are components of plasma membrane glycoproteins and glycolipids. They form a thick, fuzzy cell coating, the **glycocalyx** (Fig. 12-39), which contains numerous identity markers that function in various recognition processes. The human erythrocyte has ~100 known **blood group determinants** that comprise 15 genetically distinct blood group systems. Of these, only two—the **ABO blood group system** (discovered in 1900 by Karl Landsteiner) and the **rhesus (Rh) blood group system**—have major clinical im-

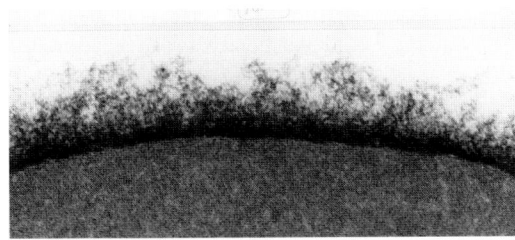

FIGURE 12-39   The erythrocyte glycocalyx as revealed by electron microscopy using special staining techniques. It is up to 1400 Å thick and composed of closely packed, 12- to 25-Å-diameter oligosaccharide filaments linked to plasma membrane–associated proteins and lipids [Courtesy of Harrison Latta, UCLA.]

portance. The various blood groups are identified by means of suitable antibodies or by specific plant lectins.

a. ABO Blood Group Substances Are Carbohydrates

The ABO system consists of three blood group substances, the A, B, and H antigens, which are components of erythrocyte surface sphingoglycolipids. [Antigens are characteristic constellations of chemical groups that elicit the production of specific antibodies when injected into an animal (Section 35-2A). Each antibody molecule can specifically bind to at least two of its corresponding antigen molecules, thereby cross-linking them.] Individuals with type A cells have A antigens on their cell surfaces and carry anti-B antibodies in their serum; those with type B cells, which bear B antigens, carry anti-A antibodies; those with type AB cells, which bear both A and B antigens, carry neither anti-A nor anti-B antibodies; and type O individuals, whose cells bear neither antigen, carry both anti-A and anti-B antibodies. Consequently, the transfusion of type A blood into a type B individual, for example, causes an anti-A antibody–A antigen reaction, which agglutinates (clumps together) the transfused erythrocytes, resulting in an often fatal blockage of blood vessels. The H antigen is discussed below.

The ABO blood group substances are not confined to erythrocytes but also occur in the plasma membranes of many tissues as glycolipids of considerable diversity. In fact, in the ~80% of the population known as secretors, these antigens are secreted as *O*-linked components of glycoproteins into various body fluids, including saliva, milk, seminal fluid, gastric juice, and urine. These diverse molecules, which are 85% carbohydrate by weight and have molecular masses ranging up to thousands of kD, consist of multiple oligosaccharides attached to a polypeptide chain.

The A, B, and H antigens differ only in the sugar residues at their nonreducing ends (Table 12-5). The H antigen occurs in type O individuals; it is also the precursor oligosaccharide of A and B antigens. Type A individuals have a 303-residue glycosyltransferase that specifically adds an *N*-acetylgalactosamine residue to the terminal position of the H antigen, whereas in type B individuals, this enzyme, which differs by four amino acid residues from that of type A individuals, instead adds a galactose residue. In type O

individuals, the enzyme is inactive because its synthesis terminates after its 115th residue.

Do the different blood groups confer any biological advantages or disadvantages? Epidemiological studies indicate that type A and B individuals are less susceptible to cholera infections than are type O individuals, with the relatively rare type AB individuals being highly resistant to this deadly disease. Apparently, type A and B oligosaccharides block a receptor for the bacterium causing cholera, *Vibrio cholera* (Section 19-2C). In addition, type O individuals, particularly nonsecretors, have a higher incidence of peptic (stomach) ulcers. However, type A individuals have a higher incidence of stomach cancer, heart disease, and pernicious anemia (Section 25-2E).

F. Gap Junctions

Most eukaryotic cells are in metabolic as well as physical contact with neighboring cells. This contact is brought about by tubular particles, named **gap junctions,** that join discrete regions of neighboring plasma membranes much like hollow rivets (Fig. 12-40). Indeed, these intercellular channels are so widespread that many whole organs are continuous from within. Thus gap junctions are important intercellular communication channels. For example, the synchronized contraction of heart muscle is brought about by flows of ions through gap junctions (heart muscle is not innervated as is skeletal muscle). Likewise, gap junctions serve as conduits for some of the substances that mediate embryonic development; blocking gap junctions with antibodies that bind to them causes developmental abnormalities in species as diverse as hydras, frogs, and mice. Gap junctions also function to nourish cells that are distant from the blood supply, such as bone and lens cells. Thus, it is not surprising that, in humans, defects in gap junctions are associated with several neurodegenerative diseases as well as developmental anomalies of the cardiovascular system.

Gap junctions consist of a single sort of protein subunit known as a **connexin.** A single gap junction consists of two

TABLE 12-5   Structures of the A, B, and H Antigenic Determinants in Erythrocytes

Type	Antigen
H	Galβ(1→4)GlcNAc ··· ↑1,2 L-Fucα
A	GalNAcα(1→3)Galβ(1→4)GlcNAc ··· ↑1,2 L-Fucα
B	Galα(1→3)Galβ(1→4)GlcNAc ··· ↑1,2 L-Fucα

Abbreviations: Gal = galactose, GalNAc = *N*-acetylgalactosamine, GlcNAc = *N*-acetylglucosamine, L-Fuc = L-fucose.

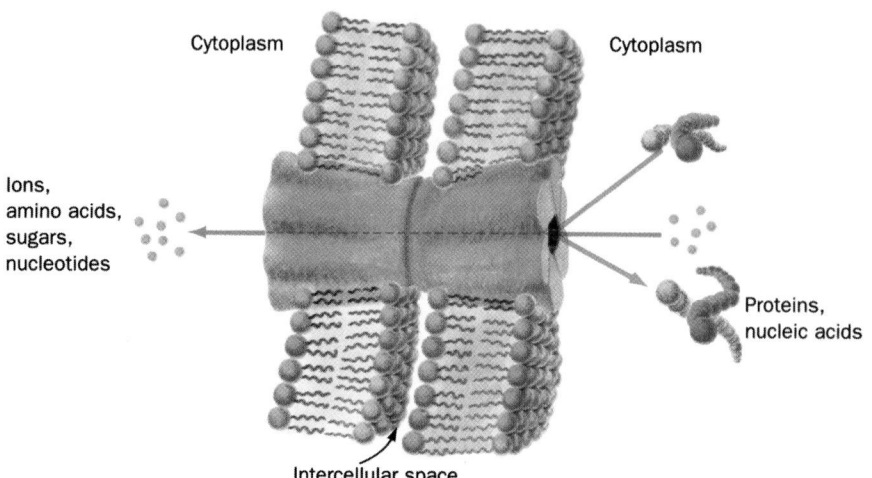

Cytoplasm

Cytoplasm

Ions,
amino acids,
sugars,
nucleotides

Proteins,
nucleic acids

Intercellular space

FIGURE 12-40 Model of a gap junction.
Gap junctions between adjacent cells
consist of two apposed plasma
membrane–embedded hexagonal studs
that bridge the gap between the cells.
Small molecules and ions, but not
macromolecules, can pass between cells
via the gap junction's central channel.

apposed hexagonal rings of connexins, called **connexons,**
one from each of the adjoining plasma membranes
(Fig. 12-40). A given animal expresses numerous genetically distinct connexins, with rodents, for example,
expressing 13 different connexins ranging in molecular
mass from 25 to 50 kD. Many types of cells simultaneously
express several different species of connexins, and in cells
that do so, there is considerable evidence that at least some
connexons may be formed from two or more species of
connexins. Moreover, the gap junctions joining two cells
may consist of two different types of connexons. These
various types of gap junctions presumably differ in their
selectivities for the substances they transmit.

Mammalian gap junction channels are 16 to 20 Å in
diameter, which Werner Loewenstein established by microinjecting single cells with fluorescent molecules of various sizes and observing with a fluorescence microscope
whether the fluorescent probe passed into neighboring
cells. The molecules and ions that can pass freely between
neighboring cells are limited in molecular mass to a maximum of ~1000 D; macromolecules such as proteins and
nucleic acids cannot leave a cell via this route.

The diameter of a gap junction channel varies with Ca^{2+}
concentration: The channels are fully open when the Ca^{2+}
level is $<10^{-7}M$ and narrow as the Ca^{2+} concentration increases until, above $5 \times 10^{-5}M$, they close. This shutter
system is thought to protect communities of interconnected
cells from the otherwise catastrophic damage that would
result from the death of even one of their members. Cells
generally maintain very low cytosolic Ca^{2+} concentrations
($<10^{-7}M$) by actively pumping Ca^{2+} out of the cell as well
as into their mitochondria and endoplasmic reticulum
(Section 20-3B; Ca^{2+} is an important intracellular messenger whose cytosolic concentration is precisely regulated).
Ca^{2+} floods back into leaky or metabolically depressed
cells, thereby inducing closure of their gap junctions and
sealing them off from their neighbors.

a. Connexins Contain Transmembrane 4-Helix Bundles

Gap junctions spontaneously associate *in vivo* to form
membranous plaques of ordered particles that are
essentially 2-dimensional crystals. This permitted Mark

Yeager to determine the structure of a recombinant cardiac
gap junction through electron crystallography in a manner
resembling that used with bacteriorhodopsin (Section
12-3A). The structure (Fig. 12-41), whose resolution is
7.5 Å in the membrane plane and 21 Å normal to the membrane, reveals a particle with D_6 symmetry, a diameter of
~70 Å, and a length of ~150 Å which encloses a central
channel whose diameter varies from ~40 Å at its mouth
to ~15 Å in its interior. The TM portions of the gap junc-

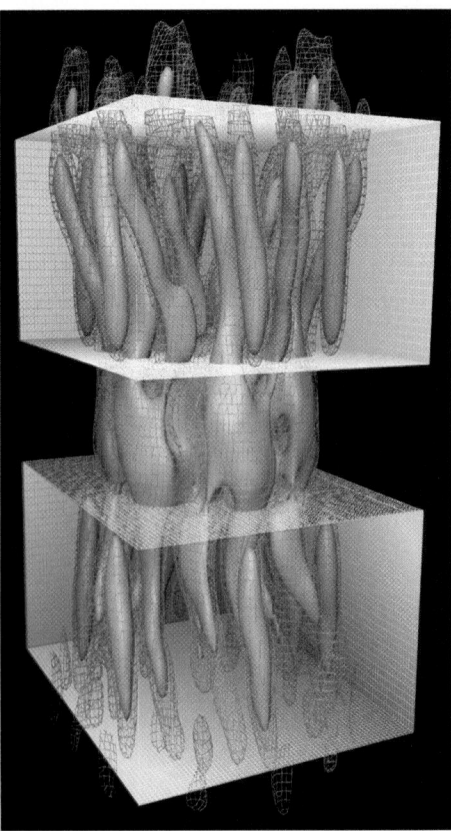

**FIGURE 12-41 Electron crystal structure of a cardiac gap junc-
tion.** The electron density, at two different levels, is represented
by the solid and mesh contours (*gold*). The white boxes indicate the positions of the cell membranes. [Courtesy of Mark
Yeager, The Scripps Research Institute, La Jolla, California.]

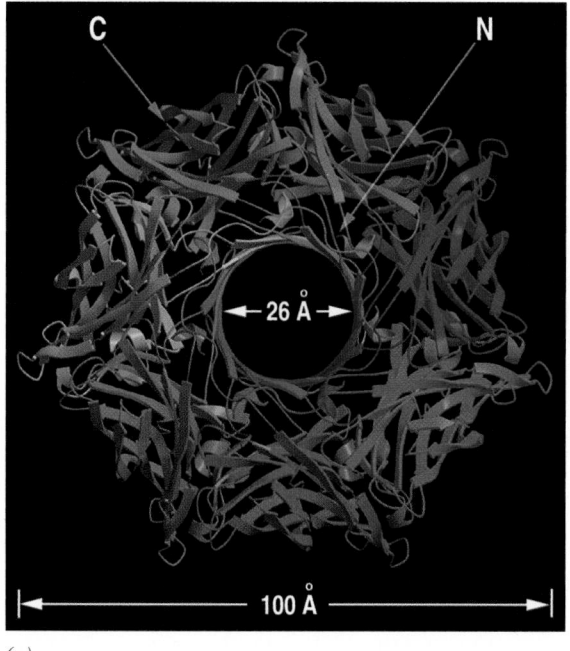

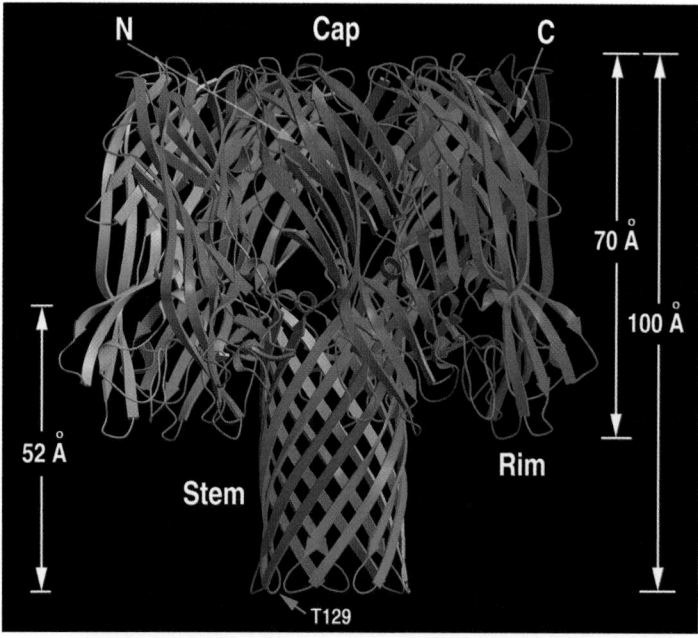

(a)

(b)

FIGURE 12-42 X-Ray structure of α-hemolysin. Views (*a*) along and (*b*) perpendicular to the heptameric transmembrane pore's 7-fold axis. Each subunit is drawn with a different color. (*c*) The monomer unit with its three domains drawn in different colors. [Courtesy of Eric Gouaux, Columbia University. PDBid 7AHL.]

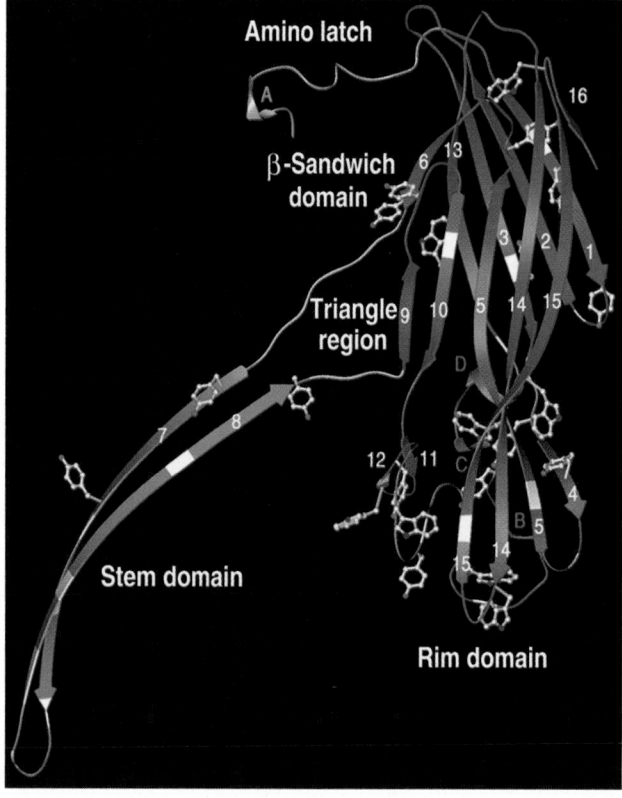

(c)

tion each contain 24 rods of electron density that are arranged with hexagonal symmetry and which extend normal to the membrane plane. This is consistent with the conclusions of hydropathy plots resembling Fig. 12-22 that each connexin contains 4 conserved TM helices. Since both the N- and C-termini of connexins are located on the cytoplasmic side of the plasma membrane, it is evident that the TM portion of a connexin consists of an up–down–up–down 4-helix bundle.

G. *Channel-Forming Proteins*

A number of bacterial toxins are synthesized as water-soluble monomers that, on interacting with their target membrane via a specific receptor protein, spontaneously insert into the membrane as a TM pore. This process, which for many such **channel-forming toxins (CFTs)** requires their oligomerization, causes the leakage of small ions and molecules from the target cell, thereby killing it through loss of osmotic balance. The formation of only one CFT-based pore is often sufficient to kill a cell.

One of the best characterized CFTs is **α-hemolysin,** which the human pathogen *Staphylococcus aureus* secretes as a water-soluble 293-residue monomer and which spontaneously inserts into the membranes of erythrocytes and several other types of cells in the form of heptameric pores. Even though the α-hemolysin monomer is water-soluble and lacks clearly hydrophobic segments, the heptamer acts as a typical TM protein in that it is not released from the

membrane by treatment with high salt, low pH, or chaotropic agents but, instead, requires treatment with detergents for this to occur.

The X-ray structure of detergent-solublized α-hemolysin, determined by Eric Gouaux, reveals a striking mushroom-shaped heptameric complex that is 100 Å in height and 100 Å in diameter (Figs. 12-42*a* and *b*). A 14- to 46-Å-diameter solvent-filled channel, which runs along the protein's 7-fold axis, forms a TM pore. The stem of the mushroom, the protein's TM segment, consists of a 52-Å-high and 26-Å-diameter, porinlike, 14-stranded, antiparallel β barrel composed of seven 2-stranded antiparallel β sheets, one from each subunit (Fig. 12-42b). The remainder of each subunit consists of a β sandwich domain and a rim domain, which together form a 70-Å-long ellipsoid (Fig. 12-42c). Seven of these ellipsoids are distributed in a ring, thereby forming the mushroom's cap and rim. The rim domain projects toward and probably interacts with the membrane's phospholipid head groups via the basic and aromatic residues that extend from the crevice between the top of the stem and rim.

A variety of experimental evidence indicates that the spontaneous formation of the heptameric TM pore occurs via several discrete steps: (1) the binding of the aqueous monomer to the membrane surface, probably through the interaction of the protein's polypeptide loops with the surface groups of the lipid bilayer; (2) the formation of the heptamer on the surface of the membrane; and (3) the insertion of the 14-stranded β barrel through the membrane to form the TM pore. The structural details of this process are as yet unknown, although it seems clear that there is little change in the monomers' secondary structure on their assembly to form the heptameric TM pore. The reason why monomers do not form heptamers in aqueous solution is probably due to differences between the strengths of the intrasubunit interactions in the monomer in aqueous solution and the intersubunit interactions in the heptamer in the membrane.

Not all CFTs form pores using β barrels. Rather, a variety of CFTs, notably several *E. coli* proteins known as **colicins,** form pores that are lined with α helices. Most such pores consist of monomers.

4 ■ MEMBRANE ASSEMBLY AND PROTEIN TARGETING

As cells grow and divide, they synthesize new membranes. How are such asymmetric membranes generated? One way in which this might occur is through self-assembly. Indeed, when the detergent used to disperse a biological membrane is removed, liposomes form in which functional integral proteins are embedded. In most cases, however, these model membranes are symmetrical, both in their lipid distribution between the inner and outer leaflets of the bilayer and in the orientations of their embedded proteins. An alternative hypothesis of membrane assembly is that *it occurs on the scaffolding of preexisting membranes; that is, membranes are generated by the expansion of old ones rather than by the creation of new ones.* In this section we shall see that this is, in fact, how biological membranes are generated. In doing so, we shall consider how proteins are inserted into and passed through membranes as well as how portions of membranes in the form of vesicles pinch off from one membrane and fuse with another, thereby transporting proteins and lipids between these membranes. These highly complex processes are indicative of the intricacies of biological processes in general.

A. *Lipid Distributions in Membranes*

The enzymes involved in the biosynthesis of membrane lipids are mostly integral proteins (Section 25-8). Their substrates and products are themselves membrane components, so that membrane lipids are fabricated on site. Eugene Kennedy and James Rothman demonstrated this to be the case in bacteria through the use of selective labeling. They gave growing bacteria a 1-min pulse of $^{32}PO_4^{3-}$ so as to label radioactively the phosphoryl groups of only the newly synthesized phospholipids. **Trinitrobenzenesulfonic acid (TNBS),** a membrane-impermeable reagent that combines with phosphatidylethanolamine (**PE;** Fig. 12-43), was then immediately added to the cell suspension. Analysis of the resulting doubly labeled membrane showed that none of the TNBS-labeled PE was radioactively labeled. This observation indicates that *newly made PE is synthesized on the cytoplasmic face of the membrane (Fig. 12-44, top right).*

a. Membrane Proteins Catalyze Phospholipid Flip-Flops

If an interval of only 3 min is allowed to elapse between the $^{32}PO_4^{3-}$ pulse and the TNBS addition, about half of the ^{32}P-labeled PE is also TNBS labeled (Fig. 12-44, *bottom*). This observation indicates that the flip-flop rate of PE in

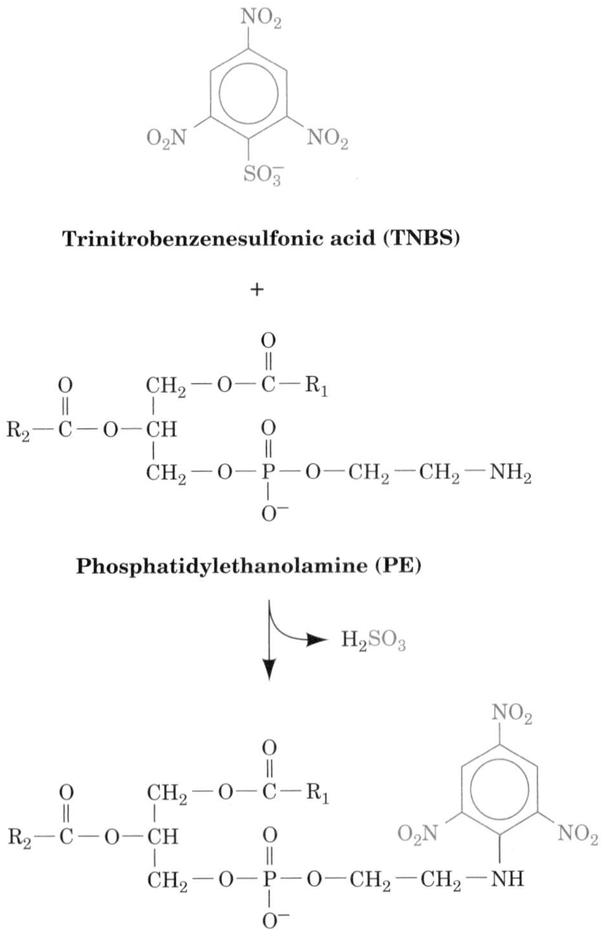

Trinitrobenzenesulfonic acid (TNBS)

+

Phosphatidylethanolamine (PE)

FIGURE 12-43 Reaction of TNBS with PE.

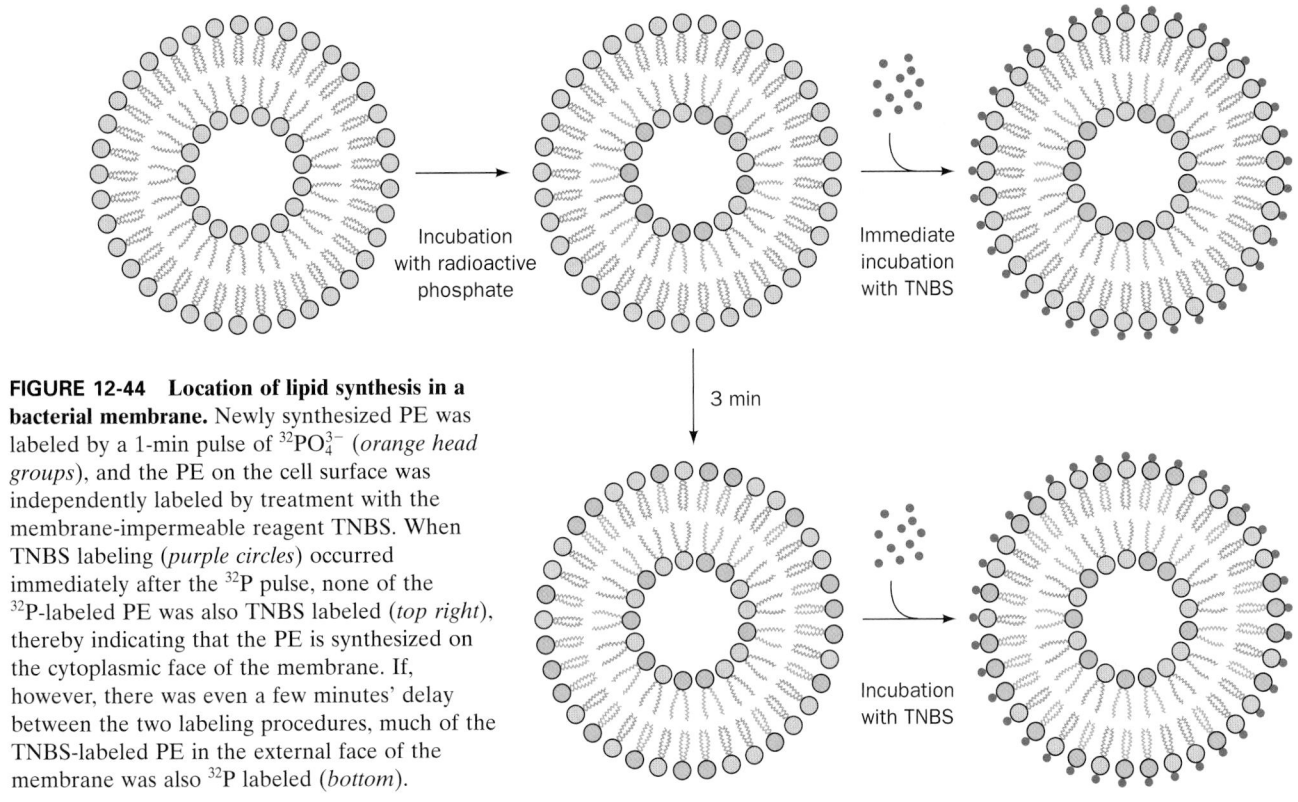

FIGURE 12-44 Location of lipid synthesis in a bacterial membrane. Newly synthesized PE was labeled by a 1-min pulse of $^{32}PO_4^{3-}$ (*orange head groups*), and the PE on the cell surface was independently labeled by treatment with the membrane-impermeable reagent TNBS. When TNBS labeling (*purple circles*) occurred immediately after the ^{32}P pulse, none of the ^{32}P-labeled PE was also TNBS labeled (*top right*), thereby indicating that the PE is synthesized on the cytoplasmic face of the membrane. If, however, there was even a few minutes' delay between the two labeling procedures, much of the TNBS-labeled PE in the external face of the membrane was also ^{32}P labeled (*bottom*).

the bacterial membrane is ~100,000-fold greater than it is in bilayers consisting of only phospholipids (where, it will be recalled, the flip-flop rates have half-times of many days).

How do phospholipids synthesized on one side of the membrane reach its other side so quickly? Phospholipid flip-flops appear to be facilitated in two ways:

1. Membranes contain proteins known as **flipases** that catalyze the flip-flops of specific phospholipids. These proteins tend to equilibrate the distribution of their corresponding phospholipids across a bilayer; that is, the net transport of a phospholipid is from the side of the bilayer with the higher concentration of the phospholipid to the opposite side. Such a process, as we shall see in Section 20-2, is a form of **facilitated diffusion.**

2. Membranes contain proteins known as **phospholipid translocases** that transport specific phospholipids across a bilayer in a process that is driven by ATP hydrolysis. These proteins can transport certain phospholipids from the side of a bilayer that has the lower concentration of the phospholipid being translocated to the opposite side, thereby establishing a nonequilibrium distribution of the phospholipid. Such a process, as we shall see in Section 20-2, is a form of **active transport.**

The observed distribution of phospholipids across membranes (e.g., Fig. 12-35) therefore appears to arise from the membrane orientations of the enzymes that synthesize phospholipids combined with the countervailing tenden-

cies of ATP-dependent phospholipid translocases to generate asymmetric phospholipid distributions and those of flipases to randomize these distributions.

b. A Membrane's Characteristic Lipid Composition Can Arise in Several Ways

In eukaryotic cells, lipids are synthesized on the cytoplasmic face of the endoplasmic reticulum, from where they are transported to other membranes. Perhaps the most important mechanism of lipid transport is the budding off of membranous vesicles from the ER and their subsequent fusion with other membranes (Sections 12-4C and 12-4D). However, this mechanism does not explain the different lipid compositions of the various membranes in a cell. Lipids may also be transported between membranes by the **phospholipid exchange proteins** present in many cells. These proteins spontaneously transfer specific phospholipids, one molecule at a time, between two membranes separated by an aqueous medium. A membrane's characteristic lipid composition may also arise through on-site remodeling and/or selective degradation of its component lipids through the action of specific enzymes (Section 25-8A).

B. *The Secretory Pathway*

Membrane proteins, as are all proteins, are ribosomally synthesized under the direction of messenger RNA templates such that each polypeptide grows from its N-terminus to its C-terminus by the stepwise addition of amino acid residues (Section 5-4B). Cytologists have long noted two classes of eukaryotic ribosomes, those free in the cytosol and those bound to the endoplasmic reticulum **(ER)** so as to form the **rough endoplasmic reticulum** (**RER,** so called because of the knobby appearance its bound ribosomes give it; Fig. 1-5). Both classes of ribosomes are nevertheless structurally identical; they differ only in the nature of the polypeptide they are synthesizing. *Free ribosomes synthesize mostly soluble and mitochondrial proteins, whereas membrane-bound ribosomes manufacture TM proteins and proteins destined for secretion, operation within the ER, or incorporation into **lysosomes*** (membranous organelles containing a battery of hydrolytic enzymes that function to degrade and recycle cell components; Section 1-2A). These latter proteins initially appear in the RER.

a. The Secretory Pathway Accounts for the Targeting of Many Secreted and Membrane Proteins

How are RER-destined proteins differentiated from other proteins? And how do these large, relatively polar molecules pass through the RER membrane? These processes occur via the **secretory pathway,** which was first described by Günter Blobel, Cesar Milstein, and David Sabatini around 1975. Since ~25% of the different species of proteins synthesized by all types of cells are integral proteins and many others are secreted, *~40% of the various types of proteins that a cell synthesizes must be processed via the secretory pathway or some other protein targeting pathway* (e.g., that which directs proteins to the mitochondrion; Section 12-4E). In this subsection, we first present an overview of the secretory pathway and then discuss aspects of it in detail. The secretory pathway is outlined in Fig. 12-45:

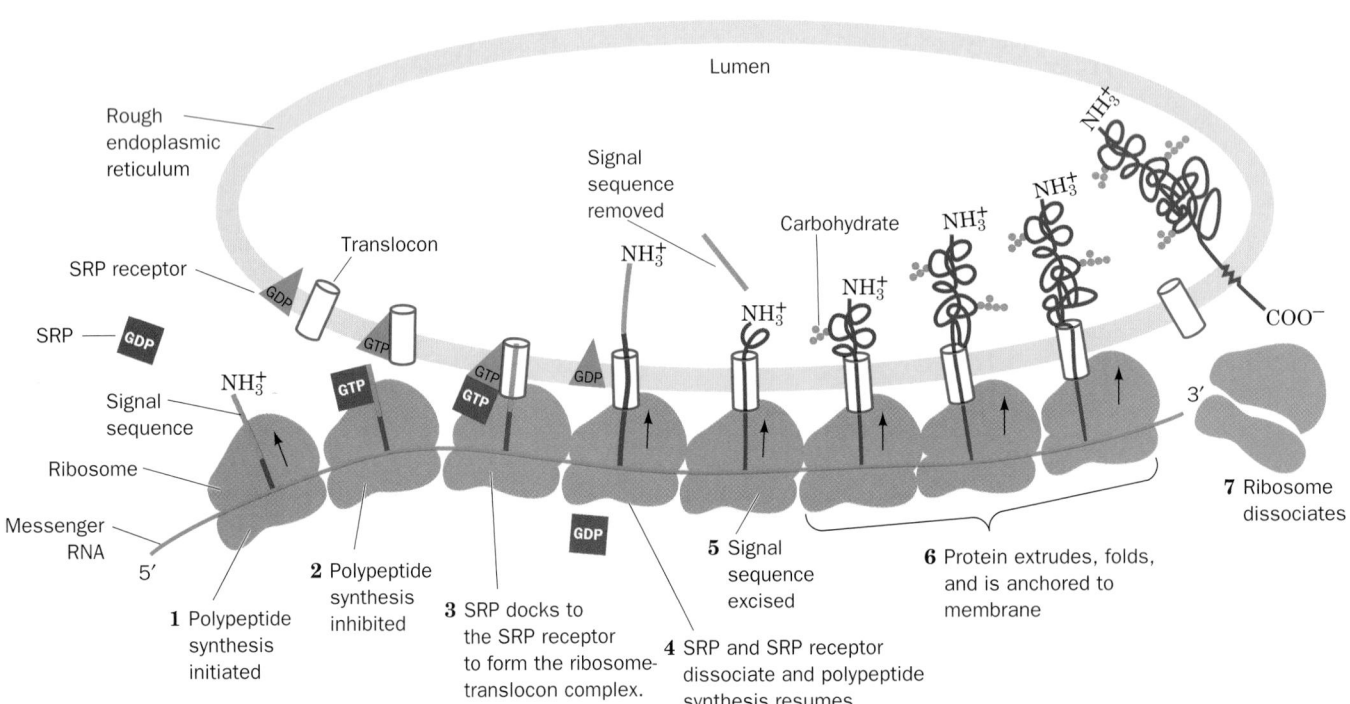

FIGURE 12-45   The ribosomal synthesis, membrane insertion, and initial glycosylation of an integral protein via the secretory pathway. (1) Protein synthesis is initiated at the N-terminus of the polypeptide, which consists of a 13- to 36-residue signal sequence. **(2)** A signal recognition particle (SRP) binds to the ribosome and the signal sequence emerging from it, thereby arresting polypeptide synthesis. **(3)** The SRP is bound by the transmembrane SRP receptor (SR) in complex with the translocon, thereby bringing together the ribosome and the translocon. **(4)** The SRP and SR hydrolyze their bound GTPs, causing them to dissociate from the ribosome–translocon complex. The ribosome then resumes the synthesis of the polypeptide, which passes through the translocon into the lumen of the ER. **(5)** Shortly after the entrance of the signal sequence into the lumen of the endoplasmic reticulum, it is proteolytically excised. **(6)** As the growing polypeptide chain passes into the lumen, it commences folding into its native conformation, a process that is facilitated by its interaction with the chaperone protein Hsp70 (not shown). Simultaneously, enzymes initiate the polypeptide's specific glycosylation. Once the protein has folded, it cannot be pulled out of the membrane. At points determined by its sequence, the polypeptide becomes anchored in the membrane (proteins destined for secretion pass completely into the ER lumen). **(7)** Once polypeptide synthesis is completed, the ribosome dissociates into its two subunits. 🎝 **See the Animated Figures**

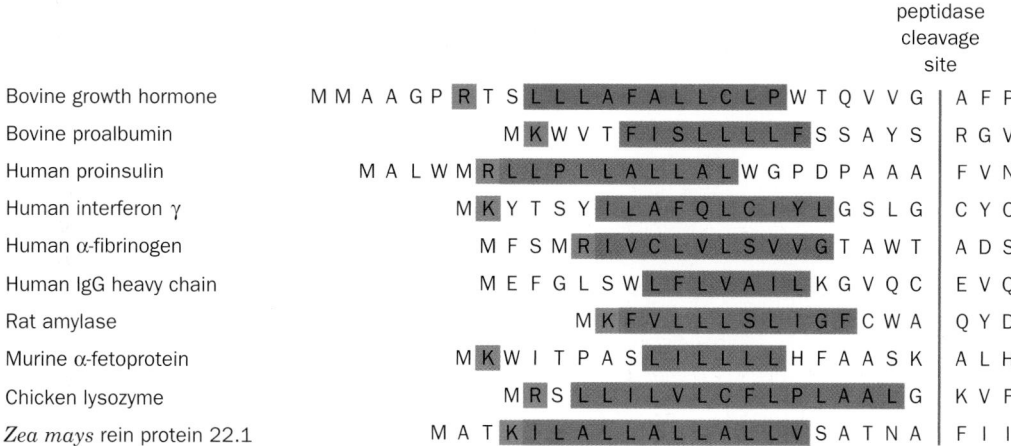

FIGURE 12-46   N-Terminal sequences of some eukaryotic secretory preproteins. The hydrophobic cores (*brown*) of most signal peptides are preceded by basic residues (*blue*). [After Watson, M.E.E., *Nucleic Acids Res.* **12**, 5147–5156 (1984).]

1. *All secreted, ER-resident, and lysosomal proteins, as well as many TM proteins, are synthesized with leading (N-terminal) 13- to 36-residue **signal peptides.*** These signal peptides consist of a 6- to 15-residue hydrophobic core flanked by several relatively hydrophilic residues that usually include one or more basic residues near the N-terminus (Fig. 12-46). Signal peptides otherwise have little sequence similarity. However, a variety of evidence indicates they form α helices in nonpolar environments.

2. When the signal peptide first protrudes beyond the ribosomal surface (when the polypeptide is at least ~40 residues long), the **signal recognition particle (SRP),** a 325-kD complex of six different polypeptides and an ~300-nucleotide RNA molecule, binds to both the signal peptide and the ribosome accompanied by replacement of the SRP's bound GDP by GTP. The SRP's resulting conformational change causes the ribosome to arrest further polypeptide growth, thereby preventing the RER-destined protein from being released into the cytosol.

3. The SRP–ribosome complex diffuses to the RER surface, where it is bound by the **SRP receptor (SR;** also called **docking protein)** in complex with the **translocon,** a protein pore in the ER membrane through which the growing polypeptide will be extruded. In forming the SR–translocon complex, the SR's bound GDP is replaced by GTP.

4. The SRP and SR stimulate each other to hydrolyze their bound GTP to GDP (which is energetically equivalent to ATP hydrolysis), resulting in conformational changes that cause them to dissociate from each other and from the ribosome–translocon complex. This permits the bound ribosome to resume polypeptide synthesis such that the growing polypeptide's N-terminus passes through the translocon into the lumen of the ER. Most ribosomal processes, as we shall see in Section 32-3, are driven by GTP hydrolysis.

5. Shortly after the signal peptide enters the ER lumen, it is specifically cleaved from the growing polypeptide by a membrane-bound **signal peptidase** (polypeptide chains with their signal peptide still attached are known as **preproteins;** signal peptides are alternatively called **presequences**).

6. The nascent (growing) polypeptide starts to fold to its native conformation, a process that is facilitated by its interaction with an ER-resident chaperone protein Hsp70 (Section 9-2C). Enzymes in the ER lumen then initiate **posttranslational modification** of the polypeptide, such as the specific attachments of "core" carbohydrates to form glycoproteins (Section 23-3B); the formation of disulfide bonds as facilitated by protein disulfide isomerase (Section 9-2A), an ER–resident protein; and the attachment of GPI anchors (Section 23-3B).

7. When polypeptide synthesis is completed, the protein is released from both the ribosome and the translocon, and the ribosome dissociates from the RER. Secretory, ER-resident, and lysosomal proteins pass completely through the RER membrane into the lumen. TM proteins, in contrast, contain one or more hydrophobic ~20-residue "membrane anchor" sequences that remain embedded in the membrane.

The secretory pathway also occurs in prokaryotes for the insertion of certain proteins into the cell membrane (whose exterior is equivalent to the ER lumen). Indeed, all forms of life yet tested have homologous SRPs and SRs.

b. The X-Ray Structure of the SRP Core Reveals How It Binds Signal Peptide

Mammalian SRPs consist of six polypeptides known as **SRP9, SRP14, SRP19, SRP54, SRP68,** and **SRP72** (where the numbers are their molecular masses in kD) and a 7S RNA [RNAs are often classified according to their sedi-

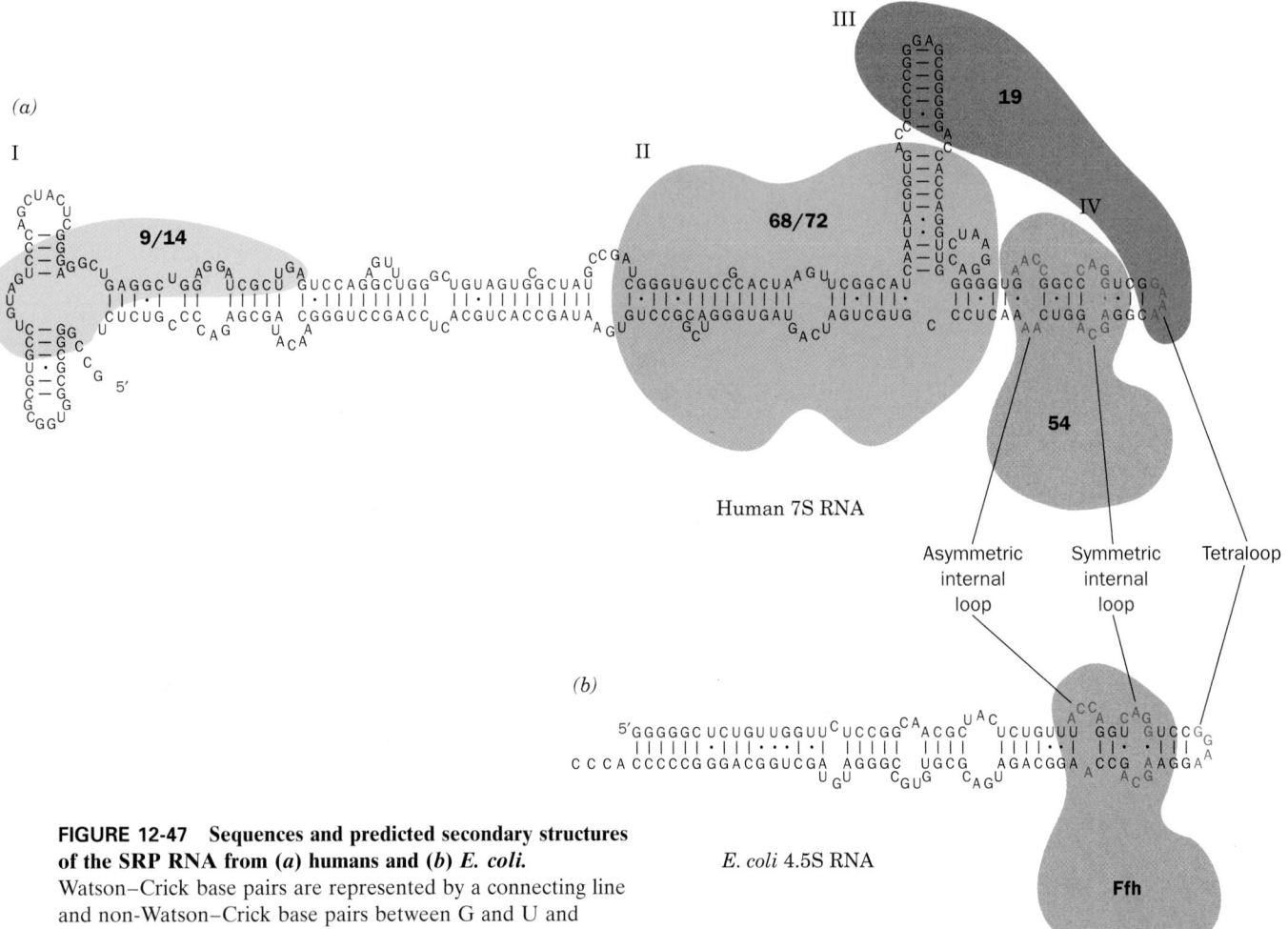

FIGURE 12-47 Sequences and predicted secondary structures of the SRP RNA from (a) humans and (b) E. coli. Watson–Crick base pairs are represented by a connecting line and non-Watson–Crick base pairs between G and U and between G and A are indicated by a dot. The three principal conserved features of domain IV in both sequences, the tetraloop, the asymmetric internal loop, and the symmetric internal loop, are highlighted in red. Note the similarities of both the sequences and positions of these conserved features. In mammals, SRP9 and SRP14 form a heterodimer named **SRP9/14** that is implicated in interacting with the ribosome, whereas SRP68 and SRP72 form a heterodimer dubbed

SRP68/72 that is required for protein translocation. The regions of the SRP RNA that these proteins protect from nuclease digestion or chemical derivatization are indicated by shading. In *E. coli,* the SRP54 homolog Ffh binds to the domain IV–like portion of its SRP RNA. [After Walter, P. and Johnson, A.E., *Annu. Rev. Cell Biol.* **10,** 94 (1994).]

mentation rate in Svedberg units (S), which increases with their molecular mass (Section 6-6D); human 7S RNA consists of 299 nucleotides **(nt)**. Electron micrographs indicate that SRP RNA has the shape of an elongated rod that is 240 Å long and 60 Å wide. This is consistent with its predicted secondary structure (Fig. 12-47a), which indicates that SRP RNA comprises 4 domains, of which only the ~50-nt domain IV is highly conserved. Many prokaryotic SRPs are much simpler; that in *E. coli* consists of a single polypeptide named **Ffh** that is homologous to SRP54 (Ffh for *F*ifty-*f*our *h*omolog) and a 4.5S RNA (114 nt; Fig. 12-47b) that, in part, is predicted to have a secondary structure similar to that of domain IV. Replacing SRP54 with Ffh or vice versa yields functional SRPs, at least *in vitro,* thereby suggesting that the Ffh–4.5S RNA complex is a structurally minimized version of the eukaryotic SRP.

The SRP54/Ffh protein consists of three domains: the N-terminal N domain, whose function is unknown; the central G domain, which contains the SRP's GTPase function and mediates its interaction with the SRP receptor; and the C-terminal M domain, which is rich in methionine (14 of its 102 residues in *E. coli* and 11 in humans, although only 4 of these Met positions are in common) and binds both the SRP RNA and the signal peptide. The X-ray structure of Ffh from *Thermus aquaticus* (Fig. 12-48), determined by Robert Stroud and Peter Walter, reveals that its N domain forms a bundle of four antiparallel α helices and that its G domain consists of an open β sheet (Section 8-3B) that structurally resembles those of other GTPases. The structure of the M domain is discussed below.

Jennifer Doudna determined the X-ray structure of the *E. coli* SRP's conserved core, that of a complex of the Ffh

M domain with domain IV RNA. This X-ray structure (Fig. 12-49*a*) reveals that the 49-nt RNA, in agreement with its predicted secondary structure (Fig. 12-47*b*), forms a 70-Å-long double helical rod in which the RNA chain doubles back on itself via a 4-nt unpaired loop (a so-called **tetraloop**; RNA, as does DNA, can form a base-paired double helix, although its conformation is distinctly different from that of B-DNA; Section 29-1B). The bases of the RNA's so-called symmetric internal loop (Figs. 12-47 and 12-49*a*), somewhat unexpectedly, associate via non–Watson–Crick base pairing interactions and thereby continue the sugar–phosphate backbone's double helical conformation as well as the stacking interactions between successive base pairs. In contrast, the asymmetric internal loop's unpaired 4-nt segment is looped out from the double helical stack, thereby leaving a large cavity in the center of the helix, which is filled by a remarkable cluster of 2 hydrated Mg^{2+} ions and 28 ordered water molecules. The RNA and protein interact mainly via a dense hydrogen bonded network involving the RNA's symmetric and asymmetric internal loops and the M domain's helices 2, 2b, and 3 (Fig. 12-49*a*).

A 33-residue segment in the complex's 102-residue M domain is disordered and hence not visible in the X-ray structure. However, the remaining segments of this protein are closely superimposable on those of the M domain in the X-ray structure of *T. aquaticus* Ffh, in which the entire domain is visible (Fig. 12-49*a*). The segment that is disordered in *E. coli* Ffh, the so-called finger loop, together with adjacent portions of the M domain, forms a deep groove

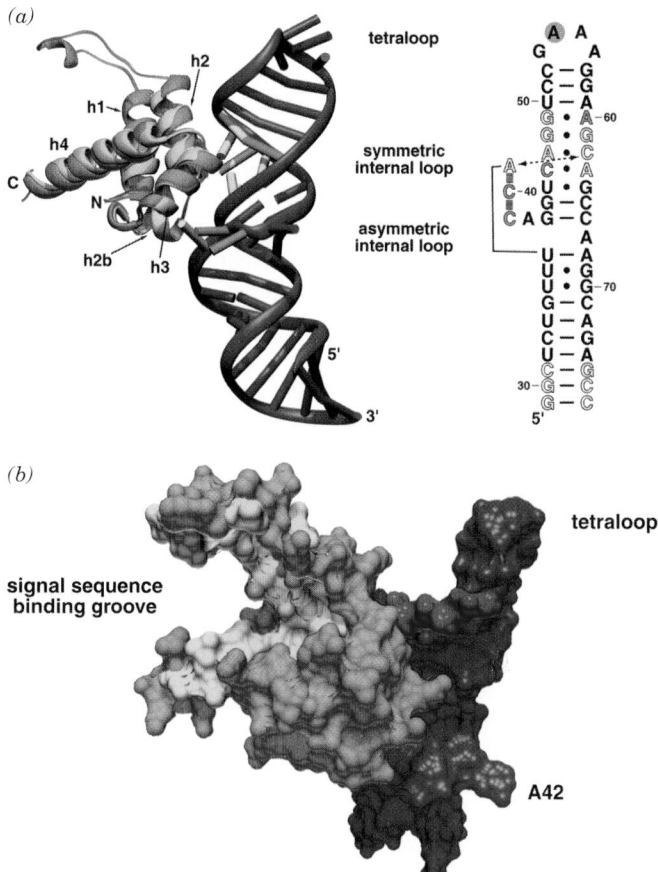

(a)

(b)

FIGURE 12-49   X-Ray structure the SRP core from *E. coli*.
(*a*) The RNA is represented in ladder form in purple, with its invariant bases yellow and its highly conserved bases green. The M domain is drawn as a cyan ribbon with the superimposed structure of the unbound M domain from *T. aquaticus* in pink. Successive α helices are named h1, h2, etc. The secondary structure of the RNA is drawn on the right so as to indicate its three-dimensional architecture, with the bases of its symmetric and asymmetric internal loops as well as those at the 3′ and 5′ ends of the RNA that are not native to *E. coli* 4.5S RNA printed in outline form. (*b*) The molecular surface of the complex oriented to show the putative signal peptide binding groove. The RNA is dark blue, the protein is pink, the hydrophobic residues lining the signal peptide binding groove are yellow, and the RNA's adjoining phosphate groups are red. [Courtesy of Robert Batey and Jennifer Doudna, Yale University. PDBid 1DUL.]

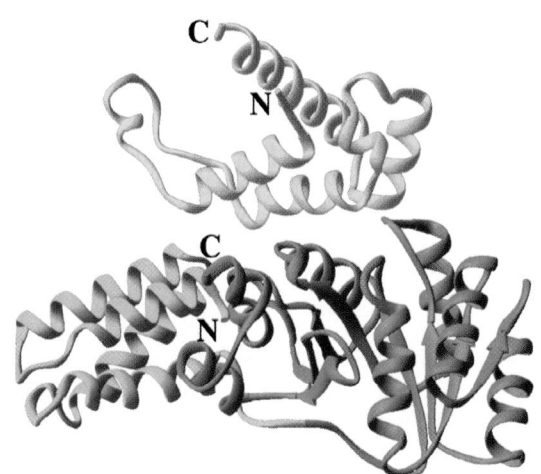

FIGURE 12-48   X-Ray structure of *T. aquaticus* Ffh. The protein is shown in ribbon form with its N domain cyan, its G domain green, and its M domain gold. The 11-residue segment connecting the G and M domains is disordered in the X-ray structure. Since Ffh crystallizes as a cyclic trimer and the disordered segment could be as long as ~40 Å, it is not possible to determine which of the three M domains in this trimer is linked to a given G domain. The M domain drawn here has therefore been arbitrarily selected from among those in the trimer. [Based on an X-ray structure determined by Peter Walter and Robert Stroud, University of California at San Francisco. PDBid 2FFH.]

in *T. aquaticus* Ffh that is lined almost entirely with hydrophobic residues including, in *E. coli*, 11 of its 14 Met residues (Fig. 12-49*b*; the Met side chain has physical properties similar to that of an *n*-butyl group). This 15-Å-wide by 25-Å-long groove apparently forms the binding site for the signal peptide's hydrophobic helix. Its flexible unbranched Met side chain "bristles" and finger loop presumably provide the groove with the plasticity to bind a variety of different signal sequences so long as they are hydrophobic and form an α helix. Indeed, in the X-ray structure of the human SRP54 M domain, the N-terminal helix, which is longer than that in the *E. coli* Ffh M domain, extends from the protein core to bind in the hydrophobic groove of an adjacent molecule.

What is the function of SRP RNA? The X-ray structure of the complex reveals that the RNA seamlessly continues the protein's hydrophobic binding groove onto a ledge formed by the RNA's sugar–phosphate backbone above the symmetric internal loop (Fig. 12-49*b*). This suggests that the helical signal peptide binds to the complex such that its N-terminal basic residues (Fig. 12-46) interact with the ledge's anionic phosphate groups.

c. Secretory Pathway Initiation Is Driven by GTP Hydrolysis

In eukaryotes, the SRP receptor is a heterodimer of subunits named **SRα** and **SRβ**. SRβ is a 271-residue integral protein that has an N-terminal TM segment, whereas SRα is a 638-residue peripheral protein that is apparently membrane-bound through the association of its N-terminal segment with SRβ. Both SRα and SRβ are GTPases.

In *E. coli*, the SR consists of a single 497-residue subunit named **FtsY,** whose C-terminal portion is homologous to that of SRα, although their N-terminal portions have no sequence similarity. Curiously, the X-ray structure of the C-terminal portion of FtsY closely resembles that of the N and G domains of SRP54 (Fig. 12-48), to which it is ~34% identical in sequence.

The targeting of the SRP–ribosome complex to the ER membrane is mediated by the GTPase functions of SRP54, SRα, and SRβ. In numerous biological systems, mainly those mediating translation (Section 32-3), vesicle transport (Sections 12-4C and 12-4D), and signal transduction (Section 19-2), *GTPases function as molecular switches that endow the system with unidirectionality and specificity.* These so-called **G proteins** have at least two stable conformations: GDP-bound and GTP-bound. Interconversion between these states only occurs in a unidirectional cycle due to the irreversibility of GTP hydrolysis. In most cases, a G protein must interact with other proteins in order to change conformational states. Thus, GTP hydrolysis often requires stimulation by a specific **GTPase activating protein (GAP),** and the exchange of bound GDP for GTP may require the action of a specific **guanine nucleotide exchange factor (GEF;** Section 19-2C). The need for these particular factors confers specificity on the system.

The GEF for the SRP is the complex of the newly emerged signal sequence with the M domain of SRP54, which induces the adjoining G domain to exchange its bound GDP for GTP (Fig. 12-45, Stage 2). The formation of the resulting SRP · GTP complex results in a conformational change that locks the SRP to the ribosome, which, in turn, induces translational arrest. The GEF for the SR appears to be an empty translocon, which thereby associates with the resulting SR · GTP complex to which the SRP · GTP–ribosome complex then binds (Fig. 12-45, Stage 3). Evidently, the SRP and SR, both in their GTP forms, act as "molecular matchmakers" to bring together an empty translocon with a ribosome synthesizing a signal sequence–bearing polypeptide. The SRP and the SR then reciprocally stimulate each other's GTPase functions (act as mutual GAPs;

neither protein alone has significant GTPase activity) followed by their dissociation, yielding free SRP · GDP and SR · GDP complexes ready to participate in a new round of the secretory pathway (Fig. 12-45, Stage 4). The release of the SRP and SR permits the now translocon-associated ribosome to recommence translation, thereby extruding the polypeptide it is synthesizing into or through the ER membrane as described below.

d. The Translocon Is a Multifunctional Transmembrane Pore

How are preproteins transported across or inserted into the RER membrane? In 1975, Blobel postulated that these processes are mediated by an aqueous TM channel. However, it was not until 1991 that he was able to show that these channels actually exist through electrophysiological measurements indicating that the RER membrane contains ion-conducting channels. These increase in number when the ribosome-bearing side of the RER is treated with **puromycin** (an antibiotic that causes the ribosome to prematurely release the growing polypeptide; Section 32-3D), thereby suggesting that the channels are usually plugged by the presence of the polypeptides. By linking fluorescent dyes whose fluorescence is sensitive to the polarity of their environment to a nascent polypeptide, Arthur Johnson demonstrated that these channels, now called translocons, enclose aqueous pores that completely span the ER membrane.

The various ER transmembrane proteins that comprise the translocon have been identified through the use of photoactivatable groups that were attached to signal sequences and mature regions of preproteins. On exposure to light of the proper wavelength, the photoactivatable groups react with nearby proteins to form covalent cross-links, thereby permitting the identification of these proteins. The major component of the translocon in mammals, dubbed **Sec61** (**SecYEG** in prokaryotes), is a heterotrimeric protein whose subunits are named **Sec61α, Sec61β,** and **Sec61γ.** Sec61α (476 residues) is predicted to have 10 TM helices and has both its N- and C-termini in the cytosol. Sec61β (96 residues) and Sec61γ (68 residues) are each predicted to have a single TM helix and each has its N-terminus in the cytosol.

Cryoelectron microscopy of yeast Sec61 in complex with a ribosome by Andrej Sali, Joachim Frank, and Blobel (Fig. 12-50) revealed that Sec61 forms a funnel-shaped toroid that has four attachments to the ribosome and whose central channel is aligned with the tunnel through which the nascent polypeptide exits the ribosome (Section 32-3D). The Sec61 complex has a thickness of 48 Å, an outer diameter of 85 to 95 Å, and an average inner diameter of 15 Å. The volume of the pore structure indicates that it consists of three Sec61 heterotrimers.

An additional component of the mammalian translocon is known as **translocating chain-associated membrane protein** (**TRAM,** 374 residues; predicted to have 8 TM helices and has both its N- and C-termini in the cytosol). Through the use of Sec61-containing liposomes that either did or did not also contain TRAM, Tom Rapoport demonstrated that TRAM is required for the translocation and membrane integration of most but not all preproteins into the

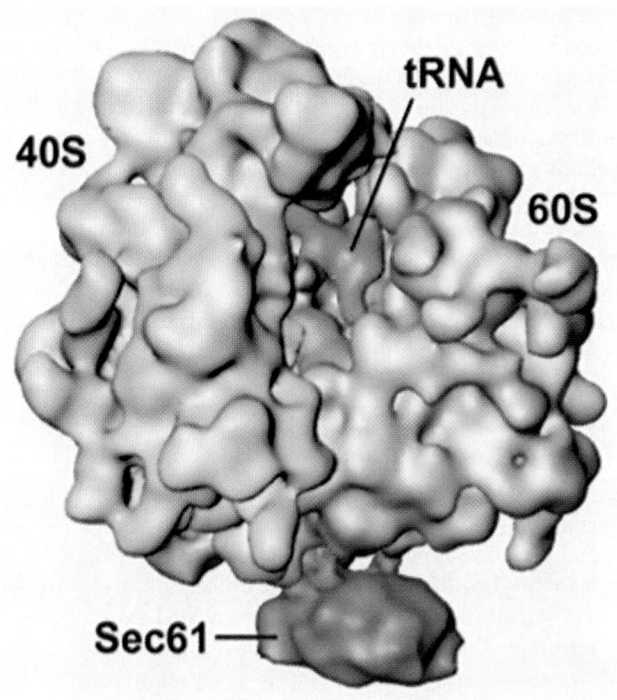

(a)

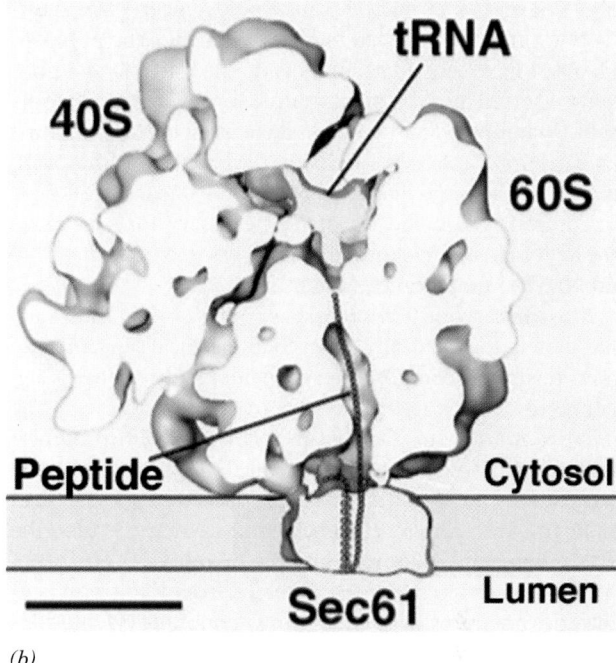

(b)

FIGURE 12-50   The structure of yeast Sec61 in its complex with a translating ribosome as determined by cryoelectron microscopy at 15 Å resolution. (*a*) View with the Sec61 oligomer in red, the ribosome's small (40S) subunit in yellow, its large (60S) subunit in blue, and a tRNA that is bound in the ribosomal P-site (Fig. 5-28) in green. (*b*) View as in Part *a* but cut away along a plane that passes through both the Sec61 pore and the tunnel through which the growing polypeptide exits the large ribosomal subunit (*black line*). The scale bar is 100 Å long. [Courtesy of Joachim Frank, State University of New York at Albany.]

liposome. Whether or not a given preprotein requires TRAM for translocation depends on its signal sequence, although no particular characteristic of this sequence appears to be critical for TRAM dependence.

How wide is the translocon pore when it is translocating a polypeptide? At a minimum, it would have to be ~7 Å across (the diameter of an extended anhydrous polypeptide), although if a TM sequence took up its α helical conformation while still in the pore, the pore would have to be at least ~12 Å wide. To determine the internal diameter of a functioning translocon pore, Johnson assembled translocation intermediates with fluorescent dyes incorporated at specific sites along the nascent polypeptide and determined their accessibility to **fluorescence quenchers** of different sizes (fluorescence quenchers are agents that, on colliding with an excited fluorophore, remove its excited state energy). These experiments indicate that a functioning translocon has a surprisingly large pore diameter of 40 to 60 Å. However, similar measurements indicate that an inactive (ribosome-free) translocon has a pore diameter of only 9 to 15 Å. Perhaps, the Sec61 pore seen in Fig. 12-50 adopts the ribosome-free conformation despite its associated ribosome.

Notwithstanding the enormous pore size of an active translocon, the ER membrane's permeability barrier is maintained because the ribosome forms a tight seal with the translocon (in contrast to the rather tenuous association seen in Fig. 12-50). Moreover, after the ribosome has dissociated from the translocon, the seal is preserved by

the binding of an ER-resident Hsp70 homolog (Section 9-2C) named **BiP** to the luminal face of the translocon. In fact, after the ribosome has bound to the translocon and translocation has commenced, BiP does not dissociate until the nascent polypeptide has reached a length of ~70 residues. This process, which is probably mediated by an interaction between the signal peptide and the translocon, presumably provides a safety mechanism that maintains the ER's permeability barrier until the ribosome can form a tight seal at the cytosolic end of the translocon.

e. The Translocon Inserts Transmembrane Helices into the ER Membrane

In addition to forming a conduit for soluble proteins to enter the ER, *the translocon must insert an integral protein's TM segments into the ER membrane.* The translocon, in concert with the ribosome, recognizes these TM segments and laterally installs them into the lipid bilayer via a largely unknown mechanism. The basis for this recognition must be more than just an ~20-residue stretch of nonpolar residues since predictions based only on sequence hydrophobicity do not always correctly identify a polypeptide's TM segments.

The signal sequences of many TM proteins are not cleaved by signal peptidase but, instead, are inserted into the ER membrane. Such so-called **signal-anchor sequences** *may be oriented with their N-termini either in the cytosol* *(e.g., Sec61α) or in the ER lumen (e.g., glycophorin A).* If the N-terminus is installed in the cytosol, then the polypeptide must have looped around inside the translocon before being inserted into the membrane. Moreover, for polytopic (multispanning) TM proteins such as Sec61α, this must have occurred for each successive TM helix. It is therefore likely that the large diameter of the active translocon pore is required to provide the space necessary for successive TM helices to reverse their direction prior to being inserted into the ER membrane.

Most eukaryotic TM proteins adopt an orientation such that their cytoplasmically exposed interhelical segments are more positively charged than their luminally exposed segments, but this is not always the case. Nevertheless, one might reasonably expect that in a polytopic TM protein it is the membrane orientation of the N-terminal TM helix that dictates the orientations of the succeeding TM helices (many of which have yet to be synthesized at the time the N-terminal helix is inserted into the membrane). However, the deletion or insertion of a TM helix from/into a polypeptide does not necessarily change the membrane orientations of the succeeding TM helices: When two successive TM helices have the same preferred orientation, one of them may be forced out of the membrane. Thus, the basis for the orientation of a TM protein in the ER membrane and the mechanism through which the translocon achieves this orientation are poorly understood, although photocrosslinking studies strongly implicate TRAM in this process.

f. Protein Folding in the ER Is Facilitated by Molecular Chaperones

The ER, as does the cytosol, contains a battery of molecular chaperones that assist in protein folding and act as agents of quality control. The best characterized of these is BiP, an Hsp70 homolog (Section 9-2C). BiP associates with many secretory and membrane proteins although, if folding proceeds normally, these interactions are weak and short-lived. However, proteins that are improperly folded, incorrectly glycosylated, or improperly assembled form stable complexes with BiP that are often exported to the cytosol via a poorly understood process involving the translocon called **retrotranslocation,** where they are proteolytically degraded (Section 32-6). Two other notable ER-resident chaperones are **calreticulin** and **calnexin,** homologous proteins that participate in facilitating and monitoring the folding and assembly of glycoproteins (Section 23-3B). The ER also contains protein disulfide isomerases (PDIs; Section 9-2A) and peptidyl prolyl cis–trans isomerases (PPIs; Section 9-2B).

Abnormalities of protein folding and assembly are emerging as important mechanisms of disease (e.g., Section 9-5). For instance, **cystic fibrosis** is the most common life-threatening recessive genetic disease in the Caucasian population (affecting one in ~2000 individuals). It occurs in homozygotes for a defective **cystic fibrosis transmembrane regulator (CFTR) protein,** a 1480-residue glycoprotein with 12 TM helices that functions as a Cl^- transporter in the plasma membrane of epithelial cells. Individuals with cystic fibrosis produce highly viscous mucus that, in its most damaging effects, blocks the small airways in the lungs. This leads to persistent infections, which cause severe progressive lung degeneration that is usually fatal by around age 30. Although cystic fibrosis is caused by any of a large number of mutations in the CFTR gene, 70% of the cases arise from the deletion of Phe 508 (ΔF508), which is positioned in a cytoplasmic loop of the CFTR protein (which initially occupies the ER lumen). The oligosaccharide chains of ΔF508 are improperly processed in the ER and consequently the defective loop folds incorrectly. Although ΔF508 retains almost full biological activity, it is nevertheless not released by calnexin in the ER, resulting in its retrotranslocation and degradation by a (in this case overly zealous) proteolytic surveillance system (Section 32-6B).

C. Vesicle Formation

Shortly after their polypeptide synthesis is completed, the partially processed transmembrane, secretory, and lysosomal proteins appear in the Golgi apparatus (Fig. 1-5), a 0.5- to 1.0-μm-diameter organelle consisting of a stack of 3 to 6 or more (depending on the species) flattened and functionally distinct membranous sacs known as **cisternae,** where further posttranslational processing, mainly glycosylation, occurs (Section 23-3B). The Golgi stack (Fig. 12-51) has two distinct faces, each comprised of a network of interconnected membranous tubules: the **cis Golgi network (CGN),** which is opposite the ER and is the port through which proteins enter the Golgi apparatus; and the **trans Golgi network (TGN),** through which processed proteins exit to their final destinations. The intervening Golgi stack contains at least three different types of sacs, the **cis, medial,** and **trans cisternae,** each of which contains different sets of glycoprotein processing enzymes.

Proteins transit from one end of the Golgi stack to the other while being modified in a stepwise manner, a process that is described in Section 23-3B. These proteins are transported via two mechanisms:

1. They are conveyed between successive Golgi compartments in the cis to trans direction as cargo within membranous vesicles that bud off of one compartment and fuse with a successive compartment, a process known as forward or **anterograde transport.**

2. They are carried as passengers in Golgi compartments that transit the Golgi stack, that is, the cis cisternae eventually become trans cisternae, a process called **cisternal progression** or **maturation.** This process is mediated through the backward or **retrograde transport** of Golgi-resident proteins from one compartment to the preceding one via membranous vesicles.

On reaching the trans Golgi network, the now mature proteins are sorted and sent to their final cellular destinations.

a. Membrane, Secretory, and Lysosomal Proteins Are Transported in Coated Vesicles

The vehicles in which proteins are transported between the RER, the Golgi apparatus, and their final destinations, as well as between the different compartments of the Golgi apparatus, are known as **coated vesicles** (Fig. 12-52). This is because these 60- to 150-nm-diameter membranous sacs

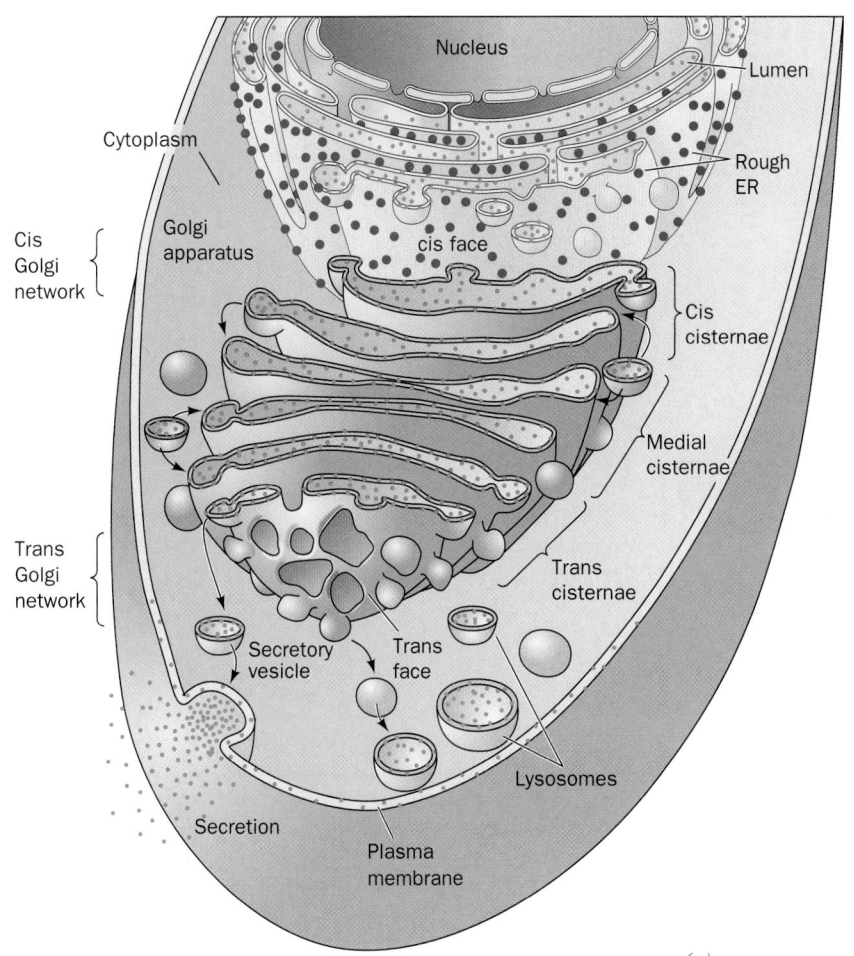

FIGURE 12-51 Posttranslational processing of proteins. Proteins destined for secretion, insertion into the plasma membrane, or transport to lysosomes are synthesized by RER-associated ribosomes (*blue dots*; *top*). As they are synthesized, the proteins (*red dots*) are either injected into the lumen of the ER or inserted into its membrane. After initial processing in the ER, the proteins are encapsulated in vesicles that bud off from the ER membrane and subsequently fuse with the cis Golgi network. The proteins are progressively processed in the cis, medial, and trans cisternae of the Golgi. Finally, in the trans Golgi network (*bottom*), the completed glycoproteins are sorted for delivery to their final destinations, the plasma membrane, **secretory vesicles,** or lysosomes, to which they are transported by yet other vesicles.

(a)

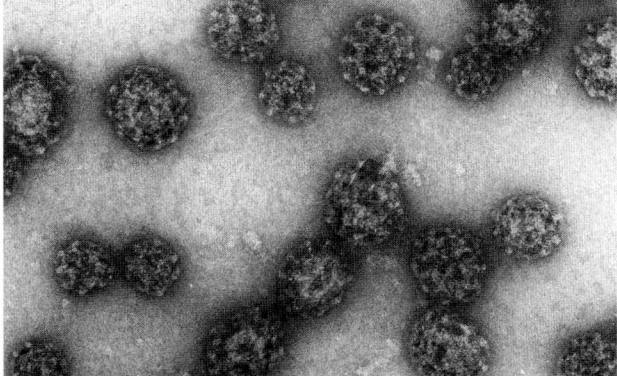

FIGURE 12-52 Electron micrographs of coated vesicles.
(*a*) Clathrin-coated vesicles. Note their polyhedral character. [Courtesy of Barbara Pearse, Medical Research Council, Cambridge, U.K.] (*b*) COPI-coated vesicles. (*c*) COPII-coated vesicles. The insets in Parts *b* and *c* show the respective vesicles at higher magnification. [Courtesy of Lelio Orci, University of Geneva, Switzerland.]

(b)

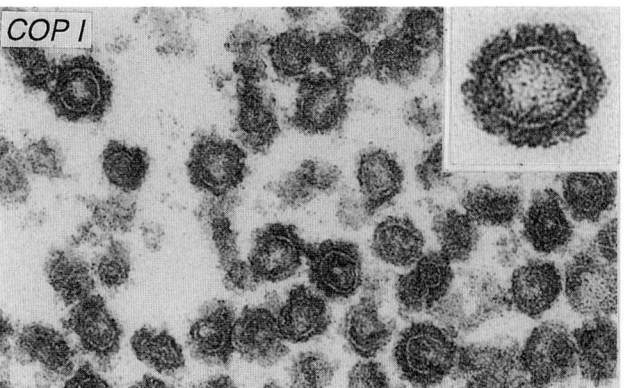

(c)

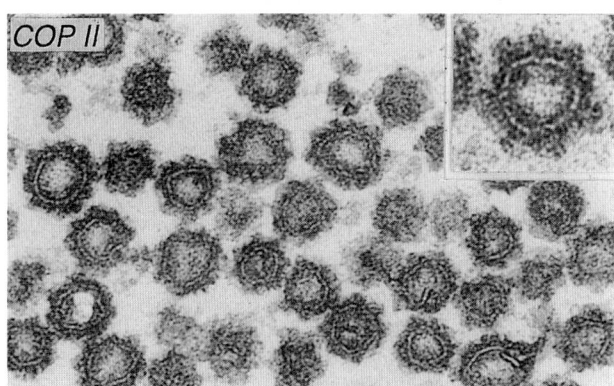

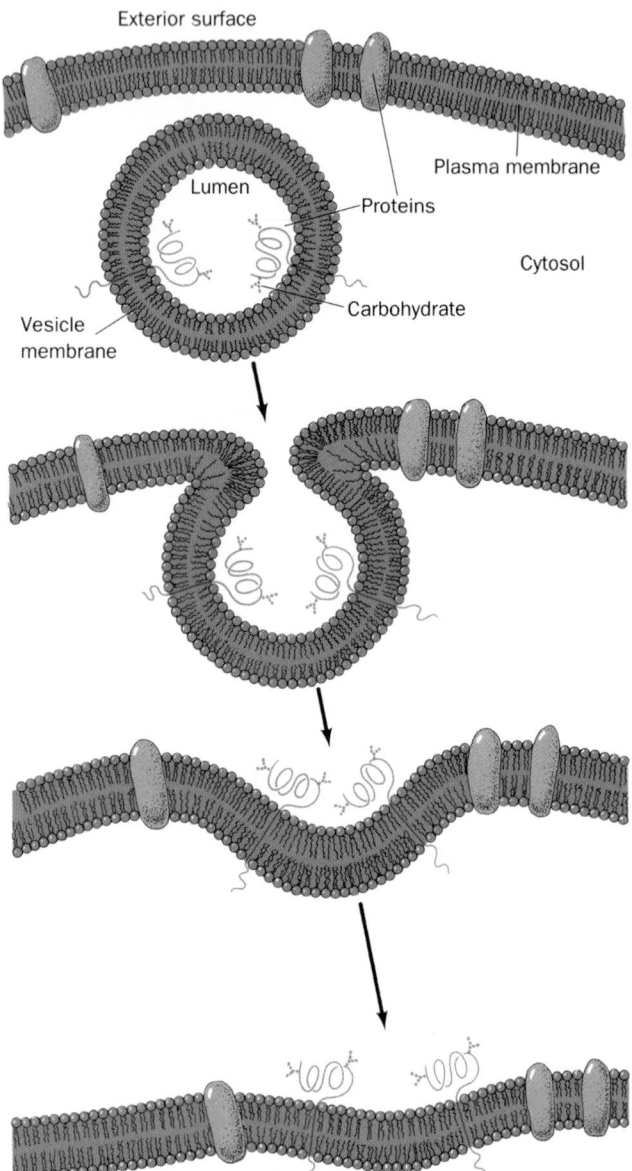

FIGURE 12-53   The fusion of a vesicle with the plasma membrane preserves the orientation of the integral proteins embedded in the vesicle bilayer. The inside of the vesicle and the exterior of the cell are topologically equivalent because the same side of the protein is always immersed in the cytosol. Note that any soluble proteins contained within the vesicle would be secreted. In fact, proteins destined for secretion are packaged in membranous secretory vesicles that subsequently fuse with the plasma membrane as shown.

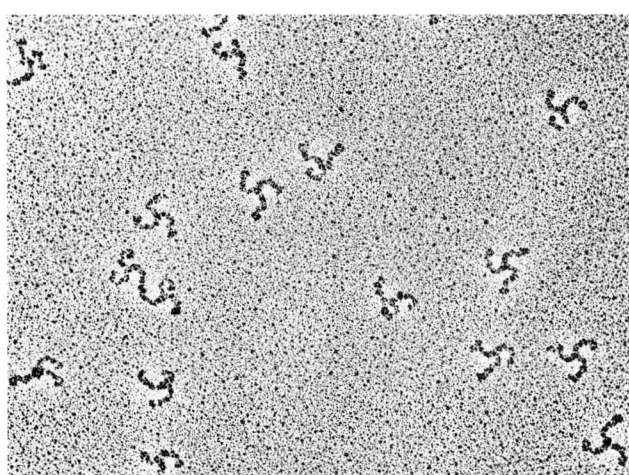

FIGURE 12-54   Electron micrograph of triskelions. The variable orientations of their legs are indicative of their flexibility. [Courtesy of Daniel Branton, Harvard University.]

cisternae are topologically equivalent to the outside of the cell. This explains why the carbohydrate moieties of integral glycoproteins and the GPI anchors of GPI-linked proteins occur only on the external surfaces of plasma membranes.

The three major types of coated vesicles that are known are characterized by their proteins coats. These are:

1. Clathrin (Fig. 12-52*a*), a protein that forms a polyhedral framework around vesicles that transport TM, GPI-linked, and secreted proteins from the Golgi to the plasma membrane. The clathrin cages can be dissociated to flexible three-legged proteins known as **triskelions** (Fig. 12-54) that consist of three so-called heavy chains (**HC,** 190 kD) that each bind one of two homologous light chains, **LCa** or **LCb** (24–27 kD), at random.

2. COPI protein (Fig. 12-52*b; COP* for *coat protein*), which forms a fuzzy rather than a polyhedral coating about vesicles that carry out both the anterograde and retrograde transport of proteins between successive Golgi compartments. In addition, COPI-coated vesicles return escaped ER-resident proteins from the Golgi to the ER (see below). COPI consists of 7 different subunits (α, 160 kD; β, 110 kD; β', 102 kD; γ, 98 kD; δ, 61 kD; ε, 31 kD; and ζ, 20 kD). The soluble complex comprising the COPI protomer is named **coatomer.**

3. COPII protein (Fig. 12-52*c*), which transports proteins from the ER to the Golgi. The COPII vesicle components are then recycled by COPI-coated vesicles for participation in another round of vesicle formation (the COPI vesicle components entering the ER are presumably recycled by COPII-coated vesicles). The COPII coat consists of two conserved protein heterodimers that in yeast are named **Sec23/24p** and **Sec13/31p.**

All of the above coated vesicles also carry receptors, which bind the proteins being transported, as well as **fusion pro-**

are initially encased on their outer (cytosolic) faces by specific proteins that act as flexible scaffolding in promoting vesicle formation. A vesicle buds off from its membrane of origin and later fuses to its target membrane. *This process preserves the orientation of the transmembrane protein (Fig. 12-53), so that the lumens of the ER and the Golgi*

teins, which mediate the fusion of these vesicles with their target membranes. We discuss these processes below and in Section 12-4D.

b. Clathrin Cages Are Formed by Overlapping Heavy Chains

Clathrin-coated vesicles **(CCVs)** are structurally better characterized than those coated with COPI or COPII. Clathrin forms polyhedral cages in which, as Barbara Pearse discovered, each vertex is the center (hub) of a triskelion, and its edges, which are ~150 Å long, are formed by the overlapping legs of four triskelions (Fig. 12-55*a*). Such polyhedra, which have 12 pentagonal faces and a variable number of hexagonal faces (for geometric reasons explained in Section 33-2A), are the most parsimonious way of enclosing spheroidal objects in polyhedral cages. The volume enclosed by a clathrin polyhedron, of course, increases with its number of hexagonal faces.

The triskelion's ~450-Å-long legs are each formed by the 1675-residue heavy chains (HCs), which trimerize via their C-terminal domains (Fig. 12-55*b*). Although the X-ray structure of an entire HC has not been determined, those of its N-terminal segment and a portion of its proximal leg have been elucidated:

1. The N-terminal segment (residues 1–494; Fig. 12-56*a, b*), whose structure was determined by Stephen Harrison and Tomas Kirchhausen, consists of two domains: (i) an N-terminal 7-bladed **β propeller** in which each structurally similar propeller blade consists of a 4-stranded an-

tiparallel β sheet (Fig. 12-56*b*); and (ii) a C-terminal linker that consists of 10 α helices of variable lengths (2–4 turns) connected by short loops and arranged in an irregular right-handed helix (a helix of helices, that is, a **superhelix**) named an **α zigzag.**

2. The proximal leg segment (residues 1210–1516; Fig. 12-56*c*), whose structure was determined by Peter Hwang and Robert Fletterick, consists of 24 linked α helices that are arranged similarly but more regularly than the above α zigzag to form a rod-shaped right-handed superhelix. The rigidity of this motif is attributed to its continuous hydrophobic core together with the efficient interdigitation of its side chains where its crossing antiparallel α helices come into contact (Section 8-3B).

The positions of these fragments in a clathrin HC is diagrammed in Fig. 12-56*d*.

Sequence and structural alignments indicate that HC residues 537 to 1566 consist of seven homologous ~145-residue **clathrin heavy chain repeats (CHCRs)** that are arranged in tandem and which each contain 10 helices (the above proximal leg segment consists of all of CHRC6 together with the C- and N-terminal portions of CHRC5 and CHRC7; Fig. 12-56*c*). Hence the entire HC leg appears to consist of an extended superhelix of linked α helices. Nevertheless, triskelion legs exhibit considerable flexibility (Fig. 12-54), a functional necessity for the formation of different sized vesicles as well as for the bud-

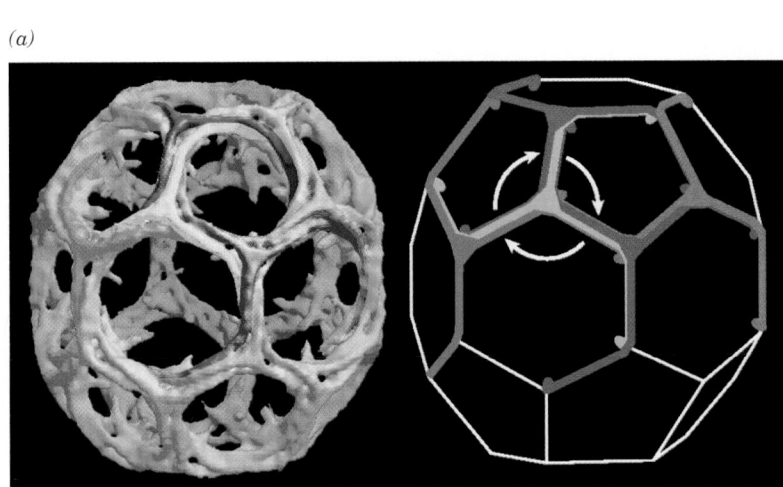

(a)

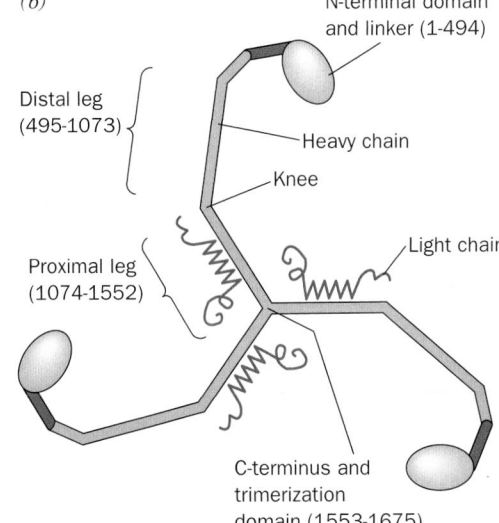

(b)

FIGURE 12-55 Anatomy of a clathrin-coated vesicle. (*a*) A cryoelectron microscopy–based image of a clathrin cage at 21 Å resolution with its triskelions differently colored. Its adaptor protein–containing core has been removed for clarity. As the accompanying diagram (*right*) indicates, a triskelion is centered on each of this polyhedral cage's 36 vertices, the cage edges are formed by the antiparallel legs of adjacent triskelions, and the linker and N-terminal domains project inward. Clathrin forms polyhedral cages with a large range of different sizes (number of hexagons): That shown here is only ~600 Å in diameter, whereas clathrin-coated membranous vesicles are typically ~1200 Å in diameter or larger. [Electron micrograph by Barbara Pearse and courtesy of H.T. McMahon, MRC Laboratory for Molecular Biology, Cambridge, U.K.] (*b*) Schematic diagram of a triskelion indicating its structural subdivisions.

(a)

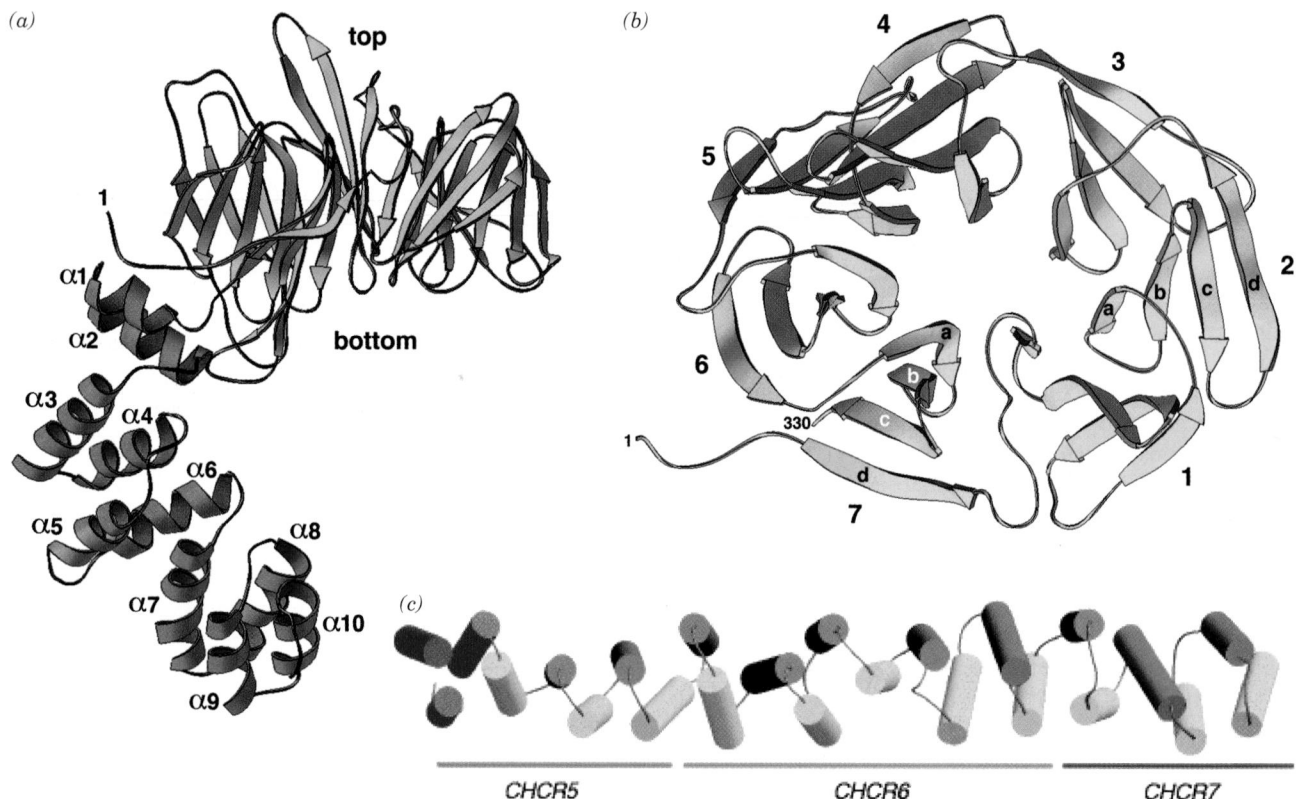

(b)

(c)

CHCR5   CHCR6   CHCR7

FIGURE 12-56   Structure of the clathrin heavy chain. (*a*) The
X-ray structure of the N-terminal domain and part of the linker
of rat HC. The N-terminal domain forms a seven-bladed β
propeller (*yellow*) that is seen here in side view, and the linker
(*red*) forms an α zigzag. (*b*) The β propeller as viewed from the
top along its pseudo-seven-fold axis. [Parts *a* and *b* courtesy of
Tomas Kirchhausen, Harvard Medical School. PDBid 1BPO.]
(*c*) The X-ray structure of bovine clathrin HC residues 1210 to
1516 as viewed with its N-terminus on the left. The helices are
alternately colored yellow and green with the exception of the
three N-terminal helices, which are colored gray to indicate that
they are poorly resolved. The red, green, and purple bars
denote the regions of CHCR5, CHCR6, and CHCR7,
respectively. [Courtesy of Peter Hwang, University of California
at San Francisco. PDBid 1B89.] (*d*) Schematic diagram of a
single clathrin heavy chain indicating the positions of its
N-terminal β propeller (*magenta*), the following α zigzag linker

(d)

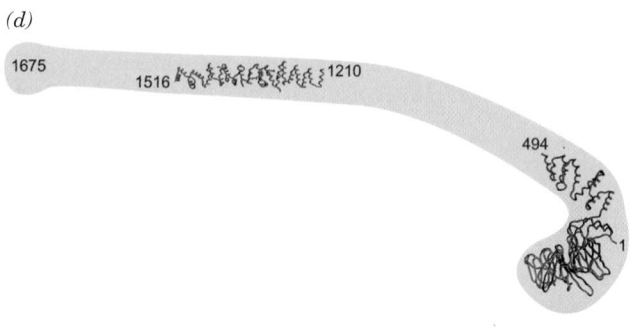

(*blue*), and the proximal leg segment (*cyan*). [Courtesy of
Barbara Pearse, MRC Laboratory of Molecular Biology,
Cambridge, U.K.]

ding of a vesicle from a membrane surface, which is accompanied by a large change in its curvature. The HC appears to flex mainly along a segment of the knee between
its proximal and distal legs that is free of contacts with
other molecules in clathrin cages.

The proximal leg segment bears extensive hydrophobic
surface patches that follow the grooves between adjacent
helices. This suggests that the lengthwise association of two
proximal legs in a clathrin cage (Fig. 12-55*a*) is stabilized
by the burial of these hydrophobic patches through the
complementary packing of the helices of one proximal leg
in the grooves on another.

Light chains (LCs) are not required for clathrin cage assembly. Indeed, LCs inhibit heavy chain polymerization *in*

vitro, which suggests that they have a regulatory role in preventing inappropriate clathrin cage assembly in the cytosol.
The X-ray structure of the HC proximal leg segment, which
encompasses the LC binding site, contains a prominent basic groove in which the highly acidic LCs are postulated to
bind. In fact, an 84% identical muscle homolog of clathrin
HC, in which three of the basic residues in this groove have
been replaced, fails to bind LCs. This suggests that LC binding interferes with the formation of salt bridges between
HCs that stabilize cage assembly. The segments of the 60%
identical LCa and LCb that differ in sequence are confined
to regions that do not participate in HC binding and hence
are likely to contain sites for the attachment of cytosolic
factors that regulate vesicle uncoating (see below).

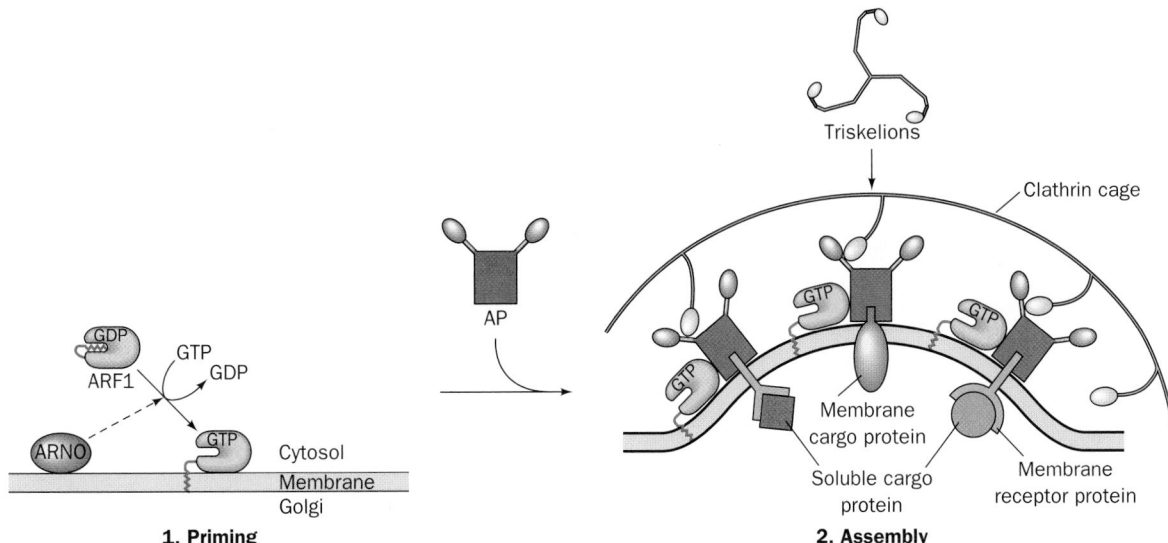

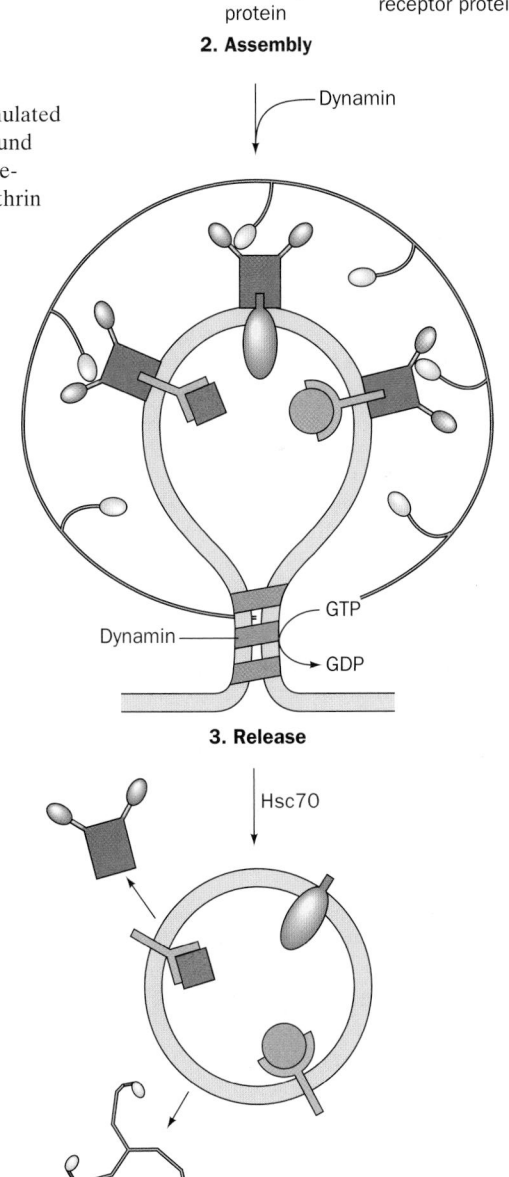

FIGURE 12-57 Formation of clathrin-coated vesicles. (1) The ARNO-stimulated exchange of ARF1's bound GDP for GTP frees ARF1 · GDP's protein-bound N-terminal myristoyl group for insertion into the membrane. **(2)** Membrane-bound ARF1 · GTP binds adapter proteins (APs). These, in turn, bind clathrin HC, thereby promoting the formation of a clathrin coat, which causes the vesicle to bud out from the membrane. In addition, APs bind the transmembrane receptors of cargo proteins as well as transmembrane cargo proteins. **(3)** The vesicle is released from the membrane through the action of the GTPase dynamin. **(4)** Shortly after the vesicle is released from the membrane, the clathrin coat and the APs dissociate from the vesicle.

c. Clathrin-Coated Vesicles Also Participate in Endocytosis

CCVs, as we have seen, transport TM and secretory proteins from the trans Golgi network (TGN) to the plasma membrane (Fig. 12-51). In addition, through a process known as **endocytosis** (discussed in Section 12-5B), they act to engulf specific proteins from the extracellular medium by the invagination of a portion of the plasma membrane and to transport them to intracellular destinations.

d. The Formation of CCVs Is a Complex Process

The formation of CCVs involves four stages (Fig. 12-57): (1) priming, (2) assembly, (3) release, and (4) uncoating. We outline these processes below.

1. Priming: The Activation of ARF1. Vesicle formation begins with the binding to the membrane of the myristoylated small (181-residue) GTPase named **ARF1** (ARF for **ADP-ribosylation factor,** because it was first described as a cofactor in the cholera toxin-catalyzed ADP-ribosylation of the GTPases known as heterotrimeric G proteins; Section 19-2). ARFs, which are members of the **Ras** superfamily (Ras is a small GTPase that participates in intracellular signaling; Section 19-3C), are

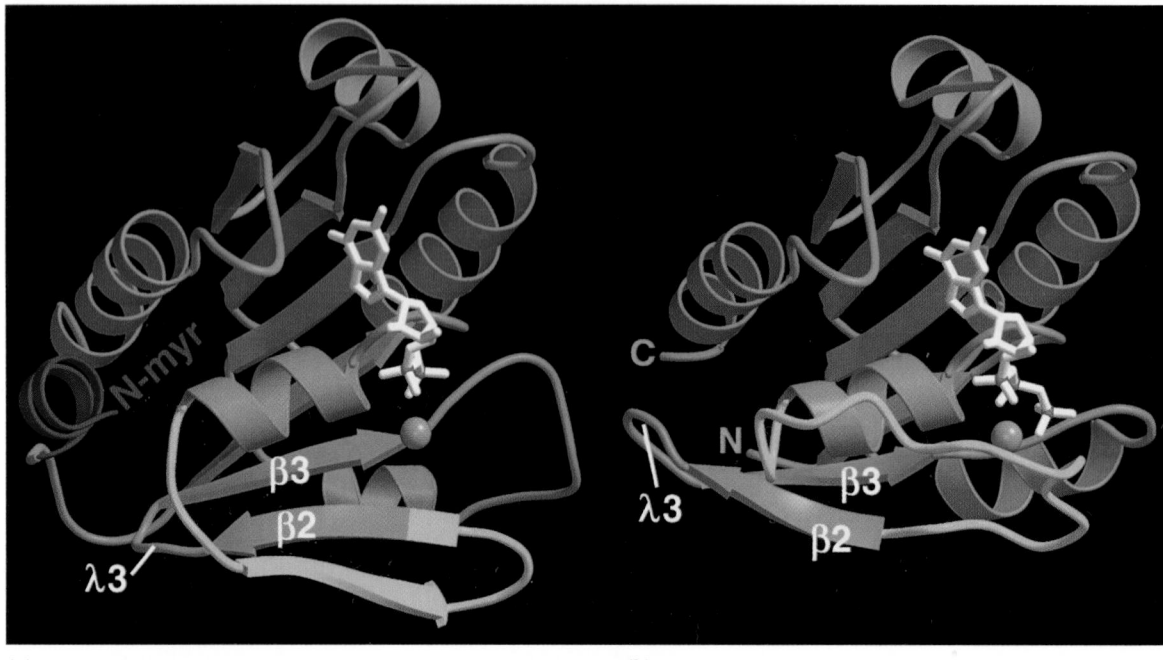

(a)   *(b)*

**FIGURE 12-58   X-Ray structures of (*a*) ARF1 · GDP and (*b*)
ARF1 · GDPNP.** (GDPNP is a nonhydrolyzable GTP analog in
which the O atom linking GTP's β- and γ-phosphorus atoms is
replaced by an NH group.) The bound nucleotides are drawn in
stick form in white with their phosphorus atoms magenta and
their bound Mg^{2+} ions shown as lavender spheres. In ARF1 ·
GDP, the protein's N-terminal helix (*red*) together with its
covalently linked myristoyl group (not present in the X-ray
structures) are bound in a shallow hydrophobic groove on the
surface of the protein formed in part by the residues of loop λ3.
However, the replacement of GDP by GDPNP (and

presumably GTP) induces a conformational change in residues
37 to 53 (*yellow*) that displaces strand β2 by two residues along
strand β3, a shift of 7 Å. The resulting movement of loop λ3
eliminates the binding site for the N-terminus, thereby making
the myristoyl group available for membrane insertion (residues
1–17 of the GDPNP complex are disordered). [Courtesy of
Jonathan Goldberg, Memorial Sloan-Kettering Cancer Center,
New York. The X-ray structure of ARF1 · GDP was
determined by Dagmar Ringe, Brandeis University. PDBid
1HUR.]

water-soluble cytosolic proteins when binding GDP, but
when binding GTP they associate with membranes through
the insertion of their N-terminal myristoyl groups into the
bilayer (Section 12-3B). The comparison of X-ray
structures of ARF1 · GDP and ARF1 · GTP, determined
by Dagmar Ringe and by Jonathan Goldberg, indicate that
this occurs because the N-terminal helix of ARF1 · GDP
together with its appended myristoyl group are bound in
a shallow groove in the protein (Fig. 12-58*a*) that is absent
in ARF1 · GTP (Fig. 12-58*b*).

The guanine nucleotide exchange factor (GEF) for
ARF1, which in humans is called **ARNO** (for **ARF
nucleotide-binding site opener;** 399 residues), contains an
~200-residue domain similar to the highly conserved yeast
protein **Sec7.** When ARNO or its isolated Sec7 domain is
incubated with myristoylated ARF1 · GDP, it fails to
catalyze nucleotide exchange unless lipid micelles are also
present, thereby suggesting that ARNO is activated only
when localized to a membrane surface. Indeed, ARNO
contains a **pleckstrin homology (PH) domain,** an ~100-
residue module occurring in numerous proteins (Section
19-3C) that binds the minor membrane phospholipid
phosphatidylinositol-4,5-bisphosphate (PIP₂),

Phosphatidylinositol-4,5-bisphosphate (PIP₂)

which is also a precursor of compounds that participate in
intracellular signaling (Section 19-4A).

**2. Assembly: Adaptor Proteins Link Cargo Proteins to
the Clathrin Coat.** Membrane-bound ARF1 · GTP acts to
recruit **adapter proteins (APs)** to the membrane surface.
APs bind clathrin HC together with TM proteins that are
either receptors that selectively bind soluble cargo proteins

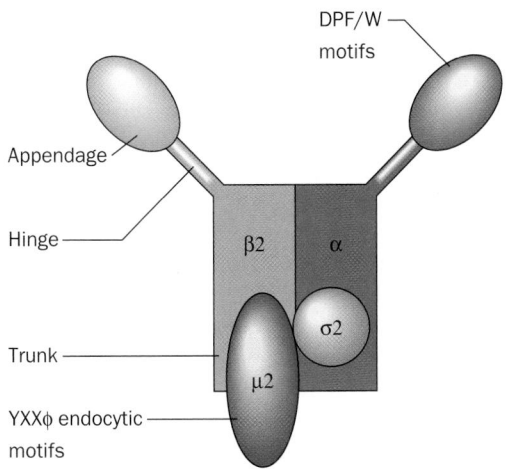

FIGURE 12-59   Schematic diagram of the AP2 heterotetramer. AP1 has a similar structure. [After Pearse, B.M., Smith, C.J., and Owen, D.J., *Curr. Opin. Struct. Biol.* **10**, 223 (2000).]

inside the vesicle or are cargo proteins themselves. APs comprise the central cores of CCVs and, in fact, are the scaffolding on which clathrin cages form. The APs bind clathrin via its N-terminal β propeller domain (Fig. 12-56*a*), which forms the knobs that project inward from clathrin cages (Fig. 12-55*a*). The grooves between the propeller blades on the top face of the β propeller (Fig. 12-56*b*) probably form the AP binding sites.

AP1 is the main AP contained in the coated vesicles originating from the TGN, whereas the homologous **AP2** predominates in endocytotic vesicles. Both APs are heterotetramers: AP1 consists of the subunits γ, β1 (~110 kD each), μ1 (~50 kD), and σ1 (~17 kD), whereas the corresponding subunits of the better characterized AP2 are α, β2, μ2, and σ2 (Fig. 12-59). Electron microscopy and X-ray studies indicate that the large subunits each consist of a trunk and an appendage domain joined by a flexible and proteolytically sensitive hinge region (Fig. 12-59). The AP2 hinge region of β2 binds to the clathrin β propeller, whereas the cytoplasmic domains of target proteins bind most commonly to μ2 via YXXϕ sequences (where ϕ is a bulky hydrophobic residue), but in some cases to the β2 trunk via D/EXXXLL sequences. This explains why the proteolytic excision of AP2's appendage domain prevents the assembly of clathrin coats, although the remaining AP2 trunk can still bind to membranes that contain proteins bearing a YXXϕ internalization signal. Additional APs, including **AP3, AP4,** and **AP180,** have been identified but are not well characterized.

3. Release: Vesicle Scission Is Mediated by Dynamin. The budding of a CCV from its parent membrane appears to be mechanically driven by the formation of the clathrin cage. However, the actual scission of the coated bud from its parent membrane to form a coated vesicle requires the participation of **dynamin**, an ~870-residue GTPase. Dynamin contains a PIP₂-binding PH domain, which re-

cruits dynamin to the membrane. On binding GTP, dynamin forms a helical oligomer that wraps about the base of the budding vesicle so as to squeeze this region down to a thin tube (Fig. 12-60). The oligomerization together with the presence of PIP₂ stimulates dynamin to hydrolyze its bound GTP (dynamin also contains a GAP domain), causing the helical oligomer to lengthen its pitch. However, the way in which this process releases the vesicle from the membrane is poorly understood.

4. Uncoating: The Recycling of Clathrin and Adapter Proteins. Shortly after the formation of a CCV, the clathrin is released as triskelions, thereby recycling them for participation in the formation of additional coated vesicles [note that a slight clockwise twist of a triskelion will detach it from its clathrin cage (e.g., arrows about the yellow triskelion in Fig. 12-55*a*)]. This process is mediated by the ATPase **Hsc70**, a homolog of the chaperone Hsp70 (Section 9-2C) present in all eukaryotic cells, which, on ATP hydroysis, forms a complex with triskelions that may then facilitate the assembly of new CCVs. Following clathrin release from newly formed vesicles, the APs are also released, although the factors that mediate this process, if any, are unknown. It would seem likely, although it has not been shown to be the case, that this vesicle uncoating process is initiated by the hydrolysis of ARF1's bound GTP to GDP, which would release ARF1 from the membrane and, presumably, from binding an AP. In any case, the coating and uncoating of vesicles by clathrin must be closely regulated processes since both occur simultaneously.

A variety of regulatory and accessory proteins of largely unknown function have also been implicated in CCV for-

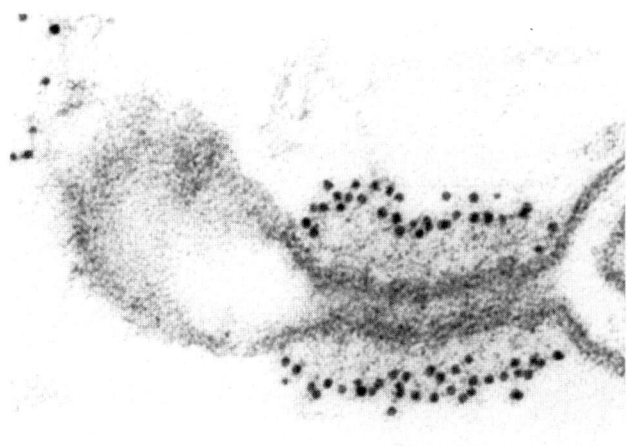

FIGURE 12-60   Electron micrograph of a budding coated vesicle. The vesicle was incubated with the nonhydrolyzable GTP analog **GTPγS** (in which a terminal O atom on the γ-phosphorus atom of GTP is replaced by S) and then treated with gold-tagged anti-dynamin antibodies (*black dots*). Note that the dynamin surrounds a long narrow tube at the base of the budding vesicle that has not pinched off from the membrane. [Courtesy of Pietro De Camilli, Yale University School of Medicine.]

mation. Moreover, many of the proteins described above are each present in several isoforms. Hence it is clear that our understanding of this process is far from complete.

e. The Assembly of COPI- and COPII-Coated Vesicles Resembles That of Clathrin-Coated Vesicles

COPI- and COPII-coated vesicles are both assembled in processes, elucidated in large part by Randy Schekman, that resemble CCV assembly:

1. Priming: COPI-coated vesicles are primed identically to CCVs: ARF1 is recruited to the membrane by the ARNO-promoted exchange of its bound GDP for GTP (Fig. 12-57, Step 1). COPII-coated vesicle assembly is similarly primed but by different proteins: **Sar1p** is the small GTPase that carries out this process, and the exchange of its GDP for GTP is mediated by the transmembrane GEF **Sec12p.**

2. Assembly: ARF1 · GTP stoichiometrically recruits intact coatomers to form COPI-coated vesicles. Most of the 7 COPI coatomer subunits have homologs in the clathrin system and function accordingly: The β-, γ-, δ-, and ζ-COPs correspond to the β, α, μ, and σ subunits of AP2, respectively (Fig. 12-59), and the α- and ε-COPs correspond to the clathrin heavy and light chains. In COPII coat formation, Sar1p · GTP recruits Sec23/24p, which in turn recruits Sec13/31p.

3. Release: Both COPI- and COPII-coated vesicles spontaneously bud off from their parent membranes; these processes appear to have no requirement for an analog of dynamin as does CCV release.

4. Uncoating: As is the case for CCVs, COPI- and COPII-coated vesicles uncoat shortly after being released from their parent membranes. These processes appear to be initiated by the hydrolysis of the GTPs bound to ARF1 and Sar1p, which thereby weaken the attachment of COPI and COPII to their respective vesicles. The GTPase activating protein (GAP) for COPI vesicles, a 415-residue protein named **ARF GAP,** appears to be a component of the COPI coat. In COPII vesicles, Sec23p has been identified as the GAP for Sar1p.

f. Proteins Are Directed to the Lysosome by Carbohydrate Recognition Markers

How are proteins in the ER selected for transport to the Golgi apparatus and from there to their respective membranous destinations? A clue as to the nature of this process is provided by the human hereditary defect known as **I-cell disease** (alternatively, **mucolipidosis II**) which, in homozygotes, is characterized by severe progressive psychomotor retardation, skeletal deformities, and death by age 10. The lysosomes in the connective tissue of I-cell disease victims contain large inclusions (after which the disease is named) of glycosaminoglycans and glycolipids as a result of the absence of several lysosomal hydrolases. These enzymes are synthesized on the RER with their correct amino acid sequences but, rather than being dis-

patched to the lysosomes, are secreted into the extracellular medium. This misdirection results from the absence of a mannose-6-phosphate recognition marker on the carbohydrate moieties of these hydrolases because an enzyme required for mannose phosphorylation fails to recognize the lysosomal proteins. The mannose-6-phosphate residues are normally bound by a receptor in the coated vesicles that transport lysosomal hydrolases from the Golgi apparatus to the lysosomes (Section 23-3B). No doubt, other glycoproteins are directed to their intracellular destinations by similar carbohydrate markers.

g. ER-Resident Proteins Have the C-Terminal Sequence KDEL

Most soluble ER-resident proteins in mammals have the C-terminal sequences KDEL (HDEL in yeast), KKXX, or KXKXXX (where X represents any amino acid residue), whose alteration results in the secretion of the resulting protein. By what means are these proteins selectively retained in the ER? Since many ER-resident proteins freely diffuse within the ER, it seems unlikely that they are immobilized by membrane-bound receptors within the ER. Rather, it has been shown that ER-resident proteins, as do secretory and lysosomal proteins, readily leave the ER via COPII-coated vesicles but that ER-resident proteins are promptly retrieved from the Golgi and returned to the ER in COPI-coated vesicles. Indeed, coatomer binds the Lys residues in the C-terminal KKXX motif of transmembrane proteins, which presumably permits it to gather these proteins into COPI-coated vesicles. Furthermore, genetically appending KDEL to the lysosomal protease **cathepsin D** causes it to accumulate in the ER, but it nevertheless acquires an *N*-acetylglucosaminyl-1-phosphate group, a modification that is made in an early Golgi compartment. Presumably, a membrane-bound receptor in a post-ER compartment binds the KDEL signal and the resulting complex is returned to the ER in a COPI-coated vesicle. **KDEL receptors** have, in fact, been identified in yeast and humans. However, the observation that former KDEL proteins whose KDEL sequences have been deleted are, nevertheless, secreted relatively slowly suggests that there are mechanisms for retaining these proteins in the ER by actively withholding them from the bulk flow of proteins through the secretory pathway.

D. *Vesicle Fusion*

Vesicles that travel only short distances (<1 μm) between their parent and target membranes (e.g., between neighboring Golgi cisternae) do so via simple diffusion, a process that typically takes from one to several minutes. However, vesicles that have longer distances to commute (e.g., from the TGN to the plasma membrane) are actively transported along cytoskeletal microtubules (Section 1-2A) by the motor proteins **dynein** and **kinesin,** which unidirectionally crawl along microtubule "tracks" in an ATP-driven process (Section 35-3F).

a. Vesicle Fusion Is Most Easily Studied in Yeast and in Synapses

On arriving at its target membrane, a vesicle fuses with it, thereby releasing its contents on the opposite side of the target membrane (Fig. 12-53). How do vesicles fuse and why do they fuse only with their target membranes and not other membranes? Progress in answering these questions has been made mainly by using two experimental approaches, the genetic dissection of this process in yeast and its biochemical analysis in **synapses,** the junctions between neurons (nerve cells) and between neurons and muscles (Fig. 12-61).

When a nerve impulse in the presynaptic cell reaches a synapse, it triggers the fusion of **neurotransmitter-** containing **synaptic vesicle** with the **presynaptic membrane** (a specialized section of the neuron's plasma membrane), thereby releasing the neurotransmitter (a small molecule) into the ~200-Å-wide **synaptic cleft** (the process whereby membranous vesicles fuse with the plasma membrane to release their contents outside the cell is called **exocytosis**). The neurotransmitter rapidly diffuses across the synaptic cleft to the postsynaptic membrane, where it binds to specific receptors that thereupon trigger the continuation of the nerve impulse in the postsynaptic cell (Section 20-5C). The homogenization of nerve tissue causes its presynaptic endings to pinch off and reseal to form **synaptosomes,** which can be readily isolated by density gradient ultracentrifugation for subsequent study.

b. Vesicle Fusion Requires the Coordinated Actions of Many Proteins

Biological membranes do not spontaneously fuse. Indeed, being negatively charged, they strongly repel one another at short distances. These repulsive forces must be overcome if biological membranes are to fuse. As we shall see below, we are just beginning to understand how this complicated process occurs.

Studies of the mechanism of vesicle fusion were pioneered by Rothman, who demonstrated that the fusion process is blocked by low concentrations of the cysteine-alkylating agent **N-ethylmaleimide (NEM),**

(a)

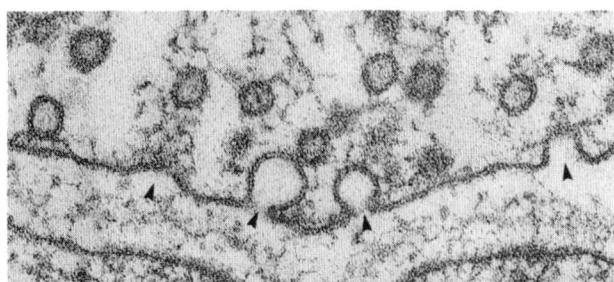

(b)

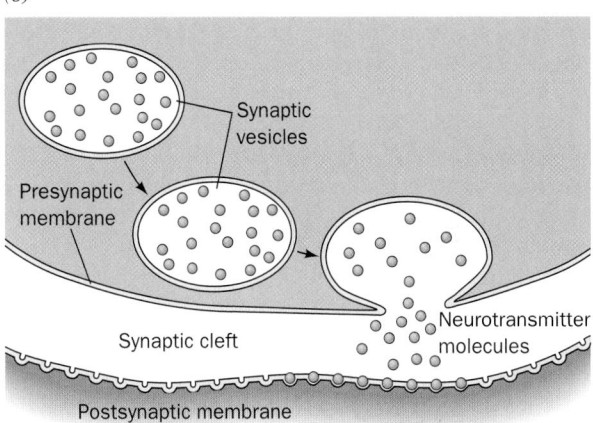

FIGURE 12-61 Transmission of nerve impulses across a synaptic cleft. (*a*) Electron micrograph of a frog neuromuscular junction in which the synaptic vesicles are undergoing exocytosis (*arrows*) with the presynaptic membrane (*top*). [Courtesy of John Heuser, Washington University School of Medicine, St. Louis, Missouri.] (*b*) The neurotransmitter, which is thereby discharged into the synaptic cleft, rapidly (in <0.1 ms) diffuses to the postsynaptic membrane, where it binds to transmembrane receptors, triggering a new nerve impulse.

indicating the presence of an **NEM-sensitive fusion (NSF) protein.** NSF is a cytosolic ATPase that does not bind to membranes unless a **soluble NSF attachment protein (SNAP)** is also present. SNAPs bind to membranes in the absence of NSF, demonstrating that SNAPs bind before NSF. SNAPs bind to alkali-extracted membranes, which indicates that **SNAP receptors (SNAREs)** are integral or lipid-linked proteins.

Three classes of proteins appear to participate in all vesicle fusion reactions:

1. Rab proteins, which are small GTPases of the Ras superfamily that, when binding GTP, are anchored to the vesicle membrane by two geranylgeranyl groups (Section 12-3B). *Rab · GTP functions to recognize the vesicle's target membrane and to form a relatively loose association with it,* a process known as "tethering," which is poorly understood. Cells express numerous Rab isoforms, 11 in yeast and 63 in humans, each localized to a specific membrane compartment. On vesicle fusion, Rab hydrolyzes its bound GTP, the resulting Rab · GDP is extracted from the membrane and transferred to a new vesicle, and its GDP is replaced by GTP, thereby recycling the system. There are indications that Rab proteins also mediate the vesicle interactions with the cytoskeleton that function in transporting vesicles to their proper destinations.

2. SNAREs, which form cognate combinations of membrane-associated proteins known as **R-SNAREs** and **Q-SNAREs** (because they contain conserved Arg and Gln residues in their cytoplasmic domains; they were originally named **v-SNAREs** and **t-SNAREs,** respectively, because they are mainly associated with the vesicle and target membranes). The best characterized SNAREs are those functioning at neuronal synapses: **Synaptobrevin** (alteratively, **VAMP** for *vesicle associated membrane protein*) is an R-SNARE, whereas **syntaxin** and **SNAP-25** (for *synaptosome associated protein of 25 kD*) are Q-SNAREs. *R-SNAREs and Q-SNAREs associate to firmly anchor the vesicle to its target membrane,* a process called "docking." The docked complexes, which are described below, are eventually disassembled by NSF in association with a SNAP protein. (Note that SNAP-25 is not a SNAP protein; by curious coincidence, the two independently characterized proteins were assigned the same acronym before it was realized that they are functionally associated.)

3. The **SM proteins** (so called because they are named **Sec1** in yeast and **Munc18** in mammals), which in synapses bind to syntaxin so as to prevent synaptobrevin and SNAP-25 from binding to it. Mutational studies indicate that these 65- to 70-kD hydrophilic proteins are essential for vesicle fusion.

c. SNAREs Form a Stable 4-Helix Bundle

The R-SNARE synaptobrevin and the Q-SNAREs syntaxin and SNAP-25 form a highly stable complex; boiling SDS solution is required to dissociate it. Synaptobrevin and syntaxin each have a C-terminal TM helix, and SNAP-25 is anchored to the membrane via palmitoyl groups that are linked to Cys residues in its central region. The X-ray structure of the associating portions of this complex (Fig. 12-62*a*), determined by Reinhard Jahn and Axel Brünger, reveals it to be a bundle of four parallel ~65-residue α helices with two of the helices formed by the N- and C-terminal segments of SNAP-25. Since synaptobrevin is anchored in the vesicle membrane and syntaxin and SNAP-25 are anchored in the target membrane, this so-called core complex firmly ties together the two membranes (Fig. 12-62*b*).

The four helices of the core complex wrap around each other with a gentle left-handed twist. For the most part, the sequence of each helix has the expected 7-residue repeat, $(a\text{-}b\text{-}c\text{-}d\text{-}e\text{-}f\text{-}g)_n$, with residues *a* and *d* hydrophobic (note that this property is characteristic of 4- and 3-helix bundles as well as of coiled coils). However, the central layer of side chains along the length of the 4-helix bundle consists of an Arg residue from synaptobrevin that is hydrogen bonded to three Gln side chains, one from syntaxin and one from each of the SNAP-25 helices. These highly

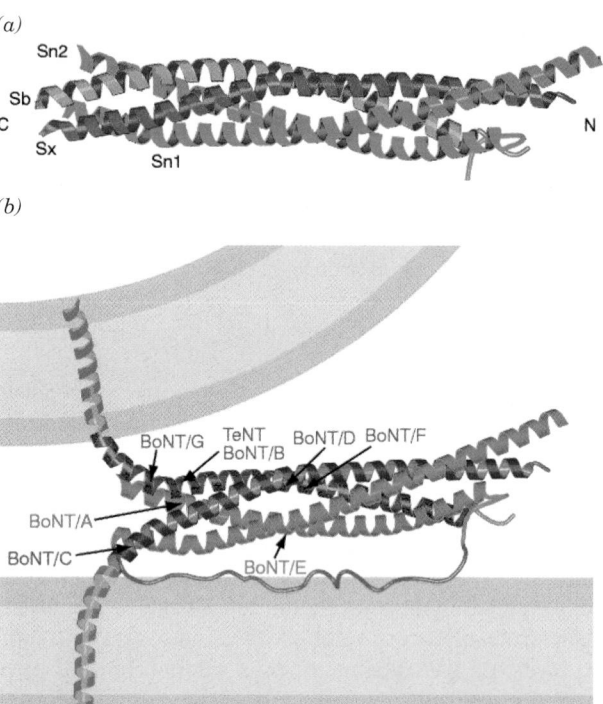

FIGURE 12-62   X-Ray structure of the syntaxin–synaptobrevin–SNAP-25 core complex. (*a*) Ribbon diagram showing the syntaxin helix (Sx) in red, the synaptobrevin helix (Sb) in blue, and the N- and C-terminal helices of SNAP-25 (Sn1 and Sn2) in green. (*b*) Model of the synaptic fusion complex linking two membranes (*gray*). The helices of the core complex are colored as in Part *a*. The transmembrane C-terminal extensions of syntaxin and synaptobrevin are modeled as helices (*yellow-green*). The loop connecting the N- and C-terminal helices of SNAP-25 is speculatively represented as an unstructured loop (*brown*). Recall that this loop is anchored to the membrane via Cys-linked palmitoyl groups (*not shown*). The cleavage sites for the various clostridial neurotoxins are indicated by the arrows. [Courtesy of Axel Brünger, Yale University. PDBid 1SFC.]

conserved polar residues are sealed off from the aqueous environment such that their interactions are enhanced by the low dielectric constant of their environment. It therefore appears that these interactions serve to bring the four helices into proper register.

Since mammalian cells contain 30 different R-SNAREs and Q-SNAREs, it would seem likely that their interactions are at least partially responsible for the specificity that vesicles exhibit in fusing with their target membranes. Indeed, Rothman has shown this to be the case by determining, *in vitro,* the rate of fusion of liposomes bearing different SNAREs. In testing all the R-SNAREs in the yeast genome against Q-SNAREs known to be localized to the yeast Golgi, vacuole, and plasma membranes, he found that liposome fusion only occurs when the combinations of R- and Q-SNAREs correspond to those mediating membrane flow *in vivo.* Nevertheless, it seems likely that the *in vivo* specificity of vesicle fusion is augmented by other mechanisms such as the localization of cognate R-

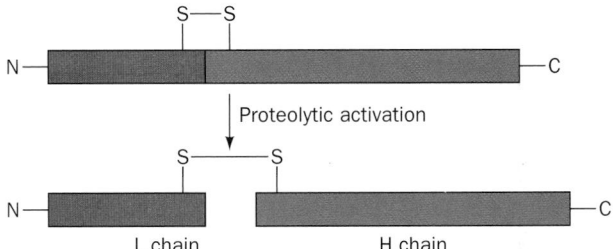

FIGURE 12-63   Model of clostridial neurotoxins and their activation by host proteases. The disulfide bond linking the L and H segments is cleaved after the neurotoxin is taken up by its target neuron.

and Q-SNAREs to particular regions in the cell and by the actions of regulatory proteins including, as is indicated above and discussed below, Rab proteins.

d. Tetanus and Botulinus Toxins Specifically Cleave SNAREs

The frequently fatal infectious diseases **tetanus** (which arises from wound contamination) and **botulism** (a type of food poisoning) are caused by certain anaerobic bacteria of the genus *Clostridium*. These bacteria produce extremely potent protein neurotoxins that inhibit the release of neurotransmitters into synapses. In fact, botulinal toxins are the most powerful known toxins; they are ~10 millionfold more toxic than cyanide (10^{-10} g · kg^{-1} will kill a mouse).

There are 7 serologically distinct types of botulinal neurotoxins, designated **BoNT/A** through **BoNT/G,** and one type of tetanus neurotoxin, **TeTx.** Each of these homologous proteins is synthesized as a single ~150-kD polypeptide chain that is cleaved by host proteases to yield an ~50-kD L chain that remains disulfide-linked to the ~100-kD H chain (Fig. 12-63). The H chains bind to specific types of neurons (via gangliosides and protein receptors), where they facilitate the uptake of the L chain by endocytosis. *The L chains are proteases, and each cleaves its target SNARE at a specific site* (Fig. 12-62b). This pre-

vents the formation of the core complex and thereby halts the exocytosis of synaptic vesicles. The H chain of TeTx specifically binds to inhibitory neurons (which function to moderate excitory nerve impulses) and is thereby responsible for the spastic paralysis characteristic of tetanus. The H chains of the BoNTs instead bind to motor neurons (which innervate muscles) and thus cause the flaccid paralysis characteristic of botulism.

The administration of carefully controlled quantities of botulinal toxin (trade name Botox) is medically useful in relieving the symptoms of certain types of chronic muscle spasms. Moreover, this toxin is being used cosmetically: Its injection into the skin relaxes the small muscles causing wrinkles and hence these wrinkles disappear for ~3 months.

e. Bilayer Fusion May Be Catalyzed by Specific Proteins

The association of Q-SNAREs on a vesicle with an R-SNARE on its target membrane brings the two bilayers into close proximity, yielding a so-called **trans-SNARE complex.** But what induces the fusion of the juxtaposed bilayers? One possibility is that the mechanical stresses arising from the formation of a ring of several trans-SNARE complexes cause lipid molecules inside the ring to leave their bilayer to form a hypothetical transient structure with lipids from the opposing bilayer, leading to bilayer fusion. Indeed, as we discussed above, liposomes containing the corresponding Q- and R-SNAREs spontaneously fuse. However, this *in vitro* process takes 30 to 40 minutes whereas, for example, the *in vivo* fusion of a synaptic vesicle with the presynaptic membrane takes <0.3 ms (Section 20-5C). This suggests that other proteins may be involved in catalyzing the actual bilayer fusion process. In fact, the fusion of yeast vacuoles, which is preceded by trans-SNARE complex formation, appears to be catalyzed by a multisubunit transmembrane pore-forming protein known as **V0.** Trans-SNARE complex formation apparently brings V0 molecules from the two membranes into apposition to form an aqueous pore between the two vesicles (Fig. 12-64a) in a manner reminiscent of the formation of a gap junction between adjacent cells (Fig. 12-40). It is

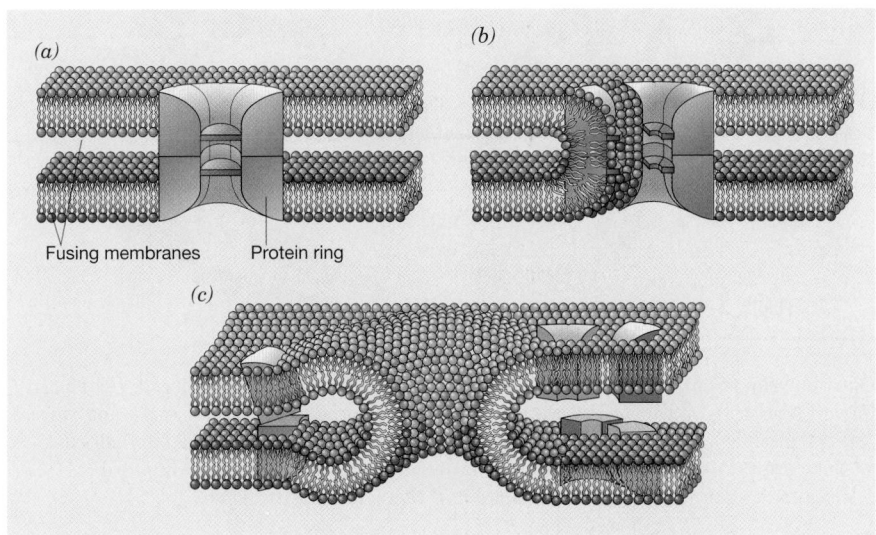

FIGURE 12-64   Model for the fusion of yeast vacuoles. (*a*) The formation of trans-SNARE complexes (not shown) induces the dimerization of two cylindrical V0 molecules (here shown in cutaway view with their hydrophobic and hydrophilic surfaces green and yellow) embedded in opposite bilayers. (*b*) The subunits of the V0 dimer separate horizontally, thereby permitting lipids to invade its enclosed aqueous pore, the so-called fusion pore. (*c*) The fusion pore expands, eventually causing the V0 subunits to also separate vertically. [From Almers, W., *Nature* **409,** 568 (2001). Reproduced with permission.]

proposed that the radial expansion of the aqueous pore by the lateral separation of its subunits permits lipids to enter the spaces between the subunits in a way that causes the two bilayers to fuse (Figs. 12-64*b*,*c*).

f. The Structure of the nSec1–Syntaxin Complex Suggests a Function for Rab Protein

The neuronal SM protein, which is named **nSec1,** binds to syntaxin with high affinity to form a complex that is mutually exclusive with the formation of the syntaxin–synaptobrevin–SNAP-25 complex. The X-ray structure of nSec1 in complex with the cytoplasmic domain of syntaxin (Fig. 12-65), determined by William Weis, reveals that this portion of the 288-residue syntaxin forms an N-terminal up–down–up–down four-helix bundle. Syntaxin's C-terminal helix (but lacking its TM portion) adopts a bent and somewhat irregular conformation, which differs from that in the core complex displayed in Fig. 12-62. In contrast, the remaining N-terminal 3-helix bundle is closely superimposable on the NMR structure of this segment alone. The 594-residue nSec1 is an arch-shaped molecule that binds syntaxin, and in particular its C-terminal helix, in the cleft of the arch (Fig. 12-65*c*).

The formation of the syntaxin–synaptobrevin–SNAP-25 complex that mediates vesicle fusion requires that the nSec1–syntaxin complex dissociate and that syntaxin's N-terminal 3-helix bundle release the C-terminal helix. Mutational studies indicate that Rab protein and/or its effectors mediate this process. It has therefore been proposed that the binding of Rab and/or its effectors to the nSec1–syntaxin complex causes nSec1 to change conformation, which in turn induces syntaxin's N-terminal 3-helix bundle to release the C-terminal helix, thereby permitting the SNARE complex to form.

g. NSF Mediates Core Complex Disassembly

The SNARE complex in the fused membranes, the so-called **cis-SNARE complex,** must eventually be dissociated in order for its component proteins to participate in a new round of vesicle fusion. This process is mediated by NSF, an ATP-dependent cytosolic protein that binds to SNAREs (SNAP receptors) through the intermediacy of adaptor proteins called SNAPs (soluble NSF attachment proteins). Although it was initially proposed that the NSF-mediated disassembly of the cis-SNARE complex drives membrane fusion, it is now clear that NSF functions to recycle SNAREs after their participation in membrane fusion, that is, *NSF functions as an ATP-driven molecular chaperone.*

NSF is a hexamer of identical 752-residue subunits. Sequence analysis and limited proteolysis studies indicate that each subunit consists of three domains:

1. An N-terminal so-called N-domain (residues 1–205), which mediates NSF's interactions with SNAPs and SNAREs.

2. A D1 domain (206–487), which binds ATP and catalyzes its hydroysis in a process that drives the disassembly of the cis-SNARE complex.

3. A C-terminal D2 domain (488–752), which is homologous to D1. D2 binds ATP with a much higher affinity than does D1 but hydrolyzes it very slowly, if at all. D2·ATP mediates the hexamerization of NSF, which is required for NSF activity.

The X-ray structure of the D2 domain of NSF was independently determined by Weis and by Jahn and Brünger. Its wedge-shaped subunits associate to form a 116-Å-diameter and 40-Å-high disk-shaped hexamer that has an ~18-Å-diameter central pore (Fig. 12-66). The ATP

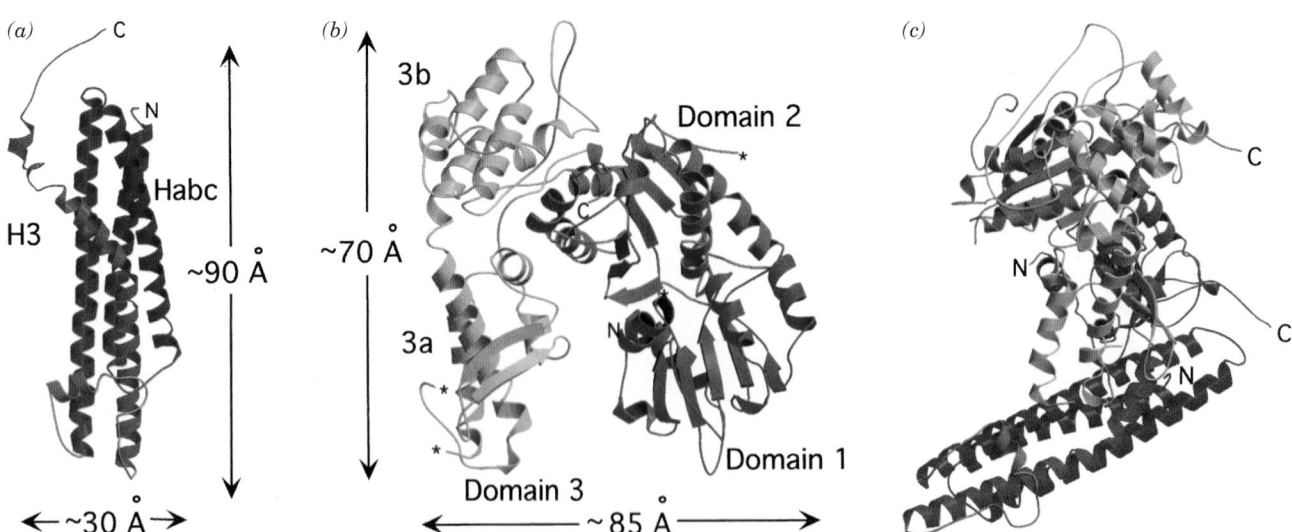

FIGURE 12-65 X-Ray structure of the complex between nSec1 and syntaxin. (*a*) Ribbon diagram of syntaxin with its N-terminal 3-helix bundle (Habc) orange and the cytoplasmic portion, its C-terminal helix (H3; the segment that forms a component of the core complex), purple. (*b*) Ribbon diagram of nSec1 with its three domains differently colored. (*c*) The nSec1–syntaxin complex colored as in Parts *a* and *b* and viewed such that the nSec1 is rotated by 90° about the vertical axis relative to Part *b*. [Courtesy of William Weis, Stanford University School of Medicine. PDBid 1DN1.]

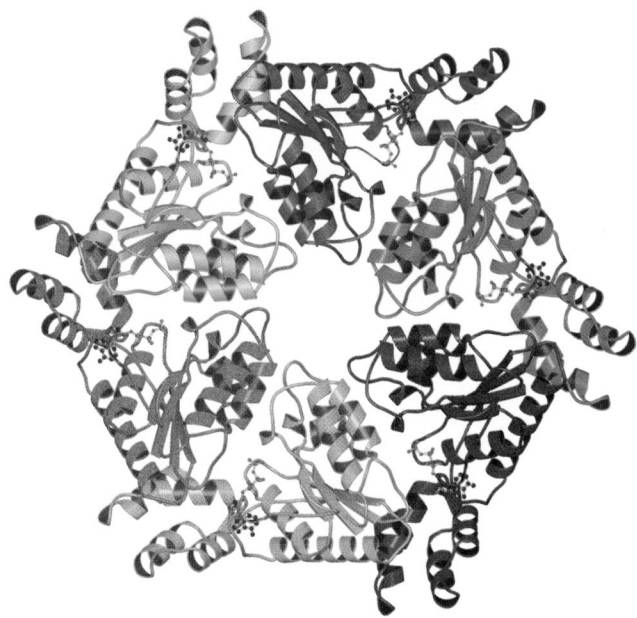

FIGURE 12-66 X-Ray structure of the NSF D2 hexamer as viewed from its N-terminal end along its 6-fold axis. Each subunit is differently colored. The bound ATPs are drawn in ball-and-stick form. [Courtesy of Axel Brünger, Yale University. PDBid 1NSF.]

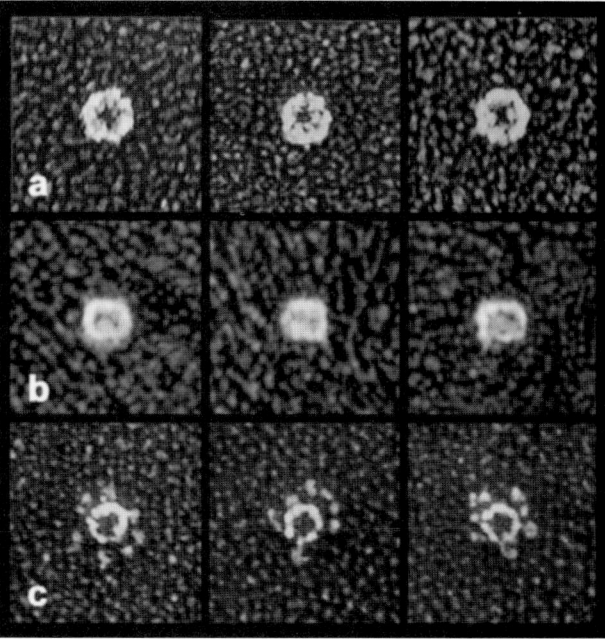

FIGURE 12-67 Quick-freeze/deep-etch electron micrographs of NSF hexamers. (*a*) Top and (*b*) side views in the presence of ATP. (*c*) Top view in the presence of ATPγS. [Courtesy of John Heuser, Washington University School of Medicine, St. Louis, Missouri.]

is bound near the interface between two subunits, where it presumably helps stabilize their association.

Electron micrographs by Jahn and John Heuser of intact NSF in the presence of ATP have the appearance of an ~120-Å-diameter hexagonal ring with a 30- to 50-Å central opening when seen in top view (Fig. 12-67*a*) and of a 120-Å by 150-Å rectangle when seen in side view (Fig. 12-67*b*). The length of the rectangle is about twice the height of the D2 disk, which suggests that D1 forms a D2-like hexagonal disk that stacks on D2. In the presence of ADP, NSF has an identical appearance, which suggests that D1 rapidly hydrolyzes its bound ATP to form ADP. However, in the presence of the nonhydrolyzable ATP analog **ATPγS** (in which a terminal O atom on the γ-phosphorus atom of ATP is replaced by S), NSF displays 6 globular feet that are tightly packed around the somewhat smaller hexagonal ring (Fig. 12-67*c*). Since the hexagonal rings but not the globules are seen when D1–D2 constructs are imaged in the presence of ATPγS, the globules must be the N domains. Evidently, the N domains are held tightly around the central disk of stacked D1 and D2 hexamers when D1 binds ADP but are released when D1 binds ATP.

The mechanism whereby NSF disassembles the cis-SNARE complex is largely unknown. The rod-shaped SNARE core complex (Fig. 12-62*a*), which is 20 to 25 Å in diameter, is too wide to fit inside the 18-Å-diameter central pore of the D2 hexamer (and presumably the similarly shaped D1 hexamer) without significant structural changes. It is therefore unlikely that the core complex

binds inside NSF's central cavity in a manner similar to the way that the GroEL–GroES chaperonin system binds its substrate proteins (Section 9-2C). Moreover, electron micrographs indicate that the complex of SNAP and the three SNARE proteins binds to one end of NSF in the presence of ATPγS (but not at all in the presence of ADP). Since NSF oligomers containing mixtures of active and inactive D1 domains are unable to disassemble SNARE complexes, it appears that the NSF subunits function in a cooperative manner.

E. Protein Targeting to Mitochondria

Although mitochondria contain functioning genetic and protein synthesizing systems, their genomes encode only a handful of inner membrane proteins (13 in humans; 8 in yeast). The vast majority of mitochondrial proteins (>98%), which comprise 10 to 20% of intracellular proteins, are encoded by nuclear genes and are synthesized by cytosolic ribosomes. They must therefore traverse one or both mitochondrial membranes (Section 1-2A) to reach their final destinations. In this subsection, we discuss how proteins are imported into mitochondria and are directed to their correct destinations [outer membrane, inner membrane, intermembrane space, or **matrix** (the space enclosed by the inner membrane)]. Our rapidly developing knowledge of this process was elucidated in large part through investigations in yeast and in the pink bread mold *Neurospora crassa* by Walter Neupert, Nikolaus Pfanner, and Gottfried Schatz. However, there is considerable evidence that this process is well conserved among all eukaryotes. Note that the transport systems we describe here and in Section 12-4B resemble those that mediate the import of proteins into chloroplasts (in which proteins must cross up to three membranes) and peroxisomes (Section 1-2A).

a. Proteins Must Be Unfolded for Import into Mitochondria

Nuclear-encoded mitochondrial proteins are fully synthesized by cytosolic ribosomes before they are imported into mitochondria; that is, they are posttranslationally im-

ported, in contrast to secretory pathway proteins, which are cotranslationally imported (Section 12-4B). One might expect, therefore, that mitochondrial proteins, many of which are integral proteins, would at least partially fold and/or nonspecifically aggregate in the cytosol before encountering the mitochondrial import system. Yet a variety of evidence indicates that *only unfolded proteins can pass through mitochondrial membranes*. For example, **dihydrofolate reductase (DHFR),** a normally cytosolic enzyme, is imported into yeast mitochondria when it is preceded by the targeting sequence of a cytosolically synthesized mitochondrial protein (see below). However, the importation

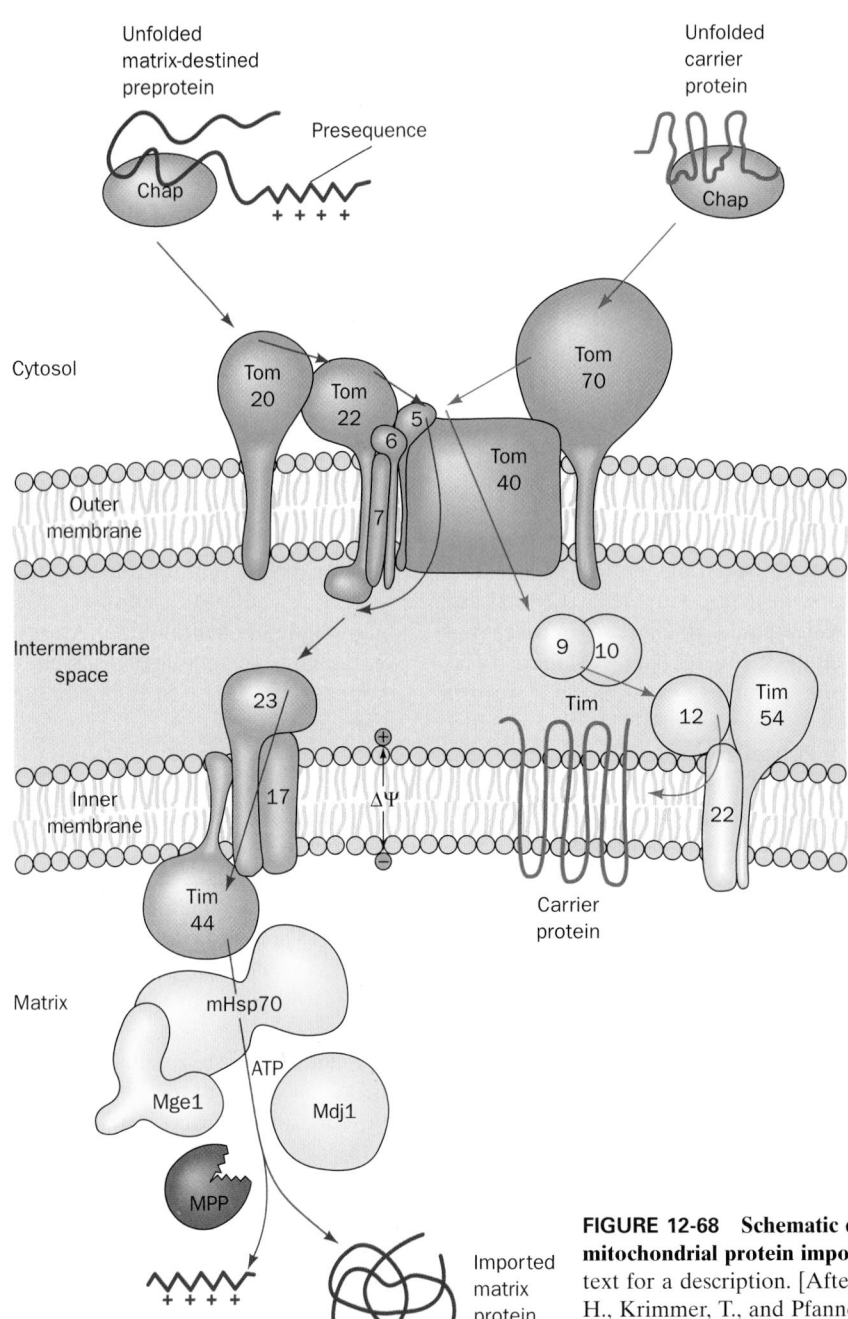

FIGURE 12-68 Schematic diagram of the mitochondrial protein import machinery. See the text for a description. [After Voos, W., Martin, H., Krimmer, T., and Pfanner, N., *Biochim. Biophys. Acta* **1422,** 237 (1999).]

of this chimeric protein is arrested by the presence of **methotrexate,** an analog of DHFR's normal substrate **dihydrofolate** (Section 28-4B), which binds to DHFR with such high affinity that it stabilizes the protein's native conformation.

The import competence of mitochondrially destined proteins is maintained in the cytosol by a variety of ATP-dependent molecular chaperones. These include members of the Hsp70 family (Section 9-2C) and, in mammals, a protein named **mitochondrial import stimulation factor (MSF).** Consequently, the genetically engineered shutdown of Hsp70 production in yeast causes the cells to cytosolically accumulate proteins that would otherwise be imported into the mitochondria. Moreover, the rate of the Hsp70-facilitated mitochondrial import of a protein is enhanced by its prior denaturation by urea. Evidently, Hsp70 functions in this process as an ATP-driven "protein unfoldase."

b. Transport of Proteins across the Outer Mitochondrial Membrane

Cytosolically synthesized matrix proteins have N-terminal targeting sequences of 20 to 60 residues that are rich in basic and hydroxylated side chains but have few, if any, acidic side chains. These presequences, which do not interact with the SRP, form amphipathic α helices in solution.

The protein subunits that participate in importing proteins across the outer mitochondrial membrane are called **TOM proteins** (for *t*ranslocase of the *o*uter *m*embrane) and are named Tomxx, where xx is the molecular mass of the subunit in kilodaltons. Likewise, the proteins involved in translocating proteins across the inner mitochondrial membrane are called **TIM proteins** (for *t*ranslocase of the *i*nner *m*embrane) and are named Timxx.

The machinery that imports proteins through the outer mitochondrial membrane does so as follows (Fig. 12-68, *top left*):

1. The signal sequence of an unfolded preprotein associates with the cytoplasmic domain of mitochondrial receptor proteins: Hsp70-associated proteins interact mainly with **Tom20** in complex with **Tom22,** whereas MSF releases its bound preproteins to **Tom70.** The NMR structure of Tom20's cytosolic domain in complex with an 11-residue segment of a presequence peptide (Fig. 12-69) was determined by Toshiya Endo and Daisuke Kohda. The Tom20 domain consists mainly of 5 helices, whose two N-terminal helices form a nonpolar surface groove in which the helical presequence binds mainly via hydrophobic interactions rather than ionic interactions. Evidently, Tom20 recognizes the presequence's amphipathic helix but not its positive charges. Thus these positive charges, which are required for mitochondrial import, must interact with other elements of the mitochondrial import machinery such as Tom22 or **Tom5** (see below).

2. Tom20 and Tom70 deliver preproteins to the **general import pore (GIP),** which consists of **Tom40,** a polytopic TM protein, which CD measurements indicate

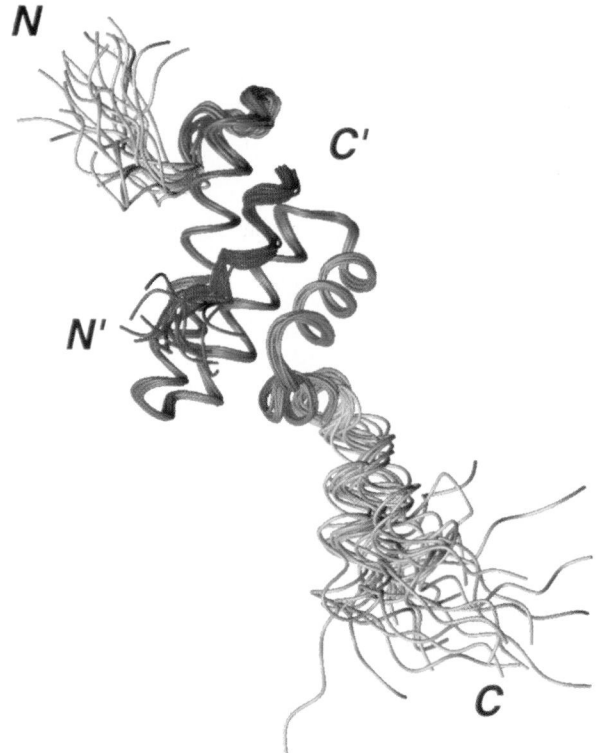

FIGURE 12-69 **NMR structure of the cytoplasmic domain of rat Tom20 in complex with the C-terminal 11-residue segment (GPRLSRLLSYA) of the 22-residue presequence of the rat mitochondrial enzyme aldehyde dehydrogenase.** The diagram is a superposition of the 20 final structures in the NMR analysis (Section 8-3A) in which the residues used to make the superposition are blue (Tom20) and red (presequence) and the remaining residues are gray (Tom20) and orange (presequence). [Courtesy of Toshiya Endo, Nagoya University, Nagoya, Japan, and Daisuke Kohda, Biomolecular Engineering Research Institute, Osaka, Japan. PDBid 1OM2.]

consists mainly of β sheets (as do mitochondrial porins, which are also outer membrane proteins; Fig. 12-27). Electrophysiological measurements demonstrate that Tom40 contains a cation-selective hydrophilic channel through which precursor proteins are presumably transported. In mitochondria, Tom40 is closely associated with Tom22 as well as with three small subunits, Tom5, **Tom6,** and **Tom7,** to form the **TOM core complex.** Tom5, whose absence greatly slows the transfer of proteins to the GIP, binds presequences and probably mediates the transfer of Tom20/Tom22-bound preproteins to the Tom40 channel. Tom6 mediates the assembly of Tom22 with Tom40, thereby facilitating preprotein transfer to the channel. In contrast, Tom7 destabilizes the TOM complex and hence is likely to function in the lateral insertion of outer membrane proteins such as porins, a process that is probably directed by internal hydrophobic "stop transfer" sequences. Electron micrographs of the *Neurospora* TOM

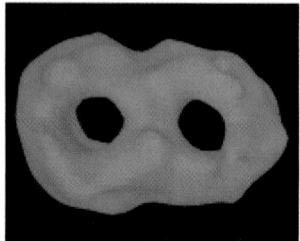 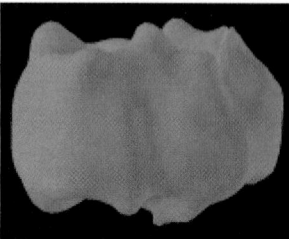

FIGURE 12-70   Electron microscopy–based image of the TOM core complex particles from *Neurospora*. The particles, which are shown in top view (*left*) and side view (*right*), contain two openings that presumably represent the mitochondrial outer membrane's protein-conducting channel. [Courtesy of Stephan Nussberger and Walter Neupert, Universität München, Germany.]

core complex (Fig. 12-70) reveal an ~70-Å-high (~20 Å larger than the thickness of the lipid bilayer) and ~120-Å-wide particle containing two ~21-Å-diameter pores that presumably are the protein-conducting channels. This agrees with permeability experiments using cations of various sizes, which indicate that the Tom40 pore is ~22 Å in diameter.

3. The forces driving the translocation of polypeptides across the outer mitochondrial membrane remain largely enigmatic. A proposed mechanism, the **acid chain hypothesis,** is that the positively charged presequence is sequentially transferred between acidic (negatively charged) patches to which it binds with successively higher affinities. Such patches have been shown to be present on the cytoplasmic faces of Tom20, Tom22, and Tom5, as well as on the intermembrane faces of Tom40 and Tom22.

c. Transport of Proteins across the Inner Mitochondrial Membrane

At this point, the mitochondrial protein import pathway bifurcates (Fig. 12-68, *middle*): Proteins that are directed to the matrix are transported by a different TIM complex than those that are being inserted into the inner mitochondrial membrane. We shall discuss these two pathways in the order they were discovered.

Matrix-destined proteins are conducted across the inner mitochondrial membrane via a protein channel formed by the TM proteins **Tim23** and **Tim17** (Fig. 12-68, *bottom left*). In the presence of methotrexate, the above DHFR chimera becomes stuck in the membrane with the spacer linking the enzyme to its N-terminal presequence simultaneously spanning the TOM and TIM complexes. Such stable translocation intermediates accumulate at sites where, as electron microscopy studies reveal, the outer and inner mitochondrial membranes approach one another more closely than elsewhere in the mitochondrion, possibly with the TOM and TIM complexes in contact. If the spacer is so short that it cannot span both membranes (less than ~40 residues), no stable translocation intermediate is formed. Thus, it appears that presequences make their way be-

tween the TOM and TIM complexes without the aid of chaperones.

The translocation of a protein across the inner mitochondrial membrane requires energy in the form of both ATP and an electrostatic potential across the inner mitochondrial membrane. This so-called **membrane potential** (Section 20-1), $\Delta\Psi$, which is metabolically generated (Section 22-3B), probably functions to electrophoretically transport the positively charged N-terminal signal sequence into the matrix (the matrix is negatively charged with respect to the cytosol). The ATP is utilized by mitochondrial Hsp70 **(mHsp70),** which binds to **Tim44** on the inner face of the inner mitochondrial membrane, where it is thought to mechanically pull the protein through the Tim17/23 pore in a ratchetlike manner. mHsp70 functions in partnership with the cochaperone protein **Mdj1,** an Hsp40 homolog (Section 9-2C), and **Mge1,** which facilitates ADP–ATP exchange in mHsp70.

Once the preprotein has entered the matrix, its N-terminal signal sequence is excised by **matrix processing peptidase (MPP),** an essential protein. The imported protein then folds/assembles to its native state via ATP-driven processes that are mediated by a battery of chaperone proteins including mHsp70 (only about 10% of which is associated with Tim44) and Hsp60/Hsp10 (homologs of the GroEL/GroES system; Section 9-2C).

d. Insertion of Carrier Proteins into the Inner Mitochondrial Membrane

The mitochondrial inner membrane is impermeable to nearly all polar substances and hence contains numerous **metabolite carrier proteins** to permit the acquisition of reactants and the delivery of products. The members of this family of polytopic TM proteins include the **ATP/ADP carrier** (alternatively, the **ATP–ADP translocator,** which exchanges the ATP synthesized in the matrix for the ADP product of cytosolic ATP hydrolysis; Section 20-4C) and the **phosphate carrier** (which returns the phosphate product of cytosolic ATP hydrolysis to the matrix; Section 22-1B).

The members of the metabolite carrier family lack the N-terminal presequences of matrix-targeted proteins, but instead contain poorly characterized internal targeting sequences. Nevertheless, they are translocated across the outer mitochondrial membrane by the TOM complex (Fig. 12-68, *top right*), although they bind to the Tom70 receptor, which preferentially interacts with proteins carrying the internal targeting signals, rather than the Tom20/Tom22 receptor, which mainly binds matrix-targeted proteins.

On passing through the TOM complex, members of the metabolite carrier family are escorted across the intermembrane space by a complex of the homologous proteins **Tim9** and **Tim10,** probably $(Tim9)_3(Tim10)_3$. The observation that carrier proteins in mitochondria depleted of Tim9 and Tim10 are not inserted into the GIP, as indicated by their failure to reach a protease-resistant state, suggests that it is the binding of the carrier protein to the Tim9–Tim10 complex that drives its translocation across the outer mitochondrial membrane. The Tim9–Tim10 complex delivers the carrier protein to the peripheral protein **Tim12** (a homolog of Tim9 and Tim10), which is associated with the integral proteins **Tim22** (which has significant homlogy to the pore-forming components of Tim17 and Tim 23 of the foregoing TIM complex) and **Tim54** to form a different TIM complex (Fig. 12-68, *bottom right*). *Tim22, possibly in concert with Tim54, then mediates the insertion of the carrier protein into the inner mitochondrial membrane, where it assembles to form homodimers.* This process occurs via an unknown but membrane potential–dependent mechanism.

e. Nonmatrix Mitochondrial Proteins May Be Targeted through a Variety of Import Mechanisms

Some of the cytosolically synthesized proteins destined for insertion into the mitochondrion's inner membrane or for residence in its intermembrane space are first imported into the matrix as described above and then exported from there to their final destinations. These proteins, for the most part, are synthesized with bipartite N-terminal targeting sequences whose inner (more C-terminal) segments, once exposed by the removal of the above-described N-terminal presequence, direct the proteins to their target compartments. This indirect routing may reflect the mitochondrion's prokaryotic origin (Section 1-2A): The primordial mitochondrion, being a gram-negative bacterium, synthesized all of its proteins in its cytoplasm (the primordial matrix) so that membrane-bound or intermembrane proteins had to be exported to these destinations.

Some inner membrane and intermembrane proteins have completely independent targeting mechanisms. For example, cytochrome *c*, a peripheral protein that is associated with the outer surface of the inner mitochondrial membrane, readily traverses the outer mitochondrial membrane as **apocytochrome *c*** (cytochrome *c* without its covalently attached heme; Figure 9-36), probably by passing through a porinlike protein named **P70.** The driving force for this process is provided by **cytochrome *c* heme lyase (CCHL),** the enzyme that covalently attaches heme to apocytochrome *c* in the intermembrane space. Heme attachment is required to retain cytochrome *c* within the intermembrane space. Note that apocytochrome *c* does not have a cleavable targeting sequence; the residues that serve to identify this protein to its importation apparatus are part of the mature protein. CCHL is likewise synthesized as the mature-length protein but, in contrast to apocytochrome *c*, is imported into the intermembrane space through the TOM complex.

5 ■ LIPOPROTEINS

*Lipids and proteins associate noncovalently to form **lipoproteins,** which function in the blood plasma as transport vehicles for triacylglycerols and cholesterol.* In this section, we discuss the structure, function, and dysfunction of lipoproteins, and how eukaryotic cells take up lipoproteins and other specific proteins from their external medium through receptor-mediated endocytosis.

A. *Lipoprotein Structure*

Lipids, such as phospholipids, triacylglycerols, and cholesterol, are but sparingly soluble in aqueous solution. Hence, *they are transported by the circulation as components of lipoproteins, globular micellelike particles that consist of a nonpolar core of triacylglycerols and cholesteryl esters surrounded by an amphiphilic coating of protein, phospholipid, and cholesterol.* Lipoproteins have been classified into five broad categories on the basis of their functional and physical properties (Table 12-6):

1. Chylomicrons, which transport exogenous (externally supplied; in this case, dietary) triacylglycerols and cholesterol from the intestines to the tissues.

2–4. Very low density lipoproteins (VLDL), intermediate density lipoproteins (IDL), and low density lipoproteins (LDL), a group of related particles that transport endogenous (internally produced) triacylglycerols and

TABLE 12-6 **Characteristics of the Major Classes of Lipoproteins in Human Plasma**

	Chylomicrons	VLDL	IDL	LDL	HDL
Density (g · cm^{-3})	<0.95	<1.006	1.006–1.019	1.019–1.063	1.063–1.210
Particle diameter (Å)	750–12,000	300–800	250–350	180–250	50–120
Particle mass (kD)	400,000	10,000–80,000	5000–10,000	2300	175–360
% Protein[a]	1.5–2.5	5–10	15–20	20–25	40–55
% Phospholipids[a]	7–9	15–20	22	15–20	20–35
% Free cholesterol[a]	1–3	5–10	8	7–10	3–4
% Triacylglycerols[b]	84–89	50–65	22	7–10	3–5
% Cholesteryl esters[b]	3–5	10–15	30	35–40	12
Major apolipoproteins	A-I, A-II, B-48, C-I, C-II, C-III, E	B-100, C-I, C-II, C-III, E	B-100, C-I, C-II, C-III, E	B-100	A-I, A-II, C-I, C-II, C-III, D, E

[a]Surface components.
[b]Core lipids.

cholesterol from the liver to the tissues (the liver synthesizes triacylglycerols from excess carbohydrates; Section 25-4).

5. High density lipoproteins (HDL), which transport endogenous cholesterol from the tissues to the liver.

Lipoprotein particles undergo continuous metabolic processing, so that they have variable properties and compositions (Table 12-6). Each contains just enough protein, phospholipid, and cholesterol to form an ~20-Å-thick monolayer of these substances on the particle surface (Fig. 12-71). Lipoprotein densities increase with decreasing particle diameter because the density of their outer coating is greater than that of their inner core.

a. Apolipoproteins Have Amphipathic Helices That Coat Lipoprotein Surfaces

The protein components of lipoproteins are known as **apolipoproteins** or just **apoproteins.** At least nine apolipoproteins are distributed in significant amounts in the different human lipoproteins (Tables 12-6 and 12-7). Most of them are water-soluble and associate rather weakly with lipoproteins. Hence, they readily transfer between lipoprotein particles via the aqueous phase. CD measurements indicate that *apolipoproteins have a high helix content, which increases when they are incorporated in lipoproteins.* Apparently the helices are stabilized by a lipid environment, presumably because helices fully satisfy the polypeptide backbone's hydrogen bonding potential in a lipoprotein's water-free interior.

b. The X-Ray Structure of ApoA-I Mimics That in HDL

Apolipoprotein A-I (apoA-I) is HDL's major apoprotein. Sequence analysis indicates that apoA-I consists mainly of repeated amphipathic α helices of 11 or 22 residues that provide the protein's lipid-binding regions. *These putative α helices, as well as similar helices that occur in most other apolipoproteins, have their hydrophobic*

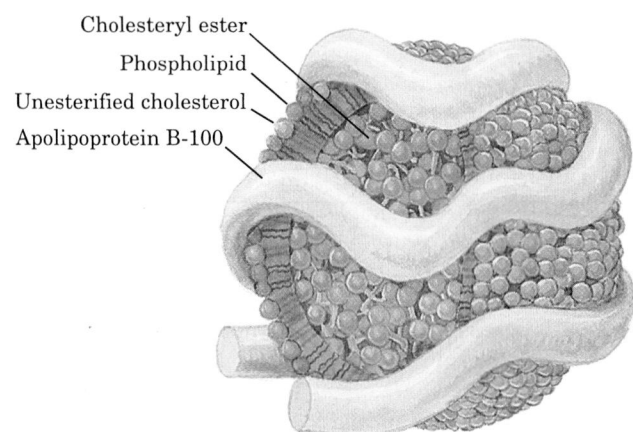

FIGURE 12-71   LDL, the major cholesterol carrier of the bloodstream. This spheroidal particle consists of some 1500 cholesteryl ester molecules surrounded by an amphiphilic coat of 800 phospholipid molecules, 500 cholesterol molecules, and a single 4536-residue molecule of apolipoprotein B-100.

and hydrophilic residues on opposite sides of the helical cylinders (Fig. 12-72). Furthermore, the polar helix face has a zwitterionic character in that its negatively charged residues project from the center of this face, whereas its positively charged residues are located at its edges. Indeed, a synthetic 22-residue polypeptide of high helix-forming propensity, which was designed by E. Thomas Kaiser to have this polarity distribution but to otherwise have minimal similarity to the repeating apoA-I sequences, behaves much like apoA-I in binding to egg lecithin liposomes. Evidently, the structural role of apoA-I, and probably most other apolipoproteins, is fulfilled by its helical segments rather than by any organized tertiary structure. This suggests that *lipoprotein α helices float on phospholipid surfaces, much like logs on water.* The phospholipids are presumably arrayed with their charged groups bound to

TABLE 12-7   Properties of the Major Species of Human Apolipoproteins

Apolipoprotein	Number of Residues	Molecular Mass[a] (kD)	Function
A-I	243	29	Activates LCAT[b]
A-II	77	17	Inhibits LCAT, activates hepatic lipase
B-48	2152	241	Cholesterol clearance
B-100	4536	513	Cholesterol clearance
C-I	56	6.6	Activates LCAT?
C-II	79	8.9	Activates LPL[c]
C-III	79	8.8	Inhibits LPL, activates LCAT?
D	169	19	Unknown
E	299	34	Cholesterol clearance

[a]All apolipoproteins are monomers but apoA-II, which is a disulfide-linked dimer.
[b]LCAT: Lecithin–cholesterol acyltransferase.
[c]LPL: Lipoprotein lipase.

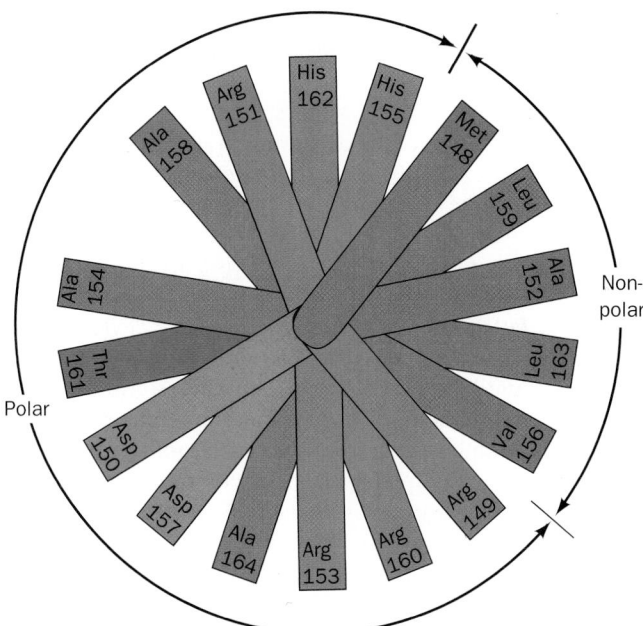

FIGURE 12-72   A helical wheel projection of the amphipathic α helix constituting residues 148 to 164 of apolipoprotein A-I. (In a helical wheel representation, the side chain positions are projected down the helix axis onto a plane.) Note the segregation of nonpolar, acidic, and basic residues to different sides of the helix. Other apolipoprotein helices have similar polarity distributions. [After Kaiser, E.T., *in* Oxender, D.L. and Fox, C.F. (Eds.), *Protein Engineering*, p. 194, Liss (1987).]

oppositely charged residues on the polar face of the helix and with the first few methylene groups of their fatty acid residues in hydrophobic association with the nonpolar face of the helix.

A variety of criteria indicate that apoA-I undergoes significant secondary structural changes on binding lipid. However, apo Δ(1–43)A-I, a truncation mutant that lacks residues 1 to 43 of the 243-residue human apoA-I, has a conformation that closely resembles that of lipid-bound apoA-I, whether or not lipid is present. Lipid-free apo Δ(1–43)A-I is therefore likely to provide a valid structural model for lipid-bound apoA-I.

The X-ray structure of apo Δ(1–43)A-I (Fig. 12-73) was determined by David Borhani and Christie Brouillette. It revealed that, over most of its length, each polypeptide chain forms a pseudocontinuous amphipathic α helix that is punctuated by kinks at Pro residues that are spaced mainly at intervals of 22 residues to form 10 helical segments arranged in the shape of a twisted horseshoe. Two such monomers (e.g., the green and cyan monomers in Fig. 12-73) associate in an antiparallel fashion along most of their lengths to form a dimer that has the shape of a twisted elliptical ring. Two such dimers, in turn, associate via their hydrophobic surfaces to form an elliptical tetramer with D_2 symmetry that has outer dimensions of 135×90 Å and an inner hole of 95×50 Å. The surface of this tetrameric ring, which consists of up–down–up–down 4-helix bundles over about three-fourths of its circumference, is hydrophilic with a uniform electrostatic potential, whereas the interior of each 4-helix bundle contains mainly Val and Leu side chains. Since, in this conformation, these hydrophobic residues are unavailable for binding to lipid, it is postulated that they associate in the lipid-free crystal so as to shelter the lipid-binding face of apo Δ(1–43)A-I dimers from contact with water (which fills the spaces in the crystal).

The sizes and shapes of the apo Δ(1–43)A-I dimer and tetramer seem ideal for wrapping around the 50- to 120-Å-diameter HDL particles. Since HDL particles often contain two or four apoA-I monomers, it is proposed that when pairs of apoA-I monomers bind to HDL, they do so as the above-described antiparallel dimer. Its exposed non-

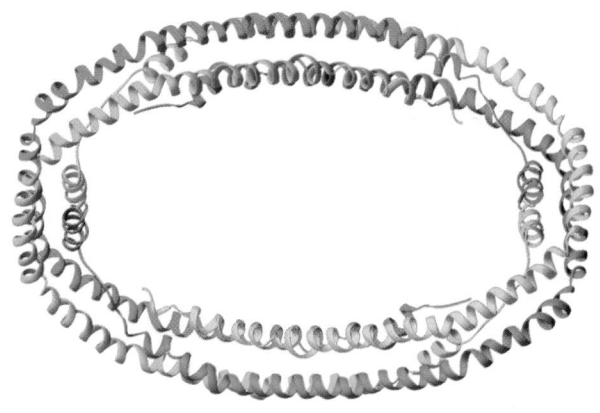

(a)

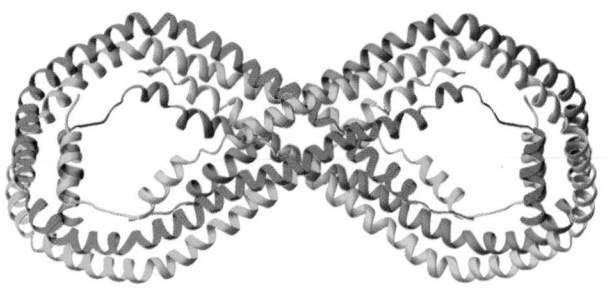

(b)

FIGURE 12-73   X-Ray structure of human apo Δ(1–43)A-I. The four monomers of the D_2-symmetric tetramer it forms are drawn in different colors. (*a*) View along the 2-fold axis relating the green and magenta subunits and the cyan and gold subunits. (*b*) View from the top of Part *a* along the 2-fold axis relating the cyan and green subunits and the gold and magenta subunits. The latter dimers, which interact along most of their

lengths, probably maintain their identities in HDL particles, whereas the other pairings, whose interactions are less extensive, are unlikely to do so. [Based on an X-ray structure by David Borhani, Southern Research Institute, Birmingham, Alabama, and Christie Brouillette, University of Alabama Medical Center, Birmingham. PDBid 1AV1.]

polar side chains could then hydrophobically interact with the HDL particle's buried nonpolar groups. Two such dimers could associate on the surface of an HDL particle to form a tetramer, although, most probably, in a different manner than is seen in the structure of apo Δ(1–43)A-I.

B. *Lipoprotein Function*

The various lipoproteins have different physiological functions, as we discuss below.

a. Chylomicrons Are Delipidated in the Capillaries of Peripheral Tissues

Chylomicrons, which are assembled by the intestinal mucosa, function to keep exogenous triacylglycerols and cholesterol suspended in aqueous solution. These lipoproteins are released into the intestinal lymph (known as **chyle**), which is transported through the lymphatic ves-

sels before draining into the large body veins via the thoracic duct. After a fatty meal, the otherwise clear chyle takes on a milky appearance.

Chylomicrons adhere to binding sites on the inner surface (endothelium) of the capillaries in skeletal muscle and adipose tissue. There, within minutes after entering the bloodstream, the chylomicron's component triacylglycerols are hydrolyzed through the action of **lipoprotein lipase (LPL),** an extracellular enzyme that is activated by **apoC-II.** The tissues then take up the liberated monoacylglycerol and fatty acid hydrolysis products. The chylomicrons shrink as their triacylglycerols are progressively hydrolyzed until they are reduced to cholesterol-enriched **chylomicron remnants.** The chylomicron remnants reenter the circulation by dissociating from the capillary endothelium and are subsequently taken up by the liver, as is explained below. *Chylomicrons therefore function to deliver dietary triacylglycerols to muscle and adipose tissue and dietary cholesterol to the liver (Fig. 12-74, left).*

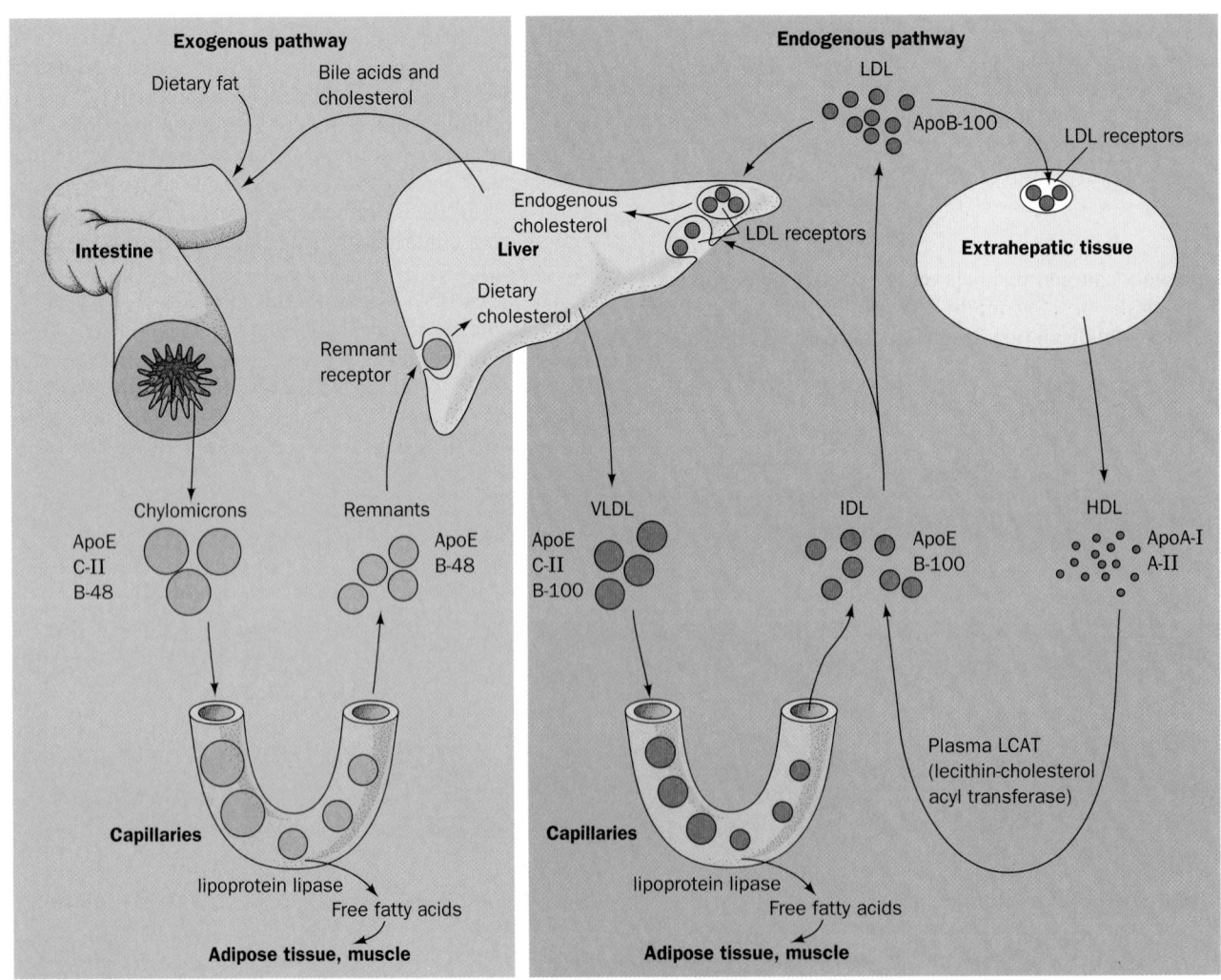

FIGURE 12-74   Model for plasma triacylglycerol and cholesterol transport in humans. [After Brown, M.S. and Goldstein, J.L., *in* Brunwald, E., Isselbacher, K.J., Petersdorf, R.G., Wilson, J.D., Martin, J.B., and Fauci, A.S. (Eds.), *Harrison's Principles of Internal Medicine* (11th ed.), *p.* 1652, McGraw-Hill (1987).] 🔁 **See the Animated Figures**

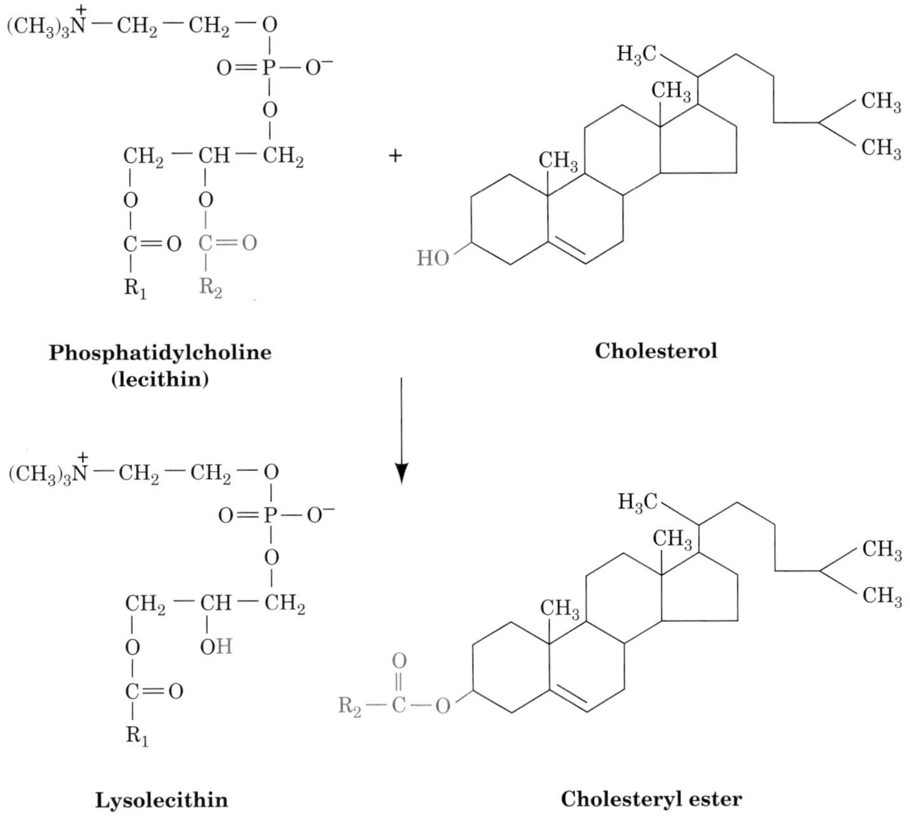

FIGURE 12-75   Reaction catalyzed by lecithin–cholesterol acyltransferase (LCAT). The transferred acyl group is most often a linoleic acid residue.

b. VLDL Are Degraded Much Like Chylomicrons

VLDL, which are synthesized in the liver as lipid transport vehicles, are also degraded by lipoprotein lipase (Fig. 12-74, *right*). The VLDL remnants appear in the circulation, first as IDL and then as LDL. In the transformation of VLDL to LDL, all their proteins but **apoB-100** are removed and much of their cholesterol is esterified by the HDL-associated enzyme **lecithin–cholesterol acyltransferase (LCAT),** as is discussed below. This enzyme transfers a fatty acid residue from atom C2 of lecithin to cholesterol with the concomitant formation of **lysolecithin** (Fig. 12-75).

ApoB-100, a 4536-residue monomeric glycoprotein (and thus one of the largest monomeric proteins known), has a hydrophobicity approaching that of integral proteins and contains relatively few amphipathic helices. Hence, in contrast to the other, less hydrophobic plasma apolipoproteins, apoB-100 is neither water-soluble nor transferred between lipoprotein particles. Each LDL particle contains but one molecule of apoB-100, which immunoelectron microscopy indicates has an extended form that covers at least half of the particle surface (Fig. 12-71). Chylomicrons, however, contain **apoB-48,** a 2152-residue protein that is identical in sequence to the N-terminal 48% of apoB-100. Indeed, both proteins are encoded by the same gene. The remarkable mechanism by which this gene expresses different length proteins in liver and intestines is discussed in Section 31-4A.

c. Cells Take Up Cholesterol through Receptor-Mediated Endocytosis of LDL

Cholesterol, as we have seen, is an essential component of animal cell membranes. The cholesterol may be externally supplied or, if this source is insufficient, internally synthesized (Section 25-6A). Michael Brown and Joseph Goldstein have demonstrated that *cells obtain exogenous cholesterol mainly through the endocytosis (engulfment) of LDL particles via the following process:* The LDL is sequestered by **LDL receptor (LDLR),** a cell-surface transmembrane glycoprotein, which specifically binds both apoB-100 and **apoE.** The LDLR is an 839-residue glycoprotein that has a 767-residue N-terminal exoplasmic (outside the plasma membrane) portion, a 22-residue TM segment that presumably forms an α helix, and a 50-residue C-terminal domain. Its 322-residue N-terminal domain, which consists of 7 tandemly repeated ~40-residue Cys-rich modules, mediates the binding of lipoproteins.

The X-ray structure of the LDLR's fifth ligand binding repeat (LR5), determined by Peter Kim and James Berger, reveals a polypeptide backbone that is essentially devoid of regular secondary structure but which wraps around to

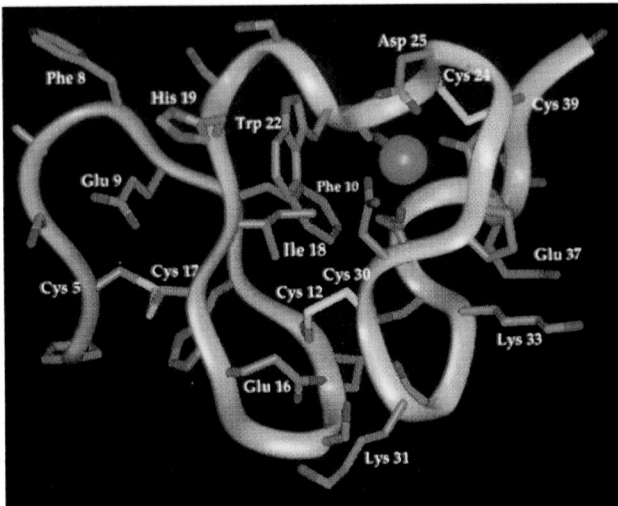

FIGURE 12-76   X-Ray structure of ligand binding repeat 5 (LR5) of the human LDL receptor. The polypeptide backbone of this 37-residue segment is represented by a silver ribbon and its side chains are shown in stick form with C green, N blue, O red, and S yellow. The bound Ca^{2+} ion is represented by a cyan sphere. [Courtesy of Peter Kim and James Berger, MIT. PDBid 1AJJ.]

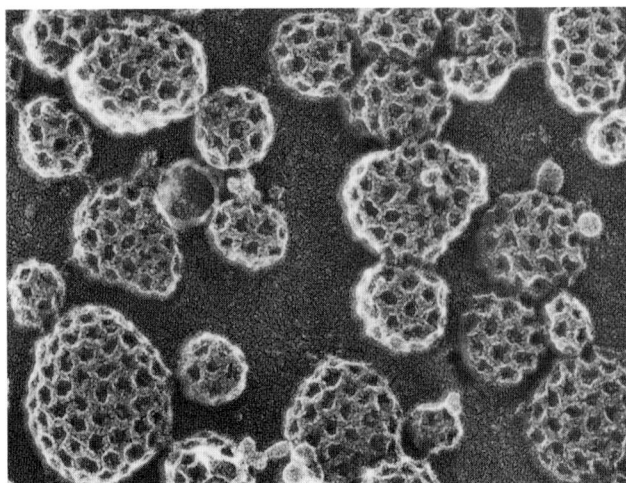

FIGURE 12-77   Freeze-etch electron micrograph of coated pits on the inner surface of a cultured fibroblast's plasma membrane. Compare this figure with that of clathrin-coated vesicles (Fig. 12-52*a*). [Courtesy of John Heuser, Washington University School of Medicine, St. Louis, Missouri.]

form a 2-turn right-handed helix (Fig. 12-76). The LR5 structure appears to be in large part organized about a Ca^{2+} ion that is octahedrally liganded mainly by conserved acidic side chains near LR5's C-terminal end. Interestingly, the overall topology of LR5 resembles those of the homologous LR1 and LR2, as determined by NMR, even though there is no indication that these latter modules bind Ca^{2+}.

LDLRs cluster into **coated pits,** which serve to gather the cell-surface receptors that are destined for endocyto-

sis while excluding other cell-surface proteins. The coated pits, which have a clathrin backing (Fig. 12-77), invaginate into the plasma membrane to form clathrin-coated vesicles (Section 12-4C) that subsequently fuse with lysosomes (Fig. 12-78). *Such **receptor-mediated endocytosis** (Fig. 12-79) is a general mechanism whereby cells take up large molecules, each through a corresponding specific receptor.* Indeed, the liver takes up chylomicron remnants in this manner through the mediation of a separate **remnant receptor** that specifically binds apoE.

In the lysosome, as demonstrated by radioactive labeling studies, the LDL's apoB-100 is rapidly degraded to its component amino acids (Fig. 12-79). The cholesteryl esters

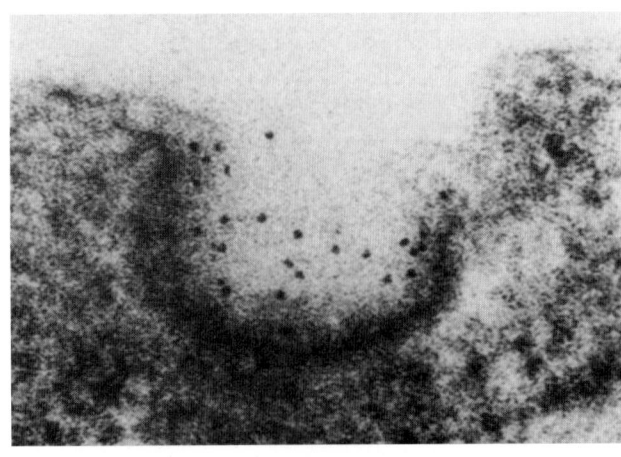

(a)

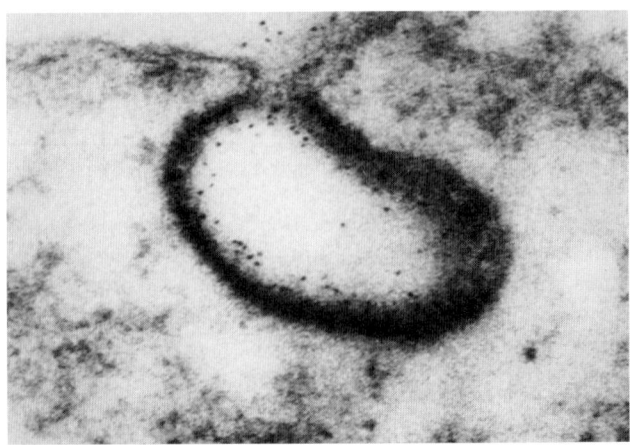

(b)

FIGURE 12-78   Electron micrographs showing the endocytosis of LDL by cultured human fibroblasts. The LDL was conjugated to ferritin so that it appears as dark dots. (*a*) LDL bound to a coated pit on the cell surface. (*b*) The coated pit invaginates and begins to pinch off from the cell membrane to form a coated vesicle enclosing the bound LDL. [From Anderson, R.G.W., Brown, M.S., and Goldstein, J.L., *Cell* **10,** 356 (1977). Copyright © 1977 by Cell Press.]

are hydrolyzed by a lysosomal lipase to yield cholesterol, which is subsequently incorporated into the cell membranes. Any excess intracellular cholesterol is reesterified for storage within the cell through the action of **acyl-CoA: cholesterol acyltransferase (ACAT).**

The overaccumulation of cellular cholesteryl esters is prevented by two feedback mechanisms:

1. High intracellular levels of cholesterol suppress the synthesis of LDLR, thus decreasing the rate of LDL accumulation by endocytosis (although LDLR cycles in and out of the cell every 10 to 20 min, it is slowly degraded by the cell such that its half-life is ~20 h).

2. Excess intracellular cholesterol inhibits the biosynthesis of cholesterol (Section 25-6B).

d. ApoE's Receptor Binding Domain Contains a Four-Helix Bundle

ApoE is a 299-residue monomeric protein that consists of two independently folded domains: an N-terminal domain that binds strongly to the LDLR but only weakly to lipid and a C-terminal domain that binds to the lipoprotein surface but lacks affinity for the LDLR. Proteolysis of apoE yields fragments corresponding to apoE's N-terminal domain (residues 1–191) and C-terminal domain (residues 216–299). Sequence analysis suggests that the C-terminal

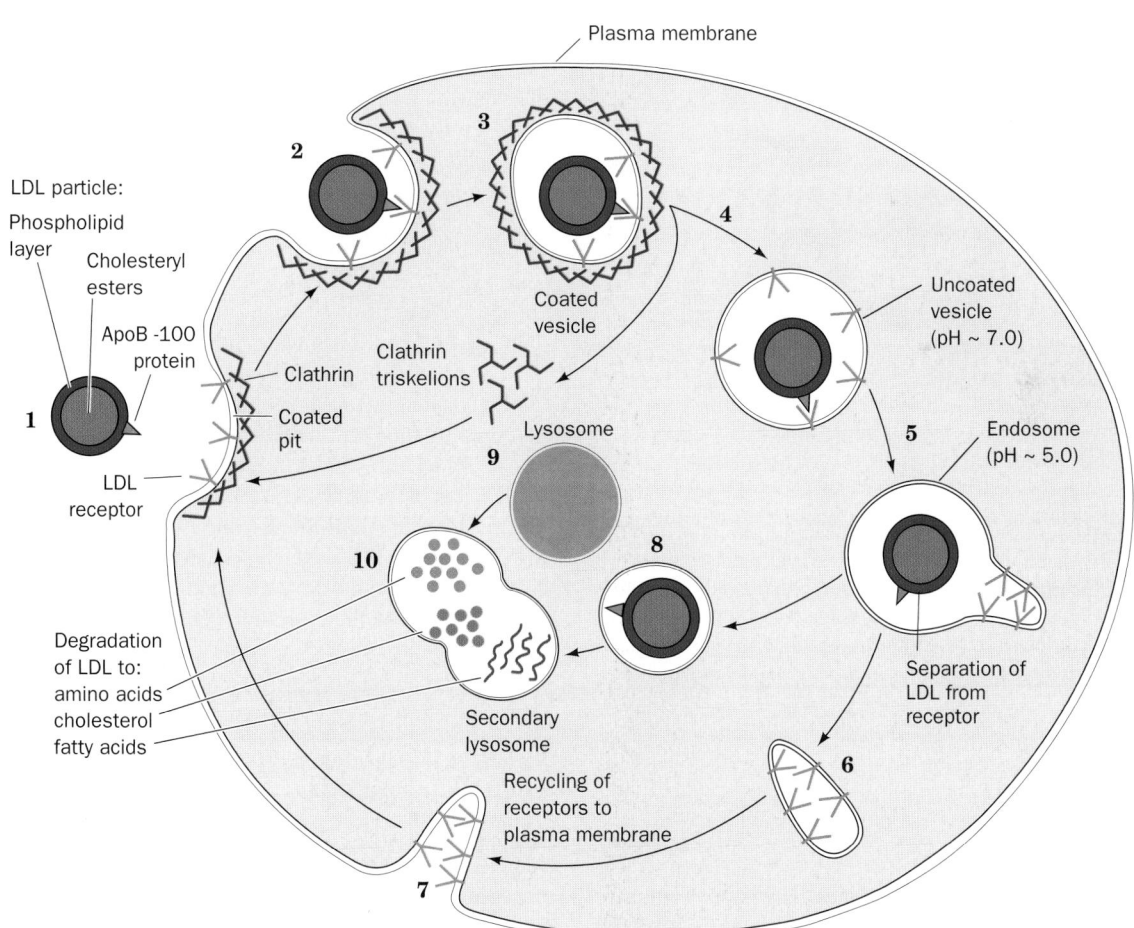

FIGURE 12-79 Sequence of events in the receptor-mediated endocytosis of LDL. The LDL specifically binds to LDL receptors on clathrin-coated pits **(1).** These bud into the cell **(2)** to form coated vesicles **(3),** whose clathrin coats depolymerize as triskelions, resulting in the formation of uncoated vesicles **(4).** These vesicles then fuse with vesicles called **endosomes (5),** which have an internal pH of ~5.0. The acidity induces the LDL to dissociate from its receptor. The LDL accumulates in the vesicular portion of the endosome, whereas the LDL receptors concentrate in the membrane of an attached tubular structure, which then separates from the endosome **(6)** and subsequently recycles the LDL receptors to the plasma membrane **(7).** The vesicular portion of the endosome **(8)** fuses with a lysosome **(9),** yielding a **secondary lysosome (10),** wherein the apoB-100 component of the LDL is degraded to its component amino acids and the cholesteryl esters are hydrolyzed to yield cholesterol and fatty acids.

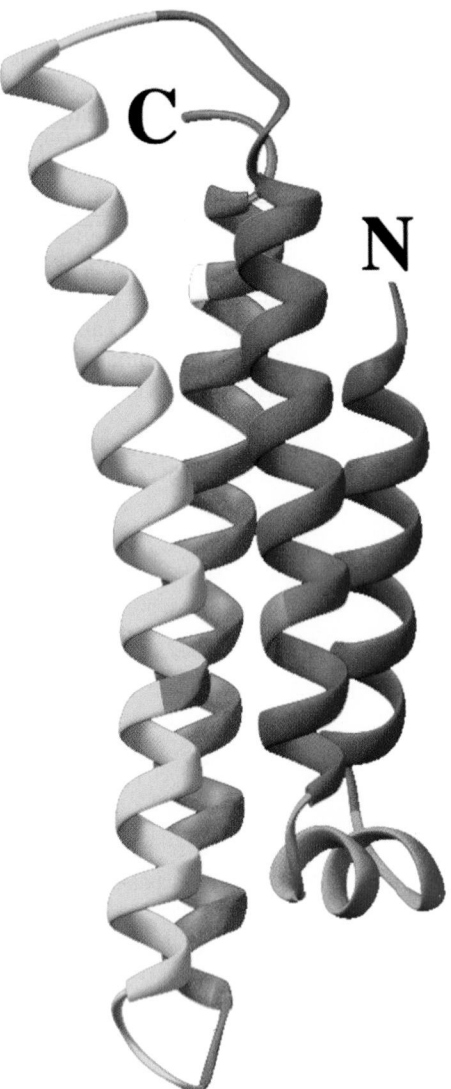

FIGURE 12-80   Ribbon diagram of the receptor-binding domain of human apolipoprotein E. The helices are colored, from N- to C-terminus, in rainbow order, blue to red. Residues 61, 112, and 158 are colored orange, magenta, and white, respectively. [Based on an X-ray structure by David Agard, University of California at San Francisco. PDBid 1LPE.]

domain is largely composed of helices. The X-ray structure of the N-terminal domain (Fig. 12-80), determined by David Agard, reveals that it consists mainly of five α helices, four of which form an elongated (65 Å) up–down–up–down four-helix bundle. The helices of the four-helix bundle, as expected, are strongly amphipathic, with their hydrophobic residues sequestered inside the protein, out of contact with solvent, whereas their hydrophilic residues are solvent-exposed. The structure appears to be further stabilized by numerous salt bridges on the protein's highly charged surface.

The C-terminal helix of the apoE N-terminal fragment contains nine closely spaced basic residues that are not involved in salt bridges, thereby producing a large positively charged patch on the protein surface. ApoE variants in which one of these basic residues is replaced by a neutral or acidic residue all display reduced affinity for the LDLR, thereby suggesting that the patch forms apoE's binding site for the LDLR. Hence this C-terminal helix has been dubbed apoE's receptor-binding helix.

The LDLR binds apoB-100 and apoE with comparable affinities. ApoB-100 (but not apoB-48) contains a conserved segment that is similar to apoE's receptor-binding helix, although the two proteins otherwise have no apparent sequence similarity. In VLDL, the receptor-binding domain of apoB-100 is unavailable for receptor binding but is exposed on transformation of the VLDL to LDL.

e. HDL Transports Cholesterol from the Tissues to the Liver

HDL has essentially the opposite function of LDL: *It removes cholesterol from the tissues.* HDL is assembled in the plasma from components obtained largely through the degradation of other lipoproteins. *Circulating HDL acquires its cholesterol by extracting it from cell-surface membranes and converts it to cholesteryl esters through the action of LCAT, an enzyme that is activated by apoA-I. HDL therefore functions as a cholesterol scavenger.*

The liver is the only organ capable of disposing of significant quantities of cholesterol (by its conversion to bile acids; Section 25-6C). This occurs through the mediation of both the LDLR and a specific HDL receptor named **SR-BI** (for *s*cavenger *r*eceptor class *B* type *I*). About half of the VLDL, after its degradation to IDL and LDL, is taken up by the liver via LDLR-mediated endocytosis (Fig. 12-74, *right*). However, hepatocytes (liver cells) take up cholesteryl esters from HDL by an entirely different mechanism: Rather than being engulfed and degraded, the SR-BI–bound HDL selectively transfers its component cholesteryl esters to the cell. The lipid-depleted HDL then dissociates from the cell and reenters the circulation.

C. *Lipoprotein Dysfunction in Atherosclerosis and Alzheimer's Disease*

Atherosclerosis, the most common form of **arteriosclerosis** (hardening of the arteries), is characterized by the presence of **atheromas** (Greek: *athera,* mush), arterial thickenings that, on sectioning, exude a pasty yellow deposit of almost pure cholesteryl esters.

Atherosclerosis is a progressive disease that begins as intracellular lipid deposits in the smooth muscle cells of the inner arterial wall. These lesions eventually become fibrous, calcified plaques that narrow and even block the arteries. The resultant roughening of the arterial wall promotes the formation of blood clots, which may also occlude the artery. A blood flow stoppage, known as an **infarction,** causes the death of the deprived tissues. Although atheromas can occur in many different arteries, they are most common in the coronary arteries, the arteries supplying the heart. This results in **myocardial infarctions** or "heart attacks," the most common cause of death in Western industrialized countries.

a. Deficient LDL Receptors Result in Atherosclerosis

The development of atherosclerosis is strongly correlated with the level of plasma cholesterol. This is particularly evident in individuals with **familial hypercholesterolemia (FH).** Homozygotes with this inherited disorder have such high levels of the cholesterol-rich LDL in their plasma that their plasma cholesterol levels are three- to fivefold greater than the average level of ~175 mg · 100 mL^{-1}. This situation results in the deposition of cholesterol in their skin and tendons as yellow nodules known as **xanthomas.** However, far greater damage is caused by the rapid formation of atheromas that, in homozygotes, cause death from myocardial infarction as early as the age of 5. Heterozygotes, which comprise ~1 person in 500, are less severely afflicted; they develop symptoms of coronary artery disease after the age of 30.

Cells taken from FH homozygotes completely lack functional LDLR, whereas those taken from heterozygotes have about half of the normal complement. Homozygotes and, to a lesser extent, heterozygotes are therefore unable to utilize the cholesterol in LDL. Rather, their cells must synthesize most of the cholesterol for their needs. The high level of plasma LDL in these individuals results from two related causes:

1. Its decreased rate of degradation because of the lack of LDLR.

2. Its increased rate of synthesis from IDL due to the failure of LDLR to take up IDL.

Many of the LDLR's point mutations that cause FH alter residues in LR5 (Fig. 12-76). Some of these eliminate or misplace Ca^{2+}-binding ligands, whereas others delete hydrogen bonds or disulfide bonds that stabilize the LR5 backbone structure. Since apoproteins must be associated with lipids in order to bind to the LDLR with high affinity, the LR5 structure suggests that LDL binds to the hydrophobic concave face that is located on the top of LR5, as shown in Fig. 12-76.

b. Scavenger Receptors Take Up Oxidized LDL

Atherosclerotic plaques in individuals with FH contain **macrophages** (a type of white blood cell that ingests and, if possible, destroys a variety of foreign and endogenous substances) that contain so much cholesterol that they are known as **foam cells.** How do macrophages take up cholesterol? Macrophages from both normal individuals and those with FH have few LDLRs and therefore take up little native LDL. However, they avidly take up, by endocytosis, LDL that has been chemically modified by the acetylation of its Lys residues (which eliminates this side chain's positive charge, thereby increasing LDL's negative charge). The macrophage cell-surface receptors that bind

acetylated LDL are known as **scavenger receptors** because they also bind certain other polyanionic molecules.

Scavenger receptors avidly take up oxidized LDL. The unsaturated fatty acids of LDL are highly susceptible to chemical oxidation but, in the blood, are protected from oxidation by antioxidants. However, these antioxidants are thought to become depleted when LDL is trapped for an extended time within the artery walls (where it is thought to gain access by an injury to the arterial lining through which plasma leaks into the arterial wall). As a consequence, oxygen radicals convert LDL's unsaturated fatty acids to aldehydes and oxides that react with its Lys residues, thereby mimicking acetylation. The ingested LDL is degraded as described above and its cholesterol converted to cholesteryl esters which accumulate as insoluble residues.

The physiological significance of this scenario has been demonstrated by the observations that antibodies against aldehyde-conjugated Lys residues stain atherosclerotic plaques, that LDL from atherosclerotic plaques binds to scavenger receptors and produce foam cells *in vitro,* and that antioxidants inhibit atherosclerosis in rabbits that have an animal counterpart of FH. It should be noted that tobacco smoke oxidizes LDL, which may explain why smoking leads to an increased incidence of atherosclerosis. High plasma levels of LDL, of course, also accelerate LDL uptake.

If this model of atheroma formation is correct, then *the optimal level of plasma LDL is the lowest concentration that can adequately supply cholesterol to the cells.* Such a level, which is thought to be ~25 mg of cholesterol · 100 mL^{-1}, occurs in various mammalian species that are not naturally susceptible to atherosclerosis as well as in newborn humans. Yet the plasma LDL level in adult Western men averages ~7-fold higher than this supposed optimal level. The reasons for such a high plasma cholesterol level are poorly understood (but see below), although it is clear that it is affected by diet and by environmental stress. Medical strategies for reducing the level of plasma cholesterol are considered in Section 25-6B.

c. Atherosclerosis Is a Multifactorial Disease

Epidemiological studies indicate that high plasma HDL levels are strongly correlated with a low incidence of cardiovascular disease. Women have HDL levels higher than men and also less heart disease. Many of the factors that decrease the incidence of heart disease also tend to increase HDL levels. These include strenuous exercise, weight loss, certain drugs such as alcohol, and female sex hormones known as **estrogens** (Section 19-1G). Conversely, cigarette smoking is inversely related to HDL concentration. Curiously, in communities that have a very low incidence of coronary artery disease, both the mean HDL and LDL concentrations are low. The reasons for these various effects are unknown.

There is also a strong inverse correlation in humans between the risk for atherosclerosis and the plasma level of apoA-I, HDL's major protein component, which is required for its assembly. To investigate whether apoA-I has a direct antiatherogenic effect, mice of a strain that develops diet-induced **fatty streak lesions** in their large blood vessels were genetically modified to express high plasma levels of human apoA-I (fatty streak lesions are the precursors of atherosclerotic plaques, which these mice have too short a lifetime to develop). These transgenic mice are significantly protected from developing fatty streak lesions. Yet transgenic mice that overexpress mouse **apoA-II,** another major HDL protein, develop more and larger fatty streak lesions then their nontransgenic counterparts. Since the plasma HDL–cholesterol levels in the latter transgenic mice are significantly elevated, it appears that both the composition and the level of plasma HDL are important atherosclerotic mediators. Similarly, transgenic mice that express high levels of human apoE or human LDLR resist the elevation in plasma LDL levels that would otherwise be brought on by a cholesterol-rich diet, whereas mice in which the gene encoding apoE has been knocked out rapidly develop atherosclerotic lesions.

Cholesteryl ester transfer protein (CETP) is a plasma protein that exchanges neutral lipids (e.g., cholesteryl esters and triacylglycerols) among lipoproteins and hence functions analogously to phospholipid exchange proteins (Section 12-4A). Since VLDL and LDL are triacylglycerol-rich whereas HDL are cholesteryl ester–rich (Table 12-6), CETP mediates the net transport of cholesteryl esters from HDL to VLDL and LDL (and of triacylglycerols in the opposite direction). Consequently, animals that express CETP have higher cholesterol levels in their VLDL and LDL and lower cholesterol levels in their HDL than animals that do not express CETP. Mice of a strain that normally have little or no CETP activity were made transgenic for CETP and fed an atherogenic (high-fat, high-cholesterol) diet. These transgenic mice developed atherosclerotic lesions far more rapidly than their similarly fed nontransgenic counterparts. Since the two types of mice had similar total plasma cholesterol levels, these results suggest that *the progression of atherosclerotic lesions is more a function of how cholesterol is partitioned between lipoproteins than it is of the total plasma cholesterol level.*

Increased risk of atherosclerosis in humans is also associated with elevated plasma levels of lipoprotein **Lp(a)**, a variant of LDL in which apoB-100 is tightly associated with the 4259-residue plasma protein **apo(a).** Rodents and most other nonprimate mammals lack the gene for apo(a). However, mice transgenic for human apo(a) rapidly develop fatty streak lesions when given a high-fat diet (approximating human diets in industrialized Western countries). Apo(a) mainly consists of repeated segments that are homologous to **plasminogen,** a plasma protein that, when activated, functions to proteolytically dismantle blood clots (Section 35-1F). The normal function of apo(a) in humans is unknown, although it has been hypothesized that it participates in healing blood vessel wounds.

d. ApoE4 Is Implicated in Both Cardiovascular Disease and Alzheimer's Disease

There are three common allelic variants of apoE in humans: **apoE2** (occurring in 15% of the population), which

has Cys at positions 112 and 158; **apoE3** (78% occurrence), in which these residues are Cys and Arg, respectively (Fig. 12-80 shows the structure of apoE3 with residues 112 and 158 magenta and white); and **apoE4** (7% occurrence), in which these residues are both Arg. These differences have medical significance: ApoE3 has a preference for binding to HDL, whereas apoE4 has a preference for binding to VLDL, which is probably why apoE4 is associated with elevated plasma concentrations of LDL and thus an increased risk of cardiovascular disease. Evidently, changes in apoE's N-terminal domain can affect the function of its C-terminal lipoprotein-binding domain.

ApoE4, as we have seen in Section 9-5B, is also associated with a greatly (16-fold) increased incidence of Alzheimer's disease (AD). This observation is perhaps less surprising when it is realized that apoE is expressed by certain nerve cells and is present in the cerebrospinal fluid, where it functions in mediating cholesterol transport, much as is does in blood plasma (cholesterol is abundant in nerve cell plasma membranes, which mediate neurotransmission; Section 20-5C).

Brain tissue from AD victims reveals numerous extracellular amyloid plaques, which consist of fibrillar deposits of amyloid β (Aβ) peptide that arises through proteolysis of the normally occurring amyloid precursor protein (Section 9-5B). Amyloid plaques appear to be AD's pathogenic agent. Immunochemical staining indicates that apoE is associated with amyloid plaques. *In vitro* experiments demonstrate that both apoE3 and apoE4 form SDS-stable complexes with Aβ peptide that, after long incubation times, aggregate and precipitate from solution as a matrix of fibrils that closely resemble those in amyloid plaques. ApoE4 forms this complex more readily than apoE3 and yields a denser, more extensive matrix.

Comparison of the X-ray structures of apoE4 and apoE3 reveals that there are only minor differences in their backbone conformations, which are restricted to the immediate vicinity of their site of difference (residue 112: Cys in ApoE3 and Arg in ApoE4). The only two side chains in apoE4 that undergo changes in conformation relative to those in apoE3 are Glu 109, which swings around in apoE4 to form a salt bridge with Arg 112, and Arg 61 (orange in Fig. 12-80), which contacts Cys 112 in apoE3 but swings away to accommodate the new salt bridge in apoE4. Thus, both Glu 109 and Arg 61 are candidates for mediating the functional differences between apoE3 and apoE4. However, the mutagenic substitution of Ala for Glu 109 in apoE3 does not significantly alter its preference for binding to HDL over VLDL. In contrast, the substitution of Thr for Arg 61 in apoE4 gives this protein an apoE3-like preference for HDL over VLDL. Evidently, the position of Arg 61 is critical in determining the HDL/VLDL preference of apoE. This hypothesis is supported by the observation that residue 61 is invariably Thr in the 10 apoEs of known sequences from other species. None of these species exhibits the complete pathology of AD, although it remains to be demonstrated that Arg 61 actually contributes to the differential binding of apoE3 and apoE4 to Aβ peptide.

e. ApoE2 Has a Low Affinity for LDL Receptor

ApoE2 binds to the LDLR with only 0.1% of the affinity of apoE3 or apoE4. Thus the presence of apoE2 is the underlying cause of **familial type III hyperlipoproteinemia,** which is characterized by elevated plasma cholesterol and triglyceride levels and hence accelerated coronary artery disease.

The defective binding of apoE2 to the LDLR is caused by the substitution of Cys for Arg 158, a position (white in Fig. 12-80) that lies outside the previously identified receptor-binding region, residues 136 to 150 (located in the bottom half of the C-terminal helix in Fig. 12-80). In apoE3, Asp 154 forms a salt bridge with Arg 158 (which is situated one turn farther along the α helix). In apoE2, since Arg 158 is replaced with Cys, this salt bridge cannot

form. Rather, as the X-ray structure of apoE2 reveals, Asp 154 forms a salt bridge with Arg 150 (which is situated one turn earlier on the α helix), thereby altering the side chain conformation of this LDLR-binding residue. In fact, the disruption of this abnormal salt bridge by the mutagenic replacement of Asp 154 in ApoE2 with Ala restores LDLR binding affinity to a nearly normal level.

Individuals with type III hyperlipoproteinemia are particularly responsive to a low-fat, low-calorie diet and to a reduction in body weight. It is therefore postulated that the altered lipid composition of lipoproteins under such a regimen causes significant amounts of apoE2 to assume a receptor-active conformation, leading to normal or near-normal rates of lipoprotein clearance from the circulation.

CHAPTER SUMMARY

1 ■ Lipid Classification Fatty acids are long-chain carboxylic acids that may have one or more double bonds that are usually cis. Their anions are amphiphilic molecules that form micelles in water. Fatty acids rarely occur free in nature but rather are components of lipids. The most abundant class of lipids, the triacylglycerols or neutral fats, are nonpolar molecules that constitute the major nutritional store of animals. The lipids that occur in membranes are the phospholipids, the sphingolipids, and, in eukaryotes, cholesterol or similar sterols. Sphingolipids such as cerebrosides and gangliosides have complex carbohydrate head groups that act as specific recognition markers in various biological processes.

2 ■ Properties of Lipid Aggregates The molecular shapes of membrane lipids cause them to aggregate in aqueous solution as bilayers. These form closed vesicles known as liposomes that are useful model membranes and drug delivery systems. Bilayers are essentially impermeable to polar molecules, except for water. Likewise, the flip-flop of a lipid in a bilayer is an extremely rare event. In contrast, bilayers above their transition temperatures behave as two-dimensional fluids in which the individual lipid molecules freely diffuse in the bilayer plane. Cholesterol decreases membrane fluidity and broadens the temperature range of its order–disorder transition by interfering with the orderly packing of the lipids' fatty acid side chains.

3 ■ Biological Membranes Biological membranes contain a high proportion of proteins. Integral proteins, for example, bacteriorhodopsin, the photosynthetic reaction center, porins, and cyclooxygenase, have nonpolar surface regions that hydrophobically associate with the bilayer core. Peripheral proteins, for example, cytochrome *c,* bind to integral proteins on the membrane surface or to phospholipid head groups via polar interactions. Specific integral proteins are invariably associated with a particular side of the membrane or, if they are transmembrane proteins, have only one orientation. Lipid-linked proteins contain covalently attached isoprenoid, fatty acyl, and/or glycosylphosphatidylinositol (GPI) groups that serve to anchor these proteins to membranes and to mediate protein–protein interactions. According to the fluid mosaic model of membrane structure, integral proteins resemble icebergs floating on a two-dimen-

sional lipid sea. These proteins, as observed by the freeze-fracture and freeze-etch techniques, are randomly distributed in the membrane. Certain lipids and/or proteins may form specific aggregates on one leaflet of a membrane.

The erythrocyte cytoskeleton is responsible for the shape, flexibility, and fluidity of the red cell. Spectrin, the major constituent of the cytoskeleton, is a wormlike $(\alpha\beta)_2$ heterotetramer that is cross-linked by actin oligomers and band 4.1 protein. The resulting protein meshwork is anchored to the membrane by the association of spectrin with ankyrin, which, in turn, binds to band 3 protein, a transmembrane protein that forms an anion channel.

The erythrocyte surface bears the various blood group antigens. The antigens of the ABO system differ in the sugar at a nonreducing end. The ABO blood group substances occur in the plasma membranes of many cells and in the secretions of many individuals.

Gap junctions are hexagonal transmembrane protein tubes that link adjoining cells. The gap junction's central channel, which closes at high intracellular levels of Ca^{2+}, allows small molecules and ions but not macromolecules to pass between cells. The connexin subunits of a gap junction's two face-to-face hexameric connexons each contain four transmembrane helices.

Bacterial channel-forming toxins such as α-hemolysin form oligomers on the outer surface of a target cell's plasma membrane. These insert themselves into the membrane to form pores through which small molecules and ions leak, thereby killing the cell.

4 ■ Membrane Assembly and Protein Targeting New membranes are generated by the expansion of old ones. Lipids are synthesized by membrane-bound enzymes and are deposited on one side of the membrane. They migrate to the other side by flip-flops that are catalyzed by membrane-bound flipases and phospholipid translocases. In eukaryotes, lipids are transported between different membranes by lipid vesicles or by phospholipid-exchange proteins.

In the secretory pathway, transmembrane proteins and proteins destined for secretion are ribosomally synthesized with an N-terminal signal sequence. A signal peptide is bound by an RNA-containing signal recognition particle (SRP), which thereupon arrests polypeptide synthesis. The SRP–ribosome complex then binds to the SRP receptor (SR) in complex with the translocon on the endoplasmic reticulum (ER) membrane and, on GTP hydrolysis by both the SRP and SR, resumes polypeptide synthesis. As a protein destined for secretion passes through the translocon into the ER lumen, its signal peptide is removed by an ER-resident signal peptidase, its folding is facilitated through interactions with ER-resident chaperone proteins such as BiP, and its posttranslational processing, mainly signal peptide excision and glycosylation, is initiated. Integral proteins, whose transmembrane (TM) segments each contain signal-anchor sequences, also enter the translocon, which laterally installs these TM segments into the ER membrane.

Proteins are transferred between the ER, the Golgi apparatus (where further posttranslational processing takes place), and their final destinations via membranous vesicles that are coated with clathrin, COPI, or COPII. Clathrin-coated vesicles also participate in endocytosis. Polyhedral clathrin cages are formed by triskelions, which are trimers of heavy chains, each of which binds a light chain. Clathrin-coated vesicle formation is primed by the action of ARNO, a guanine nucleotide exchange factor (GEF) that induces the small GTPase ARF1 to exchange its bound GDP for GTP and thereupon insert its myristoyl group into the membrane. ARF1 · GTP then recruits adapter proteins such as AP1 and AP2, which simultaneously bind clathrin heavy chains and TM proteins that are cargo proteins or are receptors for soluble cargo proteins inside the vesicle. The formation of the clathrin cage drives vesicle budding, but its actual release from its parent membrane requires the action of the GTPase dynamin. Shortly after its release, the vesicle uncoats in a process mediated by the chaperone protein Hsc70. COPI- and COPII-coated vesicles undergo similar processes, although they do not require a dynaminlike protein to bud off from their parent membranes. The receptors in coated vesicles bind their target cargo proteins through specific signals such as the mannose-6-phosphate group that directs proteins to the lysosome or the C-terminal KDEL sequence that retrieves normally ER-resident proteins from the Golgi to the ER.

The fusion of a vesicle with its target membrane is initiated when a Rab protein, a small GTPase, loosely joins (tethers) the two membranes. The vesicle is then more firmly anchored (docked) to the membrane through interactions between cognate R-SNAREs on the vesicle and Q-SNAREs on the target membrane. In neurons, synaptic vesicles dock with the presynaptic membrane through the association of the R-SNARE synaptobrevin (VAMP) with the Q-SNAREs syntaxin and SNAP-25 to form a 4-helix bundle. These neuronal SNAREs are specifically cleaved by tetanus and botulinal neurotoxins. The final steps of membrane fusion may be catalyzed by other proteins such as V0 in the fusion of yeast vacuoles. The neuronal SM protein nSec1 binds to syntaxin with high affinity so as to prevent the formation of the SNARE complex. Rab protein and/or its effectors apparently induce nSec1 to release syntaxin and thereby permit the formation of the SNARE complex. After vesicle fusion, the SNARE complex must be dissociated in order to be recycled. This occurs through the auspices of the ATP-driven molecular chaperone NSF, which binds to the SNARE complex through the intermediacy of a SNAP protein.

Nuclear-encoded mitochondrial proteins are synthesized by cytosolic ribosomes and enter the mitochondrion posttranslationally. A protein can only pass through a membrane in its unfolded form and hence must first be unfolded by ATP-driven molecular chaperones such as Hsp70 and MSF. A matrix-directed protein is passed through the mitochondrial outer membrane via the TOM complex, which recognizes the protein's positively charged N-terminal signal sequence. The N-terminal presequence then crosses the intermembrane space to encounter a TIM complex, which translocates it through the inner membrane into the matrix. This latter process is driven by both the mitochondrial membrane potential, which electrophoretically draws the positively charged presequence into the matrix, and by the ATP-driven chaperone mHsp70, which binds to Tim44 and ratchets the unfolded protein into the matrix. MPP then excises the presequence from the protein, which is subsequently folded to its native state by a battery of resident chaperones including mHsp70 and Hsp60/Hsp10. Metabolite carrier proteins, which lack N-terminal presequences but have internal targeting sequences, also enter the intermembrane space via the TOM complex. However, they are then escorted across the intermembrane space by a complex of Tim9 and Tim10, which conducts them to a different TIM complex that laterally inserts them into the inner mitochondrial membrane. Many proteins destined for insertion into the inner membrane or residence in the intermembrane space first enter the matrix. On removal of the N-terminal presequence from their bipartite signal sequence, they are translocated to their final destinations in a memento of the mitochondrion's prokaryotic origin. Apocytochrome *c* freely passes through the outer mitochondrial membrane via the porinlike protein P70 into the intermembrane space, where it is trapped by the CCHL-catalyzed attachment of its heme group.

5 ■ Lipoproteins   Lipids are transported in the circulation by plasma lipoproteins. These are essentially droplets of triacylglycerols and cholesteryl esters coated with a monolayer of phospholipids, cholesterol, and apolipoproteins. The amphiphilic apolipoprotein helices float on the lipoprotein surface in hydrophobic contact with its lipid interior. Chylomicrons and VLDL function to respectively transport triacylglycerols and cholesterol from the intestines and the liver to the tissues. HDL transports mainly cholesterol from the tissues to the liver, the only organ capable of disposing of significant quantities of cholesterol. The triacylglycerols of chylomicrons and VLDL are degraded by lipoprotein lipase that lines the capillaries. LDL, the cholesterol-containing degradation product of VLDL, binds to cell-surface LDL receptors (LDLRs) and is taken into cells by receptor-mediated endocytosis. The presence of excess intracellular cholesterol inhibits the synthesis of both LDLR and cholesterol. A major cause of atherosclerosis is an excess of plasma LDL, a phenomenon that is particularly evident in individuals with familial hypercholesterolemia, who lack functional LDLRs. The excess LDL becomes oxidized and is taken up by the macrophages that inhabit atherosclerotic plaques via their scavenger receptors. Atherosclerosis, a multifactorial disease, is also correlated with a low concentration of HDL, which functions as a cholesterol scavenger. The apoE2 and apoE4 variants of apoE are implicated in cardiovascular disease, whereas apoE4 is also implicated in Alzheimer's disease.

REFERENCES

GENERAL

Finean, J.B., Coleman, R., and Michell, R.H., *Membranes and Their Cellular Functions* (3rd ed.), Blackwell (1984).

Jain, M.K., *Introduction to Biological Membranes* (2nd ed.), Wiley (1988).

Mellman, I. and Warren, G., The road taken: Past and future foundations of membrane traffic, *Cell* **100**, 99–112 (2000). [An intellectual history and review of membrane trafficking.]

Robertson, R.N., *The Lively Membranes*, Cambridge University Press (1983).

Tanford, C., *The Hydrophobic Effect: Formation of Micelles and Biological Membranes* (2nd ed.), Wiley–Interscience (1980). [An exposition of the thermodynamic properties of micelles and membranes.]

LIPIDS AND BILAYERS

Cullis, R.R. and Hope, M.J., Physical properties and functional roles of lipids in membranes, *in* Vance, D.E. and Vance, J. (Eds.), *Biochemistry of Lipids, Lipoproteins and Membranes*, Elsevier (1991).

Giles, C.H., Franklin's teaspoon of oil, *Chem. Ind.*, 1616–1624 (1969). [A historical account of Benjamin Franklin's investigations into the effect of oil on waves.]

Gurr, M.I. and Harwood, J.L., *Lipid Biochemistry: An Introduction* (4th ed.), Chapman and Hall (1991).

Hakomori, S., Glycosphingolipids, *Sci. Am.* **254**(5), 44–53 (1986).

Harwood, J.L., Understanding liposomal properties to aid their clinical usage, *Trends Biochem. Sci.* **17**, 203–204 (1992).

Lasic, D.D., Novel applications of liposomes, *Trends Biotech.* **16**, 307–321 (1998). [An authoritative review on the uses of liposomes as delivery vehicles for drugs, vaccines, gene therapy agents, and cosmetic products.]

Lasic, D.D. and Papahadjopoulos, D. (Eds.), *Medical Applications of Liposomes*, Elsevier (1998).

Scott, L.H., Modeling the lipid component of membranes, *Curr. Opin. Struct. Biol.* **12**, 495–502 (2002).

Storch, J. and Kleinfeld, A.M., The lipid structure of biological membranes, *Trends Biochem. Sci.* **10**, 418–421 (1985).

MEMBRANE PROTEINS

Cowan, S.W., Bacterial porins: Lessons from three high-resolution structures, *Curr. Opin Struct. Biol.* **3**, 501–507 (1993).

Cowan, S.W., Schirmer, T., Rummel, G., Steiert, M., Ghosh, R., Pauptit, R.A., Jansonius, J.N., and Rosenbusch, J.P., Crystal structures explain functional properties of two *E. coli* porins, *Nature* **358**, 727–733 (1992).

Deisenhofer, J. and Michel, H., High-resolution structures of photosynthetic reaction centers, *Annu. Rev. Biophys. Biophys. Chem.* **20**, 247–266 (1991); *and* Deisenhofer, J., Epp, O., Miki, K., Huber, R., and Michel, H., Structure of the protein subunits in the photosynthetic reaction centre of *Rhodopseudomonas viridis* at 3 Å resolution, *Nature* **318**, 618–624 (1985).

Grigorieff, N., Ceska, T.A., Downing, K.H., Baldwin, J.M., and Henderson, R., Electron-crystallographic refinement of the structure of bacteriorhodopsin, *J. Mol. Biol.* **259**, 393–421 (1996); *and* Belrhali, H., Nollert, P., Royant, A., Menzel, C., Rosenbusch, J.P., Landau, E.M., and Pebay-Peyroula, E., Protein, lipid and water organization in bacteriorhodopsin crystal: A molecular view of the purple membrane at 1.9 Å resolution, *Structure* **7**, 909–917 (1999).

Haupts, U., Tittor, J., and Oesterhelt, D., Closing in on bacteriorhodopsin: Progress in understanding the molecule, *Annu. Rev. Biophys. Biomol. Struct.* **28**, 67–99 (1999).

Killian, J.A. and von Heijne, G., How proteins adapt to a membrane–water interface, *Trends Biochem. Sci.* **25**, 429–434 (2000).

Lemmon, M.A. and Engelman, D.M., Helix–helix interactions inside lipid bilayers, *Curr. Opin. Struct. Biol.* **2**, 511–518 (1992).

Pebay-Peyroula, E., Rummel, G., Rosenbusch, J.P., and Landau, E.M., X-Ray structure of bacteriorhodopsin at 2.5 angstroms from microcrystals grown in lipidic cubic phases, *Science* **277**, 1676–1681 (1997); *and* Gouaux, E., It's not just a phase: Crystallization and X-ray structure of bacteriorhodopsin in lipidic cubic phases, *Structure* **6**, 5–10 (1998).

Picot, D., Loll, P.J., and Garavito, R.M., The X-ray crystal structure of the membrane protein prostaglandin synthase H$_2$-1, *Nature* **367**, 243–249 (1994).

Popot, J.-L. and Engelman, D.M., Helical membrane protein folding, stability, and evolution, *Annu. Rev. Biochem.* **69**, 881–922 (2000).

Rees, D.C., De Antonio, L., and Eisenberg, D., Hydrophobic organization of membrane proteins, *Science* **245**, 510–512 (1989).

Schulz, G.E., β-Barrel membrane proteins, *Curr. Opin. Struct. Biol.* **10**, 443–447 (2000).

Stowell, M.H.B. and Rees, D.C., Structure and stability of membrane proteins, *Adv. Prot. Chem.* **46**, 279–311 (1995).

Subramanium, S., The structure of bacteriorhodopsin: An emerging consensus, *Curr. Opin. Struct. Biol.* **9**, 462–468 (1999). [Compares the six structures of bacteriorhodopsin that have been independently determined by electron or X-ray crystallography and finds them to be remarkably similar.]

Weiss, M.S. and Schulz, G.E., Structure of porin refined at 1.8 Å resolution, *J. Mol. Biol.* **227**, 493–509 (1992); *and* Weiss, M.S., Wacker, T., Weckesser, J., Welte, W., and Schulz, G.E., The three-dimensional structure of porin from *Rhodobacter capsulatus* at 3 Å resolution, *FEBS Lett.* **267**, 268–272 (1990).

White, S.H. and Wimley, W.C., Membrane protein folding and stability: Physical principles, *Annu. Rev. Biophys. Biomol. Struct.* **28**, 319–365 (1999).

LIPID-LINKED PROTEINS

Clarke, S., Protein isoprenylation and methylation at carboxyl-terminal cysteine residues, *Annu. Rev. Biochem.* **61**, 355–386 (1992).

Cross, G.A.M., Glycolipid anchoring of plasma membrane proteins, *Annu. Rev. Cell Biol.* **6**, 1–39 (1990).

Englund, P.T., The structure and biosynthesis of glycosyl phosphatidylinositol protein anchors, *Annu. Rev. Biochem.* **62**, 65–100 (1993).

Marshall, C.J., Protein prenylation: A mediator of protein–protein interactions, *Science* **259**, 1865–1866 (1993).

Schafer, W.R. and Rine, J., Protein prenylation: Genes, enzymes, targets, and functions, *Annu. Rev. Genet.* **30**, 209–237 (1992).

Schlesinger, M.J. (Ed.), *Lipid Modification of Proteins*, CRC Press (1993).

Tartakoff, A.M. and Singh, N., How to make a glycoinositol phospholipid anchor, *Trends Biochem. Sci.* **17**, 470–473 (1992).

Zhang, F.L. and Casey, P.J., Protein prenylation: Molecular mechanisms and functional consequences, *Annu. Rev. Biochem.* **65**, 241–269 (1996).

MEMBRANE STRUCTURE

Brown, D.A. and London, E., Structure and function of sphingolipid- and cholesterol-rich membrane rafts, *J. Biol. Chem.* **275**, 17221–17224 (2000); *and* Function of lipid rafts in biological membranes, *Annu. Rev. Cell Dev. Biol.* **14**, 111–136 (1998).

Dawidowicz, E.A., Dynamics of membrane lipid metabolism and turnover, *Annu. Rev. Biochem.* **56**, 43–61 (1987).

Edidin, M., Lipid microdomains in cell surface membranes, *Curr. Opin. Struct. Biol.* **7**, 528–532 (1997).

Frye, C.D. and Edidin, M., The rapid intermixing of cell surface antigens after formation of mouse–human heterokaryons, *J. Cell Sci.* **7**, 319–335 (1970).

Galbiati, F., Razani, B., and Lisanti, M.P., Emerging themes in rafts and caveolae, *Cell* **106**, 403–411 (2001).

Simons, K. and Ikonen, E., Functional rafts in cell membranes, *Nature* **387**, 569–572 (1997).

Singer, S.J. and Nicolson, G.L., The fluid mosaic model of the structure of cell membranes, *Science* **175**, 720–731 (1972). [A landmark paper on membrane structure.]

Webb, W.W., Luminescence measurements of macromolecular mobility, *Ann. N.Y. Acad. Sci.* **366**, 300–314 (1981). [A discussion of the fluorescence photobleaching recovery technique.]

THE RED CELL MEMBRANE

Agre, P. and Parker, J.C. (Eds.), *Red Blood Cell Membranes,* Marcel Dekker (1989). [Contains useful articles on red cell membrane components and architecture.]

Bennett, V., Ankyrins, *J. Biol. Chem.* **267**, 8703–8706 (1992).

Bretscher, A., Microfilament structure and function in the cortical cytoskeleton, *Annu. Rev. Cell Biol.* **7**, 337–374 (1991).

Davies, K.E. and Lux, S.E., Hereditary disorders of the red cell membrane, *Trends Genet.* **5**, 222–227 (1989).

Elgsaeter, A., Stokke, B.T., Mikkelsen, A., and Branton, D., The molecular basis of erythrocyte shape, *Science* **234**, 1217–1223 (1986).

Gallagher, P.G. and Benz, E.J., Jr., The erythrocyte membrane and cytoskeleton: Structure, function, and disorders, *in* Stamatoyannopoulos, G., Majerus, P.W., Perlmutter, R.M., and Varmus, H. (Eds.), *The Molecular Basis of Blood Diseases* (3rd ed.), Chapter 8, Elsevier (2001).

Gilligan, D.M. and Bennett, V., The junctional complex of the membrane skeleton, *Sem. Hematol.* **30**, 74–83 (1993).

Grum, V.L., Li, D., MacDonald, R.I., and Mondragón, A., Structures of two repeats of spectrin suggest models of flexibility, *Cell* **98**, 523–535 (1999).

Jennings, M.L., Structure and function of the red blood cell anion transport protein, *Annu. Rev. Biophys. Biophys. Chem.* **18**, 397–430 (1989).

Liu, S.-C. and Derick, L.H., Molecular anatomy of the red blood cell membrane skeleton: Structure–function relationships, *Sem. Hematol.* **29**, 231–243 (1992).

Luna, E.J. and Hitt, A.L., Cytoskeleton–plasma membrane interactions, *Science* **258**, 955–964 (1992).

Reithmeier, R.A.F., The erythrocyte anion transporter (band 3), *Curr. Opin. Struct. Biol.* **3**, 513–515 (1993).

Sedgwick, S.G. and Smerdon, S.J., The ankyrin repeat: a diversity of interactions on a common framework, *Trends Biochem. Sci.* **24**, 311–319 (1999).

Schofield, A.E., Reardon, R.M., and Tanner, M.J.A., Defective anion transport activity of the abnormal band 3 in hereditary ovalocytotic red blood cells, *Nature* **355**, 836–838 (1992).

Viel, A. and Branton, D., Spectrin: On the path from structure to function, *Curr. Opin. Cell Biol.* **8**, 49–55 (1996).

BLOOD GROUPS

Vitala, J. and Järnefelt, J., The red cell surface revisited, *Trends Biochem. Sci.* **10**, 392–395 (1985).

Watkins, H.M., Biochemistry and genetics of the ABO, Lewis and P group systems, *Adv. Human Genet.* **10**, 1–136 (1980).

Yamamoto, F., Clausen, H., White, T., Marken, J., and Hakomori, S., Molecular genetic basis of the histo-blood group ABO system, *Nature* **345**, 229–233 (1990).

GAP JUNCTIONS

Bruzzone, R., White, T.W., and Paul, D.L., Connections with connexins: The molecular basis of direct intracellular signaling, *Eur. J. Biochem.* **238**, 1–27 (1996).

Goodenough, D.A., Goliger, J.A., and Paul, D.L., Connexins, connexons, and intercellular communication, *Annu. Rev. Biochem.* **65**, 475–502 (1996).

Unger, V.M., Kumar, N.M., Gilula, N.B., and Yeager, M., Three-dimensional structure of a recombinant gap junction membrane channel, *Science* **283**, 1176–1180 (1999).

Yeager, M., Unger, V.M., and Falk, M.M., Synthesis, assembly and structure of gap junction intercellular channels, *Curr. Opin. Struct. Biol.* **8**, 517–524 *and* 810–811 (1998).

CHANNEL-FORMING PROTEINS

Gouaux, J.E., Channel-forming toxins: Tales of transformation, *Curr. Opin. Struct. Biol.* **7**, 566–573 (1997).

Song, L., Hobaugh, M.R., Shustak, C., Chesley, S., Bayley, H., and Gouaux, J.E., Structure of staphylococcal α-hemolysin, a heptameric transmembrane pore, *Science* **274**, 1859–1866 (1996).

LIPID ASYMMETRY IN MEMBRANES

Devaux, P.E., Protein involvement in transmembrane lipid asymmetry, *Annu. Rev. Biophys. Biomol. Struct.* **21**, 417–439 (1992).

Op den Kamp, J.A.F., Lipid asymmetry in membranes, *Annu. Rev. Biochem.* **48**, 47–71 (1979).

Wirtz, K.W.A., Phospholipid transfer proteins, *Annu. Rev. Biochem.* **60**, 73–99 (1991).

SECRETORY PATHWAY

Alberts, B., Johnson, A., Lewis, J., Raff, M. Roberts, K., and Walter, P., *The Molecular Biology of the Cell* (4th ed.), Chap. 12, Garland Science (2002).

Batey, R.T., Rambo, R.P., Lucast, L., Rha, B., and Doudna, J.A., Crystal structure of the ribonuclear core protein of the signal recognition particle, *Science* **287**, 1232–1239 (2000).

Beckmann, R., Spahn, C.M.T., Eswar, N., Helmers, J., Penczek, P.A., Sali, A., Frank, J., and Blobel, G., Architecture of the protein-conducting channel associated with the translating 80S ribosome, *Cell* **107**, 361–372 (2001).

Chevet, E., Cameron, P.H., Pelletier, M.F., Thomas, D.Y., and Bergeron, J.J.M., The endoplasmic reticulum: Integration of protein folding, quality control, signaling and degradation, *Curr. Opin. Struct. Biol.* **11**, 120–124 (2001).

Fewell, S.W., Travers, K.J., Weissman, J.S., and Brodsky, J.L., The action of molecular chaperones in the early secretory pathway, *Annu. Rev. Genet.* **35**, 149–191 (2001).

Johnson, A.E. and van Waes, M.A., The translocon: A dynamic gateway at the ER membrane, *Annu. Rev. Cell Dev. Biol.* **15**, 799–842 (1999).

Keenan, R.J., Freymann, D.M., Stroud, R.M., and Walter, P., The signal recognition particle, *Annu. Rev. Biochem.* **70**, 755–775 (2001).

Keenan, R.J., Freymann, D.M., Walter, P., and Stroud, R.M.,

Crystal structure of the signal sequence binding subunit of the signal recognition particle, *Cell* **94,** 181–191 (1998).

Kornfield, S. and Sly, W.S., I-cell disease and pseudo-Hurler polydystrophy disorders of liposomal enzyme phosphorylation and localization, *in* Scriver, C.R., Beaudet, A.L., Sly, W.S., and Valle, D. (Eds.), *The Metabolic & Molecular Bases of Inherited Disease* (8th ed.), *pp.* 3469–3482, McGraw-Hill (2001).

Lippincott-Schwartz, J., Roberts, R.H., and Hirschberg, K., Secretory protein trafficking and organelle dynamics in living cells, *Annu. Rev. Cell Dev. Biol.* **16,** 557–589 (2000).

Lodish, H., Berk, A., Zipursky, S.L., Matsudaira, P., Baltimore, D., and Darnell, J., *Molecular Cell Biology* (4th ed.), Chapter 17, Freeman (2000).

Matlack, K.E.S., Mothes, W., and Rapoport, T., Protein translocation: Tunnel vision, *Cell* **92,** 381–390 (1998).

Montoya, G., Svensson, C., Luirink, J., and Sinning, I., Crystal structure of the NG domain from the signal-recognition particle receptor FtsY, *Nature* **385,** 365–368 (1997).

Nakai, K., Protein sorting signals and prediction of subcellular localization, *Adv. Prot. Chem.* **54,** 277–344 (2000).

Phoenix, D.A. (Ed.), *Protein Targeting and Translocation,* Portland Press (1998).

Rapoport, T.A., Jungnickel, B., and Kutay, U., Protein transport across the eukaryotic endoplasmic reticulum and bacterial inner membranes, *Annu. Rev. Biochem.* **65,** 271–303 (1996).

Rothman, J.E. and Orci, L., Molecular dissection of the secretory pathway, *Nature* **355,** 409–415 (1992).

Stroud, R.M. and Walter, P., Signal recognition and protein targeting, *Curr. Opin. Struct. Biol.* **9,** 754–759 (1999).

Walter, P. and Johnson, A.E., Signal sequence recognition and protein targeting to the endoplasmic reticulum membrane, *Annu. Rev. Cell Biol.* **10,** 87–119 (1994).

Wild, K., Weichenrieder, O., Strub, K., Sinning, I., and Cusack, S., Towards the structure of the mammalian signal recognition particle, *Curr. Opin. Struct. Biol.* **12,** 72–81 (2002).

Zheng, N. and Gierasch, L., Signal sequences: The same yet different, *Cell* **86,** 849–852 (1996).

COATED VESICLES

Bottomley, M.J., Surdo, P.L., and Driscoll, P.C., How dynamin sets vesicles Phree! *Curr. Biol.* **9,** R301–R304 (1999).

Brodsky, F.M., Chen, C.-Y., Knuehl, C., Towler, M.C., and Wakeham, D.E., Biological basket weaving: Formation and function of clathrin-coated vesicles, *Annu. Rev. Cell Dev. Biol.* **17,** 515–568 (2001).

Collins, B.M., McCoy, A.J., Kent, H.M., Evans, P.R., and Owen, D.J., Molecular architecture and functional model of the endocytotic AP2 complex, *Cell* **109,** 523–535 (2002).

Donaldson, J.G. and Lippincott-Schwartz, J., Sorting and signaling at the Golgi complex, *Cell* **101,** 693–696 (2000).

Evans, P.R. and Owen, D.J., Endocytosis and vesicle trafficking, *Curr. Opin. Struct. Biol.* **12,** 814–821 (2002).

Hinshaw, J.E., Dynamin and its role in membrane fusion, *Annu. Rev. Cell Dev. Biol.* **16,** 483–519 (2000).

Hirst, J. and Robinson, M.S., Clathrin and adaptors, *Biochim. Biophys. Acta* **1404,** 173–193 (1998).

Kirchhausen, T., Adaptors for clathrin-mediated traffic, *Annu. Rev. Cell Dev. Biol.* **15,** 705–732 (1999).

Kirchhausen, T., Clathrin, *Annu. Rev. Biochem.* **69,** 699–727 (2000).

Kreis, T.E., Lowe, M., and Pepperkok, R., COPs regulating membrane traffic, *Annu. Rev. Cell Dev. Biol.* **11,** 677–706 (1995).

Marsh, M. (Ed.), *Endocytosis,* Oxford (2001). [Contains chapters on various aspects of endocytosis including clathrin-dependent and -independent endocytosis, mechanisms of membrane fusion, and vesicle recycling.]

Marsh, M. and McMahon, H.T., The structural era of endocytosis, *Science* **285,** 215–220 (1999). [A review of the structure and functions of clathrin-coated vesicles.]

McNiven, M.A., Cao, H., Pitts, K.R., and Yoon, Y., The dynamin family of mechanoenzymes: Pinching in new places, *Trends Biochem. Sci.* **25,** 115–120 (2000).

Neufield, E.F., Lysosomal storage diseases, *Annu. Rev. Biochem.* **60,** 257–280 (1991).

Pishavee, B. and Payne, G.S., Clathrin coats—threads laid bare, *Cell* **95,** 443–446 (1998).

Pearse, B.M.F., Smith, C.J., and Owen, D.J., Clathrin coat construction in endocytosis, *Curr. Opin. Struct. Biol.* **10,** 220–228 (2000); *and* Smith, C.J. and Pearse, B.M.F., Clathrin: Anatomy of a coat protein, *Trends Cell Biol.* **9,** 335–338 (1999). [Authoritative reviews.]

Pelham, H.R.B. and Rothman, J.E., The debate about transport in the Golgi—two sides of the same coin, *Cell* **102,** 713–719 (2000).

Roth, M.G., Snapshots of ARF1: Implications for mechanisms of activation and inactivation, *Cell* **97,** 149–152 (1999).

Schekman, R. and Orci, L., Coat proteins and vesicle budding, *Science* **271,** 1526–1533 (1996).

Schmid, S.L., Clathrin-coated vesicle formation and protein sorting: An integrated process, *Annu. Rev. Biochem.* **66,** 511–548 (1997).

Smith, C.J., Grigorieff, N., and Pearse, B.M.F., Clathrin coats at 21 Å resolution: A cellular assembly designed to recycle multiple membrane receptors, *EMBO J.* **17,** 4943–4953 (1998).

Springer, S., Spang, A., and Schekman, R., A primer on vesicle budding, *Cell* **97,** 145–148 (1999).

ter Haar, E., Musacchio, A., Harrison, S.C., and Kirchhausen, T., Atomic structure of clathrin: A β propeller terminal domain joins an α zigzag linker, *Cell* **95,** 563–573 (1998).

Ybe, J.A., Brodsky, F.M., Hofmann, K., Lin, K., Liu, S.-H., Chen, L., Earnest, T.N., Fletterick, R.J., and Hwang, P.K., Clathrin self-assembly is mediated by a tandemly repeated superhelix, *Nature* **399,** 371–375 (1999). [The X-ray structure of the clathrin heavy chain proximal leg segment.]

VESICLE FUSION

Alberts, B., Johnson, A., Lewis, J., Raff, M., Roberts, K., and Walter, P., *The Molecular Biology of the Cell* (4th ed.), Chap. 13, Garland Science (2002).

Brünger, A.T., Structure of proteins involved in synaptic vesicle fusion in neurons, *Annu. Rev. Biophys. Biomol. Struct.* **30,** 151–171 (2001).

Chan, Y.A. and Scheller, R.H., SNARE-mediated membrane fusion, *Nature Rev. Mol. Cell. Biol.,* **2,** 98–106 (2001).

Gonzalez, L., Jr. and Scheller, R.H., Regulation of membrane trafficking: Structural insights from a Rab effector complex, *Cell* **96,** 755–758 (1999).

Hanson, P.I., Roth, R., Morisaki, H., Jahn, R., and Heuser, J.E., Structure and conformational changes in NSF and its membrane receptor complexes visualized by quick freeze/deep etch electron microscopy, *Cell* **90,** 523–535 (1997).

Jahn, S. and Südhof, T.C., Membrane fusion and exocytosis, *Annu. Rev. Biochem.* **68,** 863–911 (1999).

May, A.P., Whiteheart, S.W., and Weis, W.I., Unraveling the mechanism of the vesicle transport ATPase NSF, the *N*-ethylmalemide-sensitive factor, *J. Biol. Chem.* **276,** 21991–21994 (2001).

Mayer, A., Membrane fusion in eukaryotic cells, *Annu. Rev. Cell*

Dev. Biol., **18,** 289–314 (2002).

McNew, J.A., Parlati, F., Fukuda, R., Johnston, R.J., Paz, K., Paumet, F., Söllner, T.H., and Rothman, J.E., Compartmental specificity of cellular membrane fusion encoded in SNARE proteins, *Nature* **407,** 153–159 (2000).

Misura, K.M.S., May, A.P., and Weis, W.I., Protein–protein interactions in intercellular membrane fusion, *Curr. Opin. Struct. Biol.* **10,** 662–671 (2000).

Misura, K.M.S., Scheller, R.H., and Weis, W.I., Three-dimensional structure of the neuronal-Sec1–syntaxin 1a complex, *Nature* **404,** 355–362 (2000).

Nichols, B.J. and Pelham, H.R.B., SNAREs and membrane fusion in the Golgi apparatus, *Biochim. Biophys. Acta* **1404,** 9–31 (1998).

Niemann, H., Blasi, J., and Jahn, R., Clostridial neurotoxins: New tools for dissecting exocytosis, *Trends Cell Biol.* **4,** 179–185 (1994).

Peters, C., Bayer, M.J., Bühler, S., Andersen, J.S., Mann, M., and Mayer, A., *Trans*-complex formation by proteolipid channels in the terminal phase of membrane fusion, *Nature* **409,** 581–588 (2001); *and* Mayer, A., What drives membrane fusion in eukaryotes, *Trends Biochem. Sci.* **26,** 717–723 (2001).

Sutton, R.B., Fasshauer, D., Jahn, R., and Brünger, A.T., Crystal structure of a SNARE complex involved in synaptic exocytosis at 2.4 Å resolution, *Nature* **395,** 347–353 (1998).

Yu, R.C., Hanson, P.I., Jahn, R., and Brünger, A.T., Structure of the ATP-dependent oligomerization domain of *N*-ethylmaleimide sensitive factor complexed with ATP, *Nature Struct. Biol.* **5,** 803–810 (1998); *and* Lenzen, C.U., Steinmann, D., Whiteheart, S.W., and Weis, W.I., Crystal structure of the hexamerization domain of *N*-ethylmaleimide-sensitive fusion protein, *Cell* **94,** 525–536 (1998).

Zerial, M. and McBride, H., Rab proteins as membrane organizers, *Nature Rev. Mol. Cell Biol.* **2,** 107–119 (2001).

MITOCHONDRIAL AND NUCLEAR PROTEIN TARGETING

Abe, Y., Shodai, T., Muto, T., Mihara, K., Torii, H., Nishikawa, S., Endo, T., and Kohda, D., Structural basis of presequence recognition by the mitochondrial protein import receptor Tom20, *Cell* **100,** 551–560 (2000). [An NMR structure.]

Ahting, U., Thun, C., Hegerl, R., Typke, D., Nargang, F.E., Neupert, W., and Nussberger, S., The TOM core complex: The general protein import pore of the outer membrane of mitochondria, *J. Cell Biol.* **147,** 959–968 (1999). [An electron microscopy study of the TOM core complex.]

Dalbey, R.E. and Kuhn, A., Evolutionarily related insertion pathways of bacterial, mitochondrial, and thylakoid membrane proteins, *Annu. Rev. Cell Dev. Biol.* **16,** 51–87 (2000).

Gabriel, K., Buchanan, S.K., and Lithgow, T., The alpha and the beta: Protein translocation across mitochondrial and plastid outer membranes, *Trends Biochem. Sci.* **26,** 36–40 (2001).

Koehler, C.M., Merchant, S., and Schatz, G., How membrane proteins travel across the mitochondrial intermembrane space, *Trends Biochem. Sci.* **24,** 428–432 (1999).

Neupert, W., Protein import into mitochondria, *Annu. Rev. Biochem.* **66,** 863–917 (1997).

Ryan, M.T., Wagner, R., and Pfanner, N., The transport machinery for the import of preproteins across the outer mitochondrial membrane, *Int. J. Biochem. Cell Biol.* **32,** 13–21 (2000).

Schatz, G. and Dobberstein, B., Common principles of protein translocation across membranes, *Science* **271,** 1519–1526 (1996).

Voos, W., Martin, H., Krimmer, T., and Pfanner, N., Mechanisms of protein translocation in mitochondria, *Biochim. Biophys. Acta* **1422,** 235–254 (1999); *and* Truscott, K.N. and Pfanner, N., Import of carrier proteins into mitochondria, *Biol. Chem.* **380,** 1151–1156 (1999).

LIPOPROTEINS

Borhani, D.W., Rogers, D.P., Engler, J.A., and Brouillette, C.G., Crystal structure of truncated human apolipoprotein A-I suggests a lipid-bound conformation, *Proc. Natl. Acad. Sci.* **94,** 12291–12296 (1997).

Brown, M.S. and Goldstein, J.L., A receptor-mediated pathway for cholesterol homeostasis, *Science* **232,** 34–47 (1986). [A Nobel prize address.]

Brown, M.S. and Goldstein, J.L., Koch's postulates for cholesterol, *Cell* **71,** 187–188 (1992).

Chan, L., Apolipoprotein B, the major protein component of triglyceride-rich and low density lipoproteins, *J. Biol. Chem.* **267,** 25621–25624 (1992).

Fass, D., Blacklow, S., Kim, P.S., and Berger, J.M., Molecular basis of familial hypercholesterolaemia from structure of LDL receptor module, *Nature* **388,** 691–693 (1997).

Hajjar, K.A., and Nachman, R.L., The role of lipoprotein(a) in atherogenesis and thrombosis, *Annu. Rev. Med.* **47,** 423–442 (1996).

Krieger, M., Charting the fate of the "good cholesterol": Identification and characterization of the high-density lipoprotein receptor SR-BI, *Annu. Rev. Biochem.* **68,** 523–558 (1999).

Lawn, R.M., Wade, D.P., Hammer, R.E., Chiesa, G., Verstuyft, J.G., and Rubin, E.M., Atherogenesis in transgenic mice expressing human apolipoprotein(*a*), *Nature* **360,** 670–672 (1992).

Lodish, H., Berk, A., Zipursky, S.L., Matsudaira, P., Baltimore, D., and Darnell, J., *Molecular Cell Biology* (4th ed.), *pp.* 727–733, Freeman (2000). [Discussion of receptor-mediated endocytosis.]

Marotti, K.R., Castle, C.K., Boyle, T.P., Lin, A.H., Murray, R.W., and Melchior, G.W., Severe atherosclerosis in transgenic mice expressing simian cholesteryl ester transfer protein, *Nature* **364,** 73–75 (1993).

Parthasarathy, S., Steinberg, D., and Witzum, J.L., The role of oxidized low-density lipoproteins in the pathogenesis of atherosclerosis, *Annu. Rev. Med.* **43,** 219–225 (1992).

Rosseneu, M. (Ed.), *Structure and Function of Apolipoproteins,* CRC Press (1992).

Rubin, E.M., Krauss, R.M., Spangler, E.A., Verstuyft, J.G., and Clift, S.M., Inhibition of early atherogenesis in transgenic mice by human apolipoprotein AI, *Nature* **353,** 265–266 (1991); *and* Warden, C.H., Hedrick, C.C., Qiao, J.-H., Castellani, L.W., and Lusis, A.J., Atherosclerosis in transgenic mice overexpressing apolipoprotein A-II, *Science* **261,** 469–472 (1993).

Schmid, S.L., The mechanism of receptor mediated endocytosis: More questions than answers, *BioEssays* **14,** 589–596 (1992).

Schumaker, V.N. (Ed.), *Adv. Prot. Chem.* **45** (1994). [Contains authoritative chapters on apoB and LDL structure (*pp.* 205–248), apoE (*pp.* 249–302), amphipathic α helices (*pp.* 303–369), and lipophorin (*pp.* 371–415).]

Scriver, C.R., Beaudet, A.L., Sly, W.S., and Valle, D. (Eds.), *The Metabolic & Molecular Bases of Inherited Disease* (8th ed.), Chapters 114–123, McGraw-Hill (2001). [Authoritative discussions of diseases characterized by abnormal lipid metabolism.]

Smythe, E. and Warren, G., The mechanism of receptor-mediated endocytosis, *Eur. J. Biochem.* **202,** 689–699 (1992).

Steinberg, D., Low density lipoprotein oxidation and its pathobiological significance, *J. Biol. Chem.* **272,** 20963–20966 (1997).

Weisgraber, K.H. and Mahley, R.W., Human apolipoprotein E: The Alzheimer's disease connection, *FASEB J.* **10,** 1485–1493 (1996).

Wilson, C., Wardell, M.R., Weisgraber, K.H., Mahley, R.W., and Agard, D.A., Three-dimensional structure of the LDL receptor-binding domain of human apolipoprotein E, *Science* **252,** 1817–1822 (1991).

■ PROBLEMS

1. Explain the difference in melting points between trans-oleic acid (44.5°C) and cis-oleic acid (13.4°C).

2. Why do animals that live in cold climates generally have more polyunsaturated fatty acid residues in their fats than do animals that live in warm climates?

*3. How many different isomers of phosphatidylserine, triacylglycerol, and cardiolipin can be made from four types of fatty acids?

4. Estimate the thickness of the surface layer formed by Benjamin Franklin's teaspoon of oil on Clapham pond (1 teaspoon = 5 mL and 1 acre = 4047 m^2).

5. "Hard water" contains a relatively high concentration of Ca^{2+}. Explain why soap is ineffective for washing in hard water.

6. Explain why pure hydrocarbons do not form monolayers on water.

7. Soap bubbles are inside-out bilayers; that is, the polar head groups of the amphiphiles, together with some water, are in opposition, whereas their hydrophobic tails extend into the air. Explain the physical basis of this phenomenon.

8. Describe the action of detergents in extracting integral proteins from membranes. How do they keep these proteins from precipitating? Why do mild detergents such as Triton X-100 bind only to proteins that form lipid complexes?

*9. Is the transmembrane portion of glycophorin A (Fig. 12-21) α helical (use the Chou and Fasman rules; Section 9-3A)?

10. The symmetries of oligomeric integral proteins are constrained by the requirement that their subunits must all have the same orientation with respect to the plane of the membrane. What symmetries can these proteins have? Explain. (Protein symmetry is discussed in Section 8-5B.)

11. (a) How many residues must an α helix contain in order to span the 30-Å-thick hydrocarbon core of a lipid bilayer? (b) How many residues in a β sheet are required to span this bilayer core if it is inclined by 30° with respect to the normal to the membrane plane? (c) Why do most transmembrane α helices and β strands have more than these minimum numbers?

12. Explain why antibodies against type A blood group antigens are inhibited by *N*-acetylgalactosamine, whereas anti-B antibodies are inhibited by galactose.

13. Individuals with a certain one of the ABO blood types are said to be "universal donors," whereas those with another type are said to be "universal recipients." What are these blood types? Explain.

14. Anti-H antibodies are not normally found in human blood. They may, however, be elicited in animals by the injection of human blood. How would such antibodies be expected to react with tissues from individuals with type A, type B, and type O blood groups?

15. *Thermus aquaticus* is a thermophilic bacterium that grows between the temperatures of 50 and 80°C. Although the signal peptide binding groove of its Ffh M domain is lined with hydrophobic groups, only three of them are Met side chains. In contrast, the binding grooves of mesophilic organisms (those that live at normal temperatures) are lined with numerous Met side chains (11 in *E. coli*). In addition, the finger loop forming one wall of the binding groove is disordered in the X-ray structure of the *E. coli* M domain but ordered in that of *T. aquaticus* (Fig. 12-48; both proteins were crystallized at room temperature). Suggest a reason for these evolutionary adaptations in *T. aquaticus*.

*16. In a genetically distinct form of familial hypercholesterolemia, LDL binds to the cell surface but fails to be internalized by endocytosis. Electron microscopy reveals that each mutant cell has its normal complement of coated pits but that ferritin-conjugated LDL does not bind to them. Rather, the bound LDL is uniformly distributed about the noncoated regions of the cell surfaces. Apparently the binding properties of the mutant LDL receptors are normal but they are in the wrong place. What do these data suggest about how the LDL receptor is assembled into coated pits?

17. Table 12-6 indicates that the densities of lipoproteins increase as their particle diameters decrease. Explain.

18. Certain types of animal viruses form by budding out from a cell surface much like coated pits bud into the cytoplasm during endocytosis to form coated vesicles. In both cases, the membranous vesicles form on a polyhedral protein scaffolding. Sketch the budding of an animal virus and indicate the location of its membrane relative to its protein shell.

19. Why aren't chylomicrons taken up by LDL receptors?

20. Wolman's disease is a rapidly fatal homozygous defect characterized by a severe deficiency in **cholesteryl ester hydrolase,** the enzyme that catalyzes the hydrolysis of intracellularly located cholesteryl esters. Describe the microscopic appearance of the cells of victims of Wolman's disease.

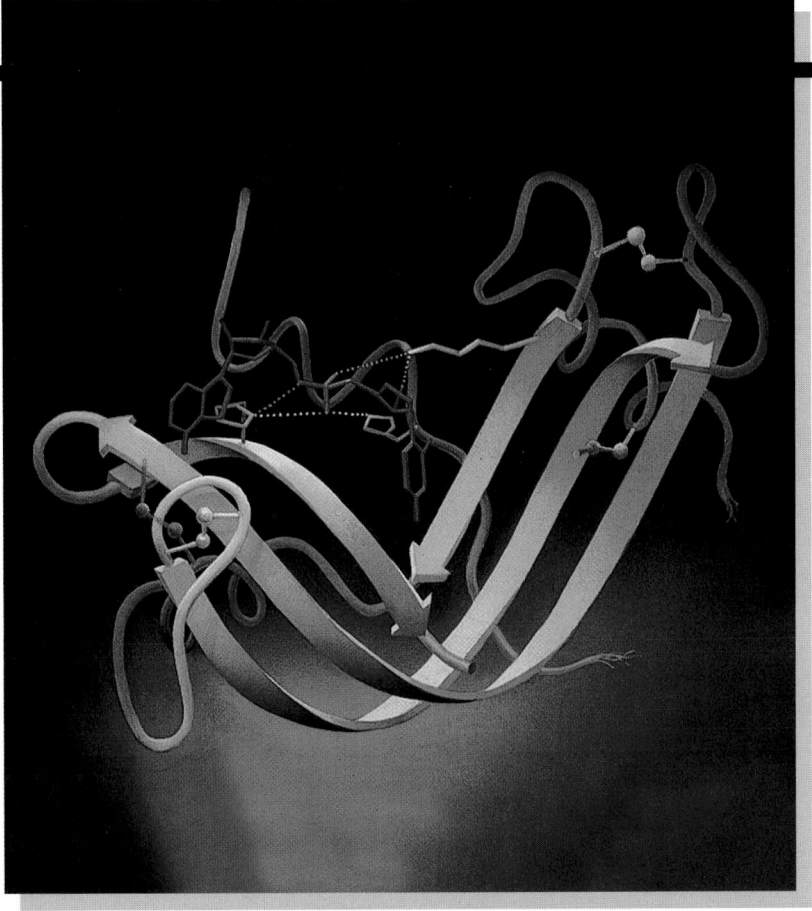

*Bovine pancreatic ribonuclease S in complex
with a nonhydrolyzable substrate analog,
the dinucleotide phosphonate UpcA.*

MECHANISMS OF
ENZYME ACTION

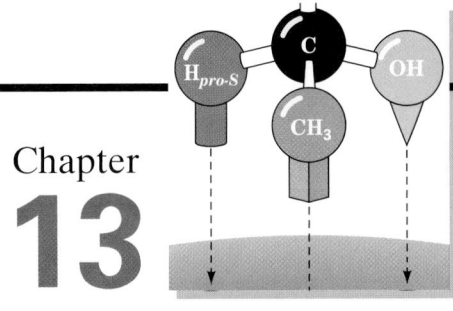

Chapter 13

Introduction to Enzymes

The enormous variety of biochemical reactions that comprise life are nearly all mediated by a series of remarkable biological catalysts known as **enzymes.** Although enzymes are subject to the same laws of nature that govern the behavior of other substances, they differ from ordinary chemical catalysts in several important respects:

1. Higher reaction rates: The rates of enzymatically catalyzed reactions are typically factors of 10^6 to 10^{12} greater than those of the corresponding uncatalyzed reactions and are at least several orders of magnitude greater than those of the corresponding chemically catalyzed reactions.

2. Milder reaction conditions: Enzymatically catalyzed reactions occur under relatively mild conditions: temperatures below 100°C, atmospheric pressure, and nearly neutral pH's. In contrast, efficient chemical catalysis often requires elevated temperatures and pressures as well as extremes of pH.

3. Greater reaction specificity: Enzymes have a vastly greater degree of specificity with respect to the identities of both their **substrates** (reactants) and their products than do chemical catalysts; that is, enzymatic reactions rarely have side products. For example, in the enzymatic synthesis of proteins on ribosomes (Section 32-3), polypeptides consisting of well over 1000 amino acid residues are made all but error free. Yet, in the chemical synthesis of polypeptides, side reactions and incomplete reactions presently limit the lengths of polypeptides that can be accurately produced in reasonable yields to ~100 residues (Section 7-5B).

4. Capacity for regulation: The catalytic activities of many enzymes vary in response to the concentrations of substances other than their substrates and products. The mechanisms of these regulatory processes include allosteric control, covalent modification of enzymes, and variation of the amounts of enzymes synthesized.

Consideration of these remarkable catalytic properties of enzymes leads to one of the central questions of biochemistry: *How do enzymes work?* We address this issue in this part of the text.

In this chapter, following a historical review, we commence our study of enzymes with a discussion of two clear instances of enzyme action: one that illustrates how enzyme specificity is manifested, and a second that exemplifies the regulation of enzyme activity. These are by no means exhaustive treatments but are intended to highlight these all-important aspects of enzyme mechanism. We shall encounter numerous other examples of these phenomena in our study of metabolism (Chapters 16–28). These two expositions are interspersed with a consideration of the role of enzymatic cofactors. The chapter ends with a short synopsis of enzyme nomenclature. In Chapter 14 we take up the formalism of enzyme kinetics because the study of the rates of enzymatically catalyzed reactions provides indispensable mechanistic information. Finally, Chapter 15 is a general discussion of the catalytic mechanisms employed by enzymes, followed by an examination of the mechanisms of several specific enzymes.

1 ■ HISTORICAL PERSPECTIVE

The early history of **enzymology,** the study of enzymes, is largely that of biochemistry itself; these disciplines evolved together from nineteenth century investigations of fermentation and digestion. Research on fermentation is widely considered to have begun in 1810 with Joseph Gay-Lussac's determination that ethanol and CO_2 are the principal products of sugar decomposition by yeast. In 1835, Jacob Berzelius, in the first general theory of chemical catalysis, pointed out that an extract of malt known as **diastase** (now known to contain the enzyme α-amylase; Section 11-2D) catalyzes the hydrolysis of starch more efficiently than does sulfuric acid. Yet, despite the ability of mineral acids to mimic the effect of diastase, it was the inability to reproduce most other biochemical reactions in the laboratory that led Louis Pasteur, in the mid-nineteenth century, to propose that the processes of fermentation could only occur in living cells. Thus, as was common in his era, Pasteur assumed that living systems were endowed with a "vital force" that permitted them to evade the laws of nature governing inanimate matter. Others, however, notably Justus von Liebig, argued that biological processes are caused by the action of chemical substances that were then known as "ferments." Indeed, the name "enzyme" (Greek: *en*, in + *zyme*, yeast) was coined in 1878 by Fredrich Wilhelm Kühne in an effort to emphasize that there is something *in* yeast, as opposed to the yeast itself, that catalyzes the reactions of fermentation. Nevertheless, it was not until 1897 that Eduard Buchner obtained a cell-free yeast extract that could carry out the synthesis of ethanol from glucose (**alcoholic fermentation;** Section 17-3B).

Emil Fischer's discovery, in 1894, that glycolytic enzymes can distinguish between stereoisomeric sugars led to the formulation of his **lock-and-key hypothesis:** *The specificity of an enzyme (the lock) for its substrate (the key) arises from their geometrically complementary shapes.* Yet the chemical composition of enzymes was not firmly established until well into the twentieth century. In 1926, James Sumner, who crystallized the first enzyme, jack bean **urease,** which catalyzes the hydrolysis of urea to NH_3 and CO_2, demonstrated that these crystals consist of protein. Since Sumner's preparations were somewhat impure, however, the protein nature of enzymes was not generally accepted until the mid-1930s, when John Northrop and Moses Kunitz showed that there is a direct correlation between the enzymatic activities of crystalline pepsin, trypsin, and chymotrypsin and the amounts of protein present. Enzymological experience since then has amply demonstrated that enzymes are proteins (although it has recently been shown that RNA can also have catalytic properties; Section 31-4B).

Although the subject of enzymology has a long history, most of our understanding of the nature and functions of enzymes is a product of the last 50 years. Only with the advent of modern techniques for separation and analysis (Chapter 6) has the isolation and characterization of an enzyme become less than a monumental task. It was not until 1963 that the first amino acid sequence of an enzyme, that of **bovine pancreatic ribonuclease A** (Section 15-1A), was reported in its entirety, and not until 1965 that the first X-ray structure of an enzyme, that of hen egg white **lysozyme** (Section 15-2A), was elucidated. In the years since then, thousands of enzymes have been purified and characterized to at least some extent, and the pace of this endeavor is rapidly accelerating.

2 ■ SUBSTRATE SPECIFICITY

The noncovalent forces through which substrates and other molecules bind to enzymes are similar in character to the forces that dictate the conformations of the proteins themselves (Section 8-4): Both involve van der Waals, electrostatic, hydrogen bonding, and hydrophobic interactions. In general, a substrate-binding site consists of an indentation or cleft on the surface of an enzyme molecule that is complementary in shape to the substrate (geometric comple-

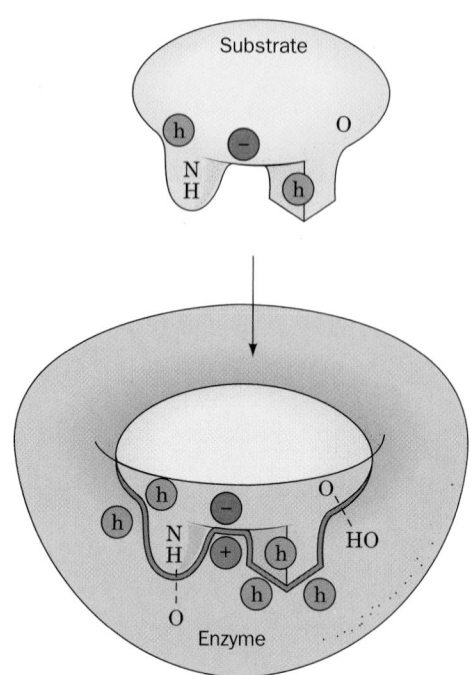

FIGURE 13-1   An enzyme–substrate complex illustrating both the geometric and the physical complementarity between enzymes and substrates. Hydrophobic groups are represented by an h in a brown circle, and dashed lines represent hydrogen bonds.

mentarity). Moreover, the amino acid residues that form the binding site are arranged to interact specifically with the substrate in an attractive manner (electronic complementarity; Fig. 13-1). Molecules that differ in shape or functional group distribution from the substrate cannot productively bind to the enzyme; that is, they cannot form enzyme–substrate complexes that lead to the formation of products. The substrate-binding site may, in accordance with the lock-and-key hypothesis, exist in the absence of bound substrate or it may, as suggested by the induced-fit hypothesis (Section 10-4C), form about the substrate as it binds to the enzyme. *X-Ray studies indicate that the substrate-binding sites of most enzymes are largely preformed but that most of them exhibit at least some degree of induced fit on binding substrate.*

A. Stereospecificity

Enzymes are highly specific both in binding chiral substrates and in catalyzing their reactions. This **stereospecificity** arises because enzymes, by virtue of their inherent chirality (proteins consist of only L-amino acids), form asymmetric active sites. For example, trypsin readily hydrolyzes polypeptides composed of L-amino acids but not those consisting of D-amino acids. Likewise, the enzymes involved with glucose metabolism (Section 17-2) are specific for D-glucose residues.

Enzymes are absolutely stereospecific in the reactions they catalyze. This was strikingly demonstrated for the case of **yeast alcohol dehydrogenase (YADH)** by Frank

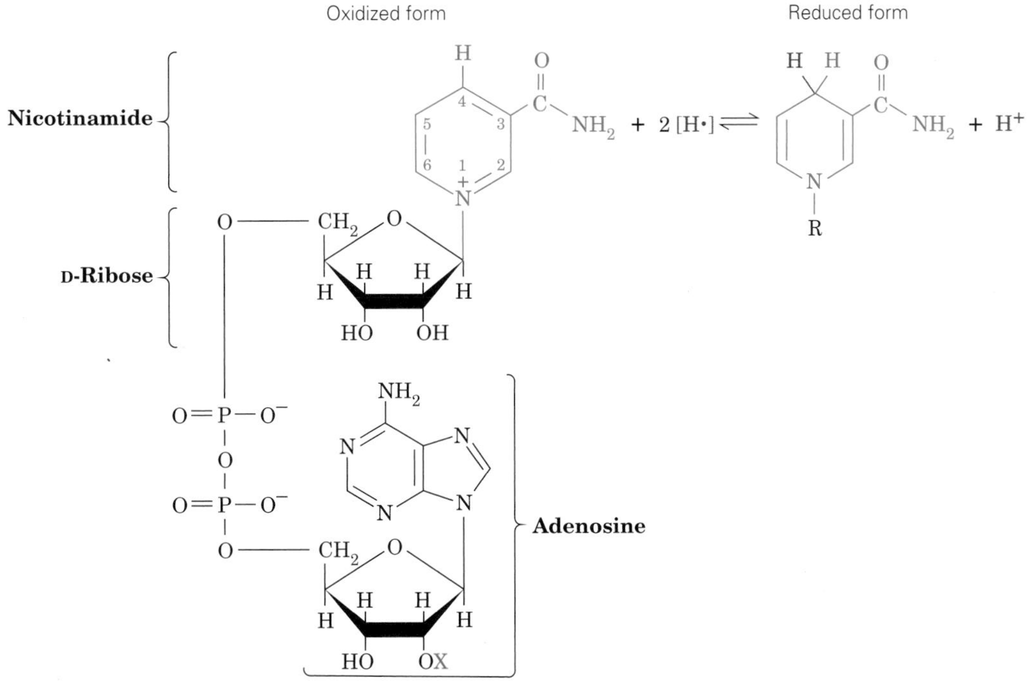

Oxidized form   Reduced form

Nicotinamide

$$+ \; 2\,[H\cdot] \rightleftharpoons \qquad\qquad + \; H^+$$

D-Ribose

Adenosine

X = H   **Nicotinamide adenine dinucleotide (NAD⁺)**
X = PO₃²⁻   **Nicotinamide adenine dinucleotide phosphate (NADP⁺)**

FIGURE 13-2 **The structures and reaction of nicotinamide adenine dinucleotide (NAD⁺) and nicotinamide adenine dinucleotide phosphate (NADP⁺).** Their reduced forms are **NADH** and **NADPH**. These substances, which are collectively referred to as the **nicotinamide coenzymes** or **pyridine nucleotides** (nicotinamide is a pyridine derivative) function, as is indicated in later chapters, as intracellular carriers of reducing equivalents (electrons). Note that only the nicotinamide ring is changed in the reaction. Reduction formally involves the transfer of two hydrogen atoms (H·), although the actual reduction may occur via a different mechanism.

Westheimer and Birgit Vennesland. Alcohol dehydrogenase catalyzes the interconversion of ethanol and acetaldehyde according to the reaction:

$$\mathrm{CH_3CH_2OH + NAD^+} \underset{}{\overset{\mathrm{YADH}}{\rightleftharpoons}} \overset{\textstyle O}{\overset{\|}{\mathrm{CH_3CH}}} + \mathrm{NADH + H^+}$$
    Ethanol   **Acetaldehyde**

The structures of **NAD⁺** and **NADH** are presented in Fig. 13-2. Ethanol, it will be recalled, is a prochiral molecule (see Section 4-2C for a discussion of prochirality):

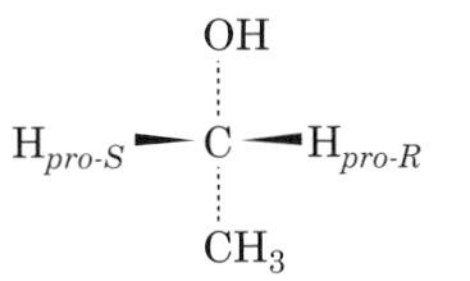

$$\mathrm{H}_{pro\text{-}S} \blacktriangleright \; \underset{\textstyle CH_3}{\overset{\textstyle OH}{C}} \; \blacktriangleleft \mathrm{H}_{pro\text{-}R}$$

Ethanol's two methylene H atoms may be distinguished if *the molecule is held in some sort of asymmetric jig (Fig. 13-3). The substrate-binding sites of enzymes are, of course, just such jigs because they immobilize the reacting groups of the substrate on the enzyme surface.*

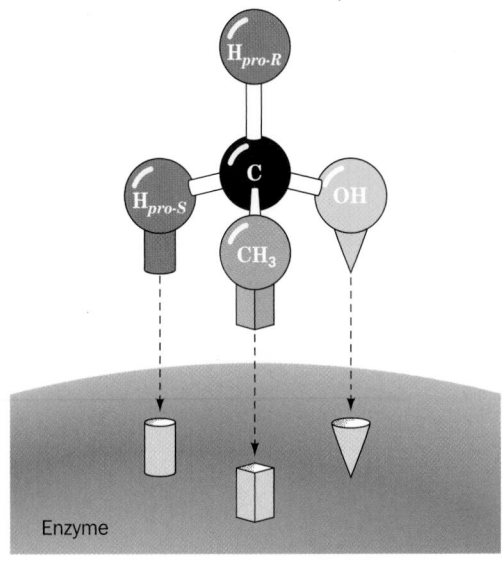

FIGURE 13-3 **Prochiral differentiation.** The specific attachment of a prochiral center to an enzyme binding site permits the enzyme to differentiate between prochiral groups. Note: if it was possible, the binding of the prochiral molecule's mirror image to the same three sites from the underside of the enzyme would still result in H$_{pro\text{-}R}$ pointing towards a different position.

Westheimer and Vennesland elucidated the stereospecific nature of the YADH reaction through the following series of experiments:

1. If the YADH reaction is carried out with deuterated ethanol, the product NADH is deuterated:

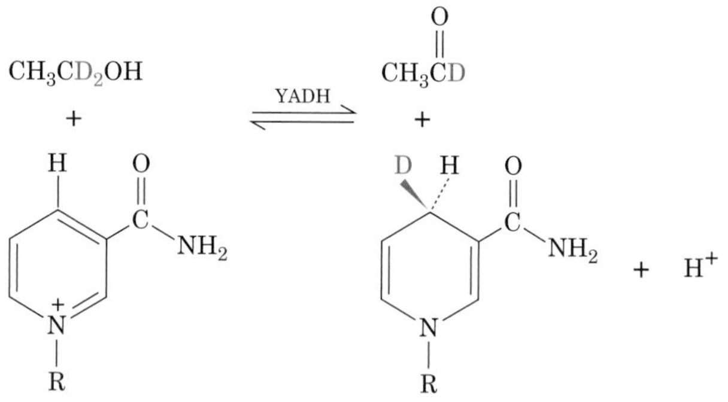

NAD⁺   **NADD**

Note that the nicotinamide ring of NAD⁺ is also prochiral.

2. On isolating this NADD and using it in the reverse reaction to reduce normal acetaldehyde, the deuterium is quantitatively transferred from the NADD to the acetaldehyde to form the product ethanol:

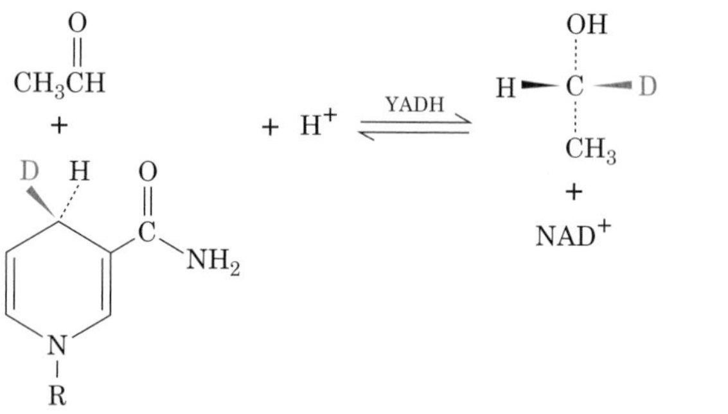

3. If the enantiomer of the foregoing CH₃CHDOH is made as follows:

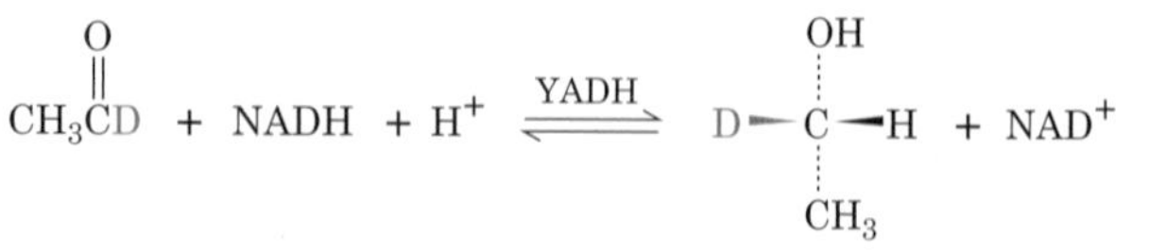

none of the deuterium is transferred from the product ethanol to NAD⁺ in the reverse reaction.

4. If, however, this ethanol is converted to its tosylate and then inverted by S_N2 hydrolysis to yield the enantiomeric ethanol,

p-Toluenesulfonyl chloride (tosyl chloride)

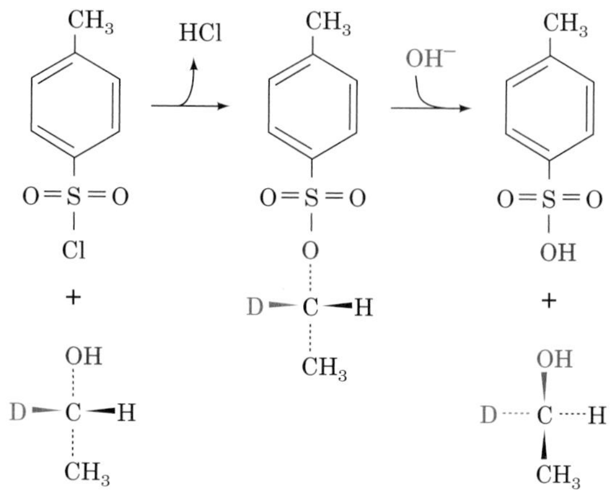

the deuterium is again quantitatively transferred to NAD⁺ in the YADH reaction.

The foregoing observations, in addition to showing that there is direct hydrogen transfer in the YADH reaction (Experiments 1 and 2), indicate that the enzyme distinguishes between the *pro-S* and *pro-R* hydrogens of ethanol as well as the *si* and *re* faces of the nicotinamide ring of NAD⁺ (Experiments 2–4). It was later demonstrated, by stereospecific syntheses, that YADH transfers the *pro-R* hydrogen of ethanol to the *re* face of the nicotinamide ring of NAD⁺ as is drawn in the preceding diagram.

The stereospecificity of YADH is by no means unusual. As we consider biochemical reactions we shall find that nearly all enzymes that participate in chiral reactions are absolutely stereospecific.

a. Stereospecificity in the NADH-Dependent Dehydrogenases May Have Functional Significance

In our exploration of metabolism, we shall encounter numerous species of NADH-dependent dehydrogenases that function to reduce (or oxidize) a great variety of substrates. These various dehydrogenases are more or less equally distributed between those transferring the *pro-R* (*re*-side) and the *pro-S* (*si*-side) hydrogens at C4 of NADH (also known as A-side and B-side transfers).

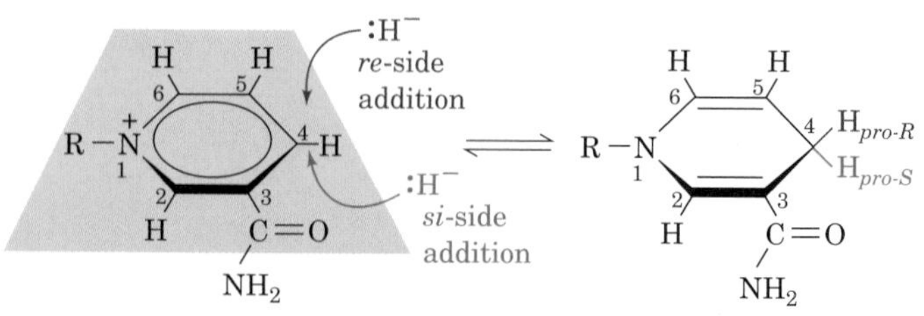

Yet, despite the fact that *si*- and *re*-side hydrogen transfers to or from the nicotinamide ring yield chemically identical products, a particular specificity of transfer is rigidly maintained within classes of dehydrogenases catalyzing similar reactions in different organisms. Indeed, dehydrogenases that catalyze reactions whose equilibrium constants with their natural substrates in the direction of reduction are $<10^{-12}M$ almost always transfer the nicotinamide's *pro-R* hydrogen, whereas those with equilibrium constants $>10^{-10}M$ generally transfer the *pro-S* hydrogen. Why has evolution so assiduously maintained this stereospecificity? Is it simply the result of a historical accident or does it serve some physiological function?

The NADH hydrogen transferred in a given enzymatic reaction is almost certainly that on the side of the nicotinamide ring facing the substrate. It was therefore widely assumed that the stereospecificity in any given class of dehydrogenases simply arose through a random choice made early in evolutionary history. Once made, this choice became "locked in," because flipping a nicotinamide ring about its glycosidic bond in NADH would result, it was presumed, in its carboxamide group obstructing catalytically essential residues on the enzyme.

In an effort to shed light on this matter, Steven Benner mutated YADH in a manner that the X-ray structure of the closely similar enzyme horse **liver alcohol dehydrogenase (LADH)** suggests permits the *si* face of nicotinamide to bind to the enzyme without interfering with catalysis. The resulting mutant enzyme (Leu 182 → Ala) makes one stereochemical "mistake" every 850,000 turnovers versus one mistake every 7 billion turnovers for wild-type (unmutated) YADH. This 8000-fold decrease in stereospecificity indicates that at least some of the side chains responsible for YADH's stereospecificity are not essential for catalysis and hence strengthens the argument that stereospecificity in the dehydrogenases has functional significance.

B. *Geometric Specificity*

The stereospecificity of enzymes is not particularly surprising in light of the complementarity of an enzymatic binding site for its substrate. A substrate of the wrong chirality will not fit into an enzymatic binding site for much the same reasons that you cannot fit your right hand into your left glove. *In addition to their stereospecificity, however, most enzymes are quite selective about the identities of the chemical groups on their substrates.* Indeed, such **geometric specificity** is often a more stringent requirement than is stereospecificity. After all, your left glove will more or less fit left hands that have somewhat different sizes and shapes than your own.

Enzymes vary considerably in their degree of geometric specificity. A few enzymes are absolutely specific for only one compound. Most enzymes, however, catalyze the reactions of a small range of related compounds. For example, YADH catalyzes the oxidation of small primary and secondary alcohols to their corresponding aldehydes or ketones but none so efficiently as that of ethanol. Even methanol and isopropanol, which differ from ethanol only

by the deletion or addition of a CH_2 group, are oxidized by YADH at rates that are, respectively, 25-fold and 2.5-fold slower than that for ethanol. Similarly, $NADP^+$, which differs from NAD^+ only by the addition of a phosphoryl group at the $2'$ position of its adenosine ribose group (Fig. 13-2), does not bind to YADH. On the other hand, there are many enzymes that bind $NADP^+$ but not NAD^+.

Some enzymes, particularly digestive enzymes, are so permissive in their ranges of acceptable substrates that their geometric specificities are more accurately described as preferences. Carboxypeptidase A, for example, catalyzes the hydrolysis of C-terminal peptide bonds to all residues except Arg, Lys, and Pro if the preceding residue is not Pro (Table 7-1). However, the rate of this enzymatic reaction varies with the identities of the residues in the vicinity of the C-terminus of the polypeptide (see Fig. 7-5). Some enzymes are not even very specific in the type of reaction they catalyze. Thus chymotrypsin, in addition to its ability to mediate peptide bond hydrolysis, also catalyzes ester bond hydrolysis.

Moreover, the acyl group acceptor in the chymotrypsin reaction need not be water; amino acids, alcohols, or ammonia can also act in this capacity. You should realize, however, that such permissiveness is much more the exception than the rule. Indeed, most intracellular enzymes function *in vivo* (in the cell) to catalyze a particular reaction on a specific substrate.

3 ■ COENZYMES

Enzymes catalyze a wide variety of chemical reactions. Their functional groups can facilely participate in acid–base reactions, form certain types of transient covalent bonds, and take part in charge–charge interactions (Section 15-1). They are, however, less suitable for catalyzing oxidation–reduction reactions and many types of group-transfer processes. Although enzymes catalyze such reactions, they mainly do so in association with small molecule **cofactors,** which essentially act as the enzymes' "chemical teeth."

Cofactors may be metal ions, such as the Zn^{2+} required for the catalytic activity of carboxypeptidase A, or organic molecules known as **coenzymes,** such as the NAD^+ in YADH (Section 13-2A). Some cofactors, for instance

TABLE 13-1   The Common Coenzymes

Coenzyme	Reaction Mediated	Section Discussed
Biotin	Carboxylation	23-1A
Cobalamin (B_{12}) coenzymes	Alkylation	25-2E
Coenzyme A	Acyl transfer	21-2A
Flavin coenzymes	Oxidation–reduction	16-5C
Lipoic acid	Acyl transfer	21-2A
Nicotinamide coenzymes	Oxidation–reduction	13-2A
Pyridoxal phosphate	Amino group transfer	26-1A
Tetrahydrofolate	One-carbon group transfer	26-4D
Thiamine pyrophosphate	Aldehyde transfer	17-3B

NAD^+, are but transiently associated with a given enzyme molecule, so that, in effect, they function as cosubstrates. Other cofactors, known as **prosthetic groups,** are essentially permanently associated with their protein, often by covalent bonds. For example, the heme prosthetic group of hemoglobin is tightly bound to its protein through extensive hydrophobic and hydrogen bonding interactions together with a covalent bond between the heme Fe^{2+} ion and His F8 (Sections 10-1A and 10-2B).

Coenzymes are chemically changed by the enzymatic reactions in which they participate. Thus, in order to complete the catalytic cycle, the coenzyme must be returned to its original state. For prosthetic groups, this can occur only in a separate phase of the enzymatic reaction sequence. For transiently bound coenzymes, such as NAD^+,

TABLE 13-2   Vitamins That Are Coenzyme Precursors

Vitamin	Coenzyme	Human Deficiency Disease
Biotin	Biocytin	*a*
Cobalamin (B_{12})	Cobalamin (B_{12}) coenzymes	Pernicious anemia
Folic acid	Tetrahydrofolate	Megaloblastic anemia
Nicotinamide	Nicotinamide coenzymes	Pellagra
Pantothenate	Coenzyme A	*a*
Pyridoxine (B_6)	Pyridoxal phosphate	*a*
Riboflavin (B_2)	Flavin coenzymes	*a*
Thiamine (B_1)	Thiamine pyrophosphate	Beriberi

*a*No specific name; deficiency in humans is rare or unobserved.

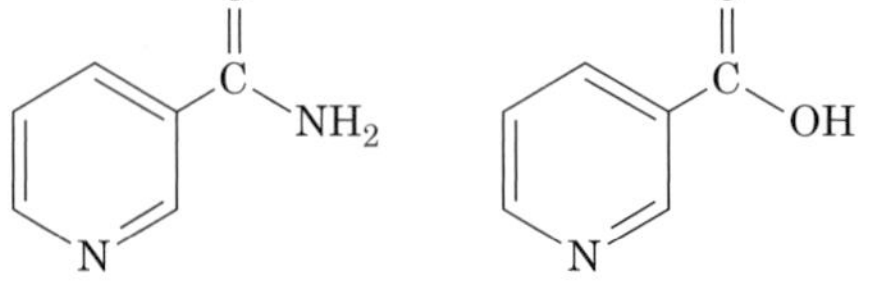

however, the regeneration reaction may be catalyzed by a different enzyme.

A catalytically active enzyme–cofactor complex is called a **holoenzyme.** The enzymatically inactive protein resulting from the removal of a holoenzyme's cofactor is referred to as an **apoenzyme;** that is,

Apoenzyme (*inactive*) + cofactor $\rightleftharpoons$

holoenzyme (*active*)

Table 13-1 lists the most common coenzymes together with the types of reactions in which they participate. We shall describe the structures of these substances and their reaction mechanisms in the appropriate sections of the textbook.

a. Many Vitamins Are Coenzyme Precursors

Many organisms are unable to synthesize certain portions of essential cofactors and therefore these substances must be present in the organism's diet; thus they are **vitamins.** In fact, many coenzymes were discovered as growth factors for microorganisms or substances that cure nutritional deficiency diseases in humans and animals. For example, the NAD^+ component **nicotinamide** (alternatively known as **niacinamide**) or its carboxylic acid analog **nicotinic acid** (**niacin;** Fig. 13-4), relieves the dietary deficiency disease in humans known as **pellagra.** Pellagra, which is characterized by diarrhea, dermatitis, and dementia, was endemic in the rural southern United States in the early twentieth century. Most animals, including humans, can synthesize nicotinamide from the amino acid tryptophan (Section 28-6A). The corn-rich diet that was prevalent in the rural South, however, contained little available nicotinamide or tryptophan from which to synthesize it. [Corn actually contains significant quantities of nicotinamide but in a form that requires treatment with base before it can be intestinally absorbed. The Mexican Indians, who are thought to have domesticated the corn plant, customarily soak corn meal in lime water—dilute $Ca(OH)_2$ solution—before using it to bake their staple food, tortillas.]

The vitamins in the human diet that are coenzyme precursors are all **water-soluble vitamins** (Table 13-2). In contrast, the **lipid-soluble vitamins,** such as **vitamins A** and **D,** are not components of coenzymes, although they are also

FIGURE 13-4   Structures of nicotinamide and nicotinic acid. These vitamins form the redox-active components of the nicotinamide coenzymes NAD^+ and $NADP^+$ (compare with Fig. 13-2).

Nicotinamide (niacinamide)   Nicotinic acid (niacin)

required in trace amounts in the diets of many higher animals. The distant ancestors of humans probably had the ability to synthesize the various vitamins, as do many modern plants and microorganisms. Yet, since vitamins are normally available in the diets of higher animals, which all eat other organisms, or are synthesized by the bacteria that normally inhabit their digestive systems, it is believed that the then superfluous cellular machinery to synthesize them was lost through evolution.

4 ■ REGULATION OF ENZYMATIC ACTIVITY

An organism must be able to regulate the catalytic activities of its component enzymes so that it can coordinate its numerous metabolic processes, respond to changes in its environment, and grow and differentiate, all in an orderly manner. There are two ways that this may occur:

1. *Control of enzyme availability:* *The amount of a given enzyme in a cell depends on both its rate of synthesis and its rate of degradation.* Each of these rates is directly controlled by the cell. For example, *E. coli* grown in the absence of the disaccharide lactose (Fig. 11-12) lack the enzymes to metabolize this sugar. Within minutes of their exposure to lactose, however, these bacteria commence synthesizing the enzymes required to utilize this nutrient (Section 31-1A). Similarly, the various tissues of a higher organism contain different sets of enzymes, although most of its cells contain identical genetic information. How cells achieve this control of enzyme synthesis is a major subject of Part V of this textbook. The degradation of proteins is discussed in Section 32-6.

2. *Control of enzyme activity:* *An enzyme's catalytic activity may be directly regulated through conformational or structural alterations.* The rate of an enzymatically catalyzed reaction is directly proportional to the concentration of its enzyme–substrate complex, which, in turn, varies with the enzyme and substrate concentrations and with the enzyme's substrate-binding affinity (Section 14-2A). The catalytic activity of an enzyme can therefore be controlled through the variation of its substrate-binding affinity. Recall that Sections 10-1 and 10-4 detail how hemoglobin's oxygen affinity is allosterically regulated by the binding of ligands such as O_2, CO_2, H^+, and BPG. These homotropic and heterotropic effects (ligand binding that, respectively, alters the binding affinity of the same or different ligands) result in cooperative (sigmoidal) O_2-binding curves such as those of Figs. 10-6 and 10-8. *An enzyme's substrate-binding affinity may likewise vary with the binding of small molecule effectors, thereby changing the enzyme's catalytic activity.* In this section we consider the allosteric control of enzymatic activity by examining one particular example: **aspartate transcarbamoylase (ATCase)** from *E. coli*. (The activities of many enzymes are similarly regulated through their reversible covalent modification, usually by the phosphorylation of a Ser residue. We study this form of enzymatic regulation in Section 18-3.)

a. The Feedback Inhibition of ATCase Regulates Pyrimidine Biosynthesis

Aspartate transcarbamoylase catalyzes the formation of **N-carbamoylaspartate** from **carbamoyl phosphate** and aspartate:

Carbamoyl phosphate   **Aspartate**

aspartate transcarbamoylase

$+$ $H_2PO_4^-$

N-Carbamoylaspartate

Arthur Pardee demonstrated that this reaction is the first step unique to the biosynthesis of pyrimidines (Section 28-2A), major components of nucleic acids.

The allosteric behavior of *E. coli* ATCase was investigated by John Gerhart and Howard Schachman, who demonstrated that this enzyme exhibits positive homotropic cooperative binding of both its substrates, namely, aspartate and carbamoyl phosphate. Moreover, ATCase is heterotropically inhibited by **cytidine triphosphate (CTP)**, a pyrimidine nucleotide, and is heterotropically activated by **adenosine triphosphate (ATP)**, a purine nucleotide. CTP therefore decreases the enzyme's catalytic rate, whereas ATP increases it (Fig. 13-5).

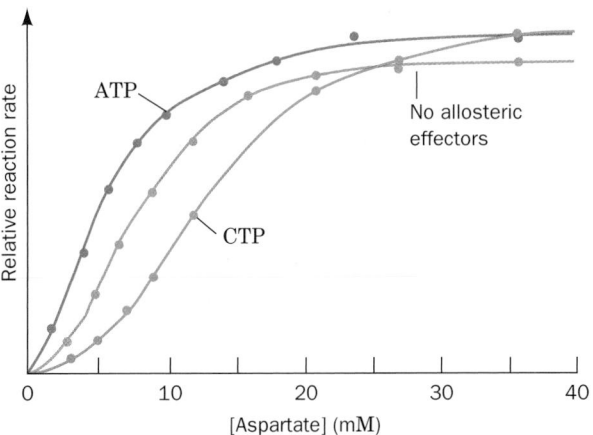

FIGURE 13-5 The rate of the reaction catalyzed by ATCase as a function of aspartate concentration. The rates were measured in the absence of allosteric effectors, in the presence of 0.4 m*M* CTP (inhibition), and in the presence of 2.0 m*M* ATP (activation). [After Kantrowitz, E.R., Pastra-Landis, S.C., and Lipscomb, W.N., *Trends Biochem. Sci.* **5**, 125 (1980).] ◈ See the Animated Figures

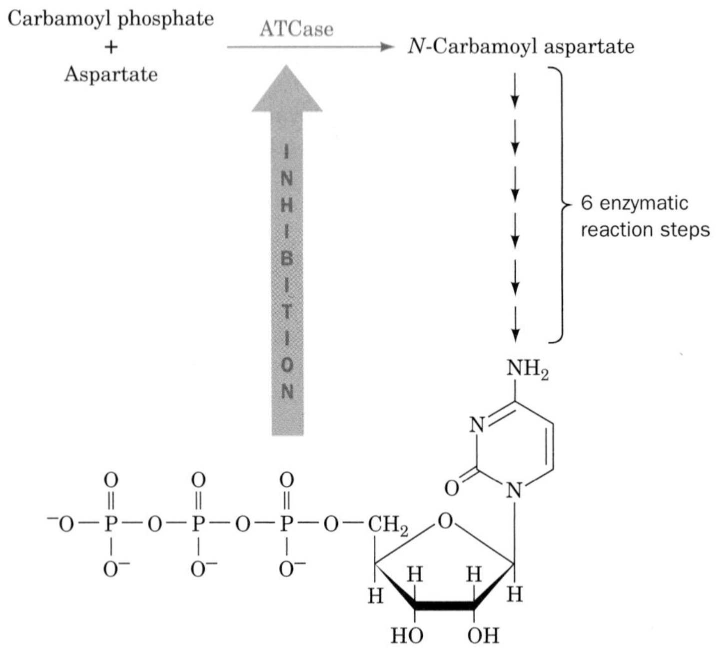

FIGURE 13-6 Schematic representation of the pyrimidine biosynthesis pathway. CTP, the end product of the pathway, inhibits ATCase, which catalyzes the pathway's first step.

CTP, a product of the pyrimidine biosynthesis pathway (Fig. 13-6), is a nucleic acid precursor (Section 5-4). Consequently, when rapid nucleic acid biosynthesis has depleted a cell's CTP pool, this effector dissociates from ATCase through mass action, thereby deinhibiting the enzyme and increasing the rate of CTP synthesis. Conversely, if the rate of CTP synthesis outstrips its rate of uptake, the resulting excess CTP inhibits ATCase, which, in turn, reduces the rate of CTP synthesis. *This is an example of* ***feedback inhibition,*** *a common mode of metabolic regulation in which the concentration of a biosynthetic pathway product controls the activity of an enzyme near the beginning of that pathway.*

The metabolic significance of the ATP activation of ATCase is that it tends to coordinate the rates of synthesis of purine and pyrimidine nucleotides for nucleic acid biosynthesis. For instance, if the ATP and CTP concentrations are out of balance with ATP in excess, ATCase is activated to synthesize pyrimidines until balance is achieved. (Note: The ATP concentration in cells is normally greater than the CTP concentration because ATP is in greater demand. Hence the ATP concentration required to activate ATCase is higher than the CTP concentration required to inhibit it by an equal amount.) Conversely, if CTP is in excess, the resulting CTP inhibition of ATCase permits purine biosynthesis to attain this balance.

b. Allosteric Changes Alter ATCase's Substrate-Binding Sites

E. coli ATCase (300 kD) has the subunit composition c_6r_6, where c and r represent its catalytic and regulatory subunits. The X-ray structure of ATCase (Fig. 13-7), determined by William Lipscomb, reveals that the catalytic subunits are arranged as two sets of trimers (c_3) in complex with three sets of regulatory dimers (r_2) to form a molecule with the rotational symmetry of a trigonal prism (D_3 symmetry; Section 8-5B). Each regulatory dimer joins two catalytic subunits in different c_3 trimers.

Dissociated catalytic trimers retain their catalytic activity, exhibit a noncooperative (hyperbolic) substrate saturation curve, have a maximum catalytic rate higher than that of intact enzyme, and are unaffected by the presence of either ATP or CTP. The isolated regulatory dimers bind these allosteric effectors but are devoid of enzymatic activity. Evidently, *the regulatory subunits allosterically reduce the activity of the catalytic subunits in the intact enzyme.*

As allosteric theory predicts (Section 10-4), the activator ATP preferentially binds to ATCase's active (R or high substrate affinity) state, whereas the inhibitor CTP preferentially binds to the enzyme's inactive (T or low substrate affinity) state. Similarly, the unreactive bisubstrate analog ***N-(phosphonacetyl)-L-aspartate (PALA)***

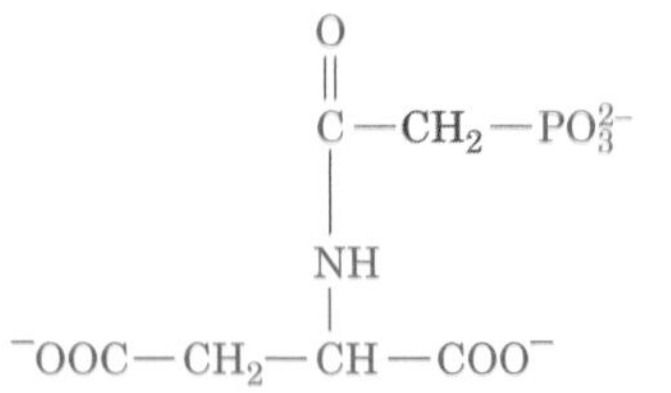

N-(Phosphonacetyl)-
L-aspartate (PALA)

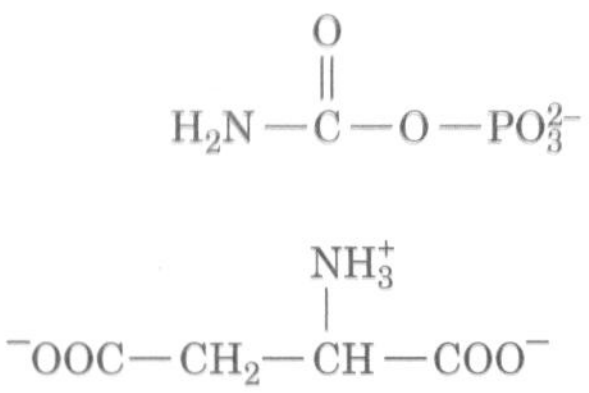

Carbamoyl phosphate
+
Aspartate

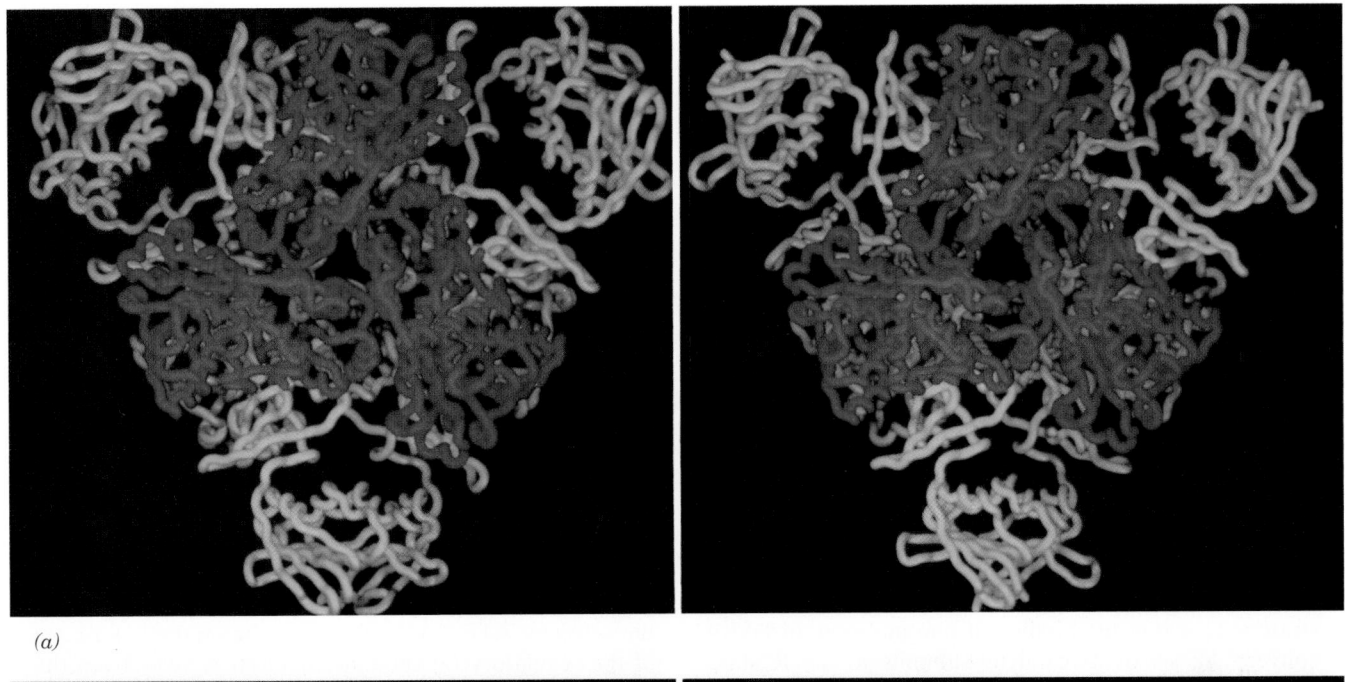

(a)

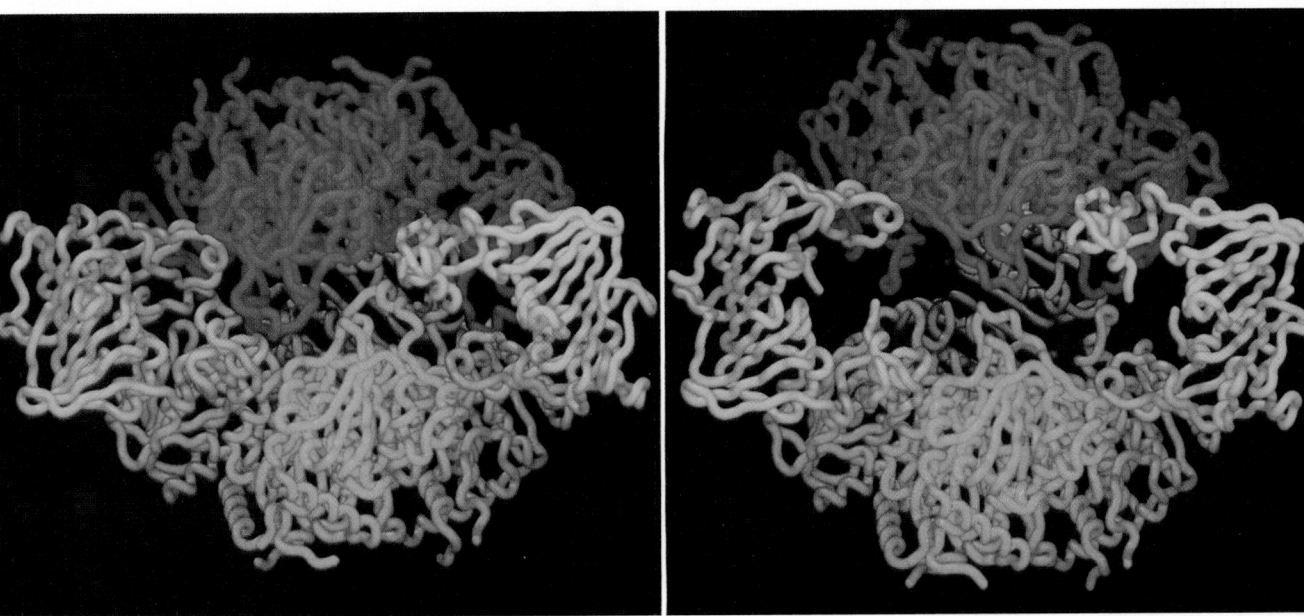

(b)

FIGURE 13-7 X-Ray structure of ATCase. The polypeptide backbones of T-state ATCase (*left*) and R-state ATCase (*right*) are viewed (*a*) along the protein's molecular threefold axis of symmetry and (*b*) along a molecular twofold axis of symmetry. The regulatory dimers (*yellow*) join the upper catalytic trimer (*red*) to the lower catalytic trimer (*blue*). [Courtesy of Michael Pique, The Scripps Research Institute, La Jolla, California. X-Ray structures by William Lipscomb, Harvard University. PDBids 4AT1 and 8ATC.] 🎷 **See Kinemage Exercise 11-1**

binds tightly to R-state but not to T-state ATCase (the use of unreactive substrate analogs is common in the study of enzyme mechanisms because they form stable complexes that are amenable to structural study rather than rapidly reacting to form products as do true substrates).

The X-ray structures of the T-state ATCase–CTP complex and the R-state ATCase–PALA complex reveal that the T → R transition maintains the protein's D_3 symmetry.

The comparison of these two structures (Fig. 13-7) indicates that in the T → R transition, the enzyme's catalytic trimers separate along the molecular threefold axis by ~11 Å and reorient about this axis relative to each other by 12° such that these trimers assume a more nearly eclipsed configuration. In addition, the regulatory dimers rotate clockwise by 15° about their twofold axes and separate by ~4 Å along the threefold axis. Such large quaternary shifts are reminiscent of those in hemoglobin (Section 10-2B).

ATCase's substrates, carbamoyl phosphate and aspartate, each bind to a separate domain of the catalytic subunit (Fig. 13-8). The binding of PALA to the enzyme, which presumably mimics the binding of both substrates, induces active site closure in a manner that would bring them together so as to promote their reaction. The resulting atomic shifts, up to 8 Å for some residues (Fig. 13-8), trigger ATCase's T → R quaternary shift. Indeed, *ATCases's tertiary and quaternary shifts are so tightly coupled through extensive intersubunit contacts (see below) that they cannot occur independently (Fig. 13-9).* The binding of substrate to one catalytic subunit therefore increases the substrate-binding affinity and catalytic activity of the other catalytic subunits and hence accounts for the enzyme's positively cooperative substrate binding, much as occurs in hemoglobin (Section 10-2C). Thus, low levels of PALA actually activate ATCase by promoting its T → R transition: ATCase has such high affinity for this unreactive bisubstrate analog that the binding of one molecule of PALA converts all six of its catalytic subunits to the R state. Evidently, *ATCase closely follows the symmetry model of allosterism (Section 10-4B).*

c. The Structural Basis of Allosterism in ATCase

What are the interactions that stabilize the T and R states of ATCase and why must their interconversion be concerted? The region of the protein that undergoes the most profound conformational rearrangement with the T → R transition is a flexible loop composed of residues 230 to 250 in the catalytic (*c*) subunit, the so-called 240s loop [the symmetry-related red and blue loops at the cen-

ters of Fig. 13-7b that lie side by side in the T state (*left*) but are vertically apposed in the R state (*right*)]. In the T state, each 240s loop forms two intersubunit hydrogen bonds with the vertically opposite *c* subunit (Fig. 13-7b, *left*), together with an intrasubunit hydrogen bond. Domain closure as a consequence of substrate binding (Figs. 13-8 and 13-9) ruptures these hydrogen bonds and replaces them, in the R state, with new intrachain hydrogen bonds. The consequent reorientation of the 240s loop is thought to be largely responsible for the quaternary shift to the R state (see below). Since the Glu 239 carboxyl group is the acceptor in all of the above T-state interchain and R-state intrachain hydrogen bonds, this hypothesis is corroborated by the observation that the mutation of Glu 239 to Gln converts ATCase to an enzyme that is devoid of both homotropic and heterotropic effects and that has a quaternary structure midway between those of the R and T states.

The structural basis for heterotropic effects in ATCase is gradually being unveiled. Both the inhibitor CTP and the activator ATP bind to the same site on the outer edge of the regulatory (*r*) subunit, about 60 Å away from the nearest catalytic site. CTP binds preferentially to the T state, increasing its stability, while ATP binds preferentially to the R state, increasing its stability. The binding of these effectors to their less favored states also has structural consequences. When CTP binds to R-state ATCase, it reorients several residues at the nucleotide binding site, which induces a contraction in the length of the regulatory dimer (r_2). This distortion, through the interactions of residues at the *r–c* interface, causes the catalytic trimers (c_3) to come together by 0.5 Å (become more T-like, that is, less active, which presumably destabilizes the R state). This, in turn,

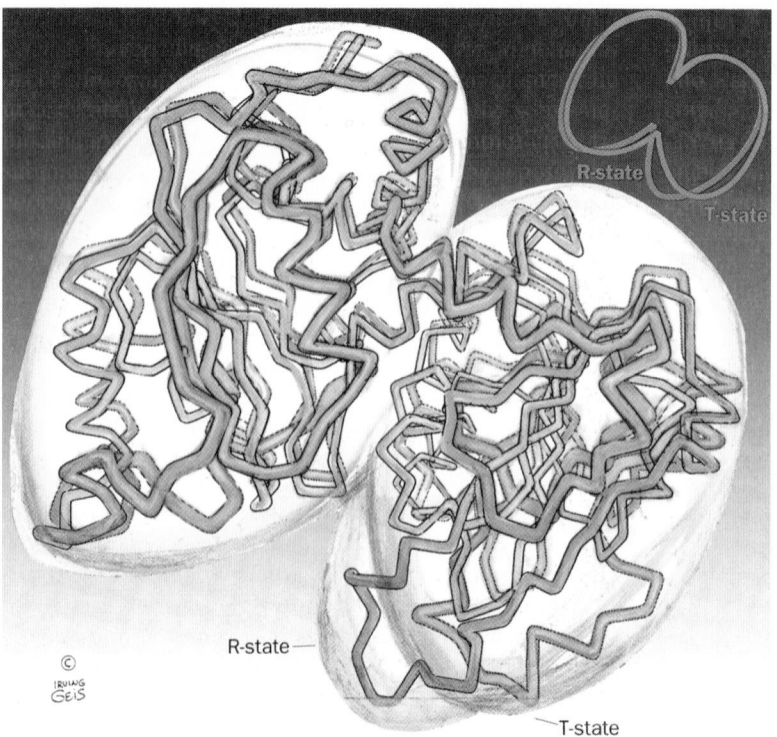

R-state

T-state

FIGURE 13-8 Comparison of the polypeptide backbones of the ATCase catalytic subunit in the T state (*orange*) and the R state (*blue*). The subunit consists of two domains, with the one on the left containing the carbamoyl phosphate binding site and that on the right forming the aspartic acid binding site. The T → R transition brings the two domains together such that their two bound substrates can react to form product. [Illustration, Irving Geis/Geis Archives Trust. Copyright Howard Hughes Medical Institute. Reproduced with permission. X-Ray structures by William Lipscomb, Harvard University.]

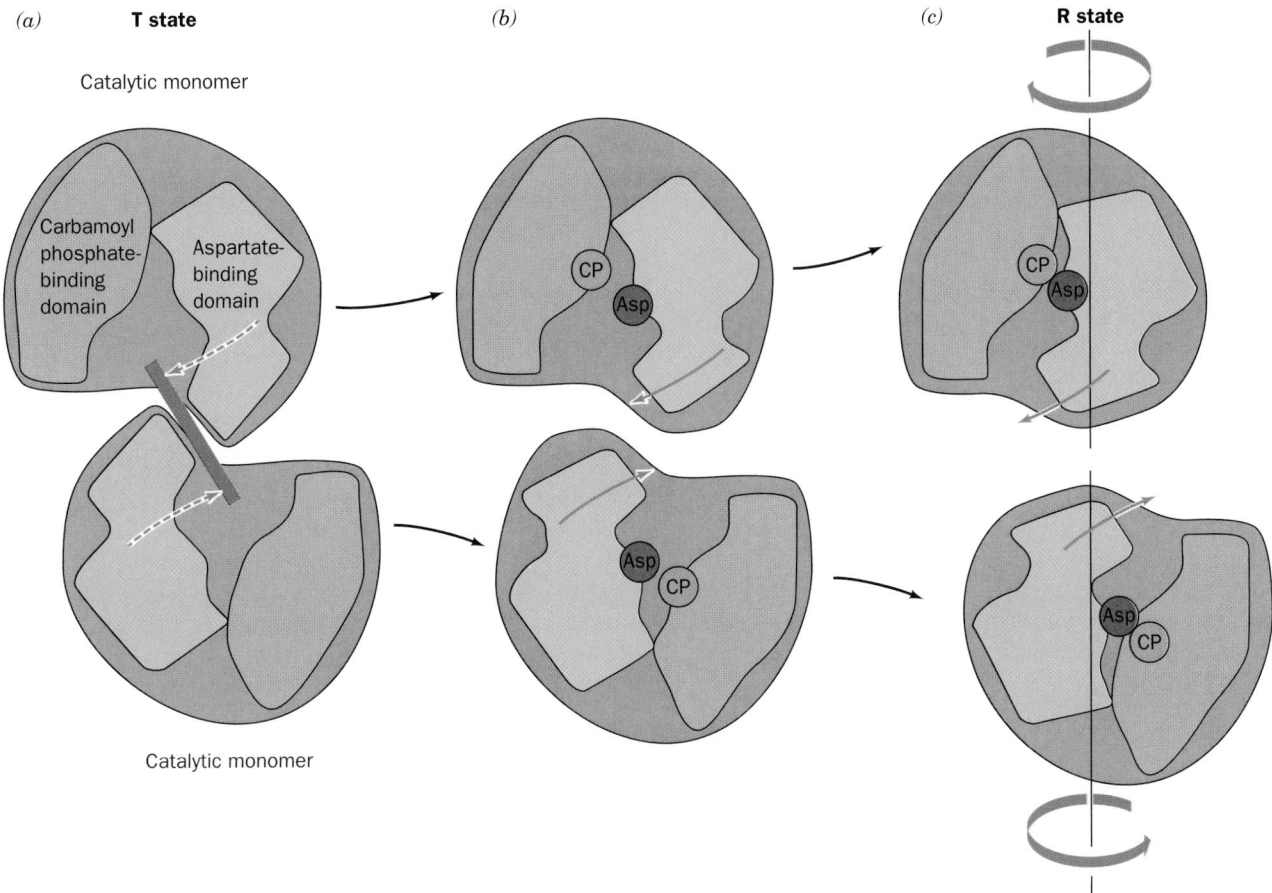

FIGURE 13-9 **Schematic diagram indicating the tertiary and quaternary conformational changes in two vertically interacting catalytic ATCase subunits.** (*a*) In the absence of bound substrate the protein is held in the T state because the motions that bring together the two domains of each subunit (*dashed arrows*) are prevented by steric interference (*purple bar*) between the contacting aspartic acid binding domains. (*b*) The binding of carbamoyl phosphate (CP) followed by aspartic acid (Asp) to their respective binding sites causes the subunits to move apart and rotate with respect to each other so as to permit the T → R transition. (*c*) In the R state, the two domains of each subunit come together so as to promote the reaction of their bound substrates to form products. [Illustration, Irving Geis/Geis Archives Trust. Copyright Howard Hughes Medical Institute. Reproduced with permission.] *See* Kinemage Exercises 11-1 and 11-2.

reorients key residues in the enzyme's active sites, thereby decreasing the enzyme's catalytic activity. ATP has essentially opposite effects when binding to the T-state enzyme: It causes the catalytic trimers to move apart by 0.4 Å (become more R-like, that is, more active, which presumably destabilizes the T state), thereby reorienting key residues in the enzyme's active sites so as to increase the enzyme's catalytic activity. The binding of CTP to T-state ATCase does not further compress the catalytic trimers but, nevertheless, perturbs active site residues in a way that further stabilizes the T state. Although the X-ray structure of ATP complexed to R-state ATCase has not yet been reported, it is expected that ATP binding perturbs the R state in a manner analogous but opposite to the binding of CTP to T-state ATCase.

d. Allosteric Transitions in Other Enzymes Resemble Those of Hemoglobin and ATCase

Allosteric enzymes are widely distributed in nature and tend to occupy key regulatory positions in metabolic path-

ways. Three such enzymes, in addition to hemoglobin and ATCase, have had their X-ray structures determined in both their R and T states: **phosphofructokinase** (Sections 17-2C and 17-4F), **fructose-1,6-bisphosphatase** (Section 17-4F), and **glycogen phosphorylase** (Section 18-1A). In all five proteins, quaternary changes, through which binding and catalytic effects are communicated among active sites, are concerted and preserve the symmetry of the protein. This is because each of these proteins has two sets of alternative contacts, which are stabilized largely by hydrogen bonds that mostly involve side chains of opposite charge. In all five proteins, the quaternary shifts are primarily rotations of subunits relative to one another with only small translations. Secondary structures are largely preserved in T → R transitions, which is probably important for mechanically transmitting heterotropic effects over the tens of Å necessary in these proteins. The ubiquity of these structural features among allosteric proteins of known structures suggests that the regulatory mechanisms of other allosteric enzymes, by and large, follow this model.

5 ■ A PRIMER OF ENZYME NOMENCLATURE

Enzymes, as we have seen throughout the text so far, are commonly named by appending the suffix *-ase* to the name of the enzyme's substrate or to a phrase describing the enzyme's catalytic action. Thus urease catalyzes the hydrolysis of urea and alcohol dehydrogenase catalyzes the oxidation of alcohols to their corresponding aldehydes. Since there were at first no systematic rules for naming enzymes, this practice occasionally resulted in two different names being used for the same enzyme or, conversely, in the same name being used for two different enzymes. Moreover, many enzymes, such as catalase, which mediates the dismutation of H_2O_2 to H_2O and O_2, were given names that provide no clue as to their function; even such atrocities as "old yellow enzyme" have crept into use. In an effort to eliminate this confusion and to provide rules for rationally naming the rapidly growing number of newly discovered enzymes, a scheme for the systematic functional classification and nomenclature of enzymes was adopted by the International Union of Biochemistry and Molecular Biology (IUBMB).

Enzymes are classified and named according to the nature of the chemical reactions they catalyze. There are six major classes of reactions that enzymes catalyze (Table 13-3), as well as subclasses and sub-subclasses within these classes. Each enzyme is assigned two names and a four-number classification. Its **recommended name** is convenient for everyday use and is often an enzyme's previously used trivial name. Its **systematic name** is used when ambiguity must be minimized; it is the name of its substrate(s) followed by a word ending in *-ase* specifying the type of reaction the enzyme catalyzes according to its major group classification. For example, the most recent version of the *Enzyme Nomenclature Database* (available from http://expasy.org/enzyme and http://www.chem.qmw.ac.uk/iubmb/enzyme/, which contained a growing list of over 4000 entries at the beginning of 2003) indicates that the

TABLE 13-3   Enzyme Classification According to Reaction Type

Classification	Type of Reaction Catalyzed
1. Oxidoreductases	Oxidation–reduction reactions
2. Transferases	Transfer of functional groups
3. Hydrolases	Hydrolysis reactions
4. Lyases	Group elimination to form double bonds
5. Isomerases	Isomerization
6. Ligases	Bond formation coupled with ATP hydrolysis

enzyme whose recommended name is carboxypeptidase A (Section 7-1A) has the systematic name **peptidyl-L-amino acid hydrolase** and the **classification number** EC 3.4.17.1. Here "EC" stands for Enzyme Commission, the first number (3) indicates the enzyme's major class (hydrolases; Table 13-3), the second number (4) denotes its subclass [acting on peptide bonds (peptide hydrolases)], the third number (17) designates its sub-subclass (metallocarboxypeptidases; carboxypeptidase A has a bound Zn^{2+} ion that is essential for its catalytic activity), and the fourth number (1) is the enzyme's arbitrarily assigned serial number in its sub-subclass. As another example, the enzyme with the recommended name alcohol dehydrogenase (Section 13-2A) has the systematic name **alcohol:NAD⁺ oxidoreductase** and the classification number EC 1.1.1.1. In this text, as in general biochemical terminology, we shall most often use the recommended names of enzymes but, when ambiguity must be minimized, we shall refer to an enzyme's systematic name. Note that the enzyme entries in the above websites contain links to databases describing the properties of the selected enzyme. The website named BRENDA (http://www.brenda.uni-koeln.de/) provides enzyme nomenclature together with a wealth of physical and functional data concerning each enzyme.

■ CHAPTER SUMMARY

2 ■ Substrate Specificity Enzymes specifically bind their substrates through geometrically and physically complementary interactions. This permits enzymes to be absolutely stereospecific, both in binding substrates and in catalyzing reactions. Enzymes vary in the more stringent requirement of geometric specificity. Some are highly specific for the identity of their substrates, whereas others can bind a wide range of substrates and catalyze a variety of related types of reactions.

3 ■ Coenzymes Enzymatic reactions involving oxidation–reduction reactions and many types of group-transfer processes are mediated by coenzymes. Many vitamins are coenzyme precursors.

4 ■ Regulation of Enzyme Activity Enzymatic activity may be regulated by the allosteric alteration of substrate-binding affinity. For example, the rate of the reaction catalyzed by *E. coli* ATCase is subject to positive homotropic

control by substrates, heterotropic inhibition by CTP, and heterotropic activation by ATP. ATCase has the subunit composition c_6r_6. Its isolated catalytic trimers are catalytically active but not subject to allosteric control. The regulatory dimers bind ATP and CTP. Substrate binding induces a tertiary conformational shift in the catalytic subunits, which increases the subunit's substrate-binding affinity and catalytic efficiency. This tertiary shift is strongly coupled to ATCase's large quaternary T → R conformational shift, thereby accounting for the enzyme's allosteric properties. Other allosteric enzymes appear to operate in a similar manner.

5 ■ A Primer of Enzyme Nomenclature Enzymes are systematically classified according to their recommended name, their systematic name, and their classification number, which is indicative of the type of reaction catalyzed by the enzyme.

REFERENCES

GENERAL
Dixon, M. and Webb, E.C., *Enzymes* (3rd ed.), Academic Press (1979). [A treatise on enzymes.]

HISTORY
Friedmann, H.C. (Ed.), *Enzymes,* Hutchinson Ross (1981). [A compendium of classic enzymological papers published between 1761 and 1974; with commentary.]

Fruton, J.S., *Molecules and Life, pp.* 22–86, Wiley (1972).

Schlenk, F., Early research on fermentation—a story of missed opportunities, *Trends Biochem. Sci.* **10,** 252–254 (1985).

SUBSTRATE SPECIFICITY
Creighton, D.J. and Murthy, N.S.R.K., Stereochemistry of enzyme-catalyzed reactions at carbon, *in* Sigman, D.S. and Boyer, P.D. (Eds.), *The Enzymes* (3rd ed.), Vol. 19, *pp.* 323–421, Academic Press (1990). [Section II discusses the stereochemistry of reactions catalyzed by nicotinamide-dependent dehydrogenases.]

Fersht, A., *Structure and Mechanism in Protein Science,* Freeman (1999).

Lamzin, V.S., Sauter, Z., and Wilson, K.S., How nature deals with stereoisomers, *Curr. Opin. Struct. Biol.* **5,** 830–836 (1995).

Mesecar, A.D. and Koshland, D.E. Jr., A new model for protein stereospecificity, *Nature* **403,** 614–615 (2000).

Ringe, D., What makes a binding site a binding site? *Curr. Opin. Struct. Biol.* **5,** 825–829 (1995).

Weinhold, E.G., Glasfeld, A., Ellington, A.D., and Benner, S.A., Structural determinants of stereospecificity in yeast alcohol dehydrogenase, *Proc. Natl. Acad. Sci.* **88,** 8420–8424 (1991).

REGULATION OF ENZYME ACTIVITY
Allewell, N.M., *Escherichia coli* aspartate transcarbamoylase: Structure, energetics, and catalytic and regulatory mechanisms, *Annu. Rev. Biophys. Biophys. Chem.* **18,** 71–92 (1989).

Evans, P.R., Structural aspects of allostery, *Curr. Opin. Struct. Biol.* **1,** 773–779 (1991).

Gouaux, J.E., Stevens, R.C., Ke, H., and Lipscomb, W.N., Crystal structure of the Glu-289 → Gln mutant of aspartate carbamoyl-transferase at 3.1-Å resolution: An intermediate quaternary structure, *Proc. Natl. Acad. Sci.* **86,** 8212–8216 (1989).

Jin, L., Stec, B., Lipscomb, W.N., and Kantrowitz, E.R., Insights into the mechanisms of catalysis and heterotropic regulation of *Escherichia coli* aspartate transcarbamoylase based upon a structure of the enzyme complexed with the bisubstrate analogue N-phosphonacetyl-L-aspartate at 2.1 Å, *Proteins* **37,** 729–742 (1999).

Kantrowitz, E.R. and Lipscomb, W.N., *Escherichia coli* aspartate transcarbamylase: The molecular basis for a concerted allosteric transition, *Trends Biochem. Sci.* **15,** 53–59 (1990).

Koshland, D.E., Jr., The key–lock theory and the induced fit theory, *Angew. Chem. Int. Ed. Engl.* **33,** 2375–2378 (1994).

Lipscomb, W.N., Structure and function of allosteric enzymes, *Chemtracts—Biochem. Mol. Biol.* **2,** 1–15 (1991).

Macol, C.P., Tsuruta, H., Stec, B., and Kantrowitz, E.R., Direct structural evidence for a concerted allosteric transition in *Escherichia coli* aspartate transcarbamoylase, *Nature Struct. Biol.* **8,** 423–426 (2001).

Schachman, H.K., Can a simple model account for the allosteric transition of aspartate transcarbamoylase? *J. Biol. Chem.* **263,** 18583–18586 (1988).

Stevens, R.C. and Lipscomb, W.N., A molecular mechanism for pyrimidine and purine nucleotide control of aspartate transcarbamoylase, *Proc. Natl. Acad. Sci.* **89,** 5281–5285 (1992).

Zhang, Y. and Kantrowitz, E.R., Probing the regulatory site of *Escherichia coli* aspartate transcarbamoylase by site specific mutagenesis, *Biochemistry* **31,** 792–798 (1992).

ENZYME NOMENCLATURE
Enzyme Nomenclature, Academic Press (1992). [Recommendations of the Nomenclature Committee of the IUBMB on the nomenclature and classification of enzymes.]

Tipton, K.F., The naming of parts, *Trends Biochem. Sci.* **18,** 113–115 (1993). [A discussion of the advantages of a consistent naming scheme for enzymes and the difficulties of formulating one.]

PROBLEMS

1. Indicate the products of the YADH reaction with normal acetaldehyde and NADH in D$_2$O solution.

2. Indicate the product(s) of the YADH-catalyzed oxidation of the chiral methanol derivative (R)-TDHCOH.

3. The enzyme **fumarase** catalyzes the hydration of the double bond of **fumarate:**

Fumarate   **L-Malate**

Predict the action of fumarase on **maleate,** the cis isomer of fumarate. Explain.

4. Write a balanced equation for the chymotrypsin-catalyzed reaction between an ester and an amino acid.

5. Hominy grits, a regional delicacy of the southern United States, is made from corn that has been soaked in a weak lye (NaOH) solution. What is the function of this unusual treatment?

6. Which of the curves in Fig. 13-5 exhibits the greatest cooperativity? Explain.

7. What are the advantages of having the final product of a multistep metabolic pathway inhibit the enzyme that catalyzes the first step?

8. Using the Web or the references, find the systematic names and classification numbers for the enzymes whose recommended names are catalase, aspartate carbamoyltransferase (aspartate transcarbamoylase), and trypsin.

Chapter
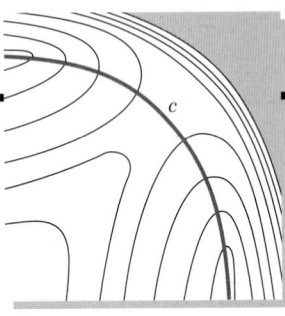

14

Rates of Enzymatic Reactions

Kinetics is the study of the rates at which chemical reactions occur. A major purpose of such a study is to gain an understanding of a reaction mechanism, that is, a detailed description of the various steps in a reaction process and the sequence with which they occur. Thermodynamics, as we saw in Chapter 3, tells us whether a given process can occur spontaneously but provides little indication as to the nature or even the existence of its component steps. In contrast, *the rate of a reaction and how this rate changes in response to different conditions is intimately related to the path followed by the reaction and is therefore indicative of its reaction mechanism.*

In this chapter, we take up the study of **enzyme kinetics,** a subject that is of enormous practical importance in biochemistry because:

1. It is through kinetic studies that the binding affinities of substrates and inhibitors to an enzyme can be determined and that the maximum catalytic rate of an enzyme can be established.

2. By observing how the rate of an enzymatic reaction varies with the reaction conditions and combining this information with that obtained from chemical and structural studies of the enzyme, the enzyme's catalytic mechanism may be elucidated.

3. Most enzymes, as we shall see in later chapters, function as members of metabolic pathways. The study of the kinetics of an enzymatic reaction leads to an understanding of that enzyme's role in an overall metabolic process.

4. Under the proper conditions, the rate of an enzymatically catalyzed reaction is proportional to the amount of the enzyme present, and therefore most enzyme assays (measurements of the amount of enzyme present) are based on kinetic studies of the enzyme. Measurements of enzymatically catalyzed reaction rates are therefore among the most commonly employed procedures in biochemical and clinical analyses.

We begin our consideration of enzyme kinetics by reviewing chemical kinetics because enzyme kinetics is based on this formalism. Following that, we derive the basic equations of enzyme kinetics, describe the effects of inhibitors on enzymes, and consider how the rates of enzymatic reactions vary with pH. We end by outlining the kinetics of complex enzymatic reactions.

Kinetics is, by and large, a mathematical subject. Although the derivations of kinetic equations are occasionally rather detailed, the level of mathematical skills it requires should not challenge anyone who has studied elementary calculus. Nevertheless, to prevent mathematical detail from obscuring the underlying enzymological principles, the derivations of all but the most important kinetic equations have been collected in the appendix to this chapter. Those who wish to cultivate a deeper understanding of enzyme kinetics are urged to consult this appendix.

1 ■ CHEMICAL KINETICS

Enzyme kinetics is a branch of chemical kinetics and, as such, shares much of the same formalism. In this section we shall therefore review the principles of chemical kinetics so that, in later sections, we can apply them to enzymatically catalyzed reactions.

A. *Elementary Reactions*

A reaction of overall stoichiometry

$$A \rightarrow P$$

may actually occur through a sequence of **elementary reactions** (simple molecular processes) such as

$$A \rightarrow I_1 \rightarrow I_2 \rightarrow P$$

Here A represents reactants, P products, and I_1 and I_2 symbolize **intermediates** in the reaction. *The characterization of the elementary reactions comprising an overall reaction process constitutes its mechanistic description.*

a. Rate Equations

At constant temperature, elementary reaction rates vary with reactant concentration in a simple manner. Consider the general elementary reaction:

$$aA + bB + \cdots + zZ \rightarrow P$$

The rate of this process is proportional to the frequency with which the reacting molecules simultaneously come together, that is, to the products of the concentrations of the reactants. This is expressed by the following **rate equation**

$$\text{Rate} = k[A]^a[B]^b \cdots [Z]^z \qquad [14.1]$$

where k is a proportionality constant known as a **rate constant**. The **order** of a reaction is defined as $(a + b + \cdots + z)$, the sum of the exponents in the rate equation. *For an elementary reaction, the order corresponds to the molecularity of the reaction, the number of molecules that must simultaneously collide in the elementary reaction.* Thus the elementary reaction $A \rightarrow P$ is an example of a **first-order** or **unimolecular** reaction, whereas the elementary reactions $2A \rightarrow P$ and $A + B \rightarrow P$ are examples of **second-order** or **bimolecular** reactions. Unimolecular and bimolecular reactions are common. **Termolecular** reactions are unusual and fourth- and higher order elementary reactions are unknown. This is because the simultaneous collision of three molecules is a rare event; that of four or more molecules essentially never occurs.

B. *Rates of Reactions*

We can experimentally determine the order of a reaction by measuring [A] or [P] as a function of time; that is,

$$v = -\frac{d[A]}{dt} = \frac{d[P]}{dt} \qquad [14.2]$$

where v is the instantaneous rate or **velocity** of the reaction. For the first-order reaction $A \rightarrow P$:

$$v = -\frac{d[A]}{dt} = k[A] \qquad [14.3a]$$

For second-order reactions such as $2A \rightarrow P$:

$$v = -\frac{d[A]}{dt} = k[A]^2 \qquad [14.3b]$$

whereas for $A + B \rightarrow P$, a second-order reaction that is first order in [A] and first order in [B],

$$v = -\frac{d[A]}{dt} = -\frac{d[B]}{dt} = k[A][B] \qquad [14.3c]$$

The rate constants of first- and second-order reactions must have different units. In terms of units, v in Eq. [14.3a] is expressed as $M \cdot s^{-1} = kM$. Therefore, k must have units of reciprocal seconds (s^{-1}) in order for Eq. [14.3a] to balance. Similarly, for second-order reactions, $M \cdot s^{-1} = kM^2$, so that k has the units $M^{-1} s^{-1}$.

The order of a specific reaction can be determined by measuring the reactant or product concentrations as a function of time and comparing the fit of these data to equations describing this behavior for reactions of various orders. To do this we must first derive these equations.

a. First-Order Rate Equation

The equation for [A] as a function of time for a first-order reaction, $A \rightarrow P$, is obtained by rearranging Eq. [14.3a]

$$\frac{d[A]}{[A]} \equiv d \ln[A] = -k \, dt$$

and integrating it from $[A]_o$, the initial concentration of A, to [A], the concentration of A at time t:

$$\int_{[A]_o}^{[A]} d \ln[A] = -k \int_0^t dt$$

This results in

$$\ln[A] = \ln[A]_o - kt \qquad [14.4a]$$

or, by taking the antilogs of both sides,

$$[A] = [A]_o e^{-kt} \qquad [14.4b]$$

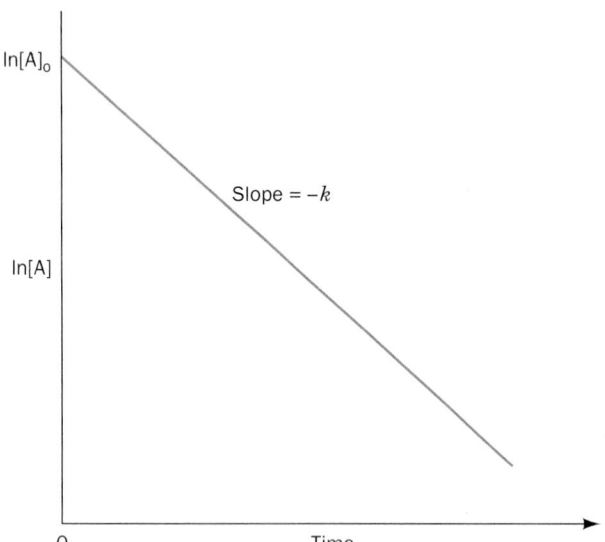

FIGURE 14-1   Plot of ln[A] versus time for a first-order reaction. This illustrates the graphical determination of the rate constant *k* using Eq. [14.4a].

Equation [14.4a] is a linear equation in terms of the variables ln[A] and *t* as is diagrammed in Fig. 14-1. Therefore, if a reaction is first order, a plot of ln[A] versus *t* will yield a straight line whose slope is −*k,* the negative of the first-order rate constant, and whose intercept on the ln[A] axis is ln[A]$_o$.

Substances that are inherently unstable, such as radioactive nuclei, decompose through first-order reactions (first-order processes are not just confined to chemical reactions). One of the hallmarks of a first-order reaction is that *the time for half of the reactant initially present to decompose, its* **half-time** *or* **half-life,** *$t_{1/2}$, is a constant and hence independent of the initial concentration of the reactant.* This is easily demonstrated by substituting the relationship [A] = [A]$_o$/2 when *t* = $t_{1/2}$ into Eq. [14.4a] and rearranging:

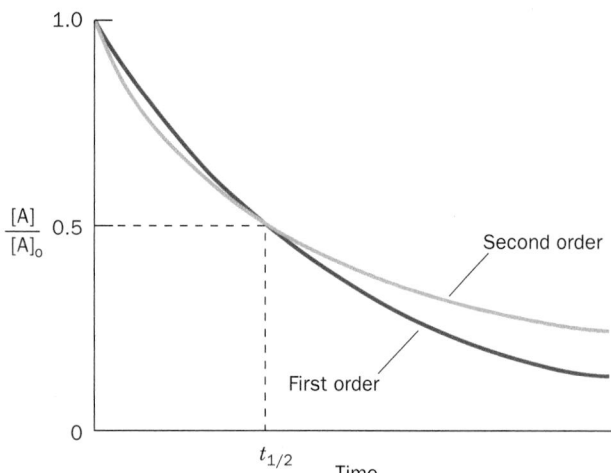

FIGURE 14-2   Comparison of the progress curves for first- and second-order reactions that have the same value of $t_{1/2}$. [After Tinoco, I., Jr., Sauer, K., and Wang, J.C., *Physical Chemistry. Principles and Applications in Biological Sciences* (2nd ed.), p. 291, Prentice-Hall (1985).]

$$\ln\left(\frac{[A]_o/2}{[A]_o}\right) = -kt_{1/2}$$

Thus

$$t_{1/2} = \frac{\ln 2}{k} = \frac{0.693}{k} \qquad [14.5]$$

In order to appreciate the course of a first-order reaction, let us consider the decomposition of ^{32}P, a radioactive isotope that is widely used in biochemical research. It has a half-life of 14 days. Thus, after 2 weeks, one-half of the ^{32}P initially present in a given sample will have decomposed; after another 2 weeks, one-half of the remainder, or three-quarters of the original sample, will have decomposed; etc. The long-term storage of waste ^{32}P therefore presents little problem, since after 1 year (26 half-lives), only 1 part in 2^{26} = 67 million of the original sample will remain. How much will remain after 2 years? In contrast, ^{14}C, another commonly employed radioactive tracer, has a half-life of 5715 years. Only a small fraction of a given quantity of ^{14}C will decompose over the course of a human lifetime.

b.   Second-Order Rate Equation for One Reactant

In a second-order reaction with one type of reactant, 2A → P, the variation of [A] with time is quite different from that in a first-order reaction. Rearranging Eq. [14.3b] and integrating it over the same limits used for the first-order reaction yields:

$$\int_{[A]_o}^{[A]} -\frac{d[A]}{[A]^2} = k\int_0^t dt$$

so that

$$\frac{1}{[A]} = \frac{1}{[A]_o} + kt \qquad [14.6]$$

Equation [14.6] is a linear equation in terms of the variables 1/[A] and *t.* Consequently, Eqs. [14.4a] and [14.6] may be used to distinguish a first-order from a second-order reaction by plotting ln[A] versus *t* and 1/[A] versus *t* and observing which, if any, of these plots is a straight line.

Figure 14-2 compares the different shapes of the progress curves describing the disappearance of A in first- and second-order reactions having the same half-times. Note that before the first half-time, the second-order progress curve descends more steeply than the first-order curve, but after this time the first-order progress curve is the more rapidly decreasing of the two. The half-time for a second-order reaction is expressed $t_{1/2}$ = 1/(k[A$_o$]) and therefore, in contrast to a first-order reaction, is dependent on the initial reactant concentration.

C.   *Transition State Theory*

The goal of kinetic theory is to describe reaction rates in terms of the physical properties of the reacting molecules. A theoretical framework for doing so, which explicitly considers the structures of the reacting molecules and how they collide, was developed in the 1930s, principally by Henry Eyring. This view of reaction processes, known as **transition state theory** or **absolute rate theory,** is the foundation of much of modern kinetics and has provided an ex-

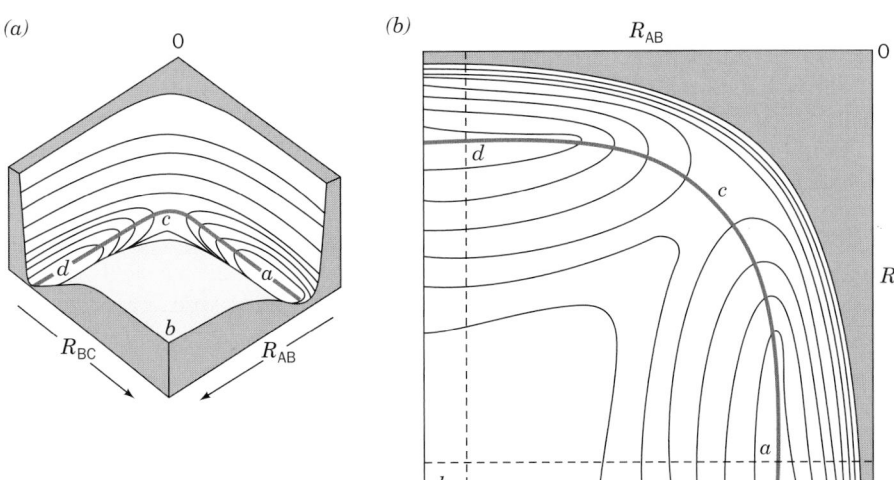

FIGURE 14-3 Potential energy of the colinear H + H₂ system as a function of its internuclear distances, R_{AB} and R_{BC}. The reaction is represented as (a) a perspective drawing and (b) the corresponding contour diagram. The points a and d are approaching potential energy minima, b is approaching a maximum, and c is a saddle point. [After Frost, A.A. and Pearson, R.G., *Kinetics and Mechanism* (2nd ed.), p. 80, Wiley (1961).]

traordinarily productive framework for understanding how enzymes catalyze reactions.

a. The Transition State

Consider a bimolecular elementary reaction involving three atoms A, B, and C:

$$A—B + C \rightarrow A + B—C$$

Clearly atom C must approach the diatomic molecule A—B so that, at some point in the reaction, a high-energy (unstable) complex represented as A ··· B ··· C exists in which the A—B covalent bond is in the process of breaking while the B—C bond is in the process of forming.

Let us consider the simplest example of this reaction: that of a hydrogen atom with diatomic hydrogen (H_2) to yield a new H_2 molecule and a different hydrogen atom:

$$H_A—H_B + H_C \rightarrow H_A + H_B—H_C$$

The potential energy of this triatomic system as a function of the relative positions of its component atoms is plotted in Fig. 14-3. Its shape is of two long and deep valleys parallel to the coordinate axes with sheer walls rising toward the axes and less steep ones rising toward a plateau where both coordinates are large (the region of point b). The two

valleys are joined by a pass or saddle near the origin of the diagram (point c). The minimum energy configuration is that of an H_2 molecule and an isolated atom, that is, with one coordinate large and the other at the H_2 covalent bond distance [near points a (the reactants) and d (the products)]. During a collision, the reactants generally approach one another with little deviation from the minimum energy reaction pathway (line a—c—d) because other trajectories would require much greater energy. As the atom and molecule come together, they increasingly repel one another (have increasing potential energy) and therefore usually fly apart. *If, however, the system has sufficient kinetic energy to continue its coalescence, it will cause the covalent bond of the H_2 molecule to weaken until ultimately, if the system reaches the saddle point (point c), there is an equal probability that either the reaction will occur or that the system will decompose back to its reactants.* Therefore, at this saddle point, the system is said to be at its **transition state** and hence to be an **activated complex.** Moreover, since the concentration of the activated complex is small, *the decomposition of the activated complex is postulated to be the rate-determining process of this reaction.*

*The minimum free energy pathway of a reaction is known as its **reaction coordinate.*** Figure 14-4a, which is

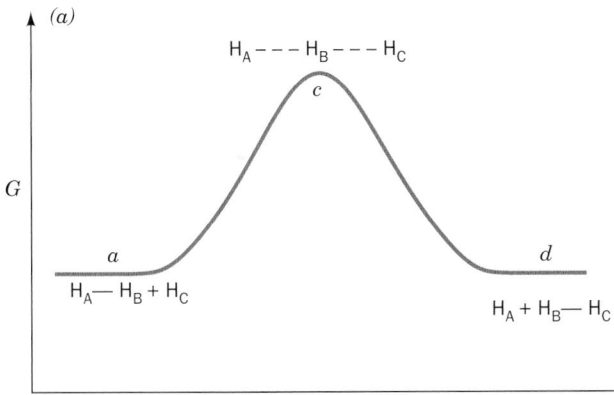

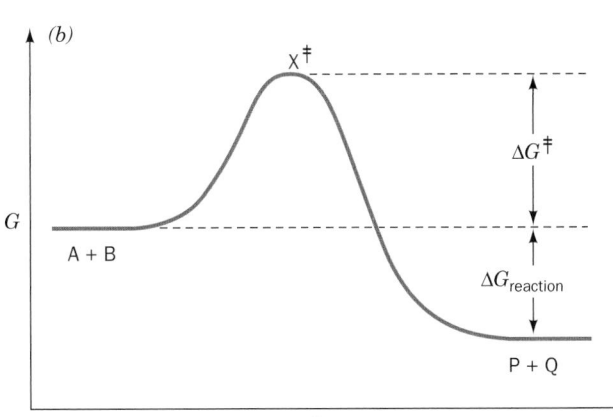

FIGURE 14-4 Transition state diagrams. (a) For the H + H₂ reaction. This is a section taken along the a—c—d line in Fig.

14-3. (b) For a spontaneous reaction, that is, one in which the free energy decreases.

called a **transition state diagram** or a **reaction coordinate diagram,** shows the free energy of the $H + H_2$ system along the reaction coordinate (line $a—c—d$ in Fig. 14-3). It can be seen that the transition state is the point of highest free energy on the reaction coordinate. If the atoms in the triatomic system are of different types, as is diagrammed in Fig. 14-4b, the transition state diagram is no longer symmetrical because there is a free energy difference between reactants and products.

b. Thermodynamics of the Transition State

The realization that the attainment of the transition state is the central requirement in any reaction process led to a detailed understanding of reaction mechanisms. For example, consider a bimolecular reaction that proceeds along the following pathway:

$$A + B \underset{}{\overset{K^{\ddagger}}{\rightleftharpoons}} X^{\ddagger} \xrightarrow{k'} P + Q$$

where $X^{\ddagger}$ represents the activated complex. Therefore, considering the preceding discussion,

$$\frac{d[P]}{dt} = k[A][B] = k'[X^{\ddagger}] \qquad [14.7]$$

where k is the ordinary rate constant of the elementary reaction and k' is the rate constant for the decomposition of $X^{\ddagger}$ to products.

In contrast to stable molecules, such as A and P, which occur at energy minima, the activated complex occurs at an energy maximum and is therefore only metastable (like a ball balanced on a pin). Transition state theory nevertheless assumes that $X^{\ddagger}$ is in rapid equilibrium with the reactants; that is,

$$K^{\ddagger} = \frac{X^{\ddagger}}{[A][B]} \qquad [14.8]$$

where $K^{\ddagger}$ is an equilibrium constant. *This central assumption of transition state theory permits the powerful formalism of thermodynamics to be applied to the theory of reaction rates.*

If $K^{\ddagger}$ is an equilibrium constant it can be expressed as:

$$-RT \ln K^{\ddagger} = \Delta G^{\ddagger} \qquad [14.9]$$

where $\Delta G^{\ddagger}$ is the Gibbs free energy of the activated complex less that of the reactants (Fig. 14-4b), T is the absolute temperature, and R ($= 8.3145 \text{ J} \cdot \text{K}^{-1} \cdot \text{mol}^{-1}$) is the gas constant (this relationship between equilibrium constants and free energy is derived in Section 3-4A). Then combining Eqs. [14.7] through [14.9] yields

$$\frac{d[P]}{dt} = k'e^{-\Delta G^{\ddagger}/RT}[A][B] \qquad [14.10]$$

This equation indicates that the rate of a reaction depends not only on the concentrations of its reactants but also decreases exponentially with $\Delta G^{\ddagger}$. *Thus, the larger the difference between the free energy of the transition state and that of the reactants, that is, the less stable the transition state, the slower the reaction proceeds.*

In order to continue, we must now evaluate k', the rate constant for passage of the activated complex over the maximum in the transition state diagram (sometimes referred to as the **activation barrier** or the **kinetic barrier** of the reaction). This transition state model permits us to do so (although the following derivation is by no means rigorous). The activated complex is held together by a bond that is associated with the reaction coordinate and that is assumed to be so weak that it flies apart during its first vibrational excursion. Therefore, k' is expressed

$$k' = \kappa\nu \qquad [14.11]$$

where ν is the vibrational frequency of the bond that breaks as the activated complex decomposes to products and κ, the **transmission coefficient,** is the probability that the breakdown of the activated complex, $X^{\ddagger}$, will be in the direction of product formation rather than back to reactants. For most spontaneous reactions in solution, κ is between 0.5 and 1.0; for the colinear $H + H_2$ reaction, we saw that it is 0.5.

We have nearly finished our job of evaluating k'. All that remains is to determine the value of ν. Planck's law states that

$$\nu = \varepsilon/h \qquad [14.12]$$

where, in this case, ε is the average energy of the vibration that leads to the decomposition of $X^{\ddagger}$, and h ($= 6.6261 \times 10^{-34} \text{ J} \cdot \text{s}$) is Planck's constant. Statistical mechanics tells us that at temperature T, the classical energy of an oscillator is

$$\varepsilon = k_B T \qquad [14.13]$$

where k_B ($= 1.3807 \times 10^{-23} \text{ J} \cdot \text{K}^{-1}$) is a constant of nature known as the **Boltzmann constant** and $k_B T$ is essentially the available thermal energy. Combining Eqs. [14.11] through [14.13]

$$k' = \frac{\kappa k_B T}{h} \qquad [14.14]$$

Then assuming, as is done for most reactions, that $\kappa = 1$ (κ can rarely be calculated with any confidence), the combination of Eqs. [14.7] and [14.10] with [14.14] yields the expression for the rate constant k of our elementary reaction:

$$k = \frac{k_B T}{h} e^{-\Delta G^{\ddagger}/RT} \qquad [14.15]$$

This equation indicates that *the rate of reaction decreases as its free energy of activation, $\Delta G^{\ddagger}$, increases.* Conversely, as the temperature rises, so that there is increased thermal energy available to drive the reacting complex over the activation barrier, the reaction speeds up. (Of course, enzymes, being proteins, are subject to thermal denaturation, so that the rate of an enzymatically catalyzed reaction falls precipitously with increasing temperature once the enzyme's denaturation temperature has been surpassed.) Keep in mind, however, that transition state theory is an ideal model; real systems behave in a more complicated, although qualitatively similar, manner.

c. Multistep Reactions Have Rate-Determining Steps

Since chemical reactions commonly consist of several elementary reaction steps, let us consider how transition state theory treats such reactions. For a multistep reaction such as

$$A \xrightarrow{k_1} I \xrightarrow{k_2} P$$

where I is an intermediate of the reaction, there is an activated complex for each elementary reaction step; the shape of the transition state diagram for such a reaction reflects the relative rates of the elementary reactions involved. For this reaction, if the first reaction step is slower than the second reaction step ($k_1 < k_2$), then the activation barrier of the first step must be higher than that of the second step, and conversely if the second reaction step is the slower (Fig. 14-5). Since the rate of formation of product P can only be as fast as the slowest elementary reaction, *if one reaction step of an overall reaction is much slower than the other, the slow step acts as a "bottleneck" and is therefore said to be the **rate-determining step** of the reaction.*

d. Catalysis Reduces $\Delta G^{\ddagger}$

Biochemistry is, of course, mainly concerned with enzyme-catalyzed reactions. *Catalysts act by lowering the activation barrier for the reaction being catalyzed (Fig.14-6).* If a catalyst lowers the activation barrier of a reaction by $\Delta\Delta G^{\ddagger}_{cat}$, then, according to Eq. [14.15], the rate of the reaction is enhanced by the factor $e^{\Delta\Delta G^{\ddagger}_{cat}/RT}$. Thus, a 10-fold rate enhancement requires that $\Delta\Delta G^{\ddagger}_{cat} = 5.71$ kJ · mol^{-1}, less than half the energy of a typical hydrogen bond; a million-fold rate acceleration occurs when $\Delta\Delta G^{\ddagger}_{cat} = 34.25$ kJ · mol^{-1}, a small fraction of the energy of most covalent bonds. The rate enhancement is therefore a sensitive function of $\Delta\Delta G^{\ddagger}_{cat}$.

Note that the kinetic barrier is lowered by the same amount for both the forward and the reverse reactions (Fig. 14-6). Consequently, a catalyst equally accelerates the forward and the reverse reactions so that the equilibrium constant for the reaction remains unchanged. The chemical mechanisms through which enzymes lower the activation barriers of reactions are the subject of Section 15-1. There we shall see that the most potent such mechanism often involves the enzymatic binding of the transition state of the catalyzed reaction in preference to the substrate.

2 ■ ENZYME KINETICS

*📖 See **Guided Exploration 12: Michaelis-Menten Kinetics, Lineweaver-Burk Plots and Enzyme Inhibition*** The chemical reactions of life are mediated by enzymes. These remarkable catalysts, as we saw in Chapter 13, are individually highly specific for particular reactions. Yet collectively they are extremely versatile in that the many thousand enzymes now known carry out such diverse reactions as hydrolysis, polymerization, functional group transfer, oxidation–reduction,

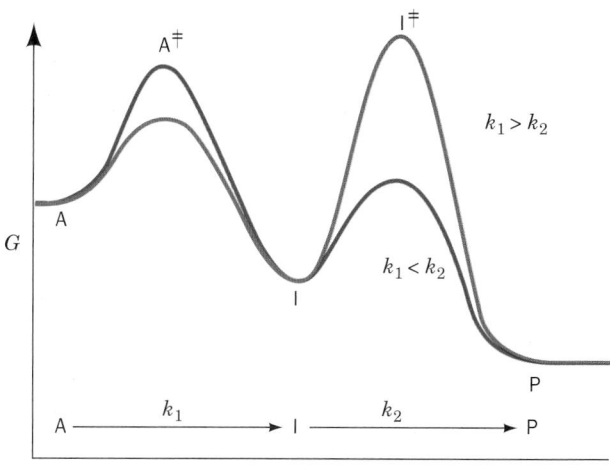

FIGURE 14-5 Transition state diagram for the two-step overall reaction $A \rightarrow I \rightarrow P$. For $k_1 < k_2$ (*green curve*), the first step is rate determining, whereas if $k_1 > k_2$ (*red curve*), the second step is rate determining.

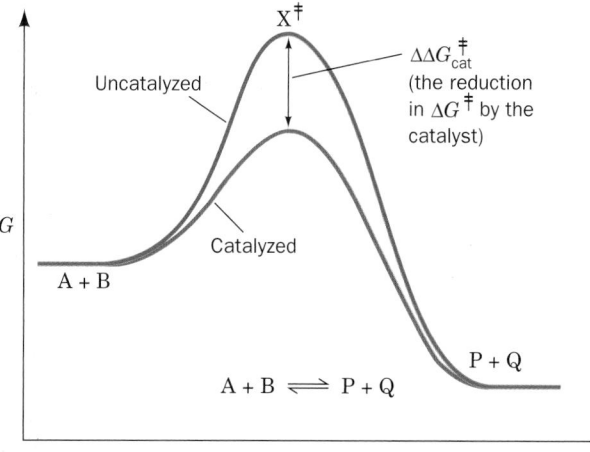

FIGURE 14-6 The effect of a catalyst on the transition state diagram of a reaction. Here $\Delta\Delta G^{\ddagger} = \Delta G^{\ddagger}_{uncat} - \Delta G^{\ddagger}_{cat}$.

dehydration, and isomerization, to mention only the most common classes of enzymatically mediated reactions. Enzymes are not passive surfaces on which reactions take place but, rather, are complex molecular machines that operate through a great diversity of mechanisms. For instance, some enzymes act on only a single substrate molecule; others act on two or more different substrate molecules whose order of binding may or may not be obligatory. Some enzymes form covalently bound intermediate complexes with their substrates; others do not.

Kinetic measurements of enzymatically catalyzed reactions are among the most powerful techniques for elucidating the catalytic mechanisms of enzymes. The remainder of this chapter is therefore largely concerned with the development of the kinetic tools that are most useful in the determination of enzymatic mechanisms. We begin, in this section, with a presentation of the basic theory of enzyme kinetics.

A. *The Michaelis–Menten Equation*

The study of enzyme kinetics began in 1902 when Adrian Brown reported an investigation of the rate of hydrolysis of sucrose as catalyzed by the yeast enzyme **invertase** (now known as **β-fructofuranosidase**):

$$\text{Sucrose} + H_2O \rightarrow \text{glucose} + \text{fructose}$$

Brown demonstrated that when the sucrose concentration is much higher than that of the enzyme, the reaction rate becomes independent of the sucrose concentration; that is, the rate is **zero order** with respect to sucrose. He therefore proposed that the overall reaction is composed of two elementary reactions in which the substrate forms a complex with the enzyme that subsequently decomposes to products and enzyme:

$$E + S \underset{k_{-1}}{\overset{k_1}{\rightleftharpoons}} ES \overset{k_2}{\longrightarrow} P + E$$

Here E, S, ES, and P symbolize the enzyme, substrate, **enzyme–substrate complex,** and products, respectively (for enzymes composed of multiple identical subunits, E refers to active sites rather than enzyme molecules). According to this model, *when the substrate concentration becomes high enough to entirely convert the enzyme to the ES form, the second step of the reaction becomes rate limiting and the overall reaction rate becomes insensitive to further increases in substrate concentration.*

The general expression for the **velocity** (rate) of this reaction is

$$v = \frac{d[P]}{dt} = k_2[ES] \qquad [14.16]$$

The overall rate of production of ES is the difference between the rates of the elementary reactions leading to its appearance and those resulting in its disappearance:

$$\frac{d[ES]}{dt} = k_1[E][S] - k_{-1}[ES] - k_2[ES] \qquad [14.17]$$

This equation cannot be explicitly integrated, however, without simplifying assumptions. Two possibilities are:

1. Assumption of equilibrium: In 1913, Leonor Michaelis and Maude Menten, building on earlier work by Victor Henri, assumed that $k_{-1} \gg k_2$, so that the first step of the reaction achieves equilibrium.

$$K_S = \frac{k_{-1}}{k_1} = \frac{[E][S]}{[ES]} \qquad [14.18]$$

Here K_S is the dissociation constant of the first step in the enzymatic reaction. With this assumption, Eq. [14.17] can be integrated. Although this assumption is not often correct, in recognition of the importance of this pioneering work, the noncovalently bound enzyme–substrate complex ES is known as the **Michaelis complex.**

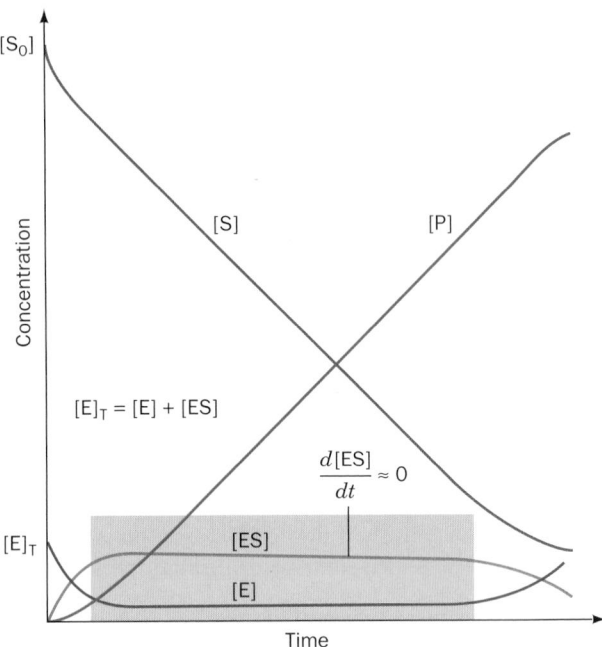

FIGURE 14-7 Progress curves for the components of a simple Michaelis–Menten reaction. Note that with the exception of the transient phase of the reaction, which occurs before the shaded block, the slopes of the progress curves for [E] and [ES] are essentially zero so long as [S] >> $[E]_T$ (within the shaded block). [After Segel, I.H., *Enzyme Kinetics*, p. 27, Wiley (1975).] ♫ See the Animated Figures

2. Assumption of steady state: Figure 14-7 illustrates the progress curves of the various participants in the preceding reaction model under the physiologically common condition that substrate is in great excess over enzyme. With the exception of the initial stage of the reaction, the so-called **transient phase,** which is usually over within milliseconds of mixing the enzyme and substrate, [ES] remains approximately constant until the substrate is nearly exhausted. Hence, the rate of synthesis of ES must equal its rate of consumption over most of the course of the reaction; that is, [ES] maintains a **steady state.** One can therefore assume with a reasonable degree of accuracy that [ES] is constant; that is,

$$\frac{d[ES]}{dt} = 0 \qquad [14.19]$$

This so-called **steady-state assumption** was first proposed in 1925 by G. E. Briggs and John B. S. Haldane.

In order to be of use, kinetic expressions for overall reactions must be formulated in terms of experimentally measurable quantities. The quantities [ES] and [E] are not, in general, directly measurable but the total enzyme concentration

$$[E]_T = [E] + [ES] \qquad [14.20]$$

is usually readily determined. The rate equation for our enzymatic reaction is then derived as follows. Combining

Eq. [14.17] with the steady-state assumption, Eq. [14.19], and the conservation condition, Eq. [14.20], yields:

$$k_1([E]_T - [ES])[S] = (k_{-1} + k_2)[ES]$$

which on rearrangement becomes

$$[ES](k_{-1} + k_2 + k_1[S]) = k_1[E]_T[S]$$

Dividing both sides by k_1 and solving for [ES],

$$[ES] = \frac{[E]_T[S]}{K_M + [S]}$$

where K_M, which is known as the **Michaelis constant,** is defined

$$K_M = \frac{k_{-1} + k_2}{k_1} \qquad [14.21]$$

The meaning of this important constant is discussed below.

The **initial velocity** of the reaction from Eq. [14.16] can then be expressed in terms of the experimentally measurable quantities $[E]_T$ and $[S]$:

$$v_o = \left(\frac{d[P]}{dt}\right)_{t=t_s} = k_2[ES] = \frac{k_2[E]_T[S]}{K_M + [S]} \quad [14.22]$$

where t_s is the time when the steady state is first achieved (usually milliseconds after $t = 0$). The use of the initial velocity (operationally taken as the velocity measured before more than ~10% of the substrate has been converted to product) rather than just the velocity minimizes such complicating factors as the effects of reversible reactions, inhibition of the enzyme by product, and progressive inactivation of the enzyme.

The **maximal velocity** of a reaction, V_{max}, occurs at high substrate concentrations when the enzyme is **saturated,** that is, when it is entirely in the ES form:

$$V_{max} = k_2[E]_T \qquad [14.23]$$

Therefore, combining Eqs. [14.22] and [14.23], we obtain

$$v_o = \frac{V_{max}[S]}{K_M + [S]} \qquad [14.24]$$

*This expression, the **Michaelis–Menten equation,** is the basic equation of enzyme kinetics.* It describes a rectangular hyperbola such as is plotted in Fig. 14-8 (although this curve is rotated by 45° and translated to the origin with respect to the examples of hyperbolas seen in most elementary algebra texts). The saturation function for oxygen binding to myoglobin, Eq. [10.4], has the same functional form.

a. Significance of the Michaelis Constant

The Michaelis constant, K_M, has a simple operational definition. At the substrate concentration where $[S]=K_M$, Eq. [14.24] yields $v_o = V_{max}/2$ so that K_M *is the substrate concentration at which the reaction velocity is half-maximal.* Therefore, if an enzyme has a small value of K_M, it achieves

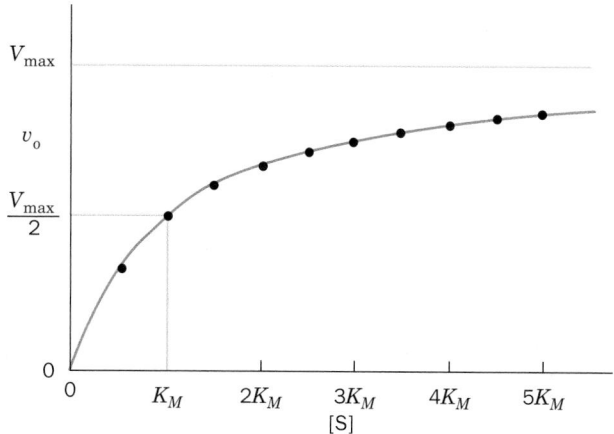

FIGURE 14-8 Plot of the initial velocity v_o of a simple Michaelis–Menten reaction versus the substrate concentration [S]. Points are plotted in 0.5-K_M intervals of substrate concentration between $0.5K_M$ and $5K_M$. *See the Animated Figures*

maximal catalytic efficiency at low substrate concentrations.

The magnitude of K_M varies widely with the identity of the enzyme and the nature of the substrate (Table 14-1). It is also a function of temperature and pH (see Section 14-4). The Michaelis constant (Eq. [14.21]) can be expressed as

$$K_M = \frac{k_{-1}}{k_1} + \frac{k_2}{k_1} = K_S + \frac{k_2}{k_1} \qquad [14.25]$$

Since K_S is the dissociation constant of the Michaelis complex, as K_S decreases, the enzyme's affinity for substrate increases. K_M is therefore also a measure of the affinity of the enzyme for its substrate providing k_2/k_1 is small compared with K_S, that is, $k_2 < k_{-1}$.

B. *Analysis of Kinetic Data*

There are several methods for determining the values of the parameters of the Michaelis–Menten equation. At very high values of [S], the initial velocity v_o asymptotically approaches V_{max}. In practice, however, it is very difficult to assess V_{max} accurately from direct plots of v_o versus [S] such as Fig. 14-8. Even at such high substrate concentrations as $[S] = 10K_M$, Eq. [14.24] indicates that v_o is only 91% of V_{max}, so that the extrapolated value of the asymptote will almost certainly be underestimated.

A better method for determining the values of V_{max} and K_M, which was formulated by Hans Lineweaver and Dean Burk, uses the reciprocal of Eq. [14.24]:

$$\frac{1}{v_o} = \left(\frac{K_M}{V_{max}}\right)\frac{1}{[S]} + \frac{1}{V_{max}} \qquad [14.26]$$

This is a linear equation in $1/v_o$ and $1/[S]$. If these quantities are plotted, in the so-called **Lineweaver–Burk** or **double-reciprocal plot,** the slope of the line is K_M/V_{max}, the $1/v_o$ intercept is $1/V_{max}$, and the extrapolated $1/[S]$ intercept is $-1/K_M$ (Fig. 14-9). A disadvantage of this plot is that most experimental measurements involve relatively high [S] and are therefore crowded onto the left side of the graph. Furthermore, for small values of [S], small errors in v_o lead to large errors in $1/v_o$ and hence to large errors in K_M and V_{max}.

Several other types of plots, each with its advantages and disadvantages, have been formulated for the determination of V_{max} and K_M from kinetic data. With the advent of conveniently available computers, however, kinetic data are commonly analyzed by mathematically sophisticated statistical treatments. Nevertheless, Lineweaver–Burk plots are valuable for the visual presentation of kinetic data as well as being useful in the analysis of kinetic data from enzymes requiring more than one substrate (Section 14-5C).

a. k_{cat}/K_M Is a Measure of Catalytic Efficiency

An enzyme's kinetic parameters provide a measure of its catalytic efficiency. We may define the **catalytic constant** of an enzyme as

$$k_{cat} = \frac{V_{max}}{[E]_T} \qquad [14.27]$$

This quantity is also known as the **turnover number** of an enzyme because it is the number of reaction processes (turnovers) that each active site catalyzes per unit time. The turnover numbers for a selection of enzymes are given in Table 14-1. Note that these quantities vary by over eight orders of magnitude depending on the identity of the enzyme as well as that of its substrate. Equation [14.23] indicates that for the Michaelis–Menten model, $k_{cat} = k_2$. For enzymes with more complicated mechanisms, k_{cat} may be a function of several rate constants.

When [S] $<< K_M$, very little ES is formed. Consequently, $[E] \approx [E]_T$, so that Eq. [14.22] reduces to a second-order rate equation:

$$v_o \approx \left(\frac{k_2}{K_M}\right)[E]_T[S] \approx \left(\frac{k_{cat}}{K_M}\right)[E][S] \qquad [14.28]$$

k_{cat}/K_M is the apparent second-order rate constant of the enzymatic reaction; the rate of the reaction varies directly

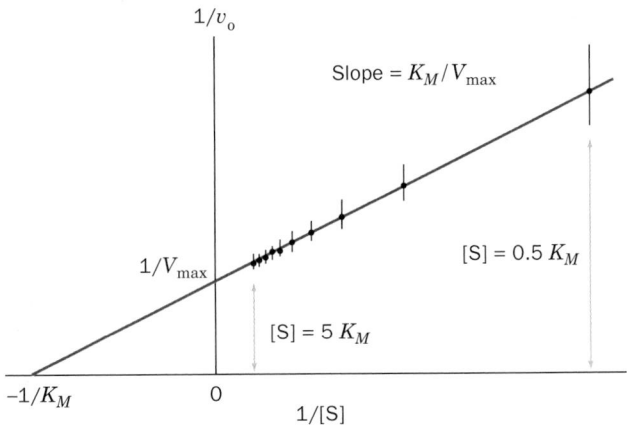

FIGURE 14-9   A double reciprocal (Lineweaver–Burk) plot. Error bars are $\pm 0.05 V_{max}$. The indicated points are the same as those in Fig. 14-8. Note the large effect of small errors at small [S] (large 1/[S]) and the crowding together of points at large [S]. *See the Animated Figures*

with how often enzyme and substrate encounter one another in solution. *The quantity k_{cat}/K_M is therefore a measure of an enzyme's catalytic efficiency.*

b. Some Enzymes Have Attained Catalytic Perfection

Is there an upper limit on enzymatic catalytic efficiency? From Eq. [14.21] we find

$$\frac{k_{cat}}{K_M} = \frac{k_2}{K_M} = \frac{k_1 k_2}{k_{-1} + k_2} \qquad [14.29]$$

This ratio is maximal when $k_2 >> k_{-1}$, that is, when the formation of product from the Michaelis complex, ES, is fast compared to its decomposition back to substrate and enzyme. Then $k_{cat}/K_M = k_1$, the second-order rate constant for the formation of ES. The term k_1, of course, can be no greater than the frequency with which enzyme and substrate molecules collide with each other in solution. This **diffusion-controlled limit** is in the range of 10^8 to 10^9 $M^{-1} \cdot s^{-1}$. Thus, enzymes with such values of k_{cat}/K_M must catalyze a reaction almost every time they encounter a substrate molecule. Table 14-1 indicates that several enzymes,

TABLE 14-1   Values of K_M, k_{cat}, and k_{cat}/K_M for Some Enzymes and Substrates

Enzyme	Substrate	$K_M (M)$	$k_{cat} (s^{-1})$	$k_{cat}/K_M (M^{-1} \cdot s^{-1})$
Acetylcholinesterase	Acetylcholine	9.5×10^{-5}	1.4×10^4	1.5×10^8
Carbonic anhydrase	CO_2	1.2×10^{-2}	1.0×10^6	8.3×10^7
	HCO_3^-	2.6×10^{-2}	4.0×10^5	1.5×10^7
Catalase	H_2O_2	2.5×10^{-2}	1.0×10^7	4.0×10^8
Chymotrypsin	N-Acetylglycine ethyl ester	4.4×10^{-1}	5.1×10^{-2}	1.2×10^{-1}
	N-Acetylvaline ethyl ester	8.8×10^{-2}	1.7×10^{-1}	1.9
	N-Acetyltyrosine ethyl ester	6.6×10^{-4}	1.9×10^2	2.9×10^5
Fumarase	Fumarate	5.0×10^{-6}	8.0×10^2	1.6×10^8
	Malate	2.5×10^{-5}	9.0×10^2	3.6×10^7
Superoxide dismutase	Superoxide ion ($O_2^{\cdot -}$)	3.6×10^{-4}	1.0×10^6	2.8×10^9
Urease	Urea	2.5×10^{-2}	1.0×10^4	4.0×10^5

FIGURE 14-10 Cross section through the active site of human superoxide dismutase (SOD). The enzyme binds both a Cu^{2+} and a Zn^{2+} ion (*orange and cyan spheres*). SOD's molecular surface is represented by a dot surface that is colored according to its electrostatic charge, with red most negative, yellow negative, green neutral, cyan positive, and blue most positive. The electrostatic field vectors are represented by similarly colored arrows. Note how this electrostatic field would draw the negatively charged superoxide ion into its binding site, which is located between the Cu^{2+} ion and Arg 143. [Courtesy of Elizabeth Getzoff, The Scripps Research Institute, La Jolla, California.]

namely, catalase, superoxide dismutase, fumarase, acetylcholinesterase, and possibly carbonic anhydrase, have achieved this state of virtual catalytic perfection.

Since the active site of an enzyme generally occupies only a small fraction of its total surface area, how can any enzyme catalyze a reaction every time it encounters a substrate molecule? In the case of **superoxide dismutase (SOD),** it appears that the arrangement of charged groups on the enzyme's surface serves to electrostatically guide the charged substrate to the enzyme's active site (Fig. 14-10). [SOD, which is present in nearly all cells, functions to inactivate the highly reactive and therefore destructive **superoxide radical** $(O_2^{-\cdot})$ by catalyzing the reaction $2O_2^{-\cdot} + 2H^+ \rightarrow H_2O_2 + O_2$.] Other enzymes, including **acetylcholinesterase** (Section 20-5C), have similar mechanisms to funnel polar substrates to their active sites.

C. *Reversible Reactions*

The Michaelis–Menten model implicitly assumes that enzymatic reverse reactions may be neglected. Yet many enzymatic reactions are highly reversible (have a small free energy of reaction) and therefore have products that back react to form substrates at a significant rate. In this section we therefore relax the Michaelis–Menten restriction of no

back reaction and, by doing so, discover some interesting and important kinetic principles.

a. The One-Intermediate Model

Modification of the Michaelis–Menten model to incorporate a back reaction yields the following reaction scheme:

$$E + S \underset{k_{-1}}{\overset{k_1}{\rightleftharpoons}} ES \underset{k_{-2}}{\overset{k_2}{\rightleftharpoons}} P + E$$

(Here ES might just as well be called EP because this model does not specify the nature of the intermediate complex.) The equation describing the kinetic behavior of this model, which is derived in Appendix A of this chapter, is expressed

$$v = \frac{\dfrac{V^f_{max}[S]}{K^S_M} - \dfrac{V^r_{max}[P]}{K^P_M}}{1 + \dfrac{[S]}{K^S_M} + \dfrac{[P]}{K^P_M}} \qquad [14.30]$$

where

$$V^f_{max} = k_2[E]_T \qquad V^r_{max} = k_{-1}[E]_T$$

$$K^S_M = \frac{k_{-1} + k_2}{k_1} \qquad K^P_M = \frac{k_{-1} + k_2}{k_{-2}}$$

and

$$[E]_T = [E] + [ES]$$

This is essentially a Michaelis–Menten equation that works backwards as well as forwards. Indeed, at $[P] = 0$, that is, when $v = v_o$, this equation becomes the Michaelis–Menten equation.

b. The Haldane Relationship

At equilibrium (which occurs after the reaction has run its course), $v = 0$, so Eq. [14.30], which holds at equilibrium as well as at steady state, can be solved to yield

$$K_{eq} = \frac{[P]_{eq}}{[S]_{eq}} = \frac{V^f_{max} K^P_M}{V^r_{max} K^S_M} \qquad [14.31]$$

where $[P]_{eq}$ and $[S]_{eq}$ are the concentrations of P and S at equilibrium. This so-called **Haldane relationship** demonstrates that *the kinetic parameters of a reversible enzymatically catalyzed reaction are not independent of one another. Rather, they are related by the equilibrium constant for the overall reaction, which, of course, is independent of the presence of the enzyme.*

c. Kinetic Data Cannot Unambiguously Establish a Reaction Mechanism

An enzyme that forms a reversible complex with its substrate should likewise form one with its product; that is, it should have a mechanism such as

$$E + S \underset{k_{-1}}{\overset{k_1}{\rightleftharpoons}} ES \underset{k_{-2}}{\overset{k_2}{\rightleftharpoons}} EP \underset{k_{-3}}{\overset{k_3}{\rightleftharpoons}} P + E$$

The equation describing the kinetic behavior of this two-intermediate model, whose derivation is analogous to that described in Appendix A for the one-intermediate model, has a form identical to that of Eq. [14.30]. However, its parameters V_{max}^f, V_{max}^r, K_M^S, and K_M^P are defined in terms of the six kinetic constants of the two-intermediate model rather than the four of the one-intermediate model. In fact, the steady-state rate equations for reversible reactions with three or more intermediates also have this same form but with yet different definitions of the four parameters.

The values of V_{max}^f, V_{max}^r, K_M^S, and K_M^P in Eq. [14.30] can be determined by suitable manipulations of the initial substrate and product concentrations under steady-state conditions. This, however, will not yield the values of the rate constants for our two-intermediate model because there are six such constants and only four equations describing their relationships. Moreover, steady-state kinetic measurements are incapable of distinguishing the number of intermediates in a reversible enzymatic reaction because the form of Eq. [14.30] does not change with the number of intermediates.

The functional identities of the equations describing these reaction schemes may be understood in terms of an analogy between our *n*-intermediate reversible reaction model and a "black box" containing a system of water pipes with one inlet and one drain:

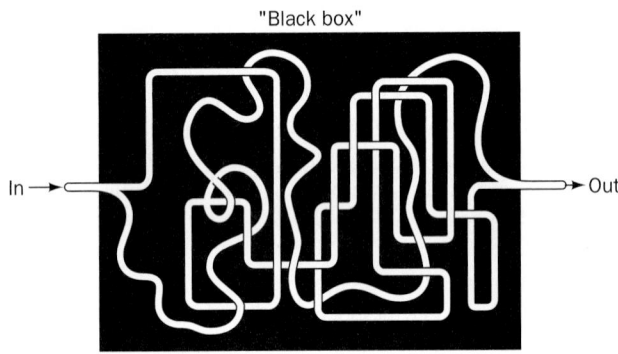

"Black box"

At steady state, that is, after the pipes have filled with water, one can measure the relationship between input pressure and output flow. However, such measurements yield no information concerning the detailed construction of the plumbing connecting the inlet to the drain. This would require additional observations such as opening the black box and tracing the pipes. *Likewise, steady-state kinetic measurements can provide a phenomenological description of enzymatic behavior, but the nature of the intermediates remains indeterminate. Rather, these intermediates must be detected and characterized by independent means such as by spectroscopic analysis.*

The foregoing discussion brings to light a central principle of kinetic analysis: *The steady-state kinetic analysis of a reaction cannot unambiguously establish its mechanism.* This is because no matter how simple, elegant, or rational a mechanism one postulates that fully accounts for kinetic data, there are an infinite number of alternative mechanisms, perhaps complicated, awkward, and seemingly irrational, that can account for these kinetic data equally well. Usually it is the simpler and more elegant mechanism that

turns out to be correct, but this is not always the case. *If, however, kinetic data are not compatible with a given mechanism, then the mechanism must be rejected.* Therefore, although kinetics cannot be used to establish a mechanism unambiguously without confirming data, such as the physical demonstration of an intermediate's existence, the steady-state kinetic analysis of a reaction is of great value because it can be used to eliminate proposed mechanisms.

3 ■ INHIBITION

Many substances alter the activity of an enzyme by combining with it in a way that influences the binding of substrate and/or its turnover number. Substances that reduce an enzyme's activity in this way are known as **inhibitors.**

Many inhibitors are substances that structurally resemble their enzyme's substrate but either do not react or react very slowly compared to substrate. Such inhibitors are commonly used to probe the chemical and conformational nature of a substrate-binding site as part of an effort to elucidate the enzyme's catalytic mechanism. In addition, many enzyme inhibitors are effective chemotherapeutic agents, since an "unnatural" substrate analog can block the action of a specific enzyme. For example, **methotrexate** (also called **amethopterin**) chemically resembles **dihydrofolate.** Methotrexate binds tightly to the enzyme **dihydrofolate reductase,** thereby preventing it from carrying out its normal function, the reduction of dihydrofolate to **tetrahydrofolate,** an essential cofactor in the biosynthesis of the DNA precursor dTMP (Section 28-3B):

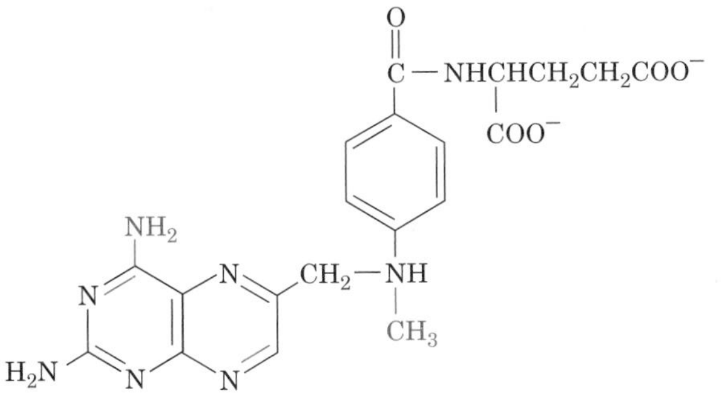

Methotrexate

$$\downarrow \text{dihydrofolate reductase}$$

NO REACTION

Rapidly dividing cells, such as cancer cells, which are actively engaged in DNA synthesis, are far more susceptible to methotrexate than are slower growing cells such as those of most normal mammalian tissues. Hence, methotrexate, when administered in proper dosage, kills cancer cells without fatally poisoning the host.

There are various mechanisms through which enzyme inhibitors can act. In this section, we discuss several of the simplest such mechanisms and their effects on the kinetic behavior of enzymes that follow the Michaelis–Menten model.

A. *Competitive Inhibition*

A substance that competes directly with a normal substrate for an enzymatic binding site is known as a **competitive inhibitor.** Such an inhibitor usually resembles the substrate to the extent that it specifically binds to the active site but differs from it so as to be unreactive. Thus methotrexate is a competitive inhibitor of dihydrofolate reductase. Similarly, **succinate dehydrogenase,** a citric acid cycle enzyme that functions to convert **succinate** to **fumarate** (Section 21-3F), is competitively inhibited by **malonate,** which structurally resembles succinate but cannot be dehydrogenated:

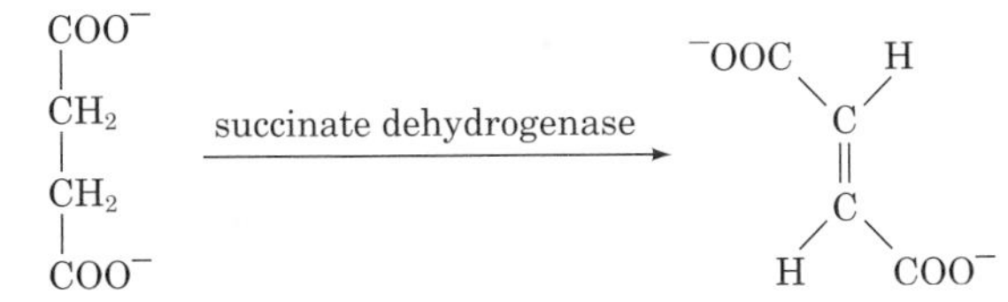

Succinate   **Fumarate**

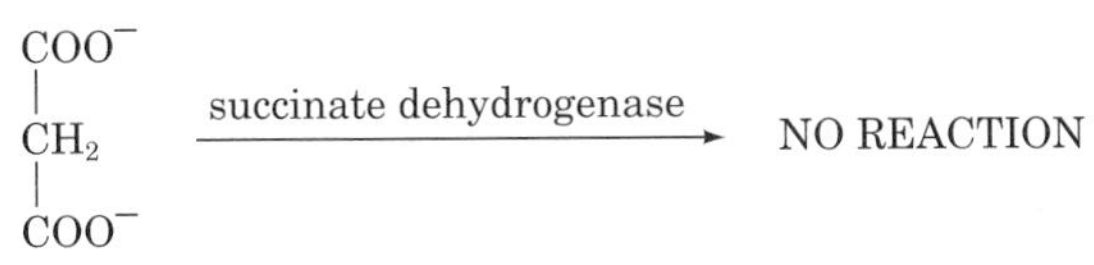

Malonate

The effectiveness of malonate in competitively inhibiting succinate dehydrogenase strongly suggests that the en-zyme's substrate-binding site is designed to bind both of the substrate's carboxylate groups, presumably through the influence of two appropriately placed positively charged residues.

The general model for competitive inhibition is given by the following reaction scheme:

$$E + S \underset{k_{-1}}{\overset{k_1}{\rightleftharpoons}} ES \overset{k_2}{\longrightarrow} P + E$$
$$+$$
$$I$$
$$K_I \Updownarrow$$
$$EI + S \longrightarrow \text{NO REACTION}$$

Here it is assumed that I, the inhibitor, binds reversibly to the enzyme and is in rapid equilibrium with it so that

$$K_I = \frac{[E][I]}{[EI]} \qquad [14.32]$$

and EI, the enzyme–inhibitor complex, is catalytically inactive. *A competitive inhibitor therefore acts by reducing the concentration of free enzyme available for substrate binding.*

Our goal, as before, is to express v_o in terms of measurable quantities, in this case $[E]_T$, $[S]$, and $[I]$. We begin, as in the derivation of the Michaelis–Menten equation, with the expression for the conservation condition, which must now take into account the existence of EI.

$$[E]_T = [E] + [EI] + [ES] \qquad [14.33]$$

The enzyme concentration can be expressed in terms of [ES] by rearranging Eq. [14.17] under the steady-state condition:

$$[E] = \frac{K_M[ES]}{[S]} \qquad [14.34]$$

That of the enzyme–inhibitor complex is found by rearranging Eq. [14.32] and substituting Eq. [14.34] into it:

$$[EI] = \frac{[E][I]}{K_I} = \frac{K_M[ES][I]}{[S]K_I} \qquad [14.35]$$

Substituting the latter two results into Eq. [14.33] yields

$$[E]_T = [ES]\left\{\frac{K_M}{[S]}\left(1 + \frac{[I]}{K_I}\right) + 1\right\}$$

which can be solved for [ES] by rearranging it to

$$[ES] = \frac{[E]_T[S]}{K_M\left(1 + \frac{[I]}{K_I}\right) + [S]}$$

so that, according to Eq. [14.22], the initial velocity is expressed

$$v_o = k_2[ES] = \frac{k_2[E]_T[S]}{K_M\left(1 + \frac{[I]}{K_I}\right) + [S]} \qquad [14.36]$$

Then defining

$$\alpha = \left(1 + \frac{[I]}{K_I}\right) \qquad [14.37]$$

and $V_{max} = k_2[E]_T$ as in Eq. [14.23],

$$v_o = \frac{V_{max}[S]}{\alpha K_M + [S]} \qquad [14.38]$$

This is the Michaelis–Menten equation with K_M modulated by α, a function of the inhibitor concentration (which, according to Eq. [14.37], must always be ≥ 1). The value of [S] at $v_o = V_{max}/2$ is therefore αK_M.

Figure 14-11 shows the hyperbolic plot of Eq. [14.38] for various values of α. Note that as $[S] \to \infty$, $v_o \to V_{max}$ for any value of α. The larger the value of α, however, the greater [S] must be to approach V_{max}. Thus, the inhibitor does not affect the turnover number of the enzyme. Rather, the presence of I has the effect of making [S] appear more dilute than it actually is, or alternatively, making K_M appear larger than it really is. Conversely, increasing [S] shifts the substrate-binding equilibrium toward ES. Hence, there is true competition between I and S for the enzyme's substrate-binding site; their binding is mutually exclusive.

Recasting Eq. [14.38] in the double-reciprocal form yields

$$\frac{1}{v_o} = \left(\frac{\alpha K_M}{V_{max}}\right)\frac{1}{[S]} + \frac{1}{V_{max}} \qquad [14.39]$$

A plot of this equation is linear and has a slope of $\alpha K_M/V_{max}$, a 1/[S] intercept of $-1/\alpha K_M$, and a $1/v_o$ intercept of $1/V_{max}$ (Fig. 14-12). *The double-reciprocal plots for a competitive inhibitor at various concentrations of I intersect at $1/V_{max}$ on the $1/v_o$ axis; this is diagnostic for competitive inhibition as compared with other types of inhibition (Sections 14-3B and 14-3C).*

By determining the values of α at different inhibitor concentrations, the value of K_I can be found from Eq. [14.37]. In this way, competitive inhibitors can be used to probe the structural nature of an active site. For example, to ascertain the importance of the various segments of an ATP molecule

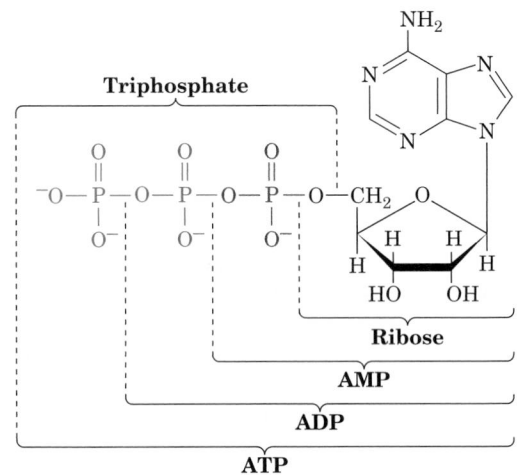

for binding to the active site of an ATP-requiring enzyme, one might determine the K_I, say, for ADP, AMP (adenosine monophosphate), ribose, triphosphate ion, etc. Since many of these ATP components are catalytically inactive, inhibition studies are the most convenient means of monitoring their binding to the enzyme.

If the inhibitor binds irreversibly to the enzyme, the inhibitor is classified as an **inactivator,** as is any agent that somehow inactivates the enzyme. Inactivators truly reduce the effective level of $[E]_T$ at all values of [S]. Reagents that modify specific amino acid residues can act in this manner.

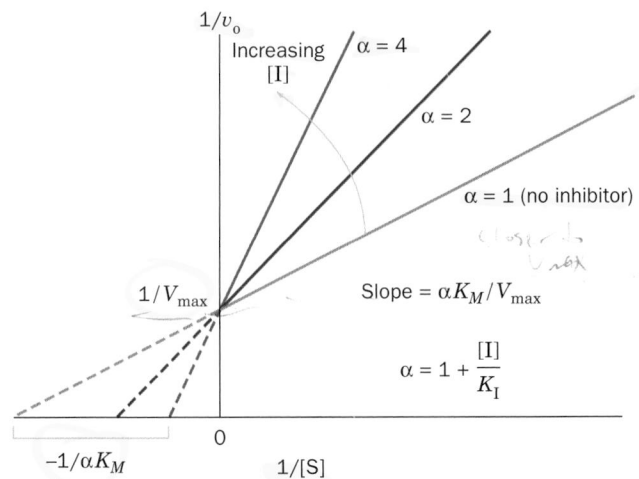

FIGURE 14-11   Competitive inhibition. Plot of the initial velocity v_o of a simple Michaelis–Menten reaction versus the substrate concentration [S] in the presence of different concentrations of a competitive inhibitor.

FIGURE 14-12   Lineweaver–Burk plot of the competitively inhibited Michaelis–Menten enzyme described by Fig. 14-11. Note that all lines intersect on the $1/v_o$ axis at $1/V_{max}$. See the Animated Figures

B. *Uncompetitive Inhibition*

In **uncompetitive inhibition,** the inhibitor binds directly to the enzyme–substrate complex but not to the free enzyme:

$$E + S \underset{k_{-1}}{\overset{k_1}{\rightleftharpoons}} ES \xrightarrow{k_2} P + E$$

$$+$$

$$I$$

$$K'_I \Big\updownarrow$$

$$ESI \longrightarrow NO\ REACTION$$

The inhibitor-binding step, which has the dissociation constant

$$K'_I = \frac{[ES][I]}{[ESI]} \qquad [14.40]$$

is assumed to be at equilibrium. The binding of the uncompetitive inhibitor, which need not resemble the substrate, is envisioned to cause structural distortion of the active site, thereby rendering the enzyme catalytically inactive. (If the inhibitor binds to enzyme alone, it does so without affecting its affinity for substrate.)

The Michaelis–Menten equation for uncompetitive inhibition, which is derived in Appendix B of this chapter, is

$$v_o = \frac{V_{max}[S]}{K_M + \alpha'[S]} \qquad [14.41]$$

where

$$\alpha' = 1 + \frac{[I]}{K'_I} \qquad [14.42]$$

Inspection of this equation indicates that *at high values of [S], v_o asymptotically approaches V_{max}/α', so that, in contrast to competitive inhibition, the effects of uncompetitive inhibition on V_{max} are not reversed by increasing the substrate concentration.* However, at low substrate concentrations, that is, when $[S] << K_M$, the effect of an uncompetitive inhibitor becomes negligible, again the opposite behavior of a competitive inhibitor.

When cast in the double-reciprocal form, Eq. [14.41] becomes

$$\frac{1}{v_o} = \left(\frac{K_M}{V_{max}}\right)\frac{1}{[S]} + \frac{\alpha'}{V_{max}} \qquad [14.43]$$

The Lineweaver–Burk plot for uncompetitive inhibition is linear with slope K_M/V_{max}, as in the uninhibited reaction, and with $1/v_o$ and $1/[S]$ intercepts of α'/V_{max} and $-\alpha'/K_M$, respectively. *A series of Lineweaver–Burk plots at various uncompetitive inhibitor concentrations consists of a family of parallel lines (Fig. 14-13). This is diagnostic for uncompetitive inhibition.*

Uncompetitive inhibition requires that the inhibitor affect the catalytic function of the enzyme but not its substrate binding. For single-substrate enzymes it is difficult to conceive of how this could happen with the exception

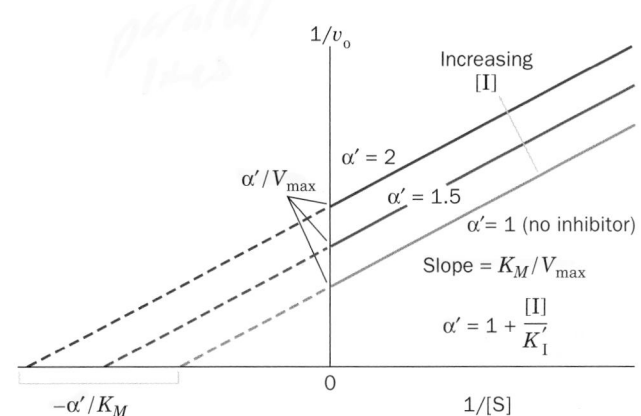

FIGURE 14-13 Lineweaver–Burk plot of a simple Michaelis–Menten enzyme in the presence of uncompetitive inhibitor. Note that all lines have identical slopes of K_M/V_{max}. See the **Animated Figures**

of small inhibitors such as protons (see Section 14-4) or metal ions. As we discuss in Section 14-5C, however, uncompetitive inhibition is important for multisubstrate enzymes.

C. *Mixed Inhibition*

If both the enzyme and the enzyme–substrate complex bind inhibitor, the following model results:

$$E + S \underset{k_{-1}}{\overset{k_1}{\rightleftharpoons}} \qquad ES \xrightarrow{k_2} P + E$$

$$+ \qquad\qquad +$$

$$I \qquad\qquad I$$

$$K_I \Big\updownarrow \qquad\quad K'_I \Big\updownarrow$$

$$EI \qquad\quad ESI \longrightarrow NO\ REACTION$$

Both of the inhibitor-binding steps are assumed to be at equilibrium but with different dissociation constants:

$$K_I = \frac{[E][I]}{[EI]} \quad and \quad K'_I = \frac{[ES][I]}{[ESI]} \qquad [14.44]$$

This phenomenon is alternatively known as **mixed inhibition** or **noncompetitive inhibition.** Presumably a mixed inhibitor binds to enzyme sites that participate in both substrate binding and catalysis.

The Michaelis–Menten equation for mixed inhibition, which is derived in Appendix C of this chapter, is

$$v_o = \frac{V_{max}[S]}{\alpha K_M + \alpha'[S]} \qquad [14.45]$$

where α and α' are defined in Eqs. [14.37] and [14.42], respectively. It can be seen from Eq. [14.45] that the name "mixed inhibition" arises from the fact that the denominator has the factor α multiplying K_M as in competitive inhibition (Eq. [14.38]) and the factor α' multiplying [S] as in uncompetitive inhibition (Eq. [14.41]). Mixed inhibitors

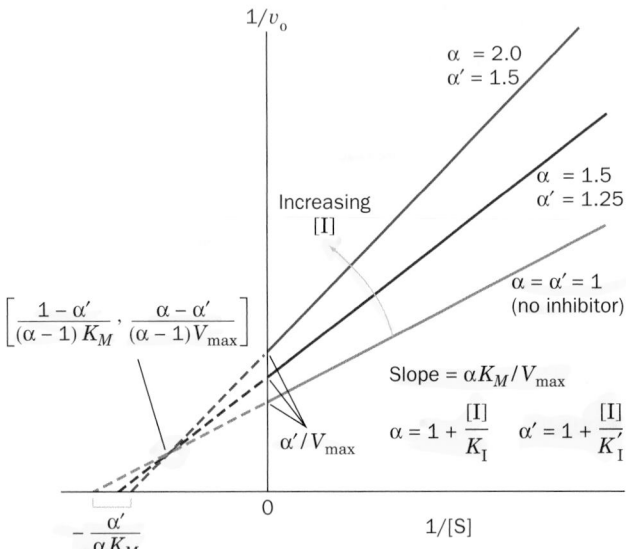

FIGURE 14-14   Lineweaver–Burk plot of a simple Michaelis–Menten enzyme in the presence of a mixed inhibitor. Note that the lines all intersect to the left of the $1/v_o$ axis. The coordinates of this intersection point are given in brackets. Note that when $K_I = K_I'$, $\alpha = \alpha'$ and the lines intersect on the $1/[S]$ axis at $-1/K_M$. 🔊 **See the Animated Figures**

are therefore effective at both high and low substrate concentrations.

The Lineweaver–Burk equation for mixed inhibition is

$$\frac{1}{v_o} = \left(\frac{\alpha K_M}{V_{max}}\right)\frac{1}{[S]} + \frac{\alpha'}{V_{max}} \qquad [14.46]$$

The plot of this equation consists of lines that have slope $\alpha K_M/V_{max}$ with a $1/v_o$ intercept of α'/V_{max} and a $1/[S]$ intercept of $-\alpha'/\alpha K_M$ (Fig. 14-14). Algebraic manipulation of Eq. [14.46] for different values of [I] reveals that this equation describes a family of lines that intersect to the left of the $1/v_o$ axis (Fig. 14-14). For the special case in which $K_I = K_I'$ ($\alpha = \alpha'$), the intersection is, in addition, on the $1/[S]$ axis, a situation which, in an ambiguity of nomenclature, is sometimes described as noncompetitive inhibition.

Table 14-2 provides a summary of the preceding results concerning the inhibition of simple Michaelis–Menten en-

zymes. The quantities K_M^{app} and V_{max}^{app} are the "apparent" values of K_M and V_{max} that would actually be observed in the presence of inhibitor for the Michaelis–Menten equation describing the inhibited enzymes.

4 ■ EFFECTS OF pH

Enzymes, being proteins, have properties that are quite pH sensitive. Most proteins, in fact, are active only within a narrow pH range, typically 5 to 9. This is a result of the effects of pH on a combination of factors: (1) the binding of substrate to enzyme, (2) the catalytic activity of the enzyme, (3) the ionization of substrate, and (4) the variation of protein structure (usually significant only at extremes of pH).

a. pH Dependence of Simple Michaelis–Menten Enzymes

The initial rates for many enzymatic reactions exhibit bell-shaped curves as a function of pH (e.g., Fig. 14-15). These curves reflect the ionizations of certain amino acid residues that must be in a specific ionization state for enzyme activity. The following model can account for such pH effects.

$$
\begin{array}{ccc}
\text{E}^- & & \text{ES}^- \\[2pt]
K_{E2}\,\Big\|\,\text{H}^+ & & K_{ES2}\,\Big\|\,\text{H}^+ \\[4pt]
\text{EH} + \text{S} & \underset{k_{-1}}{\overset{k_1}{\rightleftharpoons}} \text{ESH} & \xrightarrow{k_2} \text{P} + \text{EH} \\[4pt]
K_{E1}\,\Big\|\,\text{H}^+ & & K_{ES1}\,\Big\|\,\text{H}^+ \\[2pt]
\text{EH}_2^+ & & \text{ESH}_2^+
\end{array}
$$

In this expansion of the simple one substrate–no back reaction mechanism, it is assumed that only EH and ESH are catalytically active.

The Michaelis–Menten equation for this model, which is derived in Appendix D, is

$$v_o = \frac{V_{max}'[S]}{K_M' + [S]} \qquad [14.47]$$

TABLE 14-2   Effects of Inhibitors on the Parameters of the Michaelis–Menten Equation[a]

Type of Inhibition	V_{max}^{app}	K_M^{app}
None	V_{max}	K_M
Competitive	V_{max}	αK_M
Uncompetitive	V_{max}/α'	K_M/α'
Mixed	V_{max}/α'	$\alpha K_M/\alpha'$

[a] $\alpha = 1 + \dfrac{[I]}{K_I}$ and $\alpha' = 1 + \dfrac{[I]}{K_I'}$.

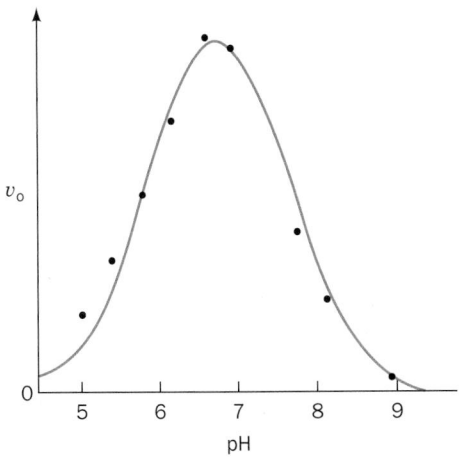

FIGURE 14-15 Effect of pH on the initial rate of the reaction catalyzed by the enzyme fumarase. [After Tanford, C., *Physical Chemistry of Macromolecules*, p. 647, Wiley (1961).]

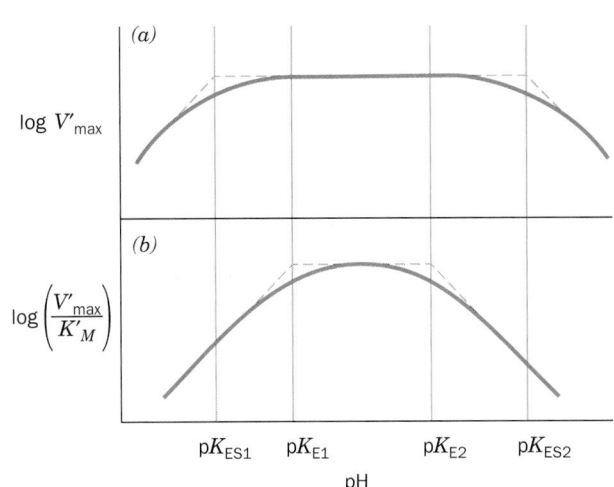

FIGURE 14-16 The pH dependence of (*a*) log V'_{max} and (*b*) log (V'_{max}/K'_M). This indicates how the values of the molecular ionization constants can be determined by graphical extrapolation.

Here the apparent Michaelis–Menten parameters are defined

$$V'_{max} = V_{max}/f_2 \qquad \text{and} \qquad K'_M = K_M(f_1/f_2)$$

where

$$f_1 = \frac{[\text{H}^+]}{K_{E1}} + 1 + \frac{K_{E2}}{[\text{H}^+]}$$

$$f_2 = \frac{[\text{H}^+]}{K_{ES1}} + 1 + \frac{K_{ES2}}{[\text{H}^+]}$$

and V_{max} and K_M refer to the active forms of the enzyme, EH and ESH. Note that at any given pH, Eq. [14.47] behaves as a simple Michaelis–Menten equation, but because of the pH dependence of f_1 and f_2, v_o varies with pH in a bell-shaped manner (e.g., Fig. 14-15).

b. Evaluation of Ionization Constants

The ionization constants of enzymes that obey Eq. [14.47] can be evaluated by the analysis of the curves of log V'_{max} versus pH, which provides values of K_{ES1} and K_{ES2} (Fig. 14-16*a*), and of log (V'_{max}/K'_M) versus pH, which yields K_{E1} and K_{E2} (Fig. 14-16*b*). This, of course, entails the determination of the enzyme's Michaelis–Menten parameters at each of a series of different pH's.

The measured pK's often provide valuable clues as to the identities of the amino acid residues essential for enzymatic activity. For example, a measured pK of ~4 suggests that an Asp or Glu residue is essential to the enzyme. Similarly, pK's of ~6 or ~10 suggest the participation of a His or a Lys residue, respectively. However, a given acid–base group may vary by as much as several pH units from its expected value as a consequence of the electrostatic influence of nearby charged groups, as well as of the proximity of regions of low polarity. For example, the car-

boxylate group of a Glu residue forming a salt bridge with a Lys residue is stabilized by the nearby positive charge and therefore has a lower pK than it would otherwise have; that is, it is more difficult to protonate. Conversely, a carboxylate group immersed in a region of low polarity is less acidic than normal because it attracts protons more strongly than if it were in a region of higher polarity. The identification of a kinetically characterized pK with a particular amino acid residue must therefore be verified by other types of measurements such as the use of group-specific reagents to inactivate a putative essential residue.

5 ■ BISUBSTRATE REACTIONS

We have heretofore been concerned with reactions involving enzymes that require only a single substrate. Yet enzymatic reactions involving two substrates and yielding two products

$$\text{A} + \text{B} \underset{}{\overset{\text{E}}{\rightleftharpoons}} \text{P} + \text{Q}$$

account for ~60% of known biochemical reactions. Almost all of these so-called **bisubstrate reactions** are either **transferase** reactions in which the enzyme catalyzes the transfer of a specific functional group, X, from one of the substrates to the other:

$$\text{P}\!-\!\text{X} + \text{B} \underset{}{\overset{\text{E}}{\rightleftharpoons}} \text{P} + \text{B}\!-\!\text{X}$$

or oxidation–reduction reactions in which reducing equivalents are transferred between the two substrates. For example, the hydrolysis of a peptide bond by trypsin (Section 7-1E) is the transfer of the peptide carbonyl group from

(a)

$$R_1 - \overset{\overset{O}{\|}}{C} - NH - R_2 \;+\; H_2O \;\xrightarrow{\text{trypsin}}\; R_1 - \overset{\overset{O}{\|}}{C} - O^- \;+\; H_3\overset{+}{N} - R_2$$

Polypeptide

(b)

$$CH_3 - \overset{\overset{\displaystyle H}{|}}{\underset{\underset{\displaystyle H}{|}}{C}} - OH \;+\; NAD^+ \xrightarrow[\;\;\;\;\;\;\;\searrow\;\;\;\;\;\;\;]{\overset{\text{alcohol}}{\text{dehydrogenase}}} CH_3 - \overset{\overset{O}{\|}}{C}H \;+\; NADH$$

$$H^+$$

FIGURE 14-17   Some bisubstrate reactions. (*a*) In the peptide hydrolysis reaction catalyzed by trypsin, the peptide carbonyl group, with its pendent polypeptide chain, is transferred from the peptide nitrogen atom to a water molecule. (*b*) In the alcohol dehydrogenase reaction, a hydride ion is formally transferred from ethanol to NAD^+.

the peptide nitrogen atom to water (Fig. 14-17*a*). Similarly, in the alcohol dehydrogenase reaction (Section 13-2A), a hydride ion is formally transferred from ethanol to NAD^+ (Fig. 14-17*b*). Although such bisubstrate reactions could, in principle, occur through a vast variety of mechanisms, only a few types are commonly observed.

A. Terminology

We shall follow the nomenclature system introduced by W. W. Cleland for representing enzymatic reactions:

1. Substrates are designated by the letters A, B, C, and D *in the order that they add to the enzyme.*

2. Products are designated P, Q, R, and S *in the order that they leave the enzyme.*

3. Stable enzyme forms are designated E, F, and G with E being the free enzyme, if such distinctions can be made. A stable enzyme form is defined as one that by itself is incapable of converting to another stable enzyme form (see below).

4. The numbers of reactants and products in a given reaction are specified, in order, by the terms **Uni** (one), **Bi** (two), **Ter** (three), and **Quad** (four). A reaction requiring one substrate and yielding three products is designated a Uni Ter reaction. In this section, we shall be concerned with reactions that require two substrates and yield two products, that is, Bi Bi reactions. Keep in mind, however, that there are numerous examples of even more complex reactions.

a. Types of Bi Bi Reactions

Enzyme-catalyzed group-transfer reactions fall under two major mechanistic classifications:

1. Sequential Reactions: *Reactions in which all substrates must combine with the enzyme before a reaction can occur and products can be released are known as **Sequential reactions.*** In such reactions, the group being transferred,

X, is directly passed from A (= P—X) to B, yielding P and Q (= B—X). Hence, such reactions are also called **single-displacement reactions.**

Sequential reactions can be subclassified into those with a compulsory order of substrate addition to the enzyme, which are said to have an **Ordered mechanism,** and those with no preference for the order of substrate addition, which are described as having a **Random mechanism.** In the Ordered mechanism, the binding of the first substrate is apparently required for the enzyme to form the binding site for the second substrate, whereas for the Random mechanism, both binding sites are present on the free enzyme.

Let us describe enzymatic reactions using Cleland's shorthand notation. The enzyme is represented by a horizontal line and successive additions of substrates and release of products are denoted by vertical arrows. Enzyme forms are placed under the line and rate constants, if given, are to the left of the arrow or on top of the line for forward reactions. An **Ordered Bi Bi** reaction is represented:

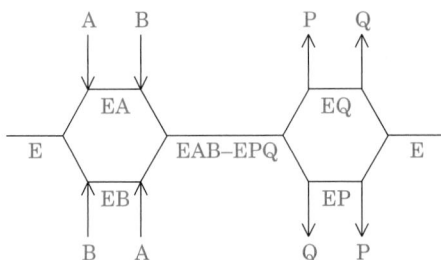

where A and B are said to be the **leading** and **following** substrates, respectively. Here, only minimal details are given concerning the interconversions of intermediate enzyme forms because, as we have seen for reversible single-substrate enzymes, steady-state kinetic measurements provide no information concerning the number of intermediates in a given reaction step. Many NAD^+- and $NADP^+$-requiring dehydrogenases follow an Ordered Bi Bi mechanism in which the coenzyme is the leading reactant.

A **Random Bi Bi** reaction is diagrammed:

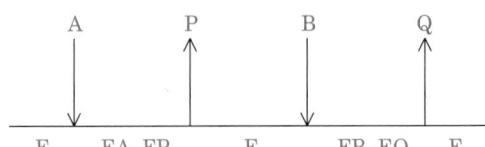

Some dehydrogenases and kinases operate through Random Bi Bi mechanisms.

2. Ping Pong Reactions: *Mechanisms in which one or more products are released before all substrates have been added are known as **Ping Pong reactions.*** The **Ping Pong Bi Bi** reaction is represented by

In it, a functional group X of the first substrate A (= P—X) is displaced from the substrate by the enzyme E to yield the first product P and a stable enzyme form F (= E—X) in which X is tightly (often covalently) bound to the enzyme (Ping). In the second stage of the reaction, X is displaced from the enzyme by the second substrate B to yield the second product Q (= B—X), thereby regenerating the original form of the enzyme, E (Pong). Such reactions are therefore also known as **double-displacement reactions**. *Note that in Ping Pong Bi Bi reactions, the substrates A and B do not encounter one another on the surface of the enzyme.* Many enzymes, including chymotrypsin (Section 15-3), transaminases (Section 26-1A), and some flavoenzymes, react with Ping Pong mechanisms.

B. *Rate Equations*

Steady-state kinetic measurements can be used to distinguish among the foregoing bisubstrate mechanisms. In order to do so, one must first derive their rate equations. This can be done in much the same manner as for single-substrate enzymes, that is, solving a set of simultaneous linear equations consisting of an equation expressing the steady-state condition for each kinetically distinct enzyme complex plus one equation representing the conservation condition for the enzyme. This, of course, is a more complex undertaking for bisubstrate enzymes than it is for single-substrate enzymes.

The rate equations for the above described bisubstrate mechanisms in the absence of products are given below in double-reciprocal form.

a. Ordered Bi Bi

$$\frac{1}{v_o} = \frac{1}{V_{\max}} + \frac{K_M^A}{V_{\max}[A]} + \frac{K_M^B}{V_{\max}[B]} + \frac{K_S^A K_M^B}{V_{\max}[A][B]} \quad [14.48]$$

b. Rapid Equilibrium Random Bi Bi

The rate equation for the general Random Bi Bi reaction is quite complicated. However, in the special case that both substrates are in rapid and independent equilibrium with the enzyme, that is, when the EAB–EPQ intercon-

version is rate determining, the initial rate equation reduces to the following relatively simple form. This mechanism is known as the **Rapid Equilibrium Random Bi Bi** mechanism:

$$\frac{1}{v_o} = \frac{1}{V_{\max}} + \frac{K_S^A K_M^B}{V_{\max} K_S^B [A]} + \frac{K_M^B}{V_{\max}[B]} + \frac{K_S^A K_M^B}{V_{\max}[A][B]}$$
$$[14.49]$$

c. Ping Pong Bi Bi

$$\frac{1}{v_o} = \frac{K_M^A}{V_{\max}[A]} + \frac{K_M^B}{V_{\max}[B]} + \frac{1}{V_{\max}} \quad [14.50]$$

d. Physical Significance of the Bisubstrate Kinetic Parameters

The kinetic parameters in the equations describing bisubstrate reactions have meanings similar to those for single-substrate reactions. $V_{\max}$ is the maximal velocity of the enzyme obtained when both A and B are present at saturating concentrations, K_M^A and K_M^B are the respective concentrations of A and B necessary to achieve $\frac{1}{2}V_{\max}$ in the presence of a saturating concentration of the other, and K_S^A and K_S^B are the respective dissociation constants of A and B from the enzyme, E.

C. *Differentiating Bisubstrate Mechanisms*

One can discriminate between Ping Pong and Sequential mechanisms from their contrasting properties in linear plots such as those of the Lineweaver–Burk type.

a. Diagnostic Plot for Ping Pong Bi Bi Reactions

A plot of $1/v_o$ versus $1/[A]$ at constant [B] for Eq. [14.50] yields a straight line of slope $K_M^A/V_{\max}$ and an intercept on the $1/v_o$ axis equal to the last two terms in Eq. [14.50]. Since the slope is independent of [B], such plots for different values of [B] yield a family of parallel lines (Fig. 14-18). A plot of $1/v_o$ versus $1/[B]$ for different values of [A] likewise yields a family of parallel lines. *Such parallel lines are diagnostic for a Ping Pong mechanism.*

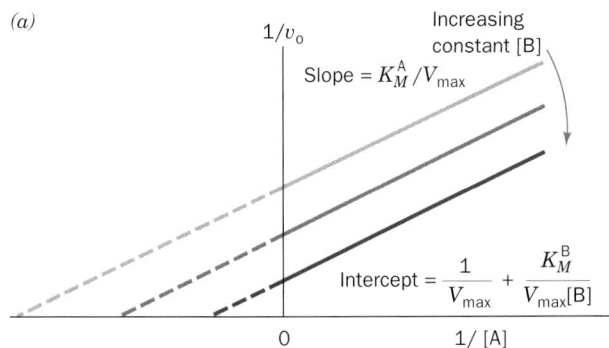

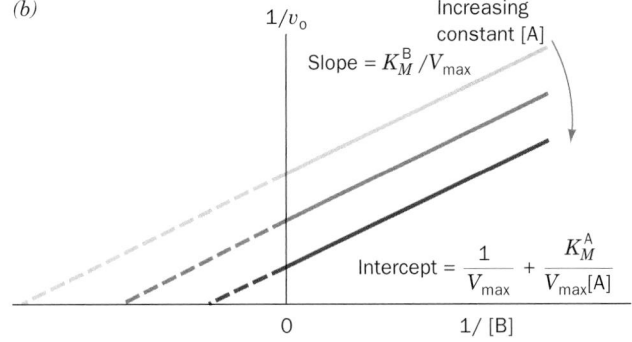

FIGURE 14-18   Double-reciprocal plots for an enzymatic reaction with a Ping Pong Bi Bi mechanism. (*a*) Plots of $1/v_o$ versus $1/[A]$ at various constant concentrations of B. (*b*) Plots of $1/v_o$ versus $1/[B]$ at various constant concentrations of A.

(a)

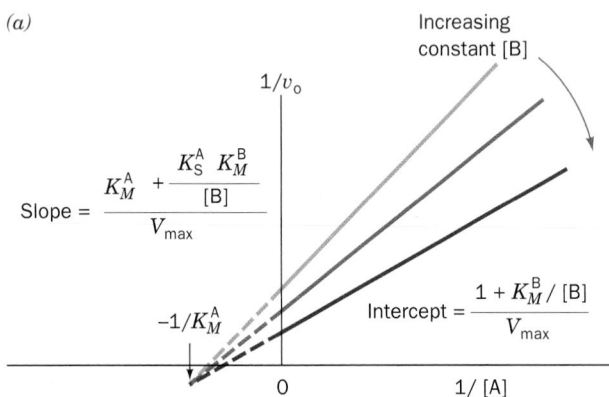

(b)

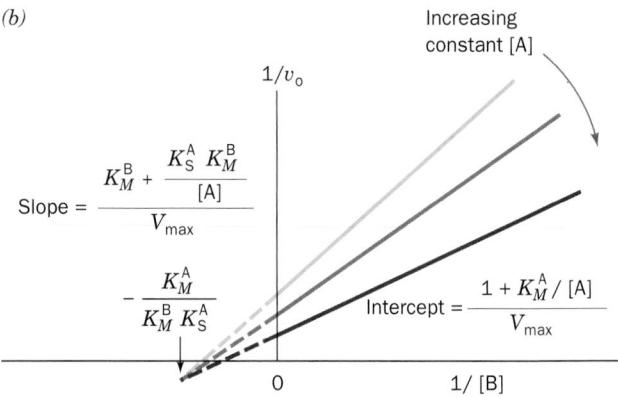

FIGURE 14-19   Double-reciprocal plots of an enzymatic reaction with a Sequential Bi Bi mechanism. (*a*) Plots of $1/v_o$ versus $1/[A]$ at various constant concentrations of B. (*b*) Plots of $1/v_o$ versus $1/[B]$ at various constant concentrations of A.

The corresponding plots for Rapid Equilibrium Random Bi Bi reactions have identical appearances; their lines all intersect to the left of the $1/v_o$ axis.

b. Diagnostic Plot for Sequential Bi Bi Reactions

The equations representing the Ordered Bi Bi mechanism (Eq. [14.48]) and the Rapid Equilibrium Random Bi Bi mechanism (Eq. [14.49]) have identical functional dependence on [A] and [B].

Equation [14.48] can be rearranged to

$$\frac{1}{v_o} = \frac{K_M^A}{V_{max}}\left(1 + \frac{K_S^A K_M^B}{K_M^A[B]}\right)\frac{1}{[A]} + \frac{1}{V_{max}}\left(1 + \frac{K_M^B}{[B]}\right) \quad [14.51]$$

Thus plotting $1/v_o$ versus $1/[A]$ for constant [B] yields a linear plot with a slope equal to the coefficient of $1/[A]$ and an intercept on the $1/v_o$ axis equal to the second term of Eq. [14.51] (Fig. 14-19*a*). Alternatively, Eq. [14.48] can be rearranged to

$$\frac{1}{v_o} = \frac{K_M^B}{V_{max}}\left(1 + \frac{K_S^A}{[A]}\right)\frac{1}{[B]} + \frac{1}{V_{max}}\left(1 + \frac{K_M^A}{[A]}\right) \quad [14.52]$$

which yields a linear plot of $1/v_o$ versus $1/[B]$ for constant [A] with a slope equal to the coefficient of $1/[B]$ and an intercept on the $1/v_o$ axis equal to the second term of Eq. [14.52] (Fig. 14-19*b*). *The characteristic feature of these plots, which is indicative of a Sequential mechanism, is that the lines intersect to the left of the $1/v_o$ axis.*

c. Differentiating Random and Ordered Sequential Mechanisms

The Ordered Bi Bi mechanism may be experimentally distinguished from the Random Bi Bi mechanism through **product inhibition studies.** If only one product of the reaction, P or Q, is added to the reaction mixture, the reverse reaction still cannot occur. Nevertheless, by binding to the enzyme, this product will inhibit the forward reaction. For an Ordered Bi Bi reaction, Q (= B—X, the second product to be released) directly competes with A (= P—X, the leading substrate) for binding to E and hence is a competitive inhibitor of A when [B] is fixed (the presence of X in Q = B—X interferes with the binding of A = P—X). However, since B combines with EA, not E, Q is a mixed inhibitor of B when [A] is fixed (Q interferes with both the binding of B to enzyme and with the catalysis of the reaction). Similarly, P, which combines only with EQ, is a mixed inhibitor of A when [B] is held constant and of B when [A] is held constant. In contrast, in a Rapid Equilibrium Bi Bi reaction, since both products as well as both substrates can combine directly with E, both P and Q are competitive inhibitors of A when [B] is constant and of B when [A] is constant. These product inhibition patterns are summarized in Table 14-3.

D. *Isotope Exchange*

Mechanistic conclusions based on kinetic analyses alone are fraught with uncertainties and are easily confounded by inaccurate experimental data. A particular mechanism for an enzyme is therefore greatly corroborated if the mechanism can be shown to conform to experimental criteria other than kinetic analysis.

TABLE 14-3   Patterns of Product Inhibition for Sequential Bisubstrate Mechanisms

Mechanism	Product Inhibitor	[A] Variable	[B] Variable
Ordered Bi Bi	P	Mixed	Mixed
	Q	Competitive	Mixed
Rapid Equilibrium Random Bi Bi	P	Competitive	Competitive
	Q	Competitive	Competitive

*Sequential (single-displacement) and Ping Pong (double-displacement) bisubstrate mechanisms may be differentiated through the use of **isotope exchange** studies.* Double-displacement reactions are capable of exchanging an isotope from the first product P back to the first substrate A in the absence of the second substrate. Consider an overall Ping Pong reaction catalyzed by the bisubstrate enzyme E

$$P\!-\!X \ + \ B \ \underset{\overset{E}{\rightleftharpoons}}{} \ P \ + \ B\!-\!X$$

in which, as usual, A = P—X, Q = B—X, and X is the group that is transferred from one substrate to the other in the course of the reaction. Only the first step of the reaction can take place in the absence of B. If a small amount of isotopically labeled P, denoted P*, is added to this reaction mixture then, in the reverse reaction, P*—X will form:

Forward reaction   E + P—X → E—X + P
Reverse reaction   E—X + P* → E + P*—X

that is, isotopic exchange will occur.

In contrast, let us consider the first step of a Sequential reaction. Here a noncovalent enzyme–substrate complex forms:

$$E \ + \ P\!-\!X \ \rightleftharpoons \ E \cdot P\!-\!X$$

Addition of P* cannot result in an exchange reaction because no covalent bonds are broken in the formation of E · P—X; that is, there is no P released from the enzyme to exchange with P*. The demonstration of isotopic exchange for a bisubstrate enzyme is therefore convincing evidence favoring a Ping Pong mechanism.

a. Isotope Exchange in Sucrose Phosphorylase and Maltose Phosphorylase

The enzymes **sucrose phosphorylase** and **maltose phosphorylase** provide two clear-cut examples of how enzymatically catalyzed isotopic exchange reactions are used to differentiate kinetic mechanisms. Sucrose phosphorylase catalyzes the overall reaction

Glucose—fructose + phosphate
Sucrose
$$\Big\Vert\, E$$
Glucose-1-phosphate + fructose

If the enzyme is incubated with sucrose and isotopically labeled fructose in the absence of phosphate, it is observed that the label passes into the sucrose:

Glucose—fructose + fructose*
Sucrose
$$\Big\Vert\, E$$
Glucose—fructose* + fructose

For the reverse reaction, if the enzyme is incubated with glucose-1-phosphate and [32P]-labeled phosphate, this label exchanges into the glucose-1-phosphate:

Glucose-1-phosphate + phosphate*
$$\Big\Vert\, E$$
Glucose-1-phosphate* + phosphate

These observations indicate that a tight glucosyl–enzyme complex is formed with the release of fructose, thereby establishing that the sucrose phosphorylase reaction occurs via a Ping Pong mechanism. This finding has been conclusively corroborated by the isolation and characterization of the glucosyl–enzyme complex.

The enzyme **maltose phosphorylase** catalyzes a similar overall reaction:

Glucose—glucose + phosphate
Maltose
$$\Big\Vert\, E$$
Glucose-1-phosphate + glucose

In contrast to sucrose phosphorylase, however, it does not catalyze isotopic exchange between glucose-1-phosphate and [32P]phosphate or between maltose and [14C]glucose. Likewise, a glucosyl–enzyme complex has not been detected. This evidence is consistent with maltose phosphorylase having a sequential mechanism.

Appendix: ■ DERIVATIONS OF MICHAELIS–MENTEN EQUATION VARIANTS

A. *The Michaelis–Menten Equation for Reversible Reactions—Equation [14.30]*

The conservation condition for the reversible reaction with one intermediate (Section 14-2C) is

$$[E]_T = [E] + [ES] \qquad [14.A1]$$

The steady-state condition (as well as the equilibrium condition) is

$$\frac{d[ES]}{dt} = k_1[E][S] + k_{-2}[E][P] - (k_{-1} + k_2)[ES] = 0 \qquad [14.A2]$$

so that

$$[E] = \left(\frac{k_{-1} + k_2}{k_1[S] + k_{-2}[P]} \right)[ES] \qquad [14.A3]$$

Substituting this result into Eq. [14.A1] yields

$$[E]_T = \left(\frac{k_{-1} + k_2}{k_1[S] + k_{-2}[P]} + 1 \right)[ES] \qquad [14.A4]$$

The velocity of the reaction is expressed

$$v = -\frac{d[S]}{dt} = k_1[E][S] - k_{-1}[ES] \quad [14.A5]$$

which can be combined with Eq. [14.A3] to give

$$v = \left(\frac{k_1[S](k_{-1} + k_2)}{k_1[S] + k_{-2}[P]} - k_{-1}\right)[ES] \quad [14.A6]$$

which, in turn, is combined with Eq. [14.A4] to yield

$$v = \left(\frac{k_1k_2[S] - k_{-1}k_{-2}[P]}{k_{-1} + k_2 + k_1[S] + k_{-2}[P]}\right)[E]_T \quad [14.A7]$$

Dividing the numerator and denominator of this equation by $(k_{-1} + k_2)$ results in

$$v = \left(\frac{k_2\left(\frac{k_1}{k_{-1} + k_2}\right)[S] - k_{-1}\left(\frac{k_{-2}}{k_{-1} + k_2}\right)[P]}{1 + \left(\frac{k_1}{k_{-1} + k_2}\right)[S] + \left(\frac{k_{-2}}{k_{-1} + k_2}\right)[P]}\right)[E]_T$$

$$[14.A8]$$

Then, if we define the following parameters analogously with the constants of the Michaelis–Menten equation (Eqs. [14.23] and [14.21]),

$$V_{max}^f = k_2[E]_T \qquad V_{max}^r = k_{-1}[E]_T$$

$$K_M^S = \frac{k_{-1} + k_2}{k_1} \qquad K_M^P = \frac{k_{-1} + k_2}{k_{-2}}$$

we obtain the Michaelis–Menten equation for a reversible one-intermediate reaction:

$$v = \frac{\dfrac{V_{max}^f[S]}{K_M^S} - \dfrac{V_{max}^r[P]}{K_M^P}}{1 + \dfrac{[S]}{K_M^S} + \dfrac{[P]}{K_M^P}} \quad [14.30]$$

B. Michaelis–Menten Equation for Uncompetitive Inhibition—Equation [14.41]

For uncompetitive inhibition (Section 14-3B), the inhibitor binds to the Michaelis complex with dissociation constant

$$K_I' = \frac{[ES][I]}{[ESI]} \quad [14.A9]$$

The conservation condition is

$$[E]_T = [E] + [ES] + [ESI] \quad [14.A10]$$

Substituting in Eqs. [14.34] and [14.A9] yields

$$[E]_T = [ES]\left(\frac{K_M}{[S]} + 1 + \frac{[I]}{K_I'}\right) \quad [14.A11]$$

Defining α' similarly to Eq. [14.37] as

$$\alpha' = 1 + \frac{[I]}{K_I'} \quad [14.A12]$$

and v_o and V_{max} as in Eqs. [14.22] and [14.23], respectively,

$$v_o = k_2[ES] = \frac{V_{max}}{\dfrac{K_M}{[S]} + \alpha'} \quad [14.A13]$$

which on rearrangement yields the Michaelis–Menten equation for uncompetitive inhibition:

$$v_o = \frac{V_{max}[S]}{K_M + \alpha'[S]} \quad [14.41]$$

C. The Michaelis–Menten Equation for Mixed Inhibition—Equation [14.45]

In mixed inhibition (Section 14-3C), the inhibitor-binding steps have different dissociation constants:

$$K_I = \frac{[E][I]}{[EI]} \quad \text{and} \quad K_I' = \frac{[ES][I]}{[ESI]} \quad [14.A14]$$

(Here, for the sake of mathematical simplicity, we are making the thermodynamically unsupportable assumption that EI does not react with S to form ESI. Inclusion of this reaction requires a more complex derivation than that given here but leads to results that are substantially the same.) The conservation condition for this reaction scheme is

$$[E]_T = [E] + [EI] + [ES] + [ESI] \quad [14.A15]$$

so that substituting in Eqs. [14.A14]

$$[E]_T = [E]\left(1 + \frac{[I]}{K_I}\right) + [ES]\left(1 + \frac{[I]}{K_I'}\right) \quad [14.A16]$$

Defining α and α' as in Eqs. [14.37] and [14.A12], respectively, Eq. [14.A16] becomes

$$[E]_T = [E]\alpha + [ES]\alpha' \quad [14.A17]$$

Then substituting in Eq. [14.34]

$$[E]_T = [ES]\left(\frac{\alpha K_M}{[S]} + \alpha'\right) \quad [14.A18]$$

Defining v_o and V_{max} as in Eqs. [14.22] and [14.23] results in the Michaelis–Menten equation for mixed inhibition:

$$v_o = \frac{V_{max}[S]}{\alpha K_M + \alpha'[S]} \quad [14.45]$$

D. The Michaelis–Menten Equation for Ionizable Enzymes—Equation [14.47]

In the model presented in Section 14-4 to account for the effect of pH on enzymes, the dissociation constants for the ionizations are

$$K_{E2} = \frac{[H^+][E^-]}{[EH]} \qquad K_{ES2} = \frac{[H^+][ES^-]}{[ESH]}$$

$$K_{E1} = \frac{[H^+][EH]}{[EH_2^+]} \qquad K_{ES1} = \frac{[H^+][ESH]}{[ESH_2^+]} \quad [14.A19]$$

Protonation and deprotonation are among the fastest known reactions, so that, with the exception of the few enzymes with extremely high turnover numbers, it can be reasonably assumed that all acid–base reactions are at equilibrium. The conservation condition is

$$[E]_T = [EH]_T + [ESH]_T \quad\quad [14.A20]$$

where $[E]_T$ is the total enzyme present in any form,

$$
\begin{aligned}
[EH]_T &= [EH_2^+] + [EH] + [E^-]\\
&= [EH]\left(\frac{[H^+]}{K_{E1}} + 1 + \frac{K_{E2}}{[H^+]}\right)\\
&= [EH]f_1 \quad\quad [14.A21]
\end{aligned}
$$

and

$$
\begin{aligned}
[ESH]_T &= [ESH_2^+] + [ESH] + [ES^-]\\
&= [ESH]\left(\frac{[H^+]}{K_{ES1}} + 1 + \frac{K_{ES2}}{[H^+]}\right)\\
&= [ESH]f_2 \quad\quad [14.A22]
\end{aligned}
$$

Then making the steady-state assumption

$$\frac{d[ESH]}{dt} = k_1[EH][S] - (k_{-1} + k_2)[ESH] = 0 \quad [14.A23]$$

and solving for [EH]

$$[EH] = \frac{(k_{-1} + k_2)[ESH]}{k_1[S]} = \frac{K_M[ESH]}{[S]} \quad\quad [14.A24]$$

Therefore, from Eq. [14.A21],

$$[EH]_T = \frac{K_M[ESH]f_1}{[S]} \quad\quad [14.A25]$$

which, together with Eqs. [14.A20] and [14.A22], yields

$$[E]_T = [ESH]\left(\frac{K_M f_1}{[S]} + f_2\right) \quad\quad [14.A26]$$

As in the simple Michaelis–Menten derivation, the initial rate is

$$v_o = k_2[ESH] = \frac{k_2[E]_T}{\left(\dfrac{K_M f_1}{[S]}\right) + f_2} = \frac{(k_2/f_2)[E]_T[S]}{K_M(f_1/f_2) + [S]}$$

$$[14.A27]$$

Then defining the "apparent" values of K_M and $V_{max} = k_2[E]_T$ at a given pH:

$$K'_M = K_M(f_1/f_2) \quad\quad [14.A28]$$

and

$$V'_{max} = V_{max}/f_2 \qu\quad [14.A29]$$

the Michaelis–Menten equation modified to account for pH effects is

$$v_o = \frac{V'_{max}[S]}{K'_M + [S]} \quad\quad [14.47]$$

CHAPTER SUMMARY

1 ■ Chemical Kinetics Complicated reaction processes occur through a series of elementary reaction steps defined as having a molecularity equal to the number of molecules that simultaneously collide to form products. The order of a reaction can be determined from the characteristic functional form of its progress curve. Transition state theory postulates that the rate of a reaction depends on the free energy of formation of its activated complex. This complex, which occurs at the free energy maximum of the reaction coordinate, is poised between reactants and products and is therefore also known as the transition state. Transition state theory explains that catalysis results from the reduction of the free energy difference between the reactants and the transition state.

2 ■ Enzyme Kinetics In the simplest enzymatic mechanism, the enzyme and substrate reversibly combine to form an enzyme–substrate complex known as the Michaelis complex, which may irreversibly decompose to form product and the regenerated enzyme. The rate of product formation is expressed by the Michaelis–Menten equation, which is derived under the assumption that the concentration of the Michaelis complex is constant, that is, at a steady state. The Michaelis–Menten equation, which has the functional form of a rectangular hyperbola, has two parameters: V_{max}, the maximal rate of the reaction, which occurs when the substrate concentration is saturating, and K_M, the Michaelis constant, which has the value of the substrate concentration at the half-maximal reaction rate.

These parameters may be graphically determined using the Lineweaver–Burk plot. Physically more realistic models of enzyme mechanisms than the Michaelis–Menten model assume the enzymatic reaction to be reversible and to have one or more intermediates. The functional form of the equations describing the reaction rates for these models is independent of their number of intermediates, so that the models cannot be differentiated using only steady-state kinetic measurements.

3 ■ Inhibition Enzymes may be inhibited by competitive inhibitors, which compete with the substrate for the enzymatic binding site. The effect of a competitive inhibitor may be reversed by increasing the substrate concentration. An uncompetitive inhibitor inactivates a Michaelis complex on binding to it. The maximal rate of an uncompetitively inhibited enzyme is a function of inhibitor concentration, and therefore the effect of an uncompetitive inhibitor cannot be reversed by increasing substrate concentration. In mixed inhibition, the inhibitor binds to both the enzyme and the enzyme–substrate complex to form a complex that is catalytically inactive. The rate equation describing this situation has characteristics of both competitive and uncompetitive reactions.

4 ■ Effects of pH The rate of an enzymatic reaction is a function of hydrogen ion concentration. At any pH, the rate of a simple enzymatic reaction can be described by the Michaelis–Menten equation. However, its parameters V_{max} and K_M vary with pH. By the evaluation of kinetic rate curves as a function of pH, the pK's of an enzyme's ionizable binding and catalytic groups can be determined, which may help identify these groups.

5 ■ Bisubstrate Reactions   The majority of enzymatic reactions are bisubstrate reactions in which two substrates react to form two products. Bisubstrate reactions may have Ordered or Random Sequential mechanisms or Ping Pong Bi Bi mechanisms, among others. The initial rate equations for any of these mechanisms involve five parameters, which are analogous to either Michaelis–Menten equation parameters or equilibrium constants. The various bisubstrate mechanisms may be experimentally differentiated according to the forms of their double-reciprocal plots and from the nature of their product inhibition patterns. Isotope exchange reactions provide an additional, nonkinetic method of differentiating bisubstrate mechanisms.

REFERENCES

CHEMICAL KINETICS

Atkins, P.W. and de Paula, J., *Physical Chemistry* (7th ed.), Chapters 25–27, Freeman (2002). [Most physical chemistry textbooks have similar coverage.]

Hammes, G.G., *Principles of Chemical Kinetics,* Academic Press (1978).

Laidler, K.J., *Chemical Kinetics* (3rd ed.), Harper & Row (1987).

ENZYME KINETICS

Biswanger, H., *Enzyme Kinetics: Principles and Methods,* Wiley–VCH (2002).

Cleland, W.W., Steady state kinetics, *in* Boyer, P.D. (Ed.), *The Enzymes* (3rd ed.), Vol. 2, *pp.* 1–65, Academic Press (1970); *and* Steady-state kinetics, *in* Sigman, D.S. and Boyer, P.D. (Eds.), *The Enzymes* (3rd ed.), Vol. 19, *pp.* 99–158, Academic Press (1990).

Cleland, W.W., Determining the mechanism of enzyme-catalyzed reactions by kinetic studies, *Adv. Enzymol.* **45,** 273 (1977).

Cornish-Bowden, A., *Fundamentals of Enzyme Kinetics* (Revised ed.), Portland Press (1995). [A lucid and detailed account of enzyme kinetics.]

Cornish-Bowden, A. and Wharton, C.W., *Enzyme Kinetics,* IRL Press (1988).

Copeland, R.A., *Enzymes,* VCH (1996).

Dixon, M. and Webb, E.C., *Enzymes* (3rd ed.), Chapter IV, Academic Press (1979). [An almost exhaustive treatment of enzyme kinetics.]

Fersht, A., *Structure and Mechanism in Protein Science,* Chapters 3–7, Freeman (1999).

Gutfreund, H., *Kinetics for the Life Sciences: Receptors, Transmitters, and Catalysts,* Cambridge University Press (1995).

Hammes, G.G., *Enzyme Catalysis and Regulation,* Chapter 3, Academic Press (1982).

Knowles, J.R., The intrinsic pK_a-values of functional groups in enzymes: Improper deductions from the pH-dependence of steady state parameters, *CRC Crit. Rev. Biochem.* **4,** 165 (1976).

Kuby, S.A., *A Study of Enzymes,* Vol. I, CRC Press (1991). [Contains several chapters on enzyme kinetics.]

Marangoni, A.G., *Enzyme Kinetics. A Modern Approach,* Wiley (2002).

Piszkiewicz, D., *Kinetics of Chemical and Enzyme Catalyzed Reactions,* Oxford University Press (1977). [A highly readable discussion of enzyme kinetics.]

Purich, D.L. (Ed.), *Contemporary Enzyme Kinetics and Mechanism* (2nd ed.), Academic Press (1996) [A collection of articles on advanced topics.]

Schulz, A.R., *Enzyme Kinetics,* Cambridge (1994).

Segel, I.H., *Enzyme Kinetics,* Wiley–Interscience (1993). [A detailed and understandable treatise providing full explanations of many aspects of enzyme kinetics.]

Tinoco, I., Jr., Sauer, K., Wang, J.C., and Puglisi, J.D., *Physical Chemistry. Principles and Applications for Biological Sciences* (4th ed.), Chapters 7 and 8, Prentice-Hall (2002).

Wood, W.B., Wilson, J.H., Benbow, R.M., and Hood, L.E., *Biochemistry. A Problems Approach* (2nd ed.), Chapter 8, Benjamin/Cummings (1981). [Contains instructive problems on enzyme kinetics with answers worked out in detail.]

PROBLEMS

1. The hydrolysis of sucrose:

$$\text{Sucrose} + H_2O \rightarrow \text{glucose} + \text{fructose}$$

takes the following time course.

Time (min)	[Sucrose] (M)
0	0.5011
30	0.4511
60	0.4038
90	0.3626
130	0.3148
180	0.2674

Determine the first-order rate constant and the half-life of the reaction. Why does this bimolecular reaction follow a first-order rate law? How long will it take to hydrolyze 99% of the sucrose initially present? How long will it take if the amount of sucrose initially present is twice that given in the table?

2. By what factor will a reaction at 25°C be accelerated if a catalyst reduces the free energy of its activated complex by $1 \, \text{kJ} \cdot \text{mol}^{-1}$; by $10 \, \text{kJ} \cdot \text{mol}^{-1}$?

3. For a Michaelis–Menten reaction, $k_1 = 5 \times 10^7 \, M^{-1} \cdot \text{s}^{-1}$, $k_{-1} = 2 \times 10^4 \, \text{s}^{-1}$, and $k_2 = 4 \times 10^2 \, \text{s}^{-1}$. Calculate K_S and K_M for this reaction. Does substrate binding achieve equilibrium or the steady state?

*4. The following table indicates the rates at which a substrate reacts as catalyzed by an enzyme that follows the Michaelis–Menten mechanism: (1) in the absence of inhibitor; (2) and (3) in the presence of 10 mM concentration, respectively, of either of two inhibitors. Assume $[E]_T$ is the same for all reactions.

$[S]$ (mM)	$(1)v_o$ ($\mu M \cdot s^{-1}$)	$(2)v_o$ ($\mu M \cdot s^{-1}$)	$(3)v_o$ ($\mu M \cdot s^{-1}$)
1	2.5	1.17	0.77
2	4.0	2.10	1.25
5	6.3	4.00	2.00
10	7.6	5.7	2.50
20	9.0	7.2	2.86

(a) Determine K_M and V_{max} for the enzyme. For each inhibitor determine the type of inhibition and K_I and/or K_I'. What additional information would be required to calculate the turnover number of the enzyme? (b) For $[S] = 5$ mM, what fraction of the enzyme molecules have a bound substrate in the absence of inhibitor, in the presence of 10 mM inhibitor of type (2), and in the presence of 10 mM inhibitor of type (3)?

*5. Ethanol in the body is oxidized to acetaldehyde (CH_3CHO) by liver alcohol dehydrogenase (LADH). Other alcohols are also oxidized by LADH. For example, methanol, which is mildly intoxicating, is oxidized by LADH to the quite toxic product formaldehyde (CH_2O). The toxic effects of ingesting methanol (a component of many commercial solvents) can be reduced by administering ethanol. The ethanol acts as a competitive inhibitor of the methanol by displacing it from LADH. This provides sufficient time for the methanol to be harmlessly excreted by the kidneys. If an individual has ingested 100 mL of methanol (a lethal dose), how much 100 proof whiskey (50% ethanol by volume) must he imbibe to reduce the activity of his LADH toward methanol to 5% of its original value? The adult human body contains ~40 L of aqueous fluids throughout which ingested alcohols are rapidly and uniformly mixed. The densities of ethanol and methanol are both 0.79 g $\cdot$ cm^{-3}. Assume the K_M values of LADH for ethanol and methanol to be 1.0×10^{-3} M and 1.0×10^{-2} M, respectively, and that $K_I = K_M$ for ethanol.

6. The K_M of a Michaelis–Menten enzyme for a substrate is 1.0×10^{-4} M. At a substrate concentration of 0.2M, $v_o = 43$ $\mu M \cdot min^{-1}$ for a certain enzyme concentration. However, with a substrate concentration of 0.02M, v_o has the same value. (a) Using numerical calculations, show that this observation is accurate. (b) What is the best range of $[S]$ for measuring K_M?

7. Why are uncompetitive and mixed inhibitors generally considered to be more effective *in vivo* than competitive inhibitors?

8. Explain why an exact fit to a kinetic model of the experimental parameters describing a reaction does not prove that the reaction follows the model.

9. An enzyme that follows the model for pH effects presented in Section 14-4 has $pK_{ES1} = 4$ and $pK_{ES2} = 8$. What is the pH at which V'_{max} is a maximum for this enzyme? What fraction of V_{max} does V'_{max} achieve at this pH?

10. Derive the initial rate equation for a Rapid Equilibrium Random Bi Bi reaction. Assume the equilibrium constants K_S^A and K_S^B for binding A and B to the enzyme are independent of whether the other substrate is bound (an assumption that constrains $K_M^B = K_S^B$ in Eq. [14.49]).

*11. Consider the following variation of a Ping Pong Bi Bi mechanism.

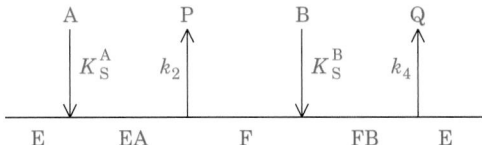

Assume that the substrate-binding reactions are in rapid equilibrium,

$$K_S^A = \frac{[E][A]}{[EA]} \quad \text{and} \quad K_S^B = \frac{[F][B]}{[FB]}$$

that both $[A]$ and $[B] >> [E]_T$, that neither product release reaction is reversible, and that the steady-state approximation is valid. (a) Derive an expression for v_o in terms of K_S^A, K_S^B, k_2, and k_4. (b) Indicate the form of the double-reciprocal plots for $1/v_o$ versus $1/[A]$ for various values of $[B]$. (c) Indicate the form of the double-reciprocal plots for $1/v_o$ versus $1/[B]$ for various values of $[A]$.

12. Creatine kinase catalyzes the reaction

$$MgADP^- + \text{phosphocreatine} \rightleftharpoons MgATP^{2-} + \text{creatine}$$

which functions to regenerate ATP in muscle. Rabbit muscle creatine kinase exhibits the following kinetic behavior. In the absence of both products, plots of $1/v_o$ versus $1/[MgADP^-]$ at different fixed concentrations of phosphocreatine yield lines that intersect to the left of the $1/v_o$ axis. Similarly, plots of $1/v_o$ versus $1/[\text{phosphocreatine}]$ in the absence of product at different fixed concentrations of MgADP$^-$ yield lines that intersect to the left of the $1/v_o$ axis. In the absence of one of the reaction products, MgATP^{2-} or creatine, plots of $1/v_o$ versus $1/[MgADP^-]$ at different concentrations of the other product intersect on the $1/v_o$ axis. The same is true of the plots of $1/v_o$ versus $1/[\text{phosphocreatine}]$. Indicate a kinetic mechanism that is consistent with this information.

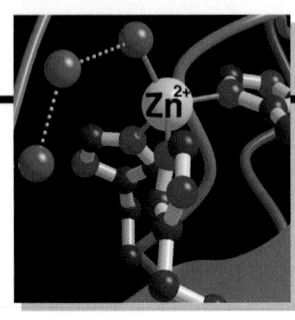

Chapter 15

Enzymatic Catalysis

1 ■ Catalytic Mechanisms
   A. Acid–Base Catalysis
   B. Covalent Catalysis
   C. Metal Ion Catalysis
   D. Electrostatic Catalysis
   E. Catalysis through Proximity and Orientation Effects
   F. Catalysis by Preferential Transition State Binding
2 ■ Lysozyme
   A. Enzyme Structure
   B. Catalytic Mechanism
   C. Testing the Phillips Mechanism
3 ■ Serine Proteases
   A. Kinetics and Catalytic Groups
   B. X-Ray Structures
   C. Catalytic Mechanism
   D. Testing the Catalytic Mechanism
   E. Zymogens
4 ■ Drug Design
   A. Techniques of Drug Discovery
   B. Introduction to Pharmacology
   C. HIV Protease and Its Inhibitors

Enzymes, as we have seen, cause rate enhancements that are orders of magnitude greater than those of the best chemical catalysts. Yet they operate under mild conditions and are highly specific as to the identities of both their substrates and their products. These catalytic properties are so remarkable that many nineteenth century scientists concluded that enzymes have characteristics that are not shared by substances of nonliving origin. To this day, there are few enzymes for which we understand in more than cursory detail how they achieve their enormous rate accelerations. Nevertheless, it is now abundantly clear that the catalytic mechanisms employed by enzymes are identical to those used by chemical catalysts. Enzymes are simply better designed.

In this chapter we consider the nature of enzymatic catalysis. We begin by discussing the underlying principles of chemical catalysis as elucidated through the study of organic reaction mechanisms. We then embark on a detailed examination of the catalytic mechanisms of several of the best characterized enzymes: **lysozyme** and the **serine proteases.** Their study should lead to an appreciation of the intracacies of these remarkably efficient catalysts as well as of the experimental methods used to elucidate their

properties. We end with a discussion of how drugs are discovered and tested, a process that depends heavily on the principles of enzymology since many drug targets are enzymes. In doing so, we consider how therapeutically effective inhibitors of **HIV-1 protease** were discovered.

1 ■ CATALYTIC MECHANISMS

Catalysis is a process that increases the rate at which a reaction approaches equilibrium. Since, as we discussed in Section 14-1C, the rate of a reaction is a function of its free energy of activation ($\Delta G^{\ddagger}$), a catalyst acts by lowering the height of this kinetic barrier; that is, a catalyst stabilizes the transition state with respect to the uncatalyzed reaction. There is, in most cases, nothing unique about enzymatic mechanisms of catalysis in comparison to nonenzymatic mechanisms. *What apparently make enzymes such powerful catalysts are two related properties: their specificity of substrate binding combined with their optimal arrangement of catalytic groups.* An enzyme's arrangement of binding and catalytic groups is, of course, the product of eons of evolution: Nature has had ample opportunity to fine-tune the performances of most enzymes.

The types of catalytic mechanisms that enzymes employ have been classified as:

   1. Acid–base catalysis.
   2. Covalent catalysis.
   3. Metal ion catalysis.
   4. Electrostatic catalysis.
   5. Proximity and orientation effects.
   6. Preferential binding of the transition state complex.

In this section, we examine these various phenomena. In doing so we shall frequently refer to the organic model compounds that have been used to characterize these catalytic mechanisms.

A. Acid–Base Catalysis

General acid catalysis is a process in which partial proton transfer from a Brønsted acid (a species that can donate protons; Section 2-2A) lowers the free energy of a reaction's transition state. For example, an uncatalyzed keto–enol tautomerization reaction occurs quite slowly as a result of

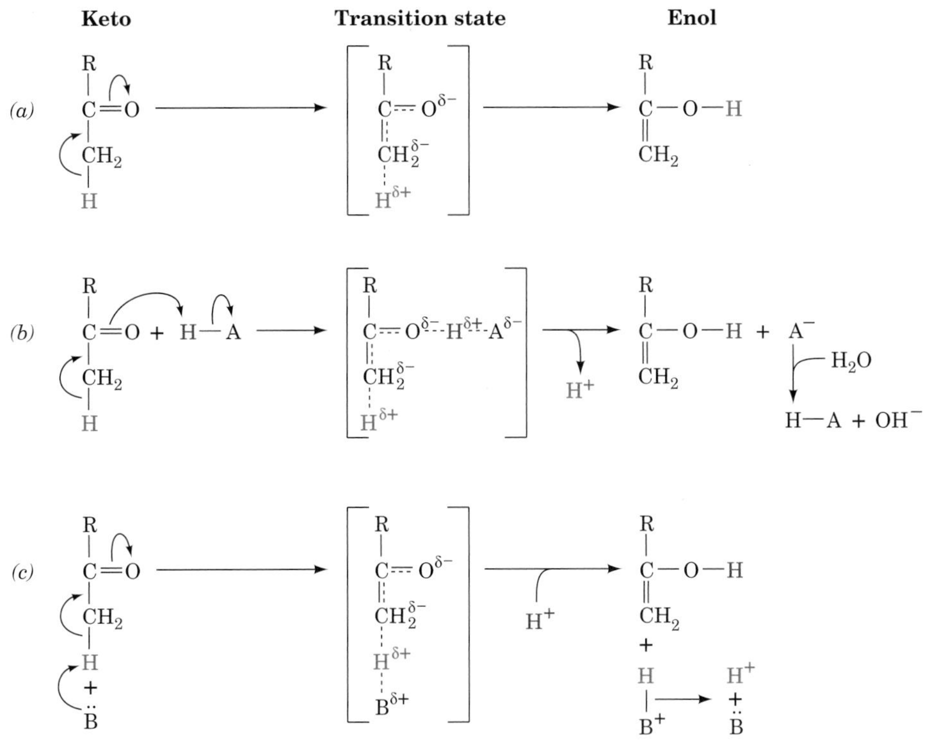

FIGURE 15-1 **Mechanisms of keto–enol tautomerization.** (*a*) Uncatalyzed, (*b*) general acid catalyzed, and (*c*) general base catalyzed.

the high energy of its carbanionlike transition state (Fig. 15-1*a*). Proton donation to the oxygen atom (Fig. 15-1*b*), however, reduces the carbanion character of the transition state, thereby catalyzing the reaction. *A reaction may also be stimulated by* **general base catalysis** *if its rate is increased by partial proton abstraction by a Brønsted base (a species that can combine with a proton; Fig. 15-1c). Some reactions may be simultaneously subject to both processes: a* **concerted general acid–base catalyzed reaction.**

a. Mutarotation Is Catalyzed by Acids and by Bases

The mutarotation of glucose provides an instructive example of acid–base catalysis. Recall that a glucose molecule can assume either of two anomeric cyclic forms through the intermediacy of its linear form (Section 11-1B):

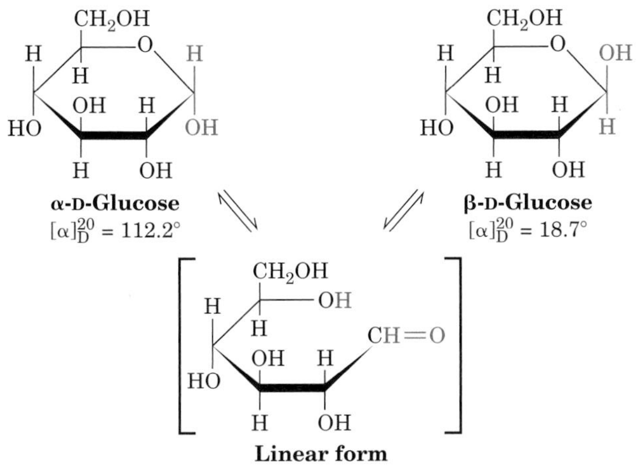

In aqueous solvents, the initial rate of mutarotation of α-D-glucose, as monitored by polarimetry (Section 4-2A), is observed to follow the relationship:

$$v = -\frac{d[\alpha\text{-D-glucose}]}{dt} = k_{obs}[\alpha\text{-D-glucose}] \quad [15.1]$$

where k_{obs} is the reaction's apparent first-order rate constant. The mutarotation rate increases with the concentrations of general acids and general bases; they are thought to catalyze mutarotation according to the mechanism:

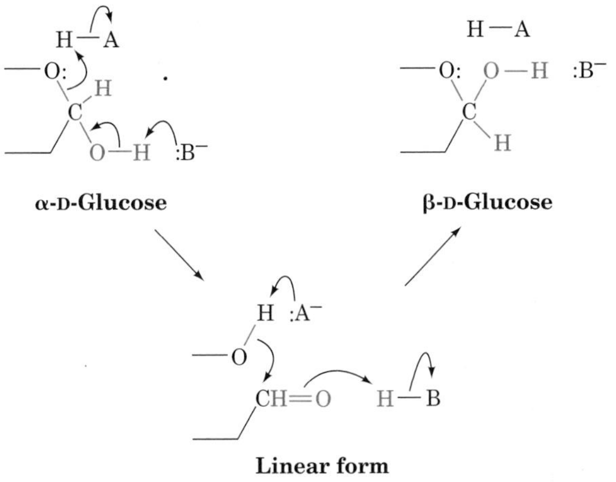

This model is consistent with the observation that in aprotic solvents such as benzene, **2,3,4,6-O-tetramethyl-α-D-glucose** (a less polar benzene-soluble analog)

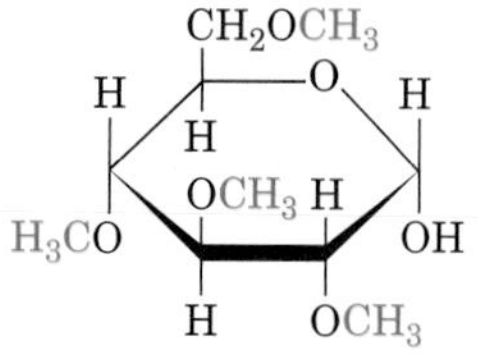

2,3,4,6-O-Tetramethyl-α-D-glucose

does not undergo mutarotation. Yet, the reaction is catalyzed by the addition of phenol, a weak benzene-soluble acid, together with pyridine, a weak benzene-soluble base, according to the rate equation:

$$v = k[\text{phenol}][\text{pyridine}][\text{tetramethyl-α-D-glucose}] \quad [15.2]$$

Moreover, in the presence of **α-pyridone,** whose acid and base groups can rapidly interconvert between two tautomeric forms and are situated so that they can simultaneously catalyze mutarotation,

α-Pyridone

Glucose

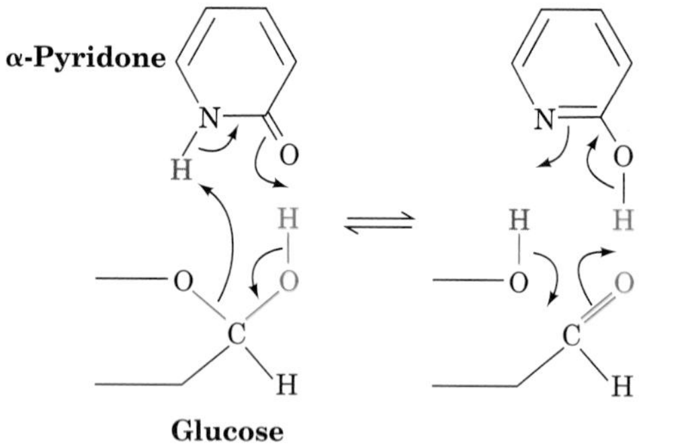

the reaction follows the rate law

$$v = k'[\text{α-pyridone}][\text{tetramethyl-α-D-glucose}] \quad [15.3]$$

where $k' = 7000M \times k$. This increased rate constant indicates that α-pyridone does, in fact, catalyze mutarotation in a concerted fashion since $1M$ α-pyridone has the same catalytic effect as impossibly high concentrations of phenol and pyridine (e.g., $70M$ phenol and $100M$ pyridine).

Many types of biochemically significant reactions are susceptible to acid and/or base catalysis. These include the hydrolysis of peptides and esters, the reactions of phosphate groups, tautomerizations, and additions to carbonyl groups. The side chains of the amino acid residues Asp, Glu, His, Cys, Tyr, and Lys have pK's in or near the physiological pH range (Table 4-1) which, we shall see, permits them to act in the enzymatic capacity of general acid and/or base catalysts in analogy with known organic mechanisms. Indeed, the ability of enzymes to arrange several catalytic groups about their substrates makes concerted acid–base catalysis a common enzymatic mechanism.

b. The RNase A Reaction Incorporates General Acid–Base Catalysis

Bovine pancreatic ribonuclease A (RNase A) provides an illuminating example of enzymatically mediated general acid–base catalysis. This digestive enzyme functions to hy-

drolyze RNA to its component nucleotides. The isolation of **2′,3′-cyclic nucleotides** from RNase A digests of RNA indicates that the enzyme mediates the following reaction sequence:

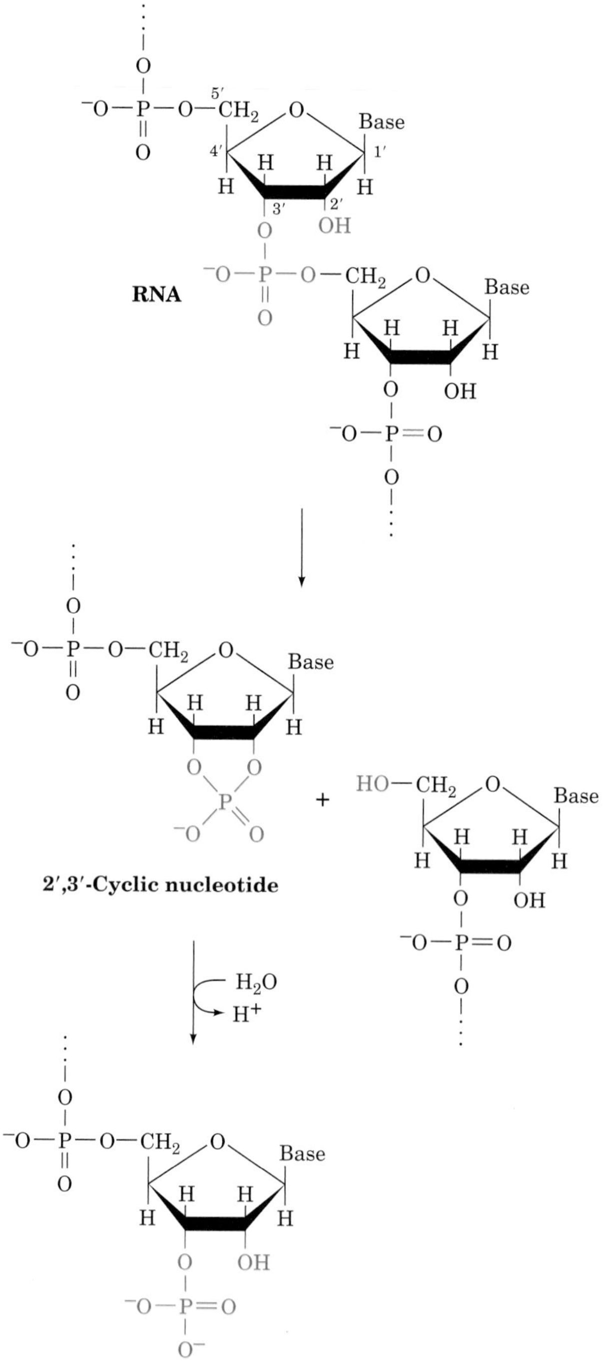

RNA

2′,3′-Cyclic nucleotide

The RNase A reaction exhibits a pH rate profile that peaks near pH 6 (Fig. 15-2). Analysis of this curve (Section 14-4), together with chemical derivatization and X-ray studies, indicates that RNase A has two essential His residues, His 12 and His 119, which act in a concerted manner as general acid and base catalysts (the structure of RNase A is sketched in Fig. 9-2). Evidently, the RNase A reaction is a two-step process (Fig. 15-3):

1. His 12, acting as a general base, abstracts a proton from an RNA 2′-OH group, thereby promoting its nucleo-

FIGURE 15-2   The pH dependence of V'_{max}/K'_M in the RNase A–catalyzed hydrolysis of cytidine-2′,3′-cyclic phosphate. V'_{max}/K'_M is given in units of $M^{-1} \cdot s^{-1}$. Analysis of this curve (Section 14-4) suggests the catalytic participation of groups with pK's of 5.4 and 6.4. [After del Rosario, E.J. and Hammes, G.G., *Biochemistry* **8,** 1887 (1969).]

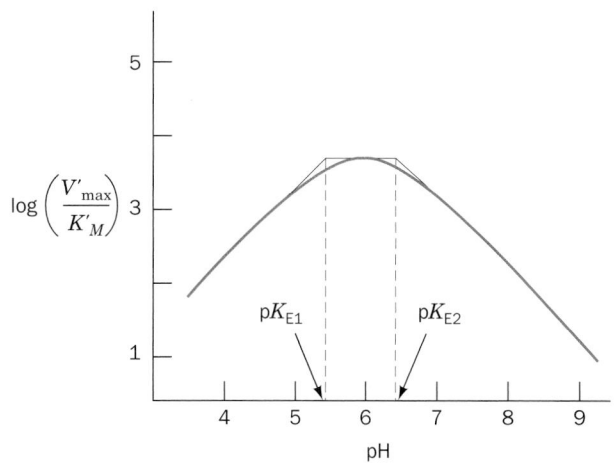

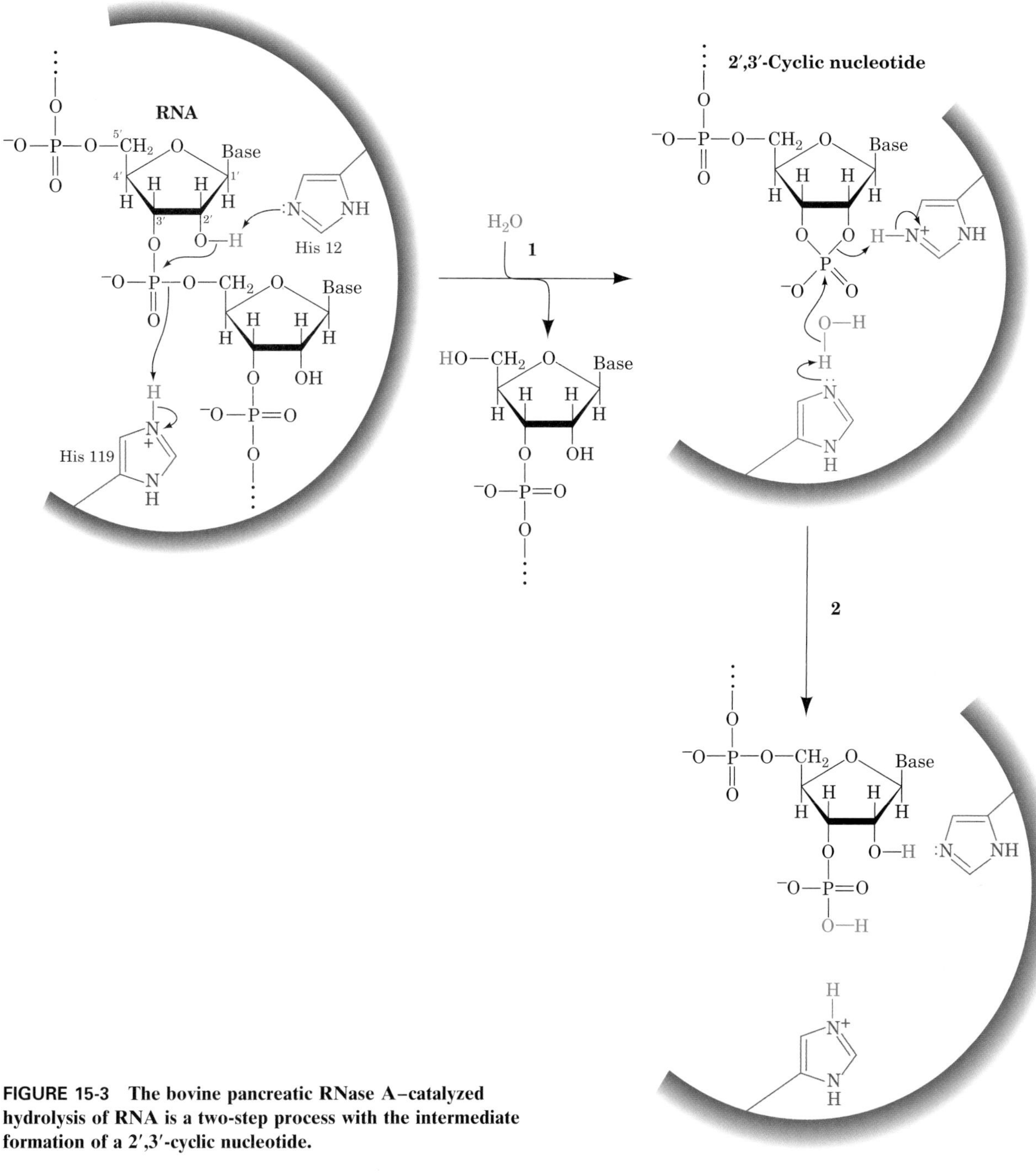

FIGURE 15-3   The bovine pancreatic RNase A–catalyzed hydrolysis of RNA is a two-step process with the intermediate formation of a 2′,3′-cyclic nucleotide.

philic attack on the adjacent phosphorus atom while His 119, acting as a general acid, promotes bond scission by protonating the leaving group.

2. The 2′,3′-cyclic intermediate is hydrolyzed through what is essentially the reverse of the first step in which water replaces the leaving group. Thus His 12 acts as a general acid and His 119 as a general base to yield the hydrolyzed RNA and the enzyme in its original state.

B. *Covalent Catalysis*

Covalent catalysis *involves rate acceleration through the transient formation of a catalyst–substrate covalent bond.* The decarboxylation of **acetoacetate,** as chemically catalyzed by primary amines, is an example of such a process (Fig. 15-4). In the first stage of this reaction, the amine nucleophilically attacks the carbonyl group of acetoacetate to form a **Schiff base** (imine bond).

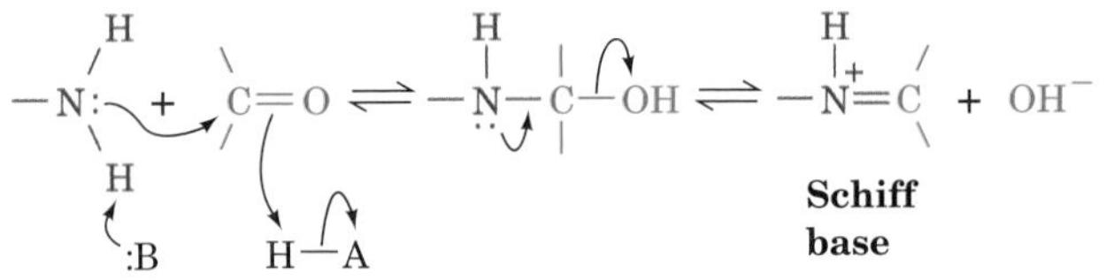

The protonated nitrogen atom of the covalent intermediate then acts as an electron sink (Fig. 15-4, bottom) so as to reduce the otherwise high-energy enolate character of the transition state. The formation and decomposition of the Schiff base occur quite rapidly, so that these steps are not rate determining in this reaction sequence.

a. Covalent Catalysis Has Both Nucelophilic and Electrophilic Stages

As the preceding example indicates, covalent catalysis may be conceptually decomposed into three stages:

1. The nucleophilic reaction between the catalyst and the substrate to form a covalent bond.

2. The withdrawal of electrons from the reaction center by the now electrophilic catalyst.

3. The elimination of the catalyst, a reaction that is essentially the reverse of stage 1.

Reaction mechanisms are somewhat arbitrarily classified as occurring with either **nucleophilic catalysis** or **electrophilic catalysis** depending on which of these effects provides the greater driving force for the reaction, that is, which catalyzes its rate-determining step. The primary amine–catalyzed decarboxylation of acetoacetate is clearly an electrophilically catalyzed reaction since its nucleophilic phase, Schiff base formation, is not its rate-determining step. In other covalently catalyzed reactions, however, the nucleophilic phase may be rate determining.

The nucleophilicity of a substance is closely related to its basicity. Indeed, the mechanism of nucleophilic catalysis resembles that of general base catalysis except that, instead of abstracting a proton from the substrate, the catalyst nucleophilically attacks it so as to form a covalent bond. Consequently, if covalent bond formation is the rate-determining step of a covalently catalyzed reaction, the reaction rate tends to increase with the covalent catalyst's basicity (pK).

An important aspect of covalent catalysis is that the more stable the covalent bond formed, the less facilely it will decompose in the final steps of a reaction. A good covalent catalyst must therefore combine the seemingly contradictory properties of high nucleophilicity and the ability to form a good leaving group, that is, to easily reverse the bond formation step. Groups with high polarizabilities (highly mobile electrons), such as imidazole and thiol functions, have these properties and hence make good covalent catalysts.

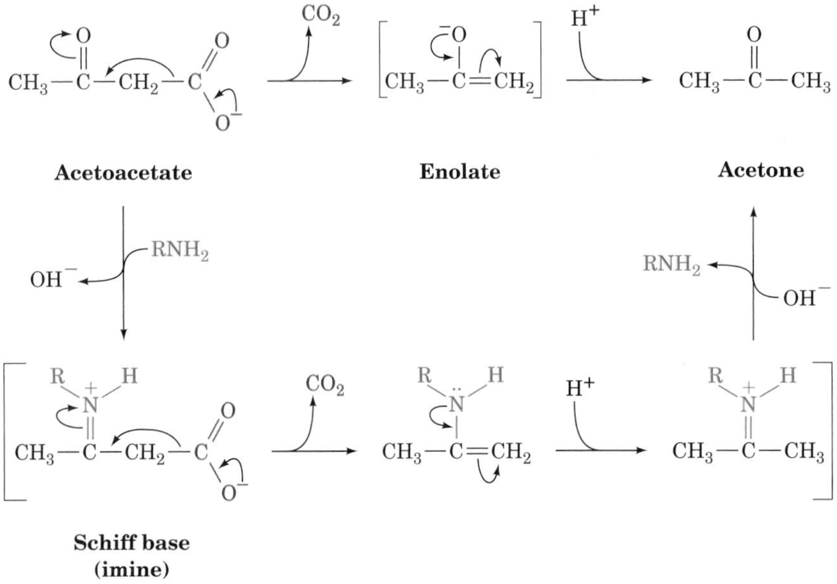

FIGURE 15-4 The decarboxylation of acetoacetate. The uncatalyzed reaction mechanism is shown at the top and the reaction mechanism as catalyzed by primary amines is shown at the bottom.

b. Certain Amino Acid Side Chains and Coenzymes Can Serve as Covalent Catalysts

Enzymes commonly employ covalent catalytic mechanisms as is indicated by the large variety of covalently linked enzyme–substrate reaction intermediates that have been isolated. For example, the enzymatic decarboxylation of acetoacetate proceeds, much as described above, through Schiff base formation with an enzyme Lys residue's ε-amino group. The covalent intermediate, in this case, has been isolated through $NaBH_4$ reduction of its imine bond to an amine, thereby irreversibly inhibiting the enzyme. Other enzyme functional groups that participate in covalent catalysis include the imidazole moiety of His, the thiol group of Cys, the carboxyl function of Asp, and the hydroxyl group of Ser. In addition, several coenzymes, most notably **thiamine pyrophosphate** (Section 17-3B) and **pyridoxal phosphate** (Section 26-1A), function in association with their apoenzymes mainly as covalent catalysts.

C. *Metal Ion Catalysis*

Nearly one-third of all known enzymes require the presence of metal ions for catalytic activity. There are two classes of metal ion–requiring enzymes that are distinguished by the strengths of their ion–protein interactions:

1. *Metalloenzymes* *contain tightly bound metal ions,* most commonly transition metal ions such as Fe^{2+}, Fe^{3+}, Cu^{2+}, Zn^{2+}, Mn^{2+}, or Co^{3+}.

2. *Metal-activated enzymes* *loosely bind metal ions from solution,* usually the alkali and alkaline earth metal ions Na^+, K^+, Mg^{2+}, or Ca^{2+}.

Metal ions participate in the catalytic process in three major ways:

1. By binding to substrates so as to orient them properly for reaction.

2. By mediating oxidation–reduction reactions through reversible changes in the metal ion's oxidation state.

3. By electrostatically stabilizing or shielding negative charges.

In this section we shall be mainly concerned with the third aspect of metal ion catalysis. The other forms of enzyme-mediated metal ion catalysis are considered in later chapters in conjunction with discussions of specific enzyme mechanisms.

a. Metal Ions Promote Catalysis through Charge Stabilization

In many metal ion–catalyzed reactions, the metal ion acts in much the same way as a proton to neutralize negative charge, that is, it acts as a Lewis acid. Yet *metal ions are often much more effective catalysts than protons because metal ions can be present in high concentrations at neutral pH's and can have charges greater than +1.* Metal ions have therefore been dubbed "superacids."

The decarboxylation of **dimethyloxaloacetate,** as catalyzed by metal ions such as Cu^{2+} and Ni^{2+}, is a nonenzymatic example of catalysis by a metal ion:

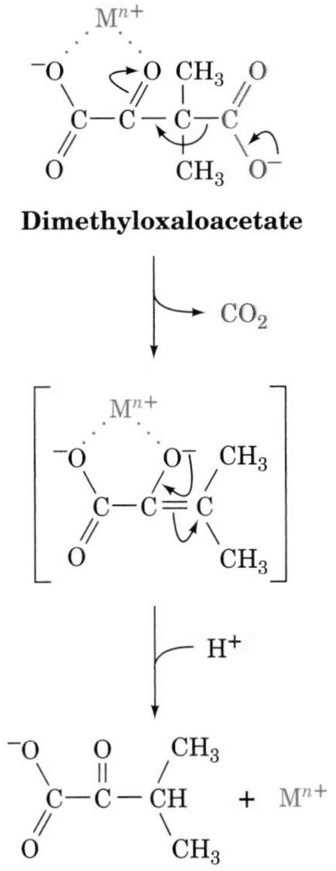

Dimethyloxaloacetate

Here the metal ion (M^{n+}), which is chelated by the dimethyloxaloacetate, electrostatically stabilizes the developing enolate ion of the transition state. This mechanism is supported by the observation that acetoacetate, which cannot form such a chelate, is not subject to metal ion–catalyzed decarboxylation. Most enzymes that decarboxylate oxaloacetate require a metal ion for activity.

b. Metal Ions Promote Nucleophilic Catalysis via Water Ionization

A metal ion's charge makes its bound water molecules more acidic than free H_2O and therefore a source of OH^- ions even below neutral pH's. For example, the water molecule of $(NH_3)_5Co^{3+}(H_2O)$ ionizes according to the reaction:

$$(NH_3)_5Co^{3+}(H_2O) \rightleftharpoons (NH_3)_5Co^{3+}(OH^-) + H^+$$

with a pK of 6.6, which is ~9 pH units below the pK of free H_2O. *The resulting metal ion–bound hydroxyl group is a potent nucleophile.*

An instructive example of this phenomenon occurs in the catalytic mechanism of **carbonic anhydrase** (Section 10-1C), a widely occurring enzyme that catalyzes the reaction:

$$CO_2 + H_2O \rightleftharpoons HCO_3^- + H^+$$

Carbonic anhydrase contains an essential Zn^{2+} ion that lies at the bottom of an ~15-Å-deep active site cleft (Fig. 8-41), where it is tetrahedrally coordinated by three evolutionarily invariant His side chains and an O atom of either an

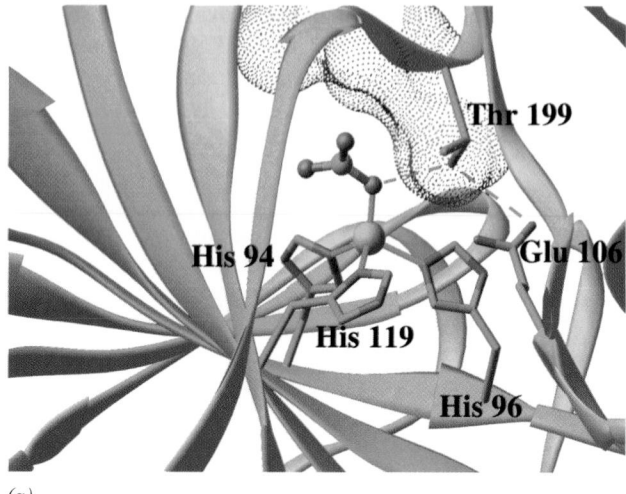

(a)

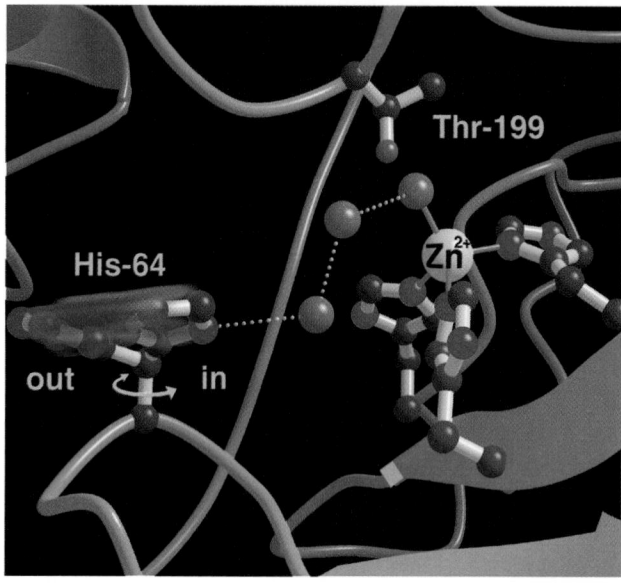

(b)

FIGURE 15-5   X-Ray structures of human carbonic anhydrase.
(*a*) Its active site in complex with bicarbonate ion. The
polypeptide is shown in ribbon form (*gold*) with its side chains
shown in stick form colored according to atom type (C green,
N blue, and O red). The protein-bound Zn^{2+} ion (*cyan sphere*)
is tetrahedally liganded (*gray bonds*) by three invariant His
side chains and the HCO_3^- ion, which is shown in ball-and-stick
form. The HCO_3^- ion also interacts with the protein via van der
Waals contacts (*dot surface colored according to atom type*) and
a hydrogen bonded network (*dashed gray lines*) involving Thr
199 and Glu 106. [Based on an X-ray structure by K. K.
Kannan, Bhabha Atomic Research Center, Bombay, India.
PDBid 1HCB.] (*b*) The active site showing the proton shuttle
through which His 64, acting as a general base, abstracts a
proton from the Zn^{2+}-bound H_2O to form an OH^- ion. The
polypeptide backbone is shown in ribbon form (*cyan*), and its
side chains and several bound solvent molecules are shown in
ball-and-stick form with C black, N blue, and O red. The
proton shuttle consists of two water molecules that form a
hydrogen bonded network (*dotted white lines*) that bridges the
Zn^{2+}-bound OH^- ion and His 64 in its "in" conformation. On
protonation, His 64 swings to the "out" conformation.
[Courtesy of David Christianson, University of Pennsylvania.]
See the Interactive Exercises

HCO_3^- ion (Fig. 15-5*a*) or a water molecule (Fig. 15-5*b*).
The enzyme has the following catalytic mechanism:

1. We begin with a water molecule bound to the protein
in the Zn^{2+} ion's fourth liganding position (Fig. 15-5*b*). This
Zn^{2+}-polarized H_2O ionizes in a process facilitated through
general base catalysis by His 64 in its "in" conformation.
Although His 64 is too far away from the Zn^{2+}-bound water to directly abstract its proton, these entities are linked
by two intervening water molecules to form a hydrogen
bonded network that is thought to act as a proton shuttle.

Im = imidazole

2. The resulting Zn^{2+}-bound OH^- ion nucleophilically
attacks the nearby enzymatically bound CO_2, thereby converting it to HCO_3^-.

Im = imidazole

In doing so, the Zn^{2+}-bound OH^- group donates a hydrogen bond to Thr 199, which in turn donates a hydrogen
bond to Glu 106 (Fig. 15-5*a*). These interactions orient the
OH^- group with the optimal geometry (see below) for
nucleophilic attack on the substrate CO_2.

3. The catalytic site is regenerated by the exchange of
the Zn^{2+}-bound HCO_3^- reaction product for H_2O together

with the deprotonation of His 64. In the latter process, His 64 swings to its "out" conformation (Fig. 15-5*b*), which may facilitate proton transfer to the bulk solvent.

c. Metal Ions Promote Reactions through Charge Shielding

Another important enzymatic function of metal ions is **charge shielding.** For example, the actual substrates of **kinases** (phosphoryl-transfer enzymes utilizing ATP) are Mg^{2+}–ATP complexes such as

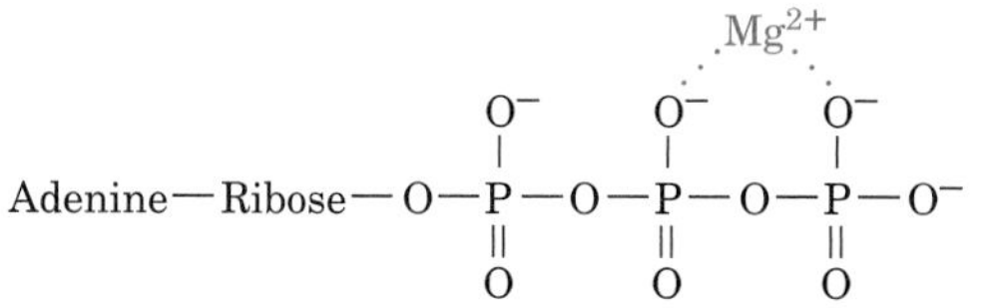

$$Adenine{-}Ribose{-}O{-}\overset{O^-}{\underset{O}{P}}{-}O{-}\overset{O^-}{\underset{O}{P}}{-}O{-}\overset{O^-}{\underset{O}{P}}{-}O^-$$

rather than just ATP. Here, the Mg^{2+} ion's role, in addition to its orienting effect, is to shield electrostatically the negative charges of the phosphate groups. Otherwise, these charges would tend to repel the electron pairs of attacking nucleophiles, especially those with anionic character.

D. *Electrostatic Catalysis*

The binding of substrate generally excludes water from an enzyme's active site. The local dielectric constant of the active site therefore resembles that in an organic solvent, where electrostatic interactions are much stronger than they are in aqueous solutions (Section 8-4A). The charge distribution in a medium of low dielectric constant can greatly influence chemical reactivity. Thus, as we have seen, the pK's of amino acid side chains in proteins may vary by several units from their nominal values (Table 4-1) because of the proximity of charged groups.

Although experimental evidence and theoretical analyses on the subject are still sparse, *there are mounting indications that the charge distributions about the active sites of enzymes are arranged so as to stabilize the transition states of the catalyzed reactions.* Such a mode of rate enhancement, which resembles the form of metal ion catalysis discussed above, is termed **electrostatic catalysis.** Moreover, in several enzymes, *these charge distributions apparently serve to guide polar substrates toward their binding sites so that the rates of these enzymatic reactions are greater than their apparent diffusion-controlled limits (Section 14-2B).*

E. *Catalysis through Proximity and Orientation Effects*

Although enzymes employ catalytic mechanisms that resemble those of organic model reactions, they are far more catalytically efficient than these models. Such efficiency must arise from the specific physical conditions at enzyme catalytic sites that promote the corresponding chemical reactions. The most obvious effects are **proximity** and **orientation:** *Reactants must come together with the proper spatial relationship for a reaction to occur.* For example, in the bimolecular reaction of imidazole with *p*-nitrophenylacetate,

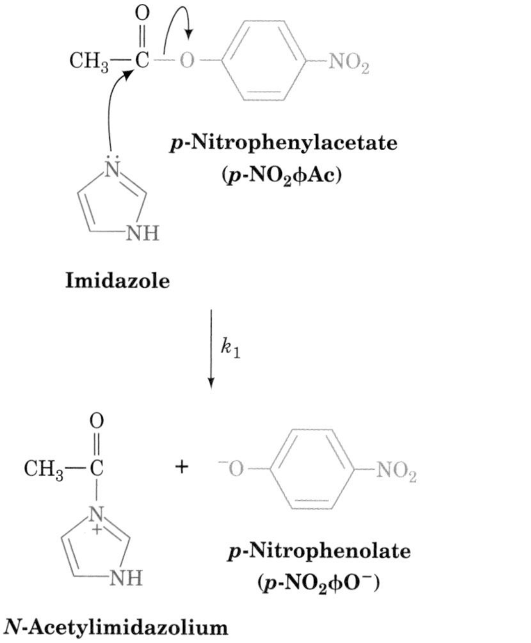

p-**Nitrophenylacetate** (*p*-$NO_2\phi Ac$)

Imidazole

k_1

N-**Acetylimidazolium**

p-**Nitrophenolate** (*p*-$NO_2\phi O^-$)

the progress of the reaction is conveniently monitored by the appearance of the intensely yellow *p*-**nitrophenolate ion:**

$$\frac{d[p\text{-}NO_2\phi O^-]}{dt} = k_1[\text{imidazole}][p\text{-}NO_2\phi Ac] \quad [15.4]$$
$$= k_1'[p\text{-}NO_2\phi Ac]$$

where ϕ = phenyl. Here k_1', the pseudo-first-order rate constant, is 0.0018 s^{-1} when [imidazole] = $1M$. However, for the intramolecular reaction

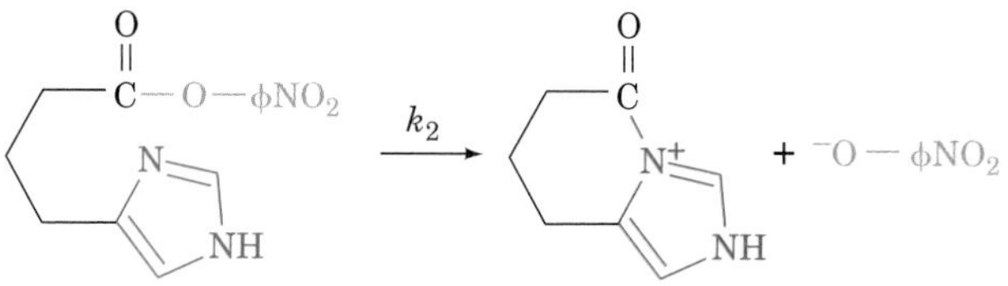

the first-order rate constant $k_2 = 0.043 \text{ s}^{-1}$; that is, $k_2 = 24k_1'$. Thus, when the $1M$ imidazole catalyst is covalently attached to the reactant, it is 24-fold more effective than when it is free in solution; that is, *the imidazole group in the intramolecular reaction behaves as if its concentration is 24M.* This rate enhancement has contributions from both proximity and orientation.

a. Proximity Alone Contributes Relatively Little to Catalysis

Let us make a rough calculation as to how the rate of a reaction is affected purely by the proximity of its reacting groups. Following Daniel Koshland's treatment, we shall make several reasonable assumptions:

1. Reactant species, that is, functional groups, are about the size of water molecules.

2. Each reactant species in solution has 12 nearest-neighbor molecules, as do packed spheres of identical size.

3. Chemical reactions occur only between reactants that are in contact.

4. The reactant concentration in solution is low enough so that the probability of any reactant species being in simultaneous contact with more than one other reactant molecule is negligible.

Then the reaction:

$$A + B \xrightarrow{k_1} A\text{—}B$$

obeys the second-order rate equation

$$v = \frac{d[A\text{—}B]}{dt} = k_1[A][B] = k_2[A, B]_{pairs} \quad [15.5]$$

where $[A,B]_{pairs}$ is the concentration of contacting molecules of A and B. The value of this quantity is

$$[A, B]_{pairs} = \frac{12[A][B]}{55.5M} \quad [15.6]$$

since there are 12 ways that A can be in contact with B, and $[A]/55.5M$ is the fraction of sites occupied by A in water solution ($[H_2O] = 55.5M$ in dilute aqueous solutions) and hence the probability that a molecule of B will be next to one of A. Combining Eqs. [15.5] and [15.6] yields

$$v = k_1\left(\frac{55.5}{12}\right)[A, B]_{pairs} = 4.6k_1[A, B]_{pairs} \quad [15.7]$$

Thus, in the absence of other effects, this model predicts that for the intramolecular reaction,

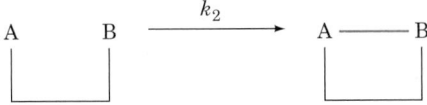

$k_2 = 4.6k_1$, which is a rather small rate enhancement. Factors that will increase this value other than proximity alone clearly must be considered.

b. Properly Orienting Reactants and Arresting Their Relative Motions Can Result in Large Catalytic Rate Enhancements

The foregoing theory is, of course, quite simple. For example, it does not take into account the relative orientations of the reacting molecules. Yet molecules are not equally reactive in all directions as Koshland's simple theory assumes. Rather, *they react most readily only if they have the proper relative orientation.* For example, in an S_N2 (bimolecular nucleophilic substitution) reaction, the incoming nucleophile optimally attacks its target C atom along the direction opposite to that of the bond to the leaving group (Fig. 15-6). The approaches of reacting atoms along a trajectory that deviates by as little as 10° from this optimum direction can reduce the reaction rate by as much as a factor of ~100. In a related phenomenon, a molecule may be maximally reactive only when it assumes a conformation that aligns its various orbitals in a way that minimizes the electronic energy of its transition state, an effect termed **stereoelectronic assistance.**

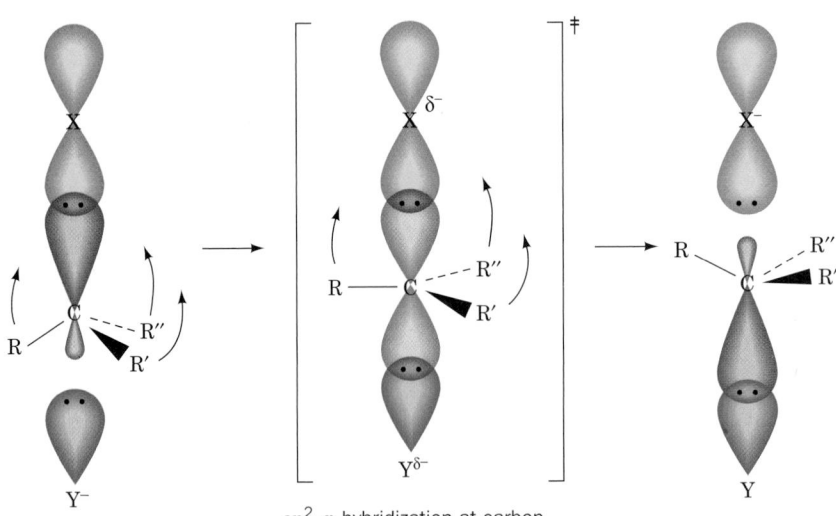

sp²–p hybridization at carbon

FIGURE 15-6   The geometry of an S_N2 reaction. The attacking nucleophile, Y^-, must approach the tetrahedrally coordinated and hence sp^3-hybridized C atom along the direction opposite that of its bond to the leaving group, X, a process called **backside attack.** In the transition state of the reaction, the C atom becomes trigonal bipyramidally coordinated and hence sp^2–p hybridized, with the p orbital (*blue*) forming partial bonds to X and Y. The three sp^2 orbitals form bonds to the C atom's three other substituents (R, R′, and R″), which have shifted their positions into the plane perpendicular to the X—C—Y axis (*curved arrows*). Any deviation from this optimal geometry would increase the free energy of the transition state, $\Delta G^\ddagger$, and hence reduce the rate of the reaction (Eq. [14.15]). The transition state then decomposes to products in which the R, R′, and R″ have inverted their positions about the C atom, which has rehybridized to sp^3, and X^- has been released.

Another effect that we have neglected in our treatment of proximity is that of motions of the reacting groups with respect to one another. Yet, in the transition state complex, the reacting groups have little relative motion. In fact, as Thomas Bruice demonstrated, the rates of intramolecular reactions are greatly increased by arresting a molecule's internal motions in a way that increases the mole fraction of the reacting groups that are in a conformation which can enter the transition state (Table 15-1). Similarly, when an enzyme brings two molecules together in a bimolecular reaction, as William Jencks pointed out, not only does it increase their proximity, but it freezes out their relative translational and rotational motions (decreases their entropy), thereby enhancing their reactivity. Theoretical studies by Bruice indicate that much of this rate enhancement can arise from the enzymatic binding of substrates in a conformation that readily enters the transition state.

Enzymes, as we shall see in Sections 15-2 and 15-3, bind substrates in a manner that both aligns and immobilizes them so as to optimize their reactivities. The free energy required to do so is derived from the specific binding free energy of substrate to enzyme.

F. Catalysis by Preferential Transition State Binding

The rate enhancements effected by enzymes are often greater than can be reasonably accounted for by the catalytic mechanisms so far discussed. However, we have not yet considered one of the most important mechanisms of enzymatic catalysis: *the binding of the transition state to an enzyme with greater affinity than the corresponding substrates or products.* When taken together with the previously described catalytic mechanisms, preferential transition state binding rationalizes the observed rates of enzymatic reactions.

The original concept of transition state binding proposed that enzymes mechanically strained their substrates toward the transition state geometry through binding sites into which undistorted substrates did not properly fit. This so-called **rack mechanism** (in analogy with the medieval torture device) was based on the extensive evidence for the role of strain in promoting organic reactions. For example, the rate of the reaction,

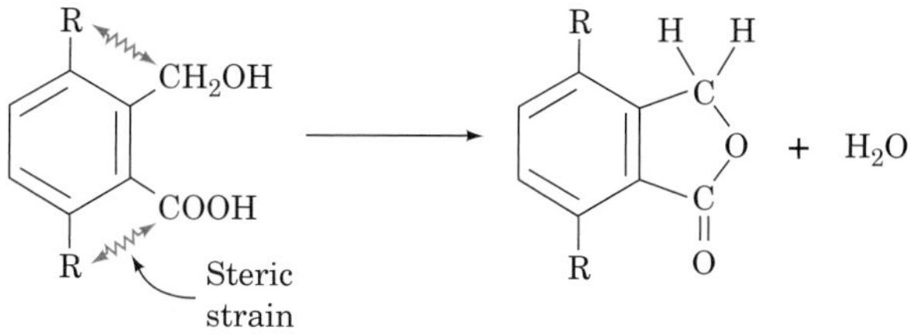

is 315 times faster when R is CH_3 rather than when it is H because of the greater steric repulsions between the CH_3 groups and the reacting groups. Similarly, ring opening reactions are considerably more facile for strained rings such

TABLE 15-1 **Relative Rates of Anhydride Formation for Esters Possessing Different Degrees of Motional Freedom in the Reaction:**

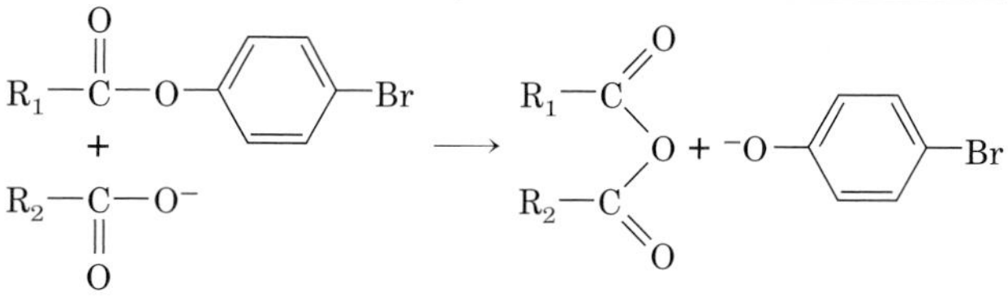

Reactants[a]	Relative Rate Constant
$CH_3COO\phi Br$ + CH_3COO^-	1.0
(see structure)	$\sim 1 \times 10^3$
(see structure)	$\sim 2.3 \times 10^5$
(see structure)	$\sim 8 \times 10^7$

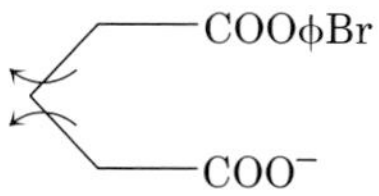

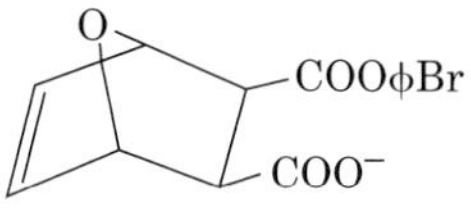

[a]Curved arrows indicate rotational degrees of freedom.

Source: Bruice, T.C. and Lightstone, F.C., *Acc. Chem. Res.* **32**, 127 (1999).

as cyclopropane than for unstrained rings such as cyclohexane. In either process, *the strained reactant more closely resembles the transition state of the reaction than does the corresponding unstrained reactant.* Thus, as was first suggested by Linus Pauling and further amplified by Richard Wolfenden and Gustav Lienhard, *interactions that preferentially bind the transition state increase its concentration and therefore proportionally increase the reaction rate.*

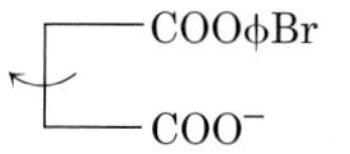

Let us quantitate this statement by considering the kinetic consequences of preferentially binding the transition state of an enzymatically catalyzed reaction involving a single substrate. The substrate S may react to form product P either spontaneously or through enzymatic catalysis:

$$S \xrightarrow{k_N} P$$
$$ES \xrightarrow{k_E} EP$$

Here k_E and k_N are the first-order rate constants for the catalyzed and uncatalyzed reactions, respectively. The relationships between the various states of these two reaction pathways are indicated in the following scheme:

$$E + S \underset{}{\overset{K_N^\ddagger}{\rightleftharpoons}} S^\ddagger + E \longrightarrow P + E$$

$$\big\Updownarrow K_R \qquad \big\Updownarrow K_T \qquad \big\Updownarrow$$

$$ES \underset{}{\overset{K_E^\ddagger}{\rightleftharpoons}} ES^\ddagger \longrightarrow EP$$

where

$$K_R = \frac{[ES]}{[E][S]} \qquad\qquad K_T = \frac{[ES^\ddagger]}{[E][S^\ddagger]}$$

$$K_N^\ddagger = \frac{[E][S^\ddagger]}{[E][S]} \quad \text{and} \quad K_E^\ddagger = \frac{[ES^\ddagger]}{[ES]}$$

are all association constants. Consequently,

$$\frac{K_T}{K_R} = \frac{[S][ES^\ddagger]}{[S^\ddagger][ES]} = \frac{K_E^\ddagger}{K_N^\ddagger} \qquad [15.8]$$

According to transition state theory, Eqs. [14.7] and [14.14], the rate of the uncatalyzed reaction can be expressed

$$v_N = k_N[S] = \left(\frac{\kappa k_B T}{h}\right)[S^\ddagger] = \left(\frac{\kappa k_B T}{h}\right)K_N^\ddagger[S] \quad [15.9]$$

Similarly, the rate of the enzymatically catalyzed reaction is

$$v_E = k_E[ES] = \left(\frac{\kappa k_B T}{h}\right)[ES^\ddagger] = \left(\frac{\kappa k_B T}{h}\right)K_E^\ddagger[ES] \quad [15.10]$$

Therefore, combining Eqs. [15.8] to [15.10],

$$\frac{k_E}{k_N} = \frac{K_E^\ddagger}{K_N^\ddagger} = \frac{K_T}{K_R} \qquad [15.11]$$

This equation indicates that *the more tightly an enzyme binds its reaction's transition state* (K_T) *relative to the substrate* (K_R), *the greater the rate of the catalyzed reaction* (k_E) *relative to that of the uncatalyzed reaction* (k_N); *that is, catalysis results from the preferential binding and therefore the stabilization of the transition state* ($S^\ddagger$) *relative to that of the substrate* (S) *(Fig. 15-7).*

According to Eq. [14.15], the ratio of the rates of the catalyzed versus the uncatalyzed reaction is expressed

$$\frac{k_E}{k_N} = \exp[(\Delta G_N^\ddagger - \Delta G_E^\ddagger)/RT] \qquad [15.12]$$

A rate enhancement factor of 10^6 therefore requires that an enzyme bind its transition state complex with 10^6-fold higher affinity than its substrate, which corresponds to a 34.2 kJ · mol^{-1} stabilization at 25°C. This is roughly the free energy of two hydrogen bonds. Consequently, *the enzymatic binding of a transition state* ($ES^\ddagger$) *by two hydrogen bonds that cannot form in the Michaelis complex* (ES) *should result in a rate enhancement of* ~10^6 *based on this effect alone.*

It is commonly observed that the specificity of an enzyme is manifested by its turnover number (k_{cat}) rather than by its substrate-binding affinity. In other words, an enzyme binds poor substrates, which have a low reaction rate, as well as or even better than good ones, which have a high reaction rate. Such enzymes apparently use a good substrate's intrinsic binding energy to stabilize the corresponding transition state; that is, *a good substrate does not necessarily bind to its enzyme with high affinity, but does so on activation to the transition state.*

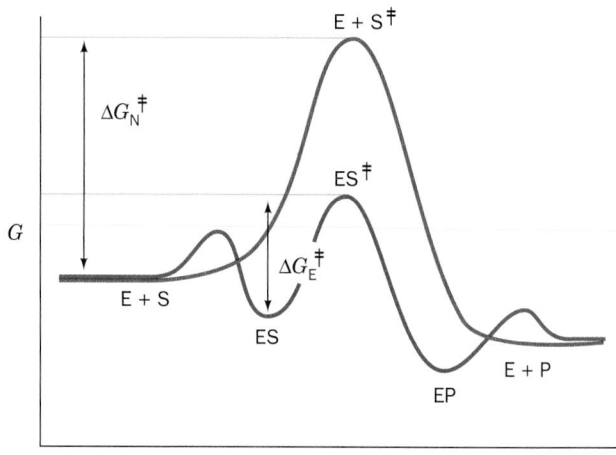

FIGURE 15-7   Reaction coordinate diagrams for a hypothetical enzymatically catalyzed reaction involving a single substrate (*blue*) and the corresponding uncatalyzed reaction (*red*).

🎞 **See the Animated Figures**

a. Transition State Analogs Are Competitive Inhibitors

If an enzyme preferentially binds its transition state, then it can be expected that **transition state analogs,** *stable molecules that resemble* $S^\ddagger$ *or one of its components, are potent competitive inhibitors of the enzyme.* For example, the reaction catalyzed by **proline racemase** from *Clostridium sticklandii* is thought to occur via a planar transition state:

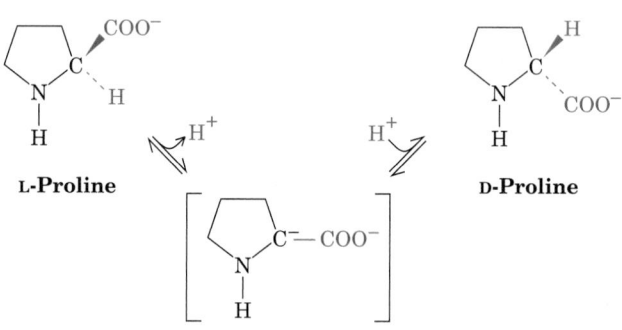

Planar transition state

Proline racemase is competitively inhibited by the planar analogs of proline, **pyrrole-2-carboxylate** and **Δ-1-pyrroline-2-carboxylate,**

Pyrrole-2-carboxylate   **Δ-1-Pyrroline-2-carboxylate**

both of which bind to the enzyme with 160-fold greater affinity than does proline. These compounds are therefore thought to be analogs of the transition state in the proline

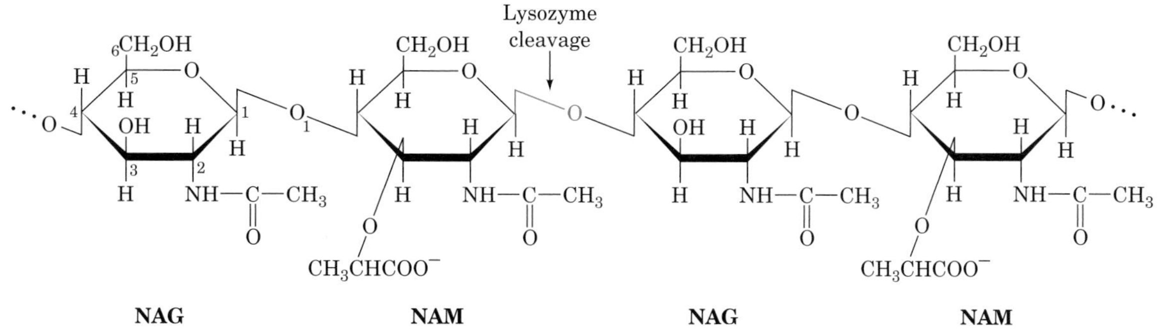

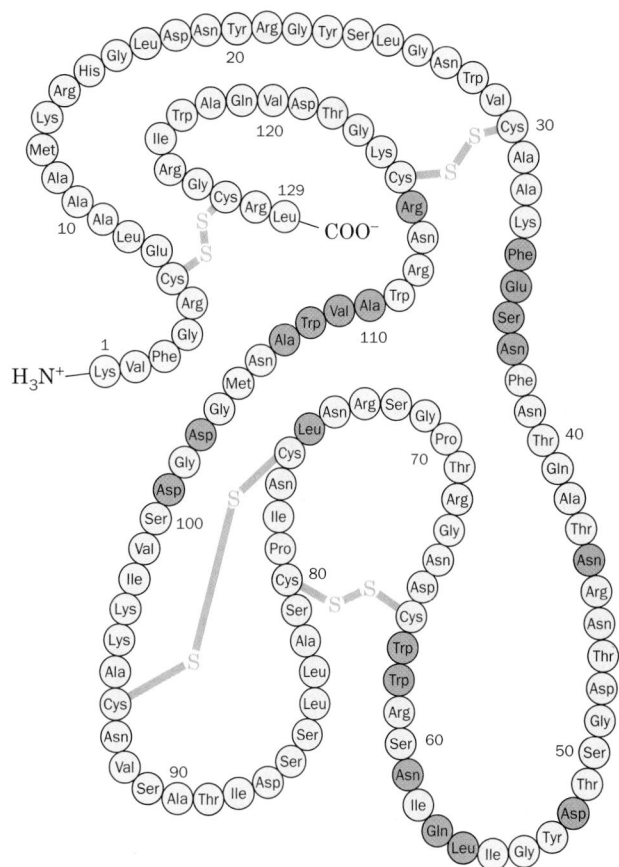

FIGURE 15-8   The alternating NAG–NAM polysaccharide component of bacterial cell walls. The position of the lysozyme cleavage site is shown.

racemase reaction. In contrast, **tetrahydrofuran-2-car-boxylate,**

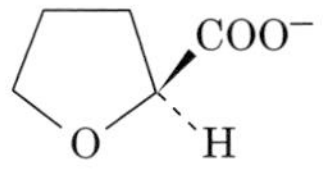

Tetrahydrofuran-2-carboxylate

which more closely resembles the tetrahedral structure of proline, is not nearly as good an inhibitor as these compounds. A 160-fold increase in binding affinity corresponds, according to Eq. [15.12], to a 12.6 kJ · mol^{-1} increase in the free energy of binding. This quantity presumably reflects the additional binding affinity that proline racemase has for proline's planar transition state over that of the undistorted molecule.

Hundreds of transition state analogs for various enzymatic reactions have been reported. Some are naturally occurring antibiotics. Others were designed to investigate the mechanisms of particular enzymes and/or to act as specific enzymatic inhibitors for therapeutic or agricultural use. Indeed, as we discuss in Section 15-4C, *the theory that enzymes bind transition states with higher affinity than substrates has led to a rational basis for drug design based on the understanding of specific enzyme reaction mechanisms.*

2 ■ LYSOZYME

In the following two sections, we shall investigate the catalytic mechanisms of several well-characterized enzymes. In doing so, we shall see how enzymes apply the catalytic principles described in Section 15-1. You should note that *the great catalytic efficiency of enzymes arises from their simultaneous use of several of these catalytic mechanisms.*

Lysozyme is an enzyme that destroys bacterial cell walls. It does so, as we saw in Section 11-3B, by hydrolyzing the β(1→4) glycosidic linkages from **N-acetylmuramic acid (NAM)** to **N-acetylglucosamine (NAG)** in the alternating NAM–NAG polysaccharide component of cell wall peptidoglycans (Fig. 15-8). It likewise hydrolyzes β(1→4)-linked poly(NAG) (chitin), a cell wall component of most fungi. Lysozyme occurs widely in the cells and secretions of vertebrates, where it may function as a bactericidal agent. However, the observation that few pathogenic bacteria are

susceptible to lysozyme alone has prompted the suggestion that this enzyme mainly helps dispose of bacteria after they have been killed by other means.

Hen egg white (HEW) lysozyme is the most widely studied species of lysozyme and is one of the mechanistically best understood enzymes. It is a rather small protein (14.7 kD) whose single polypeptide chain consists of 129 amino acid residues and is internally cross-linked by four disulfide bonds (Fig. 15-9). HEW lysozyme catalyzes the

FIGURE 15-9   Primary structure of HEW lysozyme. The amino acid residues that line the substrate-binding pocket are shown in dark purple.

hydrolysis of its substrate at a rate that is ~10^8-fold greater than that of the uncatalyzed reaction.

A. *Enzyme Structure*

The elucidation of an enzyme's mechanism of action requires a knowledge of the structure of its enzyme–substrate complex. This is because, even if the active site residues have been identified through chemical and physical means, their three-dimensional arrangements relative to the substrate as well as to each other must be known for an understanding of how the enzyme works. However, an enzyme binds its good substrates only transiently before it catalyzes a reaction and releases the products. Consequently, *most of our knowledge of enzyme–substrate complexes derives from X-ray studies of enzymes in complex with inhibitors or poor substrates* that remain stably bound to the enzyme for the several hours that are usually required to measure a protein crystal's X-ray diffraction intensities (although techniques for measuring X-ray intensities in less than 1 s have been developed). The large solvent-filled channels that occupy much of the volume of most protein crystals (Section 8-3A) often permit the formation of enzyme–inhibitor complexes by the diffusion of inhibitor molecules into crystals of the native protein.

The X-ray structure of HEW lysozyme, which was elucidated by David Phillips in 1965, was the second structure of a protein and the first of an enzyme to be determined at high resolution. The protein molecule is roughly ellipsoidal in shape with dimensions 30 × 30 × 45 Å (Fig. 15-10). *Its most striking feature is a prominent cleft, the substrate-binding site, that traverses one face of the molecule.* The polypeptide chain forms five helical segments as well as a three-stranded antiparallel β sheet that comprises much of one wall of the binding cleft (Fig. 15-10b). As expected, most of the nonpolar side chains are in the interior of the molecule, out of contact with the aqueous solvent.

a. The Nature of the Binding Site

NAG oligosaccharides of less than five residues are but very slowly hydrolyzed by HEW lysozyme (Table 15-2) although these substrate analogs bind to the enzyme's active site and are thus its competitive inhibitors. The X-ray struc-

TABLE 15-2   Rates of HEW Lysozyme-Catalyzed Hydrolysis of Selected Oligosaccharide Substrate Analogs

Compound	k_{cat} (s^{-1})
(NAG)$_2$	2.5×10^{-8}
(NAG)$_3$	8.3×10^{-6}
(NAG)$_4$	6.6×10^{-5}
(NAG)$_5$	0.033
(NAG)$_6$	0.25
(NAG–NAM)$_3$	0.5

Source: Imoto, T., Johnson, L.N., North, A.C.T., Phillips, D.C., and Rupley, J.A., *in* Boyer, P.D. (Ed.), *The Enzymes* (3rd ed.), Vol. 7, p. 842, Academic Press (1972).

ture of the (NAG)$_3$–lysozyme complex reveals that (NAG)$_3$ is bound on the right side of the enzymatic binding cleft as drawn in Fig. 15-10a for substrate residues A, B, and C. This inhibitor associates with the enzyme through strong hydrogen bonding interactions, some of which involve the acetamido groups of residues A and C, as well as through close-fitting hydrophobic contacts. In an example of induced-fit ligand binding (Section 10-4C), there is a slight (~1 Å) closure of lysozyme's binding cleft on binding (NAG)$_3$.

b. Lysozyme's Catalytic Site Was Identified through Model Building

(NAG)$_3$ takes several weeks to hydrolyze under the influence of lysozyme. It is therefore presumed that the complex revealed by X-ray analysis is unproductive; that is, the enzyme's catalytic site occurs at neither the A—B nor the B—C bonds. [Presumably, the rare occasions when (NAG)$_3$ hydrolyzes occur when it binds productively at the catalytic site.]

In order to locate lysozyme's catalytic site, Phillips used model building to investigate how a larger substrate could bind to the enzyme. Lysozyme's active site cleft is long enough to accommodate (NAG)$_6$, which the enzyme rapidly hydrolyzes (Table 15-2). However, the fourth NAG residue (residue D in Fig. 15-10a) appeared unable to bind to the enzyme because its C6 and O6 atoms too closely contact Glu 35, Trp 108, and the acetamido group of NAG residue C. This steric interference could be relieved by distorting the glucose ring from its normal chair conforma-

FIGURE 15-10   (*Opposite*) **X-Ray structure of HEW lysozyme.** (a) The polypeptide chain is shown with a bound (NAG)$_6$ substrate (*green*). The positions of the backbone C$_\alpha$ atoms are indicated together with those of the side chains that line the substrate-binding site and form disulfide bonds. The substrate's sugar rings are designated A, at its nonreducing end (*right*), through F, at its reducing end (*left*). Lysozyme catalyzes the hydrolysis of the glycosidic bond between residues D and E. Rings A, B, and C are observed in the X-ray structure of the complex of (NAG)$_3$ with lysozyme; the positions of rings D, E, and F were inferred from model building studies. [Illustration, Irving Geis/Geis Archives Trust. Copyright Howard Hughes Medical Institute. Reproduced with permission.] (*b*) A ribbon diagram of lysozyme highlighting the protein's secondary structure and indicating the positions of its catalytically important side chains, Glu 35 and Asp 52 (*red*). (*c*) A computer-generated model showing the protein's molecular envelope (*purple*) and C$_\alpha$ backbone (*blue*). The side chains of the catalytic residues, Asp 52 (*above*) and Glu 35 (*below*), are colored yellow. Note the enzyme's prominent substrate-binding cleft. [Courtesy of Arthur Olson, The Scripps Research Institute, La Jolla, California.] Parts *a, b,* and *c* have approximately the same orientation. ✎ See the Interactive Exercises and Kinemage Exercise 9

(a)

(b)

(c)

tion to that of a half-chair (Fig. 15-11). *This distortion, which renders atoms C1, C2, C5, and O5 of residue D coplanar, moves the —C6H₂OH group from its normal equatorial position to an axial position where it makes no close contacts and can hydrogen bond to the backbone carbonyl group of Gln 57 and the amido group of Val 109 (Fig. 15-12).* Continuing the model building, Phillips found that residues E and F apparently bind to the enzyme without distortion and with a number of favorable hydrogen bonding and van der Waals contacts.

We are almost in a position to identify lysozyme's catalytic site. In the enzyme's natural substrate, every second residue is an NAM. Model building, however, indicated that its lactyl side chain cannot be accommodated in the binding subsites of either residues C or E. Hence, the NAM residues must bind to the enzyme in subsites B, D, and F.

$$\cdots-\text{NAG}-\text{NAM}-\text{NAG}-\text{NAM}-\text{NAG}-\text{NAM}-\cdots\rightarrow\left(\begin{array}{c}\text{reducing}\\\text{end}\end{array}\right)$$
$$\quad\;\; \text{A}\qquad\text{B}\qquad\text{C}\qquad\text{D}\qquad\text{E}\qquad\text{F}$$

The observation that lysozyme hydrolyzes $\beta(1\rightarrow4)$ linkages from NAM to NAG implies that bond cleavage occurs either between residues B and C or between residues D and E. Since (NAG)₃ is stably bound to but not cleaved by the enzyme while spanning subsites B and C, the probable cleavage site is between residues D and E. This conclusion is supported by John Rupley's observation that lysozyme nearly quantitatively hydrolyzes (NAG)₆

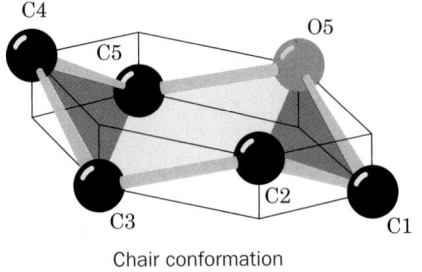

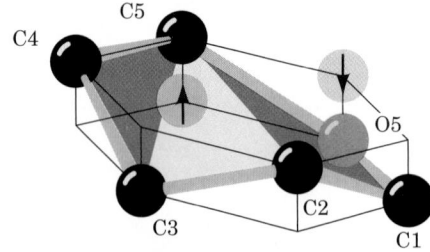

FIGURE 15-11 Chair and half-chair conformations. Hexose rings normally assume the chair conformation. It is postulated, however, that binding by lysozyme distorts the D-ring into the half-chair conformation such that atoms C1, C2, C5, and O5 are coplanar. *See the Animated Figures*

FIGURE 15-12 Interactions of lysozyme with its substrate. The view is into the binding cleft with the heavier edges of the rings facing the outside of the enzyme and the lighter ones against the bottom of the cleft. [Illustration, Irving Geis/Geis Archives Trust. Copyright Howard Hughes Medical Institute. Reproduced with permission. Based on an X-ray structure by David Phillips, Oxford University, U.K. PDBid 4LYZ.] *See Kinemage Exercise 9*

between the second and third residues from its reducing terminus (the end with a free C1—OH), just as is expected if the enzyme has six saccharide-binding subsites and cleaves its bound substrate between residues D and E.

The bond that lysozyme cleaves was identified by carrying out the lysozyme-catalyzed hydrolysis of $(NAG)_3$ in $H_2{}^{18}O$. The resulting product had ^{18}O bonded to the C1 atom of its newly liberated reducing terminus, thereby demonstrating that bond cleavage occurs between C1 and the bridge oxygen O1:

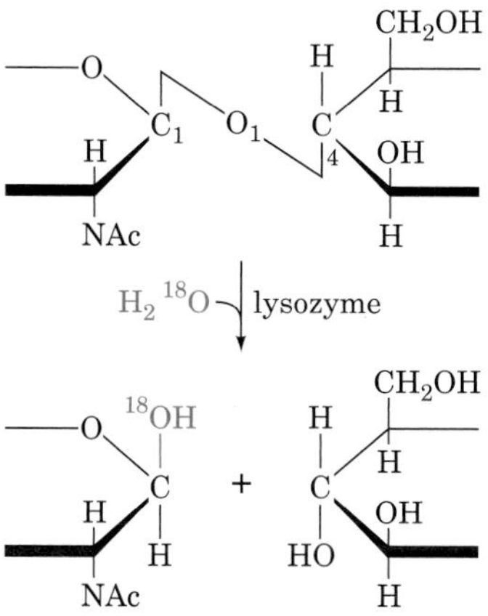

Thus, lysozyme catalyzes the hydrolysis of the C1—O1 bond of a bound substrate's D residue. Moreover, *this reaction occurs with retention of configuration, so that the D-ring product remains the β anomer.*

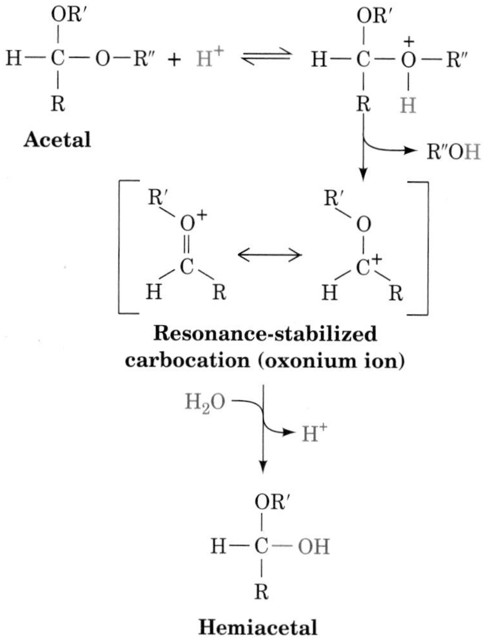

B. *Catalytic Mechanism*

It remains to identify lysozyme's catalytic groups. The reaction catalyzed by lysozyme, the hydrolysis of a glycoside, is the conversion of an acetal to a hemiacetal. Nonenzymatic acetal hydrolysis is an acid-catalyzed reaction that involves the protonation of a reactant oxygen atom followed by cleavage of its C—O bond (Fig. 15-13). This results in the formation of a resonance-stabilized carbocation that is called an **oxonium ion.** To attain maximum orbital overlap, and thus resonance stabilization, the oxonium ion's R and R′ groups must be coplanar with its C, O, and H atoms (stereoelectronic assistance). The oxonium ion then adds water to yield the hemiacetal and regenerate the acid catalyst. In searching for catalytic groups on an enzyme that mediates acetal hydrolysis, we should therefore seek a potential acid catalyst and possibly a group that could further stabilize an oxonium ion intermediate.

a. Glu 35 and Asp 52 Are Lysozyme's Catalytic Residues

The only functional groups in the immediate vicinity of lysozyme's reaction center that have the required catalytic

FIGURE 15-13 Mechanism of the nonenzymatic acid-catalyzed hydrolysis of an acetal to a hemiacetal. The reaction involves the protonation of one of the acetal's oxygen atoms followed by cleavage of its C—O bond to form an alcohol (R″OH) and a resonance-stabilized carbocation (oxonium ion). The addition of water to the oxonium ion forms the hemiacetal and regenerates the H⁺ catalyst. Note that the oxonium ion's C, O, H, R, and R′ atoms all lie in the same plane.

properties are the side chains of Glu 35 and Asp 52, residues that are invariant in the family of lysozymes of which HEW lysozyme is the prototype. These side chains, which are disposed to either side of the β(1→4) glycosidic linkage to be cleaved (Fig. 15-10), have markedly different environments. Asp 52 is surrounded by several conserved polar residues with which it forms a complex hydrogen bonded network. Asp 52 is therefore predicted to have a normal pK; that is, it should be unprotonated and hence negatively charged throughout the 3 to 8 pH range in which lysozyme is catalytically active. In contrast, *the carboxyl group of Glu 35 is nestled in a predominantly nonpolar pocket, where, as we discussed in Section 15-1D, it is likely to remain protonated at unusually high pH's for carboxyl groups.* Indeed, neutron diffraction studies, which provide similar information to X-ray diffraction studies but also reveal the positions of hydrogen atoms, indicate that Glu 35 is protonated at physiological pH's. The closest approaches in the X-ray structures between the carboxyl O atoms of both Asp 52 and Glu 35 and the C1—O1 bond of NAG residue D are ~3 Å, which makes them the prime candidates for electrostatic and acid catalysts, respectively.

b. The Phillips Mechanism

With much of the foregoing information, Phillips postulated the following enzymatic mechanism for lysozyme (Fig. 15-14):

1. Lysozyme attaches to a bacterial cell wall by binding to a hexasaccharide unit. In the process, *residue D is distorted toward the half-chair conformation* in response to the unfavorable contacts that its —C6H₂OH group would otherwise make with the protein.

2. *Glu 35 transfers its proton to the O1 of the D-ring,* the only polar group in its vicinity *(general acid catalysis). The C1—O1 bond is thereby cleaved, generating a resonance-stabilized oxonium ion at C1.*

3. The ionized carboxyl group of Asp 52 acts to *stabilize the developing oxonium ion through charge–charge interactions (electrostatic catalysis).* This carboxylate group apparently cannot form a covalent bond with the substrate because the observed ~3 Å distance between C1 and a carboxyl O atom of Asp 52 is much greater than the ~1.4 Å length of a C—O covalent bond [i.e., the reaction appears to occur via an S_N1 (unimolecular nucleophilic substitution) mechanism to yield an oxonium ion, not via a mechanism involving the transient formation of a C—O covalent bond to the enzyme; but see Section 15-2C]. The bond cleavage reaction is facilitated by the strain in the D-ring that distorts it to the planar half-chair conformation. This is a result of the oxonium ion's required planarity; that is, *the initial binding conformation of the D-ring resembles that of the reaction's transition state (transition state binding catalysis; Fig. 15-15).*

4. At this point, the enzyme releases the hydrolyzed E-ring with its attached polysaccharide (the leaving group), yielding a cationic, noncovalent, **glycosyl–enzyme intermediate.** This oxonium ion subsequently adds H₂O from solution in a reversal of the preceding steps to form product and to reprotonate Glu 35. *The reaction's retention of configuration is dictated by the shielding of one of the oxonium ion's faces by the enzymatic cleft.* The enzyme then releases the D-ring product with its attached saccharide, thereby completing the catalytic cycle.

C. Testing the Phillips Mechanism

The Phillips mechanism was formulated largely on the basis of structural investigations of lysozyme and a knowledge of the mechanism of nonenzymatic acetal hydrolysis. A variety of evidence has since been gathered that bears on the validity of this mechanism. In the remainder of this section, we discuss the highlights of these studies to illustrate how scientific models evolve.

a. Identification of the Catalytic Residues

Lysozyme's catalytically important groups have been experimentally identified through site-directed mutagenesis (Section 5-5G) and the use of group-specific reagents:

Glu 35. The mutagenesis of Glu 35 to Gln yields a protein with no detectable catalytic activity (<0.1% of wild type),

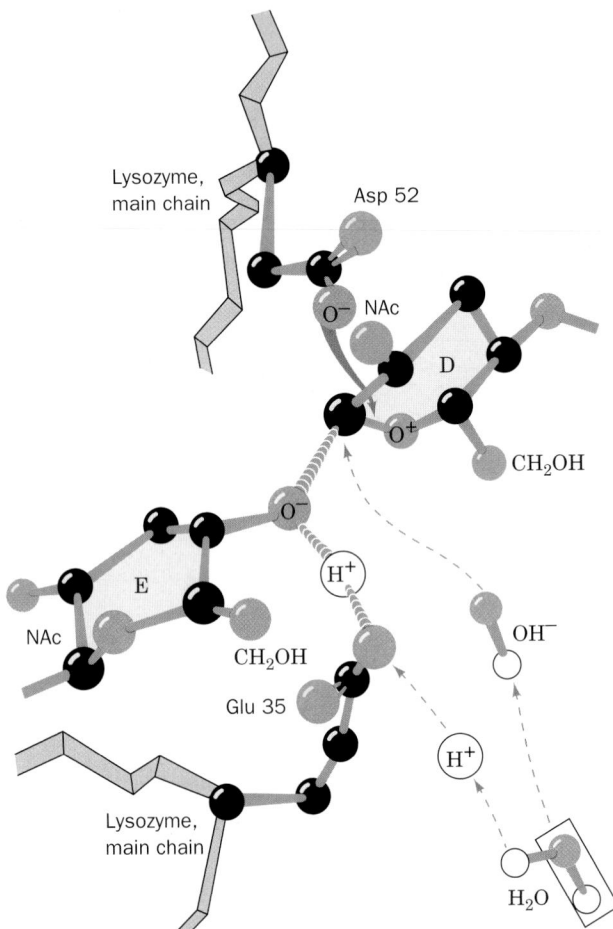

FIGURE 15-14   The Phillips mechanism for the lysozyme reaction. The cleavage of the glycosidic bond between the substrate D- and E-rings occurs through protonation of the bridge oxygen atom by Glu 35. The resulting D-ring oxonium ion is stabilized by the proximity of the Asp 52 carboxylate group and the enzyme-induced distortion of the D-ring. Once the E-ring is released, H₂O from solution provides both an OH⁻ that combines with the oxonium ion and an H⁺ that reprotonates Glu 35. NAc represents the *N*-acetylamino substituent at C2 of each glucose ring. *♪ See Kinemage Exercise 9 and the Animated Figures*

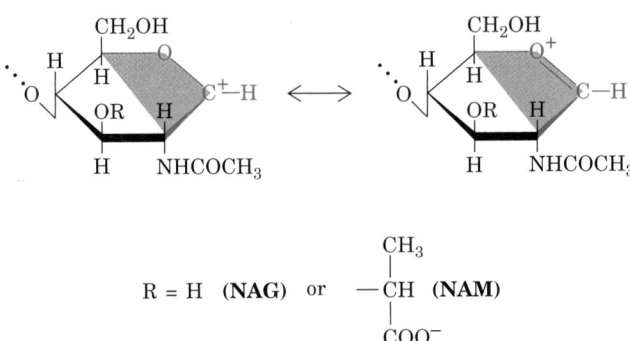

FIGURE 15-15   The D-ring oxonium ion intermediate in the Phillips mechanism is stabilized by resonance. This requires that atoms C1, C2, C5, and O5 be coplanar (*shading*); that is, the hexose ring must assume the half-chair conformation.

although it has only a ~1.5-fold decrease in substrate affinity. *Glu 35 must therefore be essential for lysozyme's catalytic activity.*

Asp 52. The mutagenesis of Asp 52 to Asn, which has a polarity comparable to that of Asp but lacks its negative charge, yields an enzyme with no more than 5% of wild-type lysozyme's catalytic activity even though this mutation causes an ~2-fold increase in the enzyme's affinity for substrate. *Asp 52 is therefore important for enzymatic activity.*

Noninvolvement of Other Amino Acid Residues. Lysozyme's other carboxyl groups besides Glu 35 and Asp 52 do not participate in the catalytic process, as was demonstrated by reacting lysozyme with carboxyl-specific reagents in the presence of substrate. This treatment yields an almost fully active enzyme in which all carboxyl groups but Glu 35 and Asp 52 are derivatized. Other group-specific reagents that modify, for instance, His, Lys, Met, or Tyr residues but induce no major protein structure disruptions cause little change in lysozyme's catalytic efficiency.

b. Role of Strain

Many of the mechanistic investigations of lysozyme have had the elusive goal of establishing the catalytic role of strain. Not all of these studies, as we shall see, have supported the Phillips mechanism, thereby stimulating a series of investigations that have only recently settled this issue.

Measurements of the binding equilibria of various oligosaccharides to lysozyme indicate that all saccharide residues except that binding to the D subsite contribute energetically toward the binding of substrate to lysozyme; binding NAM in the D subsite requires a free energy input of 12 kJ · mol^{-1} (Table 15-3). The Phillips mechanism explains this observation as being indicative of the energy penalty of straining the D-ring from its preferred chair conformation toward the half-chair form.

As we have discussed in Section 15-1F, an enzyme that catalyzes a reaction by the preferential binding of its transition state has a greater binding affinity for an inhibitor that has the transition state geometry (transition state analog) than it does for its substrate. The δ-lactone analog of (NAG)$_4$ (Fig. 15-16) is a transition state analog of lysozyme

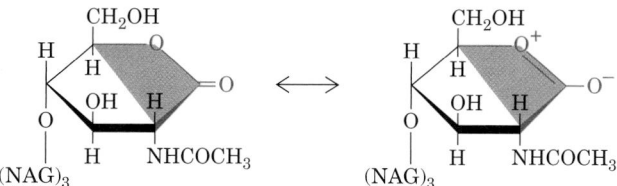

FIGURE 15-16   The δ-lactone analog of (NAG)$_4$. Its C1, O1, C2, C5, and O5 atoms are coplanar (*shading*) because of resonance, as is the D-ring in the reaction intermediate of the Phillips mechanism (compare with Fig. 15-15).

since *this compound's lactone ring has the half-chair conformation that geometrically resembles the proposed oxonium ion transition state of the substrate's D-ring.* X-Ray studies indicate, in accordance with prediction, that this inhibitor binds to lysozyme's A—B—C—D subsites such that the lactone ring occupies the D subsite in a half-chairlike conformation.

Despite the foregoing, *the role of substrate distortion in lysozyme catalysis had been questioned.* Theoretical studies by Michael Levitt and Arieh Warshel on substrate binding by lysozyme suggested that the protein is too flexible to mechanically distort the D-ring of a bound substrate. Rather, these calculations implied that transition state stabilization occurs through the displacement by substrate of several tightly bound water molecules from the D subsite. The resulting desolvation of the Asp 52 carboxylate group would significantly enhance its capacity to electrostatically stabilize the transition state oxonium ion. This study therefore concluded that "electrostatic strain" rather than steric strain is the more important factor in stabilizing lysozyme's transition state.

In an effort to obtain further experimental information bearing on the Phillips strain mechanism, Nathan Sharon and David Chipman determined the D subsite–binding affinities of several saccharides by comparing the lysozyme-binding affinities of various substrate analogs. The NAG lactone inhibitor binds to the D subsite with 9.2 kJ · mol^{-1} greater affinity than does NAG. This quantity corresponds, according to Eq. [14.15], to no more than an ~40-fold rate enhancement of the lysozyme reaction as a result of strain (recall that the difference in binding energy between a transition state analog and a substrate is indicative of the enzyme's rate enhancement arising from the preferential binding of the transition state complex). Such an enhancement is hardly a major portion of lysozyme's ~10^8-fold rate enhancement (accounting for only ~20% of the reaction's $\Delta\Delta G_{cat}^{\ddagger}$; Section 14-1C). Moreover, an **N-acetylxylosamine (XylNAc)** residue,

TABLE 15-3   Binding Free Energies of HEW Lysozyme Subsites

Site	Bound Saccharide	Binding Free Energy (kJ · mol^{-1})
A	NAG	−7.5
B	NAM	−12.3
C	NAG	−23.8
D	**NAM**	**+12.1**
E	NAG	−7.1
F	NAM	−7.1

Source: Chipman, D.M. and Sharon, N., *Science* **165,** 459 (1969).

N-Acetylxylosamine residue

which lacks the sterically hindered —C6H₂OH group of NAM and NAG, has only marginally greater binding affinity for the D subsite (-3.8 kJ · mol^{-1}) than does NAG (-2.5 kJ · mol^{-1}). Yet recall that the Phillips mechanism postulates that it is the unfavorable contacts made by this —C6H₂OH group that promotes D-ring distortion. Nevertheless, lysozyme does not hydrolyze saccharides with XylNAc in the D subsite.

The apparent inconsistencies among the foregoing experimental observations were largely rationalized by Michael James' highly accurate (1.5-Å resolution) X-ray crystal structure determination of lysozyme in complex with NAM–NAG–NAM. This trisaccharide binds, as expected, to the B, C, and D subsites of lysozyme. *The NAM in the D subsite, in agreement with the Phillips mechanism, is distorted to the half-chair conformation with its —C6H₂OH group in a nearly axial position due to steric clashes that would otherwise occur with the acetamido group of the C subsite NAG* (although, contrary to the original Phillips mechanism, Glu 35 and Trp 108 are too far away from the —C6H₂OH group to contribute to this distortion). This strained conformation is stabilized by a strong hydrogen bond between the D-ring O6 and the backbone NH of Val 109 (transition state stabilization). Indeed, the mutation of Val 109 to Pro, which lacks the NH group to make such a hydrogen bond, inactivates the enzyme. Lysozyme's lack of hydrolytic activity when XylNAc occupies its D subsite is likewise explained by the absence of this hydrogen bond and the consequent lesser stability of the XylNAc ring's half-chair transition state.

The unexpectedly small free energy differences in binding NAG, NAG lactone, and XylNAc to the D subsite are explained by the observation that undistorted NAG and XylNAc can be modeled into the D subsite as it occurs in the X-ray structure of the lysozyme ·NAM–NAG–NAM complex. NAM's bulky lactyl side chain prevents it from binding to the D subsite in this manner.

c. The Lysoyme Reaction Proceeds via a Covalent Intermediate

Alternatives to the Phillips mechanism postulate that either (1) the carboxyl group of Asp 52 displaces the leaving group to form a covalent bond to C1, thereby yielding a covalent glycosyl–enzyme ester intermediate that is subsequently displaced by water to yield product (a double-displacement mechanism); or (2) water directly displaces the leaving group (a single-displacement mechanism). A single-displacement mechanism would result in inversion of configuration between substrate and product and thus can be ruled out. A double-displacement mechanism would account for the observed retention of configuration in the lysozyme reaction (as does the Phillips mechanism). However, it is at odds with the observation that the distance between C1 in a D subsite–bound saccharide and a carboxyl O of Asp 52 (which participates in a network of hydrogen bonds that apparently hold this side chain in its position) are too long to form a covalent bond (minimally 2.3 Å in the NAM–NAG–NAM complex without significantly disrupting the protein structure vs ~1.4 Å for a

C—O single bond). Indeed, no such covalent bond had been observed in any of the numerous X-ray structures containing hen egg white (HEW) lysozyme.

Despite the foregoing, all other β-glycosidases of known structure that cleave glycosidic linkages with net retention of configuration at the anomeric carbon (as does HEW lysozyme) have been shown to do so via a covalent glycosyl–enzyme intermediate. The active sites of these so-called **retaining β-glycosidases** structurally resemble that of HEW lysozyme. Moreover, there is no direct evidence indicative of the existence of a long-lived oxonium ion at the active site of any retaining β-glycosidase, including HEW lysozyme (the lifetime of a glucosyl oxonium ion in water is ~10^{-12} s, a time only slightly longer than that of a bond vibration). Consequently, there had been a growing suspicion that the HEW lysozyme reaction also proceeds via a covalent intermediate, one between the D-ring's anomeric carbon (C1) and the side chain carboxyl group of Asp 52 to form an ester linkage:

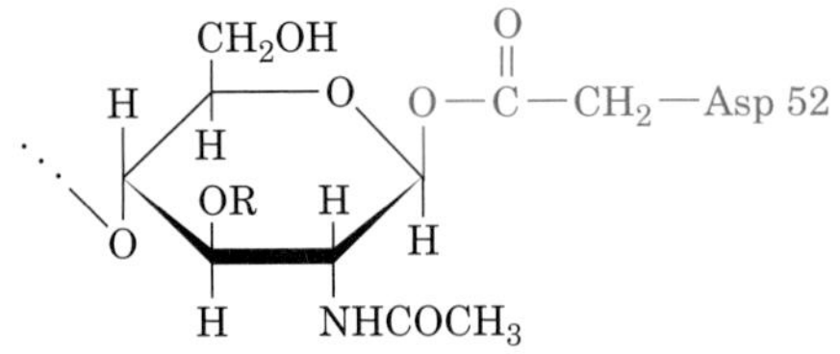

This intermediate presumably reacts with H₂O in what is essentially the reverse of the reaction leading to its formation, thereby yielding the reaction's second product (a double-displacement mechanism). In this mechanism, the oxonium ion is proposed to be the transition state on the way to forming the covalent intermediate, rather than being an intermediate itself.

If, in fact, HEW lysozyme follows this mechanism, the reason that its covalent intermediate had never been observed is that its rate of breakdown must be much faster than its rate of formation. Hence, if this intermediate is to be experimentally observed, its rate of formation must be made significantly greater than its rate of breakdown. To do so, Stephen Withers capitalized on three phenomena. First, if, as postulated, the reaction goes through an oxonium ion transition state, all steps involving its formation should be slowed by the electron withdrawing effects of substituting F (the most electronegative element) at C2 of the D-ring. Second, mutating Glu 35 to Gln (E35Q) removes the general acid–base that catalyzes the reaction, further slowing all steps involving the oxonium ion transition state. Third, substituting an additional F atom at C1 of the D-ring accelerates the formation of the intermediate because this F is a good leaving group. Making all three of these changes should increase the rate of formation of the proposed covalent intermediate relative to its breakdown and hence should result in its accumulation. Withers therefore incubated E35Q HEW lysozyme with NAG-β(1→4)-2-deoxy-2-fluoro-β-D-glucopyranosyl fluoride **(NAG2FGlcF):**

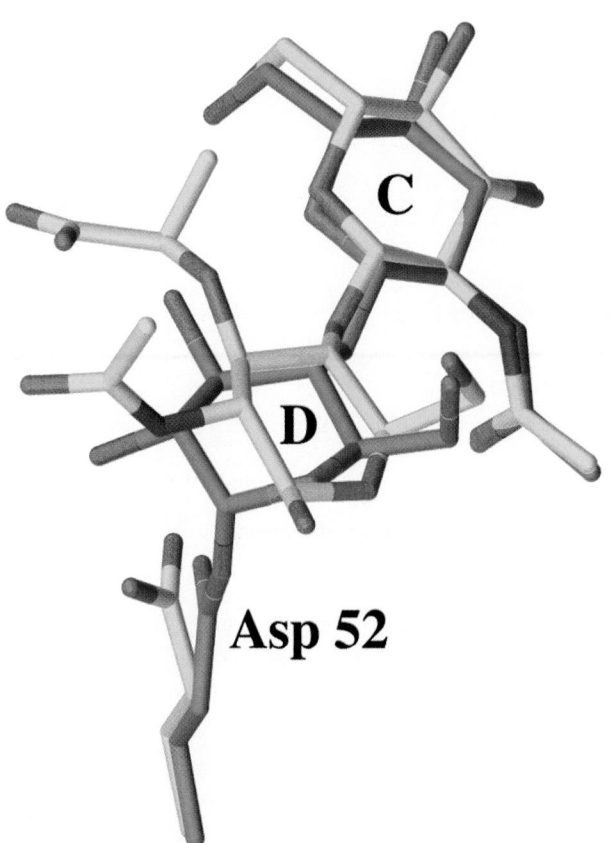

NAG2FGlcF

Electrospray ionization mass spectrometry (ESI-MS; Section 7-1J) of this reaction mixture revealed a sharp peak at 14,683 D, consistent with the formation of the proposed covalent intermediate, but no significant peak at or near the 14,314-D molecular mass of the mutant enzyme alone.

The X-ray structure of this covalent complex unambiguously reveals the expected ~1.4-Å-long covalent bond between C1 of the D-ring and a side chain carboxyl O of Asp 52 (Fig. 15-17). This D-ring adopts an undistorted chair

FIGURE 15-17   The HEW lysozyme covalent intermediate. The substrate C- and D-rings and Asp 52 are shown in the superposition of the X-ray structures of the covalent complex formed by reacting E35Q lysozyme with NAG2FGlcF (C green, N blue, O red, and F magenta) and the noncovalent complex of wild-type lysozyme with NAM–NAG–NAM (C yellow, N blue, and O red). Note that the covalent bond between Asp 52 and C1 of the D-ring forms when the D-ring in the noncovalent complex relaxes from its distorted half-chair conformation to an undistorted chair conformation and that the side chain of Asp 52 undergoes an ~45° rotation about its C_α—C_β bond. [Based on X-ray structures by David Vocadlo and Stephen Withers, University of British Columbia, Vancouver, Canada; *and* Michael James, University of Alberta, Edmonton, Canada. PDBid 1H6M.]

conformation, thus indicating that it is a reaction intermediate rather than an approximation of the transition state. The superposition of this covalent complex with that of the above described complex of NAM–NAG–NAM with wild-type HEW lysozyme reveals how this covalent bond forms (Fig. 15-17). The shortening of the 3.2-Å distance between the D-ring C1 and the Asp 52 O in the NAM–NAG–NAM complex to ~1.4 Å in the covalent complex is almost entirely a consequence of the relaxation of the D-ring from the half-chair to the chair conformation combined with an ~45° rotation of the Asp 52 side chain about its C_α—C_β bond; the positions of the D-ring O4 and O6 atoms are essentially unchanged. Hence, over 35 years after its formulation, it was shown that *the Phillips mechanism must be altered to take into account the transient formation of this covalent glycosyl–enzyme ester intermediate (covalent catalysis).* Keep in mind, however, that in order to form this covalent linkage, the D-ring must pass through an oxonium-like transition state, which requires it to transiently assume the half-chair conformation.

3 ■ SERINE PROTEASES

Our next example of enzymatic mechanisms is a diverse group of proteolytic enzymes known as the **serine proteases** (Table 15-4). These enzymes are so named because they have a common catalytic mechanism characterized by the possession of a peculiarly reactive Ser residue that is essential for their enzymatic activity. The serine proteases are the most thoroughly understood family of enzymes as a result of their extensive examination over a nearly 50-year period by kinetic, chemical, physical, and genetic techniques. In this section, we mainly study the best characterized serine proteases, **chymotrypsin, trypsin,** and **elastase.** We also consider how these three enzymes, which are synthesized in inactive forms, are physiologically activated.

A. *Kinetics and Catalytic Groups*

Chymotrypsin, trypsin, and elastase are digestive enzymes that are synthesized by the pancreatic acinar cells (Fig. 1-10*c*) and secreted, via the pancreatic duct, into the duodenum (the small intestine's upper loop). All of these enzymes catalyze the hydrolysis of peptide (amide) bonds but with different specificities for the side chains flanking the scissile (to be cleaved) peptide bond (recall that chymotrypsin is specific for a bulky hydrophobic residue preceding the scissile peptide bond, trypsin is specific for a positively charged residue, and elastase is specific for a small neutral residue; Table 7-2). Together, they form a potent digestive team.

a. **Ester Hydrolysis as a Kinetic Model**

That chymotrypsin can act as an esterase as well as a protease is not particularly surprising since the chemical mechanisms of ester and amide hydrolysis are almost identical. The study of chymotrypsin's esterase activity has led to important insights concerning this enzyme's catalytic

TABLE 15-4   A Selection of Serine Proteases

Enzyme	Source	Function
Trypsin	Pancreas	Digestion of proteins
Chymotrypsin	Pancreas	Digestion of proteins
Elastase	Pancreas	Digestion of proteins
Thrombin	Vertebrate serum	Blood clotting
Plasmin	Vertebrate serum	Dissolution of blood clots
Kallikrein	Blood and tissues	Control of blood flow
Complement C1	Serum	Cell lysis in the immune response
Acrosomal protease	Sperm acrosome	Penetration of ovum
Lysosomal protease	Animal cells	Cell protein turnover
Cocoonase	Moth larvae	Dissolution of cocoon after metamorphosis
α-Lytic protease	*Bacillus sorangium*	Possibly digestion
Proteases A and B	*Streptomyces griseus*	Possibly digestion
Subtilisin	*Bacillus subtilis*	Possibly digestion

Source: Stroud, R.M., *Sci. Am.* **231**(1), 86 (1974).

FIGURE 15-18   Time course of *p*-nitrophenylacetate hydrolysis as catalyzed by two different concentrations of chymotrypsin. The enzyme rapidly binds substrate and releases the first product, *p*-nitrophenolate ion, but the second product, acetate ion, is released more slowly. Consequently, the rate of *p*-nitrophenolate generation begins rapidly (burst phase) but slows as acyl–enzyme complex accumulates until the rate of *p*-nitrophenolate generation approaches that of acetate release (steady state). The extrapolation of the steady state curve to zero time (*dashed lines*) indicates the initial concentration of active enzyme. [After Hartley, B.S. and Kilby, B.A., *Biochem. J.* **56**, 294 (1954).]

mechanism. Kinetic measurements by Brian Hartley of the chymotrypsin-catalyzed hydrolysis of *p*-nitrophenylacetate

indicated that the reaction occurs in two phases (Fig. 15-18):

1. The "burst phase," in which the highly colored *p*-nitrophenolate ion is rapidly formed in amounts stoichiometric with the quantity of active enzyme present.

2. The "steady-state phase," in which *p*-nitrophenolate is generated at a reduced but constant rate that is independent of substrate concentration.

These observations have been interpreted in terms of a two-stage reaction sequence in which the enzyme (1) rapidly reacts with the *p*-nitrophenylacetate to release

p-nitrophenolate ion forming a covalent acyl–enzyme intermediate that (2) is slowly hydrolyzed to release acetate:

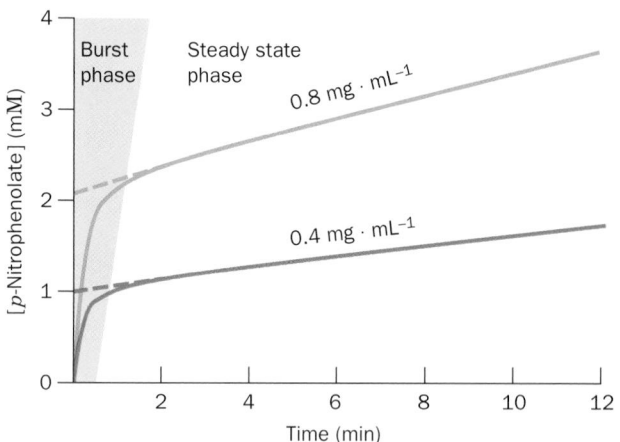

Chymotrypsin evidently follows a Ping Pong Bi Bi mechanism (Section 14-5A). Chymotrypsin-catalyzed amide hydrolysis has been shown to follow a reaction pathway similar to that of ester hydrolysis but with the first step of the reaction, enzyme acylation, being rate determining rather than the deacylation step.

b. Identification of the Catalytic Residues

Chymotrypsin's catalytically important groups were identified by chemical labeling studies. These are described below.

Ser 195. A diagnostic test for the presence of the **active Ser** of serine proteases is its reaction with **diisopropylphosphofluoridate (DIPF):**

**Diisopropylphospho-
fluoridate (DIPF)**

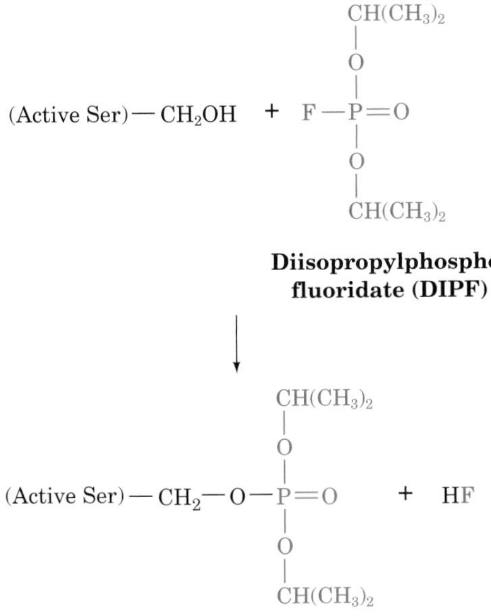

DIP–Enzyme

which irreversibly inactivates the enzyme. Other Ser residues, including those on the same protein, do not react with DIPF. *DIPF reacts only with Ser 195 of chymotrypsin, thereby demonstrating that this residue is the enzyme's active Ser.*

The use of DIPF as an enzyme inactivating agent came about through the discovery that organophosphorus compounds such as DIPF are potent nerve poisons. The neurotoxicity of DIPF arises from its ability to inactivate **acetylcholinesterase,** a serine esterase that catalyzes the hydrolysis of **acetylcholine:**

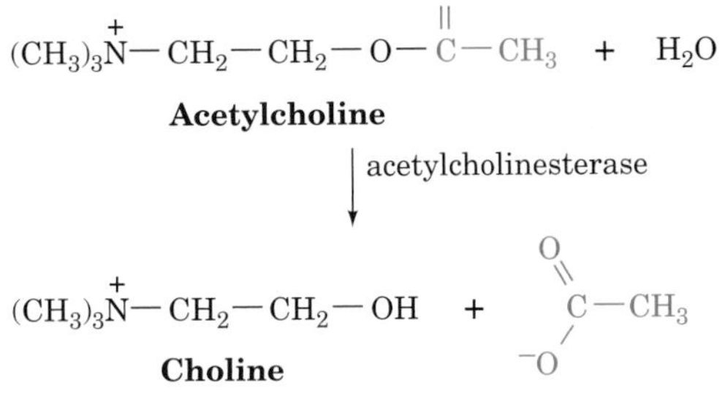

Acetylcholine

acetylcholinesterase

Choline

Acetylcholine is a **neurotransmitter:** It transmits nerve impulses across the **synapses** (junctions) between certain types of nerve cells (Sections 12-4D and 20-5C). The inactivation of acetylcholinesterase prevents the otherwise rapid hydrolysis of the acetylcholine released by a nerve impulse and thereby interferes with the regular sequence of nerve impulses. DIPF is of such great toxicity to humans that it has been used militarily as a nerve gas. Related compounds, such as **parathion** and **malathion,**

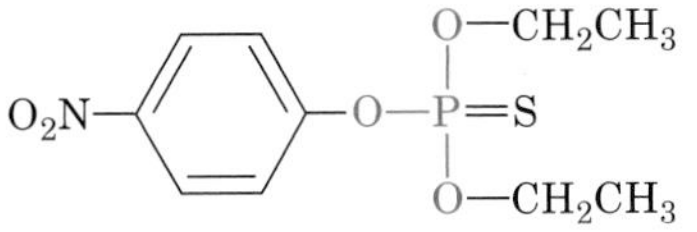

Parathion

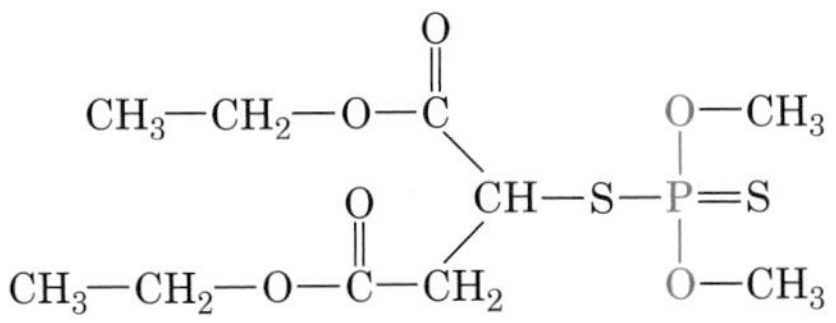

Malathion

are useful insecticides because they are far more toxic to insects than to mammals.

His 57. A second catalytically important residue was discovered through **affinity labeling.** In this technique, a substrate analog bearing a reactive group specifically binds at the enzyme's active site, where it reacts to form a stable covalent bond with a nearby susceptible group (these reactive substrate analogs have therefore been described as the "Trojan horses" of biochemistry). The affinity labeled groups can subsequently be identified by peptide mapping (Section 7-1K). Chymotrypsin specifically binds **tosyl-L-phenylalanine chloromethyl ketone (TPCK),**

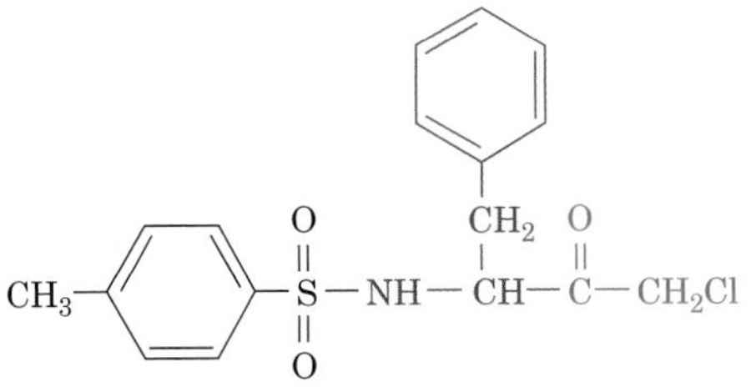

because of its resemblance to a Phe residue (one of chymotrypsin's preferred residues; Table 7-2). Active site–bound TPCK's chloromethyl ketone group is a strong alkylating agent; it reacts with His 57 (Fig. 15-19), thereby

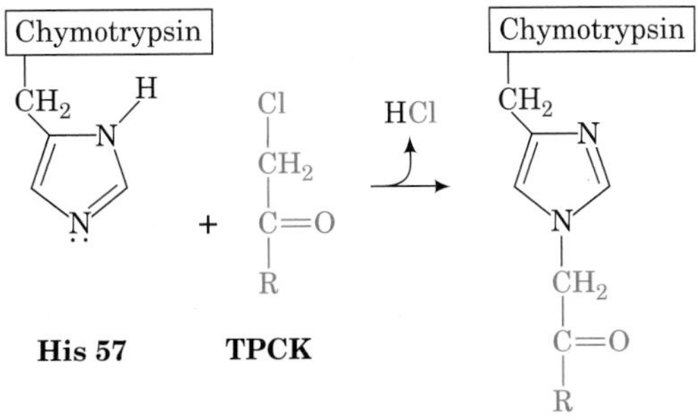

FIGURE 15-19   Reaction of TPCK with chymotrypsin to alkylate His 57.

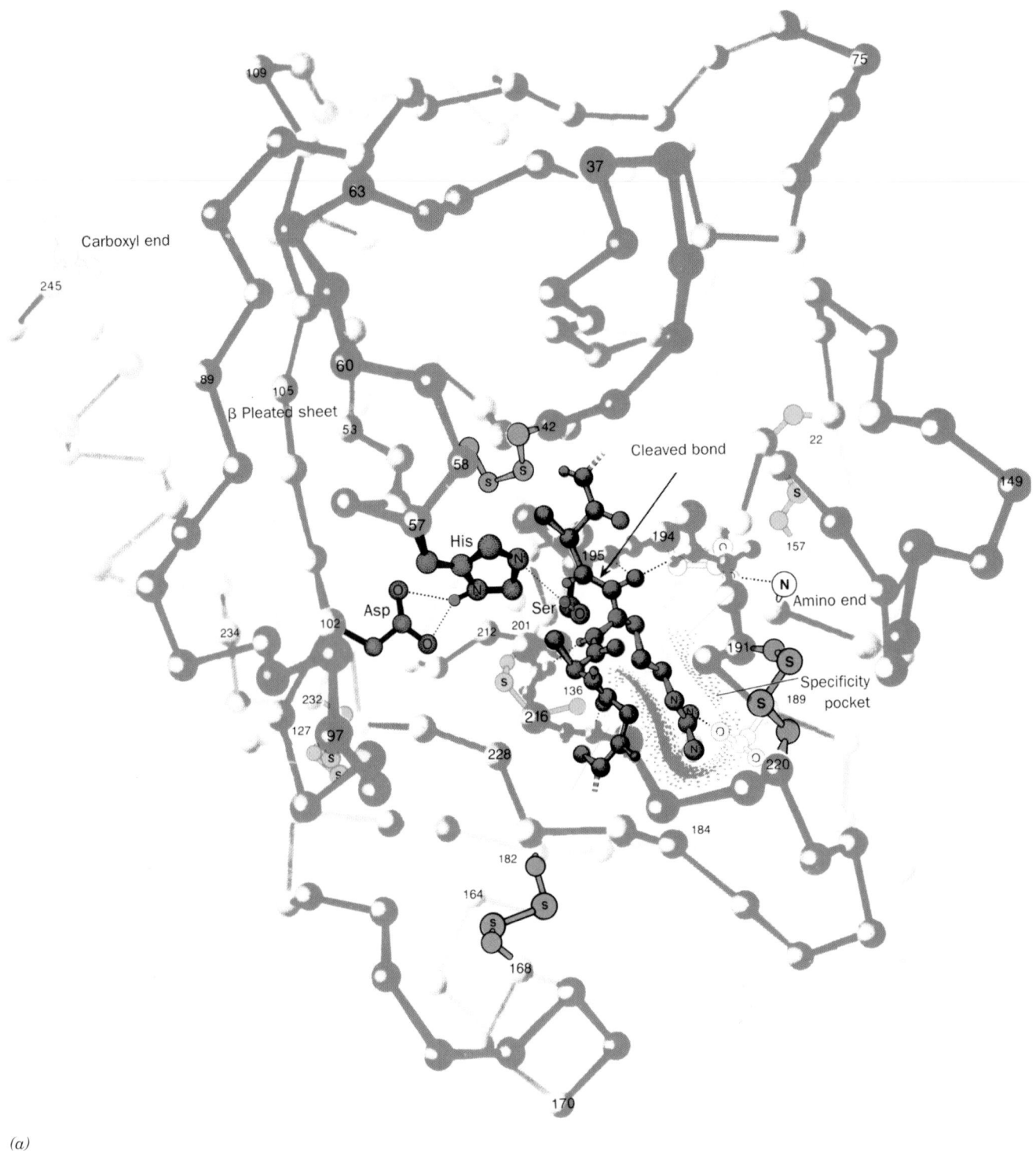

(a)

FIGURE 15-20 X-Ray structure of bovine trypsin. (*a*) A drawing of the enzyme in complex with a polypeptide substrate (*green*) that has its Arg side chain occupying the enzyme's specificity pocket (*stippling*). The C_α backbone of the enzyme is shown together with its disulfide bonds and the side chains of the catalytic triad, Ser 195, His 57, and Asp 102. The active sites of chymotrypsin and elastase contain almost identically arranged catalytic triads. [Illustration, Irving Geis/Geis Archives Trust. Copyright Howard Hughes Medical Institute. Reproduced with permission.] (*b*) A ribbon diagram of trypsin highlighting its secondary structure and indicating the arrangement of its catalytic triad. (*c*) A drawing showing the surface of trypsin (*blue*) superimposed on its polypeptide backbone (*purple*). The side chains of the catalytic triad are shown in green. [Courtesy of Arthur Olson, The Scripps Research Institute, La Jolla, California.] Parts *a, b,* and *c* have approximately the same orientation. [Based on an X-ray structure by Robert Stroud, University of California at San Francisco, PDBid 1TGN.]

🗘 **See Kinemage Exercise 10-1**

inactivating the enzyme. The TPCK reaction is inhibited by **β-phenylpropionate,**

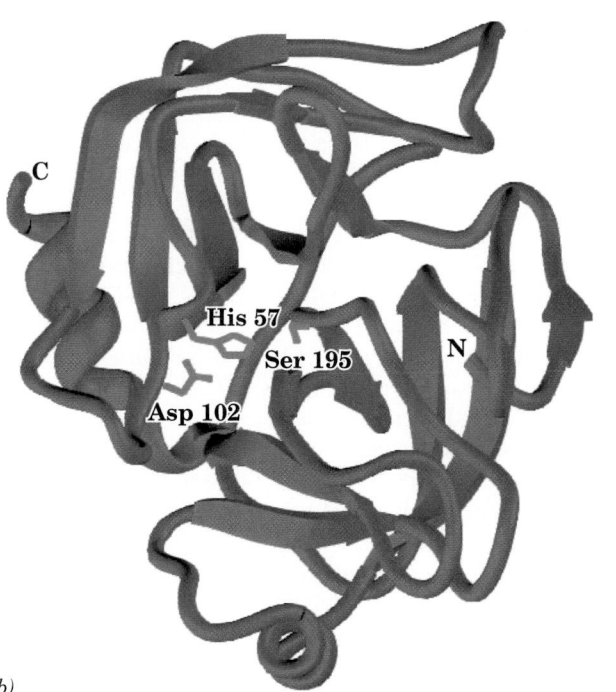

β-Phenylpropionate

a competitive inhibitor of chymotrypsin that presumably competes with TPCK for its enzymatic binding site. Moreover, the TPCK reaction does not occur in 8*M* urea, a denaturing reagent, or with DIP–chymotrypsin, in which the active site is blocked. These observations establish that *His 57 is an essential active site residue of chymotrypsin.*

B. *X-Ray Structures*

Bovine chymotrypsin, bovine trypsin, and porcine elastase are strikingly homologous: The primary structures of these ~240-residue monomeric enzymes are ~40% identical and their internal sequences are even more alike (in comparison, the α and β chains of human hemoglobin have a 44% sequence identity). Furthermore, *all of these enzymes have an active Ser and a catalytically essential His as well as similar kinetic mechanisms.* It therefore came as no surprise when their X-ray structures all proved to be closely related.

To most conveniently compare the structures of these three digestive enzymes, they have been assigned the same amino acid residue numbering scheme. Bovine chymotrypsin is synthesized as an inactive 245-residue precursor named **chymotrypsinogen** that is proteolytically converted to chymotrypsin (Section 15-3E). In what follows, the numbering of the amino acid residues in chymotrypsin, trypsin, and elastase will be that of the corresponding residues in bovine chymotrypsinogen.

The X-ray structure of bovine chymotrypsin was elucidated in 1967 by David Blow. This was followed by the determination of the structures of bovine trypsin (Fig. 15-20) by Robert Stroud and Richard Dickerson, and porcine elastase by David Shotton and Herman Watson. Each of these proteins is folded into two domains, both of which have extensive regions of antiparallel β-sheets in a barrel-like arrangement but contain little helix. *The catalytically essential His 57 and Ser 195 are located at the substrate-binding site together with the invariant (in all serine proteases) Asp 102, which is buried in a solvent-inaccessible pocket. These three residues form a hydrogen bonded constellation referred to as the* **catalytic triad** (Figs. 15-20 and 15-21).

a. The Structural Basis of Substrate Specificity Can Be Quite Complex

The X-ray structures of the above three enzymes suggest the basis for their differing substrate specificities (Table 7-2):

1. In chymotrypsin, the bulky aromatic side chain of the preferred Phe, Trp, or Tyr residue that contributes the

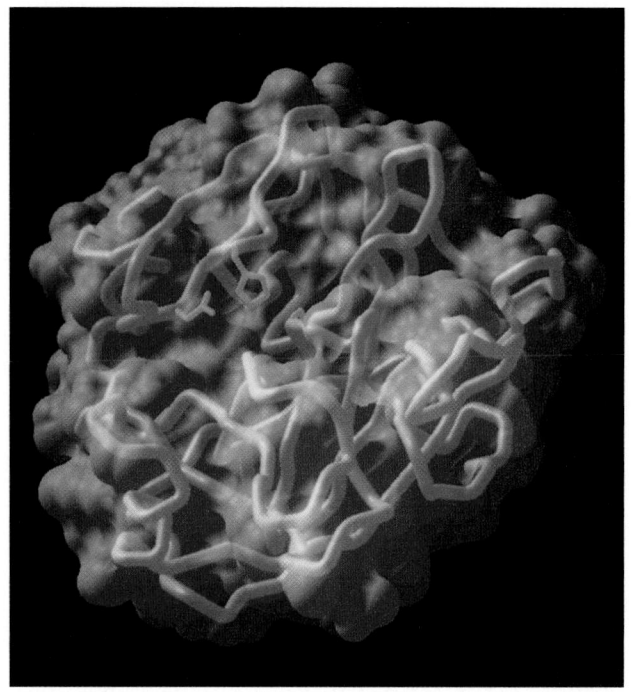

(b)

(c)

FIGURE 15-20 (*Continued*)

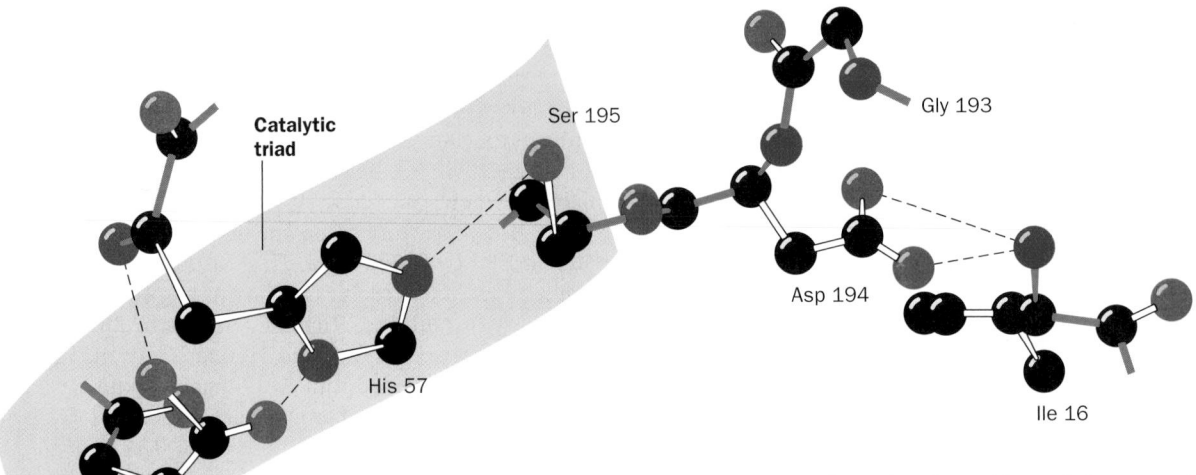

FIGURE 15-21   The active site residues of chymotrypsin. The view is in approximately the same direction as in Fig. 15-20. The catalytic triad consists of Ser 195, His 57, and Asp 102. [After Blow, D.M. and Steitz, T.A., *Annu. Rev. Biochem.* **39,** 86 (1970).]

carbonyl group of the scissile peptide fits snugly into a slit-like hydrophobic pocket, the specificity pocket, that is located near the catalytic groups (Fig. 15-20*a*).

2. In trypsin, the residue corresponding to chymotrypsin Ser 189, which lies at the back of the specificity pocket, is the anionic residue Asp. The cationic side chains of trypsin's preferred residues, Arg or Lys, can therefore form ion pairs with this Asp residue. The rest of chymotrypsin's specificity pocket is preserved in trypsin so that it can accommodate the bulky side chains of Arg and Lys.

3. Elastase is so named because it rapidly hydrolyzes the otherwise nearly indigestible Ala, Gly, and Val-rich protein **elastin** (a connective tissue protein with rubberlike elastic properties). Elastase's specificity pocket is largely occluded by the side chains of a Val and a Thr residue that replace two Gly's lining this pocket in both chymotrypsin and trypsin. Consequently elastase, whose specificity pocket is better described as a depression, specifically cleaves peptide bonds after small neutral residues, particularly Ala. In contrast, chymotrypsin and trypsin hydrolyze such peptide bonds extremely slowly because these small substrates cannot be sufficiently immobilized on the enzyme surface for efficient catalysis to occur (Section 15-1E).

Thus, for example, trypsin catalyzes the hydrolysis of peptidyl amide substrates with an Arg or Lys residue preceding the scissile bond with an efficiency, as measured by k_{cat}/K_M (Section 14-2B), that is 10^6-fold greater than that for the corresponding Phe-containing substrates. Conversely, chymotrypsin catalyzes the hydrolysis of substrates after Phe, Trp, and Tyr residues 10^4-fold more efficiently than after the corresponding Lys-containing substrates.

Despite the foregoing, the mutagenic change in trypsin of Asp 189 → Ser (D189S) by William Rutter did not switch its specificity to that of chymotrypsin but instead yielded

a poor, nonspecific protease. Moreover, even replacing the other three residues in trypsin's specificity pocket that differ from those in chymotrypsin, with those of chymotrypsin, fails to yield a significantly improved enzyme. However, trypsin is converted to a reasonably active chymotrypsin-like enzyme when, in addition to the foregoing changes (collectively designated S1), both of its two surface loops that connect the walls of the specificity pocket, L1 (residues 185–188) and L2 (residues 221–225), are replaced by those of chymotrypsin (termed Tr→Ch[S1+L1+L2]). Although this mutant enzyme still has a low substrate-binding affinity, K_S, the additional mutation Y172W in a third surface loop yields an enzyme (Tr→Ch[S1+L1+L2+Y172W]) that has 15% of chymotrypsin's catalytic efficiency. Curiously, these loops, whose sequences are largely conserved in each enzyme, are not structural components of either the specificity pocket or the extended substrate binding site in chymotrypsin or in trypsin (Fig. 15-20*a*).

Careful comparisons, by Charles Craik and Robert Fletterick, of the X-ray structures of chymotrypsin and trypsin with those of the closely similar Tr→Ch[S1+L1+L2] and Tr→Ch[S1+L1+L2+Y172W] in complex with a Phe-containing chloromethyl ketone inhibitor reveal the structural basis of substrate specificity in trypsin and chymotrypsin. Efficient catalysis in the serine proteases requires that the enzyme's active site be structurally intact and that the substrate's scissile bond be properly positioned relative to the catalytic triad and other components of the active site (see below). The above mutagenic changes do not affect the structure of the catalytic triad or those portions of the active site that bind the substrate's leaving group (that segment on the C-terminal side of the scissile bond). However, the main chain conformation of the conserved Gly 216 (which forms two hydrogen bonds to the backbone of the third residue before the substrate's scissile bond in an antiparallel β pleated sheet–like arrangement) differs in trypsin and chymotrypsin and adopts a chymotrypsin-like structure in both hybrid proteins. Evidently, if Gly 216 adopts a trypsin-like conformation, the scissile bond in Phe-containing substrates is misoriented for efficient catalysis. Thus, despite the fact that Gly 216 is conserved in trypsin and chymotrypsin, the differing structures of loop L2 in the two enzymes maintain it in distinct conformations.

Loop L1, which interacts with L2 in both trypsin and chymotrypsin, is largely disordered in the X-ray structure of Tr→Ch[S1+L1+L2]. Modeling a trypsin-like L1 into Tr→Ch[S1+L1+L2] results in severe steric clashes with the chymotrypsin-like L2. Thus, the requirement of a chymotrypsin-like L1 for the efficient catalysis by Tr→Ch[S1+L1+L2] appears to arise from the need to permit L2 to adopt a chymotrypsin-like conformation.

Residue 172 is located at the base of the specificity pocket. The improvement in substrate binding affinity of Tr→Ch[S1+L1+L2+Y172W] over Tr→Ch[S1+L1+L2] arises from structural rearrangements in this region of the enzyme caused by the increased bulk and different hydrogen bonding requirements of Trp versus Tyr. These changes appear to improve both the structural stability of residues forming the specificity pocket and their specificity for chymotrypsin-like substrates. These results therefore highlight an important caveat for genetic engineers: *Enzymes are so exquisitely tailored to their functions that they often respond to mutagenic tinkering in unexpected ways.*

b. Evolutionary Relationships among Serine Proteases

We have seen that sequence and structural homologies among proteins reveal their evolutionary relationships (Sections 7-3 and 9-6). *The great similarities among chymotrypsin, trypsin, and elastase indicate that these proteins evolved through gene duplications of an ancestral serine protease followed by the divergent evolution of the resulting enzymes (Section 7-3C).*

Several serine proteases from various sources provide further insights into the evolutionary relationships among the serine proteases. **Streptomyces griseus protease A (SGPA)** is a bacterial serine protease of chymotryptic specificity that exhibits extensive structural similarity, although only ~20% sequence identity, with the pancreatic serine proteases. The primordial trypsin gene evidently arose before the divergence of prokaryotes and eukaryotes.

There are three known serine proteases whose primary and tertiary structures bear no discernible relationship to each other or to chymotrypsin but which, nevertheless, contain catalytic triads at their active sites whose structures closely resemble that of chymotrypsin:

1. Subtilisin, an endopeptidase that was originally isolated from *Bacillus subtilis.*

2. Wheat germ **serine carboxypeptidase II,** an exopeptidase whose structure is surprisingly similar to that of

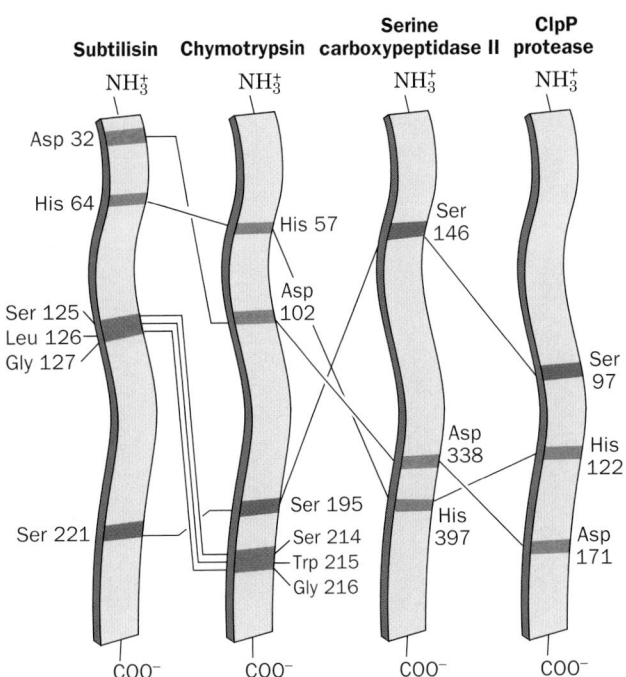

FIGURE 15-22   Relative positions of the active site residues in subtilisin, chymotrypsin, serine carboxypeptidase II, and ClpP protease. The peptide backbones of Ser 214, Trp 215, and Gly 216 in chymotrypsin, and their counterparts in subtilisin, participate in substrate-binding interactions. [After Robertus, J.D., Alden, R.A., Birktoft, J.J., Kraut, J., Powers, J.C., and Wilcox, P.E., *Biochemistry* **11,** 2449 (1972).] ♦♫ **See Kinemage Exercise 10-2**

carboxypeptidase A (Fig. 8-19*a*) even though the latter protease has an entirely different catalytic mechanism from that of the serine proteases (see Problem 3).

3. *E. coli* **ClpP,** which functions in the degradation of cellular proteins (Section 32-6B).

Since the orders of the corresponding active site residues in the amino acid sequences of the four types of serine proteases are quite different (Fig. 15-22), *it seems highly improbable that they could have evolved from a common ancestor serine protease. These proteins apparently constitute a remarkable example of* **convergent evolution:** *Nature seems to have independently discovered the same catalytic mechanism at least four times.* (In addition, **human cytomegalovirus protease,** an essential protein for virus replication that bears no resemblance to the above proteases, has active site Ser and His residues whose relative positions are similar to those in other serine proteases but lacks an active site Asp residue; it appears to have a catalytic dyad.)

C. Catalytic Mechanism

♦♫ **See Guided Exploration 12: The Catalytic Mechanism of Serine Proteases** The extensive active site homologies among the various serine proteases indicate that they all have the same catalytic mechanism. On the basis of considerable chemical and structural data gathered in many laboratories, the following catalytic mechanism has been formulated for the serine proteases, here given in terms of chymotrypsin (Fig. 15-23):

1. After chymotrypsin has bound substrate to form the Michaelis complex, *Ser 195, in the reaction's rate-determining step, nucleophilically attacks the scissile peptide's carbonyl group to form a complex known as the **tetrahedral***

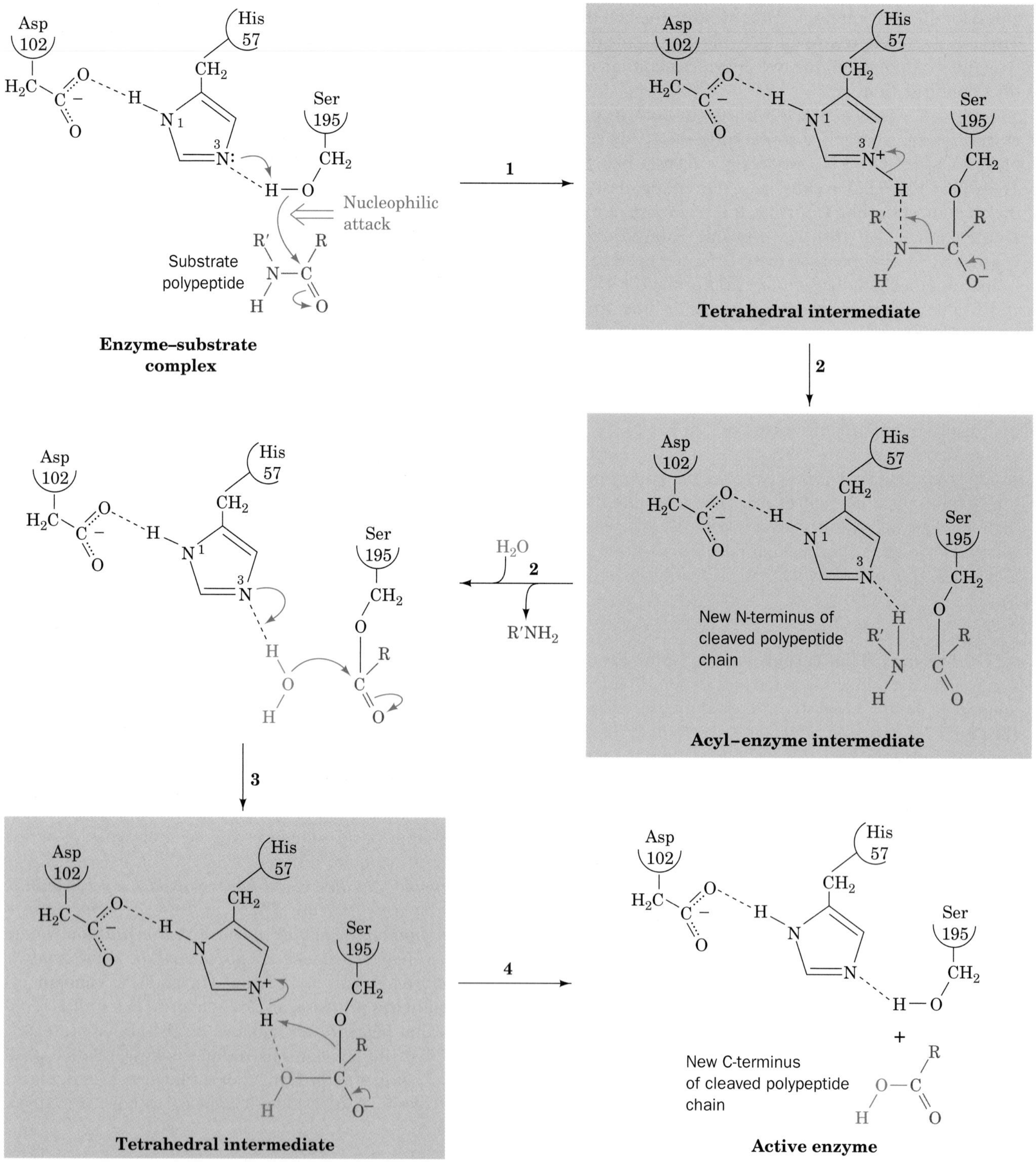

FIGURE 15-23 Catalytic mechanism of the serine proteases. The reaction involves **(1)** the nucleophilic attack of the active site Ser on the carbonyl carbon atom of the scissile peptide bond to form the tetrahedral intermediate; **(2)** the decomposition of the tetrahedral intermediate to the acyl–enzyme intermediate through general acid catalysis by the active site Asp-polarized His, followed by loss of the amine product and its replacement by a water molecule; **(3)** the reversal of Step 2 to form a second tetrahedral intermediate; and **(4)** the reversal of Step 1 to yield the reaction's carboxyl product and the active enzyme.

intermediate (covalent catalysis). X-Ray studies indicate that Ser 195 is ideally positioned to carry out this nucleophilic attack (proximity and orientation effects). The imidazole ring of His 57 takes up the liberated proton, thereby forming an imidazolium ion (general base catalysis). This process is aided by the polarizing effect of the unsolvated carboxylate ion of Asp 102, which is hydrogen bonded to His 57 (electrostatic catalysis; see Section 15-3D). Indeed, the mutagenic replacement of trypsin's Asp 102 by Asn leaves the enzyme's K_M substantially unchanged at neutral pH but reduces its k_{cat} to <0.05% of its wild-type value. Neutron diffraction studies have demonstrated that *Asp 102 remains a carboxylate ion rather than abstracting a proton from the imidazolium ion to form an uncharged carboxylic acid group.* The tetrahedral intermediate has a well-defined, although transient, existence. We shall see that *much of chymotrypsin's catalytic power derives from its preferential binding of the transition state leading to this intermediate (transition state binding catalysis).*

2. The tetrahedral intermediate decomposes to the **acyl–enzyme intermediate** under the driving force of proton donation from N3 of His 57 (general acid catalysis). The amine leaving group ($R'NH_2$, the new N-terminal portion of the cleaved polypeptide chain) is released from the enzyme and replaced by water from the solvent.

3 & 4. The acyl-enzyme intermediate (which, in the absence of enzyme, would be a stable compound) is rapidly deacylated by what is essentially the reverse of the previous steps followed by the release of the resulting carboxylate product (the new C-terminal portion of the cleaved polypeptide chain), thereby regenerating the active enzyme. In this process, water is the attacking nucleophile and Ser 195 is the leaving group.

D. Testing the Catalytic Mechanism

The formulation of the foregoing model for catalysis by serine proteases has prompted numerous investigations of its validity. In this section we discuss several of the most revealing of these studies.

a. The Tetrahedral Intermediate Is Mimicked in a Complex of Trypsin with Trypsin Inhibitor

Convincing structural evidence for the existence of the tetrahedral intermediate was provided by Robert Huber in an X-ray study of the complex between **bovine pancreatic trypsin inhibitor (BPTI)** and trypsin. The 58-residue protein BPTI, whose folding pathway we examined in Section 9-1C, binds to and inactivates trypsin, thereby preventing any trypsin that is prematurely activated in the pancreas from digesting that organ (see Section 15-3E). BPTI binds to the active site region of trypsin across a tightly packed interface that is cross-linked by a complex network of hydrogen bonds. This complex's $10^{13} M^{-1}$ association constant, among the largest of any known protein–protein interaction, emphasizes BPTI's physiological importance.

The portion of BPTI in contact with the trypsin active site resembles bound substrate. The side chain of BPTI Lys 15I (here "I" differentiates BPTI residues from trypsin residues) occupies the trypsin specificity pocket (Fig. 15-24a) and the peptide bond between Lys 15I and Ala 16I is positioned as if it were the scissile peptide bond (Fig. 15-24b). What is most remarkable about this structure is that *its active site complex assumes a conformation well along the reaction coordinate toward the tetrahedral intermediate: The side chain oxygen of trypsin Ser 195, the active Ser, is in closer-than-van der Waals contact (2.6 Å) with the pyramidally distorted carbonyl carbon of BPTI's "scissile" peptide.* Despite this close contact, the proteolytic reaction cannot proceed past this point along the reaction coordinate because of the rigidity of the active site complex and because it is so tightly sealed that the leaving group cannot leave and water cannot enter the reaction site.

Protease inhibitors are common in nature, where they have protective and regulatory functions. For example, certain

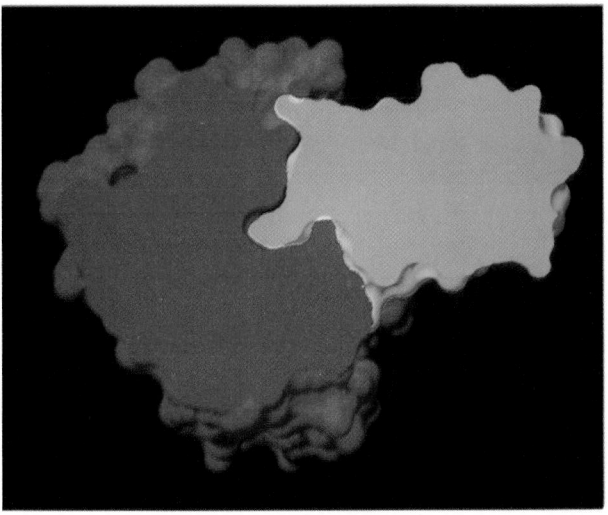

(a)

(b)

FIGURE 15-24   Trypsin–BPTI complex. (*a*) The X-ray structure shown as a cutaway surface drawing indicating how trypsin (*red*) binds BPTI (*green*). The green protrusion extending into the red cavity near the center of the figure represents the Lys 15I side chain occupying trypsin's specificity pocket. Note the close complementary fit of these two proteins. [Courtesy of Michael Connolly, New York University.] (*b*) Trypsin Ser 195, the active Ser, is in closer-than-van der Waals contact with the carbonyl carbon of BPTI's scissile peptide, which is pyramidally distorted toward Ser 195. The normal proteolytic reaction is apparently arrested somewhere along the reaction coordinate between the Michaelis complex and the tetrahedral intermediate.

plants release protease inhibitors in response to insect bites, thereby causing the offending insect to starve by inactivating its digestive enzymes. Protease inhibitors constitute ~10% of the nearly 200 proteins of blood serum. For instance, **α₁-proteinase inhibitor,** which is secreted by the liver, inhibits **leukocyte elastase** (leukocytes are a type of white blood cell; the action of leukocyte elastase is thought to be part of the inflammatory process). Pathological variants of α₁-proteinase inhibitor with reduced activity are associated with **pulmonary emphysema,** a degenerative disease of the lungs resulting from the hydrolysis of its elastic fibers. Smokers also suffer from reduced activity of their α₁-proteinase inhibitor because of the oxidation of its active site Met residue. Full activity of this inhibitor is not regained until several hours after smoking.

b. Serine Proteases Preferentially Bind the Transition State

Detailed comparisons of the X-ray structures of several serine protease–inhibitor complexes have revealed a further structural basis for catalysis in these enzymes (Fig. 15-25):

1. The conformational distortion that occurs with the formation of the tetrahedral intermediate causes the carbonyl oxygen of the scissile peptide to move deeper into the active site so as to occupy a previously unoccupied position, the **oxyanion hole.**

2. *There it forms two hydrogen bonds with the enzyme that cannot form when the carbonyl group is in its normal trigonal conformation.* These two enzymatic hydrogen bond donors were first noted by Joseph Kraut to occupy corresponding positions in chymotrypsin and subtilisin. He proposed the existence of the oxyanion hole based on the

premise that convergent evolution had made the active sites of these unrelated enzymes functionally identical.

3. The tetrahedral distortion, moreover, permits the formation of an otherwise unsatisfied hydrogen bond between the enzyme and the backbone NH group of the residue preceding the scissile peptide. Consequently, *the enzyme binds the tetrahedral intermediate in preference to either the Michaelis complex or the acyl–enzyme intermediate.*

It is this phenomenon that is responsible for much of the catalytic efficiency of serine proteases (see below). In fact, the reason that DIPF is such an effective inhibitor of serine proteases is because its tetrahedral phosphate group makes this compound a transition state analog of the enzyme.

c. The Tetrahedral Intermediate and the Water Molecule Attacking the Acyl–Enzyme Intermediate Have Been Directly Observed

Most enzymatic reactions turn over far too rapidly for their intermediate states to be studied by X-ray or NMR techniques. Consequently, much of our structural knowledge of these intermediate states derives from the study of enzyme–inhibitor complexes or complexes of substrates with inactivated enzymes. Yet the structural relevance of these complexes is subject to doubt precisely because they are catalytically unproductive.

In an effort to rectify this situation for serine proteases, Janos Hadju and Christopher Schofield searched for peptide–protease complexes that are stable at pH's at which the protease is inactive but which could be rendered active by changing the pH. To do so, they screened libraries

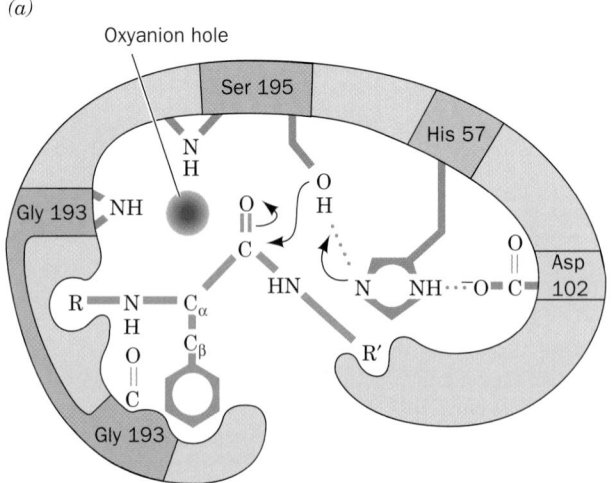

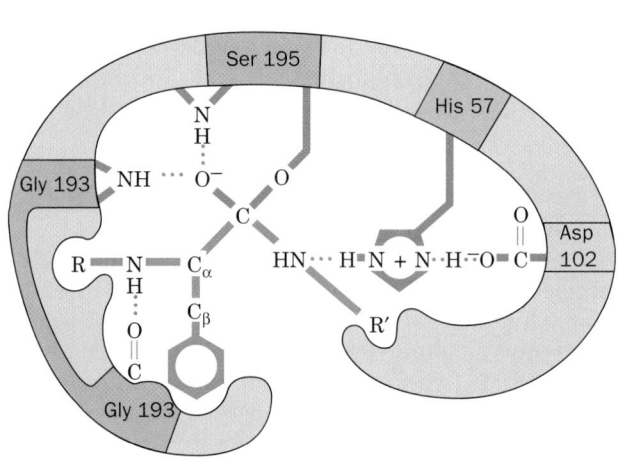

FIGURE 15-25 Transition state stabilization in the serine proteases. (*a*) In the Michaelis complex, the trigonal carbonyl carbon of the scissile peptide is conformationally constrained from binding in the oxyanion hole (*upper left*). (*b*) In the tetrahedral intermediate, the now charged carbonyl oxygen of the scissile peptide (the oxyanion) has entered the oxyanion hole, thereby hydrogen bonding to the backbone NH groups of

Gly 193 and Ser 195. The consequent conformational distortion permits the NH group of the residue preceding the scissile peptide bond to form an otherwise unsatisfied hydrogen bond to Gly 193. Serine proteases therefore preferentially bind the tetrahedral intermediate. [After Robertus, J.D., Kraut, J., Alden, R.A., and Birktoft, J.J., *Biochemistry* **11,** 4302 (1972).]

🔖 See Kinemage Exercise 10-3

of peptides for their ability to bind to porcine pancreatic elastase at pH 3.5 (at which pH His 57 is protonated and hence unable to act as a general base) through the use of ESI-MS (Section 7-1J). They thereby discovered that YPFVEPI, a heptapeptide segment of the human milk protein **β-casein** that is named **BCM7,** forms a complex with elastase, whose mass is consistent with the formation of an ester linkage between BCM7 and the enzyme. In the presence of $^{18}OH_2$ at pH 7.5 (where elastase is active), the ^{18}O label was incorporated into both BCM7 and the elastase–BCM7 complex, thereby demonstrating that the reaction of BCM7 with elastase is reversible at this pH. Fragmentation studies by fast atom bombardment–tandem mass spectroscopy (FAB–MS/MS; Section 7-1J) further revealed that BCM7 that had been incubated with elastase in the presence of $^{18}OH_2$ at pH 7.5 incorporated the ^{18}O label into only its C-terminal Ile residue.

The X-ray structure of the BCM7–elastase complex at pH 5 (Fig. 15-26a) revealed that BCM7's C-terminal carboxyl group, in fact, forms an ester linkage with elastase's Ser 195 side chain hydroxyl group to form the expected acyl–enzyme intermediate. Moreover, this X-ray structure reveals the presence of a bound water molecule that appears poised to nucleophilically attack the ester linkage (the distance from this water molecule to BCM7's C-terminal C atom is 3.1 Å and the line between them is nearly perpendicular to the plane of the acyl group). His 57, which is hydrogen bonded to this water molecule, is properly positioned to abstract one of its protons, thereby activating it for the nucleophilic attack (general base catalysis). The carbonyl O atom of the acyl group occupies the enzyme's oxyanion hole such that it is hydrogen bonded to the main chain N atoms of both Ser 195 and Gly 193. This is in agreement with spectroscopic measurements indicating that the acyl–enzyme intermediate's carbonyl group is, in fact, hydrogen bonded to the oxyanion hole. It was initially assumed that the oxyanion hole acts only to stabilize the tetrahedral oxyanion transition state that resides near the tetrahedral intermediate on the catalytic reaction coordinate. However, it now appears that the oxyanion hole also functions to polarize the carbonyl group of the acyl–enzyme intermediate toward an oxyanion (electrostatic catalysis).

The catalytic reaction was initiated in crystals of the BCM7–elastase complex by transferring them to a buffer at pH 9. After soaking in this buffer for 1 min, the crystals were rapidly frozen in liquid N$_2$ ($-196°C$), thereby arresting the enzymatic reaction (recall that the catalytically essential collective motions of proteins cease at such low temperatures; Section 9-4). The X-ray structure of such a frozen crystal (Fig. 15-26b) revealed that the above acyl–enzyme intermediate had converted to the tetrahedral in-

(a)

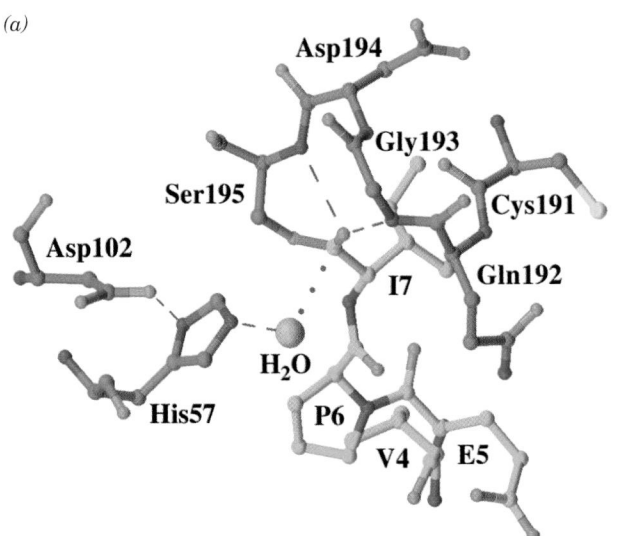

(b)

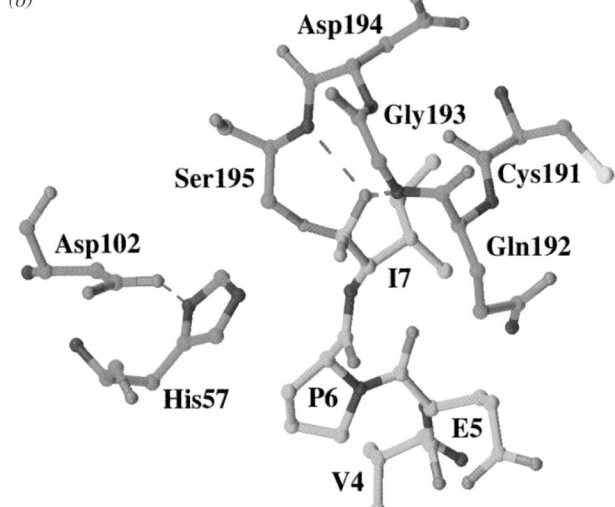

FIGURE 15-26 X-Ray structures of porcine pancreatic elastase in complex with the heptapeptide BCM7 (YPFVEPI). The residues of elastase are specified by the three-letter code and those of BCM7 are specified by the one-letter code. (a) The complex at pH 5. The enzyme's active site residues and the heptapeptide (whose N-terminal three residues are disordered) are shown in ball-and-stick form with elastase C green, BCM7 C cyan, N blue, O red, S yellow, and the bond between the Ser 195 O atom and the C-terminal C atom of BCM7 magenta. The enzyme-bound water molecule, which appears poised to nucleophilically attack the acyl–enzyme's carbonyl C atom, is represented by an orange sphere. The dashed gray lines represent the catalytically important hydrogen bonds and the dotted gray line indicates the trajectory that the bound water molecule presumably follows in nucleophilically attacking the acyl group's carbonyl C atom. (b) The complex after being brought to pH 9 for 1 min and then rapidly frozen in liquid nitrogen. The various groups in the structure are represented and colored as in Part a. Note that the water molecule in Part a has become a hydroxyl substituent (*orange*) to the carbonyl C atom, thereby yielding the tetrahedral intermediate. [Based on X-ray structures by Christopher Schofield and Janos Hadju, University of Oxford, U.K. PDBids (a) 1HAX and (b) 1HAZ.]

termediate, whose oxyanion, as expected, remained hydrogen bonded to the N atoms of Ser 195 and Gly 193. Comparison of this crystal structure with that of the acyl–enzyme intermediate reveals that the enzyme's active site residues do not significantly change their positions in the conversion from the acyl–enzyme intermediate to the tetrahedral intermediate. However, the peptide substrate must do so out of steric necessity when the trigonal planar acyl group converts to the tetrahedral oxyanion (compare Figs. 15-26*a* and 15-26*b*). In response, several enzyme residues that contact the peptide but which are distant from the active site also shift their positions (not shown in Fig. 15-26).

d. The Role of the Catalytic Triad: Low-Barrier Hydrogen Bonds

The earlier literature postulated that the Asp 102-polarized His 57 side chain directly abstracts a proton from Ser 195, thereby converting its weakly nucleophilic —CH_2OH group to a highly nucleophilic alkoxide ion, —CH_2O^-:

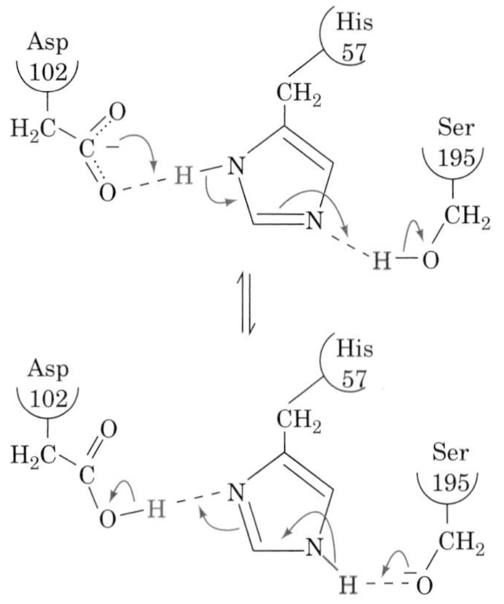

"Charge relay system"

In the process, the anionic charge of Asp 102 was thought to be transferred, via a tautomeric shift of His 57, to Ser 195. The catalytic triad was therefore originally named the **charge relay system.** It is now realized, however, that such a mechanism is implausible because an alkoxide ion ($pK \geq 15$) has far greater proton affinity than does His 57 ($pK \approx 7$, as measured by NMR techniques). How, then, can Asp 102 nucleophilically activate Ser 195?

A possible solution to this conundrum has been pointed out by W.W. Cleland and Maurice Kreevoy and, independently, by John Gerlt and Paul Gassman. Proton transfers between hydrogen bonded groups (D—H $\cdots$ A) only occur at physiologically reasonable rates when the pK of the proton donor is no more than 2 or 3 pH units greater than that of the protonated form of the proton acceptor (the height of the kinetic barrier, $\Delta G^{\ddagger}$, for the protonation of an acceptor by a more basic donor increases with the dif-

ference between the pK's of the donor and acceptor). However, when the pK's of the hydrogen bonding donor (D) and acceptor (A) groups are nearly equal, the distinction between them breaks down: *The hydrogen atom becomes more or less equally shared between them* (D $\cdots$ H $\cdots$ A). Such **low-barrier hydrogen bonds (LBHBs)** are unusually short and strong (they are also known as **short, strong hydrogen bonds**): They have, as studies of model compounds in the gas phase indicate, association free energies as high as -40 to -80 kJ $\cdot$ mol^{-1} versus the -12 to -30 kJ $\cdot$ mol^{-1} for normal hydrogen bonds (the energy of the normally covalent D—H bond is subsumed into the low-barrier hydrogen bonding system) and a D $\cdots$ A length of <2.55 Å for O—H $\cdots$ O and <2.65 Å for N—H $\cdots$ O versus 2.8 to 3.1 Å for normal hydrogen bonds.

LBHBs are unlikely to exist in dilute aqueous solution because water molecules, which are excellent hydrogen bonding donors and acceptors, effectively compete with D—H and A for hydrogen bonding sites. However, LBHBs may exist in nonaqueous solution and in the active sites of enzymes that exclude bulk solvent water. If so, an effective enzymatic "strategy" would be to convert a weak hydrogen bond in the Michaelis complex to a strong hydrogen bond in the transition state, thereby facilitating proton transfer while applying the difference in the free energy between the normal and low-barrier hydrogen bonds to preferentially binding the transition state. In fact, as Perry Frey has shown, the NMR spectrum of the proton linking His 57 to Asp 102 in chymotrypsin (which exhibits a particularly large downfield chemical shift indicative of deshielding) is consistent with the formation of an LBHB in the transition state (see Fig. 15-25*b*; the pK's of protonated His 57 and Asp 102 are nearly equal in the anhydrous environment of the active site complex). This presumably promotes proton transfer from Ser 195 to His 57 as in the charge relay mechanism. Moreover, an ultrahigh (0.78 Å) resolution X-ray structure of *Bacillus lentus* subtilisin by Richard Bott reveals that the hydrogen bond between His 64 and Asp 32 of its catalytic triad has an unusually short N $\cdots$ O distance of 2.62 ± 0.01 Å and that its H atom is nearly centered between the N and O atoms (note that this highly accurate protein X-ray structure is one of the very few in which H atoms are observed and in which short D $\cdots$ A distances are confidently measured).

Although several studies, such as the foregoing, have revealed the existence of unusually short hydrogen bonds in enzyme active sites, it is far more difficult to demonstrate experimentally that they are unusually strong, as LBHBs are predicted to be. In fact, several studies of the strengths of unusually short hydrogen bonds in organic model compounds in nonaqueous solutions suggest that these hydrogen bonds are not unusually strong. Consequently, a lively debate has ensued as to the catalytic significance of LBHBs. Yet if enzymes do not form LBHBs, it remains to be explained how, in numerous widely accepted enzymatic mechanisms that we shall encounter, the conjugate base of an acidic group can abstract a proton from a far more basic group.

e. Much of a Serine Protease's Catalytic Activity Arises from Preferential Transition State Binding

Despite the foregoing, blocking the action of the catalytic triad through the specific methylation of His 57 by treating chymotrypsin with **methyl-*p*-nitrobenzene sulfonate**

Methyl-*p*-nitrobenzene sulfonate

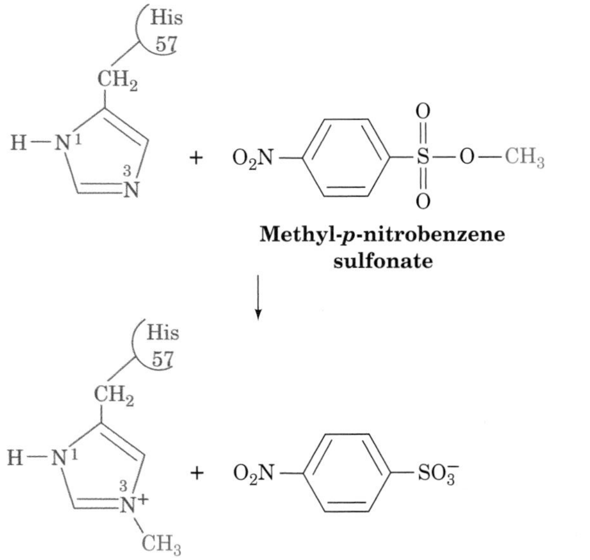

yields an enzyme that is a reasonably good catalyst: It enhances the rate of proteolysis by as much as a factor of 2×10^6 over the uncatalyzed reaction, whereas the native enzyme has a rate enhancement factor of $\sim 10^{10}$. Similarly, the mutation of Ser 195, His 57, or even all three residues of the catalytic triad yields enzymes that enhance proteolysis rates by $\sim 5 \times 10^4$-fold over that of the uncatalyzed reaction. Evidently, the catalytic triad provides a nucleophile and is an alternate source and sink of protons (general acid–base catalysis). However, *a large portion of chymotrypsin's rate enhancement must be attributed to its preferential binding of the catalyzed reaction's transition state.*

E. *Zymogens*

Most proteolytic enzymes are biosynthesized as somewhat larger inactive precursors known as **zymogens** (enzyme precursors, in general, are known as **proenzymes**). In the case of digestive enzymes, the reason for this is clear: If these enzymes were synthesized in their active forms, they would digest the tissues that synthesized them. Indeed, **acute pancreatitis,** a painful and sometimes fatal condition that can be precipitated by pancreatic trauma, is characterized by the premature activation of the digestive enzymes synthesized by this gland.

a. Serine Proteases Are Autocatalytically Activated

Trypsin, chymotrypsin, and elastase are activated according to the following pathways:

Trypsin. The activation of **trypsinogen,** the zymogen of trypsin, occurs as a two-stage process when trypsinogen enters the duodenum from the pancreas. **Enteropeptidase,** a single-pass transmembrane serine protease that is located

in the duodenal mucosa, specifically hydrolyzes trypsinogen's Lys 15—Ile 16 peptide bond, thereby excising its N-terminal hexapeptide (Fig. 15-27). This yields the active enzyme, which has Ile 16 at its N-terminus. Since this activating cleavage occurs at a trypsin-sensitive site (recall that trypsin cleaves after Arg and Lys residues), the small amount of trypsin produced by enteropeptidase also catalyzes activation, generating more trypsin, etc.; that is, trypsinogen activation is autocatalytic.

Chymotrypsin. Chymotrypsinogen is activated by the specific tryptic cleavage of its Arg 15—Ile 16 peptide bond to form **π-chymotrypsin** (Fig. 15-28). π-Chymotrypsin subsequently undergoes autolysis (self-digestion) to specifically excise two dipeptides, Ser 14–Arg 15 and Thr 147–Asn 148, thereby yielding the equally active enzyme **α-chymotrypsin** (heretofore and hereafter referred to as chymotrypsin). The biochemical significance of this latter process, if any, is unknown.

Elastase. Proelastase, the zymogen of elastase, is activated similarly to trypsinogen by a single tryptic cleavage that excises a short N-terminal polypeptide.

b. Biochemical "Strategies" That Prevent Premature Zymogen Activation

Trypsin activates pancreatic **procarboxypeptidases A** and **B** and **prophospholipase A₂** (the action of phospholipase A_2 is outlined in Section 25-1) as well as the pancreatic serine proteases. Premature trypsin activation can consequently trigger a series of events that lead to pancreatic self-digestion. Nature has therefore evolved an elaborate defense against such inappropriate trypsin activation. We have already seen (Section 15-3D) that pancreatic trypsin inhibitor binds essentially irreversibly to any trypsin formed in the pancreas so as to inactivate it. Furthermore, the trypsin-catalyzed activation of trypsinogen (Fig. 15-27) occurs quite slowly, presumably because the unusually large negative charge of its highly evolutionarily conserved N-terminal hexapeptide repels the Asp at the back of trypsin's specificity pocket. Finally, pancreatic zymogens are stored in intracellular vesicles called

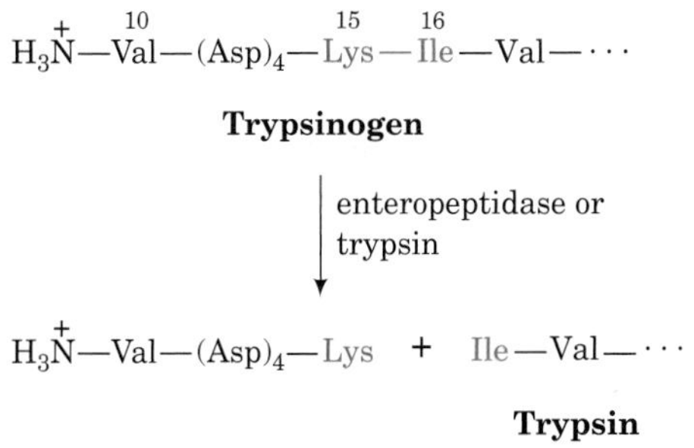

FIGURE 15-27 Activation of trypsinogen to form trypsin. Proteolytic excision of the N-terminal hexapeptide is catalyzed by either enteropeptidase or trypsin. The chymotrypsinogen residue numbering is used here; that is, Val 10 is actually trypsinogen's N-terminus and Ile 16 is trypsin's N-terminus.

FIGURE 15-28 Activation of chymotrypsinogen by proteolytic cleavage. Both π- and α-chymotrypsin are enzymatically active. ♖ **See Kinemage Exercise 10-4**

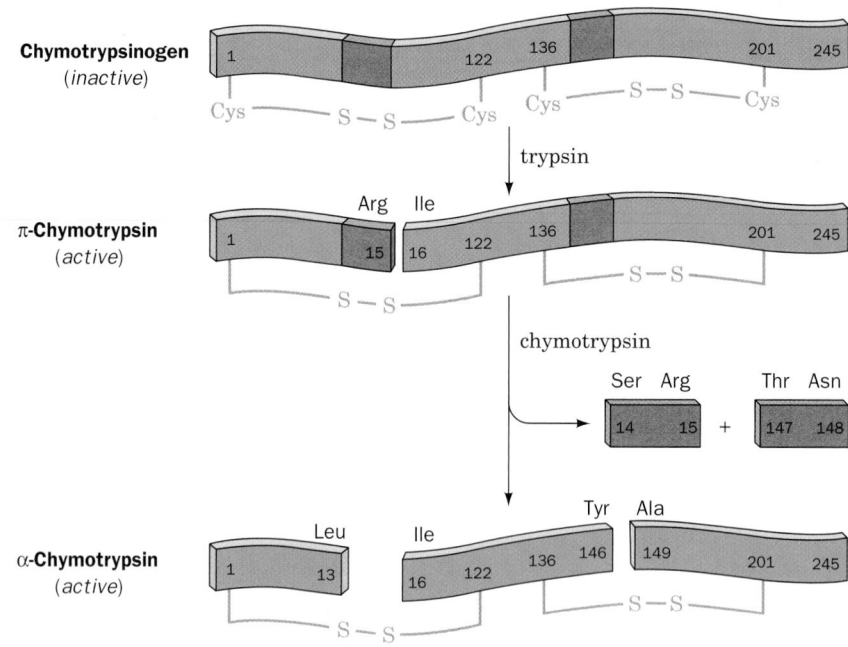

c. Zymogens Have Distorted Active Sites

Since the zymogens of trypsin, chymotrypsin, and elastase have all their catalytic residues, why aren't they enzymatically active? Comparisons of the X-ray structures of trypsinogen with that of trypsin and of chymotrypsinogen with that of chymotrypsin show that on activation, the newly liberated N-terminal Ile 16 residue moves from the surface of the protein to an internal position, where its free cationic amino group forms an ion pair with the invariant anionic Asp 194 (Fig. 15-21). Aside from this change, however, the structures of these zymogens closely resemble those of their corresponding active enzymes. Surprisingly, this resemblance includes their catalytic triads, an observation which led to the discovery that these zymogens are actually enzymatically active, albeit at a very low level. Careful comparisons of the corresponding enzyme and zymogen structures, however, revealed the reason for this low activity: *The zymogens' specificity pockets and oxyanion holes are improperly formed such that, for example, the amide NH of chymotrypsin's Gly 193 points in the wrong direction to form a hydrogen bond with the tetrahedral intermediate (see Fig. 15-25).* Hence, the zymogens' very low enzymatic activity arises from their reduced ability to bind substrate productively and to stabilize the tetrahedral intermediate. These observations provide further structural evidence favoring the role of preferred transition state binding in the catalytic mechanism of serine proteases.

4 ■ DRUG DESIGN

The improvements in medical care over the past several decades are, in large measure, attributable to the development of a huge variety of drugs, which have eliminated

zymogen granules whose membranous walls are thought to be resistant to enzymatic degradation.

or greatly relieved numerous human ailments. Such medications include antibiotics (which have enormously reduced the impact of infectious diseases), anti-inflammatory agents (which reduce the effects of inflammatory diseases such as arthritis), analgesics and anesthetics (which make modern surgical techniques possible), agents that reduce the incidence and severity of cardiovascular disease and stroke, antidepressants, antipsychotics, agents that inhibit stomach acid secretion (which prevent stomach ulcers and heartburn), agents to combat allergies and asthma, immunosuppressants (which make organ transplants possible), agents used for cancer chemotherapy, and a great variety of other substances.

Early human cultures almost certainly recognized both the beneficial and toxic effects of indigenous plant and animal products and used many of them as "medications." Unfortunately, most of these substances were useless or even harmful. Although there were sporadic attempts over the 2500 years preceding the modern era to formulate rational systems of drug discovery, they had little success because they were based mainly on unfounded theories and superstition (e.g., the doctrine of signatures stated that if a plant resembles a particular body part, it must be designed by nature to influence that body part) rather than observation and experiment. Consequently, at the beginning of the 20th century, only three known drugs, apart from folk medicines, were effective in treating specific diseases: (1) **Digitalis,** a heart stimulant extracted from the foxglove plant (Section 20-3A), was used to treat various heart conditions; (2) **quinine** (Section 26-4A), obtained from the bark and roots of the *Cinchona* tree, was used to treat malaria; and (3) mercury was used to treat syphilis (a cure that was often worse than the disease). It was not until several decades later that the rise of the scientific method coupled to the rapidly increasing knowledge of physiology, biochemistry, and chemistry led to effective methods of drug discovery. In fact, the vast majority of drugs in use today were discovered and developed in the past three decades.

In this section we discuss the elements of drug discovery and **pharmacology** (the science of drugs, including their composition, uses, and effects). The section ends with a consideration of one of the major successes of modern drug discovery methods, HIV protease inhibitors.

A. *Techniques of Drug Discovery*

Most drugs act by modifying the function of a particular **receptor** in the body or in an invading pathogen. In most cases, the receptor is a protein to which the drug specifically binds. It may be an enzyme, a transmembrane channel that transports a specific substance into or out of a cell (Chapter 20), and/or a protein that participates in an inter- or intracellular signaling pathway (Chapter 19). In all of these cases, a substance that in binding to a receptor modulates its function is known as an **agonist,** whereas a substance that binds to a receptor without affecting its function but blocks the binding of agonists is called an **antagonist.** The biochemical and physiological effects of a drug and its mechanism of action are referred to as its **pharmacodynamics.**

a. Drug Discovery Is a Complex Procedure

How are new drugs discovered? Nearly all drugs that have been in use for over a decade were discovered by screening large numbers of synthetic compounds and natural products for the desired effect. Drug candidates that are natural products are usually discovered by the fractionation of the organisms in which they occur, which are often plants used in folk remedies of the conditions of interest. Humans having the condition whose treatment is being sought cannot be used as "guinea pigs" in this initial screening process, and even guinea pigs or other laboratory animals such as mice or dogs (if they can be made to be suitable models of the condition under consideration) are too expensive to use on the many thousands of compounds that are usually tested. Thus, *in vitro* **screens** are initially used, such as the degree of binding of a drug candidate to an enzyme that is implicated in a disease of interest, toxicity toward the target bacteria in the search for a new antibiotic, or effects on a line of cultured mammalian cells. However, as the number of drug candidates is winnowed down, more sensitive screens such as testing in animals are employed.

A drug candidate that exhibits a desired effect is called a **lead compound** (or, colloquially, a lead). A good lead compound binds to its target receptor with a dissociation constant, $K_D < 1\ \mu M$. Such a high affinity is necessary to minimize a drug's less specific binding to other macromolecules in the body and to ensure that only low doses of the drug need be taken. For enzyme inhibitors, the dissociation constant is the inhibitor's K_I or K_I' (Section 14-3). Other common measures of the effect of a drug are the **IC$_{50}$,** the *i*nhibitor *c*oncentration at which an enzyme exhibits 50% of its maximal activity; the **ED$_{50}$,** the *e*ffective *d*ose of a drug required to produce a therapeutic effect in 50% of a test sample; the **TD$_{50}$,** the mean *t*oxic *d*ose

required to produce a particular toxic effect in animals; and the **LD$_{50}$,** the mean *l*ethal *d*ose required to kill 50% of a test sample.

For an inhibitor of an enzyme that follows Michaelis-Menten kinetics, the IC$_{50}$ is determined by measuring the ratio v_I/v_0 for several values of [I] at constant [S], where v_I is the initial velocity of the enzyme when the inhibitor concentration is [I]. By dividing equation [14.24] by equation [14.38] with α defined according to equation [14.37], we see that

$$\frac{v_I}{v_0} = \frac{K_M + [S]}{K_M\alpha + [S]} = \frac{K_M + [S]}{K_M\left(1 + \dfrac{[I]}{K_I}\right) + [S]} \quad [15.13]$$

When $v_I/v_0 = 0.5$ (50% inhibition),

$$[I] = [IC_{50}] = K_I\left(1 + \frac{[S]}{K_M}\right) \quad [15.14]$$

Consequently, if the measurements of v_I/v_0 are made with $[S] << K_M$, then $[IC_{50}] = K_I$.

The ratio TD$_{50}$/ED$_{50}$ is defined as a drug's **therapeutic index,** the ratio of the dose of the drug that produces toxicity to that which produces the desired effect. It is, of course, preferable that a drug have a high therapeutic index, but this is not always possible.

b. Cathepsin K Is a Drug Target for Osteoporosis

The development of genomic sequencing techniques (Section 7-2B) and hence the characterization of tens of thousands of previously unknown genes is providing an enormous number of potential drug targets. For example, **osteoporosis,** a condition that afflicts postmenopausal women and elderly men, is characterized by the progressive loss of bone mass leading to a greatly increased frequency of bone fracture, particularly of the hip, spine, and wrist. Bones consist of a protein matrix that is >90% type I collagen (Section 8-2B), in which spindle- or plate-shaped crystals of **hydroxyapatite,** $Ca_5(PO_4)_3OH$, are embedded. Bones are by no means static structures. They undergo continuous remodeling through the countervailing action of two types of bone cells: **osteoblasts,** which synthesize bone's organic matrix in which its mineral component is laid down; and **osteoclasts,** which solubilize mineralized bone matrix through the secretion of proteolytic enzymes into an extracellular bone resorption pit, which is maintained at pH 4.5. The acidic solution dissolves the bone's mineral component, thereby exposing its protein matrix to proteolytic degradation. Osteoporosis arises when bone resorption outstrips bone formation.

In the search for a drug target for osteoporosis, a cDNA library (Sections 5-5E and 5-5F) was prepared from an **osteoclastoma** (a cancer derived from osteoclasts; normally osteoclasts are very rare cells). Around 4% of these cDNAs encode a heretofore unknown protease, which was named **cathepsin K** (cathepsins are proteases that occur in the lysosome). Further studies, both at the cDNA and protein levels, indicated that cathepsin K is only expressed at high levels in osteoclasts. Microscopic examination of osteoclasts that had been stained with antibodies directed against cathepsin K revealed that this enzyme is localized at the contact site between osteoclasts and the bone resorption pit. Subsequently, it was shown that mutations in the gene encoding cathepsin K are the cause of **pycnodysostosis,** a rare hereditary disease which is characterized by hardened and fragile bones, short stature, skull deformities, and osteoclasts that demineralize bone normally but do not degrade its protein matrix. Evidently, cathepsin K functions to degrade the protein matrix of bone and hence is an attractive drug target for the treatment of osteoporosis.

c. SARs and QSARs Are Useful Tools for Drug Discovery

A lead compound is used as a point of departure to design more efficacious compounds. Experience has shown that even minor modifications to a drug candidate can result in major changes in its pharmacological properties. Thus, one might place methyl, chloro, hydroxyl, or benzyl groups at various places on a lead compound in an effort to improve its pharmacodynamics. For most drugs in use today, 5 to 10 thousand related compounds were typically synthesized in generating the medicinally useful drug. These were not random procedures but were guided by experience as medicinal chemists tested various derivatives

of a lead compound: For those compounds that had improved efficacy, derivatives were made and tested; etc. This process has been systematized through the use of **structure–activity relationships (SARs):** the determination, via synthesis and screening, of which groups on a lead compound are important for its drug function and which are not. For example, if a phenyl group on a lead compound interacts hydrophobically with a flat region of its receptor, then hydrogenating the phenyl ring to form a nonplanar cyclohexane ring will yield a compound with reduced affinity for the receptor.

A logical extension of the SAR concept is to quantify it, that is, to determine a **quantitative structure–activity relationship (QSAR).** This idea is based on the premise that there is a relatively simple mathematical relationship between the biological activity of a drug and its physicochemical properties. For instance, if the hydrophobicity of a drug is important for its biological activity, then changing the substituents on the drug so as to alter its hydrophobicity will affect its activity. A measure of the substance's hydrophobicity is its **partition coefficient,** P, between the two immiscible solvents, octanol and water, at equilibrium:

$$P = \frac{\text{concentration of drug in octanol}}{\text{concentration of drug in water}} \quad [15.15]$$

Biological activity may be expressed as $1/C$, where C is the drug concentration required to achieve a specified level of biological function (e.g., IC_{50}). Then a plot of $\log 1/C$ versus $\log P$ (the use of logarithms keeps the plot on a manageable scale) for a series of derivatives of the lead compound having a relatively small range of $\log P$ values often indicates a linear relationship (Fig. 15-29a), which can therefore be expressed:

$$\log\left(\frac{1}{C}\right) = k_1 \log P + k_2 \quad [15.16]$$

Here k_1 and k_2 are constants, whose optimum values in this QSAR can be determined by computerized curve-fitting methods. For compounds with a larger range of $\log P$ values, it is likely that a plot of $\log 1/C$ versus $\log P$ will have a maximum value (Fig. 15-29b) and hence be better described by a quadratic equation:

$$\log\left(\frac{1}{C}\right) = k_1(\log P)^2 + k_2 \log P + k_3 \quad [15.17]$$

Of course, the biological activities of few substances depend only on their hydrophobicities. A QSAR can therefore simultaneously take into account several physicochemical properties of substituents such as their pK values, van der Waals radii, hydrogen bonding energy, and conformation. The values of the constants for each of the terms in a QSAR is indicative of the contribution of that term to the drug's activity. The use of QSARs to optimize the biological activity of a lead compound has proven to be a valuable tool in drug discovery.

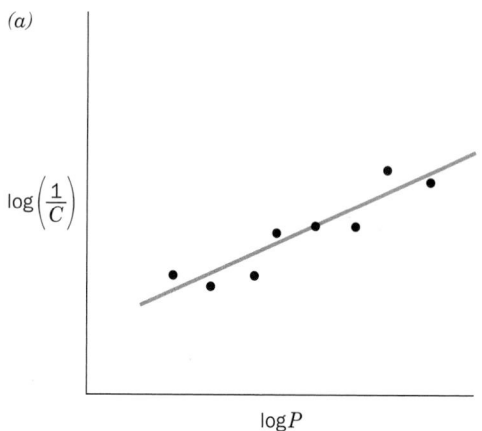

(a)

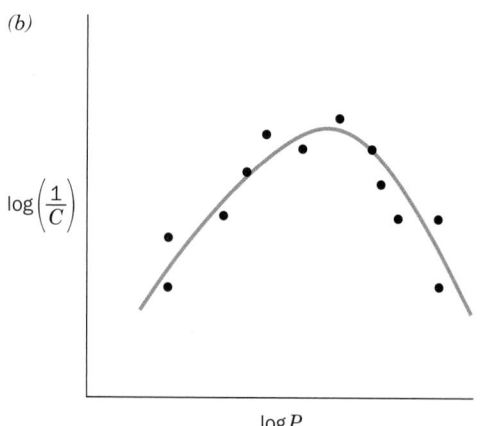

(b)

FIGURE 15-29 Hypothetical QSAR plots of log(1/C) versus log P for a series of related compounds. (*a*) A plot that is best described by a linear equation. (*b*) A plot that is best described by a quadratic equation.

d. Structure-Based Drug Design

Since the mid 1980s, dramatic advances in the speed and precision with which a macromolecular structure can be determined by X-ray crystallography and NMR (Section 8-3A) have enabled **structure-based drug design,** a process that greatly reduces the number of compounds that need be synthesized in a drug discovery program. As its name implies, structure-based drug design (also called **rational drug design**) uses the structure of a receptor in complex with a drug candidate to guide the development of more efficacious compounds. Such a structure will reveal, for example, the positions of the hydrogen bonding donors and acceptors in a receptor binding site as well as cavities in the binding site into which substituents might be placed on a drug candidate to increase its binding affinity for the receptor. These direct visualization techniques are usually supplemented with molecular modeling tools such as the computation of the minimum energy conformation of a proposed derivative, quantum mechanical calculations that determine its charge distribution and hence how it would interact electrostatically with the receptor, and docking simulations in which an inhibitor candidate is computationally modeled into the binding site on the receptor to assess potential interactions. Structure-based drug design is an iterative process: The structure of the receptor in complex with a compound with improved properties is determined in an effort to further improve its properties.

e. Combinatorial Chemistry and High-Throughput Screening

As structure-based methods were developed, it appeared that they would become the dominant mode of drug discovery. However, the recent advent of **combinatorial chemistry** techniques to rapidly and inexpensively synthesize large numbers of related compounds combined with the development of robotic **high-throughput screening** techniques has caused the drug discovery "pendulum" to again swing toward the "make-many-compounds-and-see-what-they-do" approach. A familiar example of combinatorial chemistry is the parallel synthesis of the large number of different oligonucleotides on a DNA chip (Section 7-6B). Similarly, if a lead compound can be synthesized in a stepwise manner from several smaller modules, then the substituents on each of these modules can be varied in parallel to produce a library of related compounds (e.g., Fig. 15-30).

A variety of synthetic techniques have been developed that permit the combinatorial synthesis of thousands of related compounds in a single procedure. Thus, whereas investigations into the importance of a hydrophobic group at a particular position in a lead compound might previously have prompted the individual syntheses of only the ethyl, propyl, and benzyl derivatives of the compound, the use of combinatorial synthesis would permit the generation of perhaps 100 different groups at that position. This would far more effectively map out the potential range of the substituent and possibly identify an unexpectedly active analog. Interestingly, QSAR and computational techniques have been combined in the development of "virtual combinatorial chemistry," a procedure in which libraries of compounds are computationally "synthesized" and "analyzed" to predict their efficacy, thereby again reducing the number of compounds that must actually be synthesized in order to generate an effective drug.

B. Introduction to Pharmacology

The *in vitro* development of an effective drug candidate is only the first step in the drug development process. *Besides causing the desired response in its isolated target receptor, a useful drug must be delivered in sufficiently high concentration to this receptor where it resides in the human body without causing unacceptable side effects.*

a. Pharmacokinetics Is a Multifaceted Phenomenon

The most convenient form of drug administration is orally (by mouth). In order to reach its target receptor, a drug administered in this way must surmount a series of formidable barriers: (1) It must be chemically stable in the highly acidic (pH 1) environment of the stomach and must not be degraded by the digestive enzymes in the gastrointestinal tract; (2) it must be absorbed from the gastrointestinal tract into the bloodstream, that is, it must pass through several cell membranes; (3) it must not bind too tightly to other substances in the body (e.g., lipophilic substances tend to be absorbed by certain plasma proteins and by fat tissue; anions may be bound by plasma proteins, mainly **albumin;** and cations may be bound by nucleic acids); (4) it must survive derivatization by the battery of enzymes, mainly in the liver, that function to detoxify **xenobiotics** (foreign compounds), as discussed below (note that the intestinal blood flow drains directly into the liver via the portal vein, so that the liver processes all orally ingested substances before they reach the rest of the body); (5) it must avoid rapid excretion by the kidneys; (6) it must pass from the capillaries to its target tissue; (7) if it is targeted to the brain, it must cross the **blood–brain barrier,** which blocks the passage of most polar substances; and (8) if it is targeted to an intracellular receptor, it must pass through the plasma membrane and, possibly, other intracellular membranes. The ways in which a drug interacts with these various barriers is known as its **pharmacokinetics.** Thus, the **bioavailability** of a drug (the extent to which it reaches its site of action, which is usually taken to be the systemic circulation) depends on both the dose given and its pharmacokinetics. Of course, barriers (1) and (2) can be circumvented by injecting the drug [e.g., some forms

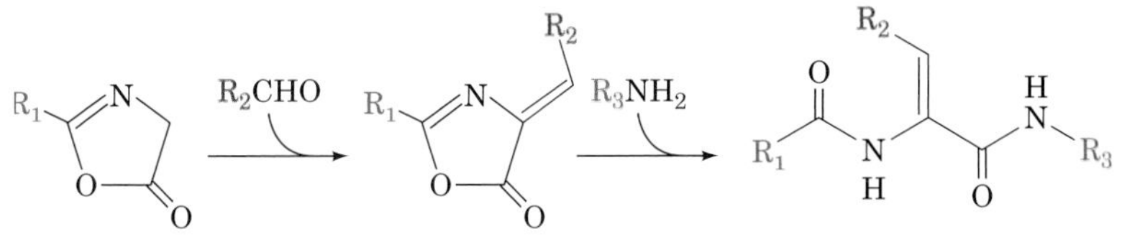

FIGURE 15-30 The combinatorial synthesis of arylidene diamides. If ten different variants of each R group are used in the synthesis, then 1000 different derivatives will be synthesized.

of penicillin (Fig. 11-25) must be injected because their functionally essential β-lactam rings are highly susceptible to acid hydrolysis], but this mode of drug delivery is undesirable for long-term use.

Since the pharmacokinetics of a drug candidate is as important to its efficacy as is its pharmacodynamics, both must be optimized in producing a medicinally useful drug. The following empirically based rules, formulated by Christopher Lipinski and known as **Lipinski's "rule of five,"** state that a compound is likely to exhibit poor absorption or permeation if:

1. Its molecular mass is greater than 500 D.

2. It has more than 5 hydrogen bond donors (expressed as the sum of its OH and NH groups).

3. It has more than 10 hydrogen bond acceptors (expressed as the sum of its N and O atoms).

4. Its value of log P is greater than 5.

Drug candidates that disobey Rule 1 are likely to have low solubilities and to only pass through cell membranes with difficulty; those that disobey Rules 2 and/or 3 are likely to be too polar to pass through cell membranes; and those that disobey Rule 4 are likely to be poorly soluble in aqueous solution and hence unable to gain access to membrane surfaces. Thus, *the most effective drugs are usually a compromise; they are neither too lipophilic nor too hydrophilic.* In addition, their pK values are usually in the range 6 to 8 so that they can readily assume both their ionized and unionized forms at physiological pH's. This permits them to cross cell membranes in their unionized form and to bind to their receptor in their ionized form. However, since the concentration of a drug at its receptor depends, as we saw, on many different factors, the pharmacokinetics of a drug candidate may be greatly affected by even small chemical changes. QSARs and other computational tools have been developed to predict these effects but they are, as yet, rather crude.

b. Toxicity and Adverse Reactions Eliminate Most Drug Candidates

The final criteria that a drug candidate must meet are that its use be safe and efficacious in humans. Tests for these properties are initially carried out in animals, but since humans and animals often react quite differently to a particular drug, the drug must ultimately be tested in humans through **clinical trials.** In the United States, clinical trials are monitored by the Food and Drug Administration (FDA) and have three increasingly detailed (and expensive) phases:

Phase I. This phase is primarily designed to test the safety of a drug candidate but is also used to determine its dosage range and the optimal dosage method (e.g., orally vs injection) and frequency. It is usually carried out on a small number (20–100) of normal, healthy volunteers, but in the case of a drug candidate known to be highly toxic (e.g., a cancer chemotherapeutic agent), it is carried out on volunteer patients with the target disease.

Phase II. This phase mainly tests the efficacy of the drug against the target disease in 100 to 500 volunteer patients but also refines the dosage range and checks for side effects. The effects of the drug candidate are usually assessed via **single blind tests,** procedures in which the patient is unaware of whether he/she has received the drug or a control substance. Usually the control substance is a **placebo** (an inert substance with the same physical appearance, taste, etc., as the drug being tested) but, in the case of a life-threatening disease, it is an ethical necessity that the control substance be the best available treatment against the disease.

Phase III. This phase monitors adverse reactions from long-term use as well as confirming efficacy in 1000 to 5000 patients. It pits the drug candidate against control substances through the statistical analysis of carefully designed **double blind tests,** procedures in which neither the patients nor the clinical investigators evaluating the patients' responses to the drug know whether a given patient has received the drug or a control substance. This is done to minimize bias in the subjective judgments the investigators must make.

Currently, only about 5 drug candidates in 5000 that enter preclinical trials reach clinical trials. Of these, only one, on average, is ultimately approved for clinical use, with ~40% of drug candidates passing Phase I trials and ~50% of those passing Phase II trials (most drug candidates that enter Phase III trials are successful). In recent years, the preclinical portion of a drug discovery process has averaged ~3 years to complete, whereas successful clinical trials have usually required an additional 7 to 10 years. These successive stages of the drug discovery process are increasingly expensive, so that to successfully bring a drug to market costs, on average, around $300 million.

The most time-consuming and expensive aspect of a drug development program is identifying a drug candidate's rare adverse reactions. Nevertheless, it is not an uncommon experience for a drug to be brought to market only to be withdrawn some months or years later when it is found to have caused unanticipated life-threatening side effects in as few as 1 in 10,000 individuals (the search for new applications of an approved drug and its postmarketing surveillance are known as its Phase IV clinical trials). For example, in 1997, the FDA withdrew its approval of the drug **fenfluramine (fen),**

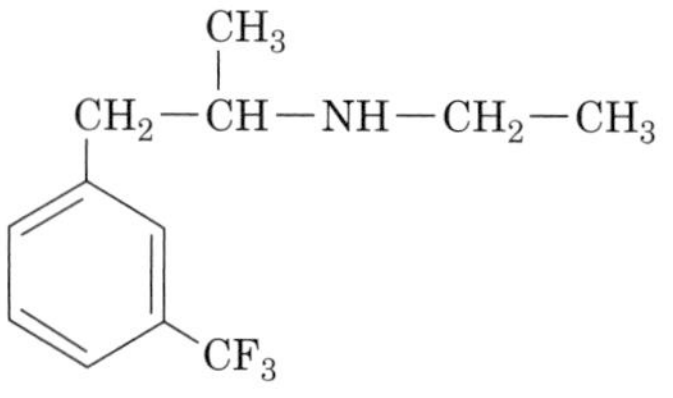

Fenfluramine

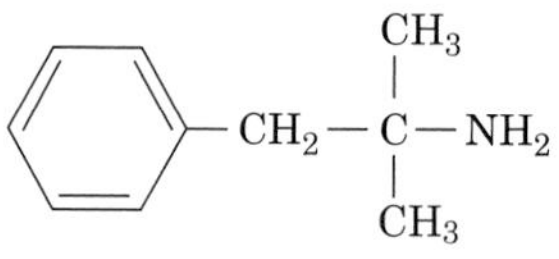

Phentermine

which it had approved in 1973 for use as an appetite suppressant in short-term (a few weeks) weight loss programs. Fenfluramine had become widely prescribed, often for extended periods, together with another appetite suppressant, **phentermine (phen;** approved in 1959), a combination known as **fen-phen** (although the FDA had not approved of the use of the two drugs in combination, once it approves a drug for any purpose, a physician may prescribe it for any other purpose). The withdrawal of fenfluramine was prompted by over 100 reports of heart valve damage in individuals (mostly woman) who had taken fen-phen for an average of 12 months (phentermine was not withdrawn because the evidence indicated that fenfluramine was the responsible agent). This rare side effect had not been observed in the clinical trials of fenfluramine, in part because, being an extremely unusual type of drug reaction, it had not been screened for.

c. The Cytochromes P450 Metabolize Most Drugs

Why is it that a drug that is well tolerated by the majority of patients can pose such a danger to others? *Differences in reactions to drugs arise from genetic differences among individuals as well as differences in their disease states, other drugs they are taking, age, sex, and environmental factors.* The **cytochromes P450,** which function in large part to detoxify xenobiotics and participate in the metabolic clearance of the majority of drugs in use, provide instructive examples of these phenomena.

The cytochromes P450 constitute a superfamily of heme-containing enzymes that occur in nearly all living organisms, from bacteria to mammals [their name arises from the characteristic 450-nm peak in their absorption spectra when reacted in their Fe(II) state with CO]. Humans express ~100 **isozymes** (catalytically and structurally similar but genetically distinct enzymes from the same organism; also called **isoforms**) of cytochromes P450, mainly in the liver but also in other tissues (its various isozymes are named by the letters "CYP" followed by a number designating its family, an uppercase letter designating its subfamily, and often another number; e.g., CYP2D6). These **monooxygenases** (Fig. 15-31), which in animals are embedded in the endoplasmic reticulum membrane, catalyze reactions of the sort

$$RH + O_2 + 2H^+ + 2e^- \rightleftharpoons ROH + H_2O$$

The electrons (e^-) are supplied by NADPH, which passes them to cytochrome P450's heme prosthetic group via the intermediacy of the enzyme **cytochrome P450 reductase.** Here RH represents a wide variety of usually lipophilic compounds for which the different cytochromes P450 are specific. They include polycyclic aromatic hydrocarbons [PAHs, frequently carcinogenic (cancer-causing) compounds that are present in tobacco smoke, broiled meats, and other pyrolysis products], polycyclic biphenyls (PCBs, which were widely used in electrical insulators and as plasticizers and are also carcinogenic), steroids (in whose syntheses cytochromes P450 participate; Sections 25-6A and 25-6C), and many different types of drugs. The xenobiotics are thereby converted to a more water-soluble form, which aids in their excretion by the kidneys. Moreover, the newly generated hydroxyl groups are often enzymatically conju-

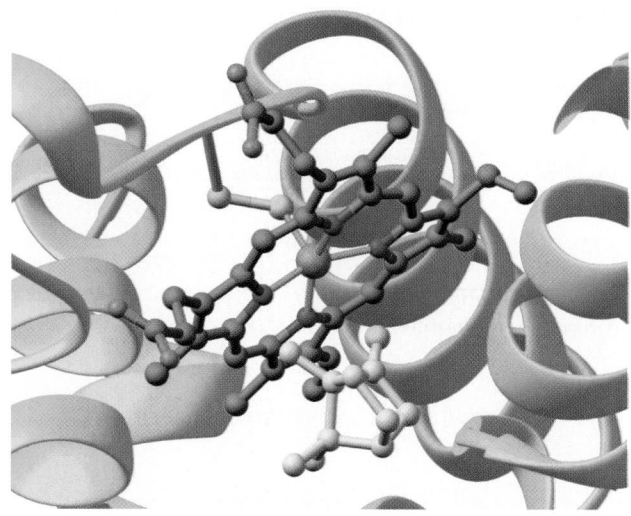

FIGURE 15-31   X-Ray structure of cytochrome P450$_{CAM}$ from *Pseudomonas putida* **showing its active site region.** The heme group, the Cys side chain that axially ligands its Fe atom, and the enzyme's lipophilic substrate **thiocamphor** are shown in ball-and-stick form with N blue, O red, S yellow, Fe orange, and the C atoms of the heme, its liganding Cys side chain, and the thiocamphor green, cyan, and pale blue-green, respectively. The bonds liganding the Fe are gray. [Based on an X-ray structure by Thomas Poulos, University of California at Irvine. PDBid 8CPP.]

gated (covalently linked) to polar substances such as glucuronic acid (Section 11-1C), glycine, sulfate, and acetate, which further enhances aqueous solubility. The many types of cytochromes P450 in animals, which have different substrate specificities (although these specificities tend to be broad and hence often overlap), are thought to have arisen in response to the numerous toxins which plants produce, presumably to discourage animals from eating them.

Drug–drug interactions are often mediated by cytochromes P450. For example, if drug A is metabolized by or otherwise inhibits a cytochrome P450 isozyme that metabolizes drug B, then coadministering drugs A and B will cause the bioavailability of drug B to increase above the value it would have had if it alone had been administered. This phenomenon is of particular concern if drug B has a low therapeutic index. Conversely, if, as is often the case, drug A induces the increased expression of the cytochrome P450 isozyme that metabolizes it and drug B, then coadministering drugs A and B will reduce drug B's bioavailability, a phenomenon that was first noted when certain antibiotics caused oral contraceptives to lose their efficacy. Moreover, if drug B is metabolized to a toxic product, its increased rate of reaction may result in an adverse reaction. Environmental pollutants such as PAHs or PCBs are also known to induce the expression of specific cytochrome P450 isozymes and thereby alter the rates at which certain drugs are metabolized. Finally, some of these same effects may occur in patients with liver disease, as well as arising from age-based, gender-based, and individual differences in liver physiology.

Although the cytochromes P450 presumably evolved to detoxify and/or help eliminate harmful substances, in sev-

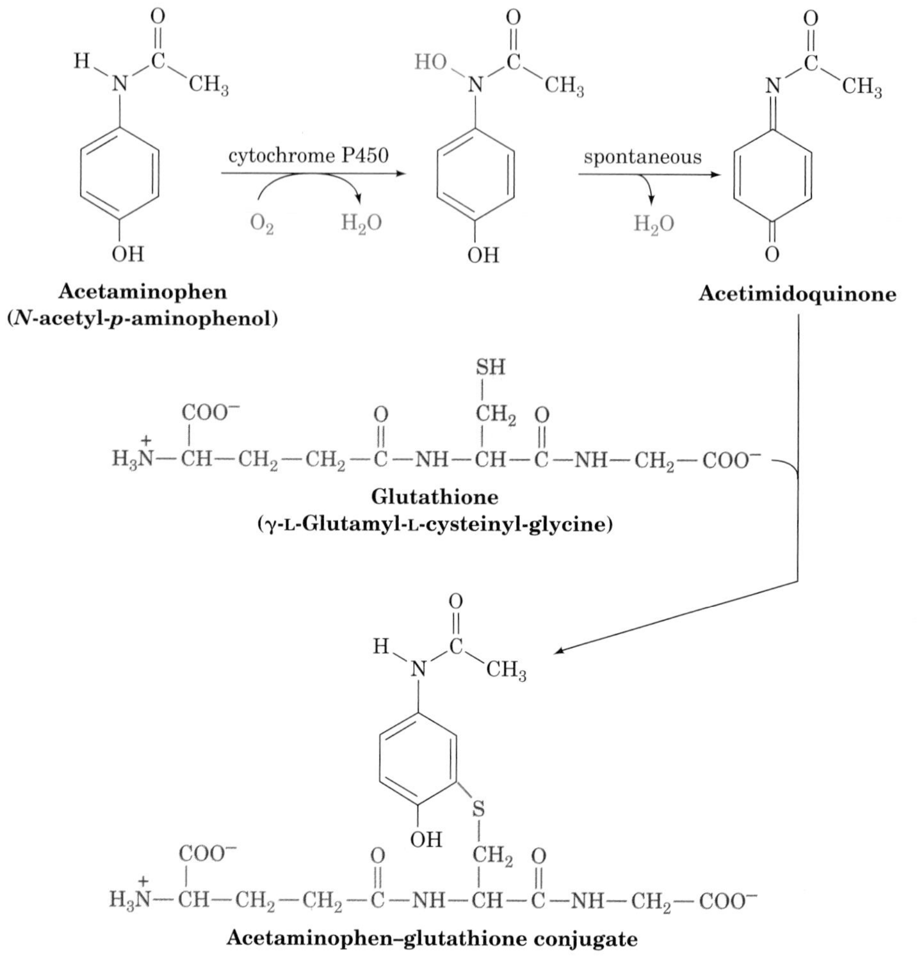

FIGURE 15-32 The metabolic reactions of acetaminophen that convert it to its conjugate with glutathione.

eral cases they have been shown to participate in converting relatively innocuous compounds to toxic agents. For example, **acetaminophen** (Fig. 15-32), a widely used analgesic and antipyretic (fever reducer), is quite safe when taken in therapeutic doses (1.2 g/day for an adult) but in large doses (>10 g) is highly toxic. This is because, in therapeutic amounts, 95% of the acetaminophen present is enzymatically glucuronidated or sulfated at its —OH group to the corresponding conjugates, which are readily excreted. The remaining 5% is converted, through the action of a cytochrome P450 (CYP2E1), to **acetimidoquinone** (Fig. 15-32), which is then conjugated with **glutathione,** a tripeptide with an unusual γ-amide bond that participates in a wide variety of metabolic processes (Section 26-4C). However, when acetaminophen is taken in large amounts, the glucuronidation and sulfation pathways become saturated and hence the cytochrome P450-mediated pathway becomes increasingly important. If hepatic (liver) glutathione is depleted faster than it can be replaced, acetimidoquinone, a reactive compound, instead conjugates with the sulfhydryl groups of cellular proteins, resulting in often fatal hepatotoxicity.

Many of the cytochromes P450 in humans are unusually **polymorphic,** that is, there are several common alleles (variants) of the genes encoding these enzymes. Alleles that cause diminished, enhanced, and qualitatively altered rates of drug metabolism have been characterized for many of the cytochromes P450. The distributions of these various alleles differ markedly among ethnic groups and hence probably arose to permit each group to cope with the toxins in its particular diet.

Polymorphism in a given cytochrome P450 results in differences between individuals in the rates at which they metabolize certain drugs. For instance, in cases that a cytochrome P450 variant has absent or diminished activity, otherwise standard doses of a drug that the enzyme normally metabolizes may cause the bioavailability of the drug to reach toxic levels. Conversely, if a particular P450 enzyme has enhanced activity (usually because the gene encoding it has been duplicated one or more times), higher than normal doses of a drug that the enzyme metabolizes would have to be administered to obtain the required therapeutic effect. However, if the drug is metabolized to a toxic product, this may result in an adverse reaction. Several known P450 variants have altered substrate specificities and hence produce unusual metabolites, which also may cause harmful side effects.

Experience has amply demonstrated that *there is no such thing as a drug that is entirely free of adverse reactions.* However, as the enzymes and their variants that participate in drug metabolism are characterized and rapid and inexpensive genotyping methods are developed, it may become possible to tailor drug treatment to an individual's genetic makeup rather than to the population as a whole.

C. HIV Protease and Its Inhibitors

Acquired immunodeficiency syndrome (AIDS), the only major epidemic attributable to a previously unknown pathogen to appear in the 20th century (it was first described in 1981), is caused by **human immunodeficiency virus type 1** (**HIV-1;** the closely related **HIV-2,** which we shall not explicitly discuss here, also causes AIDS and has a similar response to drugs). HIV-1, which was discovered in 1983, is a **retrovirus,** a family of viruses that were independently characterized in 1970 by David Baltimore and Howard Temin. The retroviral genome is a single-stranded RNA that reproduces inside its host cell by transcribing the RNA to double-stranded DNA in a process mediated by the virally encoded enzyme **reverse transcriptase** (Section 30-4C). The DNA is then inserted into the host cell's chromosomal DNA by a viral enzyme named **integrase** and is passively replicated along with the cell's DNA. However, under activating conditions (which for HIV-1 often is an infection by another pathogen), the retroviral DNA is transcribed, the proteins it encodes are expressed and inserted in or anchored to the host cell plasma membrane, and new **virions** (virus particles) are produced by the budding out of a viral protein-laden segment of plasma membrane so as to enclose viral RNA (Fig. 15-33).

HIV-1 is targeted to and specifically replicates within **helper T cells,** essential components of the immune system (Section 35-2A). Unlike most types of retroviruses, HIV-1 eventually kills the cells producing it. Although the helper T cells within which HIV-1 are actively replicating are often destroyed by the immune system, those within which the HIV-1 is latent (its DNA is not being transcribed) are not detected by the immune system and hence provide a reservoir of HIV-1 (other types of cells also harbor HIV-1). Consequently, over a several year period after the initial infection (during most of which the host exhibits no obvious symptoms), the host's immune system is steadily depleted until it has deteriorated to the point that the host regularly falls victim to and is eventually killed by opportunistic pathogens that individuals with normally functioning immune systems can readily withstand. It is this latter stage of an HIV infection that is called AIDS. In the absence of effective therapy, AIDS is almost invariably fatal. Through the year 2002, an estimated 30 million people had died of AIDS and an estimated 42 million others, largely in sub-Saharan Africa, were HIV-positive, numbers that are increasing at the rate of ~5 million per year. As a consequence of this global catastrophe, HIV has been characterized and effective countermeasures against it have been devised faster than for any other pathogen in history.

a. Reverse Transcriptase Inhibitors Are Only Partially Effective

The first drug to be approved by the FDA (in 1987) to fight AIDS was **3′-azido-3′-deoxythymidine (AZT; zidovudine),**

$$HOCH_2 \quad O \quad T$$
$$N = N = N \quad H$$

**3′-Azido-3′-deoxythymidine
(AZT; zidovudine)**

which had first been synthesized in 1964 as a possible anticancer agent (it was ineffective). AZT is a nucleoside analog that, on enzymatic conversion to its triphosphate in the cell (the plasma membrane is impermeable to nucleoside triphosphates), inhibits HIV-1 reverse transcriptase, as do the several other drugs (Section 30-4C) that the FDA had approved to treat AIDS prior to 1996. Unfortunately, these agents only slow the progression of an HIV infection but do not stop it. This is in part because they are toxic, mainly

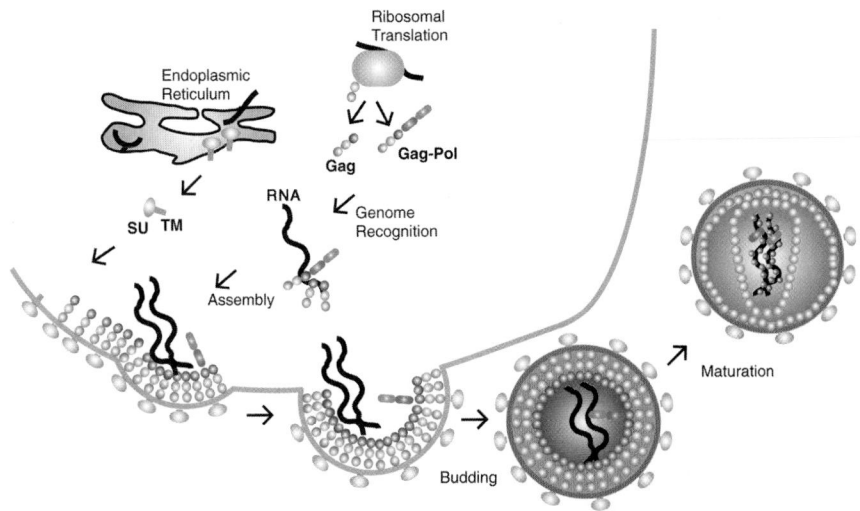

FIGURE 15-33 The assembly, budding, and maturation of HIV-1. SU is the surface glycoprotein **gp120** and TM is the transmembrane protein **gp41.** [After Turner, B.G. and Summers, M.F., *J. Mol. Biol.* **285,** 4 (1999).]

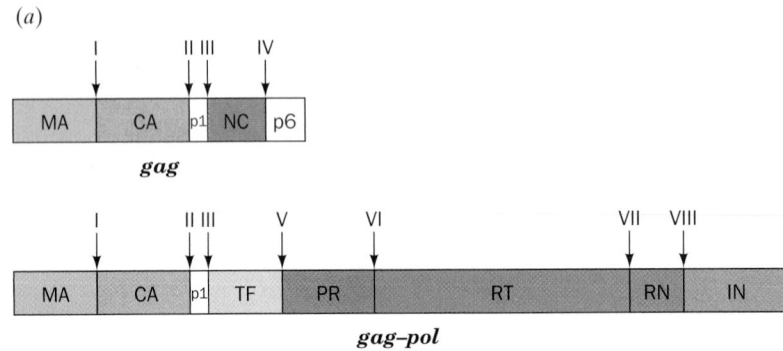

FIGURE 15-34   HIV-1 polyproteins. (*a*) The organization of
the HIV-1 gag and gag–pol polyproteins. The symbols used are
MA, matrix protein; CA, capsid protein; NC, nucleocapsid
protein; TF, transmembrane protein; PR, protease; RT, reverse
transcriptase; RN, ribonuclease; and IN, integrase. (*b*) The
sequences flanking the HIV-1 protease cleavage sites (*red
bonds*) indicated in Part *a*.

Cleavage site	Sequence
I	··· Ser - Gln - Asn - Tyr — Pro - Ile - Val - Gln ···
II	··· Ala - Arg - Val - Leu — Ala - Glu - Ala - Met ···
III	··· Ala - Thr - Ile - Met — Met - Gln - Arg - Gly ···
IV	··· Pro - Gly - Asn - Phe — Leu - Gln - Ser - Arg ···
V	··· Ser - Phe - Asn - Phe — Pro - Gln - Ile - Thr ···
VI	··· Thr - Leu - Asn - Phe — Pro - Ile - Ser - Pro ···
VII	··· Ala - Glu - Thr - Phe — Tyr - Val - Asp - Gly ···
VIII	··· Arg - Lys - Ile - Leu — Phe - Leu - Asp - Gly ···

to the bone marrow cells that are blood cell precursors,
and hence cannot be taken in large doses. More important,
however, is that reverse transcriptase, unlike most other
DNA polymerases (Section 30-2A), cannot correct its mistakes and hence frequently generates mutations (about one
per 10^4 bp and, since the viral genome consists of $\sim 10^4$ bp,
each viral genome bears, on average, one new mutation).
Consequently, *under the selective pressure of an anti-HIV
drug such as AZT, the drug's target receptor rapidly evolves
to a drug-resistant form.*

b.   HIV-1 Polyproteins Are Cleaved by HIV-1 Protease

HIV-1, as do other retroviruses, synthesizes its proteins
in the form of **polyproteins,** which each consist of several
tandemly linked proteins (Fig. 15-34). HIV-1 encodes two
polyproteins, **gag** (55 kD) and **gag–pol** (160 kD), which are
both anchored to the plasma membrane via N-terminal
myristoylation (Section 12-3B). These polyproteins are
then cleaved to their component proteins through the action of **HIV-1 protease,** but only after this enzyme has excised itself from gag–pol. This process occurs only after the
virion has budded off from the host cell and results in a
large structural reorganization of the virion (Fig. 15-33).
The virion is thereby converted from its noninfectious immature form to its pathogenic mature form. If HIV-1 protease is inactivated, either mutagenically or by an inhibitor,
the virion remains noninfectious. Hence HIV-1 protease is
an opportune drug target.

c.   Aspartic Proteases and Their Catalytic Mechanism

HIV-1 protease is a member of the **aspartic protease**
family (also known as **acid proteases**), so called because

these enzymes all contain catalytically essential Asp
residues that occur in the signature sequence Asp–
Thr/Ser–Gly. Humans have several known aspartic proteases including **pepsin,** a digestive enzyme secreted by the
stomach (its specificity is indicated in Table 7-2) that functions at pH 1 and which was the first enzyme to be recognized (named in 1825 by T. Schwann); **chymosin** (formerly
rennin), a stomach enzyme, occurring mainly in infants,
that specifically cleaves a Phe–Met peptide bond in the
milk protein κ-casein, thereby causing milk to curdle, making it easier to digest (calf stomach chymosin has been used
for millennia to make cheese); **cathepsins D** and **E,** lysosomal proteases that function to degrade cellular proteins;
renin, which participates in the regulation of blood pressure and electrolyte balance (Fig. 15-35); and **β-secretase**
(also known as **memapsin 2**), a transmembrane protein
common in brain that participates in cleaving Aβ precursor protein to yield amyloid-β protein (Aβ), which is implicated in Alzheimer's disease (Section 9-5B). In addition,
many fungi secrete aspartic proteases, presumably to aid
them in invading the tissues they colonize.

Eukaryotic aspartic proteases are ~330-residue monomeric proteins. The X-ray structure of pepsin (Fig. 15-36*a*),
which closely resembles those of other eukaryotic aspartic
proteases, reveals that this croissant-shaped protein consists of two homologous domains that are related by
approximate 2-fold symmetry (although only about 25
residues in the core β sheets of each domain are closely
related by this symmetry). Each domain contains a catalytically essential Asp in an analogous position. The X-
ray structures of enzyme–inhibitor complexes of various
aspartate proteases indicate that substrates bind in a
prominent cleft between the two domains that could ac-

¹Asp-Arg-Val-Tyr-Ile-His-Pro-Phe-His-Leu-Val-Ile-His¹³

Angiotensinogen

H_2O ⟍ renin

¹Asp-Arg-Val-Tyr-Ile-His-Pro-Phe-His-Leu¹⁰ + Val-Ile-His

Angiotensin I

H_2O ⟍ angiotensin converting enzyme (ACE)

¹Asp-Arg-Val-Tyr-Ile-His-Pro-Phe⁸ + His-Leu

Angiotensin II

FIGURE 15-35 Renin participation in blood pressure regulation. Renin proteolytically cleaves the 13-residue polypeptide **angiotensinogen** to the 10-residue polypeptide **angiotensin I.** This latter peptide is then cleaved by **angiotensin converting enzyme (ACE)** to the 8-residue polypeptide **angiotensin II,** which, on binding to its receptor, induces vasoconstriction and retention of Na^+ and water by the kidneys, resulting in increased blood pressure. Consequently there have been considerable efforts to develop both renin and ACE inhibitors for the control of **hypertension** (high blood pressure), although as yet, only ACE inhibitors have been approved as drugs.

commodate an ~8-residue polypeptide segment in an extended β sheetlike conformation. The active site Asp residues are located at the base of this cleft (Fig. 15-36a).

What is the catalytic mechanism of eukaryotic aspartic proteases? Proteolytic enzymes, in general, have three essential catalytic components:

1. A nucleophile to attack the carbonyl C atom of the scissile peptide to form a tetrahedral intermediate (Ser 195 serves this function in trypsin; Fig. 15-23).

2. An electrophile to stabilize the negative charge that develops on the carbonyl O atom of the tetrahedral intermediate (the H-bonding donors lining the oxyanion hole, Gly 193 and Ser 195, do so in trypsin; Fig. 15-25).

3. A proton donor so as to make the amide N atom of the scissile peptide a good leaving group (the imidazolium group of His 57 in trypsin; Fig. 15-23).

Pepsin's pH rate profile (Section 14-4) suggests that it has two ionizable essential residues, one with $pK \approx 1.1$ and the other with $pK \approx 4.7$, which are almost certainly the carboxyl groups of its essential Asp residues. At the pH of the stomach, the Asp residue with pK 4.7 is protonated and that with pK 1.1 is partially ionized. This suggests that the ionized carboxyl group acts as a nucleophile to form the putative tetrahedral intermediate. However, no covalent intermediate between an aspartic protease and its substrate has ever been detected.

The two active site Asp residues in eukaryotic aspartic proteases are in close proximity and both appear to form hydrogen bonds to a bridging water molecule that is present in several X-ray structures of eukaryotic aspartic proteases (Fig. 15-36b). This, together with a variety of enzymological and kinetic data, led Thomas Meeks to propose

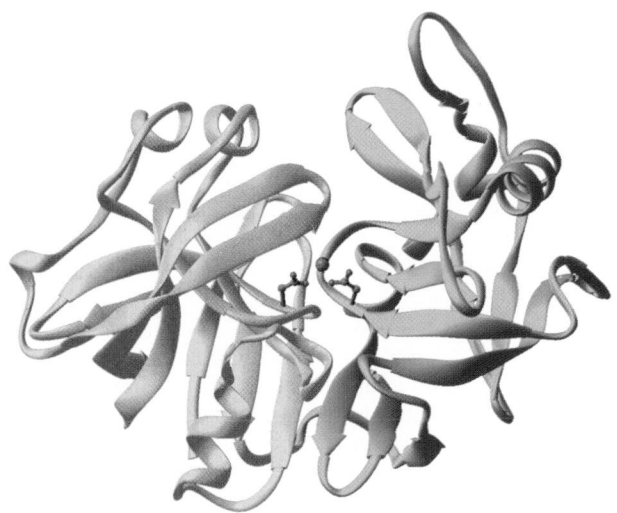

(a)

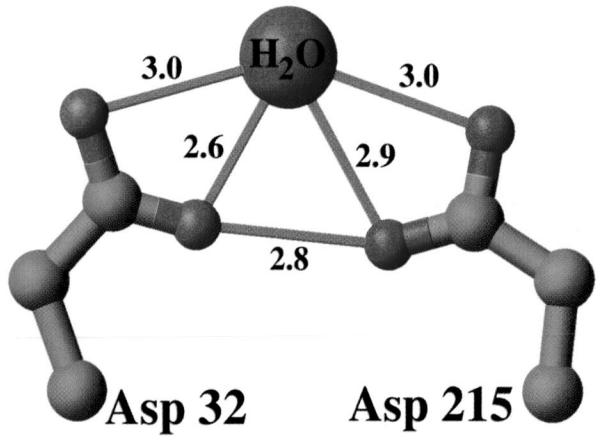

(b)

FIGURE 15-36 X-Ray structure of pepsin. (*a*) Ribbon diagram in which the N-terminal domain (residues 1–172) is gold, the C-terminal domain (residues 173–326) is cyan, the side chains of the active site Asp residues are shown in ball-and-stick form with C green and O red, and the water molecule that is bound by these Asp side chains is represented by a large red sphere. The protein is viewed with the pseudo-2-fold axis relating core portions of the two domains tipped from vertical toward the viewer. (*b*) Enlarged view of the active site Asp residues and their bound water molecule indicating the lengths (in Å) of possible hydrogen bonds (*thin gray bonds*). The X-ray structures of other aspartic proteases exhibit similar interatomic distances. [Based on an X-ray structure by Anita Sielecki and Michael James, University of Alberta, Edmonton, Canada. PDBid 4PEP.]

the following catalytic mechanism for aspartic proteases (Fig. 15-37):

1. An active site Asp carboxylate group, acting as a general base, activates the bound water molecule, the so-called lytic water, to nucleophilically attack the scissile peptide's carbonyl C as an OH⁻ ion. Proton donation (general acid catalysis) by the second, previously uncharged active site Asp stabilizes the oxyanion that would otherwise form in the resulting tetrahedral intermediate.

2. The N atom of the scissile peptide is protonated by the first Asp (general acid catalysis) resulting, through charge rearrangement and proton transfer to the second Asp (general base catalysis), in amide bond scission.

Aspartic proteases are inhibited by compounds with tetrahedral carbon atoms at a position mimicking a scissile peptide bond (see below). This strongly suggests that these enzymes preferentially bind their transition states (transition state stabilization), thereby enhancing catalysis.

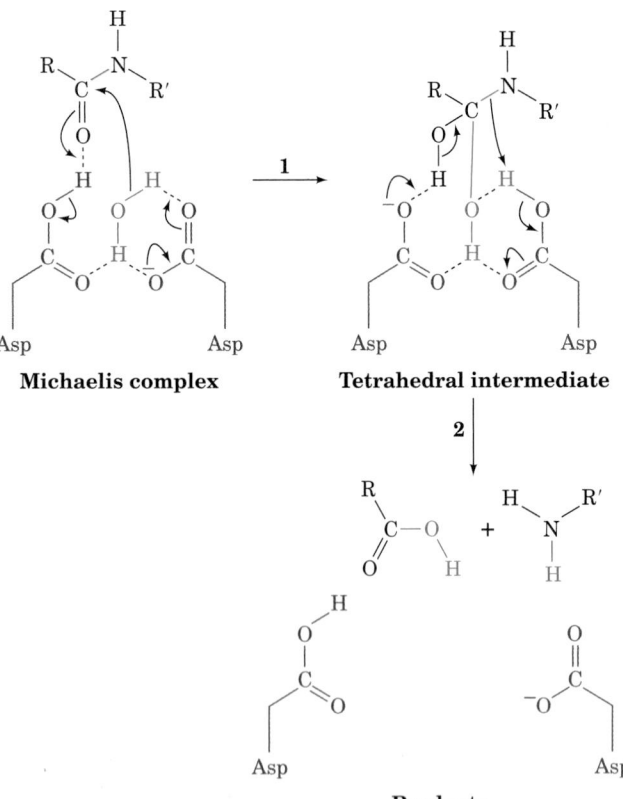

FIGURE 15-37 Catalytic mechanism of aspartic proteases.
(1) The nucleophilic attack of the enzyme-activated water molecule (*red*) on the carbonyl carbon atom of the scissile peptide bond (*green*) to form the tetrahedral intermediate. This reaction step is promoted by general base catalysis by the Asp on the right and general acid catalysis by the Asp on the left (*blue*). **(2)** The decomposition of the tetrahedral intermediate to form products via general acid catalysis by the Asp on the right and general base catalysis by the Asp on the left.

d. HIV-1 Protease Inhibitors Are Effective Anti-AIDS Agents

HIV-1 protease differs from eukaryotic aspartic proteases in that it is a homodimer of 99-residue subunits. Nevertheless, its X-ray structure (Fig. 15-38a), determined independently in 1989 by Alexander Wlodawer, by Manual Navia and Paula Fitzgerald, and by Tom Blundell, closely resembles those of eukaryotic aspartic proteases. Thus, HIV-1 protease has the enzymatically unusual property that its single active site is formed by two identical sym-

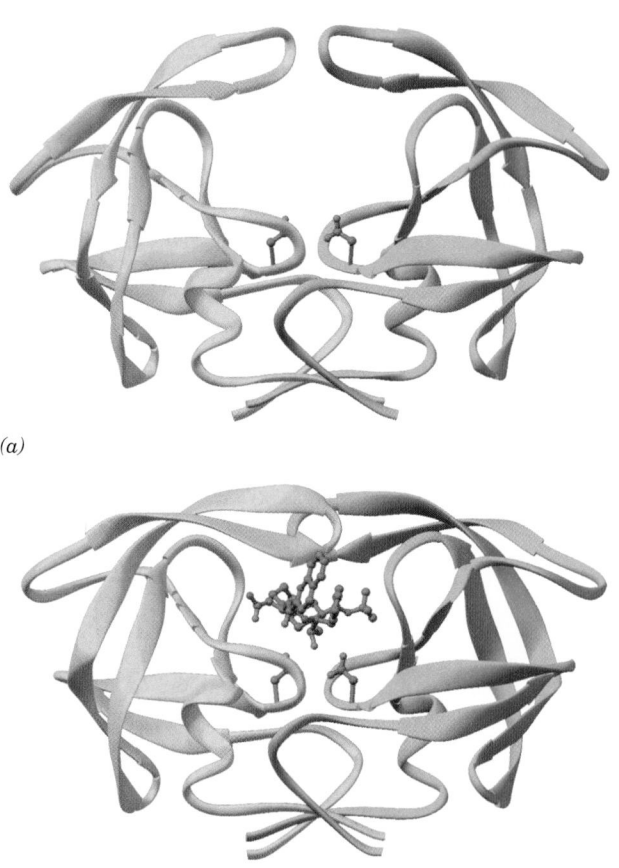

(a)

(b)

FIGURE 15-38 X-Ray structure of HIV-1 protease.
(a) Uncomplexed and (b) in complex with its inhibitor saquinavir (structural formula in Fig. 15-41). In each structure, the homodimeric protein is viewed with its 2-fold axis of symmetry vertical and is shown as a ribbon diagram with one subunit gold and the other cyan. The side chains of the active site Asp residues, Asp 25 and Asp 25′, as well as the saquinavir in Part *b*, are shown in ball-and-stick form with C green, N blue, and O red. Note how the β hairpin "flaps" at the top of the uncomplexed enzyme have folded down over the inhibitor in the saquinavir complex. Compare these structures with that of the similarly viewed pepsin in Fig. 15-36a. [Part *a* based on an X-ray structure by Tom Blundell, Birkbeck College, London, U.K., and Part *b* based on an X-ray structure by Robert Crowther, Hoffmann-LaRoche Ltd., Nutley, New Jersey. PDBids (a) 3PHV and (b) 1HXB.]

See the Interactive Exercises

metrically arranged subunits. Quite possibly HIV-1 protease resembles the putative primordial aspartic protease that, through gene duplication, evolved to form the eukaryotic enzymes (although HIV-1 protease is well suited to the limited amount of genetic information that a virus can carry).

Once the structure of HIV-1 protease became available, intensive efforts were mounted in numerous laboratories to find therapeutically effective inhibitors of this enzyme. In this process, ~200 X-ray structures and several NMR structures have been reported of HIV-1 protease, its mutants, and the proteases of other retroviruses, both alone and in their complexes with a great variety of inhibitors. Hence, HIV-1 protease is perhaps the most exhaustively structurally studied protein.

Comparison of the X-ray structure of HIV-1 protease alone (Fig. 15-38a) with that of its complexes with polypeptidelike inhibitors (e.g., Fig. 15-38b) reveals that, on binding an inhibitor, the β hairpin "flaps" covering the "top" of the substrate-binding cleft move down by as much as 7 Å to enclose the inhibitor. Such an inhibitor binds to the 2-fold symmetric enzyme in a two-fold pseudosymmetric extended conformation such that the inhibitor interacts with the enzyme much like a strand in a β sheet (Fig. 15-39). On the "floor" of the binding cleft, each signature sequence (Asp 25–Thr 26–Gly 27) is located in a loop that is stabilized by a network of hydrogen bonds similar to that observed in eukaryotic aspartic proteases. The inhibitor interacts with the enzyme via a hydrogen bond to the active site residue Asp 25. However, contrary to the case for eukaryotic aspartic proteases (Fig. 15-36b), no X-ray structure of an HIV-1 protease contains a water molecule within hydrogen bonding distance of Asp 25 or Asp 25'. On the flap side of the binding cleft, the inhibitor interacts with Gly 48 and Gly 48' and with a water molecule that is not the attacking nucleophile but which mediates the contacts between the flaps and the inhibitor backbone.

Although HIV-1 protease specifically cleaves the gag and gag–pol polyproteins at a total of 8 sites (Fig. 15-34b), these sites appear to have little in common except that their immediately flanking residues are nonpolar and mostly bulky. Indeed, binding studies indicate that HIV-1 pro-

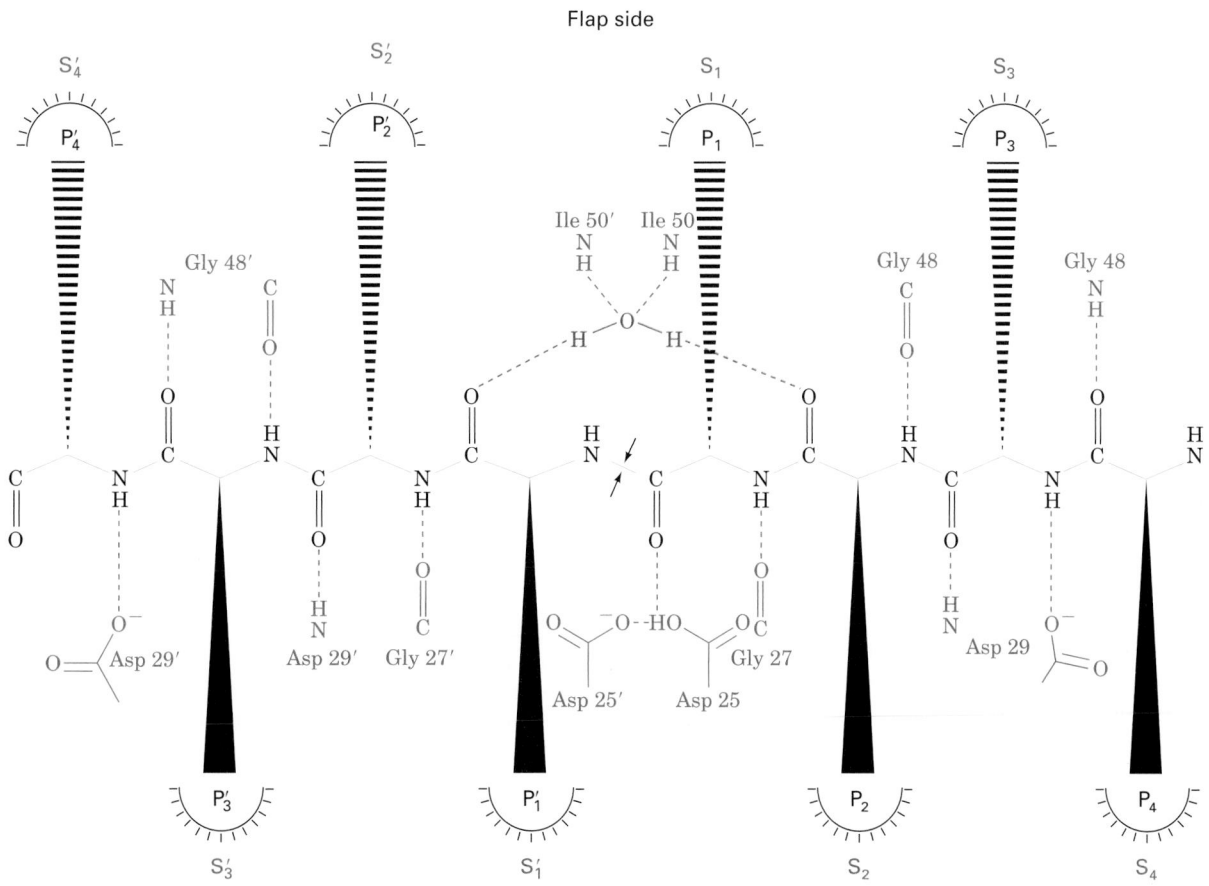

FIGURE 15-39   Arrangement of hydrogen bonds between HIV-1 protease and a modeled substrate. In the nomenclature used here, polypeptide residues in one subunit are assigned primed numbers to differentiate them from the residues of the other subunit; substrate residues on the N-terminal side of the scissile peptide bond are designated P_1, P_2, P_3, ... , counting toward the N-terminus; substrate residues on its C-terminal side are designated P_1', P_2', P_3', ... , counting toward the C-terminus; and the symbols S_1, S_2, S_3, ... , and S_1', S_2', S_3', ... , designate the enzyme's corresponding residue-binding subsites. The scissile peptide bond is marked by arrows. [After Wlodawer, A. and Vondrasek, J., *Annu. Rev. Biophys. Biomol. Struct.* **27,** 257 (1998).]

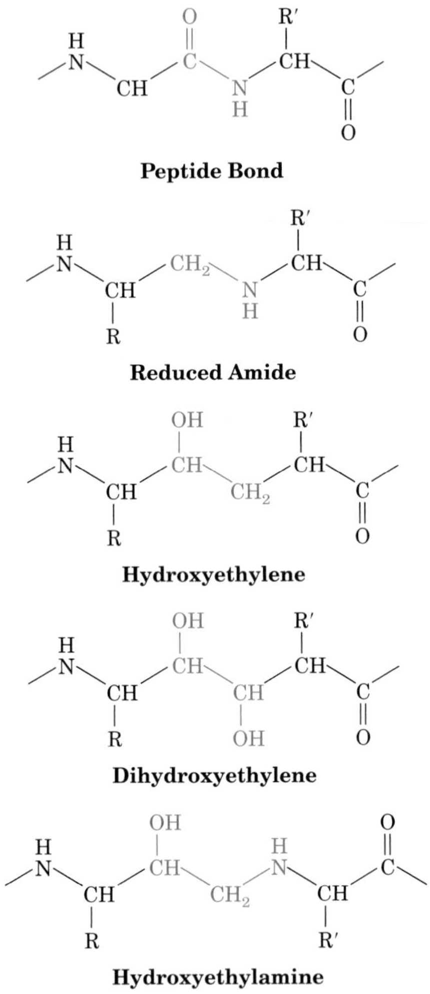

Peptide Bond

Reduced Amide

Hydroxyethylene

Dihydroxyethylene

Hydroxyethylamine

FIGURE 15-40   **Comparison of a normal peptide bond (*top*) to a selection of groups (*red*) that are isosteres (stereochemical analogs) of the tetrahedral intermediate in reactions catalyzed by aspartic proteases.**

Indinavir (Crixivan™)

Nelfinavir (Viracept™)

Ritonavir (Norvir™)

Saquinavir (Invirase™)

Amprenavir (Agenerase™)

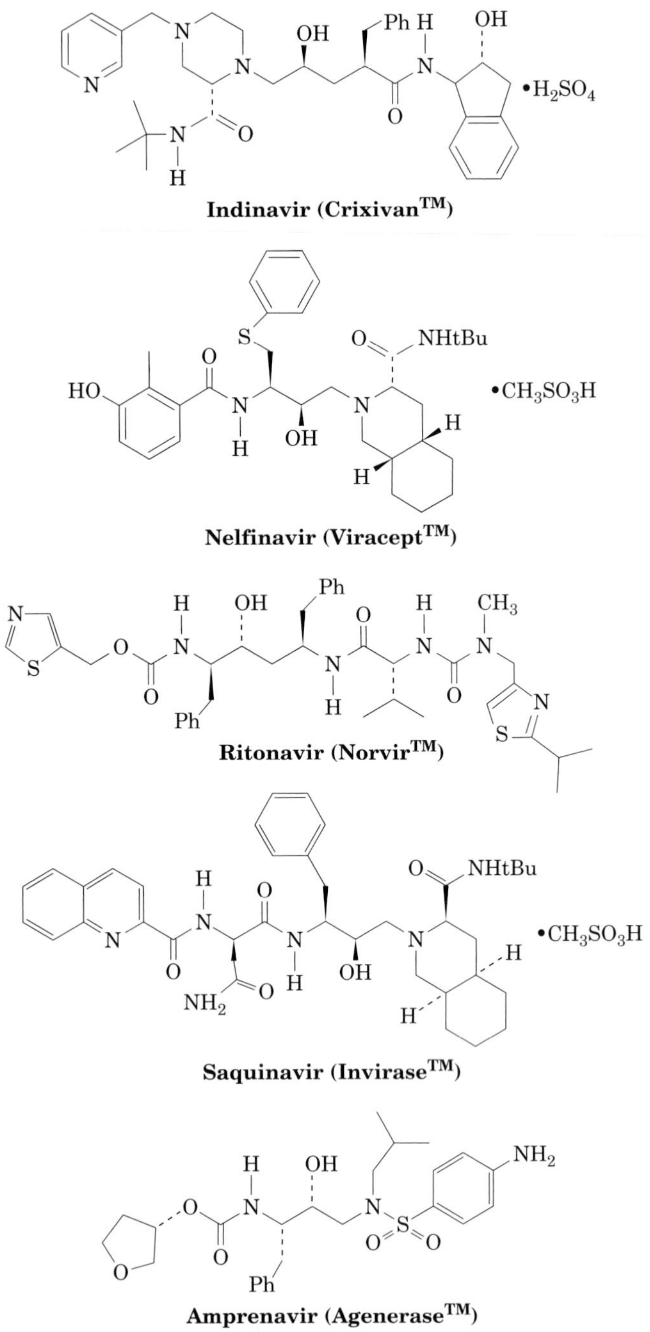

FIGURE 15-41   Some HIV-1 protease inhibitors that are in clinical use. Note that in addition to its generic (chemical) name, each drug has a proprietary trade name, here in parentheses, under which it is marketed.

tease's specificity arises from the cumulative effects of the interactions between the enzyme and the amino acids in positions P_4 through P_4'. However, three of the peptides cleaved by HIV-1 have either the sequence Phe-Pro or Tyr-Pro, which are sequences that human aspartic proteases do not cleave. Hence, HIV-1 protease inhibitors containing groups that resemble either of these dipeptides would be unlikely to inhibit essential human aspartic proteases.

An effective HIV-1 protease inhibitor should resemble a substrate with its scissile peptide replaced by a group that the enzyme cannot cleave. Such a group should, preferably, enhance the enzyme's affinity for the inhibitor. Mimics of the tetrahedral intermediate (Fig. 15-37), that is, transition state analogs, are likely to do so. Consequently, a variety of such groups (Fig. 15-40) have been investigated in efforts to synthesize therapeutically effective inhibitors of HIV-1 protease.

Although HIV-1 protease has high *in vitro* affinity for its polypeptide-based inhibitors, these substances have poor oral bioavailability (they are degraded by digestive

proteases) and pharmacokinetics (they do not readily pass through cell membranes.). Consequently, therapeutically effective HIV-1 protease inhibitors must be **peptido-mimetics** (peptide mimics), substances that sterically and perhaps physically, but not chemically, resemble polypeptides. The use of peptidomimetics also permits conformational constraints to be imposed on a drug candidate that would not be present in the corresponding polypeptide.

As of early 2003, the FDA had approved six HIV-1 protease inhibitors (Fig. 15-41), the first of which, **saquinavir,** was sanctioned in late 1995. These peptidomimetics have IC_{50}s against HIV in culture ranging from 2 to 60 nM but have little or no activity against human aspartic proteases (K_I's >10 μM). They are the first drugs to clearly prolong the lives of AIDS victims. Their development, in each case, was a complex iterative process that required the design, synthesis, and evaluation of numerous related compounds. In several cases, these investigations capitalized on the wealth of experience gained in developing peptidomimetic inhibitors of the aspartic protease renin and in the resulting stockpiles of these compounds.

All the FDA-approved HIV-1 protease inhibitors initially cause a rapid and profound decline in a patient's plasma HIV load, which is often paralleled by immune system recovery. However, as we saw with reverse transcriptase inhibitors, mutant forms of the protease that are resistant to the inhibitor being used arise, usually within 4 to 12 weeks. Moreover, such a mutant protease is likely to be resistant to other HIV-1 protease inhibitors, because all of the HIV-1 protease inhibitors are targeted to the same binding site. This has led to the use of combination therapies in which an HIV-1 protease inhibitor is administered together with one, or more often, two reverse transcriptase inhibitors. This is because any virus that gains resistance to one drug in a regimen will be suppressed by the other drug(s) in that regimen. In addition, the HIV-1 protease inhibitor **ritonavir** has been shown to be a potent inhibitor of the cytochrome P450 isoforms (CYP3A4,5,7) that metabolize other protease inhibitors and hence is usually prescribed in low dosage as an adjunct to another protease inhibitor to improve the latter's pharmacokinetics.

The plasma virus levels in many patients who were placed on combination therapy rapidly became undetectable and have remained so for several years. This, however, does not constitute a cure: If drug therapy is interrupted, the virus will reappear in the plasma because certain tissues in the body harbor latent viruses that are unaffected by and/or inaccessible to drug therapy. Thus, the presently available anti-HIV medications must be taken for a lifetime.

Current anti-HIV therapies are by no means ideal. To maximize their oral bioavailability, some of the different drugs must be taken well before or after a meal but others must be taken with a meal. To minimize the probability of resistant forms of HIV arising, the bioavailability of each drug must be maintained at a certain minimum level and hence each drug must be taken on a rigid schedule. Moreover, these drugs have significant side effects, mainly fatigue, nausea, diarrhea, tingling and numbness with ritonavir, and kidney stones with **indinavir.** Consequently, numerous AIDS patients fail to take their medications properly, which greatly increases the likelihood that they will develop resistance to these drugs and infect others with drug-resistant viruses. Finally, HIV-1 protease inhibitors, being complex molecules, are difficult to synthesize and therefore are relatively expensive, so that in the developing countries in which AIDS is most prevalent, governments and most individuals cannot afford to purchase these drugs, even if they were to be supplied at cost. It is therefore important that anti-HIV therapies be developed that are easy for patients to comply with, are inexpensive, and ideally, will totally eliminate an HIV infection.

CHAPTER SUMMARY

1 ■ Catalytic Mechanisms Most enzymatic mechanisms of catalysis have ample precedent in organic catalytic reactions. Acid- and base-catalyzed reactions occur, respectively, through the donation or abstraction of a proton to or from a reactant so as to stabilize the reaction's transition state complex. Enzymes often employ ionizable amino acid side chains as general acid–base catalysts. Covalent catalysis involves nucleophilic attack of the catalyst on the substrate to transiently form a covalent bond followed by the electrophilic stabilization of a developing negative charge in the reaction's transition state. Various protein side chains as well as certain coenzymes can act as covalent catalysts. Metal ions, which are common enzymatic components, catalyze reactions by stabilizing developing negative charges in a manner resembling general acid catalysis. Metal ion–bound water molecules are potent sources of OH^- ions at neutral pH's. Metal ions also facilitate enzymatic reactions through the charge shielding of bound substrates. The arrangement of charged groups about an enzymatic active site of low dielectric constant in a manner that stabilizes the transition state complex results in the electrostatic catalysis of the enzymatic reaction. Enzymes catalyze reactions by bringing their substrates into close proximity in reactive orientations. The enzymatic binding of the substrates in a bimolecular reaction arrests their relative motions resulting in a rate enhancement. The preferential enzymatic binding of the transition state of a catalyzed reaction over the substrate is an important rate enhancement mechanism. Transition state analogs are potent competitive inhibitors because they bind to the enzyme more tightly than does the corresponding substrate.

2 ■ Lysozyme Lysozyme catalyzes the hydrolysis of β(1→4)-linked poly(NAG–NAM), the bacterial cell wall polysaccharide, as well as that of poly(NAG). According to the Phillips mechanism, lysozyme binds a hexasaccharide so as to distort its D-ring toward the half-chair conformation of the planar oxonium ion transition state. This is followed by cleavage of the C1—O1 bond between the D- and E-rings as promoted by proton donation from Glu 35. Finally, the resulting oxonium ion transition state is electrostatically stabilized by the nearby carboxyl group of Asp 52 so that the E-ring can be replaced by OH^- to form the hydrolyzed product. The roles of Glu 35 and Asp 52 in lysozyme catalysis have been verified through mutagenesis studies. Similarly, structural and binding studies indicate that strain is of major catalytic importance in the lysozyme mechanism. However, mass spectrometry and X-ray studies have demonstrated that the lysozyme reaction proceeds via a covalent glycosyl–enzyme intermediate involving Asp52 rather than by the noncovalently bound oxonium ion intermediate postulated by the Phillips mechanism.

3 ■ Serine Proteases   Serine proteases constitute a widespread class of proteolytic enzymes that are characterized by the possession of a reactive Ser residue. The pancreatically synthesized digestive enzymes trypsin, chymotrypsin, and elastase are sequentially and structurally related but have different side chain specificities for their substrates. All have the same catalytic triad, Asp 102, His 57, and Ser 195, at their active sites. The differing side chain specificities of trypsin and chymotrypsin depend in a complex way on the structures of the loops that connect the walls of the specificity pocket, as well as on the charge of the side chain at the base of the specificity pocket. Subtilisin, serine carboxypeptidase II, and ClpP are unrelated serine proteases that have essentially the same active site geometry as do the pancreatic enzymes. Catalysis in serine proteases is initiated by the nucleophilic attack of the active Ser on the carbonyl carbon atom of the scissile peptide to form the tetrahedral intermediate, a process that may be facilitated by the formation of a low-barrier hydrogen bond between Asp 102 and His 57. The tetrahedral intermediate, which is stabilized by its preferential binding to the enzyme's active site, then decomposes to the acyl–enzyme intermediate under the impetus of proton donation from the Asp 102-polarized His 57. After the replacement of the leaving group by solvent H_2O, the catalytic process is reversed to yield the second product and the regenerated enzyme. The Asp 102–His 57 couple therefore functions in the reaction as a proton shuttle. The active Ser is not unusually reactive but is ideally situated to nucleophilically attack the activated scissile peptide. The X-ray structure of the trypsin–BPTI complex indicates the existence of the tetrahedral intermediate, whereas X-ray structures of a complex of elastase with the heptapeptide BCM7 have visualized both the acyl–enzyme intermediate and the tetrahedral intermediate.

The pancreatic serine proteases are synthesized as zymogens to prevent pancreatic self-digestion. Trypsinogen is activated by a single proteolytic cleavage by enteropeptidase. The resulting trypsin similarly activates trypsinogen as well as chymotrypsinogen, proelastase, and other pancreatic digestive enzymes. Trypsinogen's catalytic triad is structurally intact. The zymogen's low catalytic activity arises from a distortion of its specificity pocket and oxyanion hole, so that it is unable to productively bind substrate or preferentially bind the catalytic reaction's transition state.

4 ■ Drug Design   Drugs act by binding to and thereby modifying the functions of receptors. Many promising drug candidates, which are known as lead compounds, have been found by methods in which a large number of compounds are tested for drug efficacy in an assay that is a suitable surrogate of the disease/condition under consideration. Lead compounds are then chemically manipulated in the search for compounds with improved drug efficacy. Structure–activity relationships (SARs) and quantitative structure–activity relationships (QSARs) are useful tools in this endeavor. Structure-based drug design uses the X-ray and NMR structures of drug candidates in complex with their target proteins, together with a variety of molecular modeling tools, to guide the search for improved drug candidates. However, the advent of combinatorial chemistry and high-throughput screening procedures has extended the "make-many-compounds-and-see-what-they-do" approaches to drug discovery. In order to reach their target receptors, drugs must have favorable pharmacokinetics, that is, they must readily traverse numerous physical barriers in the body, avoid chemical transformation by enzymes, and not be excreted too rapidly. Most useful drugs are neither too lipophilic nor too hydrophilic so that they can both gain access to the necessary membranes and pass through them. Drug toxicity, dosage, efficacy, and the nature of rare adverse reactions are determined through extensive and carefully designed clinical trials. Most drugs are metabolically cleared through oxidative hydroxylation by one of the ~100 cytochrome P450 isozymes. This permits the hydroxylated drugs to be enzymatically conjugated to polar groups such as glucuronic acid and glycine, which increases their rates of excretion by the kidneys. Drug–drug interactions are frequently mediated by cytochromes P450. Polymorphisms among cytochromes P450 are often responsible for the variations among individuals in their response to a given drug, including adverse reactions.

The formulation of HIV-1 protease inhibitors to control HIV infections is one of the major triumphs of modern drug discovery methods. HIV are retroviruses that attack specific immune system cells and thereby degrade the immune system over a period of several years to the point that it is no longer able defend against opportunistic infections. HIV-1 protease functions to cleave the polyproteins in immature HIV-1 virions that have budded out from a host cell, thus generating the mature, infectious form. HIV-1 protease is an aspartic protease that, as eukaryotic aspartic proteases such as pepsin, uses its two active site Asp residues to activate its bound lytic water molecule as the nucelophile that attacks and thereby cleaves specific peptide bonds in the substrate polyprotein. All of the FDA-approved peptidomimetic inhibitors of HIV-1 protease cause a rapid and profound decrease in plasma HIV levels, although they do not entirely eliminate the virus. They are used in combination with reverse transcriptase inhibitors to minimize the ability of the rapidly mutating HIV to evolve drug-resistant forms.

REFERENCES

GENERAL

Bender, M.L., Bergeron, R.J., and Komiyama, M., *The Bioorganic Chemistry of Enzymatic Catalysis,* Wiley (1984).

Fersht, A., *Structure and Mechanism in Protein Science,* Freeman (1999).

Jencks, W.P., *Catalysis in Chemistry and Enzymology,* Dover (1987). [A classic and, in many ways, still current work.]

Walsh, C., *Enzymatic Reaction Mechanisms,* Freeman (1979). [A compendium of enzymatic reactions.]

CATALYTIC MECHANISMS

Atkins, W.M. and Sligar, S.G., Protein engineering for studying enzyme catalytic mechanism, *Curr. Opin. Struct. Biol.* **1,** 611–616 (1991).

Bruice, T.C., Some pertinent aspects of mechanism as determined with small molecules, *Annu. Rev. Biochem.* **45,** 331–373 (1976).

Bruice, T.C. and Benkovic, S.J., Chemical basis for enzyme catalysis, *Biochemistry* **39,** 6267–6274 (2000); *and* Bruice, T.C. and Lightstone, F.C., Ground state and transition state contributions to the rates of intramolecular and enzymatic reactions, *Acc. Chem. Res.* **32,** 127–136 (1999).

Christianson, D.W. and Cox, J.D., Catalysis by metal-activated hydroxide in zinc and manganese metalloenzymes, *Annu. Rev. Biochem.* **68,** 33–57 (1999). [Discusses the enzymatic mechanism of carbonic anhydrase.]

Glusker, J.P., Structural aspects of metal liganding to functional groups in proteins, *Adv. Protein Chem.* **42,** 1–76 (1991).

Hackney, D.D., Binding energy and catalysis, *in* Sigman, D.S. and Boyer, P.D. (Eds.), *The Enzymes* (3rd ed.), Vol. 19, *pp.* 1–36, Academic Press (1990).

Jencks, W.P., Binding energy, specificity, and enzymatic catalysis: The Circe effect, *Adv. Enzymol.* **43,** 219–410 (1975).

Kraut, J., How do enzymes work? *Science* **242,** 533–540 (1988).

Lolis, E. and Petsko, G.A., Transition-state analogues in protein crystallography: Probes of the structural source of enzyme catalysis, *Annu. Rev. Biochem.* **59,** 597–630 (1990).

Page, M.I., Entropy, binding energy, and enzyme catalysis, *Angew. Chem. Int. Ed. Engl.* **16,** 449–459 (1977).

Schramm, V.L., Enzymatic transition states and transition state analog design, *Annu. Rev. Biochem.* **67,** 693–720 (1998).

Villafranca, J.J. and Nowak, T., Metal ions at enzyme active sites, *in* Sigman, D.S. (Ed.), *The Enzymes* (3rd ed.), Vol. 20, *pp.* 63–94, Academic Press (1992).

Warshel, A., Computer simulations of enzymatic reactions, *Curr. Opin. Struct. Biol.* **2,** 230–236 (1992).

Williams, R.J.P., Are enzymes mechanical devices? *Trends Biochem. Sci.* **18,** 115–117 (1993). [Argues that the mechanical aspects of enzymes have received insufficient consideration.]

Wolfenden, R., Analogue approaches to the structure of the transition state in enzyme reactions, *Acc. Chem. Res.* **5,** 10–18 (1972).

LYSOZYME

Blake, C.C.F., Johnson, L.N., Mair, G.A., North, A.C.T., Phillips, D.C., and Sarma, V.R., Crystallographic studies of the activity of hen egg-white lysozyme, *Proc. R. Soc. London Ser. B* **167,** 378–388 (1967).

Chipman, D.M. and Sharon, N., Mechanism of lysozyme action, *Science* **165,** 454–465 (1969).

Ford, L.O., Johnson, L.N., Machin, P.A., Phillips, D.C., and Tijan, R., Crystal structure of a lysozyme–tetrasaccharide lactone complex, *J. Mol. Biol.* **88,** 349–371 (1974).

Imoto, T., Johnson, L.N., North, A.C.T., Phillips, D.C., and Rupley, J.A., Vertebrate lysozymes, *in* Boyer, P.D. (Ed.), *The Enzymes* (3rd ed.), Vol. 7, *pp.* 665–868, Academic Press (1972). [An exhaustive review.]

Johnson, L.N., Cheetham, J., McLaughlin, P.J., Acharya, K.R., Barford, D., and Phillips, D.C., Protein–oligosaccharide interactions: Lysozyme, phosphorylase, amylases, *Curr. Top. Microbiol. Immunol.* **139,** 81–134 (1988).

Jollès, P. (Ed.), *Lysozymes: Model Enzymes in Biochemistry and Biology,* Birkhaüser Verlag (1996).

Kirby, A.J., Turning lysozyme upside down, *Nature Struct. Biol.* **2,** 923–925 (1995); *and* Illuminating an ancient retainer, *Nature Struct. Biol.* **3,** 107–108 (1996). [Argues, based on organic mechanistic principles as well as the known structures and chemistry of certain glycosidases, that the reaction catalyzed by HEW lysozyme is likely to proceed via a double displacement mechanism.]

McKenzie, H.A. and White, F.H., Jr., Lysozyme and α-lactalbumin: Structure, function and interrelationships, *Adv. Protein Chem.* **41,** 173–315 (1991).

Mooser, G., Glycosidases and glycosyltransferases, *in* Sigman, D.S. (Ed.), *The Enzymes* (3rd ed.), Vol. 20, *pp.* 187–233, Academic Press (1992). [Section II discusses lysozyme.]

Phillips, D.C., The three-dimensional structure of an enzyme molecule, *Sci. Am.* **215**(5), 75–80 (1966). [A marvelously illustrated article on the structure and mechanism of lysozyme.]

Schindler, M., Assaf, Y., Sharon, N., and Chipman, D.M., Mechanism of lysozyme catalysis: Role of ground-state strain in subsite D in hen egg-white and human lysozymes, *Biochemistry* **16,** 423–431 (1977).

Secemski, I.I., Lehrer, S.S., and Lienhard, G.E., A transition state analogue for lysozyme, *J. Biol. Chem.* **247,** 4740–4748 (1972). [Binding studies on the lactone derivative of (NAG)₄.]

Strynadka, N.C.J. and James, M.N.G., Lysozyme revisited: Crystallographic evidence for distortion of an *N*-acetylmuramic acid residue bound in site D, *J. Mol. Biol.* **220,** 401–424 (1991).

Vocadlo, D.J., Davies, G.J., Laine, R., and Withers, S.G., Catalysis by hen egg-white lysozyme proceeds via a covalent intermediate, *Nature* **412,** 835–838 (2001).

Warshel, A. and Levitt, M., Theoretical studies of enzymatic reactions; dielectric, electrostatic and steric stabilization of the carbonium ion in the reaction of lysozyme, *J. Mol. Biol.* **103,** 227–249 (1976). [Theoretical indications that lysozyme catalysis occurs through electrostatic rather than steric strain.]

White, A. and Rose, D.R., Mechanism of catalysis by retaining β-glycosyl hydrolases, *Curr. Opin. Struct. Biol.* **7,** 645–651 (1997).

SERINE PROTEASES

Blow, D.M., The tortuous story of Asp...His...Ser: Structural analysis of chymotrypsin, *Trends Biochem. Sci.* **22,** 405–408 (1998). [A personal memoir of the structural determination of α-chymotrypsin in the years 1967 through 1969.]

Cleland, W.W., Frey, P.A., and Gerlt, J.A., The low barrier hydrogen bond in enzymatic catalysis, *J. Biol. Chem.* **273,** 25529–25532 (1998).

Corey, D.R. and Craik, C.S., An investigation into the minimum requirements for peptide hydrolysis by mutation of the catalytic triad of trypsin, *J. Am. Chem. Soc.* **114,** 1784–1790 (1992).

Ding, X., Rasmussen, B.F., Petsko, G.A., and Ringe, D., Direct structural observation of an acyl-enzyme intermediate in the hydrolysis of an ester substrate by elastase, *Biochemistry* **33,** 9285–9293 (1994).

Dodson, G. and Wlodawer, A., Catalytic triads and their relatives, *Trends Biochem. Sci.* **23,** 347–352 (1998).

Frey, P.A., Whitt, S.A., and Tobin, J.B., A low-barrier hydrogen bond in the catalytic triad of serine proteases, *Science* **264,** 1927–1930 (1994).

James, M.N.G., Sielecki, A.R., Brayer, G.D., Delbaere, L.T.J., and Bauer, C.A., Structure of product and inhibitor complexes of *Streptomyces griseus* protease A at 1.8 Å resolution, *J. Mol. Biol.* **144,** 45–88 (1980).

Kuhn, P., Knapp, M., Soltis, S.M., Ganshaw, G., Thoene, M., and Bott, R., The 0.78 Å structure of a serine protease: *Bacillus lentus* subtilisin, *Biochemistry* **37**, 13446–13452 (1998).

Liao, D.-I. and Remington, S.J., Structure of wheat serine carboxypeptidase II at 3.5-Å resolution, *J. Biol. Chem.* **265**, 6528–6531 (1990).

Neurath, H., Evolution of proteolytic enzymes, *Science* **224**, 350–357 (1984).

Perona, J.J. and Craik, C.S., Evolutionary divergence of substrate specificity within the chymotrypsin-like serine protease fold, *J. Biol. Chem.* **272**, 29987–29990 (1997); *and* Structural basis of substrate specificity in the serine proteases, *Protein. Sci.* **4**, 337–360 (1995).

Perrin, C.L. and Nielson, J.B., "Strong" hydrogen bonds in chemistry and biology, *Annu. Rev. Phys. Chem.* **48**, 511–544 (1997). [A detailed review which concludes that the evidence for the importance of LBHBs in enzymatic reactions is inconclusive.]

Phillips, M.A. and Fletterick, R.J., Proteases, *Curr. Opin. Struct. Biol.* **2**, 713–720 (1992).

Roberts, R.M., Mathialagan, N., Duffy, J.Y., and Smith, G.W., Regulation and regulatory role of proteinase inhibitors, *Crit. Rev. Euk. Gene Express.* **5**, 385–435 (1995).

Shan, S., Loh, S., and Herschlag, D., The energetics of hydrogen bonds in model systems: Implications for enzymatic catalysis, *Science* **272**, 97–101 (1996).

Stroud, R.M., Kossiakoff, A.A., and Chambers, J.L., Mechanism of zymogen activation, *Annu. Rev. Biophys. Bioeng.* **6**, 177–193 (1977).

Wang, J., Hartling, J.A., and Flanagan, J.M., The structure of ClpP at 2.3 Å resolution suggests a model for ATP-dependent proteolysis, *Cell* **91**, 447–456 (1997).

Wilmouth, R.C., Edman, K., Neutze, R., Wright, P.A., Clifton, I.J., Schneider, T.R., Schofield, C.J., and Hadju, J., X-Ray snapshots of serine protease catalysis reveals a tetrahedral intermediate, *Nature Struct. Biol.* **8**, 689–694 (2001); *and* Wilmouth, R.C., Clifton, I.J., Robinson, C.V., Roach, P.L., Aplin, R.T., Westwood, N.J., Hadju, J., and Schofield, C.J., Structure of a specific acyl-enzyme complex formed between β-casomorphin-7 and porcine pancreatic elastase, *Nature Struct. Biol.* **4**, 456–461 (1997).

DRUG DISCOVERY

Debouck, C. and Metcalf, B., The impact of genomics on drug discovery, *Annu. Rev. Pharmacol. Toxicol.* **40**, 193–208 (2000).

Gordon, E.M. and Kerwin, J.F., Jr. (Eds.), *Combinatorial Chemistry and Molecular Diversity in Drug Discovery*, Wiley-Liss (1998).

Gringauz, A., *Introduction to Medicinal Chemistry*, Wiley-VCH (1997).

Harman, J.G., Limbird, L.E., Molinoff, P.B., Ruddon, R.W., and Gilman, A.G. (Eds.), *Goodman & Gilman's The Pharmacologic Basis of Therapeutics* (9th ed.), McGraw-Hill (2000).

Ingelman-Sundberg, M., Oscarson, M., and McLellan, R.A., Polymorphic human cytochrome P450 enzymes: An opportunity for individualized drug treatment, *Trends Pharmacol. Sci.* **20**, 342–349 (1999).

Katzung, B.G. (Ed.), *Basic & Clinical Pharmacology* (7th ed.), Appleton & Lange (1998).

Marrone, T.J., Briggs, J.M., and McCammon, J.A., Structure-based drug design: Computational advances, *Annu. Rev. Pharmacol. Toxicol.* **37**, 71–90 (1997).

Mycek, M.J., Harvey, R.A., and Champe, P.C., *Lippincott's Illustrated Reviews: Pharmacology* (2nd ed.), Lippincott–Raven Publishers (1997).

Navia, M.A. and Murcko, M.A., Use of structural information in drug design, *Curr. Opin. Struct. Biol.* **2**, 202–210 (1992).

Ohlstein, E.H., Ruffolo, R.R., Jr., and Elliott, J.D., Drug discovery in the next millennium, *Annu. Rev. Pharmacol. Toxicol.* **40**, 177–191 (2000).

Patrick, G.L, *An Introduction to Medicinal Chemistry*, Oxford University Press (1995).

Smith, D.A. and van der Waterbeemd, H., Pharmacokinetics and metabolism in early drug design, *Curr. Opin. Chem. Biol.* **3**, 373–378 (1999).

Terrett, N.O., *Combinatorial Chemistry*, Oxford University Press (1998).

Walsh, G., *Biopharmaceuticals: Biochemistry and Biotechnology*, Wiley (1998).

White, R.E., High-throughput screening in drug metabolism and pharmacokinetic support of drug discovery, *Annu. Rev. Pharmacol. Toxicol.* **40**, 133–157 (2000).

Wong, L.-L., Cytochrome P450 monooxygenases, *Curr. Opin. Chem. Biol.* **2**, 263–268 (1998).

HIV-1 PROTEASE AND OTHER ASPARTIC PROTEASES

Davies, D.R., The structure and function of the aspartic proteases, *Annu. Rev. Biophys. Biophys. Chem.* **19**, 189–215 (1990).

Erickson, J.W. and Burt, S.K., Structural mechanisms of HIV drug resistance, *Annu. Rev. Pharmacol. Toxicol.* **36**, 545–571 (1996).

Flexner, C., Dual protease inhibitor therapy in HIV-infected patients: Pharmacological rationale and clinical benefits, *Annu. Rev. Pharmacol. Toxicol.* **40**, 649–674 (2000).

Kling, J., Blocking HIV's "scissors," *Modern Drug Discovery* **3**(2), 37–45 (2000).

Meeks, T.D., Catalytic mechanisms of the aspartic proteases, *in* Sinnott, M. (Ed.), *Comprehensive Biological Catalysis*, Vol. 1, pp. 327–344, Academic Press (1998).

Richman, D.D., HIV chemotherapy, *Nature* **410**, 995–1001 (2001).

Tomesselli, A.G., Thaisrivongs, S., and Heinrikson, R. L., Discovery and design of HIV protease inhibitors as drugs for treatment of AIDS, *Adv. Antiviral Drug Design* **2**, 173–228 (1996).

Turner, B.G. and Summers, M.F., Structural biology of HIV, *J. Mol. Biol.* **285**, 1–32 (1999). [A review.]

Wilk, T. and Fuller, S.D., Towards the structure of human immunodeficiency virus: Divide and conquer? *Curr. Opin. Struct. Biol.* **9**, 231–243 (1999).

Wlodawer, A. Rational approach to AIDs drug design through structural biology, *Annu. Rev. Med.* **53**, 595–614 (2001); *and* Wlodawer, A. and Vondrasek, J., Inhibitors of HIV-1 protease: A major success of structure-assisted drug design, *Annu. Rev. Biophys. Biomol. Struct.* **27**, 249–284 (1998).

PROBLEMS

1. Explain why γ-pyridone is not nearly as effective a catalyst for glucose mutarotation as is α-pyridone. What about β-pyridone?

2. RNA is rapidly hydrolyzed in alkaline solution to yield a mixture of nucleotides whose phosphate groups are bonded to either the 2′ or the 3′ positions of the ribose residues. DNA, which lacks RNA's 2′ OH groups, is resistant to alkaline degradation. Explain.

3. Carboxypeptidase A, a Zn^{2+}-containing enzyme, hydrolyzes the C-terminal peptide bonds of polypeptides (Section 7-1A). In the enzyme–substrate complex, the Zn^{2+} ion is coordinated to three enzyme side chains, the carbonyl oxygen of the scissile peptide bond, and a water molecule. A plausible model for the enzyme's reaction mechanism that is consistent with X-ray and enzymological data is diagrammed in Fig. 15-42. What are the roles of the Zn^{2+} ion and Glu 270 in this mechanism?

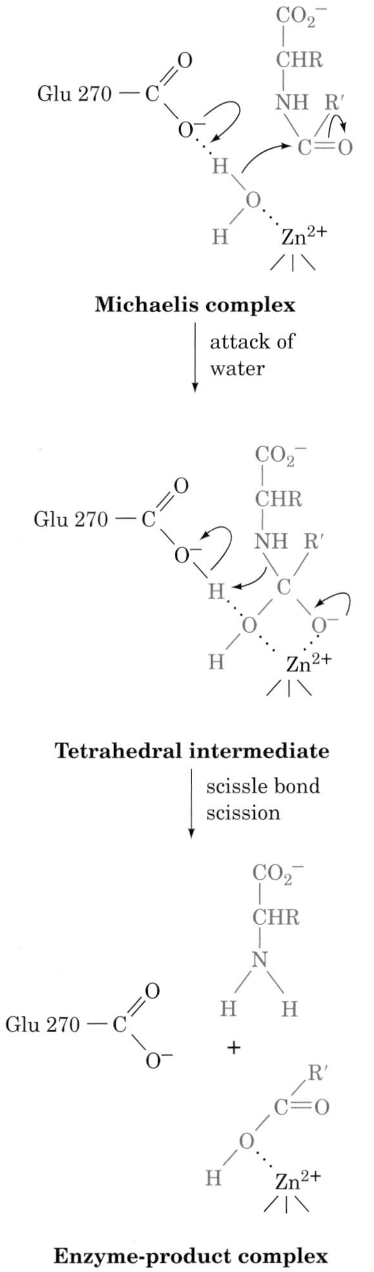

FIGURE 15-42 Mechanism of carboxypeptidase A.

4. In the following lactonization reaction,

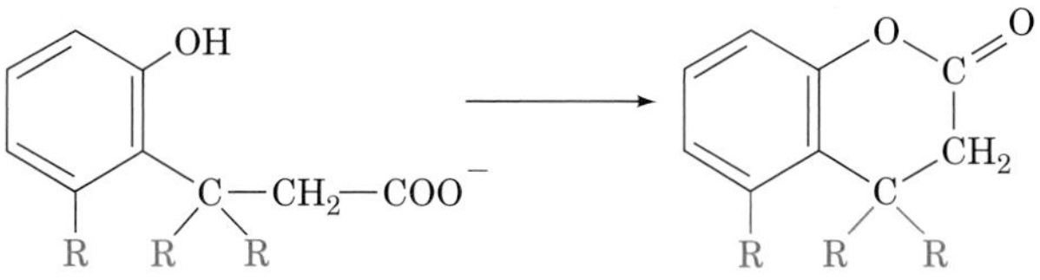

the relative reaction rate when $R = CH_3$ is 3.4×10^{11} times that when $R = H$. Explain.

*5. Derive the analog of Eq. [15.11] for an enzyme that catalyzes the reaction:

$$A + B \rightarrow P$$

Assume the enzyme must bind A before it can bind B:

$$E + A + B \rightleftharpoons EA + B \rightleftharpoons EAB \rightarrow EP$$

6. Explain, in thermodynamic terms, why an "enzyme" that stabilizes its Michaelis complex as much as its transition state does not catalyze a reaction.

7. Suggest a transition state analog for proline racemase that differs from those discussed in the text. Justify your suggestion.

8. Wolfenden has stated that it is meaningless to distinguish between the "binding sites" and the "catalytic sites" of enzymes. Explain.

9. Explain why oxalate ($^-OOCCOO^-$) is an inhibitor of oxaloacetate decarboxylase.

10. In light of the information given in this chapter, why are enzymes such large molecules? Why are active sites almost always located in clefts or depressions in enzymes rather than on protrusions?

11. Predict the effects on lysozyme catalysis of changing Phe 34, Ser 36, and Trp 108 to Arg, assuming that this change does not significantly alter the structure of the protein.

*12. The incubation of $(NAG)_4$ with lysozyme results in the slow formation of $(NAG)_6$ and $(NAG)_2$. Propose a mechanism for this reaction. What aspect of the Phillips mechanism is established by this reaction?

13. How would the lysozyme binding affinity of the following $\beta(1\rightarrow4)$-linked tetrasaccharide

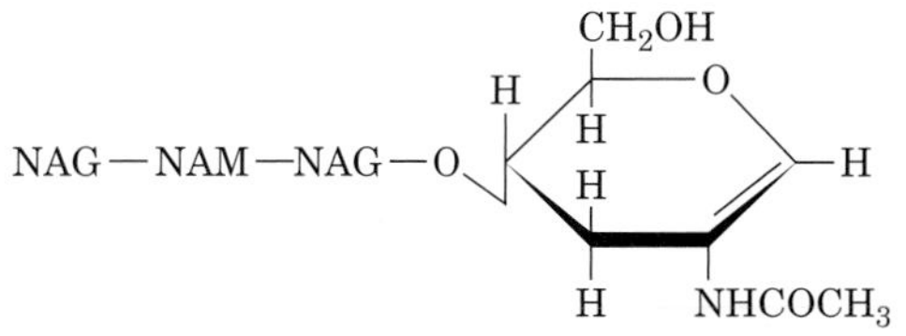

compare with that of NAG–NAM–NAG–NAM? Explain.

14. A major difficulty in investigating the properties of the pancreatic serine proteases is that these enzymes, being proteins themselves, are self-digesting. This problem is less severe, however, for solutions of chymotrypsin than it is for solutions of trypsin or elastase. Explain.

15. The comparison of the active site geometries of chymotrypsin and subtilisin under the assumption that their similarities have catalytic significance has led to greater mechanistic understanding of both these enzymes. Discuss the validity of this strategy.

16. Benzamidine ($K_I = 1.8 \times 10^{-5}M$) and **leupeptin** ($K_I = 1.8 \times 10^{-7}M$)

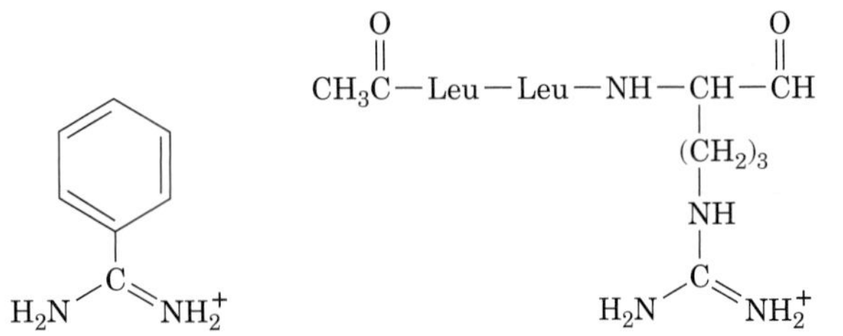

Benzamidine   **Leupeptin**

are both specific competitive inhibitors of trypsin. Explain their mechanisms of inhibition. Design leupeptin analogs that inhibit chymotrypsin and elastase.

17. Trigonal boronic acid derivatives have a high tendency to form tetrahedral adducts. **2-Phenylethyl boronic acid**

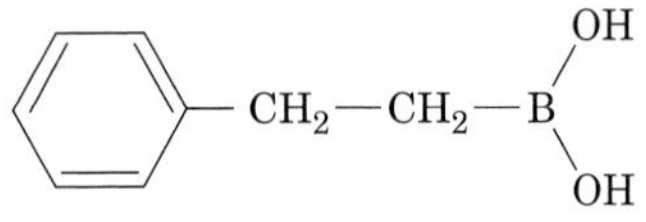

2–Phenylethyl boronic acid

is an inhibitor of subtilisin and chymotrypsin. Indicate the structure of these enzyme–inhibitor complexes.

18. Tofu (bean curd), a high-protein soybean product that is widely consumed in China and Japan, is prepared in such a way as to remove the trypsin inhibitor present in soybeans. Explain the reason(s) for this treatment.

19. Explain why mutating all three residues of trypsin's catalytic triad has essentially no greater effect on the enzyme's catalytic rate enhancement than mutating only Ser 195.

20. Explain why chymotrypsin is not self-activating as is trypsin.

21. Does Lipinski's "rule of five" predict that a hexapeptide would be a therapeutically effective drug? Explain.

22. The preferred antidote for acetaminophen overdose is *N*-acetylcysteine. Explain why the administration of this substance, which must occur within 8 to 16 hours of the overdose, is an effective treatment.

23. Why would the activation of HIV-1 protease before the virus buds from its host cell be disadvantageous to the virus? Explain.

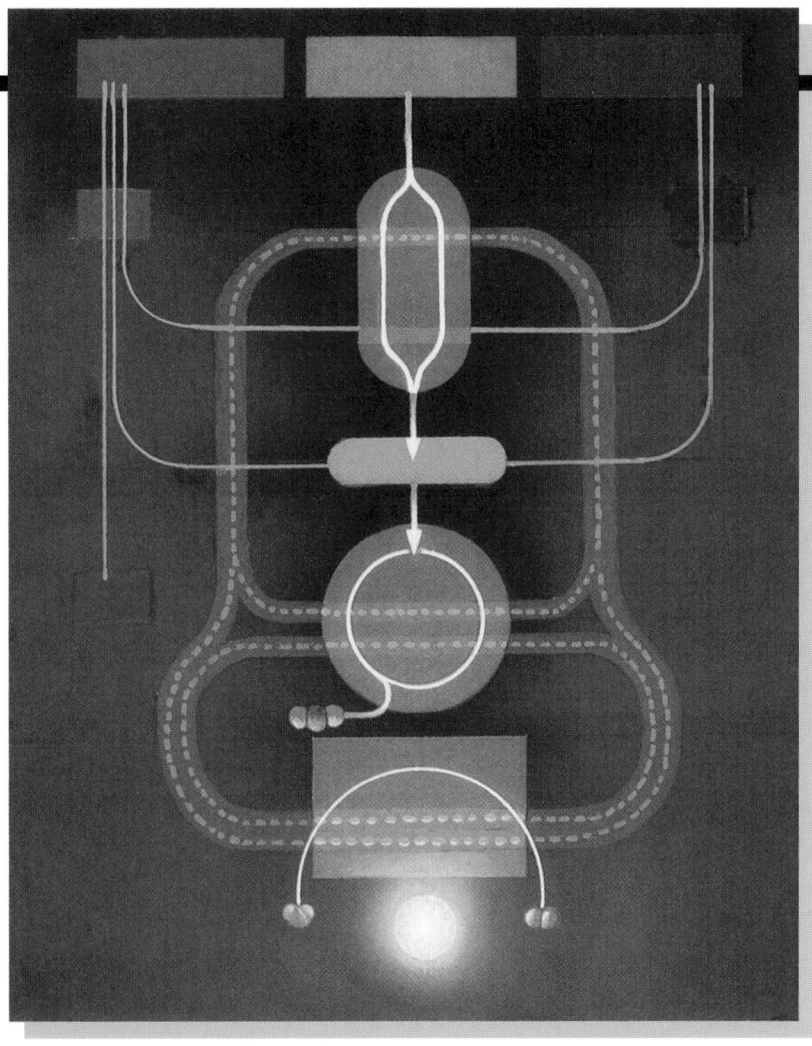

*Schematic diagram of the major pathways
of energy metabolism.*

PART

IV

METABOLISM

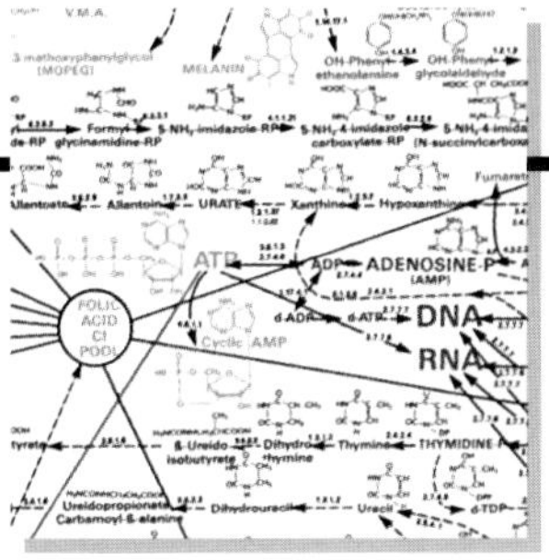

Chapter 16

Introduction to Metabolism

Living organisms are not at equilibrium. Rather, they require a continuous influx of free energy to maintain order in a universe bent on maximizing disorder. **Metabolism** is the overall process through which living systems acquire and utilize the free energy they need to carry out their various functions. *They do so by coupling the exergonic reactions of nutrient oxidation to the endergonic processes required to maintain the living state* such as the performance of mechanical work, the active transport of molecules against concentration gradients, and the biosynthesis of complex molecules. How do living things acquire this nec-

essary free energy? And what is the nature of the energy coupling process? **Phototrophs** (plants and certain bacteria; Section 1-1A) acquire free energy from the sun through **photosynthesis,** a process in which light energy powers the endergonic reaction of CO_2 and H_2O to form carbohydrates and O_2 (Chapter 24). **Chemotrophs** obtain their free energy by oxidizing organic compounds (carbohydrates, lipids, proteins) obtained from other organisms, ultimately phototrophs. *This free energy is most often coupled to endergonic reactions through the intermediate synthesis of "high-energy" phosphate compounds such as **adenosine triphosphate (ATP;** Section 16-4). In addition to being completely oxidized, nutrients are broken down in a series of metabolic reactions to common intermediates that are used as precursors in the synthesis of other biological molecules.*

A remarkable property of living systems is that, despite the complexity of their internal processes, they maintain a steady state. This is strikingly demonstrated by the observation that, over a 40-year time span, a normal human adult consumes literally tons of nutrients and imbibes over 20,000 L of water, but does so without significant weight change. This steady state is maintained by a sophisticated set of metabolic regulatory systems. In this introductory chapter to metabolism, we outline the general characteristics of metabolic pathways, study the main types of chemical reactions that comprise these pathways, and consider the experimental techniques that have been most useful in their elucidation. We then discuss the free energy changes associated with reactions of phosphate compounds and oxidation–reduction reactions. Finally we consider the thermodynamic nature of biological processes, that is, what properties of life are responsible for its self-sustaining character.

1 ■ METABOLIC PATHWAYS

Metabolic pathways are series of consecutive enzymatic reactions that produce specific products. Their reactants, intermediates, and products are referred to as **metabolites.** Since an organism utilizes many metabolites, it has many

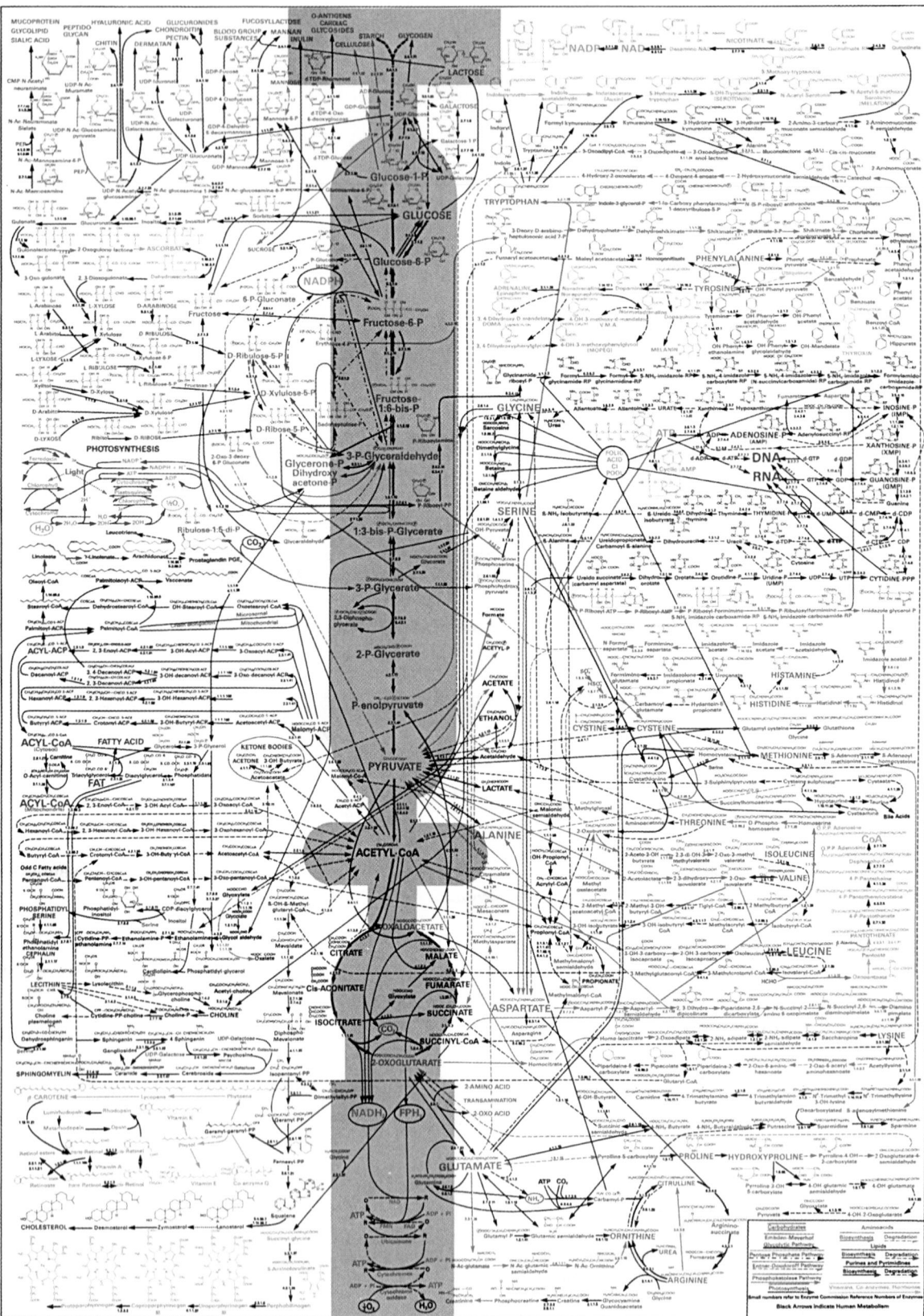

FIGURE 16-1 Map of the major metabolic pathways in a typical cell. The main pathways of glucose metabolism are shaded. [Designed by Donald Nicholson. Published by BDH Ltd., Poole 2, Dorset, England.]

metabolic pathways. Figure 16-1 shows a metabolic map for a typical cell with many of its interconnected pathways. Each reaction on the map is catalyzed by a distinct enzyme, of which there are ~4000 known. At first glance, this network seems hopelessly complex. Yet, by focusing on its major areas in the following chapters, for example, the main pathways of glucose oxidation (the shaded areas of Fig. 16-1), we shall become familiar with its most important avenues and their interrelationships. Maps of metabolic pathways in a more readable form can be found on the Web at http://www.expasy.org/cgi-bin/search-biochem-index, http://www.tcd.ie/Biochemistry/IUBMB-Nicholson, and http://www.genome.ad.jp/kegg/metabolism.html.

The reaction pathways that comprise metabolism are often divided into two categories:

1. Catabolism, or degradation, in which nutrients and cell constituents are broken down exergonically to salvage their components and/or to generate free energy.

2. Anabolism, or biosynthesis, in which biomolecules are synthesized from simpler components.

The free energy released by catabolic processes is conserved through the synthesis of ATP from ADP and phosphate (or) through the reduction of the coenzyme $NADP^+$ to NADPH (Fig. 13-2). ATP and NADPH are the major free energy sources for anabolic pathways (Fig. 16-2).

A striking characteristic of degradative metabolism is that *it converts large numbers of diverse substances (carbohydrates, lipids, and proteins) to common intermediates.* These intermediates are then further metabolized in a central oxidative pathway that terminates in a few end products. Figure 16-3 outlines the breakdown of various foodstuffs, first to their monomeric units, and then to the common intermediate, **acetyl-coenzyme A (acetyl-CoA)** (Fig. 21-2).

Biosynthesis carries out the opposite process. *Relatively few metabolites, mainly pyruvate, acetyl-CoA, and the citric acid cycle intermediates, serve as starting materials for a host of varied biosynthetic products.* In the next several chapters we discuss many degradative and biosynthetic

pathways in detail. For now, let us consider some general characteristics of these processes.

Five principal characteristics of metabolic pathways stem from their function of generating products for use by the cell:

1. Metabolic pathways are irreversible. A highly exergonic reaction (having a large negative free energy change) is irreversible; that is, it goes to completion. If such a reaction is part of a multistep pathway, it confers directionality on the pathway; that is, it makes the entire pathway irreversible.

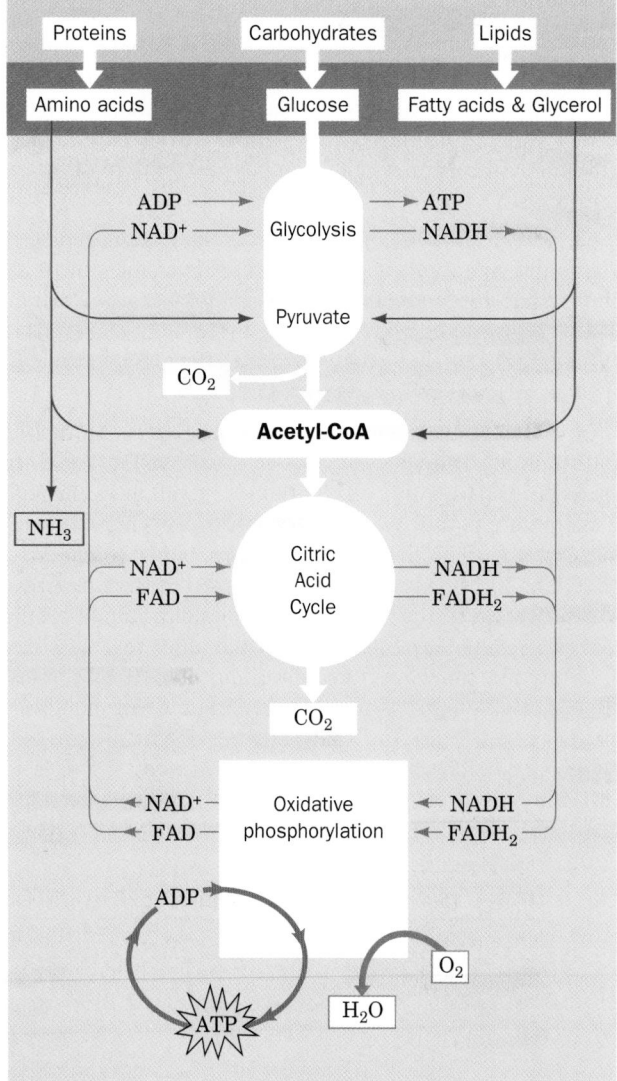

FIGURE 16-3 Overview of catabolism. Complex metabolites such as carbohydrates, proteins, and lipids are degraded first to their monomeric units, chiefly glucose, amino acids, fatty acids, and glycerol, and then to the common intermediate, acetyl coenzyme A (acetyl-CoA). The acetyl group is then oxidized to CO_2 via the citric acid cycle with concomitant reduction of NAD^+ and FAD. Reoxidation of these latter coenzymes by O_2 via the electron-transport chain and oxidative phosphorylation yields H_2O and ATP.

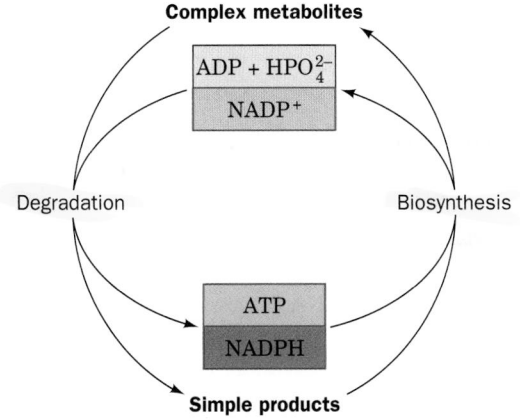

FIGURE 16-2 ATP and NADPH are the sources of free energy for biosynthetic reactions. They are generated through the degradation of complex metabolites.

2. Catabolic and anabolic pathways must differ. *If two metabolites are metabolically interconvertible, the pathway from the first to the second must differ from the pathway from the second back to the first:*

$$
\begin{array}{ccc}
 & A & \\
① \nearrow & & \searrow ② \\
\searrow & & \nearrow \\
 & Y \leftarrow X &
\end{array}
$$

This is because if metabolite 1 is converted to metabolite 2 by an exergonic process, the conversion of metabolite 2 to metabolite 1 requires that free energy be supplied in order to bring this otherwise endergonic process "back up the hill." Consequently, the two pathways must differ in at least one of their reaction steps. *The existence of independent interconversion routes, as we shall see, is an important property of metabolic pathways because it allows independent control of the two processes.* If metabolite 2 is required by the cell, it is necessary to "turn off" the pathway from 2 to 1 while "turning on" the pathway from 1 to 2. Such independent control would be impossible without different pathways.

3. Every metabolic pathway has a first committed step. Although metabolic pathways are irreversible, most of their component reactions function close to equilibrium. Early in each pathway, however, there is an irreversible (exergonic) reaction that "commits" the intermediate it produces to continue down the pathway.

4. All metabolic pathways are regulated. Metabolic pathways are regulated by laws of supply and demand. In order to exert control on the flux of metabolites through a metabolic pathway, it is necessary to regulate its rate-limiting step. The first committed step, being irreversible, functions too slowly to permit its substrates and products to equilibrate (if the reaction was at equilibrium, it would not be irreversible). Since most of the other reactions in a pathway function close to equilibrium, the first committed step is often one of its rate-limiting steps. Most metabolic pathways are therefore controlled by regulating the enzymes that catalyze their first committed step(s). This is an efficient way to exert control because it prevents the unnecessary synthesis of metabolites further along the path-

way when they are not required. Specific aspects of such flux control are discussed in Section 17-4B.

5. Metabolic pathways in eukaryotic cells occur in specific cellular locations. The compartmentation of the eukaryotic cell allows different metabolic pathways to operate in different locations, as is listed in Table 16-1 (these organelles are described in Section 1-2A). For example, ATP is mainly generated in the mitochondrion but much of it is utilized in the cytoplasm. The synthesis of metabolites in specific membrane-bounded subcellular compartments makes their transport between these compartments a vital component of eukaryotic metabolism. Biological membranes are selectively permeable to metabolites because of the presence in membranes of specific transport proteins. The transport protein that facilitates the passage of ATP through the mitochondrial membrane is discussed in Section 20-4C, along with the characteristics of membrane transport processes in general. The synthesis and utilization of acetyl-CoA are also compartmentalized. This metabolic intermediate is utilized in the cytosolic synthesis of fatty acids but is synthesized in mitochondria. Yet there is no transport protein for acetyl-CoA in the mitochondrial membrane. How cells solve this fundamental problem is discussed in Section 25-4D. In multicellular organisms, compartmentation is carried a step further to the level of tissues and organs. The mammalian liver, for example, is largely responsible for the synthesis of glucose from noncarbohydrate precursors (**gluconeogenesis; Section 23-1**) so as to maintain a relatively constant level of glucose in the circulation, whereas adipose tissue is specialized for the storage and mobilization of triacylglycerols. The metabolic interdependence of the various organs is the subject of Chapter 27.

2 ■ ORGANIC REACTION MECHANISMS

Almost all of the reactions that occur in metabolic pathways are enzymatically catalyzed organic reactions. Section 15-1 details the various mechanisms enzymes have at their disposal for catalyzing reactions: acid–base catalysis,

TABLE 16-1   Metabolic Functions of Eukaryotic Organelles

Organelle	Function
Mitochondrion	Citric acid cycle, electron transport and oxidative phosphorylation, fatty acid oxidation, amino acid breakdown
Cytosol	Glycolysis, pentose phosphate pathway, fatty acid biosynthesis, many reactions of gluconeogenesis
Lysosomes	Enzymatic digestion of cell components and ingested matter
Nucleus	DNA replication and transcription, RNA processing
Golgi apparatus	Posttranslational processing of membrane and secretory proteins; formation of plasma membrane and secretory vesicles
Rough endoplasmic reticulum	Synthesis of membrane-bound and secretory proteins
Smooth endoplasmic reticulum	Lipid and steroid biosynthesis
Peroxisomes (glyoxisomes in plants)	Oxidative reactions catalyzed by amino acid oxidases and catalase; glyoxylate cycle reactions in plants

covalent catalysis, metal ion catalysis, electrostatic catalysis, proximity and orientation effects, and transition state binding. Few enzymes alter the chemical mechanisms of these reactions, so *much can be learned about enzymatic mechanisms from the study of nonenzymatic model reactions.* We therefore begin our study of metabolic reactions by outlining the types of reactions we shall encounter and the mechanisms by which they have been observed to proceed in nonenzymatic systems.

Christopher Walsh has classified biochemical reactions into four categories: (1) **group-transfer reactions;** (2) **oxidations and reductions;** (3) **eliminations, isomerizations, and rearrangements;** and (4) **reactions that make or break carbon–carbon bonds.** Much is known about the mechanisms of these reactions and about the enzymes that catalyze them. The discussions in the next several chapters focus on these mechanisms as they apply to specific metabolic interconversions. In this section we outline the four reaction categories and discuss how our knowledge of their reaction mechanisms derives from the study of model organic reactions. We begin by briefly reviewing the chemical logic used in analyzing these reactions.

A. *Chemical Logic*

A covalent bond consists of an electron pair shared between two atoms. In breaking such a bond, the electron pair can either remain with one of the atoms **(heterolytic bond cleavage)** or separate such that one electron accompanies each of the atoms **(homolytic bond cleavage)** (Fig. 16-4). Homolytic bond cleavage, which usually produces unstable radicals, occurs mostly in oxidation–reduction reactions. Heterolytic C—H bond cleavage involves either carbanion and proton (H$^+$) formation or carbocation (carbonium ion) and hydride ion (H$^-$) formation. Since hydride ions are highly reactive species and carbon atoms are slightly more electronegative than hydrogen atoms, bond cleavage in which the electron pair remains with the carbon atom is the predominant mode

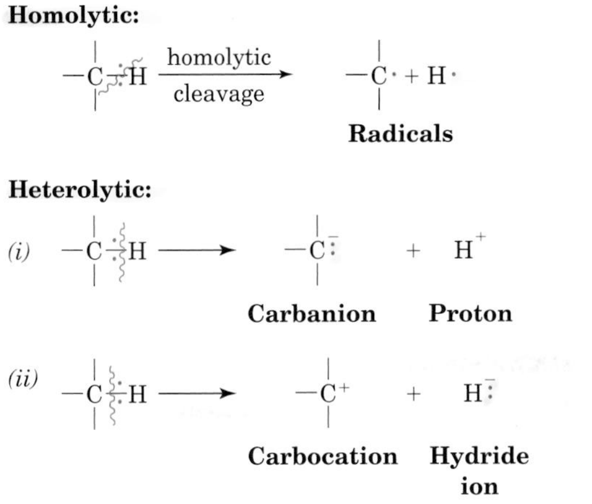

FIGURE 16-4 Modes of C—H bond breaking. Homolytic cleavage yields radicals, whereas heterolytic cleavage yields either (*i*) a carbanion and a proton or (*ii*) a carbocation and a hydride ion.

of C—H bond breaking in biochemical systems. Hydride ion abstraction occurs only if the hydride is transferred directly to an acceptor such as NAD$^+$ or NADP$^+$.

Compounds participating in reactions involving heterolytic bond cleavage and bond formation are categorized into two broad classes: electron rich and electron deficient. Electron-rich compounds, which are called **nucleophiles** (nucleus lovers), are negatively charged or contain unshared electron pairs that easily form covalent bonds with electron-deficient centers. Biologically important nucleophilic groups include amino, hydroxyl, imidazole, and sulfhydryl functions (Fig. 16-5*a*). The nucleophilic forms of these groups are also their basic forms. Indeed, nucleophilicity and basicity are closely related properties (Section 15-1B): A compound acts as a base when it forms a covalent bond with H$^+$, whereas it acts as a nucleophile when it forms a covalent bond with an electron-deficient center

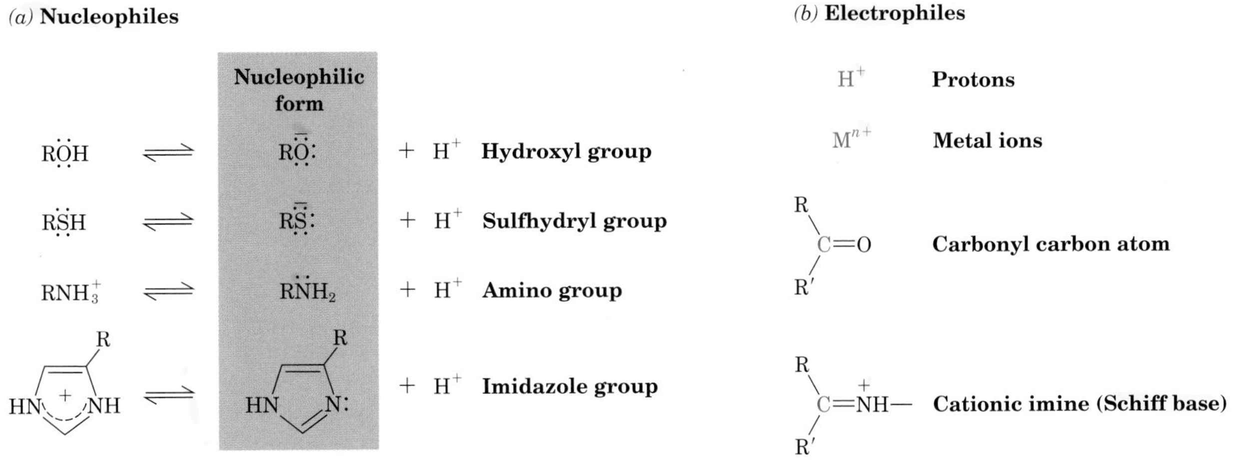

FIGURE 16-5 Biologically important nucleophilic and electrophilic groups. (*a*) Nucleophiles are the conjugate bases of weak acids such as the hydroxyl, sulfhydryl, amino, and imidazole groups. (*b*) Electrophiles contain an electron-deficient atom (*red*).

other than H$^+$, usually an electron-deficient carbon atom:

Basic reaction
of an amine   R—N̈H$_2$ + H$^+$ ⟶ R—N$^+$—H

Nucleophilic
reaction of an
amine   R—N̈H$_2$ + C=O ⟶ R—N—C—OH

tween an amine and an aldehyde or ketone, is represented:

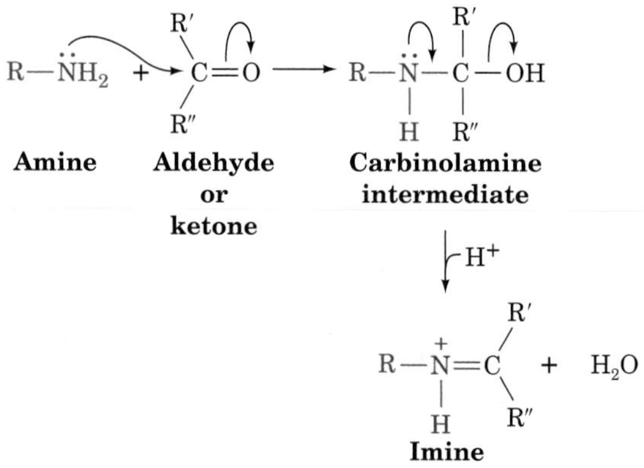

R—N̈H$_2$ + C=O ⟶ R—N̈—C—OH

Amine   **Aldehyde**   **Carbinolamine**
              or   **intermediate**
              ketone

⟶ H$^+$

R—N$^+$=C + H$_2$O

Imine

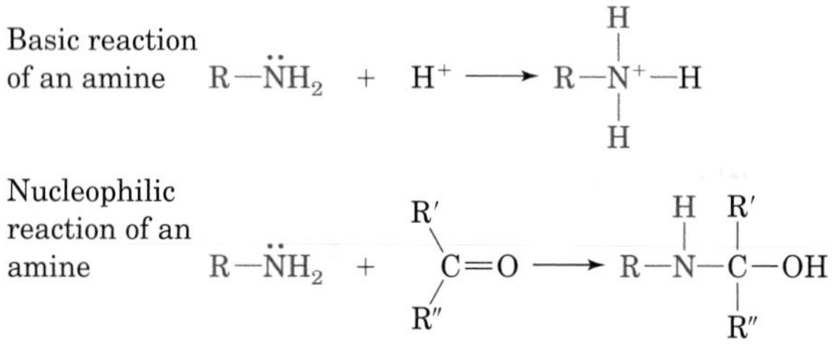

Electron-deficient compounds are called **electrophiles** (electron lovers). They may be positively charged, contain an unfilled valence electron shell, or contain an electronegative atom. The most common electrophiles in biochemical systems are H$^+$, metal ions, the carbon atoms of carbonyl groups, and cationic imines (Fig. 16-5*b*).

Reactions are best understood if the electron pair rearrangements involved in going from reactants to products can be traced. In illustrating these rearrangements we shall use the **curved arrow convention** in which the movement of an electron pair is symbolized by a curved arrow emanating from the electron pair and pointing to the electron-deficient center attracting the electron pair. For example, imine formation, a biochemically important reaction be-

In the first reaction step, the amine's unshared electron pair adds to the electron-deficient carbonyl carbon atom while one electron pair from its C=O double bond transfers to the oxygen atom. In the second step, the unshared electron pair on the nitrogen atom adds to the electron-deficient carbon atom with the elimination of water. *At all times, the rules of chemical reason prevail:* For example, there are never five bonds to a carbon atom or two bonds to a hydrogen atom.

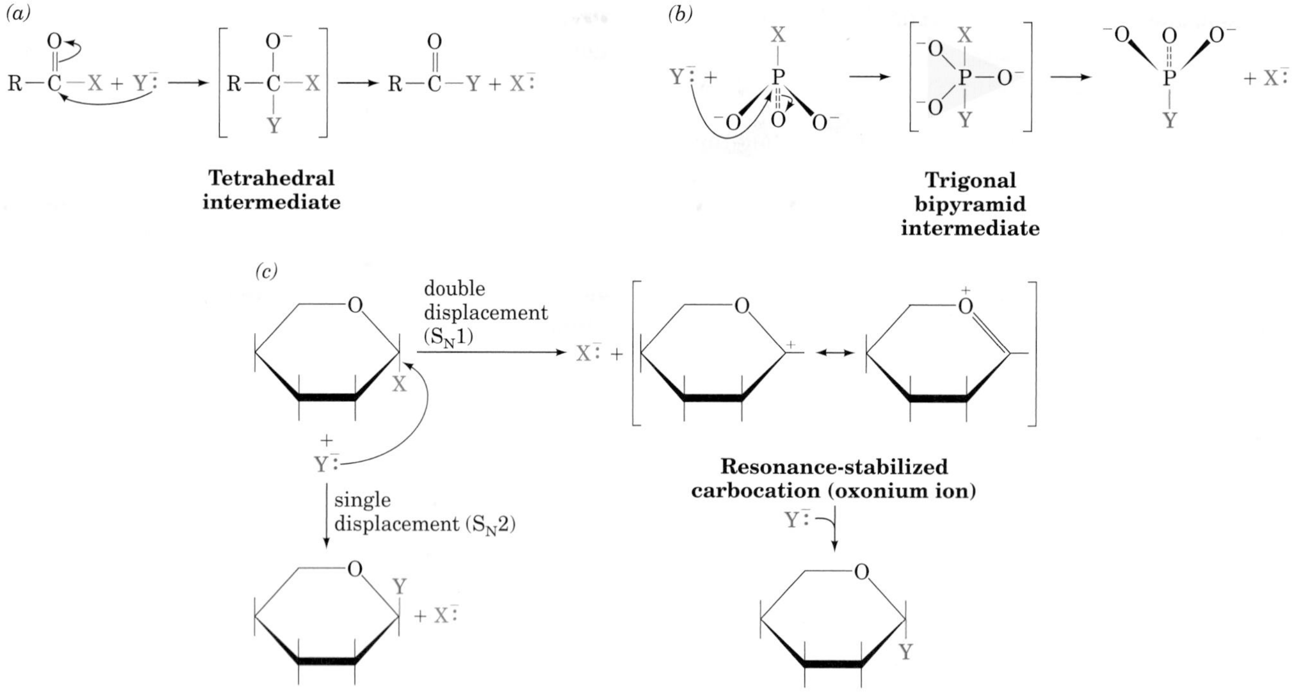

(*a*)

R—C—X + Y: ⟶ [R—C—X] ⟶ R—C—Y + X:
                    |
                    Y

**Tetrahedral
intermediate**

(*b*)

Y: + P ⟶ [P—O$^-$] ⟶ P + X:

**Trigonal
bipyramid
intermediate**

(*c*)

double
displacement
(S$_N$1)
⟶ X: +

**Resonance-stabilized
carbocation (oxonium ion)**

Y:

single
displacement (S$_N$2)

+ X:

FIGURE 16-6 Types of metabolic group-transfer reactions.
(*a*) Acyl group transfer involves addition of a nucleophile (Y) to the electrophilic carbon atom of an acyl compound to form a tetrahedral intermediate. The original acyl carrier (X) is then expelled to form a new acyl compound. (*b*) Phosphoryl group transfer involves the in-line (with the leaving group) addition of a nucleophile (Y) to the electrophilic phosphorus atom of a tetrahedral phosphoryl group. This yields a trigonal bipyramidal intermediate whose apical positions are occupied by the leaving group (X) and the attacking group (Y). Elimination of the leaving group (X) to complete the transfer reaction results in

the phosphoryl group's inversion of configuration. (*c*) Glycosyl group transfer involves the substitution of one nucleophilic group for another at C1 of a sugar ring. This reaction usually occurs via a double displacement mechanism in which the elimination of the original glycosyl carrier (X) is accompanied by the intermediate formation of a resonance-stabilized carbocation (oxonuim ion) followed by the addition of the adding nucleophile (Y). The reaction also may occur via a single displacement mechanism in which Y directly displaces X with inversion of configuration.

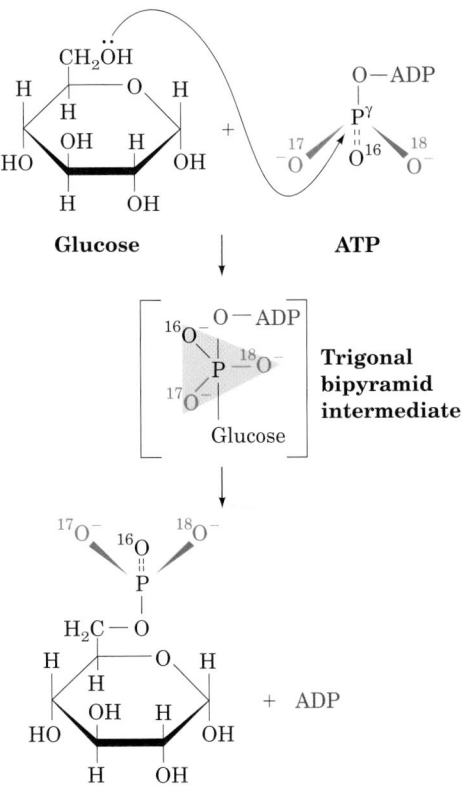

Glucose-6-phosphate

FIGURE 16-7 The phosphoryl-transfer reaction catalyzed by hexokinase. During its transfer to the 6-OH of glucose, the γ-phosphoryl group of ATP made chiral by isotopic substitution undergoes inversion of configuration via a trigonal bipyramidal intermediate.

B. *Group-Transfer Reactions*

The group transfers that occur in biochemical systems involve the transfer of an electrophilic group from one nucleophile to another:

$$Y: \quad + \quad A-X \quad \longrightarrow \quad Y-A + X:$$

| **Nucleophile** | **Electrophile–** |
| | **nucleophile** |

They could equally well be called nucleophilic substitution reactions. The most commonly transferred groups in biochemical reactions are acyl groups, phosphoryl groups, and glycosyl groups (Fig. 16-6):

1. Acyl group transfer from one nucleophile to another almost invariably involves the addition of a nucleophile to the acyl carbonyl carbon atom so as to form a tetrahedral intermediate (Fig. 16-6a). Peptide bond hydrolysis, as catalyzed, for example, by chymotrypsin (Section 15-3C), is a familiar example of such a reaction.

2. Phosphoryl group transfer proceeds via the in-line addition of a nucleophile to a phosphoryl phosphorus atom to yield a trigonal bipyramidal intermediate whose apexes are occupied by the adding and leaving groups (Fig. 16-6b). The overall reaction results in the tetrahedral phosphoryl group's inversion of configuration. Indeed, chiral phosphoryl com-

pounds have been shown to undergo just such an inversion. For example, Jeremy Knowles has synthesized ATP made chiral at its γ-phosphoryl group by isotopic substitution and demonstrated that this group is inverted on its transfer to glucose in the reaction catalyzed by **hexokinase** (Fig. 16-7).

3. Glycosyl group transfer involves the substitution of one nucleophilic group for another at C1 of a sugar ring (Fig. 16-6c). This is the central carbon atom of an acetal. Chemical models of acetal reactions generally proceed via acid-catalyzed cleavage of the first bond to form a resonance-stabilized carbocation at C1 (an oxonium ion). The lysozyme-catalyzed hydrolysis of bacterial cell wall polysaccharides (Section 15-2B) is such a reaction.

C. *Oxidations and Reductions*

Oxidation–reduction (redox) reactions involve the loss or gain of electrons. The thermodynamics of these reactions is discussed in Section 16-5. Many of the redox reactions that occur in metabolic pathways involve C—H bond cleavage with the ultimate loss of two bonding electrons by the carbon atom. These electrons are transferred to an electron acceptor such as NAD^+ (Fig. 13-2). Whether these reactions involve homolytic or heterolytic bond cleavage has not always been rigorously established. In most instances heterolytic cleavage is assumed when radical species are not observed. It is useful, however, to visualize redox C—H bond cleavage reactions as hydride transfers as diagrammed below for the oxidation of an alcohol by NAD^+:

| **General base** | **Alcohol** | NAD^+ |

| **General acid** | **Ketone** | **NADH** |

For aerobic organisms, the terminal acceptor for the electron pairs removed from metabolites by their oxidation is molecular oxygen (O_2). Recall that this molecule is a ground state diradical species whose unpaired electrons have parallel spins. The rules of electron pairing (the Pauli exclusion principle) therefore dictate that O_2 can only accept unpaired electrons; that is, electrons must be transferred to O_2 one at a time (in contrast to redox processes in which electrons are transferred in pairs). Electrons that are removed from

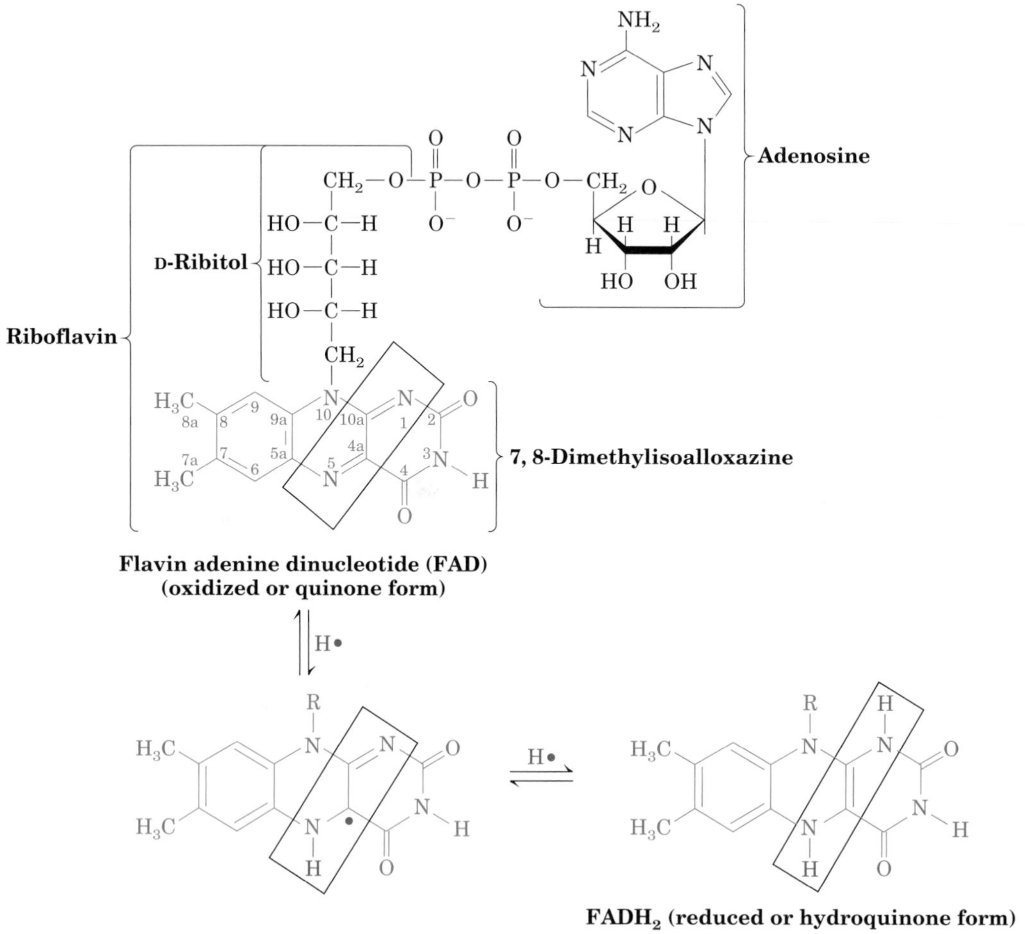

FIGURE 16-8   The molecular formula and reactions of the coenzyme flavin adenine dinucleotide (FAD). The term "flavin" is synonymous with the isoalloxazine system. The D-ribitol residue is derived from the alcohol of the sugar D-ribose. The FAD may be half-reduced to the stable radical FADH· or fully reduced to $FADH_2$ (*boxes*). Consequently, different FAD-containing enzymes cycle between different oxidation states of FAD. FAD is usually tightly bound to its enzymes, so that this coenzyme is normally a prosthetic group rather than a cosubstrate as is, for example, NAD^+. Consequently, although humans and other higher animals are unable to synthesize the isoalloxazine component of flavins and hence must obtain it in their diets [for example, in the form of **riboflavin (vitamin B_2)**], riboflavin deficiency is quite rare in humans. The symptoms of riboflavin deficiency, which are associated with general malnutrition or bizarre diets, include an inflamed tongue, lesions in the corners of the mouth, and dermatitis.

metabolites as pairs must therefore be passed to O_2 via the electron-transport chain one at a time. This is accomplished through the use of conjugated coenzymes that have stable radical oxidation states and can therefore undergo both $1e^-$ and $2e^-$ redox reactions. One such coenzyme is **flavin adenine dinucleotide (FAD;** Fig. 16-8). **Flavins** (substances that contain the **isoalloxazine** ring) can undergo two sequential one-electron transfers or a simultaneous two-electron transfer that bypasses the semiquinone state.

D. *Eliminations, Isomerizations, and Rearrangements*

a. Elimination Reactions Form Carbon–Carbon Double Bonds

Elimination reactions result in the formation of a double bond between two previously single-bonded saturated centers. The substances eliminated may be H_2O, NH_3, an alcohol (ROH), or a primary amine (RNH_2). The dehydration of an alcohol, for example, is an elimination reaction:

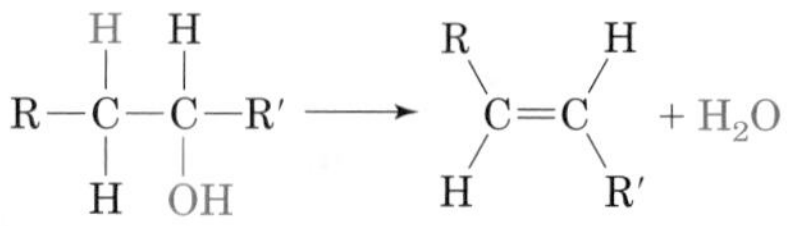

Bond breaking and bond making in this reaction may proceed via one of three mechanisms (Fig. 16-9a): (1) concerted; (2) stepwise with the C—O bond breaking first to form a carbocation; or (3) stepwise with the C—H bond breaking first to form a carbanion.

Enzymes catalyze dehydration reactions by either of two simple mechanisms: (1) protonation of the OH group by an acidic group (acid catalysis) or (2) abstraction of the proton by a basic group (base catalysis). Moreover, in a stepwise reaction, the charged intermediate may be stabilized by an oppositely charged active site group (electrostatic catalysis). The glycolytic enzyme **enolase** (Section 17-2I) and the citric acid cycle enzyme **fumarase** (Section 21-3G) catalyze such dehydration reactions.

Elimination reactions may take one of two possible stereochemical courses (Fig. 16-9*b*): (1) trans (anti) eliminations, the most prevalent biochemical mechanism, and (2) cis (syn) eliminations, which are biochemically less common.

b. Biochemical Isomerizations Involve Intramolecular Hydrogen Atom Shifts

Biochemical **isomerization reactions** involve the intramolecular shift of a hydrogen atom so as to change the location of a double bond. In such a process, a proton is removed from one carbon atom and added to another. The metabolically most prevalent isomerization reaction is the **aldose–ketose interconversion,** a base-catalyzed reaction that occurs via **enediolate anion** intermediates (Fig. 16-10).

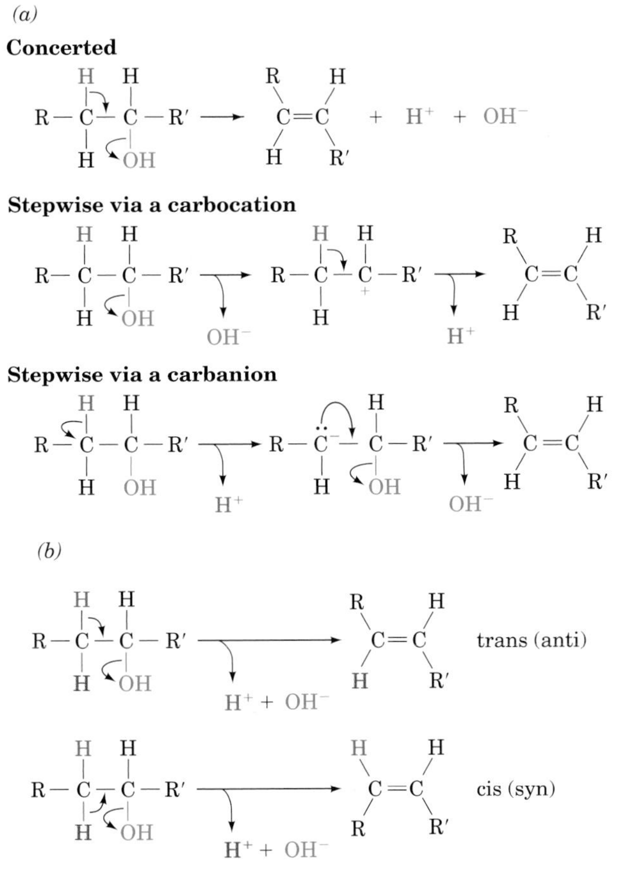

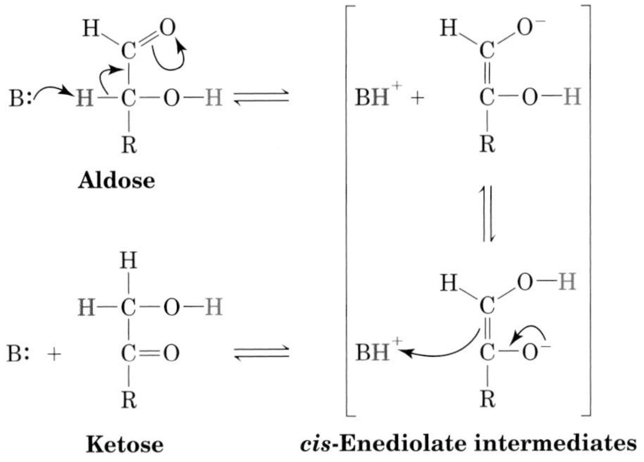

FIGURE 16-10 Mechanism of aldose–ketose isomerization. The reaction occurs with acid–base catalysis and proceeds via *cis*-enediolate intermediates.

The glycolytic enzyme **phosphoglucose isomerase** catalyzes such a reaction (Section 17-2B).

Racemization is an isomerization reaction in which a hydrogen atom shifts its stereochemical position at a molecule's only chiral center so as to invert that chiral center (e.g., the racemization of proline by proline racemase; Section 15-1F). Such an isomerization is called an **epimerization** in a molecule with more than one chiral center.

c. Rearrangements Produce Altered Carbon Skeletons

Rearrangement reactions break and reform C—C bonds so as to rearrange a molecule's carbon skeleton. There are few such metabolic reactions. One is the conversion of L-methylmalonyl-CoA to **succinyl-CoA** by **methylmalonyl-CoA mutase,** an enzyme whose prosthetic group is a **vitamin B$_{12}$** derivative:

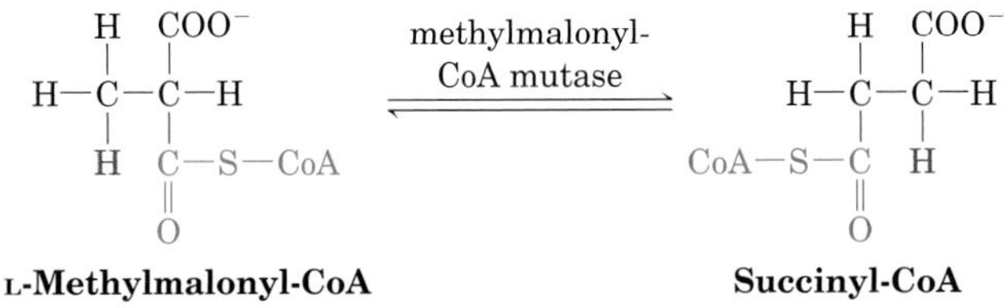

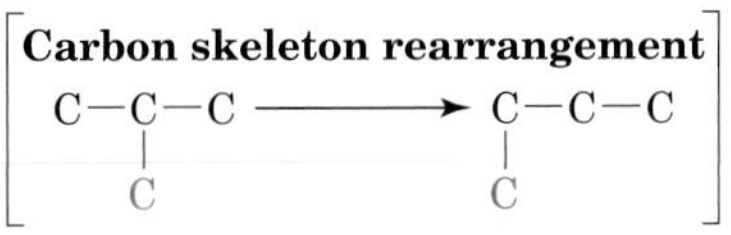

This reaction is involved in the oxidation of fatty acids with an odd number of carbon atoms (Section 25-2E) and several amino acids (Section 26-3E).

E. *Reactions That Make and Break Carbon–Carbon Bonds*

Reactions that make and break carbon–carbon bonds form the basis of both degradative and biosynthetic metabolism. The breakdown of glucose to CO$_2$ involves five such cleav-

FIGURE 16-9 Possible elimination reaction mechanisms using dehydration as an example. Reactions may be (*a*) either concerted, stepwise via a carbocation intermediate, or stepwise via a carbanion intermediate; and may occur with (*b*) either trans (anti) or cis (syn) stereochemistry.

ages, whereas its synthesis involves the reverse process. Such reactions, considered from the synthetic direction, involve addition of a nucleophilic carbanion to an electrophilic carbon atom. The most common electrophilic carbon atoms in such reactions are the sp^2-hybridized carbonyl carbon atoms of aldehydes, ketones, esters, and CO_2:

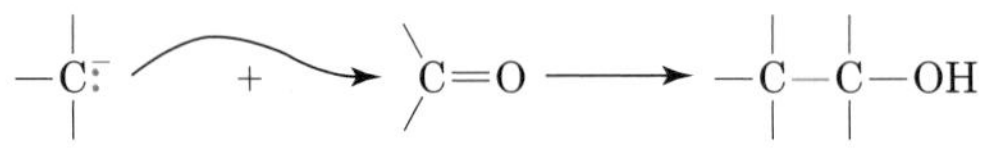

Stabilized carbanions must be generated to add to these electrophilic centers. Three examples are the **aldol condensation** (catalyzed, e.g., by **aldolase;** Section 17-2D), **Claisen ester condensation** (**citrate synthase;** Section 21-3A), and the decarboxylation of β-keto acids (**isocitrate dehydrogenase,** Section 21-3C; and **fatty acid synthase,** Section 25-4C). In nonenzymatic systems, both the aldol condensation and Claisen ester condensation involve the base-catalyzed generation of a carbanion α to a carbonyl group (Fig. 16-11a,b). The carbonyl group is electron

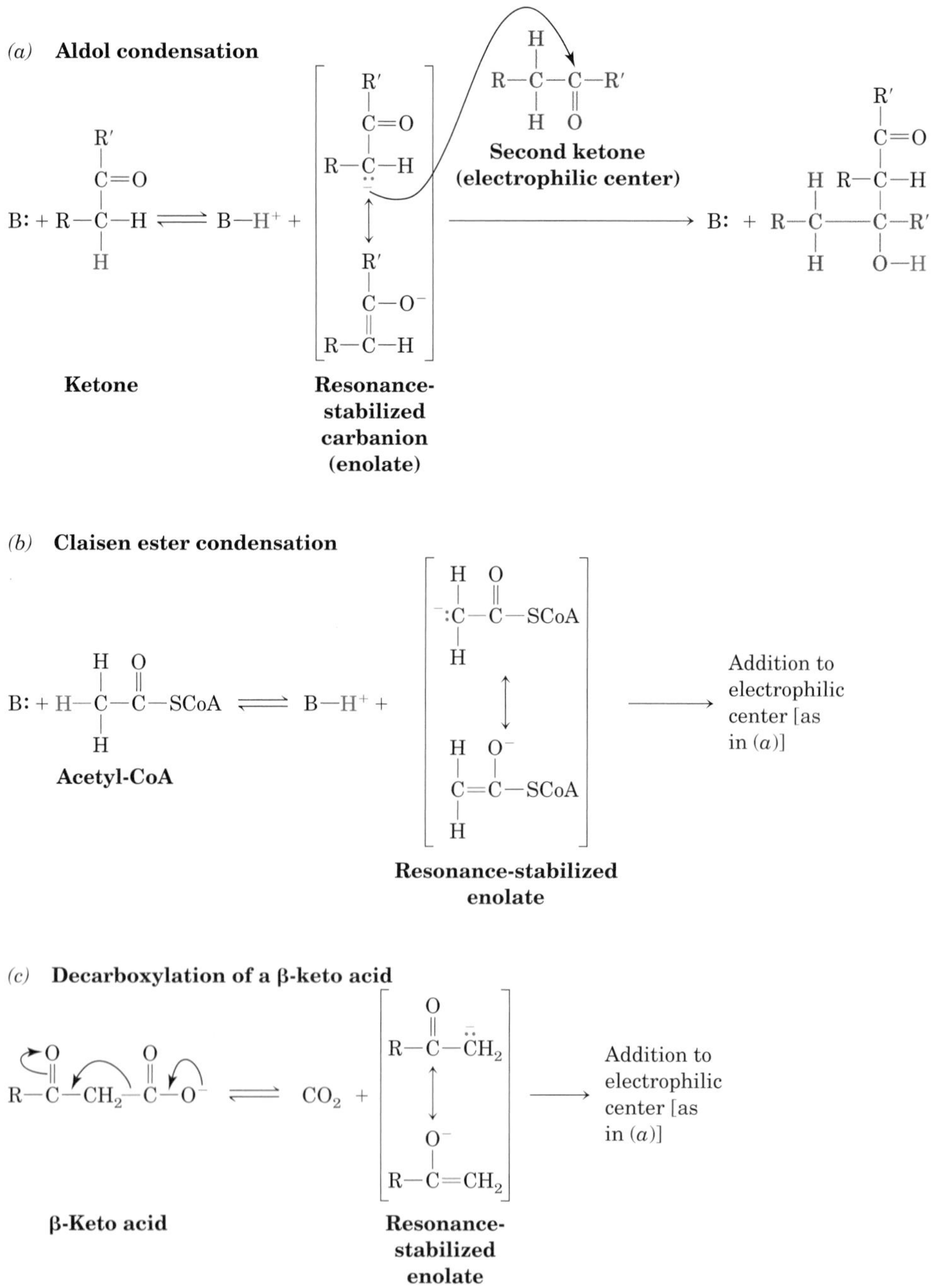

FIGURE 16-11 Examples of C—C bond formation and cleavage reactions. (a) Aldol condensation, (b) Claisen ester condensation, and (c) decarboxylation of a β-keto acid. All three types of reactions involve generation of a resonance-stabilized carbanion followed by addition of this carbanion to an electrophilic center.

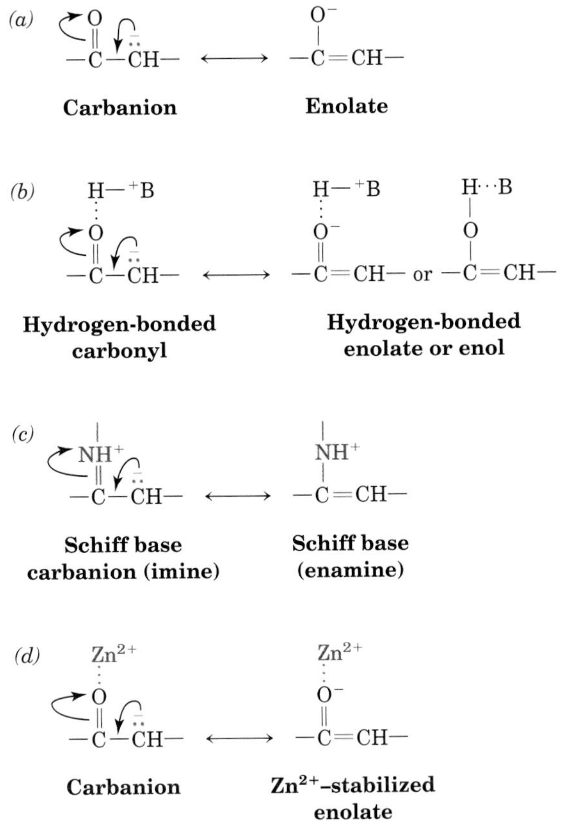

FIGURE 16-12   **Stabilization of carbanions.** (*a*) Carbanions adjacent to carbonyl groups are stabilized by the formation of enolates. (*b*) Carbanions adjacent to carbonyl groups hydrogen bonded to general acids are stabilized electrostatically or by charge neutralization. (*c*) Carbanions adjacent to protonated imines (Schiff bases) are stabilized by the formation of enamines. (*d*) Metal ions stabilize carbanions adjacent to carbonyl groups by the electrostatic stabilization of the enolate.

withdrawing and thereby provides resonance stabilization by forming an enolate (Fig. 16-12*a*). The enolate may be further stabilized by neutralizing its negative charge. Enzymes do so through hydrogen bonding or protonation (Fig. 16-12*b*), conversion of the carbonyl group to a protonated Schiff base (covalent catalysis; Fig. 16-12*c*), or by its coordination to a metal ion (metal ion catalysis; Fig. 16-12*d*). The decarboxylation of a β-keto acid does not require base catalysis for the generation of the resonance-stabilized carbanion; the highly exergonic formation of CO_2 provides its driving force (Fig. 16-11*c*).

3 ■ EXPERIMENTAL APPROACHES TO THE STUDY OF METABOLISM

A metabolic pathway can be understood at several levels:

1. In terms of the sequence of reactions by which a specific nutrient is converted to end products, and the energetics of these conversions.

2. In terms of the mechanisms by which each intermediate is converted to its successor. Such an analysis requires the isolation and characterization of the specific enzymes that catalyze each reaction.

3. In terms of the control mechanisms that regulate the flow of metabolites through the pathway. An exquisitely complex network of regulatory processes renders metabolic pathways remarkably sensitive to the needs of the organism; the output of a pathway is generally only as great as required.

As you might well imagine, the elucidation of a metabolic pathway on all of these levels is a complex process, involving contributions from a variety of disciplines. Most of the techniques used to do so involve somehow perturbing the system and observing the perturbation's effect on growth or on the production of metabolic intermediates. One such technique is the use of metabolic inhibitors that block metabolic pathways at specific enzymatic steps. Another is the study of genetic abnormalities that interrupt specific metabolic pathways. Techniques have also been developed for the dissection of organisms into their component organs, tissues, cells, and subcellular organelles, and for the purification and identification of metabolites as well as the enzymes that catalyze their interconversions. The use of isotopic tracers to follow the paths of specific atoms and molecules through the metabolic maze has become routine. Techniques utilizing NMR technology are able to trace metabolites noninvasively as they react *in vivo*. This section outlines the use of these various techniques.

A. *Metabolic Inhibitors, Growth Studies, and Biochemical Genetics*

a. Pathway Intermediates Accumulate in the Presence of Metabolic Inhibitors

The first metabolic pathway to be completely traced was the conversion of glucose to ethanol in yeast by a process known as **glycolysis** (Section 17-1A). In the course of these studies, certain substances, called **metabolic inhibitors,** were found to block the pathway at specific points, thereby causing preceding intermediates to build up. For instance, iodoacetate causes yeast extracts to accumulate fructose-1,6-bisphosphate, whereas fluoride causes the buildup of two phosphate esters, 3-phosphoglycerate and 2-phosphoglycerate. The isolation and characterization of these intermediates was vital to the elucidation of the glycolytic pathway: Chemical intuition combined with this information led to the prediction of the pathway's intervening steps. Each of the proposed reactions was eventually shown to occur *in vitro* as catalyzed by a purified enzyme.

b. Genetic Defects Also Cause Metabolic Intermediates to Accumulate

Archibald Garrod's realization, in the early 1900s, that human genetic diseases are the consequence of deficiencies in specific enzymes (Section 1-4C) also contributed to the elucidation of metabolic pathways. For example, on the ingestion of either phenylalanine or tyrosine, individuals with the largely harmless inherited condition known as **al-captonuria,** but not normal subjects, excrete **homogentisic acid** in their urine (Section 26-3H). This is because the liver of alcaptonurics lacks an enzyme that catalyzes the breakdown of homogentisic acid. Another genetic disease, **phenylketonuria** (Section 26-3H), results in the accumulation of **phenylpyruvate** in the urine (and which, if untreated, causes severe mental retardation in infants). Ingested phenylalanine and phenylpyruvate appear as phenylpyruvate in the urine of affected subjects, whereas tyrosine is metabolized normally. The effects of these two abnormalities suggested the pathway for phenylalanine metabolism diagrammed in Fig. 16-13. However, the supposition that phenylpyruvate but not tyrosine occurs on the normal pathway of phenylalanine metabolism because phenylpyruvate accumulates in the urine of phenylketonurics has proved incorrect. This indicates the pitfalls of relying solely on metabolic blocks and the consequent buildup of intermediates as indicators of a metabolic pathway. In this case, phenylpyruvate formation was later shown to arise from a normally minor pathway that becomes significant only when the phenylalanine concentration is abnormally high, as it is in phenylketonurics.

c. Metabolic Blocks Can Be Generated by Genetic Manipulation

Early metabolic studies led to the astounding discovery that *the basic metabolic pathways in most organisms are essentially identical.* This metabolic uniformity has greatly facilitated the study of metabolic reactions. A mutation that inactivates or deletes an enzyme in a pathway of interest can be readily generated in rapidly reproducing microorganisms through the use of **mutagens** (chemical agents that induce genetic changes; Section 32-1A), X-rays, or genetic engineering techniques (Sections 5-5). Desired mutants are identified by their requirement of the pathway's end product for growth. For example, George Beadle and Edward Tatum proposed a pathway of arginine biosynthesis in the mold *Neurospora crassa* based on their analysis of three arginine-requiring **auxotrophic mutants** (mutants requiring a specific nutrient for growth), which were isolated after X-irradiation (Fig. 16-14). This landmark study also conclusively demonstrated that enzymes are specified by genes (Section 1-4C).

d. Genetic Manipulations of Higher Organisms Provide Metabolic Insights

Transgenic organisms (Section 5-5H) constitute valuable resources for the study of metabolism. *They can be used to both create metabolic blocks and to express genes in tissues where they are not normally present.* For example, **creatine kinase** catalyzes the formation of **phosphocreatine** (Section 16-4C), a substance that functions to generate ATP rapidly when it is in short supply. This enzyme is normally present in many tissues, including brain and muscle, but not in liver. The introduction of the gene encoding creatinine kinase into the liver of a mouse causes the liver to synthesize phosphocreatine when the mouse is fed creatine, as demonstrated by localized *in vivo* NMR techniques (Fig. 16-15; NMR is discussed below). The presence of phosphocreatine in a transgenic mouse liver protects the animal against the sharp drop in [ATP] ordinarily caused by fructose overload (Section 17-5A). This

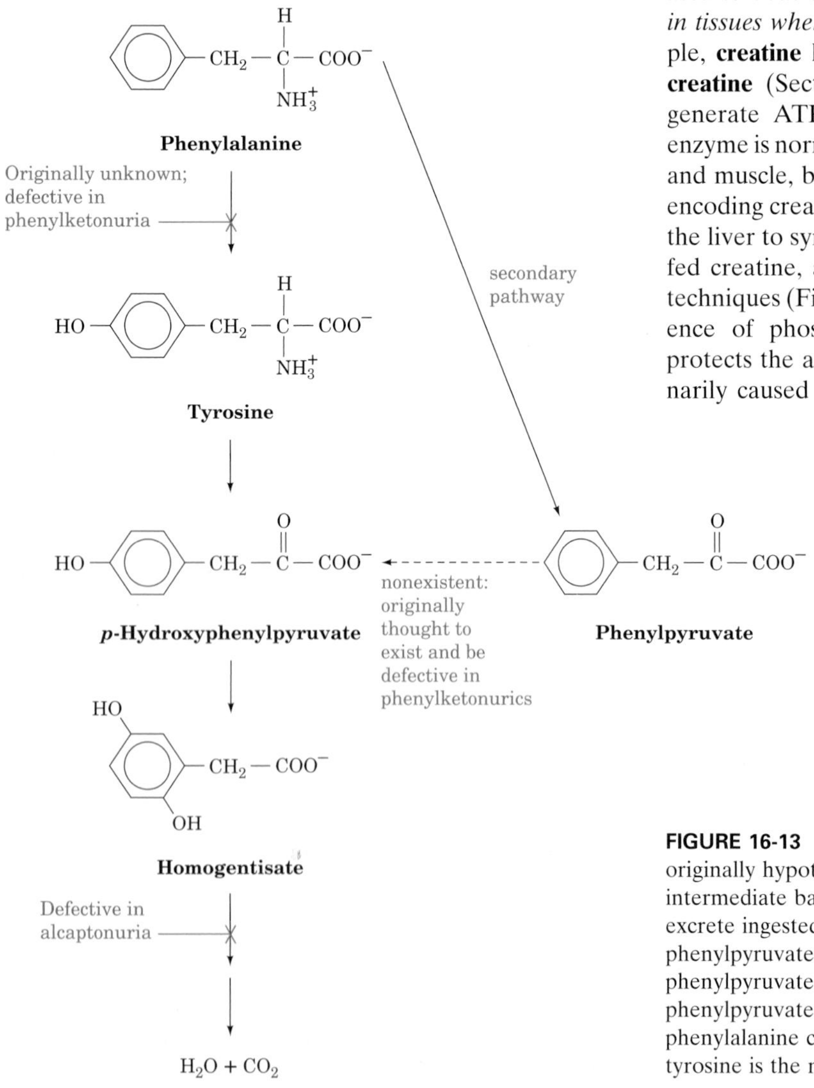

FIGURE 16-13   Pathway for phenylalanine degradation. It was originally hypothesized that phenylpyruvate was a pathway intermediate based on the observation that phenylketonurics excrete ingested phenylalanine and phenylpyruvate as phenylpyruvate. Further studies, however, demonstrated that phenylpyruvate is not a homogentisate precursor; rather, phenylpyruvate production is significant only when the phenylalanine concentration is abnormally high. Instead, tyrosine is the normal product of phenylalanine degradation.

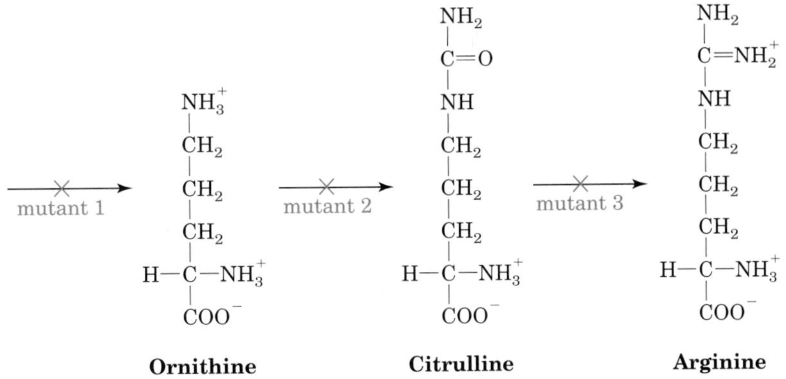

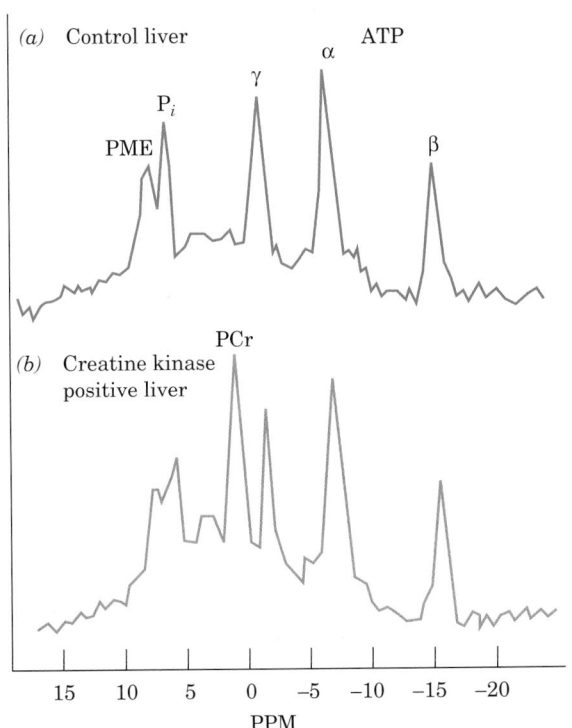

Ornithine   **Citrulline**   **Arginine**

FIGURE 16-14 Pathway of arginine biosynthesis indicating the positions of genetic blocks. All of these mutants grow in the presence of arginine, but mutant 1 also grows in the presence of the (nonstandard) α-amino acids **citrulline** or **ornithine** and mutant 2 grows in the presence of citrulline. This is because in mutant 1, an enzyme leading to the production of ornithine is absent but enzymes farther along the pathway are normal. In mutant 2, the enzyme catalyzing citrulline production is defective, whereas in mutant 3 an enzyme involved in the conversion of citrulline to arginine is lacking.

genetic manipulation technique is being used to study mechanisms of metabolic control *in vivo.*

Metabolic pathways are regulated both by controlling the activities of regulatory enzymes (Sections 17-4 and 18-3) and by controlling their concentrations at the level of gene expression (Sections 31-3, 32-4, and 34-3). The important question of how hormones and diet control metabolic processes at the level of gene expression is being addressed through the use of transgenic animals. **Reporter genes** (genes whose products are easily detected; Section 5-5G) are placed under the influence of **promoters** (genetic elements that regulate transcriptional initiation; Section 5-5A) that control the expression of specific regulatory enzymes, and the resulting composite gene is expressed in animals. The transgenic animals can then be treated with specific hormones and/or diets and the production of the reporter gene product measured. For instance, in an investigation by Richard Hanson, the promoter for the enzyme **phosphoenolpyruvate carboxykinase (PEPCK)** was attached to the structural gene encoding **growth hormone (GH).** PEPCK, an important regulatory enzyme in **gluconeogenesis** (the synthesis of glucose from noncarbohydrate precursors; Section 23-1), is normally present in liver and kidneys but not in blood. GH, however, is secreted into the blood and its presence there can be readily quantitated by an ELISA (Section 6-1D). Mice transgenic for PEPCK/GH were fed either a high-carbohydrate/low-protein diet or a high-protein/low-carbohydrate diet, which are known to decrease and increase PEPCK activity, respectively. GH in high concentrations was detected only in the serum of PEPCK/GH mice on a high-protein diet, thereby indicating that the GH was synthesized under the same dietary control as that of the PEPCK expressed by the normal gene. Thus, the activity of PEPCK in PEPCK/GH mice can be continuously monitored, albeit indirectly, through serum GH assays (the direct measurement of PEPCK in mouse liver or kidney requires the sacrifice of the animal and hence can be done only once). Such use of reporter genes has proved to be of great value in the study of the genetic control of metabolism.

Modern techniques also make it possible to insert a mutation that inactivates or deletes an enzyme or control protein in a pathway of interest in higher organisms such

FIGURE 16-15 The expression of creatine kinase in transgenic mouse liver as demonstrated by localized *in vivo* ^{31}P NMR. (*a*) The spectrum of a normal mouse liver after the mouse had been fed a diet supplemented with 2% creatine. The peaks corresponding to inorganic phosphate (P_i), the α, β, and γ phosphoryl groups of ATP, and phosphomonoesters (PME) are labeled. (*b*) The spectrum of the liver of a mouse transgenic for creatine kinase that had been fed a diet supplemented with 2% creatine. The phosphocreatine peak is labeled PCr. [After Koretsky, A.P., Brosnan, M.J., Chen, L., Chen, J., and Van Dyke, T.A., *Proc. Natl. Acad. Sci.* **87,** 3114 (1990)].

as mice (**knockout mice**; Section 5-5H). Knockout mice have proved useful for studying metabolic control mechanisms. For example, PEPCK activity is thought to be controlled exclusively by increasing or decreasing its availability. Diet affects its production, as we have seen. However, this demand-based control is superimposed on the developmental regulation of PEPCK production. The enzyme is not produced at all in early embryos and only appears near birth, when gluconeogenesis is required to supply the glucose that had been previously available *in utero*. One of the proteins thought to be responsible for the developmental regulation of PEPCK production is **CCAAT/enhancer-binding protein α (C/EBPα)**, a **transcription factor** (Section 5-4A; transcriptional regulation in eukaryotes is discussed in Section 34-3B). Newborn mice homozygous for the targeted deletion of the *c/ebpα* gene (*c/ebpα* knockout mice) do not produce C/EBPα and therefore do not produce PEPCK. Consequently, their livers cannot synthesize the glucose necessary to maintain adequate blood glucose levels once they are disconnected from the maternal circulation. Indeed, these mice become so hypoglycemic that they die within 8 hours of birth. Clearly C/EBPα has an important role in the developmental regulation of PEPCK.

B. *Isotopes in Biochemistry*

The specific labeling of metabolites such that their interconversions can be traced is an indispensable technique for elucidating metabolic pathways. Franz Knoop formulated this technique in 1904 to study fatty acid oxidation. He fed dogs fatty acids chemically labeled with phenyl groups and isolated the phenyl-substituted end products from their urine. From the differences in these products when the phenyl-substituted starting material contained odd and even numbers of carbon atoms he deduced that fatty acids are degraded in C_2 units (Section 25-2).

a. Isotopes Specifically Label Molecules without Altering Their Chemical Properties

Chemical labeling has the disadvantage that the chemical properties of labeled metabolites differ from those of normal metabolites. This problem is eliminated by labeling molecules of interest with **isotopes** (atoms with the same number of protons but a different number of neutrons in their nuclei). Recall that the chemical properties of an element are a consequence of its electron configuration which, in turn, is determined by its atomic number, not its atomic mass. The metabolic fate of a specific atom in a metabolite can therefore be elucidated by isotopically labeling that position and following its progress through the metabolic pathway of interest. The advent of isotopic labeling and tracing techniques in the 1940s therefore revolutionized the study of metabolism. (**Isotope effects,** which are changes in reaction rates arising from the mass differences between isotopes, are in most instances negligible. Where they are significant, most noticeably between hydrogen and its isotopes deuterium and tritium, they have been used to gain insight into enzymatic reaction mechanisms.)

b. NMR Can Be Used to Study Metabolism in Whole Animals

Nuclear magnetic resonance (NMR) detects specific isotopes due to their characteristic nuclear spins. Among the isotopes that NMR can detect are ^1H, ^{13}C, and ^{31}P. Since the NMR spectrum of a particular nucleus varies with its immediate environment, it is possible to identify the peaks corresponding to specific atoms even in relatively complex mixtures.

The development of magnets large enough to accommodate animals and humans, and to localize spectra to specific organs, has made it possible to study metabolic pathways noninvasively by NMR techniques. Thus, ^{31}P NMR can be used to study energy metabolism in muscle by monitoring the levels of ATP, ADP, inorganic phosphate, and phosphocreatine (Figure 16-15). Indeed, a ^{31}P NMR system has been patented to measure the muscular metabolic efficiency and maximum power of race horses while they are walking or running on a motor-driven treadmill in order to identify promising animals and to evaluate the efficacy of their training and nutritional programs.

Isotopically labeling specific atoms of metabolites with ^{13}C (which is only 1.10% naturally abundant) permits the metabolic progress of the labeled atoms to be followed by ^{13}C NMR. Figure 16-16 shows *in vivo* ^{13}C NMR spectra of a rat liver before and after an injection of D-[1-^{13}C]glucose. The ^{13}C can be seen entering the liver and then being converted to glycogen (the storage form of glucose; Chapter 18). ^1H NMR techniques are being used to determine the *in vivo* levels of a variety of metabolites in tissues such as brain and muscle.

c. The Detection of Radioactive Isotopes

All elements have isotopes. For example, the atomic mass of naturally occurring Cl is 35.45 D because, at least on Earth, it is a mixture of 55% ^{35}Cl and 45% ^{36}Cl (other isotopes of Cl are present in only trace amounts). Stable isotopes are generally identified and quantitated by mass spectrometry or NMR techniques. Many isotopes, however, are unstable; they undergo **radioactive decay,** a process that involves the emission from the radioactive nuclei of subatomic particles such as helium nuclei (**α particles**), electrons (**β particles**), and/or photons (**γ radiation**). Radioactive nuclei emit radiation with characteristic energies. For example, ^3H, ^{14}C, and ^{32}P all emit β particles but with respective energies of 0.018, 0.155, and 1.71 MeV. The radiation from ^{32}P is therefore highly penetrating, whereas that from ^3H and ^{14}C is not. (^3H and ^{14}C, as all radioactive isotopes, must, nevertheless, be handled with great caution because they can cause genetic damage on ingestion.)

Radiation can be detected by a variety of techniques. Those most commonly used in biochemical investigations are **proportional counting** (known in its simplest form as **Geiger counting**), **liquid scintillation counting,** and **autoradiography.** Proportional counters electronically detect the ionizations in a gas caused by the passage of radiation. Moreover, they can also discriminate between particles of different energies and thus simultaneously determine the amounts of two or more different isotopes present.

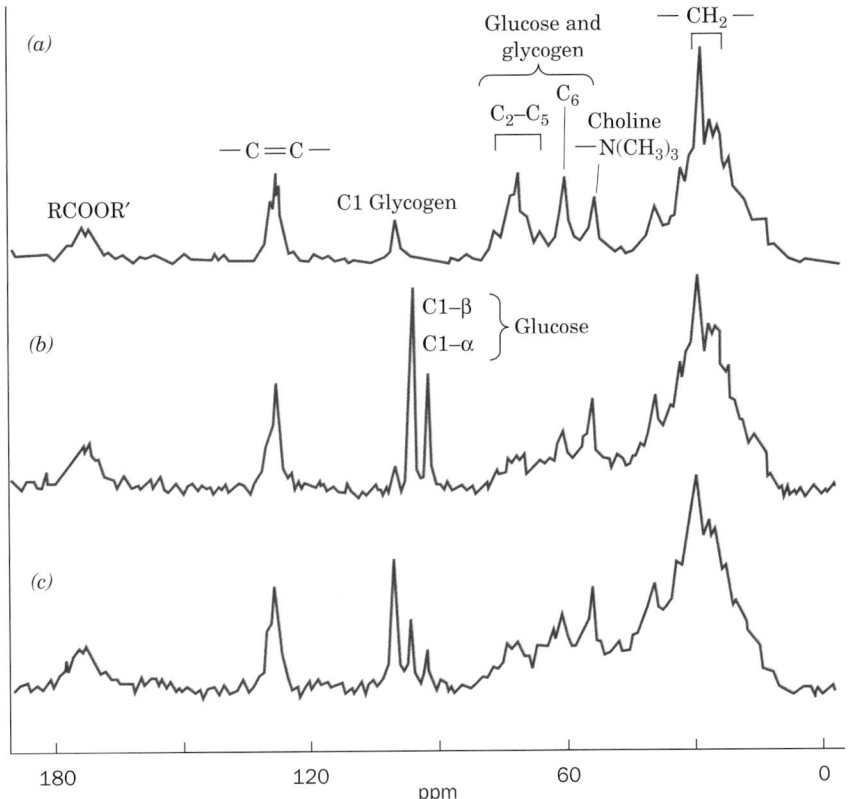

FIGURE 16-16 **The conversion of [1-¹³C]glucose to glycogen as observed by localized *in vivo* ¹³C NMR.** (*a*) The natural abundance ¹³C NMR spectrum of the liver of a live rat. Note the resonance corresponding to C1 of glycogen. (*b*) The ¹³C NMR spectrum of the liver of the same rat ~5 min after it was intravenously injected with 100 mg of [1-¹³C]glucose (90% enriched). The resonances of the C1 atom of both the α and β anomers of glucose are clearly distinguishable from each other and from the resonance of the C1 atom of glycogen. (*c*) The ¹³C NMR spectrum of the liver of the same rat ~30 min after the [1-¹³C]glucose injection. The C1 resonances of both the α and β glucose anomers are much reduced while the C1 resonance of glycogen has increased. [After Reo, N.V., Siegfried, B.A., and Acherman, J.J.H., *J. Biol. Chem.* **259**, 13665 (1984)].

Although proportional counters are quite simple to use, the radiation from two of the most widely used isotopes in biochemical analysis, ³H and ¹⁴C, have insufficient penetrating power to enter a proportional counter's detection chamber with reasonable efficiency. This limitation is circumvented through liquid scintillation counting. In this technique, a radioactive sample is dissolved or suspended in a solution containing fluorescent substances that emit a pulse of light when struck by radiation. The light is detected electronically so that the number of light pulses can be counted. The emitting nucleus can also be identified because the intensity of a light pulse is proportional to the radiation energy (the number of fluorescent molecules excited by a radioactive particle is proportional to the particle's energy).

In autoradiography, radiation is detected by its blackening of photographic film. The radioactive sample is laid on, or in some cases mixed with, the photographic emulsion and, after sufficient exposure time (from minutes to months), the film is developed. Autoradiography is widely used to locate radioactive substances in polyacrylamide gels (e.g., Fig. 6-27). Position-sensitive radiation counters (electronic film) are similarly employed.

d. Radioactive Isotopes Have Characteristic Half-Lives

Radioactive decay is a random process whose rate for a given isotope depends only on the number of radioactive atoms present. It is therefore a simple first-order process whose half-life, $t_{1/2}$, is a function only of the rate constant, k, for the decay process (Section 14-1B):

$$t_{1/2} = \frac{\ln 2}{k} = \frac{0.693}{k} \qquad [14.5]$$

Because k is different for each radioactive isotope, each has a characteristic half-life. The properties of some isotopes in common biochemical use are listed in Table 16-2.

TABLE 16-2 **Some Trace Isotopes of Biochemical Importance**

Stable Isotopes

Nucleus	Natural Abundance (%)
²H	0.015
¹³C	1.07
¹⁵N	0.37
¹⁸O	0.20

Radioactive Isotopes

Nucleus	Radiation Type	Half-Life
³H	β	12.33 years
¹⁴C	β	5715 years
²²Na	β⁺, γ	2.60 years
³²P	β	14.28 days
³⁵S	β	87.2 days
⁴⁵Ca	β	162.7 days
⁶⁰Co	β, γ	5.271 years
¹²⁵I	γ	59.4 days
¹³¹I	β, γ	8.04 days

Source: Holden, N.E., *in* Lide, D.R. (Ed.), *Handbook of Chemistry and Physics* (82nd ed.), pp. 11–51 to 197, CRC Press (2001).

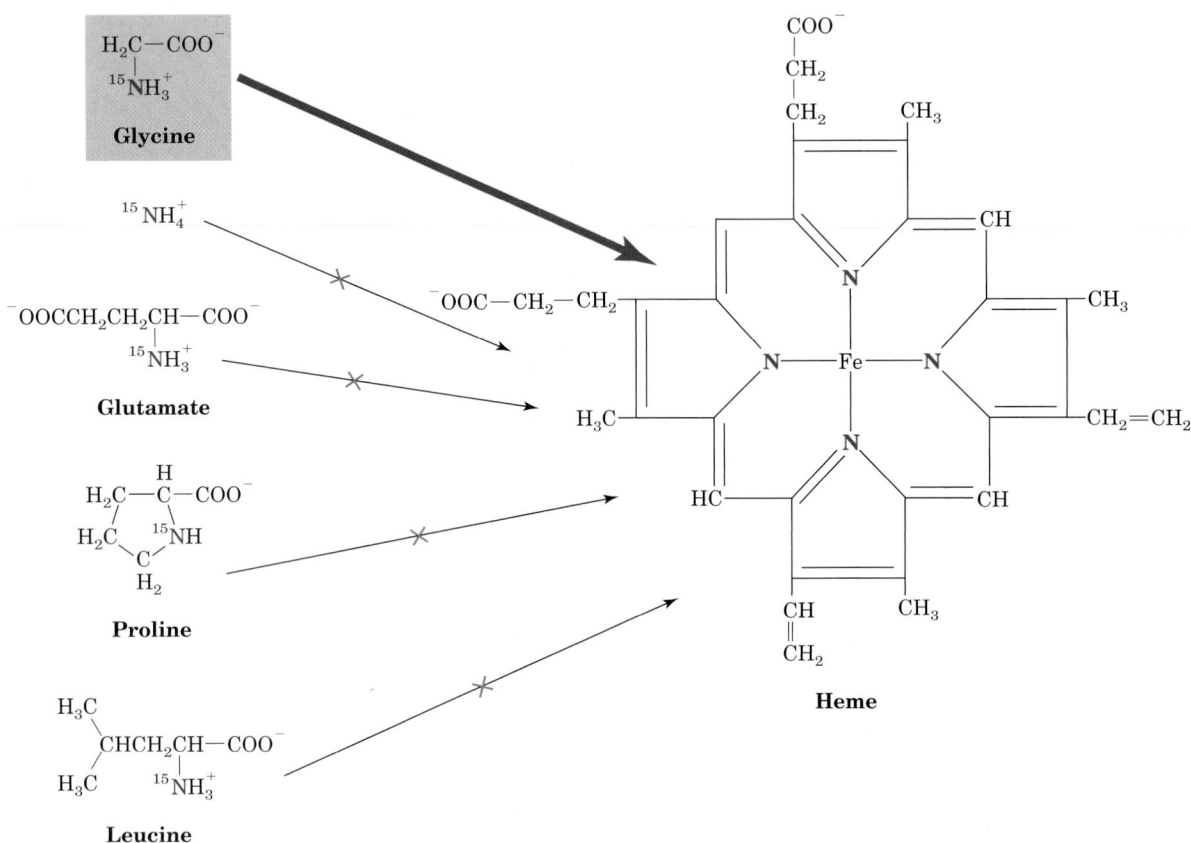

FIGURE 16-17   The metabolic origin of the nitrogen atoms in heme. Only [^{15}N]glycine, of many ^{15}N-labeled metabolites, is an ^{15}N-labeled heme precursor.

e. Isotopes Are Indispensable for Establishing the Metabolic Origins of Complex Metabolites and Precursor–Product Relationships

The metabolic origins of complex molecules such as heme, cholesterol, and phospholipids may be determined by administering isotopically labeled starting materials to animals and isolating the resulting products. One of the early advances in metabolic understanding resulting from the use of isotopic tracers was the demonstration, by David Shemin and David Rittenberg in 1945, that the nitrogen atoms of heme are derived from glycine rather than from ammonia, glutamic acid, proline, or leucine (Section 26-4A). They showed this by feeding rats these ^{15}N-labeled nutrients, isolating the heme in their blood, and analyzing it for ^{15}N content. Only when the rats were fed [^{15}N]glycine did the heme contain ^{15}N (Fig. 16-17). This technique was also used to demonstrate that all of cholesterol's carbon atoms are derived from acetyl-CoA (Section 25-6A).

Isotopic tracers are also useful in establishing the order of appearance of metabolic intermediates, their so-called **precursor–product relationships.** An example of such an analysis concerns the biosynthesis of the complex phospholipids called **plasmalogens** and **alkylacylglycerophospholipids** (Section 25-8A). Alkylacylglycerophospholipids are ethers, whereas the closely related plasmalogens are vinyl ethers. Their similar structures brings up the inter-

esting question of their biosynthetic relationship: Which is the precursor and which is the product? Two possible modes of synthesis can be envisioned (Fig. 16-18):

I. The starting material is converted to the vinyl ether (plasmalogen), which is then reduced to yield the ether (alkylacylglycerophospholipid). Accordingly, the vinyl ether would be the precursor and the ether the product.

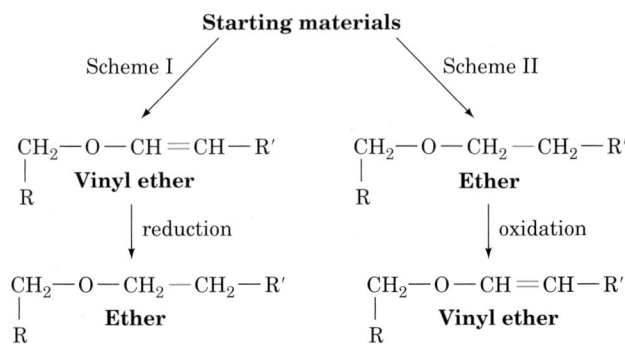

FIGURE 16-18   Two possible pathways for the biosynthesis of ether– and vinyl ether–containing phospholipids. (I) The vinyl ether is the precursor and the ether is the product. **(II)** The ether is the precursor and the vinyl ether is the product.

II. The ether is formed first and then oxidized to yield the vinyl ether. The ether would then be the precursor and the vinyl ether the product.

Precursor–product relationships can be most easily sorted out through the use of radioactive tracers. A pulse of the labeled starting material is administered to an organism and the specific radioactivities of the resulting metabolic products are followed with time (Fig. 16-19):

$$\text{Starting material*} \rightarrow \text{A*} \rightarrow \text{B*} \rightarrow \text{later products*}$$

(here the * represents the radioactive label). Metabolic pathways, as we shall see in Section 16-6B, normally operate in a steady state; that is, the throughput of metabolites in each of its reaction steps is equal. Moreover, the rates of most metabolic reactions are first order for a given substrate. Making these assumptions, we note that the rate of change of B's radioactivity, [B*], is equal to the rate of passage of label from A* to B* less the rate of passage of label from B* to the pathway's next product:

$$\frac{d[\text{B*}]}{dt} = k[\text{A*}] - k[\text{B*}] = k([\text{A*}] - [\text{B*}]) \quad [16.1]$$

where k is the pseudo-first-order rate constant for both the conversion of A to B and the conversion of B to its product, and t is time. Inspection of this equation indicates the criteria that must be met to establish that A is the precursor of B (Fig. 16-19):

1. Before the radioactivity of the product [B*] is maximal, $d[\text{B*}]/dt > 0$, so [A*] > [B*]; that is, *while the radioactivity of a product is rising, it should be less than that of its precursor.*

2. When [B*] is maximal, $d[\text{B*}]/dt = 0$, so [A*] = [B*]; that is, *when the radioactivity of a product is at its peak, it should be equal to that of its precursor.* This result also implies that *the radioactivity of a product peaks after that of its precursor.*

3. After [B*] begins to decrease, $d[\text{B*}]/dt < 0$, so [A*] < [B*]; that is, *after the radioactivity of a product has peaked, it should remain greater than that of its precursor.*

Such a determination of the precursor–product relationship between alkylacylglycerophospholipid and plasmalogen, using ^{14}C-labeled starting materials, indicated that the ether is the precursor and the vinyl ether is the product (Fig. 16-18, Scheme II).

C. *Isolated Organs, Cells, and Subcellular Organelles*

In addition to understanding the chemistry and catalytic events that occur at each step of a metabolic pathway, it is important to learn where a given pathway occurs within an organism. Early workers studied metabolism in whole animals. For example, the role of the pancreas in diabetes was established by Frederick Banting and Charles Best in 1921 by surgically removing that organ from dogs and observing that these animals then developed the disease.

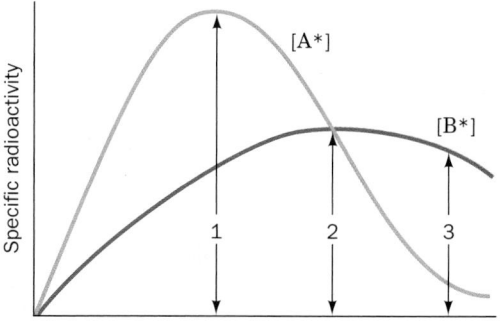

FIGURE 16-19 The flow of a pulse of radioactivity from precursor to product. At point 1, product radioactivity (B*, *purple*) is increasing and is less than that of its precursor (A*, *orange*); at point 2, product radioactivity is maximal and is equal to that of its precursor; and at point 3, product radioactivity is decreasing and is greater than that of its precursor.

The metabolic products produced by a particular organ can be studied by **organ perfusion** or in **tissue slices.** In organ perfusion, a specific organ is surgically removed from an animal and the organ's arteries and veins are connected to an artificial circulatory system. The composition of the material entering the organ can thereby be controlled and its metabolic products monitored. Metabolic processes can be similarly studied in slices of tissue thin enough to be nourished by free diffusion in an appropriate nutrient solution. Otto Warburg pioneered the tissue slice technique in the early twentieth century through his studies of respiration, in which he used a manometer to measure the changes in gas volume above tissue slices as a consequence of their O_2 consumption.

A given organ or tissue generally contains several cell types. **Cell sorters** are devices that can separate cells according to type once they have been treated with the enzymes trypsin and collagenase to destroy the intercellular matrix that binds them into a tissue. This technique allows further localization of metabolic function. A single cell type may also be grown in **tissue culture** for study. Although culturing cells often results in their loss of differentiated function, techniques have been developed for maintaining several cell types that still express their original characteristics.

As discussed in Section 16-1, metabolic pathways in eukaryotes are compartmentalized in various subcellular organelles (Table 16-1). For example, oxidative phosphorylation occurs in the mitochondrion, whereas glycolysis and fatty acid biosynthesis occur in the cytosol. Such observations are made by breaking cells open and fractionating their components by differential centrifugation (Section 6-1B), possibly followed by zonal ultracentrifugation through a sucrose density gradient or by equilibrium density gradient ultracentrifugation in a CsCl density gradient, which, respectively, separate particles according to their size and density (Section 6-5B). The cell fractions are then analyzed for biochemical function.

4 ■ THERMODYNAMICS OF PHOSPHATE COMPOUNDS

processes that require energy are driven by those that produce it.

The endergonic processes that maintain the living state are driven by the exergonic reactions of nutrient oxidation. This coupling is most often mediated through the syntheses of a few types of "high-energy" intermediates whose exergonic consumption drives endergonic processes. These intermediates therefore form a sort of universal free energy "currency" through which free energy–producing reactions "pay for" the free energy–consuming processes in biological systems.

Adenosine triphosphate (**ATP;** Fig. 16-20), which occurs in all known life-forms, is the "high-energy" intermediate that constitutes the most common cellular energy currency. Its central role in energy metabolism was first recognized in 1941 by Fritz Lipmann and Herman Kalckar. ATP consists of an **adenosine** moiety to which three **phosphoryl groups** ($-PO_3^{2-}$) are sequentially linked via a **phosphoester bond** followed by two **phosphoanhydride bonds. Adenosine diphosphate** (**ADP**) and **5′-adenosine monophosphate** (**AMP**) are similarly constituted but with only two and one phosphoryl units, respectively.

In this section we consider the nature of phosphoryl-transfer reactions, discuss why some of them are so exergonic, and outline how the cell consumes and regenerates ATP.

A. *Phosphoryl-Transfer Reactions*

Phosphoryl-transfer reactions,

$$R_1-O-PO_3^{2-} + R_2-OH \rightleftharpoons R_1-OH + R_2-O-PO_3^{2-}$$

are of enormous metabolic significance. Some of the most important reactions of this type involve the synthesis and hydrolysis of ATP:

$$ATP + H_2O \rightleftharpoons ADP + P_i$$
$$ATP + H_2O \rightleftharpoons AMP + PP_i$$

where $\mathbf{P}_i$ and $\mathbf{PP}_i$, respectively, represent **orthophosphate** (PO_4^{3-}) and **pyrophosphate** ($P_2O_7^{4-}$) in any of their ionization states. *These highly exergonic reactions are coupled to numerous endergonic biochemical processes so as to drive them to completion. Conversely, ATP is regenerated by coupling its formation to a more highly exergonic metabolic process* (the thermodynamics of coupled reactions is discussed in Section 3-4C).

To illustrate these concepts, let us consider two examples of phosphoryl-transfer reactions. The initial step in the metabolism of glucose is its conversion to glucose-6-phosphate (Section 17-2A). Yet the direct reaction of glucose and P_i is thermodynamically unfavorable (Fig. 16-21a). In biological systems, however, this reaction is coupled to the exergonic hydrolysis of ATP, so the overall reaction is thermodynamically favorable. ATP can be similarly regenerated by coupling its synthesis from ADP and P_i to the even more exergonic hydrolysis of **phosphoenolpyruvate** (Fig. 16-21b; Section 17-2J).

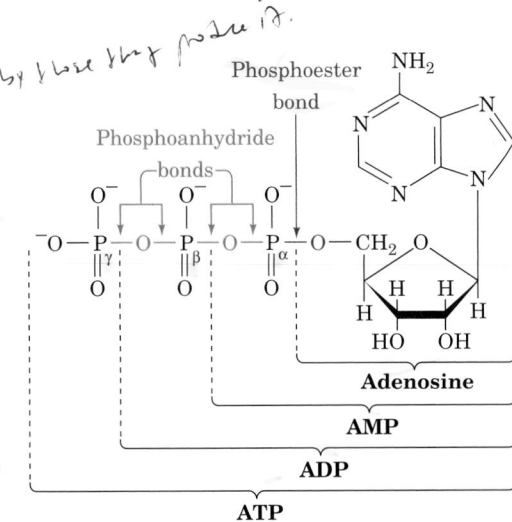

FIGURE 16-20   The structure of ATP indicating its relationship to ADP, AMP, and adenosine. The phosphoryl groups, starting with that on AMP, are referred to as the α, β, and γ phosphates. Note the difference between phosphoester and phosphoanhydride bonds.

The bioenergetic utility of phosphoryl-transfer reactions stems from their kinetic stability to hydrolysis combined with their capacity to transmit relatively large amounts of free energy. The $\Delta G^{\circ\prime}$ values of hydrolysis of several phosphorylated compounds of biochemical importance are tabulated in Table 16-3. The negatives of these values are often referred to as **phosphate group-transfer potentials;** they are a measure of the tendency of phosphorylated compounds to transfer their phosphoryl groups to water. Note that ATP has an intermediate phosphate group-transfer potential. Under standard conditions, the compounds above ATP in Table 16-3 can spontaneously transfer a phosphoryl group to ADP to form ATP, which can, in turn,

TABLE 16-3   Standard Free Energies of Phosphate Hydrolysis of Some Compounds of Biological Interest

Compound	$\Delta G^{\circ\prime}(kJ \cdot mol^{-1})$
Phosphoenolpyruvate	−61.9
1,3-Bisphosphoglycerate	−49.4
Acetyl phosphate	−43.1
Phosphocreatine	−43.1
PP_i	−33.5
ATP ($\rightarrow$ AMP + PP$_i$)	**−32.2**
ATP ($\rightarrow$ ADP + P$_i$)	**−30.5**
Glucose-1-phosphate	−20.9
Fructose-6-phosphate	−13.8
Glucose-6-phosphate	−13.8
Glycerol-3-phosphate	−9.2

Source: Jencks, W.P., *in* Fasman, G.D. (Ed.), *Handbook of Biochemistry and Molecular Biology* (3rd ed.), *Physical and Chemical Data*, Vol. I, *pp.* 296–304, CRC Press (1976).

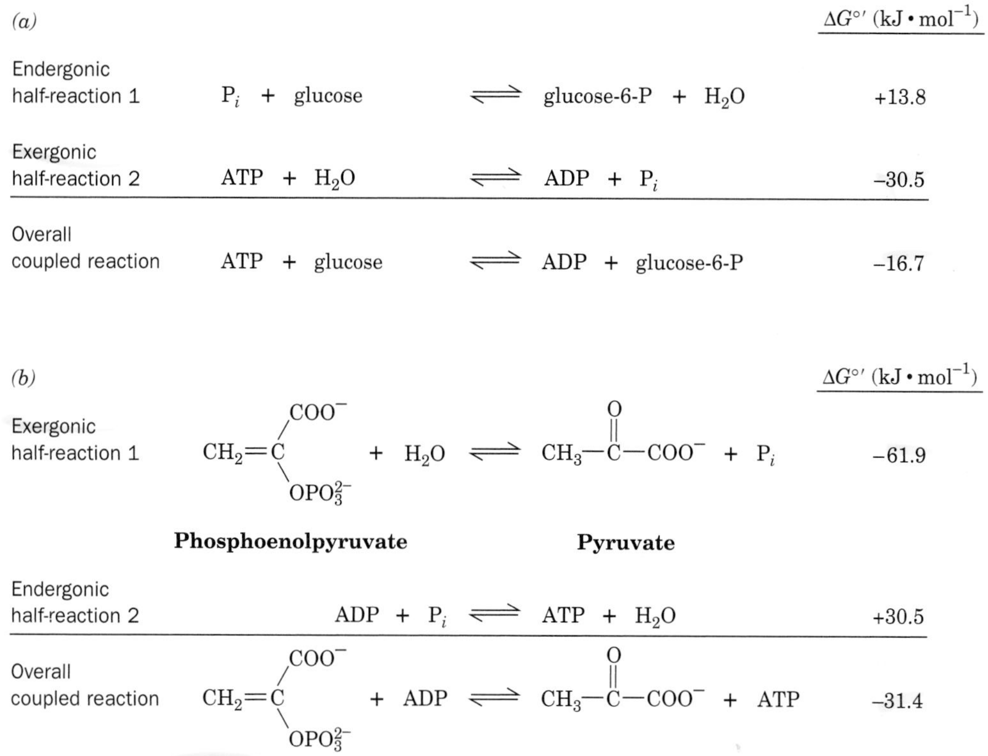

(a)
$\Delta G^{\circ\prime}$ (kJ · mol^{-1})

Endergonic half-reaction 1	P$_i$ + glucose	$\rightleftharpoons$ glucose-6-P + H$_2$O	+13.8
Exergonic half-reaction 2	ATP + H$_2$O	$\rightleftharpoons$ ADP + P$_i$	−30.5
Overall coupled reaction	ATP + glucose	$\rightleftharpoons$ ADP + glucose-6-P	−16.7

(b)
$\Delta G^{\circ\prime}$ (kJ · mol^{-1})

Exergonic half-reaction 1

$$CH_2{=}C\!\!\begin{array}{c}{}^{COO^-}\\{}\\{}_{OPO_3^{2-}}\end{array} + H_2O \rightleftharpoons CH_3{-}\overset{\displaystyle O}{\overset{\|}{C}}{-}COO^- + P_i \qquad -61.9$$

Phosphoenolpyruvate   **Pyruvate**

Endergonic half-reaction 2   ADP + P$_i$ $\rightleftharpoons$ ATP + H$_2$O   +30.5

Overall coupled reaction

$$CH_2{=}C\!\!\begin{array}{c}{}^{COO^-}\\{}\\{}_{OPO_3^{2-}}\end{array} + ADP \rightleftharpoons CH_3{-}\overset{\displaystyle O}{\overset{\|}{C}}{-}COO^- + ATP \qquad -31.4$$

FIGURE 16-21 Some overall coupled reactions involving ATP.
(*a*) The phosphorylation of glucose to form glucose-6-phosphate and ADP. (*b*) The phosphorylation of ADP by phosphoenolpyruvate to form ATP and pyruvate. Each reaction has been conceptually decomposed into a direct phosphorylation step (half-reaction 1) and a step in which ATP is hydrolyzed (half-reaction 2). Both half-reactions proceed in the direction in which the overall reaction is exergonic ($\Delta G < 0$).

spontaneously transfer a phosphoryl group to the hydrolysis products (ROH form) of the compounds below it.

a. ΔG of ATP Hydrolysis Varies with pH, Divalent Metal Ion Concentration, and Ionic Strength

The ΔG of a reaction varies with the total concentrations of its reactants and products and thus with their ionic states (Eq. [3.15]). The ΔG's of hydrolysis of phosphorylated compounds are therefore highly dependent on pH, divalent metal ion concentration (divalent metal ions such as Mg^{2+} have high phosphate-binding affinities), and ionic strength. Reasonable estimates of the intracellular values of these quantities as well as of [ATP], [ADP], and [P$_i$] (which are generally on the order of millimolar) indicate that ATP hydrolysis under physiological conditions has $\Delta G \approx -50$ kJ · mol^{-1} rather than the -30.5 kJ · mol^{-1} of its $\Delta G^{\circ\prime}$. Nevertheless, for the sake of consistency in comparing reactions, we shall usually refer to the latter value.

The above situation for ATP is not unique. It is important to keep in mind that *within a given cell, the concentrations of most substances vary both with location and time. Indeed, the concentrations of many ions, coenzymes, and metabolites commonly vary by several orders of magnitude across membranous organelle boundaries.* Unfortunately, it is usually quite difficult to obtain an accurate measurement of the concentration of any particular chemical

species in a specific cellular compartment. The ΔG's for most *in vivo* reactions are therefore little more than estimates.

B. *Rationalizing the "Energy" in "High-Energy" Compounds*

Bonds whose hydrolysis proceeds with large negative values of $\Delta G^{\circ\prime}$ (customarily more negative than -25 kJ · mol^{-1}) are often referred to as **"high-energy" bonds** or **"energy-rich" bonds** and are frequently symbolized by the squiggle ($\sim$). Thus ATP may be represented as AR—P$\sim$P$\sim$P, where A, R, and P symbolize adenyl, ribosyl, and phosphoryl groups, respectively. Yet, the phosphoester bond joining the adenosyl group of ATP to its α-phosphoryl group appears to be not greatly different in electronic character from the so-called "high-energy" bonds bridging its β and γ phosphoryl groups. In fact, none of these bonds have any unusual properties, so the term "high-energy" bond is somewhat of a misnomer. (In any case, it should not be confused with the term "bond energy," which is defined as the energy required to break, not hydrolyze, a covalent bond.) Why then, should the phosphoryl-transfer reactions of ATP be so exergonic? The answer comes from the comparison of the stabilities of the reactants and products of these reactions.

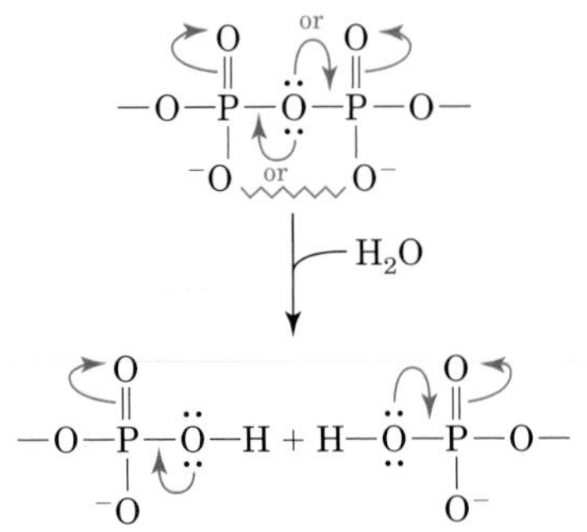

FIGURE 16-22   Resonance and electrostatic stabilization in a phosphoanhydride and its hydrolysis products. The competing resonances (*curved arrows* from central O) and charge–charge repulsions (*zigzag line*) between the phosphoryl groups of a phosphoanhydride decrease its stability relative to its hydrolysis products.

Several different factors appear to be responsible for the "high-energy" character of phosphoanhydride bonds such as those in ATP (Fig. 16-22):

1. The resonance stabilization of a phosphoanhydride bond is less than that of its hydrolysis products. This is because a phosphoanhydride's two strongly electron-withdrawing phosphoryl groups must compete for the π electrons of its bridging oxygen atom, whereas this competition is absent in the hydrolysis products. In other words, the electronic requirements of the phosphoryl groups are less satisfied in a phosphoanhydride than in its hydrolysis products.

2. Of perhaps greater importance is the destabilizing effect of the electrostatic repulsions between the charged groups of a phosphoanhydride in comparison to that of its hydrolysis products. In the physiological pH range, ATP has three to four negative charges whose mutual electrostatic repulsions are partially relieved by ATP hydrolysis.

3. Another destabilizing influence, which is difficult to assess, is the smaller solvation energy of a phosphoanhydride in comparison to that of its hydrolysis products. Some estimates suggest that this factor provides the dominant thermodynamic driving force for the hydrolysis of phosphoanhydrides.

A further property of ATP that suits it to its role as an energy intermediate stems from the relative kinetic stability of phosphoanhydride bonds to hydrolysis. Most types of anhydrides are rapidly hydrolyzed in aqueous solution. Phosphoanhydride bonds, however, have unusually large free energies of activation. Consequently, ATP is reasonably stable under physiological conditions but is readily hydrolyzed in enzymatically mediated reactions.

a.   Other "High-Energy" Compounds

The compounds in Table 16-3 with phosphate group-transfer potentials significantly greater than that of ATP have additional destabilizing influences:

1. Acyl phosphates. The hydrolysis of **acyl phosphates**

(mixed phosphoric–carboxylic anhydrides), such as **acetyl phosphate** and **1,3-bisphosphoglycerate,**

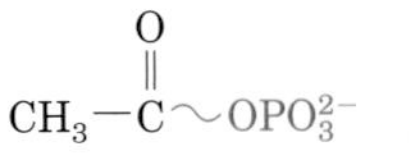

Acetyl phosphate

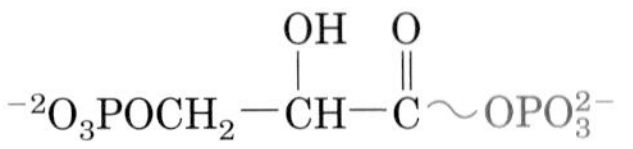

1,3-Bisphosphoglycerate

is driven by the same competing resonance and differential solvation influences that function in the hydrolysis of phosphoanhydrides. Apparently these effects are more pronounced for acyl phosphates than for phosphoanhydrides.

2. Enol phosphates. The high phosphate group-transfer potential of an **enol phosphate,** such as phosphoenolpyruvate (Fig. 16-21*b*), derives from its **enol** hydrolysis product being less stable than its **keto** tautomer. Consider the hydrolysis reaction of an enol phosphate as occurring in two steps (Fig. 16-23). The hydrolysis step is subject to the driving forces discussed above. *It is therefore the highly exergonic enol–keto conversion that provides phosphoenolpyruvate with the added thermodynamic impetus to phosphorylate ADP to form ATP.*

3. Phosphoguanidines. The high phosphate group-transfer potentials of **phosphoguanidines,** such as **phosphocreatine** and **phosphoarginine,** largely result from the competing resonances in their **guanidino** group, which are even more pronounced than they are in the phosphate group of phosphoanhydrides (Fig. 16-24). Consequently, phosphocreatine can phosphorylate ADP (see Section 16-4C).

Compounds such as **glucose-6-phosphate** or **glycerol-3-phosphate,**

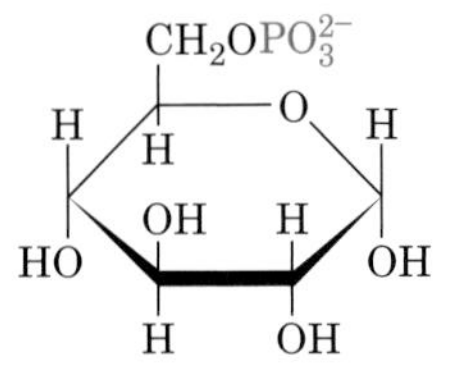

α-D-**Glucose-6-phosphate** 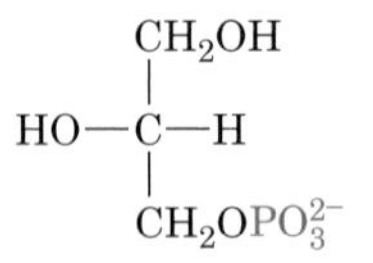   L-**Glycerol-3-phosphate**

which are below ATP in Table 16-3, have no significantly different resonance stabilization or charge separation in comparison with their hydrolysis products. Their free energies of hydrolysis are therefore much less than those of the preceding "high-energy" compounds.

C.   *The Role of ATP*

As Table 16-3 indicates, *in the thermodynamic hierarchy of phosphoryl-transfer agents, ATP occupies the middle rank.* This enables ATP to serve as an energy conduit between "high-energy" phosphate donors and "low-energy" phosphate acceptors (Fig. 16-25). Let us examine the general biochemical scheme of how this occurs.

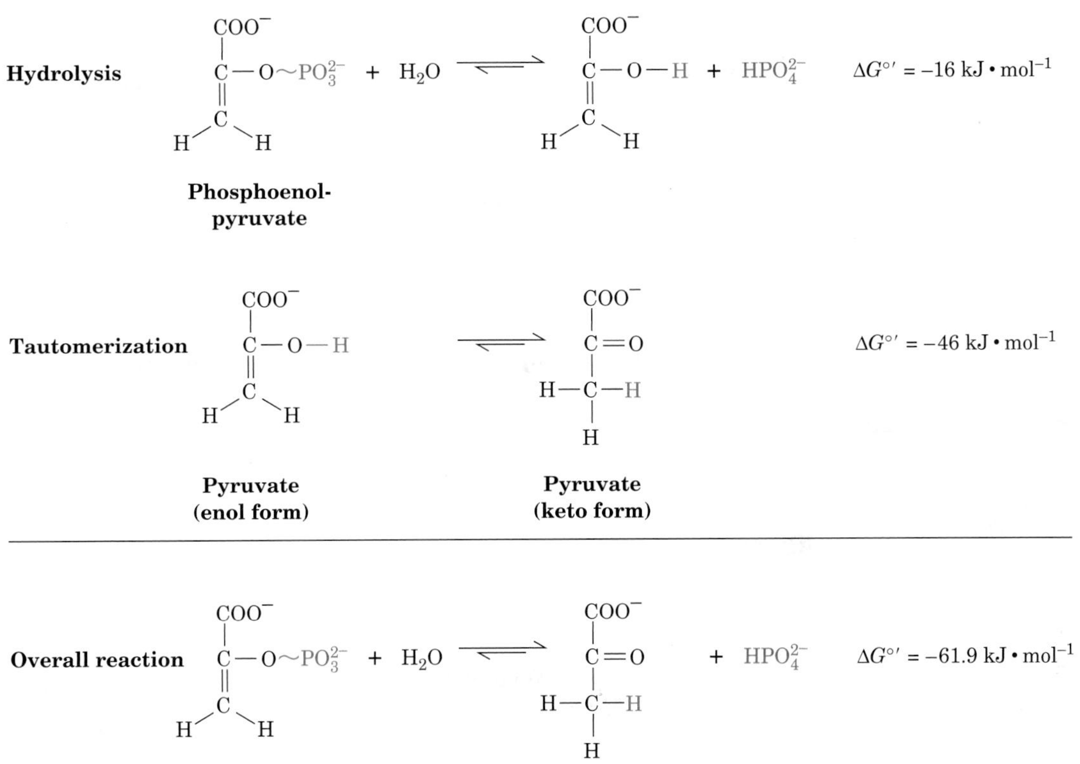

Hydrolysis

$$\begin{array}{c} COO^- \\ | \\ C-O\sim PO_3^{2-} \\ \| \\ C \\ H \quad H \end{array} + H_2O \rightleftharpoons \begin{array}{c} COO^- \\ | \\ C-O-H \\ \| \\ C \\ H \quad H \end{array} + HPO_4^{2-} \qquad \Delta G^{\circ\prime} = -16\ kJ\cdot mol^{-1}$$

**Phosphoenol-
pyruvate**

Tautomerization

$$\begin{array}{c} COO^- \\ | \\ C-O-H \\ \| \\ C \\ H \quad H \end{array} \rightleftharpoons \begin{array}{c} COO^- \\ | \\ C=O \\ | \\ H-C-H \\ | \\ H \end{array} \qquad \Delta G^{\circ\prime} = -46\ kJ\cdot mol^{-1}$$

**Pyruvate
(enol form)**   **Pyruvate
(keto form)**

Overall reaction

$$\begin{array}{c} COO^- \\ | \\ C-O\sim PO_3^{2-} \\ \| \\ C \\ H \quad H \end{array} + H_2O \rightleftharpoons \begin{array}{c} COO^- \\ | \\ C=O \\ | \\ H-C-H \\ | \\ H \end{array} + HPO_4^{2-} \qquad \Delta G^{\circ\prime} = -61.9\ kJ\cdot mol^{-1}$$

FIGURE 16-23 Hydrolysis of phosphoenolpyruvate. The reaction is broken down into two steps, hydrolysis and tautomerization.

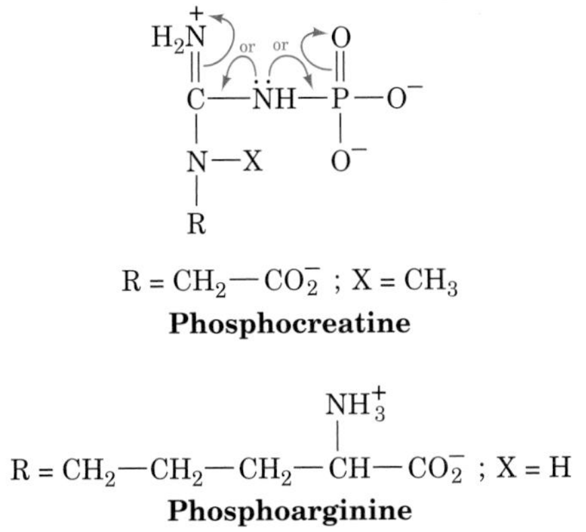

$$\begin{array}{c} H_2N^+ \quad or \quad or \quad O \\ \| \qquad \qquad \| \\ C-NH-P-O^- \\ | \qquad\quad | \\ N-X \qquad O^- \\ | \\ R \end{array}$$

$R = CH_2-CO_2^-$; $X = CH_3$
Phosphocreatine

$$R = CH_2-CH_2-CH_2-\overset{\overset{\displaystyle NH_3^+}{|}}{CH}-CO_2^-\ ;\ X = H$$
Phosphoarginine

FIGURE 16-24 Competing resonances in phosphoguanidines.

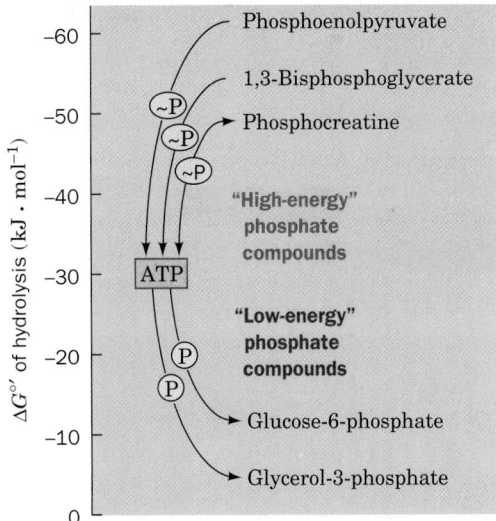

FIGURE 16-25 The flow of phosphoryl groups from "high-energy" phosphate donors, via the ATP–ADP system, to "low-energy" phosphate acceptors.

In general, the highly exergonic phosphoryl-transfer reactions of nutrient degradation are coupled to the formation of ATP from ADP and P_i through the auspices of various enzymes known as **kinases;** these enzymes catalyze the transfer of phosphoryl groups between ATP and other molecules. Consider the two reactions in Fig. 16-21*b*. If carried out independently, these reactions would not influence each other. In the cell, however, the enzyme **pyruvate kinase** couples the two reactions by catalyzing the transfer of the phosphoryl group of phosphoenolpyruvate directly to ADP to result in an overall exergonic reaction.

a. Consumption of ATP

In its role as the universal energy currency of living systems, ATP is consumed in a variety of ways:

1. Early stages of nutrient breakdown. The exergonic hydrolysis of ATP to ADP may be enzymatically coupled

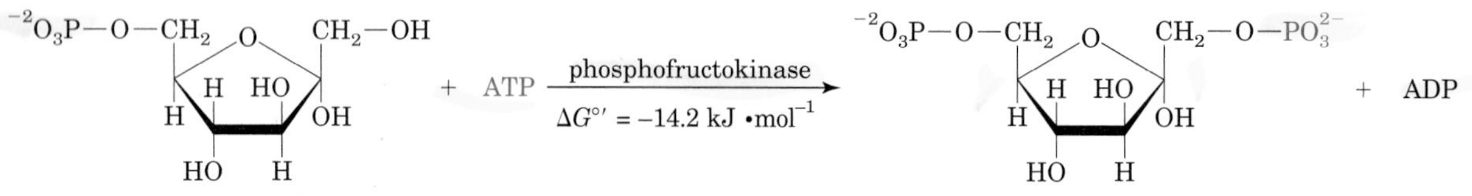

FIGURE 16-26   The phosphorylation of fructose-6-phosphate by ATP to form fructose-1,6-bisphosphate and ADP.

to an endergonic phosphorylation reaction to form "low-energy" phosphate compounds. We have seen one example of this in the hexokinase-catalyzed formation of glucose-6-phosphate (Fig. 16-21a). Another example is the **phosphofructokinase**-catalyzed phosphorylation of **fructose-6-phosphate** to form **fructose-1,6-bisphosphate** (Fig. 16-26). Both of these reactions occur in the first stage of glycolysis (Section 17-2).

2. Interconversion of nucleoside triphosphates. Many biosynthetic processes, such as the synthesis of proteins and nucleic acids, require nucleoside triphosphates other than ATP. These include the ribonucleoside triphosphates CTP, GTP, and UTP (Section 1-3C) which, together with ATP, are utilized, for example, in the biosynthesis of RNA (Section 31-2) and the deoxyribonucleoside triphosphate DNA precursors dATP, dCTP, dGTP, and dTTP (Section 5-4C). All these **nucleoside triphosphates (NTPs)** are synthesized from ATP and the corresponding **nucleoside diphosphate (NDP)** in reactions catalyzed by the nonspecific enzyme **nucleoside diphosphate kinase:**

$$ATP + NDP \rightleftharpoons ADP + NTP$$

The $\Delta G^{\circ\prime}$ values for these reactions are nearly zero, as might be expected from the structural similarities among the NTPs. These reactions are driven by the depletion of the NTPs through their exergonic hydrolysis in the biosynthetic reactions in which they participate (Section 3-4C).

3. Physiological processes. The hydrolysis of ATP to ADP and P_i energizes many essential endergonic physiological processes such as chaperone-assisted protein folding (Section 9-2C), muscle contraction, and the transport of molecules and ions against concentration gradients. In general, these processes result from conformational changes in proteins (enzymes) that occur in response to their binding of ATP. This is followed by the exergonic hydrolysis of ATP and release of ADP and P_i, thereby causing these processes to be unidirectional (irreversible).

4. Additional phosphoanhydride cleavage in highly endergonic reactions. Although many reactions involving ATP yield ADP and P_i (**orthophosphate cleavage**), others yield AMP and PP_i (**pyrophosphate cleavage**). In these latter cases, the PP_i is rapidly hydrolyzed to $2P_i$ by **inorganic pyrophosphatase** ($\Delta G^{\circ\prime} = -33.5$ kJ · mol^{-1}) so that *the pyrophosphate cleavage of ATP ultimately results in the hydrolysis of two "high-energy" phosphoanhydride bonds.* The attachment of amino acids to tRNA molecules for protein synthesis is an example of this phenomenon (Fig. 16-27 and Sections 5-4B and 32-2C). The two steps of the reaction involving the amino acid are readily reversible because the free energies of hydrolysis of the bonds formed are comparable to that of ATP hydrolysis. The overall reaction is driven to completion by the hydrolysis of PP_i, which is essentially irreversible. Nucleic acid biosynthesis from the appropriate NTPs also releases PP_i (Sections 30-2A and 31-2). The free energy changes of

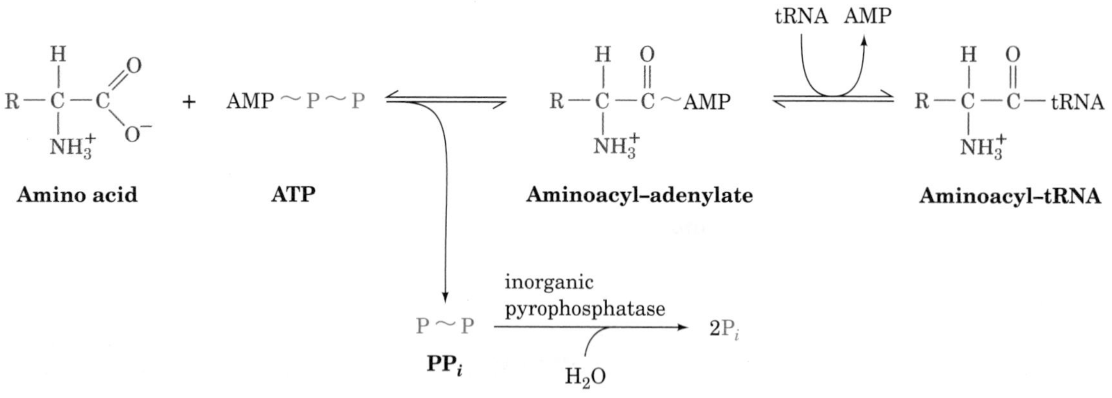

FIGURE 16-27   Pyrophosphate cleavage in the synthesis of an aminoacyl–tRNA. Here the squiggle (~) represents a "high-energy" bond. In the first reaction step, the amino acid is **adenylylated** by ATP. In the second step, a tRNA molecule displaces the AMP moiety to form an aminoacyl–tRNA. The highly exergonic hydrolysis of pyrophosphate ($\Delta G^{\circ\prime} = -33.5$ kJ · mol^{-1}) drives the reaction forward.

these vital reactions are around zero, so the subsequent hydrolysis of PP_i is essential to drive the synthesis of nucleic acids.

b. Formation of ATP

To complete its intermediary metabolic function, ATP must be replenished. This is accomplished through three types of processes:

1. Substrate-level phosphorylation. ATP may be formed, as is indicated in Fig. 16-21b, from phosphoenolpyruvate by direct transfer of a phosphoryl group from a "high-energy" compound to ADP. Such reactions, which are referred to as **substrate-level phosphorylations,** most commonly occur in the early stages of carbohydrate metabolism (Section 17-2).

2. Oxidative phosphorylation and photophosphorylation. Both oxidative metabolism and photosynthesis act to generate a proton (H^+) concentration gradient across a membrane (Sections 22-3 and 24-2D). Discharge of this gradient is enzymatically coupled to the formation of ATP from ADP and P_i (the reverse of ATP hydrolysis). In oxidative metabolism, this process is called **oxidative phosphorylation,** whereas in photosynthesis it is termed **photophosphorylation.** Most of the ATP produced by respiring and photosynthesizing organisms is generated in this manner.

3. Adenylate kinase reaction. The AMP resulting from pyrophosphate cleavage reactions of ATP is converted to ADP in a reaction catalyzed by the enzyme **adenylate kinase** (Section 17-4F):

$$AMP + ATP \rightleftharpoons 2ADP$$

The ADP is subsequently converted to ATP through substrate-level phosphorylation, oxidative phosphorylation, or photophosphorylation.

c. Rate of ATP Turnover

The cellular role of ATP is that of a free energy transmitter rather than a free energy reservoir. The amount of ATP in a cell is typically only enough to supply its free energy needs for a minute or two. Hence, ATP is continually being hydrolyzed and regenerated. Indeed, ^{32}P-labeling experiments indicate that the metabolic half-life of an ATP molecule varies from seconds to minutes depending on the cell type and its metabolic activity. For instance, brain cells have only a few seconds supply of ATP (which, in part, accounts for the rapid deterioration of brain tissue by oxygen deprivation). *An average person at rest consumes and regenerates ATP at a rate of ~3 mol (1.5 kg) · h^{-1} and as much as an order of magnitude faster during strenuous activity.*

d. Phosphocreatine Provides a "High-Energy" Reservoir for ATP Formation

Muscle and nerve cells, which have a high ATP turnover (a maximally exerting muscle has only a fraction of a second's ATP supply), have a free energy reservoir that functions to regenerate ATP rapidly. In vertebrates, phosphocreatine (Fig. 16-24) functions in this capacity. It is synthesized by the reversible phosphorylation of creatine by ATP as catalyzed by **creatine kinase:**

$$ATP + creatine \rightleftharpoons phosphocreatine + ADP$$
$$\Delta G^{\circ\prime} = +12.6 \, kJ \cdot mol^{-1}$$

Note that this reaction is endergonic under standard conditions; however, the intracellular concentrations of its reactants and products (typically 4 mM ATP and 0.013 mM ADP) are such that it operates close to equilibrium ($\Delta G \approx 0$). Accordingly, when the cell is in a resting state, so that [ATP] is relatively high, the reaction proceeds with net synthesis of phosphocreatine, whereas at times of high metabolic activity, when [ATP] is low, the equilibrium shifts so as to yield net synthesis of ATP. *Phosphocreatine thereby acts as an ATP "buffer" in cells that contain creatine kinase.* A resting vertebrate skeletal muscle normally has sufficient phosphocreatine to supply its free energy needs for several minutes (but for only a few seconds at maximum exertion). In the muscles of some invertebrates, such as lobsters, phosphoarginine performs the same function. These phosphoguanidines are collectively named **phosphagens.**

5 ■ OXIDATION–REDUCTION REACTIONS

Oxidation–reduction reactions, processes involving the transfer of electrons, are of immense biochemical significance; living things derive most of their free energy from them. In photosynthesis (Chapter 24), CO_2 is **reduced** (gains electrons) and H_2O is **oxidized** (loses electrons) to yield carbohydrates and O_2 in an otherwise endergonic process that is powered by light energy. In aerobic metabolism, which is carried out by all eukaryotes and many prokaryotes, the overall photosynthetic reaction is essentially reversed so as to harvest the free energy of oxidation of carbohydrates and other organic compounds in the form of ATP (Chapter 22). Anaerobic metabolism generates ATP, although in lower yields, through intramolecular oxidation–reductions of various organic molecules, for example, glycolysis (Chapter 17); in certain anaerobic bacteria, ATP is generated through the use of non-O_2 oxidizing agents such as sulfate or nitrate. In this section we outline the thermodynamics of oxidation–reduction reactions in order to understand the quantitative aspects of these crucial biological processes.

A. The Nernst Equation

Oxidation–reduction reactions (also known as **redox** or **oxidoreduction reactions**) resemble other types of chemical reactions in that they involve group transfer. For instance, hydrolysis transfers a functional group to water. In oxidation–reduction reactions, the "groups" transferred are electrons, which are passed from an **electron donor** (**reductant** or **reducing agent**) to an **electron acceptor** (**oxidant** or **oxidizing agent**). For example, in the reaction

$$Fe^{3+} + Cu^+ \rightleftharpoons Fe^{2+} + Cu^{2+}$$

Cu^+, the reductant, is oxidized to Cu^{2+} while Fe^{3+}, the oxidant, is reduced to Fe^{2+}.

Redox reactions may be divided into two **half-reactions** or **redox couples**, such as

$$Fe^{3+} + e^- \rightleftharpoons Fe^{2+} \text{ (reduction)}$$
$$Cu^+ \rightleftharpoons Cu^{2+} + e^- \text{ (oxidation)}$$

whose sum is the above whole reaction. These half-reactions occur during oxidative metabolism in the vital mitochondrial electron transfer mediated by **cytochrome c oxidase** (Section 22-2C). Note that for electrons to be transferred, both half-reactions must occur simultaneously. In fact, the electrons are the two half-reactions' common intermediate.

a. Electrochemical Cells

A half-reaction consists of an electron donor and its conjugate electron acceptor; in the oxidation half-reaction shown above, Cu^+ is the electron donor and Cu^{2+} is its conjugate electron acceptor. Together these constitute a **conjugate redox pair** analogous to the conjugate acid–base pair (HA and A^-) of a Brønsted acid (Section 2-2A). An important difference between redox pairs and acid–base pairs, however, is that *the two half-reactions of a redox reaction, each consisting of a conjugate redox pair, may be physically separated so as to form an* **electrochemical cell** (Fig. 16-28). In such a device, each half-reaction takes place in its separate **half-cell,** and electrons are passed between half-cells as an electric current in the wire connecting their two electrodes. A salt bridge is necessary to complete the electrical circuit by providing a conduit for ions to migrate in the maintenance of electrical neutrality.

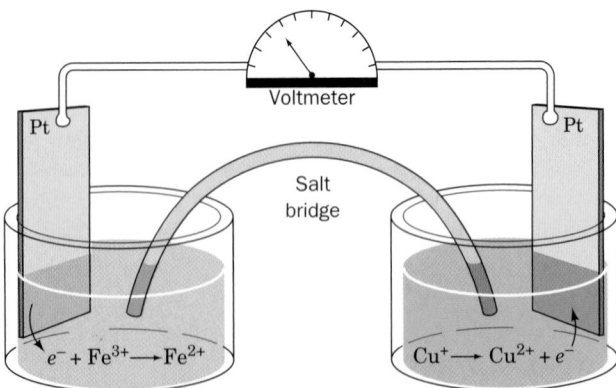

FIGURE 16-28 Example of an electrochemical cell. The half-cell undergoing oxidation (here $Cu^+ \rightarrow Cu^{2+} + e^-$) passes the liberated electrons through the wire to the half-cell undergoing reduction (here $e^- + Fe^{3+} \rightarrow Fe^{2+}$). Electroneutrality in the two half-cells is maintained by the transfer of ions through the electrolyte-containing salt bridge.

The free energy of an oxidation–reduction reaction is particularly easy to determine through a simple measurement of the voltage difference between its two half-cells. Consider the general redox reaction:

$$A_{ox}^{n+} + B_{red} \rightleftharpoons A_{red} + B_{ox}^{n+}$$

in which n electrons per mole of reactants are transferred from reductant (B_{red}) to oxidant (A_{ox}^{n+}). The free energy of this reaction is expressed, according to Eq. [3.15], as

$$\Delta G = \Delta G^\circ + RT \ln\left(\frac{[A_{red}][B_{ox}^{n+}]}{[A_{ox}^{n+}][B_{red}]}\right) \qquad [16.2]$$

Equation [3.12] indicates that, under reversible conditions,

$$\Delta G = -w' = -w_{el} \qquad [16.3]$$

where w', the non-pressure–volume work, is, in this case, w_{el}, the electrical work required to transfer the n moles of electrons through the electric potential difference $\Delta\mathscr{E}$. This, according to the laws of electrostatics, is

$$w_{el} = n\mathscr{F}\Delta\mathscr{E} \qquad [16.4]$$

where $\mathscr{F}$, the **faraday,** is the electrical charge of 1 mol of electrons ($1\mathscr{F} = 96,494$ C · mol^{-1} = 96,494 J · V^{-1} · mol^{-1}). Thus, substituting Eq. [16.4] into Eq. [16.3],

$$\Delta G = -n\mathscr{F}\Delta\mathscr{E} \qquad [16.5]$$

Combining Eqs. [16.2] and [16.5], and making the analogous substitution for ΔG°, yields the **Nernst equation:**

$$\Delta\mathscr{E} = \Delta\mathscr{E}^\circ - \frac{RT}{n\mathscr{F}} \ln\left(\frac{[A_{red}][B_{ox}^{n+}]}{[A_{ox}^{n+}][B_{red}]}\right) \qquad [16.6]$$

which was originally formulated in 1881 by Walther Nernst. Here $\Delta\mathscr{E}$, the **electromotive force (emf)** or **redox potential,** may be described as the "electron pressure" that the electrochemical cell exerts. The quantity $\Delta\mathscr{E}^\circ$, the redox potential when all components are in their standard states, is called the **standard redox potential.** If these standard states refer to biochemical standard states (Section 3-4B), then $\Delta\mathscr{E}^\circ$ is replaced by $\Delta\mathscr{E}^{\circ'}$. Note that a positive $\Delta\mathscr{E}$ in Eq. [16.5] results in a negative ΔG; in other words, *a positive $\Delta\mathscr{E}$ is indicative of a spontaneous reaction, one that can do work.*

B. Measurements of Redox Potentials

The free energy change of a redox reaction may be determined, as Eq. [16.5] indicates, by simply measuring its redox potential with a voltmeter (Fig. 16-28). Consequently, voltage measurements are commonly employed to characterize the sequence of reactions comprising a metabolic electron-transport pathway (such as mediates, e.g., oxidative metabolism; Chapter 22).

Any redox reaction can be divided into its component half-reactions:

$$A_{ox}^{n+} + ne^- \rightleftharpoons A_{red}$$
$$B_{ox}^{n+} + ne^- \rightleftharpoons B_{red}$$

TABLE 16-4 **Standard Reduction Potentials of Some Biochemically Important Half-reactions**

Half-Reaction	$\mathscr{E}°'$ (V)
$\frac{1}{2}O_2 + 2H^+ + 2e^- \rightleftharpoons H_2O$	0.815
$SO_4^{2-} + 2H^+ + 2e^- \rightleftharpoons SO_3^{2-} + H_2O$	0.48
$NO_3^- + 2H^+ + 2e^- \rightleftharpoons NO_2^- + H_2O$	0.42
Cytochrome a_3 $(Fe^{3+}) + e^- \rightleftharpoons$ cytochrome a_3 (Fe^{2+})	0.385
$O_2(g) + 2H^+ + 2e^- \rightleftharpoons H_2O_2$	0.295
Cytochrome a $(Fe^{3+}) + e^- \rightleftharpoons$ cytochrome a (Fe^{2+})	0.29
Cytochrome c $(Fe^{3+}) + e^- \rightleftharpoons$ cytochrome c (Fe^{2+})	0.235
Cytochrome c_1 $(Fe^{3+}) + e^- \rightleftharpoons$ cytochrome c_1 (Fe^{2+})	0.22
Cytochrome b $(Fe^{3+}) + e^- \rightleftharpoons$ cytochrome b (Fe^{2+}) (*mitochondrial*)	0.077
Ubiquinone $+ 2H^+ + 2e^- \rightleftharpoons$ ubiquinol	0.045
Fumarate$^- + 2H^+ + 2e^- \rightleftharpoons$ succinate$^-$	0.031
$FAD + 2H^+ + 2e^- \rightleftharpoons FADH_2$ (*in flavoproteins*)	−0.040
Oxaloacetate$^- + 2H^+ + 2e^- \rightleftharpoons$ malate$^-$	−0.166
Pyruvate$^- + 2H^+ + 2e^- \rightleftharpoons$ lactate$^-$	−0.185
Acetaldehyde $+ 2H^+ + 2e^- \rightleftharpoons$ ethanol	−0.197
$FAD + 2H^+ + 2e^- \rightleftharpoons FADH_2$ (*free coenzyme*)	−0.219
$S + 2H^+ + 2e^- \rightleftharpoons H_2S$	−0.23
Lipoic acid $+ 2H^+ + 2e^- \rightleftharpoons$ dihydrolipoic acid	−0.29
$NAD^+ + H^+ + 2e^- \rightleftharpoons NADH$	−0.315
$NADP^+ + H^+ + 2e^- \rightleftharpoons NADPH$	−0.320
Cystine $+ 2H^+ + 2e^- \rightleftharpoons$ 2 cysteine	−0.340
Acetoacetate$^- + 2H^+ + 2e^- \rightleftharpoons$ β-hydroxybutyrate$^-$	−0.346
$H^+ + e^- \rightleftharpoons \frac{1}{2}H_2$	−0.421
Acetate$^- + 3H^+ + 2e^- \rightleftharpoons$ acetaldehyde $+ H_2O$	−0.581

Source: Mostly from Loach, P.A., *in* Fasman, G.D. (Ed.), *Handbook of Biochemistry and Molecular Biology* (3rd ed.), *Physical and Chemical Data*, Vol. I, *pp.* 123–130, CRC Press (1976).

where, by convention, both half-reactions are written as reductions. These half-reactions can be assigned **reduction potentials,** $\mathscr{E}_A$ and $\mathscr{E}_B$, in accordance with the Nernst equation:

$$\mathscr{E}_A = \mathscr{E}_A° - \frac{RT}{n\mathscr{F}} \ln\left(\frac{[A_{red}]}{[A_{ox}^{n+}]}\right) \quad [16.7a]$$

$$\mathscr{E}_B = \mathscr{E}_B° - \frac{RT}{n\mathscr{F}} \ln\left(\frac{[B_{red}]}{[B_{ox}^{n+}]}\right) \quad [16.7b]$$

For the redox reaction of any two half-reactions:

$$\Delta\mathscr{E}° = \mathscr{E}°_{(e^-\ acceptor)} - \mathscr{E}°_{(e^-\ donor)} \quad [16.8]$$

Thus, when the reaction proceeds with A as the electron acceptor and B as the electron donor, $\Delta\mathscr{E}° = \mathscr{E}_A° - \mathscr{E}_B°$ and similarly for $\Delta\mathscr{E}$.

Reduction potentials, like free energies, must be defined with respect to some arbitrary standard. By convention, standard reduction potentials are defined with respect to the standard hydrogen half-reaction

$$2H^+ + 2e^- \rightleftharpoons H_2(g)$$

in which H^+ at pH 0, 25°C, and 1 atm is in equilibrium with $H_2(g)$ that is in contact with a Pt electrode. This half-cell is arbitrarily assigned a standard reduction potential of $\mathscr{E}° = 0$ V(1 V $= 1$ J $\cdot$ C^{-1}). For the biochemical convention, we likewise define the standard (pH = 0) hydrogen half-reaction as having $\mathscr{E}' = 0$ so that the hydrogen half-cell at the biochemical standard state (pH = 7) has $\mathscr{E}°' = -0.421$ V (Table 16-4). When $\Delta\mathscr{E}$ is positive, ΔG is negative (Eq. [16.5]), indicating a spontaneous process. In combining two half-reactions under standard conditions, the direction of spontaneity therefore involves the reduction of the redox couple with the more positive standard reduction potential. In other words, *the more positive the standard reduction potential, the greater the tendency for the redox couple's oxidized form to accept electrons and thus become reduced.*

a. Biochemical Half-Reactions Are Physiologically Significant

The biochemical standard reduction potentials ($\mathscr{E}°'$) of some biochemically important half-reactions are listed in Table 16-4. The oxidized form of a redox couple with a large positive standard reduction potential has a high affinity for electrons and is a strong electron acceptor (oxidizing agent), whereas its conjugate reductant is a weak electron donor (reducing agent). For example, O_2 is the strongest oxidizing agent in Table 16-4, whereas H_2O, which tightly holds its electrons, is the table's weakest reducing agent. The converse is true of half-reactions with large negative standard reduction potentials. Since electrons spontaneously flow from low to high reduction potentials, they are transferred, under standard conditions, from the reduced products in any half-reaction in Table 16-4 to the oxidized reactants of any half-reaction above it (although this may not occur at a measurable rate in the absence of a suitable enzyme). Thus, in biological systems, the approximate lower limit for a standard reduction potential is -0.421 V because reductants with a lesser value of $\mathscr{E}^{\circ\prime}$ would reduce protons to H_2. However, reducing centers in proteins that are protected from water may have lower potentials. Note that the Fe^{3+} ions of the various cytochromes tabulated in Table 16-4 have significantly different redox potentials. This indicates that *the protein components of redox enzymes play active roles in electron-transfer reactions by modulating the redox potentials of their bound redox-active centers.*

Electron-transfer reactions are of great biological importance. For example, in the mitochondrial electron-transport chain (Section 22-2), the primary source of ATP in eukaryotes, electrons are passed from NADH (Fig. 13-2) along a series of electron acceptors of increasing reduction potential (many of which are listed in Table 16-4) to O_2. ATP is generated from ADP and P_i by coupling its synthesis to this free energy cascade. *NADH thereby functions as an energy-rich electron-transfer coenzyme.* In fact, the oxidation of one NADH to NAD^+ supplies sufficient free energy to generate three ATPs. The NAD^+/NADH redox couple functions as the electron acceptor in many exergonic metabolite oxidations. In serving as the electron donor in ATP synthesis, it fulfills its cyclic role as a free energy conduit in a manner analogous to ATP. The metabolic roles of redox coenzymes are further discussed in succeeding chapters.

C. *Concentration Cells*

A concentration gradient has a lower entropy (greater order) than the corresponding uniformly mixed solution and therefore requires the input of free energy for its formation. Consequently, discharge of a concentration gradient is an exergonic process that may be harnessed to drive an endergonic reaction. For example, discharge of a proton concentration gradient (generated by the reactions of the electron-transport chain) across the inner mitochondrial membrane drives the enzymatic synthesis of ATP from ADP and P_i (Section 22-3). Likewise, nerve impulses, which require electrical energy, are transmitted through the discharge of $[Na^+]$ and $[K^+]$ gradients that nerve cells generate across their cell membranes (Section 20-5B). Quantitation of the free energy contained in a concentration gradient is accomplished by use of the concepts of electrochemical cells.

The reduction potential and free energy of a half-cell vary with the concentrations of its reactants. An electrochemical cell may therefore be constructed from two half-cells that contain the same chemical species but at different concentrations. The overall reaction for such an electrochemical cell may be represented

$$A_{ox}^{n+}(\text{half-cell 1}) + A_{red}(\text{half-cell 2}) \rightleftharpoons$$
$$A_{ox}^{n+}(\text{half-cell 2}) + A_{red}(\text{half-cell 1}) \qquad [16.9]$$

and, according to the Nernst equation, since $\Delta\mathscr{E}^{\circ} = 0$ when the same reaction occurs in both cells,

$$\Delta\mathscr{E} = \frac{RT}{n\mathscr{F}} \ln\left(\frac{[A_{ox}^{n+}(\text{half-cell 2})][A_{red}(\text{half-cell 1})]}{[A_{ox}^{n+}(\text{half-cell 1})][A_{red}(\text{half-cell 2})]}\right)$$

Such **concentration cells** are capable of generating electrical work until they reach equilibrium. This occurs when the concentration ratios in the half-cells become equal ($K_{eq} = 1$). The reaction constitutes a sort of mixing of the two half-cells; the free energy generated is a reflection of the entropy of this mixing. The thermodynamics of concentration gradients as they apply to membrane transport is discussed in Section 20-1.

6 ■ THERMODYNAMICS OF LIFE

One of the last refuges of **vitalism,** the doctrine that biological processes are not bound by the physical laws that govern inanimate objects, was the belief that living things can somehow evade the laws of thermodynamics. This view was partially refuted by elaborate calorimetric measurements on living animals that are entirely consistent with the energy conservation predictions of the first law of thermodynamics. However, the experimental verification of the second law of thermodynamics in living systems is more difficult. It has not been possible to measure the entropy of living matter because the heat, q_p, of a reaction at a constant T and P is only equal to $T\Delta S$ if the reaction is carried out reversibly (Eq. [3.8]). Obviously, the dismantling of a living organism to its component molecules for such a measurement would invariably result in its irreversible death. Consequently, the present experimentally verified state of knowledge is that the entropy of living matter is less than that of the products to which it decays.

In this section we consider the special aspects of the thermodynamics of living systems. Knowledge of these matters, which is by no means complete, has enhanced our understanding of how metabolic pathways are regulated, how cells respond to stimuli, and how organisms grow and change with time.

A. *Living Systems Cannot Be at Equilibrium*

Classical or **equilibrium thermodynamics** (Chapter 3) applies largely to reversible processes in closed systems. The fate of any isolated system, as we discussed in Section 3-4A, is that it must inevitably reach equilibrium. For example, if its reactants are in excess, the forward reaction will proceed faster than the reverse reaction until equilibrium is attained ($\Delta G = 0$). In contrast, open systems may remain in a nonequilibrium state as long as they are able to acquire free energy from their surroundings in the form of reactants, heat, or work. While classical thermodynamics provides invaluable information concerning open systems by indicating whether a given process can occur spontaneously, further thermodynamic analysis of open systems requires the application of the more recently elucidated principles of **nonequilibrium** or **irreversible thermodynamics.** In contrast to classical thermodynamics, this theory explicitly takes time into account.

Living organisms are open systems and therefore can never be at equilibrium. As indicated above, they continuously ingest high-enthalpy, low-entropy nutrients, which they convert to low-enthalpy, high-entropy waste products. The free energy resulting from this process is used to do work and to produce the high degree of organization characteristic of life. If this process is interrupted, the organism ultimately reaches equilibrium, which for living things is synonymous with death. For example, one theory of aging holds that senescence results from the random but inevitable accumulation in cells of genetic defects that interfere with and ultimately disrupt the proper functioning of living processes. [The theory does not, however, explain how single-celled organisms or the germ cells of multicellular organisms (sperm and ova), which are in effect immortal, are able to escape this so-called **error catastrophe.**]

Living systems must maintain a nonequilibrium state for several reasons:

1. Only a nonequilibrium process can perform useful work.

2. The intricate regulatory functions characteristic of life require a nonequilibrium state because a process at equilibrium cannot be directed (similarly, a ship that is dead in the water will not respond to its rudder).

3. The complex cellular and molecular systems that conduct biological processes can be maintained only in the nonequilibrium state. Living systems are inherently unstable because they are degraded by the very biochemical reactions to which they give rise. Their regeneration, which must occur almost simultaneously with their degradation, requires the continuous influx of free energy. For example, the ATP-generating consumption of glucose (Section 17-2), as has been previously mentioned, occurs with the initial consumption of ATP through its reactions with glucose to form glucose-6-phosphate and with fructose-6-phosphate to form fructose-1,6-bisphosphate. Consequently, if metabolism is suspended long enough to exhaust the available ATP supply, glucose metabolism cannot be resumed. Life therefore differs in a fundamental way from a complex machine such as a computer. Both require a throughput of free energy to be active. However, the function of the machine is based on a static structure, so that the machine can be repeatedly switched on and off. Life, in contrast, is based on a self-destructing but self-renewing process which, once interrupted, cannot be reinitiated.

B. *Nonequilibrium Thermodynamics and the Steady State*

In a nonequilibrium process, something (such as matter, electrical charge, or heat) must flow, that is, change its spatial distribution. In classical mechanics, the acceleration of mass occurs in response to force. *Similarly, flow in a thermodynamic system occurs in response to a thermodynamic force* **(driving force)**, *which results from the system's nonequilibrium state.* For example, the flow of matter in diffusion is motivated by the thermodynamic force of a concentration gradient; the migration of electrical charge (electric current) occurs in response to a gradient in an electric field (a voltage difference); the transport of heat results from a temperature gradient; and a chemical reaction results from a difference in chemical potential. Such flows are said to be **conjugate** to their thermodynamic force.

A thermodynamic force may also promote a **nonconjugate flow** under the proper conditions. For example, a gradient in the concentration of matter can give rise to an electric current (a concentration cell), heat (such as occurs on mixing H_2O and HCl), or a chemical reaction (the mitochondrial production of ATP through the dissipation of a proton gradient). Similarly, a gradient in electrical potential can motivate a flow of matter (electrophoresis), heat (resistive heating), or a chemical reaction (the charging of a battery). When a thermodynamic force stimulates a nonconjugate flow, the process is called **energy transduction.**

a. Living Things Maintain the Steady State

*Living systems are, for the most part, characterized by being in a **steady state.*** By this it is meant that all flows in the system are constant, so that the system does not change with time. Some environmental steady-state processes are schematically illustrated in Fig. 16-29. Ilya Prigogine, a pioneer in the development of irreversible thermodynamics, has shown that a steady-state system produces the maximum amount of useful work for a given energy expenditure under the prevailing conditions. *The steady state of an open system is therefore its state of maximum thermodynamic efficiency.* Furthermore, in analogy with Le Châtelier's principle, slight perturbations from the steady state give rise to changes in flows that counteract these perturbations so as to return the system to the steady state. *The steady state of an open system is therefore analogous to the equilibrium state of an isolated system; both are stable states.*

In the following chapters we shall see that many biological regulatory mechanisms function to maintain a steady state. For example, the flow of reaction intermediates through a metabolic pathway is often inhibited by an excess of final product and stimulated by an excess of starting material through the allosteric regulation of its key enzymes (Section 13-4). Living things have apparently evolved so as to take maximum thermodynamic advantage of their environments.

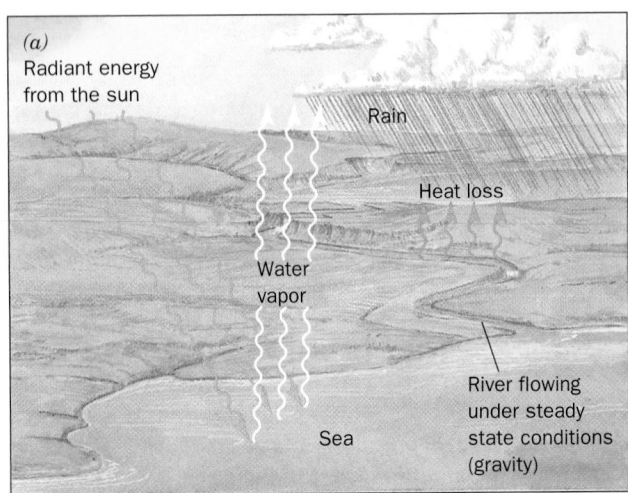

(a)
Radiant energy from the sun

Rain

Heat loss

Water vapor

River flowing under steady state conditions (gravity)

Sea

(b)
Radiant energy from the sun

Heat loss

Photosynthesis

CO_2 + H_2O

Breakdown of carbohydrates

FIGURE 16-29   Two examples of open systems in a steady state. (*a*) A constant flow of water in the river occurs under the influence of the force of gravity. The water level in the reservoir is maintained by rain, the major source of which is the evaporation of seawater. Hence the entire cycle is ultimately powered by the sun. (*b*) The steady state of the biosphere is similarly maintained by the sun. Plants harness the sun's radiant energy to synthesize carbohydrates from CO_2 and H_2O. The eventual metabolism of the carbohydrates by the plants or by the animals that eat them results in the release of their stored free energy and the return of the CO_2 and H_2O to the environment to complete the cycle.

C. *Thermodynamics of Metabolic Control*

a. Enzymes Selectively Catalyze Required Reactions

Biological reactions are highly specific; only reactions that lie on metabolic pathways take place at significant rates despite the many other thermodynamically favorable reactions that are also possible. As an example, let us consider the reactions of ATP, glucose, and water. Two thermodynamically favorable reactions that ATP can undergo are phosphoryl transfer to form ADP and glucose-6-phosphate, and hydrolysis to form ADP and P_i (Fig. 16-21*a*). The free energy profiles of these reactions are diagrammed in Fig. 16-30. ATP hydrolysis is thermodynamically favored over the phosphoryl transfer to glucose. However, their relative rates are determined by their free energies of activation to their transition states ($\Delta G^\ddagger$ values; Section 14-

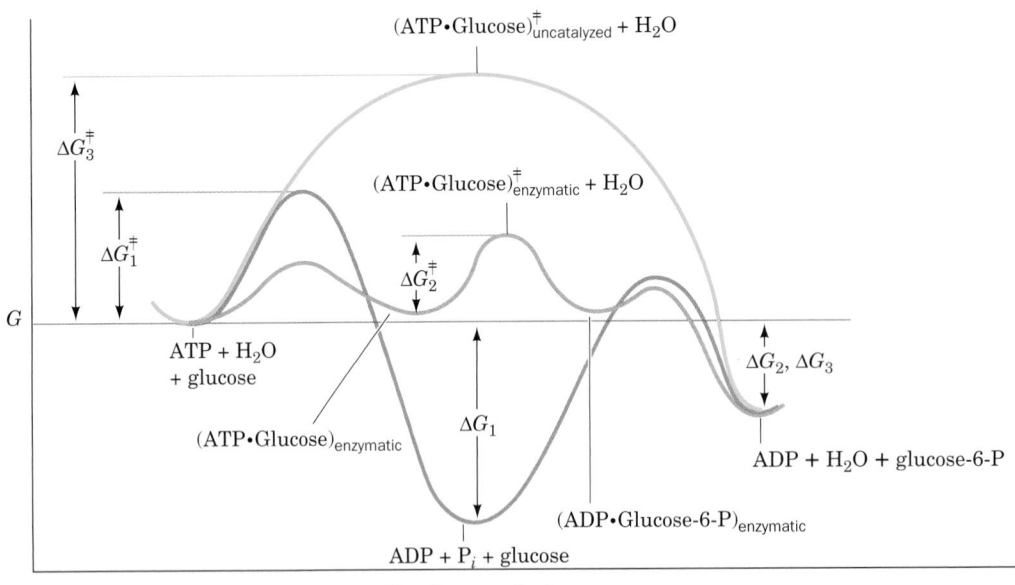

$(ATP \cdot Glucose)^{\ddagger}_{uncatalyzed} + H_2O$

$(ATP \cdot Glucose)^{\ddagger}_{enzymatic} + H_2O$

$\Delta G^{\ddagger}_3$

$\Delta G^{\ddagger}_1$

$\Delta G^{\ddagger}_2$

G

ATP + H_2O + glucose

$(ATP \cdot Glucose)_{enzymatic}$

ΔG_1

$\Delta G_2, \Delta G_3$

ADP + H_2O + glucose-6-P

$(ADP \cdot Glucose-6-P)_{enzymatic}$

ADP + P_i + glucose

Reaction coordinate

FIGURE 16-30 Reaction coordinate diagrams for (1) the reaction of ATP and water (*purple curve*), and the reaction of ATP and glucose (2) in the presence (*orange curve*) and (3) in the absence (*yellow curve*) of an appropriate enzyme. Although the hydrolysis of ATP is a more exergonic reaction than the phosphorylation of glucose (ΔG_1 is more negative than ΔG_2), the latter reaction is predominant in the presence of a suitable enzyme because it is kinetically favored ($\Delta G^{\ddagger}_2 < \Delta G^{\ddagger}_1$).

1C) and the relative concentrations of glucose and water. The larger $\Delta G^{\ddagger}$, the slower the reaction. In the absence of enzymes, $\Delta G^{\ddagger}$ for the phosphoryl-transfer reaction is greater than that for hydrolysis, so the hydrolysis reaction predominates (although neither reaction occurs at a biologically significant rate).

The free energy barriers of both of the nonenzymatic reactions are far higher than that of the enzyme-catalyzed phosphoryl transfer to glucose. Hence enzymatic formation of glucose-6-phosphate is kinetically favored over the nonenzymatic hydrolysis of ATP. *It is the role of an enzyme, in this case hexokinase, to selectively reduce the free energy of activation of a chemically coupled reaction so that it approaches equilibrium faster than the more thermodynamically favored uncoupled reaction.*

b. Many Enzymatic Reactions Are Near Equilibrium

Although metabolism as a whole is a nonequilibrium process, many of its component reactions function close to equilibrium. The reaction of ATP and creatine to form phosphocreatine (Section 16-4C) is an example of such a reaction. The ratio [creatine]/[phosphocreatine] depends on [ATP] because creatine kinase, the enzyme catalyzing this reaction, has sufficient activity to equilibrate the reaction rapidly. The net rate of such an equilibrium reaction is effectively controlled by varying the concentrations of its reactants and/or products.

c. Pathway Throughput Is Regulated by Controlling Enzymes Operating Far from Equilibrium

Other biological reactions function far from equilibrium. For example, the phosphofructokinase reaction (Fig. 16-26) has an equilibrium constant of $K'_{eq} = 300$ but under physiological conditions in rat heart muscle has the mass action ratio [fructose-1,6-bisphosphate][ADP]/[fructose-6-phosphate][ATP] = 0.03, which corresponds to $\Delta G = -25.7$ kJ $\cdot$ mol^{-1} (Eq. [3.15]). This situation arises from a buildup of reactants because there is insufficient phosphofructokinase activity to equilibrate the reaction. Changes in substrate concentrations therefore have relatively little effect on the rate of the phosphofructokinase reaction; the enzyme is close to saturation. Only changes in the activity of the enzyme, through allosteric interactions, for example, can significantly alter this rate. An enzyme such as phosphofructokinase is therefore analogous to a dam on a river. Substrate **flux** (rate of flow) is controlled by varying its activity (allosterically or by other means), much as a dam controls the flow of a river below the dam by varying the opening of its floodgates (when the water levels on the two sides of the dam are different, that is, when they are not at equilibrium).

Understanding of how reactant flux in a metabolic pathway is controlled requires knowledge of which reactions are functioning near equilibrium and which are far from it. Most enzymes in a metabolic pathway operate near equilibrium and therefore have net rates that are sensitive only to their substrate concentrations. However, as we shall see in the following chapters (particularly in Section 17-4), *certain enzymes, which are strategically located in a metabolic pathway, operate far from equilibrium. These enzymes, which are targets for metabolic regulation by allosteric interactions and other mechanisms, are responsible for the maintenance of a stable steady-state flux of metabolites through the pathway.* This situation, as we have seen, maximizes the pathway's thermodynamic efficiency.

CHAPTER SUMMARY

1 ■ Metabolic Pathways Metabolic pathways are series of consecutive enzymatically catalyzed reactions that produce specific products for use by an organism. The free energy released by degradation (catabolism) is, through the intermediacy of ATP and NADPH, used to drive the endergonic processes of biosynthesis (anabolism). Carbohydrates, lipids, and proteins are all converted to the common intermediate acetyl-CoA, whose acetyl group is then converted to CO_2 and H_2O through the action of the citric acid cycle and oxidative phosphorylation. A relatively few metabolites serve as starting materials for a host of biosynthetic products. Metabolic pathways have five principal characteristics: (1) Metabolic pathways are irreversible; (2) if two metabolites are interconvertible, the synthetic route from the first to the second must differ from the route from the second to the first; (3) every metabolic pathway has an exergonic first committed step; (4) all metabolic pathways are regulated, usually at the first committed step; and (5) metabolic pathways in eukaryotes occur in specific subcellular compartments.

2 ■ Organic Reaction Mechanisms Almost all metabolic reactions fall into four categories: (1) group-transfer reactions; (2) oxidation–reduction reactions; (3) eliminations, isomerizations, and rearrangements; and (4) reactions that make or break carbon–carbon bonds. Most of these reactions involve heterolytic bond cleavage or formation occurring through the addition of nucleophiles to electrophilic carbon atoms. Group-transfer reactions therefore involve transfer of an electrophilic group from one nucleophile to another. The main electrophilic groups transferred are acyl groups, phosphoryl groups, and glycosyl groups. The most common nucleophiles are amino, hydroxyl, imidazole, and sulfhydryl groups. Electrophiles participating in metabolic reactions are protons, metal ions, carbonyl carbon atoms, and cationic imines. Oxidation–reduction reactions involve loss or gain of electrons. Oxidation at carbon usually involves C—H bond cleavage, with the ultimate loss by C of the two bonding electrons through their transfer to an electron acceptor such as NAD$^+$. The terminal electron acceptor in aerobes is O_2. Elimination reactions are those in which a C=C double bond is created from two saturated carbon centers with the loss of H_2O, NH_3, ROH, or RNH$_2$. Dehydration reactions are the most common eliminations. Isomerizations involve shifts of double bonds within molecules. Rearrangements are biochemically uncommon reactions in which intramolecular C—C bonds are broken and reformed to produce new carbon skeletons. Reactions that make and break C—C bonds form the basis of both degradative and biosynthetic metabolism. In the synthetic direction, these reactions involve addition of a nucleophilic carbanion to an electrophilic carbon atom. The most common electrophilic carbon atom is the carbonyl carbon, whereas carbanions are usually generated by removal of a proton from a carbon atom adjacent to a carbonyl group or by decarboxylation of a β-keto acid.

3 ■ Experimental Approaches to the Study of Metabolism
Experimental approaches employed in elucidating metabolic pathways include the use of metabolic inhibitors, growth studies, and biochemical genetics. Metabolic inhibitors block pathways at specific enzymatic steps. Identification of the resulting intermediates indicates the course of the pathway. Mutations, which occur naturally in genetic diseases or can be induced by mutagens, X-rays, or genetic engineering, may also result in the absence or inactivity of an enzyme. Modern genetic techniques make it possible to express foreign genes in higher organisms (transgenic animals) or eliminate (knock out) a gene and study the effects of these changes on metabolism. When isotopic labels are incorporated into metabolites and allowed to enter a metabolic system, their paths may be traced from the distribution of label in the intermediates. NMR is a noninvasive technique that may be used to detect and study metabolites *in vivo*. Studies on isolated organs, tissue slices, cells, and subcellular organelles have contributed enormously to our knowledge of the localization of metabolic pathways.

4 ■ Thermodynamics of Phosphate Compounds   Free energy is supplied to endergonic metabolic processes by the ATP produced via exergonic metabolic processes. ATP's $-30.5 \text{ kJ} \cdot \text{mol}^{-1} \Delta G^{\circ\prime}$ of hydrolysis is intermediate between those of "high-energy" metabolites such as phosphoenolpyruvate and "low-energy" metabolites such as glucose-6-phosphate. The "high-energy" phosphoryl groups are enzymatically transferred to ADP, and the resulting ATP, in a separate reaction, phosphorylates "low-energy" compounds. ATP may also undergo pyrophosphate cleavage to yield PP_i,

whose subsequent hydrolysis adds further thermodynamic impetus to the reaction. ATP is present in too short a supply to act as an energy reservoir. This function, in vertebrate nerve and muscle cells, is carried out by phosphocreatine, which under low-ATP conditions readily transfers its phosphoryl group to ADP to form ATP.

5 ■ Oxidation–Reduction Reactions   The half-reactions of redox reactions may be physically separated to form two electrochemical half-cells. The redox potential for the reduction of A by B,

$$A_{ox}^{n+} + B_{red} \rightleftharpoons A_{red} + B_{ox}^{n+}$$

in which n electrons are transferred, is given by the Nernst equation

$$\Delta \mathscr{E} = \Delta \mathscr{E}^{\circ} - \frac{RT}{n\mathscr{F}} \ln\left(\frac{[A_{red}][B_{ox}^{n+}]}{[A_{ox}^{n+}][B_{red}]}\right)$$

The redox potential of such a reaction is related to the reduction potentials of its component half-reactions, $\mathscr{E}_A$ and $\mathscr{E}_B$, by

$$\Delta \mathscr{E} = \mathscr{E}_A - \mathscr{E}_B$$

If $\mathscr{E}_A > \mathscr{E}_B$, then A_{ox}^{n+} has a greater electron affinity than does B_{ox}^{n+}. The reduction potential scale is defined by arbitrarily setting the reduction potential of the standard hydrogen half-cell to zero. Redox reactions are of great metabolic importance. For example, the oxidation of NADH yields three ATPs through the mediation of the electron-transport chain.

6 ■ Thermodynamics of Life   Living organisms are open systems and therefore cannot be at equilibrium. They must continuously dissipate free energy in order to carry out their various functions and preserve their highly ordered structures. The study of nonequilibrium thermodynamics has indicated that the steady state, which living processes maintain, is the state of maximum efficiency under the constraints governing open systems. Control mechanisms that regulate biological processes preserve the steady state by regulating the activities of enzymes that are strategically located in metabolic pathways.

 REFERENCES

METABOLIC STUDIES
Beadle, G.W., Biochemical genetics, *Chem. Rev.* **37,** 15–96 (1945). [A classic review summarizing the "one gene–one enzyme" hypothesis.]

Cerdan, S. and Seelig, J., NMR studies of metabolism, *Annu. Rev. Biophys. Biophys. Chem.* **19,** 43–67 (1990).

Cooper, T.G., *The Tools of Biochemistry,* Chapter 3, Wiley-Interscience (1977). [A presentation of radiochemical techniques.]

Freifelder, D., *Biophysical Chemistry* (2nd ed.), Chapters 5 and 6, Freeman (1982). [A discussion of the principles of radioactive counting and autoradiography.]

Fruton, J.S. and Simmons, S., *General Biochemistry* (2nd ed.), Chapter 16, Wiley (1958). [Outlines the classic methods for the study of intermediate metabolism.]

Goodridge, A.G., The new metabolism: Molecular genetics in the

analysis of metabolic regulation, *FASEB J.* **4,** 3099–3110 (1990).

Hevesy, G., Historical sketch of the biological application of tracer elements, *Cold Spring Harbor Symp. Quant. Biol.* **13,** 129–150 (1948).

Jeffrey, F.M.H., Rajagopal, A., Malloy, C.R., and Sherry, A.D., ^{13}C-NMR: A simple yet comprehensive method for analysis of intermediary metabolism, *Trends Biochem. Sci.* **16,** 5–10 (1991).

Koretsky, A.P., Investigation of cell physiology in the animal using transgenic technology, *Am. J. Physiol.* **262,** C261–C275 (1992).

Koretsky, A.P. and Williams, D.S., Application of localized *in vivo* NMR to whole organ physiology in the animal, *Annu. Rev. Physiol.* **54,** 799–826 (1992).

McGrane, M.M., Yun, J.S., Patel, Y.M., and Hanson, R.W., Metabolic control of gene expression: *In vivo* studies with transgenic mice, *Trends Biochem. Sci.* **17**, 40–44 (1992).

Michal, G. (Ed.), *Biochemical Pathways. An Atlas of Biochemistry and Molecular Biology,* Wiley (1999). [An encyclopedic compendium of metabolic pathways.]

Shemin, D. and Rittenberg, D., The biological utilization of glycine for the synthesis of the protoporphyrin of hemoglobin, *J. Biol Chem.* **166**, 621–625 (1946).

Shulman, R.G. and Rothman, D.L., ^{13}C NMR of intermediary metabolism: Implications for systematic physiology, *Annu. Rev. Physiol.* **63**, 15–48 (2001).

Suckling, K.E. and Suckling, C.J., *Biological Chemistry,* Cambridge University Press (1980). [Presents the organic chemistry of biochemical reactions.]

Walsh, C., *Enzymatic Reaction Mechanisms,* Chapter 1, Freeman (1979). [A discussion of the types of biochemical reactions.]

Wang, N.-D., Finegold, M.J., Bradley, A., Ou, C.N., Abdelsayed, S.V., Wilde, M.D.,Taylor, L.R., Wilson, D.R., and Darlington, G.J., Impaired energy homeostasis in C/EBPα knockout mice, *Science* **269**, 1108–1112 (1995).

Westheimer, F.H., Why nature chose phosphates, *Science* **235**, 1173–1178 (1987).

BIOENERGETICS

Alberty, R.A., Standard Gibbs free energy, enthalpy and entropy changes as a function of pH and pMg for reactions involving adenosine phosphates, *J. Biol. Chem.* **244**, 3290–3302 (1969).

Alberty, R.A., Calculating apparent equilibrium constants of enzyme-catalyzed reactions at pH 7, *Biochem. Ed.* **28**, 12–17 (2000).

Caplan, S.R., Nonequilibrium thermodynamics and its application to bioenergetics, *Curr. Top. Bioenerg.* **4**, 1–79 (1971).

Crabtree, B. and Taylor, D.J., Thermodynamics and metabolism, *in* Jones, M.N. (Ed.), *Biochemical Thermodynamics*, pp. 333–378, Elsevier (1979).

Dickerson, R.E., *Molecular Thermodynamics,* Chapter 7, Benjamin (1969). [An interesting chapter on the thermodynamics of life.]

Henley, H.J.M., An introduction to nonequilibrium thermodynamics, *J. Chem. Ed.* **41**, 647–655 (1964).

Katchelsky, A. and Curran, P.F., *Nonequilibrium Thermodynamics in Biophysics,* Harvard University Press (1965).

Morowitz, H.J., *Foundations of Bioenergetics,* Academic Press (1978).

PROBLEMS

1. Glycolysis (glucose breakdown) has the overall stoichiometry:

Glucose + 2ADP + 2P$_i$ + 2NAD$^+$ →

2pyruvate + 2ATP + 2NADH + 2H$^+$ + 2H$_2$O

whereas that of gluconeogenesis (glucose synthesis) is

2Pyruvate + 4ATP + 2NADH + 2H$^+$ + 4H$_2$O →

glucose + 4ADP + 4P$_i$ + 2NAD$^+$

What is the overall stoichiometry of the glycolytic breakdown of 1 mol of glucose followed by its gluconeogenic synthesis? Explain why it is necessary that the pathways of these two processes be independently controlled and why they must differ by at least one reaction.

2. It has been postulated that a trigonal bipyrimidal pentacovalent phosphorus intermediate can undergo a vibrational deformation process known as **pseudorotation** in which its apical ligands exchange with two of its equatorial ligands via a tetragonal pyrimidal transition state:

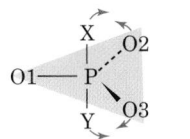

**Trigonal bipyramid
[X and Y apical]**

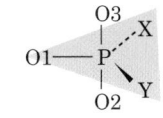

**Trigonal bipyramid
[O2 and O3 apical]**

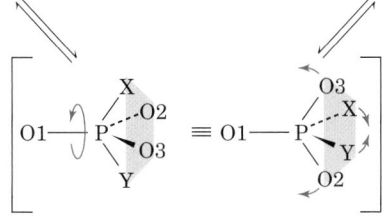

**Tetragonal pyramidal
transition state**

In a nucleophilic substitution reaction, would two cycles of pseudorotation, so as to place the leaving group (X) in an apical position and the attacking group (Y) in an equatorial position, lead to retention or inversion of configuration on the departure of the leaving group?

3. One **Curie (Ci)** of radioactivity is defined as 3.70×10^{10} disintegrations per second, the number that occurs in 1 g of pure ^{226}Ra. A sample of ^{14}CO$_2$ has a specific radioactivity of 5μCi $\cdot$ μmol^{-1}. What percentage of its C atoms are ^{14}C?

4. In the hydrolysis of ATP to ADP and P$_i$, the equilibrium concentration of ATP is too small to be measured accurately. A better way of determining K'_{eq}, and hence $\Delta G^{\circ\prime}$ of this reaction, is to break it up into two steps whose values of $\Delta G^{\circ\prime}$ can be accurately determined. This has been done using the following pair of reactions (the first being catalyzed by **glutamine synthetase**):

(1) ATP + glutamate + NH$_3^+$ ⇌

ADP + P$_i$ + glutamine + H$^+$ $\Delta G_1^{\circ\prime}$ = −16.3 kJ $\cdot$ mol^{-1}

(2) Glutamate + NH$_3^+$ ⇌ glutamine + H$_2$O + H$^+$

$\Delta G_2^{\circ\prime}$ = 14.2 kJ $\cdot$ mol^{-1}

What is the $\Delta G^{\circ\prime}$ of ATP hydrolysis according to these data?

*5. Consider the reaction catalyzed by hexokinase:

ATP + glucose ⇌ ADP + glucose-6-phosphate

A mixture containing 40 m*M* ATP and 20 m*M* glucose was incubated with hexokinase at pH 7 and 25°C. Calculate the equilibrium concentrations of the reactants and products (see Table 16-3).

6. In aerobic metabolism, glucose is completely oxidized in the reaction

Glucose + 6O$_2$ ⇌ 6CO$_2$ + 6H$_2$O

with the coupled generation of 38 ATP molecules from 38 ADP and 38 P$_i$. Assuming the ΔG for the hydrolysis of ATP to ADP and P$_i$ under intracellular conditions is −50 kJ $\cdot$ mol^{-1} and that for the combustion of glucose is −2823.2 kJ $\cdot$ mol^{-1}, what is the efficiency of the glucose oxidation reaction in terms of the free energy sequestered in the form of ATP?

7. Typical intracellular concentrations of ATP, ADP, and P_i in muscles are 5.0, 0.5, and 1.0 mM, respectively. At 25°C and pH 7: (a) What is the free energy of hydrolysis of ATP at these concentrations? (b) Calculate the equilibrium concentration ratio of phosphocreatine to creatine in the creatine kinase reaction:

$$\text{Creatine} + \text{ATP} \rightleftharpoons \text{phosphocreatine} + \text{ADP}$$

if ATP and ADP have the above concentrations. (c) What concentration ratio of ATP to ADP would be required under the foregoing conditions to yield an equilibrium concentration ratio of phosphocreatine to creatine of 1? Assuming the concentration of P_i remained 1.0 mM, what would the free energy of hydrolysis of ATP be under these latter conditions?

*8. Assuming the intracellular concentrations of ATP, ADP, and P_i, are those given in Problem 7: (a) Calculate the concentration of AMP at pH 7 and 25°C under the condition that the adenylate kinase reaction:

$$2\text{ADP} \rightleftharpoons \text{ATP} + \text{AMP}$$

is at equilibrium. (b) Calculate the equilibrium concentration of AMP when the free energy of hydrolysis of ATP to ADP and P_i is -55 kJ · mol^{-1}. Assume [P_i] and ([ATP] + [ADP]) remain constant.

9. Using the data in Table 16-4, list the following substances in order of their decreasing oxidizing power: (a) fumarate$^-$, (b) cystine, (c) O_2, (d) NADP$^+$, (e) cytochrome c (Fe^{3+}), and (f) lipoic acid.

10. Calculate the equilibrium concentrations of reactants and products for the reaction:

$$\text{Acetoacetate}^- + \text{NADH} + \text{H}^+ \rightleftharpoons$$
$$\beta\text{-hydroxybutyrate}^- + \text{NAD}^+$$

when the initial concentrations of acetoacetate$^-$ and NADH are 0.01 and 0.005M, respectively, and β-hydroxybutyrate and NAD$^+$ are initially absent. Assume the reaction takes place at 25°C and pH 7.

11. In anaerobic bacteria, the final metabolic electron acceptor is some molecule other than O_2. A major requirement for any redox pair utilized as a metabolic free energy source is that it provides sufficient free energy to generate ATP from ADP and P_i. Indicate which of the following redox pairs are sufficiently exergonic to enable a properly equipped bacterium to utilize them as a major energy source. Assume that redox reactions forming ATP require two electrons and that $\Delta\mathscr{E} = \Delta\mathscr{E}°'$.

(a) Ethanol + NO$_3^-$ (c) H$_2$ + S
(b) Fumarate$^-$ + SO$_3^{2-}$ (d) Acetaldehyde + acetaldehyde

12. Calculate $\Delta G°'$ for the following pairs of half-reactions at pH 7 and 25°C. Write a balanced equation for the overall reaction and indicate the direction in which it occurs spontaneously under standard conditions.

(a) (H$^+$/$\frac{1}{2}$H$_2$) and ($\frac{1}{2}$O$_2$ + 2H$^+$/H$_2$O)
(b) (Pyruvate$^-$ + 2H$^+$/lactate$^-$) and (NAD$^+$ + H$^+$/NADH)

*13. The chemiosmotic hypothesis (Section 22-3A) postulates that ATP is generated in the two-electron reaction:

$$\text{ADP} + \text{P}_i + 2\text{H}^+ \text{ (low pH)} \rightleftharpoons \text{ATP} + \text{H}_2\text{O} + 2\text{H}^+ \text{ (high pH)}$$

which is driven by a metabolically generated pH gradient in the mitochondria. What is the magnitude of the pH gradient required for net synthesis of ATP at 25°C and pH 7, if the steady-state concentrations of ATP, ADP, and P_i are 0.01, 10, and 10 mM, respectively?

14. Gastric juice is 0.15M HCl. The blood plasma, which is the source of this H$^+$ and Cl$^-$, is 0.10M in Cl$^-$ and has a pH of 7.4. Calculate the free energy necessary to produce the HCl in 0.1 L of gastric juice at 37°C.

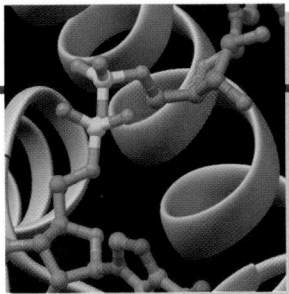

Chapter
17 Glycolysis

At this point we commence our discussions of specific metabolic pathways by considering **glycolysis** (Greek: *glykos,* sweet; *lysis,* loosening), the pathway by which **glucose** is converted via **fructose-1,6-bisphosphate** to **pyruvate** with the generation of 2 mol of ATP per mol of glucose. This sequence of 10 enzymatic reactions, which is probably the most completely understood biochemical pathway, plays a key role in energy metabolism by providing a significant portion of the energy utilized by most organisms and by preparing glucose, as well as other carbohydrates, for oxidative degradation.

In our study of glycolysis, and indeed of all of metabolism, we shall attempt to understand the pathway on four levels:

1. The chemical interconversion steps, that is, the sequence of reactions by which glucose is converted to the pathway's end products.

2. The mechanism of the enzymatic conversion of each pathway intermediate to its successor.

3. The energetics of the conversions.

4. The mechanisms controlling the **flux** (rate of flow) of metabolites through the pathway.

The flux of metabolites through a pathway is remarkably sensitive to the needs of the organism for the products of the pathway. Through an exquisitely complex network of control mechanisms, flux through a pathway is only as great as required.

1 ■ THE GLYCOLYTIC PATHWAY

An overview of glucose metabolism is diagrammed in Fig. 17-1. *Under aerobic conditions, the pyruvate formed by glycolysis is further oxidized by the citric acid cycle (Chapter 21) and oxidative phosphorylation (Chapter 22) to CO_2 and water. Under anaerobic conditions, however, the pyruvate is instead converted to a reduced end product, which is **lactate** in muscle (**homolactic fermentation;** a fermentation is an anaerobic biological reaction process) and ethanol + CO_2 in yeast (**alcoholic fermentation**).*

A. *Historical Perspective*

The fermentation of glucose to ethanol and CO_2 by yeast (Fig. 17-2) has been a useful process since before the dawn of recorded history. Winemaking and bread baking both exploit this process. Yet the scientific investigation of the mechanism of glycolysis began only in the latter half of the nineteenth century.

In the years 1854 to 1864, Louis Pasteur established that fermentation is caused by microorganisms. It was not until 1897, however, that Eduard Buchner demonstrated that cell-free yeast extracts can also carry out this process. This discovery refuted the then widely held belief that fermentation, and every other biological process, was mediated by some "vital force" inherent in living matter, and thereby brought glycolysis within the province of chemistry. This was a major step in the development of biochemistry as a science. Although, in principle, the use of cell-free extracts enabled a systematic "dissection" of the reactions involved in the pathway, the complete elucidation of the glycolytic pathway was still a long-range project because analytical techniques for the isolation and identification of intermediates and enzymes had to be developed concurrently.

In the years 1905 to 1910, Arthur Harden and William Young made two important discoveries:

1. Inorganic phosphate is required for fermentation and is incorporated into fructose-1,6-bisphosphate, an intermediate in the process.

2. A cell-free yeast extract can be separated, by dialysis, into two fractions that are both required for fermentation: A nondialyzable heat-labile fraction they named **zymase;** and a dialyzable, heat-stable fraction they called **cozymase.** It was shown later by others that zymase is a mixture of enzymes and that cozymase is a mixture of cofactors: coenzymes such as NAD^+, ATP, and ADP, as well as metal ions.

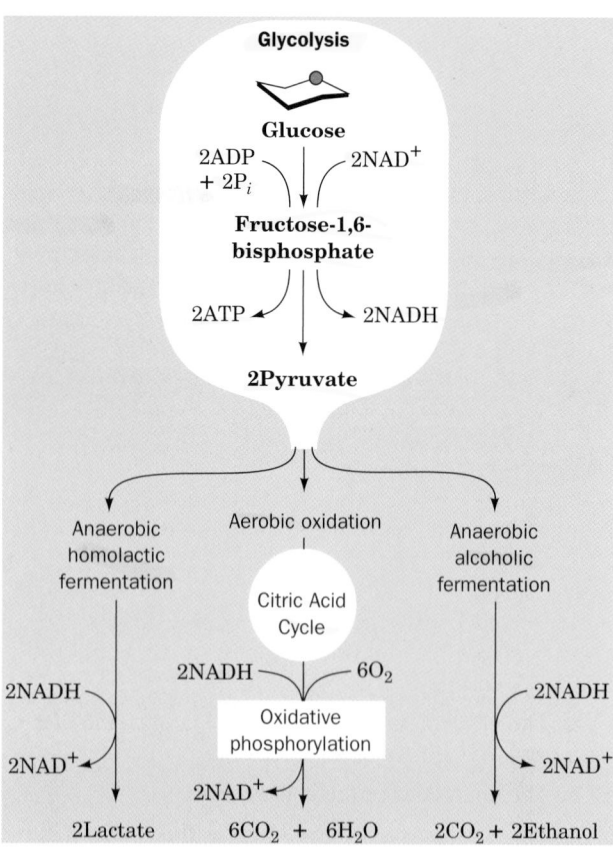

FIGURE 17-1   Glycolysis. Glycolysis converts glucose to pyruvate while generating two ATPs. Under anaerobic conditions, alcoholic fermentation of pyruvate occurs in yeast, whereas homolactic fermentation occurs in muscle. Under aerobic conditions, pyruvate is oxidized to H_2O and CO_2 via the citric acid cycle (Chapter 21) and oxidative phosphorylation (Chapter 22).

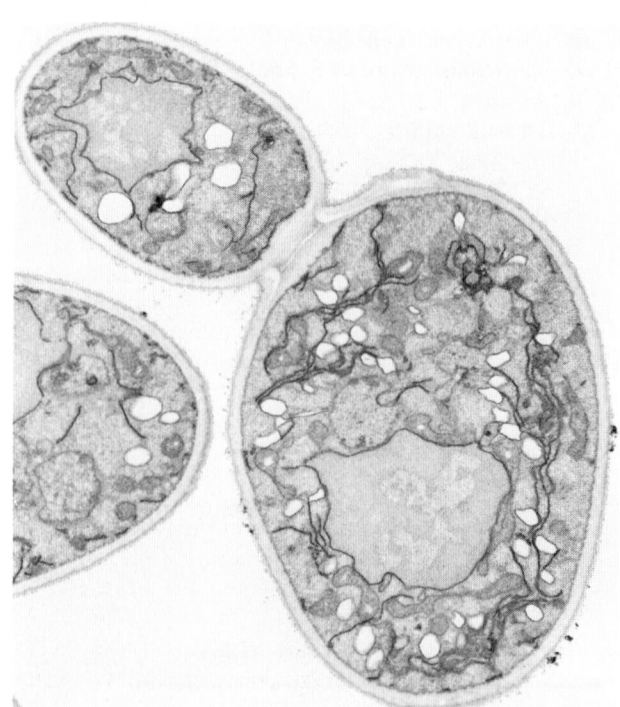

FIGURE 17-2   Electron micrograph of yeast cells. [Biophoto Associates.]

In their efforts to identify pathway intermediates, the early investigators of glycolysis developed a general technique of metabolic investigation that is still widely used today: *Reagents are found that inhibit the production of pathway products, thereby causing the buildup of metabolites that can then be identified as pathway intermediates.* Over years of investigation that attempted to identify glycolytic intermediates, various reagents were found that inhibit the production of ethanol from glucose in yeast extracts. The use of different inhibitors results in the accumulation of different intermediates. For example, the addition of iodoacetate to fermenting yeast extracts causes the buildup of fructose-1,6-bisphosphate, whereas addition of fluoride ion induces the accumulation of **3-phosphoglycerate** and **2-phosphoglycerate**:

3-Phosphoglycerate ⟶ **2-Phosphoglycerate**

The mechanisms by which these inhibitors act are discussed in Sections 17-2D and 17-2I, respectively.

One remarkable finding of these studies was that the same intermediates and enzyme activities could be isolated not only from yeast, but from a great variety of other organisms. With few exceptions (see Problem 17-10), *living things all metabolize glucose by identical pathways. In spite of their enormous diversity, they share a common biochemistry.*

By 1940, the efforts of many investigators had come to fruition with the elucidation of the complete pathway of glycolysis. The work of three of these individuals, Gustav Embden, Otto Meyerhof, and Jacob Parnas, has been commemorated in that glycolysis is alternatively known as the **Embden–Meyerhof–Parnas pathway.** Other major contributors to the elucidation of this pathway were Carl and Gerty Cori, Carl Neuberg, Robert Robison, and Otto Warburg.

B. *Pathway Overview*

Before beginning our detailed discussion of the enzymes of glycolysis, let us first take a moment to survey the overall pathway as it fits in with animal metabolism as a whole. Glucose usually arises in the blood as a result of the breakdown of higher polysaccharides (Sections 11-2B, 11-2D, and 18-1) or from its synthesis from noncarbohydrate sources (**gluconeogenesis;** Section 23-1). The fate of nonglucose hexoses is discussed in Section 17-5. Glucose enters most cells by a specific carrier that transports it from the exterior of the cell into the cytosol (Section 20-2E). *The*

enzymes of glycolysis are located in the cytosol, where they are only loosely associated, if at all, with cell structures such as membranes. However, there is considerable circumstantial evidence that successive enzymes in the glycolytic pathway loosely associate, presumably to facilitate the efficient transfer of intermediates between enzymes. Such associations of functionally related enzymes have been referred to as **metabolons.** Nevertheless, no actual complexes of glycolytic enzymes have yet been isolated.

Glycolysis converts glucose to two C_3 units (pyruvate) of lower free energy in a process that harnesses the released free energy to synthesize ATP from ADP and P_i. This process requires a pathway of chemically coupled phosphoryl-transfer reactions (Sections 16-4 and 16-6). Thus the chemical strategy of glycolysis is:

1. Add phosphoryl groups to the glucose.

2. Chemically convert phosphorylated intermediates into compounds with high phosphate group-transfer potentials.

3. Chemically couple the subsequent hydrolysis of reactive substances to ATP synthesis.

The 10 enzyme-catalyzed reactions of glycolysis are diagrammed in Fig. 17-3. Note that ATP is used early in the pathway to synthesize phosphoryl compounds (Reactions 1 and 3) but is later resynthesized (Reactions 7 and 10). Glycolysis may therefore be considered to occur in two stages:

Stage I (Reactions 1–5): A preparatory stage in which the hexose glucose is phosphorylated and cleaved to yield two molecules of the triose **glyceraldehyde-3-phosphate.** This process utilizes two ATPs in a kind of energy investment.

Stage II (Reactions 6–10): The two molecules of glyceraldehyde-3-phosphate are converted to pyruvate, with concomitant generation of four ATPs. Glycolysis therefore has a net profit of two ATPs per glucose: Stage I consumes two ATPs; Stage II produces four ATPs.

The overall reaction is

$$\text{Glucose} + 2\text{NAD}^+ + 2\text{ADP} + 2P_i \rightarrow$$
$$2\text{NADH} + 2\text{pyruvate} + 2\text{ATP} + 2\text{H}_2\text{O} + 4\text{H}^+$$

a. The Oxidizing Power of NAD⁺ Must Be Recycled

NAD^+ is the primary oxidizing agent of glycolysis. The NADH produced by this process (Fig. 17-3, Reaction 6) must be continually reoxidized to keep the pathway supplied with NAD^+. There are three common ways that this occurs (Fig. 17-1, *bottom*):

1. Under anaerobic conditions in muscle, NAD^+ is regenerated when NADH reduces pyruvate to lactate (homolactic fermentation; Section 17-3A).

2. Under anaerobic conditions in yeast, pyruvate is decarboxylated to yield CO_2 and acetaldehyde and the latter

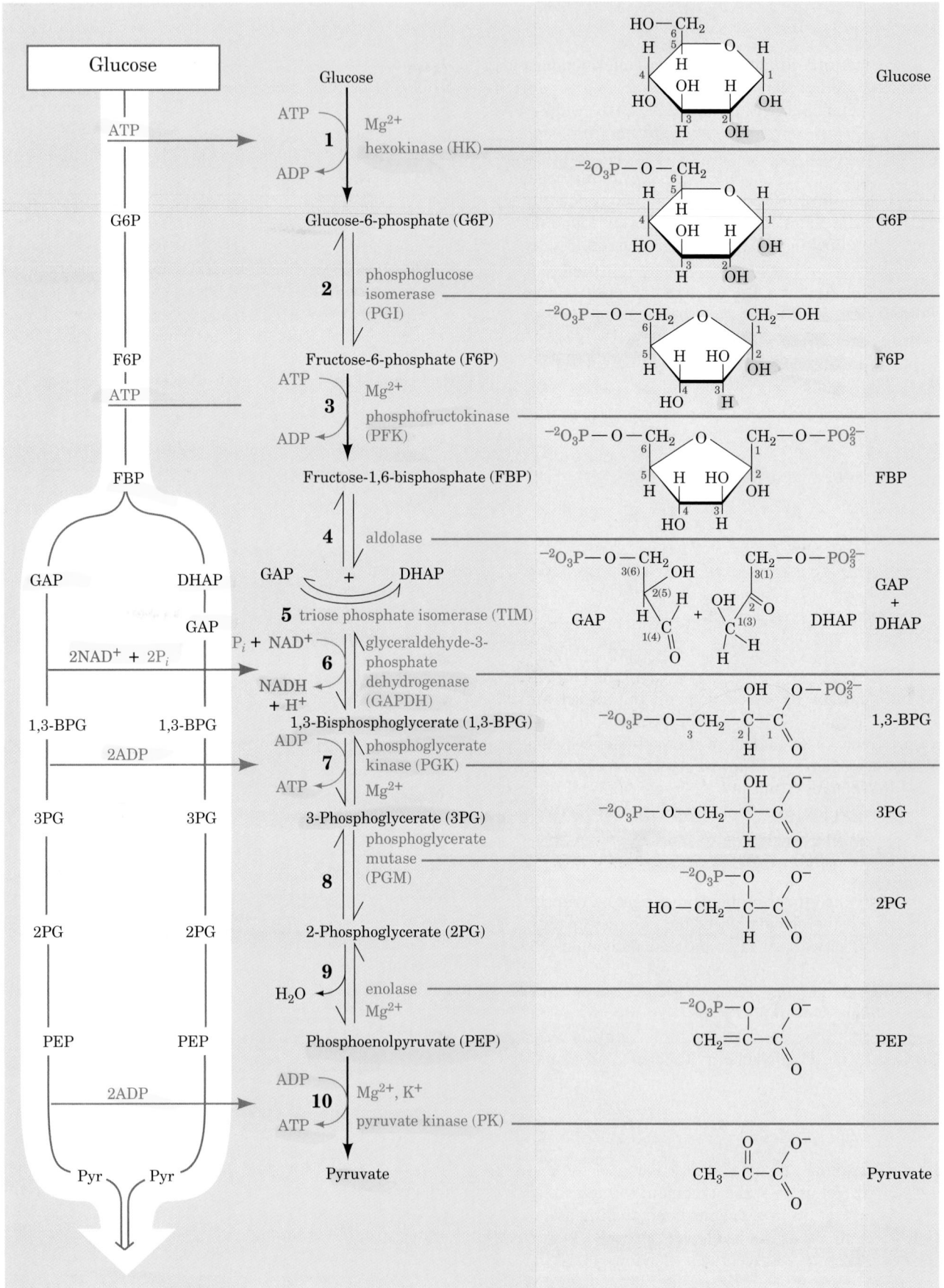

FIGURE 17-3 Degradation of glucose via the glycolytic pathway. Glycolysis may be considered to occur in two stages. Stage I (Reactions **1–5**): Glucose is phosphorylated and cleaved to form two molecules of the triose glyceraldehyde-3-phosphate. This requires the expenditure of two ATPs in an "energy investment" (Reactions **1** and **3**). Stage II (Reactions **6–10**): The two molecules of glyceraldehyde-3-phosphate are converted to pyruvate with the concomitant generation of four ATPs (Reactions **7** and **10**). 🐁 See the Animated Figures

is reduced by NADH to yield NAD$^+$ and ethanol (alcoholic fermentation; Section 17-3B).

3. Under aerobic conditions, the mitochondrial oxidation of each NADH to NAD$^+$ yields three ATPs (Section 22-2A).

Thus in aerobic glycolysis, NADH may be thought of as a "high-energy" compound, whereas in anaerobic glycolysis its free energy of oxidation is dissipated as heat.

2 ■ THE REACTIONS OF GLYCOLYSIS

See Guided Exploration 14: Glycolysis Overview In this section we examine the reactions of glycolysis more closely, describing the properties of the individual enzymes and their mechanisms. In Section 17-3 we consider the anaerobic fate of pyruvate. Finally, in Section 17-4 we consider the thermodynamics of the entire process and address the problem of how the flux of metabolites through the pathway is controlled. As we study the individual glycolytic enzymes we shall encounter many organic reaction mechanisms (Section 16-2). Indeed, the study of organic reaction mechanisms has been invaluable in understanding the mechanisms by which enzymes catalyze reactions.

Note that the X-ray structure of each of the 10 glycolytic enzymes has been reported. All of these enzymes are either homodimers or homotetramers with D_2 symmetry (Section 8-5B), whose subunits consist mainly of α/β domains (Section 8-3B).

A. *Hexokinase: First ATP Utilization*

Reaction 1 of glycolysis is the transfer of a phosphoryl group from ATP to glucose to form **glucose-6-phosphate (G6P)** in a reaction catalyzed by **hexokinase (HK):**

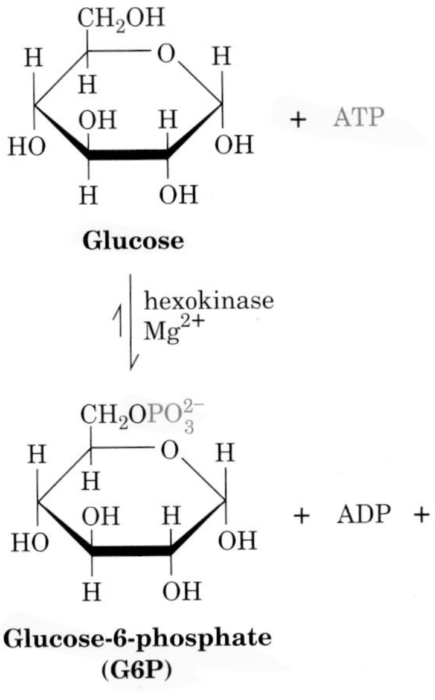

Glucose

hexokinase
Mg^{2+}

Glucose-6-phosphate (G6P)

A **kinase** is an enzyme that transfers phosphoryl groups between ATP and a metabolite (Section 16-4C). The

metabolite that serves as the phosphoryl group acceptor for a specific kinase is identified in the prefix of the kinase name. HK is a relatively nonspecific enzyme contained in all cells that catalyzes the phosphorylation of hexoses such as D-glucose, D-mannose, and D-fructose. Liver cells also contain **glucokinase,** which catalyzes the same reaction but which is primarily involved in the maintenance of blood glucose levels (Section 18-3F). The second substrate for HK, as with other kinases, is an Mg^{2+}–ATP complex. In fact, uncomplexed ATP is a potent competitive inhibitor of HK. In what follows, we shall rarely refer to this Mg^{2+} requirement, but keep in mind that it is essential for kinase enzymatic activity (other divalent metal ions such as Mn^{2+} often satisfy the metal ion requirements of kinases *in vitro*, but Mg^{2+} is the normal physiological species).

a. Kinetics and Mechanism of the Hexokinase Reaction

Hexokinase has a Random Bi Bi mechanism in which the enzyme forms a ternary complex with glucose and Mg^{2+}–ATP before the reaction occurs. The Mg^{2+}, by complexing with the phosphate oxygen atoms, is thought to shield their negative charges, making the phosphorus atom more accessible for the nucleophilic attack of the C6—OH group of glucose (Fig. 17-4).

An important mechanistic question is why does HK catalyze the transfer of a phosphoryl group from ATP to glucose to yield G6P, but not to water to yield ADP + P$_i$ (ATP hydrolysis)? Water is certainly small enough to fit into the phosphoryl acceptor group's enzymatic binding site. Furthermore, phosphoryl transfer from ATP to water is more exergonic than it is to glucose (Table 16-3), particularly since [H$_2$O] = 55.5 M and [glucose] = 5 to 10 mM *in vivo*. Yet HK catalyzes phosphoryl transfer to glucose 40,000 times faster than it does to water.

The answer was provided by Thomas Steitz via X-ray structural studies of yeast HK. Comparison of the X-ray structures of HK and the glucose–HK complex indicates that *glucose induces a large conformational change*

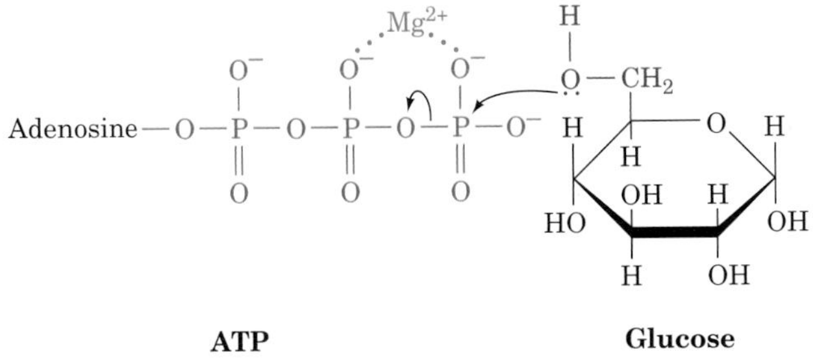

ATP   **Glucose**

FIGURE 17-4 The nucleophilic attack of the C6—OH group of glucose on the γ phosphate of an Mg^{2+}–ATP complex. The position of the Mg^{2+} ion is shown as an example; its actual position(s) has not been conclusively established. In any case, the Mg^{2+} functions to shield the negatively charged groups of ATP and thereby facilitates the nucleophilic attack.

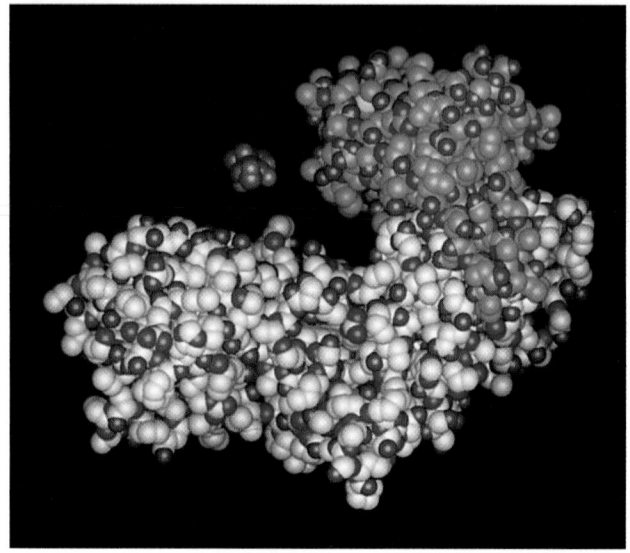

(a)

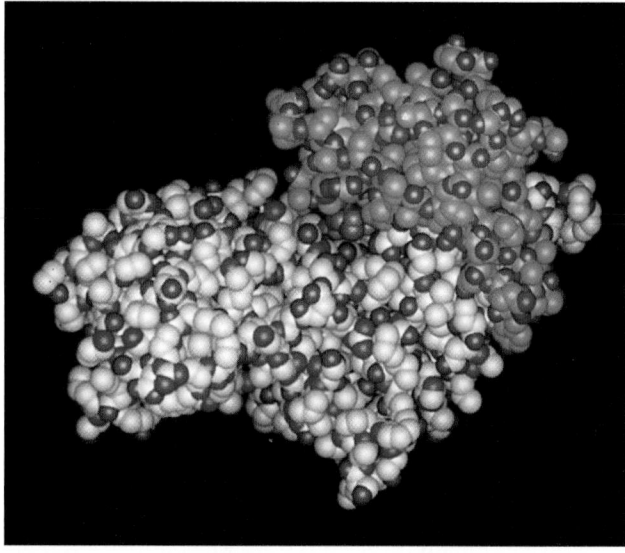

(b)

FIGURE 17-5 Conformation changes in yeast hexokinase on binding glucose. Space-filling models of a subunit of (*a*) free hexokinase and (*b*) in complex with glucose (*purple*). Note the prominent bilobal appearance of the free enzyme (the C atoms in the small lobe are shaded green, whereas those in the large lobe are light gray; the N and O atoms are blue and red). In the enzyme–substrate complex these lobes have swung together so as to engulf the substrate. [Based on X-ray structures by Thomas Steitz, Yale University. PDBids (*a*) 2YHX and (*b*) 1HKG.] 🔹 **See the Interactive Exercises**

in HK (Fig. 17-5). The two lobes that form its active site cleft swing together by up to 8 Å so as to engulf the glucose in a manner that suggests the closing of jaws. *This movement places the ATP in close proximity to the* —C6H₂OH *group of glucose and excludes water from the active site (catalysis by proximity effects; Section 15-1E).* If the catalytic and reacting groups were in the proper position for reaction while the enzyme was in the open position (Fig. 17-5*a*), ATP hydrolysis would almost certainly be the dominant reaction. This conclusion is confirmed by the observation that **xylose,** which differs from glucose only by the lack of the —C6H₂OH group,

α-D-**Xylose**

greatly enhances the rate of ATP hydrolysis by HK (presumably xylose induces the activating conformational change while water occupies the binding site of the missing hydroxymethyl group). Clearly, *this substrate-induced conformational change in HK is responsible for the enzyme's specificity.* In addition, the active site polarity is reduced by exclusion of water, thereby expediting the nucleophilic reaction process. Other kinases have the same deeply clefted structure as HK (e.g., Section 17-2G) and undergo conformational changes on binding their sub-

strates. This suggests that all kinases have similar mechanisms for maintaining specificity.

B. *Phosphoglucose Isomerase*

Reaction 2 of glycolysis is the conversion of G6P to **fructose-6-phosphate (F6P)** by **phosphoglucose isomerase (PGI;** also called **glucose-6-phosphate isomerase).** This is the isomerization of an aldose to a ketose:

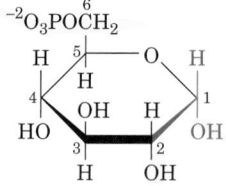

Glucose-6-phosphate (G6P)

phosphoglucose isomerase (PGI)

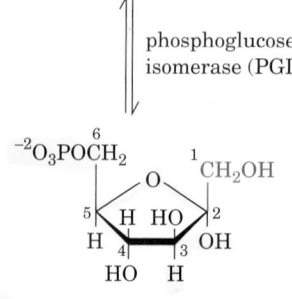

Fructose-6-phosphate (F6P)

Since G6P and F6P both exist predominantly in their cyclic forms (Fig. 11-4 shows these structures for the unphos-

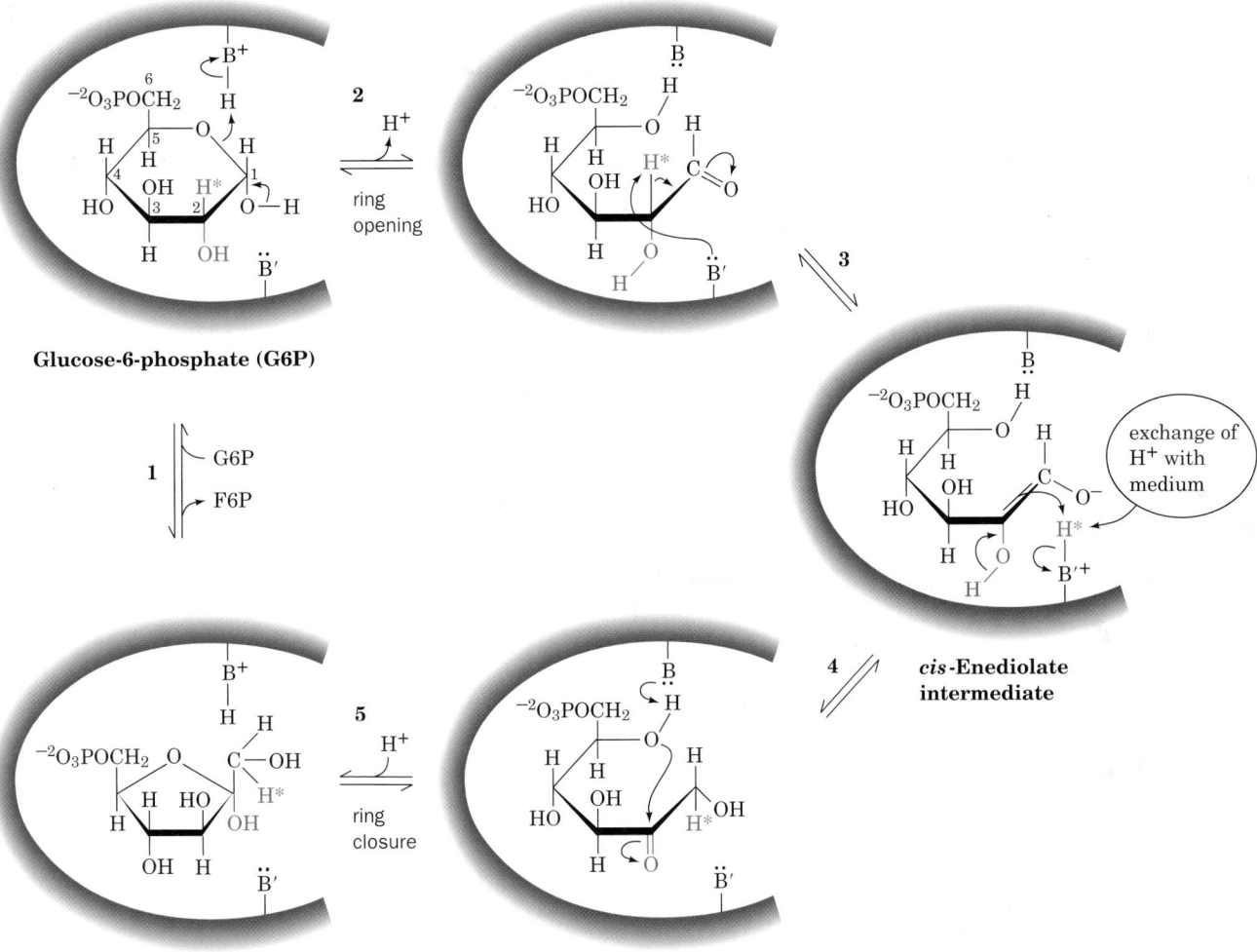

Glucose-6-phosphate (G6P)

cis-Enediolate intermediate

Fructose-6-phosphate (F6P)

FIGURE 17-6 Reaction mechanism of phosphoglucose isomerase. The active site catalytic residues (BH$^+$ and B$'$) are thought to be Lys and a His–Glu dyad, respectively.

phorylated sugars), the reaction requires ring opening, followed by isomerization, and subsequent ring closure. The determination of the enzyme's pH dependence led to a hypothesis for amino acid side chain participation in the catalytic mechanism. The catalytic rate exhibits a bell-shaped pH dependence curve with characteristic pK's of 6.7 and 9.3, which suggests the catalytic participation of both a His and a Lys (Section 14-4). Indeed, comparison of the amino acid sequences of PGI from several different organisms reveals that both a His and Lys are conserved. However, a Glu residue is also conserved, and as we have seen for lysozyme (Section 15-2C), Glu can have an unusually high pK under certain conditions. In fact, the X-ray structure of PGI reveals that Glu 216 and His 388 form a hydrogen bonded catalytic dyad (resembling the interaction of the Asp and His residues in the catalytic triad of serine proteases; Fig. 15-19), which facilitates the action of His 388 as an acid–base catalyst.

A proposed reaction mechanism for the PGI reaction involves general acid–base catalysis by the enzyme (Fig. 17-6):

Step 1 Substrate binding.

Step 2 An acid, presumably the Lys ε-amino group, catalyzes ring opening.

Step 3 A base, presumably the imidazole portion of the His–Glu dyad, abstracts the acidic proton from C2 to form a cis-enediolate intermediate (this proton is acidic because it is α to a carbonyl group).

Step 4 The proton is replaced on C1 in an overall proton transfer. Protons abstracted by bases are labile and exchange rapidly with solvent protons. Nevertheless, Irwin Rose confirmed this step by demonstrating that [2-^{3}H]G6P is occasionally converted to [1-^{3}H]F6P by intramolecular proton transfer before the ^{3}H has had a chance to exchange with the medium.

Step 5 Ring closure to form the product, which is subsequently released to yield free enzyme, thereby completing the catalytic cycle.

PGI, like most enzymes, catalyzes reactions with nearly absolute stereospecificity. To appreciate this, let us compare the proposed enzymatic reaction mechanism with that in the nonenzymatic base-catalyzed isomerization of glucose, fructose, and mannose (Fig. 17-7). Glucose and mannose are epimers of one another because they differ only with respect to their configuration at one chiral center, C2 (Section 11-1A). In the enediolate intermediate, as well as in the linear form of fructose, C2 has no chirality. Therefore, in nonenzymatic systems, base-catalyzed isomerization of glucose also results in racemization of C2 with the production of mannose. In the presence of PGI, however, ^{1}H NMR measurements indicate that the rate of the isomerization reaction is several orders of magnitude greater than that of the epimerization reaction. Evidently, PGI shields the face of the enediolate to which H^+ must be added to form mannose-6-phosphate.

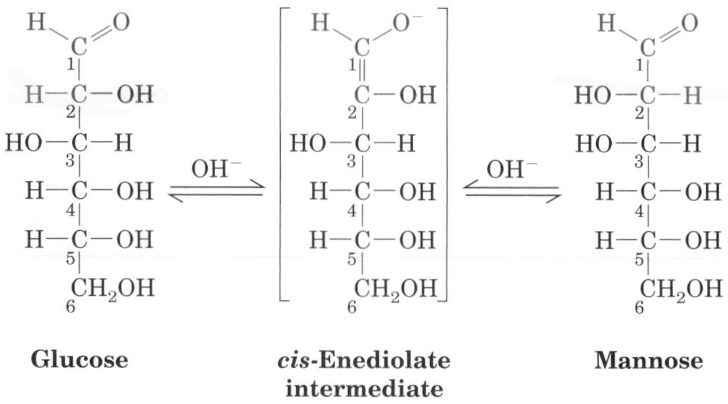

C. *Phosphofructokinase: Second ATP Utilization*

In Reaction 3 of glycolysis, **phosphofructokinase (PFK)** phosphorylates F6P to yield fructose-1,6-bisphosphate [**FBP** or **F1,6P**; previously known as **fructose-1,6-diphosphate (FDP)**]:

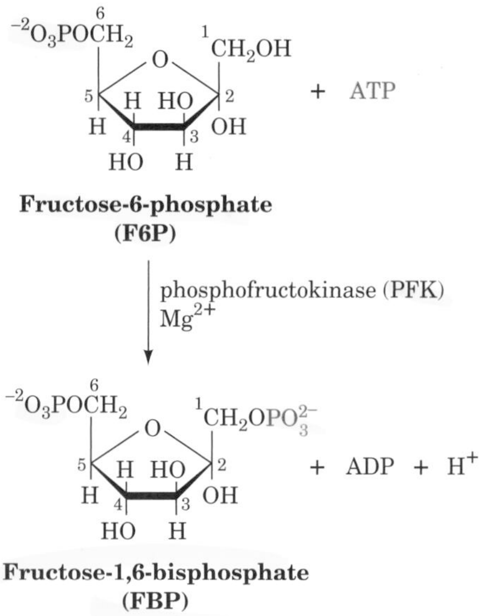

Fructose-6-phosphate (F6P)

phosphofructokinase (PFK)
Mg^{2+}

Fructose-1,6-bisphosphate (FBP)

This reaction is similar to the hexokinase reaction (Reaction 1 in Fig. 17-3; Section 17-2A). PFK catalyzes the nucleophilic attack by the C1—OH group of F6P on the electrophilic γ-phosphorus atom of the Mg^{2+}–ATP complex.

PFK plays a central role in the control of glycolysis because it catalyzes one of the pathway's rate-determining reactions. In many organisms the activity of PFK is enhanced allosterically by several substances, including AMP, and inhibited allosterically by several other substances, including ATP and citrate. The control of PFK is exquisitely complex; the mechanism by which it regulates the glycolytic pathway is examined in Section 17-4F.

FIGURE 17-7   Base-catalyzed isomerization of glucose, mannose, and fructose. In the absence of enzyme, this reaction is nonstereospecific.

D. *Aldolase*

Aldolase catalyzes Reaction 4 of glycolysis, the cleavage of FBP to form the two trioses **glyceraldehyde-3-phosphate (GAP)** and **dihydroxyacetone phosphate (DHAP)**:

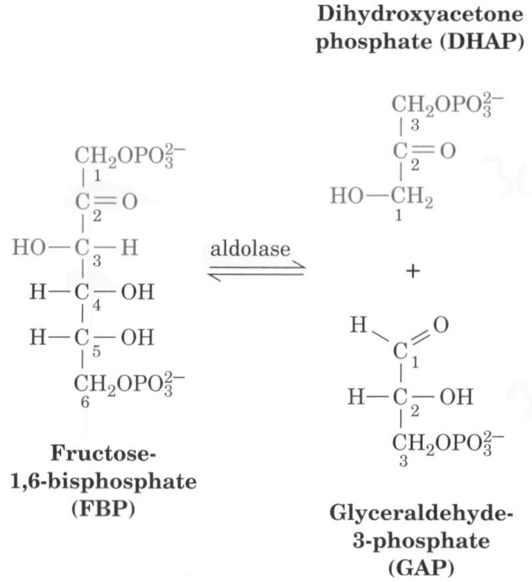

This reaction is an **aldol cleavage (retro aldol condensation)** whose nonenzymatic base-catalyzed mechanism is

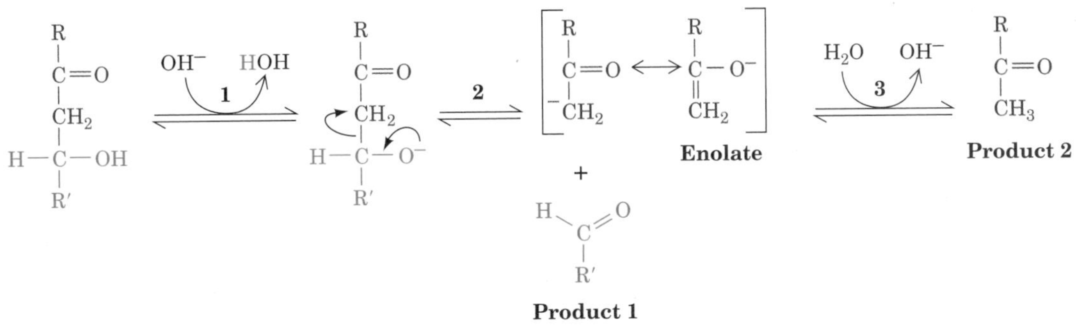

FIGURE 17-8 **Mechanism for base-catalyzed aldol cleavage.** Aldol condensation occurs by the reverse mechanism.

shown in Fig. 17-8. Note that aldol cleavage between C3 and C4 of FBP requires a carbonyl at C2 and a hydroxyl at C4. Hence, the "logic" of Reaction 2 in the glycolytic pathway, the isomerization of G6P to F6P, is clear. Aldol cleavage of G6P would have resulted in products of unequal carbon chain length, while aldol cleavage of FBP results in two interconvertible C3 compounds that can therefore enter a common degradative pathway. The enolate intermediate in the aldol cleavage reaction is stabilized by resonance, as shown, as a result of the electron-withdrawing character of the carbonyl oxygen atom.

Note that at this point in the pathway the atom numbering system changes. Atoms 1, 2, and 3 of glucose become atoms 3, 2, and 1 of DHAP, thus reversing order. Atoms 4, 5, and 6 become atoms 1, 2, and 3 of GAP (Fig. 17-3).

a. There Are Two Mechanistic Classes of Aldolases

Aldol cleavage is catalyzed by stabilizing its enolate intermediate through increased electron delocalization. There are two types of aldolases that are classified according to the chemistry they employ to stabilize the enolate. In Class I aldolases, which occur in animals and plants, the reaction occurs as follows (Fig. 17-9):

Step 1 Substrate binding.

Step 2 Reaction of the FBP carbonyl group with the ε-amino group of the active site Lys 226 to form an iminium cation, that is, a protonated Schiff base.

Step 3 C3—C4 bond cleavage resulting in enamine formation and the release of GAP. The iminium ion, as we saw in Section 16-2E, is a better electron-withdrawing group than is the oxygen atom of the precursor carbonyl group. Thus, catalysis occurs because the enamine intermediate (Fig. 17-9, Step 3) is more stable than the corresponding enolate intermediate of the base-catalyzed aldol cleavage reaction (Fig. 17-8, Step 2).

Step 4 Protonation of the enamine to an iminium cation.

Step 5 Hydrolysis of this iminium cation to release DHAP, with regeneration of the free enzyme.

Proof for the formation of the Schiff base was provided by "trapping" ^{14}C-labeled DHAP on the enzyme by reacting it with with NaBH$_4$, which reduces imines to amines:

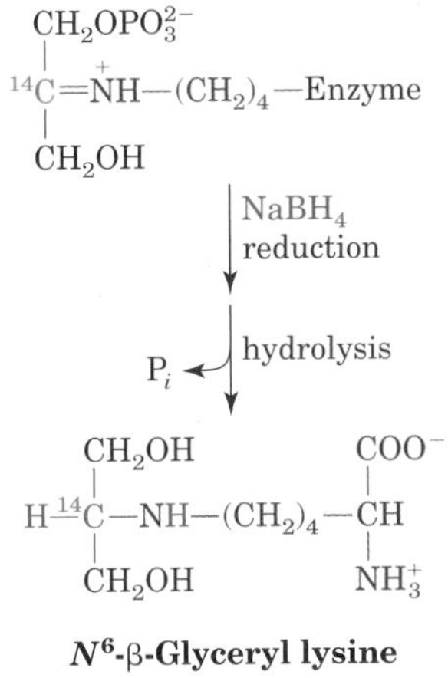

$$CH_2OPO_3^{2-}$$
$$^{14}C=\overset{+}{N}H-(CH_2)_4-Enzyme$$
$$CH_2OH$$

$$\downarrow \text{NaBH}_4 \text{ reduction}$$

$$P_i \longleftarrow \downarrow \text{hydrolysis}$$

$$\begin{array}{cc} CH_2OH & COO^- \\ H-^{14}C-NH-(CH_2)_4-CH \\ CH_2OH & NH_3^+ \end{array}$$

N^6-β-Glyceryl lysine

The radioactive product was hydrolyzed and identified as **N^6-β-glyceryl lysine.**

Cys and His residues were initially thought to act as the acid and base catalysts that facilitate the proton transfers in the aldolase reaction because the appropriate group-specific reagents inactivate the enzyme by reacting with these residues. For example, the reaction of a specific Cys residue of aldolase with iodoacetic acid inactivates the enzyme and results in the buildup of the FBP observed in the early glycolysis inhibition studies (Section 17-1A). However, site-directed mutagenesis to Ala of the Cys residue thought to be involved in the catalytic activity results in no loss of enzymatic function. Modification of this Cys residue is now thought to prevent the conformational changes required for productive substrate binding.

An early X-ray structure of aldolase suggested that a Tyr side chain was positioned to act as the active site acid–base catalyst and that the His was instead necessary for the maintenance of the Tyr's catalytically active orientation. A re-examination of the X-ray data caused yet another modification of the mechanism. The Tyr originally seen at the

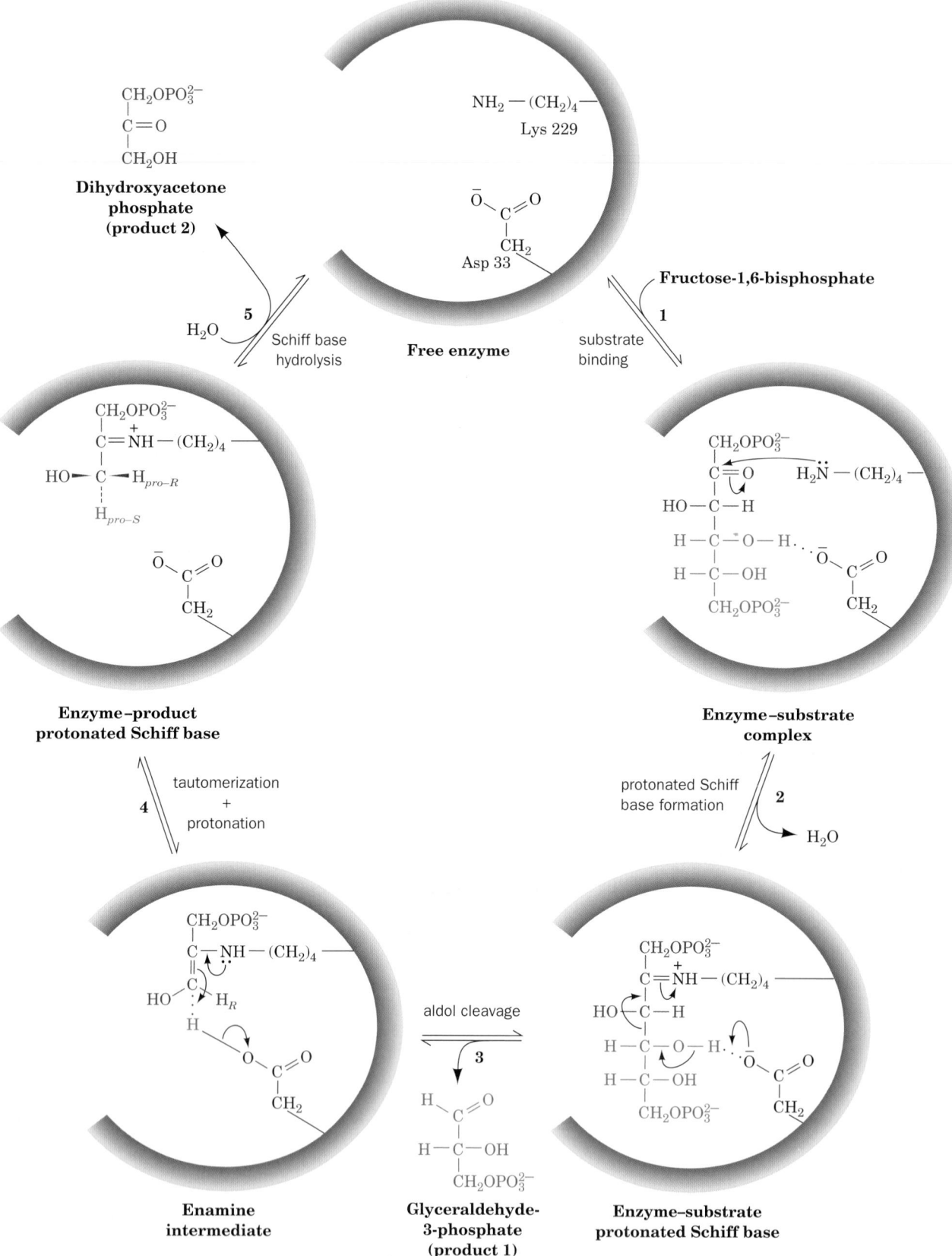

FIGURE 17-9 Enzymatic mechanism of Class I aldolase. The reaction involves **(1)** substrate binding; **(2)** Schiff base formation between the enzyme's active site Lys residue and FBP; **(3)** aldol cleavage to form an enamine intermediate of the enzyme and DHAP with release of GAP (shown with its *re* face up); **(4)** tautomerization and protonation to the iminium form of the Schiff base; and **(5)** hydrolysis of the Schiff base with release of DHAP.

active site has changed position in this new analysis, so that it is out of reach of the active site. Asp 33 and Lys 146 now appear to be acting as acid–base catalysts. These residues are evolutionarily conserved and their mutagenesis eliminates enzyme activity. This is an excellent example of the caution that must be exercised in the interpretation of chemical modification and structural data, and the power of site-directed mutagenesis in the study of enzyme mechanisms (although see Section 15-3B).

Class II aldolases, which occur in fungi, algae, and some bacteria, do not form a Schiff base with the substrate. Rather, a divalent cation, usually Zn^{2+} or Fe^{2+}, polarizes the carbonyl oxygen of the substrate to stabilize the enolate intermediate of the reaction (Fig. 16-12d):

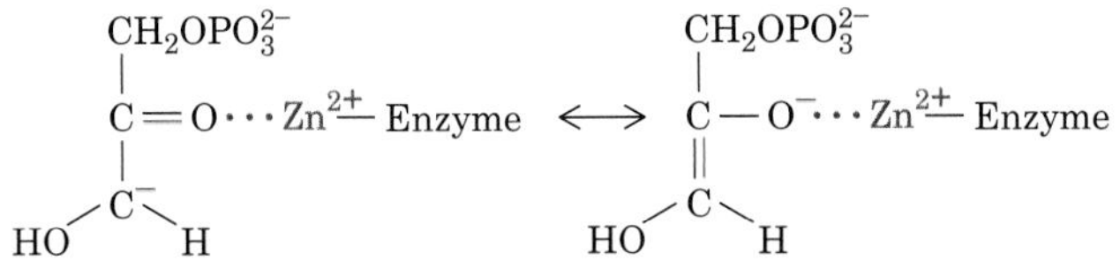

Are the two classes of aldolases related? Although both classes exhibit the Uni Bi kinetics implicit in their mechanisms, they exhibit only ~15% sequence identity, which places them in the lower end of the twilight zone for establishing homology (Section 7-4B). Nevertheless, their X-ray structures reveal that they have the same fold, the α/β barrel. The evolution of this particularly common fold is discussed in Section 8-3B.

b. Why Two Classes of Aldolase?

Since glycolysis presumably arose very early in evolutionary history, the existence of two classes of aldolase is unexpected. It had originally been postulated that, since Class I aldolases occur in higher organisms, Class II aldolases must be the more primitive enzyme form, that is, less metabolically capable than are the Class I enzymes. However, the discovery that some organisms simultaneously express both classes of aldolase suggests that both enzyme classes are evolutionarily ancient and equally adept at carrying out their metabolic functions. Thus, the expression of both classes of aldolase in some organisms probably represents an ancient metabolic redundancy that evolution has eliminated in most contemporary organisms. Whatever the reason for the occurrence of two classes of aldolase, the fact that Class II aldolases do not occur in mammals makes them an attractive target in the development of antibacterial drugs.

c. Aldolase Is Stereospecific

The aldolase reaction provides another example of the extraordinary stereospecificity of enzymes. In the nonenzymatic aldol condensation to form hexose-1,6-bisphosphate from DHAP and GAP, there are four possible products depending on whether the *pro-R* or *pro-S* hydrogen at C3 of DHAP is removed and whether the resulting carbanion attacks GAP on its *re* or its *si* face:

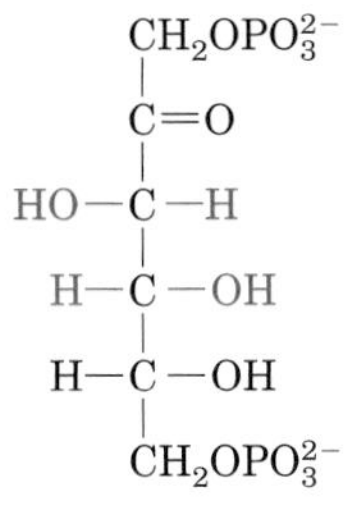

D-Fructose 1,6-bisphosphate, 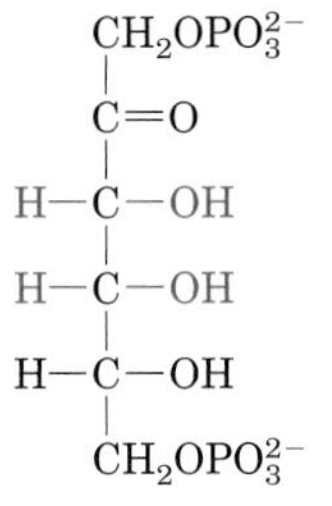 D-Psicose 1,6-bisphosphate, 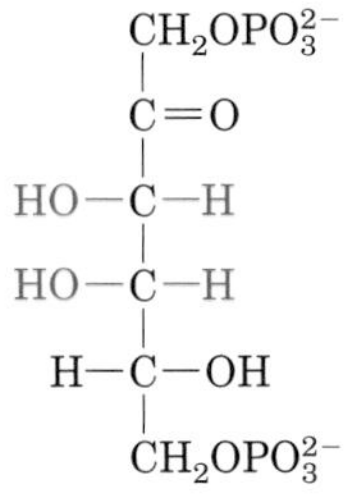 D-Tagatose 1,6-bisphosphate, 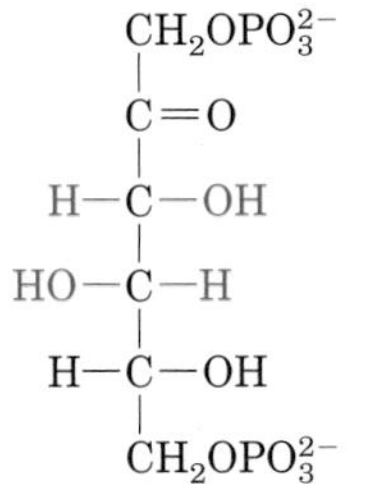 D-Sorbose 1,6-bisphosphate

In the enzymatic aldol condensation (Fig. 17-9 in reverse), carbanion formation from the enzyme–DHAP iminium ion (Fig. 17-9, Step 4 in reverse) occurs with removal of only the *pro-S* hydrogen. Attack of this carbanion occurs only on the *si* face of the enzyme-bound GAP carbonyl group, so that only FBP is formed (Fig. 17-9, Step 3 in reverse).

E. Triose Phosphate Isomerase

Only one of the products of the aldol cleavage reaction, GAP, continues along the glycolytic pathway (Fig. 17-3). However, DHAP and GAP are ketose–aldose isomers just as are F6P and G6P. Interconversion of GAP and DHAP therefore probably occurs via an enediol or enediolate intermediate in analogy with the phosphoglucose isomerase reaction (Fig. 17-6). **Triose phosphate isomerase (TIM or TPI**; Figs. 8-19b and 8-52) catalyzes this process in Reaction 5 of glycolysis, the final reaction of Stage I:

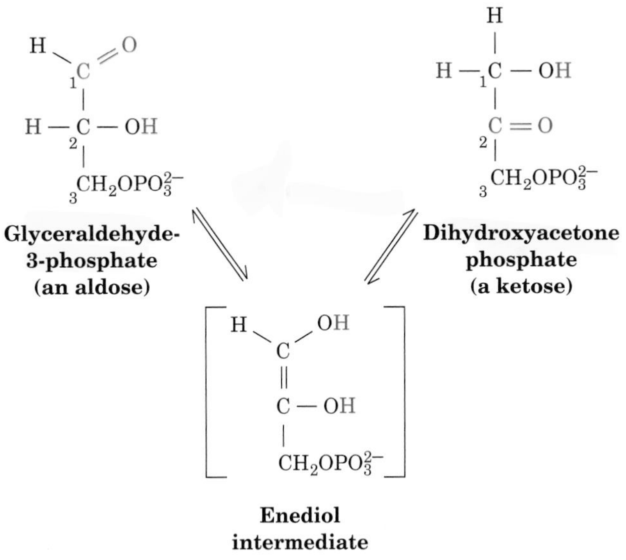

Glyceraldehyde-3-phosphate (an aldose), Dihydroxyacetone phosphate (a ketose), Enediol intermediate

Support for this reaction scheme comes from the use of the transition state analogs **phosphoglycohydroxamate** and **2-phosphoglycolate,** stable compounds whose geometric structures resemble that of the proposed enediol or enediolate intermediate:

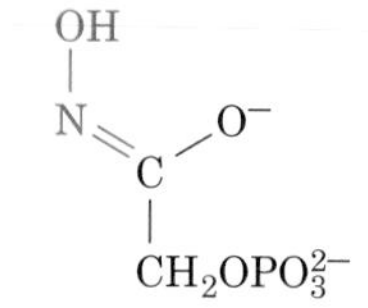

Phosphoglyco-hydroxamate

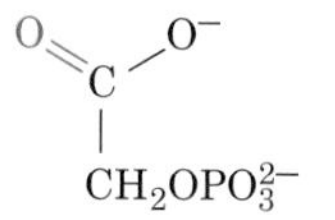

2-Phosphoglycolate

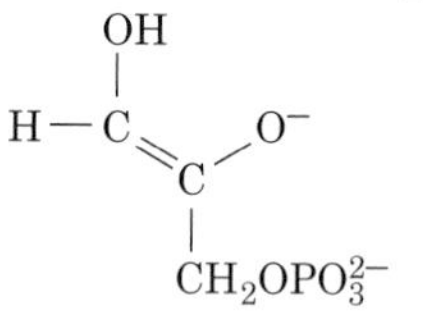

Proposed enediolate intermediate

Since enzymes catalyze reactions by binding the transition state complex more tightly than the substrate (Section 15-1F), phosphoglycohydroxamate and 2-phosphoglycolate should bind more tightly to TIM than does substrate. In fact, phosphoglycohydroxamate and 2-phosphoglycolate bind 155- and 100-fold more tightly to TIM than do either GAP or DHAP.

a. Glu 165 Functions as a General Base

The pH dependence of the TIM reaction is a bell-shaped curve with pK's of 6.5 and 9.5. The similarity of these pK's to the corresponding quantities of the phosphoglucose isomerase reaction suggests the participation of both an acid and a base in the TIM reaction as well. However, pH studies alone, as we have seen, are difficult to interpret in terms of specific amino acid residues since the active site environment may alter the pK of an acidic or basic group.

Affinity labeling reagents have been employed in an effort to identify the base at the active site of TIM. Both **bromohydroxyacetone phosphate** and **glycidol phosphate**

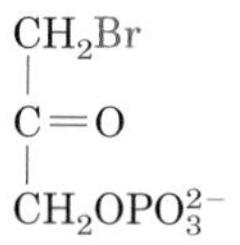

Bromohydroxyacetone phosphate

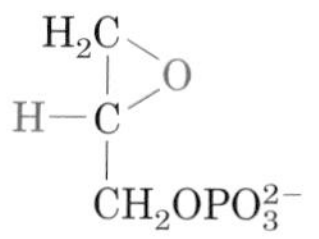

Glycidol phosphate

inactivate TIM by forming esters of Glu 165, whose carboxylate group, X-ray studies indicate, is ideally situated to abstract the C2 proton from the substrate (general base catalysis). In fact, the mutagenic replacement of Glu 165 by Asp, which X-ray studies show withdraws the carboxylate group only ~1 Å farther away from the sub-

strate than its position in the wild-type enzyme, reduces TIM's catalytic power ~1000-fold. Note that Glu 165's pK is drastically altered from the 4.1 value of the free amino acid to the observed 6.5 value. This provides yet another striking example of the effect of the environment on the properties of amino acid side chains.

b. The TIM Reaction Probably Occurs via Concerted General Acid–Base Catalysis Involving Low-Barrier Hydrogen Bonds

The X-ray structure of yeast TIM in complex with phosphoglycohydroxamate indicates that His 95 is hydrogen bonded to and hence is properly positioned to protonate the carbonyl oxygen atom of GAP (general acid catalysis):

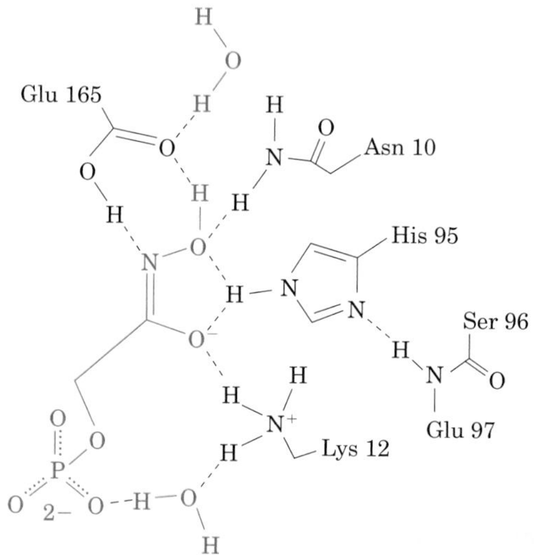

However, NMR studies indicate that His 95 is in its neutral imidazole form rather than its protonated imidazolium form. How can an imidazole N3—H group, which has a highly basic pK of ~14, protonate a carbonyl oxygen atom that, when protonated, has a very acidic pK of <0? Likewise, how can the Glu 165 carboxylate group (pK 6.5) abstract the C2 proton (pK ~17) from GAP? A plausible answer is that these proton shifts are facilitated by the formation of low-barrier hydrogen bonds (LBHBs). These unusually strong associations (-40 to -80 kJ · mol^{-1}, vs -12 to -30 kJ · mol^{-1} for normal hydrogen bonds), as we have seen in the case of the serine protease catalytic triad (Section 15-3D), form when the pK's of the hydrogen bonding donor and acceptor groups are nearly equal. They can be important contributors to rate enhancement if they form only in the transition state of an enzymatically catalyzed reaction.

In converting GAP to the enediol (or enediolate) intermediate (Fig. 17-10, *left*), the pK of the protonated form of its carbonyl oxygen, which becomes a hydroxyl group, increases to ~14, which closely matches that of neutral His 95. The resulting LBHB between this hydroxyl group and His 95 permits the neutral imidazole side chain to proto-

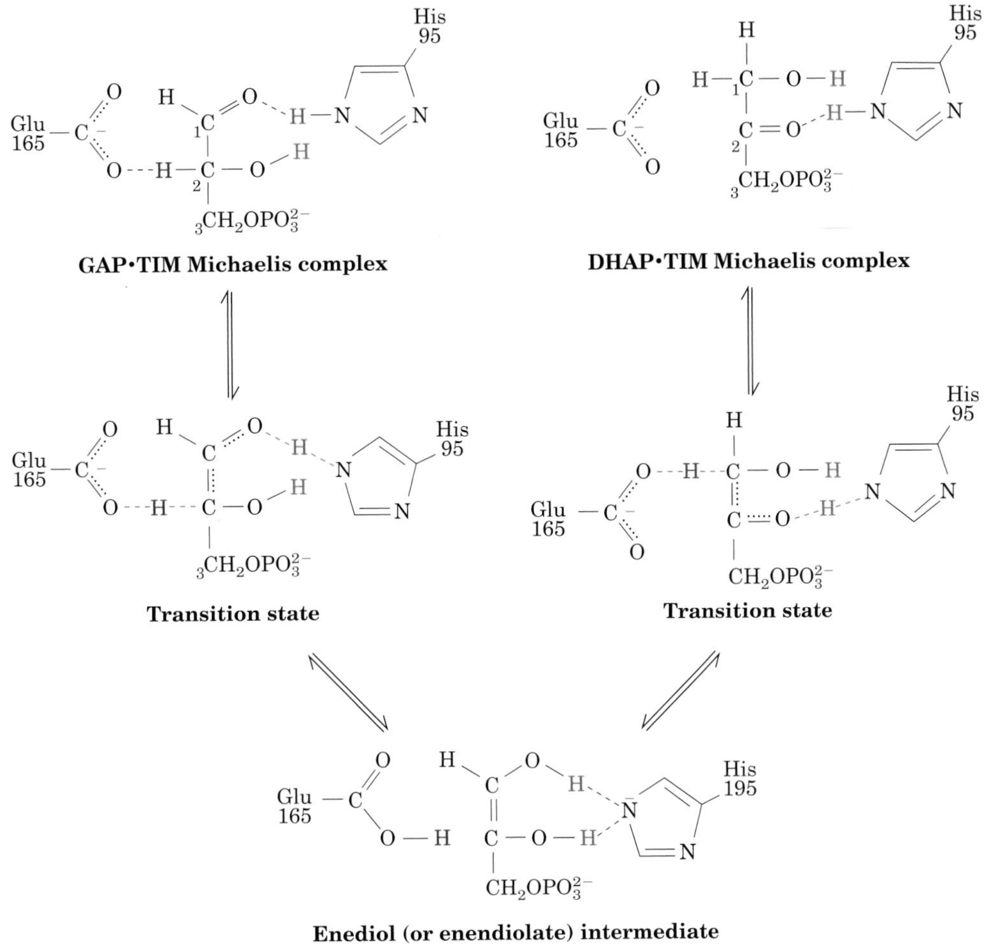

GAP·TIM Michaelis complex

DHAP·TIM Michaelis complex

Transition state

Transition state

Enediol (or enendiolate) intermediate

FIGURE 17-10 Proposed enzymatic mechanism of the TIM reaction. The reaction proceeds via the concerted abstraction of the C2—H proton of GAP by the carboxylate group of Glu 165 and the protonation of the GAP carbonyl oxygen atom by the imidazole group of His 95. The pK's of the corresponding donor and acceptor groups participating in each proton transfer process become nearly equal in the transition state and hence form low-barrier hydrogen bonds (*red dashed lines*), which act to stabilize the transition state. The resulting enediol (or possibly the electrostatically stabilized enediolate) intermediate then reacts in a similar fashion with the carboxyl group of Glu 165 protonating C1, while the deprotonated N3 atom of His 95 abstracts the proton on the 2-hydroxyl group to yield DHAP.

nate the oxygen atom. Likewise, as the carbonyl oxygen is protonated, the pK of the C2—H proton decreases to ~7, close to the pK of the Glu 165 carboxylate. It therefore appears that the reaction occurs via simultaneous proton abstraction by Glu 165 and protonation by His 95 (concerted general acid–base catalysis). The LBHBs postulated to form in the transition state, but not in the Michaelis complex, between Glu 165 and C2—H and between His 95 and the carbonyl oxygen atom are thought to provide some of the transition state stabilization necessary to catalyze the reaction. The positively charged side chain of Lys 12, which is probably responsible for the 9.5 pK observed in TIM's pH rate profile, is thought to electrostatically stabilize the negatively charged transition state. The conversion of the enediol(ate) intermediate to DHAP is likewise facilitated by the formation of transition state LBHBs (Fig. 17-10, *right*).

c. A Flexible Loop Both Preferentially Binds and Protects the Enediol Intermediate

The comparison of the X-ray structure of the TIM · phosphoglycohydroxamate complex with that of TIM alone reveals that a 10-residue loop, which is closed over the active site in the enzyme–substrate complex, is flipped up in the unoccupied active site like a hinged lid, a movement that involves main chain shifts of >7 Å (Fig. 17-11). A four-residue segment of this loop makes a hydrogen bond with the phosphate group of the substrate. Mutagenic excision of these four residues neither significantly distorts the protein nor greatly impairs substrate

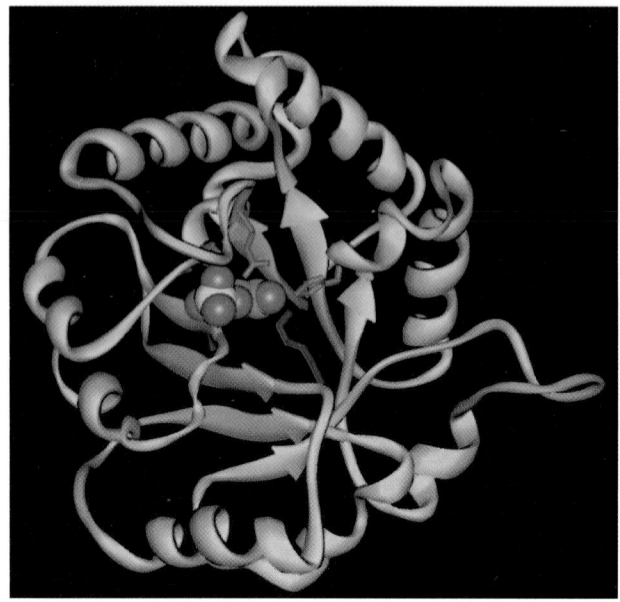

FIGURE 17-11 Ribbon diagram of yeast TIM in complex with its transition state analog 2-phosphoglycolate. A single 248-residue subunit of this homodimeric enzyme is viewed roughly along the axis of its α/β barrel. The enzyme's flexible loop, residues 168 through 177, is cyan and the side chains of Lys 12, His 95, and Glu 165 are blue, magenta, and red, respectively. The 2-phosphoglycolate is shown in space-filling form colored according to atom type (C green, O red, P yellow). [Based on an X-ray structure by Gregory Petsko, Brandeis University. PDBid 2YPI.] 🔎 See the Interactive Exercises and Kinemage Exercises 12-1 and 12-2

binding. The catalytic power of the mutant enzyme is, nevertheless, reduced 10^5-fold and it only weakly binds phosphoglycohydroxamate. Evidently, the closed loop preferentially stabilizes the enzymatic reaction's enediol-like transition state.

The closed loop conformation in the TIM reaction provides a striking example of the stereoelectronic control that

enzymes can exert on a reaction (Section 15-1E). In solution, the enediol intermediate readily breaks down with the elimination of the phosphate at C3 to form the toxic compound **methylglyoxal** (Fig. 17-12a). On the enzyme's surface, however, this reaction is prevented because the phosphate group is held by the flexible loop in the plane of the enediol, a position that disfavors phosphate elimi-

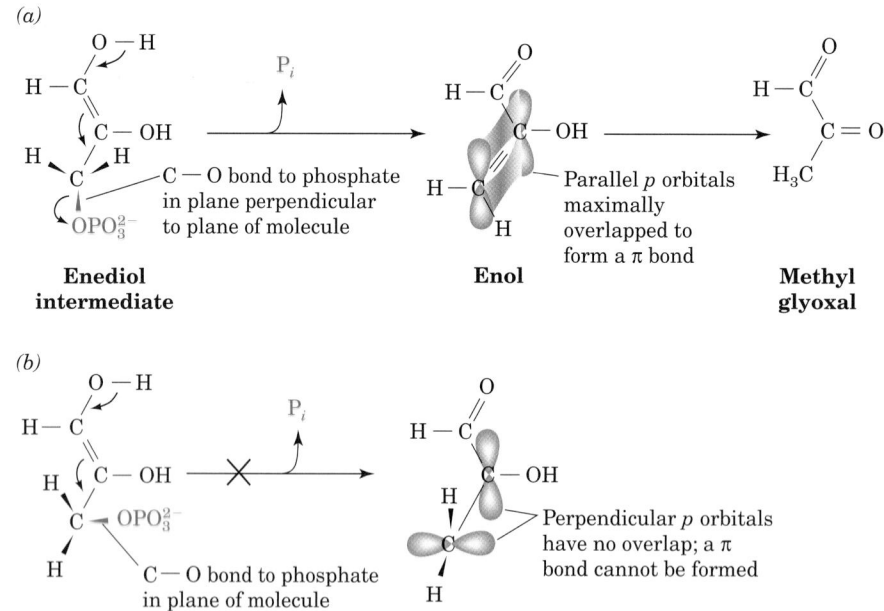

FIGURE 17-12 The spontaneous decomposition of the enediol intermediate in the TIM reaction to form methylglyoxal through the elimination of a phosphate group. (a) This reaction can occur only when the C—O bond to the phosphate group lies in a plane that is nearly perpendicular to that of the enediol so as to permit the formation of a double bond in the intermediate enol product. (b) When the C—O bond to the

phosphate group lies in a plane that is nearly parallel to that of the enediol, the p orbitals on the resulting intermediate product would be perpendicular to each other and hence lack the overlap necessary to form a π bond, that is, a double bond. The resulting unsatisfied bonding capacity greatly increases the energy of the reaction intermediate and hence makes the reaction highly unfavorable.

nation. In order for this elimination to occur, the C—O bond to the phosphate group must lie, as shown in Fig. 17-12a, in the plane perpendicular to that of the enediol. This is because, if the phosphate group were to be eliminated while this C—O bond was in the plane of the enediol as diagrammed in Fig. 17-12b, the CH_2 group of the resulting enol product would be twisted 90° out of the plane of the rest of the molecule. Such a conformation is energetically prohibitive because it prevents the formation of the enol's double bond by eliminating the overlap between its component *p* orbitals. In the mutant enzyme lacking the flexible loop, the enediol is able to escape: ~85% of the enediol intermediate is released into solution, where it rapidly decomposes to methylglyoxal and P_i. Thus, flexible loop closure also ensures that substrate is efficiently transformed to product.

On the basis of the foregoing X-ray structures, it had been widely assumed that the binding of substrate to TIM is **ligand-gated,** that is, it induces loop closure. Yet, if this were the case, the reversibilty of the TIM reaction and the chemical resemblance of its reactant and product (GAP and DHAP) make it difficult to rationalize how product could be released. However, NMR measurements by John Williams and Ann McDermott reveal that, in fact, loop motion still occurs when TIM is binding either glycerol-3-phosphate (a substrate analog) or 2-phosphoglycolate (a transition state analog) and is sufficiently fast (on a time scale of 100 μs) to account for the catalytic reaction rate (a turnover time of 230 μs). This is a clear example of how complementary information supplied by X-ray and NMR methods has yielded important insights into an enzymatic mechanism that neither technique alone could have provided.

d. TIM Is a Perfect Enzyme

TIM, as Jeremy Knowles demonstrated, has achieved catalytic perfection in that the rate of bimolecular reaction between enzyme and substrate is diffusion controlled; that is, product formation occurs as rapidly as enzyme and substrate can collide in solution, so that any increase in TIM's catalytic efficiency would not increase the reaction rate (Section 14-2B). Because of the high interconversion efficiency of GAP and DHAP, these two metabolites are maintained in equilibrium: $K = [GAP]/[DHAP] = 4.73 \times 10^{-2}$; that is, [DHAP] >> [GAP] at equilibrium. However, *as GAP is utilized in the succeeding reaction of the glycolytic pathway, more DHAP is converted to GAP, so that these compounds maintain their equilibrium ratio.* One common pathway therefore accounts for the metabolism of both products of the aldolase reaction.

Let us now take stock of where we are in our travels down the glycolytic pathway. At this stage, the glucose, which has been transformed into two GAPs, has completed the preparatory stage of glycolysis. This process has required the expenditure of two ATPs. However, this investment has resulted in the conversion of one glucose to two C_3 units, each of which has a phosphoryl group that, with a little chemical artistry, can be converted to a "high-energy" compound (Section 16-4B) whose free energy of hydrolysis can be coupled to ATP synthesis. *This energy investment is doubly repaid in the final stage of glycolysis in which the two phosphorylated C_3 units are transformed to two pyruvates with the coupled synthesis of four ATPs per glucose.*

F. *Glyceraldehyde-3-Phosphate Dehydrogenase: First "High-Energy" Intermediate Formation*

Reaction 6 of glycolysis involves the oxidation and phosphorylation of GAP by NAD^+ and P_i as catalyzed by **glyceraldehyde-3-phosphate dehydrogenase (GAPDH;** Figs. 8-45 and 8-53b):

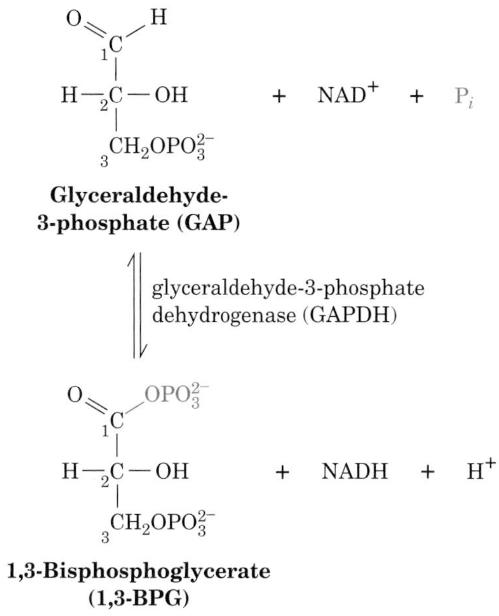

Glyceraldehyde-3-phosphate (GAP)

glyceraldehyde-3-phosphate dehydrogenase (GAPDH)

1,3-Bisphosphoglycerate (1,3-BPG)

This is the first instance of the chemical artistry alluded to above. *In this reaction, aldehyde oxidation, an exergonic reaction, drives the synthesis of the acyl phosphate **1,3-bisphosphoglycerate (1,3-BPG;** previously called **1,3-diphosphoglycerate).*** Recall that acyl phosphates are compounds with high phosphate group-transfer potential (Section 16-4B).

a. Mechanistic Studies

Several key enzymological experiments have contributed to the elucidation of the GAPDH reaction mechanism (Fig. 17-13):

1. GAPDH is inactivated by alkylation with stoichiometric amounts of iodoacetate. The presence of **carboxymethylcysteine** in the hydrolysate of the resulting alkylated enzyme (Fig. 17-13a) suggests that GAPDH has an active site Cys sulfhydryl group.

2. GAPDH quantitatively transfers ^{3}H from C1 of GAP to NAD^+ (Fig. 17-13b), thereby establishing that this reaction occurs via direct hydride transfer.

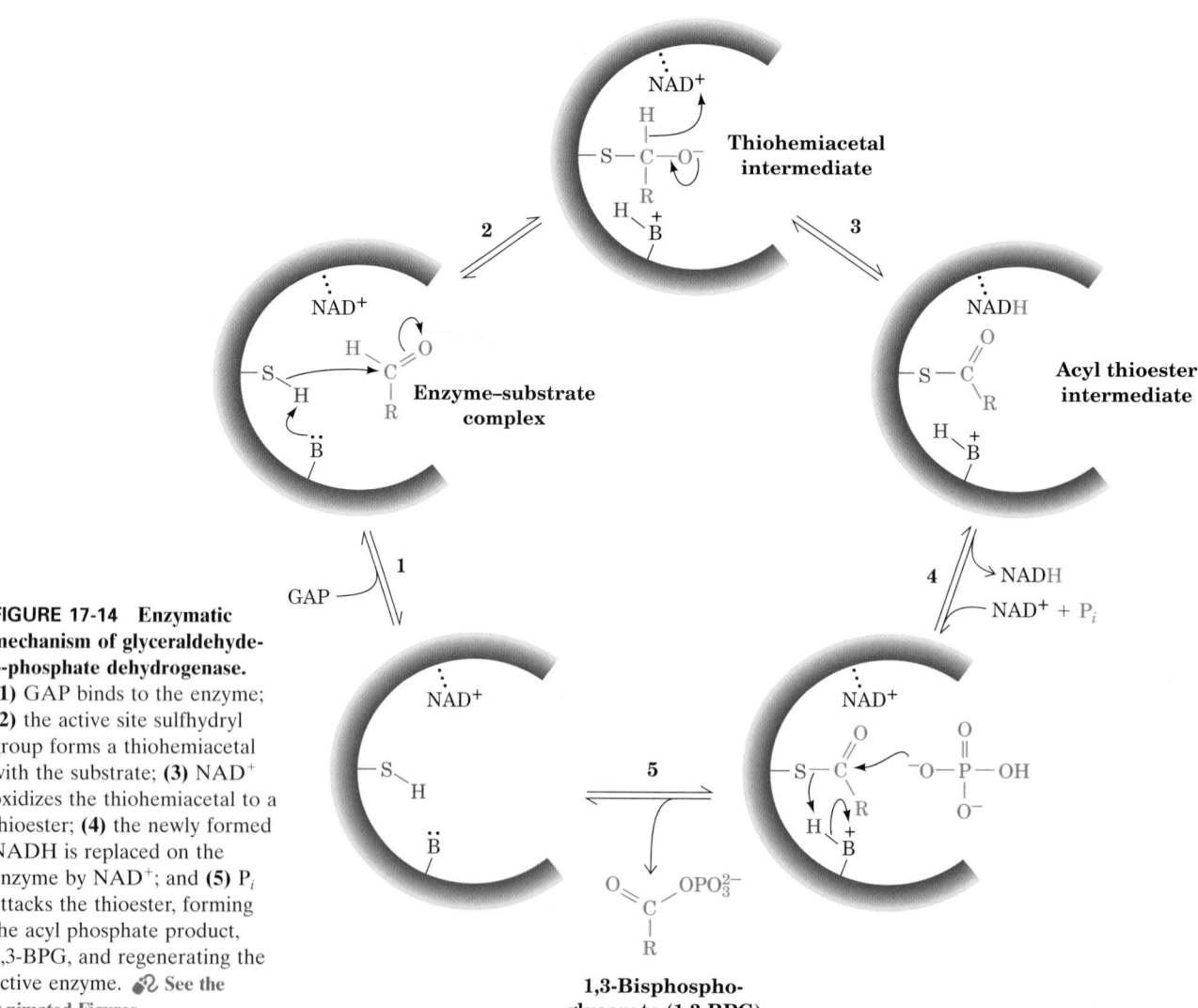

(a)

Enzyme—CH$_2$—SH + ICH$_2$COO$^-$ → Enzyme—CH$_2$—S—CH$_2$COO$^-$ —protein hydrolysis→

$$\overset{NH_3^+}{\underset{COO^-}{CH}}—CH_2—S—CH_2COO^- + \text{Other amino acids}$$

GAPDH   Active site   Iodoacetate
**   Cys**

Carboxy-methylcysteine

(b)

[1-^{3}H]GAP + NAD$^+$ + P$_i$ —GAPDH→ 1,3-Bisphosphoglycerate (1,3-BPG) + NAD^{3}H

[1-^{3}H]GAP

1,3-Bisphosphoglycerate (1,3-BPG)

(c)

Acetyl phosphate + phosphate —GAPDH→ products

Acetyl phosphate

FIGURE 17-13 Some reactions employed in elucidating the enzymatic mechanism of GAPDH. (*a*) The reaction of iodoacetate with an active site Cys residue. (*b*) Quantitative tritium transfer from substrate to NAD$^+$. (*c*) The enzyme-catalyzed exchange of ^{32}P from phosphate to acetyl phosphate.

FIGURE 17-14 Enzymatic mechanism of glyceraldehyde-3-phosphate dehydrogenase. (**1**) GAP binds to the enzyme; (**2**) the active site sulfhydryl group forms a thiohemiacetal with the substrate; (**3**) NAD$^+$ oxidizes the thiohemiacetal to a thioester; (**4**) the newly formed NADH is replaced on the enzyme by NAD$^+$; and (**5**) P$_i$ attacks the thioester, forming the acyl phosphate product, 1,3-BPG, and regenerating the active enzyme. 🔁 See the **Animated Figures**

1,3-Bisphospho-glycerate (1,3-BPG)

3. GAPDH catalyzes exchange of ^{32}P between $[^{32}P]P_i$ and the product analog **acetyl phosphate** (Fig. 17-13c). Such isotope exchange reactions are indicative of an acyl–enzyme intermediate (Section 14-5D).

David Trentham has proposed a mechanism for GAPDH based on this information and the results of kinetic studies (Fig. 17-14):

Step 1 GAP binds to the enzyme.

Step 2 The essential sulfhydryl group, acting as a nucleophile, attacks the aldehyde to form a **thiohemiacetal.**

Step 3 The thiohemiacetal undergoes oxidation to an **acyl thioester** by direct transfer of a hydride to NAD^+. This intermediate, which has been isolated, has a high group-transfer potential. *The energy of aldehyde oxidation has not been dissipated but has been conserved through the synthesis of the thioester and the reduction of NAD^+ to NADH.*

Step 4 Another molecule of NAD^+ replaces NADH.

Step 5 The thioester intermediate undergoes nucleophilic attack by P_i to regenerate free enzyme and form 1,3-BPG. This "high-energy" mixed anhydride generates ATP from ADP in the next reaction of glycolysis.

G. *Phosphoglycerate Kinase: First ATP Generation*

Reaction 7 of the glycolytic pathway results in the first formation of ATP together with **3-phosphoglycerate (3PG)** in a reaction catalyzed by **phosphoglycerate kinase (PGK):**

1,3-Bisphosphoglycerate (1,3-BPG)

phosphoglycerate kinase (PGK)

3-Phosphoglycerate (3PG)

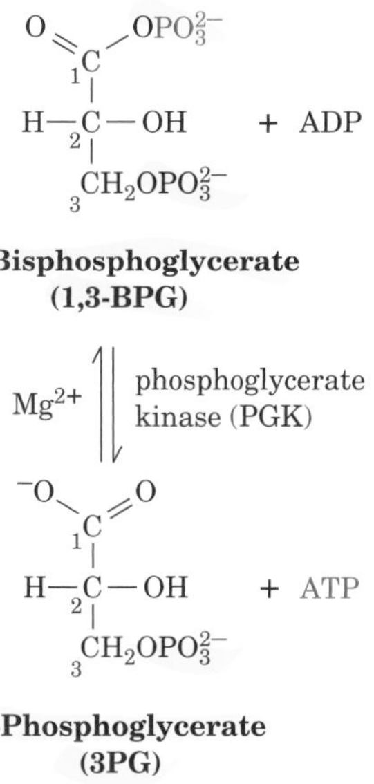

(*Note:* The name "kinase" is given to any enzyme that transfers a phosphoryl group between ATP and a metabolite. Nothing is implied as to the exergonic direction of transfer.)

PGK (Fig. 17-15) is conspicuously bilobal in appearance. The Mg^{2+}–ADP binding site is located on one domain, ~10 Å from the 1,3-BPG binding site, which is on

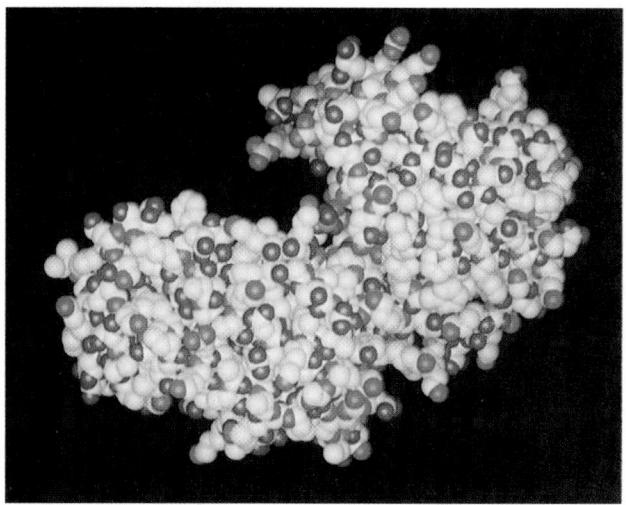

FIGURE 17-15 Space-filling model of yeast phosphoglycerate kinase showing its deeply clefted bilobal structure. The substrate binding site is at the bottom of the cleft as marked by the P atom (*magenta*) of 3PG. Compare this structure with that of hexokinase (Fig. 16-5a). [Based on an X-ray structure by Herman Watson, University of Bristol, U.K. PDBid 3PGK.]

the other domain. X-Ray structures indicate that, *on substrate binding, the two domains of PGK swing together so as to permit the substrates to react in a water-free environment as occurs with hexokinase (Section 17-2A).* Indeed, the appearance of PGK is remarkably similar to that of hexokinase (Fig. 17-5a), even though the structures of these proteins are otherwise unrelated.

Figure 17-16 indicates a reaction mechanism for PGK that is consistent with its observed sequential kinetics. The terminal phosphoryl oxygen of ADP nucleophilically attacks the C1 phosphorus atom of 1,3-BPG to form the reaction product.

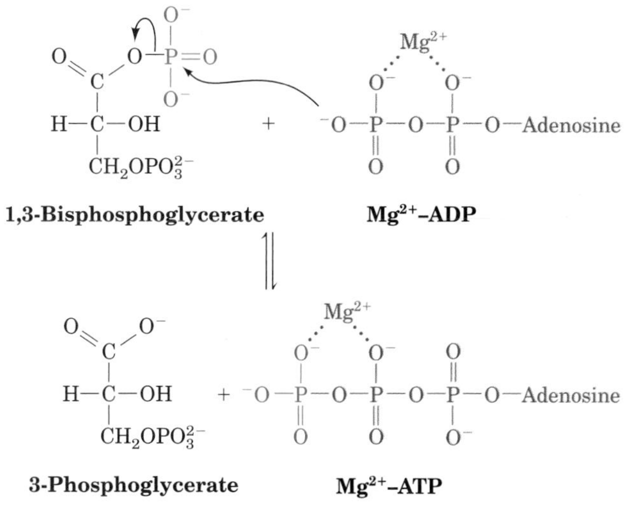

1,3-Bisphosphoglycerate　　　**Mg^{2+}–ADP**

3-Phosphoglycerate　　　**Mg^{2+}–ATP**

FIGURE 17-16 Mechanism of the PGK reaction. The Mg^{2+} positions are shown as examples; their actual binding sites are unknown.

The energetics of the overall GAPDH–PGK reaction pair are

$$GAP + P_i + NAD^+ \rightarrow 1,3\text{-}BPG + NADH$$
$$\Delta G^{\circ\prime} = +6.7 \text{ kJ} \cdot \text{mol}^{-1}$$

$$1,3\text{-}BPG + ADP \rightarrow 3PG + ATP$$
$$\Delta G^{\circ\prime} = -18.8 \text{ kJ} \cdot \text{mol}^{-1}$$

$$GAP + P_i + NAD^+ + ADP \rightarrow 3PG + NADH + ATP$$
$$\Delta G^{\circ\prime} = -12.1 \text{ kJ} \cdot \text{mol}^{-1}$$

Although the GAPDH reaction is endergonic, the strongly exergonic nature of the transfer of a phosphoryl group from 1,3-BPG to ADP makes the overall synthesis of NADH and ATP from GAP, P_i, NAD^+, and ADP favorable.

H. *Phosphoglycerate Mutase*

In Reaction 8 of glycolysis, 3PG is converted to **2-phosphoglycerate (2PG)** by **phosphoglycerate mutase (PGM):**

**3-Phosphoglycerate
(3PG)** ⇌ (phosphoglycerate mutase (PGM)) **2-Phosphoglycerate
(2PG)**

A **mutase** catalyzes the transfer of a functional group from one position to another on a molecule. This reaction is a necessary preparation for the next reaction in glycolysis, which generates a "high-energy" phosphoryl compound for use in ATP synthesis.

a. Reaction Mechanism of PGM

At first sight, the reaction catalyzed by PGM appears to be a simple intramolecular phosphoryl transfer. This is not the case, however. *The active enzyme has a phosphoryl group at its active site, which it transfers to the substrate to form a bisphospho intermediate. This intermediate then rephosphorylates the enzyme to form the product and regenerate the active phosphoenzyme.* The following experimental data permitted the elucidation of PGM's enzymatic mechanism:

1. Catalytic amounts of **2,3-bisphosphoglycerate (2,3-BPG;** previously known as **2,3-diphosphoglycerate)**

**2,3-Bisphosphoglycerate
(2,3-BPG)**

are required for enzymatic activity; that is, 2,3-BPG acts as a reaction primer.

2. Incubation of the enzyme with catalytic amounts of ^{32}P-labeled 2,3-BPG yields a ^{32}P-labeled enzyme. Zelda Rose demonstrated that this was a result of the phosphorylation of a His residue:

Phospho-His residue

3. The enzyme's X-ray structure shows His at the active site (Fig. 17-17). In the active enzyme, His 8 is phosphorylated.

These data are consistent with a mechanism in which the active enzyme contains a phospho-His residue at the active site (Fig. 17-18):

Step 1 3PG binds to the phosphoenzyme in which His 8 is phosphorylated.

Step 2 This phosphoryl group is transferred to the substrate, resulting in an intermediate 2,3-BPG · enzyme complex.

Steps 3 and 4 The complex decomposes to form the product 2PG with regeneration of the phosphoenzyme.

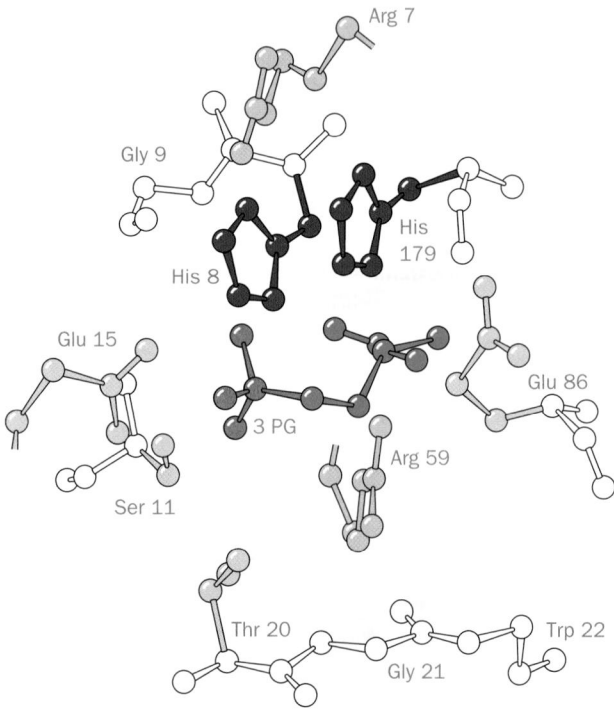

FIGURE 17-17 The active site region of yeast phosphoglycerate mutase (dephospho form) showing the substrate, 3-phosphoglycerate, and some of the side chains that approach it. His 8 is phosphorylated in the active enzyme. [After Winn, S.I., Watson, H.I., Harkins, R.N., and Fothergill, L.A., *Phil. Trans. R. Soc. London Ser. B* **293,** 126 (1981). PDBid 3PGM.]

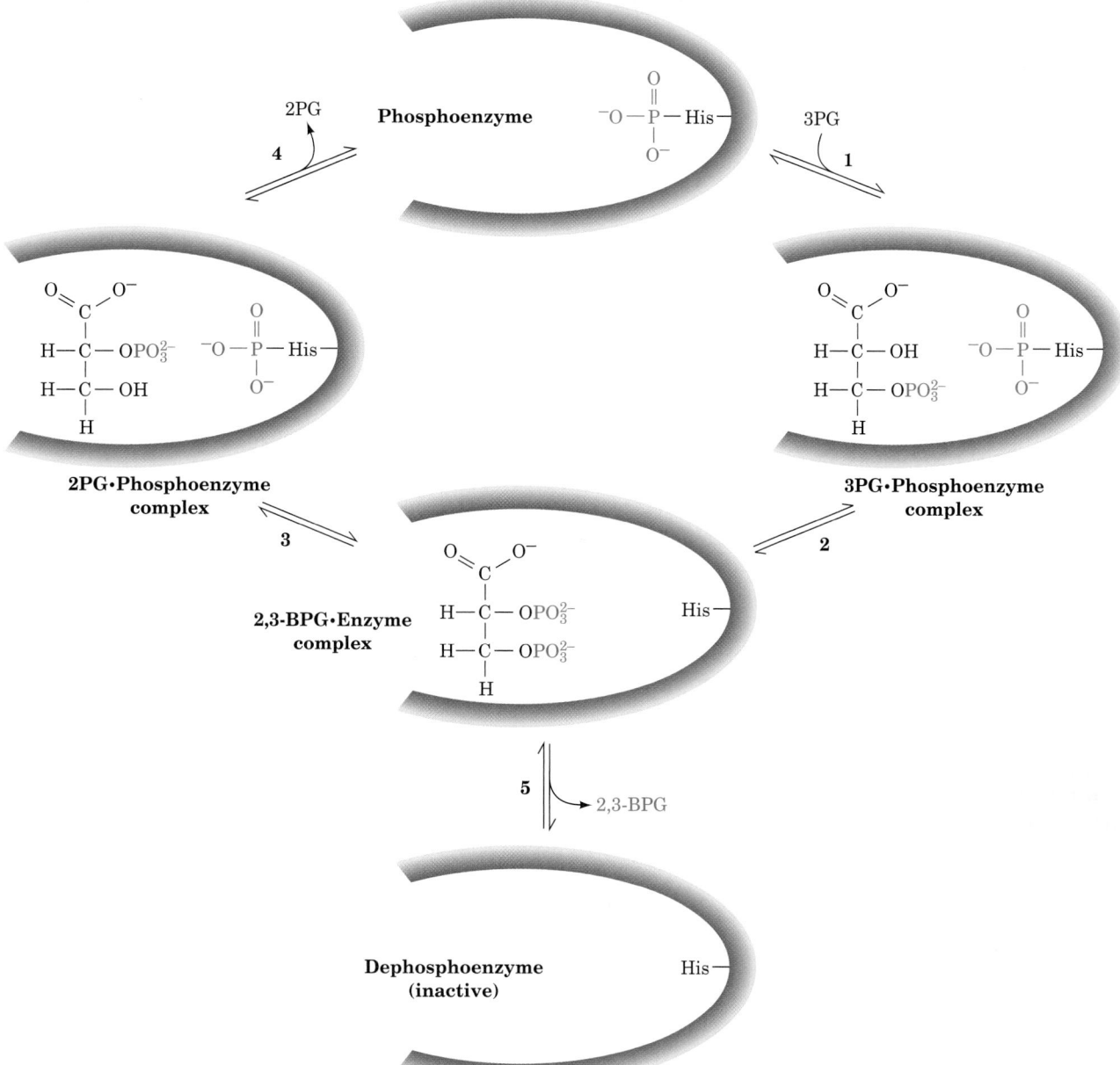

FIGURE 17-18 Proposed reaction mechanism for phosphoglycerate mutase. The active form of the enzyme contains a phospho-His residue at the active site. **(1)** Formation of an enzyme–substrate complex; **(2)** transfer of the enzyme-bound phosphoryl group to the substrate; **(3)** rephosphorylation of the enzyme by the other phosphoryl group of the substrate; and **(4)** release of product regenerating the active phosphoenzyme. **(5)** Occasionally, 2,3-BPG dissociates from the enzyme, leaving it in an inactive, dephospho form that must be rephosphorylated by the reverse reaction.

The phosphoryl group on 3PG therefore ends up on the C2 of the next 3PG to undergo reaction.

Occasionally, 2,3-BPG dissociates from the enzyme (Fig. 17-18; Step 5), leaving it in an inactive form. Trace amounts of 2,3-BPG must therefore always be available to regenerate the active phosphoenzyme by the reverse reaction.

b. Glycolysis Influences Oxygen Transport

2,3-BPG specifically binds to deoxyhemoglobin and thereby alters the oxygen affinity of hemoglobin (Section 10-1D). The concentration of 2,3-BPG in erythrocytes is much higher ($\sim$5 m*M*) than the trace amounts required for its use as a primer of PGM. Erythrocytes synthesize and degrade 2,3-BPG by a detour from the glycolytic pathway, diagrammed in Fig. 17-19. **Bisphosphoglycerate mutase** catalyzes the transfer of a phosphoryl group from C1 to C2 of 1,3-BPG. The resulting 2,3-BPG is hydrolyzed to 3PG by **2,3-bisphosphoglycerate phosphatase.** The rate of glycolysis affects the oxygen affinity of hemoglobin through the mediation of 2,3-BPG. Consequently, inherited defects

FIGURE 17-19   The pathway for the synthesis and degradation of 2,3-BPG in erythrocytes is a detour from the glycolytic pathway.

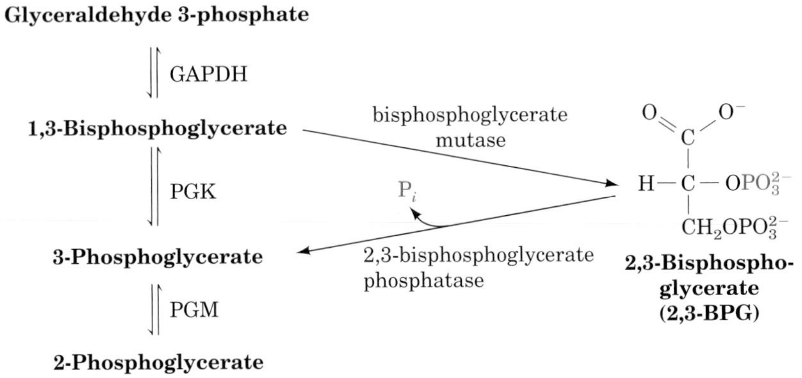

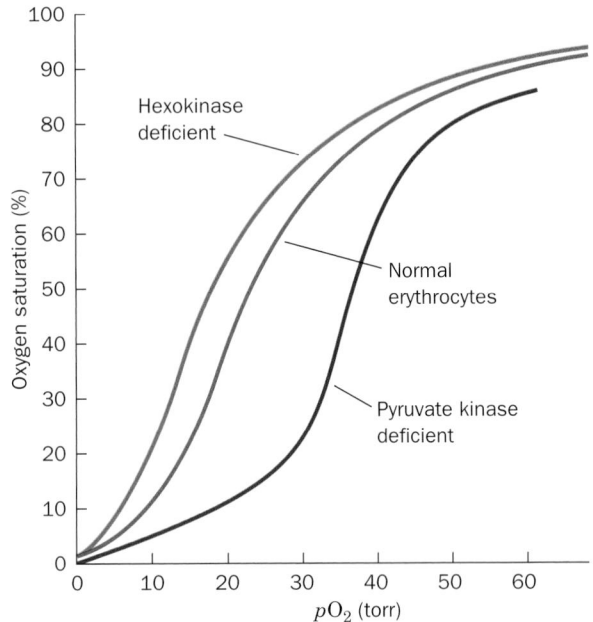

Glyceraldehyde 3-phosphate

GAPDH

1,3-Bisphosphoglycerate ⟶ bisphosphoglycerate mutase

PGK P_i

3-Phosphoglycerate ⟵ 2,3-bisphosphoglycerate phosphatase

PGM

2,3-Bisphospho-glycerate (2,3-BPG)

2-Phosphoglycerate

of glycolysis in erythrocytes alter the capacity of the blood to transport oxygen (Fig. 17-20). For example, the concentration of glycolytic intermediates in hexokinase-deficient erythrocytes is less than normal because hexokinase catalyzes the first reaction of glycolysis. This results in a diminished 2,3-BPG concentration and therefore in increased hemoglobin oxygen affinity. Conversely, pyruvate kinase deficiency decreases hemoglobin oxygen affinity through the increase of 2,3-BPG resulting from the blockade of the last reaction in glycolysis. Thus, although erythrocytes, which lack nuclei and other organelles, have but a minimal metabolism, this metabolism is physiologically significant.

I. *Enolase: Second "High-Energy" Intermediate Formation*

In Reaction 9 of glycolysis, 2PG is dehydrated to **phosphoenolpyruvate (PEP)** in a reaction catalyzed by **enolase:**

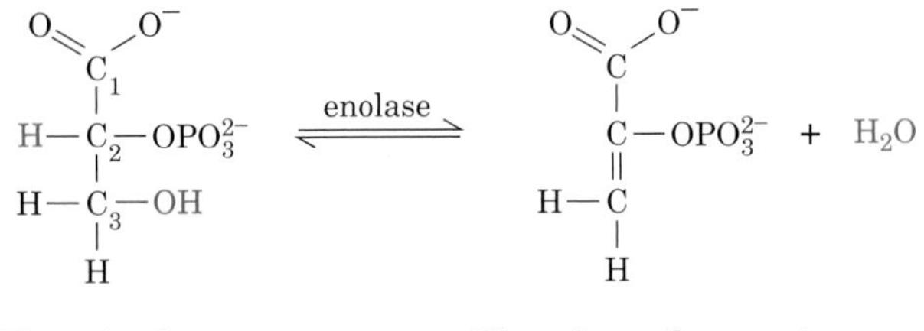

2-Phosphoglycerate (2PG) ⇌ [enolase] **Phosphoenolpyruvate (PEP)** + H_2O

The enzyme forms a complex with a divalent cation such as Mg^{2+} before the substrate is bound. A second divalent metal ion then binds to the enzyme. As is mentioned in Section 17-1A, fluoride ion inhibits glycolysis, resulting in the accumulation of 2PG and 3PG. It does so by strongly inhibiting enolase in the presence of P_i. F^- and P_i form a tightly bound complex with the Mg^{2+} at the enzyme's active site, blocking substrate binding and thereby inactivating the enzyme. Enolase's substrate, 2PG, therefore builds up and, as it does so, is equilibrated with 3PG by PGM.

a. Catalytic Mechanism of Enolase

The dehydration (elimination of H_2O) catalyzed by enolase might occur in one of three ways (Fig. 16-9a): (1) The —OH group at C3 can leave first, generating a carbocation

FIGURE 17-20   The oxygen-saturation curves of hemoglobin in normal erythrocytes (*red curve*) and those from patients with hexokinase deficiency (*green*) and with pyruvate kinase deficiency (*purple*). [After Delivoria-Papadopoulos, M., Oski, F.A., and Gottlieb, A.J., *Science* **165,** 601 (1969).]

at C3; (2) the C2 proton can leave first, generating a carbanion at C2; or (3) the reaction can be concerted. Isotope exchange studies by Paul Boyer demonstrated that the C2 proton of 2PG exchanges with solvent 12 times faster than the rate of PEP formation. However, the C3 oxygen exchanges with solvent at a rate roughly equivalent with the overall reaction rate. This suggests the following mechanism (Fig. 17-21):

Step 1 Rapid carbanion formation at C2 facilitated by a general base on the enzyme. The abstracted proton can readily exchange with the solvent, accounting for its observed rapid exchange rate.

Step 2 Rate-limiting elimination of the —OH group at C3. This is consistent with the slow rate of exchange of this hydroxyl group with solvent.

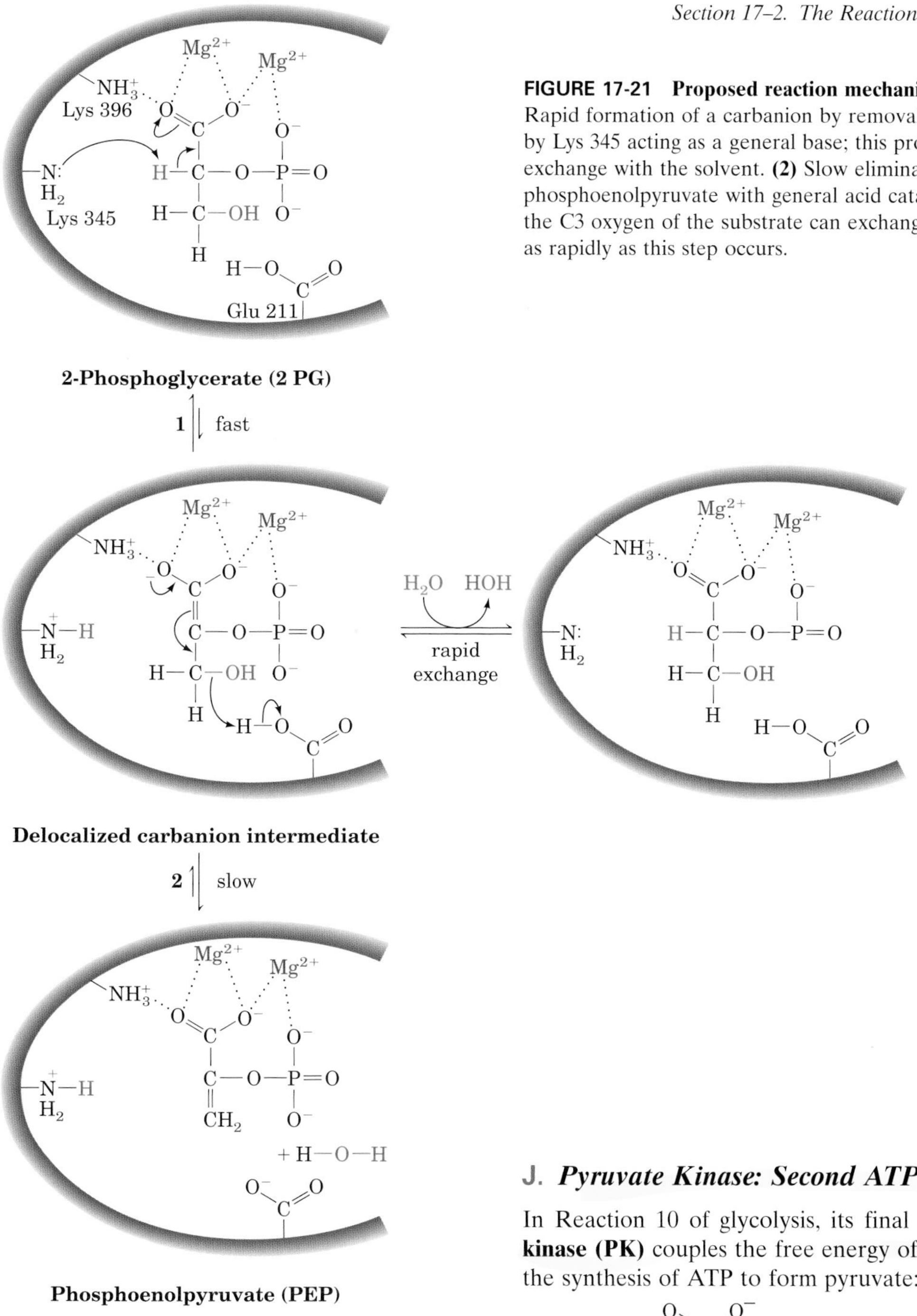

2-Phosphoglycerate (2 PG)

$$1 \Big\Updownarrow \text{ fast}$$

FIGURE 17-21 Proposed reaction mechanism of enolase. (1) Rapid formation of a carbanion by removal of a proton at C2 by Lys 345 acting as a general base; this proton can rapidly exchange with the solvent. **(2)** Slow elimination of H_2O to form phosphoenolpyruvate with general acid catalysis by Glu 211; the C3 oxygen of the substrate can exchange with solvent only as rapidly as this step occurs.

Delocalized carbanion intermediate

$$2 \Big\Updownarrow \text{ slow}$$

Phosphoenolpyruvate (PEP)

The enolase reaction (Fig. 17-21) is of mechanistic interest because it involves the abstraction of the decidedly nonacidic proton at C2 (pK > 30), followed by the elimination of an OH^- ion, which is a poor leaving group. The X-ray structure of yeast enolase in complex with two Mg^{2+} ions and an equilibrium mixture of 2PG and PEP (enolase's substrate and product), determined by George Reed and Ivan Rayment, reveals that enolase binds 2PG in an intricate complex that involves both Mg^{2+} ions. Mutagenic and enzymological studies indicate that the reaction involves the Lys 345 side chain functioning as a general base and the Glu 211 side chain functioning as a general acid. Lys 396 and the two Mg^{2+} ions are thought to stabilize the increased negative charge that develops on the carboxylate ion in the delocalized carbanion intermediate.

J. *Pyruvate Kinase: Second ATP Generation*

In Reaction 10 of glycolysis, its final reaction, **pyruvate kinase (PK)** couples the free energy of PEP hydrolysis to the synthesis of ATP to form pyruvate:

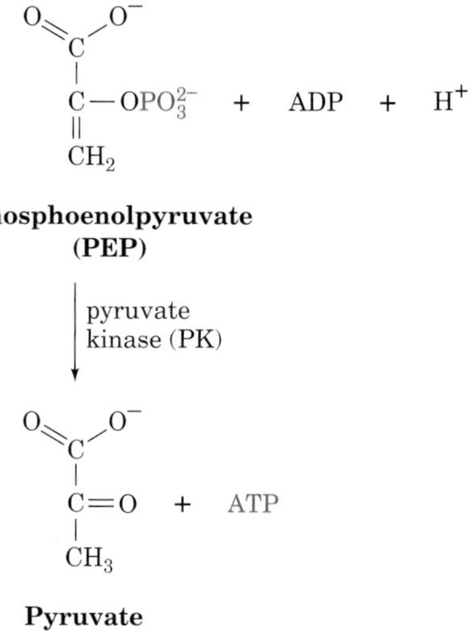

Phosphoenolpyruvate (PEP)

pyruvate kinase (PK)

Pyruvate

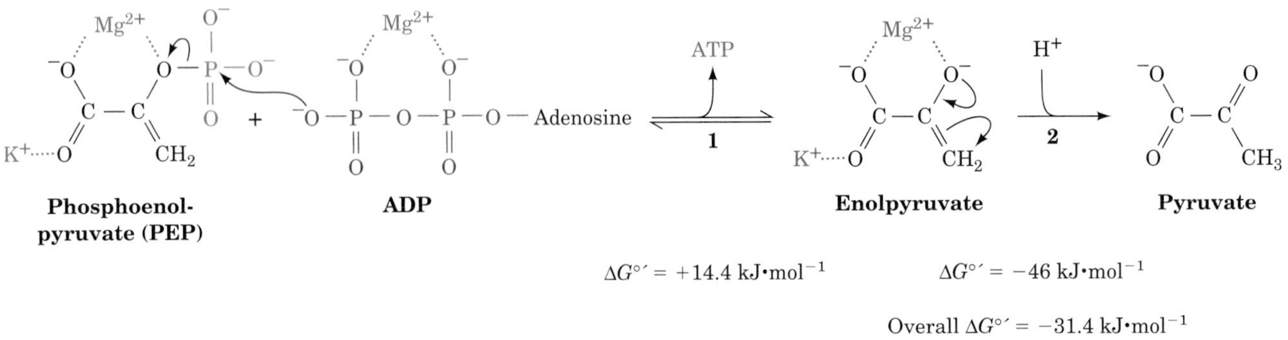

Phosphoenol-pyruvate (PEP) **ADP** **Enolpyruvate** **Pyruvate**

$\Delta G^{\circ\prime} = +14.4 \text{ kJ·mol}^{-1}$ $\Delta G^{\circ\prime} = -46 \text{ kJ·mol}^{-1}$

Overall $\Delta G^{\circ\prime} = -31.4 \text{ kJ·mol}^{-1}$

FIGURE 17-22 Mechanism of the reaction catalyzed by pyruvate kinase. (1) Nucleophilic attack of an ADP β-phosphoryl oxygen atom on the phosphorus atom of PEP to form ATP and enolpyruvate; and **(2)** tautomerization of enolpyruvate to pyruvate.

a. Catalytic Mechanism of PK

The PK reaction, which requires the participation of both monovalent (K^+) and divalent (Mg^{2+}) cations, occurs as follows (Fig. 17-22):

Step 1 A β-phosphoryl oxygen of ADP nucleophilically attacks the PEP phosphorus atom, thereby displacing enolpyruvate and forming ATP. This reaction conserves the free energy of PEP hydrolysis.

Step 2 Enolpyruvate converts to pyruvate. This enol–keto tautomerization is sufficiently exergonic to drive the coupled endergonic synthesis of ATP (Section 16-4C).

We can now see the "logic" of the enolase reaction. The standard free energy of hydrolysis of 2PG ($\Delta G^{\circ\prime}$) is only $-17.6 \text{ kJ·mol}^{-1}$, which is insufficient to drive ATP synthesis ($\Delta G^{\circ\prime} = 30.5 \text{ kJ·mol}^{-1}$ for ATP synthesis from ADP and P_i). The dehydration of 2PG results in the formation of a "high-energy" compound capable of such synthesis [the standard free energy of hydrolysis of PEP is $-61.9 \text{ kJ·mol}^{-1}$ (Fig. 16-23)]. In other words, PEP is a "high-energy" compound, 2PG is not.

3 ■ FERMENTATION: THE ANAEROBIC FATE OF PYRUVATE

For glycolysis to continue, NAD^+, which cells have in limited quantities, must be recycled after its reduction to NADH by GAPDH (Fig. 17-3; Reaction 6). In the presence of oxygen, the reducing equivalents of NADH are passed into the mitochondria for reoxidation (Chapter 22). Under anaerobic conditions, on the other hand, the NAD^+ is replenished by the reduction of pyruvate in an extension of the glycolytic pathway. Two processes for the anaerobic replenishment of NAD^+ are homolactic and alcoholic fermentation, which occur in muscle and yeast, respectively.

A. *Homolactic Fermentation*

In muscle, particularly during vigorous activity when the demand for ATP is high and oxygen has been depleted, **lactate dehydrogenase (LDH)** catalyzes the oxidation of

NADH by pyruvate to yield NAD^+ and **lactate.** This reaction is often classified as Reaction 11 of glycolysis:

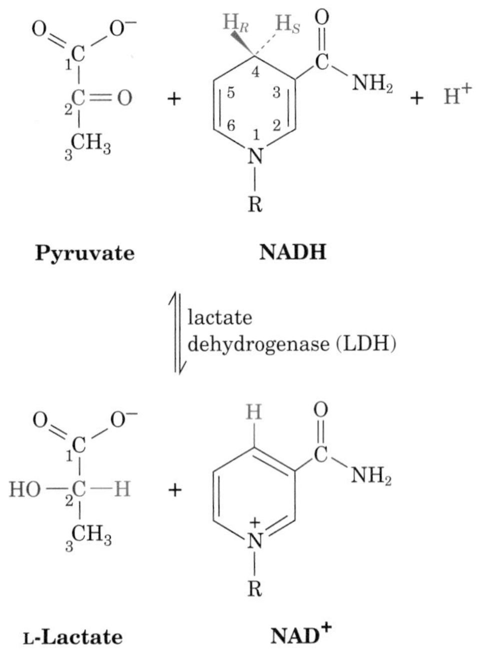

Pyruvate   **NADH**

lactate dehydrogenase (LDH)

L-Lactate   **NAD⁺**

LDH, as do other NAD^+-requiring enzymes, catalyzes its reaction with absolute stereospecificity: The *pro-R* (A-side) hydrogen at C4 of NADH is stereospecifically transferred to the *re* face of pyruvate at C2 to form L- (or S-) lactate. This regenerates NAD^+ for participation in the GAPDH reaction. The hydride transfer to pyruvate is from the same face of the nicotinamide ring as that to acetaldehyde in the alcohol dehydrogenase reaction (Section 13-2A) but from the opposite (*si*) face of the nicotinamide ring as that to GAP in the GAPDH reaction (Section 17-2F).

Mammals have two different types of LDH subunits, the M type and the H type, which together form five tetrameric isozymes: M_4, M_3H, M_2H_2, MH_3, and H_4. Although these hybrid forms occur in most tissues, the H-type subunit predominates in aerobic tissues such as

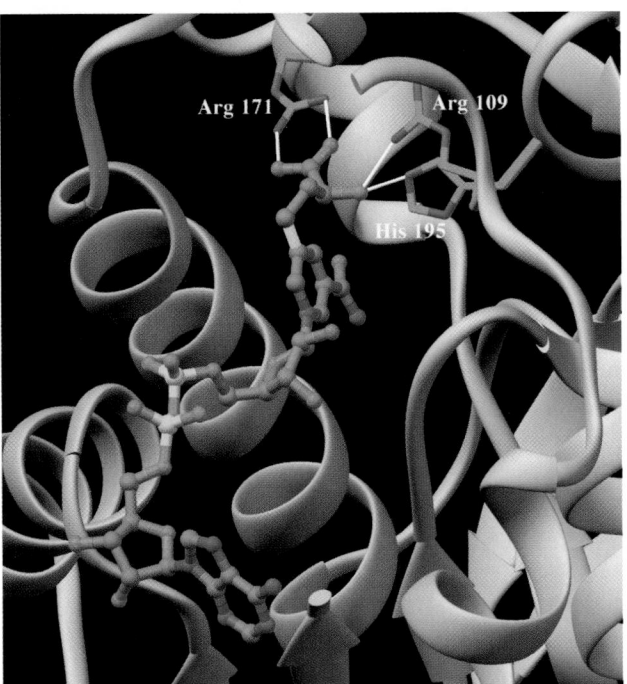

FIGURE 17-23 The active site region of porcine H₄ LDH in complex with *S*-lac-NAD⁺, a covalent adduct of lactate and NAD⁺. The adduct is shown in ball-and-stick form colored according to atom type (C green, N blue, O red, and P yellow) except for the covalent bond between lactate atom C3 and nicotinamide atom C4, which is light green. The three LDH side chains that form hydrogen bonds (*white lines*) with the pyruvate residue are shown in stick form, also colored according to atom type but with C magenta. [Based on an X-ray structure by Michael Rossmann, Purdue University. PDBid 5LDH.]

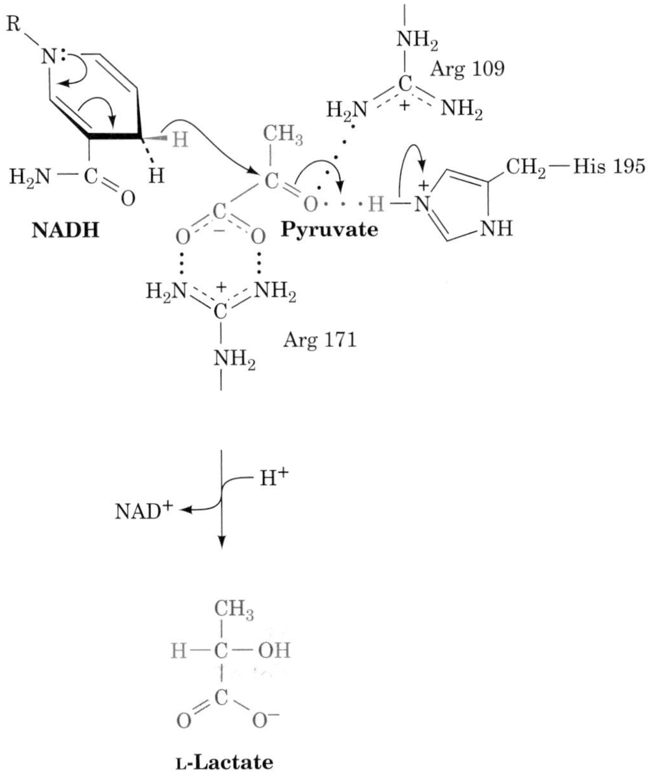

FIGURE 17-24 Reaction mechanism of lactate dehydrogenase. The reaction involves direct hydride transfer from NADH to pyruvate's carbonyl carbon atom accompanied by proton donation from the imidazolium group of His 195 to the pyruvate carbonyl oxygen atom. The latter process is facilitated by the positive charge on the nearby side chain of Arg 109.

heart muscle, while the M-type subunit predominates in tissues that are subject to anaerobic conditions such as skeletal muscle and liver. H₄ LDH has a low K_M for pyruvate and is allosterically inhibited by high levels of this metabolite, whereas the M₄ isozyme has a higher K_M for pyruvate and is not inhibited by it. The other isozymes have intermediate properties that vary with the ratio of their two types of subunits. It has therefore been proposed, although not without disagreement, that H-type LDH is better adapted to function in the oxidation of lactate to pyruvate, whereas M-type LDH is more suited to catalyze the reverse reaction.

The X-ray structure of porcine H₄ LDH in complex with **S-lac-NAD⁺** (a bisubstrate analog in which atom C3 of lactate is covalently linked to nicotinamide atom C4 of NAD⁺) was determined by Michael Rossmann (Fig. 17-23; he also determined the X-ray structure of dogfish M₄ LDH shown in Fig. 8-54). Lactate atom O2, its hydroxyl oxygen, is hydrogen bonded to the side chains of both Arg 109 and His 195, whereas the lactate carboxyl group at C1 is doubly hydrogen bonded to the side chain of Arg 171. On the basis of this structure and extensive enzymological evidence, Rossmann proposed the following mechanism for

pyruvate reduction by LDH (Fig. 17-24): *The pro-R hydride is transferred from C4 of NADH's nicotinamide ring to C2 of pyruvate with the concomitant transfer of a proton from the imidazolium moiety of His 195 to pyruvate O2, thereby yielding NAD⁺ and lactate.* The proton transfer is facilitated by repulsive interactions with the closely associated positively charged side chain of Arg 109. These interactions also serve to properly orient the pyruvate, as does the salt bridge that the pyruvate carboxyl group forms with the side chain of Arg 171.

The overall process of anaerobic glycolysis in muscle can be represented:

Glucose + 2ADP + 2P$_i$ →

$$\text{2lactate} + \text{2ATP} + 2H_2O + 2H^+$$

Much of the lactate, the end product of anaerobic glycolysis, is exported from the muscle cell via the blood to the liver, where it is reconverted to glucose (Section 23-1C).

Contrary to the widely held belief, it is not lactate buildup in the muscle *per se* that causes muscle fatigue and soreness but the accumulation of glycolytically generated acid (muscles can maintain their work load in the presence of high lactate concentrations if the pH is kept constant; but see Section 27-2B). Indeed, it is well known among hunters that the meat of an animal that has run to

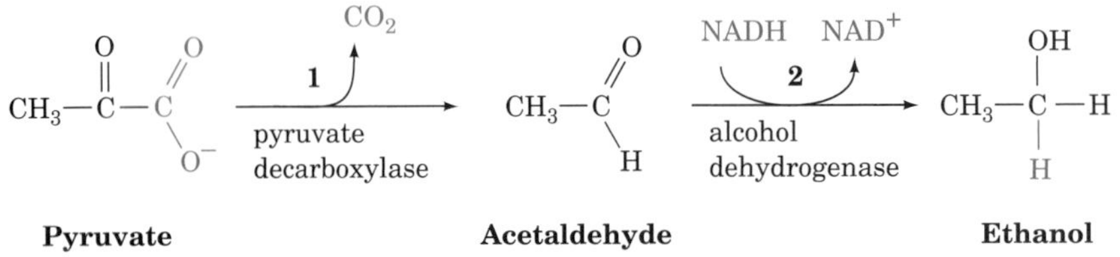

FIGURE 17-25 The two reactions of alcoholic fermentation. (1) Decarboxylation of pyruvate to form acetaldehyde is followed by **(2)** reduction of the acetaldehyde to ethanol by NADH.

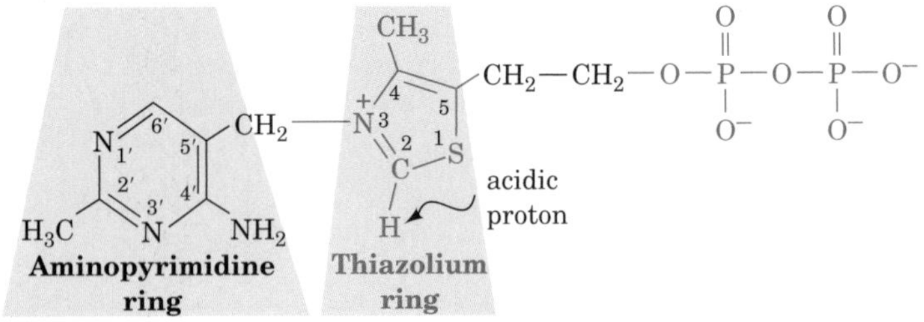

FIGURE 17-26 Thiamine pyrophosphate. The thiazolium ring constitutes its catalytically active functional group.

exhaustion before being killed has a sour taste. This is a result of lactic acid buildup in the muscles.

B. *Alcoholic Fermentation*

Under anaerobic conditions in yeast, NAD^+ is regenerated in a manner that has been of importance to mankind for thousands of years: the conversion of pyruvate to ethanol and CO_2. Ethanol is, of course, the active ingredient of wine and spirits; CO_2 so produced leavens bread. From the point of view of the yeast, however, alcoholic fermentation has a practical benefit that homolactic fermentation cannot supply: Yeast employ ethanol as a kind of antibiotic to eliminate competing organisms. This is because yeast can grow in ethanol concentrations >12% (2.5*M*), whereas few other organisms can survive in >5% ethanol (recall that ethanol is a widely used antiseptic).

a. TPP Is an Essential Cofactor of Pyruvate Decarboxylase

Yeast produces ethanol and CO_2 via two consecutive reactions (Fig. 17-25). The first reaction is the decarboxylation of pyruvate to form acetaldehyde and CO_2 as catalyzed by **pyruvate decarboxylase (PDC;** an enzyme not present in animals). PDC contains the coenzyme **thiamine pyrophosphate [TPP;** Fig. 17-26; also called **thiamin diphosphate (ThDP)],** which it binds tightly but noncovalently. The coenzyme is employed because decarboxylation of an α-keto acid such as pyruvate requires the buildup of negative charge on the carbonyl carbon atom in the transition state, an unstable situation:

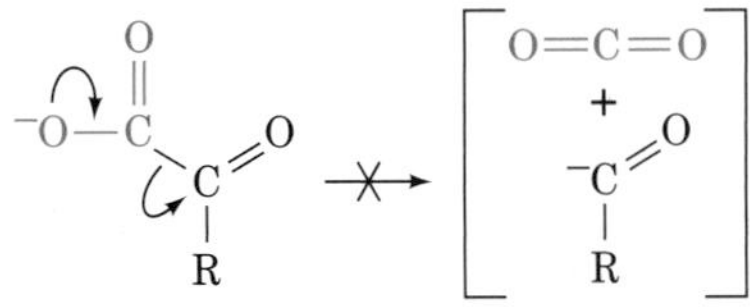

This transition state may be stabilized by delocalization of the developing negative charge into a suitable "electron sink." The amino acid residues of proteins function poorly in this capacity but TPP does so easily.

*The "business" end of TPP is the **thiazolium ring*** (Fig. 17-26). Its C2—H group is relatively acidic because of the adjacent positively charged quaternary nitrogen atom, which electrostatically stabilizes the carbanion formed on dissociation of the proton. This dipolar carbanion (or **ylid**) is the active form of the coenzyme. The mechanism of PDC catalysis is as follows (Fig. 17-27):

Step 1 Nucleophilic attack by the ylid form of TPP on the carbonyl carbon of pyruvate to form a covalent adduct.

Step 2 Departure of CO_2 to generate a resonance-stabilized carbanion adduct in which the thiazolium ring of the coenzyme acts as an electron sink.

Step 3 Protonation of the carbanion.

Step 4 Elimination of the TPP ylid to form acetaldehyde and regenerate the active enzyme.

This mechanism has been corroborated by the isolation of the **hydroxyethylthiamine pyrophosphate** intermediate (Fig. 17-27).

The X-ray structure of PDC in complex with TPP (Fig. 17-28), which was determined by William Furey and Martin Sax, has suggested a role for TPP's aminopyrimidine ring in the formation of the active ylid. Ylid formation requires a base to remove the C2 proton. Yet PDC has no basic side chain that is properly positioned to do so. The amino group of the enzyme-bound TPP's aminopyrimidine ring is suitably positioned to accept this proton; however, its p*K* is too low to do so efficiently and one of its protons sterically clashes with the C2 proton. It is therefore proposed that the aminopyrimidine is converted to its

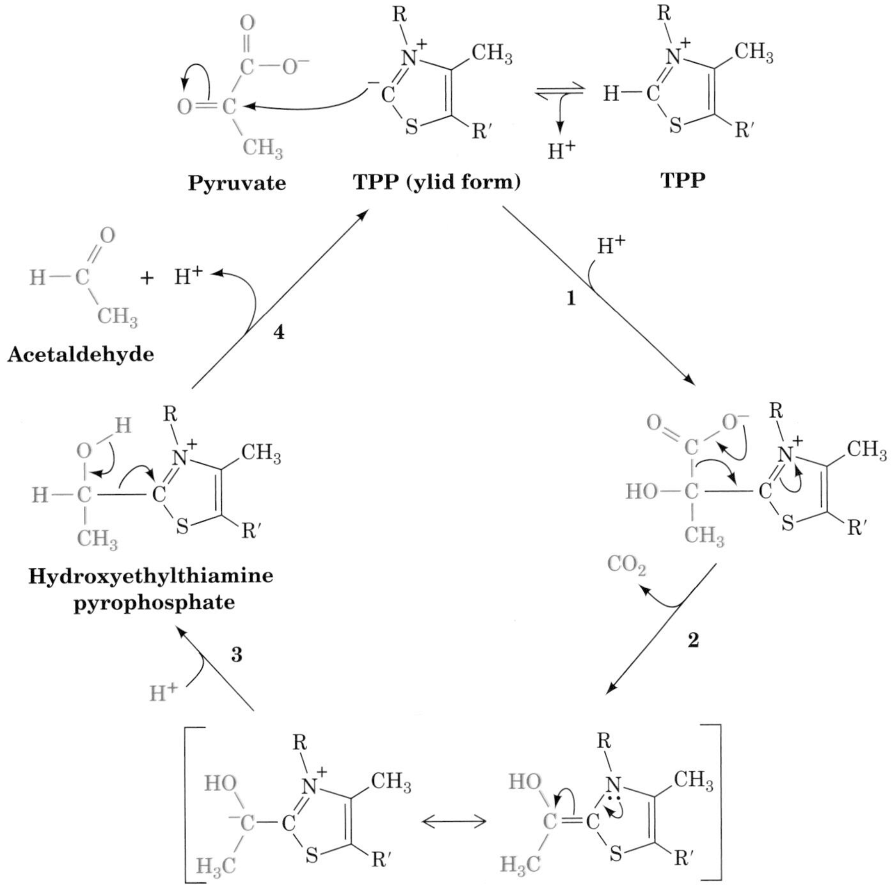

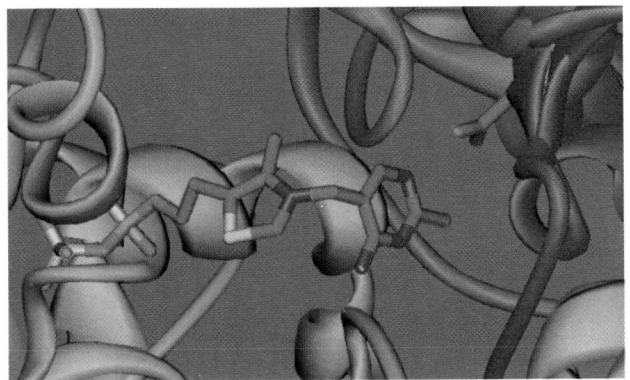

FIGURE 17-27 Reaction mechanism of pyruvate decarboxylase. (1) Nucleophilic attack by the ylid form of TPP on the carbonyl carbon of pyruvate; **(2)** departure of CO_2 to generate a resonance-stabilized carbanion; **(3)** protonation of the carbanion; and **(4)** elimination of the TPP ylid and release of product.

FIGURE 17-28 A portion of the X-ray structure of pyruvate decarboxylase from *Saccharomyces uvarum* (brewer's yeast) in complex with its TPP cofactor. The enzyme's identical 563-residue subunits form a tightly associated dimer, two of which associate loosely to form a tetramer. The TPP and the side chain of Glu 51 are shown in stick form with C green, N blue, O red, S yellow, and P gold. The TPP binds in a cavity situated between the dimer's two subunits (*cyan and magenta*), where it hydrogen bonds to Glu 51. [Based on an X-ray structure by William Furey and Martin Sax, Veterans Administration Medical Center and University of Pittsburgh, Pittsburgh, Pennsylvania. PDBid 1PYD.] 🐭 **See the Interactive Exercises**

imino tautomeric form on the enzyme's surface in a reaction involving proton donation by Glu 51 (Fig. 17-29). The imine, in turn, accepts a proton from C2, thereby forming the ylid, with tautomerization back to the amino form. The participation of N1′ and the 4′-amino group of the aminopyrimidine is supported by experiments showing that TPP analogs missing either of these functionalities are catalytically inactive. H/D exchange experiments followed by ^{1}H NMR analysis of the exchange products indicate that when TPP is bound to PDC in complex with the substrate analog **pyruvamide** (CH_3—CO—CO—NH_2), its rate of exchange to form the active species (ylid) is much greater ($>6 \times 10^2$ s^{-1}) than the enzyme's catalytic rate ($k_{cat} = 10$ s^{-1}). Moreover, the mutation of PDC's Glu 51 to Gln reduces the rate of H/D exchange to 1.7 s^{-1}, thereby supporting Glu 51's postulated function of donating a proton to N1′ of TPP's aminopyridine ring.

b. Beriberi Is a Thiamine Deficiency Disease

The ability of TPP's thiazolium ring to add to carbonyl groups and act as an "electron sink" makes it the coenzyme most utilized in α-keto acid decarboxylations. TPP is also involved in decarboxylation reactions that we shall encounter in other metabolic pathways. Consequently, thi-

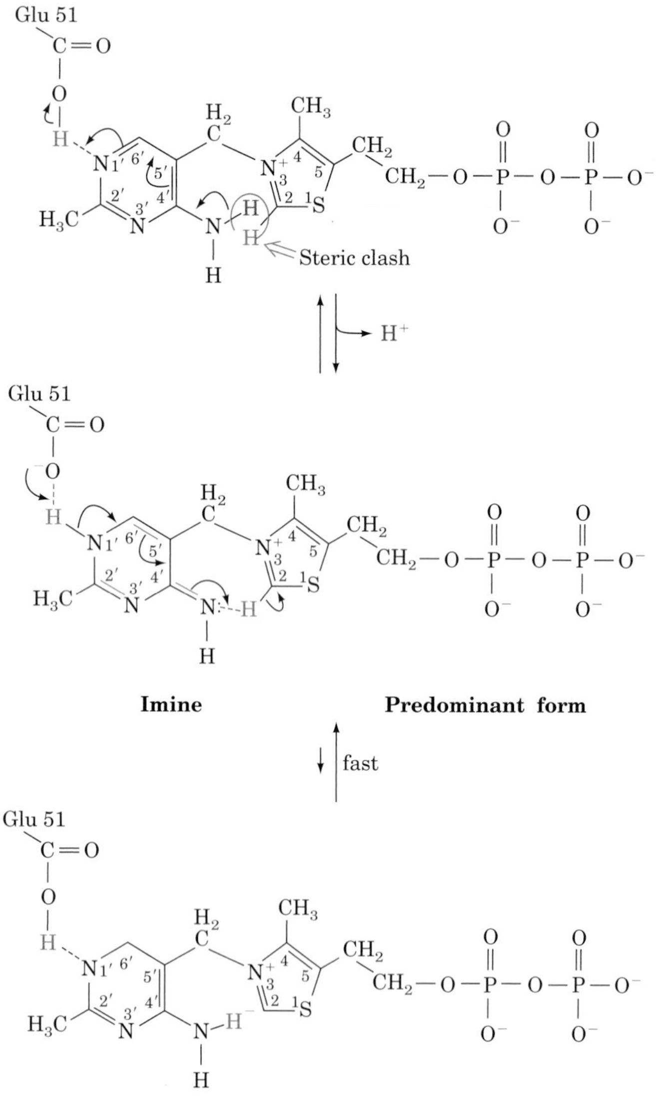

Glu 51

Imine

Predominant form

| fast

Glu 51

Ylid

FIGURE 17-29   The formation of the active ylid form of TPP in the pyruvate decarboxylase reaction. This reaction requires the participation of TPP's aminopyrimidine ring together with general acid catalysis by Glu 51. The predominant form of the cofactor on the enzyme is the imine, but the rate of formation of the active ylid is fast relative to the enzyme's catalytic rate.

amine **(vitamin B$_1$),** which is neither synthesized nor stored in significant amounts by the tissues of most vertebrates, is required in their diets. Its deficiency in humans results in an ultimately fatal condition known as **beriberi** that is characterized by neurological disturbances causing pain, paralysis and atrophy (wasting) of the limbs, and/or cardiac failure resulting in edema (the accumulation of fluid in tissues and body cavities). Beriberi was particularly prevalent in the rice-consuming areas of Asia because of the custom of polishing this staple grain to remove its coarse but thiamine-containing outer layers. Beriberi fre-

quently develops in chronic alcoholics as a consequence of their penchant for drinking but not eating.

c. Reduction of Acetaldehyde and Regeneration of NAD$^+$

The acetaldehyde formed by the decarboxylation of pyruvate is reduced to ethanol by NADH in a reaction catalyzed by **alcohol dehydrogenase (ADH).** Each subunit of the tetrameric yeast ADH (YADH) binds one NADH and one Zn^{2+} ion. The Zn^{2+} ion functions to polarize the carbonyl group of acetaldehyde (Fig. 17-30), so as to stabilize the developing negative charge in the transition state of the reaction (the role of metal ions in enzymes is discussed in Section 15-1C). This facilitates the transfer of NADH's *pro-R* hydrogen (the same atom that LDH transfers) to acetaldehyde's *re* face, forming ethanol with the transferred hydrogen in the *pro-R* position (Section 13-2A).

Both homolactic and alcoholic fermentation have the same function: the anaerobic regeneration of NAD$^+$ for continued glycolysis. Their main difference is in their metabolic products.

Mammalian liver ADH **(LADH)** functions to metabolize the alcohols anaerobically produced by intestinal flora as well as those from external sources (the direction of the ADH reaction varies with the relative concentrations of ethanol and acetaldehyde). Each subunit of this dimeric enzyme binds one NAD$^+$ and two Zn^{2+} ions, although only one of these ions participates directly in catalysis. There is significant amino acid sequence similarity between YADH and LADH, so it is quite likely that both enzymes have the same general mechanism.

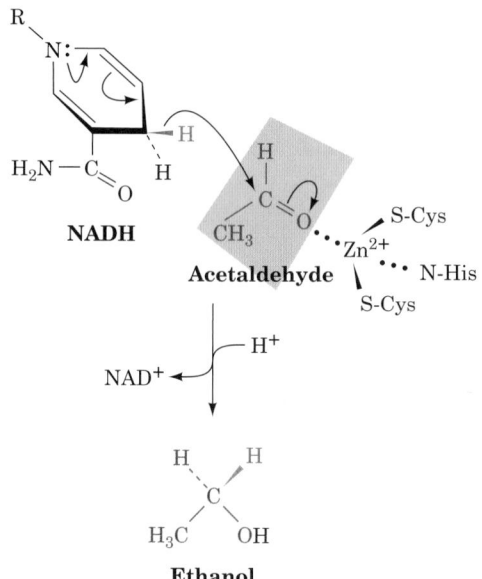

FIGURE 17-30   The reaction mechanism of alcohol dehydrogenase involves direct hydride transfer of the *pro-R* hydrogen of NADH to the *re* face of acetaldehyde.

C. *Energetics of Fermentation*

Thermodynamics permits us to dissect the process of fermentation into its component parts and to account for the free energy changes that occur. This enables us to calculate the efficiency with which the free energy of degradation of glucose is utilized in the synthesis of ATP. The overall reaction of homolactic fermentation is

$$\text{Glucose} \rightarrow 2\text{lactate} + 2\text{H}^+$$
$$\Delta G^{\circ\prime} = -196\,\text{kJ} \cdot \text{mol}^{-1} \text{ of glucose}$$

($\Delta G^{\circ\prime}$ is calculated from the data in Table 3-4 using Eqs. [3.19] and [3.21] adapted for 2H^+ ions.) For alcoholic fermentation, the overall reaction is

$$\text{Glucose} \rightarrow 2\text{CO}_2 + 2\text{ethanol}$$
$$\Delta G^{\circ\prime} = -235\,\text{kJ} \cdot \text{mol}^{-1} \text{ of glucose}$$

Each of these reactions is coupled to the net formation of two ATPs, which requires $\Delta G^{\circ\prime} = +61\,\text{kJ} \cdot \text{mol}^{-1}$ of glucose consumed (Table 16-3). Dividing the $\Delta G^{\circ\prime}$ of ATP formation by that of lactate formation indicates that homolactic fermentation is 31% "efficient"; that is, 31% of the free energy released by this process under standard biochemical conditions is sequestered in the form of ATP. The rest is dissipated as heat, thereby making the process irreversible. Likewise, alcoholic fermentation is 26% efficient under biochemical standard state conditions. Actually, *under physiological conditions, where the concentrations of reactants and products differ from those of the standard state, these reactions have free energy efficiencies of >50%.*

a. Glycolysis Is Used for Rapid ATP Production

Anaerobic fermentation utilizes glucose in a profligate manner compared to oxidative phosphorylation: Fermentation results in the production of 2 ATPs per glucose, whereas oxidative phosphorylation yields 38 ATPs per glucose (Chapter 22). This accounts for Pasteur's observation that yeast consumes far more sugar when growing anaerobically than when growing aerobically (the **Pasteur effect;** Section 22-4C). However, *the rate of ATP production by anaerobic glycolysis can be up to 100 times faster than that of oxidative phosphorylation. Consequently, when tissues such as muscle are rapidly consuming ATP, they regenerate it almost entirely by anaerobic glycolysis.* (Homolactic fermentation does not really "waste" glucose since the lactate so produced is aerobically reconverted to glucose by the liver; Section 23-1C).

Skeletal muscles consist of both **slow-twitch** (Type I) and **fast-twitch** (Type II) **fibers.** Fast-twitch fibers, so called because they predominate in muscles capable of short bursts of rapid activity, are nearly devoid of mitochondria, so that they must obtain nearly all of their ATP through anaerobic glycolysis, for which they have a particularly large capacity. Muscles designed to contract slowly and steadily, in contrast, are enriched in slow-twitch fibers that are rich in mitochondria and obtain most of their ATP through oxidative phosphorylation. (Fast- and slow-twitch

fibers were originally known as white and red fibers, respectively, because otherwise pale colored muscle tissue, when enriched with mitochondria, takes on the red color characteristic of their heme-containing cytochromes. However, fiber color has been shown to be an imperfect indicator of muscle physiology.)

In a familiar example, the flight muscles of migratory birds such as ducks and geese, which need a continuous energy supply, are rich in slow-twitch fibers and therefore such birds have dark breast meat. In contrast, the flight muscles of less ambitious fliers, such as chickens and turkeys, which are used only for short bursts (often to escape danger), consist mainly of fast-twitch fibers that form white meat. In humans, the muscles of sprinters are relatively rich in fast-twitch fibers, whereas distance runners have a greater proportion of slow-twitch fibers (although their muscles have the same color). World class distance runners have a remarkably high capacity to generate ATP aerobically. This was demonstrated by the noninvasive ^{31}P NMR monitoring of the ATP, P_i, phosphocreatine, and pH levels in their exercising but untrained forearm muscles. These observations suggest that the muscles of these athletes are better endowed genetically for endurance exercise than those of "normal" individuals.

4 ■ METABOLIC REGULATION AND CONTROL

Living organisms, as we saw in Section 16-6, are thermodynamically open systems that tend to maintain a steady state rather than reaching equilibrium (death for living things). Thus the **flux** *(rate of flow) of intermediates through a metabolic pathway is constant; that is, the rates of synthesis and breakdown of each pathway intermediate maintain it at a constant concentration.* Such a state, it will be recalled, is one of maximum thermodynamic efficiency (Section 16-6B). *Regulation of the steady state (**homeostasis**) must be maintained in the face of changes in flux through the pathway in response to changes in demand.*

The terms metabolic control and metabolic regulation are often used interchangeably. However, for our purposes we shall give them different definitions: **Metabolic regulation** is the process by which the steady-state flow of metabolites through a pathway is maintained, whereas **metabolic control** is the influence exerted on the enzymes of a pathway in response to an external signal in order to alter the flux of metabolites.

A. *Homeostasis and Metabolic Control*

There are two reasons why metabolic flow must be controlled:

I. To provide products at the rate they are needed, that is, to balance supply with demand.

II. To maintain the steady-state concentrations of the intermediates in a pathway within a narrow range (homeostasis).

Organisms maintain homeostasis for several reasons:

1. In an open system, such as metabolism, the steady state is the state of maximum thermodynamic efficiency (Section 16-5B).

2. Many intermediates participate in more than one pathway, so that changing their concentrations may disturb a delicate balance.

3. The rate at which a pathway can respond to a control signal slows if large changes in intermediate concentrations are involved.

4. Large changes in intermediate concentrations may have deleterious effects on cellular osmotic properties.

The concentrations of intermediates and the level of metabolic flux at which a pathway is maintained vary with the needs of the organism through a highly responsive system of precise controls. Such pathways are analogous to rivers that have been dammed to provide a means of generating electricity. Although water is continually flowing in and out of the lake formed by the dam, a relatively constant water level is maintained. The rate of water outflow from the lake is precisely controlled at the dam and is varied in response to the need for electrical power. In this section, we examine the mechanisms by which metabolic pathways in general, and the glycolytic pathway in particular, are controlled in response to biological energy needs.

B. *Metabolic Flux*

Since a metabolic pathway is a series of enzyme-catalyzed reactions, it is easiest to describe the flux of metabolites through the pathway by considering its reaction steps individually. The flux of metabolites, J, through each reaction step is the rate of the forward reaction, v_f, less that of the reverse reaction, v_r:

$$J = v_f - v_r \qquad [17.1]$$

At equilibrium, by definition, there is no flux ($J = 0$), although v_f and v_r may be quite large. At the other extreme, in reactions that are far from equilibrium, $v_f >> v_r$, so that the flux is essentially equal to the rate of the forward reaction, $J \approx v_f$. *The flux throughout a steady-state pathway is constant and is set (generated) by the pathway's rate-determining step (or steps). Consequently, control of flux through a metabolic pathway requires: (1) that the flux through this **flux-generating step** vary in response to the organism's metabolic requirements and (2) that this change in flux be communicated throughout the pathway to maintain a steady state.*

The classic description of metabolic control and regulation is that every metabolic pathway has a rate-limiting step and is regulated by controlling the rate of this pivotal enzyme. These so-called regulatory enzymes are almost invariably allosteric enzymes subject to feedback inhibition (Section 13-4) and are often also controlled by covalent modification (which we discuss in Section 18-3).

Several questions arise. Are these regulatory enzymes really rate limiting for the pathway? Is there really only one step in the pathway that is rate limiting, or might there be a number of enzymes contributing to the regulation of the pathway? Does controlling these enzymes really control the flux of metabolites through the pathway or is the function of feedback inhibition really to maintain a steady state? These are complicated questions with complicated answers.

C. *Metabolic Control Analysis*

While it has been common practice to assume that every metabolic pathway has a rate-limiting step, experiments suggest that the situation becomes more complex when these pathways are combined in a living organism. Hence, it is important to develop methods to quantitatively analyze metabolic systems in order to establish mechanisms of control and regulation. **Metabolic control analysis,** developed by Henrik Kacser and Jim Burns and independently by Reinhart Heinrich and Tom Rapoport, provides a framework for considering these problems. It is a way of quantitatively describing the behavior of metabolic systems in response to various perturbations.

a. The Flux Control Coefficient Measures the Sensitivity of the Flux to the Change in Enzyme Concentration

Metabolic control analysis makes no *a priori* assumption that only one step is rate limiting. Instead, it defines a **flux control coefficient,** C^J (where J is an index, not an exponent), to measure the sensitivity of flux to a change in enzyme concentration. The flux control coefficient is defined as the fractional change in flux, J, with respect to the fractional change in enzyme concentration, [E]:

$$C^J = \frac{\partial J/J}{\partial [E]/[E]} = \frac{\partial \ln J}{\partial \ln [E]} \approx \frac{\Delta J/J}{\Delta [E]/[E]} \qquad [17.2]$$

(recall that $\partial x/x = \partial \ln x$).

The flux control coefficient is the analog of the kinetic order of a reaction. If a reaction is first order in substrate concentration, [S], then doubling [S] doubles the rate of the reaction, whereas if the reaction is zero order in [S] (e.g., in a saturated enzymatic reaction), then the reaction rate is insensitive to the value of [S]. Similarly, if the flux control coefficient of an enzyme is 1, then doubling the concentration of the enzyme, [E], doubles the flux through the pathway and if it is zero, the flux is insensitive to the value of [E]. Of course, the flux control coefficient may have some intermediate value between 0 and 1. For example, if a 10% increase in the enzyme concentration increases the flux by only 7.5%, the flux control coefficient would be 0.075/0.10 = 0.75.

Of course, the flux through a metabolic pathway may be controlled by more than one enzyme. In this case, the flux control coefficient for each of the participating enzymes is the fraction of the total control on the pathway exerted by that enzyme. Thus, *the sum of all the flux control coefficients involved in controlling a pathway must equal 1.* This is the **additivity theorem of metabolic control.**

b. Recombinant DNA Technology Has Been Used to Measure Flux Control Coefficients

The flux control coefficient is a variable that has been experimentally determined *in vivo* for many enzymes that had previously been assumed to catalyze the rate-determining steps for their pathways. For example, the citric acid cycle enzyme **citrate synthase** (Sections 21-1A and 21-3A) catalyzes an irreversible reaction ($\Delta G^{o\prime} = -31.5$ kJ · mol^{-1}) and has therefore long been assumed to be one of the enzymes regulating the flow through the citric acid cycle (Section 21-4). Daniel Koshland determined how the activity of citrate synthase affected the flux through the citric acid cycle via genetic engineering techniques that permitted him to control the concentration of this enzyme *in vivo.* He constructed a plasmid (Fig. 17-31) that contained the gene for citrate synthase under the control of (directly downstream from) a modified ***lac*** **promoter,** together with the ***lacI*** gene, which encodes the ***lac*** **repressor** (in the absence of inducer, the *lac* repressor binds to the *lac* promoter and thereby prevents the transcription of the genes it controls by RNA polymerase; Section 5-4A), and the ***amp^R*** gene, which confers resistance to the antibiotic **ampicillin.** This plasmid was introduced into a mutant *E. coli* that lacked a gene for citrate synthase and was ampicillin-sensitive. These *E. coli* were grown in the presence of ampicillin (which killed any cells that had not taken up the plasmid) and of varying amounts of **isopropylthiogalactoside (IPTG),** a nonmetabolizable inducer of the *lac* operon (Section 5-5C).

Using this system, the citrate synthase concentration was measured as a function of [IPTG] and the growth rate of the *E. coli* was determined as a function of [IPTG] when glucose and/or acetate were the sole carbon sources. When acetate was the sole carbon source, the *E. coli* obtained most of their metabolic energy via the citric acid cycle and their growth varied proportionately with [citrate synthase]. The flux control coefficient of the enzyme in this case approached its maximum value of 1, that is, the flux through the citric acid cycle was almost entirely controlled by the activity of citrate synthase. However, when glucose was also available, the *E. coli* grew rapidly, even when [citrate synthase] was low, and were unaffected by changes in its value. Here, the flux control coefficient was near zero, indicating that the flux through the citric acid cycle was reduced to the point that even low concentrations of citrate synthase were in catalytic excess (evidently, when glucose is present, the citric acid cycle has a secondary role in energy production and biosynthesis in *E. coli*).

c. The Rates of Enzymatic Reactions Respond to Changes in Flux

Let us consider how a constant flux is maintained throughout a metabolic pathway by analyzing the response of an enzyme-catalyzed reaction to a change in the flux of the reaction preceding it. In the following steady-state pathway:

$$\text{S} \xrightarrow[\substack{\text{rate-determining}\\ \text{step(s)}}]{J} \text{A} \underset{v_r}{\overset{v_f}{\rightleftharpoons}} \text{B} \xrightarrow{J} \text{P}$$

the flux, J, through the reaction A $\rightleftharpoons$ B, which must be identical to the flux through the rate-determining step(s), is expressed by Eq. [17.1] ($J = v_f - v_r$). If the flux of the rate-determining step increases by the amount ΔJ, the increase must be communicated to the next reaction step in the pathway by an increase in v_f (Δv_f) in order to reestablish the steady state. Qualitatively, we can see that this occurs because an increase in J causes an increase in [A], which in turn causes an increase in v_f. The amount of increase in [A] (Δ[A]) that causes v_f to increase the appropriate amount (Δv_f) is determined as follows:

$$\Delta J = \Delta v_f \qquad [17.3]$$

Dividing Eq. [17.3] by J, multiplying the right side by v_f/v_f, and substituting in Eq. [17.1] yields

$$\frac{\Delta J}{J} = \frac{\Delta v_f}{v_f}\frac{v_f}{J} = \frac{\Delta v_f}{v_f}\frac{v_f}{(v_f - v_r)} \qquad [17.4]$$

which relates $\Delta J/J$, the fractional change in flux through the rate-determining step(s), to $\Delta v_f/v_f$, the fractional

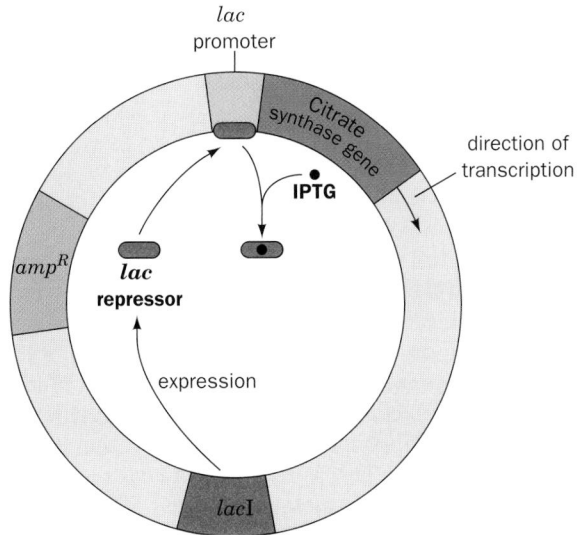

FIGURE 17-31   Schematic diagram of the plasmid constructed to control the amount of citrate synthase produced by *E. coli.* The *lacI* gene encodes the *lac* repressor, which binds to the *lac* promoter. This prevents the transcription of its immediately downstream gene, which encodes citrate synthase. On binding IPTG, the *lac* repressor releases the *lac* promoter, thereby permitting the expression of citrate synthase. Consequently, the concentration of the nonmetabolizable IPTG controls the level of expression of citrate synthase. The *amp^R* gene encodes a protein that provides resistance to the antibiotic ampicillin. Hence, in the presence of ampicillin, only those otherwise ampicillin-sensitive *E. coli* that have taken up the plasmid will survive.

change in v_f, the forward rate of the next reaction in the pathway.

In Section 14-2A, we discussed the relationship between substrate concentration and the rate of an enzymatic reaction as expressed by the Michaelis–Menten equation:

$$v_f = \frac{V_{max}^f[A]}{K_M + [A]} \qquad [14.24]$$

In the simplest and physiologically most common situation, $[A] << K_M$, so that

$$v_f = \frac{V_{max}^f[A]}{K_M} \qquad [17.5]$$

and

$$\Delta v_f = \frac{V_{max}^f \Delta[A]}{K_M} \qquad [17.6]$$

Hence,

$$\frac{\Delta v_f}{v_f} = \frac{\Delta[A]}{[A]} \qquad [17.7]$$

that is, the fractional change in forward reaction rate is equal to the fractional change in substrate concentration. Then, by substituting Eq. [17.7] into Eq. [17.4], we find that

$$\frac{\Delta J}{J} = \frac{\Delta[A]}{[A]} \frac{v_f}{(v_f - v_r)} \qquad [17.8]$$

This equation relates the fractional change in flux through a metabolic pathway's rate-determining step(s) to the fractional change in substrate concentration necessary to communicate that change to the following reaction steps. *The quantity $v_f/(v_f - v_r)$ is a measure of the sensitivity of a reaction's fractional change in flux to its fractional change in substrate concentration.* This quantity is also a measure of the reversibility of the reaction, that is, how close it is to equilibrium:

1. In an irreversible reaction, v_r approaches 0 (relative to v_f) and therefore $v_f/(v_f - v_r)$ approaches 1. The reaction therefore requires a nearly equal fractional increase in its substrate concentration in order to respond to a fractional increase in flux.

2. As a reaction approaches equilibrium, v_r approaches v_f and hence $v_f/(v_f - v_r)$ approaches infinity. The reaction's response to a fractional increase in flux therefore requires a much smaller fractional increase in its substrate concentration.

Consequently, *the ability of a reaction to communicate a change in flux increases as the reaction approaches equilibrium.* A series of sequential reactions that are all near equilibrium therefore have the same flux and maintain concentrations of intermediates in a steady state (homeostasis).

d. The Elasticity Coefficient Measures the Sensitivity of an Enzymatic Reaction to the Change in Substrate Concentration

The ratio $v_f/(v_f - v_r)$, which measures the sensitivity of an enzymatic reaction rate to the change in substrate concentration, is called, in metabolic control analysis, the **elasticity coefficient, ε**. It is the fractional change in the net rate of an enzyme reaction, v, with respect to the fractional change in the substrate concentration, [A]:

$$\varepsilon = \frac{\partial v/v}{\partial[A]/[A]} = \frac{\partial \ln v}{\partial \ln[A]} \approx \frac{v_f}{v_f - v_r} \qquad [17.9]$$

(When studying an individual enzymatic reaction, so that the fractional change in the net rate through the reaction, $\Delta v/v$, corresponds to the fractional change in the flux, $\Delta J/J$, and $[A] << K_M$, this equation is simply a rearrangement of Eq. [17.8].) The value of the elasticity coefficient depends on the kinetic characteristics of the enzyme and how close to equilibrium the enzyme is functioning. As mentioned above, if an enzyme is functioning far from equilibrium ($v_f >> v_r$), changing the substrate concentration will have a small effect on the net rate of the enzyme reaction (ε will be close to 1). However, if the enzyme is functioning very close to equilibrium so that both the forward and reverse reaction rates are much faster than the overall net rate, ε approaches infinity and it takes only a tiny change in substrate concentration to adjust to a new flux. Such large elasticity coefficients are therefore associated with maintaining homeostasis.

D. *Supply–Demand Analysis*

Early studies on control of metabolic pathways focused on individual pathways, ignoring their overall physiological functions. Control was always assumed to reside within the pathway. However, often when enzymes thought to be "rate controlling" in an individual pathway were overproduced in living organisms using genetic engineering techniques, increases in enzyme concentrations of as much as 10-fold had no effect on the flux through the specific pathway studied. The flux control coefficients of the overproduced enzymes were near zero in the *in vivo* system; they were already present in metabolic excess. The flux must somehow have been controlled from outside of the pathway. This is because, as we now realize, it is impossible to separate a pathway from the process(es) that utilizes the product(s) of that pathway (i.e., the living organisms must have reduced the activities of these enzymes in keeping with their metabolic requirements, that is, they maintained homeostasis).

Degradation pathways are inextricably linked to the biosynthetic pathways that utilize their products (Fig. 16-2). This is a **supply–demand process** and both supply and demand are involved in the two metabolic control challenges: flux control and homeostasis. Jan-Hendrik Hofmeyr and Athel Cornish-Bowden have used metabolic control analysis to explore such a system, lumping all of the reactions of the supply pathway(s) together into one block and all of the reactions of the demand pathway(s) into a second block.

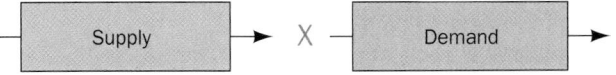

Here X is the intermediate that is produced by the supply block for use by the demand block. For the supply block, X is a product and a feedback inhibitor, so that as the concentration of X increases, the rate of flow through the supply block decreases. For the demand block, X is a substrate, so that as the concentration of X increases, the rate of flow through the demand block increases until it becomes saturated. When the flux through the supply block is equal to the flux through the demand block, the concentration of X is in a steady state, the point at which its rate of production is equal to its rate of utilization. This rate defines the actual flux through the supply–demand system and the steady-state concentration of X.

a. The Steady-State Concentration of Intermediates Responds to Changes in Supply and/or Demand

The response of the steady-state concentration of the intermediate, X, to any small change in the rate of the supply or demand block depends entirely on the elasticity coefficients of the two blocks at the steady state. Imagine that the activity of the demand block increases. This would result in a decrease of [X] and a concomitant increase in the flux through the supply block as feedback inhibition is decreased. The shift would continue until the rates through the supply and demand blocks equalize, shifting the system to a new steady state with a lower [X] and higher overall flux. Alternatively, if the activity of the supply block increases, producing a higher value of [X], the demand block would respond by increasing its rate to re-establish a new steady state at this higher [X]. *The higher the elasticity coefficient of the responding block, the smaller the change that [X] must make in order to re-establish a steady state.*

The question remains, where is the control? Is it in the supply block or in the demand block? The answer is that it is in the block for which the elasticity coefficient is lowest. Since it is the change in [X] that causes the readjustment of the steady state and the change in flux, the block for which the largest change in [X] is produced for a given change in rate is the controlling block. The change in flux for a given change in the rate in a particular block is its flux control coefficient, so the *control lies in the block that has the highest flux control coefficient and the lowest elasticity coefficient.* For example, if the supply block has a very high elasticity coefficient and the demand block has a very low elasticity coefficient, increased demand need cause very little decrease in [X] to result in a change in supply rate to reach a new steady state. However, because of the low elasticity of the demand block, there will have to be a

much larger increase in [X] due to an increase in supply to cause the rate of the demand block to increase enough to reach a new steady state. Consequently, increasing the activity of the demand block would have a much larger effect on the flux than increasing the activity of the supply block. Thus, for this case, the flux is much more sensitive to changes in demand than to changes in supply, that is, the flux control coefficient of the demand block is much greater than that of the supply block.

There is a reciprocal relationship between the flux control coefficient and the elasticity coefficient. The larger the flux control coefficient, the lower the elasticity coefficient, and vice versa. The ratio of the elasticity coefficients of the supply and demand blocks determines the distribution of flux control between supply and demand. *When the ratio of the supply elasticity coefficient to the demand elasticity coefficient is greater than 1, as in our example, flux control lies in the demand portion of the pathway, and vice versa.*

b. The Elasticity Coefficient Describes the Regulation of Steady-State Intermediate Concentrations

In addition to controlling flux through the supply–demand system, the steady-state concentrations of the intermediates are also regulated. We have seen that the larger the elasticity coefficient of a given block, the smaller the change in [X] that is needed to re-establish a steady state and change the flux. Keeping the change in [X] as small as possible while changing the flux and maintaining a steady state is very important. *The larger the elasticity coefficient, the more sensitive the regulation of homeostasis.*

Flux control requires a high flux control coefficient, which requires a low elasticity coefficient. Regulation of homeostasis requires a high elasticity coefficient, which requires a low flux control coefficient. A large difference in the elasticity coefficients of the supply and demand blocks therefore leads to the exclusive control of flux by one or the other of the blocks. *The functions of flux and concentration control are mutually exclusive. If the demand block controls the flux, the function of the supply block is to regulate homeostasis.*

c. Feedback Inhibition Is Required for Homeostasis, Not Flux Control

When the demand block is exerting flux control, an increase in demand results in a decrease in the concentration of X, thereby decreasing feedback inhibition of the supply block. Feedback inhibition might therefore appear to be an essential part of the control process. In fact, this is not the case. Feedback inhibition is not part of the control system but part of the homeostasis system. It determines the range of [X] at which there is a steady state. In the absence of feedback inhibition, the supply block will be insensitive to [X] for most of that concentration range but will become sensitive to [X] near equilibrium, where the demand block could then control the flux. However, this would require such high concentrations of X and the other metabolites in the supply pathway as to be osmotically dangerous. Feedback inhibition maintains homeostasis at physiologically reasonable metabolite concentrations.

E. *Mechanisms of Flux Control*

a. Flux through a Pathway Is Controlled at Its Rate-Determining Step(s)

The metabolic flux through an entire pathway is determined by controlling its rate-determining step(s), which by definition is much slower than the following reaction step(s). The product(s) of the rate-determining step(s) is therefore removed before it can equilibrate with the reactant, so that the rate-determining step(s) functions far from equilibrium and has a large negative free energy change. In an analogous manner, the flow of a river can only be controlled at a dam, which creates a difference in water levels between its upstream and downstream sides; this is a situation that also has a large negative free energy change, in this case resulting from the hydrostatic pressure head. Yet, as we have just seen, the elasticity coefficient, ε, of a nonequilibrium reaction ($v_f >> v_r$) is close to 1; that is, its substrate concentration must double (in the absence of other controlling effects) in order to double the reaction flux rate (Eq. [17.8]). However, some pathway fluxes vary by factors that are much greater than can be explained by changes in substrate concentrations. For example, glycolytic fluxes are known to vary by factors of 100 or more, whereas variations of substrate concentrations over such a large range are unknown. Consequently, although changes in substrate concentration can communicate a change in flux at the rate-determining step(s) to the other (near equilibrium; $v_f \approx v_r$) reaction steps of the pathway, there must be other mechanisms that control the flux of the rate-determining step(s).

The flux through a rate-determining step(s) of a pathway may be altered by several mechanisms:

1. Allosteric control: Many enzymes are allosterically controlled (Section 13-4) by effectors that are often substrates, products, or coenzymes in the pathway but not necessarily of the enzyme in question (feedback regulation). One such enzyme is PFK, an important glycolytic regulatory enzyme (Section 17-4F).

2. Covalent modification (enzymatic interconversion): Many enzymes that regulate pathway fluxes may be enzymatically phosphorylated and dephosphorylated at specific Ser, Thr, and/or Tyr residues or covalently modified in some other way. Such enzymatic modification processes, which are themselves subject to control, greatly alter the activities of the modified enzymes. This flux control mechanism is discussed in Section 18-3.

3. Substrate cycles: If v_f and v_r in Eq. [17.8] represent the rates of two opposing nonequilibrium reactions that are catalyzed by different enzymes, v_f and v_r may be inde-

pendently varied. The flux through such a substrate cycle, as we shall see in the next section, is more sensitive to the concentrations of allosteric effectors than is the flux through a single unopposed nonequilibrium reaction.

4. Genetic control: Enzyme concentrations, and hence enzyme activities, may be altered by protein synthesis in response to metabolic needs. Genetic control of enzyme concentrations is a major concern of Part V of this textbook.

Mechanisms 1 to 3 can respond rapidly (within seconds or minutes) to external stimuli and are therefore classified as "short-term" control mechanisms. Mechanism 4 responds more slowly to changing conditions (within hours or days in higher organisms) and is therefore referred to as a "long-term" control mechanism.

F. *Regulation of Glycolysis in Muscle*

Elucidation of the flux regulation mechanisms of a given pathway involves the determination of the pathway's regulatory enzymes involved in the rate-determining steps together with the identification of the modulators of these enzymes and their mechanism(s) of modulation. A hypothesis may then be formulated that can be tested *in vivo*. A common procedure for establishing regulatory mechanisms involves three steps.

1. Identification of the rate-determining step(s) of the pathway. One way to do so is to measure the *in vivo* ΔG's of all the reactions in the pathway to determine how close to equilibrium they function. Those that operate far from equilibrium are potential control points; the enzymes catalyzing them may be regulated by one or more of the mechanisms listed above. Another way of establishing the rate-determining step(s) of a pathway is to measure the effect of a known inhibitor on a specific reaction step and on the flux through the pathway as a whole. The ratio of the fractional change in the activity of the inhibited enzyme to the fractional change in the total flux (the flux control coefficient) will vary between 0 and 1. The closer the ratio is to 1, the more involved that enzyme is in the regulation of the total flux through the pathway.

2. *In vitro* identification of allosteric modifiers of the enzymes catalyzing the rate-determining reactions. The mechanisms by which these compounds act are determined from their effects on the enzyme's kinetics. From this information, a model of the allosteric mechanisms for regulating the pathway may be formulated.

3. Measurement of the *in vivo* levels of the proposed regulators under various conditions to establish whether these concentration changes are consistent with the proposed regulation mechanism.

a. Free Energy Changes in the Reactions of Glycolysis

Let us examine the thermodynamics of glycolysis with an eye toward understanding its regulatory mechanisms.

TABLE 17-1   **Standard Free Energy Changes ($\Delta G°'$), and Physiological Free Energy Changes (ΔG) in Heart Muscle, of the Reactions of Glycolysis**[a]

Reaction	Enzyme	$\Delta G°'$ (kJ·mol^{-1})	ΔG (kJ·mol^{-1})
1	HK	−20.9	−27.2
2	PGI	+2.2	−1.4
3	PFK	−17.2	−25.9
4	Aldolase	+22.8	−5.9
5	TIM	+7.9	Negative
6 + 7	GAPDH + PGK	−16.7	−1.1
8	PGM	+4.7	−0.6
9	Enolase	−3.2	−2.4
10	PK	−23.0	−13.9

[a]Calculated from data in Newsholme, E.A. and Start, C., *Regulation in Metabolism*, p. 97, Wiley (1973).

This must be done separately for each type of tissue in question because glycolysis is regulated in different tissues in different ways. We shall confine ourselves to muscle tissue. First we establish the pathway's possible regulation points through the identification of its nonequilibrium reactions. Table 17-1 lists the standard free energy change ($\Delta G°'$) and the actual physiological free energy change (ΔG) associated with each reaction in the pathway. It is important to realize that the free energy changes associated with the reactions under standard conditions may differ dramatically from those in effect under physiological conditions. For example, the $\Delta G°'$ for aldolase is +22.8 kJ·mol^{-1}, whereas under physiological conditions in heart muscle it is close to zero, indicating that the *in vivo* activity of aldolase is sufficient to equilibrate its substrates and products. The same is true of the GAPDH + PGK reaction series. Indeed, in a steady-state pathway, all the reactions must have $\Delta G < 0$. This is because if $\Delta G > 0$ for any reaction, its flux would be in the reverse direction.

In the glycolytic pathway, only three reactions, those catalyzed by hexokinase (HK), phosphofructokinase (PFK), and pyruvate kinase (PK), function with large negative free energy changes in heart muscle under physiological conditions (Table 17-1). These nonequilibrium reactions of glycolysis are the candidates for the flux-control points. The

other glycolytic reactions function near equilibrium: Their forward and reverse rates are much faster than the actual flux through the pathway (although their forward rates must be at least slightly greater than their reverse rates). Consequently, these near-equilibrium reactions are very sensitive to changes in the concentration of pathway intermediates and rapidly communicate any changes in flux generated at the rate-determining step(s) throughout the rest of the pathway, ensuring the maintenance of a steady state.

b. Phosphofructokinase Is a Major Target for Regulating the Flux of Glycolysis in Muscle

In vitro kinetic studies of HK, PFK, and PK indicate that each is controlled by a variety of compounds, some of which are listed in Table 17-2. Yet, when the G6P source for glycolysis is glycogen, rather than glucose, as is often the case in skeletal muscle (Section 18-1), the hexokinase reaction is not required. *PFK, an elaborately controlled enzyme functioning far from equilibrium, evidently is a major target for regulating glycolysis in muscle under most conditions.*

PFK (Fig. 17-32a) is a tetrameric enzyme with two conformational states, R and T, that are in equilibrium. ATP is both a substrate and an allosteric inhibitor of PFK. Each subunit has two binding sites for ATP, a substrate site and an inhibitor site. The substrate site binds ATP equally well in either conformation, but the inhibitor site binds ATP almost exclusively in the T state. The other substrate of PFK, F6P, preferentially binds to the R state. Consequently, at high concentrations, ATP acts as a heterotropic allosteric inhibitor of PFK by binding to the T state, thereby shifting the T $\rightleftharpoons$ R equilibrium in favor of the T state and thus decreasing PFK's affinity for F6P (this is similar to the action of 2,3-BPG in decreasing the affinity of hemoglobin for O$_2$; Section 10-2F). In graphical terms, at high concentrations of ATP, the hyperbolic (noncooperative) curve of PFK activity versus [F6P] is converted to the sigmoidal

TABLE 17-2   **Some Effectors of the Nonequilibrium Enzymes of Glycolysis**

Enzyme	Inhibitors	Activators[a]
HK	G6P	−
PFK	ATP, citrate, PEP	ADP, AMP, cAMP, FBP, F2,6P, F6P, NH$_4^+$, P$_i$
PK (muscle)	ATP	AMP, PEP, FBP

[a]The activators for PFK are better described as deinhibitors of ATP because they reverse the effect of inhibitory concentrations of ATP.

(a)

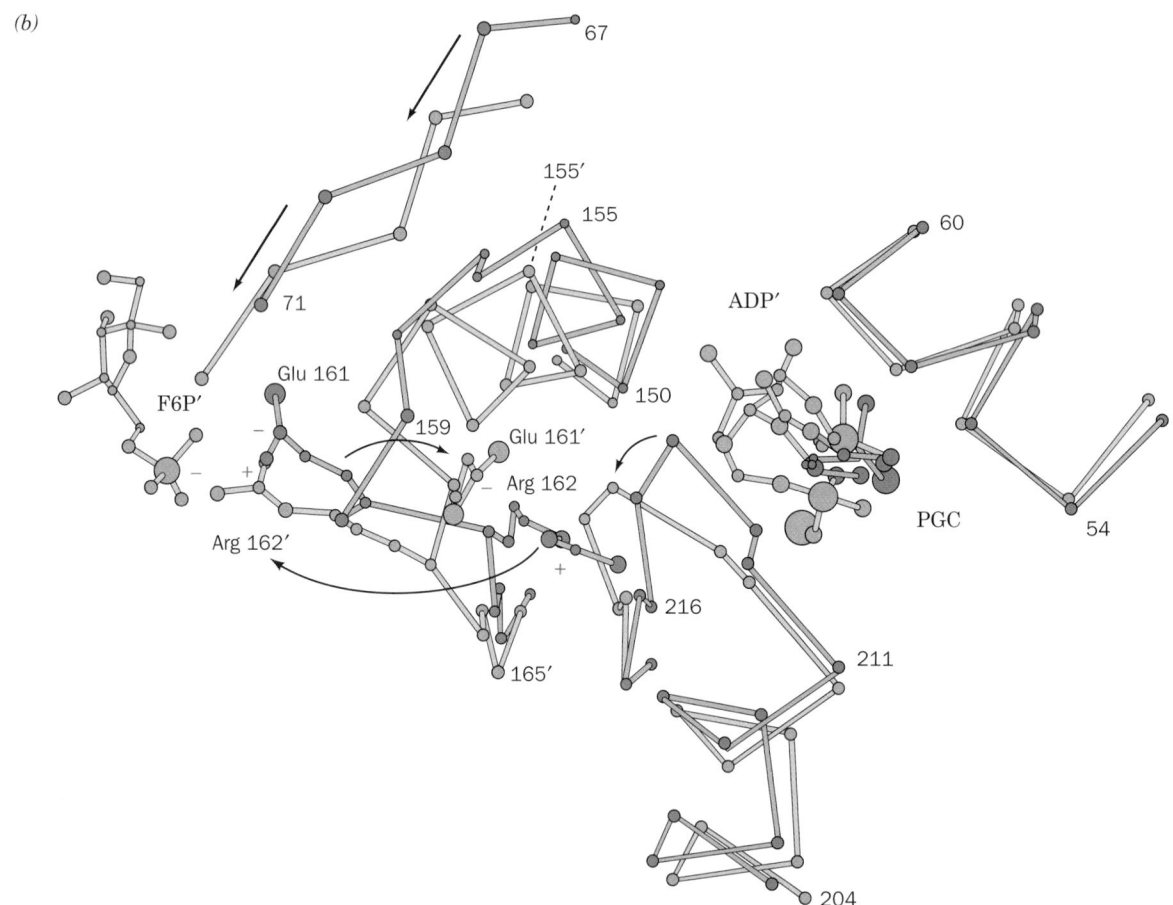

FIGURE 17-32   X-Ray structure of PFK. (*a*) A ribbon diagram showing two subunits of the tetrameric *E. coli* protein (related by a twofold axis perpendicular to the page through the center of the figure) with helices pink, β strands gray, and the remaining chain segments white. Each of the subunits in the protein is in association with its substrates F6P (*near the center of each subunit*) and ATP · Mg^{2+} (*lower right and upper left*; the green balls represent Mg^{2+}), together with the activator ADP · Mg^{2+} (*top right and lower left, in the rear*). [Courtesy of Phillip Evans, Cambridge University, U.K. PDBid 1PFK.] (*b*) A superposition of those segments of the T-state (*blue*) and R-state (*red*) enzymes that undergo a large conformational rearrangement on the T → R allosteric transition (indicated by the arrows). Residues of the R-state structure are marked by a prime. Also shown are bound ligands: the nonphysiological inhibitor 2-phosphoglycolate (**PGC;** a PEP analog) for the T state, and the cooperative substrate F6P and the activator ADP for the R state. [After Schirmer, T. and Evans, P.R., *Nature* **343,** 142 (1990). PDBids 4PFK and 6PFK.] ✷ **See Kinemage Exercises 13-1 and 13-2**

(b)

(cooperative) curve characteristic of allosteric enzymes (Fig. 17-33; cooperative and noncooperative processes are discussed in Section 10-1B). For example, when [F6P] = 0.5 m*M* (the dashed line in Fig. 17-33), the enzyme is nearly maximally active, but in the presence of 1 m*M* ATP, the activity drops to 15% of its original level (a nearly 7-fold decrease). [Actually, the most potent allosteric effector of PFK is **fructose-2,6-bisphosphate (F2,6P).** We discuss the

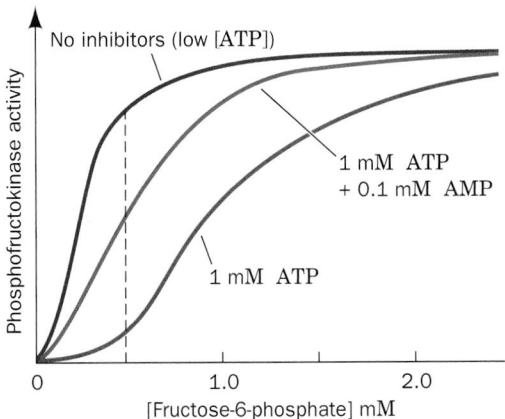

FIGURE 17-33 PFK activity versus F6P concentration. The various conditions are: blue, no inhibitors (low, noninhibitory [ATP]); green, 1 m*M* ATP (inhibitory); and red, 1 m*M* ATP + 0.1 m*M* AMP. [After data from Mansour, T.E. and Ahlfors, C.E., *J. Biol. Chem.* **243**, 2523–2533 (1968).]

role of F2,6P in regulating PFK activity when we study the mechanism by which the liver maintains blood glucose concentrations (Section 18-3F).]

c. Structural Basis for PFK's Allosteric Change in F6P Affinity

The X-ray structures of PFK from several organisms have been determined for both the R and the T states by Phillip Evans. The R state of PFK is homotropically stabilized by the binding of its substrate fructose-6-phosphate (F6P). In the R state of *Bacillus stearothermophilus* PFK, the side chain of Arg 162 forms a salt bridge with the phosphoryl group of an F6P bound in an active site of another subunit (Fig. 17-32*b*). However, Arg 162 is located at the end of a helical turn that unwinds on transition to the T state. The positively charged side chain of Arg 162 thereby swings away and is replaced by the negatively charged side chain of Glu 161. As a consequence, the doubly negative phosphoryl group of F6P has a greatly diminished affinity for the T-state enzyme. The unwinding of this helical turn, which is obligatory for the R → T transition, is prevented by the binding of the activator ADP to its effector site on PFK in the R state, and facilitated by the binding of ATP to this effector site in the T state. Evidently, the same conformational shift is responsible for both the homotropic and the heterotropic allosteric effects in PFK.

d. AMP Overcomes the ATP Inhibition of PFK

Direct allosteric control of PFK by ATP may superficially appear to be the means by which glycolytic flux is regulated. After all, when [ATP] is high as a result of low metabolic demand, PFK is inhibited and flux through the pathway is low; conversely, when [ATP] is low, flux through the pathway is high and ATP is synthesized to replenish the pool. Consideration of the physiological variation in ATP concentration, however, indicates that the situation

must be more complex. The metabolic flux through glycolysis may vary by 100-fold or more, depending on the metabolic demand for ATP. However, measurements of [ATP] *in vivo* at various levels of metabolic activity indicate that [ATP] varies <10% between rest and vigorous exertion. Yet *there is no known allosteric mechanism that can account for a 100-fold change in flux of a nonequilibrium reaction with only 10% change in effector concentration.* Thus, some other mechanism, or mechanisms, must be responsible for controlling glycolytic flux.

The inhibition of PFK by ATP is relieved by AMP. This results from AMP's preferential binding to the R state of PFK. If a PFK solution containing 1 m*M* ATP and 0.5 m*M* F6P is brought to 0.1 m*M* in AMP, the activity of PFK rises from 10 to 50% of its maximal activity, a 5-fold increase (Fig. 17-33).

[ATP] decreases by only 10% in going from a resting state to one of vigorous activity because it is buffered by the action of two enzymes: creatine kinase (Section 16-4C) and, of particular importance to this discussion, **adenylate kinase** (**AK;** also known as **myokinase**). Adenylate kinase catalyzes the reaction

$$2\text{ADP} \rightleftharpoons \text{ATP} + \text{AMP} \qquad K = \frac{[\text{ATP}][\text{AMP}]}{[\text{ADP}]^2} = 0.44$$

which rapidly equilibrates the ADP resulting from ATP hydrolysis in muscle contraction with ATP and AMP.

In muscle, [ATP] is ~50 times [AMP] and ~10 times [ADP], so that, *as a result of the adenylate kinase reaction, a 10% decrease in [ATP] will cause over a fourfold increase in [AMP]* (see Problem 11 in this chapter). Consequently, a metabolic signal consisting of a decrease in [ATP] too small to relieve PFK inhibition is amplified significantly by the adenylate kinase reaction, which increases [AMP] by an amount sufficient to produce a much larger increase in PFK activity.

e. Adenylate Kinase's Internal Motions Act as an Energetic Counterweight to Balance Substrate Binding

Adenylate kinase, like other kinases, must be specific to prevent undesirable phosphoryl-transfer reactions such as hydrolysis. However, once the reaction has occurred, the tightly bound products must be rapidly released to maintain the enzyme's catalytic efficiency. With kinases such as hexokinase and phosphoglycerate kinase, this process is accomplished by the closing of "jaws" on the bound substrates that open when product is formed (Figs. 17-5 and 17-15), a process that is presumably driven by the exergonic free energy change of the reaction the enzyme catalyzes. However, since the AK reaction is energetically neutral (it replaces one phosphodiester bond with another), AK specificity is accomplished by a somewhat different means. Comparison of the X-ray structures, determined by Georg Schulz, of unliganded AK with AK in complex with the inhibitory bisubstrate analog **Ap₅A** (two ADPs connected by a fifth phosphate) indicates that two ~30-residue domains of AK close over the Ap₅A, thereby

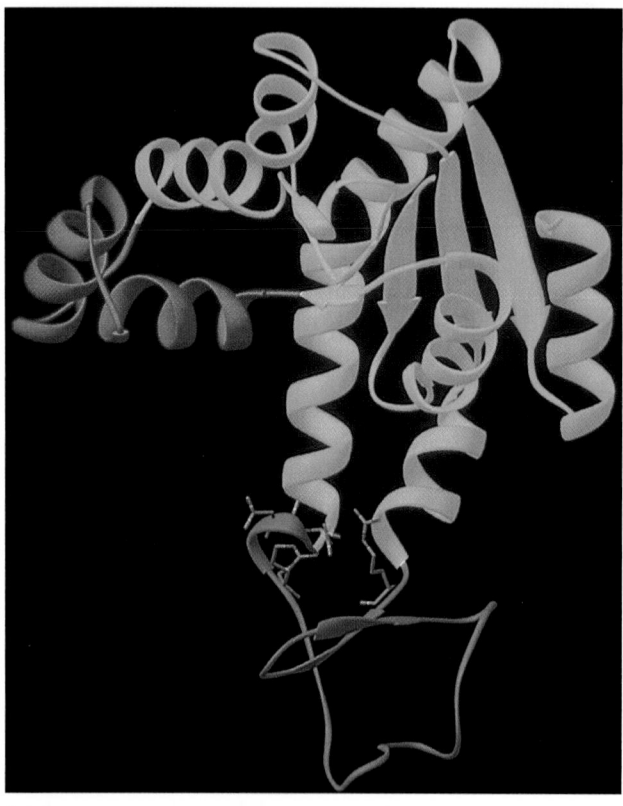

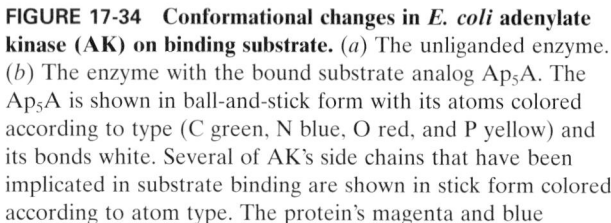

(a)

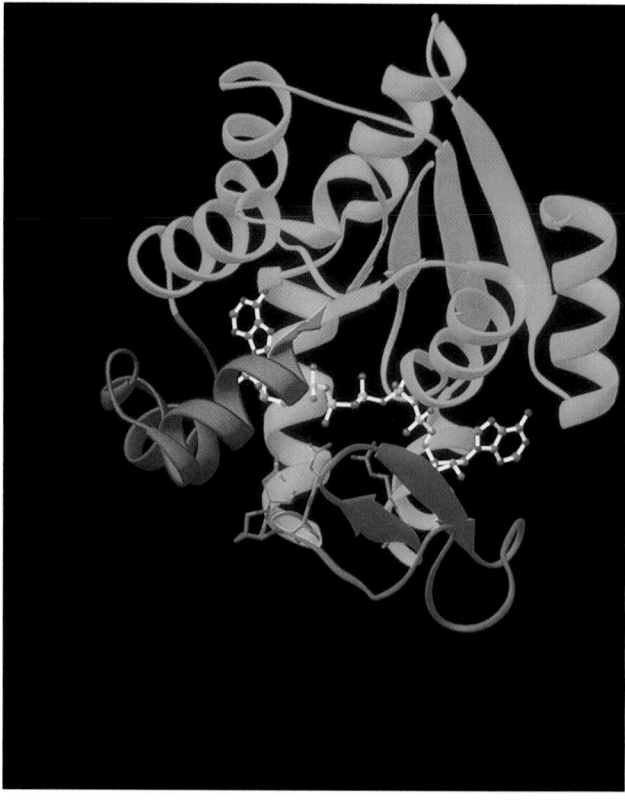

(b)

FIGURE 17-34   Conformational changes in *E. coli* adenylate kinase (AK) on binding substrate. (*a*) The unliganded enzyme. (*b*) The enzyme with the bound substrate analog Ap₅A. The Ap₅A is shown in ball-and-stick form with its atoms colored according to type (C green, N blue, O red, and P yellow) and its bonds white. Several of AK's side chains that have been implicated in substrate binding are shown in stick form colored according to atom type. The protein's magenta and blue domains undergo extensive conformational changes on ligand binding, whereas the remainder of the protein (*gold*), whose orientation is the same in Parts *a* and *b*, largely maintains its conformation. Compare these structures to that of porcine AK (Fig. 8-54*b*). [Based on X-ray structures by Georg Schulz, Institut für Organische Chemie und Biochemie, Freiburg, Germany. PDBids (*a*) 4AKE and (*b*) 1AKE.]

tightly binding it and excluding water (Fig. 17-34). This comparison also suggests how AK avoids falling into the energy well of tight-binding substrates and products. On binding substrate, a portion of the protein remote from the active site increases its chain mobility and thereby "absorbs" some of the free energy of substrate binding (recall that an X-ray structure determination reveals atomic mobilities as well as positions; Section 9-4). This region "resolidifies" on product release. This mechanism, Schulz has hypothesized, acts as an "energetic counterweight" that permits facile product release and hence maintains a high reaction rate.

f. Substrate Cycling Can Increase Flux Sensitivity

Even though a mechanism exists for amplifying the effect of a small change in [ATP] by producing a larger change in [AMP], a fourfold increase in [AMP] would allosterically increase the activity of PFK by only ~10-fold, an amount insufficient to account for the observed 100-fold increase in glycolytic flux. Small changes in effector concentration (and therefore v_f) can only cause relatively large changes in the flux through a reaction ($v_f - v_r$) if the reaction is functioning close to equilibrium. The reason for this high sensitivity is that for such reactions, the term $v_f/(v_f - v_r)$ in Eq. [17.8] (the elasticity coefficient) is large, that is, the reverse reaction contributes significantly to the value of the net flux. This is not the case for the PFK reaction.

Such equilibrium-like conditions may be imposed on a nonequilibrium reaction if a second enzyme catalyzes the regeneration of substrate from product in a thermodynamically favorable manner. Then v_r is no longer negligible compared to v_f. This situation requires that the forward process (formation of FBP from F6P) and reverse process (breakdown of FBP to F6P) be accomplished by different reactions since the laws of thermodynamics would otherwise be violated. In the following paragraphs, we discuss the nature of such **substrate cycles**.

Under physiological conditions, the reaction catalyzed by PFK:

Fructose-6-phosphate + ATP →

$$\text{fructose-1,6-bisphosphate} + \text{ADP}$$

is highly exergonic ($\Delta G = -25.9$ kJ · mol^{-1}, Table 17-1). Consequently, the back reaction has a negligible rate compared to the forward reaction. **Fructose-1,6-bisphosphatase (FBPase)**, however, which is present in many mammalian tissues (and which is an essential enzyme in gluconeogenesis; Section 23-1), catalyzes the exergonic hydrolysis of FBP ($\Delta G = -8.6$ kJ · mol^{-1}):

Fructose-1,6-bisphosphate + H$_2$O →

$$\text{fructose-6-phosphate} + \text{P}_i$$

Note that the combined reactions catalyzed by PFK and FBPase result in net ATP hydrolysis:

$$\text{ATP} + \text{H}_2\text{O} \rightleftharpoons \text{ADP} + \text{P}_i$$

Such a set of opposing reactions is known as a substrate cycle because it cycles a substrate to an intermediate and back again. When this set of reactions was discovered, it was referred to as a **futile cycle** since its net result seemed to be the useless consumption of ATP. In fact, when it was found that the PFK activators AMP and F2,6P allosterically inhibit FBPase, it was suggested that only one of these enzymes was functional in a cell under any given set of conditions. It was subsequently demonstrated, however, that both enzymes often function simultaneously at significant rates.

g. Substrate Cycling Can Account for Glycolytic Flux Variation

Eric Newsholme proposed that substrate cycles are not at all "futile" but, rather, have a regulatory function. The *in vivo* activities of enzymes and concentrations of metabolites are extremely difficult to measure, so that their values are rarely known accurately. However, let us make the physiologically reasonable assumption that a fourfold increase in [AMP], resulting from the adenylate kinase reaction, causes PFK activity (v_f) to increase from 10 to 90% of its maximum and FBPase activity (v_r) to decrease from 90 to 10% of its maximum. The maximum activity of muscle PFK is known from *in vitro* studies to be ~10-fold greater than that of muscle FBPase. Hence, if we assign full activity of PFK to be 100 arbitrary units, then full activity of FBPase is 10 such units. The flux through the PFK reaction in glycolysis under conditions of low [AMP] is

$$J_{low} = v_f(low) - v_r(low) = 10 - 9 = 1$$

where v_f is catalyzed by PFK and v_r by FBPase. The flux under conditions of high [AMP] is

$$J_{high} = v_f(high) - v_r(high) = 90 - 1 = 89$$

Substrate cycling could therefore amplify the effect of changes in [AMP] on the net rate of phosphorylation of F6P. Without the substrate cycle, a fourfold increase in [AMP] increases the net flux by about 9-fold, whereas with the cycle the same increase in [AMP] causes a $J_{high}/J_{low} = 89/1 \approx 90$-fold increase in net flux. Consequently, under the above assumptions, *a 10% change in [ATP] could stimulate a 90-fold change in flux through the glycolytic pathway by a combination of the adenylate kinase reaction and substrate cycles.*

h. Physiological Impact of Substrate Cycling

Substrate cycling, if it has a regulatory function, does not increase the maximum flux through a pathway. On the contrary, it functions to decrease its minimum flux. In a sense, the substrate is put into a "holding pattern." In the case described above, *the cycling of substrate is the energetic "price" that a muscle must pay to be able to change rapidly from a resting state, in which substrate cycling is maximal, to one of sustained high activity.* However, the rate of substrate cycling may itself be under hormonal or nervous control so as to increase the sensitivity of the metabolic system under conditions when high activity (fight or flight) is anticipated (we address the involvement of hormones in metabolic regulation in Sections 18-3E and 18-3F).

In some tissues, substrate cycles function to produce heat. For example, many insects require a thoracic temperature of 30°C to be able to fly. Yet bumblebees are capable of flight at ambient temperatures as low as 10°C. Bumblebee flight muscle FBPase has a maximal activity similar to that of its PFK (10-fold greater than our example for mammalian muscle); furthermore, unlike all other known muscle FBPases, it is not inhibited by AMP. This permits the FBPase and PFK of bumblebee flight muscle to be highly active simultaneously so as to generate heat. Since the maximal rate of FBP cycling possible in bumblebee flight muscle generates only 10 to 15% of the required heat, however, other mechanisms of thermogenesis must also be operative. Nevertheless, FBP cycling is probably significant because, unlike bumblebees, honeybees, which have no FBPase activity in their flight muscles, cannot fly when the temperature is low.

i. Substrate Cycling, Thermogenesis, and Obesity

Many animals, including adult humans, generate some of their body heat, particularly when it is cold, through substrate cycling in muscle and liver, one mechanism of a process known as **nonshivering thermogenesis** (the muscle contractions of shivering or any other movement also produce heat; another mechanism of nonshivering thermogenesis is described in Section 22-3D). Substrate cycling is stimulated by thyroid hormones (which stimulate metabolism in most tissues; Section 19-1D), as is indicated, for example, by the observation that rats lacking a functioning thyroid gland do not survive at 5°C. Chronically obese individuals tend to have lower than normal metabolic rates, which is probably due, in part, to a reduced rate of nonshivering thermogenesis. Such individuals therefore tend to be cold sensitive. Indeed, whereas normal individuals increase their rate of thyroid hormone activation on exposure to cold, genetically obese animals and obese humans fail to do so.

j. The Overexpression of PFK Does Not Increase the Rate of Glycolysis

PFK has long been thought to be the controlling enzyme of glycolysis. It was therefore expected that increasing the level of expression of PFK in yeast cells via genetic engineering techniques would increase the rate of glycolysis independent of the demand for glycolytic products. It has been amply demonstrated, however, that this is not the case. Although PFK is a major regulatory enzyme of glycolysis, its catalytic activity *in vivo* is controlled by the concentrations of the effectors that reflect the needs of the demand blocks that utilize its products.

Metabolic control analysis, in addition to helping us recognize that control can be shared by several enzymes in a pathway, has also alerted us to the difference between control and regulation. *Although PFK has a major role in regulating the flux through glycolysis, it is controlled, in vivo, by factors outside the pathway.* An increase in the *in vivo* concentration of PFK will therefore not increase the flux through the pathway because these controlling factors adjust the catalytic activity of PFK only to meet the needs of the cell.

5 ■ METABOLISM OF HEXOSES OTHER THAN GLUCOSE

While glucose is the primary end product of the digestion of starch and glycogen (Sections 11-2C and 11-2D), three other hexoses are also prominent digestion products: **Fructose,** obtained from fruits and from the hydrolysis of sucrose (table sugar); **galactose,** obtained from the hydrolysis of lactose (milk sugar); and **mannose,** obtained from the digestion of polysaccharides and glycoproteins. After digestion, these monosaccharides enter the bloodstream, which carries them to various tissues. *The metabolism of fructose, galactose, and mannose proceeds by their conversion to glycolytic intermediates, from which point they are broken down in a manner identical to glucose.*

A. *Fructose*

Fructose is a major fuel source in diets that contain large amounts of sucrose (a disaccharide of fructose and glucose; Fig. 11-12). There are two pathways for the metabolism of fructose; one occurs in muscle and the other occurs in liver. This dichotomy results from the different enzymes present in these various tissues.

Fructose metabolism in muscle differs little from that of glucose. Hexokinase (Section 17-2A), which converts glucose to G6P on entry into muscle cells, also phosphorylates fructose, yielding F6P (Fig. 17-35, *left*). The entry of fructose into glycolysis therefore involves only one reaction step.

Liver contains little hexokinase; rather, it contains glucokinase, which phosphorylates only glucose (Section 17-2A). Fructose metabolism in liver must therefore differ from that in muscle. In fact, liver converts fructose to glycolytic intermediates through a pathway that involves seven enzymes (Fig. 17-35, *right*):

1. Fructokinase catalyzes the phosphorylation of fructose by ATP at C1 to form **fructose-1-phosphate.** *Neither hexokinase nor phosphofructokinase can phosphorylate fructose-1-phosphate at C6 to form the glycolytic intermediate fructose-1,6-bisphosphate.*

2. Class I aldolase (Section 17-2D) has several isoenzyme forms. Muscle contains Type A aldolase, which is specific for fructose-1,6-bisphosphate. Liver, however, contains Type B aldolase, which also utilizes fructose-1-phosphate as a substrate (Type B aldolase is also called **fructose-1-phosphate aldolase**). In liver, fructose-1-phosphate therefore undergoes an aldol cleavage (Section 17-2D):

Fructose-1-phosphate $\rightleftharpoons$

dihydroxyacetone phosphate + glyceraldehyde

3. Direct phosphorylation of glyceraldehyde by ATP through the action of **glyceraldehyde kinase** forms the glycolytic intermediate glyceraldehyde-3-phosphate.

4–7. Alternatively, glyceraldehyde is converted to the glycolytic intermediate dihydroxyacetone phosphate by reduction to glycerol by NADH as catalyzed by alcohol dehydrogenase (Reaction 4), phosphorylation to glycerol-3-phosphate by ATP through the action of **glycerol kinase** (Reaction 5), and reoxidation by NAD^+ to dihydroxyacetone phosphate as mediated by glycerol phosphate dehydrogenase (Reaction 6). The DHAP is then converted to GAP by triose phosphate isomerase (Reaction 7).

As this complex series of reactions suggests, the liver has an enormous repertory of enzymes. This is because the liver is involved in the breakdown of a great variety of metabolites. Efficiency in metabolic processing dictates that many of these substances be converted to glycolytic intermediates. The liver, in fact, contains many of the enzymes necessary to do so.

a. Excessive Fructose Depletes Liver P_i

At one time, fructose was thought to have advantages over glucose for intravenous feeding. The liver, however, encounters metabolic problems when the blood concentration of this sugar is too high (higher than can be attained by simply eating fructose-containing foods). When the fructose concentration is high, fructose-1-phosphate may be produced faster than Type B aldolase can cleave it. Intravenous feeding of large amounts of fructose may therefore result in high enough fructose-1-phosphate accumulation to severely deplete the liver's store of P_i. Under

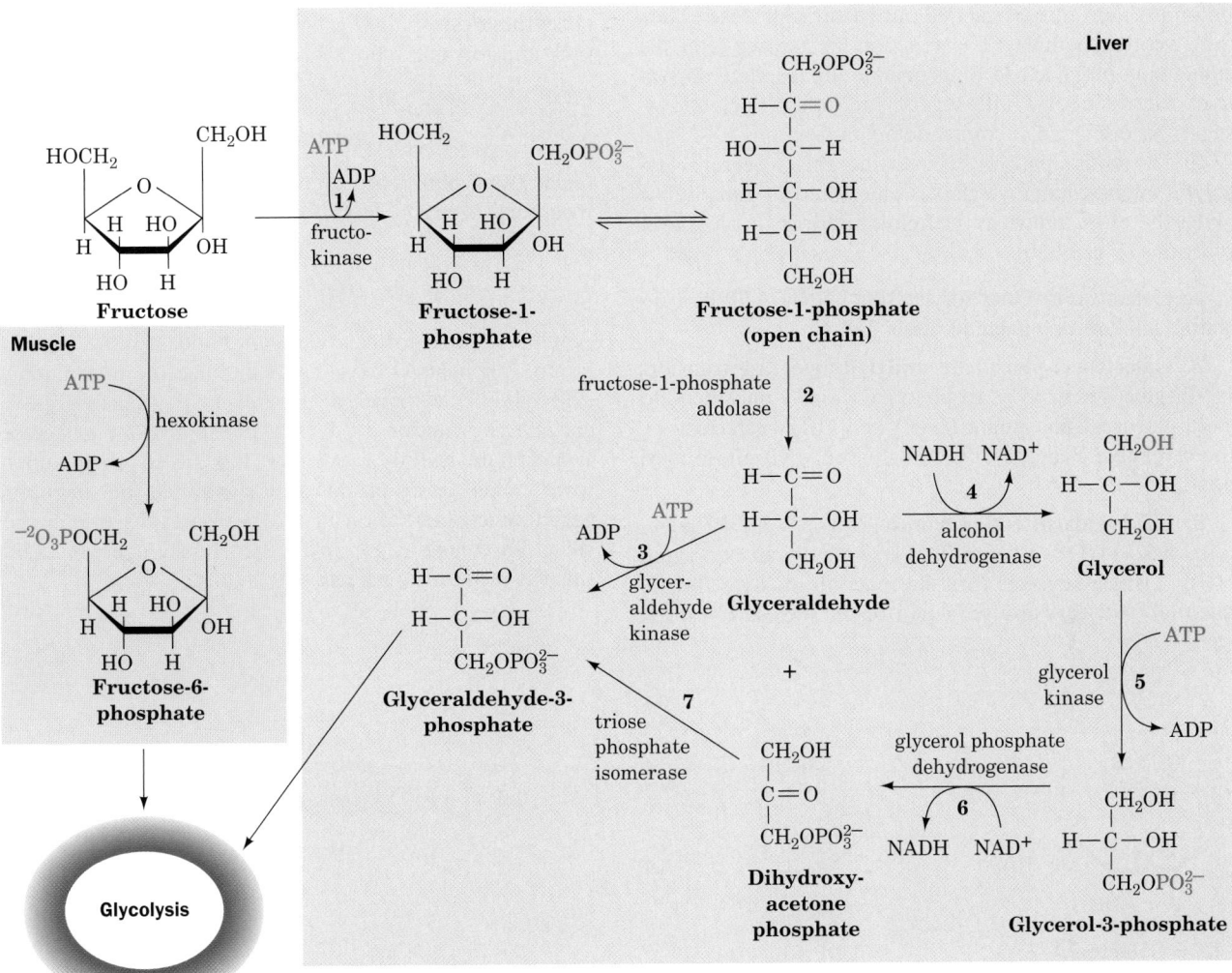

FIGURE 17-35 Metabolism of fructose. In muscle (*left*), the conversion of fructose to the glycolytic intermediate F6P involves only one enzyme, hexokinase. In liver (*right*), seven enzymes participate in the conversion of fructose to glycolytic intermediates: **(1)** fructokinase, **(2)** fructose-1-phosphate aldolase, **(3)** glyceraldehyde kinase, **(4)** alcohol dehydrogenase, **(5)** glycerol kinase, **(6)** glycerol phosphate dehydrogenase, and **(7)** triose phosphate isomerase.

these conditions, [ATP] drops, thereby activating glycolysis and lactate production. The lactate concentration in the blood and the consequent low pH under such conditions can reach life-threatening levels.

Fructose intolerance, a genetic disease in which ingestion of fructose causes the same fructose-1-phosphate accumulation as with its intravenous feeding, results from a deficiency of Type B aldolase. This condition appears to be self-limiting: Individuals with fructose intolerance rapidly develop a strong distaste for anything sweet.

B. *Galactose*

Galactose comprises half of the milk sugar lactose and is thus a major fuel constituent of dairy products. Galactose

and glucose are epimers that differ only in their configuration about C4:

α-ᴅ-**Glucose**   α-ᴅ-**Galactose**

The enzymes of glycolysis are specific; they do not recognize the galactose configuration. An epimerization reaction must therefore be carried out before galactose enters the glycolytic pathway. This reaction takes place after the conversion of galactose to its uridine diphosphate derivative. The role of UDP–sugars and other nucleotidyl–sugars is discussed in more detail in Sections 18-2 and 23-3. The entire pathway converting galactose to a glycolytic intermediate, which was elucidated by Luis Leloir and is therefore known as the **Leloir pathway,** involves four reactions (Fig. 17-36):

1. Galactose is phosphorylated at C1 by ATP in a reaction catalyzed by **galactokinase.**

2. Galactose-1-phosphate uridylyltransferase transfers **UDP–glucose's** uridylyl group to galactose-1-phosphate to yield **glucose-1-phosphate (G1P)** and **UDP–galactose** by the reversible cleavage of UDP–glucose's pyrophosphoryl bond.

3. UDP–galactose-4-epimerase converts UDP–galactose back to UDP–glucose. This enzyme has an associated NAD^+, which suggests that the reaction involves the sequential oxidation and reduction of the hexose C4 atom:

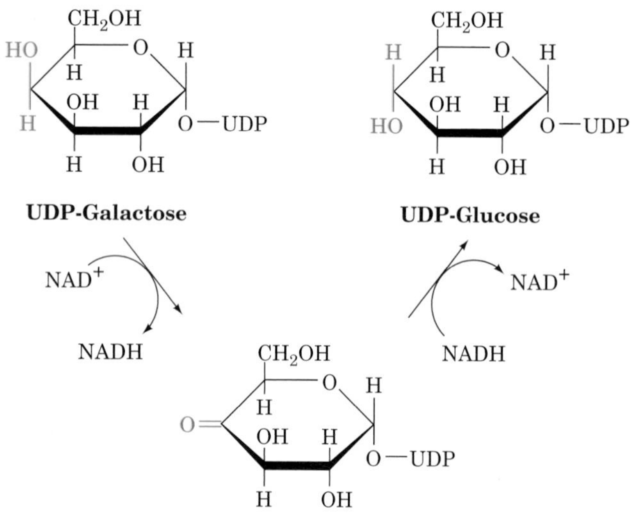

tacks the α-phosphoryl group of UDP–glucose, displacing G1P and forming a uridylyl–His intermediate:

UDP–glucose + E–His 166 $\rightleftharpoons$

glucose-1-phosphate + E–His-UMP

Galactose-1-phosphate then displaces the uridylyl group from the enzyme's His to form UDP–galactose:

Galactose-1-phosphate + E–His-UMP $\rightleftharpoons$

UDP–galactose + E–His

A Gln residue forms hydrogen bonds to the uridylyl group's phosphoryl oxygens to stabilize the intermediate uridylyl–His. Mutation of this Gln to Arg inactivates the enzyme. Formation of UDP–galactose from galactose-1-phosphate is thus prevented, leading to a buildup of toxic metabolic by-products. For example, the increased galactose concentration in the blood results in a higher galactose concentration in the lens of the eye where this sugar is reduced to **galactitol.**

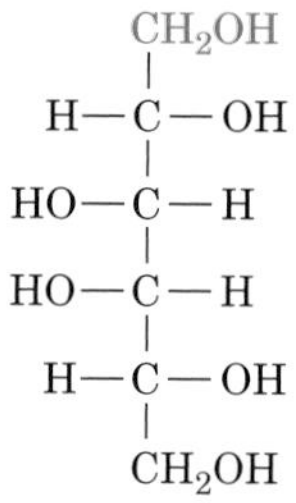

UDP-Galactose   **UDP-Glucose**

D-Galactitol

The presence of this sugar alcohol in the lens eventually causes cataract formation (clouding of the lens).

Galactosemia is treated by a galactose-free diet. Except for the mental retardation, this reverses all symptoms of the disease. The galactosyl units that are essential for the synthesis of glycoproteins (Section 11-3C) and glycolipids (Section 12-1D) may be synthesized from glucose by a reversal of the epimerase reaction. These syntheses therefore do not require dietary galactose.

4. G1P is converted to the glycolytic intermediate G6P by the action of **phosphoglucomutase** (Section 18-1B).

a. Galactosemia

Galactosemia is a genetic disease characterized by the inability to convert galactose to glucose. Its symptoms include failure to thrive, mental retardation, and in some instances death from liver damage. Most cases of galactosemia involve a mutation in the enzyme catalyzing Reaction 2 of the interconversion, galactose-1-phosphate uridylyltransferase. This reaction is a double displacement in which an enzyme His side chain first nucleophilically at-

C. Mannose

Mannose, a common component of glycoproteins (Section 11-3C), and glucose are C2 epimers:

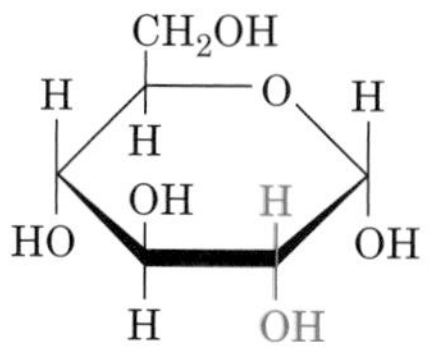

α-D-Glucose 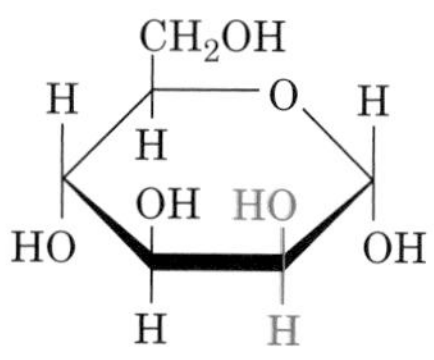   **α-D-Mannose**

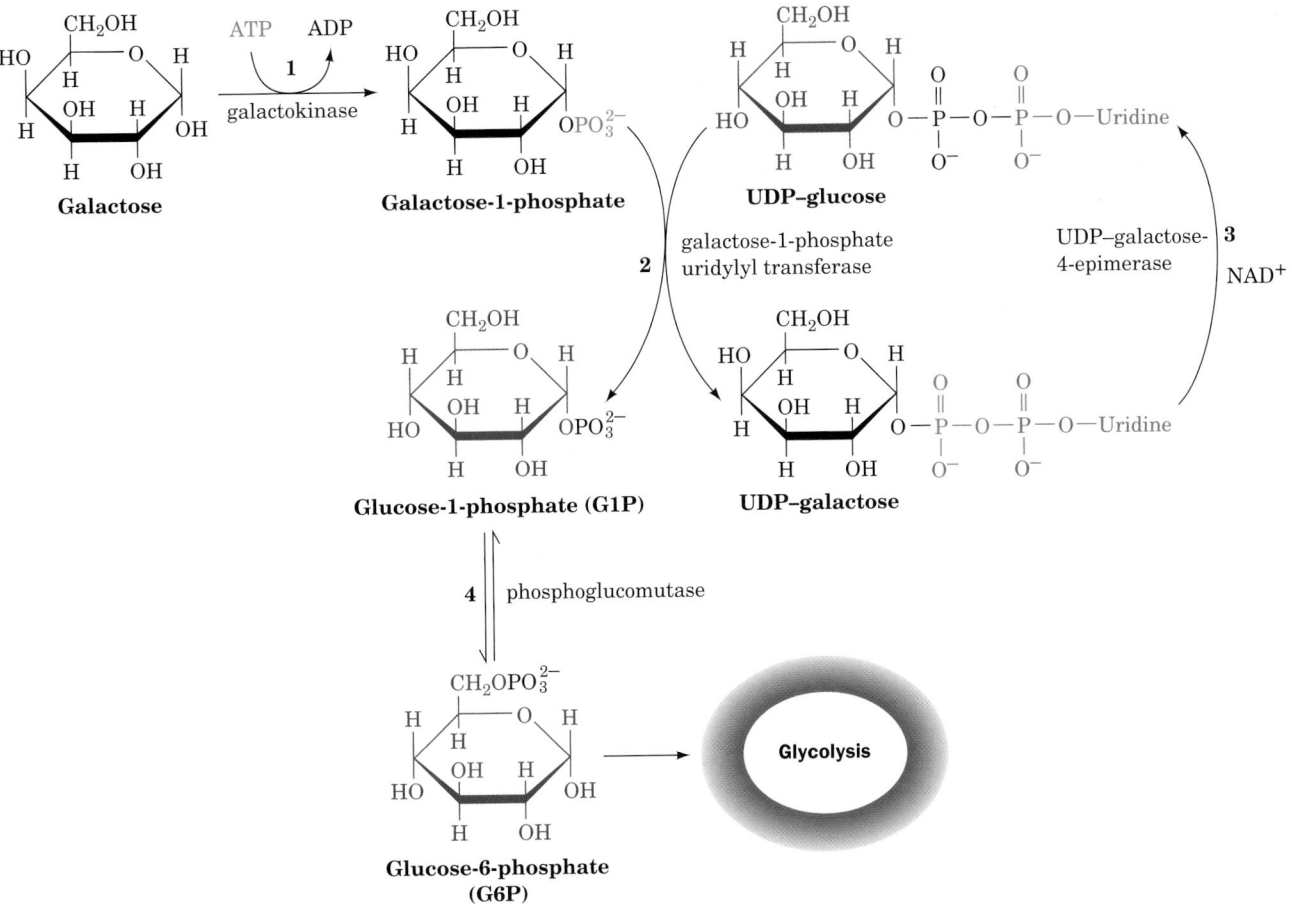

FIGURE 17-36 Metabolism of galactose. Four enzymes participate in the conversion of galactose to the glycolytic intermediate G6P: **(1)** galactokinase, **(2)** galactose-1-phosphate uridylyltransferase, **(3)** UDP–galactose-4-epimerase, and **(4)** phosphoglucomutase.

Mannose enters the glycolytic pathway after its conversion to F6P via a two-reaction pathway (Fig. 17-37):

1. Hexokinase (Section 17-2A) converts mannose to mannose-6-phosphate.

2. Phosphomannose isomerase then converts this aldose to the ketose F6P. The mechanism of the phosphomannose isomerase reaction resembles that catalyzed by phosphoglucose isomerase (Section 17-2B); it involves an enediolate intermediate.

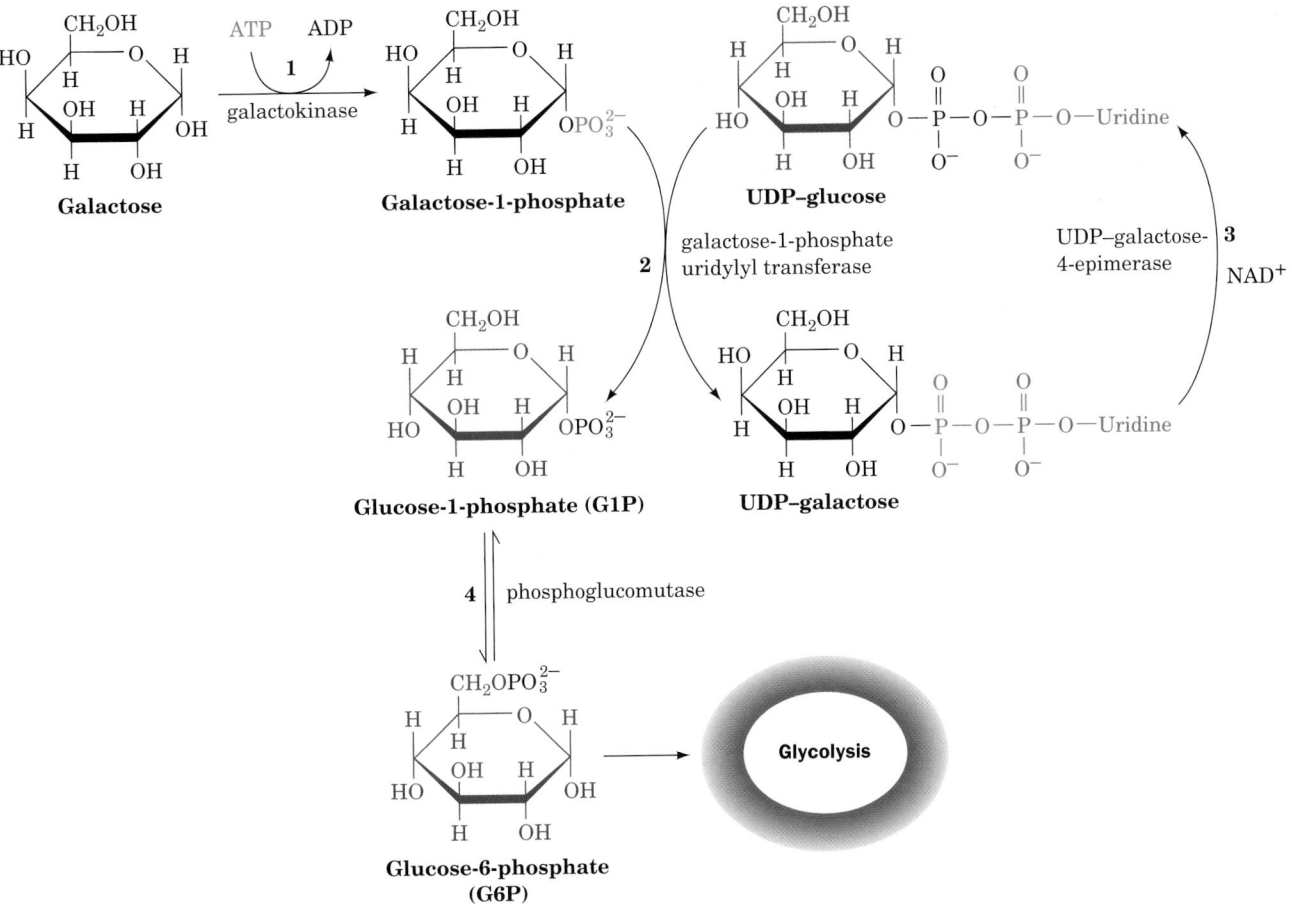

FIGURE 17-37 Metabolism of mannose. Two enzymes are required to convert mannose to the glycolytic intermediate F6P: **(1)** hexokinase and **(2)** phosphomannose isomerase.

CHAPTER SUMMARY

1 ■ The Glycolytic Pathway Glycolysis is the metabolic pathway by which most life-forms degrade glucose to two molecules of pyruvate with the concomitant net generation of two ATPs. The overall reaction:

$$\text{Glucose} + 2\text{NAD}^+ + 2\text{ADP} + 2\text{P}_i \rightarrow$$
$$2\text{NADH} + 2\text{pyruvate} + 2\text{ATP} + 2\text{H}_2\text{O} + 4\text{H}^+$$

occurs in 10 enzymatically catalyzed reactions.

2 ■ The Reactions of Glycolysis In the preparatory stage of glycolysis, which encompasses its first five reactions, glucose reacts with two ATPs, in an "energy investment," to form fructose-1,6-bisphosphate, which is subsequently converted to two molecules of glyceraldehyde-3-phosphate. In the second stage of glycolysis, the "payoff" stage, which comprises its last five reactions, glyceraldehyde-3-phosphate reacts with NAD^+ and P_i to form the "high-energy" compound 1,3-bisphosphoglycerate. This compound reacts in the last four reactions of the pathway with two ADPs to form pyruvate and two ATPs per molecule. The mechanisms of the 10 glycolytic enzymes have been elucidated through chemical and kinetic measurements combined with X-ray structural studies. The glycolytic enzymes exhibit stereospecificity in the reactions that they catalyze. In at least two kinases, phosphoryl transfer from substrate to water is prevented by substrate-induced conformational changes that form the active site and exclude water from it.

3 ■ Fermentation: The Anaerobic Fate of Pyruvate The NAD^+ consumed in the formation of 1,3-BPG must be regenerated if glycolysis is to continue. In the presence of O_2, NAD^+ is regenerated by oxidative phosphorylation in the mitochondria. Under anaerobic conditions in muscle, pyruvate is reduced by NADH, yielding lactate and NAD^+ in a reaction catalyzed by lactate dehydrogenase. In many muscles, particularly during strenuous activity, the process of homolactic fermentation is a major free energy source. In anaerobic yeast, NAD^+ is regenerated by alcoholic fermentation in two reactions. First pyruvate is decarboxylated to acetaldehyde by pyruvate decarboxylase, an enzyme that requires thiamine pyrophosphate as a cofactor. The acetaldehyde is then reduced by NADH to form ethanol and NAD^+ in a reaction catalyzed by alcohol dehydrogenase.

4 ■ Metabolic Regulation and Control Metabolic regulation is the process by which the steady-state flow of metabolites through a pathway is maintained. Metabolic control is the force exerted on the enzymes of the pathway in response to an external signal in order to increase or decrease the flow while maintaining the steady state to the extent possible. Homeostasis is the regulation of the steady state. Metabolic flow must be controlled to balance supply with demand and also to maintain homeostasis. It is possible for more than one enzyme to be rate-limiting in a metabolic pathway. Metabolic control analysis provides a framework for the study of metabolic systems *in vivo* that share control among more than one enzyme, and it quantitatively describes flux control and homeostasis. The flux control coefficient measures the sensitivity of the flux to a change in enzyme concentration. The elasticity coefficient measures the sensitivity of an enzymatic rate to the change in substrate concentration. Both supply and demand are involved in flux control and homeostasis. The response of the steady-state concentration of intermediates to changes in the supply or demand blocks depends entirely on the elasticity coefficients of the two blocks at the steady state. When the supply elasticity coefficient is greater than the demand elasticity coefficient, flux control lies in the demand portion of the pathway, and vice versa. Homeostasis control depends on large elasticity coefficients, whereas flux control requires a low elasticity coefficient and a high flux control coefficient. If the demand block controls the flux, the function of the supply block is to control homeostasis. Feedback inhibition determines the range of concentration of intermediates at which there is a steady state. It maintains homeostasis at physiologically reasonable metabolite concentrations, sometimes far from their equilibrium values.

The flux through a reaction that is close to equilibrium is very sensitive to changes in substrate concentration. Hence, the steady-state flux through a metabolic pathway can only be regulated by a nonequilibrium reaction. Nonequilibrium reactions are controlled by allosteric interactions, substrate cycles, covalent modification, and genetic (long-term) control mechanisms. In muscle glycolysis, phosphofructokinase (PFK) catalyzes one of the flux-generating steps. Although PFK is inhibited by high concentrations of one of its substrates, ATP, the 10% variation of [ATP] over the range of metabolic activity has insufficient influence on PFK activity to account for the observed 100-fold range in glycolytic flux. [AMP] has a fourfold variation in response to the 10% variation of [ATP] through the action of adenylate kinase. Although AMP relieves the ATP inhibition of PFK, its concentration variation is also insufficient to account for the observed glycolytic flux range. However, the product of the PFK reaction, fructose-1,6-bisphosphate, is hydrolyzed to F6P by FBPase, which is inhibited by AMP. The substrate cycle catalyzed by these two enzymes confers, at least in principle, the necessary sensitivity of the glycolytic flux to variations in [AMP]. Substrate cycling is an important source of nonshivering thermogenesis.

5 ■ Metabolism of Hexoses Other than Glucose Digestion of carbohydrates yields glucose as the primary product. Other prominent products are fructose, galactose, and mannose. These monosaccharides are metabolized through their conversion to glycolytic intermediates.

REFERENCES

GENERAL

Cornish-Bowden, A. (Ed.), *New Beer in an Old Bottle: Eduard Buchner and the Growth of Biochemical Knowledge*, Universitat de València (1997). [The 1897 paper by Eduard Buchner reporting the discovery of cell-free fermentation (in the original German as well as its English and Spanish translations) together with a series of essays discussing the historical context of this discovery and the modern study of multienzyme systems.]

Fersht, A., *Structure and Mechanism in Protein Science,* Freeman (1999).

Fruton, J.S., *Molecules and Life: Historical Essays on the Interplay of Chemistry and Biology,* Wiley–Interscience (1974). [Includes a detailed historical account of the elucidation of fermentation.]

Saier, M.H., Jr., *Enzymes in Metabolic Pathways,* Chapter 5, Harper & Row (1987).

Walsh, C., *Enzymatic Reaction Mechanisms,* Freeman (1979).

ENZYMES OF GLYCOLYSIS

The Enzymes of Glycolysis: Structure, Activity and Evolution, *Phil. Trans. R. Soc. London Ser. B* **293,** 1–214 (1981). [A collection of authoritative discussions on the enzymes of glycolysis.]

Allen, S.C. and Muirhead, H., Refined three-dimensional structure of cat-muscle (M1) pyruvate kinase at a resolution of 2.6 Å, *Acta Cryst.* D**52,** 499–504 (1996).

Bennett, W.S., Jr. and Steitz, T.A., Glucose-induced conformational change in yeast hexokinase, *Proc. Natl. Acad. Sci.* **75,** 4848–4852 (1978).

Berstein, B.E., Michels, P.A.M., and Hol, W.G.J., Synergistic effects of substrate-induced conformational changes in phosphoglycerate activation, *Nature* **385,** 275–278 (1997).

Biesecker, G., Harris, J.I., Thierry, J.C., Walker, J.E., and Wonacott, A.J., Sequence and structure of D-glyceraldehyde-3-phosphate dehydrogenase from *Bacillus stearothermophilus, Nature* **266,** 328–333 (1977).

Boyer, P.D. (Ed.), *The Enzymes* (3rd ed.), Vols. 5–9 and 13, Academic Press (1972–1976). [Contains early detailed reviews of the various glycolytic enzymes.]

Cleland, W.W. and Kreevoy, M.M., Low-barrier hydrogen bonds and enzymic catalysis, *Science* **264,** 1887–1890 (1994); *and* Gerlt, J.A. and Gassman, P.G., Understanding the rates of certain enzyme-catalyzed reactions: Proton abstraction from carbon acids, acyl-transfer reactions, and displacement of phosphodiesters, *Biochemistry* **32,** 11943–11952 (1993).

Dalby, A., Dauter, Z., and Littlechild, J.A., Crystal structure of human muscle aldolase complexed with fructose 1,6-bisphosphate: Mechanistic implications, *Protein Sci.* **8,** 291–297 (1999).

Davenport, R.C., Bash, P.A., Seaton, B.A., Karplus, M., Petsko, G.A., and Ringe, D., Structure of the triosephosphate isomerase–phosphoglycohydroxamate complex: An analogue of the intermediate on the reaction pathway, *Biochemistry* **30,** 5821–5826 (1991); *and* Lolis, E. and Petsko, G.A., Crystallographic analysis of the complex between triosephosphate isomerase and 2-phosphoglycolate at 2.5 Å resolution: Implications for catalysis, *Biochemistry* **29,** 6619–6625 (1990).

Evans, P.R. and Hudson, P.J., Structure and control of phosphofructokinase from *Bacillus stearothermophilus, Nature* **279,** 500–504 (1979).

Gefflaut, T., Blonski, C., Perie, J., and Willson, M., Class I aldolases: Substrate specificity, mechanism, inhibitors and structural aspects, *Prog. Biophys. Mol. Biol.* **63,** 301–340 (1995).

Hall, D.R., Leonard, G.A., Reed, C.D., Watt, C.I., Berry, A., and Hunter, W.N., The crystal structure of *Escherichia coli* class II fructose-1,6-bisphosphate aldolase in complex with phosphoglycohydroxamate reveals details of mechanism and specificity, *J. Mol. Biol.* **287,** 383–394 (1999).

Harlos, K., Vas, M., and Blake, C.C.F., Crystal structure of the binary complex of pig muscle phosphoglycerate kinase and its substrate 3-phospho-D-glycerate, *Proteins* **12,** 133–144 (1992).

Jedrzejas, M.J., Structure, function, and evolution of phospho-glycerate mutase: Comparison with fructose-2,6-bisphosphatase, acid phosphatase, and alkaline phosphatase, *Prog. Biophys. Mol. Biol.* **73,** 263–287 (2000).

Jeffrey, C.J., Bahnson, B.J., Chien, W., Ringe, D., and Petsko, G.A., Crystal structure of rabbit phosphoglucose isomerase, a glycolytic enzyme that moonlights as neuroleukin, autocrine motility factor, and differentiation mediator, *Biochemistry* **39,** 955–964 (2000).

Joseph, D., Petsko, G.A., and Karplus, M., Anatomy of a conformational change: Hinged "lid" motion of the triosephosphate isomerase loop, *Science* **249,** 1425–1428 (1990).

Knowles, J.R., Enzyme catalysis: Not different, just better, *Nature* **350,** 121–124 (1991). [A lucid discussion of TIM's catalytic mechanism.]

Kuby, S.A. (Ed.), *A Study of Enzymes,* Vol. II, CRC Press (1991). [Chapters 17, 18, 19, and 20 discuss the mechanisms of adenylate kinase, PFK, PGI and TIM, and aldolase, respectively. Chapter 4 discusses thiamine-dependent reaction mechanisms.]

Marsh, J.J. and Lebherz, H.G., Fructose-bisphosphate aldolases: An evolutionary history, *Trends Biochem. Sci.* **17,** 110–113 (1992).

Maurer, P.J. and Nowak, T., Fluoride inhibition of yeast enolase. 1. Formation of ligand complexes, *Biochemistry* **20,** 6894–6900 (1981); *and* Nowak, T. and Maurer, P.J., Fluoride inhibition of yeast enolase. 2. Structural and kinetic properties of ligand complexes determined by nuclear relaxation rate studies, *Biochemistry* **20,** 6901–6911 (1981).

Morris, A.J. and Tolan, D.R., Lysine-146 of rabbit muscle aldolase is essential for cleavage and condensation of the C3-C4 bond of fructose 1,6-bis(phosphate), *Biochemistry* **33,** 12291–12297 (1994); *and* Site-directed mutagenesis identifies aspartate 33 as a previously unidentified critical residue in the catalytic mechanism of rabbit aldolase A, *J. Biol. Chem.* **268,** 1095–1100 (1993).

Muirhead, H. and Watson, H. Glycolytic enzymes: From hexose to pyruvate, *Curr. Opin. Struct. Biol.* **2,** 870–876 (1992).

Seeholzer, S.H., Phosphoglucose isomerase: A ketol isomerase with aldol C2-epimerase activity, *Proc. Natl. Acad. Sci.* **90,** 1237–1241 (1993).

Reed, G.H., Poyner, R.R., Larsen, T.M., Wedekind, J.E., and Rayment, I., Structural and mechanistic studies on enolase, *Curr. Opin. Struct. Biol.* **6,** 736–743 (1996).

Williams, J.C. and McDermott, A.E., Dynamics of the flexible loop of triosephosphate isomerase: The loop motion is not ligand-gated, *Biochemistry* **34,** 8309–8319 (1995).

ENZYMES OF ANAEROBIC FERMENTATION

Boyer, P.D. (Ed.), *The Enzymes* (3rd ed.), Vol. 11, Academic Press (1975). [Contains authoritative reviews on alcohol dehydrogenase, lactate dehydrogenase, and the evolutionary and structural relationships among the dehydrogenases.]

Dyda, F., Furey, W., Swaminathan, S., Sax, M., Farrenkopf, B., and Jordan, F., Catalytic centers in the thiamin diphosphate dependent enzyme pyruvate decarboxylase at 2.4-Å resolution, *Biochemistry* **32,** 6165–6170 (1993).

Golbik, R., Neef, H., Hubner, G., Konig, S., Seliger, B., Meshalkina, L., Kochetov, G.A., and Schellenberger, A., Function of the aminopyridine part in thiamine pyrophosphate enzymes, *Bioinorg. Chem.* **19,** 10–17 (1991).

Park, J.H., Brown, R.L., Park, C.R., Cohn, M., and Chance, B., Energy metabolism in the untrained muscle of elite runners as observed by ^{31}P magnetic resonance spectroscopy: Evidence suggesting a genetic endowment for endurance exercise. *Proc. Natl. Acad. Sci.* **85,** 8780–8785 (1988).

CONTROL OF METABOLIC FLUX

Crabtree, B. and Newsholme, E.A., A systematic approach to describing and analyzing metabolic control systems, *Trends Biochem. Sci.* **12,** 4–12 (1987).

Fell, D.A., Metabolic control analysis: A survey of its theoretical and experimental development, *Biochem. J.* **286,** 313–330 (1992).

Fell, D., *Understanding the Control of Metabolism,* Portland Press (1997).

Hofmeyr, J.-H. S. and Cornish-Bowden, A., Regulating the cellular economy of supply and demand, *FEBS Lett.* **476,** 47–51 (2000).

Kacser, H. and Burns, J.A. (with additional comments by Kacser, H. and Fell, D.A.), The control of flux, *Biochem. Soc. Trans.* **23,** 341–366 (1995).

Kacser, H. and Porteous, J.W., Control of metabolism: What do we have to measure? *Trends Biochem. Sci.* **12,** 5–14 (1987).

Lardy, H. and Schrago, E., Biochemical aspects of obesity, *Annu. Rev. Biochem.* **59,** 689–710 (1990).

Newsholme, E.A., Challiss, R.A.J., and Crabtree, B., Substrate cycles: their role in improving sensitivity in metabolic control, *Trends Biochem. Sci.* **9,** 277–280 (1984).

Perutz, M.F., Mechanism of cooperativity and allosteric regulation in proteins, *Q. Rev. Biophys.* **22,** 139–236 (1989). [Section 6 discusses PFK.]

Schaaf, I., Heinisch, J., and Zimmermann, K., Overproduction of glycolytic enzymes in yeast, *Yeast* **5,** 285–290 (1989).

Schirmer, T. and Evans, P.R., Structural basis of the allosteric behaviour of phosphofructokinase, *Nature* **343,** 140–145 (1990).

Walsh, K. and Koshland, D.E., Jr., Characterization of rate-controlling steps *in vivo* by use of an adjustable expression vector, *Proc. Natl. Acad. Sci.* **82,** 3577–3581 (1985).

METABOLISM OF HEXOSES OTHER THAN GLUCOSE

Frey, P.A., The Leloir pathway: A mechanistic imperative for three enzymes to change the stereochemical configuration of a single carbon in galactose, *FASEB J.* **10,** 461–470 (1996).

Scriver, C.R., Beaudet, A.L., Sly, W.S., and Valle, D. (Eds.), *The Metabolic & Molecular Bases of Inherited Disease* (8th ed.), McGraw-Hill (2001). [Chapters 70 and 72 discuss fructose and galactose metabolism and their genetic disorders.]

PROBLEMS

1. Write out the reactions of the glycolytic pathway from glucose to lactate using structural formulas for all intermediates. Learn the names of these intermediates and the enzymes that catalyze the reactions.

2. $\Delta G^{\circ\prime}$ for the aldolase reaction is $+22.8$ kJ $\cdot$ mol^{-1}. In the cell, at 37°C, the mass action ratio [DHAP]/[GAP] $= 5.5$. Calculate the equilibrium ratio of [FBP]/[GAP] when [GAP] is (a) $2 \times 10^{-5}M$ and (b) $10^{-3}M$.

3. The pH dependence of the rate of the triose phosphate isomerase (TIM) reaction has characteristic pK's of 6.5 and 9.5. His 95, a catalytically essential residue, has been shown to have a pK of 4.5. Why doesn't the pH rate curve indicate the existence of this pK?

4. Arsenate, a structural analog of phosphate, can act as a substrate for any reaction in which phosphate is a substrate. Arsenate esters, unlike phosphate esters, are kinetically as well as thermodynamically unstable and hydrolyze almost instantaneously. Write a balanced overall equation for conversion of glucose to pyruvate in the presence of ATP, ADP, NAD$^+$, and either (a) phosphate or (b) arsenate. (c) Why is arsenate a poison?

5. When glucose is degraded anaerobically via glycolysis there is no overall oxidation or reduction of the substrate. The fermentation reaction is therefore said to be "balanced." The free energy required for ATP formation is nevertheless obtained from favorable electron-transfer reactions. Which metabolic intermediate is the electron donor and which is the electron acceptor when glucose is degraded by a balanced glycolytic fermentation: (a) in muscle and (b) in yeast?

6. In which carbon atoms of pyruvate would radioactivity be found if glucose metabolized by the glycolytic pathway were labeled with ^{14}C at: (a) C1 and (b) C4? (*Note:* Assume that triose phosphate isomerase is able to equilibrate dihydroxyacetone phosphate and glyceraldehyde-3-phosphate.)

*7. The following reaction is catalyzed by an enzyme very similar to Class I aldolases:

Fructose-6-phospate   **Erythrose-4-phospate**

transaldolase

Glyceraldehyde-3-phosphate   **Sedoheptulose-7-phosphate**

Write a plausible mechanism for this reaction using curved arrows to indicate the electron flow.

8. The half-reactions involved in the LDH reaction and their standard reduction potentials are:

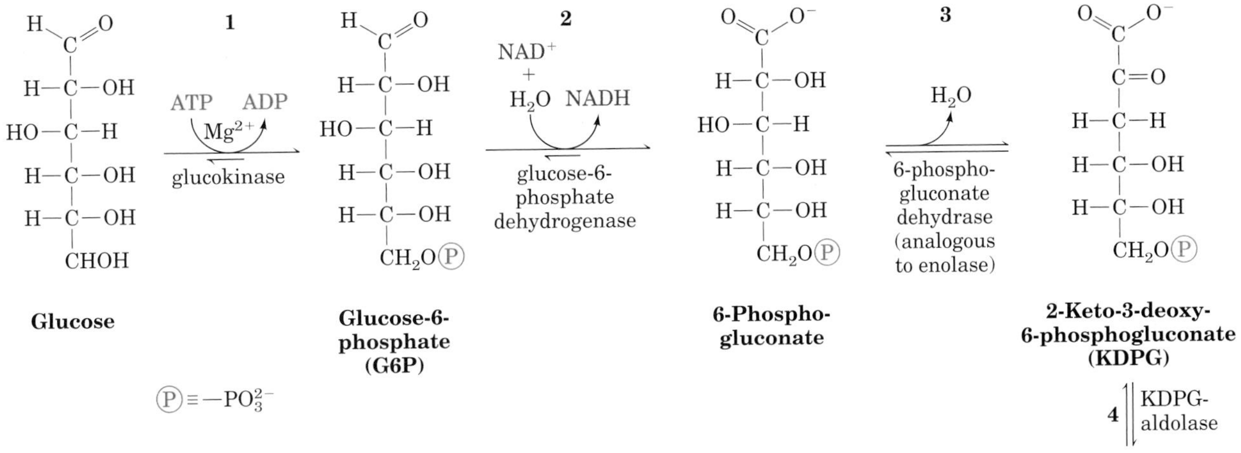

Glucose

Glucose-6-phosphate (G6P)

6-Phospho-gluconate

2-Keto-3-deoxy-6-phosphogluconate (KDPG)

$\text{(P)} \equiv -PO_3^{2-}$

ATP ADP, Mg^{2+}, glucokinase (1)

$NAD^+ + H_2O$, NADH, glucose-6-phosphate dehydrogenase (2)

H_2O, 6-phosphogluconate dehydrase (analogous to enolase) (3)

KDPG-aldolase (4)

Pyruvate

Glyceraldehyde-3-phosphate (GAP)

$CO_2 + CH_3CH_2OH$ **Ethanol**

same reactions as in glycolysis and alcoholic fermentation

NAD^+ NADH

$2ATP + H_2O$ $2ADP + P_i$

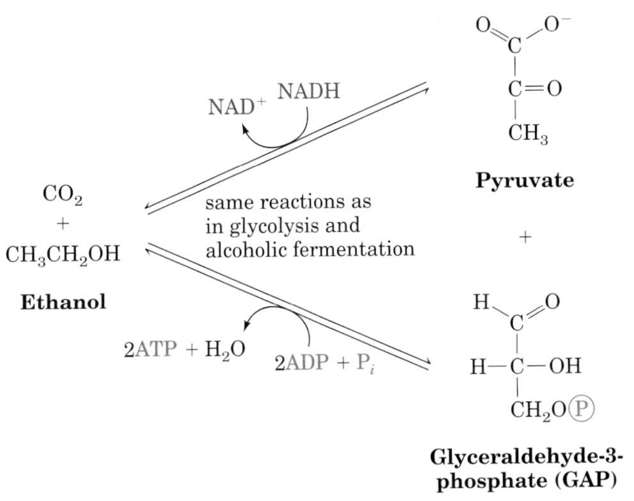

FIGURE 17-38 Entner–Doudoroff pathway for glucose breakdown.

Pyruvate $+ 2H^+ + 2e^- \rightarrow$ lactate $\mathscr{E}^{\circ\prime} = -0.185$ V

$NAD^+ + 2H^+ + 2e^- \rightarrow NADH + H^+$ $\mathscr{E}^{\circ\prime} = -0.315$ V

Calculate ΔG for the reaction under the following conditions:

(a) [lactate]/[pyruvate] = 1; $[NAD^+]/[NADH]$ = 1
(b) [lactate]/[pyruvate] = 160; $[NAD^+]/[NADH]$ = 160
(c) [lactate]/[pyruvate] = 1000; $[NAD^+]/[NADH]$ = 1000
(d) Under what conditions will the reaction spontaneously favor NADH oxidation?
(e) In order for the free energy change of the glyceraldehyde-3-phosphate dehydrogenase reaction to favor glycolysis, the $[NAD^+]/[NADH]$ ratio must be maintained close to 10^3. Under anaerobic conditions in mammalian muscle, lactate dehydrogenase performs this function. How high can the [lactate]/[pyruvate] ratio become in muscle cells before the LDH-catalyzed reaction ceases to be favorable in the direction of NAD^+ production while maintaining the foregoing $[NAD^+]/[NADH]$ ratio constant?

*9. Based on the involvement of thiamine pyrophosphate (TPP) in the pyruvate decarboxylase reaction, which of the following reactions, if any, might be expected to utilize TPP as a cofactor?

(a)

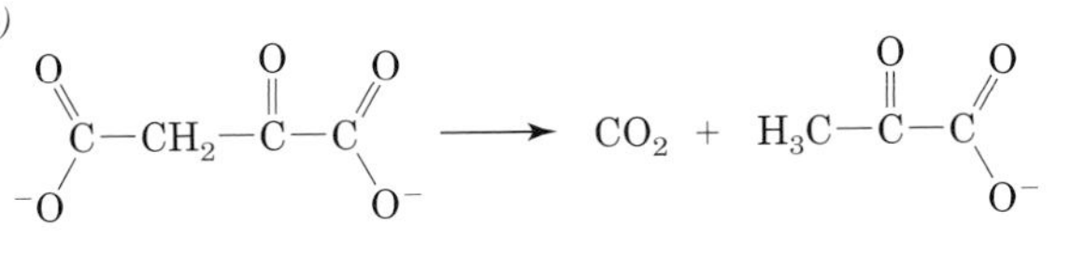

$^-O-\overset{O}{\overset{\|}{C}}-CH_2-\overset{O}{\overset{\|}{C}}-\overset{O}{\overset{\|}{C}}-O^- \longrightarrow CO_2 + H_3C-\overset{O}{\overset{\|}{C}}-\overset{O}{\overset{\|}{C}}-O^-$

(b)

$2 \ \ \overset{O\ \ \ \ O^-}{\underset{CH_3}{\overset{\diagdown\diagup}{\underset{|}{C}}}}\underset{|}{\overset{}{C=O}} \longrightarrow CO_2 + HO-\overset{}{\underset{\underset{CH_3}{\overset{|}{C=O}}}{\overset{|}{C}}}-CH_3$

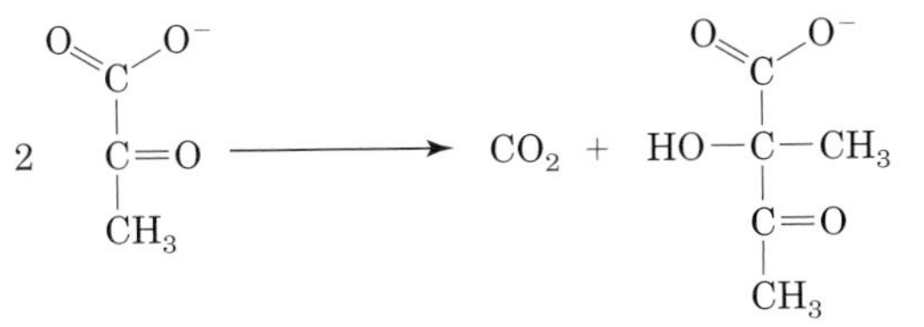

Write hypothetical mechanisms for each reaction showing where TPP is involved or why it is unnecessary.

10. The glycolytic pathway for glucose breakdown is almost universal. Some bacteria, however, utilize an alternate route called the **Entner-Doudoroff pathway** (Fig. 17-38). Like the glycolytic pathway in yeast, the final product is ethanol. (a) Write balanced equations for the conversion of glucose to ethanol and CO_2 via the Entner-Doudoroff pathway and the yeast alcoholic fermentation. (b) Infer from your stoichiometries why the glycolytic pathway rather than the Entner-Doudoroff pathway is almost universal.

*11. The hydrolysis of ATP to ADP in the cell results in a concomitant change in [AMP] as mediated by adenylate kinase. (a) Assuming that [ATP] >> [AMP] and that the total adenine nucleotide concentration in the cell, A_T = [AMP] + [ADP] + [ATP], is constant, derive an expression for [AMP] in terms of [ATP] and A_T. (b) Assuming an initial [ATP]/[ADP] of 10 and A_T = 5 m*M*, calculate the ratio of the final to initial values of [AMP] on a 10% decrease of [ATP].

Chapter 18

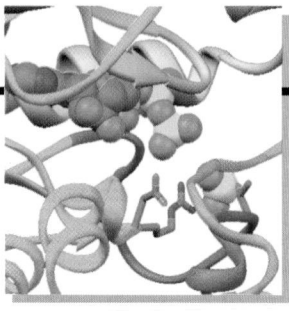

Glycogen Metabolism

Everything should be made as simple as possible, but not simpler.

Albert Einstein

Glucose, a major metabolic fuel source, is degraded via glycolysis to produce ATP (Chapter 17). Higher organisms protect themselves from potential fuel shortage by polymerizing excess glucose for storage as high molecular mass glucans (glucose polysaccharides) that may be readily mobilized in times of metabolic need. In plants, this glucose storage substance is starch, a mixture of the $\alpha(1\rightarrow4)$-linked glucan α-amylose (Fig. 11-17) and amylopectin, which differs from α-amylose by the presence of $\alpha(1\rightarrow6)$ branches every 24 to 30 residues (Fig. 11-18). In animals, the storage glucan is **glycogen** (Fig. 18-1), which differs from amylopectin only in that its branches occur every 8 to 14 residues. Glycogen occurs in 100- to 400-Å-diameter cytoplasmic granules (Figs. 11-19 and 18-1c), which contain up to 120,000 glucose units. They are especially prominent in the cells that make the greatest use of glycogen, muscle (maximally 1–2% glycogen by weight) and liver cells (maximally 10% glycogen by weight, an ~12-h energy supply

for the body). Glycogen granules also contain the enzymes that catalyze glycogen synthesis and degradation as well as some of the enzymes that regulate these processes.

As we shall see in this chapter, glycogen's glucose units are mobilized by their sequential removal from the glucan chains' nonreducing ends (ends lacking a C1—OH group). *Glycogen's highly branched structure is therefore physiologically significant: It permits glycogen's rapid degradation through the simultaneous release of the glucose units at the end of every branch.*

Why does the body go to such metabolic effort to use glycogen for energy storage when fat, which is far more abundant in the body, seemingly serves the same purpose? The answer is threefold:

   1. Muscles cannot mobilize fat as rapidly as they can glycogen.

   2. The fatty acid residues of fat cannot be metabolized anaerobically (Section 25-2).

   3. Animals cannot convert fatty acids to glucose (Section 23-1), so fat metabolism alone cannot adequately maintain essential blood glucose levels (Section 18-3F).

As with all metabolic processes, there are several levels on which glycogen metabolism may be understood. We shall examine this process in order to understand the pathway's thermodynamics and the reaction mechanisms of its individual steps but will emphasize the mechanisms by which glycogen synthesis and breakdown rates are controlled. We began our consideration of metabolic control mechanisms in Section 17-4 with a discussion of the role of allosteric interactions and substrate cycles in the regulation of glycolysis. Glycogen metabolism's more complex control systems provide us with examples of several additional control processes: covalent modification of enzymes and enzyme cascades. In addition, we shall consider glycogen metabolism as a model for the role of hormones in the overall regulatory process. We end the chapter by discussing the consequences of genetic defects in various enzymes of glycogen metabolism.

1 ■ GLYCOGEN BREAKDOWN

Liver and muscle are the two major storage tissues for glycogen. In muscle, the need for ATP results in the conversion of glycogen to glucose-6-phosphate (G6P) for

(a)

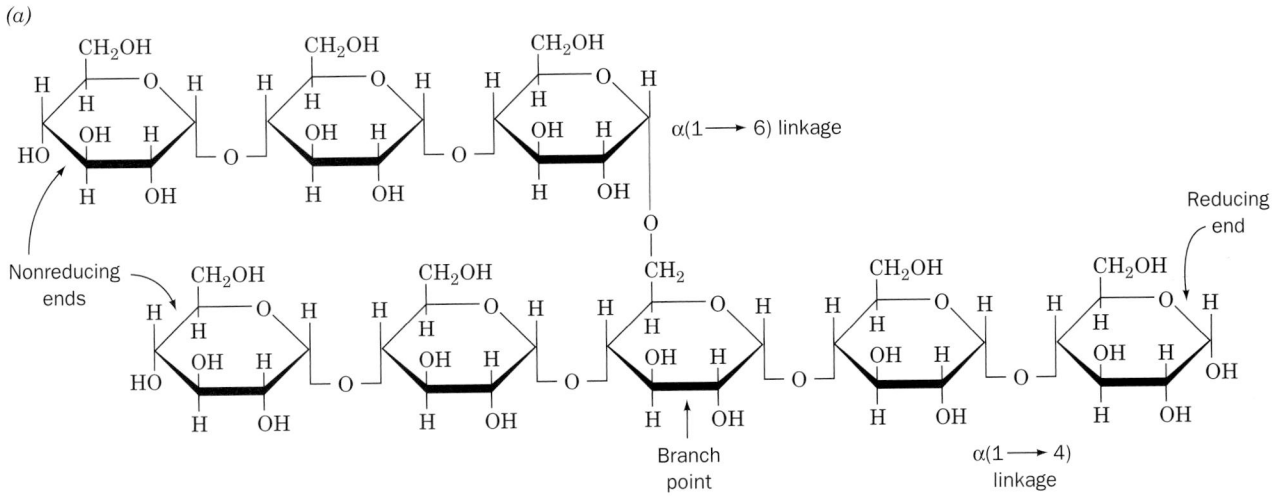

(b)

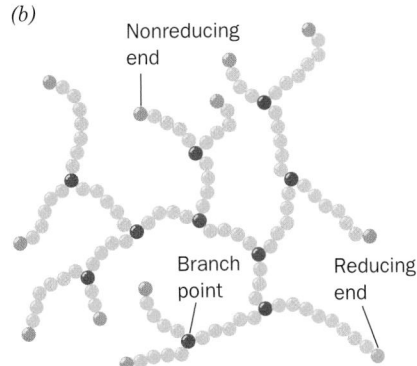

(c)

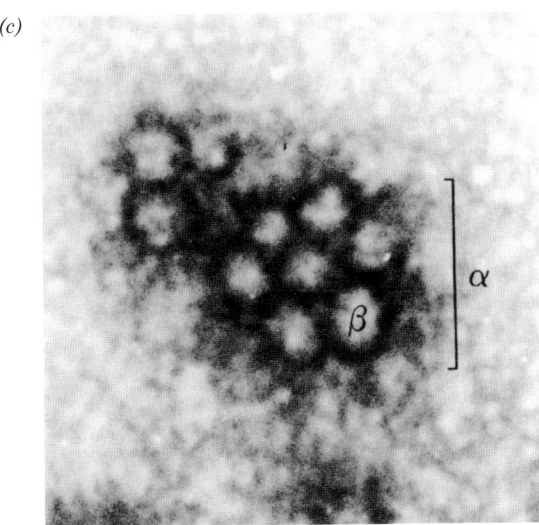

FIGURE 18-1 Structure of glycogen. (*a*) Molecular formula. In the actual molecule the chains are much longer than shown. (*b*) Schematic diagram illustrating its branched structure. Branch points in the actual molecule are separated by 8 to 14 glucosyl units. Note that the molecule, no matter how big, has but one reducing end. (*c*) Electron micrograph of a glycogen granule from rat skeletal muscle. Each granule (α) consists of several spherical glycogen molecules (β) and its associated proteins. [From Calder, P.C., *Int. J. Biochem.* **23,** 1339 (1991). Copyright Elsevier Science. Used with permission.]

entry into glycolysis. In liver, low blood glucose concentration triggers glycogen breakdown to G6P, which in this case is hydrolyzed to glucose and released into the bloodstream to reverse this situation.

Glycogen breakdown requires the actions of three enzymes:

1. Glycogen phosphorylase (or simply **phosphorylase**) catalyzes glycogen **phosphorolysis** (bond cleavage by the substitution of a phosphate group) to yield **glucose-1-phosphate (G1P).**

$$\text{Glycogen} + \text{P}_i \rightleftharpoons \text{glycogen} + \text{G1P}$$
$$(n \text{ residues}) \qquad\qquad (n-1 \text{ residues})$$

This enzyme will only release a glucose unit that is at least five units from a branch point.

2. Glycogen debranching enzyme removes glycogen's branches, thereby permitting the glycogen phosphorylase reaction to go to completion. It also hydrolyzes α(1→6)-linked glucosyl units to yield glucose. Consequently, ~92% of glycogen's glucose residues are converted to G1P. The remaining ~8%, those at the branch points, are converted to glucose.

3. Phosphoglucomutase converts G1P to G6P, which, as we have seen (Section 17-2A), is also formed in the first step of glycolysis through the action of either hexokinase or glucokinase. G6P can either continue along the glycolytic pathway (as in muscle) or be hydrolyzed to glucose (as in liver).

In this section, we discuss the structures and mechanisms of action of these three enzymes.

A. Glycogen Phosphorylase

Glycogen phosphorylase is a dimer of identical 842-residue (97-kD) subunits that catalyzes the controlling step in glycogen breakdown. It is regulated both by allosteric interactions and by covalent modification. *The enzyme-*

catalyzed modification/demodification process yields two forms of phosphorylase: **phosphorylase a,** *which has a phosphoryl group esterified to Ser 14 in each of its subunits, and* **phosphorylase b,** *which lacks these phosphoryl groups. Phosphorylase's allosteric inhibitors, ATP, G6P, and glucose, and its allosteric activator, AMP (to name only the enzyme's most prominent effectors), interact differently with the phospho- and dephosphoenzymes, resulting in an extremely sensitive control process. We study this control process in Section 18-3C.*

a. Structural Domains and Binding Sites

The high-resolution X-ray structures of phosphorylase *a* and phosphorylase *b* were determined by Robert Fletterick and Louise Johnson, respectively. The structure of phosphorylase *b*, despite its lack of a Ser-linked phosphate, is very similar to that of phosphorylase *a* (Fig. 18-2). Both structures have two domains, an N-terminal domain (residues 1–484; among the largest known domains), and a C-terminal domain (residues 485–842). The N-terminal domain is further divided into an interface subdomain (residues 1–315), which includes the covalent modification site (Ser 14), the allosteric effector site, and all the intersubunit contacts in the dimer; and a glycogen-binding subdomain (residues 316–484), which contains the "glycogen storage site" (see below). The catalytic site is located at the center of the subunit where these two subdomains come together with the C-terminal domain. In Section 18-3

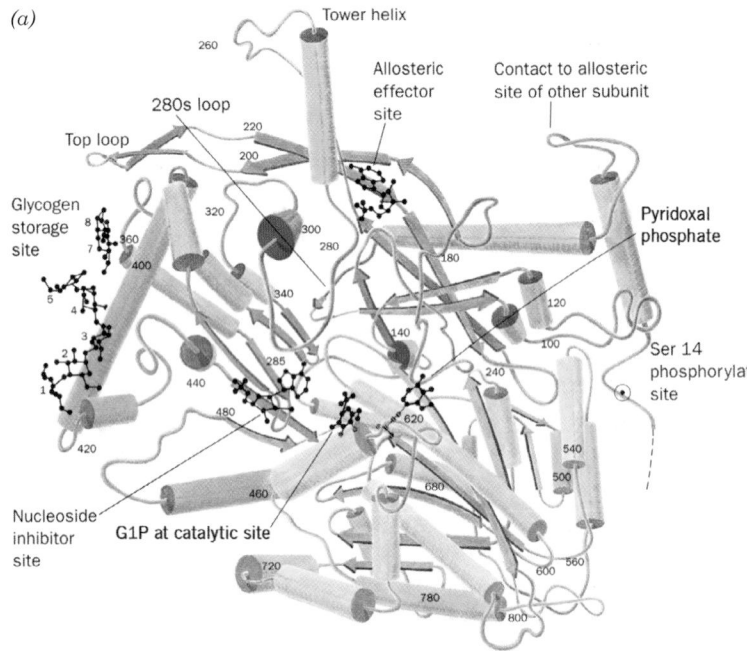

(a)

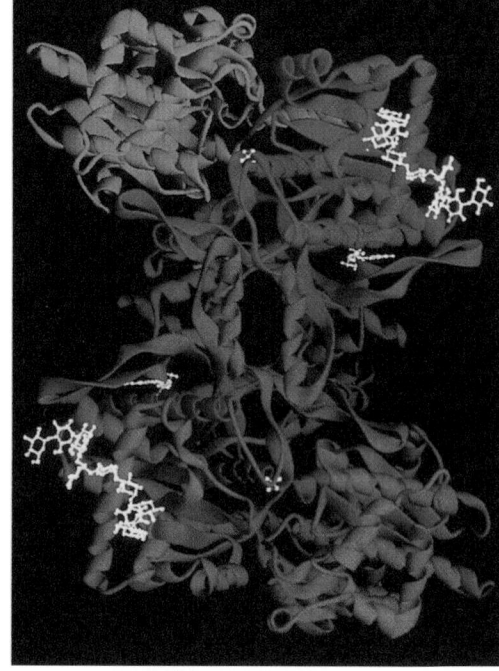

(b)

FIGURE 18-2 X-Ray structure of rabbit muscle glycogen phosphorylase. (*a*) Ribbon diagram of a phosphorylase *b* subunit. It consists of an N-terminal domain, which is subdivided into an interface subdomain (residues 1–315) and a glycogen-binding subdomain (residues 316–484), and a C-terminal domain (residues 485–842). AMP is shown bound at both the allosteric effector site and the nucleoside inhibitor site. G1P is shown bound at the catalytic site. The pyridoxal phosphate, which is partially hidden from view, is bound at Lys 678 in the C-terminal domain. **Maltoheptaose,** an $\alpha(1\rightarrow4)$-linked glucose heptamer, is bound to the glycogen storage site. Residues 1–11, which do not appear in the electron density map, are represented by a dashed line. Ser 14 is the site of enzymatic phosphorylation. [After McLaughlin, P.J., Stuart, D.I., Klein, H.W., Oikonomakos, N.G., and Johnson, L.N., *Biochemistry* **23**, 5865 (1984). PDBid 1GPB.] (*b*) A ribbon diagram of the glycogen phosphorylase *a* dimer viewed along its molecular twofold axis of symmetry (this view is related to that in Part *a* by an ~45° rotation about the vertical axis; the structural differences between the enzyme's two forms are relatively small). The bottom subunit is colored orange, whereas the top subunit's N-terminal and C-terminal domains are colored blue and green, respectively. The various bound ligands are white: The phosphate group at the center of each

(c)

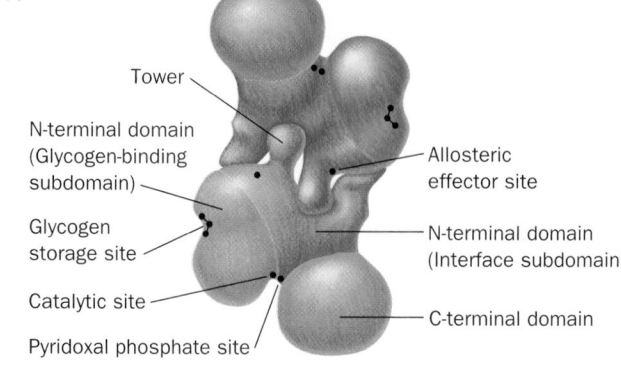

subunit marks the enzyme's catalytic sites (the Ser 14 phosphate groups in both subunits are hidden in this drawing), two chains of maltoheptaose are bound at each glycogen storage site, and the AMPs at the "back" of the protein identify the allosteric effector sites. [Courtesy of Stephen Sprang, University of Texas Southwest Medical Center.] (*c*) An interpretive "low-resolution" drawing of Part *b* showing the enzyme's various ligand-binding sites. 🔖 **See Kinemage Exercise 14-1**

we discuss the allosteric behavior of glycogen phosphorylase and the conformational differences between phosphorylases *a* and *b*.

Glycogen forms a left-handed helix with 6.5 glucose residues per turn, similar to α-amylose (Fig. 11-17*b*). An ~30-Å-long crevice on the surface of the phosphorylase monomer that has the same radius of curvature as glycogen connects the glycogen storage site, which binds glycogen, to the active site, which phosphorylizes it. *Since this crevice can accommodate four or five sugar residues in a chain but is too narrow to admit branched oligosaccharides, it provides a clear physical rationale for the inability of phosphorylase to cleave glycosyl residues closer than five units from a branch point.* Presumably the glycogen storage site increases the catalytic efficiency of phosphorylase by permitting it to phosphorylize many glucose residues on the same glycogen particle without having to dissociate and reassociate completely between catalytic cycles.

b. Pyridoxal Phosphate Is an Essential Cofactor for Phosphorylase

Phosphorylase contains **pyridoxal-5-phosphate (PLP)**

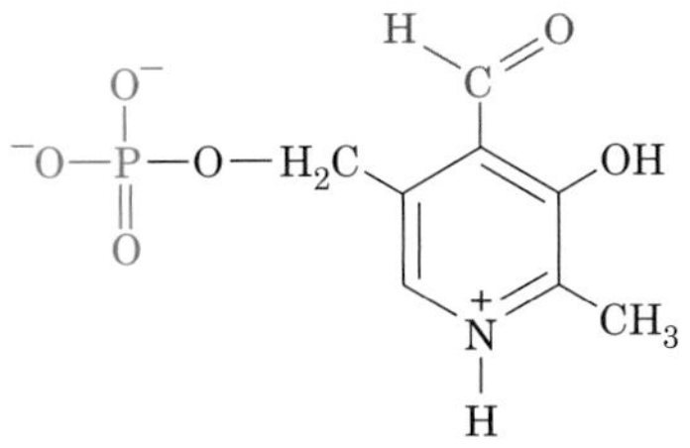

Pyridoxal phosphate (PLP)

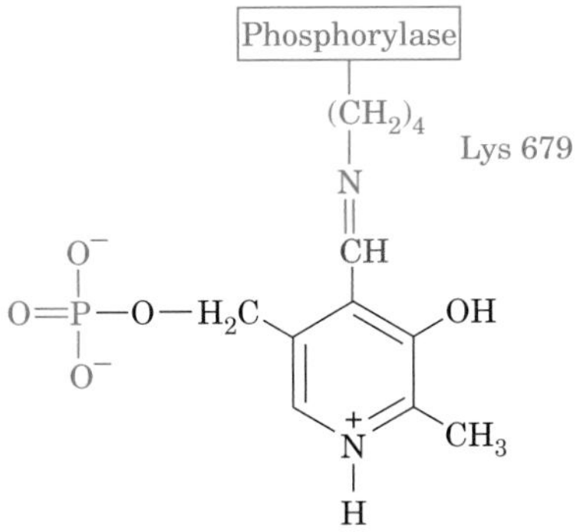

**PLP covalently bound to
phosphorylase via a
Schiff base to Lys 679**

and requires it for activity. This vitamin B_6 derivative is covalently linked to phosphorylase via a Schiff base to Lys 679. PLP is similarly linked to a variety of enzymes involved in amino acid metabolism, where it is an essential cofactor in transamination reactions (Section 26-1A). The mechanism of PLP participation in the phosphorylase reaction must differ from that in these other enzymes because, for example, reduction of the Schiff base with

$NaBH_4$ ($-HC=N- \rightarrow -H_2C-NH-$) has no effect on the activity of phosphorylase, whereas it inactivates the PLP-requiring enzymes of amino acid metabolism. This is an intriguing example of nature's opportunism in using the same cofactor to perform different chemistries.

Extensive studies on phosphorylase using PLP analogs in which various parts of this molecule are missing or modified indicate that only its phosphate group participates in the catalytic process. Indeed, the X-ray structures of phosphorylase reveal that only PLP's phosphate group is near this enzyme's active site. This phosphoryl group most probably functions as an acid–base catalyst.

c. Kinetics and Reaction Mechanism

The phosphorylase reaction results in the cleavage of the C1—O1 bond from a nonreducing terminal glucosyl unit of glycogen, yielding G1P. This reaction proceeds with retention of configuration, which suggests that the phosphorolysis occurs via a double displacement mechanism (two sequential nucleophilic substitutions, each occurring with inversion of configuration; Fig. 16-6*c*) involving a covalent glucosyl–enzyme intermediate. Yet, phosphorylase exhibits Rapid Equilibrium Random Bi Bi kinetics (Section 14-5), not Ping Pong kinetics, as would be expected for a double displacement mechanism. Furthermore, all attempts to establish the existence of the putative covalent intermediate have been unsuccessful.

An alternative mechanism (Fig. 18-3), which is compatible with the available kinetic, chemical, and structural data, commences with the formation of a ternary enzyme · P_i · glycogen complex, followed by the generation of an intermediate shielded oxonium ion similar to the transition state in the lysozyme reaction (which also involves glycosidic bond cleavage in a polysaccharide; Section 15-2B). *Bond cleavage, with its consequent oxonium ion formation, is assisted by protonation of the glycosidic oxygen by the P_i substrate (acid catalysis).* Phosphorylase has no protein nucleophilic or carboxylate groups in the vicinity of the scissile glycosidic bond and hence would be unable to form a covalent intermediate as does lysozyme. However, since the PLP phosphoryl group is within hydrogen bonding distance of the P_i, it appears that bond cleavage is facilitated by the simultaneous protonation of the reacting P_i by the PLP phosphoryl group in a kind of proton relay. The resulting oxonium ion (Fig. 18-3) is stabilized through its formation of an ion pair with the anionic P_i (electrostatic catalysis), which subsequently collapses to yield product, G1P, in a reaction step that is facilitated by the abstraction of a proton from P_i by the PLP phosphoryl group (base catalysis).

Support for the oxonium ion mechanism comes from the observation that **1,5-gluconolactone**

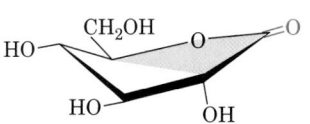

1,5-Gluconolactone

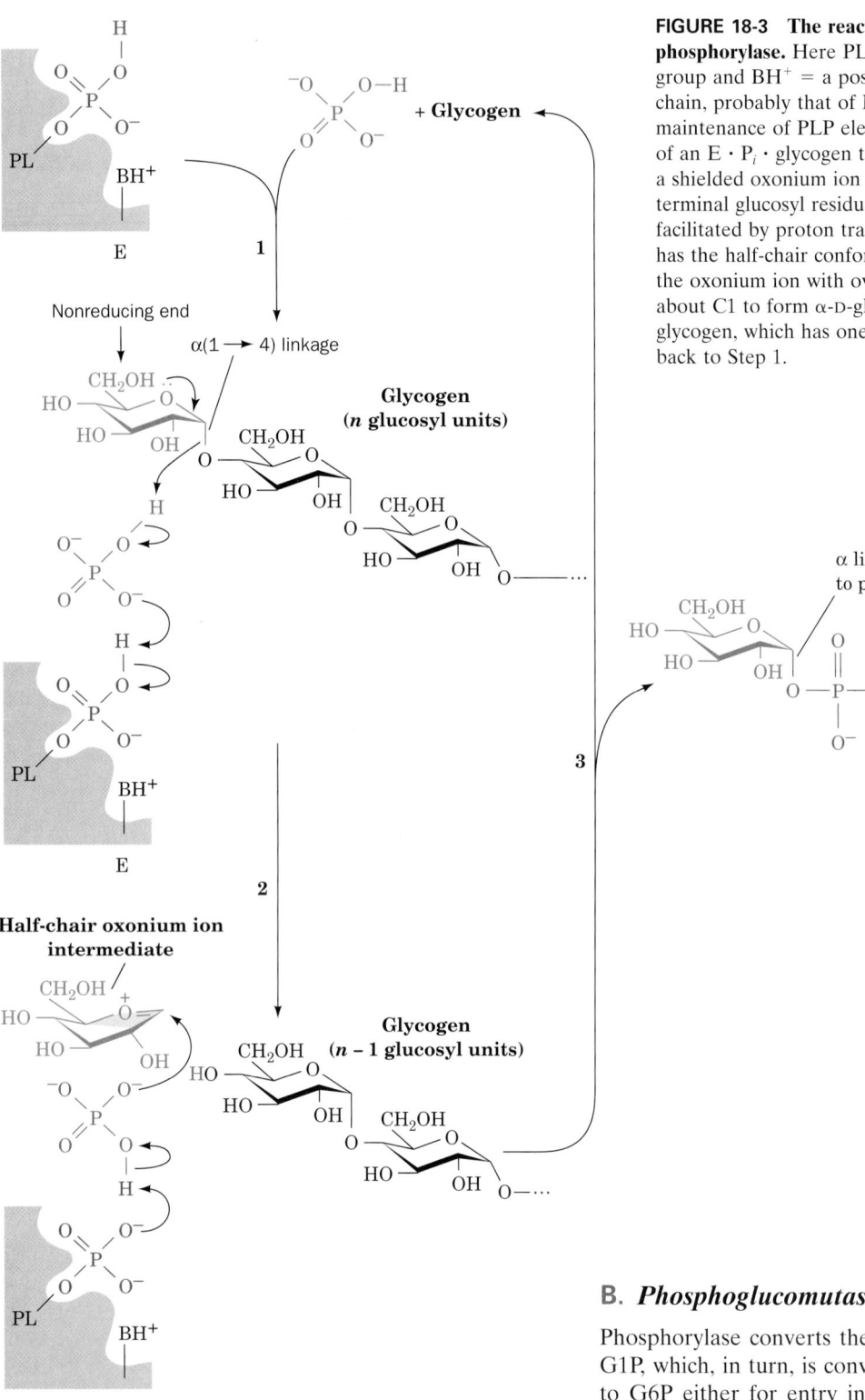

FIGURE 18-3   The reaction mechanism of glycogen phosphorylase. Here PL = an enzyme-bound pyridoxal group and BH$^+$ = a positively charged amino acid side chain, probably that of Lys 568, necessary for the maintenance of PLP electrical neutrality. **(1)** Formation of an E · P$_i$ · glycogen ternary complex. **(2)** Formation of a shielded oxonium ion intermediate from the α-linked terminal glucosyl residue involving acid catalysis by P$_i$ as facilitated by proton transfer from PLP. The oxonium ion has the half-chair conformation. **(3)** Reaction of P$_i$ with the oxonium ion with overall retention of configuration about C1 to form α-D-glucose-1-phosphate. The glycogen, which has one less residue than before, cycles back to Step 1.

B. *Phosphoglucomutase*

Phosphorylase converts the glucosyl units of glycogen to G1P, which, in turn, is converted by phosphoglucomutase to G6P either for entry into glycolysis in muscle or hydrolysis to glucose in liver. The X-ray structure of rabbit muscle phosphoglucomutase indicates that the active site of this 561-residue monomeric enzyme is largely buried at the bottom of a particularly deep crevice in the protein. The phosphoglucomutase reaction is similar to that catalyzed by phosphoglycerate mutase (Section 17-2H). A phosphoryl group is transferred from the active phosphoenzyme to G1P, forming **glucose 1,6-bisphosphate (G1,6P),** which then rephosphorylates the enzyme to yield G6P

is a potent inhibitor of phosphorylase. 1,5-Gluconolactone has the same half-chair conformation as the proposed oxonium ion, suggesting that it is a transition state analog that mimics the oxonium ion at the active site of phosphorylase (Section 15-1F).

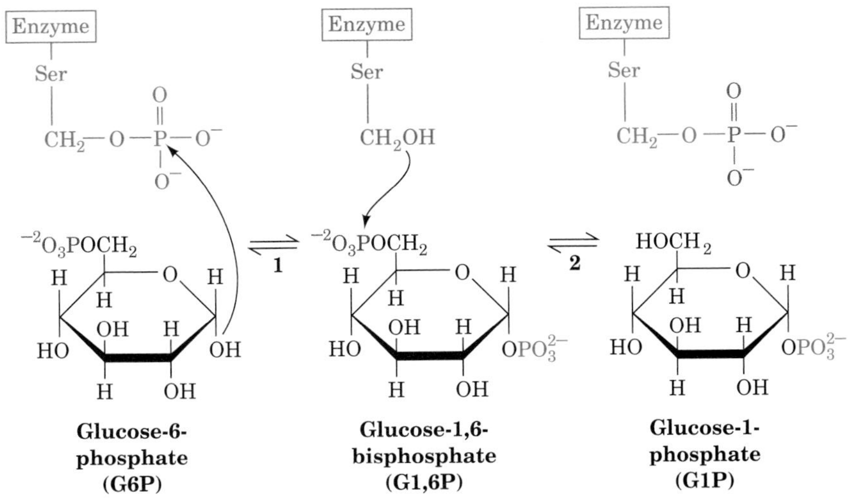

FIGURE 18-4 The mechanism of action of phosphoglucomutase. (1) The OH group at C1 of G6P attacks the phosphoenzyme to form a dephosphoenzyme–G1,6P intermediate. **(2)** The Ser—OH group on the dephosphoenzyme attacks the phosphoryl group at C6 to regenerate the phosphoenzyme with the formation of G1P.

(Fig. 18-4). An important difference between this enzyme and phosphoglycerate mutase is that the phosphoryl group in phosphoglucomutase is covalently bound to a Ser hydroxyl group rather than to a His imidazole nitrogen.

G1,6P occasionally dissociates from phosphoglucomutase, resulting in the inactivation of this enzyme. The presence of small amounts of G1,6P is therefore necessary to keep phosphoglucomutase fully active. This intermediate is provided by **phosphoglucokinase,** which catalyzes the phosphorylation of the C6—OH group of G1P by ATP.

C. *Glycogen Debranching Enzyme*

Glycogen debranching enzyme acts as an $\alpha(1\rightarrow4)$ transglycosylase (glycosyl transferase) by transferring an $\alpha(1\rightarrow4)$-linked trisaccharide unit from a "limit branch" of glycogen to the nonreducing end of another branch (Fig. 18-5). This reaction forms a new $\alpha(1\rightarrow4)$ linkage with three more units available for phosphorylase-catalyzed phosphorolysis. The $\alpha(1\rightarrow6)$ bond linking the remaining glycosyl residue in the branch to the main chain is hydrolyzed

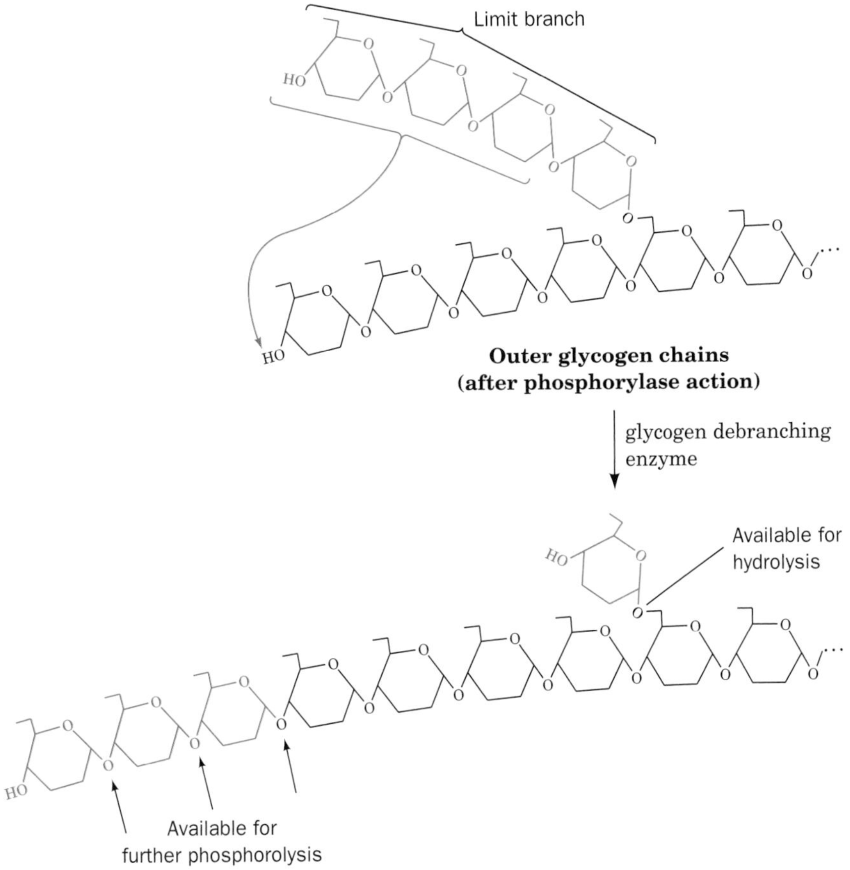

FIGURE 18-5 Reactions catalyzed by debranching enzyme. The enzyme transfers the terminal three $\alpha(1\rightarrow4)$-linked glucose residues from a "limit branch" of glycogen to the nonreducing end of another branch. The $\alpha(1\rightarrow6)$ bond of the residue remaining at the branch point is hydrolyzed by further action of debranching enzyme to yield free glucose. The newly elongated branch is subject to degradation by glycogen phosphorylase.

(not phosphorylized) by the same debranching enzyme to yield glucose and debranched glycogen. Thus *debranching enzyme has different active sites for the transferase reaction and the α(1→6)-glucosidase reaction.* The presence of two independent catalytic activities on the same enzyme no doubt improves the efficiency of the debranching process.

The maximal rate of the glycogen phosphorylase reaction is much greater than that of the glycogen debranching reaction. Consequently, the outermost branches of glycogen, which comprise nearly half of its residues, are degraded in muscle in a few seconds under conditions of high metabolic demand. Glycogen degradation beyond this point requires debranching and hence occurs more slowly. This, in part, accounts for the fact that a muscle can sustain its maximum exertion for only a few seconds.

D. *Thermodynamics of Glycogen Metabolism: The Need for Separate Pathways of Synthesis and Breakdown*

The $\Delta G°'$ (ΔG under standard biochemical conditions) for the phosphorylase reaction is $+3.1$ kJ · mol^{-1}, so, as Eq. [3.15] indicates, this reaction is at equilibrium ($\Delta G = 0$) at 25°C when $[P_i]/[G1P] = 3.5$. In the cell, however, this concentration ratio varies between 30 and 100, which places ΔG in the range -5 to -8 kJ · mol^{-1}; that is, *under physiological conditions, glycogen breakdown is exergonic.* The synthesis of glycogen from G1P under physiological conditions is therefore thermodynamically unfavorable without free energy input. Consequently, *glycogen biosynthesis and breakdown must occur by separate pathways. Thus we encounter a recurrent metabolic strategy: Biosynthetic and degradative pathways of metabolism are almost always different (Section 16-1).* There are two important reasons for this. The first, as we have seen, is that both pathways may be required under similar *in vivo* metabolite concentrations. This situation is thermodynamically impossible if one pathway is just the reverse of the other. The second reason is equally important: Reactions catalyzed by different enzymes can be independently regulated, which permits very fine flux control. We have seen this principle in operation in the glycolytic conversion of fructose-6-phosphate (F6P) to fructose-1,6-bisphosphate (F1,6P) by phosphofructokinase (PFK; Section 17-4B). The reverse process in that case (hydrolysis of F1,6P) is catalyzed by fructose bisphosphatase (FBPase). Independent control of those two enzymes provides precise regulation of glycolytic flux.

Glycogen metabolism, like glycolysis, is exquisitely regulated by the independent control of its synthetic and degradative pathways. In the next section we examine the pathway of glycogen synthesis and, in Section 18-3, we explore the regulatory process.

2 ■ GLYCOGEN SYNTHESIS

Although the thermodynamic arguments presented in Section 18-1D demonstrate that glycogen synthesis and breakdown must occur by separate pathways, it was not thermodynamic arguments that led to the general acceptance of this idea. Rather, it was the elucidation of the cause of **McArdle's disease,** a rare inherited glycogen storage disease that results in painful muscle cramps on strenuous exertion (Section 18-4). The muscle tissue from individuals with McArdle's disease exhibits no glycogen phosphorylase activity and is therefore incapable of glycogen breakdown. Their muscles, nevertheless, contain moderately high quantities of normal glycogen. Clearly, there must be separate pathways for glycogen synthesis and breakdown.

Since the direct conversion of G1P to glycogen and P_i is thermodynamically unfavorable (positive ΔG) under all physiological P_i concentrations, glycogen biosynthesis requires an additional exergonic step. This is accomplished, as Luis Leloir discovered in 1957, by combining G1P with uridine triphosphate (UTP) to form **uridine diphosphate glucose (UDP–glucose** or **UDPG):**

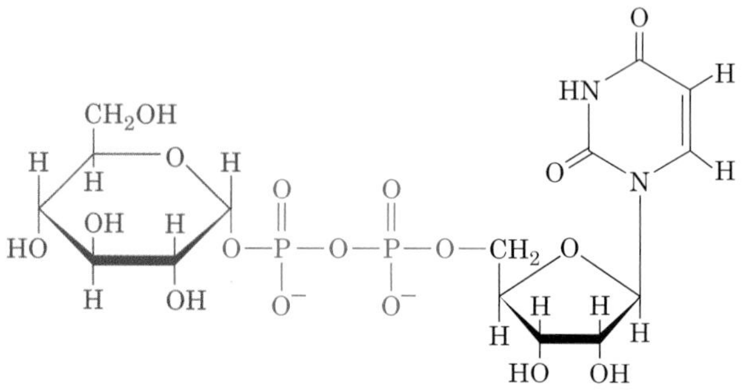

Uridine diphosphate glucose (UDPG)

UDPG's "high-energy" status permits it to spontaneously donate glucosyl units to the growing glycogen chain.

The enzymes catalyzing the three steps involved in the glycogen synthesis pathway are **UDP–glucose pyrophosphorylase, glycogen synthase,** and **glycogen branching enzyme.** In this section, we examine the reactions catalyzed by these enzymes. Discussion of how these enzymes are controlled is reserved for Section 18-3.

A. *UDP–Glucose Pyrophosphorylase*

UDP–glucose pyrophosphorylase catalyzes the reaction of UTP and G1P (Fig. 18-6). In this reaction, the phosphoryl oxygen of G1P attacks the α phosphorus atom of UTP to form UDPG and release PP_i. The $\Delta G°'$ of this phosphoanhydride exchange is, as expected, nearly zero. However, the PP_i formed is hydrolyzed in a highly exergonic reaction by the omnipresent enzyme inorganic pyrophosphatase. The overall reaction for the formation of UDPG is therefore also highly exergonic:

	$\Delta G°'$ (kJ · mol^{-1})
G1P + UTP $\rightleftharpoons$ UDPG + PP_i	~0
H_2O + PP_i $\rightleftharpoons$ $2P_i$	-33.5
Overall: G1P + UTP $\rightleftharpoons$ UDPG + $2P_i$	-33.5

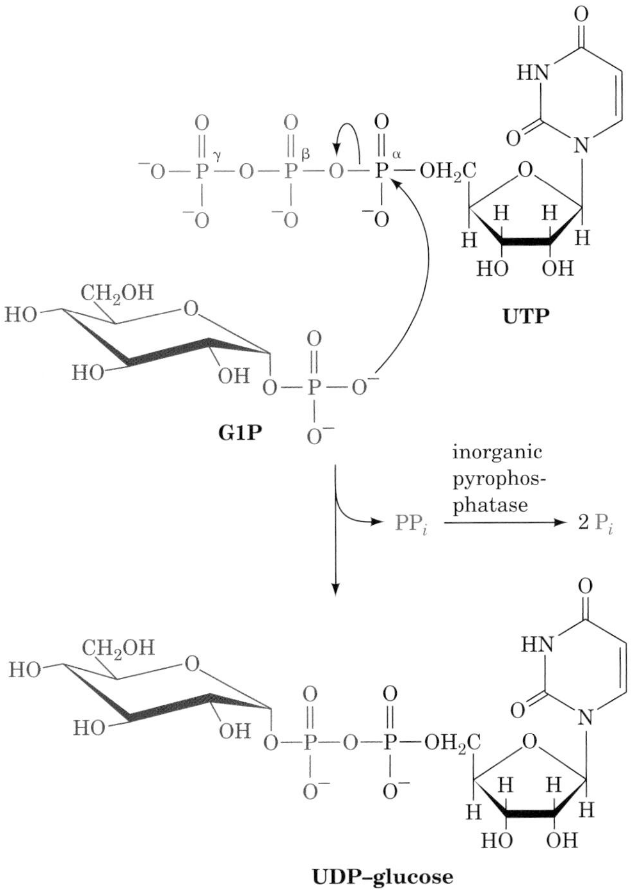

FIGURE 18-6 Reaction catalyzed by UDP–glucose pyrophosphorylase. The reaction is a phosphoanhydride exchange in which the phosphoryl oxygen of G1P attacks the α phosphorus atom of UTP to form UDPG and release PP_i. The PP_i is rapidly hydrolyzed by inorganic pyrophosphatase.

The cleavage of a nucleoside triphosphate to form PP_i is a common biosynthetic strategy. The free energy of PP_i hydrolysis can then be utilized together with the free energy of nucleoside triphosphate hydrolysis to drive an otherwise endergonic reaction to completion (Section 16-4C).

B. *Glycogen Synthase*

In the next step of glycogen synthesis, the glycogen synthase reaction, the glucosyl unit of UDPG is transferred to the C4—OH group on one of glycogen's nonreducing ends to form an $\alpha(1\rightarrow4)$-glycosidic bond (Fig. 18-7). The glycogen synthase reaction, like those of glycogen phosphorylase and lysozyme, is thought to involve a glucosyl oxonium ion intermediate or transition state since it is also inhibited by 1,5-gluconolactone, an analog that mimics the oxonium ion's half-chair geometry.

The $\Delta G°'$ for the glycogen synthase reaction is -13.4 $kJ \cdot mol^{-1}$, making the overall reaction spontaneous under the same conditions that glycogen breakdown by glycogen phosphorylase is also spontaneous. The rates of both reactions may then be independently controlled. There is, however, an energetic price for doing so. In this case, *for each molecule of G1P that is converted to glycogen and then regenerated, one molecule of UTP is hydrolyzed to UDP and P_i. The cyclic synthesis and breakdown of glycogen is therefore not a perpetual motion "machine" but, rather, is an "engine" that is powered by UTP hydrolysis.* The UTP is replenished through a phosphate-transfer reaction mediated by **nucleoside diphosphate kinase** (Section 28-1B):

$$UDP + ATP \rightleftharpoons UTP + ADP$$

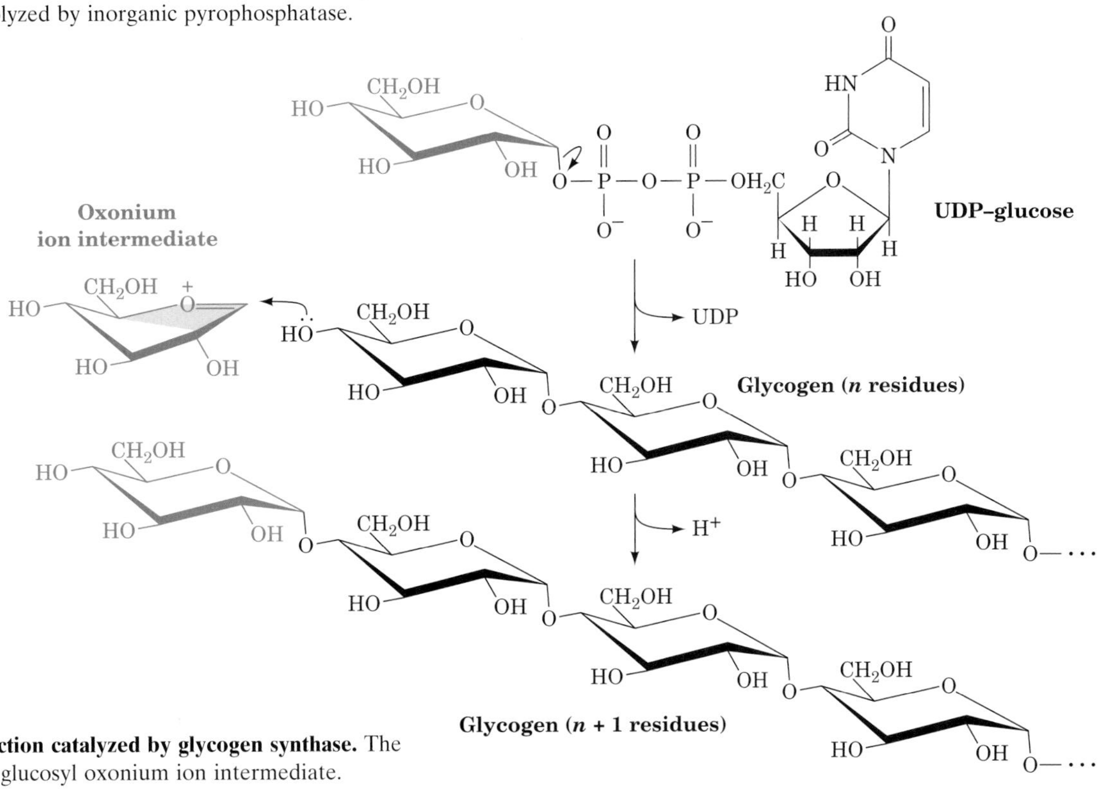

FIGURE 18-7 Reaction catalyzed by glycogen synthase. The reaction involves a glucosyl oxonium ion intermediate.

so that UTP hydrolysis is energetically equivalent to ATP hydrolysis. This reaction occurs via a Ping Pong mechanism in which an active site His residue is transiently phosphorylated at its N_δ position much as occurs in the phosphoglycerate mutase reaction of glycolysis (Section 17-2H).

Glycogen synthase cannot simply link together two glucose residues; it can only extend an already existing $\alpha(1{\rightarrow}4)$-linked glucan chain. How, then, is glycogen synthesis initiated? The answer is that the first step in glycogen synthesis is the self-catalyzed attachment of a glucose residue to the Tyr 194 OH group of a protein named **glycogenin.** Glycogenin, which was discovered by William Whelan, further extends the glucan chain by up to seven additional UDP–glucose-supplied residues, forming a "primer" for the initiation of glycogen synthesis. Only at this point does glycogen synthase commence glycogen synthesis, which it initiates on the "primer" while tightly complexed to glycogenin. However, these proteins dissociate from one another after the growing glycogen granule has reached some minimum size. Analysis of glycogen granules for glycogenin and glycogen synthase shows that they are present in a 1:1 ratio. Evidently, each glycogen molecule (Fig. 18-1*c*) is associated with one molecule each of glycogenin and glycogen synthase.

C. *Glycogen Branching*

Glycogen synthase catalyzes only $\alpha(1{\rightarrow}4)$-linkage formation to yield α-amylose. Branching to form glycogen is accomplished by a separate enzyme, **amylo-(1,4→1,6)-transglycosylase (branching enzyme),** which is distinct from glycogen debranching enzyme. Branches are created by the transfer of terminal chain segments consisting of ~7 glucosyl residues to the C6—OH groups of glucose residues on the same or another glycogen chain (Fig. 18-8). Each transferred segment must come from a chain of at least 11 residues, and the new branch point must be at least 4 residues away from other branch points.

Debranching (Section 18-1C) involves breaking and reforming $\alpha(1{\rightarrow}4)$-glycosidic bonds and only the hydrolysis of $\alpha(1{\rightarrow}6)$-glycosidic bonds; branching, on the other hand, involves breaking $\alpha(1{\rightarrow}4)$-glycosidic bonds and reforming $\alpha(1{\rightarrow}6)$ linkages. The need to hydrolyze glycogen's $\alpha(1{\rightarrow}6)$-glycosidic bonds rather than to convert them to $\alpha(1{\rightarrow}4)$ linkages is explained by the energetics of these reactions. The free energy of hydrolysis of an $\alpha(1{\rightarrow}4)$-glycosidic bond is -15.5 kJ · mol^{-1}, whereas that of an $\alpha(1{\rightarrow}6)$-glycosidic bond is only -7.1 kJ · mol^{-1}. Consequently, the hydrolysis of an $\alpha(1{\rightarrow}4)$-glycosidic bond drives the synthesis of an $\alpha(1{\rightarrow}6)$-glycosidic bond, but the reverse reaction is endergonic.

a. Glycogen Particles Are Fabricated to Optimize Glucose Mobilization

The biological function of glycogen is to maximize the density of stored glucose units consistent with the need to rapidly mobilize it under conditions of high metabolic demand. To do so, three related parameters must be opti-

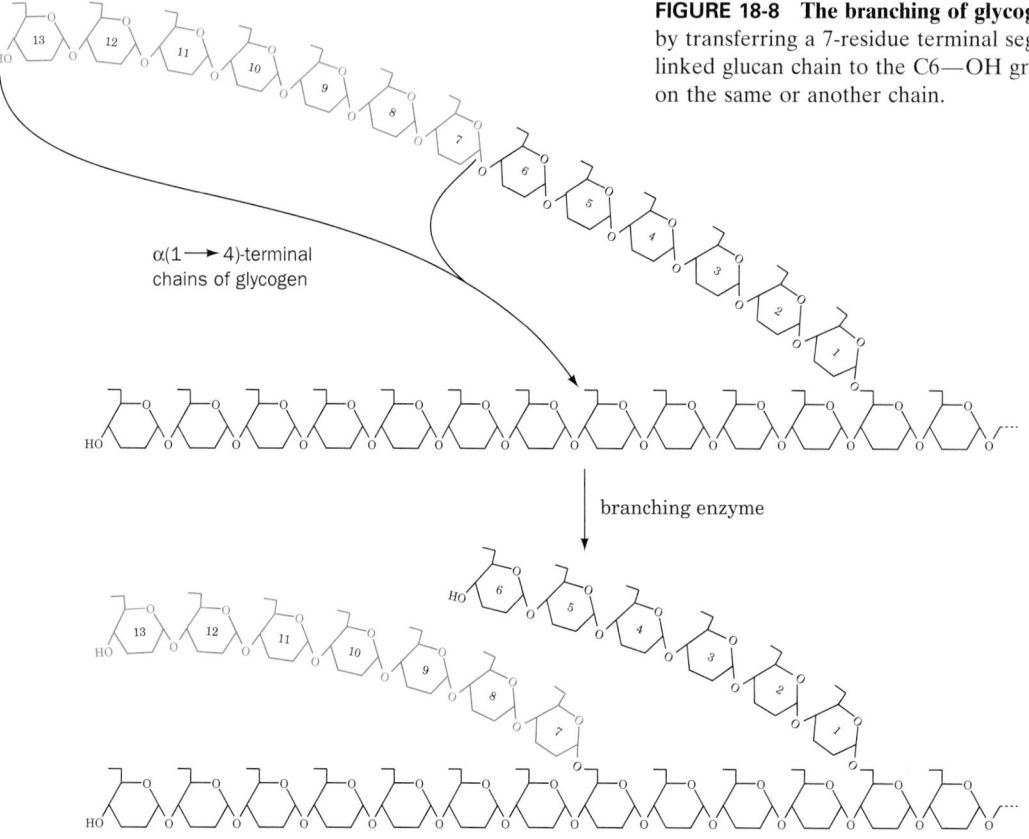

FIGURE 18-8   The branching of glycogen. Branches are formed by transferring a 7-residue terminal segment from an $\alpha(1{\rightarrow}4)$-linked glucan chain to the C6—OH group of a glucose residue on the same or another chain.

$\alpha(1{\longrightarrow}4)$-terminal chains of glycogen

branching enzyme

mized: the number of tiers of branches in a glycogen molecule, the number of branches per tier, and the average chain length per tier. For a glycogen molecule with a fixed number of residues, the number of outer branches from which glucose can be mobilized before debranching is required decreases as the average chain length increases (recall that debranching is a slower process than phosphorolysis). However, molecules with longer chains have a greater number of glucose residues that can be phosphorylized between branch points. Since the density of outermost branches is sterically limited, the maximum size of a glycogen molecule decreases as the average number of branches per tier increases. Mature glycogen particles from a variety of animals have ~12 tiers of branches, with ~2 branches per tier and branch lengths averaging ~13 residues. Mathematical analysis suggests that these values are close to optimal for mobilizing the greatest amount of glucose in the shortest possible time.

3 ■ CONTROL OF GLYCOGEN METABOLISM

We have just seen that both glycogen synthesis and breakdown are exergonic under the same physiological conditions. If both pathways operate simultaneously, however, all that is achieved is wasteful hydrolysis of UTP. This situation is similar to that of the phosphofructokinase–fructose bisphosphatase substrate cycle (Section 17-4F). Glycogen phosphorylase and glycogen synthase therefore must be under stringent control such that glycogen is either synthesized or utilized according to cellular needs. The astonishing mechanism of this control is the next topic of our discussion. It involves not only allosteric control and

substrate cycles but enzyme-catalyzed covalent modification of both glycogen synthase and glycogen phosphorylase. The covalent modification reactions are themselves ultimately under hormonal control through an enzymatic cascade.

A. *Direct Allosteric Control of Glycogen Phosphorylase and Glycogen Synthase*

As we saw in Section 17-4B, the net flux of reactants, J, through a step in a metabolic pathway is the difference between the forward and reverse reaction velocities, v_f and v_r. The variation in the flux through any step in a pathway with a change in substrate concentration approaches infinity as that reaction step approaches equilibrium ($v_f \approx v_r$; Eq. [17.4]). The flux through a near-equilibrium reaction is therefore all but uncontrollable. As we have seen for the case of PFK and FBPase, however, *precise flux control of a pathway is possible when an enzyme functioning far from equilibrium is opposed by a separately controlled enzyme. Then, v_f and v_r vary independently. In fact, under these circumstances, even the flux direction is controlled if v_r can be made larger than v_f.* Exactly this situation occurs in glycogen metabolism through the opposition of the glycogen phosphorylase and glycogen synthase reactions. The rates of both of these reactions are under allosteric control by effectors that include ATP, G6P, and AMP. In muscle, glycogen phosphorylase is activated by AMP and inhibited by ATP and G6P (Fig. 18-9, *left*). Glycogen synthase, on the other hand, is activated by G6P. When there is high demand for ATP (low [ATP], low [G6P], and high [AMP]), glycogen phosphorylase is stimulated and glycogen synthase is inhibited, so flux through this pathway favors glycogen breakdown. When [ATP] and [G6P] are high, the reverse is true and glycogen synthesis is favored.

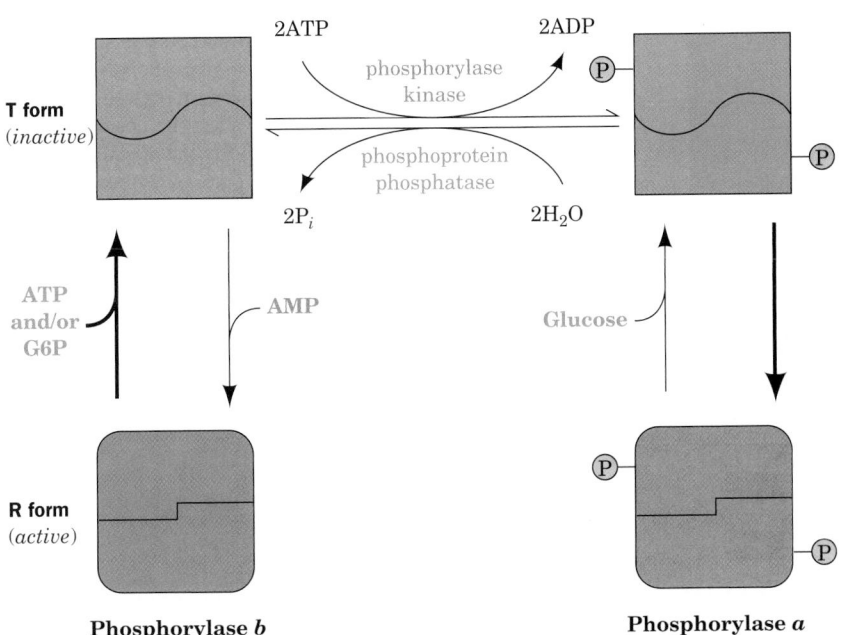

Phosphorylase *b*   **Phosphorylase *a***

FIGURE 18-9 The control of glycogen phosphorylase activity. The enzyme may assume the enzymatically inactive T conformation (*above*) or the catalytically active R form (*below*). The conformation of phosphorylase *b* is allosterically controlled by effectors such as AMP, ATP, and G6P and is mostly in the T state under physiological conditions. In contrast, the modified form of the enzyme, phosphorylase *a*, is largely unresponsive to these effectors and is mostly in the R state unless there is a high level of glucose. Under usual physiological conditions, the enzymatic activity of glycogen phosphorylase is essentially determined by its rates of modification and demodification. Note that only the T form enzyme is subject to phosphorylation and dephosphorylation, so effector binding influences the rates of these modification/demodification events.

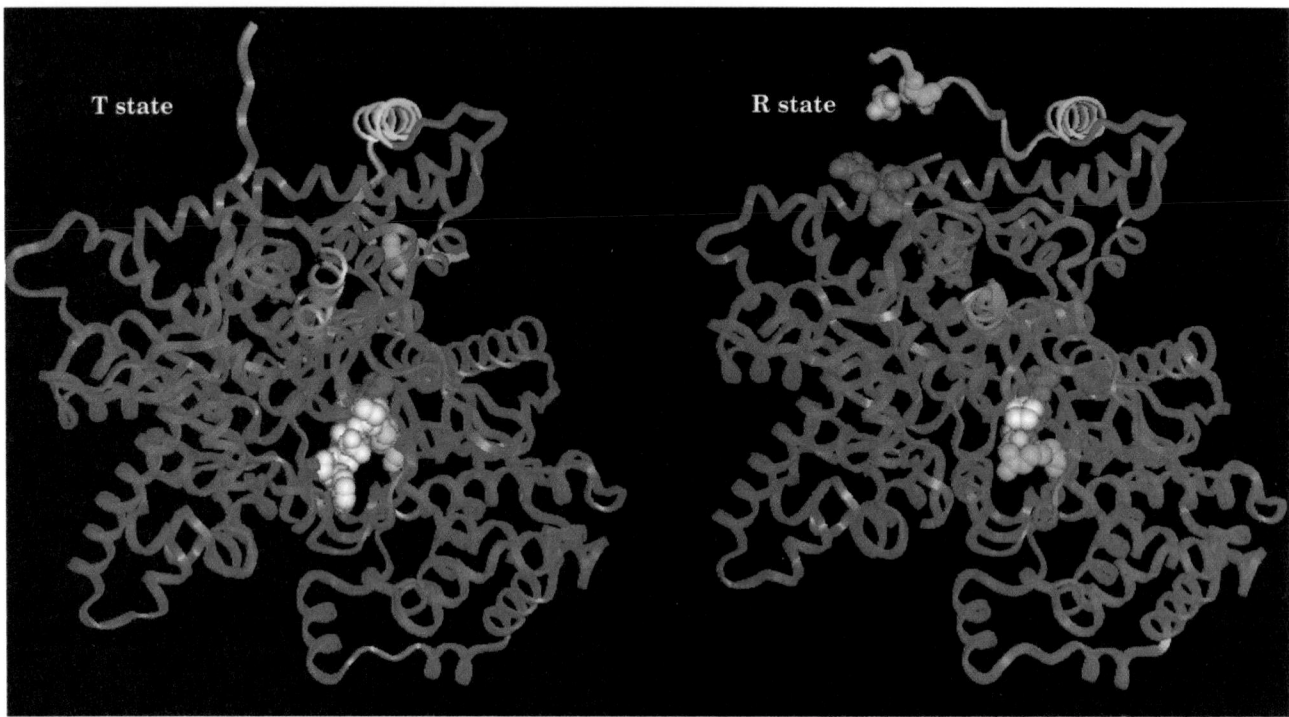

(a)

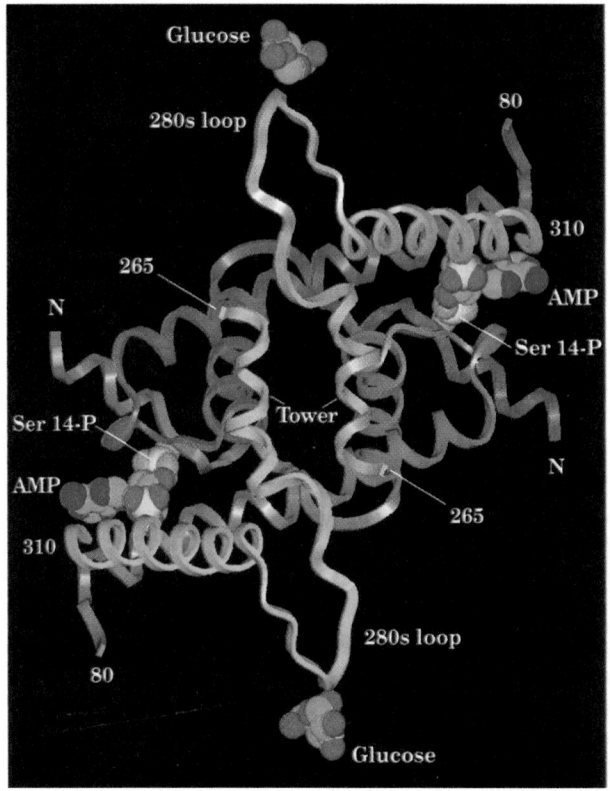

(b)

FIGURE 18-10   Conformational changes in glycogen phosphorylase. (*a*) A ribbon diagram of one subunit of the dimeric enzyme glycogen phosphorylase *b* shown (*left*) in the T state in the absence of allosteric effectors and (*right*) in the R state with bound AMP. The view is of the lower (*orange*) subunit in Fig. 18-2*b* as seen from the top of the page. The tower helix is blue, the N-terminal helix is cyan, and the N-terminal residues that change conformation on AMP binding are green. Of the groups that are shown in space-filling representation, Ser 14, the phosphorylation site, is light green; AMP is orange; the active site PLP is red; the Arg 569 side chain, which reorients in the T→R transition so as to interact with the substrate phosphate, is cyan; residues 282 to 284 of the 280s loop, which in the R state are mostly disordered and hence not seen, are white; and the phosphates, both at the active site and at the R state Ser 14 phosphorylation site (which is shown here only to mark its position in phosphorylase *a*), are yellow. (*b*) The portion of the glycogen phosphorylase *a* dimer in the vicinity of the dimer interface showing the position of the Ser 14 phosphate group, the AMP bound in the allosteric effector site, and the active site–bound glucose molecule. The view is along the molecular 2-fold axis and hence is similar to that in Fig. 18-2*b*. Residues 6 to 80 and 265 to 310 are, respectively, cyan and blue in one subunit and pink and magenta in the other. The AMP and glucose are shown in space-filling representation with C green, N dark blue, O red, and P yellow. The Ser 14 phosphate group is also shown in space-filling representation with O orange and P white. [X-Ray structure coordinates courtesy of Stephen Sprang, University of Texas Southwest Medical Center.] *See Kinemage Exercises 14-2 and 14-3*

The structural differences between the active (R) and inactive (T) conformations of glycogen phosphorylase (Fig. 18-10*a*) are fairly well understood in terms of the symmetry model of allosterism (Section 10-4B). The T-state enzyme has a buried active site and hence a low affinity for its substrates, whereas the R-state enzyme has an accessible catalytic site and a high-affinity phosphate binding site.

AMP promotes phosphorylase's T (*inactive*) → R (*active*) conformational shift by binding to the R state of the enzyme at its allosteric effector site (Fig. 18-9, *left*). In doing so, AMP's adenine, ribose, and phosphate groups bind to separate segments of the polypeptide chain so as to link the active site, the subunit interface, and the N-terminal region (Fig. 18-10*b*), the latter having undergone a large conformational shift (36 Å for Ser 14) from its position in the T-state enzyme (Fig. 18-10*a*). AMP binding also causes glycogen phosphorylase's tower helices (Figs. 18-2 and 18-10) to tilt and pull apart so as to achieve a more favorable packing. These tertiary movements trigger a concerted T→R transition, which largely consists of an ~10° relative rotation of the two subunits about an axis at the subunit interface that is perpendicular to the dimer's 2-fold axis of symmetry. The enzyme's 2-fold symmetry is thereby preserved in accordance with the symmetry model of allosterism. The movement of the tower helices also displaces and disorders a loop (the 280s loop, residues 282–286), which covers the T-state active site so as to prevent substrate access. It also causes the Arg 569 side chain, which is located in the active site near the PLP phosphoryl group and the P_i-binding site, to rotate in a way that increases the enzyme's binding affinity for its anionic P_i substrate (Figure 18-10*a*).

Curiously, ATP also binds to the allosteric effector site, but in the T state, so that it inhibits rather than promotes the T→R conformational shift. This is because, as structural analysis indicates, the β and γ phosphate groups of ATP bind to the enzyme such that its ribose and α phosphate groups are displaced relative to those of AMP, thus destabilizing the R state. The inhibitory action of ATP on phosphorylase is therefore simply understood: It competes with AMP for binding to phosphorylase and, in doing so, prevents the relative motions of the three polypeptide segments required for phosphorylase activation.

The above allosteric interactions are superimposed on an even more sophisticated control system involving covalent modifications (phosphorylation/dephosphorylation) of glycogen phosphorylase and glycogen synthase. These modifications alter the structures of the enzymes so as to change their responses to allosteric regulators. We shall therefore discuss the general concept of covalent modification and how it increases the sensitivity of a metabolic system to effector concentration changes. We subsequently consider the functions of such modifications in glycogen metabolism. Only then will we be in a position to take up the detailed consideration of allosteric control in glycogen metabolism.

B. *Covalent Modification of Enzymes by Cyclic Cascades: Effector "Signal" Amplification*

Glycogen synthase and glycogen phosphorylase can each be enzymatically interconverted between two forms with different kinetic and allosteric properties through a complex series of reactions known as a **cyclic cascade.** *The interconversion of these different enzyme forms involves distinct,* *enzyme-catalyzed* **covalent modification** *and* **demodification reactions.**

Compared with other regulatory enzymes, enzymatically interconvertible enzyme systems:

1. Can respond to a greater number of allosteric stimuli.

2. Exhibit greater flexibility in their control patterns.

3. Possess enormous amplification potential in their responses to variations in effector concentrations.

This is because *the enzymes that modify and demodify a target enzyme are themselves under allosteric control. It is therefore possible for a small change in concentration of an allosteric effector of a modifying enzyme to cause a large change in the concentration of an active, modified target enzyme.* Such a cyclic cascade is diagrammed in Fig. 18-11.

a. Description of a General Cyclic Cascade

Figure 18-11*a* shows a general scheme for a cyclic cascade where, by convention, the more active target enzyme form has the subscript *a* and the less active form has the subscript *b*. Here, modification, in this case, phosphorylation, activates the enzyme. Note that the modifying enzymes, F and R, are active only when they have bound

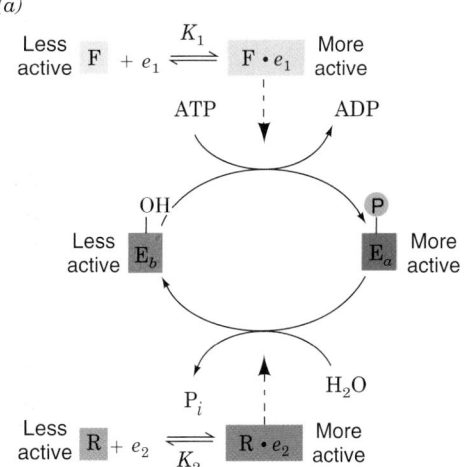

FIGURE 18-11 A monocyclic enzyme cascade. (*a*) General scheme, where F and R are, respectively, the modifying and demodifying enzymes. These are allosterically converted from their inactive to their active conformations on binding their respective effectors, e_1 and e_2. The target enzyme, E, is more active in the modified form (E_a) and less active in the unmodified form (E_b). Dashed arrows symbolize catalysis of the indicated reactions. (*b*) Chemical equations for the interconversion of the target enzyme's unmodified and modified forms E_b and E_a.

their respective allosteric effectors e_1 and e_2. The kinetic mechanisms for the interconversion of the unmodified and modified forms of the target enzyme, E_b and E_a, are indicated in Fig. 18-11*b*.

In the steady state, the fraction of E in the active form, $[E_a]/[E]_T$ (where $[E]_T = [E_a] + [E_b]$ is the total enzyme concentration), determines the rate of the reaction catalyzed by E. This fraction is a function of the total concentrations of the modifying enzymes, $[F]_T$ and $[R]_T$, the concentrations of their allosteric effectors, e_1 and e_2, the dissociation constants of these effectors, K_1 and K_2, and the substrate dissociation constants, K_f and K_r, of the target enzymes, as well as the rate constants, k_f and k_r, for the interconversions themselves (Fig. 18-11). This relationship is obviously quite complex. Nevertheless, it can be shown that, in a cyclic cascade, a relatively small change in the concentration of e_1, the allosteric effector of the modifying enzyme F, can result in a much larger change in $[E_a]/[E]_T$, the fraction of E in the active form. In other words, *the cascade functions to amplify the sensitivity of the system to an allosteric effector.*

We have so far considered the covalent modification of only one enzyme, a **monocyclic cascade.** Imagine a **bicyclic cascade** involving the covalent modification of one of the modifying enzymes (F), as well as the metabolic target enzyme (E) (Fig. 18-12). As you might expect, the amplification potential of a "signal," e_1, as well as the control flexibility of such a system, is enormous.

The activities of both glycogen phosphorylase and glycogen synthase are controlled by bicyclic cascades. Let us now examine the enzymatic interconversions involved in these bicyclic cascades. We shall specifically focus on the covalent modifications of glycogen phosphorylase and glycogen synthase, the structural effects of these covalent modifications, and how these structural changes affect the interactions of their allosteric effectors. We shall then consider the cyclic cascades as a whole, studying the various modification enzymes involved and their "ultimate" allosteric effectors. Finally, we shall see how the various cyclic cascades of glycogen metabolism function in different physiological situations.

C. *Glycogen Phosphorylase Bicyclic Cascade*

In 1938, Carl and Gerti Cori found that glycogen phosphorylase exists in two forms, the *b* form that requires AMP for activity, and the *a* form that is active without AMP. It nevertheless took 20 years for the development of the protein chemistry techniques through which Edwin Krebs and Edmund Fischer demonstrated, in 1959, that phosphorylases *a* and *b* correspond to forms of the protein in which a specific residue, Ser 14, is enzymatically phosphorylated or dephosphorylated, respectively.

a. **Glycogen Phosphorylase: The Cascade's Target Enzyme**

The activity of glycogen phosphorylase is allosterically controlled, as we saw, through AMP activation and ATP, G6P, and glucose inhibition (Section 18-3A). Super-

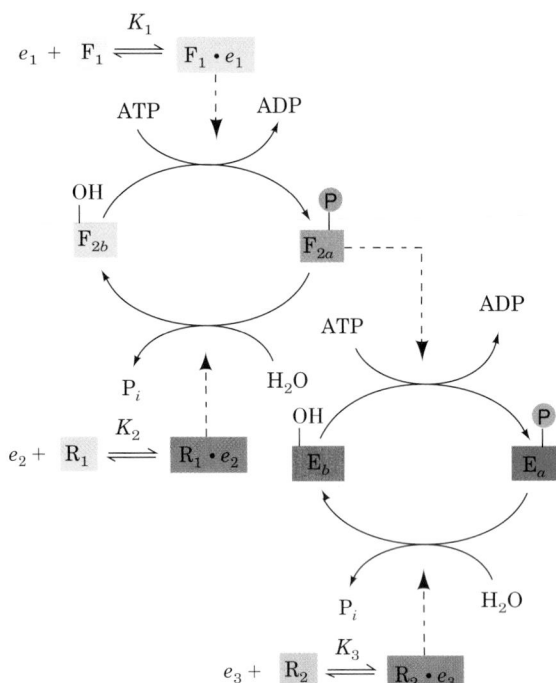

FIGURE 18-12   A bicyclic enzyme cascade. See the legend of Fig. 18-11 for symbol definitions. In a bicyclic cascade, one of the modifying enzymes (F_2) is also subject to covalent modification. It is active in the modified state (F_{2a}) and inactive in the unmodified state (F_{2b}).

imposed on this allosteric control is control by enzymatic interconversion through a bicyclic cascade involving the actions of three enzymes (Figs. 18-12 and 18-13, *left*):

1. Phosphorylase kinase, which specifically phosphorylates Ser 14 of glycogen phosphorylase *b* (Fig. 18-12, enzyme F_2).

2. Protein kinase A, which phosphorylates and thereby activates phosphorylase kinase (Fig. 18-12, enzyme F_1).

3. Phosphoprotein phosphatase-1, which dephosphorylates and thereby deactivates both glycogen phosphorylase *a* and phosphorylase kinase (Fig. 18-12, enzymes R_1 and R_2).

In an interconvertible enzyme system, the "modified" form of the enzyme bears the prefix *m* and the "original" (unmodified) form bears the prefix *o*, whereas the enzyme's most active and least active forms are identified by the suffixes *a* and *b,* respectively. In this case, *o*-phosphorylase *b* (unmodified, least active) is the form under allosteric control by AMP, ATP, and G6P (Fig. 18-9, *left*). Phosphorylation to yield *m*-phosphorylase *a* (modified, most active) all but removes the effects of these allosteric modulators. In terms of the symmetry model of allosterism (Section 10-4B), *the phosphorylation of Ser 14 shifts the enzyme's T (inactive) $\rightleftharpoons$ R (active) equilibrium in fa-*

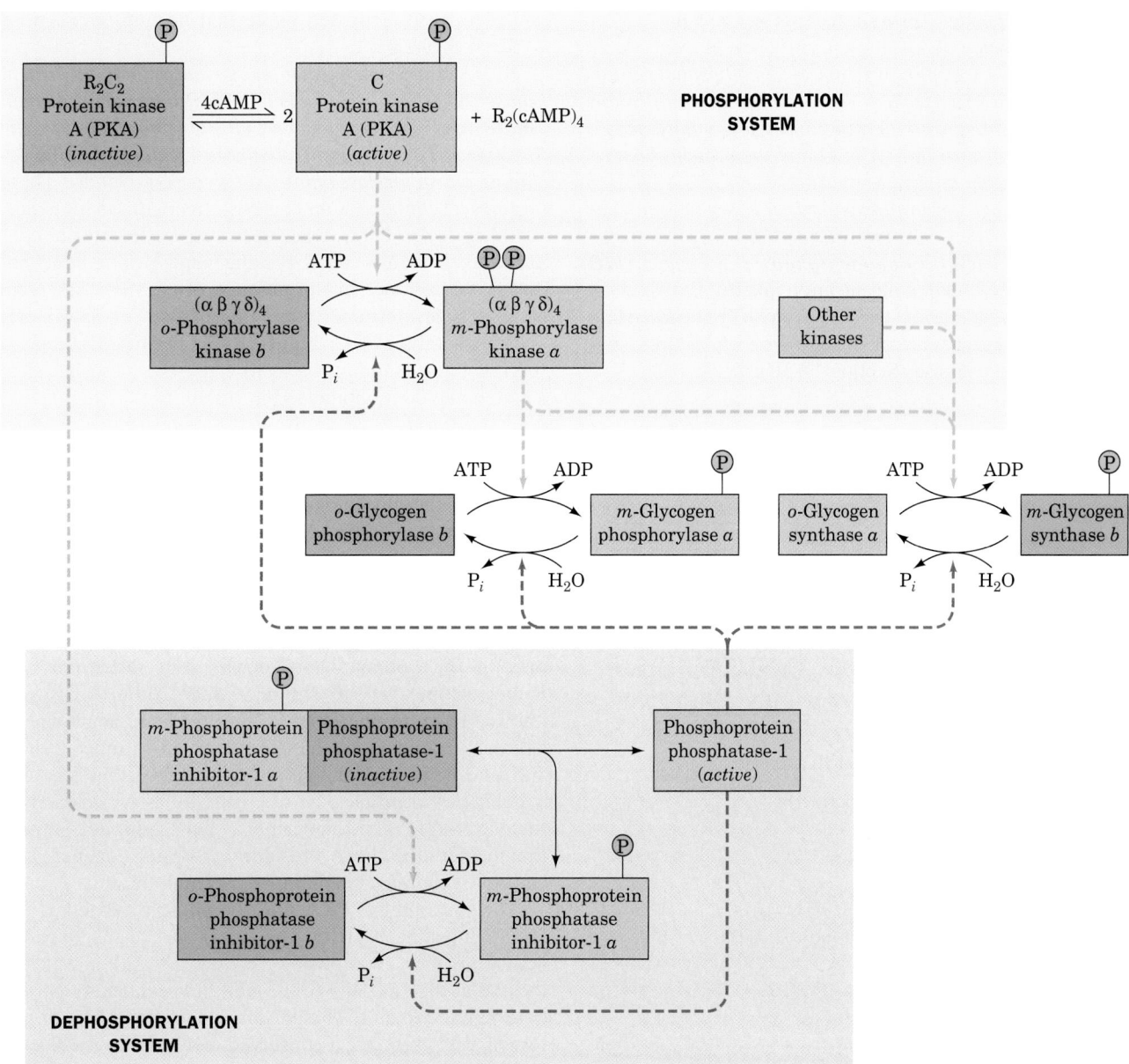

FIGURE 18-13 Schematic diagram of the major enzymatic modification/demodification systems involved in the control of glycogen metabolism in muscle. Modification (phosphorylation) systems are shaded in yellow, demodification (dephosphorylation) systems are shaded in lavender, active enzymes/inhibitors are shaded in green, and inactive enzymes/inhibitors are shaded in orange. Dashed yellow and purple arrows indicate facilitation of a modification and demodification reaction. Note that glycogen phosphorylase activity is controlled by a bicyclic enzyme cascade (*left*) and glycogen synthase activity is controlled by both a bicyclic and a monocyclic enzyme cascade (*right*). By convention, the modified form of the enzyme bears the prefix *m* and the "original" (unmodified) form bears the prefix *o*. The most active and least active forms of the enzymes are identified by the suffixes *a* and *b*, respectively. Further control of phosphoprotein phosphatase-1 covalent modification is diagrammed in Fig. 18-21. 🔊 **See Guided Exploration 15: Control of Glycogen Breakdown, and the Animated Figures**

vor of the R state (Fig. 18-9, right). Indeed, *phosphorylase a's Ser 14-phosphoryl group is analogous to an allosteric activator:* It forms ion pairs with two Arg side chains on the opposite subunit, thereby knitting the subunits together in much the same way as does AMP when it binds tightly to a site between the subunits (Fig. 18-10b).

In the resting cell, the concentrations of ATP and G6P are high enough to inhibit phosphorylase *b*. *The level of*

phosphorylase activity is therefore largely determined by the fraction of the enzyme present as phosphorylase a. The steady-state fraction of phosphorylated enzyme (E_a) depends on the relative activities of phosphorylase kinase (F_2), protein kinase A (F_1), and phosphoprotein phosphatase-1 (R_1 and R_2). This interrelationship is remarkably elaborate for glycogen phosphorylase. Let us consider the actions of these enzymes.

b. cAMP-Dependent Protein Kinase: A Crucial Regulatory Link

Phosphorylase kinase, which converts phosphorylase b to phosphorylase a, is itself subject to covalent modification (Fig. 18-13). For phosphorylase kinase to be fully active, Ca^{2+} must be present (see below) and the protein must be phosphorylated.

In both the glycogen phosphorylase and glycogen synthase cascades, the primary intracellular signal, e_1, is **adenosine-3',5'-cyclic monophosphate (3',5'-cyclic AMP or cAMP).** The cAMP concentration in a cell is a function of the ratio of its rate of synthesis from ATP by **adenylate cyclase (AC)** and its rate of breakdown to AMP by a specific **phosphodiesterase** (Section 19-2E):

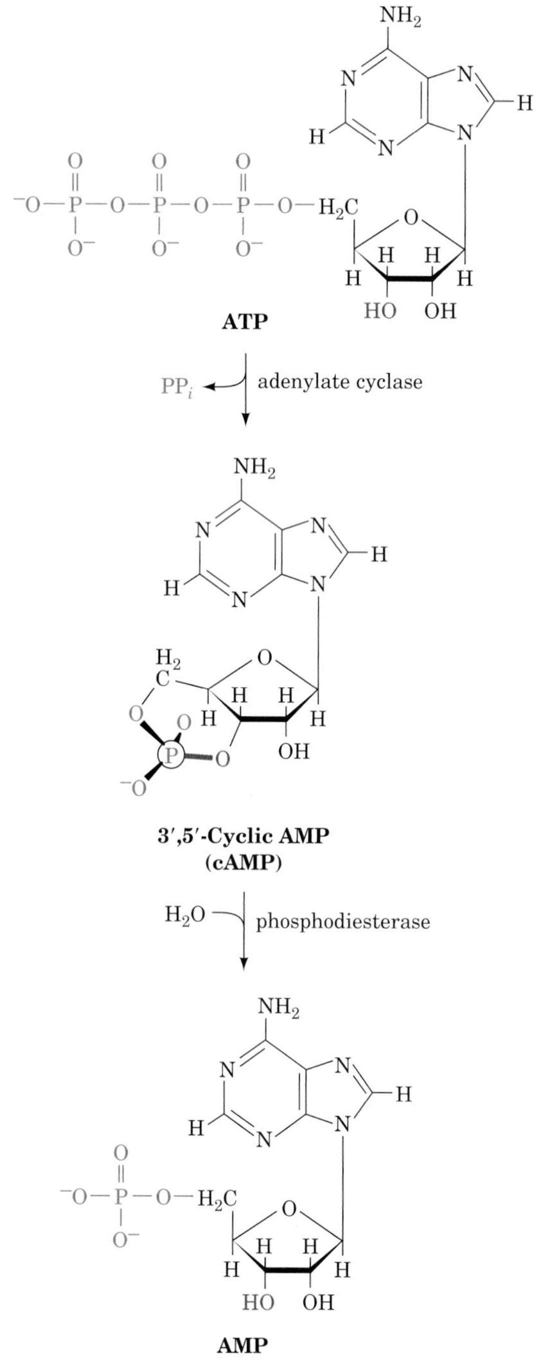

ATP

adenylate cyclase

3',5'-Cyclic AMP (cAMP)

phosphodiesterase

AMP

AC is, in turn, activated by certain hormones (Sections 18-3E and 19-2D).

cAMP is absolutely required for the activity of **protein kinase A [PKA; also called cAMP-dependent protein kinase (cAPK)],** *an enzyme that phosphorylates specific Ser and/or Thr residues of numerous cellular proteins, including phosphorylase kinase and glycogen synthase.* These proteins all contain PKA's consensus recognition sequence, Arg-Arg-X-Ser/Thr-Y, where Ser/Thr is the phosphorylation site, X is any small residue, and Y is a large hydrophobic residue. In the absence of cAMP, PKA is an inactive heterotetramer consisting of two regulatory (R) and two catalytic (C) subunits, R_2C_2. The cAMP binds to the regulatory subunits so as to cause the dissociation of active catalytic monomers (Fig. 18-13; top). *The intracellular concentration of cAMP therefore determines the fraction of PKA in its active form and thus the rate at which it phosphorylates its substrates.* In fact, in all known eukaryotic cases, the physiological effects of cAMP are exerted through the activation of specific protein kinases.

The X-ray structure of the 350-residue C subunit of mouse PKA in complex with Mg^{2+}–ATP and a 20-residue inhibitor peptide was determined by Susan Taylor and Janusz Sowadski (Fig. 18-14), and that of a similar complex of the porcine heart enzyme was determined by Robert Huber. The C subunit, as are other kinases of known structure (e.g., Figs. 17-5 and 17-15), is bilobal. It has an N-terminal lobe that consists of a 5-stranded β sheet and an α helix, and a larger C-terminal lobe that is mainly α helical. A deep cleft between the lobes is occupied by the Mg^{2+}–ATP and the segment of the inhibitor peptide that includes the above 5-residue consensus sequence. This cleft must therefore contain PKA's catalytic site.

The C subunit of PKA must be phosphorylated at Thr 197 for maximal activity. Thr 197 is part of the so-called activation loop (comprising residues 184–208), which is located at the "mouth" of the cleft between PKA's N- and C-terminal domains. The phosphoryl group at Thr 197 interacts with Arg 165, a conserved residue that is adjacent to Asp 166, the catalytic base that activates the substrate protein's target Ser/Thr hydroxyl group for phosphorylation. Thus the phosphoryl group at PKA's Thr 197 functions to properly orient its active site residues.

Protein kinases play key roles in the signaling pathways by which many hormones, growth factors, neurotransmitters, and toxins affect the functions of their target cells (Chapter 19), as well as in controlling metabolic pathways. Indeed, 581 human proteins constituting ~1.7% of human genes are predicted to be protein kinases, which accounts for the observation that ~30% of the proteins in mammalian cells are phosphorylated. The ~1000 different protein kinases that have been sequenced share a conserved catalytic core corresponding to residues 40 to 280 of PKA's C subunit. In addition to phosphorylating other proteins, many protein kinases are themselves phosphoproteins whose activities are controlled by phosphorylation, often at their activation loops. However, since PKA is normally fully phosphorylated at Thr 197, it is unclear whether its activity is regulated *in vivo* by phosphorylation/dephosphorylation.

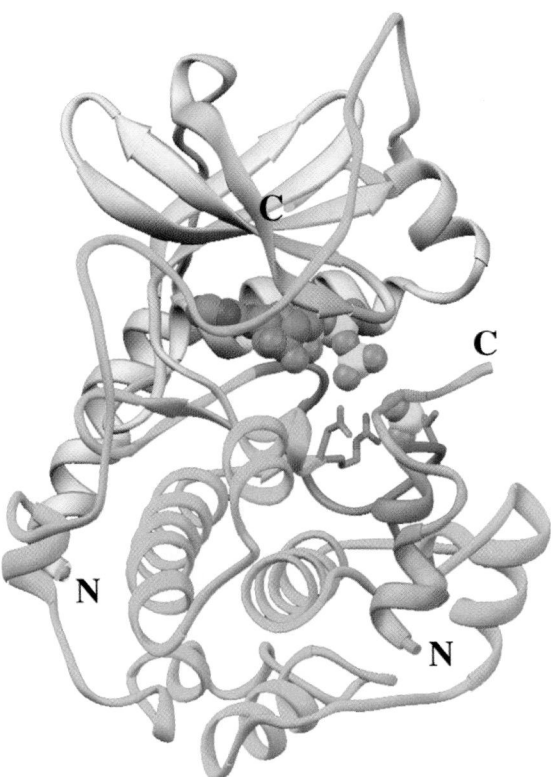

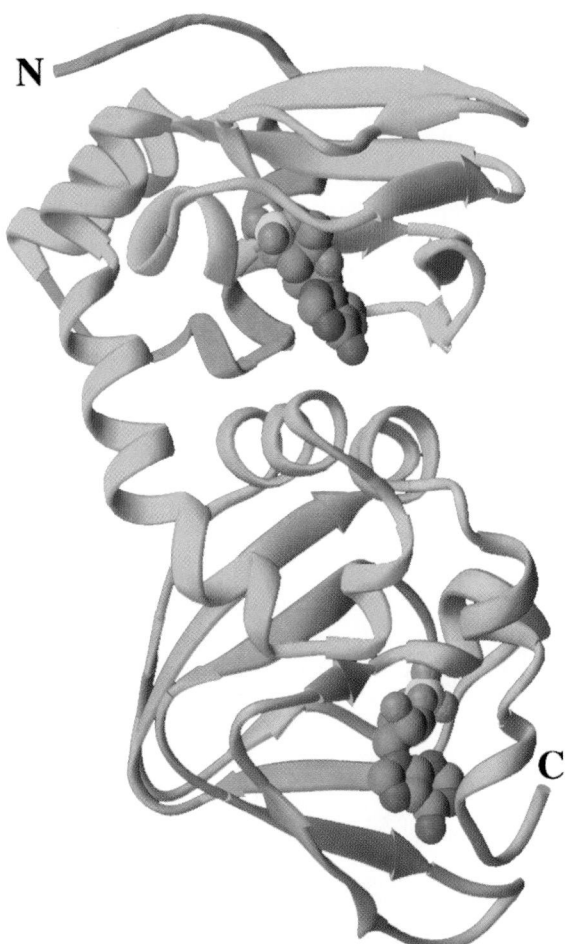

FIGURE 18-14   X-ray structure of the catalytic (C) subunit of mouse protein kinase A (PKA). The protein is in complex with ATP and a 20-residue peptide segment of a naturally occurring protein kinase inhibitor. The N-terminal domain is pink and its C-terminal domain is cyan with its activation loop light blue. The polypeptide inhibitor is orange and its pseudo–target sequence, Arg-Arg-Asn-Ala-Ile, is magenta, with the Ala replacing the Ser to be phosphorylated (note that the enzyme's true target sequence is Arg-Arg-X-Ser/Thr-Y, where X is a small residue, Y is a large hydrophobic residue, and Ser/Thr, which is replaced by Ala in the polypeptide inhibitor, is the residue that the enzyme phosphorylates). The ATP and the phosphoryl group of phosphoThr 197 are shown in space-filling form and the side chains of the catalytically essential Arg 165, Asp 166, and Thr 197 are shown in stick form, all colored according to atom type (C green, N blue, O red, and P yellow). Note that the inhibitor's pseudo–target sequence is in close proximity to the ATP's γ phosphate group, the group that the enzyme transfers. [Based on an X-ray structure by Susan Taylor and Janusz Sowadski, University of California at San Diego. PDBid 1ATP.] 🖭 **See the Interactive Exercises and Kinemage Exercise 15**

FIGURE 18-15   X-ray structure of the regulatory (R) subunit of bovine protein kinase A (PKA). The protein lacks its N-terminal 91 residues (which contain its dimerization domain) and is complexed with cAMP. The N-terminal region, which includes its autoinhibitor segment, is magenta, domain A is cyan, and domain B is orange. The cAMPs are shown in space-filling form colored according to atom type (C green, N blue, O red, and P yellow). [Based on an X-ray structure by Susan Taylor, University of California at San Diego. PDBid 1RGS.]

c. PKA's R Subunit Competitively Inhibits Its C Subunit

The R subunit of PKA has a well-defined domain structure that was first characterized by limited proteolysis. It consists of, from N- to C-terminus, a dimerization domain, an autoinhibitor segment, and two tandem homologous cAMP-binding domains, A and B. In the R_2C_2 complex, the autoinhibitor segment, which resembles the C subunit's substrate peptide, binds in the C subunit's active site (as does the inhibitory peptide in Fig. 18-14) so as to block substrate binding. Thus, the R subunit is a competitive inhibitor of PKA's substrate proteins.

Each R subunit cooperatively binds 2 cAMPs. When the B domain lacks bound cAMP, it masks the A domain so as to prevent it from binding cAMP. However, the binding of cAMP to the B domain triggers a conformational change that permits the A domain to bind cAMP, which, in turn, releases the C subunits from the complex.

Taylor determined the X-ray structure of the R subunit lacking its N-terminal 91 residues and in complex with two cAMPs (Fig. 18-15). This truncated protein is unable to dimerize but, in the absence of cAMP, forms a tight inactive complex with the C subunit, and on binding cAMP, releases active C subunits as do intact R_2 dimers. As previously predicted by sequence alignments, the A and B domains are structurally similar to each other and to the prokaryotic cAMP-binding transcriptional regulator named **catabolite gene activator protein (CAP;** Section 31-3C). Extensive interactions between the A and B domains presumably mediate the conformational change that "opens up" domain A for cAMP binding when cAMP has bound to the B domain. The autoinhibitory segment, which in the free R subunit is extremely sensitive to proteolysis, has its first 21 residues disordered in the X-ray structure.

d. Phosphorylase Kinase: Coordination of Enzyme Activation with [Ca²⁺]

Phosphorylase kinase (PhK) is activated by Ca^{2+} concentrations as low as $10^{-7}M$ as well as by covalent modification. This 1300-kD enzyme consists of four nonidentical subunits that form the active oligomer $(\alpha\beta\gamma\delta)_4$. The isolated γ subunit is capable of full catalytic activity (ability to convert phosphorylase *b* to phosphorylase *a*), whereas the α, β, and δ subunits are inhibitors of the catalytic reaction.

The δ subunit, which is known as **calmodulin (CaM),** confers Ca^{2+} sensitivity on the complex. When Ca^{2+} binds to calmodulin's four Ca^{2+}-binding sites, this ubiquitous eukaryotic regulatory protein undergoes an extensive conformational change (see below) that activates phosphorylase kinase. Glycogen phosphorylase therefore becomes phosphorylated and the rate of glycogen breakdown increases. The physiological significance of this Ca^{2+} activation process is that nerve impulses trigger muscle contraction through the release of Ca^{2+} from intracellular reservoirs (Section 35-3C). *This transient increase in cytosolic [Ca^{2+}] induces both muscle contraction and the increase in glycogen breakdown that supplies glycolysis, which in turn, generates the ATP required for muscle contraction.*

e. Calmodulin: A Ca²⁺-Activated Switch

Calmodulin is a ubiquitous eukaryotic Ca^{2+}-binding protein that participates in numerous cellular regulatory processes. In some of these, CaM functions as a monomeric protein, whereas in others (e.g., PhK) it is a subunit of a larger protein. The X-ray structure of this highly conserved 148-residue protein, determined by Charles Bugg, has a curious dumbbell-like shape in which CaM's two globular domains are connected by a seven-turn α helix (Fig. 18-16). CaM has two high-affinity Ca^{2+}-binding sites on each of its globular domains, both of which are formed by nearly superimposable helix–loop–helix motifs known as **EF hands** (Fig. 18-17) that also occur in numerous other Ca^{2+}-sensing proteins of known structure. The Ca^{2+} ion in each of these sites is octahedrally coordinated by oxygen atoms from the

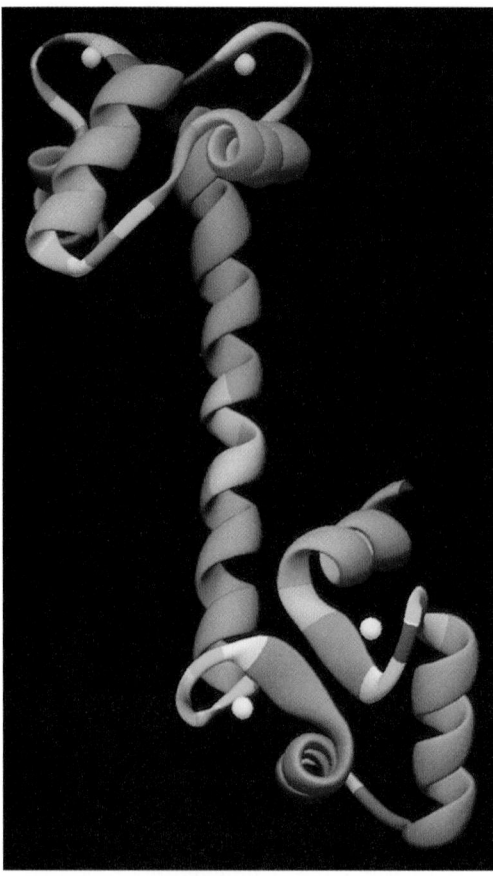

FIGURE 18-16 X-Ray structure of rat testis calmodulin. This monomeric 148-residue protein contains two remarkably similar globular domains separated by a seven-turn α helix. The residues are color coded according to their backbone conformation angles (ϕ and ψ; Fig. 8-7): cyan, α helical angles; green, β sheet angles; yellow, between helix and sheet; and purple, left-handed helix. The Gly residues are white and the N-terminus is blue. The two Ca^{2+} ions bound to each domain are represented by white spheres. [Courtesy of Mike Carson, University of Alabama at Birmingham. X-Ray structure determined by Charles Bugg, University of Alabama at Birmingham. PDBid 3CLN.] 🔎 See Kinemage Exercise 16-1

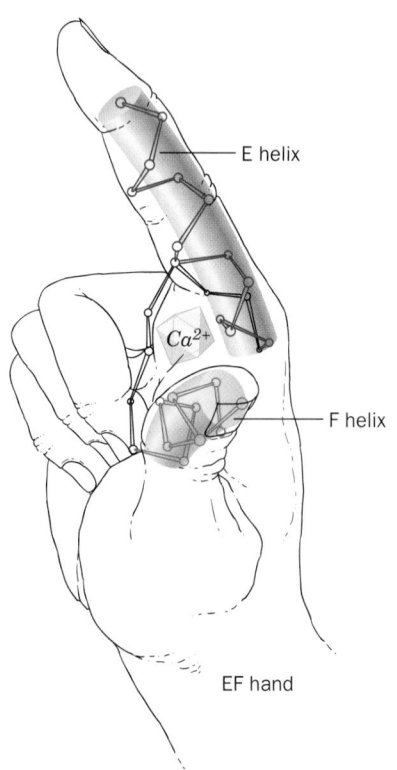

FIGURE 18-17 EF hand. The Ca^{2+}-binding sites in many proteins that function to sense the level of Ca^{2+} are formed by helix–loop–helix motifs named EF hands. [After Kretsinger, R.H., *Annu. Rev. Biochem.* **45,** 241 (1976).] 🔎 See Kinemage Exercise 16-1

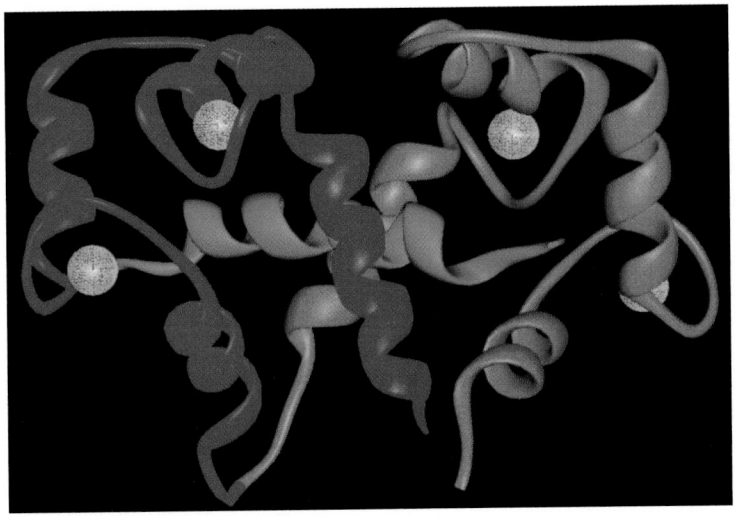

(a)

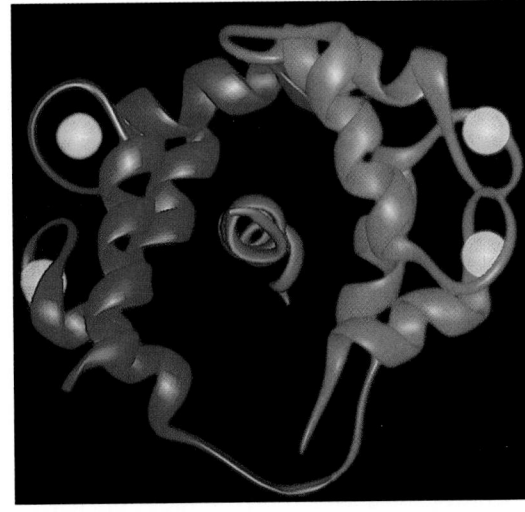

(b)

FIGURE 18-18   NMR structure of (Ca²⁺)₄–CaM from *Drosophila melanogaster* in complex with its 26-residue target polypeptide from rabbit skeletal muscle myosin light chain kinase (MLCK). The N terminal domain of CaM is blue, its C-terminal domain is red, the target polypeptide is green, and the Ca²⁺ ions are represented by cyan spheres. (*a*) A view of the complex in which the N-terminus of the target polypeptide is on the right, and (*b*) the perpendicular view as seen from the right side of Part *a*. In both views, the pseudo-twofold axis relating the N- and C-terminal domains of CaM is approximately vertical. Note how the middle segment of the long central helix in uncomplexed CaM (Fig. 18-16) has unwound and bent (bottom loop in Part *b*) such that CaM forms a globular protein that largely encloses the helical target polypeptide within a hydrophobic tunnel in a manner resembling two hands holding a rope (the target polypeptide assumes the random coil conformation in solution). However, the conformations of CaM's two globular domains are essentially unchanged by the complexation. Evidently, CaM's bound Ca²⁺ ions serve to organize and stabilize the target binding conformations of its globular domains. [Based on an NMR structure by Marius Clore, Angela Gronenborn, and Ad Bax, National Institutes of Health. PDBid 2BBM.] 🔖 See Kinemage Exercise 16-2

backbone and side chains of the loop as well as from a protein-associated water molecule.

The binding of Ca²⁺ to either domain of CaM induces a conformational change in that domain, which exposes an otherwise buried Met-rich hydrophobic patch. This patch, in turn, binds with high affinity to the CaM-binding domain of the phosphorylase kinase γ subunit, as well as to the CaM-binding domains of numerous other Ca²⁺-regulated proteins (many of which interact with CaM that is free in solution), and in doing so modulates the activities of these proteins. These CaM-binding domains have little mutual sequence homology but are all basic amphiphilic α helices. In fact, ~20-residue segments of these helices, as well as synthetic amphiphilic helices composed of only Leu, Lys, and Trp residues, are bound by Ca²⁺–CaM as tightly as the target proteins themselves.

Despite uncomplexed CaM's extended appearance in its X-ray structure (Fig. 18-16), a variety of studies indicate that both of its globular domains can simultaneously bind to a single target helix. Evidently, CaM's central α helix serves as a flexible linker rather than as a rigid spacer, a property that probably further increases the range of target sequences to which CaM can bind. This idea is confirmed by the NMR structure (Fig. 18-18) of (Ca²⁺)₄–CaM in complex with its 26-residue CaM-binding target polypeptide of skeletal muscle **myosin light chain kinase (MLCK;** a homolog of the PKA C subunit, which phosphorylates and thereby activates the light chains of the muscle protein **myosin**; Section 35-3D), which was determined by Marius Clore, Angela Gronenborn, and Ad Bax. Indeed, the extended conformation of CaM's central helix in Fig. 18-16 is probably an artifact arising from crystal packing forces, considering that this helix's central two turns contact no other portion of the protein and hence are maximally solvent-exposed (almost all other known α helices are at least partially buried in a protein). Moreover, a polypeptide with the sequence of this helix assumes a random coil conformation in aqueous solution. Nevertheless, the flexible linker is essential to the function of CaM: In the presence of Ca²⁺, CaM's individual domains (obtained by tryptic cleavage), when in high concentration, are able to bind their target proteins but fail to even marginally activate them unless present in several hundred-fold excess.

How does Ca²⁺–CaM activate its target protein kinases? MLCK contains a C-terminal segment whose sequence resembles that of MLCK's target polypeptide on the light chain of myosin but lacks a phosphorylation site. A model of MLCK, based on the X-ray structure of the 30% identical C subunit of PKA, strongly suggests that this autoinhibitor peptide inactivates MLCK by binding in its active site. Indeed, the excision of MLCK's autoinhibitor peptide by limited proteolysis permanently activates this enzyme. MLCK's CaM-binding segment overlaps this autoinhibitor peptide. Evidently, *the binding of Ca²⁺–CaM to this peptide segment extracts the autoinhibitor from MLCK's active site, thereby activating this enzyme (Fig. 18-19).*

Ca^{2+}–CaM's other target proteins, including the phosphorylase kinase γ subunit, are presumably activated in the same way. The X-ray structures of two homologous protein kinases support this so-called **intrasteric mechanism,** those of **calmodulin-dependent protein kinase I (CaMKI)** and **twitchin kinase.** While the details of binding of the autoinhibitory sequence differ for each of these protein kinases, the general mode of autoinhibition and activation by Ca^{2+}–CaM is the same.

PKA's R subunit, as we have seen, contains a similar autoinhibitory sequence adjacent to its two tandem cAMP-binding domains. In this case, however, the autoinhibitory peptide is allosterically ejected from the C subunit's active site by the binding of cAMP to the R subunit (which lacks a Ca^{2+}–CaM binding site).

f. Phosphorylase Kinase's γ Subunit Is Controlled by Multiple Autoinhibitors

Phosphorylase kinase's 386-residue γ subunit consists of an N-terminal kinase domain, which is 36% identical in sequence to the C subunit of PKA, and a C-terminal regulatory domain, which contains a CaM-binding peptide and an overlapping autoinhibitor segment. Evidently, Ca^{2+}–CaM relieves this inhibition, as is diagrammed in Fig. 18-19. This explains why the N-terminal 298-residue segment of the PhK γ subunit, termed **PhKγ_t** (t for truncated), displays catalytic activity comparable to that of fully activated PhK but is unaffected by Ca^{2+} or phosphorylation signals.

The X-ray structure of PhKγ_t in complex with ATP and a heptapeptide related to the natural substrate was determined by Johnson (Fig. 18-20). It reveals, as expected, that PhKγ_t structurally resembles PKA (Fig. 18-14) as well as other protein kinases of known structure including CaMKI and twitchin kinase. Comparisons of these various structures shed light on how the catalytic activity of PhK is regulated. Numerous protein kinases, including PKA, are activated by the phosphorylation of Ser, Thr, and/or Tyr residues in their activation loops, which, as we saw in Fig. 18-14, interacts with a conserved Arg residue that thereby correctly positions the adjacent catalytically important Asp residue. However, the PhK γ subunit is not subject to phosphorylation. Rather, its activation loop residue that might otherwise be phosphorylated is Glu 182, whose negative charge mimics the presence of a phosphate group by interacting with Arg 148 so as to correctly position Asp 149 (Fig. 18-20). Thus, the PhK catalytic site maintains an active conformation but, in the absence of Ca^{2+}, is inactivated by the binding of its C-terminal autoinhibitor segment.

Sites on both the α and β subunits of PhK are subject to phosphorylation by PKA (Fig. 18-13). This activates PhK at much lower Ca^{2+} concentrations than otherwise, and full enzyme activity is obtained in the presence of Ca^{2+} only when both these subunits are phosphorylated. The β subunit does, in fact, have an autoinhibitor sequence, sug-

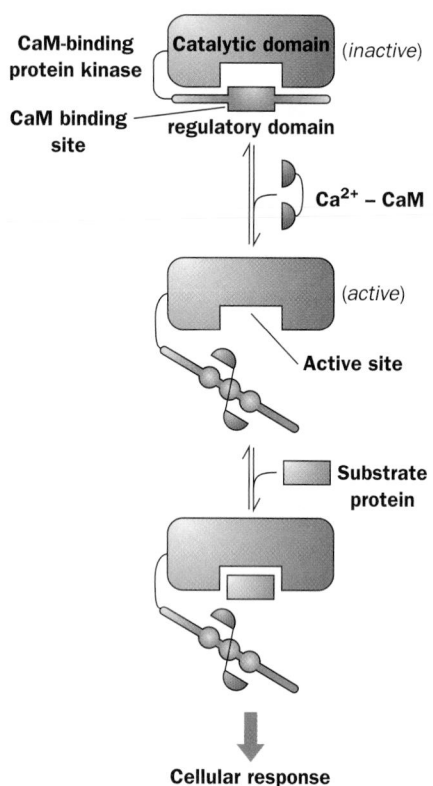

FIGURE 18-19   Schematic diagram of the Ca^{2+}–CaM-dependent activation of protein kinases. Autoinhibited kinases have an N- or C-terminal "pseudosubstrate" sequence (*red*) that binds at or near the enzyme's active site (*brown*) so as to inhibit its function. This autoinhibitory segment is in close proximity with or overlaps a Ca^{2+}–CaM binding sequence. Consequently, Ca^{2+}–CaM (*green*) binds to this sequence so as to extract it from the enzyme's active site, thereby activating the enzyme to phosphorylate other proteins (*magenta*). [After Crivici, A. and Ikura, M., *Annu. Rev. Biophys. Biomol. Struct.* **24,** 88 (1995).]

gesting that phosphorylation changes its conformation so as to make it unavailable for inhibiting the γ subunit's active site. This would explain the synergistic effect of phosphorylation and Ca^{2+} on the activity of PhK: Ca^{2+}–CaM sequesters the γ subunit's autoinhibitory segment, whereas phosphorylation of the β subunit removes yet another autoinhibitor. The way in which the phosphorylation of the α subunit modulates the activity of PhK is, as yet, unknown.

g. Phosphoprotein Phosphatase-1

The steady-state phosphorylation levels of most enzymes involved in cyclic cascades are maintained by the opposition of kinase-catalyzed phosphorylations and the hydrolytic dephosphorylations catalyzed by phosphoprotein phosphatases. The phosphatase involved in the cyclic cascades controlling glycogen metabolism is phosphoprotein phosphatase-1. This enzyme hydrolyzes the phosphoryl groups from *m*-glycogen phosphorylase *a,* both α and β subunits of phosphorylase kinase, and two other proteins involved in glycogen metabolism, as discussed below.

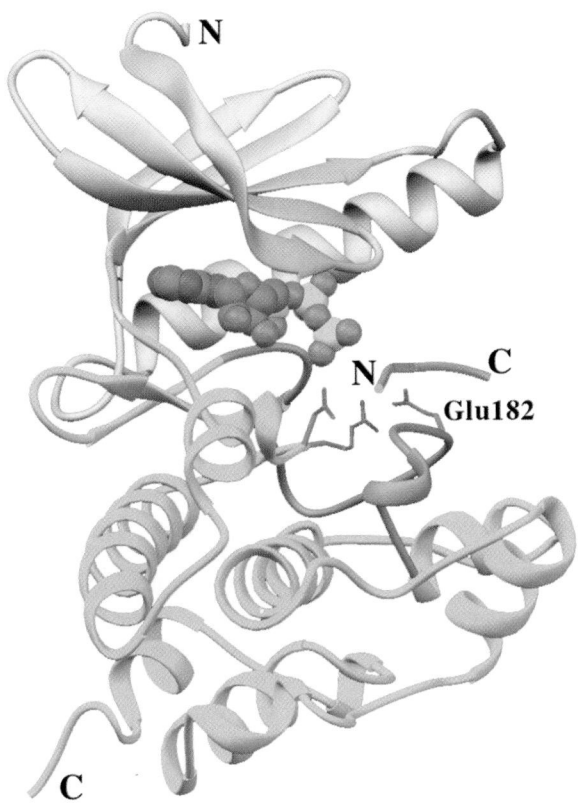

The catalytic subunit of phosphoprotein phosphatase-1 **(PP1),** which is designated **PP1c,** hydrolyzes phosphoryl groups on Ser/Thr residues via a single step mechanism.

FIGURE 18-20   X-ray structure of rabbit muscle PhKγ$_t$ in complex with ATP and a heptapeptide (RQMSFRL). This heptapeptide is related in sequence to the enzyme's natural substrate (KQISVRG). The protein is shown in the "standard" protein kinase orientation with its N terminal domain pink, its C-terminal domain cyan, and its activation loop light blue. The heptapeptide is orange, with its residue to be phosphorylated (Ser) gray. The ATP is shown in space-filling form and the side chains of the catalytically essential Arg 148, Asp 149, and Glu 182 are shown in stick form, all colored according to atom type (C green, N blue, O red, and P yellow). Note the structural similarities and differences between this protein and the homologous C subunit of PKA (Fig. 18-14). [After an X-ray structure by Louise Johnson, Oxford University, Oxford, U.K. PDBid 2PHK.]

The X-ray structure of PP1c indicates that it contains a binuclear metal ion center (both metals are Mn^{2+} in the recombinant enzyme) which, it is proposed, activates a water molecule (promotes its ionization to OH^-, Section 15-1C) for nucleophilic attack on the phosphoryl group.

PP1c binds to glycogen through the intermediacy of regulatory proteins in both muscle and liver. In muscle, PP1c is only active when it is bound to glycogen through this glycogen-binding **G$_M$ subunit.** The activity of PP1c and its affinity for the G$_M$ subunit are regulated by phosphorylation of the G$_M$ subunit at two separate sites (Fig. 18-21). Phosphorylation of site 1 by **insulin-stimulated protein kinase** activates phosphoprotein phosphatase-1, whereas phosphorylation of site 2 by PKA (which can also phosphorylate site 1) causes the enzyme to be released into the

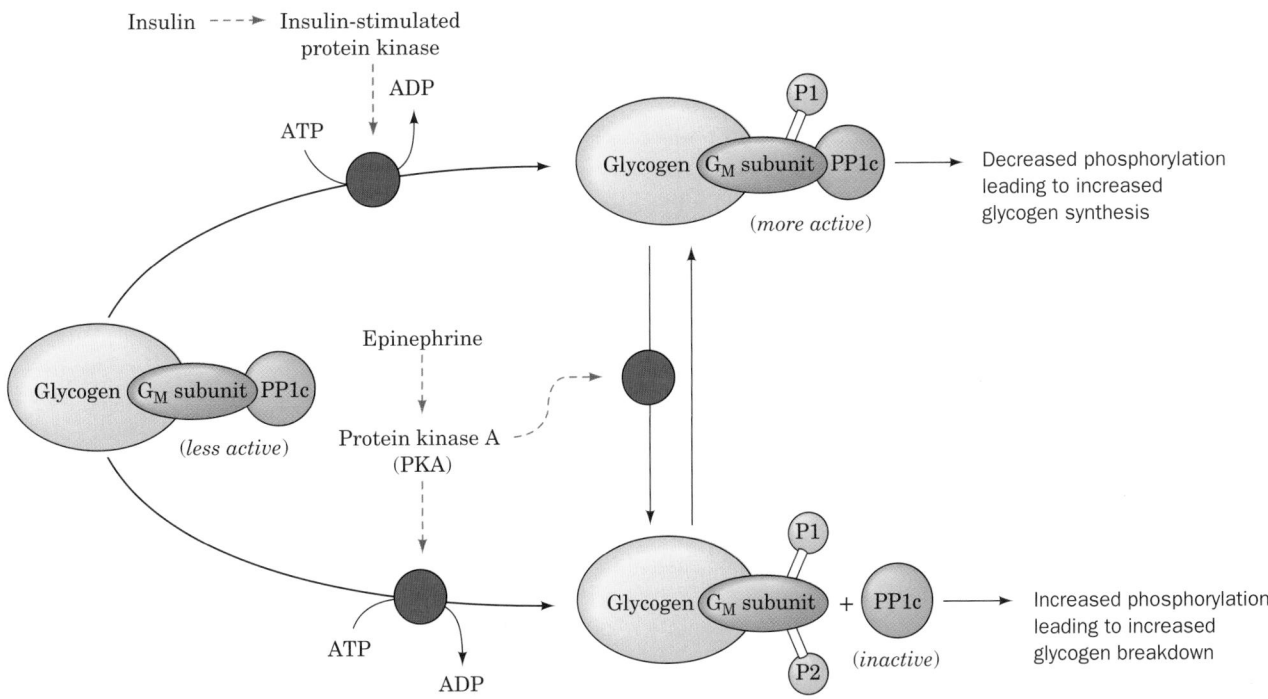

FIGURE 18-21   The antagonistic effects of insulin and epinephrine on glycogen metabolism in muscle. This occurs through their effects on the phosphoprotein phosphatase-1 catalytic subunit, PP1c, via its glycogen-bound G$_M$ subunit. Green dots and dashed arrows indicate activation.

cytoplasm, where it cannot dephosphorylate the glycogen-bound enzymes of glycogen metabolism.

In the cytosol, PP1c is also inhibited by its binding to the protein **phosphoprotein phosphatase inhibitor 1 (inhibitor-1).** This latter protein provides yet another example of control by enzymatic interconversion: It too is modified by PKA and demodified by PP1c (Fig. 18-13, *bottom*), although, in this case, a Thr, not a Ser, is phosphorylated/dephosphorylated. The protein is a functional inhibitor only when it is phosphorylated. *The concentration of cAMP therefore controls the fraction of an enzyme in its phosphorylated form, not only by increasing the rate at which it is phosphorylated, but also by decreasing the rate at which it is dephosphorylated. In the case of glycogen phosphorylase, an increase in [cAMP] results not only in an increase in this enzyme's rate of activation, but also in a decrease in its rate of deactivation.*

The activity of phosphoprotein phosphatase-1 in liver is also controlled by its binding to glycogen through the intermediacy of a glycogen-binding subunit, here named G_L. When bound to G_L, PP1c is activated toward dephosphorylation of the glycogen-bound enzymes of glycogen metabolism. However, G_L is not subject to control via phosphorylation as is G_M in the muscle. Rather, the binding of *m*-phosphorylase *a* to G_L strongly inhibits the activity of PP1c by an allosteric mechanism.

Among the major conformational changes that glycogen phosphorylase undergoes in converting from the T to the R state is the movement of the Ser 14-phosphoryl group from the surface of the T-state (inactive) enzyme to a position buried a few angstroms beneath the protein's surface at the dimer interface in the R-state (active) enzyme (Fig. 18-10). Both the R and T forms of phosphorylase *a* strongly bind the $G_L \cdot$ PP1c *complex,* but only in the T-state enzyme is the Ser 14-phosphoryl group accessible for hydrolysis by PP1c. Consequently, under the conditions that phosphorylase *a* converts to the T state (Section 18-3G), PP1c hydrolyzes its now exposed Ser 14-phosphoryl group. This converts *m*-phosphorylase *a* to *o*-phosphorylase *b*, which has only a low affinity for binding the $G_L \cdot$PP1c complex and hence does not inhibit PP1c. One effect of phosphorylase *a* demodification, therefore, is to relieve the inhibition of PP1c and thus allow it to excise the phosphoryl groups of other susceptible phosphoproteins. Since phosphorylase *a* has a high affinity for the $G_L \cdot$ PP1c complex and is in ~10-fold greater concentration, *relief of PP1c inhibition only occurs when more than ~90% of the glycogen phosphorylase is in the o-phosphorylase b form.* Glycogen synthase is among the proteins that are dephosphorylated by the $G_L \cdot$PP1c complex when it is no longer inhibited by phosphorylase. However, in contrast to phosphorylase, dephosphorylation activates glycogen synthase. This enzyme is involved in its own bicyclic cascade whose properties we shall now examine.

D. *Glycogen Synthase Bicyclic Cascade*

Like glycogen phosphorylase, glycogen synthase exists in two enzymatically interconvertible forms:

1. The modified (*m*; phosphorylated) form that is inactive under physiological conditions (the *b* form).

2. The original (*o*; dephosphorylated) form that is active (the *a* form).

m-Glycogen synthase *b* is under allosteric control; it is strongly inhibited by physiological concentrations of ATP, ADP, and P_i and hence the modified enzyme is almost totally inactive *in vivo.* The activity of the unmodified enzyme is essentially independent of these effectors, so the cell's glycogen synthase activity varies with the fraction of the enzyme in its unmodified form.

The mechanistic details of the interconversion of modified and unmodified forms of glycogen synthase are particularly complex and are therefore not as well understood as those of glycogen phosphorylase. It has been clearly established that the fraction of unmodified glycogen synthase is, in part, controlled by a bicyclic cascade involving phosphorylase kinase (PhK) and phosphoprotein phosphatase-1, enzymes that are also involved in the glycogen phosphorylase bicyclic cascade (Fig. 18-13, *right*). This demodification process is facilitated by G6P, whose binding to *m*-glycogen synthase *b* induces it to undergo a conformational change that exposes its phosphoryl groups to the surface of the protein, thereby making them available for dephosphorylation by phosphoprotein phosphatase-1.

Glycogen synthase is phosphorylated at several sites. Several protein kinases are known to at least partially deactivate human muscle glycogen synthase by phosphorylating this homotetramer at 1 or more of 9 Ser residues in the N- and C-terminal segments on its 737-residue subunits. These enzymes include PhK, PKA (so glycogen synthase deactivation may also be considered to occur via a monocyclic cascade), CaMKI (which is activated by the presence of Ca^{2+}), **protein kinase C (PKC;** which responds to the extracellular presence of certain hormones via a mechanism described in Sections 18-3G and 19-4C), **AMP-dependent protein kinase (AMPK;** which responds to ATP availability and hence acts as a fuel gauge; Sections 25-5 and 27-1), and **glycogen synthase kinase-3 [GSK3;** which is inhibited by **insulin** (Sections 18-3E and F), whose presence therefore results in the dephosphorylation and hence activation of glycogen synthase]. Why glycogen synthase deactivation is so elaborately controlled compared to its activation or the activation/deactivation of glycogen phosphorylase is unclear, although, whatever the reasons, it closely monitors the organism's metabolic state.

E. *Integration of Glycogen Metabolism Control Mechanisms*

Whether there is net synthesis or degradation of glycogen and at what rate depends on the relative balance of the active forms of glycogen synthase and glycogen phosphorylase. This, in turn, largely depends on the rates of the phosphorylation and dephosphorylation reactions of the two bicyclic cascades. These cascades, one controlling the rate of glycogen breakdown and the other controlling the rate of glycogen synthesis, are intimately related. They are linked by protein kinase A and phosphorylase kinase, which, through phosphorylation, activate glycogen phosphorylase as they inactivate glycogen synthase (Fig. 18-13).

The cascades are also linked by phosphoprotein phosphatase-1, which in liver is inhibited by phosphorylase *a* and therefore is unable to activate (dephosphorylate) glycogen synthase unless it first inactivates (also by dephosphorylation) phosphorylase *a*.

a. Hormones Are Important Regulators of Glycogen Metabolism

Glycogen metabolism is largely regulated by the peptide hormone **insulin** (Fig. 7-2) acting in opposition to **glucagon,** another peptide hormone,

^+H_3N - His - Ser - Glu - Gly - Thr - Phe - Thr - Ser - Asp - Tyr -   10
  Ser - Lys - Tyr - Leu- Asp- Ser - Arg - Arg - Ala - Gln-   20
  Asp- Phe- Val - Gln - Trp - Leu - Met - Asn- Thr - COO^- 29
Glucagon

together with the adrenal hormones **epinephrine (adrenalin)** and **norepinephrine (noradrenalin):**

X = CH_3 **Epinephrine**
X = H   **Norepinephrine**

Hormonal stimulation of cells at their plasma membranes occurs through the mediation of transmembrane proteins called **receptors.** *Different cell types have different complements of receptors and thus respond to different sets of hormones.* For example, both muscle and liver cells have abundant insulin and **adrenoreceptors** (receptors responsive to epinephrine and norepinephrine), whereas glucagon receptors are more prevalent in liver than in skelatal muscle.

b. Second Messengers Mediate Glucagon- and Epinephrine-Stimulated Glycogen Breakdown

The response to glucagon and epinephrine involves the release inside the cell of molecules known as **second messengers,** *that is, intracellular mediators of the externally received hormonal message.* Different receptors act to release different second messengers. Indeed, cAMP was identified by Earl Sutherland as the first known instance of a second messenger through his demonstration that glucagon and epinephrine act at cell surfaces to stimulate adenylyl cyclase (AC) to increase [cAMP] [the mechanism of AC activation, as well as a discussion of other second messengers, including Ca^{2+}, **inositol-1,4,5-trisphosphate (IP$_3$),** and **diacylglycerol (DAG),** is elaborated in Sections 19-2D and 19-4A]. Following this discovery, it was realized that cAMP, which is present in all forms of life, is an essential control element in many biological processes.

When hormonal stimulation by glucagon or epinephrine increases the intracellular cAMP concentration, the protein kinase A activity increases, increasing the rates of phosphorylation of many proteins and decreasing their dephosphorylation rates as well. A decrease in dephosphorylation rates, as previously noted, increases the phosphorylation level of phosphoprotein phosphatase inhibitor-1, which in turn inhibits phosphoprotein phosphatase-1. An increase in the concentration of phosphorylase *a* also contributes to the inhibition of phosphoprotein phosphatase-1.

Because of the amplifying properties of the cyclic cascades, a small change in [cAMP] results in a large change in the fraction of enzymes in their phosphorylated forms. When a large fraction of the glycogen metabolism enzymes are present in their phosphorylated forms, the metabolic flux is in the direction of glycogen breakdown, since glycogen phosphorylase is active and glycogen synthase is inactive. When [cAMP] decreases, phosphorylation rates decrease, dephosphorylation rates increase, and the fraction of enzymes in their dephospho forms increases. The resultant activation of glycogen synthase and the inhibition of glycogen phosphorylase cause a change in the flux direction toward net glycogen synthesis.

F. Maintenance of Blood Glucose Levels

An important function of the liver is to maintain the blood concentration of glucose, the brain's primary fuel source, at ~5 m*M*. When blood [glucose] decreases beneath this level, usually during exercise or well after meals have been digested, the liver releases glucose into the bloodstream. The process is mediated by the hormone glucagon as follows:

1. Glucose inhibits the pancreatic α cells from secreting glucagon into the bloodstream. When the blood glucose concentration falls, this inhibition is released causing the α cells to secrete glucagon.

2. Glucagon receptors on liver cell surfaces respond to the presence of glucagon by activating adenylate cyclase, thereby increasing the [cAMP] inside these cells.

3. The [cAMP] increase, as described above, triggers an increase in the rate of glycogen breakdown, leading to increased intracellular [G6P].

4. G6P, in contrast to glucose, cannot pass through the cell membrane. However, in liver, which does not employ glucose as a major energy source, the enzyme **glucose-6-phosphatase (G6Pase)** hydrolyzes G6P:

$$G6P + H_2O \rightarrow glucose + P_i$$

The resulting glucose enters the bloodstream, thereby increasing the blood glucose concentration. Muscle and brain cells, however, lack G6Pase so that they retain their G6P.

G6P hydrolysis requires intracellular G6P transport. G6P is produced in the cytosol, whereas G6Pase resides in endoplasmic reticulum (ER) membrane. G6P must therefore be imported into the ER by a **G6P translocase** before it can be hydrolyzed. The resulting glucose and P_i are then returned to the cytosol via specific transport proteins (Section 18-3G). A defect in any of the components of this G6P hydrolysis system results in **type I glycogen storage disease** (Section 18-4).

How does this delicately balanced system respond to an increase in blood [glucose]? When blood sugar is high, normally immediately after meals have been digested, glucagon levels decrease and insulin is released from the pancreatic β cells. *The rate of glucose transport across many cell membranes increases in response to insulin (through the insulin-dependent glucose transporter **GLUT4**; Section 20-2E), [cAMP] decreases, and glycogen metabolism therefore shifts from glycogen breakdown to glycogen synthesis.* The mechanism of insulin action is quite complex and has only recently begun to be understood (Sections 19-3 and 19-4F), but one of its target enzymes appears to be phosphoprotein phosphatase-1.

In muscle, insulin and epinephrine have antagonistic effects on glycogen metabolism. Epinephrine promotes glycogenolysis by activating the cAMP-dependent phosphorylation cascade, which stimulates glycogen breakdown while inhibiting glycogen synthesis. Insulin, as we saw in Section 18-3C, activates insulin-stimulated protein kinase to phosphorylate site 1 on the glycogen-binding G_M subunit of phosphoprotein phosphatase-1 so as to activate this protein to dephosphorylate the enzymes of glycogen metabolism (Fig. 18-21). The storage of glucose as glycogen is thereby stimulated through the inhibition of glycogen breakdown and the stimulation of glycogen synthesis.

In liver, it appears that glucose and glucose-6-phosphate themselves may be the messengers to which the glycogen metabolism system responds. *Glucose inhibits phosphorylase a by binding only to the active site of the enzyme's inactive T state, but in a manner different from that of substrate.* The presence of glucose therefore shifts phosphorylase a's T $\rightleftharpoons$ R equilibrium toward the T state (Fig. 18-9, *right*). This conformational shift, as we saw in Section 18-3C, exposes the Ser 14-phosphoryl group to phosphoprotein phosphatase-1, resulting in the demodification of phosphorylase a. An increase in glucose concentration therefore promotes inactivation of glycogen phosphorylase a through the enzyme's conversion to phosphorylase b (Fig. 18-22; i.e., phosphorylase a acts as a glucose receptor). The concomitant release of phosphoprotein phosphatase-1 inhibition (recall that it specifically binds to and is thereby inactivated by phosphorylase a), moreover, results in the activation (dephosphorylation) of m-glycogen synthase b. In addition, glucose is converted to G6P by **glucokinase** (see below), which facilitates the dephosphorylation and activation of m-glycogen synthase b to o-glycogen synthase a. Above a glucose concentration of 7 m*M*, these processes reverse the flux of glycogen metabolism from breakdown to synthesis. The liver can thereby store the excess glucose as glycogen.

a. Glucokinase Forms G6P at a Rate Proportional to the Glucose Concentration

The liver's function in "buffering" the blood [glucose] is made possible because this organ contains a variant of hexokinase (the first glycolytic enzyme; Section 17-2A) known as glucokinase **(GK; also called hexokinase D** and

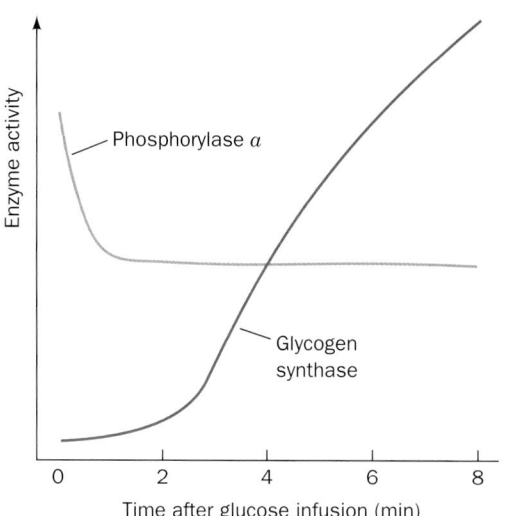

FIGURE 18-22   The enzymatic activities of phosphorylase *a* and glycogen synthase in mouse liver in response to an infusion of glucose. Phosphorylase *a* is rapidly inactivated and, somewhat later, glycogen synthase is activated. [After Stalmans, W., De Wulf, H., Hue, L., and Hers, H.-G. *Eur. J. Biochem.* **41**, 129 (1974).]

hexokinase IV). The hexokinase in most cells obeys Michaelis–Menten kinetics, has a high glucose affinity ($K_M < 0.1$ m*M*; the value of [glucose] at which the enzyme achieves half-maximal velocity; Section 14-2A), and is inhibited by its reaction product, G6P. GK, in contrast, has a much lower glucose affinity (reaching half of its maximal velocity at ~5 m*M*) and displays sigmoidal kinetics with a Hill constant (Section 10-1B) of 1.5, so *its activity increases rapidly with the blood [glucose] over the normal physiological range (Fig. 18-23; see Problem 5 of this chapter). GK, moreover, is not inhibited by physiological concentrations of G6P.* Consequently, the higher the blood [glucose], the faster the liver converts glucose to G6P [liver cells, unlike most cells, contain ample quantities of the insulin-independent glucose transporter **GLUT2** (Section 20-2E) and are therefore freely permeable to glucose; their glucose transport rate is unresponsive to insulin]. Thus at low blood [glucose], the liver does not compete with other tissues for the available glucose supply, whereas at high blood [glucose], when the glucose needs of these tissues are met, glucose in the liver is converted to G6P. The excess glucose in the liver induces the inactivation of glycogen phosphorylase and the release of phosphoprotein phosphatase-1, whereas the resulting G6P allosterically facilitates the activation of glycogen synthase via dephosphorylation. The net result is that the liver converts the excess glucose to glycogen. (Note that GK is a monomeric enzyme, so that its sigmoidal rate increase with [glucose] is a puzzling observation in light of the various allosteric models indicating that monomeric enzymes are incapable of cooperative behavior. Since GK does not exhibit Michaelis–Menten kinetics, the glucose concentration when this enzyme has half its maximal activity is known as its $K_{0.5}$ in analogy with the operational definition of K_M.)

GK is subject to metabolic controls. Emile Van Schaftingen isolated **glucokinase regulatory protein**

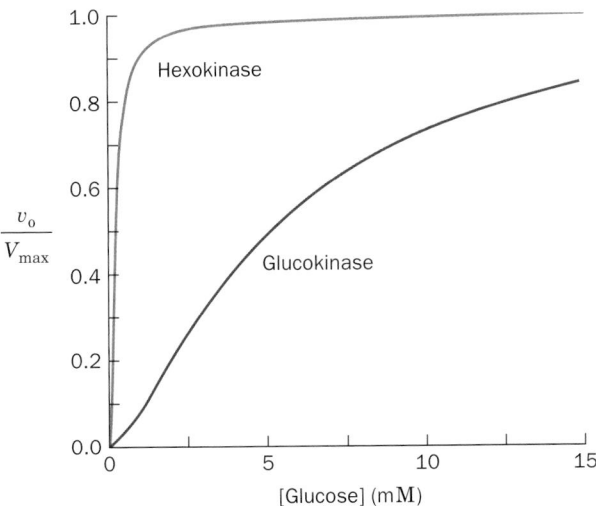

FIGURE 18-23 Comparison of the relative enzymatic activities of hexokinase and glucokinase over the physiological blood glucose range. The affinity of glucokinase for glucose ($K_{0.5}$ = 5 m*M*) is much lower than that of hexokinase (K_M = 0.1 m*M*) and exhibits sigmoid rather than hyperbolic variation with [glucose]. [The glucokinase curve was generated using the Hill equation (Eq. [10.7]) with K = 10 m*M* and n = 1.5 as determined by Cardenas, M.L., Rabajille, E., and Niemeyer, H., *Eur. J. Biochem.* **145**, 163–171 (1984).]

(GKRP; a 625-residue monomer) from rat liver, which, in the presence of the glycolytic intermediate fructose-6-phosphate (F6P), is a competitive inhibitor of glucokinase. Fructose-1-phosphate (F1P), an intermediate in liver fructose metabolism (Section 17-5A), overcomes this inhibition. Since fructose is normally available only from dietary sources (e.g., sucrose), fructose may be the signal that triggers the uptake of dietary glucose by the liver.

b. Glucokinase Regulates Glucose Homeostasis via an Intracellular Localization Mechanism

Intracellular localization plays an important role in GK inhibition by GKRP. GK translocates freely between the nucleus and cytoplasm. However, GKRP is located exclusively in the nucleus. When the glucose concentration is low, GK remains bound to GKRP in the nucleus, where it

is unavailable to phosphorylate glucose. At increased concentrations of glucose and/or F1P, GK dissociates from GKRP and translocates to the cytoplasm, where it phosphorylates glucose to G6P, starting it on its way toward glycogen synthesis. The antagonism of GK and GKRP provides a major mechanism for controlling glucose phosphorylation and glycogen metabolism in the liver. Their flux control coefficients (Section 17-4C) in liver cells are close to +1 for GK and −1 for GKRP (a negative flux control coefficient is indicative of inhibition). This countervailing relationship provides a sensitive mechanism for maintaining glucose homeostasis.

Phosphoglucomutase, which has a high enough activity to equilibrate its substrate and product and therefore functions in either direction, transforms G6P to G1P, which is then converted to glycogen. Some of the G6P is also reconverted to glucose by the action of glucose-6-phosphatase in what amounts to a "futile" cycle. This is apparently the energetic price of effective glucose "buffering" of the blood.

c. Fructose-2,6-Bisphosphate Activates Glycolysis
β-D-Fructose-2,6-bisphosphate (F2,6P)

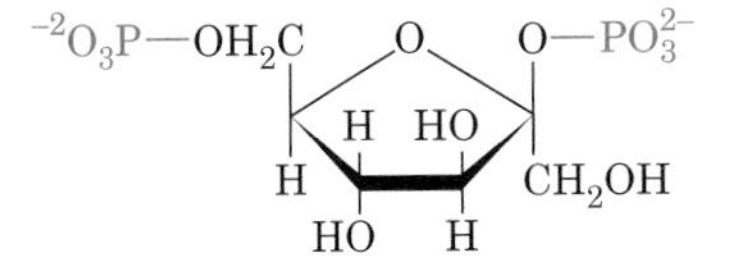

$$^{-2}O_3P-OH_2C \quad\quad O \quad\quad O-PO_3^{2-}$$

**β-D-Fructose-2,6-bisphosphate
(F2,6P)**

is also an important factor in the liver's maintenance of blood [glucose]. *F2,6P, which is not a glycolytic metabolite, is an extremely potent allosteric activator of animal phosphofructokinase (PFK) and an inhibitor of fructose bisphosphatase (FBPase).* F2,6P, which was independently discovered in 1980 by Simon Pilkis, by Van Schaftingen and Henri-Géry Hers, and by Kosaku Uyeda, therefore stimulates glycolytic flux (the F6P–FBP substrate cycle is discussed in Section 17-4F).

The concentration of F2,6P in the cell depends on the balance between its rates of synthesis and degradation by ***phosphofructokinase-2 (PFK-2; also called 6PF-2-K)*** *and* ***fructose bisphosphatase-2 (FBPase-2; also called F-2,6-Pase),*** *respectively (Fig. 18-24).* These enzyme activities are

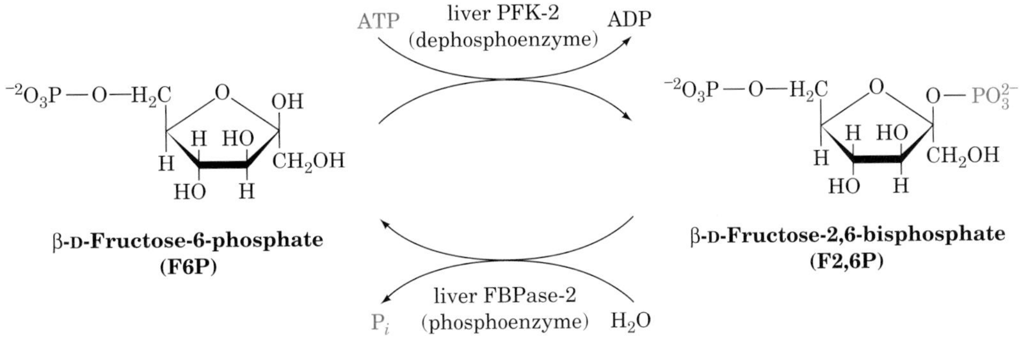

$$^{-2}O_3P-O-H_2C \quad\quad O \quad\quad OH$$

**β-D-Fructose-6-phosphate
(F6P)**

ATP → liver PFK-2 (dephosphoenzyme) → ADP

$$^{-2}O_3P-O-H_2C \quad\quad O \quad\quad O-PO_3^{2-}$$

**β-D-Fructose-2,6-bisphosphate
(F2,6P)**

P_i → liver FBPase-2 (phosphoenzyme) → H_2O

FIGURE 18-24 Formation and degradation of β-D-fructose-2,6-bisphosphate as catalyzed by PFK-2 and FBPase-2. These two enzyme activities occur on different domains of the same protein molecule. Dephosphorylation of the liver enzyme activates PFK-2 but deactivates FBPase-2.

located on different domains of a single subunit of the homodimeric protein named **PFK-2/FBPase-2.** The X-ray structure of the H256A mutant of rat testis PFK-2/FBPase-2 in complex with F6P, P$_i$, succinate, and the nonhydrolyzable ATP analog **adenosine-5'-(β,γ-imido)triphosphate (ADPNP or AMP-PNP)**

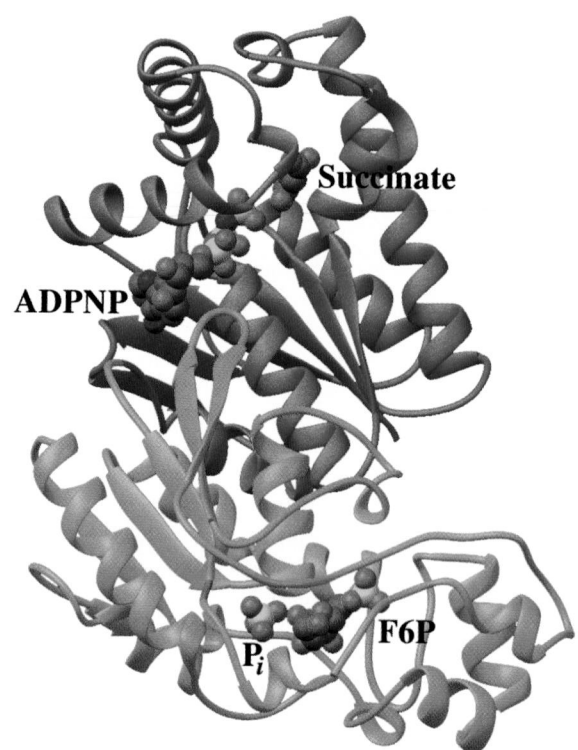

$$^-O-\underset{\underset{O^-}{|}}{\overset{\overset{O}{\|}}{P}}-NH-\underset{\underset{O^-}{|}}{\overset{\overset{O}{\|}}{P}}-O-\underset{\underset{O^-}{|}}{\overset{\overset{O}{\|}}{P}}-O-CH_2\ O$$

Adenosine-5'-(β, γ-imido)triphosphate (ADPNP)

was determined by Uyeda and Charles Hasemann (Fig. 18-25). It indicates, in agreement with a variety of studies, that the PFK-2 activity resides on each subunit's 246-residue N-terminal domain, whereas the FBPase-2 activity resides on each subunit's 213-residue C-terminal domain. The succinate, which binds in the vicinity of ADPNP's γ phosphate group, presumably occupies the F6P binding pocket in the PFK-2 active site, whereas the F6P and P$_i$ mark the F2,6P binding site in the FBPase-2 active site. The FBPase-2 domain is structurally related to the glycolytic enzyme phosphoglycerate mutase (PGM; Section 17-2H) and shares a common catalytic mechanism involving a covalent phosphoHis intermediate (His 256 in FBPase-2). The PFK-2 domain is structurally related to adenylate kinase (Section 17-4F) but not, as had been speculated, to PFK (also called **PFK-1** to distinguish this glycolytic enzyme from PFK-2).

The enzyme activities of PFK-2/FBPase-2 are subject to allosteric regulation by a variety of metabolic intermediates as well as to phosphorylation/dephosphorylation by protein kinase A and a phosphoprotein phosphatase. Phosphorylation of the liver enzyme at its Ser 32 inhibits its PFK-2 activity and activates its FBPase-2 activity. Thus, the pancreatic α cell's release of glucagon in response to low blood [glucose] results, through an increase in liver [cAMP], in a decreased liver [F2,6P]. This situation, in turn, decreases the PFK-1 activity, thereby inhibiting glycolysis. Hence, the G6P resulting from the concurrent stimulation of glycogen degradation is converted to glucose and secreted as described above rather than being metabolized. Simultaneously, the deinhibition of FBPase (also called **FBPase-1** to distinguish it from FBPase-2) by the decrease of [F2,6P] stimulates **gluconeogenesis,** the formation of glucose from nonglucose precursors such as amino acids by a pathway that effectively reverses glycolytic flux (and in which FBPase-1 is a key regulatory enzyme; Section 23-1). This process provides a second means of glucose production. Conversely, when the blood [glucose] is high, cAMP levels decrease, the liver PFK-2/FBPase-2 is dephosphorylated by phosphoprotein phosphatase-1 activating PFK-2, which, in turn, causes a rise in [F2,6P]. PFK-1

FIGURE 18-25 X-ray structure of the H256A mutant of rat testis PFK-2/FBPase-2. The N-terminal PFK-2 domain is light blue and the C-terminal FBPase-2 domain is orange. The bound ADPNP, succinate, F6P, and P$_i$ are shown in space-filling form colored according to atom type (C green, N blue, O red, and P yellow). The P$_i$, which occupies the binding site of the F2,6P's 2-phosphate group, is opposite the site that would be occupied by the side chain of the wild-type enzyme's His 256 (*magenta*), to which it would be transferred in the catalytic reaction. The succinate occupies the presumed F6P binding pocket of the PFK-2 domain. [Based on an X-ray structure by Kosaku Uyeda and Charles Hasemann, University of Texas Southwestern Medical Center. PDBid 2BIF.]

is therefore activated, FBPase-1 is inhibited, and the net glycolytic flux changes from gluconeogenesis to glycolysis.

The F2,6P control systems in skeletal muscle and in heart muscle function quite differently from that in liver due to the presence in these tissues of different PFK-2/FBPase-2 isozymes. In heart and skeletal muscles, increased glycogen breakdown is coordinated with increased glycolysis rather than increased glucose secretion. This is because phosphorylation of the heart muscle PFK-2/FBPase-2 isozyme occurs at entirely different sites (Ser 406 and Thr 475 of the 530-residue protein) from that of the liver isozyme (Ser 32 of the 470-residue protein) and activates rather than inhibits PFK-2. Consequently, hormones that stimulate glycogen breakdown also increase heart muscle [F2,6BP], thereby stimulating glycolysis as well. The skeletal muscle and testis isozymes lack phosphorylation sites altogether and are therefore not subject to cAMP-dependent phosphorylation control.

G. *Response to Stress*

Epinephrine and norepinephrine, which are often called the "fight or flight" hormones, are released into the bloodstream by the adrenal glands in response to stress. Epinephrine receptors (known as **β-adrenoreceptors;** Section 19-1F) present on the surfaces of liver and muscle cells respond to these hormones just as glucagon receptors respond to the presence of glucagon; they activate adenylate cyclase, thereby increasing intracellular [cAMP]. Indeed, epinephrine also stimulates the pancreatic α cells to release glucagon, which further increases liver [cAMP]. The G6P produced by the consequent glycogen breakdown in muscle enters the glycolytic pathway, thereby generating ATP and helping the muscles cope with the stress that triggered the epinephrine release.

The liver's response to stress, in addition to its response to the glucagon released by pancreatic epinephrine stimulation, involves response to epinephrine stimulation via two types of receptors, β-adrenoreceptors, as discussed above, and **α-adrenoreceptors.** α-Adrenoreceptors act by stimulating **phospholipase C** to release other second messengers, namely inositol-1,4,5-trisphosphate (IP$_3$), diacylglycerol (DAG), and Ca^{2+} (Fig. 18-26a), which act to reinforce the cells' response to cAMP. As we mentioned in Section 18-3C, phosphorylase kinase, which both activates phosphorylase and inactivates glycogen synthase, is fully active only when both phosphorylated and in the presence of increased [Ca^{2+}]. In addition, glycogen synthase is inactivated by phosphorylation by several other Ca^{2+}-dependent protein kinases, including protein kinase C (Section 18-3D). Protein kinase C requires both Ca^{2+} and DAG for activity (Section 19-4C). This dual stimulation of receptors in response to epinephrine causes the liver to produce G6P, which is hydrolyzed by G6Pase, resulting in the release of glucose into the bloodstream, thereby further fueling the muscles (Fig. 18-26b).

4 ■ GLYCOGEN STORAGE DISEASES

With glycogen metabolism being such a finely controlled system, it is not surprising that genetically determined enzyme deficiencies result in disease states. The study of these disease states and the enzyme deficiencies that cause them has provided insights into the system's balance. In this sense, genetic diseases are valuable research tools. Conversely, the biochemical characterization of the pathways affected by a genetic disease often leads, as we shall see, to useful strategies for its treatment. Many diseases have been characterized that result from inherited deficiencies of one or another of the enzymes of glycogen metabolism. These defects are listed in Table 18-1 and discussed in this section.

Type I: **Glucose-6-Phosphatase Deficiency (von Gierke's Disease)**

G6Pase catalyzes the final step leading to the release of glucose into the bloodstream by the liver. Deficiency of this enzyme results in an increase of intracellular [G6P], which leads to a large accumulation of glycogen of normal structure in the liver and kidney (recall that G6P inhibits glycogen phosphorylase and activates glycogen synthase) and an inability to increase blood glucose concentration in response to glucagon or epinephrine. Similar difficulties occur when there are defects in the protein that transports glucose across the liver cell plasma membrane (Section 20-2E) or in any of the proteins that transport glucose, G6P, or P$_i$ across the endoplasmic reticulum membrane (Section 18-3F; Fig. 18-26b). The symptoms of Type I glycogen storage disease include massive liver enlargement, severe **hypoglycemia** (low blood sugar) after a few hour fast, and a general failure to thrive. Treatment of the disease has included drug-induced inhibition of glucose uptake by the liver to increase blood [glucose], continuous intragastric feeding overnight,

TABLE 18-1   Hereditary Glycogen Storage Diseases

Type	Enzyme Deficiency	Tissue	Common Name	Glycogen Structure
I	Glucose-6-phosphatase	Liver	von Gierke's disease	Normal
II	α-1,4-Glucosidase	All lysosomes	Pompe's disease	Normal
III	Amylo-1,6-glucosidase (debranching enzyme)	All organs	Cori's disease	Outer chains missing or very short
IV	Amylo-(1,4→1,6)-transglycosylase (branching enzyme)	Liver, probably all organs	Andersen's disease	Very long unbranched chains
V	Glycogen phosphorylase	Muscle	McArdle's disease	Normal
VI	Glycogen phosphorylase	Liver	Hers' disease	Normal
VII	Phosphofructokinase	Muscle	Tarui's disease	Normal
VIII	Phosphorylase kinase	Liver	X-Linked phosphorylase kinase deficiency	Normal
IX	Phosphorylase kinase	All tissues		Normal
0	Glycogen synthase	Liver		Normal, deficient in quantity

(a)

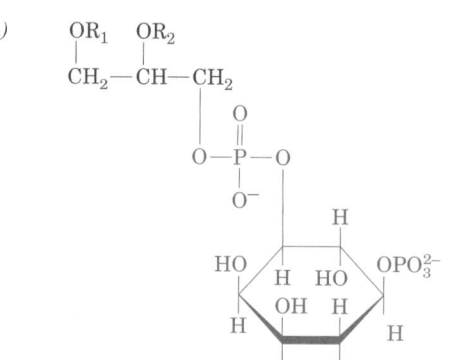

Phosphatidylinositol-4,5-bisphosphate (PIP$_2$)

phospholipase C   H$_2$O

**1,2-Diacylglycerol
(DAG)**

**Inositol-1,4,5-triphosphate
(IP$_3$)**

(b)

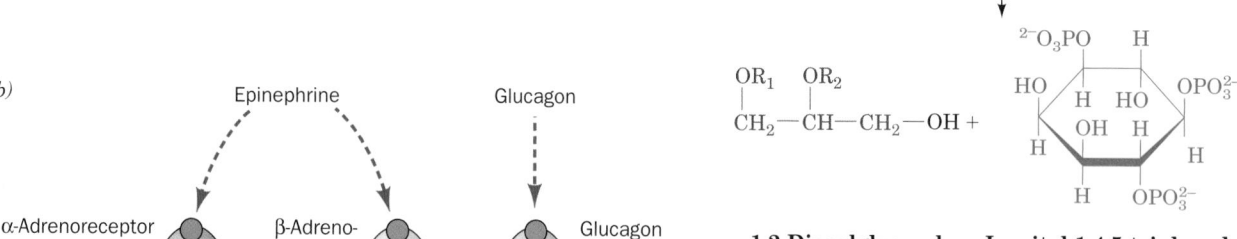

FIGURE 18-26 The liver's response to stress. (*a*) Stimulation of α-adrenoreceptors by epinephrine activates phospholipase C to hydrolyze phosphatidylinositol-4,5-bisphosphate (PIP$_2$) to inositol-1,4,5-trisphosphate (IP$_3$) and diacylglycerol (DAG). (*b*) The participation of two second messenger systems: The cAMP-mediated stimulation of glycogenolysis and inhibition of glycogen synthesis triggered by glucagon and β-adrenoreceptor activation; and the IP$_3$, DAG, and Ca^{2+}-mediated stimulation of glycogenolysis and inhibition of glycogen synthesis triggered by α-adrenoreceptor activation. IP$_3$ stimulates the release of Ca^{2+} from the endoplasmic reticulum, whereas DAG, together with Ca^{2+}, activates protein kinase C to phosphorylate and thereby inactivate glycogen synthase. G6Pase occupies the endoplasmic reticulum. Consequently, the cytosolically produced G6P is transported into the endoplasmic reticulum via the **T1 G6P translocase,** where it is hydrolyzed to glucose and P$_i$. The glucose and P$_i$ are then returned to the cytosol by the **T2 and T3 transporters,** respectively, and the glucose is exported from the cell via the GLUT2 glucose transporter.

again to increase blood [glucose], oral adminstration of uncooked corn starch (which is only slowly broken down to glucose), and surgical transposition of the portal vein, which ordinarily feeds the liver directly from the intestines, so as to allow this glucose-rich blood to reach peripheral tissues before it reaches the liver. This latter treatment has the added benefit of allowing the tissues to receive more glucose while decreasing the storage of this glucose as liver glycogen. Liver transplantation has also been successful in the few patients in which this treatment has been tried.

A gene therapy protocol (Section 5-5H) is being developed to correct type I glycogen storage disease. G6Pase-deficient knockout mice (Section 5-5H) have been treated with a viral vector containing the mouse G6Pase gene. This treatment, which delivers G6Pase to the livers of these mice, greatly increases their survival rate and corrects the metabolic abnormalities associated with this glycogen storage disease.

Type II: α-1,4-Glucosidase Deficiency (Pompe's Disease)

This is the most devastating glycogen storage disease. It results in a large accumulation of glycogen of normal structure in the lysosomes of all cells and causes death by cardiorespiratory failure, usually before the age of 1 year. We have not discussed **α-1,4-glucosidase** in the sections on the pathways of glycogen synthesis and breakdown since it is not among those enzymes. It occurs in lysosomes, where it functions to hydrolyze maltose (Section 11-2B) and linear oligosaccharides, as well as the outer branches of glycogen, thereby yielding free glucose. However, this second pathway of glycogen metabolism is not quantitatively important. The reason that lysosomes normally take up and degrade glycogen granules is unknown.

Type III: Amylo-1,6-Glucosidase (Debranching Enzyme) Deficiency (Cori's Disease)

In this disease, glycogen of abnormal structure containing very short outer chains accumulates in both liver and muscle since, in the absence of debranching enzyme, the glycogen cannot be further degraded. Its hypoglycemic symptoms are similar to, but not as severe as, those of von Gierke's disease (Type I). The low blood sugar, which in this case is a result of the decreased efficiency of glycogen breakdown, is treated with frequent feeding and a high-protein diet [in response to low blood sugar, the liver, through gluconeogenesis (Section 23-1), synthesizes glucose from amino acids]. For unknown reasons, the symptoms of Cori's disease often disappear at puberty.

Type IV: Amylo-(1,4→1,6)-Transglycosylase (Branching Enzyme) Deficiency (Andersen's Disease)

This is one of the most severe glycogen storage diseases; victims rarely survive past the age of 5 years because of liver dysfunction. Glycogen concentration in liver is not increased but its structure is abnormal, with very long unbranched chains resulting from the lack of branching enzyme. This decreased branching greatly reduces the solubility of glycogen. It has been suggested that the liver

dysfunction may be caused by a "foreign body" immune reaction to the abnormal glycogen.

Type V: Muscle Phosphorylase Deficiency (McArdle's Disease)

We have mentioned this condition in connection with the realization that glycogen synthesis and breakdown must occur by different pathways (Section 18-2). Its major symptom, which is most severely manifested in early adulthood, is painful muscle cramps on exertion. This situation is a result of the inability of the glycogen breakdown system to provide sufficient fuel for glycolysis to keep up with the metabolic demand for ATP. Studies by ^{31}P NMR on human forearm muscle have noninvasively corroborated this conclusion by demonstrating that exercise in individuals with McArdle's disease leads to elevated muscle ADP levels compared to those of normal individuals (Fig. 18-27). Curiously, if McArdle's victims continue their exertions, their cramps subside. This "second wind" effect has been attributed to vasodilation, which gives the muscles increased access to the glucose and fatty acids in the blood for use as alternative fuels to glycogen. Liver glycogen phosphorylase is normal in these individuals, implying the presence of different glycogen phosphorylase isozymes in muscle and liver.

Type VI: Liver Phosphorylase Deficiency (Hers' Disease)

Patients with a deficiency of liver phosphorylase have symptoms similar to those with mild forms of Type I glycogen storage disease. The hypoglycemia in this case results from the inability of glycogen phosphorylase to respond to the need for glucose production by the liver.

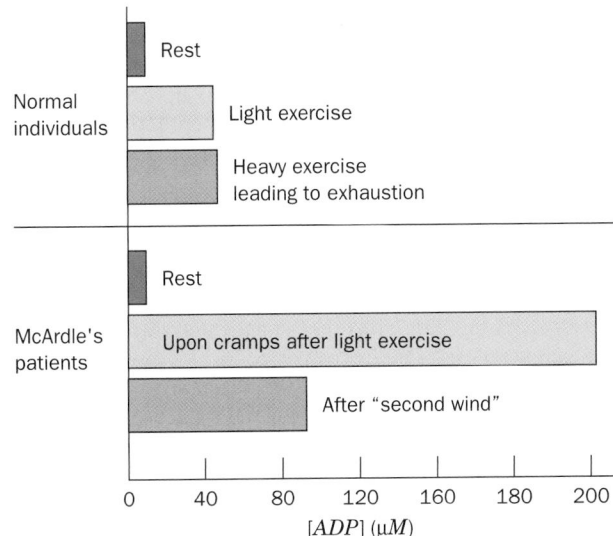

FIGURE 18-27 The ADP concentration in human forearm muscles during rest and following exertion in normal individuals and those with McArdle's disease. The ADP concentration was determined from ^{31}P NMR measurements on intact forearms. [After Radda, G.K., *Biochem. Soc. Trans.* **14,** 522 (1986).]

Type VII: Muscle Phosphofructokinase Deficiency (Tarui's Disease)

The result of a deficiency of the glycolytic enzyme PFK-1 in muscle is an abnormal buildup of the glycolytic metabolites G6P and F6P. High concentrations of G6P increase the activities of glycogen synthase (G6P activates glycogen synthase and inactivates glycogen phosphorylase) and UDP–glucose pyrophosphorylase (G6P is in equilibrium with G1P, a substrate for the enzyme) so that glycogen accumulates in muscle. Other symptoms are similar to those of Type V glycogen storage disease, muscle phosphorylase deficiency, since PFK deficiency prevents glycolysis from keeping up with the ATP demand of muscle contraction.

Type VIII: Liver Phosphorylase Kinase Deficiency (X-Linked Phosphorylase Kinase Deficiency)

Some individuals with symptoms of Type VI glycogen storage disease have liver phosphorylase of normal structure. However, they have a defective phosphorylase kinase, which results in their inability to convert phosphorylase *b* to phosphorylase *a*. The gene encoding the α subunit of phosphorylase kinase resides on the X chromosome and hence Type VIII disease is X-linked rather than autosomal, as are the other glycogen storage diseases.

Type IX: Phosphorylase Kinase Deficiency

Phosphorylase kinase deficiency, when it is autosomal recessive, is caused by a mutation in one of the genes encoding a β, γ, or δ subunit of phosphorylase kinase. Since different organs contain different isozymes of phosphorylase kinase, the symptoms and severity of Type IX disease vary with the affected organs.

Type 0: Liver Glycogen Synthase Deficiency

This is the only disease of glycogen metabolism in which there is a deficiency rather than an overabundance of glycogen. The activity of liver glycogen synthase is extremely low in individuals with this disease and hence they exhibit hyperglycemia after meals and hypoglycemia at other times. However, the primary lesion may not be in the synthase itself because other metabolic defects may lead to an imbalance of the glycogen synthase cyclic cascade. The root cause of Type 0 glycogen storage disease is still under investigation.

Since many of the glycogen storage diseases have similar clinical symptoms, these diseases are best diagnosed via genetic testing.

CHAPTER SUMMARY

1 ■ Glycogen Breakdown In animals, when glucose is not needed as a source of metabolic energy, it is stored, predominantly in liver and muscle cells, as glycogen, an $\alpha(1{\rightarrow}4)$-linked glucan with $\alpha(1{\rightarrow}6)$ branches every 8 to 14 residues. Glycogen breakdown to glucose-6-phosphate (G6P) is a two-step process. Glycogen phosphorylase catalyzes the phosphorolysis of the glycosidic linkage of a terminal glucosyl residue to form glucose-1-phosphate (G1P). Phosphoglucomutase interconverts G1P and G6P. Glycogen debranching enzyme allows complete degradation of glycogen by catalyzing the transfer of three-residue chains onto the nonreducing ends of other chains and catalyzing the hydrolysis of the remaining $\alpha(1{\rightarrow}6)$-linked glucosyl unit to glucose.

2 ■ Glycogen Synthesis Glycogen is synthesized from G6P by a pathway different from that of glycogen breakdown. G6P is converted to G1P under the influence of phosphoglucomutase. UDP–glucose pyrophosphorylase utilizes UTP to convert G1P to UDP–glucose, the activated intermediate in glycogen synthesis. The hydrolysis of the PP_i product by inorganic pyrophosphatase drives the reaction to completion. Glucosyl units are transferred from UDP–glucose to the C4—OH group of a terminal residue on a growing glycogen chain by glycogen synthase. Branching occurs through the action of a branching enzyme, which transfers ~7-residue segments of $\alpha(1{\rightarrow}4)$-linked chains to the C6—OH group of a glucosyl residue on the same or another glycogen chain.

3 ■ Control of Glycogen Metabolism The rates at which glycogen is synthesized by glycogen synthase and degraded by glycogen phosphorylase are controlled by the levels of their allosteric effectors such as ATP, AMP, G6P, and glucose. Superimposed on this allosteric control is control by the phos-

phorylation/dephosphorylation of these enzymes. The kinases and phosphatases that catalyze these modifications are part of amplifying cascades that are ultimately controlled by the hormones glucagon, insulin, and epinephrine, and by Ca^{2+}. Glucagon and epinephrine stimulate glycogen breakdown by stimulating adenylate cyclase to increase in the intracellular [cAMP]. cAMP is a "second messenger" that activates protein kinase A (PKA), which, through its activation of phosphorylase kinase, results in the phosphorylation of both glycogen phosphorylase and glycogen synthase. Phosphorylation activates glycogen phosphorylase but inactivates glycogen synthase. In addition, epinephrine causes an increase in the concentrations of other second messengers, namely, inositol-1,4,5-trisphosphate (IP_3), diacylglycerol (DAG), and Ca^{2+}, that reinforce the cAMP-dependent responses. Ca^{2+}, which is also released into muscle cytosol by nerve impulses, binds to calmodulin so as to induce this protein to activate protein kinases via an intrasteric mechanism by which Ca^{2+}–CaM extracts autoinhibitory sequences from the kinases' active sites. A decrease in [cAMP] and/or the presence of insulin leads to the activation of phosphoprotein phosphatase-1 to dephosphorylate glycogen phosphorylase and glycogen synthase. The phosphoproteins that participate in glycogen metabolism are dephosphorylated through the action of phosphoprotein phosphatase-1, which is only active when it is associated with a glycogen particle through the intermediacy of glycogen-binding G_M or G_L subunits. When [glucose] is high, the liver synthesizes glucose-6-phosphate (G6P) and ultimately glycogen from glucose via the action of glucokinase (GK), which has kinetic properties distinct from those of other hexokinases. When [glucose] is low, GK is inhibited by glucokinase regulatory protein (GKRP) and G6Pase hydrolyzes the G6P product of glycogen breakdown (which is favored at low [glucose]) for export to other tissues. The concentration in liver of F2,6P, an activator of PFK and an inhibitor of FBPase, is also dependent on the rates of cAMP-dependent phosphorylation and dephosphorylation. It is both synthesized and degraded through the action of PFK-2/FBPase-2, whose enzymatic activities are oppositely controlled by both allosteric regulation and phosphorylation/dephosphorylation.

4 ■ Glycogen Storage Diseases Glycogen storage diseases are caused by a genetic deficiency of one or another of the enzymes of glycogen metabolism. Ten different deficiencies of varying severity have been reported in humans.

■ REFERENCES

GENERAL

Boyer, P.D. and Krebs, E.G. (Eds.), *The Enzymes* (3rd ed.), Vol. 17, Academic Press (1986). [Contains detailed articles on the enzymes of glycogen metabolism and their control.]

Walsh, C., *Enzymatic Reaction Mechanisms,* Freeman (1979).

GLYCOGEN METABOLISM

Browner, M.F. and Fletterick, R.J., Phosphorylase: A biological transducer, *Trends Biochem. Sci.* **17,** 66–71 (1992).

Dai, J.-B., Liu, Y., Ray, W.J., Jr., and Konno, M., The crystal structure of muscle phosphoglucomutase refined at 2.7-angstrom resolution, *J. Biol. Chem.* **267,** 6322–6337 (1992).

Johnson, L.N., Glycogen phosphorylase: Control by phosphorylation and allosteric effectors, *FASEB J.* **6,** 2274–2282 (1992); *and* Rabbit muscle glycogen phosphorylase b. The structural basis of activation and catalysis, *in* Harding, J.J. and Crabbe, M.J.C. (Eds.), *Post-Translational Modifications of Proteins,* pp. 81–151, CRC Press (1993).

Johnson, L.N. and Barford, D., Glycogen phosphorylase, *J. Biol. Chem.* **265,** 2409–2412 (1990).

Madsen, N.B., Glycogen phosphorylase and glycogen synthetase, *in* Kuby, S.A. (Ed.), *A Study of Enzymes,* Vol. II, *pp.* 139–158, CRC Press (1991).

Meléndez-Hevia, E., Waddell, T.G., and Shelton, E.D., Optimization of molecular design in the evolution of metabolism: The glycogen molecule, *Biochem. J.* **295,** 477–483 (1993).

Palm, D., Klein, H.W., Schinzel, R.S., Bucher, M., and Helmreich, E.J.M., The role of pyridoxal 5′-phosphate in glycogen phosphorylase catalysis, *Biochemistry* **29,** 1099–1107 (1990).

Roach, P.J. and Skurat, A.V., Self-glucosylating initiator proteins and their role in glycogen biosynthesis, *Prog. Nucl. Acid Res. Mol. Biol.* **57,** 289–316 (1997). [Discusses glycogenin.]

Smythe, C. and Cohen, P., The discovery of glycogenin and the priming mechanism for glycogen biosynthesis, *Eur. J. Biochem.* **200,** 625–631 (1991).

Sprang, S.R., Acharya, K.R., Goldsmith, E.J., Stuart, D.I., Varvill, K., Fletterick, R.J., Madsen, N.B., and Johnson, L.N., Structural changes in glycogen phosphorylase induced by phosphorylation, *Nature* **336,** 215–221 (1988).

Sprang, S.R., Withers, S.G., Goldsmith, E.J., Fletterick, R.J., and Madsen, N.B., Structural basis for the activation of glycogen phosphorylase b by adenosine monophosphate. *Science* **254,** 1367–1371 (1991).

CALMODULIN AND ITS CONTROL OF GLYCOGEN METABOLISM

Babu, Y.S., Sack, J.S., Greenough, T.J., Bugg, C.E., Means, A.R., and Cook, W.J., Three-dimensional structure of calmodulin, *Nature* **315,** 37–40 (1985).

Crivici, A. and Ikura, M., Molecular and structural basis of target recognition by calmodulin, *Annu. Rev. Biophys. Biomol. Struct.* **25,** 85–116 (1995).

Ikura, M., Clore, G.M., Gronenborn, A.M., Zhu, G., and Bax, A., Solution structure of a calmodulin-target peptide complex by multidimensional NMR, *Science* **256,** 632–638 (1992); *and* Meador, W.E., Means, A.R., and Quiocho, F.A., Target enzyme recognition by calmodulin: 2.4 Å structure of a calmodulin-peptide complex, *Science* **257,** 1251–1255 (1992).

James, P., Vorherr, T., and Carafoli, E., Calmodulin-binding domains: Just two faced or multifaceted? *Trends Biochem. Sci.* **20,** 38–42 (1995).

Nakayama, S. and Kretsinger, R.H., Evolution of the EF-hand family of proteins, *Annu. Rev. Biophys. Biomol. Struct.* **23,** 473–507 (1994).

PROTEIN KINASES AND PROTEIN PHOSPHATASES

Bollen, M., Keppens, S., and Stalmans, W., Specific features of glycogen metabolism in the liver, *Biochem. J.* **336,** 19–31 (1998).

Bossemeyer, D., Engh, R.A., Kinzel, V., Ponstingl, H., and Huber, R., Phosphotransferase and substrate binding mechanism of the cAMP-dependent protein kinase catalytic subunit from porcine heart as deduced from the 2.0 Å structure of the complex with Mn^{2+} adenyl imidodiphosphate and inhibitor peptide PKI (5-24), *EMBO J.* **12,** 849–859 (1993).

Egloff, M.P., Johnson, D.F., Moorhead, G., Cohen, P.T.W., Cohen, P., and Barford, D., Structural basis for the recognition of regulatory subunits by the catalytic subunit of protein phosphatase 1, *EMBO J.* **16,** 1876–1887 (1997).

Goldberg, J., Huang, H., Kwon, Y., Greengard, P., Nairn, A.C., and Kuriyan, J., Three-dimensional structure of the catalytic subunit of protein serine/threonine phosphatase-1, *Nature* **376,** 745–753 (1995).

Johnson, L.N., Lowe, E.D., Noble, M.E.M., and Owen, D.J., The structural basis for substrate recognition and control by protein kinases, *FEBS Lett.* **430,** 1–11 (1998).

Kobe, B. and Kemp, B.E., Active site-directed protein regulation, *Nature* **402,** 373–376 (1999). [Discusses intrasteric regulation.]

Lowe, E.D., Noble, M.E.M., Skamnaki, V.T., Oikonomakos, N.G., Owen, D.J., and Johnson, L.N., The crystal structure of a phosphorylase kinase peptide substrate complex: Kinase substrate recognition, *EMBO J.* **16,** 6646–6658 (1997).

Manning, G., Whyte, D.B., Martinez, R., Hunter, T., and Sundarsanum, S., The protein kinase complement of the human genome, *Science* **298,** 1912–1934 (2002).

Nordlie, R.C., Foster, J.D., and Lange, A.J., Regulation of glucose production by the liver, *Annu. Rev. Nutr.* **19,** 379–406 (1999).

Smith, C.M., Radzio-Andzelm, E., Akamine, M.P., Madhusudan, and Taylor, S.S., The catalytic subunit of cAMP-dependent protein kinase: Prototype for an extended network of communication, *Prog. Biophys. Mol. Biol.* **71,** 313–341 (1999).

Su, Y., Dostmann, W.R.G., Herberg, F.W., Durick, K., Xuong, N., Ten Eyck, L., Taylor, S.S., and Varughese, K.I., Regulatory subunit of protein kinase A: Structure of deletion mutant with cAMP binding domains, *Science* **269,** 807–813 (1995).

Taylor, S.S., Knighton, D.R., Zheng, J., Sowadski, J.M., Gibbs, C.S., and Zoller, M.J., A template for the protein kinase family, *Trends Biochem. Sci.* **18,** 84–89 (1993); *and* Taylor, S.S., Knighton, D.R., Zheng, J., Ten Eyck, L.F., and Sowadski, J.M., Structural framework for the protein kinase family, *Annu. Rev. Cell Biol.* **8,** 429–462 (1992).

Villafranca, J.E., Kissinger, C.R., and Parge, H.E., Protein serine/threonine phosphatases, *Curr. Opin. Biotech.* **7,** 397–402 (1996).

GLUCOKINASE AND PFK-2/FBPase-2

Cornish-Bowden, A. and Cárdenas, M.L., Hexokinase and "glucokinase" in liver metabolism, *Trends Biochem. Sci.* **16,** 281–282 (1991).

de la Iglesia, N., Mukhtar, M., Seoane, J., Guinovart, J.J., and Agius, L., The role of the regulatory protein of glucokinase in the glucose sensory mechanism of the hepatocyte, *J. Biol. Chem.* **275,** 10597–10603 (2000).

Iynedjian, P.B., Mammalian glucokinase and its gene, *Biochem. J.* **293,** 1–13, (1993). [Reviews the function and control of glucokinase.]

Okar, D.A., Manzano, À., Navarro-Sabatè, A., Riera, L., Bartrons, R., and Lange, A.J., PFK-2/FBPase-2: maker and breaker of the essential biofactor fructose-2,6-bisphosphate, *Trends Biochem. Sci.* **26,** 30–35 (2001).

Pilkis, S.J., 6-Phosphofructo-2-kinase/fructose-2,6-bisphosphatase: a metabolic signaling enzyme, *Annu. Rev. Biochem.* **64,** 799–835 (1995).

Rousseau, G.G. and Hue, L., Mammalian 6-phosphofructo-2-kinase/fructose-2,6-bisphosphatase: A bifunctional enzyme that controls glycolysis, *Prog. Nucleic Acid Res. Mol. Biol.* **45,** 99–127 (1993).

Van Schaftingen, E., Vandercammen, A., Detheux, M., and Davies, D.R., The regulatory protein of liver glucokinase, *Adv. Enzyme Reg.* **32,** 133–148 (1992).

Villar-Palasi, C. and Guinovart, J. J., The role of glucose 6-phosphate in the control of glycogen synthase, *FASEB J.* **11,** 544–558 (1997).

Yuan, M.H., Mizuguchi, H., Lee, Y.-H., Cook, P.F., Uyeda, K., and Hasemann, C.A., Crystal structure of the H256A mutant of rat testis fructose-6-phosphate,2-kinase/fructose-2,6-bisphosphatase, *J. Biol. Chem.* **274,** 2176–2184 (1999); *and* Haseman, C.A., Istvan, E.S., Uyeda, K., and Deisenhofer, J., The crystal structure of the bifunctional enzyme 6-phosphofructo-2-kinase/fructose-2,6-bisphosphatase reveals distinct homologies, *Structure* **4,** 1017–1029 (1996).

GLYCOGEN STORAGE DISEASES

Bartram, C., Edwards, R.H.T., and Beynon, R.J., McArdle's disease-muscle glycogen phosphorylase deficiency, *Biochim. Biophys. Acta* **1272,** 1–13 (1995). [A review.]

Chen, Y.-T., Glycogen storage diseases, *in* Scriver, C.R., Beaudet, A.L. Sly, W.S., and Valle, D. (Eds.), *The Metabolic & Molecular Bases of Inherited Disease* (8th ed.), pp. 1521–1552, McGraw-Hill, New York (2001). [Begins with a review of glycogen metabolism.]

Radda, G.K., Control of bioenergetics: From cells to man by phosphorus nuclear-magnetic-resonance spectroscopy, *Biochem. Soc. Trans.* **14,** 517–525 (1986). [Discusses the noninvasive diagnosis of McArdle's disease by ^{31}P NMR.]

Zingone, A., Hiraiwa, H., Pan, C.-J., Lin, B., Chen, H., Ward, J.M., and Chou, J.Y., Correction of glycogen storage disease type 1a in a mouse model by gene therapy, *J. Biol. Chem.* **275,** 828–832 (2000).

PROBLEMS

1. A glycogen molecule consisting of 100,000 glucose residues is branched, on average, every 10 residues with one branch per tier. (a) How many reducing ends does it have? (b) How many tiers of branches does it have, on average?

2. A mature glycogen particle typically has 12 tiers of branches with 2 branches per tier and 13 residues per branch. How many glucose residues are in such a particle?

3. The complete metabolic oxidation of glucose to CO_2 and O_2 yields 38 ATPs (Chapter 22). What is the fractional energetic cost of storing glucose as glycogen and later metabolizing the glycogen rather than directly metabolizing the glucose? (Recall that glycogen's branched structure results in its degradation to 92% G1P and 8% glucose.)

4. What are the effects of the following on the rates of glycogen synthesis and glycogen degradation: (a) increasing the Ca^{2+} concentration, (b) increasing the ATP concentration, (c) inhibiting adenylate cyclase, (d) increasing the epinephrine concentration, and (e) increasing the AMP concentration?

5. Show that hexokinase activity but not glucokinase activity is insensitive to blood [glucose] over the physiological range. Calculate the ratio of glucokinase to hexokinase activities when [glucose] is 2 mM (hypoglycemic), 5 mM (normal), and 25 mM (diabetic). Assume that $K_M = 0.1$ mM for hexokinase and that both enzymes have the same V_{max}.

6. Compare the properties of a bicyclic cascade with those of a monocyclic cascade.

7. The V_{max} of muscle glycogen phosphorylase is much larger than that of liver. Discuss the functional significance of this phenomenon.

8. How does epinephine act on muscles to prepare them for "fight or flight"?

*9. A complication of glycogen metabolism that we have not discussed is that many protein kinases, including phosphorylase kinase, are autophosphorylating; that is, they can specifically phosphorylate and thereby activate themselves. Discuss how this phenomenon affects glycogen metabolism, taking into consideration the possibilities that phosphorylase kinase autophosphorylation may be an intramolecular or an intermolecular process.

10. Explain the symptoms of von Gierke's disease.

11. A sample of glycogen from a patient with liver disease is incubated with P_i, normal glycogen phosphorylase, and normal debranching enzyme. The ratio of glucose-1-phosphate to glucose formed in this reaction mixture is 100. What is the patient's most likely enzymatic deficiency? What is the probable structure of the patient's glycogen?

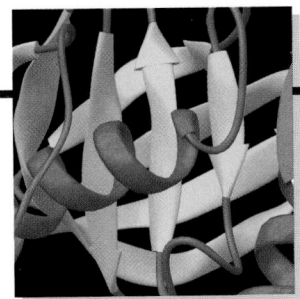

Chapter 19

Signal Transduction

Living things coordinate their activities at every level of their organization through complex chemical signaling systems. Intercellular signals occur through the mediation of chemical messengers known as **hormones** and, in higher animals, via neuronally transmitted electrochemical impulses. Intracellular communications are maintained by the synthesis or alteration of a great variety of different substances that are often integral components of the processes they control. For example, metabolic pathways, as we have seen, are regulated by the feedback control of allosteric enzymes by metabolites in those pathways or by the covalent modification of these enzymes. In this chapter we consider chemical signaling and how these signals are mediated. We begin by discussing the functions of the major human hormone systems. We then discuss the three major pathways whereby intercellular signals are transduced (converted) to intracellular signals, namely, those that utilize (1) heterotrimeric G proteins, (2) receptor tyrosine kinases, and (3) phosphoinositide cascades. Neurotransmission is discussed in Section 20-5.

1 ■ HORMONES

Hormones are classified according to the distance over which they act (Fig. 19-1):

1. Endocrine hormones act on cells distant from the site of their release. Endocrine hormones, for example, insulin and epinephrine, are synthesized and released in the bloodstream by specialized ductless **endocrine glands.**

2. Paracrine hormones (alternatively, **local mediators**) act only on cells close to the cell that released them. For example, an essential element of the immune response is initiated when a white blood cell known as a **macrophage** that has encountered a specific antigen binds a so-called **T cell** specific for that antigen. The macrophage thereupon releases a **protein growth factor** named **interleukin-1** that stimulates the bound T cell to proliferate and differentiate (Section 35-2A).

3. Autocrine hormones act on the same cell that released them. For example, the response of a T cell to interleukin-1 is enhanced by the T cell's autostimulatory release of the protein growth factor **interleukin-2.**

We are already familiar with certain aspects of hormonal control. For instance, we have considered how epinephrine, insulin, and glucagon regulate energy metabolism through the intermediacy of cAMP (Sections 18-3E and 18-3G). In this section we extend and systematize this information. Before we do so, you should note that biochemical communications are not limited to intracellular and intercellular signals. Many organisms release substances called **pheromones** that alter the behavior of other

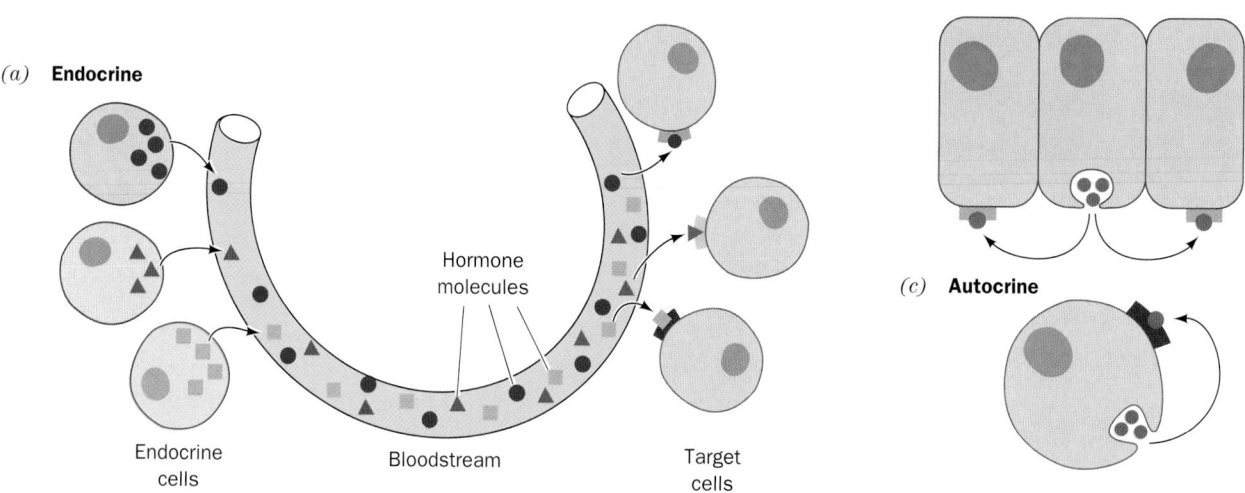

(a) **Endocrine**

Hormone molecules

Endocrine cells

Bloodstream

Target cells

(b) **Paracrine**

(c) **Autocrine**

FIGURE 19-1 Classification of hormones. Hormonal communications are classified according to the distance over which the signal acts: *(a)* endocrine signals are directed at distant cells through the intermediacy of the bloodstream, *(b)* paracrine signals are directed at nearby cells, and *(c)* autocrine signals are directed at the cell that produced them.

organisms of the same species in much the same way as hormones. Pheromones are commonly sexual attractants but some have other functions in species, such as ants, that have complex social interactions.

The human endocrine system (Fig. 19-2) secretes a wide variety of hormones (Table 19-1) that enable the body to:

1. Maintain homeostasis (e.g., insulin and glucagon maintain the blood glucose level within rigid limits during feast or famine).

2. Respond to a wide variety of external stimuli (such as the preparation for "fight or flight" engendered by epinephrine and norepinephrine).

3. Follow various cyclic and developmental programs (for instance, sex hormones regulate sexual differentiation, maturation, the menstrual cycle, and pregnancy; Sections 19-1G and 19-1I).

Most hormones are either polypeptides, amino acid derivatives, or steroids, although there are important exceptions to this generalization. In any case, *only those cells with a specific receptor for a given hormone will respond to its presence even though nearly all cells in the body may be exposed to the hormone.* Hormonal messages are therefore quite specifically addressed.

In this section, we outline the hormonal functions of the various endocrine glands. Throughout this discussion keep in mind that these glands are not just a collection of independent secretory organs but form a complex and highly interdependent control system. Indeed, as we shall see, the secretion of many hormones is under feedback control through the secretion of other hormones to which the orig-

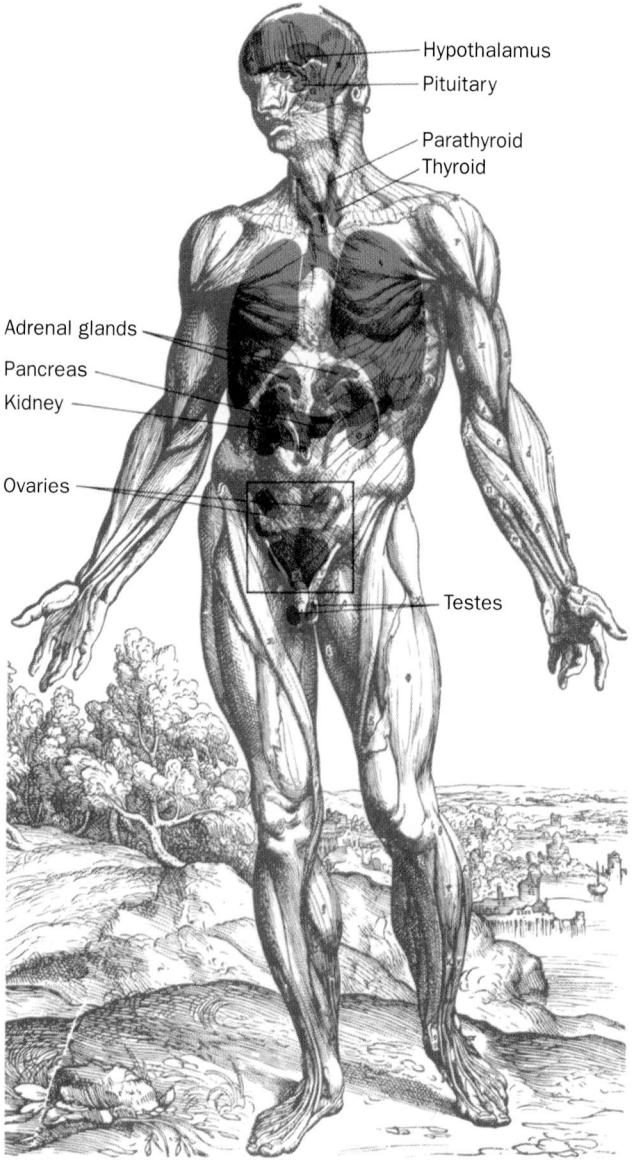

Hypothalamus
Pituitary
Parathyroid
Thyroid
Adrenal glands
Pancreas
Kidney
Ovaries
Testes

FIGURE 19-2 Major glands of the human endocrine system. Other tissues, the intestines, for example, also secrete endocrine hormones.

TABLE 19-1 Some Human Hormones

Hormone	Origin	Major Effects
Polypeptides		
Corticotropin-releasing factor (CRF)	Hypothalamus	Stimulates ACTH release
Gonadotropin-releasing factor (GnRF)	Hypothalamus	Stimulates FSH and LH release
Thyrotropin-releasing factor (TRF)	Hypothalamus	Stimulates TSH release
Growth hormone-releasing factor (GRF)	Hypothalamus	Stimulates growth hormone release
Somatostatin	Hypothalamus	Inhibits growth hormone release
Adrenocorticotropic hormone (ACTH)	Adenohypophysis	Stimulates the release of adrenocorticosteroids
Follicle-stimulating hormone (FSH)	Adenohypophysis	In ovaries, stimulates follicular development, ovulation, and estrogen synthesis; in testes, stimulates spermatogenesis
Luteinizing hormone (LH)	Adenohypophysis	In ovaries, stimulates oocyte maturation and follicular synthesis of estrogens and progesterone; in testes, stimulates androgen synthesis
Chorionic gonadotropin (CG)	Placenta	Stimulates progesterone release from the corpus luteum
Thyrotropin (TSH)	Adenohypophysis	Stimulates T_3 and T_4 release
Somatotropin (growth hormone)	Adenohypophysis	Stimulates growth and synthesis of somatomedins
Met-enkephalin	Adenohypophysis	Opioid effects on central nervous system
Leu-enkephalin	Adenohypophysis	Opioid effects on central nervous system
β-Endorphin	Adenohypophysis	Opioid effects on central nervous system
Vasopressin	Neurohypophysis	Stimulates water resorption by kidney and increases blood pressure
Oxytocin	Neurohypophysis	Stimulates uterine contractions
Glucagon	Pancreas	Stimulates glucose release through glycogenolysis and stimulates lipolysis
Insulin	Pancreas	Stimulates glucose uptake through gluconeogenesis, protein synthesis, and lipogenesis
Gastrin	Stomach	Stimulates gastric acid and pepsinogen secretion
Secretin	Intestine	Stimulates pancreatic secretion of HCO_3^-
Cholecystokinin (CCK)	Intestine	Stimulates gallbladder emptying and pancreatic secretion of digestive enzymes and HCO_3^-
Gastric inhibitory peptide (GIP)	Intestine	Inhibits gastric acid secretion and gastric emptying; stimulates pancreatic insulin release
Parathyroid hormone	Parathyroid	Stimulates Ca^{2+} uptake from bone, kidney, and intestine
Calcitonin	Thyroid	Inhibits Ca^{2+} uptake from bone and kidney
Somatomedins	Liver	Stimulate cartilage growth; have insulin-like activity
Steroids		
Glucocorticoids	Adrenal cortex	Affect metabolism in diverse ways, decrease inflammation, increase resistance to stress
Mineralocorticoids	Adrenal cortex	Maintain salt and water balance
Estrogens	Gonads	Maturation and function of secondary sex organs, particularly in females
Androgens	Gonads	Maturation and function of secondary sex organs, particularly in males; male sexual differentiation
Progestins	Ovaries and placenta	Mediate menstrual cycle and maintain pregnancy
Vitamin D	Diet and sun	Stimulates Ca^{2+} absorption from intestine, kidney, and bone
Amino Acid Derivatives		
Epinephrine	Adrenal medulla	Stimulates contraction of some smooth muscles and relaxes others, increases heart rate and blood pressure, stimulates glycogenolysis in liver and muscle, stimulates lipolysis in adipose tissue
Norepinephrine	Adrenal medulla	Stimulates arteriole contraction, decreases peripheral circulation, stimulates lipolysis in adipose tissue
Triiodothyronine (T_3)	Thyroid	General metabolic stimulation
Thyroxine (T_4)	Thyroid	General metabolic stimulation

inal hormone-secreting gland responds. Much of our understanding of hormonal function has come from careful measurements of hormone concentrations, the effects of changes of these concentrations on physiological functions, and measurements of the affinities with which hormones bind to their receptors. We begin, therefore, with a consideration of how physiological hormone concentrations are measured and how receptor–ligand interactions are quantified.

A. *Quantitative Measurements*

a. Radioimmunoassays

The serum concentrations of hormones are extremely small, generally between 10^{-12} and $10^{-7}M$, so they usually must be measured by indirect means. Biological assays were originally employed for this purpose but they are generally slow, cumbersome, and imprecise. Such assays have therefore been largely supplanted by **radioimmunoassays.** In this technique, which was developed by Rosalyn Yalow, the unknown concentration of a hormone, H, is determined by measuring how much of a known amount of the radioactively labeled hormone, H*, binds to a fixed quantity of anti-H antibody in the presence of H. This competition reaction is easily calibrated by constructing a standard curve indicating how much H* binds to the antibody as a function of [H]. The high ligand affinity and specificity that antibodies possess gives radioimmunoassays the advantages of great sensitivity and specificity.

b. Receptor Binding

Receptors, as do other proteins, bind their corresponding ligands according to the laws of mass action:

$$R + L \rightleftharpoons R \cdot L$$

Here R and L represent receptor and ligand, and the reaction's dissociation constant is expressed:

$$K_L = \frac{[R][L]}{[R \cdot L]} = \frac{([R]_T - [R \cdot L])[L]}{[R \cdot L]} \qquad [19.1]$$

where the total receptor concentration, $[R]_T = [R] + [R \cdot L]$. Equation [19.1] may be rearranged to a form analogous to the Michaelis–Menten equation of enzyme kinetics (Section 14-2A):

$$Y = \frac{[R \cdot L]}{[R]_T} = \frac{[L]}{K_L + [L]} \qquad [19.2]$$

where Y is the fractional occupation of the ligand-binding sites. Equation [19.2] represents a hyperbolic curve (Fig. 19-3a) in which K_L may be operationally defined as the ligand concentration at which the receptor is half-maximally occupied by ligand.

Although K_L and $[R]_T$ may, in principle, be determined from an analysis of a hyperbolic plot such as Fig. 19-3a, the analysis of a linear form of the equation is a simpler procedure. Equation [19.1] may be rearranged to:

$$\frac{[R \cdot L]}{[L]} = \frac{([R]_T - [R \cdot L])}{K_L} \qquad [19.3]$$

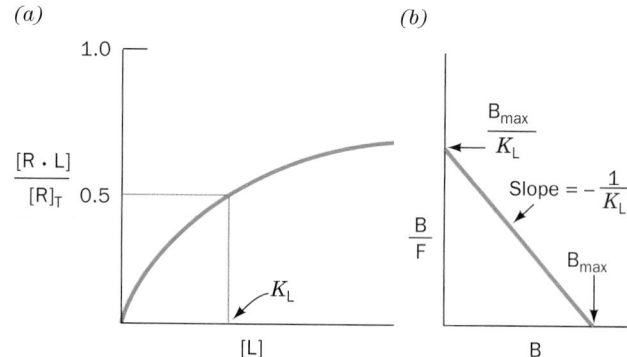

FIGURE 19-3   Binding of ligand to receptor. (*a*) A hyperbolic plot. (*b*) A Scatchard plot. Here $B \equiv [R \cdot L]$, $F \equiv [L]$, and $B_{max} \equiv [R]_T$.

Now, in keeping with customary receptor-binding nomenclature, let us redefine $[R \cdot L]$ as B (for bound ligand), [L] as F (for free ligand), and $[R]_T$ as B_{max}. Then Eq. [19.3] becomes:

$$\frac{B}{F} = \frac{(B_{max} - B)}{K_L} = -\frac{1}{K_L} \cdot B + \frac{B_{max}}{K_L} \qquad [19.4]$$

A plot of B/F versus B, which is known as a **Scatchard plot** (after George Scatchard, its originator), therefore yields a straight line of slope $-1/K_L$ whose intercept on the B axis is B_{max} (Fig. 19-3b). Here, both B and F may be determined by filter-binding assays as follows. Most receptors are insoluble membrane-bound proteins and may therefore be separated from soluble free ligand by filtration (receptors that have been solubilized may be separated from free ligand by filtration, for example, through nitrocellulose since proteins nonspecifically bind to nitrocellulose). Hence, through the use of radioactively labeled ligand, the values of B and F ($[R \cdot L]$ and [L]) may be determined, respectively, from the radioactivity on the filter and that remaining in solution. The rate of $R \cdot L$ dissociation is generally so slow (half-times of minutes to hours) as to cause insignificant errors when the filter is washed to remove residual free ligand.

c. Competitive-Binding Studies

Once the receptor-binding parameters for one ligand have been determined, the dissociation constant of other ligands for the same ligand-binding site may be determined through competitive-binding studies. The model describing this competitive binding is analogous to the competitive inhibition of a Michaelis–Menten enzyme (Section 14-3A):

$$\begin{array}{c} R + L \overset{K_L}{\rightleftharpoons} R \cdot L \\ + \\ I \\ K_I \Big\updownarrow \\ R \cdot I + L \longrightarrow \text{No binding} \end{array}$$

where I is the competing ligand whose dissociation constant with the receptor is expressed:

$$K_I = \frac{[R][I]}{[R \cdot I]} \quad [19.5]$$

Thus, in direct analogy with the derivation of the equation describing competitive inhibition:

$$[R \cdot L] = \frac{[R]_T[L]}{K_L\left(1 + \frac{[I]}{K_I}\right) + [L]} \quad [19.6]$$

The relative affinities of a ligand and an inhibitor may therefore be determined by dividing Eq. [19.6] in the presence of inhibitor with that in the absence of inhibitor:

$$\frac{[R \cdot L]_I}{[R \cdot L]_0} = \frac{K_L + [L]}{K_L\left(1 + \frac{[I]}{K_I}\right) + [L]} \quad [19.7]$$

When this ratio is 0.5 (50% inhibition), the competitor concentration is referred to as $[I_{50}]$ in analogy with the $[IC_{50}]$ of drugs that inhibit enzymes (Section 15-4A). Thus, solving Eq. [19.7] for K_I at 50% inhibition:

$$K_I = \frac{[I_{50}]}{1 + \frac{[L]}{K_L}} \quad [19.8]$$

B. *Pancreatic Islet Hormones*

The pancreas is a large glandular organ, the bulk of which is an **exocrine gland** dedicated to producing digestive enzymes such as trypsin, RNase A, α-amylase, and phospholipase A_2 that it secretes via the pancreatic duct into the small intestine. However, ~1 to 2% of pancreatic tissue consists of scattered clumps of cells known as **islets of Langerhans,** which comprise an endocrine gland that functions to maintain energy metabolite homeostasis. Pancreatic islets contain three types of cells, each of which secretes a characteristic polypeptide hormone:

1. The α cells secrete glucagon (29 residues; Section 18-3E).

2. The β cells secrete insulin (51 residues; Fig. 9-4).

3. The δ cells secrete **somatostatin** (14 residues).

Insulin, which is secreted in response to high blood glucose levels, primarily functions to stimulate muscle, liver, and adipose cells to store glucose for later use by synthesizing glycogen, protein, and fat (Section 27-2). Glucagon, which is secreted in response to low blood glucose, has essentially the opposite effects: It stimulates liver to release glucose through glycogenolysis (Section 18-3E) and gluconeogenesis (Section 23-1) and it stimulates adipose tissue to release fatty acids through lipolysis. Somatostatin, which is also secreted by the hypothalamus (Section 19-1H), inhibits the release of insulin and glucagon from their islet cells and is therefore thought to have a paracrine function in the pancreas.

Polypeptide hormones, as are other proteins destined for secretion, are ribosomally synthesized as preprohormones, processed in the rough endoplasmic reticulum and Golgi apparatus to form the mature hormone, and then packaged in secretory granules to await the signal for their release by exocytosis (Sections 12-4B, 12-4C, and 12-4D). The most potent physiological stimuli for the release of insulin and glucagon are, respectively, high and low blood glucose concentrations, so that islet cells act as the body's primary glucose sensors. However, the release of these hormones is also influenced by the autonomic (involuntary) nervous system and by hormones secreted by the gastrointestinal tract (Section 19-1C).

C. *Gastrointestinal Hormones*

The digestion and absorption of nutrients are complicated processes that are regulated by the autonomic nervous system in concert with a complex system of polypeptide hormones. Indeed, gastrointestinal hormones are secreted into the bloodstream by a system of specialized cells lining the gastrointestinal tract whose aggregate mass is greater than that of the rest of the endocrine system. The four best characterized gastrointestinal hormones are:

1. Gastrin (17 residues), which is produced by the gastric mucosa, stimulates the gastric secretion of HCl and **pepsinogen** (the zymogen of the digestive protease pepsin). Gastrin release is stimulated by amino acids and partially digested protein as well as by the vagus nerve (which innervates the stomach) in response to stomach distension. Gastrin release is inhibited by HCl and by other gastrointestinal hormones.

2. Secretin (27 residues), which is produced by the mucosa of the duodenum (upper small intestine) in response to acidification by gastric HCl, stimulates the pancreatic secretion of HCO_3^- so as to neutralize this acid.

3. Cholecystokinin (**CCK;** 8 residues), which is produced by the duodenum, stimulates gallbladder emptying, the pancreatic secretion of digestive enzymes and HCO_3^- (and thus enhances the effect of secretin), and inhibits gastric emptying. CCK is released in response to the products of lipid and protein digestion, that is, fatty acids, monoacylglycerols, amino acids, and peptides.

4. Gastric inhibitory peptide (**GIP,** also known as **glucose-dependent insulinotropic polypeptide;** 42 residues), which is produced by specialized cells lining the small intestine, is a potent inhibitor of gastric acid secretion, gastric mobility, and gastric emptying. However, GIP's major physiological function is to stimulate pancreatic insulin release. Indeed, the release of GIP is stimulated by the presence of glucose in the gut, which accounts for the observation that, after a meal, the blood insulin level increases before the blood glucose level does.

These gastrointestinal hormones form families of related polypeptides: The C-terminal pentapeptides of gastrin and CCK are identical; secretin, GIP, and glucagon are closely similar.

Several other polypeptides that affect gastrointestinal function have been isolated from the gut. However, the physiological roles of many of these substances are unclear (but see Section 27-3C). Much of this difficulty stems from the diffuse distribution of gastrointestinal hormone-secreting cells that precludes their excision, a procedure that is commonly used in controlled studies of the effects of other endocrine hormones.

D. *Thyroid Hormones*

*The thyroid gland produces two related hormones, **triiodo-thyronine (T₃)** and **thyroxine (T₄)**,*

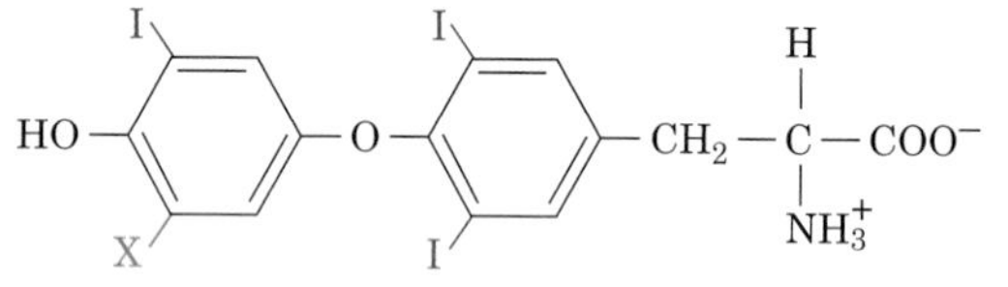

X = H   **Triiodothyronine (T₃)**
X = I   **Thyroxine (T₄)**

that stimulate metabolism in most tissues (adult brain is a conspicuous exception). The production of these unusual iodinated amino acids begins with the synthesis of **thyroglobulin,** a 2748-residue protein. Thyroglobulin is post-translationally modified in a series of biochemically unique reactions (Fig. 19-4):

1. Around 20% of thyroglobulin's 140 Tyr residues are iodinated in a **thyroperoxidase (TPO)**-catalyzed reaction forming **2,5-diiodotyrosyl** residues.

2. Two such residues are oxidatively coupled to yield T₃ and T₄ residues.

3. Mature thyroglobulin itself is hormonally inactive. However, some five or six molecules of the active hormones, T₃ and T₄, are produced by the proteolysis in the lysosome of thyroglobulin on hormonal stimulation of the thyroid (Section 19-1H).

How do thyroid hormones work? T₃ and T₄, being non-polar substances, are transported by the blood in complex

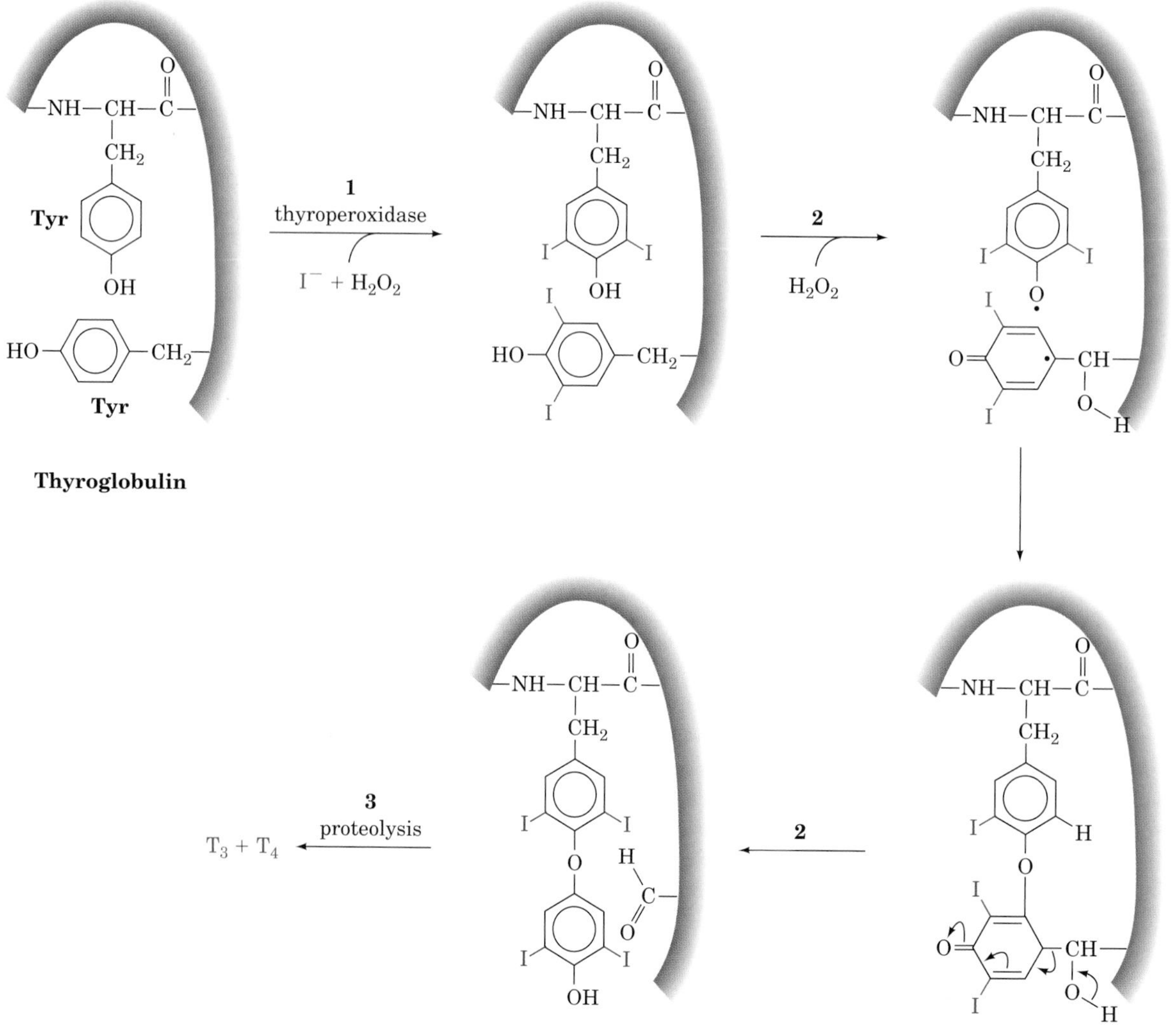

FIGURE 19-4   Biosynthesis of T₃ and T₄ in the thyroid gland. The pathway involves the iodination, coupling, and hydrolysis (proteolysis) of thyroglobulin Tyr residues. The relatively scarce I⁻ is actively sequestered by the thyroid gland.

with plasma carrier proteins, primarily **thyroxine-binding globin,** but also **prealbumin** and **albumin.** The hormones then pass through the cell membranes of their target cells into the cytosol, where they bind to a specific protein. Since the resulting hormone–protein complex does not enter the nucleus, it is thought that this complex acts to maintain an intracellular reservoir of thyroid hormones. The true **thyroid hormone receptor** is a chromosomally associated protein and therefore does not leave the nucleus. *The binding of T_3, and to a lesser extent T_4, activates this receptor as a transcription factor (Section 5-4A), resulting in increased rates of expression of numerous metabolic enzymes.* High affinity thyroid hormone–binding sites also occur on the inner mitochondrial membrane (the site of electron transport and oxidative phosphorylation; Section 22-1), suggesting that these receptors may directly regulate O_2 consumption and ATP production.

Abnormal levels of thyroid hormones are common human afflictions. **Hypothyroidism** is characterized by lethargy, obesity, and cold dry skin, whereas **hyperthyroidism** has the opposite effects. The inhabitants of areas in which the soil has a low iodine content often develop hypothyroidism accompanied by an enlarged thyroid gland, a condition known as **goiter.** The small amount of NaI often added to commercially available table salt ("iodized" salt) easily prevents this iodine deficiency disease. Young mammals require thyroid hormone for normal growth and development: Hypothyroidism during the fetal and immediate postnatal periods results in irreversible physical and mental retardation, a syndrome named **cretinism.**

E. *Control of Calcium Metabolism*

Ca^{2+} forms **hydroxyapatite,** $Ca_5(PO_4)_3OH$, the major mineral constituent of bone, and is an essential element in many biological processes including the mediation of hormonal signals as a second messenger, the triggering of muscle contraction, the transmission of nerve impulses, and blood clotting. The extracellular $[Ca^{2+}]$ must therefore be closely regulated to keep it at its normal level of ~1.2 mM. Three hormones have been implicated in maintaining Ca^{2+} homeostasis (Fig. 19-5):

1. Parathyroid hormone (PTH), an 84-residue polypeptide secreted by the parathyroid gland, which increases serum $[Ca^{2+}]$ by stimulating its resorption from bone and kidney and by increasing the dietary absorption of Ca^{2+} from the intestine.

2. Vitamin D, a group of steroidlike substances that act in a synergistic manner with PTH to increase serum $[Ca^{2+}]$.

3. Calcitonin, a 33-residue polypeptide synthesized by specialized thyroid gland cells, which decreases serum $[Ca^{2+}]$ by inhibiting the resorption of Ca^{2+} from bone and kidney.

We shall briefly discuss the functions of these hormones.

a. Parathyroid Hormone

The bones, the body's main Ca^{2+} reservoir, are by no means metabolically inert. They are continually "remodeled" through the action of two types of bone cells: **osteoblasts,** which synthesize the collagen fibrils that form the bulk of bone's organic matrix, the scaffolding on which its $Ca_5(PO_4)_3OH$ mineral phase is laid down; and **osteoclasts,** which participate in bone resorption (Section 15-4A). *PTH inhibits collagen synthesis by osteoblasts and stimulates bone resorption by osteoclasts. The main effect of PTH, however, is to increase the rate at which the kidneys excrete phosphate, the counterion of Ca^{2+} in bone.* The consequent decreased serum $[P_i]$ causes $Ca_5(PO_4)_3OH$ to leach out of bone through mass action and thus increase serum $[Ca^{2+}]$. In addition, PTH stimulates the production of the active form of vitamin D by the kidney, which, in turn, enhances the transfer of intestinal Ca^{2+} to the blood (see below).

b. Vitamin D

Vitamin D is a group of dietary substances that prevent **rickets,** a disease of children characterized by stunted growth and deformed bones stemming from insufficient bone mineralization (vitamin D deficiency in adults is known as **osteomalacia,** a condition characterized by weakened, demineralized bones). Although rickets was first described in 1645, it was not until the early twentieth century that it was discovered that animal fats, particularly fish liver oils, are effective in preventing this deficiency disease. Moreover, rickets can also be prevented by exposing children to sunlight or just UV light in the wavelength range 230 to 313 nm, regardless of their diets.

The D vitamins, which we shall see are really hormones, are sterol derivatives in which the steroid B ring (Fig. 12-9) is disrupted at its 9,10 position. The natural form of the vitamin, **vitamin D_3 (cholecalciferol),** is nonenzymatically

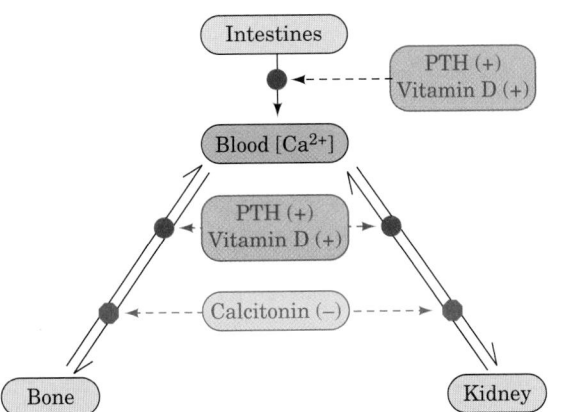

FIGURE 19-5 The roles of PTH, vitamin D, and calcitonin in controlling Ca^{2+} metabolism.

formed in the skin of animals through the photolytic action of UV light on **7-dehydrocholesterol:**

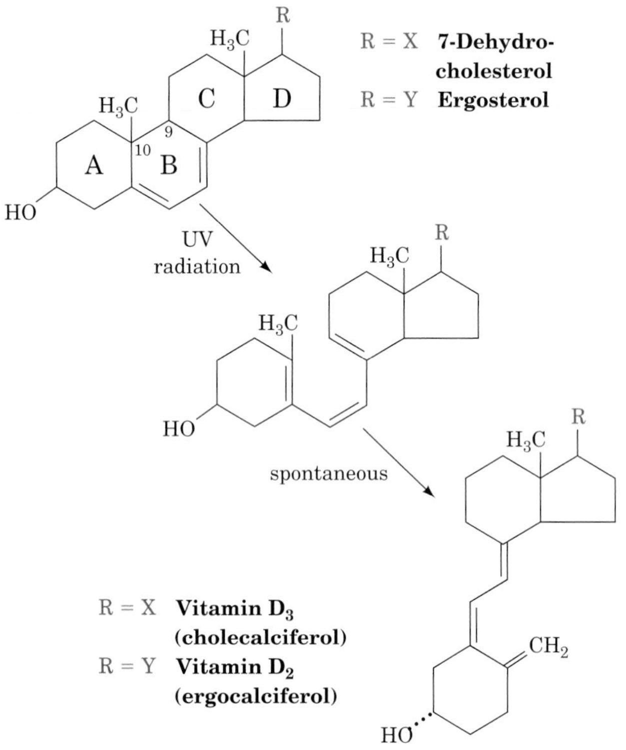

R = X   **7-Dehydro-cholesterol**
R = Y   **Ergosterol**

UV radiation

spontaneous

R = X   **Vitamin D₃ (cholecalciferol)**
R = Y   **Vitamin D₂ (ergocalciferol)**

X =

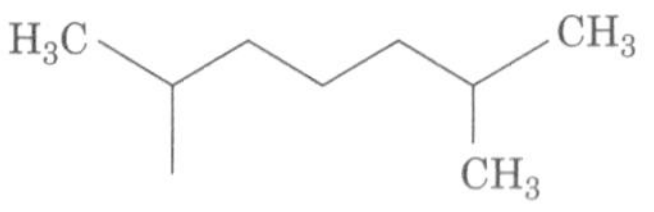

Y =

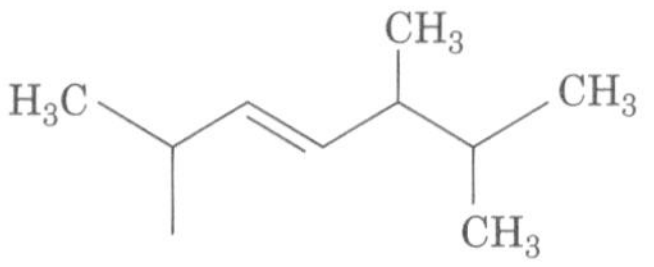

Vitamin D₂ (ergocalciferol), which differs from vitamin D_3 only by a side chain double bond and methyl group, is formed by the UV irradiation of the plant sterol **ergosterol.** Since vitamins D_2 and D_3 have essentially identical biological activities, vitamin D_2 is commonly used as a vitamin supplement, particularly in milk.

Vitamins D_2 and D_3 are hormonally inactive as such; they gain biological activity through further metabolic processing, first in the liver and then in the kidney (Fig. 19-6):

1. In the liver, vitamin D_3 is hydroxylated to form **25-hydroxycholecalciferol** in an O_2-requiring reaction catalyzed by **cholecalciferol-25-hydroxylase.**

2. The 25-hydroxycholecalciferol is transported to the kidney, where it is further hydroxylated by a mitochondrial oxygenase, **25-hydroxycholecalciferol-1α-hydroxylase,** to yield the active hormone **1α,25-dihydroxycholecalciferol [1,25(OH)₂D].** *The activity of 25-hydroxycholecalci-*

ferol-1-α-hydroxylase is regulated by PTH, so this reaction is an important control point in Ca^{2+} homeostasis.

1,25(OH)₂D acts to increase serum [Ca²⁺] by promoting the intestinal absorption of dietary Ca^{2+} and by stimulating Ca^{2+} release from bone. Intestinal Ca^{2+} absorption is stimulated through increased synthesis of a **Ca²⁺-binding protein,** which functions to transport Ca^{2+} across the intestinal mucosa. $1,25(OH)_2D$ binds to cytoplasmic receptors in intestinal epithelial cells that, on transport to the nucleus, function as transcription factors for the Ca^{2+}-binding protein. The maintenance of electroneutrality requires that Ca^{2+} transport be accompanied by that of counterions, mostly P_i, so that $1,25(OH)_2D$ also stimulates the intestinal absorption of P_i. The observation that $1,25(OH)_2D$, like PTH, stimulates the release of Ca^{2+} and P_i from bone seems paradoxical in view of the fact that low levels of $1,25(OH)_2D$ result in subnormal bone mineralization. Presumably the increased serum $[Ca^{2+}]$ resulting from $1,25(OH)_2D$-stimulated intestinal uptake of Ca^{2+} causes bone to take up more Ca^{2+} than it loses through direct hormonal stimulation.

Vitamin D, unlike the water-soluble vitamins, is retained by the body, so that excessive intake of vitamin D over long periods causes **vitamin D intoxication.** The consequent high serum $[Ca^{2+}]$ results in aberrant calcification of a wide variety of soft tissues. The kidneys are particularly prone to calcification, a process that can lead to the formation of kidney stones and ultimately kidney failure. In addition, vitamin D intoxication promotes bone demineralization to the extent that bones are easily fractured. The observation that the level of skin pigmentation in indigenous human populations tends to increase with their proximity to the equator is explained by the hypothesis that skin pigmentation functions to prevent vitamin D intoxication by filtering out excessive solar radiation.

c. Calcitonin

Calcitonin has essentially the opposite effect of PTH; it lowers serum [Ca²⁺]. It does so primarily by inhibiting osteoclastic resorption of bone. Since PTH and calcitonin both stimulate the synthesis of cAMP in their target cells (Section 19-2A), it is unclear how these hormones can oppositely affect osteoclasts. Calcitonin also inhibits kidney from resorbing Ca^{2+}, but in this case the kidney cells that calcitonin influences differ from those that PTH stimulates to resorb Ca^{2+}.

F. *Epinephrine and Norepinephrine*

The adrenal glands consist of two distinct types of tissue: the **medulla** (core), which is really an extension of the sympathetic nervous system (a part of the autonomic nervous system), and the more typically glandular **cortex** (outer layer). Here we consider the hormones of the adrenal medulla; those of the cortex are discussed in the following subsection.

The adrenal medulla synthesizes two hormonally active **catecholamines** *(amine-containing derivatives of* **catechol,**

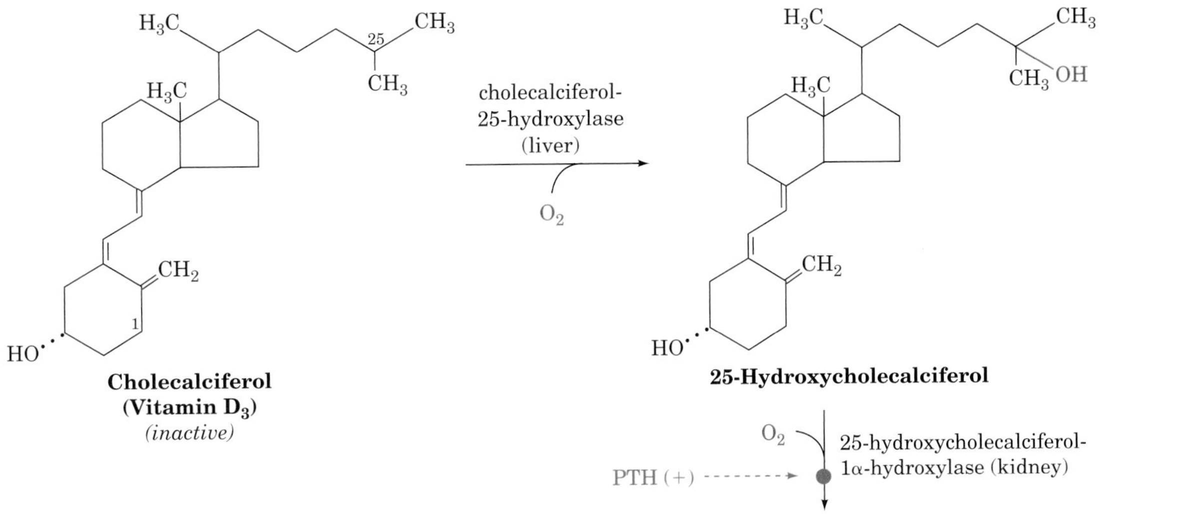

cholecalciferol-
25-hydroxylase
(liver)

O_2

**Cholecalciferol
(Vitamin D$_3$)**
(inactive)

25-Hydroxycholecalciferol

O_2 25-hydroxycholecalciferol-
1α-hydroxylase (kidney)

PTH (+)

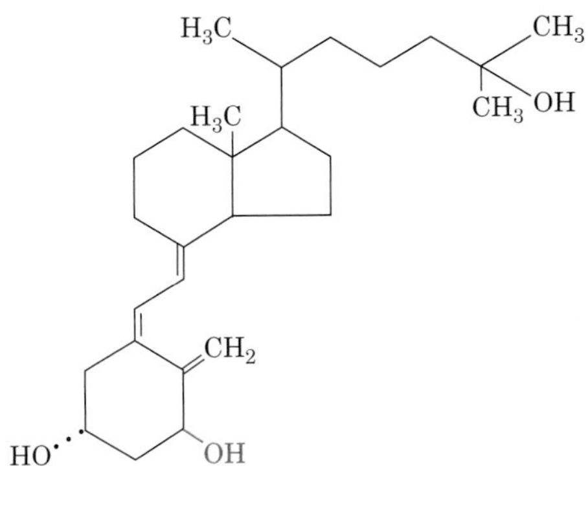

**1α,25-Dihydroxycholecalciferol
[1,25(OH)$_2$D]**
(active)

**FIGURE 19-6   Activation of vitamin D$_3$ as a hormone in liver
and kidney.** Vitamin D$_2$ (ergocalciferol) is similarly activated.

1,2-dihydroxybenzene), **norepinephrine (noradrenalin)**
and its methyl derivative **epinephrine (adrenalin):**

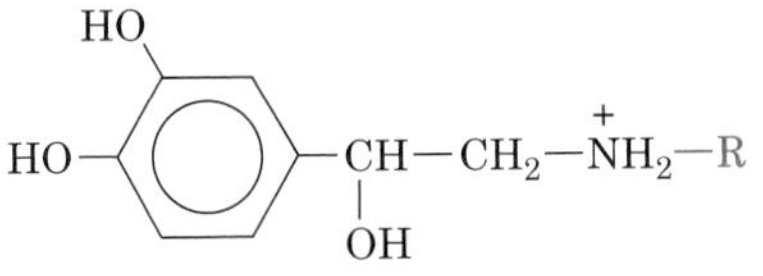

R = H   **Norepinephrine (noradrenalin)**
R = CH$_3$ **Epinephrine (adrenalin)**

These hormones are synthesized from tyrosine, as is described in Section 26-4B, and stored in granules to await
their exocytotic release under the control of the sympathetic nervous system.

   The biological effects of catecholamines are mediated
by two classes of plasma transmembrane receptors, the
α- and the β-adrenoreceptors (also known as **adrenergic
receptors**). These glycoproteins were originally identified
on the basis of their varying responses to certain **agonists**
(substances that bind to a hormone receptor so as to evoke
a hormonal response) and **antagonists** (substances that
bind to a hormone receptor but fail to elicit a hormonal
response, thereby blocking agonist action). The β- but not
the α-adrenoreceptors, for example, are stimulated by
isoproterenol but blocked by **propranolol,**

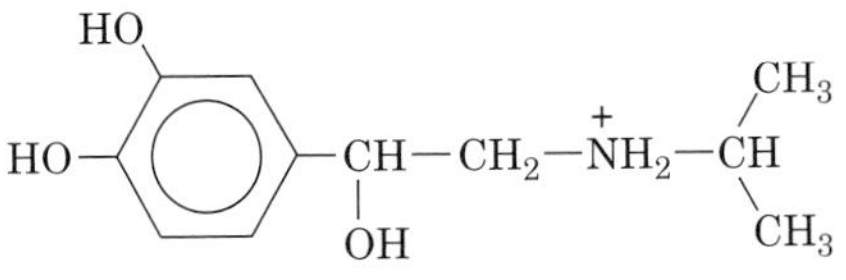

Isoproterenol

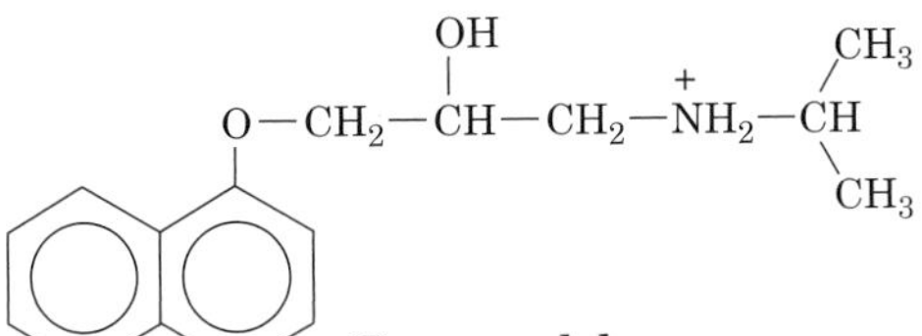

Propranolol

whereas α- but not β-adrenoreceptors are blocked by
phentolamine:

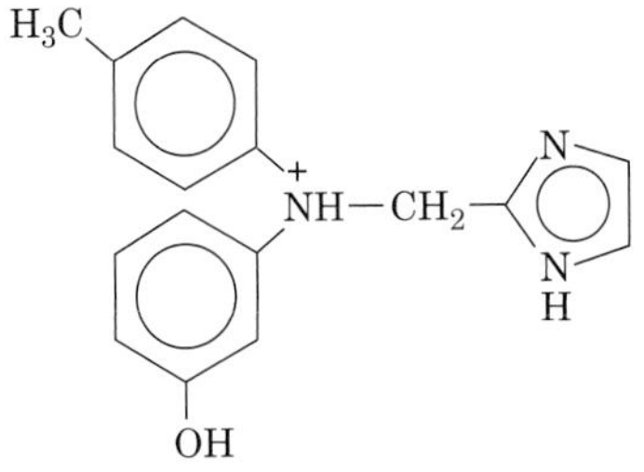

Phentolamine

   The α- and β-adrenoreceptors, which occur in separate
tissues in mammals, generally respond differently and
often oppositely to catecholamines. For instance, β-
adrenoreceptors, which activate adenylate cyclase, stimulate glycogenolysis and gluconeogenesis in liver (Sections
18-3E and 18-3G), glycogenolysis and glycolysis in skeletal muscle, lipolysis in adipose tissue, the relaxation of

smooth (involuntary) muscle in the bronchi and the blood vessels supplying the skeletal (voluntary) muscles, and increased heart action. In contrast, α-adrenoreceptors, whose intracellular effects are mediated either by the inhibition of adenylate cyclase (α_2 **receptors;** Section 19-2C) or via the phosphoinositide cascade (α_1 **receptors;** Section 19-4A), stimulate smooth muscle contraction in blood vessels supplying peripheral organs such as skin and kidney, smooth muscle relaxation in the lung and gastrointestinal tract, and blood platelet aggregation. *Most of these diverse effects are directed toward a common end: the mobilization of energy resources and their shunting to where they are most needed to prepare the body for sudden action.*

The varying responses and tissue distributions of the α- and β-adrenoreceptors and their subtypes to different agonists and antagonists have important therapeutic consequences. For example, propranolol is used for the treatment of high blood pressure and protects heart attack victims from further heart attacks, whereas epinephrine's bronchodilator effects make it clinically useful in the treatment of **asthma,** a breathing disorder caused by the inappropriate contraction of bronchial smooth muscle.

G. *Steroid Hormones*

a. The Adrenocortical Steroids Mediate a Wide Variety of Metabolic Functions

The adrenal cortex produces at least 50 different **adrenocortical steroids** *(whose synthesis is outlined in Section 25-6C).* These have been classified according to the physiological responses they evoke:

1. The **glucocorticoids** affect carbohydrate, protein, and lipid metabolism in a manner nearly opposite to that of insulin and influence a wide variety of other vital functions, including inflammatory reactions and the capacity to cope with stress.

2. The **mineralocorticoids** largely function to regulate the excretion of salt and water by the kidney.

3. The **androgens** and **estrogens** affect sexual development and function. They are made in larger quantities by the gonads.

Glucocorticoids, the most common of which are **cortisol** (also known as **hydrocortisone**)

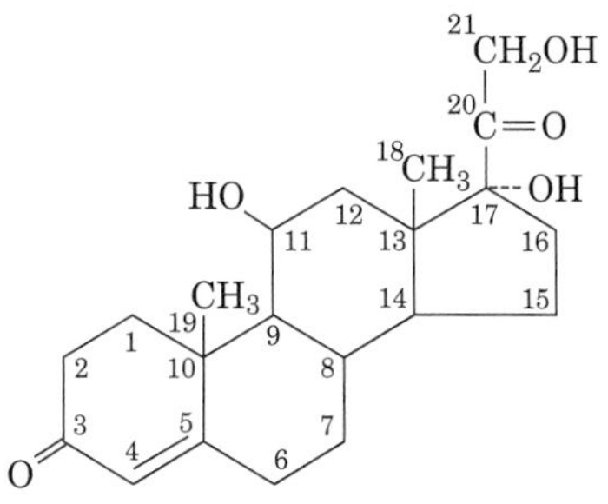

Cortisol (hydrocortisone)

and **corticosterone,** and the mineralocorticoids, the most common of which is **aldosterone,** are all C_{21} compounds:

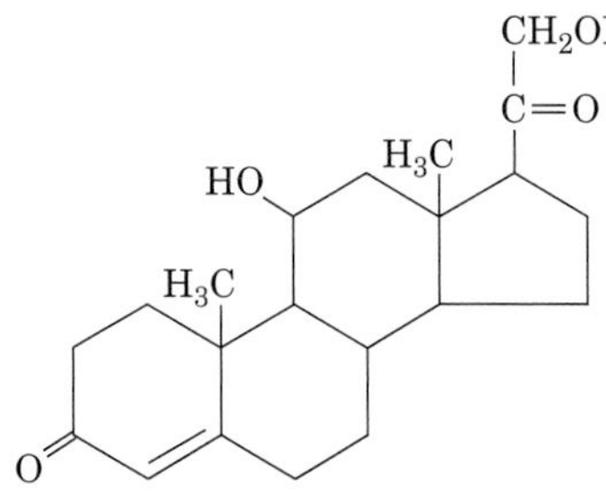

Corticosterone

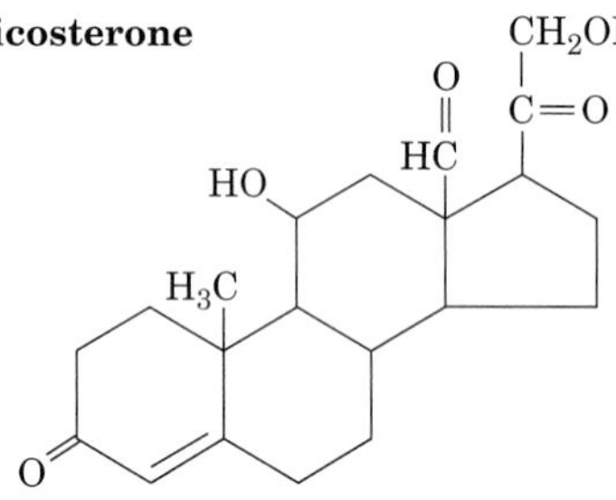

Aldosterone

Steroids, being water insoluble, are transported in the blood in complex with the glycoprotein **transcortin** and, to a lesser extent, by albumin. The steroids (including vitamin D) spontaneously pass through the membranes of their target cells to the cytosol, where they bind to their cognate receptors. The steroid–receptor complexes then migrate to the cell nucleus, where they function as transcription factors to induce, or in some cases repress, the transcription of specific genes (a process that is discussed in Section 34-3B). In this way, the glucocorticoids and the mineralocorticoids influence the expression of numerous metabolic enzymes in their respective target tissues. Thyroid hormones, which are also nonpolar, function similarly. However, as we shall see in the following sections, all other hormones act less directly in that they bind to their cognate cell-surface receptors and thereby trigger complex cascades of events within cells that ultimately influence transcription as well as other cellular processes.

Impaired adrenocortical function, either through disease or trauma, results in a condition known as **Addison's disease,** which is characterized by hypoglycemia, muscle weakness, Na^+ loss, K^+ retention, impaired cardiac function, loss of appetite, and a greatly increased susceptibility to stress. The victim, unless treated by the administration of glucocorticoids and mineralocorticoids, slowly languishes and dies without any particular pain or distress. The opposite problem, adrenocortical hyperfunction, which is usually caused by a tumor of the adrenal cortex or the pituitary gland (Section 19-1H), results in **Cushing's syndrome,** which is characterized by fatigue, hyperglycemia, edema (water retention), and a redistribution of body fat to yield a characteristic "moon face." Long-term treatments of various diseases with synthetic glucocorticoids result in similar symptoms.

b. Gonadal Steroids Mediate Sexual Development and Function

The gonads (testes in males, ovaries in females), in addition to producing sperm or ova, secrete steroid hormones (androgens and estrogens) that regulate sexual differentiation, the expression of secondary sex characteristics, and sexual behavior patterns. Although testes and ovaries both synthesize androgens and estrogens, the testes predominantly secrete androgens, which are therefore known as **male sex hormones,** whereas ovaries produce mostly estrogens, which are consequently termed **female sex hormones.**

Androgens, of which **testosterone** is prototypic,

Testosterone

β-Estradiol

Progesterone

lack the C_2 substituent at C17 present in glucocorticoids and are therefore C_{19} compounds. Estrogens, such as **β-estradiol,** resemble androgens but lack a C10 methyl group because they have an aromatic A ring and are therefore C_{18} compounds. Interestingly, testosterone is an intermediate in estrogen biosynthesis (Section 25-6C). A second class of ovarian steroids, C_{21} compounds called **progestins,** help mediate the menstrual cycle and pregnancy (Section 19-1I). **Progesterone,** the most abundant progestin, is, in fact, a precursor of glucocorticoids, mineralocorticoids, and testosterone (Section 25-6C).

c. Sexual Differentiation Is Both Hormonally and Genetically Controlled

What factors control sexual differentiation? If the gonads of an embryonic male mammal are surgically removed, that individual will become a phenotypic female. Evidently, *mammals are programmed to develop as females unless embryonically subjected to the influence of testicular*

hormones. Indeed, genetic males with absent or nonfunctional cytosolic androgen receptors are phenotypic females, a condition named **testicular feminization.** Curiously, estrogens appear to play no part in embryonic female sexual development, although they are essential for female sexual maturation and function.

Normal individuals have either the XY (male) or the XX (female) genotypes (Section 1-4C). However, those with the abnormal genotypes XXY **(Klinefelter's syndrome)** and X0 (only one sex chromosome; **Turner's syndrome)** are, respectively, phenotypic males and phenotypic females, although both are sterile. Apparently, *the normal Y chromosome confers the male phenotype, whereas its absence results in the female phenotype.* There are, however, rare (1 in 20,000) XX males and XY females. DNA hybridization studies have revealed that these XX males (who are sterile and have therefore been identified through infertility clinics) have a small segment of a normal Y chromosome translocated onto one of their X chromosomes, whereas XY females are missing this segment.

Early male and female embryos—through the sixth week of development in humans—have identical undifferentiated genitalia. Evidently, the Y chromosome contains a gene, **testes-determining factor (TDF),** that induces the differentiation of testes, whose hormonal secretions, in turn, promote male development. The misplaced chromosomal segments in XY females and XX males have a common 140-kb sequence that contains a structural gene dubbed **SRY** (for *sex-determining region of Y*) that encodes an 80-residue DNA-binding motif. Several sex-reversed XY women have a mutation in the region of their *SRY* gene encoding this DNA-binding domain that eliminates its ability to bind DNA, a mutation that is not present in their father's gene. *SRY* is expressed in embryonic gonadal cells previously shown to be responsible for testis determination. Moreover, of eleven XX mice that were made transgenic for **Sry** (the mouse analog of *SRY*), three were males. Thus, *TDF/SRY* is the first clear example of a mammalian gene that controls the development of an entire organ system (development is discussed in Section 34-4B).

H. *Control of Endocrine Function: The Hypothalamus and Pituitary Gland*

The anterior lobe of the **pituitary gland** (the **adenohypophysis**) and the **hypothalamus,** a nearby portion of the brain, constitute a functional unit that hormonally controls much of the endocrine system. *The neurons of the hypothalamus synthesize a series of polypeptide hormones known as **releasing factors** and **release-inhibiting factors** which, on delivery to the adenohypophysis via a direct circulatory connection (their half-lives are on the order of a few minutes), stimulate or inhibit the release of the corresponding **trophic hormones** into the bloodstream.* Trophic hormones, by definition, stimulate their target endocrine tissues to secrete the hormones they synthesize. Since releasing and release-inhibiting factors, trophic hormones, and endocrine hormones are largely secreted in nanogram, microgram, and milligram quantities per day, respectively, and tend to have progressively longer half-lives, these hormonal systems can be said to form amplifying cascades. Four such systems are prominent in humans (Fig. 19-7; *left*):

1. Corticotropin-releasing factor (**CRF;** 41 residues) causes the adenohypophysis to release **adrenocorticotropic hormone** (**ACTH;** 39 residues), which stimulates the release of adrenocortical steroids. *The entire system is under feedback control: ACTH inhibits the release of CRF and the adrenocortical steroids inhibit the release of both CRF and ACTH. Moreover, the hypothalamus, being part of the brain, is also subject to neuronal control, so the hypothalamus forms the interface between the nervous system and the endocrine system.*

2. Thyrotropin-releasing factor (**TRF**), a tripeptide with an N-terminal **pyroGlu** residue (a Glu derivative in which the side chain carboxyl group forms an amide bond with its amino group),

Thyrotropin-releasing factor (TRF)

stimulates the adenohypophysis to release the trophic hormone **thyrotropin** (**thyroid-stimulating hormone; TSH**) which, in turn, stimulates the thyroid to synthesize and release T_3 and T_4. TRF, as are other releasing factors, is present in the hypothalamus in only vanishingly small quantities. It was independently characterized in 1969 by Roger Guillemin and Andrew Schally using extracts of the hypothalami from over 2 million sheep and 1 million pigs.

3. Gonadotropin-releasing factor (**GnRF;** 10 residues)

$$\overset{1}{\text{pyroGlu}}\text{-His-Trp-Ser-Tyr-Gly-Leu-Arg-Pro-}\overset{10}{\text{Gly}}\text{-NH}_2$$

Gonadotropin-releasing factor (GnRF)

stimulates the adenohypophysis to release **luteinizing hormone** (**LH**) and **follicle-stimulating hormone** (**FSH**), which are collectively known as **gonadotropins.** In males, LH stimulates the testes to secrete androgens, whereas FSH promotes spermatogenesis. In females, FSH stimulates the development of ovarian follicles (which contain the immature ova), whereas LH triggers ovulation.

4. Growth hormone-releasing factor (**GRF;** 44 residues) and **somatostatin** [14 residues; also known as **growth hormone release-inhibiting factor** (**GRIF**)], stimulate/inhibit the release of **growth hormone** (**GH**) from the adenohypophysis. GH (also called **somatotropin**), in turn,

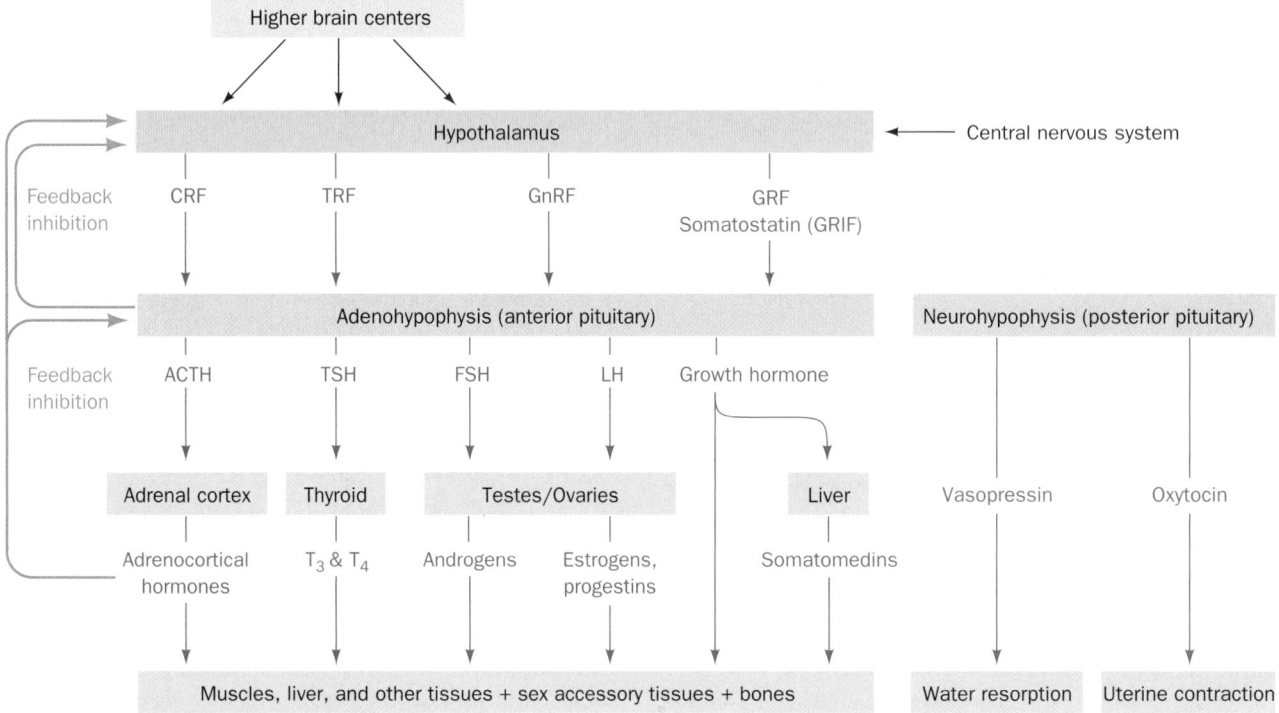

FIGURE 19-7   Hormonal control circuits, indicating the relationships between the hypothalamus, the pituitary, and the target tissues. Releasing factors and release-inhibiting factors secreted by the hypothalamus signal the adenohypophysis to secrete or stop secreting the corresponding trophic hormones, which, for the most part, stimulate the corresponding endocrine gland(s) to secrete their respective endocrine hormones. The endocrine hormones, in addition to controlling the growth, differentiation, and metabolism of their corresponding target tissues, influence the secretion of releasing factors and trophic hormones through feedback inhibition. The levels of trophic hormones likewise influence the levels of their corresponding releasing factors.

stimulates generalized growth (see Fig. 5-5 for a striking example of its effect). GH directly accelerates the growth of a variety of tissues (in contrast to TSH, LH, and FSH, which act only indirectly by activating endocrine glands) and induces the liver to synthesize a series of polypeptide growth factors termed **somatomedins** that stimulate cartilage growth and have insulinlike activities.

TSH, LH, and FSH are heterodimeric glycoproteins which, in a given species, all have the same α subunit (92 residues) and a homologous β subunit (114, 114, and 118 residues, respectively, in humans). Human GH consists of a single 191-residue polypeptide chain, which is unrelated to TSH, LH, or FSH.

a. The Neurohypophysis Secretes Oxytocin and Vasopressin

The posterior lobe of the pituitary, the **neurohypophysis,** which is anatomically distinct from the adenohypophysis, secretes two homologous nonapeptide hormones (Fig. 19-7, *right*): **vasopressin** [also known as **antidiuretic hormone (ADH)**], which increases blood pressure and stimulates the kidneys to retain water; and **oxytocin,** which causes contraction of uterine smooth muscle and therefore induces labor:

Cys-Tyr-Phe-Gln-Asn-Cys-Pro-Arg-Gly-NH$_2$
Human vasopressin

Cys-Tyr- Ile -Gln-Asn-Cys-Pro-Leu-Gly-NH$_2$
Human oxytocin

The rate of vasopressin release is largely controlled by osmoreceptors, which monitor the osmotic pressure of the blood.

I. *Control of the Menstrual Cycle*

The menstrual cycle and pregnancy are particularly illustrative of the interactions among hormonal systems. The ~28-day human menstrual cycle (Fig. 19-8) begins during menstruation with a slight increase in the FSH level that initiates the development of a new ovarian follicle. As the follicle matures, it secretes estrogens that act to sensitize the adenohypophysis to GnRF. This process culminates in a surge of LH and FSH, which triggers ovulation. The ruptured ovarian follicle, the **corpus luteum,** secretes progesterone and estrogens, which inhibit further gonadotropin secretion by the adenohypophysis and stimulate the uterine lining to prepare for the implantation of a fertilized ovum. If fertilization does not occur, the corpus luteum regresses, progesterone and estrogen levels fall, and menstruation (the sloughing off of the uterine lining) ensues. The reduced steroid levels also permit a slight increase in the FSH level, which initiates a new menstrual cycle.

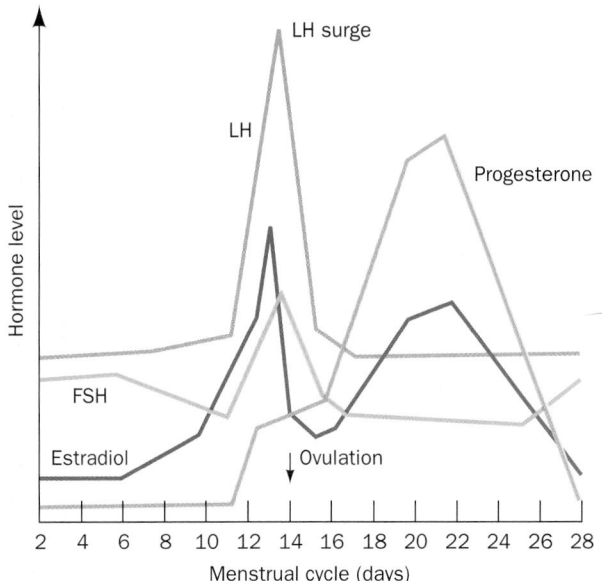

FIGURE 19-8 **Patterns of hormone secretion during the menstrual cycle in the human female.**

A fertilized ovum that has implanted into the hormonally prepared uterine lining soon commences synthesizing **chorionic gonadotropin (CG).** This heterodimeric glycoprotein hormone contains a 145-residue β subunit that has a high degree of sequence identity with those of LH (85%), FSH (45%), and TSH (36%) over their N-terminal 114 residues and the same α subunit. CG stimulates the corpus luteum to continue secreting progesterone rather than regressing and thus prevents menstruation. Pregnancy tests utilize immunoassays that can detect CG in blood or urine within a few days after embryo implantation. Most female oral contraceptives (birth control pills) contain progesterone derivatives, whose ingestion induces a state of pseudopregnancy in that they inhibit the midcycle surge of FSH and LH so as to prevent ovulation.

J. *Growth Hormone and Its Receptor*

The binding of growth hormone activates its receptor to stimulate growth and metabolism in muscle, bone, and cartilage cells. This 620-residue receptor is a member of a large family of structurally related protein growth factor receptors, which includes those for various interleukins. All of these receptors consist of an N-terminal extracellular ligand-binding domain, a single transmembrane segment that is almost certainly helical, and a C-terminal cytoplasmic domain that is not homologous within the superfamily but in many cases contains a tyrosine kinase function (Section 19-3A).

The X-ray structure of the 191-residue human growth hormone **(hGH)** in complex with the 238-residue extracellular domain of its binding protein **(hGHbp),** determined by Abraham de Vos and Anthony Kossiakoff, revealed that this complex consists of two molecules of

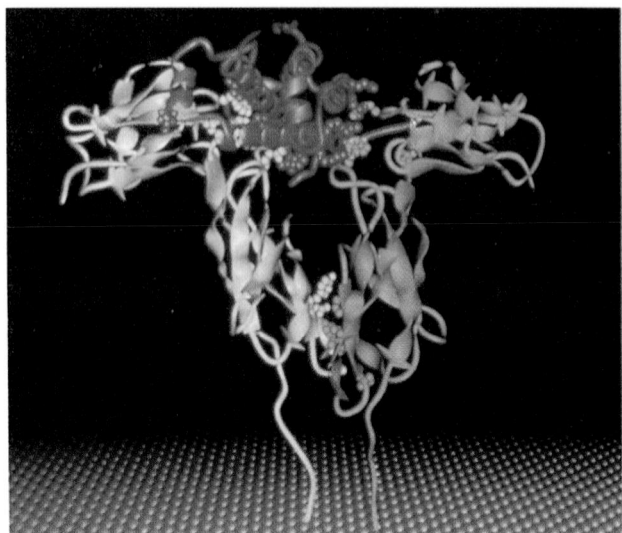

FIGURE 19-9   X-Ray structure of human growth hormone (hGH) in complex with two molecules of its receptor's extracellular domain (hGHbp). The proteins are shown in ribbon form, with the two hGHbp molecules, which together bind one molecule of hGH, green and blue and with the hGH red. The side chains involved in intersubunit interactions are shown in space-filling form. The yellow pebbled surface represents the cell membrane, through which the C-terminal ends of the hGHbp molecules are shown penetrating as they do in the intact hGH receptor. [Courtesy of Abraham de Vos and Anthony Kossiakoff, Genentech Inc., South San Francisco, California. PDBid 3HHR.] *See the Interactive Exercises*

FIGURE 19-10   Acromegaly. The characteristic enlarged features of Akhenaton, the Pharaoh who ruled Egypt in the years 1379–1362 B.C., strongly suggest that he suffered from acromegaly. [Agytisches Museum, Staadtliche Museen Preussicher Kulturbesitz, Berlin, Germany. Photo by Margarete Busing.]

hGHbp bound to a single hGH molecule (Fig. 19-9). hGH consists largely of a four-helix bundle, which closely resembles that in the previously determined X-ray structure of porcine GH, although with significant differences that may be caused by the binding of hGH to its receptor. A variety of other protein growth factors with known structures, including several interleukins, contain similar four-helix bundles. Each hGHbp molecule consists of two structurally homologous domains, each of which forms a topologically identical sandwich of a three- and a four-stranded antiparallel β sheet that resembles the immunoglobulin fold (Section 35-2B).

The two hGHbp molecules bind to hGH with near twofold symmetry about an axis that is roughly perpendicular to the helical axes of the hGH four-helix bundle and, presumably, to the plane of the cell membrane to which the intact hGH receptor is anchored (Fig. 19-9). The C-terminal domains of the two hGHbp molecules are almost parallel and in contact with one another. Intriguingly, the two hGHbp molecules use essentially the same residues to bind to sites that are on opposite sides of hGH's four-helix bundle and that have no structural similarity. The X-ray structure is largely consistent with the results of mutational studies designed to identify the hGH and hGHbp residues important for receptor binding.

The ligand-induced dimerization of hGHbp has important implications for the mechanism of signal transduction. The dimerization, which does not occur in the absence of hGH, apparently brings together the intact receptors' in-

tracellular domains in a way that activates an effector protein such as a tyrosine kinase (Section 19-3A). Indeed, hGH mutants that cannot induce receptor dimerization are biologically inactive. Numerous other protein growth factors also induce the dimerization of their receptors.

a.   Abnormal GH Production Causes Abnormal Growth

Overproduction of GH, usually a consequence of a pituitary tumor, results in excessive growth. If this condition commences while the skeleton is still growing, that is, before its growth plates have ossified, then this excessive growth is of normal proportions over the entire body, resulting in **gigantism.** Moreover, since excessive GH inhibits the testosterone production necessary for growth plate ossification, such "giants" continue growing throughout their abnormally short lives. If, however, the skeleton has already matured, GH stimulates only the growth of soft tissues, resulting in enlarged hands and feet and thickened facial features, a condition named **acromegaly** (Fig. 19-10). The opposite problem, GH deficiency, which results in insufficient growth **(dwarfism),** can be treated before skeletal maturity by regular injections of hGH (animal GH is ineffective in humans). Since hGH was, at first, available only from the pituitaries of cadavers, it was in very short supply. Now, however, hGH can be synthesized in virtually unlimited amounts via recombinant DNA techniques. Indeed, there is concern that GH will be taken in an uncontrolled manner by individuals wishing to increase their athletic prowess (it would be very difficult to prove that an individual had used exogenously supplied hGH because it is normally present in the human body, even in adults, and is rapidly degraded).

K. *Opioid Peptides*

Among the most intriguing hormones secreted by the adenohypophysis are polypeptides that have opiatelike effects on the central nervous system. These include the 31-residue **β-endorphin,** its N-terminal pentapeptide, termed **methionine-enkephalin,** and the closely similar **leucine-enkephalin** (although the enkaphalins are independently expressed).

$$
\begin{aligned}
&\overset{1}{\text{Tyr}}\text{-Gly-Gly-Phe-}\overset{5}{\text{Met}}\text{-Thr-Ser-Glu-Lys-}\overset{10}{\text{Ser}}\\
&\overset{11}{\text{Gln}}\text{-Thr-Pro-Leu-}\overset{15}{\text{Val}}\text{-Thr-Leu-Phe-Lys-}\overset{20}{\text{Asn}}\\
&\overset{21}{\text{Ala}}\text{-Ile-Val-Lys-}\overset{25}{\text{Asn}}\text{-Ala-His-Lys-Lys-}\overset{30}{\text{Gly}}\\
&\overset{31}{\text{Gln}}\text{-}
\end{aligned}
$$

β-Endorphin

Tyr-Gly-Gly-Phe-Met

Methionine-enkephalin (Met-enkephalin)

Tyr-Gly-Gly-Phe-Leu

Leucine-enkephalin (Leu-enkephalin)

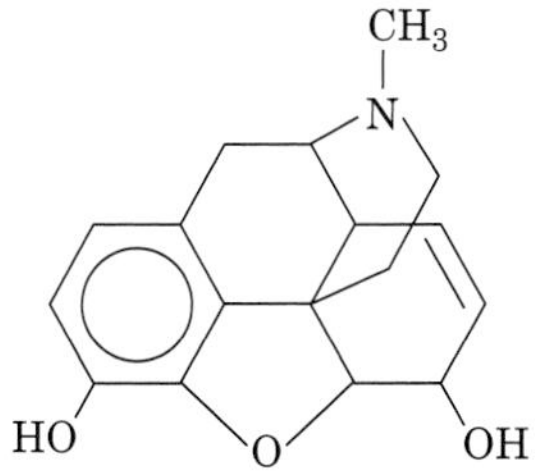

Morphine (an opiate)

These substances bind to **opiate receptors** in the brain and have been shown to be their physiological agonists. The role of these so-called **opioid peptides** has yet to be definitively established, but it appears they are important in the control of pain and emotional states. Pain relief through the use of acupuncture and placebos as well as such phenomena as "runner's high" may be mediated by opioid peptides.

L. *The Hormonal Function of Nitric Oxide*

Nitric oxide (NO) is a reactive and toxic free radical gas. Thus, it came as a great surprise that *this molecule functions as an intercellular signal in regulating blood vessel dilation and serves as a neurotransmitter. It also functions in the immune response.* The role of NO in vasodilation was discovered through the observation that substances such as acetylcholine (Section 20-5C) and bradykinin (Section 7-5B), which act through the phosphoinositide signaling system (Section 19-4) to increase the flow through blood vessels by eliciting smooth muscle relaxation, require an intact **endothelium** overlying the smooth muscle (the endothelium is a layer of cells that lines the inside of certain body cavities such as blood vessels). Evidently, endothelial cells respond to the presence of these vasodilation agents by releasing a diffusible and highly labile substance (half-life ~5 s) that induces the relaxation of smooth muscle cells. This substance was identified as NO, in part, through parallel studies identifying NO as the active metabolite that mediates the well-known vasodilating effects of antianginal organic nitrates such as **nitroglycerin**

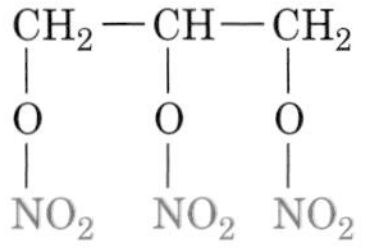

$$
\begin{array}{ccc}
CH_2 & CH & CH_2 \\
| & | & | \\
O & O & O \\
| & | & | \\
NO_2 & NO_2 & NO_2
\end{array}
$$

Nitroglycerin

(**angina pectoris** is a condition caused by insufficient blood flow to the heart muscle, leading to severe chest pain).

a. Nitric Oxide Synthase Requires Five Redox-Active Cofactors

NO is synthesized by **NO synthase (NOS),** which catalyzes the NADPH-dependent 5-electron oxidation of L-arginine by O_2 to yield NO and the amino acid citrulline with the intermediate formation of N^ω-**hydroxy-L-arginine** (Fig. 19-11). Three isozymes of NOS have been identified in mammals,

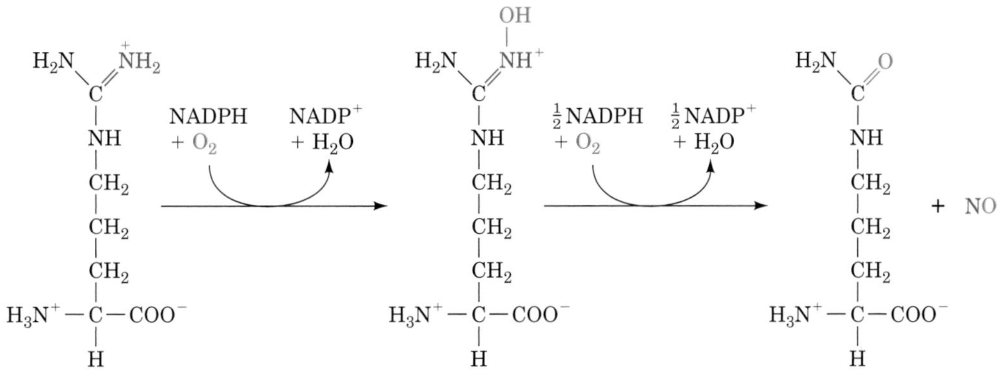

FIGURE 19-11 The NO synthase (NOS) reaction. The N^ω-hydroxy-L-arginine intermediate is tightly bound to the enzyme.

neuronal NOS (nNOS), inducible NOS (iNOS), and **endo-thelial NOS (eNOS),** which are also known as **NOS-1, -2,** and **-3,** respectively. These isozymes, which have >50% sequence identity, are all homodimeric proteins of 125- to 160-kD subunits that each consist of two domains:

1. An N-terminal, ~500-residue, oxygenase domain that catalyzes both reaction steps of Fig. 19-11 and contains the dimer interface. This domain binds the substrates O_2 and L-arginine and two redox-active prosthetic groups, Fe(III)-heme and **5,6,7,8-tetrahydrobiopterin (H_4B),**

5,6,7,8-Tetrahydrobiopterin

a compound that also functions in the hydroxylation of phenylalanine to tyrosine (Section 26-3H). The X-ray structures of the oxygenase domains of iNOS (Fig. 19-12), determined by John Tainer, and of eNOS, independently determined by Thomas Poulos and Patricia Weber, are closely similar.

2. A C-terminal, ~600-residue, reductase domain that supplies the electrons for the NOS reaction. It binds NADPH and two redox-active prosthetic groups, an FAD (Fig. 16-8) and a **flavin mononucleotide (FMN;** FAD lacking its AMP residue; Fig. 22-17a) via three nucleotide binding modules. This domain is homologous to **cytochrome P_{450} reductase,** an enzyme that participates in detoxification processes (Section 15-4B).

The NADPH bound to the reductase domain transmits its electrons via the FAD and then the FMN to the heme in the oxidase domain. Interestingly, the reductase domain transmits its electrons to the oxidase domain on the opposite subunit. This was shown by Dennis Stuehr through his construction of an NOS heterodimer in which one subunit was full length and the other consisted of only an oxygenase domain. If a mutation that disrupts the L-arginine binding site was in the full-length subunit, the enzymatic activity of this heterodimer was unaffected, but if it was in the oxygenase-only subunit, activity was abolished.

The heme Fe atom is 5-coordinated with its axial ligand supplied by a specific Cys S atom (Fig. 19-12). The L-arginine substrate binds on the opposite side of the heme from this Cys with the N atom to be hydroxylated ~4.0 Å distant from the Fe atom, a distance too large for covalent bond formation. Since O_2 is known to react with the heme Fe atom, it presumably binds between it and this N atom.

NOS requires bound H_4B to produce NO. In the absence of this prosthetic group, NOS efficiently catalyzes the oxidation of NADPH by O_2 to yield H_2O_2. Nevertheless, H_4B does not undergo net oxidation in the NOS reaction (as it does in the reaction hydroxylating phenylalanine to tyrosine; Section 26-3H) and, moreover, binds too far away from the heme and on its opposite side from the L-arginine to di-

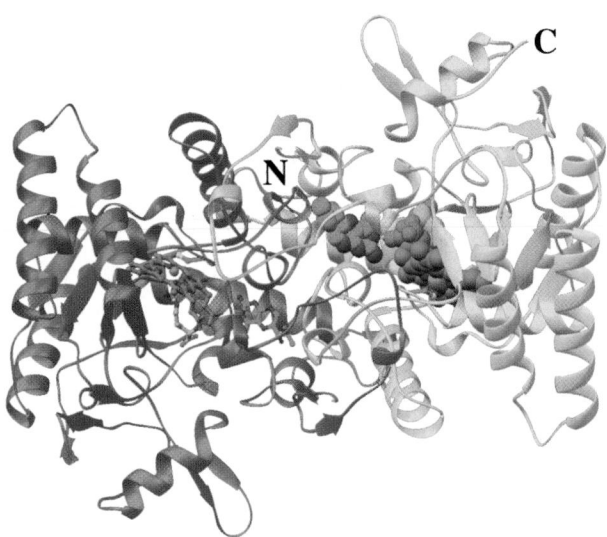

FIGURE 19-12   X-Ray structure of the oxygenase domain of iNOS. The homodimeric protein is viewed along its 2-fold axis of symmetry with one subunit gold and the other gray. The heme and H_4B prosthetic groups together with the substrate L-arginine are shown in the gray subunit in ball-and-stick form and in the gold subunit in space-filling form, both colored according to atom type with C of heme, H_4B, and L-arginine green, magenta, and cyan, respectively, and N blue, O red, and Fe orange. The side chain of Cys 194, the residue that axially ligands the heme Fe, is shown in ball-and-stick form in both subunits with C magenta and S yellow. The C-terminal 15 residues of the gray subunit have been deleted to better expose the underlying heme and L-arginine. [Based on an X-ray structure by Thomas Poulos, University of California at Irvine. PDBid 1NOD.]

rectly participate in the hydroxylation reaction (Fig. 19-12). In fact, the cytochromes P450 (Section 15-4B), which catalyze hydroxylation reactions similar to that of NOS, lack H_4B. Thus the function of H_4B in NOS is, as yet, unknown.

NO rapidly diffuses across cell membranes, although its high reactivity prevents it from traveling >1 mm from its site of synthesis (in particular, it efficiently reacts with both oxyhemoglobin and deoxyhemoglobin: NO + HbO_2 → NO_3^- + Hb; and NO + Hb → HbNO; Section 10-1A). *The physiological target of NO in smooth muscle cells is guanylate cyclase (GC),* which catalyzes the reaction of GTP to yield **3′,5′-cyclic GMP (cGMP),**

3′,5′-Cyclic GMP (cGMP)

an intracellular second messenger that resembles 3′,5′-cyclic AMP (cAMP; GC is a homolog of adenylate cyclase; Section 18-3C). cGMP causes smooth muscle relaxation through its stimulation of protein phosphorylation by **cGMP-dependent protein kinase.** NO reacts with GC's heme prosthetic group to yield **nitrosoheme,** whose presence increases GC's activity by up to 200-fold, presumably via a conformation change resembling that in hemoglobin on binding O_2 (Section 10-2B; although GC binds O_2 quite poorly).

b. eNOS and nNOS but Not iNOS Are Regulated by [Ca^{2+}]

Ca^{2+}–calmodulin activates eNOS and nNOS by binding to the ~30-residue N-terminal segments of their reductase domains. Thus, for example, the stimulatory action of vasodilatory agents on the phosphoinositide signaling system (Section 19-4A) in endothelial cells to produce an influx of Ca^{2+} results in the synthesis of NO. Hence, *NO functions to transduce hormonally induced increases in intracellular [Ca^{2+}] in endothelial cells to increased rates of production of cGMP in neighboring smooth muscle cells.*

NO produced by neuronal NOS mediates vasodilation through endothelium-independent neural stimulation of smooth muscle. In this signal transduction pathway, which is responsible for the dilation of cerebral and other arteries as well as penile erection (see Section 19-2E), nerve impulses cause an increased [Ca^{2+}] in nerve terminals, thereby stimulating neuronal NOS. The resultant NO diffuses to nearby smooth muscle cells, where it binds to guanylate cyclase and activates it to synthesize cGMP as described above.

Inducible NOS (iNOS) is unresponsive to Ca^{2+} even though it has two tightly bound calmodulin subunits. However, it is transcriptionally induced in macrophages and **neutrophils** (white blood cells that function to ingest and kill bacteria), as well as in endothelial and smooth muscle cells (in contrast, eNOS and nNOS are expressed **constitutively,** that is, at a constant rate). Several hours after exposure to **cytokines** (protein growth factors that regulate the differentiation, proliferation, and activities of the numerous types of cells, most conspicuously blood cells; interleukins are cytokines) and/or **endotoxins** (bacterial cell wall lipopolysaccharides that elicit inflammatory responses; Section 35-2F), these cells begin to produce large quantities of NO and continue to do so for many hours. Activated macrophages and neutrophils also produce superoxide radical ($O_2^-\cdot$), which chemically combines with NO to form the even more toxic **peroxynitrite** (**OONO$^-$**, which rapidly reacts with H_2O to yield the highly reactive **hydroxide radical, OH·,** and NO_2) that they use to kill ingested bacteria. Indeed, NOS inhibitors block the cytotoxic actions of macrophages.

Cytokines and endotoxins induce a long-lasting and profound vasodilation and a poor response to vasoconstrictors such as epinephrine. The sustained release of NO has been implicated in **septic shock** (an often fatal immune system overreaction to bacterial infection that results in a catastrophic reduction in blood pressure), in inflammation-related tissue damage as occurs in autoimmune diseases

such as rheumatoid arthritis, and in the damage to neurons in the vicinity of but not directly killed by a stroke (which often does greater damage than the stroke itself). Many of these conditions might be alleviated if drugs can be developed that selectively inhibit iNOS and/or nNOS, while permitting eNOS to carry out its essential function of maintaining vascular tone. Moreover, the administration of NO itself appears to be medically useful. For example, the inhalation of low levels NO has been used to reduce pulmonary hypertension (high blood pressure in the lung, an often fatal condition caused by constriction of its arteries) in newborn infants.

2 ■ HETEROTRIMERIC G PROTEINS

We have seen (Section 18-3) that hormones such as glucagon and epinephrine regulate glycogen metabolism by stimulating adenylate cyclase (AC) to synthesize the second messenger cAMP from ATP. The cAMP then binds to protein kinase A (PKA) so as to activate this enzyme to initiate cascades of phosphorylation/dephosphorylation events that ultimately control the activities of glycogen phosphorylase and glycogen synthase. Numerous other extracellular signaling molecules (known as ligands or agonists) also activate the intracellular synthesis of cAMP, thereby eliciting a cellular response. But what is the mechanism through which the binding of a ligand to an extracellular receptor induces AC to synthesize cAMP in the cytosol? In answering this question we shall see that the systems that link extracellular receptors to AC as well as other **effectors** have a surprising complexity that endows them with immense capacity for both signal amplification and regulatory flexibility.

A. *Overview*

🔎 **See Guided Exploration 16. Mechanisms of Hormone Signaling Involving the Adenylate Cyclase System** Adenylate cyclase, which is located on the plasma membrane's cytoplasmic surface, and the receptors that activate it, which are exposed to the cellular exterior, are separate proteins that do not physically interact. Rather, *they are functionally coupled by* **heterotrimeric G proteins** *(Fig. 19-13)*, so called because they specifically bind the guanine nucleotides GTP and GDP. AC is activated by a heterotrimeric G protein (often called just a G protein) but only when the G protein is complexed with GTP. However, G protein slowly hydrolyzes GTP to GDP + P_i (at the leisurely rate of 2–3 min^{-1}) and thereby deactivates itself (if G proteins were efficient enzymes, they would be unable to effectively activate AC). G protein is reactivated by a GDP–GTP exchange reaction that is catalyzed by the ligand–receptor complex but not by unoccupied receptor. *Heterotrimeric G protein therefore mediates the transduction of an extracellular signal (a ligand) to an intracellular signal (the cAMP). Moreover, the receptor–G protein–AC system amplifies the extracellular signal because each ligand–receptor complex activates many G proteins before it is inactivated by the*

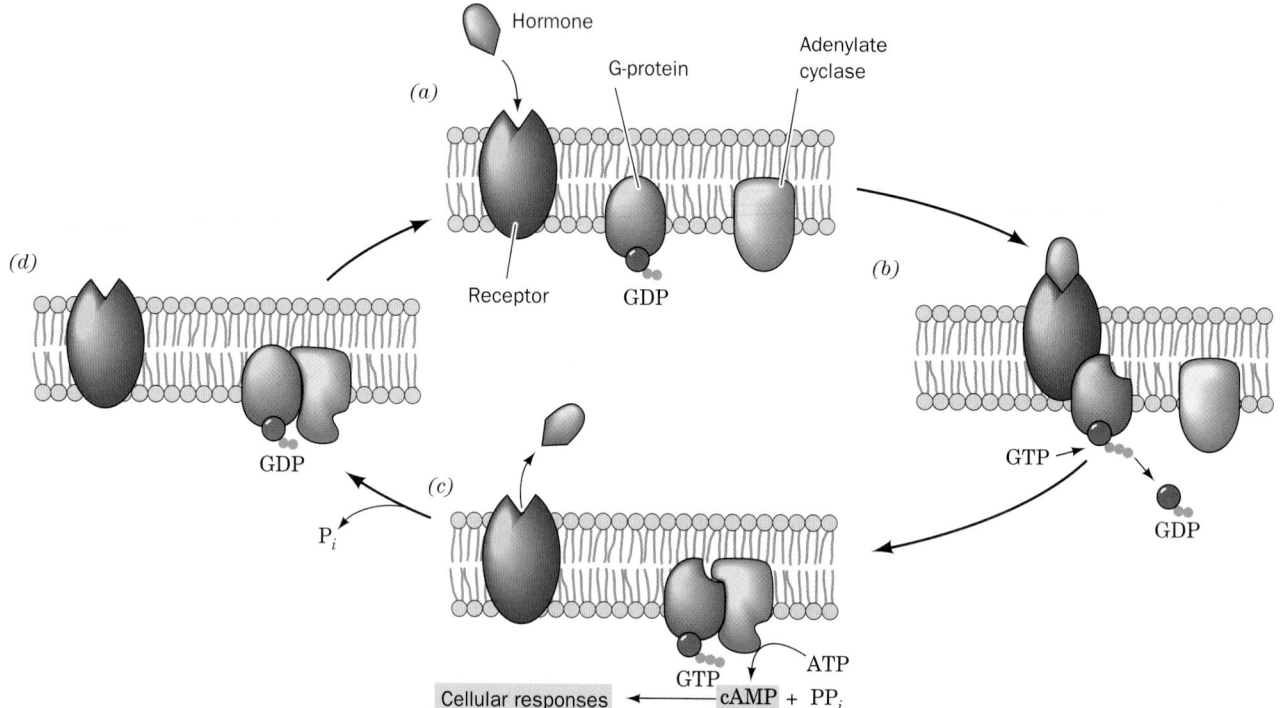

FIGURE 19-13 Activation/deactivation cycle for hormonally stimulated AC. (*a*) In the absence of hormone, heterotrimeric G protein binds GDP and AC is catalytically inactive. (*b*) The hormone–receptor complex stimulates the G protein to exchange its bound GDP for GTP. (*c*) The G protein·GTP complex, in turn, binds to and thereby activates AC to produce cAMP. (*d*) The eventual G protein-catalyzed hydrolysis of its bound GTP to GDP causes G protein to dissociate from and hence deactivate AC.

spontaneous dissociation of the ligand and, during its lifetime, each G protein·GTP–AC complex catalyzes the formation of many cAMP molecules. In this section, we discuss how this process occurs.

Heterotrimeric G proteins are members of the superfamily of regulatory GTPases that are collectively known as **G proteins** (whether one is referring to a heterotrimeric or some other species of G protein is usually clear from context). G proteins other than heterotrimeric G proteins have a wide variety of essential functions including signal transduction (e.g., **Ras;** Section 19-3C), vesicle trafficking (e.g., Arf, dynamin, and Rab; Sections 12-4C and 12-4D), the regulation of the actin cytoskeleton (by **Rho;** Section 35-3E), translation (as ribosomal accessory factors; Section 32-3), and targeting (as a component of the signal recognition particle; Section 12-4B). The many G proteins share common structural motifs that bind guanine nucleotides (GDP and GTP) and catalyze the hydrolysis of GTP to GDP + P_i (see below).

B. G Protein-Coupled Receptors

The receptors responsible for activating AC and other targets of heterotrimeric G proteins are all integral proteins with 7 transmembrane helices (Fig. 19-14). These **G protein-coupled receptors (GPCRs; also called heptahelical, 7 TM,** and **serpentine receptors)** constitute one of the largest known protein families (>1000 species in mammals, which constitutes >3% of the ~30,000 genes in the human genome). They include receptors for nucleosides, nucleotides, Ca^{2+}, catecholamines [epinephrine and norepinephrine as well as **dopamine** (Section 26-4B)] and other biogenic amines (e.g., **histamine** and **serotonin;** Section 26-4B); for **eicosanoids (prostaglandins, prostacyclins, thromboxanes, leukotrienes,** and **lipoxins,** derivatives of the C_{20} fatty acid arachidonic acid, which are potent local mediators of numerous important physiological processes; Section 25-7); and for most of the large variety of peptide and protein hormones discussed in Section 19-1. In addition, GPCRs have important sensory functions: They constitute the olfactory (odorant) and gustatory (taste) receptors (of which there are estimated to be ~500 different types in mammals), as well as the several light-sensing proteins in the retina of the eye, which are known as **rhodopsins.** Moreover, the GPCRs constitute the most important class of drug targets in the pharmaceutical

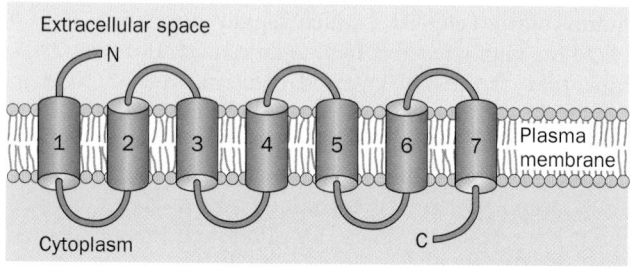

FIGURE 19-14 General structure of a G protein-coupled receptor (GPCR).

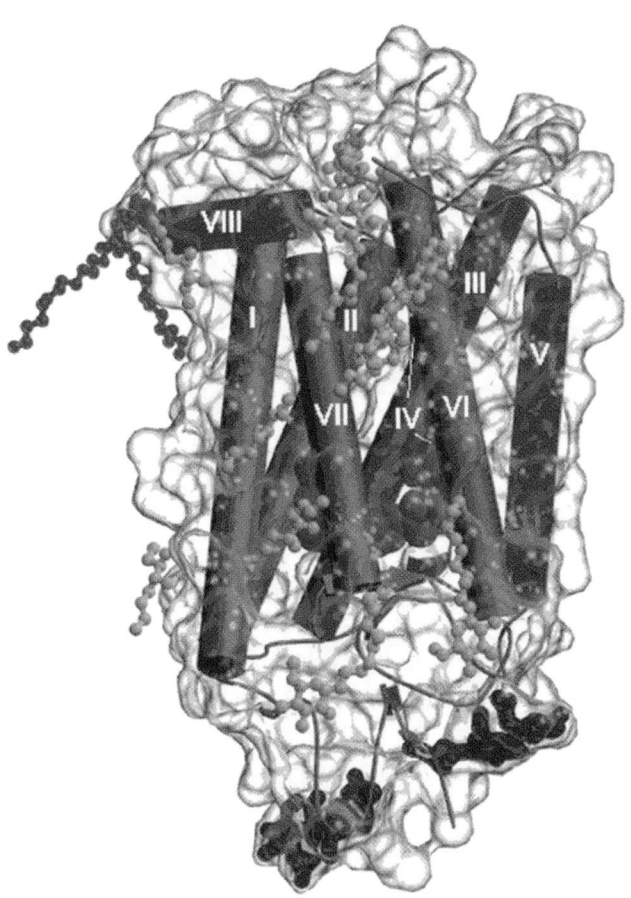

FIGURE 19-15   X-Ray structure of bovine rhodopsin. The structure is viewed parallel to the plane of the membrane with the cytoplasm above. The transparent surface represents the protein's solvent-accessible surface. The polypeptide backbone is shown in tube-and-arrow representation (*blue*). Note its bundle of 7 nearly parallel transmembrane helices. The protein's retinal prosthetic group is shown in space-filling form (*red*) and its two N-linked oligosaccharides (*dark blue*) and its two covalently attached palmitoyl groups (*green*) are shown in ball-and-stick form. Detergent molecules, which facilitated the crystallization of this integral protein and which are associated with its hydrophobic transmembrane surface, are drawn in ball-and-stick form in yellow. [Courtesy of Ronald Stenkamp, University of Washington. PDBid 1HZX.]

arsenal (Section 15-4): ~60% of approved drugs elicit their therapeutic effects by selectively interacting with a specific GPCR.

The only GPCR whose X-ray structure has yet been reported is that of bovine rhodopsin (Fig. 19-15), which was determined by Krzysztof Palczewski, Ronald Stenkamp, and Masashi Miyano. Rhodopsin consists of the 348-residue protein **opsin** that is covalently linked to the chromophore **retinal** (Fig. 12-25) via a Schiff base to Lys 296, much as occurs in the homologous bacteriorhodopsin (Section 12-3A), which is a heptahelical light-driven proton pump (Section 22-3B). The absorption of a photon causes the rhodopsin-bound retinal to isomerize from its ground state 11-cis form to its all-trans form. The isomerization causes opsin to undergo a transient conformational change before the all-*trans*-retinal is hydrolyzed and dissociated from the opsin (which is subsequently regenerated by the addition of 11-*cis*-retinal delivered from adjacent epithelial cells in the retina). It is this conformational change, which occurs mainly on retinal's cytoplasmic surface, that activates its cognate G protein. Note that the transmembrane helices of the GPCRs are generally uniform in size (20–27 residues), but that their N- and C-terminal segments and the loops connecting their transmembrane helices (which form their ligand and G protein binding sites) vary widely in length with the identity of the GPCR (7–595 residues for the N- and C-termini and 5–230 residues for the loops).

a. Receptors Are Subject to Desensitization

One of the hallmarks of biological signaling systems is that they adapt to long-term stimuli by reducing their response to them, a process named **desensitization**. *These signaling systems therefore respond to changes in stimulation levels rather than to their absolute values.* What is the mechanism of desensitization? In the case of β-adrenoreceptors, continuous exposure to epinephrine leads to the phosphorylation of one or more of the receptor's Ser residues. This phosphorylation, which is catalyzed by a specific kinase that acts on the hormone–receptor complex but not on the receptor alone, decreases the influence of hormone on heterotrimeric G protein, at least in part by reducing the receptor's epinephrine-binding affinity. The phosphorylated receptors, moreover, are endocytotically sequestered in specialized vesicles that are devoid of both heterotrimeric G protein and AC, thereby further attenuating the cell's response to epinephrine. If the epinephrine level is reduced, the receptor is slowly dephosphorylated by a phosphatase and returned to the cell surface, thereby restoring the cell's epinephrine sensitivity.

C. *Heterotrimeric G Proteins: Structure and Function*

Heterotrimeric G proteins, which were first characterized by Alfred Gilman and Martin Rodbell, are more complex than Fig. 19-13 implies: They consist, as their name indi-

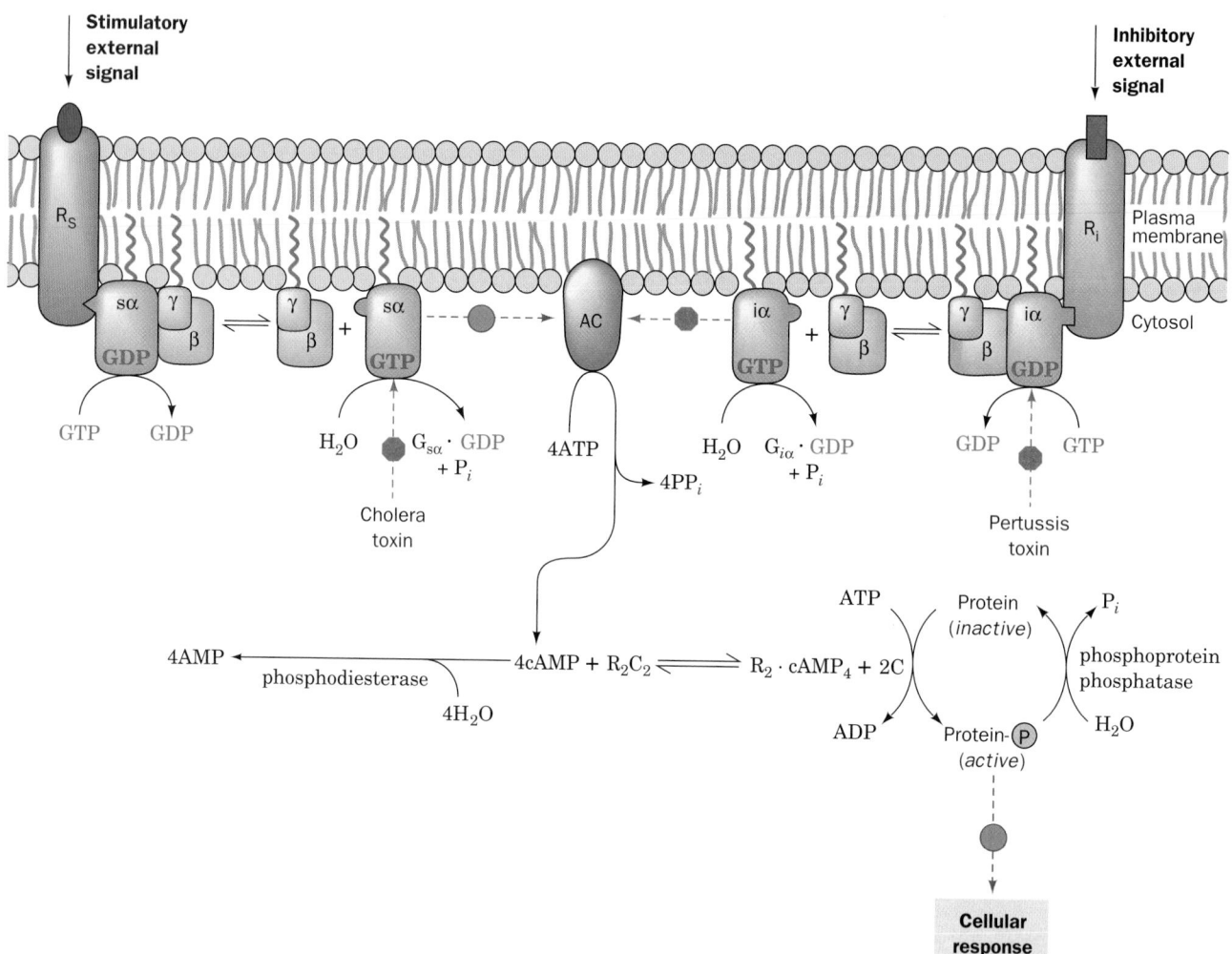

FIGURE 19-16 Mechanism of receptor-mediated activation/ inhibition of AC. The binding of hormone to a stimulatory receptor, R_s (*left*), induces it to bind G_s protein, which, in turn, stimulates the $G_{s\alpha}$ subunit of this $G_{s\alpha}G_{\beta\gamma}$ heterotrimer to exchange its bound GDP for GTP. The $G_{s\alpha}$·GTP complex then dissociates from $G_{\beta\gamma}$ and, until it catalyzes the hydrolysis of its bound GTP to GDP, stimulates adenylate cyclase to convert ATP to cAMP. The binding of hormone to the inhibitory receptor, R_i (*right*), triggers an almost identical chain of events except that the presence of $G_{i\alpha}$·GTP complex inhibits AC from synthesizing cAMP. R_2C_2 represents protein kinase A, whose catalytic subunit, C, when activated by the dissociation of the regulatory dimer as R_2·cAMP$_4$ (Section 18-3C), activates various cellular proteins by catalyzing their phosphorylation. The sites of action of cholera and pertussis toxins are indicated.

cates, of three different subunits, α, β, and γ (45, 37, and 9 kD, respectively), of which it is G_α that binds GDP and GTP (Fig. 19-16) and hence is a member of the G protein superfamily. The binding of G_α·GDP–$G_\beta G_\gamma$ to its cognate ligand–GPCR complex induces the G_α to exchange its bound GDP for GTP and, in so doing, to dissociate from $G_\beta G_\gamma$. In contrast, G_β and G_γ bind one another with such high affinity that they only dissociate under denaturing conditions. Consequently, we shall henceforth refer to their complex as $G_{\beta\gamma}$.

Both G_α and $G_{\beta\gamma}$ are membrane-anchored proteins: G_α through its myristoylation or palmitoylation or both at or near its N-terminus, and $G_{\beta\gamma}$ through the prenylation of G_γ at its C-terminus (Section 12-3B). These lipid modifications stabilize the interactions of G_α with $G_{\beta\gamma}$, as they localize both to the inner surface of the plasma membrane.

GTP binding, in addition to decreasing G_α's affinity for its cognate ligand–GPCR complex, increases its affinity for its effector, AC. Thus, *it is the binding of G_α·GTP that activates AC (Fig. 19-16, left).*

$G_{\beta\gamma}$ can also directly participate in signal transduction: It activates a wide variety of signaling proteins including several isoforms of AC (see below), certain Na$^+$, K$^+$, and Ca^{2+}-specific ion channels, various **protein tyrosine kinases**

(Section 19-3A), and **phospholipase C-β** (**PLC-β**; a component of the phosphoinositide signaling system; Section 19-4B). $G_{\beta\gamma}$ thereby provides an important source of cross talk between signaling systems.

On the eventual G_α-catalyzed hydrolysis of GTP, the resulting $G_\alpha \cdot GDP$ complex dissociates from AC and reassociates with $G_{\beta\gamma}$ to reform inactive G protein. *Since G_α hydrolyzes its bound GTP at a characteristic rate, it functions as a molecular clock that limits the length of time that both $G_\alpha \cdot GTP$ and $G_{\beta\gamma}$ can interact with their effectors.*

Several types of ligand–GPCR complexes may activate the same G protein. This occurs, for example, in liver cells in response to the binding of the corresponding hormones to glucagon receptors and to β-adrenoreceptors. In such cases, the amount of cAMP produced is the sum of that induced by the individual hormones. G proteins may also act in other ways than by activating AC: They are known, for example, to stimulate the opening of K^+ channels in heart cells and to participate in the phosphoinositide signaling system (Section 19-4A).

Some ligand–GPCR complexes inhibit rather than activate AC (Fig. 19-16, right). These include the α_2-adrenoreceptor and receptors for somatostatin and opioids. The inhibitory effect is mediated by "inhibitory" G protein, **G_i,** which may have the same β and γ subunits as does "stimulatory" G protein, **G_s,** but has a different α subunit, $G_{i\alpha}$ (41 kD). G_i acts analogously to G_s in that on binding to its corresponding ligand–GPCR complex, its $G_{i\alpha}$ subunit exchanges bound GDP for GTP and dissociates from $G_{\beta\gamma}$. However, $G_{i\alpha}$ inhibits rather than activates AC, through direct interactions and possibly because the liberated $G_{\beta\gamma}$ binds to and sequesters **$G_{s\alpha}$.** The latter mechanism is supported by the observation that liver cell membranes contain far more G_i than G_s. The activation of G_i in such cells would therefore release enough $G_{\beta\gamma}$ to bind the available $G_{s\alpha}$.

$G_{s\alpha}$ and $G_{i\alpha}$ are members of a family of related proteins, many of which have downstream effectors other than AC. This family also includes:

1. $G_{q\alpha}$, which forms a link in the phosphoinositide signaling system (Section 19-4B).

2. Transducin ($G_{t\alpha}$), a variant of $G_{i\alpha}$, which transduces visual stimuli by coupling the light-induced conformational change of rhodopsin to the activation of a specific phosphodiesterase, which thereupon hydrolyzes cGMP to GMP. This **cGMP-phosphodiesterase (cGMP-PDE)** is an $\alpha\beta\gamma_2$ heterotetramer that is activated by the displacement of its inhibitory γ subunits (**PDEγ**) by their tighter binding to $G_{t\alpha} \cdot GTP$. A cation-specific transmembrane channel (Section 20-3A) that is held open by the binding of cGMP closes on the resulting reduction in [cGMP], thereby triggering a nerve impulse (Section 20-5B) indicating that light has been detected.

3. G_{olf}, a variant of $G_{s\alpha}$, which is expressed only in olfactory sensory neurons and participates in odorant signal transduction.

4. $G_{12\alpha}$ and **$G_{13\alpha}$,** which participate in the regulation of the cytoskeleton.

This heterogeneity in G proteins occurs in the β and γ subunits as well as in the α subunits. In fact, 20 different α subunits, 6 different β subunits, and 12 different γ subunits have been identified in mammals, some of which appear to be ubiquitously expressed, whereas others are expressed only in specific cells. Thus, a cell may contain several closely related G proteins of a given type that interact with varying specificities with receptors and effectors. This complex signaling system presumably permits cells to respond in a graded manner to a variety of stimuli.

a. G Proteins Often Require Accessory Proteins to Function

The proper physiological functioning of a G protein often requires the participation of several other types of proteins:

1. A **GTPase-activating protein (GAP),** which as its name implies, stimulates its corresponding G protein to hydrolyze its bound GTP. This rate enhancement can be >2000-fold. The downstream effectors of $G_{t\alpha}$ and $G_{q\alpha}$, cGMP-PDE and PLC-β (Section 19-4B), respectively, exhibit GAP activities toward $G_{t\alpha}$ and $G_{q\alpha}$ (which otherwise would hydrolyze GTP at physiologically insignificant rates), but AC does not exhibit GAP activity toward either $G_{s\alpha}$ or $G_{i\alpha}$. However, a family of >20 **RGS proteins** (for *r*egulators of *G* protein *s*ignaling) function as GAPs for G_α subunits, although their physiological functions are as yet not well understood.

2. A **guanine nucleotide exchange factor [GEF;** alternatively **guanine nucleotide releasing factor (GRF)],** which induces its corresponding G protein to release its bound GDP. The G protein subsequently binds another guanine nucleotide (GTP or GDP, which most G proteins bind with approximately equal affinities), but since cells maintain a GTP concentration that is 10-fold higher than that of GDP, this, in effect, exchanges the bound GDP for GTP. For heterotrimeric G proteins, the ligand–GPCR complexes function as GEFs.

3. A **guanine nucleotide dissociation inhibitor (GDI).** A $G_{\beta\gamma}$ may be regarded as its associated G_α's GDI because GDP dissociates slowly from isolated G_α subunits but is essentially irreversibly bound by heterotrimers.

b. The X-Ray Structures of G_α Proteins Rationalize Their Functions

The X-ray structures of the C-terminal 325 residues of the 350-residue bovine transducin-α ($G_{t\alpha}$) in its complexes

(a)

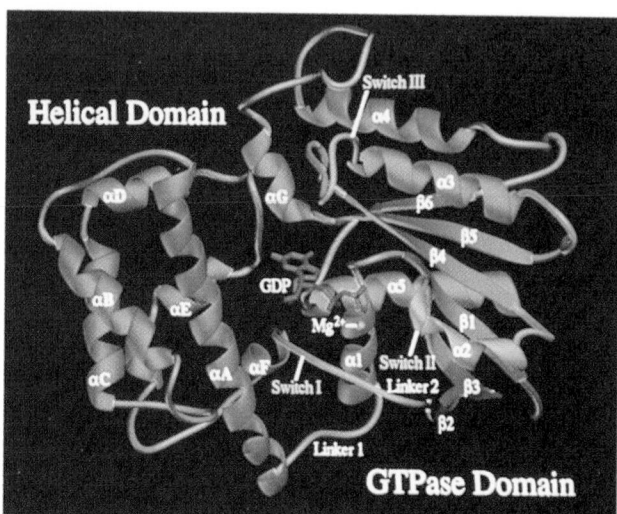

(b)

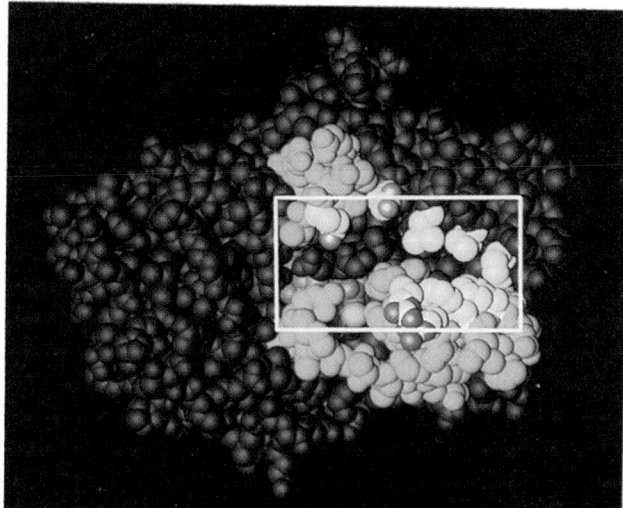

(c)

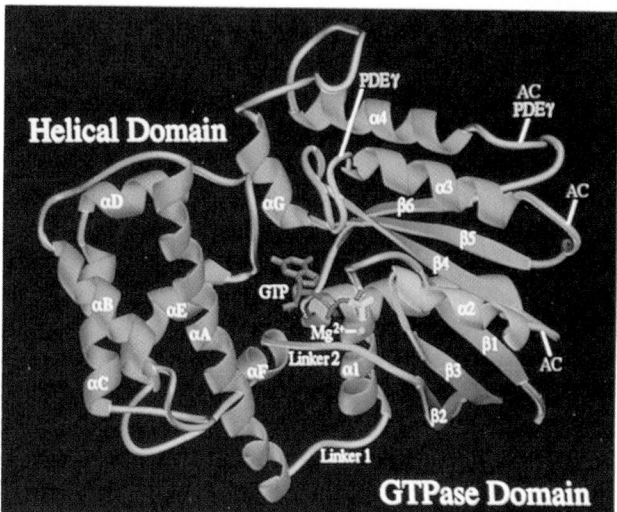

(d)

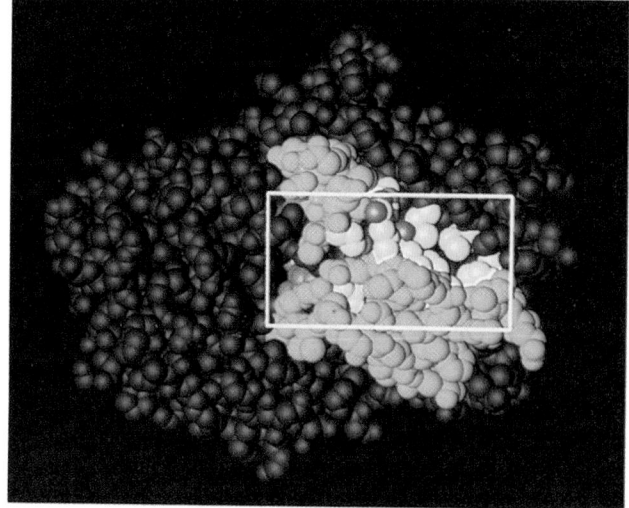

FIGURE 19-17   Structural differences between the inactive and active forms of $G_{t\alpha}$ (transducin). The change in structure is indicated by the comparison of the X-ray structure of $G_{t\alpha}\cdot$GDP in its *(a)* ribbon and *(b)* spacing-filling forms with that of $G_{t\alpha}\cdot$GTPγS in its *(c)* ribbon and *(d)* spacing-filling forms, all viewed from the same direction. In the ribbon drawings, helices and sheets are green; the segments linking them are gold; the guanine nucleotides are magenta, except for the γ phosphate of GTPγS, which is yellow; and the bound Mg^{2+} ion is represented by a blue ball. The protein's three switch regions (I, II, and III) are highlighted in cyan. In Part *c*, the two loop regions of the protein that are implicated in its interaction with the cGMP-

phosphodiesterase subunit to which it binds (PDEγ) are pointed out with yellow labels, whereas the three loop regions that are implicated in the interaction of the homologous $G_{s\alpha}$ with adenylate cyclase (AC) are indicated with pink labels. The space-filling models are colored similarly to the ribbon diagrams except for the yellow residues, which here represent those that appear to propagate or stabilize the structural transitions induced by the binding of the γ phosphate group. The box in the space-filling models outlines the cavity in $G_{t\alpha}\cdot$GDP that closes when the GDP is replaced by GTPγS and that has been implicated in modulating the affinity of $G_{t\alpha}$ for $G_{\beta\gamma}$ and for the receptor. [Courtesy of Paul Sigler, Yale University. PDBids 1TAG and 1TND.]

with GDP (Fig. 19-17*a,b*) and with the nonhydrolyzable GTP analog **GTPγS**

(Fig. 19-17*c,d*) were determined by Heidi Hamm and Paul Sigler. $G_{t\alpha}$ consists of two clearly delineated domains connected by two polypeptide linkers: (1) a highly conserved GTPase domain that is structurally similar to those in other G proteins of known structure (and hence is often described as a Ras-like domain), and (2) a helical domain that is unique to heterotrimeric G proteins. Guanine nucleotides bind to $G_{t\alpha}$ in a deep cleft that is flanked by these domains. The X-ray structures of $G_{i\alpha}\cdot$GTPγS and

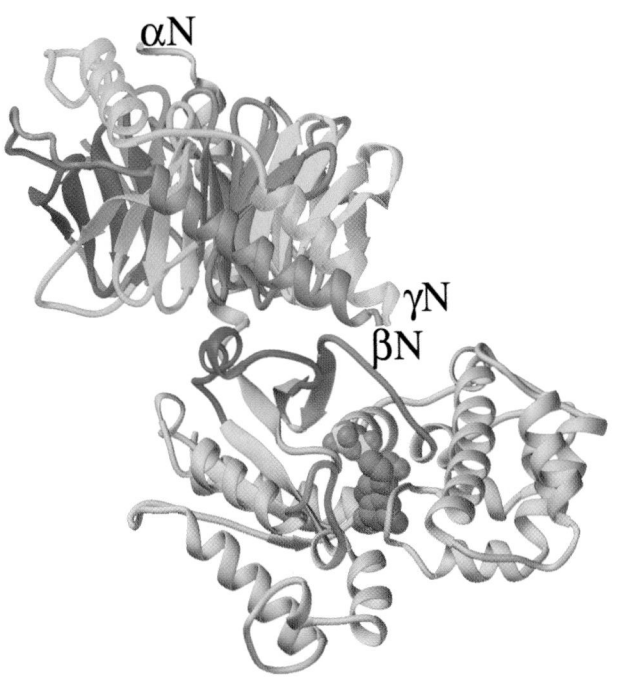

αN

γN

βN

(a)

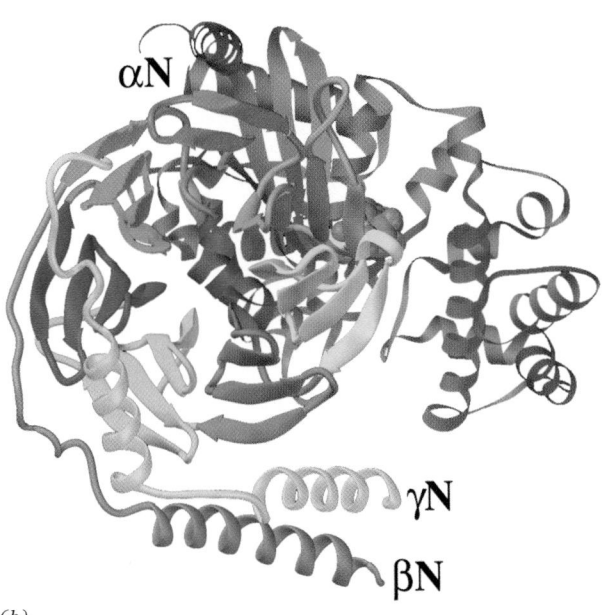

αN

γN

βN

(b)

FIGURE 19-18   X-Ray structure of the heterotrimeric G protein G$_i$. (*a*) The G$_\alpha$ subunit is violet with its Switch I, II, and III segments green, blue, and red, respectively, and with its bound GDP shown in space-filling form with C green, N blue, O red, and P yellow. The G$_\beta$ subunit's N-terminal segment is light blue and each blade of its β-propeller has a different color. The G$_\gamma$ subunit is gold. The view is perpendicular to the axis of the G$_\beta$ subunit's β-propeller. The plasma membrane would be at the top of the drawing as inferred from the positions of the N terminus of G$_\alpha$ and the neighboring C terminus of G$_\gamma$, which,

in vivo, are lipid-linked to the plasma membrane. However, the orientation of the protein relative to the plasma membrane is unknown. (*b*) View related to that in Part *a* by a 90° rotation about its horizontal axis and thus looking from the general direction of the plasma membrane. The protein is colored as in Part *a* except that the G$_\alpha$ subunit is mainly gray. [Based on an X-ray structure by Alfred Gilman and Stephan Sprang, University of Texas Southwestern Medical Center. PDBid 1GP2.] ✎ See the Interactive Exercises

G$_{s\alpha}$·GTPγS, both determined by Gilman and Stephen Sprang, closely resemble that of the G$_{t\alpha}$·GTPγS.

Comparison of the structures of the G$_{t\alpha}$·GDP and G$_{t\alpha}$·GTPγS complexes reveals that the presence of GTP's γ phosphate group promotes significant conformational shifts in only three so-called switch regions, all of which are located on the facing side of G$_{t\alpha}$ in Fig. 19-17. The γ phosphate hydrogen bonds to side chains on Switches I and II, thereby pulling these polypeptide segments in toward it and causing Switch II to contact Switch III in a way that pulls it to the right (Fig. 19-17). These concerted conformational shifts cause an extensive cavity over the GDP-binding site to largely fill in the GTPγS complex.

Switches I and II have counterparts in other G proteins of known structure. Portions of these polypeptide segments have been implicated in the interactions of G$_{t\alpha}$ with the cGMP-PDE it activates and in the interactions between the closely related G$_{s\alpha}$ with its target AC (Section 19-2D). A question that remains unanswered is how a liganded receptor induces its target G$_\alpha$ subunit to exchange its bound GDP for GTP.

c. The X-Ray Structures of Heterotrimeric G Proteins

The X-ray structures of heterotrimeric G proteins were determined by Gilman and Sprang (G$_i$·GTP; Fig. 19-18)

and by Hamm and Sigler (G$_t$·GDP). These structures reveal that the G$_\beta$ subunit (Fig. 19-18*b*) consists of an N-terminal helical domain followed by a C-terminal domain comprising seven 4-stranded antiparallel β sheets arranged like the blades of a propeller—a so-called **β-propeller**—that surround a water-filled central channel. The β-propeller's repeating β sheets are each formed by a **WD40** sequence motif, so called because they are ~40 residues in length and often contain the conserved dipeptide segment WD. The WD40 motif occurs in a functionally diverse group of 4- to 8-bladed β-propeller proteins, including the 7-bladed N-terminal domain of the clathrin heavy chain (Section 12-4C). The G$_\gamma$ subunit consists mainly of two helical segments joined by a polypeptide link (Fig. 19-18*b*). It is closely associated with G$_\beta$ along its entire extended length through mainly hydrophobic interactions and hence has no tertiary structure. The X-ray structure of isolated G$_{\beta\gamma}$ is essentially identical to that in the G$_\alpha$·GDP–G$_{\beta\gamma}$ complex, thereby indicating that the structure of G$_{\beta\gamma}$ is unchanged by its association with G$_\alpha$·GDP.

G$_\alpha$ and G$_{\beta\gamma}$ associate mainly via highly conserved contacts between the Switch I and II regions of G$_\alpha$ and the loops and turns at the bottom of G$_{\beta\gamma}$'s β-propeller (Fig. 19-18). In addition, there is a less extensive interaction between the N-terminal helix of G$_\alpha$ (which is disordered in

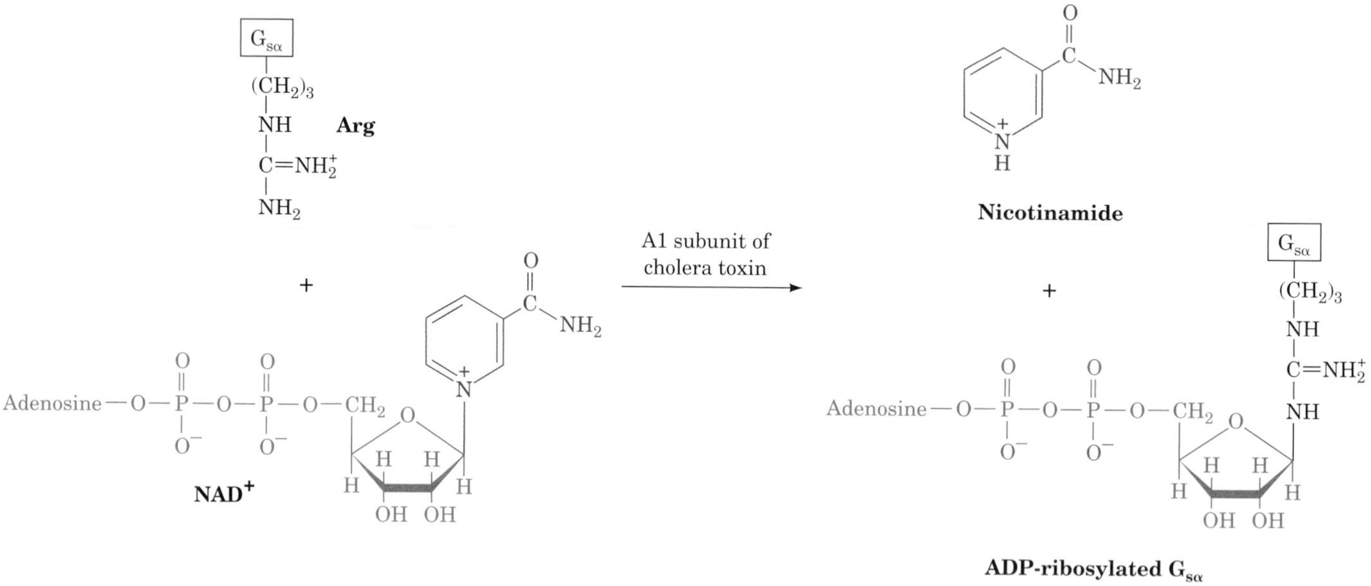

FIGURE 19-19   Mechanism of action of cholera toxin. The cholera toxin's A1 fragment catalyzes the ADP-ribosylation of a specific Arg residue on $G_{s\alpha}$ by NAD^+, thereby rendering this subunit incapable of hydrolyzing GTP.

G_α alone) and the first blade of the G_β propeller (back side of Fig. 19-18*a*). Comparison of the structure of $G_\alpha \cdot GDP$–$G_{\beta\gamma}$ with that of $G_\alpha \cdot GTP\gamma S$ reveals why G_α cannot simultaneously bind GTP and $G_{\beta\gamma}$: *In* $G_\alpha \cdot GDP$–$G_{\beta\gamma}$*, the Switch II segment of* G_α *contacts* G_β *in a way that prevents Switch II from assuming the conformation it requires to bind GTP's* γ *phosphate.* Moreover, the conformation changes in Switch II are coordinated with those in Switch I so that, together, they close over the GDP bound to G_α–$G_{\beta\gamma}$, thereby accounting for its tight binding relative to that in $G_\alpha \cdot GDP$.

d.   Cholera Toxin Stimulates Adenylate Cyclase by Permanently Activating $G_{s\alpha}$

The major symptom of **cholera,** an intestinal disorder caused by the bacterium *Vibrio cholerae,* is massive diarrhea that, if untreated, frequently results in death from dehydration. This dreaded disease is not an infection in the usual sense since the vibrio neither invades nor damages tissues but merely colonizes the intestine, as does *E. coli.* The catastrophic fluid loss that cholera induces (often over 6 liters per hour!) occurs in response to a bacterial toxin. Indeed, merely replacing cholera victims' lost water and salts enables them to survive the few days necessary to immunologically eliminate the bacterial infestation.

Cholera toxin (CT; also known as **choleragen)** is an 87-kD protein of subunit composition AB_5 in which the B subunits (103 residues each) form a pentagonal ring to which the A subunit (240 residues) is bound. Previous to CT's secretion, its A subunit is cleaved at a single site by a bacterial protease to yield two fragments, A1 (the N-terminal ~195 residues) and A2 (the C-terminal ~45 residues), that remain joined by a disulfide bond. On the binding of CT to its cell surface receptor, ganglioside G_{M1} (Sections 12-1D and 25-8C), the nicked A subunit (but not the B subunits) is taken into the cell via receptor-mediated

endocytosis and travels backward through the secretory pathway (Section 12-4B) to the Golgi apparatus. From there, it is conducted into the endoplasmic reticulum (ER) via the binding of A2's C-terminal KDEL sequence to a KDEL receptor (which normally functions to retrieve ER-resident proteins that have escaped the ER; Section 12-4C). The A1 fragment is then released from A2 and enters the cytoplasm through the translocon (which normally conducts growing and still unfolded polypeptides into the ER; Section 12-4B) via a process in which A1 is unfolded through the chaperonelike action of protein disulfide isomerase (PDI; Section 9-2A).

Once inside the cell, A1 catalyzes the irreversible transfer of the ADP–ribose unit from NAD^+ to a specific Arg side chain of $G_{s\alpha}$ (Fig. 19-19). *ADP-ribosylated* $G_{s\alpha} \cdot GTP$ *can activate AC but is incapable of hydrolyzing its bound GTP.* As a consequence, the AC remains "locked" in its active state. The epithelial cells of the small intestine normally secrete digestive fluid (an HCO_3^--rich salt solution) in response to small [cAMP] increases that activate intestinal Na^+ pumps through their phosphorylation by PKA (ion pumps are discussed in Sections 20-4 and 20-5). The ~100-fold rise in intracellular [cAMP] induced by CT causes these epithelial cells to pour out enormous quantities of digestive fluid, thereby producing the symptoms of cholera. CT also affects other tissues *in vitro* but does not do so *in vivo* because CT is not absorbed from the gut into the bloodstream.

The remarkable X-ray structure of CT (Fig. 19-20*a*), determined by Graham Shipley and Edwin Westbrook, reveals that its A2 segment forms an unusual extended helix whose C-terminal end inserts into the B_5 pentamer's solvent-filled central pore, where it is noncovalently anchored. The N-terminal segment of A2 extends beyond the B_5 pentamer so as to tether the wedge-shaped A1 segment to B_5, much like a balloon on a string. The X-ray structure

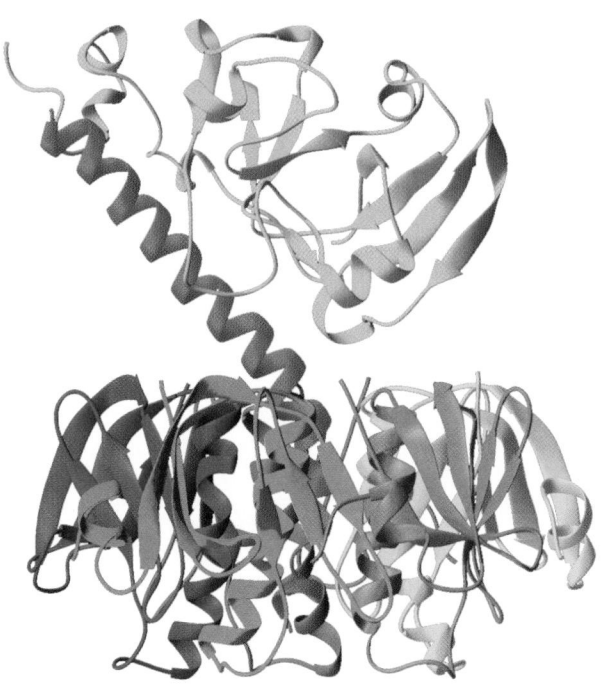

(a)

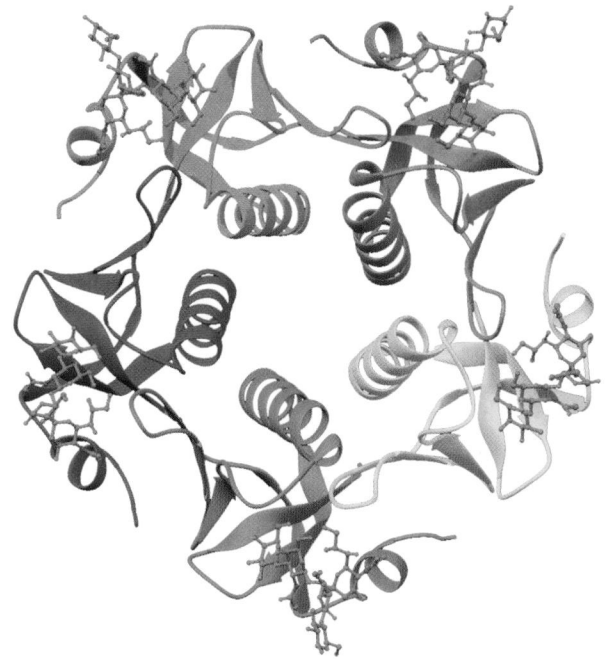

(b)

FIGURE 19-20   X-Ray structure of cholera toxin. (*a*) The entire AB$_5$ complex as viewed parallel to the presumed direction of the plane of the plasma membrane to which it binds, exterior up. The A1 segment is cyan, the A2 segment is gray, and each B subunit has a different color. Although the A1 and A2 segments in this structure form a continuous polypeptide chain, residues 193–195, which immediately precede the peptide bond that is cleaved on toxin activation, are disordered and hence not visible here (*upper left*). The C-terminal end of the A2 helix binds in the pentamer's central pore. [Based on an X-ray structure by Graham Shipley, Boston University School of Medicine, and Edwin Westbrook, Northwestern University. PDBid 1XTC.] (*b*) The structure of only the B$_5$ pentamer in which each subunit is binding CT's G$_{M1}$ receptor pentasaccharide. The structure is viewed as from the bottom of Part *a*. The subunits of the B$_5$ pentamer are colored as in Part *a* and the pentasaccharides are shown in ball-and-stick form with C green, N blue, and O red. Note the pentamer's large central pore. [Based on an X-ray structure by Wim Hol, University of Washington. PDBid 2CHB.]

of only B$_5$ in complex with the pentasaccharide from its G$_{M1}$ receptor (Fig. 19-20*b*), determined by Wim Hol, indicates that this pentasaccharide binds through an extensive hydrogen bonded network to each B subunit on the face of B$_5$ opposite that which binds A. The binding of the A subunit or the receptor pentasaccharide to B$_5$ causes only modest structural changes at their respective binding sites without altering B$_5$'s subunit interfaces. A1 contains an elongated crevice in the vicinity of a catalytically implicated residue, Glu 112, that presumably forms its active site.

Certain strains of *E. coli* cause a diarrheal disease (travelers' diarrhea) similar to, although considerably less severe, than cholera, through their production of **heat-labile enterotoxin (LT),** a protein that closely resembles CT (their A and B subunits are >80% identical and form AB$_5$ toxins that have closely similar X-ray structures) and has the same mechanism of action. The reasons for the difference in severity of these infections are unclear (cholera can be fatal within hours, whereas enterotoxic strains of *E. coli* usually only temporarily incapacitate adults). It might be due to the modest structural differences between the toxins, differences in the amounts of toxin secreted, and/or variations in microbial ecology.

The foregoing results provide a structural basis for the design of ligands that interfere with the binding of CT and LT to their receptors. Since these receptors occur on the surface of the intestinal epithelium, ligands that compete with them need not pass through any membrane. This greatly increases the usual ~500-D size limit for an effective drug candidate (Section 15-4B). Moreover, a large ligand is unlikely to enter the bloodstream and hence would have minimal side effects. Consequently, the synthesis of multivalent ligands that simultaneously bind with high affinity to all 5 receptor binding sites on an AB$_5$ molecule has yielded promising lead compounds against CT and LT.

e. Pertussis Toxin ADP-Ribosylates G$_{i\alpha}$

Bordetella pertussis, the bacterium that causes **pertussis** (whooping cough; a disease that is still responsible for ~300,000 infant deaths per year worldwide), produces an AB$_5$ protein, **pertussis toxin (PT),** that ADP-ribosylates a specific Cys residue in G$_{i\alpha}$. In doing so, it prevents G$_{i\alpha}$ from exchanging its bound GDP for GTP and therefore from inhibiting AC. PT's X-ray structure, determined by Randy Read, reveals that its A and B subunits are structurally homologous to those of CT and LT, although the A subunit of PT extends from the opposite face of its B pentamer rel-

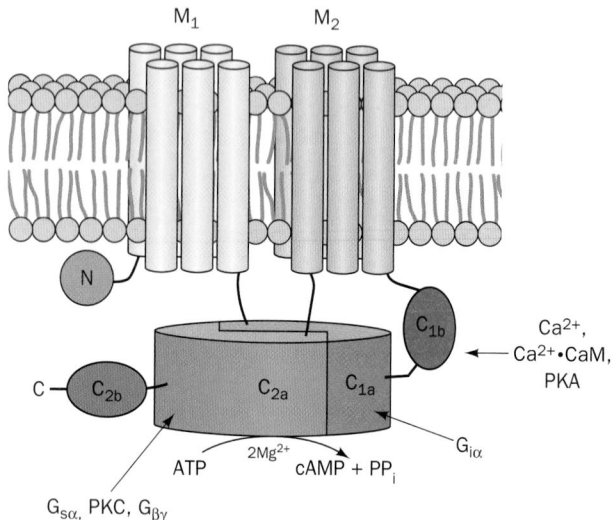

FIGURE 19-21 Schematic diagram of a typical mammalian AC. The M_1 and M_2 domains are each predicted to contain 6 transmembrane helices. C_{1a} and C_{2a} form the enzyme's pseudosymmetric catalytic core. The domains with which various regulatory proteins are known to interact are indicated. [After Tesmer, J.J.G. and Sprang, S.R., *Curr. Opin. Struct. Biol.* **8,** 713 (1998).]

ative to that in CT. Moreover, PT's B pentamer consists of 4 different subunits (one in two copies), each of which is only ~15% identical to the B subunits of CT and LT.

D. *Adenylate Cyclases*

The heterotrimeric G proteins G_s and G_i function to control the activities of adenylate cyclase (AC). In fact, mammals have 10 known isoforms of AC, AC1 through AC10 (alternatively, AC-I through AC-X), which are each expressed in a tissue-specific manner and differ in their regulatory properties. These ~120-kD transmembrane glycoproteins each consist of a small N-terminal domain (N), followed by 2 repeats of a unit consisting of a transmembrane domain (M) followed by 2 consecutive cytoplasmic domains (C), thus forming the sequence $NM_1C_{1a}C_{1b}M_2C_{2a}C_{2b}$ (Fig. 19-21). The ~40% identical C_{1a} and C_{2a} domains associate to form the AC's catalytic core, whereas C_{1b}, as well as C_{1a} and C_{2a}, bind regulatory molecules. Thus, $G_{i\alpha}$ inhibits AC1, 5, and 6 by binding to C_{1a}; $G_{s\alpha}$ activates all AC isoforms but AC9 by binding to C_{2a}; $G_{\beta\gamma}$ inhibits AC1 but activates AC2, 4, and 7 by binding to C_{2a}; and Ca^{2+}–calmodulin (Ca^{2+}–CaM; Section 18-3C) activates AC1, 3, and 8 by binding to C_{1b}. Moreover, the C_{2a} of AC2, 5, and 7 are activated by the phosphorylation of specific Ser/Thr control sites, for example, by **protein kinase C (PKC;** Section 19-4C), whereas the C_{1b} of AC5 and 6 are similarly inhibited by PKA, for example. Clearly, cells can respond to a great variety of stimuli in determining their cAMP levels.

No X-ray structure of an intact AC isoform has yet been reported. However, Sprang has determined the X-ray structure of a hybrid catalytic core consisting of the C_{1a} do-

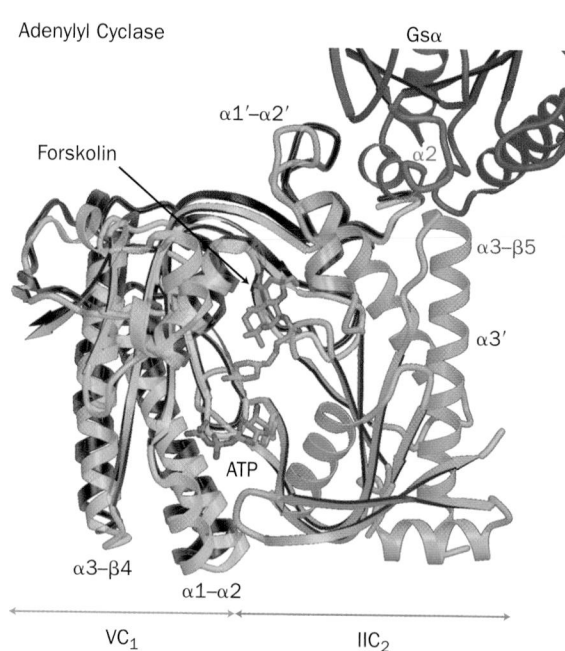

FIGURE 19-22 The X-ray structure of an AC catalytic core. This core consists of dog VC_1 and rat IIC_2 in complex with bovine $G_{s\alpha}$·GTPγS, forskolin, and is shown with a model of ATP. VC_1 is tan, IIC_2 is violet, and $G_{s\alpha}$, which is only shown in part, is gray, with its IIC_2-contacting segments, Switch II and the α3–β5 loop, highlighted in red and blue. The forskolin and the ATP are shown in stick form with C gray, N cyan, O red, and P green. The transparent maroon ribbon shows the nonoverlapping portions of the catalytically inactive rat IIC_2 homodimer in which one of its subunits is superimposed on IIC_2 in the VC_1–IIC_2 complex. [Courtesy of Heidi Hamm, Northwestern University Medical School. The X-ray structures of the VC_1–IIC_2–$Gs_{s\alpha}$ complex and the IIC_2 homodimer were determined by John Tesmer and Stephen Sprang, University of Texas Southwestern Medical Center, and by James Hurley, NIH. PDBids 1AZS and 1AB8.]

main of AC5 (VC_1) and the C_{2a} domain of AC2 (IIC_2) in complex with $G_{s\alpha}$·GTPγS, **ATPαS** (an isomer of ATPγS with the S atom on the α phosphate), and **forskolin**

Forskolin

(a product of the plant *Coleus forskohlii* that activates all ACs but AC9 and functions to lower blood pressure). The VC_1·IIC_2 catalytic core is enzymatically active and is sensitive to both $G_{s\alpha}$·GTP and forskolin. Its X-ray structure (Fig. 19-22) reveals that VC_1 and IIC_2 form a pseudosymmetric heterodimer that binds ATPγS and forskolin at pseudosymmetrically related sites of their interface.

$G_{s\alpha} \cdot GTP\gamma S$ interacts with IIC_2 mainly via its Switch II helix, which binds in a cleft on IIC_2.

The X-ray structure of the catalytically inactive C_{2a} homodimer in complex with two symmetrically arranged forskolin molecules, by James Hurley, provides a perhaps crude model for the unactivated heterodimer. Comparison of these structures (Fig. 19-22) suggests that the binding of $G_{s\alpha} \cdot GTP$ to the $C_{1a} \cdot C_{2a}$ catalytic core pries open the Switch II binding cleft on C_{2a} in a way that mechanically forces C_{1a} to rotate ~10° with respect to C_{2a}. This is postulated to reorient the complex's active site residues such that they can efficiently catalyze the conversion of ATP to cAMP. The conformation change that $G_{s\alpha}$ undergoes on hydrolyzing its bound GTP to GDP (Fig. 19-17) apparently reorients its Switch II region such that it no longer can bind to C_{2a}, thereby causing AC to revert to its inactive conformation.

VC_1 has a cleft that corresponds to the $G_{s\alpha}$-binding cleft on IIC_2. This suggests that this cleft on VC_1 provides the binding site for $G_{i\alpha}$. Indeed, mutagenesis studies on VC_1 are consistent with this hypothesis. However, the cleft on VC_1 is too narrow to accommodate the binding of a Switch II helix. This further suggests that the binding of $G_{i\alpha} \cdot GTP$ to C_{1a} pries open this cleft in a way that reorients the complex's catalytic residues so as to reduce its catalytic activity.

E. *Phosphodiesterases*

In any chemically based signaling system, the signal molecule must eventually be eliminated in order to control the amplitude and duration of the signal and to prevent interference with the reception of subsequent signals. In the case of cAMP, this second messenger is hydrolyzed to AMP by enzymes known as **cAMP-phosphodiesterases (cAMP-PDEs).**

The PDE superfamily, which includes both cAMP-PDEs and cGMP-PDEs, is encoded in mammals by at least 20 different genes grouped into 11 families (PDE1 through PDE11). Moreover, many of the mRNAs transcribed from these genes have alternative initiation sites and alternative splice sites (Sections 5-4A and 34-3C), so that mammals express ~50 PDE isoforms. These are functionally distinguished by their substrate specificities (for cAMP, cGMP, or both) and kinetic properties, their responses (or lack of them) to various activators and inhibitors (see below), and their tissue, cellular, and subcellular distributions. The PDEs have characteristic modular architectures with a conserved ~270-residue catalytic domain near their C-termini and widely divergent regulatory domains or motifs, usually in their N-terminal portions. Some PDEs are membrane-anchored, whereas others are cytosolic.

PDE activity, as might be expected, is elaborately controlled. Depending on its isoform, a PDE may be activated by one or more of a variety of agents including Ca^{2+}–CaM; phosphorylation by PKA, insulin-stimulated protein kinase (Section 20-3C), and **calmodulin-dependent protein kinase II**; and the binding of cGMP to a noncatalytic site.

However, for some PDEs, cGMP is inhibitory. Phosphorylated PDEs are dephosphorylated by a variety of protein phosphatases including Ca^{2+}–CaM-dependent phosphatase and **protein phosphatase-2A.** Thus, the PDEs provide a means for "cross talk" between cAMP-based signaling systems and those using other types of signals.

PDEs are inhibited by a variety of drug agents that influence such widely divergent disorders as asthma, congestive heart failure, depression, erectile dysfunction, inflammation, and retinal degeneration. **Sildenafil** (trade name **Viagra**),

Sildenafil (Viagra™)

a compound used to treat erectile dysfunction, specifically inhibits PDE5, which hydrolyzes only cGMP. Sexual stimulation in males causes penile nerves to release NO, which activates guanylate cyclase to produce cGMP. This induces vascular smooth muscle relaxation in the penis, thereby increasing the inflow of blood, which results in an erection. This cGMP is eventually hydrolyzed by PDE5. Sildenafil is therefore an effective treatment in men who produce insufficient NO and hence cGMP to otherwise generate a satisfactory erection.

3 ■ TYROSINE KINASE–BASED SIGNALING

📖 See Guided Exploration 17. Mechanisms of Hormone Signaling Involving the Receptor Tyrosine Kinase System We have seen that glycogen synthesis and breakdown are regulated by the phosphorylation/dephosphorylation of the enzymes that catalyze these metabolic processes as well as of many of the enzymes that catalyze these modification/demodification processes (Section 18-3). Numerous other eukaryotic processes are similarly regulated. In fact, nearly one-third of eukaryotic proteins are subject to reversible phosphorylation, and the human genome is estimated to encode ~2000 protein kinases (the Protein Kinase Resource at http://pkr.sdsc.edu/html/index.shtml is a searchable compendium on protein kinases). The vast majority of the phosphorylated amino acid residues are Ser or Thr; only about 1 in 2000 is Tyr. Nevertheless, as we discuss in this section, Tyr phosphorylation is of central importance in regulating a variety of essential cellular processes.

A. *Receptor Tyrosine Kinases*

Many protein growth factors variously control the differentiation, proliferation, migration, metabolic state, and survival of their target cells by binding to their cognate **receptor tyrosine kinases (RTKs).** The RTKs form a diverse family of over 50 transmembrane glycoproteins (Fig. 19-23) that each have a C-terminal cytoplasmic **protein tyrosine kinase (PTK)** domain and a single-pass transmembrane segment that is presumably an α helix. As their name indicates, PTKs catalyze the ATP-dependent phosphorylation of their target proteins at specific Tyr residues:

The PTK domains of RTKs are homologous to and, as we

shall see, structurally resemble the far more abundant Ser/Thr-specific protein kinases such as PKA (Fig. 18-14).

RTKs are activated by the binding of a cognate protein growth factor to one or more of their extracellular domains. It seems unlikely that the single transmembrane helix of a monomeric RTK such as **epidermal growth factor receptor (EGFR;** Fig. 19-23) has the structural complexity to transmit the ligand-binding state of its extracellular domain(s) to its cytoplasmic tyrosine kinase domain. Rather, as we have seen for human growth hormone receptor (which is not an RTK), ligand binding induces receptor dimerization (Fig. 19-9). This, in turn, activates the RTK's PTK activity, as we discuss below. For RTKs that are permanent dimers, such as the **insulin receptor (InsR;** Fig. 19-23), the PTK is thought to be activated by a ligand-induced structural change (probably a counter-rotation of the two protomers that preserves the dimer's 2-fold axis of symmetry) that is transmitted across the membrane.

a. FGF and Heparin Sulfate Are Required to Activate FGF Receptor

The mammalian **fibroblast growth factors (FGFs)** form a family of at least 21 structurally related proteins (FGF1–21) that regulate a variety of critical biological

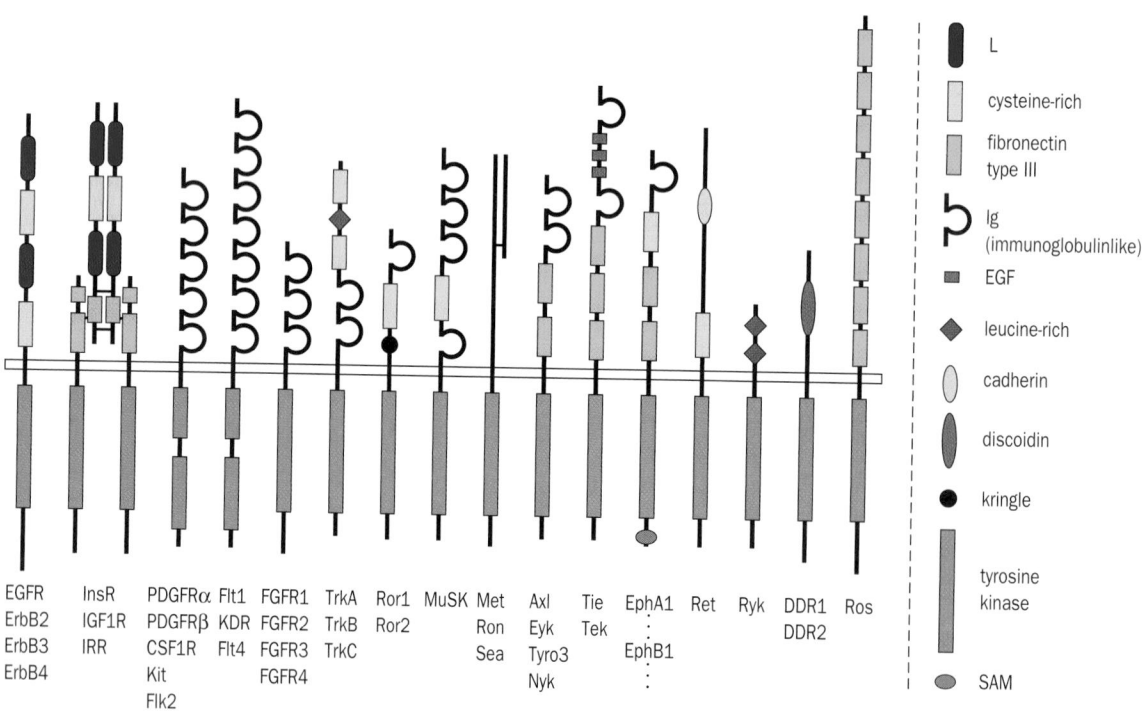

FIGURE 19-23   Domain organization in a variety of receptor tyrosine kinase (RTK) subfamilies. One to five members of each subfamily are depicted. The narrow rectangle that extends horizontally across the diagram represents the plasma membrane, with the extracellular region above it and the cytoplasm below it. The polypeptides are shown only approximately to scale, with their N-termini above. **EGFR, InsR, PDGFR,** and **FGFR** refer to **epidermal growth factor receptor, insulin receptor, platelet-derived growth factor receptor,** and **fibroblast growth factor receptor,** respectively. The RTKs' extracellular portions are modularly constructed from a variety of often repeating domains that are identified at the right of the diagram. Note that the tyrosine kinase domains of PDGFR and **Flt1** subfamiles are interrupted by ~100-residue **kinase inserts** and that the members of the InsR subfamily are α2β2 heterotetramers, whose subunits are disulfide-linked (short horizontal lines). [Courtesy of Stevan Hubbard, New York University School of Medicine.]

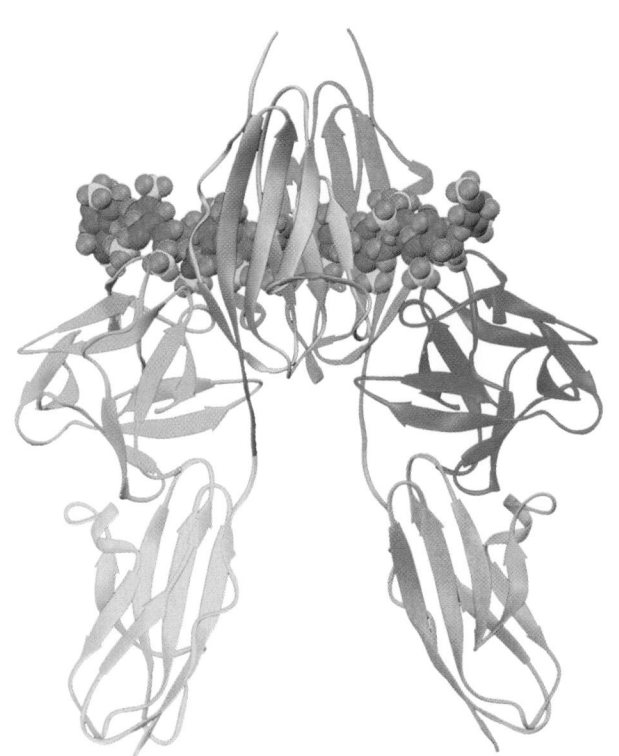

FIGURE 19-24 **The X-ray structure of the 2:2:2 complex of FGF2, the D2–D3 portion of FGFR1, and a heparin decasaccharide** as viewed normal to the dimeric complex's two-fold axis of symmetry, with the plasma membrane below. The polypeptides are represented in ribbon form with the FGF molecules cyan and light blue, the D2 and D3 domains of one FGFR monomer lavender and gold, and those of the other magenta and orange. The two heparin decasaccharides are shown in space-filling form with C green, N blue, O red, and S yellow. The D2 and D3 domains of FGFR each have the characteristic structure of the **immunoglobulin (Ig)** domain: a β sandwich comprised of a three-stranded and a 4-stranded antiparallel β sheet (Section 35-2B). [Based on an X-ray structure by Moosa Mohammadi, New York University School of Medicine. PDBid 1FQ9.]

processes including cell growth, differentiation, and migration and that are expressed in specific spatial and temporal patterns in embryos and adults. FGF-stimulated processes are mediated by four FGF receptors (FGFR1–4), which each bind a unique subset of the FGFs, thereby accounting for the diversity and tight regulation of the foregoing processes. FGFR dimerization in solution requires the presence of heparan sulfate proteoglycans (Section 11-3A) in addition to that of FGF.

FGF receptors each consist of, from N- to C-terminus (Fig. 19-23), three extracellular immunoglobulinlike domains (D1–D3 for domains 1–3), a single transmembrane helix, and a cytoplasmic domain with PTK activity. Of these, only the D2 and D3 domains are involved in FGF binding (in general, only a few of the domains in the extracellular portions of RTKs participate in ligand binding). Moosa Mohammadi determined the X-ray structure of the 2:2:2 complex of FGF2, the D2–D3 segment of FGFR1,

and a heparin decasaccharide (Fig. 11-20). It reveals (Fig. 19-24) that each FGF monomer binds to the D2 and D3 domain on one FGFR subunit and, more tenuously, to the D2 domain on the other subunit, whereas the heparin cross-links each FGF monomer to both D2 domains (whose contacts in the absence of FGF and heparin are insufficient to support appreciable FGFR dimerization).

b. RTK Dimers Are Activated by Autophosphorylation

The dimerization of an RTK (or its conformation change in the case of the insulin receptor subfamily) brings its cytoplasmically located PTK domains into apposition, such that they cross-phosphorylate each other on specific Tyr residues on their activation loops (Fig. 19-25a). This **autophosphorylation** activates the PTK in much the same way as we saw that activation loop phosphorylation induces PKA to phosphorylate its target proteins (Section 18-3C). In many cases, the activated PTK further phosphorylates the opposing RTK subunit at specific Tyr residues outside of the PTK domain (Fig. 19-25). This, as we shall see in Section 19-3C, provides binding sites for certain cytoplasmic proteins. The activated PTK may also phosphorylate specific Tyr residues on a variety of cytoplasmic proteins. In both cases, as we discuss in Section 19-3D, this causes the activation of the proteins that participate in executing the instructions implied by the extracellular presence of the protein growth factor.

c. The Insulin Receptor PTK Undergoes Major Conformation Changes on Autophosphorylation

How does autophosphorylation activate a PTK? The comparison of the X-ray structures of the PTK domain of the insulin receptor in its inactive unphosphorylated and active triphosphorylated states has done much to answer this question. The insulin receptor is expressed as a single 1382-residue precursor peptide that is proteolytically processed to yield the disulfide-linked α and β subunits (731 and 619 residues) of the mature receptor (Fig. 19-23). The X-ray structure of a 306-residue PTK-containing segment of the β subunit that was phosphorylated at its three autophosphorylation sites, Tyr residues 1158, 1162, and 1163 (using the precursor numbering system), and in complex with the nonhydrolyzable ATP analog ADPNP and an 18-residue peptide substrate, was determined by Stevan Hubbard. It reveals (Fig. 19-26a) that this PTK structurally resembles other PTKs of known structure as well as protein Ser/Thr kinases such as PKA (Fig. 18-14) and the phosphorylase kinase γ subunit (Fig. 18-20).

Comparison of this structure with that of the unphosphorylated and uncomplexed protein, determined by Hubbard and Wayne Hendrickson, reveals that, on phosphorylation and substrate binding, the PTK's N-terminal lobe undergoes a nearly rigid 21° rotation relative to the C-terminal lobe about the long axis of the protein (Fig. 19-26b). This dramatic conformational change closes the active site cleft about the ADPNP and, presumably, correctly positions critical residues for substrate binding and catalysis. All three phosphorylated Tyr residues occur on the PTK's activation loop (residues 1149–1170; recall that

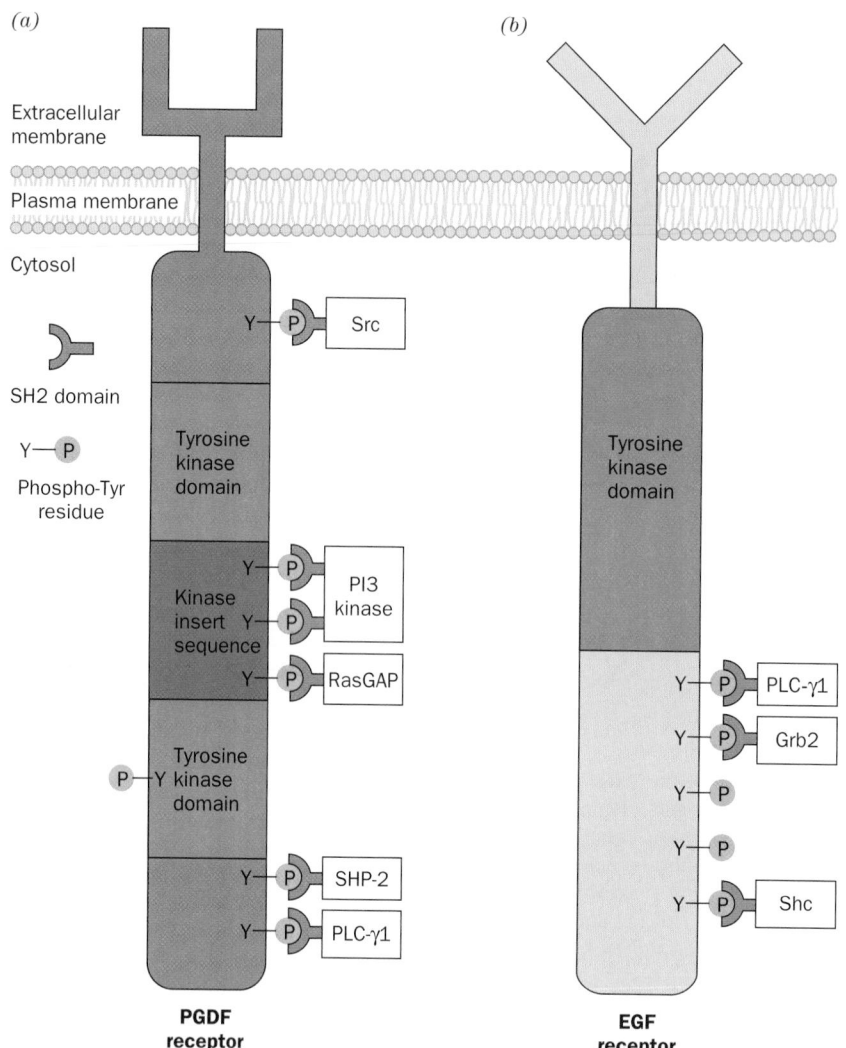

FIGURE 19-25   Schematic diagrams of RTKs. (*a*) The PGDF receptor and (*b*) the EGF receptor. Their autophosphorylation sites and the proteins that are activated by binding to these sites via their SH2 domains (all of which are discussed in this chapter) are indicated. Note that almost all of the autophosphoryated Tyr residues that bind to other proteins lie outside the tyrosine kinase domains. [After Pawson, T. and Schlessinger, J., *Curr. Biol.* **3**, 435 (1993).]

many Ser/Thr kinases are also phosphorylated on their activation loops; Section 18-3C). The unphosphorylated activation loop threads through the PTK's active site so as to prevent the binding of both ATP and protein substrates. However, on phosphorylation, the activation loop assumes a conformation that does not occlude the active site (Fig. 19-26*b*) but instead forms part of the substrate recognition site. The phosphate group on Tyr 1163 bridges the activation loop by forming a hydrogen bond with the side chain of the conserved Arg 1155 on one side of the loop and with the main chain N of Gly 1166 on its other side. The phosphate group of Tyr 1162 makes two hydrogen bonds to the side chain of the conserved Arg 1164. However, the phosphate group of Tyr 1168 makes no protein contacts, which suggests that it forms a docking site for downstream signaling proteins (see below). These observations are in agreement with experiments indicating that the tyrosine

kinase activity of the insulin receptor increases with the degree of phosphorylation at its 3 autophosphorylatable Tyr side chains and that full activity is not achieved until Tyr 1163 is phosphorylated. Indeed, nearly all known RTKs have between one and three autophosphorylatable Tyr residues in their activation loops (a major exception being the members of EGFR subfamily; Fig 19-25*b*) and, in all phosphorylated protein kinases of known structure, assume similar conformations.

Only the 6 centrally located residues, GDYMNM, of the 18-residue substrate peptide are seen in the foregoing X-ray structure. These include the YMXM sequence found in all efficient substrates of the insulin receptor. This segment associates with the kinase as a strand in a β sheet. Its Met side chains fit into adjacent hydrophobic pockets of the protein, whereas its phosphorylatable Tyr side chain extends toward the γ phosphate group of the ADPNP (Fig.

(a)

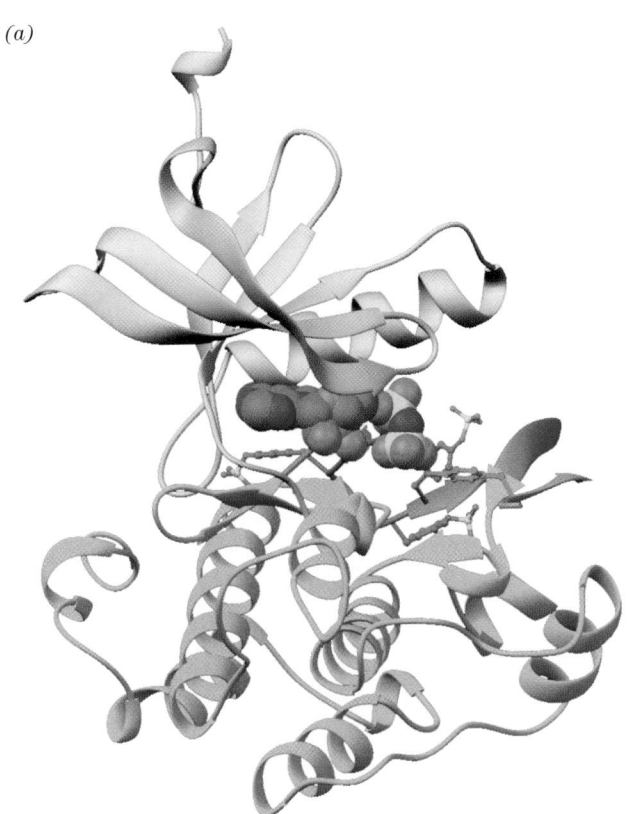

(b)

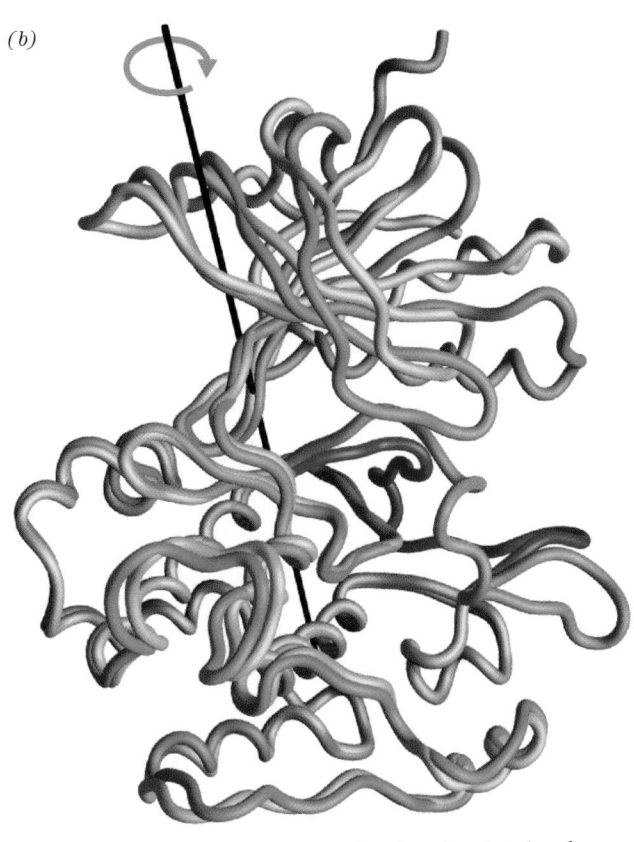

FIGURE 19-26   X-Ray structure of the PTK domain of the insulin receptor. (*a*) The PTK phosphorylated at Tyr residues 1158, 1162, and 1163 and in complex with ADPNP and an 18-residue polypeptide substrate. The PTK is shown in the "standard" protein kinase orientation with its N-terminal domain pink, its C-terminal domain cyan, and its activation loop light blue. Its three phosphorylated Tyr side chains are shown in ball-and-stick form with C green, N blue, O red, and P yellow. The ADPNP, which is identically colored, is shown in space-filling form. The substrate polypeptide is orange (only 6 of its residues are visible) and its phosphorylatable Tyr residue is shown in ball-and-stick form with C magenta and O red.

(*b*) The polypeptide backbones of the phosphorylated and unphosphorylated forms of the insulin receptor PTK domain are superimposed on their C-terminal lobes. The phosphorylated protein is green with its activation loop blue, and the unphosphorylated protein is yellow with its activation loop red. The axis (*black*) and arrow (*blue*) indicate the rotation required to align the N-terminal domain of the unphosphorylated protein with that of the phosphorylated protein. [Part *a* based on an X-ray structure by and Part *b* courtesy of Stevan Hubbard, New York University Medical School. PDBids 1IR3 for the phosphorylated protein and 1IRK for the unphosphorylated protein.] 🖱 **See the Interactive Exercises**

19-26*a*). The specificity of the protein for phosphorylating Tyr rather than Ser or Thr is explained by the observation that the side chain of Tyr, but not those of Ser or Thr, are long enough to reach the active site.

B. *Cancer: The Loss of Control of Growth*

Before we continue our discussion of signaling pathways, let us consider cancer, a group of diseases that is characterized by defects in signal transduction causing uncontrolled growth. Indeed, studies of cancer have greatly increased our understanding of signal transduction and vice versa.

The cells of the body normally remain under strict developmental control. For instance, during embryogenesis, cells must differentiate, proliferate, migrate, and even die

in the correct spatial arrangement and temporal sequence to yield a normally functioning organism. In the adult, the cells of certain tissues, such as the intestinal epithelium, the blood-forming tissues of the bone marrow, and those of hair follicles, continue to proliferate. Most adult body cells, however, have permanently ceased doing so.

Cells occasionally lose their developmental controls and commence excessive proliferation. The resulting tumors can be of two types:

1. Benign tumors, such as warts and moles, grow by simple expansion and often remain encapsulated by a layer of connective tissue. Benign tumors are rarely life threatening, although if they occur in an enclosed space such as in the brain or secrete large amounts of certain hormones, they can be lethal.

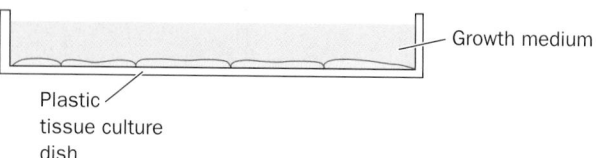

FIGURE 19-27 Growth pattern of vertebrate cells in culture. (*a*) Normal cells stop growing through contact inhibition once they have formed a confluent monolayer. (*b*) In contrast, transformed cells lack contact inhibition; they pile up to form a multilayer.

2. Malignant tumors or **cancers** grow in an invasive manner and shed cells that, in a process known as **metastasis,** colonize new sites in the body. Malignant tumors are almost invariably life threatening; they are responsible for 20% of the mortalities in the United States.

The most obvious and the medically most significant property of cancer cells is that they proliferate uncontrollably. For instance, when grown in a tissue culture dish, normal cells form a monocellular layer on the bottom of the dish and then, through a process termed **contact inhibition,** cease dividing (Fig. 19-27*a*). In contrast, the growth of malignant cells is unhampered by intercellular contacts; in culture they form multicellular layers (Fig. 19-27*b*). Moreover, even in the absence of contact inhibition, normal cells are far more limited in their capacity to reproduce than are cancer cells. Normal cells, depending on the species and age of the animal from which they were taken, will only divide in culture 20 to 60 times before they reach **senescence** (a stage at which they cease dividing; Section 30-4D) and die (a phenomenon that, no doubt, is at the heart of the aging process). *Cancer cells, on the other hand, are immortal; there is no limit to the number of times they can divide.* In fact, some cancer cell lines have been maintained in culture through thousands of divisions spanning over 5 decades. Immortal cells, however, are not necessarily malignant: *The hallmark of cancer is immortality combined with uncontrolled growth.*

a. Cancer Is Caused by Carcinogens, Radiation, and Viruses

Most cancers are caused by agents that damage DNA or interfere with its replication or repair. These include a great variety of man-made and naturally occurring substances known as **chemical carcinogens** (Section 30-5F), as well as radiation, both electromagnetic and particulate, with sufficient energy to break chemical bonds. In addition, *certain viruses induce the formation of malignant tumors in their hosts (see below).*

Almost all malignant tumors result from the **transformation** of a single cell [conversion to the cancerous state;

this term should not be confused with the acquisition of genetic information from exogenously supplied DNA (Section 5-2A)], which, then being free of its normal developmental constraints, proliferates. Yet, considering, for example, that the human body consists of around 10^{14} cells, transformation must be a very rare event. One of the major reasons for this, as the age distribution of the cancer death rate indicates (Fig. 19-28), is that *transformation requires a cell or its ancestors to have undergone several independent and presumably improbable carcinogenic changes.* Consequently, exposure to a carcinogen may prime many cells for transformation, but a malignant tumor may not form until decades later when one of these cells suffers a final transforming event.

The viral induction of cancer was first observed in 1911 by Peyton Rous, who demonstrated that cell-free filtrates from certain chicken **sarcomas** (malignant tumors arising from connective tissues) promote new sarcomas in chickens (Fig. 19-29). Although decades were to pass before the significance of this work was appreciated (Rous was awarded the Nobel Prize in 1966 at the age of 85), many other such **tumor viruses** have since been characterized. The **Rous sarcoma virus (RSV),** as are all known RNA tumor viruses, is a retrovirus (an RNA virus that replicates its chromosome by copying it to DNA in a process mediated by a virally encoded reverse transcriptase, inserting the DNA into the host's genome, and then transcribing this DNA). It contains a gene, **v-***src* ("v" for *v*iral, "*src*" for *sarcoma*), which encodes a protein named **v-Src** that mediates host cell transformation. v-*src* has therefore been termed an **oncogene** (Greek: *onkos,* mass or tumor).

What is the origin of v-*src* and what is its viral function? Hybridization studies (Section 5-3C) by Michael Bishop

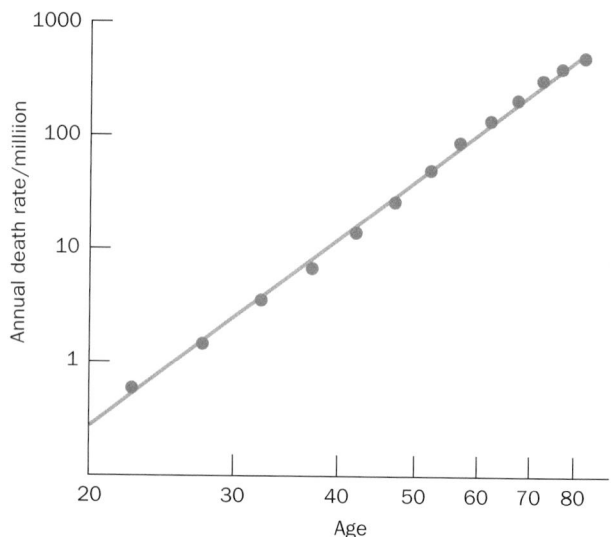

FIGURE 19-28 Variation of the cancer death rate in humans with age. The linearity of this log–log plot can be explained by the hypothesis that several randomly occurring mutations are required to generate a malignancy. The slope of the line suggests that, on the average, five such mutations are required for a malignant transformation.

(a)

(b)

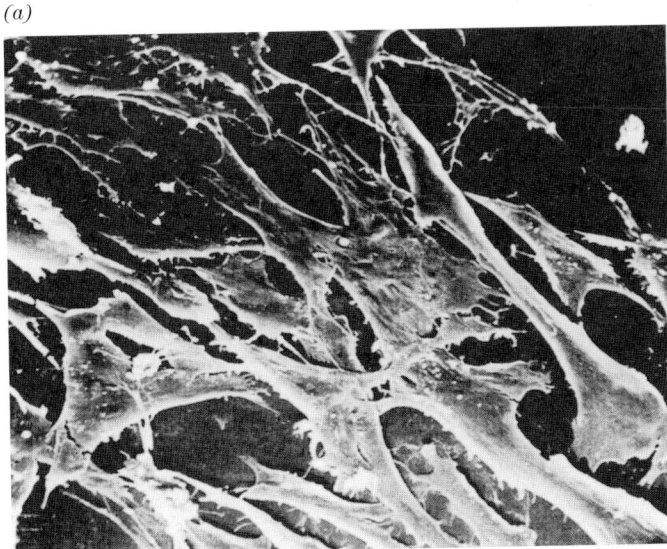

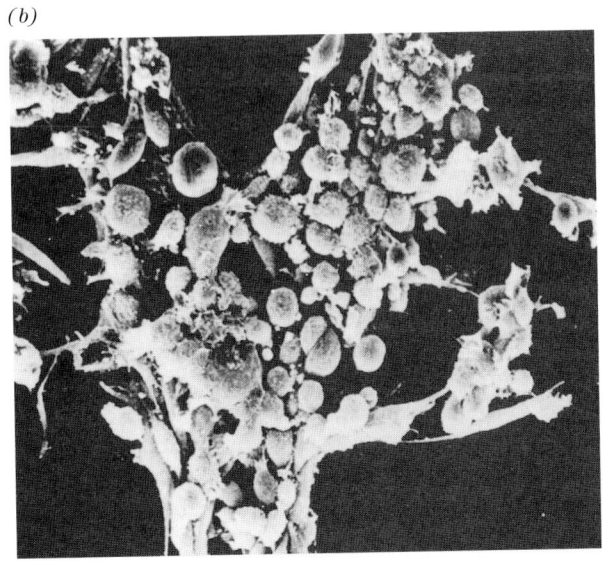

FIGURE 19-29 Transformation of cultured chicken fibroblasts by Rous sarcoma virus. (*a*) Normal cells adhere to the surface of the culture dish, where they assume a flat extended conformation. (*b*) On infection with RVS, these cells become rounded and cluster together in piles. [Courtesy of G. Steven Martin, University of California at Berkeley.]

and Harold Varmus in 1976 led to the remarkable discovery that *uninfected chicken cells contain a gene, **c-src** ("c" for cellular), that is homologous to v-src*. Moreover, c-src is highly conserved in a wide variety of eukaryotes that span the evolutionary scale from *Drosophila* to humans. This observation strongly suggests that c-src, which antibodies directed against v-Src indicated is expressed in normal cells, is an essential cellular gene. In fact, *both v-Src and its normal cellular analog, **c-Src**, function to stimulate cell proliferation* (Section 19-3C). Apparently, v-src was originally acquired from a cellular source by an initially nontransforming ancestor of RSV. By maintaining the host cell in a proliferative state (cells are usually not killed by RSV infection), v-Src presumably enhances the viral replication rate.

b. Viral Oncogene Products Mimic the Effects of Protein Growth Factors and Hormones

The proteins encoded by many viral oncogenes are analogs of various growth factor and hormone system components. For instance:

1. The **v-sis** oncogene of **simian sarcoma virus** encodes a protein secreted by infected cells that is nearly identical to PDGF. Hence, the uncontrolled growth of simian sarcoma virus–infected cells apparently results from the continuous and inappropriate presence of this PDGF homolog.

2. Nearly half of the more than 20 known retroviral oncogenes, including v-src, encode PTKs. For example, the **v-erbB** oncogene specifies a truncated version of the EGF receptor (Fig. 19-25*b*) that lacks the EGF-binding domain but retains its transmembrane segment and its protein kinase domain. *Evidently, oncogene-encoded PTKs inap-*

propriately phosphorylate the target proteins normally recognized by RTKs, thereby driving the afflicted cells to a state of unrestrained proliferation.

3. The **v-ras** oncogene encodes a protein, **v-Ras,** that functionally resembles the monomeric G-protein **c-Ras** (Section 19-3C) in that it is localized on the cytoplasmic side of the mammalian plasma membrane where, when binding GTP, it activates a variety of cellular processes by stimulating the phosphorylation of numerous proteins at specific Ser and Thr residues. Although v-Ras hydrolyzes GTP to GDP, it does so much more slowly than c-Ras. The restraint to protein phosphorylation that GTP hydrolysis would normally impose on c-Ras is thus greatly reduced in v-Ras, thereby transforming the cell.

4. Several viral oncogenes, including **v-jun** and **v-fos,** encode nuclear proteins whose corresponding normal cellular analogs are synthesized in response to growth factors such as EGF and PDGF that induce mitosis (cell division). Many such proteins, including the v-*jun* and v-*fos* gene products, bind to DNA, strongly suggesting that they influence its transcription and/or replication. Indeed, **v-jun** is 80% identical in sequence to the **proto-oncogene** (normal cellular analog of an oncogene) **c-jun,** which encodes a transcription factor named **Jun** (also called **AP-1;** Section 19-3D). Moreover, Jun/AP-1 forms a tight complex with the protein encoded by the proto-oncogene **c-fos,** which greatly increases the ability of Jun/AP-1 to stimulate transcription from Jun-responsive genes.

Oncogene products therefore appear to be functionally modified or inappropriately expressed components of elaborate control networks that regulate cell growth and differentiation. The complexity of these networks (as we shall see, cells generally respond to a variety of growth factors, hormones, and transcription factors in partially overlapping ways) is probably why malignant transformation requires several independent carcinogenic events. Note, however, that few human cancers are virally induced; nearly all of them arise from genetic alterations involving proto-oncogenes. We discuss the nature of these alterations in Section 34-4C.

C. *Relaying the Signal: Binding Modules, Adaptors, GEFs, and GAPs*

Many autophosphorylated RTKs can directly phosphorylate their target proteins. Surprisingly, however, not all RTKs do so. How, then, do they activate their target proteins? The answer, as we shall see, is through a highly diverse and complicated set of interconnected signaling pathways involving cascades of associating proteins.

a. Two-Hybrid Systems Identify Proteins That Interact *in Vivo*

Before we consider the interacting proteins that participate in RTK-mediated signal transduction, let us discuss one of the most often used methods to detect their associations *in vivo,* the **two-hybrid system.** This ingenious experimental technique, which was formulated by Stanley Fields, is based on the peculiar bipartite nature of many transcription factors (proteins that bind to the promoters and other upstream control regions of eukaryotic genes and, in doing so, influence the rate at which RNA polymerase initiates the transcription of these genes; Section 5-4A). Such transcription factors, as we further discuss in Section 34-3B, contain a DNA-binding domain (DBD) that targets the transcription factor to a specific DNA sequence and an activation domain (AD) that induces RNA polymerase to initiate transcription at a nearby transcriptional initiation site. These two domains function independently, so that a genetically engineered hybrid protein with the DBD of one transcription factor and the AD of another will activate the transcription of the gene for which the DBD is targeted. Moreover, it makes little difference as to whether the DBD is on the N-terminal or the C-terminal side of the AD, regardless of how they are arranged in their parent proteins. Evidently, as long as a DBD and an AD are held in proximity, they function as a transcription factor for the gene(s) to which the DBD is targeted.

The two-hybrid system employs two different plasmids in yeast (Fig. 19-30): One encodes a hybrid protein consisting of a DBD fused to a so-called bait or probe protein; the other encodes a hybrid protein consisting of an AD fused to a so-called fish or target protein. The DBD is targeted to a reporter gene that has been engineered into the yeast chromosome such as the *E. coli lacZ* gene, which encodes the enzyme β-galactosidase. On culture plates containing X-gal (a colorless compound that turns blue on being hydrolyzed by β-galactosidase; Section 5-5C), yeast colonies expressing β-galactosidase turn blue, thereby indicating that the bait and fish proteins they encode associate with one another. Using this technique, cells can be screened for fish proteins that specifically interact with a particular bait protein by properly inserting the various cDNAs (Section 5-5F) derived from the cells into the AD-containing plasmid. A fish protein selected in this way can then be identified by sequencing its cDNA.

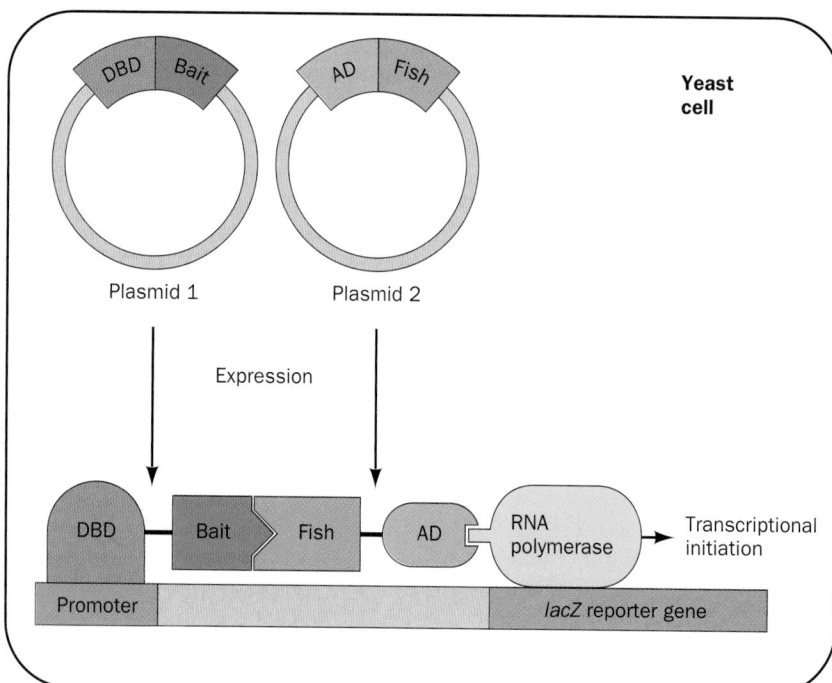

FIGURE 19-30   The two-hybrid system. It utilizes two plasmids expressed in yeast: One encodes a hybrid protein consisting of the DNA-binding domain (DBD) of a transcription factor fused to a "bait" protein, and the other encodes the activation domain (AD) of a transcription factor fused to a "fish" protein. The DBD specifically binds to the promoter of a reporter gene, here *lacZ.* If the "fish" protein associates with the "bait" protein, its attached AD induces RNA polymerase to initiate the transcription of the reporter gene. The β-galactosidase encoded by *lacZ* gene is readily detected through the use of X-gal, which turns blue when hydrolyzed by β-galactosidase. Yeast colonies expressing "bait" and "fish" proteins that do not interact remain colorless, as do colonies that do not express both plasmids.

b. SH2 Domains Mediate Signal Transduction

Now let us return to our discussion of RTK-mediated signaling. Many (>100) of the diverse cytoplasmic proteins that bind to autophosphorylated receptors, for example, the cytoplasmic PTK **c-Src** (henceforth referred as just Src), certain GTPase activating proteins (GAPs), and **phospholipase C-γ** (Section 19-4B), contain one or two conserved ~100-residue modules known as **Src homology domain 2** (**SH2;** so named because they were first noticed in tyrosine kinases related to Src; **SH1** refers to their catalytic domains). *SH2 domains specifically bind phosphoTyr residues in their target peptides with high affinity;* they bind their unphosphorylated target peptides weakly if at all. Most of the phosphoTyr residues to which SH2 binds are located in the juxtamembrane (just after the transmembrane helix), kinase insert, and C-terminal regions of RTKs; those in the activation loop function mainly to stimulate PTK activity. Indeed, RTK autophosphorylation occurs in two phases: First the activation loops of the RTK are phosphorylated and then the resulting activated PTK phosphorylates the other sites on the opposing subunit of the RTK.

The X-ray and NMR structures of SH2 domains from several proteins, both alone and in complex with phosphoTyr (pY)-containing polypeptides, have been determined. SH2 is a hemispherically shaped domain, which contains a central 5-stranded antiparallel β sheet that is sandwiched between two nearly parallel α helices. The N- and C-terminal residues of SH2 are in close proximity on the surface opposite the peptide binding site, which suggests that this domain can be inserted between any two surface residues on a protein without greatly disturbing its fold or function. Indeed, the sequences of a variety of SH2-containing proteins reveal no apparent preference for the location of this domain in a protein.

John Kuriyan determined the X-ray structure of the SH2 domain of Src in complex with an 11-residue polypeptide containing the sequence pYEEI, a tetrapeptide segment that binds to this SH2 domain with high affinity. The 11-residue peptide binds to the SH2 domain in an extended conformation, with contact being made primarily by the pYEEI tetrapeptide (Fig. 19-31a). The phosphoTyr side chain inserts into a small cleft formed, in part, by three highly conserved positively charged residues, including the side chain of an invariant Arg, which contacts the phosphate group. The Ile side chain is similarly inserted into a nearby hydrophobic pocket and the entire tetrapeptide segment interacts very tightly with SH2, although the side chains of the peptide's two central Glu residues do not project toward SH2. Thus, the peptide resembles a two-pronged plug that is inserted into a two-holed socket on SH2 (Fig. 19-31b). Comparison of this structure with that of uncomplexed Src SH2 indicates that, on binding peptide, SH2 undergoes only small conformational changes that are localized at its peptide-binding site. These structures provide a simple explanation for why SH2 does not bind the far more abundant phosphoSer- and phosphoThr-contain-

(a)

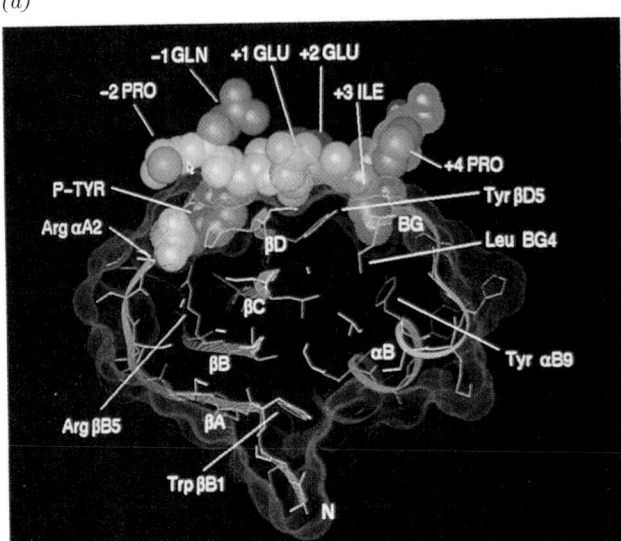

(b)

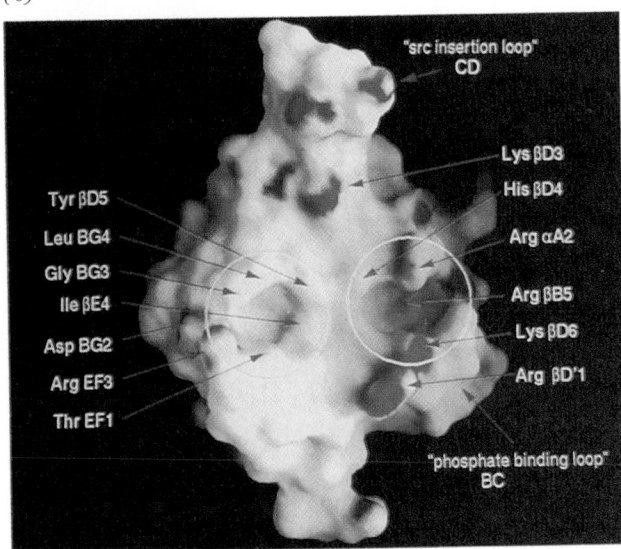

FIGURE 19-31   X-Ray structure of the 104-residue Src SH2 domain in complex with an 11-residue polypeptide (EPQpYEEIPIYL) containing the protein's pYEEI target tetrapeptide. (*a*) A cutaway view of the complex, in which its solvent-accessible surface is represented by red dots, the protein (*pink*) is shown in ribbon form with its side chains in stick form, and the N-terminal 8-residue segment of the bound polypeptide is shown in space-filling form with its backbone yellow, its side chains green, and its phosphate group white [the N-terminal Pro side chain (*left*) is largely obscured in this view and the C-terminal three residues are disordered]. (*b*) The molecular surface of the protein only, as viewed toward the peptide binding site and colored according to its local electrostatic potential with the most positive regions deep blue and the most negative regions deep red. The binding pockets for the phosphoTyr (*right*) and Ile (*left*) side chains are circled in yellow and important tyresidues are identified by red arrows. [Courtesy of John Kuriyan, The Rockefeller University.]

ing peptides: The side chains of these residues are too short for their phosphate groups to contact the invariant Arg side chain at the bottom of the phosphoTyr binding pocket.

c. PTB Domains Also Bind PhosphoTyr-Containing Peptides

A second type of motif that specifically binds to phosphoTyr-containing target peptides is known as the **phosphotyrosine-binding (PTB) domain.** PTB domains specifically bind the consensus sequence NPXpY (where X is any residue) and hence recognize the sequence on the N-terminal side of pY rather than that on its C-terminal side, as do SH2 domains.

The NMR structure of the 195-residue PTB domain of **Shc,** an adaptor protein (see below), in complex with a 12-residue target peptide containing the centrally located sequence NPQpY was determined by Stephen Fesik. The structure consists of a β sandwich comprising two nearly perpendicular antiparallel β sheets flanked by three α helices (Fig. 19-32). The N-terminal segment of the target phosphopeptide assumes an extended conformation that, in effect, forms an additional antiparallel β strand of one of the β sheets. The NPQpY segment forms a β turn in which the phosphate group contacts an Arg side chain that mutational studies indicate is essential for the binding of target peptide.

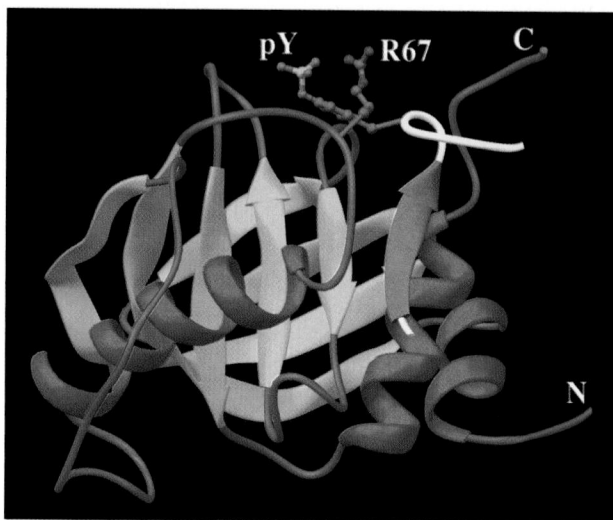

FIGURE 19-32   The NMR structure of the PTB domain of Shc in complex with a 12-residue polypeptide (HIIENPQpYFSDA) from the Shc binding site of a nerve growth factor (NGF) receptor. The PTB domain is colored according to its secondary structure (helices red, β strands gold, and coil light blue), as is the target peptide (β strand magenta and coil white). The target peptide's phosphoTyr (pY) side chain and the side chain of Arg 67, with which it closely interacts, are shown in ball-and-stick form with C green, N blue, O red, and P yellow. Note that the pY side chain extends from a β turn on the peptide ligand. [Based on an NMR structure by Stephen Fesik, Abbott Laboratories, Abbott Park, Illinois. PDBid 1SHC.]

d. SH3 Domains Bind to Proline-Rich Peptides

Many of the RTKs that contain SH2 domains also have one or more 50- to 75-residue **SH3 domains.** Moreover, SH3 is contained in several membrane-associated proteins that lack an SH2. The SH3 domain, which is unrelated to SH2, binds Pro-rich sequences of 9 or 10 residues containing the motif Pro-X-X-Pro, with the residues surrounding this motif targeting these sequences to specific SH3 domains. The physiological function of SH3 is less apparent than that of SH2 because SH3 occurs in a greater variety of proteins, including receptor and nonreceptor tyrosine kinases, adaptor proteins such as **Grb2** (see below), and structural proteins such as spectrin and myosin. However, the observation that the deletion of the SH3 domain-encoding segments from the proto-oncogenes *Src* and *Abl* (which both encode PTKs) converts them to oncogenes suggests that SH3, much like SH2 and PTB, functions to mediate the interactions between kinases and regulatory proteins. SH2, PTB, and SH3 have therefore been called "molecular velcro."

The X-ray and NMR structures of SH3 domains from several proteins indicate that the SH3 core consists of two 3-stranded, antiparallel β sheets that pack against each other with their strands nearly perpendicular. As with SH2, the close proximity of SH3's N- and C-termini suggests that this domain could be modularly inserted between two residues on the surface of another protein without greatly perturbing either structure. The X-ray structures of the SH3 domains from the tyrosine kinases Abl and **Fyn** in complex with two different 10-residue Pro-rich polypeptides to which they tightly bind were determined by Andrea Musacchio and Matti Sareste. Both decapeptides assume nearly identical conformations with their C-terminal 7 residues in the polyproline II helix conformation (Section 8-2B). The peptides bind to SH3 over their entire length in three geometrically complementary cavities (Fig. 19-33), which are mostly occupied by Pro side chains.

e. Other Binding Modules

Several other binding modules have been implicated in mediating signal transduction. These include:

1. The **WW domain** (named after its 2 highly conserved Trp residues), an ~40-residue module that binds to Pro-rich sequences on its target proteins.

2. The **pleckstrin homology (PH) domain** (so named because was first recognized in **pleckstrin** [for *pl*atelet and *leu*kocyte *C k*inase subs*tr*ate prote*in*)], an ~120-residue module that is present in >100 proteins. It structurally resembles the PTB domain (Fig. 19-32) but binds to the inositol head groups of phosphoinositides (Section 19-4). It therefore targets its attached proteins to the inner surface of the plasma membrane. In addition to its role in intracellular signaling, the PH domain participates in cytoskeletal organization, regulation of intracellular membrane transport, and modification of membrane lipids. Its structure is discussed in Section 19-4B.

3. The **PDZ domain** (named after the three proteins in

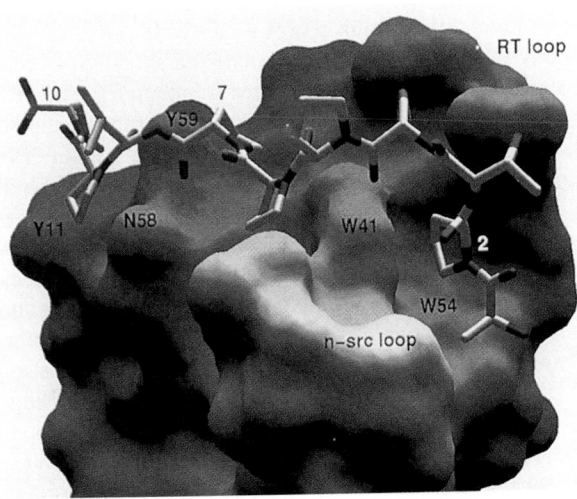

FIGURE 19-33 **X-Ray structure of the SH3 domain from Abl protein in complex with its 10-residue target Pro-rich polypeptide (APTMPPPLPP).** The protein is represented by its surface diagram and the peptide is drawn in stick form with C white, N blue, O red, and S green. Residues contacting the polypeptide are identified by their one-letter codes. [Courtesy of Andrea Musacchio, European Molecular Biology Laboratory, Heidelberg, Germany. PDBid 1ABO.]

which it was first described: *PSD-95, Dlg, and ZO-1*), an ~100-residue module that mainly binds to the C-terminal tripeptide, Ser/Thr-X-Val, of its target proteins.

Many of the proteins that participate in signal transduction consist of several modular units that also occur in several, if not many, other such proteins. These modules may have enzymatic activities (e.g., PTK activity) or bind to specific molecular motifs, such as a phosphoTyr residue within a specific sequence (e.g., SH2 domains) or another protein module (e.g., SH3 domains). Apparently, *signaling proteins have arisen through the evolutionary shuffling of these modules to generate different combinations of interactions and activities.* Indeed, we shall see that the complex behavior of these signaling proteins is a consequence of the interactions among their various modules.

f. Ras Is Activated by Phosphorylated RTKs via a Grb2–Sos Complex

c-Ras (*or just* Ras), *a proto-oncogene product, is a monomeric membrane-anchored (by prenylation) G protein that lies at the center of an intracellular signaling system: It regulates such essential cellular functions as growth and differentiation through the phosphorylation and hence activation of a variety of proteins.* In the signaling pathway described in Section 19-3D, the binding of ligand to RTKs ultimately activates a guanine nucleotide exchange factor (GEF; Section 19-2C) to exchange a Ras-bound GDP for GTP. Only Ras·GTP is capable of further relaying the signal. However, as do the homologous and structurally similar α subunits of the heterotrimeric G proteins, Ras even-

tually hydrolyzes its bound GTP to GDP, thereby halting further signal transduction and limiting the magnitude of the signal generated by the binding of ligand to the receptor. Indeed, the structure of Ras closely resembles those of the GTPase domains of the heterotrimeric G protein G_α subunits, including their Switch I and Switch II regions (Fig. 19-17). Mammalian cells express four Ras homologs: H-Ras, N-Ras, K-Ras 4A, and K-Ras 4B.

Molecular genetic analyses of signaling in a variety of distantly related organisms (notably humans, mice, *Xenopus, Drosophila,* and the nematode worm *Caenorhabditis elegans*) have revealed a remarkably conserved pathway in which *RTKs funnel the signal that they have bound ligand to Ras, which, in turn, relays the signal, via a so-called MAP kinase cascade, to the transcriptional apparatus in the nucleus (Section 19-3D).* Nevertheless, the way in which messages are passed between the RTKs and Ras remained enigmatic for several years until investigations in numerous laboratories revealed their major details. In particular, these studies demonstrated that *two previously characterized proteins,* **Grb2** *and* **Sos,** *form a complex that bridges activated RTKs and Ras in a way that induces Ras to exchange its bound GDP for GTP, thereby activating it (i.e., they act as a GEF).*

The mammalian protein Grb2, a 217-residue homolog of **drk** in *Drosophila* and **Sem-5** in *C. elegans,* consists almost entirely of an SH2 domain flanked by two SH3 domains. Sos protein (the 1596-residue product of the *Son of Sevenless* gene, so named because Sos interacts with the *Sevenless* gene product, an RTK that regulates the development of the R7 photoreceptor cell in the *Drosophila* compound eye), which is required for Ras-mediated signaling, contains a central domain homologous to known Ras-GEFs and a Pro-rich sequence in its C-terminal segment similar to known SH3-binding motifs. Moreover, mammalian homologs of Sos **(mSos)** have been shown to specifically stimulate guanine nucleotide exchange in mammalian Ras proteins. Western blotting techniques (Section 6-4B) using anti-Grb2 and anti-mSos antibodies indicate that Grb2 binds to the C-terminal segment of mSos but not when one of the Pro residues in Sos' SH3-binding motif has been replaced by Leu or in the presence of synthetic polypeptides that have these Pro-rich sequences. Similar studies indicated that in the presence of epidermal growth factor (EGF), the EGF receptor (an RTK; Figs. 19-23 and 19-25*b*) specifically binds the Grb2–mSos complex, an interaction that is blocked by the presence of a phosphopeptide with the sequence of the peptide segment containing one of the activated EGF receptor's phosphoTyr residues. Evidently, Grb2's SH2 domain binds a phosphoTyr-containing peptide segment in an activated RTK while its two SH3 domains bind the Pro-rich sequences of Sos. The GEF function of Sos is thereby stimulated to activate Ras.

g. Grb2, Shc, and IRS Are Adaptors That Recruit Sos to the Vicinity of Ras

The X-ray structure of Grb2, determined by Arnaud Ducruix, reveals that neither of its SH3 domains contacts

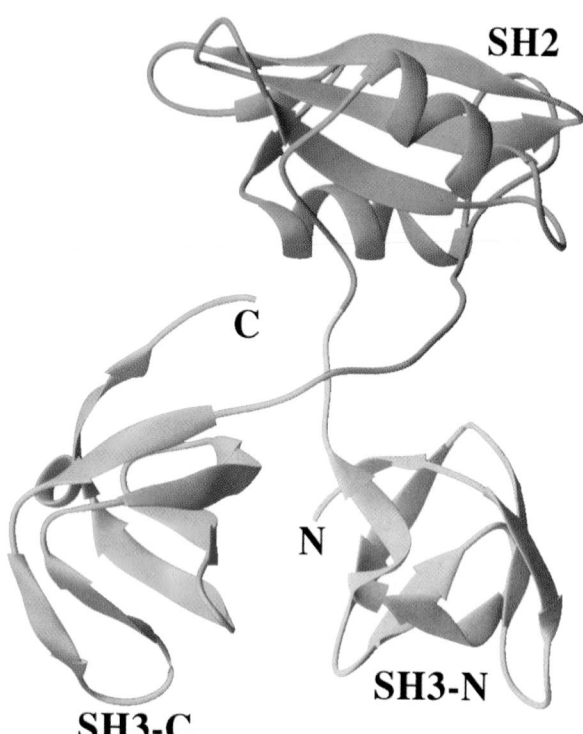

FIGURE 19-34 X-Ray structure of Grb2. Its SH2 domain (*green*) is linked to its flanking SH3 domains (*cyan and orange*) via apparently unstructured and hence flexible 4-residue linkers. [Based on an X-ray structure by Arnaud Ducruix, Université de Paris-Sud, Gif sur Yvette Cedex, France. PDBid 1GRI.]

its SH2 domain (Fig. 19-34). Moreover, although the two SH3 domains are in contact, their interface area is relatively small and hence is more likely to be an artifact of crystallization than a structural feature of Grb2 in solution. It therefore appears that Grb2's SH3 domains are flexibly linked to its SH2 domain. How does the binding of such a pliable **adaptor** (a linker that lacks enzymatic activity) to a phosphorylated RTK stimulate Sos to act as Ras's GEF? Grb2 and Sos bind one another so tightly that they are essentially permanently associated in the cell. Hence, when Grb2 binds its target phosphorylated RTK, it recruits Sos to the inner surface of the plasma membrane, where Sos's then increased local concentration causes it to more readily bind to the membrane-anchored Ras.

Shc proteins are also adaptors that link activated RTKs to Ras. Shc proteins consist of an N-terminal PTB domain (Fig. 19-32), a central effector region (CH1), and a C-terminal SH2 domain that binds to certain activated RTKs. Moreover, Shc proteins are also major targets of various RTKs, which phosphorylate them within their CH1 domains at sequences that thereupon form binding sites for the Grb2 SH2 domain. Activated RTKs may therefore bind Grb2 indirectly via Shc as well as directly. Moreover, in some cases, an Shc–Grb2–Sos complex that lacks a bound RTK may activate Ras.

The activated (autophosphorylated) insulin receptor does not directly interact with SH2 domain-containing proteins. Rather, it mainly phosphorylates an ~1300-residue

protein named the **insulin receptor substrate** (**IRS;** actually a family of four homologous proteins named IRS1–4 that are each expressed in a tissue-specific manner). The IRS proteins all have an N-terminal "targeting" region consisting of a PH domain that localizes the IRS to the interior of the plasma membrane, followed by a PTB domain that binds the IRS to a phosphoTyr residue of an activated insulin receptor (Fig. 19-35). The insulin receptor thereupon phosphorylates the IRS at one or more of its 6 to 8 Tyr residues, converting them to SH2 binding sites that then couple this system to SH2-containing proteins (Section 19-4F). Adapters with multiple SH2 binding sites such as the IRS proteins and Shc are also known as **docking proteins** because they function as platforms for the recruitment of a variety of downstream signaling molecules in response to the activation of their corresponding RTK. Thus, a docking protein increases the complexity and regulatory flexibility of its RTK-initiated signaling pathway as well as amplifying its signal.

h. Sos Functions to Pry Open Ras's Nucleotide Binding Site

The X-ray structure, determined by Kuriyan, of Ras in complex with a 506-residue GEF-containing segment of Sos reveals how Sos induces Ras to exchange its normally tightly bound GDP for GTP. This Sos segment consists of two α helical domains, of which only the C-terminal so-called catalytic domain contacts Ras. Ras is packed against the center of the elongated bowl-shaped catalytic domain (Fig. 19-36). Those portions of Ras that interact with the catalytic domain include both its Switch I and Switch II regions as well as the loop that binds the α and β phosphates of GDP and GTP, the so-called P-loop (although GMP binds to Ras with 10^6-fold less affinity than does GDP, so that the β phosphate of GDP is largely responsible for its tight binding to Ras). This interaction displaces Switch I relative to its position in the X-ray structure of Ras in complex with the nonhydrolyzable GTP analog **GDPNP:**

$$^-O-\overset{\overset{\displaystyle O}{\|}}{P}-NH-\overset{\overset{\displaystyle O}{\|}}{P}-O-\overset{\overset{\displaystyle O}{\|}}{P}-CH_2$$

Guanosine-5′-(β,γ-imido)triphosphate (GDPNP)

Ras's nucleotide binding site is thereby partially opened up and a Leu and a Glu side chain from Sos are, respectively, introduced into Ras' Mg^{2+}-binding site and the site that binds the α phosphate group of GDP/GTP. However,

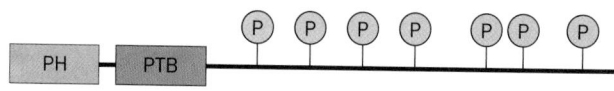

FIGURE 19-35 Structure of an insulin receptor substrate protein. An IRS contains a PH and a PTB domain at its N-terminus followed by multiple phosphoTyr-containing binding sites for the SH2 domains of downstream signaling proteins.

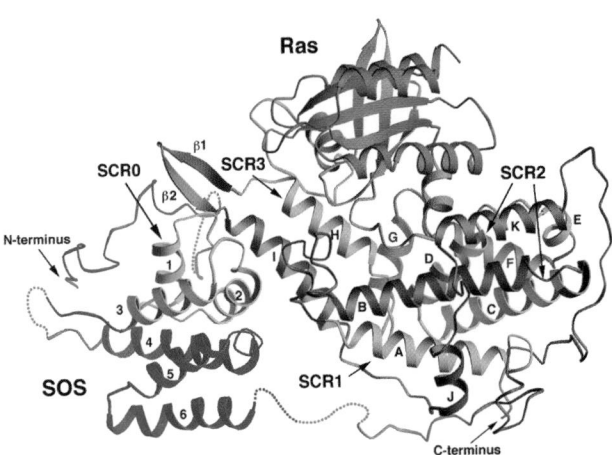

FIGURE 19-36 X-Ray structure of the complex between Ras and the GEF-containing region of Sos. The N-terminal domain of the Sos region is blue, its catalytic domain is green, and Ras is mainly gray, with its Switch I and II regions orange and its P loop red. Conserved regions (the SCRs) among the Ras family of GEFs are cyan. [Courtesy of John Kuriyan, The Rockefeller University. PDBid 1BDK.]

this interaction does not significantly occlude Ras's guanine and ribose binding sites. This rationalizes how the Sos–Ras interaction can be strong enough to displace the normally tightly bound GDP from Ras but yet weak enough so that GTP (or the 10-fold less abundant GDP) can subsequently displace Sos from Ras.

The different families of small G proteins interact with different classes of GEFs, whose catalytic domains share no sequence similarity and are structurally unrelated [e.g., rhodopsin (Fig. 19-15) is a GEF for $G_{t\alpha}$]. Nevertheless, many of these GEFs share the same general mechanism for promoting GDP–GTP exchange, which suggests that this mechanism arose on several occasions through convergent evolution.

i. GAPs Function to Turn Off Ras-Mediated Signals

Ras hydrolyzes its bound GTP with a rate constant of 0.02 min^{-1} (vs 2–3 min^{-1} for G_α subunits), too slowly for effective signal transduction. This led to the discovery of a 120-kD GTPase activating protein, **RasGAP,** that, on binding Ras·GTP, accelerates the rate of GTP hydrolysis by a factor of 10^5. RasGAP's physiological importance as a regulator of Ras-mediated signal transduction is demonstrated by the observation that the relative biological activities of Ras mutants are better correlated with their resistance to regulation by RasGAP than by their intrinsic GTPase activity.

The mechanism whereby RasGAP activates the GTPase activity of Ras was revealed by the X-ray structure, determined by Alfred Wittinghofer, of the 334-residue GTPase-activating domain of RasGAP (GAP334) bound to Ras in its complex with GDP and AlF_3 (Fig. 19-37a). GAP334, which consists of two all-helical domains, interacts with Ras over an extensive surface that includes

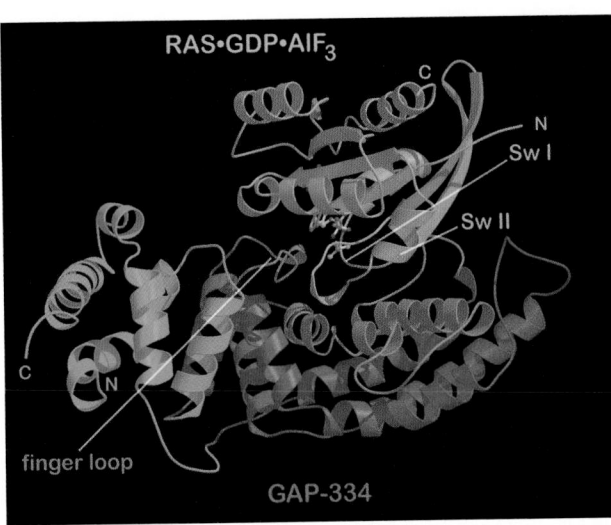

(a)

FIGURE 19-37 X-Ray structure of the GAP334·Ras·GDP· AlF_3 complex. (*a*) A ribbon diagram in which the "extra" domain of GAP334 (which does not contact Ras) is green, its catalytic domain is red with those portions contacting Ras light brown, Ras is yellow, and its bound GDP and AlF_3 are shown in ball-and-stick form. Sw I and Sw II indicate the switch regions of Ras. (*b*) The active site region of the complex in which Ras is orange, GAP334 is red, and their catalytically

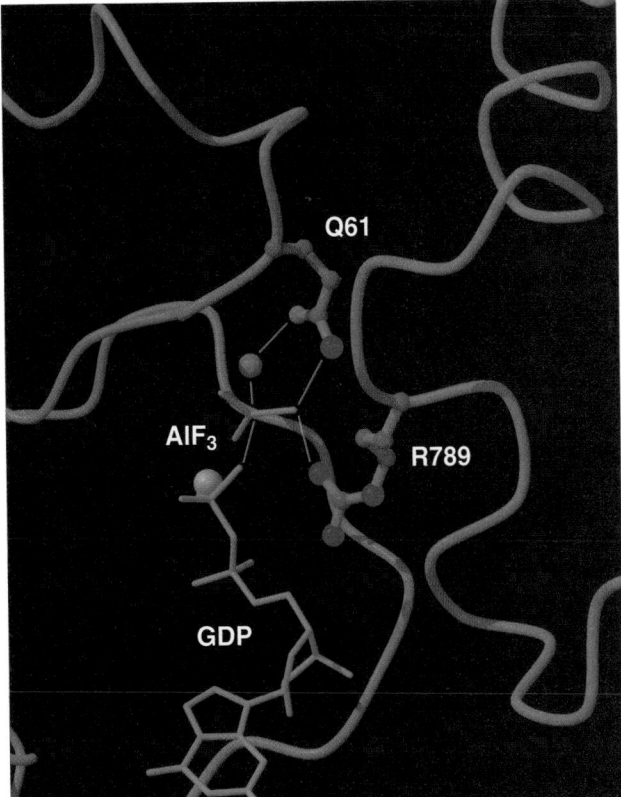

(b)

important side chains, Arg 789 of GAP334 and Gln 61 of Ras, are shown in ball-and-stick form with C gray, N blue, and O red. The GDP and AlF_3 are shown in stick form in green, the nucleophilic water molecule is represented by a red sphere, a bound Mg^{2+} ion is shown as a silver sphere, and hydrogen bonds are represented by thin gray lines. [Courtesy of Alfred Wittinghofer, Max-Planck-Institut für Molekulare Physiologie, Dortmund, Germany. PDBid 1WQ1.]

its Switch I and II regions. The AlF$_3$, which has trigonal planar symmetry, binds to Ras at the expected position of GTP's γ phosphate group, with the Al atom opposite a bound water molecule that presumably would be the at-

tacking nucleophile in the GTPase reaction. Since Al—F and P—O bonds have similar lengths and phosphoryl-transfer reactions occur via a trigonal bipyramidal transition state (Fig. 16-6b), the GDP–AlF$_3$–H$_2$O assembly

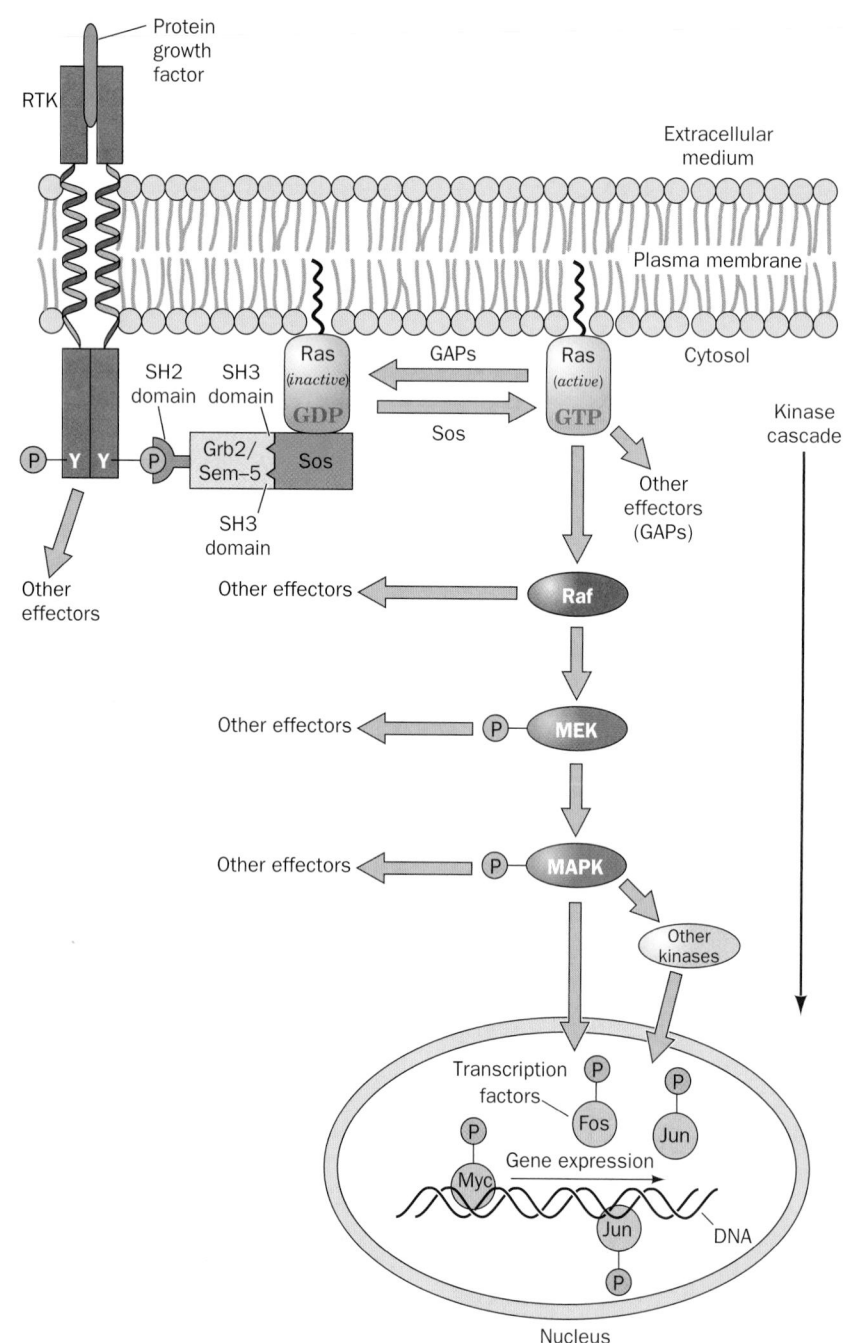

FIGURE 19-38 The Ras-activated MAP kinase cascade. This signaling cascade begins when an RTK binds its cognate growth factor, thereby inducing the autophosphorylation of this RTK's cytosolic domain. Grb2/Sem-5 binds to the resulting phosphoTyr-containing peptide segment via its SH2 domain and simultaneously binds to Pro-rich segments on Sos via its two SH3 domains. This activates Sos as a guanine nucleotide releasing factor (GRF) to exchange Ras' bound GDP for GTP, which activates Ras to bind to Raf. Then, Raf, a Ser/Thr kinase, phosphorylates MEK, which in turn phosphorylates MAPK, which then migrates to the nucleus, where it phosphorylates transcription factors such as Fos, Jun, and Myc, thereby modulating gene expression. The MAP kinase cascade eventually returns to its resting state through the actions of protein phosphatases (Section 19-3F) after a GTPase activating protein (GAP) deactivates Ras by inducing it to hydrolyze its bound GTP to GDP. [After Egan, S.E. and Weinberg, R.A., *Nature* **365**, 782 (1993).] ♨ See the Animated Figures

presumably resembles the GTPase reaction's transition state, with the AlF_3 mimicking the planar PO_3 group. Note that Ras·GDP does not by itself bind AlF_3.

GAP334 binds to Ras with GAP334's exposed so-called finger loop inserted into the Ras active site such that the finger loop's Arg 789 side chain interacts with both the Ras-bound GDP's β phosphate and the AlF_3 (Fig. 19-37b). In Ras·GTP, this Arg side chain would be in an excellent position to stabilize the developing negative charge in the GTPase reaction's transition state. Indeed, the catalytically more efficient G_α subunits contain an Arg residue (Arg 178 in $G_{i\alpha}$), whose guanidinium group occupies a nearly identical position (in $G_{s\alpha}$, this is the Arg side chain that is ADP-ribosylated by cholera toxin; Section 19-2C). The main chain carbonyl O of Arg 789 hydrogen bonds to the side chain N of Ras' catalytically important Gln 61. The O of this side chain is thereby positioned to hydrogen bond with the nucleophilic water molecule while its NH_2 group interacts with an F atom of AlF_3 (Fig. 19-37b), an arrangement that presumably stabilizes the GTPase reaction's transition state.

j. Oncogenic Mutants of Ras Are GAP-Insensitive

Mutations in Ras of Gly 12 and Gln 61 are its most common oncogenic mutations (an oncogenic form of which is found in ~30% of human cancers). These mutations prevent RasGAP from activating Ras to hydrolyze its bound GTP and hence lock Ras into its active conformation. The foregoing X-ray structure reveals why these mutants are GAP-insensitive. Gly 12 is in such close proximity to the finger loop that even the smallest possible residue change (to Ala) would sterically interfere with the geometry of the transition state through steric clashes with the main chain of Arg 789 (of RasGAP) and the side chain NH_2 of Gln 61. The observation that Gly 12 mutants of Ras bind GTP with nearly wild-type affinity therefore suggests that larger side chains can be tolerated at Ras residue 12 in the Ras–RasGAP Michaelis complex but not in the transition state. The apparent participation of Gln 61 in transition state stabilization confirms that this residue has an essential role in catalysis.

D. *MAP Kinase Signaling Cascades*

The signaling pathway downstream of Ras consists of a linear series of Ser/Thr kinases that form a so-called **MAP kinase cascade** *(Fig. 19-38). Many of the proteins that participate in MAP kinase cascades are the products of proto-oncogenes:*

1. Raf, a Ser/Thr protein kinase, is activated by direct interaction with Ras·GTP (although other signaling pathways may activate Raf by phosphorylating it at multiple Ser and Thr residues; see below). The X-ray structure of the Ras homolog **Rap1A** in complex with GDPNP and the Ras binding domain of Raf (RafRBD), determined by Wittinghofer, reveals that the two proteins associate largely by mutually extending their antiparallel β sheets across a mainly polar interface (Fig. 19-39). Although Ras·GTP has

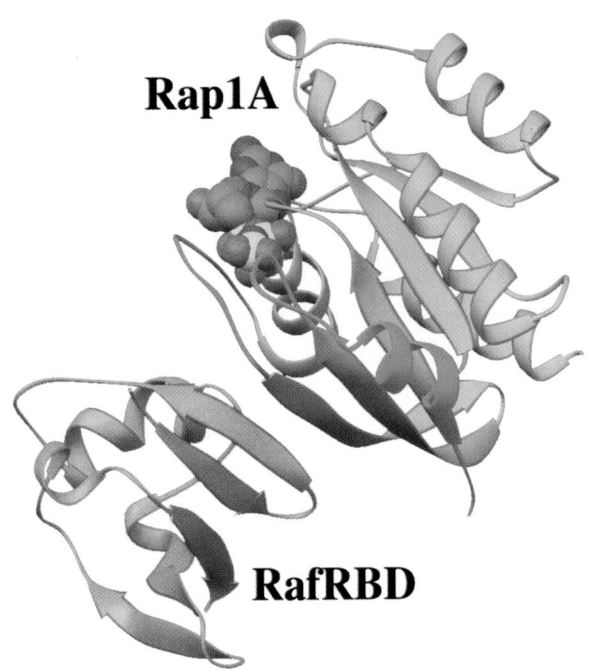

FIGURE 19-39   X-Ray structure of the Ras binding domain of Raf (RafRBD; *orange*) in complex with Rap1A·GDPNP (*light blue*). The Switch I and II regions of Rap1A are magenta and light green, and its bound GDPNP is shown in space-filling form with C green, N blue, O red, and P yellow. Rap1A·GDPNP and Ras·GDPNP have nearly identical structures. [Based on an X-ray structure by Alfred Wittinghofer, Max-Planck-Institut für Molekulare Physiologie, Dortmund, Germany. PDBid 1GUA.]

1300-fold greater affinity for binding to Raf than does Ras·GDP, it is unclear from this structure how GTP hydrolysis by Ras affects the Ras–Raf interface. Quite possibly the conformational change in Ras Switch I perturbs the Ras–Raf interface to the point that it dissociates.

2. Activated Raf phosphorylates a protein alternatively known as **MEK** and **MAP kinase kinase (MKK)** at specific Ser and Thr residues, thereby activating it as a Ser/Thr kinase. [Raf is therefore a **MAP kinase kinase kinase (MKKK)**].

3. Activated MEK phosphorylates a family of proteins named **mitogen-activated protein kinases (MAP kinases** or **MAPKs)** or **extracellular-signal-regulated kinases (ERKs)**. For more than marginal activation, a MAPK must be phosphorylated at both its Thr and Tyr residues in the sequence Thr-X-Tyr. MEK (which stands for *M*AP kinase/*E*RK-activating *k*inase) catalyzes both phosphorylations and thus has dual specificity for Ser/Thr and Tyr. The X-ray structure of the unphosphorylated MAP kinase ERK2, determined by Elizabeth Goldsmith, reveals that this protein structurally resembles other protein kinases of known structure and that its Tyr residue that becomes phosphorylated blocks the peptide-binding site in its unphosphorylated form.

4. The activated MAPKs phosphorylate a variety of cytoplasmic and membrane-associated proteins, including Sos and EGFR, at Ser/Thr-Pro motifs. In addition, the MAPKs migrate from the cytosol to the nucleus, where they phosphorylate a large variety of transcription factors including Jun/AP-1, Fos, and **Myc.** These activated transcription factors, in turn, induce the transcription of their target genes (Section 34-3B). The effects commissioned by the extracellular presence of the protein growth factor that initiated the signaling cascade are thereby produced.

MAP kinase cascades can be activated in other ways besides by liganded RTKs. For example, Raf may also be activated via its Ser/Thr phosphorylation by **protein kinase C,** which is activated via the phosphoinositide signaling system described in Section 19-4. Alternatively, Ras may be activated by subunits of certain heterotrimeric G proteins. Thus, *the MAP kinase cascade serves to integrate a variety of extracellular signals.*

a. Scaffold and Anchoring Proteins Organize and Position Protein Kinases

Eukaryotic cells contain numerous different MAPK signaling cascades, each with a characteristic set of component kinases, which in mammals comprise at least 14 MKKKs, 7 MKKs, and 12 MAPKs (Fig. 19-40). Although each MAPK is activated by a specific MKK, a given MKK can be activated by more than one MKKK. Moreover, several pathways may be activated by a single type of receptor. How then does a cell prevent inappropriate cross talk between closely related signaling pathways? One way that this occurs is through the use of **scaffold proteins,** proteins that bind some or all of the component protein kinases of a particular signaling cascade so as to ensure that the protein kinases of a given pathway interact only with one another. In addition, a scaffold protein can control the subcellular location of its associated kinases.

The first known scaffold protein was discovered through the genetic analysis of a MAP kinase cascade in yeast, which demonstrated that this protein, **Ste5p,** binds the MKKK, MKK, and MAPK components of the pathway and that, *in vivo,* the scaffold's absence inactivates the pathway. Evidently, the interactions between successive kinase components of this MAP kinase cascade are, by themselves, insufficient for signal transmission.

Perhaps the most well-characterized mammalian scaffold protein is **JIP-1** (for *J*NK *I*nteracting *P*rotein-1). JIP-1 (Fig. 19-41*a*) simultaneously binds **HPK1** (for

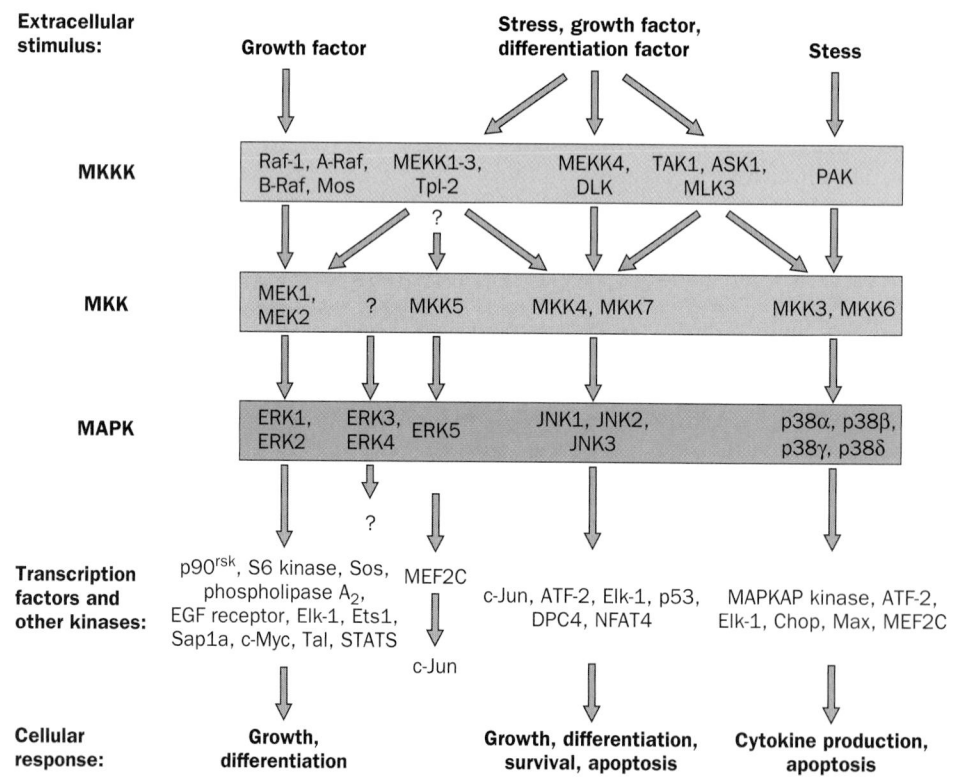

FIGURE 19-40   MAP kinase cascades in mammalian cells. Each MAP kinase cascade consists of an MKKK, an MKK, and an MAPK. Various external stimuli may each activate one or more MKKKs, which in turn, may activate one or more MKKs. However, the MKKs are relatively specific for their target MAPKs. The activated MAPKs phosphorylate specific transcription factors (e.g., **Elk-1, Ets1, p53, NFAT4, Max**) as

well as specific kinases (e.g., **p90[rsk], S6 kinase, MAPKAP kinase**). The resulting activated transcription factors and kinases then induce cellular responses such as growth, differentiation, and **apoptosis** (programmed cell death; Section 34-4E). [After Garrington, T.P. and Johnson, G.L., *Curr. Opin. Cell Biol.* **11,** 212 (1999).]

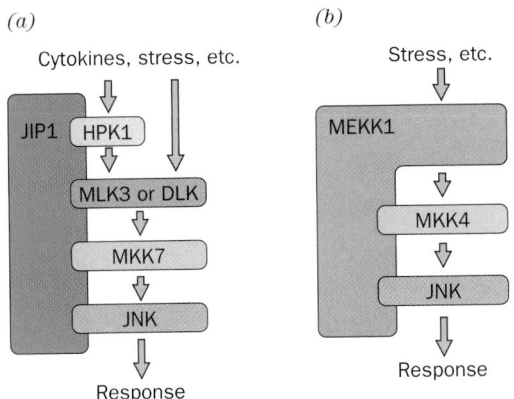

(a) *(b)*

Cytokines, stress, etc.

Stress, etc.

FIGURE 19-41 Some examples of scaffold proteins that modulate mammalian MAP kinase cascades. (*a*) JIP-1 binds all protein components of the MAP kinase cascade in which HPK1 phosphorylates MKL3 or DLK (MKKKs), which then phosphorylates MKK7, which then phosphorylates JNK (a MAPK). (*b*) MEKK1 (an MKKK) is the kinase for MKK4 and also binds JNK, MKK4's target MAPK. [After Garrington, T.P. and Johnson, G.L., *Curr. Opin. Cell Biol.* **11**, 213 (1999).]

Hematopoietic Progenitor Kinase-1), a Ras analog and hence an **MKKK kinase (MKKKK);** the MKKKs **MLK3** and **DL3; MKK7;** and the MAPK **JNK** (for *Jun N*-terminal *Kinase*). **MEKK1** is a somewhat different type of scaffold protein (Fig. 19-41*b*); this functional MKKK binds its substrate, **MKK4,** as well as the latter's substrate, JNK.

Protein Ser/Thr kinases may similarly be individually tethered to their sites of action by **anchoring proteins.** For example, protein kinase A (PKA), which participates in numerous parallel signaling pathways including that regulating glycogen metabolism (Section 18-3), associates with several unrelated so-called **A-kinase anchoring proteins (AKAPs).** The different AKAPs, which all bind the regulatory (R) subunits of PKA, target PKA to different subcellular locations (e.g, to vesicle or plasma membranes or to particular receptors) and may also bind other signaling proteins (e.g., PP1, the protein phosphatase that removes phosphate groups installed by PKA; Section 18-3C).

b. Anthrax Lethal Factor Specifically Cleaves MAPKKs

Anthrax, an infectious disease caused by the bacterium *Bacillus anthracis*, affects mainly herbivorous animals such as cattle, sheep, and goats. On rare occasions, however, it may be transmitted to humans (but not between humans), in whom it is often fatal, if untreated, through massive septic shock (Section 19-1L). Anthrax spores are significant agents of biological warfare because their inhalation results in inhalational anthrax, a form of the disease that is nearly always fatal. This is because by the time the symptoms of inhalational anthrax become apparent, the bacterial infection has already released so much toxin that

eliminating the infection through antibiotic treatment does not reverse the progress of the disease.

Anthrax toxin consists of three proteins that act in concert: **protective antigen (PA), lethal factor (LF),** and **edema factor (EF;** oedema factor in British English). PA, which is named for its use in vaccines, is a 735-residue, 4-domain protein that binds to its host cell-surface receptor (a single pass transmembrane protein) via its C-terminal domain. Most of PA's N-terminal domain is then cleaved away by a cell-surface protease, whereupon the remaining membrane-bound portions of PA form cyclic heptamers reminiscent of the cyclic pentamers formed by cholera toxin (Fig. 19-20). The heptameric PA then binds LF and/or EF by their homologous N-terminal domains and mediates their endocytotic uptake into the cell. Indeed, the intravenous administration of only PA and LF rapidly kills animals. EF is a calmodulin-activated adenylate cyclase whose action upsets water homeostasis and hence is probably responsible for the massive **edema** (abnormal buildup of intercellular fluids) seen in cutaneous anthrax infections.

LF is a 776-residue monomeric protease that has only one known cellular target: *It cleaves members of the MKK family of proteins near their N-termini so as to excise the docking sequences for their cognate downstream MAPKs.* It thereby disrupts the signal transduction pathways in which these proteins participate. However, anthrax infection targets mainly macrophages, a type of white blood cell (mice whose blood has been depleted of macrophages are resistant to anthrax). Low levels of LF, which occur in early stages of anthrax infection, cleave **MKK3,** which inhibits macrophages from releasing but not producing the inflammatory mediators NO (Section 19-1L) and **tumor necrosis factor-α (TNFα;** a cytokine that has opposite effects to most protein growth factors and is largely responsible for the wasting seen in chronic infections; Section 34-4E). This has the effect of reducing and/or delaying the immune response. In contrast, high levels of LF, which occur in late stages of infection, trigger macrophage lysis, causing the sudden release of NO and TNF-α, which presumably results in the massive septic shock that causes death.

E. Tyrosine Kinase–Associated Receptors

Many cell-surface receptors are not members of the receptor families that we have discussed and do not respond to ligand binding by autophosphorylation. These include the receptors for growth hormone (Fig. 19-9), the cytokines, the **interferons** (which defend against viral infections; Section 32-4A), and *T cell receptors* [which control the proliferation of immune system cells known as *T* lymphocytes (*T* cells); Section 35-2D]. *Ligand binding induces these tyrosine kinase–associated receptors to dimerize (and, in some cases, to trimerize), often with different types of subunits, in a way that activates associated nonreceptor tyrosine kinases (NRTKs).* The domain organization

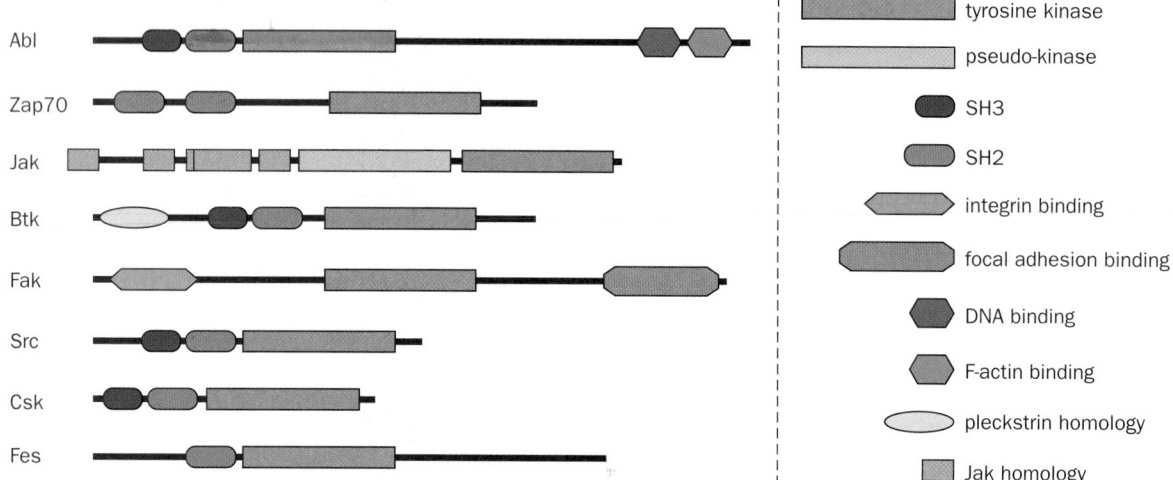

FIGURE 19-42 Domain organization of the major NRTK subfamilies. The N-termini for these polypeptides, which are drawn approximately to scale, are on the left and the domain identification is provided on the right. [Courtesy of Stevan Hubbard, New York University School of Medicine.]

for the major subfamilies of NRTKs is diagrammed in Fig. 19-42.

a. The Structure of Src Reveals Its Autoinhibitory Mechanism

Many of the NRTKs that are activated by tyrosine kinase–associated receptors belong to the **Src family,** which contains at least nine members including Src, Fyn, and **Lck.** Most of these ~530-residue membrane-anchored (by myristoylation) proteins have both an SH2 and an SH3 domain and all have a PTK domain. Hence, a Src-related kinase may also be activated by association with an autophosphorylated RTK. Although Src-related kinases are each associated with different receptors, they phosphorylate overlapping sets of target proteins. This complex web of interactions explains why different ligands often activate some of the same signaling pathways.

Src, as indicated in Fig. 19-42, consists of, from N- to C-terminus, a myristoylated N-terminal "unique" domain that differs among Src family members, an SH3 domain, an SH2 domain, a PTK domain, and a short C-terminal tail. Phosphorylation of Tyr 416 in the PTK's activation loop activates Src, whereas phosphorylation of Tyr 527 in its C-terminal tail deactivates it. *In vivo,* Src is phosphorylated at either Tyr 416 or Tyr 527, but not at both. The dephosphorylation of Tyr 527 or the binding of external ligands to the SH2 or the SH3 domain activates Src, a state that is then maintained by the autophosphorylation of Tyr 416. When Tyr 527 is phosphorylated and no activating phosphopeptides are available, Src's SH2 and SH3 domains function to deactivate its PTK domain, that is, Src is then autoinhibited.

The X-ray structure of Src·ADPNP lacking its N-terminal domain and with Tyr 527 phosphorylated, determined by Stephen Harrison and Michael Eck, reveals the structural basis of Src autoinhibition (Fig. 19-43). As biochemical studies had previously shown, the SH2 domain

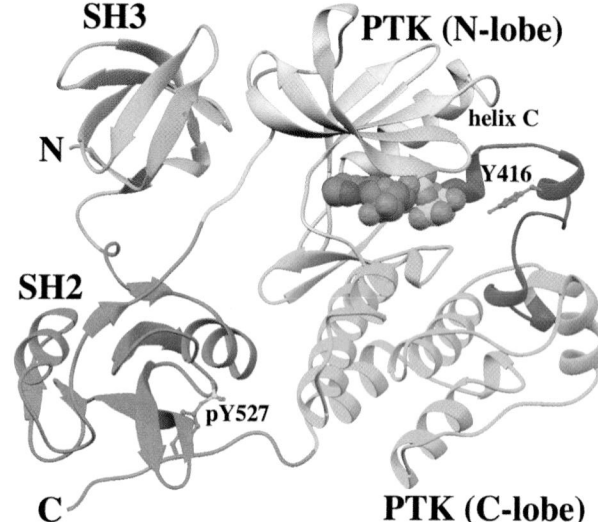

FIGURE 19-43 X-Ray structure of Src·ADPNP lacking its N-terminal domain and with Tyr 527 phosphorylated. The SH3 domain is orange, the SH2 domain is magenta, the linker joining the SH2 domain to the PTK domain is green with its 5-residue polyproline II helix gold, the N-terminal lobe of the PTK domain is pink, the C-terminal lobe is cyan with its activation loop light blue, and the C-terminal tail is red. The ADPNP is shown in space-filling form and Y416 and pY527 are shown in ball-and-stick form, all with C green, N blue, O red, and P yellow. [Based on an X-ray structure by Stephen Harrison and Michael Eck, Harvard Medical School. PDBid 2SRC.]

binds phosphoTyr 527, which occurs in the sequence pYNPG rather than the pYEEI sequence characteristic of high affinity Src SH2 target peptides. Although the pYNP segment binds to SH2 as does the pYEE segment in Fig. 19-31b, the succeeding residues are poorly ordered in the X-ray structure and, moreover, the SH2 pocket in which the Ile side chain of pYEEI binds is unoccupied. Apparently, the phosphoTyr 527-containing peptide segment binds to the Src SH2 domain with reduced affinity relative to its target peptides.

The SH3 domain binds to the linker connecting the SH3 domain to the N-terminal lobe of the PTK domain. Residues 249 to 253 of this linker form a polyproline II helix that binds to the SH3 domain in much the same way as do SH3's Pro-rich target peptides (Fig. 19-33). However, the only Pro in this segment is residue 250. The polar side chain of Gln 253, which occupies the position of the second Pro in SH3's normal Pro-X-X-Pro target sequence, does not enter the hydrophobic binding pocket that this second Pro would occupy (Fig. 19-33), and hence the path of the peptide deviates from that of Pro-rich target peptides at this point. Apparently, this interaction is also weaker than those with Src's SH3 target peptides.

Src's SH2 and SH3 domains bind on the opposite side of the PTK domain from its active site. How, then, does the conformation shown in Fig. 19-43 inhibit the PTK's activity? The two lobes of Src's PTK domain are, for the most part, closely superimposable on their counterparts in the PTK domain of phosphorylated and hence activated Lck (a Src family member) as well as the C subunit of activated PKA (Fig. 18-14). However, Src helix C (the only helix in the PTK's N-terminal lobe) is displaced from the interface between the N- and C-terminal lobes relative to its counterparts in Lck and PKA. Helix C contains the conserved residue Glu 310 (using Src numbering), which in activated Lck and PKA projects into the catalytic cleft, where it forms a salt bridge with Lys 295, an important ligand of

the substrate ATP's α and β phosphates. However, in inactive Src, Glu 310 forms an alternative salt bridge with Arg 385, whereas Lys 295 instead interacts with Asp 404. In activated Lck, Arg 385 forms a salt bridge with phosphoTyr 416.

The foregoing structural observations suggest the following scenario for Src activation (Fig. 19-44):

1. The dephosphorylation of Tyr 527 and/or the binding of the SH2 and/or SH3 domains to their target peptides (for which SH2 and SH3 have greater affinity than their internal Src binding sites) releases these domains from their PTK-bound positions shown in Fig. 19-43, thus relaxing conformational constraints on the PTK domain. This allows the PTK's active site cleft to open, thereby disrupting the structure of its partially helical activation loop (which occupies a blocking position in the active site cleft; Fig. 19-43) so as to expose Tyr 416 to autophosphorylation.

2. The resulting phosphoTyr 416 forms a salt bridge with Arg 385, which sterically requires the structural reorganization of the activation loop to its active, nonblocking conformation. The consequent rupture of the Glu 310–Arg 385 salt bridge frees helix C to assume its active orientation which, in turn, allows Glu 310 to form its catalytically important salt bridge to Lys 295, thereby activating the Src PTK activity.

The above mechanism is, perhaps unexpectedly, critically dependent on the rigidity of the 8-residue linker joining the SH2 and SH3 domains. Thus, replacing three of these linker residues with Gly (whose lack of a C_β atom

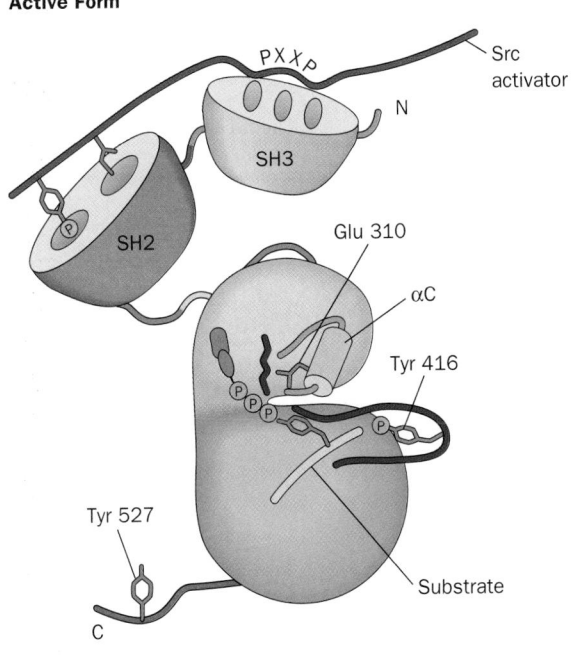

FIGURE 19-44 Schematic model of Src activation. See the text for an explanation. The coloring scheme largely matches that in Fig. 19-43, as does the viewpoint. [After Young, M.A.,

Gonfloni, F., Superti-Furga, G., Roux, B., and Kuriyan, J., *Cell* **105,** 115 (2001).]

makes it the least conformationally restricted residue) results in a protein that is no longer deactivated by the phosphorylation of Tyr 527. This is corroborated by molecular dynamics simulations (Section 9-4) indicating that the thermal motions of the SH2 and SH3 domains are highly correlated (move as a unit) when Tyr 527 is phosphorylated but that this correlation is significantly reduced when Tyr 527 is dephosphorylated or when Gly replaces the three linker residues.

b. The JAK-STAT Pathway Relays Cytokine-Based Signals

The signal that certain cytokines have been bound by their cognate receptors is transmitted within the cell, as James Darnell elucidated, by the **JAK-STAT pathway.** Cytokine receptors form complexes with proteins of the **Janus kinase (JAK)** family of NRTKs, so named because each of its four ~1150-residue members **(JAK1, JAK2, JAK3,** and **Tyk2)** has two PTK domains (Janus is the 2-faced Roman god of gates and doorways), although only

the C-terminal domain is functional (Fig. 19-42). **STATs** (for *Signal Transducers and Activators of Transcription)* comprise a family of seven ~800-residue proteins that are the only known transcription factors whose activities are regulated by Tyr phosphorylation and that have SH2 domains.

The JAK-STAT pathway functions as is diagrammed in Fig. 19-45:

1. Ligand binding induces the cytokine receptor to dimerize (or, in some cases, trimerize or even tetramerize).

2. The cytokine receptor's two associated JAKs are thereby brought into apposition, whereupon they reciprocally phosphorylate each other and then their associated receptors, a process resembling the autophosphorylation of dimerized RTKs (Section 19-3A). Note that unlike most NRTKs, JAKs lack both SH2 and SH3 domains.

3. STATs bind to the phosphoTyr group on their cognate activated receptor via their SH2 domain and are then phosphorylated on a conserved Tyr residue by the associated JAK.

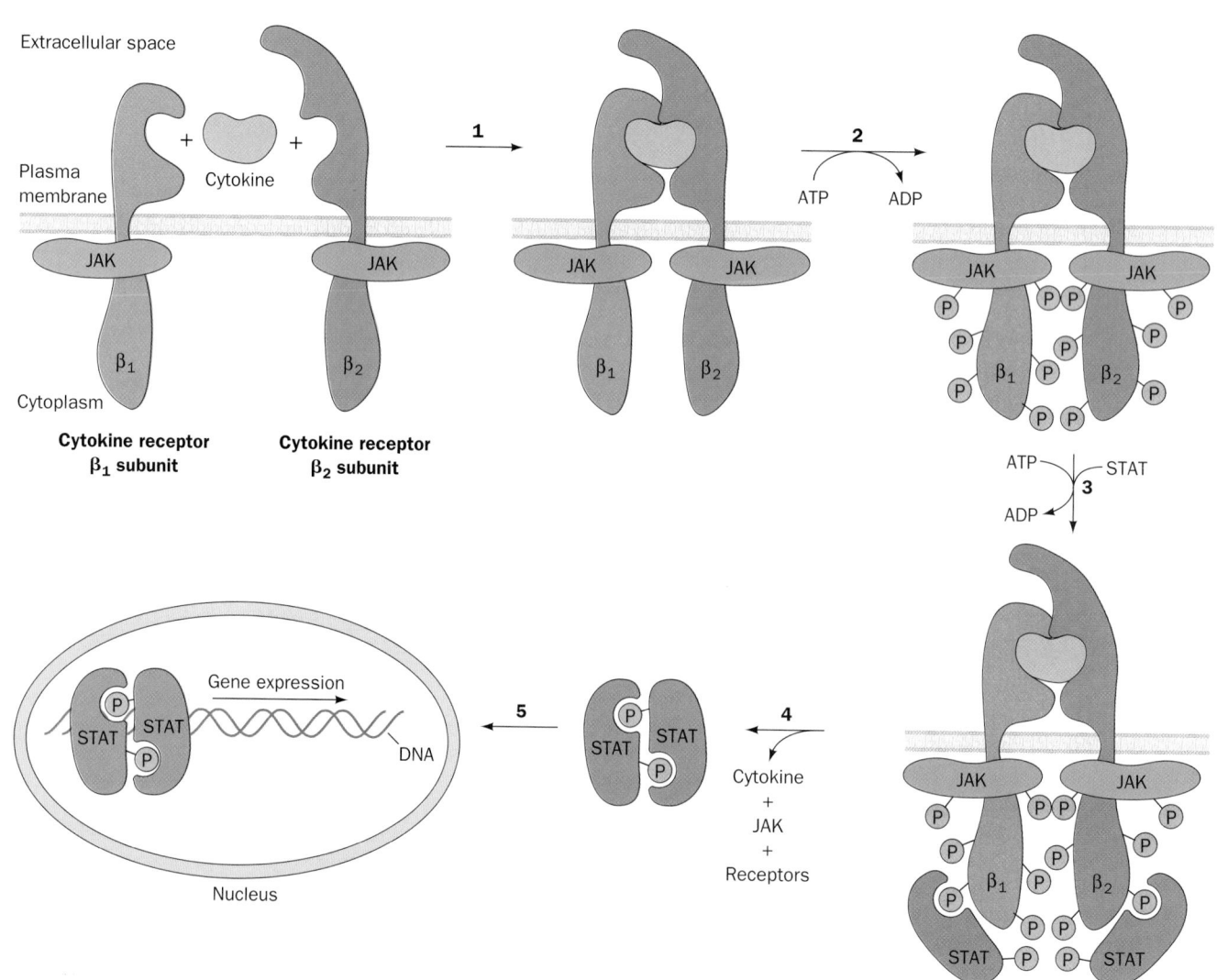

FIGURE 19-45   The JAK-STAT pathway for the intracellular relaying of cytokine signals. See the text for details. [After Carpenter, L.R., Yancopoulos, G.D., and Stahl, N., *Adv. Protein Chem.* **52,** 109 (1999).]

4. Following their dissociation from the receptor, the phosphorylated STATs homo- or heterodimerize via the association of their phosphoTyr residue with the SH2 domain on the opposing subunit.

5. The STAT dimers are then translocated to the nucleus, where these now functional transcription factors induce the expression of their target genes in much the same way as do the transcription factors that are phosphorylated by the MAPKs (Fig. 19-38).

c. PTKs Are Targets of Anticancer Drugs

The hallmark of **chronic myelogenous leukemia (CML)** is a specific chromosomal translocation (Section 34-4C) forming the so-called **Philadelphia chromosome** in which the *Abl* gene (which encodes the NRTK Abl) is fused with the *Bcr* gene (which encodes the protein Ser/Thr kinase **Bcr**). The Abl portion of this Bcr–Abl fusion protein is constitutively (continuously, without regulation) activated, probably because its Bcr portion oligomerizes. Hematopoietic stem cells (from which all blood cells are descended) bearing the Philadelphia chromosome are therefore primed to develop CML (malignancy requires several independent genetic alterations; Section 19-3B). Without a bone marrow transplant (a high-risk procedure that is unavailable to most individuals due to the lack of a suitable donor), CML is invariably fatal, with an average survival time of ~6 years.

An inhibitor of Abl would be expected to prevent the proliferation of, and even kill, CML cells. However, to be an effective anti-CML agent, such a substance must not inhibit other protein kinases because this would almost certainly cause serious side effects. Derivatives of 2-phenylaminopyrimidine bind to Abl with exceptionally high affinity and specificity. One such derivative, **gleevec** (also known as **glivec** and **STI-571**),

Gleevec (STI-571)

which was developed by Brian Druker and Nicholas Lydon, has caused the remission of symptoms in ~96% of CML patients with almost no serious side effects. This unprecedented performance occurs, in part, because gleevec is essentially inactive against other PTKs (an exception being the PDGF receptor) as well as protein Ser/Thr kinases. It is therefore the first protein kinase inhibitor that the FDA has approved for drug use (although, because the approval process took only 3 years, it has not yet been unequivocally established that gleevec actually prolongs the life of CML victims).

Abl resembles Src (Fig. 19-42) but lacks Src's C-terminal regulatory phosphorylation site (Figs. 19-43 and 19-44). The X-ray structure of Abl's PTK domain in complex with a truncated form of gleevec,

determined by Kuriyan (Fig. 19-46), reveals, as expected, that the truncated gleevec binds in Abl's ATP binding site [the piperazinyl group that this inhibitor lacks relative to

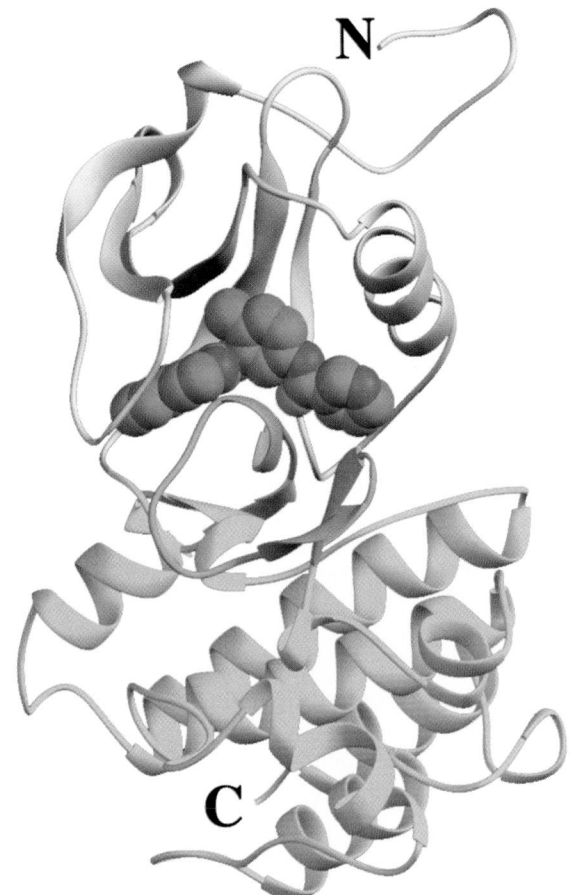

FIGURE 19-46 X-Ray structure of the Abl PTK domain in complex with a truncated derivative of gleevec. The protein is viewed from the right of the "standard" view of protein kinases (e.g., Figs. 19-26a and 19-43), with its N-terminal lobe pink, its C-terminal lobe cyan, and its activation loop light blue. The truncated gleevec, which occupies the PTK's ATP binding site, is shown in space-filling form with C green, N blue, and O red. [Based on an X-ray structure by John Kuriyan, The Rockefeller University. PDBid 1FPU.]

gleevec does not significantly alter its target discrimination but does increase gleevec's solubility and hence its bioavailability (Section 15-4B); it probably binds in a solvent accessible groove at the back of Abl]. Abl thereby adopts an inactive conformation in which its activation loop, which is not phosphorylated, appears to mimic the way in which substrate peptides bind to PTKs (such as the insulin receptor, Fig. 19-26a); that is, the activation loop assumes an autoinhibitory conformation. As a consequence, the N-terminal end of the activation loop, which has the highly conserved sequence Asp-Phe-Gly (whose Asp side chain, in the active PTK, ligates an Mg^{2+} ion that is essential for catalysis), assumes a conformation that is quite different from that observed in the X-ray structures of inactive Abl and Src (Fig. 19-43) because this latter conformation would block the binding of gleevec. This suggests that other clinically effective inhibitors can be found that exploit the subtle conformational differences among protein kinases.

F. *Protein Phosphatases*

As we previously discussed (Section 19-2E), to prevent an intracellular signaling pathway from being stuck in the "on" position, its signals must be rapidly eliminated once the message has been delivered. For proteins with phosphoTyr or phospho-Ser/Thr residues, this task is carried out by a variety of protein phosphatases, of which ~500 are encoded by the human genome (vs ~2000 protein kinases).

a. Protein Tyrosine Phosphatases Also Mediate Signal Transduction

The enzymes that dephosphorylate Tyr residues, the **protein tyrosine phosphatases (PTPs),** are not just simple housekeeping enzymes but are signal transducers in their own right. The PTPs, which were discovered by Nicholas Tonks, form a large family of diverse proteins that are present in all eukaryotes. Each PTP contains at least one conserved ~240-residue phosphatase domain that has the 11-residue signature sequence (I/V)HCXAGXGR(S/T)G, the so-called CX_5R motif, which contains the enzyme's catalytically essential Cys and Arg residues. The reaction proceeds via the nucleophilic attack of the Cys thiolate group on the P atom of the bound phosphoTyr to yield Tyr and a cysteinyl–phosphate intermediate that is subsequently hydrolyzed.

The PTPs have been subclassified into three groups: (1) receptorlike PTPs, (2) intracellular PTPs, and (3) dual-specificity PTPs, which can also dephosphorylate phospho-Ser/Thr residues. The receptorlike PTPs are constructed much like the RTKs (Fig. 19-23); that is, they have, from N- to C-terminus, an extracellular domain consisting of often multiple repeating modules that occur in other proteins, a single transmembrane helix, and a cytoplasmic domain consisting of a catalytically active PTP domain that, in most cases, is followed by a second PTP domain with little or no catalytic activity. These inactive PTP domains

are, nevertheless, highly conserved, which suggests that they have an important although as yet unknown function. Perhaps they serve as auxiliary substrate-binding sites. Biochemical and structural analyses indicate that ligand-induced dimerization of a receptorlike PTP reduces its catalytic activity, probably by blocking its active sites. Intracellular PTPs contain only one PTP domain, which is flanked by regions containing motifs, such as SH2 domains, that participate in protein–protein interactions. Structural studies reveal that the active sites of receptorlike and intracellular PTPs are too deep to bind phospho-Ser/Thr side chains—as we also saw to be the case for both PTK and SH2 domains (Sections 19-3A and 19-3C). However, the active site pockets of dual-specificity PTPs are sufficiently shallow to bind both phosphoTyr and phospho-Ser/Thr residues.

b. SHP-2 Is Inactivated by Binding Its Unliganded N-Terminal SH2 Domain

The cytoplasmic PTP **SHP-2** (also called **SH-PTP2**), which is expressed in all mammalian cells, binds to PTKs that are activated by a variety of ligands, including cytokines, growth factors, and hormones. The 591-residue SHP-2 consists of two tandem SH2 domains, followed by a PTP domain and a 66-residue C-terminal tail that contains Tyr-phosphorylation sites as well as a Pro-rich segment that may bind SH3 or WW-containing proteins. SHP-2's PTP activity is increased ~10-fold on binding peptides with a single phosphoTyr residue and ~100-fold with those having two phosphoTyr residues and at much lower peptide concentrations. SHP-2 binds to both growth factor and cytokine receptors via its SH2 domains and, when its C-terminal tail is phosphorylated, also functions as an adaptor to recruit Grb2 so as to activate MAP kinase pathways (Section 19-3D).

The X-ray structure of SHP-2 lacking its C-terminal tail (Fig. 19-47), determined by Eck and Steven Shoelson, reveals that the N-terminal SH2 domain (N-SH2) interacts extensively with the PTP domain. N-SH2 inhibits the PTP by inserting its D'E loop far into the PTP's 9-Å-deep catalytic cleft, where the loop interacts with the PTP's catalytic Arg and Cys residues and prevents the active site closure observed in the X-ray structure of a PTP in complex with a phosphopeptide. In contrast, the more C-terminal SH2 domain (C-SH2) does not have a significant interface with either the N-SH2 or the PTP domains.

The phosphopeptide-binding sites on both SH2 domains face away from the PTP domain and are therefore fully exposed on the protein surface. However, the comparison of the structure of N-SH2 complexed to a phosphopeptide with that in the above autoinhibited form of SHP-2 indicates that, in the autoinhibited form, N-SH2 adopts a conformation in which it is unable to bind phosphoTyr. Evidently, the conformations of N-SH2's PTP-binding surface and phosphopeptide binding site are allosterically linked such that its binding of PTP and phosphopeptide are mutually exclusive. The C-SH2 domain does not participate in PTP activation, although it almost

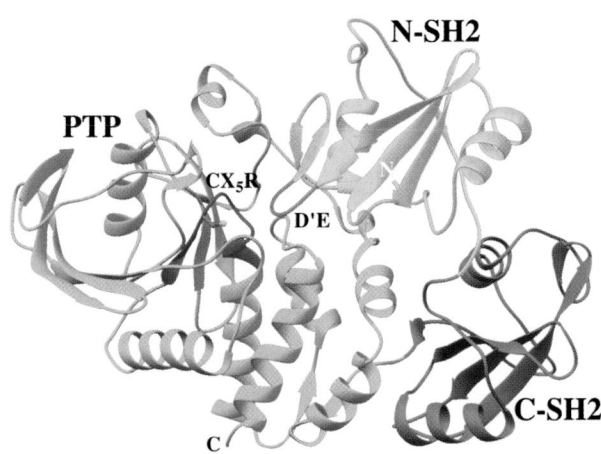

FIGURE 19-47 X-Ray structure of the protein tyrosine phosphatase SHP-2. In this structure, its N-SH2 domain is gold with its D′E loop red, its C-SH2 domain is green, and its PTP domain is cyan, with its 11-residue signature sequence, its CX_5R motif, blue and the side chain of its catalytically essential Cys residue shown in ball-and-stick form with C green and S yellow. [Based on an X-ray structure by Michael Eck and Steven Shoelson, Harvard Medical School. PDBid 2SHP.]

tein phosphorylation in controlling glycogen metabolism; Section 18-3C). The majority of these enzymes are members of two protein families: the **PPP family,** which consists of **PP1, PP2A,** and **PP2B** (PP for *p*hospho*p*rotein *p*hosphatase); and the **PPM family,** which consists of **PP2C.** The PPP and PPM families are unrelated to each other or to the PTPs. We have already considered PP1 in connection with the role of its catalytic subunit, PP1c, in dephosphorylating the proteins that regulate glycogen metabolism as well as the roles of its targeting subunits, G_M and G_L, in binding PP1c to glycogen in muscle and liver (Section 18-3C). Indeed, all PP1c's are associated with one or two regulatory (R) subunits that function to modulate the activity of their bound PP1c's, target them to substrates in specific subcellular locations, or modify their substrate specificities. It is the large variety of these R subunits that permits the limited number (1–8) of genetically distinct but closely similar (~90% sequence identity) PP1c's in a eukaryotic cell to carry out their diverse functions.

X-Ray structures have shown that PPP catalytic centers each contain an Fe^{2+} (or possibly a Fe^{3+}) ion and a Zn^{2+} (or possibly an Mn^{2+}) ion, whereas PPM catalytic centers each contain two Mn^{2+} ions. These binuclear metal ion centers nucleophilically activate water molecules to dephosphorylate substrates in a single reaction step.

certainly contributes binding energy and specificity to the binding of a phosphopeptide.

c. Bubonic Plague Virulence Requires a PTP

Bacteria lack PTKs and hence do not synthesize phosphoTyr residues. Nevertheless, PTPs are expressed by bacteria of the genus *Yersinia*, most notably *Yersinia pestis,* the pathogen that causes **bubonic plague** (the flea-transmitted "Black Death," which, since the 6th century, has been responsible for an estimated ~200 million human deaths including about one-third of the European population in the years 1347–1350). The *Y. pestis* PTP, **YopH,** which is required for bacterial virulence, is far more active than other known PTPs. Hence, when *Yersinia* injects YopH into a cell, the cell's phosphoTyr-containing proteins are catastrophically dephosphorylated. Although YopH is only ~15% identical in sequence to mammalian PTPs, it contains all of their invariant residues and their X-ray structures are closely similar. This suggests that an ancestral *Yersinia* acquired a PTP gene from a eukaryote. However, the discovery of a dual-specificity protein phosphatase in a free-living cyanobacterium raises the possibility that PTPs arose before the divergence of eukaryotes and prokaryotes.

d. Cells Contain Several Types of Protein Ser/Thr Phosphatases

The **protein Ser/Thr phosphatases** were first characterized by Earl Sutherland (who discovered the role of cAMP as a 2nd messenger; Section 18-3E) and by Edmond Fischer and Edwin Krebs (who discovered the role of pro-

e. PP2A Is Structurally Variable and Functionally Diverse

PP2A participates in a wide variety of regulatory processes including those governing metabolism, DNA replication, transcription, and development. It consists of three different subunits:

1. A catalytic subunit (C), whose N-terminal catalytic domain contains the ~280-residue catalytic core common to all PPP family members. Its C-terminal regulatory domain contains an activating binding site for Ca^{2+}–calmodulin, an inactivating Tyr phosphorylation site that is targeted by a variety of PTKs including the EGF and insulin receptors, and a C-terminal autoinhibitory tail.

2. A scaffold subunit (A; also called PR65), with which the C subunit is tightly associated in the cell.

3. One of four different regulatory subunits (B, B′, B″, and B‴), which bind to both the A and C subunits.

All of PP2A's subunits have multiple isoforms and splice variants that are expressed in a tissue-specific and developmentally specific manner, thereby generating an enormous panoply of enzymes that are targeted to different phosphoproteins in distinct subcellular sites. This complexity is a major cause of our limited understanding of how PP2A carries out its diverse cellular functions, even though it comprises between 0.3 and 1% of cellular proteins.

The X-ray structures of neither the B or C subunits of PP2A have yet been reported (although the C subunit's catalytic core is expected to resemble those in the known X-ray structures of other PPP proteins). However, the X-ray struc-

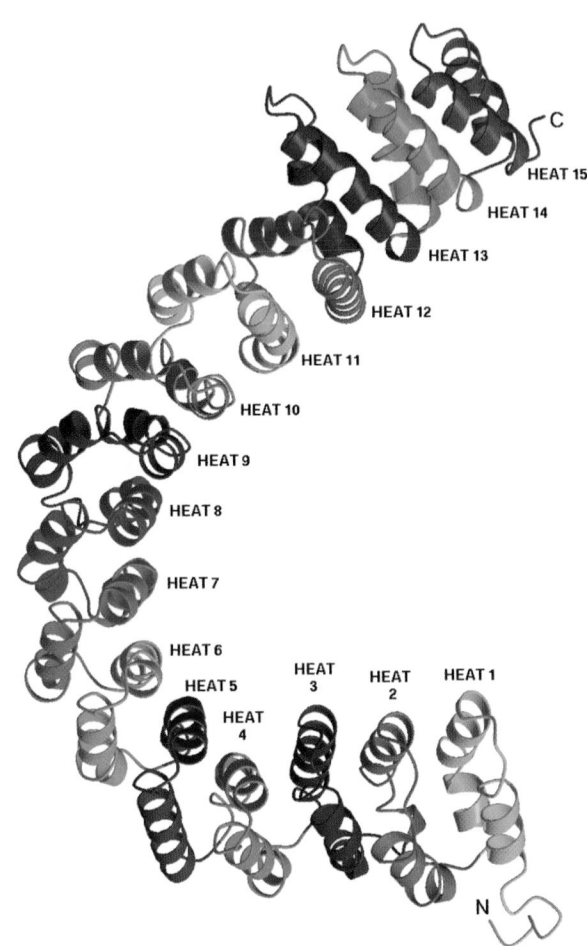

FIGURE 19-48   X-Ray structure of the A subunit of PP2A.
Each of its HEAT repeats is shown in a different color.
Compare this structure to that of IκBα(Fig. 12-38), which forms
a right-handed solenoid consisting of ankyrin repeats. [Courtesy
of Bostjan Kobe, St. Vincent's Institute of Medical Research,
Fitzroy, Victoria, Australia. X-Ray structure by David Barford,
University of Oxford, U.K. PDBid 1B3U.]

ture of its A subunit (Fig. 19-48), determined by David
Barford, reveals a remarkable solenoidal protein that consists of 15 imperfect tandem repeats of a 39-residue sequence
termed HEAT (because it occurs in proteins named
*H*untingtin, *E*F3, *A* subunit of PP2A, and *T*OR1). Successive
HEAT repeats, which each consist of two antiparallel helices
joined by a short linker, stack on one another with their corresponding helices nearly parallel so as to form an ~100-Å-
long right-handed superhelix (helix of helices) with a hook-
like shape. Mutational studies in which specific HEAT
repeats were deleted indicate that HEAT repeats 11 to 15
are necessary and sufficient to bind the PP2A C subunit,
whereas the B subunits bind to HEAT repeats 1 to 10. These
subunits are thought to bind to a ridge of conserved hydrophobic side chains on the A subunit's concave surface.

f. PP2B Is the Target of Immunosuppressant Drugs

PP2B, which is also known **calcineurin (CaN),** is unique
among protein Ser/Thr phosphatases in that it is activated
by Ca^{2+}. CaN is a heterodimer composed of a catalytic A
subunit (CaNA) and a regulatory B subunit (CaNB).

CaNA contains an N-terminal catalytic domain followed
by a CaNB-binding domain, a calmodulin (CaM)-binding
domain, and a C-terminal autoinhibitory segment. CaNB,
which has 35% sequence identity with CaM, binds four
Ca^{2+} ions via its four EF hand motifs (Section 18-3C). CaN
is activated by the binding of Ca^{2+} to CaNB and Ca^{2+}–CaM
to CaNA.

Calcineurin plays an essential role in the antigen-
induced proliferation of *T* cells. As we shall see in Section
35-2D, the binding of an antigenic peptide to a **T cell
receptor,** a tyrosine kinase–associated receptor, initiates a
complex series of signaling events involving the Src-like
PTKs Lck and Fyn, a MAP kinase cascade, and the phosphoinositide cascade (Section 19-4), which, among other
things, releases Ca^{2+} into the cytosol. The Ca^{2+}, in turn,
activates CaN to dephosphorylate the transcription factor
NFATₚ (for *n*uclear *f*actor of *a*ctivated *T* cells). NFATₚ in
complex with CaN is then translocated to the nucleus
where, in concert with other transcription factors, it induces some of the early steps in *T* cell proliferation.

As we discussed Section 9-2B, the fungal products **cyclosporin A (CsA)** and **FK506** are highly effective immunosuppressants that are in clinical use for the prevention of organ-transplant rejection and for the treatment of
autoimmune disorders (processes that are mediated by *T*
cells). CsA and FK506, respectively, bind to the peptidyl
prolyl cis–trans isomerases (rotamases) **cyclophilin** and
FK506 binding protein (FKBP12), which are therefore
collectively known as **immunophilins.** However, the observation that both CsA and FK506 are effective immunosuppressants at concentrations far below those of the
immunophilins suggests that it is the presence of the cy-
clophilin·CsA and FKBP12·FK506 complexes themselves,
rather than the inhibition of their rotamase activity, that
interferes with *T* cell proliferation. It is, in fact, the binding of either complex to CaN that prevents it from dephosphorylating NFATₚ and thereby suppresses *T* cell
proliferation.

The X-ray structures of the bovine complex
FKBP12·FK506–CaN, by Manuel Navia, and the corresponding human complex, by Ernest Villafranca, reveal
how the FKBP12·FK506 complex binds to CaN (Fig.
19-49*a*). The catalytic domain of CaNA, with its binuclear
Fe^{2+}–Zn^{2+} center marking its active site, resembles those
of other protein Ser/Thr phosphatases of known structure.
A 22-residue α helix at the C-terminal end of this phosphatase domain, which extends out from the phosphatase
domain by up to 40 Å, provides much of the CaNB binding site. Beyond this helix, the C-terminal portion of CaNA,
which contains the CaM binding site and the autoinhibitory segment, is not visible due to disorder. However,
in the X-ray structure of CaN alone (Fig. 19-49*b*), the autoinhibitory segment is seen to bind in the CaNA active site
so as to block the access of substrate phosphoproteins. The
structure of CaNB, which has four bound Ca^{2+} ions, resembles that of Ca^{2+}–CaM in complex with a helical target
peptide (Fig. 18-18) except that CaNB's two globular domains are on the same side of its bound peptide rather than
on opposite sides as are those of Ca^{2+}–CaM. CaNB thereby
forms a continuous groove in which the CaNA helix binds.

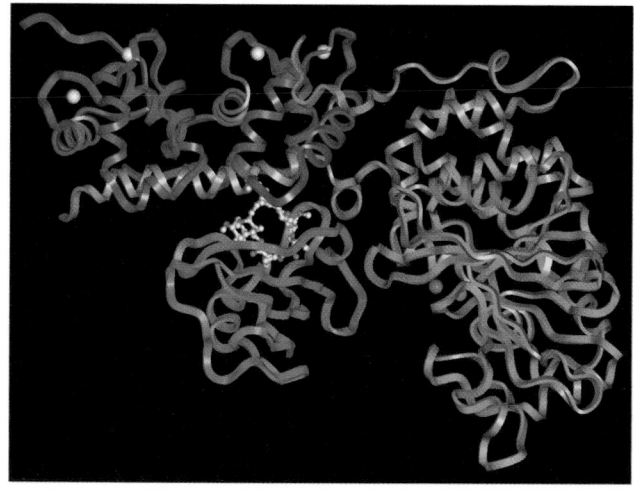

(a)

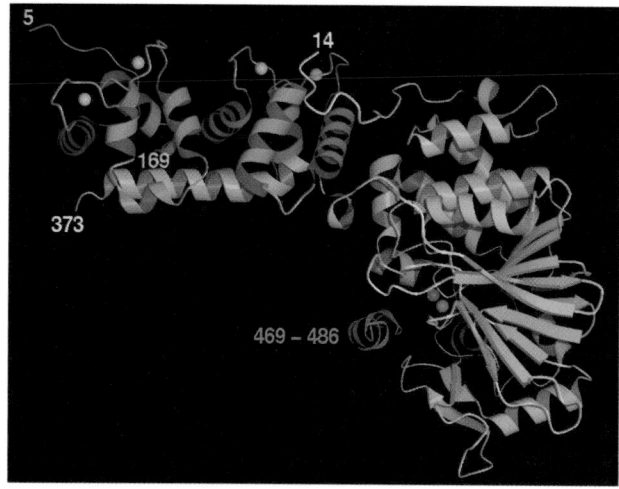

(b)

FIGURE 19-49 Calcineurin. *(a)* X-Ray structure of human FKBP12·FK506–CaN. The CaNA subunit is orange, the CaNB subunit is blue, and the FKBP12 is purple. The FK506 is shown in ball-and-stick form with C white, N blue, and O red; the Fe^{2+} and Zn^{2+} ions in the CaNA active site are, respectively, represented by red and green spheres; and the Ca^{2+} ions bound to CaNB are represented by white spheres. *(b)* X-Ray structure of human CaN with CaNA yellow, its autoinhibitory segment red, and CaNB cyan. The Fe^{2+}, Zn^{2+}, and Ca^{2+} ions are red, green, and white, respectively. [Courtesy of J. Ernest Villafranca, Blanchette Rockefeller Neurosciences Institute, Morgantown, West Virginia.]

FKBP12·FK506 binds CaN so as to contact both CaNA and CaNB, with the portion of FK506 that extends out from its FKBP12 binding site forming a significant part of this interface. The structures of FKBP12 and CaN in this complex closely resemble those in the X-ray structures of these proteins alone. It therefore appears that FK506 provides a critical component of this contact. Nevertheless, no part of the FKBP12·FK506 complex is within 10 Å of CaN's phosphatase site (although the CaN autoinhibitory segment has been displaced). This accounts for the observations that FKBP12·FK506 strongly inhibits CaN from dephosphorylating a 20-residue phosphopeptide but actually increases the rate at which CaN dephosphorylates the much smaller *p*-nitrophenylphosphate by a factor of 3.

The CaN residues that mutational studies have implicated in binding cyclosporin·CsA are largely the same as those that contact FKBP12·FK506 (Fig. 19-49*a*). Evidently, cyclosporin·CsA and FKBP12·FK506 bind to the same general region of CaN and hence have a similar mechanism of inhibition. However, the characteristics of rotamases that apparently uniquely suit them for their roles in CaN inhibition remain unknown.

4 ■ THE PHOSPHOINOSITIDE CASCADE

Extracellular signals often cause a transient rise in the cytosolic $[Ca^{2+}]$, which, in turn, activates a great variety of enzymes through the intermediacy of calmodulin and its homologs. An increase in cytosolic $[Ca^{2+}]$ triggers such diverse cellular processes as glycogenolysis (Section 18-3C) and muscle contraction (Section 35-3C). What is the source of this Ca^{2+} and how does it enter the cytosol? In certain types of cells, neurons (nerve cells; Fig. 1-10*d*), for exam-

ple, the Ca^{2+} originates in the extracellular fluid. However, the observation that the absence of extracellular Ca^{2+} does not inhibit certain Ca^{2+}-mediated processes led to the discovery that, in these cases, cytosolic Ca^{2+} is obtained from intracellular reservoirs, mostly the endoplasmic reticulum (and its equivalent in muscle, the sarcoplasmic reticulum). Extracellular stimuli leading to Ca^{2+} release must therefore be mediated by an intracellular signal.

The first clue as to the nature of this signal came from observations that the intracellular mobilization of Ca^{2+} and the turnover of **phosphatidylinositol-4,5-bisphosphate** (**PIP₂** or **PtdIns-4,5-P₂**; Fig. 19-50), which constitutes <0.05% of cellular lipids, are strongly correlated. This in-

$$\begin{array}{c}
\text{O} \\
\text{CH}_2\text{—O—C—R}_1 \\
\text{R}_2\text{—C—O—C—H} \\
\text{CH}_2\text{—O—P—O} \\
\text{O}^-
\end{array}$$

Y = H or PO_3^{2-}

FIGURE 19-50 Molecular formula of the phosphatidylinositides. The head group of these glycerophospholipids is *myo*-inositol that may be phosphorylated at its 3-, 4-, and/or 5- positions. R_1 is predominantly the hydrocarbon tail of stearic acid (an 18:0 fatty acid; Table 12-1) and R_2 is predominantly the hydrocarbon tail of arachidonic acid (a 20:4 fatty acid).

formation led Robert Michell to propose, in 1975, that PIP$_2$ hydrolysis is somehow associated with Ca^{2+} release.

A. Ca^{2+}, Inositol Trisphosphate, and Diacylglycerol Are Second Messengers

Investigations, notably by Mabel and Lowell Hokin, Michael Berridge, and Michell, eventually revealed that *PIP$_2$ is part of an important second messenger system, the phosphoinositide cascade, that mediates the transmission*

of numerous hormonal signals including those of vasopressin, CRF, TRF (Section 19-1H), acetylcholine (a neurotransmitter; Section 20-5C), epinephrine (with α_1-adrenoreceptors; Section 19-1F), EGF, and PDGF. Remarkably, this system yields up to three separate types of second messengers through the following sequence of events (Fig. 19-51):

1–3. The ligand–receptor interactions described below activate a phosphoinositide-specific **phospholipase C (PLC)** to hydrolyze PIP$_2$ to **inositol-1,4,5-trisphosphate**

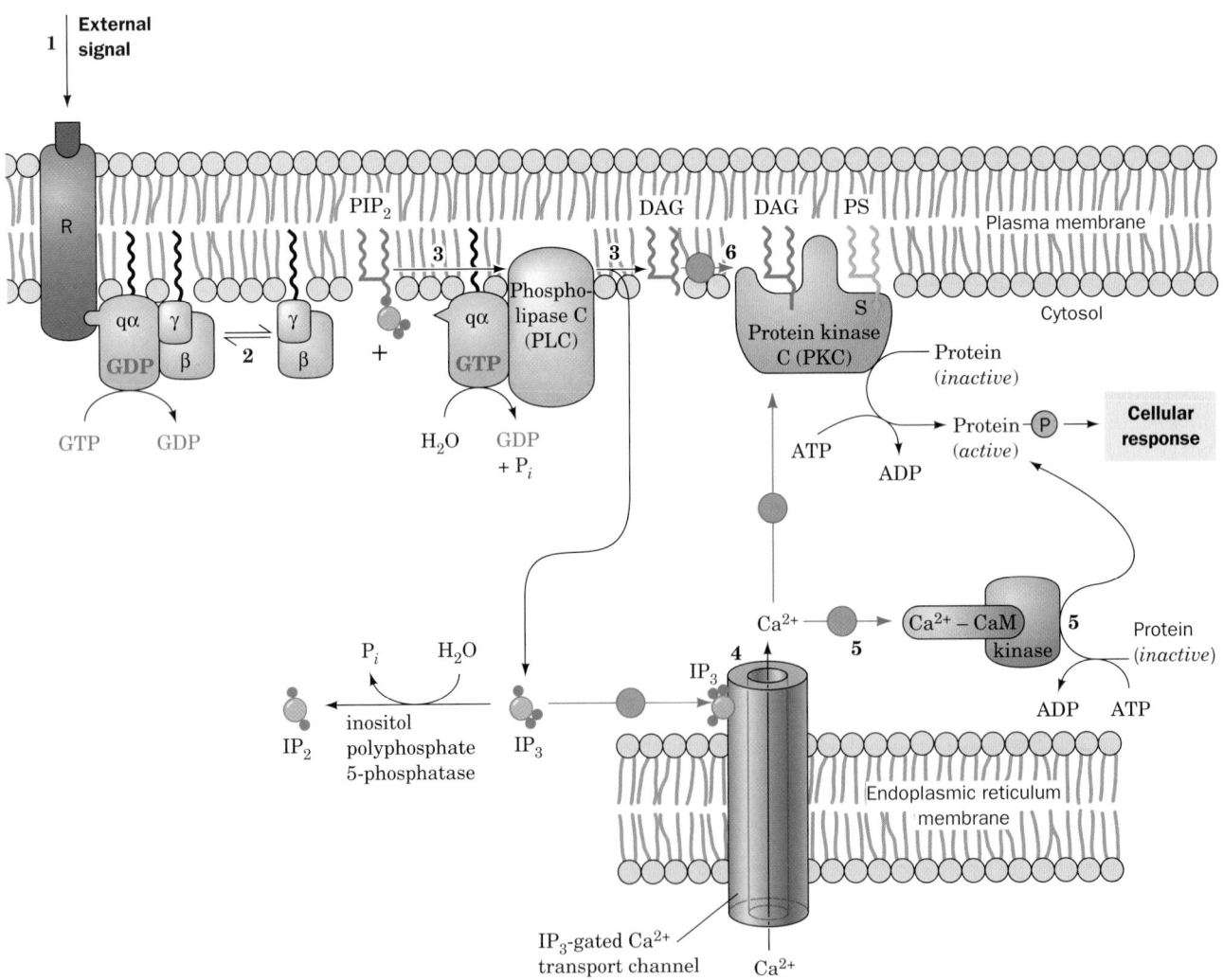

FIGURE 19-51   Role of PIP$_2$ in intracellular signaling. (1) The binding of a ligand to a cell surface receptor, R, activates a phosphoinositide-specific phospholipase C through the intermediacy of what is shown here as **(2)** a G protein (G$_q$; Fig. 19-16) but in many cases is an RTK, an NRTK, or, possibly, Ca^{2+}. Phospholipase C catalyzes the hydrolysis of PIP$_2$ to IP$_3$ and DAG **(3)**. The water-soluble IP$_3$ stimulates the release of Ca^{2+} sequestered in the endoplasmic reticulum **(4)**, which in turn activates numerous cellular processes through the intermediacy of calmodulin and its homologs **(5)**. The nonpolar DAG remains associated with the inner leaflet of the membrane, where it activates protein kinase C to phosphorylate and thereby modulate the activities of a number of cellular proteins **(6)**. This latter activation process also requires the presence of the membrane lipid phosphatidylserine (PS) and Ca^{2+}. *See the* **Animated Figures**

(IP₃ or **Ins-1,4,5-P₃)** and ***sn*-1,2-diacylglycerol (DAG** or **DG)**

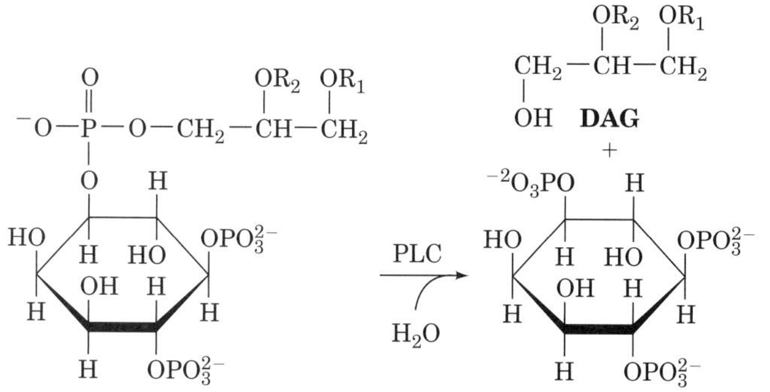

[the stereospecific numbering (*sn*) system is described in the legend of Fig. 12-3]. PLCs catalyze the hydrolysis of the bond linking a glycerophospholipid to its phosphoryl group as indicated in Fig. 19-52 (which also shows the actions of other types of phospholipases). Note that this reaction occurs at the interface between the aqueous phase and the membrane in such a way that both PIP₂ and its amphipathic hydrolysis product DAG remain associated with the membrane during the catalytic reaction.

4. The water-soluble IP₃, acting as a second messenger, diffuses through the cytoplasm to the ER, from which it stimulates the release of Ca^{2+} into the cytoplasm by binding to and thereby opening an ER-bound transmembrane Ca^{2+}-specific ion channel known as the **IP₃ receptor** (ion channels are discussed in Chapter 20).

5. The Ca^{2+}, in turn, stimulates a variety of cellular processes through the intermediacy of calmodulin and its homologs.

6. The amphipathic DAG is constrained to remain in the inner leaflet of the plasma membrane, where it nevertheless also acts as a second messenger by activating **protein kinase C (PKC)** in the presence of Ca^{2+} and phosphatidylserine (PS; which is located exclusively on the cytoplasmic face of the plasma membrane). This membrane-bound enzyme (actually a family of enzymes; Section 19-4C), in turn, phosphorylates and thereby modulates the activities of several different proteins including

glycogen synthase (Section 18-3D). DAG, which predominantly has a stearoyl group at its 1-position and an arachidonoyl group at its 2-position, is further degraded in some cells by **cytosolic phospholipase A₂ (cPLA₂)** to yield arachidonate, the major substrate for the biosynthesis of prostaglandins, prostacyclins, thromboxanes, leukotrienes, and lipoxins. These paracrine hormones, as we discuss in Section 25-7, mediate or modulate a wide variety of physiological functions.

B. *The Phospholipases C*

Mammals express four classes of phosphoinositide-specific PLCs comprising 11 different isozymes, β1–4, γ1–2, δ1–4, and ε (Fig. 19-53; the isozyme originally named PLC-α is actually a proteolytic fragment of PLC-δ1), some of which additionally have splice variants. All of these PLCs require the presence of Ca^{2+} for enzymatic activity. The PLC-δ isozymes (~760 residues), the smallest of these PLCs, consist of, from N- to C-terminus, an ~120-residue pleckstrin homology (PH) domain (Section 19-3C); an ~130-residue EF hand domain that contains four EF hand motifs (Fig. 18-17); two conserved regions known as X and Y that together form the ~250-residue PLC catalytic domain and which are separated by an ~60-residue linker; and an ~130-residue **C2 domain,** a domain that in many cases binds Ca^{2+} and which occurs in >400 proteins that mainly participate in signal transduction and membrane interactions. The PLC-β isozymes (~1200 residues) have an additional ~420-residue C-terminal tail that has been implicated in both membrane association and regulation by G proteins (see below). In contrast, the PLC-γ isozymes (~1270 residues) contain an ~420-residue insert between X and Y that consists of an additional PH domain that is split by two SH2 domains that are implicated in binding to activated PTKs (see below) as well as an SH3 domain. PLC-ε (~2300 residues) differs from other PLCs in that it lacks a PH domain and has an N-terminal RasGEF domain and two C-terminal Ras-binding **(RA)** domains. The observation that the PLCs in plants and lower eukaryotes such as yeast are δ-like suggests that the PLC-β, γ, and ε isozymes in mammals evolved from a primordial PLC-δ.

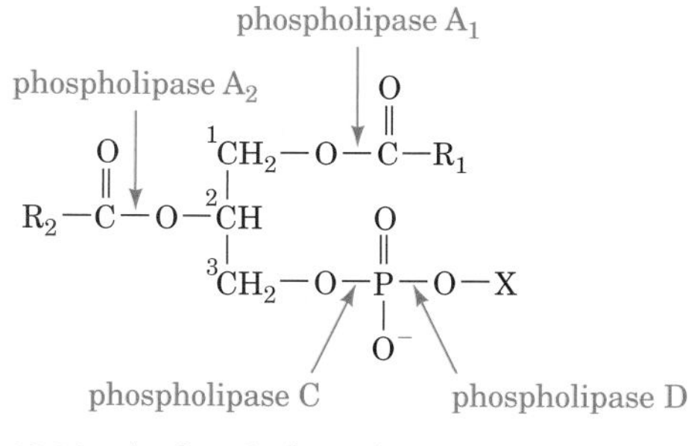

FIGURE 19-52 A phospholipase is named according to the bond that it cleaves on a glycerophospholipid. X is a phosphoinosityl group for this discussion.

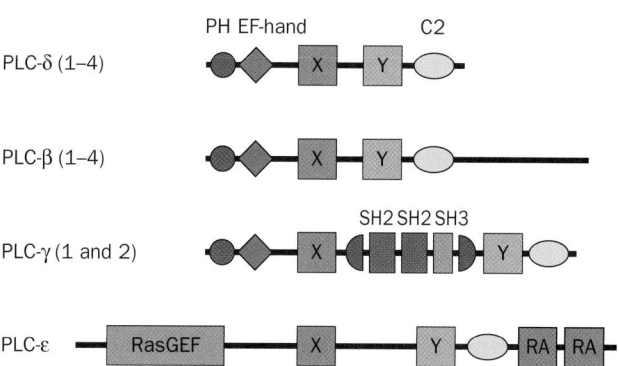

FIGURE 19-53 Domain organization of the four classes of phosphoinositide-specific PLCs. [After Rhee, S.G., *Annu. Rev. Biochem.* **70,** 284 (2001).]

a. The Structure of PLC-δ1 Explains Its Function

The X-ray structure of PLC-δ1 lacking its N-terminal PH domain (but catalytically active *in vitro*) and in complex with IP₃ and Ca²⁺ was determined by Roger Williams. It reveals (Fig. 19-54) that the enzyme's remaining three domains are linked by extended polypeptide segments that form tight interactions with these domains. EF hand motifs 1 and 2 form a lobe that is closely superimposable on the N- and C-terminal lobes of calmodulin (Figs. 18-16 and 18-18; the N-terminal helix of EF hand 1 is disordered), as do EF hands 3 and 4. Although EF hands are usually associated with Ca²⁺ ions, no Ca²⁺ is observed to bind to any of the EF hands in PLC-δ1. In fact, EF hands 3 and 4 lack typical Ca²⁺ ligands. Thus, although mutations to this domain are generally deleterious to the enzyme's catalytic activity, its function is unclear.

The X and Y regions of PLC-δ1's catalytic domain form an α/β barrel (Section 8-3B) that is seen in side view in Fig. 19-54. The X region contributes a typical βαβαβαβα motif to form half the barrel. However, the Y region, which forms the second half of the barrel, has a flexible loop in place of this motif's first α helix. The 43-residue peptide linking X and Y is disordered and hence is not visible in the X-ray structure; its deletion does not greatly affect the enzyme's activity. The IP₃ is stereospecifically bound to the enzyme in a broad depression on the C-terminal face of the 8-stranded parallel β barrel via an extensive network of hydrogen bonds and charge–charge interactions. A 6-coordinate Ca²⁺ is bound at the bottom of this active site, with one of its ligands contributed by the 2-hydroxyl group of IP₃ and the remainder formed by highly conserved Asp, Glu, and Asn side chains. The catalytic reaction is postulated to occur via a mechanism analogous to that of the RNase A–catalyzed hydrolysis of RNA (Fig. 15-3) in which PIP₂'s 2-hydroxyl group nucleophilically attacks the neighboring 1-phosphate group to form DAG and a cyclic phos-phodiester intermediate that is subsequently hydrolyzed to yield IP₃. The Ca²⁺ (rather than a His side chain as in RNase A) is properly positioned to promote the deprotonation of the 2-hydroxyl group so as to enhance its nucleophilicity and to subsequently help stabilize the developing negative charge on the pentavalent phosphorus in the catalytic reaction's transition state (Fig. 16-6*b*). This explains why **2-deoxy-PIP₂** is not hydrolyzed by mammalian phosphinositide-specific PLC's.

The C2 domain, which consists of a sandwich of two 4-stranded antiparallel β sheets, also binds a Ca²⁺ ion. However, the X-ray structure of PLC-δ1 complexed with the calcium analog lanthanum revealed two additional metal ion (Ca²⁺) binding sites in close proximity to the latter Ca²⁺ binding site. All of these metal ions lie in a crevice at one end of the β sandwich, where they are exposed on the surface of the enzyme. It therefore seems likely that, *in vivo*, they associate with anionic head groups such as that of phosphatidylserine on the surface of a membrane. Since the extensive interface between the C2 and catalytic domain appears to be rigid, this interaction probably helps bind the catalytic domain to the membrane such that it can productively interact with PIP₂ molecules. This association appears to be supplemented through the interactions of a hydrophobic ridge comprised of 3 loops from one side of the active site opening that is postulated to penetrate into the membrane's nonpolar region during catalysis. This would explain how the enzyme can catalyze the hydrolysis of PIP₂ to DAG and IP₃ while the former two compounds remain associated with the membrane.

b. The Pleckstrin Homology Domain Tethers PLC-δ1 to the Membrane

The X-ray structure of PLC-δ1's N-terminal PH domain (absent in Fig. 19-54) in complex with IP₃, determined by Joseph Schlessinger and Sigler, reveals that IP3 binds to a positively charged surface of the protein (Fig. 19-55). This is consistent with the PH domain's proposed role as a membrane anchor, as are the observations that this PH domain binds PIP₂ with much greater affinity ($K_D = 1.7 \, \mu M$) than does PLC-δ1's catalytic domain ($K_D > 0.1 \, mM$). Since the peptide segment linking the PH domain to the rest of the enzyme is probably flexible, it appears that the PH domain functions to tether the enzyme to the membrane. This accounts for kinetic measurements indicating that the enzyme catalyzes multiple cycles of PIP₂ hydrolysis without releasing the membrane.

c. The PLC-β Isozymes Are Activated by Heterotrimeric G Proteins

The PLC-βs are hormonally regulated by certain G protein-coupled receptors (e.g., those for histamine, vasopressin, TSH, thromboxane A₂, and angiotensin II) via their associated heterotrimeric G proteins, as indicated in Fig. 19-51. In particular, they are activated through their interactions with the α subunits of the G_q subfamily (Section 19-2C) in complex with GTP. G_qα·GTPγS activates the PLC-β isoforms in order of potency β1 > β3 > β2, with the position of β4 in this hierarchy indeterminate

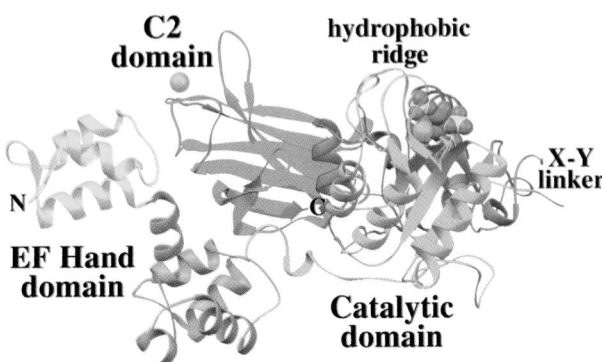

FIGURE 19-54   X-Ray structure of phospholipase C-δ1 lacking its N-terminal PH domain in complex with PIP₃ and Ca²⁺ ions. Its first two EF hand motifs are gold, the second two are orange, the X-region of the catalytic domain is cyan, its Y region is light blue, the loops forming the catalytic domain's hydrophobic ridge are tan, and the C2 domain is magenta. The PIP₃ is shown in space-filling form with C green, O red, and P yellow, and the Ca²⁺ ions are represented by silver spheres. [Based on an X-ray structure by Roger Williams, MRC Laboratory of Molecular Biology, Cambridge, U.K. PDBid 1DJX.]

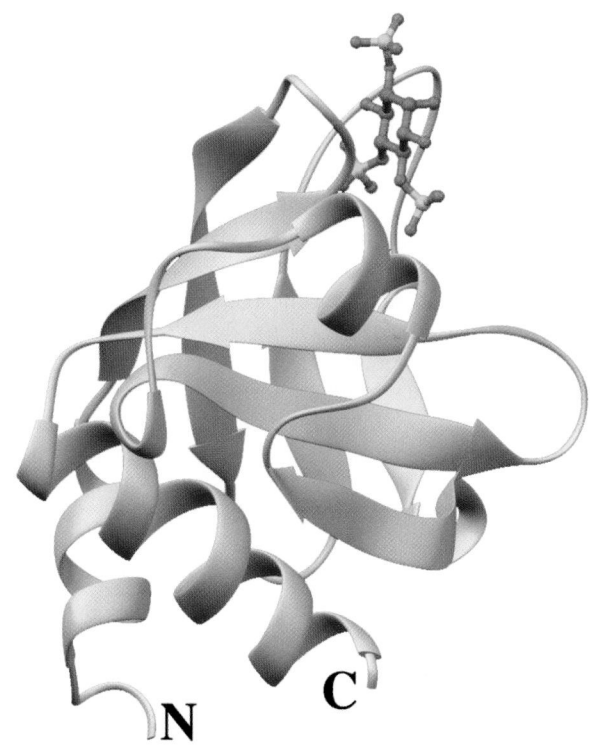

FIGURE 19-55   X-Ray structure of the pleckstrin homology domain of PLC-δ1 in complex with PIP₃. The PIP₃ is shown in ball-and-stick form with C green, O red, and P yellow. The PH domain consists largely of the β barrel/sandwich of 7 antiparallel strands and C-terminal α helix common to the numerous PH domains of known structure. [Based on an X-ray structure by Joseph Schlessinger, New York University Medical Center, and Paul Sigler, Yale University. PDBid 1MAI.]

because it is inhibited by GTPγS. The C-terminal tail that is unique to PLC-β isoforms (Fig. 19-53) has been implicated, via studies of truncated enzymes, in binding to $G_{q\alpha}$·GTP. This ~420-residue segment, which according to secondary structure predictions (Section 9-3A) is mainly α helical, contains a high proportion of basic residues that are present in clusters. These basic clusters, whose mutation results in reduced responses to $G_{q\alpha}$·GTP, are likely to interact with acidic phospholipids. These interactions are critical for the membrane localization of the PLC-β isoforms because their PH domains do not bind PIP_2. Thus, membrane localization of the PLC-β isoforms might be achieved via their binding to phospholipid head groups and/or their interactions with membrane-anchored $G_{q\alpha}$·GTP. An important aspect of the regulation of PLC-β isoforms by $G_{q\alpha}$·GTP is that the PLC-βs function as GAPs to increase the GTPase activity of $G_{q\alpha}$ by >50-fold, thereby limiting the activating function of $G_{q\alpha}$.

The PLC-β isoforms are independently activated by $G_{\beta\gamma}$ complexes, which may be supplied by the dissociation of heterotrimeric G proteins other than $G_{q\alpha}G_{\beta\gamma}$. Moreover, their order of efficacy with $G_{\beta\gamma}$ differs from that with $G_{q\alpha}$·GTP: β3 > β2 > β1, with β4 insensitive to the presence of $G_{\beta\gamma}$. Although the concentration of $G_{\beta\gamma}$ required

for maximal activation of the PLC-βs is much greater than that of $G_{q\alpha}$·GTP, their final extents of activation are similar. The sites on PLC-β2 that interact with $G_{\beta\gamma}$ are its PH domain and a 10-residue segment near the N-terminus of its Y region. This explains why C-terminally truncated PLC-β2 is activated by $G_{\beta\gamma}$ but not by $G_{q\alpha}$·GTP. The region of $G_{\beta\gamma}$ that interacts with PLC-βs overlaps the region through which it binds G_α subunits, thereby explaining why a $G_{\beta\gamma}$ cannot simultaneously bind a PLC-β and a G_α·GDP.

d. The PLC-γ Isozymes Are Activated by Protein Tyrosine Kinases

The PLC-γ isozymes in a wide variety of cells are activated by certain protein growth factors including EGF, PDGF, FGF, and NGF. These growth factors cause their corresponding receptors, which are RTKs (Section 19-3A), to autophosphorylate at particular Tyr residues. Some of these phosphoTyr sites are specifically bound, as mutational studies indicate, by the more N-terminal SH2 (N-SH2) domain on PLC-γ1 (Figs. 19-53 and 19-25a) but not its C-SH2 domain. The N-SH2 domain binds peptides containing a phosphoTyr residue followed by at least 5 predominantly hydrophobic residues, in contrast to the SH2 domain of Src, which preferentially binds pYEEI containing peptides (Section 19-3C).

The activated receptors for all four of the above growth factors phosphorylate PLC-γ1 at the same three Tyr residues, 771, 783 (located between the C-SH2 and SH3 domains), and 1254 (located in the C-terminal tail). In fact, mutating Tyr 783 to Phe completely blocks the activation of PLC-γ1 by PDGF, although this mutant PLC-γ1 still associates with the PDGF receptor. Conversely, mutating certain RTK autophosphorylation sites (e.g., the PDGF receptor's Tyr 1021) disrupts their binding of PLC-γ1 and hence its activation, even though these mutant receptors catalyze detectable levels of growth factor–dependent Tyr phosphorylation on PLC-γ1. Evidently, growth factor–induced activation of PLC-γ1 requires both the activating Tyr phosphorylation of PLC-γ1 and its association with the growth factor receptor, the latter presumably bringing PLC-γ1 into contact with its substrate, PIP_2, in the inner leaflet of the plasma membrane. The PLC-γ isozymes may also be activated by NRTKs, such as members of the Src and JAK families (all of which are membrane associated), that have been activated by tyrosine kinase–associated receptors (Section 19-3E). The function of the PLC-γ SH3 domain is unclear.

e. The Activities of PLC-δ Isozymes May Be Regulated by [Ca²⁺]

Despite the fact that the only phosphoinositide-specific PLC whose structure is known is that of PLC-δ1, the way in which the PLC-δ isozymes are regulated is poorly understood. The higher sensitivities of the PLC-δ isozymes to Ca^{2+} compared to those of the other PLCs suggests that the PLC-δ isozymes are regulated by changes in intracellular $[Ca^{2+}]$. Thus, the activation of PLC-δ isozymes may occur secondarily to the receptor-mediated activation of other PLC isozymes through their induction of the opening of Ca^{2+} channels (Fig. 19-51).

f. PLC-ε Is Activated by Ras·GTP

The presence of RasGEF and Ras-binding (RA) domains on PLC-ε suggests that PLC-ε is activated by Ras·GTP. This is, in fact, the case, as indicated by the observation that PLC-ε binds Ras·GTP with high affinity but does not bind Ras·GDP. Since Ras is membrane-anchored, this interaction brings PLC-ε into proximity with the membrane, where its PIP_2 substrate is located. Although the growth factor–induced activation of Ras is terminated by the hydrolysis of its bound GTP to GDP, the resulting Ras·GDP may be rapidly converted to Ras·GTP by the RasGEF domain of PLC-ε, thereby prolonging the receptor-mediated activation of PLC-ε. PLC-ε may also be activated by $G_{12\alpha}$.

C. *The Protein Kinases C*

Protein kinase C (PKC) is the Ser/Thr protein kinase that transduces the numerous signals mediated by the release of DAG (Fig. 19-51). In mammals, it is actually a family of eleven ~700-residue monomeric isozymes classified in 3 subfamilies: the "conventional" PKCs (α, β1, β2, and γ, of which β1 and β2 are splice variants of the same gene), the "novel" PKCs (δ, ε, η, θ, and μ), and the "atypical" PKCs (ζ and λ). The conventional PKCs, which are activated by both DAG and Ca^{2+}, each consist of an N-terminal autoinhibitory pseudosubstrate followed by four conserved domains, C1 through C4 (C for PKC homology). The DAG-binding **C1 domain,** which occurs in several other proteins including Raf (in which it does not bind DAG), consists of two tandemly repeated ~50-residue Cys-rich motifs, C1A and C1B. However, only C1B appears to bind DAG. C2, which often binds Ca^{2+}, is also a component of PLC (Fig. 19-54) as well as many other signaling proteins. C3 and C4 form the N- and C-terminal lobes of the protein kinase, which is similar in sequence to that of PKA. The protein kinase is maintained in its inactive state through its binding of the pseudosubstrate (as with MLCK; Section 18-3C). The novel PKCs, which are activated by DAG but not by Ca^{2+}, resemble the conventional PKCs except that their C2 domains do not bind Ca^{2+}. The atypical PKCs, which are unresponsive to both DAG and Ca^{2+}, have only one Cys-rich motif in their C1 domains and appear to lack C2 domains.

a. The C1 and C2 Domains Anchor PKC to the Plasma Membrane

Phorbol esters such as **12-*O*-myristoylphorbol-13-acetate** (which occurs in croton seed oil) are potent activators of protein kinase C; they structurally resemble DAG but bind to PKC with ~250-fold greater affinity. Consequently, phorbol esters are the most effective known **tumor promoters** (substances that are not in themselves carcinogenic but increase the potency of known carcinogens). The X-ray structure of the C1B motif of PKC δ in complex with phorbol-13-acetate, determined by Hurley, reveals that this 50-residue motif is largely knit together by two Zn^{2+} ions, each of which is tetrahedrally liganded by one His and three Cys side chains (Fig. 19-56). The phorbol ester binds in a narrow groove between two loops that consist mainly of nonpolar residues. Since phorbol esters are also nonpolar, the entire top third of the complex, as shown in Fig. 19-56, is hydrophobic. Very few soluble proteins have such

12-*O*-Myristoylphorbol-13-acetate

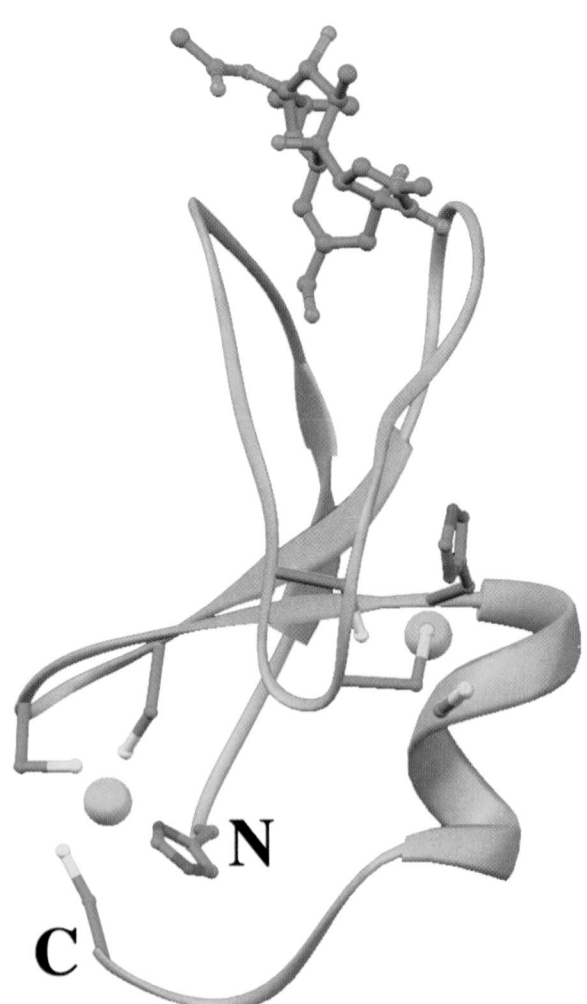

FIGURE 19-56   X-Ray structure of the C1B motif of PKC in complex with phorbol-13-acetate. The protein tetrahedrally ligands two Zn^{2+} ions, each via a His side chain and three Cys side chains. These side chains are shown in ball-and-stick form, as is the bound phorbol-13-acetate (C green, N blue, O red, and S yellow). The Zn^{2+} ions are represented by cyan spheres. [Based on an X-ray structure by James Hurley, NIH. PDBid 1PTR.]

a large fraction of their surface formed by a continuous nonpolar region. Moreover, the middle third of the protein surface, that below the nonpolar region, forms a positively charged belt about the protein. This suggests that, *in vivo*, the hydrophobic portion of the complex is inserted into the nonpolar region of its associated membrane such that the motif's positively charged belt interacts with the membrane's negatively charged head groups. This hypothesis is supported by NMR measurements indicating that residues on the ligand-binding portion of C1B interact with lipid. The fatty acyl group that is esterified to phorbol's 12-position in effective tumor promoters presumably extends into the membrane so as to help anchor the C1 domain to the membrane.

The comparison of this structure with that of C1B alone indicates that C1B does not undergo significant structural change on binding phorbol ester. Evidently, phorbol esters, and presumably DAG, activate PKC by anchoring it to the membrane rather than by an allosteric mechanism. The C2 domain, as we have seen for PLC (Section 19-4B), also does so via Ca^{2+}-mediated binding to the membrane's phosphatidylserine head groups. These interactions are synergistic in that the greater the Ca^{2+} concentration, the lower the concentration of phorbol ester or DAG necessary to activate PKC and vice versa. Nevertheless, both the C1 and C2 domains must be membrane anchored in order to activate the protein kinase. This is because the conformation required to do so extracts the N-terminal pseudosubstrate from the protein kinase active site.

b. PKC Is Primed by Phosphorylation

The activation of all mammalian PKCs but PKC μ is accompanied by their phosphorylation at three conserved Ser or Thr residues. One of these residues (Thr 500 in PKC β2) is in the protein kinase's activation loop, whereas the remaining two are in its C-terminal segment (Thr 641 and Ser 660 of the 673-residue PKC β2). The sequence of events that activate PKC, which was largely elucidated by Alexandra Newton, occurs as follows (Fig. 19-57):

1. Newly synthesized PKC binds to the membrane (or possibly to the underlying cytoskeleton), where **phosphoinositide-dependent protein kinase-1 (PDK1)** phosphorylates its activation loop. The resulting negative charge on the activation loop is postulated to properly align the active site residues of PKC for catalysis, much as we have seen for PKA (Section 18-3C; PKA's activation loop is also phosphorylated by PDK1). In fact, the mutagenic replacement of PKC α's activation loop Thr with a neutral nonphosphorylatable residue yields an inactivatable enzyme, whereas its replacement with Glu yields an enzyme that requires only DAG and Ca^{2+} for activation. A modeling study based on the structure of PKA suggests that PKC's active site–bound pseudosubstrate would mask the activation loop phosphorylation site and hence that the pseudosubstrate must be out of the active site for phosphorylation to occur.

2. The now catalytically competent PKC rapidly autophosphorylates its other two phosphorylation sites. The

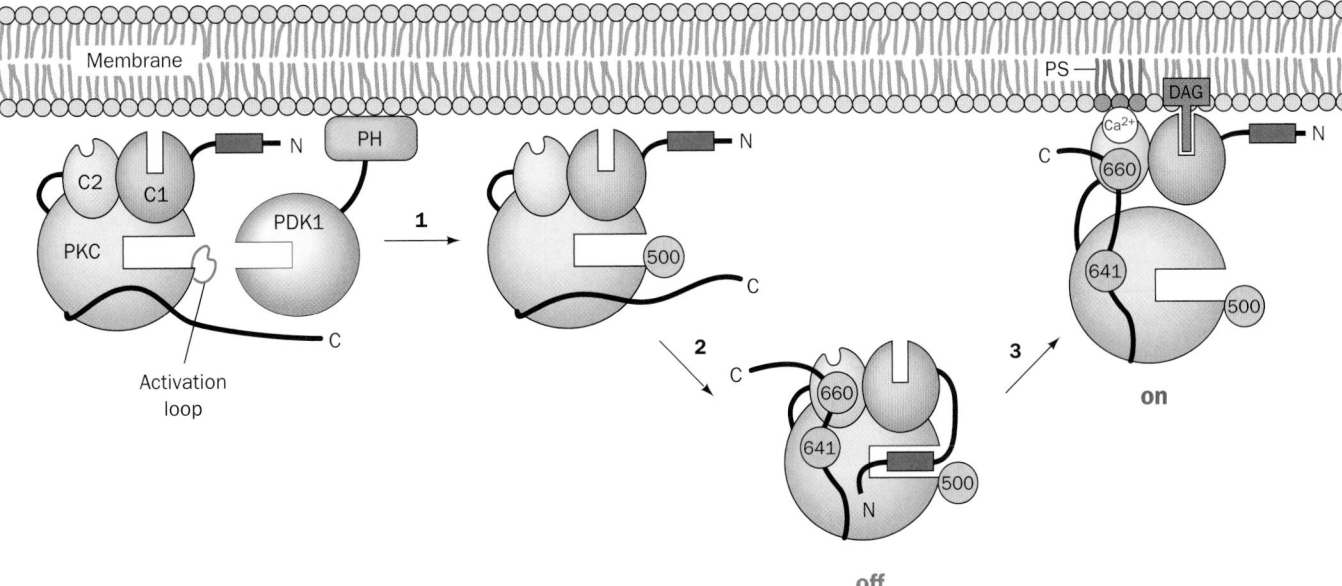

FIGURE 19-57   Activation of PKC. (1) Newly synthesized PKC is phosphorylated on its activation loop (here represented by Thr 500 of PKC β2; *yellow ball*) by phosphoinositide-dependent protein kinase-1 (PDK1), which is tethered to the membrane via its C-terminal pleckstrin homolgy domain (PH). **(2)** The now catalytically competent PKC autophosphorylates 2 sites on its C-terminal segment (here represented by Thr 641 and Ser 660 of PKC β2). However, the N-terminal pseudosubstrate segment now binds to PKC's active site, so that the enzyme remains inactive. **(3)** On the binding of PKC's C1 domain to membrane-bound DAG (the product of extracellular signals inducing phosphoinositide hydrolysis) together with the Ca^{2+}-mediated binding of the C2 domain to phosphatidylserine (PS) in the membrane, the pseudosubstrate is ejected from the PKC active site, thereby yielding active enzyme. [After a drawing by Toker, A. and Newton, A.C., *Cell* **103**, 187 (2000).]

autophosphorylation of Thr 641 appears to lock PKC into its active conformation, as suggested by the observation that in PKC β2, which has been phosphorylated at only Thr 500 and Thr 641, the selective dephosphorylation of Thr 500 yields active enzyme. The autophosphorylation of the third phosphorylation site correlates with the release of PKC into the cytosol, where PKC is maintained in its inactive state by the binding of its pseudosubstrate to its active site.

3. This autoinhibition is relieved, as described above, when PKC again binds to the membrane via DAG binding to its C1 domain and Ca^{2+}-mediated binding of its C2 domain to phosphatidylserine (PS).

D. *The Phosphoinositide 3-Kinases*

The inositol head group of phosphatidylinositol has 5 free hydroxyl groups that can be phosphorylated (Fig. 19-50). However, only its 3-, 4-, and 5-positions are known to be phosphorylated *in vivo*, and *these occur in all possible combinations (Fig. 19-58), each of which participates in signaling.*

The phosphorylations of these various phosphoinositides are catalyzed by ATP-dependent enzymes known as **phosphoinositide 3-kinases (PI3Ks), phosphoinositide 4-kinases (PIP4Ks),** and **phosphoinositide 5-kinases (PIP5Ks).** Their various products function as second messengers by recruiting the proteins that bind them to the cy-

toplasmic surface of the plasma membrane (see below). The resulting colocalization of enzymes and substrates results in further signaling activity that controls such vital functions as cell survival, proliferation, cytoskeletal rearrangement, and vesicle trafficking. In addition, the activities of these enzymes are required for vesicle trafficking via the secretory and endocytotic pathways (Sections 12-4C and 12-4D).

The PI3Ks are the presently best understood phosphoinositide kinases. Consequently, in this subsection, we discuss the PI3Ks and their products as a paradigm of all phosphoinositide kinases and the signals they produce.

a. PI3Ks Have Three Classes

Mammalian PI3Ks are divided into 3 classes according to their structures (Fig. 19-59), substrate specificities, and modes of regulation:

1. Class I PI3Ks are heterodimeric receptor-regulated enzymes that preferentially phosphorylate **phosphatidyl-inositol-4,5-bisphosphate (PtdIns-4,5-P_2).** Their ~1070-residue catalytic subunits interact with Ras·GTP via a Ras-binding domain (RBD) near their N-termini. Their regulatory subunits are adaptor proteins that link the catalytic subunits to upstream signaling events and hence form two subclasses according to the type of upstream effectors with which they interact:

(a) The class IA PI3Ks (PI3Kα, β, and δ) are activated by RTKs via the mediation of the adapter subunit **p85** (of which there are seven isoforms), which contains SH2 and SH3 domains and may be phosphorylated on specific Tyr side chains.

(b) The class IB PI3K, of which PI3Kγ is its only member, is activated by the $G_{\beta\gamma}$ dimers of heterotrimeric G proteins, with its adaptor subunit **p101** rendering it far more sensitive to $G_{\beta\gamma}$.

2. Class II PI3Ks (PI3K-C2α, β, and γ) are ~1650-residue monomers that are characterized by a C-terminal C2 domain that does not appear to bind Ca^{2+}. They preferentially phosphorylate PtdIns and **PtdIns-4-P.** Since they lack adapters, the way in which class II PI3Ks are controlled is unknown.

3. Class III PI3K, which has one known isoform, phosphorylates only PtdIns. It is a heterodimer with an 887-residue catalytic subunit and an adaptor subunit known as

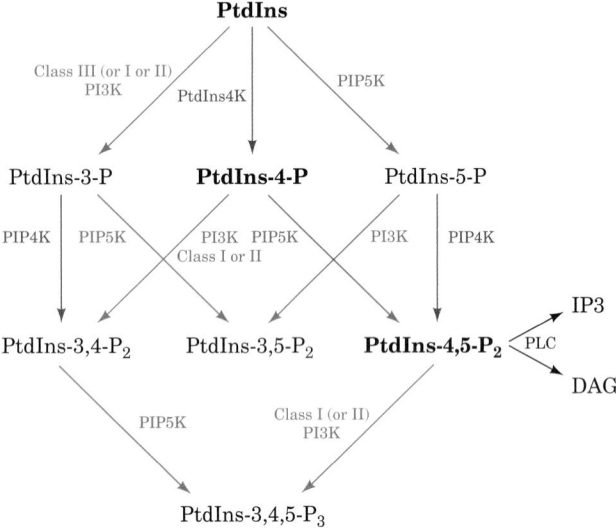

FIGURE 19-58   Flow chart of reactions in the synthesis of phosphoinositides in mammalian cells. PtdIns, PtdIns-4-P, and PtdIns-4,5-P_2 (PIP$_2$) are written in bold type to indicate their abundance: Together they comprise ~90% of the cell's total phosphoinositides. **PtdIns-3-P** and **PtdIns-5-P** each comprise 2–5% of the total, whereas the levels of **PtdIns-3,4-P_2** and **PtdIns-3,4,5-P_3** (PIP$_3$) are barely detectable in quiescent cells but rise to 1 to 3% of the total in stimulated cells. **PtdIns-3,5-P_2** comprises ~2% of the phosphoinositides in fibroblasts. [After Fruman, D.A., Meyers, R.E., and Cantley, L.C., *Annu. Rev. Biochem.* **67,** 501 (1998).]

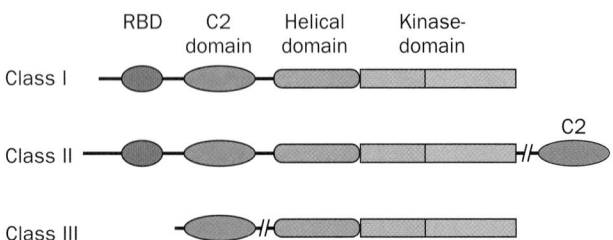

FIGURE 19-59   Domain organization of the 3 classes of PI3Ks. [After Walker, E.H., Persic, O., Ried, C., Stephens, L., and Williams, R.L., *Nature* **402,** 314 (1999).]

p150. Class III PI3K is constitutively active, that is, it is unregulated and hence is thought to be the cell's main provider of **PtdIns-3-P,** whose level is essentially unaltered by cellular stimulation. It is thought to be the evolutionary predecessor of the other classes because it is the only class of PI3K present in yeast.

In addition to their lipid kinase activities, all PI3Ks have Ser/Thr protein kinase activity, although the physiological significance of this dual specificity is unclear.

b. PI3Kγ Is a Multidomain Protein

The X-ray structure of PI3Kγ·ATP, in which the PI3Kγ lacks its N-terminal 143 residues (which are important for interaction with the p101 adaptor; the analogous portion of PI3Kα interacts with its p85 adaptor), was determined by Williams. It reveals that its RBD, C2, and helical domains form a relatively compact layer that packs against the "back" of the kinase domain (Fig. 19-60). As expected, the kinase domain is grossly similar to those of protein kinases in that it is bilobal, with its N-lobe consisting largely of a five-stranded β sheet and its C-lobe being predominantly helical. However, there are also major differences between these kinase domains, as can be seen by comparing the catalytic domain in Fig. 19-60 with that in, for example, Fig. 19-26a.

The RBD domain of PI3Kγ has the same fold as that of RafRBD (Fig. 19-39). Indeed, in the X-ray structure of PI3Kγ–Ras·GDPNP (Fig. 19-61), also determined by

Williams, the PI3K RBD interacts with Ras in a similar manner as we have seen that RafRBD interacts with the Ras homolog Rap1A (Fig. 19-39) in that they continue each other's central β sheets. However, Ras bound to PI3Kγ is rotated by 35° relative to Rap1A bound to RafRBD. Contacts between the Switch I region of Ras and the PI3Kγ stabilize this interaction and ensure its dependence on Ras·GTP. This complex also contains intermolecular contacts involving the Switch II region of Ras. Such an interaction had previously only been observed between Ras and its upstream effectors. Comparison of the structure of the PI3Kγ–Ras complex with that of PI3Kγ·ATP indicates that Ras binding induces the C-lobe of PI3Kγ's catalytic domain to pivot relative to its N-lobe in a way that substantially alters the putative binding pocket for the phosphoinositide head group. This, presumably, accounts for the ~15-fold activation of PI3Kγ on binding Ras·GTP.

The C2 domain of PI3Kγ forms the same sandwich of two 4-stranded antiparallel β sheets seen in the C2 domain of PLC-δ1 (Fig. 19-54). However, in contrast to the C2 domain of PLC-δ1 (Section 19-4B), that of PI3Kγ does not bind Ca^{2+} ions. Nevertheless, the PI3Kγ C2 domain appears to participate in membrane association, as indicated by the observation that this isolated C2 domain binds to phospholipid vesicles with an affinity similar to that of the intact enzyme. This interaction is presumably mediated by patches of basic residues on the surface of the C2 domain.

The PI3Kγ helical domain consists of 5 repeating pairs of antiparallel helices that form a superhelix, which closely

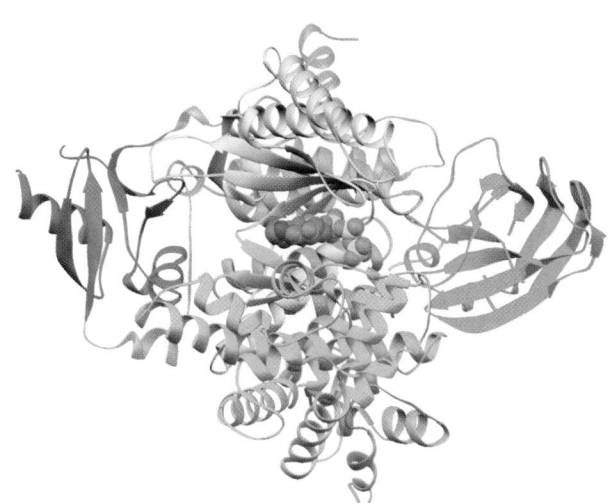

FIGURE 19-60   X-Ray structure of PI3Kγ·ATP. The protein is shown in ribbon form with its Ras-binding domain (RBD) green, its C2 domain magenta, its helical domain orange, the N- and C-lobes of its kinase domain pink and cyan, and interdomain segments gray. The ATP is shown in space-filling form with C green, N blue, O red, and P yellow. The protein is oriented such that its kinase domain is seen in "standard" view. The protein appears fragmented because several of its segments are disordered, including much of the kinase's activation loop. [Based on an X-ray structure by Roger Williams, MRC Laboratory of Molecular Biology, Cambridge, U.K. PDBid 1E8X.]

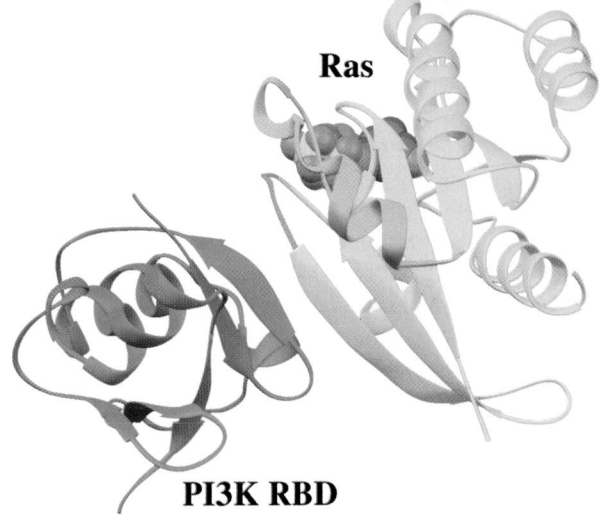

FIGURE 19-61   X-Ray structure of PI3Kγ–Ras·GDPNP. Here only the PI3Kγ RBD (*green*) and the Ras·GDPNP (*gold*) are drawn, with the Switch I and Switch II regions of Ras magenta and cyan and its bound GDPNP shown in space-filling form (C green, N blue, O red, and P yellow). The view, which is similar to that of Fig. 19-39, is related to that in Fig. 19-60 by rotating it clockwise by ~40° about its vertical axis and then turning it 180° about the axis perpendicular to the page. [Based on an X-ray structure by Roger Williams, MRC Laboratory of Molecular Biology, Cambridge, U.K. PDBid 1HE8.]

resembles that formed by the HEAT repeats in the A sub-unit of protein phosphatase 2A (PP2A; Fig. 19-48), even though PI3Kγ does not contain a HEAT sequence motif. In analogy with the function of the PP2A A subunit to bind other proteins (Section 19-4E), it is proposed that the largely solvent-exposed helical domain of PI3Kγ functions to interact with the proteins that bind PI3Kγ, such as its p101 adaptor and G$_{\beta\gamma}$.

c. Akt Activation Requires Its PH Domain-Mediated Binding to 3-Phosphoinositides

The PtdIns-3,4-P$_2$ and PtdIns-3,4,5-P$_3$ products of PI3Ks (Fig. 19-58) bind to their downstream effectors mainly via certain pleckstrin homology (PH) domains that preferentially bind the head groups of these 3-phosphoinositides rather than that of PIP$_2$, (as does the PH domain of PLC-δ; Fig. 19-55). Examples of PH domain-containing proteins that do so are the 556-residue phosphoinositide-dependent protein kinase-1 (PDK1), which, as we have seen, phosphorylates the activation loops of PKA (Section 19-4C) and PKC, and the Ser/Thr protein kinase **Akt** [also called **protein kinase B (PKB)**], a proto-oncogene product that is implicated in regulating multiple biological processes including gene expression, apoptosis, and cellular proliferation and hence is likely to phosphorylate many target proteins. The ~480-residue Akt consists of an N-terminal PH domain that binds 3-phosphoinositides and a C-terminal kinase domain that is homologous to those of PKA and PKC. Akt is present in multicellular organisms in 3 isoforms but is absent in yeast, which suggests that it evolved from the PKA/PKC family coincidentally with multicellular organisms.

Akt is not activated by the binding of target lipids to its PH domain. Rather, this requires the phosphorylation of its Thr 308, which is located in Akt's activation loop. The kinase that phosphorylates this site on Akt is the constitutively active PDK1, whose PH domain also binds 3-phosphoinositides (PDK1 also phosphorylates Akt's Ser 473, but this is not required for activation). Mutations of the residues in Akt's PH domain responsible for lipid binding block its phosphorylation *in vitro* by PDK1. However, the deletion of Akt's PH domain overcomes this enzyme's need for 3-phosphoinositides. This suggests that the binding of Akt to these membrane-bound lipids induces a conformation change that permits PDK1 to phosphorylate and hence activate Akt. It therefore appears that it is the 3-phosphoinositide-mediated colocalization of Akt and PDK1 that leads to Akt activation and hence that it is the action of PI3K that is functionally responsible for this process. In contrast, the PDK1-mediated phosphorylation of PKA, which lacks a PH domain, occurs in the absence of 3-phosphoinositides and is therefore constitutive.

d. The FYVE Domain Binds the PtdIns-3-P Head Group

The singly phosphorylated PtdIns-3-P is rarely bound by PH domains. Rather, its direct effects are mediated by **FYVE domains** [named after the 4 proteins in which it was first identified: *F*ab1p, *Y*OTB, *V*ac1p, and **Early endosome**

antigen 1 (EEA1)], which have been identified in ~60 proteins. For instance, the 1410-residue eukaryotic protein EEA1, which has a 65-residue, C-terminal FYVE domain, initiates endosome fusion in eukaryotic cells (Fig. 12-79) by recruiting the membrane-anchored small GTPase **Rab5** and the transmembrane protein **syntaxin** (Section 12-4D).

The NMR structure of the EEA1 FYVE domain, determined by Michael Overduin, reveals that it assumes similar conformations in the free state, when binding dibutanoyl-PtdIns-3-P (Fig. 19-62), and when bound to dodecylphosphocholine (DPC) micelles enriched with this PtdIns-3-P. The protein is largely held together by two bound Zn^{2+} ions, each of which is tetrahedrally liganded by 4 conserved Cys side chains. The PtdIns-3-P head group is held in its binding pocket by a network of electrostatic, hydrogen bonding, and hydrophobic interactions involving a highly conserved (R/K)(R/K)HHCR motif (RRHHCR in EEA1).

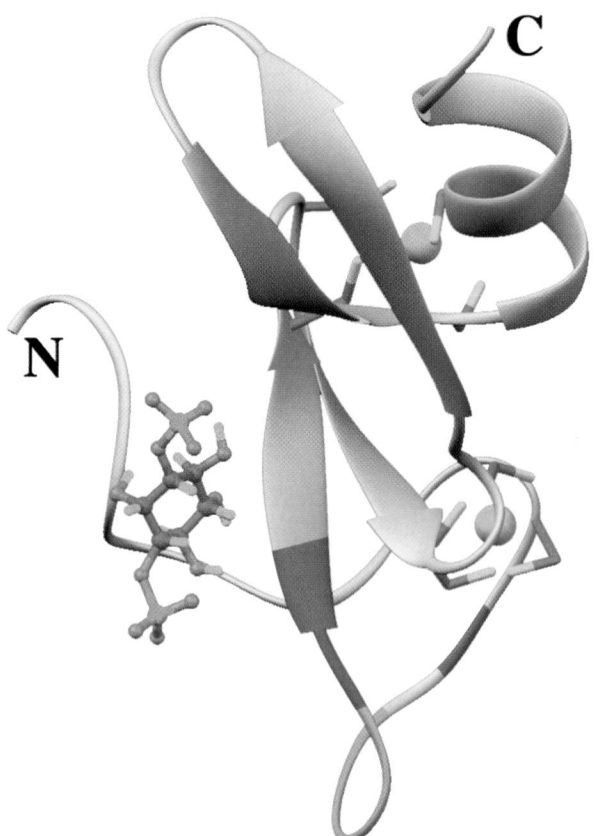

FIGURE 19-62   NMR structure of the EEA1 FYVE domain in complex with PtdIns-3-P. The head group of PtdIns-3-P is drawn in ball-and-stick form (C green, O red, P magenta, H gray). The protein binds two Zn^{2+} ions (*cyan spheres*) that are each tetrahedrally liganded by four Cys side chains that are drawn in stick form (C green and S yellow). The 5-residue loop that inserts into DPC micelles is orange and its flanking basic residues are blue. [Based on an NMR structure by Michael Overduin, University of Colorado Health Sciences Center. PDBid 1HYI.]

The NMR evidence indicates that, on the addition of DPC micelles, the FYVE domain·PtdIns-3-P complex inserts a hydrophobic 5-residue loop (FSVTV; orange in Fig. 19-62), which is flanked by basic residues (blue in Fig. 19-62), into the lipid layer. This also occurs in the absence of PtdIns-3-P but to a much lesser extent. Conversely, membrane insertion increases the binding affinity of the FYVE domain for PtdIns-3-P 20-fold (from 1 μM to 50 nM). The origin of this latter effect appears to be that the 10-residue segment preceding the membrane insertion loop, the unliganded protein's most disordered region, becomes more ordered and moves toward the binding pocket on binding PtdIns-3-P. This has led to the proposal that the FYVE domain is recruited to membranes via the insertion of its hydrophobic loop into the lipid bilayer. This, in turn, primes the protein for the recognition of PtdIns-3-P, whose binding induces the protein's otherwise mobile N-terminal segment to clamp down over the PtdIns-3-P head group.

E. *Inositol Polyphosphate Phosphatases*

Signaling via the phosphoinositide cascade is terminated through the actions of a variety of inositol phosphatases that are functionally classified as 1-, 3-, 4-, and 5-phosphatases. We end our consideration of the phosphoinositide cascade by discussing the characteristics of these essential enzymes.

a. The Inositol Polyphosphate 5-Phosphatases Act In Numerous Signaling Pathways

The first **inositol polyphosphate 5-phosphatases** that were studied hydrolyze IP$_3$ (Ins-1,4,5-P$_3$) to **IP$_2$ (Ins-1,4-P$_2$)** and thereby terminate cellular Ca^{2+} mobilization (Fig. 19-51, *bottom*). Mammals express >10 isozymes that have 5-phosphatase activity. These enzymes share a common catalytic core and have been classified according to their substrate specificities into two groups: Type I enzymes dephosphorylate inositol phosphates, whereas type II enzymes, in addition, hydrolyze the corresponding phosphoinositides.

Type I 5-phosphatases, which hydrolyze only IP$_3$ and **Ins-1,3,4,5-P$_4$,** are membrane-anchored via prenylation. That expressed in blood platelets (a type of blood cell that participates in blood clotting; Section 35-1), which is representative of this group, forms a stoichiometric complex with **pleckstrin,** a 350-residue protein that consists largely of two PH domains. When platelets are stimulated by the proteolytic clotting enzyme **thrombin** (Section 35-1B), pleckstrin is phosphorylated on Ser and Thr residues by PKC, which in turn activates its associated 5-phosphatase. Note that PKC is activated by DAG, a product of PLC, which simultaneously generates the type I 5-phosphatase substrate IP$_3$ (Fig. 19-51). Hence, the PLC product IP$_3$ activates Ca^{2+} release, whereas its coproduct DAG activates type I 5-phosphatase through pleckstrin phosphorylation to terminate the Ca^{2+} signal. This termination is apparently important for normal cell growth as a decrease in the expression of type I 5-phosphatase causes increased and even uncontrolled (malignant) cell growth.

Type II 5-phosphatases share increased similarities in their catalytic cores relative to type I enzymes and, in addition, have a so-called type II domain on the N-terminal side of their catalytic cores. They occur in three main subtypes: **GIPs, SHIPs,** and **SCIPs.** GIPs are so called because they have a C-terminal GAP domain (*G*AP-containing *i*nositol *p*hosphatase), although they have no demonstrated GAP activity. GIPs hydrolyze IP$_3$ and Ins-1,3,4,5-P$_4$ and their corresponding lipids, PtdIns-4,5-P$_2$ and PtdIns-3,4,5-P$_3$, although with varying catalytic efficiencies.

There are only two known GIPs, **5-phosphatase II** and **OCRL.** OCRL is so called because its mutation causes the X-linked hereditary disease **oculocerebrorenal dystrophy** (also called **Lowe syndrome**), which is characterized by congenital cataracts, progressive retinal degeneration, mental retardation, and renal tubule defects leading to kidney failure in early adulthood. The 901-residue OCRL occurs mainly on the surface of lysosomes, where it is anchored through prenylation. Renal tubule cells from Lowe syndrome patients are deficient in PtdIns-4,5-P$_2$ and PtdIns-3,4,5-P$_3$ hydrolytic activity, whereas the corresponding inositol phosphates are hydrolyzed normally, thereby indicating that OCRL is a lipid phosphatase. PtdIns-4,5-P$_2$ stimulates the budding of membrane vesicles from lysosomes, so that the accumulation of this lipid probably leads to abnormally increased trafficking of enzymes from the lysosome to the extracellular space. Indeed, the lysosomal enzymes in these cells appear to be missorted (as are various lysosomal hydrolases in I-cell disease; Section 12-4C). It is therefore proposed that this lifelong leakage of enzymes from the lysosomes in Lowe syndrome patients causes tissue damage that eventually results in kidney failure and blindness.

SHIPs only hydrolyze substrates that also have a phosphate in their 3-positions. The two known members of this group, **SHIP** (for *SH*2-containing *i*nositol-5-*p*hosphatases) and **SHIP2,** are ~1200-residue proteins that have an N-terminal SH2 domain. Thus, these proteins can bind to PTKs and, in fact, are phosphorylated by them to yield a consensus binding sequence for PTB domains (NPXpY; Section 19-3C). Moreover, they also contain a C-terminal Pro-rich domain that may bind to SH3-containing proteins. Thus, it appears that SHIP activity may be under the control of several systems. Indeed, SHIP1, which is expressed only in hematopoietic (blood-forming) cells, associates with the adaptor proteins Grb2 and Shc (Section 19-3C). It functions to hydrolyze PtdIns-3,4,5-P$_3$, which is implicated in activating Akt and PLC. SHIP2 functions similarly in nonhematopoietic cells, where it limits cellular responses to insulin, EGF, and PDGF.

SCIPs (*S*ac1-containing *i*nositol *p*hosphatases) are so named because they contain an N-terminal domain that is homologous to the yeast phosphatidylinositol phosphatase **Sac1.** The first SCIP to be characterized is named **synaptojanin1** because it was purified from synaptic vesicles and because the presence of two phosphatase domains is reminiscent of the two kinase domains in Janus kinases (JAKs; Section 19-3E). The 1575-residue synaptojanin1's 5-phosphatase domain hydrolyzes PIP$_3$ and PtdIns-4,5-P$_2$

and its Sac1 phosphatase domain hydrolyzes PtdIns-3-P and PtdIns-4-P. Synaptojanin1 is expressed only in neurons, where it forms complexes with the G protein dynamin (Section 12-4C) and thereby participates in synaptic vesicle recycling. The closely similar **synaptojanin2** is ubiquitously expressed but its functions are largely unknown.

b. Inositol Polyphosphate 1-Phosphatase Is Implicated in Bipolar Disorder

Mammals express only one type of **inositol polyphosphate 1-phosphatase,** a 399-residue enzyme that hydrolyzes **Ins-1,4-P_2** and Ins-1,3,4-P_3 (IP$_3$) but does not act on lipid substrates. This enzyme is inhibited by Li$^+$ ion. The therapeutic efficacy of Li$^+$ in controlling the incapacitating mood swings of manic-depressive individuals (those with **bipolar disorder**) therefore suggests that this mental illness is caused by an aberration of 1-phosphatase in the brain, possibly resulting in abnormal activation of Ca^{2+}-mobilizing receptors (Fig. 19-51, *bottom*). Indeed, *Drosophila* in which this 1-phosphatase has been deleted exhibit neurological deficits (the so-called "shaker" phenotype) that appear identical to those of wild-type *Drosophila* treated with Li$^+$.

c. The Inositol Polyphosphate 3-Phosphatase PTEN Is a Tumor Suppressor

The **inositol polyphosphate 3-phosphatases** undo the actions of the PI3Ks. The best characterized of these enzymes is the 403-residue **PTEN** (for *p*hosphatase and *ten*sin homolog; **tensin** is a cytoskeletal actin-binding protein), which *in vitro* dephosphorylates all 3-phosphorylated phosphoinositides and Ins-1,3,4,5-P_4. PTEN is a **tumor suppressor** (a protein whose loss of function results in cancer), presumably because its 3-phosphatase activity functions to downregulate the PtdIns-3,4,5-P_3-activated Akt. In fact, mutant forms of PTEN commonly occur in many types of

cancers. PTEN can also dephosphorylate Ser-, Thr-, and Tyr-phosphorylated peptides, although this activity requires the peptides to be highly acidic.

The X-ray structure of PTEN, determined by Jack Dixon and Nikola Pavletich, reveals the protein to consist of an N-terminal phosphatase domain and a C-terminal C2 domain (Fig. 19-63). The structure of its phosphatase domain resembles that common to protein tyrosine phosphatase (PTP) domains (e.g., Fig. 19-47), but with a larger active site pocket, presumably to accommodate the large size of its PtdIns-3,4,5-P_3 substrate. The C2 domain lacks bound Ca^{2+} as well as the ligands to bind it but, nevertheless, binds to phospholipid membranes, as does the C2 domain of PI3Kγ (Fig. 19-60). The phosphatase and C2 domains associate across an extensive interface, whose residues are frequently mutated in cancer. A similar tight interface between the C2 and kinase domain occurs in PLC-δ1 (Fig. 19-54). This suggests that PTEN's C2 domain functions to productively position its attached phosphatase domain at the membrane.

d. The Inositol Polyphosphate 4-Phosphatases Control the Level of PtdIns-3,4-P_2

There are two isoforms of inositol 4-phosphatases, **4-phosphatases I** and **II,** which catalyze the hydrolysis of **Ins-1,3,4-P_3, Ins-2,4-P_2,** and PtdIns-3,4-P_2. In fact, these ~940-residue proteins account for >95% of the observed PtdIns-3,4-P_2 phosphatase activity in many human tissues, thereby suggesting that they play an important role in the metabolism of this second messenger. This is supported by the observation that stimulating human platelets by thrombin or a Ca^{2+} ionophore results in the inactivation of 4-phosphatase I through its proteolytic cleavage by the Ca^{2+}-dependent protease **calpain.** This inactivation of 4-phosphatase I correlates with the Ca^{2+}- and/or aggregation-dependent accumulation of PtdIns-3,4-P_2 characteristic of human platelets (which aggregate in the initial stages of blood clot formation; Section 35-1).

F. *Epilog: Complex Systems and Emergent Properties*

Complex systems are, by definition, difficult to understand and substantiate. Familiar examples include Earth's weather system, the economies of large countries, the ecologies of even small areas, and the human brain. Biological signal transduction systems, as is amply evident from a reading of this chapter, are complex systems. Thus, as we have seen, a hormonal signal is typically transduced through several intracellular signaling pathways, each of which consists of numerous components, many of which interact with components of other signaling pathways. For example, the **insulin signaling system** (Fig. 19-64), although not yet fully elucidated, is clearly highly complex. Upon binding insulin, the insulin receptor autophosphorylates itself at several Tyr residues (Section 19-3A) and then Tyr-phosphorylates its target proteins, thereby activating several signaling pathways that control a diverse array of effects:

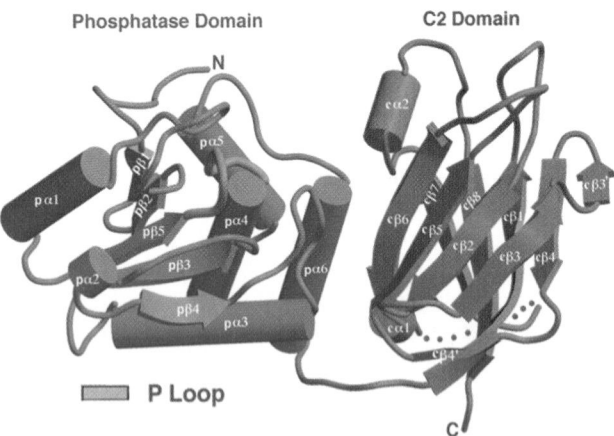

FIGURE 19-63 X-Ray structure of PTEN. The protein is shown with its phosphatase domain blue, its C2 domain red, and the P loop, which interacts with the substrate, tan. The dotted line represents a 24-residue segment that was deleted from the protein to facilitate its crystallization. [Courtesy of Nikola Pavletich, Memorial Sloan-Kettering Cancer Center, New, York, New York. PDBid 1DR5.]

1. Phosphorylation of Shc (Section 19-3C) results in stimulation of a MAP-kinase cascade (Section 19-3D), ultimately affecting growth and differentiation.

2. Phosphorylation of **Gab-1 (Grb2-associated binder-1)** similarly activates this MAP-kinase cascade.

3. Phosphorylation of insulin receptor substrate (IRS) proteins (Section 19-3C) activates a phosphoinositide cascade via a PI3K (Section 19-4D), ultimately stimulating a variety of metabolic processes including glycogen synthesis (Section 18-3E) and glucose transport (Section 20-2E), as well as cell growth and differentiation.

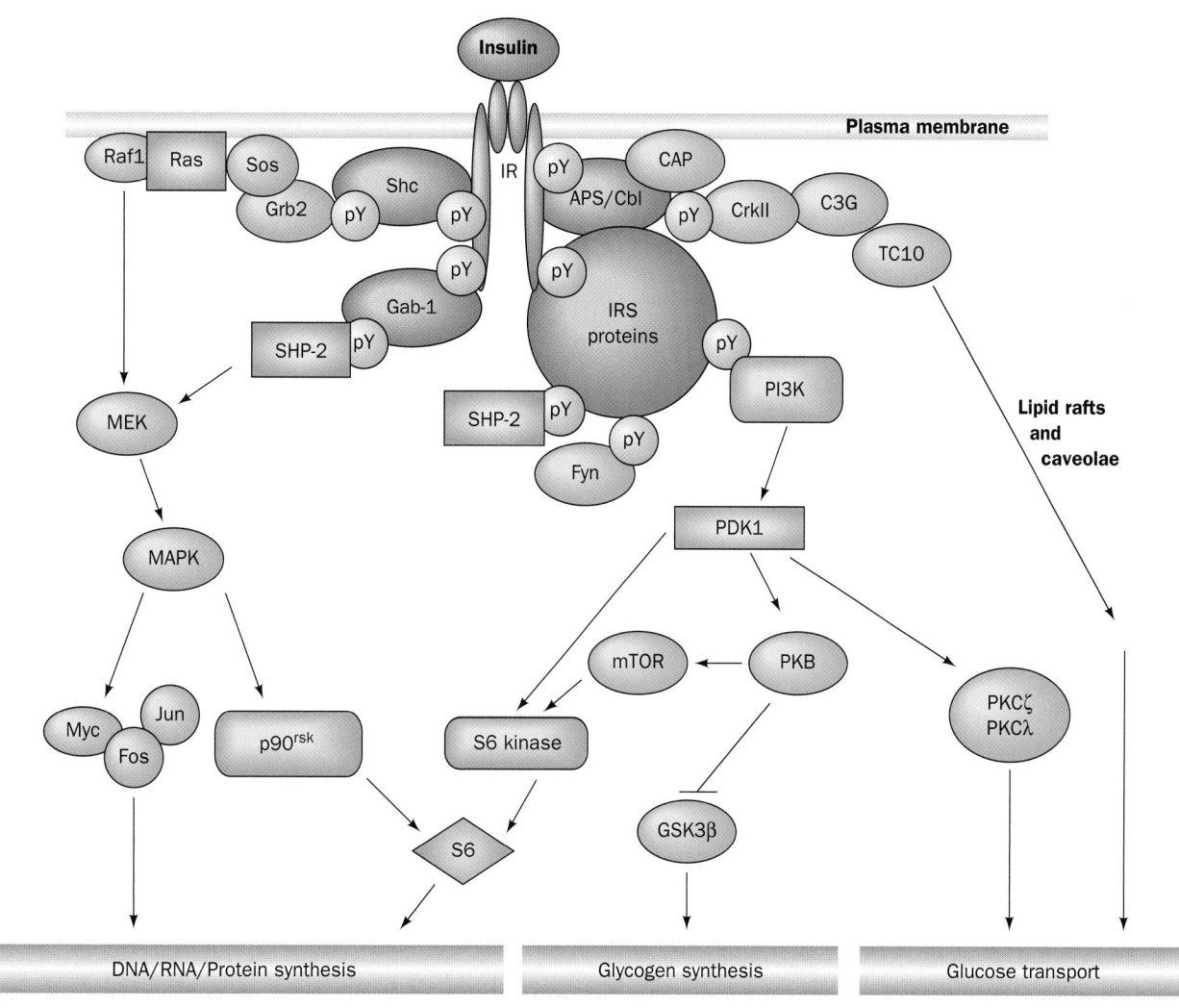

FIGURE 19-64   Insulin signal transduction. The binding of insulin to the insulin receptor **(IR)** induces its autophosphorylation at several Tyr residues on its β subunits. Several proteins, including Shc, Gab-1, the APS/Cbl complex, and IRS proteins, bind to these pY residues where they are Tyr-phosphorylated by the activated insulin receptor, thereby activating MAPK and PI3K phosphorylation cascades as well as a lipid raft and caveolae-associated regulation process. The MAPK cascade regulates the expression of genes involved in cellular growth and differentiation. The PI3K cascade leads to changes in the phosphorylation states of several enzymes, so as to stimulate glycogen synthesis, as well as other pathways. The PI3K cascade also participates in the control of vesicle trafficking, leading to the translocation of the **GLUT4** glucose transporter to the cell surface and thus increasing the rate of glucose transport into the cell (Section 20-2E). Glucose transport control is also exerted by the APS/Cbl system in a PI3K-independent manner involving lipid rafts and caveolae. Other symbols: Myc, Fos, and Jun (transcription factors; Section 19-3D), SHP-2 (an SH2-containing PTP; Section 19-3F), **CAP** (Cbl-associated protein), **C3G** [a guanine nucleotide exchange factor (GEF)], **CrkII** [an SH2/SH3-containing adapter protein), PDK1 (phosphoinositide-dependent protein kinase-1; Section 19-4C), PKB (protein kinase B, also named Akt; Section 19-4D), **mTOR** [for *mammalian target of r*apamycin, a PI3K-related protein kinase; **rapamycin** is an immunosupressant similar to FK506 (Section 9-2B); mTOR is also known as **FKBP12-rapamycin-associated protein (FRAP)**], S6 [a protein subunit of the eukaryotic ribosome's small subunit (Section 32-3A); its phosphorylation stimulates translation], and PKCζ and PKCλ (atypical isoforms of protein kinase C; Section 19-4D) [After Zick, Y., *Trends Cell Biol.* **11,** 437 (2001)].

4. Phosphorylation of the **APS/Cbl** complex (APS for *A*dapter *p*rotein containing *p*lekstrin homology and *S*rc homology-2 domains; Cbl is an SH2/SH3-binding docking protein that is a proto-oncogene product) leads to the stimulation of **TC10** [a G protein in the Rho family (Section 35-3E)], and to the PI3K-independent regulation of glucose transport involving the participation of lipid rafts and caveolae (Section 12-3C).

The predominant approach in science is reductionist: the effort to understand a system in terms of its component parts. Thus chemists and biochemists explain the properties of molecules in terms of the properties of their component atoms, cell biologists explain the nature of cells in terms of the properties of their component macromolecules, and biologists explain the characteristics of multicellular organisms in terms of the properties of their component cells. However, complex systems have **emergent properties,** properties that are not readily predicted from an understanding of their component parts (i.e., the whole is greater than the sum of its parts). Indeed, life itself is an emergent property that arises from the numerous chemical reactions that occur in a cell.

In order to elucidate the emergent properties of a complex system, an integrative approach is required. For signal transduction systems, such an approach would entail determining how each of the components of each signaling pathway in a cell interacts with all of the other such components under the conditions that each of these components experiences within its local environment. Yet techniques for doing so are not often available. Moreover, these systems are by no means static but vary, over multiple time scales, in response to cellular and organismal programs. Consequently, the means for understanding the holistic performance of cellular signal transduction systems are only in their earliest stages of development. Such an understanding is likely to have important biomedical consequences since many diseases, including cancer, diabetes, and a variety of neurological disorders, are caused by malfunctions of signal transduction systems.

CHAPTER SUMMARY

1 ■ Hormones Chemical messengers are classified as autocrine, paracrine, or endocrine hormones if they act on the same cell, cells that are nearby, or cells that are distant from the cell that secreted them, respectively. The body contains a complex endocrine system that controls many aspects of its metabolism. Hormone levels may be determined through radioimmunoassays. Receptors are membrane-bound proteins that bind their ligands according to the laws of mass action. The parameters describing the binding of a radiolabeled ligand to its receptor can be determined from Scatchard plots. The dissociation constants of additional ligands for the same receptor binding site can then be determined through competitive binding studies. The pancreatic islet cells secrete insulin and glucagon, polypeptide hormones that induce liver and adipose tissue to store or release glucose and fat, respectively. Gastrointestinal polypeptide hormones coordinate various aspects of digestion. The thyroid hormones, T_3 and T_4, are iodinated amino acid derivatives that generally stimulate metabolism by activating cellular transcription factors. Ca^{2+} metabolism is regulated by the levels of PTH, vitamin D, and calcitonin. PTH and vitamin D induce an increase in blood $[Ca^{2+}]$ by stimulating Ca^{2+} release from bone and its adsorption from kidney and intestine, whereas calcitonin has the opposite effects. Vitamin D is a steroid derivative that must be obtained in the diet or by exposure to UV radiation. Vitamin D, after being sequentially processed in the liver and kidney to $1,25(OH)_2D$, stimulates the synthesis of a Ca^{2+}-binding protein in the intestinal epithelium. The adrenal medulla secretes the catecholamines epinephrine and norepinephrine, which bind to α- and β-adrenoreceptors on a great variety of cells so as to prepare the body for "fight or flight." The adrenal cortex secretes glucocorticoid and mineralocorticoid steroids. Glucocorticoids affect metabolism in a manner opposite to that of insulin as well as mediating a wide variety of other vital functions. Mineralocorticoids regulate the excretion of salt and water by kidney. The gonads secrete steroid sex hormones, the androgens (male hormones) and estrogens (female hormones), which regulate sexual differentiation, the development of secondary sex characteristics, and sexual behavior patterns. Ovaries, in addition, secrete progestins that help mediate the menstrual cycle and pregnancy. Mammalian embryos develop as females unless subjected to the influence of the androgen testosterone. *SRY,* a gene that encodes a DNA-binding protein and that is normally located on the Y chromosome, induces the development of testes, which in turn secrete testosterone. The hypothalamus secretes a series of polypeptide releasing factors and release-inhibiting factors such as CRF, TRF, GnRF, and somatostatin that control the secretion of the corresponding trophic hormones from the pituitary gland's adenohypophysis. Most of these trophic hormones, such as ACTH, TSH, LH, and FSH, stimulate their target endocrine glands to secrete the corresponding hormones. However, growth hormone acts directly on tissues as well as stimulating liver to synthesize growth factors known as somatomedins. The pituitary gland's neurohypophysis secretes the polypeptides vasopressin, which stimulates the kidneys to retain water, and oxytocin, which stimulates uterine contraction. The menstrual cycle results from a complex interplay of hypothalamic, adenohypophyseal, and steroid sex hormones. A fertilized and implanted ovum secretes CG, which binds to the same receptor and has similar effects as LH, thus preventing menstruation. The binding of hGH to its receptor causes the receptor to dimerize, thereby providing the intracellular signal that the receptor has bound hCG. Many other hormonal signals are similarly mediated. The adenohypophysis also secretes opioid peptides that have opiatelike effects on the central nervous system. Nitric oxide (NO), a highly reactive radical gas, functions as a local mediator that regulates vasodilation, serves as neurotransmitter, and functions in the immune response. In mammals, it is synthesized by three isozymes of nitric oxide synthase (NOS), an enzyme that contains 5 redox-active prosthetic groups. eNOS and nNOS are activated by Ca^{2+} through their binding of Ca^{2+} – calmodulin; iNOS is transcriptionally controlled. NO activates guanylate cyclase to produce cGMP, which in turn activates cGMP-dependent protein kinase.

2 ■ Heterotrimeric G Proteins Ligand (hormone) binding to certain seven-transmembrane receptors activates the $G_{s\alpha}$ subunit of a stimulatory G protein to replace its bound GDP with GTP, release its associated $G_{\beta\gamma}$ subunits, and activate adenylate cyclase (AC) to synthesize cAMP. Activation continues until $G_{s\alpha}$ hydrolyzes its bound GTP to GDP and recombines with $G_{\beta\gamma}$. Several types of activated hormone receptors in a cell may stimulate the same G_s protein. There are also inhibitory G proteins, which may have the same G_β and G_γ subunits as does G_s, but which have an inhibitory $G_{i\alpha}$ subunit that deactivates adenylate cyclase. Biological signaling systems are subject to desensitization through the phosphorylation and endocytotic sequestering of the cell-surface receptors. Cholera toxin (CT) and heat-labile enterotoxin (LT), related bacterial AB_5 proteins, induce uncontrolled cAMP production by ADP-ribosylating $G_{s\alpha}$ so as to render it incapable of hydrolyzing GTP. Pertussis toxin, also an AB_5 protein, similarly ADP-ribosylates $G_{i\alpha}$. The catalytic core of the numerous isoforms of AC are pseudosymmetric heterodimers that are activated, in most cases, by the binding of the Switch II region of $G_{s\alpha}$·GTP to a cleft in an AC's C_{1a} domain. cAMP and cGMP are eliminated through the actions of numerous phosphodiesterases (PDEs), whose activities are controlled by a variety of agents, thereby providing for cross talk between signaling systems.

3 ■ Tyrosine Kinase–Based Signaling The binding of ligands such as hormones and protein growth factors activates receptor tyrosine kinases (RTKs) by inducing them to dimerize and thereupon autophosphorylate specific Tyr residues in the activation loops of their tyrosine kinase domains. This is usually followed by the autophosphorylation of Tyr residues on other cytoplasmic domains. Cancer cells' immortality and their uncontrolled proliferation endow them with the capacity to form invasive and metastatic tumors. Rous sarcoma virus, a retrovirus causing sarcomas in chickens, carries an oncogene, v-*src*, that is homologous to the normal cellular gene c-*src*. Both genes encode a protein tyrosine kinase (PTK) that stimulates cell division. Oncogene products include analogs of growth factors, growth factor receptors, nuclear proteins that stimulate transcription and/or cell division, and G proteins. An autophosphorylated RTK may activate other proteins by phosphorylating them on specific Tyr side chains. It can also modulate the activities of specific proteins through the binding of an RTK's phosphoTyr-containing peptide segment to SH2 and PTB domains on these proteins or on adaptors that bind to these proteins. Grb2, an adaptor protein, binds to certain activated RTKs in this way and simultaneously, via its SH3 domains, to Sos protein. The bound Sos, in turn, functions as a guanine nucleotide exchange factor (GEF) to induce the small G protein Ras to exchange its bound GDP for GTP. Ras is a poor GTPase but it is aided in eventually hydrolyzing its bound GTP to GDP by the GTPase activating protein (GAP) RasGAP, which insinuates a catalytically important Arg side chain into Ras's otherwise inefficient active site. Mutations that interfere with the ability of Ras–RasGAP to hydrolyze Ras's bound GTP are oncogenic. The binding of Ras·GTP to Raf, a protein Ser/Thr kinase, activates Raf to

phosphorylate MEK, a MAP kinase kinase (MKK), which in turn phosphorylates MAP kinase (MAPK). The activated MAPK phosphorylates various cytoplasmic and membrane-associated proteins and, in addition, is translocated to the nucleus where it phosphorylates certain transcription factors, which thereupon induce the transcription of their target genes. The proteins of such MAP kinase cascades are organized by their binding to scaffold proteins, which also prevents the members of different MAP kinase cascades in a cell from inappropriately phosphorylating one another. However, activated members of a MAP kinase cascade may phosphorylate other regulatory proteins, thereby eliciting cross talk between different signal transduction pathways. Tyrosine kinase–associated receptors transduce the signal that they have bound effector by activating associated nonreceptor tyrosine kinases (NRTKs), many of which are members of the Src or JAK families. Activated JAK proteins phosphorylate STAT proteins, which then dimerize and are translocated to the nucleus, where they function as transcription factors. Gleevec is a highly selective Abl inhibitor that is clinically effective in the treatment of chronic myelogenous leukemia (CML). Phosphorylated proteins are deactivated by protein phosphatases. Some protein tyrosine phosphatases (PTPs) are transmembrane receptors that are deactivated by ligand-induced dimerization. Other PTPs are cytoplasmic and are activated by their binding to activated PTKs, for example, via SH2 domains, as does SHP-2. Cells contain several types of Ser/Thr protein phosphatases: PP1 participates in the regulation of glycogen metabolism; PP2A, which participates in a wide variety of regulatory processes, is a heterotrimer with numerous variants and hence specificities and cellular locations; and calcineurin (CaN; also called PP2B) is a Ca^{2+}-activated heterodimeric phosphatase that is the target of the immunosuppressive drugs cyclosporin A and FK506 via the binding of their complexes with the rotamases cyclophilin and FKBP12 to CaN so as to prevent the binding of CaN's target phosphopeptides.

4 ■ The Phosphoinositide Cascade PIP_2, a minor phospholipid component of the plasma membrane's inner leaflet, can yield up to three types of second messengers. Hormone–receptor interactions, through the intermediacy of a G protein or an RTK, stimulate the corresponding phospholipase C (PLC) to hydrolyze PIP_2 to the water-soluble IP_3 and the membrane-bound DAG. The IP_3 stimulates the release of Ca^{2+} from the endoplasmic reticulum through ligand-gated channels. The Ca^{2+} binds to calmodulin, which in turn activates a variety of cellular processes. The DAG activates protein kinase C (PKC) to phosphorylate and thereby modulate the activities of numerous cellular proteins. DAG may also be degraded to yield arachidonate, an important intermediate in the biosynthesis of prostaglandins and related compounds. PLC-δ1 is bound to the membrane containing its PIP_2 substrate via its PH domain, which binds PIP_2, and its C2 domain, which binds to phosphatidylserine molecules in the membrane via three Ca^{2+} ions. The various classes of PLCs are activated in different ways, all of which bring the PLC into contact with its PIP_2 substrate in the membrane: PLC-β's by binding $G_{q\alpha}$·GTP and $G_{\beta\gamma}$; PLC-γ's by binding to phosphorylated PTKs via SH2 domains followed by phosphorylation of the PLC by the PTK; PLC-δ's by Ca^{2+}; and PLC-ε by binding Ras·GTP. "Conventional" PKCs are activated by both Ca^{2+}

and DAG. Phorbol esters, which are DAG mimics that activate PKC, are the most potent known tumor promoters. DAG and Ca^{2+} synergistically bind PKC to the membrane via its C1 and C2 domains, which conformationally extracts PKC's N-terminal pseudosubstrate from the kinase's active site. The kinase is catalytically activated by phosphorylation on its activation loop by PDK1 followed by autophosphorylation at two more sites. Phosphoinositides may be phosphorylated at their inositol head group's 3-, 4-, and 5-positions in all combinations, yielding membrane-bound second messengers that function by recruiting the proteins that bind them to the membrane surface. Mammalian phosphoinositide 3-kinases (PI3Ks) form three classes that differ according to their structures, substrate specificities, and modes of regulation. The $PtdIns-3,4-P_2$ and $PtdIns-3,4,5-P_3$ products of PI3Ks bind to the PH domain of the proto-oncogene product Akt (PKB), thereby colocalizing Akt with PDK1, which is also tethered to the membrane via its PH domain, so that PDK1 phosphorylates and thereby activates Akt. $PtdIns-3-P$ is bound by FYVE domains, which, like PH domains, are held together by two tetrahedrally liganded Zn^{2+} ions. The various types of inositide polyphosphate phosphatases function to terminate signaling by the phosphoinositide cascade. OCRL, a type II 5-phosphatase that participates in controlling vesicle budding from the lysosome, is mutated in oculocerebrorenal disease (Lowe syndrome). The only 1-phosphatase expressed by mammals, which hydrolyzes $Ins-1,4-P_2$ and PIP_3, is inhibited by Li^+ ion and is thereby implicated in bipolar disorder. The 3-phosphatase PTEN, a tumor suppressor whose mutant forms are common to many cancers, undoes the actions of PI3Ks. Type I 4-phosphatase in blood platelets is inactivated through proteolytic cleavage by the Ca^{2+}-activated protease calpain. Cellular signal transduction systems, such as the insulin signaling system, are complex systems with emergent properties that are, as yet, poorly understood.

REFERENCES

GENERAL

Gomperts, B.D., Tatham, P.E.R., and Kramer, I.M., *Signal Transduction,* Academic Press (2002).

Helmreich, E.J.M., *The Biochemistry of Cell Signaling,* Oxford (2001).

Krauss, G., *Biochemistry of Signal Transduction and Regulation* (2nd ed.), Wiley-VCH (2001).

Science's Signal Transduction Knowledge Environment (STKE). http://stke.sciencemag.org/ [A database on signaling molecules and their relationships to one another. This database is introduced in a series of authoritative articles in *Science* **296**, 1632–1657 (2002). To gain full access to the database requires a subscription, which many colleges and universities have through their libraries.]

HORMONES

Alderton, W.K., Cooper, C.E., and Knowles, R.G., Nitric oxide synthases: structure, function, and inhibition, *Biochem. J.* **357**, 593–615 (2002).

Capel, B., Sex in the 90s: *SRY* and the switch to the male pathway, *Annu. Rev. Physiol.* **60**, 497–523 (1998).

Crane, B.R., Arvai, A.S., Ghosh, D.K., Wu, C., Getzoff, E.D., Stuehr, D.J., and Tainer, J.A., Structure of nitric oxide synthase oxygenase dimer with pterin and substrate, *Science* **279**, 2121–2126 (1998); Raman, C.S., Li, H., Martásek, P., Král, V., Masters, B.S.S., and Poulos, T.L., Crystal structure of a constitutive endothelial nitric oxide synthase: A paradigm for function involving a novel metal center, *Cell* **95**, 939–950 (1998); *and* Fischmann, T.O., et al., Structural characterization of nitric oxide synthase isoforms reveals striking active site conservation, *Nature Struct. Biol.* **6**, 233–242 (1999).

DeGroot, L.J. and Jameson, J.L. (Eds.), *Endocrinology* (4th ed.), Saunders (2001). [A 3-volume compendium.]

Hadley, M.E., *Endocrinology* (5th ed.), Prentice-Hall (2000).

Ignarro, L.J. (Ed.), *Nitric Oxide. Biology and Pathobiology,* Academic Press (2000).

Kossiakoff, A.A. and de Vos, A.M., Structural basis for cytokine hormone–receptor recognition and receptor activation, *Adv. Protein Chem.* **52**, 67–108 (1999).

Ma, Y.-A., Sih, C.J. and Harms, A., Enzymatic mechanism of thyroxine biosynthesis. Identification of the "lost three-carbon fragment," *J. Am. Chem. Soc.* **121**, 8967–8968 (1999).

Norman, A.W. and Litwack, G., *Hormones* (2nd ed.), Academic Press (1997).

Pfeiffer, S., Mayer, B., and Hemmens, B., Nitric oxide: Chemical puzzles posed by a biological messenger, *Angew. Chem. Int. Ed.* **38**, 1714–1731 (1999).

Schafer, A.J. and Goodfellow, P.N., Sex determination in humans, *BioEssays* **18**, 955–963 (1996).

Stamler, J.S., Singel, D.J., and Loscalzo, J., Biochemistry of nitric oxide and its redox-activated forms, *Science* **258**, 1898–1902 (1992).

Stuehr, D.J., Structure-function aspects in the nitric oxide synthases, *Annu. Rev. Pharmacol.* **37**, 339–359 (1997).

HETEROTRIMERIC G PROTEINS

Bockaert, J. and Pin, J.P., Molecular tinkering of G protein-coupled receptors: An evolutionary success, *EMBO J.* **18**, 1723–1729 (1999). [Discusses the different types of GPCRs.]

Clapham, D.E. and Neer, E.J., G protein βγ subunits, *Annu. Rev. Pharmacol. Toxicol.* **37**, 167–203 (1997).

Corbin, J.D. and Francis, S.H., Cyclic GMP phosphodiesterase-5: Target of sildenafil, *J. Biol. Chem.* **274**, 13729–13732 (1999).

Fan, E., Merritt, E.A., Verlinde, C.L.M.J., and Hol, W.G.J., AB_5 toxins: Structures and inhibitor design, *Curr. Opin. Struct. Biol.* **10**, 680–686 (2000).

Hall, A. (Ed.), *GTPases,* Oxford University Press (2000).

Hamm, H.E., The many faces of G protein signaling, *J. Biol. Chem.* **273**, 669–672 (1998).

Houslay, M.D. and Milligan, G., Tailoring cAMP-signaling responses through isoform multiplicity, *Trends Biochem. Sci.* **22**, 217–224 (1997).

Hurley, J.H., The adenylyl and guanylyl cyclase superfamily, *Curr. Opin. Struct. Biol.* **8**, 770–777 (1998).

Ji, T.H., Grossmann, M., and Ji, I., G protein-coupled receptors, *J. Biol. Chem.* **273**, 17299–17302 (1998).

Noel, J.P., Hamm, H.E., and Sigler, P.B., The 2.2 Å crystal structure of transducin-α complexed with GTPγS, *Nature* **366**, 654–663 (1993); *and* Lambright, D.G., Noel, J.P., Hamm, H.E., and Sigler, P.B., Structure determinants for activation of the α-subunit of a heterotrimeric G protein, *Nature* **369**, 621–628 (1994). [The first paper describes the GTPγS complex and the second compares it with the GDP complex.]

Palczewski, K., et al., Crystal structure of rhodopsin: A G protein–coupled receptor, *Science* **289**, 739–745 (2000); Teller, D.C., Okada, T., Behnke, C.A., Palczewski, K., and Stenkamp, R.E., Advances in determination of a high-resolution three-dimensional structure of rhodopsin, a model of G-protein-coupled receptors (GCPRs), *Biochemistry* **40**, 7761–7772 (2001); *and* Okada, T. and Palczewski, K., Crystal structure of rhodopsin: Implications for vision and beyond, *Curr. Opin. Struct. Biol.* **11**, 420–426 (2001).

Soderling, S.H. and Beavo, J.A., Regulation of cAMP and cGMP signaling: New phosphodiesterases and new functions, *Curr. Opin. Cell Biol.* **12**, 174–179 (2000); *and* Beavo, J.A., Cyclic nucleotide phosphodiesterases: Functional implications of multiple isoforms, *Physiol. Rev.* **75**, 725–748 (1995).

Sprang, S.R., G protein mechanisms: Insights from structural analysis, *Annu. Rev. Biochem.* **66**, 639–678 (1997).

Stein, P.E., Boodhoo, A., Armstrong, G.D., Cockle, S.A., Klein, M.H., and Read, R.J., The crystal structure of pertussis toxin, *Structure* **2**, 45–57 (1994).

Strader, C.D., Fong, T.M., Tota, M.R., Underwood, D., and Dixon, R.A.F., Structure and function of G protein–coupled receptors, *Annu. Rev. Biochem.* **63**, 101–132 (1994).

Sunahara, R.K., Tesmer, J.J.G., Gilman, A.G., and Sprang, S.R., Crystal structure of the adenylyl cyclase activator $G_{s\alpha}$, *Science* **278**, 1943–1947 (1997).

Tesmer, J.J.G. and Sprang, S.R., The structure, catalytic mechanism and regulation of adenylyl cyclase, *Curr. Opin. Struct. Biol.* **8**, 713–719 (1998).

Tesmer, J.J.G., Sunahara, R.K., Gilman, A.G., and Sprang, S.R., Crystal structure of the catalytic domains of adenylyl cyclase in a complex with $G_{s\alpha}$·GTPγS, *Science* **278**, 1907–1916 (1997).

Vetter, I.R., and Wittinghofer, A., The guanine nucleotide–binding switch in three dimensions, *Science* **294**, 1299–1304 (2001).

Wall, M.A., Coleman, D.E., Lee, E., Iñiguez-Lluhi, J.A., Posner, B.A., Gilman, A.G., and Sprang, S.R., The structure of the G protein heteotrimer $G_{i\alpha 1}\beta_1\gamma_2$, *Cell* **83**, 1047–1058 (1995); *and* Lambright, D.G., Sondek, J., Bohm, A., Skiba, N.P., Hamm, H.E., and Sigler, P.B., The 2.0 Å crystal structure of a heterotrimeric G protein, *Nature* **379**, 311–319 (1996).

Zhang, R.-G., Scott, D.L., Westbrook, M.L., Nance, S., Spangler, B.D., Shipley, G.G., and Westbrook, E.M., The three-dimensional crystal structure of cholera toxin, *J. Mol. Biol.* **251**, 563–573 (1995); *and* Merrrit, E.A., Sarfaty, S., Jobling, M.G., Chang, T., Holmes, R.K., Hirst, T.R., and Hol, W.G.J., Structural studies of receptor binding by cholera toxin mutants, *Protein Sci.* **6**, 1516–1528 (1997).

TYROSINE KINASE–BASED SIGNALING

Angier, N., *Natural Obsessions: The Search for the Oncogene*, Houghton Mifflin (1988). [A chronicle of how science is really done.]

Barford, D., Das, A.K., and Egloff, M.-P., The structure and mechanism of protein phosphatases: Insights into catalysis and regulation, *Annu. Rev. Biophys. Biomol. Struct.* **27**, 133–164 (1998).

Blumer, K.J. and Johnson, G.L., Diversity in function and regulation of MAP kinase pathways, *Trends Biochem. Sci.* **19**, 236–240 (1994).

Bollen, M., Combinatorial control of protein phosphatase-1, *Trends Biochem. Sci.* **26**, 426–431 (2001).

Boriak-Sjodin, P.A., Margarit, S.M., Bar-Sagi, D., and Kuriyan, J., The structural basis of the activation of Ras by Sos, *Nature* **394**, 337–343 (1998).

Capdeville, R., Buchdunger, E., Zimmermann, J., and Matter, A., Glivec (STI571, ImatinIB), a rationally developed targeted anticancer drug, *Nature Rev. Drug Discov.* **1**, 493–502 (2002).

Carlisle Michel, J. J. and Scott, J.D., AKAP mediated signal transduction, *Annu. Rev. Pharmacol. Toxicol.* **42**, 235–257 (2002).

Carpenter, L.R., Yancopoulos, G.D., and Stahl, N., General mechanisms of cytokine receptor signaling, *Adv. Protein Chem.* **52**, 109–140 (1999).

Chang, L. and Karin, M., Mammalian MAP kinase signaling cascades, *Nature* **410**, 37–40 (2001).

Charbonneau, H. and Tonks, N.K., 1002 protein phosphatases? *Annu. Rev. Cell Biol.* **8**, 463–493 (1992).

Cherfils, J. and Chardin, P., GEFs: Structural basis for their activation of small GTP-binding proteins, *Trends Biochem. Sci.* **24**, 306–311 (1999).

Chien, C.-T., Bartel, P.L., Sternglanz, R., and Fields, S., The two-hybrid system: A method to identify and clone genes from proteins that interact with a protein of interest, *Proc. Natl. Acad. Sci.* **88**, 9578–9582 (1991).

Corbett, K.D. and Alber, T., The many faces of Ras: Recognition of small GTP-binding proteins, *Trends Biochem. Sci.* **26**, 710–716 (2001).

Druker, B.J. and Lydon, N.B., Lessons learned from the development of an Abl tyrosine kinase inhibitor for chronic myelogenous leukemia, *J. Clin. Invest.* **105**, 3–7 (2000).

Edwards, A.S. and Scott, J.D., A-kinase anchoring proteins: Protein kinase A and beyond, *Curr. Opin. Cell. Biol.* **12**, 217–221 (2000).

Gamblin, S.J. and Smerdon, S.J., GTPase-activating proteins and their complexes, *Curr. Opin. Struct. Biol.* **8**, 195–201 (1998).

Garrington, T.P. and Johnson, G.L., Organization and regulation of mitogen-activated protein kinase signaling pathways, *Curr. Opin. Cell Biol.* **11**, 211–218 (1999).

Goldstein, B., *Tyrosine Phosphoprotein Phosphatases* (2nd ed.), Oxford (1998). [A sourcebook.]

Groves, M.R., Hanlon, N., Turowski, P., Hemmings, B.A., and Barford, D., The structure of the protein phosphatase 2A PR65/A subunit reveals the conformation of its 15 tandemly repeated HEAT motifs, *Cell* **96**, 99–110 (1999).

Hof, P., Pluskey, S., Dhe-Paganon, S., Eck, M.J., and Shoelson, S.E., Crystal structure of tyrosine phosphatase SHP-2, *Cell* **92**, 441–450 (1998).

Hubbard, S.R., Crystal structure of the activated insulin receptor tyrosine kinase in complex with peptide substrate and ATP analog, *EMBO J.* **16**, 5572–5581 (1997); *and* Hubbard, S.R., Wei, L., Ellis, L., and Hendrickson, W.A., Crystal structure of the tyrosine kinase domain of the human insulin receptor, *Nature* **372**, 746–753 (1994).

Hubbard, S.R. and Hill, J.H., Protein tyrosine kinase structure and function, *Annu. Rev. Biochem.* **69**, 373–398 (2000).

Kissinger, C.R., et al., Crystal structures of human calcineurin and the human FKBP12–FK506–calcineurin complex, *Nature* **378**, 641–644 (1995); *and* Griffith, J.P., et al., X-Ray structure of calcineurin inhibited by the immunophilin–immunosuppressant FKBP12–FK506 complex, *Cell* **82**, 507–522 (1995).

Kolch, W., Meaningful relationships: The regulation of the Ras/Raf/MEK/ERK pathway by protein interactions, *Biochem. J.* **351**, 289–305 (2000).

Kuriyan, J. and Cowburn D., Structures of SH2 and SH3 domains, *Curr. Opin. Struct. Biol.* **3**, 828–837 (1993).

Li, L. and Dixon, J.E., Form, function, and regulation of protein tyrosine phosphatases and their involvement in human disease, *Sem. Immunol.* **12**, 75–84 (2000).

Lim, W.A., The modular logic of signaling proteins: building allosteric switches from simple binding domains, *Curr. Opin. Struct. Biol.* **12,** 61–68 (2002).

Maignan, S., Guilloteau, J.-P., Fromage, N., Arnoux, B., Becquart, J., and Ducruix, A., Crystal structure of the mammalian Grb2 adaptor, *Science* **268,** 291–293 (1995).

Margolis, B., The PTB domain: The name doesn't say it all, *Trends Endocrin. Metab.* **10,** 262–267 (1999).

Millward, T.A., Zolnierowicz, S., and Hemmings, B.A., Regulation of protein kinase cascades by protein phosphatase 2A, *Trends Biochem. Sci.* **24,** 186–191 (1999).

Musacchio, A., Sareste, M., and Wilmanns, M., High-resolution crystal structures of tyrosine kinase SH3 domains complexed with proline-rich peptides, *Nature Struct. Biol.* **1,** 546–551 (1994).

Nassar, N., Horn, G., Herrmann, C., Scherer, A., McCormack, F., and Wittinghofer, A., The 2.2 Å crystal structure of the Ras-binding domain of the serine/threonine kinase c-Raf1 in complex with Rap1A and a GTP analogue, *Nature* **375,** 554–560 (1995).

O' Shea, J.J., Gadino, M., and Schreiber, R.D., Cytokine signaling in 2002: New surprises in the Jak/Stat pathway, *Cell* **109,** S121–S131 (2002).

Pellizzari, R., Guidi-Rontani, C., Vitale, G., Mock, M., and Montecucco, C., Anthrax lethal factor cleaves MKK3 in macrophages and inhibits the LPS/IFNγ-induced release of NO and TNFα, *FEBS Lett.* **462,** 199–204 (1999).

Scheffzek, K., Ahmadian, M.R., Kabsch, W., Wiesmüller, L., Lautwein, A., Schmitz, F., and Wittinghofer, A., The Ras-RasGAP complex: Structural basis for GTPase activation and its loss in oncogenic Ras mutants, *Science* **277,** 333–338 (1997).

Schindler, C. and Darnell, J.E., Jr., Transcriptional responses to polypeptide ligands: The JAK-STAT pathway, *Annu. Rev. Biochem.* **64,** 621–651 (1995).

Schindler, T., Bornmann, W., Pellicenna, P., Miller, W.T., Clarkson, B., and Kuriyan, J., Structural mechanism for STI-571 inhibition of Abelson tyrosine kinase, *Science* **289,** 1938–1942 (2000).

Schlessinger, J., Cell signaling by receptor tyrosine kinases, *Cell* **103,** 211–225 (2000).

Schlessinger, J., Plotnikov, A.N., Ibrahimi, O.A., Eliseenkova, A.V., Yeh, B.K., Yayon, A., Linhardt, R.J., and Mohammadi, M., Crystal structure of a ternary FGF-FGFR-heparin complex reveals a dual role for heparin in FGF binding and dimerization, *Mol. Cell* **6,** 743–750 (2000).

Shuai, K., The STAT family of proteins in cytokine signaling, *Prog. Biophys. Mol. Biol* **71,** 405–422 (1999).

Sprang, S., GEFs: Master regulators of G-protein activation, *Trends Biochem. Sci.* **26,** 266–267 (2001).

Thomas, S.M. and Brugge, J.S., Cellular functions regulated by Src family kinases, *Annu. Rev. Cell Dev. Biol.* **13,** 513–609 (1997).

Tonks, N.K. and Neel, B.G., Combinitorial control of the specificity of protein tyrosine phosphatases, *Curr. Opin. Cell Biol.* **13,** 182–195 (2001).

Varmus, H. and Weinberg, R.A., *Genes and the Biology of Cancer,* Scientific American Library (1993).

Vogelstein, B. and Kinzler, K.W., The multistep nature of cancer, *Trends Genet.* **9,** 138–140 (1993).

Waksman, G., Shoelson, S.E., Pant, N., Cowburn, D., and Kuriyan, J., Binding of high affinity phosphotyrosyl peptide to the Src SH2 domain: Crystal structures of the complexed and peptide-free forms, *Cell* **72,** 779–790 (1993).

Walton, K.M. and Dixon, J.E., Protein tyrosine phosphatases, *Annu. Rev. Biochem.* **62,** 101–120 (1993).

Whitmarsh, A.J. and Davis, R.J., Structural organization of MAP-kinase signaling modules by scaffold proteins in yeast and mammals, *Trends Biochem. Sci.* **23,** 481–485 (1998).

Widman, C., Gibson, S., Jarpe, M.B., and Johnson, G.L., Mitogen-activated protein kinase: Conservation of a three-kinase module from yeast to humans, *Physiol. Rev.* **79,** 143–180 (1999).

Xu, W., Doshi, A., Lei, M., Eck, M.J., and Harrison, S.C., Crystal structures of c-Src reveal features of its autoinhibitory mechanism, *Mol. Cell* **3,** 629–638 (1999); *and* Xu, W., Harrison, S.C., and Eck, M.J., Three dimensional structure of the tyrosine kinase c-Src, *Nature* **385,** 595–602 (1995).

Yaffe, M.B., Phosphotyrosine-binding domains in tyrosine transduction, *Nature Rev. Mol. Cell Biol.* **3,** 177–186 (2002).

Young, M.A., Gonfloni, F., Superti-Furga, G., Roux, B., and Kuriyan, J., Dynamic coupling between the SH2 and SH3 domains of c-Src and Hck underlies their inactivation by C-terminal tyrosine phosphorylation, *Cell* **105,** 115–126 (2001).

Zhang, Z.-Y., Protein tyrosine phosphatases: structure and function, substrate specificity, and inhibitor development, *Annu. Rev. Pharmaccol. Toxicol.* **42,** 209–234 (2002).

Zhou, M.-M., et al., Structure and ligand recognition of the phosphotyrosine binding domain of Shc, *Nature* **378,** 584–592 (1995).

THE PHOSPHOINOSITIDE CASCADE

Berridge, M.J., Inositol trisphosphate and calcium signaling, *Nature* **361,** 315–325 (1993).

Brazil, D.P., Park, J., and Hemmings, B.A., PKB binding proteins: Getting in on the Akt, *Cell* **111,** 292–303 (2002); Brazil, D.P. and Hemmings, D.A., Ten years of protein kinase B signaling: A hard Akt to follow, *Trends Biochem. Sci.* **26,** 657–664 (2001); *and* Chan, T.O., Rittenhouse, S.E., and Tsichlis, P.N., AKT/PKB and other D3 phosphoinositide regulated kinases: Kinase action by phosphoinositide-dependent phosphorylation, *Annu. Rev. Biochem.* **68,** 965–1014 (1999).

Cho, W., Membrane targeting by C1 and C2 domains, *J. Biol. Chem.* **276,** 32407–32410 (2001).

Cockcroft, S. (Ed.), *Biology of Phosphoinositides,* Oxford (2000).

Essen, L.-O., Perisic, O., Katan, M., Wu, Y., Roberts, M.F., and Williams, R.L., Structural mapping of the catalytic mechanism for a mammalian phosphoinositide-specific phospholipase C, *Biochemistry* **36,** 1704–1718 (1997); *and* Essen, L.-O., Perisic, O., Cheung, R., Katan, M., and Williams, R.L., Crystal structure of a mammalian phosphoinositide-specific phospholipase Cδ, *Nature* **380,** 595–602 (1996).

Exton, J.H., Regulation of phosphoinositide phospholipases by hormones, neurotransmitters, and other agonists linked to G proteins, *Annu. Rev. Pharmacol. Toxicol.* **36,** 481–509 (1996).

Ferguson, K.M., Lemmon, M.A., Schlessinger, M.A., and Sigler, P.B., Structure of the high affinity complex of inositol trisphosphate with a phospholipase C pleckstrin homology domain, *Cell* **83,** 1037–1046 (1995).

Fruman, D.A., Meyers, R.E., and Cantley, L.C., Phosphoinositide kinases, *Annu. Rev. Biochem.* **67,** 481–507 (1998); *and* Anderson, R.A., Boronenkov, I.V., Doughman, S.D., Kunz, J., and Loijens, J.C., Phosphatidylinositide kinases, a multifaceted family of signaling enzymes, *J. Biol. Chem.* **274,** 9907–9910 (1999).

Hurley, J.H. and Misra, S., Signaling and subcellular targeting by membrane-binding domains, *Annu. Rev. Biophys. Biomol. Struct.* **29**, 49–79 (2000).

Katso, R., Okkenhaug, K., Ahmadi, K., White, S., Timms, J., and Waterfield, M.D., Cellular function of phosphoinositide 3-kinases: Implications for development, immunity, homeostasis, and cancer, *Annu. Rev. Cell Dev. Biol.* **17**, 615–675 (2001).

Kutateladze, T. and Overduin, M., Structural mechanism of endosome docking by the FYVE domain, *Science* **291**, 1793–1796 (2001).

Lee, J.-O., Yang, H., Georgescu, M.-M., Di Cristofano, A., Maehama, T., Shi, Y., Dixon, J.E., Pandolfi, P., and Pavletich, N.P., Crystal structure of the PTEN tumor suppressor: Implications for its phosphoinositide phosphatase activity and membrane association, *Cell* **99**, 323–344 (1999).

Leevers, S.J., Vanhaesebroeck, B., and Waterfield, M.D., Signaling though phosphoinositide 3-kinases: The lipids take centre stage, *Curr. Opin. Cell Biol.* **11**, 219–225 (1999).

Maehama, T., Taylor, G.S., and Dixon, J.E., PTEN and myotubularin: Novel phosphoinositide phosphatases, *Annu. Rev. Biochem.* **70**, 247–279 (2001).

Majerus, P.W., Kisseleva, M.V., and Norris, F.A., The role of phosphatases in inositol signaling reactions, *J. Biol. Chem.* **274**, 10669–10672 (1999).

Newton, A.C. and Johnson, J.E., Protein kinase C: A paradigm for the regulation of protein function by two membrane-targeting modules, *Biochim. Biophys. Acta* **1376**, 155–172 (1998).

Rameh, L.E. and Cantley, L.C., The role of phosphoinositide 3-kinase lipid products in cell function, *J. Biol. Chem.* **274**, 8347–8350 (1999).

Rhee, S.G., Regulation of phosphoinositide-specific phospholipase C, *Annu. Rev. Biochem.* **70**, 281–312 (2001); *and* Rhee, S.G. and Bae, Y.S., Regulation of phosphoinositide-specific phospholipase C, *J. Biol. Chem.* **272**, 15045–15048 (1997).

Ron, D. and Kazanietz, M.G., New insights into the regulation of protein kinase C and novel phorbol ester receptors, *FASEB J.* **13**, 1658–1676 (1999).

Saltiel, A.R. and Pessin, J.E., Insulin signaling pathways in time and space, *Trends Cell Biol.* **12**, 65–71 (2002).

Singer, W.D., Brown, H.A., and Sternweis, P.C., Regulation of eukaryotic phosphatidylinositol-specific phospholipase C and phospholipase D, *Annu. Rev. Biochem.* **66**, 475–509 (1997).

Vanhaesebroek, B., Leevers, S.J., Ahmadi, K., Timms, J., Katso, R., Driscoll, P.C., Woscholski, R., Parker, P.J., and Waterfield, M.D., Synthesis and function of 3-phosphorylated inositol lipids, *Annu. Rev. Biochem.* **70**, 535–632 (2001).

Walker, E.H., Persic, O., Ried, C., Stephens, L., and Williams, R.L., Structural insights into phosphoinositide 3-kinase catalysis and signaling, *Nature* **402**, 313–320 (1999); *and* Pacold, M.E., et al., Crystal structure and functional analysis of Ras binding to its effector phosphoinositide 3-kinase γ, *Cell* **103**, 931–943 (2000).

Weng, G., Bhalla, U.S., and Iyengar, R., Complexity in biological signaling systems, *Science* **284**, 92–96 (1999).

Wymann, M.P. and Pirola, L., Structure and function of phosphoinositide 3-kinases, *Biochim. Biophys. Acta* **1436**, 127–150 (1998).

Zick, Y. Insulin resistance: a phosphorylation-based uncoupling of insulin signaling, *Trends Cell Biol.* **11**, 437–441 (2001).

PROBLEMS

1. Explain the following observations: (a) Thyroidectomized rats, when deprived of food, survive for 20 days while normal rats starve to death within 7 days. (b) Cushing's syndrome, which results from excessive secretion of adrenocortical steroids, can be caused by a pituitary tumor. (c) **Diabetes insipidus,** which is characterized by unceasing urination and unquenchable thirst, results from an injury to the pituitary. (d) The growth of malignant tumors derived from sex organs may be slowed or even reversed by the surgical removal of the gonads and the adrenal glands.

2. How does the presence of the nonhydrolyzable GTP analog GDPNP affect cAMP-dependent receptor systems?

3. Explain why individuals who regularly handle dynamite (which is nitroglycerin soaked into an absorbant such as wood pulp) as part of their jobs have an unusually high incidence of heart attacks on weekends.

4. A dose-dependent side effect of sildenafil (Viagra) is the transient impairment of blue/green color discrimination. What is the biochemical basis for this phenomenon?

5. Retroviruses bearing oncogenes will infect cells from their corresponding host animal but will usually not transform them. Yet these retroviruses will readily transform immortalized cells derived from the same organism. Explain.

6. Explain why mutations of the Arg residue in $G_{s\alpha}$ that is ADP-ribosylated by cholera toxin are oncogenic mutations. Why doesn't cholera toxin cause cancer?

7. Would the following alterations to Src be oncogenic? Explain. (a) The deletion or inactivation of the SH3 domain. (b) The mutation of Tyr 416 to Phe. (c) The mutation of Tyr 527 to Phe. (d) The replacement of Src residues 249 to 253 with the sequence APTMP.

8. JIP-1 was originally so named because, when it was first characterized by overexpression in mammalian cells, it appeared to act as a "*J*NK *I*nhibitor *P*rotein." What is the basis of this observation?

9. Why does pertussis toxin appear to inhibit certain isozymes of PLC? Identify these isozymes.

10. PKC's autoinhibitory pseudosubstrate occurs at its N-terminus, whereas that of MLCK occurs at its C-terminus (Fig. 18-19). To further investigate this phenomenon, a colleague proposes to construct a PKC with its pseudosubstrate attached to the protein's C-terminus with a sufficiently long linker so that the pseudosubstrate could bind in the enzyme's active site. Would you expect this variant PKC to be activatable? Explain.

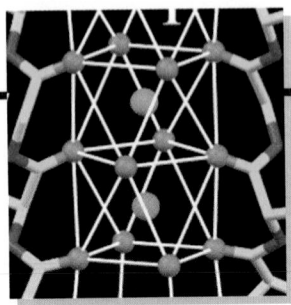

Chapter 20

Transport through Membranes

Metabolism occurs within cells that are separated from their environments by plasma membranes. Eukaryotic cells, in addition, are compartmentalized by intracellular membranes that form the boundaries and internal structures of their various organelles. The nonpolar cores of biological membranes make them highly impermeable to most ionic and polar substances, so that *these substances can traverse membranes only through the action of specific* **transport proteins.** Such proteins are therefore required to mediate all transmembrane movements of ions, such as Na^+, K^+, Ca^{2+}, and Cl^-, as well as metabolites such as pyruvate, amino acids, sugars, and nucleotides, and even water (despite its relatively high permeability in bilayers; Section 12-2B). Transport proteins are also responsible for all biological electrochemical phenomena such as neurotransmission. In this chapter, we discuss the thermodynamics, kinetics, and chemical mechanisms of these membrane

transport systems and end with a discussion of the mechanism of neurotransmission.

1 ■ THERMODYNAMICS OF TRANSPORT

As we saw in Section 3-4A, the free energy of a solute, A, varies with its concentration:

$$\overline{G}_A - \overline{G}_A^{\circ\prime} = RT \ln[A] \qquad [20.1]$$

where $\overline{G}_A$ is the **chemical potential** (partial molar free energy) of A (the bar indicates quantity per mole) and $\overline{G}_A^{\circ\prime}$ is the chemical potential of its standard state. Strictly speaking, this equation applies only to ideal solutions; for nonideal (real) solutions, molar concentrations must be replaced by activities (appendix to Chapter 3). In the dilute (millimolar) solutions that are characteristic of laboratory conditions, the activity of a substance closely approaches its molar concentration in value. However, this is not the case in the highly concentrated cellular milieu (Section 3-4C). Yet it is difficult to determine the activity of a substance in a cellular compartment. Hence, in the following derivations, we shall make the simplifying assumption that activities are equal to molar concentrations.

The diffusion of a substance between two sides of a membrane

$$A(out) \rightleftharpoons A(in)$$

thermodynamically resembles a chemical equilibration. A difference in the concentrations of the substance on two sides of a membrane generates a chemical potential difference:

$$\Delta\overline{G}_A = \overline{G}_A(in) - \overline{G}_A(out) = RT \ln\left(\frac{[A]_{in}}{[A]_{out}}\right) \qquad [20.2]$$

Consequently, if the concentration of A outside the membrane is greater than that inside, $\Delta\overline{G}_A$ for the transfer of A from outside to inside will be negative and the spontaneous net flow of A will be inward. If, however, [A] is greater inside than outside, $\Delta\overline{G}_A$ is positive and an inward net flow of A can only occur if an exergonic process, such as ATP hydrolysis, is coupled to it to make the overall free energy change negative.

a. Membrane Potentials Arise from Transmembrane Concentration Differences of Ionic Substances

The permeabilities of biological membranes to ions such as H^+, Na^+, K^+, Cl^-, and Ca^{2+} are controlled by specific membrane-embedded transport systems that we shall discuss in later sections. *The resulting charge differences across a biological membrane generate an electric potential difference, $\Delta\Psi = \Psi(in) - \Psi(out)$, where $\Delta\Psi$ is termed the* **membrane potential.** Consequently, if A is ionic, Eq. [20.2] must be amended to include the electrical work required to transfer a mole of A across the membrane from outside to inside:

$$\Delta\overline{G}_A = RT\ln\left(\frac{[A]_{in}}{[A]_{out}}\right) + Z_A\mathscr{F}\Delta\Psi \qquad [20.3]$$

where Z_A is the ionic charge of A; $\mathscr{F}$, the Faraday constant, is the charge of one mole of electrons ($96,485\ C\cdot mol^{-1}$); and $\overline{G}_A$ is now termed the **electrochemical potential** of A.

Membrane potentials in living cells can be measured directly with microelectrodes. $\Delta\Psi$ values of -100 mV (inside negative) are not uncommon (note that $1\ V = 1\ J\cdot C^{-1}$). Thus the last term of Eq. [20.3] is often significant for ionic substances.

2 ■ KINETICS AND MECHANISMS OF TRANSPORT

Thermodynamics indicates whether a given transport process will be spontaneous but, as we saw for chemical and enzymatic reactions, provides no indication of the rates of these processes. Kinetic analyses of transport processes together with mechanistic studies have nevertheless permitted these processes to be characterized. There are two types of transport processes: **nonmediated transport** and **mediated transport.** Nonmediated transport occurs through simple diffusion. In contrast, *mediated transport occurs through the action of specific carrier proteins* that are variously called **carriers, permeases, porters, translocases, translocators,** and **transporters.** Mediated transport is further classified into two categories depending on the thermodynamics of the system:

1. Passive-mediated transport or **facilitated diffusion** in which specific molecules flow from high concentration to low concentration so as to equilibrate their concentration gradients.

2. Active transport in which specific molecules are transported from low concentration to high concentration, that is, against their concentration gradients. Such an endergonic process must be coupled to a sufficiently exergonic process to make it favorable.

In this section, we consider the nature of nonmediated transport and then compare it to passive-mediated transport as exemplified by ionophores, porins, glucose

transporters, and K^+ channels. Active transport is examined in succeeding sections.

A. *Nonmediated Transport*

The driving force for the nonmediated flow of a substance A through a medium is A's electrochemical potential gradient. This relationship is expressed by the **Nernst–Planck equation:**

$$J_A = -[A]U_A(d\overline{G}_A/dx) \qquad [20.4]$$

where J_A is the **flux** (rate of passage per unit area) of A, x is distance, $d\overline{G}_A/dx$ is the electrochemical potential gradient of A, and U_A is its **mobility** (velocity per unit force) in the medium. If we assume, for simplicity, that A is an uncharged molecule so that $\overline{G}_A$ is given by Eq. [20.1], the Nernst–Planck equation reduces to

$$J_A = -D_A(d[A]/dx) \qquad [20.5]$$

where $D_A \equiv RTU_A$ is the **diffusion coefficient** of A in the medium of interest. This is **Fick's first law of diffusion,** which states that *a substance diffuses in the direction that eliminates its concentration gradient, $d[A]/dx$, at a rate proportional to the magnitude of this gradient.*

For a membrane of thickness x, Eq. [20.5] is approximated by

$$J_A = \frac{D_A}{x}([A]_{out} - [A]_{in}) = P_A([A]_{out} - [A]_{in}) \qquad [20.6]$$

where D_A is the diffusion coefficient of A inside the membrane and $P_A = D_A/x$ is termed the membrane's **permeability coefficient** for A. The permeability coefficient is indicative of the solute's tendency to transfer from the aqueous solvent to the membrane's nonpolar core. It should therefore vary with the ratio of the solute's solubility in a nonpolar solvent resembling the membrane's core (e.g., olive oil) to that in water, a quantity known as the solute's **partition coefficient** between the two solvents. Indeed, the fluxes of many nonelectrolytes across erythrocyte membranes vary linearly with their concentration differences across the membrane as predicted by Eq. [20.6] (Fig. 20-1). Moreover, their permeability coefficients, as

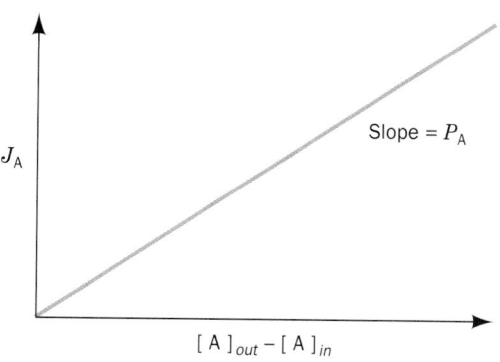

FIGURE 20-1 **Linear relationship between diffusional flux (J_A) and ($[A]_{out} - [A]_{in}$) across a semipermeable membrane.** See Eq. [20.6].

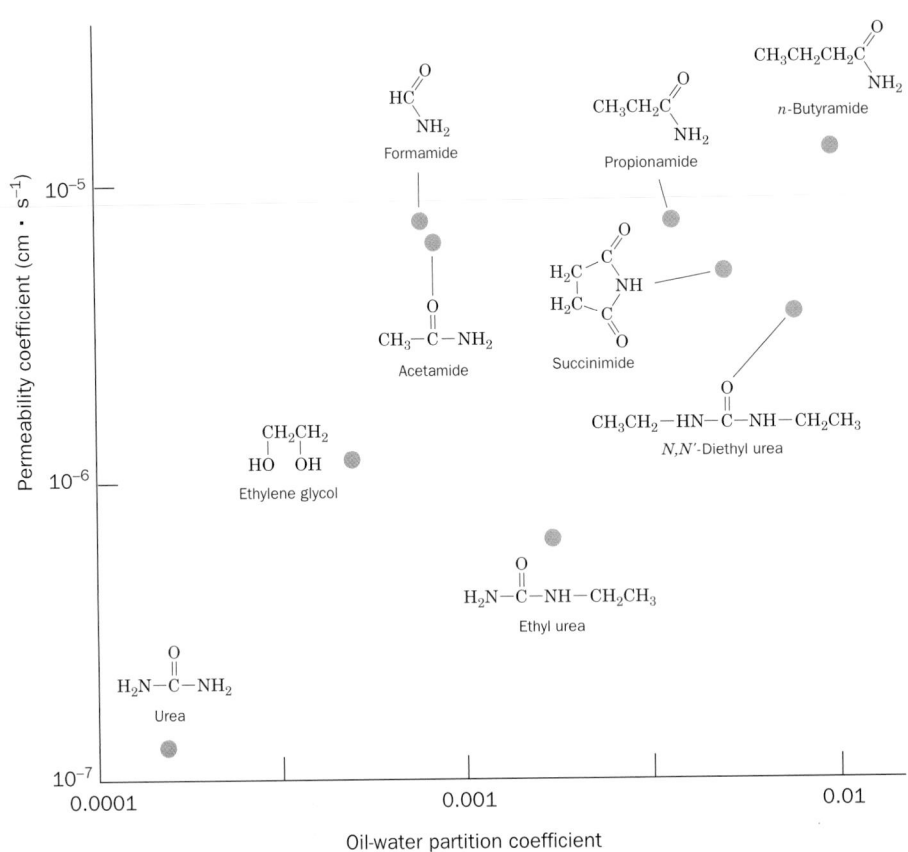

FIGURE 20-2 Permeability correlates with membrane solubility. The permeability coefficients of various organic molecules in plasma membranes from the alga *Nitella mucronata* versus their partition coefficients between olive oil and water (a measure of a molecule's polarity). This more or less linear log–log plot indicates that the rate-limiting step for the nonmediated entry of a molecule into a cell is its passage through the membrane's hydrophobic core. [Based on data from Collander, R., *Physiol. Plant.* **7**, 433–434 (1954).]

obtained from the slopes of plots such as Fig. 20-1, correlate rather well with their measured partition coefficients between nonpolar solvents and water (Fig. 20-2).

B. *Kinetics of Mediated Transport: Glucose Transport into Erythrocytes*

Despite the success of the foregoing model in predicting the rates at which many molecules pass through membranes, there are numerous combinations of solutes and membranes that do not obey Eq. [20.6]. The flux in such a system is not linear with the solute concentration difference across the corresponding membrane (Fig. 20-3) and, furthermore, the solute's permeability coefficient is much larger than is expected on the basis of its partition coefficient. Such behavior indicates that *these solutes are conveyed across membranes in complex with carrier molecules; that is, they undergo mediated transport.*

The system that transports glucose across the erythrocyte membrane provides a well-characterized example of passive-mediated transport: It invariably transports glucose down its concentration gradient but not at the rate predicted by Eq. [20.6]. Indeed, the **erythrocyte glucose transporter** exhibits four characteristics that differentiate mediated from nonmediated transport: (1) *speed and speci-*

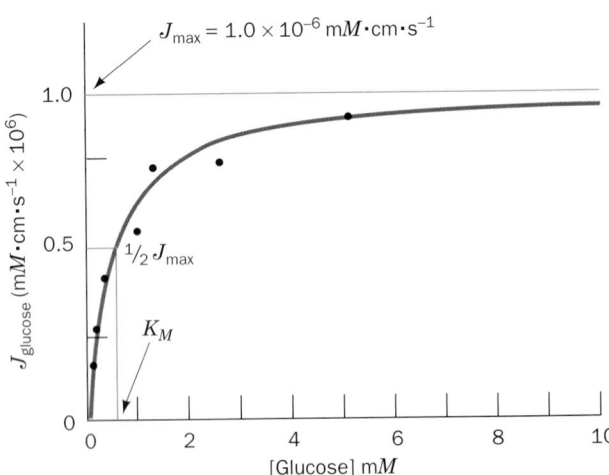

FIGURE 20-3 Variation of glucose flux into human erythrocytes with the external glucose concentration at 5°C. The black dots are experimentally determined data points, and the solid green line is computed from Eq. [20.7] with $J_{max} = 1.0 \times 10^{-6}$ mM · cm · s^{-1} and $K_M = 0.5$ mM. The nonmediated glucose flux increases linearly with [glucose] (Fig. 20-1) but would not visibly depart from the baseline on the scale of this drawing. [Based on data from Stein, W.D., *Movement of Molecules across Membranes*, p. 134, Academic Press (1967).]

ficity, (2) *saturation kinetics,* (3) *susceptibility to competitive inhibition, and* (4) *susceptibility to chemical inactivation.* In the following paragraphs we shall see how the erythrocyte glucose transporter exhibits these qualities.

a. Speed and Specificity

Table 20-1 indicates that the permeability coefficients of D-glucose and D-mannitol in synthetic bilayers, and that of D-mannitol in the erythrocyte membrane, are in reasonable agreement with the values calculated from the diffusion and partition coefficients of these sugars between water and olive oil. However, the experimentally determined permeability coefficient for D-glucose in the erythrocyte membrane is four orders of magnitude greater than its predicted value. *The erythrocyte membrane must therefore contain a system that rapidly transports glucose and that can distinguish D-glucose from D-mannitol.*

b. Saturation Kinetics

The concentration dependence of glucose transport indicates that its flux obeys the relationship:

$$J_A = \frac{J_{max}[A]}{K_M + [A]} \qquad [20.7]$$

This **saturation function** has a familiar hyperbolic form (Fig. 20-3). We have seen it in the equation describing the binding of O_2 to myoglobin (Eq. [10.4]) and in the Michaelis–Menten equation describing the rates of enzymatic reactions (Eq. [14.24]). Here, as before, K_M may be defined operationally as the concentration of glucose when the transport flux is half of its maximal rate, $J_{max}/2$. *This observation of **saturation kinetics** for glucose transport was the first evidence that a specific, saturable number of sites*

TABLE 20-1   Permeability Coefficients of Natural and Synthetic Membranes to D-Glucose and D-Mannitol at 25°C

Membrane Preparation	Permeability Coefficient ($cm \cdot s^{-1}$)	
	D-Glucose	D-Mannitol
Synthetic lipid bilayer	2.4×10^{-10}	4.4×10^{-11}
Calculated nonmediated diffusion	4×10^{-9}	3×10^{-9}
Intact human erythrocyte	2.0×10^{-4}	5×10^{-9}

Source: Jung, C.Y., *in* Surgenor, D. (Ed.), *The Red Blood Cell,* Vol. 2, p. 709, Academic Press (1975).

on the membrane were involved in the transport of any substance.

The transport process can be described by a simple four-step kinetic scheme involving binding, transport, dissociation, and recovery (Fig. 20-4). Its binding and dissociation steps are analogous to the recognition of a substrate and the release of product by an enzyme. The mechanisms of transport and recovery are discussed in Section 20-2D.

c. Susceptibility to Competitive Inhibition

Many compounds structurally similar to D-glucose inhibit glucose transport. A double-reciprocal plot (Section 14-2B) for the flux of glucose into erythrocytes in the presence or absence of 6-O-benzyl-D-galactose (Fig. 20-5) shows behavior typical of competitive inhibition of glucose transport (competitive inhibition of enzymes is discussed in Section 14-3A). *Susceptibility to competitive inhibition indicates that there is a limited number of sites available for mediated transport.*

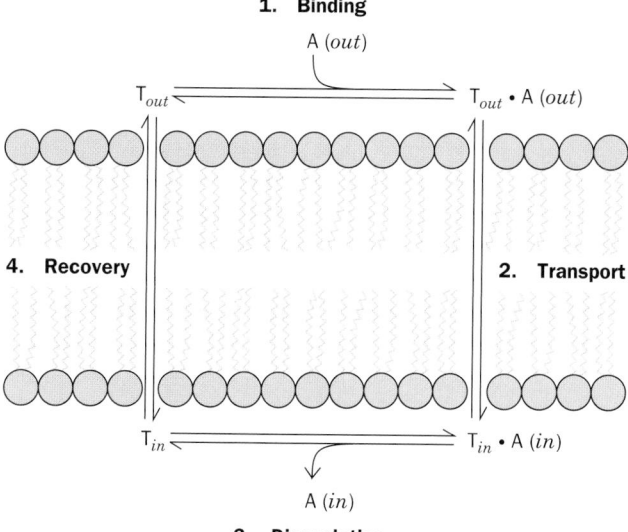

FIGURE 20-4   General kinetic scheme for membrane transport. The scheme involves four steps: binding, transport, dissociation, and recovery. T is the transport protein whose binding site for solute A is located on either the inner or the outer side of the membrane at any one time.

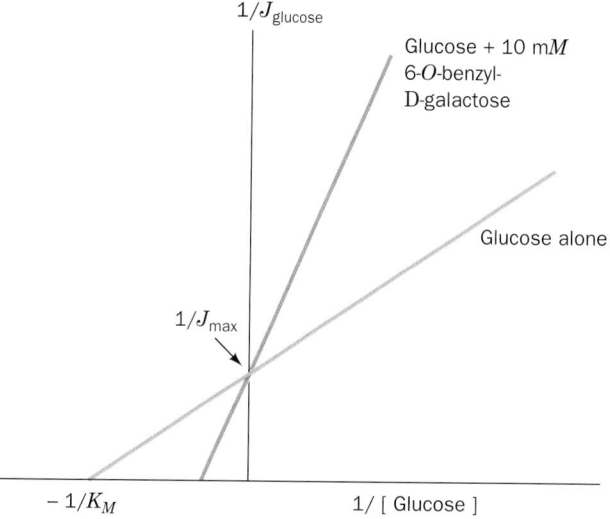

FIGURE 20-5   Double-reciprocal plots for the net flux of glucose into erythrocytes in the presence and absence of 6-O-benzyl-D-galactose. The pattern is that of competitive inhibition. [After Barnett, J.E.G., Holman, G.D., Chalkley, R.A., and Munday, K.A., *Biochem. J.* **145,** 422 (1975).]

d. Susceptibility to Chemical Inactivation

Treatment of erythrocytes with $HgCl_2$, which reacts with protein sulfhydryl groups

$$RSH + HgCl_2 \rightarrow RS—Hg—Cl + HCl$$

and thus inactivates many enzymes, causes the rapid, saturable flux of glucose to disappear, so that its permeability constant approaches that of mannitol. *The erythrocyte glucose transport system's susceptibility to such protein-modifying agents indicates that it, in fact, is a protein.*

All of the above observations indicate that *glucose transport across the erythrocyte membrane is mediated by a limited number of protein carriers.* Before we discuss the mechanism of this transport system, however, we shall examine some simpler models of facilitated diffusion.

C. *Ionophores*

Our understanding of mediated transport has been enhanced by the study of **ionophores,** substances that vastly increase the permeability of membranes to particular ions.

a. Ionophores May Be Carriers or Channel Formers

Ionophores are organic molecules of diverse types, many of which are antibiotics of bacterial origin. Cells and organelles actively maintain concentration gradients of various ions across their membranes (Section 20-3A). The antibiotic properties of ionophores arise from their tendency to discharge these vital concentration gradients.

There are two types of ionophores:

1. Carriers, *which increase the permeabilities of membranes to their selected ion by binding it, diffusing through the membrane, and releasing the ion on the other side (Fig. 20-6a).* For net transport to occur, the uncomplexed ionophore must then return to the original side of the membrane ready to repeat the process. Carriers therefore share the common property that *their ionic complexes are soluble in nonpolar solvents.*

2. Channel formers, *which form transmembrane channels or pores through which their selected ions can diffuse (Fig. 20-6b).*

Both types of ionophores transport ions at a remarkable rate. For example, a single molecule of the carrier antibiotic **valinomycin** transports up to 10^4 K^+ ions per second across a membrane. Channel formers have an even greater ion throughput; for example, each membrane channel composed of the antibiotic **gramicidin A** permits the passage of over 10^7 K^+ ions $\cdot$ s^{-1}. Clearly, the presence of either type of ionophore, even in small amounts, greatly increases the permeability of a membrane toward the specific ions transported. However, *since ionophores passively permit ions to diffuse across a membrane in either direction, their effect can only be to equilibrate the concentrations of their selected ions across the membrane.*

Carriers and channel formers are easily distinguished experimentally through differences in the temperature dependence of their action. Carriers depend on their ability

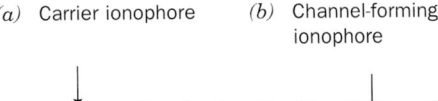

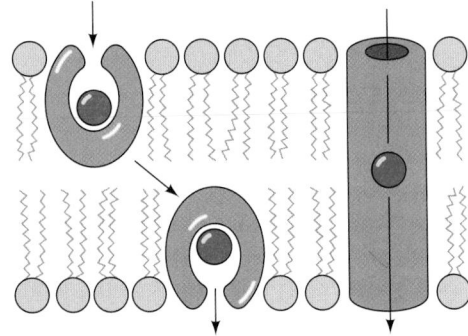

FIGURE 20-6   Ion transport modes of ionophores. (*a*) Carrier ionophores transport ions by diffusing through the lipid bilayer. (*b*) Channel-forming ionophores span the membrane with a channel through which ions can diffuse.

to diffuse freely across the membrane. Consequently, cooling a membrane below its transition temperature (the temperature below which it becomes a gel-like solid; Section 12-2C) essentially eliminates its ionic permeability in the presence of carriers. In contrast, membrane permeability in the presence of channel formers is rather insensitive to temperature because, once in place, channel formers need not move to mediate ion transport.

b. The K^+–Valinomycin Complex Has a Polar Interior and a Hydrophobic Exterior

Valinomycin, which is perhaps the best characterized carrier ionophore, specifically binds K^+ and the biologically unimportant Rb^+. It is a **cyclic depsipeptide** that contains both D- and L-amino acid residues (Fig. 20-7; a depsipeptide contains ester linkages as well as peptide bonds). The X-ray structure of valinomycin's K^+ complex (Fig. 20-8a) indicates that the K^+ is octahedrally coordinated by the carbonyl groups of its 6 Val residues, which also form its ester linkages. The cyclic, intramolecularly hydrogen bonded valinomycin backbone follows a zigzag path that surrounds the K^+ coordination shell with a sinuous molecular bracelet. *Its methyl and isopropyl side chains project*

Valinomycin

FIGURE 20-7   Valinomycin. This cyclic depsipeptide (has both ester and amide bonds) contains both D- and L-amino acids.

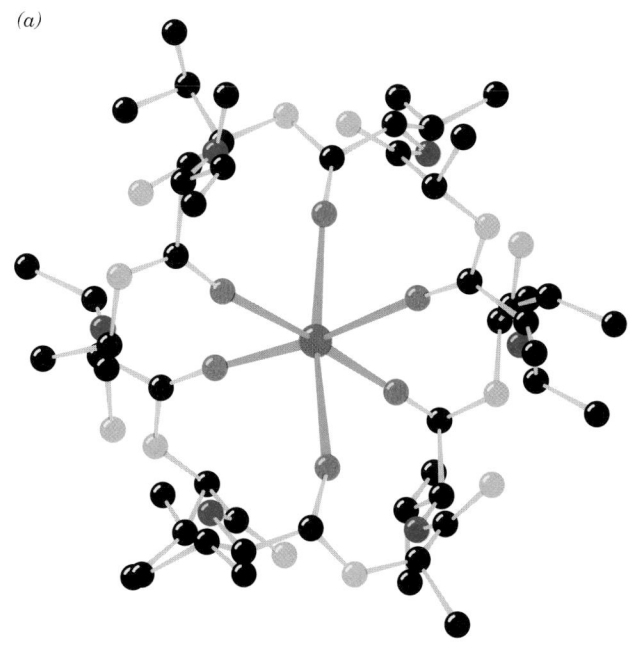

(a)

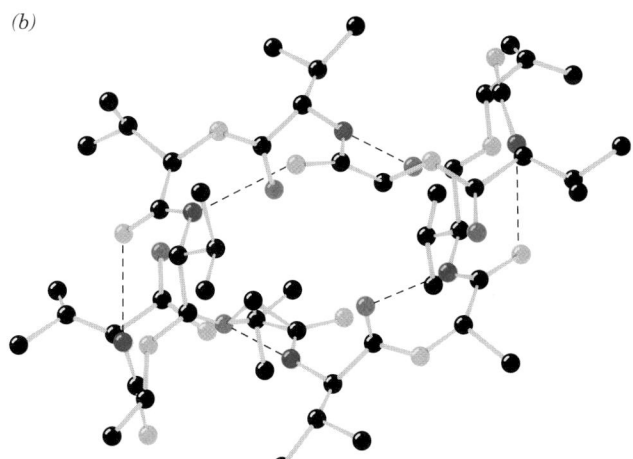

(b)

FIGURE 20-8 Valinomycin X-ray structures. (*a*) The K$^+$ complex. The six oxygen atoms that octahedrally complex the K$^+$ ion are darker red than the other oxygen atoms. [After Neupert-Laves, K. and Dobler, M., *Helv. Chim. Acta* **58,** 439 (1975).] (*b*) Uncomplexed valinomycin. [After Smith, G.D., Duax, W.L., Langs, D.A., DeTitta, G.T., Edmonds, R.C., Rohrer, D.C., and Weeks, C.M., *J. Am. Chem. Soc.* **97,** 7242 (1975).] Hydrogen atoms are not shown.

cannot simultaneously coordinate these ions. Complexes of these ions with water are therefore energetically more favorable than their complexes with valinomycin. This accounts for valinomycin's 10,000-fold greater binding affinity for K$^+$ over Na$^+$. Indeed, no other substance discriminates more acutely between Na$^+$ and K$^+$.

The Na$^+$-binding ionophore **monensin** (Fig. 20-9*a*), a linear polyether carboxylic acid, is chemically different from valinomycin. Nevertheless, X-ray analysis reveals that monensin's Na$^+$ complex has the same general features as valinomycin's K$^+$ complex in that monensin octahedrally coordinates Na$^+$ so as to wrap it in a nonpolar jacket (Fig. 20-9*b*). Other carrier ionophores have similar characteristics.

c. Gramicidin A Forms Helical Transmembrane Channels

Gramicidin A is a channel-forming ionophore from *Bacillus brevis* that permits the passage of protons and alkali metal cations but is blocked by Ca^{2+}. It is a 15-residue linear polypeptide of alternating L and D residues that is

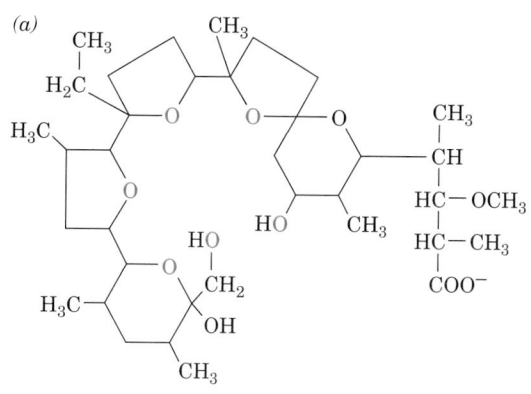

(a)

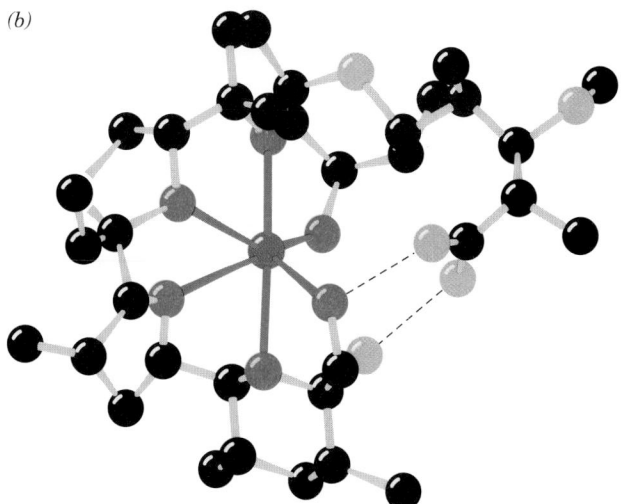

(b)

FIGURE 20-9 Monensin. (*a*) The structural formula with the six oxygen atoms that octahedrally complex Na$^+$ indicated in red. (*b*) The X-ray structure of the Na$^+$ complex (H atoms not shown). [After Duax, W.L., Smith, G.D., and Strong, P.D., *J. Am. Chem. Soc.* **102,** 6728 (1980).]

outward from the bracelet to provide the spheroidal complex with a hydrophobic exterior that makes it soluble in nonpolar solvents and in the hydrophobic cores of lipid bilayers. Uncomplexed valinomycin (Fig. 20-8*b*) has a more open conformation than its K$^+$ complex, which presumably facilitates the rapid binding of K$^+$.

K$^+$ (ionic radius, r = 1.33 Å) and Rb$^+$ (r = 1.49 Å) fit snugly into valinomycin's coordination site. However, the rigidity of the valinomycin complex makes this site too large to accommodate Na$^+$ (r = 0.95 Å) or Li$^+$ (r = 0.60 Å) properly; that is, valinomycin's six carbonyl oxygen atoms

chemically blocked at its amino terminus by formylation and at its carboxyl terminus by an amide bond with ethanolamine (Fig. 20-10). Note that all of its residues are

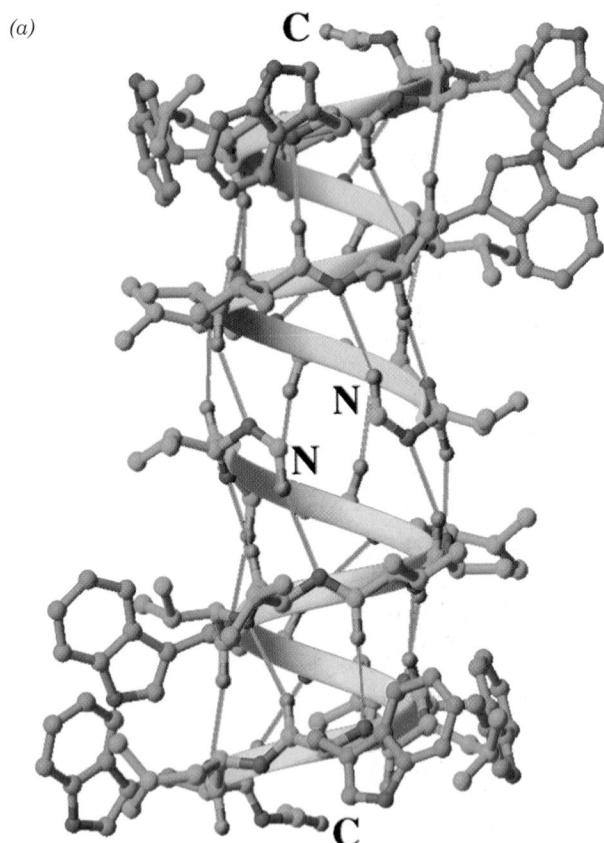

Gramicidin A

FIGURE 20-10 Gramicidin A. This polypeptide consists of 15 alternating D- and L-amino acid residues and is blocked at both its N- and C-termini.

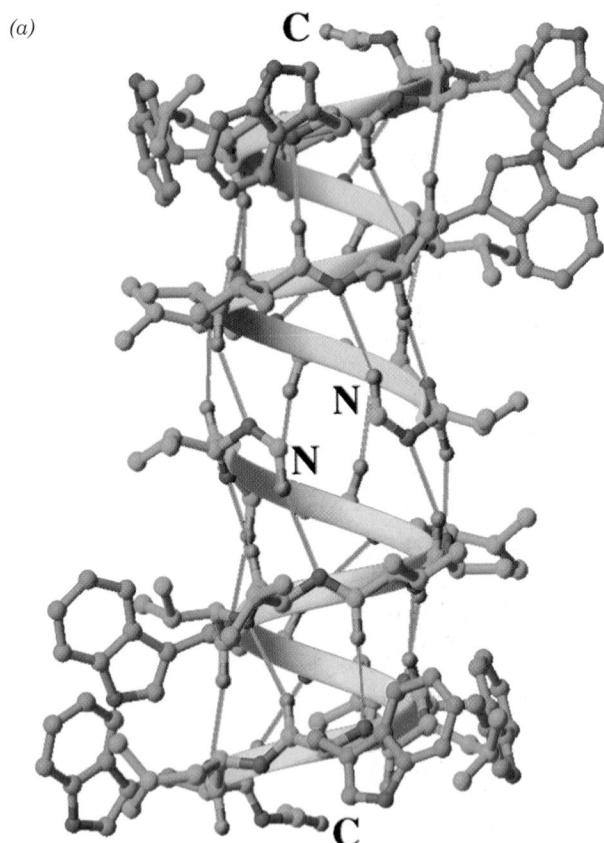

FIGURE 20-11 NMR structure of gramicidin A embedded in a dimyristoyl phosphatidylcholine bilayer. (*a*) View from within the bilayer along the homodimeric helix's twofold axis. The polypeptide is shown in ball-and-stick form colored according to atom type (N blue, O red, and C green except for the side chain C's of Trp residues, which are magenta). The cyan and gold ribbons indicate the helical paths taken by the upper and lower polypeptide backbones. Hydrogen bonds are represented by gray lines. H atoms have been omitted for clarity. The ~25-Å-long helix is right-handed with 6.5 residues per turn. Its novel hydrogen bonding arrangement is possible because the alternating D and L configurations of the amino acid residues

hydrophobic, as is expected for a small transmembrane polypeptide. Its NMR structure in a lipid bilayer, determined by Timothy Cross, indicates that *gramicidin A dimerizes in a head-to-head fashion to form a transmembrane channel* (Fig. 20-11*a*). This is corroborated by the observation that two gramicidin A molecules whose N-terminal amino groups are covalently cross-linked form a functional ion channel. Moreover, channel activity is abolished when a charged residue is introduced at gramicidin A's N-terminus, but not at its C-terminus.

The gramicidin A channel resembles a rolled up parallel β sheet and has therefore been dubbed a **β helix**. Successive backbone N—H groups in this structure alternately point up and down the helix to hydrogen bond with backbone carbonyl groups (Fig. 20-11*a*). As a consequence of its alternating D and L residues, the side chains of the β helix festoon its periphery to form the channel's required hydrophobic exterior (recall that in a β sheet of all L-amino acid residues, the side chains alternately extend to opposite sides of the sheet; Fig. 8-17). The polar backbone groups thus line the central channel (Fig. 20-11*b*) and thereby facilitate the passage of ions. The four Trp side chains in the C-terminal half of each polypeptide chain are oriented with their polar N—H groups directed toward the bilayer surface, thereby aligning the dimeric helix perpendicular to the bilayer. Indeed, the replacement of any of these Trp residues by Phe significantly reduces the β he-

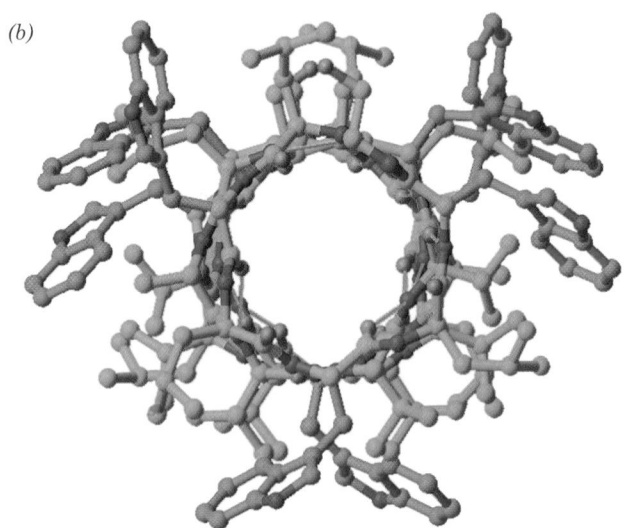

permit both successive N—H and successive C=O groups to point in opposite directions along the helix axis. Note that all backbone N—H groups not extending from the top or bottom of the helix form either intra- or intermolecular hydrogen bonds. (*b*) View perpendicular to the bilayer plane (rotated 90° about the horizontal axis relative to Part *a*). The 4-Å-diameter cylindrical pore that runs the length of the helical dimer is lined by all of the polypeptide's polar backbone groups and is wide enough to permit the passage of alkali metal cations. [Based on an NMR structure by Timothy Cross, Florida State University. PDBid 1MAG.]

lix's stability and hence the channel's conductivity. Note that gramicidin A carries out a function similar to that of the far more complex channel-forming toxins such as α-hemolysin (Section 12-3G), although the latter has a much wider pore and hence does not discriminate among ions.

D. *Maltoporin: The Structural Basis of Sugar Discrimination*

The **porins** are homotrimeric transmembrane proteins that facilitate the transport of small molecules and ions across the outer membranes of gram-negative bacteria and mitochondria. Each subunit consists mainly of a 16- to 22-stranded antiparallel β barrel that forms a solvent-accessible channel along the barrel axis (Section 12-3A). In the *E. coli* OmpF porin (Fig. 12-27), this ~50-Å-long channel is constricted near its center to an elliptical pore that has a minimum cross section of 7×11 Å. Consequently, solutes of more than ~600 D are too large to pass through this channel.

Maltoporin is a bacterial porin that facilitates the diffusion of **maltodextrins** [the $\alpha(1{\rightarrow}4)$-linked glucose oligosaccharide degradation products of starch; e.g., maltose (Fig. 11-12)]. The X-ray structure of *E. coli* maltoporin (Fig. 20-12), determined by Tilman Schirmer, reveals that maltoporin is structurally similar to OmpF porin (Fig. 12-27), but with an 18-stranded rather than a 16-stranded antiparallel β barrel enclosing each subunit's transport channel. Three long loops from the extracellular face of each maltoporin subunit fold inward into the barrel, thereby constricting the channel near the center of the membrane to a diameter of ~5 Å (which is considerably smaller than OmpF's aperture) and giving the channel an hourglass-like cross section. The channel is lined on one side with a series of six contiguous aromatic side chains arranged in a left-handed helical path that matches the left-handed helical curvature of α-amylose (Fig. 11-17). This "greasy slide" extends from the channel's vestibule floor, through its constriction, to its periplasmic outlet.

The way in which oligosaccharides interact with maltoporin was investigated by determining the X-ray structures of maltoporin in its complexes with the maltodextrins Glc_2 (maltose), Glc_3, Glc_6, and sucrose (a glucose–fructose disaccharide; Fig. 11-12). Two Glc_2 molecules, one Glc_3 molecule, and a Glc_5 segment of Glc_6 occupied the maltoporin channel in contact and conformity with the greasy slide. Thus the hydrophobic faces of the maltodextrin's glycosyl residues stack on aromatic side chains, as is often observed in complexes of sugars with proteins. The glucose hydroxyl groups, which are arranged in two strips along opposite edges of the maltodextrins, form numerous hydrogen bonds with polar side chains that line these strips. Six of these seven polar side chains are charged, which probably strengthens their hydrogen bonds, as has also been observed in complexes of sugars with proteins. Tyr 118, which protrudes into the channel opposite the greasy slide, apparently functions as a steric barrier that only permits the passage of near-planar groups such as glucosyl residues. Thus the hook-shaped sucrose, which maltoporin trans-

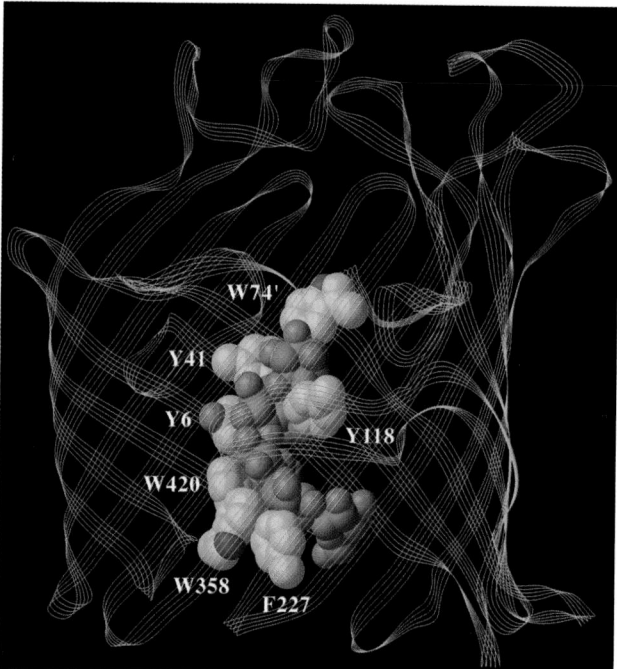

FIGURE 20-12   The X-ray structure of a subunit of *E. coli* maltoporin in complex with a maltodextrin of 6 glucosyl units (Glc_6). The structure is viewed from within the bacterial outer membrane with its extracellular surface above. The polypeptide backbone is represented by a multithreaded ribbon (*cyan*). The Glc_6 (only 5 of whose glucosyl units are observed) and the aromatic side chains lining the constricted region of the protein's centrally located transport channel are shown in space-filling form colored according to atom type (N blue, O red, protein side chain C's gold, and glucosyl C's green). Note the pronounced left-handed helical twist of the Glc_6 unit. The so-called greasy slide, which consists of the aromatic side chains of 6 residues (W74′ is contributed by an overhanging loop from an adjacent subunit), conforms closely to this shape. The side chain of Y118 protrudes into the channel opposite the greasy slide so as to allow only the transit of near planar groups such as glucosyl residues. The maltodextrin's hydroxyl groups are arranged in two strips flanking the greasy slide (only one of which is seen here) that form an extensive hydrogen bonded network with mainly charged side chains (not shown). [Based on an X-ray structure by Tilman Schirmer, University of Basel, Switzerland. PDBid 1MPO.]

ports quite slowly, binds to maltoporin with only its glucose residue inserted into the constricted part of the channel and its bulky fructose residue extending into the extracellular vestibule.

The above structures suggest a model for the selective transport of maltodextrins by maltoporin. At the start of the translocation process, the entering glucosyl residue interacts with the readily accessible end of the greasy slide in the extracellular vestibule of the channel. Further translocation along the helical channel requires the maltodextrin to follow a screwlike path that maintains the helical structure of the oligosaccharide, much like the movement of a bolt through a nut, thereby excluding molecules of comparable size that have different shapes. The translocation process is unlikely to encounter any large energy barrier due to the smooth surface of the greasy slide and the multiple polar groups at the channel constriction that would permit the essentially continuous exchange of hydrogen bonds as a maltodextrin moves through the constriction. Thus, maltoporin can be regarded as an enzyme that catalyzes the translocation of its substrate from one compartment to another.

E. *Passive-Mediated Glucose Transport*

The human erythrocyte glucose transporter is a 492-residue glycoprotein which, according to sequence hydropathy analysis (Sections 8-4C and 12-3A), has 12 membrane-spanning α helices (Fig. 20-13) that are thought to form a hydrophobic cylinder. Five of these helices (3, 5, 7, 8, and 11) are amphipathic and hence most likely form a hydrophilic channel through which glucose is transported. A highly charged 66-residue domain located between helices 6 and 7, together with the 43-residue C-terminal domain, occupy the cytoplasm, whereas a 34-residue carbohydrate-bearing domain located between helices 1 and 2 is externally located. The glucose transporter accounts for 2% of erythrocyte membrane proteins and runs as band 4.5 in SDS–PAGE gels of erythrocyte membranes (Section 12-3D; it is not visible on the gel depicted in Fig. 12-36 because the heterogeneity of its oligosaccharides makes the protein band diffuse).

a. Glucose Transport Occurs via a Gated Pore Mechanism

The erythrocyte glucose transporter has glucose binding sites on each side of the erythrocyte membrane but these have different steric requirements. Thus, John Barnett showed that 1-propylglucose will not bind to the extracellular surface of the glucose transporter but will bind to its cytoplasmic surface, whereas the converse is true of 6-propylglucose. He therefore proposed that the glucose transporter has two alternate conformations: one with the glucose binding site facing the external cell surface, requiring O1 contact and leaving O6 free, and the other with the glucose binding site facing the cytoplasm, requiring O6 contact and leaving O1 free (Fig. 20-14). *Transport apparently takes place by binding glucose to the protein on one face of the membrane, followed by a conformational change that closes the first site while exposing the other.* Glucose can then dissociate from the protein, having been translocated across the membrane. The transport cycle of this so-called **gated pore** is completed by the reversion of the glucose transporter to its initial conformation in the absence of bound glucose. Since this cycle can occur in either direction, the direction of net glucose transport is from high to low glucose concentrations. The glucose transporter thereby provides a means of equilibrating the glucose concentration across the erythrocyte membrane without any accompanying leakage of small molecules or ions.

b. Eukaryotes Express a Variety of Glucose Transporters

The erythrocyte glucose transporter, known also as **GLUT1,** has a highly conserved amino acid sequence (98% sequence identity between humans and rats), which suggests that all segments of this protein are functionally significant. GLUT1 is expressed in most tissues, although in liver and muscle, tissues that are highly active in glucose transport, it is present in only tiny amounts. Three other glucose transporters, **GLUT2, GLUT3,** and **GLUT4,** have been characterized (**GLUT5** was originally thought to be a glucose transporter but was later shown to be a fructose transporter). They are 40 to 65% identical to GLUT1 but have different tissue distributions. For example, GLUT2 is prominent in pancreatic β cells (which secrete insulin in response to increased [glucose] in blood; Section 18-3F) and liver (where its defects result in symptoms resembling Type I glycogen storage disease; Section 18-4), whereas GLUT4 occurs mainly in muscle and fat cells. Note that

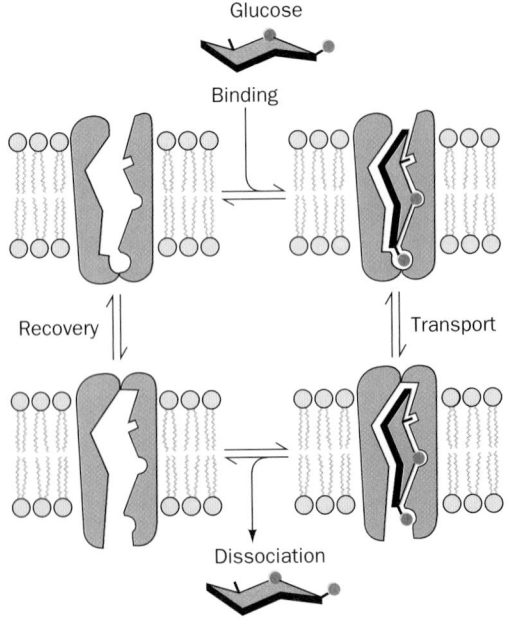

FIGURE 20-14 Alternating conformation model for glucose transport. Such a system is also known as a "gated pore." [After Baldwin, S.A. and Lienhard, G.E., *Trends Biochem. Sci.* **6,** 210 (1981).] See the Animated Figures

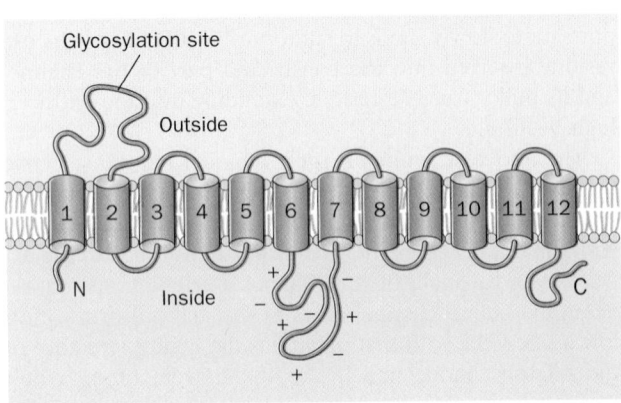

FIGURE 20-13 Predicted secondary structure and membrane orientation of the glucose transporter.

the tissue distributions of these glucose transporters correlate with the response of these tissues to insulin: Liver is unresponsive to insulin (liver functions, in part, to maintain the level of blood glucose; Section 18-3F), whereas muscle and fat cells take up glucose when stimulated by insulin. Novel members of the GLUT family from GLUT5 through GLUT12 have been identified in various tissues although they have yet to be completely characterized.

c. Cellular Glucose Uptake Is Regulated through the Insulin-Sensitive Exocytosis/Endocytosis of Glucose Transporters

Insulin stimulates fat and muscle cells to take up glucose. Within ~15 min after the administration of insulin, the J_{max} for passive-mediated glucose transport into these cells increases 6- to 12-fold, whereas the K_M remains constant. On withdrawal of the insulin, the rate of glucose uptake returns to its basal level within 20 min to 2 h depending on conditions. Neither the increase nor the decrease in the rate of glucose transport is affected by the presence of protein synthesis inhibitors, so that these observations cannot be a consequence of the synthesis of new glucose transporter or of a protein that inhibits it. How, then, does insulin regulate glucose transport?

Basal state fat and muscle cells store most of their glucose transporters in internal membranous vesicles. On insulin stimulation, these vesicles fuse with the plasma membrane in a process known as **exocytosis** *(Fig. 20-15).* The consequent increased number of cell-surface glucose transporters (GLUT4) results in a proportional increase in the cell's glucose uptake rate. On insulin withdrawal, the process is reversed through the endocytosis of plasma

membrane-embedded glucose transporters. The deletion or mutation of GLUT4's N-terminal eight residues, particularly Phe 5, causes this transporter to accumulate in the plasma membrane. Evidently, GLUT4's N-terminal segment targets it for sequestration by the cell's endocytotic machinery, although how insulin controls this system, which accounts for most of insulin's effects on muscle and fat cells, is imperfectly understood. However, research indicates that this exocytotic mechanism involves a tyrosine phosphorylation cascade that is triggered by the binding of insulin to the insulin receptor (Sections 19-3A and 19-4F), and includes the activation of a phosphoinositide 3-kinase (PI3K; Section 19-4D) as well as vesicle fusion (Section 12-4D).

F. K^+ Channels: Ion Discrimination

Potassium ions diffuse from the cytoplasm (where $[K^+] > 100$ mM) to the extracellular space (where $[K^+] < 5$ mM) through transmembrane proteins known as **K^+ channels,** a process that underlies numerous important biological processes including maintenance of cellular osmotic balance, neurotransmission (Section 20-5), and signal transduction (Chapter 19). Although there is a large diversity of K^+ channels, even within single organisms, all of them have similar sequences, exhibit comparable permeability characteristics, and most importantly, are at least 10,000-fold more permeable to K^+ than Na$^+$. Since this high selectivity (around the same as that of valinomycin; Section 20-2C) implies energetically strong interactions between K^+ and the protein, how can the K^+ channel maintain its observed nearly diffusion-limited throughput rate of up to 10^8 ions per second (a 10^4-fold greater rate than that of valinomycin)?

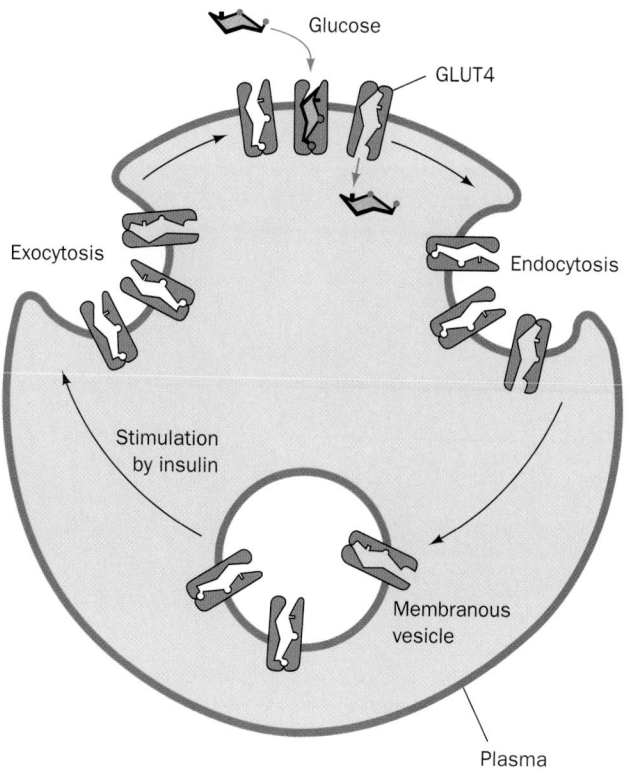

FIGURE 20-15 Regulation of glucose uptake in muscle and fat cells. Regulation is mediated by the insulin-stimulated exocytosis (the opposite of endocytosis; Section 12-5B) of membranous vesicles containing GLUT4 glucose transporters (*left*). On insulin withdrawal, the process reverses itself through endocytosis (*right*).

a. The X-Ray Structure of KcsA Reveals the Basis of K⁺ Channel Selectivity

KcsA, the K⁺ channel from *Streptomyces lividans,* is a tetramer of identical 158-residue subunits. The X-ray structure of its N-terminal 125-residue segment, determined by Roderick MacKinnon, reveals that each KcsA subunit forms two nearly parallel transmembrane helices that are inclined ~25° from the normal to the membrane plane and which are connected by an ~20-residue pore region (Fig. 20-16a). As is true of all known K⁺ channels, four such subunits associate to form a 4-fold rotationally symmetric assembly surrounding a central pore. The four inner (C-terminal) helices, which largely form the pore, pack against each other near the cytoplasmic side of the membrane much like the poles of an inverted teepee. The four outer helices, which face the lipid bilayer, buttress the inner helices but do not contact the adjacent outer helices. The pore regions, which each consist of a so-called turret, pore helix, and selectivity filter, occupy the open extracellular end of the teepee, with the pore helices fitting in between

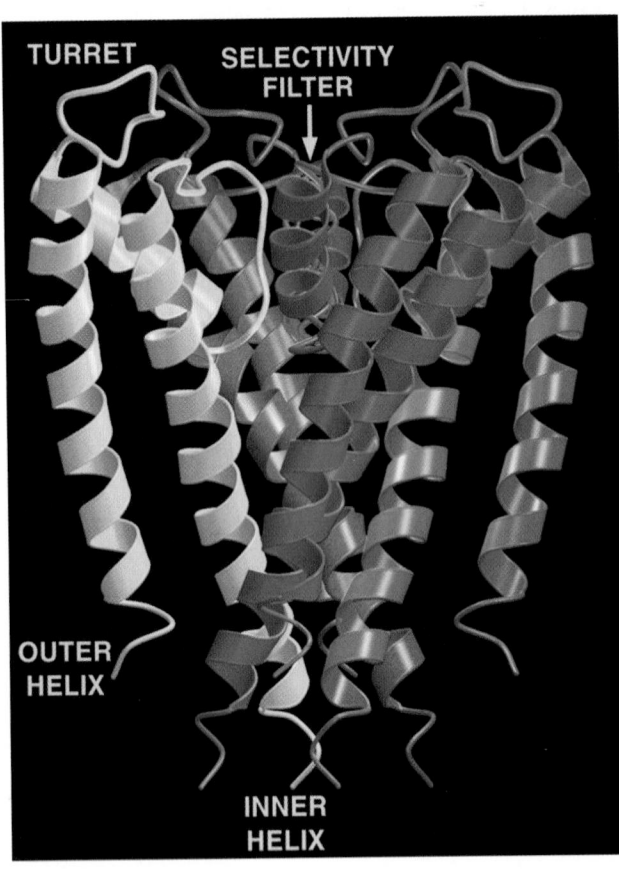

(a)

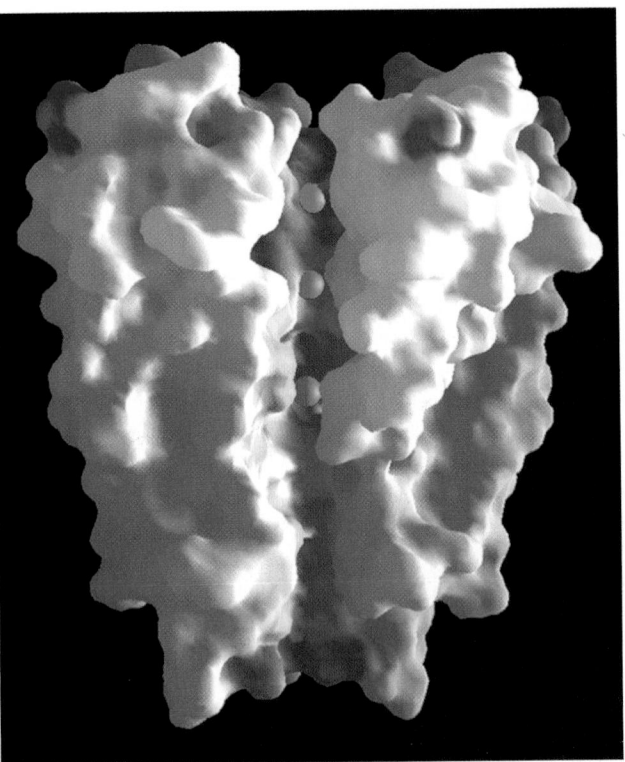

(b)

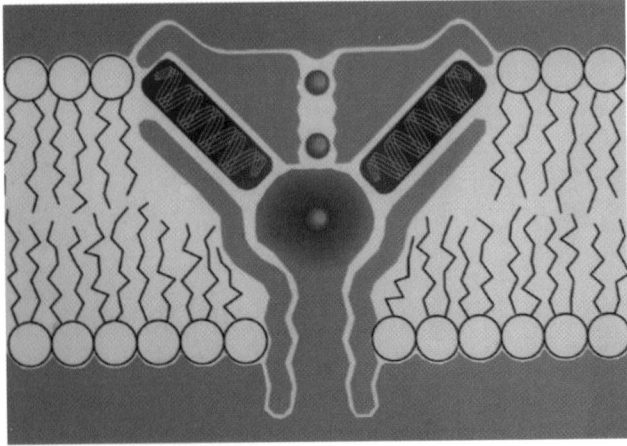

(c)

FIGURE 20-16   X-Ray structure of the KcsA K⁺ channel.
(*a*) Ribbon diagram of the tetramer as viewed from within the plane of the membrane with the cytoplasm below and the extracellular region above. The protein's 4-fold axis of rotation is vertical and each of its identical subunits is differently colored. (*b*) A cutaway diagram viewed similarly to Part *a* in which the K⁺ channel is represented by its solvent-accessible surface. The surface is colored according to its physical properties, with negatively charged areas red, uncharged areas white, positively charged areas blue, and hydrophobic areas of the central pore yellow. K⁺ ions are represented by green spheres. (*c*) A schematic diagram indicating how the K⁺ channel stabilizes a cation in the center of the membrane. The central pore's 10-Å-diameter aqueous cavity (which contains ~50 water molecules) stabilizes a K⁺ ion (*green spheres*) in the otherwise hydrophobic membrane interior. In addition, the C-terminal ends of the pore helices (*red*) all point toward the K⁺ ion, thereby electrostatically stabilizing it via their dipole moments (an α helix has a strong dipole moment with its negative end pointing toward the helix's C-terminal end because the bond dipoles of its component carbonyl and N—H groups are all

parallel to the helix axis with their negative ends pointing toward its C-terminal end; Fig. 8-11). This effect is magnified by the low dielectric constant at the center of the membrane interior. Electrostatic calculations indicate that the cavity is tuned to maximally stabilize monovalent cations. [Courtesy of Roderick MacKinnon, Rockefeller University. PDBid 1BL8]

its poles. Several K$^+$ ions and ordered water molecules are seen to occupy the central pore (Figs. 20-16b and 20-17a).

The 45-Å-long central pore has variable width: It starts at its cytoplasmic side (Fig. 20-16b, *bottom*) as an ~6-Å-diameter and 18-Å-long tunnel, the so-called internal pore, whose entrance is lined with four anionic side chains that presumably help exclude anions (red area at the bottom of Fig. 20-16b). The internal pore then widens to form a cavity ~10 Å in diameter. These regions of the central pore are both wide enough so that a K$^+$ ion could move through them in its hydrated state. However, the upper part of the pore, the so-called selectivity filter, narrows to 3 Å, thereby forcing a transiting K$^+$ ion to shed its waters of hydration. The walls of the internal pore and the cavity are lined with hydrophobic groups that interact minimally with diffusing ions (yellow area of the pore in Fig. 20-16b). However, the

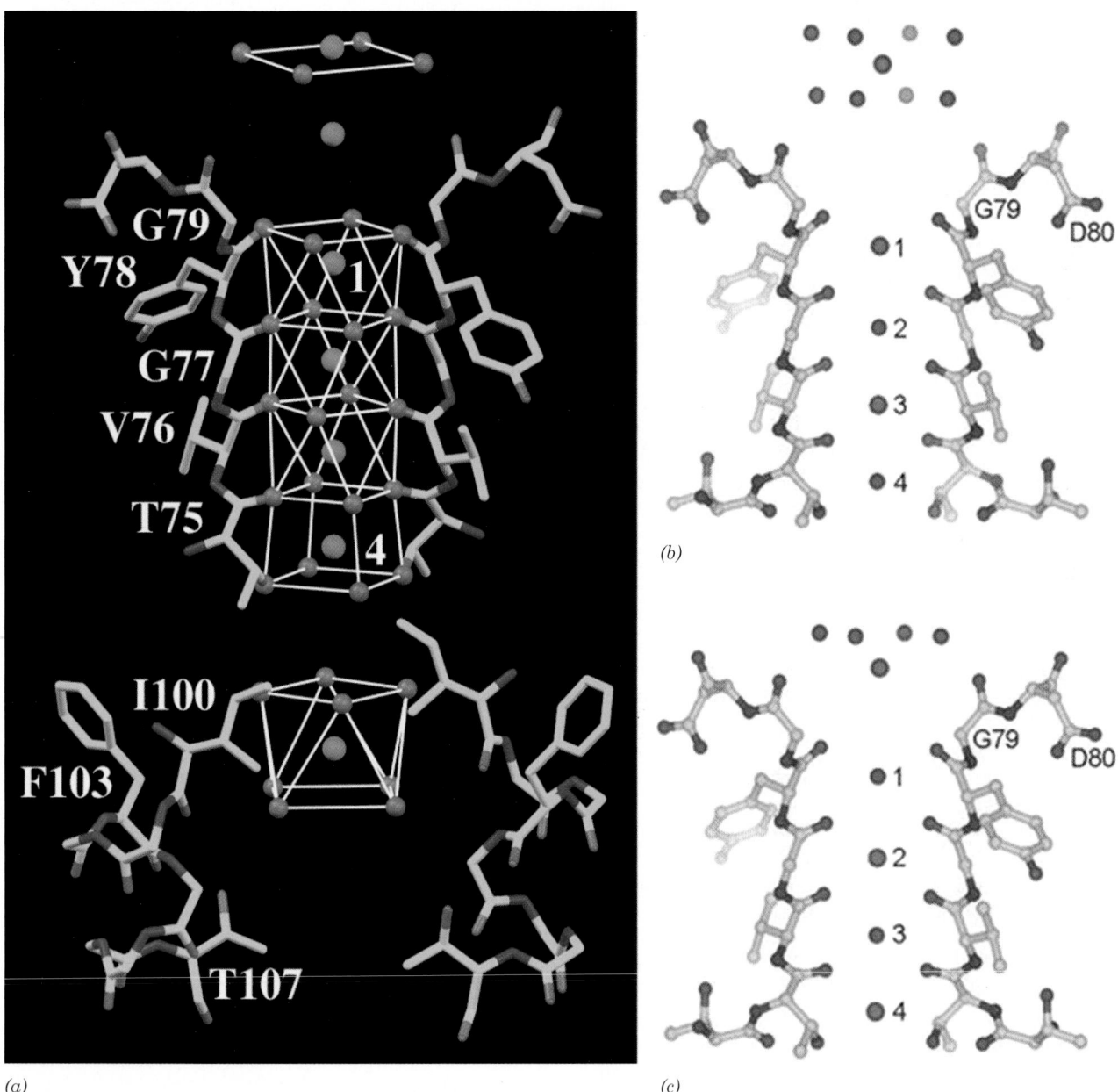

(a)

(b)

(c)

FIGURE 20-17 Portions of the KcsA K$^+$ channel responsible for its ion selectivity viewed similarly to Fig. 20-16. (*a*) The X-ray structure of the residues forming the cavity (*bottom*) and selectivity filter (*top*) but with the front and back subunits omitted for clarity. Atoms are colored according to type, with C yellow, N blue, O red, and K$^+$ ions represented by green spheres. The water and protein O atoms that ligand the K$^+$ ions, including those contributed by the front and back subunits, are represented by red spheres. The coordination polyhedra formed by these O atoms are outlined by thin white lines. (*b* and *c*) Two alternative K$^+$ binding states of the selectivity filter, whose superposition is presumed to be responsible for the electron density observed in the X-ray structure of KcsA. Atoms are colored as in Part *a*. Note that K$^+$ ions occupying the selectivity filter are interspersed with water molecules and that the K$^+$ ion immediately above the selectivity filter in Part *b* is farther above the protein than that in Part *c*. Hence these ions maintain a constant spacing while traversing the selectivity filter. [Part *a* based on an X-ray structure by, and Parts *b* and *c* courtesy of, Roderick MacKinnon, Rockefeller University. PDBid 1K4C.] 🔎 **See the Interactive Exercises**

selectivity filter (red area of the pore at the top of Fig. 20-16b) is lined with closely spaced main chain carbonyl oxygens of residues (Fig. 20-17a, top) that are highly conserved in all K^+ channels (their so-called signature sequence, TVGYG) and whose mutations disrupt the ability of the channel to discriminate between K^+ and Na^+ ions.

What is the function of the cavity? Energy calculations indicate that an ion moving through a narrow transmembrane pore must surmount an energy barrier that is maximal at the center of the membrane. The existence of the cavity reduces this electrostatic destabilization by surrounding the ion with polarizable water molecules (Fig. 20-16c). In addition, the C-terminal ends of the four pore helices point directly at the center of the cavity, so that their helix dipoles impose a negative electrostatic potential on the cavity that lowers the electrostatic barrier facing a cation crossing a lipid bilayer.

Remarkably, the K^+ ion occupying the cavity is liganded by 8 ordered water molecules located at the corners of a square antiprism (a cube with one face twisted by 45° with respect to the opposite face) in which the K^+ ion is centered (Fig. 20-17a, bottom). K^+ in aqueous solution is known to have such an inner hydration shell but it had never before been visualized. The K^+ ion is precisely centered in the cavity but yet its liganding water molecules are not in van der Waals contact with the walls of the cavity. Indeed, there is room in the cavity for ~40 additional water molecules although they are unseen in the X-ray structure because they are disordered. This disorder arises because the cavity is lined with hydrophobic groups (mainly the side chains of Ile 100 and Phe 103; Fig. 20-17a) that interact but weakly with water molecules, thus allowing them to interact freely with the K^+ ion so as to form an outer hydration shell. What, then, holds the hydrated K^+ ion in place? Apparently, it is very weak indirect hydrogen bonds involving such protein groups as the hydroxyl group of Thr 107 and possibly carbonyl O atoms from the pore and inner helices. The absence of such an ordered hydration complex when Na^+ rather than K^+ occupies the cavity is indicative of a precise geometric match between the hydrated K^+ and the cavity (the ionic radii of Na^+ and K^+ are 0.95 Å and 1.33 Å, respectively). The cavity thereby provides a high effective K^+ concentration (~2M) at the center of the membrane and positions the K^+ ion on the pore axis ready to enter the selectivity filter.

How does the K^+ channel discriminate so acutely between K^+ and Na^+ ions? The main chain O atoms lining the selectivity filter form a stack of rings (Fig. 20-17a; top) that provide a series of closely spaced sites of appropriate dimensions for coordinating dehydrated K^+ ions but not the smaller Na^+ ions. The structure of the protein surrounding the selectivity filter suggests that the diameter of the pore is rigidly maintained, thus making the energy of a dehydrated Na^+ in the selectivity filter considerably higher than that of hydrated Na^+ and thereby accounting for the K^+ channel's high selectivity for K^+ ions.

Since the selectivity filter appears designed to specifically bind K^+ ions, how does it support such a high throughput of these ions (up to 10^8 ions · s^{-1})? The structure in Fig. 20-17a shows what appear to be 4 K^+ ions in the se-

lectivity filter and two more just outside it on its extracellular side. Such closely spaced positive ions would strongly repel one another and hence represent a high energy situation. However, a variety of evidence suggests that this structure is really a superposition of two sets of K^+ ions, one with K^+ ions at the topmost position in Fig. 20-17a and at positions 1 and 3 in the selectivity filter (Fig. 20-17b) and the second with K^+ ions at the second position from the top in Fig. 17-20a and at positions 2 and 4 in the selectivity filter (Fig. 20-17c; X-ray structures can show overlapping atoms because they are averages of many unit cells). Within the selectivity filter, the positions not occupied by K^+ ions are instead occupied by water molecules that coordinate the neighboring K^+ ions.

The electron density that is represented as the topmost 4 water molecules in Fig. 20-17a is highly elongated in the vertical direction in this otherwise high resolution (2.0 Å) structure. Hence it is thought to actually arise from 8 water molecules that ligand the topmost K^+ ion in Fig. 20-17b to form an inner hydration shell similar to that of the K^+ in the central cavity (Fig. 20-17a, bottom). Moreover, the four water molecules liganding the topmost K^+ ion in Fig. 20-17c also contribute to this electron density. This latter ring of 4 waters provides half of the associated K^+ ion's 8 liganding O atoms. The others are contributed by the carbonyl O atoms of the 4 Gly 79 residues, which are properly oriented to do so. It therefore appears that a dehydrated K^+ ion transits the selectivity filter (moves to successive positions in Figs. 20-17b, c) by exchanging the properly spaced ligands extending from its walls and then exits into the extracellular solution by exchanging protein ligands for water molecules and hence again acquiring a hydration shell. These ligands are spaced and oriented such that there is little free energy change (estimated to be <12 kJ · mol^{-1}) along the reaction coordinate via which a K^+ ion transits the selectivity filter and enters the extracellular solution. The rapid dehydration of the K^+ ion entering the selectivity channel from the cavity is, presumably, similarly managed. The essentially level free energy landscape throughout this process is, of course, conducive to the rapid transit of K^+ ions through the ion channel and hence must be a product of evolutionary fine-tuning. Energy calculations indicate that mutual electrostatic repulsions between successive K^+ ions, whose movements are concerted, balances the attractive interactions holding these ions in the selectivity filter and hence further facilitates their rapid transit.

3 ■ ATP-DRIVEN ACTIVE TRANSPORT

Mediated transport is categorized according to the stoichiometry of the transport process (Fig. 20-18):

1. A **uniport** involves the movement of a single molecule at a time. The erythrocyte glucose transporter is a uniport system.

2. A **symport** simultaneously transports two different molecules in the same direction.

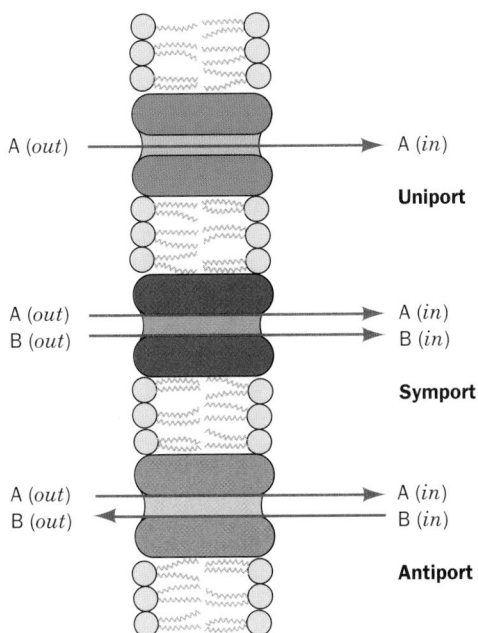

FIGURE 20-18 **Uniport, symport, and antiport translocation systems.**

3. An **antiport** simultaneously transports two different molecules in opposite directions.

The electrical character of ion transport is further specified as:

1. Electroneutral (electrically silent) if there is simultaneous charge neutralization, either by symport of oppositely charged ions or antiport of similarly charged ions.

2. Electrogenic if the transport process results in a charge separation across the membrane.

Since the glucose concentration in blood plasma is generally higher than that in cells, the erythrocyte glucose transporter normally transports glucose into the erythrocyte, where it is metabolized via glycolysis. Many substances, however, are available on one side of a membrane in lower concentrations than are required on the other side of the membrane. Such substances must be actively and selectively transported across the membrane against their concentration gradients.

Active transport is an endergonic process that is often coupled to the hydrolysis of ATP. How is this coupling accomplished? In endergonic biosynthetic reactions, it often occurs through the direct phosphorylation of a substrate by ATP; for example, the formation of UTP in the synthesis of glycogen (Section 18-2B). Membrane transport, however, is usually a physical rather than a chemical process; the transported molecule is not chemically altered. Determining the mechanism by which the free energy of ATP hydrolysis is coupled to endergonic physical processes has therefore been a challenging problem.

Three types of ATP hydrolyzing, transmembrane proteins have been identified that actively transport cations:

1. P-type ATPases are located mostly in plasma membranes and are so named because they are autophosphorylated by ATP during the transport process. P-type ATPases are known that transport H^+, Na^+, K^+, Ca^{2+}, Cu^{2+}, Cd^{2+}, and Mg^{2+} against a concentration gradient. They are distinguished from the other types of cation-translocating ATPases by their inhibition by **vanadate** (VO_4^{3-}, a phosphate analog; see Problem 4 in this chapter).

2. F-type ATPases (F_1F_0) function to translocate protons into mitochondria and bacterial cells, which in turn powers ATP synthesis. They are discussed in Section 22-3C.

3. V-type ATPases are located in plant vacuolar membranes and acidic vesicles, such as animal lysosomes, and are homologous to the F-type ATPases.

Anions are transported by a fourth type of ATPase, the so-called **A-type ATPases.** In this section, we discuss P-type ATPases. We also examine a bacterial active transport process in which the molecules transported are concomitantly phosphorylated. In the next section, we study secondary active transport systems, so called because they utilize the free energy of electrochemical gradients generated by ion-pumping ATPases to transport ions and neutral molecules against their concentration gradients.

A. (Na^+-K^+)–ATPase of Plasma Membranes

One of the most thoroughly studied active transport systems is the (Na^+-K^+)–**ATPase** of plasma membranes. This transmembrane protein was first isolated in 1957 by Jens Skou. It consists of two types of subunits: An ~1000-residue nonglycosylated α subunit that contains the enzyme's catalytic activity and ion binding sites, and an ~300-residue glycoprotein β subunit of unknown function. The sequences of the α subunits from several animal species, which are 98% identical in mammals, suggest that this subunit has around eight transmembrane α-helical segments and two large cytoplasmic domains. The β subunit is similarly predicted to have a single transmembrane helix and a large extracellular domain. The protein is thought to have the subunit composition $(\alpha\beta)_2$ (Fig. 20-19), but it is unclear

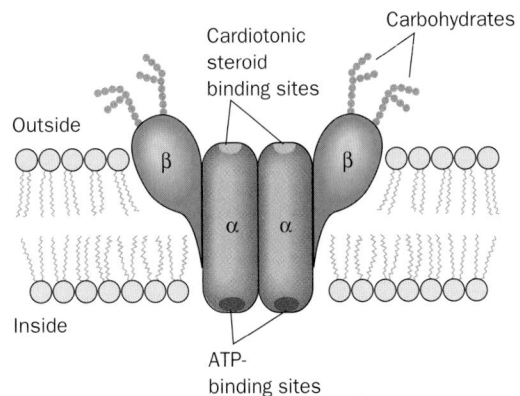

FIGURE 20-19 **Putative dimeric structure of the (Na^+-K^+)–ATPase indicating its orientation in the plasma membrane.**

whether this dimeric structure is a functional necessity. The enzyme is often called the **(Na⁺–K⁺) pump** because *it pumps Na⁺ out of and K⁺ into the cell with the concomitant hydrolysis of intracellular ATP.* The overall stoichiometry of the $(Na^+–K^+)$–ATPase reaction is

$$3Na^+(in) + 2K^+(out) + ATP + H_2O \rightleftharpoons$$
$$3Na^+(out) + 2K^+(in) + ADP + P_i$$

The $(Na^+–K^+)$–ATPase is therefore an electrogenic antiport: Three positive charges exit the cell for every two that enter. This extrusion of Na^+ enables animal cells to control their water content osmotically; without functioning $(Na^+–K^+)$ pumps, animal cells, which lack cell walls, would swell and burst (recall that lipid bilayers are permeable to H_2O; Section 12-2B). Moreover, the electrochemical potential gradient generated by the $(Na^+–K^+)$ pump is responsible for the electrical excitability of nerve cells (Section 20-5B) and provides the free energy for the active transport of glucose and amino acids into some cells (Section 20-4A). In fact, *all cells expend a large fraction of the ATP they produce (typically 30% and up to 70% in nerve cells) to maintain their required cytosolic Na⁺ and K⁺ concentrations.*

a. ATP Phosphorylates an Essential Asp during the Transport Process

The free energy of ATP hydrolysis powers the endergonic transport of Na⁺ and K⁺ against an electrochemical gradient. In coupling these two processes, a kinetic barrier must somehow be erected against the "downhill" transport of Na^+ and K^+ along their ion concentration gradients, while simultaneously facilitating their "uphill" transport. In addition, futile ATP hydrolysis must be prevented in the absence of uphill transport. How the enzyme does so is by no means well understood, although many of its mechanistic aspects have been elucidated.

A key discovery was that the protein is phosphorylated by ATP in the presence of Na^+ during the transport process. The use of chemical trapping techniques demonstrated that this phosphorylation occurs on an Asp residue to form a highly reactive **aspartyl phosphate** intermediate. For instance, sodium borohydride reduces acyl phosphates to their corresponding alcohols. In the case of an aspartyl phosphate residue, the alcohol is **homoserine.** By use of $[^3H]NaBH_4$ to reduce the phosphorylated enzyme, radioactive homoserine was, in fact, isolated from the acid hydrolysate (Fig. 20-20). The phosphorylated residue, Asp 376, begins the highly conserved sequence DKTG that occurs in the central region of the polypeptide chain.

b. The (Na⁺–K⁺)–ATPase Has Two Major Conformational States

The observations that ATP only phosphorylates the $(Na^+–K^+)$–ATPase in the presence of Na^+, while the aspartyl phosphate residue is only subject to hydrolysis in the presence of K^+, led to the realization that *the enzyme has two major conformational states, E_1 and E_2.* These states have different tertiary structures, different catalytic activities, and different ligand specificities:

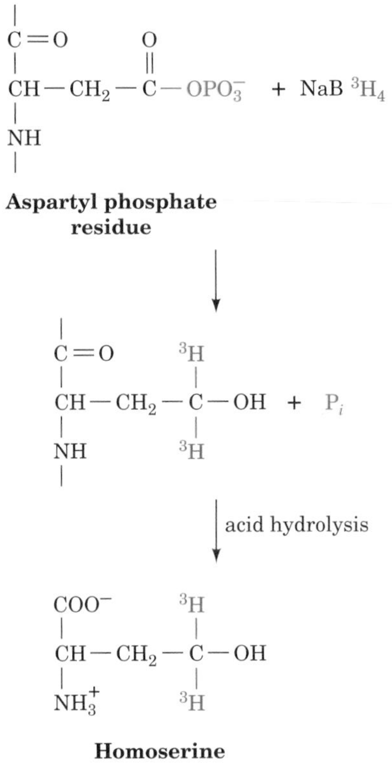

Aspartyl phosphate residue

Homoserine

FIGURE 20-20   Reaction of $[^3H]NaBH_4$ with phosphorylated $(Na^+–K^+)$–ATPase. The isolation of $[^3H]$homoserine following acid hydrolysis of the protein indicates that the original phosphorylated amino acid residue is Asp.

1. E_1 has an inward-facing high-affinity Na^+ binding site ($K_M = 0.2$ mM, well below the intracellular $[Na^+]$) and reacts with ATP to form the activated product $E_1{\sim}P$ only when Na^+ is bound.

2. $E_2{-}P$ has an outward-facing high-affinity K^+ binding site ($K_M = 0.05M$, well below the extracellular $[K^+]$) and hydrolyzes to form $P_i + E_2$ only when K^+ is bound.

c. An Ordered Sequential Kinetic Reaction Mechanism Accounts for the Coupling of Active Transport with ATP Hydrolysis

The $(Na^+–K^+)$–ATPase is thought to operate in accordance with the following ordered sequential reaction scheme (Fig. 20-21):

1. $E_1 \cdot 3Na^+$, which acquired its Na^+ inside the cell, binds ATP to yield the ternary complex $E_1 \cdot ATP \cdot 3Na^+$.

2. The ternary complex reacts to form the "high-energy" aspartyl phosphate intermediate $E_1{\sim}P \cdot 3Na^+$.

3. This "high-energy" intermediate relaxes to its "low-energy" conformation, $E_2{-}P \cdot 3Na^+$, and releases its bound Na^+ outside the cell; that is, Na^+ is transported through the membrane.

4. $E_2{-}P$ binds $2K^+$ from outside the cell to form $E_2{-}P \cdot 2K^+$.

5. The phosphate group is hydrolyzed, yielding $E_2 \cdot 2K^+$.

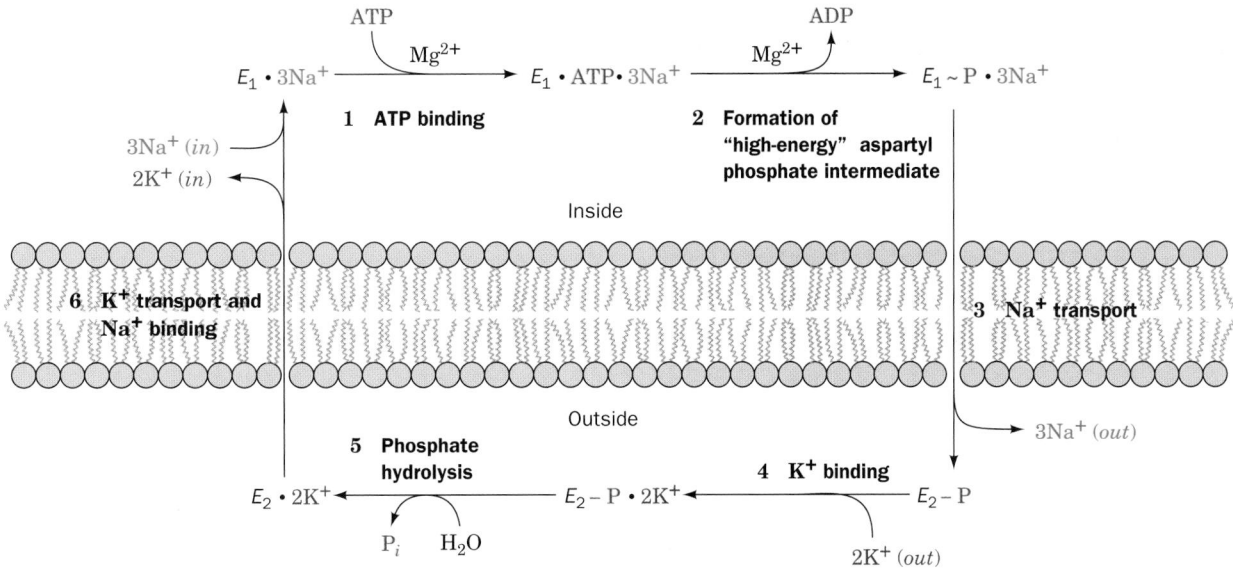

FIGURE 20-21 Kinetic scheme for the active transport of Na$^+$ and K$^+$ by (Na$^+$–K$^+$)–ATPase.

6. $E_2 \cdot 2K^+$ changes conformation, releases its 2K$^+$ inside the cell, and replaces it with 3Na$^+$, thereby completing the transport cycle.

The enzyme is thought to have only one set of cation binding sites, which apparently changes both its orientation and its specificity during the course of the transport cycle.

The obligatory order of the reaction requires that ATP can be hydrolyzed only as Na$^+$ is transported "uphill." Conversely, Na$^+$ can be transported "downhill" only if ATP is concomitantly synthesized. Consequently, although each of the above reaction steps is, in fact, individually reversible, the cycle, as is diagrammed in Fig. 20-21, circulates only in the clockwise direction under normal physiological conditions; that is, ATP hydrolysis and ion transport are coupled processes. Note that the **vectorial** (unidirectional) nature of the reaction cycle results from the alternation of some of the steps of the exergonic ATP hydrolysis reaction (Steps 1 + 2 and Step 5) with some of the steps of the endergonic ion transport process (Steps 3 + 4 and Step 6). Thus, neither reaction can go to completion unless the other one also does.

d. Mutual Destabilization Accounts for the Rate of Na$^+$ and K$^+$ Transport

The above ordered kinetic mechanism accounts only for the coupling of active transport with ATP hydrolysis. *In order to maintain a reasonable rate of transport, the free energies of all its intermediates must be roughly equal. If some intermediates were much more stable than others, the stable intermediates would accumulate, thereby severely reducing the overall transport rate.* For example, in order for Na$^+$ to be transported out of the cell, uphill, its binding must be strong to E_1 on the inside and weak to E_2 on the outside. Strong binding means greater stability and a potential bottleneck. This difficulty is counteracted by the phosphory-

lation of $E_1 \cdot 3Na^+$ and its subsequent conformational change to yield the low Na$^+$ affinity E_2—P (Steps 2 and 3, Fig. 20-21). Likewise, the strong binding of K$^+$ to E_2—P on the outside is attenuated by its dephosphorylation and conformational change to yield the low K$^+$ affinity E_1 (Steps 5 and 6, Fig. 20-21). It is these mutual destabilizations that permit Na$^+$ and K$^+$ to be transported at a rapid rate.

e. Cardiac Glycosides Specifically Inhibit the (Na$^+$–K$^+$)–ATPase

Study of the (Na$^+$–K$^+$)–ATPase has been greatly facilitated by the use of **cardiac glycosides** (also called **cardiotonic steroids**), natural products that increase the intensity of heart muscle contraction. Indeed, **digitalis**, an extract of purple foxglove leaves (Fig. 20-22a), which contains a mixture of cardiac glycosides including **digitoxin** (**digitalin**; Fig. 20-22b), has been used to treat congestive heart failure for centuries. The cardiac glycoside **ouabain** (pronounced wabane; Fig. 20-22b), a product of the East African ouabio tree, has been long used as an arrow poison. These two steroids, which are still among the most commonly prescribed cardiac drugs, inhibit the (Na$^+$–K$^+$)–ATPase by binding strongly to an externally exposed portion of the enzyme (the drugs are ineffective when injected inside cells) so as to block Step 5 in Fig. 20-21. The resultant increase in intracellular [Na$^+$] stimulates the cardiac (Na$^+$–Ca^{2+}) antiport system, which pumps Na$^+$ out of and Ca^{2+} into the cell (Section 22-1B). The increased cytosolic [Ca^{2+}] boosts the [Ca^{2+}] in other cellular organelles, principally the sarcoplasmic reticulum. Thus, the release of Ca^{2+} to trigger muscle contraction (Section 35-3C) produces a larger than normal increase in cytosolic [Ca^{2+}], thereby intensifying the force of cardiac muscle contraction. Ouabain, which was once thought to be produced only by plants, has recently been discovered also to be an ani-

(a)

Digitoxin (digitalin)

Ouabain

(b)

FIGURE 20-22   Cardiac glycosides. (*a*) The leaves of the purple foxglove plant are the source of the heart muscle stimulant digitalis. [Derek Fell.] (*b*) Digitoxin (digitalin), the major component of digitalis, and ouabain, a cardiac glycoside isolated from the East African ouabio tree, are among the most commonly prescribed cardiac drugs.

mal hormone that is secreted by the adrenal cortex and functions to regulate cell [Na^+] and overall body salt and water balance.

B. *Ca^{2+}–ATPase*

Ca^{2+} often acts as a second messenger in a manner similar to cAMP (Section 19-4). Transient increases in cytosolic [Ca^{2+}] trigger numerous cellular responses, including muscle contraction (Section 35-3C), release of neurotransmitters (Section 20-5C), and, as we have seen, glycogen breakdown (Section 18-3C). Moreover, Ca^{2+} is an important activator of oxidative metabolism (Section 22-4).

The use of phosphate as a basic energy currency requires cells to maintain a low internal [Ca^{2+}] because, for example, $Ca_3(PO_4)_2$ has a maximum aqueous solubility of 65 μ*M*. Thus, the [Ca^{2+}] in the cytosol (~0.1 μ*M*) is four orders of magnitude less than it is in the extracellular spaces (~1500 μ*M*). This large concentration gradient is maintained by the active transport of Ca^{2+} across the plasma membrane, the endoplasmic reticulum (the sarcoplasmic reticulum in muscle), and the mitochondrial inner membrane. We discuss the mitochondrial system in Section 22-1B. The plasma membrane and endoplasmic reticulum each contain a P-type **Ca^{2+}–ATPase (Ca^{2+} pump)** that actively pumps Ca^{2+} out of the cytosol at the expense of ATP hydrolysis. Their kinetic mechanisms (Fig. 20-23) are very similar to that of the (Na^+–K^+)–ATPase (Fig. 20-21).

a. The X-Ray Structure of the Ca^{2+}–ATPase Provides Mechanistic Clues

The X-ray structure of the 994-residue Ca^{2+}–ATPase from rabbit muscle sarcoplasmic reticulum in complex with

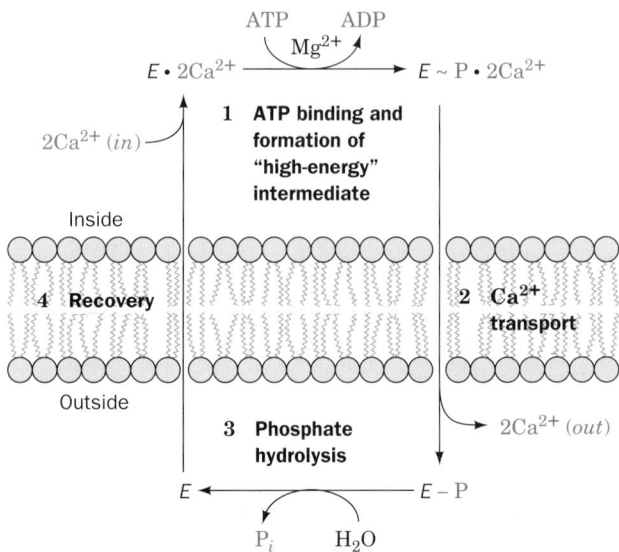

FIGURE 20-23   Kinetic mechanism of Ca^{2+}–ATPase. Here (*in*) refers to the cytosol and (*out*) refers to the outside of the cell for plasma membrane Ca^{2+}–ATPase or to the lumen of the endoplasmic reticulum (sarcoplasmic reticulum) for the Ca^{2+}–ATPase of that membrane.

two Ca^{2+} ions and the AMP analog **2′,3′-*O*-(2,4,6-trinitro-phenyl)-AMP (TNP-AMP)**

2′,3′-*O*-(2,4,6-trinitrophenyl)-AMP (TNP-AMP)

was determined by Chikashi Toyoshima. This 140-Å-long monomeric protein (Fig. 20-24) consists of a transmem-brane domain (M) composed of 10 helices (M1–M10) of varied lengths and three well-separated cytoplasmic do-mains: the actuator domain (A), so named because it par-ticipates in the transmission of major conformational changes (see below); the nucleotide-binding domain (N), which binds the TNP-AMP and presumably ATP; and the phosphorylation domain (P), which contains the phospho-rylatable Asp 351 at the beginning of the conserved sequence DKTG. The two Ca^{2+} ions are bound in close proximity in the center of the transmembrane domain. The Ca^{2+}-binding cavity is formed, in large part, by the dis-ruption (unwinding) of helices M4 and M6 in this region. Indeed, 3 of the 6 O atoms liganding one of the Ca^{2+} ions are the main chain carbonyl O atoms from 3 residues in the disrupted segment of helix M4. Although the Ca^{2+}–ATPase contains no obvious pore through which Ca^{2+} ions might be pumped from the cytoplasm to the sarcoplasmic reticulum (such as that in the K^+ channels; Fig. 20-16*b*), the unwound portions of helices M4 and M6 provide rows of main chain carbonyl groups pointing from the Ca^{2+}

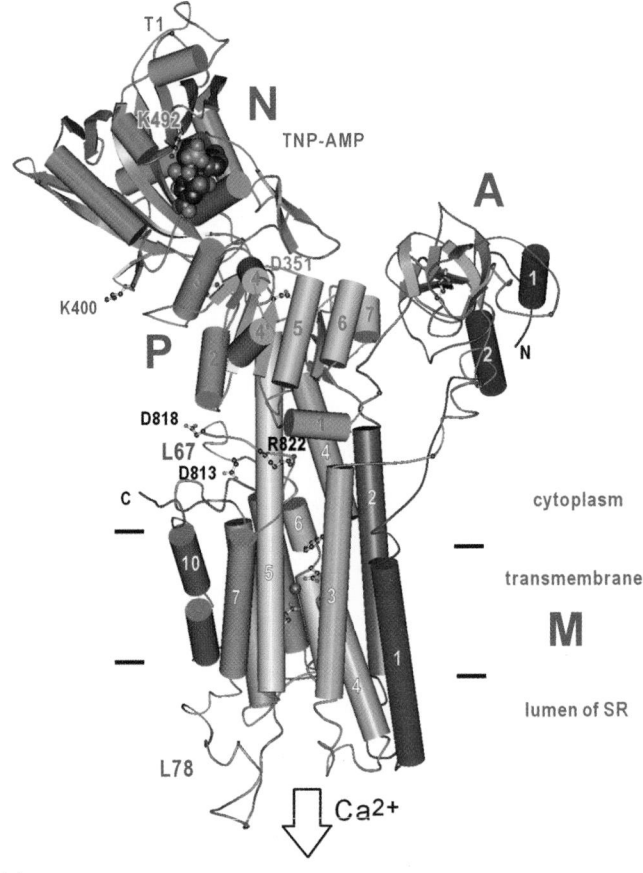

(a)

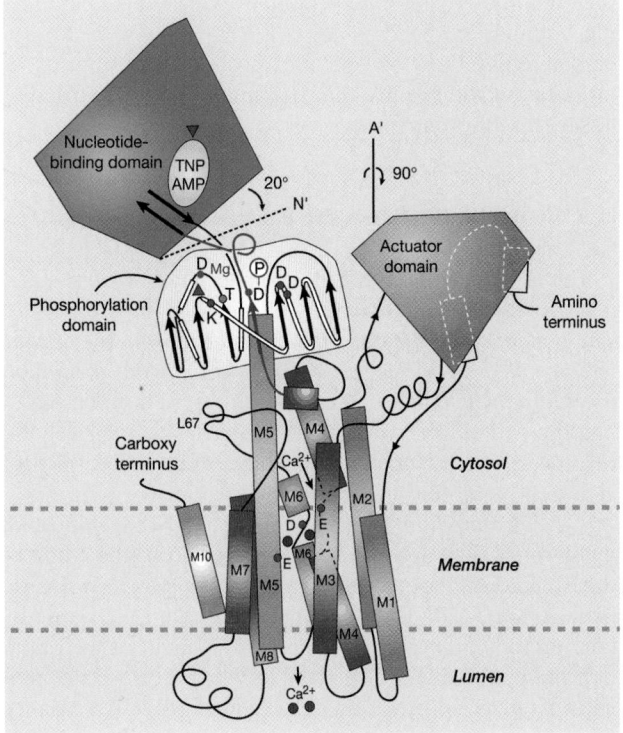

(b)

FIGURE 20-24   X-Ray structure of the Ca^{2+}–ATPase from rabbit muscle sarcoplasmic reticulum. (*a*) A tube-and-arrow diagram viewed parallel to the membrane plane in which the polypeptide chain is colored in rainbow order from N-terminus (*blue*) to C-terminus (*red*) and the transmembrane, actuator, nucleotide-binding, and phosphorylation domains are labeled M, A, N, and P. The Ca^{2+} ions bound in the transmembrane region are represented by large purple spheres, the TNP-AMP bound to the N domain is shown in space-filling form, and key side chains are drawn in ball-and-stick form. [Courtesy of Chikashi Toyoshima, University of Tokyo, Tokyo, Japan. PDBid 1EUL.] (*b*) A schematic diagram of the structure, viewed similarly to Part *a*, in which Ca^{2+} ions are represented by green spheres, the positions of residues involved in Ca^{2+} binding or catalysis are marked by small spheres and single letters, and the position of the phosphorylation site, Asp 351, is indicated by the circled P. N′ and A′ indicate the rotation of the nucleotide-binding and actuator domains required to match the electron density in the low resolution structure of calcium-free Ca^{2+}–ATPase. [From MacLennan, D.H. and Green, N.M., *Nature* **405,** 634 (2000).]

binding sites to the cytoplasm. It is postulated that these O atoms function to strip away the tightly bound waters of hydration from Ca^{2+} ions as they pass from the cytoplasm to their binding sites. The outlet for the Ca^{2+} ions is thought to be in the area surrounded by helices M3 to M5, where there is also a ring of O atoms.

The phosphorylation site (Asp 351) on the P domain is >25 Å distant from the TNP-AMP bound to the N domain. Thus, ATP hydroysis requires the closure of the "jaws" formed by the P and N domains as occurs in many kinases (e.g., Fig. 17-5). Moreover, the Ca^{2+} binding site is ~80 Å distant from the ATP binding site. How do these sites communicate their binding states to one another? The comparison of the X-ray structure in Fig. 20-24*a* with the low resolution (8 Å) electron crystallography–based structure of calcium-free Ca^{2+}–ATPase reveals that, in the absence of Ca^{2+} and bound nucleotide, the protein's A domain has rotated by 90° relative to the transmembrane domain and that the P–N "jaws" have undergone a partial (20°) closure (Fig. 20-24*b*). Toyoshima has therefore postulated that Ca^{2+} binding induces these large domain motions by shifting the positions of helices M1, M2, and M3 such that they release the A domain, which in turn induces the opening of the P–N jaws. Clearly, however, the understanding of how this complex protein carries out its vital function will require the determination of its structure in several different states.

b. Calmodulin Regulates the Plasma Membrane Ca^{2+} Pump

For a cell to maintain its proper physiological state, it must regulate the activities of its ion pumps precisely. *The regulation of the Ca^{2+} pump in the plasma membrane is controlled by the level of Ca^{2+} through the mediation of **calmodulin (CaM)**.* This ubiquitous eukaryotic Ca^{2+}-binding protein participates in numerous cellular regulatory processes including, as we have seen, the control of glycogen metabolism (Section 18-3C).

Ca^{2+}–Calmodulin activates the Ca^{2+}–ATPase of plasma membranes. The activation, as deduced from the study of the isolated ATPase, results in a decrease in its K_M for Ca^{2+} from 20 to 0.5 μM. Ca^{2+}–CaM activates the Ca^{2+} pump by binding to an inhibitory polypeptide segment of the pump in a manner similar to the way in which Ca^{2+}–CaM activates its target protein kinases (Section 18-3C). Evidence supporting this mechanism comes from proteolytically excising the Ca^{2+} pump's CaM-binding polypeptide, yielding a truncated pump that is active even in the absence of CaM. Synthetic peptides corresponding to this CaM-binding segment not only bind Ca^{2+}–CaM but inhibit the truncated pump by increasing its K_M for Ca^{2+} and decreasing its V_{max}. This suggests that, in the absence of Ca^{2+}–CaM, the CaM-binding segment of the pump interacts with the rest of the protein so as to inhibit its activity. When the Ca^{2+} concentration increases, Ca^{2+}–CaM forms and binds to the CaM-binding segment of the pump in a way that causes it to dissociate from the rest of the pump, thereby relieving the inhibition.

Now we can see how Ca^{2+} regulates its own cytoplasmic concentration: At Ca^{2+} levels below calmodulin's ~1 μM dissociation constant for Ca^{2+}, the Ca^{2+}–ATPase is relatively inactive due to autoinhibition by its CaM-binding segment. If, however, the $[Ca^{2+}]$ rises to this level, Ca^{2+} binds to calmodulin, which, in turn, binds to the CaM-binding segment so as to relieve the inhibition, thereby activating the Ca^{2+} pump:

$$Ca^{2+} + CaM \rightleftharpoons Ca^{2+}\text{-}CaM^* + pump\,(inactive) \rightleftharpoons Ca^{2+}\text{-}CaM^* \cdot pump\,(active)$$

(CaM* indicates activated calmodulin). This interaction decreases the pump's K_M for Ca^{2+} to below the ambient $[Ca^{2+}]$, thereby causing Ca^{2+} to be pumped out of the cytosol. When the $[Ca^{2+}]$ decreases sufficiently, Ca^{2+} dissociates from calmodulin and this series of events reverses itself, thereby inactivating the pump. The entire system is therefore analogous to a basement sump pump that is automatically activated by a float when the water reaches a preset level.

C. *(H⁺–K⁺)–ATPase of Gastric Mucosa*

Parietal cells of the mammalian gastric mucosa secrete HCl at a concentration of $0.15M$ (pH 0.8). Since the cytosolic pH of these cells is 7.4, this represents a pH difference of 6.6 units, the largest known in eukaryotic cells. The secreted protons are derived from the intracellular hydration of CO_2 by carbonic anhydrase:

$$CO_2 + H_2O \rightleftharpoons HCO_3^- + H^+$$

The secretion of H^+ involves the participation of an **(H⁺–K⁺)–ATPase,** an electroneutral antiport with structure and properties similar to that of Ca^{2+}–ATPase. Like the related Ca^{2+}– and (Na⁺–K⁺)–ATPases, it is phosphorylated during the transport process. In this case, however, the K^+, which enters the cell as H^+ is pumped out, is subsequently externalized by its electroneutral cotransport with Cl^-. HCl is therefore the overall transported product.

For many years, effective treatment of peptic ulcers, which was a frequently fatal condition caused by the attack of stomach acid on the gastric mucosa, often required the surgical removal of the affected portions of the stomach or sometimes the entire stomach. The discovery, by James Black, of **cimetidine,**

Cimetidine

Histamine

which inhibits stomach acid secretion, has almost entirely eliminated the need for this dangerous and debilitating surgery.

The (H^+-K^+)–ATPase of the gastric mucosa is activated by histamine stimulation of a cell-surface receptor in a process mediated by cAMP. Cimetidine (trade name Tagamet) and its analogs, which competitively inhibit the binding of histamine to this receptor, are presently among the most commonly used drugs in the United States.

D. *Group Translocation*

Group translocation is a variation of ATP-driven active transport that most bacteria use to import certain sugars. It is required for many bacterial processes, both useful and harmful (to humans), such as those that produce cheese, soy sauce, and dental cavities. *It differs from active transport in that the molecules transported are simultaneously modified chemically.* The most extensively studied example of group translocation is the **phosphoenolpyruvate-dependent phosphotransferase system (PTS)** of *E. coli* discovered by Saul Roseman in 1964. Phosphoenolpyruvate (PEP) is the phosphoryl donor for this system (recall that PEP is the "high-energy" phosphoryl donor for ATP synthesis in the pyruvate kinase reaction of glycolysis; Section 17-2J). *The PTS simultaneously transports and phosphorylates sugars. Since the cell membrane is impermeable to sugar phosphates, once they enter the cell, they remain there.* Some of the PTS-transported sugars are listed in Table 20-2.

The PTS system involves two soluble cytoplasmic proteins, **Enzyme I (EI)** and **HPr** (for *h*istidine-containing *p*hosphocarrier *p*rotein), which participate in the transport

TABLE 20-2 Some of the Sugars Transported by the *E. coli* PEP-Dependent Phosphotransferase System (PTS)

Glucose	Galactitol
Fructose	Mannitol
Mannose	Sorbitol
N-Acetylglucosamine	Xylitol

of all sugars (Fig. 20-25). In addition, for each sugar the system transports, there is a specific transmembrane transport protein **EII**, which consists of at least three functional components: two that are cytoplasmic, **EIIA** and **EIIB**, and a transmembrane channel, **EIIC**. These three components associate differently in different EII's. In *E. coli*, for example, EIIA, EIIB, and EIIC are separate subunits in cellobiose-specific EII; EIIB and EIIC are linked and EIIA is separate in glucose-specific EII; and all three components are present on a single peptide in mannitol-specific EII.

Glucose transport, which resembles that of other sugars, involves the transfer of a phosphoryl group from PEP to glucose with net inversion of configuration about the phosphorus atom. Since each phosphoryl transfer involves inversion (Section 16-2B), an odd number of transfers must be involved. Four phosphorylated protein intermediates have been identified, indicative of five phosphoryl transfers:

$$PEP \rightarrow EI \rightarrow HPr \rightarrow EIIA^{glc} \rightarrow EIIBC^{glc} \rightarrow glucose$$

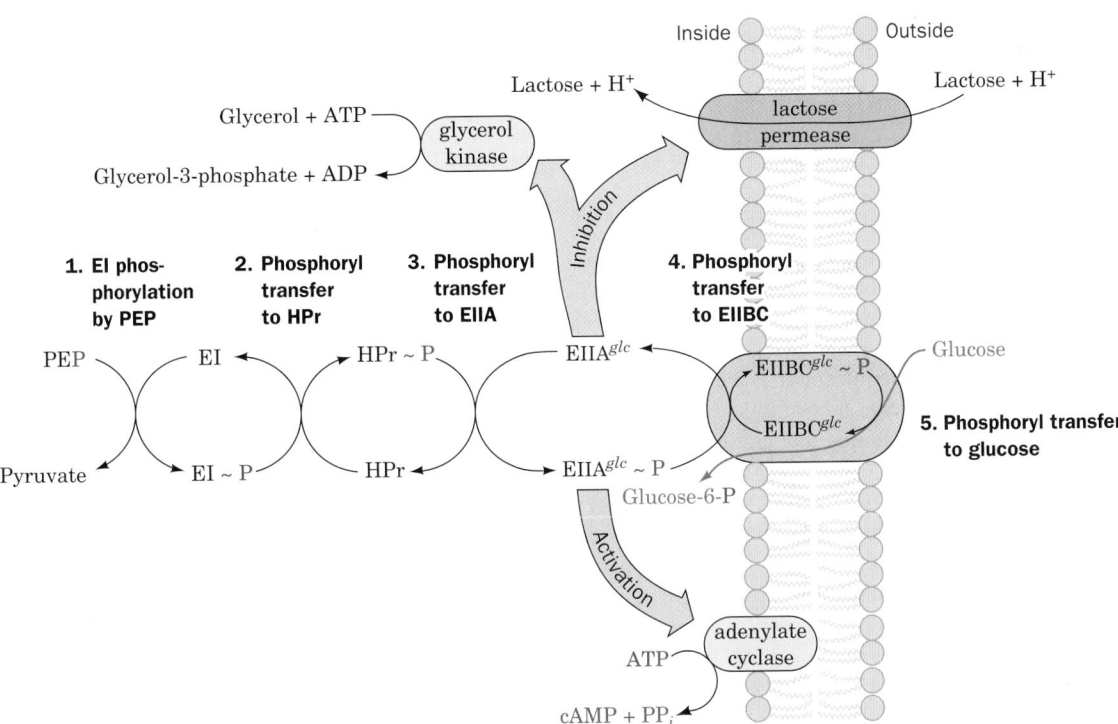

FIGURE 20-25 Transport of glucose by the PEP-dependent phosphotransferase system (PTS). HPr and EI are cytoplasmic proteins common to all sugars transported. EIIA^{glc} and EIIBC^{glc} are proteins specific for glucose. EIIA^{glc} inhibits non- PTS transport proteins such as the lactose permease (Section 20-4B) and enzymes such as glycerol kinase. Adenylate cyclase is activated by the presence of EIIA^{glc}~P (or possibly inhibited by the presence of EIIA^{glc}).

The transport process occurs as follows (Fig. 20-25):

1. PEP phosphorylates EI at N3 (N_ϵ) of His 189 to form a reactive phosphohistidine adduct.

Phosphohistidine residue

2. The phosphoryl group is transferred to N1 (N_δ) of His 15 on HPr. His is apparently a favored phosphoryl group acceptor in phosphoryl-transfer reactions. It also participates in the phosphoglycerate mutase reaction of glycolysis (Section 17-2H).

3. HPr~P continues the phosphoryl-transfer chain by phosphorylating EIIA^{glc} at N3 of His 90.

4. The fourth phosphoryl transfer is to Cys 421 of EIIB^{glc}.

5. The phosphoryl group is finally transferred from EIIB^{glc} to glucose, which, in the process, is transported across the membrane by EIIC^{glc}. Glucose is released into the cytoplasm only after it has been phosphorylated to glucose-6-phosphate (G6P).

Thus the transport of glucose is driven by its indirect, exergonic phosphorylation by PEP. The PTS is an energy-efficient system since only one ATP equivalent is required to both transport and phosphorylate glucose. When the active transport and phosphorylation steps occur separately, as they do in many cells, two ATPs are hydrolyzed per glucose processed.

a. Bacterial Sugar Transport Is Genetically Regulated

The PTS is more complex than the other transport systems we have encountered, probably because it is part of a complicated regulatory system governing sugar transport. When any of the sugars transported by the PTS is abundant, the active transport of sugars which enter the cell via other transport systems is inhibited. This inhibition, called **catabolite repression,** is mediated through the cAMP concentration (Section 31-3C). cAMP activates the transcription of genes that encode various sugar transport proteins, including **lactose permease** (Section 20-4B). The presence of glucose results in a decrease in [cAMP], which, in turn, represses the synthesis of these other sugar transport proteins. Direct inhibition of the sugar transport proteins themselves, as well as of certain enzymes, also occurs.

The mechanism for control of [cAMP] is thought to reside in EIIA^{glc}, which is transiently phosphorylated in Step 3 of the PTS transport process (Fig. 20-25). When glucose is plentiful, this enzyme is present mostly in its dephospho form since EIIA^{glc}~P rapidly transfers its phosphoryl group through EIIBC^{glc} to glucose. Under these condi-

tions, adenylate cyclase is inactive, although whether dephospho EIIA^{glc} inhibits this enzyme or EIIA^{glc}~P activates it is unclear. However, dephospho EIIA^{glc} binds to and inhibits many non-PTS transporters and enzymes that participate in the metabolism of sugars other than glucose (the metabolite of choice for many bacteria), including lactose permease and **glycerol kinase** (Section 17-5). In the absence of glucose, EIIA^{glc} is converted to EIIA^{glc}~P, thereby relieving the inhibition of non-PTS transporters. In addition, adenylate cyclase is activated to produce cAMP, which, in turn, induces the increased production of some of the non-PTS transporters and enzymes that EIIA^{glc} inhibits. This is a form of energy conservation for the cell. Why synthesize the proteins required for the transport and metabolism of all sugars when the metabolism of only one sugar at a time will do?

b. The X-Ray Structure of EIIA^{glc} in Complex with Glycerol Kinase

The X-ray structures of EIIA^{glc}, both alone and in complex with one of its regulatory targets, glycerol kinase, which were determined by James Remington and Roseman, have revealed how EIIA^{glc} inhibits at least some of its targets and why EIIA^{glc}~P does not do so. EIIA^{glc} contains two His residues, His 75 and His 90, that are required for phosphoryl transfer, although only His 90 is necessary for EIIA^{glc} to accept a phosphate from HPr. The X-ray structure of *E. coli* EIIA^{glc} alone reveals that these two His residues lie in close proximity (their N3 atoms are 3.3 Å apart) in a depression on the surface of the protein that is surrounded by a remarkable ~18-Å-diameter hydrophobic ring consisting of 11 Phe, Val, and Ile side chains.

The X-ray structure of EIIA^{glc} in complex with glycerol kinase (Fig. 20-26) confirms that this hydrophobic gasket is indeed the site of interaction between the two proteins and reveals how the phosphorylation of His 90 disrupts this interaction. The two active site His residues, which are completely buried within the hydrophobic interaction surface, coordinate a previously unanticipated Zn^{2+} ion, which is additionally coordinated to Glu 478 of glycerol kinase and a water molecule. The phosphorylation of EIIA^{glc} His 90 to yield EIIA^{glc}~P no doubt disrupts this intermolecular interaction, thereby releasing glycerol kinase and reversing its inhibition.

4 ■ ION GRADIENT–DRIVEN ACTIVE TRANSPORT

Systems such as the (Na^+–K^+)–ATPase discussed above utilize the free energy of ATP hydrolysis to generate electrochemical potential gradients across membranes. Conversely, *the free energy stored in an electrochemical potential gradient may be harnessed to power various endergonic physiological processes.* Indeed, ATP synthesis by mitochondria and chloroplasts is powered by the dissipation of proton gradients generated through electron transport and photosynthesis (Sections 22-3C and 24-2D). In this section we discuss active transport processes that are

FIGURE 20-26   The X-ray structure of *E. coli* EIIA^{glc} (*yellow,* a 168-residue monomer) in complex with one of its regulatory targets, glycerol kinase (*blue,* a tetramer of identical 501-residue subunits). The two proteins associate, in part, by tetrahedrally coordinating a Zn^{2+} ion via the side chains of His 75 and His 90 of EIIA^{glc}, a carboxylate oxygen from Glu 478 of glycerol kinase, and a water molecule. These groups are shown in ball-and-stick form with C gray, N blue, O red, and Zn^{2+} white. The Zn^{2+}-mediated interaction between EIIA^{glc} and glycerol kinase inactivates glycerol kinase, presumably through an induced-fit mechanism. The phosphorylation of EIIA^{glc} at His 90 disrupts this interaction, thereby reversing the inhibition of glycerol kinase. [Courtesy of James Remington, University of Oregon. PDBid 1GLA.]

driven by the dissipation of ion gradients. We consider three examples: intestinal uptake of glucose by the **Na$^+$–glucose symport,** uptake of lactose by *E. coli* **lactose permease,** *and the mitochondrial* **ATP–ADP translocater.**

A. Na$^+$–Glucose Symport

Nutritionally derived glucose is actively concentrated in **brush border cells** of the intestinal epithelium by an Na$^+$-dependent symport (Fig. 20-27). It is transported from these cells to the circulatory system via a passive-mediated glucose uniport located on the capillary side of the cell and which is similar to that of the erythrocyte membrane (Section 20-2B). Note that *although the immediate energy source for glucose transport from the intestine is the Na$^+$ gradient, it is really the free energy of ATP hydrolysis that powers this process through the maintenance of the Na$^+$ gradient by the (Na$^+$–K$^+$)–ATPase.* Nevertheless, since

glucose enhances Na$^+$ resorption, which in turn enhances water resorption, glucose (possibly as sucrose), in addition to salt and water, should be fed to individuals suffering from severe salt and water losses resulting from diarrhea.

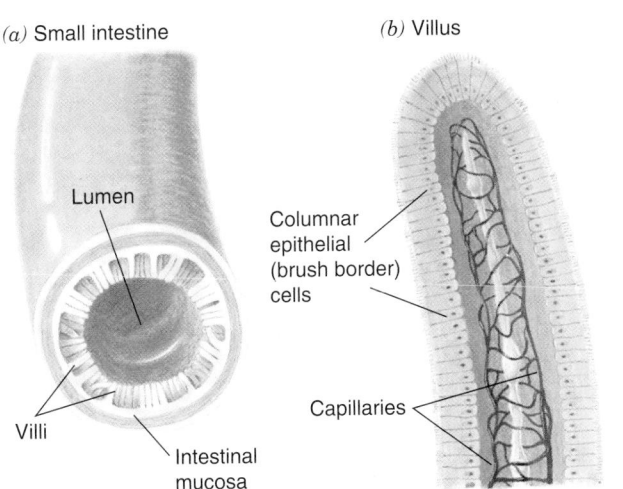

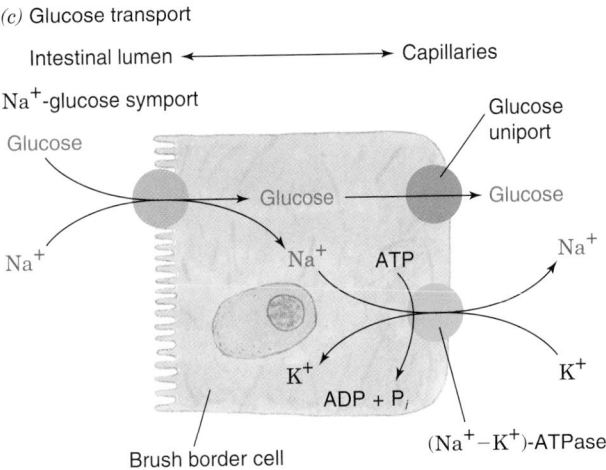

FIGURE 20-27   Glucose transport in the intestinal epithelium. The brushlike villi lining the small intestine greatly increase its surface area, thereby facilitating the absorption of nutrients. The brush border cells from which the villi are formed actively concentrate glucose from the intestinal lumen in symport with Na$^+$, a process that is driven by the (Na$^+$–K$^+$)–ATPase, which is located on the capillary side of the cell and functions to maintain a low internal [Na$^+$]. The glucose is exported to the bloodstream via a separate passive-mediated uniport system like that in the erythrocyte.

a. Active and Passive Glucose Transporters Exhibit Differential Drug Susceptibilities

The two glucose transport systems are inhibited by different drugs:

1. Phlorizin inhibits Na^+-dependent glucose transport.

2. Cytochalasin B inhibits Na^+-independent transport.

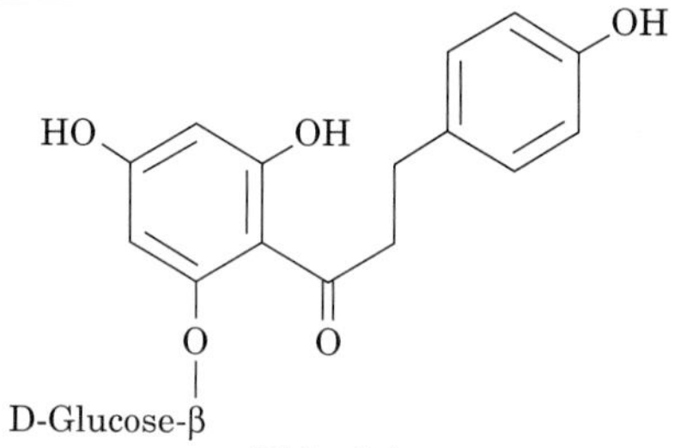

Phlorizin

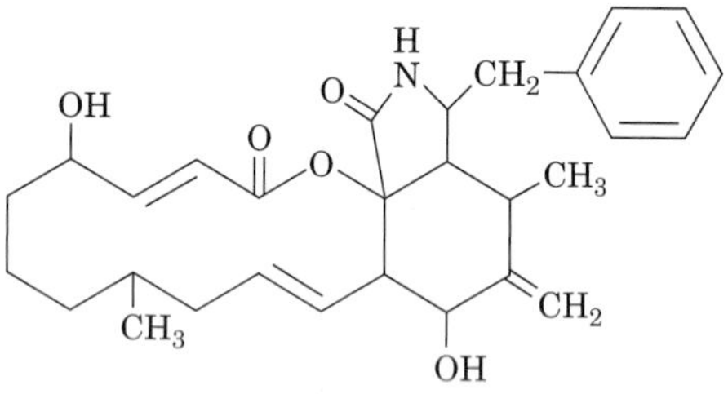

Cytochalasin B

Phlorizin binds only to the external surface of the Na^+-dependent glucose transporter, whereas cytochalasin B binds to the cytoplasmic surface of the Na^+-independent glucose transporter. This further indicates that these proteins are asymmetrically inserted into membranes. The use of these inhibitors permits the actions of the two glucose transporters to be studied separately in intact cells.

Kinetic studies indicate that the Na^+–glucose symport binds its substrates, Na^+ and glucose, in random order (Fig. 20-28), although binding of Na^+ increases the affinity of the transporter for glucose to such an extent that the upper pathway is heavily favored. Only when both substrates are bound, however, does the protein change its conformation to expose the binding sites to the inside of the cell. This requirement for concomitant Na^+ and glucose transport prevents the wasteful dissipation of the Na^+ gradient.

B. *Lactose Permease*

Gram-negative bacteria such as *E. coli* contain several active transport systems for concentrating sugars. We have already discussed the PTS system. Another extensively studied system, **lactose permease** (also known as **galactoside permease**), *utilizes the proton gradient across the bacterial cell membrane to cotransport H^+ and lactose (Fig. 20-29).* The proton gradient is metabolically generated through oxidative metabolism in a manner similar to that in mitochondria (Section 22-3B). The electrochemical potential gradient created by both these systems is used mainly to drive the synthesis of ATP.

How do we know that lactose transport requires the presence of a proton gradient? Ronald Kaback has established the requirement for this gradient through the following observations:

1. The rate of lactose transport into bacteria is increased enormously by the addition of D-lactate, an energy source for transmembrane proton gradient generation. Conversely, inhibitors of oxidative metabolism, such as cyanide, block both the formation of the proton gradient and lactose transport.

2. 2,4-Dinitrophenol, a proton ionophore that dissipates transmembrane proton gradients (Section 22-3D), inhibits lactose transport into both intact bacteria and membrane vesicles.

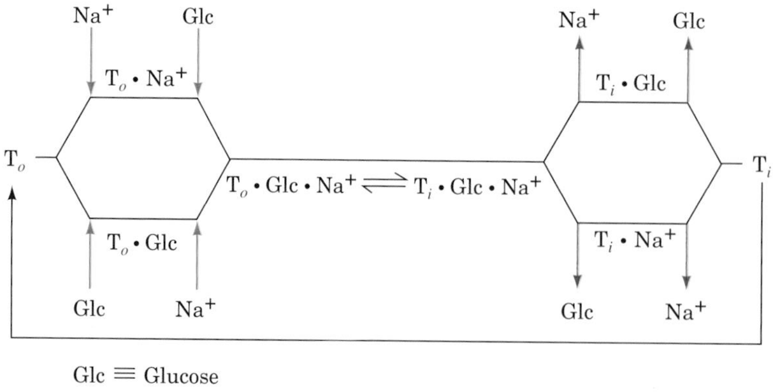

Glc $\equiv$ Glucose

FIGURE 20-28   The Na^+–glucose symport system represented as a Random Bi Bi kinetic mechanism. Binding of Na^+ increases the affinity of the transporter for glucose to such an extent that the upper pathway is heavily favored. T_o and T_i, respectively, represent the transport protein, with its binding sites exposed to the outer and inner surfaces of the membrane. [After Crane, R.K. and Dorando, F.C., in Martonosi, A.N. (Ed.), *Membranes and Transport*, Vol. 2, p. 154, Plenum Press (1982).]

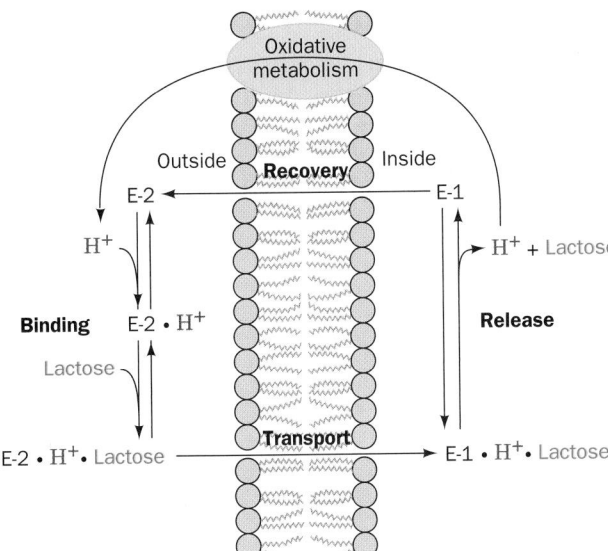

FIGURE 20-29 Kinetic mechanism of lactose permease in *E. coli.* H^+ binds first to E-2 outside the cell, followed by lactose. They are released in random order from E-1 inside the cell. E-2 must bind both lactose and H^+ in order to change conformation to E-1, thereby cotransporting these substances into the cell. E-1 changes conformation to E-2 when neither lactose nor H^+ is bound, thus completing the transport cycle.

3. The fluorescence of **dansylaminoethylthio-galactoside,**

Lactose

Dansylaminoethylthiogalactoside

a competitive inhibitor of lactose transport, is sensitive to the polarity of its environment and thus changes when it binds to lactose permease. Fluorescence measurements indicate that it does not bind to membrane vesicles that contain lactose permease in the absence of a transmembrane proton gradient.

a. Lactose Permease Has Two Major Conformational States

Lactose permease is a 417-residue monomer that, like the mammalian glucose transporters (Section 20-2E) to which it is distantly related, consists mainly of 12 trans-

membrane helices with its N- and C-termini in the cytoplasm. Like the $(Na^+–K^+)$–ATPase, it has two major conformational states (Fig. 20-29):

1. E-1, which has a low-affinity lactose binding site facing the interior of the cell.

2. E-2, which has a high-affinity lactose binding site facing the exterior of the cell.

E-1 and E-2 can only interconvert when their H^+ and lactose binding sites are either both filled or both empty. This prevents not only dissipation of the H^+ gradient without cotransport of lactose into the cell, but also transport of lactose out of the cell without cotransport of H^+ against its concentration gradient.

C. *ATP–ADP Translocator*

ATP generated in the mitochondrial matrix (its inner compartment; Section 1-2A) through oxidative phosphorylation (Section 22-3C) is largely utilized in the cytosol to drive such endergonic processes as biosynthesis, active transport, and muscle contraction. The inner mitochondrial membrane contains a system that transports ATP out of the matrix in exchange for ADP produced in the cytosol by ATP hydrolysis. This antiport, the **ATP–ADP translocator,** is electrogenic since it exchanges ADP^{3-} for ATP^{4-}.

Several natural products inhibit ATP–ADP translocation. **Atractyloside** and its derivative **carboxyatractyloside** inhibit the process only from the external surface of the inner mitochondrial membrane; **bongkrekic acid** exerts its effects only on the internal surface.

R = H **Atractyloside**
R = COOH **Carboxyatractyloside**

Bongkrekic acid

These differentially acting inhibitors have been valuable tools in the isolation of the transport protein and in the elucidation of its mechanism of action. For example, the

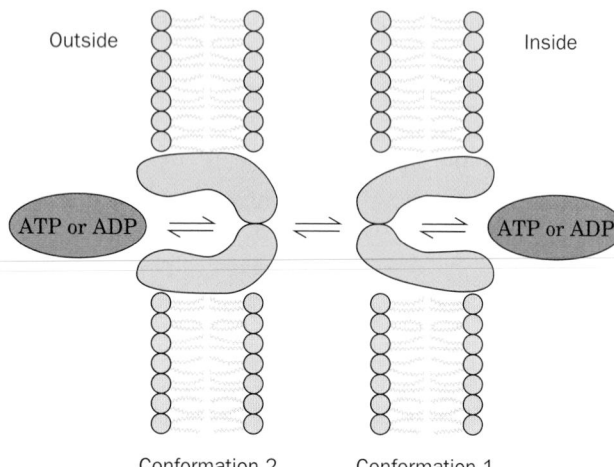

Outside   Inside

Conformation 2   Conformation 1

FIGURE 20-30   Conformational mechanism of the ATP–ADP translocator. An adenine nucleotide binding site located in the intersubunit contact area of the translocator dimer is alternately exposed to the two sides of the membrane. In contrast to the glucose transporter (Fig. 20-14), the ATP–ADP translocator can change its conformation only when bound to ADP or ATP.

translocator has been purified by affinity chromatography (Section 6-3C) using atractyloside derivatives as affinity ligands. Atractyloside binding is also a convenient means of identifying the translocator.

The ATP–ADP translocator, a dimer of identical ~300-residue subunits, has characteristics similar to those of other transport proteins. It has one binding site for which ADP and ATP compete. It has two major conformations, one with its ATP–ADP binding site facing the inside of the mitochondrion, and the other with this site facing outward (Fig. 20-30). The translocator is an antiport because it must bind ligand to change from one conformational state to the other at a physiologically reasonable rate.

The ATP–ADP translocator is not itself an active transport system. However, its electrogenic export of one negative charge per transport cycle in the direction of ATP export–ADP import is driven by the membrane potential difference, $\Delta\Psi$, across the inner mitochondrial membrane (positive outside). This results in the formation of gradients in ATP and ADP across the membrane.

5 ■ NEUROTRANSMISSION

In higher animals, the most rapid and complex intercellular communications are mediated by nerve impulses. A neuron (nerve cell; e.g., Fig. 1-10*d*) electrically transmits such a signal along its highly extended length (its **axon,** which is commonly over 1 m in larger animals) as a traveling wave of ionic currents. Signal transmission between neurons, as well as between neurons and muscles or glands, is usually chemically mediated by neurotransmitters. In this section we discuss both the electrical and chemical aspects of nerve impulse transmission.

A. *Voltage-Gated Ion Channels*

Ion gradients across cell membranes, as we have seen, are generated by specific energy-driven pumps (Section 20-3). These ion gradients are, in turn, discharged through ion channels such as K^+ channels (Section 20-2F). However, the pumps cannot keep up with the massive fluxes of ions passing through the channels. Hence ion channels are normally shut and only open transiently to perform some specific task for the cell. The opening and closing of ion channels, a process known as **gating,** occurs in response to a variety of stimuli:

1. Ligand-gated channels open in response to extracellular stimuli. For example, the KcsA K^+ channel (Section 20-2F) opens when the extracellular pH is less than ~4, whereas the ligand-gated ion channels in nerve cells open on extracellularly binding specific neurotransmitters (Section 20-5B).

2. Signal-gated channels open on intracellularly binding a second messenger such as Ca^{2+} ion or the $G_{\beta\gamma}$ subunit of a heterotrimeric G protein (Section 19-2C).

3. Voltage-gated channels open in response to a change in membrane potential. Multicellular organisms contain numerous varieties of voltage-gated channels. For example, nerve impulses arise from the sequential opening of voltage-gated channels along the length of a single nerve cell (Section 20-5B).

All voltage-gated K^+ channels are transmembrane homotetramers, each subunit of which contains an ~220-residue N-terminal cytoplasmic domain, an ~250-residue transmembrane domain consisting of six helices, S1 to S6, and an ~150-residue C-terminal cytoplasmic domain (Fig. 20-31). S5 and S6, with their intervening so-called P loop, are homologous to the KcsA K^+ channel (Section 20-2F), with the P loop containing the K^+ channel's TVGYG signature sequence.

Voltage-gated **Na^+ channels** and **Ca^{2+} channels** are ~2000-residue monomers that consist of 4 consecutive domains, each of which is homologous to the K^+ channel transmembrane domain, separated by often large cytoplasmic loops. These domains presumably assume a

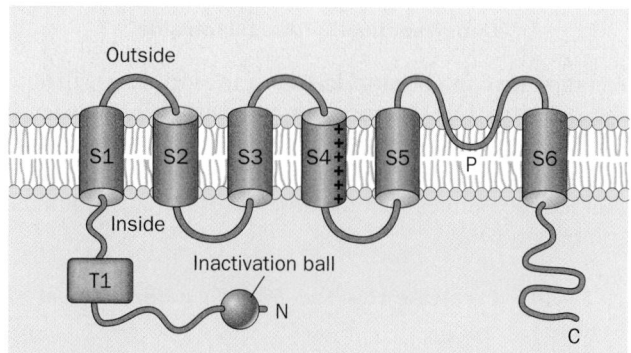

Outside

S1   S2   S3   S4   S5   P   S6

Inside

Inactivation ball

T1   N   C

FIGURE 20-31   Predicted secondary structure and membrane orientation of voltage-gated K^+ channels.

pseudotetrameric arrangement about a central pore resembling that of the subunits in voltage-gated K^+ channels. This structural homology suggests that voltage-gated ion channels share a common architecture in which differences in ion selectivity arise from precise stereochemical variations within the central pore. However, outside of their conserved transmembrane core, voltage-gated ion channels with different ion selectivities are highly divergent. For example, voltage-gated K^+ channels, which are known as **K_V channels,** have a conserved ~100-residue domain, the so-called T1 domain, that precedes the transmembrane domain and that is absent in other types of voltage-gated ion channels.

What is the nature of the gating machinery in voltage-gated ion channels? The ~19-residue S4 helix (Fig. 20-31), which contains ~5 positively charged side chains spaced

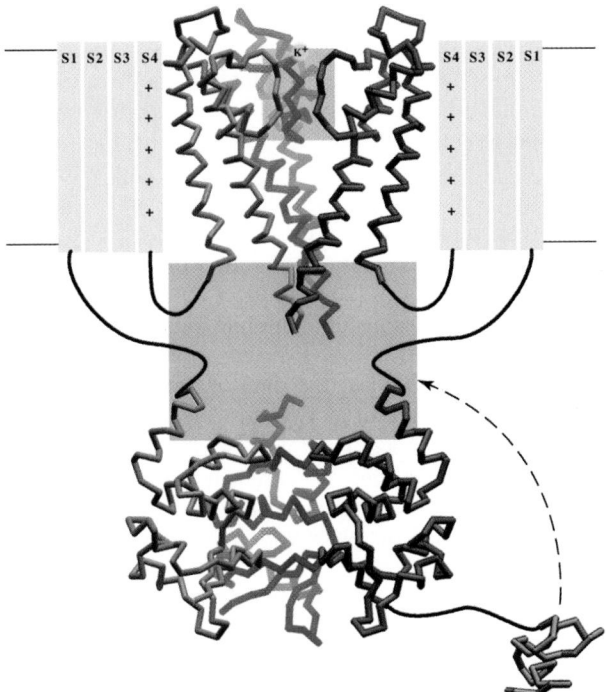

FIGURE 20-32   Composite model of the K_V channel. The channel is viewed parallel to the plane of the membrane, with its extracellular surface above. The bottom portion of the drawing shows a C_α diagram of the X-ray structure of the T1 tetramer, with each subunit in a different color (only three of the four subunits are shown for clarity). Only the inactivation peptide that is linked to the red T1 subunit is shown (*yellow*). The upper portion of the model represents its transmembrane domain, with helices S5 and S6 shown as the X-ray structure of the homologous KcsA channel (Fig. 20-16) and helices S1 to S4, whose structure is unknown, represented schematically. The C-terminal domains, which follow helices S6, have been omitted for clarity. The light green box highlights the selectivity filter formed by the four P loops. The orange box highlights the region occupied by the putative lateral windows through which both the cytoplasmic K^+ ions and the inactivation peptides gain access to the central pore. [Courtesy of Senyon Choe, The Salk Institute, La Jolla, California. PDBids for T1 and for the inactivation peptides: 1EOE and 1ZTN.]

every ~3 residues, appears to act as a voltage sensor (charged residues are unstable in a lipid environment and hence S4 is probably surrounded by protein). This was shown by covalently linking a dye whose fluorescence spectrum varies with the polarity of the environment to any of several residues in S4. Fluorescence measurements on each of these labeled ion channels revealed that when the membrane potential increases (inside becomes less negative), a stretch of at least 7 residues at the N-terminal end of S4 moves from a position within the membrane to the extracellular environment. It seems likely that this movement triggers channel opening. However, despite intensive investigations, the mechanism by which ion channels are gated open by a change in transmembrane potential is as yet unclear. Nevertheless, it seems likely that this gate is formed, at least in part, by the cytoplasmic ends of each ion channel's four S6 helices.

The K_V channel's T1 domain confers specificity in subunit oligomerization: It prevents K_V subunits of different subfamilies from co-assembling in the same tetramer. The X-ray structure of isolated T1 domains, determined by Senyon Choe and Paul Pfaffinger, reveals that this water-soluble protein forms a 4-fold rotationally symmetric homotetramer (Fig. 20-32, *bottom*). Such C_4 symmetry (Section 8-5B) is unusual for globular proteins (nearly all such homotetramers have D_2 symmetry) but, of course, is normal for homotetrameric transmembrane proteins. Presumably, the T1 tetramer hangs from the cytoplasmic face of the K_V channel's transmembrane domain such that their 4-fold axes are coincident. This further suggests that the T1 tetramer's water-filled central channel, which is ~4 Å in diameter at its narrowest point, forms the outer vestibule of the K^+ pore and perhaps even forms part of the gate. But, as we shall see below, this is not the case.

a.   Ion Channels Have Two Gates

Electrophysiological measurements indicate that the K_V channel spontaneously closes a few milliseconds after opening, a process termed **inactivation,** and does not reopen until after the membrane has repolarized (regained its resting membrane potential). Evidently, *the K_V channel contains two voltage-sensitive gates, one to open the channel on an increase in membrane potential and one to close it a short time later,* a phenomenon that has important consequences for the transmission of nerve impulses (Section 20-5B). In fact, minute "gating currents" arising from the movements of these positively charged gates in opening and closing can be detected (electrical current is the movement of charge) if the much larger currents of K^+ ions through the membrane are first blocked by plugging the K_V channel from its cytoplasmic side by high concentrations of Cs^+ or tetraethylammonium ions (which are too large to pass through the K^+ pore but apparently become stuck within it).

b.   Inactivation Occurs through the Insertion of the K_V Channel's N-Terminal Peptide into Its Central Pore

The inactivation of the K_V channel is abolished by proteolytically excising its N-terminal ~20-residue segment (its inactivation peptide), which NMR studies indicate forms a ball-like structure (Fig. 20-32, *lower right*). However, when chemically synthesized inactivation peptide is injected into the cytoplasm, the truncated K_V channel inactivates at a rate proportional to the concentration of inactivation peptide. This suggests that inactivation occurs when the inactivation ball swings around at the end of the ~65-residue peptide linking it to the T1 domain so as bind to the open K^+ pore in a way that physically blocks the passage of K^+ ions—the so-called ball-and-chain mechanism. Indeed, the time the K_V channel stays open varies with the length of this "chain," as mutationally adjusted.

Where is the inactivation peptide's binding site? Mutational analysis by MacKinnon revealed that the hydrophobic residues lining the K_V channel's internal pore and central cavity form the receptor site for the inactivation peptide. Since the ~6-Å-diameter cytoplasmic entrance to the internal pore is too narrow to admit the ball, the ball peptide must unfold in order to enter the internal pore. The first 10 residues of ball peptides are predominantly hydrophobic, whereas the succeeding 10 residues are largely hydrophilic and contain several basic residues. It therefore appears that inactivation occurs through the binding of the N-terminus of the fully extended inactivation peptide inside the internal pore via hydrophobic interactions, an association that is augmented by the binding of the basic residues in the ball peptide's C-terminal segment to the acidic residues lining the entrance of the internal pore. Thus, the inactivation peptide acts more like a snake than a ball and chain.

How does the inactivation peptide gain access to the internal pore, which is presumably covered by the T1 tetramer? The passage through the center of the T1 tetramer, as seen in its X-ray structure, is too narrow to permit the passage of the inactivation peptide. Moreover, Christopher Miller eliminated the possibility that the individual T1 domains in a tetramer can ever separate far enough to admit the inactivation peptide by showing that cross-linking adjacent T1 domains by genetically engineered disulfide bonds (whose positions were selected by referring to the T1 X-ray structure) does not significantly affect the K_V channel's gating properties. This strongly suggests that the inactivation peptide gains access to the bottom of the transmembrane pore through lateral windows between the transmembrane and T1 domains, whose sides are formed by the ~35-residue peptide segment linking these domains (Fig. 20-32, *curved dashed arrow*). Presumably, K^+ ions pass through these same windows when the K_V channel is open.

A K_V channel engineered so that only one subunit has an inactivation peptide still inactivates but at one-fourth the rate of normal K_V channels. Apparently, any of the normal K_V channel's four inactivation peptides can block the channel and it is simply a matter of chance as to which one does so. In contrast, voltage-gated Na^+ channels have only a single inactivation peptide, which is located on the segment linking the Na^+ channel's third and fourth homologous transmembrane domains. Consequently, a genetically

engineered cut of the peptide chain in this region abolishes Na^+ channel inactivation.

B. *Action Potentials*

Neurons, like other cells, generate ionic gradients across their plasma membranes through the actions of the corresponding ion-specific pumps. In particular, a (Na^+-K^+)–ATPase (Section 20-3A) pumps K^+ into and Na^+ out of the neuron to yield intracellular and extracellular concentrations of these ions similar to those listed in Table 20-3. The consequent membrane potential, $\Delta\Psi$, across a cell membrane is described by the **Goldman equation,** an extension of Eq. [20.3] that explicitly takes into account the various ions' different membrane permeabilities:

$$\Delta\Psi = \frac{RT}{\mathscr{F}}\ln\frac{\sum P_c[C(out)] + \sum P_a[A(in)]}{\sum P_c[C(in)] + \sum P_a[A(out)]} \quad [20.8]$$

Here, C and A represent cations and anions, respectively, and, for the sake of simplicity, we have made the physiologically reasonable assumption that only monovalent ions have significant concentrations. The quantities P_c and P_a, the respective **permeability coefficients** for the various cations and anions, are indicative of how readily the corresponding ions traverse the membrane (each is equal to the corresponding ion's diffusion coefficient through the membrane divided by the membrane's thickness; Section 20-2A). Note that Eq. [20.8] reduces to Eq. [20.3] if the permeability coefficients of all mobile ions are assumed to be equal.

Applying Eq. [20.8] to the data in Table 20-3 and assuming a temperature of 25°C yields $\Delta\Psi = -83$ mV (negative inside), which is in good agreement with experimentally measured membrane potentials for mammalian cells. This value is somewhat greater than the K^+ equilibrium potential, the value of $\Delta\Psi = -91$ mV obtained assuming the membrane is permeable to only K^+ ions ($P_{Na^+} = P_{Cl^-} = 0$). The membrane potential is generated by a surprisingly small imbalance in the ionic distribution across the membrane: Only ~1 ion pair per million is separated by the membrane with the anion going to the cytoplasmic side and the cation going to the external side. The resulting

TABLE 20-3 Ionic Concentrations and Membrane Permeability Coefficients in Mammals

Ion	Cell (mM)	Blood (mM)	Permeability Coefficient (cm · s⁻¹)
K^+	139	4	5×10^{-7}
Na^+	12	145	5×10^{-9}
Cl^-	4	116	1×10^{-8}
X^{-a}	138	9	0

[a] "X^- represents macromolecules that are negatively charged under physiological conditions.

Source: Darnell, J., Lodish, H., and Baltimore, D., *Molecular Cell Biology,* pp. 618 and 725, Scientific American Books (1986).

electric field is, nevertheless, enormous by macroscopic standards: Assuming a typical membrane thickness of 50 Å, it is nearly 170,000 V · cm^{-1}.

a. Nerve Impulses Are Propagated by Action Potentials

*A nerve impulse consists of a wave of transient membrane depolarization known as an **action potential** that passes along a nerve cell.* A microelectrode implanted in an axon will record that during the first ~0.5 ms of an action potential, $\Delta\Psi$ increases from its resting potential of around −60 mV to about ~30 mV (Fig. 20-33a). This de-

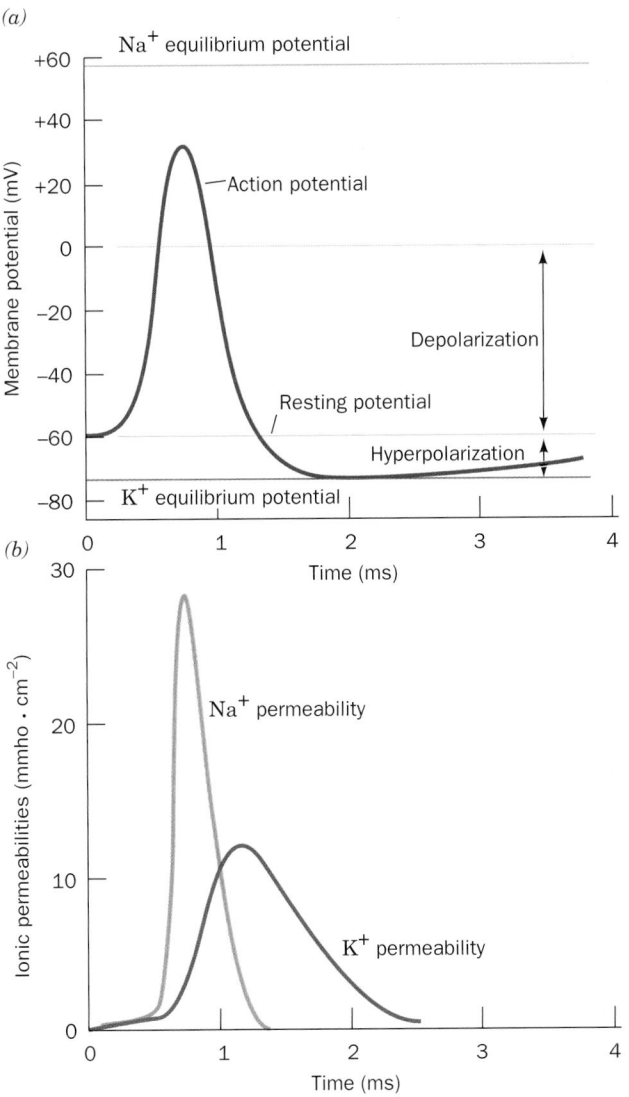

(a)

(b)

FIGURE 20-33   Time course of an action potential. (*a*) The axon membrane undergoes rapid depolarization, followed by a nearly as rapid hyperpolarization and then a slow recovery to its resting potential. (*b*) The depolarization is caused by a transient increase in Na$^+$ permeability (conductance), whereas the hyperpolarization results from a more prolonged increase in K$^+$ permeability that begins a fraction of a millisecond later. The unit of conductance, 1 mho = 1 ohm^{-1}. [After Hodgkin, A.L. and Huxley, A.F., *J. Physiol.* **117**, 530 (1952).]

polarization is followed by a nearly as rapid repolarization past the resting potential to the K$^+$ equilibrium potential (hyperpolarization) and then a slower recovery to the resting potential. What is the origin of this complicated electrical behavior? In 1952, Alan Hodgkin and Andrew Huxley demonstrated that the action potential results from a transient increase in the membrane's permeability to Na$^+$(P_{Na^+}) followed, within a fraction of a millisecond, by a transient increase in its permeability to K$^+$ (P_{K^+} Fig. 20-33b).

The ion-specific permeability changes that characterize an action potential result from the presence of Na$^+$- and K$^+$- specific voltage-gated channels. As a nerve impulse reaches a given patch of nerve cell membrane, the increased membrane potential induces the transient opening of the Na$^+$ channels, so that Na$^+$ ions diffuse into the nerve cell at the rate of ~6000 ions · ms^{-1} per channel. This increase in P_{Na^+} causes $\Delta\Psi$ to increase (Eq. [20.3]), which, in turn, induces more Na$^+$ channels to open, etc., leading to an explosive entry of Na$^+$ into the cell. Yet, before this process can equilibrate at its Na$^+$ equilibrium potential of around ~60 mV, the K$^+$ channels open (P_{K^+} increases) while the Na$^+$ channels close (inactivate; P_{Na^+} returns to its resting value). $\Delta\Psi$ therefore reverses sign and overshoots its resting potential to approach its K$^+$ equilibrium value. Eventually the K$^+$ channels also inactivate and the membrane patch regains its resting potential. The Na$^+$ channels, which remain open only 0.5 to 1.0 ms, do not reopen until the membrane has returned to its resting state, thereby limiting the axon's firing rate.

An action potential is triggered by an ~20-mV rise in $\Delta\Psi$ to about −40 mV. Action potentials therefore propagate along an axon because the initially rising value of $\Delta\Psi$ in a given patch of axonal membrane triggers the action potential in an adjacent membrane patch that does so in an adjacent membrane patch, etc. (Fig. 20-34). The nerve impulse is thereby continuously amplified so that its signal amplitude remains constant along the length of the axon (in contrast, an electrical impulse traveling down a wire dissipates as a consequence of resistive and capacitive effects). Note, however, that since the relative ion imbalance responsible for the resting membrane potential is small, only a tiny fraction of a nerve cell's Na$^+$–K$^+$ gradient is discharged by a single nerve impulse (only one K$^+$ ion per 3000–300,000 in the cytosol is exchanged for extracellular Na$^+$, as indicated by measurements with radioactive Na$^+$). An axon can therefore transmit a nerve impulse every few milliseconds without letup. This capacity to fire rapidly is an essential feature of neuronal communications: *Since nerve impulses all have the same amplitude, the magnitude of a stimulus is conveyed by the rate at which a nerve fires.*

b. The Voltage-Gated Na$^+$ Channel Is the Target of Numerous Neurotoxins

Neurotoxins have proved to be invaluable tools for dissecting the various mechanistic aspects of neurotransmission. Many neurotoxins, as we shall see, interfere with the action of neuronal voltage-gated Na$^+$ channels but, curi-

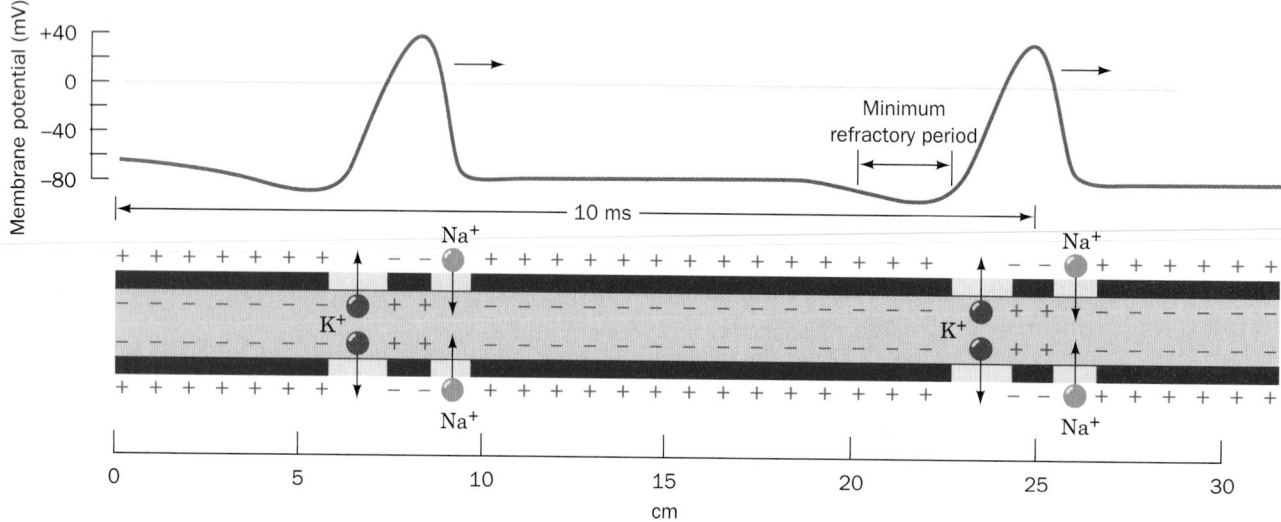

FIGURE 20-34 Action potential propagation along an axon. Membrane depolarization at the leading edge of an action potential triggers an action potential at the immediately downstream portion of the axon membrane by inducing the opening of its voltage-gated Na^+ channels. As the depolarization wave moves farther downstream, the Na^+ channels close and the K^+ channels open to hyperpolarize the membrane. After a brief refractory period, during which the K^+ channels close and the hyperpolarized membrane recovers its resting potential, a second impulse can follow. The indicated impulse propagation speed is that measured in the giant axon of the squid, which, because of its extraordinary width (~1 mm), is a favorite experimental subject of neurophysiologists. Note that the action potential in this figure appears backward from that in Fig. 20-33 because this figure shows the distribution of the membrane potential along an axon at an instant in time, whereas Fig. 20-33 shows the membrane potential's variation with time at a fixed point on the axon.

ously, few are known that affect voltage-gated K^+ channels. **Tetrodotoxin,**

Tetrodotoxin

a paralytic poison of enormous potency, which occurs mainly in the skin, ovaries, liver, and intestines of the puffer fish (known as fugu in Japan, where it is a delicacy that may be prepared only by chefs certified for their knowledge of puffer fish anatomy), acts by specifically blocking the Na^+ channel. The Na^+ channel is similarly blocked by **saxitoxin,**

Saxitoxin

a product of marine dinoflagellates (a type of plankton known as the "red tide") that is concentrated by filter-feeding shellfish to such an extent that a small mussel can contain sufficient saxitoxin to kill 50 people. Both of these neurotoxins have a cationic guanidino group, and both are effective only when applied to the external surface of a neuron (their injection into the cytoplasm elicits no response). It is therefore thought that these toxins specifically interact with an anionic carboxylate group located at the mouth of the Na^+ channel on its extracellular side.

Batrachotoxin,

Batrachotoxin

a steroidal alkaloid secreted by the skin of a Colombian arrow-poison frog, *Phyllobates aurotaenia,* is the most potent known venom (2 $\mu g \cdot kg^{-1}$ body weight is 50% lethal in mice). This substance also specifically binds to the voltage-gated Na^+ channel but, in contrast to the actions of tetrodotoxin and saxitoxin, renders the axonal membrane highly permeable to Na^+. Indeed, batrachotoxin-induced axonal depolarization is reversed by tetrodotoxin. The observation that the repeated electrical stimulation of a neuron enhances the action of batrachotoxin indicates that this toxin binds to the Na$^+$ channel in its open state.

Venoms from American scorpions contain families of 60- to 70-residue protein neurotoxins that also act to depolarize neurons by binding to their Na$^+$ channels (the different neurotoxins in the same venom appear to be specialized for binding to the Na$^+$ channels in the various species the scorpion is likely to encounter). Scorpion toxins and tetrodotoxin do not, however, compete with each other for binding to the Na$^+$ channel and therefore must bind at separate sites.

c. Nerve Impulse Velocity Is Increased by Myelination

The axons of the larger vertebrate neurons are sheathed with **myelin,** a biological "electrical insulating tape" that is

wrapped about the axon (Fig. 20-35a) so as to electrically isolate it from the extracellular medium. Impulses in myelinated nerves propagate with velocities of up to 100 m · s^{-1}, whereas those in unmyelinated nerves are no faster than 10 m · s^{-1} (imagine the coordination difficulties that, say, a giraffe would have if it had to rely on only unmyelinated nerves).

How does myelination increase the velocity of nerve impulses? Myelin sheaths are interrupted every millimeter or so along the axon by narrow unmyelinated gaps known as **nodes of Ranvier** (Fig. 20-35b), where the axon contacts the extracellular medium. Binding studies using radioactive tetrodotoxin indicate that the voltage-gated Na$^+$ channels of unmyelinated axons have rather sparse although uniform distributions in the axonal membrane of ∼20 channels · μm^{-2}. In contrast, the Na$^+$ channels of myelinated axons occur only at the nodes of Ranvier, where they are concentrated with a density of ∼10^4 channels · μm^{-2}. The action potential of a myelinated axon evidently hops between these nodes, a process named **saltatory conduction** (Latin: *saltare,* to jump). Nerve impulse transmission between the nodes must therefore occur by the passive conduction of an ionic current, a mechanism that is inherently much faster than the continuous propagation of an action potential but that is also dissipative. The nodes act as amplification stations to maintain the intensity of the electrical impulse as it travels down the axon. Without the myelin insulation, the electrical impulse would become too attenuated through transmembrane ion leakage and capacitive effects to trigger an action potential at the next node. In fact, **multiple sclerosis,** an autoimmune disease

(a)

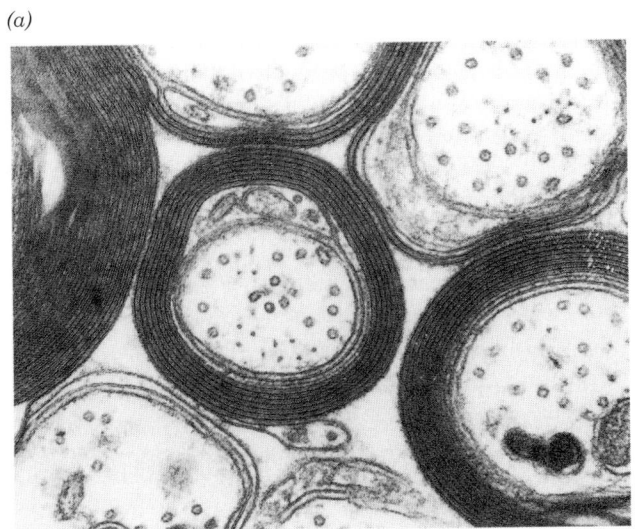

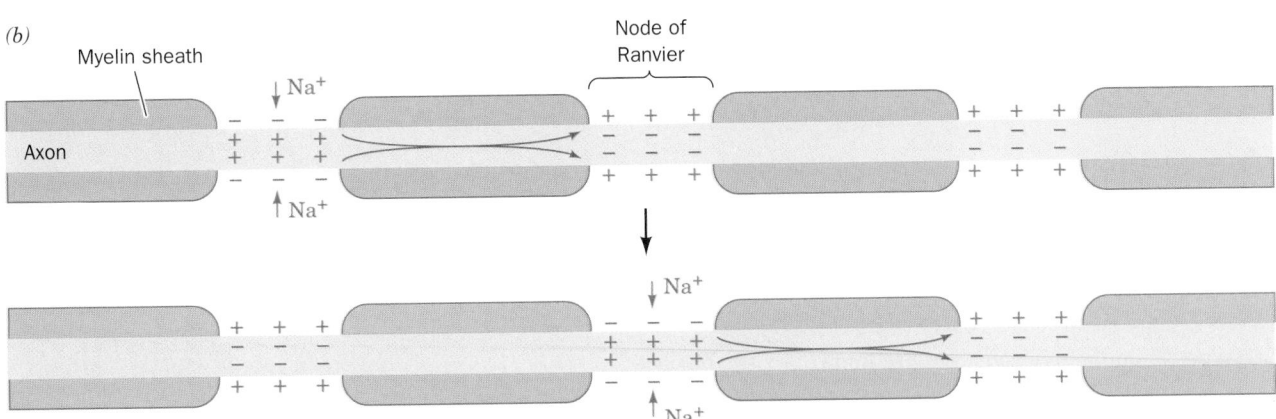

FIGURE 20-35 Myelination. (a) An electron micrograph of myelinated nerve fibers in cross section. The myelin sheath surrounding an axon is the plasma membrane of a **Schwann cell,** which, as it spirally grows around an axon, extrudes its cytoplasm from between the layers. The resulting double bilayer, which makes between 10 and 150 turns about the axon, is a good electrical insulator because of its particularly high (79%) lipid content. [Courtesy of Cedric Raine, Albert Einstein College of Medicine of Yeshiva University.] (b) A schematic diagram of a myelinated axon in longitudinal section, indicating that in the nodes of Ranvier (the relatively short gaps between adjacent myelinating cells), the axonal membrane is in contact with the extracellular medium. A depolarization generated by an action potential at one node hops, via ionic conduction, down the myelinated axon (*red arrows*), to the neighboring node, where it induces a new action potential. Nerve impulses in myelinated axons are therefore transmitted by saltatory conduction.

that demyelinates nerve fibers in the brain and spinal cord, results in serious and often fatal neurological deficiencies.

C. *Neurotransmitters and Their Receptors*

The junctions at which neurons pass signals to other neurons, muscles, or glands are called **synapses.** In **electrical synapses,** which are specialized for rapid signal transmission, the cells are separated by a gap, the **synaptic cleft,** of only 20 Å, which is spanned by gap junctions (Section 12-3F). Hence, an action potential arriving at the presynaptic side of the cleft can sufficiently depolarize the postsynaptic membrane to trigger its action potential directly. However, the >200-Å gap of most synapses is too large a distance for such direct electrical coupling. In these **chemical synapses,** the arriving action potential triggers the release from the presynaptic neuron of a specific substance known as a **neurotransmitter,** which diffuses across the cleft and binds to its corresponding receptors on the postsynaptic membrane. In **excitatory synapses,** neurotransmitter binding induces membrane depolarization, thereby triggering an action potential on the postsynaptic membrane. Conversely, neurotransmitter binding in **inhibitory synapses** alters postsynaptic membrane permeability so as to inhibit an action potential and thus attenuate excitatory signals. What is the mechanism through which an arriving action potential stimulates the release of a neurotransmitter, and by what means does its binding to a receptor alter the postsynaptic membrane's permeability? To answer these questions let us consider the workings of **cholinergic synapses;** that is, synapses that use **acetylcholine (ACh)** as a neurotransmitter:

Acetylcholine (ACh)

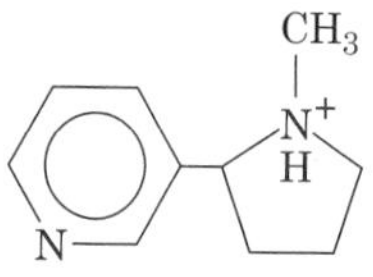

Nicotine

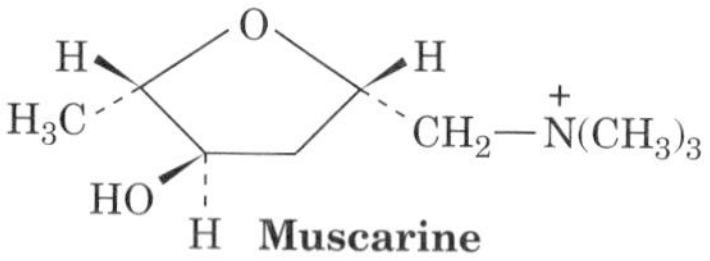

Muscarine

Two types of cholinergic synapses are known:

1. Those containing **nicotinic receptors** (receptors that respond to **nicotine**).

2. Those containing **muscarinic receptors** (receptors that respond to **muscarine,** an alkaloid produced by the poisonous mushroom *Amanita muscaria*).

In what follows, we shall focus on cholinergic synapses containing nicotinic receptors since this best characterized type of synapse occurs at all excitatory neuromuscular

junctions in vertebrates and at numerous sites in the nervous system.

a. Electric Organs of Electric Fish Are Rich Sources of Cholinergic Synapses

The study of synaptic function has been greatly facilitated by the discovery that the homogenization of nerve tissue causes its presynaptic endings to pinch off and reseal to form **synaptosomes.** The use of synaptosomes, which can be readily isolated by density gradient ultracentrifugation, has the advantage that they can be manipulated and analyzed without interference by other neuronal components.

The richest known source of cholinergic synapses is the electric organs of the freshwater electric eel *Electrophorus electricus* and saltwater electric fish of the genus *Torpedo*. Electric organs, which these organisms use to stun or kill their prey, consist of stacks of ~5000 thin flat cells called **electroplaques** that begin their development as muscle cells but ultimately lose their contractile apparatus. One side of an electroplaque is richly innervated and has high electrical resistance, whereas its opposite side lacks innervation and has low electrical resistance. Both sides maintain a resting membrane potential of around –90 mV. On neuronal stimulation, all the innervated membranes in a stack of electroplaques simultaneously depolarize to a membrane potential of around ~40 mV, yielding a potential difference across each cell of 130 mV (Fig. 20-36). Since the 5000 electroplaques in a stack are "wired" in series like the batteries in a flashlight, the total potential difference across the stack is ~5000 × 0.130 V = 650 V, enough to kill a human being.

b. Acetylcholine Is Released by the Ca^{2+}-Triggered Exocytosis of Synaptic Vesicles

ACh is synthesized near the presynaptic end of a neuron by the transfer of an acetyl group from **acetyl-CoA** [the structure of coenzyme A (CoA) is given in Fig. 21-2] to **choline** in a reaction catalyzed by **choline acetyltransferase.**

Much of this ACh is sequestered in ~400-Å-diameter membrane-enveloped **synaptic vesicles,** which typically contain ~10^4 ACh molecules each.

*The arrival of an action potential at the presynaptic membrane triggers the opening of **voltage-gated Ca^{2+} channels,***

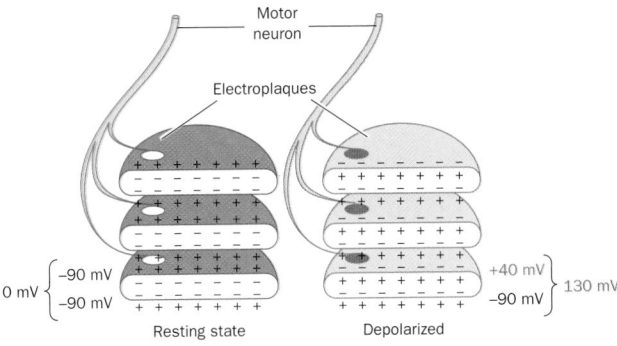

FIGURE 20-36 **The simultaneous depolarization (*red, right*) of the innervated membranes in a stack of electroplaques "wired" in series results in a large voltage difference between the two ends of the stack.** This is because the total voltage across the stack is the sum of the voltages generated by each of its electroplaques.

which transiently raises the local $[Ca^{2+}]$ from its resting level of 0.1 μM to 10 to 100 μM. The resulting influx of extracellular Ca^{2+}, in turn, stimulates the exocytosis of the synaptic vesicles in the vicinity of the Ca^{2+} channel so that they release their packets of ACh into the synaptic cleft (Fig. 12-61). The mechanism by which synaptic vesicles fuse with the presynaptic membrane is discussed in Section 12-4D.

The mechanism through which Ca^{2+} induces synaptic vesicle exocytosis is beginning to come into focus. The major Ca^{2+}-sensing protein appears to be **synaptotagmin I,** a protein with a single helix passing through the synaptic vesicle membrane, whose cytosolic domain contains four Ca^{2+} binding sites. At resting levels of Ca^{2+}, synaptotagmin I binds to the Q-SNARE syntaxin (Section 12-4D) so as to block its binding to the R-SNARE synaptobrevin and the Q-SNARE SNAP25, thereby preventing vesicle fusion. However, on binding Ca^{2+}, synaptotagmin I releases syntaxin, permitting vesicle fusion to commence.

Once triggered, the fusion of the synaptic vesicles with the presynaptic membrane occurs very rapidly (in <0.3 ms) because many synaptic vesicles are already docked with the presynaptic membrane. Each Ca^{2+} pulse triggers the exocytosis of ~10% of these docked vesicles. However, they are rapidly replaced because most of the remaining synaptic vesicles are held in reserve in a so-called active zone within 20 nm of the presynaptic membrane. Vesicles are held in the active zone by a fibrous phosphoprotein named **synapsin I** that also binds the cytoskeletal proteins actin and spectrin (Section 12-3D). Synapsin I is a substrate for calmodulin-dependent protein kinase (Section 18-3C), so that a rise in $[Ca^{2+}]$ causes its phosphorylation. This apparently releases the synaptic vesicles from the active zone, thereby permitting them to dock to the presynaptic membrane in preparation for exocytosis. The 20-nm distance between the active zone and the presynaptic membrane may be close enough for Q- and R-SNAREs to initiate the formation of their coiled coil complex (Fig. 12-62), which may thereby facilitate the docking process.

The black widow spider takes advantage of this system: Its highly neurotoxic venom protein, **α-latrotoxin** (130 kD), causes massive release of ACh at the neuromuscular junction, in part by forming homotetrameric transmembrane channels through the presynaptic membrane that act as Ca^{2+} ionophores. In contrast, **botulinus toxin,** as we have seen (Section 12-4D), interferes with the exocytosis of synaptic vesicles by proteolytically cleaving specific SNARE proteins, thereby preventing ACh release.

Exocytosed synaptic vesicle proteins are rapidly recovered from the presynaptic membrane via endocytosis mainly in clathrin-coated vesicles (Section 12-4C). However, once the resulting endocytotic vesicles lose their clathrin coats, they do not fuse with endosomes, as is usually the case (Fig. 12-79). Rather, they are immediately refilled with ACh by an H^+–ACh antiport, which is driven by the protons pumped into the vesicle by a V-type ATPase (Section 20-3), and then translocated to the active zone. This rapid recycling of the synaptic vesicles (which takes <1 min) permits neurons to fire continuously at a rate of ~50 times per second.

c. The Acetylcholine Receptor Is a Ligand-Gated Cation Channel

The **acetylcholine receptor** is a 290-kD $\alpha_2\beta\gamma\delta$ transmembrane glycoprotein whose four homologous subunits each are predicted to have 4 transmembrane helices (M1–M4) with a large N-terminal synaptic domain and a substantial cytoplasmic loop between helices M3 and M4. Electron crystallography studies, by Nigel Unwin, of the ACh receptor in its closed (unliganded) form indicate that it is an 80-Å-diameter by 125-Å-long cylinder that protrudes ~60 Å into the synaptic space and ~20 Å into the cytoplasm (Fig. 20-37). Its five rodlike subunits are arranged with pseudo-5-fold symmetry over much of their length with the clockwise order -α-β-α-δ-γ- when viewed from the synaptic space. The ACh receptor's most striking structural feature is an ~20-Å-diameter by ~65-Å-long water-filled central channel that extends from the receptor's synaptic entrance to the level of the lipid bilayer, where it forms a more constricted ~30-Å-long pore that is blocked near the middle of the bilayer. This blockage, which is presumably the channel's gate, is <6 Å thick, suggesting that it consists of one or two rings of side chains, possibly conserved Leu residues, projecting inward from the walls of the channel. The cytoplasmic end of the ACh receptor contains an ~20-Å-diameter central cavity that is ~20 Å long and that is connected to the cytoplasm via lateral openings between adjacent subunits at a level that is ~30 Å below the membrane surface. These openings, although of different sizes, are all <10 Å in diameter and hence could serve as filters to prevent the passage of cytoplasmic anions (by repulsions from anionic side chains) and large cations. Note that K_V channels appear to have similar lateral openings to the cytoplasm (Section 20-5A).

The binding of two ACh molecules, one to each α subunit, allosterically induces the opening of the channel through the bilayer to permit Na^+ and K^+ ions to diffuse,

(a)

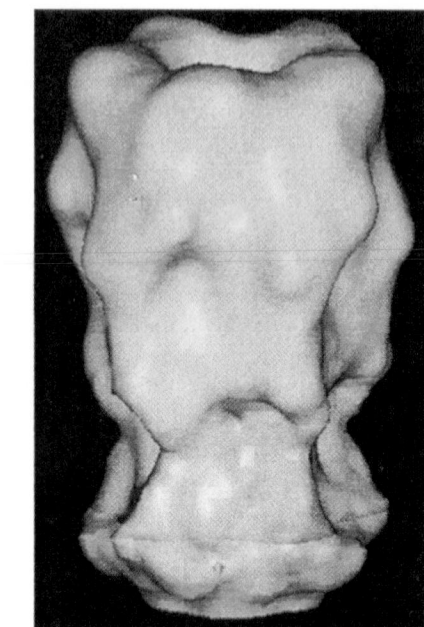

(b)

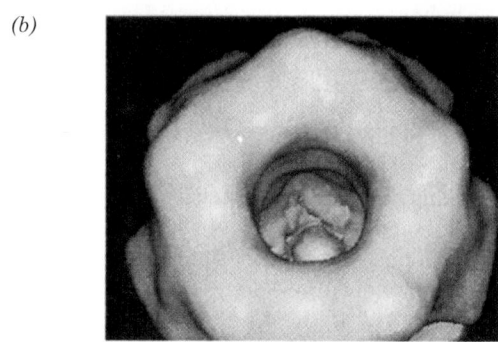

FIGURE 20-37   Electron crystal structure of the nicotinic acetylcholine receptor from the electric fish *Torpedo marmorata*. (*a*) Side view with the synaptic side up. The band across the bottom of the structure marks the position of the membrane bilayer in which the ACh receptor is embedded and separates its large extracellular portion from its smaller cytoplasmic portion. (*b*) View into the synaptic entrance of the channel. The channel narrows quite abruptly at the level of the lipid bilayer, ~60 Å below this entrance. [Courtesy of Nigel Unwin, MRC Laboratory of Molecular Biology, Cambridge, U.K.]

respectively, in and out of the cell at rates of ~20,000 of each type of ion per millisecond. The resulting depolarization of the postsynaptic membrane initiates a new action potential. After 1 to 2 ms, the ACh spontaneously dissociates from the receptor and the channel closes.

The ACh receptor is the target of some of the most deadly known neurotoxins (death occurs through respiratory arrest), whose use has greatly aided in the elucidation of receptor function. **Histrionicatoxin,** an alkaloid secreted by the skin of the Colombian arrow-poison frog *Dendrobates histrionicus*, and **d-tubocurarine,** the active ingredi-

ent of the Amazonian arrow-poison **curare** as well as a medically useful paralytic agent, are both ACh antagonists that prevent ACh receptor channel opening:

Histrionicatoxin

***d*-Tubocurarine**

Similarly, a family of homologous 7- to 8-kD venom proteins from some of the world's most poisonous snakes, including **α-bungarotoxin** from snakes of the genus *Bungarus,* **erabutoxin** from sea snakes, and **cobratoxin** from cobras, prevent ACh receptor channel opening by binding specifically and all but irreversibly to its α subunits. Indeed, detergent-solubilized ACh receptor has been purified by affinity chromatography on a column containing covalently attached cobra toxin.

d. Acetylcholine Is Rapidly Degraded by Acetylcholinesterase

*An ACh molecule that participates in the transmission of a given nerve impulse must be degraded in the few milliseconds before the potential arrival of the next nerve impulse. This essential task is accomplished by **acetylcholinesterase (AChE),*** a 75-kD fast-acting enzyme that is GPI anchored (Section 12-3B) to the surface of the postsynaptic membrane

$$H_3C-\overset{\overset{\text{O}}{\|}}{C}-O-CH_2-CH_2-\overset{+}{N}(CH_3)_3 \;+\; H_2O$$
Acetylcholine

↓ acetylcholinesterase

$$H_3C-\overset{\overset{\text{O}}{\|}}{C}-O^- +\; HO-CH_2-CH_2-\overset{+}{N}(CH_3)_3 \;+\; H^+$$
Acetate   **Choline**

(the turnover number of AChE, $k_{cat} = 14,000 \text{ s}^{-1}$; the enzyme's catalytic efficiency, $k_{cat}/K_M = 1.5 \times 10^8 \, M^{-1} \cdot \text{s}^{-1}$, is close to the diffusion-controlled limit, so that it is a nearly perfect catalyst; Section 14-2B). The resulting choline is taken up by the presynaptic cell via an Na^+–choline symport for use in the resynthesis of ACh. The operation of this transporter is similar to that of the Na^+–glucose symport of intestinal brush border cells (Section 20-4A).

AChE is a serine esterase; that is, its catalytic mechanism resembles that of serine proteases such as trypsin. These enzymes, as we have seen in Section 15-3A, are irreversibly inhibited by alkylphosphofluoridates such as diisopropylphosphofluoridate (DIPF). Indeed, related compounds such as **tabun** and **sarin**

Tabun   **Sarin**

are military nerve gases because their efficient inactivation of human AChE by reaction with the active site Ser causes paralysis stemming from cholinergic nerve impulse blockade and thus death by suffocation. **Succinylcholine,**

Succinylcholine

which is used as a muscle relaxant during surgery, is an ACh agonist that, although rapidly released by the ACh receptor, is but slowly hydrolyzed by AChE. Succinylcholine therefore produces persistent depolarization of the postsynaptic membrane. Its effects are short-lived, however, because it is rapidly hydrolyzed by the relatively nonspecific liver and plasma enzyme **butyrylcholinesterase.** Certain snake venoms, such as that of the green mamba snake, inactivate AChE, although they do so by binding to a site on AChE distinct from its active site.

e. X-Ray Structure of Acetylcholinesterase

The X-ray structure of the 537-residue AChE from the electric fish *Torpedo californica,* determined by Joel Sussman, Israel Silman, and Michal Harel, confirms that the previously identified Ser 200 and His 440 are members of AChE's catalytic triad. The structure further reveals that the third member of AChE's catalytic triad is Glu 237 rather than an Asp residue, only the second instance of a Glu in this position among the many serine proteases, lipases, and esterases of known structure. AChE's catalytic triad is arranged in what appears to be the mirror image of the catalytic triads in trypsin and subtilisin, for example (Figure 15-21), although, of course, this is not actually the case since all proteins consist of L-amino acid residues.

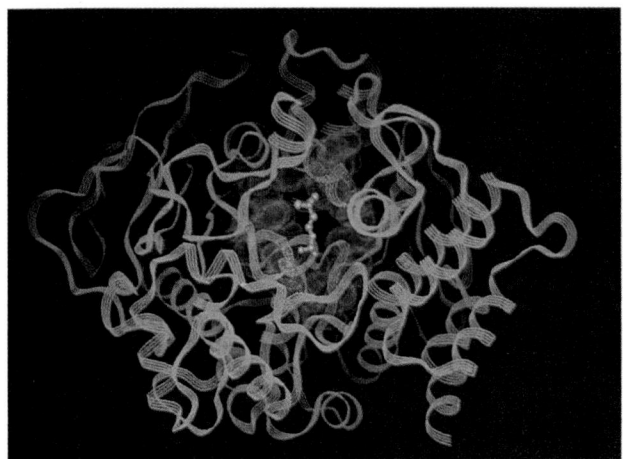

FIGURE 20-38   X-Ray structure of acetylcholinesterase (AChE). The enzyme is represented in ribbon form. The aromatic side chains lining its active site gorge (*purple*) are shown in stick form surrounded by their van der Waals dot surface. The ACh substrate, which was modeled into the active site (the enzyme was crystallized in its absence), is shown in ball-and-stick form with its atoms gold and its bonds cyan. The entrance to the gorge is at the top of the figure. [Courtesy of Joel Sussman, The Weizmann Institute of Science, Rehovot, Israel. PDBid 1ACL.]

AChE's catalytic site is near the bottom of a narrow and 20-Å-deep gorge that extends halfway through the protein and widens out near its base (Fig. 20-38). The sides of this so-called active site gorge are lined with the side chains of 14 aromatic residues that comprise 40% of its surface area. Since the side chain O atom of the active site Ser is only 4 Å from the bottom of the gorge, ACh must bind in the gorge with its positively charged trimethylammonium group surrounded by aromatic side chains. This conclusion came as a surprise since it had been understandably expected that the trimethylammonium group would be bound at an anionic site. Perhaps the weak binding provided by the interactions of the trimethylammonium group with the π electrons of the aromatic rings facilitates the rapid diffusion of ACh to the bottom of the gorge, thereby accounting for the enzyme's high turnover number. In fact, model aromatic compounds have been synthesized that also bind quaternary ammonium compounds. Intriguingly, the ACh binding site on each α subunit of the ACh receptor is also at the end of a similarly shaped gorge that opens into its central channel on its synaptic side, even though these regions of AChE and the ACh receptor exhibit no detectable sequence similarity.

f. Amino Acids and Their Derivatives Function as Neurotransmitters

The mammalian nervous system employs well over 30 substances as neurotransmitters. Some of these substances, such as glycine and glutamate, are amino acids; many others are amino acid decarboxylation products or their derivatives (often referred to as **biogenic amines**). For example, as we shall see in Section 26-4B, **dopamine,** norepinephrine, and epinephrine [which are collectively termed

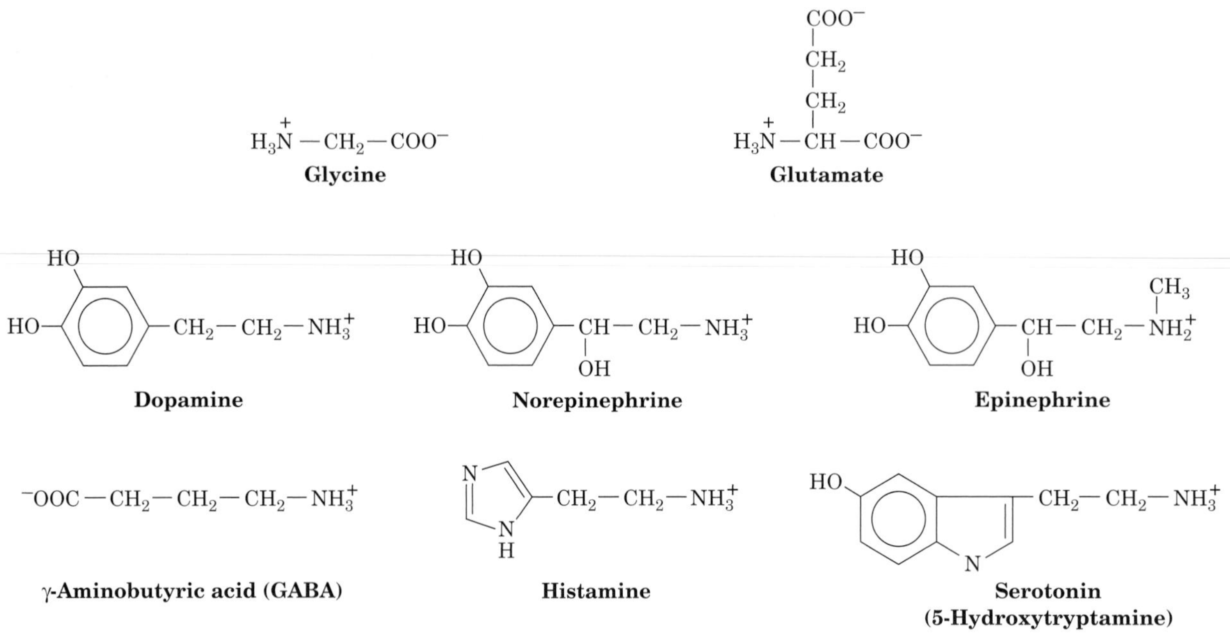

$$\underset{\textbf{Glycine}}{\overset{+}{H_3}N-CH_2-COO^-}$$

$$\underset{\textbf{Glutamate}}{\overset{COO^-}{\underset{\overset{|}{\underset{\overset{|}{CH_2}}{\underset{\overset{|}{CH_2}}{\overset{+}{H_3}N-CH-COO^-}}}}{}}}$$

Dopamine

Norepinephrine

Epinephrine

$$\underset{\textbf{γ-Aminobutyric acid (GABA)}}{^-OOC-CH_2-CH_2-CH_2-NH_3^+}$$

Histamine

Serotonin (5-Hydroxytryptamine)

FIGURE 20-39   **A selection of neurotransmitters.**

catecholamines because they are derivatives of catechol (1,2-dihydroxybenzene)] are sequentially synthesized from tyrosine, whereas **γ-aminobutyric acid (GABA), histamine,** and **serotonin** are derived from glutamate, histidine, and tryptophan, respectively (Fig. 20-39). Many of these compounds are hormonally active substances that are present in the bloodstream. However, since the brain is largely isolated from the general circulation by a selective filtration system known as the **blood–brain barrier** (Section 15-4B), the presence of these substances in the blood has no direct effect on the brain. The use of the same compounds as hormones and neurotransmitters apparently has no physiological significance but, rather, is thought to reflect evolutionary opportunism in adapting already available systems to new roles.

The use of selective staining techniques has established that each of the different neurotransmitters is used in discrete and often highly localized regions of the nervous system. The various neurotransmitters are, nevertheless, not simply functional equivalents of acetylcholine. Rather, many of them have distinctive physiological roles. For example, both GABA and glycine are inhibitory rather than excitatory neurotransmitters. The receptors for these substances are ligand-gated channels that are selectively permeable to Cl^-. Hence, their opening tends to hyperpolarize the membrane (make its membrane potential more negative) rather than depolarize it. A neuron inhibited in this manner must therefore be more intensely depolarized than otherwise to trigger an action potential (note that these neurons respond to more than one type of neurotransmitter). Thus, anion channels are inhibitory, whereas cation channels are excitatory. Ethanol, the oldest and most widely used psychoactive drug, is thought to act by inducing GABA receptors in the brain to open their Cl^- channels.

The subunits of the various neurotransmitter-gated cation channels have 20 to 40% sequence identity, as do those of the anion channels. However, the two families of channel proteins appear to be unrelated. Despite this lack of homology, the sequences of the two types of channels suggest that they have considerable structural similarity.

The actual nature of a neuron's response to a neurotransmitter depends more on the characteristics of the corresponding receptor than on the neurotransmitter's identity. Thus, as we have seen, nicotinic ACh receptors, which trigger the rapid contraction of skeletal muscles, respond to ACh within a few milliseconds by depolarizing the postsynaptic membrane. In contrast, the binding of ACh to muscarinic ACh receptors in heart muscle inhibits muscle contraction over a period of several seconds (several heartbeats). This is accomplished by hyperpolarizing the postsynaptic membrane through the closure of otherwise open K^+ channels. Slow-acting neurotransmitters may act by inducing the formation of a second messenger such as cAMP. In fact, the brain has the highest concentration of cAMP-dependent kinases in the body. The binding of catecholamines to their respective neuronal receptors, through the intermediacy of adenylate cyclase and cAMP, activates protein kinases to phosphorylate ion channels so as to alter the neuron's electrical properties. The ultimate effect of this process can be either excitatory or inhibitory. Thus catecholamines, whether acting as hormones (Section 19-1F) or as neurotransmitters, have similar mechanisms of receptor activation.

g. Neuropeptides Are Neurotransmitters

A large and growing list of hormonally active polypeptides known as **neuropeptides** also act as neurotransmitters. Not surprisingly, perhaps, the opioid peptides β-endorphin, met-enkephalin, and leu-enkephalin (Section 19-1K), as well as the hypothalamic releasing factors TRF, GnRF, and somatostatin (Section 19-1H), are in this category. What is less expected is that several gastrointestinal polypeptides, including the hormones gastrin, secretin, and cholecystokinin (CCK; Section 19-1C), may also act as neurotransmitters in discrete regions of the brain, as do the pituitary hormones oxytocin and vasopressin (Section 19-1H). Such neuropeptides differ from the simpler neu-

rotransmitters in that they seem to elicit complex behavior patterns. For example, intracranially injecting rats with a nanogram of vasopressin greatly enhances their ability to learn and remember new tasks. Similarly, injecting a male or a female rat with GnRF evokes the respective postures they require for copulation. Just how these neuropeptides operate is but one of the many enigmas of brain function and organization.

CHAPTER SUMMARY

1 ■ Thermodynamics of Transport Polar molecules and ions are transported across biological membranes by specific transmembrane transport proteins. The free energy change of the species transported depends on the ratio of its concentrations on the two sides of the membrane and, if the species is charged, on the membrane potential, $\Delta\Psi$.

2 ■ Kinetics and Mechanisms of Transport The rate of nonmediated diffusion across a membrane is a linear function of the difference in concentration of the species on the two sides of the membrane as governed by Fick's first law of diffusion. Mediated transport is characterized by rapid saturation kinetics and specificity for the substance transported. It is also subject to competitive inhibition and chemical inactivation. Ionophores transport ions through membranes. Carrier ionophores, such as valinomycin, do so by wrapping a specific ion in a hydrophobic, membrane-soluble coat that can freely diffuse through the membrane. Channel-forming ionophores, such as gramicidin A, form a transmembrane pore through which selected ions can rapidly diffuse. Maltoporin is specific for the passage of maltodextrins because its transport channel matches their left-handed helical shape and is lined with aromatic side chains that form a so-called greasy slide. Glucose transport across erythrocyte membranes is mediated by dimeric transmembrane glycoproteins that can assume two conformations: one with a glucose binding site facing the external cell surface and the other with the glucose site facing the cytosol. Transport occurs by the binding of glucose to the protein on one face of the membrane, followed by a conformational change that closes this site and exposes the other (a gated pore). The KcsA K^+ channel, a transmembrane homotetramer, permits the rapid passage of K^+ ions, for which it is highly selective. It does so, in part, because it forms an aqueous cavity surrounded by the negative ends of helix dipoles that stabilizes K^+ ions in the middle of the bilayer. K^+ ions, but not the smaller Na^+ ions, are transported because the K^+ channel's selectivity filter selectively coordinates K^+ ions by rings of O atoms in a way that allows their dehydration, passage, and subsequent hydration without significant activation barriers.

3 ■ ATP-Driven Active Transport Active transport of molecules or ions against a concentration gradient requires an input of free energy. The free energy of ATP hydrolysis is coupled to the transport of three Na^+ ions out of and two K^+ ions into the cell by the (Na^+-K^-)–ATPase. This electrogenic process involves phosphorylation of an Asp residue (by ATP) in the presence of Na^+ and its dephosphorylation (hydrolysis) in the presence of K^+. Phosphorylation and dephosphorylation are accompanied by conformational changes that ensure rapid interconversion of all intermediates along the transport

pathway. ATP-driven transport of Ca^{2+} by the Ca^{2+}–calmodulin-activated Ca^{2+}–ATPase and of H^+ by (H^+-K^+) – ATPase occur by similar phosphorylation/dephosphorylation mechanisms. The X-ray structure of the Ca^{2+}–ATPase reveals it has a 10-helix transmembrane domain that binds two Ca^{2+} ions near its center and three well-separated cytoplasmic domains that appear to undergo large rigid motions on pumping Ca^{2+} ions. Bacteria transport sugars by group translocation, a process in which the transported substance is chemically modified. The PTS system, which has important regulatory functions, phosphorylates sugars as they are transported by utilizing phosphoenolpyruvate as a phosphoryl donor.

4 ■ Ion Gradient–Driven Active Transport Active transport may be driven by the free energy stored in ion gradients (secondary active transport). Glucose is transported into intestinal epithelial cells against its concentration gradient by an Na^+–glucose symport. This process is ultimately powered by the free energy of ATP hydrolysis since the Na^+ gradient is constantly being replenished via the (Na^+-K^+)–ATPase. The system conforms to a Random Bi Bi kinetic mechanism, implying that both Na^+ and glucose must be bound for the transport-producing conformational change to occur. Lactose is transported into *E. coli* by lactose permease, an H^+ – lactose symport. This process is driven by the cell's electrochemical H^+ gradient, which is, in turn, maintained by a proton pump coupled with oxidative metabolism. The mitochondrial ATP–ADP antiport system also interacts with the membrane potential in the asymmetric transport of ATP out of and ADP into the mitochondrion.

5 ■ Neurotransmission Voltage-gated cation channels such as K_V channels open in response to the membrane potential and close a short time later through the action of a second gate that functions via a modified "ball-and-chain" mechanism. Nerve impulses are traveling waves of electrical excitation along axon plasma membranes known as action potentials that are generated by the transient opening of voltage-gated Na^+ channels to let Na^+ ions into the cell followed a short time later by the transient opening of voltage-gated K^+ channels to let K^+ ions out of the cell. Nerve impulses are chemically transmitted across most synapses by the release of neurotransmitters. Acetylcholine (ACh), the best characterized neurotransmitter, is packaged in synaptic vesicles that are exocytotically released into the synaptic cleft. This process is triggered by an increase in cytosolic $[Ca^{2+}]$ resulting from the arriving action potential's opening of voltage-gated Ca^{2+} channels. The ACh diffuses across the synaptic cleft, where it binds to the ACh receptor, a transmembrane cation channel that opens in response to ACh binding. The resultant flow of Na^+

into and K^+ out of the postsynaptic cell depolarizes the post-synaptic membrane, which, if sufficient neurotransmitter has been released, triggers a postsynaptic action potential. The ACh receptor is the target of numerous deadly neurotoxins, including histrionicatoxin, *d*-tubocurarine, and cobra toxin, which all bind to the ACh receptor so as to prevent its opening. The ACh is rapidly degraded, before the possible arrival of the next nerve impulse, through the action of acetylcholinesterase, a fast-acting serine esterase that has an unusual aromatic side chain–lined active site gorge. Nerve gases and succinylcholine inhibit acetylcholinesterase and therefore block nerve impulse transmission at cholinergic synapses.

Many specific regions of the nervous system employ neurotransmitters other than ACh. Most of these neurotransmit-

ters are amino acids, such as glycine and glutamate, or their decarboxylation products and their derivatives, including catecholamines, GABA, histamine, and serotonin. Many of these compounds are also hormonally active, but they are excluded from the brain by the blood–brain barrier. Although many neurotransmitters, such as ACh, are excitatory, others are inhibitory. The latter stimulate the opening of anion (Cl^-) channels, thereby causing the postsynaptic membrane to become hyperpolarized, so that it must be more highly depolarized than otherwise to trigger an outgoing action potential. There is also a growing list of polypeptide neurotransmitters, many of which are also polypeptide hormones, that elicit complex behavior patterns.

REFERENCES

GENERAL

Franklin, H.M., *The Vital Force: A Study of Bioenergetics*, Chapters 9 and 10, Freeman (1986).

Saier, M.H., Jr. and Reizer, J., Families and superfamilies of transport proteins common to prokaryotes and eukaryotes, *Curr. Opin. Struct. Biol.* **1**, 362–368 (1991).

Stein, W.D., *Transport and Diffusion across Cell Membranes*, Academic Press (1986).

KINETICS AND MECHANISM OF TRANSPORT

Clapham, D.E., Unlocking family secrets: K^+ channel transmembrane domains, *Cell* **97**, 547–550 (1999).

Dobler, M., *Ionophores and Their Structures*, Wiley–Interscience (1981).

Dutzler, R., Wang, Y.-F., Rizkallah, P.J., Rosenbusch, J.P., and Schirmer, T., Crystal structures of various maltooligosaccharides reveal a specific sugar translocation pathway, *Structure* **4**, 127–134 (1996); *and* Dutzler, R., Schirmer, T., Karplus, M., and Fischer, S., Translation mechanism of long sugar chains across the maltoporin membrane channel, *Structure* **10**, 1273–1284 (2002).

Kovacs, F., Quine, J., and Cross, T.A., Validation of the single-stranded channel conformation of gramicidin A by solid-state NMR, *Proc. Natl. Acad. Sci.* **96**, 7910–7915 (1999).

Rees, D.C., Chang, G., and Spencer, R.H., Crystallographic analysis of ion channels: Lessons and challenges, *J. Biol. Chem.* **275**, 713–716 (2000).

Zhou, Y., Morais-Cabral, J.H., Kaufman, A., and MacKinnon, R., Chemistry of ion coordination and hydration revealed by a K^+ channel–Fab complex at 2.0 Å resolution, *Nature* **414**, 43–48 (2001); *and* Doyle, D.A., Cabral, J.M., Pfuetzner, R.A., Kuo, A., Gulbis, J.M., Cohen, S.L., Chait, B.T., and MacKinnon, R., The structure of the potassium channel: Molecular basis of K^+ conduction and selectivity, *Science* **280**, 69–77 (1998). [High and medium resolution X-ray structures of the KcsA channel.]

GLUCOSE TRANSPORT

Barnett, J.E.G., Holman, G.D., Chalkley, R.A., and Munday, K.A., Evidence for two asymmetric conformational states in the human erythrocyte sugar transport system, *Biochem. J.* **145**, 417–429 (1975).

Czech, M.P., Clancy, B.M., Pessino, A., Woon, C.-W., and Harrison, S.A., Complex regulation of simple sugar transport in insulin-responsive cells, *Trends Biochem. Sci.* **17**, 197–200 (1992).

Elsas, L.J. and Longo, N., Glucose transporters, *Annu. Rev. Med.* **43**, 377–393 (1992).

Piper, R.C., Tai, C., Kulesza, P., Pang, S., Warnock, D., Baenziger, J., Slot, J.W., Geuze, H.J., Puri, C., and James, D.E., GLUT-4 NH_2 Terminus contains a phenylalanine-based targeting motif that regulates intracellular sequestration, *J. Cell Biol.* **121**, 1221–1232 (1993).

Silverman, M., Structure and function of hexose transporters, *Annu. Rev. Biochem.* **60**, 757–794 (1991). [Discusses both the passive facilitative glucose transporters and the Na^+–glucose symport.]

Walmsley, A.R., Barrett, M.P., Bringaud, F., and Gould, G.W., Sugar transporters from bacteria, parasites, and mammals: Structure–activity relationships, *Trends Biochem. Sci.* **22**, 476–481 (1998).

(Na^+-K^+)–ATPase

Blaustein, M.P., Physiological effects of endogenous ouabain: Control of intracellular Ca^{2+} stores and cell responsiveness, *Am. J. Physiol.* **264**, C1367–C1378 (1993).

Cantley, L.C., Carilli, C.T., Smith, R.L., and Perlman, D., Conformational changes of Na,K-ATPase necessary for transport, *Curr. Top. Membr. Transp.* **19**, 315–322 (1983).

Gadsby, D.C., The Na/K pump of cardiac cells, *Annu. Rev. Biophys. Bioeng.* **13**, 373–398 (1984).

Mercer, R.W., Schneider, J.W., and Benz, E.J., Jr., Na,K–ATPase structure, *in* Agre, P. and Parker, J.C. (Eds.), *Red Blood Cell Membranes*, pp. 135–165, Marcel Dekker (1989).

Møller, J.V., Juul, B., and le Maire, M., Structural organization, ion transport, and energy transduction of P-type ATPases, *Biochim. Biophys. Acta* **1236**, 1–51 (1996).

Pedersen, P.L. and Carafoli, E., Ion motive ATPases. I. Ubiquity, properties and significance to cell function; *and* II. Energy coupling and work output, *Trends Biochem. Sci.* **12**, 146–150, 186–189 (1987).

Ca^{2+}–ATPase

Carafoli, E., The Ca^{2+} pump of the plasma membrane, *J. Biol. Chem.* **267**, 2115–2118 (1992).

Enyedi, A., Vorherr, T., James, P., McCormick, D.J., Filoteo, A.G., Carafoli, E., and Penniston, J.T., The calmodulin binding domain of the plasma membrane Ca^{2+} pump interacts both with calmodulin and with another part of the pump, *J. Biol. Chem.* **264**, 12313–12321 (1989).

Jencks, W.P., Coupling of hydrolysis of ATP and the transport of Ca^{2+} by the calcium ATPase of sarcoplasmic reticulum,

Biochem. Soc. Trans. **20,** 555–559 (1992). [An excellent discussion of the mechanism of coupling chemical energy to the vectorial transport of ions against a concentration gradient.]

MacLennan, D.H., Rice, W.J., and Green, N.M., The mechanism of Ca^{2+} transport by sarco(endo)plasmic reticulum Ca^{2+}-ATPases, *J. Biol. Chem.* **272,** 28815–28818 (1997).

Toyoshima, C., Nakasako, M., and Ogawa, H., Crystal structure of the calcium pump of sacroplasmic reticulum at 2.6 Å resolution, *Nature* **405,** 647–655 (2000).

Wuytack, F. and Raeymaekers, L, The Ca^{2+}-transport ATPases from the plasma membrane, *J. Bioenerg. Biomembr.* **24,** 285–300 (1992).

Zhang, P., Toyoshima, C., Yonekura, K., Green, N.M., and Stokes, D.L., Structure of the calcium pump from sacroplasmic reticulum at 8-Å resolution, *Nature* **392,** 835–840 (1998).

(H^+–K^+)–ATPase

Besanèon, M., Shin, J.M., Mercier, F., Munson, K., Rabon, E., Hersey, S., and Sachs, G., Chemomechanical coupling in the gastric H,K ATPase, *Acta Physiol. Scand.* **146,** 77–88 (1992).

Sachs, G., Besanèon, M., Shin, J.M., Mercier, F., Munson, K., and Hersey, S., Structural aspects of the gastric H,K ATPase, *J. Bioenerg. Biomembr.* **24,** 301–308 (1992).

PEP-DEPENDENT PHOSPHOTRANSFERASE SYSTEM

Erni, B., Group translocation of glucose and other carbohydrates by the bacterial phosphotransferase system, *Int. Rev. Cytology* **137A,** 127–148 (1992).

Herzberg, O. and Klevit, R., Unraveling a bacterial hexose transport pathway, *Curr. Opin. Struct. Biol.* **4,** 814–822 (1994).

Hurley, J.H., Faber, H.R., Worthylake, D., Meadow, N.D., Roseman, S., Pettigrew, D.W., and Remington, S.J., Structure of the regulatory complex of *Escherichia coli* E III^{glc} with glycerol kinase, *Science* **259,** 673–677 (1993). [EIIA was previously named EIII.]

Meadow, N.D., Fox, D.K., and Roseman, S., The bacterial phosphoenolpyruvate: Glucose phosphotransferase system, *Annu. Rev. Biochem.* **59,** 497–542 (1990); *and* Feese, M., Pettigrew, D.W., Meadow, N.D., Roseman, S., and Remington, S.J., Cation promoted association (CPA) of a regulatory and target protein is controlled by protein phosphorylation, *Proc. Natl. Acad Sci.* **91,** 3544–3548 (1994).

Saier, M.H., Jr., Chauvaux, S., Deutscher, J., Reizer, J., and Ye, J.-J., Protein phosphorylation and regulation of carbon metabolism in gram-negative versus gram-positive bacteria, *Trends Biochem. Sci.* **20,** 267–271 (1995).

Na^+–GLUCOSE SYMPORT

Wright, E.M., The intestinal Na^+/glucose cotransporter, *Annu. Rev. Physiol.* **55,** 575–589 (1993).

LACTOSE PERMEASE

Barrett, M.P., Walmsley, A.R., and Gould, G.W., Structure and function of facultative sugar transporters, *Curr. Opin. Cell Biol.* **11,** 496–502 (1999).

Kaback, H.R. and Wu, J., What to do while awaiting crystals of a membrane protein and thereafter, *Acc. Chem. Res.* **32,** 805–813 (1999); *and* From membrane to molecule to the third amino acid from the left with a membrane transport protein, *Q. Rev. Biophys.* **30,** 333–364 (1997).

ATP–ADP TRANSLOCATOR

Vignais, P.V., Block, M.R., Boulay, F., Brandolin, G., Dalbon, P., and Lauquin, G.J.M., Molecular aspects of structure-function relationships in mitochondrial adenine nucleotide carrier, *in* Bengha, G. (Ed.), *Structure and Properties of Cell Membranes,* Vol. II, *pp.* 139–179, CRC Press (1985).

NEUROTRANSMISSION

Alberts, B., Johnson, A., Lewis, J., Raff, M., Roberts, K., and Walter, P., *The Molecular Biology of the Cell* (4th ed.), Chapter 11, Garland Science (2002).

Armstrong, C.M. and Hille, B., Voltage-gated ion channels and electrical excitability, *Neuron* **20,** 371–380 (1998).

Catterall, W.A., Structure and regulation of voltage-gated Ca^{2+} channels, *Annu. Rev. Cell Dev. Biol.* **16,** 521–555 (2000).

Choe, S., Kreusch, A., and Pfaffinger, P.J., Towards the three-dimensional structure of voltage-gated potassium channels, *Trends Biochem. Sci.* **24,** 345–349 (1999).

Geppert, M. and Südhof, T.C., Rab3 and synaptotagmin. The yin and yang of synaptic transmission, *Annu. Rev. Neurosci.* **21,** 75–95 (1998).

Hille, B., *Ionic Channels of Excitable Membranes* (3rd ed.), Sinauer Associates (2001).

Horn, R., Conversation between voltage sensors and gates of ion channels, *Biochemistry* **39,** 15653–15658 (2000).

Koberz, W.R., Williams, C., and Miller, C., Hanging gondola structure of the T1 domain in a voltage-gated K^+ channel, *Biochemistry* **39,** 10347–10352 (2000); *and* Koberz, W.R. and Miller, C., K^+ channels lacking the 'tetramerization' domain: Implications for core structure, *Nature Struct. Biol.* **6,** 1122–1125 (1999).

Lin, R.C. and Scheller, R.H., Mechanisms of synaptic vesicle exocytosis, *Annu. Rev. Cell Dev. Biol.* **16,** 19–49 (2000).

Lodish, H., Berk, A., Zipursky, S.L., Matsudaira, P., Baltimore, D., and Darnell, J., *Molecular Cell Biology* (4th ed.), Chapter 21, Freeman (2000).

Lynch, D.R. and Snyder, S.H., Neuropeptides: Multiple molecular forms, metabolic pathways and receptors, *Annu. Rev. Biochem.* **55,** 773–799 (1986).

Minor, D.L., Potassium channels: Life in the post-structural world, *Curr. Opin. Struct. Biol.* **11,** 408–414 (2001).

Miyazawa, A., Fujiyoshi, Y., Stowell, M., and Unwin, N., Nicotinic acetylcholine receptor at 4.6 Å resolution: Transverse tunnels in the channel wall, *J. Mol. Biol.* **288,** 765–786 (1999); Unwin, N., Acetylcholine receptor channel imaged in the open state, *Nature* **373,** 37–43 (1995); *and* Unwin, N., Nicotinic acetylcholine receptor at 9 Å resolution, *J. Mol. Biol.* **229,** 1101–1124 (1993).

Orlova, E.V., Rahman, M.A., Gowen, B., Volynski, K.E., Ashton, A.C., Manser, C., van Heel, M., and Ushkaryov, Y.A., Structure of α-latrotoxin oligomers reveals that divalent cation-dependent tetramers form membrane pores, *Nature Struct. Biol.* **7,** 48–53 (2000).

Sussman, J.L., Harel, M., Frolow, F., Oefner, C., Goldman, A., Toker, L., and Silman, I., Atomic structure of acetylcholinesterase from *Torpedo californica:* A prototypic acetylcholine-binding protein, *Science* **253,** 872–879 (1991); *and* Sussman, J.L. and Silman, I., Acetylcholinesterase: Structure and use as model for specific cation–protein interactions, *Curr. Opin. Struct. Biol.* **2,** 721–729 (1992).

Yellin, G., The voltage-gated potassium channels and their relatives, *Nature* **419,** 35–42 (2002); *and* The moving parts of voltage-gated channels, *Q. Rev. Biophys.* **31,** 239–295 (1998).

Zhou, M., Morais-Cabral, J.H., Mann, S., and MacKinnon, R., Potassium channel receptor site for the inactivation gate and quaternary amine inhibitors, *Nature* **411,** 657–661 (2001).

PROBLEMS

1. If the glucose concentration outside a cell is 10 mM but that inside a cell is 0.1 mM, what is glucose's chemical potential difference across the membrane at 37°C?

*2. If a solution of an ionic macromolecule is equilibrated with a salt solution from which it is separated by a membrane through which the salt ions but not the macromolecule can pass, a membrane potential is generated across the membrane. This so-called **Donnan equilibrium** arises because the impermeability of the membrane to some ions but not others prevents the equalization of the ionic concentrations on the two sides of the membrane. To demonstrate this effect, assume that the Cl$^-$ salt of a monocationic protein, P$^+$, is dissolved in water to the extent that [Cl$^-$] = 0.1M and is separated by a membrane impermeable to the protein but not NaCl from an equal volume of 0.1M NaCl solution. Assuming no volume change in either compartment, what are the concentrations of the various ionic species on either side of the membrane after the system has equilibrated? What is the membrane potential across the membrane? (*Hint:* Mass is conserved and the solution on each side of the membrane must be electrically neutral. At equilibrium, $\Delta G_{Na^+} + \Delta G_{Cl^-} = 0$.)

3. How long would it take one molecule of gramicidin A to transport enough Na$^+$ to change the concentration inside an erythrocyte of volume 80 μm^3 by 10 mM? Assume the erythrocyte's Na$^+$ pumps are inoperative.

4. The (Na$^+$–K$^+$)–ATPase is inhibited by nanomolar concentrations of vanadate, which forms a pentavalent ion, VO$_5^{5-}$, with trigonal bipyramidal symmetry. Explain the mechanism of this inhibition. (*Hint:* See Section 16-2B.)

5. The (H$^+$–K$^+$)–ATPase secretes H$^+$ at a concentration of 0.18M from cells that have an internal pH of 7. What is the ΔG required for the transport of 1 mol of H$^+$ under these conditions? Assuming that the ΔG for ATP hydrolysis is −31.5 kJ · mol^{-1} under these conditions, and that the membrane potential is 0.06 V, inside negative, how much ATP must be hydrolyzed per mole of H$^+$ transported in order to make this transport exergonic?

6. A 100-Å-thick membrane has a membrane potential of 100 mV. What is the magnitude of this potential difference in V · cm^{-1}? Comment on the magnitude of this potential field in macroscopic terms.

7. The resting membrane potential ($\Delta\Psi$) of a neuron at 37°C is −60 mV (inside negative). If the free energy change associated with the transport of one Na$^+$ ion from outside to inside is −11.9 kJ · mol^{-1}, and [Na$^+$] outside the cell is 260 mM, what is [Na$^+$] inside the cell?

8. You have isolated a new strain of bacteria and would like to know whether leucine and ethylene glycol enter the cells by mediated diffusion or only by a nonmediated route. To do this you measure the initial rates of uptake of these molecules as a function of external concentration and obtain the data in the following table.

Compound	Concentration (M)	Initial Uptake Rate (arbitrary units)
Leucine	1×10^{-6}	110
	2×10^{-6}	220
	5×10^{-6}	480
	1×10^{-5}	830
	3×10^{-5}	1700
	1×10^{-4}	2600
	5×10^{-4}	3100
	1×10^{-3}	3200
Ethylene glycol	1×10^{-3}	1
	5×10^{-3}	5
	0.01	10
	0.05	50
	0.1	100
	0.5	500
	1.0	1000

Which compound(s) enters by a mediated route? What criteria did you use for this decision?

9. Draw the structures of the following compounds and predict whether they can cross a membrane without mediation or will require facilitation. Indicate the criteria you used to make these predictions. (a) Ethanol, (b) glycine, (c) cholesterol, and (d) ATP.

10. Write a kinetic scheme for the (H$^+$–K$^+$)–ATPase that provides for coupled ATP hydrolysis with H$^+$ transport. Discuss the order of substrate addition required for coupling. Identify the steps in which mutual destabilization results in reasonable rates of transport.

11. What function might the synthesis of digitalis serve in the purple foxglove plant?

12. The high water permeabilities of many cells (e.g., red blood cells and those forming kidney tubules) are, in part, mediated by transmembrane channels known as **aquaporins.** The structure of an aquaporin reveals that its 8 helices form an hourglass-shaped pore that, in its narrowest region, is 3 Å in diameter. This aquaporin is impermeable to glycerol [CHOH(CH$_2$OH)$_2$]. However, a homologous and structurally similar **glycerol channel,** which is minimally 3.4 Å wide, permits the passage of glycerol but is only poorly permeable to water. Discuss the possible differences between these channels that would account for their different permeabilities.

13. What is the resting membrane potential across an axonic membrane at 25°C (a) in the presence of tetrodotoxin or (b) with a high concentration of Cs$^+$ inside the axon (use the data in Table 20-3)? How do these substances affect the axon's action potential?

14. Why don't nerve impulses propagate in the reverse direction?

15. Decamethonium ion is a synthetic muscle relaxant.

$$(H_3C)_3\overset{+}{N}-(CH_2)_{10}-\overset{+}{N}(CH_3)_3$$
Decamethonium

What is its mechanism of action?

Chapter 21

Citric Acid Cycle

In this chapter we continue our metabolic explorations by examining the **citric acid cycle,** the common mode of oxidative degradation in eukaryotes and prokaryotes. This cycle, which is alternatively known as the **tricarboxylic acid (TCA) cycle** and the **Krebs cycle,** marks the "hub" of the metabolic system: *It accounts for the major portion of carbohydrate, fatty acid, and amino acid oxidation and generates numerous biosynthetic precursors.* The citric acid cycle is therefore **amphibolic,** that is, it operates both catabolically and anabolically.

We begin our study of the citric acid cycle with an overview of its component reactions and a historical synopsis of its elucidation. Next, we explore the origin of the cycle's starting compound, **acetyl-coenzyme A (acetyl-CoA),** the common intermediate formed by the breakdown of most metabolic fuels. Then, after discussing the reaction mechanisms of the enzymes that catalyze the cycle, we consider the various means by which it is regulated. Finally, we deal with the citric acid cycle's amphibolic nature by examining its interrelationships with other metabolic pathways.

1 ■ CYCLE OVERVIEW

🕹 **See Guided Exploration 18. Citric Acid Cycle Overview** *The citric acid cycle (Fig. 21-1) is an ingenious series of reactions that oxidizes the acetyl group of acetyl-CoA to two molecules of CO_2 in a manner that conserves the liberated free energy for utilization in ATP generation.* Before we study these reactions in detail, let us consider the cycle's chemical strategy by "walking" through the cycle and noting the fate of the acetyl group at each step. Following this preview, we shall consider some of the major discoveries that led to our present understanding of the citric acid cycle.

A. *Reactions of the Cycle*

The eight enzymes of the citric acid cycle (Fig. 21-1) catalyze a series of well-known organic reactions that cumulatively oxidize an acetyl group to two CO_2 molecules with the concomitant generation of three NADHs, one FADH$_2$, and one GTP:

1. Citrate synthase catalyzes the condensation of acetyl-CoA and **oxaloacetate** to yield **citrate,** giving the cycle its name.

2. The strategy of the cycle's next two steps is to rearrange citrate to a more easily oxidized isomer and then oxidize it. **Aconitase** isomerizes citrate, a not readily oxidized tertiary alcohol, to the easily oxidized secondary alcohol **isocitrate.** The reaction sequence involves a dehydration, producing enzyme-bound *cis*-**aconitate,** followed by a hydration, so that citrate's hydroxyl group is, in effect, transferred to an adjacent carbon atom.

3. Isocitrate dehydrogenase oxidizes isocitrate to the β-keto acid intermediate **oxalosuccinate** with the coupled reduction of NAD$^+$ to NADH; oxalosuccinate is then decarboxylated, yielding **α-ketoglutarate.** This is the first step in which oxidation is coupled to NADH production and also the first CO_2-generating step.

4. The multienzyme complex **α-ketoglutarate dehydrogenase** oxidatively decarboxylates α-ketoglutarate to **succinyl-coenzyme A.** The reaction involves the reduction of a second NAD$^+$ to NADH and the generation of a second molecule of CO_2. At this point in the cycle, two molecules

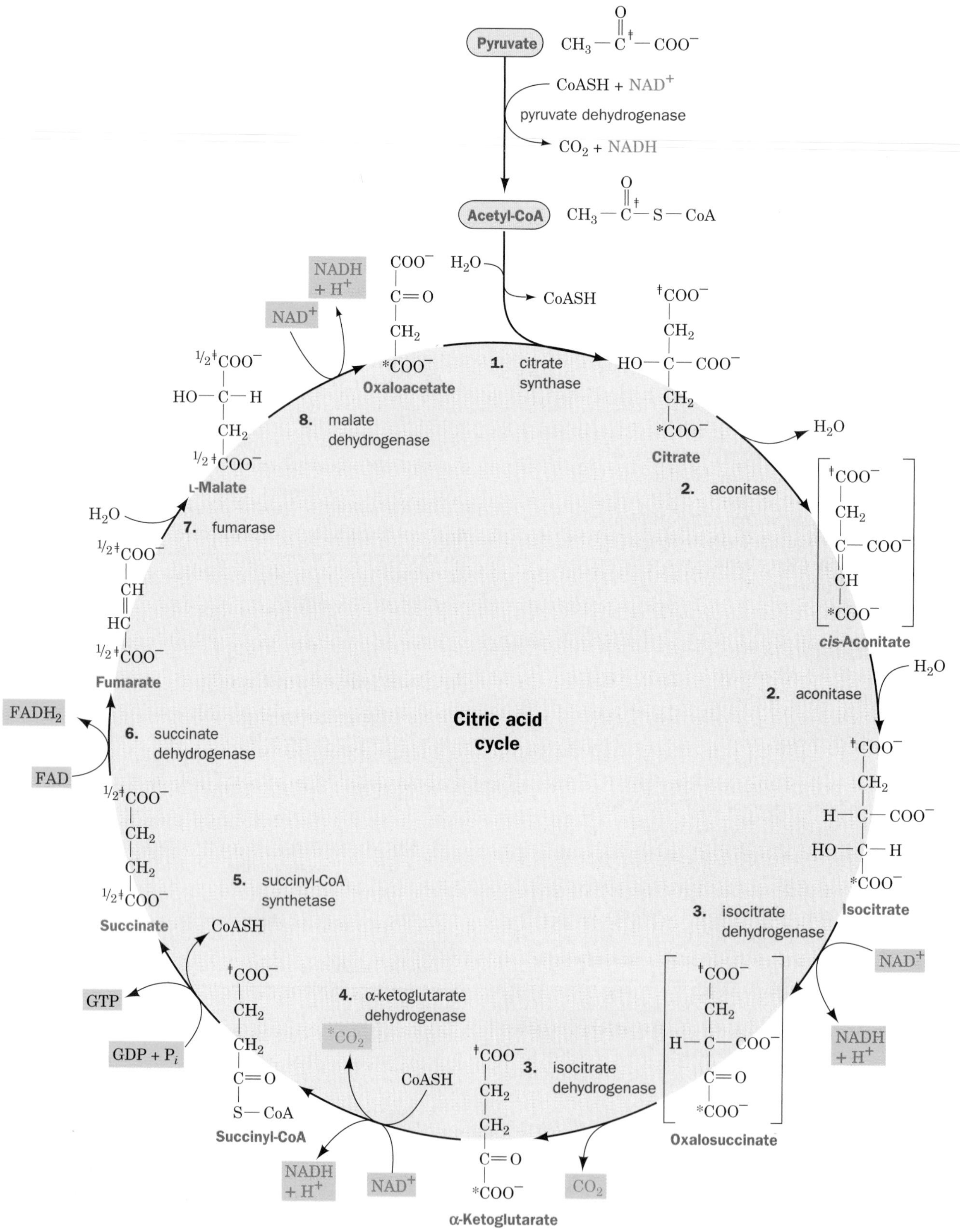

FIGURE 21-1   Reactions of the citric acid cycle. The reactants and products of this catalytic cycle are boxed. The pyruvate → acetyl-CoA reaction (*top*) supplies the cycle's substrate via carbohydrate metabolism but is not considered to be part of the cycle. The bracketed compounds are enzyme-bound intermediates. An isotopic label at C4 of oxaloacetate (*) becomes C1 of α-ketoglutarate and is released as CO_2 in Reaction 4. An isotopic label at C1 of acetyl-CoA (‡) becomes C5 of α-ketoglutarate and is scrambled in Reaction 5 between C1 and C4 of succinate (1/2‡). 🎧 **See the Animated Figures**

of CO_2 have been produced, so that the net oxidation of the acetyl group is complete. Note, however, that it is not the carbon atoms of the entering acetyl-CoA that have been oxidized.

5. Succinyl-CoA synthetase converts succinyl-coenzyme A to **succinate.** The free energy of the thioester bond is conserved in this reaction by the formation of "high-energy" GTP from GDP + P_i.

6. The remaining reactions of the cycle serve to oxidize succinate back to oxaloacetate in preparation for another round of the cycle. **Succinate dehydrogenase** catalyzes the oxidation of succinate's central single bond to a trans double bond, yielding **fumarate** with the concomitant reduction of the redox coenzyme FAD to $FADH_2$ (the molecular formulas of FAD and $FADH_2$ and the reactions through which they are interconverted are given in Fig. 16-8).

7. Fumarase then catalyzes the hydration of fumarate's double bond to yield **malate.**

8. Finally, **malate dehydrogenase** reforms oxaloacetate by oxidizing malate's secondary alcohol group to the corresponding ketone with concomitant reduction of a third NAD^+ to NADH.

Acetyl groups are thereby completely oxidized to CO_2 with the following stoichiometry:

$$3NAD^+ + FAD + GDP + P_i + acetyl\text{-}CoA \rightarrow$$
$$3NADH + FADH_2 + GTP + CoA + 2CO_2$$

The citric acid cycle functions catalytically as a consequence of its regeneration of oxaloacetate: An endless number of acetyl groups can be oxidized through the agency of a single oxaloacetate molecule.

NADH and $FADH_2$ are vital products of the citric acid cycle. Their reoxidation by O_2 through the mediation of the electron-transport chain and oxidative phosphorylation (Chapter 22) completes the breakdown of metabolic fuel in a manner that drives the synthesis of ATP. Other functions of the cycle are discussed in Section 21-5.

B. *Historical Perspective*

The citric acid cycle was proposed in 1937 by Hans Krebs, a contribution that ranks as one of the most important achievements of metabolic chemistry. We therefore outline the intellectual history of this cycle's discovery.

By the early 1930s, significant progress had been made in elucidating the glycolytic pathway (Section 17-1A). Yet the mechanism of glucose oxidation and its relationship to cellular respiration (oxygen uptake) was still a mystery. Nevertheless, the involvement of several metabolites in cellular oxidative processes was recognized. It was well known, for example, that in addition to lactate and acetate, the dicarboxylates succinate, malate, and α-ketoglutarate, as well as the tricarboxylate citrate, are rapidly oxidized by muscle tissue during respiration. It had also been shown that **malonate** (Section 21-3F), a potent inhibitor of succinate oxidation to fumarate, also inhibits cellular respiration, thereby suggesting that succinate plays a central role

in oxidative metabolism rather than being just another metabolic fuel.

In 1935, Albert Szent-Györgyi demonstrated that cellular respiration is dramatically accelerated by catalytic amounts of succinate, fumarate, malate, and oxaloacetate; that is, *the addition of any of these substances to minced muscle tissue stimulates O_2 uptake and CO_2 production far in excess of that required to oxidize the added dicarboxylic acid.* Szent-Györgyi further showed that these compounds were interconverted according to the reaction sequence:

$$Succinate \rightarrow fumarate \rightarrow malate \rightarrow oxaloacetate$$

Shortly afterward, Carl Martius and Franz Knoop demonstrated that citrate is rearranged, via cis-aconitate, to isocitrate and then dehydrogenated to α-ketoglutarate. α-Ketoglutarate was already known to undergo oxidative decarboxylation to succinate and CO_2. This extended the proposed reaction sequence to

$$Citrate \rightarrow cis\text{-}aconitate \rightarrow isocitrate \rightarrow \alpha\text{-}ketoglutarate$$
$$\rightarrow succinate \rightarrow fumarate \rightarrow malate \rightarrow oxaloacetate$$

What was necessary to close the circle so as to make the system catalytic was to establish that oxaloacetate is converted to citrate. In 1936, Martius and Knoop demonstrated that citrate could be formed nonenzymatically from oxaloacetate and pyruvate by treatment with hydrogen peroxide under basic conditions. Krebs used this chemical model as the point of departure for the biochemical experiments that led to his proposal of the citric acid cycle.

Krebs' hypothesis was based on his investigations, starting in 1936, on respiration in minced pigeon breast muscle (which has a particularly high rate of respiration). The idea of a catalytic cycle was not new to him: In 1932, he and Kurt Henseleit had elucidated the outlines of the **urea cycle,** a process in which ammonia and CO_2 are converted to urea (Section 26-2). The most important observations Krebs made in support of the existence of the citric acid cycle were as follows:

1. Succinate is formed from fumarate, malate, or oxaloacetate in the presence of the metabolic inhibitor malonate. Since malonate inhibits the direct reduction of fumarate to succinate, the succinate must be formed by an oxidative cycle.

2. Pyruvate and oxaloacetate can form citrate enzymatically. Krebs therefore suggested that the metabolic cycle is closed with the reaction:

$$Pyruvate + oxaloacetate \rightarrow citrate + CO_2$$

3. The interconversion rates of the cycle's individual steps are sufficiently rapid to account for observed respiration rates, so it must be (at least) the major pathway for pyruvate oxidation in muscle.

Although Krebs had established the existence of the citric acid cycle, some major gaps still remained in its complete elucidation. The mechanism of citrate formation did not become clear until Nathan Kaplan and Fritz Lipmann discovered **coenzyme A** in 1945 (Section 21-2), and Severo Ochoa and Feodor Lynen established, in 1951, that acetyl-CoA is the intermediate that condenses with oxaloacetate to form citrate. Oxidative decarboxylation of α-ketoglutarate to succinate was also shown to involve coenzyme A with succinyl-CoA as an intermediate.

The elucidation of the citric acid cycle was a major achievement and, like all achievements of this magnitude, required the efforts of numerous investigators. Indeed, many biochemists are still working to understand the cycle on a molecular and enzymatic level. We shall study the eight enzymes that catalyze the cycle after first discussing the cycle's major fuel, acetyl-CoA, and its formation from pyruvate.

2 ■ METABOLIC SOURCES OF ACETYL-COENZYME A

*Acetyl groups enter the citric acid cycle as **acetyl-coenzyme A (acetyl-SCoA or acetyl-CoA;** Fig. 21-2), the common product of carbohydrate, fatty acid, and amino acid breakdown.* **Coenzyme A (CoASH or CoA)** consists of a β-mercaptoethylamine group bonded through an amide linkage to the vitamin **pantothenic acid,** which, in turn, is attached to a 3′-phosphoadenosine moiety via a pyrophosphate bridge. The acetyl group of acetyl-CoA is bonded as a thioester to the sulfhydryl portion of the β-mercaptoethylamine group. *CoA thereby functions as a carrier of acetyl and other acyl groups (the A of CoA stands for "Acetylation").*

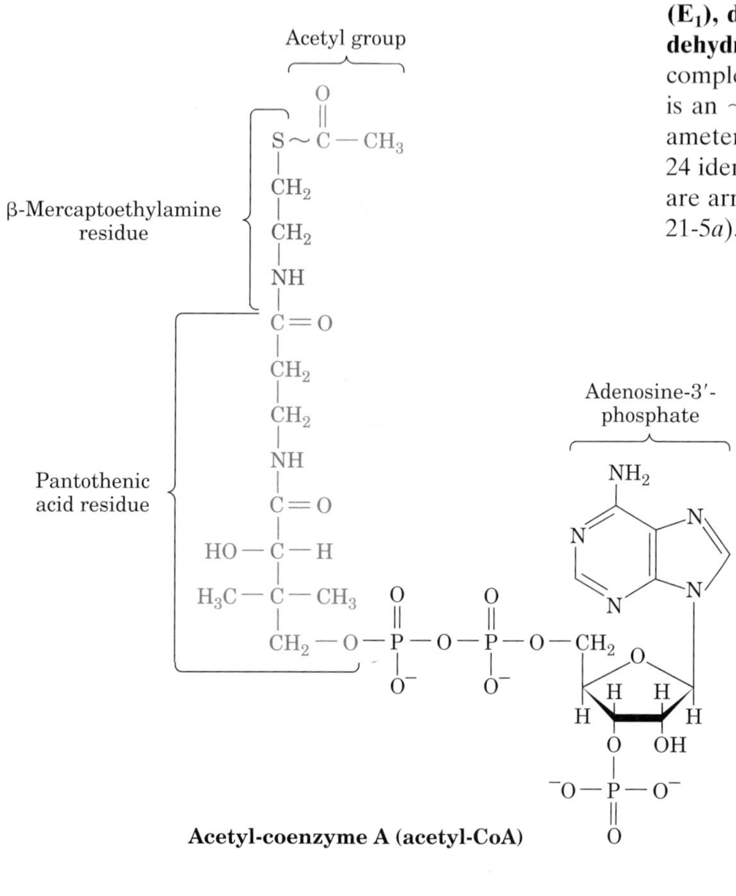

Acetyl-CoA is a "high-energy" compound: The $\Delta G°′$ for the hydrolysis of its thioester bond is -31.5 kJ · mol^{-1}, which makes this reaction slightly (1 kJ · mol^{-1}) more exergonic than that of ATP hydrolysis (Section 16-4B). The formation of this thioester bond in a metabolic intermediate therefore conserves a portion of the free energy of oxidation of a metabolic fuel.

A. *Pyruvate Dehydrogenase Multienzyme Complex (PDC)*

The immediate precursor to acetyl-CoA from carbohydrate sources is the glycolytic product pyruvate. As we saw in Section 17-3, under anaerobic conditions the NADH produced by glycolysis is reoxidized with concomitant reduction of pyruvate to lactate (in muscle) or ethanol (in yeast). Under aerobic conditions, however, NADH is reoxidized by the mitochondrial electron-transport chain (Section 22-2), so that pyruvate, which enters the mitochondrion via a specific pyruvate–H$^+$ symport (membrane transport nomenclature is discussed in Section 20-3), can undergo further oxidation. (The formation of acetyl-CoA from fatty acids and amino acids is discussed in Sections 25-2 and 26-3.)

Acetyl-CoA is formed from pyruvate through oxidative decarboxylation by a **multienzyme complex** named **pyruvate dehydrogenase.** In general, multienzyme complexes are groups of noncovalently associated enzymes that catalyze two or more sequential steps in a metabolic pathway. The **pyruvate dehydrogenase multienzyme complex (PDC)** consists of three enzymes: **pyruvate dehydrogenase (E$_1$), dihydrolipoyl transacetylase (E$_2$), and dihydrolipoyl dehydrogenase (E$_3$).** The *E. coli* pyruvate dehydrogenase complex, which was largely characterized by Lester Reed, is an ~4600-kD polyhedral particle that is ~300 Å in diameter (Fig. 21-3a). Isolated *E. coli* E$_2$ forms a particle with 24 identical subunits, which electron micrographs indicate are arranged with cubic symmetry (Figs. 21-3b, 21-4a, and 21-5a). The E$_1$ subunits form dimers that associate with the

FIGURE 21-2   Chemical structure of acetyl-CoA. The thioester bond is drawn with a $\sim$ to indicate that it is a "high-energy" bond. In CoA, the acetyl group is replaced by hydrogen.

FIGURE 21-3   Electron micrographs of the *E. coli* pyruvate dehydrogenase multienzyme complex. (*a*) The intact complex. (*b*) The dihydrolipoyl transacetylase (E$_2$) "core" complex. [Courtesy of Lester Reed, University of Texas.]

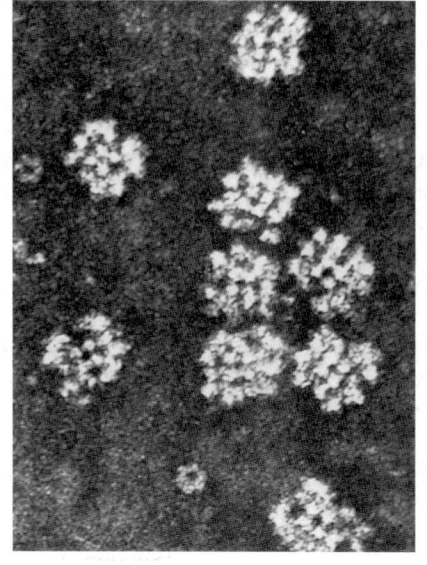
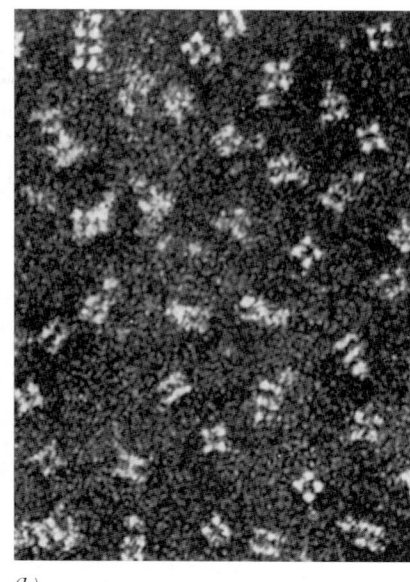

(*a*)   (*b*)

FIGURE 21-4   Structural organization of the *E. coli* PDC. (*a*) The dihydrolipoyl transacetylase (E$_2$) "core." Its 24 subunits (*green spheres*) associate as trimers located at the corners of a cube to form a particle that has cubic symmetry (*O* symmetry; Section 8-5B). (*b*) The 24 pyruvate dehydrogenase (E$_1$) subunits (*orange spheres*) form dimers that associate with the E$_2$ core (*shaded cube*) at the centers of each of its 12 edges, whereas the 12 dihydrolipoyl dehydrogenase (E$_3$) subunits (*purple spheres*) form dimers that attach to the E$_2$ cube at the centers of each of its 6 faces. (*c*) Parts *a* and *b* combined to form the entire 60-subunit complex.

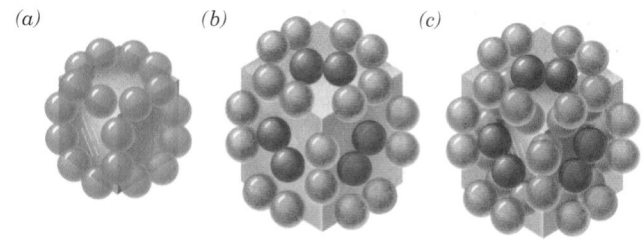

(*a*)   (*b*)   (*c*)

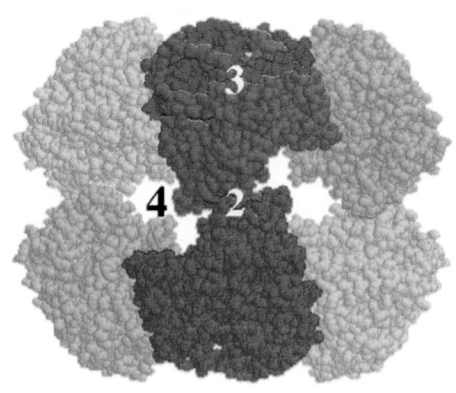

(*a*)

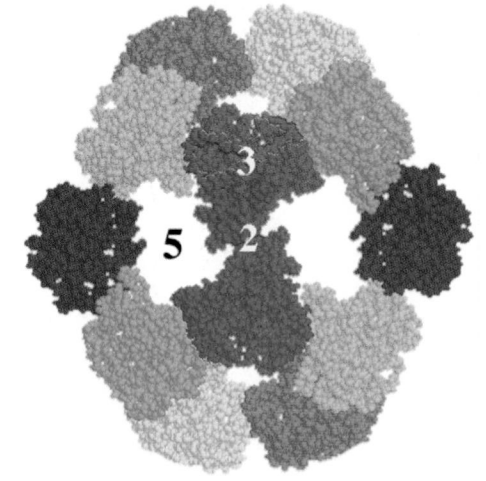

(*b*)

FIGURE 21-5   Comparison of the X-ray structures of the dihydrolipoyl transacetylase (E$_2$) cores of PDCs. The structures are shown in space-filling representation and viewed along their 2-fold axes of symmetry. The rear portion of each complex, which is almost entirely eclipsed by the forward portion, has been deleted for clarity. (*a*) The cubic *Azobacter vinelandii* E$_2$ core. It consists of 24 subunits that form 8 trimers, shown here in different colors. The positions of a 2-fold, a 3-fold, and a 4-fold axis of symmetry are indicated. Its height is ~125 Å. (*b*) The dodecahedral *B. stearothermophilus* E$_2$ core. It consists of 60 subunits that form 20 trimers, shown here in different colors. The positions of a 2-fold, a 3-fold, and a 5-fold axis of symmetry are indicated. Its outer diameter is ~237 Å. The subunits forming the trimer above the center of each drawing are individually colored. Note that these subunits are extensively associated, but that the interactions between contacting trimers in both types of complexes are relatively tenuous. Also note that these contacting trimers, respectively, form 4- and 5-membered rings that comprise the square and pentagonal faces of the cubic and dodecahedral complexes. [Based on X-ray structures by Wim Hol, University of Washington. PDBids 1EAB and 1B5S.]

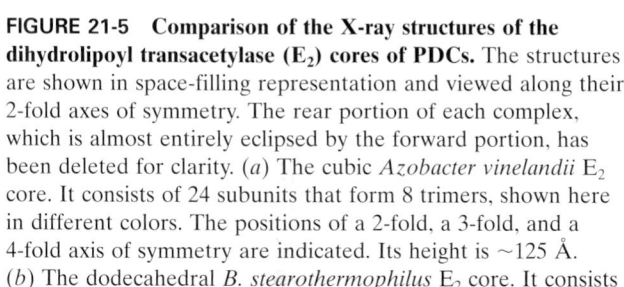

E_2 cube at the centers of the cube's 12 edges (Fig 21-4*b, c*), whereas the E_3 subunits form dimers that are located at the centers of this cube's 6 faces. We discuss the X-ray structures of the E_1, E_2, and E_3 subunits below.

a. Some PDCs Have a Dodecahedral Form

Although all PDCs catalyze the same reactions by similar mechanisms, they may have different quaternary structures. Whereas the PDCs of *E. coli* and most other gram-negative bacteria have the foregoing cubic symmetry, those of eukaryotes and some gram-positive bacteria have an analogous dodecahedral form [Fig. 21-5*b;* a dodecahedron is a regular polyhedron with *I* symmetry (Section 8-5B) that has 20 vertices, each lying on a 3-fold axis, and 12 pentagonal faces having an aggregate of 30 edges]. Thus, the mitochondrially located ~10,000-kD eukaryotic complex, the largest known multienzyme complex, consists of a dodecahedral core of 20 E_2 trimers (one centered on every vertex) surrounded by 30 E_1 $\alpha_2\beta_2$ heterotetramers (one centered on every edge) and 12 E_3 dimers (one centered in every face).

> The observation that the PDCs of both eukaryotes and gram-positive bacteria have the dodecahedral form suggests that mitochondria are descended from gram-positive bacteria. Yet mitochondria are enclosed by two membranes as are gram-negative bacteria, whereas gram-positive bacteria have only one membrane (Sections 1-2A and 11-3B). However, it has recently been demonstrated by genome sequence comparisons that mitochondria are, in fact, closely related to the obligate intracellular parasite *Rickettsia prowazekii*, the causative agent of the disease typhus. These bacteria are gram-negative but have the dodecahedral form of PDC.

b. Multienzyme Complexes Are Catalytically Efficient

Multienzyme complexes are a step forward in the evolution of catalytic efficiency. They offer the following mechanistic advantages:

1. Enzymatic reaction rates are limited by the frequency at which enzymes collide with their substrates (Section 14-2B). If a series of reactions occurs within a multienzyme complex, the distance that substrates must diffuse between active sites is minimized, thereby achieving a rate enhancement.

2. Complex formation provides the means for **channeling** (passing) metabolic intermediates between successive enzymes in a metabolic pathway, thereby minimizing side reactions.

3. The reactions catalyzed by a multienzyme complex may be coordinately controlled.

c. Acetyl-CoA Formation Occurs in Five Reactions

The PDC catalyzes five sequential reactions (Fig. 21-6) with the overall stoichiometry:

$$\text{Pyruvate} + \text{CoA} + \text{NAD}^+ \rightarrow$$
$$\text{acetyl-CoA} + \text{CO}_2 + \text{NADH}$$

The coenzymes and prosthetic groups required in this reaction sequence are thiamine pyrophosphate (TPP; Fig. 17-26), flavin adenine dinucleotide (FAD; Fig. 16-8), nicotinamide adenine dinucleotide (NAD^+; Fig. 13-2), and **lipoamide** (Fig. 21-7); their functions are listed in Table 21-1. Lipoamide consists of **lipoic acid** joined in amide linkage to the ε-amino group of a Lys residue. Reduction of its cyclic disulfide to a dithiol, **dihydrolipoamide,** and its reoxidation (Fig. 21-7) are the "business" of this prosthetic group.

The five reactions catalyzed by the PDC are as follows (Fig. 21-6):

1. Pyruvate dehydrogenase (E_1), a TPP-requiring enzyme, decarboxylates pyruvate, with the intermediate formation of hydroxyethyl-TPP. This reaction is identical with

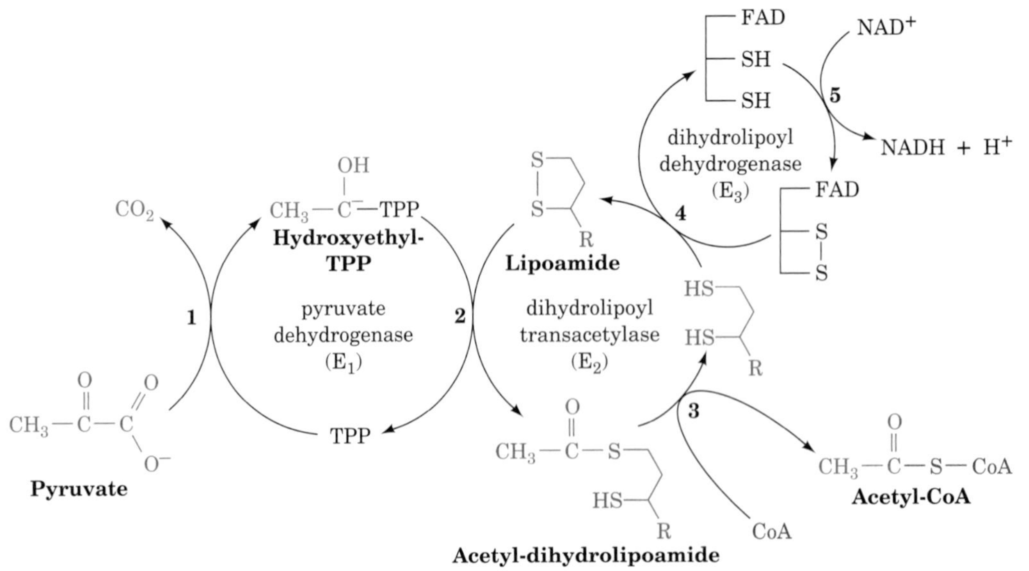

FIGURE 21-6   The five reactions of the PDC. E_1 (pyruvate dehydrogenase) contains TPP and catalyzes Reactions 1 and 2. E_2 (dihydrolipoyl transacetylase) contains lipoamide and catalyzes Reaction 3. E_3 (dihydrolipoyl dehydrogenase) contains FAD and a redox-active disulfide and catalyzes Reactions 4 and 5.

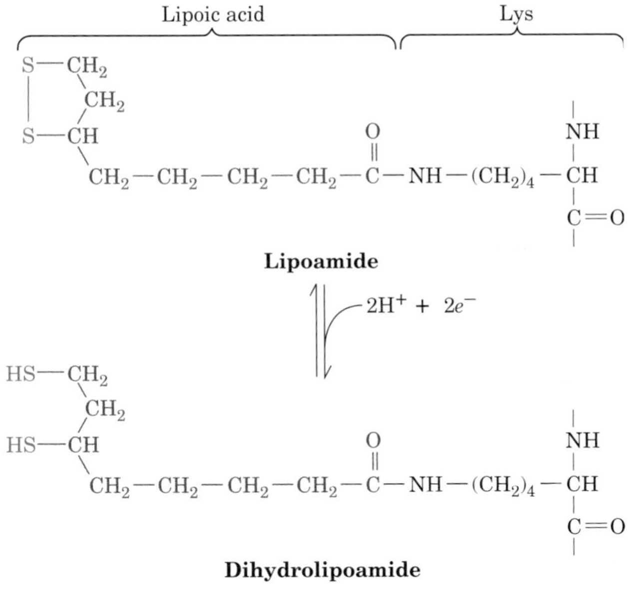

FIGURE 21-7 Interconversion of lipoamide and dihydrolipoamide. Lipoamide is lipoic acid covalently joined to the ε-amino group of a Lys residue via an amide linkage.

that catalyzed by yeast pyruvate decarboxylase (Section 17-3B):

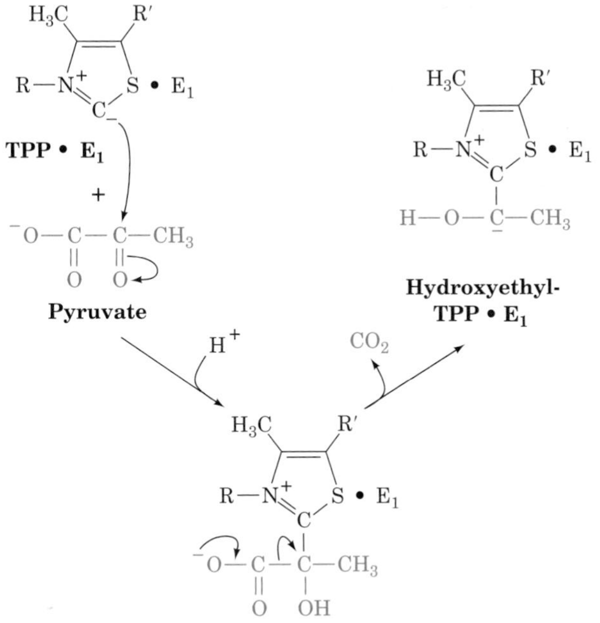

2. Unlike pyruvate decarboxylase, however, pyruvate dehydrogenase does not convert the hydroxyethyl-TPP intermediate into acetaldehyde and TPP. Instead, the hydroxyethyl group is transferred to the next enzyme in the multienzyme sequence, dihydrolipoyl transacetylase (E_2). The reaction occurs by attack of the hydroxyethyl group carbanion on the lipoamide disulfide, followed by the elimination of TPP from the intermediate adduct to form acetyl-dihydrolipoamide and regenerate active E_1. The hydroxyethyl carbanion is thereby oxidized to an acetyl group by the concomitant reduction of the lipoamide disulfide bond:

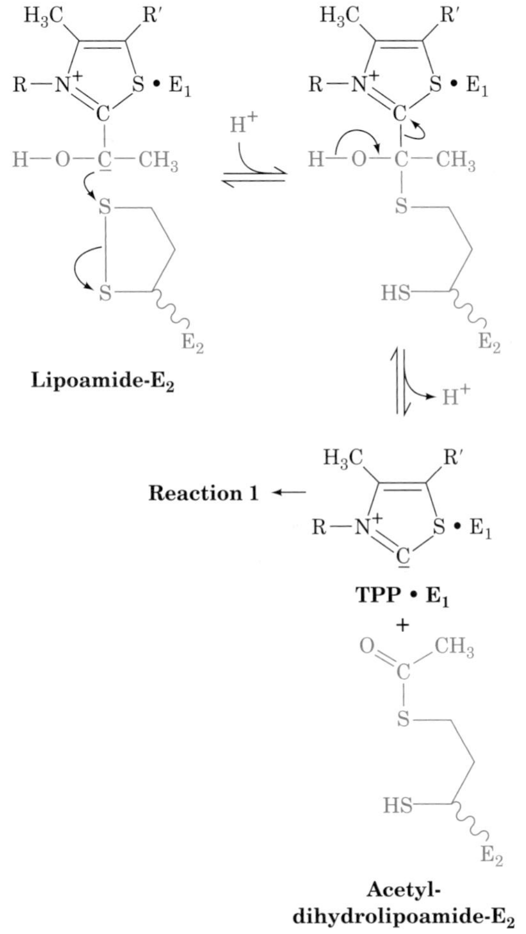

TABLE 21-1 The Coenzymes and Prosthetic Groups of Pyruvate Dehydrogenase

Cofactor	Location	Function
Thiamine pyrophosphate (TPP)	Bound to E_1	Decarboxylates pyruvate, yielding a hydroxyethyl-TPP carbanion
Lipoic acid	Covalently linked to a Lys on E_2 (lipoamide)	Accepts the hydroxyethyl carbanion from TPP as an acetyl group
Coenzyme A (CoA)	Substrate for E_2	Accepts the acetyl group from acetyl-dihydrolipoamide
Flavin adenine dinucleotide (FAD)	Bound to E_3	Reduced by dihydrolipoamide
Nicotinamide adenine dinucleotide (NAD^+)	Substrate for E_3	Reduced by $FADH_2$

3. E_2 then catalyzes the transfer of the acetyl group to CoA, yielding acetyl-CoA and dihydrolipoamide–E_2:

$$CoA—S—\overset{\overset{\displaystyle O}{\|}}{C}—CH_3$$

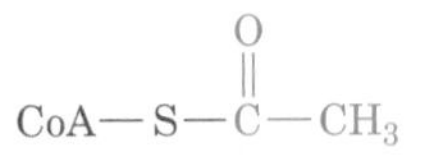

Acetyl-CoA

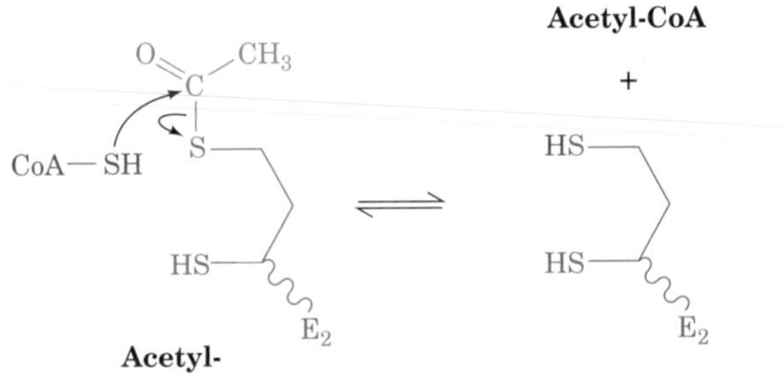

+

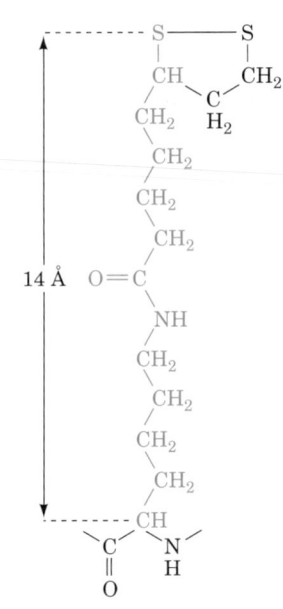

Acetyl-dihydrolipoamide-E_2   **Dihydrolipoamide-E_2**

This is a transesterification in which the sulfhydryl group of CoA attacks the acetyl group of acetyl-dihydrolipoamide–E_2 to form a tetrahedral intermediate (not shown), which decomposes to acetyl-CoA and dihydrolipoamide–E_2.

4. Dihydrolipoyl dehydrogenase (E_3; also called **lipoamide dehydrogenase**) reoxidizes dihydrolipoamide, thereby completing the catalytic cycle of E_2:

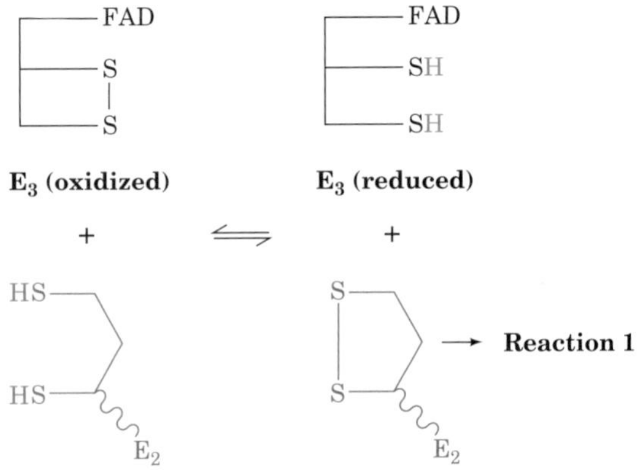

```
┌── FAD          ┌── FAD
│                │
├── S            ├── SH
│   │            │
└── S            └── SH

E₃ (oxidized)    E₃ (reduced)
```

+ ⇌ +

HS, HS ... E_2 S, S ... E_2 → **Reaction 1**

Oxidized E_3 contains a reactive disulfide group and a tightly bound FAD. The oxidation of dihydrolipoamide is a disulfide interchange reaction (Section 9-1A): The lipoamide disulfide bond forms with concomitant reduction of E_3's reactive disulfide to two sulfhydryl groups.

5. Reduced E_3 is reoxidized by NAD^+:

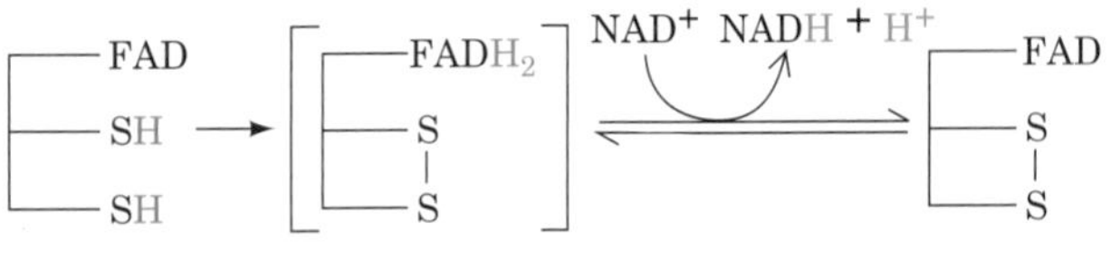

```
┌── FAD       ┌── FADH₂     NAD⁺  NADH + H⁺    ┌── FAD
│             │                                │
├── SH    →   ├── S          ⇌                 ├── S
│             │   │                            │   │
└── SH        └── S                            └── S
```

E₃ (oxidized)

↓

Reaction 4

The enzyme's active sulfhydryl groups are reoxidized by the enzyme-bound FAD, which is thereby reduced to $FADH_2$. The $FADH_2$ is then reoxidized to FAD by NAD^+, producing NADH.

d. The Lipoyllysyl Arm Transfers Intermediates between Enzyme Subunits

How are reaction intermediates channeled between the enzymes of the PDC? The group between the lipoamide

disulfide bond and the E_2 polypeptide backbone, the so-called **lipoyllysyl arm**, has a fully extended length of 14 Å:

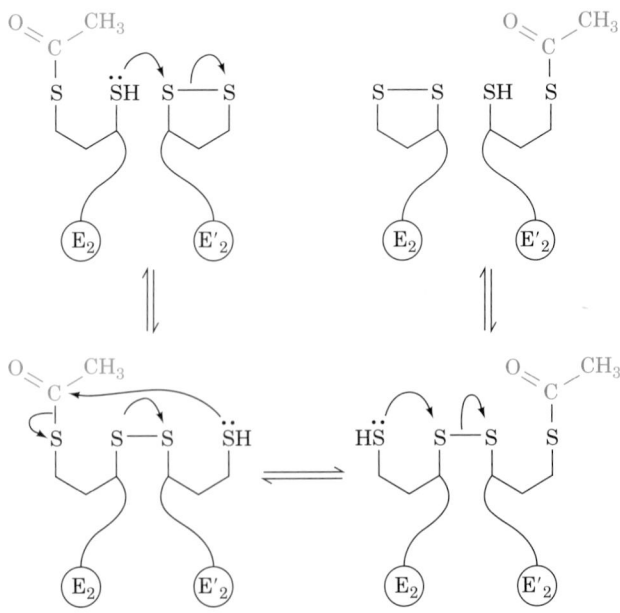

Lipoyllysyl arm
(fully extended)

This suggests that the *lipoyllysyl arm acts as a long tether that swings the lipoamide group together with its reduced acetylated reaction product (Reaction 2) between the active sites of E_1, E_2, and E_3.* Indeed, spectroscopic evidence indicates that the domain of E_2 that bears the lipoyllysyl arm is flexibly linked to the rest of the subunit. Moreover, there is rapid interchange of acetyl groups among the lipoyl groups of the E_2 core (24 lipoyl groups in *E. coli*, 60 in mammals); the tethered arms evidently also swing among themselves, exchanging both acetyl groups and disulfides:

One E_1 subunit can therefore acetylate numerous E_2 subunits and one E_3 subunit can reoxidize several dihydrolipoamide groups.

e. The Structure of E₂

Dihydrolipoyl transacetylase (E₂) consists of several domains (Fig. 21-8): one to three N-terminal lipoyl domains (~80 residues) that each covalently bind a lipoamide residue, a peripheral subunit-binding domain (~35 residues) that binds both E₁ and E₃, and a C-terminal catalytic domain (~250 residues) that contains the enzyme's catalytic center and its intersubunit binding sites. These domains are linked by 20- to 40-residue Pro- and Ala-rich segments that are highly flexible and thereby provide the lipoyl domains with the mobility they require to interact with E₁ and E₃.

The X-ray structure of the catalytic domain (residues 409–638) of E₂ from the gram-negative bacterium *Azotobacter vinelandii*, determined by Wim Hol, indicates, in

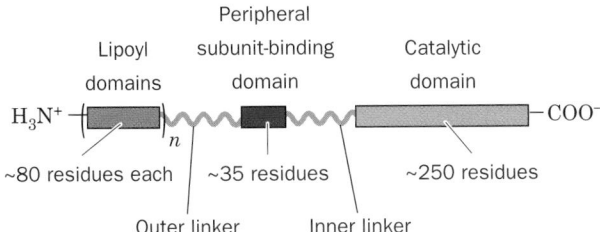

FIGURE 21-8 **Domain structure of the dihydrolipoyl transacetylase (E₂) subunit of the PDC.** The number of lipoyl domains, *n*, is species dependent: *n* = 3 for *E. coli* and *A. vinelandii*, *n* = 2 for mammals and *Streptococcus faecalis*, and *n* = 1 for *B. stearothermophilus* and yeast.

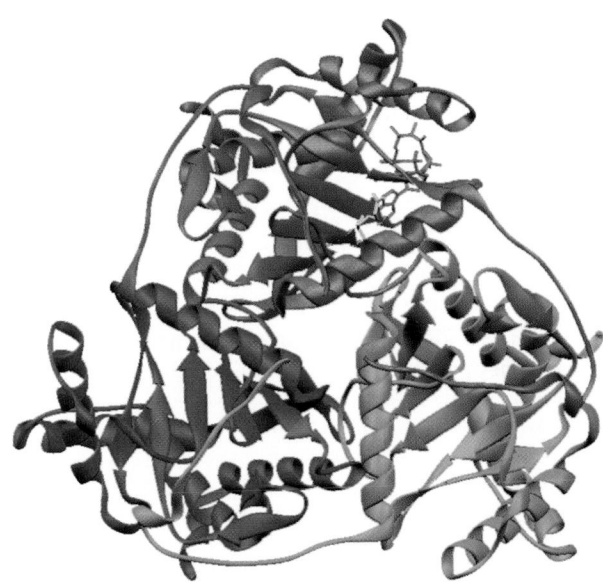

FIGURE 21-9 **X-Ray structure of a trimer of *A. vinelandii* dihydrolipoyl transacetylase (E₂) catalytic domains.** This ribbon diagram is viewed along its 3-fold axis (the cubic complex's body diagonal) from outside the complex. Coenzyme A (*magenta*) and lipoamide (*cyan*), in skeletal form, are shown bound in the active site of the red subunit. Note how the N-terminal "elbow" of each subunit extends over a neighboring subunit; its deletion greatly destabilizes the complex. Compare this drawing to that of the entire cubic complex (Fig. 21-5a). [Based on an X-ray structure by Wim Hol, University of Washington. PDBid 1EAB.]

agreement with the foregoing electron microscopic observations, that E₂ consists of eight tightly associated trimers (Fig. 21-9) arranged at the corners of a cube (Fig. 21-5a). The resulting hollow cagelike structure contains channels large enough to allow substrates to diffuse in and out. In fact, the X-ray structures of the ternary complex of the E₂ catalytic domain with coenzyme A and dihydrolipoic acid reveal that these substrates bind in extended conformations at opposite ends of a 30-Å-long channel that is located at the interface between two of the subunits in a trimer. This arrangement requires CoA to approach its binding site from inside the cube. E₂'s flexibly linked lipoyl domains are thought to protrude from the central core so as to interact with the lipoyl domains of neighboring E₂ subunits, as well as with E₁ and E₃ (see below).

The NMR structures of the lipoyl domains of E₂ from several sources verify the exposed nature of the site of the lipoyl group. They consist of a β-barrel/sandwich containing a 4-stranded and a 3-stranded antiparallel β-sheet, with the Lys to which the lipoyl group would be attached extending from an exposed position on a type I β bend linking two of the β strands in the 4-stranded sheet (Fig. 21-10a).

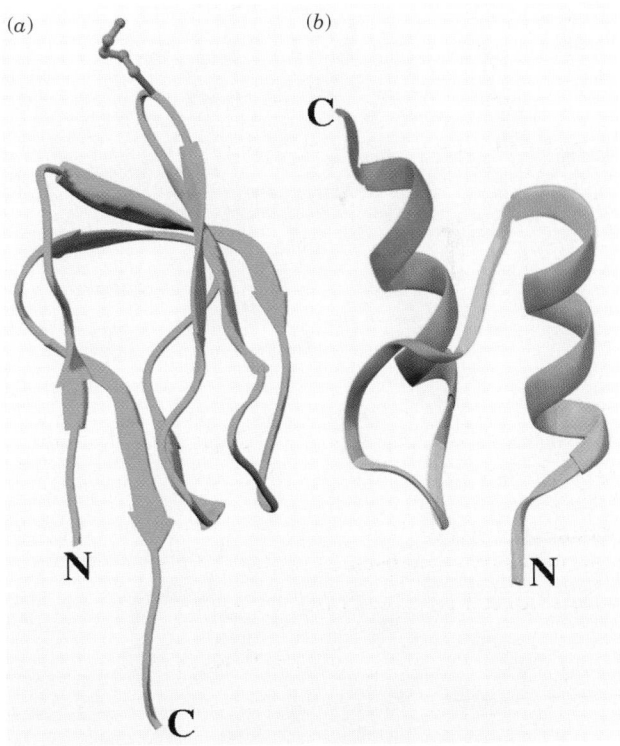

FIGURE 21-10 **NMR structures of the lipoyl and peripheral subunit-binding domains of dihydrolipoyl transacetylase (E₂).** (*a*) The NMR structure of the *A. vinelandii* dihydrolipoyl transacetylase (E₂) lipoyl domain. The Lys side chain to which the lipoyl group would be bound is shown in ball-and-stick form and is located in a type I β turn. [Based on an NMR structure by Aart de Kok, Wageningen Agricultural University, Wageningen, Netherlands. PDBid 1IYU.] (*b*) The NMR structure of the peripheral subunit-binding domain from *B. stearothermophilus* E₂. The polypeptide ribbon is color ramped, red to blue, from its N- to its C-terminus. [Based on an NMR structure by Richard Perham, Cambridge University, U.K. PDBid 2PDD.] 🖱 **See the Interactive Exercises**

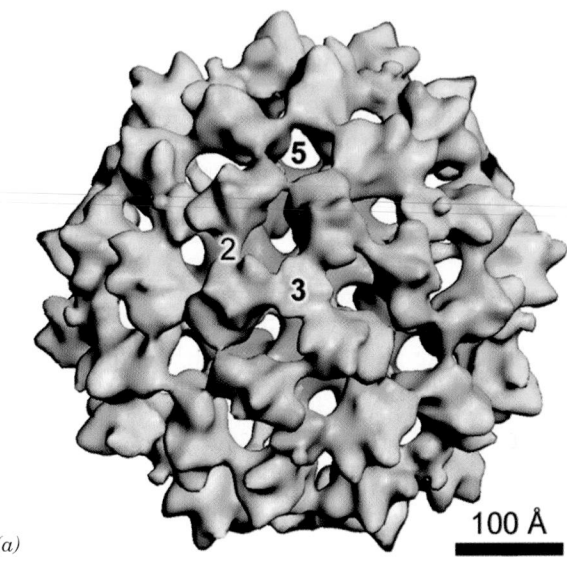

(a)

100 Å

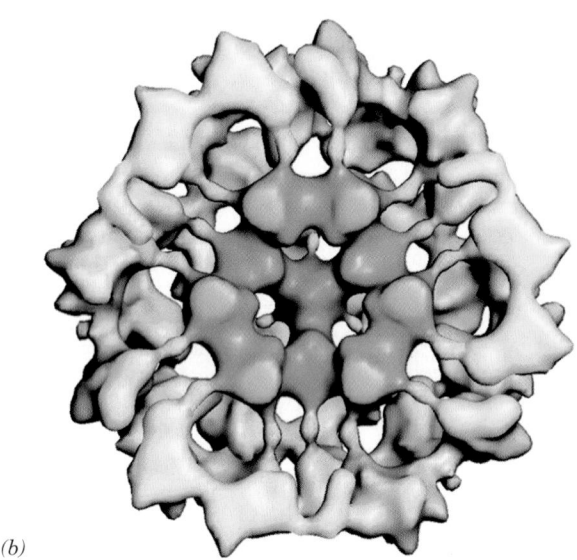

(b)

The NMR structure of the peripheral subunit-binding domain from *B. stearothermophilus,* determined by Richard Perham, reveals that its ~35-residue ordered region consists of two parallel helices separated by a loop that form a close-packed hydrophobic core (Fig. 21-10*b*). It is among the smallest known polypeptides that have an ordered structure.

f. Mammalian and Yeast PDCs Contain Additional Subunits

In mammals and yeast, the dodecahedral PDC's already complicated structure has a further level of complexity in that it contains additional subunits: About 12 copies of **E₃ binding protein (E₃BP)** facilitate the binding of E_3 to the E_2 core of the eukaryotic dodecahedral complex. E₃BP has a lipoyllysine-containing domain similar to E_2 and can accept an acetyl group, but its C-terminal domain has no catalytic activity, and the removal of its lipoyllysine domain does not diminish the catalytic activity of the complex. E₃BP's main role seems to be to aid in the binding of E_3, since limited proteolysis of E₃BP decreases E_3's binding ability.

James Stoops and Reed have determined the organization of the bovine kidney dodecahedral complex through electron microscopy (Fig. 21-11). As expected, the E_2 subunits form a dodecahedral core that is surrounded by a concentric dodecahedron of E_1 subunits (Figs. 21-11*a,b*). The E_2 subunits bind the E_1 subunits via the radially extending, ~50-Å-long inner linkers that precede the pe-

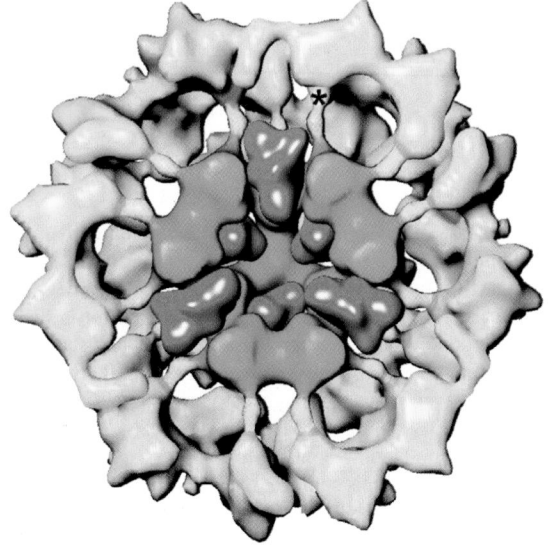

(c)

FIGURE 21-11   Electron microscopy–based images of the bovine kidney pyruvate dehydrogenase complex at ~35 Å resolution. (*a*) The entire particle as viewed along its 3-fold axis of symmetry. E_1 is yellow, the E_2 catalytic core is green, and the inner linkers that connect E_2's catalytic domains to its E_1-binding domains (Fig. 21-8) are cyan. The particle is viewed along a 3-fold axis of symmetry with the positions of a 5-fold axis and a 2-fold axis also marked. (*b*) A cutaway diagram viewed and colored as in Part *a* but with the particle's closest half removed to reveal the E_2 catalytic core and the inner linkers. Compare the green portion with the X-ray structure of a dodecahedral E_2 catalytic core as viewed along a 2-fold axis (Fig. 21-5*b*). (*c*) A cutaway diagram as in Part *b* but with E_3 dimers (Fig. 21-13*a*) shown at 20 Å resolution (*red*) modeled into the pentagonal openings of the E_2 core. The position of a peripheral subunit-binding site at the end of an E_2 inner linker is marked by an asterisk (*). [From Zhou, Z.H., McCarthy, D.B., O'Connor, C.M., Reed, L.J., and Stoops, J.K., *Proc. Natl. Acad. Sci.* **98,** 14802 (2001).]

ripheral subunit-binding domain of E_2 (Fig. 21-8; these linkers are not present in Fig. 21-5b because that X-ray structure only contains E_2's catalytic core). Although the bovine kidney PDC used in the structure determination lacked sufficient $E_3BP\cdot E_3$ to be visible, its position in the electron microscopy–based structure of yeast PDC indicates that an E_3 dimer occupies and largely fills each pentagonal opening of the E_2 core (Fig. 21-11c). Hence the E_3 subunits are not arranged with dodecahedral symmetry (similarly, the E_3 subunits in the cubic *E. coli* PDC are not arranged with cubic symmetry; Fig. 21-4b). The peripheral subunit-binding domain, which is located at the end of the E_2 inner linker (* in Fig. 21-11c), is the point about which the E_2 lipoyl domains (Fig. 21-10a) pivot. It is ~50 Å from the active sites of E_1, E_2, and E_3.

In addition to E_3BP, the mammalian PDC contains one to three copies each of **pyruvate dehydrogenase kinase** and **pyruvate dehydrogenase phosphatase.** The kinase and phosphatase function to regulate the catalytic activity of the complex (Section 21-2C).

g. Arsenic Compounds Are Poisonous because They Sequester Lipoamide

Arsenic has been known to be a poison since ancient times. As(III) compounds, such as **arsenite** (AsO_3^{3-}) and organic arsenicals, are toxic because of their ability to covalently bind sulfhydryl compounds. This is particularly true of vicinal (adjacent) sulfhydryls such as those of lipoamide because they can form bidentate adducts:

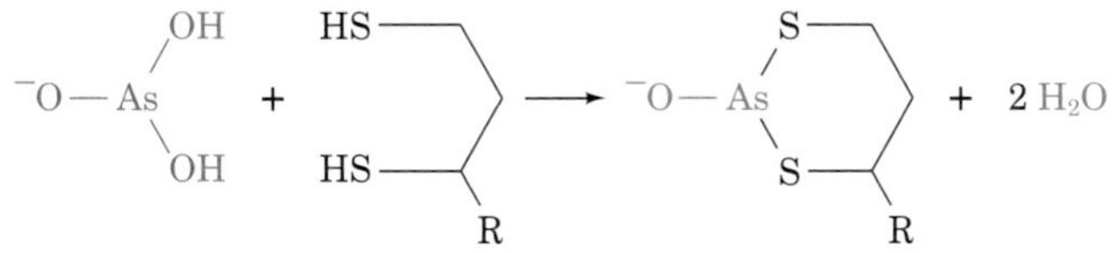

The resultant inactivation of lipoamide-containing enzymes, especially pyruvate dehydrogenase and α-ketoglutarate dehydrogenase (Section 21-3D), brings respiration to a halt.

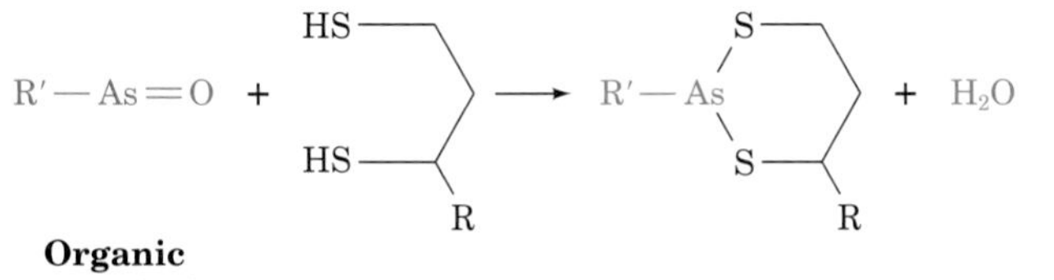

Organic arsenicals are more toxic to microorganisms than to humans, apparently because of differences in the sensitivities of their various enzymes to these compounds. This differential toxicity is the basis for the early twenti-

eth century use of organic arsenicals in the treatment of syphilis (now superseded by penicillin) and trypanosomiasis (typanosomes are parasitic protozoa that cause several diseases including **African sleeping sickness** and **Chagas-Cruz disease**). These compounds were really the first antibiotics, although, not surprisingly, they had severe side effects.

Arsenic is often suspected as a poison in untimely deaths. In fact, it has long been thought that Napoleon Bonaparte died from arsenic poisoning while in exile on St. Helena, an island in the Atlantic Ocean. This suspicion, and the chemical analyses it sparked, makes a fascinating chemical anecdote. The finding that a lock of Napoleon's hair indeed contains a high level of arsenic strongly supports the notion that arsenic poisoning at least contributed to his death. But was it murder or environmental pollution? A sample of the wallpaper from Napoleon's drawing room was found to contain the commonly used (at the time) green pigment copper arsenate ($CuHAsO_4$). It was eventually determined that in a damp climate, as occurs on St. Helena, fungi growing on the wallpaper eliminate the arsenic by converting it to the volatile and highly toxic trimethyl arsine [$(CH_3)_3As$]. Indeed, Napoleon's regular visitors also suffered from symptoms of arsenic poisoning (e.g., gastrointestinal disturbances), which appeared to moderate when they spent much of their time outdoors. Thus, Napoleon's arsenic poisoning may have been unintentional.

Retrospective detective work also suggests that Charles Darwin was a victim of chronic arsenic poisoning. For most of his life after he returned from his epic voyage, Darwin complained of numerous ailments, including eczema, vertigo, headaches, arthritis, gout, palpitations, and nausea, all symptoms of arsenic poisoning. Fowler's solution, a common nineteenth century tonic, contained 10 mg of arsenite · mL^{-1}. Many individuals, quite possibly Darwin himself, took this "medication" for years.

---

h. The Structure of E_1

In addition to the PDC, most cells contain two other closely related multienzyme complexes: the **α-ketoglutarate dehydrogenase complex** (which catalyzes Reaction 4 of the citric acid cycle; Fig. 21-1) and the **branched-chain α-keto acid dehydrogenase complex** (which participates in the degradation of isoleucine, leucine, and valine; Sections 26-3E and 26-3F). These multienzyme complexes all catalyze similar reactions: the NAD$^+$-linked oxidative decarboxylation of an α-keto acid with the transfer of the resulting acyl group to CoA. In fact, all three members of this **2-ketoacid dehydrogenase** family of multienzyme complexes share the same E_3 subunit, and their E_1 and E_2 subunits, which are specific for their corresponding substrates, are homologous and use identical cofactors.

The X-ray structure of E_1 from a PDC (pyruvate dehydrogenase) has not yet been reported. However, Hol has determined the X-ray structure of E_1 from *Pseudomonas putida* branched-chain α-keto acid dehydrogenase, which is named **2-oxoisovalerate dehydrogenase.** It consists of a 2-fold symmetric $\alpha_2\beta_2$ heterotetramer in which the β_2 dimer appears to be clamped between the two subunits of

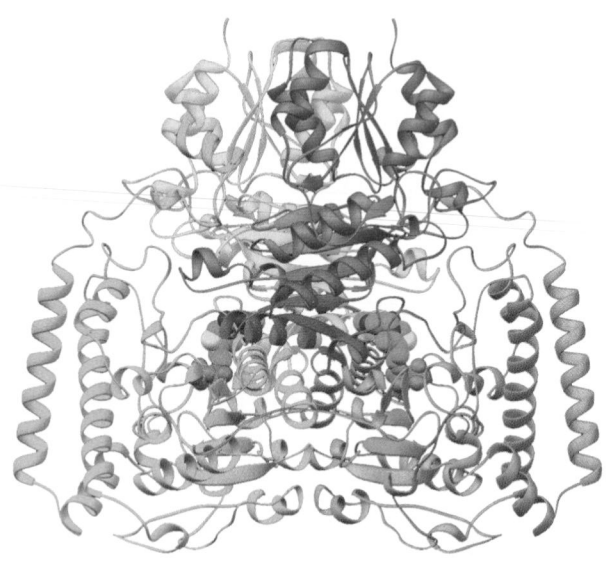

(a)

(b)

**FIGURE 21-12   X-Ray structure of E₁ from *P. putida*
branched-chain α-keto acid dehydrogenase.** (*a*) The α₂β₂
heterotetrameric protein as viewed perpendicular to its 2-fold
axis with its α subunits light blue and gold and its β subunits
cyan and orange. Its TPP cofactors, which bind at the interface
between α and β subunits, are shown in space-filling form
colored according to atom type (C green, N blue, O red, P
magenta, and S yellow). (*b*) A surface diagram of the active site
region, sliced to reveal the ~20-Å-long channel leading from

the active site to the enzyme surface. The lipoyllysyl arm of the
E₂ lipoyl domain has been modeled into the channel. The TPP–
substrate adduct [where here the adduct at TPP C2 (Fig. 17-26)
is a hydroxyisovalerate group], the lipoyllysyl arm in its
disulfide state, and two His side chains that have been
implicated in the catalytic mechanism as proton donors are
shown in stick form. [Part *a* based on an X-ray structure by,
and Part *b* courtesy of, Wim Hol, University of Washington.
PDBid 1QS0.]

the α₂ dimer (Fig. 21-12*a;* in cubic PDCs and in α-ketoglutarate dehydrogenase, which is also cubic, the E₁'s are
homodimers in which the foregoing α and β subunits are
fused). 185 residues from the 410-residue α subunit and
125 residues from the 339-residue β subunit make up a conserved structural core that occurs in other TPP-utilizing
enzymes of known structure, including pyruvate decarboxylase (Section 17-3B; which is also a homodimer). The
TPP-binding site is located at the end of an ~20-Å-long
funnel-shaped channel at the interface between an α and
a β subunit (Fig. 21-12*b*). Presumably, the lipoyllysyl arm
at the end of an E₂ lipoyl domain is inserted into this
channel, as is modeled in Fig. 21-12*b*, for the transfer of
the hydroxyacyl substrate from TPP to lipoamide.

B. *The Mechanism of Dihydrolipoyl Dehydrogenase*

The reaction catalyzed by dihydrolipoamide dehydrogenase (E₃) is more complex than Reactions 4 and 5 in Fig.
21-6 suggest. Vincent Massey demonstrated that *oxidized
dihydrolipoamide dehydrogenase contains a "redox-active"
disulfide bond, which in the enzyme's reduced form has
accepted an electron pair through bond cleavage to form a
dithiol:*

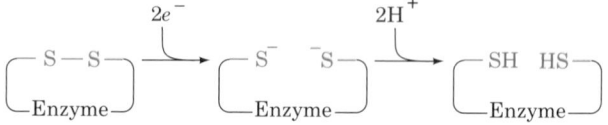

He established this through the following observations involving arsenite (which, as we saw in Section 21-2A, reacts
with vicinal sulfhydryl groups but not with disulfides).

1. The spectrum of oxidized dihydrolipoamide dehydrogenase (E) is unaffected by arsenite.

2. When NADH reacts with the oxidized enzyme in the
presence of arsenite, the resulting reduced enzyme (EH₂)
binds arsenite to form an enzymatically inactive species.

3. The oxidation state of the flavin in a **flavoprotein**
(flavin-containing protein) is readily established from its
characteristic UV–visible spectrum: FAD is an intense yellow, whereas FADH₂ is pale yellow. The spectrum of the
arsenite-inactivated EH₂ indicates that its FAD prosthetic
group is fully oxidized.

Oxidized dihydrolipoamide dehydrogenase must therefore
have a second electron acceptor in addition to FAD; arsenite's known specificity suggests that the acceptor is a
disulfide. Dihydrolipoamide dehydrogenase's amino acid
sequence indicates that its redox-active disulfide bond

forms between its Cys 43 and Cys 48, which occur on a highly conserved segment of the enzyme's polypeptide chain.

a. The X-Ray Structures of Lipoamide Dehydrogenase and Glutathione Reductase

The X-ray structures of dihydrolipoyl dehydrogenase from several microorganisms, mostly determined by Hol, reveal that each ~470-residue subunit of these homo-dimeric enzymes is folded into four domains, all of which participate in forming the subunit's catalytic center (Fig. 21-13). The flavin is almost completely buried in the protein, which prevents the surrounding solution from interfering with the electron-transfer reaction catalyzed by the enzyme (FADH$_2$, but not NADH or thiol, is rapidly oxidized by O$_2$). The redox-active disulfide, which is located on the opposite side of the flavin ring from the nicotinamide ring, links successive turns in a distorted segment of an α helix (in an undistorted helix, the C$_\alpha$ atoms of Cys 43 and Cys 48 would be too far apart to permit the disulfide bond to form).

E$_3$'s catalytic mechanism has been largely determined in analogy with that of the homologous (~33% identical) but structurally more extensively characterized **glutathione reductase (GR).** This nearly ubiquitous enzyme catalyzes the NADPH-dependent reduction of **glutathione disulfide (GSSG)** to the intracellular reducing agent **glutathione (GSH;** its physiological function is discussed in Sections 23-4E and 26-4C):

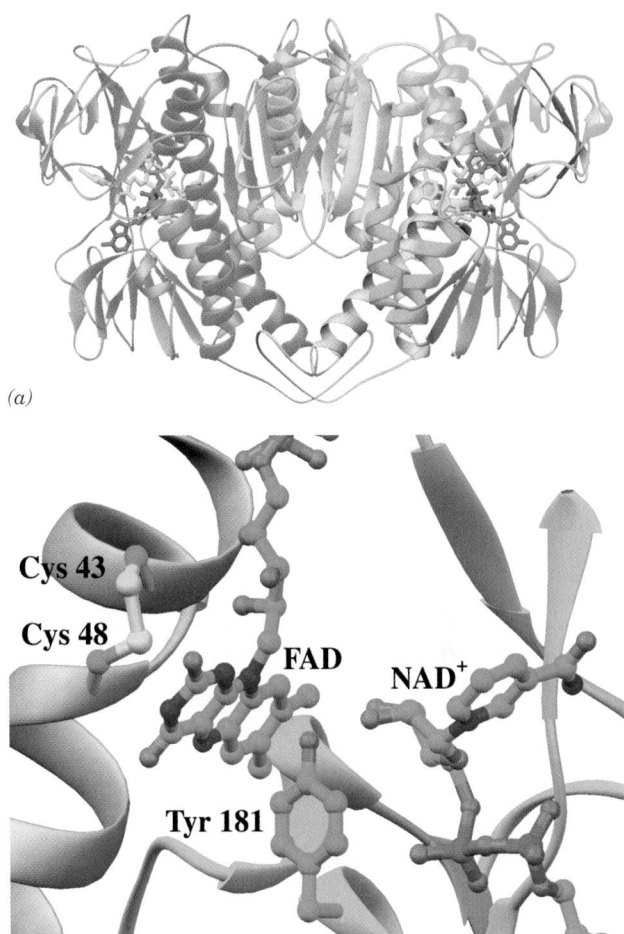

(a)

(b)

FIGURE 21-13 X-Ray structure of dihydrolipoamide dehydrogenase (E$_3$) from *P. putida* in complex with FAD and NAD$^+$. (*a*) The homodimeric enzyme viewed with its twofold axis of symmetry vertical. One subunit is gray and the other is colored according to domain, with its FAD-binding domain (residues 1–142) lavender, its NAD$^+$-binding domain (residues 143–268) cyan, its central domain (residues 269–337) yellow-green, and its interface domain (residues 338–458) gold. The NAD$^+$ and FAD in both subunits are shown in stick form, with NAD$^+$ green and FAD yellow. (*b*) The enzyme's active site region in which the central and interface domains have been deleted for clarity. The view is from roughly the same direction as for the right subunit in Part *a*. The redox-active portions of the bound NAD$^+$ and FAD cofactors, the side chains of Cys 43 and Cys 48 forming the redox-active disulfide bond, and the side chain of Tyr 181 are shown in ball-and-stick form with C green, N blue, O red, P magenta, and S yellow. Note that the side chain of Tyr 181 is interposed between the flavin and the nicotinamide rings. [Based on an X-ray structure by Wim Hol, University of Washington. PDBid 1LVL.]

Glutathione disulfide (GSSG)

Glutathione (GSH)
(γ-L-Glutamyl-L-cysteinylglycine)

Note that this reaction is analogous to that catalyzed by dihydrolipoyl dehydrogenase, but that these two reactions normally occur in opposite directions; that is, dihydrolipoyl dehydrogenase normally uses NAD$^+$ to oxidize two thiol groups to a disulfide (Fig. 21-6), whereas GR uses NADPH

to reduce a disulfide to two thiol groups. Nevertheless, the active sites of these two enzymes are closely superimposable.

The arrangement of the groups in dihydrolipoyl dehydrogenase's catalytic center and its reaction sequence are

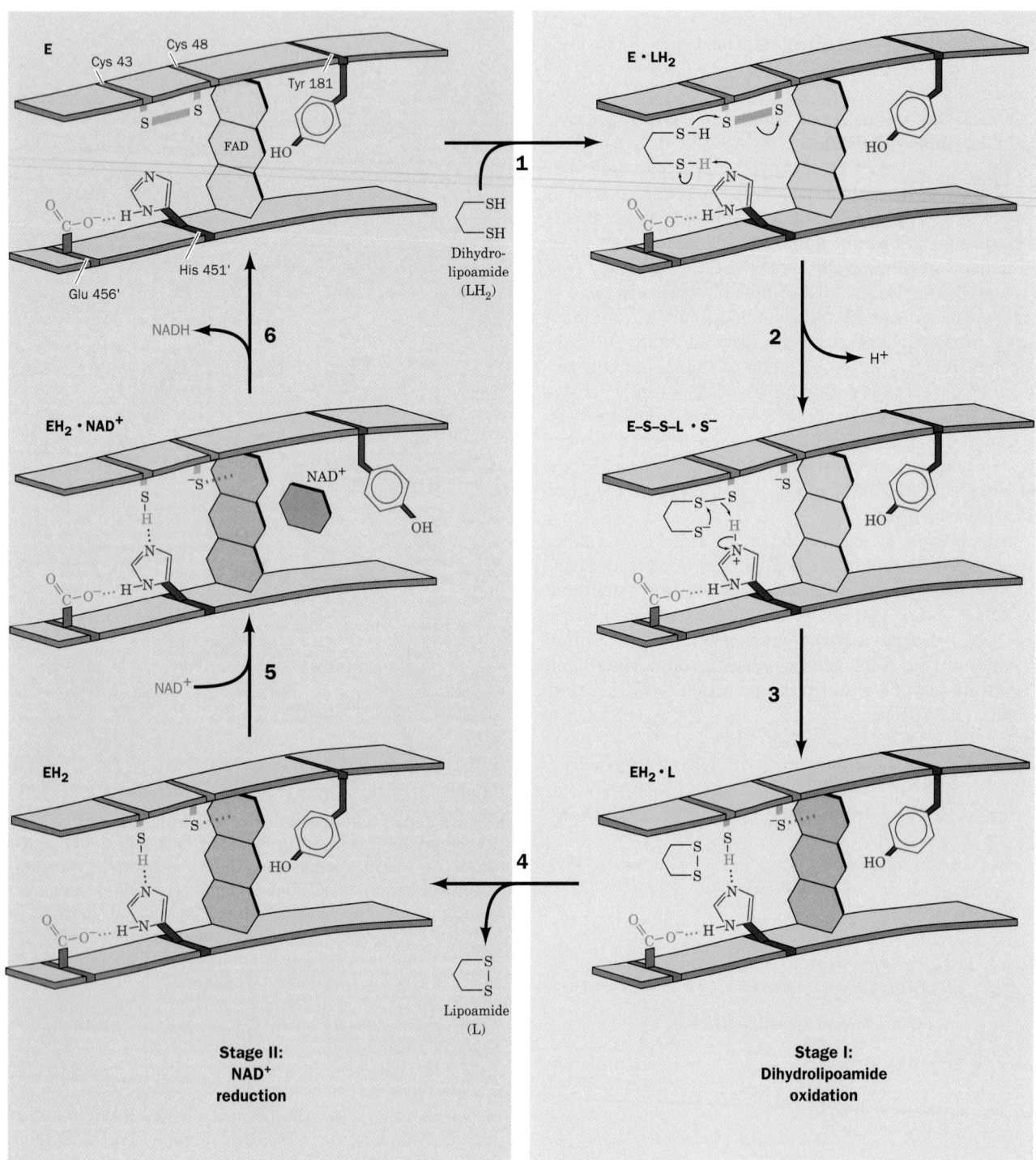

FIGURE 21-14   Catalytic reaction cycle of dihydrolipoyl dehydrogenase. The catalytic center is surrounded by protein so that the NAD$^+$ and dihydrolipoamide binding sites are in deep pockets. The catalytic cycle consists of 6 steps: **(1)** The oxidized enzyme, E, which contains a redox-active disulfide bond between Cys 43 and Cys 48, binds dihydrolipoamide, LH$_2$, the enzyme's first substrate, to form an enzyme–substrate complex, E·LH$_2$. **(2)** A substrate S atom nucleophilically attacks S$_{43}$ to yield a disulfide bond and release S$_{48}$ as a thiolate ion. The proton on the substrate's second thiol group is abstracted by His 451′ to yield a second thiolate ion, E–S–S–LS$^-$. **(3)** The substrate thiolate ion nucleophilically displaces S$_{43}$ aided by general acid catalysis by His 451′ to yield the enzyme's

lipoamide product in its complex with the stable reduced enzyme, EH$_2$·L, in which S$_{48}$ forms a charge-transfer complex with the flavin ring (*dotted red line to the red flavin ring*). **(4)** The lipoamide product is released, yielding EH$_2$. The phenol side chain of Tyr 181 continues to block access to the flavin ring of FAD, so as to prevent the enzyme's oxidation by O$_2$. **(5)** The enzyme's second substrate, NAD$^+$, binds to EH$_2$ to form EH$_2$·NAD$^+$. The phenol side chain of Tyr 181 is pushed aside by the nicotinamide ring of NAD$^+$. **(6)** The catalytic cycle is then closed by the reduction of the NAD$^+$ by EH$_2$ to reform the oxidized enzyme E and yield the enzyme's second product, NADH.

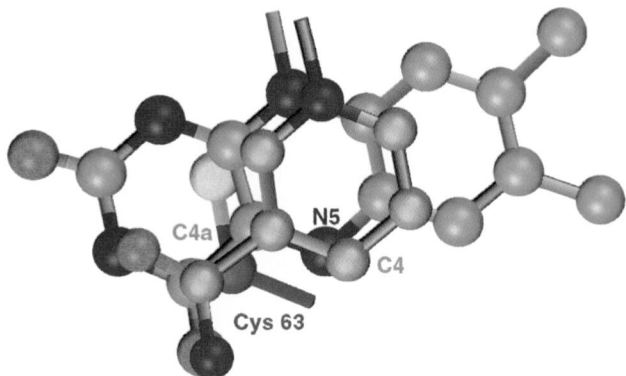

FIGURE 21-15 The complex of the flavin ring, the nicotinamide ring, and the side chain of Cys 63 observed in the X-ray structure of human erythrocyte glutathione reductase. The view is perpendicular to the flavin ring. Atoms are colored according to type with nicotinamide C cyan, flavin C green, Cys 63 C magenta, N blue, O red, and S yellow. The two planar heterocycles are parallel and in van der Waals contact. The Cys 63 S atom (equivalent to S_{48} of dihydrolipoyl dehydrogenase), a member of the redox-active disulfide, is also in van der Waals contact with the flavin ring on the opposite side of it from the nicotinamide ring. [Based on an X-ray structure by Andrew Karplus and Georg Schulz, Institut für Organische Chemie und Biochemie, Freiburg, Germany. PDBid 1GRB.]

diagrammed in Fig. 21-14. The substrate-binding site is located in the interface between the enzyme's two subunits in the vicinity of the redox-active disulfide. In the absence of NAD^+, the phenol side chain of its Tyr 181 covers the nicotinamide-binding pocket so as to shield the flavin from contact with the solution. Indeed, Fig. 21-13*b* shows an enzyme–product complex of oxidized dihydrolipoyl dehydrogenase with NAD^+ in which the side chain of Tyr 181 is interposed between the nicotinamide ring and its binding site at the enzyme's catalytic center. However, in the X-ray structure of GR in complex with NADPH, determined by Georg Schulz and Heiner Schirmer, this side chain has moved aside such that the reduced nicotinamide ring binds parallel to and in van der Waals contact with the flavin ring of the fully oxidized enzyme, E (Fig. 21-15). The H_S substituent to this reduced nicotinamide's prochiral C4 atom (that facing the flavin), the H atom that is known to be lost in the GR reaction, lies near flavin atom N5, the position through which electrons often enter a flavin ring on its reduction. This positioning is particularly significant in view of the catalytic mechanism described below.

b. Catalytic Mechanism

Protein X-ray structures neither reveal H atom positions (except in the most highly resolved structures) nor indicate pathways of electron transfer. The electron-transfer pathway in the glutathione reductase reaction has nevertheless been inferred from the X-ray structures of a series of stable enzymatic reaction intermediates as augmented with a variety of enzymological data. Here we present this mechanism in terms of the reaction catalyzed by the closely similar dihydrolipoamide dehydrogenase (Fig. 21-14).

Dihydrolipoamide dehydrogenase has a Ping Pong mechanism (Section 14-5A); each of its two substrates, dihydrolipoamide and NAD^+, react in the absence of the other (Fig. 21-6, Reactions 4 and 5). Stage I of the catalytic reaction (Fig. 21-14, Steps 1–4) involves a disulfide interchange reaction between the first substrate, dihydrolipoamide (LH_2), and the redox-active disulfide on the enzyme, a process in which His 451′ functions as a general acid–base (primed residues refer to the opposite subunit). In fact, Glu 456′, His 451′, and Cys 43 are arranged much like the catalytic triad of serine proteases (Section 15-3B), with the Cys SH replacing the Ser OH. The importance of His 451′ as an acid–base catalyst was established by Charles Williams through his observation that mutationally changing it to Gln yields an enzyme that retains only ~0.4% of the wild-type catalytic activity. The thiolate anion formed from Cys 48 in Step 2 forms a **charge-transfer complex** with the flavin in which S_{48}^- (the S atom of Cys 48) contacts the flavin ring near its 4a position (a charge-transfer complex is a noncovalent interaction in which an electron pair is partially transferred from a donor, in this case S_{48}^-, to an acceptor, in this case the oxidized flavin ring; the red color of this complex is indicative of the formation of the charge-transfer complex).

Stage II of this Ping Pong reaction (Fig. 21-14, Steps 5 and 6) involves the binding and reduction of NAD^+ followed by the release of NADH to regenerate the oxidized enzyme. The path of the electrons from the reactive disulfide in its reduced form through the FAD to NAD^+ has been elucidated by spectroscopic studies and the chemistry of model compounds. These indicate that an electron pair is rapidly transferred from S_{48}^- to the flavin ring through the transient formation of a covalent bond from S_{48} to

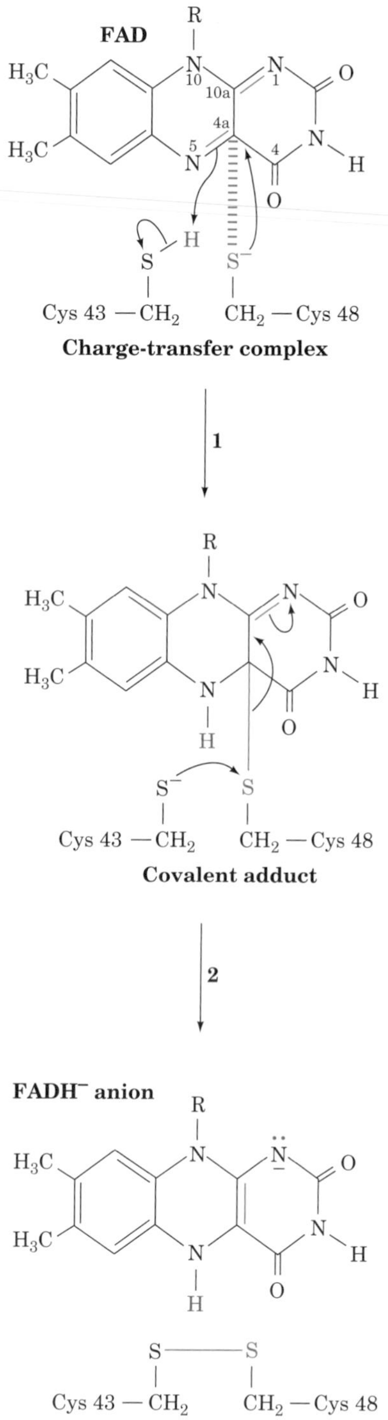

FAD

Charge-transfer complex

1

Covalent adduct

2

FADH⁻ anion

Redox-active disulfide

FIGURE 21-16   The reaction transferring an electron pair from dihydrolipoyl dehydrogenase's redox-active disulfide in its reduced form to the enzyme's bound flavin ring. (1) The collapse of the charge-transfer complex between the Cys 48 thiolate ion and the flavin ring (*dashed red line*) to form a covalent bond between S_{48} and flavin atom C4a. S_{48} is located out of the plane of the flavin, as Fig. 21-15 indicates. Flavin atom N5 acquires a proton, possibly from S_{43}, which becomes a thiolate ion. **(2)** The S_{43} thiolate atom nucleophilically attacks S_{48} to form the redox-active disulfide bond, thereby releasing the reduced flavin anion FADH⁻.

flavin atom 4a (Fig. 21-16, Step 1). His 451′ then abstracts the proton from the S_{43} thiol to form a thiolate ion, which nucleophilically attacks S_{48} to reform the redox-active disulfide group (Fig. 21-16, Step 2). The resulting reduced flavin anion (FADH⁻) has but transient existence. The H atom substituent to its N5 is immediately transferred (formally as a hydride ion) to the juxtaposed C4 atom of the nicotinamide ring (Fig. 21-15), yielding FAD and the reaction's second product, NADH, thereby completing the catalytic cycle. Thus, the FAD appears to function more like an electron conduit between the reduced form of the redox-active disulfide and NAD⁺ than as a source or sink of electrons.

C. Control of Pyruvate Dehydrogenase

The PDC regulates the entrance of acetyl units derived from carbohydrate sources into the citric acid cycle. The decarboxylation of pyruvate by E_1 is irreversible and, since there are no other pathways in mammals for the synthesis of acetyl-CoA from pyruvate, it is crucial that the reaction be carefully controlled. Two regulatory systems are employed:

1. Product inhibition by NADH and acetyl-CoA (Fig. 21-17*a*).

2. Covalent modification by phosphorylation/dephosphorylation of the pyruvate dehydrogenase (E_1) subunit (Fig. 21-17*b*; enzymatic regulation by covalent modification is discussed in Section 18-3B).

a. Control by Product Inhibition
NADH and acetyl-CoA compete with NAD⁺ and CoA for binding sites on their respective enzymes. They also drive the reversible transacetylase (E_2) and dihydrolipoyl dehydrogenase (E_3) reactions backward (Fig. 21-17a). High ratios of [NADH]/[NAD⁺] and [acetyl-CoA]/[CoA] therefore maintain E_2 in the acetylated form, incapable of accepting the hydroxyethyl group from the TPP on E_1. This, in turn, ties up the TPP on the E_1 subunit in its hydroxyethyl form, decreasing the rate of pyruvate decarboxylation.

b. Control by Phosphorylation/Dephosphorylation
Control by phosphorylation/dephosphorylation occurs only in eukaryotic enzyme complexes. These complexes contain pyruvate dehydrogenase kinase and pyruvate dehydrogenase phosphatase bound to the dihydrolipoyl transacetylase (E_2) core. *The kinase inactivates the pyruvate dehydrogenase (E_1) subunit by catalyzing the phosphorylation of a specific dehydrogenase Ser residue by ATP (Fig. 21-17b). Hydrolysis of this phosphoSer residue by the phosphatase reactivates the complex.*

Pyruvate dehydrogenase kinase is activated through its interaction with the acetylated form of E_2. Consequently, the products of the reaction, NADH and acetyl-CoA, in addition to their direct effects on the PDC, indirectly activate pyruvate dehydrogenase kinase. The resultant phosphorylation inactivates the complex just as the products

(a) **Product inhibition**

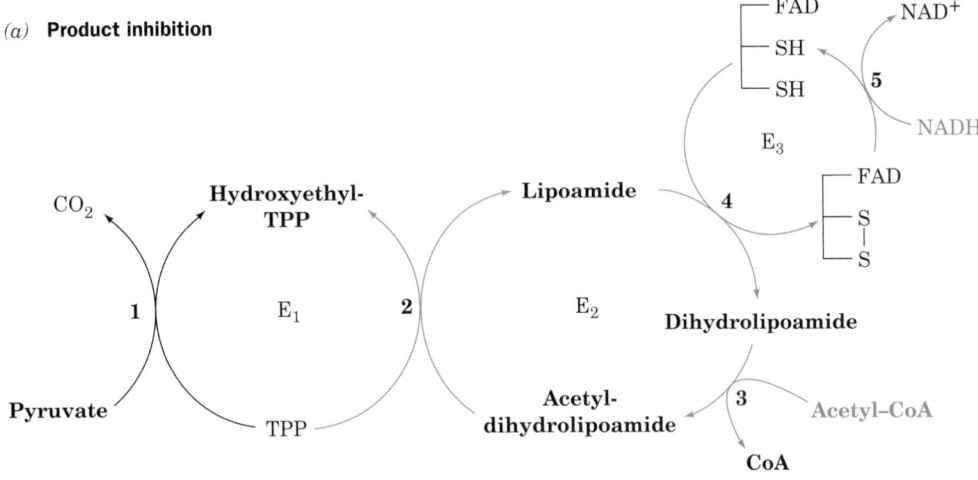

(b) **Covalent modification**

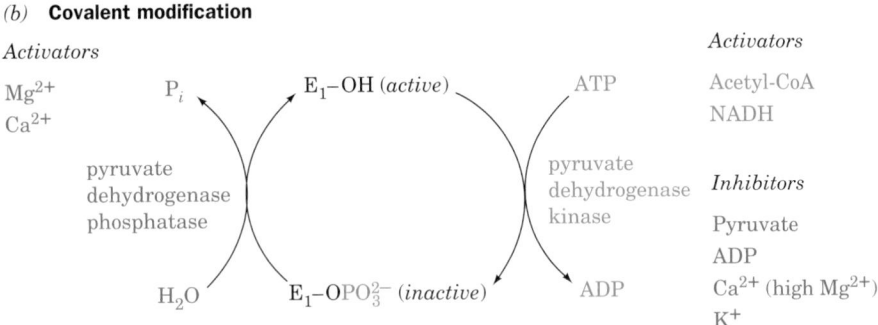

FIGURE 21-17 Factors controlling the activity of the PDC.
(*a*) Product inhibition. NADH and acetyl-CoA, respectively, compete with NAD⁺ and CoA in Reactions 3 and 5 of the pyruvate dehydrogenase reaction sequence. When the relative concentrations of NADH and acetyl-CoA are high, the reversible reactions catalyzed by E₂ and E₃ are driven backward (*red arrows*), thereby inhibiting further formation of acetyl-CoA. (*b*) Covalent modification in the eukaryotic

complex. Pyruvate dehydrogenase (E₁) is inactivated by the specific phosphorylation of one of its Ser residues in a reaction catalyzed by pyruvate dehydrogenase kinase (*right*). This phosphoryl group is hydrolyzed through the action of pyruvate dehydrogenase phosphatase (*left*), thereby reactivating E₁. The activators and inhibitors of the kinase are listed on the right and the activators of the phosphatase are listed on the left.

themselves inhibit it. Acetyl-CoA and NADH are products of fatty acid oxidation (Section 25-2) so that this inhibition of PDC serves to preserve carbohydrate stores when lipid fuels are available.

Ca²⁺ is an important second messenger signaling the need for increased energy (e.g., for muscle contraction). Increasing [Ca²⁺] enhances pyruvate dehydrogenase phosphatase activity, thus activating PDC.

Insulin is involved in the control of this system through its indirect activation of pyruvate dehydrogenase phosphatase. Recall that insulin activates glycogen synthesis as well by activating phosphoprotein phosphatase (Section 18-3C). Insulin, in response to increases in blood glucose, is now seen as promoting the synthesis of acetyl-CoA as well as glycogen. As we shall see in Section 25-4, acetyl-CoA is the precursor to fatty acids in addition to being the fuel for the citric acid cycle. Various other activators and inhibitors regulate the pyruvate dehydrogenase system (Fig. 21-17*b*); in contrast to the glycogen metabolism control system (Section 18-3), however, it is unaffected by cAMP.

3 ■ ENZYMES OF THE CITRIC ACID CYCLE

In this section we discuss the reaction mechanisms of the eight citric acid cycle enzymes. Our knowledge of these mechanisms rests on an enormous amount of experimental work; as we progress, we shall pause to examine some of these experimental details. Consideration of how this cycle is regulated and its relationship to cellular metabolism are the subjects of the following sections.

A. *Citrate Synthase*

Citrate synthase (originally named **citrate condensing enzyme***) catalyzes the condensation of acetyl-CoA and oxaloacetate (Reaction 1 of Fig. 21-1).* This initial reaction of the citric acid cycle is the point at which carbon atoms are "fed into the furnace" as acetyl-CoA. The citrate synthase reaction proceeds via an ordered sequential kinetic mechanism (Section 14-5B), with oxaloacetate adding to the enzyme before acetyl-CoA.

The X-ray structure of the free dimeric enzyme, determined by James Remington and Robert Huber, shows it to adopt an "open form" in which the two domains of each subunit form a deep cleft that contains the oxaloacetate binding site (Fig. 21-18a). On binding oxaloacetate, however, the smaller domain undergoes a remarkable 18° rotation relative to the larger domain, which closes the cleft (Fig. 21-18b).

The X-ray structures of two inhibitors of citrate synthase in ternary complex with the enzyme and oxaloacetate have also been determined. **Acetonyl-CoA,** an inhibitory analog of acetyl-CoA in the ground state, and **carboxymethyl-CoA,** a proposed transition state analog (see below),

CoAS — C(=O)—CH₃

Acetyl-CoA

CoAS — C(OH)=CH₂

Proposed enol intermediate

CoAS — CH₂ — C(=O)—CH₃

Acetonyl-CoA (ground-state analog)

CoAS — CH₂ — C(OH)=O

Carboxymethyl-CoA (transition state analog)

bind to the enzyme in its "closed" form, thereby identifying the acetyl-CoA binding site. The existence of the "open" and "closed" forms explains the enzyme's ordered sequential kinetic behavior: *The conformational change induced by oxaloacetate binding generates the acetyl-CoA binding site while sealing off the solvent's access to the bound oxaloacetate.* This is a classic example of the induced-fit model of substrate binding (Section 10-4C). Hexokinase exhibits similar behavior (Section 17-2A).

The citrate synthase reaction is a mixed aldol–Claisen ester condensation, subject to general acid–base catalysis and the intermediate participation of the enol(ate) form of acetyl-CoA. The X-ray structure of the enzyme in ternary complex with oxaloacetate and carboxymethyl-CoA reveals that three of its ionizable side chains are properly oriented to play catalytic roles: His 274, Asp 375, and His 320. The N1 atoms in both of these His side chains are hydrogen bonded to two backbone NH groups indicating that these N1 atoms are not protonated. The participation of His 274, Asp 375, and His 320 has been confirmed by kinetic studies of mutant enzymes generated by site directed mutagenesis. The following three-step mechanism, formulated mainly by Remington, takes into account these observations (Fig. 21-19):

1. The enolate of acetyl-CoA is generated in the rate-limiting step of the reaction, with the catalytic participation

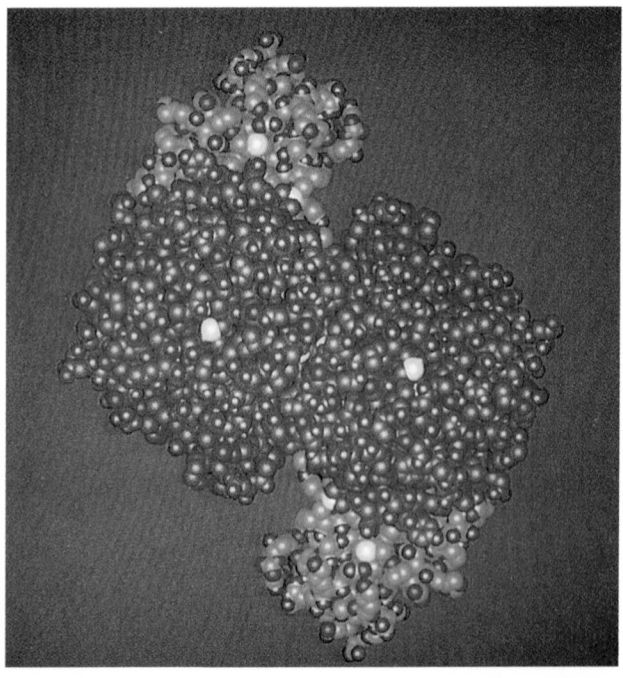

(a)

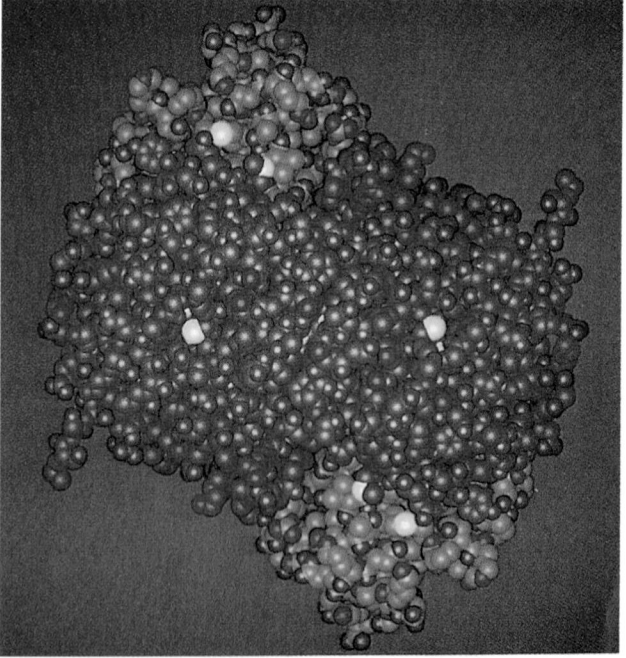

(b)

FIGURE 21-18 Conformational changes in citrate synthase.
Space-filling drawings showing citrate synthase in (a) the open conformation and (b) the closed, substrate-binding conformation. The C atoms of the small domain in each subunit of the enzyme are green and those of the large domain are purple. N, O, and S atoms in both domains are blue, red, and yellow. The view is along the homodimeric protein's twofold

rotation axis. The large conformational shift between the open and closed forms entails relative interatomic movements of up to 15 Å. [Courtesy of Anne Dallas, University of Pennsylvania, and Helen Berman, Fox Chase Cancer Center. Based on X-ray structures determined by James Remington and Robert Huber, Max-Planck-Institut für Biochemie, Martinsried, Germany. PDBids (a) 1CTS and (b) 2CTS.]

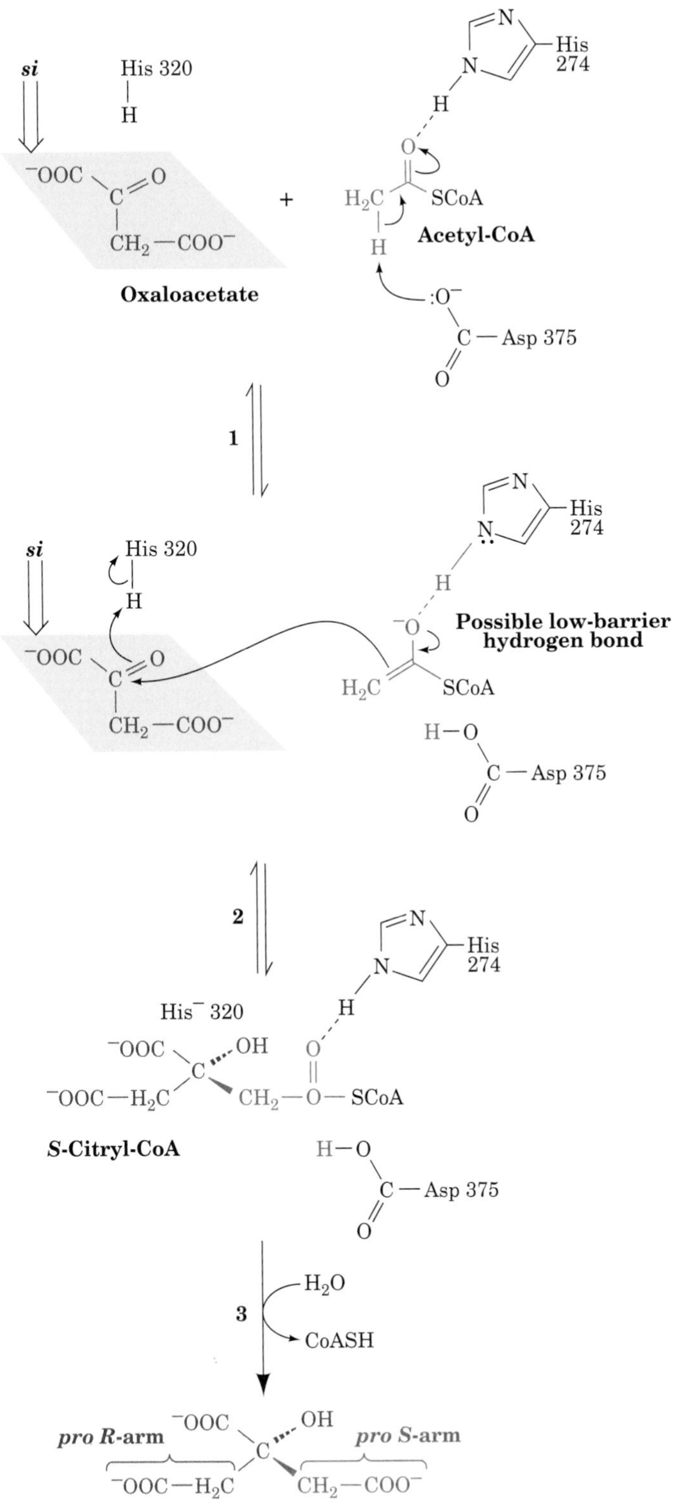

FIGURE 21-19 Mechanism and stereochemistry of the citrate synthase reaction. His 274 and His 320 in their neutral forms and Asp 375 have been implicated as general acid–base catalysts. The overall reaction's rate-limiting step is the formation of the acetyl-CoA enolate, which may be stabilized by a low-barrier hydrogen bond to His 274. The acetyl-CoA enolate then nucleophilically attacks the *si* face of oxaloacetate's carbonyl carbon. The resulting intermediate, (*S*)-citryl-CoA, is hydrolyzed to yield citrate and CoA. [Mostly after Remington, S.J., *Curr. Opin. Struct. Biol.* **2,** 732 (1992)].

of Asp 375 acting as a base to remove a proton from the methyl group and His 274, in its neutral form, H-bonding the enolate oxygen (whose protonated carbonyl form is normally far more acidic than a neutral His side chain). Much as we discussed for a similar step in the triose phosphate isomerase reaction (Section 17-2E), the pK of the protonated form of the substrate thioester carbonyl oxygen increases to ~14 on enolization, which approximates that of neutral His 274. Hence, it has been proposed that this step is facilitated through the formation of a low-barrier hydrogen bond (which, it will be recalled, is a particularly strong form of hydrogen bonding interaction in which the hydrogen atom is more or less equally shared between the "donor" and "acceptor" atoms; Section 15-3D). However, whether or not this low-barrier hydrogen bond actually forms is a subject of controversy. The presence of the thioester bond to CoA facilitates the enolization; enolization of acetate alone would require the generation of a much more highly charged and therefore less stable intermediate.

2. Citryl-CoA is formed in a second acid–base catalyzed reaction step in which the acetyl-CoA enolate nucleophilically attacks oxaloacetate while His 320, also in its neutral form, donates a proton to oxaloacetate's carbonyl group. The citryl-CoA remains bound to the enzyme.

3. Citryl-CoA is hydrolyzed to citrate and CoA, with His 320, now in the anionic form, abstracting a proton from water as it attacks the carbonyl group of the citryl-CoA and the acidic form of Asp 375 donating a proton to the CoA as it leaves. This hydrolysis provides the reaction's thermodynamic driving force ($\Delta G^{\circ\prime} = -31.5$ kJ · mol^{-1}). We shall see presently why the reaction requires such a large, seemingly wasteful, release of free energy.

Enzyme-catalyzed reactions, as we have previously noted, are stereospecific. The aldol–Claisen condensation that occurs here involves attack of the acetyl-CoA enolate exclusively at the *si* face of oxaloacetate's carbonyl carbon atom, thereby yielding (*S*)-citryl-CoA (chirality nomenclature is presented in Section 4-2C). The acetyl group of acetyl-CoA thereby only forms citrate's *pro-S* carboxymethyl group.

B. *Aconitase*

Aconitase catalyzes the reversible isomerization of citrate and isocitrate with cis-aconitate as an intermediate (Reaction 2 of Fig. 21-1). Although citrate has a plane of symmetry and is therefore not optically active, it is nevertheless prochiral; aconitase can distinguish between citrate's *pro-R* and *pro-S* carboxymethyl groups.

A combination of X-ray crystallography and site-directed mutagenesis by Helmut Beinert and David Stout has identified several amino acid residues that participate in catalysis. Formation of the *cis*-aconitate intermediate involves a dehydration in which Ser 642 alkoxide, acting as a general base, abstracts the *pro-R* proton at C2 of citrate's

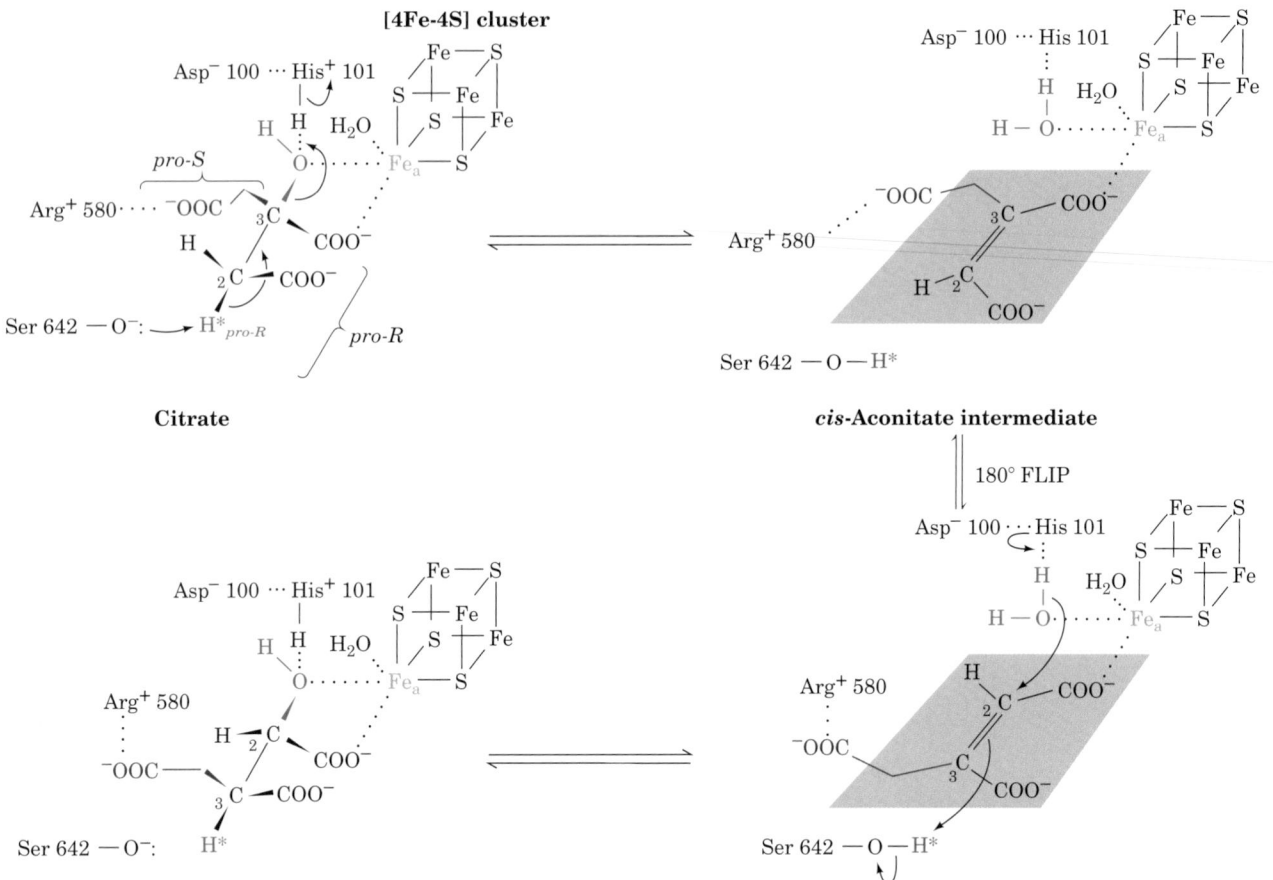

FIGURE 21-20   Mechanism and stereochemistry of the aconitase reaction. Fe$_a$ of the enzyme's [4Fe–4S] cluster coordinates the citrate hydroxyl and central carboxyl groups; Arg 580 forms a salt bridge with the *pro-S* carboxyl group; Ser 642, in its alkoxide form, acts as a general base; and the Asp 100-polarized His 101 acts as a general acid in the elimination of water to form cis-aconitate. Note the unusual 180° FLIP that cis-aconitate apparently undergoes, possibly while remaining bound to the active site; thus rehydration takes place on the opposite face of the substrate from which dehydration occurred, thereby yielding (2R,3S)-isocitrate.

pro-R carboxymethyl group (Fig. 21-20, *top*). [The O$_\gamma$ of Ser 642 occupies a sort of oxyanion hole (Section 15-3D) that apparently stabilizes its otherwise highly basic alkoxide form. The carbanion formed in the transition state of this reaction is not shown.] This is followed by the loss of the OH group at C3 in a trans elimination of H$_2$O to form the *cis*-aconitate intermediate. The latter reaction step is facilitated through general acid catalysis by His 101, whose imidazolium group is polarized through ion pairing to the side chain of Asp 100. This model is corroborated by the observations that the mutagenesis of Asp 100, His 101, or Ser 642 results in a 10^3- to 10^5-fold reduction in aconitase's catalytic activity without greatly affecting its substrate binding affinity.

Aconitase contains a covalently bound **[4Fe–4S] iron–sulfur cluster,** which is required for catalytic activity (the properties of iron–sulfur clusters are discussed in Section 22-2C). A specific Fe(II) atom in this cluster, the so-called Fe$_a$ atom, is thought to coordinate the OH group of the

substrate so as to facilitate its elimination. Iron–sulfur clusters are almost always associated with redox processes although, intriguingly, not in the case of aconitase. It has been postulated that the electronic properties of the [4Fe–4S] cluster permit the Fe$_a$ atom to expand its coordination shell from the four ligands observed in the X-ray structure of the free enzyme (three S^{2-} ions and a OH$^-$ ion) to the six octahedrally arranged ligands observed in the enzyme–substrate complex. A single metal ion, such as Zn^{2+} or Cu^{2+}, is unable to do so, and hence would require that some of its ligands be displaced on binding substrate.

The second stage of the aconitase reaction is rehydration of cis-aconitate's double bond to form isocitrate (Fig. 21-20, *bottom*). The nonenzymatic addition of H$_2$O across the double bond of cis-aconitate would yield four stereoisomers. Aconitase, however, catalyzes the stereospecific trans addition of OH$^-$ and H$^+$ across the double bond to form only (2R,3S)-isocitrate in the forward reaction and citrate in the reverse reaction.

Although citrate's OH group is lost to the solvent in the aconitase reaction, the abstracted H^+ is retained by Ser 642. Remarkably, it adds to the opposite faces of cis-aconitate's double bond in forming citrate and isocitrate. The cis-aconitate must expose a different face to the sequestered H^+ for the formation of isocitrate. This is thought to occur by a 180° flip of the intermediate on the enzyme's surface while maintaining its association with Arg 580 or else by dissociation from the enzyme and replacement by another cis-aconitate in the "flipped" orientation.

a. Fluorocitrate Inhibits Aconitase

Fluoroacetate, one of the most toxic small molecules known ($LD_{50} = 0.2$ mg · kg^{-1} of body weight in rats), occurs in the leaves of certain African, Australian, and South American poisonous plants. Interestingly, fluoroacetate itself has little toxic effect on cells; instead, cells enzymatically convert it first to fluoroacetyl-CoA and then to **(2R,3R)-2-fluorocitrate,** which specifically inhibits aconitase (see Problem 9):

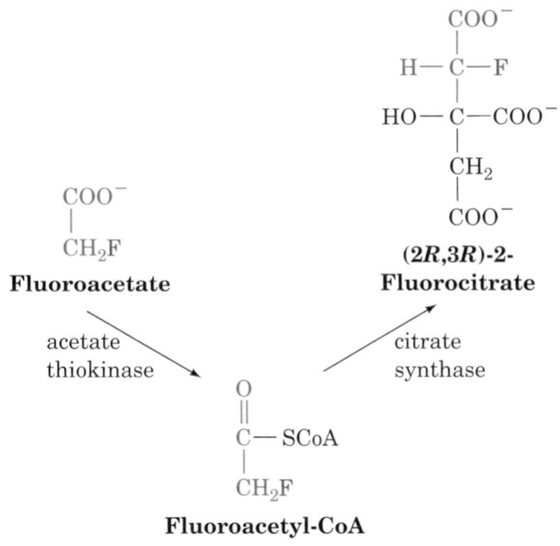

It is not clear, however, that the inhibition of aconitase fully accounts for fluorocitrate's high toxicity. Indeed, fluorocitrate also inhibits the transport of citrate across the mitochondrial membrane.

C. NAD⁺-Dependent Isocitrate Dehydrogenase

Isocitrate dehydrogenase catalyzes the oxidative decarboxylation of isocitrate to α-ketoglutarate to produce the citric acid cycle's first CO_2 and NADH (Reaction 3 of Fig. 21-1). Mammalian tissues contain two isoforms of this enzyme. Although they both catalyze the same reaction, one isoform participates in the citric acid cycle, is located entirely in the mitochondrion, and utilizes NAD^+ as a cofactor. The other isoform occurs in both the mitochondrion and the cytosol, utilizes $NADP^+$ as a cofactor, and generates NADPH for use in reductive biosynthesis.

NAD^+-dependent isocitrate dehydrogenase, which requires an Mn^{2+} or Mg^{2+} cofactor, is thought to catalyze the oxidation of a secondary alcohol (isocitrate) to a ketone (oxalosuccinate) followed by the decarboxylation of the carboxyl group β to the ketone (Fig. 21-21). In this sequence, the keto group β to the carboxyl group facilitates the decarboxylation by acting as an electron sink. The oxidation occurs with the stereospecific reduction of NAD^+ at its *re* face (A-side addition; Section 13-2A). Mn^{2+} coordinates the newly formed carbonyl group so as to polarize its electronic charge.

Although the intermediate formation of oxalosuccinate is a logical chemical prediction, evidence for it has been difficult to obtain because it has only a transient existence in reactions catalyzed by the wild-type enzyme. However, an enzymatic reaction rate can be slowed by the mutation of particular catalytically important residues, resulting in the accumulation of specific intermediates. Thus, when Lys 230, which facilitates the decarboxylation of the oxalosuccinate intermediate (Fig. 21-21) is mutated to Met in $NADP^+$-dependent isocitrate dehydrogenase, the oxalosuccinate intermediate accumulates. This accumulated intermediate was directly visualized in the X-ray structure of the mutant enzyme, in the presence of a steady-state flow of substrate, through the use of fast X-ray intensity measurements.

D. α-Ketoglutarate Dehydrogenase

α-Ketoglutarate dehydrogenase catalyzes the oxidative decarboxylation of an α-keto acid (α-ketoglutarate), releasing the citric acid cycle's second CO_2 and NADH (Reaction 4

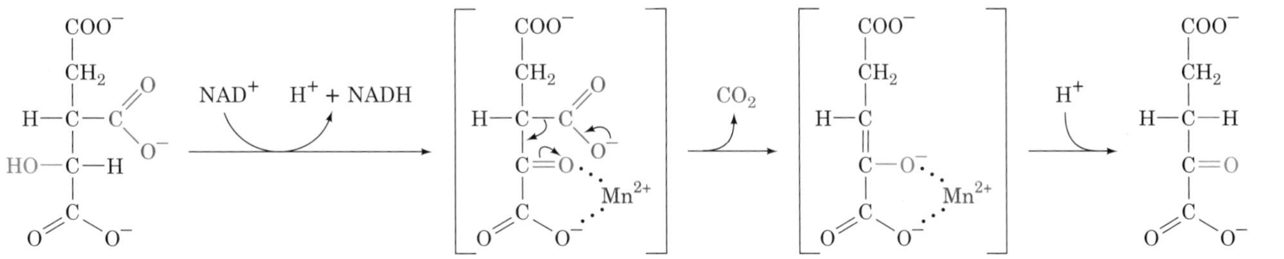

FIGURE 21-21 Probable reaction mechanism of isocitrate dehydrogenase. Oxalosuccinate is shown in brackets because it does not dissociate from the enzyme.

of Fig. 21-1). The overall reaction, which chemically resembles that catalyzed by the PDC (Fig. 21-6), is mediated by a homologous multienzyme complex consisting of **α-ketoglutarate dehydrogenase (E₁), dihydrolipoyl transsuccinylase (E₂),** and **dihydrolipoyl dehydrogenase (E₃)** in which the E_3 subunits are identical to those in the PDC (Section 21-2A).

Individual reactions catalyzed by the complex occur by mechanisms identical to those of the pyruvate dehydrogenase reaction (Section 21-2A), the product likewise being a "high-energy" thioester, in this case succinyl-CoA. There are no covalent modification enzymes in the α-ketoglutarate dehydrogenase complex, however.

E. *Succinyl-CoA Synthetase*

Succinyl-CoA synthetase (also called **succinate thiokinase***) hydrolyzes the "high-energy" compound succinyl-CoA with the coupled synthesis of a "high-energy" nucleoside triphosphate (Reaction 5 of Fig. 21-1).* (Note: Enzyme names can refer to either the forward or the reverse reaction; in this case, succinyl-CoA synthetase and succinate thiokinase refer to the reverse reaction.) GTP is synthesized from GDP + P_i by the mammalian enzyme; plant and bacterial enzymes utilize ADP + P_i to form ATP. These reactions are nevertheless equivalent since ATP and GTP are rapidly interconverted through the action of nucleoside diphosphate kinase (Section 16-4C):

$$GTP + ADP \rightleftharpoons GDP + ATP \qquad \Delta G^{\circ\prime} = 0$$

a. The Succinyl-CoA Thioester Bond Energy Is Preserved through the Formation of a Series of "High-Energy" Phosphates

How does succinyl-CoA synthetase couple the exergonic hydrolysis of succinyl-CoA ($\Delta G^{\circ\prime} = -32.6$ kJ · mol^{-1}) to the endergonic formation of a nucleoside triphosphate ($\Delta G^{\circ\prime} = 30.5$ kJ · mol^{-1})? This question was answered through the creative use of isotope tracers. In the absence of succinyl-CoA, the spinach enzyme (which utilizes adenine nucleotides) catalyzes the transfer of ATP's γ phosphoryl group to ADP as detected by ^{14}C-labeling ADP and observing the label to appear in ATP. Such an isotope exchange reaction (Section 14-5D) suggests the participation of a phosphoryl–enzyme intermediate that mediates the reaction sequence:

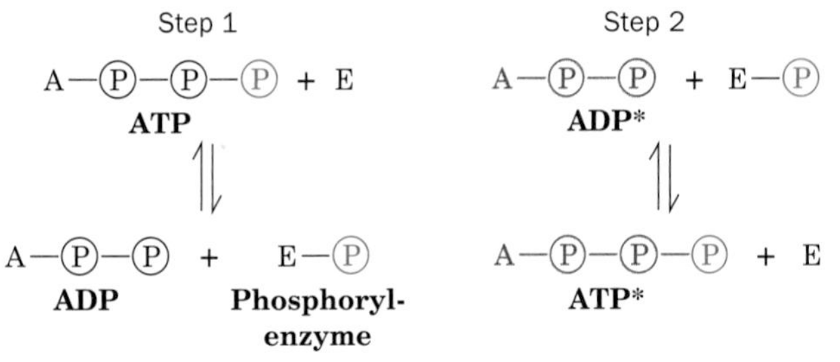

active phosphoryl–enzyme in which the phosphoryl group is covalently linked to the N3 position of a His residue.

When the succinyl-CoA synthetase reaction, which is freely reversible, is run in the direction of succinyl-CoA synthesis (opposite to its direction in the citric acid cycle) using [^{18}O]succinate as a substrate, ^{18}O is transferred from succinate to phosphate. Evidently, succinyl phosphate, a "high-energy" mixed anhydride, is transiently formed during the reaction.

These observations suggest the following three-step sequence for the mammalian succinyl-CoA synthetase reaction (Fig. 21-22):

1. Succinyl-CoA reacts with P_i to form succinyl phosphate and CoA (accounting for the ^{18}O-exchange reaction).

2. Succinyl phosphate's phosphoryl group is transferred to an enzyme His residue, releasing succinate (accounting for the 3-phosphoHis residue).

3. The phosphoryl group on the enzyme is transferred to GDP, forming GTP (accounting for the nucleoside diphosphate exchange reaction).

Note how in each of these steps *the "high-energy" succinyl-CoA's free energy of hydrolysis is conserved through the successive formation of "high-energy" compounds: First succinyl phosphate, then a 3-phosphoHis residue, and finally GTP.* The process is reminiscent of passing a hot potato.

b. A Pause for Perspective

Up to this point in the cycle, one acetyl equivalent has been completely oxidized to $2CO_2$. Two NADHs and one GTP (in equilibrium with ATP) have also been generated. In order to complete the cycle, succinate must be converted back to oxaloacetate. This is accomplished by the cycle's remaining three reactions.

F. *Succinate Dehydrogenase*

Succinate dehydrogenase catalyzes stereospecific dehydrogenation of succinate to fumarate (Reaction 6 of Fig. 21-1):

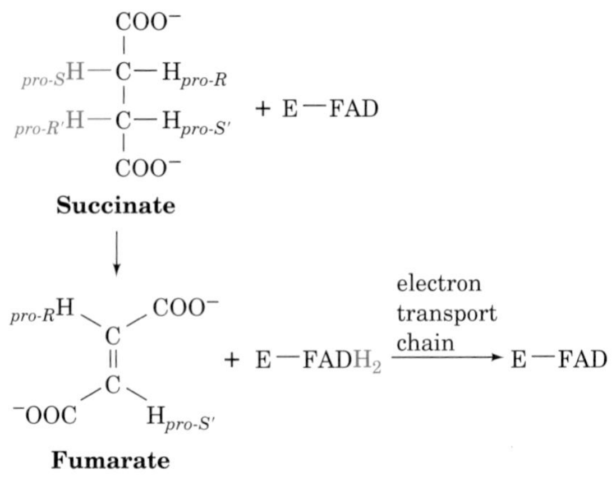

Step 1 · Step 2

Succinate

Fumarate

Indeed, this information led to the isolation of a kinetically

The enzyme is strongly inhibited by **malonate,** a structural

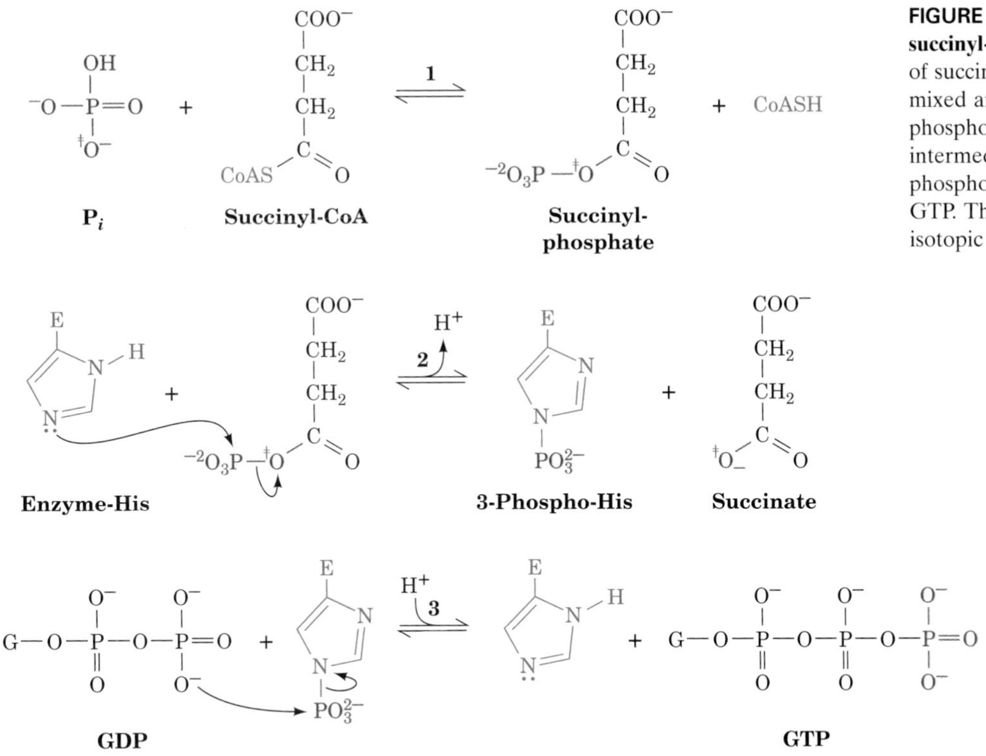

FIGURE 21-22   Reactions catalyzed by succinyl-CoA synthetase. (1) Formation of succinyl phosphate, a "high-energy" mixed anhydride. **(2)** Formation of phosphoryl–His, a "high-energy" intermediate. **(3)** Transfer of the phosphoryl group to GDP, forming GTP. The symbol ⧾ represents ^{18}O in isotopic labeling reactions.

analog of succinate and a classic example of a competitive inhibitor:

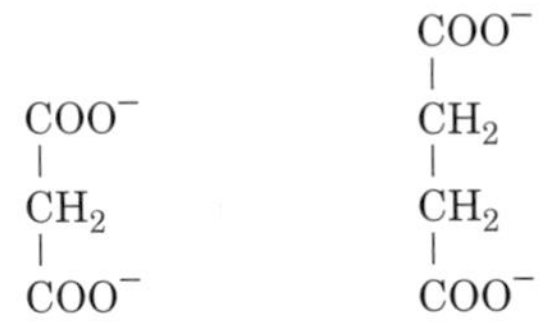

Recall that malonate inhibition of cellular respiration was one of the observations that led Krebs to hypothesize the citric acid cycle (Section 21-1B).

Succinate dehydrogenase contains an FAD, the reaction's electron acceptor. In general, FAD functions biochemically to oxidize alkanes to alkenes, whereas NAD^+ oxidizes alcohols to aldehydes or ketones. This is because the oxidation of an alkane (such as succinate) to an alkene (such as fumarate) is sufficiently exergonic to reduce FAD to $FADH_2$ but not to reduce NAD^+ to NADH. Alcohol oxidation, in contrast, can reduce NAD^+ (Table 16-4). Succinate dehydrogenase's FAD is covalently bound via its C8a atom to an enzyme His residue (Fig. 21-23). A covalent link between FAD and a protein is unusual; in most cases FAD is noncovalently although tightly bound to its

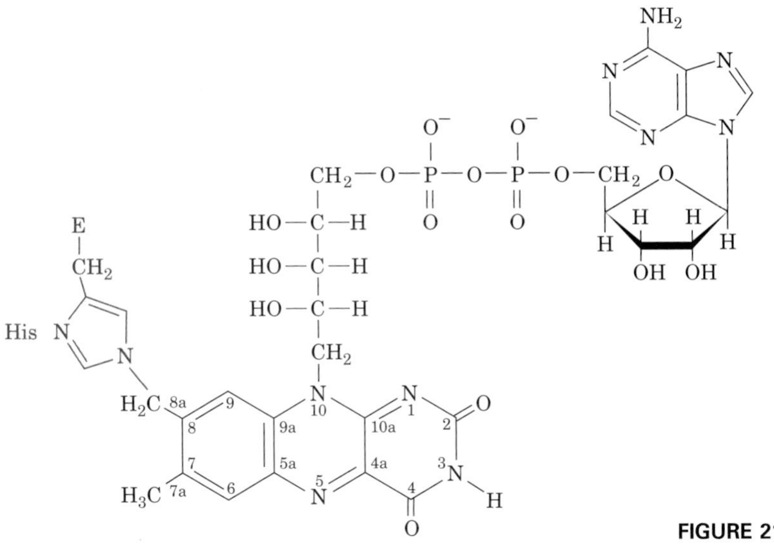

FIGURE 21-23   Covalent attachment of FAD to a His residue of succinate dehydrogenase.

associated enzyme (e.g., to dihydrolipoyl dehydrogenase; Section 21-2B).

How does succinate dehydrogenase's FADH$_2$ become reoxidized? Being permanently linked to the enzyme, this prosthetic group cannot function as a metabolite as does NADH. Rather, *succinate dehydrogenase (also known as Complex II) is reoxidized by coenzyme Q in the electron-transport chain*, an aspect of its function that we discuss in Section 22-2C. This rationalizes why succinate dehydrogenase, which is embedded in the inner mitochondrial membrane, is the only membrane-bound citric acid cycle enzyme. The others are all dissolved in the mitochondrial matrix (mitochondrial anatomy is described in Section 22-1A).

G. *Fumarase*

Fumarase (fumarate hydratase) catalyzes the hydration of fumarate's double bond to form (S)-malate (L-malate) (Reaction 7 of Fig. 21-1). Consideration of experiments that have contributed to our understanding of the fumarase mechanism illustrates the role played by independent investigations.

a. Conflicting Mechanistic Evidence: What Is the Sequence of H$^+$ and OH$^-$ Addition?

Experiments designed to establish whether the fumarase reaction occurs by a carbanion (OH$^-$ addition first) or carbocation (H$^+$ addition first) mechanism have provided contradictory information (Fig. 21-24). Evidence favoring the carbocation mechanism was obtained by studying the dehydration of (S)-malate (the fumarase reaction run in reverse) in H$_2$^{18}O. [^{18}O]Malate appears in the

reaction mixture more rapidly than it would if the ^{18}O were incorporated via a back reaction of the newly formed fumarate. This suggests the rapid formation of a carbocation intermediate at C2, from which OH$^-$ could exchange with ^{18}OH$^-$, followed by slow hydrogen removal at C3 (Fig. 21-24, *lower route*).

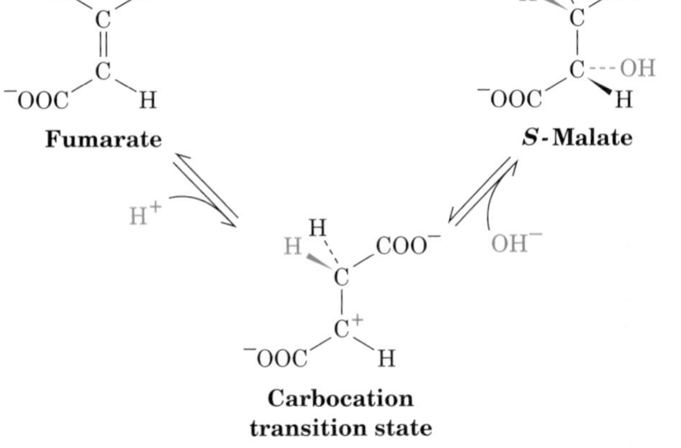

Other observations, however, indicate that the reaction occurs via the formation of a carbanion intermediate at C3 (Fig. 21-24, *upper route*). David Porter synthesized **3-nitro-2-(*S*)-hydroxypropionate,** which sterically resembles (*S*)-malate:

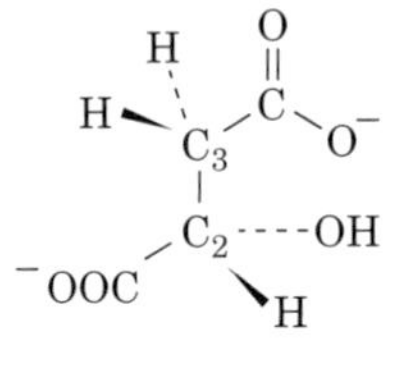

S-Malate

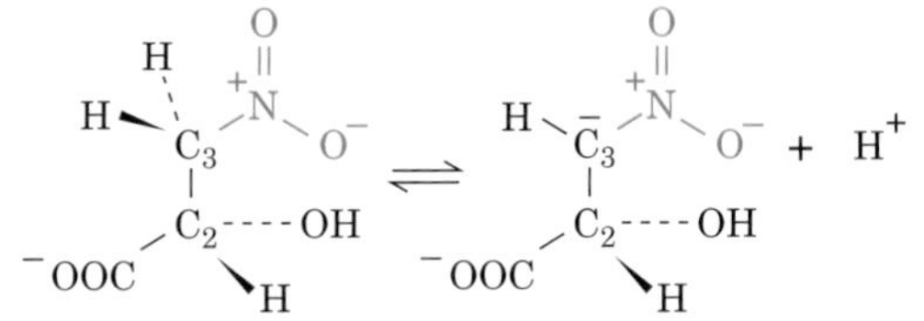

3-Nitro-2-S-hydroxypropionate

The nitro group's electron-withdrawing character renders the C3 protons relatively acidic (p$K \approx 10$). The resulting anion is an analog of the postulated C3 carbanion transition state of the fumarase reaction but not of the C2 carbocation transition state (Fig. 21-24; transition state analogs are discussed in Section 15-1F). This anion is, in fact, an excellent inhibitor of fumarase: It has an 11,000-fold higher binding affinity for the enzyme than does (*S*)-malate.

If the fumarase reaction proceeds by a carbanion mechanism, how can the rapid OH$^-$ exchange be explained? Conversely, if it has a carbocation mechanism, why is the nitro anion such an effective inhibitor? This contradictory set of observations makes only one thing clear: *When studying enzyme reaction mechanisms, it is always necessary to approach the problem from several different directions. One set of experiments should never be taken as proof, and interpretation should never be taken as fact.* Indeed, reinterpretation of the ^{18}O exchange experiment makes it consistent with a carbanion mechanism. It turns out that product release is the enzyme's rate-determining step:

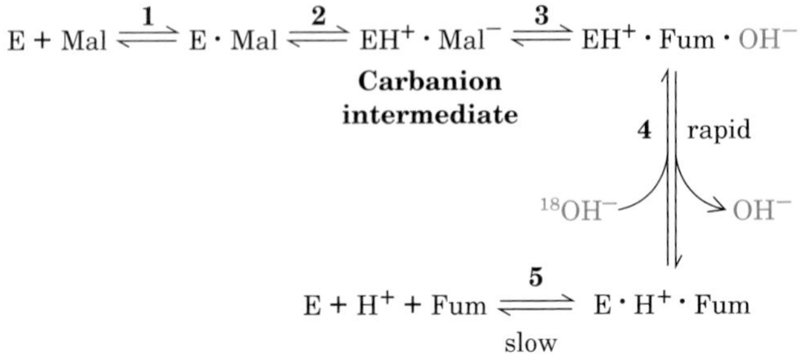

FIGURE 21-24   Possible mechanisms for the hydration of fumarate as catalyzed by fumarase.

Malate binds to fumarase **(1)**, forms a carbanion **(2)**, eliminates OH⁻ to form fumarate **(3)**, and rapidly releases OH⁻ from the enzyme surface **(4)**. Release of the other products **(5)** is slow. $^{18}OH^-$ can therefore exchange with OH⁻ to produce [^{18}O]malate more rapidly than the overall rate of the fumarase reaction.

H. *Malate Dehydrogenase*

Malate dehydrogenase catalyzes the final reaction of the citric acid cycle, the regeneration of oxaloacetate (Reaction 8 of Fig. 21-1). This occurs through the oxidation of (*S*)-malate's hydroxyl group to a ketone in an NAD⁺-dependent reaction:

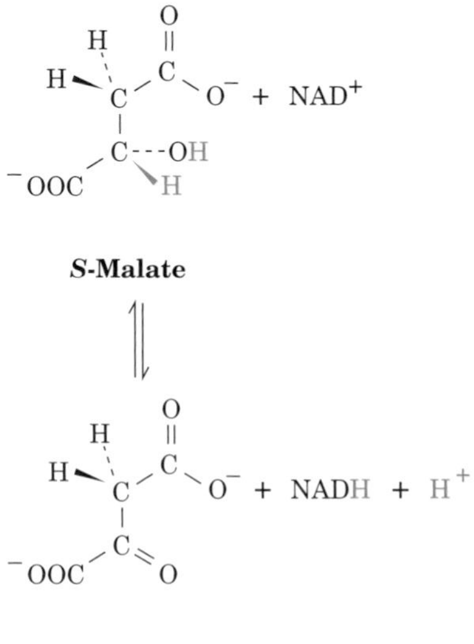

S-Malate

Oxaloacetate

The hydride ion released by the alcohol in this reaction is transferred to the *re* face of NAD⁺, the same face acted on by lactate dehydrogenase (Section 17-3A) and alcohol dehydrogenase (Section 17-3B). In fact, *X-ray crystallographic comparisons of the NAD⁺-binding domains of these three dehydrogenases indicate that they are remarkably similar and have led to the proposal that all NAD⁺-binding domains have evolved from a common ancestor (Section 9-3B).*

The $\Delta G^{\circ\prime}$ for the malate dehydrogenase reaction is $+29.7$ kJ · mol⁻¹; the concentration of oxaloacetate formed at equilibrium is consequently very low. Recall, however, that the reaction catalyzed by citrate synthase, the first enzyme in the cycle, is highly exergonic ($\Delta G^{\circ\prime} = -31.5$ kJ · mol⁻¹; Section 21-3A) because of the hydrolysis of citryl-CoA's thioester bond. We can now understand the necessity for such a seemingly wasteful process. It allows citrate formation to be exergonic at even low physiological concentrations of oxaloacetate, with the resultant initiation of another turn of the cycle.

I. *Integration of the Citric Acid Cycle*

a. The Impact of the Citric Acid Cycle on ATP Production

The foregoing discussion indicates that one turn of the citric acid cycle results in the following chemical transformations (Fig. 21-1):

1. One acetyl group is oxidized to two molecules of CO_2, a four-electron pair process (although, as we discuss below, it is not the carbon atoms of the entering acetyl group that are oxidized).

2. Three molecules of NAD⁺ are reduced to NADH, which accounts for three of the electron pairs.

3. One molecule of FAD is reduced to $FADH_2$, which accounts for the fourth electron pair.

4. One "high-energy" phosphate group is produced as GTP (or ATP).

The eight electrons abstracted from the acetyl group in the citric acid cycle subsequently pass into the electron-transport chain, where they ultimately reduce two molecules of O_2 to H_2O. The three electron pairs from NADH each produce about 3 ATPs via oxidative phosphorylation, whereas the one electron pair of $FADH_2$ produces about 2 ATPs. One turn of the citric acid cycle therefore results in the ultimate generation of about 12 ATPs. Electron transport and oxidative phosphorylation are the subject of Chapter 22.

b. Isotopic Tests of Citric Acid Cycle Stereochemistry

The reactions of the citric acid cycle have been confirmed through the use of radioactive tracer experiments that became possible in the late 1930s and early 1940s. At that time, compounds could be synthesized enriched with the stable isotope ^{13}C (detectable at the time by mass spectrometry and now also by NMR) or with the radioactive isotope ^{11}C, which has a half-life of only 20 min. ^{14}C, whose use was pioneered in the late 1940s by Samuel Ruben and Martin Kamen, has the advantage over ^{11}C that its half-life is 5715 years.

In one landmark experiment, [4-^{11}C]oxaloacetate was generated from $^{11}CO_2$ and pyruvate,

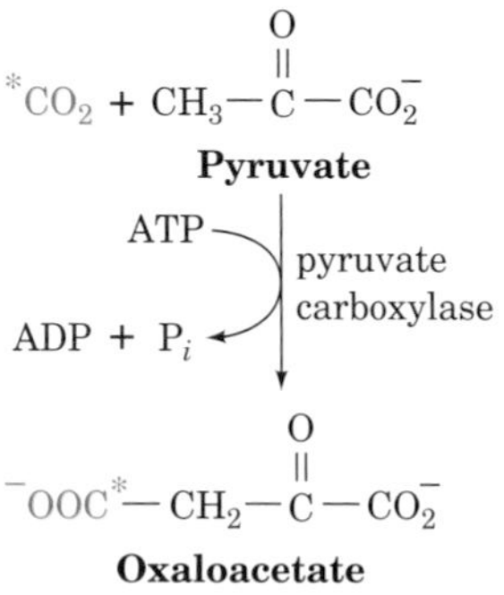

Pyruvate

Oxaloacetate

reacted in the citric acid cycle of metabolizing muscle cells, and the resulting cycle intermediates were isolated. Identification of the labeled position in the isolated α-ketoglutarate caused a furor. As we have seen in Sections 21-3A and 21-3B, citrate synthase and aconitase catalyze stereospecific reactions in which citrate synthase can distinguish between the two faces of oxaloacetate's carbonyl group and aconitase can distinguish between citrate's *pro-R* and *pro-S* carboxymethyl groups. In the early 1940s, however, the concept of prochirality had not been established; it was assumed that the two halves of citrate were indistinguishable (in nonenzymatic systems, this is, in effect, the case). It was therefore assumed that the radioactivity originally located at C4 in oxaloacetate (* in Fig. 21-1) would be scrambled in citrate so that its C1 and C6 atoms would be equally labeled, resulting in α-ketoglutarate labeled at both C1 and C5. In fact, only C1, the carboxyl group α to the keto group of α-ketoglutarate, was found to be radioactive (Fig. 21-1). This result threw the identity of the condensation product of oxaloacetate and acetyl-CoA into doubt. How could it be the symmetrical citrate molecule in light of such "conclusive" labeling experiments? This problem of which tricarboxylic acid was the cycle's original condensation product resulted in a name change from the citric acid cycle (proposed by Krebs) to the tricarboxylic acid (TCA) cycle.

In 1948, Alexander Ogston pointed out that citric acid, while symmetrical, is prochiral and thus can interact asymmetrically with the surface of aconitase (Section 13-2A). Even though citrate is now accepted as a cycle intermediate, the duality of the cycle's name persists.

While the net reaction of the cycle is oxidation of the carbon atoms of acetyl units to CO_2, the CO_2 lost in a given turn of the cycle is derived from the carbon skeleton of oxaloacetate. This can be shown by following the fate of isotopically labeled C4 of oxaloacetate (* in Fig. 21-1) through the citric acid cycle. We can see (Fig. 21-1) that it is lost as CO_2 at the α-ketoglutarate dehydrogenase reaction.

Experiments have been performed using [1-^{14}C]acetate, which cells convert to [1-^{14}C]acetyl-CoA:

$$CH_3 - {}^{*}COO^- + CoASH \xrightarrow[\text{thiokinase}]{\text{acetate} \quad ATP \quad AMP + PP_i} CH_3 - \overset{O}{\overset{\|}{C}} {}^{*}- SCoA$$

Acetate   **Acetyl-CoA**

Tracing the path of this label (‡ in Fig. 21-1) allows the reader to ascertain that the label is not lost as CO_2. It becomes scrambled (1/2‡ in Fig. 21-1) during the cycle, but not until the formation of succinate, the cycle's first rotationally symmetric (nonprochiral) intermediate.

4 ■ REGULATION OF THE CITRIC ACID CYCLE

In this section we consider how metabolic flux through the citric acid cycle is regulated. In our discussions of metabolic flux control (Section 17-4), we established that to understand how a metabolic pathway is controlled, we must identify the enzyme(s) that catalyzes its rate-determining step(s), the *in vitro* effectors of these enzymes, and the *in vivo* concentrations of these substances. *A proposed mechanism of flux control must demonstrate that an increase or decrease in flux is correlated with an increase or decrease in the concentration of the proposed effector.*

a. Citrate Synthase, Isocitrate Dehydrogenase, and α-Ketoglutarate Dehydrogenase Are the Citric Acid Cycle's Rate-Controlling Enzymes

Establishing the rate-determining steps of the citric acid cycle is more difficult than it is for glycolysis because most

TABLE 21-2   Standard Free Energy Changes ($\Delta G^{\circ\prime}$) and Physiological Free Energy Changes (ΔG) of Citric Acid Cycle Reactions

Reaction	Enzyme	$\Delta G^{\circ\prime}$ (kJ · mol^{-1})	ΔG (kJ · mol^{-1})
1	Citrate synthase	−31.5	Negative
2	Aconitase	~5	~0
3	Isocitrate dehydrogenase	−21	Negative
4	α-Ketoglutarate dehydrogenase multienzyme complex	−33	Negative
5	Succinyl-CoA synthetase	−2.1	~0
6	Succinate dehydrogenase	+6	~0
7	Fumarase	−3.4	~0
8	Malate dehydrogenase	+29.7	~0

of the cycle's metabolites are present in both mitochondria and cytosol and we do not know their distribution between these two compartments [recall that identifying a pathway's rate-determining step(s) requires determining the ΔG of each of its reactions from the concentrations of its substrates and products]. If, however, we assume equilibrium between the two compartments, we can use the total cell contents of these substances to estimate their mitochondrial concentrations. Table 21-2 gives the standard free energy changes for the eight citric acid cycle enzymes and estimates of the physiological free energy changes for the reactions in heart muscle or liver tissue. We can see that three of the enzymes are likely to function far from equilibrium under physiological conditions (negative ΔG): citrate synthase, NAD$^+$-dependent isocitrate dehydrogenase, and α-ketoglutarate dehydrogenase. We shall therefore focus our discussion on how these enzymes are regulated (Fig. 21-25).

b. The Citric Acid Cycle Is Largely Regulated by Substrate Availability, Product Inhibition, and Inhibition by Other Cycle Intermediates

In heart muscle, where the citric acid cycle functions mainly to generate ATP for use in muscle contraction, the enzymes of the cycle almost always act as a functional unit, with their metabolic flux proportional to the rate of cellular oxygen consumption. *Since oxygen consumption, NADH reoxidation, and ATP production are tightly coupled (Section 22-4), the citric acid cycle must be regulated by feedback mechanisms that coordinate its NADH production with energy expenditure.* Unlike the rate-limiting enzymes of glycolysis and glycogen metabolism, which utilize elaborate systems of allosteric control, substrate cycles, and covalent modification as flux control mechanisms, the

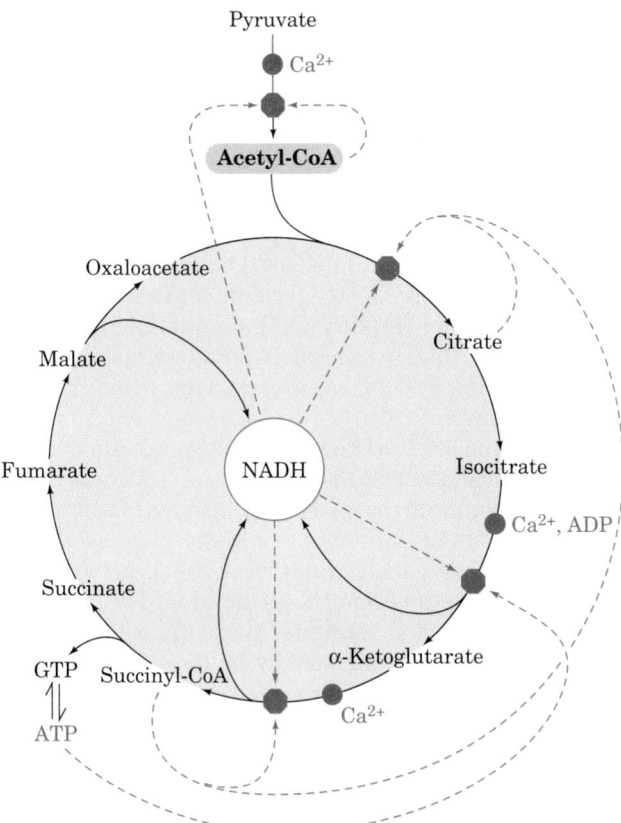

FIGURE 21-25 Regulation of the citric acid cycle. This diagram of the citric acid cycle and the pyruvate dehydrogenase reaction indicates their points of inhibition (*red octagons*) and the pathway intermediates that function as inhibitors (*dashed red arrows*). ADP and Ca^{2+} (*green dots*) are activators. 🐾 **See the Animated Figures**

regulatory enzymes of the citric acid cycle seem to be controlled almost entirely in three simple ways: (1) substrate availability, (2) product inhibition, and (3) competitive feedback inhibition by intermediates farther along the cycle. We shall encounter several examples of these straightforward mechanisms in the following discussion.

Perhaps the most crucial regulators of the citric acid cycle are its substrates, acetyl-CoA and oxaloacetate, and its product NADH. Both acetyl-CoA and oxaloacetate are present in mitochondria at concentrations that do not saturate citrate synthase. The metabolic flux through the enzyme therefore varies with substrate concentration and is subject to control by substrate availability. The production of acetyl-CoA from pyruvate is regulated by the activity of pyruvate dehydrogenase (Section 21-2B). Oxaloacetate is in equilibrium with malate, its concentration fluctuating with the [NADH]/[NAD$^+$] ratio according to the equilibrium expression

$$K = \frac{[\text{oxaloacetate}][\text{NADH}]}{[\text{malate}][\text{NAD}^+]}$$

In the transition from low to high work and respiration rates, mitochondrial [NADH] decreases. The consequent

increase in [oxaloacetate] stimulates the citrate synthase reaction, which controls the rate of citrate formation.

The observation that [citrate] invariably falls as the workload increases indicates that the rate of citrate removal increases more than its rate of formation. The rate of citrate removal is governed by NAD$^+$-dependent isocitrate dehydrogenase (aconitase functions close to equilibrium), which is strongly inhibited *in vitro* by NADH (product inhibition). Citrate synthase is also inhibited by NADH. Evidently, NAD$^+$-dependent isocitrate dehydrogenase is more sensitive to [NADH] changes than citrate synthase.

The decrease in [citrate] that occurs on transition from low to high work and respiration rates results in a domino effect:

1. Citrate is a competitive inhibitor of oxaloacetate for citrate synthase (product inhibition); the fall in [citrate] caused by increased isocitrate dehydrogenase activity increases the rate of citrate formation.

2. α-Ketoglutarate dehydrogenase is also strongly inhibited by its products, NADH and succinyl-CoA. Its activity therefore increases when [NADH] decreases.

3. Succinyl-CoA also competes with acetyl-CoA in the citrate synthase reaction (competitive feedback inhibition).

This interlocking system serves to keep the citric acid cycle coordinately regulated.

c. ADP, ATP, and Ca^{2+} Are Allosteric Regulators of Citric Acid Cycle Enzymes

In vitro studies on the enzymes of the citric acid cycle have identified a few allosteric activators and inhibitors. Increased workload is accompanied by increased [ADP] resulting from the consequent increased rate of ATP hydrolysis. ADP acts as an allosteric activator of isocitrate dehydrogenase by decreasing its apparent K_M for isocitrate. ATP, which builds up when muscle is at rest, inhibits this enzyme.

Ca^{2+}, among its many biological functions, is an essential metabolic regulator. It stimulates glycogen breakdown (Section 18-3C), triggers muscle contraction (Section 35-3C), and mediates many hormonal signals as a second messenger (Section 19-4A). Ca^{2+} also plays an important role in the regulation of the citric acid cycle (Fig. 21-25). It activates pyruvate dehydrogenase phosphatase and inhibits pyruvate dehydrogenase kinase, thereby activating the PDC to produce acetyl-CoA (Fig. 21-17*b*). In addition, Ca^{2+} activates both isocitrate dehydrogenase and α-ketoglutarate dehydrogenase. Thus, the same signal stimulates muscle contraction and the production of the ATP to fuel it.

In the liver, the role of the citric acid cycle is more complex than in heart muscle. The liver synthesizes many substances required by the body including glucose, fatty acids, cholesterol, amino acids, and porphyrins. Reactions of the citric acid cycle play a part in many of these biosynthetic pathways in addition to their role in energy metabolism.

In the next section, we discuss the contribution of the citric acid cycle to these processes.

d. Are the Enzymes of the Citric Acid Cycle Organized into a Metabolon?

Considerable efficiency can be gained by organizing the enzymes of a metabolic pathway such that the enzymes catalyzing its sequential steps interact to channel intermediates between them. Indeed, we saw this to be the case with the PDC (Section 21-2A). The advantages of such an assembly, termed a **metabolon,** include the protection of labile intermediates and the increase of their local concentration for more efficient catalysis. Considerable effort has been expended to obtain evidence for such interactions in the major metabolic pathways including glycolysis and the citric acid cycle. However, since citric acid cycle intermediates must be available for use in other metabolic pathways, any complexes between citric acid cycle enzymes are likely to be weak and therefore unable to withstand the laboratory manipulations necessary to isolate them.

Despite the foregoing, specific interactions have been demonstrated *in vivo* between the members of several pairs of citric acid cycle enzymes, including those between citrate synthase and malate dehydrogenase. For example, Paul Srere isolated the gene for a mutant citrate synthase in yeast that he termed an "assembly mutant" because it has normal enzymatic activity *in vitro* but nevertheless causes a citric acid cycle deficiency *in vivo*. This mutation occurs in a highly conserved 13-residue segment of the enzyme (Pro 354–Pro 366 in yeast), which forms a solvent-exposed loop that could interact with other proteins. To further investigate this phenomenon, Srere constructed a plasmid expressing a fusion protein (Section 5-5G) consisting of the wild-type citrate synthase peptide grafted to the C-terminal end of green fluorescent protein (GFP; Section 5-5H). If the peptide actually binds to malate dehydrogenase, the expression of this enzymatically inactive fusion protein in yeast would be expected to inhibit the citric acid cycle. This is, in fact, the case as judged by the severely decreased ability of such yeast to grow on acetate, a metabolite that can only be metabolized via the citric acid cycle. Moreover, replacing the citrate synthase peptide in the fusion protein with an unrelated peptide only slightly decreases the normal growth rate on acetate, thereby demonstrating that the specific sequence of the citrate synthase peptide is the major cause of this growth inhibition. If the decrease in growth rate is actually caused by competition of the peptide with citrate synthase for an interacting partner in a metabolon, then this should be overcome by the overexpression of either citrate synthase or its interaction partner. Indeed, when either citrate synthase or malate dehydrogenase was overexpressed in yeast expressing the GFP–citrate synthase peptide fusion protein, the growth rate on acetate was restored. However, the overexpression of aconitase did not overcome this growth inhibition. These observations support the hypothesis that malate dehydrogenase and citrate synthase must interact for optimal citric acid cycle function and identify the in-

teraction site for malate dehydrogenase on citrate synthase as the 13-residue Pro 354–Pro 366 peptide.

e. A Bacterial Isocitrate Dehydrogenase Is Regulated by Phosphorylation

Escherichia coli isocitrate dehydrogenase is a dimer of identical 416-residue subunits that is inactivated by phosphorylation of its Ser 113, an active site residue. In contrast, most other enzymes that are known to be subject to covalent modification/demodification, for example, glycogen phosphorylase (Section 18-3), are phosphorylated at allosteric sites. In the case of isocitrate dehydrogenase, phosphorylation renders the enzyme unable to bind its substrate, isocitrate.

Comparison of the X-ray structures, determined by Daniel Koshland and Robert Stroud, of isocitrate dehydrogenase alone, in its phosphorylated form, and with bound isocitrate reveals only small conformational differences, suggesting that electrostatic repulsions between the anionic isocitrate and Ser phosphate groups prevent the enzyme from binding substrate. Evidently, phosphorylation can regulate enzyme activity by directly interfering with active site ligand binding as well as by inducing a conformational change from an allosteric site.

5 ■ THE AMPHIBOLIC NATURE OF THE CITRIC ACID CYCLE

Ordinarily one thinks of a metabolic pathway as being either catabolic with the release (and conservation) of free energy, or anabolic with a requirement for free energy. The citric acid cycle is, of course, catabolic because it involves degradation and is a major free energy conservation system in most organisms. Cycle intermediates are only required in catalytic amounts to maintain the degradative function of the cycle. However, several biosynthetic pathways utilize citric acid cycle intermediates as starting materials (anabolism). The citric acid cycle is therefore **amphibolic** (both anabolic and catabolic).

All of the biosynthetic pathways that utilize citric acid cycle intermediates also require free energy. Consequently, the catabolic function of the cycle cannot be interrupted; *cycle intermediates that have been siphoned off must be replaced.* Although the mechanistic aspects of the enzymes involved in the pathways that utilize and replenish citric acid cycle intermediates are discussed in subsequent chapters, it is useful to briefly mention these metabolic interconnections here (Fig. 21-26).

a. Pathways That Utilize Citric Acid Cycle Intermediates

Reactions that utilize and therefore drain citric acid cycle intermediates are called **cataplerotic reactions** (emptying; Greek: *cata*, down + *plerotikos*, to fill). These reactions serve not only to synthesize important products but also to avoid the inappropriate buildup of citric acid cycle intermediates in the mitochondrion, for example, when there is a high rate of breakdown of amino acids to citric

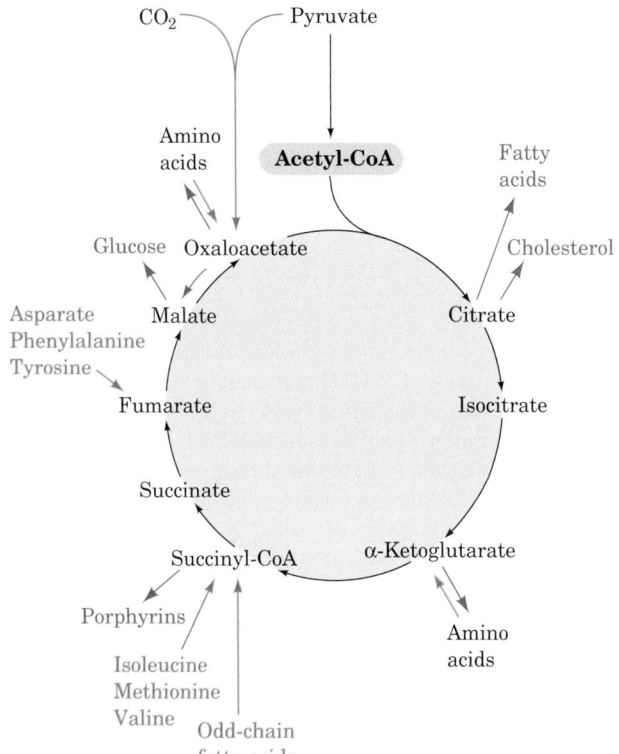

FIGURE 21-26 Amphibolic functions of the citric acid cycle.
This diagram indicates the positions at which intermediates are
cataplerotically drawn off for use in anabolic pathways (*red
arrows*) and the points where anaplerotic reactions replenish
depleted cycle intermediates (*green arrows*). Reactions
involving amino acid transamination and deamination are
reversible, so their direction varies with metabolic demand.
🔁 See the Animated Figures

acid cycle intermediates (Section 26-3). Cataplerotic reactions occur in the following pathways:

1. Glucose biosynthesis (gluconeogenesis; Section
23-1), which occurs in the cytosol, utilizes oxaloacetate as
its starting material. Oxaloacetate is not transported across
the mitochondrial membrane, but malate is. Malate that
has been transported across the mitochondrial membrane
is converted to oxaloacetate in the cytosol for gluconeogenesis.

2. Lipid biosynthesis, which includes **fatty acid biosynthesis** (Section 25-4) and **cholesterol biosynthesis** (Section
25-6A), is a cytosolic process that requires acetyl-CoA.
Acetyl-CoA is generated in the mitochondrion and is not
transported across the inner mitochondrial membrane.
Cytosolic acetyl-CoA is therefore generated by the breakdown of citrate, which can cross the inner mitochondrial
membrane, in a reaction catalyzed by **ATP-citrate lyase**
(Section 25-4D):

$$ATP + citrate + CoA \rightleftharpoons$$
$$ADP + P_i + oxaloacetate + acetyl\text{-}CoA$$

3. Amino acid biosynthesis, which utilizes citric acid cycle intermediates in two ways. α-Ketoglutarate is con-

verted to glutamate in a reductive amination reaction
involving either NAD^+ or $NADP^+$ catalyzed by **glutamate
dehydrogenase** (Section 26-1):

$$\alpha\text{-Ketoglutarate} + NAD(P)H + NH_4^+ \rightleftharpoons$$
$$glutamate + NAD(P)^+ + H_2O$$

α-Ketoglutarate and oxaloacetate are also used to synthesize glutamate and aspartate in transamination reactions
(Section 26-1):

$$\alpha\text{-Ketoglutarate} + alanine \rightleftharpoons glutamate + pyruvate$$

and

$$Oxaloacetate + alanine \rightleftharpoons aspartate + pyruvate$$

4. Porphyrin biosynthesis (Section 26-4A), which utilizes succinyl-CoA as a starting material.

5. Complete oxidation of amino acids, which requires
that the citric acid cycle intermediates to which the amino
acids are degraded be converted first to PEP [in a reaction
catalyzed by **phosphoenolpyruvate carboxykinase (PEPCK;**
Section 23-1): Oxaloacetate + GTP $\rightleftharpoons$ PEP + GDP],
then to pyruvate by pyruvate kinase (Section 17-2J), and
finally to acetyl-CoA by pyruvate dehydrogenase (Section
21-2A).

**b. Reactions That Replenish Citric Acid Cycle
Intermediates**

Reactions that replenish citric acid cycle intermediates
are called **anaplerotic reactions** (filling up, Greek: *ana,* up).
The main reaction of this type is catalyzed by **pyruvate carboxylase,** which produces oxaloacetate (Section 23-1A):

$$Pyruvate + CO_2 + ATP + H_2O \rightleftharpoons$$
$$oxaloacetate + ADP + P_i$$

This enzyme "senses" the need for more citric acid cycle
intermediates through its activator, acetyl-CoA. Any
decrease in the rate of the cycle caused by insufficient
oxaloacetate or other cycle intermediates results in an increased level of acetyl-CoA because of its underutilization.
This activates pyruvate carboxylase, which replenishes oxaloacetate, increasing the rate of the cycle. Of course, if
the citric acid cycle is inhibited at some other step, by high
NADH concentration, for example, increased oxaloacetate concentration will not activate the cycle. The excess
oxaloacetate instead equilibrates with malate, which is
transported out of the mitochondria for use in gluconeogenesis.

Degradative pathways generate citric acid cycle intermediates:

1. Oxidation of odd-chain fatty acids (Section 25-2E)
leads to the production of succinyl-CoA.

2. Breakdown of the amino acids isoleucine, methionine, and valine (Section 26-3E) also leads to the production of succinyl-CoA.

3. Transamination and deamination of amino acids lead
to the production of α-ketoglutarate and oxaloacetate.

These reactions are reversible and, depending on metabolic demand, serve to remove or replenish these citric acid cycle intermediates.

The citric acid cycle is truly at the center of metabolism (see Fig. 16-1). Its reduced products, NADH and FADH$_2$, are reoxidized by the electron-transport chain during oxidative phosphorylation and the free energy released is coupled to the biosynthesis of ATP. Citric acid cycle intermediates are utilized in the biosynthesis of many vital cellular constituents. In the next few chapters we shall explore the interrelationships of these pathways in more detail.

CHAPTER SUMMARY

1 ■ Cycle Overview The citric acid cycle, the common mode of oxidative metabolism in most organisms, is mediated by eight enzymes that collectively convert 1 acetyl-CoA to 2 CO$_2$ molecules so as to yield 3 NADHs, 1 FADH$_2$, and 1 GTP (or ATP). The NADH and FADH$_2$ are oxidized by O$_2$ in the electron-transport chain with the concomitant synthesis of around 11 more ATPs, yielding a total of about 12 ATPs for one turn of the citric acid cycle.

2 ■ Metabolic Sources of Acetyl-Coenzyme A Pyruvate, the end product of glycolysis under aerobic conditions, is converted to acetyl-CoA by the pyruvate dehydrogenase multienzyme complex (PDC), a cubic or dodecahedral cluster of three enzymes: pyruvate dehydrogenase, dihydrolipoyl transacetylase, and dihydrolipoyl dehydrogenase. The pyruvate dehydrogenase subunit catalyzes the conversion of pyruvate to CO$_2$ and a hydroxyethyl-TPP intermediate. The latter is channeled to dihydrolipoyl transacetylase, which oxidizes the hydroxyethyl group to acetate and transfers it to CoA to form acetyl-CoA. The lipoamide prosthetic group, which is reduced to the dihydro form in the process, is reoxidized by dihydrolipoamide dehydrogenase in a reaction involving bound FAD that reduces NAD$^+$ to NADH. Dihydrolipoamide transacetylase is inactivated by the formation of a covalent adduct between lipoamide and As(III) compounds. Dihydrolipoamide dehydrogenase, which closely resembles glutathione reductase, catalyzes a 2-stage reaction. In the first stage, dihydrolipoamide reduces the enzyme's redox-active disulfide group, yielding the reaction's first product, lipoamide. In the second stage, NAD$^+$ reoxidizes the reduced enzyme through the intermediacy of the enzyme's FAD prosthetic group, thereby closing the catalytic cycle and yielding the enzyme's second product, NADH. The activity of the pyruvate dehydrogenase complex varies with the [NADH]/[NAD$^+$] and [acetyl-CoA]/[CoA] ratios. In eukaryotes, the pyruvate dehydrogenase subunit is also inactivated by phosphorylation of a specific Ser residue and is reactivated by its removal. These modifications are mediated, respectively, by pyruvate dehydrogenase kinase and pyruvate dehydrogenase phosphatase, which are components of the multienzyme complex and respond to the levels of metabolic intermediates such as NADH and acetyl-CoA.

3 ■ Enzymes of the Citric Acid Cycle Citrate is formed by the condensation of acetyl-CoA and oxaloacetate by citrate synthase. The citrate is dehydrated to *cis*-aconitate and then rehydrated to isocitrate in a stereospecific reaction catalyzed by aconitase. This enzyme is specifically inhibited by (2*R*,3*R*)-2-fluorocitrate, which is enzymatically synthesized from fluoroacetate and oxaloacetate. Isocitrate is oxidatively decarboxylated to α-ketoglutarate by isocitrate dehydrogenase, which produces NADH and CO$_2$. The α-ketoglutarate, in turn, is oxidatively decarboxylated by α-ketoglutarate dehydrogenase, a multienzyme complex homologous to the PDC. This reaction generates the second NADH and CO$_2$. The resulting succinyl-CoA is converted to succinate with the generation of GTP (ATP in plants and bacteria) by succinyl-CoA synthetase. The succinate is stereospecifically dehydrogenated to fumarate by succinate dehydrogenase in a reaction that generates FADH$_2$. The final two reactions of the citric acid cycle, which are catalyzed by fumarase and malate dehydrogenase, in turn hydrate fumarate to (*S*)-malate and oxidize this alcohol to its corresponding ketone, oxaloacetate, with concomitant production of the pathway's third and final NADH.

4 ■ Regulation of the Citric Acid Cycle The enzymes of the citric acid cycle act as a functional unit that keeps pace with the metabolic demands of the cell. The flux-controlling enzymes appear to be citrate synthase, isocitrate dehydrogenase, and α-ketoglutarate dehydrogenase. Their activities are controlled by substrate availability, product inhibition, inhibition by cycle intermediates, and activation by Ca^{2+}. The enzymes of the citric acid cycle may be organized, *in vivo*, into a metabolon for channeling the products of one enzyme to the next enzyme in the cycle.

5 ■ The Amphibolic Nature of the Citric Acid Cycle Several anabolic pathways cataplerotically utilize citric acid cycle intermediates as starting materials. These essential substances are replaced by anaplerotic reactions of which the major one is synthesis of oxaloacetate from pyruvate and CO$_2$ by pyruvate carboxylase.

 # REFERENCES

HISTORY

Holmes, F.L., *Hans Krebs:* Vol. 1: *The Formation of a Scientific Life, 1900–1933;* and Vol. 2: *Architect of Intermediary Metabolism, 1933–1937,* Oxford University Press (1991 and 1993). [The biography of the discoverer of the citric acid cycle through the time of its discovery.]

Kornberg, H.L., Tricarboxylic acid cycles, *BioEssays* **7,** 236–238 (1987). [A historical synopsis of the intellectual background leading to the discovery of the citric acid cycle.]

Krebs, H.A., The history of the tricarboxylic acid cycle, *Perspect. Biol. Med.* **14,** 154–170 (1970).

PYRUVATE DEHYDROGENASE MULTIENZYME COMPLEX

Izard, T., Ævarsson, A., Allen, M.D., Westphal, A.H., Perham, R.N., de Kok, A., and Hol, W.G.J., Principles of quasi-equivalence and euclidean geometry govern the assembly of cubic and dodecahedral cores of pyruvate dehydrogenase, *Proc. Natl. Acad. Sci.* **96,** 1240–1245 (1999).

Karplus, P.A. and Schulz, G.E., Refined structure of glutathione reductase at 1.54 Å resolution, *J. Mol. Biol.* **195,** 701–729 (1987).

Karplus, P.A. and Schulz, G.E., Substrate binding and catalysis by glutathione reductase derived from refined enzyme: Substrate crystal structures at 2Å resolution, *J. Mol. Biol.* **210,** 163–180 (1989).

Mattevi, A., Obmolova, G., Sokatch, J.R., Betzel, C., and Hol, W.G.J., The refined crystal structure of *Pseudomonas putida* lipoamide dehydrogenase complexed with NAD$^+$ at 2.45 Å resolution, *Proteins* **13,** 336–351 (1992); *and* Mattevi, A., Schierbeek, A.J., and Hol, W.G.J., Refined crystal structure of lipoamide dehydrogenase from *Azotobacter vinelandii* at 2.2 Å resolution. A comparison with the structure of glutathione reductase, *J. Mol. Biol.* **220,** 975–994 (1991).

Patel, M.S. and Korotchkina, L.G., The biochemistry of the pyruvate dehydrogenase complex, *Biochem. Mol. Biol. Educ.* **31,** 5–15 (2003).

Patel, M.S., Roche, T.E., and Harris, R.A. (Eds.), *Alpha-Keto Acid Dehydrogenase Complexes,* Birkhäuser (1996).

Perham, R.N., Swinging arms and swinging domains in multifunctional enzymes: Catalytic machines for multistep reactions, *Annu. Rev. Biochem.* **69,** 961–1004 (2000). [An authoritative review on multienzyme complexes.]

Reed, L.J., A trail of research from lipoic acid to α-keto acid dehydrogenase complexes, *J. Biol. Chem.* **276,** 38329–38336 (2001). [A scientific memoir.]

Reed, L.J. and Hackert, M.L., Structure-function relationships in dihydrolipoamide acyltransferases, *J. Biol. Chem.* **265,** 8971–8974 (1990).

Roche, T.E., Baker, J.C., Yan, X., Hiromasa, Y., Gong, X., Peng, T., Dong, J., Turkan, A., and Kasten, S.E., Distinct regulatory properties of pyruvate dehydrogenase kinase and phosphatase isoforms, *Prog. Nucl. Acid Res. Mol. Biol.* **70,** 33–75 (2001).

Wegenknecht, T., Grassucci, R., Radke, G.A., and Roche, T.E., Cryoelectron microscopy of mammalian pyruvate dehydrogenase complex, *J. Biol. Chem.* **266,** 24650–24656 (1991).

Williams, C.H., Jr., Lipoamide dehydrogenase, glutathione reductase, thioredoxin reductase, and mercuric ion reductase—A family of flavoenzyme transhydrogenases, *in* Müller, F. (Ed.), *Chemistry and Biochemistry of Flavoenzymes,* Vol. III, pp. 121–211, CRC Press (2000).

Zhou, Z.H., McCarthy, D.B., O'Connor, C.M., Reed, L.J., and Stoops, J.K., The remarkable structural and functional organization of the eukaryotic pyruvate dehydrogenase complexes, *Proc. Natl. Acad. Sci.* **98,** 14802–14807 (2001); *and* Stoops, J.K., Cheng, R.H., Yazdi, M.A., Maeng, C.-Y., Schroeter, J.P., Klueppelberg, U., Kolodziej, S.J., Baker, T.S., and Reed, L.J., On the unique structural organization of the *Saccharomyces cerevisiae* pyruvate dehydrogenase complex, *J. Biol. Chem.* **272,** 5757–5764 (1997).

ENZYMES OF THE CITRIC ACID CYCLE

Beinert, H., Kennedy, M.C., and Stout, D.C., Aconitase as ironsulfur protein, enzyme, and iron-regulatory protein, *Chem. Rev.* **96,** 2335–2374 (1996).

Bolduc, J.M., Dyer, D.H., Scott, W.G., Singer, P., Sweet, R.M., Koshland, D.E., Jr., and Stoddard, B.L., Mutagenesis and Laue structures of enzyme intermediates: Isocitrate dehydrogenase, *Science* **268,** 1312–1318 (1995).

Cleland, W.W. and Kreevoy, M.M., Low-barrier hydrogen bonds and enzymic catalysis, *Science* **264,** 1887–1890 (1994).

Karpusas, M., Branchaud, B., and Remington, S.J., Proposed mechanism for the condensation reaction of citrate synthase: 1.9-Å structure of the ternary complex with oxaloacetate and carboxymethyl coenzyme A, *Biochemistry* **29,** 2213–2219 (1990).

Kurz, L.C., Nakra, T., Stein, R., Plungkhen, W., Riley, M., Hsu, F., and Drysdale, G.R., Effects of changes in three catalytic residues on the relative stabilities of some of the intermediates and transition states in the citrate synthase reaction, *Biochemistry* **37,** 9724–9737 (1998).

Lauble, H., Kennedy, M.C., Beinert, H., and Stout, D.C., Crystal structures of aconitase with isocitrate and nitroisocitrate bound, *Biochemistry* **31,** 2735–2748 (1992).

Mulholland, A.J., Lyne, P.D., and Karplus, M., Ab initio QM/MM study of the citrate synthase mechanism. A low-barrier hydrogen bond is not involved, *J. Am. Chem. Soc.* **122,** 534–535 (2000).

Porter, D.J.T. and Bright, H.J., 3-Carbanionic substrate analogues bind very tightly to fumarase and aspartase, *J. Biol. Chem.* **255,** 4772–4780 (1980).

Remington, S.J., Structure and mechanism of citrate synthase, *Curr. Top. Cell Regul.* **33,** 202–229 (1992); *and* Mechanisms of citrate synthase and related enzymes (triose phosphate isomerase and mandelate racemase), *Curr. Opin. Struct. Biol.* **2,** 730–735 (1992).

Walsh, C., *Enzymatic Reaction Mechanisms,* Freeman (1979). [Contains discussions of the mechanisms of various citric acid cycle enzymes.]

Wolodk, W.T., Fraser, M.E., James, M.N.G., and Bridger, W.A., The crystal structure of succinyl-CoA synthetase from *Escherichia coli* at 2.5Å resolution, *J. Biol. Chem.* **269,** 10883–10890 (1994).

Zheng, L., Kennedy, M.C., Beinert, H., and Zalkin, H. Mutational analysis of active site residues in pig heart aconitase, *J. Biol. Chem.* **267,** 7895–7903 (1992).

METABOLIC POISONS

Committee on the Medical and Biological Effects of Environmental Pollutants, Subcommittee on Arsenic, *Arsenic,* National Research Council, National Academy of Sciences (1977).

Gibble, G.W., Fluoroacetate toxicity, *J. Chem. Ed.* **50,** 460–462 (1973).

Jones, D.E.H. and Ledingham, K.W.D., Arsenic in Napoleon's wallpaper, *Nature* **299,** 626–627 (1982).

Lauble, H., Kennedy, M.C., Emptage, M.H., Beinert, H., and Stout, C.D., The reaction of fluorocitrate with aconitase and the crystal structure of the enzyme-inhibitor complex, *Proc. Natl. Acad. Sci.* **93,** 13699–13703 (1996).

Winslow, J.H., *Darwin's Victorian Malady,* American Philosophical Society (1971).

CONTROL MECHANISMS

Hurley, J.H., Dean, A.M., Sohl, J.L., Koshland, D.E., Jr., and Stroud, R.M., Regulation of an enzyme by phosphorylation at the active site, *Science* **249,** 1012–1016 (1990).

Owen, O.E., Kalhan, S.C., and Hanson, R.W., The key role of anaplerosis and cataplerosis for citric acid cycle function, *J. Biol. Chem.* **277,** 30409–30412 (2002).

Reed, L.J., Damuni, Z., and Merryfield, M.L., Regulation of mammalian pyruvate and branched-chain α-keto-acid dehydrogenase complexes by phosphorylation and dephosphorylation, *Curr. Top. Cell. Regul.* **27**, 41–49 (1985).

Srere, P.A., Sherry, A.D., Malloy, C.R., and Sumegi, B., Channelling in the Krebs tricarboxylic acid cycle, *in* Agius, L. and Sherratt, H.S.A. (Eds.), *Channelling in Intermediary Metabolism*, pp. 201–217, Portland Press (1997).

Stroud, R.M., Mechanisms of biological control by phosphorylation, *Curr. Opin. Struct. Biol.* **1**, 826–835 (1991). [Reviews, among other things, the inactivation of isocitrate dehydrogenase by phosphorylation.]

Vélot, C., Mixon, M.B., Teige, M., and Srere, P.A., Model of a quinary structure between Krebs TCA cycle enzymes: A model for the metabolon, *Biochemistry* **36**, 14271–14276 (1997).

Vélot, C. and Srere, P.A., Reversible transdominant inhibition of a metabolic pathway. In vivo evidence of interaction between two sequential tricarboxylic acid cycle enzymes in yeast. *J. Biol. Chem.* **275**, 12926–12933 (2000).

PROBLEMS

1. Trace the course of the radioactive label in [2-^{14}C]glucose through glycolysis and the citric acid cycle. At what point(s) in the cycle will the radioactivity be released as $^{14}CO_2$? How many turns of the cycle will be required for complete conversion of the radioactivity to CO_2? Repeat this problem for pyruvate that is ^{14}C-labeled at its methyl group.

2. The reaction of glutathione reductase with an excess of NADPH in the presence of arsenite yields a nonphysiological four-electron reduced form of the enzyme. What is the chemical nature of this catalytically inactive species?

3. Two-electron reduced dihydrolipoamide dehydrogenase (EH$_2$), but not the oxidized enzyme (E), reacts with iodoacetate (ICH$_2$COO$^-$) to yield an inactive enzyme. Explain.

4. Given the following information, calculate the physiological ΔG of the isocitrate dehydrogenase reaction at 25°C and pH 7.0: [NAD$^+$]/[NADH] = 8; [α-ketoglutarate] = 0.1 mM; [isocitrate] = 0.02 mM; assume standard conditions for CO_2 ($\Delta G°'$ is given in Table 21-2). Is this reaction a likely site for metabolic control? Explain.

5. The oxidation of acetyl-CoA to two molecules of CO_2 involves the transfer of four electron pairs to redox coenzymes. In which of the cycle's reactions do these electron transfers occur? Identify the redox coenzyme in each case. For each reaction, draw the structural formulas of the reactants, intermediates, and products and show, using curved arrows, how the electrons are transferred.

6. The citrate synthase reaction has been proposed to proceed via the formation of the enol(ate) form of acetyl-CoA. How, then, would you account for the observation that ^{3}H is not incorporated into acetyl-CoA when acetyl-CoA is incubated with citrate synthase in ^{3}H$_2$O?

7. Malonate is a competitive inhibitor of succinate in the succinate dehydrogenase reaction. Sketch the graphs that would be obtained on plotting 1/v versus 1/[succinate] at three different malonate concentrations. Label the lines for low, medium, and high [malonate].

8. Krebs found that malonate inhibition of the citric acid cycle could be overcome by raising the oxaloacetate concentration. Explain the mechanism of this process in light of your findings in Problem 7.

*9. (2*R*,3*R*)-2-Fluorocitrate contains F in the *pro-S* carboxymethyl arm of citrate [note that the rules of organic nomenclature require that atom C2 in citrate (Fig. 21-20) be renumbered as C4 in (2*R*,3*R*)-2-fluorocitrate]. This compound, but not its diastereomer, is a potent inhibitor of aconitase. (a) Draw the aconitase-catalyzed reaction pathway of (2*R*,3*R*)-2-fluorocitrate

assuming it follows the same reaction pathway as citrate (Fig. 21-20). (b) Aconitase, in fact, does not catalyze the foregoing reaction with (2*R*,3*R*)-2-fluorocitrate but, rather, yields the following tight-binding inhibitor:

Draw an alternative aconitase-catalyzed reaction that would generate this inhibitor. (c) Draw the aconitase-catalyzed reaction of (2*S*,3*R*)-3-fluorocitrate, the diastereomer of (2*R*,3*R*)-2-fluorocitrate (fluorocitrate containing F in the *pro-R* carboxymethyl arm of citrate; here the atom numbering scheme is the same as that in Fig. 21-20). Would a tight-binding inhibitor be formed?

10. Which of the following metabolites undergo net oxidation by the citric acid cycle: (a) α-ketoglutarate, (b) succinate, (c) citrate, and (d) acetyl-CoA?

11. Although there is no net synthesis of intermediates by the citric acid cycle, citric acid cycle intermediates are used in biosynthetic reactions such as the synthesis of porphyrins from succinyl-CoA. Give a reaction for the net synthesis of succinyl-CoA from pyruvate.

12. Oxaloacetate and α-ketoglutarate are precursors of the amino acids aspartate and glutamate as well as being catalytic intermediates in the citric acid cycle. Describe the net synthesis of α-ketoglutarate from pyruvate in which no citric acid cycle intermediates are depleted.

13. Lipoic acid is bound to enzymes that catalyze oxidative decarboxylation of α-keto acids. (a) What is the chemical mode of attachment of lipoic acid to enzymes? (b) Using chemical structures, show how lipoic acid participates in the oxidative decarboxylation of α-keto acids.

14. British anti-lewisite (BAL), which was designed to counter the effects of the arsenical war gas **lewisite,** is useful in treating arsenic poisoning. Explain.

British anti-lewisite (BAL)   Lewisite

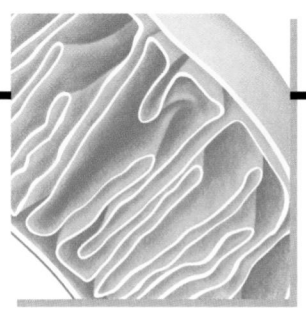

Electron Transport and Oxidative Phosphorylation

In 1789, Armand Séguin and Antoine Lavoisier (the father of modern chemistry) wrote:

> . . . in general, respiration is nothing but a slow combustion of carbon and hydrogen, which is entirely similar to that which occurs in a lamp or lighted candle, and that, from this point of view, animals that respire are true combustible bodies that burn and consume themselves.

Lavoisier had by this time demonstrated that living animals consume oxygen and generate carbon dioxide. It was

not until the early twentieth century, however, after the rise of enzymology, that it was established, largely through the work of Otto Warburg, that biological oxidations are catalyzed by intracellular enzymes. As we have seen, glucose is completely oxidized to CO_2 through the enzymatic reactions of glycolysis and the citric acid cycle. In this chapter we shall examine the fate of the electrons that are removed from glucose by this oxidation process.

The complete oxidation of glucose by molecular oxygen is described by the following redox equation:

$$C_6H_{12}O_6 + 6\,O_2 \rightarrow 6CO_2 + 6H_2O$$
$$\Delta G^{\circ\prime} = -2823 \text{ kJ} \cdot \text{mol}^{-1}$$

To see more clearly the transfer of electrons, let us break this equation down into two half-reactions. In the first half-reaction the glucose carbon atoms are oxidized:

$$C_6H_{12}O_6 + 6H_2O \rightarrow 6CO_2 + 24H^+ + 24e^-$$

and in the second, molecular oxygen is reduced:

$$6O_2 + 24H^+ + 24e^- \rightarrow 12H_2O$$

In living systems, the electron-transfer process connecting these half-reactions occurs through a multistep pathway that harnesses the liberated free energy to form ATP.

The 12 electron pairs involved in glucose oxidation are not transferred directly to O_2. Rather, as we have seen, *they are transferred to the coenzymes NAD^+ and FAD to form*

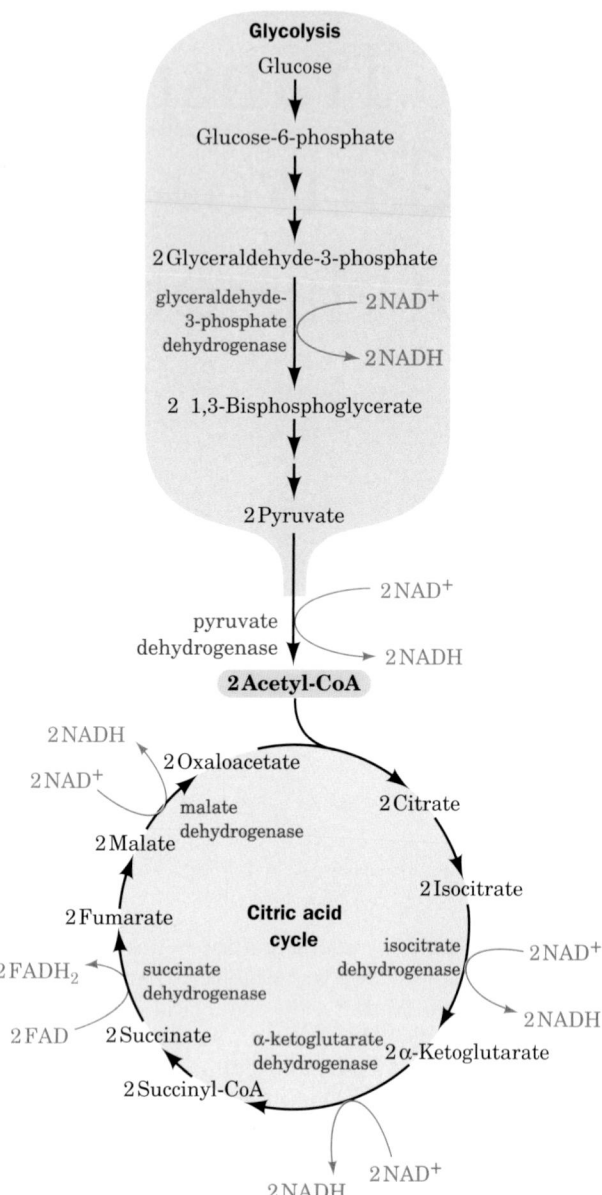

FIGURE 22-1 **The sites of electron transfer that form NADH and FADH$_2$ in glycolysis and the citric acid cycle.**

10 NADH and 2 FADH$_2$ (Fig. 22-1) in the reactions catalyzed by the glycolytic enzyme glyceraldehyde-3-phosphate dehydrogenase (Section 17-2F), pyruvate dehydrogenase (Section 21-2A), and the citric acid cycle enzymes isocitrate dehydrogenase, α-ketoglutarate dehydrogenase, succinate dehydrogenase (the only FAD reduction), and malate dehydrogenase (Section 21-3). *The electrons then pass into the **electron-transport chain**, where, through reoxidation of NADH and FADH$_2$, they participate in the sequential oxidation–reduction of over 10 redox centers before reducing O$_2$ to H$_2$O. In this process, protons are expelled from the mitochondrion. The free energy stored in the resulting pH gradient drives the synthesis of ATP from ADP and P$_i$ through **oxidative phosphorylation.*** Reoxidation of each NADH results in the synthesis of ~3 ATP, and reoxidation of FADH$_2$ yields ~2 ATP for a

total of ~38 ATP for each glucose completely oxidized to CO$_2$ and H$_2$O (including the 2 ATP made in glycolysis and the 2 ATP made in the citric acid cycle).

In this chapter we explore the mechanisms of electron transport and oxidative phosphorylation and their regulation. We begin with a discussion of mitochondrial structure and transport systems.

1 ■ THE MITOCHONDRION

The mitochondrion (Section 1-2A) is the site of eukaryotic oxidative metabolism. It contains, as Albert Lehninger and Eugene Kennedy demonstrated in 1948, the enzymes that mediate this process, including pyruvate dehydrogenase, the citric acid cycle enzymes, the enzymes catalyzing fatty acid oxidation (Section 25-2C), and the enzymes and redox proteins involved in electron transport and oxidative phosphorylation. It is therefore with good reason that the mitochondrion is described as the cell's "power plant."

A. *Mitochondrial Anatomy*

Mitochondria vary considerably in size and shape depending on their source and metabolic state. They are typically ellipsoids ~0.5 μm in diameter and 1 μm in length (about the size of a bacterium; Fig. 22-2). The mitochondrion is bounded by a smooth outer membrane and contains an extensively invaginated inner membrane. The number of invaginations, called **cristae,** varies with the respiratory activity of the particular type of cell. This is because the proteins mediating electron transport and oxidative phosphorylation are bound to the inner mitochondrial membrane, so that the respiration rate varies with membrane surface area. Liver, for example, which has a relatively low respiration rate, contains mitochondria with relatively few cristae, whereas those of heart muscle contain many. Nevertheless, the aggregate area of the inner mitochondrial membranes in a liver cell is ~15-fold greater than that of its plasma membrane.

The inner mitochondrial compartment consists of a gel-like substance of <50% water, named the **matrix,** which contains remarkably high concentrations of the soluble enzymes of oxidative metabolism (e.g., citric acid cycle enzymes), as well as substrates, nucleotide cofactors, and inorganic ions. The matrix also contains the mitochondrial genetic machinery—DNA, RNA, and ribosomes—that expresses only a few mitochondrial inner membrane proteins (Section 12-4E).

a. The Inner Mitochondrial Membrane and Cristae Compartmentalize Metabolic Functions

The outer mitochondrial membrane contains **porin,** a protein that forms nonspecific pores that permit free diffusion of up to 10-kD molecules (the X-ray structures of bacterial porins are discussed in Sections 12-3A and 20-2D). The inner membrane, which is ~75% protein by mass, is considerably richer in proteins than the outer membrane (Fig. 22-3). It is freely permeable only to O$_2$, CO$_2$, and H$_2$O and contains, in addition to respiratory chain proteins, numerous transport proteins that control

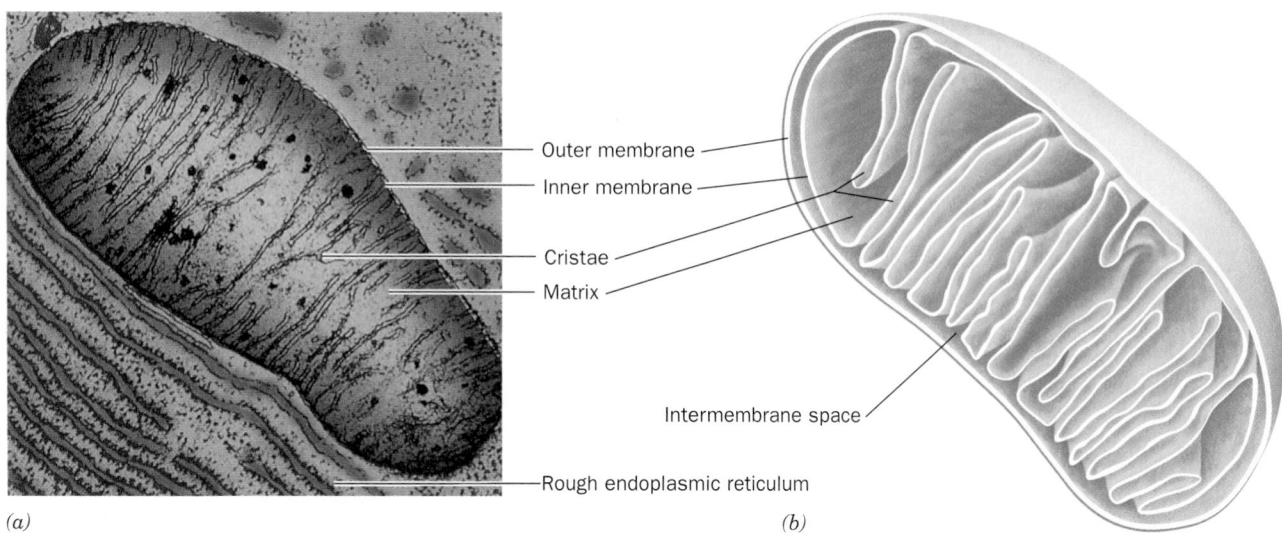

(a)   *(b)*

FIGURE 22-2   **Mitochondria.** (*a*) An electron micrograph of an animal mitochondrion. [K.R. Porter/Photo Researchers, Inc.] (*b*) Cutaway diagram of a mitochondrion.

the passage of metabolites such as ATP, ADP, pyruvate, Ca^{2+}, and phosphate (see below). *This controlled impermeability of the inner mitochondrial membrane to most ions, metabolites, and low molecular mass compounds permits the generation of ionic gradients across this barrier and results in the compartmentalization of metabolic functions between cytosol and mitochondria.*

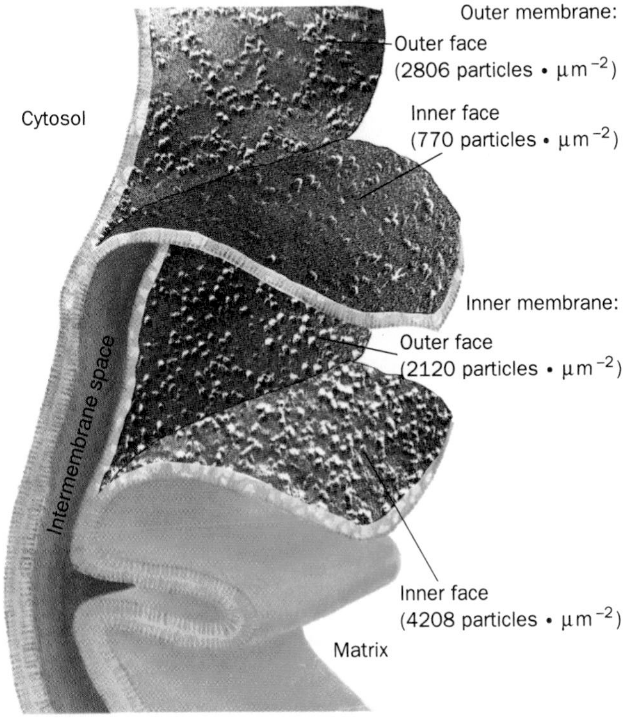

Two-dimensional electron micrographs of mitochondria such as Fig. 22-2*a* suggest that cristae resemble baffles and that the intercristal spaces communicate freely with the mitochondrion's intermembrane space, as Fig. 22-2*b* implies. However, electron microscopy–based three-dimensional image reconstruction methods have revealed that cristae can vary in shape from simple tubular entities to more complicated lamellar assemblies that merge with the inner membrane via narrow tubular structures (Fig. 22-4).

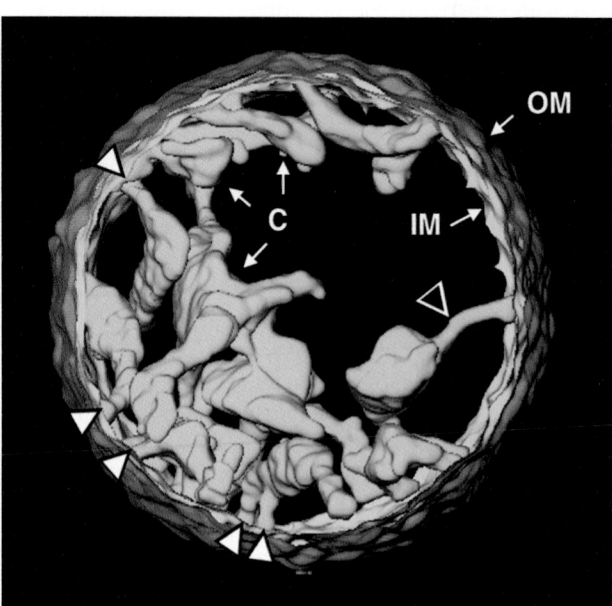

FIGURE 22-4   **Electron microscopy–based three-dimensional image reconstruction of a rat liver mitochondrion.** The outer membrane (OM) is red, the inner membrane (IM) is yellow, and the cristae (C) are green. The arrowheads point to tubular regions of the cristae that connect them to the inner membrane and to each other. [Courtesy of Carmen Mannella, Wadsworth Center, Albany, New York.]

FIGURE 22-3   **Freeze-fracture and freeze-etch electron micrographs of the inner and outer mitochondrial membranes.** The inner membrane contains about twice the density of embedded particles as does the outer membrane. [Courtesy of Lester Packer, University of California at Berkeley.]

Evidently, cristae form microcompartments that restrict the diffusion of substrates and ions between the intercristal and intermembrane spaces. This has important functional implications because it would result in a locally greater pH gradient across cristal membranes than across inner membranes that are not part of cristae, thereby significantly influencing the rate of oxidative phosphorylation (Section 22-3).

B. *Mitochondrial Transport Systems*

The inner mitochondrial membrane is impermeable to most hydrophilic substances. It must therefore contain specific transport systems to permit the following processes:

1. Glycolytically produced cytosolic NADH must gain access to the electron-transport chain for aerobic oxidation.

2. Mitochondrially produced metabolites such as oxaloacetate and acetyl-CoA, the respective precursors for cytosolic glucose and fatty acid biosynthesis, must reach their metabolic destinations.

3. Mitochondrially produced ATP must reach the cytosol, where most ATP-utilizing reactions take place, whereas ADP and P_i, the substrates for oxidative phosphorylation, must enter the mitochondrion.

We have already studied the ADP–ATP translocator and its dependence on $\Delta\Psi$, the electric potential difference across the mitochondrial membrane (Section 20-4C). The export mechanisms of oxaloacetate and acetyl-CoA from the mitochondrion are, respectively, discussed in Sections 23-1A and 25-4D. In the remainder of this section we examine the mitochondrial transport systems for P_i and Ca^{2+} and the shuttle systems for NADH.

a. P_i Transport

ATP is generated from ADP + P_i in the mitochondrion but is utilized in the cytosol. The P_i produced is returned to the mitochondrion by the **phosphate carrier,** an electroneutral P_i–H^+ symport that is driven by ΔpH. The proton that accompanies the P_i into the mitochondrion had, in effect, been previously expelled from the mitochondrion by the redox-driven pumps of the electron-transport chain (Section 22-3B). The electrochemical potential gradient generated by these proton pumps is therefore responsible for maintaining high mitochondrial ADP and P_i concentrations in addition to providing the free energy for ATP synthesis.

b. Ca^{2+} Transport

Since Ca^{2+}, like cAMP, functions as a second messenger (Section 18-3C), its concentrations in the various cellular compartments must be precisely controlled. The mitochondrion, endoplasmic reticulum, and extracellular spaces act as Ca^{2+} storage tanks. We studied the Ca^{2+}–ATPases of the plasma membrane, endoplasmic reticulum, and sarcoplasmic reticulum in Section 20-3B. Here we consider the mitochondrial Ca^{2+} transport systems.

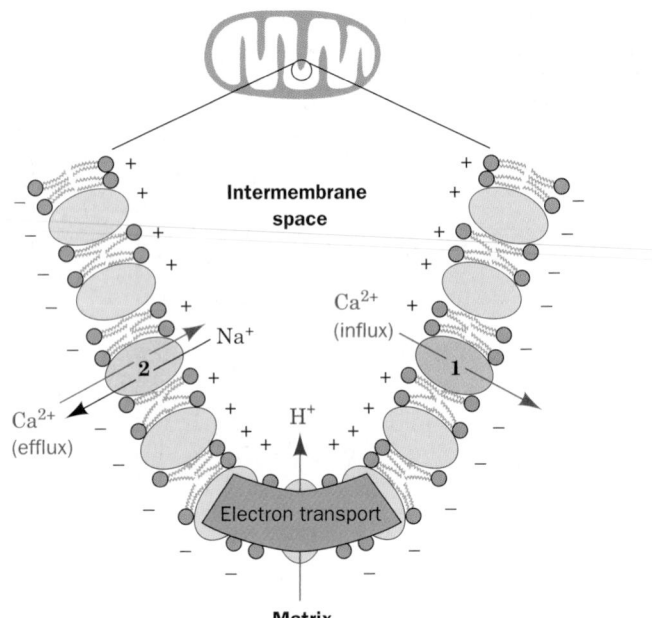

FIGURE 22-5   The two mitochondrial Ca^{2+} transport systems. System 1 mediates Ca^{2+} influx to the matrix in response to the membrane potential (negative inside). System 2 mediates Ca^{2+} efflux in exchange for Na^+.

Mitochondrial inner membrane systems separately mediate the influx and the efflux of Ca^{2+} (Fig. 22-5). The Ca^{2+} influx is driven by the inner mitochondrial membrane's membrane potential ($\Delta\Psi$, inside negative), which attracts positively charged ions. The rate of influx varies with the external $[Ca^{2+}]$ because the K_M for Ca^{2+} transport by this system is greater than the cytosolic Ca^{2+} concentration.

In heart, brain, and skeletal muscle mitochondria especially, Ca^{2+} efflux is independently driven by the Na^+ gradient across the inner mitochondrial membrane. Ca^{2+} exits the matrix only in exchange for Na^+, so that this system is an antiport. This exchange process normally operates at its maximal velocity. *Mitochondria (as well as the endoplasmic and sarcoplasmic reticulum) therefore can act as a "buffer" for cytosolic Ca^{2+} (Fig. 22-6):* If cytosolic $[Ca^{2+}]$ rises, the rate of mitochondrial Ca^{2+} influx increases while that of Ca^{2+} efflux remains constant, causing the mitochondrial $[Ca^{2+}]$ to increase while the cytosolic $[Ca^{2+}]$ decreases to its original level (its set point). Conversely, a decrease in cytosolic $[Ca^{2+}]$ reduces the influx rate, causing net efflux of $[Ca^{2+}]$ and an increase of cytosolic $[Ca^{2+}]$ back to the set point.

Oxidation carried out by the citric acid cycle in the mitochondrial matrix is controlled by the matrix $[Ca^{2+}]$ (Section 21-4). It is interesting to note, therefore, that in response to increases in cytosolic $[Ca^{2+}]$ caused by increased muscle activity, the matrix $[Ca^{2+}]$ increases, thereby activating the enzymes of the citric acid cycle. This leads to an increase in [NADH], whose reoxidation by oxidative phosphorylation (as we study in this chapter) generates the ATP needed for this increased muscle activity.

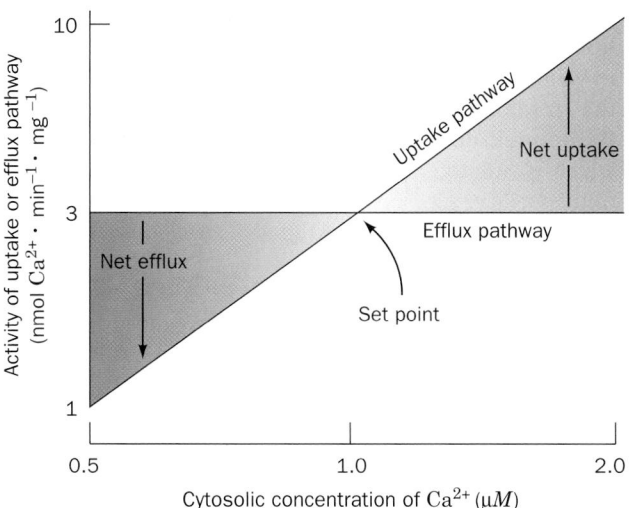

FIGURE 22-6 The regulation of cytosolic [Ca^{2+}]. The efflux pathway operates at a constant rate independent of [Ca^{2+}], whereas the activity of the influx pathway varies with [Ca^{2+}]. At the set point, the activities of the two pathways are equal and there is no net Ca^{2+} flux. An increase in cytosolic [Ca^{2+}] results in net mitochondrial influx, and a decrease in cytosolic [Ca^{2+}] results in net mitochondrial efflux. Both effects lead to the restoration of the cytosolic [Ca^{2+}]. [After Nicholls, D., *Trends Biochem. Sci.* **6**, 37 (1981).]

c. Cytoplasmic Shuttle Systems "Transport" NADH across the Inner Mitochondrial Membrane

Although most of the NADH generated by glucose oxidation is formed in the mitochondrial matrix via the citric acid cycle, that generated by glycolysis occurs in the cytosol. Yet the inner mitochondrial membrane lacks an NADH transport protein. *Only the electrons from cytosolic NADH are transported into the mitochondrion by one of several ingenious "shuttle" systems.* In the **malate–aspartate shuttle** (Fig. 22-7), which functions in heart, liver, and kidney, mitochondrial NAD$^+$ is reduced by cytosolic NADH through the intermediate reduction and subsequent regeneration of oxaloacetate. This process occurs in two phases of three reactions each:

Phase A (transport of electrons into the matrix):

1. In the cytosol, NADH reduces oxaloacetate to yield NAD$^+$ and malate in a reaction catalyzed by cytosolic malate dehydrogenase.

2. The **malate–α-ketoglutarate carrier** transports malate from the cytosol to the mitochondrial matrix in exchange for α-ketoglutarate from the matrix.

3. In the matrix, NAD$^+$ reoxidizes malate to yield NADH and oxaloacetate in a reaction catalyzed by mitochondrial malate dehydrogenase (Section 21-3H).

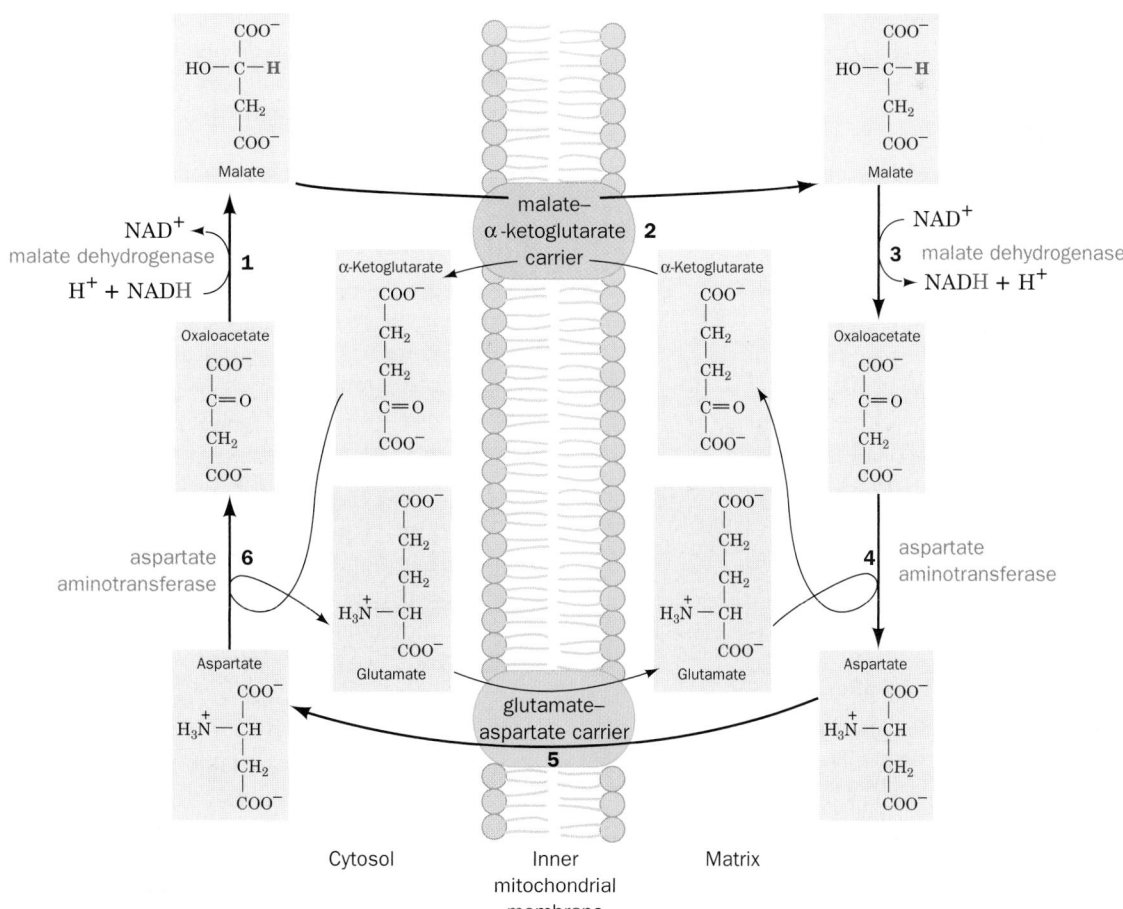

FIGURE 22-7 The malate–aspartate shuttle. The electrons of cytosolic NADH are transported to mitochondrial NADH (shown in red as hydride transfers) in Steps **1** to **3**. Steps **4** to **6** then serve to regenerate cytosolic oxaloacetate.

Phase B (regeneration of cytosolic oxaloacetate):

4. In the matrix, a transaminase (Section 26-1A) converts oxaloacetate to aspartate with the concomitant conversion of glutamate to α-ketoglutarate.

5. The **glutamate–aspartate carrier** transports aspartate from the matrix to the cytosol in exchange for cytosolic glutamate.

6. In the cytosol, a transaminase converts aspartate to oxaloacetate with the concomitant conversion of α-ketoglutarate to glutamate.

The electrons of cytosolic NADH are thereby transferred to mitochondrial NAD$^+$ to form NADH, which is subject to reoxidation via the electron-transport chain. *The malate–aspartate shuttle yields about three ATPs for every cytosolic NADH.* Note, however, that every NADH that enters the matrix is accompanied by a proton which, as we shall see (Section 22-3C), would otherwise be used to generate ~0.3 ATP. Consequently, *every cytosolic NADH that is translocated to the matrix by the malate–aspartate shuttle yields ~2.7 ATPs.*

The **glycerophosphate shuttle** (Fig. 22-8), which is simpler but less energy efficient than the malate–aspartate shuttle, occurs in brain and skeletal muscle and is particularly prominent in insect flight muscle (the tissue with the largest known sustained power output—about the same power-to-weight ratio as a small automobile engine). In it,

glycerol-3-phosphate dehydrogenase catalyzes the oxidation of cytosolic NADH by dihydroxyacetone phosphate to yield NAD$^+$, which reenters glycolysis. The electrons of the resulting **glycerol-3-phosphate** are transferred to **flavoprotein dehydrogenase** to form FADH$_2$. This enzyme, which is situated on the inner mitochondrial membrane's outer surface, supplies electrons to the electron-transport chain in a manner similar to that of succinate dehydrogenase (Section 22-2C). *The glycerophosphate shuttle therefore results in the synthesis of ~2 ATPs for every cytoplasmic NADH reoxidized, ~0.7 ATP less than the malate–aspartate shuttle.* However, the advantage of the glycerophosphate shuttle is that, being essentially irreversible, it operates efficiently even when the cytoplasmic NADH concentration is low relative to that of NAD$^+$, as occurs in rapidly metabolizing tissues. In contrast, the malate–aspartate shuttle is reversible and hence is driven by concentration gradients.

2 ■ ELECTRON TRANSPORT

In the electron-transport process, the free energy of electron transfer from NADH and FADH$_2$ to O$_2$ via protein-bound redox centers is coupled to ATP synthesis. We begin our study of this process by considering its thermodynamics. We then examine the path of electrons through the redox centers of the system and discuss the experiments used to unravel this pathway. Finally, we study the four complexes that make up the electron-transport chain. In the next section we discuss how the free energy released by the electron-transport process is coupled to ATP synthesis.

A. *Thermodynamics of Electron Transport*

We can estimate the thermodynamic efficiency of electron transport through knowledge of standard reduction potentials. As we have seen in our thermodynamic considerations of oxidation–reduction reactions (Section 16-5), an oxidized substrate's affinity for electrons increases with its standard reduction potential, $\mathscr{E}°'$ [the voltage generated by the reaction of the half-cell under standard biochemical conditions (1M reactants and products with [H$^+$] defined as 1 at pH 7) relative to the standard hydrogen electrode; Table 16-4 lists the standard reduction potentials of several half-reactions of biochemical interest]. The standard reduction potential difference, $\Delta\mathscr{E}°'$, for a redox reaction involving any two half-reactions is therefore expressed:

$$\Delta\mathscr{E}°' = \mathscr{E}°'_{(e^- \text{ acceptor})} - \mathscr{E}°'_{(e^- \text{ donor})}$$

a. NADH Oxidation Is a Highly Exergonic Reaction

The half-reactions for O$_2$ oxidation of NADH are (Table 16-4)

$$\text{NAD}^+ + \text{H}^+ + 2e^- \rightleftharpoons \text{NADH} \qquad \mathscr{E}°' = -0.315$$

and

$$\tfrac{1}{2}\text{O}_2 + 2\text{H}^+ + 2e^- \rightleftharpoons \text{H}_2\text{O} \qquad \mathscr{E}°' = 0.815 \text{ V}$$

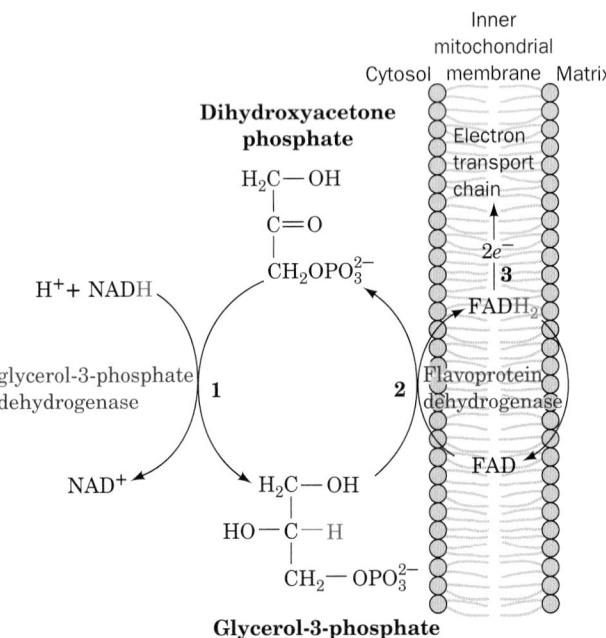

Dihydroxyacetone phosphate

H$_2$C—OH
|
C=O
|
CH$_2$OPO$_3^{2-}$

H$^+$+ NADH

glycerol-3-phosphate dehydrogenase

NAD$^+$

H$_2$C—OH
|
HO—C—H
|
CH$_2$—OPO$_3^{2-}$

Glycerol-3-phosphate

Inner mitochondrial membrane

Cytosol / Matrix

Electron transport chain

2e^-

FADH$_2$

Flavoprotein dehydrogenase

FAD

FIGURE 22-8 The glycerophosphate shuttle. The electrons of cytosolic NADH are transported to the mitochondrial electron-transport chain in three steps (shown in red as hydride transfers): **(1)** Cytosolic oxidation of NADH by dihydroxyacetone phosphate catalyzed by glycerol-3-phosphate dehydrogenase. **(2)** Oxidation of glycerol-3-phosphate by flavoprotein dehydrogenase with the reduction of FAD to FADH$_2$. **(3)** Reoxidation of FADH$_2$ with the passage of electrons into the electron-transport chain. Note that the glycerophosphate shuttle is not a membrane transport system.

Since the O_2/H_2O half-reaction has the greater standard reduction potential and therefore the higher electron affinity, the NADH half-reaction is reversed, so that NADH is the electron donor in this couple and O_2 the electron acceptor. The overall reaction is

$$\tfrac{1}{2}O_2 + NADH + H^+ \rightleftharpoons H_2O + NAD^+$$

so that

$$\Delta\mathscr{E}^{\circ\prime} = 0.815 - (-0.315) = 1.130 \text{ V}$$

The standard free energy change for the reaction can then be calculated from Eq. [16.7]:

$$\Delta G^{\circ\prime} = -n\mathscr{F}\Delta\mathscr{E}^{\circ\prime}$$

where $\mathscr{F}$, the Faraday constant, is $96{,}485 \text{ C} \cdot \text{mol}^{-1}$ of electrons and n is the number of electrons transferred per mole of reactants. Thus, since $1 \text{ V} = 1 \text{ J} \cdot \text{C}^{-1}$, for NADH oxidation:

$$\Delta G^{\circ\prime} = -2\frac{\text{mol } e^-}{\text{mol reactant}} \times 96{,}485\frac{\text{C}}{\text{mol } e^-} \times 1.13 \text{ J} \cdot \text{C}^{-1}$$
$$= -218 \text{ kJ} \cdot \text{mol}^{-1}$$

In other words, the oxidation of 1 mol of NADH by O_2 (the transfer of $2e^-$) under standard biochemical conditions is associated with the release of 218 kJ of free energy.

b. Electron Transport Is Thermodynamically Efficient

The standard free energy required to synthesize 1 mol of ATP from ADP + P_i is 30.5 kJ. The standard free energy of oxidation of NADH by O_2, if coupled to ATP synthe-sis, is therefore sufficient to drive the formation of several moles of ATP. This coupling, as we shall see, is achieved by an electron-transport chain in which electrons are passed through three protein complexes containing redox centers with progressively greater affinity for electrons (increasing standard reduction potentials) instead of directly to O_2. *This allows the large overall free energy change to be broken up into three smaller packets, each of which is coupled with ATP synthesis in a process called* **oxidative phosphorylation.** *Oxidation of 1 NADH therefore results in the synthesis of ~3 ATP.* (Oxidation of $FADH_2$, whose entrance into the electron-transport chain is regulated by a fourth protein complex, is similarly coupled to the synthesis of ~2 ATP.) The thermodynamic efficiency of oxidative phosphorylation is therefore $3 \times 30.5 \text{ kJ} \cdot \text{mol}^{-1} \times 100/218 \text{ kJ} \cdot \text{mol}^{-1} = 42\%$ under standard biochemical conditions. However, under physiological conditions in active mitochondria (where the reactant and product concentrations as well as the pH deviate from standard conditions), this thermodynamic efficiency is thought to be ~70%. In comparison, the energy efficiency of a typical automobile engine is <30%.

B. *The Sequence of Electron Transport*

⌨ **See Guided Exploration 19. Electron Transport and Oxidative Phosphorylation Overview** *The free energy necessary to generate ATP is extracted from the oxidation of NADH and $FADH_2$ by the electron-transport chain, a series of four protein complexes through which electrons pass from lower to higher standard reduction potentials (Fig. 22-9).* Electrons

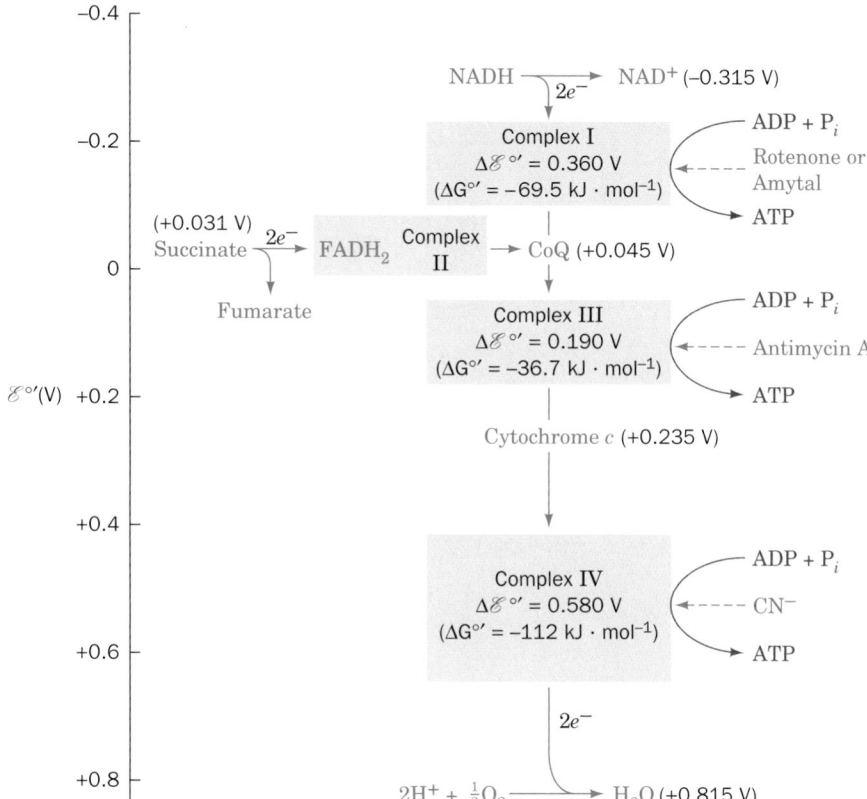

FIGURE 22-9 The mitochondrial electron-transport chain. The standard reduction potentials of its most mobile components (*green*) are indicated, as are the points where sufficient free energy is harvested to synthesize ATP (*blue*) and the sites of action of several respiratory inhibitors (*red*). (Note that Complexes I, III, and IV do not directly synthesize ATP but, rather, sequester the free energy necessary to do so by pumping protons outside the mitochondrion to form a proton gradient; Section 22-3.)

are carried from **Complexes I** and **II** to **Complex III** by **coenzyme Q** (**CoQ** or **ubiquinone;** so named because of its ubiquity in respiring organisms), and from Complex III to **Complex IV** by the peripheral membrane protein **cytochrome c** (Sections 7-3B and 9-6A).

Complex I catalyzes oxidation of NADH by CoQ:

$$NADH + CoQ \ (oxidized) \rightarrow NAD^+ + CoQ \ (reduced)$$
$$\Delta \mathscr{E}^{\circ\prime} = 0.360 \ V \qquad \Delta G^{\circ\prime} = -69.5 \ kJ \cdot mol^{-1}$$

Complex III catalyzes oxidation of CoQ (reduced) by cytochrome c:

$$CoQ \ (reduced) + 2cytochrome \ c \ (oxidized) \rightarrow$$
$$CoQ \ (oxidized) + 2cytochrome \ c \ (reduced)$$
$$\Delta \mathscr{E}^{\circ\prime} = 0.190 \ V \qquad \Delta G^{\circ\prime} = -36.7 \ kJ \cdot mol^{-1}$$

Complex IV catalyzes oxidation of cytochrome c (reduced) by O_2, the terminal electron acceptor of the electron-transport process:

$$2Cytochrome \ c \ (reduced) + \tfrac{1}{2}O_2 \rightarrow$$
$$2cytochrome \ c \ (oxidized) + H_2O$$
$$\Delta \mathscr{E}^{\circ\prime} = 0.580 \ V \qquad \Delta G^{\circ\prime} = -112 \ kJ \cdot mol^{-1}$$

The changes in standard reduction potential of an electron pair as it successively traverses Complexes I, III, and IV correspond, at each stage, to sufficient free energy to power the synthesis of an ATP molecule.

Complex II catalyzes the oxidation of $FADH_2$ by CoQ.

$$FADH_2 + CoQ \ (oxidized) \rightarrow FAD + CoQ \ (reduced)$$
$$\Delta \mathscr{E}^{\circ\prime} = 0.085 \ V \qquad \Delta G^{\circ\prime} = -16.4 \ kJ \cdot mol^{-1}$$

This redox reaction does not release sufficient free energy to synthesize ATP; it functions only to inject the electrons from $FADH_2$ into the electron-transport chain.

a. The Workings of the Electron-Transport Chain Have Been Elucidated through the Use of Inhibitors

Our understanding of the sequence of events in electron transport is largely based on the use of specific inhibitors. This sequence has been corroborated by measurements of the standard reduction potentials of the redox components of each of the complexes as well as by determining the stoichiometry of electron transport and the coupled ATP synthesis.

The rate at which O_2 is consumed by a suspension of mitochondria is a sensitive measure of the functioning of the electron-transport chain. It is conveniently measured with an **oxygen electrode** (Fig. 22-10). Compounds that inhibit electron transport, as judged by their effect on O_2 disappearance in such an experimental system, have been invaluable experimental probes in tracing the path of electrons through the electron-transport chain and in determining the points of entry of electrons from various substrates. Among the most useful such substances are **rotenone** (a plant toxin used by Amazonian Indians to poison fish and which is also used as an insecticide), **amytal** (a barbiturate), **antimycin** (an antibiotic), and **cyanide:**

Rotenone

Amytal

Cyanide

Antimycin

The following experiment illustrates the use of these inhibitors:

A buffered solution containing excess ADP and P_i is equilibrated in the reaction vessel of an oxygen electrode.

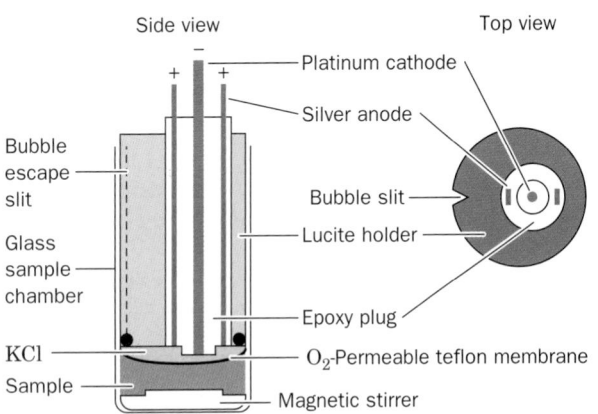

FIGURE 22-10 The oxygen electrode. This electrode consists of an Ag/AgCl reference electrode and a Pt electrode, both immersed in a KCl solution and in contact with the sample chamber through an O_2-permeable Teflon membrane. O_2 is reduced to H_2O at the Pt electrode, thereby generating a voltage with respect to the Ag/AgCl electrode that is proportional to the O_2 concentration in the sealed sample chamber. [After Cooper, T.G., *The Tools of Biochemistry, p. 69,* Wiley (1977).]

Side view

Top view

Platinum cathode
Silver anode
Bubble escape slit
Bubble slit
Lucite holder
Glass sample chamber
Epoxy plug
KCl
O_2-Permeable teflon membrane
Sample
Magnetic stirrer

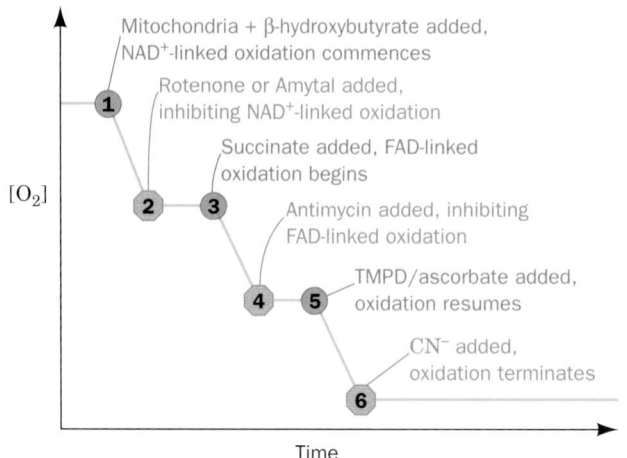

FIGURE 22-11   Effect of inhibitors on electron transport. This diagram shows an idealized oxygen electrode trace of a mitochondrial suspension containing excess ADP and P_i. At the numbered points, the indicated reagents are injected into the sample chamber and the resulting changes in $[O_2]$ are recorded. The numbers refer to the discussion in the text. [After Nicholls, D.G., *Bioenergetics, p.* 110, Academic Press (1982).]

Reagents are then injected into the chamber and the O_2 consumption recorded (Fig. 22-11):

1. Mitochondria and **β-hydroxybutyrate** are injected into the chamber. Mitochondria mediate the NAD^+-linked oxidation of β-hydroxybutyrate (Section 25-3):

$$CH_3—\underset{\underset{\text{OH}}{|}}{CH}—CH_2—CO_2^-$$

β-Hydroxybutyrate

$$NAD^+ \searrow \quad \text{β-hydroxybutyrate}$$
$$NADH + H^+ \nearrow \quad \text{dehydrogenase}$$

$$CH_3—\underset{\underset{\text{O}}{||}}{C}—CH_2—CO_2^-$$

Acetoacetate

As the resulting NADH is oxidized by the electron-transport chain with O_2 as the terminal electron acceptor, the O_2 concentration in the reaction mixture decreases.

2. Addition of rotenone or amytal completely stops the β-hydroxybutyrate oxidation.

3. Addition of succinate, which undergoes FAD-linked oxidation, causes the $[O_2]$ to resume its decrease. Electrons from $FADH_2$ are therefore still able to reduce O_2 in the presence of rotenone; that is, *electrons from $FADH_2$ enter the electron-transport chain after the rotenone-blocked step.*

4. Addition of antimycin inhibits electron transport from $FADH_2$.

5. Although NADH and $FADH_2$ are the electron-transport chain's two physiological electron donors, non-

physiological reducing agents can also be used to probe the flow of electrons. **Tetramethyl-*p*-phenylenediamine (TMPD)** is an ascorbate-reducible redox carrier that transfers electrons directly to cytochrome *c:*

Tetramethyl-*p*-phenylenediamine (TMPD), oxidized form   **Ascorbic acid**

TMPD, reduced form   **Dehydroascorbic acid**

Addition of TMPD and ascorbate to the antimycin-inhibited reaction mixture results in resumption of oxygen consumption; evidently *there is a third point at which electrons can enter the electron-transport chain.*

6. The addition of CN^- completely inhibits oxidation of all three electron donors, indicating that it blocks the electron-transport chain after the third point of entry of electrons.

Experiments such as these established the order of electron flow through the electron-transport chain complexes and the positions blocked by various electron-transport inhibitors (Fig. 22-9). This order was confirmed and extended by observations that the standard reduction potentials of the redox carriers forming the electron-transport chain complexes are very close to the standard reduction potentials of their electron donor substrates (Table 22-1). *The three jumps in reduction potential between NADH, CoQ, cytochrome c, and O_2 are each of sufficient magnitude to drive ATP synthesis.* Indeed, these redox potential jumps correspond to the points of inhibition of rotenone (or amytal), antimycin, and CN^-.

b.   Phosphorylation and Oxidation Are Rigidly Coupled

The foregoing thermodynamic studies suggest that oxidation of NADH, $FADH_2$, and ascorbate by O_2 is associated with the synthesis of about 3, 2, and 1 ATP, respectively. This stoichiometry, called the **P/O ratio** [the ratio of ATP synthesized to O atoms reduced (electron pairs taken up)], has been confirmed experimentally through measurements of O_2 uptake by resting and active mitochondria. An example of a typical experiment used to determine the P/O ratio is as follows: A suspension of mitochondria (isolated by differential centrifugation after cell disruption; Section 6-1B) containing an excess of P_i but no ADP is incubated in an oxygen electrode reaction chamber. *Oxidation and phosphorylation are closely coupled in*

TABLE 22-1   Reduction Potentials of Electron-Transport Chain Components in Resting Mitochondria

Component	$\mathscr{E}^{\circ\prime}$(V)
NADH	−0.315
Complex I (NADH:CoQ oxidoreductase; ~900 kD, 43 subunits):	
FMN	?
(Fe–S)N-1a	−0.380
(Fe–S)N-1b	−0.250
(Fe–S)N-2	−0.030
(Fe–S)N-3,4	−0.245
(Fe–S)N-5,6	−0.270
Succinate	0.031
Complex II (succinate:CoQ oxidoreductase; 127 kD, 4 subunits):	
FAD	−0.040
[2Fe–2S]	−0.030
[4Fe–4S]	−0.245
[3Fe–4S]	−0.060
Heme b_{560}	−0.080
Coenzyme Q	0.045
Complex III (CoQ:cytochrome c oxidoreductase; 243 kD, 11 subunits):	
Heme b_H (b_{562})	0.030
Heme b_L (b_{566})	−0.030
(Fe–S)	0.280
Heme c_1	0.215
Cytochrome c	0.235
Complex IV (cytochrome c oxidase; ~200 kD, 8–13 subunits):	
Heme a	0.210
Cu_A	0.245
Cu_B	0.340
Heme a_3	0.385
O_2	0.815

Source: Mainly Wilson, D.F., Erecinska, M., and Dutton, P.L., *Annu. Rev. Biophys. Bioeng.* **3,** 205 and 208 (1974); *and* Wilson, D.F., *In* Bittar, E.E. (Ed.), *Membrane Structure and Function,* Vol. 1, p. 160, Wiley (1980).

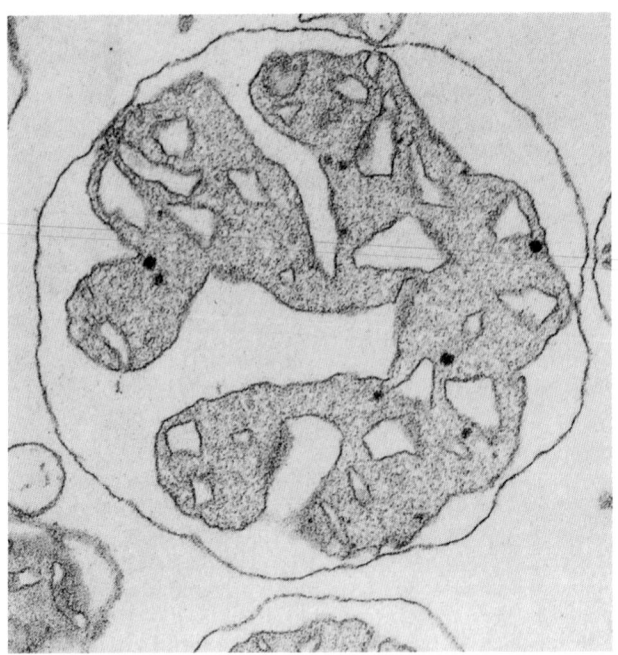

(a)

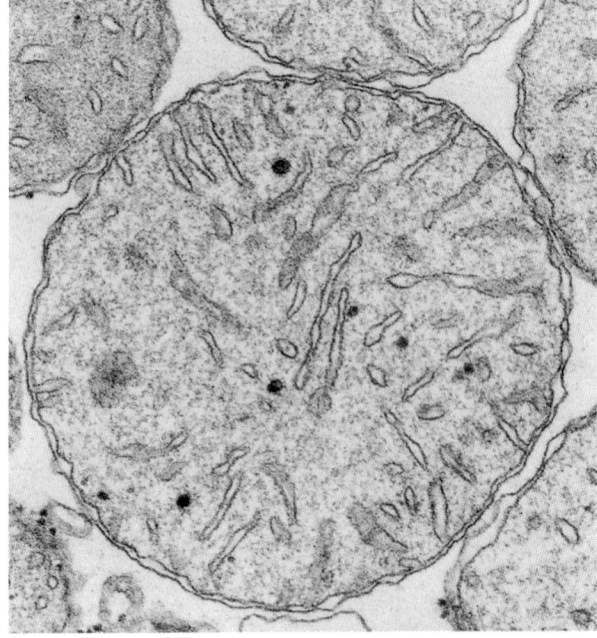

(b)

FIGURE 22-12   Electron micrographs of mouse liver mitochondria. (*a*) In the actively respiring state and (*b*) in the resting state. The cristae in actively respiring mitochondria are far more condensed than they are in resting mitochondria. [Courtesy of Charles Hackenbrock, University of North Carolina Medical School.]

well-functioning mitochondria, so electron transport can occur only if ADP is being phosphorylated (Section 22-3). Indeed, mitochondrial metabolism is so tightly regulated that even the appearances of actively respiring and resting mitochondria are greatly different (Fig. 22-12). Since no ADP is present in the reaction mixture, the mitochondria are resting and the O_2 consumption rate is minimal (Fig. 22-13; Region 1). The system is then manipulated as follows:

(a) ADP (90 μmol) and an excess of β-hydroxybutyrate (an NAD^+-linked substrate) are added. The mitochondria immediately enter the active state and the rate of oxygen consumption increases (Fig. 22-13, Region 2) and is maintained at this elevated level until all the ADP is phosphorylated. The mitochondria then return to the resting state

(Fig. 22-13, Region 3). Phosphorylation of 90 μmol of ADP under these conditions consumes 15 μmol of O_2. Since the oxidation of NADH by O_2 consumes twice as many moles of NADH as of O_2, the P/O ratio for NADH reoxidation at Region 2 is 90 μmol of ADP/(2 × 15 μmol of O_2) = 3;

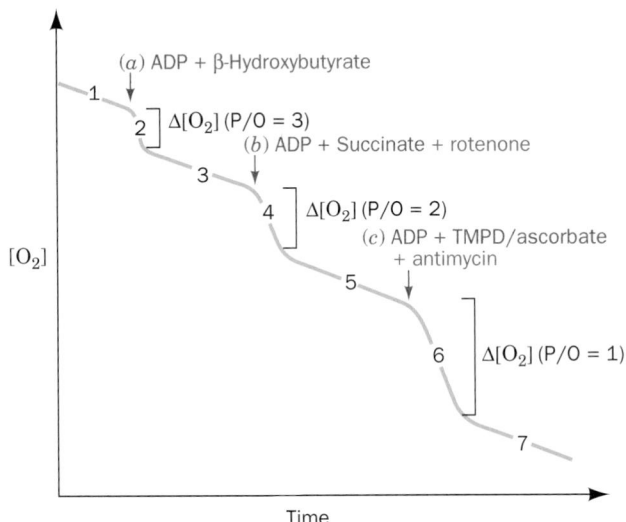

FIGURE 22-13 Determination of the stoichiometry of coupled oxidation and phosphorylation (the P/O ratio) with different electron donors. Mitochondria are incubated in excess phosphate buffer in the sample chamber of an oxygen electrode. (a) Next, 90 μmol of ADP and excess β-hydroxybutyrate are added. Respiration continues until all the ADP is phosphorylated. ΔO_2 in Region 2 is 15 μmol, corresponding to 30 μmol of NADH oxidized; thus, P/O = 90/30 = 3. (b) Then 90 μmol of ADP and excess succinate are added together with rotenone to inhibit electron transfer from NADH. ΔO_2 in Region 4 is 22.5 μmol, corresponding to 45 μmol of $FADH_2$ oxidized; P/O = 90/45 = 2. (c) Finally, 90 μmol of ADP and excess TMPD/ascorbate are added with antimycin to inhibit electron transfer from $FADH_2$. ΔO_2 in Region 6 is 45 μmol, corresponding to 90 μmol of ascorbate oxidized; P/O = 90/90 = 1.

that is, *3 mol of ADP are phosphorylated per mole of NADH oxidized.*

(b) The experiment is continued by inhibiting electron transfer from NADH by rotenone and adding an additional 90 μmol of ADP (Fig. 22-13, Region 4), this time together with an excess of the FAD-linked substrate succinate. Oxygen consumption again continues until all the ADP is phosphorylated, and the system again returns to the resting state (Fig. 22-13, Region 5). Calculation of the P/O ratio for $FADH_2$ oxidation yields the value 2; that is, *2 mol of ADP are phosphorylated per mole of $FADH_2$ oxidized.*

(c) In the same manner, *the oxidation of ascorbate/TMPD yields a P/O ratio of 1 (Fig. 22-13, Regions 6 and 7).*

These conclusions agree with the inhibitor studies indicating that there are three entry points for electrons into the electron-transport chain and with the standard reduction potential measurements exhibiting three potential jumps, each sufficient to provide the free energy for ATP synthesis (Fig. 22-9).

c. The P/O Ratios May Be Subject to Revision

Measurements of P/O ratios are subject to systematic experimental errors for which it is difficult to correct, such as inaccuracies in the measurement of the oxygen concentration, the presence of AMP, and proton leakage through the inner mitochondrial membrane. Thus, the widely accepted P/O values of 3, 2, and 1 associated with NADH-, $FADH_2$-, and ascorbate/TMPD-linked oxidation may well be in error. Indeed, measurements by Peter Hinkle have yielded values close to 2.5, 1.5, and 1 for these quantities (we shall see in Section 22-3 that the mechanism of oxidative phosphorylation does not require that P/O ratios have integer values). If these values are correct, then the number of ATP molecules that are synthesized per molecule of glucose oxidized is 2.5 ATP/NADH × 10 NADH/glucose + 1.5 ATP/$FADH_2$ × 2 $FADH_2$/glucose + 2 ATP/glucose from the citric acid cycle + 2 ATP/glucose from glycolysis = 32 ATP/glucose rather than the conventional value of 38 ATP/glucose implied by P/O ratios of 3, 2, and 1. However, since there is significant disagreement as to the validity of the revised P/O ratios, we shall, for the sake of consistency, use the more established values of 3, 2, and 1 throughout this textbook. You should nevertheless keep in mind that these values are disputed.

The determination of the *in vivo* yield of ATP/glucose is even more problematic. The reducing equivalents from the GAPDH-generated NADH may be imported into the mitochondrion via the glycerophosphate shuttle, which yields $FADH_2$, or via the malate–aspartate shuttle, which yields NADH but passes a proton into the matrix (Section 22-1B). The mix of these two shuttle systems varies from tissue to tissue and hence, so does the ATP yield. In addition, the rate of proton leakage back across the inner mitochondrial membrane, which is significant, may vary with conditions and cellular identity. Hence, the *in vivo* yield of ATP/glucose is likely to be significantly less than the above figures of 38 or 32 ATP/glucose.

How the free energy of electron transport is actually coupled to ATP synthesis, a subject of active research, is discussed in Section 22-3. We first examine the structures of the four respiratory complexes in order to understand how they are related to the function of the electron-transport chain. Keep in mind, however, that as in most areas of biochemistry, this field is under intense scrutiny and much of the information we need for a complete understanding of these relationships has yet to be elucidated.

C. *Components of the Electron-Transport Chain*

Many of the proteins embedded in the inner mitochondrial membrane are organized into the four respiratory complexes of the electron-transport chain. Each complex consists of several protein components that are associated with a variety of redox-active prosthetic groups with successively increasing reduction potentials (Table 22-1). The complexes are all laterally mobile within the inner mitochondrial membrane; they do not appear to form any stable higher structures. Indeed, they are not present in

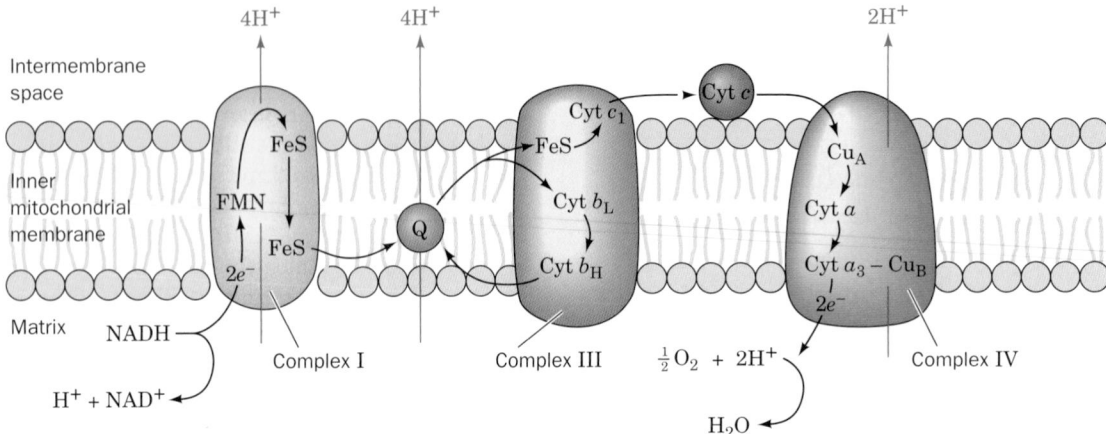

FIGURE 22-14   The mitochondrial electron-transport chain.
The pathways of electron transfer (*black*) and proton pumping
(*red*) are indicated. Electrons are transferred between
Complexes I and III by membrane-soluble CoQ (Q) and
between Complexes III and IV by the peripheral membrane
protein cytochrome *c* (Cyt *c*). Complex II (not shown) transfers
electrons from succinate to CoQ. 🔊 **See the Animated Figures**

equimolar ratios. In the following paragraphs, we examine
their structures and the agents that transfer electrons
between them. Their relationships are summarized in
Fig. 22-14.

1. Complex I (NADH:Coenzyme Q Oxidoreductase)

*Complex I (also called **NADH dehydrogenase**) passes
electrons from NADH to CoQ.* This probably largest protein component of the inner mitochondrial membrane
(~900 kD in mammals and ~700 kD in *Neurospora crassa*)
contains one molecule of **flavin mononucleotide (FMN;** a
redox-active prosthetic group that differs from FAD only
by the absence of the AMP group) and six to seven **iron–
sulfur clusters** that participate in the electron-transport
process (Table 22-1). In mammals, 7 of its 43 subunits are
encoded by mitochondrial genes, with the remainder encoded by nuclear genes.

a. Iron–Sulfur Clusters Are Redox Active

Iron–sulfur clusters, first discovered by Helmut Beinert,
commonly occur as prosthetic groups in **iron–sulfur proteins.** There are four common types of iron–sulfur clusters
(Fig. 22-15). Those designated **[2Fe–2S]** and **[4Fe–4S]**
clusters consist of equal numbers of iron and sulfide ions
and are both coordinated to four protein Cys sulfhydryl
groups. The **[3Fe–4S] cluster** is essentially a [4Fe–4S] cluster that lacks one Fe atom. One means of identifying these

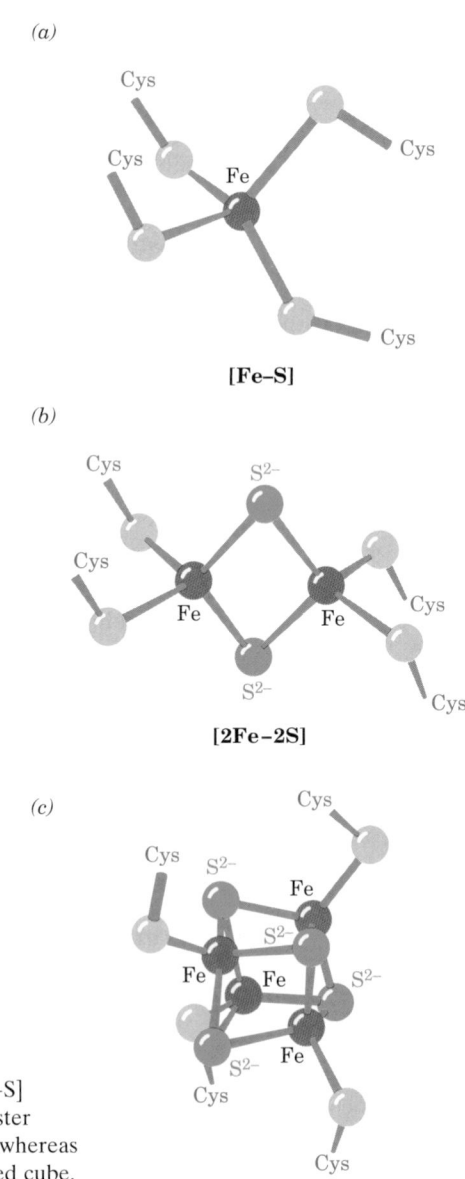

(a)

[Fe–S]

(b)

[2Fe–2S]

(c)

[4Fe–4S]

FIGURE 22-15   Structures of the common iron–sulfur clusters. (*a*) An [Fe–S]
cluster, (*b*) a [2Fe–2S] cluster, and (*c*) a [4Fe–4S] cluster. The [3Fe–4S] cluster
resembles the [4Fe–4S] cluster with one of its Fe ions removed. Note that, whereas
the Fe and S^{2-} ions of [4Fe–4S] clusters form what appears to be a distorted cube,
the structure is really two interpenetrating tetrahedra of Fe ions and S^{2-} ions.

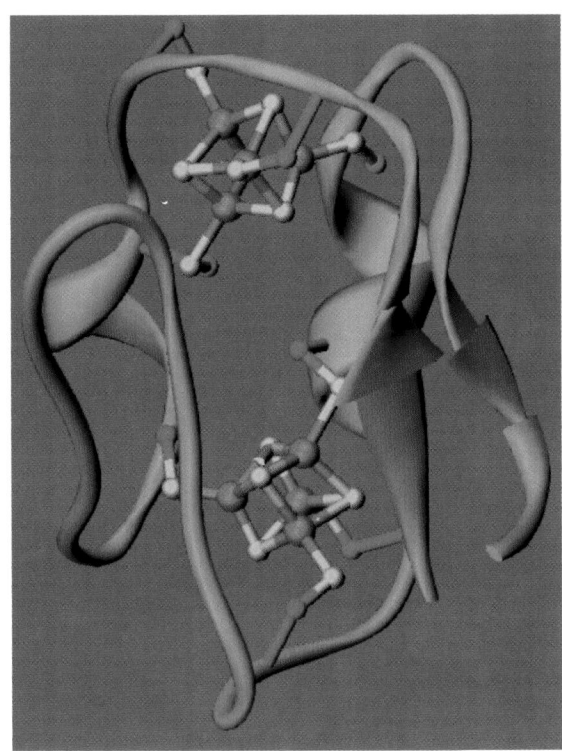

FIGURE 22-16 **X-Ray structure of ferredoxin from**
Peptococcus aerogenes. This monomeric 54-residue protein
contains two [4Fe–4S] clusters. The C_β atoms of the four Cys
residues liganding each [4Fe–4S] cluster are green, the Fe
atoms are orange, and the S atoms are yellow. [Based on an
X-ray structure by Elinor Adman, Larry Sieker, and Lyle
Jensen, University of Washington. PDBid 1FDX.] **See the
Interactive Exercises**

clusters utilizes the fact that their sulfide ions are acid labile: They are released as H_2S near pH 1. The **[Fe–S] cluster,** which occurs only in bacteria, consists of a single Fe
atom liganded to four Cys residues. Note that the Fe ions
in all four types of clusters are each coordinated by four S
atoms, which are more or less tetrahedrally disposed
around the Fe. However, in **Rieske iron–sulfur proteins**
(named after their discoverer, John Rieske), one of the Fe
atoms in a [2Fe–2S] cluster is coordinated by two His
residues rather than two Cys residues. *The oxidized and
reduced states of all iron–sulfur clusters differ by one formal charge regardless of their number of Fe ions.* This is
because the Fe ions in each cluster form a conjugated system and thus can have oxidation states between the +2
and +3 values possible for individual Fe ions. For example, each of the two [4Fe–4S] clusters in the protein **ferredoxin** (Fig. 22-16) contains one Fe(II) and three Fe(III) in
its oxidized form and two Fe(II) and two Fe(III) in its reduced form. The standard reduction potential of a given
type of iron–sulfur cluster depends on its interaction with
its associated protein as well as on its oxidation state. Iron–
sulfur proteins also occur in the photosynthetic electron-
transport chains of plants and bacteria (Section 24-2);
indeed, photosynthetic electron-transport chains are

thought to be the evolutionary precursors of oxidative
electron-transport chains (Section 1-5C).

b. The Coenzymes of Complex I
FMN and ubiquinone (CoQ), the coenzymes of
Complex I, can each adopt three oxidation states (Fig.
22-17). Although NADH can only participate in a two-
electron transfer, both FMN and CoQ are capable of accepting and donating either one or two electrons because
their semiquinone forms are stable. In contrast, the cytochromes of Complex III (see below), to which reduced
CoQ passes its electrons, are only capable of one-electron
reductions. *FMN and CoQ therefore provide an electron
conduit between the two-electron donor NADH and the
one-electron acceptors, the cytochromes.*

CoQ's hydrophobic tail makes it soluble in the inner mitochondrial membrane's lipid bilayer. In mammals, this tail
consists of 10 C_5 isoprenoid units and hence the coenzyme
is designated $Q_{10.}$ In other organisms, CoQ may have only
6 (Q_6) or 8 (Q_8) isoprenoid units.

**c. The Low Resolution Structure of Complex I Reveals
an L-Shaped Protein**
Although the X-ray structure of Complex I has not yet
been determined, low resolution electron microscopy–
based structures have been reported for Complex I from
bovine heart, *Neurospora crassa,* and *E. coli* (Fig. 22-18).
All of these structures reveal an L-shaped protein with one
domain (arm of the L) immersed in the inner mitochondrial membrane (the plasma membrane for *E. coli*) and
the other extending into the matrix (the cytosol for *E. coli*).
The *E. coli* and bovine complexes both exhibit a pronounced narrowing between the two domains. This narrow portion appears to house iron–sulfur cluster N-2
(Table 22-1) and the ubiquinone binding site.

2. Complex II (Succinate:Coenzyme Q Oxidoreductase)
*Complex II, which contains the citric acid cycle enzyme
succinate dehydrogenase (Section 21-3F) together with three
other subunits (all of which are encoded by nuclear genes),
passes electrons from succinate to CoQ.* It does so with the
participation of a covalently bound FAD, a [2Fe–2S] cluster, a [4Fe–4S] cluster, a [3Fe–4S] cluster, and one cytochrome b_{560} (Table 22-1). We discuss the structures of
the cytochromes in connection with that of Complex III
below.

The standard redox potential for electron transfer from
succinate to CoQ (Fig. 22-9) is insufficient to provide the
free energy necessary to drive ATP synthesis. Complex II
is, nevertheless, important because it injects these relatively high-potential electrons into the electron-transport
chain. Two other enzymes also synthesize and release
$CoQH_2$ in the inner mitochondrial membrane and thereby
power oxidative phosphorylation via the actions of
Complexes III and IV. These are glycerol-3-phosphate dehydrogenase of the glycerophosphate shuttle (Fig. 22-8)
and **ETF:ubiquinone oxidoreductase,** which participates in
fatty acid oxidation (Section 25-2C; ETF stands for
electron-transfer flavoprotein).

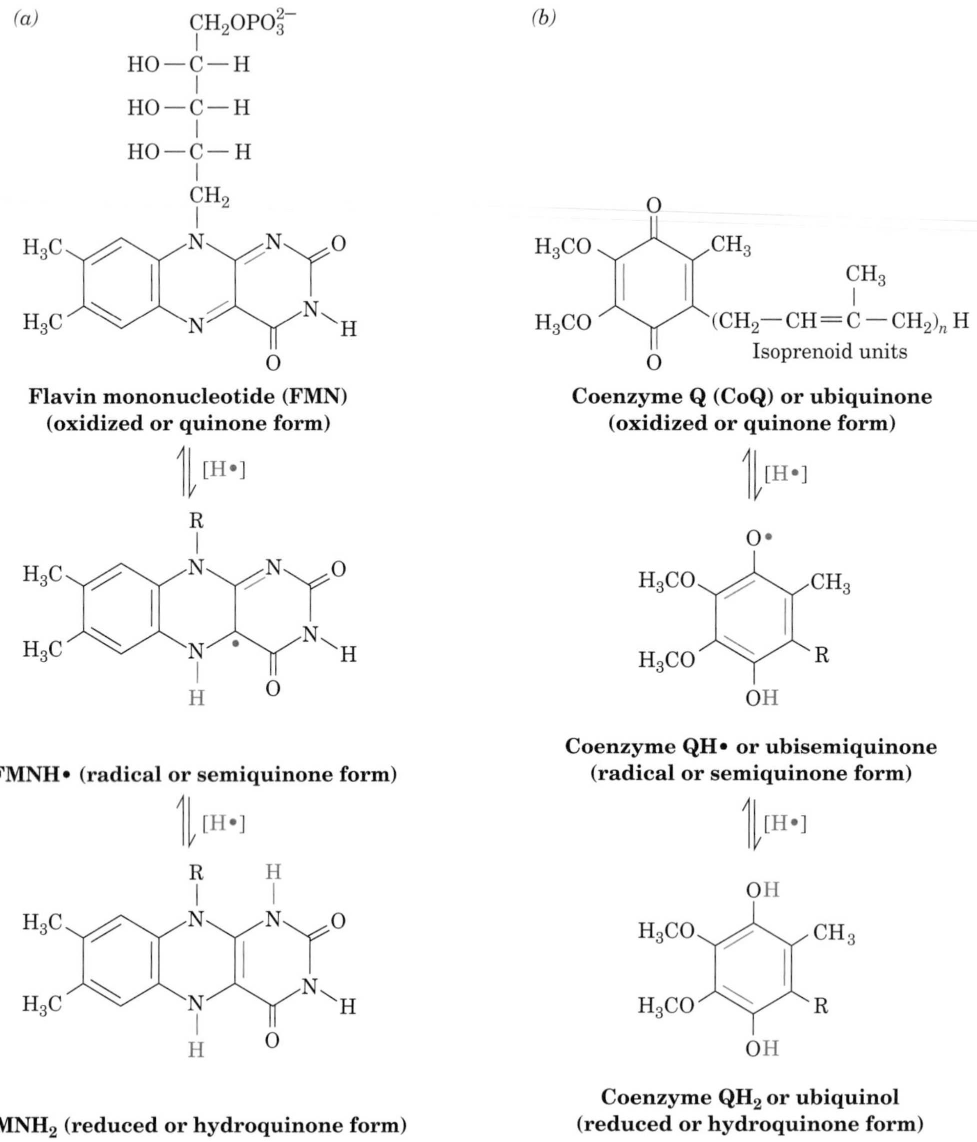

(a)

CH$_2$OPO$_3^{2-}$

HO—C—H

HO—C—H

HO—C—H

CH$_2$

Flavin mononucleotide (FMN)
(oxidized or quinone form)

⇅ [H•]

R

FMNH• (radical or semiquinone form)

⇅ [H•]

R

FMNH$_2$ (reduced or hydroquinone form)

(b)

(CH$_2$—CH=C—CH$_2$)$_n$ H

CH$_3$

Isoprenoid units

Coenzyme Q (CoQ) or ubiquinone
(oxidized or quinone form)

⇅ [H•]

O•

Coenzyme QH• or ubisemiquinone
(radical or semiquinone form)

⇅ [H•]

Coenzyme QH$_2$ or ubiquinol
(reduced or hydroquinone form)

FIGURE 22-17 Oxidation states of the coenzymes of complex I. (*a*) FMN and (*b*) CoQ. Both coenzymes form stable semiquinone free radical states.

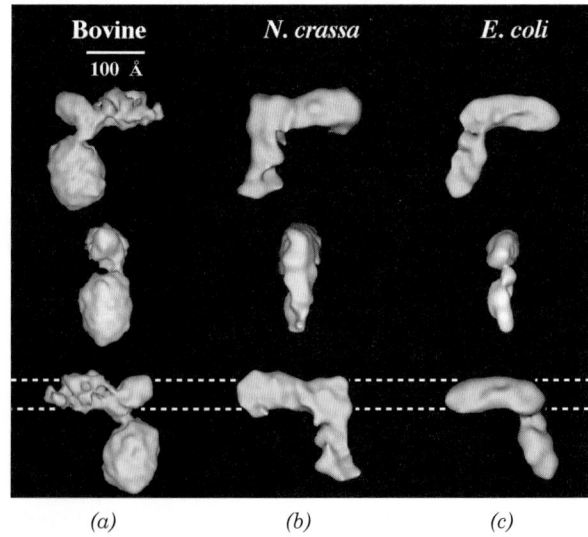

FIGURE 22-18 Electron microscopy–based three-dimensional structures of Complex I. The complexes from (*a*) bovine heart, (*b*) *N. crassa,* and (*c*) *E. coli* were determined at 22, 28, and 34 Å resolution, respectively. Successive views, top to bottom, are rotated by 90° about their vertical axes. The boundaries of the lipid bilayers in which these proteins are immersed are indicated by the dashed lines. The vertical arm protrudes into the mitochondrial matrix (bacterial cytoplasm). [Courtesy of Nikolaus Grigorieff, Brandeis University. The *N. crassa* and *E. coli* structures were determined by Vincent Guénebaut and Kevin Leonard, European Molecular Biology Laboratory, Heidleberg, Germany.]

a. Quinol–Fumarate Reductase Is a Homolog of Complex II

Although the X-ray structure of mitochondrial Complex II is unknown, that of the homologous *E. coli* respiratory complex **quinol–fumarate reductase (QFR)** has been determined. QFR functions in anaerobic organisms that use fumarate as a terminal electron acceptor, where it catalyzes the same reaction as does Complex II but in the opposite direction, that is, it uses a quinol to reduce fumarate to succinate.

The *E. coli* QFR is a 121-kD heterotetramer that has two highly conserved water-soluble subunits, a flavoprotein (Fp; 601 residues), and an iron–sulfur protein (Ip; 243 residues), together with two transmembrane subunits (130 and 118 residues) whose sequences vary among different organisms. QFR's X-ray structure in complex with its inhibitor oxaloacetate (Fig. 22-19a), determined by Douglas Rees, has the shape of the letter "b", with its bottom lobe containing Fp and Ip and its tail composed of the transmembrane subunits. The complex is oriented in the bacterial cell membrane such that Fp and Ip extend into the cytoplasm (the equivalent of the mitochondrial matrix for Complex II). Fp binds both the oxaloacetate and the FAD prosthetic group, which is covalently linked to a specific His side chain on the protein via a bond from FAD atom C8a, as likewise occurs in succinate dehydrogenase (Fig. 21-22). Ip binds the complex's three iron–sulfur clusters,

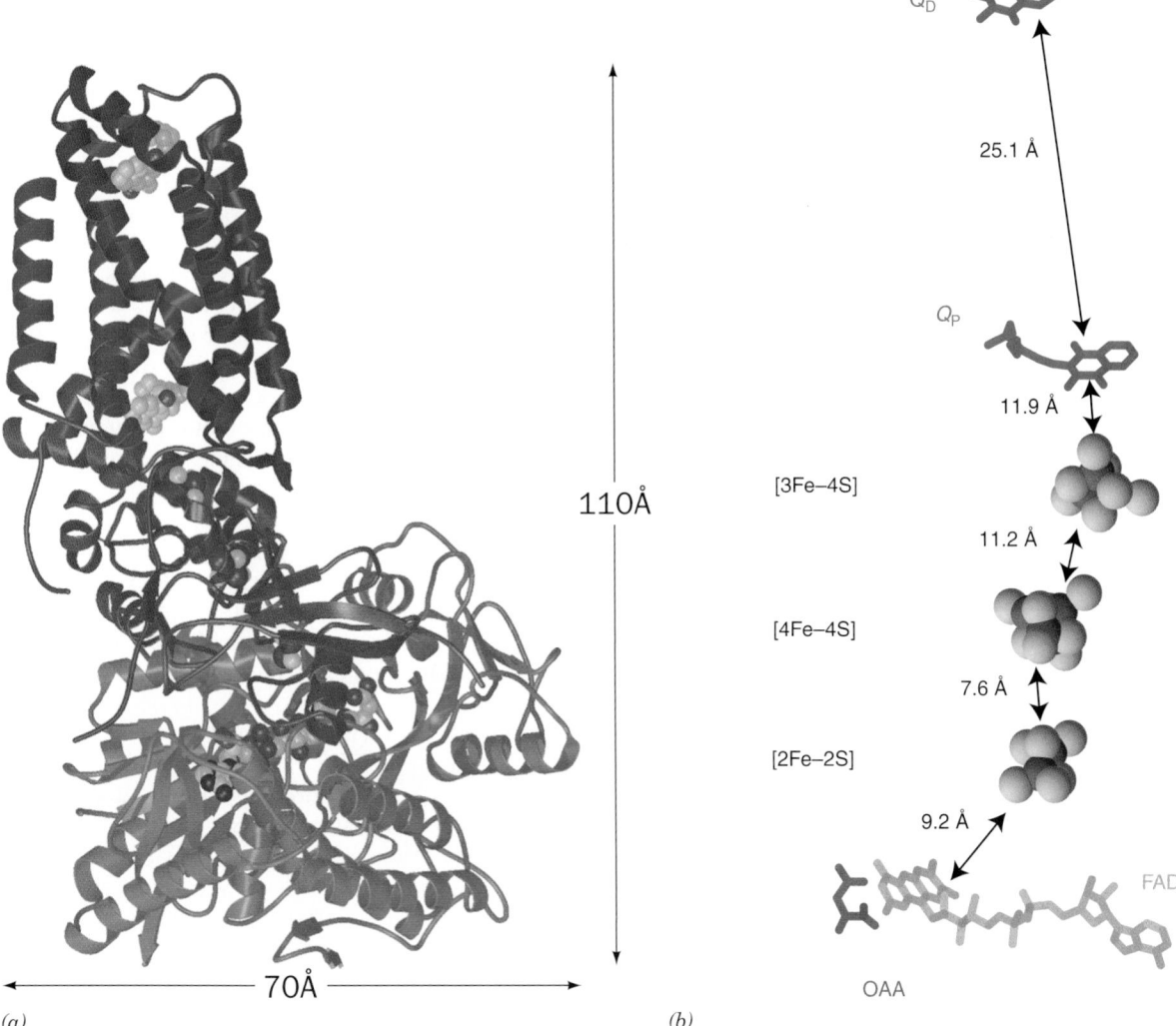

(a)

(b)

FIGURE 22-19 X-Ray structure of *E. coli* quinol–fumarate reductase (QFR) in complex with its inhibitor oxaloacetic acid (OAA). The view is parallel to the cell membrane with the cytosol (the equivalent of the mitochondrial matrix) at the bottom. (a) Ribbon diagram in which the flavoprotein (Fp) is blue, the iron–sulfur protein (Ip) is red, and the transmembrane subunits are green and purple. The OAA and its redox cofactors are shown in space-filling form with atoms colored according to type (C and S yellow; N and Fe purple, and O red). [Courtesy of Douglas Rees, California Institute of Technology. PDBid 1L0V.] (b) The arrangement of QFR's redox cofactors with their edge-to-edge distances indicated. The colors of the redox cofactors are OAA purple, FAD orange, Fe red, S yellow, and menaquinone green. [Courtesy of Tomoko Ohnishi, University of Pennsylvania. Based on an X-ray structure by Douglas Rees, Caltech. PDBid 1L0V.]

and the transmembrane subunits bind two molecules of the ubiquinone-like **menaquinone,**

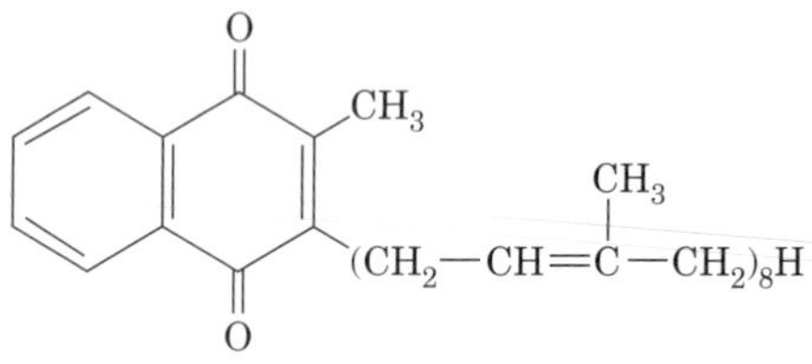

Menaquinone

which are designated Q_P and Q_D (ubiquinone can substitute for menaquinone *in vitro*). Note that the transmembrane portions of the QFRs from some species bind one or two *b*-type hemes (Complex II binds one; Table 22-1).

QFR's six redox cofactors are organized into a nearly linear chain with sequence FAD—[2Fe–2S]—[4Fe–4S]—[3Fe–4S]—Q_P—Q_D (bottom to top in Fig. 22-19*b*). These cofactors, with the exception of Q_P and Q_D, are separated by 7.6 to 11.9 Å, which are common cofactor separations in electron-transfer chains (see below). Thus, despite the measured −0.245 V reduction potential of the [4Fe–4S] cluster (Table 22-1), which would appear to be too low for it to accept electrons from succinate in the succinate → fumarate reaction, the [4Fe–4S] cluster probably participates in the electron-transfer process.

The 25.1-Å separation between Q_P and Q_D is too large for electron transfer to occur between these two redox centers at a physiologically significant rate. Rees has therefore proposed that either the Q_D site is catalytically irrelevant or that there is a third cofactor binding site approximately

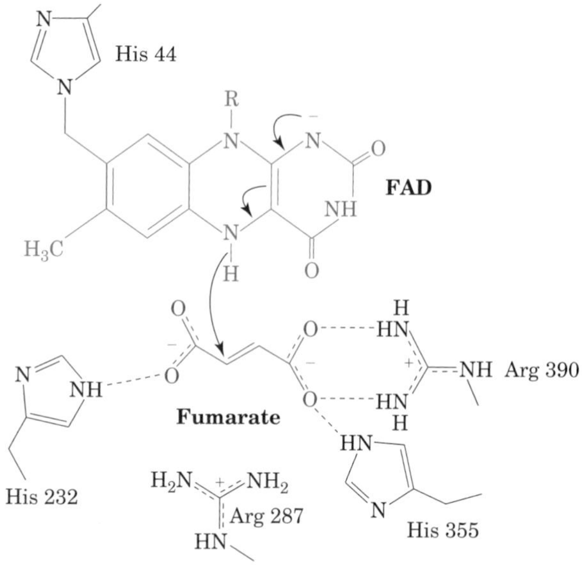

FIGURE 22-20 Active site interactions in the proposed mechanism of the QFR-catalyzed reduction of fumarate to succinate. The initial step of the reaction is hydride transfer from flavin atom N5 to the double bond of fumarate (*curved arrows*). This is followed by proton transfer to the substrate to yield succinate. [After Iverson, T.M., Luna-Chavez, C., Schröder, I., Cecchini, G., and Rees, D.C., *Curr. Opin. Struct. Biol.* **10,** 451 (2000).]

midway between Q_P and Q_D. The latter possibility is supported by the presence of unidentified electron density in a cavity between the transmembrane helices, and by mutagenic experiments that indicate that a cluster of residues in the vicinity of this cavity is essential for enzymatic activity.

The QFR·oxaloacetate structure suggests a mechanism for fumarate reduction in which a hydride ion is transferred from N5 of the flavin to the fumarate double bond (Fig. 22-20), followed by proton transfer from a nearby side chain. This mechanism is supported by the observations that four residues that contact the inhibitor or substrate, His 232, Arg 287, Arg 390, and His 355, are absolutely conserved in all known QFR and succinate dehydrogenase sequences and that the mutation of any of the first three residues inactivates these enzymes.

Even though QFR and Complex II are predicted to have similar structures, organisms such as *E. coli* that are capable of both aerobic and anaerobic metabolism employ the two different complexes for these different purposes. This appears to be because under aerobic conditions, QFR produces 25 times, as much superoxide ion as does *E. coli* Complex II as well as H_2O_2, which Complex II does not produce. These **reactive oxygen species (ROS)** are highly destructive. Comparison of the X-ray structure of QFR with that of the closely similar *E. coli* Complex II, determined by So Iwata, indicates that the electron distributions about their various redox centers greatly favors the ROS-generating side reactions of O_2 with the flavin ring of QFR relative to that of Complex II. This suggests that mutations of the genes encoding eukaryotic Complex II, which cause a wide variety of disorders including tumor formation, neurological defects, and premature aging, may result from ROS generation.

3. Complex III (Coenzyme Q:Cytochrome *c* Oxidoreductase or Cytochrome *bc₁* Complex)

Complex III passes electrons from reduced CoQ to cytochrome c. It contains four redox cofactors: two *b*-type hemes, a *c*-type heme, and one [2Fe–2S] cluster (Table 22-1).

a. Cytochromes Are Heme Proteins That Transport Electrons

Cytochromes, whose function was elucidated in 1925 by David Keilin, are redox-active proteins that occur in all organisms except a few types of obligate anaerobes. These proteins contain heme groups that reversibly alternate between their Fe(II) and Fe(III) oxidation states during electron transport.

The heme groups of the reduced [Fe(II)] cytochromes have prominent visible absorption spectra consisting of three peaks: the α, β, and γ **(Soret)** bands (Fig. 22-21*a*). The wavelength of the α peak, which varies characteristically with the particular reduced cytochrome species (it is absent in oxidized cytochromes), is useful for differentiating the various cytochromes. Accordingly, the spectra of mitochondrial membranes (Fig. 22-21*b*) indicate that they contain three cytochrome types, **cytochromes *a*, *b*,** and ***c.***

Within each type of cytochrome, different heme group environments may be characterized by slightly different α

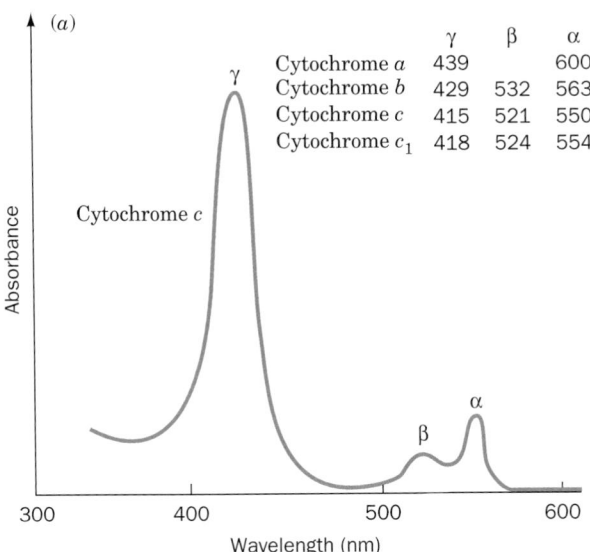
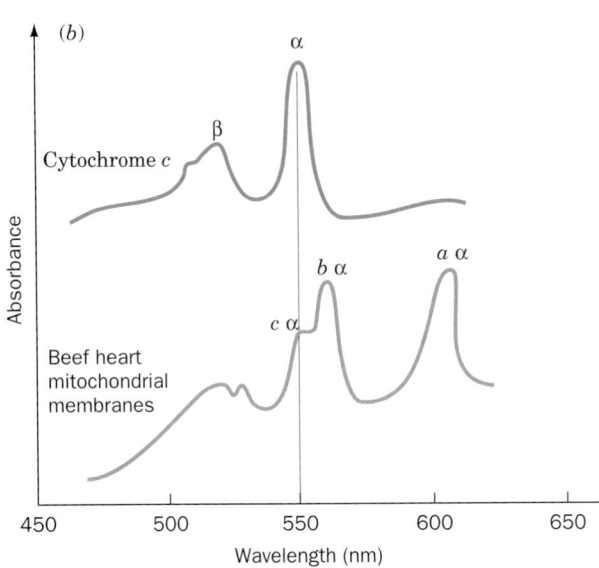

FIGURE 22-21 Visible absorption spectra of cytochromes.
(*a*) Absorption spectrum of reduced cytochrome *c* showing its characteristic α, β, and γ (Soret) absorption bands. The absorption maxima for cytochromes *a*, *b*, *c*, and *c₁* are listed. (*b*) The three separate α bands in the visible absorption

spectrum of beef heart mitochondrial membranes (*below*) indicate the presence of cytochromes *a*, *b*, and *c*. The spectrum of purified cytochrome *c* (*above*) is provided for reference. [After Nicholls, D.G. and Ferguson, S.J., *Bioenergetics 3*, p. 96, Academic Press (2002).]

Heme *a*   **Heme *b* (iron–protoporphyrin IX)**   **Heme *c***

FIGURE 22-22 Porphyrin rings in cytochromes. The (*a*) chemical structures and (*b*) axial liganding of the heme groups contained in cytochromes *a*, *b*, and *c* are shown.

Hemes *a* and *b*

Heme *c*

peak wavelengths. For example, Complex III has two *b*-type hemes: That absorbing maximally at 562 nm is referred to as b_{562} or b_H (for *high* potential; formerly called b_K), whereas that absorbing maximally at 566 nm is referred to as b_{566} or b_L (for *low* potential; formerly called b_T).

Each type of cytochrome contains a differently substituted porphyrin ring (Fig. 22-22*a*) coordinated with the redox-active iron atom. A *b*-type cytochrome contains **protoporphyrin IX,** which also occurs in hemoglobin and myoglobin (Section 10-1A). The heme group of a *c*-type

FIGURE 22-23 X-ray structures of cytochrome bc_1. (*a*) The dimeric bovine complex is viewed perpendicular to its 2-fold axis and parallel to the membrane with the matrix below. Its 11 subunits and 4 redox cofactors are colored as indicated. [Courtesy of So Iwata, Uppsala University, Uppsala, Sweden and Bing Jap, Lawrence Berkeley National Laboratory, University of California at Berkeley. PDBid 1BE3.] *See the Interactive Exercises* (*b*) The yeast enzyme in complex with cytochrome *c* and the inhibitor stigmatellin viewed with a ~90° rotation about its 2-fold axis relative to Part *a* and colored similarly. The various heme groups are differently colored but their Fe atoms are all orange. Note that only one molecule of cytochrome *c* is bound to the dimeric cytochrome bc_1. [Based on an X-ray structure by Carola Hunte, Max Planck Institute for Biophysics, Frankfurt am Main, Germany. PDBid 1KYO.]

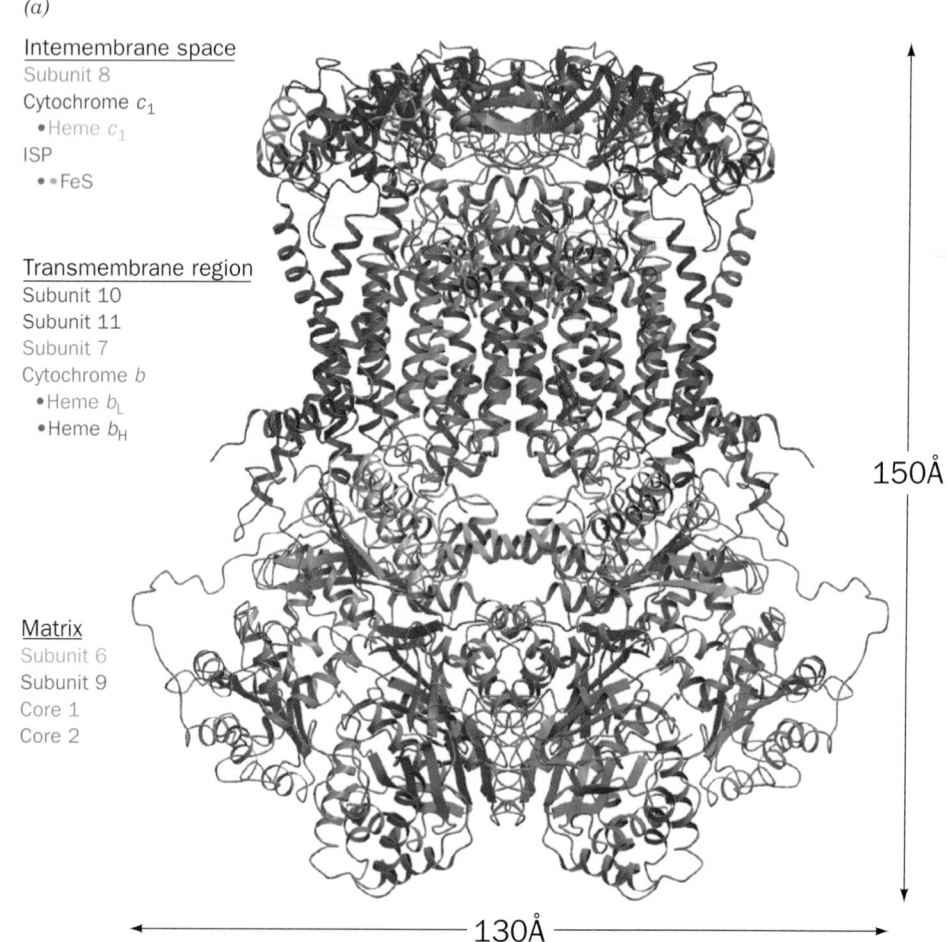

(*a*)

Intemembrane space
Subunit 8
Cytochrome c_1
• Heme c_1
ISP
• • FeS

Transmembrane region
Subunit 10
Subunit 11
Subunit 7
Cytochrome *b*
• Heme b_L
• Heme b_H

Matrix
Subunit 6
Subunit 9
Core 1
Core 2

150Å

130Å

cytochrome differs from protoporphyrin IX in that its vinyl groups have added Cys sulfhydryls across their double bonds to form thioether linkages to the protein. Heme *a* contains a long hydrophobic tail of three isoprene units (a **farnesyl** group) linked to the porphyrin via a hydroxyethyl group, as well as a formyl group in place of a methyl substituent. The axial ligands of the heme iron also vary with the cytochrome type. In cytochromes *a* and *b*, both ligands are His residues, whereas in cytochrome *c*, one is His and the other is Met (Fig. 22-22*b*). Note that the designation of a given cytochrome type refers only to the identity of the cytochrome's heme prosthetic group(s); a given cytochrome type may have any of several unrelated protein folds.

b. X-Ray Structure of the Cytochrome bc_1 Complex

All known cytochrome bc_1 complexes contain three common subunits: **cytochrome *b*,** which binds both the b_H and b_L hemes, **cytochrome c_1,** which contains a single *c*-type heme, and a Rieske iron–sulfur protein **(ISP),** which contains a [2Fe–2S] cluster. The bovine bc_1 complex contains 8 additional subunits for a total of 11 different subunits that combine to form a 2166-residue (243-kD) protomer that dimerizes. Of these, only cytochrome *b* is encoded by a mitochondrial gene.

The X-ray structures of bovine (Fig. 22-23*a*), chicken, and yeast (Fig. 22-23*b*) cytochrome bc_1 complexes have

(*b*)

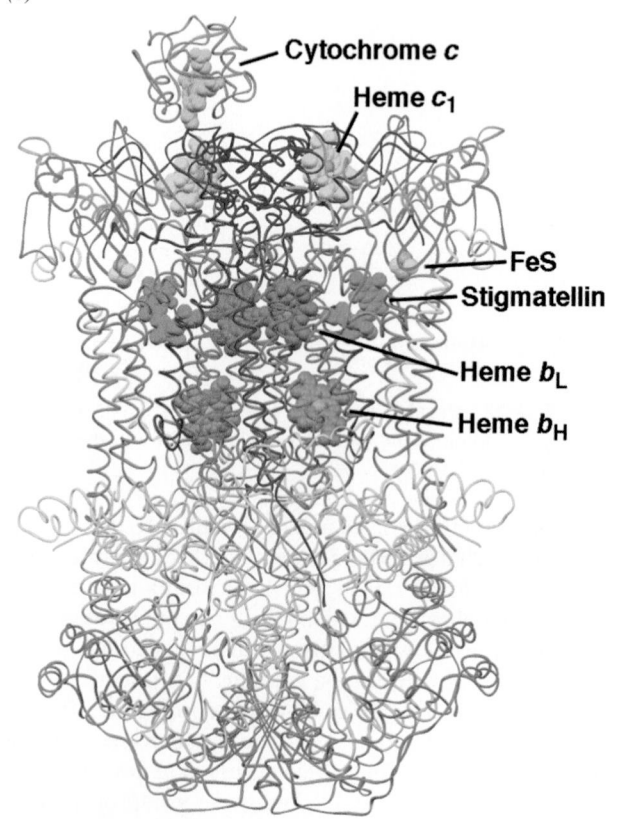

Cytochrome *c*
Heme c_1
FeS
Stigmatellin
Heme b_L
Heme b_H

been independently determined by Johann Diesenhofer, by Iwata and Bing Jap, by Edward Berry, Antony Crofts, and Sung-Hou Kim, and by Hartmut Michel. All of these structures reveal a 2-fold symmetric pear-shaped molecule of maximum diameter ~130 Å and height ~150 Å, whose wide end extends ~75 Å into the matrix space and whose narrow end extends ~35 Å into the intermembrane space. Its membrane-spanning region is ~40 Å thick and consists of 13 transmembrane helices in each protomer (12 in yeast, which consists of 9 different subunits). Eight of these transmembrane helices are contributed by the cytochrome *b* subunit, which binds both heme b_H and heme b_L within its transmembrane region, with heme b_L being closest to the intermembrane space. One of the remaining transmembrane helices is the membrane anchor of cytochrome c_1, the rest of which is a globular domain that extends into the intermembrane space. This is the portion of the complex that contains heme c_1 and to which cytochrome *c* docks (Figure 22-23*b*). The ISP is likewise anchored by a single transmembrane helix and extends into the intermembrane space. Remarkably, the two ISPs of the dimeric complex are intertwined such that the [2Fe–2S] cluster-containing domain of one protomer interacts with the cytochrome *b* and cytochrome c_1 subunits of the other protomer. The distances between the various metal centers are all quite large, ranging from 21 to 34 Å. The portion of the complex that occupies the matrix, which accounts for more than half the mass of cytochrome bc_1, consists of the structurally homologous **core 1** and **core 2** proteins together with **subunit 6** and **subunit 9.**

The route of electrons through the cytochrome bc_1 complex is discussed in Section 22-3B, together with the mechanism by which the complex preserves the free energy of electron transfer from $CoQH_2$ to cytochrome *c* for ATP synthesis.

4. Cytochrome *c*

Cytochrome *c* is a peripheral membrane protein of known crystal structure (Figs. 8-42 and 9-38*c*) that is loosely bound to the outer surface of the inner mitochondrial membrane. *It alternately binds to cytochrome c_1 (of Complex III) and to cytochrome c oxidase (Complex IV) and thereby functions to shuttle electrons between them.*

Cytochrome *c*'s binding site contains several invariant Lys residues that lie in a ring around the exposed edge of its otherwise buried heme group (Fig. 22-24). This binding site has been identified by **differential labeling:** Treatment of cytochrome *c* with acetic anhydride (which acetylates Lys residues) in the presence and absence of cytochrome c_1 demonstrated that cytochrome c_1 completely shields these cytochrome *c* Lys residues. The reactivities of other cytochrome *c* Lys residues that are distant from the exposed heme edge are unaffected by complex formation. Nearly identical results were obtained when cytochrome c_1 was replaced by cytochrome *c* oxidase. This suggests that both these proteins have negatively charged sites that are complementary to the ring of positively charged Lys residues on cytochrome *c* (see below).

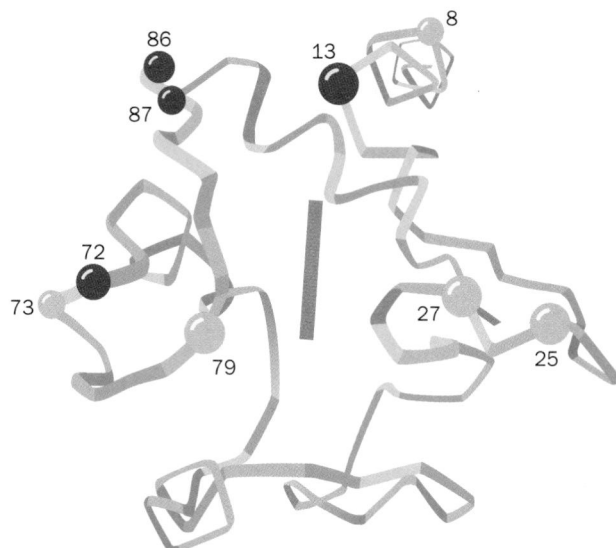

FIGURE 22-24 Ribbon diagram of cytochrome *c* showing the Lys residues involved in intermolecular complex formation. Cytochrome *c* forms complexes with cytochrome *c* oxidase and cytochrome c_1 as inferred from chemical modification studies. Dark and light blue balls, respectively, mark the positions of Lys residues whose ε-amino groups are strongly and less strongly protected by cytochrome *c* oxidase and cytochrome c_1 against acetylation by acetic anhydride. Note that these Lys residues form a ring around the heme (*solid bar*) on one face of the protein. [After Mathews, F.S., *Prog. Biophys. Mol. Biol.* **45,** 45 (1986).] *See the Interactive Exercises and Kinemage Exercise 5*

The X-ray structure of yeast cytochrome bc_1 in complex with cytochrome *c*, determined by Carola Hunte, reveals, as expected, that cytochrome *c* binds to the cytochrome c_1 subunit of cytochrome bc_1 (Fig. 22-23*b*). This association appears to be particularly tenuous because its interfacial area ($880Å^2$) is significantly less than that exhibited by protein-protein complexes known to have low stability (typically $<1600 Å^2$). Such a small interface is well suited for fast binding and release. This interface involves only two cytochrome *c* Lys residues, Lys 86 and Lys 79, which respectively contact Glu 235 and Ala 164 of cytochrome c_1. Other pairs of charged and often conserved residues surround the contact site but they are not close enough for direct polar interactions. Perhaps these interactions are mediated by water molecules that are not seen in the X-ray structure. The closest approach between the heme groups of the contacting proteins is 4.5 Å between atoms of their respective vinyl side chains, and their Fe–Fe distance is 17.4 Å. This accounts for the $8.3 \times 10^6 \, s^{-1}$ rate of electron transfer between these two redox centers (see below).

a. The Influence of Protein Structure on the Rate of Electron Transfer

Reduced hemes are highly reactive entities; they can transfer electrons over distances of 10 to 20 Å at physiologically significant rates. Hence cytochromes, in a sense, have the opposite function of enzymes: Instead of persuading unreactive substrates to react, they must prevent

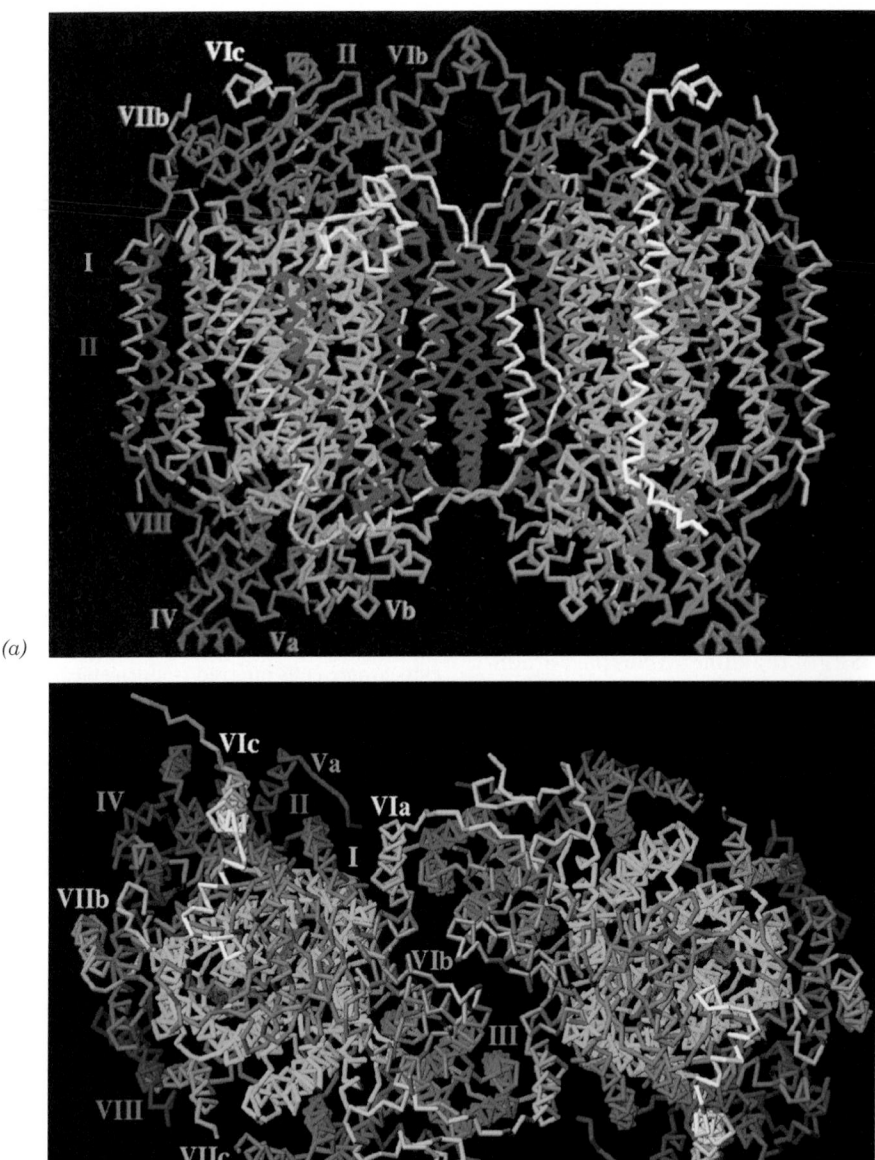

(a)

(b)

FIGURE 22-25   X-Ray structure of fully oxidized bovine heart cytochrome *c* oxidase. (*a*) View parallel to the membrane with the matrix below. The 2-fold axis of the homodimeric complex is vertical. Each of its 13 unique subunits is represented as a C_α trace that is given the same color as its name. (*b*) The complex as viewed from the top of Part *a*, along its 2-fold axis. Note the tenuous nature of the contacts between the protomers forming the dimeric complex. (*c*) A protomer viewed similarly to Part *a* showing the positions of the complex's redox centers. The protein surface is shown as a cage with its hydrophobic membrane-spanning portion in yellow and its hydrophilic portions that protrude into the intermembrane space (*top*) and matrix (*bottom*) in cyan. Heme *a* (*left*) and heme a_3 (*right*) are red, the Cu ions are green spheres, and the amino acid side chains that ligand the metal ions are green or gray. Complex IV has three additional metal ions that appear to have structural but not catalytic functions: an Mg^{2+} ion (*orange sphere*) that is liganded by a water molecule (*blue sphere*), a Zn^{2+} ion (*red sphere*) that is tetrahedrally coordinated by four Cys side chains, and an Na^+ or possibly a Ca^{2+} ion (not shown) that is near the intermembrane space. [Courtesy of Shinya Yoshikawa, Himeji Institute of Technology, Hyogo, Japan. PDBid 1OCC.]   **See the Interactive Exercises**   *(c)*

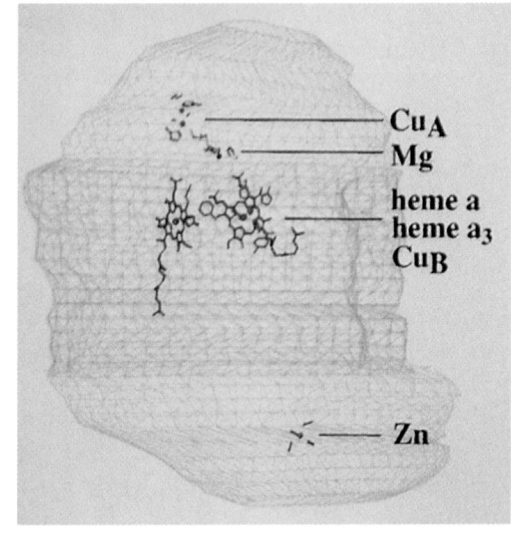

their hemes from transferring electrons nonspecifically to other cellular components. This, no doubt, is why these hemes are almost entirely enveloped by protein. However, cytochromes must also provide a path for electron transfer to an appropriate partner.

In proteins, electron transfers occur between redox-active cofactors such as hemes. However, a survey of proteins of known structure that function in electron transfer reveals that electrons travel no more then 14 Å between protein-embedded redox centers and that transfers over longer distances always involve chains of redox-active cofactors (e.g., Fig. 22-19b). Electron transfers occur far more efficiently through bonds than through space via a quantum mechanical process known as **electron tunneling.** Thus, as Harry Gray has experimentally demonstrated, electron tunneling within proteins occurs largely through the polypeptide chains between bound redox-active groups and that the rate of electron transfer varies with the structure of the intervening polypeptide. Moreover, tunneling across protein–protein interfaces is largely mediated by van der Waals interactions and water-bridged hydrogen bonds. Nevertheless, Leslie Dutton has shown that the experimentally measured electron transfer rates within proteins vary only with the distance between the electron donor and the electron acceptor and fall off with an ~10-fold decrease in rate for each 1.7 Å increase in this distance.

5. Complex IV (Cytochrome *c* Oxidase)

*Cytochrome c oxidase (**COX**), the terminal enzyme of the electron transport chain, catalyzes the one-electron oxidations of four consecutive reduced cytochrome c molecules and the concomitant four-electron reduction of one O_2 molecule to yield H_2O:*

$$4\text{Cytochrome } c^{2+} + 4H^+ + O_2 \rightarrow$$
$$4\text{cytochrome } c^{3+} + 2H_2O$$

Eukaryotic COX is an ~200-kD transmembrane protein composed of 8 to 13 subunits, whose largest and most hydrophobic chains, Subunits I, II, and III, are encoded by mitochondrial DNA. Eukaryotic COX exists in membranes as a dimer. Subunits I and II of this complex contain all four of its redox-active centers: two *a*-type hemes, *a* and a_3, and two Cu-containing centers, Cu_A and Cu_B. Heme *a* and the Cu_A center are of low potential (0.210 and 0.245 V; Table 22-1), whereas heme a_3 and Cu_B are of higher potential (0.340 and 0.385 V). Spectroscopic studies indicate that electrons are passed from cytochrome *c* to the Cu_A center, then to the heme *a*, and finally to a binuclear complex of heme a_3 and Cu_B. O_2 binds to this binuclear complex and is reduced to H_2O in a complex 4-electron reaction (see below).

a. X-Ray Structures of Cytochrome *c* Oxidase

The X-ray structures of two species of cytochrome *c* oxidase have been determined: a relatively simple form from the soil bacterium *Paracoccus denitrificans* (1106 residues) by Michel and a more complex form from bovine heart (1806 residues) by Shinya Yoshikawa. Each protomer of the 2-fold symmetric dimeric bovine COX has an ellipsoidal (potato-

like) shape comprised of a 48-Å-thick transmembrane portion and hydrophilic portions that protrude 32 and 37 Å into the mitochondrial matrix and the intermembrane space, respectively (Fig. 22-25). These protomers consist of 13 different subunits that mainly form 28 transmembrane helices. The protomer surfaces that comprise the dimer interface are concave (Fig. 22-25b) and hence their relatively tenuous contacts enclose a lipid-filled cavity. Thus a mechanistic role for dimer formation seems unlikely. *Paracoccus* COX is a monomeric complex consisting of only 4 subunits that collectively contain 22 transmembrane helices.

The structures of bovine COX Subunits I (12 transmembrane helices) and II (2 transmembrane helices), which bind all of the complex's four redox centers, are closely similar to those of *Paracoccus* COX. These are the subunits that carry out the main functions of the complex: transporting electrons from cytochrome *c* to O_2 to yield water, while pumping protons from inside (as we shall refer to the mitochondrial matrix and the bacterial cytoplasm) to outside (as we shall refer to the mitochondrial intermembrane space and the bacterial periplasmic space). Bovine Subunit III (7 transmembrane helices), whose structure also resembles that of its *Paracoccus* counterpart, does not appear to directly participate in electron transfer or proton translocation. Indeed, a complex consisting of only *Paracoccus* Subunits I and II can actively transport electrons and pump protons. Thus the function of Subunit III is unknown, although there is some evidence that it facilitates the assembly of Subunits I and II to form the active complex. Note, however, that Subunit III does not contact Subunit II. None of the 10 nuclear-encoded subunits of bovine COX resemble Subunit IV of *Paracoccus* COX (1 transmembrane helix). Seven of these bovine subunits each have one transmembrane helix, all oriented with their N-terminal ends on the matrix side of the membrane. These helices are distributed about the periphery of the dimeric core formed by Subunits I, II, and III (Fig. 22-25b). The remaining three bovine subunits are globular and associate entirely with the extramembrane portions of the complex. The X-ray structure of bovine COX provides little indication of the function of any of its nuclear-encoded subunits. Perhaps they have regulatory roles.

Subunit I binds heme *a* and the heme a_3–Cu_B binuclear center (Fig. 22-26), whose metal ions are all located ~13 Å below the membrane surface on its intermembrane/periplasmic side (Fig. 22-25c). The heme a_3 Fe has one axial His ligand, the heme *a* Fe has two axial His ligands (as in Fig. 22-22b, *top*), and the Cu_B atom has three His ligands, whose coordinating N atoms are arranged in an equilateral triangle that is centered on Cu_B and is parallel to heme a_3. The X-ray structure of fully oxidized bovine COX reveals a peroxide group (O_2^{2-}; Fig. 22-26) that bridges the heme a_3 Fe atom and Cu_B (which are separated by 4.9 Å), thereby liganding Cu_B via a distorted square-planar arrangement, a stable coordination geometry for Cu(II). However, in the X-ray structure of the fully reduced form of bovine COX (in which the heme a_3 Fe—Cu_B distance is 5.2 Å), this ligand is absent, so that Cu_B is trigonally liganded, a stable coordination geometry

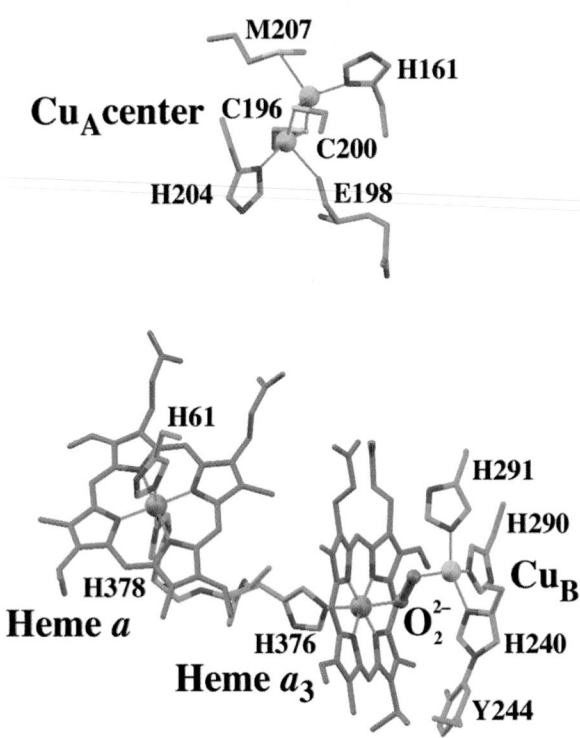

FIGURE 22-26   The redox centers in the X-Ray structure of bovine heart cytochrome *c* oxidase. The Fe and Cu ions are represented by orange and cyan spheres. Their liganding heme and protein groups (from subunit II for the Cu$_B$ center and subunit I for the others) are drawn in stick form colored according to atom type (heme C magenta, protein C green, N blue, O red, and S yellow). The peroxy group that bridges the Cu$_B$ and heme a_3 Fe ions is shown in ball-and-stick form in red. Coordination bonds are drawn as gray lines. Note that the side chains of His 240 and Tyr 244 are joined by a covalent bond (*lower right*). [Based on an X-ray structure by Shinya Yoshikawa, Himeji Institute of Technology, Hyogo, Japan. PDBid 2OCC.]

for Cu(I). The closest approach of the two heme groups is 4 Å, and the distance between their Fe atoms is 13.2 Å.

Subunit II, in addition to its two transmembrane helices, has a globular domain on the outside surface that binds the Cu$_A$ center and largely consists of a 10-stranded β barrel. The Cu$_A$ center is located ~8 Å above the outside membrane surface. Although the Cu$_A$ center was, for many years, widely believed to contain only one Cu atom, the X-ray structures of COX clearly indicate that Cu$_A$ contains two Cu atoms (Fig. 22-26). These are bridged by two Cys S atom ligands and have two additional protein ligands each to form an arrangement similar to that of a [2Fe–2S] cluster (Fig. 22-15*b*) in which the two Cu atoms are 2.4 Å apart. Spectroscopic measurements indicate that in the Cu$_A$ center's reduced form, both of its Cu atoms are in their Cu(I) states, whereas in its fully oxidized form, the newly acquired electron appears to be delocalized between the two Cu atoms such that they assume the [Cu$^{1.5+}$ · · · Cu$^{1.5+}$] state.

b. Electron and Proton Acquisition

COX's cytochrome *c* binding site is postulated to be in a corner formed by the globular domain of Subunit II and the outside surface of Subunit I, since this region is close to the Cu$_A$ site and contains 10 acidic side chains that could interact with the ring of Lys side chains surrounding cytochrome *c*'s heme crevice (Fig. 22-24). Indeed, differential labeling of cytochrome *c* oxidase's carboxyl groups in the presence and absence of cytochrome *c* demonstrated that cytochrome *c* shields the invariant residues Asp 112, Glu 114, and Glu 198 of Subunit II (bovine numbering). Glu 198 is located between the two Cys residues of Subunit II that ligand Cu$_A$ (Fig. 22-26). This observation supports the spectroscopic evidence that places the cytochrome *c* binding site on Subunit II in close proximity to Cu$_A$. Cross-linking studies have additionally shown that the cytochrome *c* surface opposite the electron-transferring site interacts with Subunit III, suggesting that Subunit III also participates in binding cytochrome *c*.

Time-resolved spectroscopic studies indicate that an electron obtained from cytochrome *c* is first acquired by the Cu$_A$ center and is then transferred to heme *a* rather than heme a_3, probably because the shortest Cu$_A$ · · · heme *a* distance of 11.7 Å is less than the shortest Cu$_A$ · · · heme a_3 distance of 14.7 Å. The electron is then rapidly transferred to the heme a_3–Cu$_B$ binuclear center, where it participates in reducing the bound O$_2$ to H$_2$O. Note that the fifth ligand of heme *a*, His 378, is separated from the fifth ligand of heme a_3, His 376, by only one residue, and hence there is a relatively short through-the-bonds electron transfer pathway between heme *a* and a_3. In addition, the closest approach of the two hemes is 4 Å.

COX must acquire four so-called **chemical** or **scalar protons** from the inside for every molecule of O$_2$ it reduces to H$_2$O. This 4-electron process is coupled to the translocation of up to four so-called **pumped** or **vectorial protons** from the inside to the outside, thereby contributing to the proton gradient that powers ATP synthesis (Section 22-2C). Note that for each turnover of the enzyme,

$$8H_{inside}^+ + 4\,\text{cyt } c^{2+} + O_2 \rightarrow 4\,\text{cyt } c^{3+} + 2H_2O + 4H_{outside}^+$$

a total of eight positive charges are transported across the membrane, thereby contributing to its membrane potential.

c. Reaction Sequence for the Reduction of O$_2$ by Cytochrome *c* Oxidase

The reduction of O$_2$ to 2 H$_2$O by cytochrome *c* oxidase takes place on the cytochrome a_3–Cu$_B$ binuclear complex (Fig. 22-26). Indeed, a synthetic model of this binuclear complex (Fig. 22-27), synthesized by James Collman, efficiently catalyzes the reduction of O$_2$ to H$_2$O when attached to an electrode.

The COX-mediated reduction of O$_2$ requires, as we shall see, the nearly simultaneous input of four electrons. However, the fully reduced a_3^{2+}–Cu$_B^{1+}$ binuclear complex can readily contribute only three electrons to its bound O$_2$ in reaching its fully oxidized a_3^{4+}–Cu$_B^{2+}$ state [cytochrome

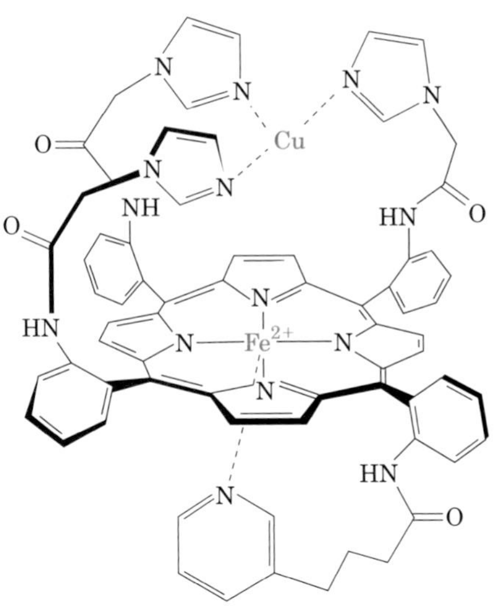

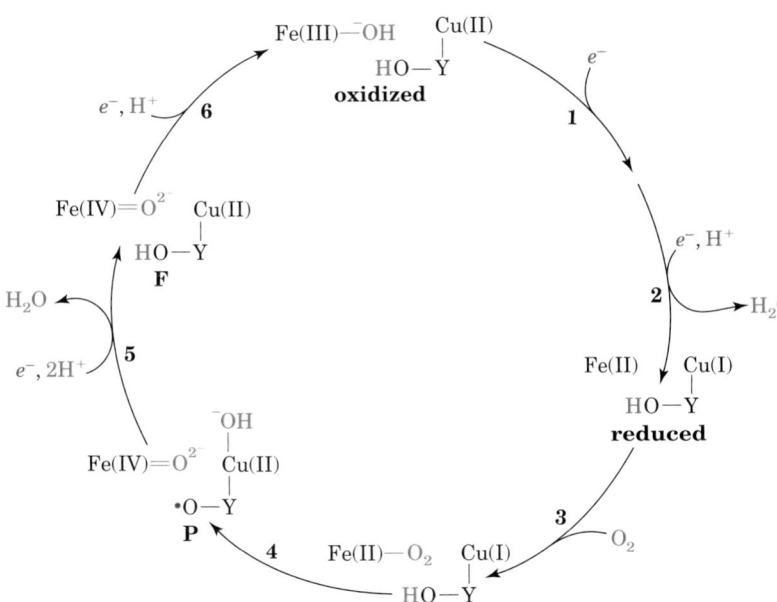

FIGURE 22-27 Synthetic model of the cytochrome a_3–Cu_B binuclear complex. This structure efficiently reduces O_2 to H_2O when attached to an electrode. The pyridine group that axially ligands the Fe ion (*bottom*) can be replaced by an imidazole group.

a_3 transiently assumes its Fe(IV) or **ferryl** oxidation state during the reduction of O_2; see below]. What is the source of the fourth electron?

The X-ray structures of both bovine and *Paracoccus* COX clearly indicate that the His 240 ligand of Cu_B (bovine numbering) is covalently cross-linked to the side chain of the conserved Tyr 244 (Fig. 22-26, *lower right*). This places Tyr 244's phenolic —OH group in close proximity to the heme a_3-ligated O_2 such that Tyr 244 can supply the fourth electron by transiently forming a tyrosyl radical (TyrO·). In fact, adding peroxide to the resting enzyme generates a tyrosyl radical, whereas mutating Tyr 244 to

Phe inactivates the enzyme. Moreover, tyrosyl radicals have been implicated in several enzymatically mediated redox processes, including the generation of O_2 from H_2O in photosynthesis (in a sense, the reverse of the COX reaction; Section 24-2C), and in the **ribonucleotide reductase** reaction (which converts NDP to dNDP; Section 28-3A). Tyr 244's phenolic —OH group is within hydrogen bonding distance of the COX-bound O_2 and hence is a likely H^+ donor during O—O bond cleavage. The formation of the covalent cross-link is expected to lower both the reduction potential and the pK of Tyr 244, thereby facilitating both radical formation and proton donation (the synthetic binuclear complex in Fig. 22-27 can function without an associated tyrosyl radical, presumably because its associated electrode can supply it with electrons much faster than cytochrome *c* can supply them to COX).

The COX reaction, elucidated in large part by Mårten Wikström and Gerald Babcock using a variety of spectroscopic techniques, involves four consecutive one-electron transfers from the Cu_A and cytochrome *a* sites and occurs as follows (Fig. 22-28):

1 and 2. The oxidized binuclear complex [Fe(III)$_{a3}$— OH^- Cu(II)$_B$] is reduced to its [Fe(II)$_{a3}$ Cu(I)$_B$] state by two consecutive one-electron transfers from cytochrome *c* via cytochrome *a* and Cu_A. A proton from the matrix is concomitantly acquired and an H_2O is released in this process. Tyr 244 (Y—OH) is in its phenolic state.

3. O_2 binds to the reduced binuclear complex so as to ligand its Fe(II)$_{a3}$ atom. It binds to the heme with much the same configuration it has in oxymyoglobin (Fig. 10-12).

4. Internal electron redistribution rapidly yields the oxyferryl complex [Fe(IV)$=O^{2-}$ HO^-—Cu(II)] in which Tyr 244 has donated an electron and a proton to the complex and thereby assumed its neutral radical state (Y—O·). This is known as compound P because it was

FIGURE 22-28 Proposed reaction sequence for the reduction of O_2 by the cytochrome a_3–Cu_B binuclear complex of cytochrome *c* oxidase. The numbered steps are discussed in the text. The entire reaction is extremely fast; it goes to completion in ~1 ms at room temperature. [Modified from Babcock, G.T., *Proc. Natl. Acad. Sci.* **96,** 12971 (1999).]

originally thought that this spectroscopically identified state was a peroxy complex. However, it has since been shown that a peroxy compound is not on the reaction pathway. The peroxy complex displayed in Fig. 22-26 is a two-electron reduced "mixed valence" state of the enzyme that cannot reduce O_2 past its peroxy form.

5. A third one-electron transfer from cytochrome *c* together with the acquisition of two protons reconverts Tyr 244 to its phenolic state, yielding compound F (for ferryl) and releasing an H_2O.

6. A fourth and final electron transfer and proton acquisition yields the oxidized [Fe(III)$_{a3}$—OH$^-$ Cu(II)$_B$] complex, thereby completing the catalytic cycle.

Note that the COX reaction proceeds without the release of the destructive partially reduced oxygen intermediates from its active site. The positions in this proposed catalytic cycle at which protons appear to be pumped from the matrix (bacterial cytoplasm) to the intermembrane (periplasmic) space are discussed in Section 22-3B. Keep in mind, however, that aspects of this cycle are uncertain and/or disputed and hence it remains under investigation.

3 ■ OXIDATIVE PHOSPHORYLATION

The endergonic synthesis of ATP from ADP and P_i in mitochondria, which, as we shall see, is catalyzed by **proton-translocating ATP synthase (Complex V),** is driven by the electron-transport process. Yet, since Complex V is physically distinct from the proteins mediating electron transport (Complexes I–IV), *the free energy released by electron transport must be conserved in a form that ATP synthase can utilize.* Such energy conservation is referred to as **energy coupling** or **energy transduction.**

The physical characterization of energy coupling has proved to be surprisingly elusive; many sensible and often ingenious ideas have failed to withstand the test of experimental scrutiny. In this section we first examine some of the hypotheses that have been formulated to explain the coupling of electron transport and ATP synthesis. We shall then explore the coupling mechanism that has garnered the most experimental support, analyze the mechanism by which ATP is synthesized by ATP synthase, and, finally, discuss how electron transport and ATP synthesis can be uncoupled.

A. Energy Coupling Hypotheses

In the more than 60 years that electron transport and oxidative phosphorylation have been studied, numerous mechanisms have been proposed to explain how these processes are coupled. In the following paragraphs, we examine the mechanisms that have received the greatest experimental attention.

1. The chemical coupling hypothesis. In 1953, Edward Slater formulated the **chemical coupling hypothesis,** in which he proposed that electron transport yielded reactive intermediates whose subsequent breakdown drove oxidative phosphorylation. We have seen, for example, that such a mechanism is responsible for ATP synthesis in glycolysis (Sections 17-2F and 17-2G). Thus, the exergonic oxidation of glyceraldehyde-3-phosphate by NAD$^+$ yields 1,3-bisphosphoglycerate, a reactive ("high-energy") acyl phosphate whose phosphoryl group is then transferred to ADP to form ATP in the phosphoglycerate kinase reaction. The difficulty with such a mechanism for oxidative phosphorylation, which has caused it to be abandoned, is that despite intensive efforts in numerous laboratories over many years, no appropriate reactive intermediates have been identified.

2. The conformational coupling hypothesis. The **conformational coupling hypothesis,** which Paul Boyer formulated in 1964, proposes that electron transport causes proteins of the inner mitochondrial membrane to assume "activated" or "energized" conformational states. These proteins are somehow associated with ATP synthase such that their relaxation back to the deactivated conformation drives ATP synthesis. As with the chemical coupling hypothesis, the conformational coupling hypothesis has found little experimental support. However, conformational coupling of a different sort appears to be involved in ATP synthesis (Section 22-3C).

3. The chemiosmotic hypothesis. The **chemiosmotic hypothesis,** proposed in 1961 by Peter Mitchell, has spurred considerable controversy, as well as much research, and is now the model most consistent with the experimental evidence. It postulates that *the free energy of electron transport is conserved by pumping H^+ from the mitochondrial matrix to the intermembrane space so as to create an electrochemical H^+ gradient across the inner mitochondrial membrane. The electrochemical potential of this gradient is harnessed to synthesize ATP (Fig. 22-29).*

Several key observations are explained by the chemiosmotic hypothesis:

(a) Oxidative phosphorylation requires an intact inner mitochondrial membrane.

(b) The inner mitochondrial membrane is impermeable to ions such as H^+, OH^-, K^+, and Cl^-, whose free diffusion would discharge an electrochemical gradient.

(c) Electron transport results in the transport of H^+ out of intact mitochondria, thereby creating a measurable electrochemical gradient across the inner mitochondrial membrane.

(d) Compounds that increase the permeability of the inner mitochondrial membrane to protons, and thereby dissipate the electrochemical gradient, allow electron transport (from NADH and succinate oxidation) to continue but inhibit ATP synthesis; that is, they "uncouple" electron transport from oxidative phosphorylation. Conversely, increasing the acidity outside the inner mitochondrial membrane stimulates ATP synthesis.

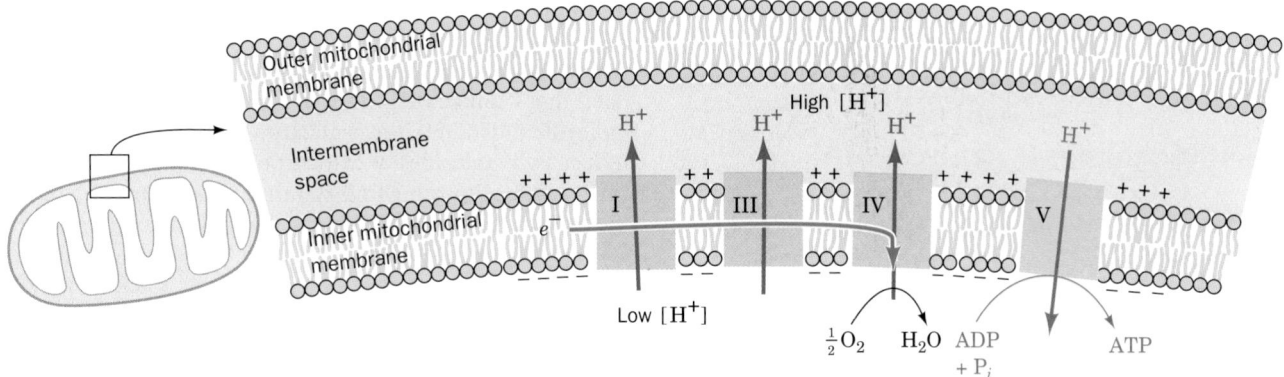

FIGURE 22-29 Coupling of electron transport (*green arrow*) and ATP synthesis. H^+ is pumped out of the mitochondrion by Complexes I, III, and IV of the electron-transport chain (*blue arrows*), thereby generating an electrochemical gradient across the inner mitochondrial membrane. The exergonic return of these protons to the matrix powers the synthesis of ATP (*red arrow*). Note that the outer mitochondrial membrane is permeable to small molecules and ions, including H^+. 🎞 **See the Animated Figures**

In the remainder of this section we examine the mechanisms through which electron transport can result in proton translocation and how an electrochemical gradient can interact with ATP synthase to drive ATP synthesis.

B. *Proton Gradient Generation*

Electron transport, as we shall see, causes Complexes I, III, and IV to transport protons across the inner mitochondrial membrane from the matrix, a region of low [H^+] and negative electrical potential, to the intermembrane space (which is in contact with the cytosol), a region of high [H^+] and positive electrical potential (Fig. 22-14). The free energy sequestered by the resulting electrochemical gradient [which, in analogy to the term electromotive force (emf), is called **proton-motive force (pmf)**] *powers ATP synthesis.*

a. Proton Pumping Is an Endergonic Process

The free energy change of transporting a proton out of the mitochondrion against an electrochemical gradient is expressed by Eq. [20.3], which, in terms of pH, is

$$\Delta G = 2.3RT\,[\text{pH}(in) - \text{pH}(out)] + Z\mathscr{F}\Delta\Psi \qquad [22.1]$$

where Z is the charge on the proton (including sign), $\mathscr{F}$ is the Faraday constant, and $\Delta\Psi$ is the membrane potential. The sign convention for $\Delta\Psi$ is that when a positive ion is transported from negative to positive, $\Delta\Psi$ is positive. Since pH(*out*) is less than pH(*in*), the export of protons from the mitochondrial matrix (against the proton gradient) is an endergonic process. In addition, *proton transport out of the matrix makes the inner membrane's internal surface more negative than its external surface.* Outward transport of a positive ion is consequently associated with a positive $\Delta\Psi$ and an increase in free energy (endergonic process), whereas the outward transport of a negative ion yields the opposite result. Clearly, it is always necessary to describe membrane polarity when specifying a membrane potential.

The measured membrane potential across the inner membrane of a liver mitochondrion, for example, is 0.168 V (inside negative; which corresponds to an ~210,000 V · cm^{-1} electric field across its ~80 Å thickness). The pH of the matrix is 0.75 units higher than that of the intermembrane space. ΔG for proton transport out of this mitochondrial matrix is therefore 21.5 kJ · mol^{-1}.

b. The Passage of About Three Protons Is Required to Synthesize One ATP

An ATP molecule's estimated physiological free energy of synthesis, around +40 to +50 kJ · mol^{-1}, is too large to be driven by the passage of a single proton back into the mitochondrial matrix; at least two protons are required. This number is difficult to measure precisely, in part because transported protons tend to leak back across the mitochondrial membrane. However, most estimates indicate that around three protons are passed per ATP synthesized.

c. Two Mechanisms of Proton Transport Have Been Proposed

Three of the four electron-transport complexes, Complexes I, III, and IV, are involved in proton translocation. Two mechanisms have been entertained that would couple the free energy of electron transport with the active transport of protons: the **redox loop mechanism** and the **proton pump mechanism.**

d. The Redox Loop Mechanism

This mechanism, proposed by Mitchell, requires that the redox centers of the respiratory chain (FMN, CoQ, cytochromes, and iron–sulfur clusters) be so arranged in the membrane that reduction would involve a redox center simultaneously accepting e^- and H^+ from the matrix side of the membrane. Reoxidation of this redox center by the next center in the chain would involve release of H^+ on the cytosolic side of the membrane together with the trans-

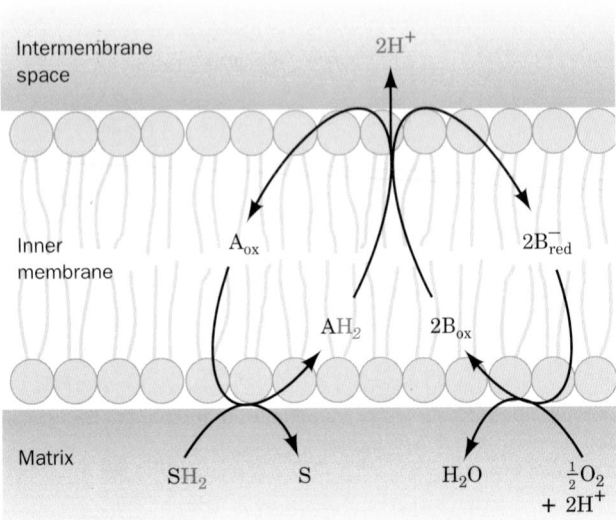

Intermembrane space

$2H^+$

Inner membrane

A_{ox}

$2B_{red}^-$

AH_2 $2B_{ox}$

Matrix

SH_2 S H_2O $\frac{1}{2}O_2 + 2H^+$

FIGURE 22-30   The redox loop mechanism for electron transport–linked H$^+$ translocation. AH_2 represents $(H^+ + e^-)$ carriers such as $FMNH_2$ and $CoQH_2$, whereas B represents pure e^- carriers such as iron–sulfur clusters and the cytochromes. These components are so arranged as to require that electron transport be accompanied by H$^+$ translocation.

fer of electrons back to the matrix side (Fig. 22-30). Electron flow from one center to the next would therefore yield net translocation of H$^+$ and the creation of an electrochemical gradient ($\Delta\Psi$ and ΔpH).

The redox loop mechanism requires that the first redox carrier contain more hydrogen atoms in its reduced state than in its oxidized state and that the second redox carrier have no difference in its hydrogen atom content between its reduced and oxidized states. Are these requirements met in the electron-transport chain? Some of the redox carriers, FMN and CoQ, in fact, contain more hydrogen atoms in their reduced state than in their oxidized state and thus can qualify as proton carriers as well as electron carriers. If these centers were spatially alternated with pure electron carriers (cytochromes and iron–sulfur clusters), such a mechanism could well be accommodated.

The main difficulty with the redox loop mechanism involves the deficiency of $(H^+ + e^-)$ carriers that can alternate with pure e^- carriers. Whereas the electron-transport chain has as many as 15 pure e^- carriers (up to 8 iron–sulfur proteins, 5 cytochromes, and 2 Cu centers), it has only 2 $(H^+ + e^-)$ carriers. The fact that there are 3 complexes with standard reduction potential changes large enough to provide free energy for ATP synthesis suggests the need for at least 3 proton-transport redox carriers. As we shall see, however, there are, in fact, 3 proton-transport sites but only two proton-transport redox carriers: Both the redox loop mechanism and the proton pump mechanism (discussed below) are employed.

e. Complex III Pumps Protons via the Q Cycle, a Type of Redox Loop

♪ **See Guided Exploration 20. The Q Cycle** Mitchell postulated that Complex III functions in a way that permits one molecule of CoQH$_2$, the 2-electron carrier, to sequentially reduce two molecules of cytochrome c, a one-electron carrier, while transporting four protons. This occurs via a modified redox loop mechanism involving a remarkable bifurcation of the flow of electrons from CoQH$_2$ to cytochrome c_1 and to cytochrome b. It is through this so-called **Q cycle** that Complex III pumps protons from the matrix to the intermembrane space.

The essence of the Q cycle is that *CoQH$_2$ undergoes a two-cycle reoxidation in which the semiquinone CoQ$^{\cdot-}$ is a stable intermediate.* This involves two independent binding sites for coenzyme Q: Q_o, which binds CoQH$_2$ and is located between the ISP and heme b_L in proximity to the intermembrane space (Fig. 22-23); and Q_i, which binds both CoQ$^{\cdot-}$ and CoQ and is located near heme b_H in proximity to the matrix. In the first cycle (Fig. 22-31a), CoQH$_2$, which is supplied by Complexes I or II on the matrix side of the inner mitochondrial membrane (**1**), diffuses through the membrane to its cytoplasmic side, where it binds to the Q_o site (**2**). There it transfers one of its electrons to the ISP (**3**), releasing its two protons into the intermembrane space and yielding CoQ$^{\cdot-}$. The ISP then reduces cytochrome c_1, whereas the CoQ$^{\cdot-}$ transfers its remaining electron to heme b_L (**4**), yielding fully oxidized CoQ. Heme b_L then reduces heme b_H (**6**). The CoQ from Step 4 is released from the Q_o site and diffuses back through the membrane to rebind at the Q_i site (**5**), where it picks up the electron from heme b_H (**7**), reverting to the semiquinone form, CoQ$^{\cdot-}$. Thus, the reaction for this first cycle is

$$CoQH_2 + \text{cytochrome } c_1(Fe^{3+}) \rightarrow$$
$$CoQ^{\cdot-} + \text{cytochrome } c_1(Fe^{2+}) + 2H^+(outside)$$

In the second cycle (Fig. 22-31b), another CoQH$_2$ repeats Steps 1 through 6: One electron reduces the ISP and then cytochrome c_1, and the other electron sequentially reduces heme b_L and then heme b_H. This second electron then reduces the CoQ$^{\cdot-}$ at the Q_i site produced in the first cycle (**8**), yielding CoQH$_2$. The protons taken up in this last step originate in the mitochondrial matrix. The reaction for the second cycle is therefore

$$CoQH_2 + CoQ^{\cdot-} + \text{cytochrome } c_1(Fe^{3+}) + 2H^+(matrix) \rightarrow$$
$$CoQ + CoQH_2 + \text{cytochrome } c_1(Fe^{2+}) + 2H^+(outside)$$

For every two CoQH$_2$ that enter the Q cycle, one CoQH$_2$ is regenerated. The combination of both cycles, in which two electrons are transferred from CoQH$_2$ to cytochrome c_1, results in the overall reaction

$$CoQH_2 + 2\text{cytochrome } c_1(Fe^{3+}) + 2H^+(matrix) \rightarrow$$
$$CoQ + 2\text{cytochrome } c_1(Fe^{2+}) + 4H^+(outside)$$

X-Ray studies of Complex III provide direct evidence for the independent existence of the Q_o and Q_i sites. The

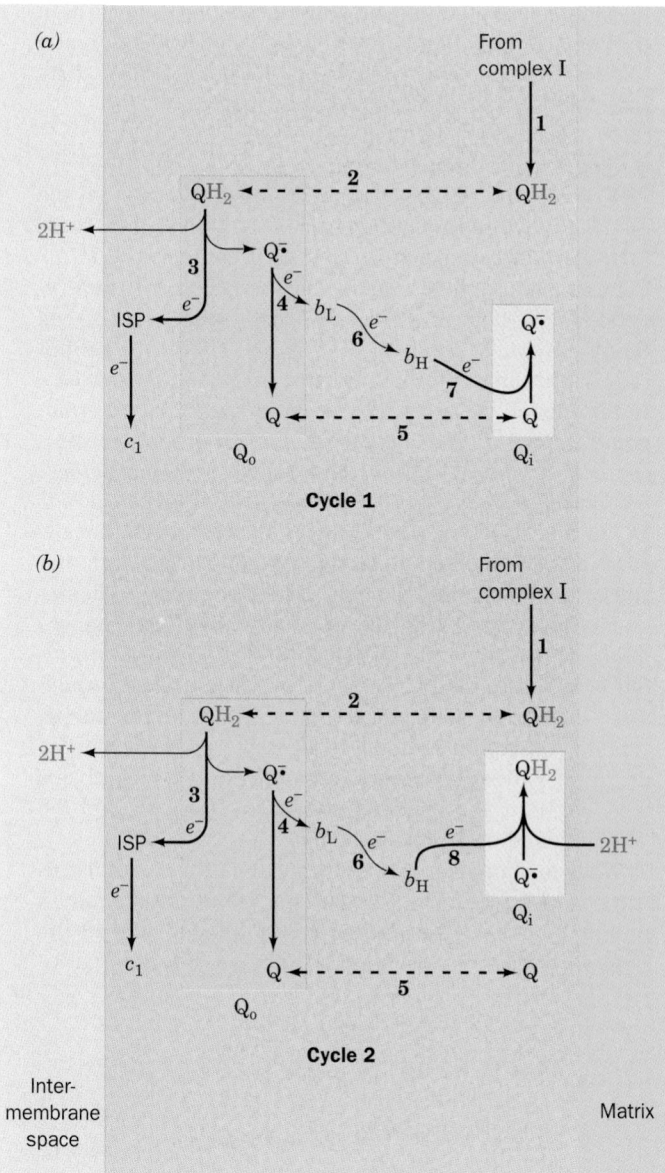

FIGURE 22-31 The Q cycle. The Q cycle is an electron-transport cycle in Complex III that accounts for H^+ translocation during the transport of electrons from cytochrome b to cytochrome c: The overall cycle is actually two cycles, the first (*a*) requiring reactions 1 through 7 and the second (*b*) requiring reactions 1 through 6 and 8. **(1)** Coenzyme QH_2 is supplied by Complex I on the matrix side of the membrane. **(2)** QH_2 diffuses to the outside of the membrane. **(3)** QH_2 reduces the Rieske iron–sulfur protein (ISP) forming $Q^{\cdot -}$ semiquinone and releasing $2H^+$. The ISP goes on to reduce cytochrome c_1. **(4)** $Q^{\cdot -}$ reduces heme b_L to form coenzyme Q. **(5)** Q diffuses to the matrix side. **(6)** Heme b_L reduces heme b_H. **(7, cycle 1 only)** Q is reduced to $Q^{\cdot -}$ by heme b_H. **(8, cycle 2 only)** $Q^{\cdot -}$ is reduced to QH_2 by heme b_H. [After Trumpower, B.L., *J. Biol. Chem.* **265**, 11410 (1990).]

antifungal agents **myxothiazol** and **stigmatellin,**

Myxothiazol

Stigmatellin

which both block electron flow from $CoQH_2$ to the ISP and to heme b_L (Steps 3 and 4 of both cycles), bind in a pocket within cytochrome b between the ISP and heme b_L (Fig. 22-23*b*). Evidently, this binding pocket overlaps the Q_o site. Similarly, antimycin (Section 22-2B), which blocks electron flow from heme b_H to CoQ and $CoQ^{\cdot -}$ (Step 7 of Cycle 1 and Step 8 of Cycle 2), binds in a pocket near heme b_H, thereby identifying this pocket as site Q_i.

The circuitous route of electron transfer in Complex III is tied to the ability of coenzyme Q to diffuse within the hydrophobic core of the membrane in order to bind to both the Q_o and Q_i sites. This process is facilitated by an indentation in the surface of cytochrome b's transmembrane region that contains Q_o from one protomer and Q_i from the other. *When $CoQH_2$ is oxidized, two reduced cytochrome c molecules and four protons appear on the outer side of the membrane.* Proton transport by the Q cycle thus follows the redox loop mechanism of proton transport in which a redox center itself (CoQ) is the proton carrier. As we shall see below, however, Complexes I and IV follow a different mechanism of proton transport, the proton pump mechanism.

f. The Bifurcation of the Q Cycle's Electron Flow Occurs via Domain Movement

Why does Q_o-bound $CoQH^{\cdot -}$ exclusively reduce heme b_L (Step 4 of the Q cycle) rather than the Rieske [2Fe–2S] cluster of the ISP, despite the greater reduction potential difference ($\Delta\mathscr{E}$) favoring the latter reaction (Table 22-1)? The remarkable answer to this question provides a fascinating insight into the inner workings of Complex III. Although the binding of stigmatellin and myxothiazol to Q_o are mutually exclusive, these inhibitors affect this site differently: Stigmatellin perturbs the spectrum and redox properties of the ISP's Rieske [2Fe–2S] cluster as well as prevents it from oxidizing cytochrome c_1, whereas myxothiazol does not interact with the ISP but, instead, shifts the spectrum of heme b_L. Evidently, stigmatellin is a $CoQH_2$ analog and myxothiazol is a $CoQH^-$ mimic.

X-Ray structures of Complex III (Section 22-2C) reveal that its Q_o site is a bifurcated pocket in which stigmatellin binds close to the ISP docking interface (see below), whereas myxothiazol binds in the vicinity of heme b_L (Fig. 22-32; their binding to Q_o is mutually exclusive because their hydrophobic tails would overlap). Furthermore, the globular Rieske [2Fe–2S] cluster–containing domain of the ISP (Fig. 22-23, *top*) is conformationally mobile and assumes a conformational state that is controlled by the ligand-binding state of Q_o: It binds to cytochrome b near the heme b_L site when stigmatellin is bound to the Q_o site, but has swung around by ~20 Å (via an $\sim57°$ hinge motion that leaves intact its tertiary structure) to bind to the cytochrome c_1 near its heme c when myxothiazol is bound to Q_o. Apparently the globular domain of the ISP functions to shuttle an electron from $CoQH_2$, which is bound to Q_o near the ISP docking interface, to heme c_1 by mechanically swinging the reduced Rieske [2Fe–2S] cluster between these sites. The resulting $CoQH^-$ shifts to the position near heme b_L (probably via a rotation about a bond connecting the semiquinone ring to its nonpolar tail), which it then reduces. Thus, $CoQH^-$ is unable to reduce the ISP (after it has reduced cytochrome c_1) because it is too far away to do so. This novel mechanism is supported by the observation that mutagenically inserting a disulfide bond either within the hinge of the ISP or between it and cytochrome b greatly diminishes the activity of Complex III but that its activity is restored on exposure to reducing agents.

g. The Proton Pump Mechanism

Complex IV (COX) transports four protons from the matrix to the intermembrane space for each O_2 it reduces ($2H^+$ per electron pair; Fig. 22-14). It contains no ($H^+ + e^-$) carriers and hence cannot do so via a redox loop (Q cycle-like) system. Rather, as we shall see, it does so via the proton pump mechanism (Fig. 22-33), which does not require that the redox centers themselves be H^+ carriers. In this model, *the transfer of electrons results in conformational changes to the complex. The unidirectional translocation of protons occurs as a result of the influence of these conformational changes on the pK's of amino acid side chains and their alternate exposure to the internal and external side of the membrane.* We have previously seen that conformation can influence pK. The Bohr effect in hemoglobin, for example, results from conformational changes induced by O_2 binding, which causes pK changes in protein acid–base groups (Section 10-2E). If such a protein were located in a membrane and if, in addition to pK changes, the conformational changes altered the side of the membrane to which the affected amino acid side chains were exposed, the result would be H^+ transport and the system would be a proton pump.

Keep in mind that protons, being atomic nuclei, must always be associated with molecules or ions. Consequently, a proton cannot be transported across a membrane in the same way that, say, a K^+ ion is. Rather, protons are translo-

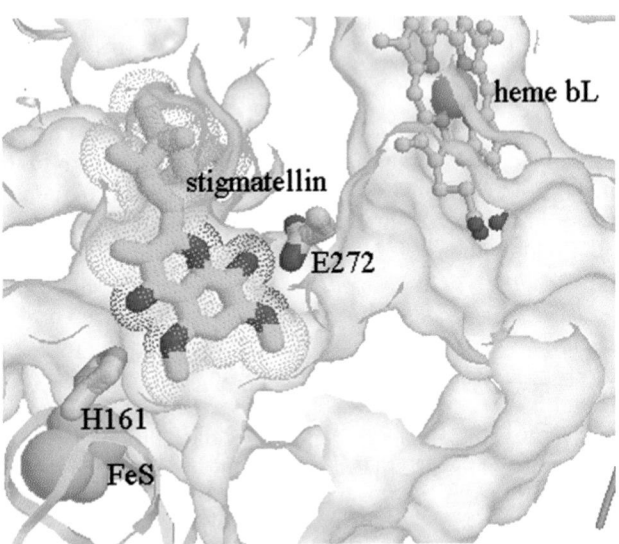

(a)

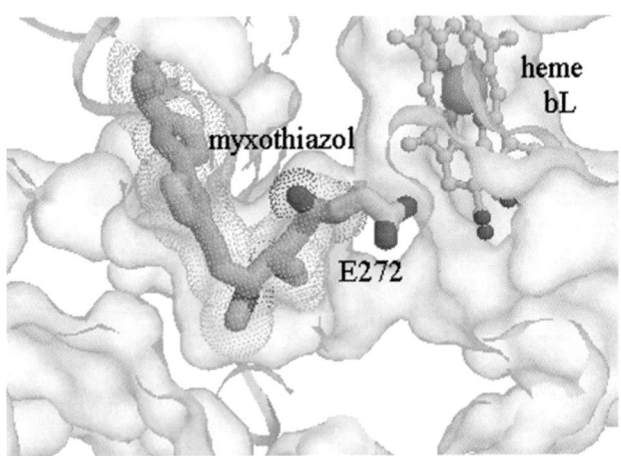

(b)

FIGURE 22-32 X-Ray structures of the Q_o binding site of the chicken cytochrome bc_1 complex occupied by inhibitors. The structures show (*a*) its complex with stigmatellin and (*b*) its complex with myxothiazol. The protein surface (*white*) has been cut away to show the Q_o pocket. Heme b_L (*upper right*) is shown in ball-and-stick form with C gray, N blue, O red, and Fe tan. In Part *a*, stigmatellin is drawn as a stick model with C yellow and its volume represented by the dotted surface. The Rieske [2Fe–2S] cluster (*lower left*) is represented by gold spheres, and the ISP domain to which it is bound is drawn as a cyan ribbon. Note that

His 61, which is a ligand of the Rieske [2Fe–2S] cluster, forms a hydrogen bond to stigmatellin. In Part *b*, the myxothiazol is drawn as a stick model with C orange. Note that the semiquinone-mimicking portion of myxothiazol binds to Q_o in proximity to heme b_L, whereas its hydrophobic tail occupies the same position as that of stigmatellin. Also note that the [2Fe–2S] cluster–containing ISP domain is not visible in this diagram; it has rotated into proximity with cytochrome c_1. [Courtesy of Antony Crofts, University of Illinois at Urbana–Champagne, and Edward Berry, University of California at Berkeley. PDBid 3BCC.]

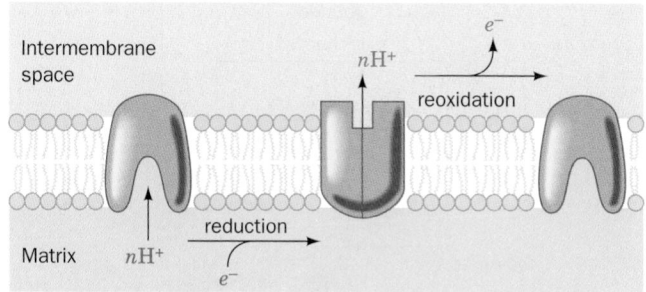

FIGURE 22-33 Proton pump mechanism of electron transport–linked proton translocation. At each H^+ translocation site, n protons bind to amino acid side chains on the matrix side (inside) of the membrane. Reduction causes a conformational change that decreases the pK's of these side chains and exposes them to the cytosolic side (outside) of the membrane, where the protons dissociate. Reoxidation results in a conformational change that restores the pump to its original conformation.

cated by hopping along chains of hydrogen bonded groups in the transport protein in much the same way that hydronium ions migrate through an aqueous solution (Fig. 2-9). Such an arrangement of hydrogen bonded groups has been described as a "proton wire." However, unlike a wire in an electrical circuit, all elements of a proton wire need not be connected at the same time and, moreover, internal water molecules, which are not always apparent in protein X-ray structures, are likely to be integral parts of the proton wire. Hence, elucidating the precise pathway of proton transport through a protein is a difficult and uncertain task.

h. Bacteriorhodopsin Is a Light-Driven Proton Pump

The simplest known and best characterized proton pump is the intrinsic membrane protein **bacteriorhodopsin** of *Halobacterium halobium*. It consists mainly of seven transmembrane helices, A through G, that form a central polar channel (Section 12-3A). This channel contains a retinal prosthetic group that is covalently linked, via a protonated Schiff base, to Lys 216, which extends from helix G (Fig. 12-24). The protein obtains the free energy required for unidirectionally pumping protons from the absorption of a photon by the retinal. This initiates a sequence of events in which the protein conformationally adjusts through successive spectroscopically characterized intermediates, designated J, K, L, M, N, and O, as the system decays back to its ground state over a period of ∼10 ms. The net result of this cycle is the translocation of one proton from the cytoplasm to the extracellular medium, thereby converting light energy to proton-motive force. The as yet incompletely understood mechanism of this process, which was elucidated through detailed structural, mutational, and time-resolved spectroscopic studies carried out in several laboratories, is outlined in Fig. 22-34:

1. On absorbing a photon, the ground state all-*trans*-retinal photoisomerizes to its 13-cis form. This is a multistep process that rapidly passes (in ∼3 ps) through the J and K states. The free end of the retinal, which is now twisted around the newly cis double bond, moves relative to the protein scaffold such that retinal's C13 methyl group

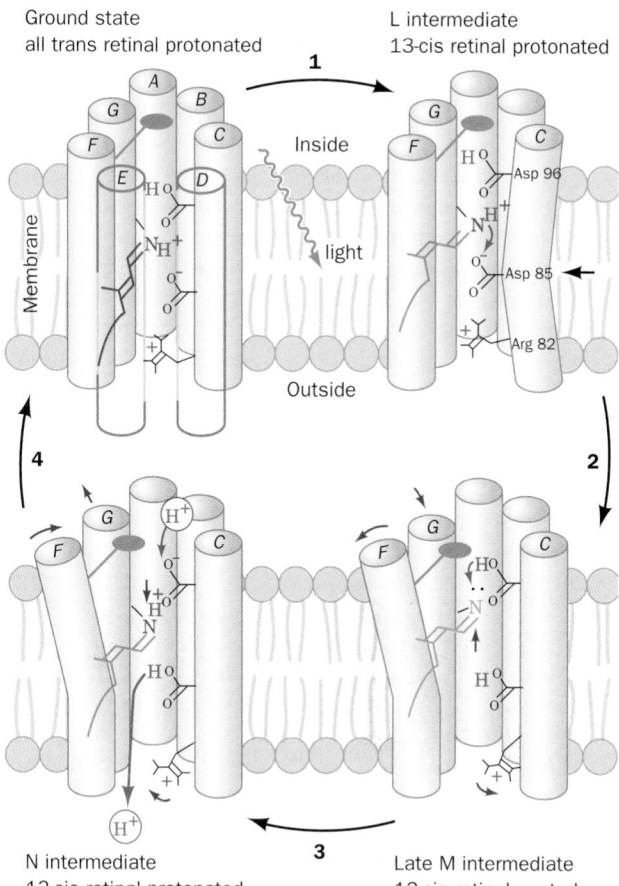

FIGURE 22-34 Proton pump of bacteriorhodopsin. See the text for a description of this mechanism. The protein is represented by its seven transmembrane helices, A through G (with helices D and E omitted for clarity in all but the upper left panel), and several mechanistically important side chains. The retinal is drawn with the approximate color of the complex in its various spectroscopically characterized states. Red arrows indicate proton movements, blue arrows indicate the movements of groups of atoms, and the "paddle" attached to helix F represents the bulky side chains that must move aside to open the cytoplasmic channel. [After Kühlbrandt, W., *Nature* **406**, 569 (2000).]

and its C14 atom shift toward the inside by 1.3 and 1.7 Å, respectively. This yields the L state.

2. Further conformational adjustments yield the M state. Here the N atom of the Schiff base has rotated and shifted from its ground state position, in which it is hydrogen bonded to an internal water molecule, to one in which it points toward the inside face of the protein in the vicinity of the hydrophobic side chains of Val 49 and Leu 93. This reduces the pK of the protonated Schiff base. In contrast, the pK of Asp 85 increases. This is because, in the ground state, Asp 85 effectively serves as the counterion of the protonated Schiff base and participates in a hydrogen bonded network with three internal water molecules, but in the M state, it is only associated with a single water molecule. Consequently, the Schiff base protonates Asp 85. This process is facilitated by a slight movement of helix C that brings Asp 85 closer to the Schiff base N atom.

The deprotonated retinal straightens and, in doing so, moves upward (toward the inside) by 0.7 to 1.0 Å. It thereby pushes against the F helix, causing its inside (cytoplasmic) end to tilt outward from the channel by ~3.5 Å and the G helix to partially replace it.

3. The movement of the F helix opens up the central channel on the inside of the membrane, admitting several water molecules that form a hydrogen bonded chain between Asp 96 and the Schiff base. One of these water molecules hydrogen bonds with Asp 96 so as to lower its p*K*. This permits Asp 96 to protonate the Schiff base via the intermediacy of the chain of hydrogen bonded water molecules, thereby yielding the N state.

4. Asp 96 is reprotonated by the cytoplasmic solution. The loops forming bacteriorhodopsin's inside surface bear numerous charged residues and hence it appears that they function as "antennas" to capture protons from the alkaline cytoplasmic medium. Asp 85 transfers its proton to the extracellular medium via a hydrogen bonded network that includes several bound water molecules. This process is facilitated by a preceding 1.6-Å displacement of the Arg 82 side chain toward a complex of residues that includes Glu 194 and Glu 204, which reduces the p*K* of this complex. The retinal then relaxes, via the O state, to its original all-trans form and helices F and G return to their original positions, thereby reforming the ground state of the protein and completing the catalytic cycle.

The retinal, which occupies the center of the protein channel, thereby acts as a one-way proton valve. The vectorial nature of this process arises from the unidirectional series of conformational changes made by the photoexcited retinal as it relaxes to its ground state. The protein's principal proton pumping motions are remarkably small, involving group movements of ~1 Å or less in response to the light-induced flexing of the retinal. These, nevertheless, cause p*K* changes in various residues that facilitate proton transfer as well as making and breaking hydrogen bonded networks of protein groups and water molecules in the proper sequence to transport a proton. A similar mechanism appears to operate in COX, but with its conformational changes motivated by redox reactions rather than by photoexcitation.

i. COX Has Two Proton Translocating Channels

Two channels that are candidates for translocating protons from the inside to the vicinity of the O_2-reducing center have been described in both bovine and *Paracoccus* COX (Fig. 22-35). These channels, which are both contained in Subunit I, are named the K- and D-channels after their respective key residues (K319 and D91 in the bovine numbering scheme, which we shall use in the subsequent discussion). Both putative channels are similar in character to that present in bacteriorhodopsin in that they consist of chains of hydrogen bonded and potentially hydrogen bonded protein groups, bound water molecules, and water-filled cavities.

The K-channel leads from K319, which is exposed to the inside, to Y244, the residue that is the proposed substrate electron and proton donor in the reaction forming the P

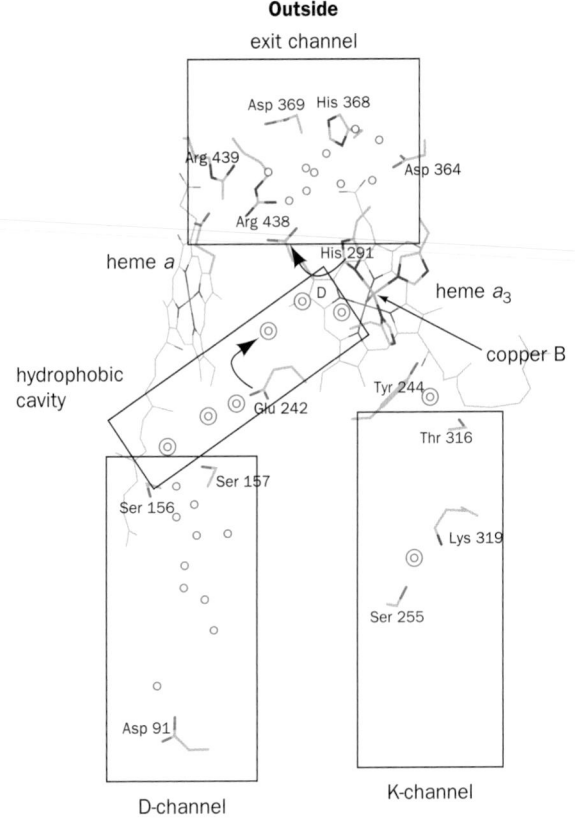

FIGURE 22-35   The proton-translocating channels in bovine COX. The enzyme is viewed parallel to the membrane with the matrix below. The four rectangles delineate the proposed proton entry and exit conduits. Single and double circles, respectively, represent water molecules that are observed in X-ray structures or that theoretical methods suggest are likely to be present. [After a drawing by Mårten Wikström, University of Helsinki, Helsinki, Finland.]

state (Step 4 of Fig. 22-28). The K319M mutant has an extremely low activity (<0.05% of wild type), which is not increased by supplying additional protons from the outside. Hence it appears that the K-channel is not connected to the putative exit channel (Fig. 22-35) that leads to the outside. It therefore seems likely that the K-channel only functions to supply chemical protons to the O_2-reduction center.

The entrance to the D-channel is within a region on the protein surface that appears likely to act as a proton-gathering antenna. The mutation of D91 to any noncarboxylate residue eliminates proton pumping but reduces the rate of O_2 reduction to only 45% of wild type (in *E. coli*). Evidently, the D-channel, in series with the exit channel, is the proton pumping channel. Moreover, there are growing indications that, in addition, the D-channel, which extends to the vicinity of the heme a_3–Cu_B binuclear center, is also a conduit for the chemical protons required for the second part of the reaction cycle (Steps 5 and 6 of Fig. 22-28).

What is the mechanism that couples O_2 reduction to proton pumping in COX? Unfortunately, X-ray crystallography has provided little guidance in answering this question because the X-ray structures of only a few different states of COX have as yet been determined and because the resolution of these structures is too low to reliably reveal small structural differences between them (keep in mind that the COX structure is among the largest determined). Consequently, the mechanisms that have been proposed to explain how COX pumps protons are largely inferences based on limited structural information, interpretation of site-directed mutagenesis experiments, spectroscopic data, theoretical considerations, and chemical intuition. Several ingenious although largely phenomenological models of how COX pumps protons have been proposed. These models agree that at least one proton is pumped during each of Steps 5 and 6 in the COX reaction cycle (Fig. 22-28) but disagree as to where in the reaction cycle the remaining two protons are pumped and how the protein actuates this process. Clearly there is still much to learn about how COX carries out its function.

C. *Mechanism of ATP Synthesis*

🐚 **See Guided Exploration 21. F_1F_0-ATPsynthase and the binding change mechanism** *The free energy of the electrochemical proton gradient across the mitochondrial membrane is harnessed in the synthesis of ATP by **proton-translocating ATP synthase** (also known as **F_1F_0–ATPase, Complex V,** and **F-type H^+–ATPase**). In the following subsections we discuss the location and structure of this ATP synthase and the mechanism by which it harnesses proton flux to drive ATP synthesis.*

a. Proton-Translocating ATP Synthase Is a Multisubunit Transmembrane Protein

Proton-translocating ATP synthase consists of two major substructures comprising 8 to 13 different subunits. Electron micrographs of mitochondria (Fig. 22-36) show lollipop-

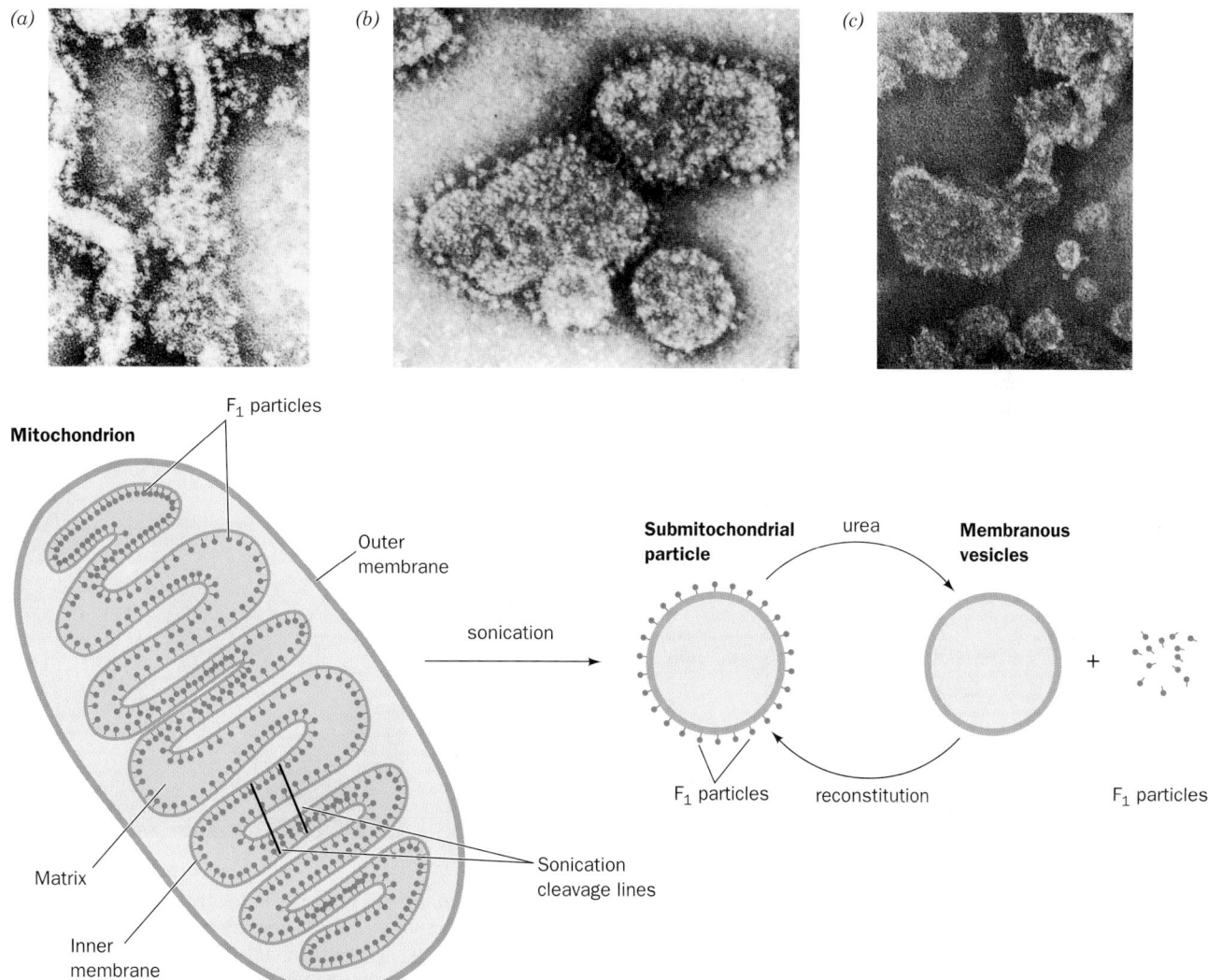

FIGURE 22-36   Electron micrographs and interpretive drawings of the mitochondrial membrane at various stages of dissection. (*a*) Cristae from intact mitochondria showing their F_1 "lollipops" projecting into the matrix. [From Parsons, D.F., *Science* **140**, 985 (1963). Copyright © 1963 American Association for the Advancement of Science. Used by permission.] (*b*) Submitochondrial particles, showing their outwardly projecting F_1 lollipops. Submitochondrial particles are prepared by the sonication (ultrasonic disruption) of inner mitochondrial membranes. [Courtesy of Peter Hinkle, Cornell University.] (*c*) Submitochondrial particles after treatment with urea. [Courtesy of Efraim Racker, Cornell University.]

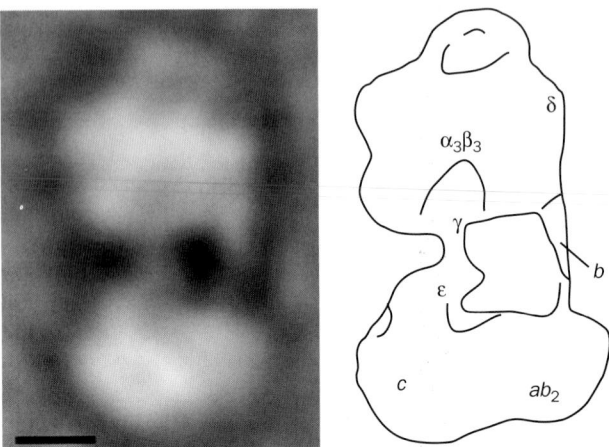

FIGURE 22-37   Electron microscopy–based image of *E. coli* F₁F₀–ATPase. The accompanying interpretive drawing indicates the positions of its component subunits, which are described in the text. [Courtesy of Roderick Capaldi, University of Oregon.]

shaped structures studding the matrix surface of the inner mitochondrial membrane (Fig. 22-36a). Similar entities have been observed lining the inner surface of the bacterial plasma membrane and in chloroplasts (Section 24-2D). Sonication of the inner mitochondrial membrane yields sealed vesicles, **submitochondrial particles,** from which the "lollipops" project (Fig. 22-36b) and which can carry out ATP synthesis.

Efraim Racker discovered that the proton-translocating ATP synthase from submitochondrial particles is comprised of two functional units, **F₀** and **F₁**. F₀ is a water-insoluble

transmembrane protein composed of as many as eight different types of subunits (although only three in *E. coli*) that contains a proton translocation channel. F₁ is a water-soluble peripheral membrane protein, composed of five types of subunits, that is easily dissociated from F₀ by treatment with urea. Solubilized F₁ can hydrolyze ATP but cannot synthesize it (hence the name ATPase). Submitochondrial particles from which F₁ has been removed by urea treatment no longer exhibit the lollipops in their electron micrographs (Fig. 22-36c) and lack the ability to synthesize ATP. If, however, F₁ is added back to these F₀-containing submitochondrial particles, their ability to synthesize ATP is restored and their electron micrographs again exhibit the lollipops. Thus *the lollipops are the F₁ particles.* Electron micrographs of F₁F₀ particles from *E. coli,* by Roderick Capaldi, clearly show their dumbbell-shaped structure in which F₀ and F₁ are joined by both an ~45-Å-long central stalk and a less substantial peripherally located connector (Fig. 22-37).

b. The X-Ray Structure of F₁ Reveals the Basis of Its Lollipoplike Structure

The F₁ subunit of the mitochondrial F₁F₀–ATPase is an $\alpha_3\beta_3\gamma\delta\varepsilon$ nonamer in which the β subunit contains the catalytic site for ATP synthesis and the δ subunit is required for binding of F₁ to F₀. The X-ray structure of F₁ from bovine heart mitochondria, determined by John Walker and Andrew Leslie, reveals that this 3440-residue (371-kD) protein is an 80-Å-high and 100-Å-wide spheroid that is mounted on a 30-Å-long stem (Fig. 22-38a). F₁'s α and β subunits, which are 20% identical in sequence and have

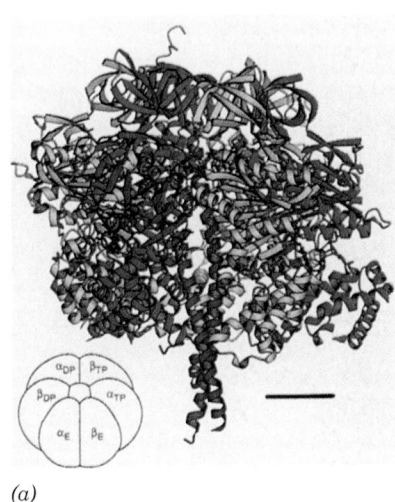

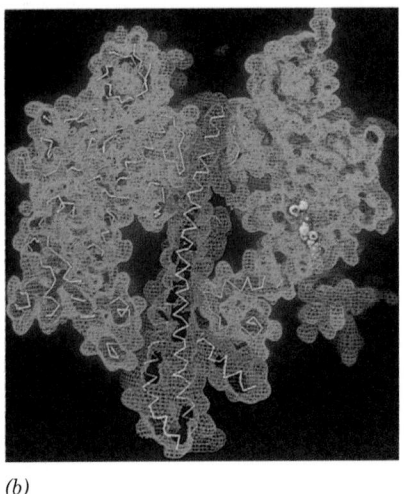

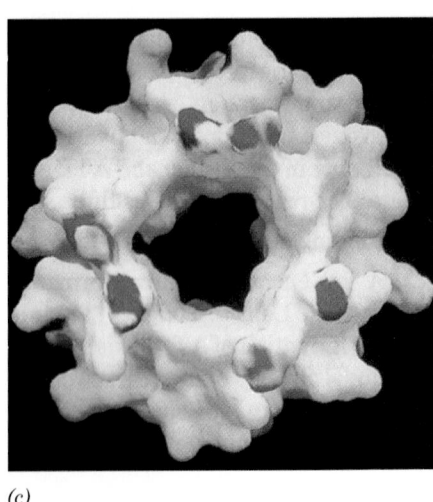

(a)   *(b)*   *(c)*

FIGURE 22-38   X-Ray structure of F₁–ATPase from bovine heart mitochondria. (*a*) A ribbon diagram in which the α, β, and γ subunits are red, yellow, and blue, respectively, and the nucleotides are black in ball-and-stick representation. The inset drawing indicates the orientation of these subunits in this view. The bar is 20 Å long. (*b*) Cross section through the electron density map of the protein in which the density for the α and β subunits is blue and that for the γ subunit is orange. The superimposed C_α backbones for these subunits are yellow, and a bound ADPNP is represented in space-filling form (C yellow, N blue, O red). Note the large central cavity surrounding the γ

subunit between the two regions where it contacts the $\alpha_3\beta_3$ assembly. (*c*) The surface of the inner portion of the $\alpha_3\beta_3$ assembly through which the C-terminal helix of the γ subunit penetrates as viewed from the top of Parts *a* and *b*. The surface is colored according to its electrical potential, with positive potentials blue and negative potentials red. Note the absence of charge on the inner surface of this sleeve. The portion of the γ subunit's C-terminal helix that contacts this sleeve is similarly devoid of charge. [From Abrahams, J.P., Leslie, A.G.W., Lutter, R., and Walker, J.E., *Nature* **370,** 623 and 627 (1994). PDBid 1BMF.] 🔗 **See the Interactive Exercises**

nearly identical folds, are arranged alternately, like the segments of an orange, about the upper portion of a 90-Å-long α helix formed by the C-terminal segment of the γ subunit. The C-terminus of this helix protrudes into a 15-Å-deep dimple that is centrally located at the top of the spheroid (Fig. 22-38*b*). The lower half of the helix forms a bent left-handed antiparallel coiled coil with the N-terminal segment of the γ subunit. This coiled coil is almost certainly a part of the ~45-Å-long stalk seen in electron micrographs of the F_1F_0–ATPase such as Fig. 22-37.

The cyclical arrangement and structural similarities of F_1's α and β subunits gives it both pseudo-threefold and pseudo-sixfold rotational symmetry. Nevertheless, the protein is asymmetric. This is in part due to the presence of the γ subunit but, more importantly, because each of the α and β subunits takes up a somewhat different conformation. Thus, one β subunit (designated β_{TP}) binds a molecule of the nonhydrolyzable ATP analog ADPNP (Section 18-3F), the second (β_{DP}) binds ADP, and the third (β_E) has an empty and distorted binding site. The α subunits, however, all bind ADPNP, although they also differ conformationally from one another. The ADPNP and ADP binding sites each lie at a radius of ~20 Å near an interface between adjacent α and β subunits and, in fact, all incorporate a few residues from the adjacent subunit.

Although 55% of the γ subunit comprising three segments as well as the entire δ and ε subunits are not visible in the X-ray structure shown in Fig. 22-38, they are seen in their entirety in the X-ray structure of bovine F_1–ATPase determined under different conditions. This latter structure (Fig. 22-39), also determined by Leslie and Walker, reveals that the C-terminal helix of the γ subunit has approximately four turns at its N-terminal end beyond that seen at the bottom of Fig. 22-38*a* (extending its length to 114 Å) and that the remaining "missing" portions of the γ subunit as well as the δ and ε subunits are wrapped about the base of the γ subunit's coiled coil. Note that, in an unfortunate confusion of nomenclature, the *E. coli* ε subunit is the homolog of the mitochondrial δ subunit; the *E. coli* δ subunit is the counterpart of the mitochondrial **oligomycin-sensitivity conferral protein (OSCP;** see Problem 9), and the mitochondrial ε subunit has no counterpart in either bacterial or chloroplast ATP synthases.

c. The *c* Subunits of F_0 Form a Transmembrane Ring That Contacts the F_1 Stalk

The F_0 component of the F_1F_0–ATPase from *E. coli* consists of three transmembrane subunits, *a*, *b*, and *c*, that form an $a_1b_2c_{9-12}$ complex. Mitochondrial F_0 additionally contains one copy each of three different subunits, *d*, F_6, and OSCP (which is the functional equivalent of the *E. coli* δ protein), as well as several "minor" subunits, *e, f, g,* and A6L, of unknown function. A variety of evidence indicates that the hydrophobic *c* subunits associate to form a ring with the ab_2 unit located at its periphery (see below). The sequence of the *a* subunit suggests this highly hydrophobic 271-residue peptide forms five transmembrane helices. The 156-residue *b* subunits each consist of a single transmembrane helix anchoring a polar inside do-

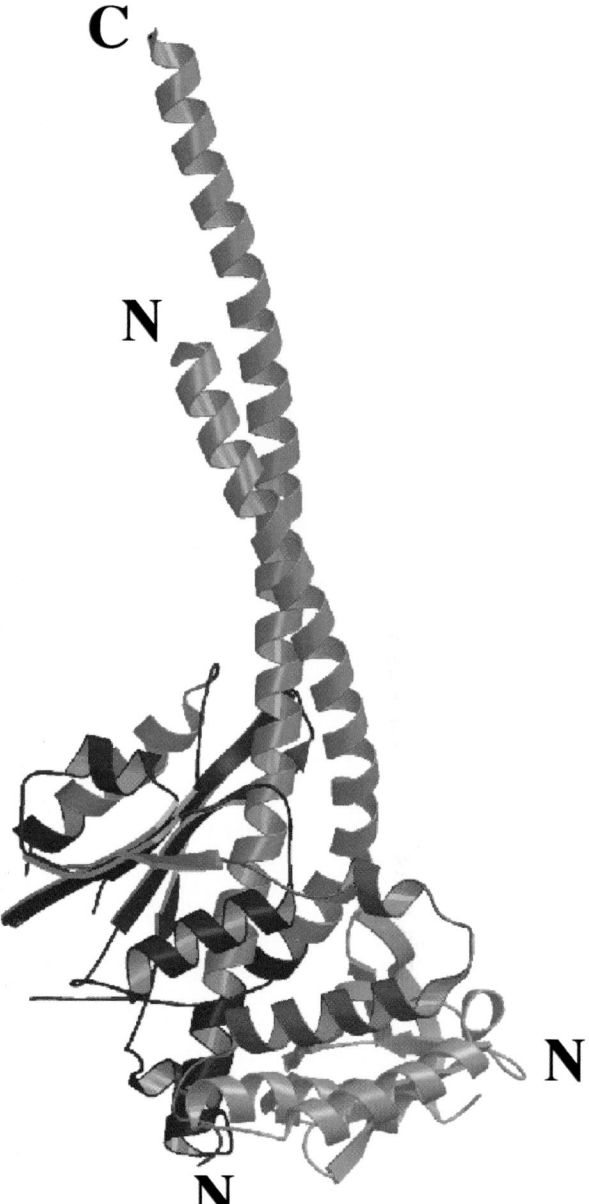

FIGURE 22-39 The γ, δ, and ε subunits in the X-ray structure of bovine F_1–ATPase. The γ (*cyan and blue*), δ (*green*), and ε (*magenta*) subunits are visible in this X-ray structure of bovine F_1–ATPase, which was determined under different conditions from that shown in Fig. 22-38. The cyan regions of the γ subunit are visible in the latter structure, whereas the blue regions as well as the entire δ and ε subunits are not. [Courtesy of Andrew Leslie and John Walker, MRC Laboratory of Molecular Biology, Cambridge, U.K. PDBid 1E79.]

main that homodimerizes to form a parallel α-helical coiled coil.

Mark Girvin and Robert Fillingame determined the NMR structure of the 79-residue *E. coli c* subunit at both pH 5 and pH 8 (Fig. 22-40). These reveal that the *c* subunit consists of two α helices that are connected by a 4-residue polar loop and that are arranged in a banana-shaped antiparallel coiled coil. The significance of the different protonation states is discussed below.

(a)   (b)

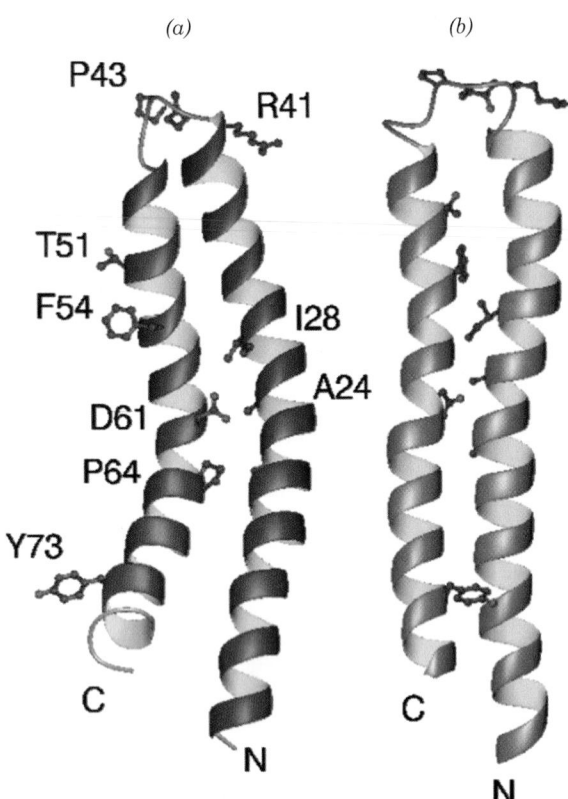

FIGURE 22-40   NMR structures of the c subunit of E. coli F$_1$F$_0$–ATPase. The structures were determined in chloroform–methanol–water (4:4:1) solution at (a) pH 8 (at which D61 is deprotonated) and (b) pH 5 (at which D61 is protonated). Selected side chains are shown to aid in the comparison of the two structures. Note that the C-terminal helix in the pH 8 structure has rotated by 140° clockwise, as viewed from the top of the drawing, relative to that in the pH 5 structure. [Courtesy of Mark Girvin, Albert Einstein College of Medicine. PDBids (a) 1C99 and (b) 1C0V.]

The low resolution X-ray structure of yeast mitochondrial F$_1$ in complex with its c-ring oligomer (Fig. 22-41) was determined by Leslie and Walker. The α and β subunits of yeast F$_1$ have similar conformations and bound nucleotides to those in bovine F$_1$ (Fig. 22-38a). The structure of the N-terminal domain of the bovine δ subunit (Fig. 22-39) fits into a region of the yeast F$_1$ electron density, where it contacts both the base of the γ subunit and the c-ring, thereby providing a footlike interface between F$_0$ and F$_1$.

The yeast c oligomer consists of 10 subunits (a number which may differ from that in E. coli) that each resemble the NMR structure of the isolated E. coli c subunit (Fig. 22-40). The c subunits associate front to back so as to form two concentric rings of α helices. Although this X-ray structure's low resolution precludes the identification of its side chains, the different lengths of the 76-residue yeast c subunit's two helices permits their identification in the c-ring: The 58- and 47-Å-long inner and outer helices are, respectively, consistent in length with the yeast c subunit's 39- and 31-residue N- and C-terminal helices. The polar loops in 5 of the 10 c subunits contact the δ subunit. The

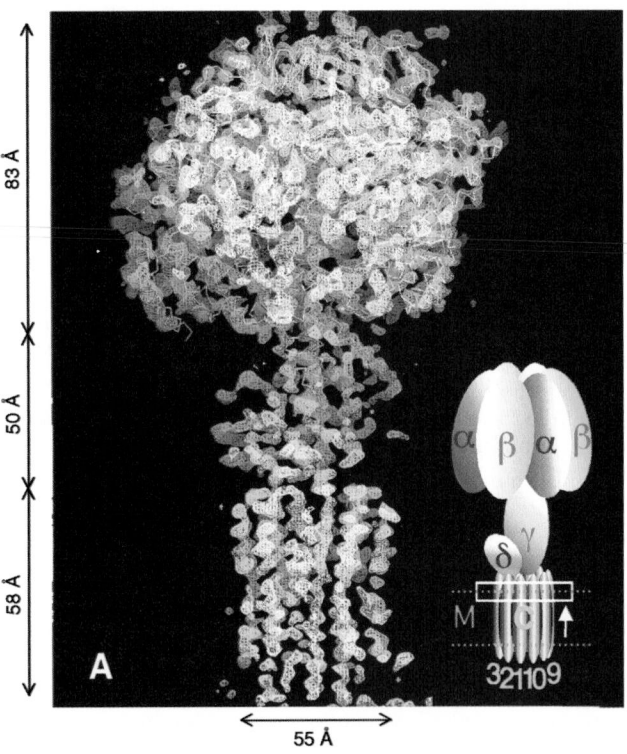

(a)

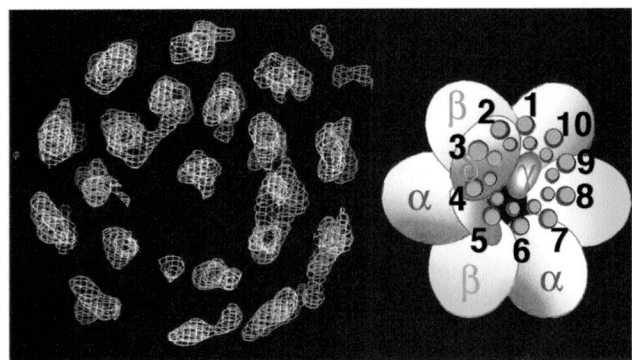

(b)

FIGURE 22-41   Low (3.9 Å) resolution electron density map of the yeast mitochondrial F$_1$–c$_{10}$ complex. (a) A view from within the inner mitochondrial membrane with the matrix above. The C$_α$ backbone of bovine F$_1$ (with α orange, β yellow, and γ green) is superimposed on the electron density map. The inset indicates the location of the subunits of the complex, with the dashed lines indicating the presumed position of the inner mitochondrial membrane (M) and with the c subunits numbered. (b) View from the intermembrane space of the boxed section of the c$_{10}$ ring in the inset of Part a. The inset indicates the positions of F$_1$'s subunits relative to the c subunits. The blue circles represent the α helices of the numbered c subunits, with the larger outer circles accounting for the larger side chains in the C-terminal helix. [Courtesy of Andrew Leslie and John Walker, Medical Research Council, Cambridge, U.K. PDBid 1QO1.]

γ subunit appears to associate with 1 or 2 other c subunits so that ~2/3 of the top surface of the c-ring contacts the F$_1$ stalk.

d. The Binding Change Mechanism: Proton-Translocating ATP Synthase Is Driven by Conformational Changes

The mechanism of ATP synthesis by proton-translocating ATP synthase can be conceptually broken down into three phases:

1. Translocation of protons carried out by F_0.

2. Catalysis of formation of the phosphoanhydride bond of ATP carried out by F_1.

3. Coupling of the dissipation of the proton gradient with ATP synthesis, which requires interaction of F_1 and F_0.

The available evidence supports a mechanism for ATP formation, proposed by Boyer, that resembles the conformational coupling hypothesis of oxidative phosphorylation (Section 22-3A). However, the conformational changes in the ATP synthase that power ATP formation are generated by proton translocation rather than by direct electron transfer, as proposed in the original formulation of the conformational coupling hypothesis.

F_1 is proposed to have three interacting catalytic protomers, each in a different conformational state: one that binds substrates and products loosely (L state), one that binds them tightly (T state), and one that does not bind them at all (open or O state). The free energy released on proton translocation is harnessed to interconvert these three states. The phosphoanhydride bond of ATP is synthesized only in the T state and ATP is released only in the O state. The reaction involves three steps (Fig. 22-42):

1. Binding of ADP and P_i to the "loose" (L) binding site.

2. A free energy–driven conformational change that converts the L site to a "tight" (T) binding site that catalyzes the formation of ATP. This step also involves conformational changes of the other two subunits that convert the ATP-containing T site to an "open" (O) site and convert the O site to an L site.

3. ATP is synthesized at the T site on one subunit while ATP dissociates from the O site on another subunit. On the surface of the active site, the formation of ATP from

ADP and P_i entails little free energy change, that is, the reaction is essentially at equilibrium. Consequently, the free energy supplied by the proton flow primarily facilitates the release of the newly synthesized ATP from the enzyme; that is, it drives the T → O transition, thereby disrupting the enzyme–ATP interactions that had previously promoted the spontaneous formation of ATP from ADP + P_i in the T site.

How is the free energy of proton transfer coupled to the synthesis of ATP? Boyer proposed that *the binding changes are driven by the rotation of the catalytic assembly, $\alpha_3\beta_3$, with respect to other portions of the F_1F_0–ATPase.* This hypothesis is supported by the X-ray structure of F_1. Thus, the closely fitting nearly circular arrangement of the α and β subunits' inner surface about the γ subunit's helical C-terminus is reminiscent of a cylindrical bearing rotating in a sleeve (Figs. 20-38*b,c*). Indeed, the contacting hydrophobic surfaces in this assembly are devoid of the hydrogen bonding and ionic interactions that would interfere with their free rotation (Fig. 22-38*c*); that is, the bearing and sleeve appear to be "lubricated." Moreover, the central cavity in the $\alpha_3\beta_3$ assembly (Fig. 22-38*b*) would permit the passage of the γ subunit's N-terminal helix within the core of this particle during rotation. Finally, the conformational differences between F_1's three catalytic sites appear to be correlated with the position of the γ subunit. *Apparently the γ subunit, which is thought to rotate within the fixed $\alpha_3\beta_3$ assembly, acts as a molecular cam shaft in linking the proton gradient–driven rotational motor to the conformational changes in the catalytic sites of F_1.* This concept is also supported by molecular dynamics simulations (Section 9-4) by Leslie, Walker, and Martin Karplus, which indicate that the conformational changes in the β subunits arise from both steric and electrostatic interactions with the rotating γ subunit.

Rotating assemblies are not unprecedented in biological systems. Bacterial flagella, which function as propellers, had previously been shown to be membrane-mounted rotary engines that are driven by the discharge of a proton gradient (Section 35-3G).

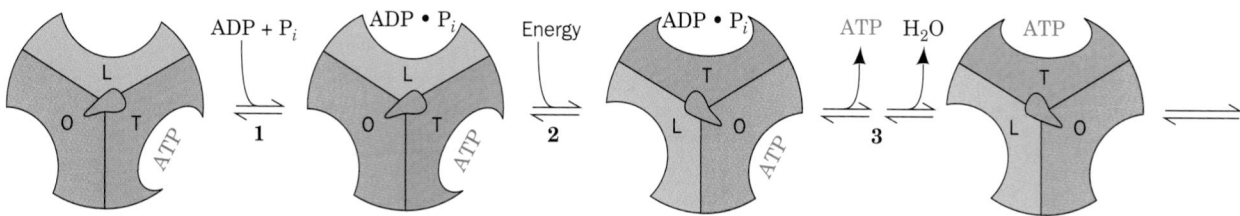

FIGURE 22-42 Energy-dependent binding change mechanism for ATP synthesis by proton-translocating ATP synthase. F_1 has three chemically identical but conformationally distinct interacting $\alpha\beta$ protomers: O, the open conformation, has very low affinity for ligands and is catalytically inactive; L has loose binding for ligands and is catalytically inactive; T has tight binding for ligands and is catalytically active. ATP synthesis occurs in three steps. **(1)** Binding of ADP and P_i to site L. **(2)** Energy-dependent conformational change converting binding site L to T, T to O, and O to L. **(3)** Synthesis of ATP at site T and release of ATP from site O. The enzyme returns to its initial state after two more passes of this reaction sequence. The energy that drives the conformational change is apparently transmitted to the catalytic $\alpha_3\beta_3$ assembly via the rotation of the $\gamma\varepsilon$ assembly (in *E. coli;* $\gamma\delta$ in mitochondria), here represented by the centrally located asymmetric object (*green*). [After Cross, R.L., *Annu. Rev. Biochem.* **50,** 687 (1980).] ⟡ See the Animated Figures

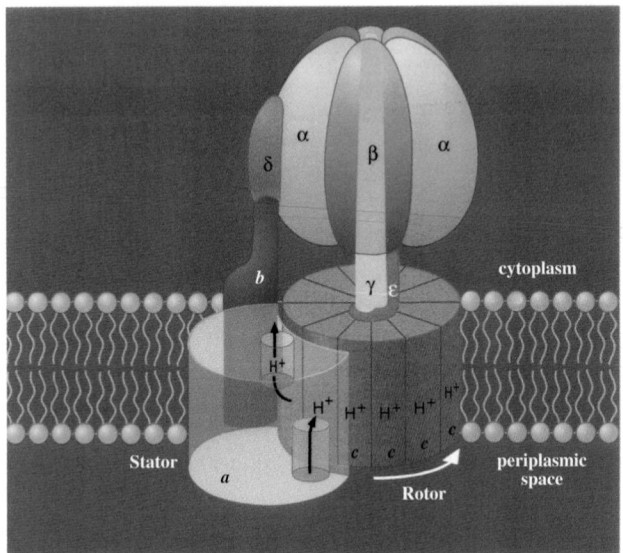

FIGURE 22-43 Model of the *E. coli* F_1F_0–ATPase. The $\gamma\varepsilon$–c_{12} ring complex is the rotor and the ab_2–$\alpha_3\beta_3\delta$ complex is the stator. Rotational motion is imparted to the rotor by the passage of protons from the outside (periplasm) to the inside (cytoplasm). Protons entering from the outside bind to a *c* subunit where it interacts with the *a* subunit, and exit to the inside after the *c* ring has made a nearly full rotation as indicated (*white arrow*), so that the *c* subunit again contacts the *a* subunit. The $b_2\delta$ complex presumably functions to prevent the $\alpha_3\beta_3$ assembly from rotating with the γ subunit. [Courtesy of Richard Cross, State University of New York, Syracuse, New York.]

e. The F_1F_0–ATPase Is a Rotary Engine

The proposed rotation of the $\alpha_3\beta_3$ assembly with respect to the γ subunit engendered by the binding change mechanism has led to the model of the F_1F_0–ATPase diagrammed in Fig. 22-43. A rotational engine must have a rotor, which rotates with respect to a stationary stator. In the F_1F_0–ATPase, the rotor is proposed to be an assembly of the *c*-ring with the γ and (*E. coli*) ε subunits, whereas the ab_2 unit and the (*E. coli*) δ subunit together with the $\alpha_3\beta_3$ spheroid form the stator. The rotation of the *c*-ring in the membrane relative to the stationary *a* subunit is driven by the migration of protons from the outside to the inside, as we discuss below. The $b_2\delta$ assembly presumably functions to hold the $\alpha_3\beta_3$ spheroid in position while the γ subunit rotates inside it.

The rotation of the *E. coli* $\gamma\varepsilon$–*c*-ring rotor with respect to the ab_2–$\alpha_3\beta_3\delta$ stator has been ingeniously demonstrated by Masamitsu Futai using techniques developed by Kazuhiko Kinosita Jr. and Masasuke Yoshida (Fig. 22-44*a*). The $\alpha_3\beta_3$ spheroid of *E. coli* F_1F_0–ATPase was fixed, head down, to a glass surface as follows: Six consecutive His residues (a so-called **His tag**) were mutagenically appended to the N-terminus of the α subunit, which is located at the top of the $\alpha_3\beta_3$ spheroid as it is drawn in Fig. 20-38*a*. The His-tagged assembly was applied to a glass surface coated with horseradish peroxidase (which, like most proteins, sticks to glass) conjugated with Ni^{2+}-**nitriloacetic acid** $[N(CH_2COOH)_3$, which tightly binds His tags], thereby

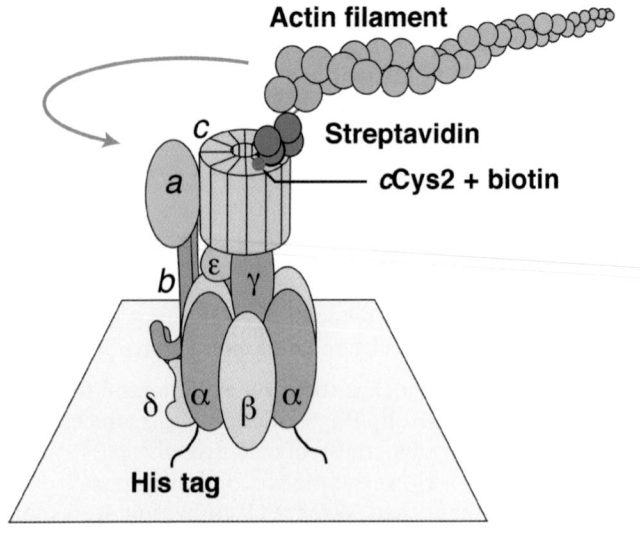

(*a*)

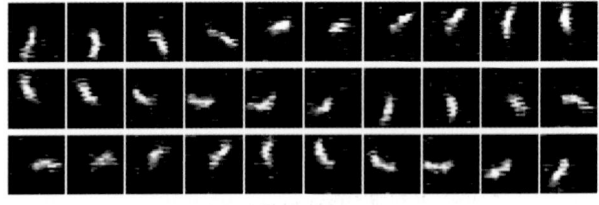

(*b*)

FIGURE 22-44 Rotation of the *c*-ring in *E. coli* F_1F_0–ATPase. (*a*) The experimental system used to observe the rotation. See the text for details. The blue arrow indicates the observed direction of rotation of the fluorescently labeled actin filament that was linked to the *c*-ring. (*b*) The rotation of a 3.6-μm-long actin filament in the presence of 5 m*M* MgATP as seen in successive video images taken through a fluorescence microscope. [Courtesy of Masamitsu Futai, Osaka University, Osaka, Japan.]

binding the F_1F_0–ATPase to the surface with its F_0 side facing away from the surface. The Glu 2 residues of this assembly's *c* subunits, which are located on the side of the *c*-ring facing away from F_1, had been mutagenically replaced by Cys residues, which were then covalently linked to **biotin** (a coenzyme that normally participates in carboxylation reactions; Section 23-1A). A fluorescently labeled and biotinylated (at one end) filament of the muscle protein **actin** (Section 35-3A) was then attached to the *c* subunit through the addition of a bridging molecule of **streptavidin**, a protein that avidly binds biotin to each of four binding sites (Cys 193 of the γ subunit, the only other Cys residue in the rotor, was mutagenically replaced by Ala to prevent it from being linked to an actin filament).

The *E. coli* F_1F_0–ATPase can work in reverse, that is, it can pump protons from the inside (cytoplasm) to the outside (periplasm) at the expense of ATP hydrolysis (this enables the bacterium to maintain its proton gradient under anaerobic conditions, which it uses to drive various processes such as flagellar rotation). Thus, the foregoing preparation was observed under a fluorescence microscope as a 5 m*M* MgATP solution was infused over it. *Many of the actin filaments were seen to rotate (Fig. 22-44b), and al-*

ways in a counterclockwise direction when viewed looking down on the glass surface (from the outside). This would permit the γ subunit to sequentially interact with the β subunits in the direction

$$\beta_E(\text{O state}) \rightarrow \beta_{DP}(\text{L state}) \rightarrow \beta_{TP}(\text{T state})$$

(Figs. 22-38*a* and 22-42), the direction expected for ATP hydrolysis.

In a variation of the above experiment, the γ subunit of the $\alpha_3\beta_3\gamma$ complex was directly cross-linked, via its Cys 193, to a fluorescently labeled actin filament and, in this case, the α subunits were immobilized by appended His tags. At very low ATP concentrations (e.g., 0.02 μ*M*), video images (Fig. 22-45) revealed that the fluorescent actin filament rotated counterclockwise in discrete steps of 120°, as the binding change mechanism predicts. Moreover, the calculated frictional work done in each rotational step is very nearly equal to the energy available from the hydrolysis of one ATP molecule, that is, *the F_1F_0–ATPase converts chemical to mechanical energy with nearly 100% efficiency.*

f. *c*-Ring Rotation Is Impelled by H$^+$-Induced Conformational Changes

The foregoing structural and biochemical information has led to the model for proton-driven rotation of the F_0 subunit that is diagrammed in Fig. 22-43. Protons from the outside enter a hydrophilic channel between the *a* subunit and the *c*-ring, where they bind to a *c* subunit. The *c*-ring then rotates nearly a full turn (while protons bind to successive *c* subunits as they pass this input channel) until the subunit reaches a second hydrophilic channel between the *a* subunit and the *c*-ring that opens into the inside. There the proton is released. Thus, the F_1F_0-ATPase, which generates 3ATP per turn and (at least in yeast) has 10 *c* subunits in its F_0 assembly, ideally forms 3/10 = 0.3 ATP for every proton it passes from the intermembrane space (outside) to the matrix (inside).

But how does the passage of protons through this system induce the rotation of the *c*-ring and hence the synthesis of ATP? The mutation of the *c* subunit's conserved Asp 61 to Asn inactivates *E. coli* F_1F_0–ATPase. The *a* subunit's invariant Arg 210 (*E. coli* numbering) has been similarly implicated in proton translocation. Through the mutagenic conversion of selected residues on the *a* and *c* subunits to Cys, Fillingame has shown that the C-terminal helix of *E. coli* subunit *c* (which contains Asp 61) can be disulfide-cross-linked to the putative fourth helix of subunit *a* (which contains Arg 210). Evidently, these helices are juxtaposed at some point in the *c*-ring's rotation cycle. Thus, it is postulated that the protonation of Asp 61 releases its attraction to Arg 210, thereby permitting the *c*-ring to rotate.

Comparison of the NMR structures of subunit *c* at pH 8 and pH 5 (Fig. 22-40), at which Asp 61 is, respectively, deprotonated and protonated, reveals that its main conformational change on protonation is an ~140° clockwise rotation (as viewed from F_1) of its Asp 61-containing C-terminal helix with respect to its N-terminal helix. Since the C-terminal helix is the *c*-ring's outer helix (Fig. 22-41*b*),

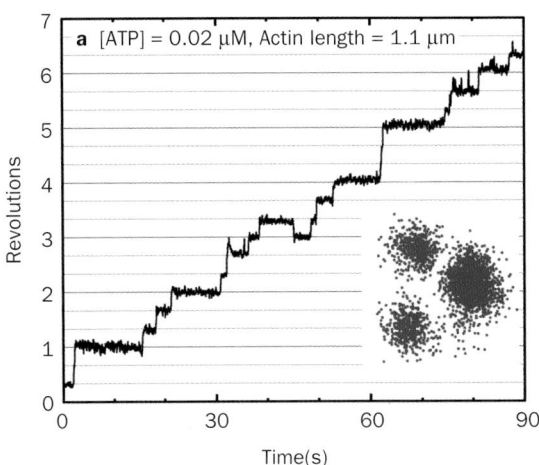

FIGURE 22-45 Stepwise rotation of the γ subunit of F_1 relative to an immobilized $\alpha_3\beta_3$ unit at low ATP concentration as observed by fluorescence microscopy. The graph plots the number of rotations made by a fluorescently labeled actin filament that was linked at one end to the γ subunit in a preparation similar to that diagrammed in Fig. 22-44*a* (but lacking F_0, δ, and ε). Note that the actin filament rotates in increments of 120°. This is also evident in the inset, which shows the superposition of the centers of the actin images (the $\delta_3\beta_3\gamma$ assembly is fixed in the center). [Courtesy of Kazuhiko Kinosita Jr., Keio University, Yokohama, Japan.]

this suggests that, on protonation, the rotation of the C-terminal helix mechanically pushes against the juxtaposed *a* subunit so as to rotate the *c*-ring in the direction indicated in Fig. 22-43.

D. *Uncoupling of Oxidative Phosphorylation*

Electron transport (the oxidation of NADH and FADH$_2$ by O$_2$) and oxidative phosphorylation (the synthesis of ATP) are normally tightly coupled due to the impermeability of the inner mitochondrial membrane to the passage of protons. Thus the only way for H$^+$ to reenter the matrix is through the F_0 portion of the proton-translocating ATP synthase. In the resting state, when oxidative phosphorylation is minimal, the electrochemical gradient across the inner mitochondrial membrane builds up to the extent that the free energy to pump additional protons is greater than the electron-transport chain can muster, thereby inhibiting further electron transport. However, many compounds, including **2,4-dinitrophenol (DNP)** and **carbonyl-cyanide-*p*-trifluoromethoxyphenylhydrazone (FCCP),** have been found to "uncouple" these processes. The chemiosmotic hypothesis has provided a rationale for understanding the mechanism by which these uncouplers act.

The presence in the inner mitochondrial membrane of an agent that renders it permeable to H$^+$ uncouples oxidative phosphorylation from electron transport by providing a route for the dissipation of the proton electrochemical gradient that does not require ATP synthesis. Uncoupling therefore allows electron transport to proceed unchecked even when ATP synthesis is inhibited. DNP and FCCP are lipophilic weak acids that therefore readily pass through

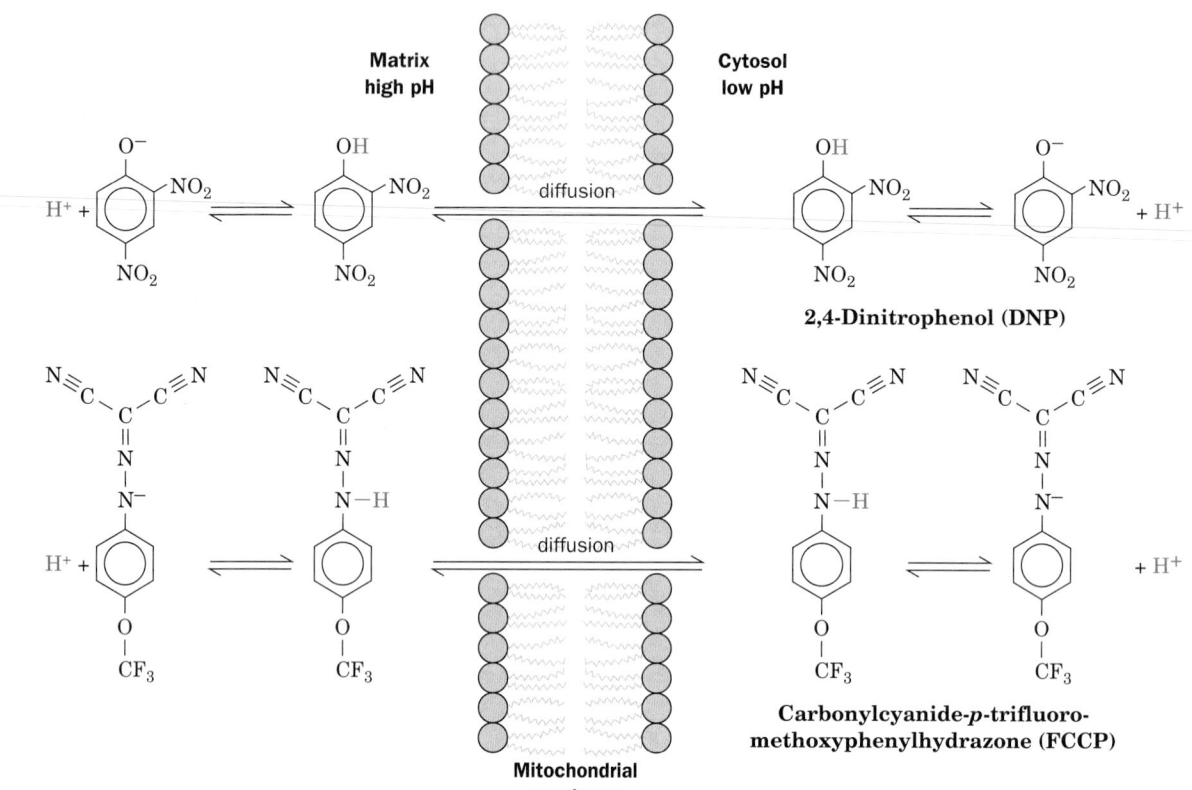

FIGURE 22-46   Uncoupling of oxidative phosphorylation. The proton-transporting ionophores DNP and FCCP uncouple oxida-

tive phosphorylation from electron transport by discharging the electrochemical proton gradient generated by electron transport.

membranes. In a pH gradient, they bind protons on the acidic side of the membrane, diffuse through, and release them on the alkaline side, thereby dissipating the gradient (Fig. 22-46). Thus, *such uncouplers are proton-transporting ionophores* (Section 20-2C).

Even before the mechanism of uncoupling was known, it was recognized that metabolic rates were increased by such compounds. Studies at Stanford University in the early part of the twentieth century documented an increase in respiration and weight loss caused by DNP. The compound was even used as a "diet pill" for several years. In the words of Efraim Racker (*A New Look at Mechanisms in Bioenergetics,* p. 155):

> *In spite of warnings from the Stanford scientists, some enterprising physicians started to administer dinitrophenol to obese patients without proper precautions. The results were striking. Unfortunately in some cases the treatment eliminated not only the fat but also the patients, and several fatalities were reported in the Journal of the American Medical Association in 1929. This discouraged physicians for a while. . . .*

a. Hormonally Controlled Uncoupling in Brown Adipose Tissue Functions to Generate Heat

The dissipation of an electrochemical H$^+$ gradient, which is generated by electron transport and uncoupled from ATP synthesis, produces heat. Heat generation is the physiological function of **brown adipose tissue (brown fat).** This tis-

sue is unlike typical (white) adipose tissue in that, besides containing large amounts of triacylglycerols, it contains numerous mitochondria whose cytochromes color it brown. Newborn mammals that lack fur, such as humans, as well as hibernating mammals, contain brown fat in their neck and upper back that functions in **nonshivering thermogenesis,** that is, as a "biological heating pad." (The ATP hydrolysis that occurs during the muscle contractions of shivering—or any other movement—also produces heat. Nonshivering thermogenesis through substrate cycling is discussed in Section 17-4F.)

The mechanism of heat generation in brown fat involves the regulated uncoupling of oxidative phosphorylation in their mitochondria. These mitochondria contain **thermogenin** [also called **uncoupling protein (UCP)**], a protein homodimer of 307-residue subunits that acts as a channel to control the permeability of the inner mitochondrial membrane to protons. In cold-adapted animals, thermogenin constitutes up to 15% of brown fat inner mitochondrial membrane proteins. The flow of protons through this channel protein is inhibited by physiological concentrations of purine nucleotides (ADP, ATP, GDP, GTP), but this inhibition can be overcome by free fatty acids. The components of this system interact under hormonal control.

Thermogenesis in brown fat mitochondria is activated by free fatty acids. These counteract the inhibitory effects of purine nucleotides, thereby stimulating the flux through the proton channel and uncoupling electron transport from oxidative phosphorylation. *The concentration of fatty acids in brown adipose tissue is controlled by the hormone **norepinephrine (noradrenaline)***

Norepinephrine

with cAMP acting as a second messenger (Section 18-3). Under norepinephrine stimulation (Fig. 22-47), the adenylate cyclase component of the norepinephrine receptor system synthesizes cAMP as described in Section 19-2. The

cAMP, in turn, allosterically activates protein kinase A (PKA), which activates **hormone-sensitive triacylglycerol lipase** by phosphorylating it (Section 25-5). Finally, the activated lipase hydrolyzes triacylglycerols to yield the free fatty acids that open the proton channel.

b. Other Tissues Contain UCP Homologs

Although it originally seemed that only brown fat mitochondria contain an uncoupling protein, it is now apparent that other tissues contain homologs of UCP1. Thus, **UCP2** is expressed in many tissues including white adipose tissue, whereas **UCP3** occurs in both brown and white adipose tissues as well as in muscle. The metabolic roles of UCP2 and UCP3 are unclear, although there are indications that UCP2 participates in diet-induced (rather than cold-induced) thermogenesis (Section 27-3E).

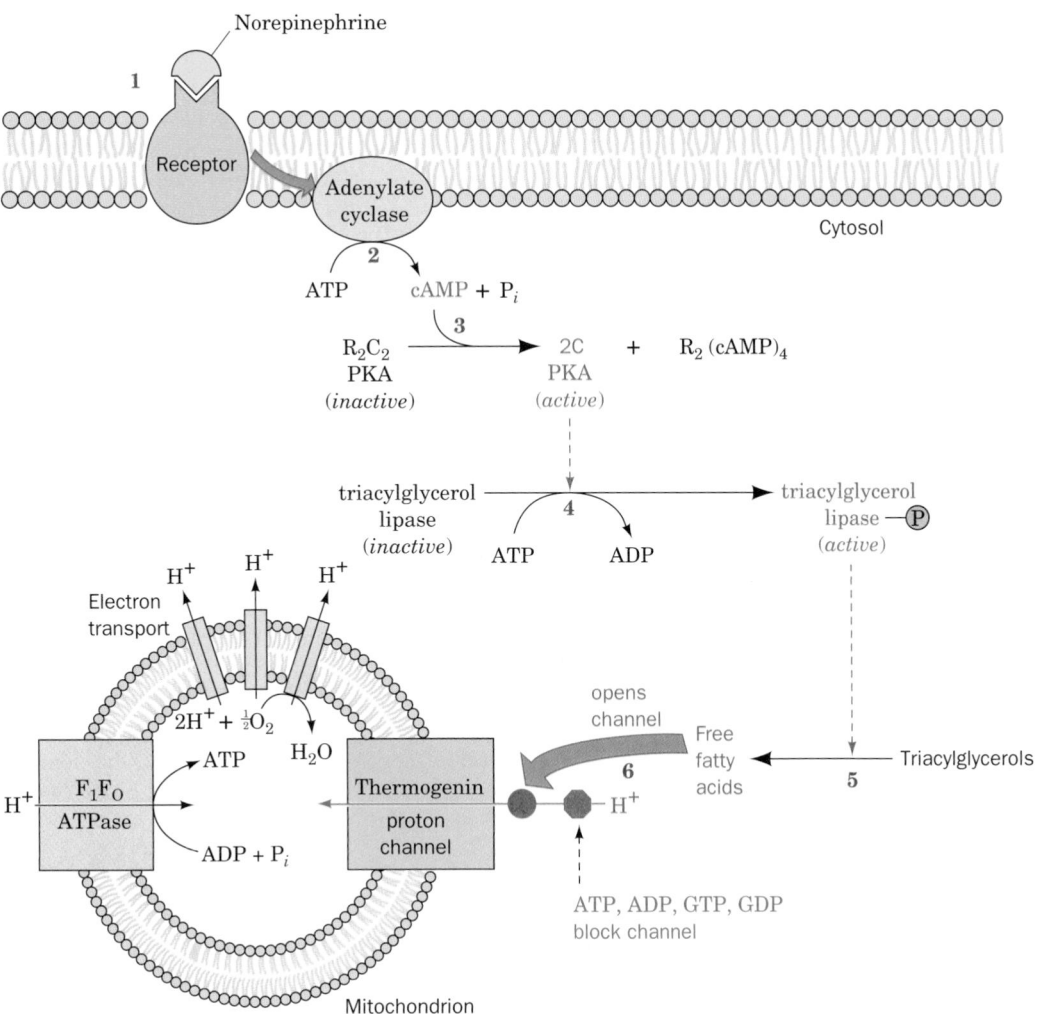

FIGURE 22-47 Mechanism of hormonally induced uncoupling of oxidative phosphorylation in brown fat mitochondria. **(1)** Norepinephrine binds to its cell surface receptor. **(2)** The norepinephrine–receptor complex stimulates adenylate cyclase, thereby causing cAMP levels to rise. **(3)** cAMP binding activates protein kinase A (PKA). **(4)** PKA phosphorylates

hormone-sensitive triacylglycerol lipase, thereby activating it. **(5)** Triacylglycerols are hydrolyzed, yielding free fatty acids. **(6)** Free fatty acids overcome the purine nucleotide block of the proton channel formed by thermogenin, allowing H$^+$ to enter the mitochondrion uncoupled from ATP synthesis.

4 ■ CONTROL OF ATP PRODUCTION

A typical adult woman requires some 1500 to 1800 kcal (6300–7500 kJ) of metabolic energy per day. This corresponds to the free energy of hydrolysis of over 200 mol of ATP to ADP and P_i. Yet the total amount of ATP present in the body at any one time is <0.1 mol; obviously, this sparse supply of ATP must be continually recycled. As we have seen, when carbohydrates serve as the energy supply and aerobic conditions prevail, this recycling involves glycogenolysis, glycolysis, the citric acid cycle, and oxidative phosphorylation.

Of course the need for ATP is not constant. There is a 100-fold change in ATP utilization between sleep and vigorous activity. *The activities of the pathways that produce ATP are under strict coordinated control so that ATP is never produced more rapidly than necessary.* We have already discussed the control mechanisms of glycolysis, glycogenolysis, and the citric acid cycle (Sections 17-4, 18-3, and 21-4). In this section we discuss the mechanisms through which oxidative phosphorylation is controlled and observe how all four systems are synchronized to produce ATP at precisely the rate required at any particular moment.

A. *Control of Oxidative Phosphorylation*

In our discussion of the control of glycolysis, we saw that most of the reactions in a metabolic pathway function close to equilibrium. *The few irreversible reactions constitute the potential control points of the pathway and usually are catalyzed by regulatory enzymes that are under allosteric control.* In the case of oxidative phosphorylation, the pathway from NADH to cytochrome c functions near equilibrium ($\Delta G' \approx 0$):

$$\tfrac{1}{2}NADH + cytochrome\ c^{3+} + ADP + P_i \rightleftharpoons$$
$$\tfrac{1}{2}NAD^+ + cytochrome\ c^{2+} + ATP$$

for which

$$K_{eq} = \left(\frac{[NAD^+]}{[NADH]}\right)^{\!1/2}\frac{[c^{2+}]}{[c^{3+}]}\frac{[ATP]}{[ADP][P_i]} \quad [22.2]$$

This pathway is therefore readily reversed by the addition of ATP. *In the cytochrome c oxidase reaction, however, the terminal step of the electron-transport chain is irreversible and is thus one of the important regulatory sites of the pathway.* Cytochrome c oxidase, in contrast to most regulatory enzyme systems, appears to be controlled exclusively by the availability of one of its substrates, reduced cytochrome c (c^{2+}). Since this substrate is in equilibrium with the rest of the coupled oxidative phosphorylation system (Eq. [22.2]), its concentration ultimately depends on the intramitochondrial [NADH]/[NAD$^+$] ratio and the **ATP mass action ratio** ([ATP]/[ADP][P_i]). By rearranging Eq. [22.2], the ratio of reduced to oxidized cytochrome c is expressed

$$\frac{[c^{2+}]}{[c^{3+}]} = \left(\frac{[NADH]}{[NAD^+]}\right)^{\!1/2}\left(\frac{[ADP][P_i]}{[ATP]}\right)K_{eq} \quad [22.3]$$

Consequently, the higher the [NADH]/[NAD$^+$] ratio and the lower the ATP mass action ratio, the higher $[c^{2+}]$ (re-

duced cytochrome c) and thus the higher the cytochrome c oxidase activity.

How is this system affected by changes in physical activity? In an individual at rest, ATP hydrolysis to ADP and P_i is minimal and the ATP mass action ratio is high; the concentration of reduced cytochrome c is therefore low and oxidative phosphorylation is minimal. Increased activity results in hydrolysis of ATP to ADP and P_i, thereby decreasing the ATP mass action ratio and increasing the concentration of reduced cytochrome c. This results in an increase in the electron-transport rate and its coupled phosphorylation. Such control of oxidative phosphorylation by the ATP mass action ratio is called **acceptor control** because the rate of oxidative phosphorylation increases with the concentration of ADP, the phosphoryl group acceptor. In terms of a supply–demand system (Section 17-4D), acceptor control is understood as control by the demand block.

The compartmentalization of the cell into mitochondria, where ATP is synthesized, and cytoplasm, where ATP is utilized, presents an interesting control problem: Is it the ATP mass action ratio in the cytosol or in the mitochondrial matrix that ultimately controls oxidative phosphorylation? Clearly the ATP mass action ratio that exerts direct control must be that of the mitochondrial matrix where ATP is synthesized. However, the inner mitochondrial membrane, which is impermeable to adenine nucleotides and P_i, depends on specific transport systems to maintain communication between the two compartments (Section 20-4C). This organization makes it possible for the transport of adenine nucleotides or P_i to participate in the control of oxidative phosphorylation.

Considerable research effort has been aimed at determining how oxidative phosphorylation is controlled in terms of metabolic control analysis. For example, Hans Westerhoff and Martin Kushmerick employed ^{31}P NMR to measure the ATP/ADP ratios in human forearm muscle at rest and during twitch contractions caused by external electrical contraction (the ^{31}P NMR spectrum of ATP is shown in Fig. 16-15). Under conditions of low to moderate ATP demand, the cytosolic mass action ratio as controlled by the demand block of the system appears to be the major control factor for mitochondrial oxidation. However, as other laboratories have shown, as the demand for ATP increases, the ADP–ATP translocator exerts greater control until finally, when the demand for ATP is high, control shifts to the supply block of the system, oxidative phosphorylation itself.

B. *Coordinated Control of ATP Production*

Glycolysis, the citric acid cycle, and oxidative phosphorylation constitute the major pathways for cellular ATP production. Control of oxidative phosphorylation by the ATP mass action ratio depends, of course, on an adequate supply of electrons to fuel the electron-transport chain. This aspect of the system's control is, in turn, dependent on the [NADH]/[NAD$^+$] ratio (Eq. [22.3]), which is maintained high by the combined action of glycolysis and the citric acid cycle in converting 10 molecules of NAD$^+$ to NADH per molecule of glucose oxidized (Fig. 22-1). It is clear, therefore, that coordinated control is necessary for the three processes. This is provided by the regulation of each of the control points of glycolysis [hexokinase, phosphofructokinase (PFK), and pyruvate kinase] and the citric acid cycle (pyruvate dehydrogenase, citrate synthase, isocitrate dehydrogenase, and α-ketoglutarate dehydrogenase) by adenine nucleotides or NADH or both as well as by certain metabolites (Fig. 22-48).

a. Citrate Inhibits Glycolysis

The main control points of glycolysis and the citric acid cycle are regulated by several effectors besides adenine nucleotides or NADH (Fig. 22-48). This is an extremely complex system with complex demands. Its many effectors, which are involved in various aspects of metabolism, increase its regulatory sensitivity. One particularly interesting regulatory effect is the inhibition of PFK by citrate. When demand for ATP decreases, [ATP] increases and [ADP] decreases. The citric acid cycle slows down at its isocitrate dehydrogenase (activated by ADP) and α-ketoglutarate dehydrogenase (inhibited by ATP) steps, thereby causing the citrate concentration to build up. Citrate can leave the mitochondrion via a specific transport system and, *once in the cytosol, acts to restrain further carbohydrate breakdown by inhibiting PFK.*

b. Fatty Acid Oxidation Inhibits Glycolysis

As we shall see in Section 25-1, the oxidation of fatty acids is an aerobic process that produces acetyl-CoA, which enters the citric acid cycle, thereby increasing both the mitochondrial and cytoplasmic concentrations of citrate. The increased [acetyl-CoA] inhibits the pyruvate dehydrogenase complex, whereas the increased [citrate] inhibits phosphofructokinase, leading to a buildup of glucose-6-phosphate, which inhibits hexokinase (Fig. 22-48). This inhibition of glycolysis by fatty acid oxidation is called the **glucose–fatty acid cycle** or **Randle cycle** (after its discoverer, Phillip Randle), although it is not, in fact, a cycle. The Randle cycle allows fatty acids to be utilized as the major fuel for oxidative metabolism in heart muscle, while conserving glucose for organs such as the brain, which require it.

C. *Physiological Implications of Aerobic versus Anaerobic Metabolism*

In 1861, Louis Pasteur observed that *when yeast are exposed to aerobic conditions, their glucose consumption and ethanol production drop precipitously* (the **Pasteur effect;** alcoholic fermentation in yeast to produce ATP, CO_2, and ethanol are discussed in Section 17-3B). An analogous effect is observed in mammalian muscle; the concentration of lactic acid, the anaerobic product of muscle glycolysis, drops dramatically when cells switch to aerobic metabolism.

FIGURE 22-48 Schematic diagram depicting the coordinated control of glycolysis and the citric acid cycle by ATP, ADP, AMP, P_i, Ca^{2+}, and the $[NADH]/[NAD^+]$ ratio (the vertical arrows indicate increases in this ratio). Here a green dot signifies activation and a red octagon represents inhibition. [After Newsholme, E.A. and Leech, A.R., *Biochemistry for the Medical Sciences, pp.* 316 and 320, Wiley (1983).] 🎵 **See the Animated Figures**

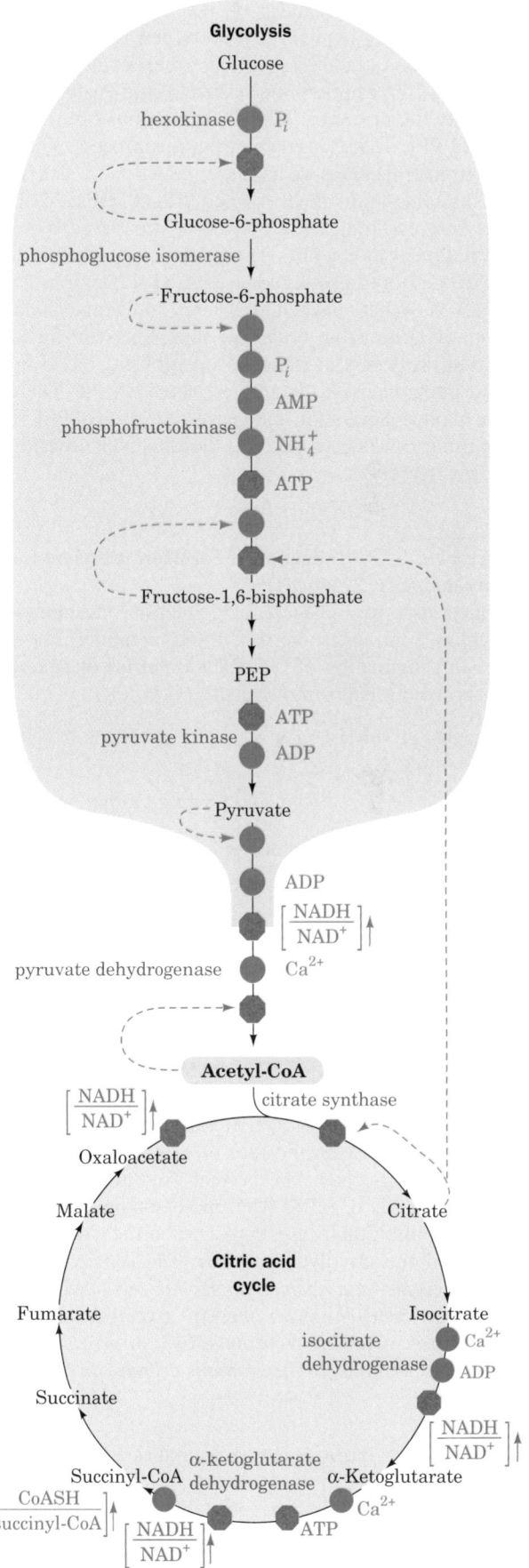

a. Hypoxia Causes an Increase in Glycolysis

In the presence of sufficient oxygen, oxidative phosphorylation supplies most of the body's ATP needs. However, during **hypoxia** (when oxygen is limiting), glycolysis must be stimulated (with its inherent increased rate of glucose consumption; the reverse of the Pasteur effect) to supply the necessary ATP. F2,6P, the most potent activator of PFK-1, also participates in this process. The concentration of F2,6P, as we have seen (Section 18-3F), is regulated by the bifunctional enzyme PFK-2/FBPase-2. In its heart isozyme, the PFK-2 activity is stimulated by phosphorylation at its Ser 466. Among the enzymes that do so is **AMP-activated protein kinase** (AMPK; Sections 25-4B and 25-5). When oxygen deficiency prevents oxidative phosphorylation from providing sufficient ATP for heart function, as occurs in **ischemia** (insufficient blood flow), the resulting increased [AMP] activates AMPK. The consequent phosphorylation and hence activation of PFK-2 results in an increase of [F2,6P], thereby activating PFK-1 and thus glycolysis.

b. Aerobic ATP Production Is Far More Efficient than Anaerobic ATP Production

One reason for the decrease in glucose consumption on switching from anaerobic to aerobic metabolism is clear from an examination of the stoichiometries of anaerobic and aerobic breakdown of glucose ($C_6H_{12}O_6$).

Anaerobic glycolysis:

$$C_6H_{12}O_6 + 2ADP + 2P_i \rightarrow$$
$$2lactate + 2H^+ + 2H_2O + 2ATP$$

Aerobic metabolism of glucose:

$$C_6H_{12}O_6 + 38ADP + 38P_i + 6O_2 \rightarrow$$
$$6CO_2 + 44H_2O + 38ATP$$

(3 ATP for each of the 10 NADH generated per glucose oxidized, 2 ATP for each of the 2 $FADH_2$ generated, 2 ATP produced in glycolysis, and 2GTP $\rightleftharpoons$ 2ATP produced in the citric acid cycle.) Thus *aerobic metabolism is 19 times more efficient than anaerobic glycolysis in producing ATP.* The switch to aerobic metabolism therefore rapidly increases the ATP mass action ratio. As the ATP mass action ratio increases, the rate of electron transport decreases, which has the effect of increasing the [NADH]/[NAD$^+$] ratio. The increases in [ATP] and [NADH] inhibit their target enzymes in the citric acid cycle and in the glycolytic pathway. *The activity of PFK, which is citrate- and adenine nucleotide-regulated and one of the rate-controlling enzymes of glycolysis, decreases manyfold on switching from anaerobic to aerobic metabolism. This accounts for the dramatic decrease in glycolysis.*

c. Anaerobic Glycolysis Has Advantages as Well as Limitations

Animals can sustain anaerobic glycolysis for only short periods of time. This is because PFK, which cannot function effectively much below pH 7, is inhibited by the acid-ification arising from lactic acid production (but see Section 27-2B). Despite this limitation and the low efficiency of glycolytic ATP production, *the enzymes of glycolysis are present in such great concentrations that when they are not inhibited, ATP can be produced much more rapidly than through oxidative phosphorylation.*

The different characteristics of aerobic and anaerobic metabolism permit us to understand certain aspects of cancer cell metabolism and cardiovascular disease.

d. Cancer Cell Metabolism

As Warburg first noted in 1926, certain cancer cells produce more lactic acid under aerobic conditions than do normal cells. This is because the glycolytic pathway in these cells produces pyruvate more rapidly than the citric acid cycle can accommodate. How can this happen given the interlocking controls on the system? One explanation is that these controls have broken down in cancer cells. Another is that their ATP utilization occurs at rates too rapid to be replenished by oxidative phosphorylation. This would alter the ratios of adenine nucleotides so as to relieve the inhibition of PFK-1. In addition, many cancer cell lines have a much larger [F2,6P] than do normal cells. These cells contain an inducible isozyme of PFK-2/FBPase-2 that has an AMPK-phosphorylatable site for activating PFK-2. Consequently, an [AMP] increase in these cells results in an increase in their [F2,6P], which further activates PFK-1 and glycolysis. Efforts to understand the metabolic differences between cancer cells and normal cells may eventually lead to a treatment of certain forms of this devastating disease.

e. Cardiovascular Disease

Oxygen deprivation of certain tissues resulting from cardiovascular disease is of major medical concern. For example, two of the most common causes of human death, **myocardial infarction** (heart attack) and **stroke,** are caused by interruption of the blood (O_2) supply to a portion of the heart or the brain, respectively. It seems obvious why this should result in a cessation of cellular activity, but why does it cause cell death?

In the absence of O_2, a cell, which must then rely only on glycolysis for ATP production, rapidly depletes its stores of phosphocreatine (a source of rapid ATP production; Section 16-4C) and glycogen. As the rate of ATP production falls below the level required by membrane ion pumps for the maintenance of proper intracellular ionic concentrations, the osmotic balance of the system is disrupted, so that the cell and its membrane-enveloped organelles begin to swell. The resulting overstretched membranes become permeable, thereby leaking their enclosed contents. [In fact, a useful diagnostic criterion for myocardial infarction is the presence in the blood of heart-specific enzymes, such as the H-type isozyme of lactate dehydrogenase (vs the M-type isozyme, which predominates in skeletal muscle; Section 17-3A), which leak out of necrotic (dead) heart tissue.] Moreover, the decreased intracellular pH that accompanies anaerobic glycolysis (because of lactic acid production) permits the released lysosomal enzymes (which are active only at acidic pH's) to degrade the cell contents. Thus, the cessation of metabolic activity results in irreversible cell damage. Rapidly respiring tissues, such as those of heart and brain, are particularly susceptible to such damage.

◼ CHAPTER SUMMARY

1 ◼ The Mitochondrion Oxidative phosphorylation is the process through which the NADH and $FADH_2$ produced by nutrient oxidation are oxidized with the concomitant formation of ATP. The process takes place in the mitochondrion, an ellipsoidal organelle that is bounded by a permeable outer membrane and contains an impermeable and highly invaginated inner membrane that encloses the matrix. Enzymes of oxidative phosphorylation are embedded in the inner mitochondrial membrane. P_i is imported into the mitochondrion by a specific transport protein. Ca^{2+} import and Ca^{2+} export proteins operate to maintain a constant cytosolic $[Ca^{2+}]$. NADH's electrons are imported into the mitochondrion by one of several shuttle systems such as the glycerophosphate shuttle or the malate–aspartate shuttle.

2 ◼ Electron Transport The standard free energy change for the oxidation of NADH by O_2 is $\Delta G^{\circ\prime} = -218 \text{ kJ} \cdot \text{mol}^{-1}$, whereas that for the synthesis of ATP from ADP and P_i is $\Delta G^{\circ\prime} = 30.5 \text{ kJ} \cdot \text{mol}^{-1}$. Consequently, the molar free energy of oxidation of NADH by O_2 is sufficient to power the synthesis of several moles of ATP under standard conditions. The electrons generated by oxidation of NADH and $FADH_2$ pass through four protein complexes, the electron-transport chain, with the coupled synthesis of ATP. Complexes I, III, and IV participate in the oxidation of NADH, producing about three ATPs per NADH, whereas $FADH_2$ oxidation, which involves Complexes II, III, and IV, produces only about two ATPs per $FADH_2$. Thus, the ratio of moles of ATP produced per mole of coenzyme oxidized by O_2, the P/O ratio, is 3 for NADH oxidation and 2 for $FADH_2$ oxidation (although some measurements indicate that these quantities are 2.5 and 1.5). The route taken by electrons through the electron-transport chain was elucidated, in part, through the use of electron-transport inhibitors. Rotenone and amytal inhibit Complex I, antimycin inhibits Complex III, and CN^- inhibits Complex IV. Also involved were measurements of the reduction potentials of the electron-carrying prosthetic groups contained in the electron-transport complexes.

Complex I contains FMN and six to seven iron–sulfur clusters in a 43-subunit (in mammals) transmembrane protein complex. This L-shaped complex passes electrons from NADH to CoQ, a nonpolar small molecule that diffuses freely within the membrane. Complex II contains the citric acid cycle enzyme succinate dehydrogenase and also passes electrons to CoQ, in this case from succinate through FAD and three iron–sulfur clusters. The X-ray structure of the *E. coli* enzyme quinol–fumarate reductase, a homolog of Complex II, indicates that its redox cofactors are arranged in a linear chain. $CoQH_2$ passes electrons to Complex III (cytochrome bc_1), a homodimeric complex whose protomers each contain two *b*-type hemes bound to a cytochrome *b* subunit, a Rieske iron–sulfur protein (ISP), and a cytochrome c_1. An electron from cytochrome c_1 of Complex III is passed to the Cu_A center of Complex IV (cytochrome *c* oxidase) via the peripheral membrane protein cytochrome *c*. This electron is then passed to cytochrome *a*, which, in turn, passes it to a binuclear center composed of heme a_3 and Cu_B, which reduces O_2 to H_2O. This process occurs in four 1-electron steps that pump four protons from the mitochondrial matrix/bacterial cytoplasm to the intermembrane space/periplasm.

3 ◼ Oxidative Phosphorylation The mechanism by which the free energy released by the electron-transport chain is stored and utilized in ATP synthesis is described by the chemiosmotic hypothesis. This hypothesis states that the free energy released by electron transport is conserved by the generation of an electrochemical proton gradient across the inner mitochondrial membrane (bacterial cell membrane; outside positive and acidic), which is harnessed to synthesize ATP. The proton gradient is created and maintained by the obligatory outward translocation of H^+ across the inner mitochondrial membrane as electrons travel through Complexes I, III, and IV.

Complex III pumps protons via a redox loop mechanism called the Q cycle, a bifurcated double cycle in which one molecule of $CoQH_2$ is oxidized to CoQ and then is re-reduced to $CoQH_2$ by a second molecule of $CoQH_2$ in a process that collectively transfers four protons from the inside to the outside while oxidizing one molecule of $CoQH_2$ to CoQ. Electrons are transferred between the two CoQ's, which are bound at different sites, Q_o and Q_i, as well as between the $CoQH_2$ bound at Q_o and cytochrome c_1 via the ISP, which undergoes a conformational change in doing so. Complex IV contains no $(H^+ + e^-)$ carriers such as $CoQH_2$ and hence translocates proteins via a proton pump mechanism. Bacteriorhodopsin, the best characterized proton pump, translocates protons in a light driven process. This involves a trans to cis isomerization of bacteriorhodopsin's retinal prosthetic group on absorbing a photon, followed by the translocation of a proton through the hydrophilic central channel of this transmembrane protein via a process that involves conformational and pK changes of the polar groups lining the channel as the retinal relaxes to its ground state. Complex IV is thought to pump protons via a similar mechanism that is driven by the changes in the redox state of its heme a_3–Cu_B binuclear center as it reduces O_2 to H_2O.

The energy stored in the electrochemical proton gradient is utilized by proton-translocating ATP synthase (Complex V, F_1F_0-ATPase) in the synthesis of ATP via the binding change mechanism, by coupling this process to the exergonic transport of H^+ back to the inside. Proton-translocating ATP synthase contains two oligomeric components: F_1 ($\alpha_3\beta_3\gamma\delta\varepsilon$), a peripheral membrane protein that appears as "lollipops" in electron micrographs of the inner mitochondrial membrane, and F_0 (ab_2c_{9-12} in *E. coli*), an integral membrane protein that contains the proton channel. The conformational changes that promote the synthesis of ATP from ADP + P_i arise through the demonstrated rotation of the γ subunit relative to the catalytic $\alpha_3\beta_3$ assembly that contains the enzyme's three active sites. The γ subunit is attached to a ring of *c* subunits in F_0, whose rotation is driven by the passage of protons between it and the *a* subunit.

Compounds such as 2,4-dinitrophenol are uncouplers of oxidative phosphorylation because they carry H^+ across the mitochondrial membrane, thereby dissipating the proton gradient and allowing electron transport to continue without concomitant ATP synthesis. Brown fat mitochondria contain a regulated uncoupling system that, under hormonal control, generates heat instead of ATP.

4 ■ Control of ATP Production   Under aerobic conditions, the rate of ATP synthesis by oxidative phosphorylation

is regulated, in a phenomenon known as acceptor control, by the ATP mass action ratio. ATP synthesis is tightly coupled to the oxidation of NADH and $FADH_2$ by the electron-transport chain. Glycolysis and the citric acid cycle are coordinately controlled so as to produce NADH and $FADH_2$ only at a rate required to meet the system's demand for ATP.

REFERENCES

HISTORICAL OVERVIEW

Ernster, L. and Schatz, G., Mitochondria: A historical review, *J. Cell Biol.* **91,** 227s–255s (1981).

Fruton, J.S., *Molecules and Life, pp.* 262–396, Wiley–Interscience (1972).

Krebs, H., *Otto Warburg. Cell Physiologist, Biochemist, and Eccentric,* Clarendon Press (1981). [A biography of one of the pioneers in the biochemical study of respiration, by a distinguished student.]

Prebble, J., Peter Mitchell and the ox phos wars, *Trends Biochem. Sci.* **27,** 209–212 (2002).

Racker, E., *A New Look at Mechanisms in Bioenergetics,* Academic Press (1976). [A fascinating personal account by one of the outstanding contributors to the field.]

GENERAL

Ernster, L. (Ed.), *Bioenergetics,* Elsevier (1984).

Harold, F.M., *The Vital Force: A Study of Bioenergetics,* Chapter 7, Freeman (1986).

Hatefi, Y., The mitochondrial electron transport chain and oxidative phosphorylation system, *Annu. Rev. Biochem.* **54,** 1015–1069 (1985).

Martonosi, A.N. (Ed.), *The Enzymes of Biological Membranes* (2nd ed.), Vol. 4, *Bioenergetics of Electron and Proton Transport,* Plenum Press (1985).

Newsholme, E. and Leech, T., *The Runner,* Fitness Books (1983). [A delightful book on the physiology and biochemistry of running.]

Nicholls, D.G. and Ferguson, S.J., *Bioenergetics* (3rd ed.), Academic Press (2002). [An authoritative monograph devoted almost entirely to the mechanism of oxidative phosphorylation and the techniques used to elucidate it.]

Schultz, B.E. and Chan, S.I., Structures and proton-pumping strategies of mitochondrial respiratory enzymes, *Annu. Rev. Biophys. Biomol. Struct.* **30,** 23–65 (2001).

MITOCHONDRIA

Frey, T.G. and Mannella, C.A., The internal structure of mitochondria, *Trends Biochem. Sci.* **23,** 319–324 (2000).

Science **283,** 1476–1497 (1999). [A series of four review articles on mitochondria.]

ELECTRON TRANSPORT

Babcock, G.T., How oxygen is activated and reduced in respiration, *Proc. Natl. Acad. Sci.* **96,** 12971–12973 (1999); *and* Proshlyakov, D.E., Pressler, M.A., and Babcock, G.T., Dioxygen activation and bond cleavage by mixed-valence cytochrome *c* oxidase, *Proc. Natl. Acad. Sci.* **95,** 8020–8025 (1998).

Beinert, H., Holm, R.H., and Münck, E., Iron-sulfur clusters: Nature's modular, multipurpose structures, *Science* **277,** 653–659 (1997).

Berry, E.A., Guergova-Kuras, M., Huang, L., and Crofts, A.R., Structure and function of cytochrome *bc* complexes, *Annu. Rev. Biochem.* **69,** 1005–1075 (2000).

Calhoun, M.W., Thomas, J.W., and Gennis, R.B., The cytochrome superfamily of redox-driven proton pumps, *Trends Biochem. Sci.* **19,** 325–330 (1994).

Collman, J.P., Rapta, M., Bröring, M., Raptova, L., Schwenninger, R., Boitrel, B., Fu, L., and L'Her, M., Close structural analogues of the cytochrome *c* oxidase Fe_{a3}/Cu_B center show clean $4e^-$ electroreduction of O_2 to H_2O at physiological pH, *J. Amer. Chem. Soc.* **121,** 1387–1388 (1999).

Crofts, A.R. and Berry, E.A., Structure and function of the cytochrome bc_1 complex of mitochondria and photosynthetic bacteria, *Curr. Opin. Struct. Biol.* **8,** 501–509 (1998).

Darrouzet, E., Moser, C.C., Dutton, P.L., and Daldal, F., Large scale domain movement in cytochrome bc_1: A new device for electron transfer in proteins. *Trends Biochem. Sci.* **26,** 445–451 (2001).

Hinkle, P.C., Kumar, M.A., Resetar, A., and Harris, D.L., Mechanistic stoichiometry of mitochondrial oxidative phosphorylation, *Biochemistry* **30,** 3576–3582 (1991). [Describes measurements of the P/O ratios indicating that their values are 2.5, 1.5, and 1.]

Hunte, C., Koepke, J., Lange, C., Rossmanith, T., and Michel, H., Structure at 2.3 Å resolution of the cytochrome bc_1 complex from the yeast *Saccharomyces cerevisiae* co-crystallized with an antibody Fv fragment, *Structure* **8,** 669–684 (2000).

Iverson, T.M., Luna-Chavez, C., Cecchini, G., and Rees, D.C., Structure of the *Escherichia coli* fumarate reductase respiratory complex, *Science* **284,** 1961–1966 (1999); *and* Iverson, T.M., Luna-Chavez, C., Schröder, I., Cecchini, G., and Rees, D.C., Analyzing your complexes: Structure of the quinol reductase respiratory complex, *Curr. Opin. Struct. Biol.* **10,** 448–455 (2000).

Iwata, S., Ostermeier, C., Ludwig, B., and Michel, H., Structure at 2.8 Å resolution of cytochrome *c* oxidase from *Paracoccus denitrificans, Nature* **376,** 660–669 (1995).

Iwata, S., Lee, J.W., Okada, K., Lee, J.K., Iwata, M., Rasmussen, B., Link, T.A., Ramaswamy, S., and Jap, B.K., Complete structure of the 11-subunit bovine mitochondrial cytochrome bc_1 complex, *Science* **281,** 64–71 (1998).

Kim, H., Xia, D., Yu, C.-A., Xia, J.-Z., Kachurin, A.M., Zhang, L., Yu, L., and Deisenhofer, J., Inhibition binding changes domain mobility in the iron–sulfur protein of mitochondrial bc_1 complex from bovine heart, *Proc. Natl. Acad. Sci.* **95,** 8026–8033 (1998); *and* Xia, D., Yu, C.-A., Kim, H., Xia, J.-Z., Kachurin, A.M., Zhang, L., Yu, L., and Deisenhofer, J., Crystal structure of the cytochrome bc_1 complex from heart mitochondria, *Science* **277,** 60–66 (1997).

Lange, C. and Hunte, C., Crystal structure of the yeast cytochrome bc_1 complex with its bound substrate cytochrome *c, Proc. Natl. Acad. Sci.* **99,** 2800–2805 (2002).

Michel, H., Behr, J., Harrenga, A., and Kannt, A., Cytochrome *c* oxidase: Structure and spectroscopy, *Annu. Rev. Biophys. Biomol. Struct.* **27,** 329–356 (1998).

Moore, G.R. and Pettigrew, G.W., *Cytochromes c. Evolutionary, Structural and Physicochemical Aspects,* Springer-Verlag (1990).

Moser, C.C., Keske, J.M., Warncke, K., Farid, R.S., and Dutton, L.S., Nature of biological electron transfer, *Nature* **355,** 796–802 (1992).

Ohnishi, T., Moser, C.C., Page, C.C., Dutton, P.L., and Yano, T., Simple redox-linked proton-transfer design: New insights from structures of quinol-fumarate reductase, *Structure* **8,** R23–R32 (2000).

Trumpower, B.L., The protonmotive Q cycle, *J. Biol. Chem.* **265,** 11409–11412 (1990).

Tsukihara, T., Aoyama, H., Yamashita, E., Tomizaki, T., Yamaguchi, H., Shinzawa-Itoh, K., Nakashima, R., Yaono, R., and Yoshikawa, S., The whole structure of the 13-subunit oxidized cytochrome *c* oxidase at 2.8 Å, *Science* **272,** 1136–1144 (1996).

Walker, J.E., The NADH:ubiquinone oxidoreductase (complex I) of respiratory chains, *Q. Rev. Biophys.* **25,** 253–324 (1992). [An exhaustive review.]

Yankovskaya, V., Horsefield, R., Törnroth, S., Luna-Chavez, C., Miyoshi, H., Légar, C., Byrne, B., Cecchini, G., and Iwata, S., Architecture of succinate dehydrogenase and reactive oxygen species generation, *Science* **299,** 700–704 (2003).

Yoshikawa, S., et al., Redox-coupled crystal structural changes in bovine heart cytochrome *c* oxidase, *Science* **280,** 1723–1729 (1998); *and* Yoshikawa, S., Beef heart cytochrome *c* oxidase, *Curr. Opin. Struct. Biol.* **7,** 574–579 (1997).

Zhang, Z., Huang, L., Shulmeister, V.M., Chi, Y.-I., Kim, K.K., Huang, L.-W., Crofts, A.R., Berry, E.A., and Kim, S.H., Electron transfer by domain movement in cytochrome bc_1, *Nature* **392,** 677–684 (1998).

BACTERIORHODOPSIN

Gennis, R.B. and Ebray, T.G., Proton pump caught in the act, *Science* **286,** 252–253 (1999).

Haupts, U., Tittor, J., and Oesterhelt, D., Closing in on bacteriorhodopsin: Progress in understanding the molecule, *Annu. Rev. Biophys. Biomol. Struct.* **28,** 367–399 (1999).

Heberle, J., Proton transfer reactions across bacteriorhodopsin and along the membrane, *Biochim. Biophys. Acta* **1458,** 135–147 (2000).

Kühlbrandt, W., Bacteriorhodopsin—the movie, *Nature* **406,** 569–570 (2000).

Lanyi, J.K., Progress toward an explicit mechanistic model for the light-driven pump, bacteriorhodopsin, *FEBS Lett.* **464,** 103–107 (1999); *and* Lanyi, J.K. and Luecke, H., Bacteriorhodopsin, *Curr. Opin. Struct. Biol.* **11,** 415–419 (2001)

OXIDATIVE PHOSPHORYLATION

Abrahams, J.P., Leslie, A.G.W., Lutter, R., and Walker, J.E., Structure at 2.8 Å resolution of F_1-ATPase from bovine heart mitochondria, *Nature* **370,** 621–628 (1994).

Boyer, P.D., The binding change mechanism for ATP synthase—some probabilities and possibilities, *Biochim. Biophys. Acta* **1140,** 215–250 (1993).

Boyer, P.D., The ATP synthase—a splendid molecular machine, *Annu. Rev. Biochem.* **66,** 717–749 (1997).

Capaldi, R. and Aggeler, R., Mechanism of F_1F_0-type ATP synthase, a biological rotary motor, *Trends Biochem. Sci.* **27,** 154–160 (2002).

Gennis, R.B., Multiple proton-conducting pathways in cytochrome oxidase and a proposed role for the active-site tyrosine, *Biochim. Biophys. Acta* **1458,** 241–248 (2000).

Gibbons, C., Montgomery, M.G., Leslie, A.G.W., and Walker, J.E., The structure of the central stalk in bovine F_1-ATPase at 2.4 Å resolution, *Nature Struct. Biol.* **7,** 1055–1061 (2000).

Klingenberg, M., Mechanism and evolution of the uncoupling protein of brown adipose tissue, *Trends Biochem. Sci.* **15,** 108–112 (1990).

Ma, J., Flynn, T.C., Cui, Q., Leslie, A.G.W., Walker, J.E., and Karplus, M. A dynamic analysis of the rotation mechanism for the conformational change in F_1-ATPase, *Structure* **10,** 921–931 (2002).

Mills, D.A., Florens, L., Hiser, C., Qian, J., and Ferguson-Miller, S., Where is outside in cytochrome *c* oxidase and how and when do protons get there? *Biochim. Biophys. Acta* **1458,** 180–187 (2000).

Mitchell, P., Vectorial chemistry and the molecular mechanics of chemiosmotic coupling: Power transmission by proticity, *Biochem. Soc. Trans.* **4,** 398–430 (1976).

Nicholls, D.G. and Rial, E., Brown fat mitochondria, *Trends Biochem. Sci.* **9,** 489–491 (1984).

Noji, H. and Yoshida, M., The rotary engine in cell ATP synthase, *J. Biol. Chem.* **276,** 1665–1668 (2001).

Rastogi, V.K. and Girvin, M.E., Structural changes linked to proton translocation by subunit *c* of the ATP synthase, *Nature* **402,** 262–268 (1999); *and* Girvin, M.E., Rastogi, V.K., Abildgaard, F., Markley, J.L., and Fillingame, R.H., Solution structure of the transmembrane H^+-transporting subunit c of the ATP synthase, *Biochemistry* **37,** 8817–8824 (1998).

Sambongi, Y., Iko, Y., Tanabe, M., Omote, H., Iwamoto-Kihara, A., Ueda, I., Yanagida, T., Wada, Y., and Futai, M., Mechanical rotation of the c subunit oligomer in ATP synthase (F_0F_1): Direct observation, *Science* **286,** 1722–1724 (1999).

Stock, D., Leslie, A.G.W., and Walker, J.E., Molecular architecture of the rotary motor in ATP synthase, *Science* **286,** 1700–1705 (1999). [The X-ray structure of the F_1-*c*-ring complex.]

Stock, D., Gibbons, C., Arechaga, I., Leslie, A.G.W., and Walker, J.E., The rotary mechanism of ATP synthase, *Curr. Opin. Struct. Biol.* **10,** 672–679 (2000).

Verkhovsky, M.I., Jasaitis, A., Verkhovskaya, M.L., Morgan, J.E., and Wikström, M., Proton translocation by cytochrome *c* oxidase, *Nature* **400,** 480–483 (1999).

Walker, J.E. (Ed.), *The Mechanism of F_1F_0-ATPase, Biochim. Biophys. Acta* **1458,** 221–514 (2000). [A series of authoritative reviews.]

Wilkens, S. and Capaldi, R.A., ATP synthase's second stalk comes into focus, *Nature* **393,** 29 (1998).

Yasuda, R., Noji, H., Kinosita, K., Jr., and Yoshida, M., F_1-ATPase is a highly efficient molecular motor that rotates with discrete 120° steps, *Cell* **93,** 1117–1124 (1998).

Yoshida, M. Muneyuki, E., and Hisabori, T., ATP synthase—a marvelous rotary engine of the cell, *Nature Rev. Mol. Cell. Biol.* **2,** 669–677 (2001); *and* Noji, H. and Yoshida, M., The rotary machine in the cell ATP synthase, *J. Biol. Chem.* **276,** 1665–1668 (2001).

Zaslavsky, D. and Gennis, R.B., Proton pumping by cytochrome oxidase: Progress, problems, and postulates, *Biochim. Biophys. Acta* **1458,** 164–179 (2000).

CONTROL OF ATP PRODUCTION

Brown, G.C., Control of respiration and ATP synthesis in mammalian mitochondria and cells, *Biochem. J.* **284,** 1–13 (1992).

Chesney J., Mitchell, R., Benigni, F., Bacher, M., Spiegel, L., Al-Abed, Y., Han, J.H., Metz, C., and Bucala, R., An inducible gene product for 6-phosphofructo-2-kinase with an AU-rich instability element: role in tumor cell glycolysis and the Warburg effect, *Proc. Natl. Acad. Sci.* **96,** 3047–3052 (2000).

Harris, D.A. and Das, A.M., Control of mitochondrial ATP synthesis in the heart, *Biochem. J.* **280,** 561–573 (1991).

Jeneson, J.A.L., Westerhoff, H.V., and Kushmerick, M.J., A metabolic control analysis of kinetic controls in ATP free energy metabolism in contracting skeletal muscle, *Am. J. Physiol. Cell Physiol.* **279**, C813–C832 (2000).

Marsin, A.-S., Bertrand, L., Rider, M.H., Deprez, J., Beauloye, C., Vincent, M.F., Van den Berghe, G., Carling, D., and Hue, L., Phosphorylation and activation of heart PFK-2 by AMPK has a role in the stimulation of glycolysis during ischaemia, *Curr. Biol.* **10**, 1247–1255 (2000).

Marsin, A.-S., Bouzin, C., Bertrand, L., and Hue, L., The stimulation of glycolysis by hypoxia in activated monocytes is me-

diated by AMP-activated protein kinase and inducible 6-phosphofructo-3-kinase, *J. Biol. Chem.* **277**, 30778–30783 (2002).

Randle, P.J., Regulatory interactions between lipids and carbohydrates: The glucose fatty acid cycle after 35 years. *Diabetes/Metab. Rev.* **14**, 263–283 (1998).

Ricquier, D. and Bouillaud, F., The mitochondrial uncoupling protein: Structural and genetic studies, *Prog. Nucl. Acid Res. Mol. Biol.* **56**, 83–108 (1997).

PROBLEMS

1. Rank the following redox-active coenzymes and prosthetic groups of the electron-transport chain in order of increasing affinity for electrons: cytochrome *a*, CoQ, FAD, cytochrome *c*, NAD$^+$.

2. Why is the oxidation of succinate to fumarate only associated with the production of two ATPs during oxidative phosphorylation, whereas the oxidation of malate to oxaloacetate is associated with the production of three ATPs?

3. What is the thermodynamic efficiency of oxidizing FADH$_2$ so as to synthesize two ATPs under standard biochemical conditions?

4. Sublethal cyanide poisoning may be reversed by the administration of nitrites. These substances oxidize hemoglobin, which has a relatively low affinity for CN$^-$, to methemoglobin, which has a relatively high affinity for CN$^-$. Why is this treatment effective?

5. Match the compound with its behavior: (1) rotenone, (2) dinitrophenol, and (3) antimycin. (a) Inhibits oxidative phosphorylation when the substrate is pyruvate but not when the substrate is succinate. (b) Inhibits oxidative phosphorylation when the substrate is either pyruvate or succinate. (c) Allows pyruvate to be oxidized by mitochondria even in the absence of ADP.

6. Nigericin is an ionophore (Section 20-2C) that exchanges K$^+$ for H$^+$ across membranes. Explain how the treatment of functioning mitochondria with nigericin uncouples electron transport from oxidative phosphorylation. Does valinomycin, an ionophore that transports K$^+$ but not H$^+$, do the same? Explain.

7. The difference in pH between the internal and external surfaces of the inner mitochondrial membrane is 1.4 pH units (external side acidic). If the membrane potential is 0.06 V (inside negative), what is the free energy released on transporting 1 mol of protons back across the membrane? How many protons must be transported to provide enough free energy for the synthesis of 1 mol of ATP (assume standard biochemical conditions)?

*8. (a) A simplistic interpretation of the Q cycle would predict that the proton pumping efficiency of cytochrome bc_1 would be reduced by no more than 50% in the presence of saturating amounts of antimycin. Explain. (b) Indicate why cytochrome bc_1 is nearly 100% inhibited by antimycin.

9. The antibiotic oligomycin B

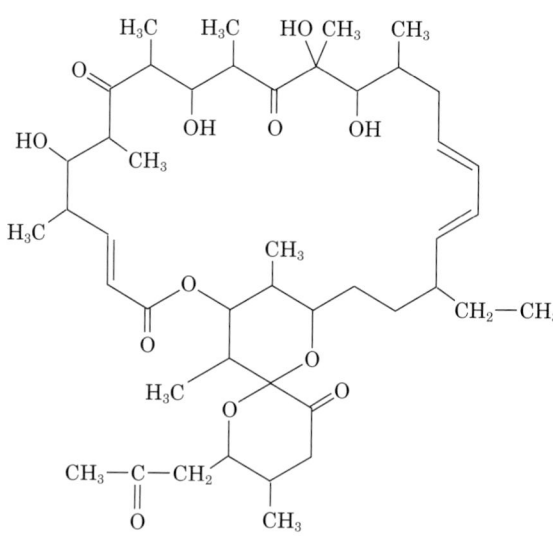

Oligomycin B

binds to the F$_0$ subunit of the mitochondrial F$_1$F$_0$ – ATPase and thereby prevents it from synthesizing ATP [note that oligomycin-sensitivity conferral protein (OSCP), the mitochondrial counterpart of the *E. coli* δ subunit (Fig. 22-43), does not bind oligomycin B.] Explain why: (a) Submitochondrial particles from which F$_1$ has been removed are permeable to protons. (b) Addition of oligomycin B to F$_1$-depleted submitochondrial particles decreases this permeability severalfold.

10. Oligomycin B (see Problem 9) and cyanide both inhibit oxidative phosphorylation when the substrate is either pyruvate or succinate. Dinitrophenol can be used to distinguish between these inhibitors. Explain.

11. The *E. coli* F$_1$F$_0$–ATPase cannot synthesize ATP when Met 23 of its γ subunit is mutated to Lys. Yet the F$_1$ component of this complex still exhibits rotation of its γ subunit relative to its α$_3$β$_3$ spheroid when it is supplied with ATP. Suggest a reason for these effects.

12. For the oxidation of a given amount of glucose, does nonshivering thermogenesis by brown fat or shivering thermogenesis by muscle produce more heat?

13. How does atractyloside affect mitochondrial respiration? (*Hint:* See Section 20-4C.)

14. Certain unscrupulous operators offer, for a fee, to freeze recently deceased individuals in liquid nitrogen until medical science can cure the disease from which they died. What is the biochemical fallacy of this procedure?

Chapter 23

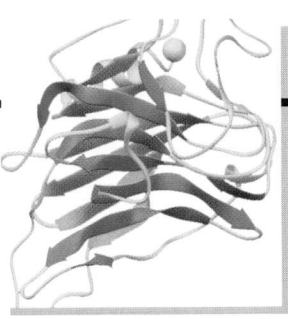

Other Pathways of Carbohydrate Metabolism

Heretofore, we have dealt with many aspects of carbohydrate metabolism. We have seen how the free energy of glucose oxidation is sequestered in ATP through glycolysis, the citric acid cycle, and oxidative phosphorylation. We have also studied the mechanism by which glucose is stored as glycogen for future use and how glycogen metabolism is controlled in response to the needs of the organism. In this chapter, we examine several other carbohydrate metabolism pathways of importance:

1. Gluconeogenesis, through which noncarbohydrate precursors such as lactate, pyruvate, glycerol, and amino acids are converted to glucose.

2. The **glyoxylate cycle,** through which plants convert acetyl-CoA to glucose.

3. Oligosaccharide and glycoprotein biosynthesis, through which oligosaccharides are synthesized and added to specific amino acid residues of proteins.

4. The **pentose phosphate pathway,** an alternate pathway of glucose degradation, which generates **NADPH,** the source of reducing equivalents in reductive biosynthesis, and **ribose-5-phosphate,** the sugar precursor of the nucleic acids.

This chapter completes our study of carbohydrate metabolism in animals; photosynthesis, which occurs only in plants and in certain bacteria, is the subject of Chapter 24.

1 ■ GLUCONEOGENESIS

Glucose occupies a central role in metabolism, both as a fuel and as a precursor of essential structural carbohydrates and other biomolecules. The brain and red blood cells are almost completely dependent on glucose as an energy source. Yet the liver's capacity to store glycogen is only sufficient to supply the brain with glucose for about half a day under fasting or starvation conditions. Thus, *when fasting, most of the body's glucose needs must be met by gluconeogenesis (literally, new glucose synthesis), the biosynthesis of glucose from noncarbohydrate precursors.* Indeed, isotopic labeling studies determining the source of glucose in the blood during a fast showed that gluconeogenesis is responsible for 64% of total glucose production over the first 22 hours of the fast and accounts for almost all the glucose production by 46 hours. Thus, gluconeogenesis provides a substantial fraction of the glucose produced in fasting humans, even after a few hours fast. Gluconeogenesis occurs in liver and, to a smaller extent, in kidney.

The noncarbohydrate precursors that can be converted to glucose include the glycolysis products lactate and pyruvate, citric acid cycle intermediates, and the carbon skeletons of most amino acids. First, however, all these substances must be converted to oxaloacetate, the starting

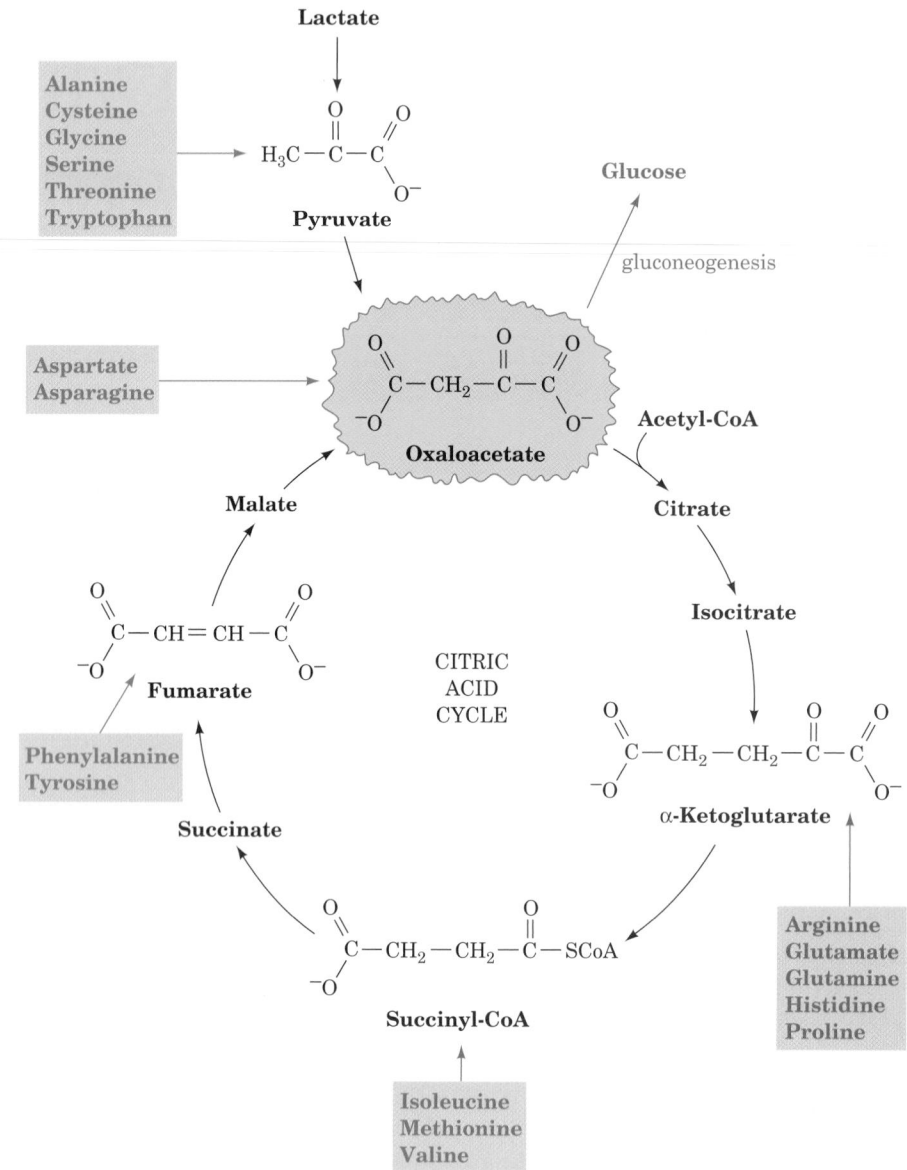

FIGURE 23-1   Pathways converting lactate, pyruvate, and citric acid cycle intermediates to oxaloacetate. The carbon skeletons of all amino acids but leucine and lysine may be, at least in part, converted to oxaloacetate and thus to glucose by these reactions.

material for gluconeogenesis (Fig. 23-1). The only amino acids that cannot be converted to oxaloacetate in animals are leucine and lysine because their breakdown yields only acetyl-CoA (Section 26-3F). *There is no pathway in animals for the net conversion of acetyl-CoA to oxaloacetate.* Likewise, fatty acids cannot serve as glucose precursors in animals because most fatty acids are degraded completely to acetyl-CoA (Section 25-2C). Unlike animals, however, plants do contain a pathway for the conversion of acetyl-CoA to oxaloacetate, the **glyoxylate cycle** (Section 23-2), so that fatty acids can serve as a plant cell's only carbon source. Glycerol, a triacylglycerol breakdown product, is converted to glucose via synthesis of the glycolytic intermediate dihydroxyacetone phosphate, as described in Section 25-1.

A. *The Gluconeogenesis Pathway*

Gluconeogenesis utilizes glycolytic enzymes. Yet three of these enzymes, hexokinase, phosphofructokinase (PFK), and pyruvate kinase, catalyze reactions with large negative free energy changes in the direction of glycolysis. These reactions must therefore be replaced in gluconeogenesis by reactions that make glucose synthesis thermodynamically favorable. Here, as in glycogen metabolism (Section 18-1D), we see the recurrent theme that *biosynthetic and degradative pathways differ in at least one reaction. This not only permits both directions to be thermodynamically favorable under the same physiological conditions but allows the pathways to be independently controlled so that one direction can be activated while the other is inhibited.*

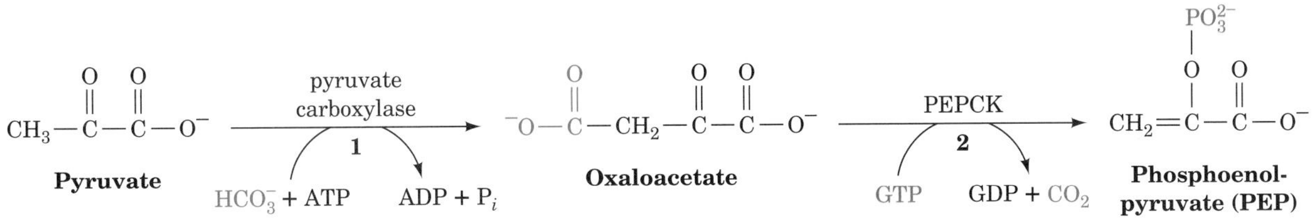

FIGURE 23-2 Conversion of pyruvate to oxaloacetate and then to phosphoenolpyruvate. The enzymes involved are **(1)** pyruvate carboxylase and **(2)** PEP carboxykinase (PEPCK).

a. Pyruvate Is Converted to Oxaloacetate before Conversion to Phosphoenolpyruvate

The formation of phosphoenolpyruvate (PEP) from pyruvate, the reverse of the pyruvate kinase reaction, is endergonic and therefore requires free energy input. This is accomplished by first converting the pyruvate to oxaloacetate. Oxaloacetate is a "high-energy" intermediate whose exergonic decarboxylation provides the free energy necessary for PEP synthesis. The process requires the participation of two enzymes (Fig. 23-2):

1. Pyruvate carboxylase catalyzes the ATP-driven formation of oxaloacetate from pyruvate and HCO_3^-.

2. PEP carboxykinase (PEPCK) converts oxaloacetate to PEP in a reaction that uses GTP as a phosphorylating agent.

b. Pyruvate Carboxylase Has a Biotin Prosthetic Group

Pyruvate carboxylase, discovered in 1959 by Merton Utter, is a tetrameric protein of identical 1158-residue subunits, each of which has a **biotin** prosthetic group. *Biotin (Fig. 23-3a) functions as a CO_2 carrier by forming a carboxyl substituent at its **ureido group** (Fig. 23-3b).* Biotin is covalently bound to the enzyme by an amide linkage between the carboxyl group of its valerate side chain and the ε-amino group of an enzyme Lys residue to form a **biocytin** (alternatively, **biotinyllysine**) residue (Fig. 23-3b). The biotin ring system is therefore at the end of a 14-Å-long flexible arm, much like that of the lipoic acid prosthetic group in the pyruvate dehydrogenase multienzyme complex (Section 21-2A).

Biotin, which was first identified in 1935 as a growth factor in yeast, is an essential human nutrient. Its nutritional deficiency is rare, however, because it occurs in many foods and is synthesized by intestinal bacteria. Human biotin deficiency almost always results from the consumption of large amounts of raw eggs. This is because egg whites contain a protein, **avidin,** that binds biotin so tightly (dissociation constant, $K = 10^{-15}M$) as to prevent its intestinal absorption (cooked eggs do not cause this problem because cooking denatures avidin). The presence of avidin in eggs is thought to inhibit the growth of microorganisms in this highly nutritious environment. The avidin homolog **streptavidin,** which is secreted by *Streptomyces avidinii,* is used as a linking agent in numerous biotechnological applications (e.g., Section 22-3C) because of its particularly high affinity for biotin.

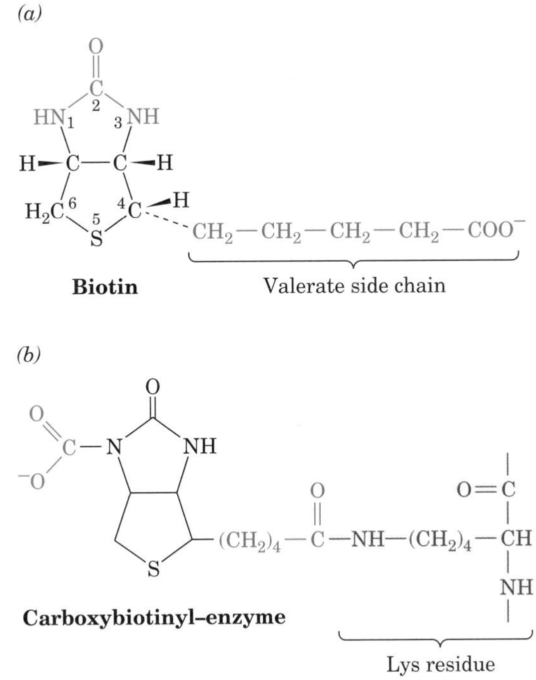

FIGURE 23-3 Biotin and carboxybiotinyl–enzyme. (*a*) Biotin consists of an imidazoline ring that is cis-fused to a tetrahydrothiophene ring bearing a valerate side chain. The chirality at each of its three asymmetric centers is indicated. Positions 1, 2, and 3 constitute a ureido group. (*b*) In carboxybiotinyl–enzyme, N1 of the biotin ureido group is the carboxylation site. Biotin is covalently attached to carboxylases by an amide linkage between its valeryl carboxyl group and the ε-amino group of an enzyme Lys side chain.

c. The Pyruvate Carboxylase Reaction

The pyruvate carboxylase reaction occurs in two phases (Fig. 23-4):

Phase I Biotin is carboxylated at its N1 atom by bicarbonate ion in a three-step reaction in which the hydrolysis of ATP to ADP + P_i functions, via the intermediate formation of **carboxyphosphate,** to dehydrate bicarbonate. This yields free CO_2, which has sufficient free energy to carboxylate biotin. The resulting carboxyl group is activated relative to bicarbonate ($\Delta G^{\circ\prime}$ for its cleavage is -19.7 kJ·mol^{-1}) and can therefore be transferred without further free energy input.

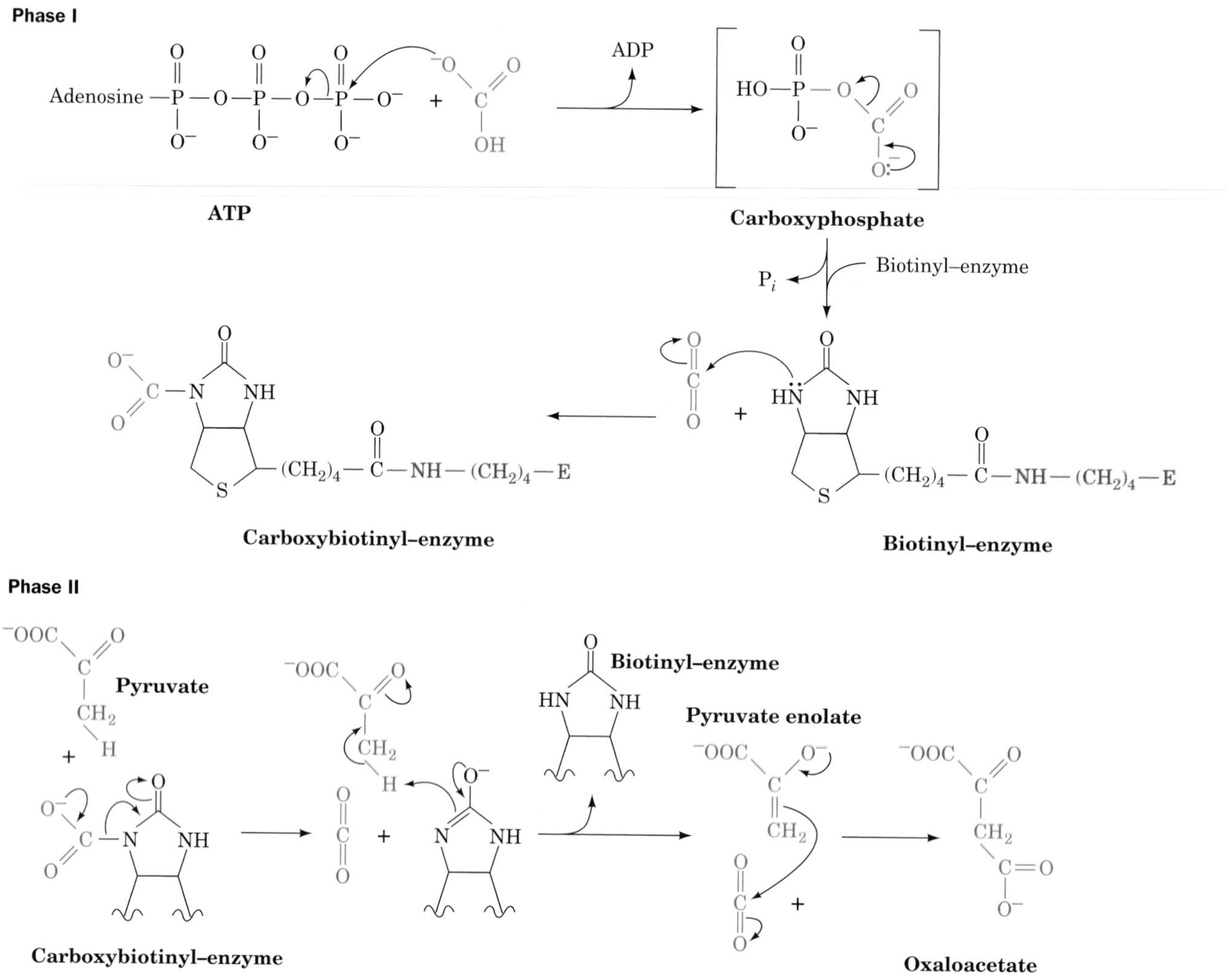

Phase I

Phase II

FIGURE 23-4   Two-phase reaction mechanism of pyruvate carboxylase. Phase I is a three-step reaction in which carboxyphosphate is formed from bicarbonate and ATP, followed by the generation of CO_2 on the enzyme, which then carboxylates biotin. **Phase II** is a three-step reaction in which CO_2 is produced at the active site via the elimination of the biotinyl enzyme, which accepts a proton from pyruvate to generate pyruvate enolate. This, in turn, nucleophilically attacks the CO_2, yielding oxaloacetate. [After Knowles, J.R., *Annu. Rev. Biochem.* **58**, 217 (1989).]

Phase II The activated carboxyl group is transferred from carboxybiotin to pyruvate in a three-step reaction to form oxaloacetate.

These two reaction phases occur on different subsites of the same enzyme; the 14-Å arm of biocytin serves to transfer the biotin ring between the two sites.

d. Acetyl-CoA Regulates Pyruvate Carboxylase

Oxaloacetate synthesis is an anaplerotic (filling up) reaction that increases citric acid cycle activity (Section 21-4). Accumulation of the citric acid cycle substrate acetyl-CoA is therefore indicative of the need for more oxaloacetate. Indeed, acetyl-CoA is a powerful allosteric activator of pyruvate carboxylase; the enzyme is all but inactive without bound acetyl-CoA. *If, however, the citric acid cycle is inhibited (by ATP and NADH, whose presence in high con-* centrations indicates a satisfied demand for oxidative phosphorylation; Section 21-4), oxaloacetate instead undergoes gluconeogenesis.*

e. PEP Carboxykinase

PEPCK, a monomeric 608-residue enzyme, catalyzes the GTP-driven decarboxylation of oxaloacetate to form PEP and GDP (Fig. 23-5). Note that the CO_2 that carboxylates pyruvate to yield oxaloacetate is eliminated in the formation of PEP. Oxaloacetate may therefore be considered to be "activated" pyruvate, with CO_2 and biotin facilitating the activation at the expense of ATP hydrolysis. Acetyl-CoA is similarly activated for fatty acid biosynthesis through such a carboxylation–decarboxylation process (forming malonyl-CoA; Section 25-4B). In general, β-keto acids may be considered "high-energy" compounds because of the high free energy of decarboxylation of the β-carboxyl group. The enolates they generate are used to form carbon–carbon bonds in fatty acid biosynthesis or phosphoenolpyruvate here in gluconeogenesis.

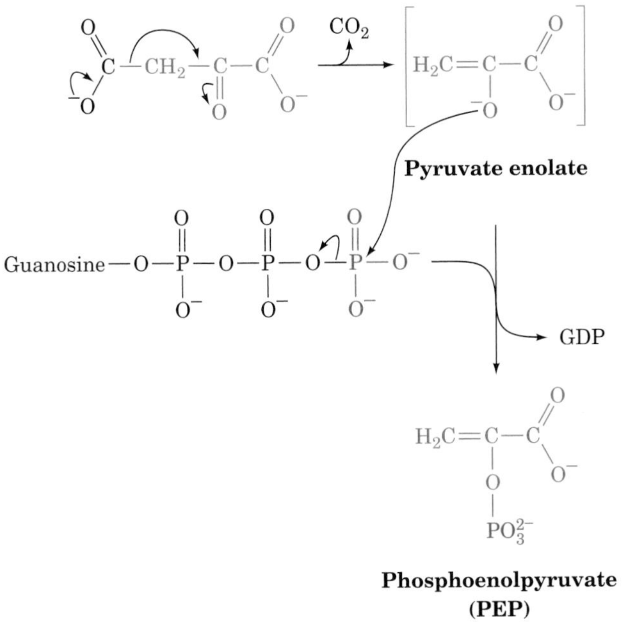

f. Gluconeogenesis Requires Metabolite Transport between Mitochondria and Cytosol

The generation of oxaloacetate from pyruvate or citric acid cycle intermediates occurs only in the mitochondrion, whereas the enzymes that convert PEP to glucose are cytosolic. The cellular location of PEPCK varies with the species. In mouse and rat liver it is located almost exclusively in the cytosol, in pigeon and rabbit liver it is mitochondrial, and in guinea pig and humans it is more or less equally distributed between both compartments. In order for gluconeogenesis to occur, either oxaloacetate must leave the mitochondrion for conversion to PEP or the PEP formed there must enter the cytosol.

PEP is transported across the mitochondrial membrane by specific membrane transport proteins. There is, however, no such transport system for oxaloacetate. It must first be converted either to aspartate (Fig. 23-6, Route 1) or to malate (Fig. 23-6, Route 2), for which mitochondrial transport systems exist (Section 22-1B). The difference between these two routes involves the transport of NADH

FIGURE 23-5   The PEPCK mechanism. Decarboxylation of oxaloacetate (a β-keto acid) forms a resonance-stabilized enolate anion whose oxygen atom attacks the γ phosphoryl group of GTP forming PEP and GDP.

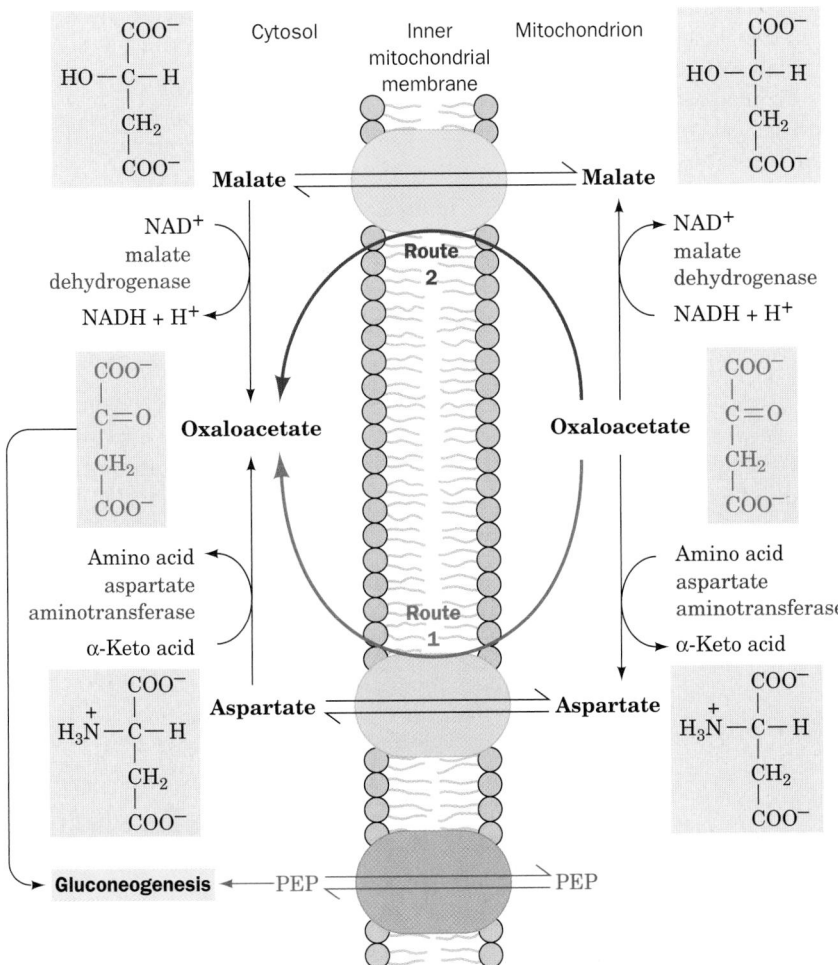

FIGURE 23-6   Transport of PEP and oxaloacetate from the mitochondrion to the cytosol. PEP is directly transported between these compartments. Oxaloacetate, however, must first be converted to either aspartate through the action of **aspartate aminotransferase** (Route 1) or to malate by malate dehydrogenase (Route 2). Route 2 involves the mitochondrial oxidation of NADH followed by the cytosolic reduction of NAD$^+$ and therefore also transfers NADH reducing equivalents from the mitochondrion to the cytosol. ✷ **See the Animated Figures**

reducing equivalents. The **malate dehydrogenase** route (Route 2) results in the transport of reducing equivalents from the mitochondrion to the cytosol, since it utilizes mitochondrial NADH and produces cytosolic NADH. The **aspartate aminotransferase** route (Route 1) does not involve NADH. Cytosolic NADH is required for gluconeogenesis so, under most conditions, the route through malate is a necessity. If the gluconeogenic precursor is lactate, however (Section 23-1C), its oxidation to pyruvate generates cytosolic NADH, so that either transport route may then be used. Of course, as we have seen, during oxidative metabolism the two routes may also alternate (with Route 2 reversed) to form the malate–aspartate shuttle, which transports NADH reducing equivalents into the mitochondrion (Section 22-1B).

In the liver, where the urea cycle occurs (Section 26-2), a third route, a modification of Route 1, may be followed for transporting oxaloacetate into the cytosol. The aspartate that enters the cytosol by Route 1 may be converted to fumarate as part of the urea cycle (Fig. 26-7), instead of being transaminated. Fumarate is then hydrated to malate and dehydrogenated to oxaloacetate by cytosolic equivalents of citric acid cycle enzymes. This third route generates cytosolic NADH in the same way as does Route 2.

g. Hydrolytic Reactions Bypass PFK and Hexokinase

The opposing pathways of gluconeogenesis and glycolysis utilize many of the same enzymes (Fig. 23-7). However, the free energy change is highly unfavorable in the gluconeogenic direction at two other points in the pathway in addition to the pyruvate kinase reaction: the PFK reaction and the hexokinase reaction. At these points, instead of generating ATP by reversing the glycolytic reactions, FBP and G6P are hydrolyzed, releasing P_i in exergonic processes catalyzed by **fructose-1,6-bisphosphatase (FBPase)** and **glucose-6-phosphatase,** respectively. *Glucose-6-phosphatase is unique to liver and kidney, permitting them to supply glucose to other tissues.*

Because of the presence of separate gluconeogenic enzymes at the three irreversible steps in the glycolytic conversion of glucose to pyruvate, both glycolysis and gluconeogenesis are rendered thermodynamically favorable. This is accomplished at the expense of the free energy of hydrolysis of two molecules each of ATP and GTP per molecule of glucose synthesized by gluconeogenesis in addition to that which would be consumed by the direct reversal of glycolysis.

Glycolysis:

Glucose + 2NAD$^+$ + 2ADP + 2P$_i$ →
  2pyruvate + 2NADH + 4H$^+$ + 2ATP + 2H$_2$O

Gluconeogenesis:

2Pyruvate + 2NADH + 4H$^+$ + **4ATP** + **2GTP** + 6H$_2$O
  → glucose + 2NAD$^+$ + 4ADP + 2GDP + 6P$_i$

Overall:

2ATP + 2GTP + 4H$_2$O → 2ADP + 2GDP + 4P$_i$

Such free energy losses in a cyclic process are thermodynamically inescapable. They are the price that must be paid to maintain independent regulation of the two pathways.

B. *Regulation of Gluconeogenesis*

If both glycolysis and gluconeogenesis were to proceed in an uncontrolled manner, the net effect would be a futile cycle wastefully hydrolyzing ATP and GTP. This does not occur. Rather, *these pathways are reciprocally regulated so*

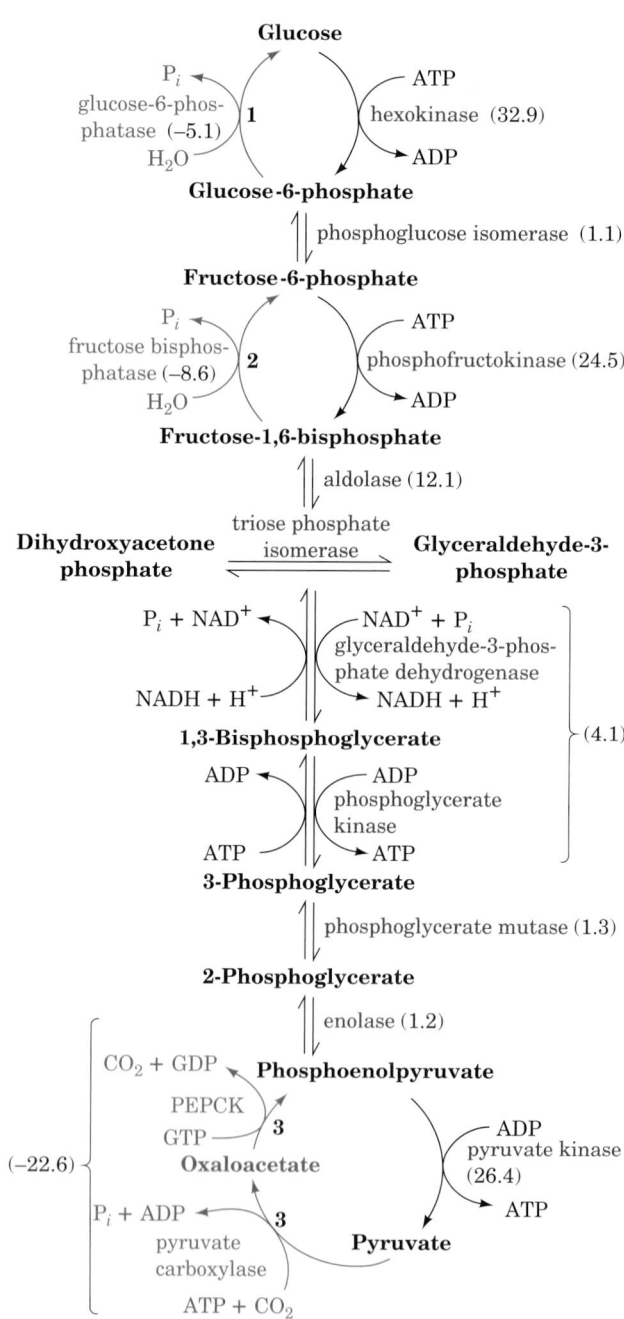

FIGURE 23-7 Pathways of gluconeogenesis and glycolysis. The three numbered steps, which are catalyzed by different enzymes in gluconeogenesis, have red arrows. The ΔG's for the reactions in the direction of gluconeogenesis under physiological conditions in liver are given in parentheses in kJ · mol^{-1}. [ΔG's obtained from Newsholme, E.A. and Leech, A.R., *Biochemistry for the Medical Sciences*, p. 448, Wiley (1983).] 🐭 See the Animated Figures

TABLE 23-1   Regulators of Gluconeogenic Enzyme Activity

Enzyme	Allosteric Inhibitors	Allosteric Activators	Enzyme Phosphorylation	Protein Synthesis
PFK	ATP, citrate	AMP, F2,6P		
FBPase	AMP, F2,6P			
Pyruvate kinase	Alanine	F1,6P	Inactivates	
Pyruvate carboxylase		Acetyl-CoA		
PEPCK				Stimulated by glucagon, thyroid hormone, and glucocorticoids, and inhibited by insulin
PFK-2	Citrate	AMP, F6P, P_i	Inactivates	
FBPase-2	F6P	Glycerol-3-P	Activates	

as to meet the needs of the organism. In the fed state, when the blood glucose level is high, the liver is geared toward fuel conservation: Glycogen is synthesized and the glycolytic pathway and pyruvate dehydrogenase are activated, breaking down glucose to acetyl-CoA for fatty acid biosynthesis and fat storage. In the fasted state, however, the liver maintains the blood glucose level both by glycogen breakdown and by reversing the flux through glycolysis toward gluconeogenesis [using mainly protein degradation products via the **glucose–alanine cycle** (Section 26-1A) and glycerol from triacylglycerol hydrolysis (Section 25-1)].

a.   Glycolysis and Gluconeogenesis Are Controlled by Allosteric Interactions and Covalent Modifications

The rate and direction of glycolysis and gluconeogenesis are controlled at the points in these pathways where the forward and reverse directions can be independently regulated: the reactions catalyzed by (1) hexokinase/glucose-6-phosphatase, (2) PFK/FBPase, and (3) pyruvate kinase/pyruvate carboxylase–PEPCK (Fig. 23-7). Table 23-1 lists these regulatory enzymes and their regulators. The dominant mechanisms are allosteric interactions and cAMP-dependent covalent modifications (phosphorylation/dephosphorylation; Section 18-3). cAMP-dependent covalent modification renders this system sensitive to control by glucagon and other hormones that alter cAMP levels.

One of the most important allosteric effectors involved in the regulation of glycolysis and gluconeogenesis is fructose-2,6-bisphosphate (F2, 6P), which activates PFK and inhibits FBPase (Section 18-3F). The concentration of F2,6P is controlled by its rates of synthesis and breakdown by phosphofructokinase-2 (PFK-2) and fructose bisphosphatase-2 (FBPase-2), respectively. Control of the activities of PFK-2 and FBPase-2 is therefore an important aspect of gluconeogenic regulation even though these enzymes do not catalyze reactions of the pathway. PFK-2 and FBPase-2 activities, which occur on separate domains of the same bifunctional enzyme, are subject to allosteric regulation as well as control by covalent modifications (Table 23-1). Low levels of blood glucose result in hormonal activation of gluconeogenesis through regulation of [F2,6P] (Fig. 23-8).

Activation of gluconeogenesis in liver also involves inhibition of glycolysis at the level of pyruvate kinase. *Liver pyruvate kinase is inhibited both allosterically by alanine (a pyruvate precursor; Section 26-1A) and by phosphoryla-*

tion. Glycogen breakdown, in contrast, is stimulated by phosphorylation (Section 18-3C). Both pathways then flow toward G6P, which is converted to glucose for export to muscle and brain. Muscle pyruvate kinase, an isozyme of the liver enzyme, is not subject to these controls. Indeed, such controls would be counterproductive in muscle since this tissue lacks glucose-6-phosphatase and thus the ability to synthesize glucose via gluconeogenesis.

b.   PEPCK Concentration Is Transcriptionally Controlled

PEPCK is the enzyme that catalyzes the first committed reaction of gluconeogenesis. It is therefore of interest (Table 23-1) that PEPCK's activity is controlled solely through the transcriptional regulation of the gene encoding it (transcriptional regulation is outlined in Section 5-4A and discussed in detail in Sections 31-3 and 34-3). In particular, the transcription of the PEPCK gene is stimulated by glucagon, glucocorticoids, and thyroid hormones, and is inhibited by insulin. For instance, the cAMP that is produced in response to stimulation of the liver by glucagon, in addition to its initiation of phosphorylation cascades (Section 18-3), induces the transcription of the PEPCK gene. Richard Hanson has

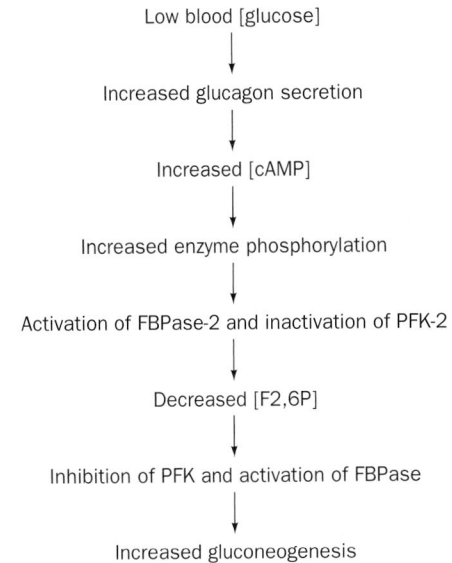

FIGURE 23-8   Hormonal regulation of [F2,6P]. This process activates gluconeogenesis in liver in response to low blood [glucose].

shown that this occurs because the PEPCK gene **promoter** (a control region that precedes the transcriptional initiation site of genes encoding proteins; Section 5-4A) contains a specific DNA sequence called the **cAMP response element (CRE)** that is bound by a **transcription factor** named **CRE binding protein (CREB),** but only when CREB is also binding cAMP (recall that a transcription factor is a protein that binds to a specific segment of its target promoter and, in doing so, activates RNA polymerase to initiate the transcription of the associated gene; Section 5-4A). However, the PEPCK gene promoter contains numerous other binding sites for specific transcription factors. Among them are the **thyroid hormone response element (TSE),** which is bound by thyroid hormone receptor in complex with thyroid hormone (Section 19-1D), and the **glucocorticoid hormone response element (GRE),** which is bound by the **glucocorticoid receptor** in complex with a glucocorticoid hormone (Sections 19-1G and 34-3B). In contrast, PEPCK gene transcription is strongly repressed by protein factors phosphorylated by the PI3K signaling cascade initiated by the binding of insulin to the insulin receptor (these protein factors may repress transcription by interfering with the binding of the above transcription factors; the mechanism of insulin signaling is discussed in Sections 19-3A, 19-3B, 19-4D, and 19-4F). The rate of PEPCK mRNA production is determined by the integration of these various interactions and hence of the signals that caused them.

C. *The Cori Cycle*

Muscle contraction is powered by hydrolysis of ATP, which is then regenerated through oxidative phosphorylation in the mitochondria of slow-twitch (red) muscle fibers and by glycolysis yielding lactate in fast-twitch (white) muscle fibers. Slow-twitch fibers also produce lactate when ATP demand exceeds oxidative flux. The lactate is transferred, via the bloodstream, to the liver, where it is reconverted to pyruvate by lactate dehydrogenase and then to glucose by gluconeogenesis. Thus, through the intermediacy of the bloodstream, liver and muscle participate in a metabolic cycle known as the **Cori cycle** (Fig. 23-9) in honor of Carl and

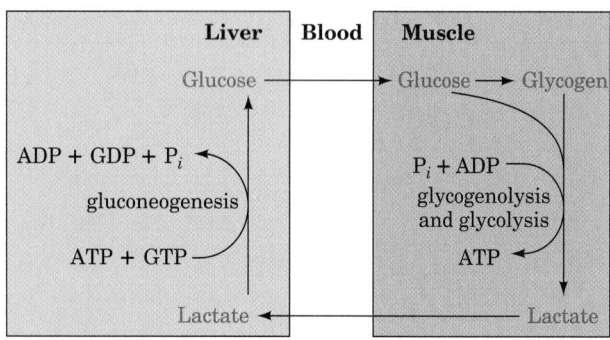

FIGURE 23-9 The Cori cycle. Lactate produced by muscle glycolysis is transported by the bloodstream to the liver, where it is converted to glucose by gluconeogenesis. The bloodstream carries the glucose back to the muscles, where it may be stored as glycogen. ✣ See the Animated Figures

Gerty Cori, who first described it. This is the same ATP-consuming glycolysis/gluconeogenesis "futile cycle" we discussed above. Here, however, instead of occurring in the same cell, the two pathways occur in different organs. Liver ATP is used to resynthesize glucose from lactate produced in muscle. The resynthesized glucose is returned to the muscle, where it is stored as glycogen and used, on demand, to generate ATP for muscle contraction. The ATP utilized by the liver for this process is regenerated by oxidative phosphorylation. After vigorous exertion, it often takes at least 30 min for all of the lactate so produced to be converted to glycogen and the oxygen consumption rate to return to its resting level, a phenomenon known as **oxygen debt.**

2 ■ THE GLYOXYLATE CYCLE

Plants, but not animals, possess enzymes that mediate the net conversion of acetyl-CoA to succinate, which is then converted, via malate, to oxaloacetate. This is accomplished via the **glyoxylate cycle** (Fig. 23-10), a pathway involving enzymes of the **glyoxysome** (a membranous plant organelle; Section 1-2A). The glyoxylate cycle involves five enzymes, three of which also participate in the citric acid cycle: citrate synthase, aconitase, and malate dehydrogenase. The two other enzymes, isocitrate lyase and malate synthase, are unique to the cycle.

The glyoxalate cycle consists of five reactions (Fig. 23-10):

Reactions 1 and 2. Glyoxysomal oxaloacetate is condensed with acetyl-CoA to form citrate, which is isomerized to isocitrate as in the citric acid cycle. Since the glyoxysome contains no aconitase, Reaction 2 presumably takes place in the cytosol.

Reaction 3. Glyoxysomal **isocitrate lyase** cleaves the isocitrate to succinate and **glyoxylate** (hence the cycle's name).

Reaction 4. Malate synthase, a glyoxysomal enzyme, condenses glyoxylate with a second molecule of acetyl-CoA to form malate.

Reaction 5. Glyoxysomal malate dehydrogenase catalyzes the oxidation of malate to oxaloacetate by NAD^+, thereby completing the cycle.

The glyoxylate cycle therefore results in the net conversion of two acetyl-CoA to succinate instead of to four molecules of CO_2, as would occur in the citric acid cycle. The succinate produced in Reaction 3 is transported to the mitochondrion, where it enters the citric acid cycle and is converted to malate, which has two alternative fates: (1) It can be converted to oxaloacetate in the mitochondrion, continuing the citric acid cycle and thereby making the glyoxylate pathway an anaplerotic process (Section 21-5); or (2) it can be transported to the cytosol, where it is converted to oxaloacetate for entry into gluconeogenesis.

The overall reaction of the glyoxylate cycle can be considered to be the formation of oxaloacetate from two molecules of acetyl-CoA.

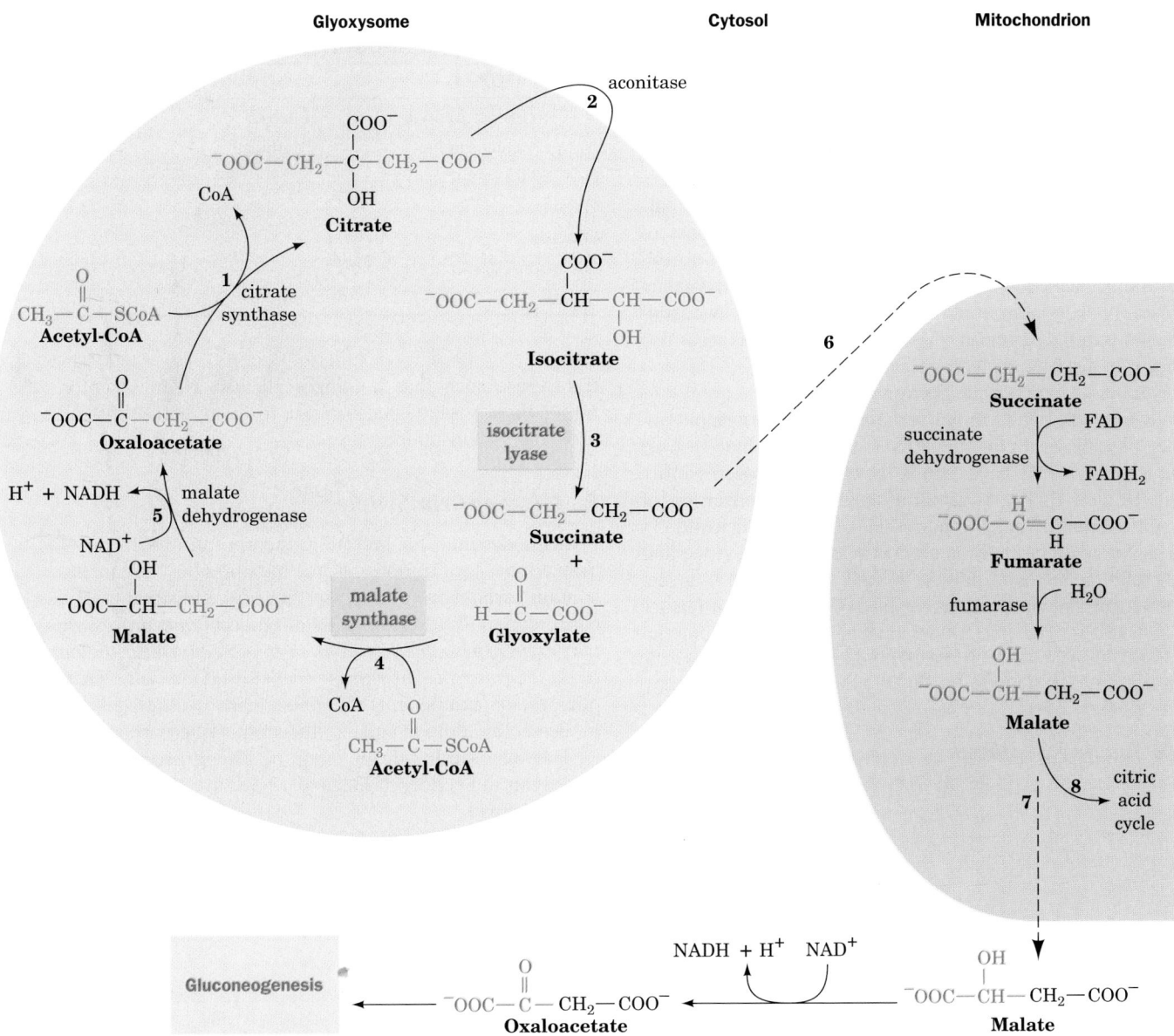

FIGURE 23-10 The glyoxylate cycle. The cycle results in the net conversion of two acetyl-CoA to succinate in the glyoxysome, which can be converted to malate in the mitochondrion for use in gluconeogenesis. Isocitrate lyase and malate synthase, enzymes unique to glyoxysomes (which occur only in plants), are boxed in blue. (1) Glyoxysomal citrate synthase catalyzes the condensation of oxaloacetate with acetyl-CoA to form citrate. (2) Cytosolic aconitase catalyzes the conversion of citrate to isocitrate. (3) Isocitrate lyase catalyzes the cleavage of isocitrate to succinate and glyoxylate. (4) Malate synthase catalyzes the condensation of glyoxylate with acetyl-CoA to form malate. (5) Glyoxysomal malate dehydrogenase catalyzes the oxidation of malate to oxaloacetate, completing the cycle. (6) Succinate is transported to the mitochondrion, where it is converted to malate via the citric acid cycle. (7) Malate is transported to the cytosol, where malate dehydrogenase catalyzes its oxidation to oxaloacetate, which can then be used in gluconeogenesis. (8) Alternatively, malate can continue in the citric acid cycle, making the glyoxylate cycle anaplerotic.

$$2\text{Acetyl-CoA} + 2\text{NAD}^+ + \text{FAD} \rightarrow$$
$$\text{oxaloacetate} + 2\text{CoA} + 2\text{NADH} + \text{FADH}_2 + 2\text{H}^+$$

Isocitrate lyase and malate synthase, the only enzymes of the glyoxylate pathway unique to plants, enable germinating seeds to convert their stored triacylglycerols, through acetyl-CoA, to glucose. It had long been assumed that this was a requirement of germination. However, a mutant of *Arabidopsis thaliana* (an oilseed plant) lacking isocitrate lyase, and hence unable to convert lipids to carbohydrate, nevertheless germinated. This process was only inhibited when the mutant plants were subjected to low light conditions. Therefore, it now appears that the glyoxylate cycle's importance in seedling growth is its anaplerotic function in providing 4-carbon units to the citric acid cycle, which can then oxidize the triacylglycerol-derived acetyl-CoA.

3 ■ BIOSYNTHESIS OF OLIGOSACCHARIDES AND GLYCOPROTEINS

Oligosaccharides consist of monosaccharide units joined together by glycosidic bonds (linkages between C1, the anomeric carbon, of one unit and an OH group of a second unit; Section 11-1C). About 80 different kinds of naturally occurring glycosidic linkages are known, most of which involve mannose, N-acetylglucosamine, N-acetylmuramic acid, glucose, galactose, fucose (6-deoxygalactose), N-acetylneuraminic acid (sialic acid), and N-acetylgalactosamine (Section 11-1C). Glycosidic linkages also occur to lipids (e.g. glycosphingolipids; Section 12-1D) and proteins (glycoproteins; Section 11-3C).

Glycosidic bond formation requires free energy input under physiological conditions ($\Delta G^{\circ\prime} = 16$ kJ · mol^{-1}). This free energy, as we have seen in the case of glycogen synthesis (Section 18-2B), is acquired through the conversion of monosaccharide units to nucleotide sugars. A nucleotide at a sugar's anomeric carbon atom is a good leaving group and thereby facilitates formation of a glycosidic bond to a second sugar unit via reactions catalyzed by **glycosyltransferases** (Fig. 23-11). The nucleotides that participate in monosaccharide transfers are UDP, GDP, and CMP; a given sugar is associated with only one of these nucleotides (Table 23-2).

A. *Lactose Synthesis*

Several disaccharides are synthesized for future use as metabolic fuels. In plants, the major fuel disaccharide is sucrose (Section 11-2B), whose synthesis is discussed in Section 24-3A. Typical of mammalian disaccharides is lactose [β-galactosyl-(1→4)-glucose; milk sugar], which is synthesized in the mammary gland by **lactose synthase** (Fig. 23-12). The donor sugar is UDP–galactose, which is formed by epimerization of UDP–glucose (Section 17-5B). The acceptor sugar is glucose.

Lactose synthase consists of two subunits:

1. Galactosyltransferase, the catalytic subunit, which occurs in many tissues, where it catalyzes the reaction of UDP–galactose and N-acetylglucosamine to yield N-acetyllactosamine, a constituent of many complex oligosaccharides (see, e.g., Fig. 23-19, Reaction 6).

2. α-Lactalbumin, a mammary gland protein with no catalytic activity, which alters the specificity of galactosyl-

TABLE 23-2   **Sugar Nucleotides and Their Corresponding Monosaccharides in Glycosyltransferase Reactions**

UDP	GDP	CMP
N-Acetylgalactosamine	Fucose	Sialic acid
N-Acetylglucosamine	Mannose	
N-Acetylmuramic acid		
Galactose		
Glucose		
Glucuronic acid		
Xylose		

transferase such that it utilizes glucose as an acceptor, rather than N-acetylglucosamine, to form lactose instead of N-acetyllactosamine.

B. *Glycoprotein Synthesis*

Proteins destined for secretion, incorporation into membranes, or localization inside membranous organelles contain carbohydrates and are therefore classified as glycoproteins. *Glycosylation and oligosaccharide processing play an indispensable role in the sorting and the distribution of these proteins to their proper cellular destinations.* Their polypeptide components are ribosomally synthesized and processed by addition and modification of oligosaccharides.

The oligosaccharide portions of glycoproteins, as we have seen in Sections 11-3C and 12-3B, are classified into three groups:

1. N-Linked oligosaccharides, which are attached to their polypeptide chain by a β-N-glycosidic bond to the side chain N of an Asn residue in the sequence Asn-X-Ser or Asn-X-Thr, where X is any amino acid residue except Pro or perhaps Asp (Fig. 23-13a).

2. O-Linked oligosaccharides, which are attached to their polypeptide chain through an α-O-glycosidic bond to the side chain O of a Ser or Thr residue (Fig. 23-13b) or, only in collagens (Section 8-2B), to that of a 5-hydroxylysine (Hyl) residue (Fig. 23-13c).

3. Glycosylphosphatidylinositol (GPI) membrane anchors, which are attached to their polypeptide chain through an amide bond between mannose-6-phosphoethanolamine and the C-terminal carboxyl group (Fig. 23-13d).

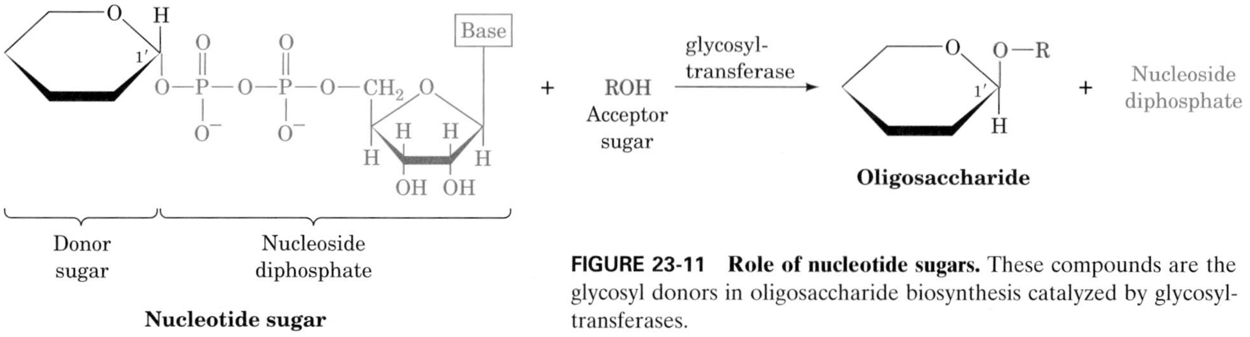

FIGURE 23-11   **Role of nucleotide sugars.** These compounds are the glycosyl donors in oligosaccharide biosynthesis catalyzed by glycosyltransferases.

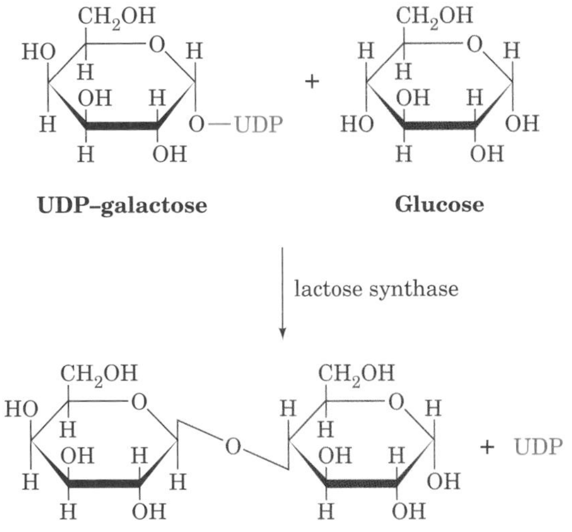

UDP–galactose   **Glucose**

lactose synthase

Lactose
[β-galactosyl-(1 ⟶ 4)-glucose]

FIGURE 23-12   Lactose Synthase. This enzyme catalyzes the formation of lactose from UDP–galactose and glucose.

We shall consider the synthesis of these three types of oligosaccharides in turn.

a. N-Linked Glycoproteins Are Synthesized in Four Stages

N-Linked glycoproteins are formed in the endoplasmic reticulum and further processed in the Golgi apparatus. Synthesis of their carbohydrate moieties occurs in four stages:

1. Synthesis of a lipid-linked oligosaccharide precursor.

2. Transfer of this precursor to the side chain N of an Asn residue on a growing polypeptide.

3. Removal of some of the precursor's sugar units.

4. Addition of sugar residues to the remaining core oligosaccharide.

We shall discuss these stages in order.

b. N-Linked Oligosaccharides Are Constructed on Dolichol Carriers

N-Linked oligosaccharides are initially synthesized as lipid-linked precursors. The lipid component in this process is **dolichol,** a long-chain polyisoprenol of 14 to 24 isoprene units (17–21 units in animals and 14–24 units in fungi and plants; isoprene units are C_5 units with the carbon skeleton

(a)

Asn

(b)

Ser (Thr)

(c)

5-Hydroxylysine

(d)

Phosphoethanolamine   C-terminal residue

Mannose

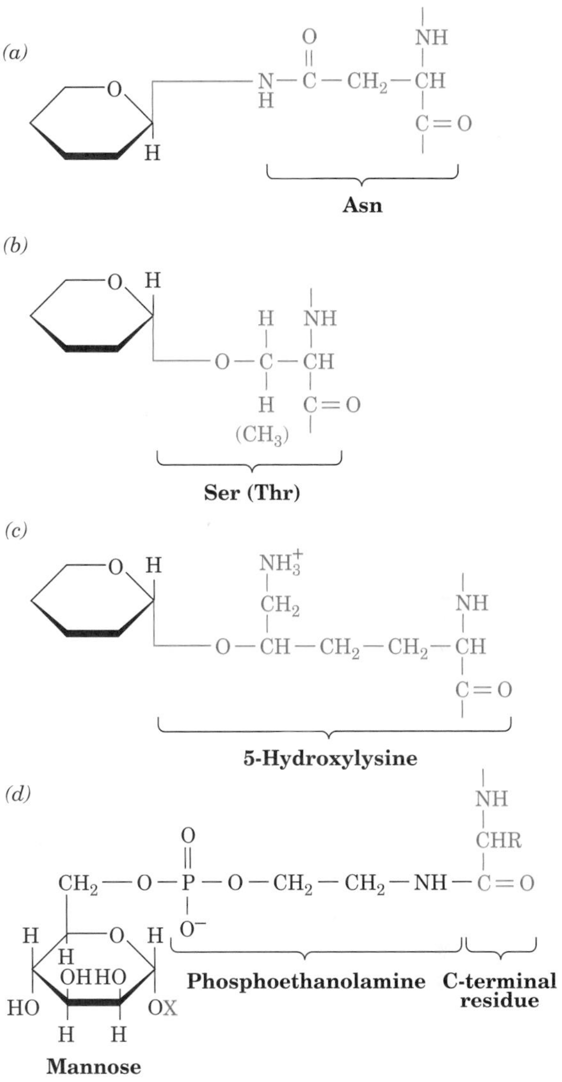

FIGURE 23-13   Types of saccharide–polypeptide linkages in glycoproteins. (*a*) An *N*-linked glycosidic bond to an Asn residue in the sequence Asn-X-Ser/Thr. (*b*) An *O*-linked glycosidic bond to a Ser (or Thr) residue. (*c*) An *O*-linked glycosidic bond to a 5-hydroxylysine residue in collagen. (*d*) An amide bond between the C-terminal amino acid of a protein and the phosphoethanolamine bridge to the 6 position of mannose in the glycophosphoinositol (GPI) anchor. The X group (*green*) denotes the rest of the GPI anchor (Fig. 12-30).

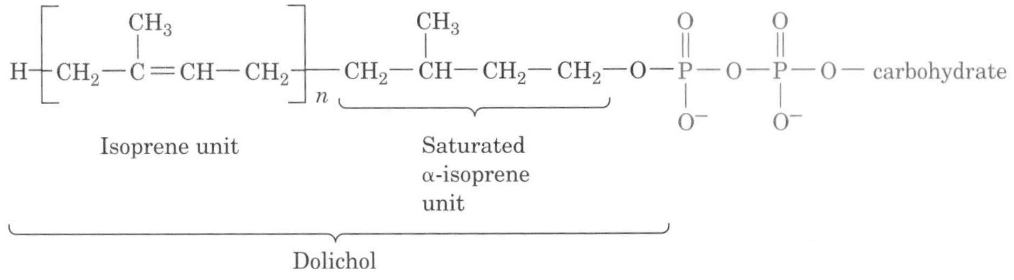

of isoprene; Section 25-6A), which is linked to the oligosaccharide precursor via a pyrophosphate bridge (Fig. 23-14). Dolichol apparently anchors the growing oligosaccharide to the endoplasmic reticulum membrane. Involvement of lipid-

Isoprene unit

Saturated α-isoprene unit

Dolichol

FIGURE 23-14   Dolichol pyrophosphate glycoside. The carbohydrate precursors of *N*-linked glycosides are synthesized as dolichol pyrophosphate glycosides. Dolichols are long-chain polyisoprenols ($n = 14–24$) in which the α-isoprene unit is saturated.

linked oligosaccharides in *N*-linked glycoprotein synthesis was first demonstrated in 1972 by Armando Parodi and Luis Leloir, who showed that, when a lipid-linked oligosaccharide containing [^{14}C]glucose is incubated with rat liver **microsomes** (vesicular fragments of isolated endoplasmic reticulum), the radioactivity becomes associated with protein.

c. *N*-Linked Glycoproteins Have a Common Oligosaccharide Core

The pathway of dolichol-PP-oligosaccharide synthesis involves stepwise addition of monosaccharide units to the growing glycolipid by specific glycosyltransferases to form a common "core" structure. Each monosaccharide unit is added by a unique glycosyltransferase (Fig. 23-15). For

example, in Reaction 2 of Fig. 23-15, five mannosyl units are added through the action of five different mannosyltransferases, each with a different oligosaccharide-acceptor specificity. The oligosaccharide core, the product of Reaction 9 in Fig. 23-15, has the composition (*N*-acetylglucosamine)$_2$(mannose)$_9$(glucose)$_3$.

Although nucleotide sugars are the most common monosaccharide donors in glycosyltransferase reactions, *several mannosyl and glucosyl residues are transferred to the growing dolichol-PP-oligosaccharide from their corresponding dolichol-P derivatives.* This requirement for **dolichol-P-mannose** was discovered by Stuart Kornfeld, who found that mutant mouse lymphoma cells (lymphoma is a type of cancer) that are unable to synthesize the nor-

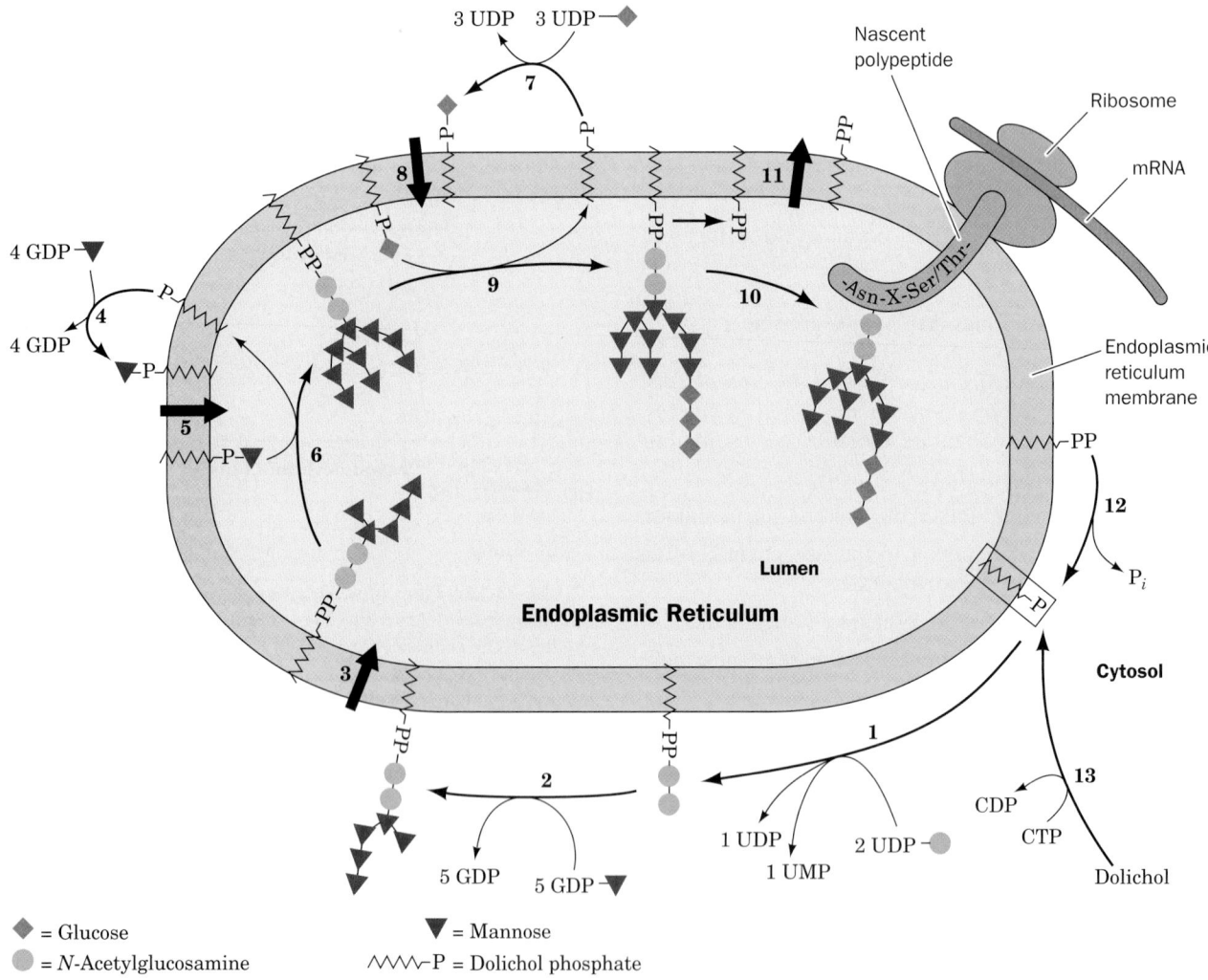

♦ = Glucose ▼ = Mannose
● = *N*-Acetylglucosamine ∿∿∿–P = Dolichol phosphate

FIGURE 23-15 Pathway of dolichol-PP-oligosaccharide synthesis.
(1) Addition of *N*-acetylglucosamine-1-P and a second *N*-acetylglucosamine to dolichol-P. **(2)** Addition of five mannosyl residues from GDP–mannose in reactions catalyzed by five different mannosyltransferases. **(3)** Membrane translocation of dolichol-PP-(*N*-acetylglucosamine)$_2$(mannose)$_5$ to the lumen of the endoplasmic reticulum (ER). **(4)** Cytosolic synthesis of dolichol-P-mannose from GDP–mannose and dolichol-P. **(5)** Membrane translocation of dolichol-P-mannose to the lumen of the ER. **(6)** Addition of four mannosyl residues from dolichol-P-mannose in reactions catalyzed by four different mannosyltransferases. **(7)** Cytosolic

synthesis of dolichol-P-glucose from UDP–glucose and dolichol-P. **(8)** Membrane translocation of dolichol-P-glucose to the lumen of the ER. **(9)** Addition of three glucosyl residues from dolichol-P-glucose. **(10)** Transfer of the oligosaccharide from dolichol-PP to the polypeptide chain at an Asn residue in the sequence Asn-X-Ser/Thr, releasing dolichol-PP. **(11)** Translocation of dolichol-PP to the cytoplasmic surface of the ER membrane. **(12)** Hydrolysis of dolichol-PP to dolichol-P. **(13)** Dolichol-P can also be formed by phosphorylation of dolichol by CTP. [Modified from Abeijon, C. and Hirschberg, C.B., *Trends Biochem. Sci.* **17**, 34 (1992).] ✷ **See the Animated Figures**

mal lipid-linked oligosaccharides formed a defective, smaller glycolipid. These cells contain all the requisite glycosyltransferases but are unable to synthesize dolichol-P-mannose (Reaction 4 in Fig. 23-15 is blocked). When this substance is supplied to the mutant cells, mannosyl units are added to the defective dolichol-PP-oligosaccharide.

d. Dolichol-PP-Oligosaccharide Synthesis Involves Topological Changes of the Intermediates

Reactions 1, 2, 4, and 7 of Fig. 23-15 all occur on the cytoplasmic side of the endoplasmic reticulum (ER) membrane. This was determined by using "right-side-out" rough ER vesicles and showing that various membrane-impermeant reagents can disrupt one or another of these reactions. Reactions 6, 9, and 10 occur in the lumen of the ER as judged by the inability of concanavalin A, a **lectin** (carbohydrate-binding protein), to bind to the products of these reactions until the membrane is permeabilized. The (mannose)$_5$(N-acetylglucosamine)$_2$-PP-dolichol product of Reaction 2, the dolichol-P-mannose product of Reaction 4, and the dolichol-P-glucose product of Reaction 7 must therefore be translocated across the ER membrane (Reactions 3, 5, and 8) such that they extend from its luminal surface in order for the synthesis of N-linked oligosaccharides to continue. The mechanisms of these various translocation processes are unknown.

e. N-Linked Oligosaccharides Are Cotranslationally Added to Proteins

Vesicular stomatitis virus (VSV), which infects cattle, producing influenza-like symptoms, provides an excellent model system for studying N-linked glycoprotein processing. The VSV coat consists of host-cell membrane in which a single viral glycoprotein, the **VSV G-protein** (not to be confused with the GTPases involved in signal transduction; Chapter 19), is embedded. Since a viral infection almost totally usurps an infected cell's protein synthesizing machinery, a VSV-infected cell's Golgi apparatus, which normally contains hundreds of different types of glycoproteins, contains virtually no other glycoprotein but G-protein. Consequently, the maturation of the G-protein is relatively easy to follow.

70 to 90% of the Asn-X-Ser/Thr sites in mature eukaryotic proteins, are N-glycosylated. Studies of VSV-infected cells indicate that the *transfer of the lipid-linked oligosaccharide to a polypeptide chain occurs while the polypeptide chain is still being synthesized.* Structural predictions (Section 9-3A), together with glycosylation studies of model polypeptides, suggest that the amino acid sequences flanking known N-glycosylation sites occur at β turns or loops in which Asn's backbone N—H group is hydrogen bonded to the Ser/Thr hydroxyl O atom (Fig. 23-16a). This explains why Pro cannot occupy the X posi-

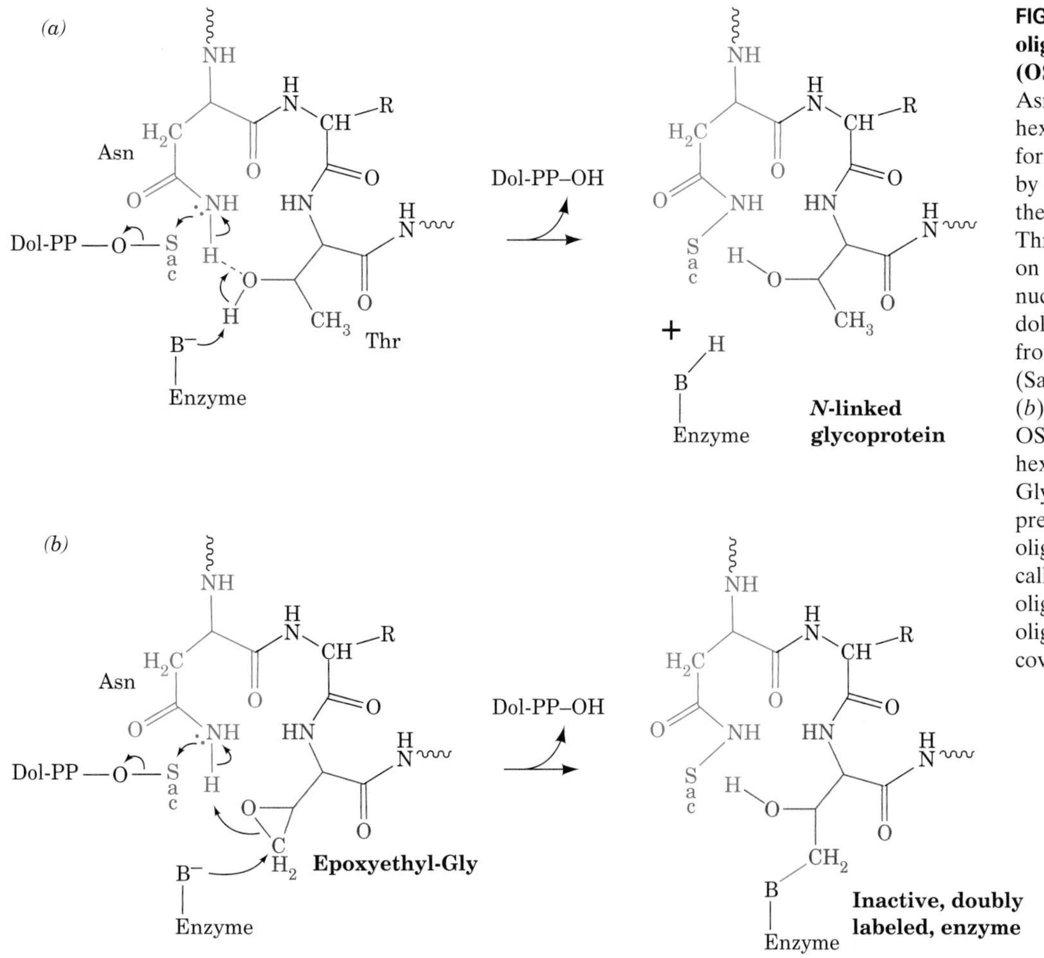

FIGURE 23-16 The oligosaccharyltransferase (OST) reaction. (*a*) The Asn-X-Thr component of a hexapeptide model substrate forms a ring that is closed by a hydrogen bond from the Asn amide group to the Thr hydroxyl group. A base on the enzyme facilitates the nucleophilic displacement of dolichol pyrophosphate from the oligosaccharide (Sac) by the amide nitrogen. (*b*) The inactivation of the OST by reacting it with a hexapeptide containing Asn-Gly-epoxyethylGly in the presence of dolichol-PP-oligosaccharide. This chemically labels the base with the oligopeptide to which the oligosaccharide has become covalently linked.

tion; it would prevent Asn-X-Ser/Thr from assuming the putative required hydrogen bonded conformation.

VSV G-protein is *N*-glycosylated by **oligosaccharyl-transferase (OST),** a membrane-bound, ~300-kD, multi-subunit enzyme that recognizes the amino acid sequence Asn-X-Ser/Thr (Fig. 23-15, Reaction 10). Ernst Bause has proposed a catalytic mechanism for OST in which an enzyme base abstracts a proton from the Ser/Thr hydroxyl group, which in turn abstracts a proton from the Asn NH_2 group, thereby promoting its nucleophilic attack on the oligosaccharide (Sac), which then displaces the dolichol pyrophosphate (Fig. 23-16*a*). This mechanism is supported by the observation that reacting the OST with dolichol-PP-oligosaccharide and a hexapeptide model substrate containing the sequence Asn-X-epoxyethylGly (rather than Asn-X-Ser/Thr) irreversibly inactivates the enzyme by covalently linking it to the now glycosylated hexapeptide (Fig. 23-16*b*).

f. The Calnexin/Calreticulin Cycle Facilitates Glycoprotein Folding

The processing of a glycoprotein-linked core oligosaccharide begins in the endoplasmic reticulum by the enzymatic trimming (removal) of its three glucose residues (Fig. 23-17, Reactions 2 and 3) and one of its mannose residues (Fig. 23-17, Reaction 4) before the protein has folded to its native conformation. This is not a straightforward

process, however, because **UDP–glucose:glycoprotein glucosyltransferase (GT),** a 1513-residue soluble protein, reglucosylates the oligosaccharides of partially folded glycoproteins, a reaction that reverses the removal of the last of the three glucose residues by **glucosidase II** (Fig. 23-17, Reaction 3). This futile cycle (most glycoproteins undergo reglucosylation at least once) is part of a chaperone-mediated glycoprotein folding process called the **calnexin/calreticulin cycle. Calnexin (CNX; ~570 residues),** which is membrane bound, and **calreticulin (CRT; ~400** residues), its soluble homolog, are ER-resident lectins that bind partially folded glycoproteins bearing a monoglucosylated oligosaccharide in a way that protects the glycoprotein from degradation and premature transfer to the Golgi apparatus. If the glycoprotein is released and deglucosylated before it has correctly folded, GT, which recognizes only non-native glycoproteins, reglucosylates it so that the CNX/CRT cycle can repeat. CNX and CRT both also bind **ERp57,** a 481-residue thiol oxidoreductase homologous to protein disulfide isomerase (PDI; Section 9-2A). While the partially folded glycoprotein is bound to the complex, ERp57 catalyzes disulfide interchange reactions to facilitate the formation of the correctly paired disulfide bonds. The CNX/ERp57 and CRT/ERp57 complexes are therefore responsible for the correct folding and disulfide bond formation of the glycoproteins in the ER. The importance of this process is demonstrated by the ob-

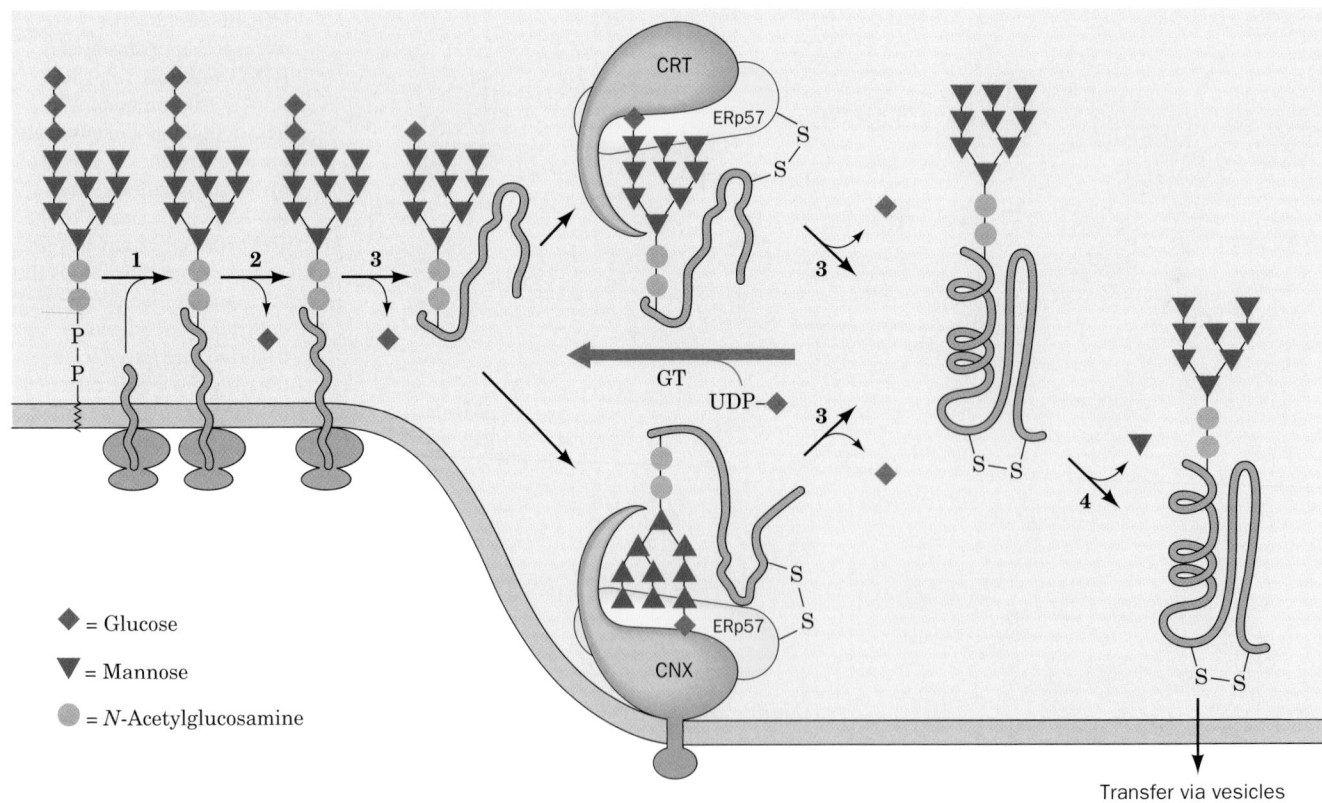

FIGURE 23-17   The calnexin/calreticulin cycle for glycoprotein folding in the endoplasmic reticulum. The reactions are catalyzed by: **(1)** oligosaccharyltransferase (OST); **(2)** α-glucosidase I; **(3)** α-glucosidase II, UDP–glucose:glycoprotein glucosyltransferase (GT), calreticulin (CRT), calnexin (CNX), and the thiol oxidoreductase ERp57; and **(4)** ER α-1,2-mannosidase. [After Helenius, A. and Aebi, M., *Science* **291,** 2367 (2001).]

servation that knockout mice lacking the gene for CRT die *in utero*.

The X-ray structure of the luminal domain of calnexin (residues 61–458), determined by Miroslaw Cygler, reveals a most unusual structure (Fig. 23-18): a compact globular domain (residues 61–262 and 415–458) from which extends a 145-Å-long arm (residues 270–414). The globular domain forms a sandwich of a 6-stranded and a 7-stranded antiparallel β sheet that binds a Ca^{2+} ion and which resembles legume lectins such as concanavalin A (Fig. 8-40). This domain binds glucose on its concave (*blue*) surface, which is lined by hydrogen bonding groups that model building suggests binds the (glucose)$_1$(mannose)$_3$ portion of calnexin's natural (glucose)$_1$(mannose)$_9$ substrate. The long arm, which consists of an extended hairpin, is known as the P domain because it has four copies each of two different Pro-rich motifs arranged in the sequence 11112222, with each ~18-residue motif 1 in antiparallel association with an ~14-residue motif 2 on the opposite strand of the hairpin. Each of these motif pairs has a similar structure, with its conserved residues maintaining identical interactions in each pair. The P domain has been shown to form the binding site for ERp57 in both calnexin and calreticulin.

g. Glycoprotein Processing is Completed in the Golgi Apparatus

Once a glycoprotein has folded to its native conformation and ER α-1,2-mannosidase has removed one of its mannosyl residues (Fig. 23-17, Step 4), the glycoprotein is transported, in membranous vesicles, to the Golgi apparatus, where it is further processed (Fig. 23-19). The Golgi apparatus (Fig. 12-51), as we discussed in Section 12-4C, consists of, from opposite the ER outward, the cis Golgi network, through which glycoproteins enter the Golgi apparatus; a stack of at least three different types of sacs, the cis, medial, and trans cisternae; and the trans Golgi net-

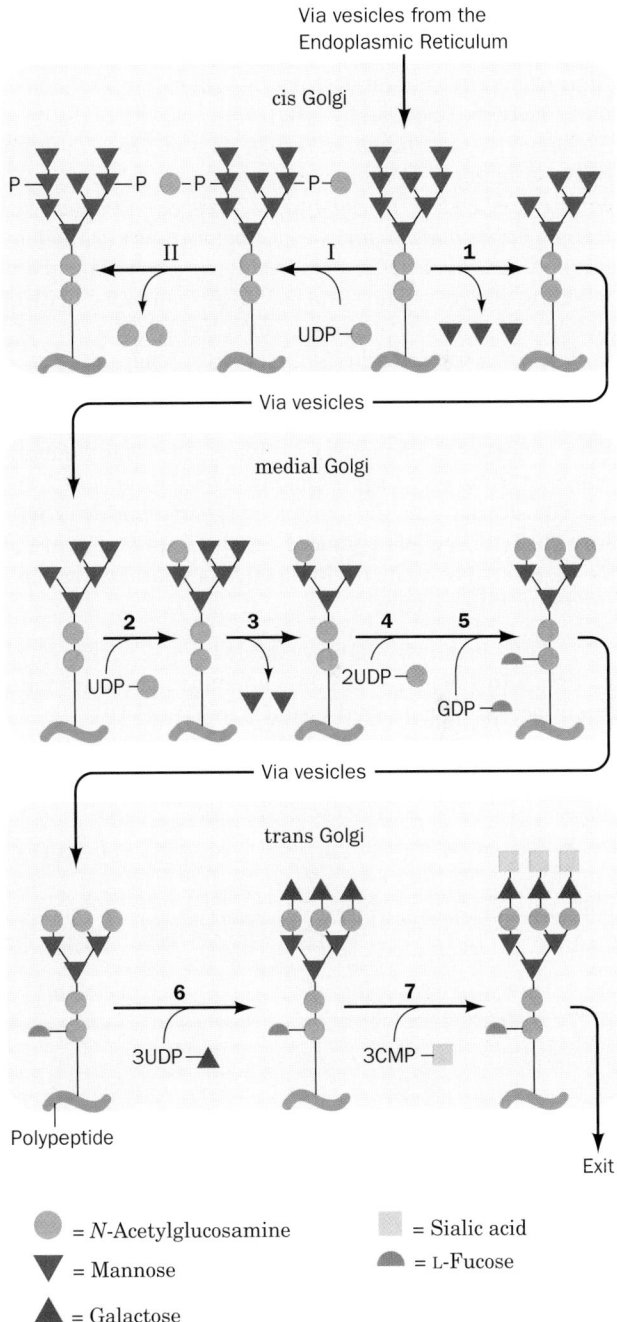

FIGURE 23-19   Oligosaccharide processing of VSV G-protein in the Golgi network. The reactions are catalyzed by: **(1)** Golgi α-mannosidase I, **(2)** *N*-acetylglucosaminyltransferase I, **(3)** Golgi α-mannosidase II, **(4)** *N*-acetylglucosaminyltransferase II, **(5)** fucosyltransferase, **(6)** galactosyltransferase, and **(7)** sialyltransferase. Lysosomal proteins are modified by: **(I)** *N*-acetylglucosaminyl phosphotransferase and **(II)** *N*-acetylglucosamine-1-phosphodiester α-*N*-acetylglucosaminidase. [Modified from Kornfeld, R. and Kornfeld, S., *Annu. Rev. Biochem.* **54,** 640 (1985).]

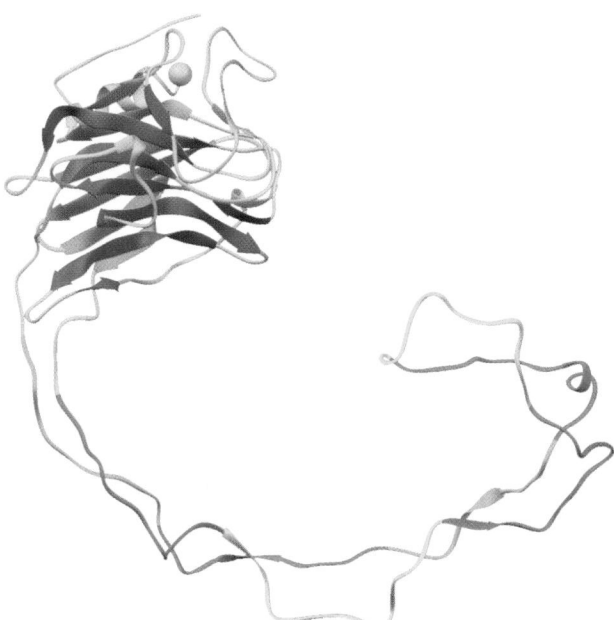

FIGURE 23-18   X-Ray structure of the luminal portion of canine calnexin. The 6- and 7-stranded antiparallel β sheets of its globular domain are colored orange and blue, with its remaining portions gray and its bound Ca^{2+} ion represented by a light green sphere. In the P domain, motifs 1 are alternately colored green and yellow and motifs 2 are alternately colored magenta and cyan. [Based on an X-ray structure by Miroslaw Cygler, Biotechnology Research Institute, NRC, Montreal, Quebec, Canada. PDBid 1JHN.]

work, through which proteins exit the Golgi apparatus. Glycoproteins traverse the Golgi stack, from the cis to the medial to the trans cisternae, each of which, as shown by James Rothman and Kornfeld, contains different sets of glycoprotein processing enzymes. As this occurs, mannose residues are trimmed from each oligosaccharide group and *N*-acetylglucosamine, galactose, fucose, and/or sialic acid residues are added to complete the processing of the glycoprotein (Fig. 23-19; Reactions 1–7). The glycoproteins are then sorted in the trans Golgi network for transport to their respective cellular destinations via membranous vesicles (Sections 12-4C and 12-4D).

There is enormous diversity among the different oligosaccharides of *N*-linked glycoproteins, as is indicated, for example, in Fig. 11-29c. Indeed, *even glycoproteins with a given polypeptide chain exhibit considerable microheterogeneity* (Section 11-3C), presumably as a consequence of incomplete glycosylation and lack of absolute specificity on the part of glycosyltransferases and glycosylases.

The processing of all *N*-linked oligosaccharides is identical through Reaction 4 of Fig. 23-17, so that all of them have a common (*N*-acetylglucosamine)$_2$(mannose)$_3$ core (five "noncore" mannose residues are subsequently trimmed from VSV G-protein; Fig. 23-19, Reactions 1 and 3). The diversity of the *N*-linked oligosaccharides therefore arises through divergence from this sequence after Fig. 23-19, Reaction 3. The resulting oligosaccharides are classified into three groups:

1. High-mannose oligosaccharides (Fig. 23-20a), which contain 2 to 9 mannose residues appended to the common pentasaccharide core (red residues in Fig. 23-20).

2. Complex oligosaccharides (Fig. 23-20b), which contain variable numbers of *N*-acetyllactosamine units as well as sialic acid and/or fucose residues linked to the core.

3. Hybrid oligosaccharides (Fig. 23-20c), which contain elements of both high-mannose and complex chains.

It is unclear how different types of oligosaccharides are related to the functions and/or final cellular locations of their glycoproteins. Lysosomal glycoproteins, however, appear to be of the high-mannose variety.

h. Inhibitors Have Aided the Study of *N*-Linked Glycosylation

Elucidation of the events in the glycosylation process has been greatly facilitated through the use of inhibitors that block specific glycosylation enzymes. Two of the most useful are the antibiotics **tunicamycin** (Fig. 23-21a), a hydrophobic analog of UDP–*N*-acetylglucosamine, and **bacitracin** (Fig. 23-22), a cyclic polypeptide. Both were discovered because of their ability to inhibit bacterial cell wall biosynthesis, a process that also involves the participation of lipid-linked oligosaccharides. Tunicamycin blocks the formation of dolichol-PP-oligosaccharides by inhibiting the synthesis of dolichol-PP-*N*-acetylglucosamine from dolichol-P and UDP–*N*-acetylglucosamine (Fig. 23-15, Reaction 1). Tunicamycin resembles an adduct of these reactants (Fig. 23-21b) and, in fact, binds to the enzyme with a dissociation constant of $7 \times 10^{-9}M$.

Bacitracin forms a complex with dolichol-PP that inhibits its dephosphorylation (Fig. 23-15, Reaction 12), thereby preventing glycoprotein synthesis from lipid-

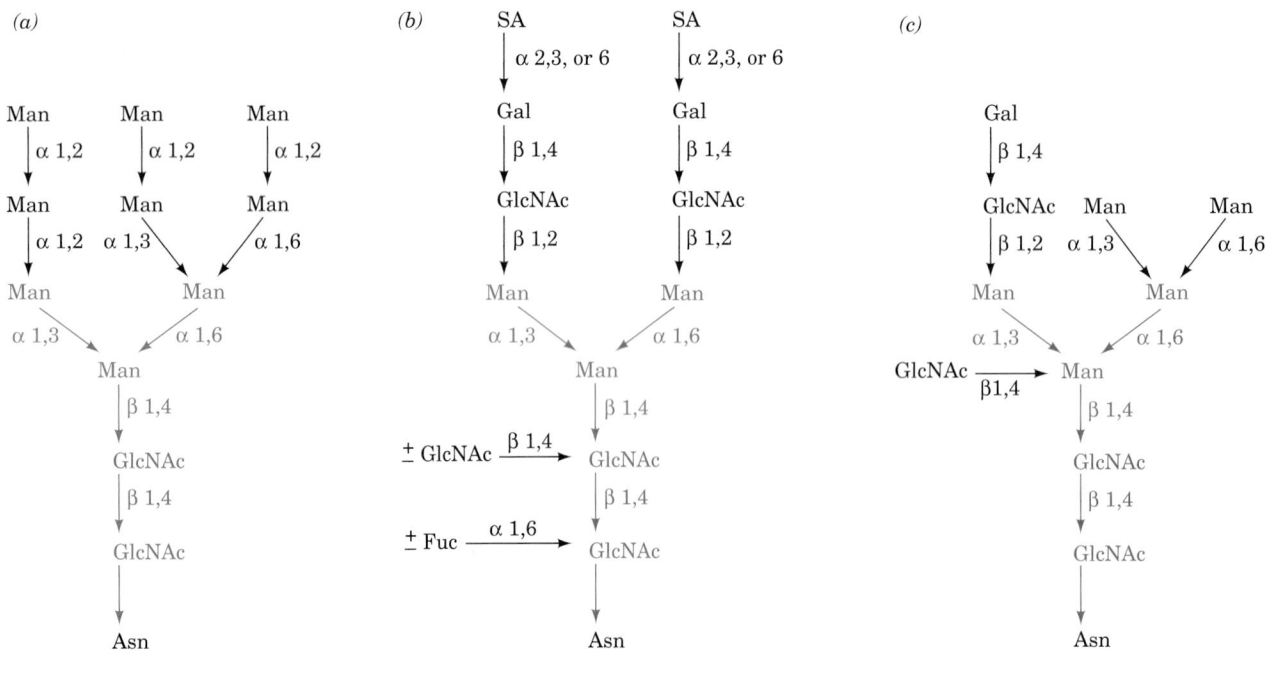

FIGURE 23-20   Types of *N*-linked oligosaccharides. Typical primary structures of (*a*) high-mannose, (*b*) complex, and (*c*) hybrid *N*-linked oligosaccharides. The pentasaccharide core common to all *N*-linked oligosaccharides is indicated in red. [After Kornfeld, R. and Kornfeld, S., *Annu. Rev. Biochem.* **54,** 633 (1985).]

FIGURE 23-21 Chemical structure of tunicamycin. The structure of (*a*) the glycosylation inhibitor tunicamycin is compared to that of (*b*) dolichol-P + UDP–*N*-acetylglucosamine.

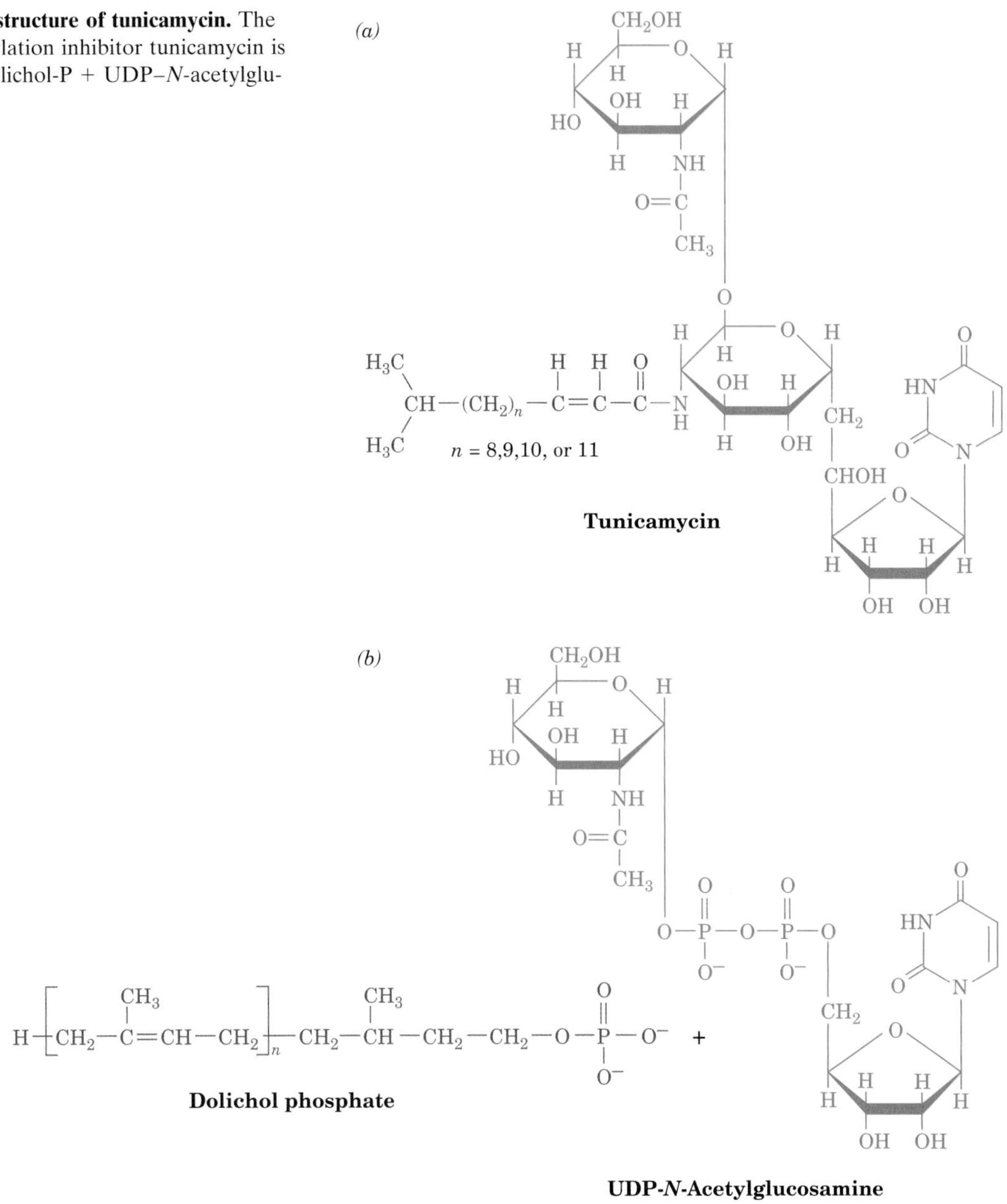

(*a*)

n = 8,9,10, or 11

Tunicamycin

(*b*)

Dolichol phosphate

UDP-*N*-Acetylglucosamine

linked oligosaccharide precursors. Bacitracin is clinically useful because it destroys bacterial cell walls but does not affect animal cells because it cannot cross cell membranes (bacterial cell wall biosynthesis is an extracellular process).

i. *O*-Linked Oligosaccharides Are Posttranslationally Formed

The study of the biosynthesis of **mucin,** an *O*-linked glycoprotein secreted by the submaxillary salivary gland, in-

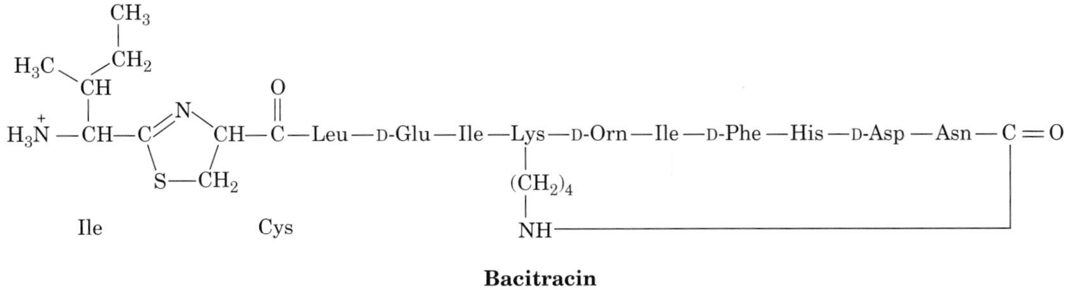

Bacitracin

FIGURE 23-22 Chemical structure of bacitracin. Note that this dodecapeptide has four D-amino acid residues and two unusual intrachain linkages. "Orn" represents the nonstandard amino acid residue ornithine (Fig. 4-22).

dicates that *O-linked oligosaccharides are synthesized in the Golgi apparatus by serial addition of monosaccharide units to a completed polypeptide chain (Fig. 23-23).* Synthesis starts with the transfer of *N*-acetylgalactosamine (GalNAc) from UDP–GalNAc to a Ser or Thr residue on the polypeptide by **GalNAc transferase.** In contrast to *N*-linked oligosaccharides, which are transferred to an Asn in a specific amino acid sequence, the *O*-glycosylated Ser and Thr residues are not members of any common sequence. Rather, it is thought that the location of glycosylation sites is specified only by the secondary or tertiary structure of the polypeptide. Glycosylation continues with stepwise addition of galactose, sialic acid, *N*-acetylglucosamine, and/or fucose by the corresponding glycosyltransferases.

j. Oligosaccharides on Glycoproteins Act as Recognition Sites

Glycoproteins that are synthesized in the endoplasmic reticulum and processed in the Golgi apparatus are targeted for secretion, insertion into cell membranes, or incorporation into cellular organelles such as lysosomes. This suggests that *oligosaccharides serve as recognition markers for this sorting process.* For example, the study of I-cell disease (Section 12-4C) demonstrated that in glycoprotein enzymes destined for the lysosome, a mannose residue is converted to mannose-6-phosphate (M6P) in the cis cisternae of the Golgi. The process involves two enzymes (Fig. 23-19, Reactions I and II), which are thought to recognize lysosomal protein precursors by certain structural features on these proteins rather than a specific amino acid sequence. In the trans Golgi network, M6P-bearing glycoproteins are sorted into lysosome-bound coated vesicles through their specific binding to one of two M6P receptors, one of which is a 275-kD membrane glycoprotein called the **M6P/IGF-II receptor** (because it has been found that this M6P receptor and the **insulinlike growth factor II receptor** are the same protein). Individuals with I-cell disease lack the enzyme catalyzing mannose phosphorylation (Fig. 23-19, Reaction I), resulting in the secretion of the normally lysosome-resident enzymes.

ABO blood group antigens (Section 12-3E) are *O*-linked glycoproteins. Their characteristic oligosaccharides are components of both cell-surface lipids and of proteins that occur in various secretions such as saliva. These oligosaccharides form antibody recognition sites.

Glycoproteins are believed to mediate cell–cell recognition. For example, an *O*-linked oligosaccharide on a glycoprotein that coats the mouse ovum surface (zona pellucida) acts as the sperm receptor. Even when this oligosaccharide is separated from its protein, it retains the ability to bind mouse sperm.

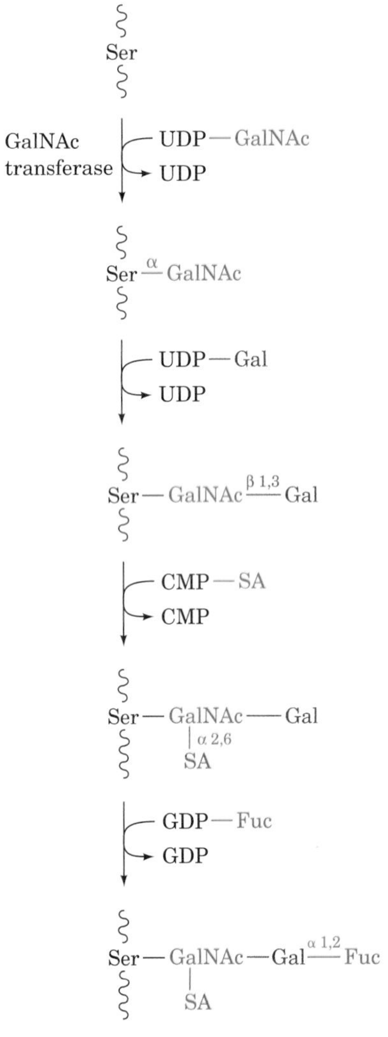

k. GPI-Linked Proteins

Glycosylphosphatidylinositol (GPI) groups function to anchor a wide variety of proteins to the exterior surface of the eukaryotic plasma membrane, thus providing an alternative to transmembrane polypeptide domains (Section 12-3B; Fig. 12-30). This anchoring results from transamidation of a preformed GPI glycolipid within 1 min of the synthesis and transfer of a target protein to the ER. Biosynthesis of the GPI core structure (Fig. 23-24a) begins on the cytoplasmic side of the ER with the transfer of *N*-acetylglucosamine from UDP–*N*-acetylglucosamine (UDP–GlcNAc) to the 6 hydroxyl of the inositol of phosphatidylinositol, followed by the removal of the acetyl group. The mammalian pathway then continues with the

FIGURE 23-23   Proposed synthesis pathway for the carbohydrate moiety of an *O*-linked oligosaccharide chain of canine submaxillary mucin. SA and Fuc represent sialic acid and fucose.

FIGURE 23-24   GPI anchors. (*Opposite*) (*a*) The pathway of synthesis of the tetrasaccharide core of glycophosphatidylinositol (GPI). The following enzymes and steps are involved: **(1)** UDP–GlcNAc:PI α1→6 *N*-acetylglucosaminyltransferase complex, **(2)** GlcNAc–PI de-*N*-acetylase, **(3)** inositol acyltransferase, **(4)** Dol-P-Man:GlcN–PI/GlcN–(acyl)PI α1→4 mannosyltransferase (MT-I), **(5)** an ethanolamine phosphotransferase, **(6)** Dol-P-Man:Man₁GlcN–(acyl)PI α1→6 mannosyltransferase (MT-II), **(7)** Dol-P–Man:Man₂GlcN–(acyl)PI α1→2 mannosyltransferase (MT-III), **(8)** lipid remodeling (replacement of the fatty acyl groups on PI), and **(9)** transfer of phosphoethanolamine from phosphatidylethanolamine to the 6-hydroxyl group of the terminal mannose residue of the core tetrasaccharide by an ethanolamine phosphotransferase. (*b*) Transamidation of the target protein, resulting in a C-terminal amide link to the GPI anchor.

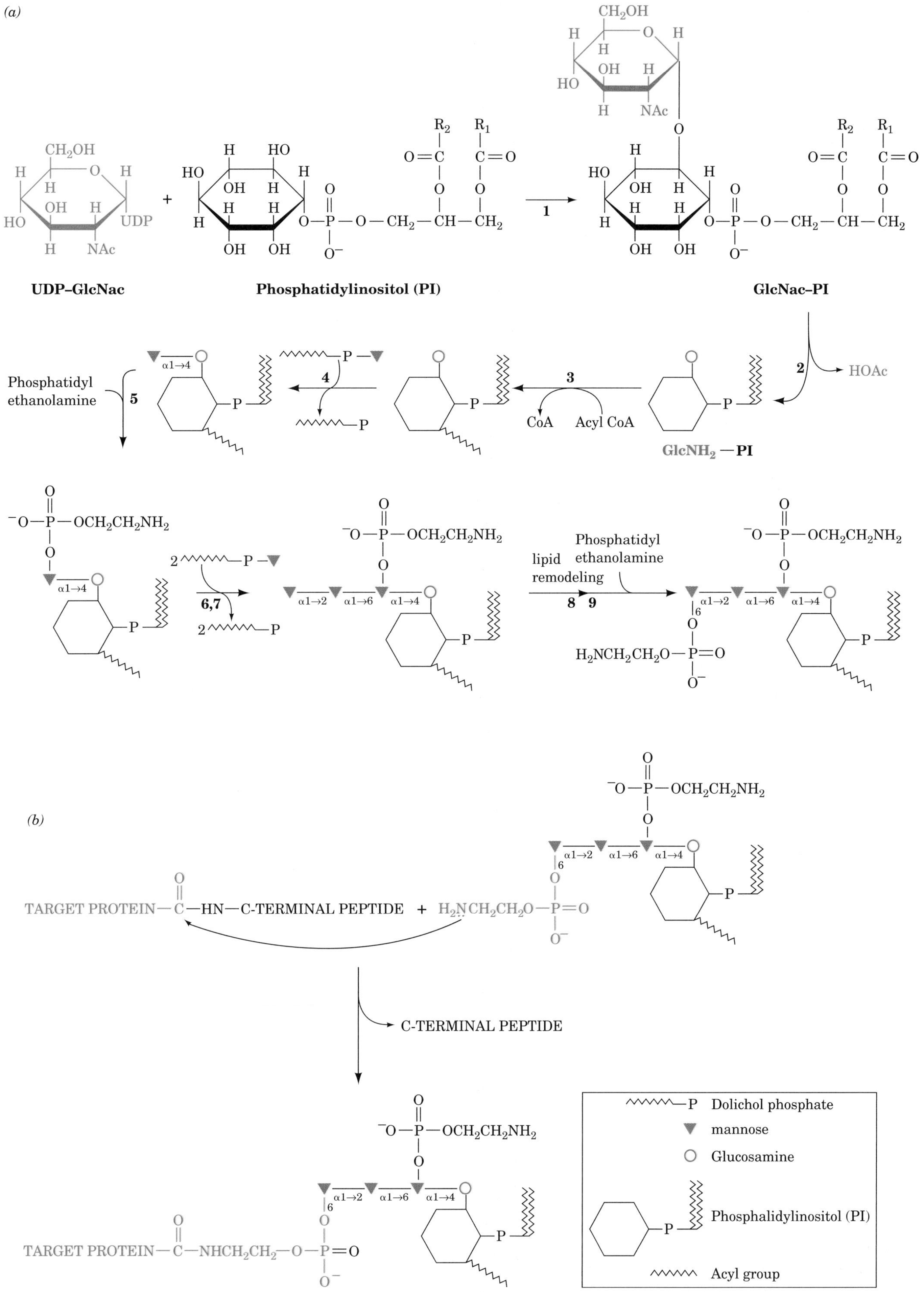

(a)
UDP–GlcNac
Phosphatidylinositol (PI)
GlcNac–PI
HOAc
GlcNH2 — PI
CoA
Acyl CoA
Phosphatidyl ethanolamine
lipid remodeling
(b)
TARGET PROTEIN — C — HN — C-TERMINAL PEPTIDE
C-TERMINAL PEPTIDE
TARGET PROTEIN — C — NHCH2CH2 — O — P = O
Dolichol phosphate
mannose
Glucosamine
Phosphalidylinositol (PI)
Acyl group

2-acylation of inositol, translocation to the luminal side of the ER membrane, and the addition of mannose from dolichol-P-mannose (Dol-P-Man; Fig. 23-15) and phosphoethanolamine from phosphatidylethanolamine (Table 12-2), as indicated in Fig. 23-24*a*. This core is modified with a variety of additional sugar residues, depending on the species and the protein to which it is attached. There is considerable diversity in the fatty acid residues of GPI anchors due to the extensive lipid remodeling that occurs during anchor synthesis. Target proteins become anchored to the membrane surface when the amino group of the GPI phosphoethanolamine nucleophilically attacks a specific amino acyl group of the protein near its C-terminus, resulting in a transamidation that releases a 20- to 30-residue hydrophobic C-terminal signal peptide (Fig. 23-24*b*). Since GPI groups are appended to proteins on the luminal surface of the RER, GPI-anchored proteins occur on the exterior surface of the plasma membrane (Fig. 12-53). However, they are distributed unevenly in the outer leaflet of the plasma membrane because they prefer to associate with sphingolipid–cholesterol rafts (Section 12-3C).

The core GPI structure is evolutionarily conserved among all eukaryotes, although there are differences between species in its synthesis. For example, the cell surface of the trypanosomes that cause African sleeping sickness (a debilitating and often fatal disease that afflicts millions of people in sub-Saharan Africa) has a dense coating of **variant surface glycoprotein (VSG)** that is GPI-anchored to its plasma membrane. The VSG coating conceals the trypanosome's plasma membrane from the host's immune system although it recognizes and attacks the VSG itself. The parasite is nevertheless able to evade the host's immunological defenses because it has a genetic repertoire of about a thousand immunologically distinct VSGs. An individual trypanosome expresses only one of its VSG genes and hence the host can mount an effective immunological attack against the prevailing population of VSGs, a process that takes around 1 week (Section 35-2A). However, by switching VSG genes, a new population of trypanosomes arises that replicates unchecked until the host can mount a new immune response, a cycle that repeats until the death of the host. The comparison of the GPI biosynthetic pathway in trypanosomes with that in mammalian systems has revealed several differences in the pathway order. For example, Steps 3 and 4 of Fig. 23-23*a* are reversed in trypanosomes. This and other differences in the substrate specificities of the enzymes catalyzing this pathway have brought to light several promising drug targets for the treatment of African sleeping sickness.

4 ■ THE PENTOSE PHOSPHATE PATHWAY

ATP is the cell's "energy currency"; its exergonic hydrolysis is coupled to many otherwise endergonic cell functions. *Cells have a second currency, reducing power.* Many endergonic reactions, notably the reductive biosynthesis of fatty acids (Section 25-4) and cholesterol (Section 25-6A),

as well as photosynthesis (Section 24-3A), require NADPH in addition to ATP. Despite their close chemical resemblance, *NADPH and NADH are not metabolically interchangeable* (recall that these coenzymes differ only by a phosphate group at the 2′-OH group of NADPH's adenosine moiety; Fig. 13-2). Whereas NADH participates in utilizing the free energy of metabolite oxidation to synthesize ATP (oxidative phosphorylation), *NADPH is involved in utilizing the free energy of metabolite oxidation for otherwise endergonic reductive biosynthesis.* This differentiation is possible because the dehydrogenase enzymes involved in oxidative and reductive metabolism exhibit a high degree of specificity toward their respective coenzymes. Indeed, cells normally maintain their $[NAD^+]/[NADH]$ ratio near 1000, which favors metabolite oxidation, while keeping their $[NADP^+]/[NADPH]$ ratio near 0.01, which favors metabolite reduction.

NADPH is generated by the oxidation of G6P via an alternative pathway to glycolysis, the **pentose phosphate pathway** *[also called the* **hexose monophosphate (HMP) shunt** *and the* **phosphogluconate pathway;** *Fig. 23-25]. The pathway also produces ribose-5-phosphate (R5P), an essential precursor in nucleotide biosynthesis (Sections 28-1, 28-2, and 28-5).* The first evidence of this pathway's existence was obtained in the 1930s by Otto Warburg, who discovered $NADP^+$ through his studies on the oxidation of G6P to 6-phosphogluconate. Further indications came from the observation that tissues continue to respire in the presence of high concentrations of fluoride ion, which, it will be recalled, blocks glycolysis by inhibiting enolase (Section 17-2I). It was not until the 1950s, however, that the pentose phosphate pathway was elucidated by Frank Dickens, Bernard Horecker, Fritz Lipmann, and Efraim Racker. Tissues most heavily involved in fatty acid and cholesterol biosynthesis (liver, mammary gland, adipose tissue, and adrenal cortex) are rich in pentose phosphate pathway enzymes. Indeed, some 30% of the glucose oxidation in liver occurs via the pentose phosphate pathway.

The overall reaction of the pentose phosphate pathway is

$$3G6P + 6NADP^+ + 3H_2O \rightleftharpoons$$
$$6NADPH + 6H^+ + 3CO_2 + 2F6P + GAP$$

However, the pathway may be considered to have three stages:

1. Oxidative reactions (Fig. 23-25, Reactions 1–3), which yield NADPH and **ribulose-5-phosphate (Ru5P).**

$$3G6P + 6NADP^+ + 3H_2O \rightarrow$$
$$6NADPH + 6H^+ + 3CO_2 + 3Ru5P$$

2. Isomerization and epimerization reactions (Fig. 23-25, Reactions 4 and 5), which transform Ru5P either to **ribose-5-phosphate (R5P)** or to **xylulose-5-phosphate (Xu5P).**

$$3Ru5P \rightleftharpoons R5P + 2Xu5P$$

3. A series of C—C bond cleavage and formation reactions (Fig. 23-25, Reactions 6–8) that convert two molecules of Xu5P and one molecule of R5P to two molecules

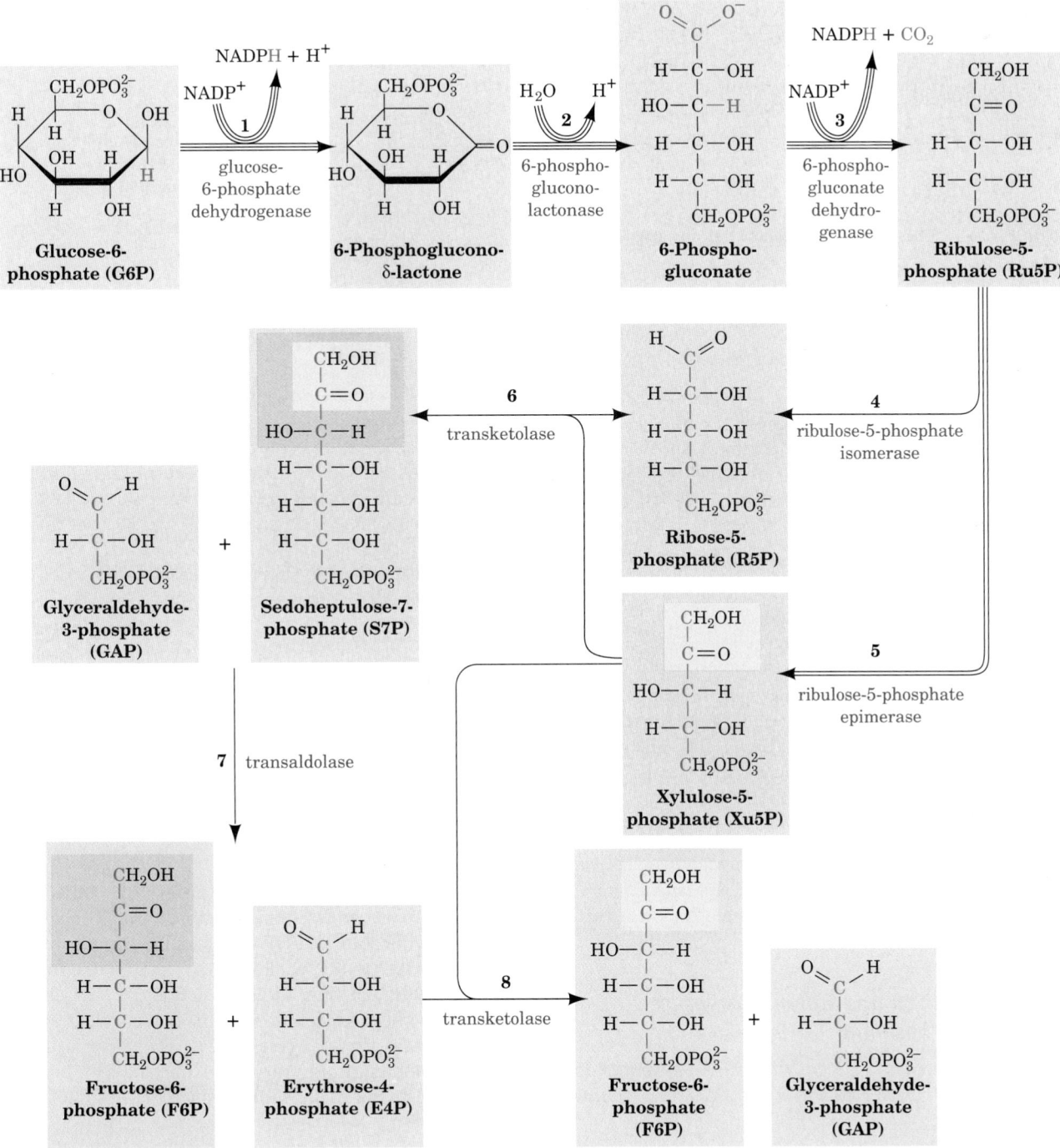

FIGURE 23-25 The pentose phosphate pathway. The number of lines in an arrow represents the number of molecules reacting in one turn of the pathway so as to convert three G6Ps to three CO_2s, two F6Ps, and one GAP. For the sake of clarity, sugars from Reaction 3 onward are shown in their linear forms. The carbon skeleton of R5P and the atoms derived from it are drawn in red and those from Xu5P are drawn in green. The C_2 units transferred by transketolase are shaded in green and the C_3 units transferred by transaldolase are shaded in blue. Double-headed arrows indicate reversible reactions.

of fructose-6-phosphate (F6P) and one of glyceraldehyde-3-phosphate (GAP).

$$R5P + 2Xu5P \rightleftharpoons 2F6P + GAP$$

The reactions of Stages 2 and 3 are freely reversible so that the products of the pathway vary with the needs of the cell.

For example, when R5P is required for nucleotide biosynthesis, Stage 3 works in reverse, producing R5P from F6P and GAP nonoxidatively. In this section, we discuss the three stages of the pentose phosphate pathway and how this pathway is controlled. We close by considering the consequences of one of its abnormalities.

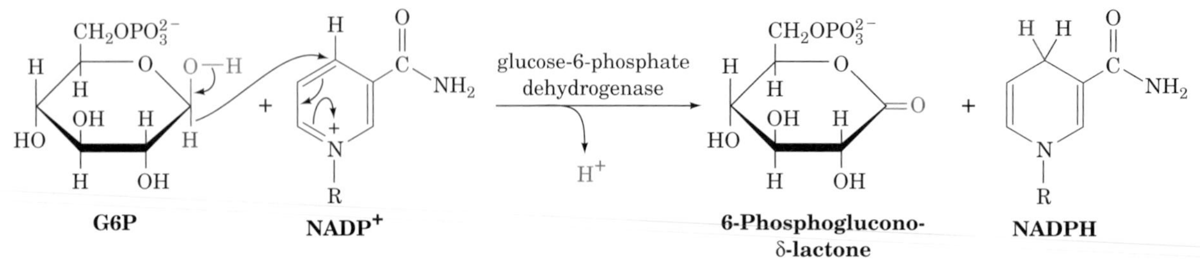

FIGURE 23-26   The glucose-6-phosphate dehydrogenase reaction.

A. *Oxidative Reactions of NADPH Production*

Only the first three reactions of the pentose phosphate pathway are involved in NADPH production.

1. Glucose-6-phosphate dehydrogenase (G6PD) catalyzes net transfer of a hydride ion to $NADP^+$ from C1 of G6P to form **6-phosphoglucono-δ-lactone** (Fig. 23-26). G6P, a cyclic hemiacetal with C1 in the aldehyde oxidation state, is thereby oxidized to a cyclic ester (lactone). The enzyme is specific for $NADP^+$ and is strongly inhibited by NADPH.

2. 6-Phosphogluconolactonase increases the rate of hydrolysis of 6-phosphoglucono-δ-lactone to **6-phosphogluconate** (the nonenzymatic reaction occurs at a significant rate), the substrate of the next oxidative enzyme in the pathway.

3. 6-Phosphogluconate dehydrogenase catalyzes the oxidative decarboxylation of 6-phosphogluconate, a β-hydroxy acid, to Ru5P and CO_2 (Fig. 23-27). The reaction is similar to that catalyzed by the citric acid cycle enzyme isocitrate dehydrogenase (Section 21-3C).

Formation of Ru5P completes the oxidative portion of the pentose phosphate pathway. *It generates two molecules of NADPH for each molecule of G6P that enters the pathway.* The product Ru5P must subsequently be converted to R5P or Xu5P for further use.

B. *Isomerization and Epimerization of Ribulose-5-Phosphate*

*Ru5P is converted to R5P by **ribulose-5-phosphate isomerase** (Fig. 23-25, Reaction 4) and to Xu5P by **ribulose-***

*5-phosphate epimerase** (Fig. 23-25, Reaction 5). These isomerization and epimerization reactions, as discussed in Section 16-2D, are both thought to occur via enediolate intermediates (Fig. 23-28).*

R5P is an essential precursor in the biosynthesis of nucleotides (Sections 28-1, 28-2, and 28-8). If, however, more R5P is formed than the cell needs, the excess, along with Xu5P, is converted to the glycolytic intermediates F6P and GAP as described below.

C. *Carbon–Carbon Bond Cleavage and Formation Reactions*

The conversion of three C_5 sugars to two C_6 sugars and one C_3 sugar involves a remarkable "juggling act" catalyzed by two enzymes, **transaldolase** and **transketolase.** As we discussed in Section 16-2E, enzymatic reactions that make or break carbon–carbon bonds usually have mechanisms that involve generation of a stabilized carbanion and its addition to an electrophilic center such as an aldehyde. This is the dominant theme of both the transaldolase and the transketolase reactions.

a. Transketolase Catalyzes the Transfer of C_2 Units

Transketolase, which has a thiamine pyrophosphate cofactor (TPP; Section 17-3B), catalyzes the transfer of a C_2 unit from Xu5P to R5P, yielding GAP and **sedoheptulose-7-phosphate (S7P)** (Fig. 23-25, Reaction 6). The reaction involves the intermediate formation of a covalent adduct between Xu5P and TPP (Fig. 23-29). The X-ray structure of this homodimeric enzyme shows that the TPP binds in a deep cleft between the subunits such that residues from

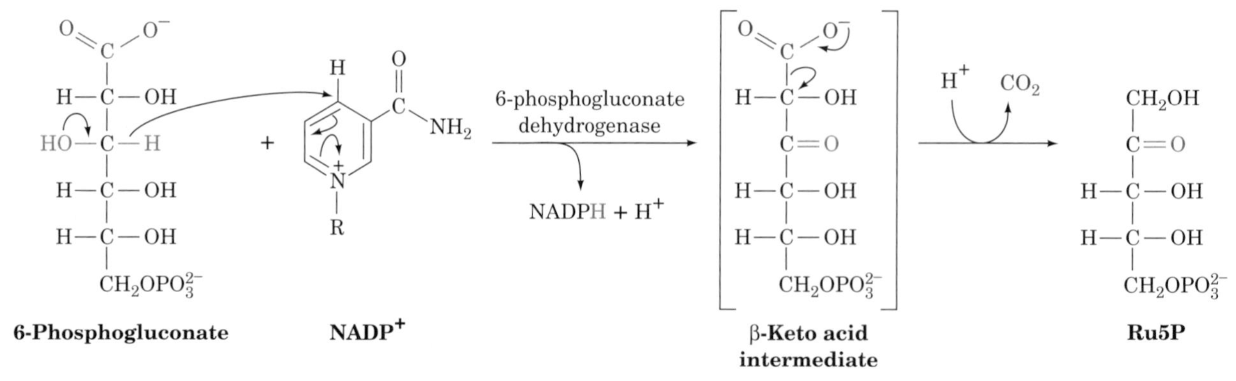

FIGURE 23-27   The phosphogluconate dehydrogenase reaction. Oxidation of the OH group forms an easily decarboxylated

β-keto acid (although the proposed intermediate has not been isolated).

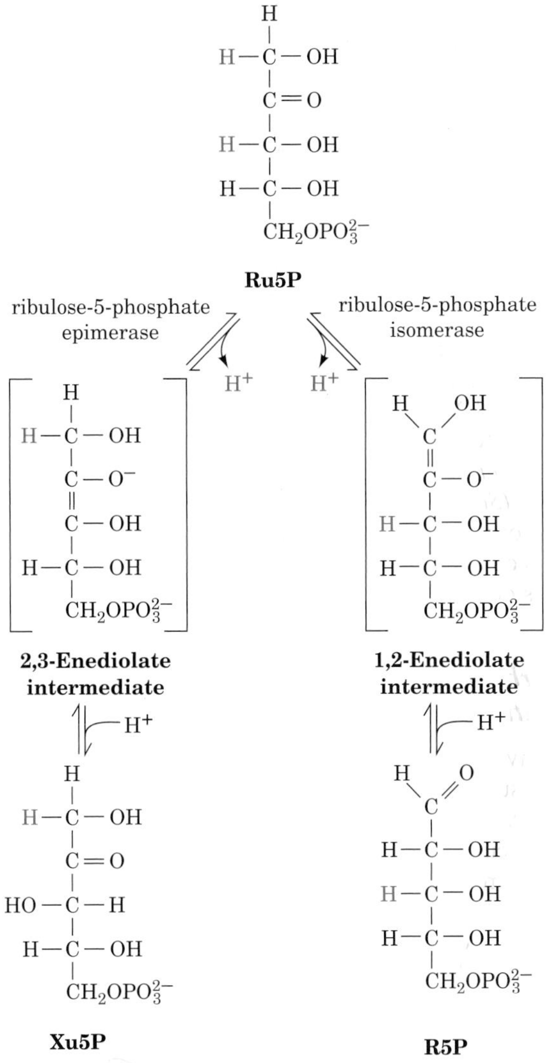

FIGURE 23-28 Ribulose-5-phosphate isomerase and ribulose-5-phosphate epimerase. The reactions catalyzed by both these enzymes involve enediolate intermediates. In the isomerase reaction (*right*), a base on the enzyme removes a proton from C1 of Ru5P to form a 1,2-enediolate and then adds a proton at C2 to form R5P. In the epimerase reaction (*left*), a base on the enzyme removes a C3 proton to form a 2,3-enediolate. A proton is then added to the same carbon atom but with inversion of configuration to yield Xu5P.

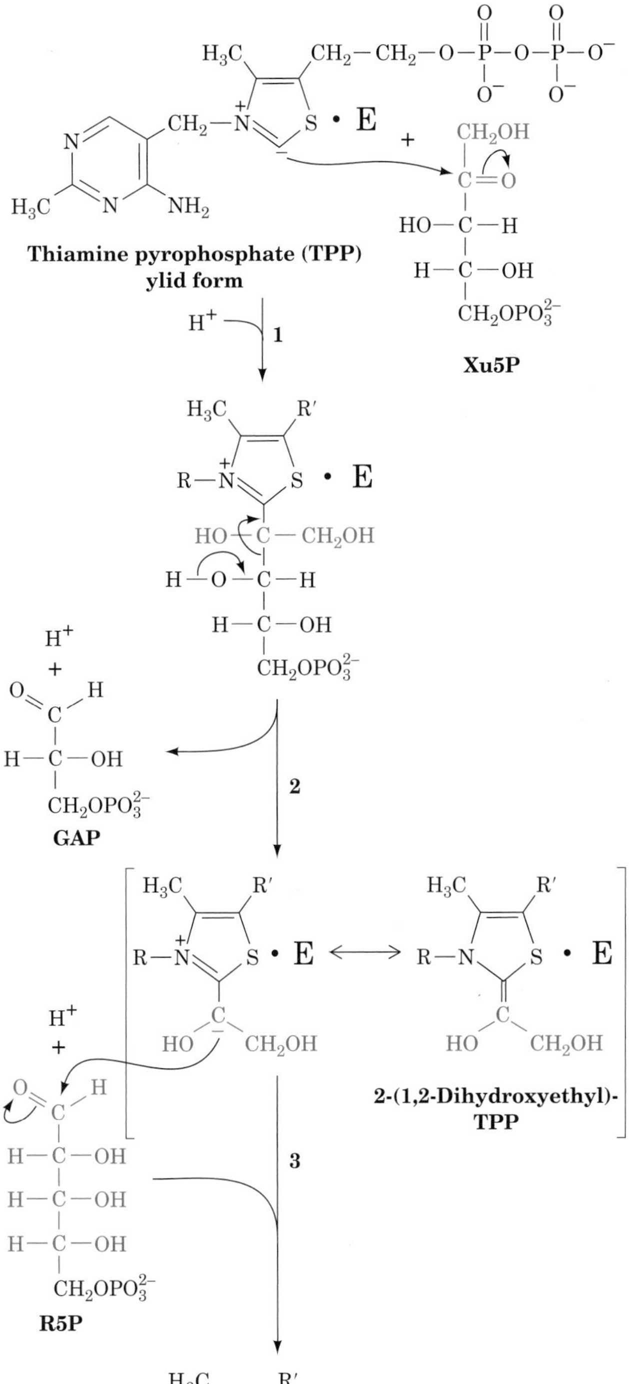

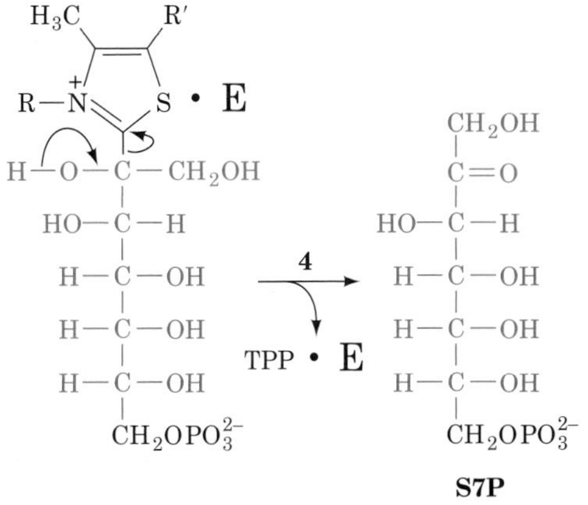

FIGURE 23-29 Mechanism of transketolase. Transketolase utilizes the coenzyme thiamine pyrophosphate to stabilize the carbanion formed on cleavage of the C2—C3 bond of Xu5P. The reaction occurs as follows: **(1)** The TPP ylid attacks the carbonyl group of Xu5P. **(2)** C2—C3 bond cleavage yields GAP and enzyme-bound 2-(1,2-dihydroxyethyl)-TPP, a resonance stabilized carbanion. **(3)** The C2 carbanion attacks the aldehyde carbon of R5P forming an S7P–TPP adduct. **(4)** TPP is eliminated yielding S7P and the regenerated TPP–enzyme.

both subunits participate in its binding, just as in pyruvate decarboxylase (another TPP-requiring enzyme; Figure 17-28). In fact the structures are so similar that it is likely that they diverged from a common ancestor.

b. Transaldolase Catalyzes the Transfer of C₃ Units

Transaldolase catalyzes the transfer of a C_3 unit from S7P to GAP, yielding **erythrose-4-phosphate (E4P)** and F6P (Fig. 23-25, Reaction 7). The reaction occurs by aldol cleavage, which begins with the formation of a Schiff base between an ε-amino group of an essential enzyme Lys residue and the carbonyl group of S7P (Fig. 23-30). Transaldolase and Class I aldolase (Section 17-2D) share a common reaction mechanism and may also share a common ancestor, despite their lack of significant sequence identity. Both are α/β barrel proteins (Section 8-3B), but while the Schiff base–forming Lys is on β4 (the fourth β strand from the N-terminus) of transaldolase, it is on β6 of Class I aldolase. Superimposing the barrel structures of these two enzymes while maintaining the alignment of the β strands bearing the Schiff base–forming Lys residues results in a significantly better fit than doing so while maintaining the alignment of their entire α/β barrels. Moreover, five of the pairs of matched active site residues in the former superposition are identical. This suggests that, during evolution, the DNA sequence for two α/β units was transferred from the N-terminus to the C-terminus of the evolving Class I aldolase, moving the active site Lys from β6 to β4. Such a circular permutation of an α/β barrel's structural elements does not greatly change its structure.

c. A Second Transketolase Reaction Yields GAP and a Second F6P Molecule

In a second transketolase reaction, a C_2 unit is transferred from a second molecule of Xu5P to E4P to form GAP and another molecule of F6P (Fig. 23-25, Reaction 8). The third phase of the pentose phosphate pathway thus transforms two molecules of Xu5P and one of R5P to two molecules of F6P and one molecule of GAP. These carbon skeleton transformations (Fig. 23-25, Reactions 6–8) are summarized in Fig. 23-31.

D. *Control of the Pentose Phosphate Pathway*

The principal products of the pentose phosphate pathway are R5P and NADPH. The transaldolase and transketolase reactions serve to convert excess R5P to glycolytic in-

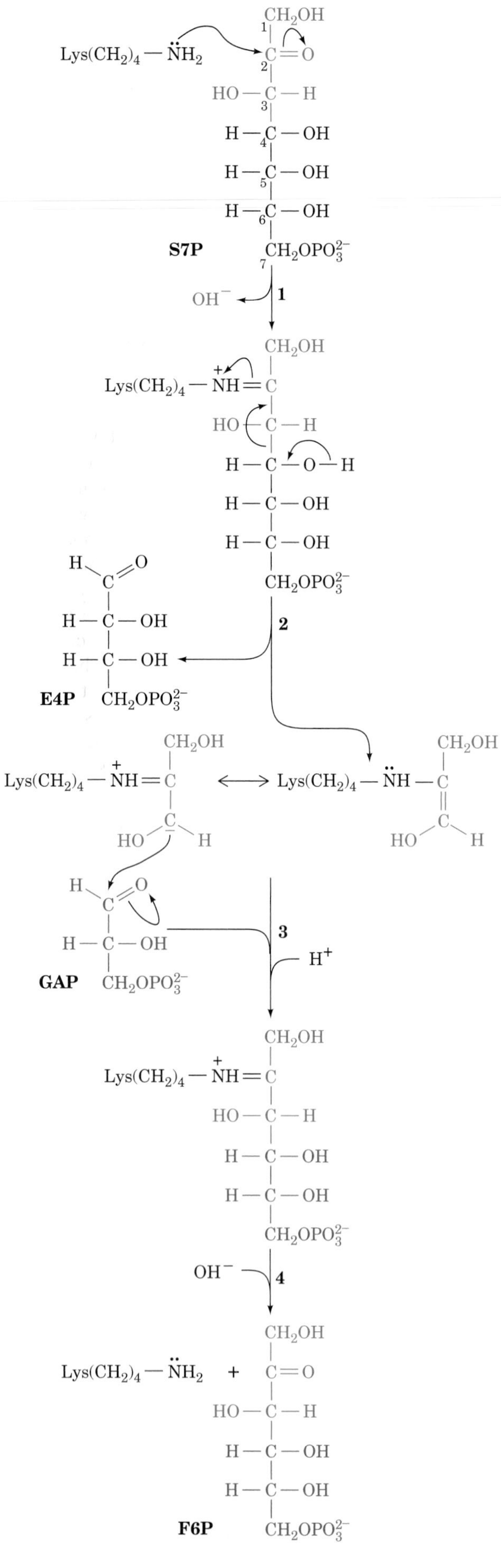

FIGURE 23-30   Mechanism of transaldolase. Transaldolase contains an essential Lys residue that forms a Schiff base with S7P to facilitate an aldol cleavage reaction. The reaction occurs as follows: **(1)** The ε-amino group of an essential Lys residue forms a Schiff base with the carbonyl group of S7P. **(2)** A Schiff base–stabilized C3 carbanion is formed in an aldol cleavage reaction between C3 and C4 that eliminates E4P. **(3)** The enzyme-bound resonance-stabilized carbanion adds to the carbonyl C atom of GAP, forming F6P linked to the enzyme via a Schiff base. **(4)** The Schiff base hydrolyzes, regenerating active enzyme and releasing F6P.

(6) $C_5 + C_5 \rightleftharpoons C_7 + C_3$

(7) $C_7 + C_3 \rightleftharpoons C_6 + C_4$

(8) $\underline{C_5 + C_4 \rightleftharpoons C_6 + C_3}$

(Sum) $3\,C_5 \rightleftharpoons 2\,C_6 + C_3$

FIGURE 23-31   Summary of carbon skeleton rearrangements in the pentose phosphate pathway. A series of carbon–carbon bond formations and cleavages convert three C_5 sugars to two C_6 and one C_3 sugar. The number to the left of each reaction is keyed to the corresponding reaction in Fig. 23-25.

termediates when the metabolic need for NADPH exceeds that of R5P in nucleotide biosynthesis. The resulting GAP and F6P can be consumed through glycolysis and oxidative phosphorylation or recycled by gluconeogenesis to form G6P. *In the latter case, 1 molecule of G6P can be converted, via six cycles of the pentose phosphate pathway and gluconeogenesis, to 6 CO_2 molecules with the concomitant generation of 12 NADPH molecules.* When the need for R5P outstrips that for NADPH, F6P and GAP can be diverted from the glycolytic pathway for use in the synthesis of R5P by reversal of the transaldolase and transketolase reactions. In fact, mass spectral analysis of the ^{13}C-labeled carbons from $[1,2-^{13}C]$glucose incorporated into RNA in rapidly proliferating cancer cells has shown that more than ~70% of the *de novo* ribose synthesis arises through this nonoxidative reversal of the pentose phosphate pathway (rather than its forward direction).

Flux through the oxidative pentose phosphate pathway and thus the rate of NADPH production is controlled by the rate of the glucose-6-phosphate dehydrogenase reaction (Fig. 23-25, Reaction 1). The activity of this enzyme, which catalyzes the pentose phosphate pathway's first committed step ($\Delta G = -17.6$ kJ · mol^{-1} in liver), is regulated by the NADP$^+$ concentration (substrate availability). When the cell consumes NADPH, the NADP$^+$ concentration rises, increasing the rate of the glucose-6-phosphate dehydrogenase reaction, thereby stimulating NADPH regeneration.

E. *Glucose-6-Phosphate Dehydrogenase Deficiency*

NADPH is required for several reductive processes in addition to biosynthesis. For example, erythrocyte membrane integrity requires a plentiful supply of reduced glutathione (GSH), a Cys-containing tripeptide (Sections 21-2B and 26-4C). A major function of GSH in the erythrocyte is to eliminate H_2O_2 and organic hydroperoxides. H_2O_2, a toxic product of various oxidative processes, reacts with double bonds in the fatty acid residues of the erythrocyte cell membrane to form organic hydroperoxides. These, in turn, react to cleave fatty acid C—C bonds, thereby damaging the membrane. In erythrocytes, the unchecked buildup of peroxides results in premature cell lysis. Peroxides are eliminated through the action of **glutathione peroxidase,**

one of the handful of enzymes with a selenium cofactor, yielding glutathione disulfide (GSSG).

$$2GSH + R-O-O-H \xrightarrow{\text{glutathione peroxidase}} GSSG + ROH + H_2O$$
Organic hydroperoxide

GSH is subsequently regenerated by the NADPH reduction of GSSG catalyzed by glutathione reductase (Section 21-2B).

$$GSSG + NADPH + H^+ \xrightarrow{\text{glutathione reductase}} 2GSH + NADP^+$$

A steady supply of NADPH is therefore vital for erythrocyte integrity.

a. Primaquine Causes Hemolytic Anemia in Glucose-6-Phosphate Dehydrogenase Mutants

A genetic defect, common in African, Asian, and Mediterranean populations, results in severe hemolytic anemia on infection or on the administration of certain drugs including the antimalarial agent **primaquine.**

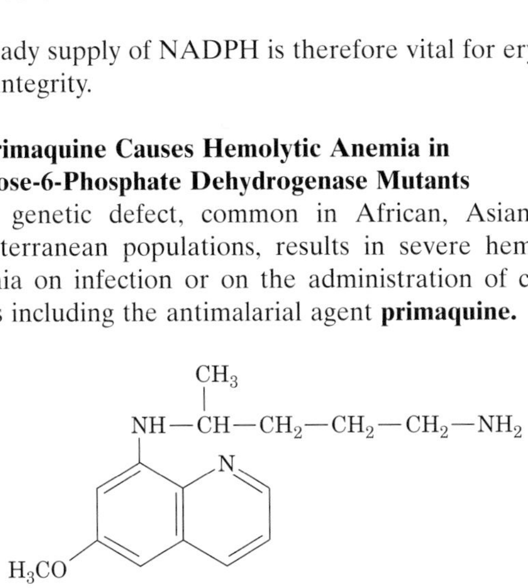

Primaquine

Similar effects, which go by the name of **favism,** occur when individuals bearing this trait eat **fava beans (broad beans,** *Vicia faba*), a staple Middle Eastern vegetable that contains small quantities of toxic glycosides. This trait has been traced to an altered gene for glucose-6-phosphate dehydrogenase (G6PD). Under most conditions, mutant erythrocytes have sufficient enzyme activity for normal function. Agents such as primaquine and fava beans, however, stimulate peroxide formation, thereby increasing the demand for NADPH to a level that mutant cells cannot meet.

The major reason for low enzymatic activity in affected cells appears to be an accelerated rate of breakdown of the mutant enzyme (protein degradation is discussed in Section 32-6). This explains why patients with G6PD deficiency react to primaquine with hemolytic anemia but recover within a week despite continued primaquine treatment. Mature erythrocytes lack a nucleus and protein synthesizing machinery and therefore cannot synthesize new enzyme molecules to replace degraded ones (they likewise cannot synthesize new membrane components, which is why they are so sensitive to membrane damage in the first place). The initial primaquine treatments result in the lysis of old red blood cells whose defective G6PD has been largely degraded. Lysis products stimulate the release of young cells that contain more enzyme and are therefore better able to cope with primaquine stress.

It is estimated that over 400 million people are deficient in G6PD, which makes this condition the most common human enzymopathy. Indeed, ~400 G6PD variants have been reported and at least 125 of them have been characterized at the molecular level. G6PD is active in a dimer–tetramer equilibrium. Many of the mutation sites in individuals with the most severe G6PD deficiency are at the dimer interface, shifting the equilibrium toward the inactive and unstable monomer.

Several G6PD variants occur with high incidence. For example, the so-called type A⁻ deficiency, which exhibits ~10% of the normal G6PD activity, has an incidence of 11% among black Americans. This variant is also the most common form of G6PD deficiency in sub-Saharan Africa. The variant "Mediterranean" is found throughout the Mediterranean and Middle East regions, and occurs in 65% of Kurdish Jews, the population with the highest

known incidence of this trait. The high prevalence of defective G6PD in malarial areas of the world suggests that such mutations confer resistance to the malarial parasite, *Plasmodium falciparum* (as we likewise saw to be the case for the sickle-cell trait; Section 7-3A). Indeed, two epidemiological studies involving over 2000 African children with A⁻ G6PD deficiency indicate that this form is associated with an ~50% reduction in the risk of severe malaria for both female heterozygotes and male hemizygotes (G6PD deficiency is an X-linked trait).

In vitro studies indicate that erythrocytes with G6PD deficiency are less suitable hosts for plasmodia than are normal cells. This is presumably because the parasite requires the products of the pentose phosphate pathway and/or because the erythrocyte is lysed before the parasite has had a chance to mature. Thus, like the sickle-cell trait (Section 7-3A), *a defective G6PD confers a selective advantage on individuals living where malaria is endemic.*

CHAPTER SUMMARY

1 ■ Gluconeogenesis Lactate, pyruvate, citric acid cycle intermediates, and many amino acids may be converted, by gluconeogenesis, to glucose via the formation of oxaloacetate. For this to occur, the three irreversible steps of glycolysis must be bypassed. The pyruvate kinase reaction is bypassed by converting pyruvate to oxaloacetate in an ATP-driven reaction catalyzed by the biotinyl-containing enzyme pyruvate carboxylase. The oxaloacetate is subsequently decarboxylated and phosphorylated by GTP to form PEP in a reaction catalyzed by PEPCK. For this to happen in species in which PEPCK is a cytosolic enzyme, the oxaloacetate must be transported from the mitochondrion to the cytosol via its interim conversion to either malate or aspartate. Conversion to malate concomitantly transports reducing equivalents to the cytosol in the form of NADH. The two other irreversible steps of glycolysis, the PFK reaction and the hexokinase reaction, are bypassed by simply hydrolyzing their products, FBP and G6P, by FBPase and glucose-6-phosphatase, respectively. A glucose molecule may therefore be synthesized from pyruvate at the expense of four ATPs more than are generated by the reverse process. Glycolysis and gluconeogenesis are reciprocally regulated so as to consume glucose when the demand for ATP is high and synthesize it when the demand is low. The control points in these processes are at pyruvate kinase/pyruvate carboxylase–PEPCK, PFK/FBPase, and hexokinase/glucose-6-phosphatase. Regulation of these enzymes is exerted largely through allosteric interactions, cAMP-dependent enzyme modifications, and, for PEPCK, gene expression. Muscle, which is incapable of gluconeogenesis, transfers much of the lactate it produces to the liver via the blood for conversion to glucose and return to the muscle. This Cori cycle shifts the metabolic burden of oxidative ATP generation for gluconeogenesis from muscle to liver.

2 ■ The Glyoxylate Cycle Animals cannot convert fatty acids to glucose because they lack the enzymes necessary to synthesize oxaloacetate from acetyl-CoA. Plants, however,

can do so via the glyoxylate cycle, a glyoxysomal process that converts two molecules of acetyl-CoA to one molecule of succinate via the intermediate formation of glyoxylate. Succinate is converted to oxaloacetate for use in gluconeogensis or the citric acid cycle.

3 ■ Biosynthesis of Oligosaccharides and Glycoproteins Glycosidic bonds are formed by transfer of the monosaccharide unit of a sugar nucleotide to a second sugar unit. Such reactions occur in the synthesis of disaccharides such as lactose and in the synthesis of the carbohydrate components of glycoproteins. In *N*-linked glycoproteins, the carbohydrate component is attached to the protein via an *N*-glycosidic bond to an Asn residue in the sequence Asn-X-Ser/Thr. In *O*-linked glycoproteins, the carbohydrate attachment is an *O*-glycosidic bond to Ser or Thr, or in collagens, to 5-hydroxylysine. In GPI-anchored proteins a glycosylphosphatidylinositol group is linked to the protein through an intermediary phosphoethanolamine bridge, which forms an amide bond to the protein's C-terminal amino acid residue. Synthesis of *N*-linked oligosaccharides begins in the endoplasmic reticulum with the multistep formation of a lipid-linked precursor consisting of dolichol pyrophosphate bonded to a common 14-residue core oligosaccharide. The carbohydrate is then transferred to an Asn residue of a growing polypeptide chain. The correct folding of the immature *N*-linked glycoprotein is assisted via the calnexin/calreticulin cycle and it is subsequently transferred, via a membranous vesicle, to the cis Golgi network of the Golgi apparatus. Processing is completed by the trimming of mannose residues followed by attachment of a variety of other monosaccharides as catalyzed by specific enzymes in the cis, medial, and trans Golgi cisternae. Completed *N*-linked glycoproteins are sorted in the trans Golgi network according to the identities of their carbohydrate components for transport, via membranous vesicles, to their final cellular destinations. Three major types of *N*-linked oligosaccharides have been identified, high mannose, complex, and hybrid oligosaccharides, all of which contain a common pentasaccharide core. Studies of glycoprotein formation have been facilitated by the use of antibiotics, such as tunicamycin and bacitracin, which inhibit specific enzymes involved in the synthesis of these oligosaccharides. *O*-Linked oligosaccharides are synthesized in the Golgi apparatus by sequential attachments of specific monosaccharide units to certain Ser or Thr residues. Carbohydrate components of glycoproteins are thought to act as recognition markers for the transport of glycoproteins to their proper cellular destinations and for cell–cell and antibody recognition. The GPI membrane anchor is appended to proteins on the luminal surface of the endoplasmic reticulum, thereby targeting GPI-anchored proteins to the external surface of the plasma membrane.

4 ■ The Pentose Phosphate Pathway The cell uses NAD^+ in oxidative reactions and employs NADPH in reductive biosynthesis. NADPH is synthesized by the pentose phosphate pathway, an alternate mode of glucose oxidation. This pathway also synthesizes R5P for use in nucleotide biosynthesis. The first three reactions of the pentose phosphate pathway involve oxidation of G6P to Ru5P with release

of CO_2 and formation of two NADPH molecules. This is followed by reactions that either isomerize Ru5P to R5P or epimerize it to Xu5P. Each molecule of R5P not required for nucleotide biosynthesis, together with two Xu5P, is converted to two molecules of F6P and one molecule of GAP via the sequential actions of transketolase, transaldolase, and, again, transketolase. The products of the pentose phosphate pathway depend on the needs of the cell. The F6P and GAP may be metabolized through glycolysis and the citric acid cycle or recycled via gluconeogenesis. If NADPH is in excess, the latter portion of the pentose phosphate pathway may be reversed to synthesize R5P from glycolytic intermediates. The pentose phosphate pathway is controlled at its first committed step, the glucose-6-phosphate dehydrogenase reaction, by the $NADP^+$ concentration. A genetic deficiency in glucose-6-phosphate dehydrogenase leads to hemolytic anemia on administration of the antimalarial drug primaquine. This X-linked deficiency, which results from the accelerated degradation of the mutant enzyme, provides resistance against severe malaria to female heterozygotes and male hemizygotes for this trait.

REFERENCES

GLUCONEOGENESIS

Croniger, C.M., Olswang, Y., Reshef, L., Kalhan, S.C., Tilghman, S.M., and Hanson, R.W., Phosphoenolpyruvate carboxykinase revisited. Insights into its metabolic role, *Biochem. Mol. Biol. Educ.* **30,** 14–20 (2002); *and* Croniger, C.M., Chakravarty, K., Olswang, Y., Cassuto, H., Reshef, L., and Hanson, R.W., Phosphoenolpyruvate carboxykinase revisited. II. Control of PEPCK-C gene expression, *Biochem. Mol. Biol. Educ.* **30,** 353–362 (2002).

Hanson, R.W. and Reshef, L., Regulation of phosphoenolpyruvate carboxykinase (GTP) gene expression, *Annu. Rev. Biochem.* **66,** 581–611 (1997).

Knowles, J.R., The mechanism of biotin-dependent enzymes, *Annu. Rev. Biochem.* **58,** 195–221 (1989).

Krauss-Friedman, N., *Hormonal Control of Gluconeogenesis,* Vols. I and II, CRC Press (1986).

Matte, A., Tari, L.W., Goldie, H., and Delbaere, T.J., Structure and mechanism of phosphoenolpyruvate carboxykinase, *J. Biol. Chem.* **272,** 8105–8108 (1997).

Pilkis, S.J., Mahgrabi, M.R., and Claus, T.H., Hormonal regulation of hepatic gluconeogenesis and glycolysis, *Annu. Rev. Biochem.* **57,** 755–783 (1988).

Rothman, D.L., Magnusson, I., Katz, L.D., Shulman, R.G., and Shulman, G.I., Quantitation of hepatic gluconeogenesis in fasting humans with ^{13}C NMR, *Science* **254,** 573–576 (1991).

Van Schaftingen, E., and Gerin, I., The glucose-6-phosphatase system, *Biochem. J.* **362,** 513–532 (2002).

THE GLYOXYLATE CYCLE

Eastmond, P.J. and Graham, I.A., Re-examining the role of the glyoxylate cycle in oilseeds, *Trends Plant Sci.* **6,** 72–77 (2001).

OLIGOSACCHARIDE BIOSYNTHESIS

Abeijon, C. and Hirschberg, C.B., Topography of glycosylation reactions in the endoplasmic reticulum, *Trends Biochem. Sci.* **17,** 32–36 (1992).

Bause, E., Wesemann, M., Bartoschek, A., and Breuer, W., Epoxyethylglycyl peptides as inhibitors of oligosaccharyl-

transferase: double-labeling of the active site, *Biochem. J.* **322,** 95–102 (1997).

Burda, P. and Aebi, M., The dolichol pathway of *N*-linked glycosylation, *Biochim. Biophys. Acta* **1426,** 239–257 (1999).

Elbein, A.D., Inhibitors of the biosynthesis and processing of N linked oligosaccharide chains, *Annu. Rev. Biochem.* **56,** 497–534 (1987).

Englund, P.T., The structure and biosynthesis of glycosyl phosphatidylinositol protein anchors, *Annu. Rev. Biochem.* **62,** 65–100 (1993).

Ferguson, M.A.J., Brimacombe, J.S., Brown, J.R., Crossman, A., Dix, A., Field, R.A., Güther, M.L.S., Milne, K.G., Sharma, D.K., and Smith, T.K., The GPI biosynthetic pathway as a therapeutic target for African sleeping sickness, *Biochim. Biophys. Acta* **1455,** 327–340 (1999).

Florman, H.M. and Wasserman, P.M., *O*-Linked oligosaccharides of mouse egg ZP3 account for its sperm receptor activity, *Cell* **41,** 313–324 (1985).

Helenius, A. and Aebi, M., Intracellular functions of N-linked glycans, *Science* **291,** 2364–2369 (2001).

Helenius, A., Trombetta, E.S., Hebert, J.N., and Simons, J.F., Calnexin, calreticulin and the folding of glycoproteins, *Trends Cell Biol.* **7,** 193–200 (1997).

Hirschberg, C.B. and Snider, M.D., Topography of glycosylation in the rough endoplasmic reticulum and the Golgi apparatus, *Annu. Rev. Biochem.* **56,** 63–87 (1987).

Kornfeld, R. and Kornfeld, S., Assembly of asparagine-linked oligosaccharides, *Annu. Rev. Biochem.* **54,** 631–664 (1985).

Maeda, Y., Watanabe, R., Harris, C.L., Hong, Y., Ohishi, K., Kinoshita, K., and Kinoshita, T., PIG-M transfers the first mannose to glycosylphosphatidylinositol on the lumenal side of the ER, *EMBO J.* **20,** 250–261 (2001).

Parodi, A.J., Role of N-oligosaccharide endoplasmic reticulum processing reactions in glycoprotein folding and degradation, *Biochem. J.* **348,** 1–13 (2000); *and* Protein glucosylation and its role in protein folding, *Annu. Rev. Biochem.* **69,** 69–93 (2000).

Schachter, H., Enzymes associated with glycosylation, *Curr. Opin. Struct. Biol.* **1,** 755–765 (1991).

Schrag, J.D., Bereron, J.J.M., Li, Y., Borisova, S., Hahn, M., Thomas, D.Y., and Cygler, M., The structure of calnexin, an ER chaperone involved in quality control of protein folding, *Mol. Cell* **8**, 633–644 (2001).

Schwartz, R.T. and Datema, R., Inhibitors of trimming: New tools in glycoprotein research, *Trends Biochem. Sci.* **9**, 32–34 (1984).

Shaper, J.H. and Shaper, N.L., Enzymes associated with glycosylation, *Curr. Opin. Struct. Biol.* **2**, 701–709 (1992).

Tartakoff, A.M. and Singh, N., How to make a glycoinositol phospholipid anchor, *Trends Biochem. Sci.* **17**, 470–473 (1992).

von Figura, K. and Hasilik, A., Lysosomal enzymes and their receptors, *Annu. Rev. Biochem.* **55**, 167–193 (1986).

THE PENTOSE PHOSPHATE PATHWAY

Adams, M.J., Ellis, G.H., Gover, S., Naylor, C.E., and Phillips, C., Crystallographic study of coenzyme, coenzyme analogue and substrate binding in 6-phosphogluconate dehydrogenase: Implications for NADP specificity and enzyme mechanism, *Structure* **2**, 651–668 (1994).

Au, S.W.N., Gover, S., Lam, V.M.S., and Adams, M.J., Human glucose-6-phosphate dehydrogenase: The crystal structure reveals a structural NADP$^+$ molecule and provides insights into enzyme deficiency, *Structure* **8**, 293–303 (2000).

Beutler, E., The molecular biology of G6PD variants and other red cell enzyme defects, *Annu. Rev. Med.* **43**, 47–59 (1992).

Boros, L.G., et al. Oxythiamine and dehydroepiandrosterone inhibit the nonoxidative synthesis of ribose and tumor cell proliferation, *Cancer Res.* **57**, 4242–4248 (1997).

Jia, J., Huang, W., Schörken, U., Sahm, H., Sprenger, G.A., Lindqvist, Y., and Schneider, G., Crystal structure of transaldolase B from Escherichia coli suggests a circular permutation of the α/β barrel within the class I aldolase family, *Structure* **4**, 715–724 (1996).

Lindqvist, Y. and Schneider, G., Thiamin diphosphate dependent enzymes: transketolase, pyruvate oxidase and pyruvate decarboxylase, *Curr. Opin. Struct. Biol.* **3**, 896–901 (1993); *and* Muller, Y.A., Lindqvist, Y., Furey, W., Schulz, G.E., Jordan, F., and Schneider, G., A thiamin diphosphate binding fold revealed by comparison of the crystal structures of transketolase, pyruvate oxidase and pyruvate decarboxylase, *Structure* **1**, 95–103 (1993).

Luzzato, L., Mehta, A., and Vulliamy, T., Glucose-6-phosphate dehydrogenase deficiency, *in* Scriver, C.R., Beaudet, A., Sly, W.S., and Valle, D. (Eds.), *The Metabolic & Molecular Bases of Inherited Disease* (8th ed.), *pp.* 4517–4553, McGraw-Hill (2001).

Ruwende, C., et al., Natural selection of hemi- and heterozygotes for G6PD deficiency in Africa by resistance to severe malaria, *Nature* **376**, 246–249 (1995).

Wood, T., *The Pentose Phosphate Pathway,* Academic Press (1985).

PROBLEMS

1. Compare the relative energetic efficiencies, in ATPs per mole of glucose oxidized, of glucose oxidation via glycolysis + the citric acid cycle versus glucose oxidation via the pentose phosphate pathway + gluconeogenesis. Assume that NADH and NADPH are each energetically equivalent to three ATPs.

2. Although animals cannot synthesize glucose from acetyl-CoA, if a rat is fed ^{14}C-labeled acetate, some of the label will appear in the glycogen extracted from its muscles. Explain.

3. Substances that inhibit specific trimming steps in the processing of *N*-linked glycoproteins have been useful tools in elucidating the pathway of this process. Explain.

4. Through clever genetic engineering you have developed an unregulatable enzyme that can interchangeably use NAD$^+$ or NADP$^+$ in a redox reaction. What would be the physiological consequence(s) on an organism of having such an enzyme?

5. What is the free energy change of the reaction

$$NADH + NADP^+ \rightleftharpoons NAD^+ + NADPH$$

under physiological conditions (p. 862)? Assume that $\Delta G^{\circ\prime} = 0$ for this reaction and that $T = 37°C$.

6. If G6P is ^{14}C-labeled at its C2 position, what is the distribution of the radioactive label in the products of the pentose phosphate pathway after one turnover of the pathway? What is the distribution of the label after passage of these products through gluconeogenesis followed by a second round of the pentose phosphate pathway?

7. After feeding rapidly growing and proliferating cells [1,2-^{13}C]glucose and isolating the RNA, you find that both the C1 and C2 atoms of the ribosyl units are labeled. Show, using chemical structures and the appropriate enzymes, how the pentose phosphate pathway can yield this distribution of the label.

8. The relative metabolic activities in an organism of glycolysis + the citric acid cycle versus the pentose phosphate pathway + gluconeogenesis can be measured by comparing the rates of $^{14}CO_2$ generation on administration of glucose labeled with ^{14}C at C1 with that of glucose labeled at C6. Explain.

9. In light of the finding that an otherwise benign or even advantageous mutation leads to abnormal primaquine sensitivity combined with the fact that human beings have enormous genetic complexity, comment on the possibility of developing drugs that exhibit no atypical side effects in any individual.

Chapter
24

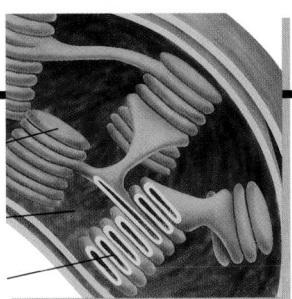

Photosynthesis

Life on Earth depends on the sun. *Plants and cyanobacteria chemically sequester light energy through photosynthesis, a light-driven process in which CO_2 is "fixed" to yield carbohydrates (CH_2O).*

$$CO_2 + H_2O \xrightarrow{\text{light}} (CH_2O) + O_2$$

This process, in which CO_2 is reduced and H_2O is oxidized to yield carbohydrates and O_2, is essentially the reverse of oxidative carbohydrate metabolism. Photosynthetically produced carbohydrates therefore serve as an energy source for the organism that produced them as well as for nonphotosynthetic organisms that directly or indirectly consume photosynthetic organisms. In fact, even modern industry is highly dependent on the products of photosynthesis because coal, oil, and gas (the so-called fossil fuels) are thought to be the remains of ancient organisms. It is estimated that photosynthesis annually fixes $\sim 10^{11}$ tons of carbon, which represents the storage of over 10^{18} kJ of energy. Moreover, photosynthesis, over the eons, has produced the O_2 in Earth's atmosphere (Section 1-5C).

The notion that plants obtain nourishment from such insubstantial things as light and air took nearly two centuries to develop. In 1648, the Flemish physician Jean-Baptiste von Helmont reported that growing a potted willow tree from a shoot caused an insignificant change in the weight of the soil in which the tree had been rooted. Although another century was to pass before the law of conservation of matter was formulated, van Helmont attributed the tree's weight gain to the water it had taken up. This idea was extended in 1727 by Stephen Hales, who proposed that plants extract some of their matter from the air.

The first indication that plants produce oxygen was found by the English clergyman and pioneering chemist Joseph Priestley, who reported:

Finding that candles burn very well in air in which plants had grown a long time, and having some reason to think, that there was something attending vegetation, which restored air that had been injured by respiration, I thought it was possible that the same process might also restore the air that had been injured by the burning of candles. Accordingly, on the 17th of August, 1771, I put a sprig of mint into a quantity of air, in which a wax candle had burned out, and found that, on the 27th of the same month, another candle burned perfectly well in it.

Although Priestley later discovered oxygen, which he named "dephlogisticated air," it was Antoine Lavoisier who elucidated its role in combustion and respiration. Nevertheless, Priestley's work inspired the Dutch physician Jan Ingen-Housz, who in 1779 demonstrated that the "purifying" power of plants resides in the influence of sunlight on their green parts. In 1782, the Swiss pastor Jean Senebier showed that CO_2, which he called "fixed air," is taken up during photosynthesis. His compatriot Théodore de Saussure found, in 1804, that the combined weights of the organic matter produced by plants and the oxygen they evolve is greater than the weight of the CO_2 they consume. He therefore concluded that water, the only other substance he added to his system, was also necessary for photosynthesis. The final ingredient in the overall photosynthetic recipe was established in 1842 by the German physiologist Robert Mayer, one of the formulators of the first law of thermodynamics, who concluded that plants convert light energy to chemical energy.

1 ■ CHLOROPLASTS

*The site of photosynthesis in eukaryotes (algae and higher plants) is the **chloroplast*** (Section 1-2A), a member of the membranous subcellular organelles peculiar to plants known as **plastids.** The first indication that chloroplasts have a photosynthetic function was Theodor Engelmann's observation, in 1882, that small, motile, O_2-seeking bacteria congregate at the surface of the alga *Spirogyra,* overlying its single chloroplast, but only while the chloroplast is illuminated. Chloroplasts must therefore be the site of light-induced O_2 evolution, that is, photosynthesis. Chloroplasts, of which there are 1 to 1000 per cell, vary considerably in size and shape but are typically ~5-μm-long ellipsoids. Like mitochondria, which they resemble in many ways, chloroplasts have a highly permeable outer membrane and a nearly impermeable inner membrane separated by a narrow intermembrane space (Fig. 24-1). The inner membrane encloses the **stroma,** a concentrated solution of enzymes much like the mitochondrial matrix, that also contains the DNA, RNA, and ribosomes involved in the synthesis of several chloroplast proteins. The stroma, in turn, surrounds a third membranous compartment, the **thylakoid** (Greek: *thylakos,* a sac or pouch). The thylakoid is probably a single highly folded vesicle, although in most organisms it appears to consist of stacks of disklike sacs named **grana,** which are interconnected by unstacked **stromal lamellae.** A chloroplast usually contains 10 to 100 grana. Thylakoid membranes arise from invaginations in the inner membrane of developing chloroplasts and therefore resemble mitochondrial cristae.

The lipids of the thylakoid membrane have a distinctive composition. They consist of only ~10% phospholipids; the majority, ~80%, are uncharged **mono-** and **digalactosyl diacylglycerols,** and the remaining ~10% are the sulfolipids **sulfoquinovosyl diacylglycerols (quinovose** is 6-deoxyglucose):

X = OH   **Galactosyl diacylglycerol**

Digalactosyl diacylglycerol

$X = SO_3^-$
* = C atom inverted   **Sulfoquinovosyl diacylglycerol**

The acyl chains of these lipids have a high degree of unsaturation, which gives the thylakoid membrane a highly fluid character.

Photosynthesis occurs in two distinct phases:

1. The **light reactions,** which use light energy to generate NADPH and ATP.

2. The **dark reactions,** actually light-independent reactions, which use NADPH and ATP to drive the synthesis of carbohydrate from CO_2 and H_2O.

The light reactions occur in the thylakoid membrane and involve processes that resemble mitochondrial electron transport and oxidative phosphorylation (Sections 22-2 and 22-3). In photosynthetic prokaryotes, which lack chloroplasts, the light reactions take place in the cell's plasma (inner) membrane or in highly invaginated structures derived from it called **chromatophores** (e.g., Fig. 24-2;

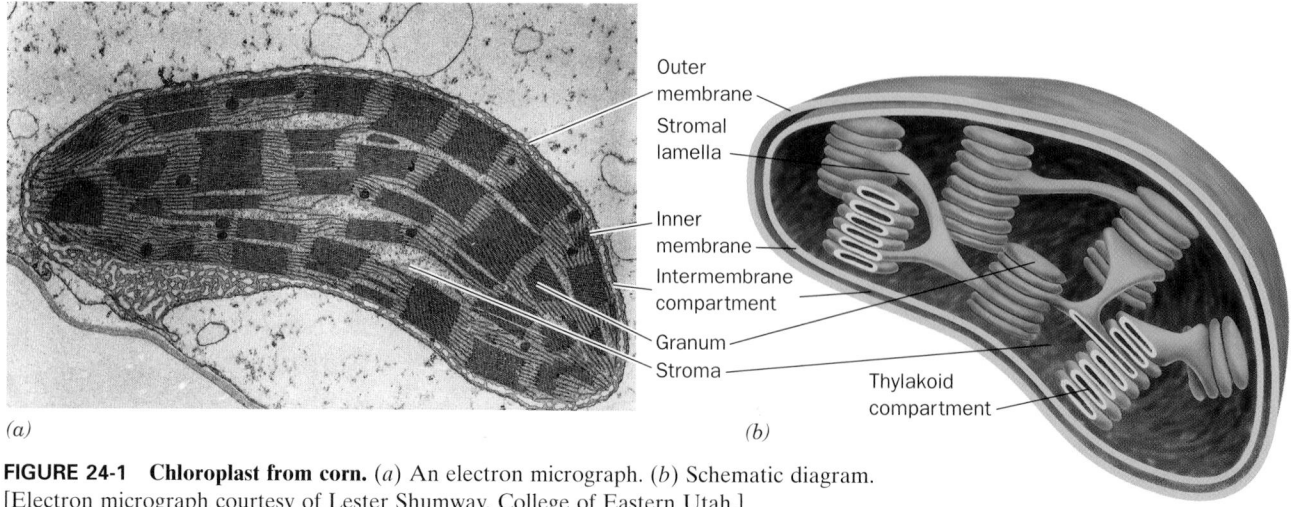

(a)

(b)

FIGURE 24-1   Chloroplast from corn. (*a*) An electron micrograph. (*b*) Schematic diagram.
[Electron micrograph courtesy of Lester Shumway, College of Eastern Utah.]

Labels: Outer membrane / Stromal lamella / Inner membrane / Intermembrane compartment / Granum / Stroma / Thylakoid compartment

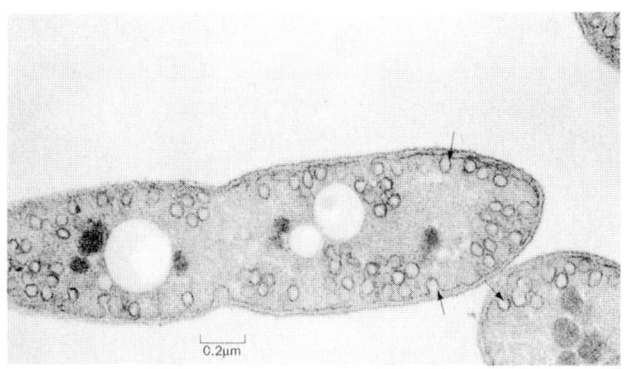

FIGURE 24-2 Electron micrograph of a section through the purple photosynthetic bacterium *Rhodobacter sphaeroides.* Its plasma membrane invaginates to form externally connected tubules known as chromatophores (*arrows;* seen here in circular cross section) that are the sites of photosynthesis. [Courtesy of Gerald A. Peters, Virginia Commonwealth University.]

recall that chloroplasts evolved from cyanobacteria that assumed a symbiotic relationship with a nonphotosynthetic eukaryote; Section 1-2A). In eukaryotes, the dark reactions occur in the stroma through a cyclic series of enzyme-catalyzed reactions. In the following sections, we consider the light and dark reactions in detail.

2 ■ LIGHT REACTIONS

In the first decades of the twentieth century, it was generally assumed that light, as absorbed by photosynthetic pigments, directly reduced CO_2, which, in turn, combined with water to form carbohydrate. In this view, CO_2 is the source of the O_2 generated by photosynthesis. In 1931, however, Cornelis van Niel showed that green photosynthetic bacteria, anaerobes that use H_2S in photosynthesis, generate sulfur:

$$CO_2 + 2H_2S \xrightarrow{light} (CH_2O) + 2S + H_2O$$

The chemical similarity between H_2S and H_2O led van Niel to propose that the general photosynthetic reaction is

$$CO_2 + 2H_2A \xrightarrow{light} (CH_2O) + 2A + H_2O$$

where H_2A is H_2O in green plants and cyanobacteria and H_2S in photosynthetic sulfur bacteria. This suggests that photosynthesis is a two-stage process in which light energy is harnessed to oxidize H_2A (the light reactions):

$$2H_2A \xrightarrow{light} 2A + 4[H]$$

and the resulting reducing agent [H] subsequently reduces CO_2 (the dark reactions):

$$4[H] + CO_2 \rightarrow (CH_2O) + H_2O$$

Thus, in aerobic photosynthesis, H_2O, not CO_2, is photolyzed (split by light).

The validity of van Niel's hypothesis was established unequivocally by two experiments. In 1937, Robert Hill

discovered that when isolated chloroplasts that lack CO_2 are illuminated in the presence of an artificial electron acceptor such as ferricyanide $[Fe(CN)_6^{3-}]$, O_2 is evolved with concomitant reduction of the acceptor [to ferrocyanide, $Fe(CN)_6^{4-}$, in our example]. This so-called **Hill reaction** demonstrates that CO_2 does not participate directly in the O_2-producing reaction. It was discovered eventually that the natural photosynthetic electron acceptor is $NADP^+$ (Fig. 13-2), whose reduction product, NADPH, is utilized in the dark reactions to reduce CO_2 to carbohydrate (Section 24-3A). In 1941, when the oxygen isotope ^{18}O became available, Samuel Ruben and Martin Kamen directly demonstrated that the source of the O_2 formed in photosynthesis is H_2O:

$$H_2^{18}O + CO_2 \xrightarrow{light} (CH_2O) + {}^{18}O_2$$

This section is a discussion of the major aspects of the light reactions.

A. *Absorption of Light*

The principal photoreceptor in photosynthesis is **chlorophyll**. This cyclic tetrapyrrole, like the heme group of globins and cytochromes (Sections 10-1A and 22-2C), is derived biosynthetically from protoporphyrin IX. Chlorophyll, however, differs from heme in four major respects (Fig. 24-3):

1. Its central metal ion is Mg^{2+} rather than Fe(II) or Fe(III).

2. It has a cyclopentenone ring, Ring V, fused to pyrrole Ring III.

3. Pyrrole Ring IV is partially reduced in **chlorophyll *a* (Chl *a*)** and **chlorophyll *b* (Chl *b*)**, the two major chlorophyll varieties in eukaryotes and cyanobacteria, whereas in **bacteriochlorophyll *a* (BChl *a*)** and **bacteriochlorophyll *b* (BChl *b*)**, the principal chlorophylls of photosynthetic bacteria, Rings II and IV are partially reduced.

4. The propionyl side chain of Ring IV is esterified to a tetraisoprenoid alcohol. In Chl *a* and *b* as well as in BChl *b* it is **phytol** but in BChl *a* it is either phytol or **geranylgeraniol,** depending on the bacterial species.

In addition, Chl *b* has a formyl group in place of the methyl substitutent to atom C3 of Ring II of Chl *a*. Similarly, BChl *a* and BChl *b* have different substituents to atom C4.

a. Light and Matter Interact in Complex Ways

As photosynthesis is a light-driven process, it is worthwhile reviewing how light and matter interact. Electromagnetic radiation is propagated as discrete **quanta (photons)** whose energy E is given by **Planck's law:**

$$E = h\nu = \frac{hc}{\lambda} \qquad [24.1]$$

where h is **Planck's constant** (6.626×10^{-34} J · s), c is the speed of light (2.998×10^8 m · s^{-1} in a vacuum), ν is the

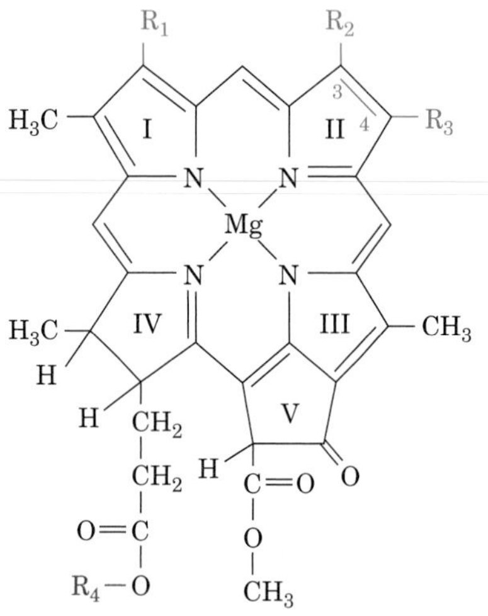

Chlorophyll 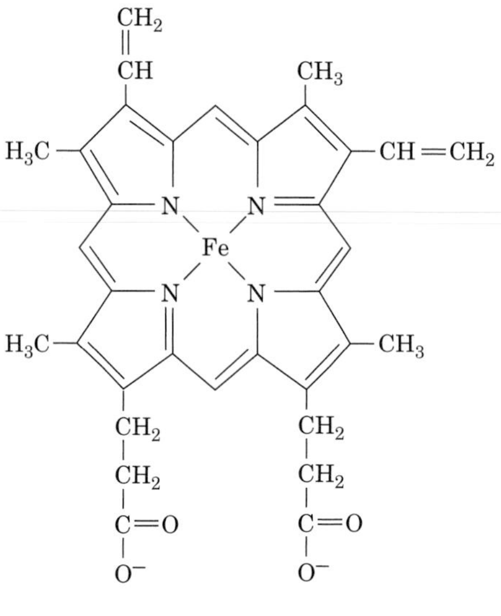   **Iron–protoporphyrin IX**

	R₁	R₂	R₃	R₄
Chlorophyll *a*	$-CH=CH_2$	$-CH_3$	$-CH_2-CH_3$	P
Chlorophyll *b*	$-CH=CH_2$	$\overset{O}{\overset{\|}{-C}}-H$	$-CH_2-CH_3$	P
Bacteriochlorophyll *a*	$\overset{O}{\overset{\|}{-C}}-CH_3$	$-CH_3^{\ a}$	$-CH_2-CH_3^{\ a}$	P or G
Bacteriochlorophyll *b*	$\overset{O}{\overset{\|}{-C}}-CH_3$	$-CH_3^{\ a}$	$=CH-CH_3^{\ a}$	P

a No double bond between positions C3 and C4.

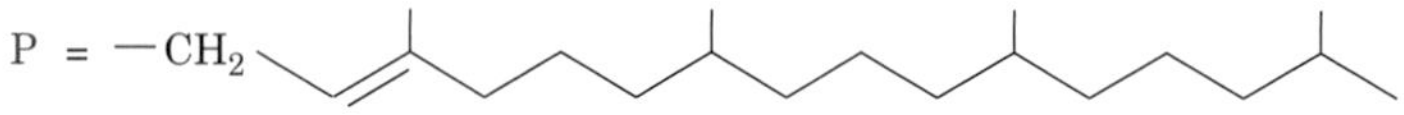

P = $-CH_2$ ~~~~~~~~~~

Phytyl side chain

G = $-CH_2$ ~~~~~~~~~~

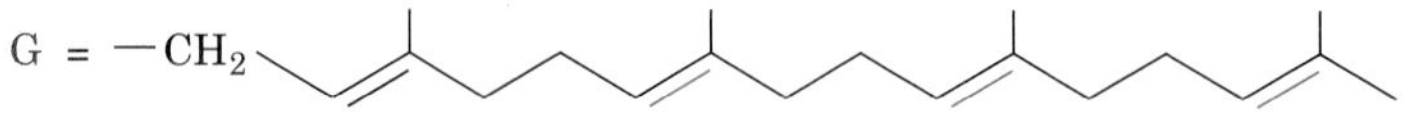

Geranylgeranyl side chain

FIGURE 24-3 Chlorophyll structures. The molecular formulas of chlorophylls *a* and *b* and bacteriochlorophylls *a* and *b* are compared to that of iron protoporphyrin IX (heme). The isoprenoid phytyl and geranylgeranyl tails presumably increase the chlorophylls' solubility in nonpolar media.

frequency of the radiation, and λ is its wavelength (visible light ranges in wavelength from 400 to 700 nm). Thus red light with λ = 680 nm has an energy of 176 kJ · einstein^{-1} (an **einstein** is a mole of photons).

Molecules, like atoms, have numerous electronic quantum states of differing energies. Moreover, because molecules contain more than one nucleus, each of their electronic states has an associated series of vibrational and rotational substates that are closely spaced in energy (Fig. 24-4). Absorption of light by a molecule usually occurs through the promotion of an electron from its ground (lowest energy) state molecular orbital to one of higher energy. However, *a given molecule can only absorb photons of certain wavelengths because, as is required by the law of con-servation of energy, the energy difference between the two states must exactly match the energy of the absorbed photon.*

The amount of light absorbed by a substance at a given wavelength is described by the **Beer–Lambert** law:

$$A = \log\frac{I_0}{I} = \varepsilon cl \qquad [24.2]$$

where *A* is the absorbance, I_0 and *I* are, respectively, the intensities of the incident and transmitted light, *c* is the molar concentration of the sample, *l* is the length of the light path through the sample in cm, and ε is the molecule's **molar extinction coefficient.** Consequently, a plot of *A* versus λ for a given molecule, its **absorption spectrum** (Fig. 24-5), is indicative of its electronic structure.

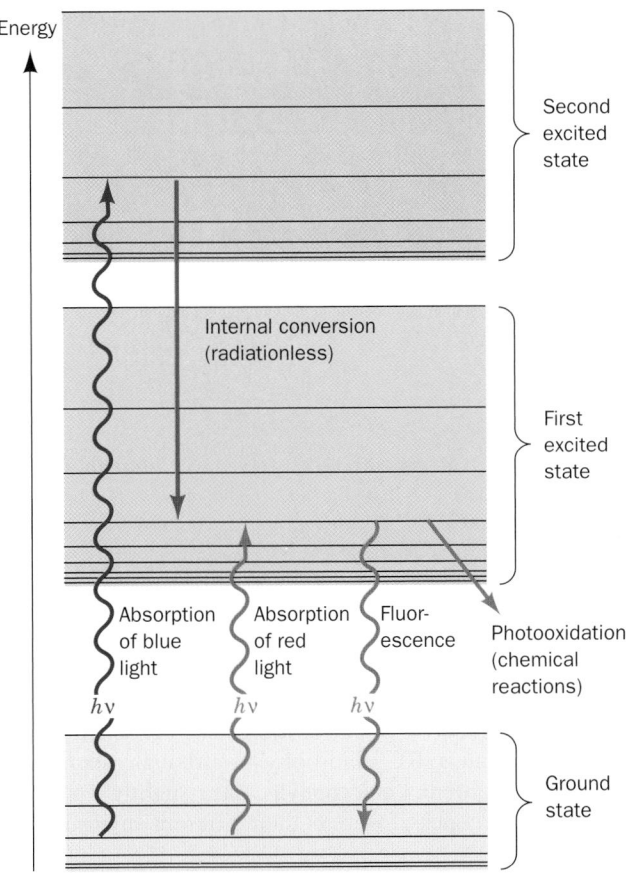

FIGURE 24-4 Energy diagram indicating the electronic states of chlorophyll and their most important modes of interconversion. The wiggly arrows represent the absorption of photons or their fluorescent emission. Excitation energy may also be dissipated in radiationless processes such as internal conversion (heat production) or chemical reactions. ☙ **See the Animated Figures**

The various chlorophylls are highly conjugated molecules (Fig. 24-3). It is just such molecules that strongly absorb visible light (the spectral band in which the solar radiation reaching Earth's surface is of peak intensity; Fig. 24-5). In fact, the peak molar extinction coefficients of the various chlorophylls, over $10^5 \, M^{-1} \cdot cm^{-1}$, are among the highest known for organic molecules. Yet the relatively small chemical differences among the various chlorophylls greatly affect their absorption spectra. These spectral differences, as we shall see, are functionally significant.

An electronically excited molecule can dissipate its excitation energy in many ways. Those modes with the greatest photosynthetic significance are as follows (Fig. 24-4):

1. Internal conversion, a common mode of decay in which electronic energy is converted to the kinetic energy of molecular motion, that is, to heat. This process occurs very rapidly, being complete in $<10^{-11}$ s. Many molecules relax in this manner to their ground states. Chlorophyll molecules, however, usually relax only to their lowest excited states. Therefore, *the photosynthetically applicable excitation energy of a chlorophyll molecule that has absorbed a photon in its short-wavelength band, which corresponds to its second excited state, is no different than if it had absorbed a photon in its less energetic long-wavelength band.*

2. Fluorescence, in which an electronically excited molecule decays to its ground state by emitting a photon. Such a process requires $\sim 10^{-8}$ s, so it occurs much more slowly than internal conversion. Consequently, a fluorescently emitted photon generally has a longer wavelength (lower energy) than that initially absorbed. Fluorescence accounts for the dissipation of only 3 to 6% of the light energy absorbed by living plants. However, chlorophyll in solution,

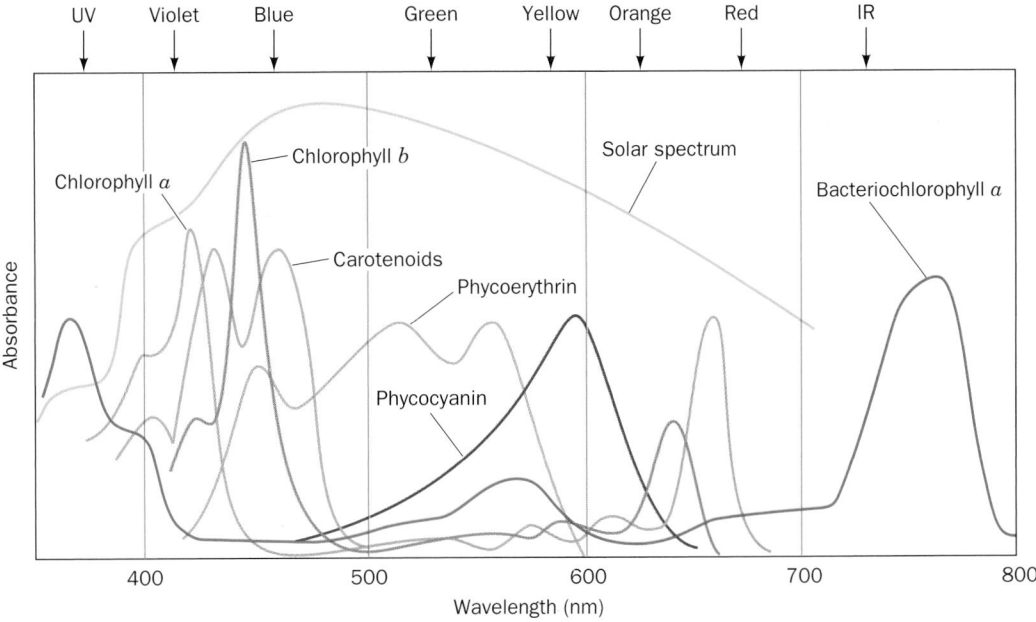

FIGURE 24-5 Absorption spectra of various photosynthetic pigments. The chlorophylls each have two absorption bands, one in the red and one in the blue. Phycoerythrin absorbs blue and green light, whereas phycocyanin absorbs yellow light. Together, these pigments absorb most of the visible light in the solar spectrum. [After a drawing by Govindjee, University of Illinois.]

where of course the photosynthetic uptake of this energy cannot occur, has an intense red fluorescence.

3. Exciton transfer (also known as **resonance energy transfer**), in which an excited molecule directly transfers its excitation energy to nearby unexcited molecules with similar electronic properties. This process occurs through interactions between the molecular orbitals of the participating molecules in a manner analogous to the interactions between mechanically coupled pendulums of similar frequencies. An exciton (excitation) may be serially transferred between members of a group of molecules or, if their electronic coupling is strong enough, the entire group may act as a single excited "supermolecule." We shall see that *exciton transfer is of particular importance in funneling light energy to photosynthetic reaction centers.*

4. Photooxidation, in which a light-excited donor molecule is oxidized by transferring an electron to an acceptor molecule, which is thereby reduced. This process occurs because the transferred electron is less tightly bound to the donor in its excited state than it is in the ground state. In photosynthesis, excited chlorophyll (Chl^*) is such a donor. *The energy of the absorbed photon is thereby chemically transferred to the photosynthetic reaction system.* Photooxidized chlorophyll, Chl^+, a cationic free radical, eventually returns to its ground state by oxidizing some other molecule.

b. Light Absorbed by Antenna Chlorophylls Is Transferred to Photosynthetic Reaction Centers

The primary reactions of photosynthesis, as is explained in Sections 24-2B and 24-2C, take place at **photosynthetic reaction centers (RCs).** Yet *photosynthetic organelles contain far more chlorophyll molecules than RCs.* This was demonstrated in 1932 by Robert Emerson and William Arnold in their studies of O_2 production by the green alga *Chlorella* (a favorite experimental subject), which had been exposed to repeated brief (10-μs) flashes of light. The amount of O_2 generated per flash was maximal when the interval between flashes was at least 20 ms. Evidently, this is the time required for a single turnover of the photosynthetic reaction cycle. Emerson and Arnold then measured the variation of O_2 yield with flash intensity when the flash interval was the optimal 20 ms. With weak flashes, the O_2 increased linearly with flash intensity such that about one molecule of O_2 was generated per eight photons absorbed (Fig. 24-6). With increasing flash intensity the efficiency of this process fell off, no doubt because the number of photons began to approach the number of photochemical units. What was unanticipated, however, was that each flash of saturating intensity produced only one molecule of O_2 per ~2400 molecules of chlorophyll present. Since at least eight photons must be sequentially absorbed to liberate one O_2 molecule (Section 24-2C), these results suggest that the photosynthetic apparatus contains ~2400/8 = 300 chlorophyll molecules per RC.

With such a great excess of chlorophyll molecules per RC, it seems unlikely that all participate directly in photochemical reactions. Rather, as subsequent experiments

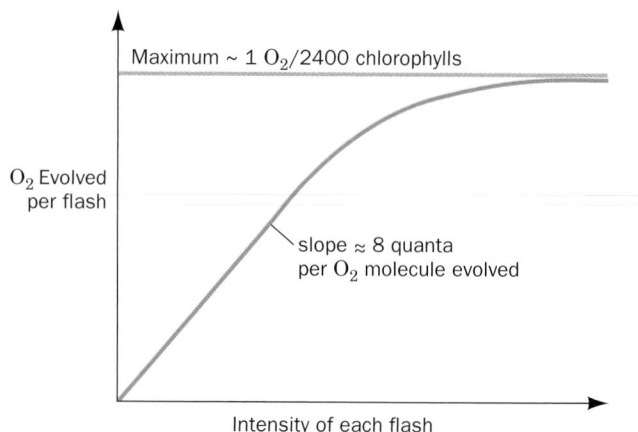

FIGURE 24-6 The amount of O_2 evolved by *Chlorella* algae versus the intensity of light flashes. Flashes are separated by dark intervals of >20 ms.

have shown, *most chlorophylls function to gather light; that is, they act as light-harvesting antennas.* These **antenna chlorophylls** pass the energy of an absorbed photon, by exciton transfer, from molecule to molecule until the excitation reaches an RC (Fig. 24-7a). There, the excitation is trapped because RC chlorophylls, although chemically identical to antenna chlorophylls, have slightly lower excited state energies because of their different environments (Fig. 24-7b).

Transfer of energy from the antenna system to an RC occurs in $<10^{-10}$ s with an efficiency of >90%. This high efficiency depends on the chlorophyll molecules having appropriate spacings and relative orientations. Even in bright sunlight, an RC intercepts only ~1 photon per second, a metabolically insignificant rate, and hence, these **light-harvesting complexes (LHCs)** serve an essential function.

c. The LHCs of Purple Photosynthetic Bacteria Contain Multiple Symmetrically Arranged Light-Absorbing Molecules

Most **purple photosynthetic bacteria,** which are among the simplest photosynthetic organisms, have two types of LHCs, **LH1** and **LH2,** that are transmembrane proteins but have different spectral and biochemical properties. LH2, which absorbs light at shorter wavelengths than LH1, rapidly passes the energy from the photons it absorbs to LH1, which, in turn, passes it to the RC. The X-ray structure of LH2 from the purple photosynthetic bacterium *Rhodospirillum (Rs.) molischianum* (Fig. 24-8), determined by Hartmut Michel, reveals that this protein is an eightfold rotationally symmetric $\alpha_8\beta_8$ 16-mer that binds 24 bacteriochlorophyll a (BChl a) molecules and 8 **lycopene** molecules (a **carotenoid;** see below):

Lycopene

The α and β subunits (56 and 45 residues, respectively) both consist largely of single helices that are aligned nearly

(a)

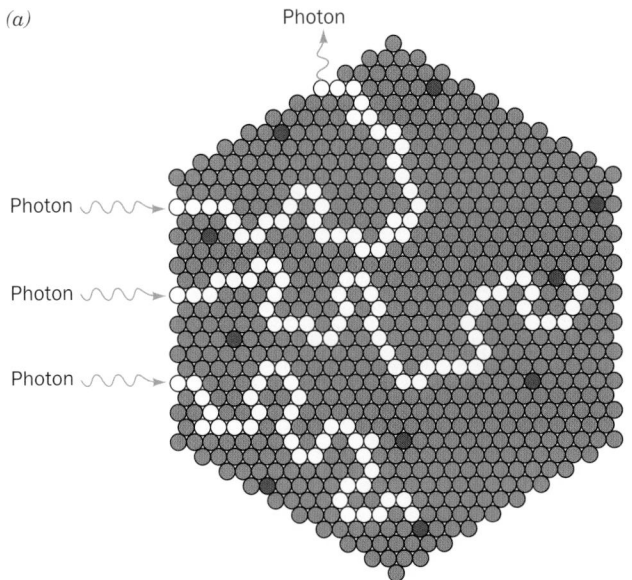

(b)

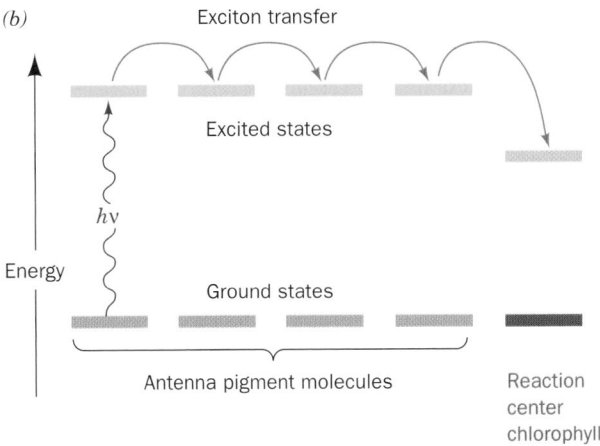

FIGURE 24-7 Flow of energy through a photosynthetic antenna complex. (*a*) The excitation resulting from photon absorption randomly migrates by exciton transfer among the molecules of the antenna complex (*light green circles*) until it is either trapped

by an RC chlorophyll (*dark green circles*) or, less frequently, fluorescently reemitted. (*b*) The excitation is trapped by the RC chlorophyll because its lowest excited state has a lower energy than those of the antenna pigment molecules.

perpendicularly to the plane of the membrane in which they are embedded. The eight α subunits pack side by side to form a hollow cylinder of diameter ~31-Å (as measured between helix axes). Each of the eight β subunits occupies

a position radially outward from an α subunit to form a concentric cylinder of diameter ~62 Å. Sixteen of the BChl *a* molecules are packed between these rings of helices in an arrangement resembling a 16-bladed turbine: Successive nearly parallel BChl *a* ring systems are in partial van der Waals contact (their Mg^{2+} ions are ~9 Å apart) with their

(a)

FIGURE 24-8 X-Ray structure of LH2 from *Rs. molischianum*. The α subunits are purple and the β subunits are pink. The bound chromophores are drawn in stick form with the BChl *a*'s green and the lycopenes yellow. The phytyl tails of the BChl *a*'s have been truncated for clarity. (*a*) View perpendicular to the bacterial membrane from the cytoplasm. The polypeptide chains are represented by tubes tracing their

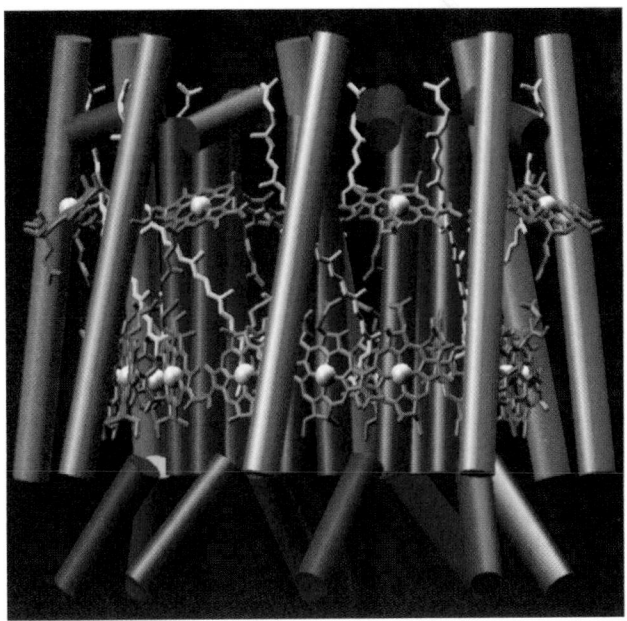

(b)

C_α atoms. (*b*) View parallel to the membrane with the cytoplasm above. The protein subunits are represented by only their helices, which are drawn as cylinders. The Mg^{2+} ions are drawn as white spheres. [Courtesy of Juergen Koepke and Hartmut Michel, Max-Planck-Institut für Biochemie, Frankfurt, Germany. PDBid 1LGH.] *See the* **Interactive Exercises**

planes perpendicular to the plane of the membrane. Their Mg^{2+} atoms are each singly axially liganded by His side chains [much like the Fe(II) in deoxyhemoglobin] that alternately extend from an α and a β subunit around the lower end of the cylinder. The remaining eight BChl *a* molecules, which are each singly axially liganded by a side chain of Asp 6α near the upper end of the cylinder, are arranged in an 8-fold symmetric ring between successive β subunit helices and are oriented with the planes of their ring systems tilted by ~35° relative to the plane of the membrane. The eight lycopene molecules are sandwiched between the α and β subunits and extend along much of their lengths, thereby contacting both sets of BChl *a* molecules. The LH2 from *Rhodopseudomonas (Rps.) acidophila*, another purple photosynthetic bacterium, is an $\alpha_9\beta_9$ 18-mer but otherwise has a similar structure in its transmembrane region to that of *Rs. molischianum*, even though their α and β subunits are only 26 and 31% identical.

Spectroscopic measurements indicate that an LH2's His-liganded and closely associated BChl *a* molecules maximally absorb radiation at a wavelength of 850 nm (and hence are called B850) and are strongly coupled, that is, they absorb radiation almost as a unit. The other, more loosely associated BChl *a* molecules (B800) maximally absorb radiation at 800 nm, largely as individual molecules (BChl *a*'s local environment in the protein alters its spectrum from that in solution; Fig. 24-5). When a B800 BChl *a* absorbs a photon, the excitation is rapidly [in ~700 femtoseconds (fs); 1 fs = 10^{-15} s] transferred to a lower energy B850 BCl *a* (which may independently absorb a photon), which even more rapidly (in ~100 fs) exchanges the excitation among the other B850 BChl *a* molecules. Hence, the B850 system acts as a kind of energy storage ring that delocalizes the excitation over a large region. The carotenoid molecules in this system absorb visible (<800 nm) light and may also facilitate the transmission of excitation between the rather distantly separated (19 Å between Mg atoms) nearest-neighbor B850 and B800 BChl *a* molecules.

LH1, like LH2, has α and β subunits of ~50 residues each. The low (8.5-Å) resolution structure of LH1 from *Rs. rubrum*, as determined by electron crystallography, reveals that it resembles LH2 but with 16-fold rotational symmetry, and forms a 116-Å-diameter cylinder with a 68-Å-diameter hole down its center. This hole is of sufficient size to contain an RC (see below), as electron microscopic studies indicate is, in fact, the case (Fig. 24-9). LH1's BChl *a* molecules absorb radiation at a longer wavelength than those of LH2 and consequently, when these two assemblies are in contact, excitation is rapidly [in 1–5 picoseconds (ps); 1 ps = 10^{-12} s] transferred from LH2 to LH1 and then (in 20–40 ps) to LH1's enclosed RC. Excitations may also be rapidly exchanged between contacting LH2s. Thus, this antenna system transfers virtually all of the radiation energy it absorbs to the RC in far less than the few nanoseconds (ns; 1 ns = 10^{-9} s) over which these excitations would otherwise decay. It should be noted that this complicated arrangement of **chromophores** (light-absorbing molecules) is among the simplest known; those

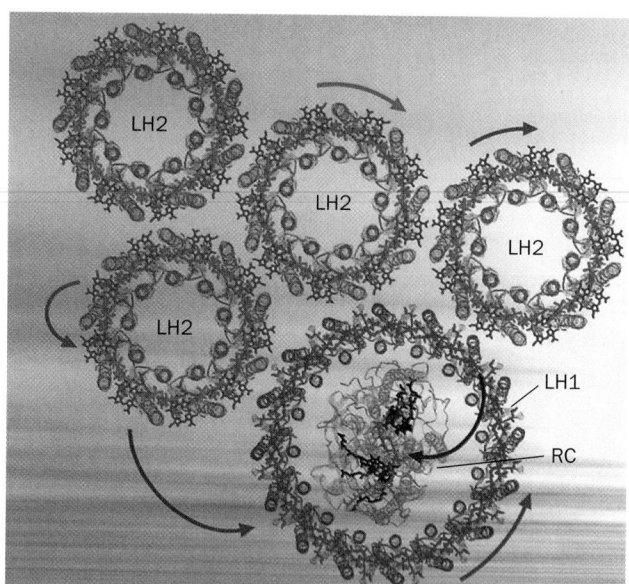

FIGURE 24-9   Model of the light-absorbing antenna system of purple photosynthetic bacteria. Several LH2s associate with each other and with LH1, which surrounds the photosynthetic reaction center (RC). The BChl *a*'s of LH2 B850 and LH1 are green, those of LH2 B800 are purple, and the light-absorbing pigments of the RC (see below) are red and black. Light absorbed by the Bchl *a* and lycopene molecules of an LH2 is rapidly transferred (*curved arrows*), often via other contacting LH2s, to LH1, which, in turn, transfers the excitation to its enclosed RC. [From Bhattarchardee, Y., *Nature* **412**, 474 (2001).]

of the light-harvesting systems of plants are even more elaborate (see below).

d. LHCs Contain Accessory Pigments

Most LHCs contain organized arrays of other light-absorbing substances in addition to chlorophyll. These **accessory pigments** function to fill in the absorption spectra of the antenna complexes in spectral regions where chlorophylls do not absorb strongly (Fig. 24-5). **Carotenoids,** which are C_{40} largely linear polyenes such as lycopene and **β-carotene,**

β-Carotene

are components of all green plants and many photosynthetic bacteria and are therefore the most common accessory pigments (they are largely responsible for the brilliant fall colors of deciduous trees as well as for the orange color of carrots, after which carotenoids are named):

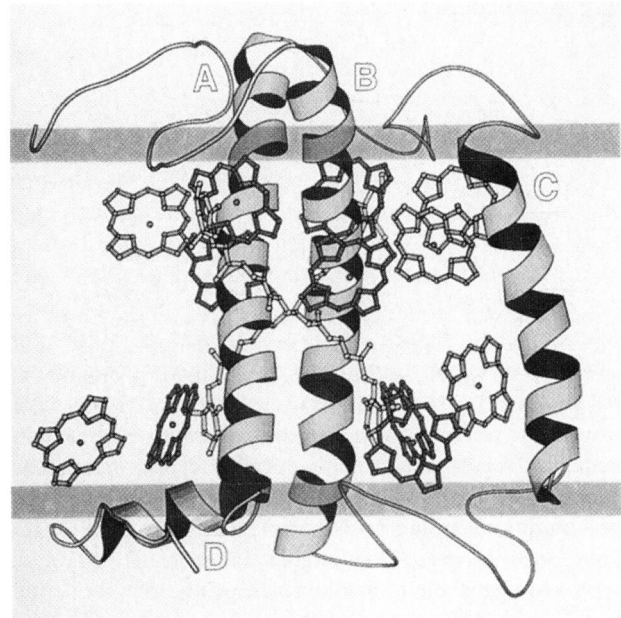

FIGURE 24-10 Structure of a subunit of the trimeric protein LHC-II from pea chloroplasts as determined by electron crystallography. This highly conserved transmembrane protein's seven Chl *a*, five Chl *b* (as represented by only their porphyrin rings), and two carotenoid molecules are dark green, light green, and yellow, respectively, and its chlorophyll-bound Mg^{2+} ions are pink. The blue bands indicate the approximate boundaries of the thylakoid membrane in which the protein is normally embedded with its upper side, as drawn, facing the stroma. Note that much of the protein, including its A and B helices and most of its pigment molecules, exhibits approximate twofold symmetry with the pseudo-twofold axis perpendicular to the plane of the membrane. [Courtesy of Werner Kühlbrandt, Max Planck Institute of Biophysics, Frankfort, Germany.]

LHC-II, the most abundant membrane protein in the chloroplasts of green plants, is a 232-residue transmembrane protein that binds at least seven Chl *a*'s, five Chl *b*'s, and two carotenoids (Fig. 24-10), thereby accounting for around half the chlorophyll in the biosphere. Carotenoids serve an additional function besides that of light-gathering antennas: Through electronic interactions, they prevent their associated light-excited chlorophyll molecules from transferring this excitation to O_2, which would otherwise yield a highly reactive and hence destructive form of O_2.

Aquatic photosynthetic organisms, which are responsible for nearly half of the photosynthesis on Earth, additionally contain other types of accessory pigments. This is because light outside the wavelengths 450 to 550 nm (blue and green light) is absorbed almost completely by passage through more than 10 m of water. In red algae and cyanobacteria, Chl *a* is therefore replaced as an antenna pigment by a series of linear tetrapyrroles, notably the red **phycoerythrobilin** and the blue **phycocyanobilin:**

Peptide-linked Phycoerythrobilin and Phycocyanobilin

The lowest excited states of these so-called **bilins** have higher energies than those of the chlorophylls, thereby facilitating energy transfer to the RC. The bilins are covalently linked via Cys S atoms to **phycobiliproteins** to form **phycoerythrin** and **phycocyanin** (spectra in Fig. 24-5). These, in turn, are organized in high molecular mass particles called **phycobilisomes** that are bound to the outer faces of photosynthetic membranes so as to funnel excitation energy to RCs over long distances with >90% efficiency.

B. *Electron Transport in Purple Photosynthetic Bacteria*

Photosynthesis is a process in which electrons from excited chlorophyll molecules are passed through a series of acceptors that convert electronic energy to chemical energy. Thus two questions arise: (1) What is the mechanism of energy transduction; and (2) how do photooxidized chlorophyll molecules regain their lost electrons? We shall see that photosynthetic bacteria solve these problems somewhat differently from cyanobacteria and plants. We first discuss these mechanisms in photosynthetic bacteria, where they are simpler and better understood. Electron transport in cyanobacteria and plants is the subject of Section 24-2C.

a. The Photosynthetic Reaction Center Is a Transmembrane Protein Containing a Variety of Chromophores

The first indication that chlorophyll undergoes direct photooxidation during photosynthesis was obtained by Louis Duysens in 1952. He observed that illumination of membrane preparations from the purple photosynthetic bacterium *Rs. rubrum* caused a slight (~2%) bleaching of their absorbance at 870 nm, which returned to their original levels in the dark. Duysens suggested that this bleaching is caused by photooxidation of a bacteriochlorophyll complex that he named **P870** (P for pigment and 870 nm for the position of the major long-wavelength absorption band of BChl *a*; photosynthetic bacteria tend to inhabit murky stagnant ponds, so that they require an infrared-absorbing species of chlorophyll). The ability to detect the presence of P870 eventually led to the purification and characterization of the RC to which it is bound.

RC particles from several species of purple photosynthetic bacteria **(PbRCs)** have similar compositions. That from *Rps. viridis* consists of three hydrophobic subunits: H (258 residues), L (273 residues), and M (323 residues). The L and M subunits of this membrane-spanning particle collectively bind four molecules of BChl *b* (which maximally absorbs light at 960 nm), two molecules of **bacteriopheophytin *b*** (**BPheo *b***; BChl *b* in which the Mg^{2+} is replaced by two protons), one nonheme/non-Fe–S Fe(II) ion, one molecule of the redox coenzyme ubiquinone (Section 22-2B), and one molecule of the related **menaquinone**

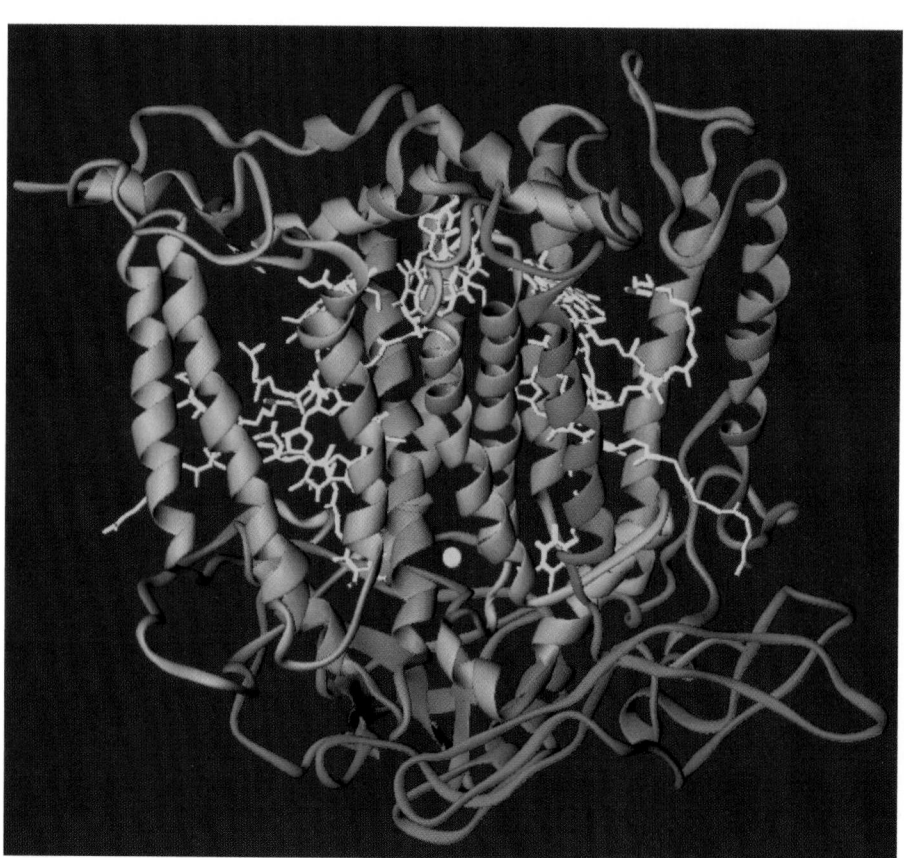

Menaquinone

(**vitamin K₂**, a substance required for proper blood clotting; Section 35-1B). In many PbRCs, however, the BChl *b*, BPheo *b*, and menaquinone are replaced by BChl *a*, BPheo *a*, and a second ubiquinone, respectively.

The RC of *Rps. viridis*, whose X-ray structure was determined by Johann Deisenhofer, Robert Huber, and Hartmut Michel in 1984, was the first transmembrane protein to be described in atomic detail (Fig. 12-26). *The protein's membrane-spanning portion consists of 11 α helices that form a 45-Å-long flattened cylinder with the expected hydrophobic surface.* A *c*-type cytochrome containing four hemes, which is an integral constituent of the RC complex in only some photosynthetic bacteria, binds to the RC on the external side of the plasma membrane. In fact, the RC from another bacterial species, *Rhodobacter (Rb.) sphaeroides*, whose X-ray structure (Fig. 24-11) was independently determined by Marianne Schiffer and by Douglas Rees and George Feher, is nearly identical to that of *Rps. viridis* but lacks such a bound cytochrome.

b. Two BChl Molecules Form a "Special Pair"

The most striking aspect of the RC is that its chromophoric prosthetic groups are arranged with nearly perfect twofold symmetry (Fig. 24-12*a*). This symmetry arises because the L and M subunits, with which these prosthetic groups are exclusively associated, have homologous sequences and similar folds. Two of the BChl *b* molecules in the *Rps. viridis* RC, the so-called **special pair,** are closely

FIGURE 24-11 A ribbon diagram of the photosynthetic reaction center (RC) from *Rb. sphaeroides*. The H, M, and L subunits, as viewed from within the plane of the plasma membrane with the cytoplasm below, are magenta, cyan and orange, respectively. The prosthetic groups are yellow and are shown in skeletal form with the exception of the Fe(II) atom, which is represented by a sphere. The 11 largely vertical helices that form the central portion of the protein constitute its transmembrane region. Compare this structure with that of the RC from *Rps. viridis* (Fig. 12-26), whose H, M, and L subunits are 39, 50, and 59% identical to those of *Rb. sphaeroides*. Note that the *Rb. sphaeroides* protein lacks the four-heme *c*-type cytochrome (*green* in Fig. 12-26) on its periplasmic surface and that the Q_A prosthetic group, whose quinone ring lies to the right of the Fe(II), is ubiquinone in *Rb. sphaeroides* but menaquinone in *Rps. viridis*. [Based on an X-ray structure by Marianne Schiffer, Argonne National Laboratory. PDBid 2RCR.] *See the* **Interactive Exercises and Kinemage Exercise 8-2**

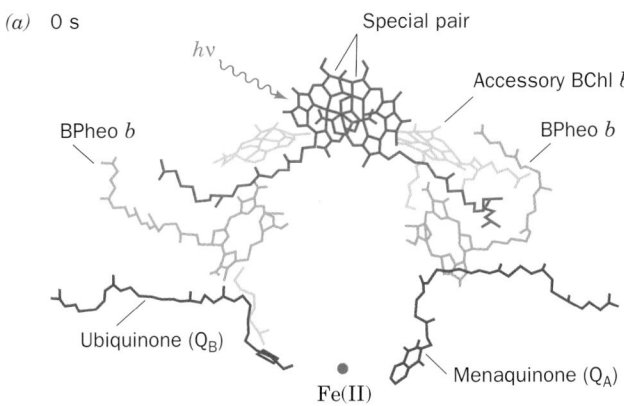

(*a*) 0 s

Special pair

hν

BPheo *b*

Accessory BChl *b*

BPheo *b*

Ubiquinone (Q_B)

Fe(II)

Menaquinone (Q_A)

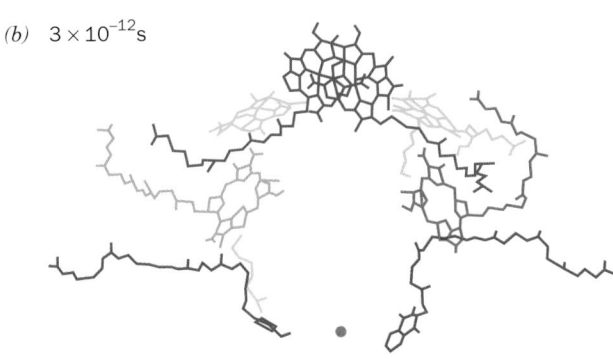

(*b*) 3×10^{-12} s

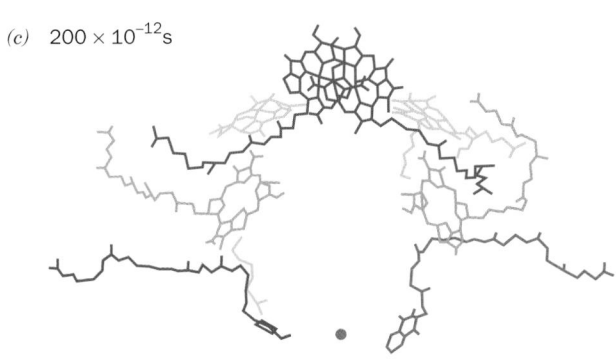

(*c*) 200×10^{-12} s

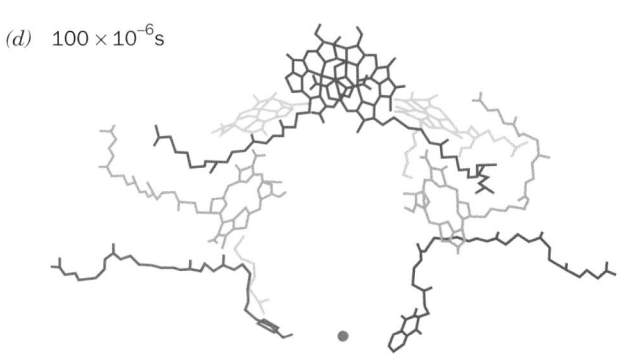

(*d*) 100×10^{-6} s

FIGURE 24-12 Sequence of excitations in the bacterial RC of *Rps. viridis.* The RC chromophores are shown in the same view as in Fig. 12-26, which resembles that in Fig. 24-11. Note that their rings, but not their aliphatic side chains, are arranged with close to twofold symmetry. (*a*) At zero time, a photon is absorbed by the "special pair" of BChl *b* molecules, thereby collectively raising them to an excited state [in each step, the excited molecule(s) is shown in red]. (*b*) Within 3 ps, an excited electron has passed to the BPheo *b* of the L subunit (right arm of the system) without becoming closely associated with the accessory BChl *b*. The special pair is thereby left with a positive charge. (*c*) Some 200 ps later, the excited electron has transferred to the menaquinone (Q_A, which is ubiquinone in *Rb. sphaeroides*). (*d*) Within the next 100 μs, the special pair has been reduced (via an electron transport chain discussed in the text), thereby eliminating its positive charge while the excited electron migrates to the ubiquinone (Q_B). After a second such electron has been transferred to Q_B, it picks up two protons from solution and exchanges with the membrane-bound ubiquinone pool. ✍ **See Kinemage Exercise 8-2**

liganded BChl *b* molecule, which, in turn, is associated with a BPheo *b* molecule. The menaquinone is in close association with the L subunit BPheo *b* (Fig. 24-12*a, right*), whereas the ubiquinone, which is but loosely bound to the protein, associates with the M subunit BPheo *b* (Fig. 24-12*a, left*). These various chromophores are closely associated with a number of protein aromatic rings, which are therefore also thought to participate in the electron-transfer process described below. The Fe(II) is positioned between the menaquinone and ubiquinone rings and is octahedrally liganded by four His side chains and the two carboxyl oxygen atoms of a Glu side chain. Curiously, the two symmetry related groups of chromophores are not functionally equivalent; electrons, as we shall see, are almost exclusively transferred through the L subunit (the right sides of Figs. 24-11 and 24-12). This effect is generally attributed to subtle structural and electronic differences between the L and M subunits.

c. The Electronic States of Molecules Undergoing Fast Reactions Can Be Monitored by EPR and Laser Spectroscopy Techniques

The turnover time of a photosynthetic reaction cycle, as we have seen, is only a few milliseconds. Its sequence of reactions can therefore only be traced by measurements that can follow extremely rapid electronic changes in molecules. Two techniques are well suited to this task:

1. Electron paramagnetic resonance (EPR) spectroscopy [also called **electron spin resonance (ESR) spectroscopy**], which detects the spins of unpaired electrons in a manner analogous to the detection of nuclear spins in NMR spectroscopy. A molecular species with unpaired electrons, such as an organic radical or a transition metal ion, has a characteristic EPR spectrum because its unpaired electrons interact with the magnetic fields generated by the nuclei and the other electrons of the molecule.

associated; they are nearly parallel and have an Mg—Mg distance of ~7 Å. The special pair occupies a predominantly hydrophobic region of the protein and each of its Mg^{2+} ions has a His side chain as a fifth ligand. Each member of the special pair is in contact with another His-

Paramagnetic species as short lived as 10 ps can exhibit definitive EPR spectra.

2. Optical spectroscopy using pulsed lasers. Laser flashes of <1 fs have been generated. By monitoring the bleaching (disappearance) of certain absorption bands and the emergence of others, laser spectroscopy can track the time course of a fast reaction process.

d. Photon Absorption Rapidly Photooxidizes the Special Pair

The sequence of photochemical events mediated by the photosynthetic reaction center is diagrammed in Fig. 24-12:

(a) The primary photochemical event of bacterial photosynthesis is absorption of a photon by the special pair (P870 or **P960** depending on whether it consists of BChl a or b; here, for argument's sake, we assume it to be P960). This event is nearly instantaneous; it occupies the ~3-fs oscillation time of a light wave. EPR measurements established that P960 is, in fact, a pair of BChl b molecules and indicated that the excited electron is delocalized over both of them.

(b) P960*, the excited state of P960, has but a fleeting existence. Laser spectroscopy has demonstrated that within ~3 ps after its formation, P960* has transferred an electron to the BPheo b on the right in Fig. 24-12b to yield P960$^+$ BPheo b^-. In forming this radical pair, the transferred electron must pass near but seems not to reduce the intervening BChl b (which is therefore termed an accessory chlorophyll), although its position strongly suggests that it has an important role in conveying electrons.

(c) By some 200 ps later, the electron has further migrated to the menaquinone (or, in many species, the second ubiquinone), designated Q_A, to form the anionic semiquinone radical Q_A^- All these electron transfers, as diagrammed in Fig. 24-13, are to progressively lower energy states, which makes this process all but irreversible.

Rapid removal of the excited electron from the vicinity of P960$^+$ is an essential feature of the RC; this prevents back reactions that would return the electron to P960$^+$ so as to provide the time required for the wasteful internal conversion of its excitation energy to heat. In fact, *this sequence of electron transfers is so efficient that its overall* **quantum yield** *(ratio of molecules reacted to photons absorbed) is virtually 100%.* No man-made device has yet approached this level of efficiency.

e. Electrons Are Returned to the Photooxidized Special Pair via an Electron-Transport Chain

The remainder of the photosynthetic electron-transport process occurs on a much slower timescale. Within ~100 μs after its formation, Q_A^-, which occupies a hydrophobic pocket in the protein, transfers its excited electron to the more solvent-exposed ubiquinone, Q_B, to form Q_B^- (Fig. 24-12d). The nonheme Fe(II) is not reduced in this process and, in fact, its removal only slightly affects the electron transfer rate, so that the Fe(II) probably functions to fine-

tune the RC's electronic character. Q_A never becomes fully reduced; it shuttles between its oxidized and semiquinone forms. Moreover, the lifetime of Q_A^- is so short that it never becomes protonated. In contrast, once the RC again becomes excited, it transfers a second electron to Q_B^- to form the fully reduced Q_B^{2-}. This anionic quinol takes up two protons from the solution on the cytoplasmic side of the plasma membrane to form Q_BH_2. Thus Q_B is a molecular transducer that converts two light-driven one-electron excitations to a two-electron chemical reduction.

The electrons taken up by Q_BH_2 are eventually returned to P960$^+$ via a complex electron-transport chain (Fig. 24-13). The details of this process are more species dependent than the preceding and are not so well understood. The available redox carriers include a membrane-bound pool of ubiquinone molecules, **cytochrome bc_1,** and **cytochrome c_2.** Cytochrome bc_1 is a transmembrane protein complex composed of a [2Fe–2S] cluster–containing subunit; a heme c-containing cytochrome c_1; a cytochrome b that contains two functionally inequivalent heme b's, b_H and b_L (H and L for high and low potential); and, in some species, a fourth subunit. Note that cytochrome bc_1 is strikingly similar to the proton-translocating Complex III of mitochondria (Section 22-2C), which is also called cytochrome bc_1. The electron-transport pathway leads from Q_BH_2 on the cytoplasmic side of the plasma membrane, through the ubiquinone pool, with which Q_BH_2 exchanges, to cytochrome bc_1, and then to cytochrome c_2 on the external (periplasmic) side of the plasma membrane. The reduced cytochrome c_2, which, as its name implies, closely resembles mitochondrial cytochrome c, diffuses along the external membrane surface until it reacts with the membrane-spanning RC to transfer an electron to P960$^+$ (the structures of several c-type cytochromes, including that of cytochrome c_2 from *Rs. rubrum*, are diagrammed in Fig. 9-38). In *Rps. viridis*, the four-heme c-type cytochrome bound to the RC complex on the external side of the plasma membrane (Fig. 12-26) is interposed between cytochrome c_2 and P960$^+$. Note that one of this c-type cytochrome's hemes is positioned to reduce the photooxidized special pair. The RC is thereby prepared to absorb another photon.

f. Photosynthetic Electron Transport Drives the Formation of a Proton Gradient

Since electron transport in PbRCs is a cyclic process (Fig. 24-13), *it results in no net oxidation–reduction. Rather, it functions to translocate the cytoplasmic protons acquired by Q_BH_2 across the plasma membrane, thereby making the cell alkaline relative to its environment.* The mechanism of this process is essentially identical to that of proton transport in mitochondrial Complex III (Section 22-3B); that is, in addition to the translocation of the two H$^+$ resulting from the two-electron reduction of Q_B to QH$_2$, a Q cycle mediated by cytochrome bc_1 translocates two H$^+$ for a total of four H$^+$ translocated per two photons absorbed (Fig. 24-13a; also see Fig. 22-31). *Synthesis of ATP, a process known as* **photophosphorylation,** *is driven by the*

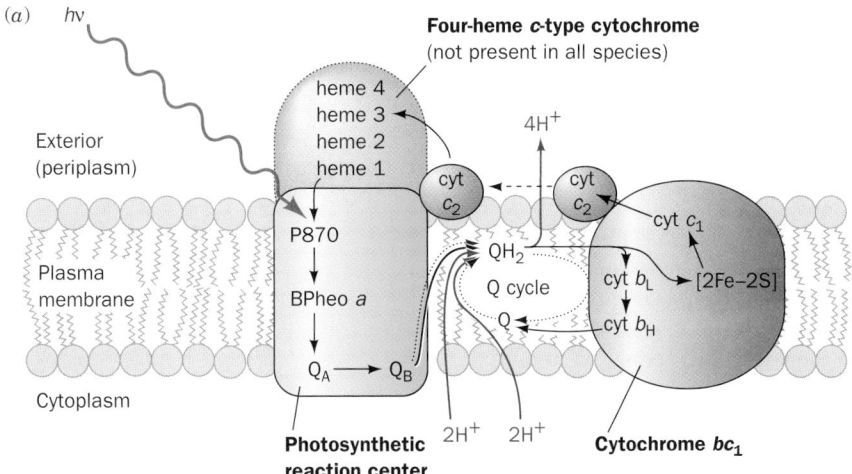

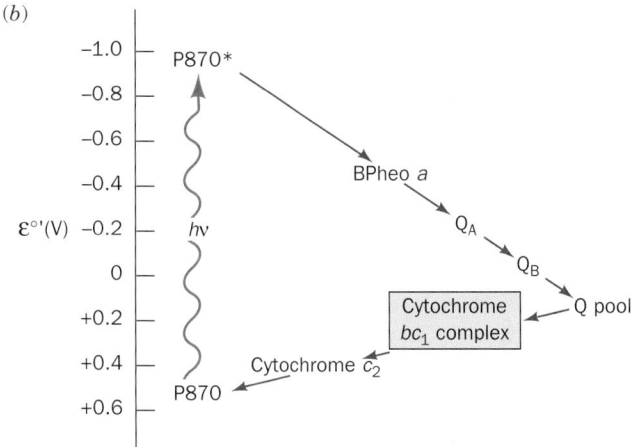

FIGURE 24-13   Photosynthetic electron-transport system of purple photosynthetic bacteria. (*a*) A schematic diagram indicating the arrangement of the system components in the bacterial plasma membrane and the flows of electrons (*black arrows*) and protons (*blue arrows*) that photon (*hv*) absorption promotes through them. The system contains two protein complexes, the RC and cytochrome bc_1. Two electrons liberated from the special pair, here P870 (as in *Rb. sphaeroides*), by the consecutive absorption of two photons are taken up by ubiquinone (Q_B) together with two protons from the cytoplasm to yield ubiquinol (QH_2). The QH_2 is released from the RC and diffuses (*dotted arrows*) through the membrane to cytochrome bc_1, which, in a two-electron reaction, oxidizes it to ubiquinone with the concomitant liberation of its two protons to the external medium. One of the two electrons is passed, via the [2Fe–2S] cluster and cytochrome c_1, to cytochrome c_2, a peripheral membrane protein that then diffuses across the external surface of the membrane so as to return the electron to P870 of the RC. The second electron from QH_2 passes, via a Q cycle, through hemes b_L and b_H of cytochrome bc_1 and then contributes to the reduction of a molecule of ubiquinone (Q) with the concomitant uptake of two more cytoplasmic protons (two rounds of a Q cycle are required for the reduction of one molecule of Q to QH_2; Fig. 22-31). The resulting QH_2 diffuses back to cytochrome bc_1. There it is again oxidized, with the liberation of its two protons to the exterior and the return of one of its two electrons, via cytochrome c_2, to P870, thereby completing the electrical circuit. Note that in every turn of a Q cycle, half the electrons liberated by the oxidation of QH_2 to Q are used to reduce Q to QH_2, so that, after a large number of turns, an electron that enters the Q cycle, on average, passes through it twice before being returned to P870. Thus, the net result of the absorption of two photons by the RC is the translocation of four H^+ from the cytoplasm to the external medium. (*b*) The approximate standard reduction potentials of the photosynthetic electron-transport system's various components.

dissipation of the resulting pH gradient in a manner that closely resembles ATP synthesis in oxidative phosphorylation (Section 22-3C). We further discuss the mechanism of photophosphorylation in Section 24-2D.

Photosynthetic bacteria use photophosphorylation-generated ATP to drive their various endergonic processes. However, unlike cyanobacteria and plants, which generate their required reducing equivalents by the light-driven oxidation of H_2O (see below), photosynthetic bacteria must obtain their reducing equivalents from the environment. Various substances, such as H_2S, S, $S_2O_3^{2-}$, H_2, and many organic compounds, function in this capacity depending on the bacterial species.

Modern photosynthetic bacteria are thought to resemble the original photosynthetic organisms. These presumably arose very early in the history of cellular life when environmentally supplied sources of "high-energy" compounds were dwindling but reducing agents were still plentiful (Section 1-5C). During this era, photosynthetic bacteria were no doubt the dominant form of life. However, their very success eventually caused them to exhaust the available reductive resources. The ancestors of modern cyanobacteria adapted to this situation by evolving a photosynthetic system with sufficient electromotive force to abstract electrons from H_2O. The gradual accumulation of the resulting toxic waste product, O_2, forced photosynthetic bacteria, which cannot photosynthesize in the presence of O_2 (although some species have evolved the ability to respire), into the narrow ecological niches to which they are presently confined (Section 1-1A).

C. *Two-Center Electron Transport*

✏ See Guided Exploration 22. Two-Center Photosynthesis (Z-scheme) Overview *Plants and cyanobacteria use the reducing power generated by the light-driven oxidation of H_2O to produce NADPH.* The component half-reactions of this process, together with their standard reduction potentials, are

$$O_2 + 4e^- + 4H^+ \rightleftharpoons 2H_2O \qquad \mathscr{E}^{\circ\prime} = +0.815 \text{ V}$$

and

$$NADP^+ + H^+ + 2e^- \rightleftharpoons NADPH \qquad \mathscr{E}^{\circ\prime} = -0.320 \text{ V}$$

Hence, the overall four-electron reaction and its standard redox potential is

$$2NADP^+ + 2H_2O \rightleftharpoons 2NADPH + O_2 + 2H^+$$
$$\Delta\mathscr{E}^{\circ\prime} = -1.135 \text{ V}$$

This latter quantity corresponds (Eq. [16.5]) to a standard free energy change of $\Delta G^{\circ\prime} = 438 \text{ kJ} \cdot \text{mol}^{-1}$, which Eq. [24.1] indicates is the energy of one einstein of 223-nm photons (UV light). Clearly, *even if photosynthesis were 100% efficient, which it is not, it would require more than one photon of visible light to generate a molecule of O_2. In fact, experimental measurements indicate that algae minimally require 8 to 10 photons of visible light to produce one molecule of O_2.* In the following subsections, we discuss how plants and cyanobacteria manage this multiphoton process.

a. Photosynthetic O_2 Production Requires Two Sequential Photosystems

Two seminal observations led to the elucidation of the basic mechanism of photosynthesis in plants:

1. The quantum yield for O_2 evolution by *Chlorella pyrenoidosa* varies little with the wavelength of the illuminating light between 400 and 675 nm but decreases precipitously above 680 nm (Fig. 24-14, lower curve). This phenomenon, the "red drop," was unexpected because Chl *a* absorbs such far-red light (Fig. 24-5).

2. Shorter wavelength light, such as yellow-green light, enhances the photosynthetic efficiency of 700-nm light well in excess of the energy content of the shorter wavelength light; that is, *the rate of O_2 evolution by both lights is greater than the sum of the rates for each light acting alone (Fig. 24-14, upper curve).* Moreover, this enhancement still occurs if the yellow-green light is switched off several seconds before the red light is turned on and vice versa.

These observations clearly indicate that two processes are involved. They are explained by a mechanistic model, the **Z-scheme,** which postulates that *O_2-producing photosynthesis occurs through the actions of two photosynthetic RCs that are connected essentially in series (Fig. 22-15).*

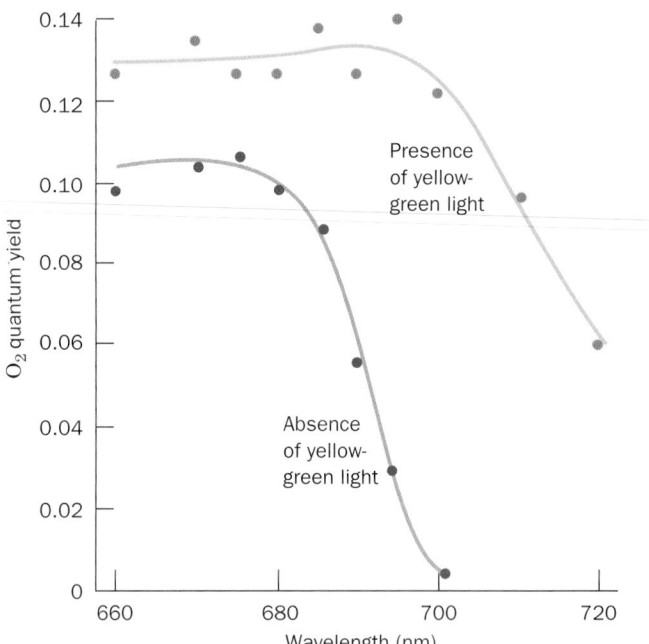

FIGURE 24-14   Quantum yield for O_2 production by *Chlorella* algae as a function of the wavelength of the incident light. The experiment was conducted in the absence (*lower curve*) and the presence (*upper curve*) of supplementary yellow-green light. The upper curve has been corrected for the amount of O_2 production stimulated by the supplementary light alone. Note that the lower curve falls off precipitously above 680 nm (the red drop). However, the supplementary light greatly increases the quantum yield in the wavelength range above 680 nm (far-red) in which the algae absorb light. [After Emerson, R., Chalmers, R., and Cederstrand, C., *Proc. Natl. Acad. Sci.* **49,** 137 (1957).]

1. Photosystem I (PSI) generates a strong reductant capable of reducing $NADP^+$ and, concomitantly, a weak oxidant.

2. Photosystem II (PSII) generates a strong oxidant capable of oxidizing H_2O and, concomitantly, a weak reductant.

The weak reductant reduces the weak oxidant, so that *PSI and PSII form a two-stage electron "energizer." Both photosystems must therefore function for photosynthesis (electron transfer from H_2O to $NADP^+$, forming O_2 and NADPH) to occur.*

The red drop is explained in terms of the Z-scheme by the observation that PSII is only poorly activated by 680-nm light. In the presence of only this far-red light, PSI is activated but is unable to obtain more than a few of the electrons it is capable of energizing. Yellow-green light, however, efficiently stimulates PSII to supply these electrons. The observation that the far-red and yellow-green lights can be alternated indicates that both photosystems remain activated for a time after the light is switched off.

The validity of the Z-scheme was established as follows. The oxidation state of **cytochrome *f,*** a *c*-type cytochrome of the electron-transport chain connecting PSI and PSII (see below), can be spectroscopically monitored. Illu-

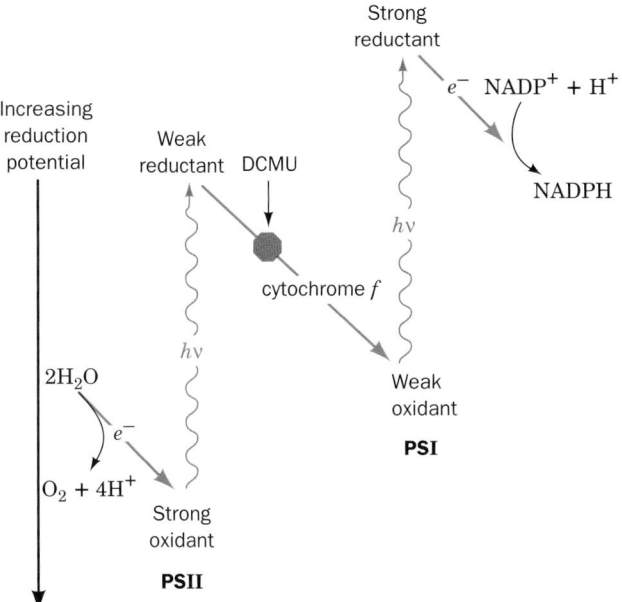

FIGURE 24-15 **The Z-scheme for photosynthesis in plants and cyanobacteria.** Two photosystems, PSI and PSII, function to drive electrons from H_2O to NADPH. The reduction potential increases downward so that electron flow occurs spontaneously in this direction. The herbicide DCMU (see text) blocks photosynthetic electron transport from PSII to cytochrome *f*.

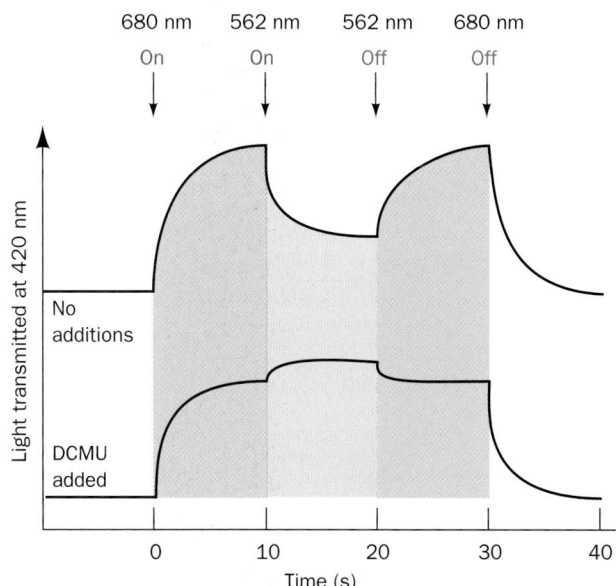

FIGURE 24-16 **The oxidation state of cytochrome *f* in *Porphyridium cruentum* algae as monitored by a weak beam of 420-nm (blue-violet) light.** An increase in the transmitted light signals the oxidation of cytochrome *f*. In the upper curve, strong light at 680 nm (far-red) causes the oxidation of the cytochrome *f* but the superposition of 562-nm (yellow-green) light causes its partial rereduction. In the lower curve, the presence of the herbicide DCMU, which inhibits photosynthetic electron transport, causes 562-nm light to further oxidize, rather than reduce, the cytochrome *f*.

mination of algae with 680-nm (far-red) light results in the oxidation of cytochrome *f* (Fig. 24-16). However, the additional imposition of a 562-nm (yellow-green) light results in this protein's partial rereduction. In the presence of the herbicide **3-(3,4-dichlorophenyl)-1,1-dimethylurea (DCMU),**

3-(3,4-Dichlorophenyl)-1,1-dimethylurea (DCMU)

which abolishes photosynthetic oxygen production, 680-nm light still oxidizes cytochrome *f* but simultaneous 562-nm light only oxidizes it further. The explanation for these effects is that 680-nm light, which efficiently activates only PSI, causes it to withdraw electrons from (oxidize) cytochrome *f*. The 562-nm light also activates PSII, which thereby transfers electrons to (reduces) cytochrome *f*. DCMU blocks electron flow from PSII to cytochrome *f* (Fig. 24-15), so an increased intensity of light, whatever its wavelength, only serves to activate PSI further.

b. O₂-Producing Photosynthesis Is Mediated by Three Transmembrane Protein Complexes Linked by Mobile Electron Carriers

The components of the Z-scheme, which mediate electron transport from H_2O to NADPH, are largely organized

into three thylakoid membrane-bound particles (Fig. 24-17): (1) PSII, (2) the **cytochrome b₆f complex,** *and (3) PSI. As in oxidative phosphorylation, electrons are transferred between these complexes via mobile electron carriers. The ubiquinone analog* **plastoquinone (Q),** *via its reduction to* **plastoquinol (QH₂),**

Plastoquinone

$2 [H \cdot]$

Plastoquinol

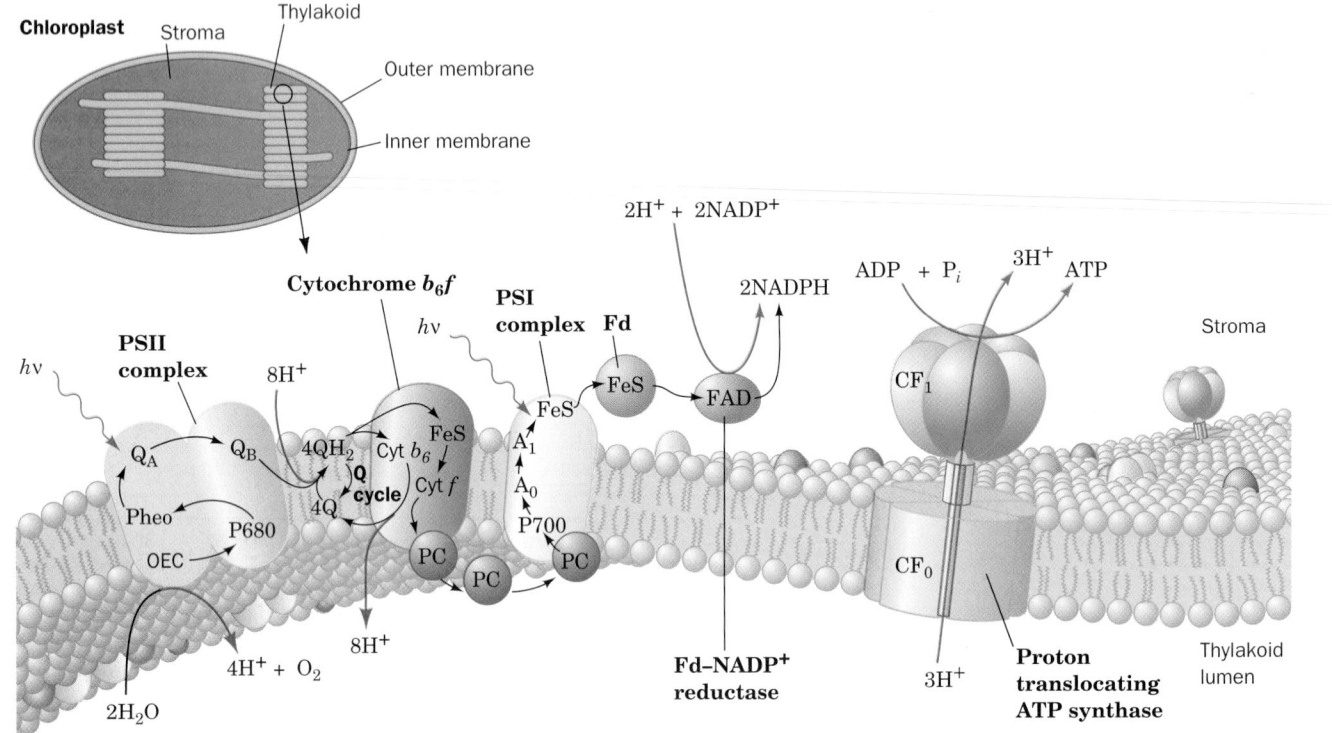

FIGURE 24-17 Schematic representation of the thylakoid membrane showing the components of its electron-transport chain. The system consists of three protein complexes: PSII, the cytochrome $b_6 f$ complex, and PSI, which are electrically "connected" by the diffusion of the electron carriers plastoquinol (Q) and plastocyanin (PC). Light-driven transport of electrons (*black arrows*) from H_2O to $NADP^+$ forming NADPH motivates the transport of protons (*red arrows*) into the thylakoid space (Fd is ferredoxin). Additional protons are split off from water by the oxygen-evolving complex (OEC), yielding O_2. The resulting proton gradient powers the synthesis of ATP by the CF_1CF_0 proton-translocating ATP synthase [CF_1 and CF_0 are chloroplast (C) analogs of mitochondrial F_1 and F_0]. The membrane also contains light-harvesting complexes whose component chlorophylls and other chromophores transfer their excitations to PSI and PSII. [After Ort, D.R. and Good, N.E., *Trends Biochem. Sci.* **13**, 469 (1988).]

links PSII to the cytochrome $b_6 f$ complex, which, in turn, interacts with PSI through the mobile Cu-containing redox protein **plastocyanin (PC).** In what follows, we trace the electron pathway through this chloroplast system from H_2O to $NADP^+$ (Fig. 24-18).

c. PSII Resembles the PbRC

PSII from the thermophilic cyanobacterium *Synechococcus elongatus* consists of at least 17 subunits, 14 of which occupy the photosynthetic membrane. These transmembrane subunits include the RC proteins **D1 (PsbA)** and **D2 (PsbD),** the chlorophyll-containing inner-antenna subunits **CP43 (PsbC)** and **CP47 (PsbB),** and the α and β subunits of **cytochrome b_{559}.** The X-ray structure of this PSII was determined by Norbert Krauss, Wolfram Saenger, and Horst-Tobias Witt at such a low resolution (3.8 Å) that only its major features such as helices and chlorophyll rings could be reliably traced. It reveals (Fig. 24-19) that this ~340-kD protein is a symmetric dimer, whose protomeric units each contain 36 transmembrane helices, 22 of which are portions of D1, D2, CP43, and CP47. Each protomer binds 32 Chl a's, 2 **pheophytin a's** (**Pheo a's;** Chl a with its

Mg^{2+} replaced by two protons), 2 hemes, 2 plastoquinones, a nonheme Fe, and a cluster of four Mn ions. (In higher plants, PSII contains ~25 subunits and forms an ~1000-kD transmembrane supercomplex with LHC-II and several other antenna proteins.)

D1 is related to D2 via a pseudo-2-fold axis of symmetry that is parallel to the dimer axis and that passes through the nonheme Fe. The arrangement of the five transmembrane helices in both D1 and D2 resembles that in the L and M subunits of the PbRC (Fig. 24-11). Indeed, these two sets of subunits have similar sequences, thereby indicating that they arose from a common ancestor. CP43 and CP47, which are also pseudosymmetrically related, each contain six helices arranged as a trimer of dimers and, respectively, bind 12 and 14 antenna Chl a's.

The cofactors of PSII's RC (Fig. 24-20) are organized similarly to those of the bacterial system (Fig. 24-12): They have essentially the same components (with Chl a, Pheo a, and plastoquinone replacing BChl b, BPheo b, and menaquinone, respectively) and are symmetrically organized along the complex's pseudo-twofold axis. The two Chl a rings labeled P_{D1} and P_{D2} in Fig. 24-20 are positioned

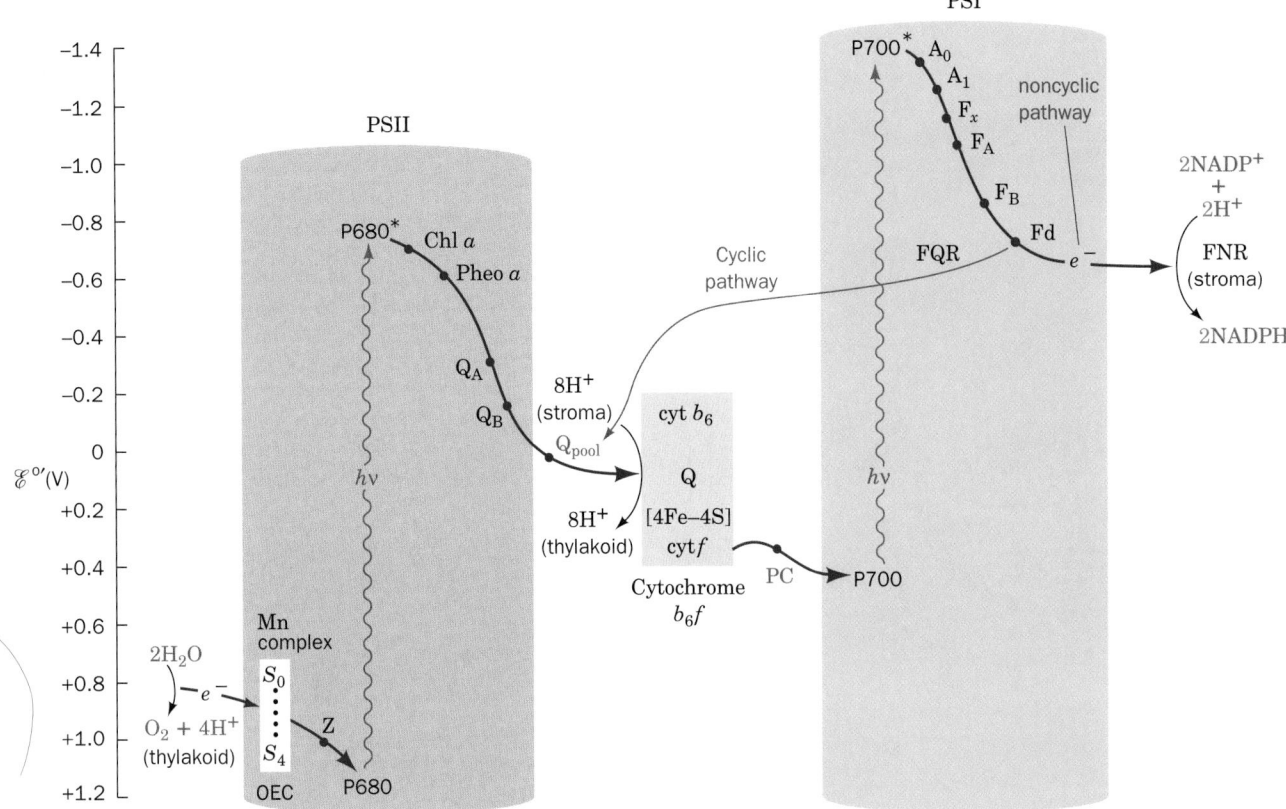

FIGURE 24-18   Detailed diagram of the Z-scheme of photosynthesis. Electrons ejected from P680 by the absorption of photons are replaced with electrons abstracted from H_2O by an Mn complex (OEC), thereby forming O_2 and four H^+. Each ejected electron is passed through a chain of electron carriers to a pool of plastoquinone molecules (Q). The resulting plastoquinol, in turn, reduces the cytochrome b_6f particle (*yellow box*) that transfers electrons with the concomitant translocation of protons, via a Q cycle, into the thylakoid lumen. Cytochrome b_6f then transfers the electrons to plastocyanin (PC). The plastocyanin regenerates photooxidized P700. The electron ejected from P700, through the intermediacy of a chain of electron carriers (A_0, A_1, F_X, F_A, F_B, and Fd), reduces $NADP^+$ to NADPH in noncyclic electron transport. Alternatively, the electron may be returned to the cytochrome b_6f complex in a cyclic process that only translocates protons into the thylakoid lumen.

analogously to the BChl *b*'s of P960's special pair and are therefore presumed to form PSII's primary electron donor, **P680** (named after the wavelength of its absorption maximum). However, the parallel rings in P680 are further apart than those in P960 (with an Mg—Mg distance of 10 Å in P680 vs 7 Å in P960), so that they are only weakly electronically coupled. Consequently, the unpaired electron in the photooxidized $P680^+$ is likely to reside on only one of these rings, presumably P_{D1} since it is closer to Tyr_Z than is P_{D2} (see below). The electron ejected from P680 follows a similar course as that in the bacterial system even though the two systems operate over different ranges of reduction potentials (compare Figs. 22-13*b* and 22-18). As indicated in the central part of Fig. 24-18, the electron is transferred to a molecule of Pheo *a* ($Pheo_{D1}$ in Fig. 24-20), probably via a Chl *a* molecule (Chl_{D1}), and then to a bound plastoquinone (Q_A). Subsequently, two electrons are transferred, one at a time, to a second plastoquinone molecule, Q_B, which takes up two protons at the stromal (cytoplasmic) surface of the thylakoid membrane. The resulting plastoquinol, Q_BH_2, then exchanges with a membrane-bound pool of plastoquinone molecules (the putative binding site for this relatively mobile plastoquinone is unoccupied in the X-ray structure of PSII). DCMU, as well as many other commonly used herbicides, competes with plastoquinone for the Q_B binding site on PSII, which explains how they inhibit photosynthesis.

Two "extra" Chl *a* molecules, $Chlz_{D1}$ and $Chlz_{D2}$, lie on the periphery of the RC, where they are postulated to function in the transfer of excitation from the antenna systems to P680. Cytochrome b_{559}, whose function is unclear, breaks the pseudosymmetry of the PSII protomer as does the Mn cluster, whose function we now discuss.

d. O_2 Is Generated in a Five-Stage Water-Splitting Reaction Mediated by an Mn-Containing Protein Complex

The oxidation of two molecules of H_2O to form one molecule of O_2 requires four electrons. Since transfer of a single electron from H_2O to $NADP^+$ requires two photo-

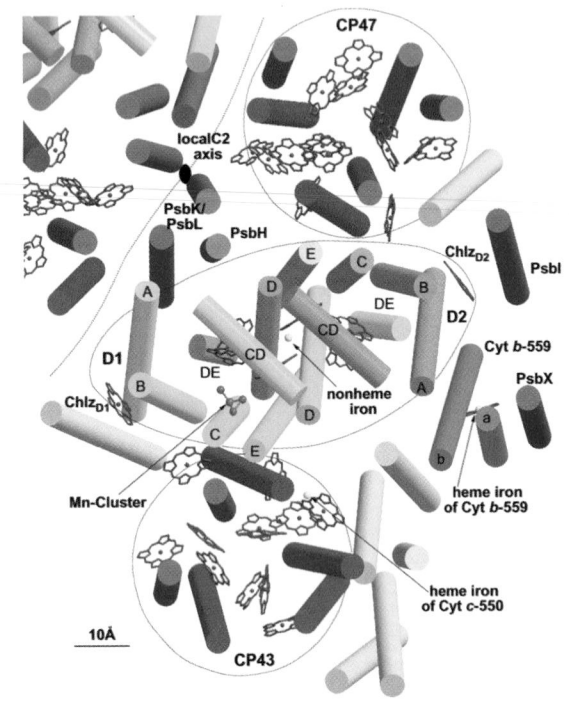

(a)

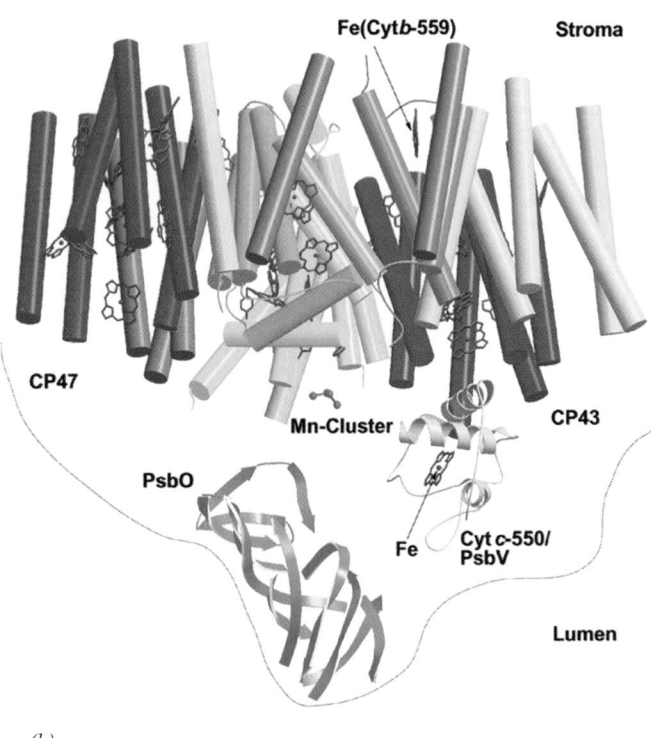

(b)

FIGURE 24-19   X-Ray structure of PSII from *S. elongatus* showing its arrangement of helices (*cylinders*) and cofactors. (*a*) View perpendicular to the membrane from the thylakoid lumen showing only the transmembrane portions of the complex. One protomer of the dimeric complex is shown in its entirety with part of the second protomer related by a local twofold axis (*black ellipse on the dotted interface*) shown at the upper left. The helices from different subunits are drawn in different colors with the seven unassigned helices light blue. The Chl *a* rings and hemes are drawn in stick form (*blue-green*). D1 and D2 are highlighted by a surrounding ellipse, as are CP43 and CP47 by circles. The Mn cluster is drawn in ball-and-stick form (*magenta*), although the actual positions of its Mn atoms are uncertain. (*b*) View from the right side of Part *a* with the lumen below and the membrane plane slightly tilted. The luminal proteins **PsbO** and **cytochrome *c*₅₅₀ (PsbV)** are, respectively, shown as a β-sheet structure (*green*) and a helical model (*light blue*). [Courtesy of Wolfram Saenger, Freie Universität Berlin, Germany. PDBid 1FE1.]

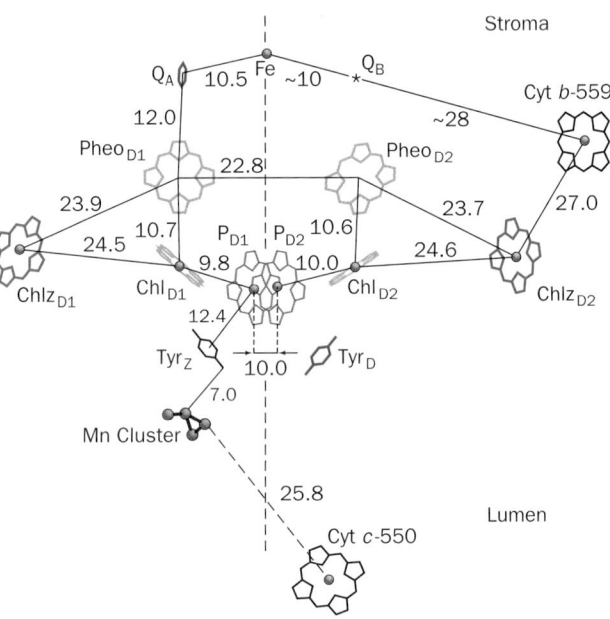

FIGURE 24-20   Arrangement of cofactors in the RC subunits (D1 and D2) of PSII. The complex is viewed along the membrane plane with the thylakoid lumen below. The dashed line that runs through the nonheme Fe is the RC's 2-fold axis of pseudosymmetry, which is parallel to the local twofold axis relating the two protomers of the PSII dimer (Fig. 24-19*a*). Thin lines indicate the center-to-center distances, in Å, between cofactors. The asterisk marks the putative binding site for Q_B. Compare this figure with Fig. 24-12 (which is shown upside down relative to this figure). [After Zouni, A., Witt, H.-T., Kern, J., Fromme, P., Krauss, N., Saenger, W., and Orth, P., *Nature* **409,** 741 (2001).]

chemical events, this accounts for the observed minimum of 8 to 10 photons absorbed per molecule of O_2 produced.

Must the four electrons necessary to produce a given O_2 molecule be removed by a single photosystem or can they be extracted by several different photosystems? Pierre

Joliet and Bessel Kok answered this question by analyzing the rate at which dark-adapted chloroplasts produce O_2 when exposed to a series of short flashes. O_2 was evolved with a peculiar oscillatory pattern (Fig. 24-21). There is virtually no O_2 evolved by the first two flashes. The third flash

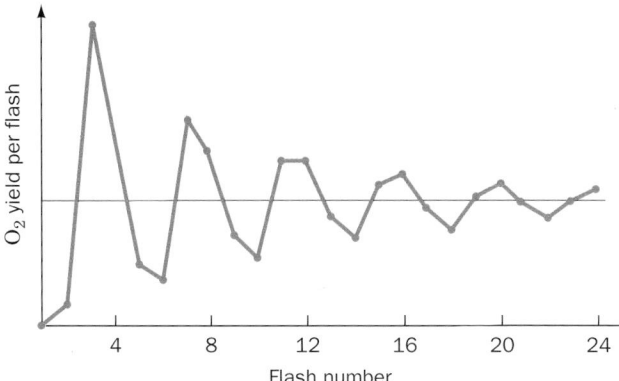

FIGURE 24-21   The O$_2$ yield per flash in dark-adapted spinach chloroplasts. Note that the yield peaks on the third flash and then on every fourth flash thereafter until the curve eventually damps out to its average value. [After Forbush, B., Kok, B., and McGloin, M.P., *Photochem. Photobiol.* **14,** 309 (1971).]

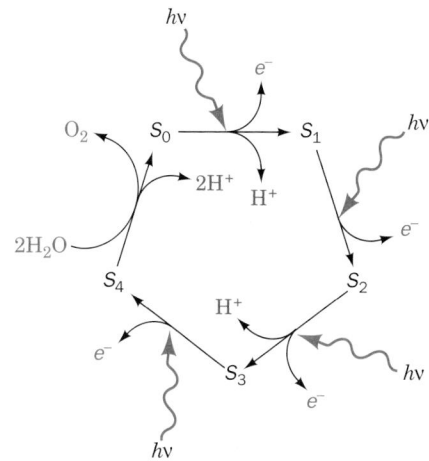

FIGURE 24-22   Schematic mechanism of O$_2$ generation in chloroplasts. Four electrons are stripped, one at a time in light-driven reactions ($S_0 \rightarrow S_4$), from two bound H$_2$O molecules. In the recovery step ($S_4 \rightarrow S_0$), which is light independent, O$_2$ is released and two more H$_2$O molecules are bound. Three of these five steps release protons into the thylakoid lumen.

results in the maximal O$_2$ yield. Thereafter, the amount of O$_2$ produced peaks with every fourth flash until the oscillations damp out to a steady state. This periodicity indicates that each O$_2$-evolving center cycles through five different states, S_0 through S_4 (Fig. 24-22). Each of the transitions between S_0 and S_4 is a photon-driven redox reaction; that from S_4 to S_0 results in the release of O$_2$. Thus, *each O$_2$ molecule must be produced by a single photosystem.* The observation that O$_2$ evolution peaks at the third rather than the fourth flash indicates that the oxygen-evolving center's resting state is predominantly S_1 rather than S_0. The oscillations gradually damp out because a small fraction of the RCs fail to be excited or become doubly excited by a given flash of light, so that they eventually lose synchrony. The five reaction steps release a total of four water-derived protons into the inner thylakoid space (lumen) in a stepwise manner (Fig. 24-22).

Since the S states function to abstract electrons from H$_2$O, their standard reduction potentials must average more than the 0.815-V value of the O$_2$/H$_2$O half-reaction (which makes P680$^+$, which abstracts these electrons, among the most powerful biological oxidizing agents known). PSII has the remarkable capacity of stabilizing these highly reactive intermediates for extended periods (typically minutes) in close proximity to water. We are only beginning to understand how this occurs. PSII contains four protein-bound Mn ions which, on excitation of chloroplasts with short flashes of light, exhibit EPR signals that have a four-flash periodicity similar to that of O$_2$ production (Fig. 24-21). These Mn ions, together with a Ca^{2+} ion and one or two Cl$^-$ ions, form a catalytically active complex, the **oxygen-evolving complex (OEC),** which binds two H$_2$O molecules so as to facilitate O$_2$ formation. The OEC cycles through a series of oxidation states [the S states, which appear to involve various combinations of Mn(II), Mn(III), Mn(IV), and Mn(V)] while abstracting protons and electrons from the H$_2$O molecules, and finally releases O$_2$ into the inner thylakoid space. Although the OEC's Mn cluster has been located in the X-ray structure

of PSII (Figs. 24-19 and 24-20), the arrangement of its Mn atoms could not be definitively established nor could its ligands be identified. However, a variety of spectroscopic evidence has led to the proposal of several structural models for the OEC including that in Fig. 24-23. Nevertheless, in the absence of a definitive structure, the mechanism through which the OEC catalyzes the oxidation of H$_2$O to O$_2$ remains speculative.

The next link in the PSII electron transport chain is an entity, originally named Z (Fig. 24-18), which relays electrons from the OEC to P680. The existence of Z is signaled by a transient EPR spectrum of illuminated chloroplasts that parallels the S-state transitions. The change in this spectrum on supplying deuterated tyrosine to cyanobacteria indicates that Z$^+$ is a tyrosyl radical (TyrO·; EPR spec-

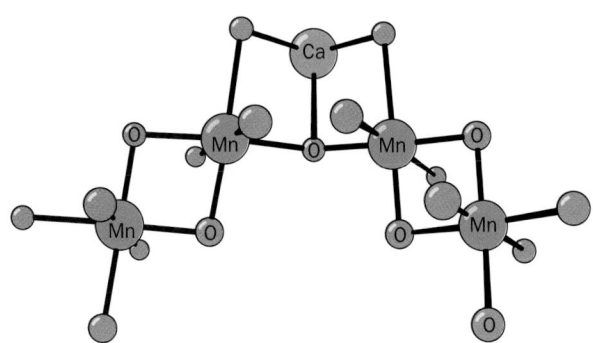

FIGURE 24-23   Proposed structural model for the OEC. This model is consistent with its spectroscopic characteristics (as are several others). Note that in this model, neighboring pairs of Mn ions are each bridged by two O atoms, a structural feature for which the spectroscopic evidence is particularly strong. [After Robblee, J.H., Cinco, R.M., and Yachindra, V.K., *Biochim. Biophys. Acta* **1503,** 16 (2001).]

tra reflect the nuclear spins of the atoms with which the unpaired electrons interact). It has been identified as Tyr$_Z$ in PSII (Fig. 24-20) due to its position between the Mn cluster and P680's chlorophyll P$_{D1}$.

e. Electron Transport through the Cytochrome b_6f Complex Generates a Proton Gradient

From the plastoquinone pool, electrons pass through the **cytochrome b_6f complex.** This integral membrane assembly closely resembles its bacterial counterpart, cytochrome bc_1 (Section 24-2B), as well as Complex III of the mitochondrial electron-transport chain (also called cytochrome bc_1; Section 22-2C). Cytochrome b_6f, which is almost certainly a functional dimer as Complex III is known to be, consists of as many as seven different subunits, with its four "large" subunits being **cytochrome f**

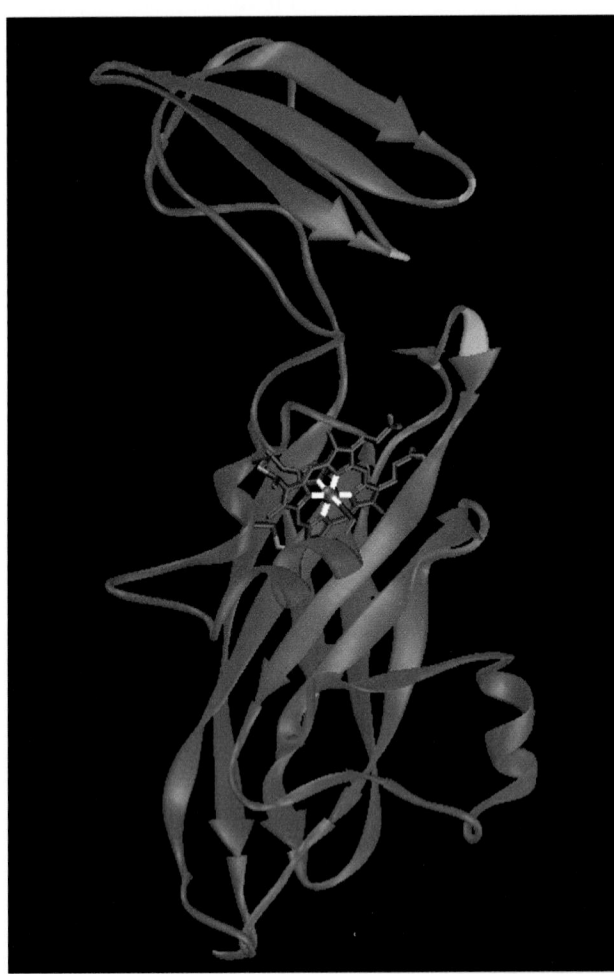

FIGURE 24-24   X-Ray structure of turnip cytochrome f. The heme group and the groups that covalently link it to the protein (Cys 21, Cys 24, His 25, and the N-terminal amino group) are shown in stick form with their C, N, O, and S atoms colored green, blue, red, and yellow, and the heme's Fe atom is represented by an orange sphere. The five Lys and Arg residues that form a positively charged patch on the small domain's surface are cyan. [Based on an X-ray structure by Janet Smith, Purdue University. PDBid 1CTM.]

(which contains one c-type heme), **cytochrome b_6** (which contains two b-type hemes), subunit IV, and a Rieske iron–sulfur protein (which contains one [2Fe–2S] cluster; Section 22-2C). The cytochrome b of mitochondrial Complex III is a fusion of chloroplast cytochrome b_6 and subunit IV, and the Rieske iron–sulfur proteins in the two complexes are homologous and structurally similar. However, cytochrome f (f for *feuille,* French for leaf) is unrelated to its counterpart in Complex III, cytochrome c_1, although both are c-type cytochromes.

The cytochrome b_6f complex transports protons as well as electrons from the outside to the inside of the thylakoid membrane. This proton translocation occurs through a Q cycle (Section 22-3B and Fig. 24-13a) in which plastoquinone is the (H$^+$ + e^-) carrier. The Q cycle mechanism predicts that two protons are translocated across the thylakoid membrane for every electron transported, but the experimental difficulties of measuring this ratio have precluded its unambiguous determination. It is, nevertheless, clear that *electron transport, via the cytochrome b_6f complex, generates much of the electrochemical proton gradient that drives the synthesis of ATP in the chloroplast (see below).*

The 285-residue cytochrome f from turnip, the largest polypeptide in the cytochrome b_6f complex, contains a single transmembrane segment near its C-terminus (residues 251–270, which presumably forms an α helix) oriented such that the protein's N-terminal 250 residues extend into the thylakoid lumen. The X-ray structure of the 252-residue N-terminal segment of cytochrome f, determined by Janet Smith, reveals an elongated two-domain protein that is dominated by β sheets (Fig. 24-24) and thus has an entirely different fold from those of other c-type cytochromes of known structure (e.g., Figs. 9-38 and 12-26a). Cytochrome f's single heme c group is nevertheless covalently linked to the larger domain of the protein via the two Cys residues in a Cys-X-Y-Cys-His sequence that is characteristic of c-type cytochromes and whose His residue forms one of the Fe(III)'s two axial ligands (Fig. 9-36). Intriguingly, however, the second axial ligand is the protein's N-terminal amino group, a group that has previously not been observed to be a heme ligand.

f. Plastocyanin Transports Electrons from Cytochrome b_6f to PSI

Electron transfer between cytochrome f, the terminal electron carrier of the cytochrome b_6f complex, and PSI is mediated by **plastocyanin (PC),** a 99-residue, monomeric, Cu-containing, peripheral membrane protein located on the thylakoid luminal surface (Fig. 24-17). Thus PC is the functional analog of cytochrome c, which transfers electrons from Complex III to Complex IV in the mitochondrial electron-transport chain (Section 22-2C).

PC's redox center cycles between its Cu(I) and Cu(II) oxidation states. The X-ray structure of PC from poplar leaves, determined by Hans Freeman, shows that its single Cu atom is coordinated with distorted tetrahedral geometry by a Cys, a Met, and two His residues (Fig. 24-25).

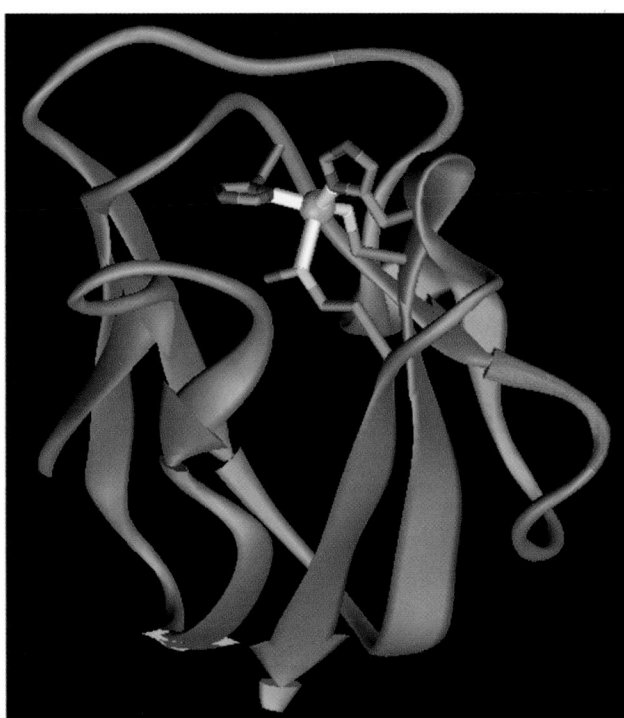

FIGURE 24-25   X-Ray structure of plastocyanin (PC) from poplar leaves. This protein, a member of the family of **blue copper proteins** (as is the globular domain of Complex IV's Subunit II, which binds the Cu_A center), folds into a β sandwich. Its Cu atom (*orange sphere*), which alternates between its Cu(I) and Cu(II) oxidation states, is tetrahedrally liganded by the side chains of His 37, Cys 84, His 87, and Met 92, which are shown in stick form with their C, N, and S atoms green, blue, and yellow. Six conserved Asp and Glu residues that form a negatively charged patch on the protein's surface are red. [Based on an X-ray structure by Mitchell Guss and Hans Freeman, University of Sydney, Australia. PDBid 1PLC.]

Cu(II) complexes with four ligands normally adopt a square planar coordination geometry, whereas those of Cu(I) are generally tetrahedral. Evidently, the strain of Cu(II)'s protein-imposed tetrahedral coordination in PC promotes its reduction to Cu(I). This hypothesis accounts for PC's high standard reduction potential (0.370 V) compared to that of the normal Cu(II)/Cu(I) half-reaction (0.158 V). This is an example of how proteins modulate the reduction potentials of their redox centers so as to match them to their function—in the case of plastocyanin, the efficient transfer of electrons from the cytochrome b_6f complex to PSI.

The structures of cytochrome f and PC suggest how these proteins associate. Cytochrome f's Lys 187, a member of a conserved group of five positively charged residues on the surface of the protein's small domain, can be cross-linked to Asp 44 on PC, which similarly occupies a conserved negatively charged surface patch. Quite possibly, the two proteins associate through electrostatic interactions, much like cytochrome c interacts with its redox

partners in the mitochondrial electron-transport chain, Complex III and Complex IV (Section 22-2C). This suggests that cytochrome b_6f contains cytochrome f rather than cytochrome c_1, its functional analog in Complex III, because cytochrome f is better suited to interact with PC, whereas cytochrome c_1 is better suited to interact with cytochrome c.

g. PSI Resembles Both PSII and the PbRC

Cyanobacterial PSIs are trimers of protomers that each consist of at least 11 different protein subunits coordinating >100 cofactors. The X-ray structure of PSI from *S. elongatus* (Fig. 24-26), determined at 2.5-Å resolution by Krauss, Saenger, and Petra Fromme, reveals that each of its 356-kD protomers contains nine transmembrane subunits (**PsaA, PsbB, PsaF, PsaI–M, and PsaX**) and three stromal (cytoplasmic) subunits (**PsaC–E**), which collectively bind 127 cofactors that comprise 30% of PSI's mass. The cofactors forming the PSI RC are all bound by the homologous subunits PsaA (755 residues) and PsaB (740 residues), whose 11 transmembrane helices each are arranged in a manner resembling those in the L and M subunits of the PbRC (Fig. 24-11) and the D1 and D2 subunits of PSII (Fig. 24-19), thus supporting the hypothesis that all RCs arose from a common ancestor. PsaA and PsaB, together with other transmembrane subunits, also bind the cofactors of the core antenna system (see below).

Figure 24-27 indicates that PSI's RC consists of six Chl a's and two molecules of **phylloquinone (vitamin K_1;** note that it has the same phytyl side chain as chlorophylls; Fig. 24-3),

Phylloquinone

all arranged in two pseudosymmetrically related branches, followed by three [4Fe–4S] clusters. The primary electron donor of this system, **P700,** consists of a pair of parallel Chl a's, A1 and B1, whose Mg^{2+} ions are separated by 6.3 Å, and thus resembles the special pair in the PbRC. However, EPR studies indicate that ~80% of the unpaired electron associated with photooxidized $P700^+$ resides on Chl a B1. A1 is followed in the left branch of Fig. 24-27 by two more Chl a rings, B2 and A3, and B1 is followed by A2 and B3 in the right branch. One or both of the third pair of Chl a molecules, A3 and B3, probably form the spectroscopically identified primary electron acceptor A_0 (right side of Fig. 24-18). The Mg^{2+} ions of A3 and B3 are each axially liganded by the S atom of a Met residue rather

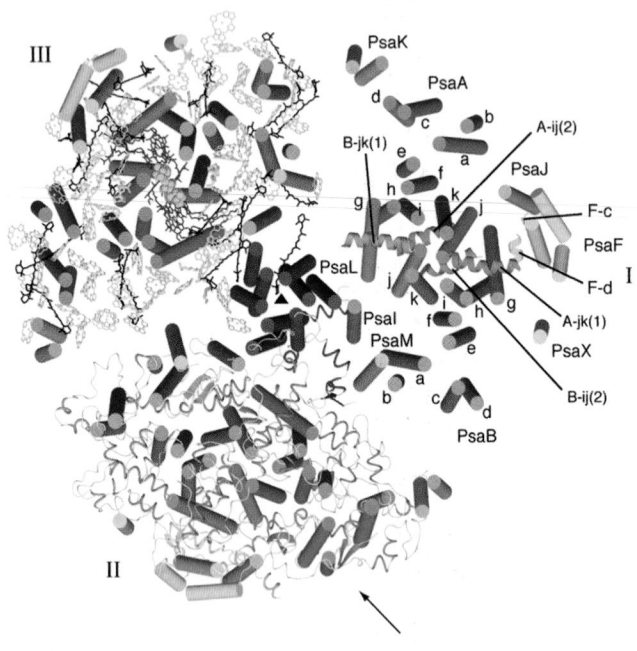

(a)

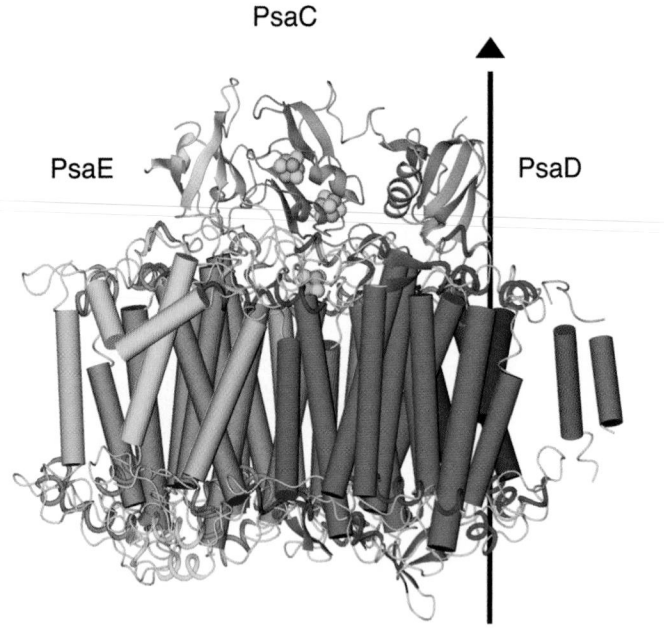

(b)

FIGURE 24-26 X-Ray structure of PSI from *S. elongatus.* (a) View of the trimer perpendicular to the membrane from its stromal side. The stromal subunits have been removed for clarity. PSI's threefold axis of symmetry is represented by the small black triangle. Different structural elements are shown for each of the three protomers (I, II, and III). I shows the arrangement of transmembrane helices (*cylinders*), which are differently colored for each subunit. The transmembrane helices of both PsaA (*blue*) and PsaB (*red*) are named a through k from their N- to C-termini. The six helices in extramembranous loop regions are drawn as spirals. II shows the transmembrane helices as cylinders with the stromal and lumenal loop regions drawn in ribbon form. III shows the transmembrane helices as cylin- ders together with all cofactors. The RC Chl *a*'s and quinones, drawn in stick form, are purple, the Fe and S atoms of the [4Fe–4S] clusters are drawn as orange and yellow spheres, the antenna system Chl *a*'s (whose side chains have been removed for clarity) are yellow, the carotenoids are black, and the bound lipids are light green. (b) One protomer as viewed parallel to the membrane along the arrow in Part *a* with the stroma above. The transmembrane subunits are colored as in Part *a* with the stromal subunits PsaC, PsaD, and PsaE pink, cyan, and light green. The vertical line and triangle mark the trimer's threefold axis of symmetry. [Courtesy of Wolfram Saenger, Freie Universität Berlin, Germany. PDBid 1JB0.]

than by a His side chain (thereby forming the only known biological examples of Mg^{2+}—S coordination). All of the residues involved in Mg^{2+} coordination and hydrogen bonding to these second and third Chl *a*'s are strictly con- served in PSI's, from cyanobacteria to higher plants, thereby suggesting that all of these interactions are important for fine-tuning their redox potentials. Electrons are passed from A3 and B3 to the phylloquinones, Q_K-A

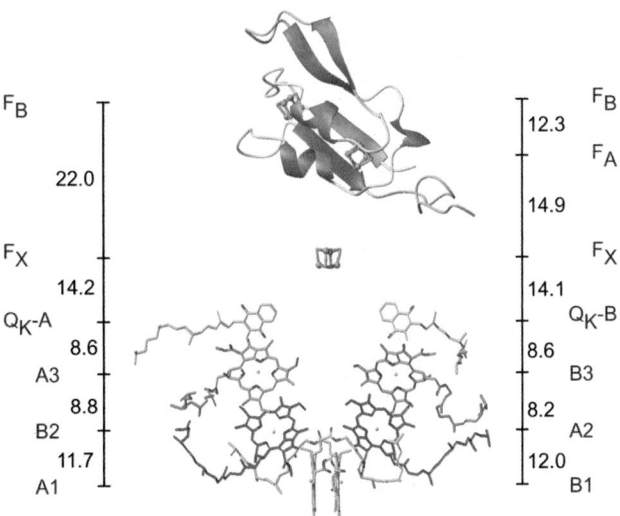

FIGURE 24-27 Cofactors of the PSI RC and PsaC. The structure is viewed parallel to the membrane plane with the stroma above. The Chl *a* and phylloquinone molecules are arranged in two branches that are related by PSI's 2-fold axis of pseudosymmetry, which is vertical in this drawing. The Chl *a*'s are labeled A or B to indicate that their Mg^{2+} ions are liganded by the side chains of PsaA or PsaB, respectively, and, from the luminal side upward, by different colors and numbers, 1 to 3. The phylloquinones are named Q_K-A and Q_K-B. PsaC is shown in ribbon form with those portions resembling segments in bacterial 2[4Fe–4S] ferredoxins pink and with insertions and extensions green. The three [4Fe–4S] clusters are shown in ball-and-stick form and labeled according to their spectroscopic identities F_X, F_A, and F_B. The center-to-center distances between cofactors (*vertical black lines*) are given in Å. Compare this figure with Figs. 24-19 and 24-12. [Courtesy of Wolfram Saenger, Freie Universität Berlin, Germany. PDBid 1JB0.]

and Q_K-B, which almost certainly correspond to the spectroscopically identified electron acceptor A_1. Spectroscopically based kinetic investigations indicate that, in contrast to the case for the PbRC, electrons pass through both branches of the PSI RC, although at different rates: $35 \times 10^6 \text{ s}^{-1}$ for the branch ending in Q_K-B and 4.4×10^6 s^{-1} for that ending in Q_K-A. Indeed, the PSI RC is most closely related to the RC of **green sulfur bacteria** (a second class of photosynthetic bacteria), which is a true homodimer.

Up until this point, PSI's RC resembles those of PSII and purple photosynthetic bacteria. However, rather than the reduced forms of either Q_K-A or Q_K-B dissociating from PSI, both of these quinones directly pass their photoexcited electron to a chain of three spectroscopically identified [4Fe–4S] clusters designated F_X, F_A, and F_B (right side of Fig. 24-18). F_X, which lies on the pseudo-twofold axis relating PsaA and PsaB, is coordinated by two Cys residues from each of these subunits. F_A and F_B are bound to the stromal subunit PsaC, which structurally resembles bacterial 2[4Fe–4S] ferredoxins (e.g., Fig. 22-16). Mutational studies on the Cys residues of PsaC that coordinate its two [4Fe–4S] clusters indicate that the cluster that lies closer to F_X is F_A and the more distant cluster is F_B (Fig. 24-27). The observation that both branches of PSI's electron-transfer pathways are active, in contrast to only one active branch in PSII and the PbRC, is rationalized by the fact that the two quinones at the ends of each branch are functionally equivalent in PSI but functionally different in PSII and the PbRC.

PSI's core antenna system consists of 90 Chl *a* molecules and 22 carotenoids (Fig. 24-26*a*). The Mg^{2+} ions of 79 of these Chl *a* molecules are axially liganded by residues of PsaA and PsaB (mostly His side chains or protein-bound water molecules), whereas the remaining 11 are so liganded by the smaller subunits PsaJ through M and PsaX. The spatial distribution of these antenna Chl *a*'s resembles that in the core antenna subunits CP43 and CP47 of PSII. Indeed, the N-terminal domains of PsaA and PsaB are similar in sequence to those of CP43 and CP47 and fold into similar structures containing six transmembrane helices each. The carotenoids, which are mostly β-carotenes, are deeply buried in the membrane, where they are in van der Waals contact with Chl *a* rings. This permits efficient energy transfer from photoexcited carotenoids to Chl *a* as well as protects PSI from photooxidative damage. PSI also tightly binds four lipid molecules such that their fatty acyl groups are embedded among the complex's transmembrane helices. This strongly suggests that these lipids have specific structural and/or functional roles rather than being artifacts of preparation. Indeed, the head group of one of them, a phospholipid, coordinates the Mg^{2+} of an antenna Chl *a*, an unprecedented interaction.

h. PSI-Activated Electrons May Reduce NADP⁺ or Motivate Proton Gradient Formation

Electrons ejected from F_B in PSI may follow either of two alternative pathways (Fig. 24-18):

1. Most electrons follow a noncyclic pathway by passing to an ~100-residue, [2Fe–2S]-containing, soluble ferredoxin **(Fd)** that is located in the stroma. Reduced Fd, in turn, reduces NADP⁺ in a reaction mediated by the ~310-residue, monomeric, FAD-containing **ferredoxin–NADP⁺ reductase (FNR,** Fig. 24-28*a*), to yield the final product of the chloroplast light reaction, NADPH. Two reduced Fd molecules successively deliver one electron each to the FAD of FNR, which thereby sequentially assumes the neutral semiquinone and fully reduced states before transferring the two electrons and a proton to the NADP⁺ via what is formally a hydride ion transfer. The X-ray structure of the complex between Fd and FNR from maize leaf (Fig. 24-28*b*), determined by Genji Kurisu, reveals that the shortest interatomic approach between Fd's [2Fe–2S] cluster and FNR's FAD is the 6.0 Å between an Fe atom and FAD atom C8a (the methyl C closest to its ribitol residue; Fig. 16-8). This is sufficiently close for direct electron transfer through space between these prosthetic groups. The complex is stabilized by five salt bridges, as similarly appears to be the case for the interaction between cytochrome *f* and PC.

2. Some electrons are returned from PSI, via cytochrome b_6, to the plastoquinone pool, thereby traversing a cyclic pathway that translocates protons across the thylakoid membrane. The likely mechanism for this process is that Fd reduces **ferredoxin–plastoquinone reductase (FQR),** which, in turn, reduces plastoquinone, which then reduces cytochrome $b_6 f$, thereby translocating protons via a Q cycle into the thylakoid lumen. However, FQR has not been isolated. Note that the cyclic pathway is independent of the action of PSII and hence does not result in the evolution of O_2. This accounts for the observation that chloroplasts absorb more than eight photons per O_2 molecule evolved.

The cyclic electron flow presumably functions to increase the amount of ATP produced relative to that of NADPH and thus permits the cell to adjust the relative amounts of these two substances produced according to its needs.

i. PSI and PSII Occupy Different Parts of the Thylakoid Membrane

Freeze-fracture electron microscopy (Section 12-3C) has revealed that the protein complexes of the thylakoid membrane have characteristic distributions (Fig. 24-29):

1. PSI occurs mainly in the unstacked stroma lamellae, in contact with the stroma, where it has access to NADP⁺.

2. PSII is located almost exclusively between the closely stacked grana, out of direct contact with the stroma.

3. Cytochrome $b_6 f$ is uniformly distributed throughout the membrane.

The high mobilities of plastoquinone and plastocyanin, the electron carriers that shuttle electrons between these

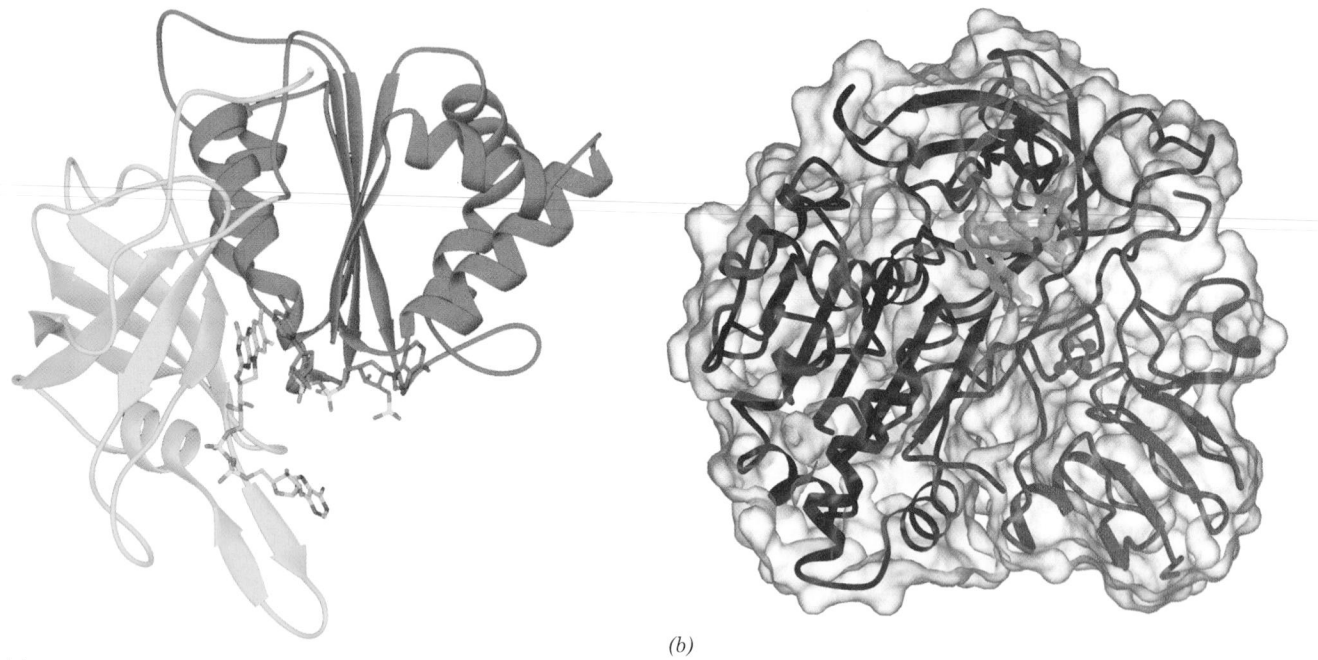

(a)

(b)

FIGURE 24-28 Ferredoxin–NADP⁺ reductase. (*a*) The X-ray
structure of the Y308S mutant form of pea ferredoxin–NADP⁺
reductase (FNR) in complex with FAD and NADP⁺. This
308-residue protein has two domains: The N-terminal domain
(*gold*), which forms the FAD binding site, folds into an
antiparallel β barrel, whereas the C-terminal domain (*magenta*),
which provides the NADP⁺ binding site, forms a dinucleotide-
binding fold (Section 8-3B). The FAD and NADP⁺ are shown
in stick form with NADP⁺ C green, FAD C cyan, N blue,
O red, and P yellow. The flavin and nicotinamide rings are in
opposition with C4 of the nicotinamide ring and C5 of the
flavin ring 3.0 Å apart, an arrangement that is consistent with
direct hydride transfer as also occurs in glutathione reductase
and dihydrolipoyl dehydrogenase (Section 21-2B). However, in

contrast to these latter enzymes, whose bound flavin and
nicotinamide rings are parallel, those in FNR are inclined by
~30°, a heretofore unobserved binding mode. [Based on an
X-ray structure by Andrew Karplus, Cornell University. PDBid
1QFY.] (*b*) The X-ray structure of the complex between Fd
(*red*) and FNR (*blue*) from maize leaf with both proteins drawn
in ribbon form embedded in their solvent-accessible surfaces.
The [2Fe–2S] cluster of Fd (*green*) and the FAD of FNR
(*yellow*) are drawn in ball-and-stick form. The Fd binds in a
hollow between FNR's two domains (Part *a*) such that the line
joining the two Fe's of the [2Fe–2S] cluster lies roughly in the
plane of the flavin ring. [Courtesy of Genji Kurisu, Osaka
University, Osaka, Japan. PDBid 1GAQ.] **See the
Interactive Exercises**

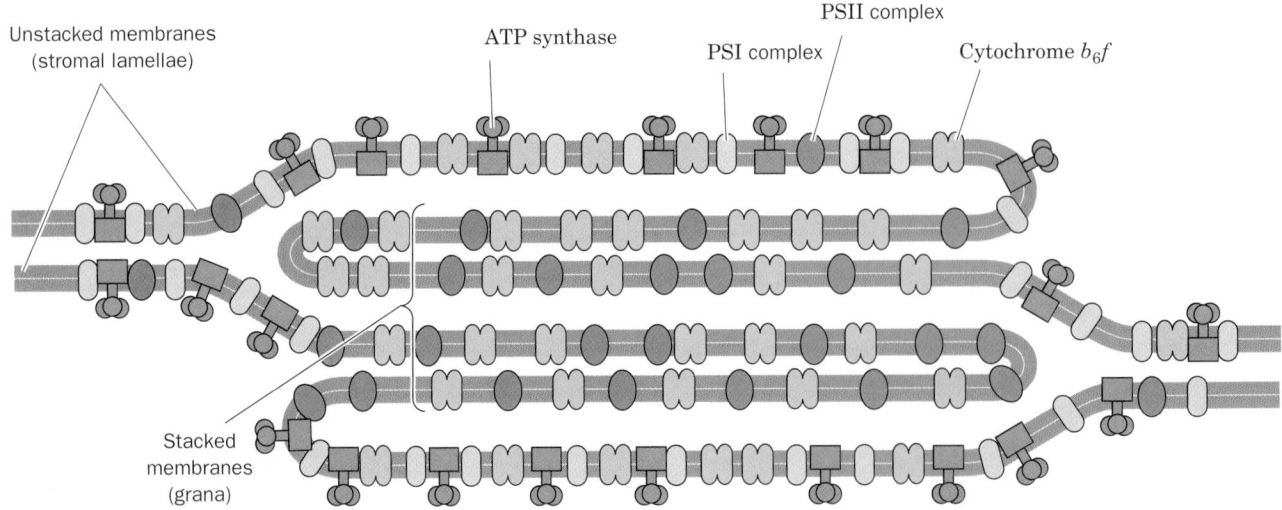

FIGURE 24-29 Segregation of PSI and PSII. The distribution
of photosynthetic protein complexes between the stacked
(grana) and the unstacked (stroma exposed) regions of the

thylakoid membrane is shown. [After Anderson, J.M. and
Anderson, B., *Trends Biochem. Sci.* **7,** 291 (1982).]

complexes, permits photosynthesis to proceed at a reasonable rate.

What function is served by the segregation of PSI and PSII, which are typically present in chloroplasts in equimolar amounts? If these two photosystems were in close proximity, the higher excitation energy of PSII (P680 vs P700) would cause it to pass a large fraction of its absorbed photons to PSI via exciton transfer; that is, PSII would act as a light-harvesting antenna for PSI (Fig. 24-7b). The separation of these particles by around 100 Å eliminates this difficulty.

The physical separation of PSI and PSII also permits the chloroplast to respond to changes in illumination. The relative amounts of light absorbed by the two photosystems vary with how the light-harvesting complexes (LHCs) are distributed between the stacked and unstacked portions of the thylakoid membrane. Under high illumination (normally direct sunlight, which contains a high proportion of short-wavelength blue light), all else being equal, PSII absorbs more light than PSI. PSI is then unable to take up electrons as fast as PSII can supply them, so the plastoquinone is predominantly in its reduced state. The reduced plastoquinone activates a protein kinase to phosphorylate specific Thr residues of the LHCs, which, in response, migrate to the unstacked regions of the thylakoid membrane, where they bind to PSI. A greater fraction of the incident light is thereby funneled to PSI. Under low illumination (normally shady light, which contains a high proportion of long-wavelength red light), PSI takes up electrons faster than PSII can provide them so that plastoquinone predominantly assumes its oxidized form. The LHCs are consequently dephosphorylated and migrate to the stacked portions of the thylakoid membrane, where they drive PSII. The chloroplast therefore maintains the balance between its two photosystems by a light-activated feedback mechanism.

D. *Photophosphorylation*

Chloroplasts generate ATP in much the same way as mitochondria, that is, by coupling the dissipation of a proton gradient to the enzymatic synthesis of ATP (Section 22-3C). This was clearly demonstrated by the imposition of an artificially produced pH gradient across the thylakoid membrane. Chloroplasts were soaked, in the dark, for several hours in a succinic acid solution at pH 4 so as to bring the thylakoid lumen to this pH (the thylakoid membrane is permeable to un-ionized succinic acid). The abrupt transfer of these chloroplasts to an ADP + P_i-containing buffer at pH 8 resulted in an impressive burst of ATP synthesis: About 100 ATPs were synthesized per molecule of cytochrome *f* present. Moreover, the amount of ATP synthesized was unaffected by the presence of electron-transport inhibitors such as DCMU. This, together with the observations that photophosphorylation requires an intact thylakoid membrane and that proton translocators such as 2,4-dinitrophenol (Section 22-3D) uncouple photophosphorylation from light-driven electron transport, provides

convincing evidence favoring the chemiosmotic hypothesis (Section 22-3A).

a. Chloroplast Proton-Translocating ATP Synthase Resembles That of Mitochondria

Electron micrographs of thylakoid membrane stromal surfaces and bacterial plasma membrane inner surfaces reveal lollipop-shaped structures (Fig. 24-30). These closely resemble the F_1 units of the proton-translocating ATP synthase studding the matrix surfaces of inner mitochondrial membranes (Fig. 22-36a). In fact, the chloroplast ATP synthase, which is termed the **CF_1CF_0 complex** (C for chloroplast), has remarkably similar properties to the mitochondrial F_1F_0 complex (Section 22-3C). For example,

1. Both F_0 and CF_0 units are hydrophobic transmembrane proteins that contain a proton-translocating channel.

2. Both F_1 and CF_1 are hydrophilic peripheral membrane proteins of subunit composition $\alpha_3\beta_3\gamma\delta\varepsilon$, of which β is a reversible ATPase.

3. Both ATP synthases are inhibited by oligomycin.

4. Chloroplast ATP synthase translocates protons out of the thylakoid lumen into the stroma (Fig. 24-17), and mitochondrial ATP synthase conducts them into the matrix space (the mitochondrial equivalent of the stroma) from the intermembrane space (Section 22-3A).

Clearly, proton-translocating ATP synthases must have evolved very early in the history of cellular life. Chloroplast ATP synthase is located in the unstacked portions of the

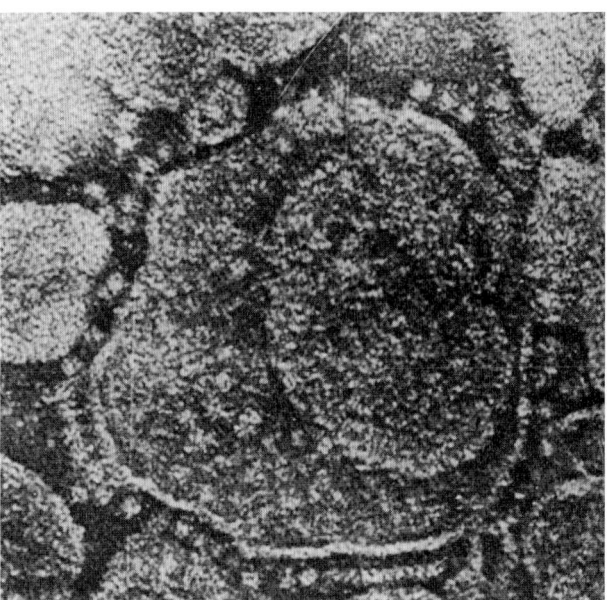

FIGURE 24-30 Electron micrograph of thylakoids. The CF_1 "lollipops" of their ATP synthases project from their stromal surfaces. Compare this with Fig. 22-36a,b. [Courtesy of Peter Hinkle, Cornell University.]

thylakoid membrane, in contact with the stroma, where there is room for the bulky CF_1 globule and access to ADP (Fig. 24-29).

b. Photosynthesis with Noncyclic Electron Transport Produces Around 1.25 ATP Equivalents per Absorbed Photon

At saturating light intensities, chloroplasts generate proton gradients of ~ 3.5 pH units across their thylakoid membranes. This, as we have seen (Figs. 24-17 and 24-18), arises from two sources:

1. The evolution of a molecule of O_2 from two H_2O molecules releases four protons into the thylakoid lumen. These protons should be considered as being supplied from the stroma by the protons and H atoms taken up in the synthesis of NADPH.

2. The transport of the liberated four electrons through the cytochrome b_6f complex occurs with the translocation of what is estimated to be eight protons from the stroma to the thylakoid lumen.

Altogether ~ 12 protons are translocated per molecule of O_2 produced by noncyclic electron transport.

The thylakoid membrane, in contrast to the inner mitochondrial membrane, is permeable to ions such as Mg^{2+} and Cl^-. Translocation of protons and electrons across the thylakoid membrane is consequently accompanied by the passage of these ions so as to maintain electrical neutrality (Mg^{2+} out and Cl^- in). This all but eliminates the membrane potential, $\Delta\Psi$ (Eq. [22.1]). *The electrochemical gradient in chloroplasts is therefore almost entirely a result of the pH gradient.*

Chloroplast ATP synthase, according to most estimates, produces one ATP for every three protons it transports out of the thylakoid lumen. Noncyclic electron transport in chloroplasts therefore results in the production of $\sim 12/3 = 4$ molecules of ATP per molecule of O_2 evolved (although this quantity is subject to revision) or around half an ATP per photon absorbed. Cyclic electron transport is a more productive ATP generator since it yields two-thirds of an ATP (two protons transported) per absorbed photon. The noncyclic process, of course, also yields NADPH, each molecule of which has the free energy to produce three ATPs (Section 22-2A; although this does not normally occur), for a total of six more ATP equivalents per O_2 produced. Consequently, the energetic efficiency of the noncyclic process is $4/8 + 6/8 = 1.25$ ATP equivalents per absorbed photon.

3 ■ DARK REACTIONS

In the previous section we saw how light energy is harnessed to generate ATP and NADPH. In this section we discuss how these products are used to synthesize carbohydrates and other substances from CO_2.

A. *The Calvin Cycle*

The metabolic pathway by which plants incorporate CO_2 into carbohydrates was elucidated between 1946 and 1953 by Melvin Calvin, James Bassham, and Andrew Benson. They did so by tracing the metabolic fate of the radioactive label from $^{14}CO_2$ as it passed through a series of photosynthetic intermediates. The basic experimental strategy they used was to expose growing cultures of algae, such as *Chlorella*, to $^{14}CO_2$ for varying times and under differing illumination conditions and then to drop the cells into boiling alcohol so as to disrupt them while preserving their labeling pattern. The radioactive products were subsequently separated and identified (an often difficult task) through the use of the then recently developed technique of two-dimensional paper chromatography (Section 6-3D) coupled with autoradiography. The overall pathway, diagrammed in Fig. 24-31, is known as the **Calvin cycle** or the **reductive pentose phosphate cycle.**

Some of Calvin's earliest experiments indicated that algae exposed to $^{14}CO_2$ for a minute or more had synthesized a complex mixture of labeled metabolic products, including sugars and amino acids. By inactivating the algae within 5 s of their exposure to $^{14}CO_2$, however, it was shown that *the first stable radioactively labeled compound formed is 3-phosphoglycerate (3PG), which is initially labeled only in its carboxyl group.* This result immediately suggested, in analogy with most biochemical experience, that the 3PG was formed by the carboxylation of a C_2 compound. Yet the failure to find any such precursor eventually forced this hypothesis to be abandoned. The actual carboxylation reaction was discovered through an experiment in which illuminated algae had been exposed to $^{14}CO_2$ for ~ 10 min so that the levels of their labeled photosynthetic intermediates had reached a steady state. The CO_2 was then withdrawn. As expected, the carboxylation product, 3PG,

FIGURE 24-31 (*Opposite*) **The Calvin cycle.** The number of lines in an arrow indicates the number of molecules reacting in that step for a single turn of the cycle that converts three CO_2 molecules to one GAP molecule. For the sake of clarity, the sugars are all shown in their linear forms, although the hexoses and heptoses predominantly exist in their cyclic forms (Section 11-1B). The ^{14}C-labeling patterns generated in one turn of the cycle through the use of $^{14}CO_2$ are indicated in red. Note that two of the Ru5Ps are labeled only at C3, whereas the third Ru5P is equally labeled at C1, C2, and C3. ♫ **See the Animated Figures**

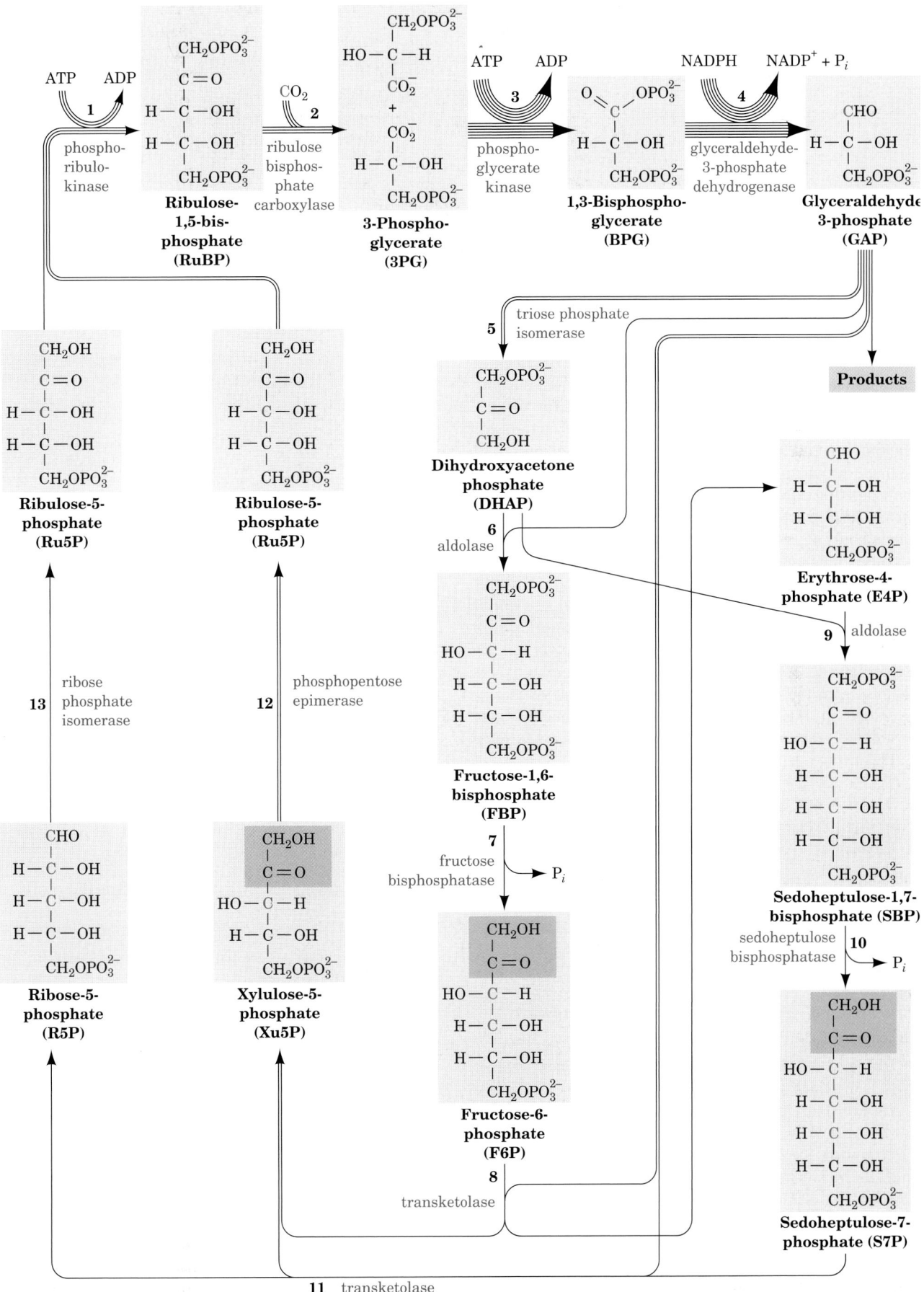

decreased in concentration (Fig. 24-32) because it was depleted by reactions farther along the pathway. The concentration of **ribulose-5-phosphate (Ru5P),**

$$
\begin{array}{c}
CH_2OH \\
| \\
C{=}O \\
| \\
H{-}C{-}OH \\
| \\
H{-}C{-}OH \\
| \\
CH_2OPO_3^{2-}
\end{array}
$$

Ribulose-5-phosphate (Ru5P)

however, simultaneously increased. Evidently, Ru5P is the Calvin cycle's carboxylation substrate. If so, the resulting C_6 carboxylation product must split into two C_3 compounds, one of which is 3PG (Fig. 24-31, Reaction 2). A consideration of the oxidation states of Ru5P and CO_2 indicates that, in fact, both C_3 compounds must be 3PG and that the carboxylation reaction requires no external redox source.

While the search for the carboxylation substrate was going on, several other photosynthetic intermediates had been identified and, through chemical degradation studies, their labeling patterns had been elucidated. For example, the hexose fructose-1,6-bisphosphate (FBP) is initially labeled only at its C3 and C4 positions (Fig. 24-31) but later becomes labeled to a lesser degree at its other atoms. Similarly, a series of tetrose, pentose, hexose, and heptose phosphates were isolated that had the identities and initial labeling patterns indicated in Fig. 24-31. A consideration of the flow of the labeled atoms through these various intermediates led, in what was a milestone of metabolic biochemistry, to the deduction of the Calvin cycle as is diagrammed in Fig. 24-31. The existence of many of its postulated reactions was eventually confirmed by *in vitro* studies using purified enzymes.

a. The Calvin Cycle Generates GAP from CO_2 via a Two-Stage Process

The Calvin cycle may be considered to have two stages:

Stage 1 The production phase (top line of Fig. 24-31), in which three molecules of Ru5P react with three molecules of CO_2 to yield six molecules of glyceraldehyde-3-phosphate (GAP) at the expense of nine ATP and six NADPH molecules. *The cyclic nature of the pathway makes this process equivalent to the synthesis of one GAP from three CO_2 molecules.* Indeed, at this point, one GAP can be bled off from the cycle for use in biosynthesis (see Stage 2).

Stage 2 The recovery phase (bottom lines of Fig. 24-31), in which the carbon atoms of the remaining five GAPs are shuffled in a remarkable series of reactions, similar to those of the pentose phosphate pathway (Section 23-4), to reform the three Ru5Ps with which the cycle began. Indeed, the elucidation of the pentose phosphate pathway at about the same time that the Calvin cycle was being worked out provided much of the biochemical evidence in support of the Calvin cycle. This stage can be conceptually decomposed into four sets of reactions (with the numbers keyed to the corresponding reactions in Fig. 24-31):

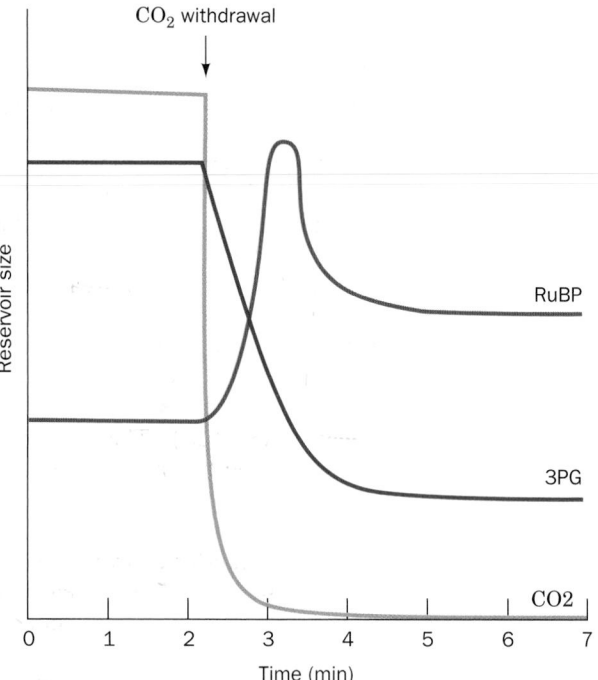

FIGURE 24-32   Algal 3BPG and RuBP levels on removal of CO_2. The time course of the levels of 3PG (*purple curve*) and RuBP (*green curve*) in steady-state $^{14}CO_2$-labeled, illuminated algae is shown during a period in which the CO_2 (*orange curve*) is abruptly withdrawn. In the absence of CO_2, the 3PG concentration rapidly decreases because it is taken up by the reactions of the Calvin cycle but cannot be replenished by them. Conversely, the RuBP concentration transiently increases as it is synthesized from the residual pool of Calvin cycle intermediates but, in the absence of CO_2, cannot be used for their regeneration.

6. $C_3 + C_3 \rightarrow C_6$

8. $C_3 + C_6 \rightarrow C_4 + C_5$

9. $C_3 + C_4 \rightarrow C_7$

11. $C_3 + C_7 \rightarrow C_5 + C_5$

The overall stoichiometry for this process is therefore

$$5\,C_3 \rightarrow 3\,C_5$$

Note that this stage of the Calvin cycle occurs without further input of free energy (ATP) or reducing power (NADPH).

b. Most Calvin Cycle Reactions Also Occur in Other Metabolic Pathways

The types of reactions in the Calvin cycle are all familiar (Section 23-4), with the exception of the carboxylation reaction. This first stage of the Calvin cycle begins with the phosphorylation of Ru5P by **phosphoribulokinase** to form **ribulose-1,5-bisphosphate (RuBP).** Following the carboxylation step, which is discussed below, the resulting 3PG is converted first to 1,3-bisphosphoglycerate (BPG) and then to GAP. This latter sequence is the reverse of two consecutive glycolytic reactions (Sections 17-2G and 17-2F) except that the Calvin cycle reaction involves NADPH rather than NADH.

The second stage of the Calvin cycle begins with the reverse of a familiar glycolytic reaction, the isomerization of GAP to dihydroxyacetone phosphate (DHAP) by triose phosphate isomerase (Section 17-2E). Following this, DHAP is directed along two analogous paths (Fig. 24-31): Reactions 6–8 or Reactions 9–11. Reactions 6 and 9 are aldolase-catalyzed aldol condensations in which DHAP is linked to an aldehyde (aldolase is specific for DHAP but accepts a variety of aldehydes). Reaction 6 is also the reverse of a glycolytic reaction (Section 17-2D). Reactions 7 and 10 are phosphate hydrolysis reactions that are catalyzed, respectively, by fructose bisphosphatase (FBPase, which we previously encountered in our discussion of glycolytic futile cycles and gluconeogenesis; Sections 17-4B and 23-1A) and **sedoheptulose bisphosphatase (SBPase).** The remaining Calvin cycle reactions are catalyzed by enzymes that also participate in the pentose phosphate pathway. In Reactions 8 and 11, both catalyzed by **transketolase,** a C_2 keto unit (shaded in green in Fig. 24-31) is transferred from a ketose to GAP to form **xylulose-5-phosphate (Xu5P)** and leave the aldoses **erythrose-4-phosphate (E4P)** in Reaction 8 and **ribose-5-phosphate (R5P)** in Reaction 11. The E4P produced by Reaction 8 feeds into Reaction 9. The Xu5Ps produced by Reactions 8 and 11 are converted to Ru5P by **phosphopentose epimerase** in Reaction 12. The R5P from Reaction 11 is also converted to Ru5P by **ribose phosphate isomerase** in Reaction 13, thereby completing a turn of the Calvin cycle. Thus only 3 of the 11 Calvin cycle enzymes, phos-phoribulokinase, the carboxylation enzyme **ribulose bisphosphate carboxylase,** and SBPase, have no equivalents in animal tissues.

c. RuBP Carboxylase Catalyzes CO_2 Fixation in an Exergonic Process

The enzyme that catalyzes CO_2 fixation, ribulose bisphosphate carboxylase **(RuBP carboxylase),** is arguably the world's most important enzyme, since nearly all life on Earth ultimately depends on its action. This protein, presumably as a consequence of its particularly low catalytic efficiency ($k_{cat} = \sim3$ s^{-1}), comprises up to 50% of leaf proteins and is therefore the most abundant protein in the biosphere (it is estimated to be synthesized at the rate of $\sim4 \times 10^9$ tons/year, which fixes $\sim10^{11}$ tons of CO_2/year; in comparison crude oil is consumed at the rate of $\sim3 \times 10^9$ tons/year). RuBP carboxylase from higher plants and most photosynthetic microorganisms consists of eight large (L) subunits (477 residues in tobacco leaves) encoded by chloroplast DNA and eight small (S) subunits (123 residues) specified by a nuclear gene (the RuBP carboxylase from certain photosynthetic bacteria is an L_2 dimer whose L subunit has 28% sequence identity with and is structurally similar to that of the L_8S_8 enzyme). X-Ray studies by Carl-Ivar Brändén and by David Eisenberg demonstrated that the L_8S_8 enzyme has the symmetry of a square prism (Fig. 24-33a). The L subunit contains the enzyme's catalytic site, as is demonstrated by its enzymatic

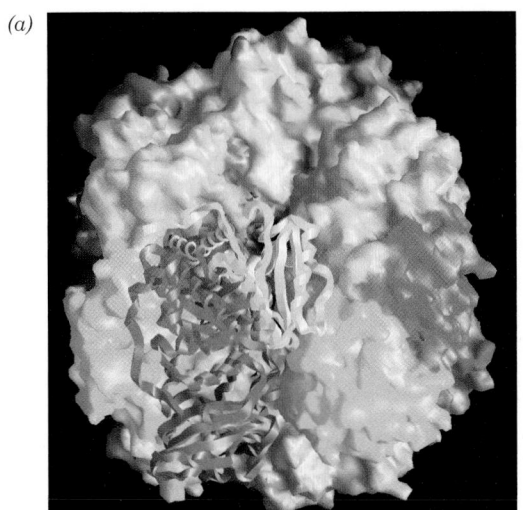

(a)

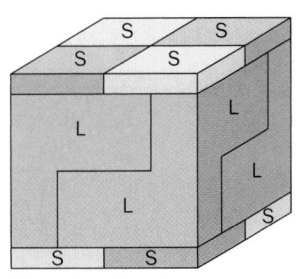

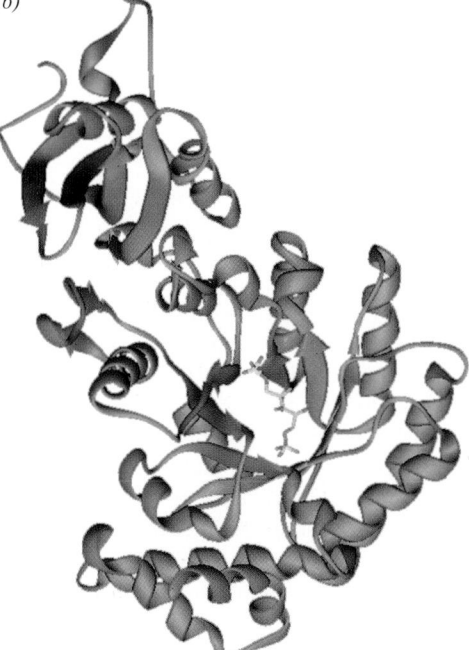

(b)

FIGURE 24-33 X-Ray structure of tobacco RuBP carboxylase. (*a*) The quaternary structure of the L_8S_8 protein. One L and one S subunit are drawn as ribbons with the remainder represented by their van der Waals surfaces. The protein, which has D_4 symmetry (the rotational symmetry of a square prism; Fig. 8-64*b*), is shown with its fourfold axis tipped toward the viewer. As the accompanying diagram indicates, the elongated L subunits (six are clearly visible in the structural drawing) can be considered to associate as two interdigitated tetramers, with that extending from the top green and that extending from the bottom cyan. The members of the S_4 tetramers that cap the top and bottom of the complex are alternately yellow and orange (only one subunit of the lower S_4 tetramer is visible). [Based on an X-ray structure by Yasushi Kai, Osaka University, Osaka, Japan. PDBid 1BUR]. (*b*) An L subunit in complex with the transition state inhibitor **2-carboxyarabinitol-1,5-bisphosphate (CABP;** see text) as drawn in stick form with C cyan, O red, and P yellow. Note that the CABP is bound in the mouth of the enzyme's α/β barrel. The subunit is oriented by a rotation about the vertical axis relative to that drawn in ribbon form in Part *a*. [Based on an X-ray structure by David Eisenberg, UCLA. PDBid 1RLC.]

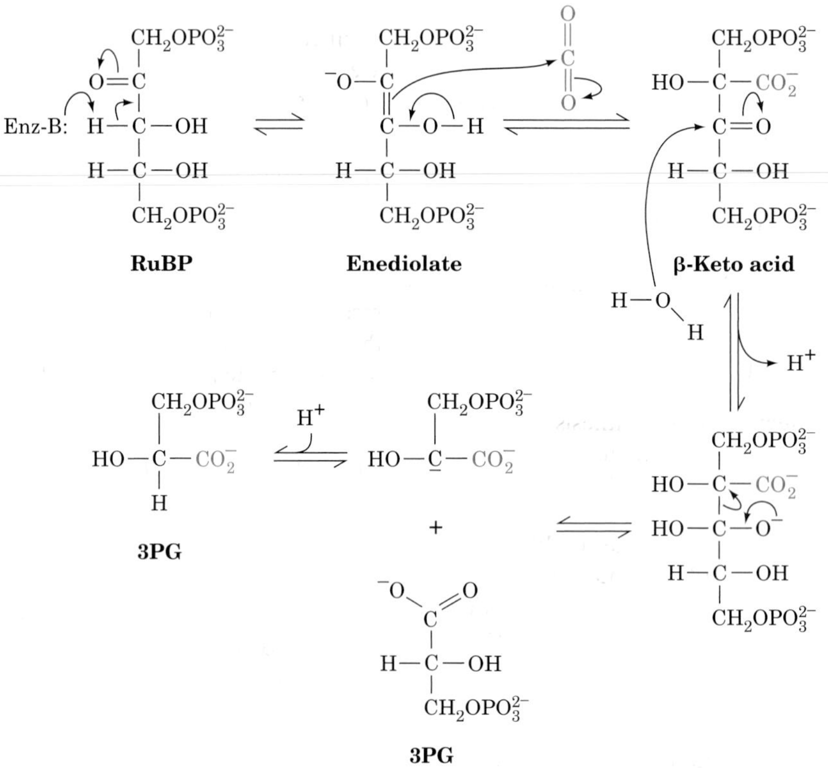

FIGURE 24-34   Probable reaction mechanism of the carboxylation reaction catalyzed by RuBP carboxylase. The reaction proceeds via an enediolate intermediate that nucleophilically attacks CO_2 to form a β-keto acid. This intermediate reacts with water to yield two molecules of 3PG. *See the Animated Figures*

activity in the absence of the S subunit. It consists of two domains (Fig. 24-33b): Residues 1 to 150 form a mixed five-stranded β sheet and residues 151 to 475 fold into an α/β barrel (Fig. 8-19b) which, as do nearly all known α/β barrel enzymes (Section 8-3B), contains the enzyme's active site at the mouth of the barrel near the C-terminus of its β strands. The function of the S subunit is unknown; attempts to show that it has a regulatory role, in analogy with other enzymes, have been unsuccessful.

The accepted mechanism of RuBP carboxylase, which was largely formulated by Calvin, is indicated in Fig. 24-34. Abstraction of the C3 proton of RuBP, the reaction's rate-determining step, generates an enediolate that nucleophilically attacks CO_2 (not HCO_3^-). The resulting β-keto acid is rapidly attacked at its C3 position by H_2O to yield an adduct that splits, by a reaction similar to aldol cleavage, to yield the two product 3PG molecules. The following evidence favors this mechanism:

1. The C3 proton of enzyme-bound RuBP exchanges with solvent, an observation compatible with the existence of the enediolate intermediate.

2. The C2 and C3 oxygen atoms remain attached to their respective C atoms, which eliminates mechanisms involving a covalent adduct such as a Schiff base between RuBP and the enzyme.

3. The trapping of the proposed β-keto acid interme-

diate by borohydride reduction and the tight enzymatic binding of its analogs, such as **2-carboxyarabinitol-1-phosphate (CA1P)** and **2-carboxyarabinitol-1,5-bisphosphate (CABP),**

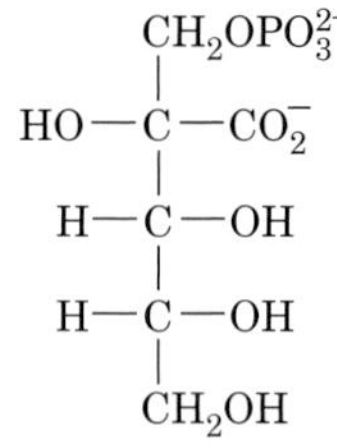

2-Carboxyarabinitol-1-phosphate (CA1P)

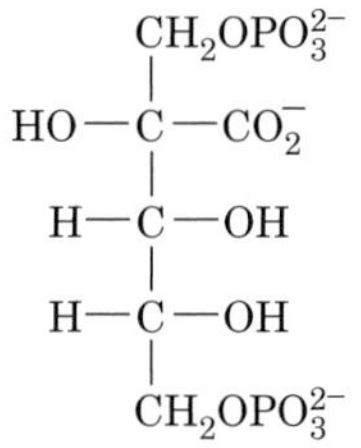

2-Carboxyarabinitol-1,5-bisphosphate (CABP)

provide strong evidence for the existence of this intermediate.

The driving force for the overall reaction, which is highly exergonic ($\Delta G°' = -35.1 \ kJ \cdot mol^{-1}$), *is provided by the cleavage of the β-keto acid intermediate to yield an additional resonance-stabilized carboxylate group.*

RuBP carboxylase activity requires a bound divalent metal ion, physiologically Mg^{2+}, which probably acts to stabilize developing negative charges during catalysis. The Mg^{2+} is, in part, bound to the enzyme by a catalytically essential carbamate group that is generated by the reaction of a nonsubstrate CO_2 with the ε-amino group of Lys 201. Although the *in vitro* activation reaction occurs spontaneously in the presence of Mg^{2+} and HCO_3^-, it is catalyzed *in vivo* by the enzyme **RuBP carboxylase activase** in an ATP-driven process.

d. GAP Is the Precursor of Glucose-1-Phosphate and Other Biosynthetic Products

The overall stoichiometry of the Calvin cycle is

$$3CO_2 + 9ATP + 6NADPH \rightarrow$$
$$GAP + 9ADP + 8P_i + 6NADP^+$$

GAP, the primary product of photosynthesis, is used in a variety of biosynthetic pathways, both inside and outside the chloroplast. For example, it can be converted to fructose-6-phosphate by the further action of Calvin cycle enzymes and then to glucose-1-phosphate (G1P) by phosphoglucose isomerase (Section 17-2B) and phosphoglucomutase (Section 18-1B). *G1P is the precursor of the higher carbohydrates characteristic of plants.* These most notably include sucrose (Section 11-2B), their major transport sugar for delivering carbohydrates to nonphotosynthesizing cells; starch (Section 11-2D), their chief storage polysaccharide; and cellulose (Section 11-2C), the primary structural component of their cell walls. In the synthesis of all these substances, G1P is activated by the formation of either ADP–, CDP–, GDP–, or UDP–glucose (Section 18-2), depending on the species and the pathway. Its glucose unit is then transferred to the nonreducing end of a growing polysaccharide chain much as occurs in the synthesis of glycogen (Section 18-2B). In the case of sucrose synthesis, the acceptor is the reducing end of F6P, with the resulting **sucrose-6-phosphate** being hydrolyzed to sucrose by a phosphatase. Fatty acids and amino acids are synthesized from GAP as is described, respectively, in Sections 25-4 and 26-5.

B. *Control of the Calvin Cycle*

During the day, plants satisfy their energy needs via the light and dark reactions of photosynthesis. At night, however, like other organisms, they must use their nutritional reserves to generate their required ATP and NADPH through glycolysis, oxidative phosphorylation, and the pentose phosphate pathway. Since the stroma contains the enzymes of glycolysis and the pentose phosphate pathway as well as those of the Calvin cycle, *plants must have a light-sensitive control mechanism to prevent the Calvin cycle from consuming this catabolically produced ATP and NADPH in a wasteful futile cycle.*

As we saw in Section 17-4F, the control of flux in a metabolic pathway occurs at enzymatic steps that are far from equilibrium; that is, those that have a large negative value of ΔG. Inspection of Table 24-1 indicates that the three best candidates for flux control in the Calvin cycle are the reactions catalyzed by RuBP carboxylase, FBPase, and SBPase (Reactions 2, 7, and 10, Fig. 24-31). In fact, the catalytic efficiencies of these three enzymes all vary, *in vivo*, with the level of illumination.

The activity of RuBP carboxylase responds to three light-dependent factors:

1. It varies with pH. On illumination, the pH of the stroma increases from around 7.0 to about 8.0 as protons are pumped from the stroma into the thylakoid lumen. RuBP carboxylase has a sharp pH optimum near pH 8.0.

TABLE 24-1 **Standard and Physiological Free Energy Changes for the Reactions of the Calvin Cycle**

Step[a]	Enzyme	$\Delta G^{\circ\prime}$ (kJ · mol^{-1})	ΔG (kJ · mol^{-1})
1	Phosphoribulokinase	−21.8	−15.9
2	Ribulose bisphosphate carboxylase	−35.1	−41.0
3 + 4	Phosphoglycerate kinase + glyceraldehyde-3-phosphate dehydrogenase	+18.0	−6.7
5	Triose phosphate isomerase	−7.5	−0.8
6	Aldolase	−21.8	−1.7
7	Fructose bisphosphatase	−14.2	−27.2
8	Transketolase	+6.3	−3.8
9	Aldolase	−23.4	−0.8
10	Sedoheptulose bisphosphatase	−14.2	−29.7
11	Transketolase	+0.4	−5.9
12	Phosphopentose epimerase	+0.8	−0.4
13	Ribose phosphate isomerase	+2.1	−0.4

[a]Refer to Fig. 24-31.

Source: Bassham, J.A. and Buchanan, B.B., *in* Govindjee (Ed.), *Photosynthesis*, Vol. II, p. 155, Academic Press (1982).

2. It is stimulated by Mg^{2+}. Recall that the light-induced influx of protons to the thylakoid lumen is accompanied by the efflux of Mg^{2+} to the stroma (Section 24-2D).

3. It is strongly inhibited by its transition state analog 2-carboxyarabinitol-1-phosphate (CA1P; Section 24-3A), which many plants synthesize only in the dark. **RuBP carboxylase activase** facilitates the release of the tight-binding CA1P from RuBP carboxylase as well as catalyzing its carbamoylation (Section 24-3A).

FBPase and SBPase are also activated by increased pH and Mg^{2+}, and by NADPH as well. The action of these factors is complemented by a second regulatory system that responds to the redox potential of the stroma. **Thioredoxin,** an ~105-residue protein that occurs in many types of cells, contains a redox-active disulfide group. Reduced thioredoxin activates both FBPase and SBPase by a disulfide interchange reaction (Fig. 24-35). This explains why these Calvin cycle enzymes are activated by reduced disulfide reagents such as dithiothreitol. The redox level of thioredoxin is maintained by **ferredoxin–thioredoxin reductase (FTR),** which contains a redox-active disulfide that is closely associated with a [4Fe–4S] cluster through which the protein directly responds to the redox state of soluble ferredoxin (Fd) in the stroma. This in turn varies with the illumination level. The thioredoxin system also deactivates phosphofructokinase (PFK), the main flux-generating enzyme of glycolysis (Section 17-4F). Thus in plants, *light stimulates the Calvin cycle while deactivating glycolysis, whereas darkness has the opposite effect* (that is, the so-called dark reactions do not occur in the dark).

We have seen that ferredoxin reduces ferredoxin–$NADP^+$ reductase (Section 24-2C) and FTR, as well as supplying electrons to the cyclic pathway of chloroplast photosynthesis, perhaps by reducing ferredoxin–plastoquinone reductase (Section 24-2C). In addition,

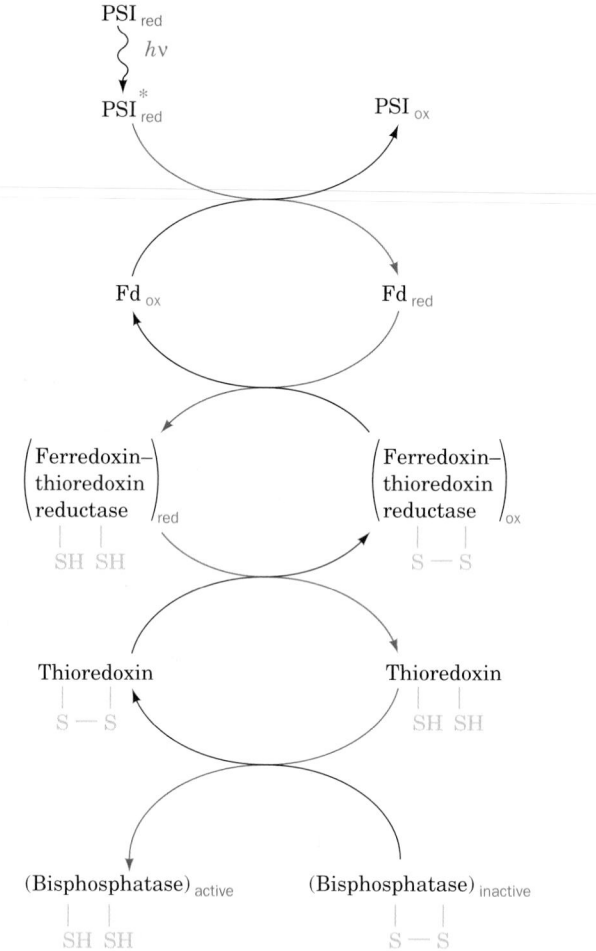

FIGURE 24-35   Light-activation mechanism of FBPase and SBPase. Photoactivated PSI reduces soluble ferredoxin (Fd), which reduces ferredoxin–thioredoxin reductase, which, in turn, reduces the disulfide linkage of thioredoxin. Reduced thioredoxin reacts with the inactive bisphosphatases by disulfide interchange, thereby activating these flux-generating Calvin cycle enzymes.

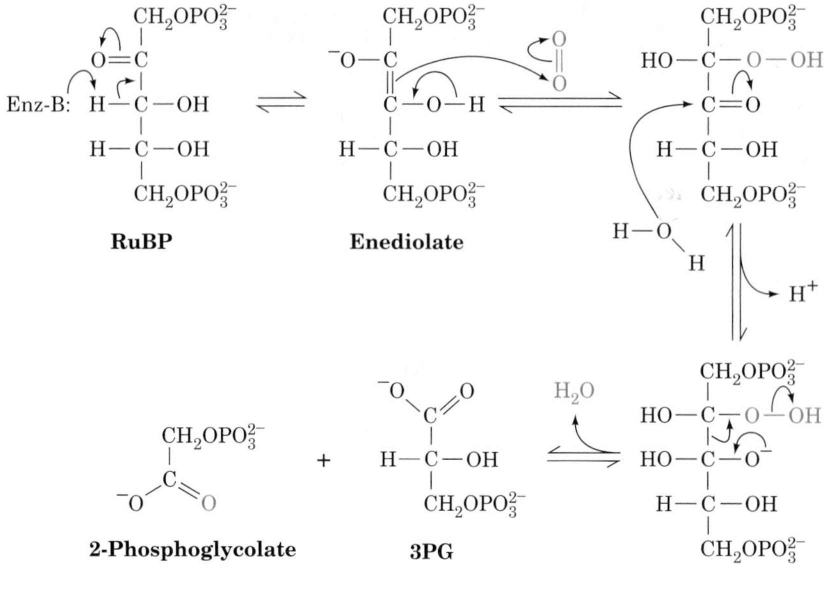

FIGURE 24-36   Probable mechanism of the oxygenase reaction catalyzed by RuBP carboxylase–oxygenase. Note the similarity of this mechanism to that of the carboxylase reaction catalyzed by the same enzyme (Fig. 24-34).

RuBP   **Enediolate**

2-Phosphoglycolate   **3PG**

ferredoxin is the reducing agent for three metabolically essential chloroplast enzymes: **sulfite reductase** (which reduces SO_3^{2-} to S^{2-}), **nitrite reductase** (which reduces NO_2^- to NH_4^+), and **glutamate synthase** (which catalyzes the reaction of α-ketoglutarate and NH_4^+ to form glutamate; Section 26-5A). Thus Fd stands at the center of a complex web of enzymatic and regulatory processes.

C. *Photorespiration and the C$_4$ Cycle*

It has been known since the 1960s that *illuminated plants consume O_2 and evolve CO_2 in a pathway distinct from oxidative phosphorylation. In fact, at low CO_2 and high O_2 levels, this **photorespiration** process can outstrip photosynthetic CO_2 fixation.* The basis of photorespiration was unexpected: *O_2 competes with CO_2 as a substrate for RuBP carboxylase* (RuBP carboxylase is therefore also called **RuBP carboxylase–oxygenase** or **RuBisCO**). In the oxygenase reaction, O_2 reacts with RuBisCO's second substrate, RuBP, to form 3PG and **2-phosphoglycolate** (Fig. 24-36). The 2-phosphoglycolate is hydrolyzed to **glycolate** by **glycolate phosphatase** and, as described below, is partially oxidized to yield CO_2 by a series of enzymatic reactions that occur in the peroxisome and the mitochondrion. Thus photorespiration is a seemingly wasteful process that undoes some of the work of photosynthesis. In the following subsections we discuss the biochemical basis of photorespiration, its significance, and how certain plants manage to evade its deleterious effects.

a. Photorespiration Dissipates ATP and NADPH
The photorespiration pathway is outlined in Fig. 24-37. Glycolate is exported from the chloroplast to the peroxisome (also called the glyoxisome, Section 1-2A), where it is oxidized by **glycolate oxidase** to **glyoxylate** and H_2O_2. The H_2O_2, a powerful and potentially harmful oxidizing agent, is disproportionated to H_2O and O_2 in the peroxisome by the heme-containing enzyme **catalase.** Some of the glyoxylate is further oxidized by glycolate oxidase to oxalate. The remainder is converted to glycine in a **transamination reaction,** as discussed in Section 26-1A, and exported to the mitochondrion. There, two molecules of glycine are converted to one molecule of serine and one of CO_2 by a reaction described in Section 26-3B. *This is the origin of the CO_2 generated by photorespiration.* The serine is transported back to the peroxisome, where a transamination reaction converts it to **hydroxypyruvate.** This substance is reduced to **glycerate** and phosphorylated in the cytosol to 3PG, which reenters the chloroplast, where it is reconverted to RuBP in the Calvin cycle. *The net result of this complex photorespiration cycle is that some of the ATP and NADPH generated by the light reactions is uselessly dissipated.*

Although photorespiration has no known metabolic function, the RuBisCOs from the great variety of photosynthetic organisms so far tested all exhibit oxygenase activity. Yet, over the eons, the forces of evolution must have optimized the function of this important enzyme. It is thought that photosynthesis evolved at a time when Earth's

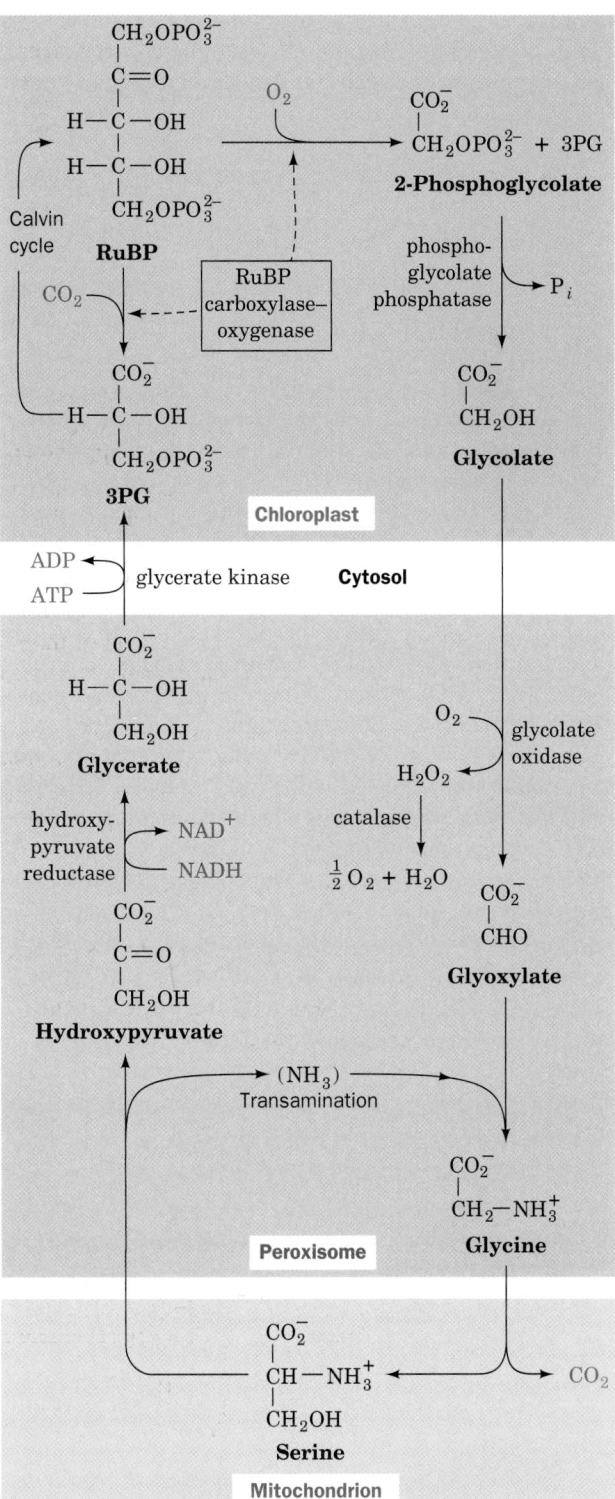

FIGURE 24-37 Photorespiration. This pathway metabolizes the phosphoglycolate produced by the RuBP carboxylase-catalyzed oxidation of RuBP. The reactions occur, as indicated, in the chloroplast, the peroxisome, the mitochondrion, and the cytosol. Note that two glycines are required to form serine + CO_2 (Section 26-3B).

atmosphere contained large quantities of CO_2 and very little O_2, so that photorespiration was of no consequence. It has therefore been suggested that the RuBisCO reaction has an obligate intermediate that is inherently autooxidizable. Another possibility is that photorespiration protects the photosynthetic apparatus from photooxidative damage when insufficient CO_2 is available to otherwise dissipate its absorbed light energy. This hypothesis is supported by the observation that when chloroplasts or leaf cells are brightly illuminated in the absence of both CO_2 and O_2, their photosynthetic capacity is rapidly and irreversibly lost.

b. Photorespiration Limits the Growth Rates of Plants

The steady-state CO_2 concentration attained when a photosynthetic organism is illuminated in a sealed system is named its **CO_2 compensation point.** For healthy plants, this is the CO_2 concentration at which the rates of photosynthesis and photorespiration are equal. For many species it is ~40 to 70 ppm (parts per million) CO_2 (the normal atmospheric concentration of CO_2 is 330 ppm), so their photosynthetic CO_2 fixation usually dominates their photorespiratory CO_2 release. However, the CO_2 compensation point increases with temperature because the oxygenase activity of RuBisCO increases more rapidly with temperature than does its carboxylase activity. Thus, *on a hot bright day, when photosynthesis has depleted the level of CO_2 at the chloroplast and raised that of O_2, the rate of photorespiration may approach that of photosynthesis. This phenomenon is, in fact, a major limiting factor in the growth of many plants.* Indeed, plants possessing a RuBisCO with

significantly less oxygenase activity would not only have increased photosynthetic efficiency but would need less water because they could spend less time with their **stomata** (the pores leading to their internal leaf spaces) open acquiring CO_2 and would have a reduced need for fertilizer because they would require less RuBisCO. The control of photorespiration is therefore an important unsolved agricultural problem that is presently being attacked through genetic engineering studies (Section 5-5).

c. C₄ Plants Concentrate CO₂

Certain species of plants, such as sugarcane, corn, and most important weeds, have a metabolic cycle that concentrates CO_2 in their photosynthetic cells, thereby almost totally preventing photorespiration (their CO_2 compensation points are in the range 2 to 5 ppm). The leaves of plants that have this so-called **C₄ cycle** have a characteristic anatomy. Their fine veins are concentrically surrounded by a single layer of so-called **bundle-sheath cells,** which in turn are surrounded by **mesophyll cells.**

The C₄ cycle (Fig. 24-38) was elucidated in the 1960s by Marshall Hatch and Rodger Slack. It begins with the uptake of atmospheric CO_2 by the mesophyll cells, which, lacking RuBisCO in their chloroplasts, do so by condensing it as HCO_3^- with phosphoenolpyruvate (PEP) to yield oxaloacetate. The oxaloacetate is reduced by NADPH to **malate,** which is exported to the bundle-sheath cells (the name C₄ refers to these four-carbon acids). There the malate is oxidatively decarboxylated by $NADP^+$ to form CO_2, pyruvate, and NADPH. The CO_2, which has been concentrated by this process, enters the Calvin cycle. The

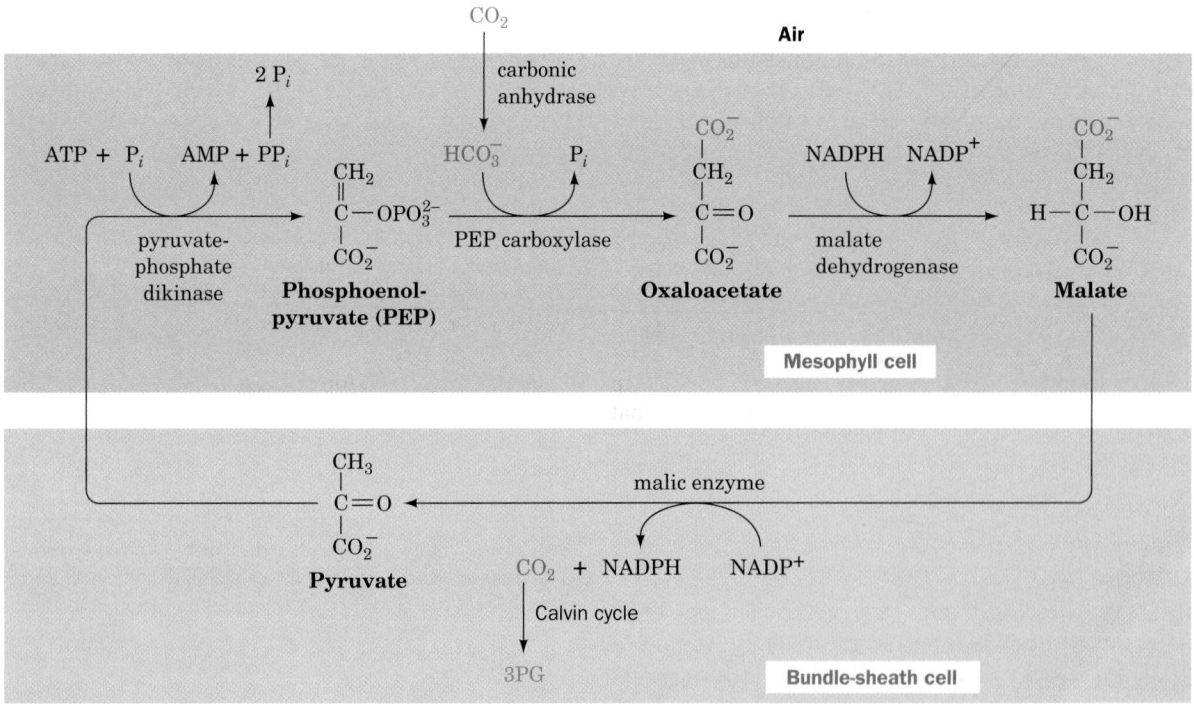

FIGURE 24-38 The C₄ pathway. CO_2 is concentrated in the mesophyll cells and transported to the bundle-sheath cells for entry into the Calvin cycle.

pyruvate is returned to the mesophyll cells, where it is phosphorylated to again form PEP. The enzyme that mediates this reaction, **pyruvate-phosphate dikinase,** has the unusual action of activating a phosphate group through the hydrolysis of ATP to AMP + PP_i. This PP_i is further hydrolyzed to two P_i, which is tantamount to the consumption of a second ATP. *CO_2 is thereby concentrated in the bundle-sheath cells at the expense of two ATPs per CO_2. The dark reactions of photosynthesis in C_4 plants therefore consume a total of five ATPs per CO_2 fixed versus the three ATPs required by the Calvin cycle alone.* The additional ATP is presumably generated through the cyclic flow of electrons in the light reactions (Section 24-2C).

C_4 plants, which comprise ~5% of terrestrial plants, occur largely in unshaded areas of tropical regions because they grow faster under hot and sunny conditions than other, so-called **C_3 plants** (so named because they initially fix CO_2 in the form of three-carbon acids). In cooler climates, where photorespiration is less of a burden, C_3 plants have the advantage because they require less energy to fix CO_2.

d. CAM Plants Store CO_2 through a Variant of the C_4 Cycle

A variant of the C_4 cycle that separates CO_2 acquisition and the Calvin cycle in time rather than in space occurs in many desert-dwelling succulent plants. If, as most plants, they opened their stomata by day to acquire CO_2, they would simultaneously transpire (lose by evaporation) what for them would be unacceptable amounts of water. To minimize this loss, these succulents only absorb CO_2 at night when the temperature is relatively cool. They store this CO_2, in a process known as **crassulacean acid metabolism (CAM;** so named because it was first discovered in plants of the family Crassulaceae), by the synthesis of malate through the reactions of the C_4 pathway (Fig. 24-38). The large amount of PEP necessary to store a day's supply of CO_2 is obtained by the breakdown of starch via glycolysis. During the course of the day, this malate is broken down to CO_2, which enters the Calvin cycle, and pyruvate, which is used to resynthesize starch. CAM plants are able, in this way, to carry out photosynthesis with minimal water loss.

CHAPTER SUMMARY

1 ■ Chloroplasts Photosynthesis is the light-driven fixation of CO_2 to form carbohydrates and other biological molecules. In plants, photosynthesis takes place in the chloroplast, which consists of an inner and an outer membrane surrounding the stroma, a concentrated enzyme solution in which the thylakoid membrane system is immersed. Photosynthesis occurs in two stages, the so-called light reactions in which light energy is harnessed to synthesize ATP and NADPH, and the dark reactions in which these products are used to drive the synthesis of carbohydrates from CO_2 and H_2O. The thylakoid membrane is the site of the photosynthetic light reactions, whereas the dark reactions take place in the stroma. The counterpart of the thylakoid in photosynthetic bacteria is a specialized portion of the plasma membrane named the chromatophore.

2 ■ Light Reactions Chlorophyll is the principal photoreceptor of photosynthesis. Light is absorbed initially by light-harvesting complexes (LHCs) that contain chlorophyll and accessory pigments such as carotenoids. The resulting excitation then migrates via exciton transfer until it reaches the reaction center chlorophyll, where it is trapped. LH2 from purple photosynthetic bacteria is a transmembrane protein that consists of eight or nine rotationally related subunits that each bind three BChl *a* molecules and one carotenoid. LH1, which is similarly arranged but 16-fold symmetric, contains a central hole that binds a photosynthetic reaction center (RC). Light energy absorbed by LH2 is transmitted to LH1, which, in turn, transmits it to the RC.

The purple bacterial photosynthetic RC (PbRC) is a particle that consists of three subunits and several redox-active small molecules that are arranged as two pseudosymmetrically related chains of electron carriers. The primary photon absorbing species of the *Rps. viridis* bacterial reaction center is a special pair of BChl *b* molecules known as P960. By rapid measurement techniques it has been determined that the electron ejected by P960* passes by a third BChl *b* to a BPheo *b* molecule located in only one of the two chains (the other is apparently nonfunctional) and then sequentially to a menaquinone (Q_A) and a ubiquinone (Q_B). The resulting Q_B^- is subsequently further reduced in a second one-electron transfer process and then takes up two protons from the cytosol to form Q_BH_2. The electrons taken up by this species are returned to P960 via a cytochrome bc_1 complex, cytochrome c_2, and, in some bacteria, a four-heme *c*-type cytochrome associated with the photosynthetic reaction center. This cyclic electron-transport process functions to translocate protons, via a Q cycle mediated by the cytochrome bc_1, from the cytoplasm to the outside of the cell. The resulting proton gradient, in a process known as photophosphorylation, drives the synthesis of ATP. Since bacterial photosynthesis does not generate the reducing equivalents needed in many biosynthetic processes, photosynthetic bacteria require an outside source of reducing agents such as H_2S.

In plants and cyanobacteria, the light reactions occur in two reaction centers, those of PSI and PSII, which are electrically "connected" in series. This enables the system to generate sufficient electromotive force to form NADPH by oxidizing H_2O in a noncyclic pathway known as the Z-scheme. PSI and PSII both contain core antenna systems and their RCs are evolutionarily related to each other and to the PbRC. PSII contains an Mn_4 complex that oxidizes two H_2O molecules to four H^+ and O_2 in four one-electron steps. The electrons are passed singly, through a Tyr side chain named Z, to photooxidized P680, the reaction center's photon-absorbing species, which consists of two Chl *a* molecules that are less closely associated than the special pair of BChl *a* molecules in the PbRC. The electron previously ejected from P680* passes through a series of carriers resembling those of the PbRC to a pool of plastoquinone molecules. The electrons then enter the cytochrome b_6f complex, which transports protons, via a Q cycle, from the stroma to the thylakoid space. These electrons are transferred individually, by a plastocyanin carrier, directly to PSI's photooxidized photon-absorbing pigment, P700, a pair of Chl *a*'s that resembles the PbRC's special pair. The electron that had been previously ejected from P700* migrates through both sides of a bifurcated chain of Chl *a* molecules and then through a chain of three [4Fe–4S] clusters to a soluble ferredoxin (Fd) that contains a [2Fe–2S] cluster. The electron then reduces $NADP^+$ in a noncyclic process mediated by ferredoxin–$NADP^+$ reductase. Alternatively, it may be returned, presumably via ferredoxin–plastoquinone reductase, to the plastoquinone pool in a cyclic process that does not require electron input from PSII and only translocates protons across the thylakoid membrane. ATP is synthesized by the CF_1CF_0-ATP synthase, which closely resembles the analogous mitochondrial complex, in a reaction driven by the dissipation of the proton gradient across the thylakoid membrane.

3 ■ **Dark Reactions** CO_2 is fixed in the photosynthetic dark reactions of plants and cyanobacteria by the reactions of the Calvin cycle. The first stage of the Calvin cycle, in sum, mediates the reaction $3RuBP + 3CO_2 \rightarrow 6GAP$ with the consumption of 9 ATP and 6 NADPH generated by the light reactions. The second stage reshuffles the atoms of five GAPs to reform the three RuBPs with which the cycle began, a process that requires no further input of free energy or reduction equivalents. The sixth GAP, the product of the Calvin cycle, is used to synthesize carbohydrates, amino acids, and fatty acids. The flux-controlling enzymes of the Calvin cycle are activated in the light through variations in the pH and the Mg^{2+} and NADPH concentrations, and by the redox level of thioredoxin. The central enzyme of the Calvin cycle, RuBP carboxylase, catalyzes both a carboxylase and an oxygenase reaction with RuBP. The latter reaction is the first step in the photorespiration cycle that liberates CO_2. The rate of photorespiration increases with temperature and decreases with CO_2 concentration, so photorespiration constitutes a significant energetic drain on most plants on hot bright days. C_4 plants, which are most common in the tropics, have a system for concentrating CO_2 in their photosynthetic cells so as to minimize the effects of photorespiration but at the cost of 2 ATP per CO_2 fixed. Certain desert plants conserve water by absorbing CO_2 at night and releasing it to the Calvin cycle by day. This crassulacean acid metabolism occurs through a process similar to the C_4 cycle.

REFERENCES

GENERAL

Blankenship, R.E., *Molecular Mechanisms of Photosynthesis,* Blackwell Science (2002).

Buchanan, B.B., Gruissem, W., and Jones, R.L. (Eds.), *Biochemistry and Molecular Biology of Plants,* American Society of Plant Physiologists (2000).

Deisenhofer, J. and Norris, J.R. (Eds.), *The Photosynthetic Reaction Center,* Vols. I and II, Academic Press (1993). [Vol. I covers chemical and biochemical aspects of photosynthesis and Vol. II is oriented toward its physical principles.]

Hall, D.O. and Rao, K.K., *Photosynthesis* (6th ed.), Cambridge (1999).

Lawlor, D.W., *Photosynthesis* (3rd ed.), BIOS Scientific Publishers Ltd. (2001).

Nicholls, D.G. and Ferguson, S.J., *Bioenergetics 3,* Chapter 6, Academic Press (2002).

CHLOROPLASTS

Bogorad, L. and Vasil, I.K. (Eds.), *The Molecular Biology of Plastids,* Academic Press (1991).

Hoober, J.K., *Chloroplasts,* Plenum Press (1984).

LIGHT REACTIONS

Allen, J.F., How does protein phosphorylation regulate photosynthesis? *Trends Biochem. Sci.* **17,** 12–17 (1992).

Anderson, J.M., Photoregulation of the composition, function and structure of thylakoid membranes, *Annu. Rev. Plant Physiol.* **37,** 93–136 (1986).

Barber, J., Photosystem II: a multisubunit membrane protein that oxidises water, *Curr. Opin. Struct. Biol.* **12,** 523–530 (2002).

Barber, J. and Anderson, B., Revealing the blueprint of photosynthesis, *Nature* **370,** 31–34 (1994).

Bendall, D.S. and Manasse, R.S., Cyclic photophosphorylation and electron transport, *Biochim. Biophys. Acta* **1229,** 23–38 (1995).

Chitnis, P.R., Photosystem I: Function and physiology, *Annu. Rev. Plant Physiol. Plant Biol.* **52,** 593–626 (2001).

Cramer, W.A., Soriano, G.M., Ponomarev, M., Huang, D., Zhang, H., Martinez, S.E., and Smith, J.L., Some new structural aspects of old controversies concerning the cytochrome $b_6 f$ complex of oxygenic photosynthesis, *Annu. Rev. Plant Physiol. Plant Mol. Biol.* **47,** 477–508 (1996).

Debus, R.J., The manganese and calcium ions of photosynthetic oxygen evolution, *Biochim. Biophys. Acta* **1102,** 269–352 (1992).

Deisenhofer, J., Epp, O., Sinning, I., and Michel, H., Crystallographic refinement at 2.3 Å resolution and refined model of the photosynthetic reaction centre from *Rhodopseudomonas viridis, J. Mol. Biol.* **246,** 429–457 (1995).

Deisenhofer, J. and Michel, H., High-resolution structures of photosynthetic reaction centers, *Annu. Rev. Biophys. Biophys. Chem.* **20,** 247–266 (1991); *and* Structures of bacterial photosynthetic reaction centers, *Annu. Rev. Cell Biol.* **7,** 1–23 (1991).

Deng, Z., Aliverti, A., Zanetti, G., Arakaki, A.K., Ottado, J., Orellano, E.G., Calcaterra, N.B., Ceccarelli, E.A., Carrillo, N., and Karplus, P.A., A productive $NADP^+$ binding mode of ferredoxin–$NADP^+$ reductase revealed by protein engineering and crystallographic studies, *Nature Struct. Biol.* **6,** 847–853 (1999); *and* Bruns, C.M. and Karplus, P.A., Refined crystal structure of spinach ferredoxin reductase at 1.7 Å resolution: Oxidized, reduced, and 2′-phospho-5′-AMP bound states, *J. Mol. Biol.* **247,** 125–145 (1995).

DiMagno, T.J., Wang, Z., and Norris, J.R., Initial electron-transfer events in photosynthetic bacteria, *Curr. Opin. Struct. Biol.* **2,** 836–842 (1992).

Diner, B.A. and Rappaport, F., Structure, dynamics, and energetics of the primary photochemistry of photosystem II of oxygenic photosynthesis, *Annu. Rev. Plant Biol.* **53,** 551–580 (2002).

El-Kabbani, O., Chang, C.-H., Tiede, D., Norris, J., and Schiffer, M., Comparison of reaction centers from *Rhodobacter sphaeroides* and *Rhodopseudomonas viridis:* Overall architecture and protein-pigment interactions, *Biochemistry* **30,** 5361–5369 (1991).

Fleming, G.R. and van Grondelle, R., Femtosecond spectroscopy of photosynthetic light-harvesting systems, *Curr. Opin. Struct. Biol.* **7,** 738–748 (1997).

Gennis, R.B., Barquera, B., Hacker, B., Van Doren, S.R., Arnaud, S., Crofts, A.R., Davidson, E., Gray, K.A., and Daldal, F., The *bc₁* complexes of *Rhodobacter sphaeroides* and *Rhodobacter capsulatus*, *J. Bioenerg. Biomembr.* **25,** 195–209 (1993).

Green, B.R. and Durnford, B.G., The chlorophyll-carotenoid proteins of oxygenic photosynthesis, *Annu. Rev. Plant Physiol. Plant Mol. Biol.* **47,** 685–714 (1996).

Heathcote, P., Fyfe, P.K., and Jones, M.R., Reaction centers: The structure and evolution of biological solar power, *Trends Biochem. Sci.* **27,** 79–87 (2002).

Horton, P., Ruban, A.V., and Walters, R.G., Regulation of light harvesting in green plants, *Annu. Rev. Plant Physiol. Plant Mol. Biol.* **47,** 655–684 (1996).

Jordan, P., Fromme, P., Witt, H.T., Klukas, O., Saenger, W., and Krauss, N., Three-dimensional structure of cyanobacterial photosystem I at 2.5 Å resolution, *Nature* **411,** 909–917 (2001).

Karrasch, S., Bullough, P.A., and Ghosh, R., The 8.5 Å projection map of the light harvesting complex I from *Rhodospirillum rubrum* reveals a ring composed of 16 subunits, *EMBO J.* **14,** 631–638 (1995).

Knaff, D.B. and Hirasawa, M., Ferredoxin-dependent chloroplast enzymes, *Biochim. Biophys. Acta* **1056,** 93–125 (1991).

Koepke, J., Hu, X., Muenke, C., Schulen, K., and Michel, H., The crystal structure of the light-harvesting complex II (B800–850) from *Rhodospirillum molischianum*, *Structure* **4,** 581–597 (1996); *and* McDermott, G., Prince, S.M., Freer, A.A., Horthornthwaite-Lawless, A.M., Papiz, M.Z., Cogdell, R.J., and Isaacs, N.W., Crystal structure of an integral membrane light-harvesting complex from photosynthetic bacteria, *Nature* **374,** 517–521 (1995). [The X-ray structures of LH2s.]

Kühlbrandt, W., Wang, D.N., and Fujiyoshi, Y., Atomic model of plant light-harvesting complex by electron crystallography, *Nature* **367,** 614–621 (1994); *and* Kühlbrandt, W., Structure and function of the plant light-harvesting complex, LHC-II, *Curr. Opin. Struct. Biol.* **4,** 519–528 (1994).

Kurisu, G., Kusunoki, M., Katoh, E., Yamazaki, T., Teshima, K., Onda, Y., Kimata-Ariga, Y., and Hase, T., Structure of the electron transfer complex between ferredoxin and ferredoxin-NADP⁺ reductase, *Nature Struct. Biol.* **8,** 117–121 (2001).

Martinez, S.E., Huang, D., Szczepaniak, A., Cramer, W.A., and Smith, J.L., Crystal structure of chloroplast cytochrome *f* reveals a novel cytochrome fold and unexpected heme ligation, *Structure* **2,** 95–105 (1994).

Okamura, M.Y. and Feher, G., Proton transfer in reaction centers from photosynthetic bacteria, *Annu. Rev. Biochem.* **61,** 861–896 (1992).

Ort, D.R. and Yocum, C.F. (Eds.), *Oxygenic Photosynthesis: The Light Reactions,* Kluwer Academic Publishers (1996).

Pullerits, T. and Sundström, V., Photosynthetic light-harvesting pigment–protein complexes: Toward understanding how and why, *Acc. Chem. Res.* **29,** 381–389 (1996).

Stanier, R.Y., Ingraham, J., Wheelis, M.L., and Painter, P.R., *The Microbial World* (5th ed.), Chapter 15, Prentice-Hall (1986). [The biology of photosynthetic eubacteria.]

Strotmann, H. and Bickel-Sandkötter, S., Structure, function, and regulation of chloroplast ATPase, *Annu. Rev. Plant Physiol.* **35,** 97–120 (1984).

Yachandra, V.K., Sauer, K., and Klein, M.P., Manganese cluster in photosynthesis: Where plants oxidize water to dioxygen, *Chem. Rev.* **96,** 2927–2950 (1996). [An authoritative review.]

Zouni, A., Witt, H.-T., Kern, J., Fromme, P., Krauss, N., Saenger, W., and Orth, P., Crystal structure of photosystem II from *Synechococcus elongatus* at 3.8 Å resolution, *Nature* **409,** 739–743 (2001).

DARK REACTIONS

Brändén, C.-I., Lindqvist, Y., and Schneider, G., Protein engineering of rubisco, *Acta Crystallogr.* **B47,** 824–835 (1991); *and* Schneider, G., Lindqvist, Y., and Branden, C.-I., RUBISCO: Structure and mechanism, *Annu. Rev. Biophys. Biomol. Struct.* **21,** 119–143 (1992).

Cushman, J.C. and Bohnert, H.J., Crassalacean acid metabolism: Molecular genetics, *Annu. Rev. Plant Physiol. Plant Mol. Biol.* **50,** 305–332 (1999).

Dai, S., Schwendtmayer, C., Schürmann, P., Ramaswamy, S., and Eklund, H., Redox signaling in chloroplasts: Cleavage of disulfides by an iron-sulfur cluster, *Science* **287,** 655–658 (2000).

Gutteridge, S., Limitations of the primary events of CO_2 fixation in photosynthetic organisms: The structure and mechanism of rubisco, *Biochim. Biophys. Acta* **1015,** 1–14 (1990).

Hartman, F.C. and Harpel, M.R., Chemical and genetic probes of the active site of D-ribulose-1,5-bisphosphate carboxylase/-oxygenase: A retrospective based on the three-dimensional structure, *Adv. Enzymol. Relat. Areas Mol. Biol.* **67,** 1–75 (1993).

Hatch, M.D., C_4 photosynthesis: A unique blend of modified biochemistry, anatomy, and ultrastructure, *Biochim. Biophys. Acta* **895,** 81–106 (1987).

Ogren, W.L., Photorespiration: Pathways, regulation, and modification, *Annu. Rev. Plant Physiol.* **35,** 415–442 (1984).

Portis, A.R., Jr., Regulation of ribulose 1,5-bisphosphate carboxylase/oxygenase activity, *Annu. Rev. Plant Physiol. Plant Mol. Biol.* **43,** 415–437 (1992); *and* Rubisco activase, *Biochim. Biophys. Acta* **1015,** 15–28 (1990).

Schreuder, H.A., Knight, S., Curmi, P.M.G., Andersson, I., Cascio, D., Sweet, R.M., Brändén, C.-I., and Eisenberg, D., Crystal structure of activated tobacco rubisco complexed with the reaction-intermediate analogue 2 carboxy-arabinitol 1,5-bisphosphate, *Protein Sci.* **2,** 1136–1146 (1993).

Spreitzer, R.J. and Salvucci, M.E., Rubisco: structure, regulatory interactions, and possibilities for a better enzyme, *Annu. Rev. Plant Biol.* **53,** 449–475 (2001).

Taylor, T.C. and Andersson, I., The structure of the complex between rubisco and its natural substrate ribulose 1,5-bisphosphate, *J. Mol. Biol.* **265,** 432–444 (1997).

PROBLEMS

1. Why is chlorophyll green in color when it absorbs in the red and the blue regions of the spectrum (Fig. 24-5)?

2. The "red tide" is a massive proliferation of certain algal species that cause seawater to become visibly red. Describe the spectral characteristics of the dominant photosynthetic pigments in these algae.

3. $H_2^{18}O$ is added to a suspension of chloroplasts capable of photosynthesis. Where does the label appear when the suspension is exposed to light?

4. Indicate, where appropriate, the analogous components in the photosynthetic electron-transport chains of purple photosynthetic bacteria and chloroplasts.

5. Antimycin inhibits photosynthesis in chloroplasts. Indicate its most likely site of action and explain your reasoning.

6. Calculate the energy efficiency of cyclic and noncyclic photosynthesis in chloroplasts using 680-nm light. What would this efficiency be with 500-nm light? Assume that ATP formation requires 59 kJ · mol^{-1} under physiological conditions.

*7. What is the minimum pH gradient required to synthesize ATP from ADP + P$_i$? Assume [ATP]/([ADP][P$_i$]) = 10^3, T = 25°C, and that three protons must be translocated per ATP generated. (See Table 16-3 for useful thermodynamic information.)

8. Indicate the average Calvin cycle labeling pattern in ribulose-5-phosphate after two rounds of exposure to $^{14}CO_2$.

9. Chloroplasts are illuminated until the levels of their Calvin cycle intermediates reach a steady state. The light is then turned off. How do the levels of RuBP and 3PG vary after this time?

10. What is the energy efficiency of the Calvin cycle combined with glycolysis and oxidative phosphorylation; that is, what percentage of the input energy can be metabolically recovered in synthesizing starch from CO_2 using photosynthetically produced NADPH and ATP rather than somehow directly storing these "high-energy" intermediates? Assume that each NADPH is energetically equivalent to three ATPs and that starch synthesis and breakdown are energetically equivalent to glycogen synthesis and breakdown.

11. If a C$_3$ plant and a C$_4$ plant are placed together in a sealed illuminated box with sufficient moisture, the C$_4$ plant thrives while the C$_3$ plant sickens and eventually dies. Explain.

12. The leaves of some species of desert plants taste sour in the early morning but, as the day wears on, they become tasteless and then bitter. Explain.

Chapter 25

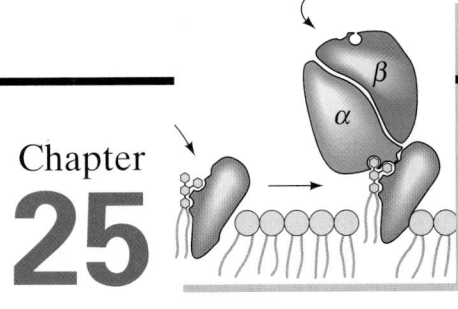

Lipid Metabolism

Lipids play indispensable roles in cell structure and metabolism. For example, triacylglycerols are the major storage form of metabolic energy in animals; cholesterol is a vital component of cell membranes and a precursor of the steroid hormones and bile salts; arachidonate, a C_{20} unsaturated fatty acid, is the precursor of the prostaglandins, prostacyclins, thromboxanes, leukotrienes, and lipoxins, potent intercellular mediators that control a variety of complex processes; and complex glycolipids and phospholipids are major components of biological membranes. We discussed the structures of simple and complex lipids in Section 12-1. In the first half of this chapter, we consider the metabolism of fatty acids and triacylglycerols, including their digestion, oxidation, and biosynthesis. We then consider how cholesterol is synthesized and utilized, and how prostaglandins, prostacyclins, thromboxanes, leukotrienes, and lipoxins are synthesized. We end by studying how complex glycolipids and phospholipids are synthesized from their simpler lipid and carbohydrate components.

1 ■ LIPID DIGESTION, ABSORPTION, AND TRANSPORT

***Triacylglycerols** (also called **fats** or **triglycerides**) constitute ~90% of the dietary lipid and are the major form of metabolic energy storage in humans.* Triacylglycerols consist of glycerol triesters of fatty acids such as palmitic and oleic acids

1-Palmitoyl-2,3-dioleoyl-glycerol

(the names and structural formulas of some biologically common fatty acids are listed in Table 12-1). Like glucose, they are metabolically oxidized to CO_2 and H_2O. Yet, since most carbon atoms of triacylglycerols have lower oxidation states than those of glucose, *the oxidative metabolism of fats yields over twice the energy of an equal weight of dry*

TABLE 25-1   Energy Content of Food Constituents

Constituent	ΔH(kJ · g^{-1} dry weight)
Carbohydrate	16
Fat	37
Protein	17

Source: Newsholme, E.A. and Leech, A.R., *Biochemistry for the Medical Sciences,* p. 16, Wiley (1983).

carbohydrate or protein (Table 25-1). Moreover, fats, being nonpolar, are stored in an anhydrous state, whereas glycogen, the storage form of glucose, is polar and is consequently stored in a hydrated form that contains about twice its dry weight of water. Fats therefore provide up to six times the metabolic energy of an equal weight of hydrated glycogen.

a. Lipid Digestion Occurs at Lipid–Water Interfaces

Since triacylglycerols are water insoluble, whereas digestive enzymes are water soluble, *triacylglycerol digestion takes place at lipid–water interfaces.* The rate of triacylglycerol digestion therefore depends on the surface area of the interface, a quantity that is greatly increased by the churning peristaltic movements of the intestine combined with the emulsifying action of **bile salts** (also called **bile acids**). The bile salts are powerful digestive detergents that, as we shall see in Section 25-6C, are synthesized by the liver and secreted via the gallbladder into the small intestine where lipid digestion and absorption mainly take place.

b. Pancreatic Lipase Requires Activation and Has a Catalytic Triad

Pancreatic **lipase (triacylglycerol lipase)** catalyzes the hydrolysis of triacylglycerols at their 1 and 3 positions to form sequentially **1,2-diacylglycerols** and **2-acylglycerols,** together with the Na$^+$ and K$^+$ salts of fatty acids (soaps). These soaps, being amphipathic, aid in the lipid emulsification process.

The enzymatic activity of pancreatic lipase greatly increases when it forms a complex with pancreatic **colipase,** a protein that forms a 1:1 complex with lipase, in the presence of mixed micelles of phosphatidylcholine (Fig. 12-4) and bile salts. This complex aids in the adsorption of the enzyme to emulsified oil droplets as well as stabilizes the enzyme in an active conformation. The X-ray structures, determined by Christian Cambillau, of pancreatic lipase–procolipase complexes, alone and cocrystallized with mixed micelles of phosphatidylcholine and bile salts, have revealed the structural basis of the activation of lipase as well as how colipase and micelles aid lipase in binding to the lipid–water interface (Fig. 25-1).

The active site of the 449-residue pancreatic lipase, which is contained in the enzyme's N-terminal domain (residues 1–336), contains a catalytic triad that closely resembles that in the serine proteases (Section 15-3B; recall that ester hydrolysis is mechanistically similar to peptide hydrolysis). In the absence of mixed micelles, lipase's active

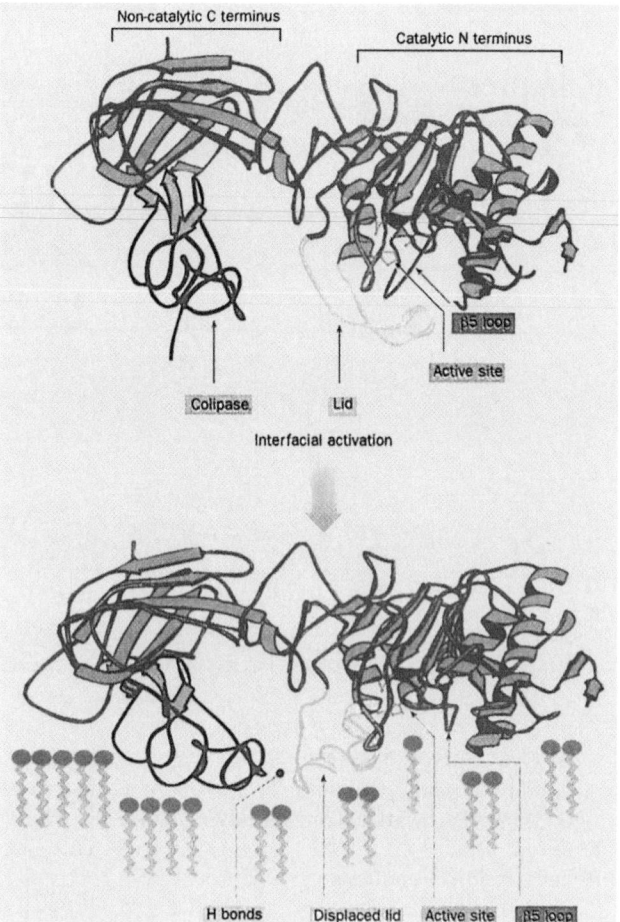

FIGURE 25-1   Mechanism of interfacial activation of triacylglycerol lipase in complex with procolipase. On binding to a phospholipid micelle (*green*), the 25-residue lid (*yellow*) covering the enzyme's active site (*tan*) changes conformation so as to expose its hydrophobic residues, thereby uncovering the active site. This causes the 10-residue β5 loop (*brown*) to move aside in a way that forms the enzyme's oxyanion hole. The procolipase (*magenta*) also changes its conformation so as to hydrogen bond to the "open" lid, thereby stabilizing it in this conformation and, together with lipase, forming an extended hydrophobic surface. [From *Nature* **362,** 793 (1993). Reproduced with permission. PDBid 1LPA.]

site is covered by a 25-residue helical lid. However, in the presence of the mixed micelles, the lid undergoes a complex structural reorganization that exposes the active site; causes a contacting 10-residue loop, the β5 loop, to change conformation in a way that forms the active enzyme's oxyanion hole; and generates a hydrophobic surface about the entrance to the active site. Indeed, the active site of the mixed micelle–containing complex contains a long rod of electron density that contacts the catalytic triad's Ser residue and appears to be a phosphatidylcholine molecule.

Procolipase binds to the C-terminal domain of lipase (residues 337–449) such that the hydrophobic tips of the three loops that comprise much of this 90-residue protein extend from the complex on the same face as lipase's active

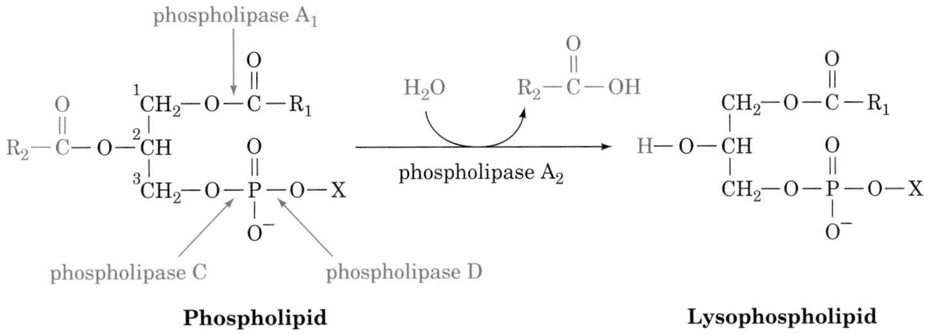

phospholipase A$_1$

$$R_2-\overset{\overset{\displaystyle O}{\|}}{C}-O-\overset{2}{\underset{|}{C}H}\begin{matrix} \overset{1}{C}H_2-O-\overset{\overset{\displaystyle O}{\|}}{C}-R_1 \\[2pt] \end{matrix} \qquad \overset{\overset{\displaystyle O}{\|}}{\underset{O^-}{P}} $$

phospholipase C   phospholipase D

Phospholipid

FIGURE 25-2   Catalytic action of phospholipase A$_2$.
Phospholipase A$_2$ hydrolytically excises the C2 fatty acid residue from a phospholipid to yield the corresponding

Lysophospholipid

lysophospholipid. The bonds hydrolyzed by other types of phospholipases, which are named according to their specificities, are also indicated.

site. A continuous hydrophobic plateau is thereby created that extends over a distance of >50 Å past the active site and that, presumably, helps bind the complex to the lipid surface. The procolipase also forms three hydrogen bonds to the opened lid, thereby stabilizing it in this conformation.

The phenomenon that lipase activity greatly increases at the lipid–water interface is called **interfacial activation.** This, it appears, requires the unmasking and restructuring of the lipase's active site through conformational changes induced by interactions with the lipid–water interface. However, neutron diffraction studies, by Juan Fontecilla-Camps, of crystals of a lipase–colipase–micelle complex, in which the lipase is in its active conformation, reveal that the activating micelle interacts, not with the substrate site, but with the concave face of colipase and the adjacent tip of the lipase's C-terminal domain (Fig. 25-1, *top left*). Apparently, micelle binding and substrate binding involve different regions of the lipase–colipase complex. Hence, strictly speaking, lipase activation appears not to be interfacial but, instead, occurs in the aqueous phase and is facilitated by the binding of colipase and a micelle.

c. Pancreatic Phospholipase A$_2$ Has a Modified Catalytic Triad

Phospholipids are degraded by pancreatic **phospholipase A$_2$**, which hydrolytically excises the fatty acid residue at C2 to yield the corresponding **lysophospholipids** (Fig. 25-2), which are also powerful detergents. Indeed, the phospholipid lecithin (phosphatidylcholine) is secreted in the bile, presumably to aid in lipid digestion.

Phospholipase A$_2$, as does triacylglycerol lipase, preferentially catalyzes reactions at interfaces. However, as Paul Sigler's determinations of the X-ray structures of the phospholipases A$_2$ from cobra venom and bee venom revealed, its mechanism of interfacial activation differs from that of triacylglycerol lipase in that it does not change its conformation. Instead, phospholipase A$_2$ contains a hydrophobic channel that provides the substrate with direct access from the phospholipid aggregate (micelle or membrane) surface to the bound enzyme's active site. Hence, on leaving its micelle to bind to the enzyme, the substrate need not become solvated and then desolvated (Fig. 25-3). In contrast, soluble and dispersed phospholipids must first

(a)

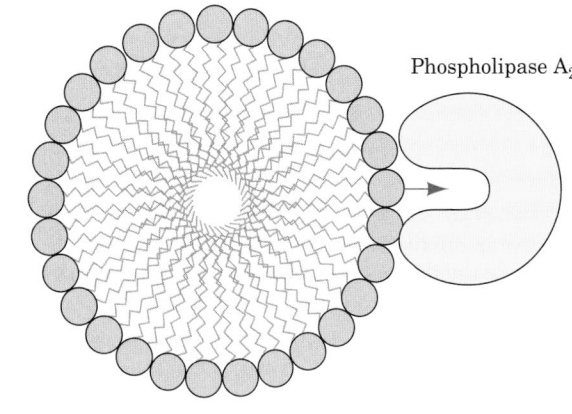

(b)

FIGURE 25-3   Substrate binding to phospholipase A$_2$. (*a*) A hypothetical model of phospholipase A$_2$ in complex with a micelle of lysophosphatidylethanolamine as shown in cross section. The protein is drawn in cyan, the phospholipid head groups are yellow, and their hydrocarbon tails are blue. The

calculated atomic motions of the assembly are indicated through a series of superimposed images taken at 5-ps intervals. [Courtesy of Raymond Salemme, E.I. du Pont de Nemours & Company.] (*b*) Schematic diagram of a productive interaction between phospholipase A$_2$ and a phospholipid contained in a micelle.

(a)

(b)

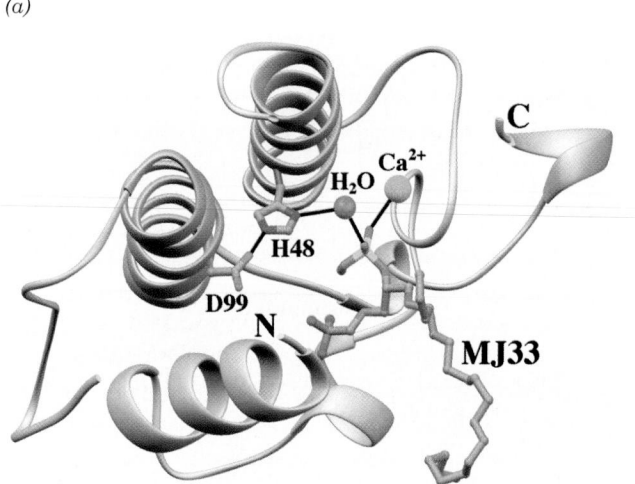

FIGURE 25-4   Structure and mechanism of phospholipase A$_2$. (*a*) The X-ray structure of the 124-residue monomeric porcine phospholipase A$_2$ (*lavender*) in complex with the tetrahedral intermediate mimic MJ33. The enzyme's active site contains a catalytic triad similar to those of the serine proteases (Fig. 15-21) with a water molecule replacing the catalytic Ser. The His 48 and Asp 99 side chains of the catalytic triad together with MJ33 are drawn in stick form colored according to atom type (C gray, N blue, O red, F green, and P yellow). The water molecule of the catalytic triad and the catalytically important Ca^{2+} ion are represented by red and cyan spheres. Catalytically important hydrogen bonds and Ca^{2+} liganding interactions are represented by thin black lines. The tetrahedral phosphoryl group of MJ33 presumably occupies the site of the unobserved H$_2$O. Residues 65 to 74 of the protein have been deleted for clarity. [Based on an X-ray structure by Mahendra Jain and Brian Bahnson, University of Delaware. PDBid 1FXF.] (*b*) The catalytic mechanism of phospholipase A$_2$. **(1)** The catalytic triad activates a second water molecule to attack the scissile carbonyl carbon with Ca^{2+} coordinating the activated water molecule as well as electrostatically stabilizing the resulting tetrahedral intermediate (rather than doing so via nucleophilic catalysis as occurs in the serine proteases; Fig. 15-23). **(2)** The tetrahedral intermediate decomposes to yield products. [After Berg, O.G., Gelb, M.H., Tsai, M.-D., and Jain, M.K., *Chem. Rev.* **101**, 2638 (2001).]

surmount these significant kinetic barriers in order to bind to the enzyme.

The catalytic mechanism of phospholipase A$_2$ also differs substantially from that of triacylglycerol lipase. Although the phospholipase A$_2$ active site contains the His and Asp components of a catalytic triad, an enzyme-bound water molecule occupies the position expected for an active site Ser. Moreover, the active site contains a bound Ca^{2+} ion and does not form an acyl–enzyme intermediate. Sigler therefore proposed that phospholipase A$_2$ catalyzes the direct hydrolysis of phospholipid with a His–Asp "catalytic dyad" activating an active site water molecule for nucleophilic attack on the ester, and with the Ca^{2+} ion stabilizing the oxyanion transition state. However, the subsequently determined X-ray structure, by Mahendra Jain

and Brian Bahnson, of phospholipase A$_2$ in complex with the tetrahedral intermediate mimic **MJ33**

MJ33 [1-Hexadecyl-3-(trifluoroethyl)-*sn*-glycero-2-phosphomethanol]

suggests that a second, previously unobserved water molecule, which is liganded by the Ca^{2+} ion, is the attacking nucleophile (Fig. 25-4*a*). This has led to the formulation of a reaction mechanism (Fig. 25-4*b*) in which the Asp–His–water catalytic triad and the Ca^{2+} ion both activate the second water molecule, with the Ca^{2+} ion also stabilizing the resulting tetrahedral intermediate.

d. Bile Salts and Fatty Acid–Binding Protein Facilitate the Intestinal Absorption of Lipids

The mixture of fatty acids and mono- and diacylglycerols produced by lipid digestion is absorbed by the cells lining the small intestine (the intestinal mucosa) in a process facilitated by bile salts. The micelles formed by the bile salts take up the nonpolar lipid degradation products so as to permit their transport across the unstirred aqueous boundary layer at the intestinal wall. The importance of this process is demonstrated in individuals with obstructed bile ducts: They absorb little of their dietary lipids but, rather, eliminate them in hydrolyzed form in their feces **(steatorrhea).** Evidently, *bile salts are not only an aid to lipid digestion but are essential for the absorption of lipid digestion products.* Bile salts are likewise required for the efficient intestinal absorption of the lipid-soluble vitamins A, D, E, and K.

Inside the intestinal cells, fatty acids form complexes with **intestinal fatty acid–binding protein (I-FABP),** a cytoplasmic protein, which serves to increase the effective solubility of these water-insoluble substances and also to protect the cell from their detergent-like effects (recall that soaps are fatty acid salts). The X-ray structures of rat I-FABP, both alone and in complex with a single molecule of palmitate, were determined by James Sacchettini. This monomeric, 131-residue protein consists largely of 10 antiparallel β strands organized into a stack of two approximately orthogonal β sheets (Fig. 25-5). The palmitate occupies a gap between two of the β strands such that it lies between the β sheets with an orientation that, over much of its length, is more or less parallel to the gapped β strands (this structure has therefore been described as forming a "β-clam"). The palmitate's carboxyl group interacts with Arg 106, Gln 115, and two bound water molecules, whereas the methylene chain is encased by the side chains of several hydrophobic, mostly aromatic, residues.

e. Lipids Are Transported in Lipoprotein Complexes

The lipid digestion products absorbed by the intestinal mucosa are converted by these tissues to triacylglycerols (Section 25-4F) and then packaged into lipoprotein particles called **chylomicrons.** These, in turn, are released into the bloodstream via the lymph system for delivery to the tissues. Similarly, triacylglycerols synthesized by the liver

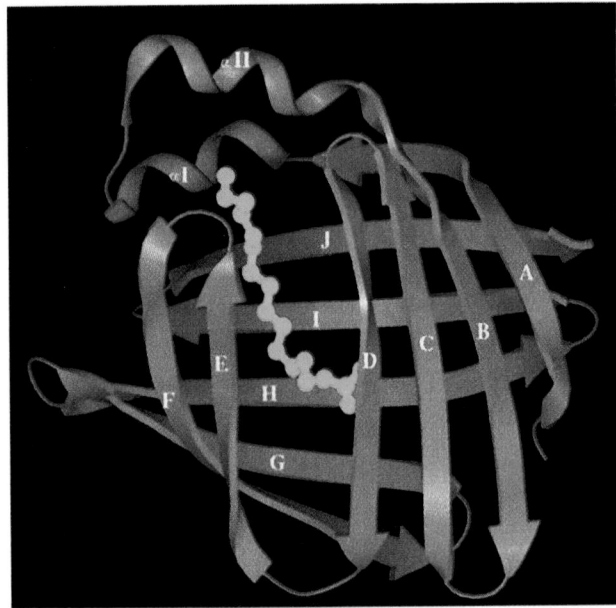

FIGURE 25-5 X-Ray structure of rat intestinal fatty acid–binding protein. The structure is shown in ribbon form (*blue*) in complex with palmitate, shown in ball-and-stick form (*yellow*). [Courtesy of James Sacchettini, Albert Einstein College of Medicine.]

are packaged into **very low density lipoproteins (VLDL)** and released directly into the blood. These lipoproteins, whose origins, structures, and functions are discussed in Section 12-5, maintain their otherwise insoluble lipid components in aqueous solution.

The triacylglycerol components of chylomicrons and VLDL are hydrolyzed to free fatty acids and glycerol in the capillaries of adipose tissue and skeletal muscle by **lipoprotein lipase** (Section 12-5B). The resulting free fatty acids are taken up by these tissues while the glycerol is transported to the liver or kidneys. There it is converted to the glycolytic intermediate dihydroxyacetone phosphate by the sequential actions of **glycerol kinase** and **glycerol-3-phosphate dehydrogenase** (Fig. 25-6).

Mobilization of triacylglycerols stored in adipose tissue involves their hydrolysis to glycerol and free fatty acids by **hormone-sensitive triacylglycerol lipase** (or just **hormone-sensitive lipase;** Section 25-5). The free fatty acids are released into the bloodstream, where they bind to **serum albumin** (or just **albumin**), a soluble 585-residue monomeric protein that comprises about half of the blood serum pro-

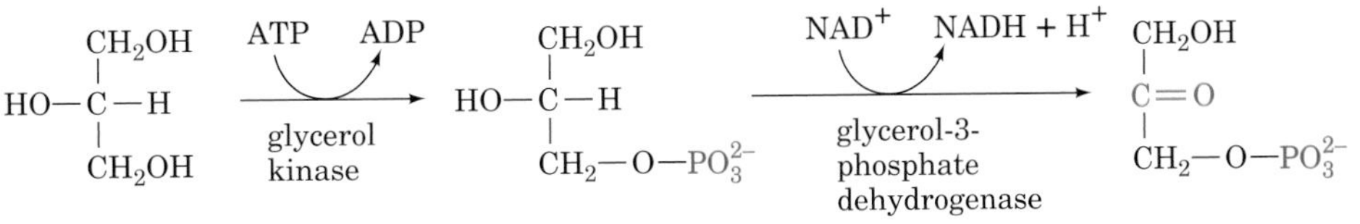

FIGURE 25-6 Conversion of glycerol to the glycolytic intermediate dihydroxyacetone phosphate.

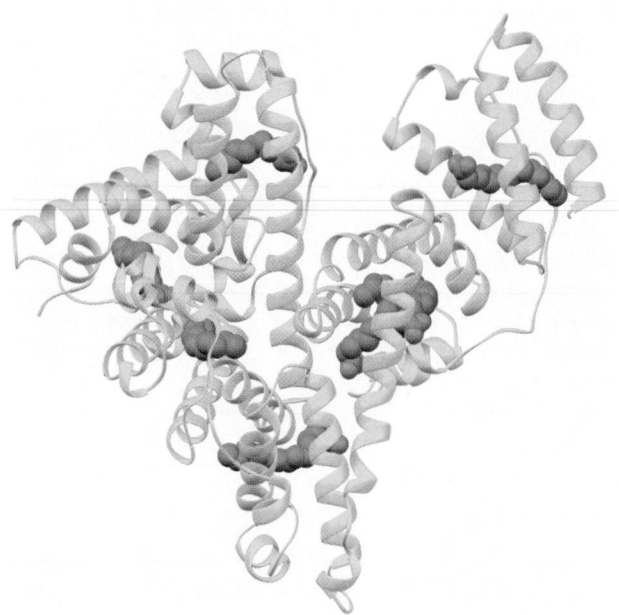

FIGURE 25-7   X-Ray structure of human serum albumin in complex with 7 molecules of palmitic acid. The protein is gold and the fatty acids are drawn in space-filling form with C green and O red. [Based on an X-ray structure by Stephen Curry, Imperial College of Science, Technology, and Medicine, London, U.K. PDBid 1E7H.]

tein. In the absence of albumin, the maximum solubility of free fatty acids is $\sim 10^{-6} M$. Above this concentration, free fatty acids form micelles that act as detergents to disrupt protein and membrane structure and would therefore be toxic. However, the effective solubility of fatty acids in fatty

acid–albumin complexes is as much as 2 mM. Nevertheless, those rare individuals with **analbuminemia** (severely depressed levels of albumin) suffer no apparent adverse symptoms; evidently, their fatty acids are transported in complex with other serum proteins.

The X-ray structure of human serum albumin in its complexes with a variety of common fatty acids, determined by Stephen Curry, reveals that each albumin molecule can bind up to seven fatty acid molecules (Fig. 25-7). However, these binding sites have different fatty acid–binding affinities so that, under normal physiological conditions, albumin carries between 0.1 and 2 fatty acid molecules per protein molecule. Albumin also binds an extraordinarily broad range of drugs and is thereby a major and usually unpredictable influence on their pharmacokinetics (Section 15-4B). Indeed, the large amounts of fatty acids in the blood after meals can significantly affect the pharmacokinetics of a drug through competitive and/or cooperative interactions.

2 ■ FATTY ACID OXIDATION

The biochemical strategy of fatty acid oxidation was understood long before the advent of modern biochemical techniques involving enzyme purification or the use of radioactive tracers. In 1904, Franz Knoop, in the first use of chemical labels to trace metabolic pathways, fed dogs fatty acids labeled at their ω (last) carbon atom by a benzene ring and isolated the phenyl-containing metabolic products from their urine. Dogs fed labeled odd-chain fatty acids excreted **hippuric acid,** the glycine amide of **benzoic acid,** whereas those fed labeled even-chain fatty acids excreted **phenylaceturic acid,** the glycine amide of **phenylacetic acid** (Fig. 25-8). Knoop therefore deduced that the oxidation of

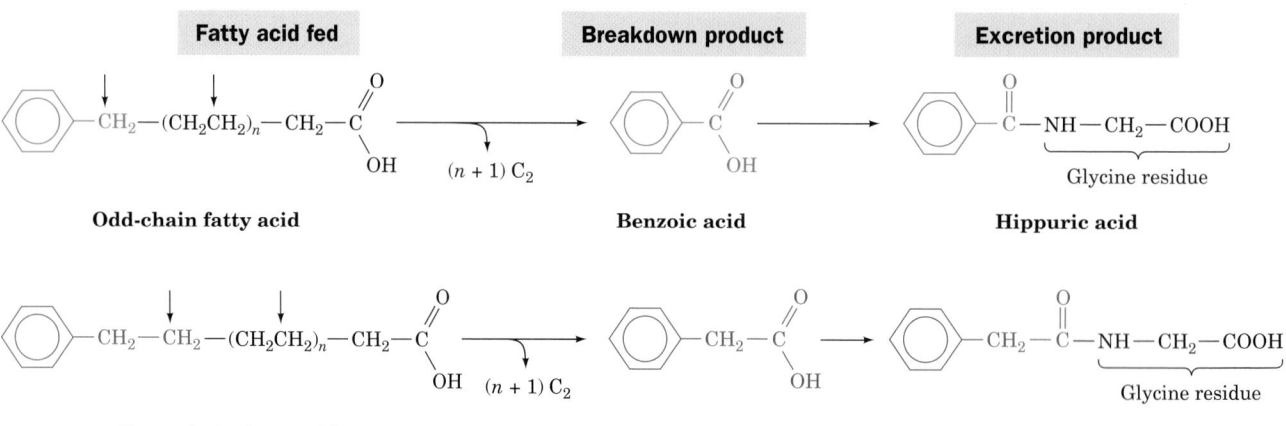

FIGURE 25-8   Franz Knoop's classic experiment indicating that fatty acids are metabolically oxidized at their β-carbon atom. ω-Phenyl-labeled fatty acids containing an odd number of carbon atoms are oxidized to the phenyl-labeled C_1 product, benzoic acid, whereas those with an even number of carbon atoms are oxidized

to the phenyl-labeled C_2 product, phenylacetic acid. These products are excreted as their respective glycine amides, hippuric and phenylaceturic acids. The vertical arrows indicate the deduced sites of carbon oxidation. The intermediate C_2 products are oxidized to CO_2 and H_2O and were therefore not isolated.

the carbon atom β to the carboxyl group is involved in fatty acid breakdown. Otherwise, the phenylacetic acid would be further oxidized to benzoic acid. Knoop proposed that this breakdown occurs by a mechanism known as **β oxidation** in which the fatty acid's C_β atom is oxidized. It was not until after 1950, following the discovery of coenzyme A, that the enzymes of fatty acid oxidation were isolated and their reaction mechanisms elucidated. This work confirmed Knoop's hypothesis.

A. *Fatty Acid Activation*

Before fatty acids can be oxidized, they must be "primed" for reaction in an ATP-dependent acylation reaction to form fatty acyl-CoA. This activation process is catalyzed by a family of at least three **acyl-CoA synthetases** (also called **thiokinases**) that differ according to their chain-length specificities. These enzymes, which are associated with either the endoplasmic reticulum (ER) or the outer mitochondrial membrane, all catalyze the reaction

Fatty acid + CoA + ATP $\rightleftharpoons$ acyl-CoA + AMP + PP$_i$

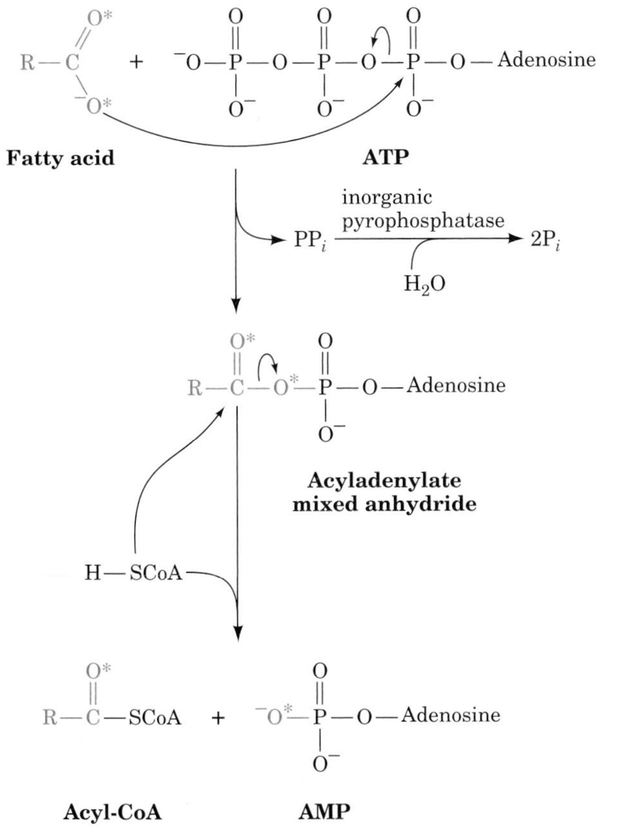

In the activation of ^{18}O-labeled palmitate by a long-chain acyl-CoA synthetase, both the AMP and the acyl-CoA products become ^{18}O labeled. This observation indicates that the reaction has an acyladenylate mixed anhydride intermediate that is attacked by the sulfhydryl group of CoA to form the thioester product (Fig. 25-9). The reaction involves both the cleavage and the synthesis of bonds with large negative free energies of hydrolysis so that the free energy change associated with the overall reaction is close to zero. The reaction is driven to completion in the cell by the highly exergonic hydrolysis of the product pyrophosphate (PP$_i$) catalyzed by the ubiquitous **inorganic pyrophosphatase.** Thus, as commonly occurs in metabolic pathways, *a reaction forming a "high-energy" bond through the hydrolysis of one of ATP's phosphoanhydride bonds is driven to completion by the hydrolysis of its second such bond.*

B. *Transport across the Mitochondrial Membrane*

Although fatty acids are activated for oxidation in the cytosol, they are oxidized in the mitochondrion as Eugene Kennedy and Albert Lehninger established in 1950. We must therefore consider how fatty acyl-CoA is transported across the inner mitochondrial membrane. A long-chain fatty acyl-CoA cannot directly cross the inner mitochondrial membrane. Rather, its acyl portion is first transferred to **carnitine** (Fig. 25-10), a compound that occurs in both plant and animal tissues. This transesterification reaction has an equilibrium constant close to 1, which indicates that the *O*-acyl bond of **acyl-carnitine** has a free energy of hydrolysis similar to that of the thioester. **Carnitine palmitoyltransferases I** and **II,** which can transfer a variety of acyl groups, are located, respectively, on the external and internal surfaces of the inner mitochondrial membrane. The translocation process itself is mediated by a specific carrier protein that transports acyl-carnitine into the mitochondrion while transporting free carnitine in the oppo-

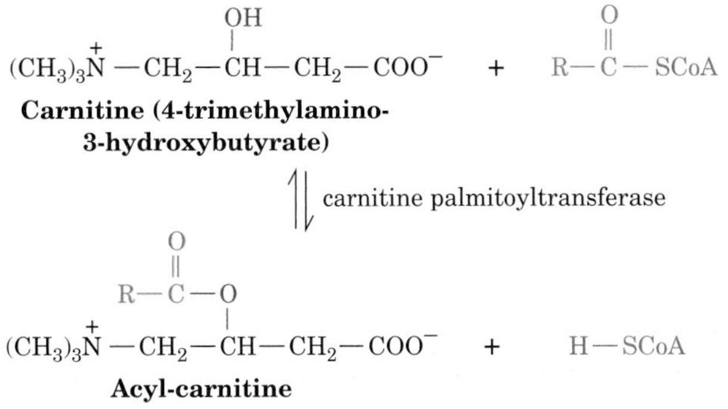

FIGURE 25-9 Mechanism of fatty acid activation catalyzed by acyl-CoA synthetase. Experiments utilizing ^{18}O-labeled fatty acids (*) demonstrate that the formation of acyl-CoA involves an intermediate acyladenylate mixed anhydride.

FIGURE 25-10 Acylation of carnitine catalyzed by carnitine palmitoyltransferase.

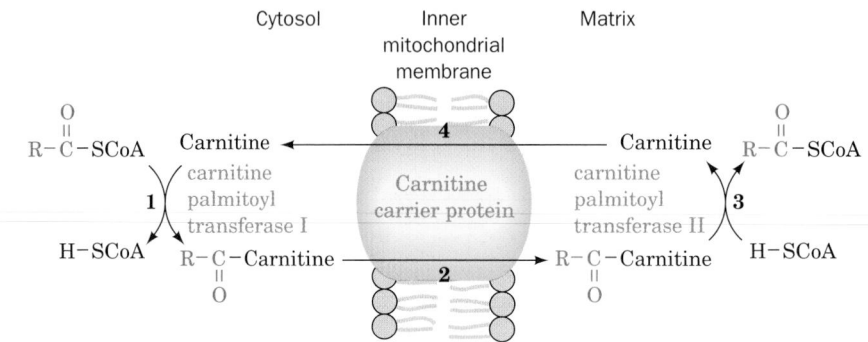

FIGURE 25-11   Transport of fatty acids into the mitochondrion.

site direction. Acyl-CoA transport therefore occurs via four reactions (Fig. 25-11):

1. The acyl group of a cytosolic acyl-CoA is transferred to carnitine, thereby releasing the CoA to its cytosolic pool.

2. The resulting acyl-carnitine is transported into the mitochondrial matrix by the transport system.

3. The acyl group is transferred to a CoA molecule from the mitochondrial pool.

4. The product carnitine is returned to the cytosol.

The cell thereby maintains separate cytosolic and mitochondrial pools of CoA. The mitochondrial pool functions in the oxidative degradation of pyruvate (Section 21-2A) and certain amino acids (Sections 26-3E–G) as well as fatty acids, whereas the cytosolic pool supplies fatty acid biosynthesis (Section 25-4). The cell similarly maintains separate cytosolic and mitochondrial pools of ATP and NAD^+.

C. β Oxidation

Fatty acids are dismembered through the β oxidation of fatty acyl-CoA, a process that occurs in four reactions (Fig. 25-12):

1. Formation of a trans-α,β double bond through dehydrogeation by the flavoenzyme **acyl-CoA dehydrogenase (AD).**

2. Hydration of the double bond by **enoyl-CoA hydratase (EH)** to form a **3-L-hydroxyacyl-CoA.**

3. NAD^+-dependent dehydrogenation of this β-hydroxyacyl-CoA by **3-L-hydroxyacyl-CoA dehydrogenase (HAD)** to form the corresponding β-ketoacyl-CoA.

4. C_α–C_β cleavage in a thiolysis reaction with CoA as catalyzed by **β-ketoacyl-CoA thiolase** (**KT;** also called just **thiolase**) to form acetyl-CoA and a new acyl-CoA containing two less C atoms than the original one.

The first three steps of this process chemically resemble

the citric acid cycle reactions that convert succinate to oxaloacetate (Sections 21-3F–H):

Succinate   **Fumarate**   **L-Malate**   →   **Oxaloacetate**

Mitochondria contain four acyl-CoA dehydrogenases, with specificities for short- (C_4 to C_6), medium- (C_6 to C_{10}), long- (between medium and very long), and very long-chain (C_{12} to C_{18}) fatty acyl-CoAs. The reaction catalyzed by these enzymes is thought to involve removal of a proton at C_α and transfer of a hydride ion equivalent from C_β to FAD (Fig. 25-12, Reaction 1). The X-ray structure of the **medium-chain acyl-CoA dehydrogenase (MCAD)** in complex with **octanoyl-CoA,** determined by Jung-Ja Kim, clearly shows how the enzyme orients the enzyme's base (Glu 376), the substrate C_α—C_β bond, and the FAD prosthetic group for reaction (Fig. 25-13).

a. Acyl-CoA Dehydrogenase Is Reoxidized via the Electron-Transport Chain

The $FADH_2$ resulting from the oxidation of the fatty acyl-CoA substrate is reoxidized by the mitochondrial electron-transport chain through the intermediacy of a

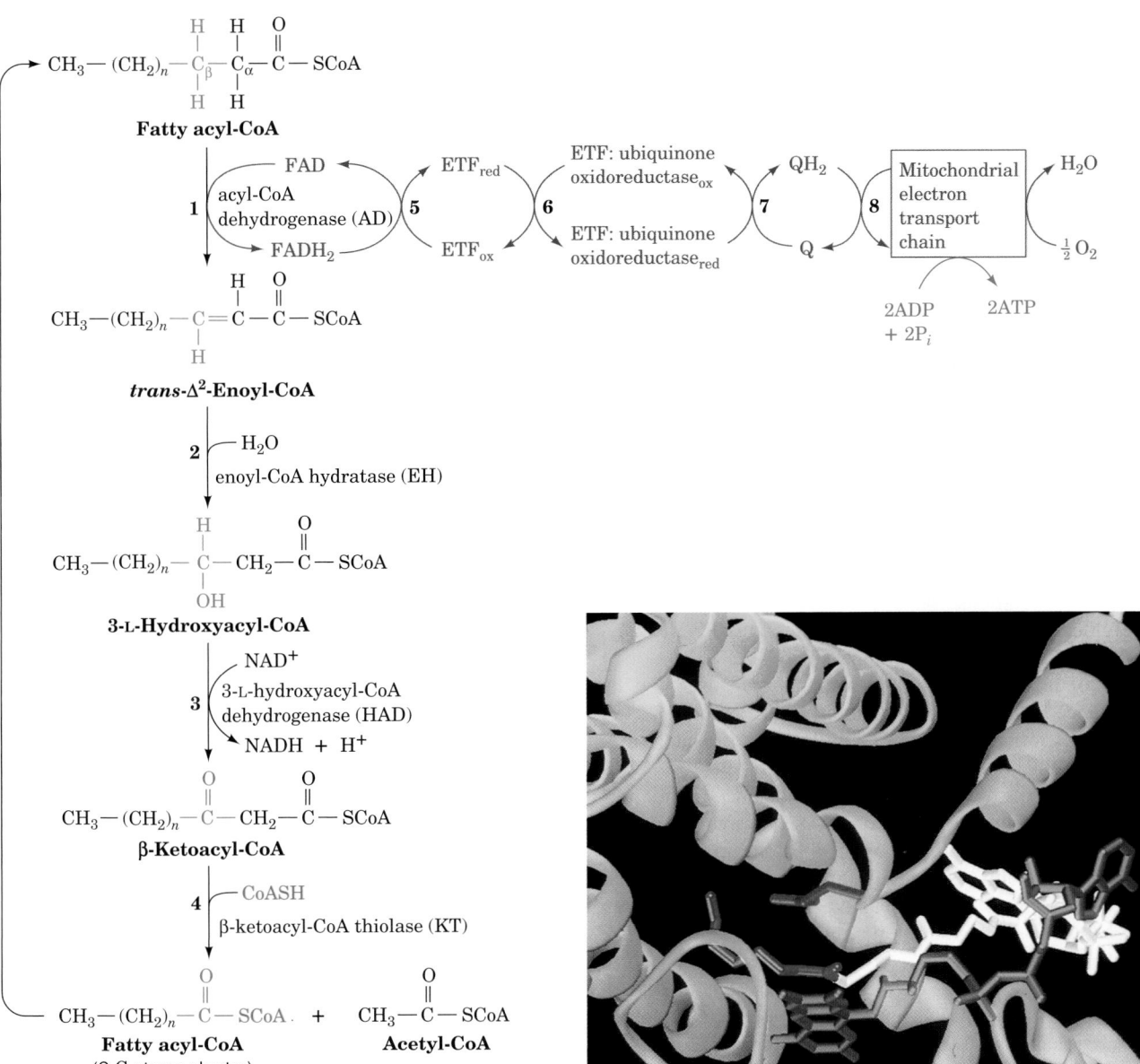

FIGURE 25-12   The β-oxidation pathway of fatty acyl-CoA.
🔊 See the Animated Figures

FIGURE 25-13   Ribbon diagram of the active site region in a subunit of medium-chain acyl-CoA dehydrogenase from pig liver mitochondria in complex with octanoyl-CoA. The enzyme is a tetramer of identical 385-residue subunits, each of which binds an FAD prosthetic group (*green*) and its octanoyl-CoA substrate (whose octanoyl and CoA moieties are blue and white) in largely extended conformations. The octanoyl-CoA binds such that its C_α—C_β bond is sandwiched between the carboxylate group of Glu 376 (*red*) and the flavin ring (*green*), consistent with the proposal that Glu 376 is the general base that abstracts the α proton in the α,β dehydrogenation reaction catalyzed by the enzyme. [Based on an X-ray structure by Jung-Ja Kim, Medical College of Wisconsin. PDBid 3MDE.]
🔊 See the Interactive Exercises

series of electron-transfer reactions. **Electron-transfer flavoprotein (ETF)** transfers two electrons from $FADH_2$ to the flavoiron–sulfur protein **ETF:ubiquinone oxidoreductase,** which in turn transfers two electrons to the mitochondrial electron-transport chain by reducing coenzyme Q (CoQ; Fig. 25-12, Reactions 5–8). Reduction of O_2 to H_2O by the electron-transport chain beginning at the CoQ stage results in the synthesis of two ATPs per two electrons transferred (Section 22-2B).

b. Acyl-CoA Dehydrogenase Deficiency Has Fatal Consequences

The unexpected death of an apparently healthy infant, often overnight, has been, for lack of any real explanation, termed **sudden infant death syndrome (SIDS).** MCAD has

been shown to be deficient in up to 10% of these infants, making this genetic disease more prevalent than **phenylketonuria (PKU)** (Section 26-3H), a genetic defect in phenylalanine degradation for which babies born in the

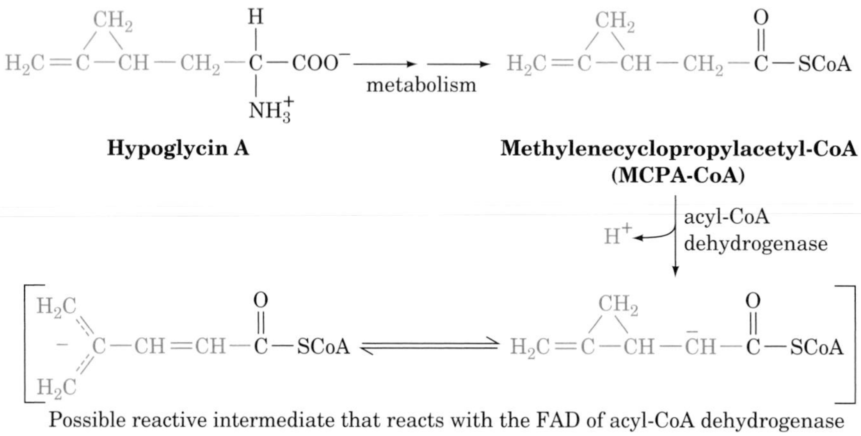

Possible reactive intermediate that reacts with the FAD of acyl-CoA dehydrogenase

FIGURE 25-14 Metabolic conversions of hypoglycin A to yield a product that inactivates acyl-CoA dehydrogenase. Spectral changes suggest that the enzyme's FAD prosthetic group has been modified.

United States are routinely tested. Glucose is the principal energy metabolism substrate just after eating, but when the glucose level later decreases, the rate of fatty acid oxidation must correspondingly increase. The sudden death of infants lacking MCAD may be caused by the imbalance between glucose and fatty acid oxidation.

Lys 304, which becomes Glu in the most prevalent mutation among individuals with MCAD deficiency, is ~20 Å distant from the enzyme's active site and hence cannot participate in binding substrate or FAD. However, since the side chains of Asp 300 and Asp 346 lie within 6 Å of Glu 304, near a subunit–subunit interface, it seems likely that the high concentration of negative charges resulting from the Lys 304 → Glu mutation structurally destabilizes the enzyme.

Deficiency of acyl-CoA dehydrogenase has also been implicated in **Jamaican vomiting sickness,** whose victims suffer violent vomiting followed by convulsions, coma, and death. Severe hypoglycemia is observed in most cases. This condition results from eating unripe **ackee fruit,** which contains **hypoglycin A,** an unusual amino acid, which is metabolized to **methylenecyclopropylacetyl-CoA (MCPA-CoA;** Fig. 25-14). MCPA-CoA, a substrate for acyl-CoA dehydrogenase, is thought to undergo the first step of the reaction that this enzyme catalyzes, removal of a proton from C_α, to form a reactive intermediate that covalently modifies the enzyme's FAD prosthetic group (Fig. 25-14). Since a normal step in the enzyme's reaction mechanism generates the reactive intermediate, MCPA-CoA is said to be a **mechanism-based inhibitor.**

c. Long-Chain Enoyl-CoAs Are Converted to Acetyl-CoA and a Shorter Acyl-CoA by Mitochondrial Trifunctional Protein

The products of acyl-CoA dehydrogenases are 2-enoyl-CoAs. Depending on their chain lengths their processing is continued by one of three systems (Fig. 25-12): the short-chain, medium-chain, or long-chain 2-enoyl-CoA hydratases (EHs), hydroxyacyl-CoA dehydrogenases (HADs), and β-ketoacyl-CoA thiolases (KTs). The long-chain (LC) versions of these enzymes are contained on one $\alpha_4\beta_4$ octameric protein, **mitochondrial trifunctional protein,** located in the inner mitochondrial membrane. LCEH and LCHAD are contained on the α subunits while LCKT is located on the β subunits. The protein is therefore a combination multifunctional protein (more than one enzyme activity on a single polypeptide chain)–multienzyme complex (a complex of polypeptides catalyzing more than one reaction). The advantage of such a trifunctional enzyme is the ability to channel the intermediates toward the final product. Indeed, no long-chain hydroxyacyl-CoA or ketoacyl-CoA intermediates are released into solution by this system.

d. The Thiolase Reaction Occurs via Claisen Ester Cleavage

The final stage of the fatty acid β-oxidation process, the thiolase reaction, forms acetyl-CoA and a new acyl-CoA, which is two carbon atoms shorter than the one that began the cycle. This occurs in five reaction steps (Fig. 25-15):

1. An active site thiol is added to the substrate β-keto group.

2. Carbon–carbon bond cleavage forms an acetyl-CoA carbanion intermediate that is stabilized by electron withdrawal into this thioester's carbonyl group. Such a reaction is known as a Claisen ester cleavage (the reverse of a Claisen condensation). The citric acid cycle enzyme citrate synthase also catalyzes a reaction that involves a stabilized acetyl-CoA carbanion intermediate (Section 21-3A).

3. The acetyl-CoA carbanion intermediate is protonated by an enzyme acid group, yielding acetyl-CoA.

4 and 5. Finally, CoA displaces the enzyme thiol group from the enzyme–thioester intermediate, yielding acyl-CoA.

The formation of an enzyme–thioester intermediate

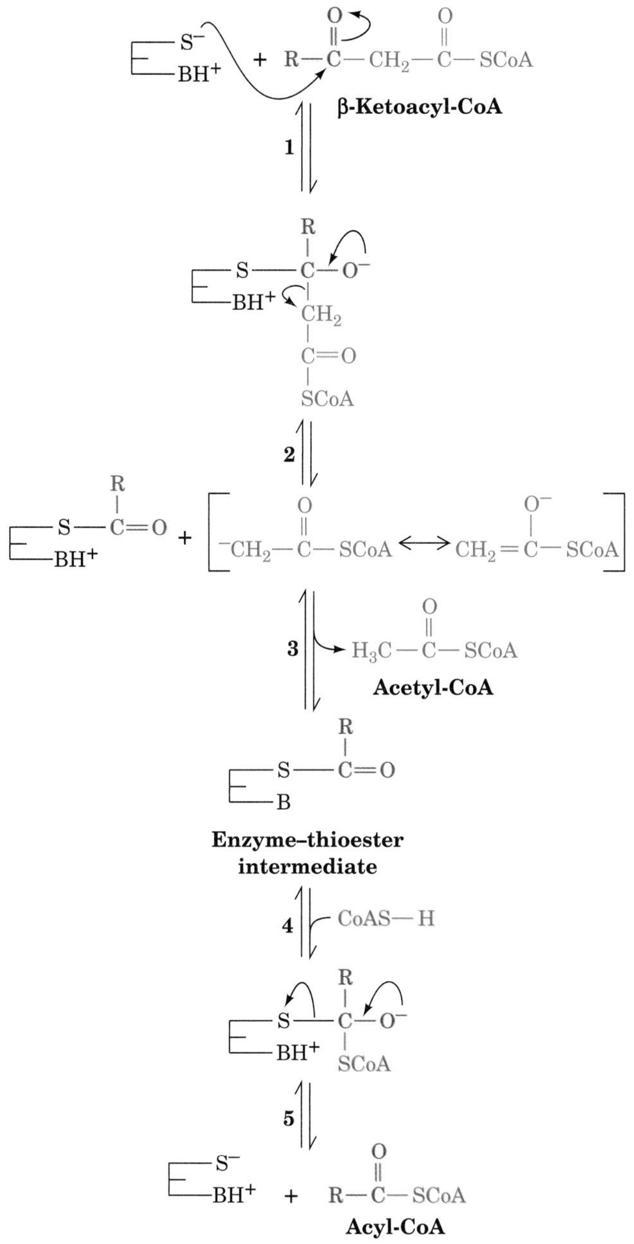

FIGURE 25-15 Mechanism of action of β-ketoacyl-CoA thiolase. An active site Cys residue participates in the formation of an enzyme thioester intermediate.

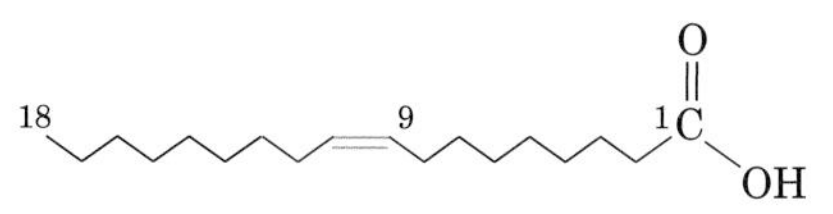

**Oleic acid
(9-*cis*-Octadecenoic acid)**

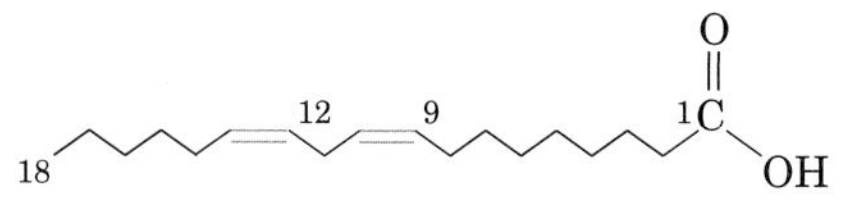

**Linoleic acid
(9,12-*cis*-Octadecadienoic acid)**

involving an active site thiol group is based on the observation that incubation of the enzyme with [^{14}C]acetyl-CoA yields a specifically labeled enzyme Cys residue (the reverse of steps 4 and 5):

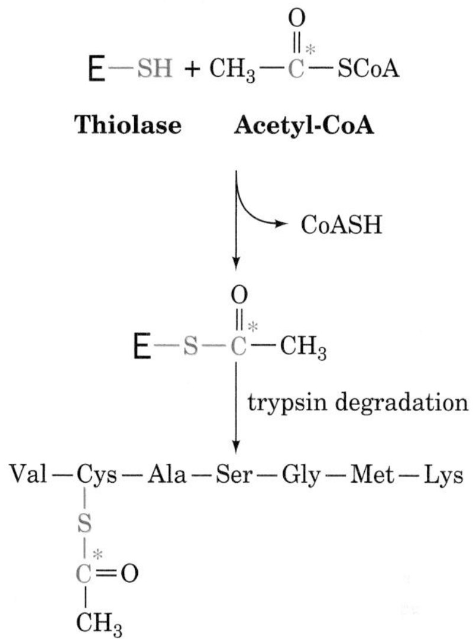

e. Fatty Acid Oxidation Is Highly Exergonic

The function of fatty acid oxidation is, of course, to generate metabolic energy. Each round of β oxidation produces one NADH, one FADH$_2$, and one acetyl-CoA. Oxidation of acetyl-CoA via the citric acid cycle generates additional FADH$_2$ and NADH, which are reoxidized through oxidative phosphorylation to form ATP. Complete oxidation of a fatty acid molecule is therefore a highly exergonic process, which yields numerous ATPs. For example, oxidation of palmitoyl-CoA (which has a C$_{16}$ fatty acyl group) involves seven rounds of β oxidation, yielding 7 FADH$_2$, 7 NADH, and 8 acetyl-CoA. Oxidation of the 8 acetyl-CoA, in turn, yields 8 GTP, 24 NADH, and 8 FADH$_2$. Since oxidative phosphorylation of the 31 NADH molecules yields 93 ATP and that of the 15 FADH$_2$ yields 30 ATPs, subtracting the 2 ATP equivalents required for fatty acyl-CoA formation (Section 25-2A), *the oxidation of one palmitate molecule has a net yield of 129 ATP.*

D. *Oxidation of Unsaturated Fatty Acids*

Almost all unsaturated fatty acids of biological origin (Section 12-1A) contain only cis double bonds, which most often begin between C9 and C10 (referred to as a Δ^9 or 9-double bond; Table 12-1). Additional double bonds, if any, occur at three-carbon intervals and are therefore never conjugated. Two examples of unsaturated fatty acids are oleic acid and linoleic acid (Fig. 25-16). Note that one of the double bonds in linoleic acid is at an odd-numbered

FIGURE 25-16 Structures of two common unsaturated fatty acids. Most unsaturated fatty acids contain unconjugated cis double bonds.

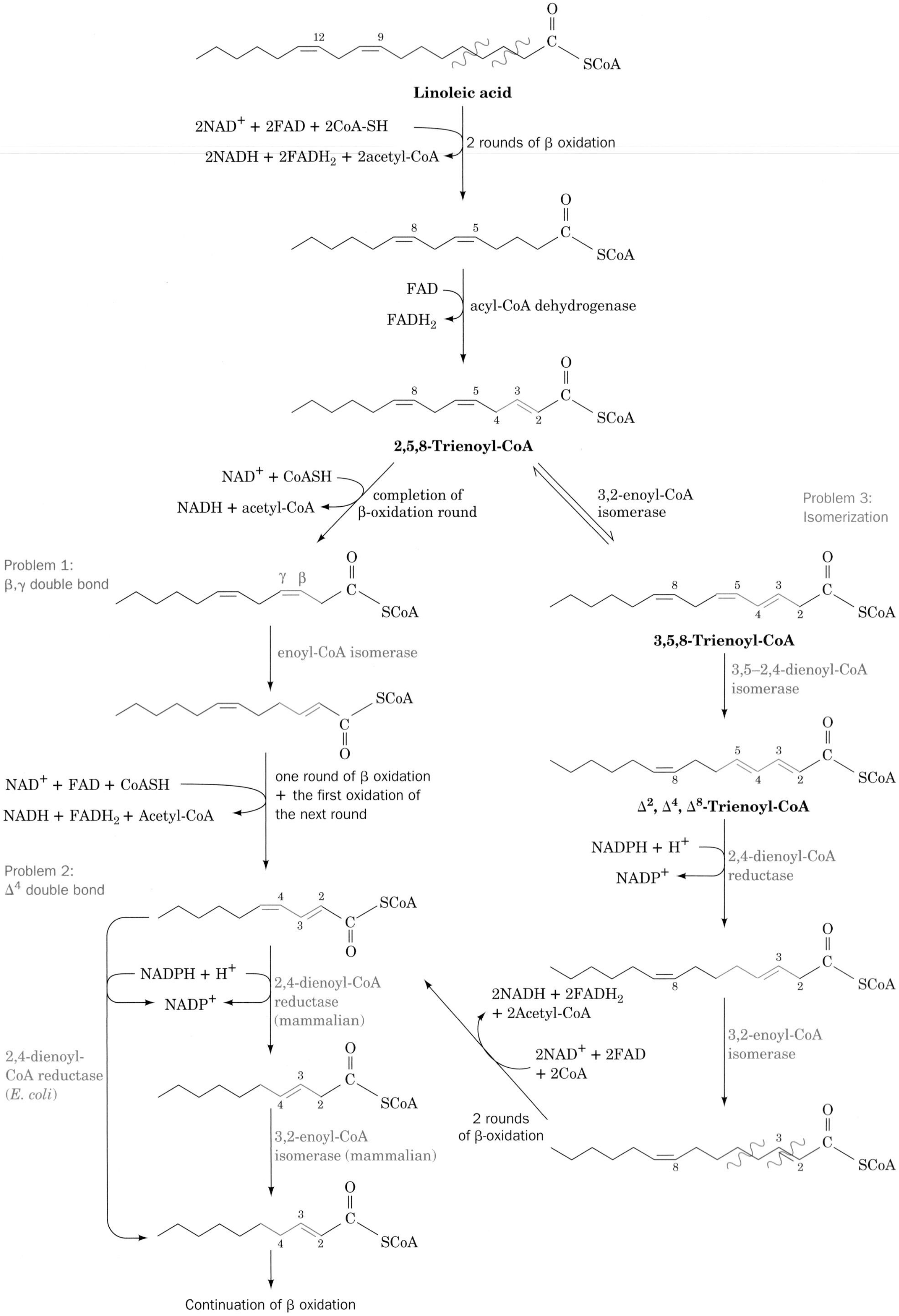

Linoleic acid

$2NAD^+ + 2FAD + 2CoA\text{-}SH$

$2NADH + 2FADH_2 + 2acetyl\text{-}CoA$ — 2 rounds of β oxidation

FAD

$FADH_2$ — acyl-CoA dehydrogenase

2,5,8-Trienoyl-CoA

$NAD^+ + CoASH$ — completion of β-oxidation round

$NADH + acetyl\text{-}CoA$

3,2-enoyl-CoA isomerase

Problem 3: Isomerization

Problem 1: β,γ double bond

enoyl-CoA isomerase

$NAD^+ + FAD + CoASH$

$NADH + FADH_2 + Acetyl\text{-}CoA$ — one round of β oxidation + the first oxidation of the next round

Problem 2: Δ^4 double bond

$NADPH + H^+$

$NADP^+$ — 2,4-dienoyl-CoA reductase (mammalian)

2,4-dienoyl-CoA reductase (*E. coli*)

3,2-enoyl-CoA isomerase (mammalian)

Continuation of β oxidation

3,5,8-Trienoyl-CoA

3,5–2,4-dienoyl-CoA isomerase

Δ^2, Δ^4, Δ^8-Trienoyl-CoA

$NADPH + H^+$

$NADP^+$ — 2,4-dienoyl-CoA reductase

$2NADH + 2FADH_2 + 2Acetyl\text{-}CoA$

$2NAD^+ + 2FAD + 2CoA$ — 2 rounds of β-oxidation

3,2-enoyl-CoA isomerase

FIGURE 25-17 (*Opposite*) **Problems in the oxidation of unsaturated fatty acids and their solutions.** Linoleic acid is used as an example. The first problem, the presence of a β,γ double bond seen in the left-hand pathway, is solved by the bond's enoyl-CoA isomerase–catalyzed conversion to a trans-α,β double bond. The second problem in the left-hand pathway, that a 2,4-dienoyl-CoA is not a substrate for enoyl-CoA hydratase, is eliminated by the NADPH-dependent reduction of the Δ^4 bond by 2,4-dienoyl-CoA reductase to yield the β-oxidation substrate *trans*-2-enoyl-CoA in *E. coli* but *trans*-3-enoyl-CoA in mammals. Mammals therefore also have 3,2-enoyl-CoA isomerase, which converts the *trans*-3-enoyl-CoA to *trans*-2-enoyl-CoA. The third problem, the isomerization of 2,5-dienoyl-CoA (originating from the oxidation of unsaturated fatty acids with double bonds at odd-numbered C atoms) to 3,5-dienoyl-CoA by 3,2-enoyl-CoA isomerase, is solved by 3,5–2,4-dienoyl-CoA isomerase, which converts the 3,5-dienoyl-CoA to 2,4-dienoyl-CoA, a substrate for 2,4-dienoyl-CoA reductase.

carbon atom and the other is at an even-numbered carbon atom. Double bonds at these positions in fatty acids pose three problems for the β-oxidation pathway that are solved through the actions of four additional enzymes (Fig. 25-17):

Problem 1: A β,γ Double Bond

The first enzymatic difficulty occurs on the left-hand pathway in Fig. 25-17 after the third round of β oxidation: The resulting cis-β,γ double bond–containing enoyl-CoA is not a substrate for enoyl-CoA hydratase. **Enoyl-CoA isomerase,** however, mediates conversion of the cis-Δ^3 double bond to the more stable, ester-conjugated trans-Δ^2 form:

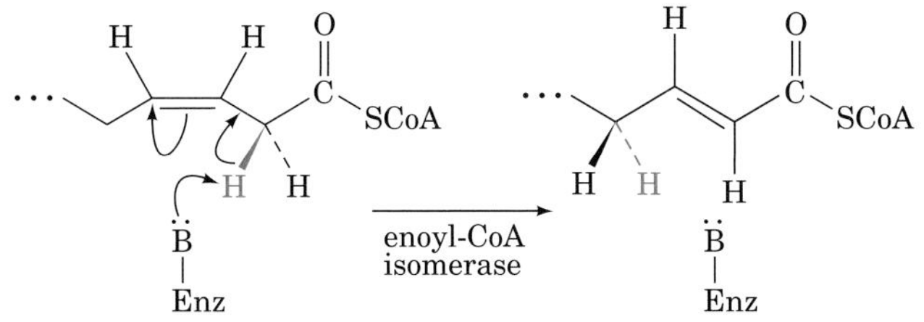

Such compounds are normal substrates of enoyl-CoA hydratase so that β oxidation can then continue.

Problem 2: A Δ^4 Double Bond Inhibits Hydratase Action

The next difficulty arises on the left-hand pathway in Fig. 25-17 in the fifth round of β oxidation. The presence of a double bond at an even-numbered carbon atom results in the formation of 2,4-dienoyl-CoA, which is a poor substrate for enoyl-CoA hydratase. However, NADPH-dependent **2,4-dienoyl-CoA reductase** reduces the Δ^4 double bond. The *E. coli* reductase produces *trans*-2-enoyl-CoA, a normal substrate of β oxidation. The mammalian reductase, however, yields *trans*-3-enoyl-CoA, which, to proceed along the β-oxidation pathway, must first be isomerized to *trans*-2-enoyl-CoA by **3,2-enoyl-CoA isomerase.**

Problem 3: The Unanticipated Isomerization of 2,5-enoyl-CoA by 3,2-enoyl-CoA Isomerase

Mammalian 3,2-enoyl-CoA isomerase catalyzes a reversible reaction that interconverts Δ^2 and Δ^3 double bonds. A carbonyl group is stabilized by being conjugated to a Δ^2 double bond. However, the presence of a Δ^5 double bond (originating from an unsaturated fatty acid with a double bond at an odd-numbered C atom such as the Δ^9 double bond of linoleic acid) is likewise stabilized by being conjugated with a Δ^3 double bond. If a 2,5-enoyl-CoA is converted by 3,2-enoyl-CoA isomerase to 3,5-enoyl CoA, which occurs up to 20% of the time, another enzyme is necessary to continue the oxidation: **3,5–2,4-Dienoyl-CoA isomerase** isomerizes the 3,5-diene to a 2,4-diene, which is then reduced by 2,4-dienoyl-CoA reductase and isomerized by 3,2-enoyl-CoA isomerase as in Problem 2 above. After two more rounds of β-oxidation, the cis-Δ^4 double bond originating from the cis-Δ^{12} double bond of linoleic acid is also dealt with as in Problem 2.

E. *Oxidation of Odd-Chain Fatty Acids*

Most fatty acids have even numbers of carbon atoms and are therefore completely converted to acetyl-CoA. Some plants and marine organisms, however, synthesize fatty acids with an odd number of carbon atoms. *The final round of β oxidation of these fatty acids forms propionyl-CoA, which, as we shall see, is converted to succinyl-CoA for entry into the citric acid cycle.* Propionate or propionyl-CoA is also produced by oxidation of the amino acids isoleucine, valine, and methionine (Section 26-3E). Furthermore, ruminant animals such as cattle derive most of their caloric intake from the acetate and propionate produced in their rumen (stomach) by bacterial fermentation of carbohydrates. These products are absorbed by the animal and metabolized after conversion to the corresponding acyl-CoA.

a. Propionyl-CoA Carboxylase Has a Biotin Prosthetic Group

The conversion of propionyl-CoA to succinyl-CoA involves three enzymes (Fig. 25-18). The first reaction is that of **propionyl-CoA carboxylase,** a tetrameric enzyme that contains a biotin prosthetic group (Section 23-1A). The reaction occurs in two steps (Fig. 25-19):

1. Carboxylation of biotin at N1′ by bicarbonate ion as in the reaction catalyzed by pyruvate carboxylase (Fig. 23-4). This step, which is driven by the concomitant hydrolysis of ATP to ADP and P_i, activates the resulting carboxyl group for transfer without further free energy input.

2. Stereospecific transfer of the activated carboxyl group from carboxybiotin to propionyl-CoA to form **(*S*)-methylmalonyl-CoA.** This step occurs via nucleophilic attack on carboxybiotin by a carbanion at C2 of propionyl-CoA (see below).

These two reaction steps occur at different catalytic sites on propionyl-CoA carboxylase. It has therefore been proposed that the biotinyllysine linkage attaching the biotin

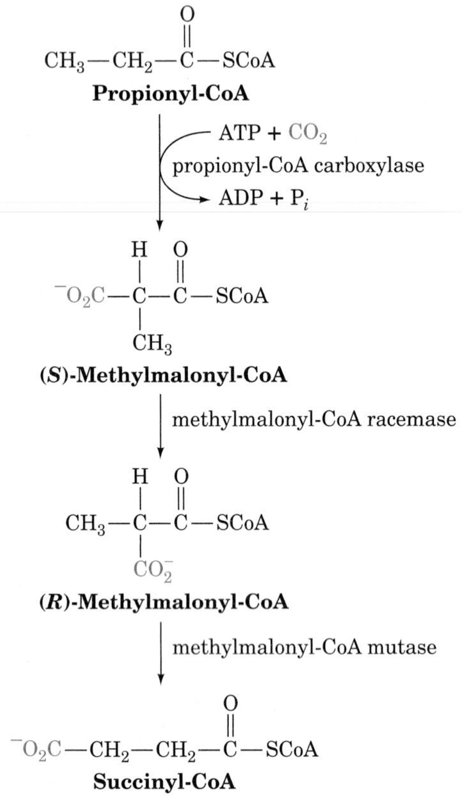

FIGURE 25-18 **Conversion of propionyl-CoA to succinyl-CoA.**

ring to the enzyme forms a flexible tether that permits the efficient transfer of the biotin ring between these two active sites as occurs in the biotin enzyme pyruvate carboxylase (Section 23-1A).

Formation of the C2 carbanion in the second stage of the propionyl-CoA carboxylase reaction involves removal of a proton α to a thioester. This proton is relatively acidic since, as we have seen in Section 25-2C, the negative charge on a carbanion α to a thioester can be delocalized into the thioester's carbonyl group. This may explain the relatively convoluted path taken in the conversion of propionyl-CoA to succinyl-CoA (Fig. 25-18). It would seem simpler, at least on paper, for this process to occur in one step, with carboxylation occurring on C3 of propionyl-CoA so as to form succinyl-CoA directly. Yet, the C3 carbanion required for such a carboxylation would be extremely unstable. Nature has instead chosen a more facile, albeit less direct route, which carboxylates propionyl-CoA at a more reactive position and then rearranges the C_4 skeleton to form the desired product.

b. Methylmalonyl-CoA Mutase Contains a Coenzyme B_{12} Prosthetic Group

Methylmalonyl-CoA mutase, which catalyzes the third reaction of the propionyl-CoA to succinyl-CoA conversion (Fig. 25-18), is specific for (*R*)-methylmalonyl-CoA even though propionyl-CoA carboxylase stereospecifically syn-

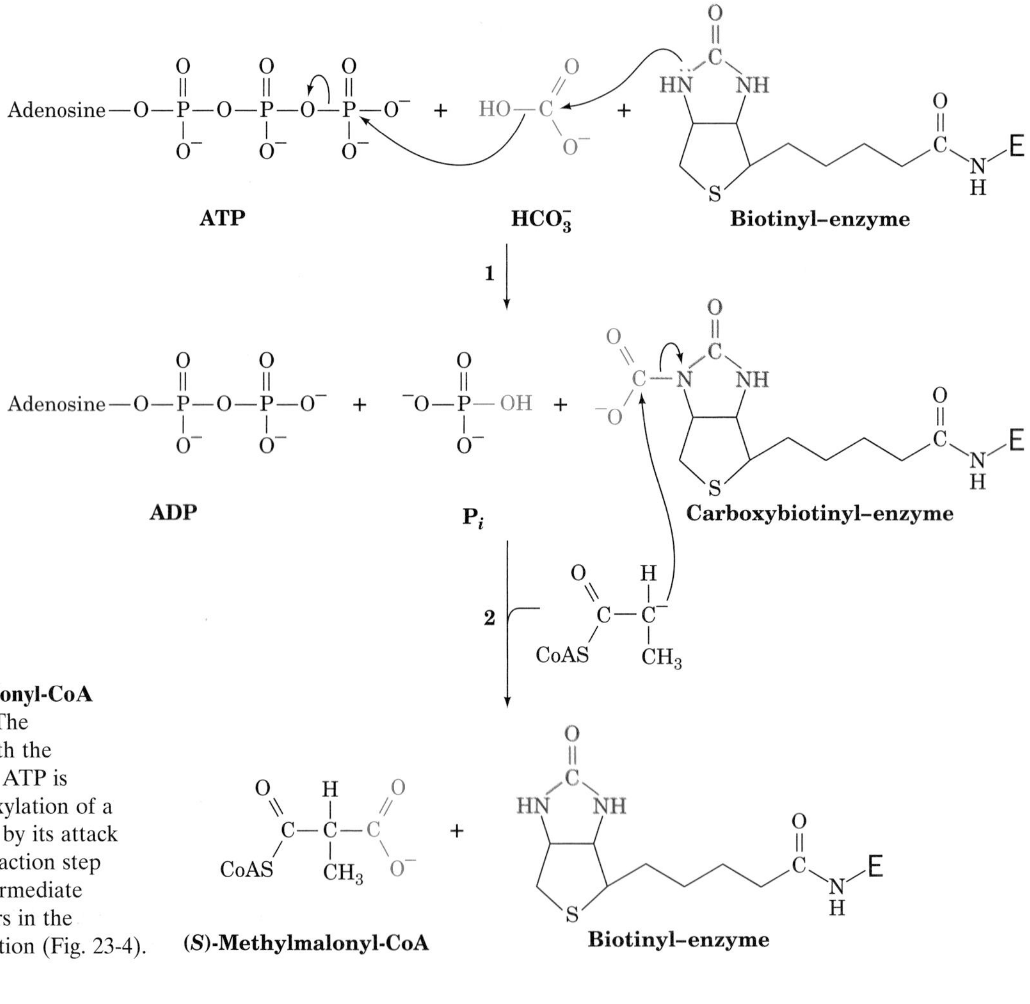

FIGURE 25-19 The propionyl-CoA carboxylase reaction. (1) The carboxylation of biotin with the concomitant hydrolysis of ATP is followed by **(2)** the carboxylation of a propionyl-CoA carbanion by its attack on carboxybiotin. Each reaction step probably involves the intermediate formation of CO_2 as occurs in the pyruvate carboxylase reaction (Fig. 23-4).

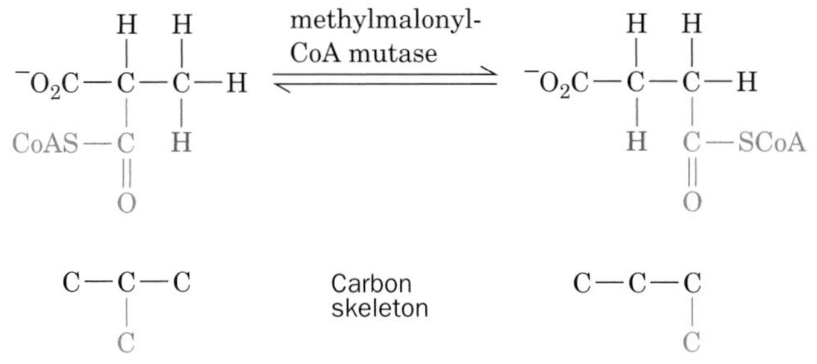

FIGURE 25-20 The rearrangement catalyzed by methylmalonyl-CoA mutase.

thesizes (*S*)-methylmalonyl-CoA. This diversion is rectified by **methylmalonyl-CoA racemase,** which interconverts the (*R*) and (*S*) configurations of methylmalonyl-CoA, presumably by promoting the reversible dissociation of its acidic α-H via formation of a resonance-stabilized carbanion intermediate:

Resonance-stabilized carbanion intermediate

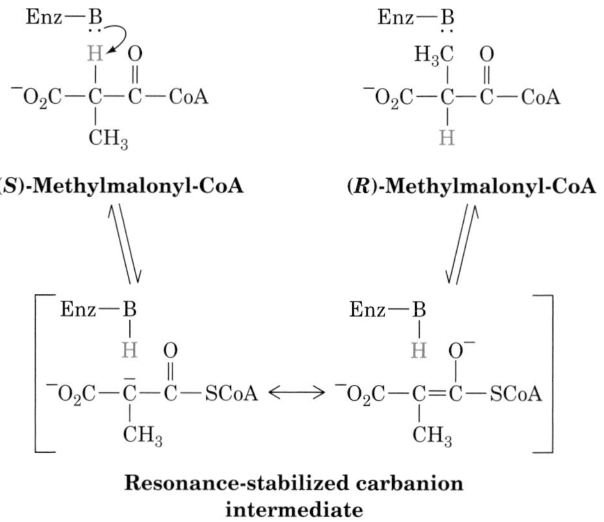

Methylmalonyl-CoA mutase, which catalyzes an unusual carbon skeleton rearrangement (Fig. 25-20), utilizes a **5′-deoxyadenosylcobalamin (AdoCbl)** prosthetic group (also called **coenzyme B$_{12}$**). Dorothy Hodgkin determined the structure of this complex molecule (Fig. 25-21) in 1956,

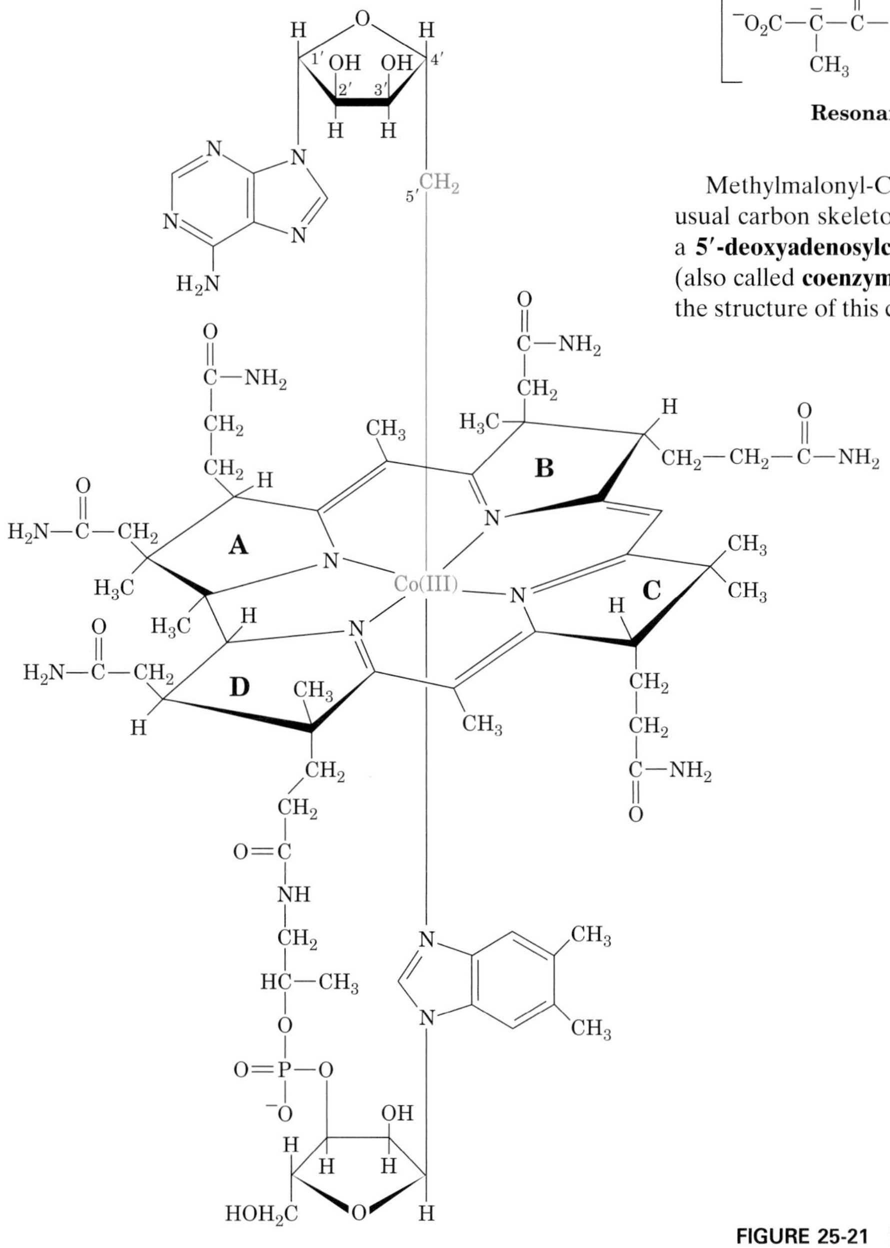

5′-Deoxyadenosylcobalamin (coenzyme B$_{12}$)

FIGURE 25-21 Structure of 5′-deoxyadenosylcobalamin (coenzyme B$_{12}$).

a landmark achievement, through X-ray crystallographic analysis combined with chemical degradation studies. AdoCbl contains a hemelike **corrin** ring whose four pyrrole N atoms each ligand a 6-coordinate Co ion. The fifth Co ligand in the free coenzyme is an N atom of a **5,6-dimethylbenzimidazole (DMB)** nucleotide that is covalently linked to the corrin D ring. The sixth ligand is a 5′-deoxyadenosyl group in which the deoxyribose C5′ atom forms a covalent C—Co bond, *one of only two known carbon–metal bonds in biology* (the other being a C—Ni bond in the bacterial enzyme **carbon monoxide dehydrogenase**). In some enzymes, the sixth ligand instead is a CH_3 group that likewise forms a C—Co bond.

AdoCbl's reactive C—Co bond participates in two types of enzyme-catalyzed reactions:

1. Rearrangements in which a hydrogen atom is directly transferred between two adjacent carbon atoms with concomitant exchange of the second substituent, X:

$$\underset{C_1}{\overset{H}{|}}{-}\underset{C_2}{\overset{X}{|}}{-} \longrightarrow \underset{C_1}{\overset{X}{|}}{-}\underset{C_2}{\overset{H}{|}}{-}$$

where X may be a carbon atom with substituents, an oxygen atom of an alcohol, or an amine.

2. Methyl group transfers between two molecules.

There are about a dozen known cobalamin-dependent enzymes. However, only two occur in mammalian systems: (1) methylmalonyl-CoA mutase, which catalyzes a carbon skeleton rearrangement (the X group in the rearrangement is —COSCoA; Fig. 25-20) and is the only B_{12}-containing enzyme that occurs in both eukaryotes and prokaryotes; and (2) **methionine synthase,** a methyl transfer enzyme that participates in methionine biosynthesis (Sections 26-3E and 26-5B). Defects in methylmalonyl-CoA mutase result in **methylmalonic aciduria,** a condition that is often fatal in infancy due to **acidosis** (low blood pH) without a diet devoid of odd-chain fatty acids and low in the amino acid residues that are degraded to propionyl-CoA (Ile, Leu, and Met; Section 26-3E).

c. The Methylmalonyl-CoA Mutase Reaction Occurs via a Free Radical Mechanism

Methylmalonyl-CoA mutase from *Propionibacterium shermanii* is an αβ heterodimer whose catalytically active 728-residue α subunit is 24% identical to its catalytically inactive 638-residue β subunit. In contrast, the human enzyme is a homodimer whose subunits are 60% identical in sequence to *P. shermanii*'s α subunit. Hence *P. shermanii*'s β subunit is thought to be an evolutionary fossil.

The X-ray structure of methylmalonyl-CoA mutase from *P. shermanii* in complex with the substrate analog **2-carboxypropyl-CoA** (which lacks methylmalonyl-CoA's thioester oxygen atom) was determined by Philip Evans. Its AdoCbl cofactor is sandwiched between the α subunit's

two domains: a 559-residue N-terminal α/β barrel (TIM barrel, the most common enzymatic motif; Section 8-3B) and a 169-residue C-terminal α/β domain that resembles a Rossmann fold (Section 8-3B). The structure of the α/β barrel contains several surprising features (Fig. 25-22):

1. The active sites of nearly all α/β barrel enzymes are located at the C-terminal ends of the barrel's β strands. However, in methylmalonyl-CoA mutase, the AdoCbl is packed against the N-terminal ends of the barrel's β strands.

2. In free AdoCbl, the Co atom is axially liganded by an N atom of its DMB group and by the adenosyl residue's 5′-CH_2 group (Fig. 25-21). However, in the enzyme, the DMB has swung aside to bind in a separate pocket and has been replaced by the side chain of His 610 from the C-terminal domain. The adenosyl group is not visible in the structure due to disorder and hence has probably also swung aside.

3. In nearly all other α/β barrel–containing enzymes, the center of the barrel is occluded by large, often branched, hydrophobic side chains. However, in methylmalonyl-CoA mutase, the 2-carboxypropyl-CoA's pantetheine group binds in a narrow tunnel through the center of the α/β barrel so as to put the methylmalonyl group of an intact substrate in close proximity to the unliganded face of the cobalamin ring. This tunnel provides the only direct access to the active site cavity, thereby protecting the reactive free radical intermediates that are produced in the catalytic reaction from side reactions (see below). The tunnel is lined by small hydrophilic residues (Ser and Thr).

Methylmalonyl-CoA mutase's substrate binding mode resembles those of several other AdoCbl-containing enzymes of known structure, which are collectively unique among α/β barrel-containing proteins.

The proposed methylmalonyl-CoA mutase reaction mechanism (Fig. 25-23) begins with **homolytic cleavage** of the cobalamin C—Co(III) bond (the C and Co atoms each acquire one of the electrons that formed the cleaved electron pair bond). The Co ion therefore fluctuates between its Co(III) and Co(II) oxidation states [the two states are spectroscopically distinguishable: Co(III) is red and diamagnetic (no unpaired electrons), whereas Co(II) is yellow and paramagnetic (unpaired electrons)]. Note that a homolytic cleavage reaction is unusual in biology; most other biological bond cleavage reactions occur via **heterolytic cleavage** (in which the electron pair forming the cleaved bond is fully acquired by one of the separating atoms).

The role of AdoCbl in the catalytic process is that of a reversible free radical generator. The C—Co(III) bond is well suited to this function because it is inherently weak (dissociation energy = 109 kJ · mol^{-1}) and appears to be further weakened through steric interactions with the enzyme. Indeed, as Fig. 25-22 indicates, the Co atom in methylmalonyl-CoA mutase has no sixth ligand and hence,

(a)

(b)

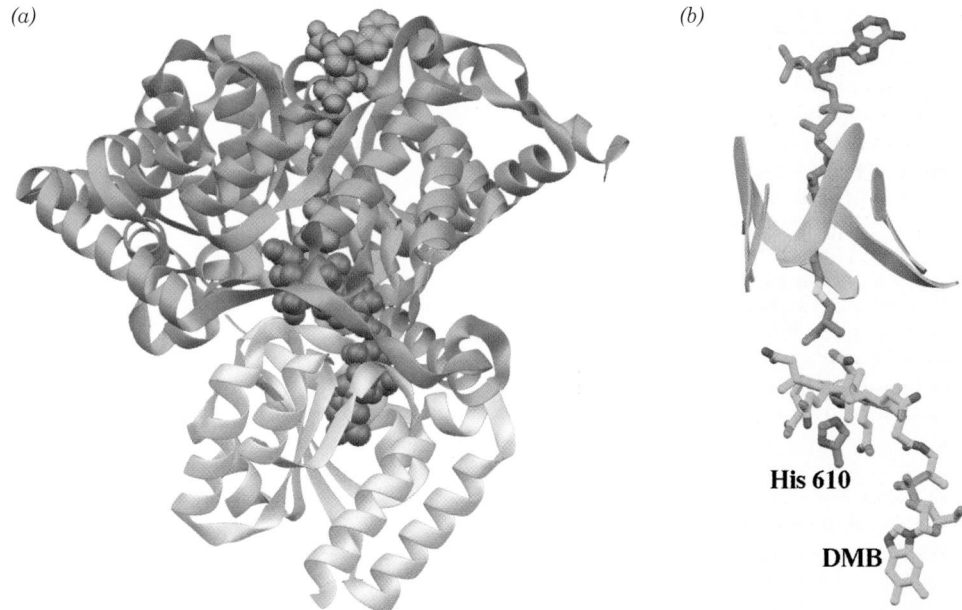

His 610

DMB

FIGURE 25-22   X-Ray structure of *P. shermanii* methylmalonyl-CoA mutase in complex with 2-carboxypropyl-CoA and AdoCbl. (*a*) The catalytically active α subunit in which the N-terminal domain is cyan with the β strands of its α/β barrel orange, and the C-terminal domain is pink. The 2-carboxypropyl-CoA (*magenta*) and AdoCbl (*green*) are drawn in space-filling form. The 2-carboxypropyl-CoA passes through the center of the α/β barrel and is oriented such that the methylmalonyl group of methylmalonyl-CoA would contact the corrin ring of the AdoCbl, which is sandwiched between the enzyme's N- and C-terminal domains. (*b*) The arrangement of the AdoCbl and 2-carboxypropyl-CoA molecules which, together with the side chain of His 610, are represented in stick form colored according to atom type (2-carboxypropyl-CoA and His C green, AdoCbl C cyan, N blue, O red, P magenta, and S yellow). The corrin ring's Co atom is represented by a lavender sphere and the α/β barrel's β strands are represented by orange ribbons. The view is similar to that in Part *a.* Note that the DMB group (*bottom*) has swung away from the corrin ring (seen edgewise) to be replaced by a side chain of His 610 from the C-terminal domain and that the 5′-deoxyadenosyl group is unseen (due to disorder). [Based on an X-ray structure by Philip Evans, MRC Laboratory of Molecular Biology, Cambridge, U.K. PDBid 7REQ.] &2 See the Interactive Exercises

as confirmed by spectroscopic measurements, is in its Co(II) state. The His N—Co bond is extremely long (2.5 Å vs 1.9–2.0 Å in various other B_{12}-containing structures). It is proposed that this strained and hence weakened bond stabilizes the Co(II) state with respect to the Co(III) state, thus favoring the formation of the adenosyl radical and facilitating the homolytic cleavage through which the catalyzed reaction occurs (Fig. 23-23). The adenosyl radical presumably abstracts a hydrogen atom from the substrate, thereby facilitating the rearrangement reaction through the intermediate formation of a cyclopropyloxy radical.

d. Succinyl-CoA Cannot Be Directly Consumed by the Citric Acid Cycle

Methylmalonyl-CoA mutase catalyzes the conversion of a metabolite to a C_4 citric acid cycle intermediate, not acetyl-CoA. The route of succinyl-CoA oxidation is therefore not as simple as it may first appear. The citric acid cycle regenerates all of its C_4 intermediates so that these compounds are really catalysts, not substrates. Consequently, succinyl-CoA cannot undergo net degradation by citric acid cycle enzymes alone. Rather, *in order for a metabolite to undergo net oxidation by the citric acid cycle, it must first*

be converted either to pyruvate or directly to acetyl-CoA. Net degradation of succinyl-CoA begins with its conversion, via the citric acid cycle, to malate. At high concentrations, malate is transported, by a specific transport protein, to the cytosol, where it may be oxidatively decarboxylated to pyruvate and CO_2 by **malic enzyme (malate dehydrogenase, decarboxylating):**

$$
\begin{array}{c}
\underset{\textbf{Malate}}{
\begin{array}{c}
CO_2^- \\
|\\
HO-C-H \\
|\\
CH_2 \\
|\\
C \\
O \quad O^-
\end{array}}
\xrightarrow[\text{NADP}^+ \;\; \text{NADPH}]{\text{H}^+ +}
\left[
\begin{array}{c}
CO_2^- \\
|\\
C=O \\
|\\
CH_2 \\
|\\
C \\
O \quad O^-
\end{array}
\right]
\xrightarrow{CO_2}
\underset{\textbf{Pyruvate}}{
\begin{array}{c}
CO_2^- \\
|\\
C=O \\
|\\
CH_3
\end{array}}
\end{array}
$$

(We previously encountered this enzyme in the C_4 cycle of photosynthesis; Fig. 24-38.) Pyruvate is then completely oxidized via pyruvate dehydrogenase and the citric acid cycle.

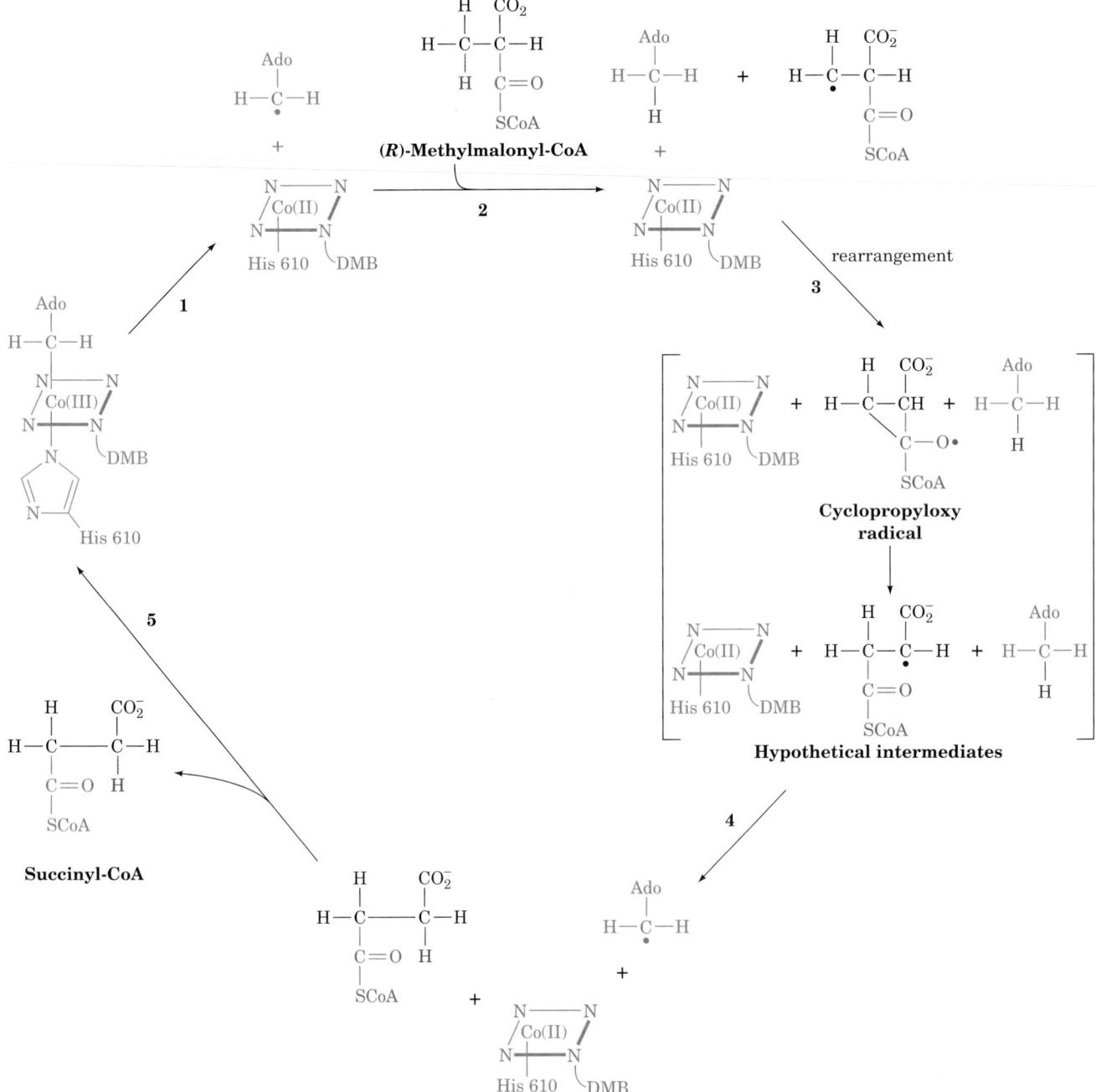

FIGURE 25-23   Proposed mechanism of methylmalonyl-CoA mutase. (1) The homolytic cleavage of the C—Co(III) bond yielding a 5′-deoxyadenosyl radical and cobalamin in its Co(II) oxidation state. **(2)** Abstraction of a hydrogen atom from the methylmalonyl-CoA by the 5′-deoxyadenosyl radical, thereby generating a methylmalonyl-CoA radical. **(3)** Carbon skeleton rearrangement to form a succinyl-CoA radical via a proposed cyclopropyloxy radical intermediate. **(4)** Abstraction of a hydrogen atom from 5′-deoxyadenosine by the succinyl-CoA radical to form succinyl-CoA and regenerate the 5′-deoxyadenosyl radical. **(5)** Release of succinyl-CoA and reformation of the coenzyme.

e. Pernicious Anemia Results from Vitamin B$_{12}$ Deficiency

The existence of **vitamin B$_{12}$** came to light in 1926 when George Minot and William Murphy discovered that **pernicious anemia,** an often fatal disease of the elderly characterized by decreased numbers of red blood cells, low hemoglobin levels, and progressive neurological deterioration, can be treated by the daily consumption of large amounts of raw liver (a treatment that some patients considered worse than the disease). It was not until 1948, however, after a bacterial assay for antipernicious anemia factor had been developed, that vitamin B$_{12}$ was isolated.

Vitamin B_{12} is synthesized by neither plants nor animals but only by a few species of bacteria. Herbivores obtain their vitamin B_{12} from the bacteria that inhabit their gut (in fact, some animals, such as rabbits, must periodically eat some of their feces to obtain sufficient amounts of this essential substance). Humans, however, obtain almost all their vitamin B_{12} directly from their diet, particularly from meat. The vitamin is specifically bound in the intestine by the glycoprotein **intrinsic factor** that is secreted by the stomach. This complex is absorbed by a specific receptor in the intestinal mucosa, where the complex is dissociated and the liberated vitamin B_{12} transported to the bloodstream. There it is bound by at least three different plasma globulins, called **transcobalamins,** which facilitate its uptake by the tissues.

Pernicious anemia is not usually a dietary deficiency disease but, rather, results from insufficient secretion of intrinsic factor. The normal human requirement for cobalamin is very small, $\sim 3 \; \mu g \cdot day^{-1}$, and the liver stores a 3- to 5-year supply of this vitamin. This accounts for the insidious onset of pernicious anemia and the fact that true dietary deficiency of vitamin B_{12}, even among strict vegetarians, is extremely rare.

F. *Peroxisomal β Oxidation*

The β oxidation of fatty acids occurs in the peroxisome as well as in the mitochondrion. Peroxisomal β oxidation in animals functions to shorten very long chain fatty acids (>22 C atoms) so as to facilitate their degradation by the mitochondrial β-oxidation system. In yeast and plants, fatty acid oxidation occurs exclusively in the peroxisomes and glyoxysomes (specialized peroxisomes, Sections 23-2 and 1-2A).

The peroxisomal pathway results in the same chemical changes to fatty acids as does the mitochondrial pathway, although the enzymes in these two organelles are different. The protein that transports very long chain fatty acids into the peroxisome, **ALD protein** (see below), does not have a carnitine requirement. The very long chain fatty acids that enter this compartment are activated by a peroxisomal very long chain acyl-CoA synthetase to form their CoA esters, and are oxidized directly. The shorter chain acyl products of this β-oxidation process are then linked to carnitine for transport into mitochondria for further oxidation.

a. X-Adrenoleukodystrophy Is Caused by a Defect in ALD Protein

X-Adrenoleukodystrophy (X-ALD), a rare X-linked inherited disease, causes very long chain saturated fatty acids to accumulate in the blood and destroy myelin, the insulating sheath surrounding the axons of many neurons (Section 20-5B). Its varied neurological symptoms present (become evident) between the ages of 4 and 10 years and are usually fatal within 1 to 10 years (except after a successful bone marrow transplant). X-ALD is caused by a defective ALD protein. Thus in X-ALD patients, **lignoceric acid** (24:0; recall that the symbol *n:m* indicates a C_n fatty acid with *m* double bonds) is converted to lignoceroyl-CoA at only 13% of the normal rate, although once formed, it undergoes β oxidation at the normal rate.

b. Peroxisomal β Oxidation Differs in Detail from Mitochondrial β Oxidation

The β-oxidation pathway in peroxisomes differs from that in mitochondria as follows:

1. The first enzyme in the peroxisomal pathway, **acyl-CoA oxidase,** catalyzes the reaction:

$$\text{Fatty acyl-CoA} + O_2 \rightarrow \textit{trans-}\Delta^2\text{-enoyl-CoA} + H_2O_2$$

This reaction involves participation of an FAD cofactor but differs from its mitochondrial counterpart in that the abstracted electrons are transferred directly to O_2 rather than passing through the electron-transport chain with its concomitant oxidative phosphorylation (Fig. 25-12). Peroxisomal fatty acid oxidation is therefore less efficient than the mitochondrial process by two ATPs for each C_2 cycle. The H_2O_2 produced is disproportionated to H_2O and O_2 through the action of peroxisomal catalase (Section 1-2A).

2. Peroxisomal enoyl-CoA hydratase and 3-L-hydroxyacyl-CoA dehydrogenase are activities that occur on a single polypeptide and therefore join the growing list of multifunctional enzymes. The reactions catalyzed are identical to those of the mitochondrial system (Fig. 25-12).

3. Peroxisomal thiolase has a different chain-length specificity than its mitochondrial counterpart. It is almost inactive with acyl-CoAs of length C_8 or less so that fatty acids are incompletely oxidized by peroxisomes.

Although peroxisomal β oxidation is not dependent on the transport of acyl groups into the peroxisome as their carnitine esters, the peroxisome contains carnitine acyltransferases. Acyl-CoAs that have been chain-shortened by peroxisomal β oxidation are thereby converted to their carnitine esters. These substances, for the most part, passively diffuse out of the peroxisome to the mitochondrion, where they are oxidized further.

G. *Minor Pathways of Fatty Acid Oxidation*

β Oxidation is blocked by an alkyl group at the C_β of a fatty acid, and thus at any odd-numbered carbon atom. One such branched-chain fatty acid, a common dietary component, is **phytanic acid.** This metabolic breakdown product of chlorophyll's phytyl side chain (Section 24-2A) is present in dairy products, ruminant fats, and fish although, surprisingly, chlorophyll itself is but a poor dietary source of phytanic acid for humans. The oxidation of branched-chain fatty acids such as phytanic acid is facilitated by **α oxida-**

tion (Fig. 25-24). In this process, the fatty acid is converted to its CoA thioester and its C_α is hydroxylated by the Fe^{2+}-containing **phytanoyl-CoA hydroxylase.** The resulting CoA thioester is, in effect, oxidatively decarboxylated to yield a

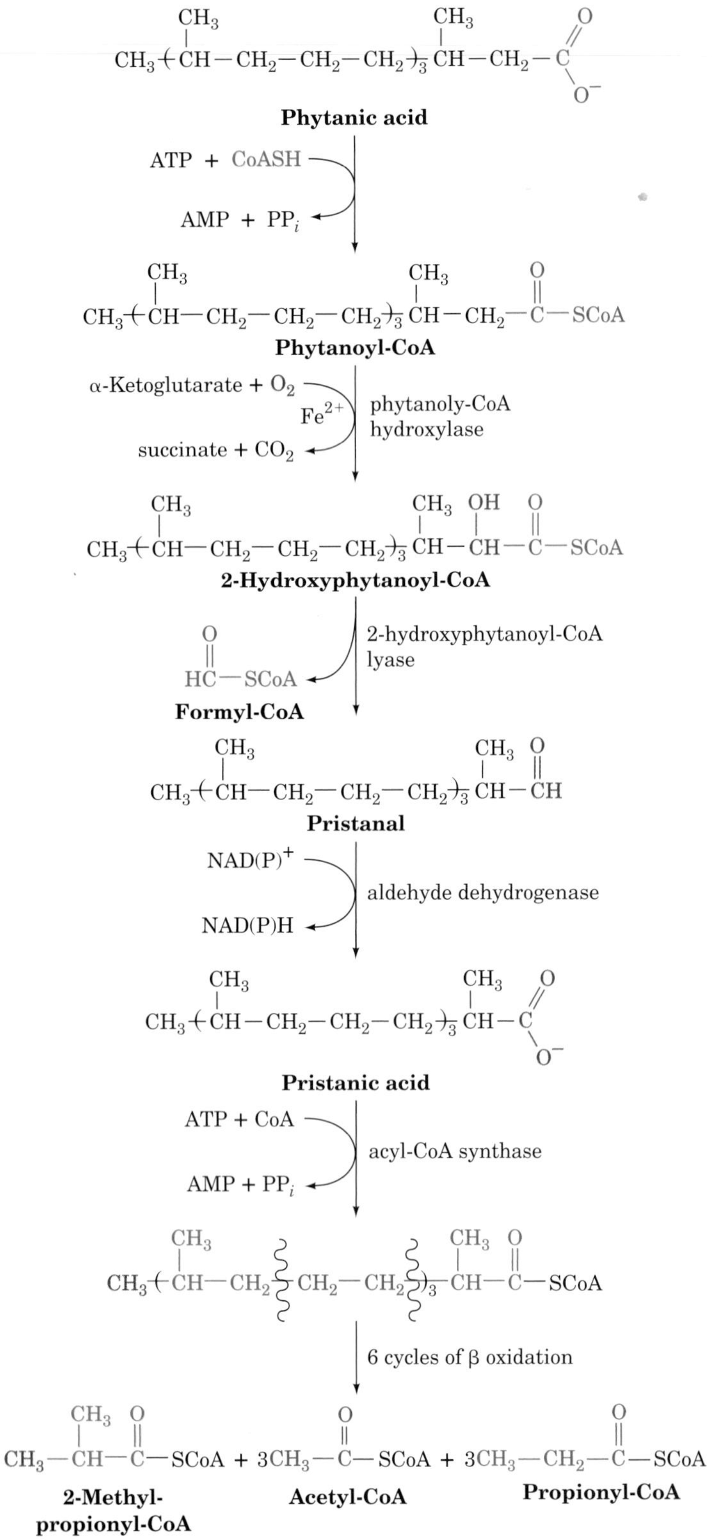

FIGURE 25-24   Pathway of α oxidation of fatty acids. Phytanic acid, a degradation product of the phytol side chain of chlorophyll, is metabolized through α oxidation to **pristanic acid** followed by β oxidation.

new fatty acid with an unsubstituted C_β. Further degradation of the molecule can then continue via six cycles of normal β oxidation to yield three propionyl-CoAs, three acetyl-CoAs, and one 2-methylpropionyl-CoA (which is converted to succinyl-CoA).

A rare genetic defect, **Refsum's disease** or **phytanic acid storage syndrome,** results from the accumulation of this metabolite throughout the body. The disease, which is characterized by progressive neurological difficulties such as tremors, unsteady gait, and poor night vision, results from a defective phytanoyl-CoA hydroxylase. Its symptoms can therefore be attenuated by a diet that restricts the intake of phytanic acid–containing foods.

Medium- and long-chain fatty acids are converted to dicarboxylic acids through **ω oxidation** (oxidation of the last carbon atom). This process, which is catalyzed by enzymes of the ER, involves hydroxylation of a fatty acid's C_ω atom by **cytochrome P450,** a monooxygenase that utilizes NADPH and O_2 (Section 15-4B). This CH_2—OH group is then oxidized to a carboxyl group, converted to a CoA derivative at either end, and oxidized via the β-oxidation pathway. ω Oxidation is probably of only minor significance in fatty acid oxidation.

3 ■ KETONE BODIES

Acetyl-CoA produced by oxidation of fatty acids in liver mitochondria can be further oxidized via the citric acid cycle as is discussed in Chapter 21. A significant fraction of this acetyl-CoA has another fate, however. *By a process known as* **ketogenesis,** *which occurs primarily in liver mitochondria, acetyl-CoA is converted to* **acetoacetate** *or* D-**β-hydroxybu-tyrate.** *These compounds, which together with* **acetone** *are somewhat inaccurately referred to as* **ketone bodies,**

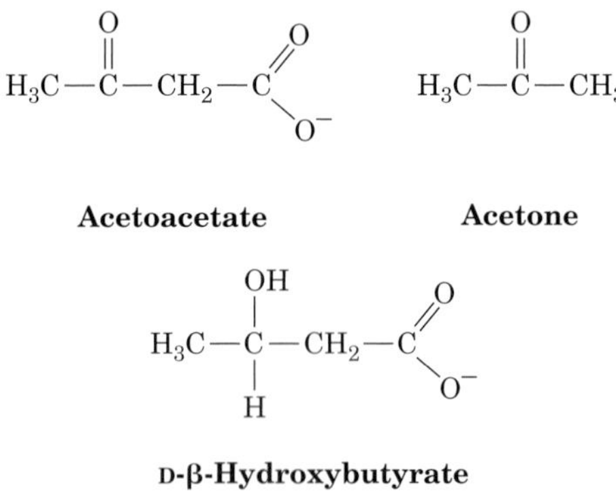

serve as important metabolic fuels for many peripheral tissues, particularly heart and skeletal muscle. The brain, under normal circumstances, uses only glucose as its energy source (fatty acids are unable to pass the blood–brain barrier), but during starvation, ketone bodies become the brain's major fuel source (Section 27-4A). *Ketone bodies are water-soluble equivalents of fatty acids.*

Acetoacetate formation occurs in three reactions (Fig. 25-25):

1. Two molecules of acetyl-CoA are condensed to **acetoacetyl-CoA** by thiolase (also called **acetyl-CoA acetyltransferase**) working in the reverse direction from the way it does in the final step of β oxidation (Section 25-2C).

2. Condensation of the acetoacetyl-CoA with a third acetyl-CoA by **HMG-CoA synthase** forms **β-hydroxy-β-methylglutaryl-CoA (HMG-CoA)**. The mechanism of this reaction resembles the reverse of the thiolase reaction (Fig. 25-15) in that an active site thiol group forms an acyl–thioester intermediate.

3. Degradation of HMG-CoA to acetoacetate and acetyl-CoA in a mixed aldol–Claisen ester cleavage is catalyzed by **HMG-CoA lyase.** The mechanism of this reaction is analogous to the reverse of the citrate synthase reaction (Section 21-3A). (HMG-CoA is also a precursor in cholesterol biosynthesis and hence may be diverted to this purpose as is discussed in Section 25-6A.)

The overall reaction catalyzed by HMG-CoA synthase and HMG-CoA lyase is

Acetoacetyl-CoA + H_2O → acetoacetate + CoA

One may well ask why this apparently simple hydrolysis reaction occurs in such an indirect manner. The answer is unclear but may lie in the regulation of the process.

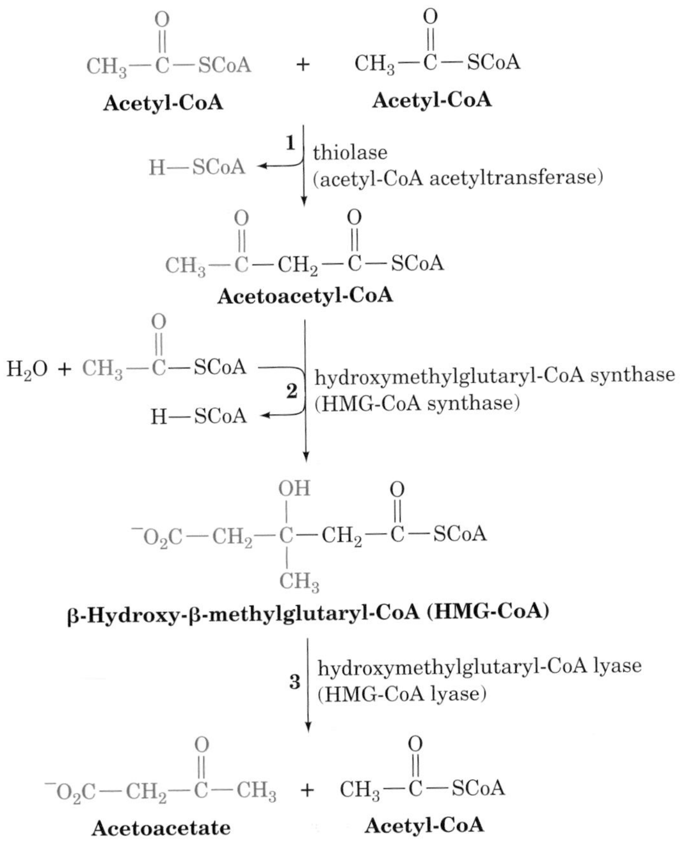

Acetoacetate may be reduced to D-β-hydroxybutyrate by **β-hydroxybutyrate dehydrogenase:**

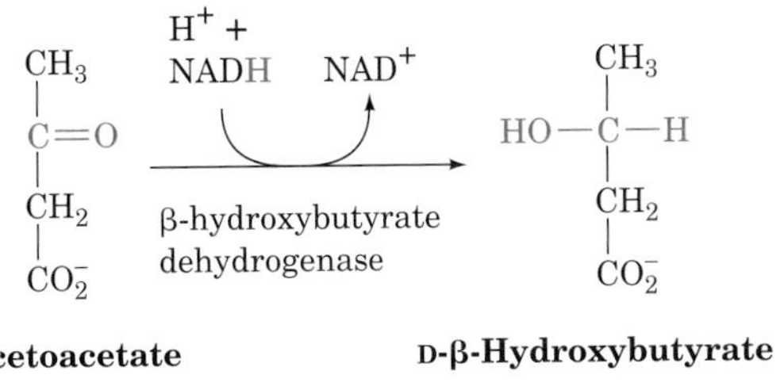

Acetoacetate   **D-β-Hydroxybutyrate**

Note that this product is the stereoisomer of the L-β-hydroxyacyl-CoA that occurs in the β-oxidation pathway. Acetoacetate, being a β-keto acid, also undergoes relatively facile nonenzymatic decarboxylation to acetone and CO_2. Indeed, the breath of individuals with **ketosis** (also called **ketoacidosis**), a potentially pathological condition in which acetoacetate is produced faster than it can be metabolized (a symptom of diabetes; Section 27-4B), has the characteristic sweet smell of acetone.

The liver releases acetoacetate and β-hydroxybutyrate, which are carried by the bloodstream to the peripheral tissues for use as alternative fuels. There, these products are converted to acetyl-CoA as is diagrammed in Fig. 25-26.

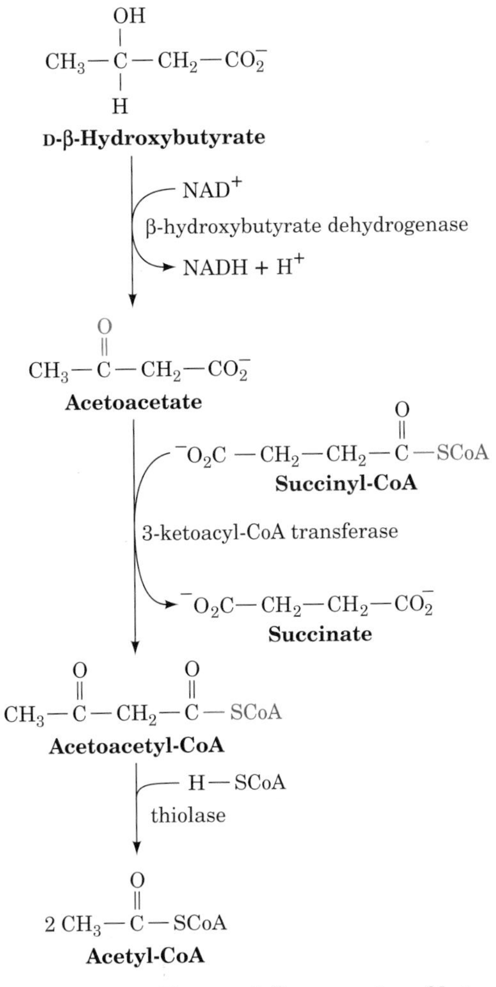

FIGURE 25-25 Ketogenesis: the enzymatic reactions forming acetoacetate from acetyl-CoA. (1) Two molecules of acetyl-CoA condense to form acetoacetyl-CoA in a thiolase-catalyzed reaction. **(2)** A Claisen ester condensation of the acetoacetyl-CoA with a third acetyl-CoA to form β-hydroxy-β-methylglutaryl-CoA (HMG-CoA) as catalyzed by HMG-CoA synthase. **(3)** The degradation of HMG-CoA to acetoacetate and acetyl-CoA in a mixed aldol–Claisen ester cleavage catalyzed by HMG-CoA lyase.

FIGURE 25-26 The metabolic conversion of ketone bodies to acetyl-CoA.

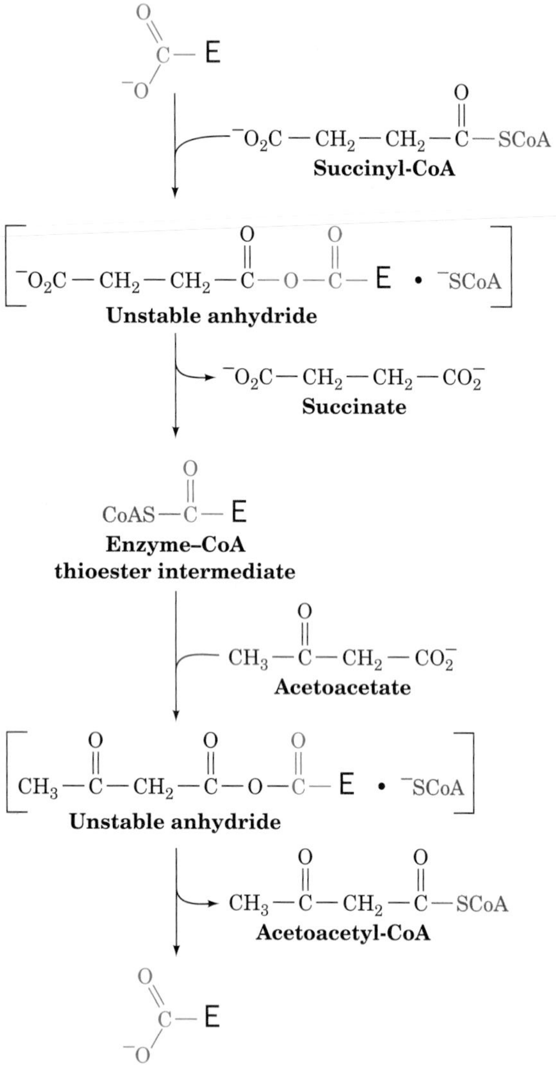

FIGURE 25-27   Proposed mechanism of 3-ketoacyl-CoA transferase involving an enzyme–CoA thioester intermediate.

The proposed reaction mechanism of **3-ketoacyl-CoA transferase** (Fig. 25-27), which catalyzes this pathway's second step, involves the participation of an active site carboxyl group both in an enzyme–CoA thioester intermediate and in an unstable anhydride. Succinyl-CoA, which acts as the CoA donor in this reaction, can also be converted to succinate with the coupled synthesis of GTP in the succinyl-CoA synthase reaction of the citric acid cycle (Section 21-3E). The "activation" of acetoacetate bypasses this step and therefore "costs" the free energy of GTP hydrolysis. The liver lacks 3-ketoacyl-CoA transferase, which permits it to supply ketone bodies to other tissues.

4 ■ FATTY ACID BIOSYNTHESIS

Fatty acid biosynthesis occurs through condensation of C_2 units, the reverse of the β-oxidation process. Through isotopic labeling techniques, David Rittenberg and Konrad

Bloch demonstrated, in 1945, that these condensation units are derived from acetic acid. Acetyl-CoA was soon proven to be a precursor of the condensation reaction, but its mechanism remained obscure until the late 1950s when Salih Wakil discovered a requirement for bicarbonate in fatty acid biosynthesis and malonyl-CoA was shown to be an intermediate. In this section, we discuss the reactions of fatty acid biosynthesis.

A. *Pathway Overview*

The pathway of fatty acid synthesis differs from that of fatty acid oxidation. This situation, as we saw in Section 18-1D, is typically the case of opposing biosynthetic and degradative pathways because it permits them both to be thermodynamically favorable and independently regulated under similar physiological conditions. Figure 25-28 outlines fatty acid oxidation and synthesis with emphasis on the differences between these pathways. Whereas fatty acid oxidation occurs in the mitochondrion and utilizes fatty acyl-CoA esters, fatty acid biosynthesis occurs in the cytosol with, as Roy Vagelos discovered, the growing fatty acids esterified to **acyl-carrier protein** (**ACP;** Fig. 25-29). ACP, like CoA, contains a phosphopantetheine group that forms thioesters with acyl groups. The phosphopantetheine phosphoryl group is esterified to a Ser OH group of ACP, whereas in CoA it is esterified to AMP. In animals ACP is part of a large multifunctional protein (Type I ACP; see below), whereas in *E. coli* it is a 125-residue polypeptide (Type II ACP). The phosphopantetheine group is transferred from CoA to apo-ACP to form the active holo-ACP by **ACP synthase.**

The redox coenzymes of the animal fatty acid oxidative and biosynthetic pathways differ (NAD^+ and FAD for oxidation; NADPH for biosynthesis) as does the stereochemistry of their intermediate steps, but their main difference is the manner in which C_2 units are removed from or added to the fatty acyl thioester chain. In the oxidative pathway, β-ketothiolase catalyzes the cleavage of the C_α—C_β bond of β-ketoacyl-CoA so as to produce acetyl-CoA and a new fatty acyl-CoA, which is shorter by a C_2 unit. The $\Delta G^{\circ\prime}$ of this reaction is very close to zero so it can also function in the reverse direction (ketone body formation). In the biosynthetic pathway, the condensation reaction is coupled to the hydrolysis of ATP, thereby driving the reaction to completion. This process involves two steps: (1) the ATP-dependent carboxylation of acetyl-CoA by **acetyl-CoA carboxylase** to form **malonyl-CoA,** and (2) the exergonic decarboxylation of the malonyl group in the condensation reaction catalyzed by **fatty acid synthase.** The mechanisms of these enzymes are described below.

B. *Acetyl-CoA Carboxylase*

*Acetyl-CoA carboxylase (**ACC**) catalyzes the first committed step of fatty acid biosynthesis and one of its rate-controlling steps.* The mechanism of this biotin-dependent enzyme is very similar to those of propionyl-CoA carboxylase (Section 25-2E) and pyruvate carboxylase (Fig.

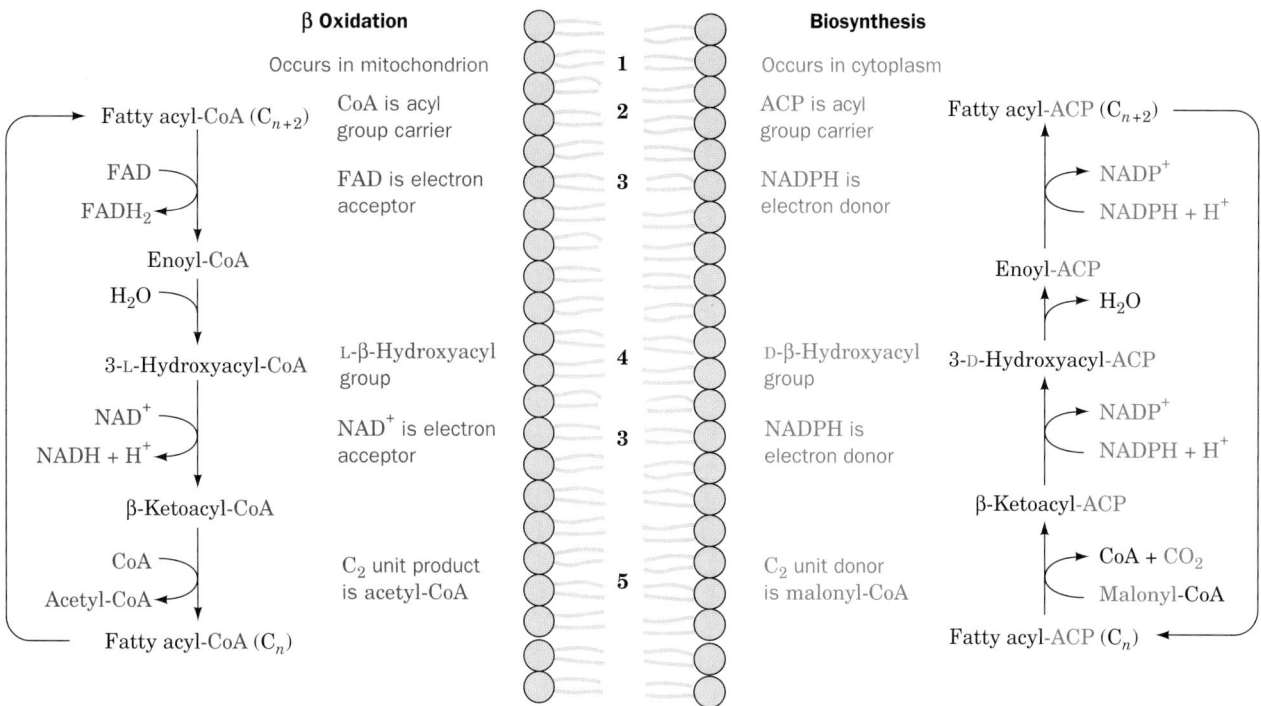

FIGURE 25-28 A comparison of fatty acid β oxidation and fatty acid biosynthesis. Differences occur in (1) cellular location, (2) acyl group carrier, (3) electron acceptor/donor, (4) stereochemistry of the hydration/dehydration reaction, and (5) the form in which C_2 units are produced/donated. 🔊 **See the Animated Figures**

23-4) in that it occurs in two steps, a CO_2 activation and a carboxylation:

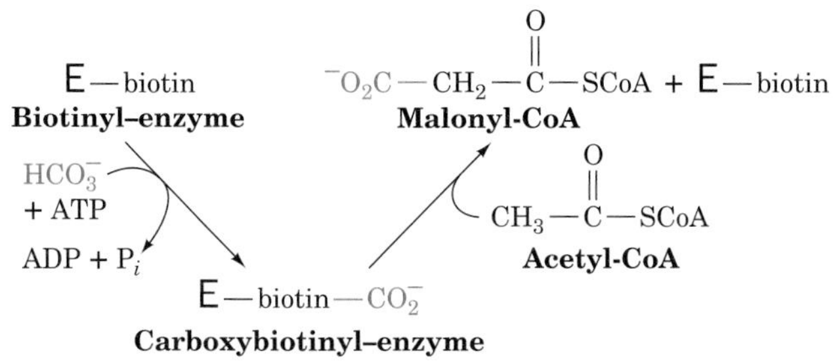

$$\text{E—biotin}$$
Biotinyl–enzyme

$$^-O_2C—CH_2—\overset{\overset{\displaystyle O}{\|}}{C}—SCoA + \text{E—biotin}$$
Malonyl-CoA

$$HCO_3^- + ATP$$
$$ADP + P_i$$

$$CH_3—\overset{\overset{\displaystyle O}{\|}}{C}—SCoA$$
Acetyl-CoA

$$\text{E—biotin—}CO_2^-$$
Carboxybiotinyl–enzyme

In *E. coli*, these steps are catalyzed by separate subunits, known as **biotin carboxylase** and **transcarboxylase,** respectively. In addition, the biotin is bound as a biocytin residue to a third subunit, termed **biotin carboxyl-carrier protein.** The mammalian and avian enzymes contain both enzymatic activities as well as the biotin carboxyl carrier on a single 2346-residue polypeptide chain.

a. Acetyl-CoA Carboxylase Is Regulated by Hormonally Controlled Reversible Phosphorylation

ACC is subject to hormonal regulation. Glucagon as well as epinephrine and norepinephrine (adrenaline and noradrenaline; Section 18-3E) trigger the enzyme's cAMP-dependent increase in phosphorylation, which inactivates the enzyme. Insulin, on the other hand, stimulates enzyme dephosphorylation and thus its activation.

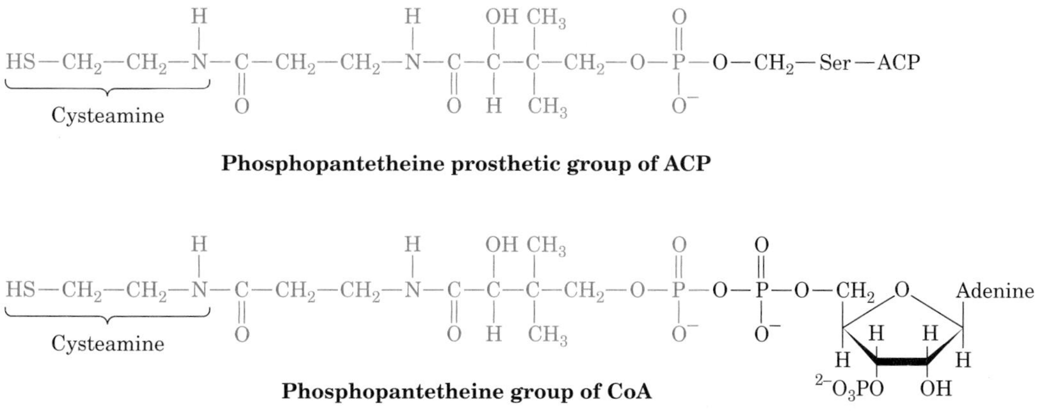

$$HS—CH_2—CH_2—\overset{\overset{\displaystyle H}{|}}{N}—\overset{\overset{\displaystyle O}{\|}}{C}—CH_2—CH_2—\overset{\overset{\displaystyle H}{|}}{N}—\overset{\overset{\displaystyle OH}{|}}{\underset{\underset{\displaystyle H}{|}}{C}}—\overset{\overset{\displaystyle CH_3}{|}}{\underset{\underset{\displaystyle CH_3}{|}}{C}}—CH_2—O—\overset{\overset{\displaystyle O}{\|}}{\underset{\underset{\displaystyle O^-}{|}}{P}}—O—CH_2—Ser—ACP$$

$$\underbrace{\qquad\qquad\qquad}_{\text{Cysteamine}}$$

Phosphopantetheine prosthetic group of ACP

$$HS—CH_2—CH_2—\overset{\overset{\displaystyle H}{|}}{N}—\overset{\overset{\displaystyle O}{\|}}{C}—CH_2—CH_2—\overset{\overset{\displaystyle H}{|}}{N}—\overset{\overset{\displaystyle OH}{|}}{\underset{\underset{\displaystyle H}{|}}{C}}—\overset{\overset{\displaystyle CH_3}{|}}{\underset{\underset{\displaystyle CH_3}{|}}{C}}—CH_2—O—\overset{\overset{\displaystyle O}{\|}}{\underset{\underset{\displaystyle O^-}{|}}{P}}—O—\overset{\overset{\displaystyle O}{\|}}{\underset{\underset{\displaystyle O^-}{|}}{P}}—O—CH_2$$

Adenine

$$^{2-}O_3PO \quad OH$$

$$\underbrace{\qquad\qquad\qquad}_{\text{Cysteamine}}$$

Phosphopantetheine group of CoA

FIGURE 25-29 The phosphopantetheine group in acyl-carrier protein (ACP) and in CoA.

The mechanism by which cAMP causes an increase in the phosphorylation state of ACC is interesting. ACC is phosphorylated, *in vitro*, by two different kinases, the cAMP-dependent protein kinase A (PKA; Section 18-3C) at Ser 77 and **AMP-dependent protein kinase (AMPK;** Sections 25-5 and 27-1**)** (which is cAMP independent) at Ser 79. Yet, when liver cells are incubated with cAMP-elevating hormones in the presence of ^{32}P-ATP, only Ser 79 is found to be labeled. Evidently, a [cAMP] increase results in a phosphorylation increase at sites modified by AMPK rather than by PKA. How can this be? It appears that, *in vivo*, the cAMP-dependent increase in phosphorylation occurs not through the phosphorylation of new sites but, rather, through the inhibition of dephosphorylation of previously phosphorylated positions. We have already seen such a mechanism in operation in the control of glycogen metabolism, where the cAMP-dependent phosphorylation of phosphoprotein phosphatase inhibitor-1 causes the inhibition of dephosphorylation (Section 18-3C). In the case of ACC, however, dephosphorylation is catalyzed by **phosphoprotein phosphatase-2A,** which is not affected by phosphoprotein phosphatase inhibitor-1. The mechanism by which PKA causes the increase in phosphorylation associated with AMPK activity is an active subject of investigation.

b. Avian and Mammalian Acetyl-CoA Carboxylases Undergo Enzyme Polymerization on Activation

Electron microscopy reveals that the flat rectangular protomers of both avian and mammalian acetyl-CoA carboxylases associate to form long filaments with molecular masses in the range 4000 to 8000 kD (Fig. 25-30). *This polymeric form of the enzyme is catalytically active but the protomer is not.* The rate of fatty acid biosynthesis is therefore controlled by the position of the equilibrium between these forms:

$$\text{Protomer}(\textit{inactive}) \rightleftharpoons \text{polymer}(\textit{active})$$

Phosphorylation favors the inactive protomer while dephosphorylation favors the active polymer. Several metabolites also affect the activity of acetyl-CoA carboxylase. Citrate binds to and increases the activity of the dephosphoenzyme. Citrate also binds to and partially activates the phosphoenzyme, whereas palmitoyl-CoA inhibits the enzyme. Thus, cytosolic citrate, whose concentration increases when the mitochondrial acetyl-CoA concentration builds up (Section 25-4D), activates fatty acid biosynthesis, whereas palmitoyl-CoA, the pathway product, is a feedback inhibitor.

c. Mammalian Acetyl-CoA Carboxylase Has Two Major Isoforms

There are two major isoforms of ACC. **α-ACC** occurs in adipose tissue and **β-ACC** occurs in tissues that oxidize but do not synthesize fatty acids, such as heart muscle. Tissues that both synthesize and oxidize fatty acids, such as liver, contain both isoforms, which are homologous although the genes encoding them are located on different

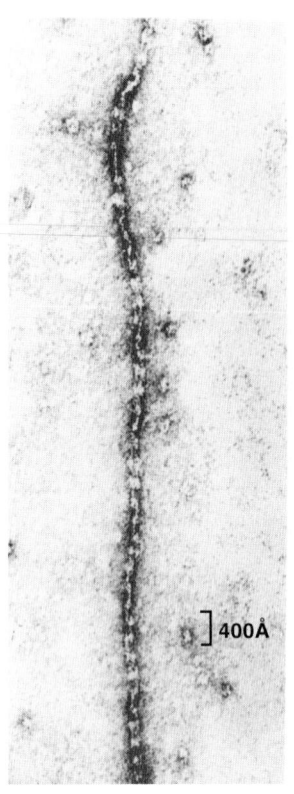
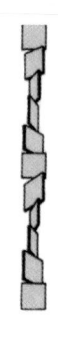

FIGURE 25-30   Association of acetyl-CoA carboxylase protomers. An electron micrograph with an accompanying interpretive drawing indicates that filaments of avian liver acetyl-CoA carboxylase consist of linear chains of flat rectangular protomers. [Courtesy of Malcolm Lane, The Johns Hopkins University School of Medicine.]

chromosomes. What is the function of β-ACC? The product of the ACC-catalyzed reaction, malonyl-CoA, strongly inhibits the mitochondrial import of fatty acyl-CoA for fatty acid oxidation, the major control point for this process. Thus it appears that β-ACC has a regulatory function (Section 25-5).

Prokaryotic acetyl-CoA carboxylases are not subject to any of these controls. This is because fatty acids in these organisms are not stored as fats but function largely as phospholipid precursors. The *E. coli* enzyme is instead regulated by guanine nucleotides so that fatty acids are synthesized in response to the cell's growth requirements.

C. Fatty Acid Synthase

The synthesis of fatty acids, mainly palmitic acid, from acetyl-CoA and malonyl-CoA involves seven enzymatic reactions. These reactions were first studied in cell-free extracts of *E. coli*, in which they are catalyzed by seven independent enzymes together with ACP. Individual enzymes with these activities also occur in chloroplasts (the only site of fatty acid synthesis in plants). In yeast, however, fatty acids are synthesized by **fatty acid synthase (FAS),** a 2500-kD $\alpha_6\beta_6$ multifunctional enzyme, whereas in animals

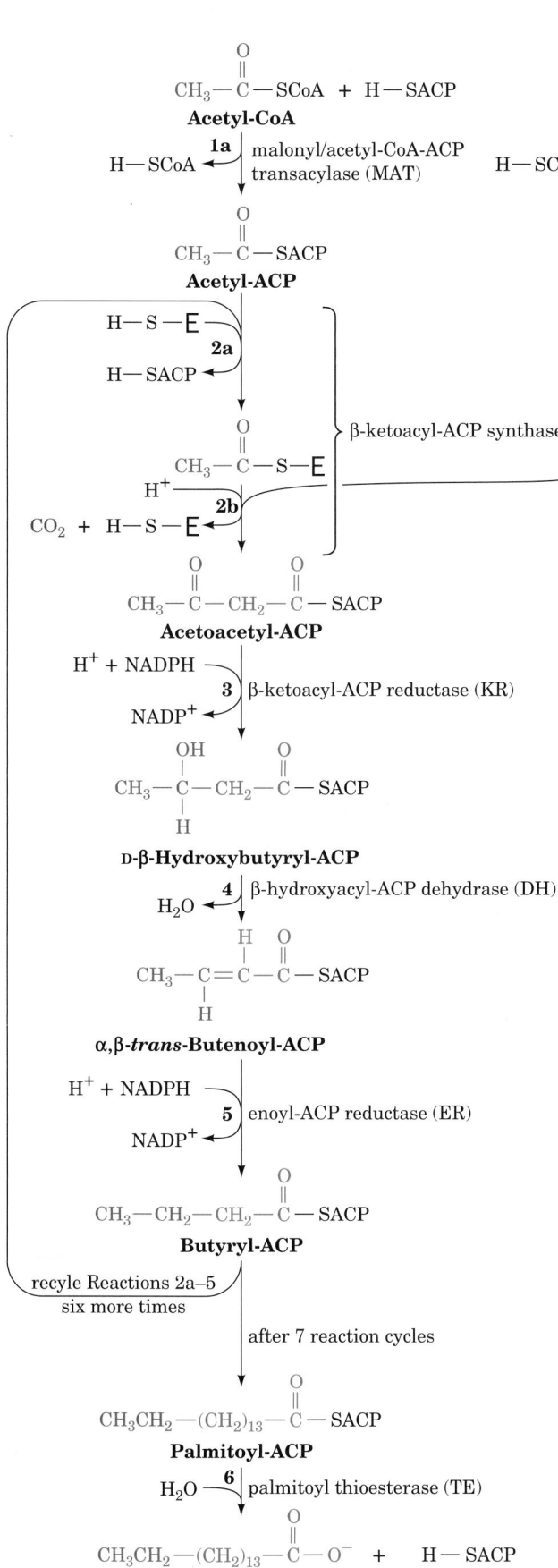

FIGURE 25-31 Reaction cycle for the biosynthesis of fatty acids. The biosynthesis of palmitate requires seven cycles of C_2 elongation followed by a final hydrolysis step. 🎵 See the Animated Figures

FAS is a multifunctional enzyme consisting of two identical 272-kD polypeptide chains, each containing all seven activities plus ACP.

The amino acid sequences of several fatty acid synthases have been deduced from their gene sequences. The sequence of the 2511-residue chicken liver enzyme is 67% identical with that of rat, with many of the mismatches arising from conservative substitutions. The regions of highest homology encompass the polypeptide segments comprising the enzymatic active sites, thereby supporting the contention that this multifunctional enzyme evolved by the joining of what were previously independent enzymes.

The reactions catalyzed by the mammalian multifunctional enzyme to synthesize palmitate are diagrammed in Fig. 25-31 with the long flexible phosphopantetheine chain of ACP (Fig. 25-29) functioning to transport the substrate between the protein's various enzymatic domains:

1a. The transfer of the acetyl group from acetyl-CoA to ACP to yield acetyl-ACP as catalyzed by **malonyl/ acetyl-CoA-ACP transacylase (MAT).**

2a. The loading of **β-ketoacyl-ACP synthase (KS;** also known as **condensing enzyme)** by the transfer of the acetyl group from ACP to a KS Cys residue, thus maintaining the acetyl group's thioester linkage.

1b. The formation of malonyl-ACP in a reaction analogous to that of Reaction 1a, which in animals is catalyzed by the same enzyme, MAT.

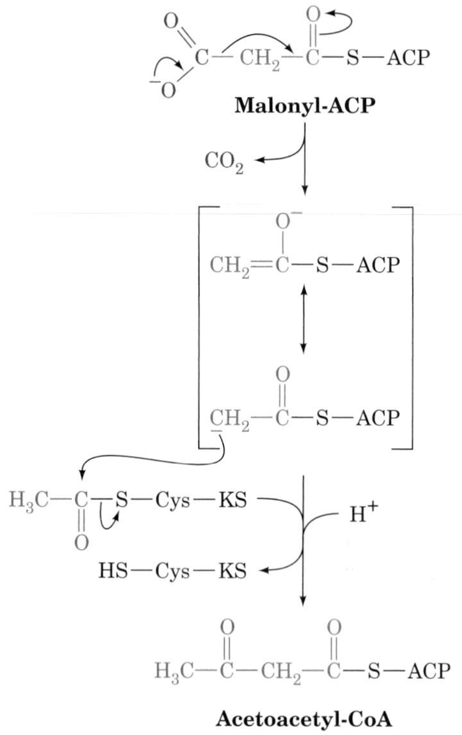

Malonyl-ACP

Acetoacetyl-CoA

FIGURE 25-32 The mechanism of carbon–carbon bond formation in fatty acid biosynthesis. The condensation of an acetyl group on the active site Cys of β-ketoacyl-ACP synthase (KS) with a malonyl group on the phosphopantetheine arm of ACP forms a β-ketoacyl-ACP. The reaction is driven by the elimination of CO_2 from the malonyl group to generate a resonance-stabilized acetyl-ACP carbanion intermediate.

2b. The coupling of the acetyl group to the C_β of the malonyl group on the ACP of either the same or the other subunit with the malonyl group's accompanying decarboxylation so as to form acetoacetyl-ACP and free the KS active site Cys-SH group (Fig. 25-32). *Consequently, the CO_2 taken up in the acetyl-CoA carboxylase reaction (Section 25-4B) does not appear in the product fatty acid. Rather, the decarboxylation functions to drive carbon–carbon bond formation in the condensation reaction which, through the acetyl-CoA carboxylase reaction, is coupled to ATP hydrolysis.*

3–5. The reduction, dehydration, and further reduction of acetoacetyl-ACP so as to form **butyryl-ACP** as sequentially catalyzed by **β-ketoacyl-ACP reductase (KR)**, **β-hydroxyacyl-ACP dehydrase (DH)**, and **enoyl-ACP reductase (ER)**. The coenzyme in both reductive steps is NADPH, whereas in β oxidation, the analogs of Reactions 3 and 5, respectively, use NAD^+ and FAD (Fig. 25-28; although in yeast, the NADPH in Reaction 5 reduces FMN to $FMNH_2$ which, in turn, reduces the C=C double bond). Moreover, Reaction 3 produces and Reaction 4 requires a D-β-hydroxyacyl group, whereas the analogous reactions in β oxidation involve the corresponding L isomer.

2a to 5 Repeat. The butyryl group from the butyryl-ACP is transferred to the Cys-SH of KS. Thus the acetyl group with which the system was initially loaded has been elongated by a C_2 unit. The ACP is "reloaded" with a malonyl group (Step 1b), and another cycle of C_2 elongation occurs. This process occurs altogether seven times to yield **palmitoyl-ACP.**

6. The palmitoyl-ACP thioester bond is hydrolyzed by **palmitoyl thioesterase (TE),** yielding palmitate, the normal product of the fatty acid synthase pathway, and regenerating the enzyme for a new round of synthesis.

The stoichiometry of palmitate synthesis therefore is

Acetyl-CoA + 7malonyl-CoA + 14NADPH + $7H^+$ → palmitate + $7CO_2$ + $14NADP^+$ + 8CoA + $6H_2O$

Since the 7 malonyl-CoA are derived from acetyl-CoA as follows:

7Acetyl-CoA + $7CO_2$ + 7ATP → 7malonyl-CoA + 7ADP + $7P_i$ + $7H^+$

the overall stoichiometry for palmitate biosynthesis is

8Acetyl-CoA + 14NADPH + 7ATP → palmitate + $14NADP^+$ + 8CoA + $6H_2O$ + 7ADP + $7P_i$

a. The Organization of the Fatty Acid Synthase Dimer

Most but not all of the enzyme activities remain functional when the native dimeric animal FAS is dissociated into monomers. Electron microscopy of these monomers indicates that they consist of a linear chain of at least four 50-Å-diameter lobes. Moreover, fragments resulting from the limited proteolysis of fatty acid synthase exhibit many of the enzymatic activities of the intact protein. Thus, *contiguous stretches of its polypeptide chain fold to form a series of autonomous domains, each with a specific but different catalytic activity.* Several other enzymes, such as mammalian acetyl-CoA carboxylase (Section 25-4B), exhibit similar multifunctionality but none has as many separate catalytic activities as does animal FAS. The order of the catalytic activities along the FAS polypeptide chain is indicated in Fig. 25-33.

The condensation reaction requires the juxtaposition of the sulfhydryl groups of an ACP phosphopantetheine and an active site Cys residue of KS. However, the monomeric form of FAS is inactive in the overall FAS reaction, which suggests that these two sulfhydryl groups are on opposite subunits and hence that these subunits interact in a head-to-tail manner. Indeed, these two sulfhydryl groups on opposite subunits are cross-linked by **1,3-dibromo-2-propanone,**

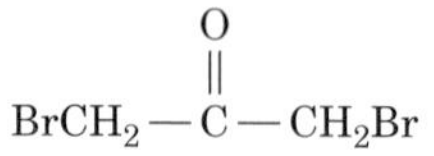

$$BrCH_2 - \overset{\overset{\displaystyle O}{\|}}{C} - CH_2Br$$

1,3-Dibromo-2-propanone

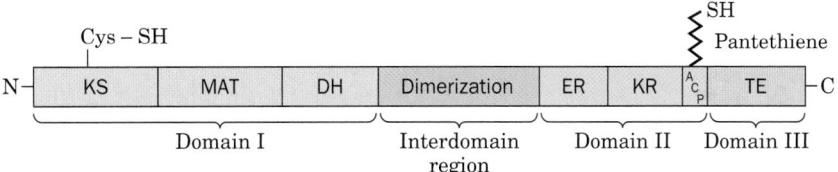

FIGURE 25-33 **Schematic diagram of the order of the enzymatic activities along the polypeptide chain of a monomer of fatty acid synthase (FAS).**

thereby demonstrating that they are in close proximity. Moreover, EM-based images of human FAS, elucidated by Wakil and Wah Chiu, reveal that the protomers in this dimer are arranged in an antiparallel orientation (Fig. 25-34). Nevertheless, Stuart Smith generated a mutant FAS consisting of a wild-type protomer and a protomer that lacks its ACP module. This heterodimer could carry out the full catalytic reaction, albeit at a reduced rate, thereby demonstrating that the active site thiol of KS is able to interact with the phosphopantetheine of ACP on the same monomer. In fact, a reinvestigation of the 1,3-dibromo-2-propanone cross-linking experiments using FAS mutants lacking the Cys or pantetheine sulfhydryl groups on one subunit revealed that head-to-tail cross-linking could occur within a single subunit. Evidently, the FAS polypeptide is sufficiently flexible for the ACP on one subunit to interact with the KS on either subunit.

b. Variations on a Theme: Polyketide Biosynthesis

Polyketides are a family of natural products that are made by the modular condensation of acyl-CoA monomers such as acetyl-CoA and propionyl-CoA with malonyl-CoA and methylmalonyl-CoA extender units whose decarboxylation drives the condensation reaction. The name polyketide comes from the fact that the primary condensation products have β-keto functional groups.

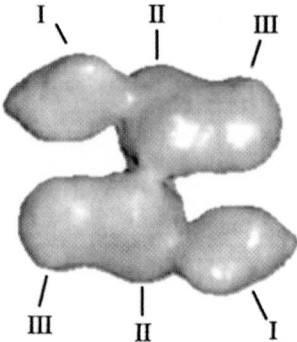

FIGURE 25-34 **EM-based image of the human FAS dimer as viewed along its 2-fold axis.** Domains I, II, and III are labeled. Note the antiparallel orientation of the monomers. [Courtesy of Salih Wakil and Wah Chiu, Baylor College of Medicine.]

Long-chain fatty acids are examples of polyketides since they are formed by the condensation of one acetyl-CoA primer and seven malonyl-CoA extender units. Following each condensation reaction, the new β-keto group may be reduced, hydrated, and reduced again as with fatty acids, or may undergo only partial modification.

All polyketide products are synthesized on very large multifunctional enzymes. We have already seen that FAS contains seven enzymatic activities as well as ACP. Another example of a polyketide is **6-deoxyerythronolide B (6dEB),** the parent **macrolactone** of the antibiotic **erythromycin A** (Section 32-3G), which is synthesized by the soil bacterium *Saccharopolyspora erythraea* from one propionyl-CoA primer and six (*S*)-methylmalonyl-CoA extenders, by **deoxyerythronolide B synthase (DEBS;** Fig. 25-35). DEBS is a 2000-kD, $\alpha_2\beta_2\gamma_2$ complex with each of its three homodimeric units catalyzing two elongation/modification cycles. Unlike FAS, which catalyzes several cycles of elongation/modification with the same active sites, DEBS catalyzes each elongation/modification cycle on a different module, allowing for differences in the modifications that occur at each cycle. Module 4, as Fig. 25-35 indicates, is almost identical in function to FAS, containing KS, **acyltransferase (AT;** a homolog of MAT), ACP, KR, DH, and ER, and reducing its primary β-ketone condensation product to a methylene group. However, it does not contain TE because the elongation process is not complete after this phase. Module 3 contains only ACP, KS, and AT and passes its β-ketone condensation product to module 4 without further modification. Modules 1, 2, 5, and 6 contain only ACP, AT, KS, and KR, the sites necessary for the condensation and ketone reduction steps, thereby generating hydroxy products. The overall organization of the modules therefore creates a polyhydroxy product containing one keto group and one methylene group in the chain. The DEBS final product, 6dEB, is a lactone produced by the reaction of the terminal hydroxyl group with the thioester anchoring the growing chain to the synthase. Various polyketide synthases have different organizations of modules, and thereby synthesize a multitude of natural products.

We next consider the means of transport of mitochondrial acetyl-CoA to the cytosol, the site of fatty acid synthesis. Following that, we examine the reactions by which fatty acids are elongated and desaturated.

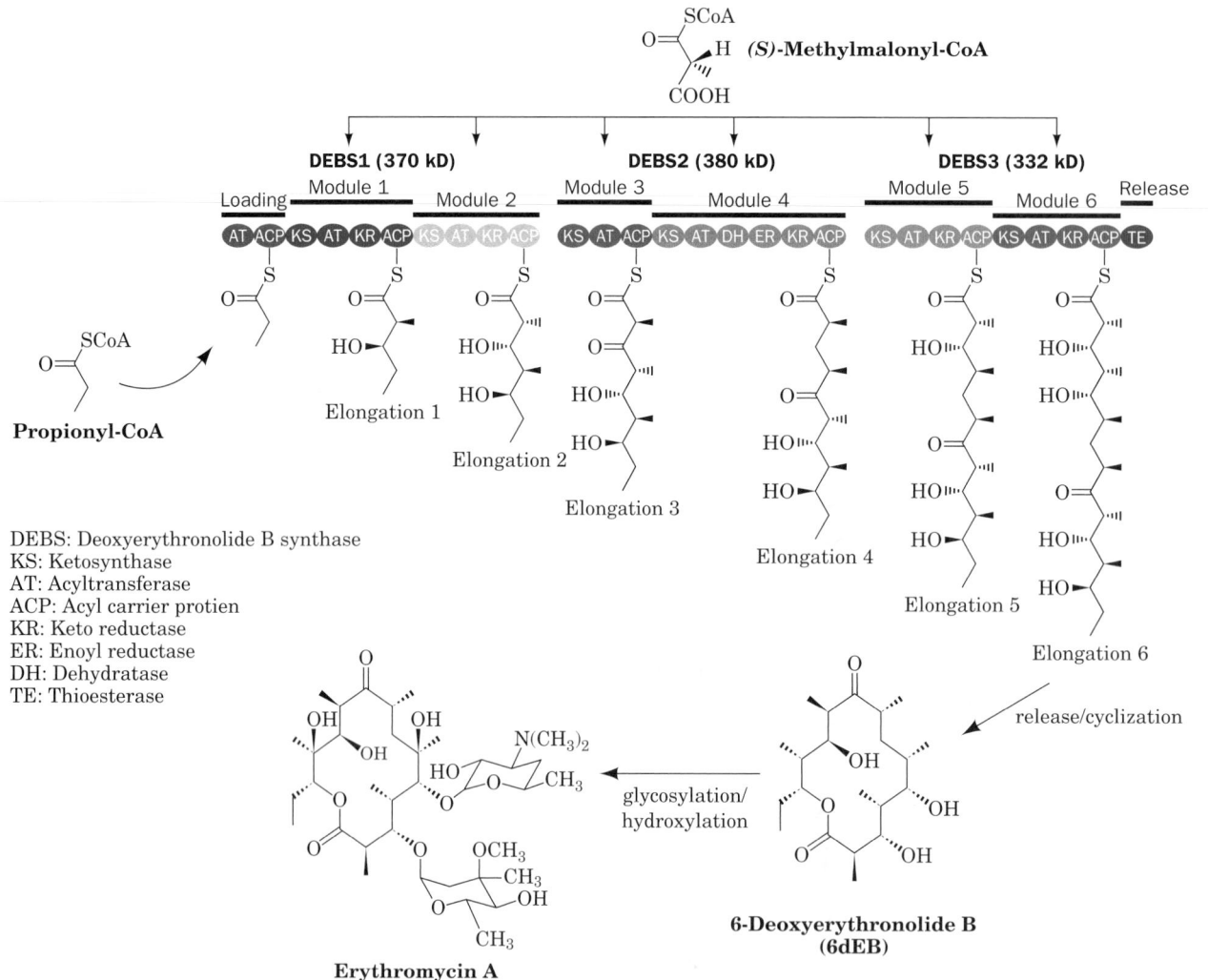

FIGURE 25-35 **An example of polyketide biosynthesis: the synthesis of erythromycin A.** [After Pfeifer, B.A., Admiraal, S.J., Gramajo, H., Cane, D.E., and Khosla, C., *Science* **291,** 1790 (2001).]

D. *Transport of Mitochondrial Acetyl-CoA into the Cytosol*

Acetyl-CoA is generated in the mitochondrion by the oxidative decarboxylation of pyruvate as catalyzed by pyruvate dehydrogenase (Section 21-2A) as well as by the oxidation of fatty acids. When the need for ATP synthesis is low, so that the oxidation of acetyl-CoA via the citric acid cycle and oxidative phosphorylation is minimal, this mitochondrial acetyl-CoA may be stored for future use as fat. Fatty acid biosynthesis occurs in the cytosol but the mitochondrial membrane is essentially impermeable to acetyl-CoA. *Acetyl-CoA enters the cytosol in the form of citrate via the* **tricarboxylate transport system** *(Fig. 25-36).* Cytosolic **ATP-citrate lyase** then catalyzes the reaction

Citrate + CoA + ATP $\rightleftharpoons$
$$\text{acetyl-CoA} + \text{oxaloacetate} + \text{ADP} + P_i$$

which resembles the reverse of the citrate synthase reaction (Section 21-3A) except that ATP hydrolysis is required to drive the intermediate synthesis of the "high-energy" cit-

ryl-CoA, whose hydrolysis drives the citrate synthase reaction to completion. ATP hydrolysis is therefore required in the ATP-citrate lyase reaction to power the resynthesis of this thioester bond. Oxaloacetate is reduced to malate by malate dehydrogenase. Malate may be oxidatively decarboxylated to pyruvate by malic enzyme (Section 25-2E) and be returned in this form to the mitochondrion. The malic enzyme reaction resembles that of isocitrate dehydrogenase in which a β-hydroxy acid is oxidized to a β-keto acid, whose decarboxylation is strongly favored (Section 21-3C). Malic enzyme's coenzyme is NADP$^+$, so when this route is used NADPH is produced for use in the reductive reactions of fatty acid biosynthesis.

Citrate transport out of the mitochondrion must be balanced by anion transport into the mitochondrion. Malate, pyruvate, and P$_i$ can act in this capacity. Malate may therefore also be transported directly back to the mitochondrion without generating NADPH. As we have seen in Section 25-4C, synthesis of each palmitate ion requires 8 molecules of acetyl-CoA and 14 molecules of NADPH. As many as 8 of these NADPH molecules may be supplied with the 8

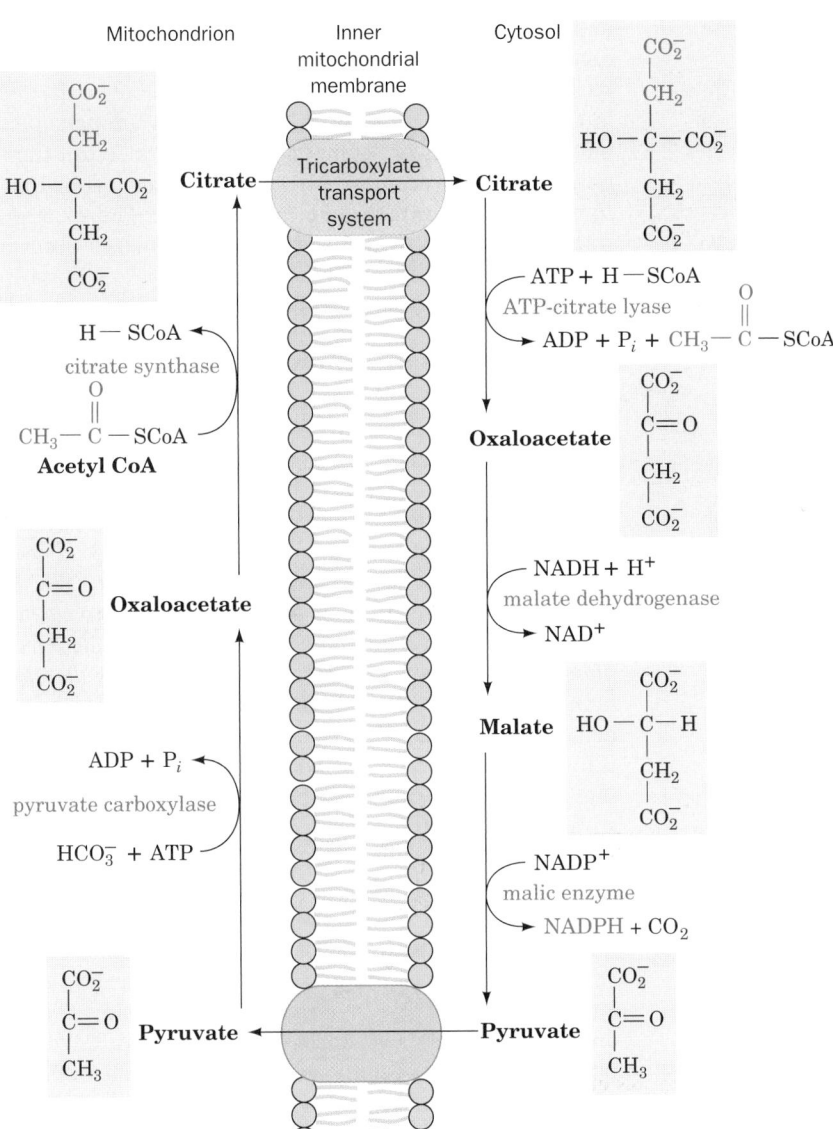

molecules of acetyl-CoA if all the malate produced in the cytosol is oxidatively decarboxylated. The remaining NADPH is provided through the pentose phosphate pathway (Section 23-4).

E. *Elongases and Desaturases*

*Palmitate (16:0), the normal product of the fatty acid synthase pathway, is the precursor of longer chain saturated and unsaturated fatty acids through the actions of **elongases** and **desaturases**.* Elongases are present in both the mitochondrion and the ER but the mechanisms of elongation at the two sites differ. Mitochondrial elongation (a process independent of the fatty acid synthase pathway) occurs by successive addition and reduction of acetyl units in a reversal of fatty acid oxidation; the only chemical difference between these two pathways occurs in the final reduction step in which NADPH takes the place of $FADH_2$ as the terminal redox coenzyme (Fig. 25-37). Elongation in the ER involves the successive condensations of malonyl-CoA with acyl-

CoA. These reactions are each followed by NADPH-associated reductions similar to those catalyzed by fatty acid synthase, the only difference being that the fatty acid is elongated as its CoA derivative rather than as its ACP derivative.

Unsaturated fatty acids are produced by **terminal desaturases**. Mammalian systems contain four terminal desaturases of broad chain-length specificities designated **Δ^9-, Δ^6-, Δ^5-,** and **Δ^4-fatty acyl-CoA desaturases.** These membrane-bound, nonheme iron–containing enzymes catalyze the general reaction

$$CH_3-(CH_2)_x-\overset{\overset{\displaystyle H}{|}}{\underset{\underset{\displaystyle H}{|}}{C}}-\overset{\overset{\displaystyle H}{|}}{\underset{\underset{\displaystyle H}{|}}{C}}-(CH_2)_y-\overset{\overset{\displaystyle O}{\|}}{C}-SCoA + NADH + H^+ + O_2$$

$$\downarrow$$

$$CH_3-(CH_2)_x-\underset{\underset{\displaystyle H}{|}}{C}=\underset{\underset{\displaystyle H}{|}}{C}-(CH_2)_y-\overset{\overset{\displaystyle O}{\|}}{C}-SCoA + 2H_2O + NAD^+$$

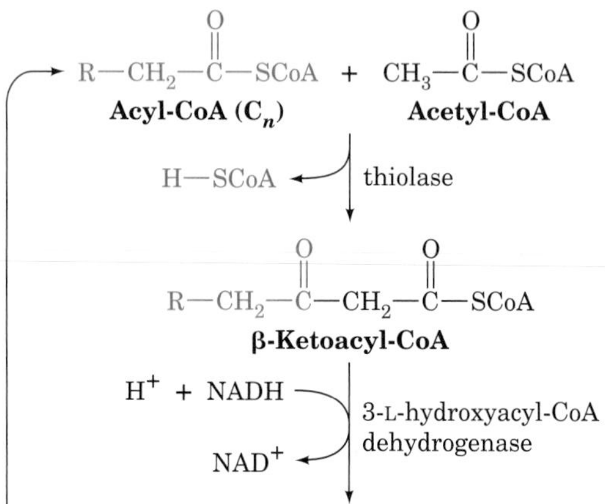

$$R-CH_2-\overset{\overset{\displaystyle O}{\|}}{C}-SCoA \quad + \quad CH_3-\overset{\overset{\displaystyle O}{\|}}{C}-SCoA$$

Acyl-CoA (C$_n$)   **Acetyl-CoA**

H—SCoA ⟵⟍ thiolase

$$R-CH_2-\overset{\overset{\displaystyle O}{\|}}{C}-CH_2-\overset{\overset{\displaystyle O}{\|}}{C}-SCoA$$

β-Ketoacyl-CoA

H$^+$ + NADH ⟍
     3-L-hydroxyacyl-CoA dehydrogenase
NAD$^+$ ⟋

$$R-CH_2-\overset{\overset{\displaystyle H}{|}}{\underset{\underset{\displaystyle OH}{|}}{C}}-CH_2-\overset{\overset{\displaystyle O}{\|}}{C}-SCoA$$

L-β-Hydroxyacyl-CoA

H$_2$O ⟵⟍ enoyl-CoA hydratase

$$R-CH_2-\overset{\overset{\displaystyle H}{|}}{C}=\overset{\overset{\displaystyle O}{\|}}{\underset{\underset{\displaystyle H}{|}}{C}}-C-SCoA$$

α,β-*trans*-Enoyl-CoA

H$^+$ + NADPH ⟍
     enoyl-CoA reductase
NADP$^+$ ⟋

$$R-CH_2-CH_2-CH_2-\overset{\overset{\displaystyle O}{\|}}{C}-SCoA$$

Acyl-CoA (C$_{n+2}$)

FIGURE 25-37   Mitochondrial fatty acid elongation. Elongation occurs by the reversal of fatty acid oxidation with the exception that the final reaction employs NADPH rather than FADH$_2$ as its redox coenzyme.

where x is at least 5 and where $(CH_2)_x$ can contain one or more double bonds. The $(CH_2)_y$ portion of the substrate is always saturated. Double bonds are inserted between existing double bonds in the $(CH_2)_x$ portion of the substrate

and the CoA group such that the new double bond is three carbon atoms closer to the CoA group than the next double bond (not conjugated to an existing double bond) and, in animals, never at positions beyond C9. Mammalian terminal desaturases are components of mini-electron-transport systems that contain two other proteins: **cytochrome b_5** and **NADH–cytochrome b_5 reductase.** The electron-transfer reactions mediated by these complexes occur at the inner surface of the ER membrane (Fig. 25-38) and are therefore not associated with oxidative phosphorylation.

a. Some Unsaturated Fatty Acids Must Be Obtained in the Diet

A variety of unsaturated fatty acids may be synthesized by combinations of elongation and desaturation reactions. However, since palmitic acid is the shortest available fatty acid in animals, the above rules preclude the formation of the Δ^{12} double bond of linoleic acid [$\Delta^{9,12}$-octadecadienoic acid; 18:2n–6 (this nomenclature is explained in Table 12-1)], a required precursor of **prostaglandins.** *Linoleic acid must consequently be obtained in the diet (ultimately from plants that have Δ^{12}- and Δ^{15}-desaturases) and is therefore termed an **essential fatty acid.*** Indeed, animals maintained on a fat-free diet develop an ultimately fatal condition that is initially characterized by poor growth, poor wound healing, and dermatitis. Linoleic acid is also an important constituent of epidermal sphingolipids that function as the skin's water-permeability barrier.

Because of the inability of animal desaturases to add double bonds to positions beyond C9, another essential fatty acid is **α-linolenic acid [ALA; $\Delta^{9,12,15}$**-octadecatrienoic acid (18:3n–3, an ω–3 fatty acid)]. This fatty acid is a precursor to **EPA ($\Delta^{5,8,11,14,17}$**-eicosapentaenoic acid; 20:5n–3) and **DHA ($\Delta^{4,7,10,13,16,19}$**-docosahexaenoic acid; 22:6n–3), polyunsaturated ω–3 fatty acids recently found to be important dietary constituents (present in fish oils) that improve cognitive function and vision, and contribute to protection against inflammation and cardiovascular disease. DHA is, among other things, the predominant fatty acid in the phospholipids of retinal rod outer segments. Substitution of DHA with otherwise identical ω–6 fatty acids in phospholipids results in impaired visual acuity. Deficiency of ω–3 polyunsaturated fatty acids in brain phospholipids is associated with memory loss and diminished cognitive function.

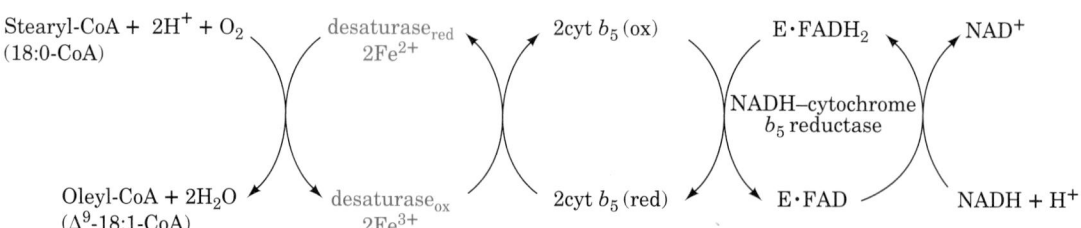

FIGURE 25-38   The electron-transfer reactions mediated by the Δ^9-fatty acyl-CoA desaturase complex. Its three proteins, desaturase, cytochrome b_5, and NADH–cytochrome b_5 reductase, are situated in the endoplasmic reticulum membrane. [After Jeffcoat, R., *Essays Biochem.* **15,** 19 (1979).]

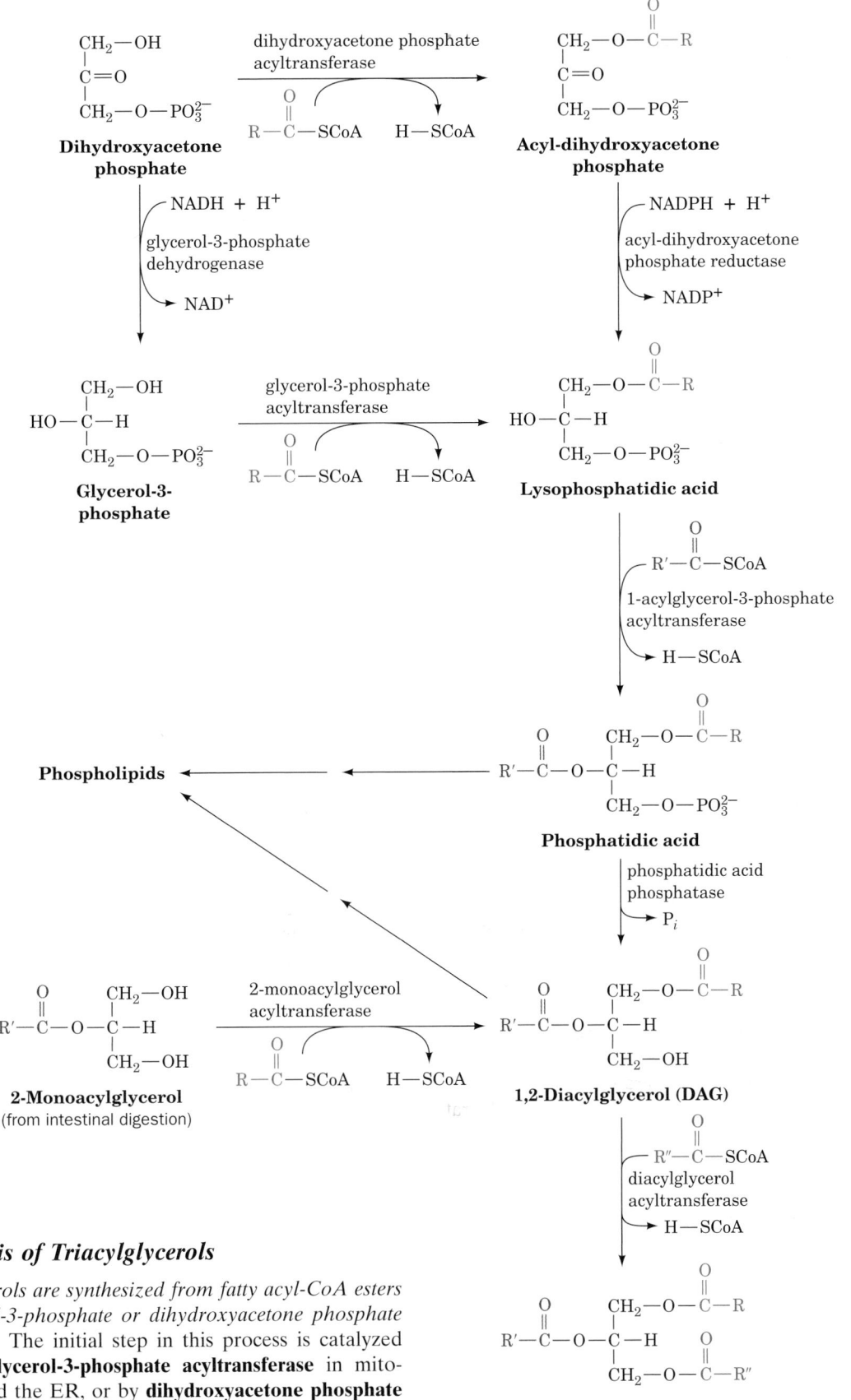

F. Synthesis of Triacylglycerols

Triacylglycerols are synthesized from fatty acyl-CoA esters and glycerol-3-phosphate or dihydroxyacetone phosphate (Fig. 25-39). The initial step in this process is catalyzed either by **glycerol-3-phosphate acyltransferase** in mitochondria and the ER, or by **dihydroxyacetone phosphate acyltransferase** in the ER or peroxisomes. In the latter case, the product **acyl-dihydroxyacetone phosphate** is reduced to the corresponding **lysophosphatidic acid** by an

FIGURE 25-39   The reactions of triacylglycerol biosynthesis.

NADPH-dependent reductase. The lysophosphatidic acid is converted to a triacylglycerol by the successive actions of **1-acylglycerol-3-phosphate acyltransferase, phosphatidic acid phosphatase,** and **diacylglycerol acyltransferase.** The intermediate phosphatidic acid and 1,2-diacylglycerol (DAG) can also be converted to phospholipids by the pathways described in Section 25-8. The acyltransferases are not completely specific for particular fatty acyl-CoAs, either in chain length or in degree of unsaturation, but in triacylglycerols of human adipose tissue, palmitate tends to be concentrated at position 1 and oleate at position 2.

a. Glyceroneogenesis is Important for Triacylglycerol Biosynthesis

The dihydroxyacetone phosphate used to make glycerol-3-phosphate for triacylglycerol synthesis comes either from glucose via the glycolytic pathway (Fig. 17-3) or from oxaloacetate via an abbreviated version of gluconeogenesis (Fig. 23-7) termed **glyceroneogenesis.** Glyceroneogenesis is necessary in times of starvation, since approximately 30% of the fatty acids that enter the liver during a fast are reesterified to triacylglycerol and exported as VLDL (Section 25-1 and 25-6A). Adipocytes also carry out glyceroneogenesis in times of starvation. They do not carry out gluconeogenesis, but contain the gluconeogenic enzyme phosphoenolpyruvate carboxykinase (PEPCK), which is upregulated when glucose concentration is low, and participates in the glyceroneogenesis required for triacylglycerol biosynthesis.

5 ■ REGULATION OF FATTY ACID METABOLISM

Discussions of metabolic control are usually concerned with the regulation of metabolite flow through a pathway in response to the differing energy needs and dietary states of an organism. For example, the difference in the energy requirement of muscle between rest and vigorous exertion may be as much as 100-fold. Such varying demands may be placed on the body when it is in either a fed or a fasted state. For instance, Eric Newsholme, an authority on the biochemistry of exercise, enjoys a 2-hour run before breakfast. Others might wish for no greater exertion than the motion of hand to mouth. In both individuals, glycogen and triacylglycerols serve as primary fuels for energy-requiring processes and are synthesized in times of quiet plenty for future use.

Hormones Regulate Fatty Acid Metabolism

Synthesis and breakdown of glycogen and triacylglycerols, as detailed in Chapter 18 and above, are processes that concern the whole organism, with its organs and tissues forming an interdependent network connected by the bloodstream. The blood carries the metabolites responsible for energy production: triacylglycerols in the form of chylomicrons and VLDL (Section 12-5A), fatty acids as their albumin complexes (Section 25-1), ketone bodies, amino acids, lactate, and glucose. The pancreatic α and β cells sense the organism's dietary and energetic state mainly through the glucose concentration in the blood. The α cells respond to the low blood glucose concentration of the fasting and energy-demanding states by secreting glucagon. The β cells respond to the high blood glucose concentration of the fed and resting states by secreting insulin. We have previously discussed (Sections 18-3E and 18-3F) how these hormones are involved in glycogen metabolism. *They also regulate the rates of the opposing pathways of lipid metabolism and therefore control whether fatty acids will be oxidized or synthesized.* Their targets are the regulatory (flux-generating) enzymes of fatty acid synthesis and breakdown in specific tissues (Fig. 25-40).

We are already familiar with most of the mechanisms by which the catalytic activities of regulatory enzymes may be controlled: substrate availability, allosteric interactions, and covalent modification (phosphorylation). These are examples of **short-term regulation,** regulation that occurs with a response time of minutes or less. *Fatty acid synthesis is controlled, in part, by short-term regulation.* Acetyl-CoA carboxylase, which catalyzes the first committed step of this pathway, is inhibited by palmitoyl-CoA and by the glucagon-stimulated cAMP-dependent increase in phosphorylation, and is activated by citrate and by insulin-stimulated dephosphorylation (Section 25-4B).

Another mechanism exists for controlling a pathway's regulatory enzymes: alteration of the amount of enzyme present by changes in the rates of protein synthesis and/or breakdown. This process requires hours or days and is therefore called **long-term regulation** (the control of protein synthesis and breakdown is discussed in Chapters 31 and 32). *Lipid biosynthesis is also controlled by long-term regulation,* with insulin stimulating and starvation inhibiting the synthesis of acetyl-CoA carboxylase and fatty acid synthase. The presence in the diet of polyunsaturated fatty acids also decreases the concentrations of these enzymes. The amount of adipose tissue lipoprotein lipase, the enzyme that initiates the entry of lipoprotein-packaged fatty acids into adipose tissue for storage (Section 12-5B), is also increased by insulin and decreased by starvation. In contrast, the concentration of heart lipoprotein lipase, which controls the entry of fatty acids from lipoproteins into heart tissue for oxidation rather than storage, is decreased by insulin and increased by starvation. *Starvation and/or regular exercise, by decreasing the glucose concentration in the blood, change the body's hormone balance. This situation results in long-term changes in gene expression that increase the levels of fatty acid oxidation enzymes and decrease those of lipid biosynthesis.*

Fatty acid oxidation is regulated largely by the concentration of fatty acids in the blood, which is, in turn, controlled by the hydrolysis rate of triacylglycerols in adipose tissue by **hormone-sensitive triacylglycerol lipase.** This enzyme is so named because it is susceptible to regulation by phosphorylation and dephosphorylation in response to hormonally controlled cAMP levels. Epinephrine and norepinephrine, as does glucagon, act to increase adipose tissue cAMP concentrations. cAMP allosterically activates

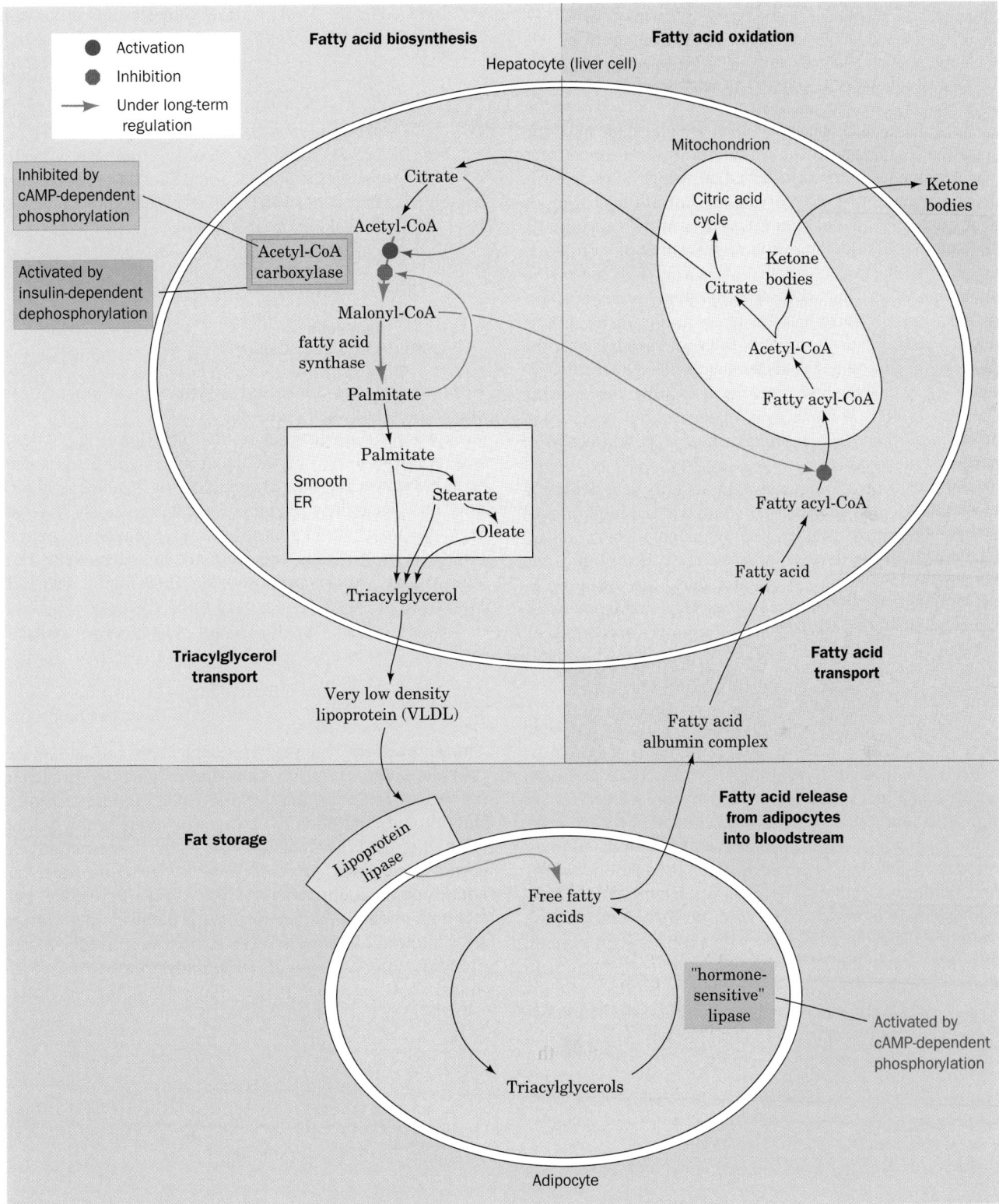

FIGURE 25-40   Sites of regulation of fatty acid metabolism.

protein kinase A (PKA) which, in turn, increases the phosphorylation levels of susceptible enzymes. Phosphorylation activates hormone-sensitive triacylglycerol lipase, thereby stimulating lipolysis in adipose tissue, raising blood fatty acid levels, and ultimately activating the β-oxidation path-

way in other tissues such as liver and muscle. In liver, this process leads to the production of ketone bodies that are secreted into the bloodstream for use by peripheral tissues as an alternative fuel to glucose. PKA, acting in concert with AMP-dependent protein kinase (AMPK), also causes the inactivation of acetyl-CoA carboxylase (Section 25-4B), one of the rate-determining enzymes of fatty acid synthesis, so that *cAMP-dependent phosphorylation simultaneously stimulates fatty acid oxidation and inhibits fatty acid synthesis.*

Insulin has the opposite effect of glucagon and epinephrine: It stimulates the formation of glycogen and triacylglycerols. This protein hormone, which is secreted in response to high blood glucose concentrations, triggers a highly complex signal transduction network (Section 19-4F) that induces the long-term regulation of numerous enzymes as well as decreasing cAMP levels. This latter situation leads to the dephosphorylation and thus the inactivation of hormone-sensitive triacylglycerol lipase, thereby reducing the amount of fatty acid available for oxidation. Insulin also stimulates the dephosphorylation of acetyl-CoA carboxylase, thereby activating this enzyme (Section 25-4B). *The glucagon–insulin ratio is therefore of prime importance in determining the rate and direction of fatty acid metabolism.*

Another control point that inhibits fatty acid oxidation when fatty acid synthesis is stimulated is the inhibition of carnitine palmitoyltransferase I by malonyl-CoA. This inhibition keeps the newly synthesized fatty acids out of the mitochondrion (Section 25-1) and thus away from the β-oxidation system. In fact, heart muscle, an oxidative tissue that does not carry out fatty acid biosynthesis, contains an isoform of acetyl-CoA carboxylase, β-ACC, whose sole function appears to be the synthesis of malonyl-CoA to regulate fatty acid oxidation.

AMPK may itself be an important regulator of fatty acid metabolism. This phosphorylating enzyme is activated by AMP and inhibited by ATP and thus has been proposed to serve as a fuel gauge for the cell. When ATP levels are high, signaling the fed and rested state, this kinase is inhibited, allowing ACC to become dephosphorylated (activated) so as to stimulate malonyl-CoA production for fatty acid synthesis in adipose tissue and for inhibition of fatty acid oxidation in muscle cells. When activity levels increase causing ATP levels to decrease with a concomitant increase in AMP levels, AMPK is activated to phosphorylate (inactivate) ACC. The resulting decrease in malonyl-CoA levels causes fatty acid biosynthesis to decrease in adipose tissue while fatty acid oxidation increases in muscle to provide the ATP for continued activity.

6 ■ CHOLESTEROL METABOLISM

Cholesterol is a vital constituent of cell membranes and the precursor of steroid hormones and bile salts. It is clearly essential to life, yet its deposition in arteries has been associated with cardiovascular disease and stroke, two leading causes of death in humans. In a healthy organism, an intricate balance is maintained between the biosynthesis, utilization, and transport of cholesterol, keeping its harmful deposition to a minimum. In this section, we study the pathways of cholesterol biosynthesis and transport and how they are

controlled. We also examine how cholesterol is utilized in the biosynthesis of steroid hormones and bile salts.

A. *Cholesterol Biosynthesis*

All of the carbon atoms of cholesterol are derived from acetate (Fig. 25-41). Observation of their pattern of incorporation into cholesterol led Konrad Bloch to propose that acetate was first converted to **isoprene units,** C_5 units that have the carbon skeleton of **isoprene:**

Isoprene
(2-Methyl-1,3-butadiene)

An isoprene unit

Isoprene units are condensed to form a linear precursor to cholesterol, and then cyclized.

Squalene, a polyisoprenoid hydrocarbon (Fig. 25-42*a*), was demonstrated to be the linear intermediate in cholesterol biosynthesis by the observation that feeding isotopically labeled squalene to animals yields labeled cholesterol. Squalene may be folded in several ways that would enable it to cyclize to the four-ring sterol nucleus (Section 12-1E). The folding pattern proposed by Bloch and Robert B. Woodward (Fig. 25-42*b*) proved to be correct.

Bloch's outline for the major stages of cholesterol biosynthesis was

Acetate → isoprenoid intermediate → squalene →
cyclization product → cholesterol

This pathway has been experimentally verified and its details elaborated. It is now known to be part of a branched pathway (Fig. 25-43) that produces several other essential isoprenoids in addition to cholesterol, namely, ubiquinone (CoQ; Fig. 22-17*b*), dolichol (Fig. 23-14), farnesylated and geranylgeranylated proteins (Fig. 12-29), and **isopentenyladenosine** (a modified base of tRNA; Fig. 32-10). We shall examine in detail the portion of this pathway that synthesizes cholesterol. Note, however, that at least 25,000 isoprenoids (also known as **terpenoids**), mostly of plant, fungal, and bacterial origin, have been characterized.

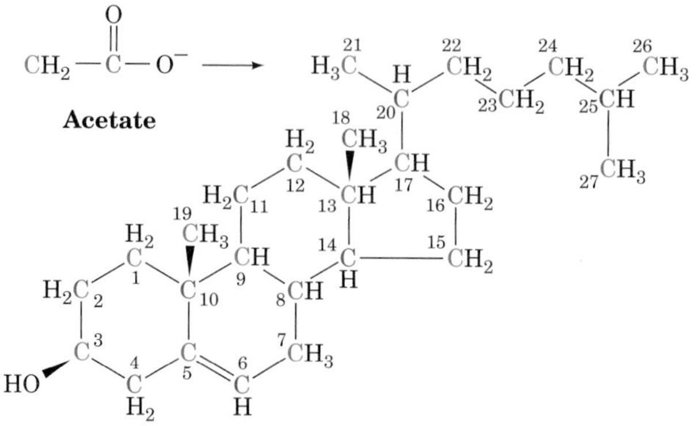

FIGURE 25-41   All of cholesterol's carbon atoms are derived from acetate.

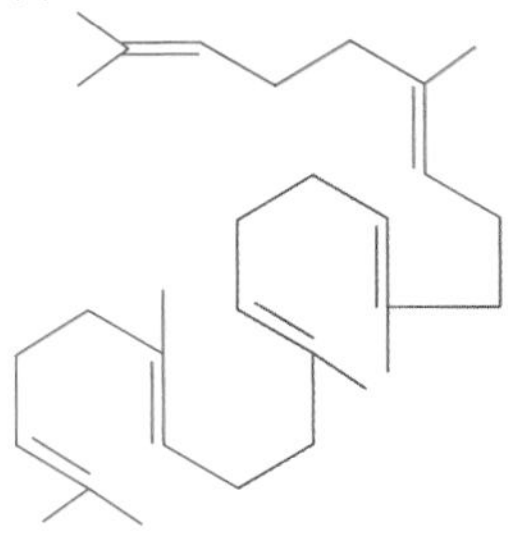

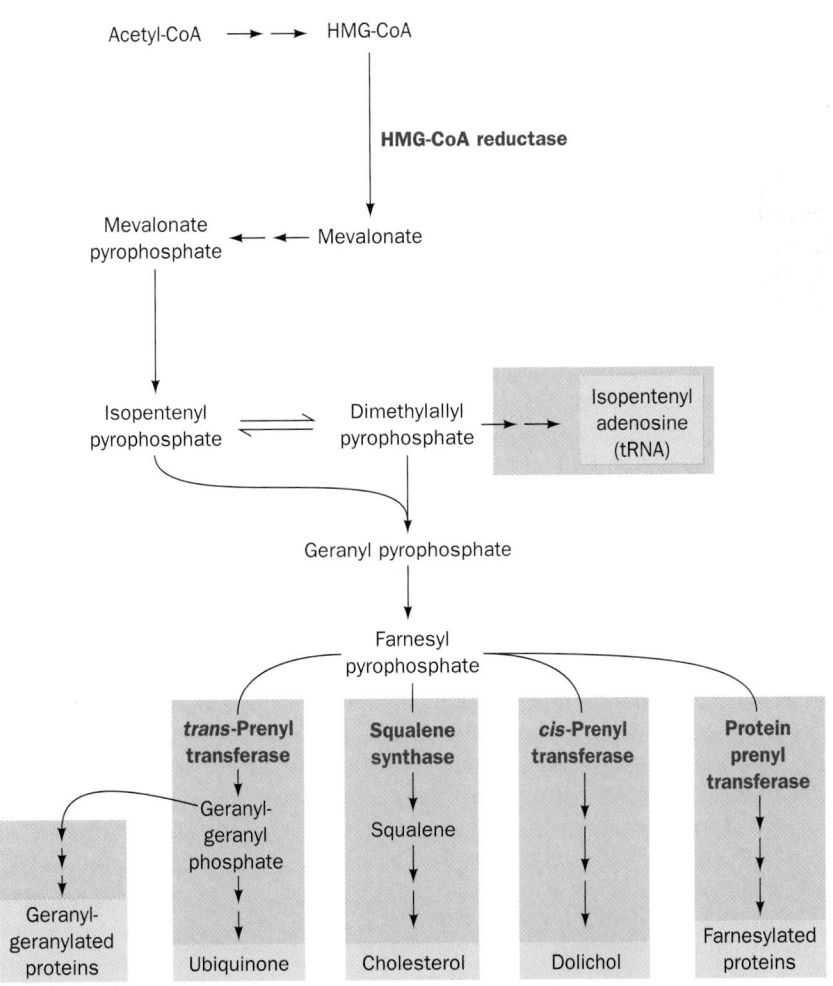

FIGURE 25-42 Squalene. (*a*) Extended conformation. Each box contains one isoprene unit. (*b*) Folded in preparation for cyclization as predicted by Bloch and Woodward.

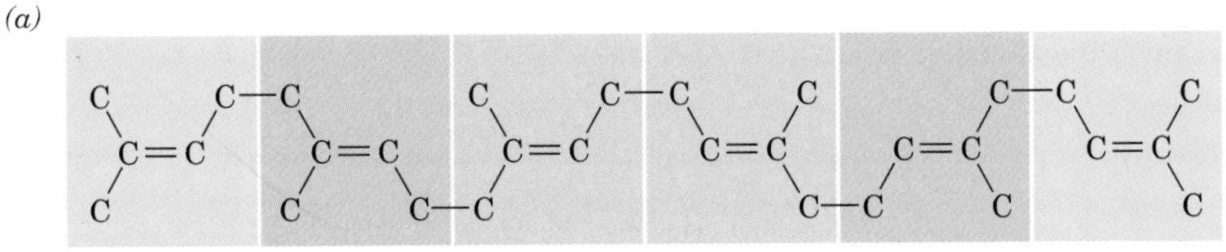

These serve as membrane constituents (e.g., cholesterol), hormones (steroids), pheromones, defensive agents, photoprotective agents (e.g., β-carotene; Section 24-2A), and visual pigments (e.g., retinal; Section 12-3A), to name only a few of their many biological functions.

a. HMG-CoA Is a Key Cholesterol Precursor

Acetyl-CoA is converted to isoprene units by a series of reactions that begins with formation of hydroxymethyl-

glutaryl-CoA (HMG-CoA; Fig. 25-25), a compound we previously encountered as an intermediate in ketone body biosynthesis (Section 25-3). HMG-CoA synthesis requires the participation of two enzymes: thiolase and HMG-CoA synthase. The enzymes forming the HMG-CoA leading to ketone bodies occur in the mitochondria, whereas those responsible for the synthesis of the HMG-CoA that is destined for cholesterol biosynthesis are located in the cytosol. Their catalytic mechanisms, however, are identical.

FIGURE 25-43 The branched pathway of isoprenoid metabolism in mammalian cells. The pathway produces ubiquinone, dolichol, farnesylated and geranylgeranylated proteins, and isopentenyl adenosine, a modified tRNA base, in addition to cholesterol.

HMG-CoA is the precursor of two isoprenoid intermediates, **isopentenyl pyrophosphate** and **dimethylallyl pyrophosphate**:

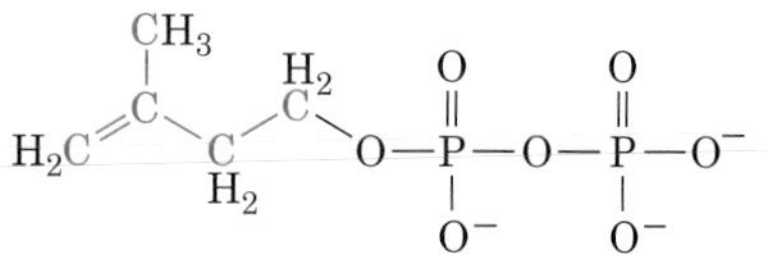

Isopentenyl pyrophosphate

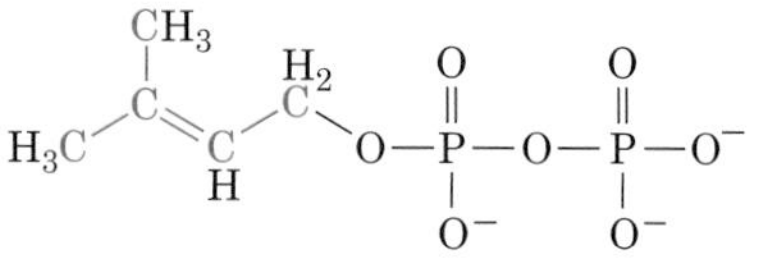

Dimethylallyl pyrophosphate

The formation of isopentenyl pyrophosphate involves four reactions (Fig. 25-44):

1. The CoA thioester group of HMG-CoA is reduced to an alcohol in an NADPH-dependent four-electron reduction catalyzed by **HMG-CoA reductase**, yielding **mevalonate**.

2. The new OH group is phosphorylated by **mevalonate-5-phosphotransferase**.

3. The phosphate group is converted to a pyrophosphate by **phosphomevalonate kinase**.

4. The molecule is decarboxylated and the resulting alcohol dehydrated by **pyrophosphomevalonate decarboxylase**.

HMG-CoA reductase mediates the rate-determining step of cholesterol biosynthesis and is the most elaborately regulated enzyme of this pathway. This 888-residue ER membrane-bound enzyme is regulated, as we shall see in Section 25-6B, by competitive and allosteric mechanisms, phosphorylation/dephosphorylation, and long-term regulation. Cholesterol itself is an important feedback regulator of the enzyme.

b. Pyrophosphomevalonate Decarboxylase Catalyzes an Apparently Concerted Reaction

5-Pyrophosphomevalonate is converted to isopentenyl pyrophosphate by an ATP-dependent dehydration–decarboxylation reaction catalyzed by **pyrophosphomevalonate decarboxylase** (Fig. 25-45). When [3-^{18}O]-5-pyrophosphomevalonate (*O in Fig. 25-45) is used as a substrate, the labeled oxygen appears in P_i. This observation suggests that 3-phospho-5-pyrophosphomevalonate is a reaction intermediate. Since all attempts to isolate this intermediate have failed, however, it has been proposed that phosphorylation, the α,β elimination of CO_2, and the elimination of P_i occur in a concerted reaction.

The equilibration between isopentenyl pyrophosphate and dimethylallyl pyrophosphate is catalyzed by **isopentenyl pyrophosphate isomerase**. The reaction is thought to occur via a protonation/deprotonation reaction involving

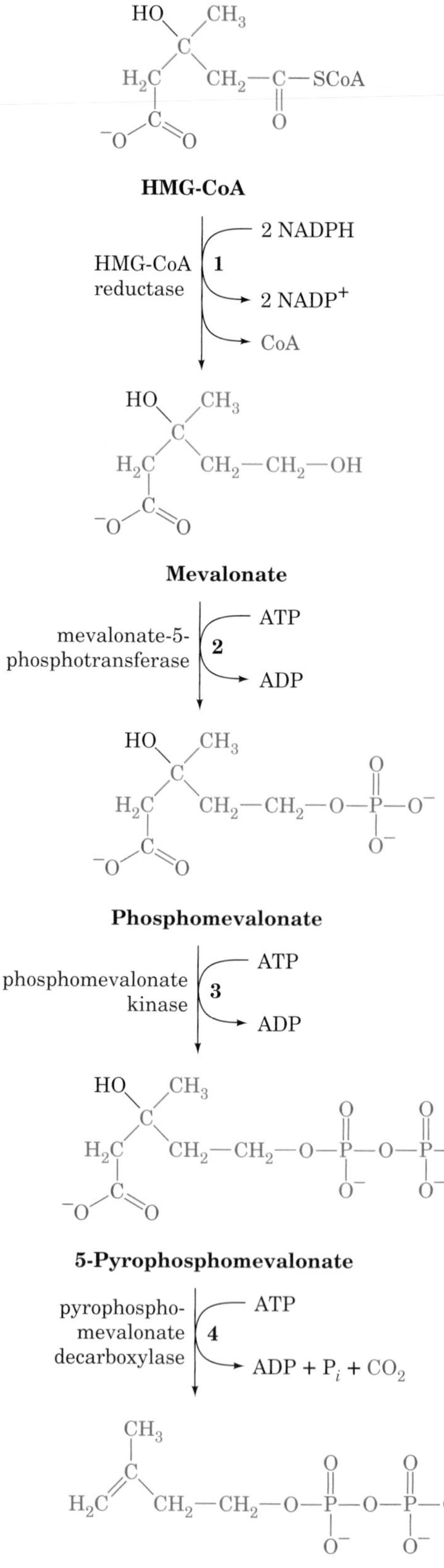

FIGURE 25-44   Formation of isopentenyl pyrophosphate from HMG-CoA.

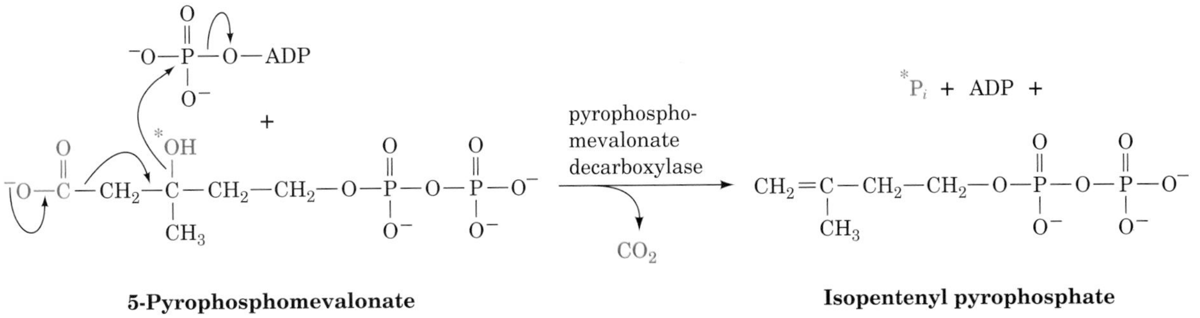

FIGURE 25-45   Action of pyrophosphomevalonate decarboxylase. The enzyme catalyzes an ATP-dependent concerted dehydration–decarboxylation of pyrophosphomevalonate, yielding isopentenyl pyrophosphate.

the intermediacy of a tertiary carbocation intermediate. Cys and Glu residues have been implicated as the general acid and base catalysts, respectively (Fig. 25-46), as supported

by site-directed mutagenesis and the X-ray structure of the enzyme. The carbocation is thought to be stabilized through interactions with the aromatic π cloud of an adjacent Trp residue. Aromatic residues provide electron-rich interactions with positively charged groups without forming covalent bonds that would destroy the intermediate.

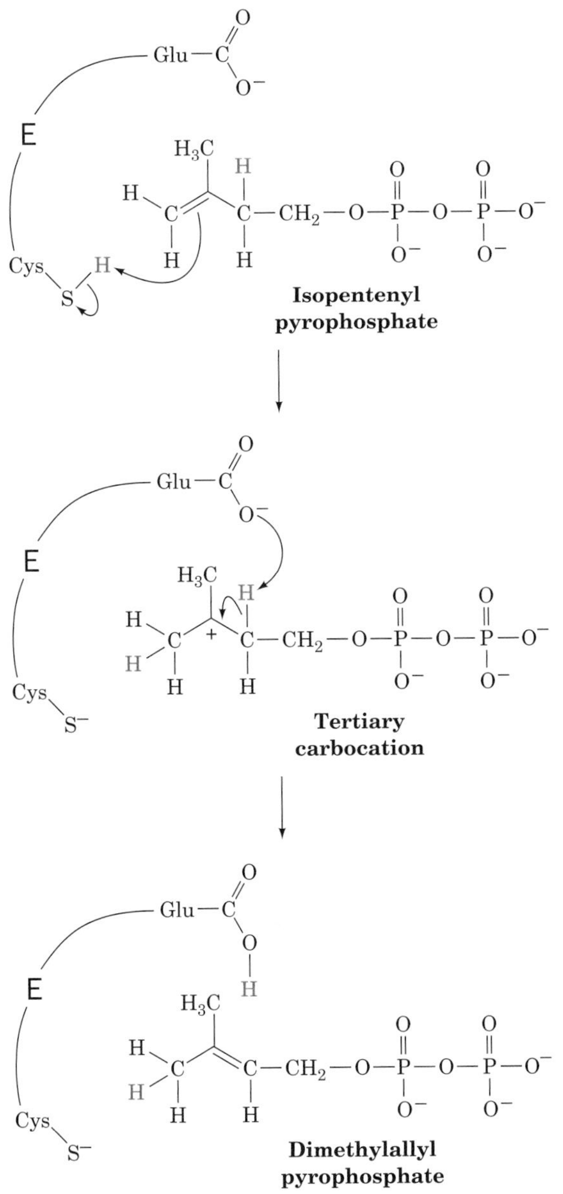

c. Squalene Is Formed by the Condensation of Six Isoprene Units

Four isopentenyl pyrophosphates and two dimethylallyl pyrophosphates condense to form the C_{30} cholesterol precursor squalene in three reactions catalyzed by two enzymes (Fig. 25-47):

1. Prenyltransferase (farnesyl pyrophosphate synthase) catalyzes the head-to-tail (1′–4) condensation of dimethylallyl pyrophosphate and isopentenyl pyrophosphate to yield **geranyl pyrophosphate.**

2. Prenyltransferase catalyzes a second head-to-tail condensation of geranyl pyrophosphate and isopentenyl pyrophosphate to yield **farnesyl pyrophosphate (FPP).**

3. Squalene synthase (SQS) then catalyzes the head-to-head (1–1′) condensation of two farnesyl pyrophosphate molecules to form squalene. Farnesyl pyrophosphate is also a precursor to dolichol, farnesylated and geranylgeranylated proteins, and ubiquinone (Fig. 25-43).

Prenyltransferase catalyzes the condensation of isopentenyl pyrophosphate with an allylic (conjugated to a C=C double bond) pyrophosphate. It is specific for isopentenyl pyrophosphate but can use either the 5-carbon dimethylallyl pyrophosphate or the 10-carbon **geranyl pyrophosphate** as its allylic substrate. The prenyltransferase–catalyzed condensation mechanism is particularly interesting

FIGURE 25-46   Mechanism of isopentenyl pyrophosphate isomerase. The enzyme interconverts isopentenyl pyrophosphate and dimethylallyl pyrophosphate by a protonation/deprotonation reaction involving a carbocation intermediate in which a Cys and a Glu residue act as a proton donor and acceptor. The carbocation intermediate is thought to be stabilized by π interactions with a nearby Trp side chain.

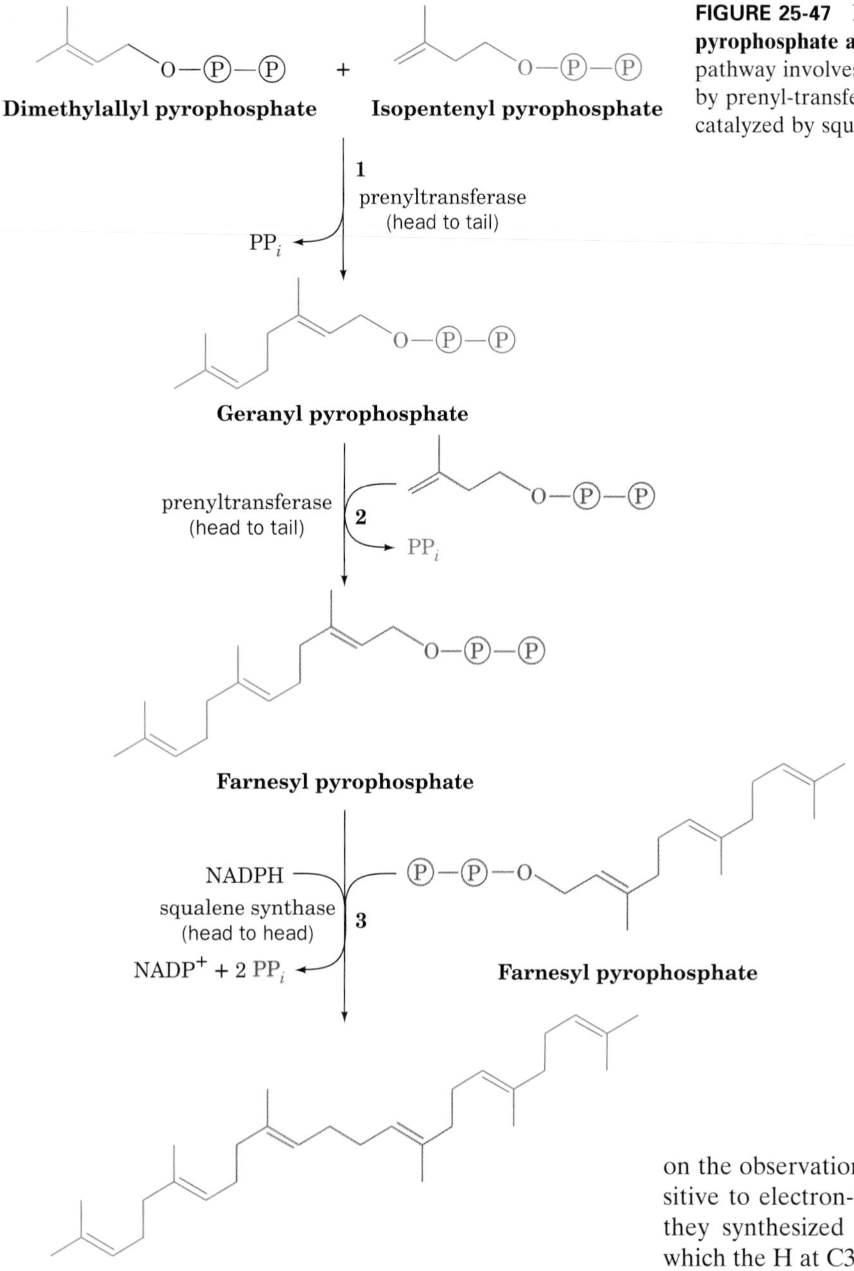

FIGURE 25-47   Formation of squalene from isopentenyl pyrophosphate and dimethylallyl pyrophosphate. The pathway involves two head-to-tail condensations catalyzed by prenyl-transferase and a head-to-head condensation catalyzed by squalene synthase.

since it is one of the few known enzyme-catalyzed reactions that proceed via a carbocation intermediate. Two possible condensation mechanisms can be envisioned (Fig. 25-48):

Scheme I   An S_N1 mechanism in which an allylic carbocation forms by the elimination of PP_i. Isopentenyl pyrophosphate then condenses with this carbocation, forming a new carbocation that eliminates a proton to form product.

Scheme II   An S_N2 reaction in which the allylic PP_i is displaced in a concerted manner. In this case, an enzyme nucleophile, X, assists in the reaction. This group is eliminated in the second step with the loss of a proton to form product.

Dale Poulter and Hans Rilling used chemical logic to differentiate between these two mechanisms. Capitalizing

on the observation that S_N1 reactions are much more sensitive to electron-withdrawing groups than S_N2 reactions, they synthesized a geranyl pyrophosphate derivative in which the H at C3 is replaced by the electron-withdrawing group F. This allylic substrate for the second $(1'-4)$ condensation catalyzed by prenyltransferase, not surprisingly, has the same K_M as the natural substrate (F and H have similar atomic radii):

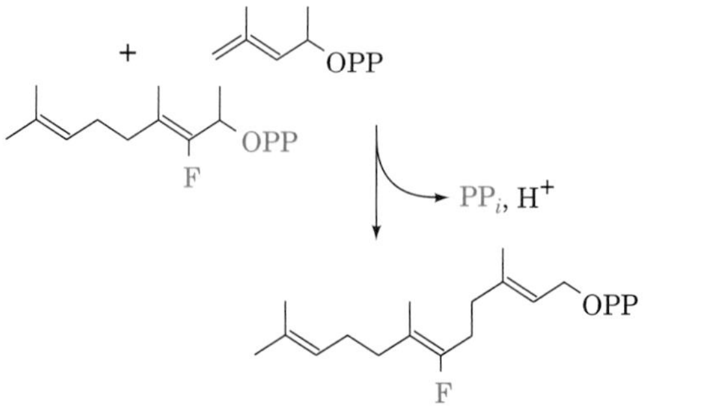

It is, however, the V_{max} of this reaction that tells the story. If the reaction is an S_N2 displacement, the fluoro

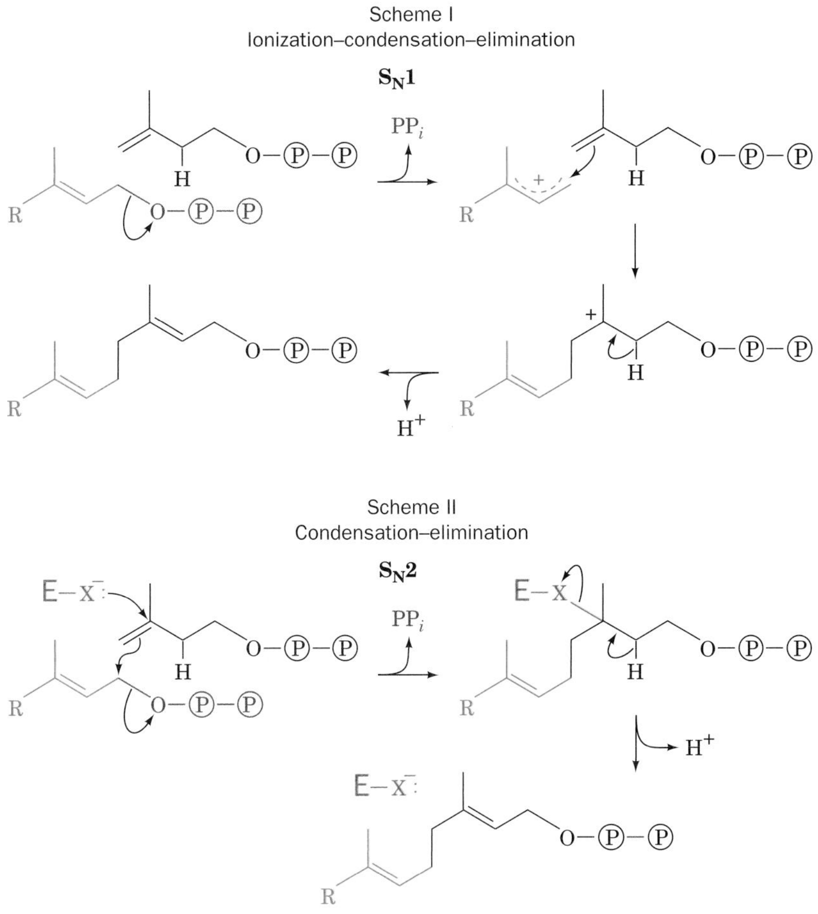

Scheme I
Ionization–condensation–elimination

S$_N$1

FIGURE 25-48 Two possible mechanisms for the prenyltransferase reaction. Scheme I involves the formation of a carbocation intermediate, whereas Scheme II involves the participation of an enzyme nucleophile, X.

Scheme II
Condensation–elimination

S$_N$2

derivative should react at a rate similar to that of the natural substrate. If, instead, the reaction has an S$_N$1 mechanism, the fluoro derivative should react orders of magnitude more slowly than the natural substrate. In fact, 3-fluorogeranyl pyrophosphate forms product at <1% of the rate of the natural substrate, strongly supporting an S$_N$1 mechanism with a carbocation intermediate.

Carbocations are now known to participate in several reactions of isoprenoid biosynthesis. The enzymes are classified according to how they generate these carbocations. Class I enzymes do so via the release of pyrophosphate, as we have seen for prenyltransferase. Class II enzymes do so by protonating a double bond, as does isopentenyl pyrophosphate isomerase (Fig. 25-46), or an epoxide, as we shall see below for oxidosqualene cyclase.

Squalene, the immediate sterol precursor, is formed by the head-to-head condensation of two FPP molecules by SQS. Although the enzyme is a Class I enzyme that is structurally related to prenyltransferase and generates carbocations by the release of pyrophosphate, the reaction is not a simple head-to-tail condensation, as might be expected, but, rather, proceeds via a complex two-step mechanism with each step catalyzed by different active site on the enzyme (Fig. 25-49):

Step I The reaction of two FPP molecules to yield the stable intermediate **presqualene pyrophosphate**. This reaction is initiated by the elimination of PP$_i$ from one

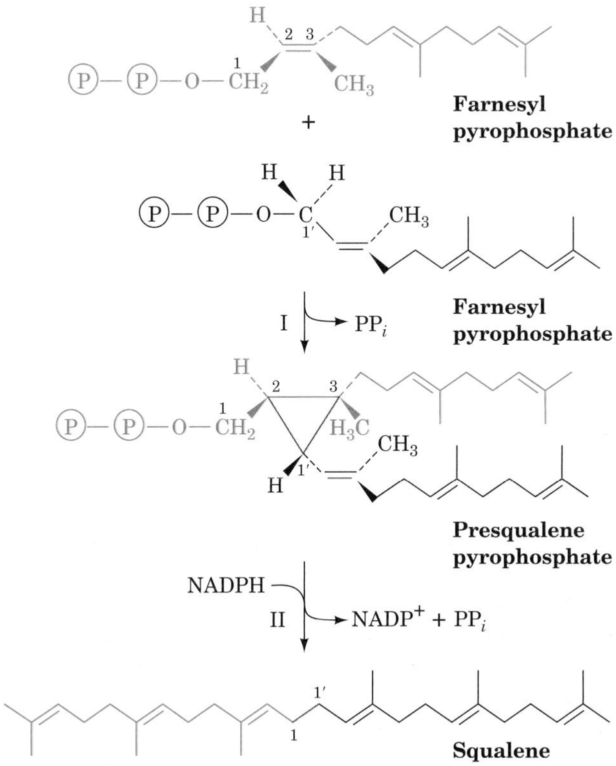

FIGURE 25-49 Action of squalene synthase. The enzyme catalyzes the head-to-head condensation of two farnesyl pyrophosphate molecules to form squalene.

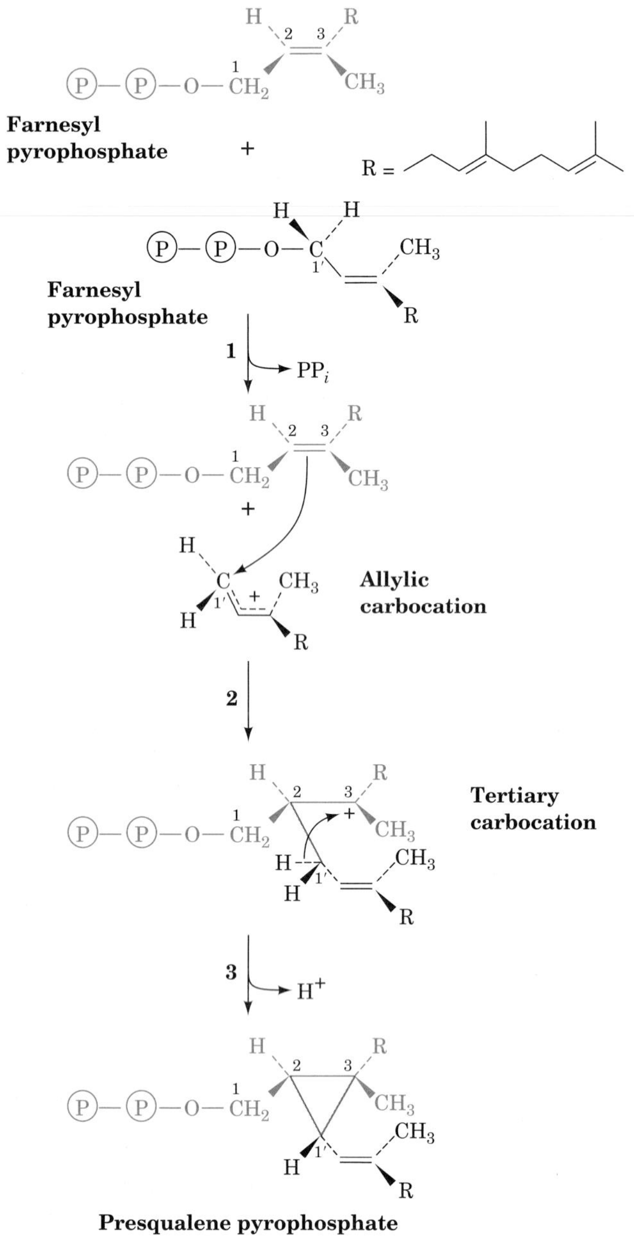

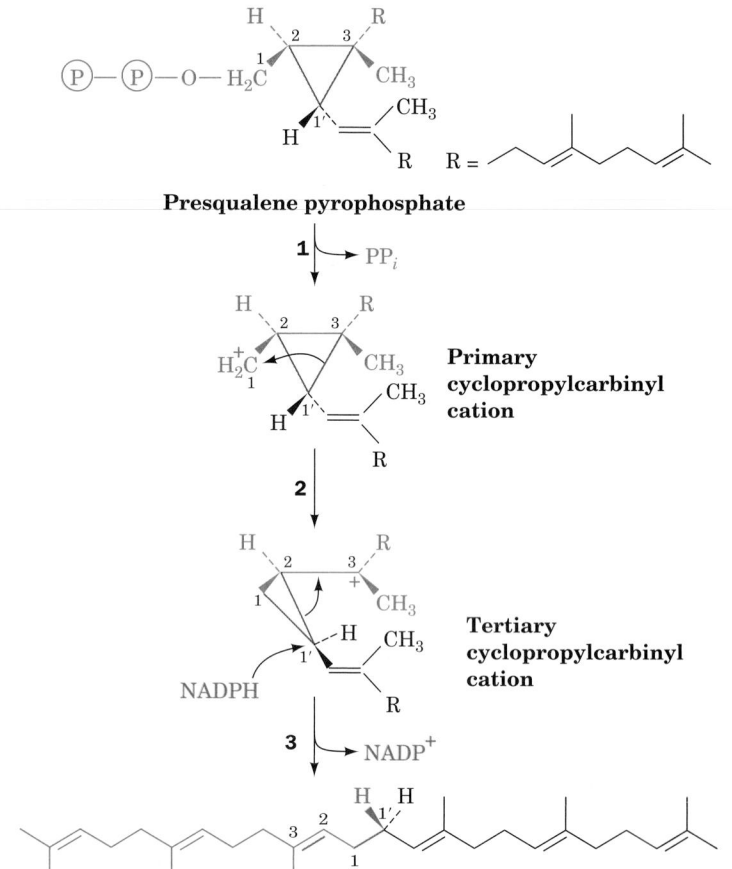

FIGURE 25-51   Mechanism of rearrangement and reduction of presqualene pyrophosphate to squalene as catalyzed by squalene synthase (Fig. 25-49, Step II). (1) Presqualene's pyrophosphate group leaves, yielding a primary carbocation at C1. **(2)** The electrons forming the C1′—C3 bond migrate to C1, forming squalene's C1—C1′ bond and a tertiary carbocation at C3. **(3)** The process is completed by the addition of an NADPH-supplied hydride ion to C1′ and the formation of the C2═C3 double bond.

FIGURE 25-50   Proposed mechanism for the formation of presqualene pyrophosphate from two farnesyl pyrophosphate molecules by squalene synthase (Fig. 25-49, Step I). (1) The pyrophosphate group on one farnesyl pyrophosphate leaves, yielding an allylic carbocation. This reaction step is facilitated by proton donation from the side chain of an essential Tyr residue, which then stabilizes the allylic cation via π–cation interactions. **(2)** The C2═C3 double bond of the second farnesyl pyrophosphate nucleophilically attacks the allylic carbocation to form a tertiary carbocation at C3. **(3)** The abstraction of the pro-*S* proton at C1′ by the phenolate group of the essential Tyr residue results in the formation of a C1′—C3 bond yielding presqualene pyrophosphate.

farnesyl pyrophosphate molecule to form an allylic carbocation at C1 that is stabilized by a π interaction with an essential Tyr residue (Fig. 25-50). The highly reactive

electron-deficient carbocation inserts into the electron-rich C2═C3 double bond of the second molecule, yielding presqualene pyrophosphate, a cyclopropylcarbinyl pyrophosphate.

Step II   The rearrangement and reduction of presqualene pyrophosphate by NADPH to form squalene. This reaction involves the formation and rearrangement of a cyclopropylcarbinyl cation in a complex reaction sequence called a **1′–2–3 process** (Fig. 25-51).

SQS, a monomeric protein, is anchored to the ER membrane via a short C-terminal membrane-spanning domain, with its active site facing the cytosol. This allows it to accept its water-soluble substrates, farnesyl pyrophosphate and NADPH, from the cytosol and release its hydrophobic product, squalene, in the ER membrane.

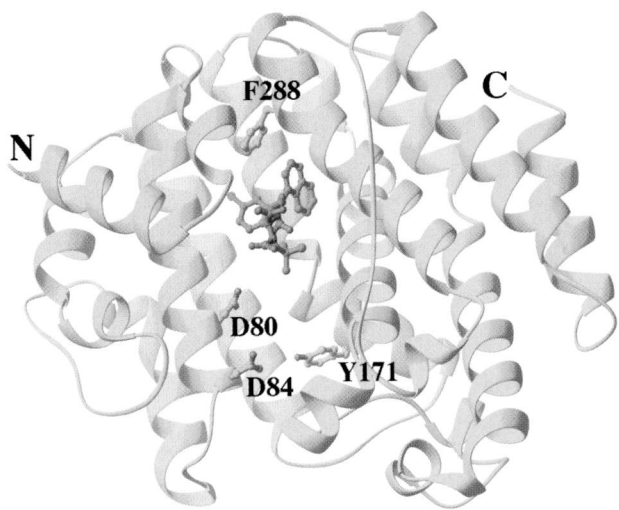

FIGURE 25-52 X-Ray structure of human squalene synthase (SQS) in complex with the inhibitor CP-320473. This inhibitor together with the side chains of D80, D84, Y171, and F288 are drawn in ball-and-stick form colored according to atom type (inhibitor C green, protein C cyan, N blue, O red, and Cl magenta). The protein is viewed looking into its central channel with the putative active sites for Steps I and II of the catalyzed reaction at the bottom and top of the cleft, respectively. The protein's C-terminal transmembrane segment (residues 371–417) together with its N-terminal 30 residues were excised to facilitate its crystallization, which did not affect its *in vitro* catalytic activity. [Based on an X-ray structure by Jayvardhan Pandit, Pfizer Central Research, Groton, Connecticut. PDBid 1EZF.]

The X-ray structure of the 417-residue human SQS, in complex with the inhibitor **CP-320473,**

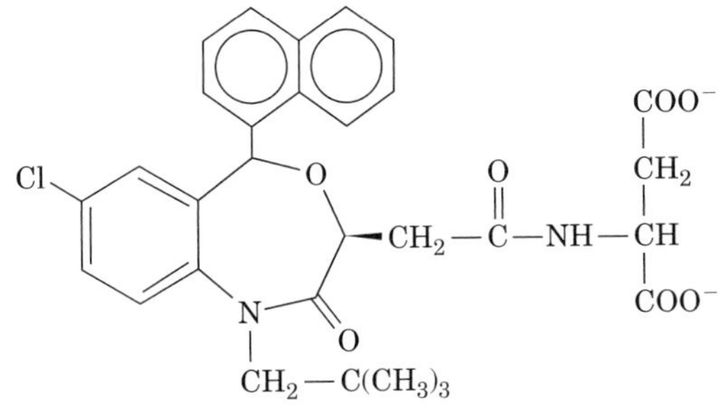

CP-320473

determined by Jayvardhan Pandit, reveals that the protein folds as a single domain with a large channel across one face into which CP-320473 binds (Fig. 25-52). The channel is lined with Asp and Arg residues that mutagenesis studies indicate are involved in FPP binding. Of these, the conserved Asp 80 and Asp 84 are implicated in binding Mg^{2+} ions that ligand an FPP pyrophosphate group. These Asp residues are adjacent to Tyr 171, which forms the base of the channel and which mutagenesis studies have identified

as the essential Tyr that is implicated in stabilizing the allylic carbocation intermediate in Step I of the SQS reaction. Step II of the SQS reaction requires that its highly reactive carbocation intermediates be shielded from contact with the aqueous solvent to prevent it from quenching the reaction. This suggests that for Step II, the presqualene pyrophosphate product of Step I moves deeper into the channel into a pocket that is lined with hydrophobic groups, including Phe 288, whose mutation inactivates the enzyme. This further suggests that Phe 288 functions to stabilize one of the cationic intermediates in Step II (Fig. 25-51) through π–cation interactions.

d. Lanosterol Is Produced by Squalene Cyclization

Squalene, an open-chain C_{30} hydrocarbon, is cyclized to form the tetracyclic steroid skeleton in two steps. **Squalene epoxidase** catalyzes the oxidation of squalene to form **2,3-oxidosqualene** (Fig. 25-53). **Oxidosqualene cyclase (lanosterol synthase)** converts this epoxide to **lanosterol,** the sterol precursor of cholesterol. The reaction is a complex process involving cyclization of 2,3-oxidosqualene to a **protosterol** cation, via a Class II mechanism involving protonation of the epoxide, and rearrangement of this cation

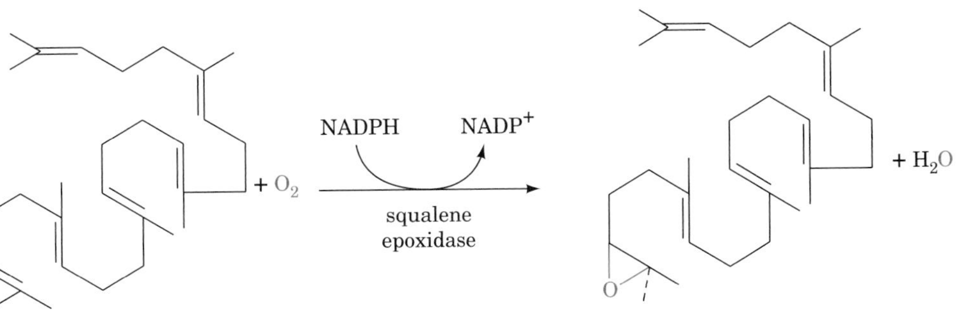

Squalene $+ O_2 \xrightarrow[\text{squalene epoxidase}]{\text{NADPH} \quad \text{NADP}^+}$ **2,3-Oxidosqualene** $+ H_2O$

FIGURE 25-53 The squalene epoxidase reaction.

to lanosterol by a series of 1,2 hydride and methyl shifts (Fig. 25-54).

The interactions of oxidosqualene with lanosterol synthase cause it to fold and react such that it only forms lanosterol. Several aromatic residues are thought to line the active site where they can stabilize intermediate carbocations. The structure of the active site presumably shields the cationic intermediates from premature quenching by either enzyme nucleophiles or water. The importance of the proper placement of residues for the formation of the correct product is demonstrated by site-directed mutagenesis: The conversion of Thr 384 in lanosterol synthase to Tyr causes some misplacement of the C8=C9 double bond, with 11% of the product having a C9=C11 double bond and 10% having a 9-hydroxy group; the double mutation, T384Y/V454I, increases the yield of the C9=C11 double bond to 64%. This, at least in part, explains how different enzymes cause squalene and oxidosqualene to form different products.

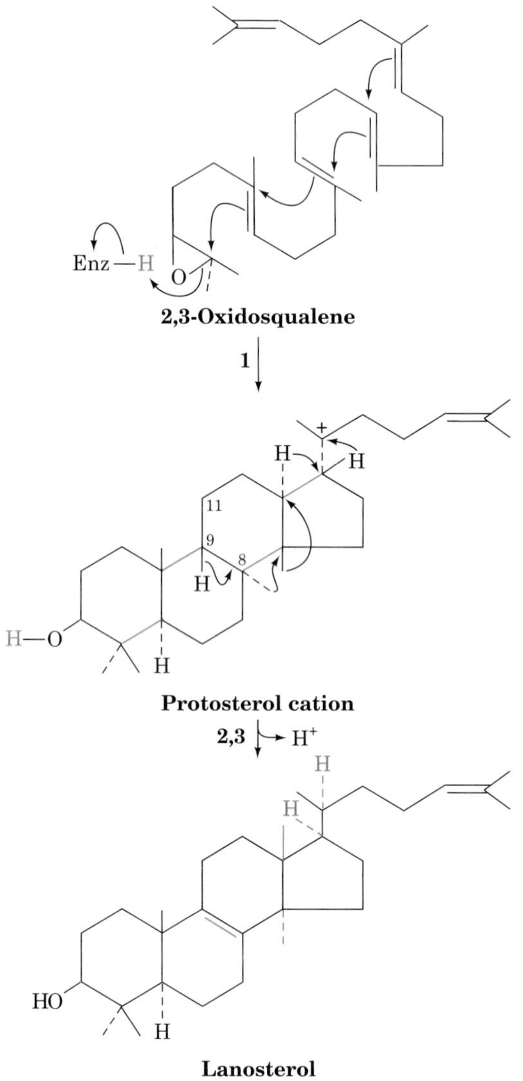

The X-ray structure of oxidosqualene cyclase has not been determined. However, Georg Schulz has determined the X-ray structure of the related **squalene–hopene cyclase** from *Alicyclobacillus acidocalderius,* which catalyzes the conversion of squalene to **hopene**

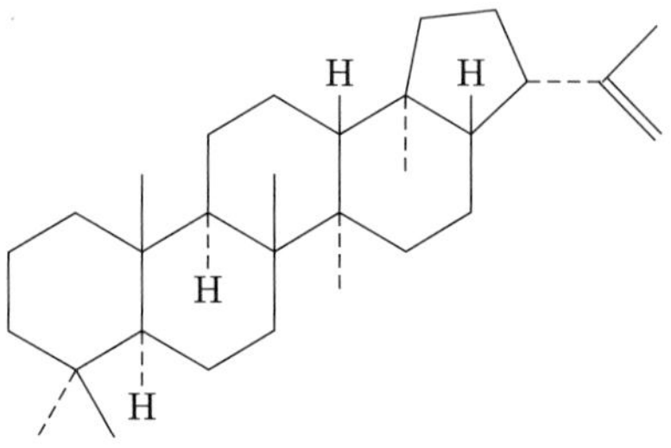

FIGURE 25-54 The oxidosqualene cyclase reaction.
(1) 2,3-Oxidosqualene is cyclized to the protosterol cation in a process that is initiated by the enzyme-mediated protonation of the squalene epoxide oxygen while this extended molecule is folded in the manner predicted by Bloch and Woodward. The opening of the epoxide leaves an electron-deficient center whose migration drives the series of cyclizations that ultimately form the protosterol cation. **(2)** A series of methyl and hydride migrations yields a hypothesized intermediate carbocation at C8, which then eliminates a proton from C9 to form the C8=C9 double bond of lanosterol.

via a cationic cyclization cascade similar to that catalyzed by oxidosqualene cyclase (although initiated by the protonation of a C=C double bond rather than an epoxide). Each 631-residue subunit of this homodimeric monotopic (integral but not transmembrane) protein consists of two structurally similar domains named **α/α barrels** (Fig. 25-55*a*). An α/α barrel consists of two concentric barrels of 6 helices each with the helices of the inner barrel largely parallel to each other and antiparallel to those of the outer barrel (much like an α/β barrel with the β strands of the inner barrel replaced by helices but with only 6 α/α units rather than 8 α/β units). This dumbbell-shaped subunit's active site is located inside a remarkably large central cavity (Fig. 25-55*b*) that presumably induces squalene to fold in such a way that it yields only hopene and which is lined with several conserved aromatic side chains that appear suitably positioned to stabilize the catalyzed reaction's intermediate carbocations. The reaction proceeds via a cationic cascade similar to that in Step 1 of the oxidosqualene cyclase cascade (Fig. 25-54), which is initiated by an enzymatic acid, probably Asp 376 at the top of the cavity, and quenched by a base, most likely a water molecule bound to Tyr 495 at the bottom of the cavity. The active site cavity is accessible from the membrane via a nonpolar channel through the enzyme's membrane-immersed portion.

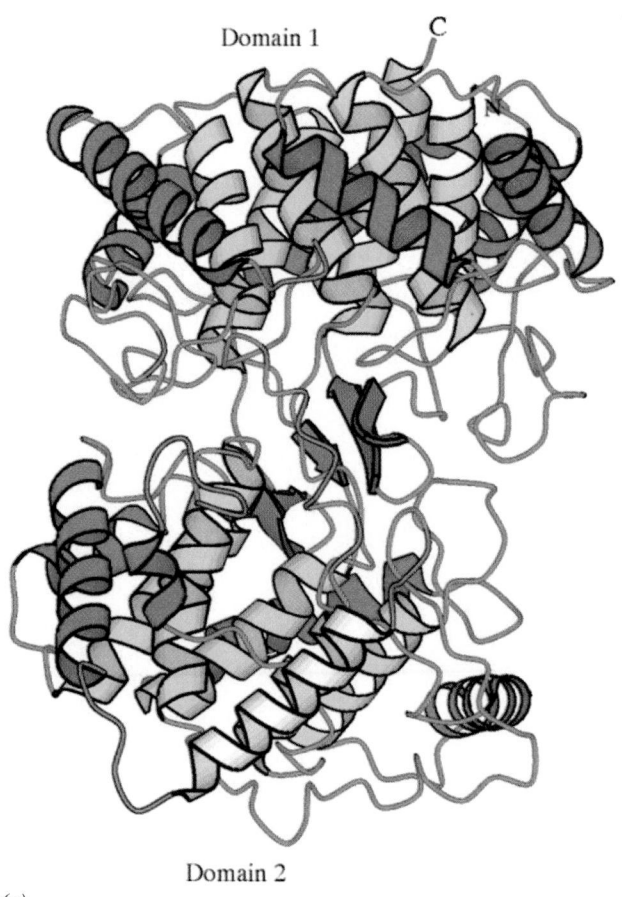

Domain 1

C

N

Domain 2

(a)

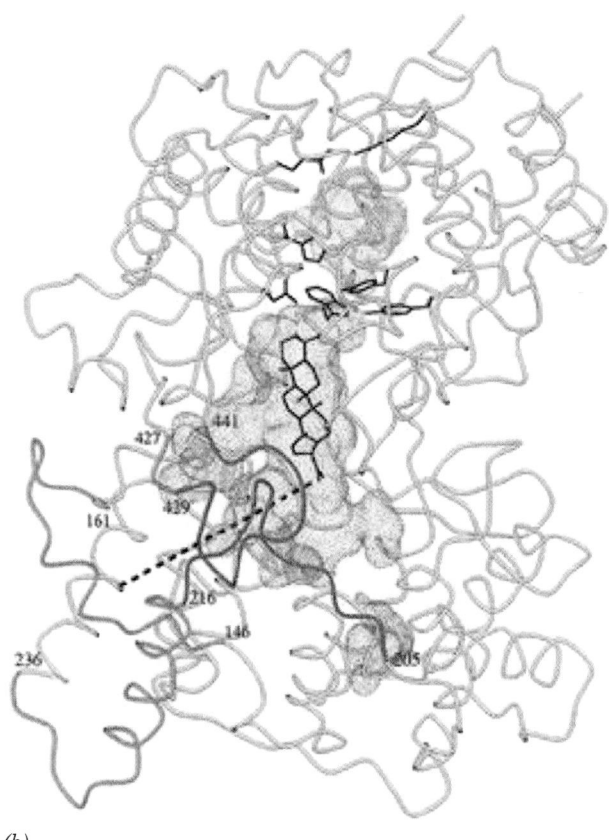

(b)

FIGURE 25-55   X-Ray structure of *A. acidocalderius* squalene–hopene cyclase. (*a*) A single subunit of the homodimeric protein. The inner helices of its two α/α barrels are yellow, its outer helices are red, β strands are green, and the nonpolar membrane-immersed portion is white. (*b*) A C_α chain trace of the subunit viewed similarly to that in Part *a* with its membrane-bound region brown. The three largest cavities in

the protein are shown, with the active site cavity blue, the upper cavity red, and the lower cavity green. Hopene, drawn in stick form, has been modeled into the active site cavity. The channel connecting the active site cavity to the membrane is indicated by the dashed line. [Courtesy of Georg Schulz, Institut für Organische Chemie und Biochemie, Freiburg im Breisgau, Germany. PDBids (*a*) 1SQC and (*b*) 2SQC.]

e. Cholesterol Is Synthesized from Lanosterol

Conversion of lanosterol to cholesterol (Fig. 25-56) is a 19-step process that we shall not explore in detail. It involves an oxidation and loss of three methyl groups. The first methyl group is removed as formate and the other two are eliminated as CO_2 in reactions that all require NADPH and O_2. The enzymes involved in this process are embedded in the ER membrane.

f. Cholesterol Is Transported in the Blood and Taken Up by Cells in Lipoprotein Complexes

Transport and cellular uptake of cholesterol are described in Section 12-5. To recapitulate, cholesterol synthesized by the liver is either converted to bile salts for use in the digestive process (Section 25-1) or esterified by **acyl-CoA:cholesterol acyltransferase (ACAT)** to form **cholesteryl esters**

Cholesteryl ester

which are secreted into the bloodstream as part of the lipoprotein complexes called **very low density lipoproteins (VLDL).** As the VLDL circulate, their component triacylglycerols and most types of their **apolipoproteins** (Table 12-6) are removed in the capillaries of muscle and adipose tissues, sequentially converting the VLDL to **intermediate-density lipoproteins (IDL)** and then to **low-density**

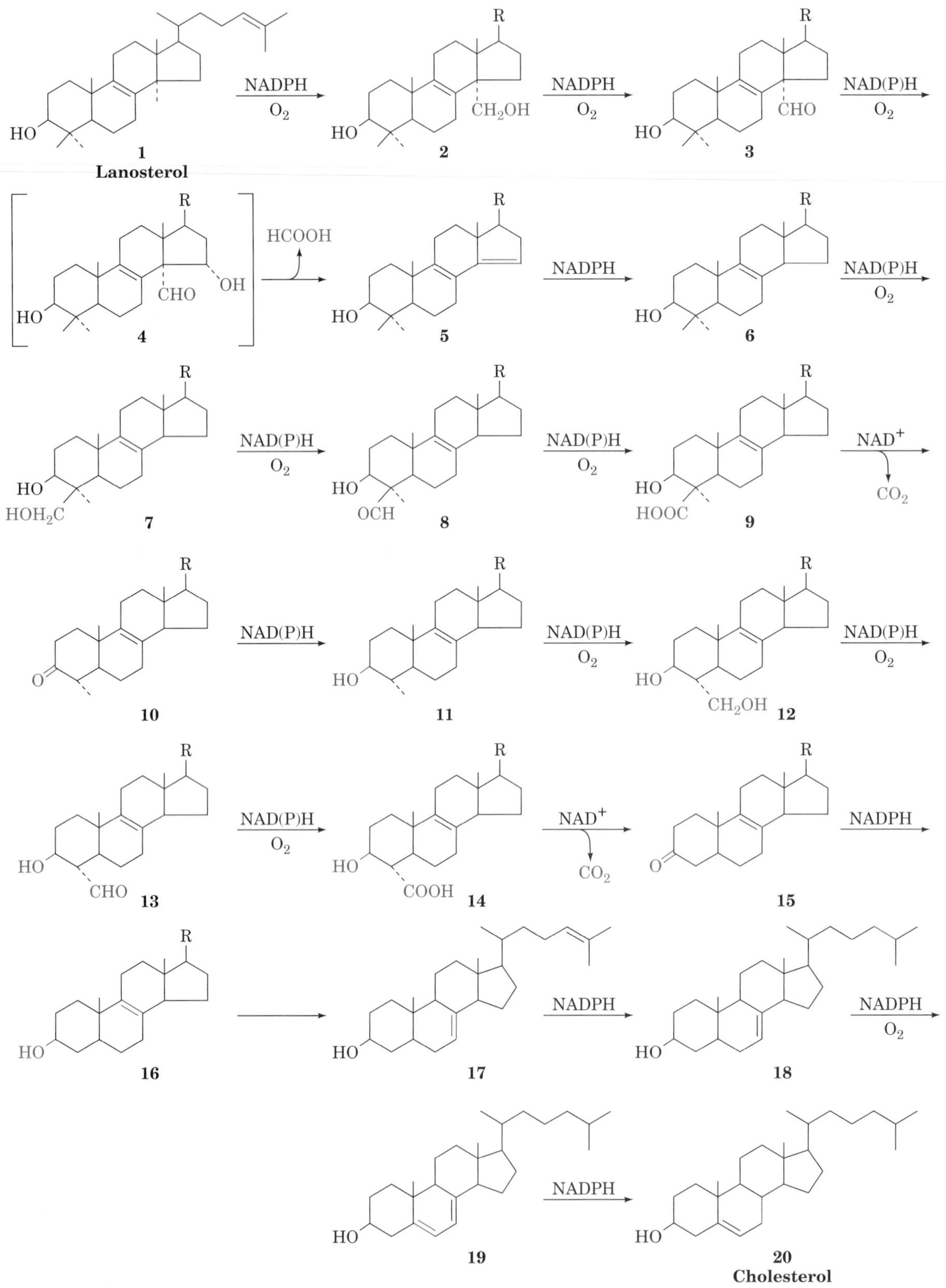

FIGURE 25-56 The 19-reaction conversion of lanosterol to cholesterol. [After Rilling, H.C. and Chayet, L.T., *in* Danielsson, H. and Sjövall, J. (Eds.), *Sterols and Bile Acids,* p. 33, Elsevier (1985), as modified by Bae, S.-H. and Paik, Y.-K., *Biochem. J.* **326,** 609–616 (1997).]

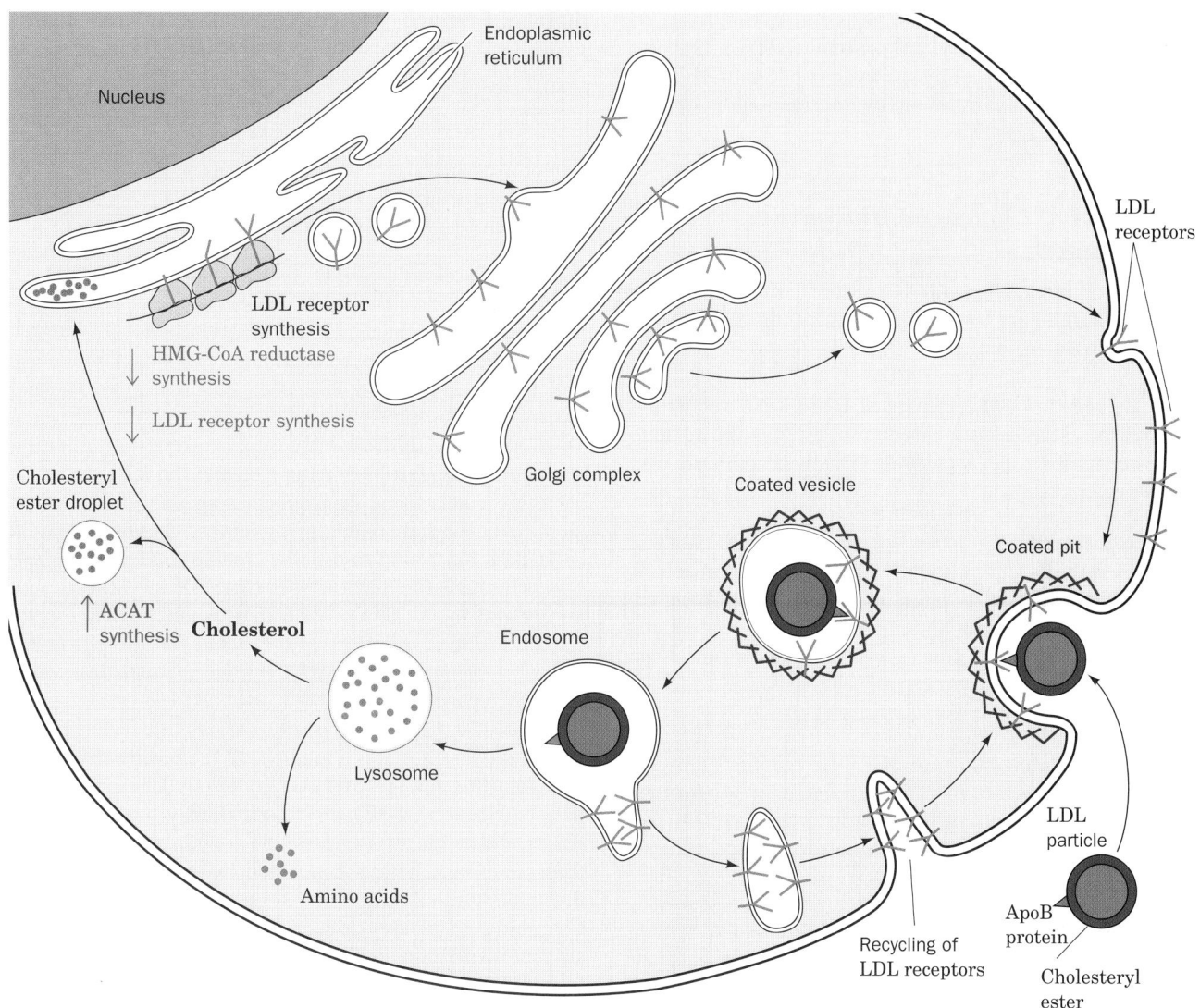

FIGURE 25-57   LDL receptor-mediated endocytosis in mammalian cells. LDL receptor is synthesized on the endoplasmic reticulum, processed in the Golgi complex, and inserted into the plasma membrane as a component of coated pits. LDL is specifically bound by the receptor on the coated pit and brought into the cell in endosomes that deliver LDL to lysosomes while recycling LDL receptor to the plasma membrane (Section 12-5B). Lysosomal degradation of LDL releases cholesterol, whose presence decreases the rate of synthesis of HMG-CoA reductase and LDL receptors (*down arrows*) while increasing that of acyl-CoA:cholesterol acyltransferase (ACAT; *up arrow*). [After Brown, M.S. and Goldstein, J.L., *Curr. Top. Cell. Reg.* **26,** 7 (1985).] 🔊 **See the Animated Figures**

lipoproteins (LDL). Peripheral tissues normally obtain most of their exogenous cholesterol from LDL by receptor-mediated endocytosis (Fig. 25-57; Section 12-5B). Inside the cell, cholesteryl esters are hydrolyzed by a lysosomal lipase to free cholesterol, which is either incorporated into cell membranes or reesterified by ACAT for storage as cholesteryl ester droplets.

Dietary cholesterol, cholesteryl esters, and triacylglycerols are transported in the blood by intestinally synthesized lipoprotein complexes called **chylomicrons.** After removal of their triacylglycerols at the peripheral tissues, the resulting **chylomicron remnants** bind to specific liver cell remnant receptors and are taken up by receptor-mediated endocytosis in a manner similar to that of LDL. In the liver, dietary cholesterol is either used in bile salt biosynthesis (Section 25-6C) or packaged into VLDL for export. *Liver and peripheral tissues therefore have two ways of obtaining cholesterol: They may either synthesize it from acetyl-CoA by the de novo pathway we have just discussed, or they may obtain it from the bloodstream by receptor-mediated endocytosis.* A small amount of cholesterol also enters cells by a non-receptor-mediated pathway.

Cholesterol actually circulates back and forth between the liver and peripheral tissues. While LDL transports cholesterol from the liver, cholesterol is transported back to the liver by **high-density lipoproteins (HDL).** Surplus cholesterol is disposed of by the liver as bile salts, thereby protecting the body from an overaccumulation of this water-insoluble substance.

B. *Control of Cholesterol Biosynthesis and Transport*

Cholesterol biosynthesis and transport must be tightly regulated. There are three ways in which the cellular cholesterol supply is maintained:

1. By regulating the activity of HMG-CoA reductase, the enzyme catalyzing the rate-limiting step in the *de novo* cholesterol biosynthesis pathway. This is accomplished in two ways:

(i) Short-term regulation of the enzyme's catalytic activity by (a) competitive inhibition, (b) allosteric effects, and (c) covalent modification involving reversible phosphorylation.

(ii) Long-term regulation of the enzyme's concentration by modulating its rates of synthesis and degradation.

2. By regulating the rate of LDL receptor synthesis, and therefore the rate of cholesterol uptake. High intracellular concentrations of cholesterol suppress LDL receptor synthesis, whereas low cholesterol concentrations stimulate it.

3. By regulating the rate of esterification and hence the removal of free cholesterol. ACAT, the enzyme that catalyzes intracellular cholesterol esterification, is regulated by reversible phosphorylation and by long-term control.

a. HMG-CoA Reductase Is the Primary Control Site for Cholesterol Biosynthesis

HMG-CoA reductase is the rate-limiting enzyme in cholesterol biosynthesis and, as therefore might be expected, constitutes the pathway's main regulatory site. The pathway branches after this reaction, however (Fig. 25-43); ubiquinone, dolichol, farnesylated and geranylgeranylated proteins, and isopentenyl adenosine are also essential, albeit minor, products. HMG-CoA is therefore subject to "multivalent" control, both long-term and short-term, in order to coordinate the synthesis of all of these products.

b. Long-Term Feedback Regulation of HMG-CoA Reductase Is Its Primary Means of Control

The main way in which HMG-CoA reductase is controlled is by long-term feedback control of the amount of enzyme present in the cell. When either LDL–cholesterol or mevalonate levels fall, the amount of HMG-CoA reductase present in the cell can rise as much as 200-fold, due to an increase in enzyme synthesis combined with a decrease in its degradation. When LDL–cholesterol or

mevalonolactone (an internal ester of mevalonate that is hydrolyzed to mevalonate and metabolized in the cell)

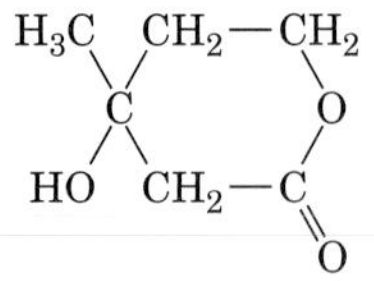

Mevalonolactone

are added back to a cell, these effects are reversed.

The mechanism by which cholesterol serves to control the expression of the >20 genes involved in its biosynthesis and uptake, such as those encoding HMG-CoA reductase and the LDL receptor, has been elucidated by Michael Brown and Joseph Goldstein. These genes all contain a DNA sequence upstream from the transcription initiation site called the **sterol regulatory element (SRE).** In order for these genes to be transcribed, a specific transcription factor, the **sterol regulatory element binding protein (SREBP),** must bind to the SRE (eukaryotic gene expression is discussed in Section 34-3). SREBP is synthesized as an integral membrane protein that, when the cholesterol concentration is sufficiently high, resides in the ER membrane in complex with **SREBP cleavage-activating protein (SCAP).** SREBP (~1160 residues) consists of three domains (Fig. 25-58): (1) an ~480-residue cytosolic N-terminal domain that is a member of the **basic helix–loop–helix/leucine zipper (bHLH/Zip)** family of transcription factors (Section 34-3B), which specifically binds to SREs; (2) an ~90-residue transmembrane domain consisting of two transmembrane helices connected by an ~30-residue hydrophilic luminal loop; and (3) an ~590-residue cytosolic C-terminal regulatory domain. SCAP (1276 residues) consists of two domains (Fig. 25-58): (1) a 730-residue N-terminal domain that contains eight transmembrane helices; and (2) a 546-residue cytosolic C-terminal domain that contains five copies of the protein–protein interaction motif known as a **WD repeat** (also called a WD40 sequence motif because it is ~40 resides long; Section 19-2C) and which presumably forms a 5-bladed β-propeller similar to the 7-bladed β-propeller of the G_β subunit (Fig. 19-18b). SCAP and SREBP associate through the interaction of the SCAP's regulatory domain with SREBP's WD domain (Fig. 25-58).

SCAP functions as a sterol sensor. An ~170-residue segment of its transmembrane domain, the **sterol-sensing domain,** interacts with sterols although how it does so is unknown. When the cholesterol in the ER membrane is depleted, SCAP changes conformation and thereupon escorts its bound SREBP to the Golgi apparatus via membranous vesicles (Sections 12-4C and 12-4D). In the Golgi apparatus, SREBP is sequentially cleaved by two membrane-bound proteases (Fig. 25-58). **Site-1 protease (S1P),** a serine protease of the subtilisin family, cleaves SREBP in the luminal loop that connects its two transmembrane helices but only when it is associated with SCAP. This cleavage exposes a peptide bond located 3 residues from the beginning of SREBP's N-terminal transmembrane helix to cleavage by **site-2 protease (S2P),** a zinc metalloprotease. This releases the bHLH/Zip domain to migrate

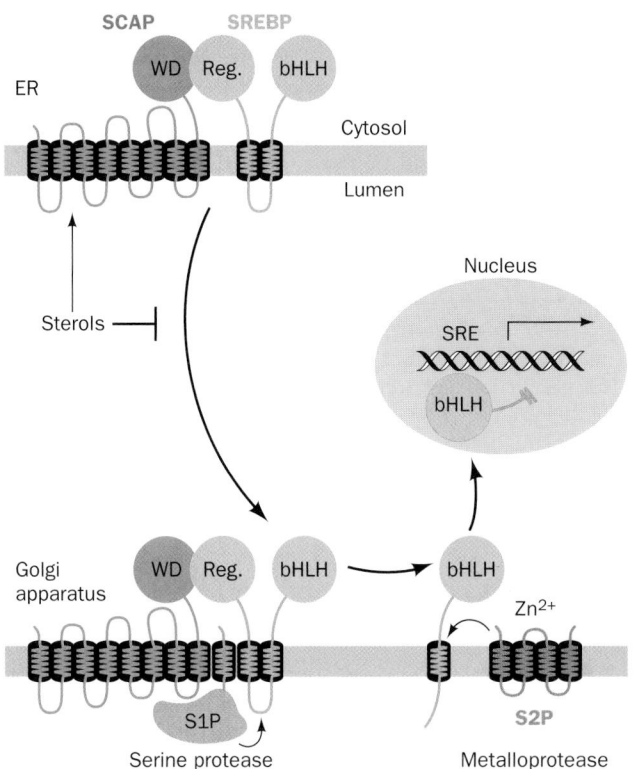

FIGURE 25-58 Model for the cholesterol-mediated proteolytic activation of SREBP. When cholesterol levels in the cell are high, the SREBP–SCAP complex resides in the ER. When cholesterol levels are low, SCAP transports SREBP via membranous vesicles to the Golgi apparatus, where SREBP undergoes sequential proteolytic cleavage by the membrane-bound proteases S1P and S2P. This releases SREBP's bHLH/Zip-containing N-terminal domain, which enters the nucleus where it binds to the SREs of its target genes, thereby inducing their transcription. [After Goldstein, J., Rawson, R.B., and Brown, M., *Arch. Biochem. Biophys.* **397,** 139 (2002).]

to the nucleus where it activates the transcription of its target genes. The cholesterol level in the cell thereby rises until SCAP no longer induces the translocation of SREBP to the Golgi, a classic case of feedback inhibition.

This complex regulatory pathway was elucidated, in part, through the generation of several lines of transgenic mice that overexpress one or another of the foregoing proteins, and knockout mice that lack one or another of these proteins. For example, knockout mice lacking either SCAP or S1P in their livers have decreased expression of both HMG-CoA reductase and LDL receptor proteins, even when fed a cholesterol-deficient diet. In contrast, transgenic mice that overexpress SREBP or SCAP have greatly increased expression of these enzymes. In fact, animals overproducing only the bHLH/Zip domain of SREBP have massively enlarged livers (up to 4-fold larger than normal) due to engorgement with triacylglycerols and cholesteryl esters and yet they continue to transcribe SREBP's target genes such that their mRNA levels are up to 75-fold greater than normal. Many individuals suffering from obesity or diabetes caused by insulin resistance (type II diabetes; Section 27-4B) have fatty livers, which in some cases leads to liver failure. Fatty livers due to insulin resistance appear to be caused by elevated levels of SREBP in response to elevated insulin levels.

c. Regulation of HMG-CoA Reductase by Covalent Modification Is a Means of Cellular Energy Conservation

HMG-CoA reductase exists in interconvertible more active and less active forms, as do glycogen phosphorylase *(Section 18-3C), glycogen synthase (Section 18-3D), pyruvate dehydrogenase (Section 21-2C), and acetyl-CoA carboxylase (Section 25-4B), among others.* The unmodified form of HMG-CoA reductase is more active and the phosphorylated form is less active. HMG-CoA reductase is phosphorylated (inactivated) at its Ser 871 in a bicyclic cascade system by the covalently modifiable enzyme AMP-dependent protein kinase (AMPK), which, as we saw in Section 25-4B, also acts on acetyl-CoA carboxylase [in this context, this enzyme was originally named **HMG-CoA reductase kinase (RK),** until it was found to be identical to AMPK]. It appears that this control is exerted to conserve energy when ATP levels fall and AMP levels rise, by inhibiting biosynthetic pathways. This hypothesis was tested by Brown and Goldstein, who used genetic engineering techniques to produce hamster cells containing a mutant HMG-CoA reductase with Ala replacing Ser 871 and therefore incapable of phosphorylation control. These cells respond normally to feedback regulation of cholesterol biosynthesis by LDL–cholesterol and mevalonate but, unlike normal cells, do not decrease their synthesis of cholesterol on ATP depletion, supporting the idea that control of HMG-CoA reductase by phosphorylation is involved in energy conservation.

d. LDL Receptor Activity Controls Cholesterol Homeostasis

LDL receptors clearly play an important role in the maintenance of plasma LDL–cholesterol levels. In normal individuals, about half of the IDL formed from the VLDL reenters the liver through LDL receptor–mediated endo-

(a) **Normal**

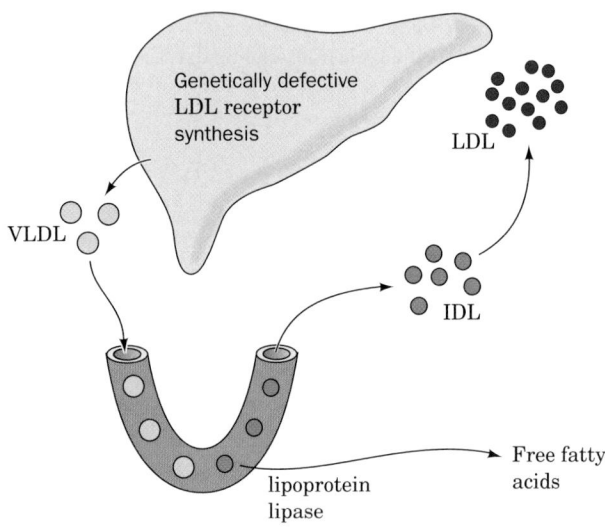

(b) **Familial hypercholesterolemia**

(c) **High cholesterol diet**

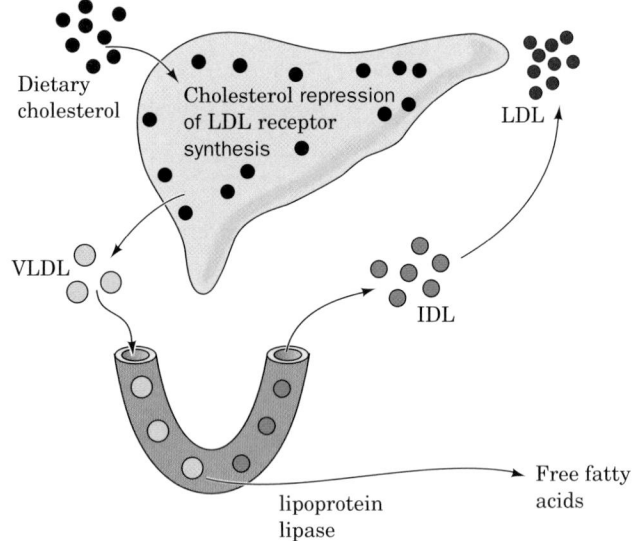

FIGURE 25-59 Control of plasma LDL production and uptake by liver LDL receptors. (*a*) In normal human subjects, VLDL is secreted by the liver and converted to IDL in the capillaries of the peripheral tissues. About half of the plasma IDL particles bind to the LDL receptor and are taken up by the liver. The remainder are converted to LDL at the peripheral tissues. (*b*) In individuals with familial hypercholesterolemia (FH), liver LDL receptors are diminished or eliminated because of a genetic defect. (*c*) In normal individuals who have a long-term high-cholesterol diet, the liver is filled with cholesterol, which represses the rate of LDL receptor production. Receptor deficiency, whether of genetic or dietary cause, raises the plasma LDL level by increasing the rate of LDL production and decreasing the rate of LDL uptake. [After Goldstein, J.L. and Brown, M.S., *J. Lipid Res.* **25,** 1457 (1984).]

cytosis (IDL and LDL both contain apolipoproteins that specifically bind to the LDL receptor; Section 12-5B). The remaining IDL are converted to LDL (Fig. 25-59*a*). *The serum concentration of LDL therefore depends on the rate at which liver removes IDL from the circulation, which, in turn, depends on the number of functioning LDL receptors on the liver cell surface.*

High blood cholesterol **(hypercholesterolemia),** which results from the overproduction and/or underutilization of LDL, is known to be caused by either of two metabolic irregularities: (1) the genetic disease **familial hypercholesterolemia (FH)** or (2) the consumption of a high-cholesterol diet. FH is a dominant genetic defect that results in a deficiency of functional LDL receptors (Section 12-5C). Homozygotes for this disorder lack functional LDL receptors, so their cells can absorb neither IDL nor LDL by receptor-mediated endocytosis. The increased concentration of IDL in the bloodstream leads to a corresponding

increase in LDL, which is, of course, underutilized since it cannot be taken up by the cells (Fig. 25-59*b*). FH homozygotes therefore have plasma LDL–cholesterol levels three to five times higher than average. FH heterozygotes, which are far more common, have about half of the normal number of functional LDL receptors and plasma LDL–cholesterol levels of about twice the average.

The long-term ingestion of a high-cholesterol diet has an effect similar, although not as extreme, as FH (Fig. 25-59c). Excessive dietary cholesterol enters the liver cells in chylomicron remnants and represses the synthesis of LDL receptor protein. The resulting insufficiency of LDL receptors on the liver cell surface has consequences similar to those of FH.

LDL receptor deficiency, whether of genetic or dietary origin, raises the LDL level by two mechanisms: (1) increased LDL production resulting from decreased IDL uptake and (2) decreased LDL uptake. Two strategies for reversing these conditions (besides maintaining a low-cholesterol diet) are being used in humans:

1. *Ingestion of anion exchange resins (Section 6-3A) that bind bile salts, thereby preventing their intestinal absorption* (resins are insoluble in water). Bile salts, which are derived from cholesterol, are normally efficiently recycled by the liver (Section 25-6C). Elimination of resin-bound bile salts in the feces forces the liver to convert more cholesterol to bile salts than otherwise. The consequent decrease in the serum cholesterol concentration induces synthesis of LDL receptors (of course, not in FH homozygotes). Unfortunately, the decreased serum cholesterol level also induces the synthesis of HMG-CoA reductase, which increases the rate of cholesterol biosynthesis. Ingestion of bile salt–binding resins such as **cholestyramine** (sold as **Questran**) therefore provides only a 15 to 20% drop in serum cholesterol levels.

2. *Treatment with competitive inhibitors of HMG-CoA reductase.* These include (Fig. 25-60) the fungal derivatives **lovastatin** (also called **mevinolin** and sold as **Mevacor**), **pravastatin (Pravachol),** and **simvastatin (Zocor)** as well as the synthetic inhibitor **atorvastatin (Lipitor),** compounds that are collectively known as **statins.** Indeed, Lipitor is presently one of the most widely prescribed drugs in the United States. The initial decreased cholesterol supply in the cell caused by the presence of statins is again met by induction of LDL receptors and HMG-CoA reductase so that, at the new steady state, the HMG-CoA reductase level is almost that of the pre-drug state. However, the increased number of LDL receptors causes increased removal of both LDL and IDL (the apoB-containing precursor to LDL), decreasing serum LDL levels appreciably. Lipitor-treated FH heterozygotes routinely show a serum cholesterol decrease of 40–50%.

The combined use of these agents, moreover, results in a clinically dramatic 50 to 60% decrease in serum cholesterol levels.

e. Overexpression of LDL Receptor Prevents Diet-Induced Hypercholesterolemia

Experiments are well underway toward the treatment of hypercholesterolemic individuals by **gene therapy** (Section 5-5H). A line of transgenic mice has been developed that overproduce the human LDL receptor. When fed a diet high in cholesterol, fat, and bile salts, these transgenic animals did not develop a detectable increase in plasma LDL. In contrast, normal mice fed the same diet exhibited large increases in plasma LDL levels. Evidently,

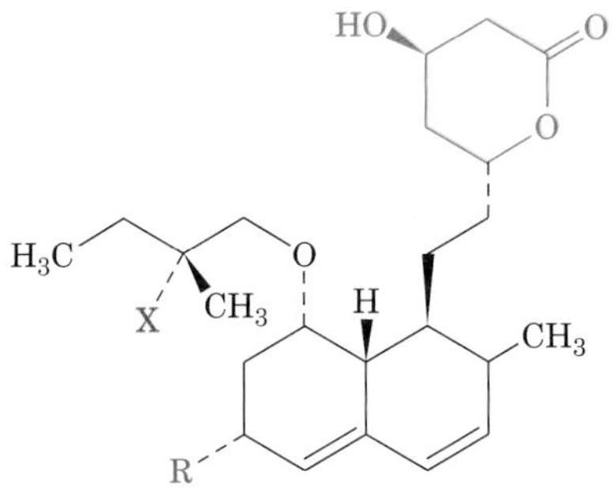

X = H   R = CH$_3$ **Lovastatin (Mevacor)**
X = H   R = OH   **Pravastatin (Pravachol)**
X = CH$_3$ R = CH$_3$ **Simvastatin (Zocor)**

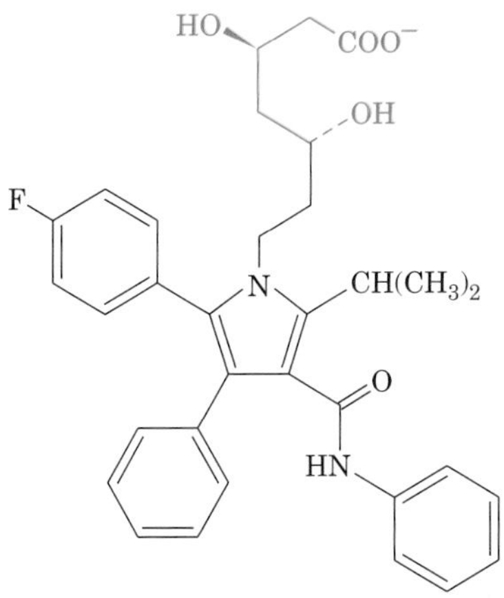

Atorvastatin (Lipitor)

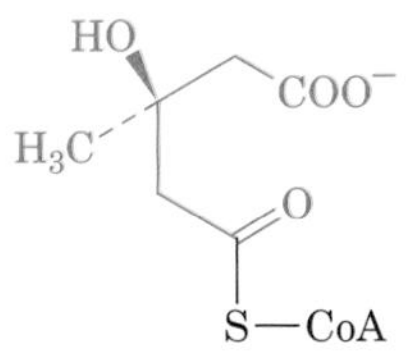

HMG-CoA

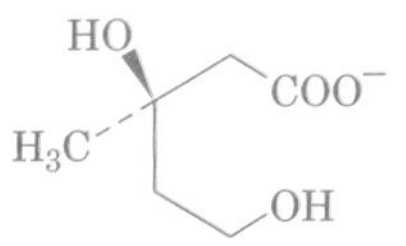

Mevalonate

FIGURE 25-60 Competitive inhibitors of HMG-CoA reductase used for the treatment of hypercholesterolemia. The molecular formulas of lovastatin (Mevacor), pravastatin (Pravachol), simvastatin (Zocor), and atorvastatin (Lipitor), all of which are potent competitive inhibitors of HMG-CoA reductase, are given. The structures of HMG-CoA and mevalonate are shown for comparison. Note that lovastatin, pravastatin, and simvastatin are lactones, whereas atorvastatin and mevalonate are hydroxy acids. The lactones are hydrolyzed enzymatically *in vivo* to their active hydroxy-acid forms.

the unregulated overexpression of LDL receptors can prevent diet-induced hypercholesterolemia, at least in mice.

C. *Cholesterol Utilization*

Cholesterol is the precursor of steroid hormones and bile salts. Steroid hormones, which are grouped into five categories, **progestins, glucocorticoids, mineralocorticoids, an-**drogens, and **estrogens,** mediate a wide variety of vital physiological functions (Section 19-1G). All contain the four-ring structure of the sterol nucleus and are remarkably similar in structure, considering the enormous differences in their physiological effects. A simplified biosynthetic scheme (Fig. 25-61) indicates their structural similarities and differences. We shall not discuss the details of these pathways.

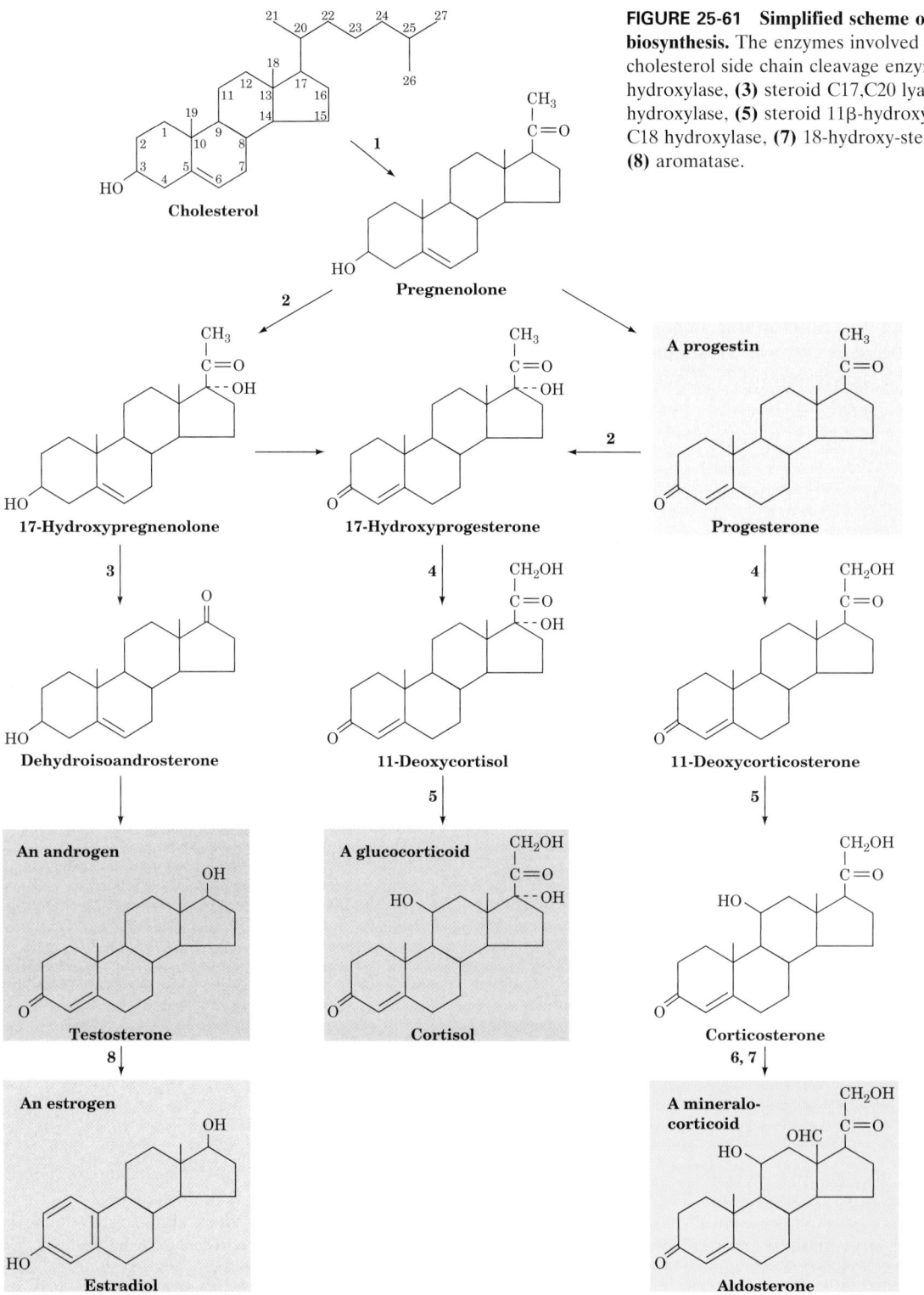

FIGURE 25-61 Simplified scheme of steroid biosynthesis. The enzymes involved are **(1)** the cholesterol side chain cleavage enzyme, **(2)** steroid C17 hydroxylase, **(3)** steroid C17,C20 lyase, **(4)** steroid C21 hydroxylase, **(5)** steroid 11β-hydroxylase, **(6)** steroid C18 hydroxylase, **(7)** 18-hydroxy-steroid oxidase, and **(8)** aromatase.

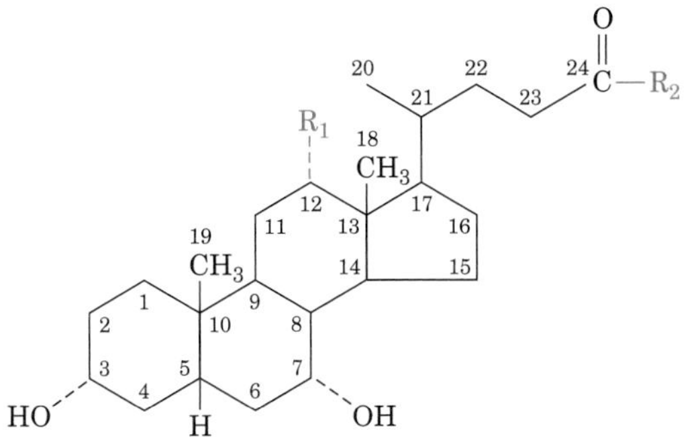

	R_1 = OH	R_1 = H
R_2 = OH	**Cholic acid**	**Chenodeoxycholic acid**
R_2 = NH—CH$_2$—COOH	**Glycocholic acid**	**Glycochenodeoxycholic acid**
R_2 = NH—CH$_2$—CH$_2$—SO$_3$H	**Taurocholic acid**	**Taurochenodeoxycholic acid**

FIGURE 25-62   Structures of the major bile acids and their glycine and taurine conjugates.

*The quantitatively most important pathway for the excretion of cholesterol in mammals is the formation of bile salts (the conjugate bases of **bile acids**).* The major bile salts, **cholate** and **chenodeoxycholate,** are synthesized in the liver and secreted as glycine or **taurine** conjugates (Fig. 25-62) into the gallbladder. From there, they are secreted into the small intestine, where they act as emulsifying agents in the digestion and absorption of fats and fat-soluble vitamins (Section 25-1). An efficient recycling system allows the bile salts to reenter the bloodstream and return to the liver for reuse several times each day. The <1 g · day^{-1} of bile salts that normally escape this recycling system are further metabolized by microorganisms in the large intestine and excreted. *This is the body's only route for cholesterol excretion.*

Comparison of the structures of cholesterol and the bile acids (Figs. 25-41 and 25-62) indicates that biosynthesis of bile acids from cholesterol involves (1) saturation of the 5,6 double bond, (2) epimerization of the 3β-OH group, (3) introduction of OH groups into the 7α and 12α positions, (4) oxidation of C24 to a carboxylic acid, and (5) conjugation of this side chain carboxylic acid with glycine or taurine. **Cholesterol 7α-hydroxylase** catalyzes the first and rate-limiting step in bile acid synthesis and is closely regulated.

7 ■ EICOSANOID METABOLISM: PROSTAGLANDINS, PROSTACYCLINS, THROMBOXANES, LEUKOTRIENES, AND LIPOXINS

Prostaglandins (PGs) were first identified in human semen by Ulf von Euler in the early 1930s through their ability to stimulate uterine contractions and lower blood pressure. von Euler thought that these compounds originated in the prostate gland (hence their name) but they were later shown to be synthesized in the seminal vesicles. By the time the mistake was realized, the name was firmly entrenched. In the mid-1950s, crystalline materials were isolated from biological fluids and called PGE (ether-soluble) and PGF (phosphate buffer–soluble; *fosfat* in Swedish). This began an explosion of research on these potent substances.

*Almost all mammalian cells except red blood cells produce prostaglandins and their related compounds, the **prostacyclins, thromboxanes, leukotrienes** and **lipoxins,** known collectively as **eicosanoids** since they are all C_{20} compounds (Greek: eikosi, twenty). The eicosanoids, like hormones, have profound physiological effects at extremely low concentrations.* For example, they mediate: (1) the inflammatory response, notably as it involves the joints (rheumatoid arthritis), skin (psoriasis), and eyes; (2) the production of pain and fever; (3) the regulation of blood pressure; (4) the induction of blood clotting; (5) the control of several reproductive functions such as the induction of labor; and (6) the regulation of the sleep/wake cycle. The enzymes that synthesize these compounds and the receptors to which they bind are therefore the targets of intensive pharmacological research.

The eicosanoids are also hormonelike in that they bind to G-protein-coupled receptors (Section 19-2B), and many of their effects are intracellularly mediated by cAMP. Unlike hormones, however, they are not transported in the bloodstream to their sites of action. Rather, these chemically and biologically unstable substances (some decompose within minutes or less *in vitro*) are local mediators (paracrine hormones; Section 19-1); that is, *they act in the same environment in which they are synthesized.*

In this section, we discuss the structures of the eicosanoids and outline their biosynthetic pathways and modes of action. As we do so, note the great diversity of their structures and functions, a phenomenon that makes the elucidation of the physiological roles of these potent substances a challenging research area.

A. Background

*Prostaglandins are all derivatives of the hypothetical C_{20} fatty acid **prostanoic acid** in which carbon atoms 8 to 12*

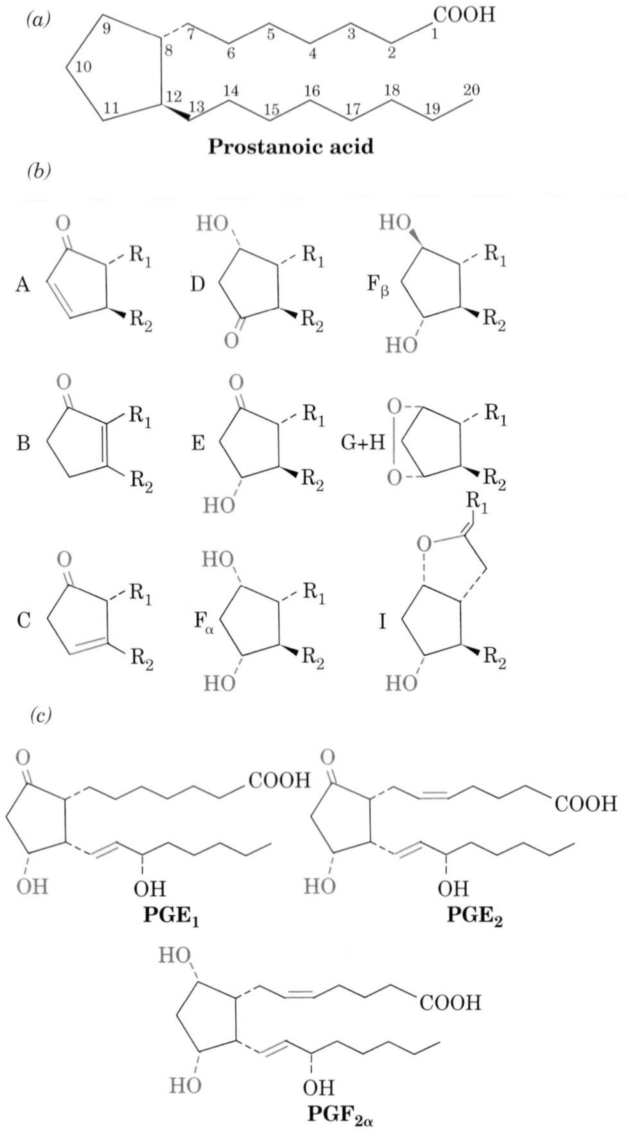

(a)

Prostanoic acid

(b)

(c)

PGE₁

PGE₂

PGF₂α

FIGURE 25-63 Prostaglandin structures. (*a*) The carbon skeleton of prostanoic acid, the prostaglandin parent compound. (*b*) Structures of prostaglandins A through I. (*c*) Structures of prostaglandins E_1, E_2, and $F_{2\alpha}$ (the first prostaglandins to be identified).

comprise a cyclopentane ring (Fig. 25-63a). Prostaglandins A through I differ in the substituents on the cyclopentane ring (Fig. 25-63*b*): **PGAs** are α,β-unsaturated ketones, **PGEs** are β-hydroxy ketones, **PGFs** are 1,3-diols, etc. In **PGF$_\alpha$,** the C9 OH group is on the same side of the ring as R_1; it is on the opposite side in **PGF$_\beta$.** The numerical subscript in the name refers to the number of double bonds contained on the side chains of the cyclopentane ring (Fig. 25-63*c*).

*In humans, the most prevalent prostaglandin precursor is **arachidonic acid (5,8,11,14-eicosatetraenoic acid),** a C_{20} polyunsaturated fatty acid that has four nonconjugated double bonds.* The double bond at C14 is six carbon atoms

from the terminal carbon atom (the ω carbon atom), making arachidonic acid an ω−6 fatty acid. Arachidonic acid is synthesized from the essential fatty acid linoleic acid (also an ω−6 fatty acid) by desaturation with a Δ^6-desaturase to yield **γ-linolenic acid (GLA),** followed by elongation and a second desaturation, this time with a Δ^5-desaturase (Fig. 25-64; Section 25-4E). Prostaglandins with the subscript 1 (the "series-1" prostaglandins) are synthesized from **dihomo-γ-linolenic acid (DGLA; 8,11,14-eicosatrienoic** acid), whereas "series-2" prostaglandins are synthesized from arachidonic acid. α-Linolenic acid (ALA), another essential fatty acid since the Δ^{15}-desaturase required for its synthesis occurs only in plants, is a precursor of 5,8,11,14,17-eicosapentaenoic acid (EPA) and the "series-3" prostaglandins. Since arachidonate is the primary prostaglandin precursor in humans, we shall mostly refer to the series-2 prostaglandins in our examples. Note, however, that when dietary linoleic acid and α-linolenic acid are equally available, the relative activities of the Δ^5- and Δ^6-desaturases are important in determining the relative amounts of these prostaglandin precursors.

a. Arachidonate Is Generated by Phospholipid Hydrolysis

Arachidonate is stored in cell membranes esterified at glycerol C2 of phosphatidylinositol and other phospholipids. The production of arachidonate metabolites is controlled by the rate of arachidonate release from these phospholipids through three alternative pathways (Fig. 25-65):

1. Phospholipase A₂ hydrolyzes acyl groups at C2 of phospholipids (Fig. 25-65*b, left*).

2. Phospholipase C (Section 19-4B) specifically hydrolyzes the phosphatidylinositol head group to yield a **1,2-diacylglycerol (DAG)** and **phosphoinositol. DAG** is phosphorylated by **diacylglycerol kinase** to phosphatidic acid, a phospholipase A₂ substrate (Fig. 25-65*b, center*). (Recall that DAG and the various phosphorylated forms of phosphoinositol are also important signaling molecules in that they mediate the phosphoinositide cascade; Section 19-4.)

3. The DAG also may be hydrolyzed directly by **diacylglycerol lipase** (Fig. 25-65*b, right*).

Corticosteroids are used as anti-inflammatory agents because they inhibit phospholipase A₂, reducing the rate of arachidonate production.

b. Aspirin Inhibits Prostaglandin Synthesis

The use of **aspirin** as an analgesic (pain-relieving), antipyretic (fever-reducing), and anti-inflammatory agent has been widespread since the nineteenth century. Yet, it was not until 1971 that John Vane discovered its mechanism of action. *Aspirin, as do other **nonsteroidal anti-inflammatory drugs (NSAIDs),** inhibits the synthesis of prostaglandins from eicosanoid precursors (Section 25-7B).* These inhibitors have therefore proved to be valuable tools in the elucidation of prostaglandin biosynthesis pathways and have provided a starting point for the rational synthesis of new anti-inflammatory drugs.

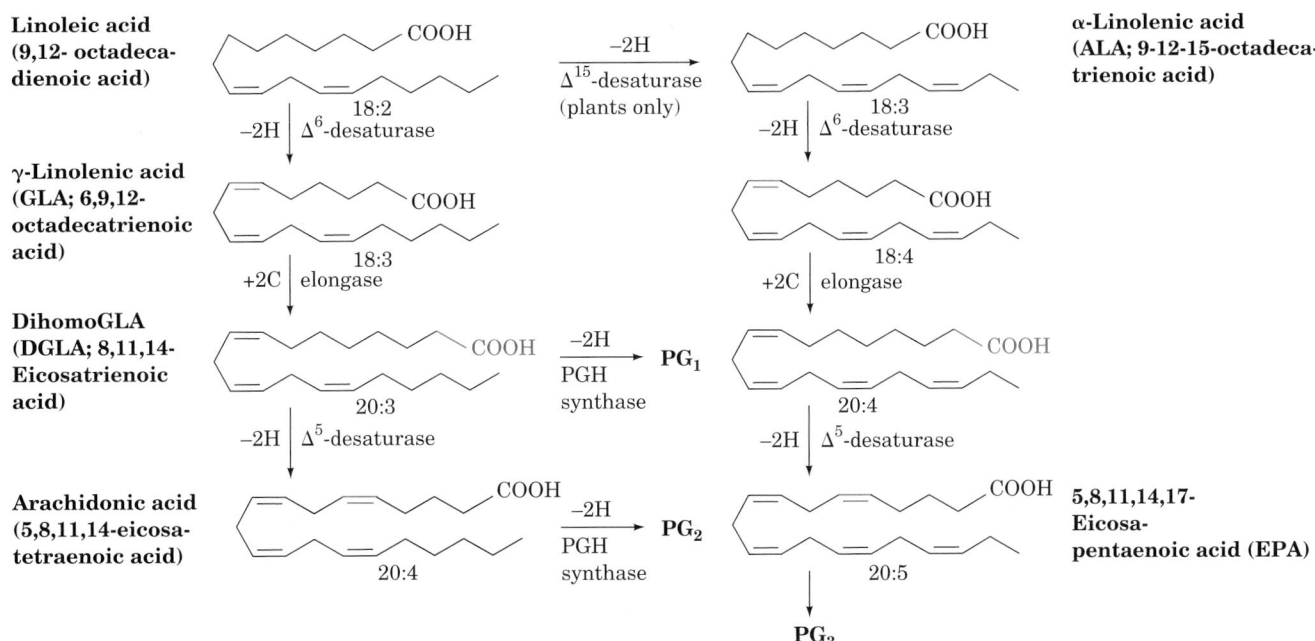

FIGURE 25-64   Synthesis of prostaglandin precursors. The linoleic acid derivatives dihomoGLA (DGLA), arachidonic acid, and 5,8,11,14,17-eicosapentaenoic acid (EPA) are the respective precursors of the series-1, series-2, and series-3 prostaglandins.

c. Arachidonic Acid Is a Precursor of Leukotrienes, Thromboxanes, and Prostacyclins

Arachidonic acid also serves as a precursor to compounds whose synthesis is not inhibited by aspirin. In fact, there are two main pathways of eicosanoid metabolism. The so-called cyclic pathway, which is inhibited by NSAIDs, forms prostaglandin's characteristic cyclopentane ring, whereas the so-called linear pathway, which is

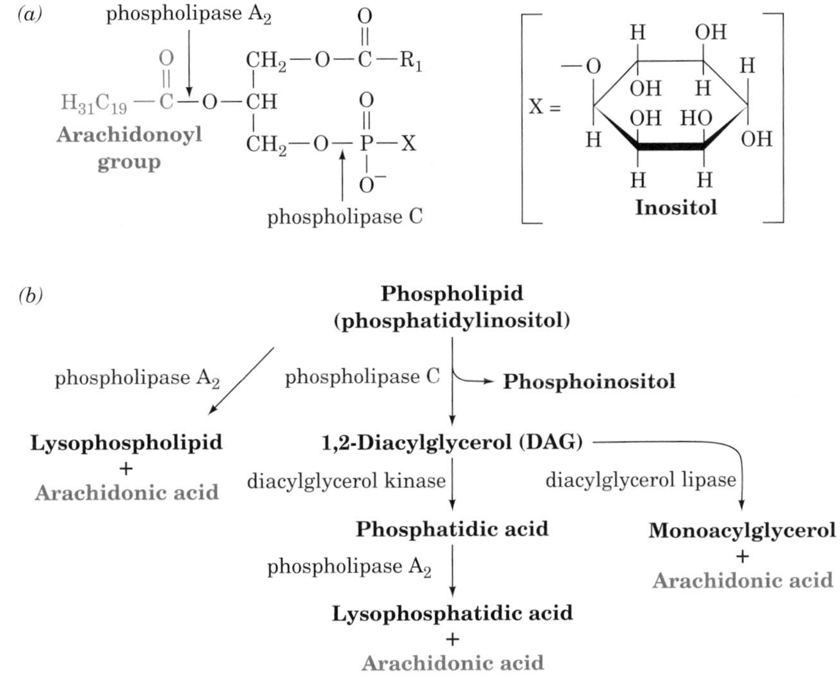

(a)

Arachidonoyl group

phospholipase A$_2$

phospholipase C

$$X = \text{Inositol}$$

(b)

Phospholipid (phosphatidylinositol)

phospholipase A$_2$ → **Lysophospholipid + Arachidonic acid**

phospholipase C → **Phosphoinositol**

1,2-Diacylglycerol (DAG)

diacylglycerol kinase → **Phosphatidic acid**

phospholipase A$_2$ → **Lysophosphatidic acid + Arachidonic acid**

diacylglycerol lipase → **Monoacylglycerol + Arachidonic acid**

FIGURE 25-65   Release of arachidonic acid by phospholipid hydrolysis. (*a*) The sites of hydrolytic cleavage mediated by phospholipases A$_2$ and C. The polar head group, X, is often inositol and its various phosphorylated forms (Section 19-4D). (*b*) Pathways of arachidonic acid liberation from phospholipids.

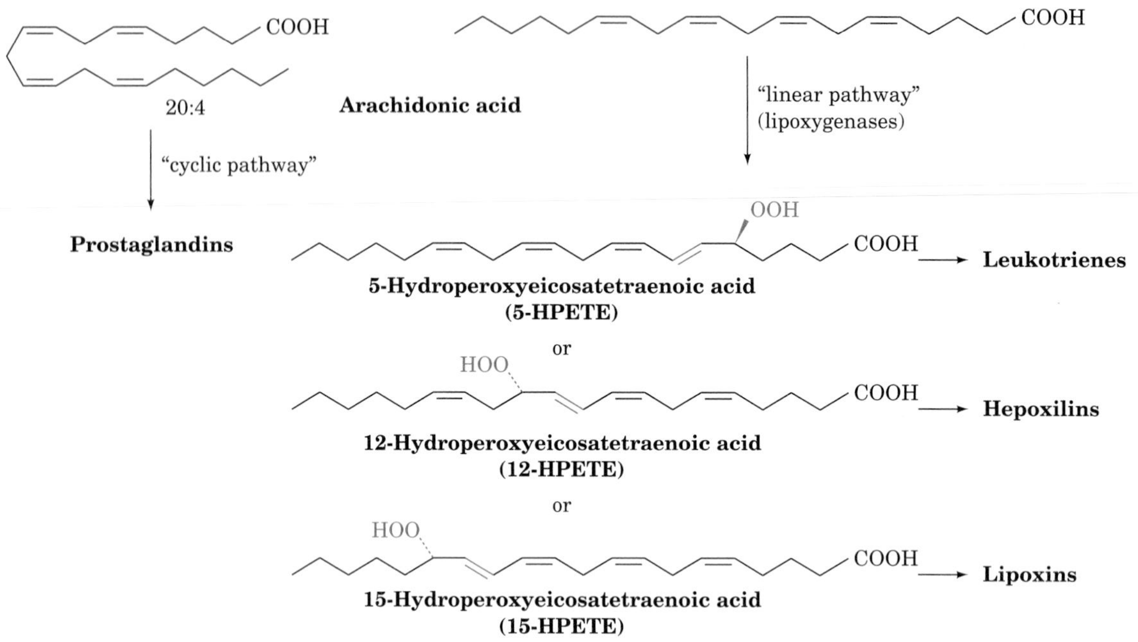

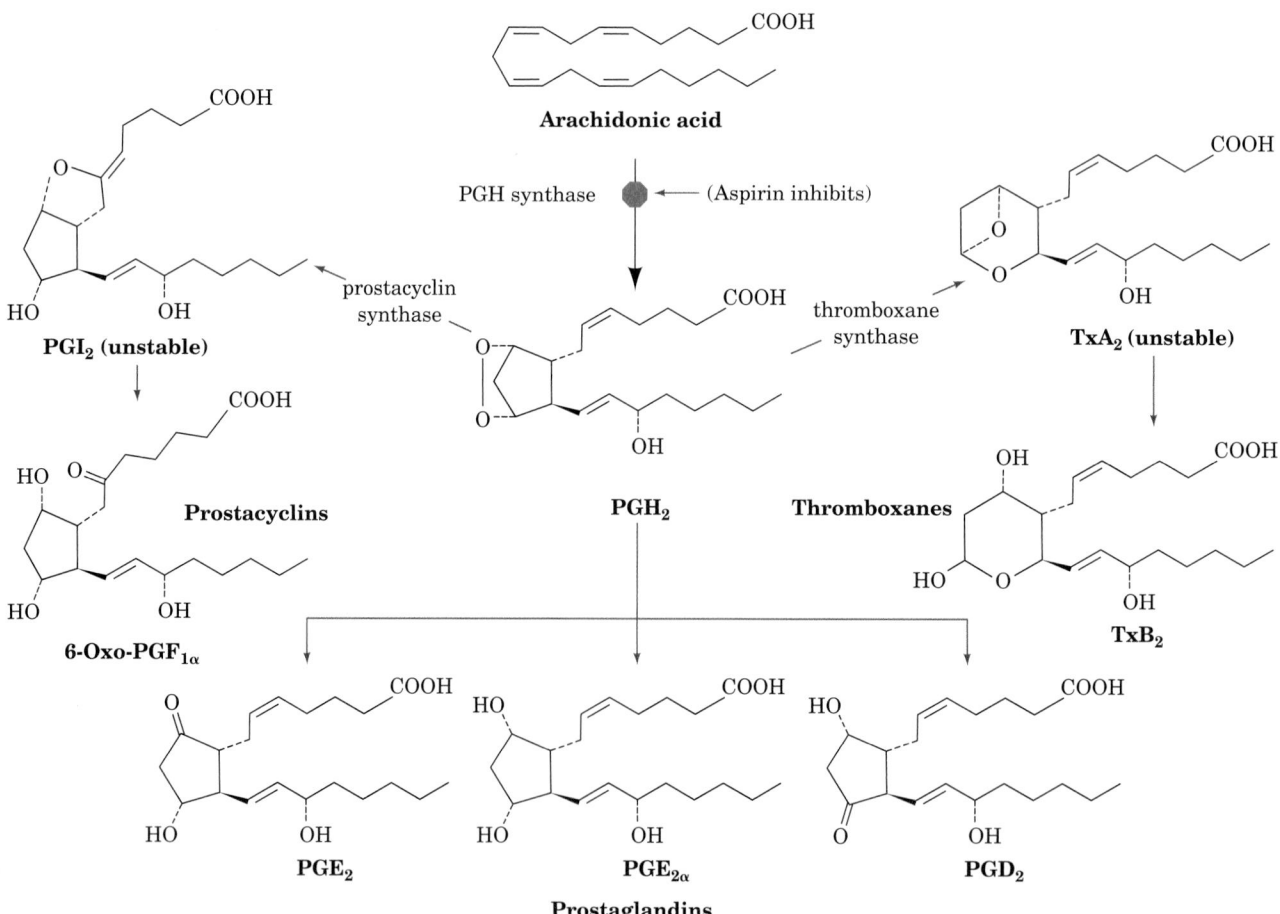

FIGURE 25-66 The cyclic and linear pathways of arachidonic acid metabolism.

FIGURE 25-67 The cyclic pathway of arachidonic acid metabolism. This pathway's branches lead to prostaglandins, prostacyclins, and thromboxanes.

not inhibited by these agents, leads to the formation of the **leukotrienes** and **HPETEs** (Fig. 25-66; Section 25-7C).

Studies using NSAIDs helped demonstrate that two structurally related and highly short-lived classes of compounds, the prostacyclins and the thromboxanes (Fig. 25-67), are also products of the cyclic pathway of eicosanoid metabolism. The specific products produced by this branched pathway depend on the tissue involved. For example, blood platelets (thrombocytes) produce thromboxanes almost exclusively; vascular endothelial cells, which make up the walls of veins and arteries, predominantly synthesize the prostacyclins; and heart muscle makes PGI_2, PGE_2, and $PGF_{2\alpha}$ in more or less equal quantities. In the remainder of this section, we study the cyclic and the linear pathways of eicosanoid metabolism.

B. *The Cyclic Pathway of Eicosanoid Metabolism: Prostaglandins, Prostacyclins, and Thromboxanes*

The first step in the cyclic pathway of eicosanoid metabolism is catalyzed by **PGH synthase** (**PGHS**; also called **prostaglandin H synthase** and **prostaglandin endoperoxide synthase**; Fig. 25-68). This heme-containing enzyme contains two catalytic activities: a cyclooxygenase activity and a peroxidase activity. The former catalyzes the tyrosyl radical-mediated addition of two molecules of O_2 to arachidonic acid, forming **PGG_2**. The latter converts the hydroperoxy function of PGG_2 to an OH group, yielding **PGH_2**. *PGH_2 is the immediate precursor of all series-2 prostaglandins, prostacyclins, and thromboxanes (Fig. 25-67).* The cyclooxygenase activity of the enzyme gives it its common name, **COX** [not to be confused with cytochrome *c* oxidase, which is also called COX (Section 22-2C)].

PGHS, a homodimeric glycoprotein of 576-residue subunits, is an integral membrane protein that extends into the lumen of the endoplasmic reticulum. Its X-ray structure, determined by Michael Garavito, reveals that each of its subunits folds into three domains (Fig. 25-69*a*): an N-terminal module that structurally resembles **epidermal growth factor** (**EGF**; a hormonally active polypeptide that stimulates cell proliferation; Section 19-3B); a central membrane-binding motif; and a C-terminal enzymatic domain. The 44-residue membrane-binding motif has a hydrophobic surface that faces away from the body of the protein but is of insufficient depth to penetrate more than one leaflet of a lipid bilayer [as is also true of squalene–hopene cyclase (Section 25-6A), the only other monotopic membrane protein of known structure].

The peroxidase active site of PGHS occurs at the interface between the large and small lobes of the catalytic domain, in a shallow cleft that contains the enzyme's Fe(III)–heme prosthetic group. The cleft exposes a large portion of the heme to solvent and is therefore thought to comprise the substrate binding site.

The cyclooxygenase active site lies on the opposite side of the heme at the end of a long narrow hydrophobic channel (~8 × 25 Å) extending from the outer surface of the

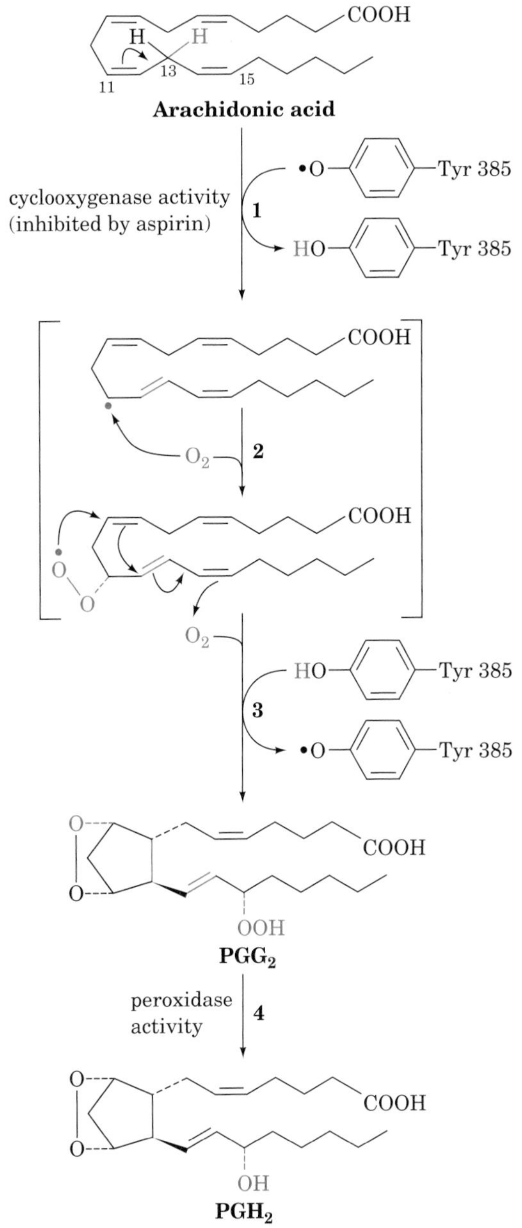

FIGURE 25-68   The reactions catalyzed by PGH synthase (PGHS). The enzyme contains two activities: a cyclooxygenase, which catalyzes Steps 1 to 3 and is inhibited by aspirin, and a peroxidase, which catalyzes Step 4. **(1)** A radical at Tyr 385 that is generated by the enzyme's heme cofactor stereospecifically abstracts a hydrogen atom from C13 of arachidonic acid, which then rearranges so that the radical is on C11. **(2)** The radical reacts with O_2 to yield a hydroperoxide radical. **(3)** The radical cyclizes and reacts with a second O_2 molecule at C15 to yield a peroxide in a process that regenerates the Tyr radical. **(4)** The enzyme's peroxidase activity converts the peroxide at C15 to a hydroxyl group.

membrane-binding motif to the center of each subunit (Fig. 25-69*b*). This channel allows access of the membrane-associated substrate to the active site. Tyr 385, which lies near the top of the channel, just beneath the heme, has

ER lumen

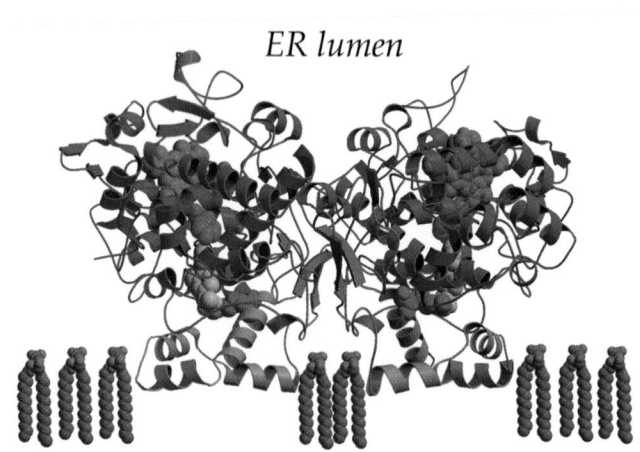

(a)

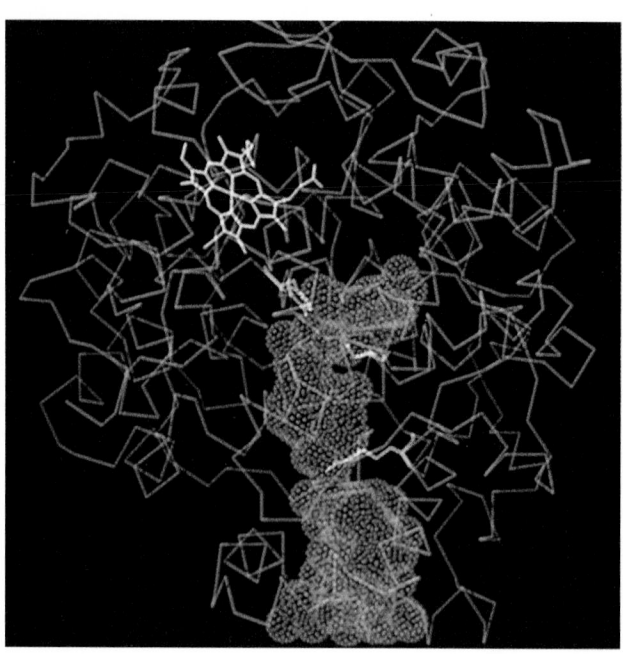

(b)

FIGURE 25-69   X-Ray structure of PGH synthase (PGHS) from sheep seminal vesicles in complex with the NSAID flurbiprofen. (*a*) This homodimeric monotopic membrane protein is viewed from in the plane of the ER membrane with its 2-fold axis of symmetry vertical. The EGF-like module is green, the membrane-binding motif is orange, and the catalytic domain is blue. The heme (*red*); fluriprofen (*yellow*); Tyr 385 (*magenta*), which forms a transient radical during the cyclooxygenase reaction; and Arg 120 (*green*), which forms an ion pair with flurbiprofen, are drawn in space-filling form.

(*b*) A C_α diagram of a PGHS subunit (*green*), the left subunit in Part *a* as viewed from 30° to the left. The peroxidase active site is located above the heme (*pink*). The hydrophobic channel, which penetrates the subunit from the membrane-binding motif at the bottom of the figure to the cyclooxygenase active site below the heme, is represented by its van der Waals surface (*blue dots*). The three residues in the channel that are shown in stick form in orange are, from top to bottom: Tyr 385, Ser 530, which is acetylated by aspirin, and Arg 120. [Courtesy of Michael Garavito, Michigan State University. PDBid 1CQE.]

been shown to form a transient radical during the cyclooxygenase reaction as does, for example, Tyr 122 in Class I ribonucleotide reductase (Section 28-3A). Indeed, the mutagenic replacement of PGHS's Tyr 385 by Phe abolishes its cyclooxygenase activity. The Tyr 385 radical is generated via an intramolecular oxidation by the heme cofactor.

The fate of PGH_2 depends on the relative activities of the enzymes catalyzing the specific interconversions (Fig. 25-67). Platelets contain **thromboxane synthase,** which mediates the formation of **thromboxane A_2 (TxA_2),** a vasoconstrictor and stimulator of platelet aggregation (an initial step in blood clotting; Section 35-1). Vascular endothelial cells contain **prostacyclin synthase,** which catalyzes the synthesis of **prostacyclin I_2 (PGI_2),** a vasodilator and inhibitor of platelet aggregation. These two substances act in opposition, maintaining a balance in the cardiovascular system.

a. NSAIDs Inhibit PGH Synthase

Nonsteroidal anti-inflammatory drugs (NSAIDs; Fig. 25-70) inhibit the synthesis of the prostaglandins,

prostacyclins, and thromboxanes by inhibiting or inactivating the cyclooxygenase activity of PGHS. Aspirin **(acetylsalicylic acid),** for example, acetylates this enzyme: If [^{14}C-*acetyl*]acetylsalicylic acid is incubated with the enzyme, radioactivity becomes irreversibly associated with the inactive enzyme as Ser 530 becomes acetylated (Fig. 25-71). The X-ray structure of PGHS reveals that Ser 530, which is not implicated in catalysis, extends into the cyclooxygenase channel just below Tyr 385 such that its acetylation would block arachidonic acid's access to the active site (Fig. 25-69*b*). The structure of PGHS, which was crystallized with the NSAID **flurbiprofen** (Fig. 25-70), indicates that this drug binds in the cyclooxygenase channel, with its carboxyl group forming an ion pair with Arg 120 (Fig. 25-69*a*). Evidently, flurbiprofen, and by implication other NSAIDs, inhibits the cyclooxygenase activity of PGHS by blocking its active site channel.

Low doses of aspirin, ~75 mg every 1 or 2 days, significantly reduce the long-term incidence of heart attacks and strokes. Such low doses selectively inhibit platelet aggregation and thus blood clot formation because these enucleated cells, which have a lifetime in the circulation of ~10

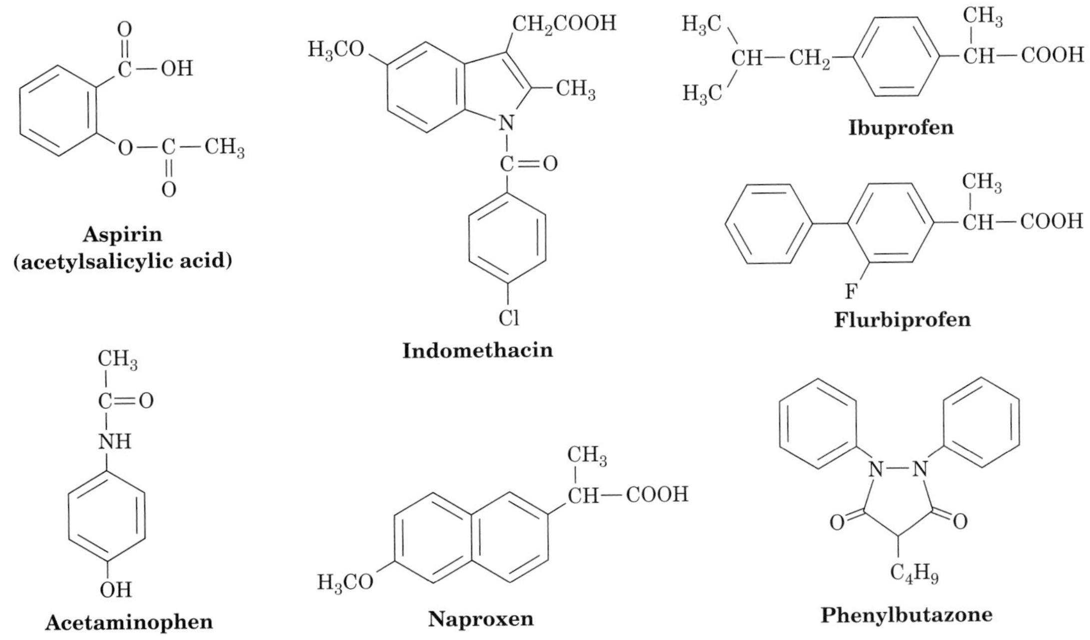

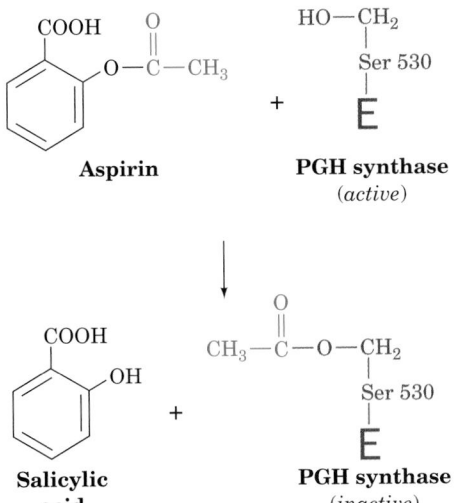

FIGURE 25-70   Some nonsteroidal anti-inflammatory drugs (NSAIDs).

FIGURE 25-71   Inactivation of PGH synthase by aspirin.
Aspirin acetylates Ser 530 of PGH synthase, thereby blocking the enzyme's cyclooxygenase activity.

days, cannot resynthesize their inactivated enzymes. Vascular endothelial cells are not so drastically affected since, for the most part, they are far from the site where aspirin is absorbed, are exposed to lesser concentrations of aspirin and, in any case, can synthesize additional PGHS.

b. COX-2 Inhibitors Lack the Side Effects of Other NSAIDs

PGHS has two isoforms, **COX-1** and **COX-2,** that share a high degree (60%) of sequence identity and structural homology. COX-1 is constitutively (without regulation) expressed in most, if not all, mammalian tissues, thereby supporting levels of prostaglandin synthesis necessary to maintain organ and tissue homeostasis such as that of the gastrointestinal mucosa. In contrast, COX-2 is only expressed in certain tissues in response to inflammatory stimuli such as cytokines, protein growth factors, and endotoxins, and hence is responsible for the elevated prostaglandin levels that cause inflammation. The NSAIDs in Fig. 25-70 are relatively nonspecific and therefore can have adverse side effects, most notably gastrointestinal ulceration, when used to treat inflammation or fever. A structure-based drug design program (Section 15-4A) was therefore instituted to create inhibitors that would target COX-2 but not COX-1. The three-dimensional structures of COX-1 and COX-2 are almost identical. However, their amino acid differences, specifically I523V, I434V, and H513R (COX-1 amino acid on the left and COX-2 amino acid on the right), make COX-2's active site channel ~20% larger in volume than that of COX-1. In addition, the fourth helix of the membrane-binding domain is oriented slightly differently so as to provide a larger opening to the channel. Medicinal chemists therefore synthesized inhibitors, collectively known as **coxibs,** that could enter the COX-2 channel but are excluded from that of COX-1. Two of these inhibitors, **rofecoxib (Vioxx)** and **celecoxib**

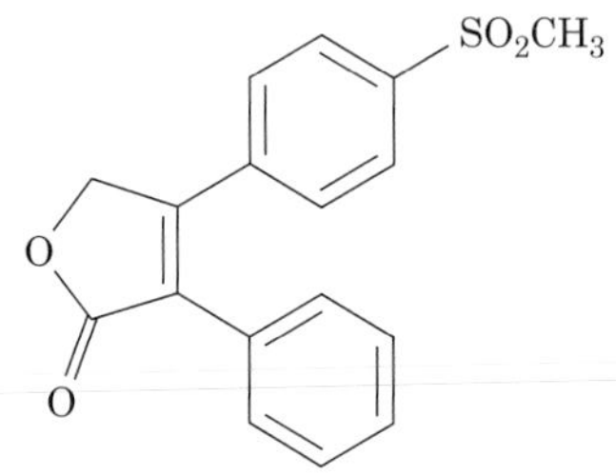

Rofecoxib (Vioxx)

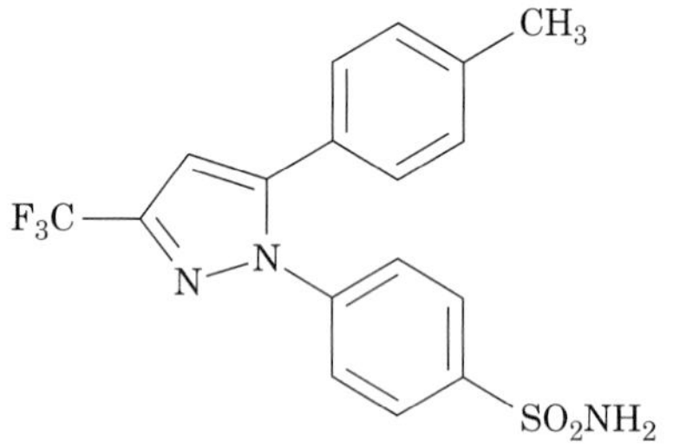

Celecoxib (Celebrex)

FIGURE 25-72   COX-2 inhibitors. Rofecoxib and celecoxib are specific inhibitors of COX-2 **(PGH synthase-2).**

(**Celebrex;** Fig. 25-72), have become major drugs for the treatment of inflammatory diseases such as arthritis because they lack the major side effects of the nonspecific NSAIDs.

c. COX-3 May Be the Target of Acetaminophen

Acetaminophen, which is among the most widely used analgesic/antipyretic drugs (but possesses little anti-inflammatory activity, so that it is not really an NSAID), does not significantly bind to either COX-1 or COX-2. Thus its mechanism of action remained a mystery until the recent discovery by Daniel Simmons of a third COX isozyme, **COX-3,** that is selectively inhibited by acetaminophen as well as by certain NSAIDS. This suggests that COX-3 is the primary target of drugs that decrease pain and fever.

C. *The Linear Pathway of Eicosanoid Metabolism: Leukotrienes and Lipoxins*

Arachidonic acid can be converted by a linear pathway to several different **hydroperoxyeicosatetraenoic acids (HPETEs)** by the **5-, 12-, and 15-lipoxygenases (5-, 12-, and 15-LOs;** Fig. 25-66). **Hepoxilins** are hydroxy epoxy derivatives of **12-HPETE** whose functions are not as yet well understood. **Lipoxins,** the products of a second lipoxygenase acting on **15-HPETE,** act as anti-inflammatory substances. Leukotrienes, derived from the 5-LO reaction, are synthesized by a variety of white blood cells, mast cells (connective tissue cells derived from the blood-forming tissues that secrete substances which mediate inflammatory and allergic reactions), as well as lung, spleen, brain, and heart. **Peptidoleukotrienes (LTC₄, LTD₄, and LTE₄)** are now recognized to be the components of the **slow reacting sub-**

stances of anaphylaxis (**SRS-A;** anaphalaxis is a violent and potentially fatal allergic reaction) released from sensitized lung after immunological challenge. These substances act at very low concentrations (as little as $10^{-10}M$) to contract vascular, respiratory, and intestinal smooth muscle. Peptidoleukotrienes, for example, are ~10,000-fold more potent than histamine, a well-known stimulant of allergic reactions. In the respiratory system, they constrict bronchi, especially the smaller airways; increase mucus secretion; and are thought to be the mediators in asthma. They are also implicated in immediate hypersensitivity (allergic) reactions, inflammatory reactions, and heart attacks.

a. Leukotriene Synthesis

The first two reactions in the conversion of arachidonic acid to leukotrienes are both catalyzed by 5-LO, which contains a nonheme, non-[Fe–S] cluster iron atom that must be in its Fe(III) state to be active. These reactions occur as follows (Fig. 25-73):

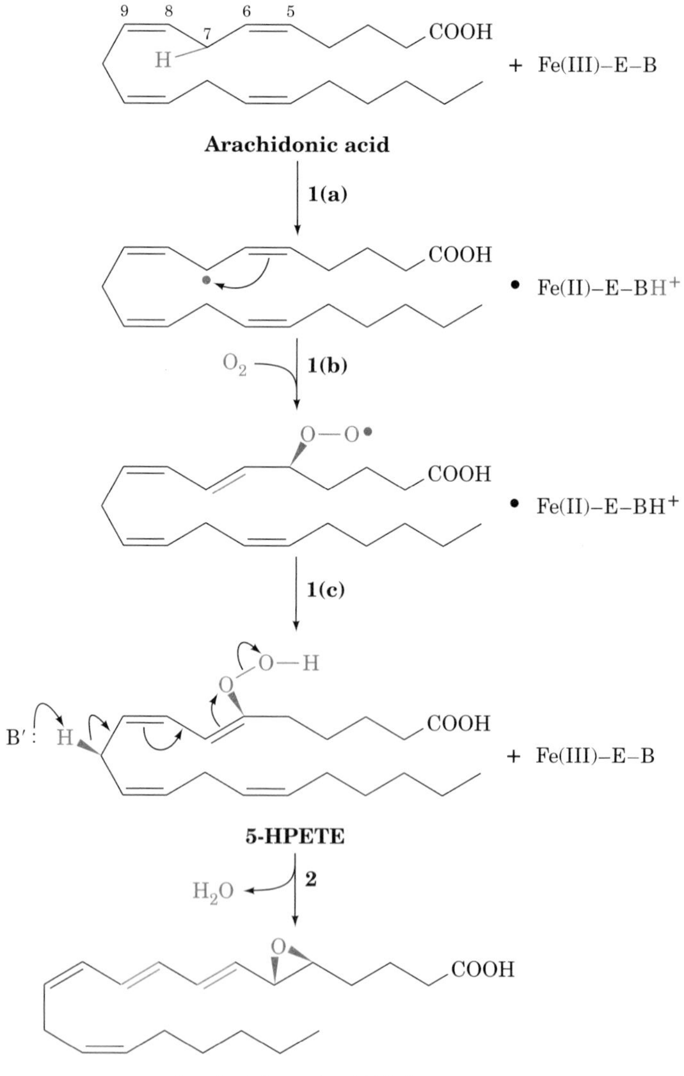

FIGURE 25-73   The 5-LO–catalyzed oxidation of arachidonic acid to LTA₄ via the intermediate 5-HPETE.

1. The oxidation of arachidonic acid to form 5-HPETE, a substance that, in itself, is not a physiological mediator. This reaction occurs in three steps:

   (a) The active site iron atom, in its active Fe(III) state, abstracts an electron from the central methylene group of the 5,8-pentadiene moiety of arachidonate and the resulting free radical loses a proton to an enzymatic base.

   (b) The free radical rearranges and adds O_2 to form a hydroperoxide radical.

   (c) The hydroperoxide radical reacts with the active site iron, now in its Fe(II) form, to yield the hydroperoxide in its anionic form, which the enzyme then protonates to yield the hydroperoxide product, regenerating the active Fe(III) enzyme.

2. The base-catalyzed elimination of water to form the unstable epoxide **leukotriene A_4** (**LTA_4**; the subscript indicates the number of carbon–carbon double bonds in the molecule, which is also its series number).

The X-ray structure of the rabbit reticulocyte 15-LO, a homolog of 5-LO, in complex with the competitive inhibitor **RS75091**

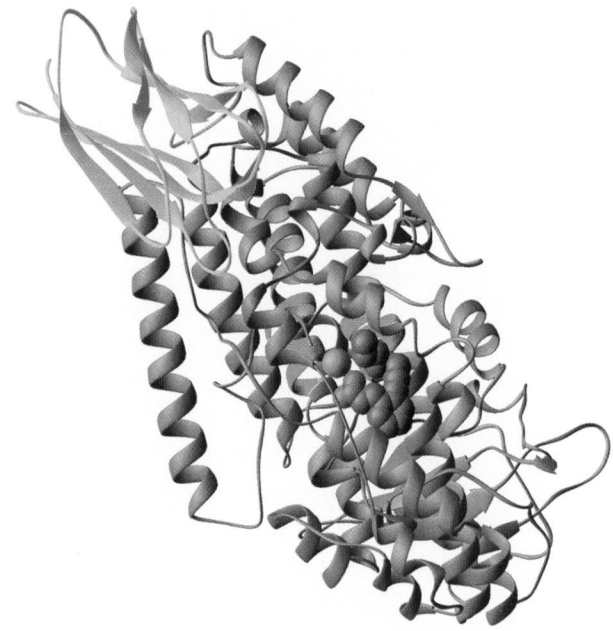

RS75091

FIGURE 25-74   X-Ray structure of rabbit reticulocyte 15-lipoxygenase (15-LO) in complex with its competitive inhibitor RS75091. The N-terminal β barrel domain is gold and the C-terminal catalytic domain is light blue with its two Fe-liganding π helical segments magenta. The Fe is represented by an orange sphere and the RS75091 is drawn in space-filling form with C green and O red. [Based on an X-ray structure by Michelle Browner, Roche Bioscience, Palo Alto, California. PDBid 1LOX.]

was determined by Michelle Browner. This 663-residue monomeric protein consists of an N-terminal 8-stranded β barrel domain and a C-terminal catalytic domain (Fig. 25-74). Its active site Fe atom is coordinated by four invariant His residues and by a C-terminal carboxylate oxygen in a liganding arrangement that is best described as a distorted octahedron with one of its six vertices unoccupied. The Fe, which is well below the protein surface, faces an internal cavity occupied by RS75091. This identifies the substrate binding cavity, which is lined with mostly hydrophobic residues and follows an irregular pathway past the Fe atom to the protein surface. Intriguingly, 15-LO (as does soybean **lipoxygenase-1,** the only other lipoxygenase of known structure) contains two rarely observed π helices (Fig. 8-14*c*), each of which contains two of the Fe-liganding His residues. Each of these π helices is embedded in a longer helix rather than being at the end of an α helix as is the case for all previously observed π helices.

The sizes of the substrate-binding cavities of the 5- and 12-LOs have been predicted through their homology modeling (Section 9-3B) with 15-LO. 5-LO and 12-LO have smaller amino acids substituted for those in 15-LO, such that, for example, 5-LO is predicted to have a cavity with ~20% greater volume than that of 15-LO. The mutagenesis of 5-LO by Harmut Kuhn so as to decrease the size of

its cavity yielded an enzyme with the specificity of 15-LO, thus supporting the proposal that it is the size of the cavity that determines lipoxygenase specificity.

b. Peptidoleukotrienes

   LTA_4 is converted to peptidoleukotrienes by reaction with **LTC_4 synthase,** a **glutathione-S-transferase** that catalyzes the addition of the glutathione sulfhydryl group to the LTA_4 epoxide, forming the first of the peptidoleukotrienes, **leukotriene C_4** (**LTC_4,** Fig. 25-75). **γ-Glutamyltransferase** removes glutamic acid, converting LTC_4 to **leukotriene D_4** (**LTD_4**). LTD_4 is converted to **leukotriene E_4** (**LTE_4**) by a dipeptidase that removes glycine. LTA_4 can also be hydrolyzed to **leukotriene B_4** (**LTB_4**), a potent chemotactic agent (a substance that attracts motile cells) involved in attracting certain types of white blood cells to fight infection.

Various inflammatory and hypersensitivity disorders (such as asthma) are associated with elevated levels of leukotrienes. The development of drugs that inhibit leukotriene synthesis has therefore been an active field of research. 5-LO activity requires the presence of **5-lipoxygenase-activating protein (FLAP),** a 161-residue integral membrane protein. FLAP binds the arachidonic acid substrate of 5-LO and facilitates enzyme–substrate binding as well as 5-LO's interaction with the membrane. Several

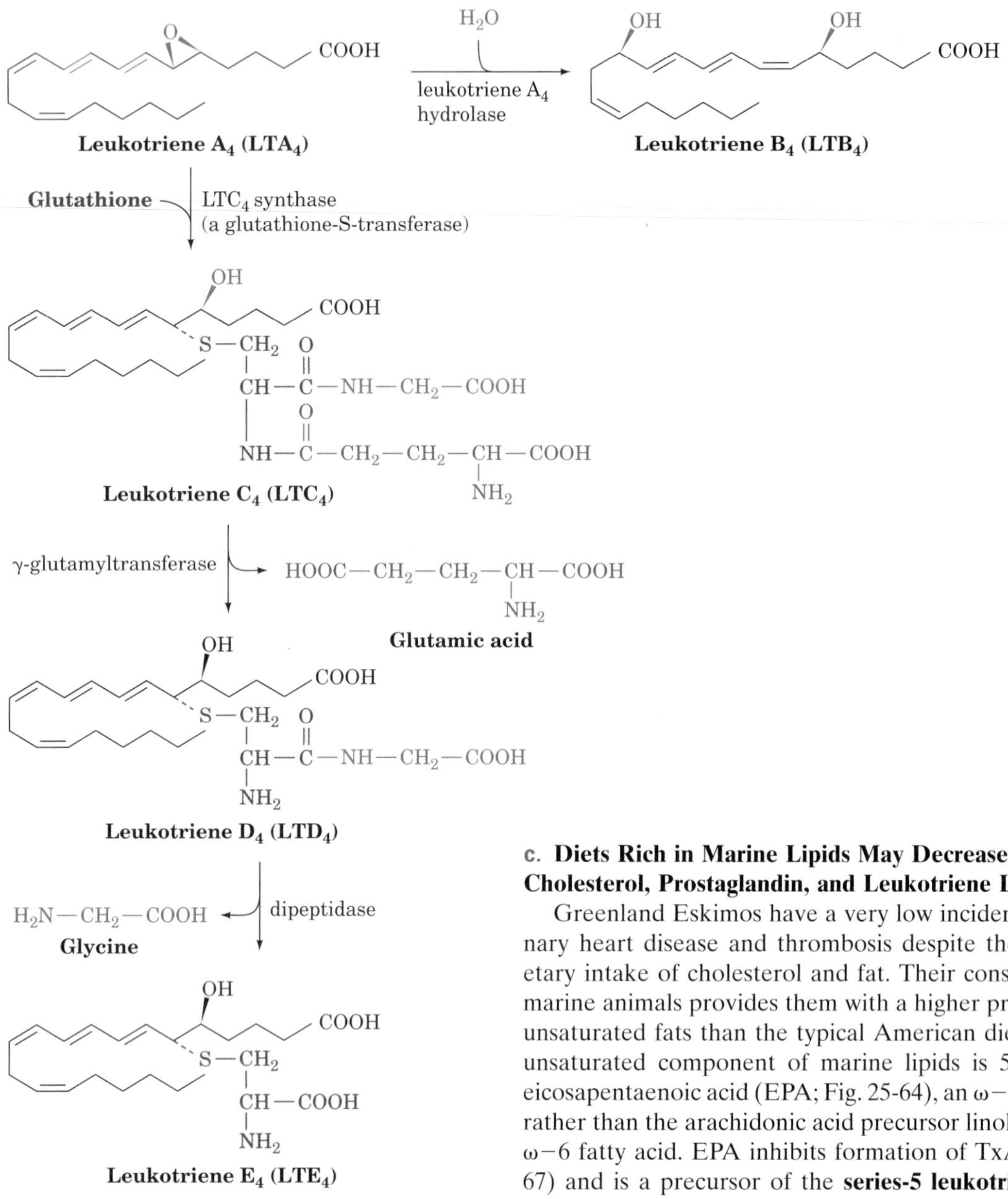

FIGURE 25-75   Formation of the leukotrienes from LTA$_4$.

inhibitors of leukotriene synthesis, such as **MK0886,**

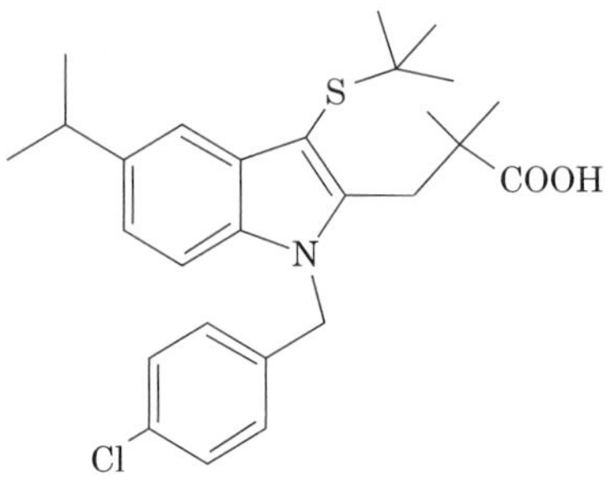

MK0886

bind to FLAP so as to inhibit both of its functions.

c. Diets Rich in Marine Lipids May Decrease Cholesterol, Prostaglandin, and Leukotriene Levels

Greenland Eskimos have a very low incidence of coronary heart disease and thrombosis despite their high dietary intake of cholesterol and fat. Their consumption of marine animals provides them with a higher proportion of unsaturated fats than the typical American diet. A major unsaturated component of marine lipids is 5,8,11,14,17-eicosapentaenoic acid (EPA; Fig. 25-64), an ω−3 fatty acid, rather than the arachidonic acid precursor linoleic acid, an ω−6 fatty acid. EPA inhibits formation of TxA$_2$ (Fig. 25-67) and is a precursor of the **series-5 leukotrienes,** compounds with substantially lower physiological activities than their arachidonate-derived (series-4) counterparts. This suggests that a diet containing marine lipids should decrease the extent of prostaglandin- and leukotriene-mediated inflammatory responses. Indeed, dietary enrichment with EPA inhibits the *in vitro* chemotactic and aggregating activities of neutrophils (a type of white blood cell). Moreover, an EPA-rich diet decreases the cholesterol and triacylglycerol levels in the plasma of hypertriacylglycerolemic patients.

d. Lipoxins and Aspirin-Induced *epi*-Lipoxins Have Anti-Inflammatory Properties

Eicosanoids are usually associated with the inflammatory response. However, some eicosanoids have anti-inflammatory properties. The **lipoxins (LXs),** products of the 12- and 15-LO pathways (sometimes also involving 5-LO), are so named because they are synthesized through *lipoxygenase interactions. Their activities appear to inhibit those of leukotrienes and hence they are anti-inflamma-

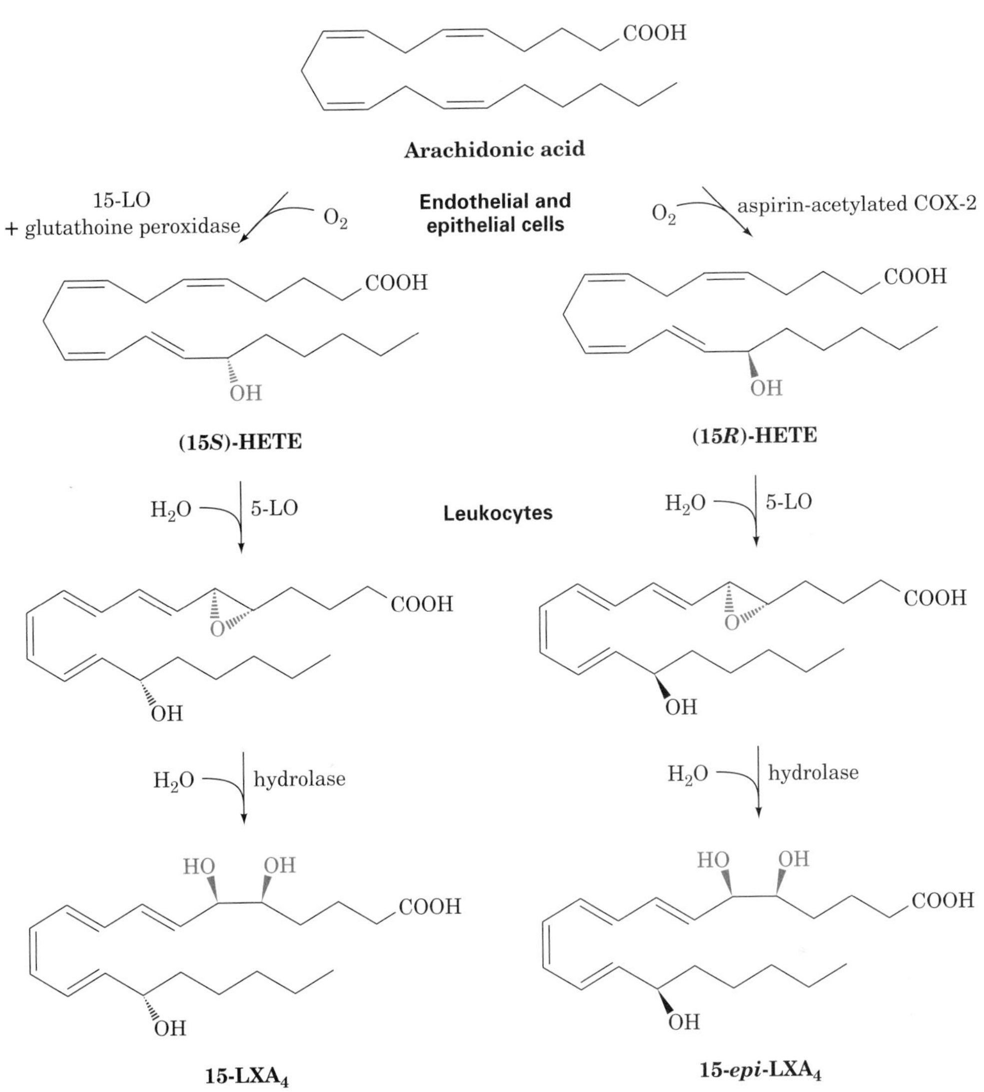

FIGURE 25-76 Lipoxin biosynthesis. The biosynthesis of the lipoxin LXA$_4$ (*left*) and the aspirin-triggered *epi*-lipoxin (ATL) 15-*epi*-LXA$_4$ (*right*). In endothelial and epithelial cells, arachidonic acid is converted by 15-LO and glutathione peroxidase to (15S)-HETE or by aspirin-acetylated COX-2 to (15R)-HETE. After transfer to leukocytes, 5-LO and a hydrolase convert these intermediate products to LXA$_4$ and 15-*epi*-LXA$_4$.

tory. There are many ways for LXs to be synthesized by combinations of the actions of 5-, 12-, and 15-LOs. Here we discuss only one such route (Fig. 25-76, *left*). **Lipoxin A$_4$ (LXA$_4$)** synthesis from arachidonic acid begins in endothelial and epithelial cells by the 15-LO–catalyzed synthesis of **(15S)-hydroperoxyeicosatetraenoic acid [(15S)-HPETE]**, which is reduced by **glutathione peroxidase** to **(15S)-hydroxyeicosatetraenoic acid [(15S)-HETE].** The (15S)-HETE then makes its way to leukocytes where it is converted to LXA$_4$ by 5-LO and a hydrolase.

Charles Serhan discovered an additional pathway for the anti-inflammatory action of aspirin that also involves lipoxin production. As we have previously seen (Fig. 25-71), aspirin covalently inhibits the cyclooxygenase activity of PGHS (COX). However, aspirin-acetylated COX-2 retains a residual 15-LO activity (Steps 3 and 4 in Fig. 25-68) through which it initiates a pathway that converts arachidonic acid to the anti-inflammatory agents called **aspirin-triggered *epi*-lipoxins** (ATLs; Fig. 25-76, *right*). This pathway begins in endothelial and epithelial cells with

the aspirin-acetylated COX-2–catalyzed conversion of arachidonic acid to **(15R)-hydroxyeicosatetraenoic acid [(15R)-HETE],** the epimer of (15S)-HETE. In leukocytes, 5-LO and a hydrolase then convert (15R)-HETE to the anti-inflammatory agent **15-*epi*-lipoxin A$_4$ (15-*epi*-LXA$_4$).**

These are indeed exciting times in the study of eicosanoid metabolism and its physiological manifestations. As the mechanisms of action of the prostaglandins, prostacyclins, thromboxanes, leukotrienes, and lipoxins are becoming better understood, they are providing the insights required for the development of new and improved therapeutic agents.

8 ■ PHOSPHOLIPID AND GLYCOLIPID METABOLISM

The "complex lipids" are dual-tailed amphipathic molecules composed of either 1,2-diacyl-sn-glycerol or N-acylsphingosine (ceramide) linked to a polar head group that is ei-

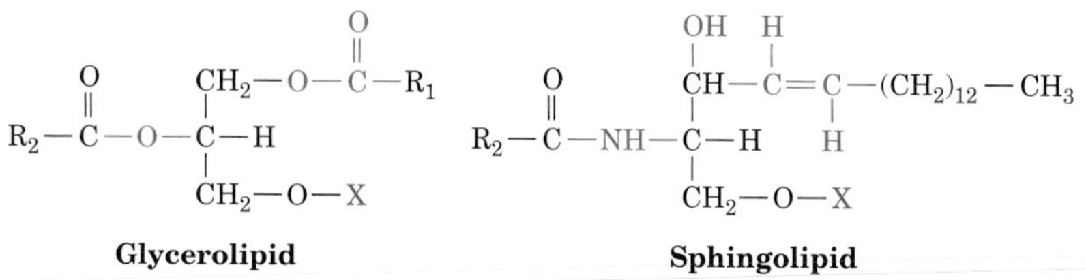

FIGURE 25-77   The glycerolipids and sphingolipids. The structures of the common head groups, X, are presented in Table 12-2.

X = H	**1,2-Diacylglycerol**
X = Carbohydrate	**Glyceroglycolipid**
X = Phosphate ester	**Glycerophospholipid**

N-**Acylsphingosine (ceramide)**
Sphingoglycolipid (glycosphingolipid)
Sphingophospholipid

ther a carbohydrate or a phosphate ester (Fig. 25-77; Sections 12-1C and 12-1D; *sn* stands for stereospecific numbering, which assigns the 1 position to the group occupying the *pro-S* position of a prochiral center). Hence, there are two categories of phospholipids, **glycerophospholipids** and **sphingophospholipids,** and two categories of glycolipids, **glyceroglycolipids** and **sphingoglycolipids** (also called **glycosphingolipids; GSLs**). In this section we describe the biosynthesis of the complex lipids from their simpler components. We shall see that the great variety of these substances is matched by the numerous enzymes required for their specific syntheses. Note also that these substances are synthesized in membranes, mostly on the cytosolic face of the endoplasmic reticulum, and from there are transported to their final cellular destinations as indicated in Sections 12-4B–D.

A. *Glycerophospholipids*

Glycerophospholipids have significant asymmetry in their C1- and C2-linked fatty acyl groups: C1 substituents are mostly saturated fatty acids, whereas those at C2 are by and large unsaturated fatty acids. We shall examine the major pathways of biosynthesis and metabolism of the glycerophospholipids with an eye toward understanding the origin of this asymmetry.

a. Biosynthesis of Diacylglycerophospholipids

The triacyglycerol precursors 1,2-diacyl-sn-glycerol and phosphatidic acid are also the precursors of certain glycerophospholipids (Figs. 25-39 and 25-77). Activated phosphate esters of the polar head groups (Table 12-2) react with the C3–OH group of 1,2-diacyl-*sn*-glycerol to form the phospholipid's phosphodiester bond. In some cases the phosphoryl group of phosphatidic acid is activated and reacts with the unactivated polar head group.

The mechanism of activated phosphate ester formation is the same for both the polar head groups **ethanolamine** and **choline** (Fig. 25-78):

1. ATP first phosphorylates the OH group of choline or ethanolamine.

2. The phosphoryl group of the resulting **phospho-ethanolamine** or **phosphocholine** then attacks CTP, dis-

placing PP*i*, to form the corresponding CDP derivatives, which are activated phosphate esters of the polar head group.

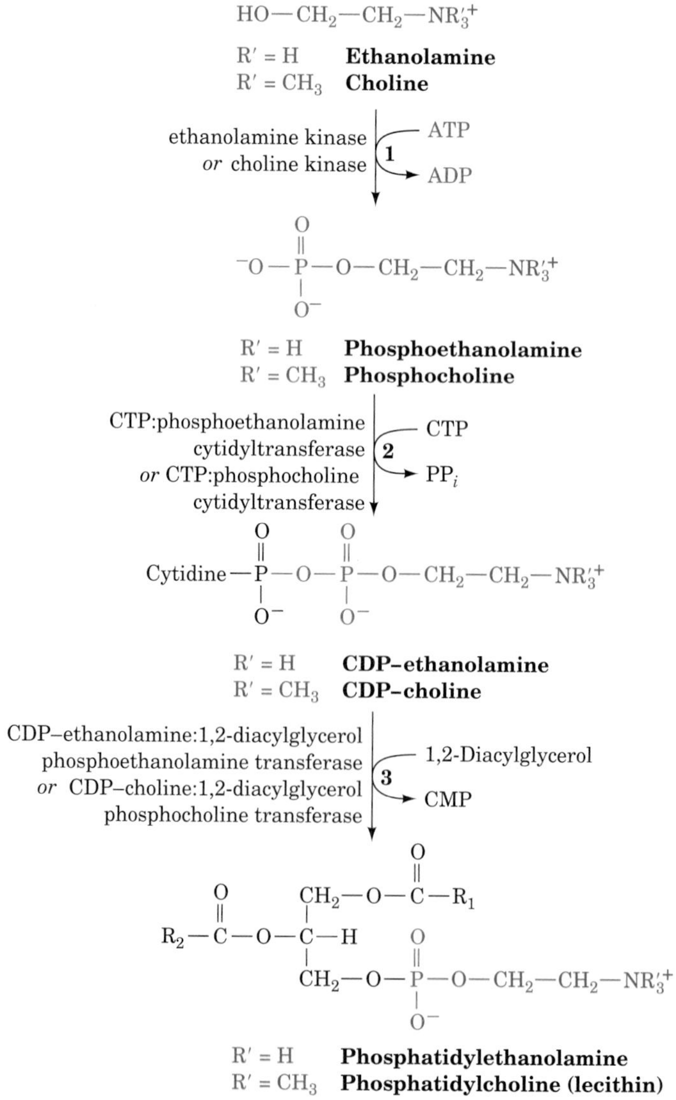

FIGURE 25-78   The biosynthesis of phosphatidylethanolamine and phosphatidylcholine. In mammals, CDP–ethanolamine and CDP–choline are the precursors of the head groups.

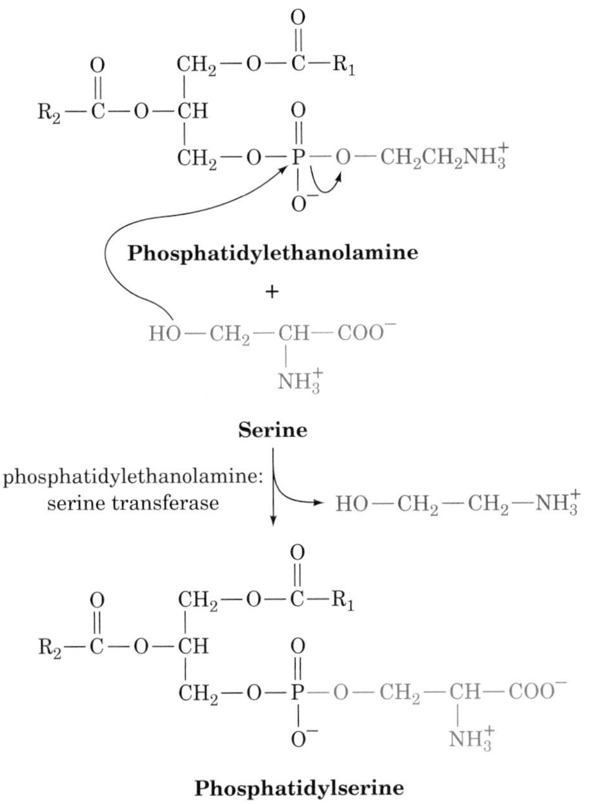

Phosphatidylethanolamine

+

Serine

phosphatidylethanolamine:
serine transferase

Phosphatidylserine

FIGURE 25-79 Phosphatidylserine synthesis. Serine replaces ethanolamine in phosphatidylethanolamine by a head group exchange reaction.

3. The C3–OH group of 1,2-diacyl-*sn*-glycerol attacks the phosphoryl group of the activated CDP–ethanolamine or CDP–choline, displacing CMP to yield the corresponding glycerophospholipid.

The liver also converts phosphatidylethanolamine to phosphatidylcholine by trimethylating its amino group, using **S-adenosylmethionine** (Section 26-3E) as the methyl donor.

Phosphatidylserine is synthesized from phosphatidylethanolamine by a head group exchange reaction catalyzed by **phosphatidylethanolamine:serine transferase** in which serine's OH group attacks the donor's phosphoryl group (Fig. 25-79). The original head group is then eliminated, forming phosphatidylserine.

In the synthesis of **phosphatidylinositol** and **phosphatidylglycerol,** the hydrophobic tail is activated rather than the polar head group. Phosphatidic acid, the precursor of 1,2-diacyl-*sn*-glycerol (Fig. 25-39), attacks the α-phosphoryl group of CTP to form the activated **CDP–diacylglycerol** and PP$_i$ (Fig. 25-80). Phosphatidylinositol results from the attack of inositol on CDP–diacylglycerol. Phosphatidylglycerol is formed in two reactions: (1) attack of the C1–OH group of *sn*-glycerol-3-phosphate on CDP–diacylglycerol, yielding **phosphatidylglycerol phosphate;** and (2) hydrolysis of the phosphoryl group to form phosphatidylglycerol.

Cardiolipin, an important phospholipid first isolated from heart tissue, is synthesized from two molecules of phosphatidylglycerol (Fig. 25-81). The reaction occurs by the attack of the C1–OH group of one of the phosphatidylglycerol molecules on the phosphoryl group of the other, displacing a molecule of glycerol.

Enzymes that synthesize phosphatidic acid have a general preference for saturated fatty acids at C1 and for unsaturated fatty acids at C2. Yet, this general preference cannot account, for example, for the observations that ~80% of brain phosphatidylinositol has a stearoyl group (18:0) at C1 and an arachidonoyl group (20:4) at C2, and that ~40% of lung phosphatidylcholine has palmitoyl groups (16:0) at both positions (this latter substance is the major component of the surfactant that prevents the lung from collapsing when air is expelled; its deficiency is responsible for **respiratory distress syndrome** in premature infants). William Lands showed *that such side chain specificity results from "remodeling" reactions in which specific acyl groups of individual glycerophospholipids are exchanged by specific phospholipases and acyltransferases.*

b. Biosynthesis of Plasmalogens and Alkylacylglycerophospholipids

Eukaryotic membranes contain significant amounts of two other types of glycerophospholipids:

1. Plasmalogens, which contain a hydrocarbon chain linked to glycerol C1 via a vinyl ether linkage:

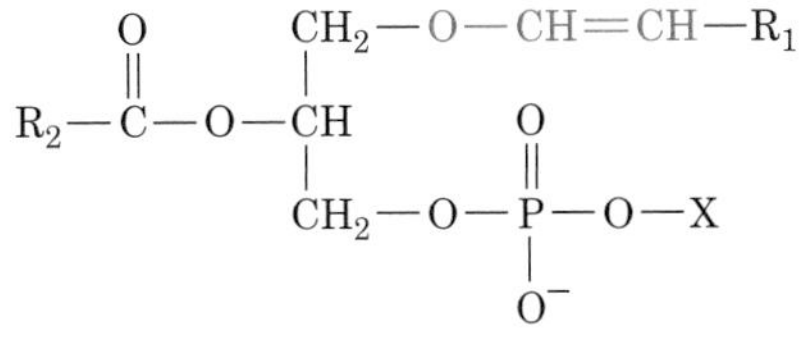

A plasmalogen

2. Alkylacylglycerophospholipids, in which the alkyl substituent at glycerol C1 is attached via an ether linkage:

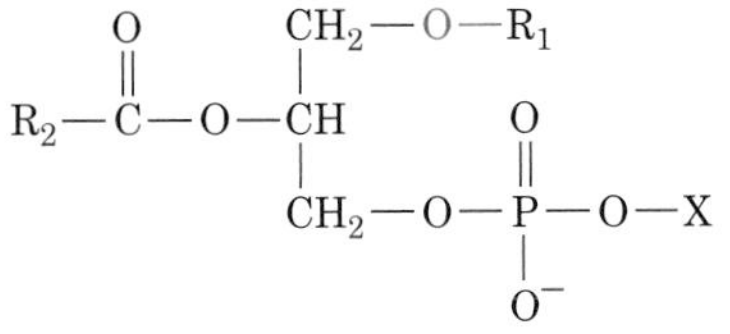

**An alkylacyl-
glycerophospholipid**

About 20% of mammalian glycerophospholipids are plasmalogens. The exact percentage varies both from species to species and from tissue to tissue within a given organism. While plasmalogens comprise only 0.8% of the phospholipids in human liver, they account for 23% of those in human nervous tissue. The alkylacylglycerophospholipids are less abundant than the plasmalogens; for instance, 59% of the ethanolamine glycerophospholipids of

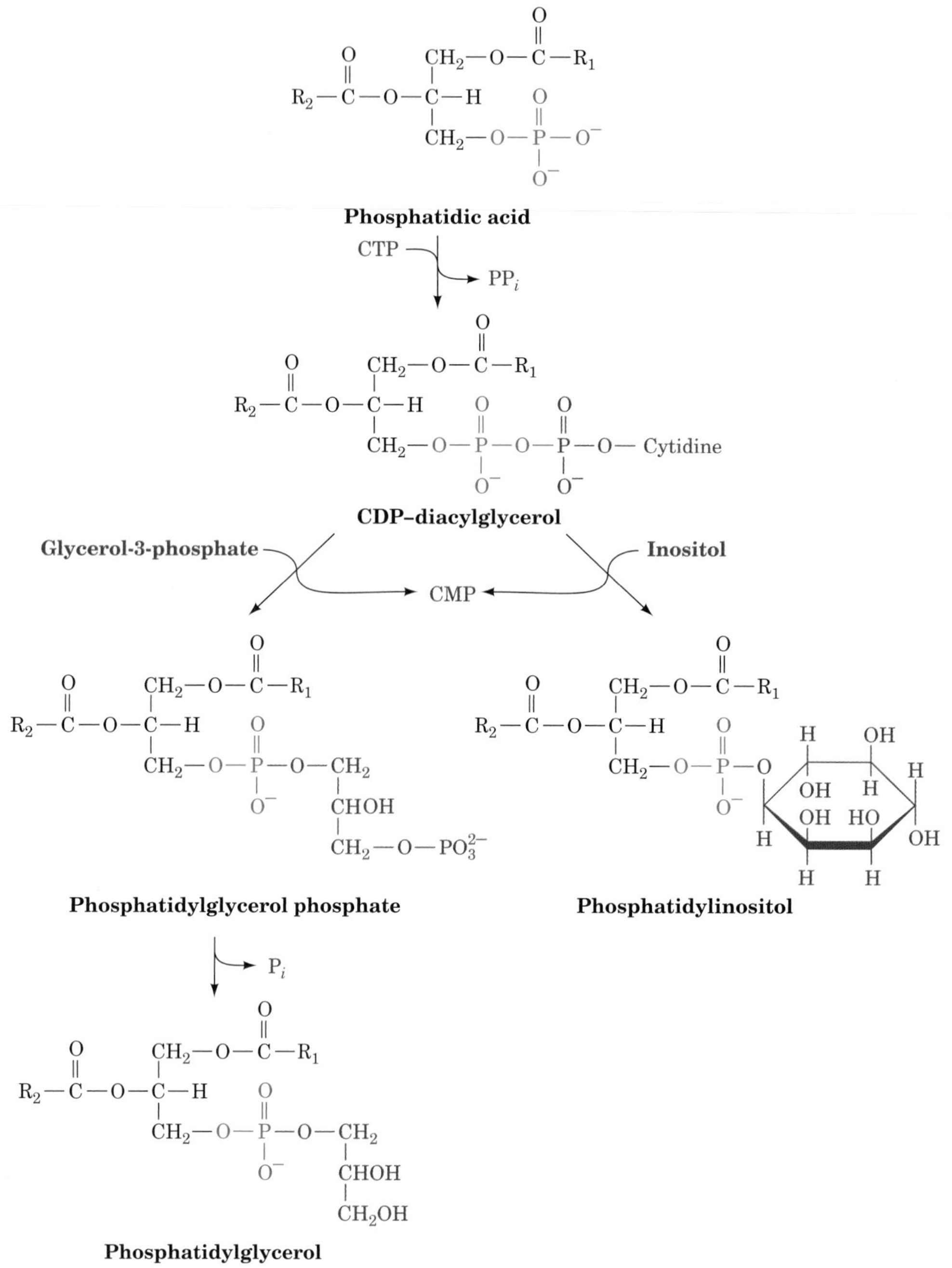

Phosphatidic acid

CTP ⟶ PP$_i$

CDP–diacylglycerol

Glycerol-3-phosphate ⟶ CMP ⟵ Inositol

Phosphatidylglycerol phosphate

Phosphatidylinositol

⟶ P$_i$

Phosphatidylglycerol

FIGURE 25-80   The biosynthesis of phosphatidylinositol and phosphatidylglycerol. In mammals, this process involves a CDP–diacylglycerol intermediate.

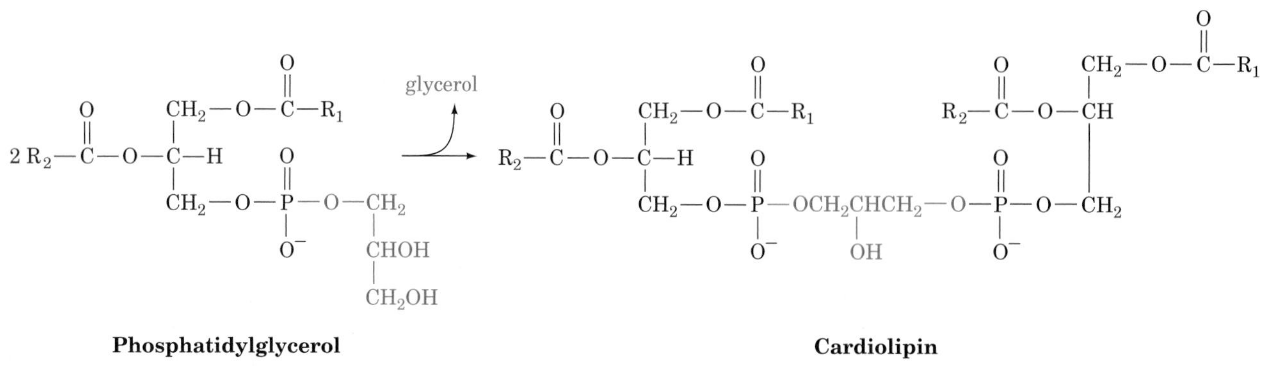

glycerol

Phosphatidylglycerol

Cardiolipin

FIGURE 25-81   The formation of cardiolipin.

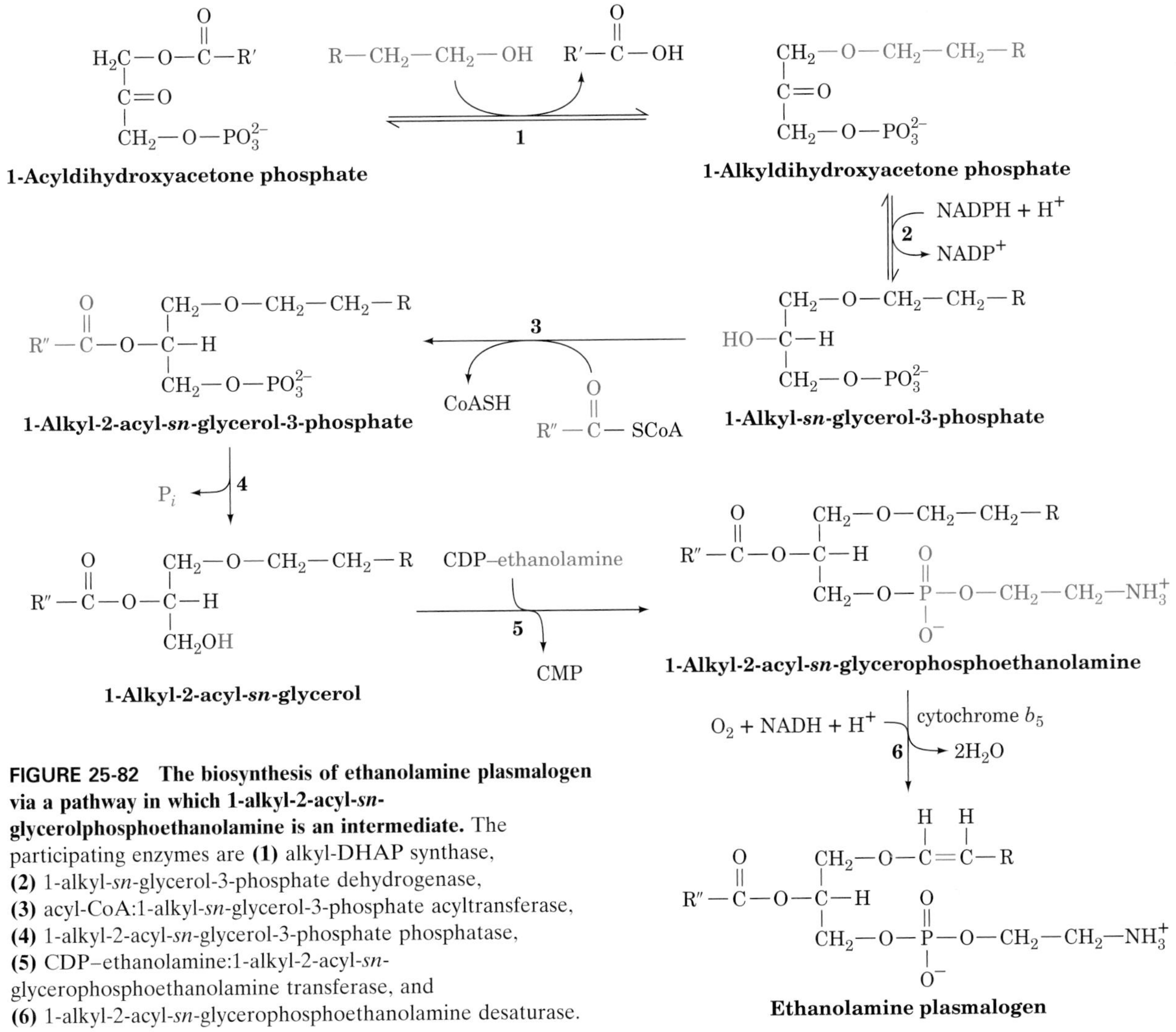

FIGURE 25-82 **The biosynthesis of ethanolamine plasmalogen via a pathway in which 1-alkyl-2-acyl-*sn*-glycerolphosphoethanolamine is an intermediate.** The participating enzymes are **(1)** alkyl-DHAP synthase, **(2)** 1-alkyl-*sn*-glycerol-3-phosphate dehydrogenase, **(3)** acyl-CoA:1-alkyl-*sn*-glycerol-3-phosphate acyltransferase, **(4)** 1-alkyl-2-acyl-*sn*-glycerol-3-phosphate phosphatase, **(5)** CDP–ethanolamine:1-alkyl-2-acyl-*sn*-glycerophosphoethanolamine transferase, and **(6)** 1-alkyl-2-acyl-*sn*-glycerophosphoethanolamine desaturase.

human heart are plasmalogens, whereas only 3.6% are alkylacylglycerophospholipids. However, in bovine erythrocytes, 75% of the ethanolamine glycerophospholipids are of the alkylacyl type.

The pathway forming ethanolamine plasmalogens and alkylacylglycerophospholipids involves several reactions (Fig. 25-82):

1. Exchange of the acyl group of **1-acyldihydroxyacetone phosphate** for an alcohol.

2. Reduction of the ketone to **1-alkyl-*sn*-glycerol-3-phosphate.**

3. Acylation of the resulting C2–OH group by acyl-CoA.

4. Hydrolysis of the phosphoryl group to yield an alkylacylglycerol.

5. Attack by the new OH group of alkylacylglycerol on CDP–ethanolamine to yield **1-alkyl-2-acyl-*sn*-glycerophosphoethanolamine.**

6. Introduction of a double bond into the alkyl group to form the plasmalogen by a desaturase having the same

cofactor requirements as the fatty acid desaturases (Section 25-4E).

Recall that the precursor–product relationship between the alkylacylglycerophospholipid and the plasmalogen was established through studies using [^{14}C]ethanolamine (Section 16-3B).

The plasmalogen with an acetyl group at R_2 and a choline polar head group (X), **1-*O*-hexadec-1'-enyl-2-acetyl-*sn*-glycero-3-phosphocholine,** is known as **platelet-activating factor (PAF).** This molecule has diverse functions and acts at very low concentrations ($10^{-10}M$) to lower blood pressure and to cause blood platelets to aggregate.

B. *Sphingophospholipids*

Only one major phospholipid contains ceramide (*N*-acylsphingosine) as its hydrophobic tail: **sphingomyelin (*N*-acylsphingosine phosphocholine;** Section 12-1D), an important structural lipid of nerve cell membranes. The molecule was once thought to be synthesized from *N*-acyl-

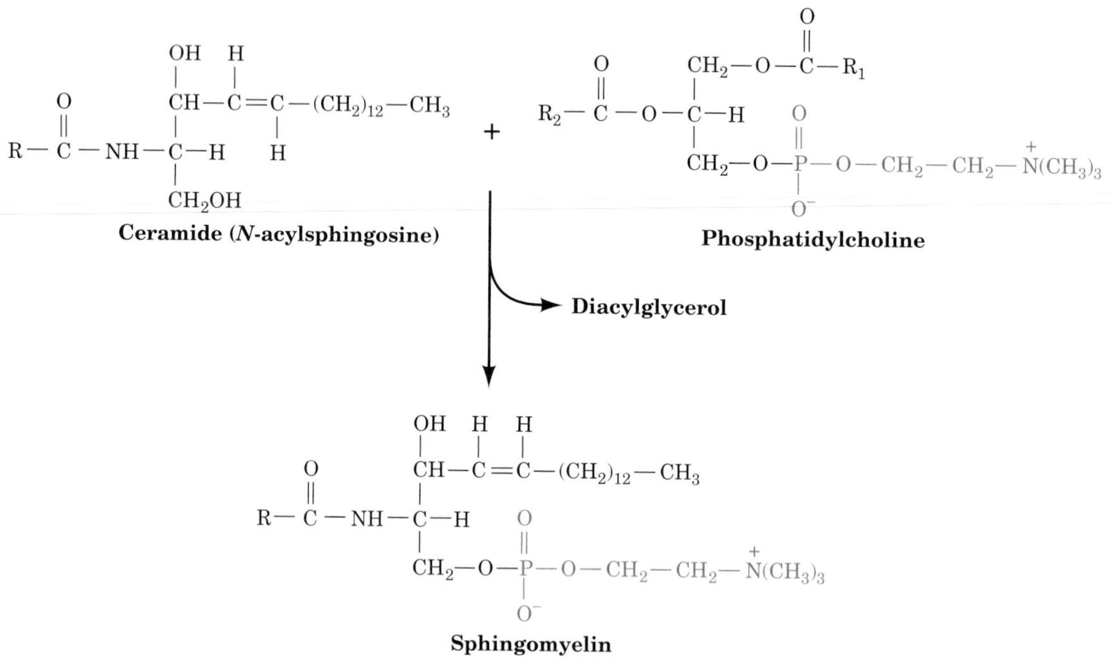

FIGURE 25-83   **The synthesis of sphingomyelin from *N*-acylsphingosine and phosphatidylcholine.**

sphingosine and CDP–choline. Recent evidence has shown, however, that the main route of sphingomyelin synthesis occurs through donation of the phosphocholine group of phosphatidylcholine to *N*-acylsphingosine (Fig. 25-83). These pathways were differentiated by establishing the precursor–product relationships between CDP–choline, phosphatidylcholine, and sphingomyelin (Section 16-3B). Mouse liver microsomes were isolated and incubated for a short time with [^{3}H]choline. Radioactivity appeared in sphingomyelin only after first appearing in both CDP–choline and phosphatidylcholine, ruling out the direct transfer of phosphocholine from CDP–choline to *N*-acylsphingosine.

The most prevalent acyl groups of sphingomyelin are palmitoyl (16:0) and stearoyl (18:0) groups. Longer chain fatty acids such as nervonic acid (24:1) and behenic acid (22:0) occur with lesser frequency in sphingomyelins.

C. *Sphingoglycolipids*

Most sphingolipids are sphingoglycolipids, that is, their polar head groups consist of carbohydrate units (Section 12-1D). The principal classes of sphingoglycolipids, as indicated in Fig. 25-84, are **cerebrosides** (ceramide monosaccharides), **sulfatides** (ceramide monosaccharide sulfates), **globosides** (neutral ceramide oligosaccharides), and **gangliosides** (acidic, sialic acid–containing ceramide oligosaccharides). The carbohydrate unit is glycosidically attached to the *N*-acylsphingosine at its C1–OH group (Fig. 25-77).

The lipids providing the carbohydrate that covers the external surfaces of eukaryotic cells are sphingoglycolipids.

Along with glycoproteins (Section 23-3), they are biosynthesized on the luminal surfaces of the endoplasmic reticulum and the Golgi apparatus and reach the plasma membrane through vesicle flow (Sections 12-4C and 12-4D), where membrane fusion results in their facing the external surface of the lipid bilayer (Fig. 12-53). Degradation of sphingoglycolipids occurs in the lysosomes after endocytosis from the plasma membrane.

In the following subsections, we discuss the biosynthesis and breakdown of *N*-acylsphingosine and sphingoglycolipids and consider the diseases caused by deficiencies in their degradative enzymes.

a. Biosynthesis of Ceramide (*N*-Acylsphingosine)

Biosynthesis of *N*-acylsphingosine occurs in four reactions from the precursors palmitoyl-CoA and serine (Fig. 25-85):

1. 3-Ketosphinganine synthase (serine palmitoyltransferase), a pyridoxal phosphate–dependent enzyme, catalyzes the condensation of palmitoyl-CoA with serine yielding **3-ketosphinganine** (pyridoxal phosphate–dependent reactions are discussed in Section 26-1A).

2. 3-Ketosphinganine reductase catalyzes the NADPH-dependent reduction of 3-ketosphinganine's keto group to form **sphinganine (dihydrosphingosine).**

3. Dihydroceramide is formed by transfer of an acyl group from an acyl-CoA to the sphinganine's 2-amino group, forming an amide bond.

4. Dihydroceramide reductase converts dihydroceramide to ceramide by an FAD-dependent oxidation reaction.

Cerebrosides

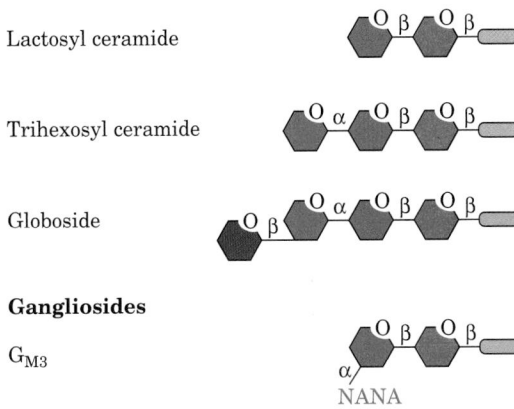

Glucocerebroside

Galactocerebroside

Sulfatide

Globosides

Lactosyl ceramide

Trihexosyl ceramide

Globoside

Gangliosides

G_{M3}

G_{M2}

G_{M1}

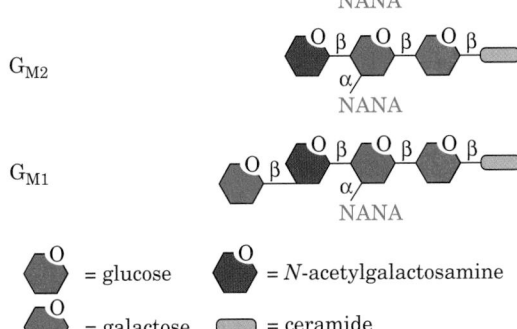

NANA = *N*-acetylneuraminic acid (sialic acid)

FIGURE 25-84 Diagrammatic representation of the principal classes of sphingoglycolipids. The G_M ganglioside structures are presented in greater detail in Fig. 12-7.

b. Biosynthesis of Cerebrosides (Glycosylceramides)

Galactocerebroside (1-β-galactosylceramide) and **glucocerebroside (1-β-glucosylceramide)** are the two most common cerebrosides. In fact, the term cerebroside is often used synonymously with galactocerebroside. Both are synthesized from ceramide by addition of a glycosyl unit from the corresponding UDP–hexose (Fig. 25-86). Galactocerebroside is a common component of brain lipids. Glucocerebroside, although relatively uncommon, is the precursor of globosides and gangliosides.

c. Biosynthesis of Sulfatides

Sulfatides (galactocerebroside-3-sulfate) account for 15% of the lipids of white matter in the brain. They are

Palmitoyl-CoA   **Serine**

1 | 3-ketosphinganine synthase
$\rightarrow$ CO_2 + CoASH

**3-Ketosphinganine
(3-ketodihydrosphingosine)**

2 | NADPH + H$^+$
3-ketosphinganine reductase
$\rightarrow$ NADP$^+$

**Sphinganine
(dihydrosphingosine)**

3 | R—C—SCoA
acyl-CoA transferase
$\rightarrow$ CoASH

**Dihydroceramide
(*N*-acylsphinganine)**

4 | FAD
dihydroceramide reductase
$\rightarrow$ FADH$_2$

**Ceramide
(*N*-acylsphingosine)**

FIGURE 25-85 The biosynthesis of ceramide (*N*-acylsphingosine).

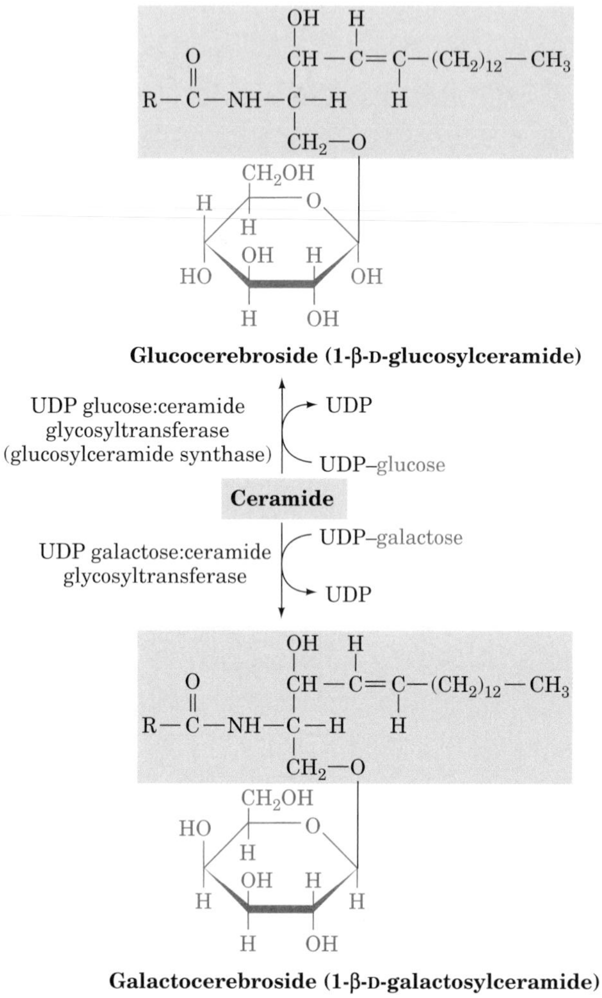

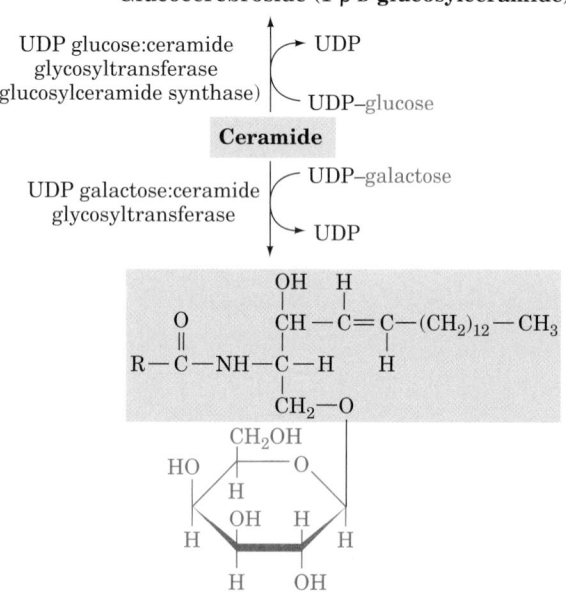

Glucocerebroside (1-β-D-glucosylceramide)

UDP glucose:ceramide
glycosyltransferase
(glucosylceramide synthase)

→ UDP

← UDP–glucose

Ceramide

← UDP–galactose

UDP galactose:ceramide
glycosyltransferase

→ UDP

Galactocerebroside (1-β-D-galactosylceramide)

FIGURE 25-86   The biosynthesis of cerebrosides.

formed by transfer of an "activated" sulfate group from **3′-phosphoadenosine-5′-phosphosulfate (PAPS)** to the C3–OH group of galactose in galactocerebroside (Fig. 25-87).

d. Biosynthesis of Globosides and Gangliosides

Biosynthesis of both globosides (neutral ceramide oligosaccharides) and gangliosides (acidic, sialic acid–containing ceramide oligosaccharides) is catalyzed by a series of **glycosyltransferases.** While the reactions are chemically similar, each is catalyzed by a specific enzyme. The pathways begin with transfer of a galactosyl unit from UDP–Gal to glucocerebroside to form a $\beta(1 \rightarrow 4)$ linkage (Fig. 25-88). Since this bond is the same as that linking glucose and galactose in lactose, this glycolipid is often referred to as **lactosyl ceramide.** Lactosyl ceramide is the precursor of both globosides and gangliosides. To form a globoside, one galactosyl and one *N*-acetylgalactosaminyl unit are sequentially added to lactosyl ceramide from UDP–Gal and UDP–GalNAc, respectively. The G_M gangliosides are formed by addition of *N*-acetylneuraminic acid (NANA, sialic acid)

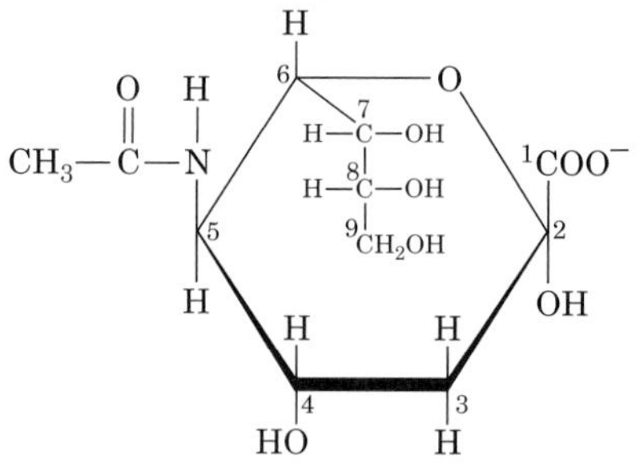

**N-Acetylneuraminic acid
(NANA, sialic acid)**

from CMP–NANA to lactosyl ceramide in $\alpha(2 \rightarrow 3)$ linkage yielding G_{M3}. The sequential additions to G_{M3} of the *N*-acetylgalactosamine and galactose units from UDP–GalNAc and UDP–Gal yield gangliosides G_{M2} and G_{M1}. Other gangliosides are formed by adding a second NANA group to G_{M3}, forming **G_{D3},** or by adding an *N*-acetylglu-

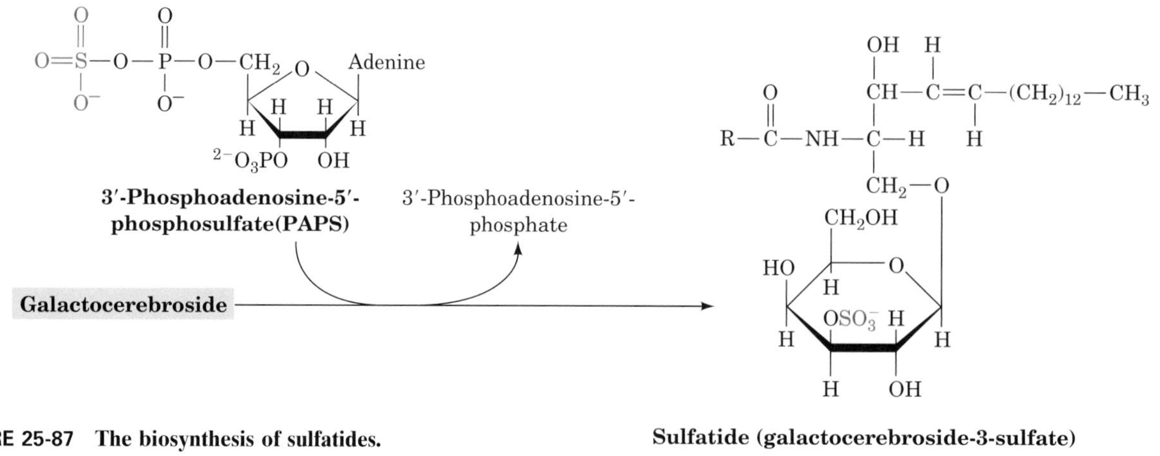

**3′-Phosphoadenosine-5′-
phosphosulfate(PAPS)**

3′-Phosphoadenosine-5′-
phosphate

Galactocerebroside

FIGURE 25-87   The biosynthesis of sulfatides.

Sulfatide (galactocerebroside-3-sulfate)

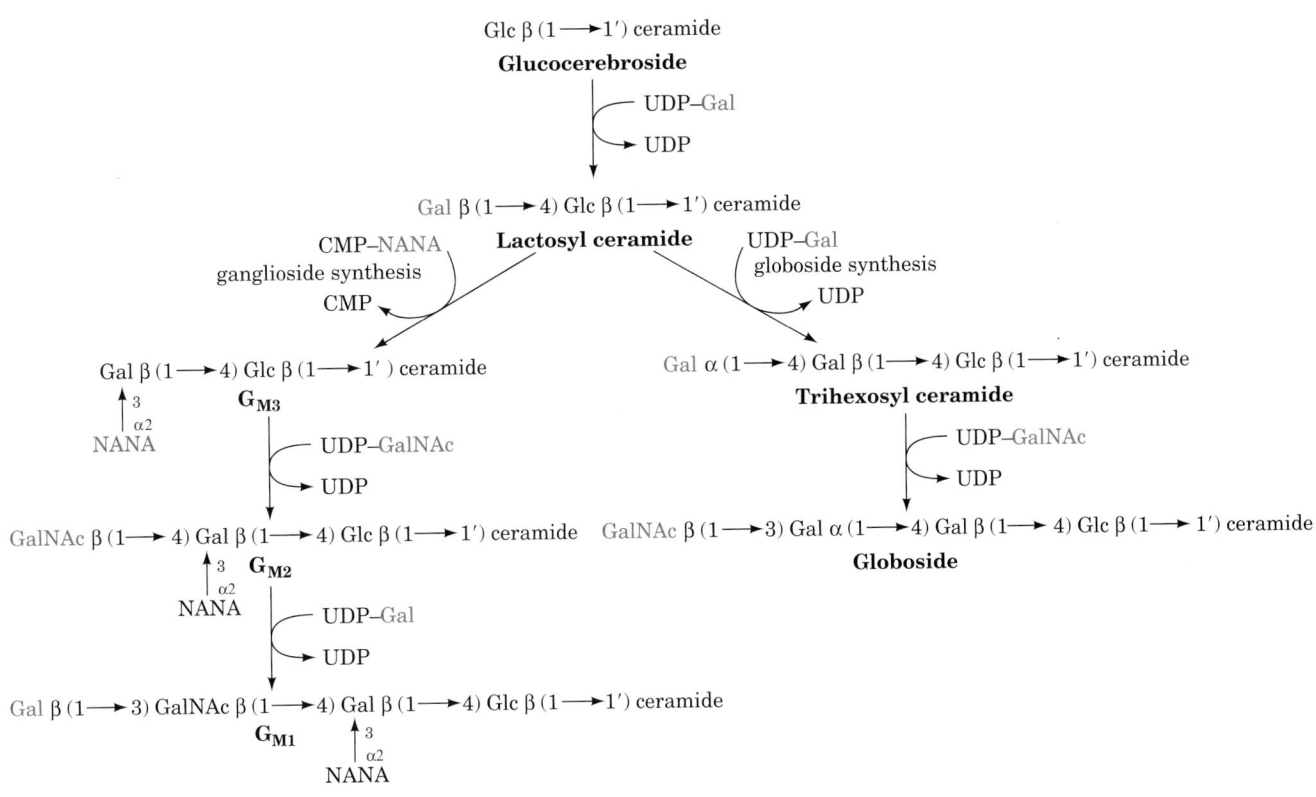

FIGURE 25-88 The biosynthesis of globosides and G_M gangliosides.

cosamine unit to lactosyl ceramide before NANA addition, forming G_{A2}. Over 60 different gangliosides are known.

e. Sphingoglycolipid Degradation and Lipid Storage Diseases

Sphingoglycolipids are lysosomally degraded by a series of enzymatically mediated hydrolytic reactions (Fig. 25-89). These reactions are catalyzed at the lipid–water interface by soluble enzymes, often with the aid of **sphingolipid activator proteins (SAPs; including saposins, G_{M2}-activator protein, and SAP-A through SAP-D).** These nonenzymatic ancillary proteins are thought to increase the accessibility of the carbohydrate moiety of the sphingolipid to the degradation enzyme. For example, G_{M2}-activator binds G_{M2} and helps expose it to the surface of the membrane. The G_{M2}-activator–G_{M2} complex can then bind **hexosaminidase A,** an αβ dimer that hydrolyzes *N*-acetylgalactosamine from G_{M2} at the lipid–water interface (Fig. 25-90).

The hereditary absence of one of the sphingolipid hydrolases or a SAP results in a **sphingolipid storage disease** (Table 25-2). One of the most common such conditions is **Tay–Sachs disease,** an autosomal recessive deficiency in hexosaminidase A. The absence of hexosaminidase A activity results in the neuronal accumulation of G_{M2} as shell-like inclusions (Fig. 25-91).

Although infants born with Tay–Sachs disease at first appear normal, by ~1 year of age, when sufficient G_{M2} has accumulated to interfere with neuronal function, they be-

come progressively weaker, retarded, and blinded until they die, usually by the age of 3 years. It is possible, however, to screen potential carriers of this disease by a simple serum assay. It is also possible to detect the disease *in utero* by assay of amniotic fluid or amniotic cells obtained by amniocentesis. The assay involves use of an artificial hexosaminidase substrate, **4-methylumbelliferyl-β-D-N-acetylglucosamine,** which yields a fluorescent product on hydrolysis:

4-Methylumbelliferyl-β-D-N-acetylglucosamine

4-Methylumbelliferone (fluorescent in alkaline medium)

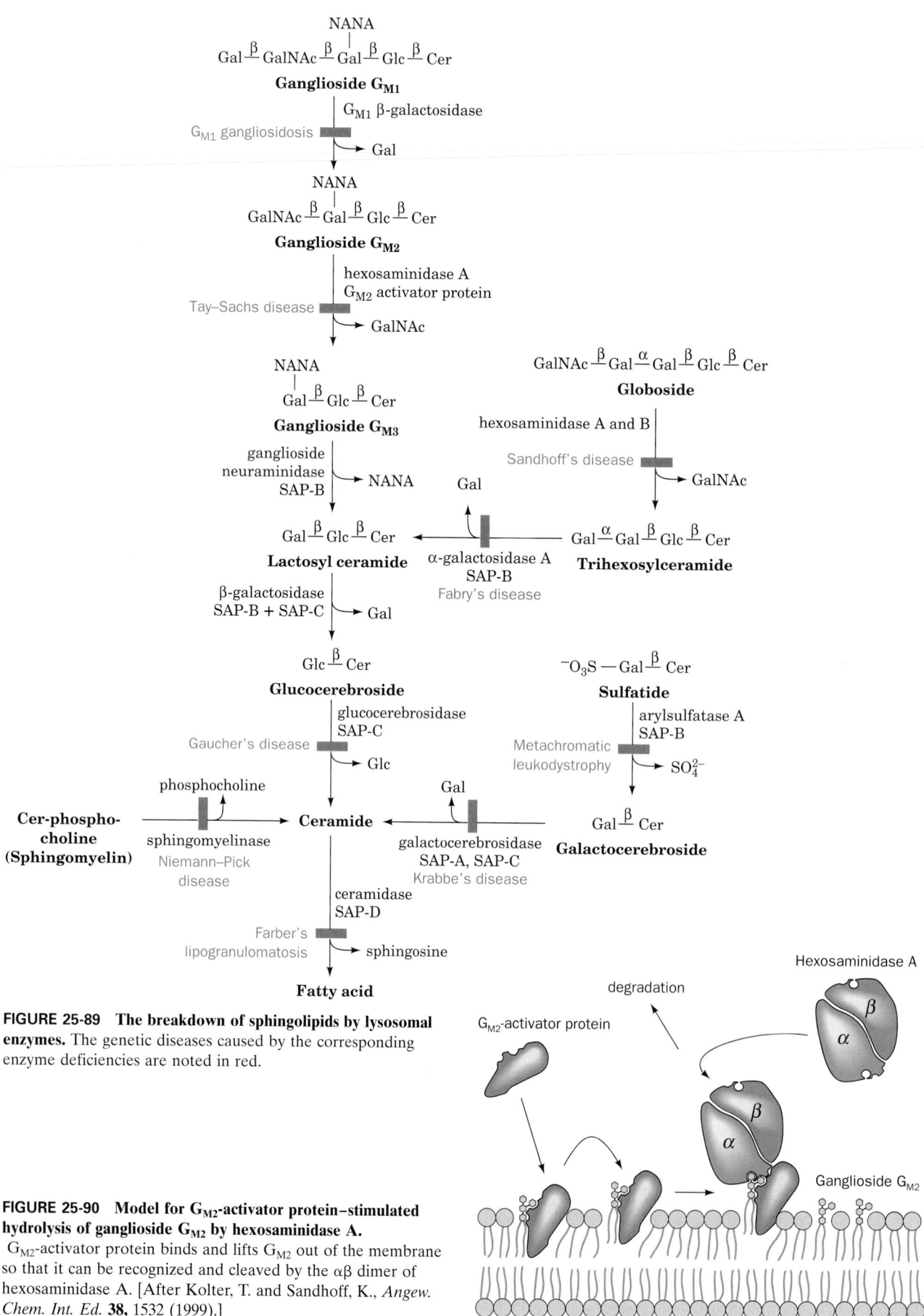

FIGURE 25-89   The breakdown of sphingolipids by lysosomal enzymes. The genetic diseases caused by the corresponding enzyme deficiencies are noted in red.

FIGURE 25-90   Model for G_{M2}-activator protein–stimulated hydrolysis of ganglioside G_{M2} by hexosaminidase A.
G_{M2}-activator protein binds and lifts G_{M2} out of the membrane so that it can be recognized and cleaved by the $\alpha\beta$ dimer of hexosaminidase A. [After Kolter, T. and Sandhoff, K., *Angew. Chem. Int. Ed.* **38,** 1532 (1999).]

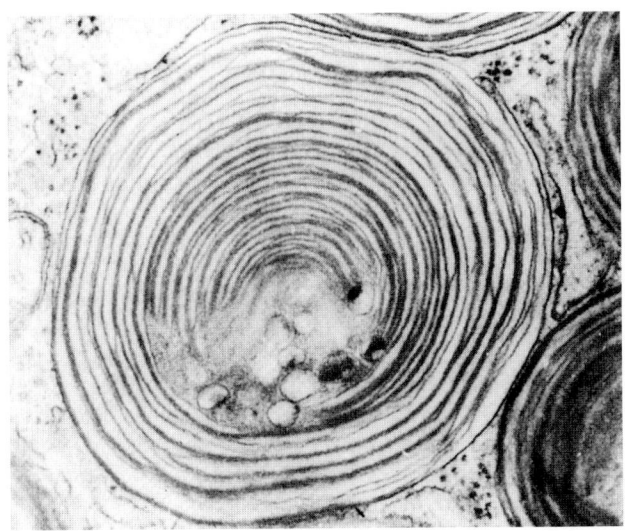

FIGURE 25-91 Cytoplasmic membranous body in a neuron affected by Tay–Sachs disease. [Courtesy of John S. O'Brien, University of California at San Diego Medical School.]

Since this substrate is also recognized by **hexosaminidase B,** which is unaffected in Tay–Sachs disease, the hexosaminidase B is first heat inactivated since it is more heat labile than hexosaminidase A. As a result of mass screening efforts, the tragic consequences of this genetic enzyme deficiency are being averted. The other sphingolipid storage diseases, although less common, have similar consequences (Table 25-2).

Sphingolipid storage diseases result when the synthetic capacity of a cell outstrips its degradative capacity leading to sphingolipid accumulation. A promising new general therapy for many of these diseases is **substrate deprivation therapy,** the inhibition of **glucosylceramide synthase,** the enzyme catalyzing the first committed step in the biosynthesis of globosides and gangliosides (Fig. 25-86). Several inhibitors of this enzyme have been developed and are in clinical trials. In a mouse model of Tay–Sachs disease, the oral administration of *N*-butyldeoxynojirimycin **(NB-DNJ)**

$$CH_2OH \quad CH_2CH_2CH_2CH_3$$

***N*-Butyldeoxynojirimycin (NB-DNJ)**

results in a 50% decrease in G_{M2} ganglioside accumulation in the brains of treated as compared to untreated Tay–Sachs mice. NB-DNJ has also been used in clinical trials with patients who suffer from non-neuropathic Gaucher's disease. Since this disease results in spleen enlargement (Table 25-2), the volume of patients' spleens were measured and shown to decrease 12 to 19% after 12 months of treatment. While much more work remains to be done in the development of drugs to treat sphingolipid storage diseases, it is clear that an understanding of the enzymes involved on the molecular level can lead to great strides at the clinical level.

TABLE 25-2 Sphingolipid Storage Diseases

Disease	Enzyme Deficiency	Principal Storage Substance	Major Symptoms
G_{M1} Gangliosidosis	G_{M1} β-galactosidase	Ganglioside G_{M1}	Mental retardation, liver enlargement, skeletal involvement, death by age 2
Tay–Sachs disease	Hexosaminidase A	Ganglioside G_{M2}	Mental retardation, blindness, death by age 3
Fabry's disease	α-Galactosidase A	Trihexosylceramide	Skin rash, kidney failure, pain in lower extremities
Sandhoff's disease	Hexosaminidases A and B	Ganglioside G_{M2} and globoside	Similar to Tay–Sachs disease but more rapidly progressing
Gaucher's disease	Glucocerebrosidase	Glucocerebroside	Liver and spleen enlargement, erosion of long bones, mental retardation in infantile form only
Niemann–Pick disease	Sphingomyelinase	Sphingomyelin	Liver and spleen enlargement, mental retardation
Farber's lipogranulomatosis	Ceramidase	Ceramide	Painful and progressively deformed joints, skin nodules, death within a few years
Krabbe's disease	Galactocerebrosidase	Deacylated galactocerebroside	Loss of myelin, mental retardation, death by age 2
Metachromatic leukodistrophy (Sulfatide lipidosis)	Arylsulfatase A	Sulfatide	Mental retardation, death in first decade

CHAPTER SUMMARY

1 ■ Lipid Digestion, Absorption, and Transport Triacylglycerols, the storage form of metabolic energy in animals, provide up to six times the metabolic energy of an equal weight of hydrated glycogen. Dietary lipids are digested by pancreatic digestive enzymes such as lipase and phospholipase A_2 that are active at the lipid–water interface of bile salt–stabilized emulsions. Bile salts are also essential for the intestinal absorption of dietary lipids, as is fatty acid–binding protein. Dietary triacylglycerols and those synthesized by the liver are transported in the blood as chylomicrons and VLDL, respectively. Triacylglycerols present in these lipoproteins are hydrolyzed by lipoprotein lipase outside the cells and enter them as free fatty acids. Fatty acids resulting from hydrolysis of adipose tissue triacylglycerols by hormone-sensitive triacylglycerol lipase are transported in the bloodstream as fatty acid–albumin complexes.

2 ■ Fatty Acid Oxidation Before fatty acids are oxidized, they are converted to their acyl-CoA derivatives by acyl-CoA synthase in an ATP-requiring process, transported into mitochondria as carnitine esters, and reconverted inside the mitochondrial matrix to acyl-CoA. β Oxidation of fatty acyl-CoA occurs in 2-carbon increments so as to convert even-chain fatty acyl-CoAs completely to acetyl-CoA. The pathway involves FAD-dependent dehydrogenation of an alkyl group, hydration of the resulting double bond, NAD$^+$-dependent oxidation of this alcohol to a ketone, and C—C bond cleavage to form acetyl-CoA and a new fatty acyl-CoA with two fewer carbon atoms. The process then repeats itself. Complete oxidation of the acetyl-CoA, NADH, and FADH$_2$ is achieved by the citric acid cycle and oxidative phosphorylation. Oxidation of unsaturated fatty acids and odd-chain fatty acids also occurs by β oxidation but requires the participation of additional enzymes. Odd-chain fatty acid oxidation generates propionyl-CoA, whose further metabolism requires the participation of (1) propionyl-CoA carboxylase, which has a biotin prosthetic group, (2) methylmalonyl-CoA racemase, and (3) methylmalonyl-CoA mutase, which contains AdoCbl (coenzyme B$_{12}$). Methylmalonyl-CoA mutase catalyzes a carbon skeleton rearrangement reaction via a free radical mechanism in which the free radical is generated by the homolytic cleavage of AdoCbl's C—Co(III) bond. β Oxidation of fatty acids takes place in the peroxisomes in addition to the mitochondrion. The peroxisomal pathway differs from the mitochondrial pathway in that the FADH$_2$ produced in the first step, rather than generating ATP by oxidative phosphorylation, is directly oxidized by O_2 to produce H_2O_2. Peroxisomal enzymes are specific for long-chain fatty acids and function in a chain-shortening process via β oxidation. The resultant intermediate chain-length products are transferred to the mitochondrion for complete oxidation.

3 ■ Ketone Bodies A significant fraction of the acetyl-CoA produced by fatty acid oxidation in the liver is converted to acetoacetate and D-β-hydroxybutyrate, which, together with acetone, are referred to as ketone bodies. The first two compounds serve as important fuels for the peripheral tissues.

4 ■ Fatty Acid Biosynthesis Fatty acid biosynthesis differs from fatty acid oxidation in several respects. Whereas fatty acid oxidation occurs in the mitochondrion utilizing fatty acyl-CoA esters, fatty acid biosynthesis occurs in the cytosol with the growing fatty acids esterified to acyl-carrier protein (ACP). The redox coenzymes differ (FAD and NAD$^+$ for oxidation; NADPH for biosynthesis), as does the stereochemistry of the pathway's intermediate steps. Oxidation produces acetyl-CoA, whereas malonyl-CoA is the immediate precursor in biosynthesis. HCO$_3^-$ is required for the activation of biosynthesis but is not incorporated in the fatty acid product. In mammals, fatty acid synthesis occurs on a single protein that contains all six activities required to do so on separate modules that are serviced by ACP. Similar but more extensive systems are used in the synthesis of the various polyketides. Acetyl-CoA is transferred from the mitochondrion to the cytosol as citrate via the tricarboxylate transport system and released by citrate cleavage to yield acetyl-CoA and oxaloacetate. Oxaloacetate is converted to malate and then to pyruvate for transport back to the mitochondrion, a process that also generates some of the NADPH required for biosynthesis. Palmitate is the primary product of fatty acid biosynthesis in animals. Longer chain fatty acids and unsaturated fatty acids are synthesized from palmitate by elongation and desaturation reactions. Triacylglycerols are synthesized from fatty acyl-CoA esters and glycerol-3-phosphate.

5 ■ Regulation of Fatty Acid Metabolism Fatty acid metabolism is regulated through the allosteric control of hormone-sensitive triacylglycerol lipase and acetyl-CoA carboxylase, phosphorylation/dephosphorylation, and/or changes in the rates of protein synthesis and breakdown. This regulation is mediated by the hormones glucagon, epinephrine, and norepinephrine, which activate degradation, and by insulin, which activates biosynthesis. These hormones interact to control the cAMP concentration, which in turn controls phosphorylation/dephosphorylation ratios via PKA. AMPK, which senses the level of ATP, is also an important regulator of fatty acid metabolism.

6 ■ Cholesterol Metabolism Cholesterol is a vital constituent of cell membranes and is the precursor of the steroid hormones and bile salts. Its biosynthesis, transport, and utilization are rigidly controlled. Cholesterol is synthesized in the liver from acetate in a pathway that involves formation of HMG-CoA from three molecules of acetate followed by reduction, phosphorylation, decarboxylation, and dehydration to the isoprene units isopentenyl pyrophosphate and dimethylallyl pyrophosphate. Four of these isoprene units are then condensed via cationic mechanisms to form squalene, which, in turn, undergoes a cyclization reaction, via a cationic cascade, to form lanosterol, the sterol precursor to cholesterol. The pathway's major control point is at HMG-CoA reductase. This enzyme is regulated by competitive and allosteric mechanisms, by phosphorylation/dephosphorylation, and, most importantly, by long-term control of the rates of enzyme synthesis and degradation. Long-term control is mediated by the integral membrane protein SREBP, which when the cholesterol level is low, is escorted by SCAP to the Golgi apparatus. There it is sequentially cleaved by the proteases S1P and S2P, thereby releasing its soluble bHLH/Zip domain to travel to the nucleus where it induces the transcription of SRE-containing genes such as those encoding HMG-CoA reductase and the LDL receptor. The liver secretes cholesterol into the bloodstream in esterified form as part of the VLDL. This complex is sequentially converted to IDL and then to LDL. LDL, which is brought into the cells by receptor-mediated endocytosis, carries the major portion of cholesterol to peripheral tissues for utilization. Excess cholesterol is returned to the liver from peripheral tissues by HDL. The cellular supply of cholesterol is controlled by three mechanisms: (1) long- and short-term regulation of HMG-CoA reductase; (2) control of LDL receptor synthesis by cholesterol concentration; and (3) long- and short-term regulation of acyl-CoA:cholesterol acyltransferase (ACAT), which mediates cholesterol esterification. Cholesterol is the precursor to the steroid hormones, which are classified as progestins, glucocorticoids, mineralocorticoids, androgens, and estrogens. The quantitatively most important pathway for the excretion of cholesterol in mammals is the formation of bile salts.

7 ■ Eicosanoid Metabolism: Prostaglandins, Prostacyclins, Thromboxanes, Leukotrienes, and Lipoxins Prostaglandins, prostacyclins, thromboxanes, leukotrienes, and lipoxins are eicosanoid products produced largely by the metabolism of arachidonate. These highly unstable compounds have profound physiological effects at extremely low concentrations. They are involved in the inflammatory response, the production of pain and fever, the regulation of blood pressure, and many other important physiological processes. Arachidonate is synthesized from linoleic acid, an essential fatty acid, and stored as phosphatidylinositol and other phospholipids. Prostaglandins, prostacyclins, and thromboxanes are synthesized via the "cyclic pathway," whereas leukotrienes and lipoxins are synthesized via the "linear pathway." Aspirin and other nonsteroidal anti-inflammatory drugs (NSAIDs) inhibit the cyclic pathway but not the linear pathway. COX-2 inhibitors are NSAIDs that bind to COX-2 but not COX-1, thereby eliminating the side effects of other NSAIDs. Peptidoleukotrienes have been identified as the slow reacting substances of anaphylaxis (SRS-A) released from sensitized lung after immunological challenge. Lipoxins and aspirin-induced *epi*-lipoxins have anti-inflammatory properties.

8 ■ Phospholipid and Glycolipid Metabolism Complex lipids have either a phosphate ester or a carbohydrate as their polar head group and either 1,2-diacyl-*sn*-glycerol or ceramide (*N*-acylsphingosine) as their hydrophobic tail. Phospholipids are either glycerophospholipids or sphingophospholipids, whereas glycolipids are either glyceroglycolipids or sphingoglycolipids. The polar head groups of glycerophospholipids, which are phosphate esters of either ethanolamine, serine, choline, inositol, or glycerol, are attached to 1,2-diacyl-*sn*-glycerol's C3–OH group by means of CTP-linked transferase reactions. The specific long-chain fatty acids found at the C1 and C2 positions are incorporated by "remodeling reactions" after the addition of the polar head group. Plasmalogens and alkylacylglycerophospholipids, respectively, contain a long-chain alkyl group in a vinyl ether linkage or in an ether linkage to glycerol's C1–OH group. Platelet-activating factor (PAF) is an important plasmalogen. The only major sphingophospholipid is sphingomyelin (*N*-acylsphingosine phosphocholine), an important structural lipid of nerve cell membranes. Most sphingolipids contain polar head groups composed of carbohydrate units and are therefore referred to as sphingoglycolipids. The principal classes of sphingoglycolipids are cerebrosides, sulfatides, globosides, and gangliosides. Their carbohydrate units, which are attached to *N*-acylsphingosine's C1–OH group by glycosidic linkages, are formed by stepwise addition of activated monosaccharide units. Several lysosomal sphingolipid storage diseases, including Tay–Sachs disease, result from deficiencies in the enzymes that degrade sphingoglycolipids.

REFERENCES

GENERAL

Boyer, P.D. (Ed.), *The Enzymes* (3rd ed.), Vol. 16, Academic Press (1983). [A collection of reviews on lipid enzymology. Section 1 deals with fatty acid biosynthesis; Section 2 covers glyceride synthesis and degradation; Sections 3–5 review phospholipid, sphingolipid, and glycolipid metabolism; and Section 6 deals with aspects of cholesterol metabolism.]

Newsholme, E.A. and Leech, A.R., *Biochemistry for the Medical Sciences,* Wiley (1983). [Chapters 6–8 contain a wealth of information on the control of fatty acid metabolism and its integration into the overall scheme of metabolism.]

Scriver, C.R., Beaudet, A.C., Sly, W.S., and Valle, D. (Eds.), *The Metabolic & Molecular Bases of Inherited Disease* (8th ed.), McGraw-Hill (2001). [Volumes 2 and 3 contain numerous chapters on defects in lipid metabolism.]

Thompson, G.A., *The Regulation of Membrane Lipid Metabolism* (2nd ed.), CRC Press (1992).

Vance, D.E. and Vance, J.E. (Eds.), *Biochemistry of Lipids, Lipoproteins and Membranes* (3rd ed.), Elsevier (1996).

LIPID DIGESTION

Berg, O.G., Gelb, M.H., Tsai, M.-D., and Jain, M.K., Interfacial enzymology: the secreted phospholipase A$_2$-paradigm, *Chem. Rev.* **101,** 2613–2653 (2001).

Bhattacharya, A.A., Grüne, T., and Curry, S., Crystallographic analysis reveals common modes of binding of medium and long-chain fatty acids to human serum albumin, *J. Mol. Biol.* **303,** 721–732 (2002).

Borgström, B., Barrowman, J.A., and Lindström, M., Roles of bile acids in intestinal lipid digestion and absorption, *in* Danielsson, H. and Sjövall, J. (Eds.), *Sterols and Bile Acids,* pp. 405–425, Elsevier (1985).

Brady, L., Brzozowski, A.M., Derewenda, Z.S., Dodson, E., Dodson, G., Tolley, S., Turkenburg, J.P., Christiansen, L., Huge-Jensen, B., Norskov, L., Thim, L., and Menge, U., A serine protease triad forms the catalytic centre of a triacylglycerol lipase, *Nature* **343,** 767–770 (1990).

Derewenda, Z.S., Structure and function of lipases, *Adv. Prot. Chem.* **45,** 1–52 (1994).

Hermoso, J., Pignol, D., Penel, S., Roth, M., Chapus, C., and Fontecilla-Camps, J.C., Neutron crystallographic evidence of lipase–colipase complex activation by a micelle, *EMBO J.* **16,** 5531–5536 (1997).

Sacchettini, J.C., Gordon, J.I., and Banaszak, L.J., Crystal structure of rat intestinal fatty-acid binding-protein, *J. Mol. Biol.* **208,** 327–339 (1989) [The structure of I-FABP in complex with palmitate]; *and* Scapin, G., Gordon, J.I., and Sacchettini, J.C., Refinement of the structure of recombinant rat intestinal fatty acid-binding apoprotein at 1.2-Å resolution, *J. Biol. Chem.* **267,** 4253–4269 (1992).

van Tilbeurgh, H., Bezzine, S., Cambillau, C., Verger, R., and Carriére, F., Colipase: structure and interaction with pancreatic lipase, *Biochim. Biophys. Acta* **1441,** 173–184 (1999); van Tilbeurgh, H., Egloff, M.-P., Martinez, C., Rugani, N., Verger, R., and Cambillau, C., Interfacial activation of the lipase–colipase complex by mixed micelles revealed by X-ray crystallography, *Nature* **362,** 814–820 (1993); *and* van Tilbeurgh, H., Sarda, L., Verger, R., and Cambillau, C., Structure of pancreatic lipase–colipase complex, *Nature* **359,** 159–162 (1992).

FATTY ACID OXIDATION

Bannerjee, R., Radical peregrinations catalyzed by coenzyme B$_{12}$-dependent enzymes, *Biochem.* **40,** 6191–6198 (2001).

Bieber, L.L., Carnitine, *Annu. Rev. Biochem.* **88,** 261–283 (1988).

Kim, J.-J.P. and Battaile, K.P., Burning fat: The structural basis of fatty acid β-oxidation, *Curr. Opin. Struct. Biol.* **12,** 721–728 (2002).

Kim, J.-J.P., Wang, M., and Pashke, R., Crystal structures of medium-chain acyl-CoA dehydrogenase from pig liver mitochondria with and without substrate, *Proc. Natl. Acad. Sci.* **90,** 7523–7527 (1993).

Kindl, H., Fatty acid degradation in plant peroxisomes: Function and biosynthesis of the enzymes involved, *Biochimie* **75,** 225–230 (1993).

Mancia, R., Smith, G.A., and Evans, P.R., Crystal structure of substrate complexes of methylmalonyl-CoA mutase, *Biochem.* **38,** 7999–8005 (1999).

Marsh, E.N.G. and Drennan, C.L., Adenosylcobalamin-dependent isomerases: new insights into structure and mechanism, *Curr. Opin. Chem. Biol.* **5,** 499–505 (2001).

Rinaldo, P., Matern, D., and Bennett, M.J., Fatty acid oxidation disorders, *Annu. Rev. Physiol.* **64,** 477–502 (2002).

Shoukry, K. and Schulz, H., Significance of the reductase-dependent pathway for the β-oxidation of unsaturated fatty acids with odd-numbered double bonds: mitochondrial metabolism of 2-*trans*-5-*cis*-octadienoyl-CoA, *J. Biol. Chem.* **273,** 6892–6899 (1998).

Sudden infant death and inherited disorders of fat oxidation, *Lancet,* 1073–1075, Nov. 8, 1986.

Thorpe, C., Green enzymes and suicide substrates: a look at acyl-CoA dehydrogenases in fatty acid oxidation, *Trends Biochem. Sci.* **14,** 148–151 (1989).

van den Bosch, H., Schutgens, R.B.H., Wanders, R.J.A., and Tager, J.M., Biochemistry of peroxisomes, *Annu. Rev. Biochem.* **61,** 157–197 (1992).

FATTY ACID BIOSYNTHESIS

Brink, J., Ludtke, S.J., Yang, C.-Y., Gu, Z.-W., Wakil, S.J., and Chiu, W., Quaternary structure of human fatty acid synthase by electron cryomicroscopy, *Proc. Natl. Acad. Sci.* **99,** 138–143 (2002).

Brownsey, R.W. and Denton, R.M., Acetyl-coenzyme A carboxylase, *in* Boyer, P.D. and Krebs, E.G. (Eds.), *The Enzymes* (3rd ed.), Vol. 18, pp. 123–146, Academic Press (1987).

Brownsey, R.W., Zhande, R., and Boone, A.N., Isoforms of acetyl-CoA carboxylase: structures, regulatory properties and metabolic functions, *Biochem. Soc. Trans.* **25,** 1232–1238 (1997).

Cane, D.E., Walsh, C.T., and Khosla, C., Harnessing the biosynthetic code: combinations, permutations and mutations, *Science* **282,** 63–68 (1998).

Jump, D.B., The biochemistry of $n-3$ polyunsaturated fatty acids, *J. Biol. Chem.* **277,** 8755–8758 (2002).

Los, D.A. and Murata, N., Structure and expression of fatty acid desaturases, *Biochim. Biophys. Acta* **1394,** 3–15 (1998).

Pfeifer, B.A., Admiraal, S.J., Gramajo, H., Cane, D.E., and Khosla, C., Biosynthesis of complex polyketides in a metabolically engineered strain of *E. coli, Science* **291,** 1790–1792 (2001).

Rangan, V.S., Joshi, A.K., and Smith, S., Mapping the functional topology of the animal fatty acid synthase by mutant complementation in vitro, *Biochemistry* **40,** 10792–10799 (2001).

Wakil, S.J., Fatty acid synthase, a proficient multifunctional enzyme, *Biochemistry* **28,** 4523–4530 (1989).

REGULATION OF FATTY ACID METABOLISM

Eaton, S., Control of mitochondrial β-oxidation flux, *Prog. Lipid Res.* **41,** 197–239 (2002).

Hardie, D.G., Carling, D., and Carlson, M., The AMP-Activated/SNF1 protein kinase subfamily: metabolic sensors of the eukaryotic cell? *Annu. Rev. Biochem.* **67,** 821–855 (1998).

Hardie, D.G. and Carling, D., The AMP-activated protein kinase: fuel gauge of the mammalian cell? *Eur. J. Biochem.* **246,** 259–273 (1997).

Munday M.R. and Hemingway C.J., The regulation of acetyl-CoA carboxylase–A potential target for the action of hypolipidemic agents, *Adv. Enzym. Regul.* **39,** 205–234 (1999).

Stralfors, P., Olsson, H., and Belfrage, P., Hormone-sensitive lipase, *in* Boyer, P.D. and Krebs, E.G. (Eds.), *The Enzymes* (3rd ed.), Vol. 18, pp. 147–177, Academic Press (1987).

Witters, L.A., Watts, T.D., Daniels, D.L., and Evans, J.L., Insulin stimulates the dephosphorylation and activation of acetyl-CoA carboxylase, *Proc. Natl. Acad. Sci.* **85,** 5473–5477 (1988).

CHOLESTEROL METABOLISM

Bloch, K., The biological synthesis of cholesterol, *Science* **150,** 19–28 (1965).

Chang, T.Y., Chang, C.C.Y., and Cheng, D., Acyl-coenzyme A:cholesterol acyltransferase, *Annu. Rev. Biochem.* **66,** 613–638 (1997).

Durbecq, V., et al., Crystal structure of isopentenyl diphosphate:dimethylallyl diphosphate isomerase, *EMBO J.* **20,** 1530–1537 (2001).

Edwards, P.A., Sterols and isoprenoids: signaling molecules derived from the cholesterol biosynthetic pathway, *Annu. Rev. Biochem.* **68,** 157–185 (1999).

Gibson, D.M. and Parker, R.A., Hydroxymethylglutaryl-coenzyme A reductase, *in* Boyer, P.D. and Krebs, E.G. (Eds.), *The Enzymes* (3rd ed.), Vol. 18, pp. 179–215, Academic Press (1987).

Goldstein, J.L. and Brown, M.S., Regulation of the mevalonate pathway, *Nature* **343,** 425–430 (1990).

Goldstein, J.L., Rawson, R.B., and Brown, M.S., Mutant mammalian cells as tools to delineate the sterol regulatory element binding protein pathway for feedback regulation of lipid synthesis, *Arch. Biochem. Biophys.* **397,** 139–148 (2002).

Istvan, E.S., Bacterial and mammalian HMG-CoA reductases: related enzymes with distinct architectures, *Curr. Opin. Struct. Biol.* **11,** 746–751 (2001).

Istvan, E.S. and Deisenhofer, J., Structural mechanism for statin inhibition of HMG-CoA reductase, *Science* **292,** 1160–1164 (2001).

Knopp, R.H., Drug therapy: drug treatment of lipid disorders, *New Engl. J. Med.* **341,** 498–511 (1999).

Meyer, M.M., Segura, M.J.R., Wilson, W.K., and Matsuda, S.P.T., Oxidosqualene cyclase residues that promote formation of cycloartenol, lanosterol and parkeol, *Angew. Chem. Int. Ed.* **39,** 4090–4092 (2000).

Rilling, H.C. and Chayet, L.T., Biosynthesis of cholesterol, *in* Danielsson, H. and Sjövall, J. (Eds.), *Sterols and Bile Acids,* pp. 1–40, Elsevier (1985).

Russell, D.W. and Setchell, K.D.R., Bile acid biosynthesis, *Biochemistry* **31,** 4737–4749 (1992).

Sato, R., Goldstein, J.L., and Brown, M.S., Replacement of serine-871 of hamster 3-hydroxy-3-methylglutaryl-CoA reductase prevents phosphorylation by AMP-activated kinase and blocks inhibition of sterol synthesis induced by ATP depletion, *Proc. Natl. Acad. Sci.* **90,** 9261–9265 (1993).

Sinensky, M. and Lutz, R.J., The prenylation of proteins, *BioEssays* **14,** 25–31 (1992).

Tansey, T.R. and Shechter, I. Structure and regulation of mammalian squalene synthase, *Biochim. Biophys. Acta* **1529,** 49–62 (2000).

Yokode, M., Hammer, R.E., Ishibashi, S., Brown, M.S., and Goldstein, J.L., Diet-induced hypercholesterolemia in mice: Prevention by overexpression of LDL receptors, *Science* **250,** 1273–1275 (1990).

Wang, K.C. and Ohnuma, S.-I., Isoprenyl diphosphate synthases, *Biochim. Biophys. Acta* **1529,** 33–48 (2000).

Wendt, K.U., Lenhart, A., and Schulz, G.E., Structure and function of squalene cyclase, *Science* **277,** 1811–1815 (1997); *and* Wendt, K.U., Poralla, K., and Schulz, G.E., The structure of the membrane protein squalene-hopene cyclase at 2.0 Å resolution, *J. Mol. Biol.* **286,** 175–187 (1998).

Wendt, K.U., Schulz, G.E., Corey, E.J. and Liu, D.R., Enzyme mechanisms for polycyclic triterpene formation, *Angew. Chem. Int. Ed.* **39,** 2812–2833 (2000).

EICOSANOID METABOLISM

Abramovitz, M., Wong, E., Cox, M.E., Richardson, C.D., Li, C., and Vickers, P.J., 5-Lipoxygenase-activating protein stimulates the utilization of arachidonic acid by 5-lipoxygenase, *Eur. J Biochem.* **215,** 105–111 (1993).

Chandrasekharan, N.V., Dai, H., Roos, K.L.T., Evanson, N.K., Tomsik, J., Elton, T.S., and Simmons, D.L., COX-3, a cyclooxygenase-1 variant inhibited by acetaminophen and other analgesic/antipyretic drugs: Cloning, structure, and expression, *Proc. Natl. Acad. Sci.* **99,** 13926–13931 (2002).

Ford-Huchinson, A.W., FLAP: a novel drug target for inhibiting the synthesis of leukotrienes, *Trends Pharm. Sci.* **12,** 68–70 (1991).

Ford-Huchinson, A.W., Gresser, M., and Young, R.N., 5-Lipoxygenase, *Annu. Rev. Biochem.* **63,** 383–417 (1994).

Gillmor, S.A., Villaseñor, A., Fletterick, R., Sigal, E., and Browner, M.F., The structure of mammalian 15-lipoxygenase reveals similarity to the lipases and the determinants of substrate specificity, *Nature Struct. Biol.* **4,** 1003–1009 (1997).

Hayaishi, O., Sleep–wake regulation by prostaglandins D$_2$ and E$_2$, *J. Biol. Chem.* **263,** 14593–14596 (1988).

Kurumbail, R.G., Kiefer, J.R., and Marnett, L.J., Cyclooxygenase enzymes: catalysis and inhibition, *Curr. Opin. Struct. Biol.* **11,** 752–760 (2001).

Lewis, R.A., Austen, F., and Soberman, R.J., Leukotrienes and other products of the 5-lipoxygenase pathway, *New Engl. J. Med.* **323,** 645–655 (1990).

Phillipson, B.E., Rothrock, D.W., Conner, W.E., Harris, W.S., and Illingworth, D.R., Reduction of plasma lipids, lipoproteins and apoproteins by dietary fish oils in patients with hypertriglyceridemia, *New Engl. J. Med.* **312,** 1210–1216 (1985).

Picot, D., Loll, P.J., and Garavito, R.M., The X-ray crystal structure of the membrane protein prostaglandin H$_2$ synthase-1, *Nature* **367,** 243–249 (1994).

Samuelsson, B. and Funk, C.D., Enzymes involved in the biosynthesis of leukotriene B4, *J. Biol. Chem.* **264,** 19469–19472 (1989).

Schwarz, K., Walther, M., Anton, M., Gerth, C., Feussner, I., and Kuhn, H., Structural basis for lipoxygenase specificity: conversion of the human leukocyte 5-lipoxygenase to a 15-lipoxygenating enzyme species by site-directed mutagenesis, *J. Biol. Chem.* **276,** 773–339 (2001).

Serhan, C.N., Lipoxins and novel aspirin-triggered 15-*epi*-lipoxins (ATL): a jungle of cell-cell interactions or a therapeutic opportunity? *Prostaglandins* **53,** 107–137 (1997).

Smith, W.L., DeWitt, D.L., and Garavito, R.M., Cyclooxygenases: structural, cellular and molecular biology, *Annu. Rev. Biochem.* **69,** 145–182 (2000).

Tsai, A.-L., Hsi, L.C., Kulmacz, R.J., Palmer, G., and Smith, W.L., Characterization of the tyrosyl radicals in ovine prostaglandin H synthase-1 by isotope replacement and site-directed mutagenesis, *J. Biol. Chem.* **269,** 5085–5091 (1994).

Turini, M.E. and DuBois, R.N., Cyclooxygenase-2: a therapeutic target, *Annu. Rev. Med.* **53,** 35–57 (2002).

Vane, J.R., Bakhle, Y.S., and Botting, R.M., Cyclooxygenases 1 and 2, *Annu. Rev. Pharmacol. Toxicol.* **38,** 97–120 (1998).

Weissman, G., Aspirin, *Sci. Am.* **264**(1), 84–90 (1991).

PHOSPHOLIPID AND GLYCOLIPID METABOLISM

Conzelmann, E. and Sandhoff, K., Glycolipid and glycoprotein degradation, *Adv. Enzymol.* **60,** 89–216 (1987).

Dowhan, W., Molecular basis for membrane diversity: Why are there so many lipids? *Annu. Rev. Biochem.* **66,** 199–232 (1997).

Kent, C., Eukaryotic phospholipid synthesis, *Annu. Rev. Biochem.* **64,** 315–342 (1995).

Kolter, T. and Sandhoff, K., Sphingolipids—their metabolic pathways and the pathobiochemistry of neurodegenerative diseases, *Angew. Chem. Int. Ed.* **38,** 1532–1568 (1999).

Neufield, E.F., Natural history and inherited disorders of a lysosomal enzyme, β-hexosaminidase, *J. Biol. Chem.* **264,** 10927–10930 (1989).

Prescott, S.M., Zimmerman, G.A., and McIntire, T.M., Platelet-activating factor, *Biol. Chem.* **265,** 17381–17384 (1990).

Tifft, C.J. and Proila, R.L., Stemming the tide: glycosphingloipid synthesis inhibitors as therapy for storage diseases, *Glycobiology* **10,** 1249–1258 (2000).

van Echten, G. and Sandhoff, K., Ganglioside metabolism, *J. Biol. Chem.* **268,** 5341–5344 (1993).

▪ PROBLEMS

1. The venoms of many poisonous snakes, including rattlesnakes, contain a phospholipase A$_2$ that causes tissue damage that is seemingly far out of proportion to the small amount of enzyme injected. Explain.

2. Why are the livers of Jamaican vomiting sickness victims usually depleted of glycogen?

3. Compare the metabolic efficiencies, in moles of ATP produced per gram, of completely oxidized fat (tripalmitoyl glycerol) versus glucose derived from glycogen. Assume that the fat is anhydrous and the glycogen is stored with twice its weight of water.

4. Methylmalonyl-CoA mutase is incubated with deuterated methylmalonyl-CoA. The coenzyme B_{12} extracted from this mutase is found to contain deuterium at its 5'-methylene group. Account for the transfer of label from substrate to coenzyme.

5. What is the energetic price, in ATP units, of converting acetoacetyl-CoA to acetoacetate and then resynthesizing acetoacetyl-CoA?

6. A fasting animal is fed palmitic acid that has a ^{14}C-labeled carboxyl group. (a) After allowing sufficient time for fatty acid breakdown and resynthesis, what would be the ^{14}C-labeling pattern in the animal's palmitic acid residues? (b) The animal's liver glycogen becomes ^{14}C labeled although there is no net increase in the amount of this substance present. Indicate the sequence of reactions whereby the glycogen becomes labeled. Why is there no net glycogen synthesis?

7. What is the ATP yield from the complete oxidation of a molecule of (a) α-linolenic acid (9,12,15-octadecatrienoic acid, 18:3n−3) and (b) **margaric acid** (heptadecanoic acid, 17:0)? Which has the greater amount of available biological energy on a per carbon basis?

*8. The role of coenzyme B_{12} in mediating hydrogen transfer was established using the coenzyme B_{12}-dependent bacterial enzyme **dioldehydrase,** which catalyzes the reaction:

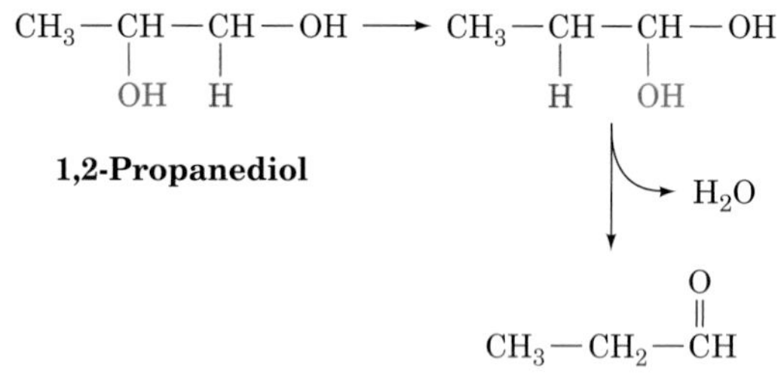

$$CH_3-CH-CH-OH \longrightarrow CH_3-CH-CH-OH$$
$$\quad\quad\; | \quad\; | \quad\quad\quad\quad\quad\quad\quad\; | \quad\; |$$
$$\quad\quad OH \;\; H \quad\quad\quad\quad\quad\quad\quad H \;\; OH$$

1,2-Propanediol

$$\downarrow \rightarrow H_2O$$

$$\quad\quad\quad O$$
$$\quad\quad\quad ||$$
$$CH_3-CH_2-CH$$

Propionaldehyde

The enzyme converts [1-^{3}H$_2$]1,2-propanediol to [1,2-^{3}H]propionaldehyde with the incorporation of tritium into both C5' positions of 5'-deoxyadenosylcobalamin's 5'-deoxyadenosyl residue. Suggest the mechanism of this reaction. What would be the products of the dioldehydrase reaction if the enzyme was supplied with [5'-^{3}H]deoxyadenosylcobalamin and unlabeled 1,2-propanediol?

9. What is the energetic price, in ATP equivalents, of breaking down palmitic acid to acetyl-CoA and then resynthesizing it?

10. What is the energetic price, in ATP equivalents, of synthesizing cholesterol from acetyl-CoA?

11. What would be the ^{14}C-labeling pattern in cholesterol if it were synthesized from HMG-CoA that was ^{14}C labeled (a) at C5, its carboxyl carbon atom, or (b) C1, its thioester carbon atom?

*12. A child suffering from severe abdominal pain is admitted to the hospital several hours after eating a meal consisting of hamburgers, fried potatoes, and ice cream. Her blood has the appearance of "creamed tomato soup" and on analysis is found to contain massive quantities of chylomicrons. As attending physician, what is your diagnosis of the patient's difficulty (the cause of the abdominal pain is unclear)? What treatment would you prescribe to alleviate the symptoms of this inherited disease?

13. Although linoleic acid is an essential fatty acid in animals, it is not required by animal cells in tissue culture. Explain.

14. The inactivation of the peroxidase function of prostaglandin H synthase (PGHS) also inactivates its cyclooxygenase function but not vice versa. Explain.

Chapter 26

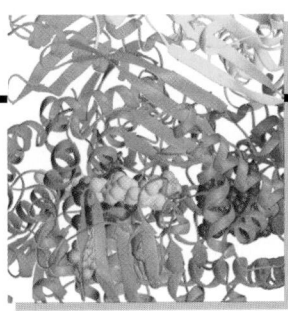

Amino Acid Metabolism

α-Amino acids, in addition to their role as protein monomeric units, are energy metabolites and precursors of many biologically important nitrogen-containing compounds, notably heme, physiologically active amines, glutathione, nucleotides, and nucleotide coenzymes. Amino acids are classified into two groups: **essential** and **nonessential.** Mammals synthesize the nonessential amino acids from metabolic precursors but must obtain the essential amino acids from their diet. Excess dietary amino acids are neither stored for future use nor excreted. Rather, they are converted to common metabolic intermediates such as pyruvate, oxaloacetate, acetyl-CoA and α-ketoglutarate. Consequently, *amino acids are also precursors of glucose, fatty acids, and ketone bodies and are therefore metabolic fuels.*

In this chapter, we consider the pathways of amino acid breakdown, synthesis, and utilization. We begin by examining the three common stages of amino acid breakdown:

1. Deamination (amino group removal), whereby amino groups are converted either to ammonia or to the amino group of aspartate.

2. Incorporation of ammonia and aspartate nitrogen atoms into urea for excretion.

3. Conversion of amino acid carbon skeletons (the α-keto acids produced by deamination) to common metabolic intermediates.

Many of these reactions are similar to those we have considered in other pathways. Others employ enzyme cofactors we have not previously encountered. One of our goals in studying amino acid metabolism is to understand the mechanisms of action of these cofactors.

After our discussion of amino acid breakdown, we examine the pathways by which amino acids are utilized in the biosynthesis of heme, physiologically active amines, and glutathione (the synthesis of nucleotides and nucleotide coenzymes is the subject of Chapter 28). Next, we study amino acid biosynthesis pathways. The chapter ends with a discussion of nitrogen fixation, a process that converts atmospheric N_2 to ammonia and is therefore the ultimate biological source of metabolically useful nitrogen.

1 ■ AMINO ACID DEAMINATION

The first reaction in the breakdown of an amino acid is almost always removal of its α-amino group with the object of excreting excess nitrogen and degrading the remaining carbon skeleton or converting it to glucose. Urea, the predominant nitrogen excretion product in terrestrial mammals, is synthesized from ammonia and aspartate. Both of these latter substances are derived mainly from glutamate, a product of most deamination reactions. In this section we examine the routes by which α-amino groups are incorporated into glutamate and then into aspartate and ammonia. In Section 26-2, we discuss urea biosynthesis from these precursors.

Most amino acids are deaminated by **transamination,** the transfer of their amino group to an α-keto acid to yield the α-keto acid of the original amino acid and a new amino acid, in reactions catalyzed by **aminotransferases** (alternatively, **transaminases**). The predominant amino group acceptor is α-ketoglutarate, producing glutamate as the new amino acid:

Amino acid + α-ketoglutarate $\rightleftharpoons$
$$\text{α-keto acid + glutamate}$$

Glutamate's amino group, in turn, is transferred to oxaloacetate in a second transamination reaction, yielding aspartate:

Glutamate + oxaloacetate $\rightleftharpoons$
$$\text{α-ketoglutarate + aspartate}$$

Transamination, of course, does not result in any net deamination. Deamination occurs largely through the oxidative deamination of glutamate by **glutamate dehydrogenase (GDH),** yielding ammonia. The reaction requires NAD^+ or $NADP^+$ as an oxidizing agent and regenerates α-ketoglutarate for use in additional transamination reactions:

Glutamate + $NAD(P)^+$ + H_2O $\rightleftharpoons$
$$\text{α-ketoglutarate} + NH_4^+ + NAD(P)H$$

The mechanisms of transamination and oxidative deamination are the subjects of this section. We also consider other means of amino group removal from specific amino acids.

A. *Transamination*

a. **Aminotransferase Reactions Occur in Two Stages:**

1. The amino group of an amino acid is transferred to the enzyme, producing the corresponding keto acid and the aminated enzyme.

Amino acid + enzyme $\rightleftharpoons$
$$\text{α-keto acid + enzyme} - NH_2$$

2. The amino group is transferred to the keto acid acceptor (e.g., α-ketoglutarate), forming the amino acid product (e.g., glutamate) and regenerating the enzyme.

α-Ketoglutarate + enzyme$-NH_2$ $\rightleftharpoons$
$$\text{enzyme + glutamate}$$

To carry the amino group, aminotransferases require participation of an aldehyde-containing coenzyme, pyridoxal-5′-phosphate (PLP), a derivative of pyridoxine (vitamin B_6; Fig. 26-1a,b). The amino group is accommodated by conversion of this coenzyme to **pyridoxamine-5′-phosphate (PMP;** Fig. 26-1c). PLP is covalently attached to the enzyme via a Schiff base (imine) linkage formed by

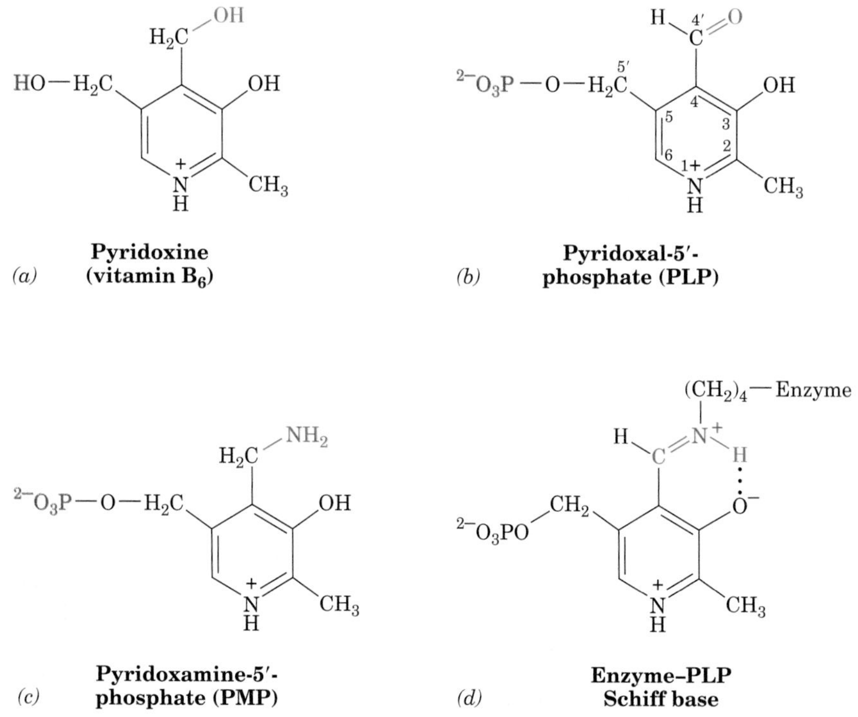

(a) **Pyridoxine (vitamin B_6)**

(b) **Pyridoxal-5′-phosphate (PLP)**

(c) **Pyridoxamine-5′-phosphate (PMP)**

(d) **Enzyme–PLP Schiff base**

FIGURE 26-1 **Forms of pyridoxal-5′-phosphate.** (*a*) Pyridoxine (vitamin B_6). (*b*) Pyridoxal-5′-phosphate (PLP). (*c*) Pyridoxamine-5′-phosphate (PMP). (*d*) The Schiff base that forms between PLP and an enzyme ε-amino group.

the condensation of its aldehyde group with the ε-amino group of an enzymatic Lys residue (Fig. 26-1*d*). This Schiff base, which is conjugated to the coenzyme's pyridinium ring, is the focus of the coenzyme's activity.

Esmond Snell, Alexander Braunstein, and David Metzler demonstrated that the aminotransferase reaction occurs via a Ping Pong Bi Bi mechanism whose two stages consist of three steps each (Fig. 26-2):

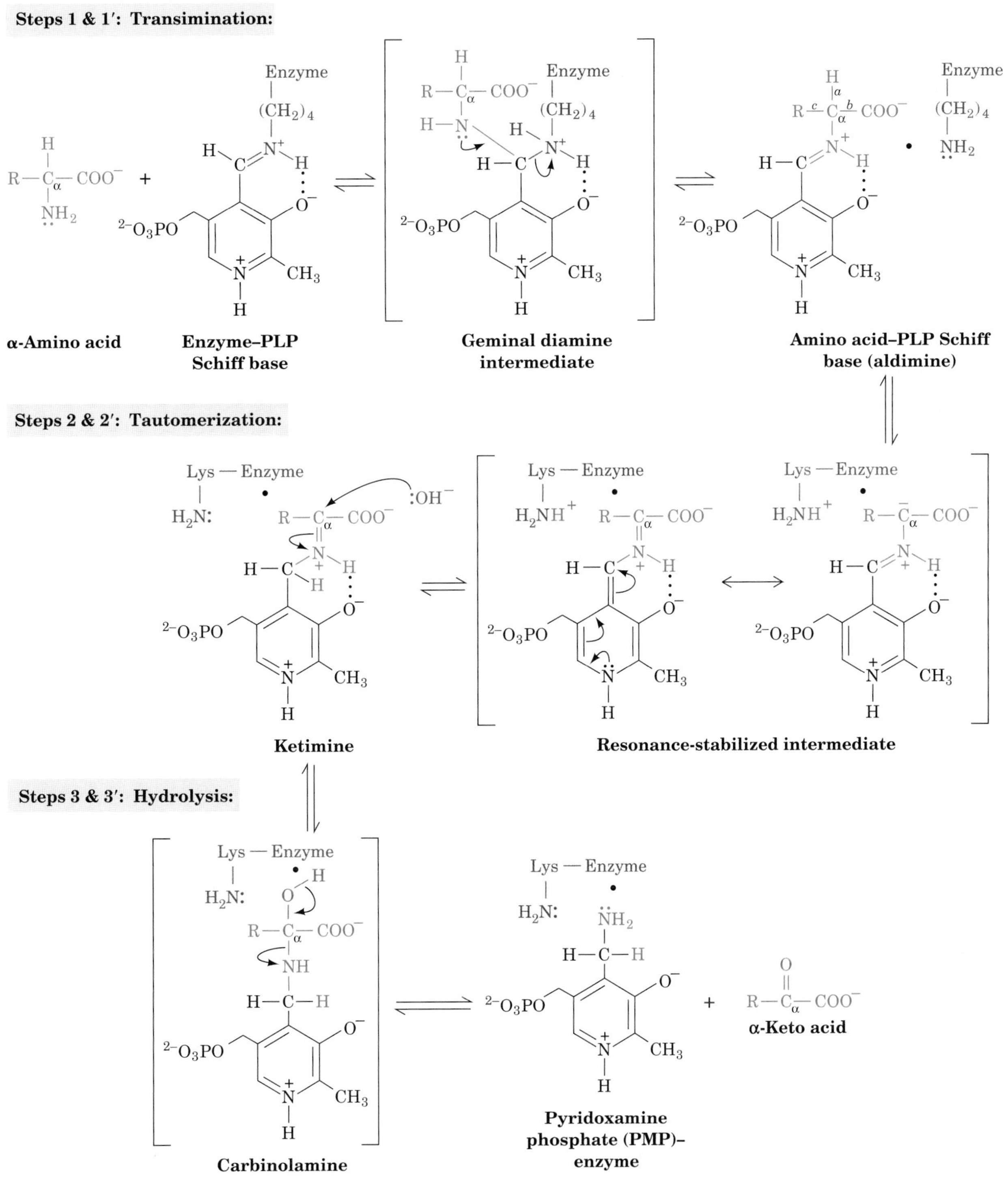

Steps 1 & 1′: Transimination:

α-Amino acid   Enzyme–PLP Schiff base   Geminal diamine intermediate   Amino acid–PLP Schiff base (aldimine)

Steps 2 & 2′: Tautomerization:

Ketimine   Resonance-stabilized intermediate

Steps 3 & 3′: Hydrolysis:

Carbinolamine   Pyridoxamine phosphate (PMP)–enzyme   α-Keto acid

FIGURE 26-2   The mechanism of PLP-dependent enzyme-catalyzed transamination. The first stage of the reaction, in which the α-amino group of an amino acid is transferred to PLP yielding an α-keto acid and PMP, consists of three steps: **(1)** transimination; **(2)** tautomerization, in which the Lys released during the transimination reaction acts as a general acid–base catalyst; and **(3)** hydrolysis. The second stage of the reaction, in which the amino group of PMP is transferred to a different α-keto acid to yield a new α-amino acid and PLP, is essentially the reverse of the first stage: Steps 3′, 2′, and 1′ are, respectively, the reverse of Steps 3, 2, and 1. *See the* **Animated Figures**

b. Stage I: Conversion of an Amino Acid to an α-Keto Acid

Step 1. The amino acid's nucleophilic amino group attacks the enzyme–PLP Schiff base carbon atom in a **transimination (trans-Schiffization)** reaction to form an amino acid–PLP Schiff base (aldimine), with concomitant release of the enzyme's Lys amino group. This Lys is then free to act as a general base at the active site.

Step 2. The amino acid–PLP Schiff base tautomerizes to an α-keto acid–PMP Schiff base by the active site Lys–catalyzed removal of the amino acid α hydrogen and protonation of PLP atom C4′ via a resonance-stabilized carbanion intermediate. This resonance stabilization facilitates the cleavage of the C_α—H bond.

Step 3. The α-keto acid–PMP Schiff base is hydrolyzed to PMP and an α-keto acid.

c. Stage II: Conversion of an α-Keto Acid to an Amino Acid

To complete the aminotransferase's catalytic cycle, the coenzyme must be converted from PMP back to the enzyme–PLP Schiff base. This involves the same three steps as above, but in reverse order:

Step 3′. PMP reacts with an α-keto acid to form a Schiff base.

Step 2′. The α-keto acid–PMP Schiff base tautomerizes to form an amino acid–PLP Schiff base.

Step 1′. The ε-amino group of the active site Lys residue attacks the amino acid–PLP Schiff base in a transimination reaction to regenerate the active enzyme–PLP Schiff base, with release of the newly formed amino acid.

The reaction's overall stoichiometry therefore is

Amino acid 1 + α-keto acid 2 ⇌
α-keto acid 1 + amino acid 2

Examination of the amino acid–PLP Schiff base's structure (Fig. 26-2, Step 1) reveals why this system is called "an electron-pusher's delight." *Cleavage of any of the amino acid C_α atom's three bonds (labeled a, b, and c) produces a resonance-stabilized C_α carbanion whose electrons are delocalized all the way to the coenzyme's protonated pyridinium nitrogen atom; that is, PLP functions as an electron sink.* For transamination reactions, this electron-withdraw-

ing capacity facilitates removal of the α proton (*a* bond cleavage) in the tautomerization of the Schiff base. PLP-dependent reactions involving *b* bond cleavage (amino acid decarboxylation) and *c* bond labilization are discussed in Section 26-4B and Sections 26-3B and 26-3G, respectively.

Aminotransferases differ in their specificity for amino acid substrates in the first stage of the transamination reaction, thereby producing the correspondingly different α-keto acid products. Most aminotransferases, however, accept only α-ketoglutarate or (to a lesser extent) oxaloacetate as the α-keto acid substrate in the second stage of the reaction, thereby yielding glutamate or aspartate as their only amino acid products. *The amino groups of most amino acids are consequently funneled into the formation of glutamate or aspartate, which are themselves interconverted by* **glutamate–aspartate aminotransferase:**

Glutamate + oxaloacetate ⇌
α-ketoglutarate + aspartate

Oxidative deamination of glutamate (Section 26-1B) yields ammonia and regenerates α-ketoglutarate for another round of transamination reactions. Ammonia and aspartate are the two amino group donors in the synthesis of urea.

d. The Glucose–Alanine Cycle Transports Nitrogen to the Liver

An important exception to the foregoing is a group of muscle aminotransferases that accept pyruvate as their α-keto acid substrate. The product amino acid, alanine, is released into the bloodstream and transported to the liver, where it undergoes transamination to yield pyruvate for use in gluconeogenesis (Section 23-1A). The resulting glucose is returned to the muscles, where it is glycolytically degraded to pyruvate. This is the **glucose–alanine cycle** (Fig. 26-3). The amino group ends up in either ammonia or aspartate for urea biosynthesis. Evidently, the glucose–alanine cycle functions to transport nitrogen from muscle to liver.

During starvation the glucose formed in the liver by this route is also used by the other peripheral tissues, breaking the cycle. Under these conditions both the amino group and the pyruvate originate from muscle protein degradation, providing a pathway yielding glucose for other tissue use (recall that muscle is not a gluconeogenic tissue; Section 23-1).

Nitrogen is also transported to the liver in the form of glutamine, synthesized from glutamate and ammonia in a

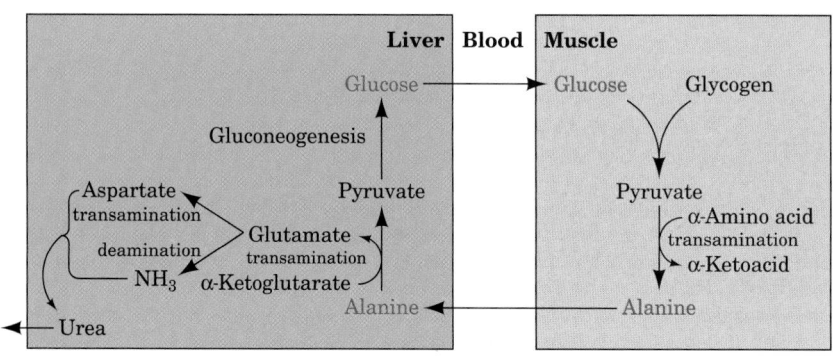

FIGURE 26-3 The glucose–alanine cycle.
See the Animated Figures

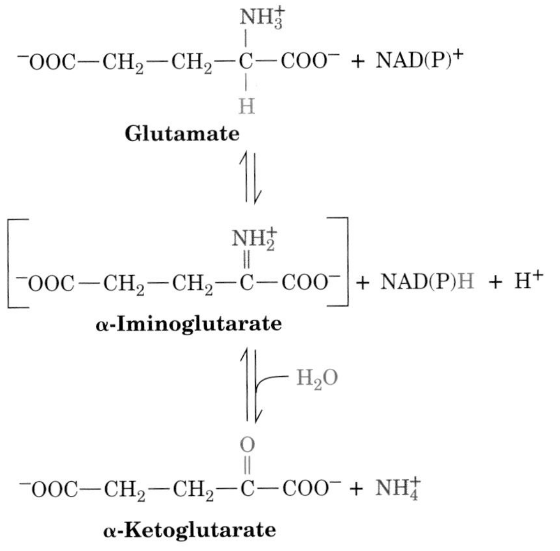

$$^-OOC-CH_2-CH_2-\overset{\overset{\displaystyle NH_3^+}{|}}{\underset{\underset{\displaystyle H}{|}}{C}}-COO^- + NAD(P)^+$$

Glutamate

$$\left[^-OOC-CH_2-CH_2-\overset{\overset{\displaystyle NH_2^+}{\|}}{C}-COO^- \right] + NAD(P)H + H^+$$

α-Iminoglutarate

$$- H_2O$$

$$^-OOC-CH_2-CH_2-\overset{\overset{\displaystyle O}{\|}}{C}-COO^- + NH_4^+$$

α-Ketoglutarate

FIGURE 26-4   The oxidative deamination of glutamate by glutamate dehydrogenase. This reaction involves the intermediate formation of α-iminoglutarate.

reaction catalyzed by **glutamine synthetase** (Section 26-5A). The ammonia is released for urea synthesis in liver mitochondria or for excretion in the kidney through the action of **glutaminase** (Section 26-3D).

B. *Oxidative Deamination: Glutamate Dehydrogenase*

Glutamate is oxidatively deaminated in the mitochondrion by glutamate dehydrogenase (GDH), the only known enzyme that, in at least some organisms, can accept either NAD^+ or $NADP^+$ as its redox coenzyme. Oxidation is thought to occur with transfer of a hydride ion from glutamate's C_α to $NAD(P)^+$, thereby forming α-iminoglutarate, which is hydrolyzed to α-ketoglutarate and ammonia (Fig. 26-4). GDH is allosterically inhibited by GTP and NADH and activated by ADP, leucine, and NAD^+ *in vitro,* suggesting that these effectors regulate the enzyme *in vivo.*

a. The X-ray Structures of GDH Reveal its Allosteric Mechanism

The X-ray structures of homohexameric GDH from bovine and human liver mitochondria, determined by Thomas Smith, reveal that each monomer has three domains, a substrate domain, a coenzyme domain, and an antenna domain. The protein, which has D_3 symmetry, can be considered to be a dimer of trimers, with the antenna domains of each trimer wrapping around each other about the 3-fold axis (Fig. 26-5a). Structural comparison of a 501-residue monomer of the bovine **GDH–glutamate–NADH–GTP** complex (Fig. 26-5b) with that of the 96% identical human **apoenzyme** (no active site or regulatory ligands bound; Fig. 26-5c) reveals that, on binding ligands, the coenzyme binding domain rotates about the so-called pivot helix so as to close the cleft between the coenzyme

and substrate domains. Simultaneously, the antenna domain twists in a way that unwinds one turn of the antenna helix that is connected to the pivot helix. Although the closed form is required for catalysis, the open form favors the association and dissociation of substrates and products. In the open state, Arg 463 (human numbering) in the center of the pivot helix interacts with the activator ADP (whose binding site in the bovine complex is occupied by the ADP moiety of an NADH; Fig. 26-5b), whereas in the closed state, the side chain of His 454 hydrogen bonds to the γ-phosphate of the inhibitor GTP. The GTP binding site is distorted and blocked in the open state so that GTP binding favors the closed form of the enzyme. This results in tight binding of substrates and products and hence inhibition of the enzyme. ADP binding favors the open form, allowing product dissociation, and therefore activates the enzyme. Allosteric interactions appear to be communicated between subunits through the interactions of the antenna domains. In fact, bacterial GDHs, which lack allosteric regulation, differ from mammalian GDHs mainly by the absence of antenna domains.

b. Hyperinsulinism/Hyperammonemia (HI/HA) Is Caused by Uncontrolled GDH Activity

Charles Stanley has reported a new form of congenital hyperinsulinism that is characterized by hypoglycemia and **hyperammonemia** (**HI/HA;** hyperammonemia is elevated levels of ammonia in the blood) and has shown that it is caused by mutations in GDH at the N-terminal end of its pivot helix in the GTP binding site or in the antenna domain near its joint with the pivot helix. The mutant enzymes have reduced sensitivity to GTP inhibition but retain their ability to be activated by ADP. The GDH mutants S448P, H454Y, and R463A, which were respectively designed to affect the antenna region, the GTP binding site, and the ADP binding site (Fig. 26-5b), all have decreased sensitivity to GTP inhibition (Fig. 26-6), with H454Y and S448P, which were previously known to be associated with HI/HA, conferring the most resistance to GTP inhibition. The hypoglycemia and hyperammonemia in HI/HA patients arises from the increased activity of the GDH mutants in the breakdown direction, producing increased amounts of α-ketoglutarate and NH_3. The increased levels of α-ketoglutarate stimulate the citric acid cycle and oxidative phosphorylation, which has been shown to lead to increased insulin secretion and hypoglycemia, thereby producing the symptoms of the disease. The NH_3 produced is usually converted to urea (Section 26-2) but can also be exported to the bloodstream.

If this scenario for the cause of HI/HA is correct, it requires a reassessment of the role of GDH in ammonia homeostasis. The equilibrium position of the GDH reaction greatly favors the synthesis of Glu ($\Delta G^{\circ\prime} \approx 30$ kJ · mol^{-1} for the reaction as written in Fig. 26-4), but studies of cellular substrate and product concentrations suggested that the enzyme functions close to equilibrium ($\Delta G \approx 0$) *in vivo.* It was therefore widely accepted that increases in $[NH_3]$, high levels of which are toxic, would cause GDH to act in reverse, removing NH_3 and hence preventing its

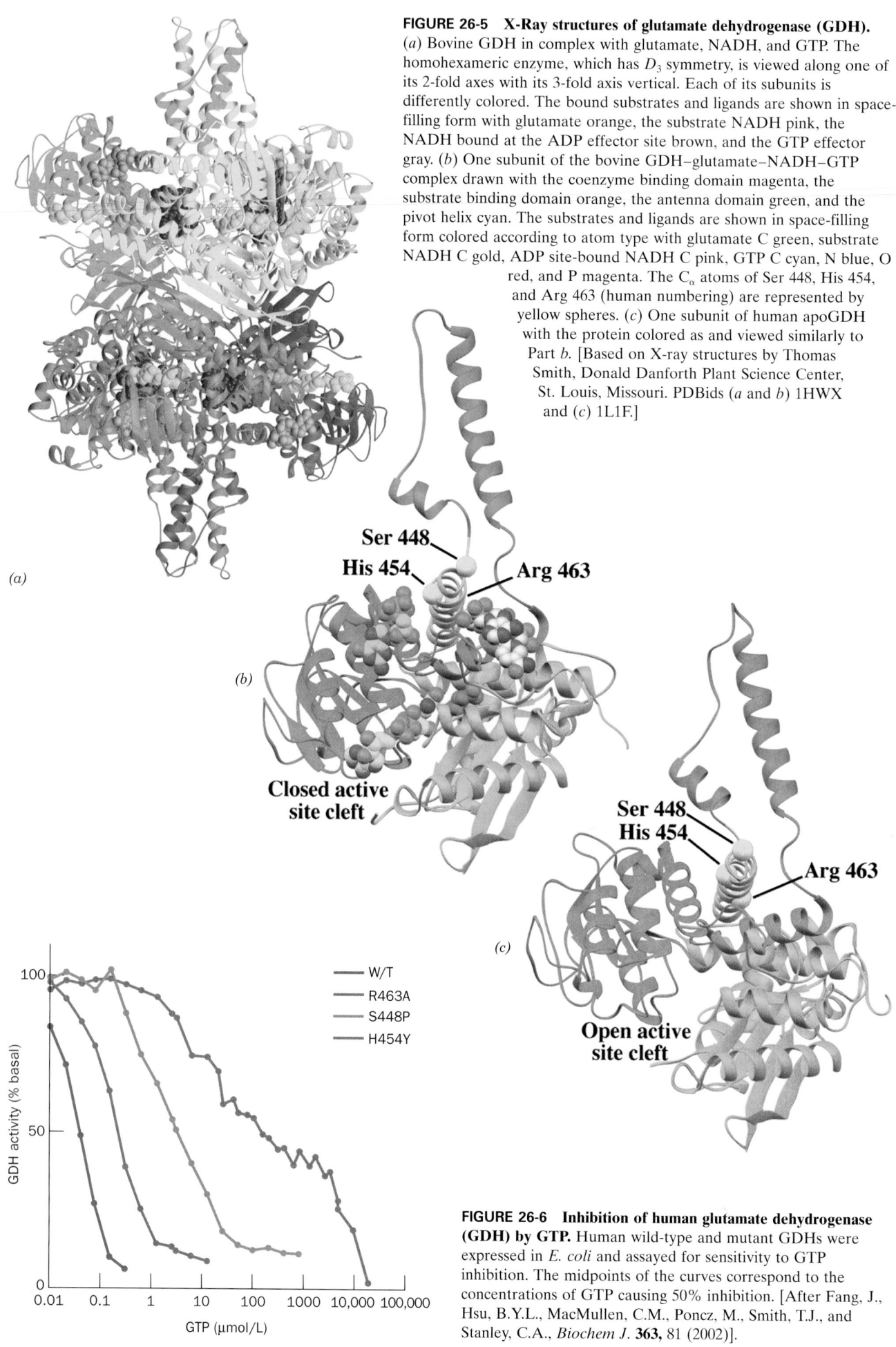

FIGURE 26-5   X-Ray structures of glutamate dehydrogenase (GDH).
(*a*) Bovine GDH in complex with glutamate, NADH, and GTP. The homohexameric enzyme, which has D_3 symmetry, is viewed along one of its 2-fold axes with its 3-fold axis vertical. Each of its subunits is differently colored. The bound substrates and ligands are shown in space-filling form with glutamate orange, the substrate NADH pink, the NADH bound at the ADP effector site brown, and the GTP effector gray. (*b*) One subunit of the bovine GDH–glutamate–NADH–GTP complex drawn with the coenzyme binding domain magenta, the substrate binding domain orange, the antenna domain green, and the pivot helix cyan. The substrates and ligands are shown in space-filling form colored according to atom type with glutamate C green, substrate NADH C gold, ADP site-bound NADH C pink, GTP C cyan, N blue, O red, and P magenta. The C_α atoms of Ser 448, His 454, and Arg 463 (human numbering) are represented by yellow spheres. (*c*) One subunit of human apoGDH with the protein colored as and viewed similarly to Part *b*. [Based on X-ray structures by Thomas Smith, Donald Danforth Plant Science Center, St. Louis, Missouri. PDBids (*a* and *b*) 1HWX and (*c*) 1L1F.]

FIGURE 26-6   Inhibition of human glutamate dehydrogenase (GDH) by GTP. Human wild-type and mutant GDHs were expressed in *E. coli* and assayed for sensitivity to GTP inhibition. The midpoints of the curves correspond to the concentrations of GTP causing 50% inhibition. [After Fang, J., Hsu, B.Y.L., MacMullen, C.M., Poncz, M., Smith, T.J., and Stanley, C.A., *Biochem J.* **363**, 81 (2002)].

buildup to toxic levels. However, since HI/HA patients have increased GDH activity yet have higher levels of NH_3 than normal, this accepted role of GDH cannot be correct. Indeed, if GDH functioned close to equilibrium, changes in its activity resulting from allosteric interactions would not result in significant flux changes.

C. *Other Deamination Mechanisms*

Two nonspecific amino acid oxidases, **L-amino acid oxidase** and **D-amino acid oxidase,** catalyze the oxidation of L- and D-amino acids, utilizing FAD as their redox coenzyme [rather than $NAD(P)^+$]. The resulting $FADH_2$ is reoxidized by O_2.

Amino acid + FAD + $H_2O \rightarrow$
$$\alpha\text{-keto acid} + NH_3 + FADH_2$$
$$FADH_2 + O_2 \rightarrow FAD + H_2O_2$$

D-Amino acid oxidase occurs mainly in kidney. Its function is an enigma since D-amino acids are associated mostly with bacterial cell walls (Section 11-3B). A few amino acids, such as serine and histidine, are deaminated nonoxidatively (Sections 26-3B and 26-3D).

2 ■ THE UREA CYCLE

Living organisms excrete the excess nitrogen resulting from the metabolic breakdown of amino acids in one of three ways. Many aquatic animals simply excrete ammonia. Where water is less plentiful, however, processes have evolved that convert ammonia to less toxic waste products that therefore require less water for excretion. One such product is urea, which is excreted by most terrestrial vertebrates; another is **uric acid,** which is excreted by birds and terrestrial reptiles:

| Ammonia | Urea | Uric acid |

Accordingly, living organisms are classified as being either **ammonotelic** (ammonia excreting), **ureotelic** (urea excreting), or **uricotelic** (uric acid excreting). Some animals can shift from ammonotelism to ureotelism or uricotelism if their water supply becomes restricted. Here we focus our attention on urea formation. Uric acid biosynthesis is discussed in Section 28-4A.

*Urea is synthesized in the liver by the enzymes of the **urea cycle.*** It is then secreted into the bloodstream and sequestered by the kidneys for excretion in the urine. The urea cycle was elucidated in outline in 1932 by Hans Krebs and Kurt Henseleit (the first known metabolic cycle; Krebs did not elucidate the citric acid cycle until 1937). Its individual

reactions were later described in detail by Sarah Ratner and Philip Cohen. The overall urea cycle reaction is

Thus, the two urea nitrogen atoms are contributed by NH_3 and aspartate, whereas the carbon atom comes from HCO_3^-. Five enzymatic reactions are involved in the urea cycle, two of which are mitochondrial and three cytosolic (Fig. 26-7). In this section, we examine the mechanisms of these reactions and their regulation.

A. *Carbamoyl Phosphate Synthetase: Acquisition of the First Urea Nitrogen Atom*

Carbamoyl phosphate synthetase (CPS) is technically not a member of the urea cycle. It catalyzes the condensation and activation of NH_3 and HCO_3^- to form **carbamoyl phosphate,** the first of the cycle's two nitrogen-containing substrates, with the concomitant hydrolysis of two ATPs. Eukaryotes have two forms of CPS:

1. Mitochondrial **CPS I** uses NH_3 as its nitrogen donor and participates in urea biosynthesis.

2. Cytosolic **CPS II** uses glutamine as its nitrogen donor and is involved in pyrimidine biosynthesis (Section 28-2A).

The reaction catalyzed by CPS I involves three steps (Fig. 26-8):

1. Activation of HCO_3^- by ATP to form **carboxyphosphate** and ADP.

2. Nucleophilic attack of NH_3 on carboxyphosphate, displacing the phosphate to form **carbamate** and P_i.

3. Phosphorylation of carbamate by the second ATP to form carbamoyl phosphate and ADP.

The reaction is essentially irreversible and is the rate-limiting step of the urea cycle. CPS I is subject to allosteric activation by **N-acetylglutamate** as is discussed in Section 26-2F.

E. coli contains only one type of CPS, which is homologous to both CPS I and CPS II. The enzyme is a heterodimer but when allosterically activated by ornithine (a urea cycle intemediate), it forms a tetramer of heterodimers, $(\alpha\beta)_4$. Its small subunit (382 residues) functions to hydrolyze glutamine and deliver the resulting NH_3 to its large subunit (1073 residues). However, if the enzyme's **glutaminase (glutamine amidotransferase)** activity is eliminated (e.g., by site-directed mutagenesis), the large subunit can still produce carbamoyl phosphate if NH_3 is supplied

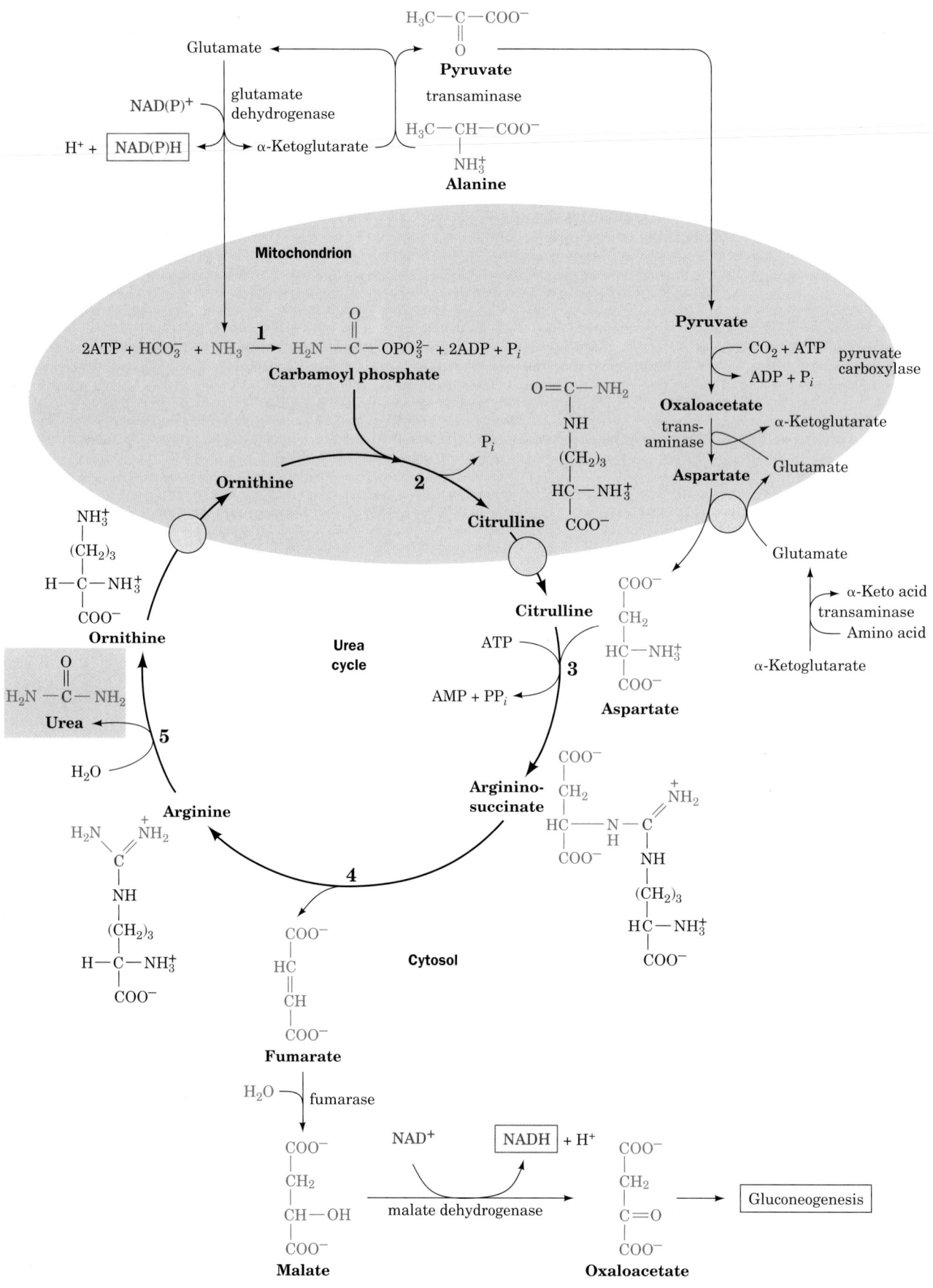

FIGURE 26-7 (*Opposite*) **The urea cycle.** Its five enzymes are: **(1)** carbamoyl phosphate synthetase, **(2)** ornithine transcarbamoylase, **(3)** argininosuccinate synthetase, **(4)** argininosuccinase, and **(5)** arginase. The reactions occur in part in the mitochondrion and in part in the cytosol with ornithine and citrulline being transported across the mitochondrial membrane by specific transport systems (*yellow circles*). One of the urea amino groups (*green*) originates as the NH_3 product of the glutamate dehydrogenase reaction (*top*). The other amino group (*red*) is obtained from aspartate through the transfer of an amino acid to oxaloacetate via transamination (*right*). The fumarate product of the argininosuccinase reaction is converted to oxaloacetate for entry into gluconeogenesis via the same reactions that occur in the citric acid cycle but take place in the cytosol (*bottom*). The ATP utilized in Reactions 1 and 3 of the cycle can be regenerated by oxidative phosphorylation from the NAD(P)H produced in the glutamate dehydrogenase (*top*) and malate dehydrogenase (*bottom*) reactions. ♨ **See the Animated Figures**

in high enough concentration. The large subunit is composed of two nearly superimposable halves that have 40% sequence identity. The N-terminal half contains the carboxyphosphate synthetic component and an oligomer-

ization domain while the C-terminal half contains the carbamoyl phosphate synthetic component and an allosteric binding domain.

a. *E. coli* CPS Contains an Extraordinarily Long Tunnel

The X-ray structure of *E. coli* CPS in complex with Mn^{2+}, ADP, P_i, and ornithine, determined by Hazel Holden and Ivan Rayment, reveals that the active site for synthesis of the carboxyphosphate intermediate is ~45 Å away from the ammonia synthesis site and also ~35 Å away from the carbamoyl phosphate synthesis active site. Astonishingly, the three sites are connected by a narrow 96-Å-long molecular tunnel that runs nearly the length of the elongated protein molecule (Fig. 26-9). It therefore appears that CPS guides its intermediate products from the active site in which they are formed to that in which they are utilized. This phenomenon, in which the intermediate of two reactions is directly transferred from one enzyme active site to another, is called **channeling** (the term "tunneling" is reserved for certain quantum mechanical phenomena).

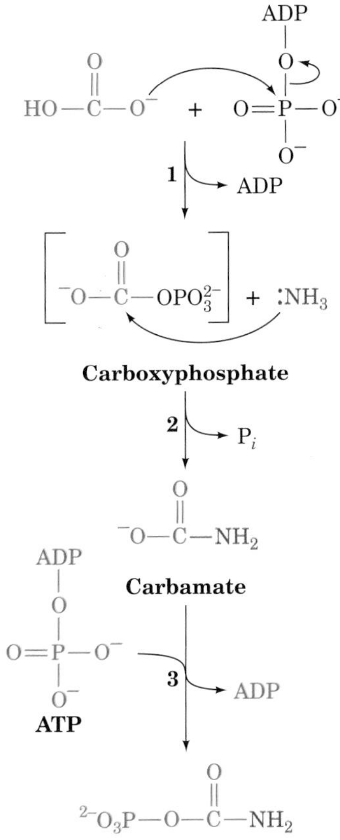

FIGURE 26-8 **The mechanism of action of CPS I.** **(1)** Activation of HCO_3^- by phosphorylation forms the intermediate, carboxyphosphate; **(2)** nucleophilic attack on carboxyphosphate by NH_3 forms the reaction's second intermediate, carbamate; and **(3)** phosphorylation of carbamate by ATP yields the reaction product carbamoyl phosphate.

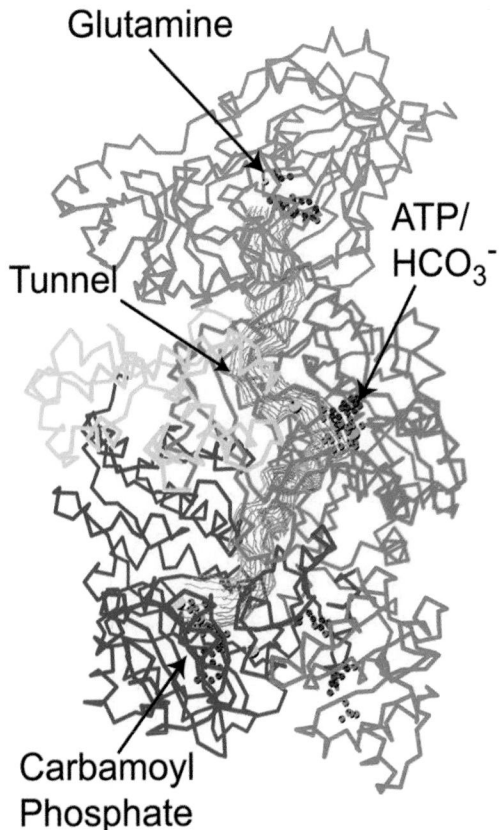

FIGURE 26-9 **X-Ray structure of *E. coli* carbamoyl phosphate synthetase (CPS).** The protein is represented by its C_α backbone. The small subunit (*magenta*) contains the glutamine binding site where NH_3 is produced or bound. The large subunit consists of the carboxyphosphate domain (*green*), the oligomerization domain (*yellow*), the carbamoyl phosphate domain (*blue*), and the allosteric binding domain (*orange*). The 96-Å-long tunnel connecting the three active sites is outlined in red. [Courtesy of Hazel Holden and Ivan Rayment, University of Wisconsin. PDBid 1JDB.]

Channeling increases the rate of a metabolic pathway by preventing the loss of its intermediate products as well as protecting the intermediate from degradation. NH_3 must travel ~45 Å down the CPS tunnel to react with carboxyphosphate to form the next intermediate, carbamate. The carbamate, in turn, must travel an additional ~35 Å to the site where it is phosphorylated by ATP to form the final product carbamoyl phosphate. The NH_3 transfer tunnel is lined with polar groups capable of forming hydrogen bonds with NH_3, whereas the tunnel through which carbamate travels is lined with backbone atoms and lacks charged groups that might induce its hydrolysis as it diffuses between active sites. Shielding and channeling are necessary because the intermediates carboxyphosphate and carbamate are extremely reactive, having half-lives of 28 and 70 ms, respectively, at neutral pH. Also, channeling allows the local concentration of NH_3 to reach a higher value than is present in the cellular medium. We shall encounter several other examples of channeling in our studies of metabolic enzymes, but the CPS tunnel is far longer than that in any other known enzyme.

B. Ornithine Transcarbamoylase

Ornithine transcarbamoylase transfers the carbamoyl group of carbamoyl phosphate to **ornithine,** yielding **citrulline** (Fig. 26-7, Reaction 2; note that both of these compounds are "nonstandard" α-amino acids in that they do not occur in proteins). The reaction occurs in the mitochondrion so that ornithine, which is produced in the cytosol, must enter the mitochondrion via a specific transport system. Likewise, since the remaining urea cycle reactions occur in the cytosol, citrulline must be exported from the mitochondrion.

C. Argininosuccinate Synthetase: Acquisition of the Second Urea Nitrogen Atom

Urea's second nitrogen atom is introduced in the urea cycle's third reaction by the condensation of citrulline's ureido group with an aspartate amino group by **argininosuccinate synthetase** (Fig. 26-10). The ureido oxygen atom is activated as a leaving group through formation of a citrullyl–AMP intermediate, which is subsequently displaced by

the aspartate amino group. Support for the existence of the citrullyl–AMP intermediate comes from experiments using ^{18}O-labeled citrulline (* in Fig. 26-10). The label was isolated in the AMP produced by the reaction, demonstrating that at some stage of the reaction, AMP and citrulline are linked covalently through the ureido oxygen atom.

D. Argininosuccinase

With formation of argininosuccinate, all of the urea molecule components have been assembled. However, the amino group donated by aspartate is still attached to the aspartate carbon skeleton. This situation is remedied by the **argininosuccinase**-catalyzed elimination of arginine from the aspartate carbon skeleton forming fumarate (Fig. 26-7, Reaction 4). Arginine is urea's immediate precursor. The fumarate produced in the argininosuccinase reaction reacts via the fumarase and malate dehydrogenase reactions to form oxaloacetate (Fig. 26-7, *bottom*), which is then used in gluconeogenesis (Section 23-1).

E. Arginase

The urea cycle's fifth and final reaction is the **arginase**-catalyzed hydrolysis of arginine to yield urea and regenerate ornithine (Fig. 26-7, Reaction 5). Ornithine is then returned to the mitochondrion for another round of the cycle. The urea cycle thereby converts two amino groups, one from NH_3 and one from aspartate, and a carbon atom from HCO_3^- to the relatively nontoxic excretion product urea at the cost of four "high-energy" phosphate bonds (three ATP hydrolyzed to two ADP, two P_i, AMP, and PP_i, followed by rapid PP_i hydrolysis). This energetic cost, together with that of gluconeogenesis, is supplied by the oxidation of the acetyl-CoA formed by the breakdown of amino acid carbon skeletons (e.g., threonine, Fig. 26-12). Indeed, half the oxygen that the liver consumes is used to provide this energy.

F. Regulation of the Urea Cycle

Carbamoyl phosphate synthetase I, the mitochondrial enzyme that catalyzes the first committed reaction of the urea cycle, is allosterically activated by **N-acetylglutamate:**

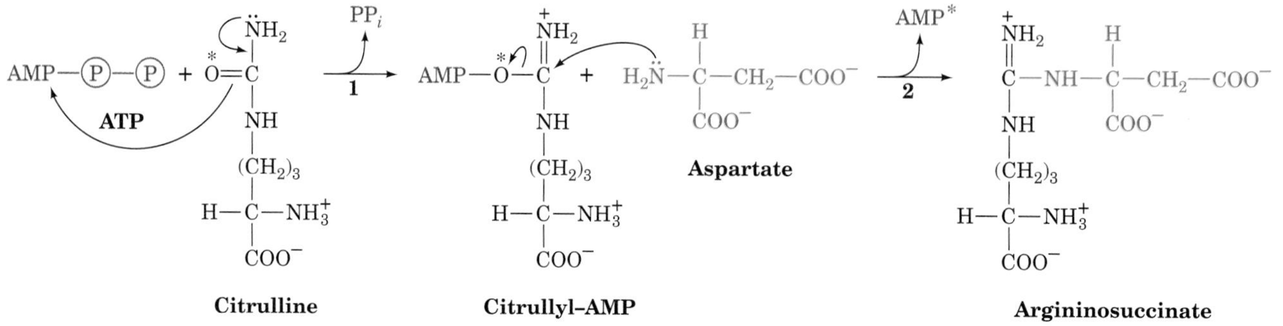

FIGURE 26-10   The mechanism of action of argininosuccinate synthetase. The steps involved are **(1)** activation of the ureido oxygen of citrulline through the formation of citrullyl–AMP and **(2)** displacement of AMP by the α-amino group of aspartate. The asterisk (*) traces the fate of ^{18}O originating in citrulline's ureido group.

$$COO^-$$
$$|$$
$$(CH_2)_2 \quad O$$
$$| \qquad \parallel$$
$$H-C-N-C-CH_3$$
$$| \quad |$$
$$^-OOC \quad H$$

N-Acetylglutamate

This metabolite is synthesized from glutamate and acetyl-CoA by **N-acetylglutamate synthase** and hydrolyzed by a specific hydrolase. The rate of urea production by the liver is, in fact, correlated with the N-acetylglutamate concentration. Increased urea synthesis is required when amino acid breakdown rates increase, generating excess nitrogen that must be excreted. Increases in these breakdown rates are signaled by an increase in glutamate concentration through transamination reactions (Section 26-1). This situation, in turn, causes an increase in N-acetylglutamate synthesis, stimulating carbamoyl phosphate synthetase and thus the entire urea cycle.

The remaining enzymes of the urea cycle are controlled by the concentrations of their substrates. Thus, inherited deficiencies in urea cycle enzymes other than arginase do not result in significant decreases in urea production (the total lack of any urea cycle enzyme results in death shortly after birth). Rather, the deficient enzyme's substrate builds up, increasing the rate of the deficient reaction to normal. The anomalous substrate buildup is not without cost, however. The substrate concentrations become elevated all the way back up the cycle to NH_3, resulting in hyperammonemia. Although the root cause of NH_3 toxicity is not completely understood, high [NH_3] puts an enormous strain on the NH_3-clearing system, especially in the brain (symptoms of urea cycle enzyme deficiencies include mental retardation and lethargy). This clearing system has been proposed to involve glutamate dehydrogenase (working in reverse) and **glutamine synthetase,** which decrease the α-ketoglutarate and glutamate pools (Sections 26-1 and 26-5A). The brain is most sensitive to the depletion of these pools. Depletion of α-ketoglutarate decreases the rate of the energy-generating citric acid cycle, whereas decreasing the glutamate concentration disturbs neuronal function, since it is both a neurotransmitter and a precursor to γ-aminobutyrate (GABA), another neurotransmitter (Section 20-5C). Glutamate depletion would also decrease the functioning of the urea cycle, since it is also the precursor to N-acetylglutamate, the major regulator of the cycle. The involvement of GDH in NH_3 clearance is a subject of debate in light of the observation that HI/HA involves deinhibition of GDH (Section 26-1B), suggesting that increased GDH activity increases the NH_3 concentration rather than decreasing it.

3 ■ METABOLIC BREAKDOWN OF INDIVIDUAL AMINO ACIDS

The degradation of amino acids converts them to citric acid cycle intermediates or their precursors so that they can be metabolized to CO_2 and H_2O or used in gluconeogenesis. Indeed, oxidative breakdown of amino acids typically accounts for 10 to 15% of the metabolic energy generated by animals. In this section we consider how amino acid carbon skeletons are catabolized. The 20 "standard" amino acids (the amino acids of proteins) have widely differing carbon skeletons, so their conversions to citric acid cycle intermediates follow correspondingly diverse pathways. We shall not describe all of the many reactions involved in detail. Rather, we shall consider how these pathways are organized and focus on a few reactions of chemical and/or medical interest.

A. Amino Acids Can Be Glucogenic, Ketogenic, or Both

"Standard" amino acids are degraded to one of seven metabolic intermediates: pyruvate, α-ketoglutarate, succinyl-CoA, fumarate, oxaloacetate, acetyl-CoA, or acetoacetate (Fig. 26-11). The amino acids may therefore be divided into two groups based on their catabolic pathways (Fig. 26-11):

1. Glucogenic amino acids, whose carbon skeletons are degraded to pyruvate, α-ketoglutarate, succinyl-CoA, fu-

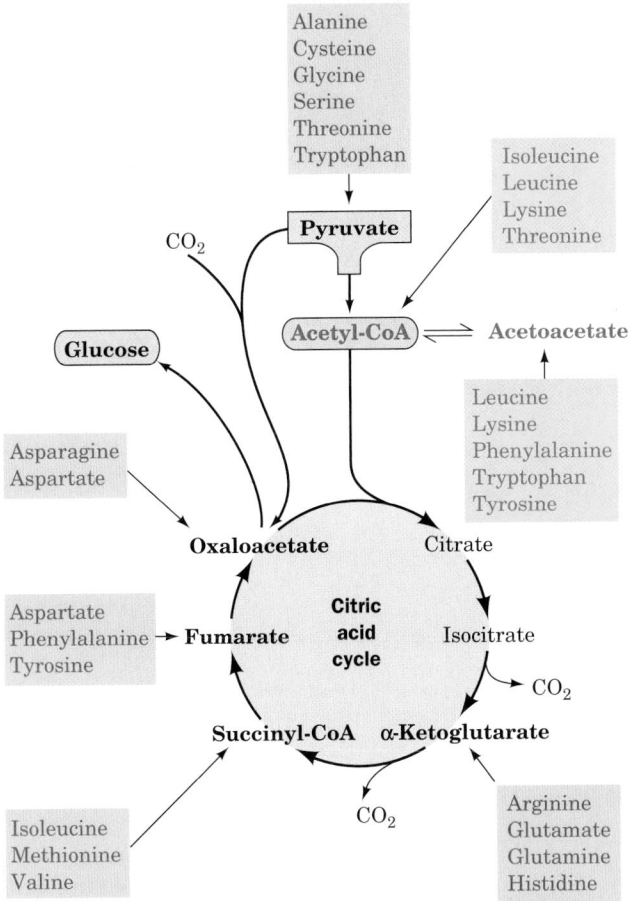

FIGURE 26-11 Degradation of amino acids to one of seven common metabolic intermediates. Glucogenic and ketogenic degradations are indicated in green and red, respectively.

marate, or oxaloacetate and are therefore glucose precursors (Section 23-1A).

2. Ketogenic amino acids, whose carbon skeletons are broken down to acetyl-CoA or acetoacetate and can thus be converted to fatty acids or ketone bodies (Section 25-3).

For example, alanine is glucogenic because its transamination product, pyruvate (Section 26-1A), can be converted to glucose via gluconeogenesis (Section 23-1A). Leucine, on the other hand, is ketogenic; its carbon skeleton is converted to acetyl-CoA and acetoacetate (Section 26-3F). Since animals lack any metabolic pathway for the net conversion of acetyl-CoA or acetoacetate to gluconeogenic precursors, no net synthesis of carbohydrates is possible from leucine, or from lysine, the only other purely ketogenic amino acid. Isoleucine, phenylalanine, threonine, tryptophan, and tyrosine, however, are both glucogenic and ketogenic; isoleucine, for example, is broken down to succinyl-CoA and acetyl-CoA and hence is a precursor of both carbohydrates and ketone bodies (Section 26-3E). The remaining 13 amino acids are purely glucogenic.

In studying the specific pathways of amino acid breakdown, we shall organize the amino acids into groups that are degraded into each of the seven metabolic intermediates mentioned above: pyruvate, oxaloacetate, α-ketoglutarate, succinyl-CoA, fumarate, acetyl-CoA, and acetoacetate. When acetoacetyl-CoA is a product in amino acid degradation, it can, of course, be directly converted to acetyl-CoA (Section 25-2). We also discuss the pathway by which, in liver, it is converted instead to acetoacetate for use as an alternative fuel source in peripheral tissues (Section 25-3).

B. *Alanine, Cysteine, Glycine, Serine, and Threonine Are Degraded to Pyruvate*

Five amino acids, alanine, cysteine, glycine, serine, and threonine, are broken down to yield pyruvate (Fig. 26-12). Tryptophan should also be included in this group since one of its breakdown products is alanine (Section 26-3G), which, as we have seen (Section 26-1A), is transaminated to pyruvate.

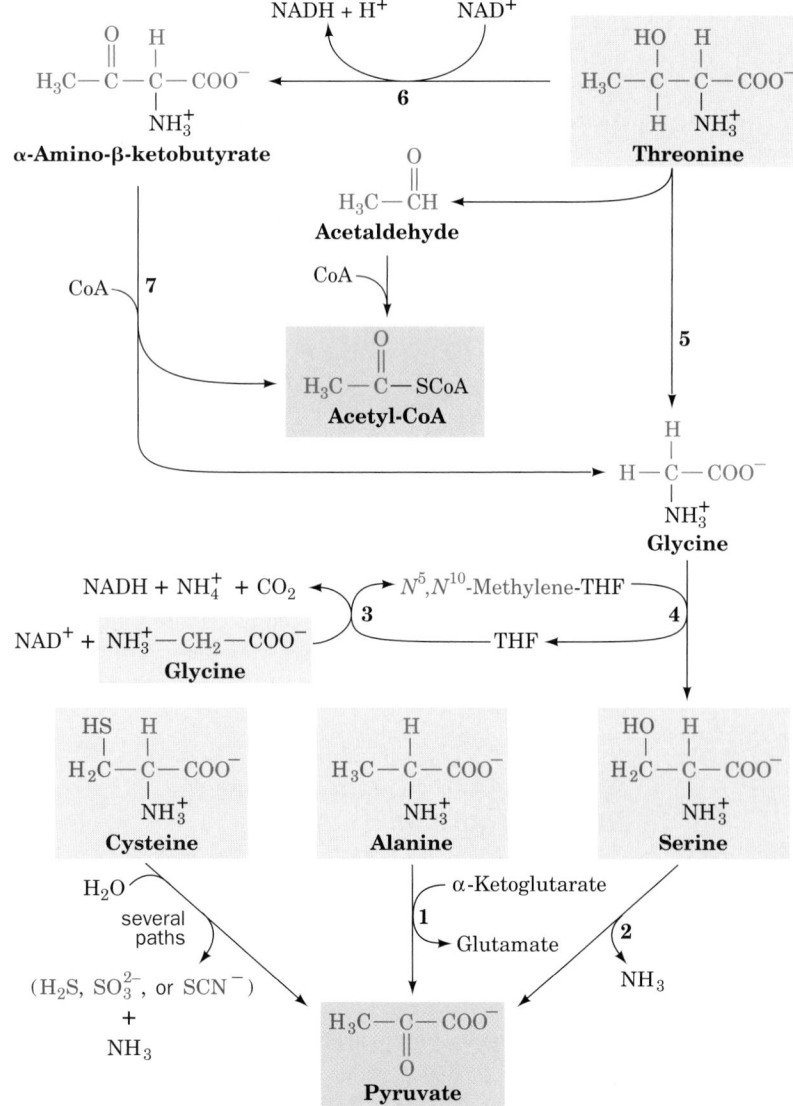

FIGURE 26-12   The pathways converting alanine, cysteine, glycine, serine, and threonine to pyruvate. The enzymes involved are **(1)** alanine aminotransferase, **(2)** serine dehydratase, **(3)** glycine cleavage system, **(4)** and **(5)** serine hydroxymethyltransferase, **(6)** threonine dehydrogenase, and **(7)** α-amino-β-ketobutyrate lyase.

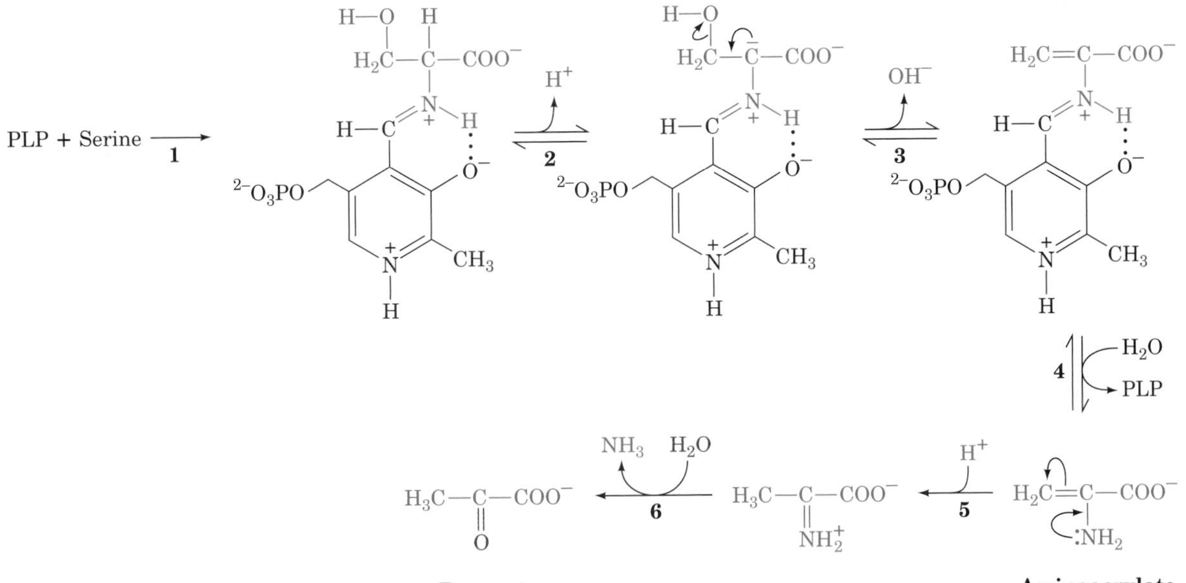

FIGURE 26-13 The serine dehydratase reaction. This PLP-dependent enzyme catalyzes the elimination of water from serine. The steps in the reaction are (1) formation of a serine–PLP Schiff base, (2) removal of the α-H atom of serine to form a resonance-stabilized carbanion, (3) β elimination of OH^-, (4) hydrolysis of the Schiff base to yield the PLP–enzyme and aminoacrylate, (5) nonenzymatic tautomerization to the imine, and (6) nonenzymatic hydrolysis to form pyruvate and ammonia.

Serine is converted to pyruvate through dehydration by **serine dehydratase.** This PLP–enzyme, like the aminotransferases (Section 26-1), functions by forming a PLP–amino acid Schiff base, which facilitates the removal of the amino acid's α-hydrogen atom. In the serine dehydratase reaction, however, the C_α carbanion breaks down with the elimination of the amino acid's C_β OH, rather than with tautomerization (Fig. 26-2, Step 2), so that the substrate undergoes α,β elimination of H_2O rather than deamination (Fig. 26-13). The product of the dehydration, the enamine **aminoacrylate,** tautomerizes nonenzymatically to the corresponding imine, which spontaneously hydrolyzes to pyruvate and ammonia.

Cysteine may be converted to pyruvate via several routes in which the sulfhydryl group is released as H_2S, SO_3^{2-}, or SCN^-.

Glycine is converted to serine by the enzyme **serine hydroxymethyltransferase,** another PLP-containing enzyme (Fig. 26-12, Reaction 4). This enzyme utilizes N^5,N^{10}-**methylene-tetrahydrofolate** (N^5,N^{10}-**methylene-THF**) as a cofactor to provide the C_1 unit necessary for this conversion. We shall defer a detailed discussion of THF cofactors until Section 26-4D.

a. The Glycine Cleavage System Is a Multienzyme Complex

The methylene group of the N^5,N^{10}-methylene-THF utilized in the conversion of glycine to serine is obtained from the methylene group of a second glycine via a reaction in which this glycine's remaining atoms are released as CO_2 and NH_4^+ (Fig. 26-12, Reaction 3). This reaction is catalyzed by the **glycine cleavage system** (also called the **glycine decarboxylase multienzyme system** in plants and **glycine synthase** when acting in the reverse direction; Section 26-5A), a complex resembling the pyruvate dehydrogenase complex (Section 21-2A) that consists of four proteins (Fig. 26-14):

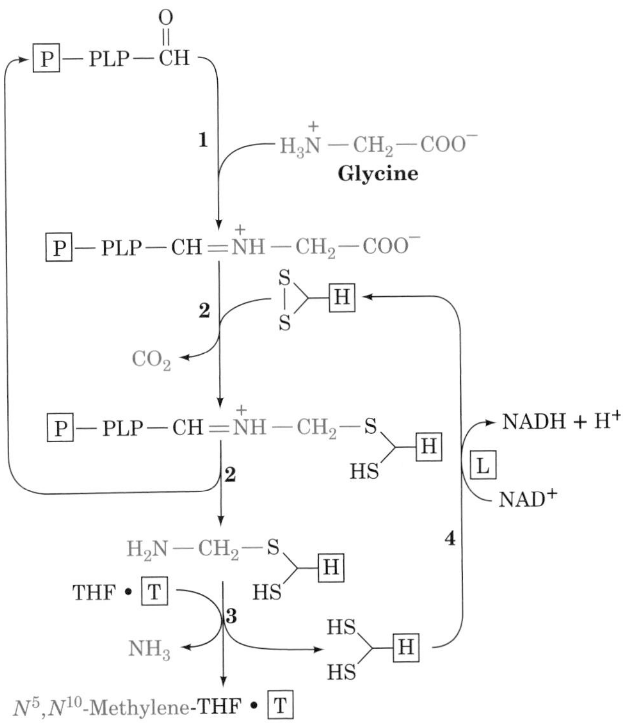

FIGURE 26-14 The reactions catalyzed by the glycine cleavage system, a multienzyme complex. The enzymes involved are (1) a PLP-dependent glycine decarboxylase (P-protein), (2) a lipoamide-containing protein (H-protein), (3) a THF-requiring enzyme (T-protein), and (4) an NAD^+-dependent, FAD-requiring dihydrolipoyl dehydrogenase (L-protein).

1. A PLP-dependent glycine decarboxylase **(P-protein).**

2. A lipoamide-containing aminomethyl carrier **(H-protein),** which carries the aminomethyl group remaining after glycine decarboxylation.

3. An N^5,N^{10}-methylene-THF synthesizing enzyme **[T-protein;** alternatively **aminomethyltransferase (AMT)],** which accepts a methylene group from the aminomethyl carrier (H-protein; the amino group is released as ammonia).

4. An NAD^+-dependent, FAD-requiring dihydrolipoyl dehydrogenase **(L-protein),** a protein shared by and known as E_3 in the pyruvate dehydrogenase complex (Section 21-2A).

Unlike the pyruvate dehydrogenase complex, the glycine cleavage system components are only loosely associated and hence are isolated as individual proteins. Nevertheless, H-protein has the central role in this multienzyme system: Its oxidized lipoyllysyl arm (Section 21-2A) is reduced as it accepts an aminomethyl group from P-protein (Fig. 26-14, Step 2), it donates the methylene group to THF in complex with T-protein as ammonia is released (Fig. 26-14, Step 3), and is then reoxidized by L-protein (Fig. 26-14, Step 4). The X-ray structure of pea leaf H-protein (Fig. 26-15), determined by Roland Douce, reveals that it is largely composed of a sandwich of 3-stranded and 6-stranded antiparallel β sheets that structurally resembles the lipoyl domain of E2 in the pyruvate dehydrogenase complex (Fig. 21-10a).

The aminomethylthio group is unstable and ordinarily is rapidly hydrolyzed to formaldehyde and NH_4^+. However, on its aminomethylation, the previously exposed lipoyl group (Fig. 26-15a) inserts into a hydrophobic cleft in the H-protein where its amino group hydrogen bonds to Glu 14 (Fig. 26-15b), thereby shielding the aminomethyl group from hydrolysis. Indeed, the replacement of Glu 14 by Ala results in the rapid hydrolysis of the aminomethyl group. It therefore appears that the T-protein in the THF · T · H complex functions to release the lipoyl group from the H-protein cleft and to orient the THF for approach to the methylene C of the aminomethyl group for reaction.

Two observations indicate that the above pathway is the major route of glycine degradation in mammalian tissues:

1. The serine isolated from an animal that has been fed [2-^{14}C]glycine is ^{14}C labeled at both C2 and C3. This observation indicates that the methylene group of the N^5,N^{10}-methylene-THF utilized by serine hydroxymethyltransferase is obtained from glycine C2.

2. The inherited human disease **nonketotic hyperglycinemia,** which is characterized by mental retardation and accumulation of large amounts of glycine in body fluids, results from the absence of one of the components of the glycine cleavage system.

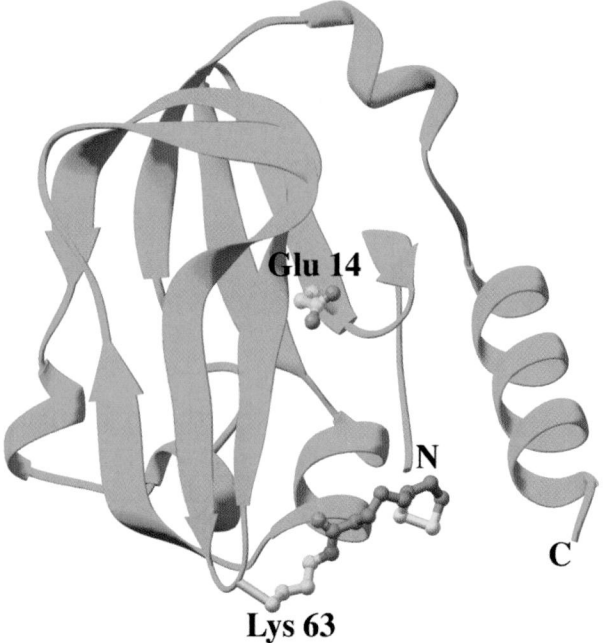

(a)

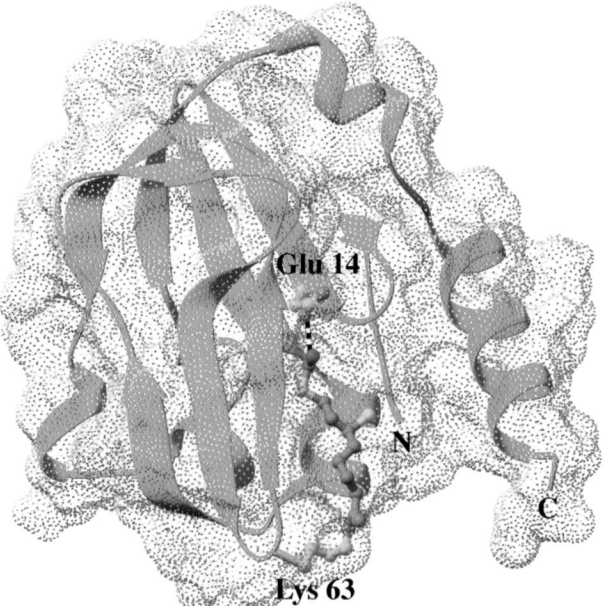

(b)

FIGURE 26-15   X-Ray structure of H-protein from the pea leaf glycine cleavage system. (a) The oxidized lipoamide-containing form in which the side chain of Glu 14 together with that of Lys 63 with its covalently linked lipoyl group are represented in ball-and-stick form colored according to atom type (Glu 14 C gold, Lys 63 C cyan, lipoyl C green, N blue, O red, and S yellow). (b) The reduced aminomethyl-dihydrolipoamide form of H-protein viewed and colored as in Part a. The dot surface represents the protein's solvent-accessible surface. Note how the aminomethyl-dihydrolipoamide has changed conformation relative to the lipoamide in Part a so as to bind in a hydrophobic cleft in the protein where its amino group is hydrogen bonded to Glu 14 (*dashed black bond*). This protects the aminomethyl group from hydrolysis. [Based on X-ray structures by Roland Douce, Centre National de la Recherche Scientifique et Commisariat à l'Energy Atomique, Grenoble, France. PDBids (a) 1HPC and (b) 1HTP.]

The glycine cleavage system and serine hydroxymethyltransferase occupy a vital role in green leaves, catalyzing the rapid destruction of the huge amounts of glycine produced by photorespiration (Section 24-3C). In fact, these enzymes comprise about half the proteins present in the mitochondria from pea and spinach leaves.

Threonine is both glucogenic and ketogenic, since one of its degradation routes produces both pyruvate and acetyl-CoA (Fig. 26-12, Reactions 6 and 7). Its major route of breakdown is through **threonine dehydrogenase,** producing **α-amino-β-ketobutyrate,** which is converted to acetyl-CoA and glycine by **α-amino-β-ketobutyrate lyase.** The glycine may be converted, through serine, to pyruvate.

b. Serine Hydroxymethyltransferase Catalyzes PLP-Dependent C_α—C_β Bond Cleavage

Threonine may also be converted directly to glycine and acetaldehyde (the latter being subsequently oxidized to acetyl-CoA), at least *in vitro,* via Reaction 5 of Fig. 26-12. Surprisingly, this reaction is catalyzed by serine hydroxymethyltransferase. We have heretofore considered PLP-catalyzed reactions that begin with the cleavage of an amino acid's C_α—H bond (Fig. 26-2). Degradation of threonine to glycine and acetaldehyde by serine hydroxymethyltransferase demonstrates that PLP also facilitates cleavage of an amino acid's C_α—C_β bond by delocalizing the electrons of the resulting carbanion into the conjugated PLP ring:

c. PLP Facilitates the Cleavage of Different Bonds in Different Enzymes

How can the same amino acid–PLP Schiff base be involved in the cleavage of the different bonds to an amino acid C_α in different enzymes? The answer to this conundrum was suggested by Harmon Dunathan. For electrons to be withdrawn into the conjugated ring system of PLP, the π-orbital system of PLP must overlap with the bonding orbital containing the electron pair being delocalized. This is possible only if the bond being broken lies in the plane perpendicular to the plane of the PLP π-orbital system (Fig. 26-16a). Different bonds to C_α can be placed in

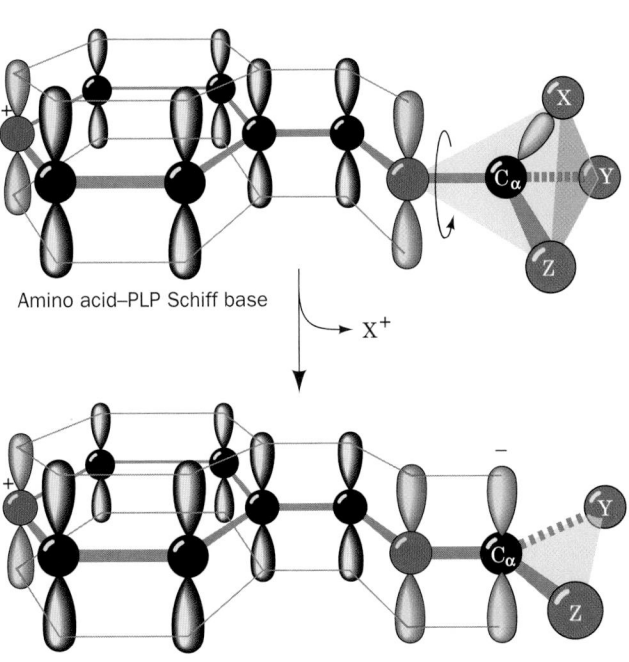

Amino acid–PLP Schiff base

Delocalized α carbanion

(a)

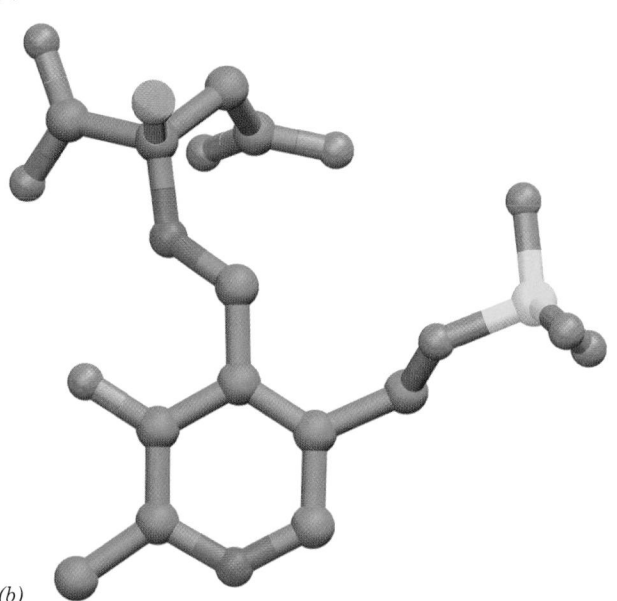

(b)

FIGURE 26-16 Bond orientation in a PLP–amino acid Schiff base. (*a*) The π-orbital framework of a PLP–amino acid Schiff base. The bond to C_α in the plane perpendicular to the PLP π-orbital system (from X in the illustration) is labile as a consequence of its overlap with the π system, which permits the broken bond's electron pair to be delocalized over the conjugated molecule. (*b*) The Schiff base complex of the inhibitor **α-methylaspartate** with PLP in the X-ray structure of porcine aspartate aminotransferase as viewed normal to the pyridoxal ring. This inhibitor is drawn in ball-and-stick form with C green, N blue, O red, and P gold, with the exception that the methyl C atom and the bond linking it to the aspartate residue are magenta. Here the methyl C occupies the position of the H atom that the enzyme normally excises from aspartate. Note that the bond linking the methyl C to aspartate is in the plane perpendicular to the pyridoxal ring and is thus ideally oriented for bond cleavage. [Part *b* based on an X-ray structure by David Metzler and Arthur Arnone, University of Iowa. PDBid 1AJS.]

this plane by rotation about the C_α—N bond. Indeed, the X-ray structure of aspartate aminotransferase reveals that the C_α—H of its aspartate substrate assumes just this conformation (Fig. 26-16*b*). Evidently, *each enzyme specifically cleaves its corresponding bond because the enzyme binds the amino acid–PLP Schiff base adduct with this bond in the plane perpendicular to that of the PLP ring.* This is an example of stereoelectronic assistance (Section 15-1E): *The enzyme binds substrate in a conformation that minimizes the electronic energy of the transition state.*

C. Asparagine and Aspartate Are Degraded to Oxaloacetate

Transamination of aspartate leads directly to oxaloacetate:

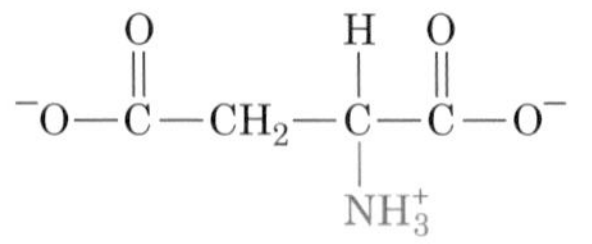

Aspartate

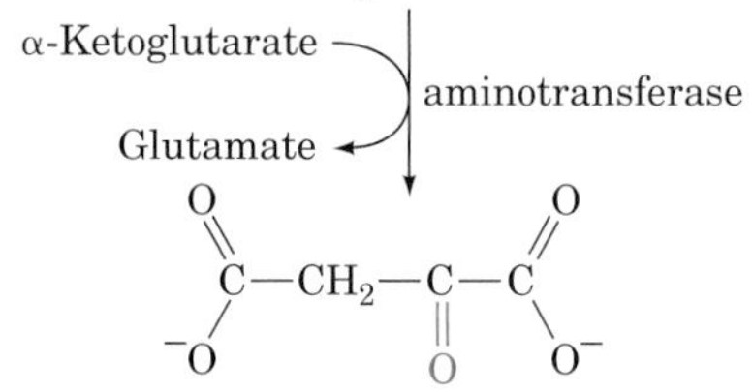

α-Ketoglutarate —⟍
 ⟩ aminotransferase
Glutamate ⟞

Oxaloacetate

Asparagine is also converted to oxaloacetate in this manner after its hydrolysis to aspartate by L-**asparaginase**:

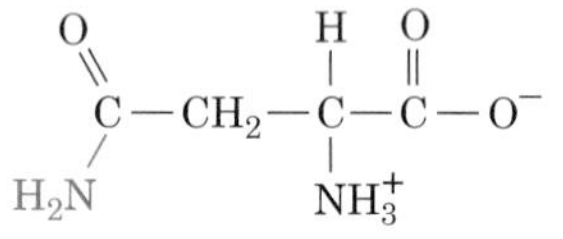

Asparagine

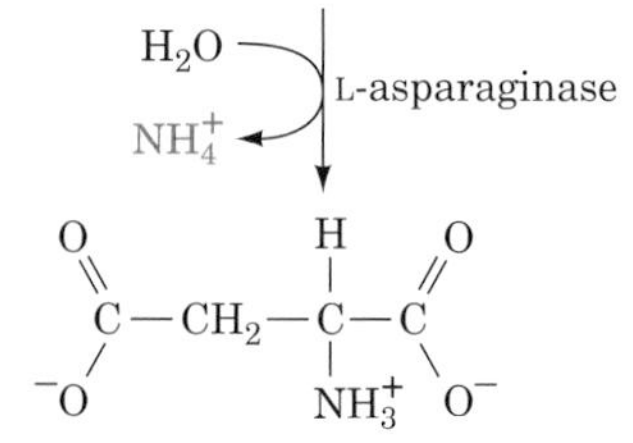

H_2O —⟍
 ⟩ L-asparaginase
NH_4^+ ⟞

Aspartate

Interestingly, L-asparaginase is an effective chemotherapeutic agent in the treatment of cancers that must obtain asparagine from the blood, particularly **acute lymphoblastic leukemia.** The cancerous cells express particularly low levels of the enzyme asparagine synthetase (Section 26-5A) and hence die without an external source of asparagine. However, L-asparaginase treatment may select for cells with increased levels of asparagine synthetase expression, and hence, in these cases, the surviving cancer cells are resistant to this treatment.

D. Arginine, Glutamate, Glutamine, Histidine, and Proline Are Degraded to α-Ketoglutarate

Arginine, glutamine, histidine, and proline are all degraded by conversion to glutamate (Fig. 26-17), which in turn is oxidized to α-ketoglutarate by glutamate dehydrogenase (Section 26-1). Conversion of glutamine to glutamate involves only one reaction: hydrolysis by **glutaminase.** Histidine's conversion to glutamate is more complicated: It is nonoxidatively deaminated, then it is hydrated, and its imidazole ring is cleaved to form ***N*-formiminoglutamate.** The formimino group is then transferred to tetrahydrofolate forming glutamate and N^5-**formimino-tetrahydrofolate** (Section 26-4D). Both arginine and proline are converted to glutamate through the intermediate formation of **glutamate-5-semialdehyde.**

E. Isoleucine, Methionine, and Valine Are Degraded to Succinyl-CoA

Isoleucine, methionine, and valine have complex degradative pathways that all yield propionyl-CoA. Propionyl-CoA, which is also a product of odd-chain fatty acid degradation, is converted, as we have seen, to succinyl-CoA by a series of reactions involving the participation of biotin and coenzyme B_{12} (Section 25-2E).

a. Methionine Breakdown Involves Synthesis of *S*-Adenosylmethionine and Cysteine

Methionine degradation (Fig. 26-18) begins with its reaction with ATP to form **S-adenosylmethionine (SAM;** alternatively **AdoMet).** *This sulfonium ion's highly reactive methyl group makes it an important biological methylating agent.* For instance, we have already seen that SAM is the methyl donor in the synthesis of phosphatidylcholine from phosphatidylethanolamine (Section 25-8A). It is also the methyl donor in the conversion of norepinephrine to epinephrine (Section 26-4B).

Methylation reactions involving SAM yield **S-adenosylhomocysteine** in addition to the methylated acceptor. The former product is hydrolyzed to adenosine and **homocysteine** in the next reaction of the methionine degradation pathway. The homocysteine may be methylated to form methionine via a B_{12}-requiring reaction in which N^5-**methyl-THF** is the methyl donor. Alternatively, the homocysteine may combine with serine to yield **cystathionine** in a PLP-requiring reaction, which subsequently forms cysteine (cysteine biosynthesis) and **α-ketobutyrate.** The α-ketobutyrate continues along the degradative pathway to propionyl-CoA and then succinyl-CoA.

b. Hyperhomocysteinemia Is Associated with Disease

Imbalance between the rate of production of homocysteine through methylation reactions utilizing SAM (Fig. 26-18, Reactions 2 and 3) and its rate of breakdown by either remethylation to form methionine (Fig. 26-18, Reaction 4) or reaction with serine to form cystathionine in the cysteine biosynthesis pathway (Fig. 26-18, Reaction 5) can result in an increase in the release of homocysteine

FIGURE 26-17 Degradation pathways of arginine, glutamate, glutamine, histidine, and proline to α-ketoglutarate. The enzymes catalyzing the reactions are **(1)** glutamate dehydrogenase, **(2)** glutaminase, **(3)** arginase, **(4)** ornithine-δ-aminotransferase, **(5)** glutamate-5-semialdehyde dehydrogenase, **(6)** proline oxidase, **(7)** spontaneous, **(8)** histidine ammonia lyase, **(9)** urocanate hydratase, **(10)** imidazolone propionase, and **(11)** glutamate formiminotransferase.

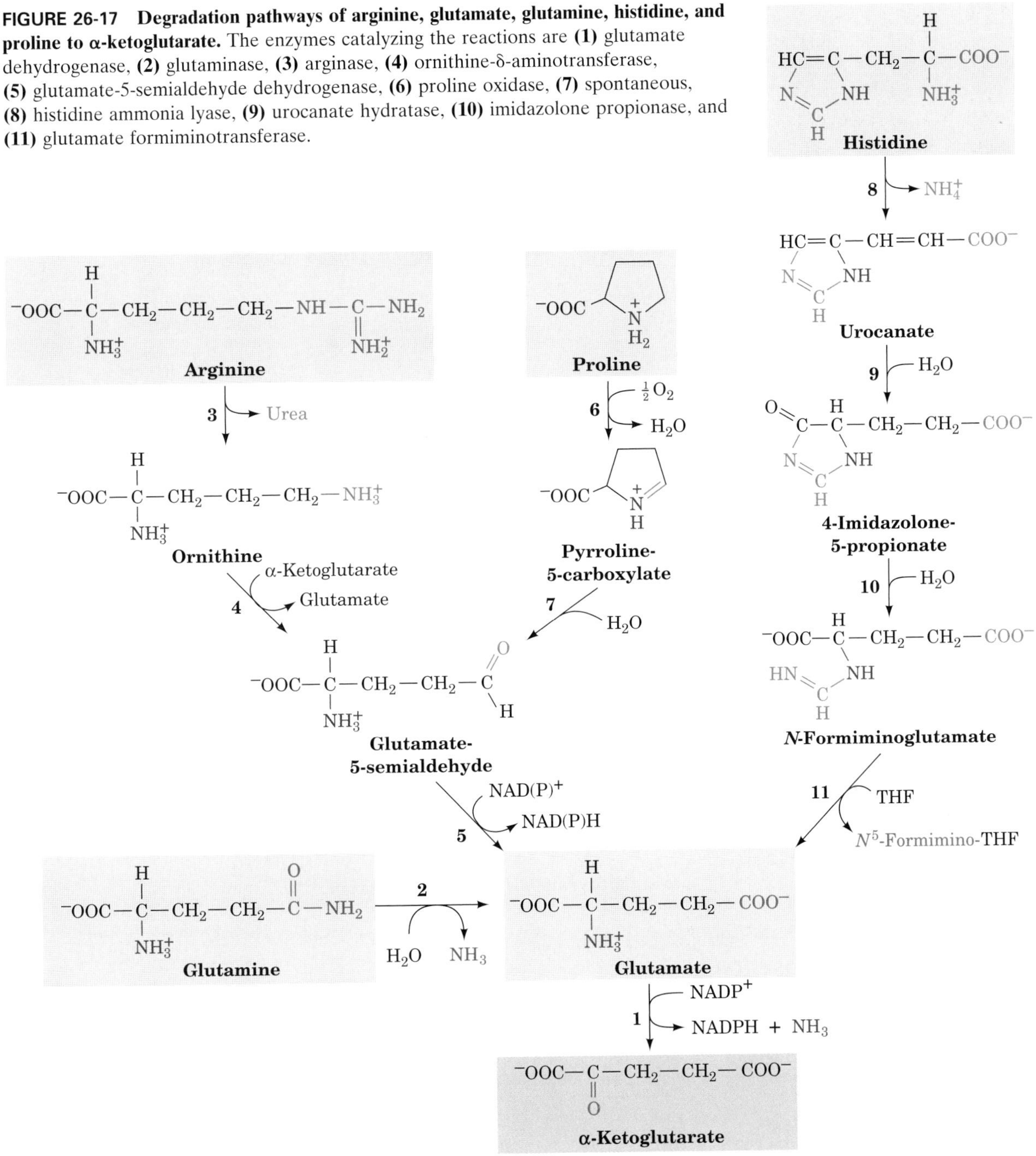

to the extracellular medium and ultimately the plasma and urine. Moderately elevated concentrations of homocysteine in the plasma, **hyperhomocysteinemia,** for reasons that are poorly understood, are closely associated with cardiovascular disease, cognitive impairment, and **neural tube defects** [the cause of a variety of severe birth defects including **spina bifida** (defects in the spinal column that often result in paralysis) and **anencephaly** (the invariably fatal failure of the brain to develop, which is the leading cause of infant death due to congenital anomalies)]. Hyperhomocysteinemia is readily controlled by ingesting the vitamin precursors of the coenzymes that participate

in homocysteine breakdown, namely, B_6 (pyridoxine, the PLP precursor; Fig. 26-1), B_{12} (Fig. 25-21), and folate (Section 26-4D). Folate, especially, appears to alleviate hyperhomocysteinemia; its administration to pregnant women dramatically reduces the incidence of neural tube defects in newborns. This has led to the discovery that 10% of the population is homozygous for the A222V mutation in N^5,N^{10}-**methylene-tetrahydrofolate reductase** (**MTHFR;** Fig. 26-18, Reaction 12; Section 26-4D), the enzyme that generates N^5-methyl-THF for the methionine synthase reaction (Fig. 26-18, Reaction 4). This mutation does not affect this homotetrameric enzyme's reaction kinetics but

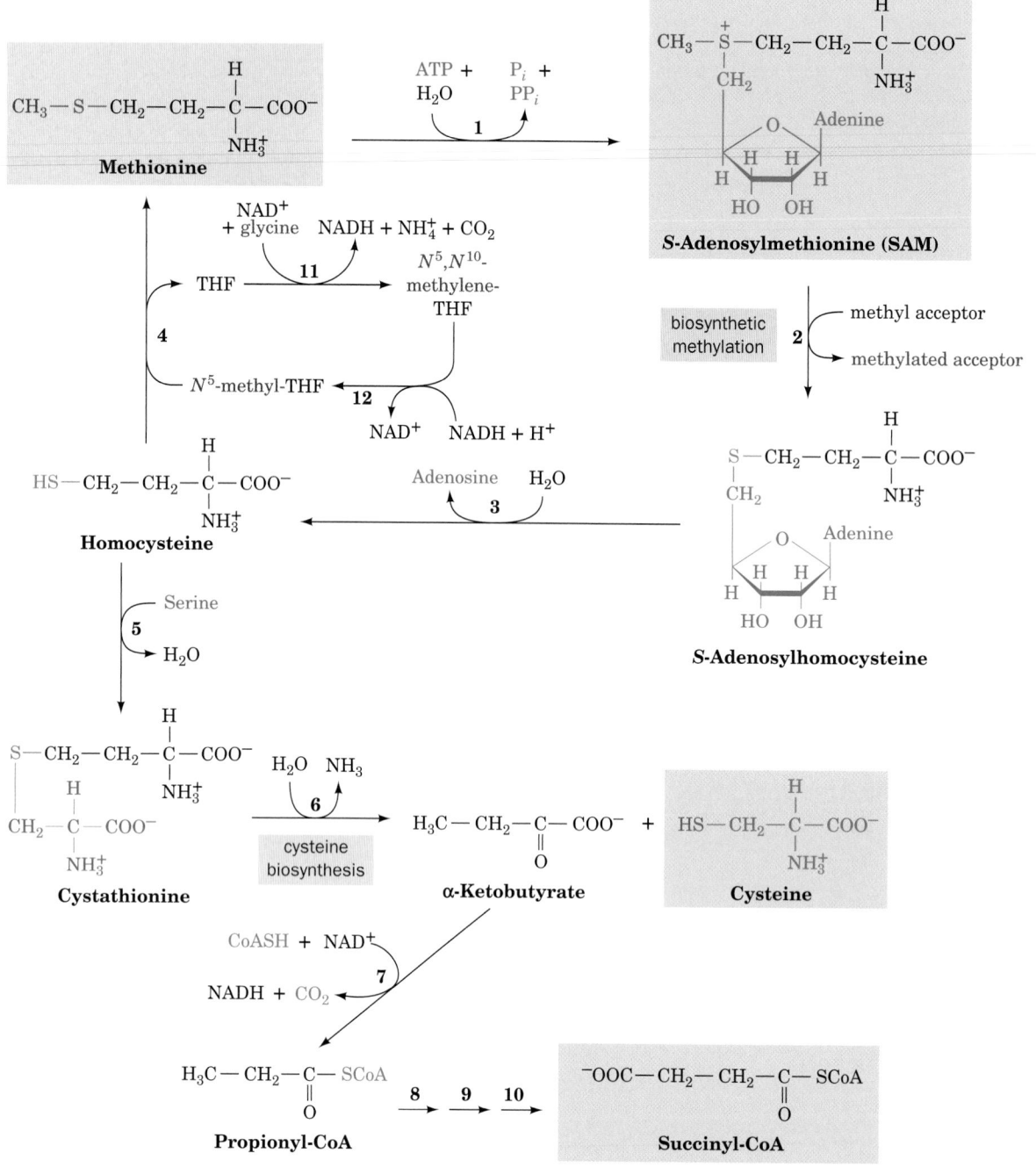

FIGURE 26-18   The pathway of methionine degradation, yielding cysteine and succinyl-CoA as products. The enzymes involved are **(1)** methionine adenosyltransferase in a reaction that yields the biological methylating agent *S*-adenosylmethionine (SAM), **(2)** methyltransferase, **(3)** adenosylhomocysteinase, **(4)** methionine synthase (a coenzyme B₁₂-dependent enzyme), **(5)** cystathionine β-synthase (a PLP-dependent enzyme), **(6)** cystathionine γ-lyase (a PLP-dependent enzyme), **(7)** α-keto acid dehydrogenase, **(8)** propionyl-CoA carboxylase, **(9)** methylmalonyl-CoA racemase, **(10)** methylmalonyl-CoA mutase (a coenzyme B₁₂-dependent enzyme; Reactions 8–10 are discussed in Section 25-2E), **(11)** glycine cleavage system (Figs. 26-12 and 26-14) or serine hydroxymethyltransferase (Fig. 26-12), **(12)** N^5,N^{10}-methylene-tetrahydrofolate reductase (a coenzyme B₁₂- and FAD-dependent enzyme; Figs. 26-19 and 26-49).

instead increases the rate at which it dissociates into dimers that readily lose their essential flavin cofactor. Folate derivatives that bind to the enzyme decrease its rate of dissociation and flavin loss, thus increasing the mutant enzyme's overall activity and decreasing the homocysteine concentration.

The X-ray structure of *E. coli* MTHFR (which is 30% identical with the catalytic domain of human MTHFR), determined by Rowena Matthews and Martha Ludwig, reveals that this 296-residue enzyme forms an α/β barrel. The FAD cofactor binds at the C-terminal ends of barrel strands β3, β4, and β5 and along helix α5 (Fig. 26-19). Ala

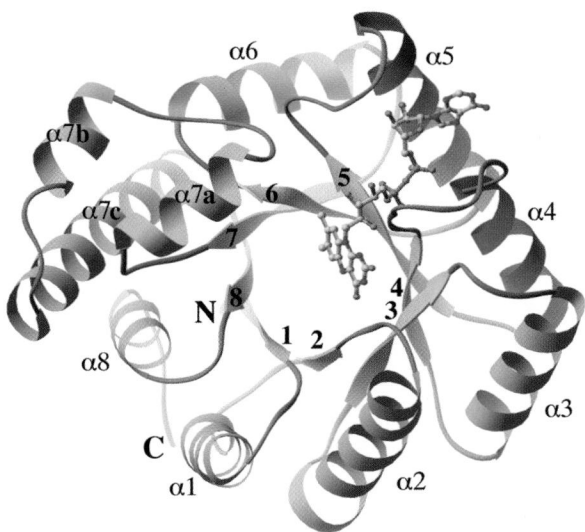

FIGURE 26-19   X-Ray structure of *E. coli* N^5,N^{10}-methylene-tetrahydrofolate reductase (MTHFR). The structure is viewed along the axis of its α/β barrel looking toward the C-terminal ends of its β strands. The protein is colored according to its secondary structure with β strands yellow and α helices cyan except for helix $\alpha5$, which is red. The enzyme's bound FAD is drawn in ball-and-stick form with C yellow, N blue, O red, and P green. Note that the AMP moiety of the FAD is in contact with helix $\alpha5$. [Courtesy of Rowena Matthews and Martha Ludwig, University of Michigan. PDBid 1B5T.]

177, which corresponds to Ala 222 in the mammalian enzyme, does not interact directly with the active site FAD. Instead it occupies a position flush against helix $\alpha5$ (which ends with residue 176). It is postulated that the replacement of Ala 177 by a bulkier Val residue would force helix $\alpha5$ to reorient. Since this helix appears to be involved in the subunit interface as well as in FAD binding, its reorientation is likely to decrease the strength of subunit and FAD interactions.

Why should this mutation be so prevalent in the human population? What selective advantage, if any, might it confer? We have seen that the gene for sickle-cell anemia provides a selective advantage against malaria (Section 10-3B). However, the selective advantage of the A222V mutation in human MTHFR is as yet a matter of speculation.

c. Methionine Synthase Is a Coenzyme B_{12}-Dependent Enzyme

Methionine synthase (alternatively **homocysteine methyltransferase),** the enzyme that catalyzes Reaction 4 in Fig. 26-18, is the only coenzyme B_{12}-associated enzyme in mammals besides methylmalonyl-CoA mutase (Section 25-2E). However, in methionine synthase, the cobalamin Co ion is axially liganded by a methyl group forming **methylcobalamin** rather than by a 5′-deoxyadenosyl group as occurs in methylmalonyl-CoA mutase (Fig. 25-21). This is because the cobalamin functions to accept the methyl group from N^5-methyl-THF to yield methylcobalamin (and THF), which, in turn, donates the methyl group to homocysteine to yield methionine.

The X-ray structure of the 246-residue methylcobalamin-binding portion of the 1227-residue monomeric *E. coli* methionine synthase, also determined by Matthews and Ludwig, reveals that it consists of two domains, an N-terminal helical domain and a C-terminal Rossmann fold-like α/β domain, with the corrin ring sandwiched between them (Fig. 26-20). The α/β domain resembles the corrin-binding α/β domain in methylmalonyl-CoA mutase (Fig. 25-22) and, in fact, sequence homologies suggest that this domain is a common binding motif in B_{12}-associated enzymes. The Co ion's second axial ligand is a His side chain as is also the case in methylmalonyl-CoA mutase; the coenzyme's 5,6-dimethylbenzamidazole (DMB) moiety, which ligands the Co ion in free methylcobalamin, has swung aside to become anchored to the protein at some distance from the corrin ring.

d. Branched-Chain Amino Acid Degradation Pathways Contain Themes Common to All Acyl-CoA Oxidations

Degradation of the branched-chain amino acids isoleucine, leucine, and valine begins with three reactions

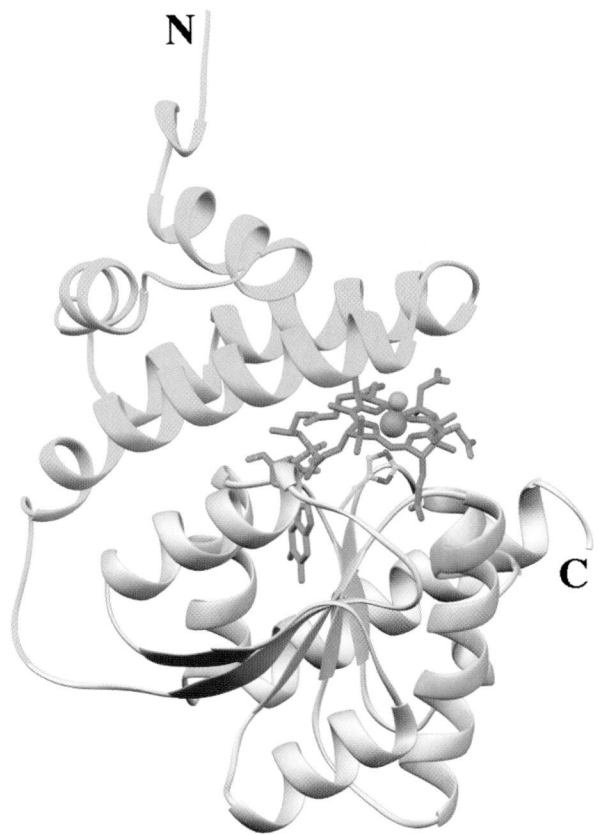

FIGURE 26-20   X-Ray structure of the B_{12}-binding domains of *E. coli* methionine synthase. Its N-terminal helical domain (residues 651–743) is cyan and its C-terminal α/β domain (residues 744–896) is pink. The methylcobalamin cofactor and its axially liganded His 759 side chain are drawn in stick form with cobalamin C green, His C gold, N blue, O red, and the Co ion and its axially liganded methyl group represented by light blue and orange spheres, respectively. [Based on an X-ray structure by Rowena Matthews and Martha Ludwig, University of Michigan. PDBid 1BMT.]

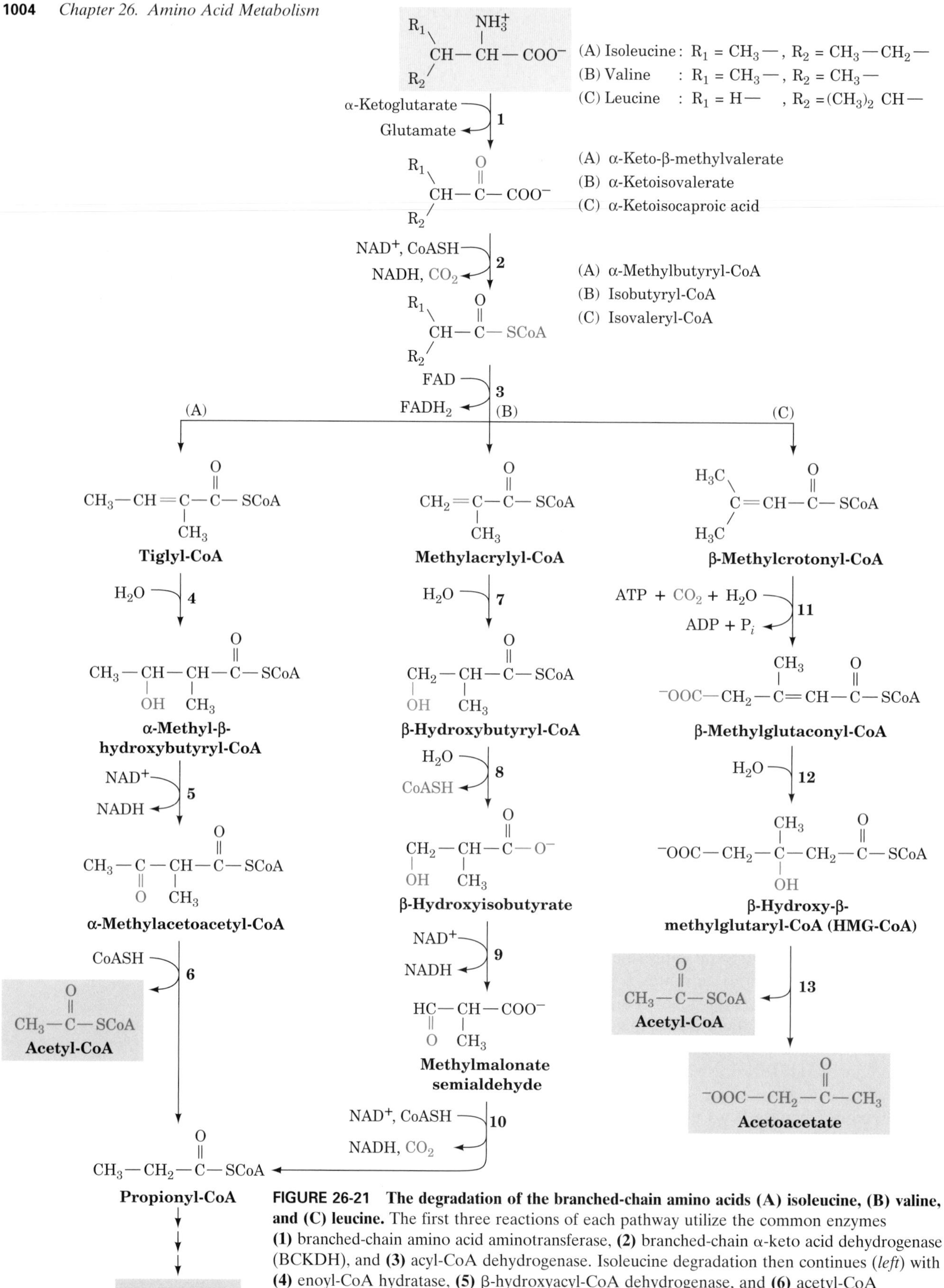

(A) Isoleucine : $R_1 = CH_3-$, $R_2 = CH_3-CH_2-$
(B) Valine   : $R_1 = CH_3-$, $R_2 = CH_3-$
(C) Leucine : $R_1 = H-$, $R_2 = (CH_3)_2 \; CH-$

(A) α-Keto-β-methylvalerate
(B) α-Ketoisovalerate
(C) α-Ketoisocaproic acid

(A) α-Methylbutyryl-CoA
(B) Isobutyryl-CoA
(C) Isovaleryl-CoA

Tiglyl-CoA

Methylacrylyl-CoA

β-Methylcrotonyl-CoA

α-Methyl-β-hydroxybutyryl-CoA

β-Hydroxybutyryl-CoA

β-Methylglutaconyl-CoA

α-Methylacetoacetyl-CoA

β-Hydroxyisobutyrate

β-Hydroxy-β-methylglutaryl-CoA (HMG-CoA)

Acetyl-CoA

Methylmalonate semialdehyde

Acetyl-CoA

Acetoacetate

Propionyl-CoA

Succinyl-CoA

FIGURE 26-21   The degradation of the branched-chain amino acids (A) isoleucine, (B) valine, and (C) leucine. The first three reactions of each pathway utilize the common enzymes **(1)** branched-chain amino acid aminotransferase, **(2)** branched-chain α-keto acid dehydrogenase (BCKDH), and **(3)** acyl-CoA dehydrogenase. Isoleucine degradation then continues (*left*) with **(4)** enoyl-CoA hydratase, **(5)** β-hydroxyacyl-CoA dehydrogenase, and **(6)** acetyl-CoA acetyltransferase to yield acetyl-CoA and the succinyl-CoA precursor propionyl-CoA. Valine degradation (*center*) continues with **(7)** enoyl-CoA hydratase, **(8)** β-hydroxy-isobutyryl-CoA hydrolase, **(9)** β-hydroxyisobutyrate dehydrogenase, and **(10)** methylmalonate semialdehyde dehydrogenase to also yield propionyl-CoA. Leucine degradation (*right*) continues with **(11)** β-methylcrotonyl-CoA carboxylase (a biotin-dependent enzyme), **(12)** β-methylglutaconyl-CoA hydratase, and **(13)** HMG-CoA lyase to yield acetyl-CoA and acetoacetate.

that employ common enzymes (Fig. 26-21, *top*): (1) transamination to the corresponding α-keto acid, (2) oxidative decarboxylation to the corresponding acyl-CoA, and (3) dehydrogenation by FAD to form a double bond.

The remainder of the isoleucine degradation pathway (Fig. 26-21, *left*) is identical to that of fatty acid oxidation (Section 25-2C): (4) double-bond hydration, (5) dehydrogenation by NAD⁺, and (6) thiolytic cleavage yielding acetyl-CoA and propionyl-CoA, which is subsequently converted to succinyl-CoA. Valine degradation is a variation on this theme (Fig. 26-21, *center*): Following (7) double-bond hydration, (8) the CoA thioester bond is hydrolyzed before (9) the second dehydrogenation reaction. The thioester bond is then regenerated as propionyl-CoA in the sequence's last reaction (10), an oxidative decarboxylation rather than a thiolytic cleavage.

e. Maple Syrup Urine Disease Results from a Defect in Branched-Chain Amino Acid Degradation

Branched-chain α-keto acid dehydrogenase (BCKDH; also known as α-ketoisovalerate dehydrogenase), which catalyzes Reaction 2 of branched-chain amino acid degradation (Fig. 26-21), is a multienzyme complex containing three enzymatic components, E1, E2, and E3, together with **BCKDH kinase** (phosphorylation inactivates) and **BCKDH phosphatase** (dephosphorylation activates), which impart control by covalent modification. This complex closely resembles the pyruvate dehydrogenase and α-ketoglutarate dehydrogenase multienzyme complexes (Sections 21-2A and 21-3D). Indeed, all three of these multienzyme complexes share a common protein component,

E₃ (dihydrolipoyl dehydrogenase), and employ the coenzymes thiamine pyrophosphate (TPP), lipoamide, and FAD in addition to their terminal oxidizing agent, NAD⁺.

A genetic deficiency in BCKDH causes **maple syrup urine disease (MSUD),** so named because the consequent buildup of branched-chain α-keto acids imparts the urine with the characteristic odor of maple syrup. Unless promptly treated by a diet low in branched-chain amino acids (but not too low because they are essential amino acids; Section 26-5), MSUD is rapidly fatal.

MSUD is an autosomal recessive disorder that is caused by defects in any of four of the complex's six subunits, E1α, E1β, E2, or E3 (E1 is an α₂β₂ heterotetramer). The determination of the X-ray structure of human BCKDH E1 by Wim Hol (Fig. 26-22) has enabled the interpretation of several of the mutations causing MSUD. The most common mutation is Y393N-α, the so-called Mennonite mutation, which occurs once in every 176 live births in the Old Order Mennonite population (versus 1 in 185,000 worldwide). This mutation is so common among Old Order Mennonites that it has been attributed to a founder effect, that is, a mutation that originated in one of the handful of founders of this isolated community. The E1 tetramer can be considered to be a dimer of αβ heterodimers with a TPP cofactor at the interface between an α subunit and a β subunit and with each α subunit contacting both the β and β′ subunits (Fig. 26-22a). The amino acid change in the Mennonite mutation occurs at the α–β′ interface: Tyr 393α is hydrogen bonded to both His 385α and Asp 328β′ (Fig. 26-22b). Its mutation to Asn disrupts these interactions and thereby impedes tetramerization.

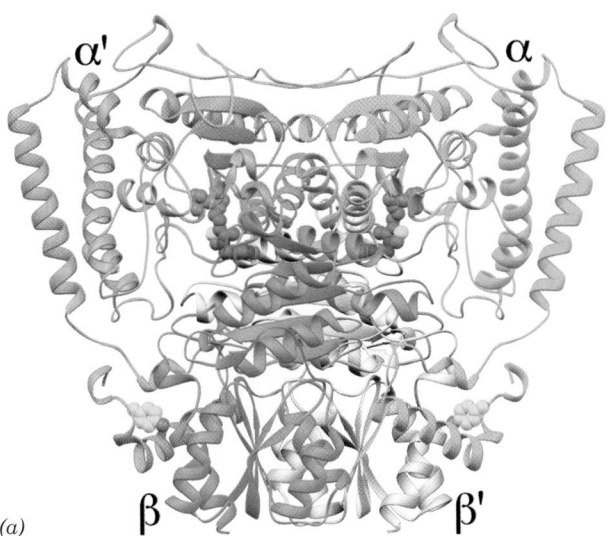

(a)

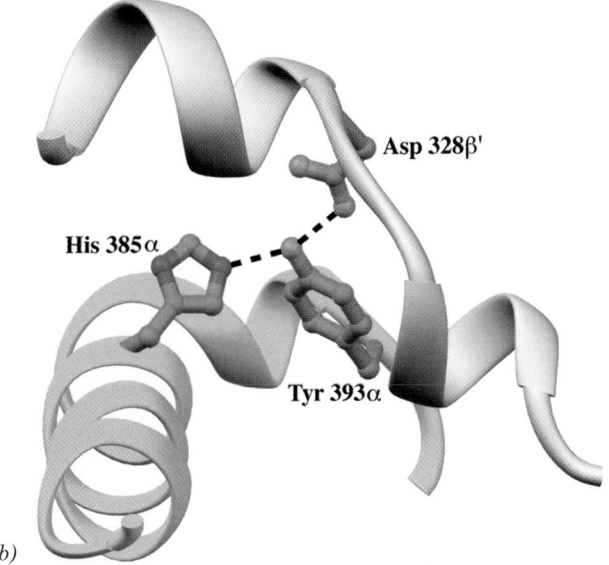

(b)

FIGURE 26-22 X-Ray structure of the E1 component of the human branched-chain α-keto acid dehydrogenase multienzyme complex. (*a*) The α₂β₂ heterotetramer. The α subunits are colored cyan and orange, and the β subunits are light blue and pink. The thiamine pyrophosphate (TPP) cofactor and Tyr 393α (which is mutated to Asn in the Mennonite mutation, causing maple syrup urine disease) are shown in space-filling form with TPP C green, Tyr 393α C gold, N blue, O red, S yellow, and P

magenta. (*b*) The α–β′ interface colored as in Part *a* and showing the interactions of Tyr 393α with His 385α and Asp 328β′. The side chains of these residues are drawn in ball-and-stick form with C green, N blue, and O red and with the hydrogen bonds between them represented by dashed lines. [Based on an X-ray structure by Wim Hol, University of Washington. PDBid 1DTW.]

F. *Leucine and Lysine Are Degraded to Acetoacetate and/or Acetyl-CoA*

Leucine is oxidized by a combination of reactions used in β oxidation and ketone body synthesis (Fig. 26-21, *right*). The first dehydrogenation and the hydration reactions are interspersed by (11) a carboxylation reaction catalyzed by a biotin-containing enzyme. The hydration reaction (12) then produces **β-hydroxy-β-methylglutaryl-CoA (HMG-CoA),** which is cleaved by HMG-CoA lyase to form acetyl-

CoA and the ketone body acetoacetate (13) (which, in turn, may be converted to 2 acetyl-CoA; Section 25-3).

Although there are several pathways for lysine degradation, the one that proceeds via formation of the α-ketoglutarate–lysine adduct **saccharopine** predominates in mammalian liver (Fig. 26-23). This pathway is of interest because we have encountered 7 of its 11 reactions in other pathways. Reaction 4 is a PLP-dependent transamination. Reaction 5 is the oxidative decarboxylation of an α-keto acid by a multienzyme complex similar to pyruvate

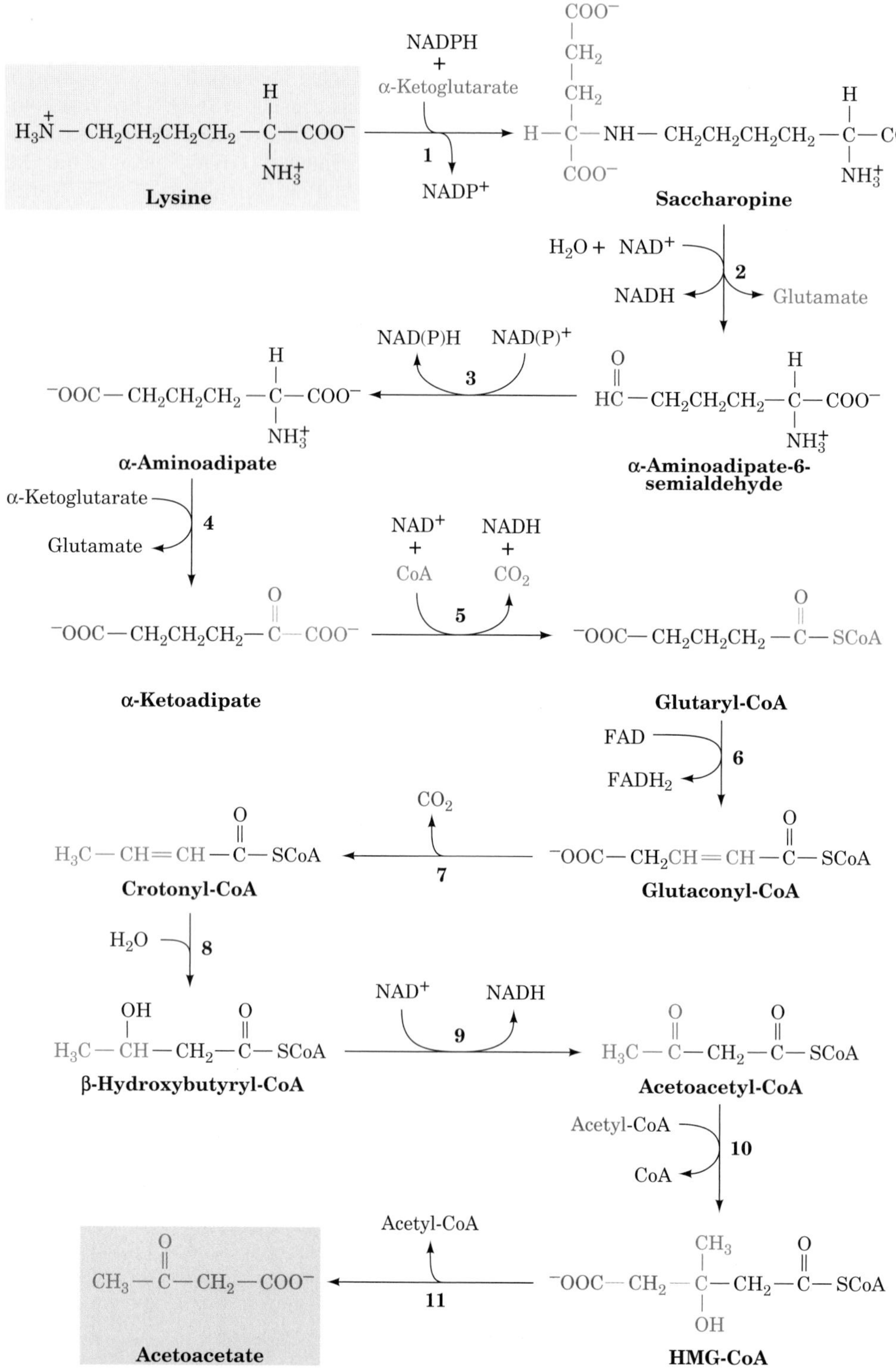

FIGURE 26-23 The pathway of lysine degradation in mammalian liver. The enzymes involved are (1) saccharopine dehydrogenase (NADP⁺, lysine forming), (2) saccharopine dehydrogenase (NAD⁺, glutamate forming), (3) aminoadipate semialdehyde dehydrogenase, (4) aminoadipate aminotransferase (a PLP enzyme), (5) α-keto acid dehydrogenase, (6) glutaryl-CoA dehydrogenase, (7) decarboxylase, (8) enoyl-CoA hydratase, (9) β-hydroxyacyl-CoA dehydrogenase, (10) HMG-CoA synthase, and (11) HMG-CoA lyase. Reactions 10 and 11 are discussed in Section 25-3.

dehydrogenase and α-ketoglutarate dehydrogenase (Sections 21-2A and 21-3D). Reactions 6, 8, and 9 are standard reactions of fatty acyl-CoA oxidation: dehydrogenation by FAD, hydration, and dehydrogenation by NAD^+. Reactions 10 and 11 are standard reactions in ketone body formation. Two molecules of CO_2 are produced at Reactions 5 and 7 of the pathway.

The saccharopine pathway is thought to predominate in mammals because a genetic defect in the enzyme that catalyzes Reaction 1 in the sequence results in **hyperlysinemia** and **hyperlysinuria** (elevated levels of lysine in the blood and urine, respectively) along with mental and physical retardation. This is yet another example of how the study of rare inherited disorders has helped to trace metabolic pathways.

Leucine's carbon skeleton, as we have seen, is converted to one molecule each of acetoacetate and acetyl-CoA, whereas that of lysine is converted to one molecule of acetoacetate and two of CO_2. Since neither acetoacetate nor acetyl-CoA can be converted to glucose in animals, leucine and lysine are purely ketogenic amino acids.

G. *Tryptophan Is Degraded to Alanine and Acetoacetate*

The complexity of the major tryptophan degradation pathway (Fig. 26-24) precludes detailed discussion of all of its reactions. However, one reaction in the pathway is of par-

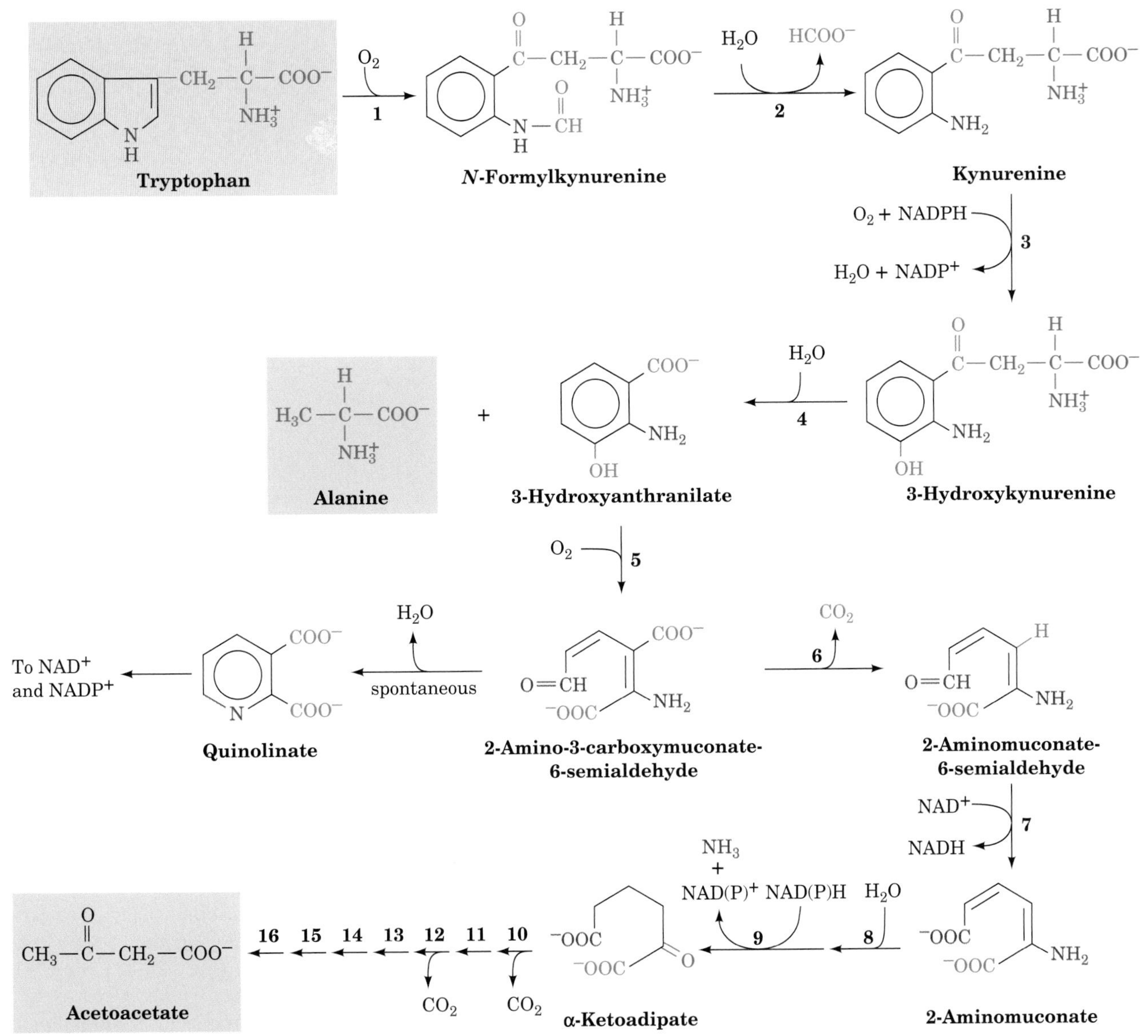

FIGURE 26-24 The pathway of tryptophan degradation. The enzymes involved are **(1)** tryptophan-2,3-dioxygenase, **(2)** formamidase, **(3)** kynurenine-3-monooxygenase, **(4)** kynureninase (PLP dependent), **(5)** 3-hydroxyanthranilate-3,4-dioxygenase, **(6)** amino carboxymuconate semialdehyde decarboxylase, **(7)** aminomuconate semialdehyde dehydrogenase, **(8)** hydratase, **(9)** dehydrogenase, and **(10–16)** enzymes of Reactions 5 through 11 in lysine degradation (Fig. 26-23). 2-Amino-3-carboxymuconate-6-semialdehyde, in addition to undergoing Reaction 6, spontaneously forms **quinolinate,** an NAD^+ and $NADP^+$ precursor (Section 28-5A).

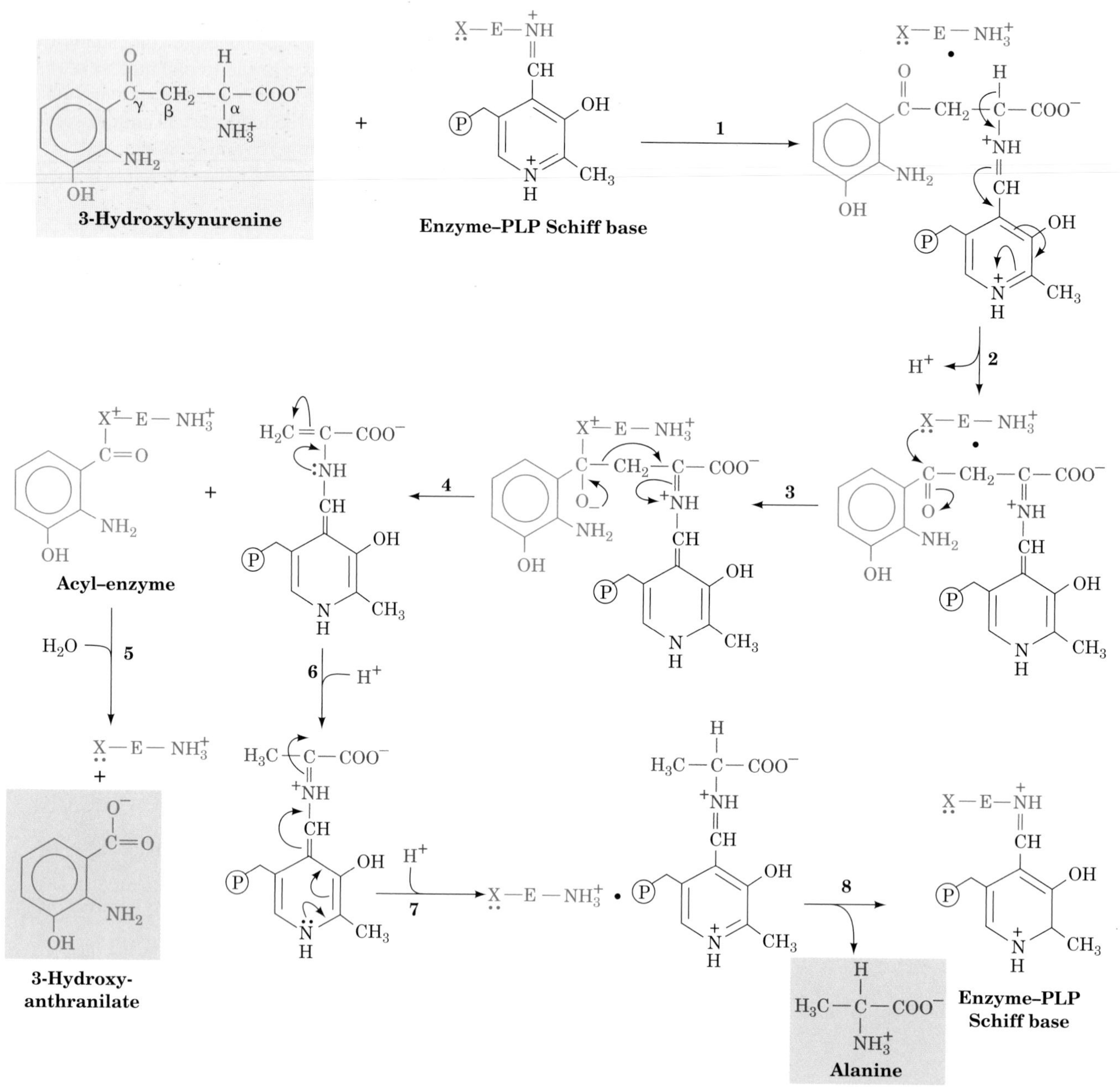

FIGURE 26-25 Proposed mechanism for the PLP-dependent kynureninase-catalyzed C_β—C_γ bond cleavage of 3-hydroxykynurenine. The reaction occurs in eight steps: **(1)** transimination, **(2)** tautomerization, **(3)** attack of an enzyme nucleophile, **(4)** C_β—C_γ bond cleavage with formation of an acyl–enzyme intermediate, **(5)** acyl–enzyme hydrolysis, **(6)** and **(7)** tautomerization, and **(8)** transimination.

ticular interest. Reaction 4, cleavage of **3-hydroxykynurenine** to alanine and **3-hydroxyanthranilate,** is catalyzed by **kynureninase,** a PLP-dependent enzyme. The reaction further demonstrates the enormous versatility of PLP. We have seen how PLP can labilize an α-amino acid's C_α—H and C_α—C_β bonds (Figs. 26-16). Here we see the facilitation of C_β—C_γ bond cleavage. The reaction follows the same steps as transamination reactions but does not hydrolyze the tautomerized Schiff base (Fig. 26-25). The proposed reaction mechanism involves an attack of an enzyme nucleophile on the carbonyl carbon (C_γ) of the tautomer-

ized 3-hydroxykynurenine–PLP Schiff base (Fig. 26-25, Step 3). This is followed by C_β—C_γ bond cleavage to generate an acyl–enzyme intermediate together with a tautomerized alanine–PLP adduct (Fig. 26-25, Step 4). Hydrolysis of the acyl–enzyme then yields 3-hydroxyanthranilate, whose further degradation yields **α-ketoadipate** (Fig. 26-24, Reactions 5–9). α-Ketoadipate is also an intermediate in lysine breakdown (Fig. 26-23, Reaction 4) so that the last seven reactions in the degradation of both these amino acids are identical, forming acetoacetate and two molecules of CO_2.

H. *Phenylalanine and Tyrosine Are Degraded to Fumarate and Acetoacetate*

Since the first reaction in phenylalanine degradation is its hydroxylation to tyrosine, a single pathway (Fig. 26-26) is responsible for the breakdown of both of these amino acids. The final products of the 6-reaction degradation are fumarate, a citric acid cycle intermediate, and acetoacetate, a ketone body.

a. Pterins Are Redox Cofactors

The hydroxylation of phenylalanine by the non-heme-iron-containing homotetrameric enzyme **phenylalanine hydroxylase (PAH)** requires O_2 and that the iron be in the Fe(II) state. The enzyme also requires the participation of **biopterin**, a pterin derivative. Pterins are compounds that contain the **pteridine** ring (Fig. 26-27). Note the resemblance between the pteridine ring and the isoalloxazine ring of the flavin coenzymes; the positions of the nitrogen atoms in pteridine are identical with those of the B and C rings of isoalloxazine. Folate derivatives also contain the pterin ring (Section 26-4D). Pterins, like flavins, participate in biological oxidations. The active form of biopterin is the fully reduced form, **5,6,7,8-tetrahydrobiopterin (BH$_4$)**. It is produced from **7,8-dihydrobiopterin** and

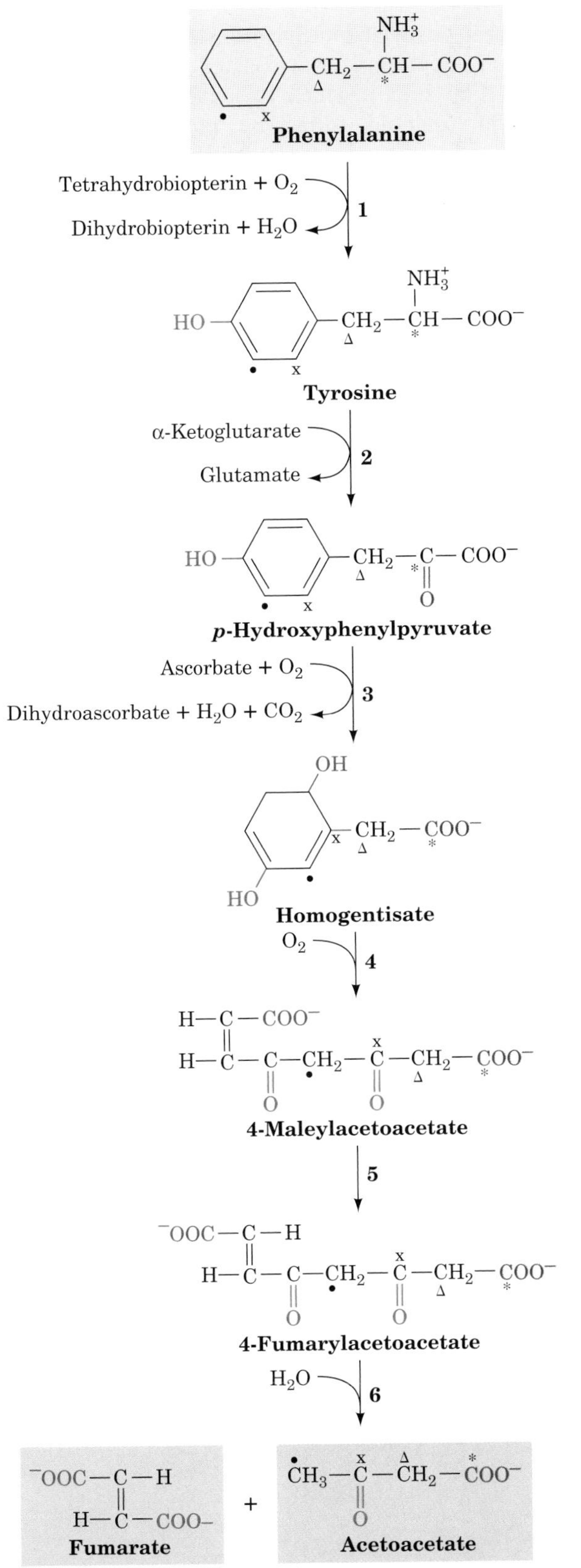

FIGURE 26-26 The pathway of phenylalanine degradation. The enzymes involved are **(1)** phenylalanine hydroxylase, **(2)** aminotransferase, **(3)** *p*-hydroxyphenylpyruvate dioxygenase, **(4)** homogentisate dioxygenase, **(5)** maleylacetoacetate isomerase, and **(6)** fumarylacetoacetase. The symbols labeling the various carbon atoms serve to indicate the group migration that occurs in Reaction 3 of the pathway (see Fig. 26-31).

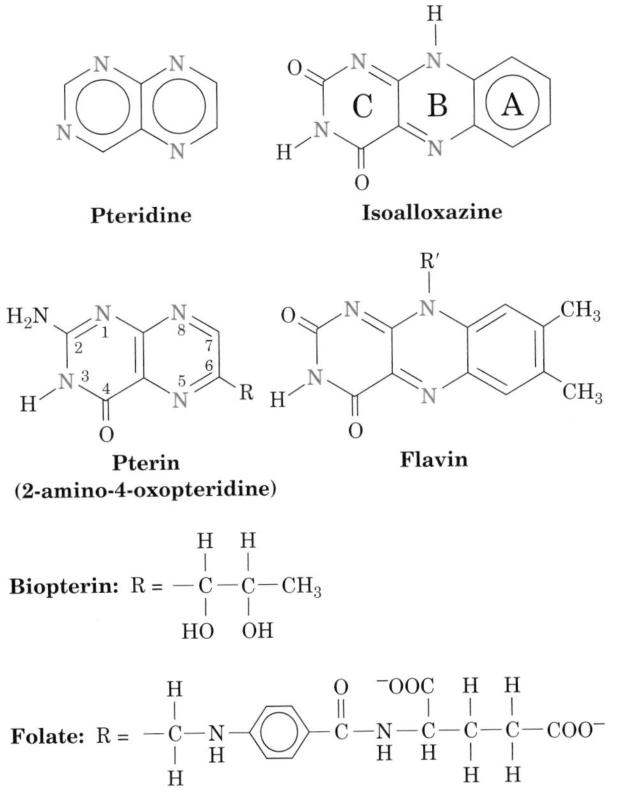

FIGURE 26-27 The pteridine ring, the nucleus of biopterin and folate. Note the similar structures of pteridine and the isoalloxazine ring of flavin coenzymes.

NADPH, in what may be considered a priming reaction, by **dihydrofolate reductase** (Fig. 26-28).

Each 452-residue subunit of the PAH homotetramer contains three domains, an N-terminal regulatory domain, a catalytic domain, and a C-terminal tetramerization domain. However, the 325-residue catalytic domain alone forms catalytically competent dimers. The X-ray structure of the catalytic domain of PAH in its Fe(II) state in complex with BH$_4$, determined by Edward Hough, reveals that the Fe(II) is octahedrally coordinated by His 285, His 290, Glu 330, and three water molecules, and that atom O4 of BH$_4$ is hydrogen bonded to two of these waters (Fig. 26-29).

In the phenylalanine hydroxylase reaction, 5,6,7,8-tetrahydrobiopterin is hydroxylated to **pterin-4a-carbinolamine** (Fig. 26-28), which is converted to **7,8-dihydrobiopterin (quinoid form)** by **pterin-4a-carbinolamine dehydratase.** The quinoid is subsequently reduced by the NAD(P)H-requiring enzyme **dihydropteridine reductase** to regenerate the active cofactor. Note that although dihydrofolate reductase and dihydropteridine reductase produce the same product, they utilize different tautomers of the substrate. Although this suggests that these enzymes may be evolutionarily related, the comparison of their X-ray structures indicates that this is not the case: Dihydropteridine reductase resembles nicotinamide

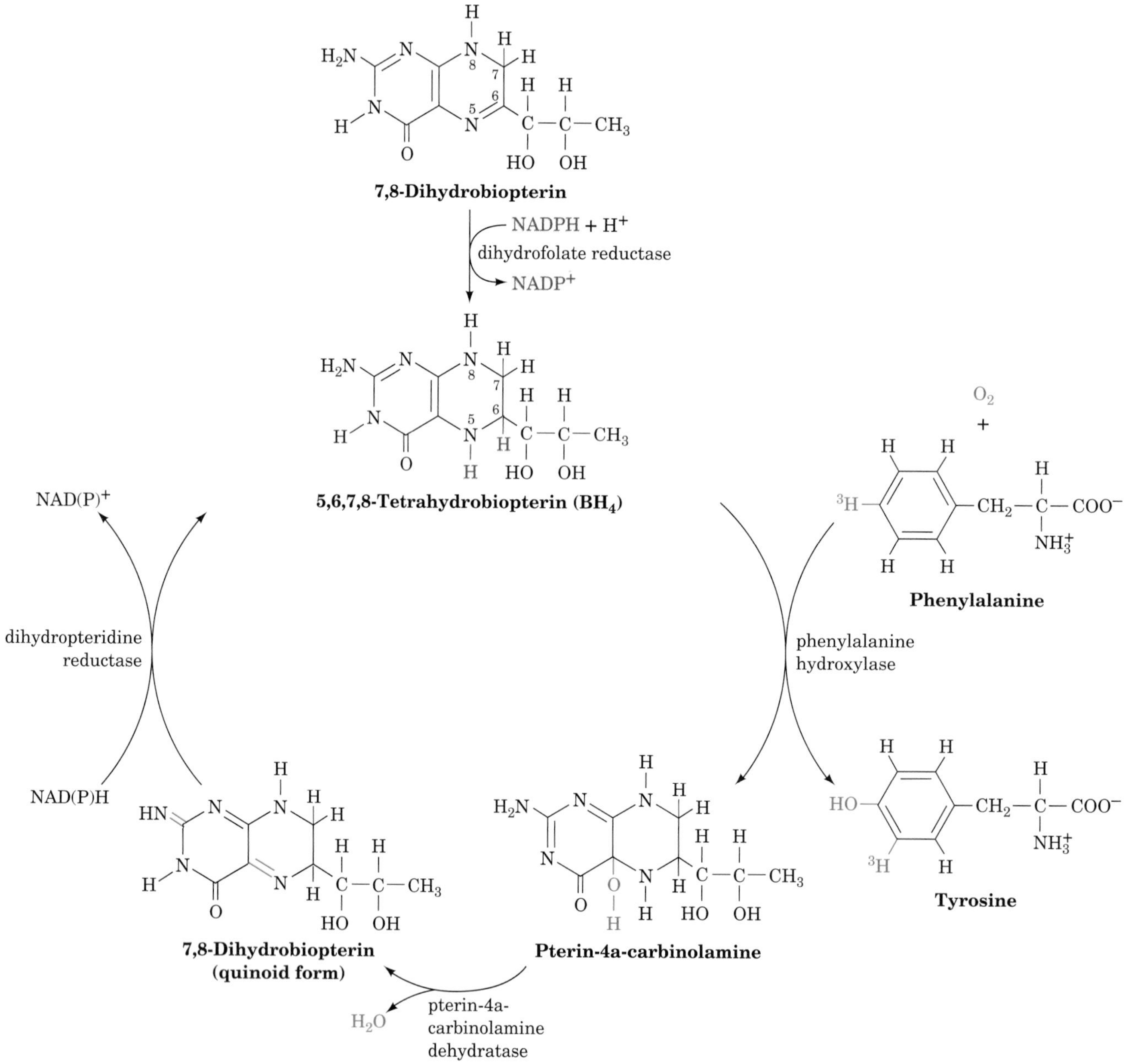

FIGURE 26-28 The formation, utilization, and regeneration of 5,6,7,8-tetrahydrobiopterin (BH$_4$) in the phenylalanine hydroxylase reaction.

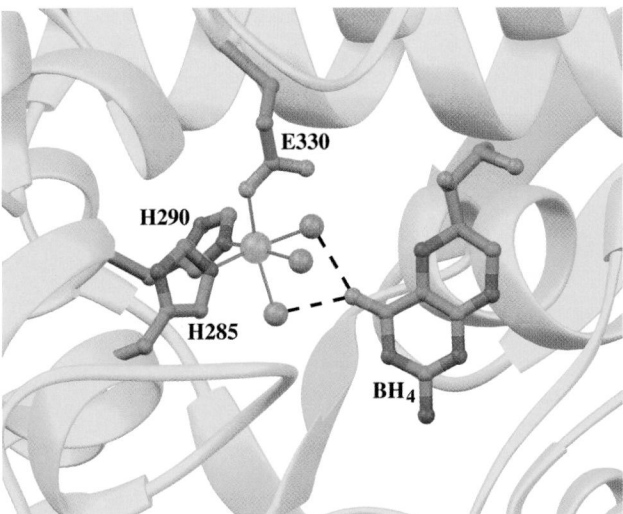

FIGURE 26-29   The active site of the Fe(II) form of phenylalanine hydroxylase (PAH) in complex with 5,6,7,8-tetrahydrobiopterin (BH₄). The Fe(II) (*orange sphere*) is octahedrally coordinated (*gray lines*) by His 285, His 290, and Glu 330 (C green, N blue, and O red) and three water molecules (*red spheres*). BH₄ atom O4 is hydrogen bonded (*black dashed lines*) to two of these water molecules. [Based on an X-ray structure by Edward Hough, University of Tromsø, Norway. PDBid 1J8U.]

coenzyme-requiring flavin-dependent enzymes such as glutathione reductase and dihydrolipoyl dehydrogenase (Section 21-2B).

b. Phenylalanine Hydroxylase Is Controled by Phosphorylation and by Allosteric Interactions

PAH initiates the detoxification of high concentrations of phenylalanine as well as the synthesis of the catecholamine hormones and neurotransmitters (Section 26-4B). It is allosterically activated by its substrate, phenylalanine, and by phosphorylation at its Ser 16 by the cAMP-dependent protein kinase A (PKA; Section 18-3C). Its second substrate, BH₄, allosterically inhibits the enzyme.

c. The NIH Shift

An unexpected aspect of the PAH reaction is that a ³H atom, which begins on C4 of phenylalanine's phenyl ring, ends up on C3 of this ring in tyrosine (Fig. 26-28, *right*) rather than being lost to the solvent by replacement with the OH group. The mechanism postulated to account for this **NIH shift** (so called because it was first characterized by chemists at the National Institutes of Health) involves the activation of oxygen by the pterin and Fe cofactors to form the pterin-4a-carbinolamine and a reactive oxyferryl group [Fe(IV)=O²⁻; Fig. 26-30, Steps 1 and 2] that reacts with the substrate to form an epoxide across the phenyl ring's 3,4 bond (Fig. 26-30, Step 3). This is followed by epoxide opening to form a carbocation at C3 (Fig. 26-30, Step 4). Migration of a hydride from C4 to C3 forms a more stable carbocation (an oxonium ion; Fig. 26-30, Step 5). This migration is followed by ring aromatization to form

tyrosine (Fig. 26-30, Step 6). **Tyrosine hydroxylase** and **tryptophan hydroxylase** (Section 26-4B) are both homologous to phenylalanine hydroxylase and utilize this same NIH shift reaction mechanism, although there may not be an epoxide intermediate in these cases.

Reaction 3 in the phenylalanine degradation pathway (Fig. 26-26) provides another example of an NIH shift. This reaction, which is catalyzed by the Fe(II)-containing **p-hydroxyphenylpyruvate dioxygenase,** involves the oxidative decarboxylation of an α-keto acid as well as ring hydroxylation. In this case, the NIH shift involves migration of an alkyl group rather than of a hydride ion to form a more stable carbocation (Fig. 26-31). This shift, which has been demonstrated through isotope-labeling studies (represented by the different symbols in Figs. 26-26 and 26-31), accounts for the observation that C3 is bonded to C4 in **p-hydroxyphenylpyruvate** but to C5 in **homogentisate.**

d. Alkaptonuria and Phenylketonuria Result from Defects in Phenylalanine Degradation

Archibald Garrod realized in the early 1900s that human genetic diseases result from specific enzyme deficiencies. We have repeatedly seen how this realization has contributed to the elucidation of metabolic pathways. The first such disease to be recognized was **alkaptonuria,** which, Garrod observed, resulted in the excretion of large quantities of homogentisic acid. This condition results from deficiency of **homogentisate dioxygenase** (Fig. 26-26, Reaction 4). Alkaptonurics suffer no ill effects other than arthritis later in life (although their urine darkens alarmingly because of the rapid air oxidation of the homogentisate they excrete).

Individuals suffering from **phenylketonuria (PKU)** are not so fortunate. Severe mental retardation occurs within a few months of birth if the disease is not detected and treated immediately (see below). Indeed, ~1% of the patients in mental institutions were, at one time (before routine screening), phenylketonurics. PKU is caused by the inability to hydroxylate phenylalanine (Fig. 26-26, Reaction 1) and therefore results in increased blood levels of phenylalanine **(hyperphenylalaninemia).** The excess phenylalanine is transaminated to **phenylpyruvate**

$$\text{C}_6\text{H}_5\text{—CH}_2\text{—}\overset{\displaystyle O}{\overset{\displaystyle \|}{\text{C}}}\text{—COO}^-$$

Phenylpyruvate

by an otherwise minor pathway. The "spillover" of phenylpyruvate (a phenylketone) into the urine was the first observation connected with the disease and gave the disease its name, although it has since been demonstrated that it is the high concentration of phenylalanine itself that gives rise to brain dysfunction. All babies born in the United States are now screened for PKU immediately after birth by testing for elevated levels of phenylalanine in the blood.

Classic PKU results from a deficiency in phenylalanine hydroxylase (PAH). When this was established in 1947, it was the first human inborn error of metabolism whose basic biochemical defect had been identified. Since then, over

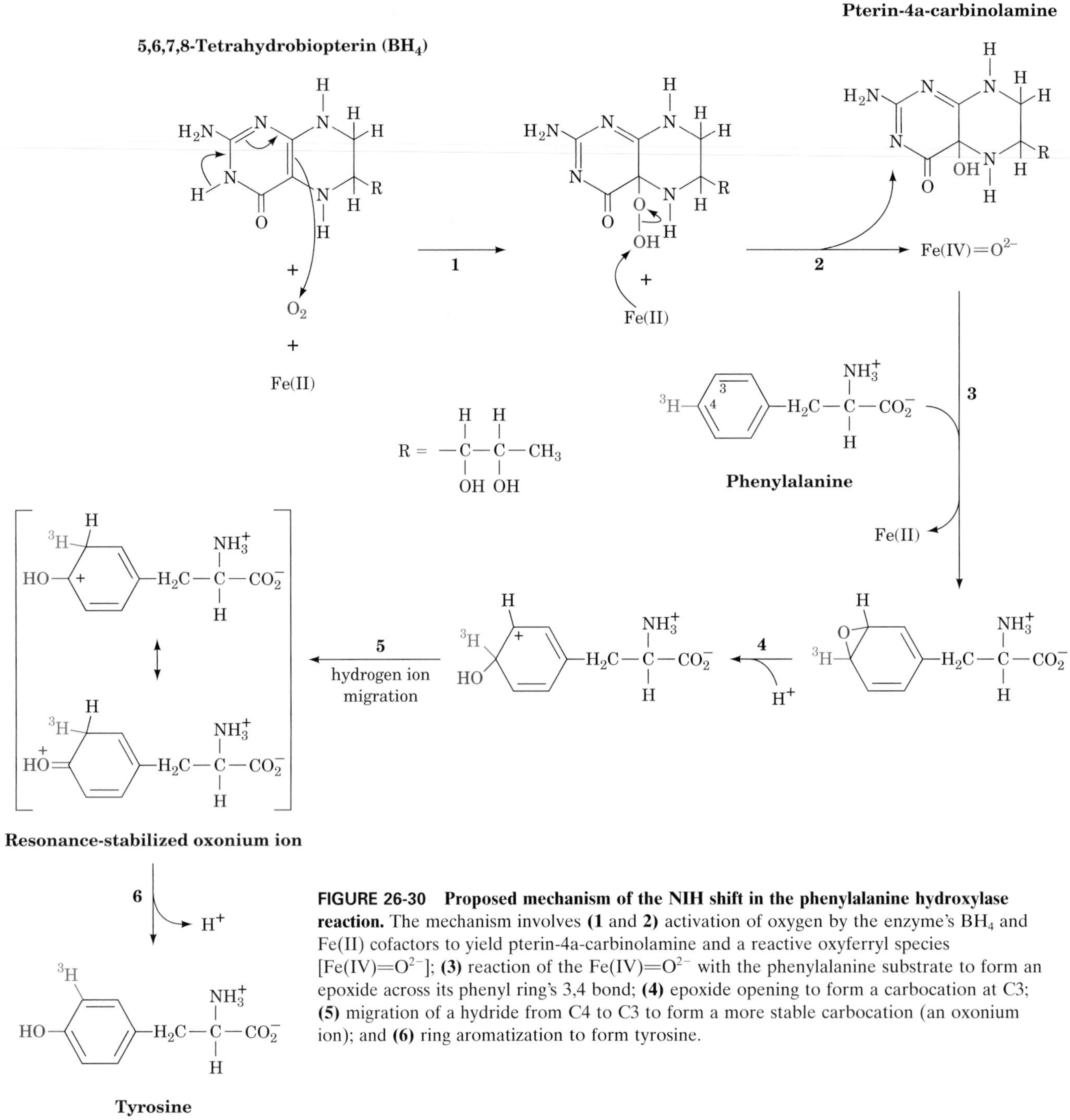

FIGURE 26-30 Proposed mechanism of the NIH shift in the phenylalanine hydroxylase reaction. The mechanism involves (**1** and **2**) activation of oxygen by the enzyme's BH_4 and Fe(II) cofactors to yield pterin-4a-carbinolamine and a reactive oxyferryl species $[Fe(IV)=O^{2-}]$; (**3**) reaction of the $Fe(IV)=O^{2-}$ with the phenylalanine substrate to form an epoxide across its phenyl ring's 3,4 bond; (**4**) epoxide opening to form a carbocation at C3; (**5**) migration of a hydride from C4 to C3 to form a more stable carbocation (an oxonium ion); and (**6**) ring aromatization to form tyrosine.

400 mutations have been identified in PAH. Because all of the tyrosine breakdown enzymes are normal, treatment consists in providing the patient with a low-phenylalanine diet and monitoring the blood level of phenylalanine to ensure that it remains within normal limits for the first 5 to 10 years of life (the adverse effects of hyperphenylalaninemia seem to disappear after that age). PAH deficiency also accounts for another common symptom of PKU: Its victims have lighter hair and skin color than their siblings. This is because tyrosine hydroxylation, the first reaction in the formation of the black skin pigment **melanin** (Section 26-4B), is inhibited by elevated phenylalanine levels.

Other causes of hyperphenylalaninemia have been discovered since the introduction of infant screening techniques. These result from deficiencies in the enzymes catalyzing the formation or regeneration of 5,6,7,8-tetrahydrobiopterin (BH_4), the PAH cofactor (Fig. 26-28). In such cases, patients must also be supplied with L-**3,4-dihydroxyphenylalanine (L-DOPA)** and **5-hydroxytryptophan,** metabolic precursors of the neurotransmitters **norepinephrine** and **serotonin,** respectively, since tyrosine hydroxylase and tryptophan hydroxylase, the PAH homologs that produce these physiologically active amines, also require 5,6,7,8-tetrahydrobiopterin (Section 26-4B). Un-

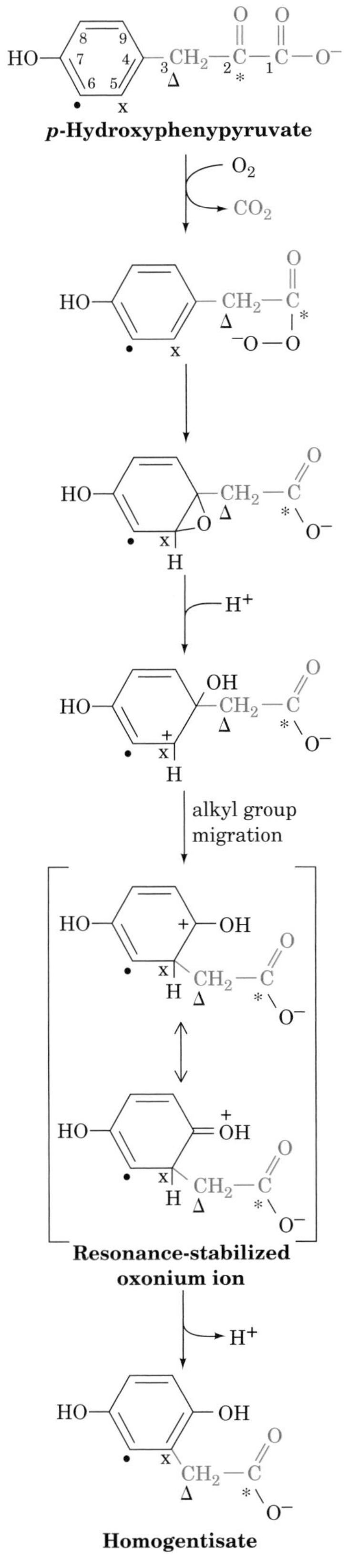

p-Hydroxyphenypyruvate

Resonance-stabilized oxonium ion

Homogentisate

FIGURE 26-31 **The NIH shift in the *p*-hydroxyphenylpyruvate dioxygenase reaction.** Carbon atoms are labeled as an aid to following the group migration constituting the shift.

fortunately, simply adding BH_4 to the diet of an affected individual is not an effective treatment because BH_4 is unstable and cannot cross the blood–brain barrier.

4 ■ AMINO ACIDS AS BIOSYNTHETIC PRECURSORS

Certain amino acids, in addition to their major function as protein building blocks, are essential precursors of a variety of important biomolecules, including nucleotides and nucleotide coenzymes, heme, various hormones and neurotransmitters, and glutathione. In this section, we therefore consider the pathways producing some of these substances. We begin by discussing the biosynthesis of heme from glycine and succinyl-CoA. We then examine the pathways by which tyrosine, tryptophan, glutamate, and histidine are converted to various neurotransmitters and study certain aspects of glutathione biosynthesis and the involvement of this tripeptide in amino acid transport and other processes. Finally, we consider the role of folate derivatives in the biosynthetic transfer of C_1 units. The biosynthesis of nucleotides and nucleotide coenzymes is the subject of Chapter 28.

A. *Heme Biosynthesis and Degradation*

Heme (Fig. 26-32), as we have seen, is an Fe-containing prosthetic group that is an essential component of many proteins, notably hemoglobin, myoglobin, and the cytochromes. The initial reactions of heme biosynthesis are

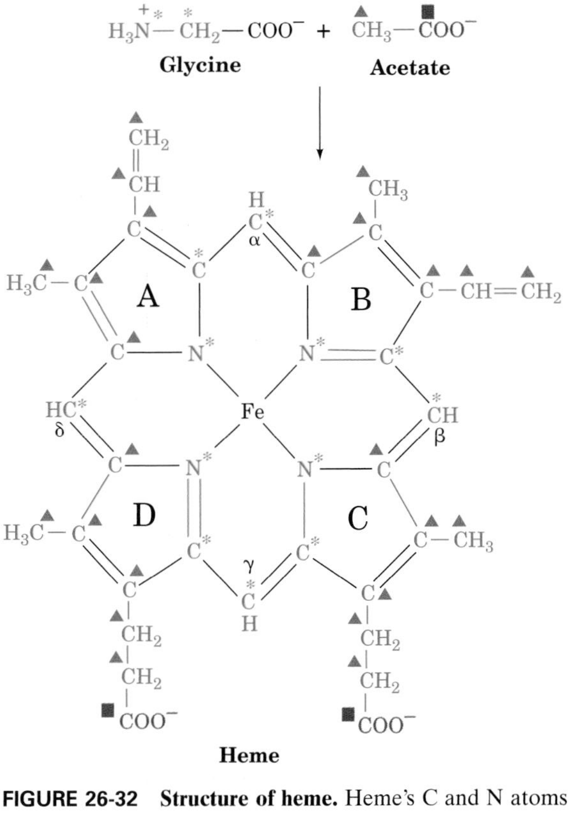

Heme

FIGURE 26-32 **Structure of heme.** Heme's C and N atoms are derived from those of glycine and acetate.

common to the formation of other tetrapyrroles including chlorophyll in plants and bacteria (Section 24-2A) and coenzyme B$_{12}$ in bacteria (Section 25-2E).

a. Porphyrins Are Derived from Succinyl-CoA and Glycine

Elucidation of the heme biosynthesis pathway involved some interesting detective work. David Shemin and David Rittenberg, who were among the first to use isotopic tracers in the elucidation of metabolic pathways, demonstrated, in 1945, that *all of heme's C and N atoms can be derived from acetate and glycine.* Only glycine, out of a variety of ^{15}N-labeled metabolites they tested (including ammonia, glutamate, leucine, and proline), yielded ^{15}N-labeled heme in the hemoglobin of experimental subjects to whom these metabolites were administered. Similar experiments, using acetate labeled with ^{14}C in its methyl or carboxyl groups, or [^{14}C$_\alpha$]glycine, demonstrated that 24 of heme's 34 carbon atoms are derived from acetate's methyl carbon, 2 from acetate's carboxyl carbon, and 8 from glycine's C$_\alpha$ atom (Fig. 26-32). None of the heme atoms is derived from glycine's carboxyl carbon atom.

Figure 26-32 indicates that heme C atoms derived from acetate methyl groups occur in groups of three linked atoms. Evidently, acetate is first converted to some other metabolite that has this labeling pattern. Shemin and Rittenberg postulated that this metabolite is succinyl-CoA based on the following reasoning (Fig. 26-33):

1. Acetate is metabolized via the citric acid cycle (Section 21-3I).

2. Labeling studies indicate that atom C3 of the citric acid cycle intermediate succinyl-CoA is derived from acetate's methyl C atom, whereas atom C4 comes from acetate's carboxyl C atom.

3. After many turns of the citric acid cycle, C1 and C2 of succinyl-CoA likewise become fully derived from acetate's methyl C atom.

We shall see that this labeling pattern indeed leads to that of heme.

In the mitochondria of yeast and animals as well as in some bacteria, the first phase of heme biosynthesis is a condensation of succinyl-CoA with glycine followed by decarboxylation to form **δ-aminolevulinic acid (ALA)** as catalyzed by the PLP-dependent enzyme **δ-aminolevulinate synthase** (Fig. 26-34). The carboxyl group lost in the decarboxylation (Fig. 26-34, Reaction 5) originates in glycine, which is why heme contains no label from this group.

b. The Pyrrole Ring Is the Product of Two ALA Molecules

The pyrrole ring is formed in the next phase of the pathway through linkage of two molecules of ALA to yield **porphobilinogen (PBG).** The reaction is catalyzed by **porphobilinogen synthase [PBGS;** alternatively, **δ-aminolevulinic acid dehydratase (ALAD)]** which, in yeast and mammals, is Zn^{2+}-dependent and involves Schiff base formation of one of the substrate molecules with an enzyme amine

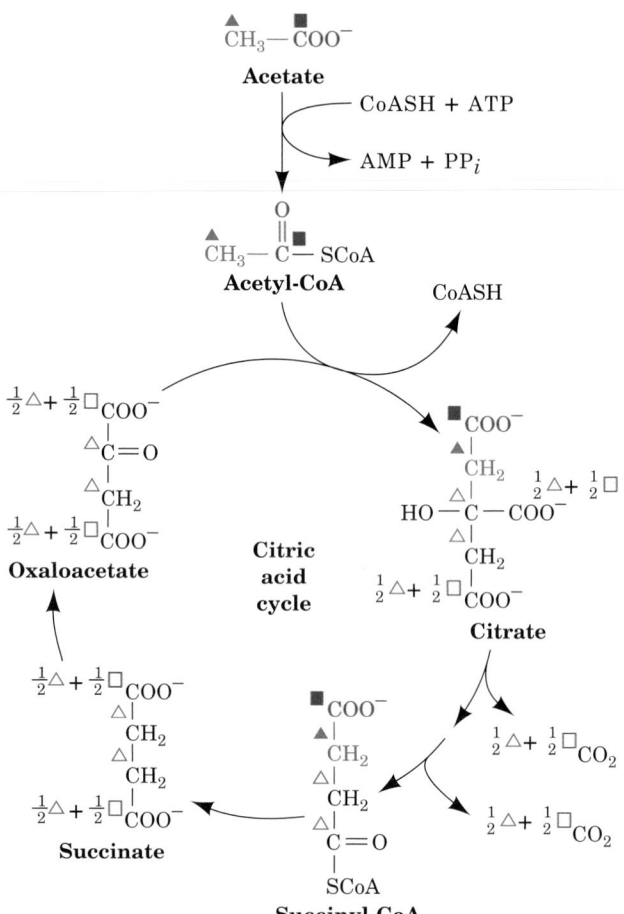

FIGURE 26-33   The origin of the C atoms of succinyl-CoA as derived from acetate via the citric acid cycle. C atoms labeled with triangles and squares are derived, respectively, from acetate's methyl and carboxyl C atoms. Filled symbols label atoms derived from acetate in the present round of the citric acid cycle, whereas open symbols label atoms derived from acetate in previous rounds of the citric acid cycle. Note that the C1 and C4 atoms of succinyl-CoA are scrambled on forming the 2-fold symmetric succinate.

group (in some bacteria and all plants, Mg^{2+} substitutes for Zn^{2+}). One possible mechanism of this condensation–elimination reaction involves formation of a second Schiff base between the ALA–enzyme Schiff base and the second ALA molecule (Fig. 26-35). At this point, if we continue tracing the acetate and glycine labels through the PBG synthase reaction (Fig. 26-35), we can begin to see how heme's labeling pattern arises.

The X-ray structure of human PBGS in covalent complex with its product PBG, determined by Jonathan Cooper, indicates that this enzyme is a homooctamer with D_4 symmetry. Each of its 330-residue subunits consists of an α/β barrel and a 39-residue N-terminal tail that wraps around a neighboring monomer (related to it by 2-fold symmetry) so that the protein is better described as a relatively loosely organized tetramer of compact dimers. As

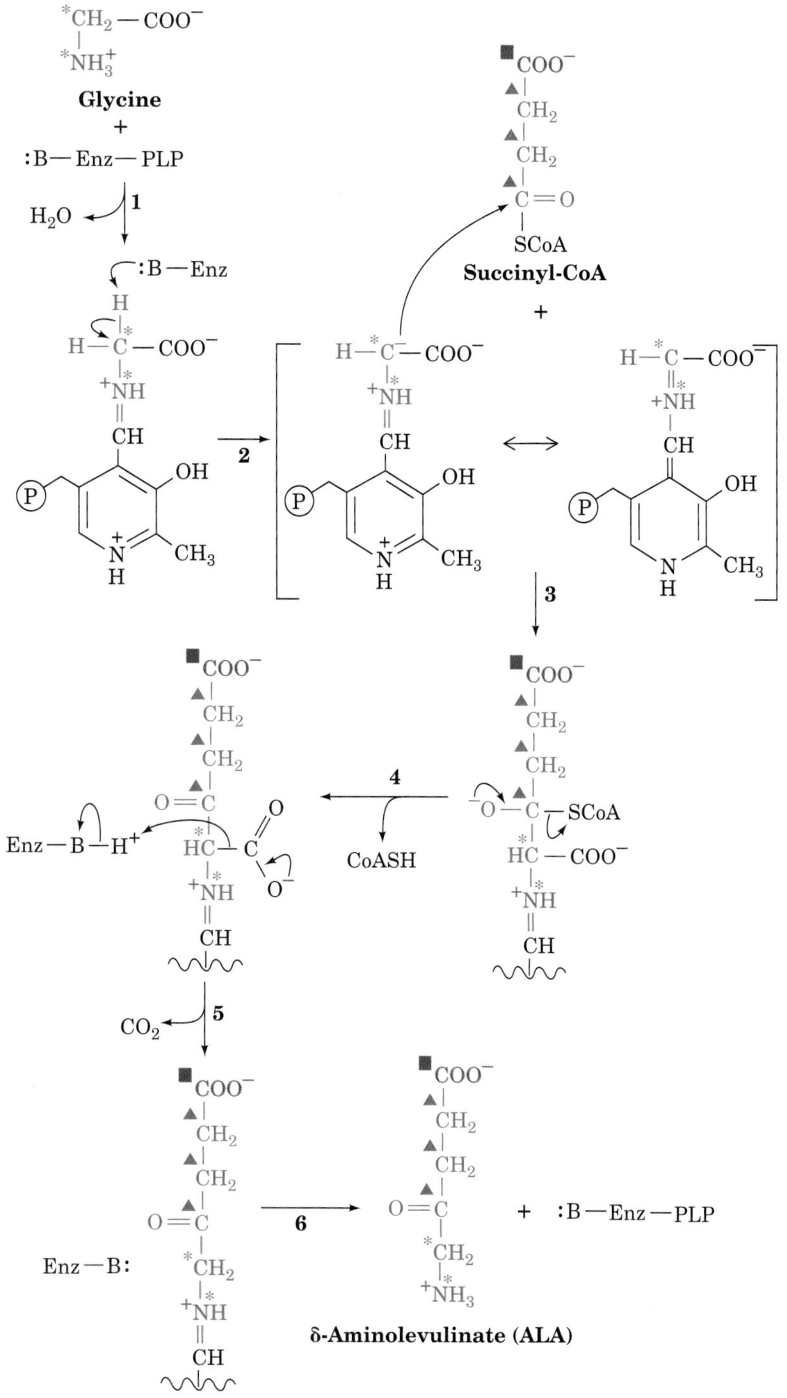

FIGURE 26-34 The mechanism of action of the PLP-dependent enzyme δ-aminolevulinate synthase. The reaction steps are **(1)** transimination, **(2)** PLP-stabilized carbanion formation, **(3)** C—C bond formation, **(4)** CoA elimination, **(5)** decarboxylation facilitated by the PLP–Schiff base, and **(6)** transimination yielding ALA and regenerating the PLP–enzyme.

is the case with nearly all α/β barrel enzymes, PBGS's active site (Fig. 26-36) lies at the mouth of the barrel at the C-terminal ends of its β strands. The active site is covered by a loop that comparison with other PBGS structures indicates forms a flexible lid over the substrate, an arrangement that is reminiscent of the glycolytic enzyme triose phosphate isomerase (TIM; Fig. 17-11). PBG is covalently bound to Lys 252 and its free amino group is coordinated to the active site Zn^{2+} ion. Lys 199 appears to be properly positioned to act as a general acid–base catalyst.

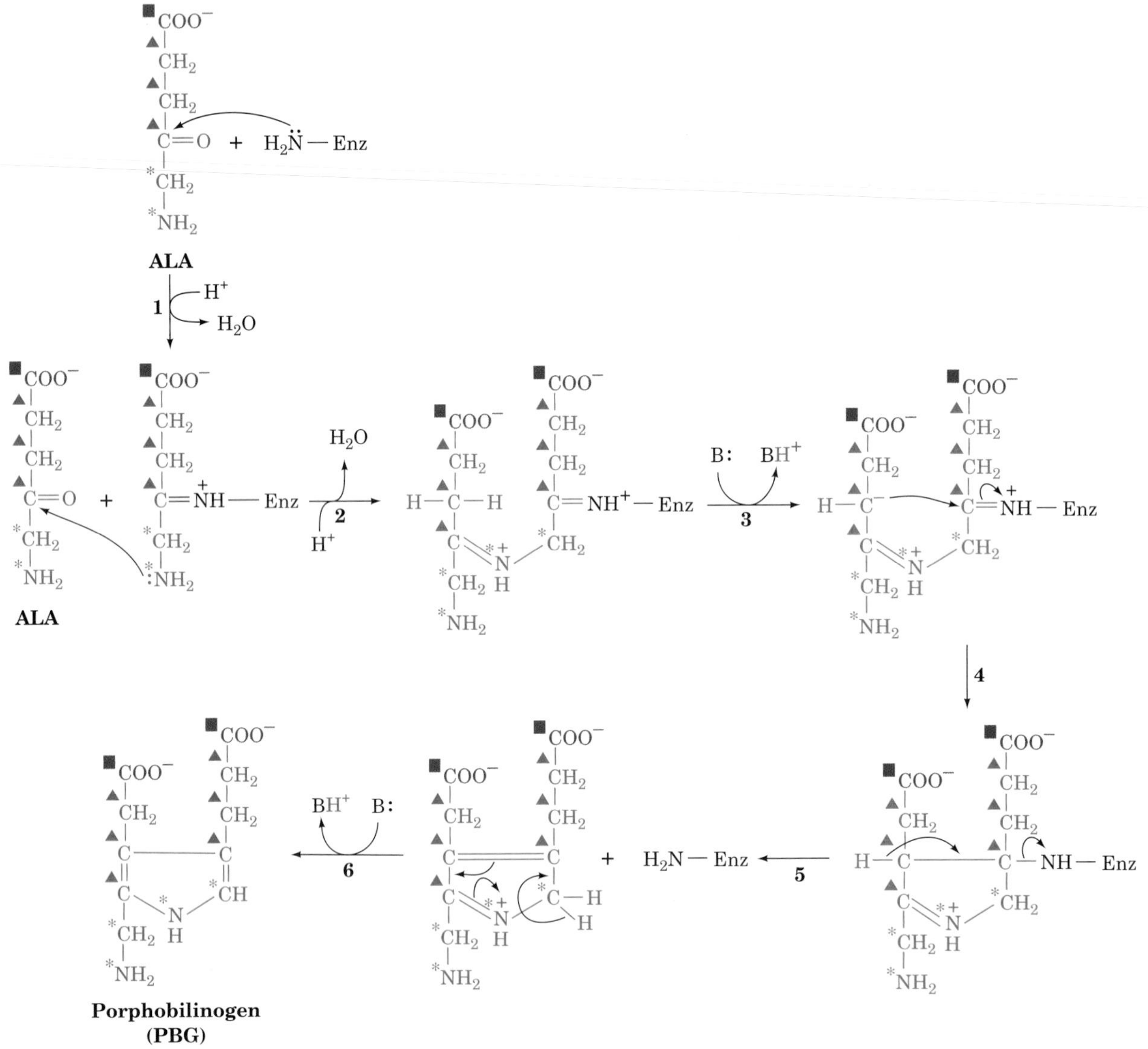

FIGURE 26-35   A possible mechanism for porphobilinogen synthase. The reaction involves (1) Schiff base formation, (2) second Schiff base formation, (3) formation of a carbanion α to a Schiff base, (4) cyclization by an aldol-type condensation, (5) elimination of the enzyme—NH₂ group, and (6) tautomerization.

Inhibition of PBG synthase by Pb^{2+} (a competitor of its active site Zn^{2+} ion) is one of the major manifestations of lead poisoning, which is among the most common acquired environmental diseases. Indeed, it has been suggested that the accumulation, in the blood, of ALA, which resembles the neurotransmitter **γ-aminobutyric acid** (Section 26-4B), is responsible for the psychosis that often accompanies lead poisoning.

c. The Porphyrin Ring Is Formed from Four PBG Molecules

The next phase of heme biosynthesis is the condensation of four PBG molecules to form **uroporphyrinogen III,** the porphyrin nucleus, in a series of reactions catalyzed by **porphobilinogen deaminase** (alternatively, **hydroxy-**

methylbilane synthase or **uroporphyrinogen synthase**) and **uroporphyrinogen III synthase.** The reaction (Fig. 26-37) begins with the enzyme's displacement of the amino group in PBG to form a covalent adduct. A second, third, and fourth PBG then sequentially add through the displacement of the primary amino group on one PBG by a carbon atom on the pyrrole ring of the succeeding PBG to yield a linear tetrapyrrole that is hydrolyzed and released from the enzyme as **hydroxymethylbilane** (also called **preuroporphyrinogen**).

d. Porphobilinogen Deaminase Has a Dipyrromethane Cofactor

Peter Shoolingin-Jordan and Alan Battersby independently showed that porphobilinogen deaminase contains a

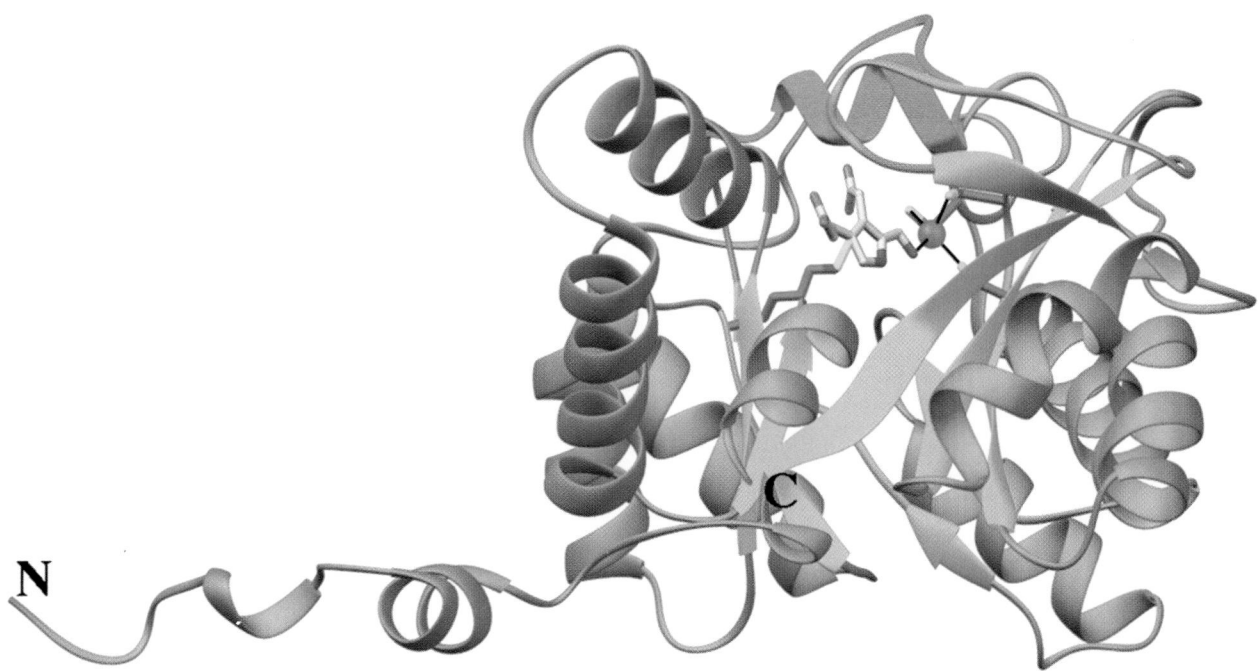

FIGURE 26-36 X-Ray structure of human porphobilinogen synthase (PBGS). A monomer of this homooctameric protein is viewed perpendicular to the axis of its α/β barrel and is drawn in gray with its β strands cyan and the loop forming its flexible lid (residues 201–222) magenta. PBGS's porphobilinogen (PBG) product, Lys 252 to which it is covalently linked, and the three Cys side chains that ligand the active site Zn^{2+} ion (*blue sphere*) are shown in stick form with PBG C pink, side chain C green, N blue, O red, S yellow, and the N—C bond linking Lys 252 to PBG gold. The active site Zn^{2+} ion is liganded (*black lines*) by the S atoms of Cys 122, Cys 124, Cys 132, and the PBG amino group. Lys 199, which lies directly behind Lys 252 in this view, appears to be properly positioned to act as an acid–base catalyst. [Based on an X-ray structure by Jonathan Cooper, University of Southampton, U.K. PDBid 1E51.]

unique **dipyrromethane** cofactor (two pyrroles linked by a methylene bridge; rings C_1 and C_2 in Fig. 26-37), which is covalently linked to the enzyme via a C—S bond to an enzyme Cys residue. Thus, the methylbilane–enzyme complex really contains a linear hexapyrrole. The subsequent reaction step, also catalyzed by porphobilinogen deaminase (Step 5 in Fig. 26-37), is the hydrolysis of the bond linking the second and third pyrrole units of the hexapyrrole to yield hydroxymethylbilane and the dipyrromethane cofactor. This cofactor is still linked to the enzyme, which is therefore ready to catalyze a new round of hydroxymethylbilane synthesis.

How is the dipyrromethane cofactor assembled? Shoolingin-Jordan has shown that porphobilinogen deaminase synthesizes its own cofactor from two PBG units using, it appears, the same catalytic machinery with which it synthesizes methylbilane. However, the enzyme Cys reacts much more rapidly with presynthesized hydroxymethylbilane to form a reaction intermediate (the product of Step 2 in Fig. 26-37) that continues to add two more PBG units. When hydroxymethylbilane is released, the enzyme retains its dipyrromethane cofactor.

The X-ray structure of *E. coli* porphobilinogen deaminase (whose sequence is >45% identical to those of mammalian enzymes), in covalent complex with its dipyrromethane cofactor, indicates that this monomeric, 307-residue protein folds into three nearly equal sized domains (Fig 26-38). The dipyrromethane cofactor lies deep in a cleft between domains 1 and 2 such that there is still considerable unoccupied space in the cleft. Although the enzyme sequentially appends four PBG residues to the cofactor, it has only one catalytic site.

If the enzyme has only one catalytic site, how does it reposition the polypyrrole chain after each catalytic cycle so that it can further extend this chain? One possibility is that the polypyrrole chain fills the cavity next to the cofactor. This model provides a simple steric rationale for why the length of the polypyrrole chain is limited to six residues (the final four of which are hydrolytically cleaved away by the enzyme to yield the hydroxymethylbilane product and regenerate the dipyrromethane cofactor).

e. Protoporphyrin IX Biosynthesis Requires Four More Reactions

Cyclization of the hydroxymethylbilane product requires **uroporphyrinogen III synthase** (Fig. 26-37). In the absence of this enzyme, hydroxymethylbilane is released from the synthase and rapidly cyclizes nonenzymatically to

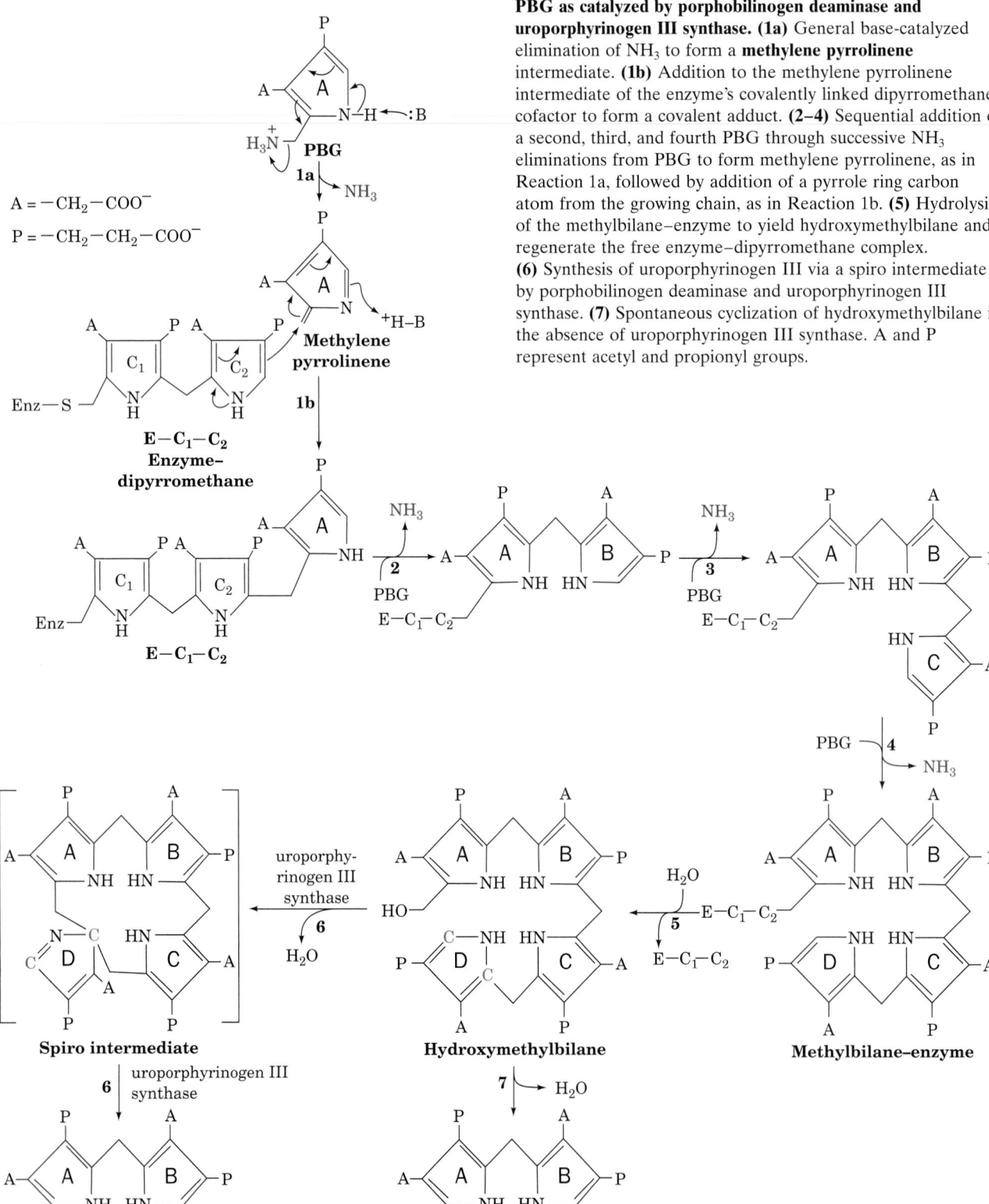

FIGURE 26-37   The synthesis of uroporphyrinogen III from PBG as catalyzed by porphobilinogen deaminase and uroporphyrinogen III synthase. (1a) General base-catalyzed elimination of NH_3 to form a **methylene pyrrolinene** intermediate. **(1b)** Addition to the methylene pyrrolinene intermediate of the enzyme's covalently linked dipyrromethane cofactor to form a covalent adduct. **(2–4)** Sequential addition of a second, third, and fourth PBG through successive NH_3 eliminations from PBG to form methylene pyrrolinene, as in Reaction 1a, followed by addition of a pyrrole ring carbon atom from the growing chain, as in Reaction 1b. **(5)** Hydrolysis of the methylbilane–enzyme to yield hydroxymethylbilane and regenerate the free enzyme–dipyrromethane complex. **(6)** Synthesis of uroporphyrinogen III via a spiro intermediate by porphobilinogen deaminase and uroporphyrinogen III synthase. **(7)** Spontaneous cyclization of hydroxymethylbilane in the absence of uroporphyrinogen III synthase. A and P represent acetyl and propionyl groups.

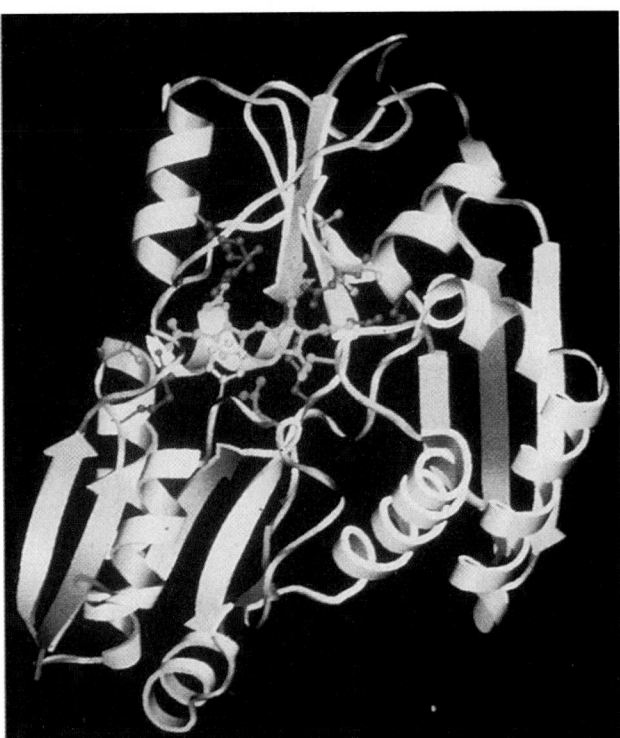

FIGURE 26-38   X-Ray structure of *E. coli* porphobilinogen deaminase in covalent complex with its dipyrromethane cofactor. The protein is shown in ribbon form and the dipyrromethane cofactor (*yellow*) together with the side chains that it contacts are shown in ball-and-stick form. [Courtesy of Gordon Louie, Stephan Wood, Peter Shoolingin-Jordan, and Tom Blundell, Birkbeck College, London, U.K. PDBid 1PDA.]

the symmetric **uroporphyrinogen I.** Heme, however, is an asymmetric molecule; the methyl substituent of pyrrole ring D has an inverted placement compared to those of rings A, B, and C (Fig. 26-32). This ring reversal to yield uroporphyrinogen III has been shown by Battersby to proceed through attachment of the methylenes from rings A and C to the same carbon of ring D so as to form a spiro compound (a bicyclic compound with a carbon atom common to both rings; Fig. 26-37).

Heme biosynthesis takes place partly in the mitochondrion and partly in the cytosol (Fig. 26-39). ALA is mitochondrially synthesized and is transported to the cytosol for conversion to PBG and then to uroporphyrinogen III. **Protoporphyrin IX,** to which Fe is added to form heme, is produced from uroporphyrinogen III in a series of reactions catalyzed by (1) **uroporphyrinogen decarboxylase,** which decarboxylates all four acetate side chains (A) to form methyl groups (M); (2) **coproporphyrinogen oxidase,** which oxidatively decarboxylates two of the propionate side chains (P) to vinyl groups (V); and (3) **protoporphyrinogen oxidase,** which oxidizes the methylene groups linking the pyrrole rings to methenyl groups. Altogether, six carboxyl groups originally from carboxyl-labeled ac-

etate are lost as CO_2. The only remaining C atoms from carboxyl-labeled acetate are the carboxyl groups of heme's two propionate side chains (P). During the coproporphyrinogen oxidase reaction, the macrocycle is transported back into the mitochondrion for the pathway's final reactions.

f. Ferrochelatase Catalyzes the Insertion of Fe(II) into Protoporphyrin IX to Form Heme

Protoporphyrin IX is converted to heme by the insertion of Fe(II) into the tetrapyrrole nucleus by **ferrochelatase,** a protein that is associated with the inner mitochondrial membrane on the matrix side. The X-ray structure of human ferrochelatase, determined by Harry Dailey and Bi-Cheng Wang, reveals that the 361-residue subunits of this homodimeric protein consist of two structurally similar domains and a C-terminal extension that occurs only in animal ferrochelatases. This C-terminal extension participates in hydrogen bonding between the monomers; bacterial ferrochelatases, which lack this extension, are monomeric. In addition, the C-terminal extension is bound to the N-terminal domain by an unusual [2Fe–2S] cluster that is coordinated by C196 of the N-terminal domain and C403, C406, and C411 of the C-terminal extension. The function of this [2Fe–2S] cluster, which is distant from the active site, is unclear although it appears likely that it has a structural role. Three mutations, C406Y, C406S, and C411G, that inactivate the enzyme and thereby cause the rare inherited disease **erythropoietic protoporphyria** (see below) demonstrate the importance of the [2Fe–2S] cluster for activity.

The ferrochelatase active site (Fig. 26-40) consists of two hydrophobic lips that, it has been proposed, participate in the enzyme's association with the membrane. The ferrochelatase reaction follows an ordered mechanism in which the Fe(II) binds to the enzyme before the porphyrin. The reaction requires that the two pyrrole NH protons be removed from the porphyrin prior to the binding of the Fe(II) (Fig. 26-39). The invariant H263 appears properly positioned to abstract these protons from the porphyrin, and the conserved and closely spaced acidic residues E343, H341, and D340 appear to form a proton conduit from H263 to the enzyme surface (Fig. 26-40), a hypothesis that is supported by mutagenesis studies. The conserved residues R164 and Y165 are located on the opposite side of the active site from H263 (and presumably on the opposite side of the bound protoporphyrin IX substrate). Their mutagenesis reduces the affinity of ferrochelatase for Fe(II) but not the porphyrin, suggesting that they have a catalytic role in the metalation reaction. Spectroscopic studies indicate that this metalation reaction is facilitated by the distortion of the porphyrin to a nonplanar conformation (a doming or ruffling).

g. Heme Biosynthesis Is Regulated Differently in Erythroid and Liver Cells

The two major sites of heme biosynthesis are erythroid cells, which synthesize ~85% of the body's heme groups, and

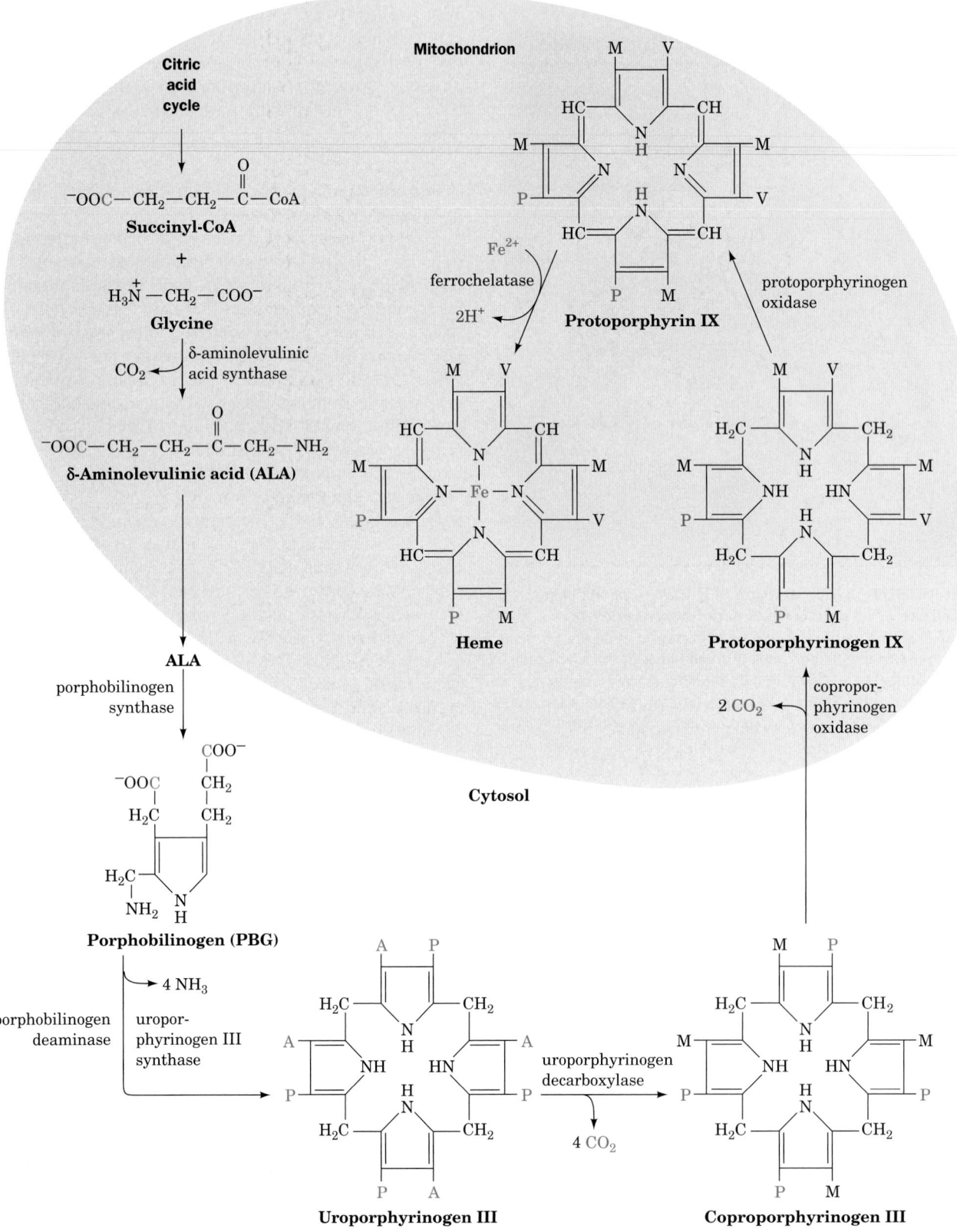

FIGURE 26-39 The overall pathway of heme biosynthesis.
δ-Aminolevulinic acid (ALA) is synthesized in the
mitochondrion by ALA synthase. ALA (*left*) leaves the
mitochondrion and is converted to PBG, four molecules of
which condense to form a porphyrin ring. The next three
reactions involve oxidation of the pyrrole ring substituents
yielding protoporphyrinogen IX whose formation is
accompanied by its transport back into the mitochondrion.
After oxidation of the methylene groups linking the pyrroles to
yield protoporphyrin IX, ferrochelatase catalyzes the insertion
of Fe^{2+} to yield heme. A, P, M, and V, respectively, represent
acetyl, propionyl, methyl, and vinyl ($-CH_2=CH_2$) groups. C
atoms originating as the carboxyl group of acetate are red.

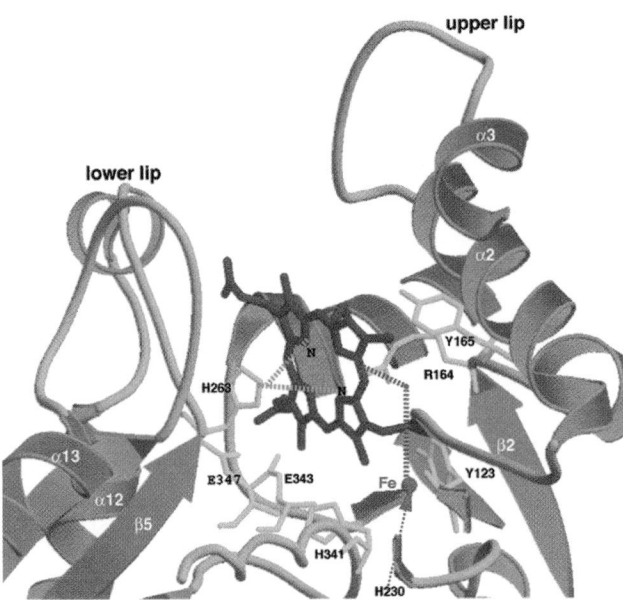

FIGURE 26-40 The active site pocket of human ferrochelatase showing the key residues in its proposed catalytic mechanism. The protoporphyrin IX substrate (*blue*), which has been model-built into the active site cleft, enters the active site from the membrane (*top*), whereas the Fe(II) enters from the matrix (*dashed red line*). The side chain of H263 is proposed to abstract protons from the two pyrrole NH groups (*dashed cyan lines*) and to pass them to the matrix via a series of closely spaced acidic groups consisting of the side chains of E343, H341, and D340. The side chains of R164 and Y165, which are on the opposite side of the active site pocket from H263, appear to participate in the metalation reaction, which is also facilitated by the enzyme-induced doming or ruffling of the porphyrin. [Courtesy of Harry Dailey, University of Georgia. PDBid 1HRK.]

the liver, which synthesizes most of the remainder. An important function of heme in liver is as the prosthetic group of **cytochrome P450,** an oxidative enzyme involved in detoxification (Section 15-4B), which is required throughout the liver cell's lifetime in amounts that vary with conditions. In contrast, erythroid cells, in which heme is, of course, a hemoglobin component, engage in heme synthesis only on differentiation, when they synthesize hemoglobin in vast quantities. This is a one-time synthesis; the heme must last the erythrocyte's lifetime (normally 120 days) since heme and hemoglobin synthesis stop on red cell maturation (protein synthesis stops on the loss of nuclei and ribosomes). The different ways in which heme biosynthesis is regulated in liver and in erythroid cells reflect these different demands: In liver, heme biosynthesis must really be "controlled," whereas in erythroid cells, the process is more like breaking a dam.

In liver, the main control target in heme biosynthesis is ALA synthase, the enzyme catalyzing the pathway's first committed step. Heme, or its Fe(III) oxidation product **hemin,** controls this enzyme's activity through three mechanisms: (1) feedback inhibition, (2) inhibition of the transport of ALA synthase from its site of synthesis in the cytosol to its reaction site in the mitochondrion (Fig. 26-39), and (3) repression of ALA synthase synthesis.

In erythroid cells, heme exerts quite a different effect on its biosynthesis. Heme stimulates, rather than represses, protein synthesis in reticulocytes (immature erythrocytes). Although the vast majority of the protein synthesized by reticulocytes is globin, there is evidence that heme also induces these cells to synthesize the enzymes of the heme biosynthesis pathway. Moreover, the rate-determining step of heme biosynthesis in erythroid cells may not be the ALA synthase reaction. Experiments on various systems of differentiating erythroid cells implicate ferrochelatase and porphobilinogen deaminase in the control of heme biosynthesis in these cells. There are also indications that cellular uptake of iron may be rate limiting. Iron is transported in the plasma complexed with the iron transport protein **transferrin.** The rate at which the iron–transferrin complex enters most cells, including those of liver, is controlled by receptor-mediated endocytosis (Section 12-4B). However, lipid-soluble iron complexes that diffuse directly into reticulocytes stimulate *in vitro* heme biosynthesis. The existence of several control points supports the supposition that when erythroid heme biosynthesis is "switched on," all of its steps function at their maximal rates rather than any one step limiting the flow through the pathway. Heme-stimulated synthesis of globin also ensures that heme and globin are synthesized in the correct ratio for assembly into hemoglobin (Section 32-4A).

h. Porphyrias Have Bizarre Symptoms

Several genetic defects in heme biosynthesis, in liver or erythroid cells, are recognized. All involve the accumulation of porphyrin and/or its precursors and are therefore known as **porphyrias** (Greek: *porphyra,* purple). Two such defects are known to affect erythroid cells: uroporphyrinogen III synthase deficiency **(congenital erythropoietic porphyria)** and ferrochelatase deficiency **(erythropoietic protoporphyria).** The former results in accumulation of uroporphyrinogen I and its decarboxylation product **coproporphyrinogen I.** Excretion of these compounds colors the urine red, their deposition in the teeth turns them a fluorescent reddish brown, and their accumulation in the skin renders it extremely photosensitive such that it ulcerates and forms disfiguring scars. Increased hair growth is also observed in afflicted individuals such that fine hair may cover much of their faces and extremities. These symptoms have prompted speculation that the werewolf legend has a biochemical basis.

The most common porphyria that primarily affects liver is porphobilinogen deaminase deficiency **(acute intermittent porphyria).** This disease is marked by intermittent attacks of abdominal pain and neurological dysfunction. Excessive amounts of ALA and PBG are excreted in the urine during and after such attacks. The urine may become red resulting from the excretion of excess porphyrins synthesized from PBG in nonhepatic cells although the skin does not become unusually photosensitive. King George III, who ruled England during the American Revolution, and who has been widely portrayed as being mad, in fact had attacks characteristic of acute intermittent porphyria, was reported to have urine the color of port wine, and had several descendants who were diagnosed as having this disease. American history might have been quite different had George III not inherited this metabolic defect.

I. Heme Is Degraded to Bile Pigments

At the end of their lifetime, red cells are removed from the circulation and their components degraded. Heme catabolism (Fig. 26-41) begins with oxidative cleavage, by

heme oxygenase, of the porphyrin between rings A and B to form **biliverdin,** a green linear tetrapyrrole. Biliverdin's central methenyl bridge (between rings C and D) is then reduced to form the red-orange **bilirubin.** The changing colors of a healing bruise are a visible manifestation of heme degradation.

The highly lipophilic bilirubin is insoluble in aqueous solutions. Like other lipophilic metabolites, such as free

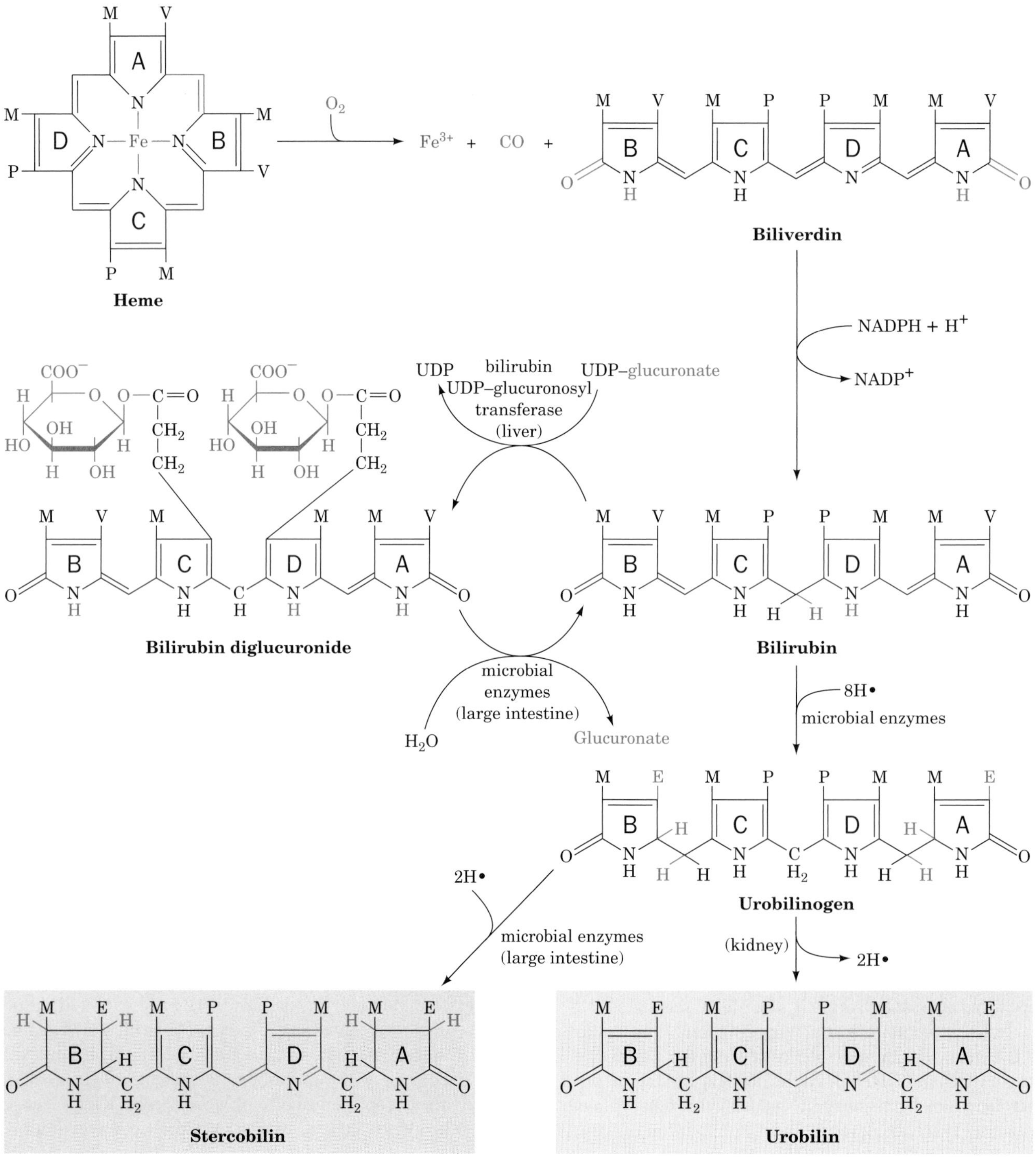

FIGURE 26-41   The heme degradation pathway. M, V, P, and E, respectively, represent methyl, vinyl, propionyl, and ethyl groups.

fatty acids, it is transported in the blood in complex with serum albumin. In the liver, its aqueous solubility is increased by esterification of its two propionate side groups with glucuronic acid, yielding **bilirubin diglucuronide,** which is secreted into the bile. Bacterial enzymes in the large intestine hydrolyze the glucuronic acid groups and, in a multistep process, convert bilirubin to several products, most notably **urobilinogen.** Some urobilinogen is reabsorbed and transported via the bloodstream to the kidney, where it is converted to the yellow **urobilin** and excreted, thus giving urine its characteristic color. Most of the urobilinogen, however, is microbially converted to the deeply red-brown **stercobilin,** the major pigment of feces.

When the blood contains excessive amounts of bilirubin, the deposition of this highly insoluble substance colors the skin and the whites of the eyes yellow. This condition, called **jaundice** (French: *jaune,* yellow), signals either an abnormally high rate of red cell destruction, liver dysfunction, or bile duct obstruction. Newborn infants, particularly when premature, often become jaundiced because their livers do not yet make sufficient **bilirubin UDP-glucuronosyltransferase** to glucuronidate the incoming bilirubin. Jaundiced infants are treated by bathing them with light from a fluorescent lamp; this photochemically converts bilirubin to more soluble isomers that the infant can degrade and excrete.

j. Hemoglobin's Reduced Affinity for CO Prevents Asphyxiation

In the reaction forming biliverdin, the methenyl bridge carbon between porphyrin rings A and B is released as CO (Fig. 26-41, *top*), which, we have seen, is a tenacious heme ligand (with 200-fold greater affinity for hemoglobin and myoglobin than O_2; Section 10-1A). Consequently, ~1% of hemoglobin's O_2-binding sites are blocked by CO, even in the absence of air pollution. However, free heme in solution binds CO with 20,000-fold greater affinity than it binds O_2. Thus, the globin (protein) portion of hemoglobin (and likewise myoglobin) somehow lowers the affinity of its bound heme for CO, thereby making O_2 transport possible. How does the globin do so?

Early X-ray structures of **carboxymyoglobin** (myoglobin with a CO ligand) indicated that the bound CO was inclined from the normal to the heme plane by 40 to 60° (the Fe—C—O bond angle appeared to be 120 to 140°), approximately the same angle with which O_2 binds to heme (Fig. 10-12). Yet, in complexes of CO with porphyrins in the absence of protein, the CO is normal to the heme plane. This suggested that the globin (in both myoglobin and hemoglobin) sterically bends the bound CO away from its preferred linear geometry, thereby reducing its affinity for CO and hence permitting the CO to be slowly exhaled. However, a variety of spectroscopic investigations together with highly accurate X-ray structures of carboxymyoglobin revealed that the bound CO is, in fact, inclined from the normal to the heme plane by ~7°, a distortion that is too small to explain the reduced affinity of myoglobin for CO. Of course, this reduced affinity might instead be explained by the distortions that the upright CO ligand im-

poses on the globin, presumably via the distal His (E7, the His residue that hydrogen bonds to the bound O_2; Section 10-2). However, studies of the energetics of binding of CO and O_2 to myoglobins in which His E7 has been mutated to nonpolar residues of comparable bulk (e.g., Leu) indicate that this is not the main determinant of the ligand affinity changes. Rather, the reduction in affinity of myoglobin for CO relative to that for O_2 has been shown to arise from the greater hydrogen bonding affinity that His E7 has for O_2 relative to CO together with electrostatic effects due to the differing charge distributions in the O_2 and CO ligands.

k. Chloroquine Prevents Malaria by Inhibiting Plasmodial Heme Sequestration

Malaria is caused by the mosquito-borne parasite *Plasmodium falciparum* (Section 7-3A), which multiplies within and destroys red blood cells in a 2-day cycle. During the intraerythrocytic stages of its life cycle, the parasite partially meets its nutritional needs by proteolyzing up to ~80% of the host cell's hemoglobin in its so-called acid food vacuole. This process releases heme which, in its soluble form, is toxic to the parasite because it damages cell membranes and inhibits a variety of enzymes. Since, unlike their human hosts, plasmodia cannot degrade heme, they sequester it within their food vacuoles in the form of harmless dark brown granules known as **hemozoin,** which consist of crystals of dimerized hemes linked together by reciprocal iron–carboxylate bonds between the ferric ions and the propionate side chains of adjacent molecules. Hemozoin has been found to be identical to **β-hematin,**

β-Hematin (hemozoin)

whose crystal structure has been determined. Dimers interact in the crystals through hydrogen bonds between the remaining carboxyl groups.

Chloroquine,

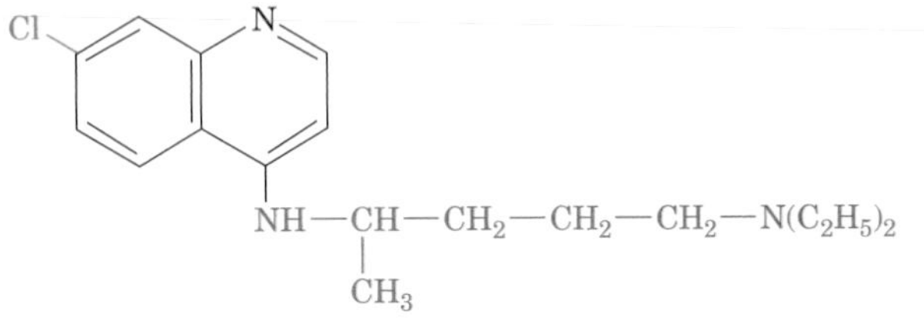

Chloroquine

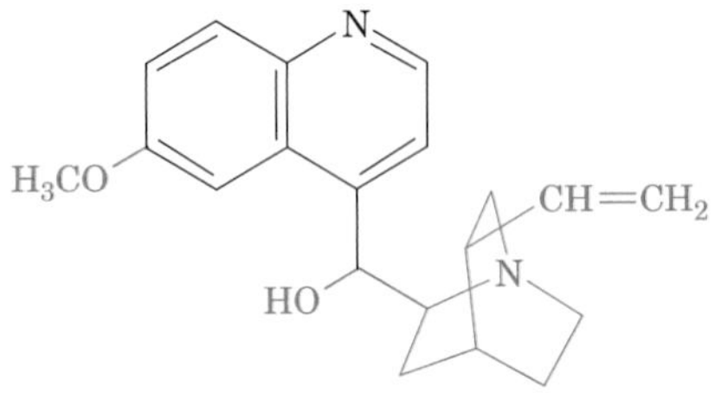

Quinine

a member of the quinoline ring-containing family of antimalarials, which includes **quinine,** is one of the most successful antimicrobial agents that has been produced. It is effective against plasmodia only during their intraerythrocytic stages. This drug, being a weak base that can readily pass through membranes in its uncharged form, accumulates in the plasmodial acid food vacuole in its acidic (charged) form in millimolar concentrations. Chloroquine and several other quinoline-containing antimalarials inhibit the crystallization of hemes to form hemozoin. This inhibition *in vivo* is almost certainly responsible for the antimalarial properties of these drugs. The mechanism of inhibition is as yet unclear although a plausible hypothesis is that the drug adsorbs onto crystallized hemozoin, inhibiting further crystallization.

The massive use of chloroquine has, unfortunately, led to the appearance of chloroquine-resistant plasmodia in nearly every malarial region of the world. Resistant plasmodia do not concentrate chloroquine in their food vacuoles to the high levels found in sensitive parasites. Rather, they export this drug out of their food vacuoles at an ~50-fold higher rate than do sensitive organisms. Since chloroquine activity and chloroquine resistance have different mechanisms, it should be possible to modify existing quinoline-containing structures or to develop new hemozoin crystallization inhibitors that are effective antimalarial agents but to which plasmodia are not (yet) resistant.

B. Biosynthesis of Physiologically Active Amines

Epinephrine, norepinephrine, dopamine, serotonin (5-hydroxytryptamine), γ-aminobutyric acid (GABA),

*and **histamine***

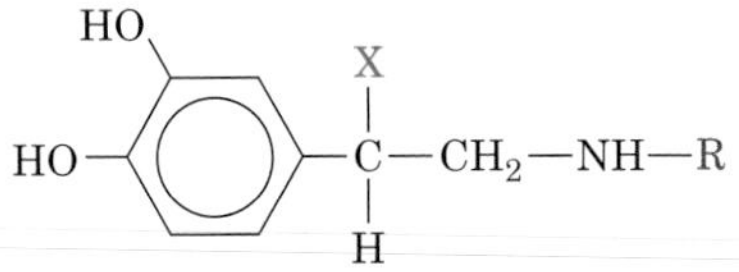

X = OH,	R = CH$_3$	**Epinephrine (Adrenalin)**
X = OH,	R = H	**Norepinephrine**
X = H,	R = H	**Dopamine**

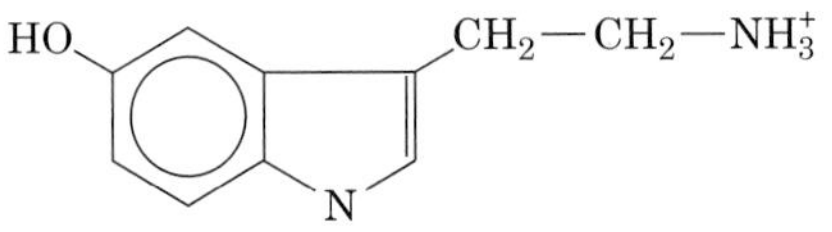

Serotonin
(5-hydroxytryptamine)

$$^-OOC-CH_2-CH_2-CH_2-NH_3^+$$

γ-Aminobutyric acid (GABA)

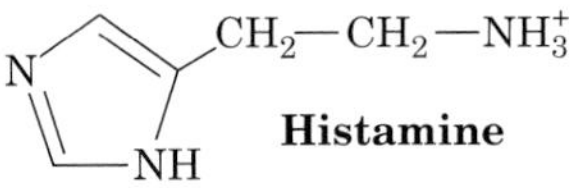

Histamine

are hormones and/or neurotransmitters derived from amino acids. For instance, epinephrine, as we have seen, activates muscle adenylate cyclase, thereby stimulating glycogen breakdown (Section 18-3E); deficiency in dopamine production is associated with **Parkinson's disease,** a degenerative condition causing "shaking palsy"; serotonin causes smooth muscle contraction; GABA is one of the brain's major inhibitory neurotransmitters (Section 20-5C), being released at 30% of its synapses; and histamine is involved in allergic responses (as allergy sufferers who take antihistamines will realize), as well as in the control of acid secretion by the stomach (Section 20-3C).

The biosynthesis of each of these physiologically active amines involves decarboxylation of the corresponding precursor amino acid. Amino acid decarboxylases are PLP-dependent enzymes that form a PLP–Schiff base with the substrate so as to stabilize the C$_\alpha$ carbanion formed on C$_\alpha$—COO$^-$ bond cleavage (Section 26-1A):

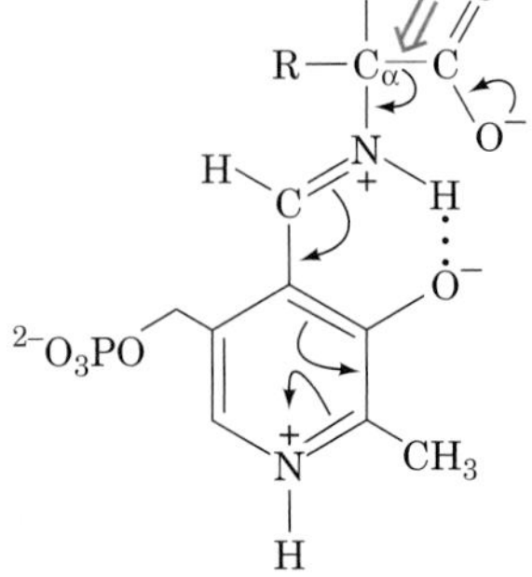

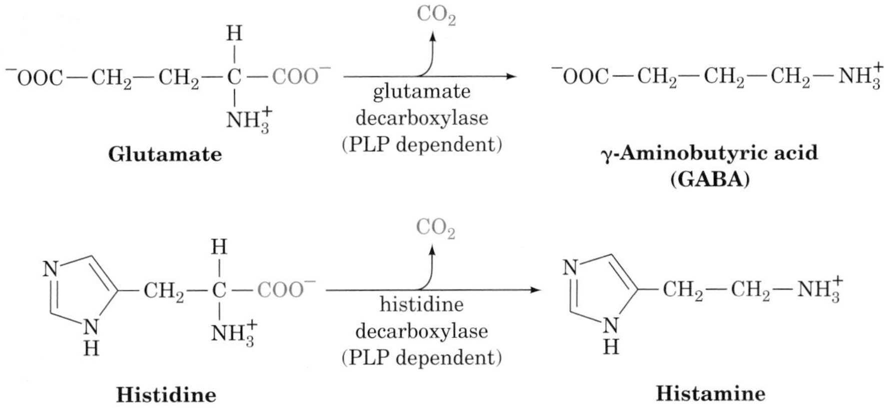

FIGURE 26-42   The formation of γ-aminobutyric acid (GABA) and histamine. The reactions involve the decarboxylations of glutamate to form GABA and of histidine to form histamine.

Formation of histamine and GABA are one-step processes (Fig. 26-42). In the synthesis of serotonin from tryptophan, the decarboxylation is preceded by a hydroxylation (Fig. 26-43) by **tryptophan hydroxylase,** one of three mammalian enzymes that has a 5,6,7,8-tetrahydrobiopterin cofactor (Section 26-3H). This hydroxylation involves an NIH shift similar to that occurring in phenylalanine hydroxylase (Fig. 26-30), although no epoxide intermediate has been observed in this case. Dopamine, norepinephrine, and epinephrine are all termed **catecholamines** because they are amine derivatives of **catechol:**

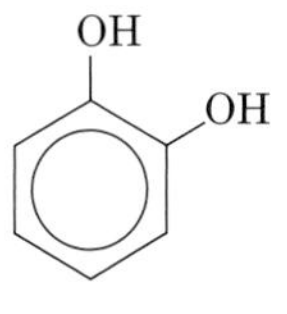

Catechol

The conversion of tyrosine to these various catecholamines occurs as follows (Fig. 26-44):

1. Tyrosine is hydroxylated to **3,4-dihydroxyphenylalanine (L-DOPA)** by **tyrosine hydroxylase,** another 5,6,7,8-tetrahydrobiopterin-requiring enzyme.

2. L-DOPA is decarboxylated to dopamine.

3. A second hydroxylation yields norepinephrine.

4. Methylation of norepinephrine's amino group by *S*-adenosylmethionine (SAM; Section 26-3E) produces epinephrine.

The specific catecholamine that a cell produces depends on which enzymes of the pathway are present. In adrenal medulla, which functions to produce hormones (Section 19-1F), epinephrine is the predominant product. In some areas of the brain, norepinephrine is more common. In other areas, most prominently the **substantia nigra,** the

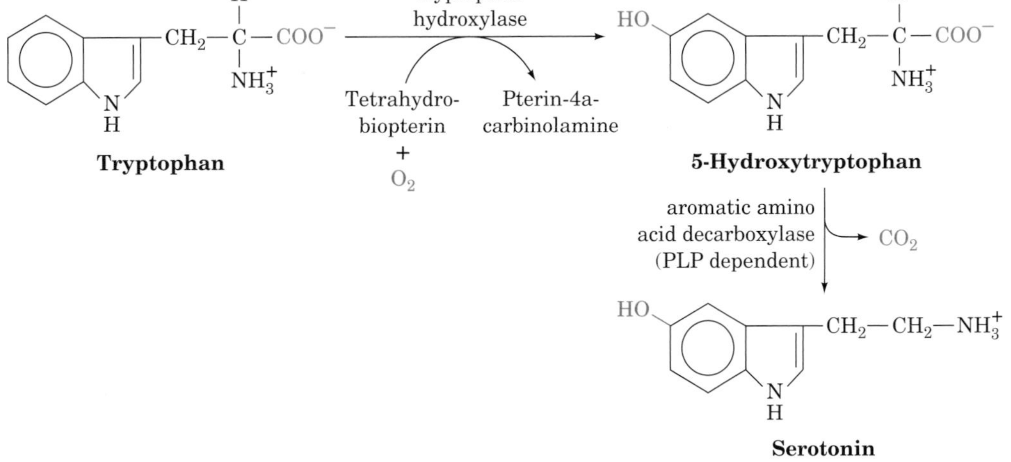

FIGURE 26-43   The formation of serotonin. The biosynthesis involves the hydroxylation and subsequent decarboxylation of tryptophan.

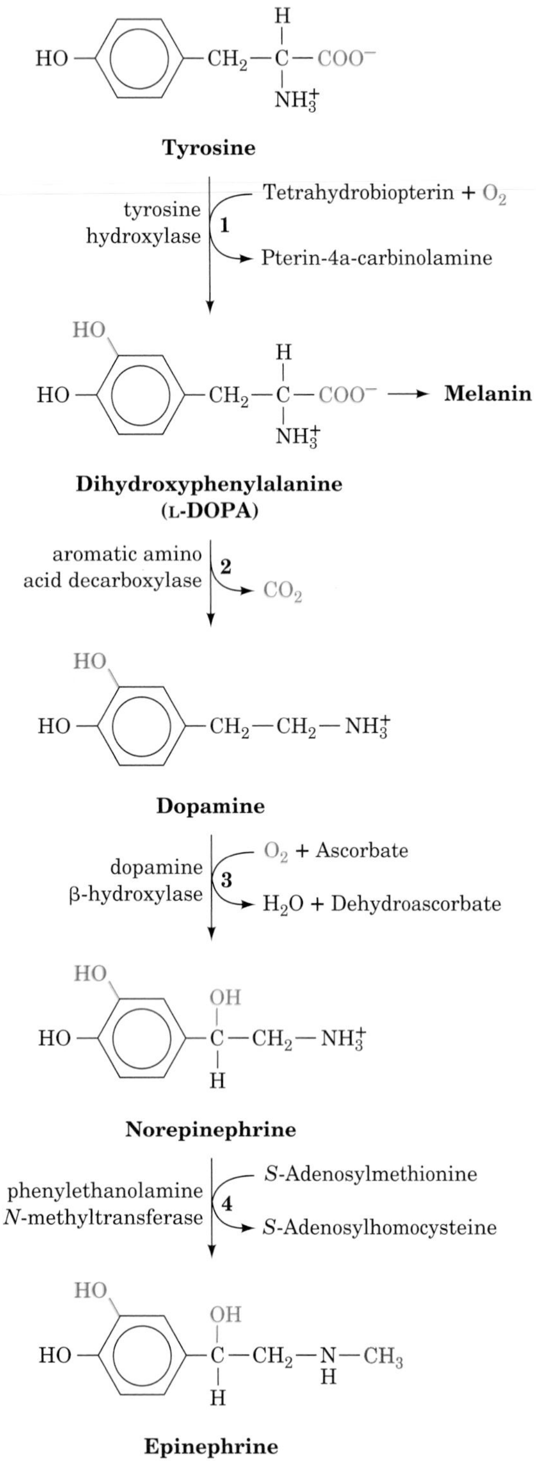

Tyrosine

tyrosine
hydroxylase **1**

Tetrahydrobiopterin + O_2

Pterin-4a-carbinolamine

**Dihydroxyphenylalanine
(L-DOPA)** → **Melanin**

aromatic amino
acid decarboxylase **2**

CO_2

Dopamine

dopamine
β-hydroxylase **3**

O_2 + Ascorbate

H_2O + Dehydroascorbate

Norepinephrine

phenylethanolamine
N-methyltransferase **4**

S-Adenosylmethionine

S-Adenosylhomocysteine

Epinephrine

**FIGURE 26-44   The sequential synthesis of L-DOPA,
dopamine, norepinephrine, and epinephrine from tyrosine.**
L-DOPA is also the precursor of the black skin pigment
melanin, an oxidized polymeric material.

pathway stops at dopamine synthesis. Indeed, Parkinson's
disease, which is caused by degeneration of the substantia
nigra, has been treated with some success by the administration of L-DOPA, dopamine's immediate precursor.

Dopamine itself is ineffective because it cannot cross the
blood–brain barrier. L-DOPA, however, is able to get to
its sites of action where it is decarboxylated to dopamine.
The enzyme catalyzing this reaction, **aromatic amino acid
decarboxylase,** decarboxylates all aromatic amino acids
and is therefore also responsible for serotonin formation.
In one approach to the treatment of Parkinson's disease,
a portion of the patient's adrenal medulla is surgically
transplanted to his or her brain. Presumably, the dopamine
and L-DOPA released by this tissue serve to replace that
lost via degeneration of the substantia nigra. L-DOPA is
also a precursor of the black skin pigment melanin.

C. Glutathione

Glutathione (GSH; γ-glutamylcysteinylglycine),

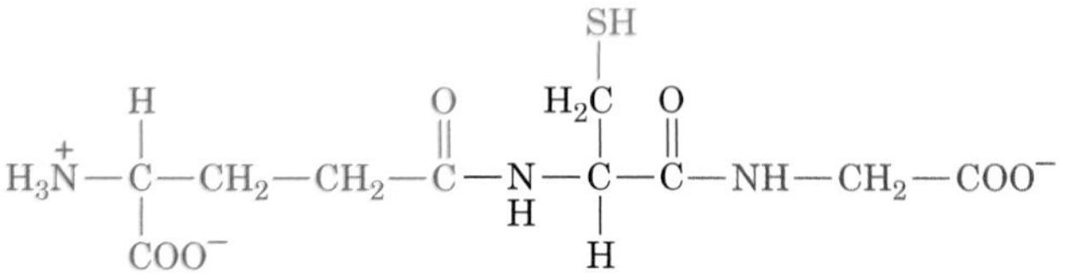

**Glutathione
(GSH; γ-glutamylcysteinylglycine)**

a tripeptide that contains an unusual γ-amide bond, participates in a variety of detoxification, transport, and metabolic processes (Fig. 26-45). For instance, it is a substrate
for peroxidase reactions, helping to destroy peroxides generated by oxidases; it is involved in leukotriene biosynthesis (Section 25-7C); and the balance between its reduced
(GSH) and oxidized (GSSG) forms maintains the
sulfhydryl groups of intracellular proteins in their correct
oxidation states.

The **γ-glutamyl cycle,** which was elucidated by Alton
Meister, *provides a vehicle for the energy-driven transport
of amino acids into cells through the synthesis and breakdown of GSH (Fig. 26-46).* GSH is synthesized from

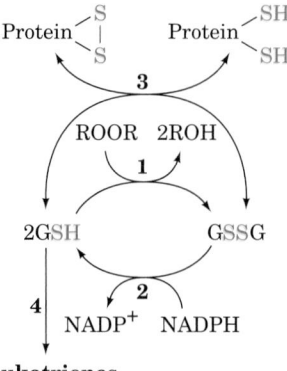

FIGURE 26-45   Some reactions involving glutathione. The
reactions and enzymes are **(1)** peroxide detoxification by
glutathione peroxidase, (2) regeneration of GSH from GSSG
by glutathione reductase (Section 21-2B), **(3)** thiol transferase
modulation of protein thiol–disulfide balance, and
(4) leukotriene biosynthesis by a glutathione-*S*-transferase.

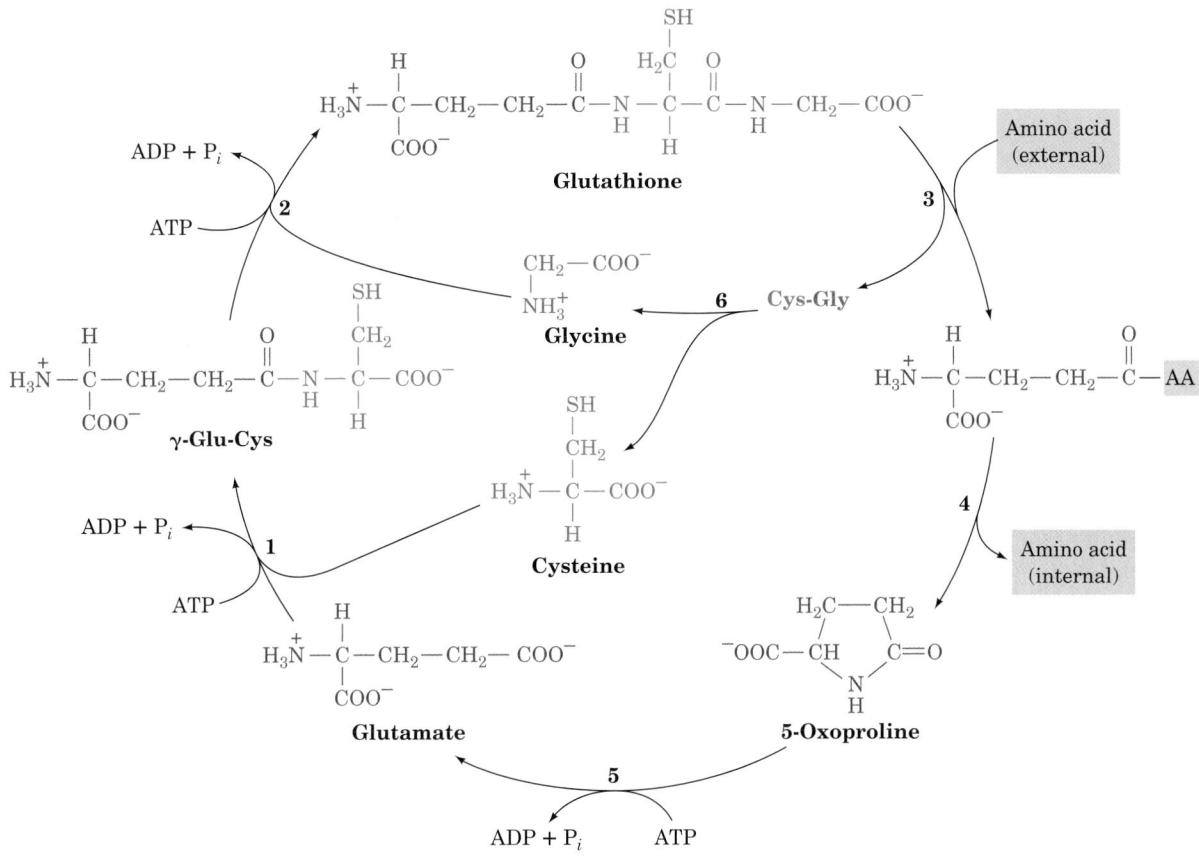

FIGURE 26-46 Glutathione synthesis as part of the γ-glutamyl cycle of glutathione metabolism. The cycle's reactions are catalyzed by **(1)** γ-glutamylcysteine synthetase, **(2)** glutathione synthetase, **(3)** γ-glutamyl transpeptidase, **(4)** γ-glutamyl cyclotransferase, **(5)** 5-oxoprolinase, and **(6)** an intracellular protease.

glutamate, cysteine, and glycine by the consecutive action of **γ-glutamylcysteine synthetase** and **GSH synthetase** (Fig. 26-46, Reactions 1 and 2). ATP hydrolysis provides the free energy for each reaction. The carboxyl group is activated for peptide bond synthesis by formation of an acyl phosphate intermediate:

$$R-\overset{\overset{O}{\|}}{C}-O^- + ATP \xrightarrow{ADP} R-\overset{\overset{O}{\|}}{C}-OPO_3^{2-}$$
$$\xrightarrow[P_i]{NH_2-R'}$$
$$R-\overset{\overset{O}{\|}}{C}-\underset{H}{N}-R'$$

The breakdown of GSH is catalyzed by **γ-glutamyl transpeptidase, γ-glutamyl cyclotransferase, 5-oxoprolinase,** and an intracellular protease (Fig. 26-46, Reactions 3–6).

Amino acid transport occurs because, whereas GSH is synthesized intracellularly and is located largely within the cell, γ-glutamyl transpeptidase, which catalyzes GSH breakdown (Fig. 26-46, Reaction 3), is situated on the cell membrane's external surface and accepts amino acids, notably cysteine and methionine. GSH is first transported to the external surface of the cell membrane, where the transfer of the γ-glutamyl group from GSH to an external amino acid occurs. The γ-glutamyl amino acid is then transported back into the cell and converted to glutamate by a two-step process in which the transported amino acid is released and **5-oxoproline** is formed as an intermediate. The last step in the cycle, the hydrolysis of 5-oxoproline, requires ATP hydrolysis. This surprising observation (amide bond hydrolysis is almost always an exergonic process) is a consequence of 5-oxoproline's unusually stable internal amide bond.

D. *Tetrahydrofolate Cofactors: The Metabolism of C_1 Units*

Many biosynthetic processes involve the addition of a C_1 unit to a metabolic precursor. A familiar example is carboxylation. For instance, gluconeogenesis from pyruvate begins with the addition of a carboxyl group to form oxaloacetate (Section 23-1A). The coenzyme involved in this and most other carboxylation reactions is biotin (Section 23-1A). In contrast, *S*-adenosylmethionine functions as a methylating agent (Section 26-3E).

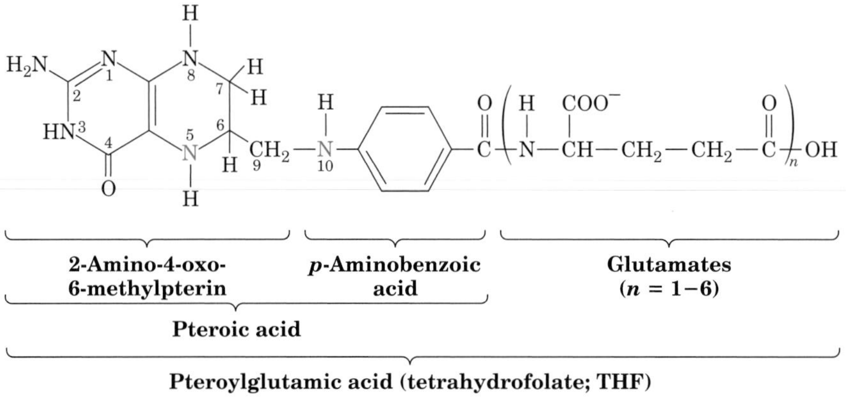

FIGURE 26-47   Tetrahydrofolate (THF).

Tetrahydrofolate (THF) is more versatile than the above cofactors in that it functions to transfer C_1 units in several oxidation states. THF is a 6-methylpterin derivative linked in sequence to *p*-aminobenzoic acid and Glu residues (Fig. 26-47). Up to five additional Glu residues may be linked to the first glutamate via isopeptide bonds to form a polyglutamyl tail.

THF is derived from **folic acid** (Latin: *folium,* leaf), a doubly oxidized form of THF that must be enzymatically reduced before it becomes an active coenzyme (Fig. 26-48). Both reductions are catalyzed by **dihydrofolate reductase (DHFR).** Mammals cannot synthesize folic acid so it must be provided in the diet or by intestinal microorganisms.

C_1 units are covalently attached to THF at its positions N5, N10, or both N5 and N10. These C_1 units, which may be at the oxidation levels of formate, formaldehyde, or

methanol (Table 26-1), are all interconvertible by enzymatic redox reactions (Fig. 26-49).

The main entry of C_1 units into the THF pool is as N^5,N^{10}-**methylene-THF** through the conversion of serine to glycine by serine hydroxymethyltransferase (Sections 26-3B and 26-5A) and the cleavage of glycine by glycine synthase (the glycine cleavage system; Section 26-3B, Fig. 26-14). Histidine also contributes C_1 units through its degradation with the formation of N^5-**formimino-THF** (Fig. 26-17, Reaction 11).

A C_1 unit in the THF pool can have several fates (Fig. 26-50):

1. It may be used directly as N^5,N^{10}-methylene-THF in the conversion of the deoxynucleotide dUMP to dTMP by **thymidylate synthase** (Section 28-3B).

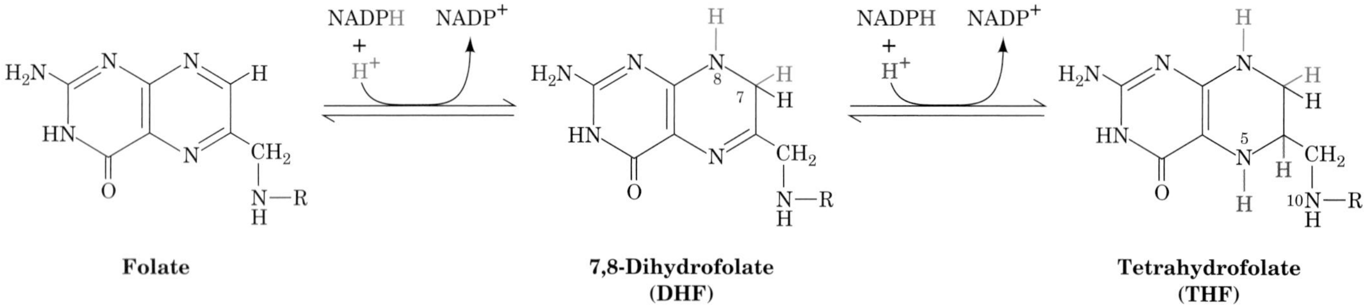

FIGURE 26-48   The two-stage reduction of folate to THF. Both reactions are catalyzed by dihydrofolate reductase (DHFR).

TABLE 26-1   Oxidation Levels of C_1 Groups Carried by THF

Oxidation Level	Group Carried	THF Derivative(s)
Methanol	Methyl ($-CH_3$)	N^5-Methyl-THF
Formaldehyde	Methylene ($-CH_2-$)	N^5,N^{10}-Methylene-THF
Formate	Formyl ($-CH=O$)	N^5-Formyl-THF, N^{10}-formyl-THF
	Formimino ($-CH=NH$)	N^5-Formimino-THF
	Methenyl ($-CH=$)	N^5,N^{10}-Methenyl-THF

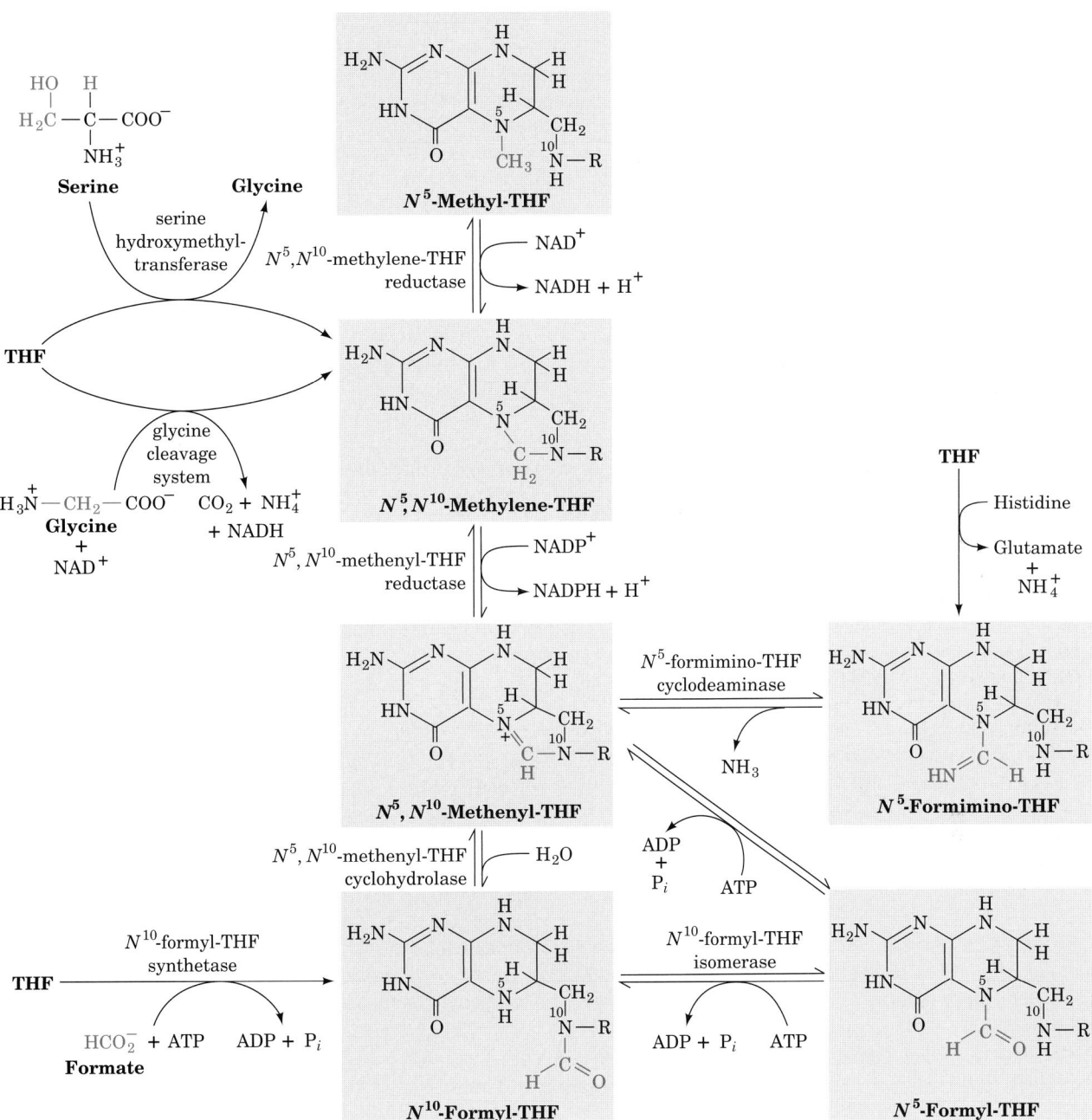

FIGURE 26-49 Interconversion of the C_1 units carried by THF.

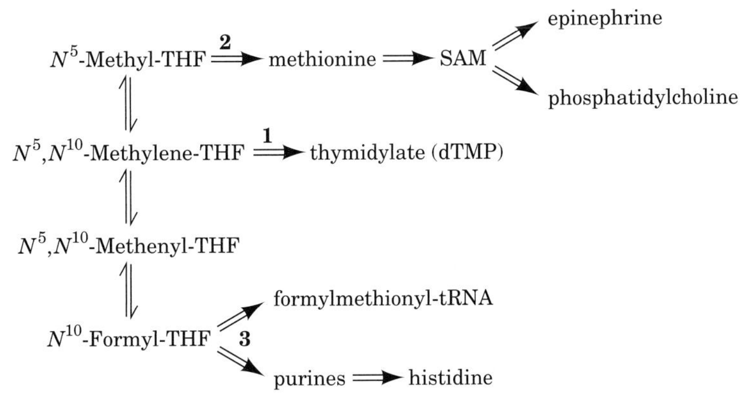

FIGURE 26-50 The biosynthetic fates of the C_1 units in the THF pool.

2. It may be reduced to N^5-**methyl-THF** for the synthesis of methionine from homocysteine (Section 26-3E).

3. It may be oxidized through N^5,N^{10}-methenyl-THF to N^{10}-**formyl-THF** for use in the synthesis of purines (Section 28-1A). Since the purine ring of ATP is involved in histidine biosynthesis in microorganisms and plants (Section 26-5B), N^{10}-formyl-THF is indirectly involved in this pathway as well. Prokaryotes use N^{10}-formyl-THF in a formylation reaction yielding **formylmethionyl-tRNA,** which they require for the initiation of protein synthesis (Section 32-3C).

Sulfonamides (sulfa drugs) such as **sulfanilamide** are antibiotics that are structural analogs of the *p*-aminobenzoic acid constituent of THF:

Sulfonamides
(R = H, sulfanilamide)

***p*-Aminobenzoic acid**

They competitively inhibit bacterial synthesis of THF at the *p*-aminobenzoic acid incorporation step, thereby blocking the above THF-requiring reactions. The inability of mammals to synthesize folic acid leaves them unaffected by sulfonamides, which accounts for the medical utility of these widely used antibacterial agents.

5 ■ AMINO ACID BIOSYNTHESIS

Many amino acids are synthesized by pathways that are present only in plants and microorganisms. Since mammals must obtain these amino acids in their diets, these substances are known as **essential amino acids.** The other amino acids, which can be synthesized by mammals from common intermediates, are termed **nonessential amino acids.** Their α-keto acid carbon skeletons are converted to amino acids by transamination reactions (Section 26-1A) utilizing the preformed α-amino nitrogen of another amino acid, usually glutamate. Yet, although it was originally presumed that glutamate can be synthesized from ammonia and α-ketoglutarate by glutamate dehydrogenase acting in reverse, it now appears that the predominant physiological direction of this enzyme is glutamate breakdown (Section 26-1B). Consequently, *preformed α-amino nitrogen should also be considered to be an essential nutrient.* In this context, it is interesting to note that, in addition to the four well-known taste receptors, those for sweet, sour, salty, and bitter tastes, a fifth taste receptor has recently been characterized, that for the meaty taste of **monosodium glutamate (MSG),** which is known as **umami** (a Japanese name).

The essential and nonessential amino acids for humans are listed in Table 26-2. Arginine is classified as essential,

even though it is synthesized by the urea cycle (Section 26-2D), because it is required in greater amounts than can be produced by this route during the normal growth and development of children (but not adults). The essential amino acids occur in animal and vegetable proteins. Different proteins, however, contain different proportions of the essential amino acids. Milk proteins, for example, contain them all in the proportions required for proper human nutrition. Bean protein, on the other hand, contains an abundance of lysine but is deficient in methionine, whereas wheat is deficient in lysine but contains ample methionine. A balanced protein diet therefore must contain a variety of different protein sources that complement each other to supply the proper proportions of all the essential amino acids.

In this section we study the pathways involved in the formation of the nonessential amino acids. We also briefly consider such pathways for the essential amino acids as they occur in plants and microorganisms. You should note, however, that although we discuss some of the most common pathways for amino acid biosynthesis, there is considerable variation in these pathways among different species. In contrast, as we have seen, the basic pathways of carbohydrate and lipid metabolism are all but universal.

A. *Biosynthesis of the Nonessential Amino Acids*

All the nonessential amino acids except tyrosine are synthesized by simple pathways leading from one of four common metabolic intermediates: pyruvate, oxaloacetate, α-ketoglutarate, and 3-phosphoglycerate. Tyrosine, which is really misclassified as nonessential, is synthesized by the one-step hydroxylation of the essential amino acid phenylalanine (Section 26-3H). Indeed, the dietary requirement for phenylalanine reflects the need for tyrosine as well. The presence of dietary tyrosine therefore decreases the need for phenylalanine. Since preformed α-amino nitrogen in

TABLE 26-2   Essential and Nonessential Amino Acids in Humans

Essential	Nonessential
Arginine[a]	Alanine
Histidine	Asparagine
Isoleucine	Aspartate
Leucine	Cysteine
Lysine	Glutamate
Methionine	Glutamine
Phenylalanine	Glycine
Threonine	Proline
Tryptophan	Serine
Valine	Tyrosine

[a]Although mammals synthesize arginine, they cleave most of it to form urea (Sections 26-2D and 26-2E).

the form of glutamate is an essential nutrient for nonessential amino acid biosynthesis, we first discuss its production by plants and microorganisms.

a. Glutamate Is Synthesized by Glutamate Synthase

Glutamate synthase, an enzyme that occurs only in microorganisms, plants, and lower animals, converts α-ketoglutarate and ammonia originating in glutamine to glutamate. The electrons required for this reductive amination come from NADPH or ferredoxin, depending on the organism. The NADPH-dependent glutamate synthase from the nitrogen-fixing bacterium *Azospirillum brasilense,* the best characterized such enzyme, is an $\alpha_2\beta_2$ heterotetramer that binds an FAD and two [4Fe–4S] clusters on each β subunit, and an FMN and a [3Fe–4S] cluster on each α subunit. The overall reaction is

$$NADPH + H^+ + \text{glutamine} + \alpha\text{-ketoglutarate}$$
$$\rightarrow 2\text{glutamate} + NADP^+$$

and involves five steps that occur at three distinct active sites (Fig. 26-51):

1. Electrons are transferred from NADPH to FAD at active site 1 on the β subunit to yield $FADH_2$.

2. The electrons are transferred from the $FADH_2$ to FMN at site 2 on a specific α subunit through the iron–sulfur clusters to yield $FMNH_2$.

3. Glutamine is hydrolyzed to α-glutamate and ammonia at site 3 on the α subunit.

4. The ammonia produced is transferred to site 2 where it reacts with α-ketoglutarate to form α-iminoglutarate.

5. The α-iminoglutarate is reduced by $FMNH_2$ to form glutamate.

In the absence of the β subunit, the α subunit can synthesize glutamate from glutamine and α-ketoglutarate using an artificial electron donor; moreover, it is homologous and functionally similar to ferredoxin-dependent glutamine synthases. The *A. brasilense* α subunit is therefore considered to be the enzyme's catalytic core.

The X-ray structure of the 1479-residue α subunit of *A. brasilense* glutamate synthase in complex with a [3Fe–4S] cluster, an FMN, an α-ketoglutarate substrate, and a

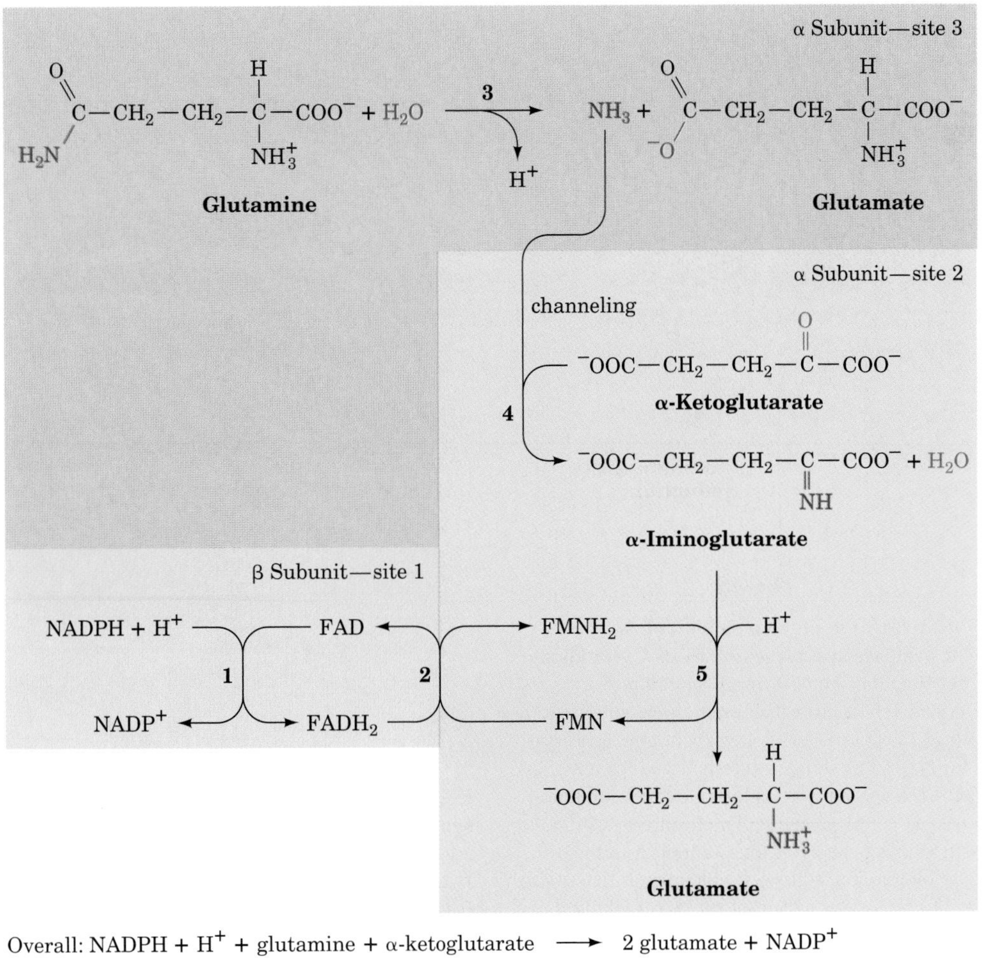

Overall: $NADPH + H^+ + \text{glutamine} + \alpha\text{-ketoglutarate} \longrightarrow 2 \text{ glutamate} + NADP^+$

FIGURE 26-51   The sequence of reactions catalyzed by glutamate synthase.

methionine sulfone inhibitor (a tetrahedral transition state analog)

Glutamine

Methionine sulfone

was determined by Andrea Mattevi (Fig. 26-52). The subunit has four domains, an N-terminal glutamine amidotransferase domain, a central domain, an FMN-binding domain, and a C-terminal **β-helix** domain (see below). The methionine sulfone binds in the N-terminal amidotrans-

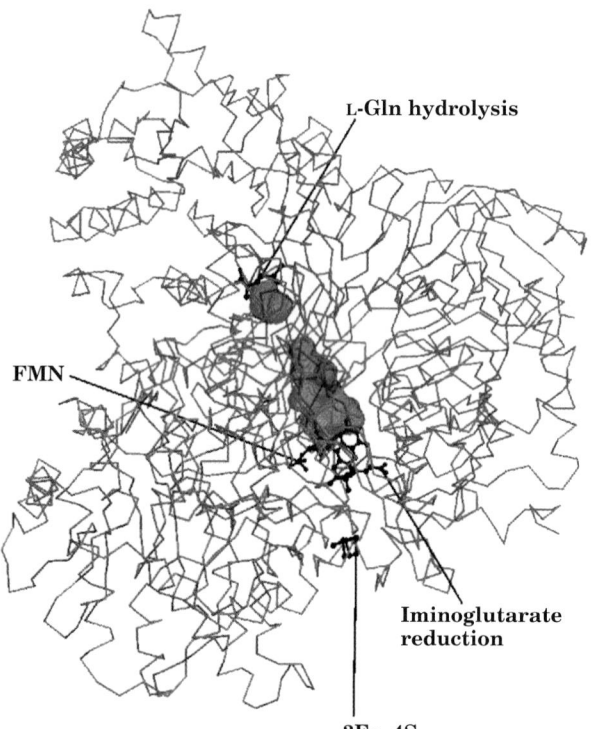

FIGURE 26-52   X-Ray structure of the α subunit of *A. brasilense* glutamate synthase as represented by its Cα backbone. The subunit has four domains, an N-terminal glutamine amidotransferase domain (*blue*) to which methionine sulfone (a glutamine tetrahedral transition state analog) is bound; a central domain (*red*); an FMN-binding domain (*green*) to which an FMN, a [3Fe–4S] cluster, and an α-ketoglutarate are bound; and a β helix domain (*purple*). The foregoing ligands are drawn in black in ball-and-stick form. The 31-Å-long tunnel from the methyl group on the methionine sulfone (analogous to the amido group of glutamine) to the α-keto group of the α-ketoglutarate is outlined by a gray surface. The tunnel is blocked in this structure (it is divided into two cavities) by the main chain atoms of four residues that protrude into the tunnel. [Courtesy of Andrea Mattevi, Universitá degli Studi di Pavia, Italy. PDBid 1EA0.]

ferase domain (site 3), where glutamine is normally hydrolyzed via the nucleophilic attack of the Cys 1 sulfhydryl group on the glutamine C_γ atom to transiently form a tetrahedral intermediate that the tetrahedral sulfone group mimics. The FMN-binding domain consists in large part of an α/β barrel, at the mouth of which the α-ketoglutarate and the [3Fe–4S] cluster bind. The distance between the methyl group on methionine sulfone (the analog of the glutamine amido group) and the α-keto group of the α-ketoglutarate is 31 Å. These two sites are connected by a tunnel through which ammonia must diffuse in order to react with α-ketoglutarate (Reaction 4, Fig. 26-51). However, the tunnel is blocked by the main chain atoms of four residues that protrude into the tunnel. Consequently, there must be at least a 2- to 3-Å shift in the structure during the reaction to permit ammonia to diffuse between these sites. Such gating of the tunnel may well be crucial for the control of enzyme function so as to avoid the wasteful hydrolysis of glutamine. In fact, glutamine is hydrolyzed only when the enzyme has bound α-ketoglutarate and reducing equivalents are available for iminoglutarate reduction.

The C-terminal domain of glutamate synthase consists largely of a 7-turn, right-handed β helix (Fig. 26-53). In this unusual motif, the polypeptide chain is wrapped in a wide helix such that neighboring turns of chain interact as do the strands of a parallel β sheet. The 43-Å-long glutamate synthase β helix has an elliptical cross section of 16 by 23 Å. β Helices have been observed in only a handful of mainly bacterial enzymes. The β helix of glutamate synthase does not contain a residue that is involved in catal-

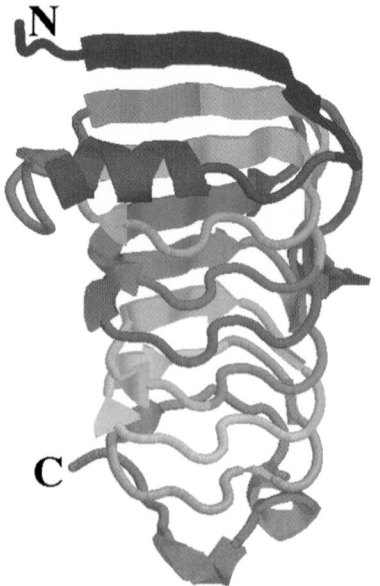

FIGURE 26-53   The β helix of *A. brasilense* glutamate synthase. The polypeptide backbone (residues 1225–1416) is color ramped from blue at its N-terminus to red at its C-terminus. Neighboring turns of polypeptide chain within the β helix interact as do the strands of parallel β sheets. Nevertheless, the conformations of many of these segments lie outside the normal range for β strands, and hence they are drawn in coil form. [Based on an X-ray structure by Andrea Mattevi, Universitá degli Studi di Pavia, Italy. PDBid 1EA0.]

ysis or electron transfer. However, it does appear to have an important structural role because some of its residues line the tunnel through which ammonia passes.

Glutamine amidotransferase domains or subunits are part of several protein structures in which glutamine is the ammonia donor for a further reaction. In glutamate synthase, it is a domain within the α subunit (Fig. 26-52). In *E. coli* carbamoyl phosphate synthetase (CPS; Section 26-2A), it is a complete subunit within a heterodimer. These domains or subunits occur in one of two families that are differentiated by their active site structures. In CPS, it belongs to the **triad family,** so called because it contains an active site Cys in a catalytic triad reminiscent of the catalytic triad in serine proteases (Section 15-3). The glutamine amidotransferase domain of glutamate synthase belongs to the **N-terminal nucleophile (Ntn) family** which, as we have seen, has an N-terminal Cys that acts as the active site nucleophile. Other enzymes involved in amino acid biosynthesis and having a glutamine amidotransferase domain include asparagine synthetase (Fig. 26-54, Reaction 4; see below), a member of the Ntn family, and imidazole glycerol phosphate synthase (Fig. 26-65, Reaction

5), which belongs to the triad family. All of these enzymes have an ammonia-channeling tunnel that connects the amidotransferase site with the ammonia-utilizing site.

b. Alanine, Asparagine, Aspartate, Glutamate, and Glutamine Are Synthesized from Pyruvate, Oxaloacetate, and α-Ketoglutarate

Pyruvate, oxaloacetate, and α-ketoglutarate are the keto acids that correspond to alanine, aspartate, and glutamate, respectively. Indeed, as we have seen (Section 26-1), the synthesis of each of these amino acids is a one-step transamination reaction (Fig. 26-54, Reactions 1–3). Asparagine and glutamine are, respectively, synthesized from aspartate and glutamate by amidation (Fig. 26-54, Reactions 4 and 5). **Glutamine synthetase** catalyzes the formation of glutamine in a reaction in which ATP is hydrolyzed to ADP and P_i via the intermediacy of **γ-glutamylphosphate** and NH_4^+ is the amino group donor (Fig. 26-54, Reaction 5). Curiously, aspartate amidation by **asparagine synthetase** to form asparagine follows a different route; it utilizes glutamine as its amino group donor and hydrolyzes ATP to AMP + PP_i (Fig. 26-54, Reaction

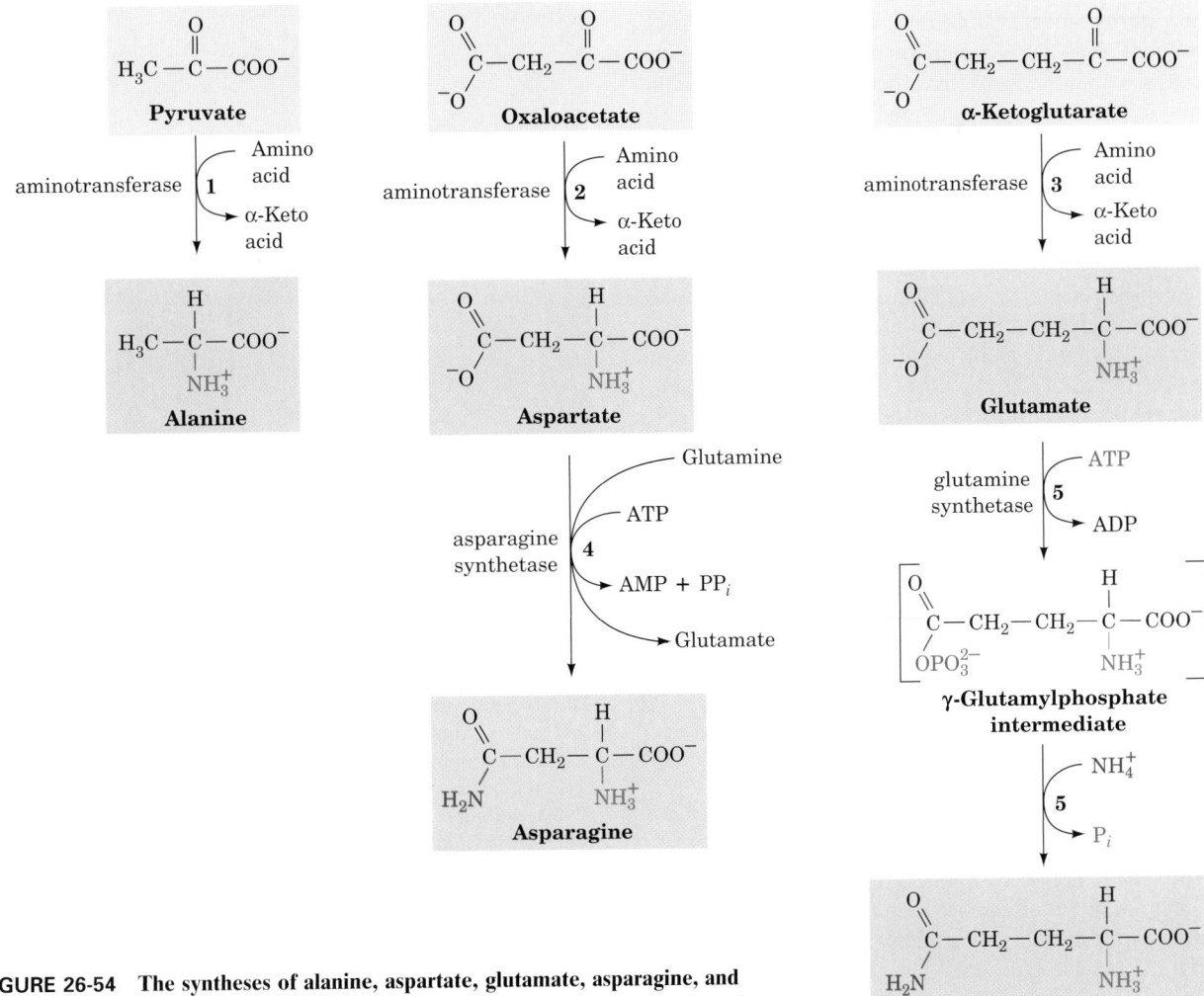

FIGURE 26-54 The syntheses of alanine, aspartate, glutamate, asparagine, and glutamine. These reactions involve the transaminations of (**1**) pyruvate, (**2**) oxaloacetate, and (**3**) α-ketoglutarate, and the amidations of (**4**) aspartate and (**5**) glutamate.

4). This enzyme is composed of a glutamine amidotransferase domain of the Ntn family (see above) and a second domain in which β-aspartyl-AMP

β-Aspartyl-AMP

is synthesized from Asp and ATP and then reacts with ammonia to form Asn. As with other glutamine amidotransferase-containing enzymes, the two domains are connected by a tunnel that channels ammonia between their two active sites.

c. Glutamine Synthetase Is a Central Control Point in Nitrogen Metabolism

Glutamine, as we have seen, is the amino group donor in the formation of many biosynthetic products, as well as being a storage form of ammonia. Glutamine synthetase's consequent pivotal position in nitrogen metabolism makes its control a vital aspect of this process. In fact, mammalian glutamine synthetases are activated by α-ketoglutarate, the product of glutamate's oxidative deamination. This control presumably prevents the accumulation of the ammonia produced by that reaction.

Bacterial glutamine synthetase, as Earl Stadtman showed, has a much more elaborate control system. This enzyme, which consists of 12 identical 469-residue subunits arranged with D_6 symmetry (Fig. 26-55), is regulated by several effectors as well as by covalent modification. Although a complete description of this complex enzyme is not given here, several aspects of its catalytic and control systems bear note.

The X-ray structure of *Salmonella typhimurium* glutamine synthetase in complex with the glutamate structural analog **phosphinothricin,**

Phosphinothricin   **Glutamate**

determined by David Eisenberg, reveals that its catalytic sites occur at the interface between the C-terminal domain of one subunit and the N-terminal domain of an adjacent subunit. These catalytic sites have a shape described as a "bifunnel" that opens at both the exposed top (ATP binding) and bottom (Glu and NH_4^+ binding) of the molecule

(between the two hexameric rings) and is narrow in the plane of its essential metal ions (two per subunit). Nucleotide binding induces conformational changes that

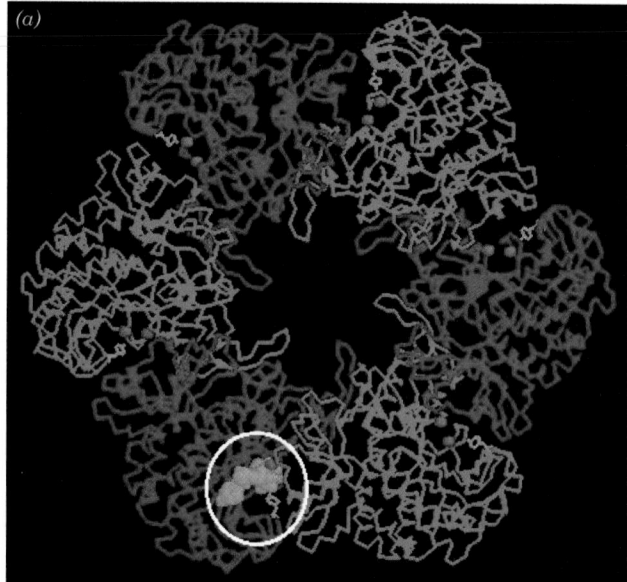

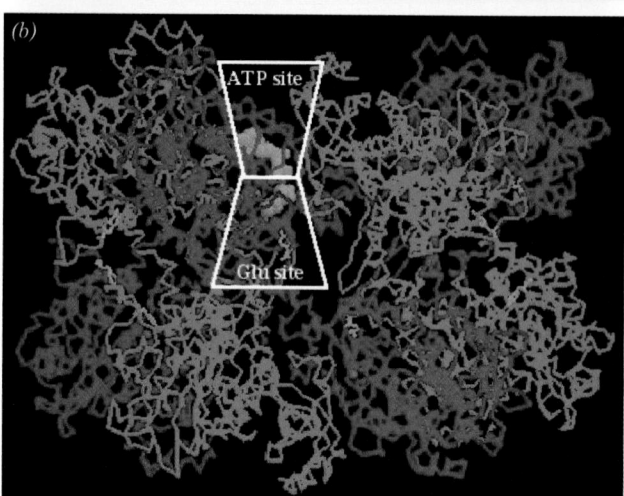

FIGURE 26-55   X-Ray structure of S. *typhimurium* glutamine synthetase. The enzyme consists of 12 identical subunits, here represented by their C_α backbones, arranged with D_6 symmetry (the symmetry of a hexagonal prism). (*a*) View down the 6-fold axis of symmetry showing only the six subunits of the upper ring in alternating blue and green. The subunits of the lower ring are roughly directly below those of the upper ring. The protein, including its side chains (not shown), has a diameter of 143 Å. The six active sites shown are marked by the pairs of Mn^{2+} ions (*red spheres*; divalent metal ions, physiologically Mg^{2+}, are required for enzymatic activity). Each adenylylation site, Tyr 397 (*yellow*), lies between two subunits at a higher radius than the corresponding active site. Also drawn in one active site are ADP (*cyan*) and phosphinothricin (*orange*), a competitive inhibitor of glutamate. The top opening to this active site, through which ATP and ADP bind, is marked with a white circle. (*b*) Side view along one of the enzyme's 2-fold axes showing only the eight nearest subunits. The molecule extends 103 Å along the 6-fold axis, which is vertical in this view. The bifunnel-shaped active site is outlined in white. [Based on an X-ray structure by David Eisenberg, UCLA. PDBid 1FPY.]

increase the enzyme's affinity for glutamate and ammonium ion, leading to an ordered sequential mechanism.

Nine feedback inhibitors cumulatively control the activity of bacterial glutamine synthetase: histidine, tryptophan, carbamoyl phosphate (as synthesized by carbamoyl phosphate synthetase II), glucosamine-6-phosphate, AMP, and CTP are all end products of pathways leading from glutamine, whereas alanine, serine, and glycine reflect the cell's nitrogen level. Several of these inhibitors act in a competitive manner, binding either to the glutamate binding site (serine, glycine, and alanine) or to the ATP binding site (AMP and CTP).

E. coli glutamine synthetase is covalently modified by adenylylation of a specific Tyr residue (Fig. 26-56). The enzyme's susceptibility to cumulative feedback inhibition increases, and its activity therefore decreases, with its degree of adenylylation. The level of adenylylation is controlled by a complex metabolic cascade that is conceptually similar to the one controlling glycogen phosphorylase (although the type of covalent modification differs in that glycogen phosphorylase is phosphorylated at a specific Ser residue; Sections 18-3B and 18-3C). Both adenylylation and deadenylylation of glutamine synthetase are catalyzed by **adenylyltransferase** in complex with a tetrameric regula-

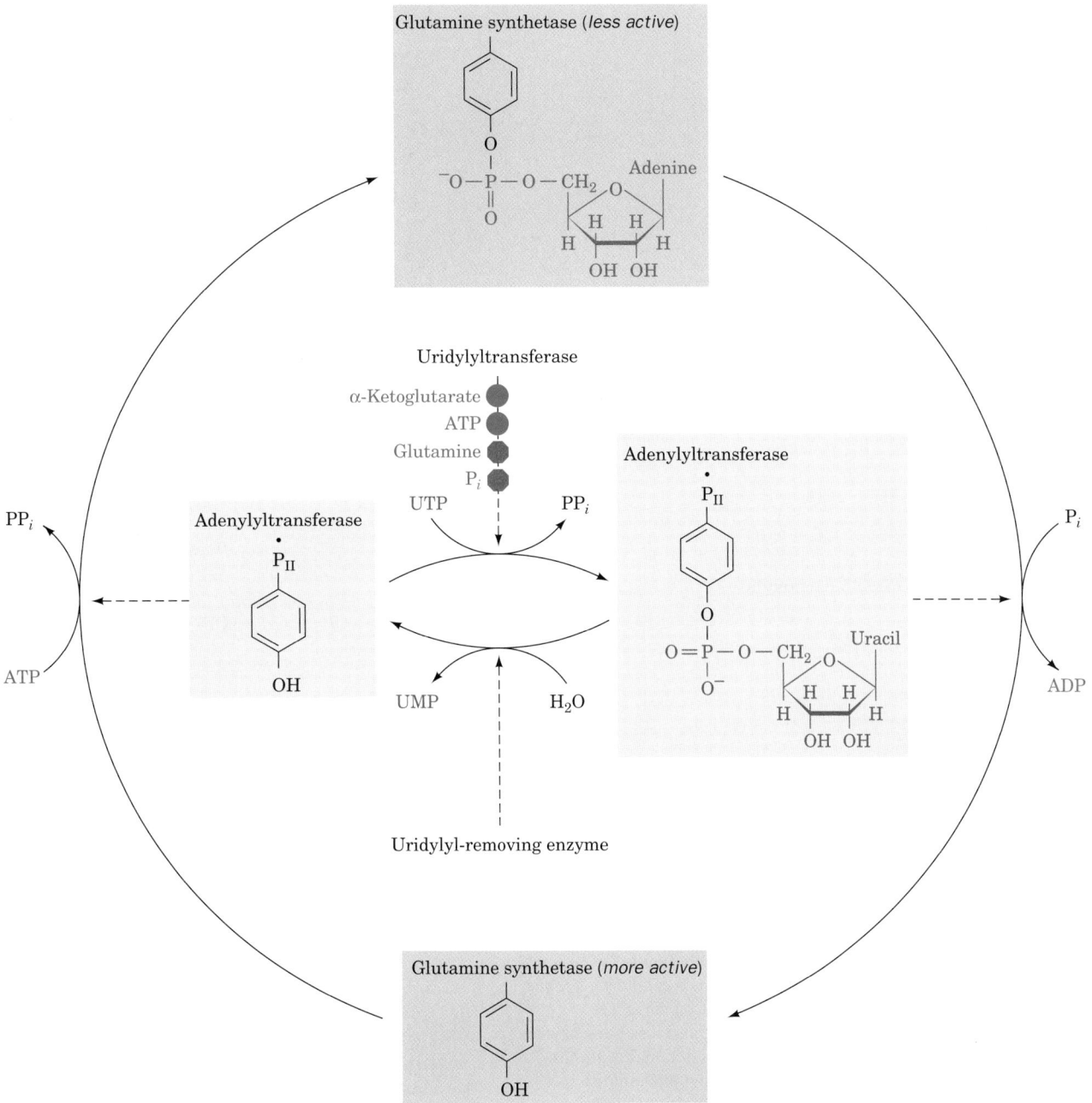

FIGURE 26-56 The regulation of bacterial glutamine synthetase. The adenylylation/deadenylylation of a specific Tyr residue is controlled by the level of uridylylation of a specific adenylyltransferase · P_{II} Tyr residue. This uridylylation level, in turn, is controlled by the relative activities of uridylyltransferase, which is sensitive to the levels of a variety of nitrogen metabolites, and uridylyl-removing enzyme, whose activity is independent of these metabolite levels.

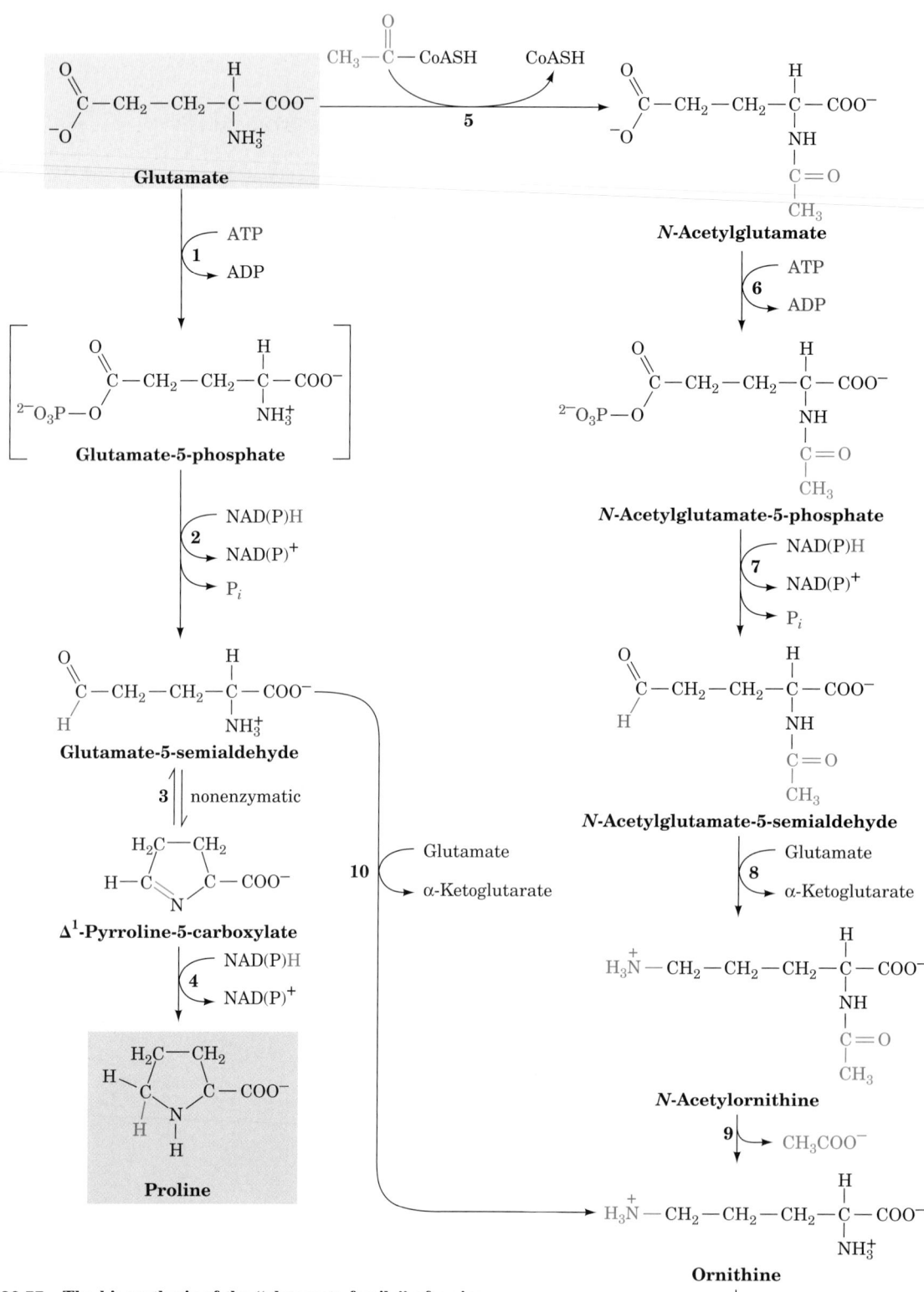

FIGURE 26-57   The biosynthesis of the "glutamate family" of amino acids: arginine, ornithine, and proline. The enzymes catalyzing proline biosynthesis are **(1)** γ-glutamyl kinase, **(2)** dehydrogenase, **(3)** nonenzymatic, and **(4)** pyrroline-5-carboxylate reductase. The enzymes catalyzing ornithine biosynthesis are **(5)** *N*-acetylglutamate synthase, **(6)** acetylglutamate kinase, **(7)** *N*-acetyl-γ-glutamyl phosphate dehydrogenase, **(8)** *N*-acetylornithine-δ-aminotransferase, and **(9)** acetylornithine deacetylase. An alternate pathway to ornithine is through Reaction **10**, catalyzed by ornithine-δ-aminotransferase. Ornithine is converted to arginine **(11)** via the urea cycle (Fig. 26-7, Reactions 2–4).

tory protein, **P$_{II}$.** This complex deadenylylates glutamine synthetase when P$_{II}$ is uridylylated (also at a Tyr residue) and adenylylates glutamine synthetase when P$_{II}$ lacks UMP residues. The level of P$_{II}$ uridylylation, in turn, depends on the relative activities of two enzymatic activities located on the same protein: a **uridylyltransferase** that uridylylates P$_{II}$ and a **uridylyl-removing enzyme** that hydrolytically excises the attached UMP groups of P$_{II}$ (Fig. 26-56). The uridylyltransferase is activated by α-ketoglutarate and ATP and inhibited by glutamine and P$_i$, whereas uridylyl-removing enzyme is insensitive to these metabolites. This complex metabolic cascade therefore renders the activity of *E. coli* glutamine synthetase extremely responsive to the cell's nitrogen requirements.

d. Glutamate Is the Precursor of Proline, Ornithine, and Arginine

Conversion of glutamate to proline (Fig. 26-57, Reactions 1–4) involves the reduction of the γ-carboxyl group to an aldehyde followed by the formation of an internal Schiff base whose further reduction yields proline. Reduction of the glutamate γ-carboxyl group to an aldehyde is an endergonic process that is facilitated by the carboxyl group's prior phosphorylation by **γ-glutamyl kinase.** The unstable product, **glutamate-5-phosphate,** has not been isolated from reaction mixtures but is presumed to be the substrate for the reduction that follows. The resulting **glutamate-5-semialdehyde** cyclizes spontaneously to form the internal Schiff base **Δ^1-pyrroline-5-carboxylate.** The final reduction to proline is catalyzed by **pyrroline-5-carboxylate reductase.** Whether the enzyme requires NADH or NADPH is unclear.

The *E. coli* pathway from glutamate to ornithine and hence to arginine likewise involves the ATP-driven reduction of the glutamate γ-carboxyl group to an aldehyde (Fig. 26-57, Reactions 6 and 7). Spontaneous cyclization of this intermediate, **N-acetylglutamate-5-semialdehyde,** is prevented by prior acetylation of its amino group by **N-acetylglutamate synthase** to form **N-acetylglutamate** (Fig. 26-57, Reaction 5). N-Acetylglutamate-5-semialdehyde, in turn, is converted to the corresponding amine by transamination (Fig. 26-57, Reaction 8). Hydrolysis of the acetyl protecting group finally yields ornithine, which, as we have seen (Section 26-2), is converted to arginine via the urea cycle. In humans, however, the pathway to ornithine is more direct. The N-acetylation of glutamate that protects it from cyclization does not occur. Rather, glutamate-5-semialdehyde, which is in equilibrium with Δ^1-pyrroline-5-carboxylate, is directly transaminated to yield ornithine in a reaction catalyzed by **ornithine-δ-aminotransferase** (Fig. 26-57, Reaction 10).

e. Serine, Cysteine, and Glycine Are Derived from 3-Phosphoglycerate

Serine is formed from the glycolytic intermediate 3-phosphoglycerate in a three-reaction pathway (Fig. 26-58):

1. Conversion of 3-phosphoglycerate's 2-OH group to a ketone yielding **3-phosphohydroxypyruvate,** serine's phosphorylated keto acid analog.

2. Transamination of 3-phosphohydroxypyruvate to phosphoserine.

3. Hydrolysis of phosphoserine to yield serine.

Serine participates in glycine synthesis in two ways (Section 26-3B):

1. Direct conversion of serine to glycine by serine hydroxymethyl transferase in a reaction that also yields N^5,N^{10}-methylene-THF (Fig. 26-12, Reaction 4 in reverse).

2. Condensation of the N^5,N^{10}-methylene-THF with CO_2 and NH_4^+ by the glycine cleavage system (Fig. 26-12, Reaction 3 in reverse).

We have already discussed the synthesis, in animals, of cysteine from serine and homocysteine, a breakdown product of methionine (Section 26-3E). Homocysteine combines with serine to yield cystathionine, which subsequently forms cysteine and α-ketobutyrate (Fig. 26-18, Reactions 5 and 6). Since cysteine's sulfhydryl group is derived from the essential amino acid methionine, cysteine is really an essential amino acid. In plants and microorganisms, however, cysteine is synthesized from serine in a two-step reaction involving the activation of the serine —OH by converting it to **O-acetylserine** followed by the displacement of acetate by sulfide (Fig. 26-59a). The sulfide required is produced from sulfate in an 8-electron reduction

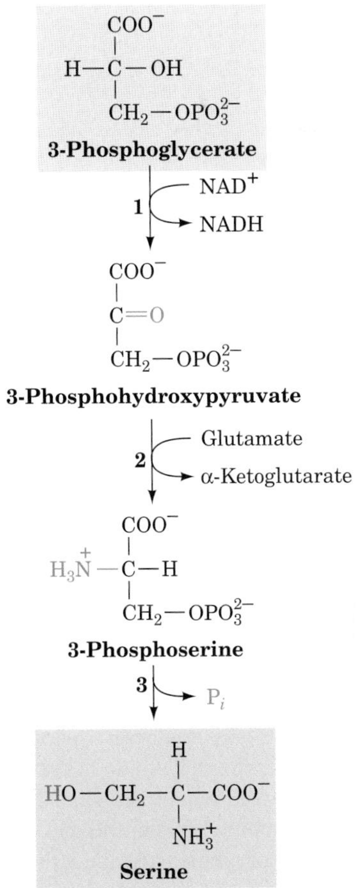

FIGURE 26-58 The conversion of 3-phosphoglycerate to serine. The pathway enzymes are **(1)** 3-phosphoglycerate dehydrogenase, **(2)** a PLP-dependent aminotransferase, and **(3)** phosphoserine phosphatase.

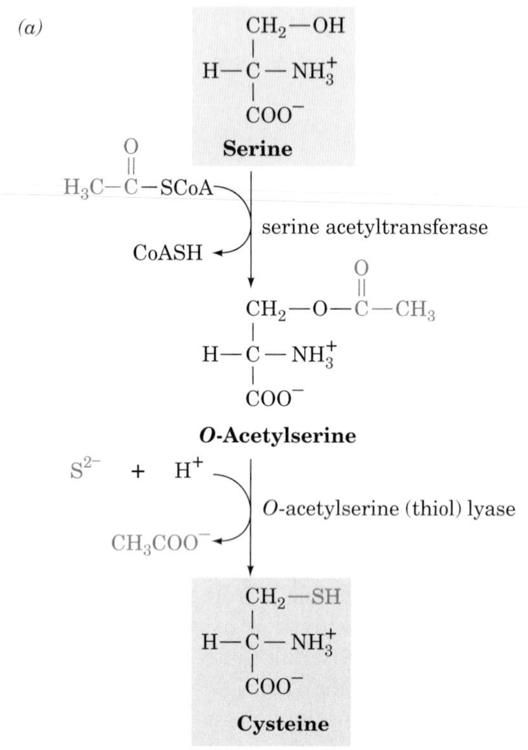

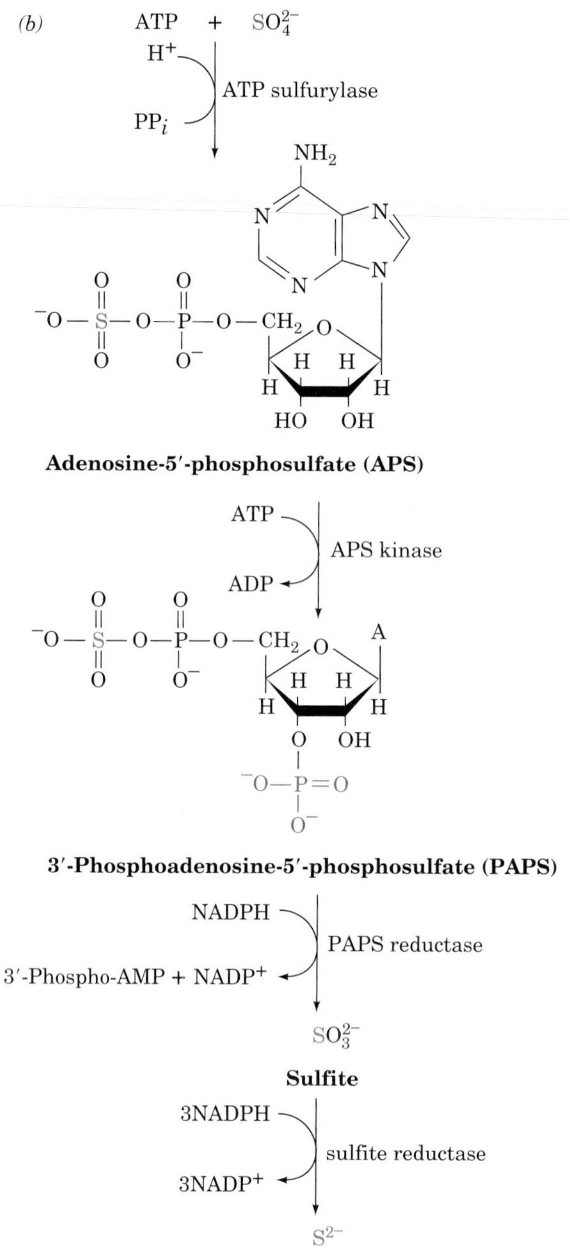

(a)

$$CH_2-OH$$
$$H-C-NH_3^+$$
$$COO^-$$
Serine

$$H_3C-\overset{O}{\overset{\|}{C}}-SCoA$$

serine acetyltransferase

CoASH

$$CH_2-O-\overset{O}{\overset{\|}{C}}-CH_3$$
$$H-C-NH_3^+$$
$$COO^-$$

O-Acetylserine

$$S^{2-} + H^+$$

O-acetylserine (thiol) lyase

$$CH_3COO^-$$

$$CH_2-SH$$
$$H-C-NH_3^+$$
$$COO^-$$
Cysteine

FIGURE 26-59 Cysteine biosynthesis. (*a*) The synthesis of cysteine from serine in plants and microorganisms. (*b*) The 8-electron reduction of sulfate to sulfide in *E. coli.*

(b)

$$ATP + SO_4^{2-}$$
$$H^+$$

ATP sulfurylase

$$PP_i$$

Adenosine-5′-phosphosulfate (APS)

ATP

APS kinase

ADP

3′-Phosphoadenosine-5′-phosphosulfate (PAPS)

NADPH

PAPS reductase

3′-Phospho-AMP + NADP$^+$

$$SO_3^{2-}$$
Sulfite

3NADPH

sulfite reductase

3NADP$^+$

$$S^{2-}$$

that occurs in *E. coli* as shown in Fig. 26-59*b*. Sulfate is first activated by the enzymes **ATP sulfurylase** and **adenosine-5′-phosphosulfate (APS) kinase.** The activated sulfate is then reduced to sulfite by **3′-phosphoadenosine-5′-phosphosulfate (PAPS) reductase** and to sulfide by **sulfite reductase.**

B. Biosynthesis of the Essential Amino Acids

Essential amino acids, like nonessential amino acids, are synthesized from familiar metabolic precursors. Their synthetic pathways are present only in microorganisms and plants, however, and usually involve more steps than those of the nonessential amino acids. For example, lysine, methionine, and threonine are all synthesized from aspartate in pathways whose common first reaction is catalyzed by **aspartokinase,** an enzyme that is present only in plants and microorganisms. Similarly, valine and leucine are formed from pyruvate; isoleucine is formed from pyruvate and α-ketobutyrate; and tryptophan, phenylalanine, and tyrosine are formed from phosphoenolpyruvate and erythrose-4-phosphate. The enzymes that synthesize essential amino acids were apparently lost early in animal evolution, possibly because of the ready availability of these amino acids in the diet.

Time and space prevent a detailed discussion of the many interesting reactions that occur in these pathways. The biosynthetic pathways of the aspartate family of amino acids, the pyruvate family, the aromatic family, and histidine are presented in Figs. 26-60 through 26-63 and 26-65

FIGURE 26-60 (*Opposite*)   **The biosynthesis of the "aspartate family" of amino acids: lysine, methionine, and threonine.** The pathway enzymes are **(1)** aspartokinase, **(2)** β-aspartate semialdehyde dehydrogenase, **(3)** homoserine dehydrogenase, **(4)** homoserine kinase, **(5)** threonine synthase (a PLP enzyme), **(6)** homoserine acyltransferase, **(7)** cystathionine γ-synthase, **(8)** cystathionine β-lyase, **(9)** methionine synthase (alternatively homocysteine methyltransferase, which also occurs in mammals; Section 26-3E), **(10)** dihydrodipicolinate synthase, **(11)** dihydrodipicolinate reductase, **(12)** *N*-succinyl-2-amino-6-ketopimelate synthase, **(13)** succinyl-diaminopimelate aminotransferase (a PLP enzyme), **(14)** succinyl-diaminopimelate desuccinylase, **(15)** diaminopimelate epimerase, and **(16)** diaminopimelate decarboxylase.

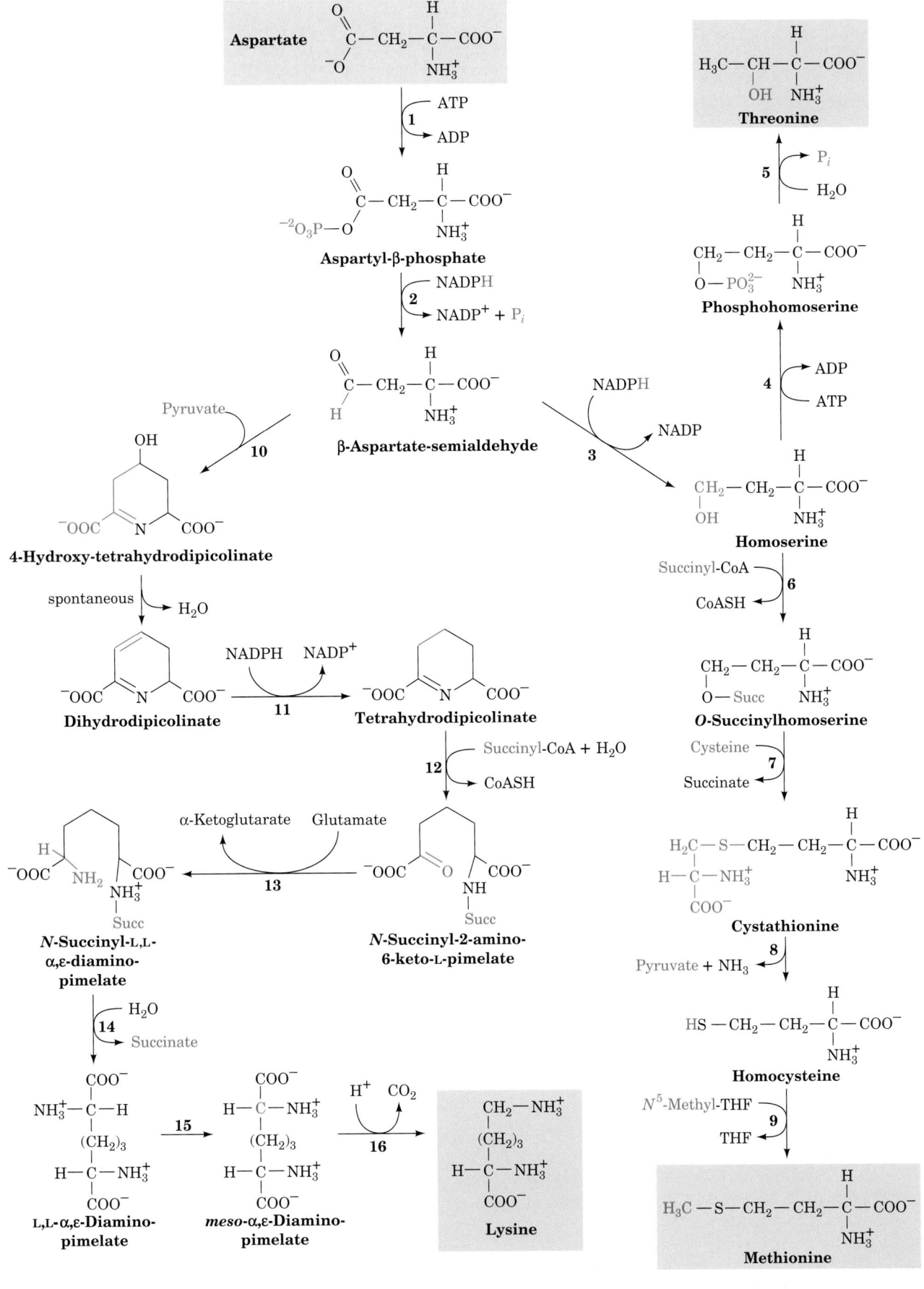

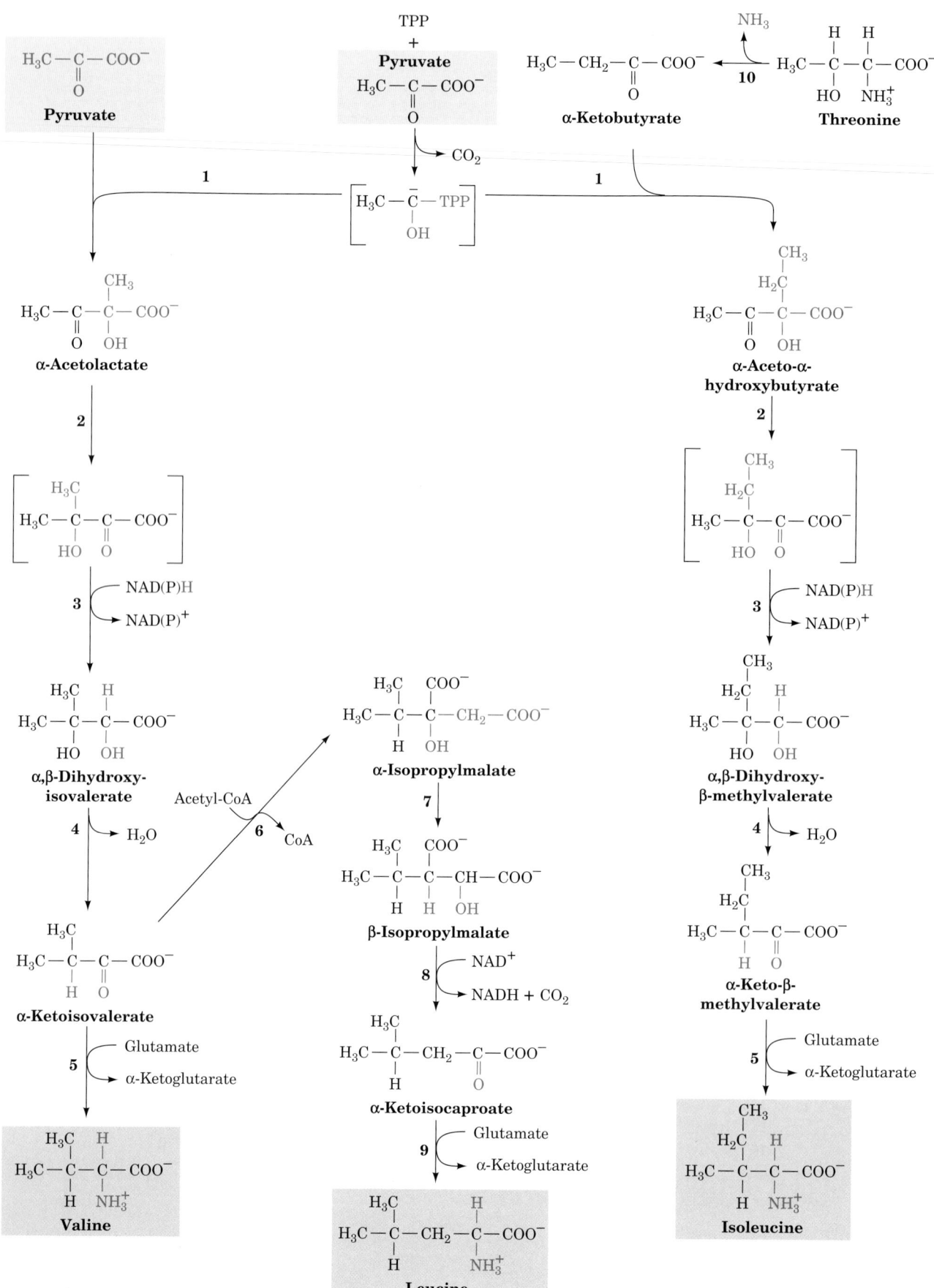

together with lists of the enzymes involved. Several agriculturally useful herbicides are specific inhibitors of some of these enzymes. Such herbicides have little toxicity toward animals and hence pose minimal risk to human health and the environment.

a. The Aspartate Family: Lysine, Methionine, and Threonine

In bacteria, aspartate is the common precursor of lysine, methionine, and threonine (Fig. 26-60). The biosynthesis of these essential amino acids all begin with the aspartokinase-catalyzed phosphorylation of aspartate to yield **aspartyl-β-phosphate.** We have seen that the control of metabolic pathways commonly occurs at the first committed step of the pathway. One might therefore expect lysine, methionine, and threonine biosynthesis to be controlled as a group. Each of these pathways is, in fact, independently controlled. *E. coli* does so via three isozymes of aspartokinase that respond differently to the three amino acids in terms both of feedback inhibition of enzyme activity and repression of enzyme synthesis. Table 26-3 summarizes this differential control. In addition, the pathway direction is controlled by feedback inhibition at the branch points by the individual amino acids. Thus methionine inhibits the *O*-acylation of homoserine (Fig. 26-60, Reaction 6), and lysine inhibits dihydrodipicolinate synthase (Fig. 26-60, Reaction 10).

b. The Pyruvate Family: Leucine, Isoleucine, and Valine

Valine and isoleucine are both synthesized via the same five-step pathway (Fig. 26-61), the only difference being in the first step of the series. In this TPP-dependent reaction, which resembles those catalyzed by pyruvate decarboxylase (Section 17-3B) and transketolase (Section 23-4C), pyruvate forms an adduct with TPP, which is decarboxylated to hydroxyethyl-TPP. This resonance-stabilized carbanion adds either to the keto group of a second pyruvate to form **acetolactate** on the way to valine, or to the keto group of threonine-derived **α-ketobutyrate** to form **α-aceto-α-hydroxybutyrate** on the way to isoleucine. The leucine biosynthetic pathway branches off from the valine pathway at α-ketoisovalerate (Fig. 26-61, Reaction 6). Reactions 6 to 8 in Fig. 26-61 are reminiscent of the first three reactions of the citric acid cycle (Sections 21-3A–C). Here, acetyl-CoA condenses with **α-ketoisovalerate** to form **α-isopropylmalate,** which then undergoes a dehy-

TABLE 26-3 Differential Control of Aspartokinase Isoenzymes in *E. Coli*

Enzyme	Feedback Inhibitor	Corepressor(s)[a]
Aspartokinase I	Threonine	Threonine and isoleucine
Aspartokinase II	None	Methionine
Aspartokinase III	Lysine	Lysine

[a]Compounds whose presence results in the transcriptional repression of enzyme synthesis (Section 31-3G).

dration/hydration reaction, followed by oxidative decarboxylation and transamination, to yield leucine.

c. The Aromatic Amino Acids: Phenylalanine, Tyrosine, and Tryptophan

The precursors to the aromatic amino acids are the glycolytic intermediate phosphoenolpyruvate (PEP) and erythrose-4-phosphate (an intermediate in the pentose phosphate pathway; Section 23-4C). Their condensation forms **2-keto-3-deoxy-D-arabinoheptulosonate-7-phosphate,** a C_7 compound that cyclizes and is ultimately converted to **chorismate** (Fig. 26-62), the branch point for tryptophan biosynthesis. Chorismate is converted either to anthranilate and then on to tryptophan, or to **prephenate** and on to either tyrosine or phenylalanine (Fig. 26-63). Although mammals synthesize tyrosine by the hydroxylation of phenylalanine (Section 26-3H), many microorganisms synthesize it directly from prephenate.

Since the synthesis of aromatic amino acids only occurs in plants and microorganisms, this pathway is a natural target for herbicides that will not be toxic to animals. For example, **glyphosate,**

$$^{-2}O_3P{-}CH_2{-}NH{-}CH_2{-}COO^-$$

Glyphosate

the active ingredient in one of the most widely used weed killers, Round-up, is a competitive inhibitor with respect to PEP in the **5-enolpyruvylshikimate-3-phosphate synthase (EPSP synthase)** reaction (Reaction 6 of Fig. 26-62).

d. A Protein Tunnel Channels the Intermediate Product of Tryptophan Synthase between Two Active Sites

The final two reactions in tryptophan biosynthesis, Reactions 5 and 6 in Fig. 26-63, are both catalyzed by **tryptophan synthase:**

1. The α subunit (268 residues) of this $\alpha_2\beta_2$ bifunctional enzyme cleaves **indole-3-glycerol phosphate,** yielding **indole** and glyceraldehyde-3-phosphate (Reaction 5).

2. The β subunit (396 residues) joins indole with L-serine in a PLP-dependent reaction to form L-tryptophan (Reaction 6).

Either subunit alone is enzymatically active, but when they are joined in the $\alpha_2\beta_2$ tetramer, the rates of both reactions

FIGURE 26-61 (*Opposite*) **The biosynthesis of the "pyruvate family" of amino acids: isoleucine, leucine, and valine.** The pathway enzymes are **(1)** acetolactate synthase (a TPP enzyme), **(2)** acetolactate mutase, **(3)** reductase, **(4)** dihydroxy acid dehydratase, **(5)** valine aminotransferase (a PLP enzyme), **(6)** α-isopropylmalate synthase, **(7)** α-isopropylmalate dehydratase, **(8)** isopropylmalate dehydrogenase, **(9)** leucine aminotransferase (a PLP enzyme), and **(10)** threonine deaminase (serine dehydratase, a PLP enzyme).

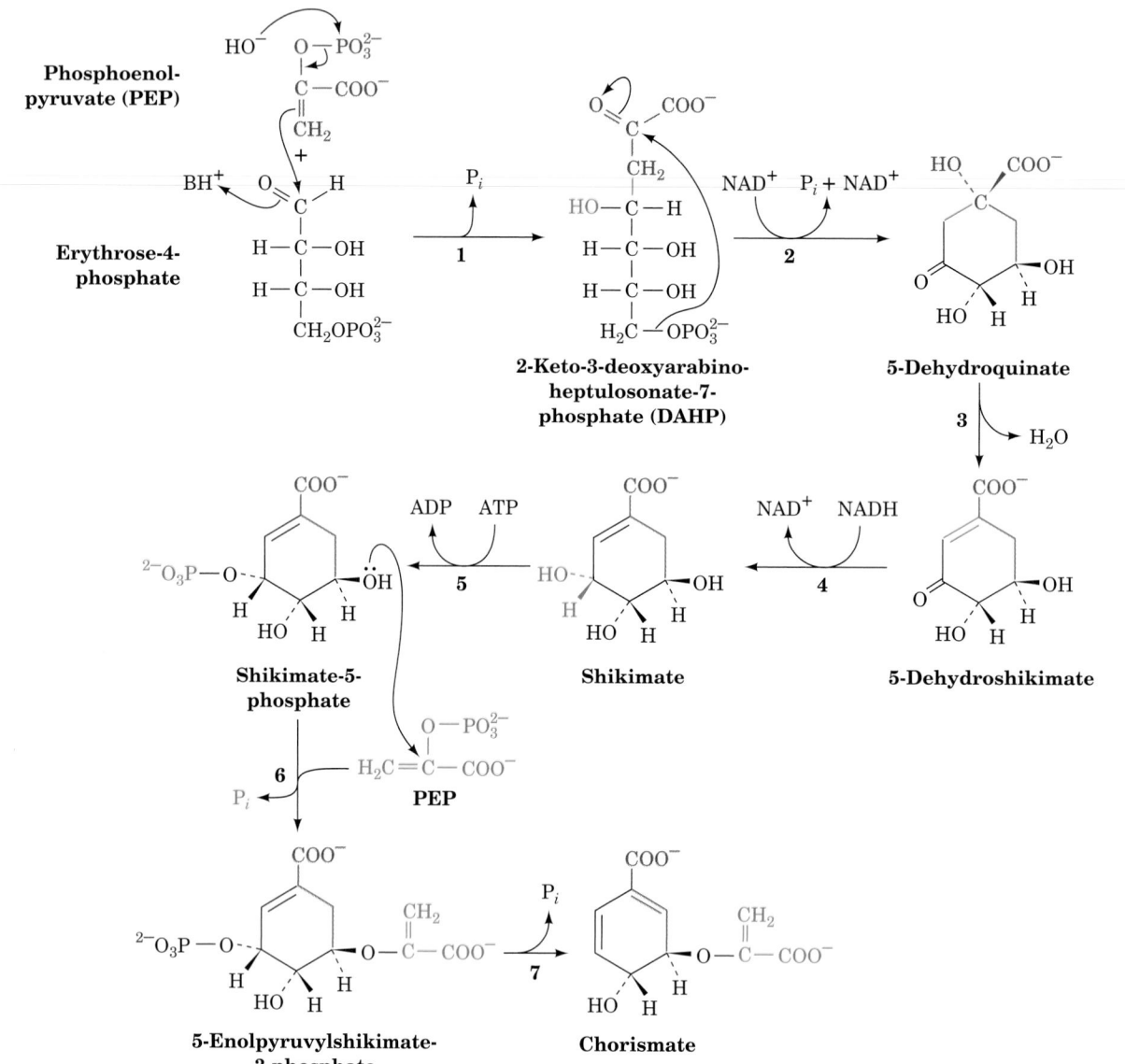

FIGURE 26-62   The biosynthesis of chorismate, the aromatic amino acid precursor. The pathway enzymes are **(1)** 2-keto-3-deoxy-ᴅ-arabinoheptulosonate-7-phosphate synthase, **(2)** dehydroquinate synthase (an NAD⁺-requiring reaction that yields an unchanged NAD⁺ product and is thereby indicative of an oxidized intermediate as similarly occurs in the UDP–galactose-4-epimerase reaction; Section 17-5B), **(3)** 5-dehydroquinate dehydratase, **(4)** shikimate dehydrogenase, **(5)** shikimate kinase, **(6)** 5-enolpyruvylshikimate-3-phosphate synthase, and **(7)** chorismate synthase.

and their substrate affinities are increased by 1 to 2 orders of magnitude. Indole, the intermediate product, does not appear free in solution; the enzyme apparently sequesters it.

The X-ray structure of tryptophan synthase from *Salmonella typhimurium*, determined by Craig Hyde, Edith Miles, and David Davies, explains the latter observation. The protein forms a 150-Å-long, 2-fold symmetric α–β–β–α complex (Fig. 26-64) in which the active sites of neighboring α and β subunits are separated by ~25 Å. *These active sites are joined by a solvent-filled tunnel that is wide enough to permit the passage of the intermediate substrate, indole.* This structure, the first in which the presence of a tunnel between active sites was observed, suggests the following series of events. The indole-3-glycerol phosphate substrate binds to the α subunit through an opening into its active site, its "front door," and the glyceraldehyde-3-phosphate product leaves via the same route. Similarly, the β subunit active site has a "front door" opening to the solvent through which serine enters and tryptophan leaves. Both active sites also have "back doors" that

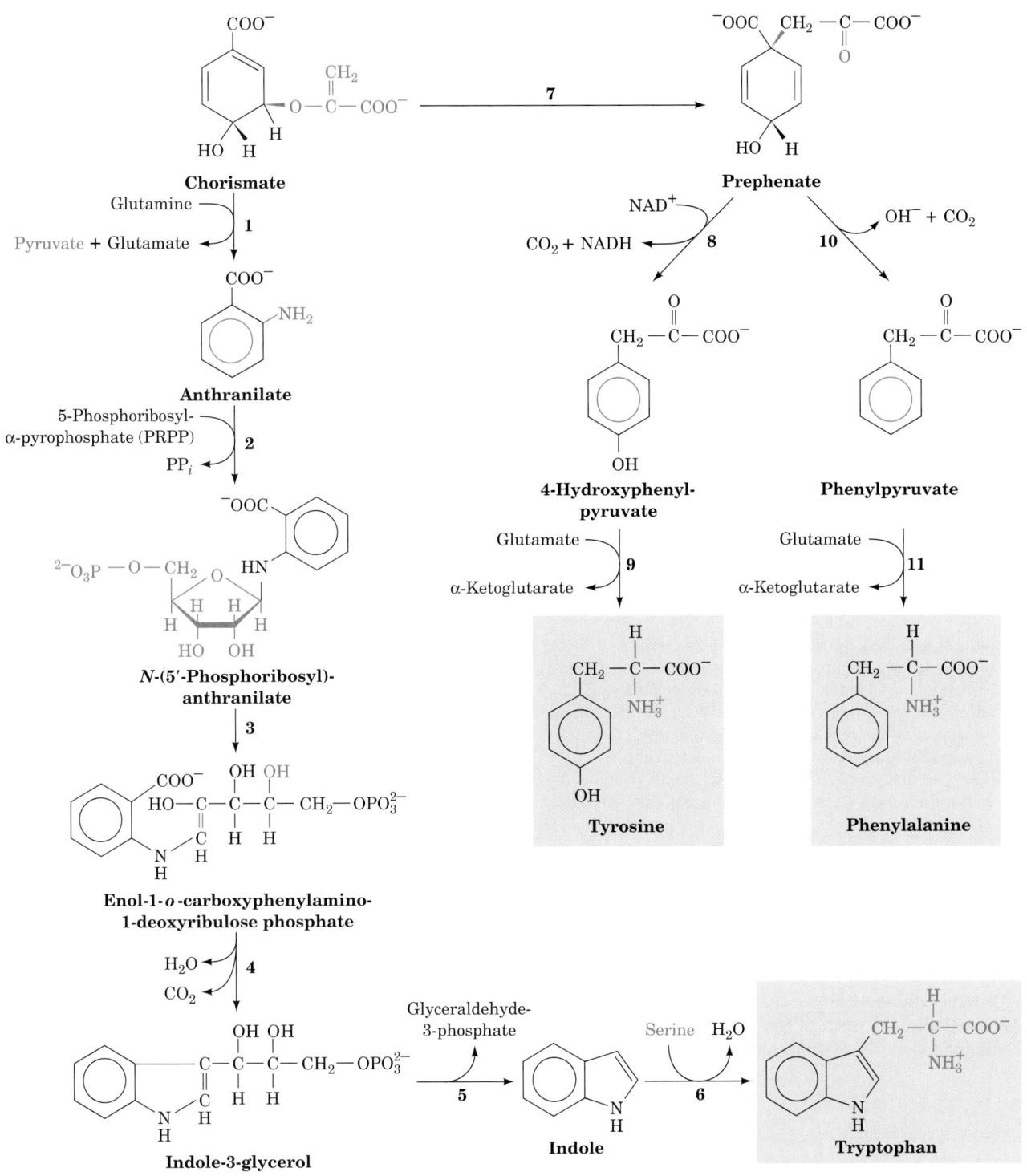

FIGURE 26-63 The biosynthesis of phenylalanine, tryptophan, and tyrosine from chorismate. The pathway enzymes are **(1)** anthranilate synthase, **(2)** anthranilate phosphoribosyltransferase, **(3)** *N*-(5′-phosphoribosyl) anthranilate isomerase, **(4)** indole-3-glycerol phosphate synthase, **(5)** tryptophan synthase, α subunit, **(6)** tryptophan synthase, β subunit, **(7)** chorismate mutase, **(8)** prephenate dehydrogenase, **(9)** aminotransferase, **(10)** prephenate dehydratase, and **(11)** aminotransferase.

are connected by the tunnel. The indole intermediate presumably diffuses between the two active sites via the tunnel and hence does not escape to the solvent.

Allosteric interactions between the subunits to control the activity of the α subunit also serve to ensure that indole is only released when the β subunit is ready to accept

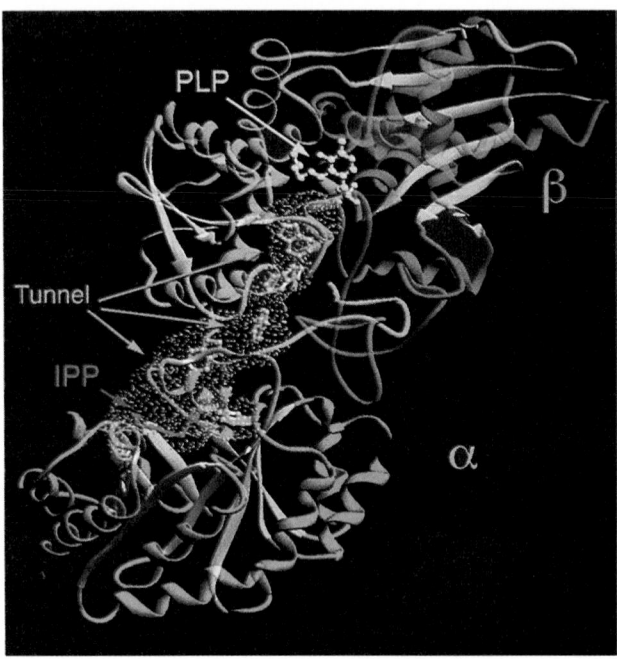

FIGURE 26-64   A ribbon diagram of the bifunctional enzyme tryptophan synthase from S. typhimurium. Only one αβ protomer of this 2-fold symmetric αββα heterotetramer is shown. The α subunit is blue, the β subunit's N-terminal domain is orange, its C-terminal domain is red, and all β sheets are tan. The active site of the α subunit is located by its bound competitive inhibitor, **indolepropanol phosphate (IPP;** *red ball-and-stick model*), whereas that of the β subunit is marked by its PLP coenzyme (*yellow ball-and-stick model*). The solvent-accessible surface of the ~25-Å-long "tunnel" connecting the active sites of the α and β subunits is outlined by a yellow dot surface. Several indole molecules (*green ball-and-stick models*) have been modeled into the tunnel in head to tail fashion, thereby demonstrating that the tunnel has sufficient width to permit the indole product of the α subunit to pass through the tunnel to the β subunit's active site. [Courtesy of Craig Hyde, National Institutes of Health.] *☞ See the Interactive Exercises*

it. Michael Dunn has shown that the elimination of water from the serine–PLP Schiff base on the β subunit to form an aminoacrylate–PLP Schiff base intermediate

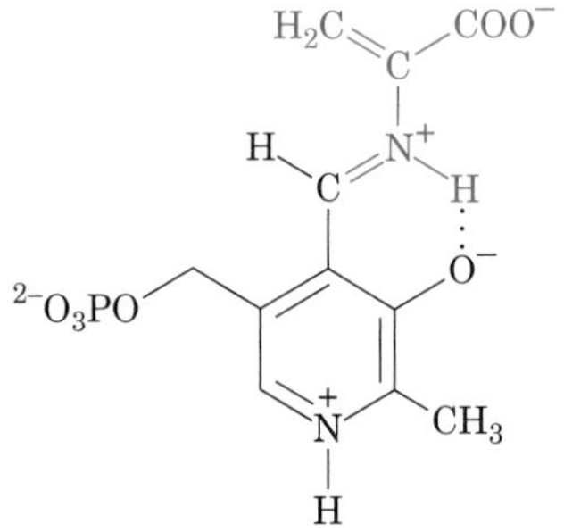

Aminoacrylate–PLP Schiff base

triggers a conformation change that activates the α subunit to produce indole. The diffusion of the indole to the β sub-

FIGURE 26-65   (*Opposite*)   **The biosynthesis of histidine.** The pathway enzymes are **(1)** ATP phosphoribosyltransferase, **(2)** pyrophosphohydrolase, **(3)** phosphoribosyl–AMP cyclohydrolase, **(4)** phosphoribosylformimino-5-aminoimidazole carboxamide ribonucleotide isomerase, **(5)** imidazole glycerol phosphate synthase (a glutamine amidotransferase), **(6)** imidazole glycerol phosphate dehydratase, **(7)** L-histidinol phosphate aminotransferase, **(8)** histidinol phosphate phosphatase, and **(9)** histidinol dehydrogenase.

unit to react with this intermediate then results in the formation of tryptophan.

Channeling may be particularly important for indole since this nonpolar molecule otherwise can escape the bacterial cell by diffusing through its plasma and outer membranes. We have seen similar phenomena in reactions involving glutamine amidotransferases (Sections 26-2A and 26-5A), as well as in the series of reactions catalyzed by fatty acid synthase, in which the growing product is kept in the vicinity of the multifuntional enzyme's active site by covalent attachment to the enzyme's flexible phospho-pantetheine arm (Section 25-4C). Channeling is also implicated in the multistep biosyntheses of purines and pyrimidines (Sections 28-1A and 28-2A).

e. Histidine Biosynthesis

Five of histidine's six C atoms are derived from **5-phos-phoribosyl-α-pyrophosphate (PRPP;** Fig. 26-65), an intermediate also involved in the biosynthesis of tryptophan (Fig. 26-63, Reaction 2), purine nucleotides (Section 28-1A), and pyrimidine nucleotides (Section 28-2A). The histidine's sixth carbon originates from ATP. The ATP atoms that are not incorporated into histidine are eliminated as **5-aminoimidazole-4-carboxamide ribonucleotide** (Fig. 26-65, Reaction 5), which is also an intermediate in purine biosynthesis (Section 28-1A).

The unusual biosynthesis of histidine from a purine has been cited as evidence supporting the hypothesis that life was originally RNA based (Section 1-5C). His residues, as we have seen, are often components of enzyme active sites, where they act as nucleophiles and/or general acid–base catalysts. The discovery that RNA can have catalytic properties (Section 31-4A) therefore suggests that the imidazole moiety of purines plays a similar role in these RNA enzymes **(ribozymes).** This further suggests that the histidine biosynthesis pathway is a "fossil" of the transition to more efficient protein-based life-forms.

6 ■ NITROGEN FIXATION

The most prominent chemical elements in living systems are O, H, C, N, and P. The elements O, H, and P occur widely in metabolically available forms (e.g., H_2O, O_2, and P_i). However, the major available forms of C and N, CO_2 and N_2, are extremely stable (unreactive); for example, the N≡N triple bond has a bond energy of 945 kJ · mol^{-1} (ver-

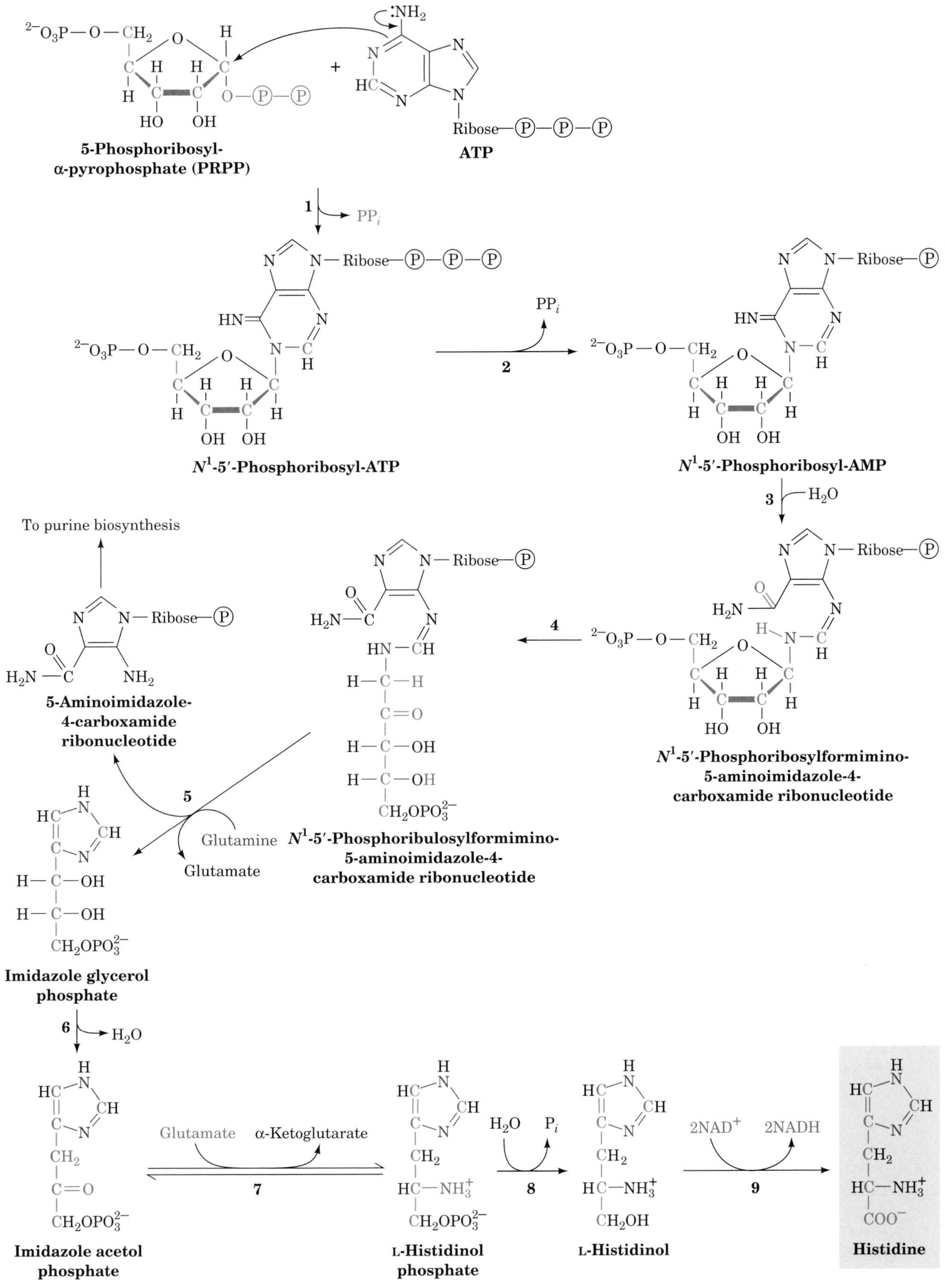

$^{2-}O_3P$—O—CH_2 ... **5-Phosphoribosyl-α-pyrophosphate (PRPP)**

$:NH_2$ ... **ATP** ... Ribose—P—P—P

1 → PP_i

N—Ribose—P—P—P ... $^{2-}O_3P$—O—CH_2 ... **N^1-5′-Phosphoribosyl-ATP**

2 → PP_i

N—Ribose—P ... $^{2-}O_3P$—O—CH_2 ... **N^1-5′-Phosphoribosyl-AMP**

3 → H_2O

N—Ribose—P ... H_2N ... $^{2-}O_3P$—O—CH_2 ... **N^1-5′-Phosphoribosylformimino-5-aminoimidazole-4-carboxamide ribonucleotide**

4 ←

N—Ribose—P ... H_2N—C ... HN—CH ... H—C—H ... C=O ... H—C—OH ... H—C—OH ... $CH_2OPO_3^{2-}$ ... **N^1-5′-Phosphoribulosylformimino-5-aminoimidazole-4-carboxamide ribonucleotide**

To purine biosynthesis

N—Ribose—P ... H_2N—C ... NH_2 ... **5-Aminoimidazole-4-carboxamide ribonucleotide**

5 ... Glutamine → Glutamate

HC ... N ... CH ... N ... H—C—OH ... H—C—OH ... $CH_2OPO_3^{2-}$ ... **Imidazole glycerol phosphate**

6 → H_2O

HC ... N ... CH ... N ... CH_2 ... C=O ... $CH_2OPO_3^{2-}$ ... **Imidazole acetol phosphate**

Glutamate → α-Ketoglutarate ... **7** ⇌

HC ... N ... CH ... N ... CH_2 ... HC—NH_3^+ ... $CH_2OPO_3^{2-}$ ... **L-Histidinol phosphate**

H_2O → P_i ... **8**

HC ... N ... CH ... N ... CH_2 ... HC—NH_3^+ ... CH_2OH ... **L-Histidinol**

$2NAD^+$ → $2NADH$ ... **9**

HC ... N ... CH ... N ... CH_2 ... HC—NH_3^+ ... COO^- ... **Histidine**

sus 351 kJ · mol⁻¹ for a C—O single bond). CO_2, with only minor exceptions, is metabolized (fixed) only by photosynthetic organisms (Chapter 24). *N_2 fixation is even less common; this element is converted to metabolically useful forms by only a few strains of bacteria, named* **diazatrophs.**

Diazatrophs of the genus *Rhizobium* live in symbiotic relationship with root nodule cells of legumes (plants belonging to the pea family, including beans, clover, and alfalfa; Fig. 26-66) where they convert N_2 to NH_3:

$$N_2 + 8H^+ + 16ATP + 16H_2O + 8e^- \rightarrow$$
$$2NH_3 + H_2 + 16ADP + 16P_i$$

The NH_3 thus formed can be incorporated either into glutamate by glutamate dehydrogenase (Section 26-1B) or into glutamine by glutamine synthetase (Section 26-5A). This nitrogen-fixing system produces more metabolically useful nitrogen than the legume needs; the excess is excreted into the soil, enriching it. It is therefore common agricultural practice to plant a field with alfalfa every few years to build up the supply of usable nitrogen in the soil for later use in growing other crops.

a. Nitrogenase Contains Novel Redox Centers

Nitrogenase, which catalyzes the reduction of N_2 to NH_3, is a complex of two proteins:

1. The **Fe-protein,** a homodimer that contains one [4Fe–4S] cluster and two ATP binding sites.

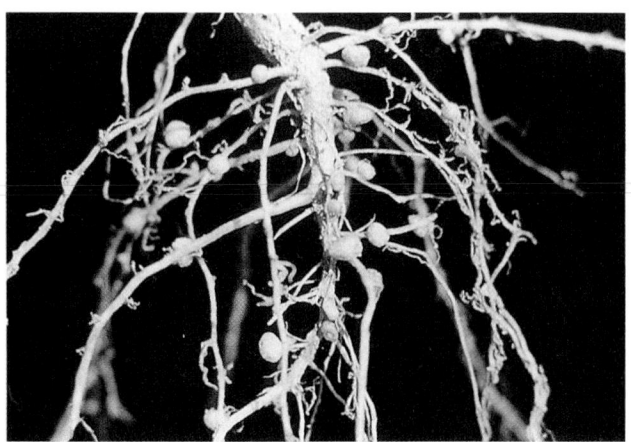

FIGURE 26-66   Photograph showing the root nodules of the legume bird's foot trefoil. [Vu/Cabisco/Visuals Unlimited.]

2. The **MoFe-protein,** an $\alpha_2\beta_2$ heterotetramer that contains Fe and Mo.

The X-ray structure of *Azotobacter vinelandii* nitrogenase in complex with the inhibitor ADP · AlF₄⁻ (which mimics the transition state in ATP hydrolysis), determined by Douglas Rees, reveals that each MoFe-protein associates with two molecules of Fe-protein (Fig. 26-67).

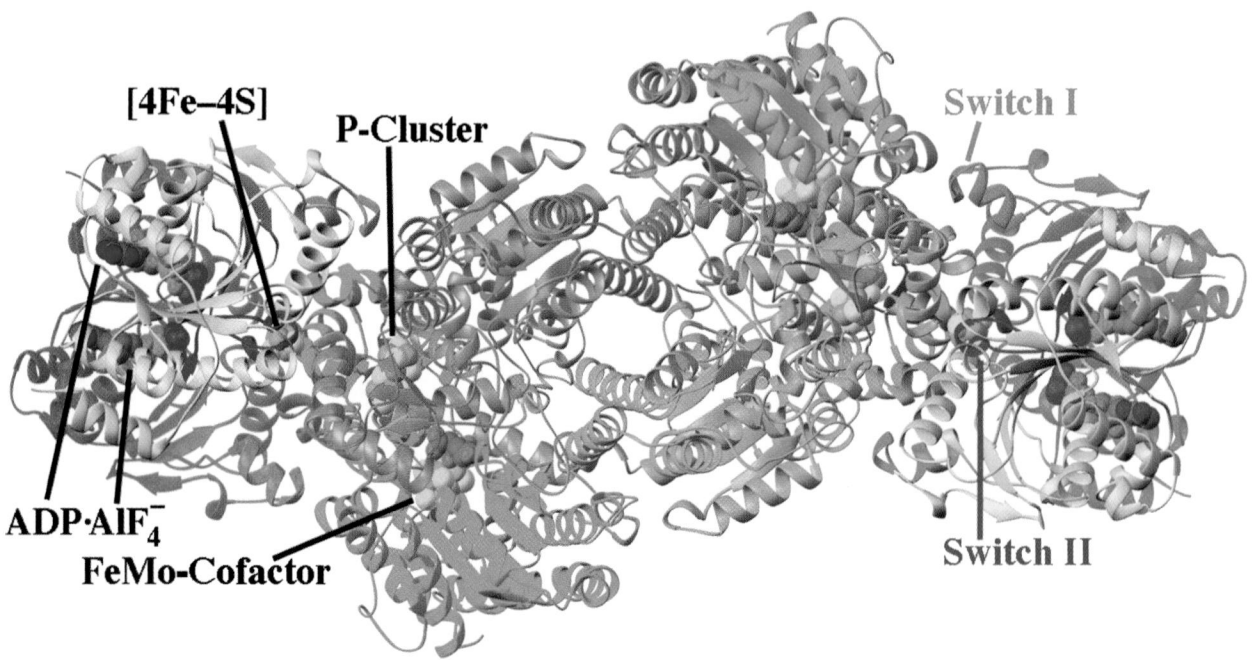

FIGURE 26-67   X-Ray structure of the *A. vinelandii* nitrogenase in complex with ADP · AlF₄⁻. The enzyme, which is viewed along its molecular 2-fold axis, is an $(\alpha\beta\gamma_2)_2$ heterooctamer in which the β–α–α–β assembly, the MoFe-protein, is flanked by two γ_2 Fe-proteins whose 289-residue subunits are related by local 2-fold symmetry. The homologous α subunits (*cyan and magenta;* 491 residues) and β subunits (*light red and light blue;* 522 residues) are related by pseudo-2-fold symmetry. The two γ subunits forming each Fe-protein (*pink and light green with their Switch I and Switch II segments red and blue*) bind to the MoFe-protein with the 2-fold axis relating them coincident with the pseudo-2-fold axis relating the MoFe-protein's α and β subunits. The ADP · AlF₄⁻, [4Fe–4S] cluster, FeMo-cofactor, and P-cluster are drawn in space-filling form with C green, N blue, O red, S yellow, Fe orange, Mo pink, and the AlF₄⁻ ion purple. [Based on an X-ray structure by Douglas Rees, California Institute of Technology. PDBid 1N2C.]

Each Fe-protein dimer's single [4Fe–4S] cluster is located in a solvent-exposed cleft between the two subunits and is symmetrically linked to Cys 97 and Cys 132 from both subunits such that an Fe-protein resembles an "iron butterfly" with the [4Fe–4S] cluster at its head. Its nucleotide binding sites are located at the interface between its two subunits.

The MoFe-protein's α and β subunits assume similar folds and extensively associate to form a pseudo-2-fold symmetric αβ dimer, two of which more loosely associate to form the 2-fold symmetric $\alpha_2\beta_2$ tetramer (Fig. 26-67). Each αβ dimer has two bound redox centers:

1. The **P-cluster** (Fig. 26-68*a,b*), which consists of two [4Fe–3S] clusters linked through an additional sulfide ion forming the eighth corner of each of the clusters to make cubane-like structures, and bridged by two Cys thiol ligands, each coordinating one Fe from each cluster. Four additional Cys thiols coordinate the remaining four Fe atoms. The positions of two of the Fe atoms in one of the [4Fe–3S] clusters change on oxidation, rupturing the bonds from these Fe atoms to the linking sulfide ion. These bonds are replaced in the oxidized state by a Ser oxygen ligand to

one of the Fe atoms, and by a bond to the amide N of a Cys from the other Fe atom.

2. The **FeMo-cofactor** (Fig. 26-68*c*), which consists of a [4Fe–3S] cluster and a [1Mo–3Fe–3S] cluster bridged by three sulfide ions. The FeMo-cofactor's Mo atom is approximately octahedrally coordinated by three cofactor sulfide ions, a His imidazole nitrogen, and two oxygens from a bound **homocitrate** ion:

$$
\begin{array}{c}
COO^- \\
| \\
CH_2 \\
| \\
CH_2 \\
| \\
HO-C-COO^- \\
| \\
CH_2 \\
| \\
COO^-
\end{array}
$$

Homocitrate

(an essential component of the FeMo-cofactor). The FeMo-cofactor contains a central cavity that a high resolution (1.16 Å) X-ray structure of *A. vinelandii* MoFe-pro-

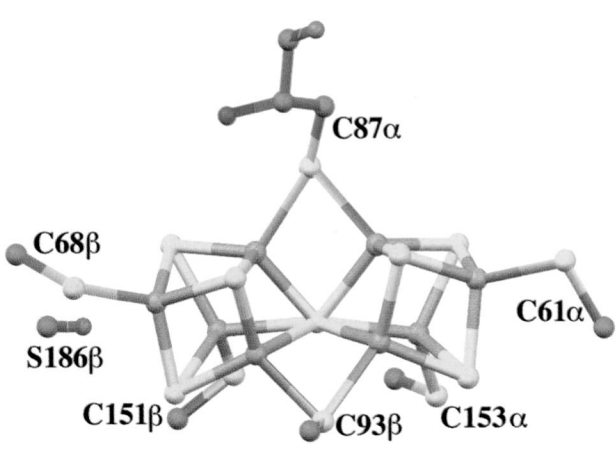

(a)

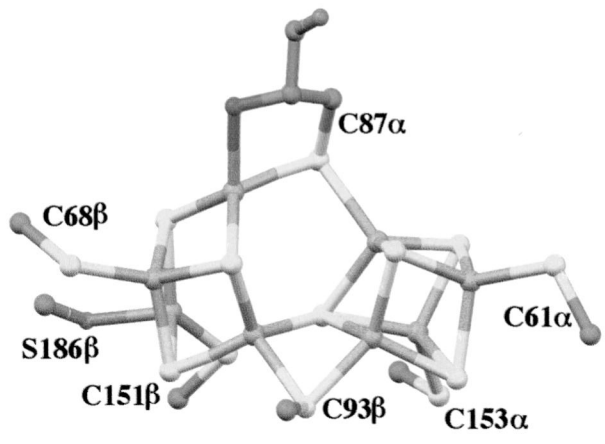

(b)

FIGURE 26-68 The prosthetic groups of the nitrogenase MoFe-protein. The molecules are drawn in ball-and-stick form with C green, N blue, O red, S yellow, Fe orange, and Mo pink. (*a*) The reduced *Klebsiella pneumoniae* P-cluster. It consists of two [4Fe–3S] complexes linked by an additional sulfide ion forming the eighth corner of each cubane-like structure, and bridged by two Cys thiol ligands, each coordinating one Fe from each cluster. Four additional Cys thiols coordinate the remaining 4 Fe atoms. (*b*) The 2-electron-oxidized *K. pneumoniae* P-cluster. In comparison with the reduced complex in Part *a*, two of the Fe—S bonds from the centrally located sulfide ion that bridges the two [4Fe–3S] clusters have been replaced by ligands from the Cys 87α amide N and the Ser 186β side chain O yielding a [4Fe–3S] cluster (*left*) and a [4Fe–4S] cluster (*right*) that remain linked by a direct Fe—S bond and two bridging Cys thiols. (*c*) The *A. vinelandi* FeMo-cofactor. It consists of a [4Fe–3S] cluster and a [1Mo–3Fe–3S] cluster that are bridged by three sulfide ions. The FeMo-cofactor is linked to the protein by only two ligands at its opposite ends, one from His 442α to the Mo atom and the other from Cys 275α to an Fe atom. The Mo atom is

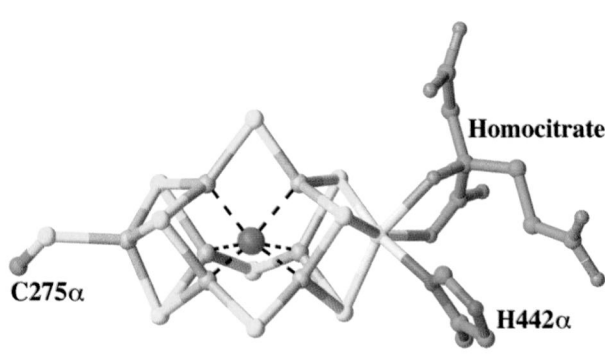

(c)

additionally doubly liganded by homocitrate. What is most likely an N atom (*blue sphere*) is liganded to the FeMo-cluster's six central Fe atoms (*dashed black lines*). [Parts *a* and *b* based on X-ray structures by David Lawson, John Innes Centre, Norwich, U.K. Part *c* based on an X-ray structure by Douglas Rees, California Institute of Technology. PDBids (*a*) 1QGU, (*b*) 1QH1, and (*c*) 1M1N.]

tein, also determined by Rees, reveals contains what most probably is a nitrogen atom (although a C or an O atom cannot be ruled out). This putative N atom is liganded to the FeMo-cofactor's central six Fe atoms such that it completes the approximate tetrahedral coordination environment of each of these Fe atoms.

The FeMo-cofactor is located ~10 Å below the α subunit surface, and hence the N_2 is thought to gain access to its binding site through conformational fluctuations of the protein (recall that myoglobin and hemoglobin likewise have no clear path for O_2 to approach its heme binding sites in these proteins; Section 10-2). The P-cluster, which is also ~10 Å below the protein surface, is at the interface between the α and β subunits on the pseudo-2-fold axis that roughly relates these two subunits. The 2-fold axis of the Fe-protein and the pseudo-2-fold axis of the MoFe-proteins coincide in their complex.

Nitrogenase hydrolyzes two ATP molecules for each electron it transfers. Since the nucleotide binding sites and the [4Fe–4S] cluster on the Fe-protein are separated by ~20 Å, a distance too large for direct coupling between electron transfer and ATP hydrolysis, it appears that these processes are allosterically coupled through conformational changes at the subunit interface. Indeed, portions of the Fe-protein resemble those of G-proteins, in which nucleotide hydrolysis is coupled to conformational changes controlling the protein's actions (Sections 19-2C and 19-3C). Specifically, two regions of the Fe-protein, designated Switch I and Switch II (Fig. 26-67), are homologous with those of Ras (Section 19-3C). The binding of ADP · AlF_4^- to Fe-protein induces conformational changes in Switch I that affect the interactions between the Fe-protein and the MoFe-protein, and in Switch II that affect the environment of the [4Fe–4S] cluster.

In nitrogenase, the [4Fe–4S] cluster of the Fe-protein approaches within ~14 Å of the P-cluster in the MoFe-protein, whereas the P-cluster and the FeMo-cofactor are ~13 Å apart. Hence, the sequence of the electron transfer steps in the nitrogenase reaction appears to be

[4Fe–4S] cluster → P-cluster → FeMo-cofactor → N_2

It therefore seems that the role of ATP hydrolysis is to stabilize a conformation in the Fe-protein that it cannot achieve on its own and which facilitates electron transfer from the [4Fe–4S] cluster on the Fe-protein to the P-cluster on the MoFe-protein.

b. N_2 Reduction Is Energetically Costly

Nitrogen fixation requires two participants in addition to N_2 and nitrogenase: (1) a source of electrons and (2) ATP. Electrons are generated either oxidatively or photosynthetically, depending on the organism. These electrons are transferred to ferredoxin (Section 22-2C), a [4Fe–4S]-containing electron carrier that transfers an electron to the Fe-protein of nitrogenase, beginning the nitrogen fixation process (Fig. 26-69). Two molecules of ATP bind to the reduced Fe-protein and are hydrolyzed as each electron is passed from the Fe-protein to the MoFe-protein. The ATP hydrolysis-induced conformational change in the Fe-protein alters its redox potential from –0.29 to –0.40 V, making the electron capable of N_2 reduction ($\mathscr{E}^{\circ\prime}$ = –0.34 V for the half-cell $N_2 + 6H^+ + 6e^- \rightleftharpoons 2NH_3$).

The actual reduction of N_2 occurs on the MoFe-protein in three discrete steps, each involving an electron pair:

$$2H^+ + 2e^- \qquad 2H^+ + 2e^- \qquad 2H^+ + 2e^-$$

$$N\equiv N \xrightarrow{\quad} H-N=N-H \xrightarrow{\quad} \underset{H}{\overset{H}{\underset{\diagdown}{N}}}-\underset{H}{\overset{H}{\underset{\diagup}{N}}} \xrightarrow{\quad} 2NH_3$$

Diimine　　　　**Hydrazine**

An electron transfer must occur six times per N_2 molecule fixed so that a total of 12 ATPs are required to fix one N_2 molecule. Although the N_2 binding site is almost certainly the FeMo-cofactor, exactly how the N_2 is bound and reduced are largely a matter of speculation. Theoretical studies suggest that the FeMo-cofactor's prismatically arranged Fe atoms provide favorable interaction sites for N_2 and its reduction products. Indeed, it seems highly likely that the putative N atom that is liganded to the FeMo-cofactor (Fig. 26-68c) participates in N_2 reduction.

Nitrogenase also reduces H_2O to H_2, which in turn reacts with **diimine** to reform N_2:

$$HN=NH + H_2 \rightarrow N_2 + 2H_2$$

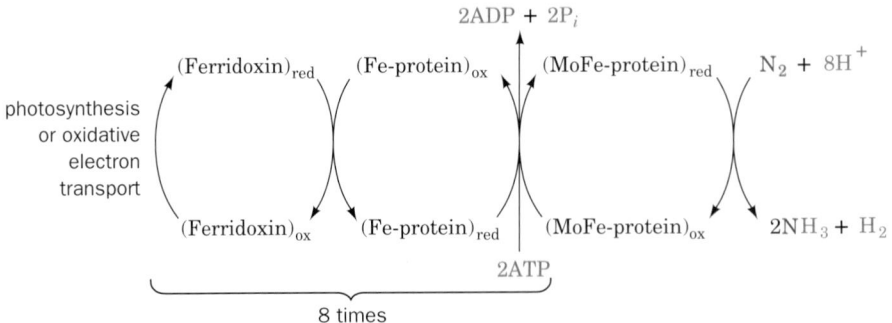

FIGURE 26-69 **The flow of electrons in the nitrogenase-catalyzed reduction of N_2.**

The resulting futile cycle is favored when the ATP level is low and/or the reduction of the Fe-protein is sluggish. Even when ATP is plentiful, however, the cycle cannot be suppressed beyond about one H_2 molecule produced per N_2 reduced and hence appears to be a requirement of the nitrogenase reaction. The total cost of N_2 reduction is therefore 8 electrons transferred and 16 ATPs hydrolyzed (physiologically, 20–30 ATPs). Hence nitrogen fixation is an energetically expensive process; indeed, the nitrogen-fixing bacteria in the root nodules of pea plants consume nearly 20% of the ATP that the plant produces.

c. Leghemoglobin Protects Nitrogenase from Oxygen Inactivation

Nitrogenase is rapidly inactivated by O_2, so the enzyme must be protected from this reactive substance. Cyanobacteria (photosynthetic oxygen-evolving bacteria; Section 1-1A) provide protection by carrying out nitrogen fixation in specialized nonphotosynthetic cells called **heterocysts,** which have Photosystem I but lack Photosystem II (Section 24-2C). In the root nodules of legumes (Fig. 26-66), however, protection is afforded by the symbiotic synthesis of **leghemoglobin.** The globin portion of this ~145-residue monomeric oxygen-binding protein is synthesized by the plant (an evolutionary curiosity since globins are otherwise known to occur only in animals), whereas the heme is synthesized by the *Rhizobium.* Leghemoglobin has a very high O_2 affinity, thus keeping the pO_2 low enough to protect the nitrogenase while providing passive O_2 transport for the aerobic bacterium.

d. Installing the Nitrogen Fixation Machinery in Nonleguminous Plants Would Revolutionize Agriculture

Although atmospheric N_2 is the ultimate nitrogen source for all living things, most plants do not support the symbiotic growth of nitrogen-fixing bacteria. They must therefore depend on a source of "prefixed" nitrogen such as nitrate or ammonia. These nutrients come from lightning discharges (the source of ~10% of naturally fixed N_2), decaying organic matter in the soil, or from fertilizer applied to it. The Haber process, which was invented by Fritz Haber in 1910, is a chemical process for N_2 fixation that is still widely used in fertilizer manufacture. This direct reduction of N_2 by H_2 to form NH_3 requires temperatures of 300 to 500°C, pressures of >300 atm, and an Fe catalyst. Intriguingly, the spacing of the Fe atoms on the surface of this catalyst resembles that of the FeMo-cofactor's central Fe atoms (Fig. 26-68c).

One of the major long-term goals of genetic engineering is to induce agriculturally useful nonleguminous plants to fix their own nitrogen, a complex undertaking in which the plant must be made to provide a hospitable environment for nitrogen fixation as well as to acquire the enzymatic machinery to do so. This would free farmers, particularly those in developing countries, from the need for either purchasing fertilizers, periodically letting their fields lie fallow (giving legumes the opportunity to grow), or following the slash-and-burn techniques that are rapidly destroying the world's tropical forests and contributing significantly to the greenhouse effect (atmospheric CO_2 pollution causing long-term global warming).

■ CHAPTER SUMMARY

1 ■ Amino Acid Deamination Amino acids are the precursors for numerous nitrogen-containing compounds such as heme, physiologically active amines, and glutathione. Excess amino acids are converted to common metabolic intermediates for use as fuels. The first step in amino acid breakdown is removal of the α-amino group by transamination. Transaminases require pyridoxal phosphate (PLP) and convert amino acids to their corresponding α-keto acids. The amino group is transferred to α-ketoglutarate to form glutamate, oxaloacetate to form aspartate, or pyruvate to form alanine. Glutamate is subsequently oxidatively deaminated by glutamate dehydrogenase (GDH) to form ammonia and regenerate α-ketoglutarate. Hyperinsulinism/hyperammonemia (HI/HA), a genetic disease, is caused by a mutation of the GDH gene that decreases GTP's ability to inhibit GDH.

2 ■ The Urea Cycle In the urea cycle, amino groups from NH_3 and aspartate combine with HCO_3^- to form urea. This pathway takes place in the liver, partially in the mitochondrion and partially in the cytosol. It begins with the ATP-dependent condensation of NH_3 and HCO_3^- by carbamoyl phosphate synthetase, an enzyme with a 96-Å-long tunnel connecting its three active sites through which its highly reactive intermediate products are channeled. The resulting carbamoyl phosphate then combines with ornithine to yield citrulline, which combines with aspartate to form argininosuccinate, which in turn is cleaved to fumarate and arginine. The arginine is then hydrolyzed to urea, which is excreted, and ornithine, which reenters the urea cycle. *N*-Acetylglutamate regulates the urea cycle by activating carbamoyl phosphate synthetase allosterically.

3 ■ Metabolic Breakdown of Individual Amino Acids The α-keto acid products of transamination reactions are degraded to citric acid cycle intermediates or their precursors. The amino acids leucine and lysine are ketogenic in that they are converted only to the ketone body precursors acetyl-CoA and acetoacetate. The remaining amino acids are, at least in part, glucogenic in that they are converted to the glucose precursors pyruvate, oxaloacetate, α-ketoglutarate, succinyl-CoA, or fumarate. Alanine, cysteine, glycine, serine, and threonine are converted to pyruvate. Serine hydroxymethyltransferase catalyzes the PLP-dependent C_α—C_β bond cleavage of serine to form glycine. This reaction requires the transfer of a methylene group from N^5,N^{10}-methylenetetrahydrofolate, which the tetrahydrofolate (THF) obtains from the glycine cleavage system, a multienzyme system. Asparagine and aspartate are converted to oxaloacetate. α-Ketoglutarate is a product of arginine, glutamate, glutamine, histidine, and proline degradation. Methionine, isoleucine, and valine are degraded to succinyl-CoA. Methionine breakdown involves the synthesis of *S*-adenosylmethionine (SAM), a sulfonium ion that acts as a methyl donor in many biosynthetic reactions. Hyperhomocysteinemia, a risk factor for cardiovascular disease, cognitive impairment, and neural tube defects, is caused by a deficiency in its folate-dependent degradation. Maple syrup urine disease (MSUD) is caused by an inherited defect in branched-chain amino acid degradation. Branched-chain amino acid degradation pathways contain reactions common to all acyl-CoA oxidations. Tryptophan is degraded to alanine and acetoacetate. Phenylalanine and tyrosine are degraded to fumarate and acetoacetate. Most individuals with the hereditary disease phenylketonuria lack phenylalanine hydroxylase (PAH), which converts phenylalanine to tyrosine.

4 ■ Amino Acids as Biosynthetic Precursors Heme is synthesized from glycine and succinyl-CoA. These precursors condense to form δ-aminolevulinic acid (ALA), which cyclizes to form the pyrrole porphobilinogen (PBG). Four molecules of PBG condense to form uroporphyrinogen III, which then goes on to form heme, with the final reaction, the insertion of Fe(II) into protoporphyrin IX, catalyzed by ferrochelatase. Defects in heme biosynthesis, which are known as porphyrias, have a variety of bizarre symptoms. Heme is degraded to form linear tetrapyrroles, which are subsequently excreted as bile pigments. The hormones and neurotransmitters L-DOPA, epinephrine, norepinephrine, serotonin, γ-aminobutyric acid (GABA), and histamine are all synthesized from amino acid precursors. Glutathione, a tripeptide that is synthesized from glutamate, cysteine, and glycine, is involved in a variety of protective, transport, and metabolic processes. Tetrahydrofolate is a coenzyme that participates in the transfer of C_1 units.

5 ■ Amino Acid Biosynthesis Amino acids are required for many vital functions of an organism. Those amino acids that mammals can synthesize from common α-keto acid carbon skeletons and preformed α-amino nitrogen such as that of glutamate are known as nonessential amino acids; those that mammals must obtain from their diet are called essential amino acids. The biosynthesis of nonessential amino acids involves relatively simple pathways, whereas those forming the essential amino acids are generally more complex.

6 ■ Nitrogen Fixation Although the ultimate source of nitrogen for amino acid biosynthesis is atmospheric N_2, this nearly inert gas must first be reduced to a metabolically useful form, NH_3, by nitrogen fixation. This process occurs only in certain types of bacteria, one genus of which occurs in symbiotic relationship with legumes. N_2 is fixed in these organisms by an oxygen-sensitive enzyme, nitrogenase, that consists of two proteins: the Fe-protein dimer, which contains one [4Fe–4S] cluster and two ATP binding sites, and the MoFe-protein $\alpha_2\beta_2$ tetramer, which contains one P-cluster (consisting of two [4Fe–3S] clusters linked by a sulfide ion) and one FeMo-cofactor (a [4Fe–3S] cluster and a [1Mo–3Fe–3S] cluster bridged by three sulfide ions and coordinated with homocitrate) in each αβ dimer. These cofactors each function as two-electron carriers for the ATP-driven reduction of N_2 to NH_3.

REFERENCES

GENERAL

Bender, D.A., *Amino Acid Metabolism*, Wiley (1985).

Scriver, C.R., Beaudet, A.C., Sly, W.S., and Valle, D. (Eds.), *The Metabolic & Molecular Bases of Inherited Disease* (8th ed.), McGraw-Hill (2001). [Volume 2, Part 8 contains numerous chapters on defects in amino acid metabolism.]

Walsh, C., *Enzymatic Reaction Mechanisms*, Chapters 24 and 25, Freeman (1979). [Discusses reactions involving PLP, THF, and SAM cofactors.]

AMINO ACID DEAMINATION AND THE UREA CYCLE

Baker, P.J., Britton, K.L., Engel, P.C., Farrants, G.W., Lilley, K.S., Rice, D.W., and Stillman, T.J., Subunit assembly and active site location in the structure of glutamate dehydrogenase, *Proteins* **12,** 75–86 (1992). [The X-ray structure of the *Clostridium symbiosum* enzyme.]

Cohen, P.P., The ornithine–urea cycle: biosynthesis and regulation of carbamyl phosphate synthetase I and ornithine transcarbamylase, *Curr. Topics Cell. Reg.* **18,** 1–19 (1981). [An interesting historical review of the discovery of urea and the urea cycle, as well as a discussion of the cycle's regulation.]

Fang, J., Hsu, B.Y.L., MacMullen, M., Poncz, M., Smith, T.J., and Stanley, C.A., Expression, purification and characterization of human glutamate dehydrogenase (GDH) allosteric regulatory mutations, *Biochem. J.* **363,** 81–87 (2002).

Holden, H.M., Thoden, J.B., and Raushel, F.M., Carbamoyl phosphate synthetase: a tunnel runs through it, *Curr. Opin. Struct. Biol.* **8,** 679–685 (1998); *and* Thoden, J.B., Holden, H.M., Wesenberg, G., Raushel, F.M., and Rayment, I., Structure of carbamoyl phosphate synthetase: A journey of 96 Å from substrate to product, *Biochemistry* **36,** 6305–6316 (1997).

Jansonius, J.N., Structure, evolution and action of vitamin B_6-dependent enzymes, *Curr. Opin. Struct. Biol.* **8,** 759–769 (1998).

Jungas, R.L., Halperin, M.L., and Brosnan, J.T., Quantitative analysis of amino acid oxidation and related gluconeogenesis in humans, *Physiol. Rev.* **72,** 419–448 (1992).

Martell, A.E., Vitamin B_6 catalyzed reactions of α-amino and α-keto acids: model systems, *Acc. Chem. Res.* **22,** 115–124 (1989).

Meijer, A.J., Lamers, W.H., and Chamuleau, R.A.F.M., Nitrogen metabolism and ornithine cycle function, *Physiol. Rev.* **70,** 701–748 (1990).

Miles, E.W., Rhee, S., and Davies, D.R., The molecular basis of substrate channeling, *Biochemistry* **274,** 12193–12196 (1999).

Saeed-Kothe, A. and Powers-Lee, S.G., Specificity determining residues in ammonia- and glutamine-dependent carbamoyl phosphate synthetases, *J. Biol. Chem.* **277,** 7231–7238 (2002).

Smith, T.J., Peterson, P.E., Schmidt, T., Fang, J., and Stanley, C.A., Structures of bovine glutamate dehydrogenase complexes elucidate the mechanism of purine regulation, *J. Mol. Biol.* **307,** 707–720 (2001).

Smith, T.J., Schmidt, T., Fang, J., Wu, J., Siuzdak, G., and Stanley, C.A., The structure of apo human glutamate dehydrogenase details subunit communication and allostery, *J. Mol. Biol.* **318,** 765–777 (2002).

Stanley, C.A., Fang, J., Kutyna, K., Hsu, B.Y.L., Ming, J.E., Glaser, B., and Poncz, M., Molecular basis and characterization of the hyperinsulinism/hyperammonemia syndrome: predominance of mutations in exons 11 and 12 of the glutamate dehydrogenase gene, *Diabetes* **49,** 667–673 (2000).

Stipanuk, M.H. and Watford, M., Amino acid metabolism, Chap. 11, *in* Stipanuk, M.H. (Ed.), *Biochemical and Physiological Basis of Nutrition,* Saunders (2000).

Torchinsky, Yu.M., Transamination: its discovery, biological and chemical aspects (1937–1987), *Trends Biochem. Sci.* **12**, 115–117 (1987).

METABOLIC BREAKDOWN OF INDIVIDUAL AMINO ACIDS

Anderson, O.A., Flatmark, T., and Hough, E., High resolution crystal structures of the catalytic domain of human phenylalanine hydroxylase in its catalytically active Fe(II) form and binary complex with tetrahydrobiopterin, *J. Mol. Biol.* **314**, 279–291 (2001).

Ævarsson, A., Chuang, J.L., Wynn, R.M., Turley, S., Chuang, D.T., and Hol, W.G.J., Crystal structure of human branched-chain α-ketoacid dehydrogenase and the molecular basis of multienzyme complex deficiency in maple syrup urine disease, *Structure* **8**, 277–291 (2000); *and* Wynn, R.M., Davie, J.R., Chuang, J.L., Cote, C.D., and Chuang, D.T., Impaired assembly of E1 decarboxylase of the branched-chain α-ketoacid dehydrogenase complex in type IA maple syrup urine disease, *J. Biol. Chem.* **273**, 13110–13110 (1998).

Binda, C., Bossi, R.T., Wakatsuki, S., Arzt, S., Coda, A., Curti, B., Vanoni, M.A., and Mattevi, A., Cross-talk and ammonia channeling between active centers in the unexpected domain arrangement of glutamate synthase, *Structure* **8**, 1299–1308 (2000).

Douce, R., Bourguignon, J., Neuburger, M., and Rébeillé, F., The glycine decarboxylase system: a fascinating complex, *Trends Plant Sci.* **6**, 167–176 (2001).

Drennen, C.L., Huang, S., Drumond, J.T., Matthews, R., and Ludwig, M.L., How a protein binds B$_{12}$: A 3.0 Å X-ray structure of B$_{12}$-binding domains of methionine synthase, *Science* **266**, 1669–1674 (1994).

Faure, M., Rourguignon, J., Neuburger, M., Macherel, D., Sieker, L., Ober, R., Kahn, R., Cohen-Addad, C., and Douce, R., Interaction between the lipoamide-containing H-protein and the lipoamide dehydrogenase (L-protein) of the glycine decarboxylase multienzyme system, 2. Crystal structures of H- and L-proteins, *Eur. J. Biochem.* **267**, 2890–2898 (2000).

Guenther, B.D., Sheppard, C.A., Tran, P., Rozen, R., Matthews, R.G., and Ludwig, M.L., The structure and properties of methylenetetrahydrofolate reductase from *Escherichia coli* suggest how folate ameliorates human hyperhomocysteinemia, *Nature, Struct. Biol.* **6**, 359–365 (1999).

Guilhaudis, L., Simorre, J.-P., Blackledge, M., Marion, D., Gans, P., Neuburger, M., and Douce, R., Combined structural and biochemical analysis of the H–T complex in the glycine decarboxylase cycle: evidence for a destabilization mechanism of the H-protein, *Biochemistry* **39**, 4259–4266 (2000).

Huang, X., Holden, H.M., and Raushel, F.M., Channeling of substrates and intermediates in enzyme-catalyzed reactions, *Annu. Rev. Biochem.* **70**, 149–180 (2001).

Jansonius, J.N. and Vincent, M.G., Structural basis for catalysis by aspartate aminotransferase, *in* Jurnak, F.A. and McPherson, A. (Eds.), *Biological Macromolecules and Assemblies,* Vol. 3, *pp.* 187–285, Wiley (1987).

Kelly, A. and Stanley, C.A., Disorders of glutamate metabolism, *Mental Retard. Devel. Dis. Res. Rev.* **7**, 287–295 (2001).

Ludwig, M.L. and Matthews, R.G., Structure-based perspectives on B$_{12}$-dependent enzymes, *Annu. Rev. Biochem.* **66**, 269–313 (1997).

Medina, M.Á., Urdiales, J.L., and Amores-Sánchez, M.I., Roles of homocysteine in cell metabolism: Old and new functions, *Eur. J. Biochem.* **268**, 3871–3882 (2001).

Nichol, C.A., Smith, G.K., and Duch, D.S., Biosynthesis and metabolism of tetrahydrobiopterin and molybdopterin, *Annu. Rev. Biochem.* **54**, 729–764 (1985).

Spiro, T.G. and Kozlowski, P.M., Is the CO adduct of myoglobin bent, and does it matter?, *Acc. Chem. Res.* **34**, 137–144 (2001).

Swain, A.L., Jaskólski, M., Housset, D., Rao, J.K.M., and Wladower, A., Crystal structure of *Escherichia coli* L-asparaginase, an enzyme used in cancer therapy, *Proc. Natl. Acad. Sci.* **90**, 1474–1478 (1993).

Varughese, K.I., Skinner, M.M., Whiteley, J.M., Matthews, D.A., and Xuong, N.H., Crystal structure of rat liver dihydropteridine reductase, *Proc. Natl. Acad. Sci.* **89**, 6080–6084 (1992).

Zalkin, H. and Smith, J.L. Enzymes utilizing glutamine as an amide donor, *Adv. Enzymol.* **72**, 87–144 (1998).

AMINO ACIDS AS BIOSYNTHETIC PRECURSORS

Battersby, A.R., Tetrapyrroles: the pigments of life, *Nat. Prod. Rep.* **17**, 507–526 (2000).

Beru, N. and Goldwasser, E., The regulation of heme biosynthesis during erythropoietin-induced erythroid differentiation, *J. Biol. Chem.* **260**, 9251–9257 (1985).

Erskine, P.T., Newbold, R., Brindley, A.A., Wood, S.P., Shoolingin-Jordan, P.M., Warren, M.J., and Cooper J.B., The X-ray structure of yeast 5-aminolaevulinic acid dehydratase complexed with substrate and three inhibitors, *J. Mol. Biol.* **312**, 133–141 (2001).

Fitzpatrick, P.F., Tetrahydropterin-dependent amino acid hydroxylases, *Annu. Rev. Biochem.* **68**, 355–381 (1999).

Grandchamp, B., Beaumont, C., de Verneuil, H., and Nordmann, Y., Accumulation of porphobilinogen deaminase, uroporphyrinogen decarboxylase, and α- and β-globin mRNAs during differentiation of mouse erythroleukemic cells: effects of succinylacetone, *J. Biol. Chem.* **260**, 9630–9635 (1985).

Jaffe, E.K., Martins, J., Li, J., Kervinen, J., and Dunbrack, R.L., Jr., The molecular mechanism of lead inhibition of human porphobilinogen synthase, *J. Biol. Chem.* **276**, 1531–1537 (2001).

Jaffe, E.K., Kervinen, J., Martins, J., Stauffer, F., Neier, R., Wlodawer, A., and Zdanov, A., Species-specific inhibition of porphobilinogen synthase by 4-oxosebacic acid, *J. Biol. Chem.* **277**, 19792–19799 (2002).

Jordan, P.M. and Gibbs, P.N.B., Mechanism of action of 5-aminolevulinate dehydratase from human erythrocytes, *Biochem. J.* **227**, 1015–1021 (1985).

Louie, G.V., Brownlie, P.D., Lambert, R., Cooper, J.B., Blundell, T.L., Wood, S.P., Warren, M.J., Woodcock, S.C., and Jordan, P.M., Structure of porphobilinogen deaminase reveals a flexible multidomain polymerase with a single catalytic site, *Nature* **359**, 33–39 (1992).

Macalpine, I. and Hunter, R., Porphyria and King George III, *Sci. Am.* **221**(1), 38–46 (1969).

Meister, A., Glutathione metabolism and its selective modification, *J. Biol. Chem.* **263**, 17205–17208 (1988).

Padmanaban, G., Venkateswar, V., and Rangarajan, P.N., Haem as a multifunctional regulator, *Trends Biochem. Sci.* **14**, 492–496 (1989).

Pagola, S., Stephens, P.W., Bohle, D.S., Kosar, A.D., and Madsen, S.K., The structure of malaria pigment β-haematin, *Nature* **404**, 307–310 (2000).

Ponka, P. and Schulman, H.M., Acquisition of iron from transferrin regulates reticulocyte heme synthesis, *J. Biol. Chem.* **260**, 14717–14721 (1985).

Schneider-Yin, X., Gouya, L., Dorsey, M., Rüfenacht, U., Deybach, J.-C., and Ferreira, G.C., Mutations in the iron-sulfur cluster ligands of the human ferrochelatase lead to erythropoietic protoporphyria, *Blood* **96**, 1545–1549 (2000).

Sellers, V.M., Wu, K.-T., Dailey, T.A., and Dailey, H.A., Heme ferrochelatase: Characterization of substrate-iron binding and proton-abstracting residues, *Biochemistry* **40,** 9821–9827 (2001); *and* Wu, C.-K., Dailey, H.A., Rose, J.P., Burden, A., Sellers, V.M., and Wang, B.-C., The 2.0 Å structure of human ferrochelatase, the terminal enzyme of heme biosynthesis, *Nature Struct. Biol.* **8,** 156–160 (2001).

Shoolingin-Jordan, P.M., Warren, M.J., and Awan, S.J., Discovery that the assembly of the dipyrromethane cofactor of porpho-bilinogen deaminase holoenzyme proceeds initially by the reaction of preuroporphyrinogen with the apoenzyme, *Biochem. J.* **316,** 373–376 (1996).

Thunell, S., Porphyrins, porphyrin metabolism and porphyrias. I. Update, *Scand. J. Clin. Lab. Invest.* **60,** 509–540 (2000).

Warren, M.J. and Scott, A.I., Tetrapyrrole assembly and modification into the ligands of biologically functional cofactors, *Trends Biochem. Sci.* **15,** 486–491 (1990).

Wellems, T.E., How chloroquine works, *Nature* **355,** 108–109 (1992).

AMINO ACIDS BIOSYNTHESIS

Adams, E. and Frand, L., Metabolism of proline and the hydroxyprolines, *Annu. Rev. Biochem.* **49,** 1005–1061 (1980).

Chaudhuri, B.N., Lange, S.C., Myers, R.S., Chittur, S.V., Davisson, V.J., and Smith, J.L., Crystal structure of imidazole glycerol phosphate synthase: a tunnel through a $(\beta/\alpha)_8$ barrel joins two active sites, *Structure* **9,** 987–997 (2001).

Cooper, A.J.L., Biochemistry of sulfur-containing amino acids, *Annu. Rev. Biochem.* **52,** 187–222 (1983).

Eisenberg, D., Gill, H.S., Pfluegl, M.U., and Rotstein, S.H., Structure-function relationships of glutamine synthetases, *Biochim. Biophys. Acta* **1477,** 122–145 (2000); *and* Gill, H.S. and Eisenberg, D., The crystal structure of phosphinothricin in the active site of glutamine synthetase illuminates the mechanism of enzymatic inhibition, *Biochemistry* **40,** 1903–1912 (2001).

Herrmann, K.M. and Somerville, R.L. (Eds.), *Amino Acids: Biosynthesis and Genetic Regulation,* Addison–Wesley (1983).

Katagiri, M. and Nakamura, M., Animals are dependent on preformed α-amino nitrogen as an essential nutrient, *Life* **53,** 125–129 (2002).

Kishore, G.M. and Shah, D.M., Amino acid biosynthesis inhibitors as herbicides, *Annu. Rev. Biochem.* **57,** 627–663 (1988). [Discusses the biosynthesis of the essential amino acids.]

Larsen, T.M., Boehlein, S.K., Schuster, S.M., Richards, N.G.J., Thoden, J.B., Holden, H.M., and Rayment, I., Three-dimensional structure of *Escherichia coli* asparagine synthetase B: a short journey from substrate to product, *Biochemistry* **38,** 16146–16167 (1999).

Miles, E.W., Structural basis for catalysis by tryptophan synthase, *Adv. Enzymol.* **64,** 93–172 (1991); Hyde, C.C. and Miles, E.W., The tryptophan synthase multienzyme complex: Exploring structure-function relationships with X-ray crystallography and mutagenesis, *Biotechnology* **8,** 27–32 (1990); *and* Hyde, C.C., Ahmed, S.A., Padlan, E.A., Miles, E.W., and Davies, D.R., Three-dimensional structure of the tryptophan synthase

α₂β₂ multienzyme complex from *Salmonella typhimurium, J. Biol. Chem.* **263,** 17857–17871 (1988).

Pan, P., Woehl, E., and Dunn, M.F., Protein architecture, dynamics and allostery in tryptophan synthase channeling, *Trends Biochem. Sci.* **22,** 22–27 (1997).

Stadtman, E.R., The story of glutamine synthetase regulation, *J. Biol. Chem.* **276,** 44357–44364 (2001).

Stallings, W.C., Abdel-Meguid, S.S., Lim, L.W., Shieh, H.-S., Dayringer, H.E., Leimgruber, N.K., Stegeman, R.A., Anderson, K.S., Sikorski, J.A., Padgette, S.R., and Kishore, G.M., Structure and topological symmetry of the glyphosate target 5-*enol*-pyruvylshikimate-3-phosphate synthase: A distinctive protein fold, *Proc. Natl. Acad. Sci.* **88,** 5046–5050 (1991). [The enzyme that catalyzes Reaction 6 of Fig. 26-62 in complex with glyphosate, an inhibitor that is a broad-spectrum herbicide.]

Wellner, D. and Meister, A., A survey of inborn errors of amino acid metabolism and transport in man, *Annu. Rev. Biochem.* **50,** 911–968 (1981).

NITROGEN FIXATION

Christiansen, J., Dean, D.R., and Seefeldt, L.C., Mechanistic features of the Mo-containing nitrogenase, *Annu. Rev. Plant Physiol. Plant Mol. Biol.* **52,** 269–295 (2001).

Einsle, O., Tezcan, F.A., Andrade, A.L.A., Schmidt, B., Yoshida, M., Howard, J.B., and Rees, D.C., Nitrogense MoFe-protein at 1.16 Å resolution: A central ligand in the FeMo-cofactor, *Science* **297,** 1696–1700 (2002).

Fisher, R.F. and Long, S.R., *Rhizobium*–plant signal exchange, *Nature* **357,** 655–660 (1992). [Discusses the signals through which Rhizobiaceae and legumes communicate to symbiotically generate the root nodules in which nitrogen fixation occurs.]

Jang, S.B., Seefeldt, L.C., and Peters, J.W., Insights into nucleotide signal transduction in nitrogenase: Structure of an iron protein with MgADP bound, *Biochemistry* **39,** 14745–14752 (2000).

Lawson, D.M. and Smith, B.E., Molybdenum nitrogenases: a crystallographic and mechanistic view, *Metal Ions Biol. Sys.* **39,** 75–120 (2002).

Mayer, S.M., Lawson, D.M., Gormal, C.A., Roe, S.M., and Smith, B.E., New insights into structure-function relationships in nitrogenase: A 1.6 Å resolution X-ray crystallographic study of *Klebsiella pneumoniae* MoFe-protein, *J. Mol. Biol.* **292,** 871–891 (1999).

Peters, J.W., Stowell, M.H.B., Soltis, S.M., Finnegan, M.G., Johnson, M.K., and Rees, D., Redox-dependent structural changes in the nitrogenase P-cluster, *Biochemistry* **36,** 1181–1187 (1997).

Rees, D.C. and Howard, J.B., Nitrogenase: standing at the crossroads, *Curr. Opin. Chem. Biol.* **4,** 559–566 (2000); *and* Howard, J.B. and Rees, D.C., Structural basis of biological nitrogen fixation, *Chem. Rev.* **96,** 2965–2982 (1996).

Schindelin, H., Kisker, C., Schlessman, J.L., Howard, J.B., and Rees., D.C., Structure of ADP · AlF₄⁻ stabilized nitrogenase complex and its implications for signal transduction, *Nature* **387,** 370–376 (1997).

PROBLEMS

1. Write the reaction for the transamination of an amino acid in terms of Cleland notation (Section 14-5A).

2. The symptoms of the partial deficiency of a urea cycle enzyme may be attenuated by a low-protein diet. Explain.

3. Why are people on a high-protein diet instructed to drink lots of water?

4. A student on a particular diet expends $10,000 \text{ kJ} \cdot \text{day}^{-1}$ while excreting 40 g of urea. Assuming that protein is 16% N by weight and that its metabolism yields $18 \text{ kJ} \cdot \text{g}^{-1}$, what percentage of the student's energy requirement is met by protein?

5. Why are phenylketonurics warned against eating products containing the artificial sweetener **aspartame** (**NutraSweet®**; chemical name L-aspartyl-L-phenylalanine methyl ester)?

6. Demonstrate that the synthesis of heme from PBG as labeled in Fig. 26-35 results in the heme-labeling pattern given in Fig. 26-32.

7. Explain why certain drugs and other chemicals can precipitate an attack of acute intermittent porphyria.

8. Heterozygotes for erythropoietic protoporphyria show only 20 to 30% residual ferrochelatase activity rather than the 50% that is normally expected for an autosomal dominant inherited disease. Provide a plausible explanation for this observation.

9. One of the symptoms of **kwashiorkor,** the dietary protein deficiency disease in children, is the depigmentation of the skin and hair. Explain the biochemical basis of this symptom.

10. What are the metabolic consequences of a defective uridylyl-removing enzyme in *E. coli*?

11. Figure 26-60, Reaction 9, indicates that methionine is synthesized in microorganisms by the methylation of homocysteine in a reaction in which N^5-methyl-THF is the methyl donor. Yet, in the breakdown of methionine (Fig. 26-18), its demethylation occurs in three steps in which SAM is an intermediate. Discuss why this reaction does not occur via the simpler one-step reversal of the methylation reaction.

*12. In the glucose–alanine cycle (Fig. 26-3), glycolytically derived pyruvate is transaminated to alanine and exported to the liver for conversion to glucose and return to the cell. Explain how a muscle cell is able to participate in this cycle under anaerobic (vigorously contracting) conditions. (*Hint:* The breakdown of many amino acids yields NH_3.)

13. Draw the activated intermediates involved in (a) glutamine and (b) asparagine biosynthesis from glutamate and aspartate, respectively. (c) Provide an example of another metabolic activation of a carboxylic acid group analogous to each of these reactions.

14. The $\alpha_2\beta_2$ tetramer of tryptophan synthase catalyzes the PLP-dependent reaction of indole-3-glycerol phosphate and serine to form tryptophan (Fig. 26-63, Reactions 5 and 6). Draw the chemical reactions involved in this synthesis, including the participation of PLP, and use curved arrows to show the flow of electrons. What role does PLP play in the reaction?

15. Suggest a reason why the nitrogen-fixing heterocysts of cyanobacteria have lost Photosystem II but retain Photosystem I.

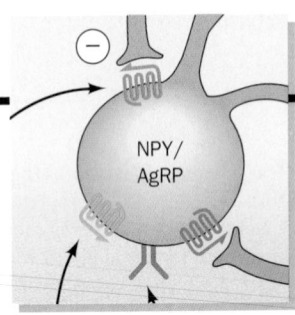

Chapter 27

Energy Metabolism: Integration and Organ Specialization

At this point in our narrative we have studied all of the major pathways of energy metabolism. Consequently, we are now in a position to consider how organisms, mammals in particular, orchestrate the metabolic symphony to meet their energy needs. This chapter therefore begins with a recapitulation of the major metabolic pathways and their control systems, then considers how these processes are apportioned among the various organs of the body, and ends with a discussion of metabolic adaptation, including how the body maintains energy balance (homeostasis), how it deals with the metabolic challenge of starvation, and how it responds to the loss of control resulting from diabetes mellitus.

1 ■ MAJOR PATHWAYS AND STRATEGIES OF ENERGY METABOLISM: A SUMMARY

Figure 27-1 indicates the interrelationships among the major pathways involved in energy metabolism. Let us review these pathways and their control mechanisms.

1. Glycolysis (Chapter 17) The metabolic degradation of glucose begins with its conversion to two molecules of pyruvate with the net generation of two molecules each of ATP and NADH. Under anaerobic conditions, pyruvate is converted to lactate (or, in yeast, to ethanol) so as to recycle the NADH. Under aerobic conditions, however, when glycolysis serves to prepare glucose for further oxidation, the NAD^+ is regenerated through oxidative phosphorylation (see below). The flow of metabolites through the glycolytic pathway is largely controlled by the activity of phosphofructokinase (PFK). This enzyme is activated by AMP and ADP, whose concentrations rise as the need for energy metabolism increases, and is inhibited by ATP and citrate, whose concentrations increase when the demand for energy metabolism has slackened. Citrate, a citric acid cycle intermediate, also inhibits PFK and glycolysis when aerobic metabolism takes over from anaerobic metabolism, making glucose oxidation more efficient (the Pasteur effect; Section 22-4C), and when fatty acid and/or ketone body oxidation (which are also aerobic pathways) are providing for energy needs (the glucose–fatty acid or Randle cycle; Section 22-4B). PFK is also activated by fructose-2,6-bisphosphate, whose concentration is regulated by the levels of glucagon, epinephrine, and norepinephrine through the intermediacy of cAMP (Section 18-3F). Liver and heart muscle F2,6P levels are regulated oppositely: A [cAMP] increase causes an [F2,6P] decrease in liver and an [F2,6P] increase in heart muscle. However, skeletal muscle [F2,6P] does not respond to changes in [cAMP].

2. Gluconeogenesis (Section 23-1) Mammals can synthesize glucose from a variety of precursors, including pyruvate, lactate, glycerol, and glucogenic amino acids (but not fatty acids), through pathways that occur mainly in liver and kidney. Many of these precursors are converted to oxaloacetate which, in turn, is converted to phosphoenolpyruvate and then, through a series of reactions that largely reverse the path of glycolysis, to glucose. The irreversible steps of glycolysis, those catalyzed by PFK and hexokinase, are bypassed in gluconeogenesis by hydrolytic reactions catalyzed, respectively, by fructose-1,6-bisphosphatase (FBPase) and glucose-6-phosphatase. FBPase and PFK may both be at least partially active simultaneously,

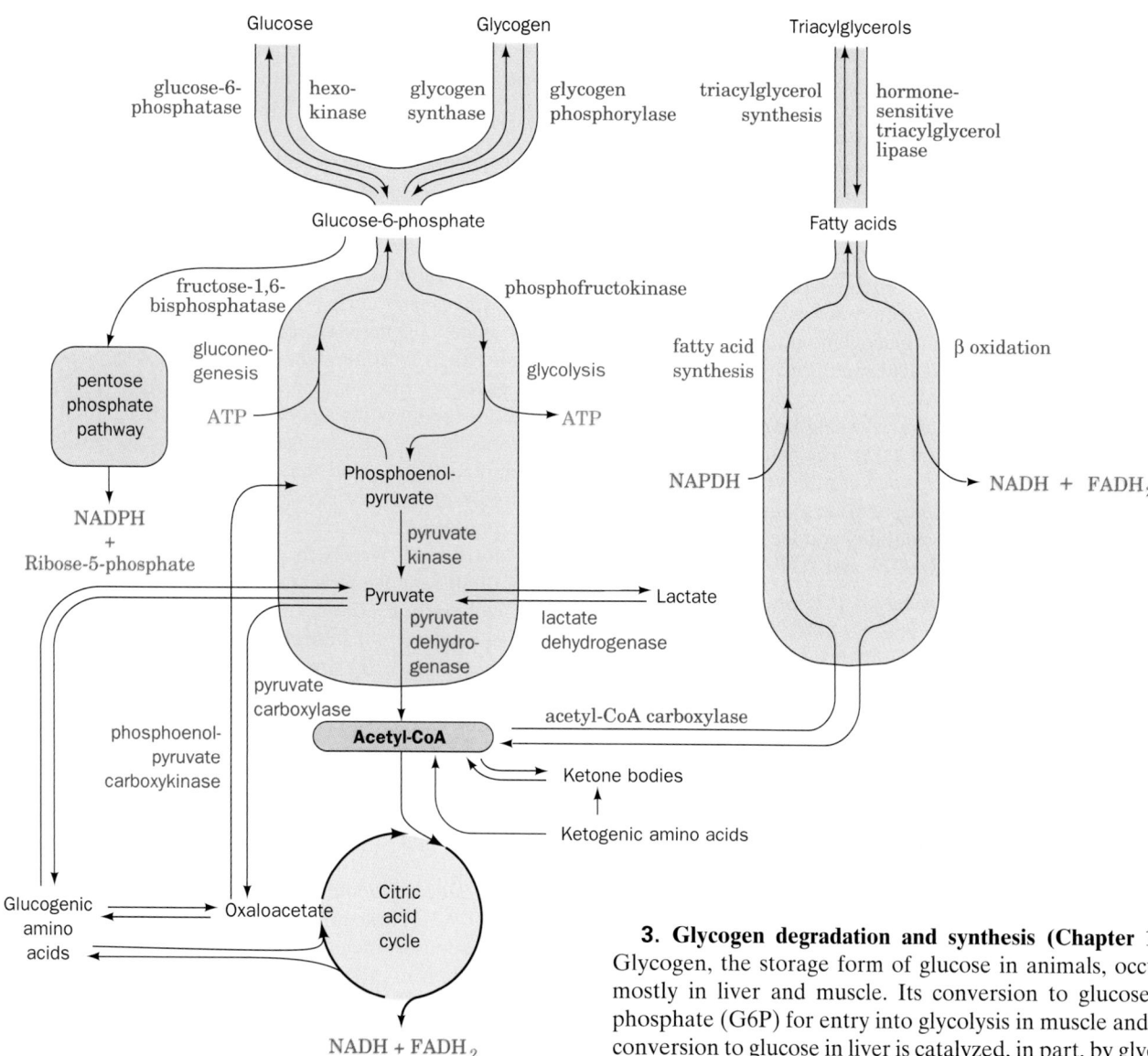

FIGURE 27-1 The major energy metabolism pathways.

creating a substrate cycle. This cycle, and the reciprocal regulation of PFK and FBPase, are important in regulating both the rate and direction of flux through glycolysis and gluconeogenesis (Sections 17-4F and 23-1B). Fatty acid and ketone body oxidation can increase the rate of gluconeogenesis in liver by decreasing the concentration of F2,6P (Section 18-3F). This occurs because the increased citrate concentration accompanying activation of the citric acid cycle during fatty acid oxidation inhibits PFK-2 as well as PFK (Table 23-1). Phosphoenolpyruvate carboxykinase (PEPCK) bypasses the third irreversible reaction of glycolysis, that catalyzed by pyruvate kinase (PK), and is controlled exclusively by long-term transcriptional regulation.

3. Glycogen degradation and synthesis (Chapter 18) Glycogen, the storage form of glucose in animals, occurs mostly in liver and muscle. Its conversion to glucose-6-phosphate (G6P) for entry into glycolysis in muscle and its conversion to glucose in liver is catalyzed, in part, by glycogen phosphorylase, whereas the opposing synthetic pathway is mediated by glycogen synthase. These enzymes are reciprocally regulated through phosphorylation/dephosphorylation reactions as catalyzed by amplifying cascades that respond to the levels of the hormones glucagon and epinephrine through the intermediacy of cAMP, and by insulin (Sections 18-3E and 19-4F). *The glucagon–insulin ratio is therefore a crucial factor in determining the rate and direction of glycogen metabolism.*

4. Fatty acid degradation and synthesis (Sections 25-1 through 25-5) Fatty acids are broken down in increments of C_2 units through β oxidation to form acetyl-CoA. They are synthesized from this compound via a separate pathway. The activity of the β-oxidation pathway varies with the fatty acid concentration. This, in turn, depends on the activity of "hormone-sensitive" triacylglycerol lipase in adipose tissue that is stimulated, through cAMP-regulated phosphorylation/dephosphorylation reactions, by glucagon and epinephrine but inhibited by insulin. The fatty acid synthesis rate varies with the activity of acetyl-CoA carboxylase, which is activated by citrate and insulin-dependent dephosphorylation, and inhibited by the pathway product palmitoyl-CoA and by cAMP- and AMP-dependent phosphorylation. Fatty acid synthesis is also subject to long-term regulation through alterations in the rates of synthesis of the enzymes mediating this process as stimulated by insulin and inhibited by fasting. *The glucagon–insulin ratio is therefore of prime importance in determining the rate and direction of fatty acid metabolism.*

5. Citric acid cycle (Chapter 21)   The citric acid cycle oxidizes acetyl-CoA, the common degradation product of glucose, fatty acids, ketone bodies, and ketogenic amino acids, to CO_2 and H_2O with the concomitant production of NADH and $FADH_2$. Many glucogenic amino acids can also be oxidized via the citric acid cycle through their breakdown, ultimately to pyruvate and then to acetyl-CoA, sometimes via the **cataplerosis** (using up) of a citric acid cycle intermediate (Section 21-5). The activities of the citric acid cycle regulatory enzymes citrate synthase, isocitrate dehydrogenase, and α-ketoglutarate dehydrogenase, are controlled by substrate availability and feedback inhibition by cycle intermediates, NADH, and ATP.

6. Oxidative phosphorylation (Chapter 22)   This mitochondrial pathway oxidizes NADH and $FADH_2$ to NAD^+ and FAD with the coupled synthesis of ATP. The rate of oxidative phosphorylation, which is tightly coordinated with the metabolic fluxes through glycolysis and the citric acid cycle, is largely dependent on the concentrations of ATP, ADP, and P_i, as well as O_2.

7. Pentose phosphate pathway (Section 23-4)   This pathway functions to generate NADPH for use in reductive biosynthesis, as well as the nucleotide precursor ribose-5-phosphate, through the oxidation of G6P. Its flux-generating step is catalyzed by glucose-6-phosphate dehydrogenase, which is controlled by the level of $NADP^+$. *The ability of enzymes to distinguish between NADH, which is mainly utilized in energy metabolism, and NADPH permits energy metabolism and biosynthesis to be regulated independently.*

8. Amino acid degradation and synthesis (Sections 26-1 through 26-5)   Excess amino acids may be degraded to common metabolic intermediates. Most of these pathways begin with an amino acid's transamination to its corresponding α-keto acid with the eventual transfer of the amino group to urea via the urea cycle. Leucine and lysine are ketogenic amino acids in that they can be converted only to acetyl-CoA or acetoacetate and hence cannot be glucose precursors. The other amino acids are glucogenic in that they may be, at least in part, converted to one of the glucose precursors pyruvate, oxaloacetate, α-ketoglutarate, succinyl-CoA, or fumarate. Five amino acids are both ketogenic and glucogenic. Essential amino acids are those that an animal cannot synthesize itself; they must be obtained from plant and microbial sources. Nonessential amino acids can be synthesized by animals utilizing preformed amino groups, via pathways that are generally simpler than those synthesizing essential amino acids.

Two compounds lie at the crossroads of the foregoing metabolic pathways: acetyl-CoA and pyruvate (Fig. 27-1). Acetyl-CoA is the common degradation product of most metabolic fuels, including polysaccharides, lipids, and proteins. Its acetyl group may be oxidized to CO_2 and H_2O via the citric acid cycle and oxidative phosphorylation or used to synthesize fatty acids. Pyruvate is the product of glycolysis, the dehydrogenation of lactate, and the breakdown of certain glucogenic amino acids. It may be oxidatively decarboxylated to yield acetyl-CoA, thereby committing its atoms either to oxidation or to the biosynthesis of fatty acids. Alternatively, it may be carboxylated via the pyruvate carboxylase reaction to form oxaloacetate which, in turn, either replenishes citric acid cycle intermediates or enters gluconeogenesis via phosphoenolpyruvate, thereby bypassing an irreversible step in glycolysis. Pyruvate is therefore a precursor of several amino acids as well as of glucose.

The foregoing pathways occur in specific cellular compartments. Glycolysis, glycogen synthesis and degradation, fatty acid synthesis, and the pentose phosphate pathway are largely or entirely cytosolically based, whereas fatty acid degradation, the citric acid cycle, and oxidative phosphorylation occur largely in the mitochondrion. Different phases of gluconeogenesis and amino acid degradation occur in each of these compartments. *The flow of metabolites across compartment membranes is mediated, in most cases, by specific carriers that are also subject to regulation.*

The enormous number of enzymatic reactions that simultaneously occur in every cell (Fig. 16-1) must be coordinated and strictly controlled to meet the cell's needs. Such regulation occurs on many levels. Intercellular communications regulating metabolism occur via certain hormones, including epinephrine, norepinephrine, glucagon, and insulin, as well as through a series of steroid hormones known as **glucocorticoids** (whose effects are discussed in Section 19-1G). These hormonal signals trigger a variety of cellular responses, including the synthesis of second messengers such as cAMP in the short term and the modulation of protein synthesis rates in the long term. On the molecular level, the enzymatic reaction rates are controlled by phosphorylation/dephosphorylation via amplifying reaction cascades, by allosteric responses to the presence of effectors, which are usually precursors or products of the reaction pathway being controlled, and by substrate availability. The regulatory machinery of opposing catabolic and anabolic pathways is generally arranged such that these pathways are reciprocally regulated.

All of the above pathways are affected in one way or another by the need for ATP, as is indicated by a cell's AMP level. Several enzymes are either activated or inhibited allosterically by AMP, and several others are phosphorylated by AMP-dependent protein kinase (AMPK, which is known as the cell's fuel gauge). AMPK's targets include the heart and inducible isozymes of the bifunctional enzyme PFK-2/FBPase-2, which control the F2,6P concentration (Sections 18-3F and 22-4C). The phosphorylation of these isozymes activates their PFK-2 activity, increasing [F2,6P], which in turn activates PFK and glycolysis. Consequently, when there is insufficient oxygen for oxidative phosphorylation to maintain adequate concentrations of ATP, the resulting AMP buildup causes the cell to turn to anaerobic glycolysis for increased ATP production.

AMPK also phosphorylates and thereby activates hormone-sensitive triacylglycerol lipase (Section 25-5). This increases the amount of fatty acid available for oxidation and hence provides another means of generating ATP. AMPK-mediated phosphorylation also inhibits certain enzymes including acetyl-CoA carboxylase (ACC; which catalyzes the first committed step of fatty acid synthesis; Section 25-4B), hydroxymethylglutaryl-CoA reductase (HMG-CoA reductase; which catalyzes the rate-determining step in cholesterol biosynthesis; Section 25-6B), and glycogen synthase (which catalyzes the rate-limiting reaction in glycogen synthesis; Section 18-3D). Consequently, when the rate of ATP production is inade-

quate, these biosynthetic pathways are turned off, thereby conserving ATP for the most vital cellular functions.

2 ■ ORGAN SPECIALIZATION

Different organs have different metabolic functions and capabilities. In this section we consider how the special needs of the mammalian body organs are met and how their metabolic capabilities are coordinated to meet these needs. In particular, we discuss brain, muscle, adipose tissue, liver, and kidney (Fig. 27-2).

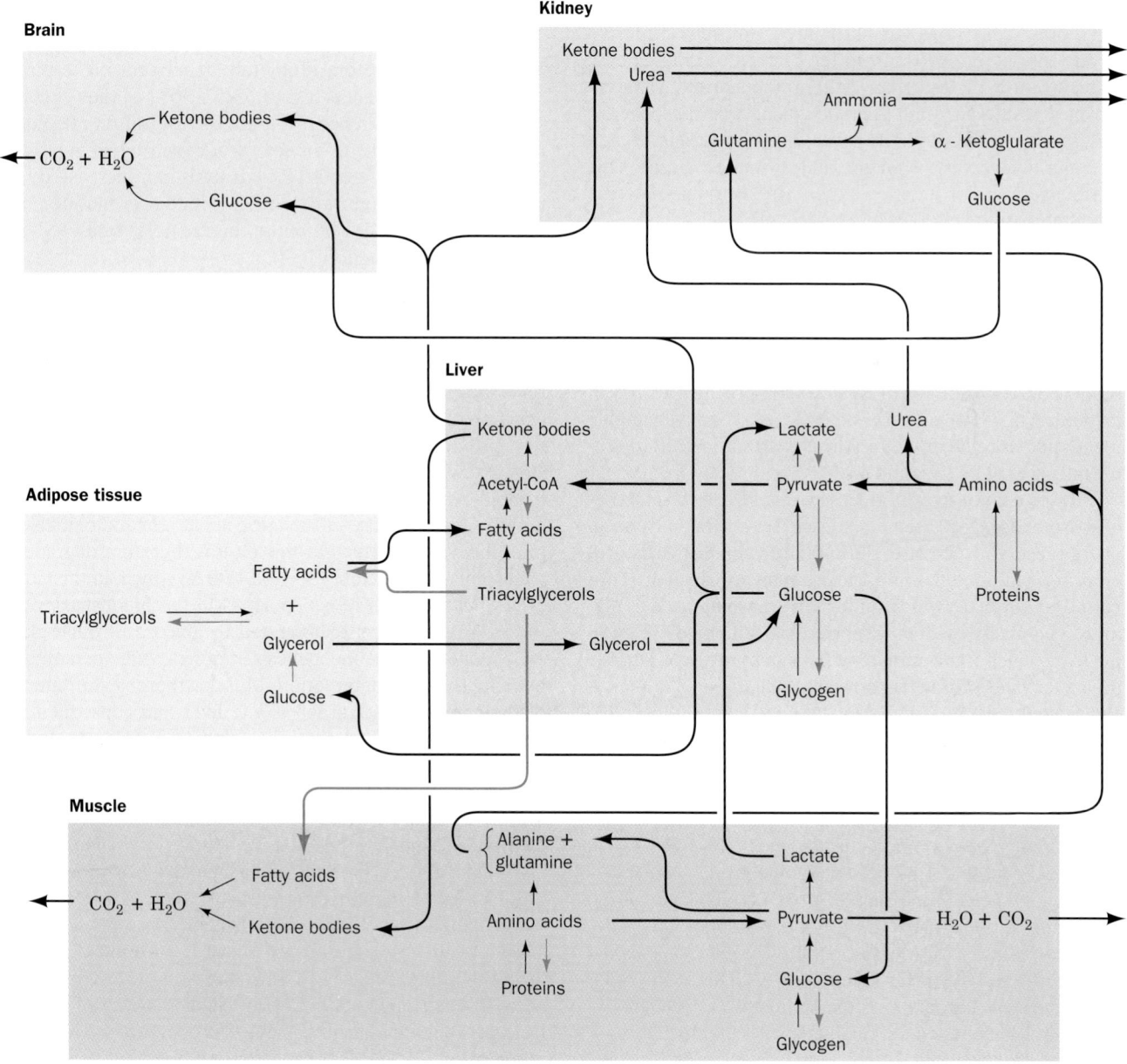

FIGURE 27-2   The metabolic interrelationships among brain, adipose tissue, muscle, liver, and kidney. The red arrows indicate pathways that predominate in the well-fed state when glucose, amino acids, and fatty acids are directly available from the intestines.

A. *Brain*

Brain tissue has a remarkably high respiration rate. For instance, the human brain only constitutes ~2% of the adult body mass but is responsible for ~20% of its resting O_2 consumption. This consumption, moreover, is independent of the state of mental activity; it varies little between sleep and the intense concentration required of, say, the study of biochemistry. Most of the brain's energy production serves to power the plasma membrane $(Na^+–K^+)$–ATPase (Section 20-3A), which maintains the membrane potential required for nerve impulse transmission (Section 20-5). In fact, the respiration of brain slices is over 50% reduced by the $(Na^+–K^+)$–ATPase inhibitor ouabain (Section 20-3A).

Under usual conditions, glucose serves as the brain's only fuel (although, with extended fasting, the brain gradually switches to ketone bodies; Section 27-4A). Indeed, since brain cells store very little glycogen, they require a steady supply of glucose from the blood. A blood glucose concentration of less than half of the normal value of ~5 mM results in brain dysfunction. Levels much below this, for example, caused by severe insulin overdose, result in coma, irreversible damage, and ultimately death. One of the liver's major functions, therefore, is to maintain the blood glucose level (Sections 18-3F and 27-2D).

B. *Muscle*

Muscle's major fuels are glucose from glycogen, fatty acids, and ketone bodies. Rested, well-fed muscle, in contrast to brain, synthesizes a glycogen store comprising 1 to 2% of its mass. The glycogen serves muscle as a readily available fuel depot since it can be rapidly converted to G6P for entry into glycolysis (Section 18-1).

Muscle cannot export glucose because it lacks glucose-6-phosphatase. Nevertheless, muscle serves the body as an energy reservoir because, during the fasting state, its proteins are degraded to amino acids, many of which are converted to pyruvate which, in turn, is transaminated to alanine. The alanine is then exported via the bloodstream to the liver, which transaminates it back to pyruvate, a glucose precursor. This process is known as the glucose–alanine cycle (Section 26-1A).

Since muscle does not participate in gluconeogenesis, it lacks the machinery that regulates this process in such gluconeogenic organs as liver and kidney. Muscle does not have receptors for glucagon, which, it will be recalled, stimulates an increase in blood glucose levels (Section 18-3F). However, muscle possesses epinephrine receptors (β-adrenergic receptors; Section 19-1F), which through the intermediacy of cAMP control the phosphorylation/dephosphorylation cascade system that regulates glycogen breakdown and synthesis (Section 18-3). This is the same cascade system that controls the competition between glycolysis and gluconeogenesis in liver in response to glucagon.

Heart muscle and skeletal muscle each contain a different isozyme of PFK-2/FBPase-2. The heart muscle isozyme is controlled by phosphorylation oppositely to that in liver, whereas skeletal muscle PFK-2/FBPase-2 is not controlled by phosphorylation at all (Section 18-3F). Thus the concentration of F2,6P rises in heart muscle but falls in liver in response to an increase in [cAMP]. Moreover the muscle isozyme of pyruvate kinase, which, it will be recalled, catalyzes the final step of glycolysis, is not subject to phosphorylation/dephosphorylation as is the liver isozyme (Section 23-1B). Thus, *whereas an increase in liver cAMP stimulates glycogen breakdown and gluconeogenesis, resulting in glucose export, an increase in heart muscle cAMP activates glycogen breakdown and glycolysis, resulting in glucose consumption. Consequently, epinephrine, which prepares the organism for action (fight or flight), acts independently of glucagon which, acting reciprocally with insulin, regulates the general level of blood glucose.*

Muscle Contraction Is Anaerobic Under Conditions of High Exertion

Muscle contraction is driven by ATP hydrolysis (Section 35-3B) and is therefore ultimately dependent on respiration. Skeletal muscle at rest utilizes ~30% of the O_2 consumed by the human body. A muscle's respiration rate may increase in response to a heavy work load by as much as 25-fold. Yet, its rate of ATP hydrolysis can increase by a much greater amount. The ATP is initially regenerated by the reaction of ADP with phosphocreatine as catalyzed by creatine kinase (Section 16-4C):

$$\text{Phosphocreatine} + \text{ADP} \rightleftharpoons \text{creatine} + \text{ATP}$$

(phosphocreatine is resynthesized in resting muscle by the reversal of this reaction). Under conditions of maximum exertion, however, such as occurs in a sprint, a muscle has only about a 4-s supply of phosphocreatine. It must then shift to ATP production via glycolysis of G6P resulting from glycogen breakdown, a process whose maximum flux greatly exceeds those of the citric acid cycle and oxidative phosphorylation. Much of this G6P is therefore degraded anaerobically to lactate (Section 17-3A) which, in the Cori cycle (Section 23-1C), is exported via the bloodstream to the liver, where it is reconverted to glucose through gluconeogenesis. Gluconeogenesis requires ATP generated by oxidative phosphorylation. Muscles thereby shift much of their respiratory burden to the liver and consequently also delay the O_2-consumption process, a phenomenon known as oxygen debt.

Muscle Fatigue Has a Protective Function

Muscle fatigue, defined as the inability of a muscle to maintain a given power output, occurs in ~20 s under conditions of maximum exertion. Such fatigue is not caused by the exhaustion of the muscle's glycogen supply. Rather, it may result from glycolytic proton generation that can drop the intramuscular pH from its resting value of 7.0 to as low as 6.4 (fatigue does not, as is widely believed, result from the buildup of lactate itself, as is demonstrated by the observation that muscles can sustain a large power output under high lactate concentrations if the pH is maintained near 7.0). Nevertheless, how acidification might cause muscle fatigue is unclear. Two other proposed causes for muscle fatigue are (1) the increased [P$_i$] arising largely from the utilization of ATP may precipitate Ca^{2+} as calcium phosphate (which is highly insoluble), thereby decreasing contractile force (muscle contraction is triggered by the release of Ca^{2+} ion; Section 35-3C); and (2) the K$^+$ ion known to be released from contracting muscle cells may result in their depolarization (Section 20-5B) and hence a reduction in their contraction. Whatever its cause(s), it seems likely that muscle fatigue is an adaptation that prevents muscle cells from committing suicide by exhausting their ATP supply (recall that glycolysis and other ATP-generating pathways must be primed by ATP).

The Heart Is a Largely Aerobic Organ

The heart is a muscular organ but one that must maintain continuous rather than intermittent activity. Thus heart muscle, except for short periods of extreme exertion, relies entirely on aerobic metabolism. It is therefore richly endowed with mitochondria; they comprise up to 40% of its cytoplasmic space, whereas some types of skeletal muscle are nearly devoid of mitochondria. The heart can metabolize fatty acids, ketone bodies, glucose, pyruvate, and lactate. Fatty acids are the resting heart's fuel of choice but, on the imposition of a heavy work load, the heart greatly increases its rate of consumption of glucose, which is derived mostly from its relatively limited glycogen store.

C. *Adipose Tissue*

Adipose tissue, which consists of cells known as adipocytes (Fig. 12-2), is widely distributed about the body but occurs most prominently under the skin, in the abdominal cavity, in skeletal muscle, around blood vessels, and in mammary gland. The adipose tissue of a normal 70-kg man contains ~15 kg of fat. This amount represents some 590,000 kJ of energy (141,000 dieter's Calories), which is sufficient to maintain life for ~3 months. Yet, adipose tissue is by no means just a passive storage depot. In fact, it is second in importance only to liver in the maintenance of metabolic homeostasis (Section 27-3).

Adipose tissue obtains most of its fatty acids from the liver or from the diet as described in Section 25-1. Fatty acids are activated by the formation of the corresponding fatty acyl-CoA and then esterified with glycerol-3-phosphate to form the stored triacylglycerols (Section 25-4F). The glycerol-3-phosphate arises from the reduction of dihydroxyacetone phosphate, which must be glycolytically generated from glucose or gluconeogenically generated from pyruvate or oxaloacetate (a process called **glyceroneogenesis;** Section 25-4F) because adipocytes lack a kinase that phosphorylates endogenous glycerol.

Adipocytes hydrolyze triacylglycerols to fatty acids and glycerol in response to the levels of glucagon, epinephrine, and insulin through a reaction catalyzed by hormone-sensitive triacylglycerol lipase (Section 25-5). If glycerol-3-phosphate is abundant, many of the fatty acids so formed are reesterified to triacylglycerols. Indeed, the average turnover time for triacylglycerols in adipocytes is only a few days. If, however, glycerol-3-phosphate is in short supply, the fatty acids are released into the bloodstream. *The rate of glucose uptake by adipocytes, which is regulated by insulin as well as by glucose availability, is therefore also an important factor in triacylglycerol formation and mobilization.* However, glycerol-3-phosphate is also produced via glyceroneogenesis under the control of PEPCK, allowing triacylglycerol turnover even when glucose concentration is low.

Obesity Results from Aberrant Metabolic Control

Obesity is one of the major health-related problems in industrial countries. Most obese people (those who are at least 20% above their desirable weights) find it inordinately difficult to lose weight or, having done so, to keep it off. Yet most animals, including humans, tend to have stable weights; that is, if they are given free access to food, they eat just enough to maintain this so-called set point weight. The nature of the regulatory machinery that controls the set point, which in obese individuals seems to be aberrantly high, is just beginning to come to light (see Section 27-3).

Formerly grossly obese individuals who have lost at least 100 kg to reach their normal weights exhibit some of the metabolic symptoms of starvation: they are obsessed with food, have low heart rates, are cold intolerant, and require 25% less caloric intake than normal individuals of similar heights and weights. In both normal and obese individuals, some 50% of the fatty acids liberated by the hydrolysis of triacylglycerols are reesterified before they can leave the adipocytes. In formerly obese subjects, this reesterification rate is only 35 to 40%, a level similar to that observed in normal individuals after a several day fast. The fat cells in normal and obese individuals, moreover, are of roughly the same size; obese people just have more of them. In fact, adipocyte precursor cells from massively obese individuals proliferate excessively in tissue culture compared to those from normal or even moderately obese subjects (adipocytes themselves do not replicate). Since fat cells, once gained, are never lost, this suggests that adipocytes, although highly elastic in size, tend to maintain a certain fixed volume and in doing so influence the metabolism and thus the appetite. This insight, unfortunately, has not yet led to a method for lowering the set points of individuals with a tendency toward obesity.

D. *Liver*

The liver is the body's central metabolic clearing house. It functions to maintain the proper levels of nutrients in the blood for use by the brain, muscles, and other tissues. The liver is uniquely situated to carry out this task because all the nutrients absorbed by the intestines except fatty acids are released into the portal vein, which drains directly into the liver.

One of the liver's major functions is to act as a blood glucose "buffer." It does so by taking up or releasing glucose in response to the levels of glucagon, epinephrine, and insulin as well as to the concentration of glucose itself. After a carbohydrate-containing meal, when the blood glucose concentration reaches ~6 m*M*, the liver takes up glucose by converting it to G6P. The process is catalyzed by glucokinase (Section 18-3F), which differs from hexokinase, the analogous glycolytic enzyme in other cells, in that glucokinase has a much lower affinity for glucose (glucokinase reaches half-maximal velocity at ~5 mM glucose vs <0.1 mM glucose for hexokinase) and is not inhibited by G6P. Liver cells, in contrast to muscle and adipose cells, are permeable to glucose, and thus insulin has no direct effect on their glucose uptake. Since the blood glucose concentration is normally less than glucokinase's K_M, the rate of glucose phosphorylation in the liver is more or less proportional to the blood glucose concentration. The other intestinally absorbed sugars, mostly fructose, galactose, and mannose, are also converted to G6P in the liver (Section 17-5). After an overnight fast, the blood glucose level drops to ~4 mM. The liver keeps it from dropping below this level by releasing glucose into the blood as is described below. In addition, lactate, the product of anaerobic glucose metabolism in the muscle, is taken up by the liver for use in gluconeogenesis and lipogenesis as well as in oxidative phosphorylation (the Cori cycle; Section 23-1C). Alanine produced in the muscle is taken up by the liver and converted to pyruvate for gluconeogenesis as well (the glucose–alanine cycle; Section 26-1A).

The Fate of Glucose-6-Phosphate Varies with Metabolic Requirements

G6P is at the crossroads of carbohydrate metabolism; it can have several alternative fates depending on the glucose demand (Fig. 27-1):

1. G6P can be converted to glucose by the action of glucose-6-phosphatase for transport via the bloodstream to the peripheral organs.

2. G6P can be converted to glycogen (Section 18-2) when the body's demand for glucose is low. Yet, increased glucose demand, as signaled by higher levels of glucagon and/or epinephrine, reverses this process (Section 18-1).

3. G6P can be converted to acetyl-CoA via glycolysis and the action of pyruvate dehydrogenase (Chapter 17 and Section 21-2). Most of this glucose-derived acetyl-CoA is used in the synthesis of fatty acids (Section 25-4), whose fate is described below, and in the synthesis of phospholipids (Section 25-8) and cholesterol (Section 25-6A). Cholesterol, in turn, is a precursor of bile salts, which are produced by the liver (Section 25-6C) for use as emulsifying agents in the intestinal digestion and absorption of fats (Section 25-1).

4. G6P can be degraded via the pentose phosphate pathway (Section 23-4) to generate the NADPH required for fatty acid biosynthesis and the liver's many other biosynthetic functions, as well as ribose-5-phosphate (R5P) for nucleotide biosynthesis (Sections 28-1A and 28-2A).

The Liver Can Synthesize or Degrade Triacylglycerols

Fatty acids are also subject to alternative metabolic fates in the liver (Fig. 27-1):

1. When the demand for metabolic fuels is high, fatty acids are degraded to acetyl-CoA and then to ketone bodies (Section 25-3) for export via the bloodstream to the peripheral tissues.

2. When the demand for metabolic fuels is low, fatty acids are used to synthesize triacylglycerols that are secreted into the bloodstream as VLDL for uptake by adipose tissue. Fatty acids may also be incorporated into phospholipids (Section 25-8).

Since the rate of fatty acid oxidation varies only with fatty acid concentration (Section 25-5), fatty acids produced by the liver might be expected to be subject to reoxidation before they can be exported. Such a futile cycle is prevented by the compartmentation of fatty acid oxidation in the mitochondrion and fatty acid synthesis in the cytosol. Carnitine palmitoyltransferase I, a component of the system that transports fatty acids into the mitochondrion (Section 25-2B), is inhibited by malonyl-CoA, the key intermediate in fatty acid biosynthesis (Section 25-4A). Hence, when the demand for metabolic fuels is low so that fatty acids are being synthesized, they cannot enter the mitochondrion for conversion to acetyl-CoA. Rather, the liver's biosynthetic demand for acetyl-CoA is met through the degradation of glucose.

When the demand for metabolic fuel rises so as to inhibit fatty acid biosynthesis, however, fatty acids are transported into the liver mitochondria for conversion to ketone bodies. Under such conditions of low blood glucose concentrations, glucokinase has reduced activity so that there is net glucose export (there is, however, always a futile cycle between the reactions catalyzed by glucokinase and glucose-6-phosphatase; Section 18-3F). The liver cannot use ketone bodies for its own metabolic purposes because liver cells lack 3-ketoacyl-CoA transferase (Section 25-3). Fatty acids rather than glucose or ketone bodies are therefore the liver's major acetyl-CoA source under conditions of high metabolic demand. The liver generates its ATP from this acetyl-CoA through the citric acid cycle and oxidative phosphorylation. The aerobic oxidation of fatty acids inhibits glucose utilization since activation of the citric acid cycle and oxidative phosphorylation increases the concentration of citrate, which inhibits glycolysis (the glucose–fatty acid or Randle cycle; Section 22-4B).

Amino Acids Are Important Metabolic Fuels

The liver degrades amino acids to a variety of metabolic intermediates (Section 26-3). These pathways mostly begin with amino acid transamination to yield the corresponding α-keto acid (Section 26-1A) with the amino group being ultimately converted, via the urea cycle (Section 26-2), to the subsequently excreted urea. Glucogenic amino acids can be converted in this manner to pyruvate or citric acid cycle intermediates such as oxaloacetate and are thereby gluconeogenic precursors (Section 23-1). Ketogenic amino acids, many of which are also glucogenic, may be converted to ketone bodies.

The liver's glycogen store is insufficient to supply the body's glucose needs for more than ~6 h after a meal. After that, glucose is supplied through gluconeogenesis from amino acids arising mostly from muscle protein degradation to alanine (the glucose-alanine cycle; Sections 26-1A and 27-2B) and glutamine (the transport form of ammonia; Section 26-1B). Thus proteins, in addition to their structural and functional roles, are important fuel resources. (Animals cannot convert fat to glucose because they lack a pathway for the net conversion of acetyl-CoA to oxaloacetate; Section 23-2).

The Liver Is the Body's Major Metabolic Processing Unit

The liver has numerous specialized biochemical functions in addition to those already mentioned. Prominent among them are the synthesis of blood plasma proteins, the degradation of porphyrins (Section 26-4A) and nucleic acid bases (Section 28-5), the storage of iron, and the detoxification of biologically active substances such as drugs, poisons, and hormones by a variety of oxidation (e.g., by cytochromes P450; Section 15-4B), reduction, hydrolysis, conjugation, and methylation reactions.

E. *Kidney*

The kidney functions to filter out the waste product urea from the blood and concentrate it for excretion, to recover important metabolites such as glucose, and to maintain the blood's pH. Blood pH is maintained by regenerating depleted blood buffers such as bicarbonate (lost by the exhalation of CO_2) and by removing for excretion excess H^+ together with the conjugate bases of excess metabolic acids such as the ketone bodies α-ketobutyrate, acetoacetate, and α-hydroxybutyrate. Phosphate, the major buffer in urine for moderate acid excretion, is accompanied by equivalent quantities of cations such as Na^+ and K^+. However, large losses of Na^+ and K^+ would upset the body's electrolyte balance, so on the production of large amounts of acids such as lactic acid or ketone bodies, the kidney produces NH_4^+ to aid in the excretion of the excess H^+ (utilizing Cl^- or the conjugate base of a metabolic acid as the counterion). This NH_4^+ is generated from glutamine, which is converted first to glutamate and then to α-ketoglutarate by glutaminase and glutamate dehydrogenase. The overall reaction is

$$\text{Glutamine} \rightarrow \text{α-ketoglutarate} + 2NH_4^+$$

The α-ketoglutarate is converted to malate by the citric acid cycle and then is exported from the mitochondria and converted either to pyruvate, which is oxidized completely to CO_2, or via oxaloacetate to PEP and then to glucose via gluconeogenesis. High fat diets, which produce high blood concentrations of free fatty acids and ketone bodies and hence high acidic loads, cause α-ketoglutarate to be converted completely to CO_2, and then to bicarbonate, thereby increasing the blood's buffering capacity. During starvation, the α-ketoglutarate enters gluconeogenesis, to the extent that the kidneys generate as much as 50% of the body's glucose supply.

3 ■ METABOLIC HOMEOSTASIS: THE REGULATION OF APPETITE, ENERGY EXPENDITURE, AND BODY WEIGHT

When a normal animal overeats, the resulting additional fat somehow signals the brain to induce the animal to eat less and to expend more energy. Conversely, the loss of fat stimulates increased eating until the lost fat is replaced. Evidently, animals have a "lipostat" that can keep the amount of body fat constant to within 1% over many years. At least a portion of the lipostat resides in the hypothalamus (a part of the brain that hormonally controls numerous physiological functions; Section 19-1H), since damaging it can yield a grossly obese animal.

Despite this obvious set of controls in animals, there has been an explosion of obesity in the human population. It has, in fact, become a world health problem, leading to diabetes and heart disease. As a result of numerous studies in recent years, researchers have been able to outline the mechanisms involved in **metabolic homeostasis,** the balance between energy influx and energy expenditure, and to identify some of the irregularities that lead to obesity. A variety of mutant strains of rodents have been generated that cause obesity. The study of these mutants has resulted in the identification of several hormones that act in a coordinated manner to regulate appetite.

A. *Leptin*

Two of the genes whose mutations cause obesity in mice are known as *obese* (*ob*) and *diabetes* (*db;* the wild-type genes are designated *OB* and *DB*). Homozygotes for defects in either of these recessive genes, *ob/ob* and *db/db*, are grossly obese and have nearly identical phenotypes (Fig. 27-3). Indeed, the way in which these phenotypes were distinguished was by surgically linking the circulation of a mutant mouse to that of a normal (*OB/OB*) mouse, a phenomenon named **parabiosis.** *ob/ob* mice so linked exhibit normalization of body weight and reduced food intake, whereas *db/db* mice do not do so. This suggests that *ob/ob* mice are deficient in a circulating factor that regulates ap-

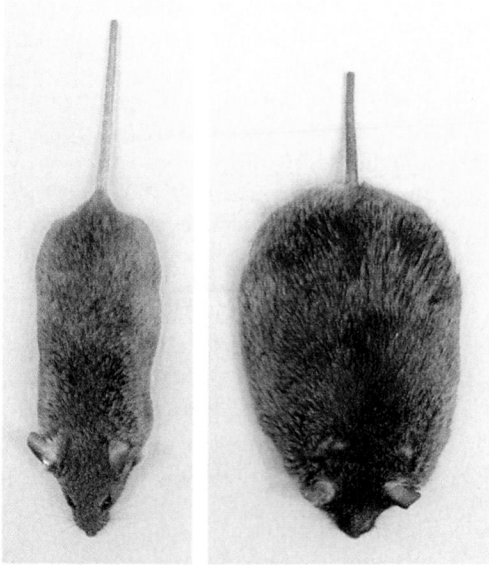

FIGURE 27-3   Normal (*OB/OB, left*) and obese (*ob/ob, right*) mice. [Courtesy of Richard D. Palmiter, University of Washington.]

petite and metabolism, whereas *db/db* mice are defective in the receptor for this circulating factor.

The mouse *OB* gene encodes a 146-residue monomeric protein named **leptin** (Greek: *leptos*, thin; Fig. 27-4), that has no apparent homology with proteins of known sequence. Leptin, which was discovered by Jeffrey Friedman, is expressed only by adipocytes, which in doing so appear to inform the brain of how much fat the body carries. Thus, injecting leptin into *ob/ob* mice causes them to eat less and to lose weight. In fact, leptin-treated *ob/ob* mice on a restricted diet lost 50% more weight than untreated *ob/ob* mice on the same diet, which suggests that reduced food intake alone is insufficient to account for leptin-induced weight loss. Leptin appears to control energy expenditure as well.

Leptin injection has no effect on *db/db* mice, which suggests that the *db* gene encodes a defective leptin receptor. The leptin receptor gene was identified by making a cDNA library from mouse brain tissue that specifically bound leptin and then identifying a receptor-expressing clone by its ability to bind leptin (gene cloning techniques are discussed in Section 5-5). This gene, which has been shown to be the *DB* gene, encodes a protein named **OB-R** (for *OB* receptor) that appears to have a single transmembrane segment and an extracellular domain that resembles the receptors for certain cytokines (proteins that regulate the differentiation, proliferation, and activities of various cells; Section 19-3E).

OB-R protein, which was discovered by Louis Tartaglia, has at least six alternatively spliced forms that appear to be expressed in a tissue-specific manner (alternative gene splicing is discussed in Section 31-4A). In normal mice, the hypothalamus expresses high levels of a splice variant of OB-R that has a 302-residue cytoplasmic segment. However, in *db/db* mice, this segment has an abnormal splice site that truncates it to only 34 residues, which almost certainly renders this OB-R variant unable to transmit leptin signals. Thus, it appears that leptin's weight-controlling effects are mediated by signal transduction resulting from its binding to the OB-R protein in the hypothalamus (signal transduction is discussed in Chapter 19).

Human leptin is 84% identical in sequence to that of mice. The use of a radioimmunoassay (Section 19-1A) to measure the serum levels of leptin in normal-weight and obese humans established that in both groups serum leptin concentrations increase with their percentage of body fat as does the *ob* mRNA content of their adipocytes. Moreover, after obese individuals had lost weight, their serum leptin concentrations and adipocyte *ob* mRNA content declined. This suggests that most obese persons produce sufficient amounts of leptin but have developed "leptin resistance." Since leptin must cross the blood–brain barrier in order to exert its effects on the hypothalamus, it has been suggested that this crossing is somehow saturable, thus limiting the concentration of leptin in the brain. The high concentration of leptin in obese individuals is not without affect, however. OB-R is also expressed in peripheral tissue where leptin has been shown to function as well. While not preventing obesity, the hor-

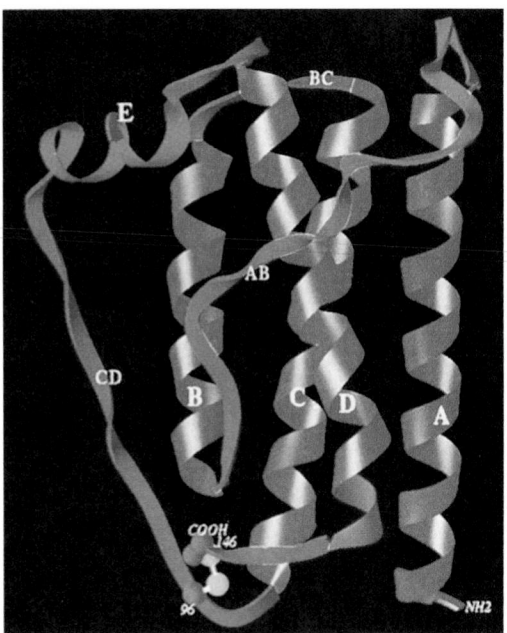

FIGURE 27-4 X-Ray structure of human leptin-E100. This mutant protein (W100E) has comparable activity to the wild-type protein but more readily crystallizes. Note that this monomeric, 146-residue, single-domain protein forms an up-up-down-down four-helix bundle, as do many cytokines and growth hormones (Fig. 19-9). A disulfide bond links the side chains of Cys 96 and Cys 146 (C green and S yellow). [Courtesy of Faming Zhang, Eli Lilly & Co., Indianapolis, Indiana.] *See the Interactive Exercises*

mone has been shown to directly stimulate the oxidation of fatty acids as well as to inhibit the accumulation of lipids in non-adipose tissue. It does so by activating AMP-dependent protein kinase (AMPK), which in turn phosphorylates and thereby inactivates acetyl-CoA carboxylase (ACC). This reduces the malonyl-CoA concentration, thereby decreasing its inhibition of carnitine palmitoyltransferase I, which then transports fatty acyl-CoA into the mitochondrion for oxidation (Section 25-5). We discuss the function of leptin in peripheral tissues in Section 27-3F.

A small minority of obese individuals have been found to be leptin deficient in a manner similar to *ob/ob* mice. Two grossly obese children who are members of the same highly consanguineous (descended from the same ancestors) family (they are cousins and both sets of parents are cousins) have been shown to be homozygous for a defective OB gene. The children, at the ages of 8 and 2 years old, respectively, weighed 86 and 29 kg and were noted to have remarkably large appetites. Their OB genes have a deletion of a single guanine nucleotide in codon 133, thereby causing a frameshift mutation that, it is likely, renders the mutant leptin biologically inactive. Moreover, their leptin serum levels were only ~10% of normal. Leptin injections have relieved their symptoms.

B. *Insulin*

We have discussed the insulin signaling cascade (Section 19-4F) and the role of insulin in peripheral tissues such as muscle and adipose tissue in stimulating the uptake of glucose (Fig. 20-15) and its storage as glycogen (Section 18-3) or fat (Section 25-5). Insulin receptors also occur in the hypothalamus. Consequently, the infusion of insulin into rats with insulin-deficient diabetes inhibits food intake, reversing the overeating behavior characteristic of the disease. Knock-out mice have been developed with a central nervous system-specific disruption of the insulin receptor gene. These mice have no alteration in brain development or survival but become obese, with increased body fat, increased leptin levels, increased serum triacylglycerol, and the elevated plasma insulin levels characteristic of insulin resistance (Section 27-4B). Evidently, insulin also plays a role in the neuronal regulation of food intake and body weight. As we discuss in Section 27-3D, insulin and leptin both act through receptors in the hypothalamus to decrease food intake.

C. *Ghrelin and PYY$_{3-36}$*

Ghrelin and PYY$_{3-36}$ Act As Short Term Regulators of Appetite

Ghrelin, which was discovered by Masayasu Kojima and Kenji Kanagawa, is an appetite-stimulating gastric peptide that is secreted by the empty stomach. This 28-residue peptide was first discovered and named for its function as a growth hormone–releasing peptide (ghrelin is an abbreviation for *growth-hormone-rel*ease). Octanoylation of its Ser 3 is required for activity.

$$\overset{10}{\text{GSXFLSPEHQ}} \; \overset{20}{\text{RVQQRKESKK}} \; \overset{28}{\text{PPAKLQPR}}$$

Human ghrelin
X = Ser modified with *n*-octanoic acid

Injection of ghrelin has been shown to induce adiposity (increased adipose tissue) in rodents by stimulating an increase in food intake while reducing fat utilization. In humans in states of positive energy balance such as obesity or high caloric intake, circulating ghrelin levels are decreased, whereas during fasting, circulating ghrelin levels increase.

PYY$_{3-36}$

$$\overset{3}{\text{I}}\text{KPEAPG}\overset{10}{\text{E}} \; \text{DASPEELNR}\overset{20}{\text{Y}} \; \text{YASLRHTLN}\overset{30}{\text{L}} \; \text{VTRQR}\overset{36}{\text{Y}}$$

Human PYY$_{3-36}$

is a peptide secreted by the gastrointestinal tract in proportion to the caloric intake of a meal, which acts to inhibit further food intake. Both rodents and humans have been shown to respond to the presence of this peptide by decreasing their food intake for up to 12 hours. Human subjects receiving a 90 minute infusion of PYY$_{3-36}$ ate only 1500 kcal of food during the next 24 hour period, whereas those receiving saline controls ate 2200 kcal during the same period.

D. *Hypothalamic Integration of Hormonal Signals*

Neurons of the Arcuate Nucleus Region of the Hypothalamus Integrate and Transmit Hunger Signals

About half of the length of the hypothalamus is taken up by the **arcuate nucleus,** a collection of neuronal cell bodies consisting of two cell types: the **NPY/AgRP** cell type and the **POMC/CART** cell type. These cell types are named after the neuropeptides they secrete. **Neuropeptide Y (NPY)**

$$\overset{1}{\text{Y}}\text{PSKPDNPG}\overset{10}{\text{E}} \; \text{DAPAGAMAR}\overset{20}{\text{Y}} \; \text{YSALRHYIN}\overset{30}{\text{L}} \; \text{ITRQR}\overset{36}{\text{Y}}\text{–NH}_2$$

Neuropeptide Y
The C-terminal carboxyl is amidated

is a potent stimulator of food intake and an inhibitor of energy expenditure, as is **Agouti related peptide (AgRP).** **Pro-opiomelanocortin (POMC)** is posttranslationally processed in the hypothalamus to release **α-melanocyte stimulating hormone (α-MSH;** Section 34-3C). **Cocaine and amphetamine-regulated transcript (CART)** and α-MSH are both inhibitors of food intake and stimulators of energy expenditure.

The balance of the secretions from these two cell types is controlled by leptin, insulin, ghrelin, and PYY$_{3-36}$ (Fig. 27-5). Leptin and insulin signal satiety and therefore decrease appetite by diffusing across the blood–brain barrier to the arcuate nucleus, where they stimulate POMC/CART neurons to produce CART and α-MSH, while inhibiting the production of NPY from NPY/AgRP neurons. Leptin receptors act through the JAK–STAT signal transduction pathway (Section 19-3E). Ghrelin has receptors on NPY/AgRP neurons that stimulate the secretion of NPY and AgRP to increase appetite. Interestingly, PYY$_{3-36}$, a peptide that is homologous to NPY, binds specifically to NPY receptor subtype Y2R on NPY/AgRP neurons. This subtype is an inhibitory receptor, however, so binding of PYY$_{3-36}$ causes a decrease in secretion from NPY/AgRP neurons. The integrated stimuli of all these secretions from the arcuate nucleus control appetite.

E. *Control of Energy Expenditure by Adaptive Thermogenesis*

The energy content of food is utilized by an organism either in the performance of work or the generation of heat. Excess energy is stored as glycogen or fat for future use. In well-balanced individuals, the storage of excess fuel remains constant over many years. However, when energy consumed is consistently greater than energy expended, obesity results. The body has several mechanisms for preventing obesity. One of them, as discussed above, is appetite control. The other is **diet-induced thermogenesis,** a form of **adaptive thermogenesis** (heat production in response to environmental stress). We have previously discussed adaptive thermogenesis in response to cold, which

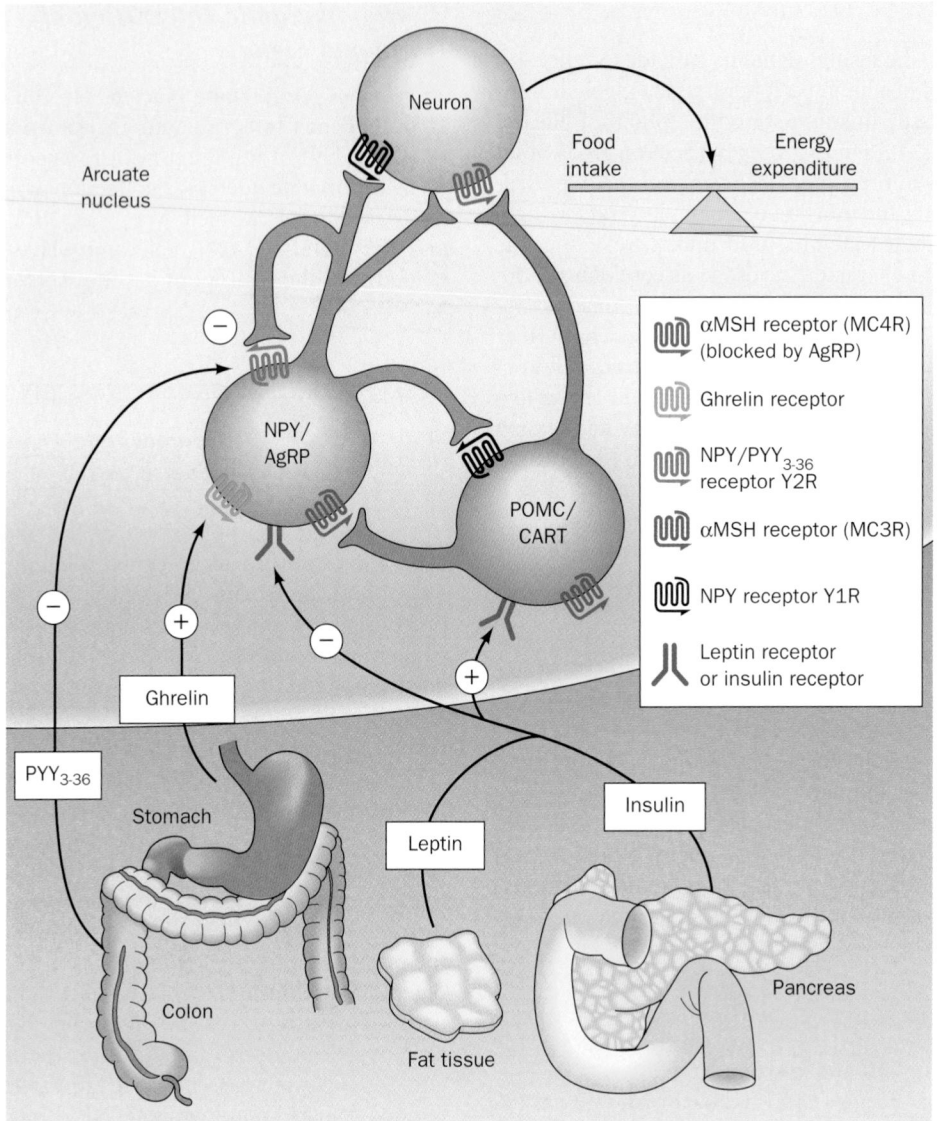

FIGURE 27-5 Hormones that control the appetite. Leptin and insulin (*lower part of the figure*) circulate in the blood at concentrations proportional to body-fat mass. They decrease appetite by inhibiting NPY/AgRP neurons (*center*) while stimulating melanocortin-producing neurons in the arcuate nucleus region of the hypothalamus. NPY and AgRP increase the appetite, and melanocortins decrease the appetite, via other neurons (*top*). Activation of NPY/AgRP-expressing neurons inhibits melanocortin-producing neurons. The gastric hormone ghrelin stimulates appetite by activating the NPY/AgRP-expressing neurons. PYY_{3-36}, released from the gastrointestinal tract, inhibits NPY/AgRP-expressing neurons and thereby decreases the appetite. PYY_{3-36} works in part through the autoinhibitory NPY receptor subtype Y2R. [After Schwartz, M.W. and Morton, G.J., *Nature* **418,** 596 (2002).]

occurs in rodents and newborn humans through the uncoupling of oxidative phosphorylation in brown adipose tissue (Section 22-3D). The mechanism of this thermogenesis involves the release of norepinephrine from the brain in response to cold, its binding to β-adrenergic receptors on brown adipose tissue inducing an increase in [cAMP], which in turn initiates an enzymatic phosphorylation cascade that activates hormone-sensitive triacylglycerol lipase. The resulting increase in the concentration of free fatty acids provides fuel for oxidation as well as inducing the opening of a proton channel, called uncoupling pro-

tein-1 (UCP1) or thermogenin, in the inner mitochondrial membrane. The opening of UCP1 discharges the proton gradient across the inner mitochondrial membrane, thus uncoupling electron transport from ATP production. The energy that would otherwise have been used to drive ATP synthesis is thereby released as heat.

Although metabolic measurements in adult humans clearly demonstrate that an increase in energy intake causes an increase in metabolic rate and thermogenesis, the cause of this increase is unclear. Adult humans have little brown adipose tissue. However, skeletal muscle represents up to 40% of their total body weight and has high mitochondrial capacity. Homologs of UCP1 have been identified: **UCP2** occurs in many tissues including white adipose tissue, whereas **UCP3** occurs in brown adipose tissue, white adipose tissue, and muscle. Leptin has been shown to upregulate UCP2. However, it has yet to be demonstrated that UCP3 in muscle participates in diet-induced thermogenesis. ATP-hydrolyzing substrate cycles such as that between fatty acids and triacylglycerol in adipose tissue (Section 27-2C) may also be involved.

F. *Did Leptin Evolve As a Thrifty Gene?*

The unusual behavior of leptin, which serves to control weight in normal-weight individuals while its concentration continues to climb without apparent effect in obese individuals, has led to the proposal that leptin evolved as a "thrifty gene." In hunter-gatherer societies, it was a distinct advantage to be able to survive intermittent famines. In order to do this, fat must be stored in adipose tissue in times of plenty, making short-term obesity advantageous. However, the accumulation of fatty acids and lipids in non-adipose tissue results in coronary artery disease, insulin resistance, and diabetes (Section 27-4B). Leptin, by directly stimulating the oxidation of fatty acids as well as inhibiting the accumulation of lipids in non-adipose tissue, is thought to protect against these diseases during short-term obesity, thereby providing an evolutionary advantage. However, in recent times in industrialized nations, the unprecedented availability of food and lack of famine has made obesity a long-term rather than a short-term condition, which is a liability rather than a benefit.

4 ■ METABOLIC ADAPTATION

In this section we consider the body's responses to two metabolically abnormal situations: (1) starvation and (2) the disease diabetes mellitus.

A. *Starvation*

Glucose is the metabolite of choice of both brain and working muscle. Yet, the body stores less than a day's supply of carbohydrate (Table 27-1). Thus, the low blood sugar resulting from even an overnight fast results, through an increase in glucagon secretion and a decrease in insulin secretion, in the mobilization of fatty acids from adipose tissue (Section 25-5). The diminished insulin level also inhibits glucose uptake by muscle tissue. Muscles therefore switch from glucose to fatty acid metabolism for energy production. The brain, however, still remains heavily dependent on glucose.

In animals, glucose cannot be synthesized from fatty acids. This is because neither pyruvate nor oxaloacetate, the precursors of glucose in gluconeogenesis (Section 23-1), can be synthesized from acetyl-CoA (the oxaloacetate in the citric acid cycle is derived from acetyl-CoA but the cyclic nature of this process requires that the oxaloacetate be consumed as fast as it is synthesized; Section 21-1A).

TABLE 27-1 Fuel Reserves for a Normal 70-kg Man

Fuel	Mass (kg)	Calories[a]
Tissues		
Fat (adipose triacyglycerols)	15	141,000
Protein (mainly muscle)	6	24,000
Glycogen (muscle)	0.150	600
Glycogen (liver)	0.075	300
Circulating fuels		
Glucose (extracellular fluid)	0.020	80
Free fatty acids (plasma)	0.0003	3
Triacylglycerols (plasma)	0.003	30
Total		166,000

[a] One (dieter's) Calorie = 1 kcal = 4.184 kJ.

Source: Cahill, G.F., Jr., *New Engl. J. Med.* **282**, 669 (1970).

During starvation, glucose must therefore be synthesized from the glycerol product of triacylglycerol breakdown and, more importantly, from the amino acids derived from the proteolytic degradation of proteins, the major source of which is muscle. Yet, the continued breakdown of muscle during prolonged starvation would ensure that this process became irreversible since a large muscle mass is essential for an animal to move about in search of food. The organism must therefore make alternate metabolic arrangements.

After several days of starvation, gluconeogenesis has so depleted the liver's oxaloacetate supply that this organ's ability to metabolize acetyl-CoA via the citric acid cycle is greatly diminished. Rather, the liver converts the acetyl-CoA to ketone bodies (Section 25-3), which it releases into the blood. The brain gradually adapts to using ketone bodies as fuel through the synthesis of the appropriate enzymes: After a 3-day fast, only about one-third of the brain's energy requirements are satisfied by ketone bodies but after 40 days of starvation, ~70% of its energy needs are so met. The rate of muscle breakdown during prolonged starvation consequently decreases to ~25% of its rate after a several-day fast. The survival time of a starving individual is therefore much more dependent on the size of his or her fat reserves than it is on his or her muscle mass. Indeed, highly obese individuals can survive for over a year without eating (and have occasionally done so in clinically supervised weight reduction programs).

B. *Diabetes Mellitus*

The polypeptide hormone insulin acts mainly on muscle, liver, and adipose tissue cells to stimulate the synthesis of glycogen, fats, and proteins while inhibiting the breakdown of these metabolic fuels. In addition, insulin stimulates the uptake of glucose by most cells, with the notable exception of brain and liver cells. Together with glucagon, which has largely opposite effects, insulin acts to maintain the proper level of blood glucose.

In the disease **diabetes mellitus,** which is the third leading cause of death in the United States after heart disease and cancer, insulin either is not secreted in sufficient amounts or does not efficiently stimulate its target cells. As a consequence, blood glucose levels become so elevated that the glucose "spills over" into the urine, providing a convenient diagnostic test for the disease. Yet, despite these high blood glucose levels, cells "starve" since insulin-stimulated glucose entry into cells is impaired. Triacylglycerol hydrolysis, fatty acid oxidation, gluconeogenesis, and ketone body formation are accelerated and, in a condition termed **ketoacidosis,** ketone body levels in the blood become abnormally high. Since ketone bodies are acids, their high concentration puts a strain on the buffering capacity of the blood and on the kidney, which controls blood pH by excreting the excess H^+ into the urine (Section 27-2E). This unusually high excess H^+ excretion is accompanied by NH_4^+, Na^+, K^+, P_i, and H_2O excretion, causing severe dehydration (which compounds the dehydration resulting from the osmotic effect of the high glucose concentration in the blood; excessive thirst is a classic symptom of diabetes) and a decrease in blood volume—ultimately life-threatening situations.

There are two major forms of diabetes mellitus:

1. Insulin-dependent, type 1, or **juvenile-onset diabetes mellitus,** which most often strikes suddenly in childhood.

2. Noninsulin-dependent, type 2, or **maturity-onset diabetes mellitus,** which usually develops rather gradually after the age of 40.

Insulin-Dependent Diabetes Is Caused by a Deficiency of Pancreatic β Cells

In insulin-dependent (type 1) diabetes mellitus, insulin is absent or nearly so because the pancreas lacks or has defective β cells. This condition results, in genetically susceptible individuals (see below), from an autoimmune response that selectively destroys their β cells. Individuals with insulin-dependent diabetes, as Frederick Banting and George Best first demonstrated in 1921, require daily insulin injections to survive and must follow carefully balanced diet and exercise regimens. Their lifespans are, nevertheless, reduced by up to one-third as a result of degenerative complications such as kidney malfunction, nerve impairment, and cardiovascular disease, as well as blindness, which apparently arise from the imprecise metabolic control provided by periodic insulin injections. Perhaps newly developed systems that monitor blood glucose levels and continuously deliver insulin in the required amounts will rectify this situation.

The usually rapid onset of the symptoms of insulin-dependent diabetes had suggested that the autoimmune attack on the pancreatic β cells responsible for this disease is one of short duration. Typically, however, the disease "brews" for several years as the aberrantly aroused immune system slowly destroys the β cells. Only when >80% of these cells have been eliminated do the classic symptoms of diabetes suddenly emerge.

Why does the immune system attack the pancreatic β cells? It has long been known that certain alleles (genetic variants) of the **Class II major histocompatibility complex (MHC) proteins** are particularly common in insulin-dependent diabetics [MHC proteins are highly polymorphic (variable within a species) immune system components to

which cell-generated antigens such as viral proteins must bind in order to be recognized as foreign; Sections 35-2A and 35-2E]. It is thought that autoimmunity against β cells is induced in a susceptible individual by a foreign antigen, perhaps a virus, which immunologically resembles some β cell component. The Class II MHC protein that binds this antigen does so with such tenacity that it stimulates the immune system to launch an unusually vigorous and prolonged attack on the antigen. Some of the activated immune system cells eventually make their way to the pancreas, where they initiate an attack on the β cells due to the close resemblance of the β cell component to the foreign antigen.

Noninsulin-Dependent Diabetes Is Characterized by Insulin Resistance as Well as Impaired Insulin Secretion

Noninsulin-dependent (type 2) diabetes mellitus **(NIDDM),** which accounts for over 90% of the diagnosed cases of diabetes and affects 18% of the population over 65 years of age, usually occurs in obese individuals with a genetic predisposition for this condition (although one that differs from that associated with insulin-dependent diabetes). These individuals may have normal or even greatly elevated insulin levels. Their symptoms arise from **insulin resistance,** an apparent lack of sensitivity to insulin in normally insulin-responsive cells. Insulin resistance, which may precede NIDDM by as much as 10 to 20 years, appears to be caused by an interruption in the insulin signaling pathway (Section 19-4F). Gerald Shulman has proposed that this interruption is caused by a Ser/Thr kinase cascade that phosphorylates proteins known as **insulin receptor substrates (IRSs;** Section 19-3C) so as to decrease their ability to be phosphorylated on their Tyr residues by activated insulin receptor. Tyrosine phosphorylation is required for IRS activation and communication with phosphoinositide 3-kinase (PI3K; Section 19-4D), which subsequently activates the translocation of GLUT4-containing vesicles to the cell surface for increased glucose transport into cells (Section 20-2E). The original Ser/Thr kinase cascade is triggered by the activation of an isoform of protein kinase C (PKC; Section 19-4C) caused by an increase in fatty acyl-CoA, diacylglycerol, and ceramides (Section 12-1D) resulting from elevated free fatty acids (Fig. 27-6). This hypothesis accounts for the observation that diet alone is often sufficient to control this type of diabetes.

DNA Chip Technology Permits the Integrated Study of Metabolic Regulation

Our ability to understand the integrated nature of metabolism and its genetic regulation in health and disease has taken a giant step forward with the advent of DNA chips (microarrays; Section 7-6B). For example, Ronald Kahn has used this technology to study the genetic basis of the metabolic abnormalities underlying both obesity and diabetes. To do so, he isolated the mRNA from the skeletal muscle of normal, diabetic and insulin-treated diabetic mice, reverse-transcribed it to cDNA (Section 5-5F), which was then hydridized to oligonucleotide microarrays that represented 14,288 mouse genes. 129 up-regulated and 106 down-regulated genes were thereby identified in diabetic mice. Not suprisingly, the expression of the mRNAs en-

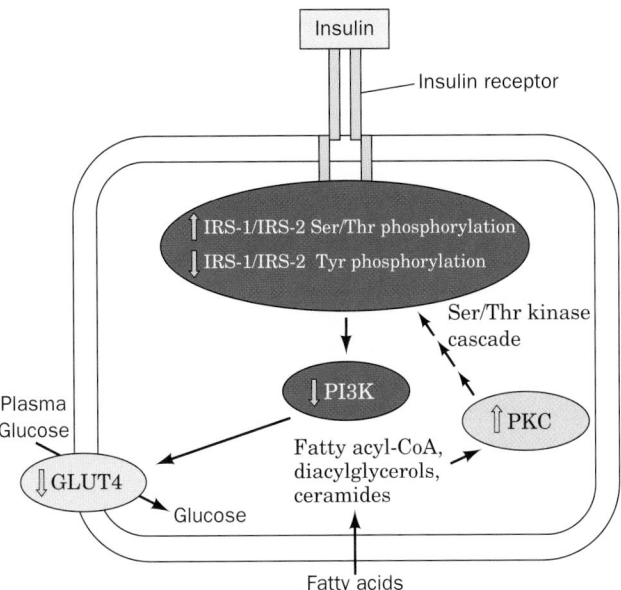

FIGURE 27-6 The mechanism through which high concentrations of free fatty acids cause insulin resistance. Elevated concentrations of free fatty acids in the blood diffuse into muscle cells where they are converted to fatty acyl-CoA, diacylglycerols, and ceramides. These lipotoxic substances activate an isoform of protein kinase C (PKC), triggering a Ser/Thr kinase cascade that results in the phosphorylation of IRS-1 and IRS-2. This phosphorylation inhibits the Tyr phosphorylation required for transmission of the insulin signal, thereby decreasing the activation of PI3K, which decreases the rate of fusion of GLUT4-containing vesicles with the plasma membrane and hence the amount of glucose entering the cell. [Modified from Shulman, G.I., *J. Clin. Invest.* **106,** 173 (2000).]

coding the fatty acid β oxidation pathway were increased, whereas those for GLUT4, glucokinase, the E1 component of the pyruvate dehydrogenase multienzyme complex, and the subunits of all four mitochondrial electron transport chain complexes were coordinately decreased. Intriguingly, only about half of these changes in gene expression could be reversed by insulin treatment. Thus, the post-genomic era will almost certainly witness an explosion in our knowledge of metabolic regulation that should yield major health benefits. Nevertheless, our ability to sensibly interpret this huge influx of information may prove to be the greatest challenge.

CHAPTER SUMMARY

1 ■ Major Pathways and Strategies of Energy Metabolism: A Summary The complex network of processes involved in energy metabolism are distributed among different compartments within cells and in different organs of the body. These processes function to generate ATP "on demand," to generate and store glucose, triacylglycerols, and proteins in times of plenty for use when needed, and to keep the concentration of glucose in the blood at the proper level for use by organs such as the brain, whose sole fuel source, under normal conditions, is glucose. The major energy metabolism pathways include glycolysis, glycogen degradation and synthesis, gluconeogenesis, the pentose phosphate pathway, and triacylglycerol and fatty acid synthesis, which are cytosolically based, and fatty acid oxidation, the citric acid cycle, and oxidative phosphorylation, which are confined to the mitochondrion. Amino acid degradation occurs, in part, in both compartments. The mediated membrane transport of metabolites therefore also plays an essential metabolic role.

2 ■ Organ Specialization The brain normally consumes large amounts of glucose. Muscle, under intense ATP demand such as in sprinting, degrades glucose and glycogen anaerobically, thereby producing lactate, which is exported via the blood to the liver for reconversion to glucose through gluconeogenesis. During moderate activity, muscle generates ATP by oxidizing glucose from glycogen, fatty acids, and ketone bodies completely to CO_2 and H_2O via the citric acid cycle and oxidative phosphorylation. Adipose tissue stores triacylglycerols and releases fatty acids into the bloodstream in response to the organism's metabolic needs. These metabolic needs are communicated to adipose tissue by means of the hormones insulin, which indicates a fed state in which storage is appropriate, and glucagon, epinephrine, and norepinephrine, which signal a need for fatty acid release to provide fuel for other tissues. The liver, the body's central metabolic clearinghouse, maintains blood glucose concentrations by storing glucose as glycogen in times of plenty and releasing glucose in times of need both by glycogen breakdown and by gluconeogenesis. It also converts fatty acids to ketone bodies for use by peripheral tissues. During a fast, it breaks down amino acids resulting from protein degradation to metabolic intermediates that can be used to generate glucose. The kidney filters out urea from the blood, recovers important metabolites, and maintains pH balance. To do so, glutamine is broken down to produce NH_4^+ for H^+ excretion. The resulting α-ketoglutarate product is converted to CO_2 to resupply HCO_3^- to the blood to maintain its buffering capacity. During starvation, the kidney uses the α-ketoglutarate from glutamine breakdown for gluconeogenesis.

3 ■ Metabolic Homeostasis: Regulation of Appetite, Energy Expenditure, and Body Weight Appetite is suppressed by the actions of leptin, a hormone produced by adipose tissue, insulin, produced by the β cells of the pancreas, and PYY_{3-36}, produced by the gastrointestinal tract, which act in the hypothalamus to inhibit the secretion of neuropeptide Y (NPY) and stimulate the secretion of α-MSH and CART. This decreases the appetite and hence food intake. Ghrelin, a hormone secreted by the empty stomach, opposes the actions of leptin, insulin, and PYY_{3-36}, stimulating appetite and food intake. Leptin also acts in peripheral tissue to stimulate energy expenditure by fatty acid oxidation and thermogenesis.

4 ■ Metabolic Adaptation During prolonged starvation, the brain slowly adapts from the use of glucose as its sole fuel source to the use of ketone bodies, thereby shifting the metabolic burden from protein breakdown to fat breakdown.

Diabetes mellitus is a disease in which insulin either is not secreted or does not efficiently stimulate its target tissues, leading to high concentrations of glucose in the blood and urine. Cells "starve" in the midst of plenty since they cannot absorb blood glucose and their hormonal signals remain those of starvation. Abnormally high production of ketone bodies is one of the most dangerous effects of uncontrolled diabetes.

REFERENCES

Chapters 17 to 26 of this text.

Baskin, D.G., Lattemann, D.F., Seeley, R.J., Woods, S.C., Porte, D., Jr., and Schwartz, M.W., Insulin and leptin: dual adiposity signals to the brain for the regulation of food intake and body weight, *Brain Res.* **848,** 114–123 (1999).

Batterham, R.L., *et al.,* Gut hormone PYY_{3-36} physiologically inhibits food intake, *Nature* **418,** 650–654 (2002).

Brüning, J.C., Gautam, D., Burks, D.J., Gillette, J., Schubert, M., Orban, P.C., Klein, R., Krone, W., Müller-Weiland, D., and Kahn, C.R., Role of brain insulin receptor in control of body weight and reproduction, *Science* **289,** 2122–2125 (2000).

Chen, H., et al., Evidence that the diabetes gene encodes the leptin receptor: Identification of a mutation in the leptin receptor gene in *db/db* mice, *Cell* **84,** 491–495 (1996); *and* Chua, S.C., Jr., Chung, W.K., Wu-Peng, S., Zhang, Y., Liu, S.-M., Tartaglia, L., and Leibel, R.L., Phenotypes of mouse *diabetes* and rat *fatty* due to mutations in the OB (leptin) receptor, *Science* **271,** 994–996 (1996).

Considine, R.V., et al., Serum immunoreactive-leptin concentrations in normal-weight and obese humans, *New Engl. J. Med.* **334,** 292–295 (1996).

Evans, J.L., Goldfine, I.D., Maddux, B.A., and Grodsky, G.M., Oxidative stress and stress-activated signaling pathways: a unifying hypothesis of type 2 diabetes, *Endocrine Rev.* **23,** 599–622 (2002).

Kristensen, P., et al., Hypothalamic CART is a new anorectic peptide regulated by leptin, *Nature* **393,** 72–76 (1998).

Lee, G.-H., Proenca, R., Montez, J.M., Carroll, K.M., Darvishzadah, J.G., Lee, J.I., and Friedman, J.M., Abnormal splicing of the leptin receptor in *diabetic* mice, *Nature* **379,** 632–635 (1996).

Lowell, B.B. and Spiegelman, B.M., Towards a molecular understanding of adaptive thermogenesis, *Nature* **404,** 652–660 (2000).

Montague, C.T., et al., Congenital leptin deficiency is associated with severe early-onset obesity in humans, *Nature* **387,** 903–908 (1997).

Moreno-Aliaga, M.J., Marti, A., García-Foncillas, J. and Martínes, J.A., DNA hybridization arrays: a powerful technology for nutritional and obesity research, *British Journal of Nutrition* **86,** 119–122 (2001).

Nakazato, M., Murakami, N., Date, Y., Kojima, M., Matsuo, H., Kangawa, K., and Matsukara, S., A role for ghrelin in the central regulation of feeding, *Nature* **409,** 194–198 (2001).

Newgard, C.B. and McGarry, J.D., Metabolic factors in pancreatic β-cell signal transduction, *Annu. Rev. Biochem.* **64,** 689–719 (1995). [Reviews the biochemical mechanisms that mediate glucose-stimulated insulin secretion by pancreatic β cells.]

Obesity, *Science* **299,** 845–860 (2003). [A series of informative articles on the origins of obesity.]

Schwartz, M.W. and Morton, G.J., Keeping hunger at bay, *Nature* **418,** 595–597 (2002).

Schwartz, M.W., Woods, S.C., Porte, D. Jr., Seeley, R.J., and Baskin, D.G., Central nervous system control of food intake, *Nature* **404,** 661–671 (2000).

Shulman, G.I., Cellular mechanisms of insulin resistance, *J. Clin. Invest.* **106,** 171–176 (2000).

Spiegelman, B.M. and Flier, J.S., Obesity and the regulation of energy balance, *Cell* **104,** 531–543 (2001).

Tartaglia, L.A., et al., Identification and expression cloning of a leptin receptor, OB-R, *Cell* **83,** 1263–1271 (1995).

Tshöp, M., Smiley, D.L., and Heiman, M.L., Ghrelin induces adiposity in rodents, *Nature* **407,** 908–913 (2000).

Unger, R.H., Leptin physiology: a second look, *Regulatory Peptides* **92,** 87–95 (2000).

Yechoor, V.K., Patti, M.-E., Saccone, R., and Kahn, C.R., Coordinated patterns of gene expression for substrate and energy metabolism in skeletal muscle of diabetic mice, *Proc. Natl. Acad. Sci.* **99,** 10587–10592 (2002).

Zhang, F., et al., Crystal structure of the *obese* protein leptin-E100, *Nature* **387,** 206–209 (1997).

Zhang, Y., Proenca, R., Maffei, M., Barone, M., Leopold, L., and Friedman, J.M., Positional cloning of the mouse *obese* gene and its human homologue, *Nature* **372,** 425–432 (1994).

Zick, Y., Insulin resistance: a phosphorylation-based uncoupling of insulin signaling, *Trends Cell Biol.* **11,** 437–441 (2001).

PROBLEMS

1. Describe the metabolic effects of liver failure.

2. What is the basis of the hypothesis that athletes' muscles are more heavily buffered than those of normal individuals?

3. Experienced runners know that it is poor practice to ingest large amounts of glucose prior to running a long-distance race such as a marathon. What is the metabolic basis of this apparent paradox?

4. Explain why urea output is vastly decreased during starvation.

5. Explain why people survive longer by total fasting than on a diet consisting only of carbohydrates.

6. Explain why the breath of an untreated diabetic smells of acetone.

7. Among the many eat-all-you-want-and-lose-weight diets that have been popular for a time is one that eliminates all carbohydrates but permits the consumption of all the protein and fat desired. Would such a diet be effective? (*Hint:* Individuals on such a diet often complain that they have bad breath.)

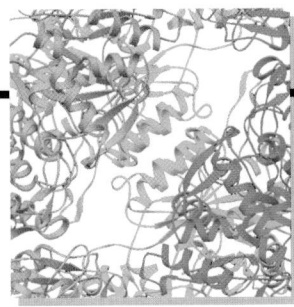

Chapter 28

Nucleotide Metabolism

Nucleotides, as we have seen, are biologically ubiquitous substances that participate in nearly all biochemical processes: They are the monomeric units of DNA and RNA; the hydrolysis of ATP and GTP drives many free energy–requiring processes; the levels of ATP, ADP, and AMP regulate numerous metabolic pathways; cAMP and cGMP mediate hormonal signals; and NAD$^+$, NADP$^+$, FMN, FAD, and coenzyme A are essential coenzymes in a great variety of enzymatic reactions. The importance of nucleotides in cellular metabolism is indicated by the observation that nearly all cells can synthesize them both *de novo* (anew) and from the degradation products of nucleic acids. In this chapter, we consider the nature of these biosynthetic pathways. In doing so, we shall examine how they are regulated and the consequences of their blockade, both by genetic defects and through the administration of chemotherapeutic agents. We then discuss how nucleotides are degraded. Finally, we outline the biosynthesis of the nucleotide coenzymes.

1 ■ SYNTHESIS OF PURINE RIBONUCLEOTIDES

In this section we commence our considerations of how nucleic acids and their components are synthesized by describing the synthesis of purine ribonucleotides. In 1948, John Buchanan obtained the first clues as to how this process occurs *de novo* by feeding a variety of isotopically labeled compounds to pigeons and chemically determining the positions of the labeled atoms in their excreted **uric acid** (a purine).

Uric acid

He used birds in these experiments because they excrete waste nitrogen almost entirely as uric acid, a water-insoluble and therefore easily isolated substance. The results of his studies, which are summarized in Fig. 28-1, demonstrated that N1 of purines arises from the amine group of aspartate; C2 and C8 originate from formate; N3

FIGURE 28-1 The biosynthetic origins of purine ring atoms. Note that C4, C5, and N7 come from a single glycine molecule but each of the other atoms is derived from an independent precursor.

and N9 are contributed by the amide group of glutamine; C4, C5, and N7 are derived from glycine (strongly suggesting that this molecule is wholly incorporated into the purine ring); and C6 comes from HCO_3^- (CO_2).

The actual pathway by which these precursors are incorporated into the purine ring, the subject of Section 28-1A, was elucidated in subsequent investigations performed largely by Buchanan and by G. Robert Greenberg. These investigations showed that the initially synthesized purine derivative is **inosine monophosphate (IMP),**

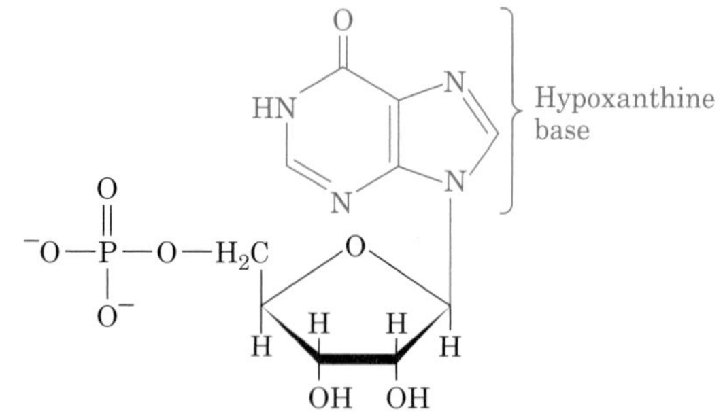

Inosine monophosphate (IMP)

the nucleotide of the base **hypoxanthine.** AMP and GMP are subsequently synthesized from this intermediate via separate pathways (Section 28-1B). Thus, contrary to naive expectation, purines are initially formed as ribonucleotides rather than as free bases. Additional studies have demonstrated that such widely divergent organisms as *E. coli*, yeast, pigeons, and humans have virtually identical pathways for the biosynthesis of purine nucleotides, thereby further demonstrating the biochemical unity of life.

A. Synthesis of Inosine Monophosphate

IMP is synthesized in a pathway comprising 11 reactions (Fig. 28-2):

1. Activation of ribose-5-phosphate. The starting material for purine biosynthesis is α-D-ribose-5-phosphate (R5P), a product of the pentose phosphate pathway (Section 23-4). In the first step of *de novo* purine biosynthesis, **ribose phosphate pyrophosphokinase** (also known as **phosphoribosylpyrophosphate synthetase**) activates R5P by reacting it with ATP to form **5-phosphoribosyl-α-pyrophosphate (PRPP).** This reaction, which occurs via the nucleophilic attack of the R5P's C1—OH group on the P_β of ATP, is unusual in that a pyrophosphoryl group is directly transferred from ATP to C1 of R5P and that the product has the α anomeric configuration. PRPP is also a precursor in the biosynthesis of pyrimidines (Section 28-2A) and the amino acids histidine and tryptophan (Section 26-5B). Thus, as is expected for an enzyme at such an important biosynthetic crossroads, the activity of ribose phosphate pyrophosphokinase varies with the concentrations of numerous metabolites, including PP_i and 2,3-bisphosphoglycerate, which are activators, and ADP and GDP, which are mixed inhibitors (Section 14-3C). The

regulation of purine nucleotide biosynthesis is further discussed in Section 28-1C.

2. Acquisition of purine atom N9. Amidophosphoribosyltransferase (alternatively, **glutamine PRPP aminotransferse** or **PurF;** the latter being named for the *E. coli* gene encoding it, *purF*) catalyzes the displacement of PRPP's pyrophosphate group by glutamine's amide nitrogen to yield **β-5-phosphoribosylamine (PRA).** This is the first reaction in the pathway that is unique to *de novo* purine biosynthesis (and hence some sources refer to it as the first reaction of the pathway, which is then said to consist of 10 reactions). This process occurs in two consecutive reactions that take place on separate active sites on the enzyme:

1. Glutamine + H_2O → glutamic acid + NH_3

2. NH_3 + PRPP → PRA + PP_i

Step 1 is catalyzed by a member of the N-terminal nucleophile (Ntn) amidotransferase family (Section 26-5A).

FIGURE 28-2 (*Opposite*) **The metabolic pathway for the** *de novo* **biosynthesis of IMP.** The purine residue is built up on a ribose ring in 11 enzymatically catalyzed reactions. The X-ray structures for all enzymes but that catalyzing Reaction 5 are shown to the outside of the corresponding reaction arrow. The peptide chains of monomeric enzymes are color-ramped from N-terminus (*blue*) to C-terminus (*red*). The oligomeric enzymes, all of which consist of identical polypeptide chains, are viewed along a rotation axis with their various chains differently colored. Bound ligands are shown in space-filling form. Enzyme 1, determined by Sine Larson, University of Copenhagen, Denmark, is a D_3 hexamer from *B. subtilis* that binds **α,β-methylene-ADP** at its catalytic (*red*) and allosteric (*blue*) sites; PDBid 1DKU. Enzyme 2, determined by Janet Smith, Purdue University, is a D_2 tetramer from *B. subtilis* that binds GMP (*blue*), ADP (*red*), and a [4Fe–4S] cluster (*orange*, which appears to have a regulatory rather than a redox function); PDBid 1AO0. Enzymes 3 and 6, both from *E. coli*, were determined by JoAnne Stubbe, MIT, and Steven Ealick, Cornell University; PDBids 1GSO and 1CLI. Enzyme 4, from *E. coli*, determined by Robert Almassy, Agouron Pharmaceuticals, San Diego, California, binds GAR (*cyan*) and 5-deazatetrahydrofolate (*red*); PDBid 1CDE. Reaction 7, in *E. coli*, is catalyzed by two sequentially acting enzymes, Class I PurE (*above*) and PurK (*below*). Class I PurE, determined by JoAnne Stubbe, MIT, and Steven Ealick, Cornell University, is a D_4 octamer that binds AIR (*red*); PDBid 1D7A. PurK, determined by JoAnne Stubbe, MIT, and Hazel Holden, University of Wisconsin, is a C_2 dimer that binds ADP (*red*); PDBid 1B6S. Enzyme 8, from yeast, was determined by Victor Lamzin, Academy of Sciences, Moscow, Russia, and Keith Wilson, EMBL, Hamburg, Germany; PDBid 1A48. Enzyme 9, from *Thermatoga maritima*, determined by Todd Yeates, UCLA, is a D_2 tetramer; PDBid 1C3U. Reactions 10 and 11 in chicken are catalyzed by a bifunctional enzyme that was determined by Stephen Benkovic, Pennsylvania State University, and Ian Wilson, The Scripps Research Institute, La Jolla, California. It forms a C_2 dimer shown with its AICAR transformylase function above and its IMP cyclohydrolase function, which binds GMP (*purple*), below; PDBid 1G8M. ✍ **See the Animated Figures**

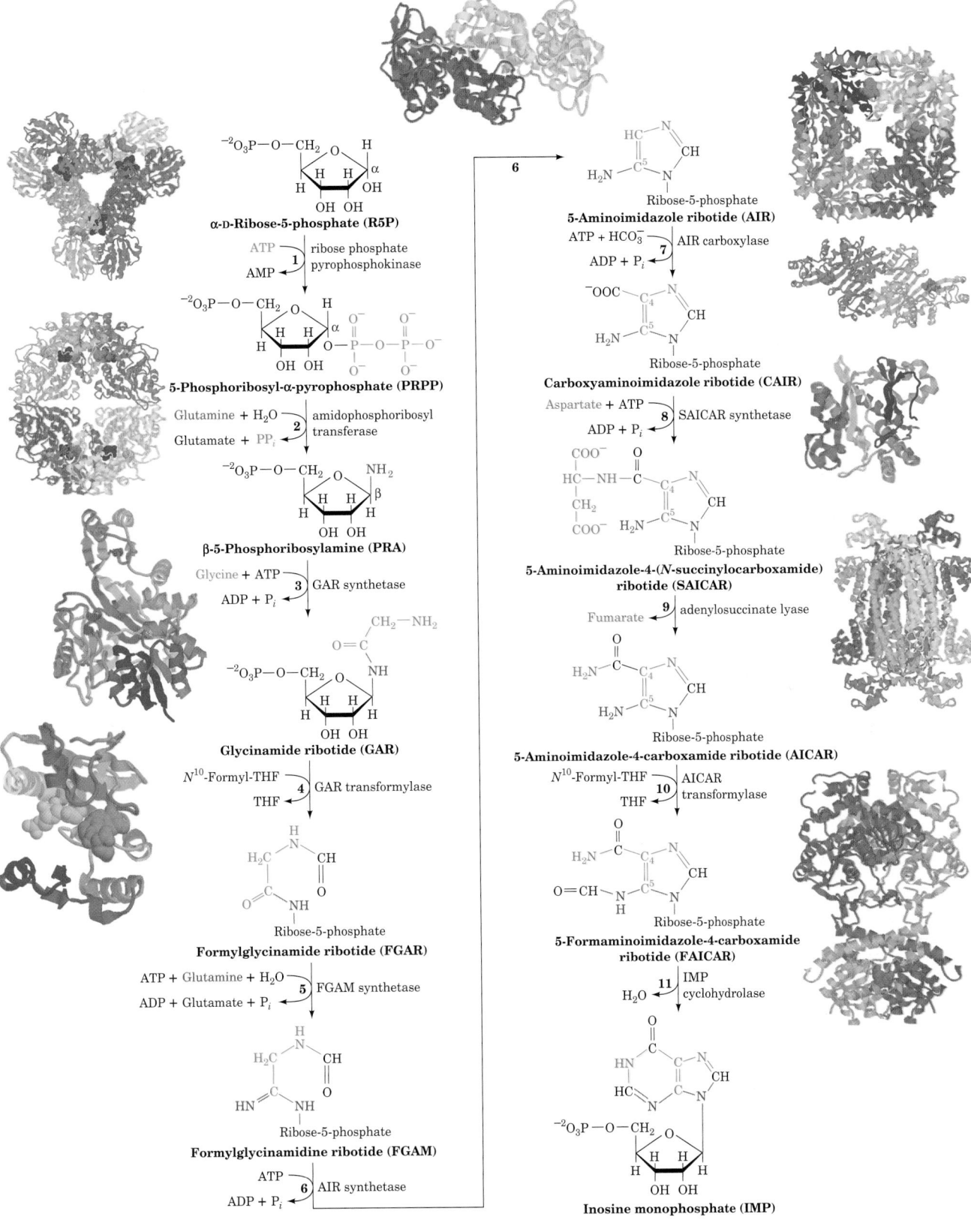

α-D-Ribose-5-phosphate (R5P)

ATP, ribose phosphate pyrophosphokinase **1**, AMP

5-Phosphoribosyl-α-pyrophosphate (PRPP)

Glutamine + H_2O, amidophosphoribosyl transferase **2**, Glutamate + PP_i

β-5-Phosphoribosylamine (PRA)

Glycine + ATP, GAR synthetase **3**, ADP + P_i

Glycinamide ribotide (GAR)

N^{10}-Formyl-THF, GAR transformylase **4**, THF

Formylglycinamide ribotide (FGAR)

ATP + Glutamine + H_2O, FGAM synthetase **5**, ADP + Glutamate + P_i

Formylglycinamidine ribotide (FGAM)

ATP, AIR synthetase **6**, ADP + P_i

5-Aminoimidazole ribotide (AIR)

ATP + HCO_3^-, AIR carboxylase **7**, ADP + P_i

Carboxyaminoimidazole ribotide (CAIR)

Aspartate + ATP, SAICAR synthetase **8**, ADP + P_i

5-Aminoimidazole-4-(N-succinylocarboxamide) ribotide (SAICAR)

Fumarate, adenylosuccinate lyase **9**

5-Aminoimidazole-4-carboxamide ribotide (AICAR)

N^{10}-Formyl-THF, AICAR transformylase **10**, THF

5-Formaminoimidazole-4-carboxamide ribotide (FAICAR)

IMP cyclohydrolase **11**, H_2O

Inosine monophosphate (IMP)

Step 2 occurs with inversion of configuration about ribose C1 and hence establishes the anomeric form of the future nucleotide. The NH_3 passes between the two active sites through a 20-Å-long tunnel that is lined with conserved nonpolar residues that lack hydrogen bonding groups and hence do not impede the diffusion of the NH_3 [we have seen that NH_3 generated by glutamine hydrolysis is similarly channeled to the active site that uses it in carbamoyl phosphate synthetase (Section 26-2A) and glutamate synthetase (Section 26-5A)]. These reactions, which are driven to completion by the subsequent hydrolysis of the released PP_i, constitute the pathway's flux-generating step. Not surprisingly, therefore, amidophosphoribosyltransferase is subject to feedback inhibition by purine nucleotides (Section 28-1C).

3. Acquisition of purine atoms C4, C5, and N7. Glycine's carboxyl group forms an amide with the amino group of PRA, yielding **glycinamide ribotide (GAR)** in a reaction, catalyzed by **GAR synthetase (PurD),** that occurs via the intermediate phosphorylation of glycine's carboxyl group. The reaction, which is reversible despite its concomitant hydrolysis of ATP to ADP + P_i, is the only step of the purine biosynthesis pathway in which more than one purine ring atom is acquired. The observation that PRA is chemically unstable (it is hydrolyzed to R5P and NH_3 with a half-life of 5 s at 37°C) suggests that GAR synthetase and amidophosphoribosyltransferase associate in a way that channels PRA between them. Indeed, a sterically and electrostatically plausible model of such a complex has been built based on the X-ray structures of these two enzymes.

4. Acquisition of purine atom C8. GAR's free α-amino group is formylated by **GAR transformylase (PurN)** to yield **formylglycinamide ribotide (FGAR).** The formyl donor in this reaction is N^{10}-formyltetrahydrofolate (N^{10}-formyl-THF), a cofactor that transfers C_1 units from such donors as serine, glycine, and formate to various acceptors in biosynthetic reactions (Section 26-4D). Structural and enzymological studies indicate that the reaction proceeds via the nucleophilic attack of the GAR amine group on the formyl carbon of N^{10}-formyl-THF to yield a tetrahedral intermediate.

5. Acquisition of purine atom N3. The amide amino group of a second glutamine is transferred to the growing purine ring to form **formylglycinamidine ribotide (FGAM).** This reaction, which is catalyzed by **FGAM synthetase (PurL),** is driven by the coupled hydrolysis of ATP to ADP + P_i. It is thought to proceed by the mechanism diagrammed in Fig. 28-3. Here the oxygen of the FGAR isoamide form reacts with ATP to yield a phosphoryl ester intermediate. This intermediate then reacts with NH_3 (the glutamine amide nitrogen as labilized through the transient formation of an enzyme thioester) to form a tetrahedral adduct. The adduct then eliminates P_i to yield the imine product, FGAM. Such reactions, in which a carboxamide oxygen is replaced by an imino group, are common in the biosynthesis of nucleotides. For example, Reaction 6 of this pathway and the reactions converting IMP to

AMP (Section 28-1B) and UTP to CTP (Section 28-2B) follow similar mechanisms, that is, conversion of a carboxamide oxygen to a phosphoryl ester that is nucleophilically attacked by an amine nitrogen atom to yield a tetrahedral adduct that, in turn, expels P_i to form product.

6. Formation of the purine imidazole ring. The purine imidazole ring is closed in an ATP-requiring intramolecular condensation that yields **5-aminoimidazole ribotide**

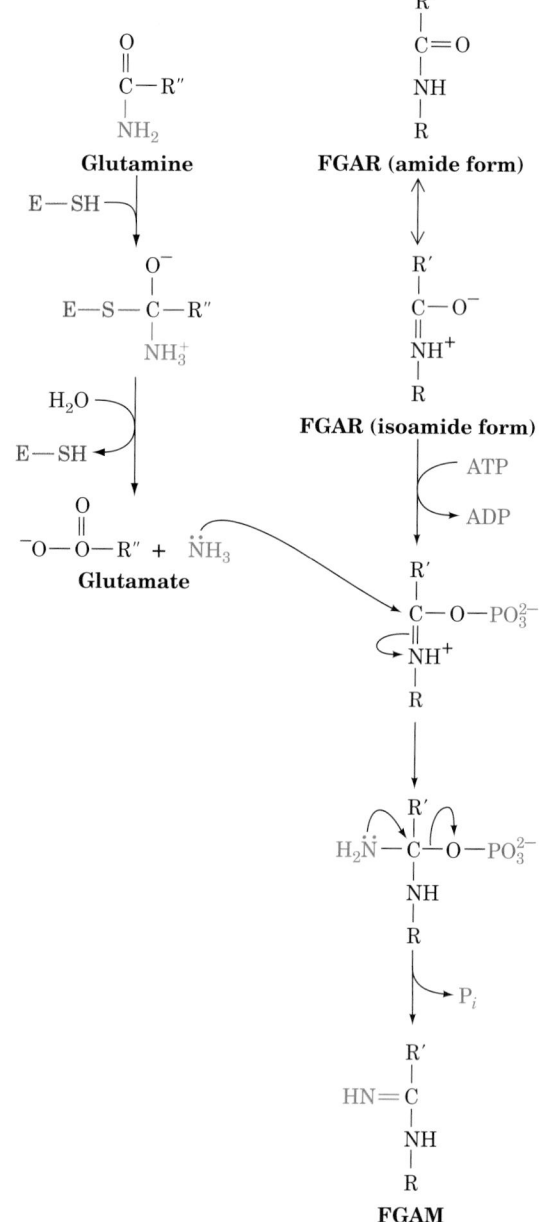

FIGURE 28-3   The proposed mechanism of formylglycinamide ribotide (FGAM) synthetase. The glutaminase domain of the enzyme contains an active site Cys residue that catalyzes the release of NH_3 with the transient formation of an enzyme thioester (not shown) whose hydrolysis produces glutamate. The isoamide form of FGAR is phosphorylated by ATP and then reacts with "NH_3" to form a tetrahedral intermediate whose collapse yields FGAM + P_i.

(AIR) in a reaction catalyzed by **AIR synthetase (PurM).** The aromatization of the imidazole ring is facilitated by the tautomeric shift of the reactant from its imine to its enamine form.

7. Acquisition of C6. In higher eukaryotes, purine C6 is introduced as HCO_3^- (CO_2) in a reaction catalyzed by **AIR carboxylase** that yields **carboxyaminoimidazole ribotide (CAIR).** However, in yeast, plants, and most prokaryotes (including *E. coli*), this overall reaction occurs in two steps that are mediated by separate enzymatic activities: **PurK** and Class I **PurE.**

PurK catalyzes the ATP-dependent carboxylation of AIR to yield N^5**-CAIR,** which Class I PurE rearranges to yield CAIR. Class I PurE is homologous to AIR carboxylase, which is therefore also called Class II PurE. Class I PurE alone can catalyze the AIR carboxylase reaction but since its K_M for HCO_3^- is 110 mM, it requires an unphysiologically high ($\sim$100 mM) HCO_3^- concentration to do so at a significant rate. However, the action of PurK decreases the HCO_3^- concentration required for the PurE-catalyzed reaction by >1000-fold, presumably through the ATP-driven formation of carbonyl phosphate, as is also postulated to occur in the carbamoyl phosphate synthetase reaction (Section 26-2A). The observation that N^5-CAIR is chemically unstable (it decomposes to AIR with a half-life of 15 s at pH 7.5 and 25°C) suggests that N^5-CAIR is channeled between PurK and Class I PurE. In fact, in yeast and plants, the N-terminus of Class I PurE is fused to the C-terminus of PurK. However, in *E. coli*, these two enzymatic activities occur on separate proteins for which there is no evidence of association.

8. Acquisition of N1. Purine atom N1 is contributed by aspartate in an amide-forming condensation reaction yielding **5-aminoimidazole-4-(N-succinylocarboxamide) ribotide (SAICAR)** that is catalyzed by **SAICAR synthetase (PurC).** The reaction, which is driven by the hydrolysis of ATP to ADP + P_i, chemically resembles Reaction 3.

9. Elimination of fumarate. SAICAR is cleaved with the release of fumarate, yielding **5-aminoimidazole-4-carboxamide ribotide (AICAR)** in a reaction catalyzed by **adenylosuccinate lyase (PurB).** Reactions 8 and 9 chemi-

cally resemble the reactions in the urea cycle in which citrulline is aminated to form arginine (Sections 26-2C and 26-2D). In both pathways, aspartate's amino group is transferred to an acceptor through an ATP-driven coupling reaction followed by the elimination of the aspartate carbon skeleton as fumarate. In plants and microorganisms, AICAR is also formed in the biosynthesis of histidine (Section 26-5B) but since in that process the AICAR is derived from ATP, it provides for no net purine biosynthesis.

10. Acquisition of C2. The final purine ring atom is acquired through formylation by N^{10}-formyltetrahydrofolate, yielding **5-formaminoimidazole-4-carboxamide ribotide (FAICAR)** in a reaction catalyzed by **AICAR transformylase (PurH).** In bacteria, this reaction and that of Reaction 4 are indirectly inhibited by sulfonamides, which, it will be recalled, prevent the synthesis of folate by competing with its *p*-aminobenzoate component (Section 26-4D). Animals, including humans, must acquire folate through the diet, since they are incapable of synthesizing it. They are therefore unaffected by sulfonamides. The antibiotic properties of sulfonamides are therefore largely a result of their inhibition of nucleic acid biosynthesis in susceptible bacteria.

11. Cyclization to form IMP. The final reaction in the pathway, ring closure to form IMP, occurs through the elimination of water as catalyzed by **IMP cyclohydrolase (PurJ).** In contrast to Reaction 6, the cyclization that forms the imidazole ring, this reaction does not entail ATP hydrolysis.

In animals, the activities catalyzing Reactions 3, 4, and 6, Reactions 7 and 8, and Reactions 10 and 11 occur on single polypeptides. The intermediate products of these multifunctional enzymes are not readily released to the medium but are channeled to the succeeding enzymatic activities of the pathway, thereby increasing the overall rates of these multistep processes and protecting the intermediates from degradation by other cellular enzymes. We have previously seen, for example, that the formation of acetyl-CoA from pyruvate takes place on the pyruvate dehydrogenase multienzyme complex, which contains three enzymes catalyzing five consecutive reactions (Section 21-2A); that all seven enzymatic activities catalyzing fatty acid synthesis in animals occur on a single protein molecule (Section 25-4C); and that the multifunctional enzymes carbamoyl phosphate synthase I (Section 26-2A), tryptophan synthase (Section 26-5B), and amidophosphoribosyltransferase pass reactive intermediate products between their active sites via protein tunnels. It is becoming increasingly apparent that the association of functionally related enzymes is a widespread phenomenon.

B. Synthesis of Adenine and Guanine Ribonucleotides

IMP does not accumulate in the cell but is rapidly converted to AMP and GMP. AMP, which differs from IMP only in the replacement of its 6-keto group by an amino

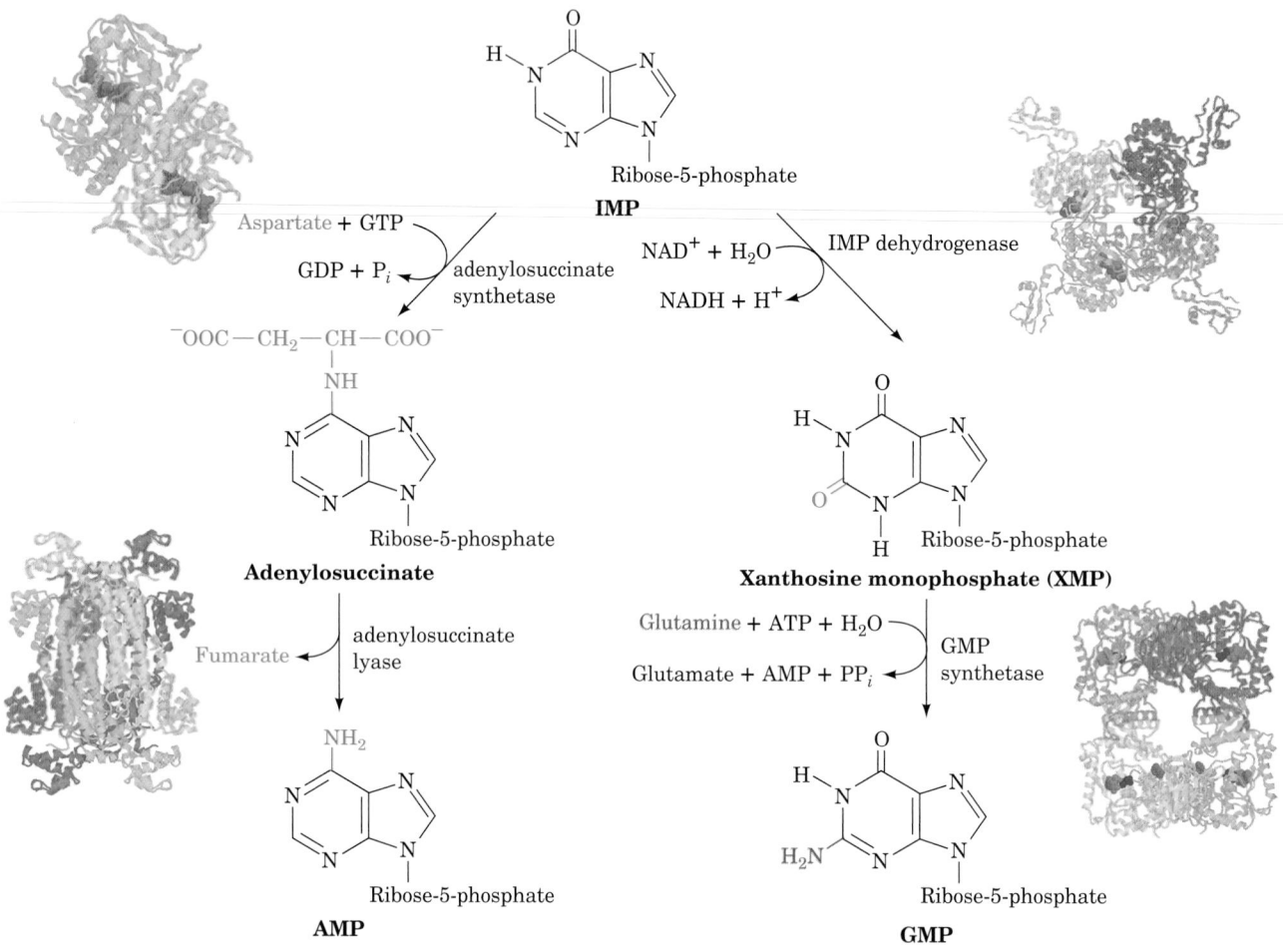

FIGURE 28-4   IMP is converted to AMP or GMP in separate two-reaction pathways. The X-ray structures for all of the enzymes catalyzing these reactions are shown to the outside of the corresponding reaction arrow. The X-ray structures for these homooligomers are shown as described in the legend to Fig. 28-2. Adenylosuccinate synthetase from *E. coli,* determined by Herbert Fromm and Richard Honzatko, Iowa State University, is a C_2 dimer in complex with IMP (*green*), GDP (*red*), and **hadacidin** (*magenta*; a competitive inhibitor of aspartate);

PDBid 1GIM. Adenylosuccinate lyase, from *Thermatoga maritima,* determined by Todd Yeates, UCLA, is a D_2 tetramer; PDBid 1C3U. IMP dehydrogenase from Chinese hamsters, determined by Keith Wilson, Vertex Pharmaceuticals, Cambridge, Massachusetts, is a C_4 tetramer in complex with oxidized IMP (*red*) and MPA (*purple*); PDBid 1JR1. GMP synthetase from *E. coli,* determined by Janet Smith, Purdue University, is a D_2 tetramer in complex with AMP (*red*), pyrophosphate (*blue*), and citrate (*purple*); PDBid 1GPM.

group, is synthesized in a two-reaction pathway (Fig. 28-4, *left*). In the first reaction, aspartate's amino group is linked to IMP in a reaction powered by the hydrolysis of GTP to GDP + P_i to yield **adenylosuccinate.** In the second reaction, **adenylosuccinate lyase** eliminates fumarate from adenylosuccinate to form AMP. This enzyme also catalyzes Reaction 9 of the IMP pathway (Fig. 28-2).

GMP is also synthesized from IMP in a two-reaction pathway (Fig. 28-4, *right*). In the first reaction, **IMP dehydrogenase** catalyzes the NAD^+-dependent oxidation of IMP to form **xanthosine monophosphate (XMP; the ribonucleotide of the base xanthine).** XMP is then converted to GMP by the replacement of its 2-keto group with glutamine's amide nitrogen in a reaction driven by the hydrolysis of ATP to AMP + PP_i (and subsequently to 2 P_i).

IMP dehydrogenase, a homotetramer of 514-residue

subunits, was incubated with IMP, NAD^+, and the fungally produced inhibitor **mycophenolic acid (MPA).**

Mycophenolic acid (MPA)

The X-ray structure of the resulting complex, determined by Keith Wilson, reveals that the enzyme had bound MPA together with a reaction intermediate in which IMP atom C2 had become covalently linked to the Cys 331 S atom and then dehydrogenated by NAD^+ to yield a thioimidate ester:

Enzyme–product thioimidate ester

The mutagenic replacement of Cys 331 by Ala inactivates the enzyme. These observations strongly support a catalytic mechanism in which the Cys 331 thiol group nucleophilically attacks IMP's C2 atom, followed by hydride transfer to NAD^+ to yield the above covalently bound intermediate, which is subsequently hydrolyzed to yield XMP. The MPA binds to the enzyme with its bicyclic ring stacked on the purine ring (as would be expected for NAD^+'s nicotinamide ring) and with its phenolic hydroxyl group in the proposed hydrolytic water site. This blocks the hydrolysis of the thioimidate ester, thereby inactivating the enzyme.

IMP dehydrogenase activity is essential to the immune response (Section 35-2) because it is required by the immune system cells known as *B* and *T* lymphocytes to generate the guanosine nucleotides they need to proliferate. Moreover, certain cancer cells have increased IMP dehydrogenase activity. Hence, IMP dehydrogenase is a target for both immunosuppressive therapy and cancer chemotherapy. Indeed, MPA is in clinical use to prevent the rejection of transplanted kidneys.

a. Nucleoside Diphosphates and Triphosphates Are Synthesized by the Phosphorylation of Nucleoside Monophosphates

In order to participate in nucleic acid synthesis, nucleoside monophosphates must first be converted to the corresponding nucleoside triphosphates. In the first of the two sequential phosphorylation reactions that do so, nucleoside diphosphates are synthesized from the corresponding nucleoside monophosphates by base-specific **nucleoside monophosphate kinases.** For example, adenylate kinase (Section 17-4F) catalyzes the phosphorylation of AMP to ADP:

$$AMP + ATP \rightleftharpoons 2ADP$$

Similarly, GDP is produced by a guanine-specific enzyme:

$$GMP + ATP \rightleftharpoons GDP + ADP$$

These nucleoside monophosphate kinases do not discriminate between ribose and deoxyribose in the substrate.

Nucleoside diphosphates are converted to the corresponding triphosphates by **nucleoside diphosphate kinase;** for instance,

$$ATP + GDP \rightleftharpoons ADP + GTP$$

Although this reaction is written with ATP as the phosphoryl donor and GDP as the acceptor, nucleoside diphosphate

kinase is nonspecific as to the bases on either of its substrates and as to whether their sugar residues are ribose or deoxyribose. The reaction occurs via a Ping Pong mechanism in which the substrate NTP phosphorylates an enzyme His residue which, in turn, phosphorylates the substrate NDP. The phosphoglycerate mutase reaction of glycolysis also has a phospho-His intermediate (Section 17-2H). The nucleoside diphosphate kinase reaction, as might be expected from the nearly identical structures of its substrates and products, normally operates close to equilibrium ($\Delta G \approx 0$). ADP is, of course, also converted to ATP by a variety of energy-releasing reactions such as those of glycolysis and oxidative phosphorylation. Indeed, it is these reactions that ultimately drive the foregoing kinase reactions.

C. Regulation of Purine Nucleotide Biosynthesis

The pathways involved in nucleic acid metabolism are tightly regulated, as is evidenced, for example, by the increased rates of nucleotide synthesis during cell proliferation. In fact, the pathways synthesizing IMP, ATP, and GTP are individually regulated in most cells so as not only to control the total amounts of purine nucleotides produced but also to coordinate the relative amounts of ATP and GTP. This control network is diagrammed in Fig. 28-5.

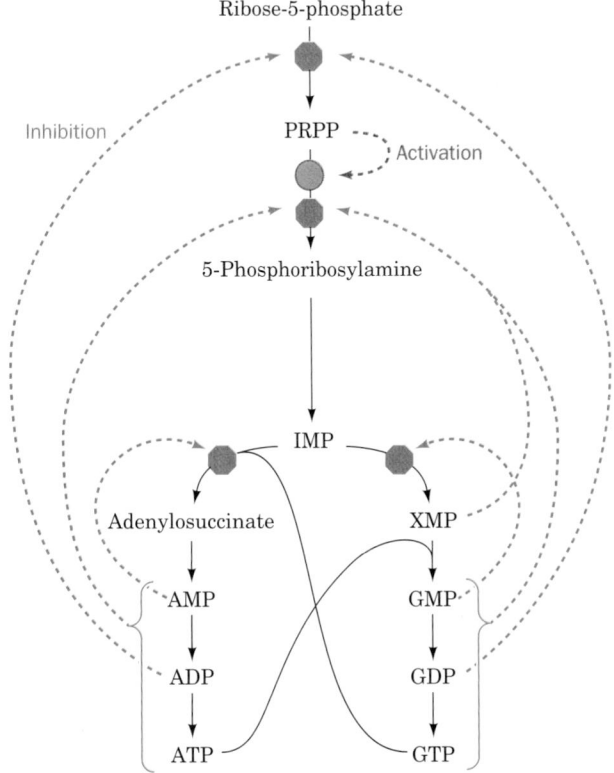

FIGURE 28-5 Control network for the purine biosynthesis pathway. Red octagons and green dots indicate control points. Feedback inhibition is indicated by dashed red arrows and feedforward activation is represented by dashed green arrows.
🐁 **See the Animated Figures**

The IMP pathway is regulated at its first two reactions: those catalyzing the synthesis of PRPP and 5-phosphoribosylamine. We have already seen that ribose phosphate pyrophosphokinase, the enzyme catalyzing Reaction 1 of the IMP pathway, is inhibited by both ADP and GDP (Section 28-1A). Amidophosphoribosyltransferase, the enzyme catalyzing the first committed step of the IMP pathway (Reaction 2), is likewise subject to feedback inhibition. In this case, however, the enzyme binds ATP, ADP, and AMP at one inhibitory site and GTP, GDP, and GMP at another. *The rate of IMP production is consequently independently but synergistically controlled by the levels of adenine nucleotides and guanine nucleotides.* Moreover, amidophosphoribosyltransferase is allosterically stimulated by PRPP (feedforward activation).

A second level of regulation occurs immediately below the branch point leading from IMP to AMP and GMP (Fig. 28-4). AMP and GMP are each competitive inhibitors of IMP in their own synthesis, so that excessive buildup of these products is impeded. In addition, the synthesis rates of adenine and guanine nucleotides are coordinated because GTP powers the synthesis of AMP from IMP, whereas ATP powers the synthesis of GMP from IMP. This reciprocity serves to balance the production of AMP and GMP (which are required in roughly equal amounts in nucleic acid biosynthesis): *The rate of synthesis of GMP increases with [ATP], whereas that of AMP increases with [GTP].*

D. *Salvage of Purines*

Most cells have an active turnover of many of their nucleic acids (particularly some types of RNA) which, through degradative processes described in Section 28-4A, result in the release of adenine, guanine, and hypoxanthine. These free purines are reconverted to their corresponding nucleotides through **salvage pathways.** In contrast to the *de novo* purine nucleotide synthesis pathway, which is virtually identical in all cells, salvage pathways are diverse in character and distribution. In mammals, purines are, for the most part, salvaged by two different enzymes. **Adenine phosphoribosyltransferase (APRT)** mediates AMP formation through the transfer of adenine to PRPP with the release of PP_i:

$$\text{Adenine} + \text{PRPP} \rightleftharpoons \text{AMP} + PP_i$$

Hypoxanthine–guanine phosphoribosyltransferase (HGPRT) catalyzes the analogous reaction for both hypoxanthine and guanine:

$$\text{Hypoxanthine} + \text{PRPP} \rightleftharpoons \text{IMP} + PP_i$$
$$\text{Guanine} + \text{PRPP} \rightleftharpoons \text{GMP} + PP_i$$

a. Lesch–Nyhan Syndrome Results from HGPRT Deficiency

The symptoms of **Lesch–Nyhan syndrome,** which is caused by a severe HGPRT deficiency, indicate that purine salvage reactions have functions other than conservation of the energy required for *de novo* purine biosynthesis.

This sex-linked congenital defect (affects almost only males) results in excessive uric acid production (uric acid is a purine degradation product; Section 28-4A) and neurological abnormalities such as spasticity, mental retardation, and highly aggressive and destructive behavior, including a bizarre compulsion toward self-mutilation. For example, many children with Lesch–Nyhan syndrome have such an irresistible urge to bite their lips and fingers that they must be restrained. If the restraints are removed, communicative patients will plead that the restraints be replaced even as they attempt to injure themselves.

The excessive uric acid production in patients with Lesch–Nyhan syndrome is readily explained. The lack of HGPRT activity leads to an accumulation of the PRPP that would normally be used in the salvage of hypoxanthine and guanine. The excess PRPP activates amidophosphoribosyltransferase (which catalyzes Reaction 2 of the IMP biosynthesis pathway; Fig. 28-2), thereby greatly increasing the rate of synthesis of purine nucleotides and consequently that of their degradation product, uric acid. Yet the physiological basis of the associated neurological abnormalities remains obscure. That a defect in a single enzyme can cause such profound but well-defined behavioral changes nevertheless has important psychiatric implications.

2 ■ SYNTHESIS OF PYRIMIDINE RIBONUCLEOTIDES

The biosynthesis of pyrimidines is a simpler process than that of purines. Isotopic labeling experiments have shown that atoms N1, C4, C5, and C6 of the pyrimidine ring are all derived from aspartic acid, C2 arises from HCO_3^-, and N3 is contributed by glutamine (Fig. 28-6). In this section we discuss the pathways for pyrimidine ribonucleotide biosynthesis and how these processes are regulated.

A. *Synthesis of UMP*

The major breakthrough in the determination of the pathway for the *de novo* biosynthesis of pyrimidine ribonucleotides was the observation that mutants of the bread mold *Neurospora crassa*, which are unable to synthesize pyrimidines and therefore require both cytosine and uracil in their growth medium, grow normally when supplied instead with the pyrimidine **orotic acid** (uracil-6-carboxylic acid).

Orotic acid (uracil-6-carboxylic acid)

This observation led to the elucidation of the following six-reaction pathway for the biosynthesis of UMP (Fig. 28-7). Note that, in contrast to the case for purine nucleotides,

Glutamine amide → N₃ — C⁴ — C⁵ ← Aspartate

HCO₃⁻ → C²₁ — C⁶ / N

FIGURE 28-6   The biosynthetic origins of pyrimidine ring atoms.

the pyrimidine ring is coupled to the ribose-5-phosphate moiety *after* the ring has been synthesized.

1. Synthesis of carbamoyl phosphate. The first reaction of pyrimidine biosynthesis is synthesis of **carbamoyl phosphate** from HCO_3^- and the amide nitrogen of glutamine by the cytosolic enzyme **carbamoyl phosphate synthetase II (CPS II).** This reaction is unusual in that it does not use biotin and consumes two molecules of ATP: One provides

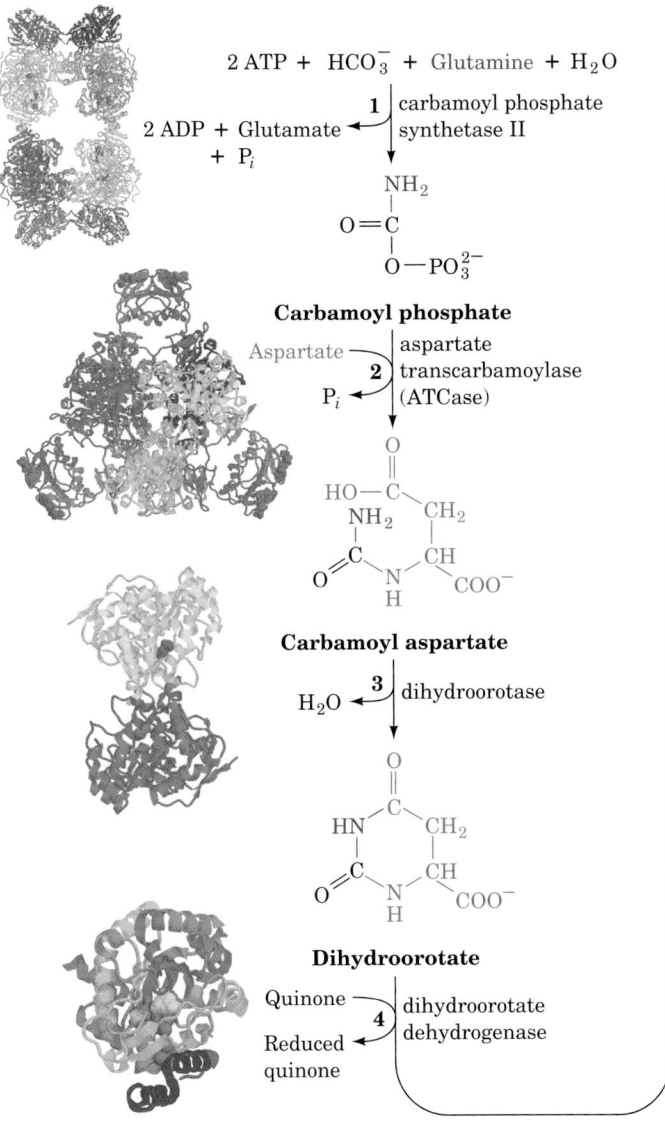

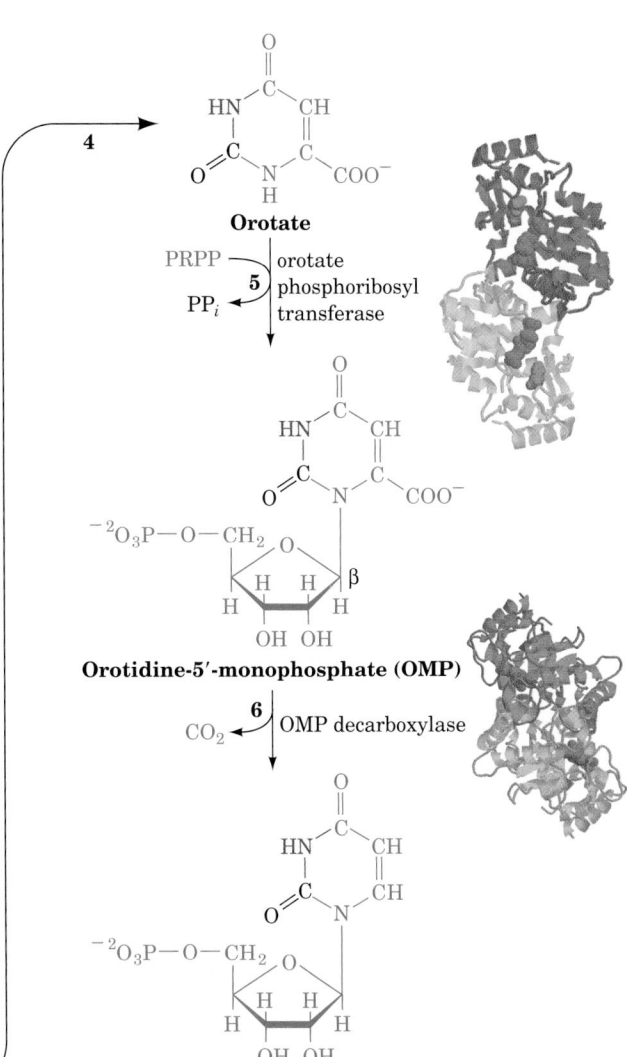

FIGURE 28-7   Metabolic pathway for the *de novo* synthesis of UMP. The pathway consists of six enzymatically catalyzed reactions. Note that, in contrast to the case for purine biosynthesis (Fig. 28-2), the pyrimidine ring is formed before its attachment to a ribose ring. The X-ray structures for the enzymes are shown as described in the legend to Fig. 28-2. Enzyme 1, from *E. coli,* determined by Hazel Holden, University of Wisconsin, is an $\alpha_4\beta_4$ heterooctamer with D_2 symmetry, whose large subunits each bind two ADPNPs (*red*) and one ornithine (*purple*); PDBid 1D3H. Enzyme 2, from *E. coli,* determined by William Lipscomb, Harvard University, is a c_6r_6 heterododecamer with D_3 symmetry, whose regulatory (*r*) subunits each bind a CTP (*green*); PDBid 5AT1. Enzyme 3 from

E. coli, determined by Hazel Holden, University of Wisconsin, is a C_2 dimer that binds carbamoyl aspartate (*purple*) in one subunit and orotate (*red*) in the other; PDBid 1J79. Enzyme 4, from humans, determined by Jon Clardy, Cornell University, is a monomer that binds orotate (*yellow*), FMN (*magenta*), and A77 1726 (*green*); PDBid 1D3H. Enzyme 5, from *Salmonella typhimurium,* determined by James Sacchettini, Albert Einstein College of Medicine, is a C_2 dimer that binds orotate (*orange*) and PRPP (*purple*); PDBid 1OPR. Enzyme 6, from *B. subtilis,* determined by Stephen Ealick, Cornell University, is a C_2 dimer that binds UMP (*green*); PDBid 1DBT.

🖙 **See the Animated Figures**

a phosphate group and the other energizes the reaction. We have previously discussed the synthesis of carbamoyl phosphate in connection with the formation of arginine (Section 26-2A). The carbamoyl phosphate that is used to synthesize arginine via the urea cycle is synthesized by a separate mitochondrial enzyme, **carbamoyl phosphate synthetase I (CPS I),** which uses ammonia as its nitrogen source. Prokaryotes only have one carbamoyl phosphate synthetase, which supplies both pyrimidine and arginine biosynthesis and utilizes glutamine. This latter enzyme, as we have seen, contains three different active sites that are connected by a remarkable 96-Å-long tunnel through which intermediate products diffuse (Fig. 26-9).

2. Synthesis of carbamoyl aspartate. Condensation of carbamoyl phosphate with aspartate to form **carbamoyl aspartate** is catalyzed by **aspartate transcarbamylase (ATCase).** This reaction, the pathway's flux-generating step, occurs without need of ATP because carbamoyl phosphate is intrinsically activated. The structure and regulation of *E. coli* ATCase is discussed in Section 13-4.

3. Ring closure to form dihydroorotate. The third reaction of the pathway was elucidated by Arthur Kornberg following his observation that microorganisms made to utilize orotic acid as a carbon source first reduce it to **dihydroorotate.** The reaction forming the pyrimidine ring yields dihydroorotate in an intramolecular condensation catalyzed by **dihydroorotase.**

4. Oxidation of dihydroorotate. Dihydroorotate is irreversibly oxidized to orotate by **dihydroorotate dehydrogenase (DHODH).** The eukaryotic enzyme, which contains FMN, is located on the outer surface of the inner mitochondrial membrane, where ubiquinone supplies its oxidizing power. The other five enzymes of pyrimidine nucleotide biosynthesis are cytosolic in animal cells. Many bacterial dihydroorotate dehydrogenases are NAD^+-linked flavoproteins that contain FMN, FAD, and a [2Fe–2S] cluster. These enzymes normally function degradatively, that is, in the direction orotate → dihydroorotate, thereby permitting these bacteria to metabolize orotate and accounting for Kornberg's observation. The reaction mediated by eukaryotic DHODH involves two redox steps, as is indicated in Fig. 28-8. The X-ray structure of human DHODH in complex with orotate, determined by Jon Clardy, reveals that the pyrimidine ring of orotate is stacked over the FMN's flavin ring with the orotate C6 and FMN N5 separated by 3.6 Å, a distance that is compatible with direct hydride transfer between these two centers. A tunnel leads from the opposite side of the flavin ring to a hydrophobic region on the enzyme surface. The enzyme presumably binds to the mitochondrial membrane surface via this hydrophobic patch, thereby permitting ubiquinone, which readily dif-

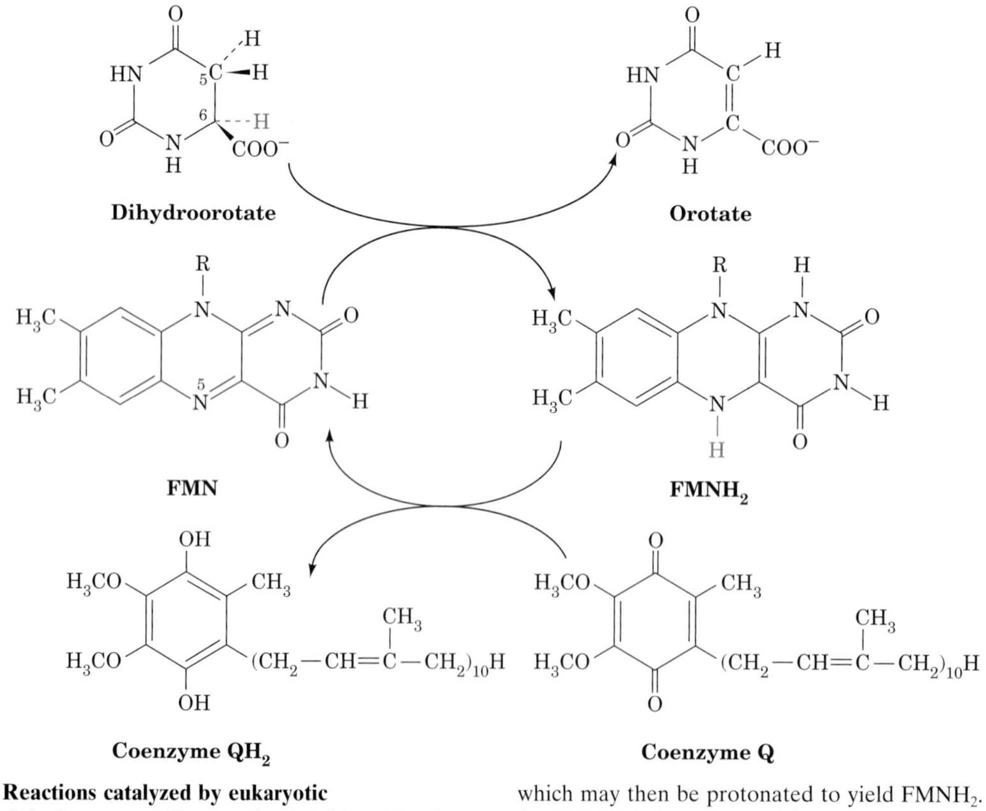

FIGURE 28-8   Reactions catalyzed by eukaryotic dihydroorotate dehydrogenase. The reaction is initiated by the enzyme-mediated abstraction of a proton from C5 of dihydroorotate followed by the direct hydride transfer from C6 of dihydroorotate to N5 of FMN to yield orotate and $FMNH^-$,

which may then be protonated to yield $FMNH_2$. The $FMNH_2$ (or $FMNH^-$) then reacts with coenzyme Q acquired from the inner mitochondrial membrane to regenerate the enzyme in its FMN form and yield coenzyme QH_2, which then re-enters the inner mitochondrial membrane.

fuses within the mitochondrial membrane, to approach and reoxidize the enzyme's bound FMNH$_2$. In the X-ray structure, this tunnel contains a tightly bound molecule named **A77 1726,** which is the primary metabolite of **leflunomide** (trade name **Arava**),

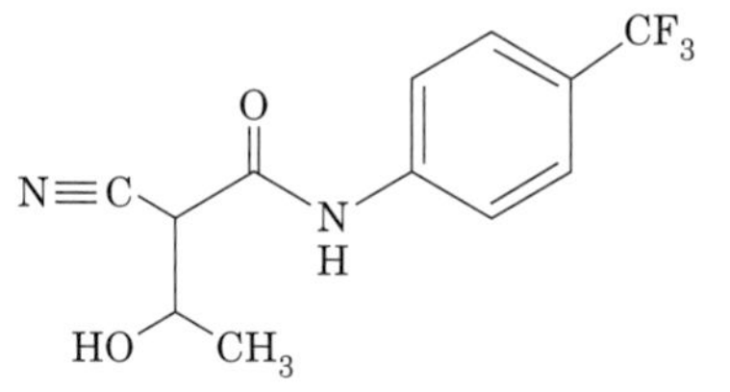

A77 1726

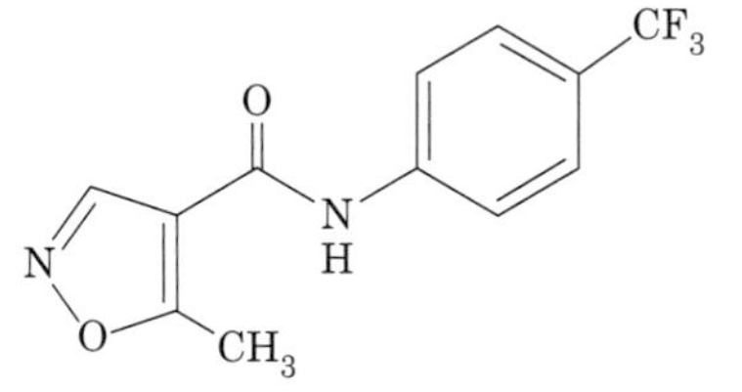

Leflunomide

a compound that is in clinical use for the treatment of rheumatoid arthritis. A77 1726 attenuates this autoimmune disease by blocking pyrimidine biosynthesis in *T* lymphocytes, thereby reducing their inappropriate proliferation. However, A77 1726 does not inhibit bacterial DHODHs.

5. Acquisition of the ribose phosphate moiety. Orotate reacts with PRPP to yield **orotidine-5′-monophosphate (OMP)** in a reaction catalyzed by **orotate phosphoribosyltransferase** and driven by hydrolysis of the eliminated PP$_i$. This reaction fixes the anomeric form of pyrimidine nucleotides in the β configuration. Orotate phosphoribosyltransferase also acts to salvage other pyrimidine bases, such as uracil and cytosine, by converting them to their corresponding nucleotides. Although the various phosphoribosyltransferases, including HGPRT, exhibit little sequence similarity, their X-ray structures indicate that they contain a common structural core that resembles the dinucleotide binding fold (Section 8-3B) but lacks one of its β strands.

6. Decarboxylation to form UMP. The final reaction of the pathway is the decarboxylation of OMP by **OMP decarboxylase (ODCase)** to form UMP. ODCase enhances the rate (k_{cat}/K_M) of OMP decarboxylation by a factor of

2×10^{23} over that of the uncatalyzed reaction, making it the most catalytically proficient enzyme known. Yet ODCase has no cofactors to help stabilize the reaction's putative carbanion intermediate. How is it able to do so? The X-ray structure, by Stephen Ealick, of ODCase from *B. subtilis* in complex with UMP indicates that a bound OMP's C6 carboxyl group that is coplanar with its pyrimidine ring would be in close proximity to the side chains of both Asp 60 and Lys 62. Ealick has therefore proposed a mechanism (Fig. 28-9) in which the electrostatic interactions between the closely spaced carboxyl groups of OMP and Asp 60 destabilize OMP's ground state. This destabilization would be reduced in the transition state by the shift of OMP's negative charge from its carboxyl group toward C6, where it would be stabilized by the adjacent positively charged side chain of Lys 62. This side chain is also proposed to protonate the fragmenting C—C bond when it becomes sufficiently basic to accept the proton, thus avoiding the formation of a high energy carbanion intermediate. The unfavorable electrostatic interaction between OMP and Asp 60 occurs because the enzyme tightly binds OMP through extensive interactions with its other functional groups. Indeed, the removal of OMP's phosphate group, which is quite distant from the C6 carboxyl group, decreases the catalytic reaction's k_{cat}/K_M by a factor of 7×10^7, thus providing a striking example of how binding energy can be applied to catalysis (preferential transition state binding).

In bacteria, the six enzyme activities mediating UMP biosynthesis occur on independent proteins. In animals, however, as Mary Ellen Jones demonstrated, the first three enzymatic activities of the pathway, carbamoyl phosphate synthetase II, ATCase, and dihydroorotase, occur on a single 210-kD polypeptide chain. Similarly, Reactions 5 and 6 of the animal pyrimidine pathway are catalyzed by a single polypeptide.

B. *Synthesis of UTP and CTP*

The synthesis of UTP from UMP is analogous to the synthesis of purine nucleotide triphosphates (Section 28-1B). The process occurs by the sequential actions of a nucleoside monophosphate kinase and nucleoside diphosphate kinase:

$$UMP + ATP \rightleftharpoons UDP + ADP$$
$$UDP + ATP \rightleftharpoons UTP + ADP$$

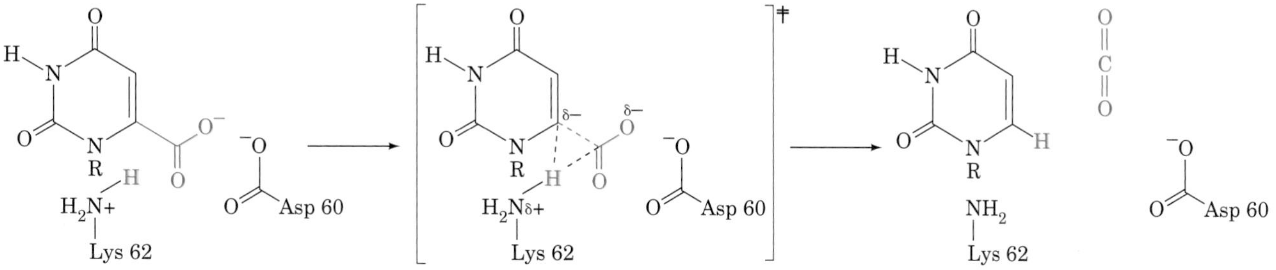

FIGURE 28-9 Proposed catalytic mechanism for OMP decarboxylase. [After Appleby, T.C., Kinsland, C., Begley, T.P., and Ealick, S.E., *Proc. Natl. Acad. Sci.* **97**, 2005 (2000).]

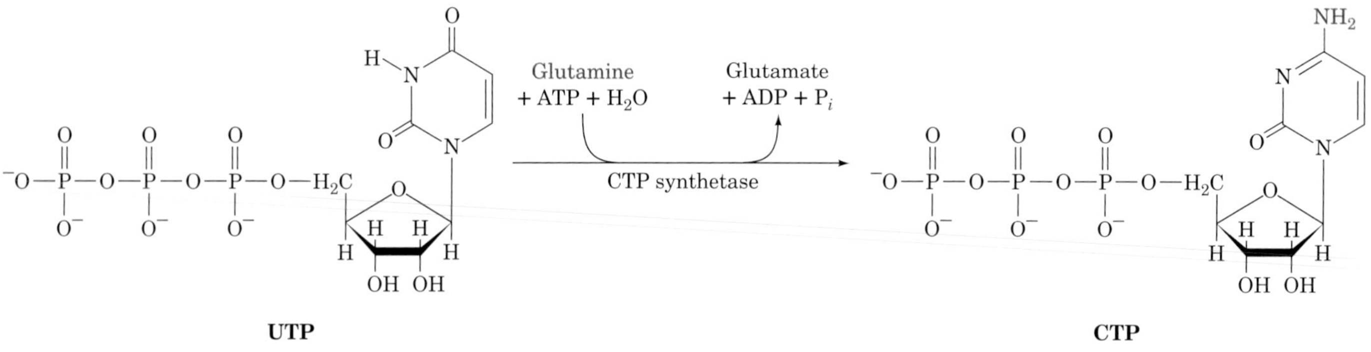

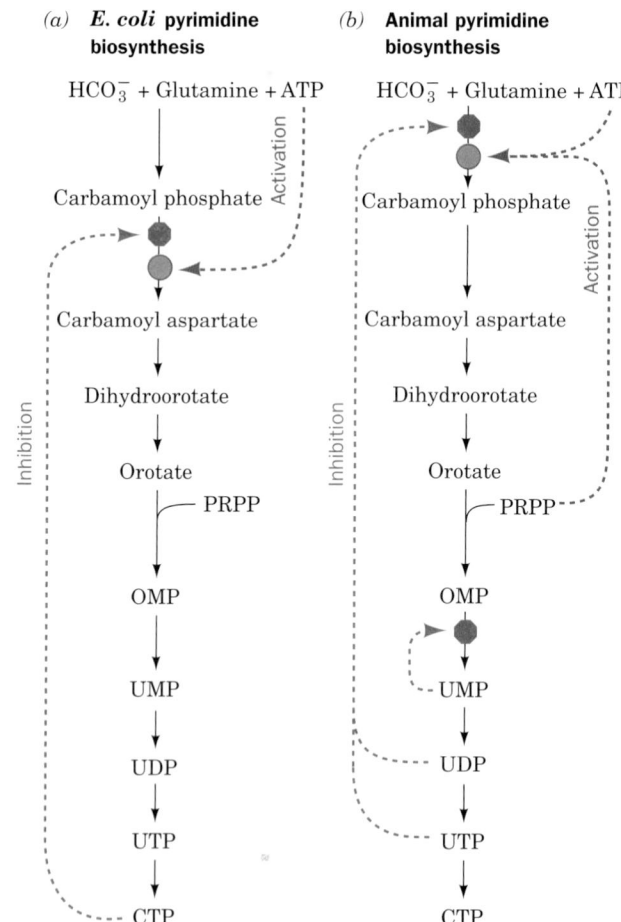

FIGURE 28-10   Synthesis of CTP from UTP.

CTP is formed by amination of UTP by **CTP synthetase** (Fig. 28-10). In animals, the amino group is donated by glutamine, whereas in bacteria it is supplied directly by ammonia.

C. *Regulation of Pyrimidine Nucleotide Biosynthesis*

In bacteria, the pyrimidine biosynthesis pathway is primarily regulated at Reaction 2, the ATCase reaction (Fig. 28-11a). In *E. coli,* control is exerted there through the allosteric stimulation of ATCase by ATP and its inhibition by CTP (Section 13-4). In many bacteria, however, UTP is the major ATCase inhibitor.

In animals, ATCase is not a regulatory enzyme. Rather, pyrimidine biosynthesis is controlled by the activity of carbamoyl phosphate synthetase II, which is inhibited by UDP and UTP and activated by ATP and PRPP (Fig. 28-11b). A second level of control in the mammalian pathway occurs at OMP decarboxylase, for which UMP and to a lesser extent CMP are competitive inhibitors. In all organisms, the rate of OMP production varies with the availability of its precursor, PRPP. The PRPP level, it will be recalled, depends on the activity of ribose phosphate pyrophosphokinase, which is inhibited by ADP and GDP (Section 28-1A).

a. Orotic Aciduria Results from an Inherited Enzyme Deficiency

Orotic aciduria, an inherited human disease, is characterized by the excretion of large amounts of orotic acid in the urine, retarded growth, and severe anemia. It results from a deficiency in the bifunctional enzyme catalyzing Reactions 5 and 6 of pyrimidine nucleotide biosynthesis. Consideration of the biochemistry of this situation led to its effective treatment: the administration of uridine and/or cytidine. The UMP formed through the phosphorylation of these nucleosides, besides replacing that normally synthesized, inhibits carbamoyl phosphate synthetase II so as to attenuate the rate of orotic acid synthesis. Few other genetic deficiencies in pyrimidine nucleotide biosynthesis are known in humans, presumably because most such defects are lethal *in utero.*

3 ■ FORMATION OF DEOXYRIBONUCLEOTIDES

DNA differs chemically from RNA in two major respects: (1) its nucleotides contain 2′-deoxyribose residues rather than ribose residues, and (2) it contains the base thymine

FIGURE 28-11   Regulation of pyrimidine biosynthesis. The control networks are shown for (*a*) *E. coli* and (*b*) animals. Red octagons and green dots indicate control points. Feedback inhibition is represented by dashed red arrows and activation is indicated by dashed green arrows. *See the Animated Figures*

(5-methyluracil) rather than uracil. In this section we consider the biosynthesis of these DNA components.

A. *Production of Deoxyribose Residues*

Deoxyribonucleotides are synthesized from their corresponding ribonucleotides by the reduction of their C2' position rather than by their de novo synthesis from deoxyribose-containing precursors.

NDP

dNDP

This pathway was established through Irwin Rose's study of how rats metabolize cytidine that is ^{14}C-labeled in both its base and ribose components. The dCMP recovered from the rats' DNA had the same labeling ratio in its cytosine and deoxyribose residues as had the original cytidine, indicating that the DNA's components remained linked during DNA synthesis. If the cytosine and the ribose residues had become separated, dilution of the labeled cytosine and ribose residues with unlabeled residues, which are present in rat tissues in different amounts, would have altered this ratio.

The enzymes that catalyze the formation of deoxyribonucleotides by the reduction of the corresponding ribonucleotides are named **ribonucleotide reductases (RNRs).** Three classes of RNRs are known that differ in their substrates (NDP or NTP), the cofactors they employ, and in the way they obtain reducing equivalents (see below). Class I and II RNRs are widely distributed among prokaryotes; some species have a Class I RNR, whereas other, sometimes related species have a Class II RNR. However, all eukaryotes except a few unicellular species have Class I RNRs. Class III RNRs occur in prokaryotes that can grow anaerobically. (Class III RNRs are O_2-sensitive whereas Class I RNRs require O_2 for activation; see below.) In fact, *E. coli*, which can grow both aerobically and anaerobically, contains a Class I and a Class III RNR. In what follows, we shall mainly discuss the mechanism of Class I RNRs but end with a consideration of the evolutionary relationships among the different classes of RNRs.

a. Class I Ribonucleotide Reductase: Structure and Mechanism

The *E. coli* Class I RNR, as Peter Reichard demonstrated, is mainly present *in vitro* as a heterotetramer that

can be decomposed to two catalytically inactive homodimers, R1$_2$ (761-residue subunits) and R2$_2$ (375-residue subunits), which together form the enzyme's two active sites (Fig. 28-12*a*). Each R1 subunit contains a substrate binding site as well as three independent effector binding sites that control both the enzyme's catalytic activity and its substrate specificity (see below). R1's catalytic residues include several redox-active thiol groups.

The X-ray structure of R2$_2$ (Fig. 28-12*b*), which was determined by Hans Eklund, reveals that each of its subunits contains a novel binuclear Fe(III) prosthetic group whose two Fe(III) ions are bridged by both an O^{2-} ion (a μ-oxo bridge) and the carboxyl group of Glu 115 (Fig. 28-12*c*). Each Fe(III) is further liganded by two carboxyl O atoms from Asp or Glu residues, a His N$_\delta$ atom, and a water molecule. The Fe(III) complex interacts with Tyr 122 to form, as EPR measurements indicate, an unusual tyrosyl free radical (TyrO·) that is 5 Å from the closest Fe atom and is buried 10 Å beneath the surface of the protein, where it is out of contact with solvent and any oxidizable side chain [tyrosyl radicals have also been observed in cytochrome *c* oxidase (Section 22-2C) and in photosystem II (Section 24-2C)].

The *E. coli* RNR is inhibited by **hydroxyurea,** which specifically quenches (destroys) the tyrosyl radical, and by **8-hydroxyquinoline,** which chelates Fe^{3+} ions.

Hydroxyurea   **8-Hydroxyquinoline**

Mammalian RNRs have similar characteristics to the *E. coli* enzyme. Indeed, hydroxyurea is in clinical use as an antitumor agent.

If *E. coli* RNR is incubated with [3'-^{3}H]UDP, a small but reproducible fraction of the ^{3}H is released as ^3H_2O. This observation, together with kinetic, spectroscopic, and site-directed mutagenesis studies, led JoAnne Stubbe to formulate the following catalytic mechanism for *E. coli* RNR (Fig. 28-13):

1. RNR's free radical (X·) abstracts an H atom from C3' of the substrate in the reaction's rate-determining step.

2 and 3. Acid-catalyzed cleavage of the C2'—OH bond releases H_2O to yield a radical–cation intermediate. The radical mediates the stabilization of the C2' cation by the 3'—OH group's unshared electron pair, thereby accounting for the radical's catalytic role.

4. The radical–cation intermediate is reduced by the enzyme's redox-active sulfhydryl pair to yield a 3'-deoxynucleotide radical and a protein disulfide group.

5. The 3' radical reabstracts an H atom from the protein to yield the product deoxynucleoside diphosphate and restore the enzyme to its radical state. A small fraction of the originally abstracted H atom exchanges with solvent before it can be replaced, thus accounting for the release of ^{3}H on reduction of [3'-^{3}H]UDP.

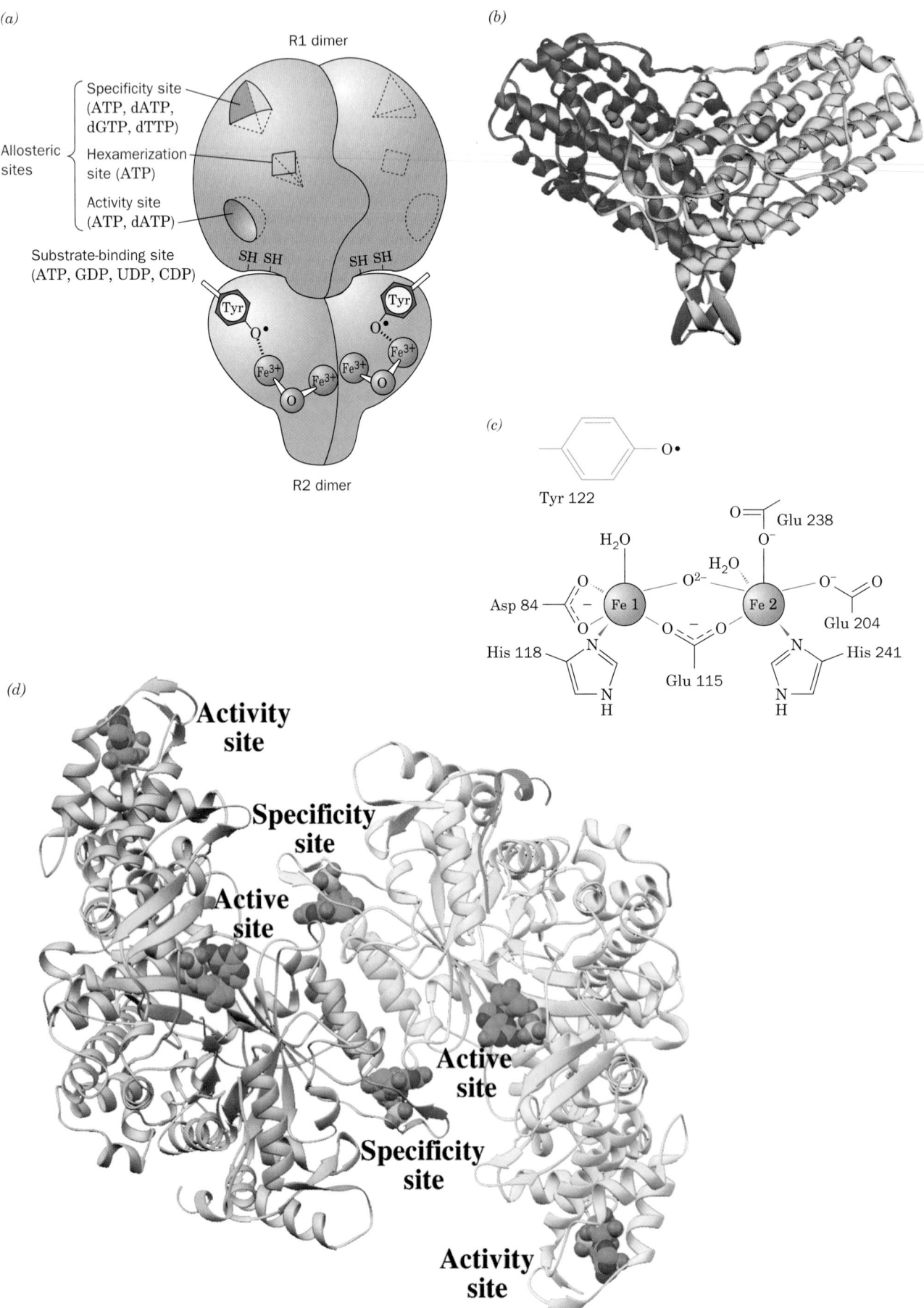

(a)

R1 dimer

Allosteric sites

Specificity site (ATP, dATP, dGTP, dTTP)

Hexamerization site (ATP)

Activity site (ATP, dATP)

Substrate-binding site (ATP, GDP, UDP, CDP)

SH SH   SH SH

Tyr   Tyr

O•   O•

Fe^{3+} Fe^{3+} Fe^{3+} Fe^{3+}

O   O

R2 dimer

(b)

(c)

Tyr 122

H_2O H_2O Glu 238

O

Asp 84   O O^{2-} Glu 204

Fe 1   Fe 2

O   O

His 118   N   N   His 241

O   O

Glu 115

N   N
H   H

(d)

Activity site

Specificity site

Active site

Active site

Specificity site

Activity site

FIGURE 28-12 (*Opposite*) **Class I ribonucleotide reductase from *E. coli*.** (*a*) A schematic diagram of its quaternary structure. The enzyme consists of two pairs of identical subunits, $R1_2$ and $R2_2$. Each R2 subunit contains a binuclear Fe(III) complex that generates a phenoxy radical at its Tyr 122. The R1 subunits each contain three different allosteric effector sites and five catalytically important Cys residues. The enzyme's two active sites are located near the interface between neighboring R1 and R2 subunits. (*b*) The X-ray structure of $R2_2$ as viewed perpendicular to its twofold axis with the dimer's longest dimension in the horizontal plane. One subunit of the homodimeric protein is shown in blue and the other in yellow. The Fe(III) ions of its binuclear Fe complexes are represented by orange spheres and the radical-harboring Tyr 122 side chains are shown in space-filling form with their C and O atoms green and red. Note that each subunit consists mainly of a bundle of eight unusually long helices. (*c*) The binuclear Fe(III) complex of R2. Each Fe(III) ion is octahedrally coordinated by a His N_δ atom and five O atoms, including those of the O^{2-} ion and the Glu carboxyl group that bridge the two Fe(III). (*d*) The X-ray structure of the R1 dimer, each subunit of which is in complex with the 20-residue C-terminal peptide of R2 together with GDP in the active site and dTTP in the specificity site. The ATP analog ADPNP bound in the activity site of the closely similar complex of R1 with the 20-residue peptide and ADPNP has been superimposed on this structure. The structure is viewed along its twofold axis with its two subunits lavender and light green, the two R2 peptides cyan and magenta, and the GDP, dTTP, and ATP shown in space-filling form colored according to atom type (C green, N blue, O red, and P gold). [Parts *b* and *d* based on X-ray structures by Hans Eklund, Swedish University of Agricultural Sciences, Uppsala, Sweden. PDBids (*b*) 1RIB and (*d*) 3R1R and 4R1R.] **See the Interactive Exercises**

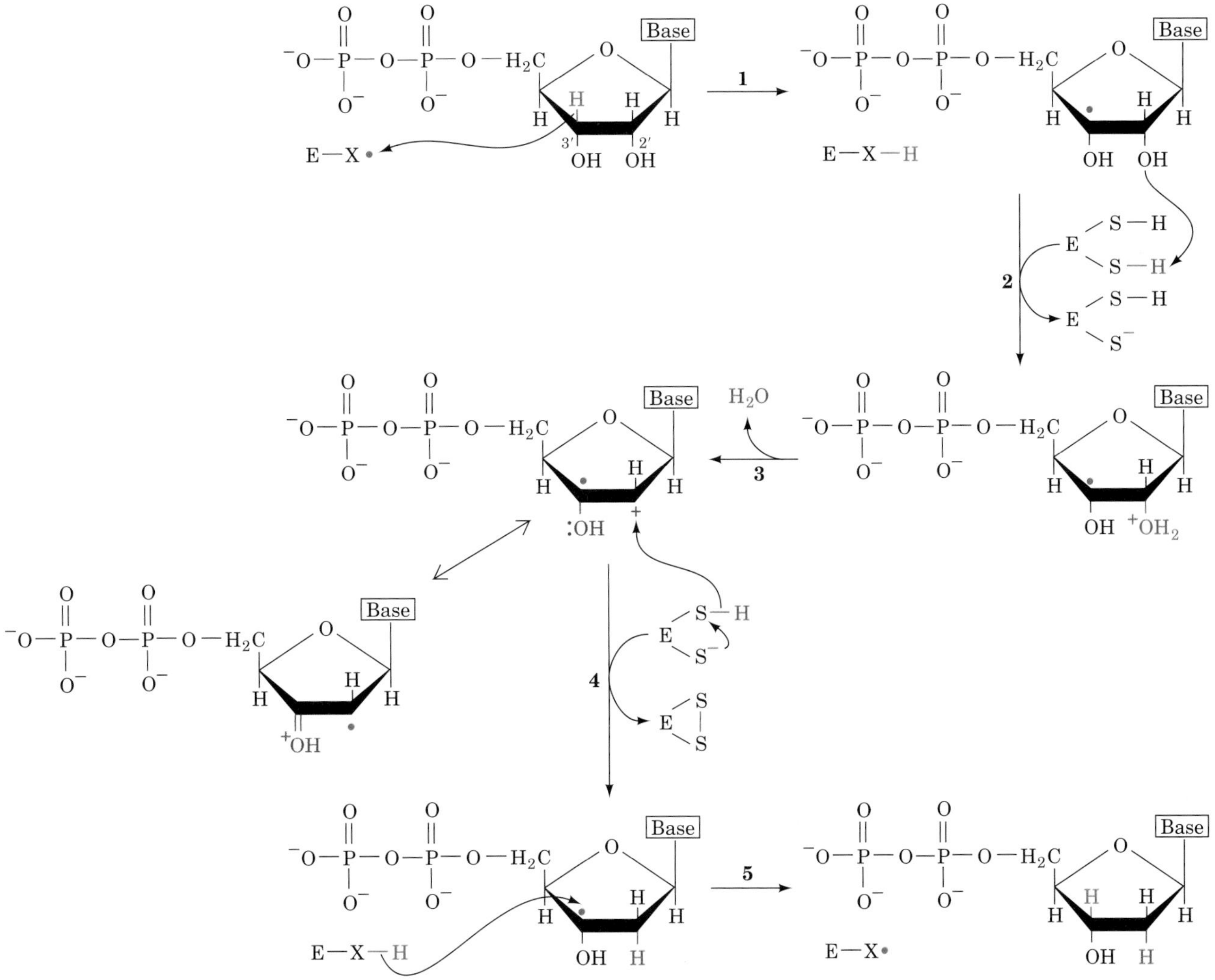

FIGURE 28-13 Enzymatic mechanism of ribonucleotide reductase. The reaction occurs via a free radical–mediated process in which reducing equivalents are supplied by the formation of an enzyme disulfide bond. [After Stubbe, J.A., *Biol. Chem.* **265**, 5330 (1990).]

The Tyr 122 radical in R2 is too far away (>10 Å) from the enzyme's catalytic site to abstract an electron directly from the substrate. Evidently, the protein mediates electron transfer from this tyrosyl radical to some other group (X· in Fig. 28-13) that is in close proximity to the substrate C3′—H group. Site-directed mutagenesis studies suggest that Cys 439 of R1, in its thiyl radical form (—S·), is the most plausible candidate for X· (which makes RNR the only enzyme in which a Cys residue is known to reduce a carbohydrate substrate). Similar studies suggest that Cys 225 and Cys 462 of R1 form the redox-active sulfhydryl pair that directly reduces substrate. Moreover, the resulting disulfide bond is subsequently reduced to regenerate active enzyme via disulfide interchange with Cys 754 and Cys 759 on R1, which are apparently positioned to accept electrons from external reducing agents (see below). Thus, each R1 subunit contains at least five Cys residues that chemically participate in nucleotide reduction.

These observations are confirmed by the X-ray structure of R1 in complex with R2's 20-residue C-terminal polypeptide (R1 does not crystallize satisfactorily in the absence of this polypeptide), also determined by Eklund (Fig. 28-12*d*). The central domain of the three-domain R1 monomer consists of a novel 10-stranded α/β barrel that is formed by the antiparallel joining of two topologically similar half-barrels, each comprising five parallel β strands connected by four α helices. As with the similar 8-stranded α/β barrels that form the active sites of numerous enzymes (Section 8-3B), R1's active site Cys residues (439, 225, 462) are located in the mouth of the 10-stranded α/β barrel.

The two R1 Cys residues, 754 and 759, that are implicated in the regeneration of the active enzyme are components of R1's C-terminal segment, which is not visible in the X-ray structure of R1 and is presumably disordered. This observation supports the hypothesis that this C-terminal segment acts to flexibly shuttle reducing equivalents from the enzyme surface to its active site.

b. Radical Generation in Class I RNR Requires the Presence of O_2

One of the most remarkable aspects of Class I RNR is its ability to stablize its normally highly reactive TyrO· radical (its half-life is 4 days in the protein vs milliseconds in solution). Yet quenching the radical, say, by hydroxyurea, inactivates the enzyme. How, then, is the radical generated in the first place? The radical may be restored *in vitro* by simply treating the inactive enzyme with Fe(II) and a reducing agent in the presence of O_2.

This is a four-electron reduction of O_2 in which the reducing agent that supplies the electron represented by e^- may be ascorbate or even excess Fe^{2+}.

c. The Inability of Oxidized RNR to Bind Substrate Serves an Important Protective Function

Comparison of the X-ray structures of reduced R1 (in which the redox-active Cys 225 and Cys 462 residues are in their SH forms) with that of oxidized R1 (in which Cys 225 and Cys 462 are disulfide-linked) reveals that Cys 462 in reduced R1 has rotated away from its position in oxidized R1 to become buried in a hydrophobic pocket, whereas Cys 225 moves into the region formerly occupied by Cys 462. The distance between the formerly disulfide-linked S atoms thereby increases from 2.0 to 5.7 Å. These movements are accompanied by small shifts of the surrounding polypeptide chain. Oxidized RNR does not bind substrate because its R1 Cys 225 would prevent the binding of substrate through steric interference of its S atom with the substrate dNDP's O2′ atom.

The inability of oxidized RNR to bind substrate has functional significance. In the absence of substrate, the enzyme's free radical is stored in the interior of the R2 protein, close to its dinuclear iron center. When substrate is bound, the radical is presumably transferred to it via a series of protein side chains in both R2 and R1. If the substrate is unable to properly react after accepting this free radical, as would be the case if RNR were in its oxidized state, this could result in the destruction of the substrate and/or the enzyme. Indeed, the mutation of the redox-active Cys 225 to Ser results in an enzyme that permits the formation of the substrate radical (Fig. 28-13); however, since the mutant enzyme is incapable of reducing it, the substrate radical instead decomposes followed by the release of its base and phosphate moieties. More importantly, a transient peptide radical forms, which cleaves and inactivates the R1 polypeptide chain while consuming the radical and thereby inactivating R2. Thus, an important role of the enzyme is to control the release of the radical's powerful oxidizing capability. It does so in part by preventing the binding of substrate while the enzyme is in its oxidized form.

d. Ribonucleotide Reductase Is Regulated by Effector-Induced Oligomerization

The synthesis of the four dNTPs in the amounts required for DNA synthesis is accomplished through feedback control. The maintenance of the proper intracellular ratios of dNTPs is essential for normal growth. Indeed, *a deficiency of any dNTP is lethal, whereas an excess is mutagenic because the probability that a given dNTP will be erroneously incorporated into a growing DNA strand increases with its concentration relative to those of the other dNTPs.*

The activities of both *E. coli* and mammalian Class I RNRs are allosterically responsive to the concentrations of various (d)NTPs. Thus, as Reichard has shown, ATP induces the reduction of CDP and UDP; dTTP induces the reduction of GDP and inhibits the reduction of CDP and

UDP; dGTP induces the reduction of ADP and, in mammals but not *E. coli*, inhibits the reduction of CDP and UDP; and dATP inhibits the reduction of all NDPs.

Barry Cooperman has shown that the catalytic activity of mouse RNR varies with its state of oligomerization, which in turn is governed by the binding of nucleotide effectors to three independent allosteric sites on R1: (1) the specificity site, which binds ATP, dATP, dGTP, and dTTP; (2) the activity site, which binds ATP and dATP; and (3) the hexamerization site, which binds only ATP. On the basis of molecular mass, ligand binding, and activity studies on mouse RNR, Cooperman formulated a model that quantitatively accounts for the allosteric regulation of Class I RNR. It has the following features (Fig. 28-14*a*):

1. The binding of ATP, dATP, dGTP, or dTTP to the specificity site induces the catalytically inactive R1 monomers to form a catalytically active dimer, $R1_2$.

2. The binding of dATP or ATP to the activity site causes the dimers to form catalytically active tetramers,

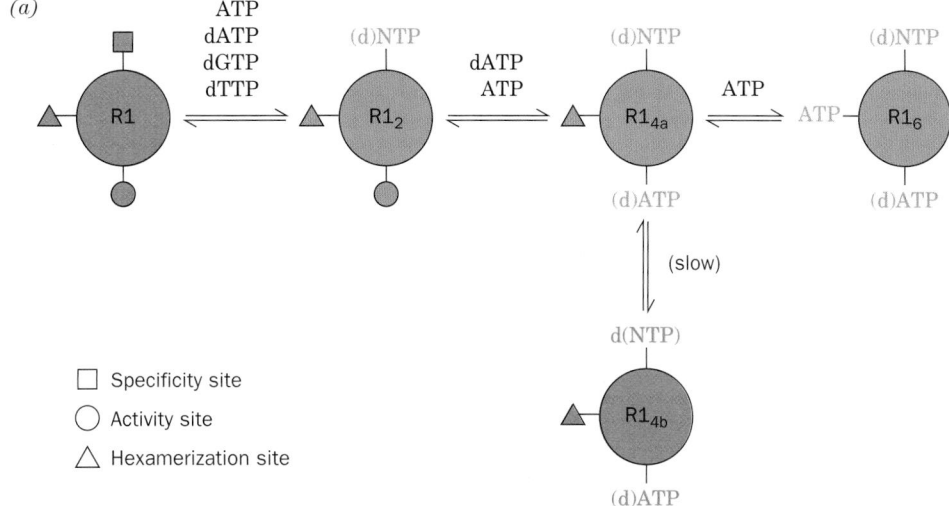

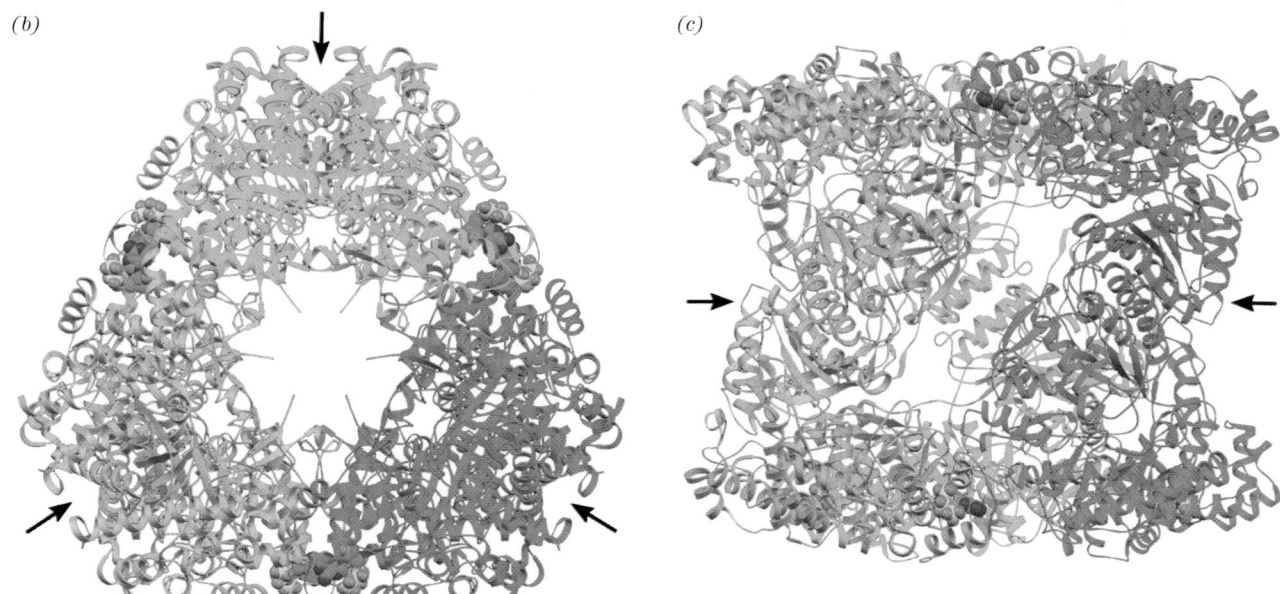

FIGURE 28-14 Ribonucleotide reductase regulation. (*a*) A model for the allosteric regulation of Class I RNR via its oligomerization. States shown in green have high activity and those shown in red have little or no activity. R2 has been omitted for simplicity. [After Kashlan, O.B., Scott, C.P., Lear, J.D., and Cooperman, B.S., *Biochemistry* **41,** 461 (2002).] (*b*) The X-ray structure of the R1 hexamer, which has D_3 symmetry, in complex with ADPNP as viewed along its 3-fold axis. Each of its three dimers are differently colored (the X-ray structure of a dimer is shown in Fig. 28-12*d*). The ADPNP, which binds to the enzyme's activity sites, is drawn in space-filling form with C green, N blue, O red, and P gold. The black arrows point along the R1 dimers' 2-fold axes and indicate the probable docking sites for the binding of R2 dimers. (*c*) The R1·ADPNP hexamer as viewed along the vertical 2-fold axis in Part *b*. [Parts *b* and *c* based on an X-ray structure by Hans Eklund, Swedish University of Agricultural Sciences, Uppsala, Sweden. PDBid 3R1R.]

R1$_{4a}$, that slowly but reversibly change conformation to a catalytically inactive state, R1$_{4b}$.

3. The binding of ATP to the hexamerization site induces the tetramers to further aggregate to form catalytically active hexamers, R1$_6$, RNR's major active form.

The concentration of ATP in a cell is such that, *in vivo*, R1 is almost entirely in its tetrameric or hexameric forms. As a consequence, ATP couples the overall rate of DNA synthesis to the cell's energy state.

The specificity and activity sites have been located in X-ray structures of *E. coli* R1 (Fig. 28-12*d*); the hexamerization site has not yet been identified. The R1 hexamer had, in fact, been previously observed in the X-ray structures of R1 (Fig. 28-14*b,c*), but the interactions between its contacting dimers are so tenuous that it was assumed that they are merely artifacts of crystallization with no physiological significance. Yet, since the activity site is located at this contact site, it now seems likely that its binding of (d)ATP induces R1 oligomerization through local conformational changes.

The foregoing model has, for simplicity, neglected the presence of R2 subunits although, of course, R1 and R2 must be present in equimolar amounts in the active enzyme. Presumably, the R1 and R2 dimers bind to one another such that their twofold axes coincide. The lack of space on the inside of the R1 hexamer dictates that the R2 dimers must contact the R1 dimers from outside the hexamer (Fig. 28-14*b*).

dCTP is not an effector of RNR. This is presumably because the intracellular balance between dCTP and dTTP is not controlled by RNR but, rather, is maintained by **deoxycytidine deaminase,** which converts dCTP to dUMP, the precursor of dTTP. This enzyme is activated by dCTP and inhibited by dTTP.

e. Thioredoxin and Glutaredoxin Are Class I Ribonucleotide Reductase's Physiological Reducing Agents

The final step in the RNR catalytic cycle is the reduction of the enzyme's newly formed disulfide bond to reform its redox-active sulfhydryl pair. Dithiols such as dithiothreitol (Section 7-1B) can serve as the reducing agent for this process *in vitro* through a disulfide interchange reaction. One of the enzyme's physiological reducing agents, however, is **thioredoxin (Trx),** a ubiquitous monomeric 108-residue protein that has a pair of closely proximal redox-active Cys residues, Cys 32 and Cys 35 (we have previously encountered thioredoxin in our study of the light-induced activation of the Calvin cycle; Section 24-3B). Thioredoxin reduces oxidized RNR via disulfide interchange.

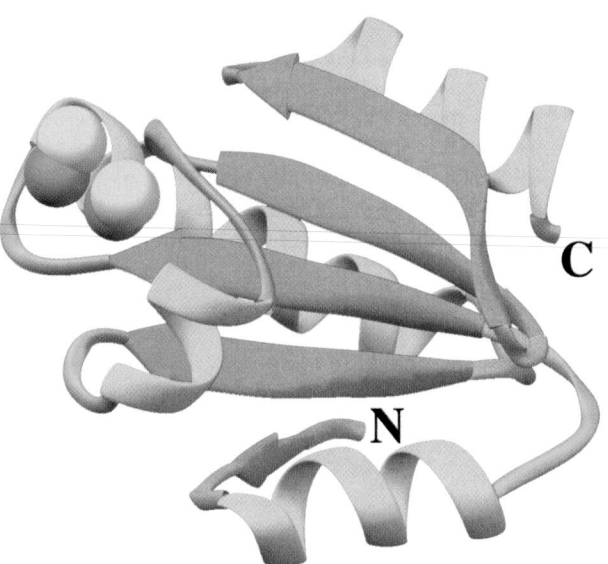

FIGURE 28-15   X-Ray structure of human thioredoxin in its reduced (sulfhydryl) state. The backbone of this 105-residue polypeptide is colored according to its secondary structure with helices cyan, sheets magenta, and the remaining portions orange. The side chains of the redox-active residues, Cys 32 and Cys 35, are shown in space-filling form with C green and S yellow. This structure closely resembles that of the homologous domains of protein disulfide isomerase (PDI, Fig. 9-15*a*). [Based on an X-ray structure by William Montfort, University of Arizona. PDBid 1ERT.]

The X-ray structure of reduced *E. coli* Trx (Fig. 28-15) reveals that the side chain of the redox-active Cys 32 is exposed on the protein's surface, where it is available for oxidation. Oxidized thioredoxin is, in turn, reduced by NADPH in a reaction mediated by the flavoprotein **thioredoxin reductase.** NADPH therefore serves as the terminal reducing agent in the RNR-mediated reduction of NDPs to dNDPs (Fig. 28-16).

The existence of a viable *E. coli* mutant devoid of thioredoxin indicates that this protein is not the only substance capable of reducing oxidized RNR *in vivo*. This observation led to the discovery of **glutaredoxin,** a disulfide-containing, monomeric, 85-residue protein that can also reduce RNR (mutants devoid of both thioredoxin and glutaredoxin are nonviable). Oxidized glutaredoxin is reduced, via disulfide interchange, by the Cys-containing tripeptide glutathione which, in turn, is reduced by NADPH as catalyzed by glutathione reductase (GR; Section 21-2B). The relative importance of thioredoxin and glutaredoxin in the reduction of RNRs remains to be established.

f. Thioredoxin Reductase Alternates Its Conformation with its Redox State

Thioredoxin reductase **(TrxR),** a homodimer of 316-residue subunits, is a homolog of GR that catalyzes a similar reaction: the reduction of a substrate disulfide bond by NADPH as mediated by an FAD prosthetic group and

$$
\begin{array}{c}
\text{Thioredoxin} \overset{\diagup \text{SH}}{\underset{\diagdown \text{SH}}{}} + \overset{\text{S} \diagdown}{\underset{\text{S} \diagup}{|}} \text{Ribonucleotide} \\
(\textit{reduced}) \qquad\qquad \text{reductase} \\
(\textit{oxidized}) \\
\big\downarrow \\
\text{Thioredoxin} \overset{\diagup \text{S}}{\underset{\diagdown \text{S}}{|}} + \overset{\text{HS} \diagdown}{\underset{\text{HS} \diagup}{}} \text{Ribonucleotide} \\
(\textit{oxidized}) \qquad\qquad \text{reductase} \\
(\textit{reduced})
\end{array}
$$

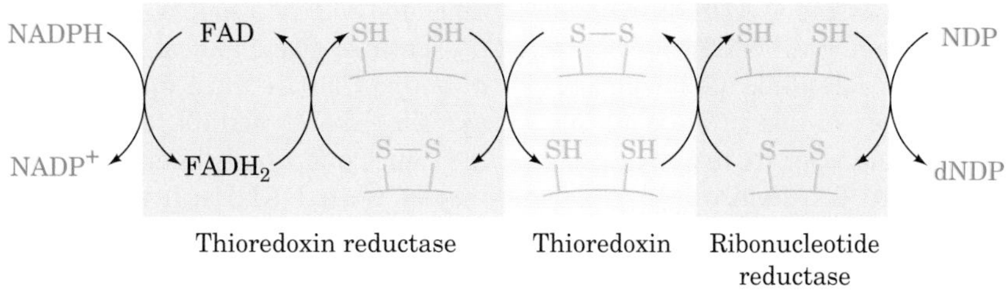

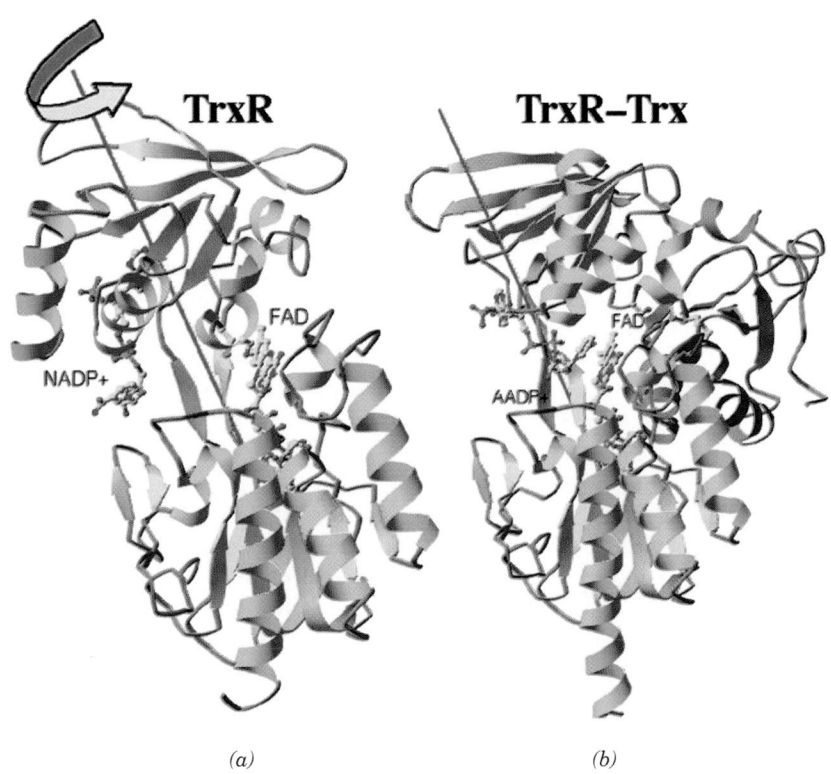

FIGURE 28-16 Electron-transfer pathway for nucleoside diphosphate (NDP) reduction. NADPH provides the reducing equivalents for this process through the intermediacy of thioredoxin reductase, thioredoxin, and ribonucleotide reductase.

a redox active sulfhydryl pair (Cys 135 and Cys 138). However, the X-ray structure of the C138S mutant of *E. coli* TrxR in complex with NADP$^+$ (Fig. 28-17*a*), determined by Charles Williams and John Kuriyan, reveals that TrxR and GR differ in their active site arrangements such that their redox-active sulfhydryl pairs are on opposite sides of the flavin rings in the two enzymes. Nevertheless, TrxR's redox-active sulfhydryl pair appears properly positioned to reduce the flavin ring. However, the NADP$^+$'s nicotinamide ring is >17 Å from the flavin ring and the redox-active sulfhydryl pair is buried such that it could not react with the enzyme's Trx substrate. How then

does TrxR manage to transfer an electron pair from its bound NADPH via its flavin ring and redox-active sulfhydryl pair to Trx?

This question was answered by Williams and Martha Ludwig through their X-ray structure determination of the C135S mutant of TrxR, whose Cys 138 is disulfide-linked to Cys 32 of the C35S mutant of *E. coli* Trx (probably the physiologically relevant disulfide bond) and which is in complex with the NADP$^+$ analog **3-aminopyridine adenine dinucleotide phosphate (AADP$^+$).** In this complex (Fig. 28-17*b*), TrxR's NADP$^+$-binding domain has rotated by 67° relative to the rest of the protein compared to its position in TrxR

(a)   *(b)*

FIGURE 28-17 X-Ray structures of *E. coli* thioredoxin reductase (TrxR). (*a*) The C138S mutant TrxR in complex with NADP$^+$. The protein is shown in ribbon form colored according to its secondary structure. The NADP$^+$, the FAD, and the side chains of Cys 135 and Ser 138 are drawn in ball-and-stick form with C yellow, N blue, O red, S green, and P magenta. (*b*) The C135S mutant TrxR in complex with AADP$^+$ and covalently linked to the C35S mutant of Trx via a disulfide bond between TrxR Cys 138 and Trx Cys 32. The TrxR is represented as in Part *a*, the Trx ribbon is gray, and its Cys 32 and Ser 35 side chains are drawn in ball-and-stick form. Comparison of these two structures reveals that TrxR's NADP$^+$-binding domain (residues 120–243) undergoes a 67° rotation about the axis drawn in blue relative to the rest of the protein, which is shown in the same orientation in both structures. [Courtesy of Martha Ludwig, University of Michigan. PDBids (*a*) 1TDF and (*b*) 1F6M.]

alone (Fig. 28-17a). This positions the AADP$^+$'s pyridine ring to react with the flavin ring and positions TrxR's redox-active sulfhydryl pair to undergo a disulfide interchange reaction with that of Trx. Moreover, in this latter conformation, the NADP$^+$-binding domain appears to provide the recognition site for the substrate Trx. Evidently, TrxR alternates its conformation with each successive step in the process of transferring an electron pair from NADPH to the flavin to its redox-active sulfhydryl pair to its bound Trx substrate. This added mechanistic complication relative to that of GR, which does not undergo a significant conformational change in reducing glutathione disulfide (Section 21-2B), has apparently evolved to permit TrxR to reduce its protein substrate: Trx would be too large for its redox-active sulfhydryl pair to properly approach the active site sulfhydryl pair in a GR-like enzyme.

g. The Three Classes of Ribonucleotide Reductases Are Evolutionarily Related

We have seen that the active forms of Class I RNRs are $R1_2R2_2$, $R1_4R2_4$, and $R1_6R2_6$ oligomers that have mechanistically essential tyrosyl radicals that are stabilized by oxo-bridged binuclear Fe(III) complexes, have NDPs for substrates, and obtain their reducing equivalents from thioredoxin and glutaredoxin. In contrast, Class II RNRs, which are α or α_2 monomers or dimers, utilize a 5'-deoxyadenosylcobalamin cofactor (coenzyme B$_{12}$; Section 25-2E) for radical generation, have NDPs for substrates, and are reduced by thioredoxin and glutaredoxin; whereas Class III RNRs, which are α_2 dimers that interact with a radical-generating protein β_2 that contains a [4Fe–4S] cluster and requires S-adenosylmethionine (SAM; Section 26-3C) and NADPH for activity, have NTPs for substrates, and their reducing equivalents are provided by the oxidation of formate to CO_2.

Since all known cellular life synthesizes its deoxyribonucleotides from ribonucleotides, the rise of an RNR must have preceded the evolutionary transition from the RNA world (Section 1-5C) to DNA-based life-forms. Did the three classes of RNRs arise independently or are they evolutionarily related? Despite the seemingly large differences between these different classes of RNRs, the reactions they catalyze are surprisingly similar. All replace the 2' OH group of ribose with H via a free radical mechanism involving a thiyl radical with the reducing equivalents provided by a Cys sulfhydryl group (Fig. 28-13; the second Cys residue of the redox-active sulfhydryl pair in Class I and II RNRs is replaced by formate in Class III RNRs). They differ mainly in the way they generate the free radical. [In Class II RNRs, the radical is generated by the homolytic cleavage of its 5'-deoxyadenosylcobalamin cofactor's C—Co(III) bond (Section 25-2E). In Class III RNRs, it is generated by the NADPH-supplied and [4Fe–4S] cluster-mediated one-electron reductive cleavage of SAM by the β_2 protein to yield methionine and the 5'-deoxyadenosyl radical (the same radical generated by the homolytic cleavage of 5'-deoxyadenosylcobalamin), which then abstracts the H atom from a C$_\alpha$—H group of a specific Gly on the α sub-

unit to yield 5'-deoxyadenosine and a stable but O$_2$-sensitive glycyl radical.] Moreover, the X-ray structures of both a Class II and a Class III RNR reveal that their active sites are formed by 10-stranded α/β barrels that have the same connectivity as and are closely superimposable on that of Class I RNRs. It therefore appears that all three classes of RNRs are evolutionarily related. Reichard has proposed that, since life arose under anaerobic conditions and that formate, one of simplest organic reductants, was probably widely available on primitive Earth (Section 1-5B), the primordial RNR was a Class III–like enzyme. The rise of photosynthetic organisms that generated O$_2$ then promoted the evolution of Class II RNRs, which can function under both anaerobic and aerobic conditions. Class I RNRs, which require the presence of O$_2$ for activation, evolved last, presumably from a Class II RNR.

h. dNTPs Are Produced by Phosphorylation of dNDPs

In pathways involving Class I and Class II RNRs, the final step in the production of dNTPs is the phosphorylation of the corresponding dNDPs:

$$dNDP + ATP \rightleftharpoons dNTP + ADP$$

This reaction is catalyzed by nucleoside diphosphate kinase, the same enzyme that phosphorylates NDPs (Section 28-1B). As before, the reaction is written with ATP as the phosphoryl donor, although any NTP or dNTP can function in this capacity. In pathways involving Class III RNRs, the production of NTPs from NDPs precedes the reduction of NTPs to dNTPs.

B. Origin of Thymine

a. dUTP Diphosphohydrolase

The dTMP component of DNA is synthesized, as we discuss below, by methylation of dUMP. The dUMP is generated through the hydrolysis of dUTP by **dUTP diphosphohydrolase (dUTPase**; also called **dUTP pyrophosphatase)**:

$$dUTP + H_2O \rightleftharpoons dUMP + PP_i$$

The reason for this apparently energetically wasteful process (dTMP, once formed, is rephosphorylated to dTTP) is that cells must minimize their concentration of dUTP in order to prevent incorporation of uracil into their DNA. This is because, as we discuss in Section 30-5B, DNA polymerase does not discriminate between dUTP and dTTP.

The X-ray structures of human dUTPases, determined by John Tainer, reveal the basis for this enzyme's exquisite specificity for dUTP. This homotrimer of 141-residue subunits binds dUTP in a snug-fitting cavity that sterically excludes thymine's C5 methyl group via the side chains of conserved residues (Fig. 28-18a). It differentiates uracil from cytosine via a set of hydrogen bonds from the protein backbone that mimic adenine's base-pairing interaction (Fig. 28-18b), and it differentiates dUTP from UTP by the steric exclusion of ribose's 2'-OH group by the side chain of a conserved Tyr.

(a)

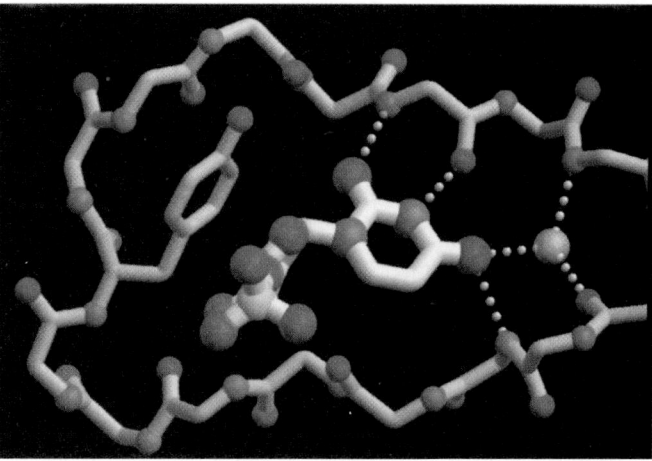

(b)

FIGURE 28-18   X-Ray structure of human dUTPase. (*a*) The molecular surface at the substrate binding site showing how the enzyme differentiates uracil from thymine. Bound dUTP is drawn in stick form with its N, O, and P atoms represented by blue, red, and yellow spheres. Mg^{2+} ions that were modeled into the structure are represented by green spheres. The protein's molecular surface is colored according to its electrostatic potential with positive, negative, and near neutral regions blue, red, and white, respectively. Note how the snug fit of the uracil ring into its binding site would sterically exclude thymine's C5 methyl group. (*b*) The substrate binding site indicating how the enzyme differentiates uracil from cytosine

and 2′-deoxyribose from ribose. dUMP bound at the active site is drawn as in Part *a*. The protein, mainly the backbone of a β hairpin motif, is similarly drawn but with thinner gray bonds. Hydrogen bonds are shown as dotted white lines, and a tightly bound, conserved water molecule is represented by a pink sphere. The pattern of hydrogen bonding donors and acceptors on the protein would prevent cytosine from binding in the active site pocket. The conserved Tyr side chain sterically excludes ribose's 2′ OH group. [Courtesy of John Tainer, The Scripps Research Institute, La Jolla, California.]

b. Thymidylate Synthase

*dTMP is synthesized from dUMP by **thymidylate synthase (TS)** with N^5,N^{10}-methylenetetrahydrofolate (N^5,N^{10}-methylene-THF) as the methyl donor:*

dUMP N^5,N^{10}**-Methylenetetrahydrofolate**

dTMP   **Dihydrofolate**

$$R = -\!\!\left\langle\!\!\bigcirc\!\!\right\rangle\!\!-\!\overset{\displaystyle O}{\underset{}{C}}\!\!\left(\!N\!-\!\overset{\displaystyle COO^-}{\underset{}{CH}}\!-\!CH_2\!-\!CH_2\!-\!\overset{\displaystyle O}{\underset{}{C}}\!\right)_{\!n}\!\!O^-;\quad n = 1\text{–}6$$

(THF cofactors are discussed in Section 26-4D). Note that the transferred methylene group (in which the carbon has the oxidation state of formaldehyde) is reduced to a methyl group (which has the oxidation state of methanol) at the expense of the oxidation of the THF cofactor to dihydrofolate (DHF).

The catalytic mechanism of TS, a highly conserved 70-kD dimeric protein, has been extensively investigated. On incubation of the enzyme with N^5,N^{10}-methylene-[6-^{3}H]THF and dUMP, the ^{3}H is quantitatively transferred to the methyl group of the product dTMP. When [5-^{3}H]dUMP is the substrate, however, the ^{3}H is released into the aqueous solvent. Such information, together with the knowledge that uracil C6, being the β position of an α,β-unsaturated ketone, is susceptible to nucleophilic attack, led Daniel Santi to propose the following mechanistic scheme for the TS reaction (Fig. 28-19):

1. An enzyme nucleophile, identified as the thiolate group of Cys 146, attacks C6 of dUMP to form a covalent adduct.

2. C5 of the resulting enolate ion attacks the CH_2 group of the iminium cation in equilibrium with N^5,N^{10}-methylene-THF to form an enzyme–dUMP–THF ternary covalent complex.

3. An enzyme base abstracts the acidic proton at the C5 position of the enzyme-bound dUMP, forming an exocyclic methylene group and eliminating the THF cofactor. The abstracted proton subsequently exchanges with solvent.

4. The redox change occurs via the migration of the N6—H atom of THF as a hydride ion to the exocyclic methylene group, converting it to a methyl group (thus accounting for the above described transfer of ^3H) and yielding DHF. This reduction promotes displacement of the Cys thiolate group from the intermediate so as to release product, dTMP, and reform active enzyme.

c. 5-Fluorodeoxyuridylate Is a Potent Antitumor Agent

The above mechanism is supported by the observation that **5-fluorodeoxyuridylate (FdUMP)**

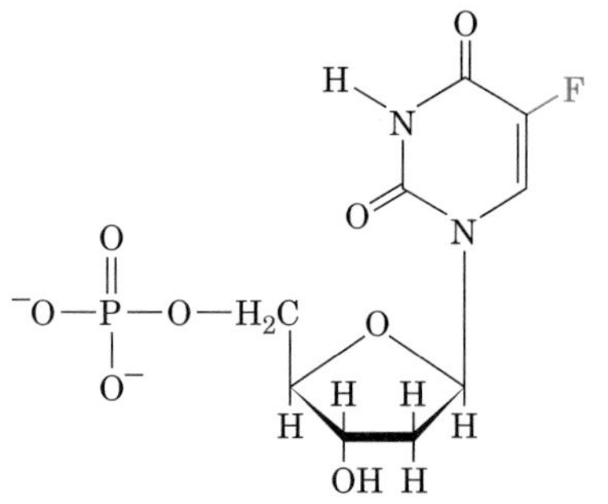

5-Fluorodeoxyuridylate (FdUMP)

is an irreversible inhibitor of TS. This substance, like dUMP, binds to the enzyme (an F atom has approximately the same radius as an H atom) and undergoes the first two steps of the normal enzymatic reaction. In Step 3, however, the enzyme cannot abstract the F atom as F^+ (recall that F is the most electronegative element), so that the enzyme is all but permanently immobilized as the enzyme–FdUMP–THF ternary covalent complex analogous to that after Step 2 in Fig. 28-19. Indeed, X-ray structural analysis by David Matthews and Jesus Villafranca revealed that crystals of *E. coli* TS that had been soaked in a solution containing FdUMP and N^5,N^{10}-methylene-THF contain precisely this complex (Fig. 28-20). Enzymatic inhibitors such as FdUMP, which inactivate an enzyme only after undergoing part or all of its normal catalytic reaction, are called **mechanism-based inhibitors** (alternatively, **suicide substrates** because they cause the enzyme to "commit suicide"). *Mechanism-based inhibitors, being targeted for particular enzymes, are among the most powerful, specific, and therefore useful enzyme inactivators.*

The strategic position of thymidylate synthase in DNA biosynthesis has led to the clinical use of FdUMP as an antitumor agent. Rapidly proliferating cells, such as cancer cells, require a steady supply of dTMP in order to survive and are therefore killed by treatment with FdUMP. In contrast, most normal mammalian cells, which grow slowly if at all, have a lesser requirement for dTMP, so that they are relatively insensitive to FdUMP (some exceptions are the bone marrow cells that comprise the blood-forming tissues and much of the immune system, the intestinal mucosa, and hair follicles). **5-Fluorouracil** and **5-fluorodeoxyuridine** are also effective antitumor agents since they are converted to FdUMP through salvage reactions.

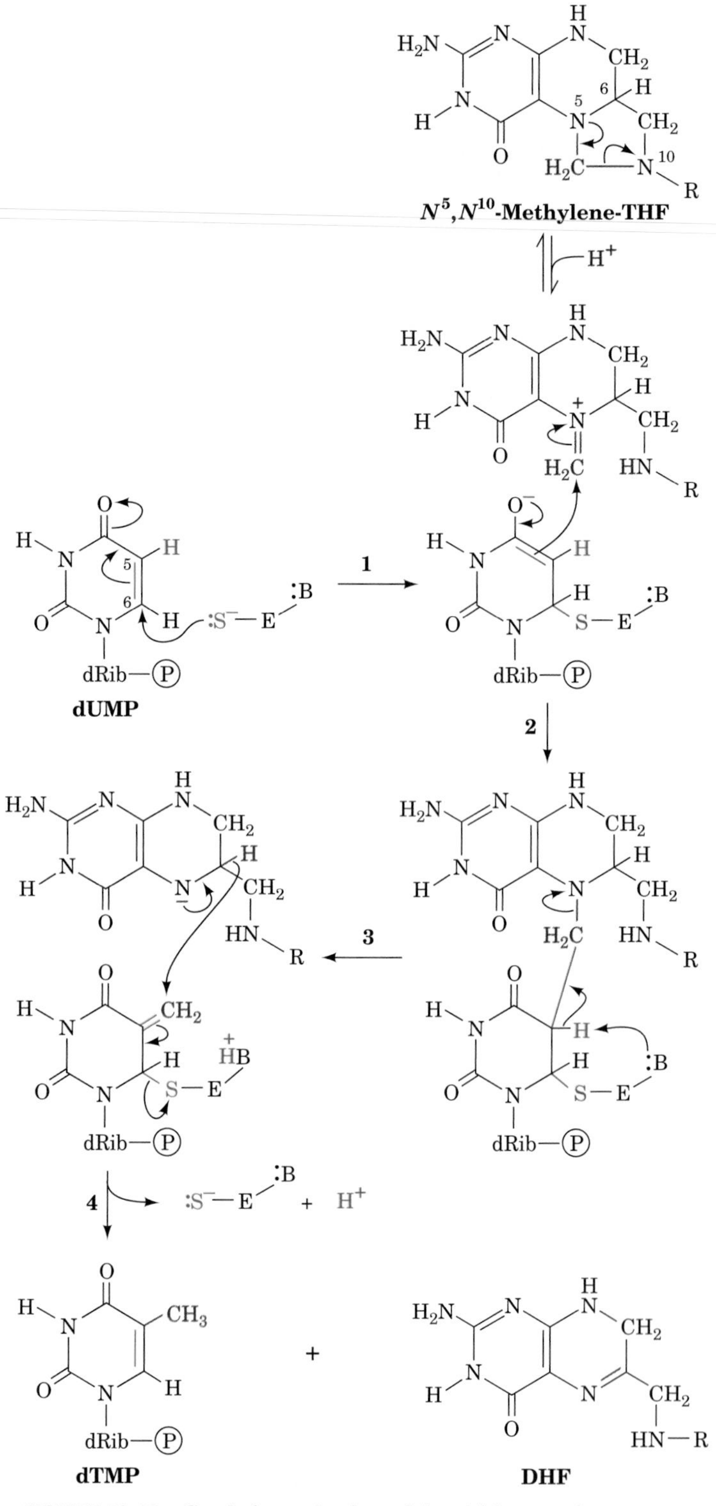

FIGURE 28-19   Catalytic mechanism of thymidylate synthase. The methyl group is supplied by N^5,N^{10}-methylene-THF, which is concomitantly oxidized to dihydrofolate.

d. N^5,N^{10}-Methylene-THF Is Regenerated in Two Reactions

The thymidylate synthase reaction is biochemically unique in that it oxidizes THF to DHF; no other enzymatic reaction employing a THF cofactor alters this coenzyme's

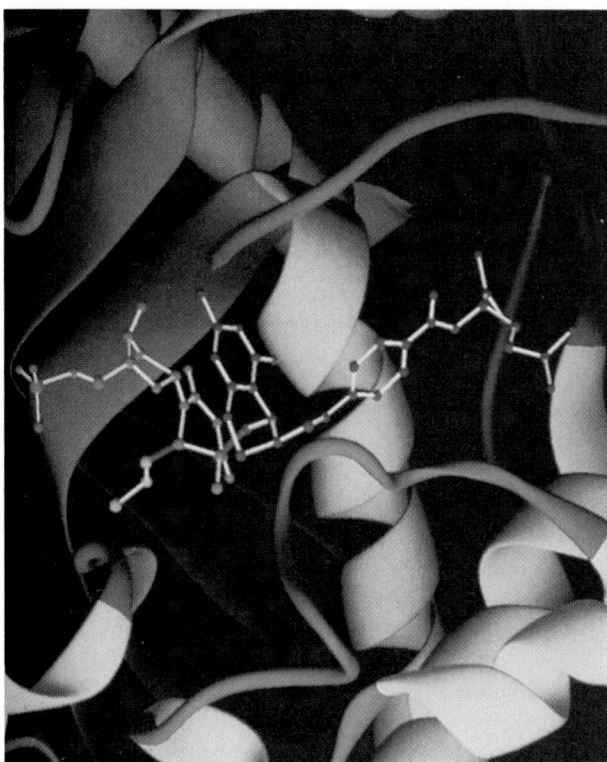

FIGURE 28-20   The X-ray structure of the *E. coli* thymidylate synthase–FdUMP–THF ternary complex. The active site region of one subunit of this dimeric enzyme (colored according to secondary structure with helices yellow, β strands orange, and other polypeptide segments blue) is shown in covalent complex with FdUMP (*green spheres*) and N^5,N^{10}-methylene-THF (*blue spheres*). The C5 and C6 atoms of FdUMP form covalent bonds (*red*) with the CH_2 group substituent to N5 of THF and the S atom of Cys 146 (*yellow spheres*). [Courtesy of Jesus Villafranca and David Matthews, Agouron Pharmaceuticals, La Jolla, California.]

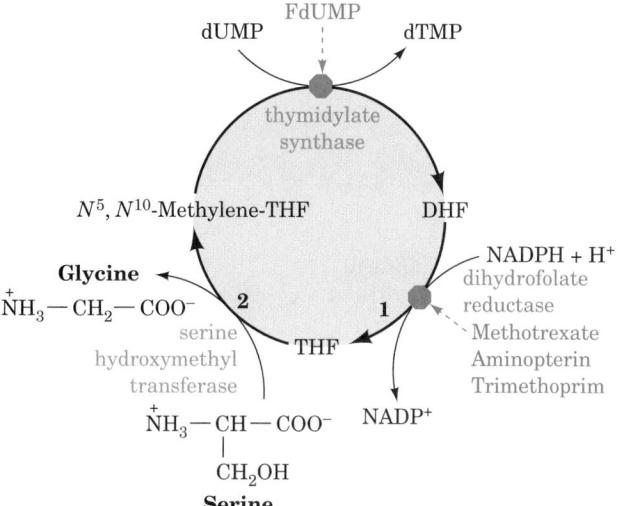

FIGURE 28-21   Regeneration of N^5,N^{10}-methylenetetra-hydrofolate. The DHF product of the thymidylate synthase reaction is converted back to N^5,N^{10}-methylene-THF by the sequential actions of (1) dihydrofolate reductase and (2) serine hydroxymethyltransferase. Thymidylate synthase is inhibited by FdUMP, whereas dihydrofolate reductase is inhibited by the antifolates methotrexate, aminopterin, and trimethoprim.

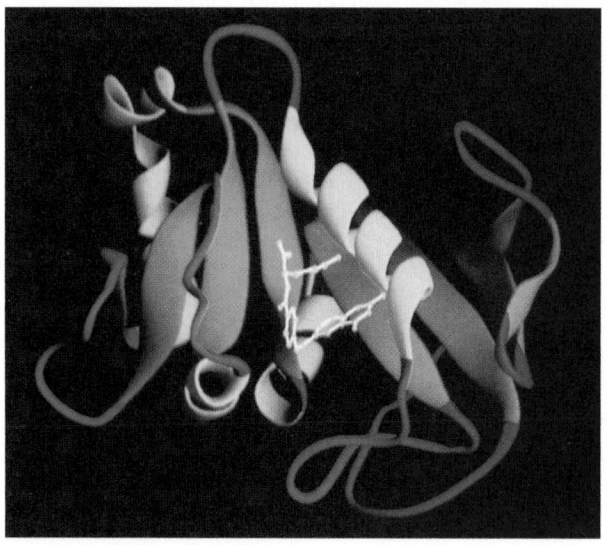

FIGURE 28-22   Ribbon diagram of human dihydrofolate reductase in complex with folate. This monomeric enzyme is colored according to secondary structure with helices yellow, β strands orange, and other polypeptide segments blue. [Courtesy of Jay F. Davies II and Joseph Kraut, University of California at San Diego.] ᛞ **See the Interactive Exercises**

net oxidation state. The DHF product of the thymidylate synthase reaction is recycled to the enzyme's N^5,N^{10}-methylene-THF cofactor through two sequential reactions (Fig. 28-21):

1. DHF is reduced to THF by NADPH as catalyzed by **dihydrofolate reductase** (**DHFR;** Section 26-4D). Although, in most organisms, DHFR is a monomeric monofunctional enzyme, in protozoa and at least some plants, DHFR and TS occur on the same polypeptide chain to form a bifunctional enzyme that has been shown to channel DHF from its TS to its DHFR active sites.

2. Serine hydroxymethyltransferase (Section 26-3B) transfers the hydroxymethyl group of serine to THF, yielding N^5,N^{10}-methylene-THF and glycine.

e. Antifolates Are Anticancer Agents
Inhibition of DHFR quickly results in all of a cell's limited supply of THF being converted to DHF by the thymidylate synthase reaction. Inhibition of DHFR there-

fore not only prevents dTMP synthesis (Fig. 28-21), but also blocks all other THF-dependent biological reactions such as the synthesis of purines (Section 28-1A), histidine, and methionine (Section 26-5B). DHFR (Fig. 28-22) therefore offers an attractive target for chemotherapy.

Methotrexate (amethopterin), aminopterin, and **trimethoprim**

R = H   **Aminopterin**
R = CH$_3$ **Methotrexate (amethopterin)**

Trimethoprim

are DHF analogs that competitively although all but irreversibly bind to DHFR with an ~1000-fold greater affinity than does DHF. These **antifolates** (substances that interfere with the action of folate cofactors) are effective anticancer agents, particularly against childhood leukemias. In fact, a successful chemotherapeutic strategy is to treat a cancer victim with a lethal dose of methotrexate and some hours later "rescue" the patient (but hopefully not the cancer) by administering massive doses of 5-formyl-THF and/or thymidine. A low dose of methotrexate is also effective in the treatment of rheumatoid arthritis, inhibiting immune system activity and thus decreasing inflammation. Trimethoprim, which was discovered by George Hitchings and Gertrude Elion, binds much more tightly to bacterial DHFRs than to those of mammals and is therefore a clinically useful antibiotic.

4 ■ NUCLEOTIDE DEGRADATION

Most foodstuffs, being of cellular origin, contain nucleic acids. Dietary nucleic acids survive the acid medium of the stomach; they are degraded to their component nucleotides, mainly in the duodenum, by pancreatic nucleases and intestinal phosphodiesterases. These ionic compounds, which cannot pass through cell membranes, are then hydrolyzed to nucleosides by a variety of group-specific nucleotidases and nonspecific phosphatases. Nucleosides may be directly absorbed by the intestinal mucosa or first undergo further degradation to free bases and ribose or ribose-1-phosphate through the action of **nucleosidases** and **nucleoside phosphorylases:**

$$\text{Nucleoside} + \text{H}_2\text{O} \xrightarrow{\text{nucleosidase}} \text{base} + \text{ribose}$$

$$\text{Nucleoside} + \text{P}_i \xrightarrow[\text{phosphorylase}]{\text{nucleoside}} \text{base} + \text{ribose-1-P}$$

Radioactive labeling experiments have demonstrated that only a small fraction of the bases of ingested nucleic acids are incorporated into tissue nucleic acids. Evidently, the *de novo* pathways of nucleotide biosynthesis largely sat-

isfy an organism's need for nucleotides. Consequently, ingested bases, for the most part, are degraded and excreted. Cellular nucleic acids are also subject to degradation as part of the continual turnover of nearly all cellular components. In this section we outline these catabolic pathways and discuss the consequences of several of their inherited defects.

A. *Catabolism of Purines*

The major pathways of purine nucleotide and deoxynucleotide catabolism in animals are diagrammed in Fig. 28-23. Other organisms may have somewhat different pathways among these various intermediates (including adenine), but all of these pathways lead to uric acid. Of course, the intermediates in these processes may instead be reused to form nucleotides via salvage reactions. In addition, ribose-1-phosphate, a product of the reaction catalyzed by **purine nucleoside phosphorylase (PNP),** is isomerized by **phosphoribomutase** to the PRPP precursor ribose-5-phosphate.

Adenosine and deoxyadenosine are not degraded by mammalian PNP. Rather, adenine nucleosides and nucleotides are deaminated by **adenosine deaminase (ADA)** and **AMP deaminase** to their corresponding inosine derivatives, which, in turn, may be further degraded. The X-ray structure of murine ADA that was crystallized in the presence of its inhibitor **purine ribonucleoside** was determined by Florante Quiocho (Fig. 28-24*a*). The enzyme forms an eight-stranded α/β barrel with its active site in a pocket at the C-terminal end of the β barrel, as occurs in nearly all known α/β barrel enzymes (Section 8-3B). Purine ribonucleoside binds to ADA in a normally rare hydrated form, **6-hydroxy-1,6-dihydropurine ribonucleoside (HDPR),**

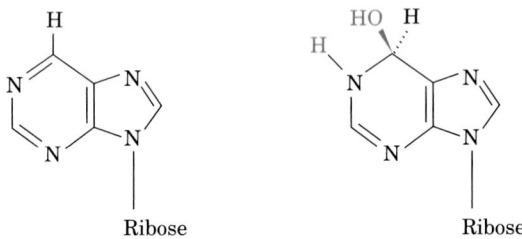

Purine ribonucleoside   **6-Hydroxy-1,6-dihydropurine ribonucleoside (HDPR)**

a nearly ideal transition state analog of the ADA reaction. Although it had been previously reported that ADA does not require a cofactor, its X-ray structure clearly reveals that a zinc ion is bound in the deepest part of the active site pocket, where it is pentacoordinated by three His side chains, a carboxyl oxygen of Asp 295, and the O6 atom of HDPR. ADA's active site complex suggests a catalytic mechanism (Fig. 28-24*b*) reminiscent of that of carbonic anhydrase (Section 15-1C): His 238, which is properly positioned to act as a general base, abstracts a proton from a bound Zn^{2+}-activated water molecule, which nucleophilically attacks the adenine C6 atom to form a tetrahedral intermediate. Products are then formed by the elimination of ammonia.

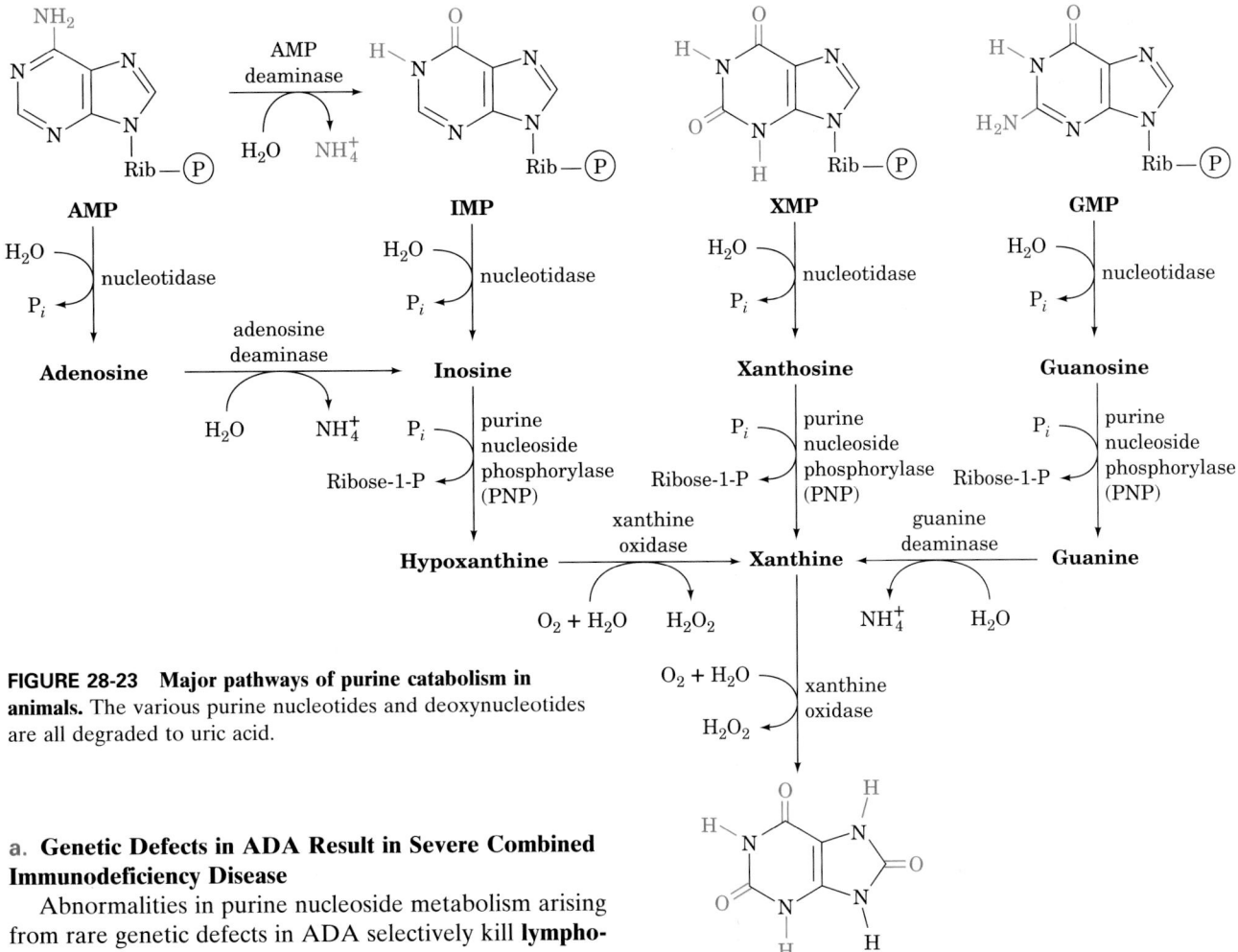

FIGURE 28-23   Major pathways of purine catabolism in animals. The various purine nucleotides and deoxynucleotides are all degraded to uric acid.

a. Genetic Defects in ADA Result in Severe Combined Immunodeficiency Disease

Abnormalities in purine nucleoside metabolism arising from rare genetic defects in ADA selectively kill **lymphocytes** (a type of white blood cell). Since lymphocytes mediate much of the immune response (Section 35-2A), ADA deficiency results in **severe combined immunodeficiency disease (SCID)** that, without special protective measures, is invariably fatal in infancy due to overwhelming infection. The mutations in all eight known ADA variants obtained from SCID patients appear to structurally perturb the active site of ADA.

Biochemical considerations provide a plausible explanation of SCID's etiology (causes). In the absence of active ADA, deoxyadenosine is phosphorylated to yield levels of dATP that are 50-fold greater than normal. This high concentration of dATP inhibits ribonucleotide reductase (Section 28-3A), thereby preventing the synthesis of the other dNTPs, choking off DNA synthesis and thus cell proliferation. The tissue-specific effect of ADA deficiency on the immune system may be explained by the observation that lymphoid tissue is particularly active in deoxyadenosine phosphorylation.

SCID caused by ADA defects does not respond to treatment by the intravenous injection of ADA because the liver clears this enzyme from the bloodstream within minutes. If, however, several molecules of the biologically inert polymer **polyethylene glycol (PEG)**

$$\text{HO} \text{--} [\text{CH}_2 \text{--} \text{CH}_2 \text{--} \text{O}]_n \text{H}$$
Polyethylene glycol

are covalently linked to surface groups on ADA, the resulting **PEG–ADA** remains in the blood for 1 to 2 weeks, thereby largely resuscitating the SCID victim's immune system. The protein-linked PEG only reduces the catalytic activity of ADA by ~40% but, evidently, masks it from the receptors that filter it out of the blood. SCID can therefore be treated effectively by PEG–ADA. This treatment, however, is expensive and not entirely satisfactory. Consequently, ADA deficiency was selected as one of the first genetic diseases to be treated by gene therapy (Section 5-5H): Lymphocytes were extracted from the blood of an ADA-deficient child and grown in the laboratory, had a normal ADA gene inserted into them via genetic engineering techniques (Section 5-5), and were then returned to the child. After 12 years, 20 to 25% of the patient's lymphocytes contained the introduced ADA gene. However, ethical considerations have mandated that the patient continue receiving injections of PEG–ADA so that the efficacy of this gene therapy protocol is unclear.

(a)

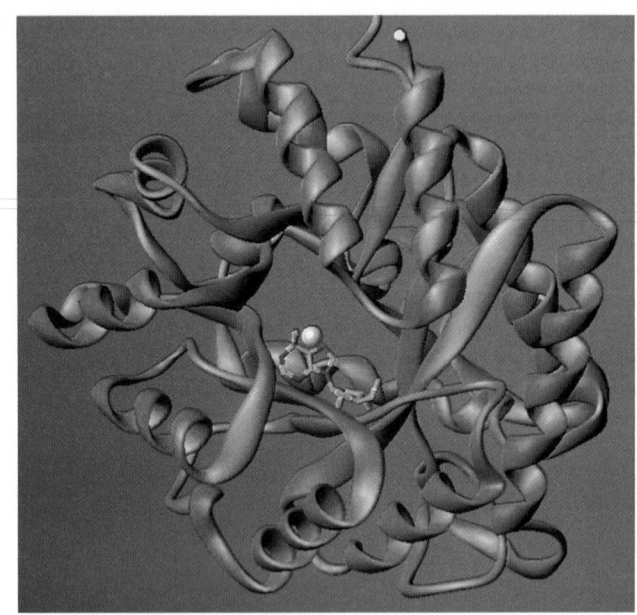

(b)

His 238

Asp 295 — C–

Glu 217 — C

Adenosine

His 238

Asp 295 — C–

Glu 217 — C–

Tetrahedral Intermediate

NH₃

His 238

Asp 295 — C–

Glu 217 — C

Inosine (enol tautomer)

FIGURE 28-24   Structure and mechanism of adenosine deaminase. (*a*) A ribbon diagram of murine adenosine deaminase in complex with its transition state analog HDPR as viewed approximately down the axis of the enzyme's α/β barrel from the N-terminal ends of its β strands. The HDPR is shown in skeletal form with its C, N, and O atoms green, blue, and red. The enzyme-bound Zn²⁺ ion, which is coordinated by HDPR's 6-hydroxyl group, is represented by a silver sphere. [Based on an X-ray structure by Florante Quiocho, Baylor College of Medicine. PDBid 1ADA.] (*b*) The proposed catalytic mechanism of adenosine deaminase. A Zn²⁺-polarized H₂O molecule (Section 15-1C) nucleophilically attacks C6 of the enzyme-bound adenosine molecule in a process that is facilitated by His 238 acting as a general base, Glu 217 acting as a general acid, and Asp 295 acting to orient the water molecule via hydrogen bonding. The resulting tetrahedral intermediate decomposes by the elimination of ammonia in a reaction that is aided by the now imidazolium and carboxyl side chains of His 238 and Glu 217 acting as a general acid and a general base, respectively. This yields inosine in its enol tautomeric form, which, on its release from the enzyme, largely assumes its dominant keto form. The Zn²⁺ is coordinated by three His side chains that are not shown. [After Wilson, D.K. and Quiocho, F.A., *Biochemistry* **32**, 1692 (1993).] ✌ See the Interactive Exercises

b. The Purine Nucleotide Cycle

The deamination of AMP to IMP, when combined with the synthesis of AMP from IMP (Fig. 28-4, *left*), has the effect of deaminating aspartate to yield fumarate (Fig. 28-25). John Lowenstein demonstrated that this **purine nucleotide cycle** has an important metabolic role in skeletal muscle. An increase in muscle activity requires an increase in the activity of the citric acid cycle. This process usually occurs through the generation of additional citric acid cycle intermediates (Section 21-4). Muscles, however, lack most of the enzymes that catalyze these anaplerotic (filling up) reactions in other tissues. Rather, muscle replenishes its citric acid cycle intermediates as fumarate generated in the purine nucleotide cycle. The importance of the purine nucleotide cycle in muscle metabolism is indicated by the observation that the activities of the three enzymes involved are all severalfold higher in muscle than in other tissues. In fact, individuals with an inherited deficiency in muscle AMP deaminase **(myoadenylate deaminase deficiency)** are easily fatigued and usually suffer from cramps after exercise.

c. Xanthine Oxidase Is a Mini-Electron-Transport Protein

Xanthine oxidase (XO) converts hypoxanthine to xanthine, and xanthine to uric acid (Fig. 28-23, *bottom*). In mammals, this enzyme occurs mainly in the liver and the small intestinal mucosa. XO is a homodimer of ~1330-

residue subunits, each of which binds a variety of electron-transfer agents: an FAD, two spectroscopically distinct [2Fe–2S] clusters, and a **molybdopterin complex (Mo-pt)**

Molybdopterin complex (Mo-pt)

in which the Mo atom cycles between its Mo(VI) and Mo(IV) oxidation states. The final electron acceptor is O_2, which is converted to H_2O_2, a potentially harmful oxidizing agent that is subsequently disproportionated to H_2O and O_2 by catalase (Section 1-2A). In XO, the polypeptide has been proteolytically cleaved into three segments (the uncleaved enzyme, which is known as **xanthine dehydrogenase,** preferably uses NAD^+ as its electron acceptor, whereas XO does not react with NAD^+).

The X-ray structure of XO from cow's milk in complex with the competitive inhibitor salicylic acid (Section 25-7B), determined by Emil Pai, reveals that the FAD and the molybdopterin complex are interposed by the two [2Fe–2S] clusters to form a mini-electron-transport chain

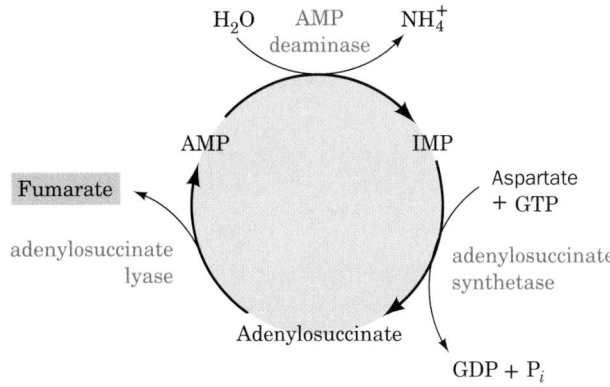

Net: H_2O + Aspartate + GTP $\rightarrow$ NH_4^+ + GDP + P_i + fumarate

FIGURE 28-25 The purine nucleotide cycle. This pathway functions, in muscle, to prime the citric acid cycle by generating fumarate.

(Fig. 28-26). Each of its three peptide segments forms a separate domain with the N-terminal domain binding the two [2Fe–2S] clusters, the central domain binding the FAD, and the C-terminal domain binding the Mo-pt complex. Although the salicylic acid does not contact the Mo-pt complex, it binds to XO in a way that blocks the approach of substrates to the metal center.

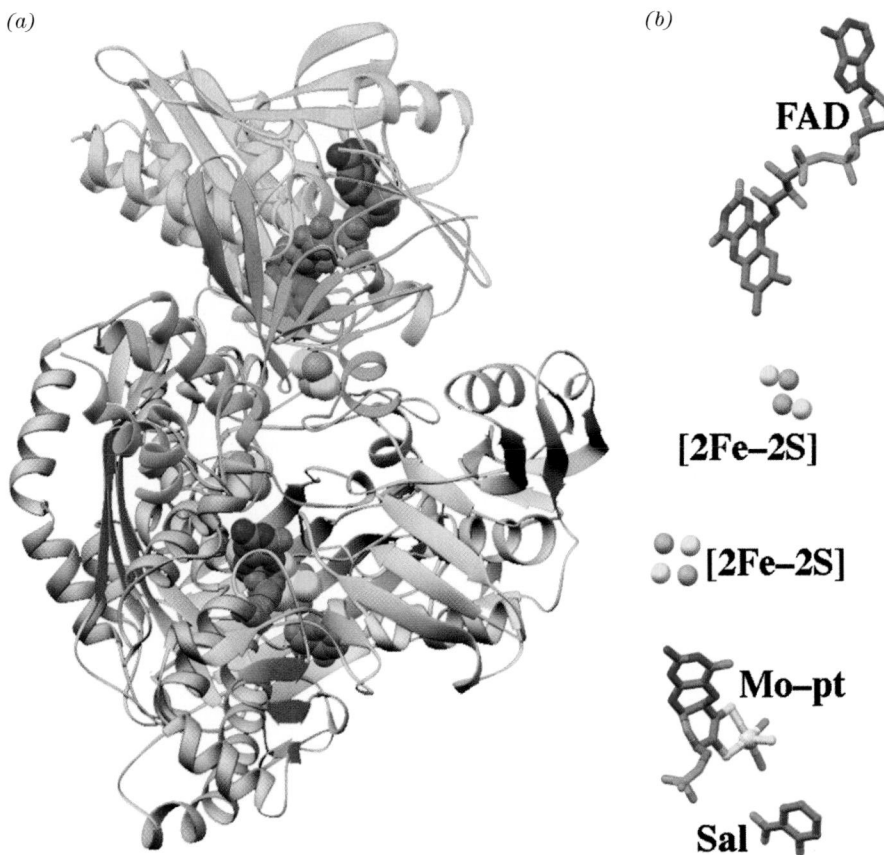

FIGURE 28-26 X-Ray structure of xanthine oxidase from cow's milk in complex with salicylic acid. (*a*) Ribbon diagram of its 1332-residue subunit in which the N-terminal domain (residues 2–165) is cyan, the central domain (residues 224–528) is gold, and the C-terminal domain (residues 571–1315) is lavender. The enzyme's redox cofactors and bound salicylic acid are shown in space-filling form with C green, N blue, O red, S yellow, P magenta, Fe orange, and Mo light blue. The ~50-residue peptide segments spanning successive pairs of domains are disordered and are apparently highly flexible. (*b*) The enzyme's redox cofactors and salicylic acid (Sal) drawn in stick form with their S, Fe, and Mo atoms represented by spheres. The atoms are colored as in Part *a* and viewed from the same direction but with greater magnification. [Based on an X-ray structure by Emil Pai, University of Toronto, Toronto, Ontario, Canada. PDBid 1FIQ.]

XO hydroxylates xanthine at its C8 position (and hypo-xanthine at its C2 position), yielding uric acid in its enol form that tautomerizes to the more stable keto form:

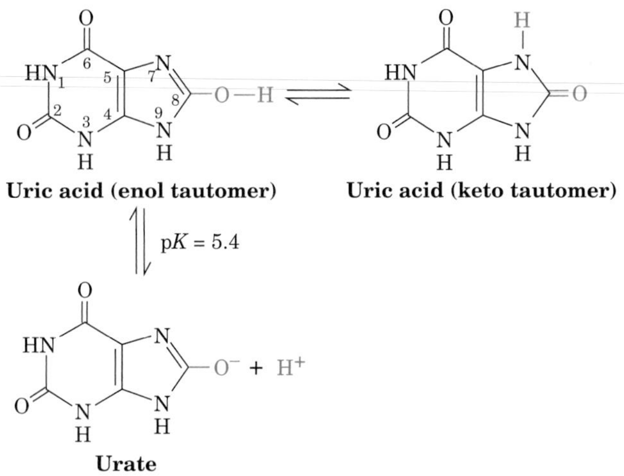

Uric acid (enol tautomer)   **Uric acid (keto tautomer)**

$$pK = 5.4$$

Urate

(its enol form ionizes with a pK of 5.4; hence, the name uric *acid*). ^{18}O-labeling experiments have demonstrated that the C8 keto oxygen of uric acid is derived from H_2O, whereas the oxygen atoms of H_2O_2 come from O_2. Chemical and spectroscopic studies suggest that the enzyme has the following mechanism (Fig. 28-27):

1. The reaction is initiated by the attack of an enzyme nucleophile, X, on the C8 position of xanthine oxidase.

2. The C8—H atom is eliminated as a hydride ion that combines with the Mo(VI) complex, thereby reducing it to the Mo(IV) state.

3. Water displaces the enzyme nucleophile producing uric acid.

In the second stage of the reaction, the now reduced enzyme is reoxidized to its original Mo(VI) state by reaction with O_2. This complex process, not surprisingly, is but poorly understood. EPR measurements indicate that electrons are funneled from the Mo(IV) through the two [2Fe–2S] clusters to the flavin and ultimately to O_2, yielding H_2O_2 and regenerated enzyme.

B. *Fate of Uric Acid*

In humans and other primates, the final product of purine degradation is uric acid, which is excreted in the urine. The same is true of birds, terrestrial reptiles, and many insects, but these organisms, which do not excrete urea, also catabolize their excess amino acid nitrogen to uric acid via purine biosynthesis. This complicated system of nitrogen excretion has a straightforward function: *It conserves water.* Uric acid is only sparingly soluble in water, so that its excretion as a paste of uric acid crystals is accompanied by very little water. In contrast, the excretion of an equivalent amount of the much more water-soluble urea osmotically sequesters a significant amount of water.

In all other organisms, uric acid is further processed before excretion (Fig. 28-28). Mammals other than primates oxidize it to their excretory product, **allantoin,** in a reaction catalyzed by the Cu-containing enzyme **urate oxidase.** A further degradation product, **allantoic acid,** is excreted by teleost (bony) fish. Cartilaginous fish and amphibia further degrade allantoic acid to urea prior to excretion. Finally, marine invertebrates decompose urea to their nitrogen excretory product, NH_4^+.

a. Gout Is Caused by an Excess of Uric Acid

Gout is a disease characterized by elevated levels of uric acid in body fluids. Its most common manifestation is excrutiatingly painful arthritic joint inflammation of sudden onset, most often in the big toe (Fig. 28-29), caused by deposition of nearly insoluble crystals of sodium urate. Sodium urate and/or uric acid may also precipitate in the

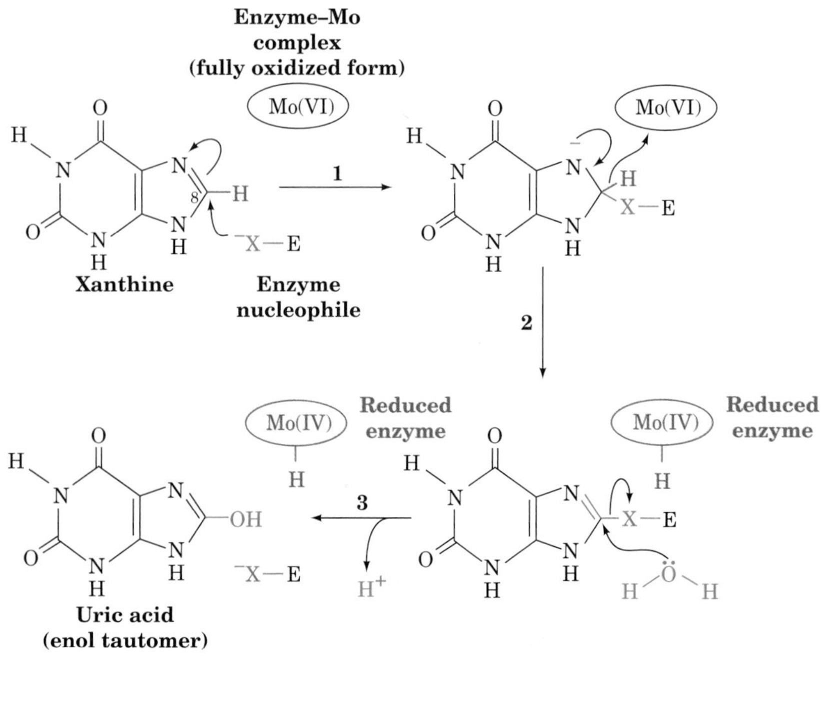

FIGURE 28-27   Mechanism of xanthine oxidase. The reduced enzyme is subsequently reoxidized by O_2, yielding H_2O_2.

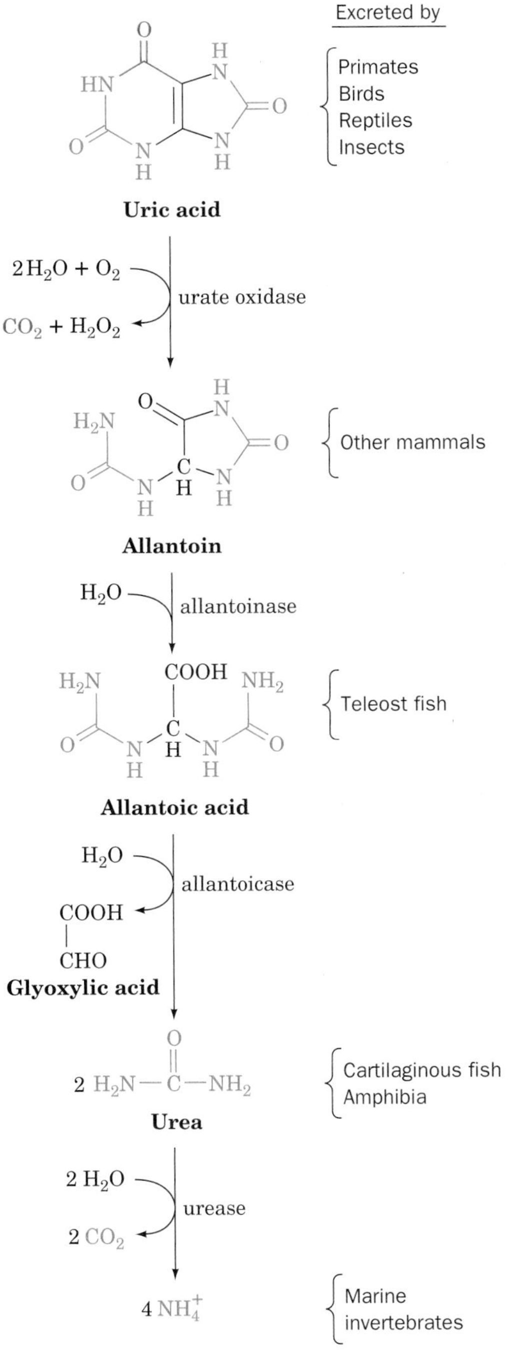

Uric acid

$2H_2O + O_2$
$CO_2 + H_2O_2$ urate oxidase

Excreted by
{ Primates
Birds
Reptiles
Insects

Allantoin   } Other mammals

H_2O allantoinase

Allantoic acid   } Teleost fish

H_2O allantoicase

Glyoxylic acid

$2 H_2N-\overset{O}{\overset{\|}{C}}-NH_2$

Urea   { Cartilaginous fish
Amphibia

$2H_2O$ urease
$2CO_2$

$4NH_4^+$ { Marine
invertebrates

FIGURE 28-28 Degradation of uric acid to ammonia. The process is arrested at different stages in the indicated species and the resulting nitrogen-containing product is excreted.

kidneys and ureters as stones, resulting in renal damage and urinary tract obstruction. Gout, which affects ~3 per 1000 persons, predominantly males, has been traditionally, although inaccurately, associated with overindulgent eating and drinking. The probable origin of this association is that in previous centuries, when wine was often contaminated with lead during its manufacture and storage, heavy drinking resulted in chronic lead poisoning, which, among other things, decreases the kidney's ability to excrete uric acid.

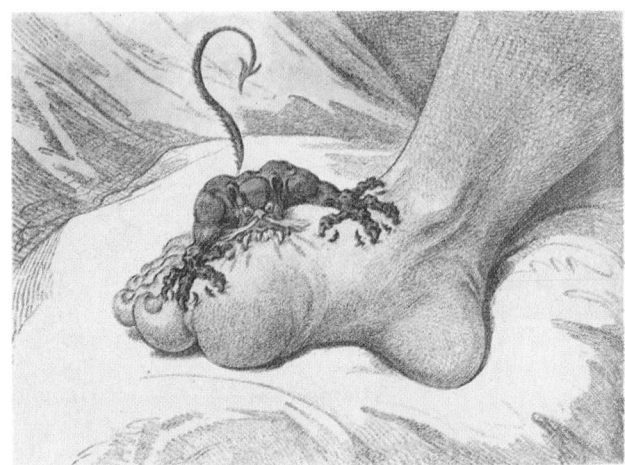

FIGURE 28-29 *The Gout*, a cartoon by James Gilroy (1799). [Yale University Medical Historical Library.]

The most prevalent cause of gout is impaired uric acid excretion (although usually for other reasons than lead poisoning). Gout may also result from a number of metabolic insufficiencies, most of which are not well characterized. One well-understood cause is HGPRT deficiency (Lesch–Nyhan syndrome in severe cases), which leads to excessive uric acid production through PRPP accumulation (Section 28-1D). Uric acid overproduction is also caused by glucose-6-phosphatase deficiency (von Gierke's glycogen storage disease; Section 18-4): The increased availability of glucose-6-phosphate stimulates the pentose phosphate pathway (Section 23-4), increasing the rate of ribose-5-phosphate production and consequently that of PRPP, which in turn stimulates purine biosynthesis.

Gout may be treated by administration of the xanthine oxidase inhibitor **allopurinol,** a hypoxanthine analog with interchanged N7 and C8 positions.

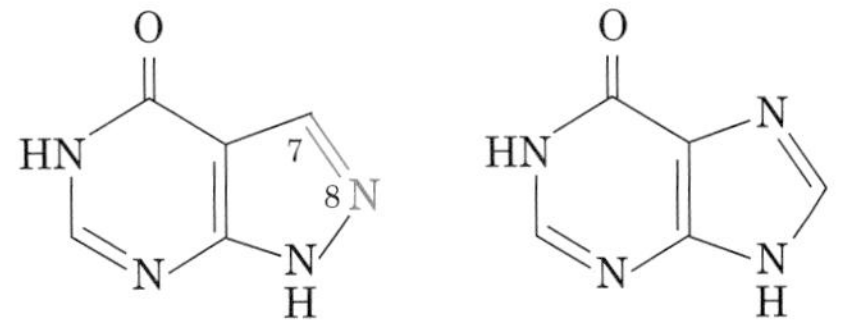

Allopurinol   **Hypoxanthine**

Xanthine oxidase hydroxylates allopurinol, as it does hypoxanthine, yielding **alloxanthine,**

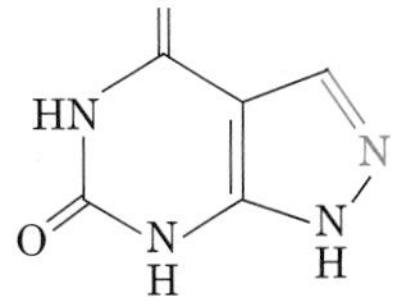

Alloxanthine

which remains tightly bound to the reduced form of the enzyme, thereby inactivating it. Allopurinol consequently alleviates the symptoms of gout by decreasing the rate of uric acid production while increasing the levels of the more

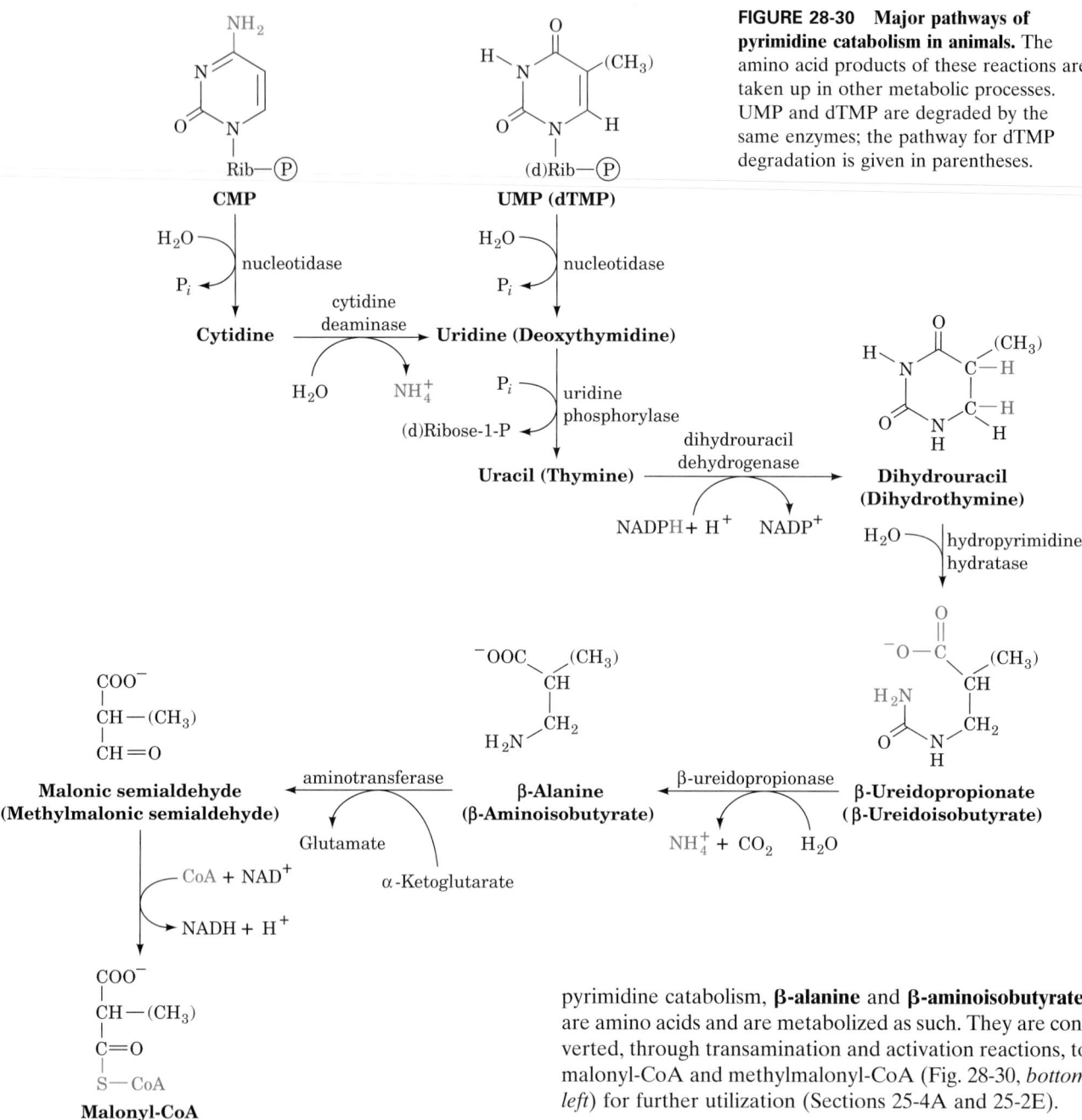

FIGURE 28-30 Major pathways of pyrimidine catabolism in animals. The amino acid products of these reactions are taken up in other metabolic processes. UMP and dTMP are degraded by the same enzymes; the pathway for dTMP degradation is given in parentheses.

soluble hypoxanthine and xanthine. Although allopurinol controls the gouty symptoms of Lesch–Nyhan syndrome, it has no effect on its neurological symptoms.

C. *Catabolism of Pyrimidines*

Animal cells degrade pyrimidine nucleotides to their component bases (Fig. 28-30, *top*). These reactions, like those of purine nucleotides, occur through dephosphorylation, deamination, and glycosidic bond cleavages. The resulting uracil and thymine are then broken down in the liver through reduction (Fig. 28-30, *middle*) rather than by oxidation, as occurs in purine catabolism. The end products of

pyrimidine catabolism, **β-alanine** and **β-aminoisobutyrate,** are amino acids and are metabolized as such. They are converted, through transamination and activation reactions, to malonyl-CoA and methylmalonyl-CoA (Fig. 28-30, *bottom left*) for further utilization (Sections 25-4A and 25-2E).

5 ■ BIOSYNTHESIS OF NUCLEOTIDE COENZYMES

In this section we outline the assembly, in animals, of the nucleotide coenzymes NAD^+ and $NADP^+$, FMN and FAD, and coenzyme A, from their vitamin precursors. These vitamins are synthesized *de novo* only by plants and microorganisms.

A. *Nicotinamide Coenzymes*

The nicotinamide moiety of the nicotinamide coenzymes (NAD^+ and $NADP^+$) is derived, in humans, from dietary nicotinamide, nicotinic acid, or the essential amino acid tryptophan (Fig. 28-31). **Nicotinate phosphoribosyltrans-**

FIGURE 28-31 Pathways for the biosynthesis of NAD$^+$ and NADP$^+$. These nicotinamide coenzymes are synthesized from their vitamin precursors, nicotinate and nicotinamide, and from the tryptophan degradation product, quinolinate.

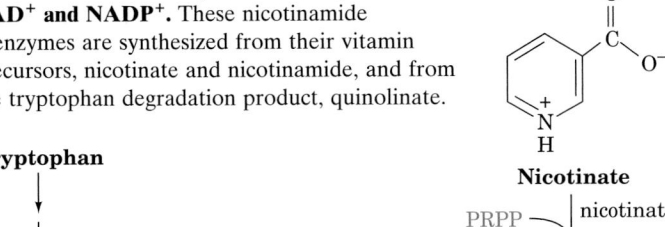

Tryptophan

quinolinate phosphoribosyl transferase

PRPP → PP$_i$ + CO$_2$

Quinolinate

Nicotinate

PRPP → PP$_i$

nicotinate phosphoribosyl transferase

Nicotinamide

PRPP → PP$_i$

nicotinamide phosphoribosyl transferase

$^{-2}$O$_3$P—O—CH$_2$

OH OH

Nicotinate mononucleotide

$^{-2}$O$_3$P—O—CH$_2$

OH OH

Nicotinamide mononucleotide (NMN)

ATP → PP$_i$

NAD$^+$ pyrophosphorylase

ATP → PP$_i$

NAD$^+$ pyrophosphorylase

Ribose —(P)—(P)— Ribose
Adenine

Nicotinate adenine dinucleotide

ATP + Glutamine + H$_2$O
ATP + Glutamate

NAD$^+$ synthetase

Ribose —(P)—(P)— Ribose
Adenine

Nicotinamide adenine dinucleotide (NAD$^+$)

ATP → ADP

NAD$^+$ kinase

$^-$O—P—O—CH$_2$

OH OH

$^-$O—P—O—CH$_2$

OH O—PO$_3^{2-}$

NH$_2$

Nicotinamide adenine dinucleotide phosphate (NADP$^+$)

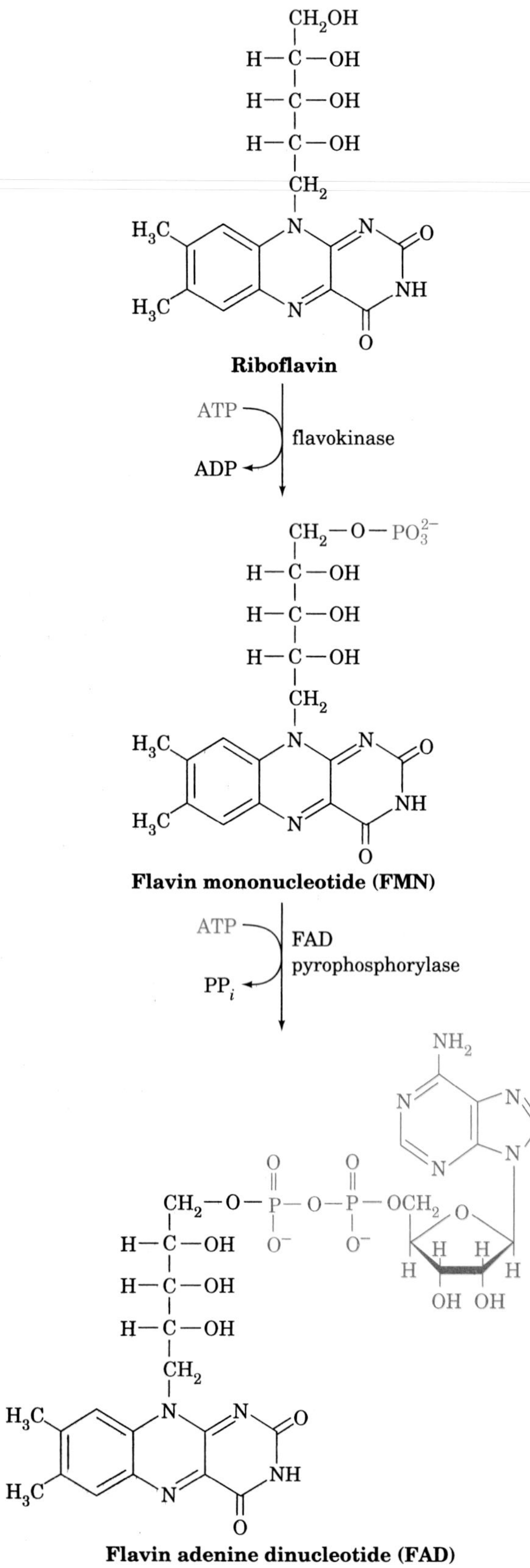

FIGURE 28-32 **Biosynthesis of FMN and FAD from the vitamin precursor riboflavin.**

ferase, which occurs in most mammalian tissues, catalyzes the formation of **nicotinate mononucleotide** from nicotinate and PRPP. This intermediate may also be synthesized from **quinolinate,** a degradation product of tryptophan (Section 26-3G), in a reaction mediated by **quinolinate phosphoribosyltransferase,** which occurs mainly in liver and kidney. A poor diet, nevertheless, may result in pellagra (nicotinic acid deficiency; Section 13-3), since, under such conditions, tryptophan will be almost entirely utilized in protein biosynthesis. Nicotinate mononucleotide is joined via a pyrophosphate linkage to an ATP-derived AMP residue by **NAD⁺ pyrophosphorylase** to yield **nicotinate adenine dinucleotide (desamido NAD⁺).** Finally, **NAD⁺ synthetase** converts this intermediate to NAD⁺ by a transamidation reaction in which glutamine is the NH_2 donor.

NAD⁺ may also be synthesized from nicotinamide. This vitamin is converted to **nicotinamide mononucleotide (NMN)** by **nicotinamide phosphoribosyltransferase,** a widely occurring enzyme distinct from nicotinate phosphoribosyltransferase. However, NAD⁺ is synthesized from NMN and ATP by NAD pyrophosphorylase, the same enzyme that synthesizes nicotinate adenine dinucleotide.

NADP⁺ is formed via the phosphorylation of the NAD⁺ adenosine residue's C2'-OH group by **NAD⁺ kinase.**

B. *Flavin Coenzymes*

FAD is synthesized from riboflavin in a two-reaction pathway (Fig. 28-32). First, the 5'-OH group of riboflavin's ribityl side chain is phosphorylated by **flavokinase,** yielding flavin mononucleotide (FMN; not a true nucleotide since its ribityl residue is not a true sugar). FAD may then be formed by the coupling of FMN and ATP-derived AMP in a pyrophosphate linkage in a reaction catalyzed by **FAD pyrophosphorylase.** Both of these enzymes are widely distributed in nature.

C. *Coenzyme A*

Coenzyme A is synthesized in mammalian cells according to the pathway diagrammed in Fig. 28-33. Pantothenate, an essential vitamin, is phosphorylated by **pantothenate kinase** and then coupled to cysteine, the future business end of CoA, by **phosphopantothenoylcysteine synthetase.** After decarboxylation by **phosphopantothenoylcysteine decarboxylase,** the resulting **4'-phosphopantethiene** is coupled to AMP in a pyrophosphate linkage by **dephospho-CoA pyrophosphorylase** and then phosphorylated at its adenosine 3'-OH group by **dephospho-CoA kinase** to form CoA. The latter two enzymatic activities occur on a single protein.

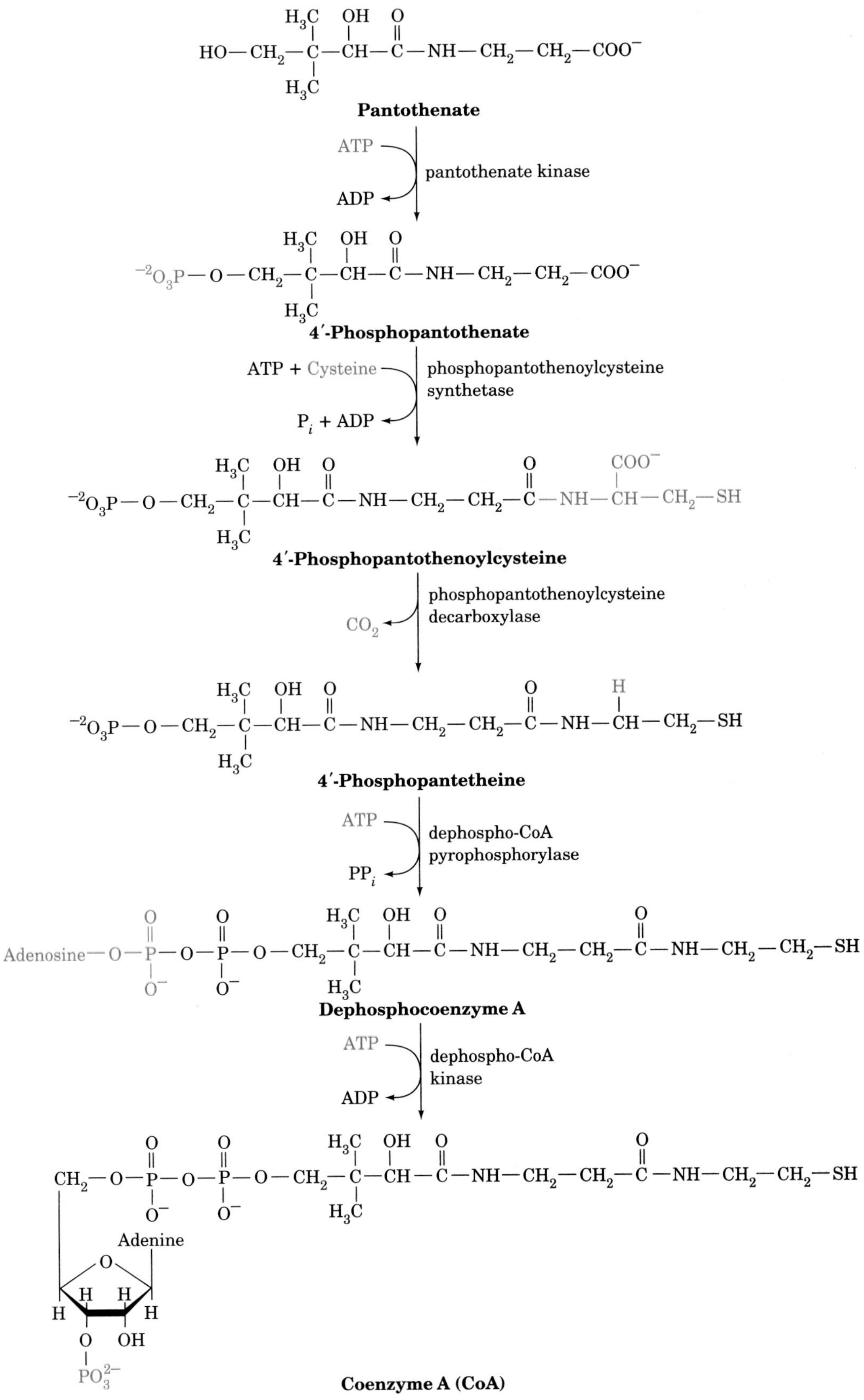

FIGURE 28-33 **Biosynthesis of coenzyme A from pantothenate, its vitamin precursor.**

CHAPTER SUMMARY

1 ■ **Synthesis of Purine Ribonucleotides** Almost all cells synthesize purine nucleotides *de novo* via similar metabolic pathways. The purine ring is constructed in an 11-step reaction sequence that yields IMP. AMP and GMP are then synthesized from IMP in separate pathways. Nucleoside diphosphates and triphosphates are sequentially formed from these products via phosphorylation reactions. The rates of synthesis of these various nucleotides are interrelated through feedback inhibition mechanisms that monitor their concentrations. Purine nucleotides may also be synthesized from free purines salvaged from nucleic acid degradation processes. The importance of these salvage reactions is demonstrated, for example, by the devastating and bizarre consequences of Lesch–Nyhan syndrome.

2 ■ **Synthesis of Pyrimidine Ribonucleotides** Cells also synthesize pyrimidines *de novo* but, in this six-step process, a free base is formed before it is converted to a nucleotide, UMP. UTP is then formed by phosphorylation of UMP, and CTP is synthesized by the amination of UTP. Pyrimidine biosynthesis is regulated by feedback inhibition as well as by the concentrations of purine nucleotides.

3 ■ **Formation of Deoxyribonucleotides** Deoxyribonucleotides are formed by reduction of the corresponding ribonucleotides. Three classes of ribonucleotide reductase (RNR) have been characterized: Class I RNR, which occurs in nearly all eukaryotes and many prokaryotes, contains an $Fe(III) — O^{2-} — Fe(III)$ group and a tyrosyl free radical; Class II and III RNRs, which occur only in prokaryotes, contain, respectively, a coenzyme B_{12} cofactor, and a [4Fe–4S] cluster together with a glycyl radical. All of them catalyze free radical–based reductions. The substrates for Class I and II RNRs are NDPs, whereas those for Class III RNRs are NTPs. Class I RNR has three independent regulatory sites that control its substrate specificity and its catalytic activity in part via its oligomerization state, thereby generating deoxynucleotides in the amounts required for DNA synthesis. The *E. coli* Class I RNR is reduced to its original state by electron-transport chains involving either thioredoxin, thioredoxin reductase, and NADPH; or glutaredoxin, glutathione, glutathione reductase, and NADPH. Thymine is synthesized by the methylation of dUMP by thymidylate synthase to form dTMP. The reaction's methyl source, N^5,N^{10}-methylene-THF, is oxidized in the reaction to yield dihydrofolate. N^5,N^{10}-Methylene-THF is subsequently regenerated through the sequential actions of dihydrofolate reductase and serine hydroxymethyltransferase. Since this sequence of reactions is required for DNA biosynthesis, it presents an excellent target for chemotherapy. FdUMP, a mechanism-based inhibitor of thymidylate synthase, and methotrexate, an antifolate that essentially irreversibly inhibits dihydrofolate reductase, are both highly effective anticancer agents.

4 ■ **Nucleotide Degradation** Purine nucleotides are catabolized to yield uric acid. Depending on the species, the uric acid is either directly excreted or first degraded to simpler nitrogen-containing substances. Overproduction or underexcretion of uric acid in humans causes gout. Pyrimidines are catabolized in animal cells to amino acids.

5 ■ **Biosynthesis of Nucleotide Coenzymes** The nucleotide coenzymes NAD^+ and $NADP^+$, FMN and FAD, and coenzyme A are synthesized in animals from vitamin precursors.

REFERENCES

GENERAL

Kornberg, A. and Baker, T.A., *DNA Replication* (2nd ed.), Chapter 2, Freeman (1992).

Scriver, C.R., Beaudet, A.L., Sly, W.S., and Valle, D. (Eds.), *The Metabolic & Molecular Bases of Inherited Disease* (8th ed.), Chapters 106–113, McGraw-Hill (2001).

PURINE NUCLEOTIDE BIOSYNTHESIS

Almassey, R.J., Janson, C.A., Kan, C.-C., and Hostomska, Z., Structures of the apo and complexed *Escherichia coli* glycinamide ribonucleotide transformylase, *Proc. Natl. Acad. Sci.* **89**, 6114–6118 (1992).

Eriksen, T.A., Kadziola, A., Bentsen, A.-K., Harlow, K.W., and Larsen, S., Structural basis for the function of *Bacillus subtilis* phosphoribosylpyrophosphate synthetase, *Nature Struct. Biol.* **7**, 303–308 (2000).

Greasley, S.E., Horton, P., Ramcharan, J., Beardsley, G.P., Benkovic, S.J., and Wilson, I.A., Crystal structure of a bifunctional transformylase and cyclohydrolase enzyme in purine biosynthesis, *Nature Struct. Biol.* **8**, 402–406 (2001).

Kappock, T.J., Ealick, S.E., and Stubbe, J., Modular evolution of the purine biosynthetic pathway, *Curr. Opin. Chem. Biol.* **4**, 567–572 (2000).

Levdikov, V.M., Barynin, V.V., Grebenko, A.I., Melik-Adamyan, W.R., Lamzin, V.S., and Wilson, K.S., The structure of

SAICAR synthase: An enzyme in the *de novo* pathway of purine biosynthesis, *Structure* **6**, 363–376 (1998).

Li, C., Kappock, T.J., Stubbe, J., Weaver, T.M., and Ealick, S.E., X-Ray crystal structure of aminoimidazole ribonucleotide synthetase (PurM) from the *Escherichia coli* purine biosynthetic pathway at 2.5 Å resolution, *Structure* **7**, 1155–1166 (1999).

Mathews, I.I., Kappock, T.J., Stubbe, J., and Ealick, S.E., Crystal structure of *Escherichia coli* PurE, an unusual mutase in the purine biosynthetic pathway, *Structure* **7**, 1395–1406 (1999).

Poland, B.W., Fromm, H.J., and Honzatko, R.B., Crystal structures of adenylosuccinate synthetase from *Escherichia coli* complexed with GDP, IMP, hadacidin, NO_3^-, and Mg^{2+}, *J. Mol. Biol.* **264**, 1013–1027 (1996).

Sintchak, M.D., Fleming, M.A., Futer, O., Raybuck, S.A., Chambers, S.P., Caron, P.R., Murcko, M.A., and Wilson, K.P., Structure and mechanism of inosine monophosphate dehydrogenase in complex with the immunosuppressant mycophenolic acid, *Cell* **85**, 921–930 (1996).

Smith, J.L., Glutamine PRPP amidotransferase: Snapshots of an enzyme in action, *Curr. Opin. Struct. Biol.* **8**, 686–694 (1998).

Tesmer, J.J., Klem, T.J., Deras, M.L., Davisson, V.J., and Smith, J.L., The crystal structure of GMP synthetase reveals a novel catalytic triad and is a structural paradigm for two enzyme families, *Nature Struct. Biol.* **3**, 74–86 (1996).

Thoden, J.B., Kappock, T.J., Stubbe, J.A., and Holden, H.M.,

Three-dimensional structure of N^5-carboxyaminoimidazole ribonucleotide synthetase: A member of the ATP grasp protein superfamily, *Biochemistry* **38,** 15480–15492 (1999). [X-ray structure of PurK.]

Toth, E.A. and Yeates, T.O., The structure of adenylosuccinate lyase, an enzyme with dual activity in the *de novo* purine biosynthetic pathway, *Structure* **8,** 163–174 (2000).

Wang, W., Kappock, T.J., Stubbe, J.A., and Ealick, S.E., X-Ray structure of glycinamide ribonucleotide synthetase from *Escherichia coli, Biochemistry* **37,** 15647–15662 (1998).

Zalkin, H. and Dixon, J.E., *De novo* purine nucleotide biosynthesis, *Prog. Nucleic Acid Res. Mol. Biol.* **42,** 259–285 (1992).

PYRIMIDINE NUCLEOTIDE BIOSYNTHESIS

Begley, T.P., Appleby, T.C., and Ealick, S.E., The structural basis for the remarkable catalytic proficiency of orotidine 5'-monophosphate decarboxylase, *Curr. Opin. Struct. Biol.* **10,** 711–718 (2000).

Jones, M.E., Orotidylate decarboxylase of yeast and man, *Curr. Top. Cell Regul.* **33,** 331–342 (1992).

Liu, S., Neidhardt, E.A., Grossman, T.H., Ocain, T., and Clardy, J., Structures of human dihydroorotate dehydrogenase in complex with antiproliferative agents, *Structure* **8,** 25–33 (1999).

Miller, B.G. and Wolfenden, R., Catalytic proficiency: The unusual case of OMP decarboxylase, *Annu. Rev. Biochem.* **71,** 847–885 (2002).

Scapin, G., Ozturk, D.H., Grubmeyer, C., and Sacchettini, J.C., The crystal structure of the orotate phosphoribosyltransferase complexed with orotate and α-D-5-phosphoribosyl-1-pyrophosphate, *Biochemistry* **34,** 10744–10754 (1995).

Thoden, J.B., Phillips, G.N., Jr., Neal, T.M., Raushel, F.M., and Holden, H.M., Molecular structure of dihydroorotase: A paradigm for catalysis through the use of a binuclear center, *Biochemistry* **40,** 6989–6997 (2001).

Traut, T.W. and Jones, M.E., Uracil metabolism—UMP synthesis from orotic acid or uridine and conversion of uracil to β-alanine: Enzymes and cDNAs, *Prog. Nucleic Acid Res. Mol. Biol.* **53,** 1–78 (1996).

SYNTHESIS OF DEOXYNUCLEOTIDES

Carreras, C.W. and Santi, D.V., The catalytic mechanism and structure of thymidylate synthase, *Annu. Rev. Biochem.* **64,** 721–762 (1995).

Eriksson, M., Uhlin, U., Ramaswamy, S., Ekberg, M., Regnström, K., Sjöberg, B.-M., and Eklund, H., Binding of allosteric effectors to ribonucleotide reductase protein R1: Reduction of active site cysteines promotes substrate binding, *Structure* **5,** 1077–1092 (1997).

Hardy, L.W., Finer-Moore, J.S., Montfort, W.R., Jones, M.O., Santi, D.V., and Stroud, R.M., Atomic structure of thymidylate synthase: Target for rational drug design, *Science* **235,** 448–455 (1987). [The X-ray structure of the *Lactobacillus casei* enzyme.]

Jordan, A. and Reichard, P., Ribonucleotide reductases, *Annu. Rev. Biochem.* **67,** 71–98 (1998).

Kashlan, O.B., Scott, C.P., Lear, J.D., and Cooperman, B.S., A comprehensive model for the allosteric regulation of mammalian ribonuclease reductase. Functional consequences of ATP- and dATP-induced oligomerization of the large subunit, *Biochemistry* **41,** 462–474 (2002).

Knighton, D.R., Kan, C.-C., Howland, E., Janson, C.A., Hostomska, Z., Welsh, K.M., and Matthews, D.A., Structure of and kinetic channeling in bifunctional dihydrofolate reductase-thymidylate synthase, *Nature Struct. Biol.* **1,** 186–194 (1994).

Kraut, J. and Matthews, D.A., Dihydrofolate reductase, *in* Jurnak,

F.A. and McPherson, A. (Eds.), *Biological Macromolecules and Assemblies,* Vol. 3, *pp.* 1–71, Wiley (1987).

Lennon, B.W., Williams, J.R., Jr., and Ludwig, M.L., Twists in catalysis: Alternating conformations in *Escherichia coli* thioredoxin reductase, *Science* **289,** 1190–1194 (2000); *and* Waksman, G., Krishna, T.S.R., Williams, C.H., Jr., and Kuriyan, J., Crystal structure of *Escherichia coli* thioredoxin reductase refined at 2 Å resolution, *J. Mol. Biol.* **236,** 800–816 (1994).

Logan, D.T., Andersson, J., Sjöberg, B.-M., and Nordlund, P., A glycyl radical site in the crystal structure of a Class III ribonucleotide reductase, *Science* **283,** 1499–1504 (1999).

Matthews, D.A., Villafranca, J.E., Janson, C.A., Smith, W.W., Welsh, K., and Freer, S., Stereochemical mechanisms of action for thymidylate synthase based on the X-ray structure of the covalent inhibitory ternary complex with 5-fluoro-2'-deoxyuridylate and 5,10-methylenetetrahydrofolate, *J. Mol. Biol.* **214,** 937–948 (1990); *and* Hyatt, D.C., Maley, F., and Montfort, W.R., Use of strain in a stereospecific catalytic mechanism: Crystal structure of *Escherichia coli* thymidylate synthase bound to FdUMP and methylenetetrahydrofolate, *Biochemistry* **36,** 4585–4594 (1997).

Mol, C.D., Harris, J.M., McIntosh, E.M., and Tainer, J.A., Human dUTP pyrophosphatase: Uracil recognition by a β hairpin and active sites formed by three separate subunits, *Structure* **4,** 1077–1092 (1996).

Nordlund, P. and Eklund, H., Structure and function of the *Escherichia coli* ribonucleotide reductase protein R2, *J. Mol. Biol.* **232,** 123–164 (1993).

Powis, G. and Montfort, W.R., Properties and biological activities of thioredoxins, *Annu. Rev. Biophys. Biomol. Struct.* **30,** 421–455 (2001).

Sintchak, M.D., Arjara, G., Kellog, B.A., Stubbe, J., and Drennan, C.L., The crystal structure of class II ribonucleotide reductase reveals how an allosterically regulated monomer mimics a dimer, *Nature Struct. Biol.* **9,** 293–300 (2002).

Stubbe, J. and Riggs-Gelasco, P., Harnessing free radicals: Formation and function of the tyrosyl radical in ribonucleotide reductase, *Trends Biochem. Sci.* **23,** 438–443 (1998).

Stubbe, J., Ge, J., and Yee, C.S., The evolution of ribonucleotide reduction revisited, *Trends Biochem. Sci.* **26,** 93–99 (2001); *and* Stubbe, J., Ribonucleotide reductases: The link between an RNA and a DNA world, *Curr. Opin. Struct. Biol.* **10,** 731–736 (2000).

Uhlin, U. and Eklund, H., Structure of ribonucleotide reductase protein R1, *Nature* **370,** 533–539 (1994).

NUCLEOTIDE DEGRADATION

Enroth, C., Eger, B.T., Okamoto, K., Nishino, T., Nishino, T., and Pai, E., Crystal structure of bovine milk xanthine dehydrogenase and xanthine oxidase: Structure based mechanism of conversion, *Proc. Natl. Acad. Sci.* **97,** 10723–10728 (2000).

Parkman, R., Weinberg, K., Crooks, G., Nolta, I., Kapoor, N., and Kohn, D., Gene therapy for adenosine deaminase deficiency, *Annu. Rev. Med.* **51,** 33–47 (2000).

Wilson, D.K., Rudolph, F.B., and Quiocho, F.A., Atomic structure of adenosine deaminase complexed with a transition-state analog: Understanding catalysis and immunodeficiency mutations, *Science* **252,** 1278–1284 (1991); Wilson, D.K. and Quiocho, F.A., A pre-transition-state mimic of an enzyme: X-ray structure of adenosine deaminase with bound 1-deaza-adenosine and zinc-activated water, *Biochemistry* **32,** 1689–1694 (1993); *and* Crystallographic observation of a trapped tetrahedral intermediate in a metalloenzyme, *Nature Struct. Biol.* **1,** 691–694 (1994).

PROBLEMS

1. Azaserine **(O-diazoacetyl-L-serine)** and **6-diazo-5-oxo-L-norleucine (DON)**

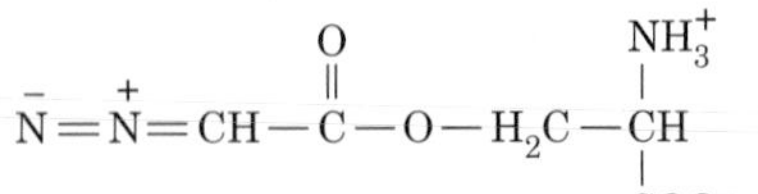

$$\overset{-}{N}=\overset{+}{N}=CH-\overset{\overset{O}{\|}}{C}-O-H_2C-\overset{\overset{NH_3^+}{|}}{\underset{|}{CH}}$$
Azaserine COO^-

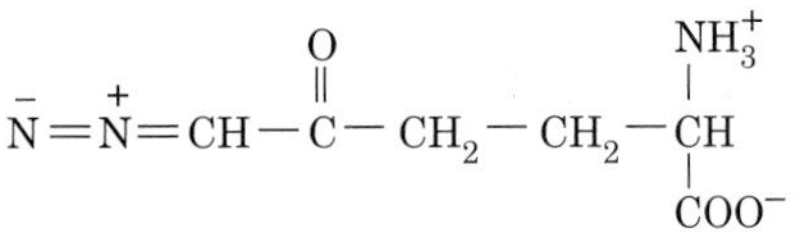

$$\overset{-}{N}=\overset{+}{N}=CH-\overset{\overset{O}{\|}}{C}-CH_2-CH_2-\overset{\overset{NH_3^+}{|}}{\underset{|}{CH}}$$
       COO^-
6-Diazo-5-oxo-L-norleucine (DON)

are glutamine analogs. They form covalent bonds to nucleophiles at the active sites of enzymes that bind glutamine, thereby irreversibly inactivating these enzymes. Identify the nucleotide biosynthesis intermediates that accumulate in the presence of either of these glutamine antagonists.

2. Suggest a mechanism for the AIR synthetase reaction (Fig. 28-2, Reaction 6).

*3. What is the energetic price, in ATPs, of synthesizing the hypoxanthine residue of IMP from CO_2 and NH_4^+?

4. Why is deoxyadenosine toxic to mammalian cells?

5. Indicate which of the following substances are mechanism-based inhibitors and explain your reasoning. (a) Tosyl-L-phenylalanine chloromethylketone with chymotrypsin (Section 15-3A). (b) Trimethoprim with bacterial dihydrofolate reductase. (c) The δ-lactone analog of $(NAG)_4$ with lysozyme (Section 15-2C). (d) Allopurinol with xanthine oxidase.

6. Why do individuals who are undergoing chemotherapy with cytotoxic (cell killing) agents such as FdUMP or methotrexate temporarily go bald?

7. Normal cells die in a nutrient medium containing thymidine and methotrexate that supports the growth of mutant cells defective in thymidylate synthase. Explain.

8. FdUMP and methotrexate, when taken together, are less effective chemotherapeutic agents than when either drug is taken alone. Explain.

9. Why is gout more prevalent in populations that eat relatively large amounts of meat?

10. Gout resulting from the *de novo* overproduction of purines can be distinguished from gout caused by impaired excretion of uric acid by feeding a patient ^{15}N-labeled glycine and determining the distribution of ^{15}N in his or her excreted uric acid. What isotopic distributions are expected for each type of defect?

11. 6-Mercaptopurine,

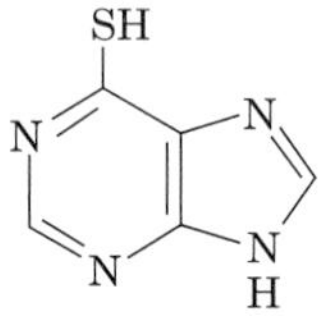

SH
6-Mercaptopurine

after conversion to the corresponding nucleotide through salvage reactions, is a potent competitive inhibitor of IMP in the pathways for AMP and GMP biosynthesis. It is therefore a clinically useful anticancer agent. The chemotherapeutic effectiveness of 6-mercaptopurine is enhanced when it is administered with allopurinol. Explain the mechanism of this enhancement.

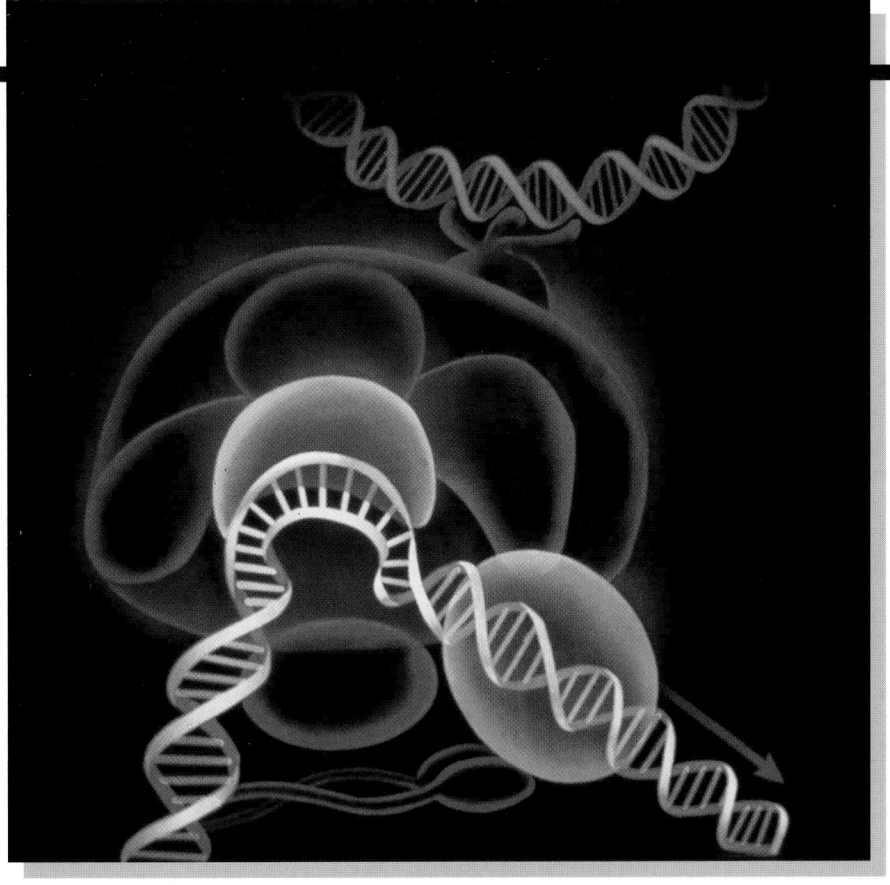

Schematic diagram of the eukaryotic preinitiation complex that is required for the transcription of DNA to messenger RNA. The TATA-box binding protein is shown in orange.

PART

V

EXPRESSION AND TRANSMISSION OF GENETIC INFORMATION

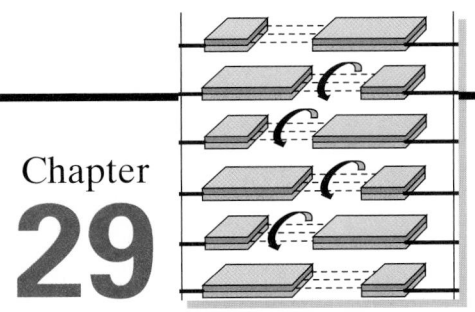

Chapter 29

Nucleic Acid Structures

*There are two classes of nucleic acids, **deoxyribonucleic acid (DNA)** and **ribonucleic acid (RNA).** DNA is the hereditary molecule in all cellular life-forms, as well as in many viruses.* It has but two functions:

1. To direct its own **replication** during cell division.

2. To direct the **transcription** of complementary molecules of RNA.

RNA, in contrast, has more varied biological functions:

1. The RNA transcripts of DNA sequences that specify polypeptides, **messenger RNAs (mRNAs),** direct the ribosomal synthesis of these polypeptides in a process known as **translation.**

2. The RNAs of ribosomes, which are about two-thirds

RNA and one-third protein, have functional as well as structural roles.

3. During protein synthesis, amino acids are delivered to the ribosome by molecules of **transfer RNA (tRNA).**

4. Certain RNAs are associated with specific proteins to form **ribonucleoproteins** that participate in the post-transcriptional processing of other RNAs.

5. In many viruses, RNA, not DNA, is the carrier of hereditary information.

The structure and properties of nucleic acids are introduced in Chapter 5. In this chapter we extend this discussion with emphasis on DNA; the structures of RNAs are detailed in Sections 31-4A and 32-2B. Methods of purifying, sequencing, and chemically synthesizing nucleic acids are discussed in Sections 6-6, 7-2, and 7-6, and recombinant DNA techniques are discussed in Section 5-5. Bioinformatics, as it concerns nucleic acids, is outlined in Section 7-4, and the Nucleic Acid Database is described in Section 8-3C.

1 ■ DOUBLE HELICAL STRUCTURES

🔎 **See Guided Exploration 23. DNA Structures** Double helical DNA has three major helical forms, B-DNA, A-DNA, and Z-DNA, whose structures are depicted in Figs. 29-1, 29-2, and 29-3. In this section we discuss the major characteristics of each of these helical forms as well as those of double helical RNA and DNA–RNA hybrid helices.

(a)

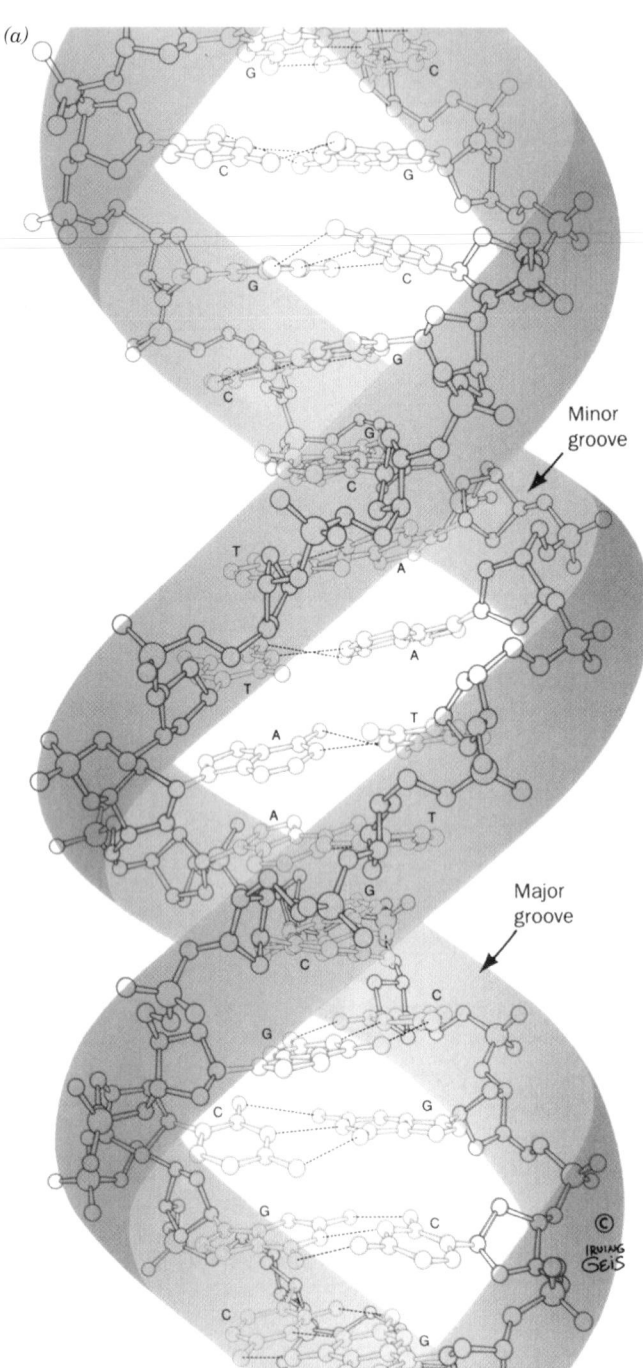

Minor groove

Major groove

FIGURE 29-1 Structure of B-DNA. The structure is represented by ball-and-stick drawings and the corresponding computer-generated space-filling models. The repeating helix is based on the X-ray structure of the self-complementary dodecamer d(CGCGAATTCGCG) determined by Richard Dickerson and Horace Drew, California Institute of Technology (PDBid 1BNA). (*a*) View perpendicular to the helix axis. In the drawing, the sugar–phosphate backbones, which wind about the periphery of the molecule, are blue, and the bases, which occupy its core, are red. In the space-filling model, C, N, O,

and P atoms are white, blue, red, and green, respectively. H atoms have been omitted for clarity in both drawings. Note that the two sugar–phosphate chains run in opposite directions. (*b*) (*Opposite*) View down the helix axis. In the drawing, the ribose ring O atoms are red and the nearest base pair is white. Note that the helix axis passes through the base pairs so that the helix has a solid core. [Illustration, Irving Geis/Geis Archives Trust. Copyright Howard Hughes Medical Institute. Reproduced with permission.] ✑ **See Kinemage Exercises 17-1 and 17-4**

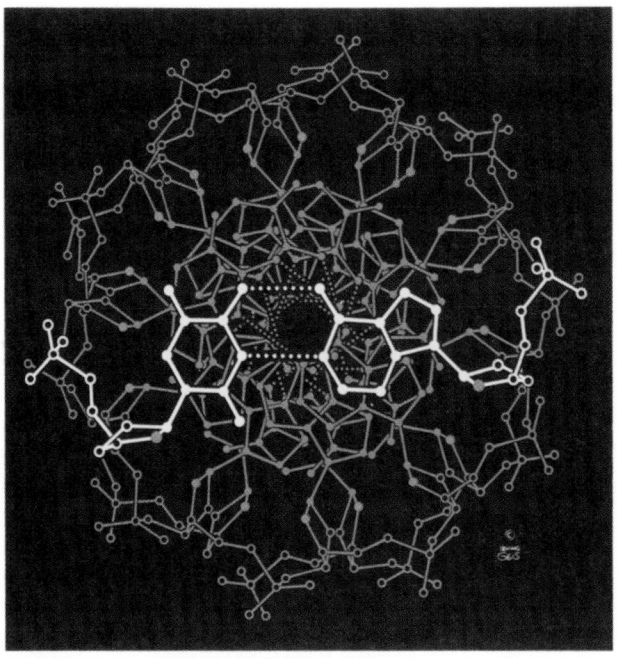 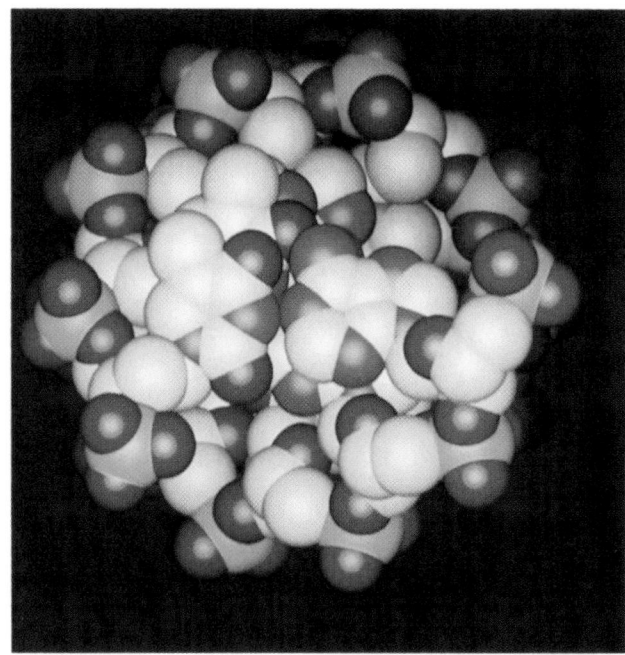

FIGURE 29-1 *(b)*

A. *B-DNA*

The structure of **B-DNA** (Fig. 29-1), the biologically predominant form of DNA, is described in Section 5-3A. To recapitulate (Table 29-1), B-DNA consists of a right-handed double helix whose two antiparallel sugar–phosphate chains wrap around the periphery of the helix. Its aromatic bases (A, T, G, and C), which occupy the core of the helix, form complementary A · T and G · C Watson–Crick base pairs (Fig. 5-12), whose planes are nearly perpendicular to the axis of the double helix. Neighboring base pairs, whose aromatic rings are 3.4 Å thick, are stacked in van der Waals contact, with the helix axis passing through the middle of each base pair. B-DNA is ~20 Å in diameter and has two deep grooves between its sugar–phosphate chains: the relatively narrow **minor groove**, which exposes that edge of the base pairs from which the glycosidic bonds (the bonds from the base N to the ribose

TABLE 29-1 Structural Features of Ideal A-, B-, and Z-DNA

	A-DNA	B-DNA	Z-DNA
Helical sense	Right-handed	Right-handed	Left-handed
Diameter	~26 Å	~20 Å	~18 Å
Base pairs per helical turn	11.6	10	12 (6 dimers)
Helical twist per base pair	31°	36°	9° for pyrimidine–purine steps; 51° for purine–pyrimidine steps
Helix pitch (rise per turn)	34 Å	34 Å	44 Å
Helix rise per base pair	2.9 Å	3.4 Å	7.4 Å per dimer
Base tilt normal to the helix axis	20°	6°	7°
Major groove	Narrow and deep	Wide and deep	Flat
Minor groove	Wide and shallow	Narrow and deep	Narrow and deep
Sugar pucker	C3'-*endo*	C2'-*endo*	C2'-*endo* for pyrimidines; C3'-*endo* for purines
Glycosidic bond	Anti	Anti	Anti for pyrimidines; syn for purines

Source: Mainly Arnott, S., *in* Neidle, S. (Ed.), *Oxford Handbook of Nucleic Acid Structure*, p. 35, Oxford University Press (1999).

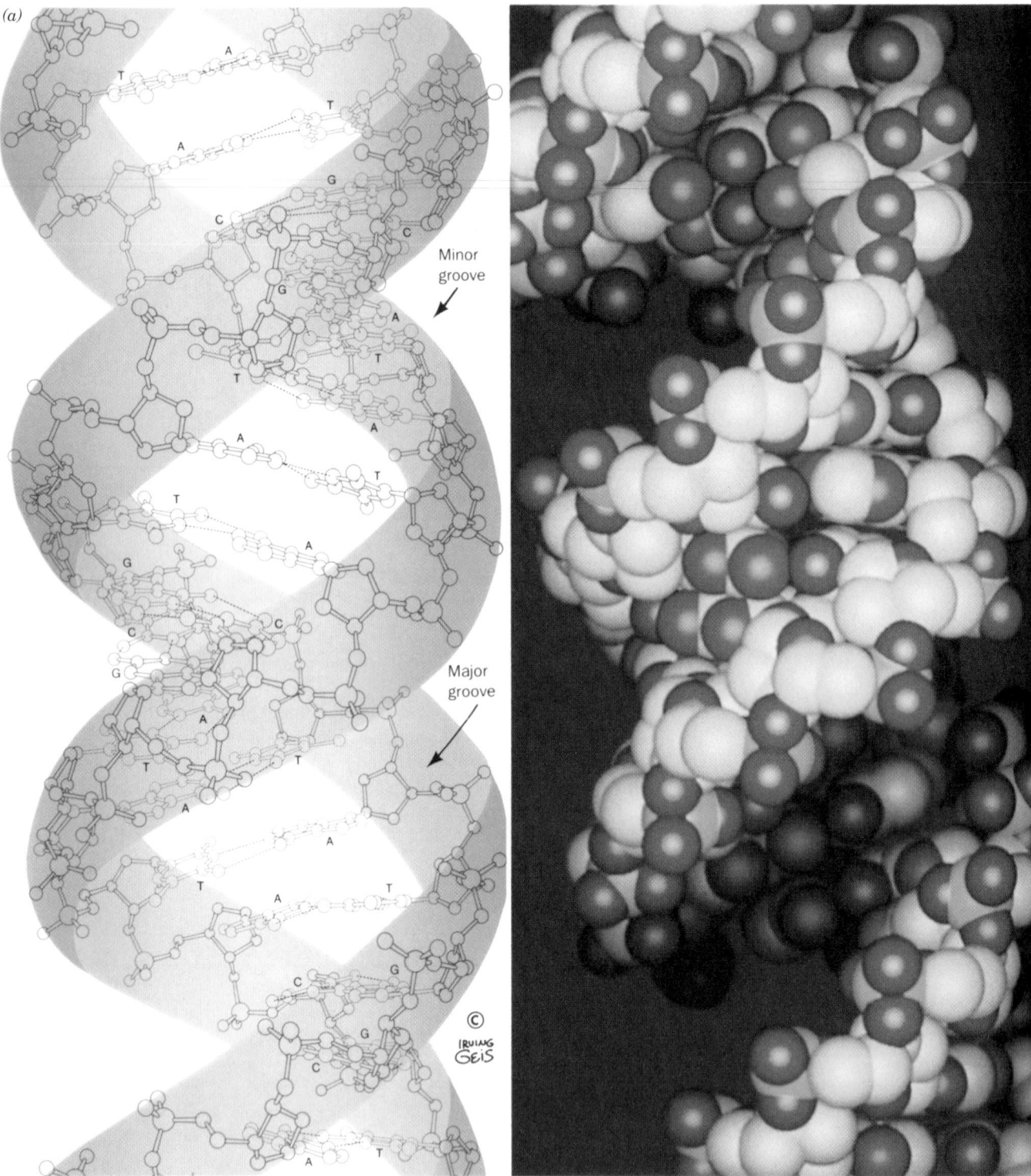

(a)

Minor groove

Major groove

IRVING GEIS

FIGURE 29-2 Structure of A-DNA. Ball-and-stick drawings and the corresponding space-filling models of A-DNA are viewed (*a*) perpendicular to the helix axis and (*b*) (*Opposite*) down the helix axis. The color codes are given in the legend to Fig. 29-1. The repeating helix was generated by Richard Dickerson based on the X-ray structure of the self-complementary octamer d(GGTATACC) determined by Olga Kennard, Dov Rabinovitch, Zippora Shakked, and Mysore Viswamitra, Cambridge University, U.K. (Nucleic Acid Data Base ID ADH010). Note that the base pairs are inclined to the helix axis and that the helix has a hollow core. Compare this figure with Fig. 29-1. [Illustration, Irving Geis/Geis Archives Trust. Copyright Howard Hughes Medical Institute. Reproduced with permission.] 🔖 **See Kinemage Exercises 17-1 and 17-5**

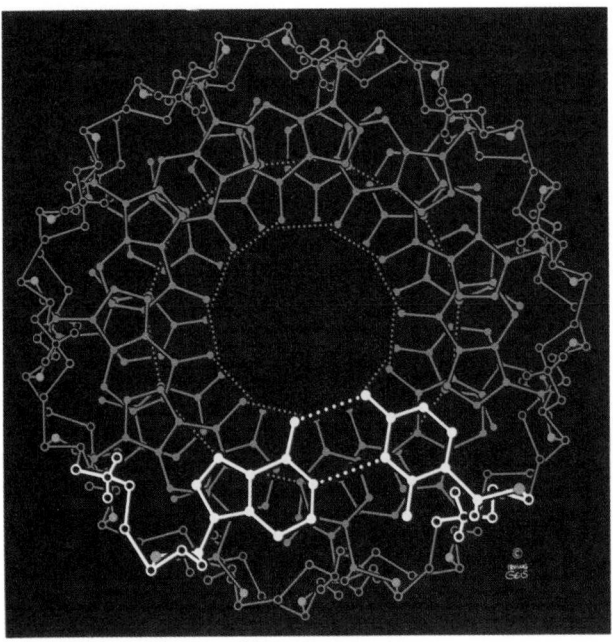

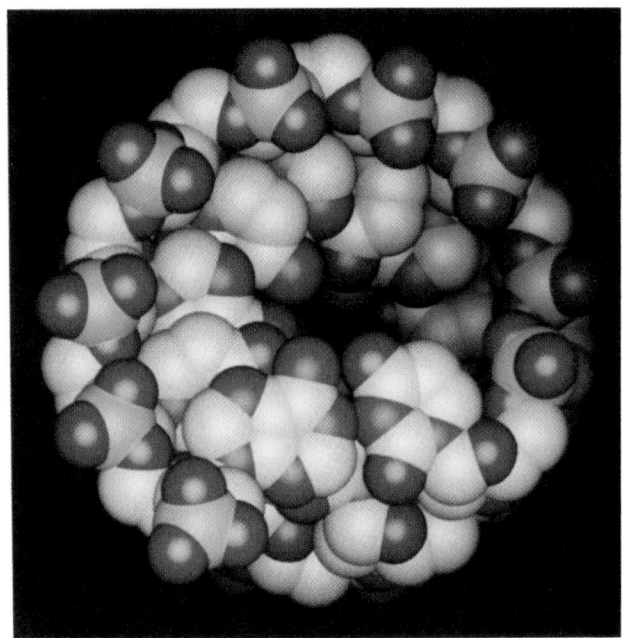

FIGURE 29-2 *(b)*

C1′) extend (toward the bottom of Fig. 5-12), and the relatively wide **major groove,** which exposes the opposite edge of each base pair (toward the top of Fig. 5-12). Canonical (ideal) DNA has a helical twist of 10 base pairs (bp) per turn and hence a pitch (rise per turn) of 34 Å.

The Watson–Crick base pairs in either orientation are structurally interchangeable, that is, A · T, T · A, G · C, and C · G can replace each other in the double helix without altering the positions of the sugar–phosphate backbones' C1 atoms. In contrast, any other combination of bases would significantly distort the double helix since the formation of a non-Watson–Crick base pair would require considerable reorientation of the DNA's sugar–phosphate backbones.

a. Real DNA Deviates from the Ideal Watson–Crick Structure

The DNA samples that were available when James Watson and Francis Crick formulated the Watson–Crick structure in 1953 were extracted from cells and hence consisted of molecules of heterogeneous lengths and base sequences. Such elongated molecules do not crystallize, but can be drawn into threadlike fibers in which the helix axes of the DNA molecules are all approximately parallel to the fiber axis but are poorly aligned, if at all, in any other way. The X-ray diffraction patterns of such fibers provide only crude, low-resolution images in which the base pair electron density is the average electron density of all the base pairs in the fiber. The Watson–Crick structure was based, in part, on the X-ray fiber diffraction pattern of B-DNA (Fig. 5-10).

By the late 1970s, advances in nucleic acid chemistry permitted the synthesis and crystallization of ever longer oligonucleotides of defined sequences (Section 7-6A),

many of which could be crystallized. Consequently, some 25 years after the Watson–Crick structure was formulated, its X-ray crystal structure was clearly visualized for the first time when Richard Dickerson and Horace Drew determined the first X-ray crystal structure of a B-DNA, that of the self-complementary dodecamer d(CGCGAATTCGCG), at near-atomic (1.9 Å) resolution. This molecule, whose structure was subsequently determined at significantly higher (1.4 Å) resolution by Loren Williams, has an average rise per residue of 3.3 Å and has 10.1 bp per turn (a helical twist of 35.5° per bp), values that are nearly equal to those of canonical B-DNA. However, individual residues depart significantly from this average conformation (Fig. 29-1). For example, the helical twist per base pair in this dodecamer ranges from 26° to 43°. Each base pair further deviates from its ideal conformation by such distortions as propeller twisting (the opposite rotation of paired bases about the base pair's long axis; in the 1.4-Å resolution structure, this quantity ranges from −23° to −7°) and base pair roll (the tilting of a base pair as a whole about its long axis; this quantity ranges from −14° to 17°).

X-Ray and NMR studies of numerous other double helical DNA oligomers have amply demonstrated that *the conformation of DNA, particularly B-DNA, is irregular in a sequence-specific manner,* although the rules specifying how sequence governs conformation have proved to be surprisingly elusive. This is because *base sequence does not so much confer a fixed conformation on a double helix as it establishes the deformability of the helix.* Thus, 5′-R–Y-3′ steps (where R and Y are the abbreviations for purines and pyrimidines, respectively) in B-DNA are easily bent because they exhibit relatively little ring–ring overlap between adjacent base pairs. In contrast, both Y–R steps and R–R steps (the latter, due to base pairing, are equiv-

(a)

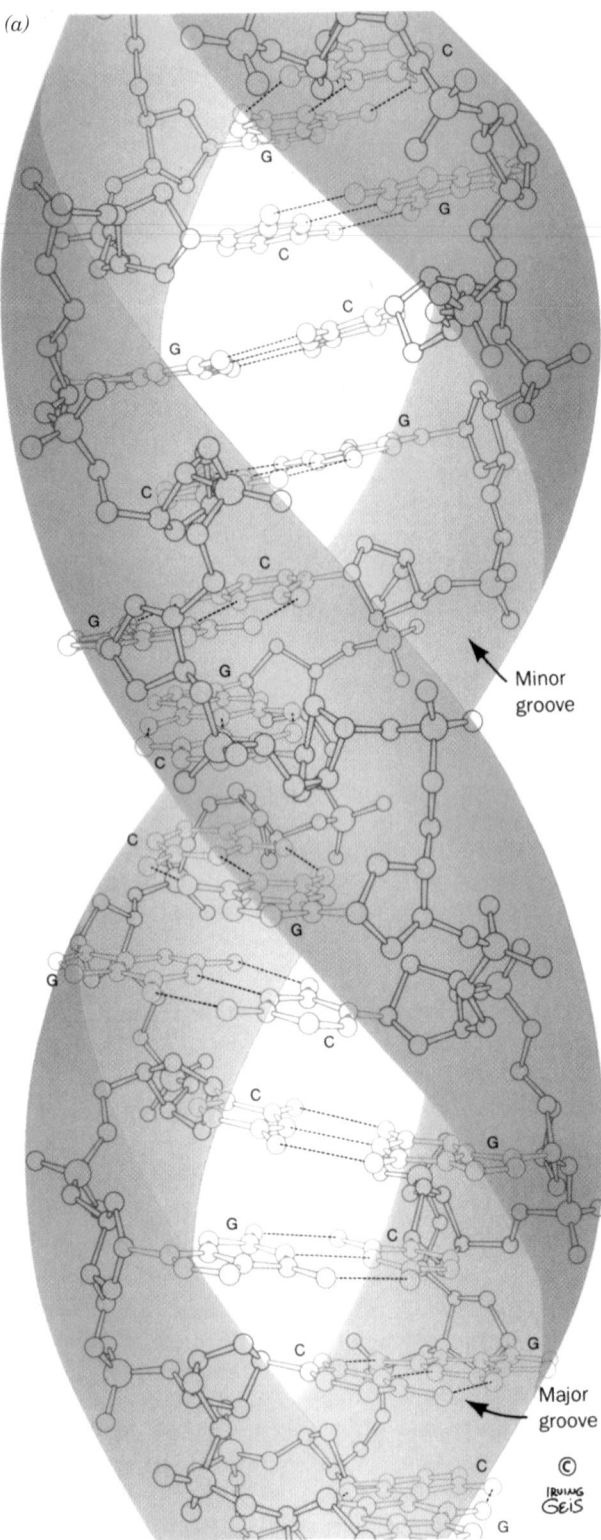

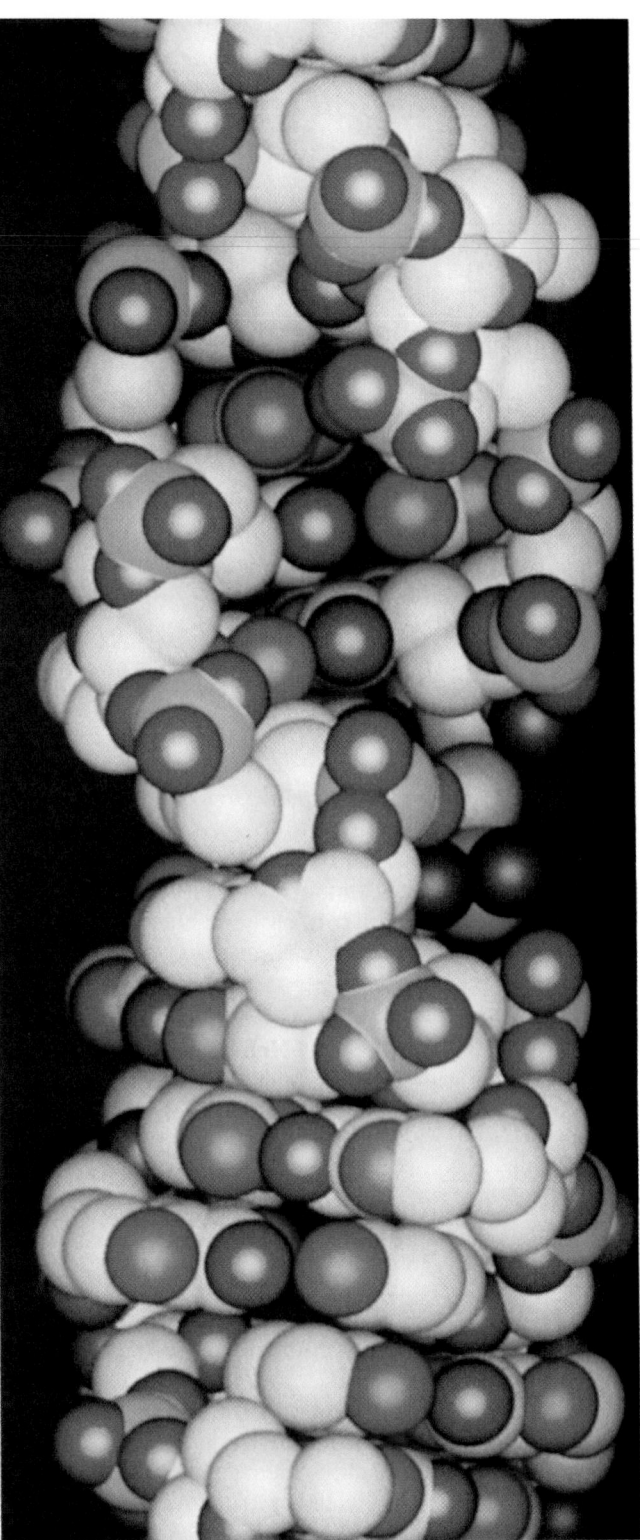

FIGURE 29-3   Structure of Z-DNA. Ball-and-stick drawings and the corresponding space-filling models of Z-DNA are viewed (*a*) perpendicular to the helix axis and (*b*) (*Opposite*) down the helix axis. The color codes are given in the legend to Fig. 29-1. The repeating helix was generated by Richard Dickerson based on the X-ray structure of the self-complementary hexamer d(CGCGCG) determined by Andrew Wang and Alexander Rich, MIT (PDBid 2DCG). Note that the helix is left handed and that the sugar–phosphate chains follow a zigzag course (alternate ribose residues lie at different radii in Part *b*), indicating that the Z-DNA's repeating motif is a dinucleotide. Compare this figure with Figs. 29-1 and 29-2. [Illustration, Irving Geis/Geis Archives Trust. Copyright Howard Hughes Medical Institute. Reproduced with permission.] **☙ See Kinemage Exercises 17-1 and 17-6**

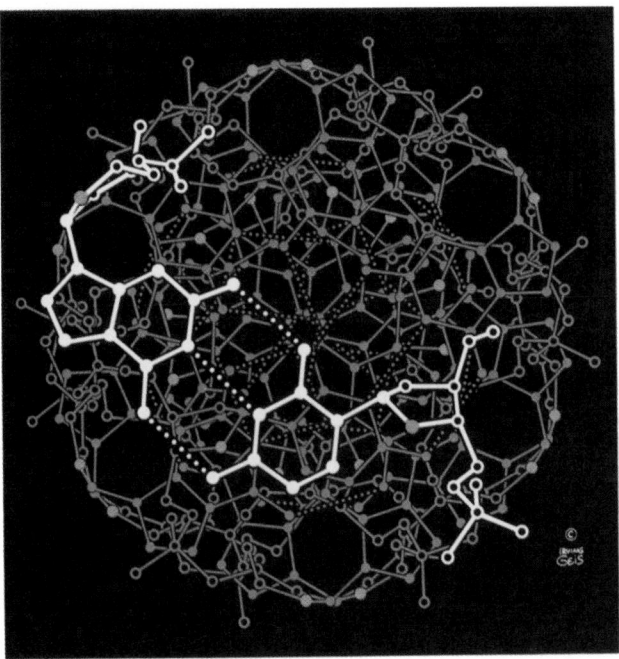 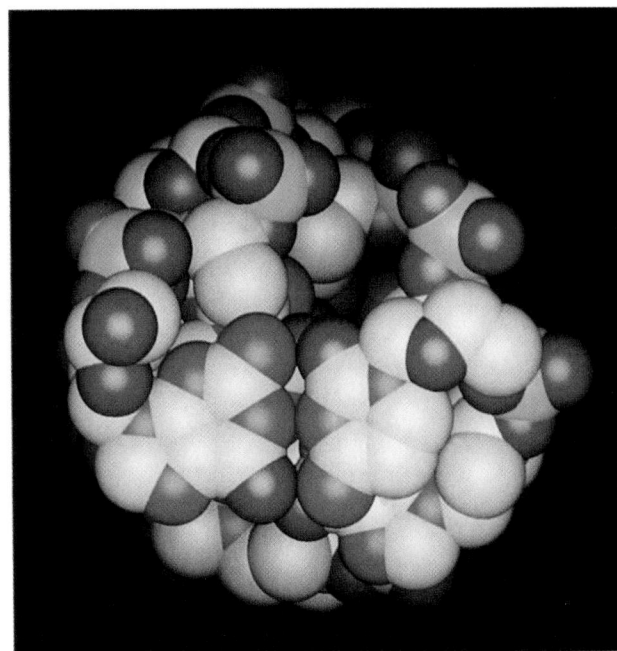

FIGURE 29-3 *(b)*

alent to Y–Y steps), and most notably A–A steps, are more rigid because the extensive ring–ring overlap between their adjacent base pairs tends to keep these base pairs parallel. *This phenomenon, as we shall see, is important for the sequence-specific binding of DNA to proteins that process genetic information.* This is because many of these proteins wrap their target DNAs around them, in many cases by bending them by well over 90°. DNAs with different sequences than the target DNA would not bind so readily to the protein because they would resist deformation to the required conformation more than the target DNA.

B. *Other Nucleic Acid Helices*

X-Ray fiber diffraction studies, starting in the mid-1940s, revealed that *nucleic acids are conformationally variable molecules.* Indeed, double helical DNA and RNA can assume several distinct structures that vary with such factors as the humidity and the identities of the cations present, as well as with base sequence. For example, fibers of B-DNA form in the presence of alkali metal ions such as Na$^+$ when the relative humidity is 92%. In this subsection, we describe the other major conformational states of double-stranded DNA as well as those of double-stranded RNA and RNA–DNA hybrid helices.

a. A-DNA's Base Pairs Are Inclined to the Helix Axis

When the relative humidity is reduced to 75%, B-DNA undergoes a reversible conformational change to the so-called A form. Fiber X-ray studies indicate that *A-DNA forms a wider and flatter right-handed helix than does B-DNA* (Fig. 29-2; Table 29-1). A-DNA has 11.6 bp per turn and a pitch of 34 Å, which gives A-DNA an axial hole

(Fig. 29-2*b*). A-DNA's most striking feature, however, is that the planes of its base pairs are tilted 20° with respect to the helix axis. Since its helix axis passes "above" the major groove side of the base pairs (Fig. 29-2*b*) rather than through them as in B-DNA, A-DNA has a deep major groove and a very shallow minor groove; it can be described as a flat ribbon wound around a 6-Å-diameter cylindrical hole. Most self-complementary oligonucleotides of <10 base pairs, for example, d(GGCCGGCC) and d(GGTATACC), crystallize in the A-DNA conformation. Like B-DNA, these molecules exhibit considerable sequence-specific conformational variation although the degree of variation is less than that in B-DNA.

A-DNA has, so far, been observed in only two biological contexts. A ~3-bp segment of A-DNA is present at the active site of DNA polymerase (Section 30-2A). In addition, Gram-positive bacteria undergoing **sporulation** (the formation, under environmental stress, of resistant although dormant cell types known as **spores;** a sort of biological lifeboat) contain a high proportion (20%) of **small acid-soluble spore proteins (SASPs).** Some of these SASPs induce B-DNA to assume the A form, at least *in vitro*. The DNA in bacterial spores exhibits a resistance to UV-induced damage that is abolished in mutants that lack these SASPs. This occurs because the B→A conformation change inhibits the UV-induced covalent cross-linking of pyrimidine bases (Section 30-5A), in part by increasing the distance between successive pyrimidines.

b. Z-DNA Forms a Left-Handed Helix

Occasionally, a seemingly well-understood or at least familiar system exhibits quite unexpected properties. Over 25 years after the discovery of the Watson–Crick

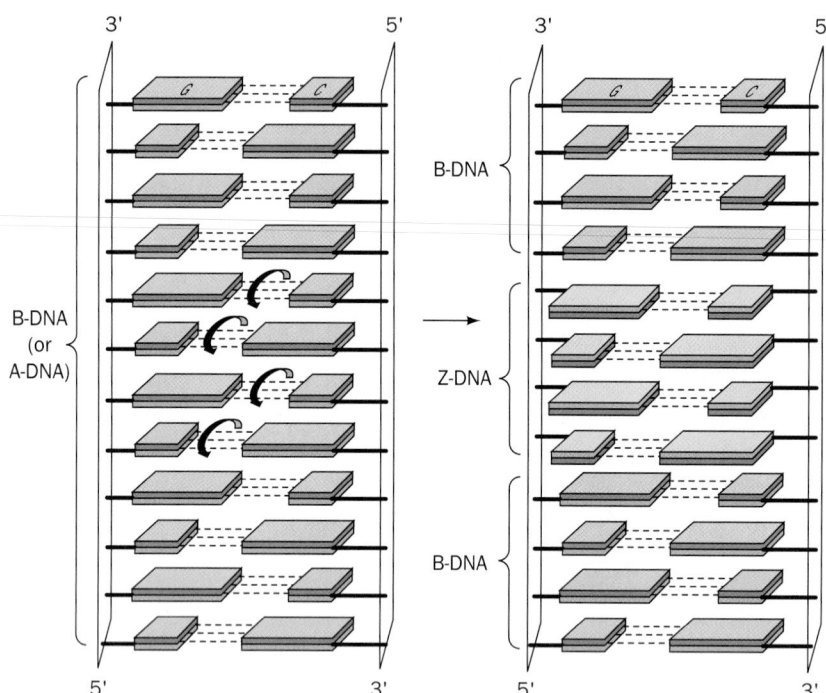

FIGURE 29-4 Conversion of B-DNA to Z-DNA. The conversion, here represented by a 4-bp DNA segment, involves a 180° flip of each base pair (*curved arrows*) relative to the sugar–phosphate chains. Here, the different faces of the base pairs are colored red and green. Note that if the drawing on the left is taken as looking into the minor groove of unwound A- or B-DNA, then in the drawing on the right, we are looking into the major groove of the unwound Z-DNA segment. [After Rich, A., Nordheim, A., and Wang, A.H.-J., *Annu. Rev. Biochem.* **53,** 799 (1984).]

structure, the crystal structure determination of the self-complementary hexanucleotide d(CGCGCG) by Andrew Wang and Alexander Rich revealed, quite surprisingly, *a left-handed double helix (Fig. 29-3; Table 29-1).* A similar helix is formed by d(CGCATGCG). *This helix, which has been dubbed* **Z-DNA,** *has 12 Watson–Crick base pairs per turn, a pitch of 44 Å, and, in contrast to A-DNA, a deep minor groove and no discernible major groove* (its helix axis passes "below" the minor groove side of its base pairs; Fig. 29-3b). Z-DNA therefore resembles a left-handed drill bit in appearance. The base pairs in Z-DNA are flipped 180° relative to those in B-DNA (Fig. 29-4) through conformational changes discussed in Section 29-2A. As a consequence, the repeating unit of Z-DNA is a dinucleotide, d(XpYp), rather than a single nucleotide as it is in the other DNA helices. The line joining successive phosphate groups on a polynucleotide strand of Z-DNA therefore follows a zigzag path around the helix (Fig. 29-3a; hence the name Z-DNA) rather than a smooth curve as it does in A- and B-DNAs (Figs. 29-1a and 29-2a).

Fiber diffraction and NMR studies have shown that complementary polynucleotides with alternating purines and pyrimidines, such as poly d(GC) · poly d(GC) and poly d(AC) · poly d(GT), take up the Z-DNA conformation at high salt concentrations. Evidently, *the Z-DNA conformation is most readily assumed by DNA segments with alternating purine–pyrimidine base sequences (for structural reasons explained in Section 29-2A).* A high salt concen-

tration stabilizes Z-DNA relative to B-DNA by reducing the otherwise increased electrostatic repulsions between closest approaching phosphate groups on opposite strands (8 Å in Z-DNA vs 12 Å in B-DNA). The methylation of cytosine residues at C5, a common biological modification (Section 30-7), also promotes Z-DNA formation since a hydrophobic methyl group in this position is less exposed to solvent in Z-DNA than it is in B-DNA.

Does Z-DNA have any biological function? Rich has proposed that the reversible conversion of specific segments of B-DNA to Z-DNA under appropriate circumstances acts as a kind of switch in regulating genetic expression, and there are indications that it transiently forms behind actively transcribing RNA polymerase (Section 31-4B). It has nevertheless been surprisingly difficult to prove the *in vivo* existence of Z-DNA. A major difficulty is demonstrating that a particular probe for detecting Z-DNA, for example, a Z-DNA-specific antibody, does not in itself cause what would otherwise be B-DNA to assume the Z conformation—a kind of biological uncertainty principle (the act of measurement inevitably disturbs the system being measured). Recently, however, Rich has discovered a family of Z-DNA-binding protein domains named **Zα,** whose existence strongly suggests that Z-DNA does, in fact, exist *in vivo.* The X-ray structure of the 81-residue Zα domain from the RNA editing enzyme **ADAR1** (Section 31-4A) in complex with d(TCGCGCG) has been determined (Fig. 29-5). The CGCGCG segment of this heptanucleotide is self-complementary, and therefore forms a twofold symmetric, 6-bp segment of Z-DNA with an overhanging dT at the 5′ end of each strand (although these dT's are disordered in the X-ray structure). A monomeric unit of Zα binds to each strand of the Z-DNA, out of contact with the Zα that binds to the opposite strand. The protein primarily interacts with Z-DNA via hydrogen bonds and salt bridges between polar and basic protein side chains and the Z-DNA's sugar–phosphate backbone. Note that none of the DNA's bases participate in these associations. The protein's DNA-binding surface, which is complementary in shape to the Z-DNA, is positively charged, as is expected for a protein that interacts with several closely spaced, anionic phosphate groups. It is postulated that ADAR1's Zα domain targets it to the Z-DNA upstream of actively transcribing genes (for reasons discussed in Section 31-4A).

c. RNA-11 and RNA–DNA Hybrids Have an A-DNA-Like Conformation

Double helical RNA is unable to assume a B-DNA-like conformation because of steric clashes involving its 2′-OH groups. Rather, it usually assumes a conformation resembling A-DNA (Fig. 29-2), known as **A-RNA** or **RNA-11,** which ideally has 11.0 bp per helical turn, a pitch of 30.9 Å, and its base pairs inclined to the helix axis by 16.7°. Many RNAs, for example, transfer and ribosomal RNAs (whose structures are detailed in Sections 32-2A and 32-3A), contain complementary sequences that form double helical stems.

Hybrid double helices, which consist of one strand each of DNA and RNA, are also predicted to have A-RNA-like conformations. In fact, the X-ray structure, by Barry Finzel, of a 10-bp complex of the DNA oligonucleotide d(GGCGCCCGAA) with the complementary RNA oligonucleotide r(UUCGGGCGCC) reveals (Fig. 29-6) that it forms a double helix with A-RNA-like character (Table 29-1) in that it has 10.9 bp per turn, a pitch of 31.3 Å, and its base pairs are, on average, inclined to the helix axis by 13.9°. Nevertheless, this hybrid helix also has B-DNA-like

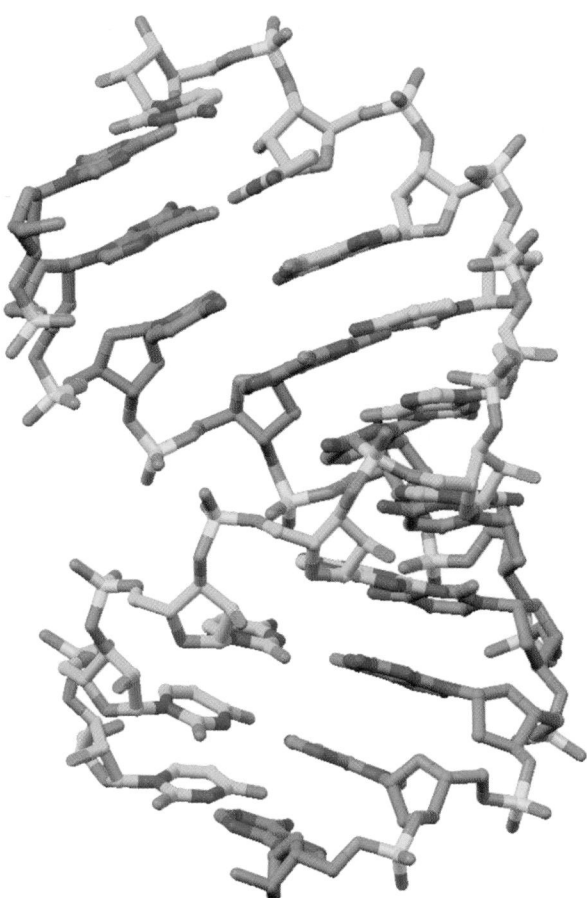

FIGURE 29-6   X-Ray structure of a 10-bp RNA–DNA hybrid helix consisting of d(GGCGCCCGAA) in complex with r(UUCGGGCGCC). The structure is shown in stick form with RNA C atoms cyan, DNA C atoms green, N blue, O red except for RNA O2′ atoms, which are magenta, and P gold. [Based on an X-ray structure by Barry Finzel, Pharmacia & Upjohn, Inc., Kalamazoo, Michigan. PDBid 1FIX.] *See the Interactive Exercises*

qualities in that the width of its minor groove (9.5 Å) is intermediate between those for canonical B-DNA (7.4 Å) and A-DNA (11 Å) and in that some of the ribose rings of its DNA strand have conformations characteristic of B-DNA (Section 29-2A), whereas others have conformations characteristic of A-RNA. Note that this structure is of biological significance because short segments of RNA · DNA hybrid helices occur in both the transcription of RNA on DNA templates (Section 31-2C) and in the initiation of DNA replication by short lengths of RNA (Section 30-1D). The RNA component of this helix is a substrate for **RNase H,** which specifically hydrolyzes the RNA strands of RNA · DNA hybrid helices *in vivo* (Section 30-4C).

2 ■ FORCES STABILIZING NUCLEIC ACID STRUCTURES

Double-stranded DNA does not exhibit the structural complexity of proteins because it has only a limited repertoire of secondary structures and no comparable tertiary

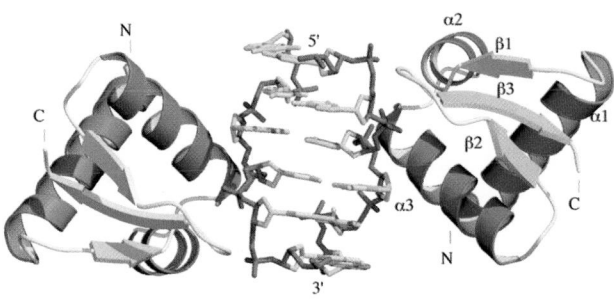

FIGURE 29-5   X-Ray structure of two ADAR1 Zα domains in complex with Z-DNA. The complex is viewed along its 2-fold axis of symmetry. The duplex of self-complementary d(CGCGCG) hexamers is shown in stick form with its backbones red and its remaining portions pink. The Zα domains are drawn in ribbon form with helices blue and sheets cyan. Note that each Zα domain contacts only one strand of the Z-DNA. [Courtesy of Alexander Rich, MIT. PDBid 1QBJ.]

or quaternary structures (although see Section 29-3). This is perhaps to be expected since there is a far greater range of chemical and physical properties among the 20 amino acid residues of proteins than there is among the four DNA bases. However, many RNAs have well-defined tertiary structures (Sections 31-4A, 32-2B, and 32-3A).

In this section we examine the forces that give rise to the structures of nucleic acids. These forces are, of course, much the same as those that are responsible for the structures of proteins (Section 8-4) but, as we shall see, the way they combine gives nucleic acids properties that are quite different from those of proteins.

A. *Sugar–Phosphate Chain Conformations*

The conformation of a nucleotide unit, as Fig. 29-7 indicates, is specified by the six torsion angles of the sugar–phosphate backbone and the torsion angle describing the orientation of the base about the glycosidic bond. It would seem that these seven degrees of freedom per nucleotide would render polynucleotides highly flexible. Yet, as we shall see, these torsion angles are subject to a variety of internal constraints that greatly restrict their conformational freedom.

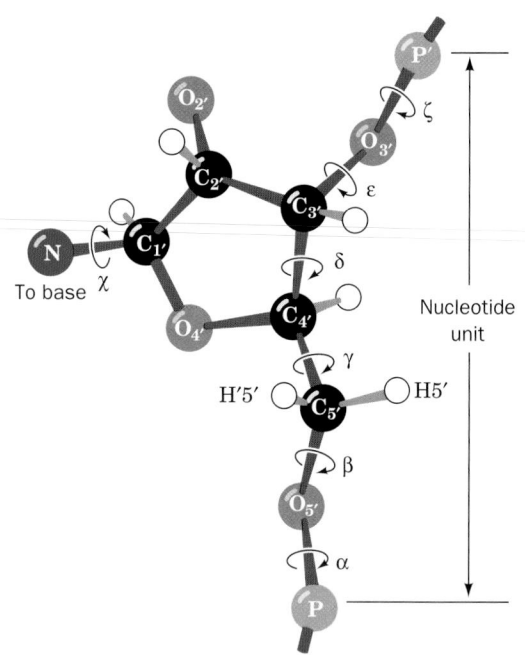

FIGURE 29-7   The conformation of a nucleotide unit is determined by the seven indicated torsion angles.

a. Torsion Angles about Glycosidic Bonds Have Only One or Two Stable Positions

The rotation of a base about its glycosidic bond is greatly hindered, as is best seen by the manipulation of a space-filling molecular model. Purine residues have two sterically permissible orientations relative to the sugar known as the **syn** (Greek: with) and **anti** (Greek: against) conformations (Fig. 29-8). For pyrimidines, only the anti conformation is easily formed because, in the syn conformation, the sugar residue sterically interferes with the pyrimidine's C2 substituent. In most double helical nucleic acids, all bases are in the anti conformation (e.g., Figs. 29-1*b* and 29-2*b*). The exception is Z-DNA (Section 29-1B), in which the alternating pyrimidine and purine residues are anti and syn (Fig. 29-3*b*). *This explains Z-DNA's pyrimidine–purine alternation.* Indeed, the base pair flips that convert B-DNA to Z-DNA (Fig. 29-4) are brought about by rotating each purine base about its gly-

cosidic bond from the anti to syn conformation, whereas it is the sugars that rotate in the pyrimidine nucleotides, thereby maintaining them in their anti conformations.

b. Sugar Ring Pucker Is Largely Limited to Only a Few of Its Possible Arrangements

The ribose ring has a certain amount of flexibility that significantly affects the conformation of the sugar–phosphate backbone. The vertex angles of a regular pentagon are 108°, a value quite close to the tetrahedral angle (109.5°), so that one might expect the ribofuranose ring to be nearly flat. However, the ring substituents are eclipsed when the ring is planar. To relieve the resultant crowding, which even occurs between hydrogen atoms, the ring **puckers;** that is, it becomes slightly nonplanar, so as to reorient the ring substituents (Fig. 29-9; this is readily observed by the manipulation of a skeletal molecular model).

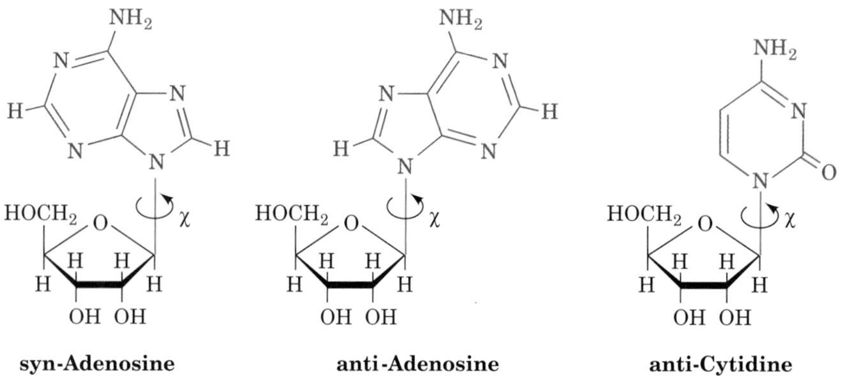

syn-Adenosine   anti-Adenosine   anti-Cytidine

FIGURE 29-8   The sterically allowed orientations of purine and pyrimidine bases with respect to their attached ribose units.

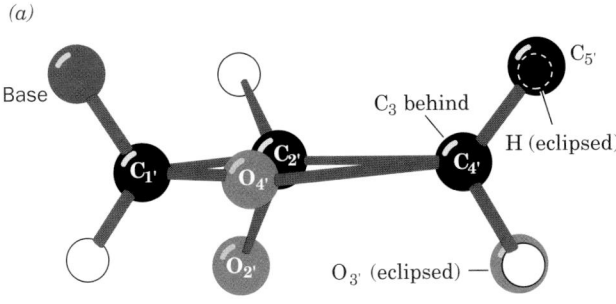

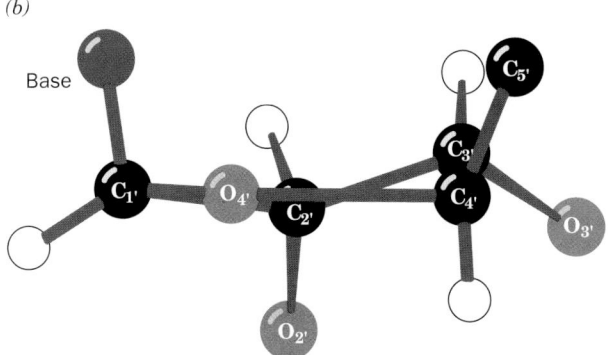

FIGURE 29-9 Sugar ring pucker. The substituents to (*a*) a planar ribose ring (here viewed down the C3′—C4′ bond) are all eclipsed. The resulting steric strain is partially relieved by ring puckering such as in (*b*), a half-chair conformation in which C3′ is the out-of-plane atom.

One would, in general, expect only three of a ribose ring's five atoms to be coplanar since three points define a plane. Nevertheless, in the great majority of the >50 nucleoside and nucleotide crystal structures that have been reported, four of the ring atoms are coplanar to within a few hundreths of an angstrom and the remaining atom is out of this plane by several tenths of an angstrom (the **half-**

chair conformation). If the out-of-plane atom is displaced to the same side of the ring as atom C5′, it is said to have the **endo** conformation (Greek: *endon,* within), whereas displacement to the opposite side of the ring from C5′ is known as the **exo** conformation (Greek: *exo,* out of). In the great majority of known nucleoside and nucleotide structures (molecules that are subject to few of the conformational constraints of double helices), the out-of-plane atom is either C2′ or C3′ (Fig. 29-10). C2′-*endo* is the most frequently occurring ribose pucker with C3′-*endo* and C3′-*exo* also being common. Other ribose conformations are rare.

The ribose pucker is conformationally important in nucleic acids because it governs the relative orientations of the phosphate substituents to each ribose residue (Fig. 29-10). For instance, it is difficult to build a regularly repeating model of a double helical nucleic acid unless the sugars are either C2′-*endo* or C3′-*endo.* In fact, canonical B-DNA has the C2′-*endo* conformation, whereas canonical A-DNA and RNA-11 are C3′-*endo.* In canonical Z-DNA, the purine nucleotides are all C3′-*endo* and the pyrimidine nucleotides are C2′-*endo,* which is another reason that the repeating unit of Z-DNA is a dinucleotide. The sugar puckers observed in the X-ray structures of A-DNA are, in fact, almost entirely C3′-*endo.* However, those of B-DNAs, although predominantly C2′-*endo,* exhibit significant variation including C4′-*exo,* O4′-*endo,* C1′-*exo,* and C3′-*exo.* This variation in B-DNA's sugar pucker is probably indicative of its greater flexibility relative to other types of DNA helices.

c. The Sugar–Phosphate Backbone Is Conformationally Constrained

If the torsion angles of the sugar–phosphate chain (Fig. 29-7) were completely free to rotate, there could probably be no stable nucleic acid structure. However, the comparison, by Muttaiya Sundaralingam, of some 40 nucleoside and nucleotide crystal structures has revealed that these angles are really quite restricted. For example, the torsion

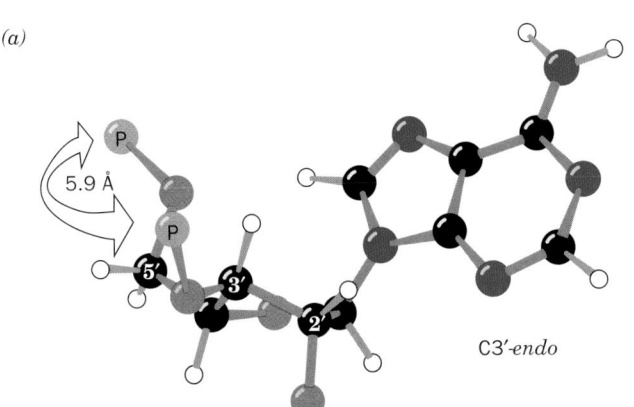

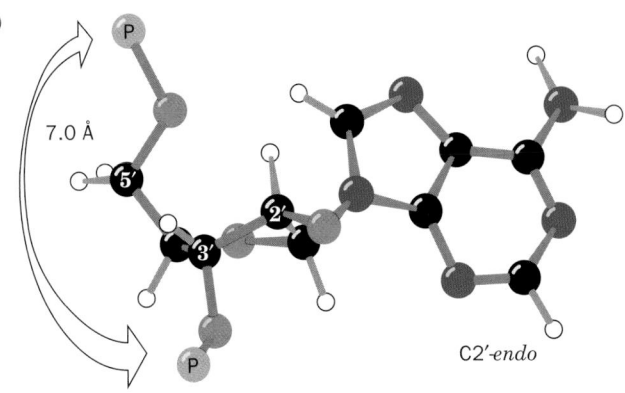

FIGURE 29-10 Nucleotide sugar conformations. (*a*) The C3′-*endo* conformation (on the same side of the sugar ring as C5′), which occurs in A-RNA and RNA-11. (*b*) The C2′-*endo* conformation, which occurs in B-DNA. The distances between adjacent P atoms in the sugar–phosphate backbone are indicated. [After Saenger, W., *Principles of Nucleic Acid Structure,* p. 237, Springer-Verlag (1983).] **See Kinemage Exercises 17-3**

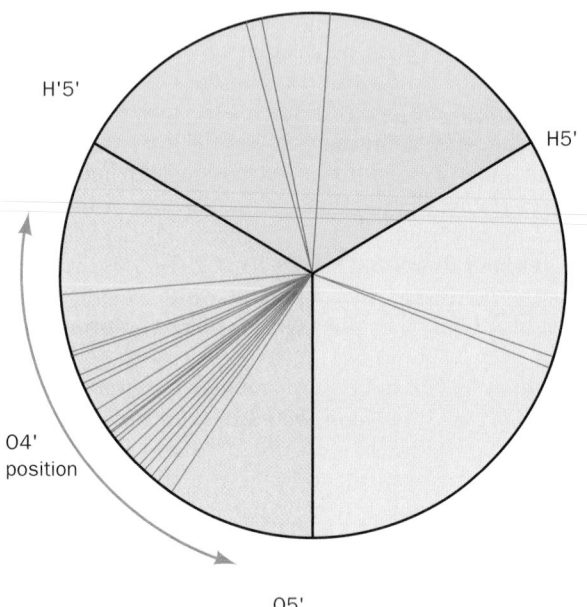

FIGURE 29-11 Conformational wheel showing the distribution of the torsion angle about the C4'—C5' bond. The torsion angle (γ in Fig. 29-7) was measured in 33 X-ray structures of nucleosides, nucleotides, and polynucleotides. Each radial line represents the position of the C4'—O4' bond in a single structure relative to the substituents of C5' as viewed from C5' to C4'. Note that most of the observed torsion angles fall within a relatively narrow range. [After Sundaralingam, M., *Biopolymers* **7,** 838 (1969).]

angle about the C4'—C5' bond (γ in Fig. 29-7) is rather narrowly distributed such that O4' usually has a gauche conformation with respect to O5' (Fig. 29-11). This is because the presence of the ribose ring together with certain noncovalent interactions of the phosphate group stiffens the sugar–phosphate chain by restricting its range of torsion angles. These restrictions are even greater in polynucleotides because of steric interference between residues.

The sugar–phosphate conformational angles of the various double helices are all reasonably strain free. *Double helices are therefore conformationally relaxed arrangements of the sugar–phosphate backbone.* Nevertheless, the sugar–phosphate backbone is by no means a rigid structure, so, on strand separation, it assumes a random coil conformation.

B. *Base Pairing*

Base pairing is apparently a "glue" that holds together double-stranded nucleic acids. Only Watson–Crick pairs occur in the crystal structures of self-complementary oligonucleotides. It is therefore important to understand how Watson–Crick base pairs differ from other doubly hydrogen bonded arrangements of the bases that have reasonable geometries (e.g., Fig. 29-12).

a. Unconstrained A · T Base Pairs Assume Hoogsteen Geometry

When monomeric adenine and thymine derivatives are cocrystallized, the A · T base pairs that form invariably have adenine N7 as the hydrogen bonding acceptor (**Hoogsteen geometry;** Fig. 29-12*b*) rather than N1 (Watson–Crick geometry; Fig. 5-12). This suggests that Hoogsteen geometry is inherently more stable for A · T pairs than is Watson–Crick geometry. Apparently steric and other environmental influences make Watson–Crick geometry the preferred mode of base pairing in double helices. A · T pairs with Hoogsteen geometry are nevertheless of biological importance; for example, they help stabilize the tertiary structures of tRNAs (Section 32-2B). In contrast, monomeric G · C pairs always cocrystallize with Watson–Crick geometry as a consequence of their triply hydrogen bonded structures.

b. Non-Watson–Crick Base Pairs Are of Low Stability

The bases of a double helix, as we have seen (Section 5-3A), associate such that any base pair position may interchangeably be A · T, T · A, G · C, or C · G without

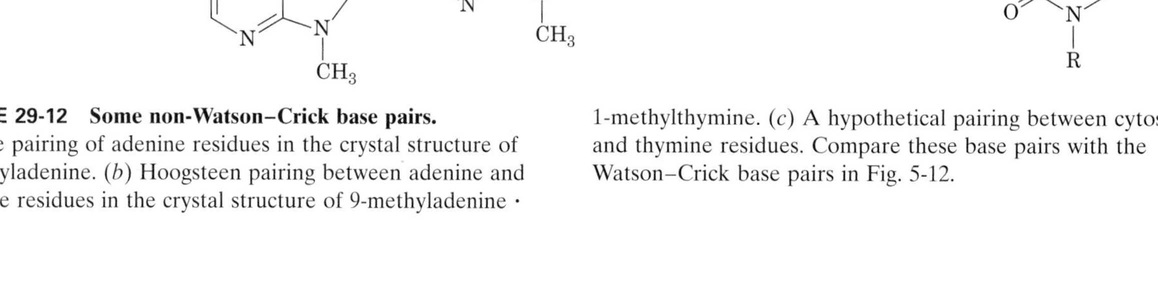

FIGURE 29-12 Some non-Watson–Crick base pairs.
(*a*) The pairing of adenine residues in the crystal structure of 9-methyladenine. (*b*) Hoogsteen pairing between adenine and thymine residues in the crystal structure of 9-methyladenine ·

1-methylthymine. (*c*) A hypothetical pairing between cytosine and thymine residues. Compare these base pairs with the Watson–Crick base pairs in Fig. 5-12.

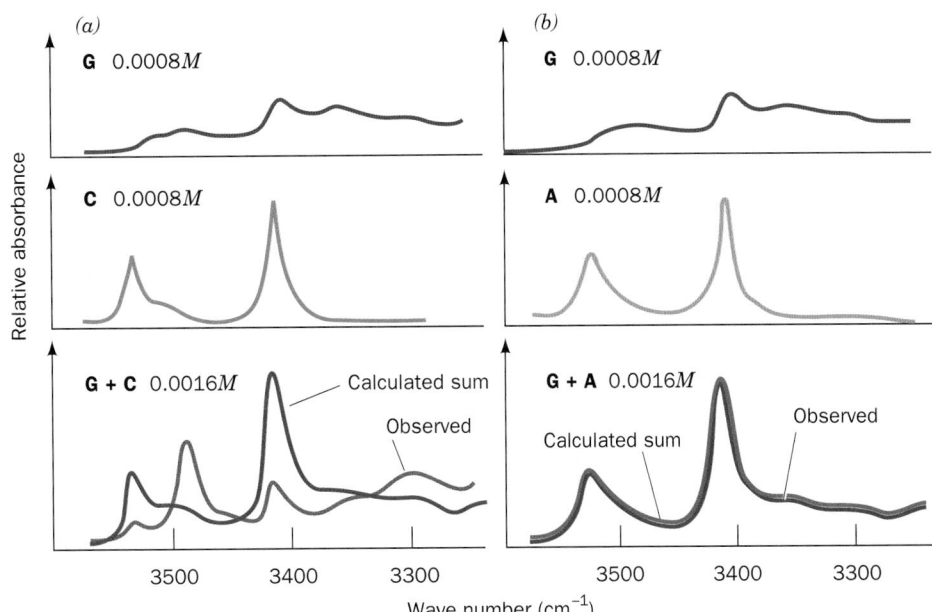

FIGURE 29-13   The IR spectra, in the N—H stretch region, of guanine, cytosine, and adenine derivatives. The derivatives were analyzed both separately and in the indicated mixtures. The solvent, CDCl$_3$, does not hydrogen bond with the bases and is relatively transparent in the frequency range of interest. (*a*) G + C. The brown curve in the lower panel, which is the sum of the spectra in the two upper panels, is the calculated spectrum of G + C for noninteracting molecules. The band near 3500 cm^{-1} in the observed G + C spectrum (*purple*) is indicative of a specific hydrogen bonding association between G and C. (*b*) G + A. The close match between the calculated and observed spectra of the G + A mixture indicates that G and A do not significantly interact. [After Kyogoku, Y., Lord, R.C., and Rich, A., *Science* **154**, 5109 (1966).]

affecting the conformations of the sugar–phosphate chains. One might reasonably suppose that this requirement of **geometric complementarity** of the Watson–Crick base pairs, A with T and G with C, is the only reason that other base pairs do not occur in a double helical environment. In fact, this was precisely what was believed for many years after the DNA double helix was discovered.

Eventually, the failure to detect pairs of different bases in nonhelical environments other than A with T (or U) and G with C led Richard Lord and Rich to demonstrate, through spectroscopic studies, that *only the bases of Watson–Crick pairs have a high mutual affinity.* Figure 29-13*a* shows the infrared (IR) spectrum in the N—H stretch region of guanine and cytosine derivatives, both separately and in a mixture. The band in the spectrum of the G + C mixture that is not present in the spectra of either of its components is indicative of a specific hydrogen bonding interaction between G and C. Such an association, which can occur between like as well as unlike molecules, may be described by ordinary mass action equations.

$$B_1 + B_2 \rightleftharpoons B_1 \cdot B_2 \qquad K = \frac{[B_1 \cdot B_2]}{[B_1][B_2]} \qquad [29.1]$$

From analyses of IR spectra such as Fig. 29-13, the values of K for the various base pairs have been determined. The self-association constants of the Watson–Crick bases are given in the top of Table 29-2 (the hydrogen bonded association of like molecules is indicated by the

appearance of new IR bands as the concentration of the molecule is increased). The bottom of Table 29-2 lists the association constants of the Watson–Crick pairs. Note that each of these latter quantities is larger than the self-association constants of either of their component bases, so that Watson–Crick base pairs preferentially form from their constituents. In contrast, the non-Watson–Crick base pairs, A·C, A·G, C·U, and G·U, whatever their geometries, have association constants that are negligible compared with the self-pairing association constants of

TABLE 29-2   Association Constants for Base Pair Formation

Base Pair	$K\ (M^{-1})^a$
Self-association	
A · A	3.1
U · U	6.1
C · C	28
G · G	10^3–10^4
Watson–Crick Base Pairs	
A · U	100
G · C	10^4–10^5

^{a}Data measured in deuterochloroform at 25°C.

Source: Kyogoku, Y., Lord, R.C., and Rich, A., *Biochim. Biophys. Acta* **179**, 10 (1969).

their constituents (e.g., Fig. 29-13*b*). *Evidently, a second reason that non-Watson–Crick base pairs do not occur in DNA double helices is that they have relatively little stability.* Conversely, the exclusive presence of Watson–Crick base pairs in DNA results, in part, from an **electronic complementarity** matching A to T and G to C. The theoretical basis of this electronic complementarity, which is an experimental observation, is obscure. This is because the approximations inherent in theoretical treatments make them unable to accurately account for the minor (few $kJ \cdot mol^{-1}$) energy differences between specific and nonspecific hydrogen bonding associations. The double helical segments of many RNAs, however, contain occasional non-Watson–Crick base pairs, most often G · U, which have functional as well as structural significance (e.g., Sections 32-2B and 32-2D).

c. Hydrogen Bonds Only Weakly Stabilize DNA

It is clear that hydrogen bonding is required for the specificity of base pairing in DNA that is ultimately responsible for the enormous fidelity required to replicate DNA with almost no error (Section 30-3D). Yet, as is also true for proteins (Section 8-4B), *hydrogen bonding contributes little to the stability of the double helix.* For instance, adding the relatively nonpolar ethanol to an aqueous DNA solution, which strengthens hydrogen bonds, destabilizes the double helix, as is indicated by its decreased melting temperature (T_m; Section 5-3C). This is because hydrophobic forces, which are largely responsible for DNA's stability (Section 29-2C), are disrupted by nonpolar solvents. In contrast, *the hydrogen bonds between the base pairs of native DNA are replaced in denatured DNA by energetically nearly equivalent hydrogen bonds between the bases and water.* This accounts for the thermodynamic observation that hydrogen bonding contributes only 2 to 8 kJ/mol of hydrogen bonds to base pairing stability.

C. *Base Stacking and Hydrophobic Interactions*

Purines and pyrimidines tend to form extended stacks of planar parallel molecules. This has been observed in the structures of nucleic acids (Figs. 29-1, 29-2, and 29-3) and in the several hundred reported X-ray crystal structures that contain nucleic acid bases. The bases in these structures are usually partially overlapped (e.g., Fig. 29-14). In fact, crystal structures of chemically related bases often exhibit similar stacking patterns. Apparently stacking interactions, which in the solid state are a form of van der Waals interaction (Section 8-4A), have some specificity, although certainly not as much as base pairing.

a. Nucleic Acid Bases Stack in Aqueous Solution

Bases aggregate in aqueous solution, as has been demonstrated by the variation of osmotic pressure with concentration. The van't Hoff law of osmotic pressure is

$$\pi = RTm \qquad [29.2]$$

where π is the osmotic pressure, m is the molality of the solute (mol solute/kg solvent), R is the gas constant, and

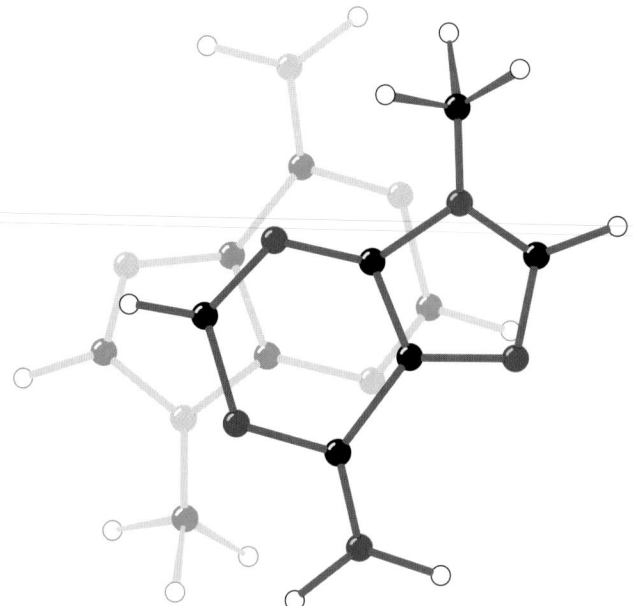

FIGURE 29-14   Stacking of adenine rings in the crystal structure of 9-methyladenine. The partial overlap of the rings is typical of the association between bases in crystal structures and in double helical nucleic acids. [After Stewart, R.F. and Jensen, L.H., *J. Chem. Phys.* **40**, 2071 (1964).]

T is the temperature. The molecular mass, *M,* of an ideal solute can be determined from its osmotic pressure since $M = c/m$, where c = g solute/kg solvent.

If the species under investigation is of known molecular mass but aggregates in solution, Eq. [29.2] must be rewritten:

$$\pi = \phi RTm \qquad [29.3]$$

where ϕ, the **osmotic coefficient,** indicates the solute's degree of association. ϕ varies from 1 (no association) to 0 (infinite association). The variation of ϕ with m for nucleic acid bases in aqueous solution (e.g., Fig. 29-15) is consistent with a model in which the bases aggregate in successive steps:

$$A + A \rightleftharpoons A_2 + A \rightleftharpoons A_3 + A \rightleftharpoons \cdots \rightleftharpoons A_n$$

where *n* is at least 5 (if the reaction goes to completion, $\phi = 1/n$). This association cannot be a result of hydrogen bonding since N^6,N^6-**dimethyladenosine,**

N^6,N^6-**Dimethyladenosine**

which cannot form interbase hydrogen bonds, has a greater degree of association than does adenosine (Fig. 29-15). Apparently *the aggregation arises from the formation of*

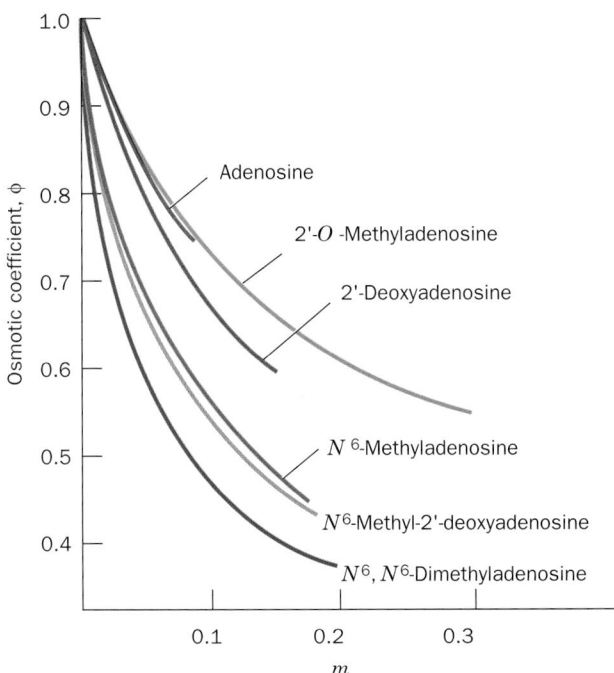

FIGURE 29-15 Variation of the osmotic coefficient φ with the molal concentrations *m* of adenosine derivatives in H₂O. The decrease of φ with increasing *m* indicates that these derivatives aggregate in solution. [After Broom, A.D., Schweizer, M.P., and Ts'o, P.O.P., *J. Am. Chem. Soc.* **89,** 3613 (1967).]

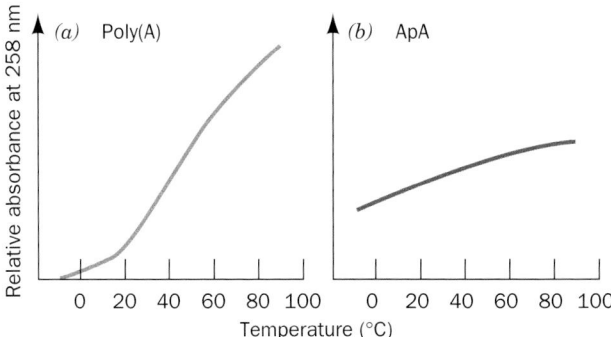

FIGURE 29-16 Melting curves for poly(A) and ApA. The broad temperature range of hyperchromic shifts at 258 nm of (*a*) poly(A) and (*b*) ApA is indicative of noncooperative conformational changes in these substances. Compare this figure with Fig. 5-16. [After Leng, M. and Felsenfeld, G., *J. Mol. Biol.* **15,** 457 (1966).]

stacks of planar molecules. This model is corroborated by proton NMR studies: The directions of the aggregates' chemical shifts are compatible with a stacked but not a hydrogen bonded model. The stacking associations of monomeric bases are not observed in nonaqueous solutions.

Single-stranded polynucleotides also exhibit stacking interactions. For example, poly(A) shows a broad increase of UV absorbance with temperature (Fig. 29-16*a*). This hyperchromism (which is indicative of nucleic acid denaturation; Section 5-3C) is independent of poly(A) concentration, so that it cannot be a consequence of intermolecular disaggregation. Likewise, it is not due to a reduction in intramolecular hydrogen bonding because poly(N^6,N^6-dimethyl A) exhibits a greater degree of hyperchromism than does poly(A). The hyperchromism must therefore arise from some sort of stacking associations within a single strand that melt out with increasing temperature. This is not a very cooperative process, as is indicated by the broadness of the melting curve and the observation that short polynucleotides, including dinucleoside phosphates such as ApA, exhibit similar melting curves (Fig. 29-16*b*).

b. Nucleic Acid Structures Are Stabilized by Hydrophobic Forces

Stacking associations in aqueous solutions are largely stabilized by hydrophobic forces. One might reasonably suppose that hydrophobic interactions in nucleic acids are similar in character to those that stabilize protein structures. However, closer examination reveals that these two types of interactions are qualitatively different in character. Thermodynamic analysis of dinucleoside phosphate melting curves in terms of the reaction

Dinucleoside phosphate (*unstacked*) ⇌
  dinucleoside phosphate (*stacked*)

(Table 29-3) indicates that *base stacking is enthalpically driven and entropically opposed. Thus the hydrophobic interactions responsible for the stability of base stacking associations in nucleic acids are diametrically opposite in character to those that stabilize protein structures* (which are enthalpically opposed and entropically driven; Section 8-4C). This is reflected in the differing structural properties of these interactions. For example, the aromatic side chains of proteins are almost never stacked and the crystal structures of aromatic hydrocarbons such as benzene, which resemble these side chains, are characteristically devoid of stacking interactions.

Hydrophobic forces in nucleic acids are but poorly understood. The observation that they are different in character from the hydrophobic forces that stabilize proteins is nevertheless not surprising because the nitrogenous bases are considerably more polar than the hydrocarbon residues of proteins that participate in hydrophobic inter-

TABLE 29-3 Thermodynamic Parameters for the Reaction

Dinucleoside phosphate ⇌ dinucleoside phosphate
(*unstacked*)   (*stacked*)

Dinucleoside Phosphate	$\Delta H_{stacking}$ (kJ · mol⁻¹)	$-T\Delta S_{stacking}$ (kJ · mol⁻¹ at 25°C)
ApA	−22.2	24.9
ApU	−35.1	39.9
GpC	−32.6	34.9
CpG	−20.1	21.2
UpU	−32.6	36.2

Source: Davis, R.C. and Tinoco, I., Jr., *Biopolymers* **6,** 230 (1968).

actions. There is, however, no theory available that adequately explains the nature of hydrophobic forces in nucleic acids (our understanding of hydrophobic forces in proteins, it will be recalled, is similarly incomplete). They are complex interactions of which base stacking is probably a significant component. Whatever their origins, hydrophobic forces are of central importance in determining nucleic acid structures.

D. *Ionic Interactions*

Any theory of the stability of nucleic acid structures must take into account the electrostatic interactions of their charged phosphate groups. Polyelectrolyte theory approximates the electrostatic interactions of DNA by considering the anionic double helix to be a homogeneously charged line or cylinder. We shall not discuss the details of this theory here, but note that it is often in reasonable agreement with experimental observations.

The melting temperature of duplex DNA increases with the cation concentration because these ions bind more tightly to duplex DNA than to single-stranded DNA due to the duplex DNA's higher anionic charge density. An increased salt concentration therefore shifts the equilibrium toward the duplex form, thus increasing the DNA's T_m. The observed relationship for Na^+ is

$$T_m = 41.1X_{G+C} + 16.6 \log[Na^+] + 81.5 \quad [29.4]$$

where X_{G+C} is the mole fraction of $G \cdot C$ base pairs (recall that T_m increases with the G + C content; Fig. 5-17); the equation is valid in the ranges $0.3 < X_{G+C} < 0.7$ and $10^{-3}M < [Na^+] < 1.0M$. Other monovalent cations such as Li^+ and K^+ have similar nonspecific interactions with phosphate groups. Divalent cations, such as Mg^{2+}, Mn^{2+}, and Co^{2+}, in contrast, specifically bind to phosphate groups, so that *divalent cations are far more effective shielding agents for nucleic acids than are monovalent cations.* For example, an Mg^{2+} ion has an influence on the DNA double helix comparable to that of 100 to 1000 Na^+ ions. Indeed, enzymes that mediate reactions with nucleic acids or just nucleotides (e.g., ATP) usually require Mg^{2+} for activity. Moreover, Mg^{2+} ions play an essential role in stabilizing

the complex structures assumed by many RNAs such as transfer RNAs (tRNAs; Section 31-2B) and ribosomal RNAs (Section 31-3A).

3 ■ SUPERCOILED DNA

🖉 **See Guided Exploration 24. DNA Supercoiling** Genetic analyses indicate that numerous viruses and bacteria have circular genetic maps, which implies that their chromosomes are likewise circular. This conclusion has been confirmed by electron micrographs in which circular DNAs are seen (Fig. 29-17). Some of these circular DNAs have a peculiar twisted appearance, a phenomenon that is known equivalently as **supercoiling, supertwisting,** and **superhelicity.** Supercoiling arises from a biologically important topological property of covalently closed circular duplex DNA that is the subject of this section. It is occasionally referred to as DNA's tertiary structure.

A. *Superhelix Topology*

Consider a double helical DNA molecule in which both strands are covalently joined to form a circular duplex molecule as is diagrammed in Fig. 29-18 (each strand can be joined only to itself because the strands are antiparallel). *A geometric property of such an assembly is that the number of times one strand wraps about the other cannot be altered without first cleaving at least one of its polynucleotide strands.* You can easily demonstrate this to yourself with a buckled belt in which each edge of the belt represents a strand of DNA. The number of times the belt is twisted before it is buckled cannot be changed without unbuckling or cutting the belt (cutting a polynucleotide strand).

This phenomenon is mathematically expressed

$$L = T + W \quad [29.5]$$

in which:

1. *L*, the **linking number** (also symbolized *Lk*), is the number of times that one DNA strand winds about the other. This integer quantity is most easily counted when the molecule's duplex axis is constrained to lie in a plane (see below).

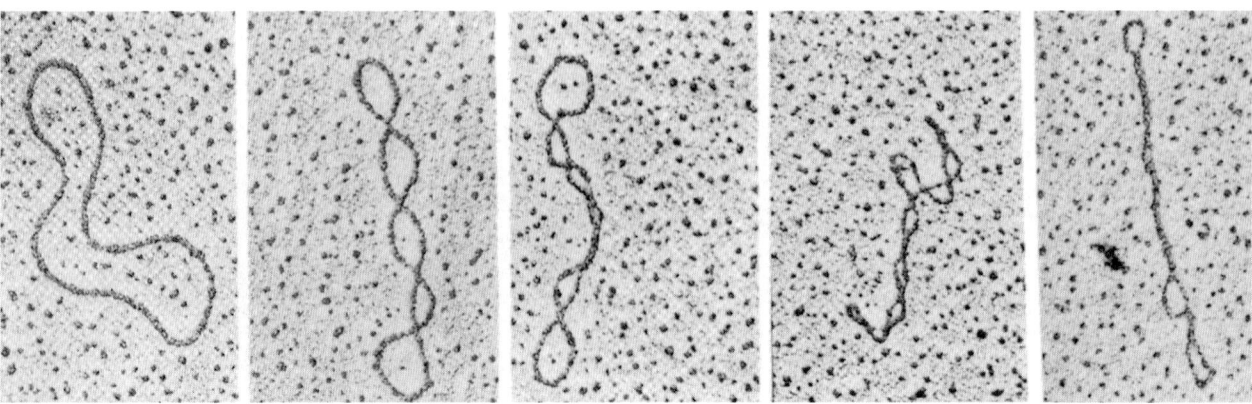

FIGURE 29-17   Electron micrographs of circular duplex DNAs. Their conformations vary from no supercoiling (*left*) to tightly supercoiled (*right*). [Electron micrographs by Laurien Polder.

From Kornberg, A. and Baker, T.A., *DNA Replication* (2nd ed.), p. 36, W.H. Freeman (1992). Used with permission.]

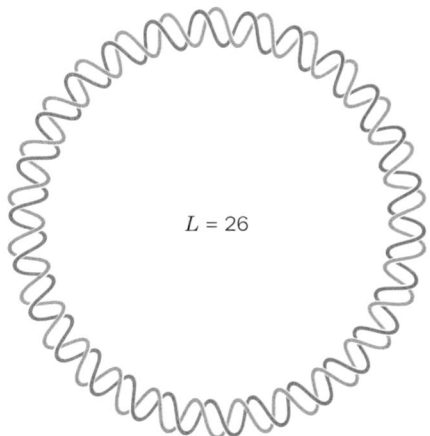

FIGURE 29-18 Schematic diagram of covalently closed circular duplex DNA that has 26 double helical turns. Its two polynucleotide strands are said to be **topologically bonded** to each other because, although they are not covalently linked, they cannot be separated without breaking covalent bonds.

However, *the linking number is invariant no matter how the circular molecule is twisted or distorted so long as both its polynucleotide strands remain covalently intact; the linking number is therefore a topological property of the molecule.*

2. *T*, the **twist** (also symbolized *Tw*), is the number of complete revolutions that one polynucleotide strand makes about the duplex axis in the particular conformation under consideration. By convention, *T* is positive for right-handed duplex turns, so that, for B-DNA in solution, the twist is normally the number of base pairs divided by 10.4 (the number of base pairs per turn of the B-DNA double helix under physiological conditions; see Section 29-3B).

3. *W*, the **writhing number** (also symbolized *Wr*), is the number of turns that the duplex axis makes about the su-perhelix axis in the conformation of interest. *It is a measure of the DNA's superhelicity.* The difference between writhing and twisting is illustrated by the familiar example in Fig. 29-19. *W* = 0 when the DNA's duplex axis is constrained to lie in a plane (e.g., Fig. 29-18); then *L = T*, so *L* may be evaluated by counting the DNA's duplex turns.

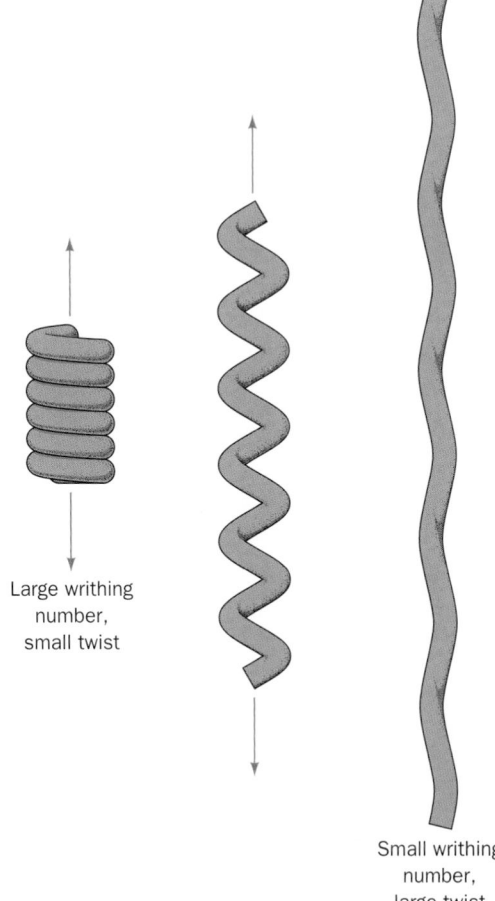

Large writhing number, small twist

Small writhing number, large twist

FIGURE 29-19 The difference between writhing and twist as demonstrated by a coiled telephone cord. In its relaxed state (*left*), the cord is in a helical form that has a large writhing number and a small twist. As the coil is pulled out (*middle*) until it is nearly straight (*right*), its writhing number becomes small as its twist becomes large.

The two DNA conformations diagrammed on the right of Fig. 29-20 are topologically equivalent; that is, they have the same linking number, *L*, but differ in their twists and

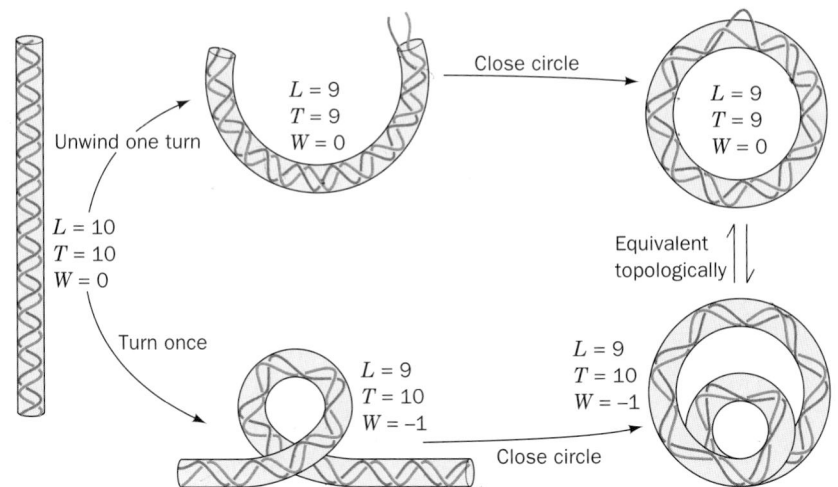

FIGURE 29-20 Two ways of introducing one supercoil into a DNA with 10 duplex turns. The two closed circular forms shown (*right*) are topologically equivalent; that is, they are interconvertible without breaking any covalent bonds. The linking number *L*, twist *T*, and writhing number *W* are indicated for each form. Strictly speaking, the linking number is only defined for a covalently closed circle.

writhing numbers. Note that T and W need not be integers, only L.

Since L is constant in an intact duplex DNA circle, for every new double helical twist, ΔT, there must be an equal and opposite superhelical twist, that is, $\Delta W = -\Delta T$. For example, a closed circular DNA without supercoils (Fig. 29-20, *upper right*) can be converted to a negatively supercoiled conformation (Fig. 29-20, *lower right*) by winding the duplex helix the same number of positive (right-handed) turns.

a. Supercoils May Be Toroidal or Interwound

A supercoiled duplex may assume two topologically equivalent forms:

1. A toroidal helix in which the duplex axis is wound as if about a cylinder (Fig. 29-21*a*).

2. An **interwound helix** in which the duplex axis is twisted around itself (Fig. 29-21*b*).

Note that these two interconvertible superhelical forms have opposite handedness. Since left-handed toroidal turns may be converted to left-handed duplex turns (see Fig. 29-19), left-handed toroidal turns and right-handed interwound turns both have negative writhing numbers. Thus an underwound duplex ($T <$ number of bp/10.4), for example, will tend to develop right-handed interwound or left-handed toroidal superhelical turns when the constraints causing it to be underwound are released (the molecular forces in a DNA double helix promote its winding to its normal number of helical turns).

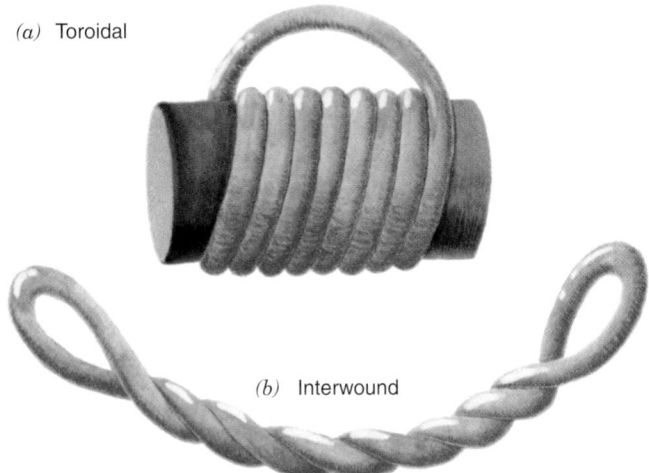

(a) Toroidal

(b) Interwound

FIGURE 29-21   Toroidal and interwound supercoils. A rubber tube that has been (*a*) toroidally coiled in a left-handed helix around a cylinder with its ends joined such that it has no twist jumps to (*b*) an interwound helix with the opposite handedness when the cylinder is removed. Neither the linking number, the twist, nor the writhing number are changed in this transformation.

b. Supercoiled DNA Is Relaxed by Nicking One Strand

Supercoiled DNA may be converted to **relaxed circles** (as appears in the leftmost panel of Fig. 29-17) by treatment with **pancreatic DNase I,** an **endonuclease** (an enzyme that cleaves phosphodiester bonds within a polynucleotide strand) that cleaves only one strand of a duplex DNA. *One single-strand nick is sufficient to relax a supercoiled DNA.* This is because the sugar–phosphate chain opposite the nick is free to swivel about its backbone bonds (Fig. 29-7) so as to change the molecule's linking number and thereby alter its superhelicity. Supercoiling builds up elastic strain in a DNA circle, much as it does in a rubber band. This is why the relaxed state of a DNA circle is not supercoiled.

B. *Measurements of Supercoiling*

Supercoiled DNA, far from being just a mathematical curiosity, has been widely observed in nature. In fact, its discovery in polyoma virus DNA by Jerome Vinograd stimulated the elucidation of the topological properties of superhelices rather than *vice versa.*

a. Intercalating Agents Control Supercoiling by Unwinding DNA

All naturally occurring DNA circles are underwound; that is, their linking numbers are less than those of their corresponding relaxed circles. This phenomenon has been established by observing the effect of ethidium ion binding on the sedimentation rate of circular DNA (Fig. 29-22). Intercalating agents such as ethidium (a planar aromatic cation; Section 6-6C) alter a circular DNA's degree of superhelicity because they cause the DNA double helix to unwind (untwist) by ~26° at the site of the intercalated molecule (Fig. 29-23). $W < 0$ in an unconstrained underwound circle because of the tendency of a duplex DNA to maintain its normal twist of 1 turn per 10.4 bp. The titration of a DNA circle by ethidium unwinds the duplex (decreases T), which must be accompanied by a compensating increase in W. This, at first, lessens the superhelicity of an underwound circle. However, as the circle binds more and more ethidium, its value of W passes through zero (relaxed circles) and then becomes positive, so that the circle again becomes superhelical. Thus the sedimentation rate of underwound DNAs, which is a measure of their compactness and therefore their superhelicity, passes through a minimum as the ethidium concentration increases. This is what is observed with native DNAs (Fig. 29-22). In contrast, the sedimentation rate of an overwound circle would only increase with increasing ethidium concentration.

b. DNAs Are Separated According to Their Linking Number by Gel Electrophoresis

Gel electrophoresis (Sections 6-4 and 6-6C) also separates similar molecules on the basis of their compactness, so that the rate of migration of a circular duplex DNA increases with its degree of superhelicity. The agarose gel electrophoresis pattern of a population of chemically iden-

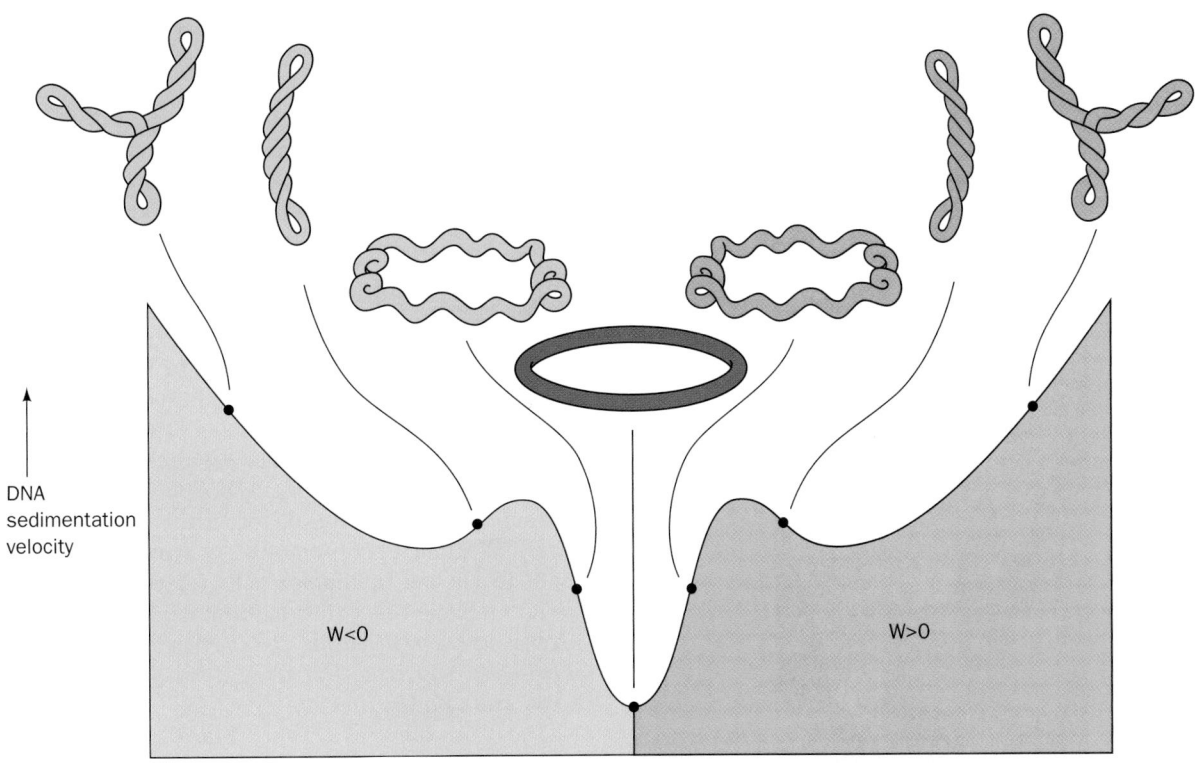

FIGURE 29-22   Sedimentation rate of underwound closed circular duplex DNA as a function of ethidium bromide concentration. The intercalation of ethidium between the base pairs locally untwists the double helix (Fig. 29-23) which, since the linking number of the circle is constant, is accompanied by an equivalent increase in the writhing number. As the negatively coiled superhelix untwists, it becomes less compact and sediments more slowly. At the low point on the curve, the DNA circles have bound sufficient ethidium to become fully relaxed. As the ethidium concentration is further increased, the DNA supercoils in the opposite direction, yielding a positively coiled superhelix. The supertwisted appearances of the depicted DNAs have been verified by electron microscopy. [After Bauer, W.R., Crick, F.H.C., and White, J.H., *Sci. Am.* **243**(1): 129 (1980). Copyright © 1981 by Scientific American, Inc.]

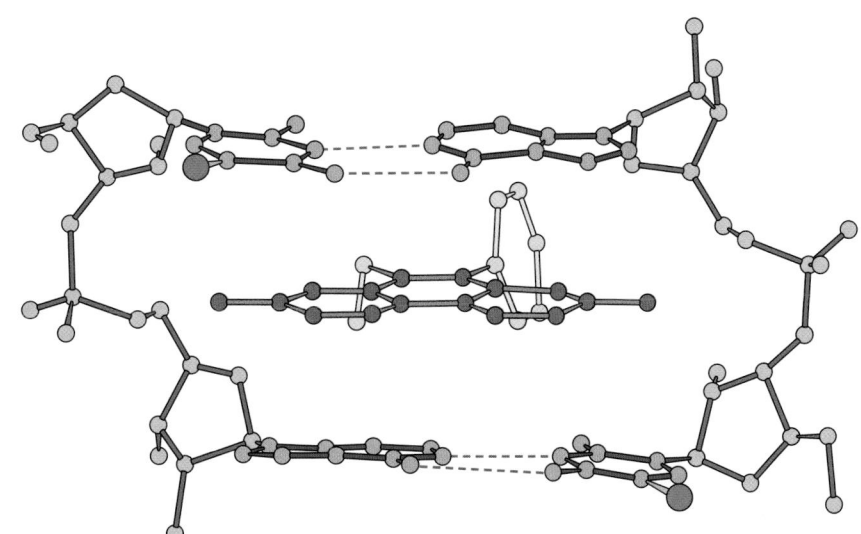

FIGURE 29-23   X-Ray structure of a complex of ethidium with 5-iodo-UpA. Ethidium (*red*) intercalates between the base pairs of the double helically paired dinucleoside phosphate and thereby provides a model for the binding of ethidium to duplex DNA. [After Tsai, C.-C., Jain, S.C., and Sobell, H.M., *Proc. Natl. Acad. Sci.* **72**, 629 (1975).]

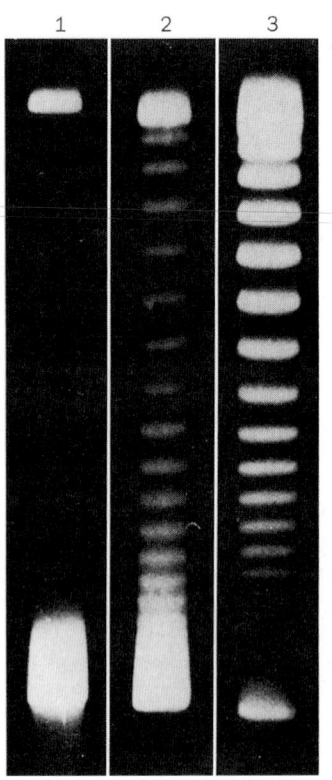

FIGURE 29-24 **Agarose gel electrophoresis pattern of SV40 DNA.** Lane 1 contains the negatively supercoiled native DNA (*lower band;* the DNA was applied to the top of the gel). In lanes 2 and 3, the DNA has been exposed for 5 and 30 min, respectively, to an enzyme, known as a type IA topoisomerase (Section 29-3C), that relaxes negative supercoils one at a time by increasing the DNA's linking number (L). The DNAs in consecutively higher bands of a given gel have successively increasing linking numbers ($\Delta L = +1$). [From Keller, W., *Proc. Natl. Acad. Sci.* **72,** 2553 (1975).]

tical DNA molecules with different linking numbers therefore consists of a series of discrete bands (Fig. 29-24). The molecules in a given band all have the same linking number and differ from those in adjacent bands by $\Delta L \pm 1$.

Comparison of the electrophoretic band patterns of **simian virus 40 (SV40)** DNA that had been enzymatically relaxed to varying degrees and then resealed (Fig. 29-24) reveals that 26 bands separate native from fully relaxed SV40 DNAs. Native SV40 DNA therefore has $W = -26$ (although it is somewhat heterogeneous in this quantity). Since SV40 DNA consists of 5243 bp, it has 1 negative superhelical turn per ~19 duplex turns. Such a **superhelix density** (W/T) is typical of circular DNAs from various biological sources.

c. DNA in Physiological Solution Has 10.4 Base Pairs per Turn

The insertion, using genetic engineering techniques (Section 5-5C), of an additional x base pairs into a superhelical DNA with a given linking number will increase the

DNA's twist and hence decrease its writhing number by x/h°, where h° is the number of base pairs per duplex turn. Such an insertion shifts the position of each band in the DNA's gel electrophoretic pattern by x/h° of the spacing between bands. By measuring the effects of several such insertions, James Wang established that $h^\circ = 10.4 \pm 0.1$ bp for B-DNA in solution under physiological conditions.

C. Topoisomerases

The normal biological functioning of DNA occurs only if it is in the proper topological state. In such basic biological processes as RNA transcription and DNA replication, the recognition of a base sequence requires the local separation of complementary polynucleotide strands. The negative supercoiling of naturally occurring DNAs results in a torsional strain that promotes such separations since it tends to unwind the duplex helix (an increase in T must be accompanied by a decrease in W). *If DNA lacks the proper superhelical tension, the above vital processes (which themselves supercoil DNA; Sections 30-2C and 31-2C) occur quite slowly, if at all.*

The supercoiling of DNA is controlled by a remarkable group of enzymes known as **topoisomerases.** They are so named because they alter the topological state (linking number) of circular DNA but not its covalent structure. There are two classes of topoisomerases:

1. Type I topoisomerases act by creating transient single strand breaks in DNA. Type I enzymes are further classified into **type IA** and **type IB topoisomerases** on the basis of their amino acid sequences and reaction mechanisms (see below).

2. Type II topoisomerases act by making transient double strand breaks in DNA.

a. Type I Topoisomerases Incrementally Relax Supercoiled DNA

Type I topoisomerases *catalyze the relaxation of supercoils in DNA by changing their linking number in increments of one turn until the supercoil is entirely relaxed.* Type IA enzymes, which are present in all cells, relax only negatively supercoiled DNA, whereas type IB enzymes, which are widely present in prokaryotes (but not *E. coli*) and eukaryotes, relax both negatively and positively coiled DNA. Although types IA and IB topoisomerases are both monomeric, ~100-kD enzymes, they share no apparent sequence or structural similities and function, as we shall see, via different enzymatic mechanisms.

A clue to the mechanism of type IA topoisomerase was provided by the observation that it reversibly **catenates** (interlinks) single-stranded circles (Fig. 29-25*a*). Apparently the enzyme operates by cutting a single strand, passing a single-strand loop through the resulting gap, and then resealing the break (Fig. 29-25*b*), thereby twisting double helical DNA by one turn. In support of this **strand passage** mechanism, the denaturation of type IA enzyme that has been incubated with single-stranded circular DNA yields a linear DNA that has its 5′-terminal phosphoryl group linked to the enzyme via a phosphoTyr diester linkage.

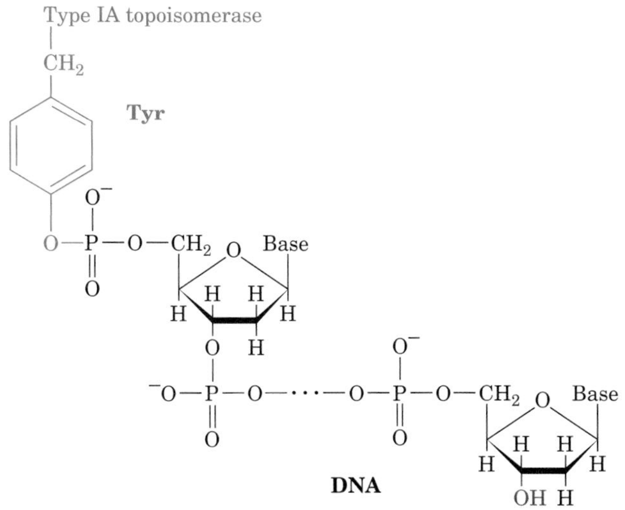

Type IA topoisomerase
|
CH_2

Tyr

DNA

In contrast, denatured type IB enzyme is linked to the 3′ end of DNA via a phosphoTyr linkage. *By forming such covalent enzyme–DNA intermediates, the free energy of the* *cleaved phosphodiester bond is preserved, so that no energy input is required to reseal the nick.*

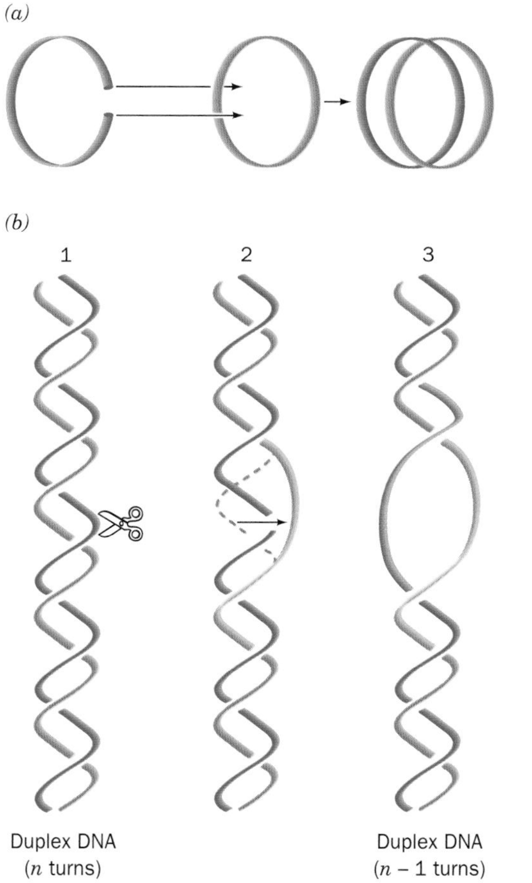

b. Type IA Topoisomerase Probably Functions via a Strand Passage Mechanism

Cells of *E. coli* contain two type IA topoisomerases named **topoisomerase I** and **topoisomerase III.** Topoisomerase III's Tyr 328 is the active site residue that forms a 5′-phosphoTyr linkage with the cleaved DNA. The X-ray structure of the inactive Y328F mutant of topoisomerase III in complex with the single-stranded octanucleotide d(CGCAACTT), determined by Alfonso Mondragón (Fig. 29-26), reveals that this 659-residue monomer folds into four domains which enclose an ~20 by 28 Å hole that is large enough to contain a duplex DNA and which is lined with numerous Arg and Lys side chains. The octanucleotide binds in a groove that is also lined with Arg and Lys side chains with its sugar–phosphate backbone in contact with the protein and with most of its bases exposed for possible base pairing. Curiously, this single-stranded DNA assumes a B-DNA-like conformation even though its complementary strand would be sterically excluded from the groove. The DNA strand is oriented with its 3′ end near the active site, where, if the mutant Phe 328 were the wild-type Tyr, its side chain would be properly positioned to nucleophilically attack the phosphate group bridging the DNA's C6 and T7 to form a 5′-phosphoTyr linkage with T7 and release C6 with a free 3′-OH. This structure and that of the homologous and structurally sim-

(a)

(b)

| 1 | 2 | 3 |

Duplex DNA
(*n* turns)

Duplex DNA
(*n* − 1 turns)

FIGURE 29-25 Type IA topoisomerase action. By cutting a single-stranded DNA, passing a loop of a second strand through the break, and then resealing the break, a type IA topoisomerase can (*a*) catenate two single-stranded circles or (*b*) unwind duplex DNA by one turn.

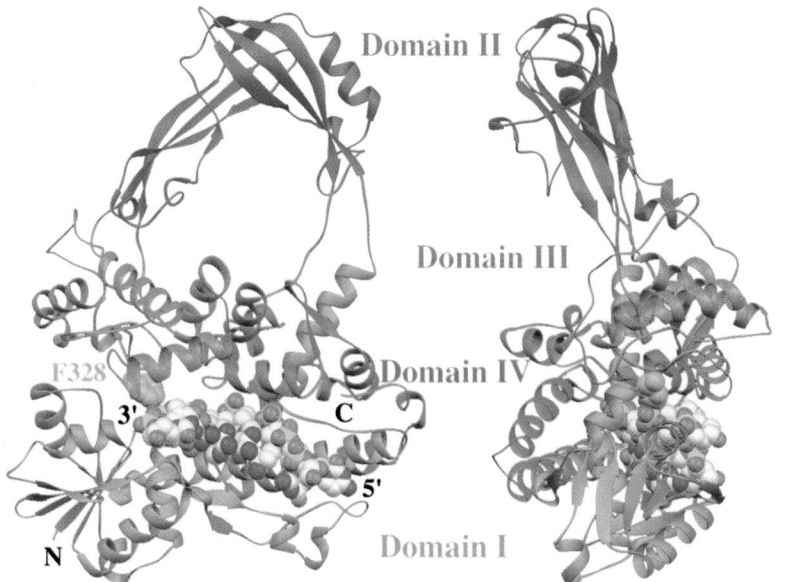

FIGURE 29-26   X-Ray structure of the Y328F mutant of *E. coli* topoisomerase III, a type IA topoisomerase, in complex with the single-stranded octanucleotide d(CGCAACTT). The two views shown are related by a 90° rotation about a vertical axis. The DNA is drawn in space-filling form with C gray, N blue, O red, and P yellow. The enzyme's active site is marked by the side chain of Phe 328, which is shown in space-filling form in yellow-green. [Based on an X-ray structure by Alfonso Mondragón, Northwestern University. PDBid 1I7D.]

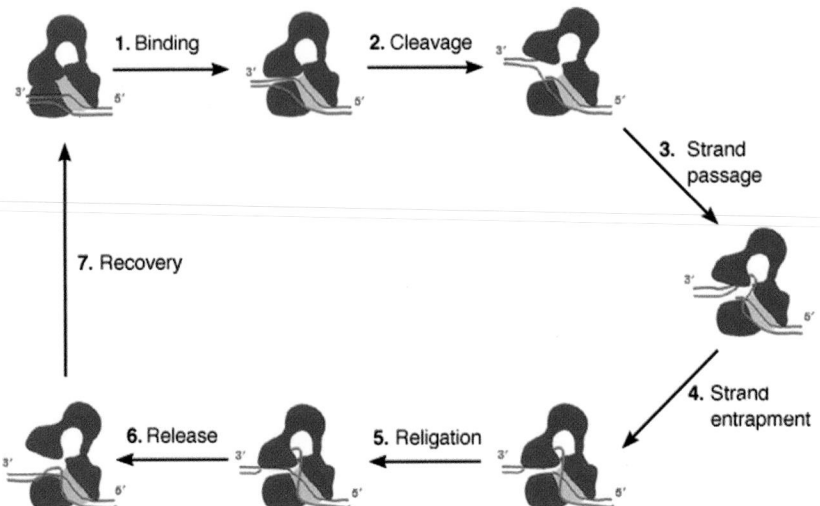

FIGURE 29-27 Proposed mechanism for the strand passage reaction catalyzed by type IA topoisomerases. The enzyme is shown in blue with the yellow patch representing the binding groove for single-stranded (ss) DNA. The two DNA strands, which are drawn in red and green, could represent the two strands of a covalently closed circular duplex or two ss circles. **(1)** The protein recognizes a ss region of the DNA, here the red strand, and binds it in its binding groove. This is followed by or occurs simultaneously with the opening of a gap between domains I and III. **(2)** The DNA is cleaved with the newly formed 5′ end, becoming covalently linked to the active site Tyr and the segment with the newly formed 3′ end remaining tightly but noncovalently bound in the binding groove. **(3)** The unbroken (*green*) strand is passed through the opening or gate formed by the cleaved (*red*) strand to enter the protein's central hole. **(4)** The unbroken strand is trapped by the partial closing of the gap. **(5)** The two cleaved ends of the red strand are rejoined in what is probably a reversal of the cleavage reaction. **(6)** The gap between domains I and III reopens to permit the escape of the red strand, yielding the reaction product in which the green strand has been passed through a transient break in the red strand. **(7)** The enzyme returns to its initial state. If the two strands form a negatively supercoiled duplex DNA, its linking number, *L,* has increased by 1; if they are separate ss circles, they have been catenated or decatenated. For duplex DNA, this process can be repeated until all of its supercoils have been removed ($W = 0$). [After a drawing by Alfonso Mondragón, Northwestern University.]

ilar *E. coli* topoisomerase I suggest the mechanism for the type IA topoisomerase-catalyzed strand passage reaction that is diagrammed in Fig. 29-27.

c. Type IB Topoisomerase Appears to Function via a Controlled Rotation Mechanism

Human **topoisomerase I (topo I)** is a 765-residue type IB topoisomerase (and hence is unrelated to *E. coli* topoisomerase I). It mediates the transient cleavage of one strand of a duplex DNA through the nucleophilic attack of Tyr 723 on a DNA P atom to yield a 3′-linked phosphoTyr diester bond and a free 5′-OH group on the succeeding nucleotide. Limited proteolysis studies revealed that topo I consists of four major regions: its N-terminal, core, linker, and C-terminal domains. The ~210-residue, highly polar, N-terminal domain, which is poorly conserved, contains several nuclear targeting signals and is dispensable for enzymatic activity.

The X-ray structure of the catalytically inactive Y723F mutant of topo I lacking its N-terminal 214 residues and in complex with a 22-bp palindromic duplex DNA was determined by Wim Hol (Fig. 29-28). The core domain of this bilobal protein is wrapped around the DNA in a tight embrace. If the mutant Phe 723 were the wild-type Tyr, its OH group would be colinear with the scissile P—O5′ bond and hence ideally positioned to nucleophilically attack this P atom so as to form a covalent linkage with the 3′ end of

the cleaved strand. As expected, the protein interacts with the DNA in a largely sequence independent manner: Of the 41 direct contacts that the protein makes to the DNA, 37 are protein–phosphate interactions and only one is base-specific. The protein interacts to a much greater extent with the five base pairs of the DNA's downstream segment (which would contain the cleaved strand's newly formed 5′ end; 29 of the 41 contacts) than it does with the base pairs of the DNA's upstream segment (to which Tyr 723 would be covalently linked; 12 of the 41 contacts).

Topo I does not seem sterically capable of unwinding supercoiled DNA via the strand passage mechanism that type IA topoisomerases appear to follow (Fig. 29-27). Rather, as is diagrammed in Fig. 29-29, it is likely that topo I relaxes DNA supercoils by permitting the cleaved duplex DNA's loosely held downstream segment to rotate relative to the tightly held upstream segment. This rotation can only occur about the sugar–phosphate bonds in the uncleaved strand (α, β, γ, ε, and ζ in Fig. 29-7) that are opposite the cleavage site because the cleavage frees these bonds to rotate. In support of this mechanism, the protein region surrounding the downstream segment contains 16 conserved, positively charged residues that form a ring about this duplex DNA, which would presumably hold the DNA in the ring but not in any specific orientation. Nevertheless, the downstream segment is unlikely to rotate freely because the cavity containing it is shaped so as to interact with the

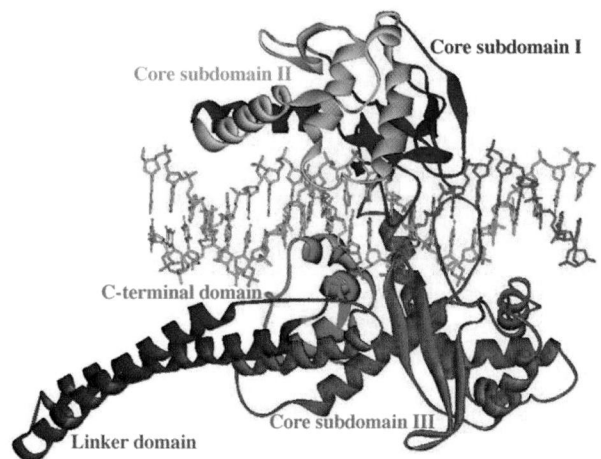

FIGURE 29-28 X-Ray structure of the N-terminally truncated, Y723F mutant of human topoisomerase I in complex with a 22-bp duplex DNA. The protein's various domains and subdomains are drawn in different colors. The DNA's uncleaved strand is cyan, and the upstream and downstream portions of the scissile strand are magenta and pink, respectively. [Courtesy of Wim Hol, University of Washington. PDBid 1A36.]

downstream segment during some portions of its rotation. Hence, type IB topoisomerases are said to mediate a **controlled rotation** mechanism in relaxing supercoiled DNA. This unwinding is driven by the superhelical tension in the DNA and hence requires no other energy input. Eventually, the DNA is religated by a reversal of the cleavage reaction and the now less supercoiled DNA is released.

d. Type II Topoisomerases Function via a Strand Passage Mechanism

The prokaryotic type II topoisomerase known as **DNA gyrase** is an ~375-kD A_2B_2 heterotetramer in which the A and B subunits are named **GyrA** and **GyrB**. *This enzyme catalyzes the stepwise negative supercoiling of DNA with the concomitant hydrolysis of an ATP to ADP + P_i.* It can also catenate and decatenate double-stranded circles as well as tie knots in them. All other type II topoisomerases, both eukaryotic and prokaryotic, only relax supercoils, although they hydrolyze ATP in doing so (DNA supercoiling in eukaryotes is generated differently from that in prokaryotes; Section 34-1B).

DNA gyrases are inhibited by a variety of substances including **novobiocin,** a member of the *Streptomyces-*

FIGURE 29-29 Controlled rotation mechanism for type IB topoisomerses. A highly negatively supercoiled DNA (*red, with a right-handed writhe*) is converted, via stages (*a*) through (*g*), to a less supercoiled form (*green*). Topo I is drawn as a bilobal space-filling structure, in which the cyan lobe is formed by core subdomains I and II (Fig. 29-28) and the magenta lobe is formed by core subdomain III, the linker domain, and the C-terminal domain. The structure shown in (*d*), which is expanded by a factor of 2, shows the downstream portion of the rotating DNA (that containing the cleaved strand's new 5′ end) at 30° intervals, all differently colored. Since the enzyme is not always in direct contact with the rotating DNA, small rocking motions of the protein (*small curved arrows*) may accompany the controlled rotation. [Courtesy of Wim Hol, University of Washington.]

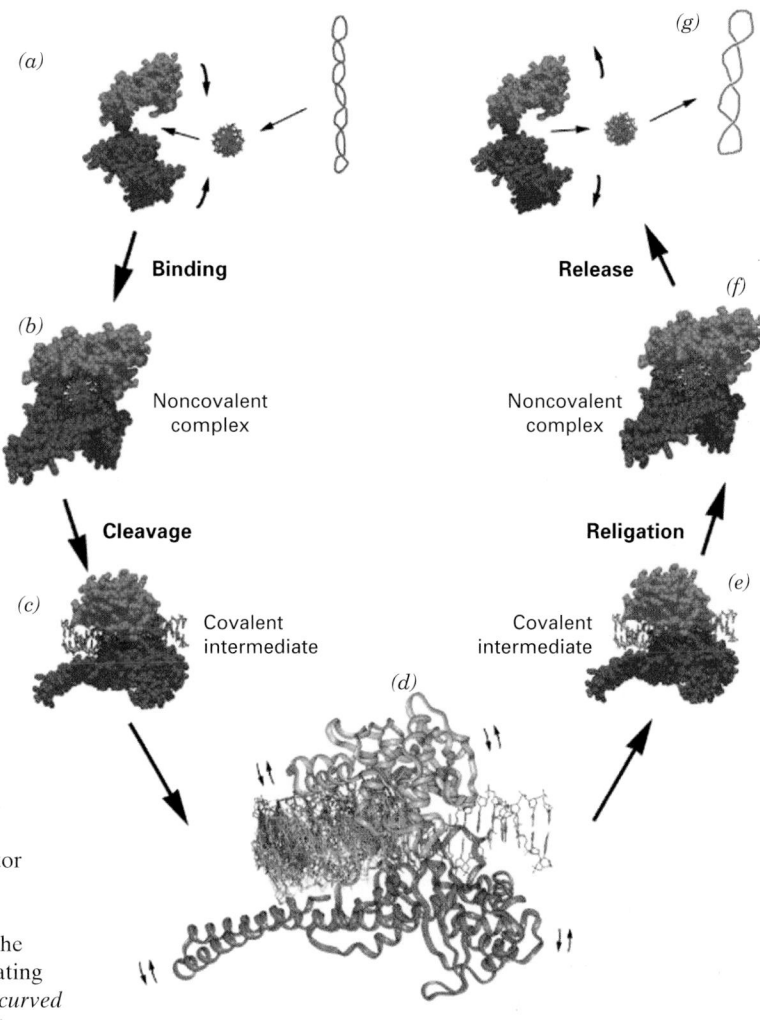

Controlled rotation

derived **coumarin** family of antibiotics, and **ciprofloxacin** (trade name **Cipro**), a member of the synthetically generated **quinolone** family of antibiotics (their coumarin and quinolone groups are drawn in red):

Novobiocin

Ciprofloxacin

These agents profoundly inhibit bacterial DNA replication and RNA transcription, thereby demonstrating the importance of properly supercoiled DNA in these processes. Studies using *E. coli* DNA gyrase mutants resistant to these substances have demonstrated that ciprofloxacin associates with GyrA and novobiocin binds to GyrB.

The gel electrophoretic pattern of duplex circles that have been exposed to DNA gyrase shows a band pattern in which the linking numbers differ by increments of 2 rather than 1, as occurs with type I topoisomerases. *Evidently, DNA gyrase acts by cutting both strands of a duplex, passing the duplex through the break, and resealing it* (Fig. 29-30). This hypothesis is corroborated by the observation that when DNA gyrase is incubated with DNA and ciprofloxacin, and subsequently denatured with guanidinium chloride, a GyrA subunit remains covalently linked to the 5' end of each of the two cut strands through a phosphoTyr linkage. These cleavage sites are staggered by 4 bp, thereby yielding sticky ends.

Saccharomyces cerevisiae (baker's yeast) **topoisomerase II (topo II)**, a type II topoisomerase, is a homodimer of subunits whose N- and C-terminal segments are homologous to DNA gyrase's B and A subunits, respectively. Hence these subfragments are designated B' and A'. The 92-kD segment encompassing residues 410 to 1202 of this 1429-residue protein can cleave duplex DNA but cannot transport it through the break because it lacks the enzyme's ATPase domain (residues 1–409). However, the C-terminal fragment (residues 1203–1429) appears to be dispensable.

The X-ray structure of the 92-kD segment (Fig. 29-31*a*), determined by James Berger, Stephen Harrison, and Wang, reveals that its two crescent-shaped monomers associate to form a heart-shaped dimer with its two B' subfragments (residues 410–633) associating at the top of the heart and its two A' subfragments (residues 683–1202) coming together at its base (point). The 49-residue segment between these two subfragments is disordered. The dimer encloses a large triangular central hole (55 Å wide and 60 Å in height). Tyr 783, the residue that forms a transient phosphoTyr covalent link with the 5' end of a cleaved DNA strand, is located at the interface between the A' and B' subfragments of the same subunit, at the end of a narrow tunnel that opens up into the central hole. Here, the A' subfragment forms a positively charged semicircular groove that funnels into this active site tunnel. B-DNA can be modeled into this groove with a 4-nt overhang of its 5'-ending strand extending into the active site tunnel. The dimer's two active site Tyr residues are located 27 Å apart such that they must move 35 to 40 Å toward and past each other to achieve positions that are properly staggered to link to the 5' ends of a cleaved duplex DNA.

The X-ray structure of an *E. coli* GyrB fragment comprising residues 2 to 393 of the 804-residue subunit in complex with the nonhydrolyzable ATP analog ADPNP was determined by Guy Dodson and Eleanor Dodson (Fig. 29-31*b*). This protein fragment, which dimerizes in solution in the presence of ADPNP, consists of two domains. The N-terminal domain, which has been implicated in ATP hydrolysis, binds Mg^{2+}–ADPNP. The C-terminal domains form the walls of a 20-Å-diameter hole through the dimer, the same diameter as that of the B-DNA double helix. All

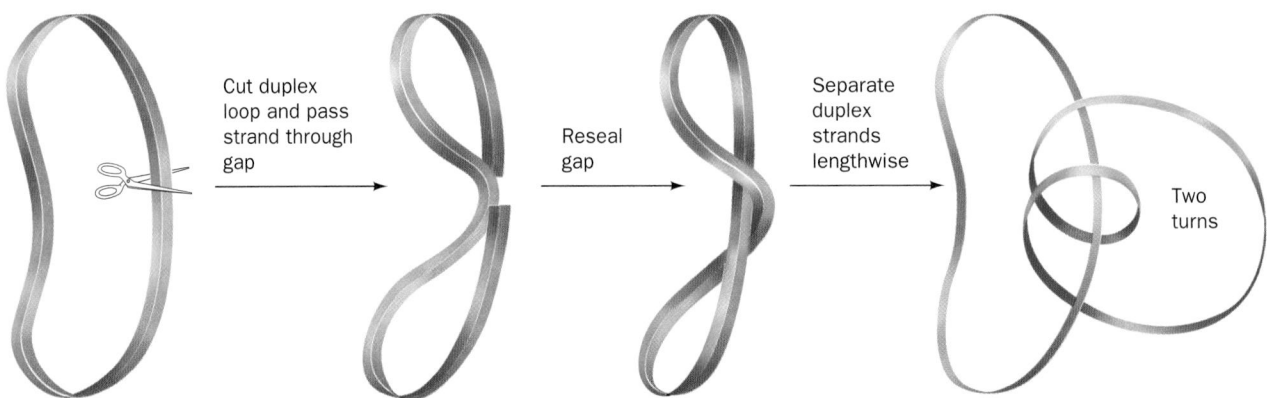

FIGURE 29-30 A demonstration, in which DNA is represented by a ribbon, that cutting a duplex circle, passing the double helix through the resulting gap, and then resealing the break changes the linking number by 2. Separating the resulting single strands (slitting the ribbon along its length; *right*) indicates that one single strand makes two complete revolutions about the other.

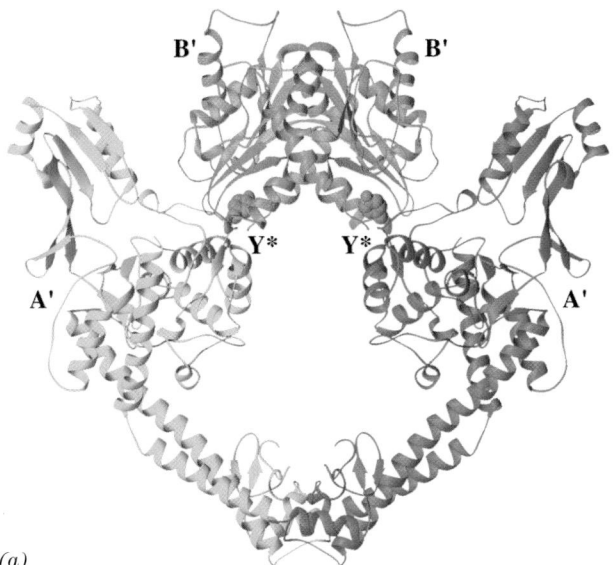

(a)

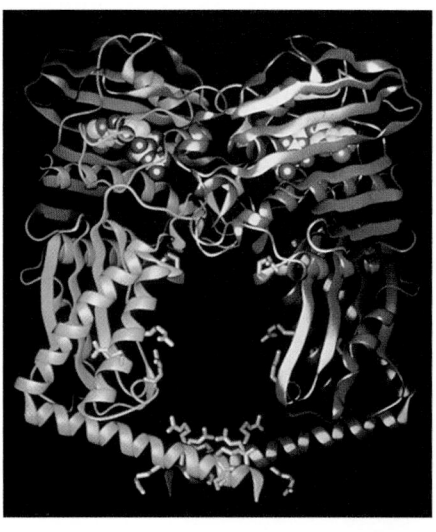

(b)

FIGURE 29-31 Structures of topoisomerase II. (*a*) X-Ray structure of the 92-kD segment of the yeast topoisomerase II (residues 410–1202) dimer as viewed with its twofold axis vertical. The A′ and B′ subfragments of one subunit are blue and red and those of the other subunit are cyan and orange. The active site Tyr 783 side chains are shown in space-filling form (C green and O red) and labeled Y*. [Based on an X-ray structure by James Berger, Stephen Harrison, and James Wang, Harvard University. PDBid 1BGW.] (*b*) X-Ray structure of a dimer of the

N-terminal fragment of *E. coli* GyrB (residues 2–393) in complex with ADPNP as viewed with its twofold axis vertical. The two identical subunits, which are colored red and green, each fold into two domains, which are represented by lighter and darker shades of color. The side chains of the Arg residues lining the 20-Å-diameter hole through the protein are shown in stick form (*blue*) and the bound ADPNP molecules are shown in space-filling form. [Courtesy of Eleanor Dodson and Guy Dodson, University of York, U.K.] *See the Interactive Exercises*

of this domain's numerous Arg residues line the walls of the cavity, as might be expected for a DNA-binding surface.

Consideration of the foregoing two structures led to a type of strand passage model for the mechanism of type II topoisomerases (Fig. 29-32) in which the DNA duplex to be

cleaved binds in the above-described groove across the top of the heart. ATP binding to the ATP-binding domain (which is absent in the 92-kD fragment) then induces a series of conformational changes in which the DNA's so-called G-segment (G for gate) is cleaved and the resulting two frag-

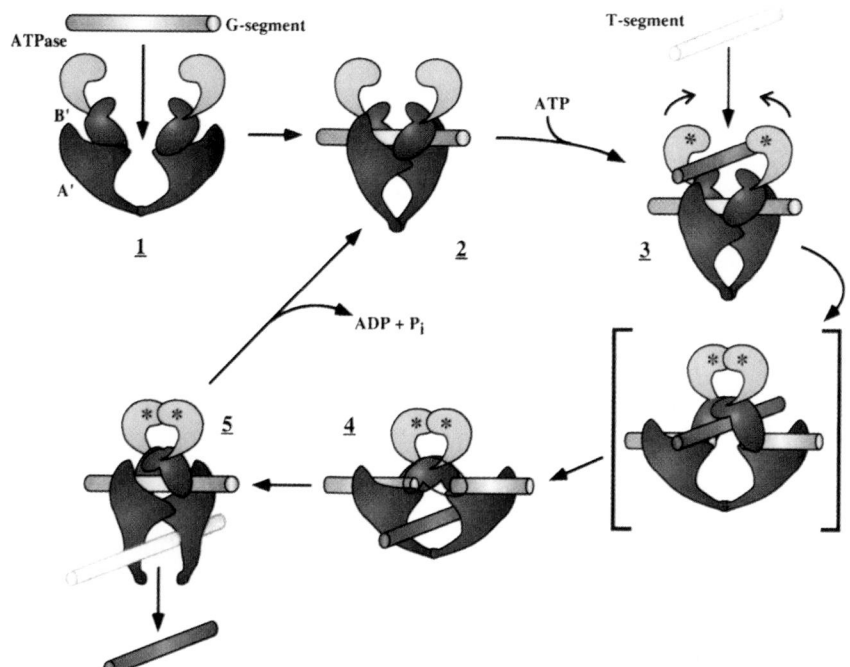

FIGURE 29-32 Model for the enzymatic mechanism of type II topoisomerases. The protein's ATPase, B′, and A′ domains are colored yellow, red, and purple, respectively, and the DNA's G- and T-segments are colored gray and green, respectively. In **1**, the G-segment binds to the enzyme, thereby inducing the conformational change drawn in **2**. The binding of ATP (represented by asterisks) and a T-segment (**3**) induces a series of conformational changes in which the G-segment is cleaved by the A′ subfragments as they separate from one another. The ATPase domains simultaneously dimerize, and the T-segment is transported through the break into the central hole (**4**; here the B′ subfragment in front is transparent for clarity). The DNA transport step is shown as proceeding through the hypothetical intermediate shown in square brackets. The G-segments are then resealed and the T-segment is released through the separation of the A′ subfragments at their dimer interface (**5**). This interface then reforms and the ATP is hydrolyzed and released to yield the enzyme in its starting state (**2**). [Courtesy of James Wang, Harvard University.]

ments are spread apart by at least 20 Å through the action of the protein. This permits the passage of the DNA's so-called T-segment (T for transported) from the top of the heart through the break and into the central hole, thereby incrementing the DNA's linking number by 2. Then, in a process that is accompanied by ATP hydrolysis, the two B′ subfragments come together to reseal the cleaved DNA, and the DNA occupying the central hole is released from the bottom of the heart by the spreading apart of the two contacting A′ subfragments (or GyrA subunits). Finally, the re-

sulting ADP and P_i are released and the A′ subfragments rejoin to yield recycled enzyme. Two independent X-ray structures support this model: that of the 92-kD segment of topo II crystallized under different conditions than the structure in Fig. 29-31a and that of a 59-kD fragment of GyrA (Fig. 29-33). The conformations that these latter proteins appear to be representative of some of the conformations predicted by the model depicted in Fig. 29-32.

e. Topoisomerse Inhibitors Are Effective Antibiotics and Cancer Chemotherapy Agents

Coumarin derivatives such as novobiocin, and quinolone derivatives such as ciprofloxacin, specifically inhibit DNA gyrases and are therefore antibiotics. In fact, ciprofloxacin is the most efficacious oral antibiotic against gram-negative bacteria presently in clinical use (novobiocin's adverse side effects and the rapid generation of bacterial resistance to it have resulted in the discontinuation of its use in the treatment of human infections). A number of substances, including **doxorubicin** (also called **adriamycin;** a product of *Streptomyces peucetius*) and **etoposide** (a synthetic derivative),

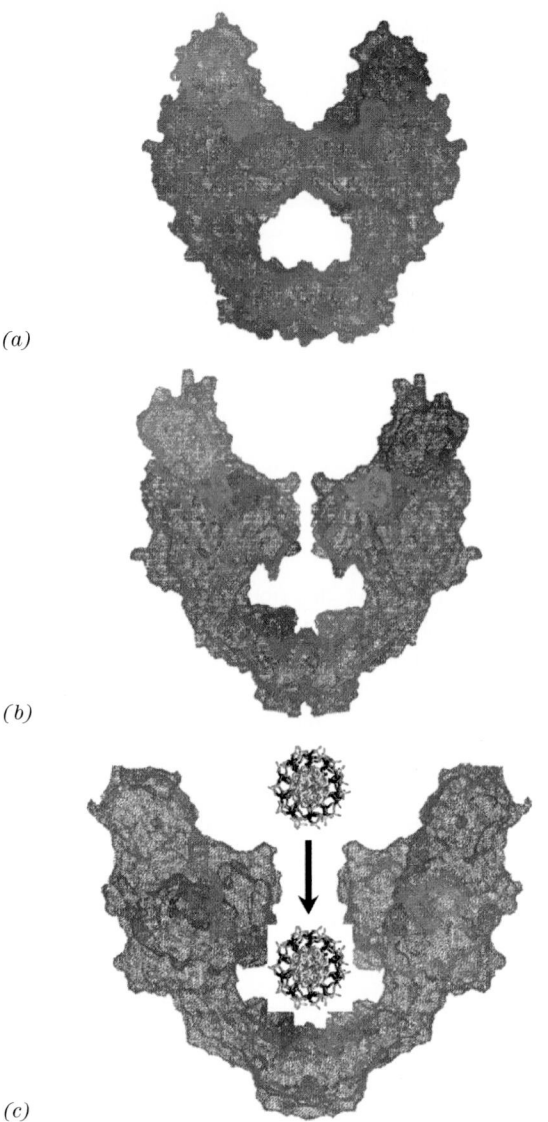

(a)

(b)

(c)

FIGURE 29-33   Surface representations of X-ray structures of type II topoisomerase A′ subfragment dimers. The proteins are viewed with their 2-fold axes vertical. (*a*) *E. coli* GyrA subunits (residues 2–523), the minimal fragments which, when complexed to Gyr B, have DNA cleavage activity. (*b*) The A′ subfragments in the X-ray structure of the 92-kD segment of topo II crystallized under different conditions from that in Fig. 29-31a. (*c*) The A′ subfragments in the X-ray structure of topo II (the blue and cyan portions of Fig. 29-31a) shown with the modeled passage of DNA into the enzyme's central hole. [Courtesy of James Berger, University of California at Berkeley PDBids (*a*) 1AB4, (*b*) 1BJT, and (*c*) 1BJW.]

Doxorubicin (Adriamycin)

Etoposide

inhibit eukaryotic type II topoisomerases and are therefore widely used in cancer chemotherapy.

Type II topoisomerase inhibitors act in either of two ways. Many of them, including novobiocin, inhibit their target enzyme's ATPase activity (novobiocin is a competitive inhibitor of ATP because it tightly binds to GyrB in a way that prevents the binding of ATP's adenine ring). They therefore kill cells by blocking topoisomerase activity, which results in the arrest of DNA replication and RNA transcription. However, other substances, including ciprofloxacin, doxorubicin, and etoposide, enhance the rate at which their target type II topoisomerases cleave double-stranded DNA and/or reduce the rate at which these enzymes reseal these breaks. Consequently, these agents induce higher than normal levels of transient protein-bridged breaks in the DNA of treated cells. These protein bridges are easily ruptured by the passage of the replication and transcription machinery, thereby rendering the breaks permanent. Although all cells have extensive enzymatic machinery to repair damaged DNA (Section 30-5), a sufficiently high level of DNA damage results in

cell death. Consequently, since rapidly replicating cells such as cancer cells have elevated levels of type II topoisomerases, they are far more likely to incur lethal DNA damage through the inhibition of their type II topoisomerases than are slow-growing or quiescent cells.

Type IB topoisomerases are specifically inhibited by the quinoline-based alkaloid **camptothecin**

Camptothecin

(a product of the Chinese tree *Camptotheca acuminata*) and its derivatives, which act by stabilizing the covalent topoisomerase I–DNA complex. These compounds, the only known naturally occurring topoisomerase IB inhibitors, are therefore potent anticancer agents.

CHAPTER SUMMARY

1 ■ Double Helical Structures B-DNA consists of a right-handed double helix of antiparallel sugar–phosphate chains with ~10 bp per turn of 34 Å and with its bases nearly perpendicular to the helix axis. Bases on opposite strands hydrogen bond in a geometrically complementary manner to form A · T and G · C Watson–Crick base pairs. At low humidity, B-DNA undergoes a reversible transformation to a wider, flatter right-handed double helix known as A-DNA. Z-DNA, which is formed at high salt concentrations by polynucleotides of alternating purine and pyrimidine base sequences, is a left-handed double helix. Double helical RNA and RNA · DNA hybrids have A-DNA-like structures. The conformation of DNA, particularly that of B-DNA, varies with its base sequence largely because DNA's deformability varies with its base sequence.

2 ■ Forces Stabilizing Nucleic Acid Structures The orientations about the glycosidic bond and the various torsion angles in the sugar–phosphate chain are sterically constrained in nucleic acids. Likewise, only a few of the possible sugar pucker conformations are commonly observed. Watson–Crick base pairing is both geometrically and electronically complementary. Yet hydrogen bonding interactions do not greatly stabilize nucleic acid structures. Rather, the structures are largely stabilized by hydrophobic interactions. Nevertheless, the hydrophobic forces in nucleic acids are qualitatively different in character from those that stabilize proteins. Electrostatic interactions between charged phosphate

groups are also important structural determinants of nucleic acids.

3 ■ Supercoiled DNA The linking number (L) of a covalently closed circular DNA is topologically invariant. Consequently, any change in the twist (T) of a circular duplex must be balanced by an equal and opposite change in its writhing number (W), which indicates its degree of supercoiling. Supercoiling can be induced by intercalation agents. The gel electrophoretic mobility of DNA increases with its degree of superhelicity. Naturally occurring DNAs are all negatively supercoiled and must be so in order to participate in DNA replication and RNA transcription. Type IA topoisomerases relax negatively supercoiled DNAs via a strand passage mechanism in which they cleave a single-strand of DNA to form a 5′-phosphoTyr bond, pass a single-strand DNA segment through the gap, and then reseal the gap. Type IB topoisomerases relax both negatively and positively supercoiled DNAs via a controlled rotation mechanism involving a single-strand cleavage in which a transient phosphoTyr bond is formed with the newly generated 3′ end. Type II topoisomerases relax duplex DNA in increments of two supertwists at the expense of ATP hydrolysis by making a double-strand scission in the DNA so as to form two transient 5′-phosphoTyr linkages, passing the duplex through the break, and resealing it. DNA gyrase also generates negative supertwists in an ATP-dependent manner. Topoisomerases are the targets of various antibiotics and chemotherapeutic agents.

REFERENCES

GENERAL

Bloomfield, V.A., Crothers, D.M., and Tinoco, I., Jr., *Nucleic Acids: Structures, Properties, and Functions*, University Science Books (2000).

Calladine, C.R. and Drew, H.R., *Understanding DNA*, Academic Press (1992). [The molecule and how it works.]

Neidle, S. (Ed.), *Oxford Handbook of Nucleic Acid Structure*, Oxford University Press (1999).

Saenger, W., *Principles of Nucleic Acid Structure,* Springer-Verlag (1984). [A detailed and authoritative exposition.]

Sinden, R.R., *DNA Structure and Function,* Academic Press (1994).

The double helix–50 years, *Nature* **421,** 395–453 (2003). [A supplement containing a series of articles on the historical, cultural, and scientific influences of the DNA double helix celebrating the fiftieth anniversary of its discovery.]

Travers, A. and Buckle, M. (Eds.) *DNA–Protein Interactions. A Practical Approach,* Oxford University Press (2000). [A laboratory manual for numerous physicochemical methods that are used to probe the interactions of DNA and proteins.]

STRUCTURES AND STABILITIES OF NUCLEIC ACIDS

Dickerson, R.E., Sequence-dependent B-DNA conformation in crystals and in protein complexes, *in* Sarma, R.H. and Sarma, M.H. (Eds.), *Structure, Motion, Interaction and Expression in Biological Molecules, pp.* 17–35, Adenine Press (1998); *and* DNA bending: the prevalence of kinkiness and the virtues of normality, *Nucleic Acids Res.* **26,** 1906–1926 (1998).

Fairhead, H., Setlow, B., and Setlow, P., Prevention of DNA damage in spores and in vitro by small, acid-soluble proteins from *Bacillus* species, *J. Bacteriol.* **175,** 1367–1374 (1993).

Joshua-Tor, L. and Sussman, J.L., The coming of age of DNA crystallography, *Curr. Opin. Struct. Biol.* **3,** 323–335 (1993).

Rich, A., Nordheim, A., and Wang, A.H.-J., The chemistry and biology of left-handed Z-DNA, *Annu. Rev. Biochem.* **53,** 791–846 (1984).

Schwartz, T., Rould, M.A., Lowenhaupt, K., Herbert, A., and Rich, A., Crystal structure of the Zα domain of the human editing enzyme ADAR1 bound to left-handed Z-DNA, *Science* **284,** 1841–1845 (1999).

Sundaralingam, M., Stereochemistry of nucleic acids and their constituents. IV. Allowed and preferred conformations of nucleosides, nucleoside mono-, di-, tri-, and tetraphosphates, nucleic acids and polynucleotides, *Biopolymers* **7,** 821–860 (1969).

Voet, D. and Rich, A., The crystal structures of purines, pyrimidines and their intermolecular structures, *Prog. Nucleic Acid Res. Mol. Biol.* **10,** 183–265 (1970).

Wing, R., Drew, H., Takano, T., Broka, C., Tanaka, S., Itakura, K., and Dickerson, R.E., Crystal structure analysis of a complete turn of B-DNA, *Nature* **287,** 755–758 (1980); and Shui, X., McFail-Isom, L., Hu, G.G., and Williams, L.D., The B-DNA decamer at high resolution reveals a spine of sodium, *Biochemistry* **37,** 8341–8355 (1998). [The Dickerson dodecamer at its original 2.5 Å resolution and at its later-determined 1.4 Å resolution.]

SUPERCOILED DNA

Bates, A.D. and Maxwell, A. *DNA Topology,* IRL Press (1993). [A monograph.]

Berger, J.M., Type II DNA topoisomerases, *Curr. Opin. Struct. Biol.* **8,** 26–32 (1998).

Berger, J.M., Gamblin, S.J., Harrison, S.C., and Wang, J.C., Structure and mechanism of DNA topoisomerase II, *Nature*

379, 225–232 (1996); Morais Cabral, J.H., Jackson, A.P., Smith, C.V., Shikotra, N., Maxwell, A., and Liddington, R.C., Crystal structure of the breakage–reunion domain of DNA gyrase, *Nature* **388,** 903–906 (1997); *and* Fass, D., Bogden, C.E., and Berger, J.M., Quaternary changes in topoisomerase II may direct orthogonal movement of two DNA strands, *Nature Struct. Biol.* **6,** 322–326 (1999).

Champoux, J.J., DNA topoisomerases: Structure, function, and mechanism, *Annu. Rev. Biochem.* **70,** 369–413 (2001).

Changela, A., DiGate, R., and Mondragón, A., Crystal structure of a complex of a type IA DNA topoisomerase with a single-stranded DNA, *Nature* **411,** 1077–1081 (2001); Mondragón, A. and DiGate, R., The structure of *Escherichia coli* DNA topoisomerase III, *Structure* **7,** 1373–1383 (1999); *and* Lima, C.D., Wang, J.C., and Mondragón, A., Three-dimensional structure of the 67K N-terminal fragment of *E. coli* DNA topoisomerase I, *Nature* **367,** 138–146 (1994).

Froelich-Ammon, S.J. and Osheroff, N., Topoisomerase poisons: Harnessing the dark side of enzyme mechanism, *J. Biol. Chem.* **270,** 21429–21432 (1995).

Horton, N.C. and Finzel, B.C., The structure of an RNA/DNA hybrid: A substrate of the ribonuclease activity of HIV-1 reverse transcriptase, *J. Mol. Biol.* **264,** 521–533 (1996).

Kanaar, R. and Cozarelli, N.R., Roles of supercoiled DNA structure in DNA transactions, *Curr. Opin. Struct. Biol.* **2,** 369–379 (1992).

Lebowitz, J., Through the looking glass: The discovery of supercoiled DNA, *Trends Biochem. Sci.* **15,** 202–207 (1990). [An informative eyewitness account of how DNA supercoiling was discovered.]

Li, T.-K. and Liu, L.F., Tumor cell death induced by topoisomerase-targeting drugs, *Annu. Rev. Pharmacol. Toxicol.* **41,** 53–77 (2001).

Maxwell, A., DNA gyrase as a drug target, *Biochem. Soc. Trans.* **27,** 48–53 (1999).

Redinbo, M.R., Stewart, L., Kuhn, P., Champoux, J.J., and Hol, W.G.J., Crystal structures of human topoisomerase I in covalent and noncovalent complexes with DNA, *Science* **279,** 1504–1513 (1998); Stewart, L., Redinbo, M.R., Qiu, X., Hol, W.G.J., and Champoux, J.J., A model for the mechanism of human topoisomerase I, *Science* **279,** 1534–1541 (1998); *and* Redinbo, M.R., Champoux, J.J., and Hol, W.G.J., Structural insights into the function of type IB topoisomerases, *Curr. Opin. Struct. Biol.* **9,** 29–36 (1999).

Wang, J.C., DNA topoisomerases, *Annu. Rev. Biochem.* **65,** 635–692 (1996).

Wang, J.C., Moving one DNA double helix through another by a type II DNA topoisomerase: The story of a simple molecular machine. *Q. Rev. Biophys.* **31,** 107–144 (1998).

Wang, J.C., Cellular roles of DNA topoisomerases: a molecular perspective, *Nature Rev. Mol. Cell Biol.* **3,** 430–440 (2002).

Wigley, D.B., Davies, G.J., Dodson, E.J., Maxwell, A., and Dodson, G., Crystal structure of an N-terminal fragment of DNA gyrase B, *Nature* **351,** 624–629 (1991).

◼ PROBLEMS

1. A · T base pairs in DNA exhibit greater variability in their propeller twisting than do G · C base pairs. Suggest the structural basis of this phenomenon.

*2. At Na$^+$ concentrations >5M, the T_m of DNA decreases with increasing [Na$^+$]. Explain this behavior. (*Hint:* Consider the solvation requirements of Na$^+$.)

*3. Why are the most commonly observed conformations of the ribose ring those in which either atom C2′ or atom C3′ is out of the plane of the other four ring atoms? (*Hint:* In puckering a planar ring such that one atom is out of the plane of the other four, the substituents about the bond opposite the out-of-plane atom remain eclipsed. This is best observed with a ball-and-stick model.)

4. Polyoma virus DNA can be separated by sedimentation at neutral pH into three components that have sedimentation coefficients of 20, 16, and 14.5S and that are known as Types I, II, and III DNAs, respectively. These DNAs all have identical base compositions and molecular masses. In 0.15*M* NaCl, both Types II and III DNA have melting curves of normal cooperativity and a T_m of 88°C. Type I DNA, however, exhibits a very broad melting curve and a T_m of 107°C. At pH 13, Types I and III DNAs have sedimentation coefficients of 53 and 16S, respectively, and Type II separates into two components with sedimentation coefficients of 16 and 18S. How do Types I, II, and III DNAs differ from one another? Explain their different physical properties.

5. When the helix axis of a closed circular duplex DNA of 2340 bp is constrained to lie in a plane, the DNA has a twist (T) of 212. When released, the DNA takes up its normal twist of 10.4 bp per turn. Indicate the values of the linking number (L), writhing number (W), and twist for both the constrained and unconstrained conformational states of this DNA circle. What is the superhelix density, σ, of both the constrained and unconstrained DNA circles?

6. A closed circular duplex DNA has a 100-bp segment of alternating C and G residues. On transfer to a solution containing a high salt concentration, this segment undergoes a transition from the B conformation to the Z conformation. What is the accompanying change in its linking number, writhing number, and twist?

7. You have discovered an enzyme secreted by a particularly virulent bacterium that cleaves the C2′—C3′ bond in the deoxyribose residues of duplex DNA. What is the effect of this enzyme on supercoiled DNA?

8. A bacterial chromosome consists of a protein–DNA complex in which its single DNA molecule appears to be supercoiled, as demonstrated by ethidium bromide titration. However, in contrast to the case with naked circular duplex DNA, the light single-strand nicking of chromosomal DNA does not abolish this supercoiling. What does this indicate about the structure of the bacterial chromosome, that is, how do its proteins constrain its DNA?

9. Although types IA and II topoisomerases exhibit no significant sequence similarity, it has been suggested that they are distantly related based on the similarites of certain aspects of their enzymatic mechanisms. What are these similarities?

10. Draw the mechanism of DNA strand cleavage and rejoining mediated by topoisomerase IA.

Chapter 29 also appeared in Volume 1 of *Biochemistry,* 3rd edition, by Donald Voet and Judith G. Voet. It is included in this Volume 2 as well, as part of the Expression and Transmission of Genetic Information.

Chapter 30

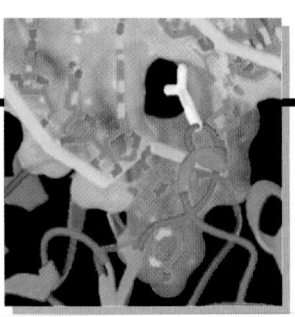

DNA Replication, Repair, and Recombination

People are DNA's way of making more DNA.

Anon.

Here we begin a three-chapter series on the basic processes of gene expression: DNA replication (this chapter), transcription (Chapter 31), and translation (Chapter 32). These processes have been outlined in Section 5-4. We shall now discuss them in greater depth with an emphasis on how we have come to know what we know.

1 ■ DNA REPLICATION: AN OVERVIEW

Watson and Crick's seminal paper describing the DNA double helix ended with the statement: "It has not escaped our notice that the specific pairing we have postulated immediately suggests a possible copying mechanism for the genetic material." In a succeeding paper they expanded on this rather cryptic remark by pointing out that a DNA strand could act as a template to direct the synthesis of its complementary strand. Although Meselson and Stahl demonstrated, in 1958, that DNA is, in fact, semiconservatively replicated (Section 5-3B), it was not until some 20 years later that the mechanism of DNA replication in prokaryotes was understood in reasonable detail. This is because, as we shall see in this chapter, the DNA replication process rivals translation in its complexity but is mediated by often loosely associated protein assemblies that are present in only a few copies per cell. *The surprising intricacy of DNA replication compared to the chemically similar transcription process (Section 31-2) arises from the need for extreme accuracy in DNA replication so as to preserve the integrity of the genome from generation to generation.*

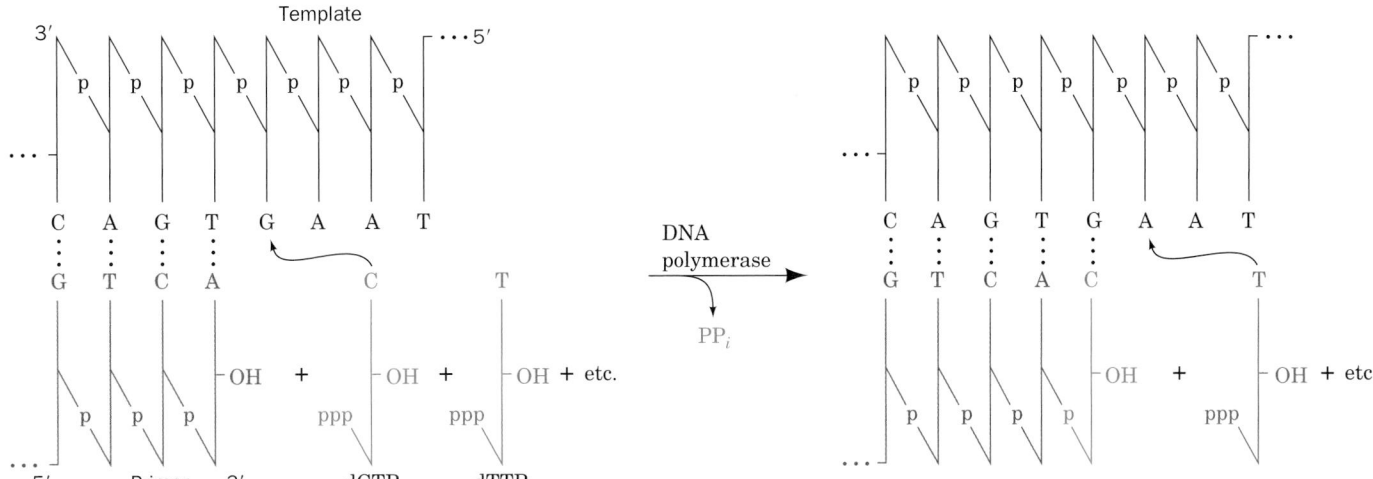

FIGURE 30-1 Action of DNA polymerase. DNA polymerases assemble incoming deoxynucleoside triphosphates on single-stranded DNA templates such that the growing strand is elongated in its 5′ → 3′ direction.

A. *Replication Forks*

*DNA is replicated by enzymes known as **DNA-directed DNA polymerases** or simply **DNA polymerases.*** These enzymes utilize single-stranded DNA as templates on which to catalyze the synthesis of the complementary strand from the appropriate deoxynucleoside triphosphates (Fig. 30-1). The incoming nucleotides are selected by their ability to form Watson–Crick base pairs with the template DNA so that the newly synthesized DNA strand forms a double helix with the template strand. *Nearly all known DNA polymerases can only add a nucleotide donated by a nucleoside triphosphate to the free 3′-OH group of a base paired polynucleotide so that DNA chains are extended only in the 5′ → 3′ direction.* DNA polymerases are discussed further in Sections 30-2A, 30-2B, and 30-4B.

a. Duplex DNA Replicates Semiconservatively at Replication Forks

John Cairns obtained the earliest indications of how chromosomes replicate through the autoradiography of replicating DNA. Autoradiograms of circular chromo-

somes grown in a medium containing [^{3}H]thymidine show the presence of replication "eyes" or "bubbles" (Fig. 30-2). These so-called **θ structures** (after their resemblance to the Greek letter theta) indicate that *double-stranded DNA (dsDNA) replicates by the progressive separation of its two parental strands accompanied by the synthesis of their complementary strands to yield two semiconservatively replicated duplex daughter strands (Fig. 30-3).* DNA replication involving θ structures is known as **θ replication.**

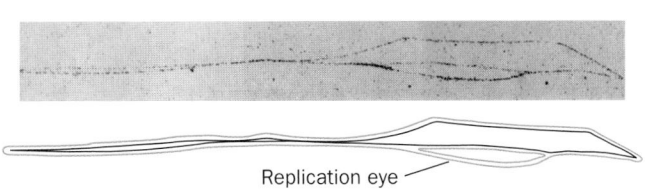

FIGURE 30-2 Autoradiogram and its interpretive drawing of a replicating *E. coli* chromosome. The bacterium had been grown for somewhat more than one generation in a medium containing [^{3}H]thymidine, thereby labeling the subsequently synthesized DNA so that it appears as a line of dark grains in the photographic emulsion (*red lines in the interpretive drawing*). The size of the replication eye indicates that the circular chromosome is about one-sixth duplicated in the present round of replication. [Courtesy of John Cairns, Cold Spring Harbor Laboratory.]

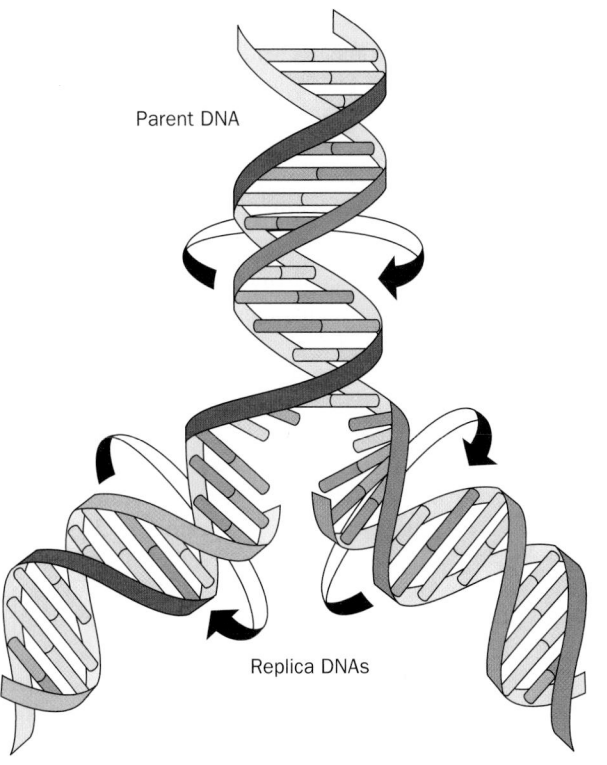

FIGURE 30-3 Replication of DNA.

A branch point in a replication eye at which DNA synthesis occurs is called a **replication fork.** A replication bubble may contain one or two replication forks **(unidirectional** or **bidirectional replication).** Autoradiographic studies have demonstrated that θ replication is almost always bidirectional (Fig. 30-4). Moreover, such experiments, together with genetic evidence, have established that prokaryotic and bacteriophage DNAs have but one **replication origin** (point where DNA synthesis is initiated).

B. *Role of DNA Gyrase*

The requirement that the parent DNA unwind at the replication fork (Fig. 30-3) presents a formidable topological obstacle. For instance, *E. coli* DNA is replicated at a rate of ~1000 nucleotides/s. If its 1300-μm-long chromosome were linear, it would have to flail around within the confines of a 3-μm-long *E. coli* cell at ~100 revolutions/s (recall that B-DNA has ~10 bp per turn). But since the *E. coli* chromosome is, in fact, circular, even this could not occur. Rather, the DNA molecule would accumulate +100 supercoils/s (see Section 29-3A for a discussion of supercoiling) until it became too tightly coiled to permit further unwinding. Naturally occurring DNA's negative supercoiling promotes DNA unwinding but only to the extent of ~5% of its duplex turns (recall that naturally occurring DNAs are typically underwound by one supercoil per ~20 duplex turns; Section 29-3B). In prokaryotes, however, negative supercoils may be introduced into DNA through the action of a Type II topoisomerase (DNA gyrase; Section 29-3C) at the expense of ATP hydrolysis. This process is essential for prokaryotic DNA replication as is demonstrated by the observation that DNA gyrase inhibitors, such as novobiocin, arrest DNA replication except in mutants whose DNA gyrase does not bind these antibiotics.

C. *Semidiscontinuous Replication*

The low-resolution images provided by autoradiograms such as Figs. 30-2 and 30-4*b* suggest that dsDNA's two antiparallel strands are simultaneously replicated at an advancing replication fork. Yet, all known DNA polymerases can only extend DNA strands in the $5' \rightarrow 3'$ direction. How, then, does DNA polymerase copy the parent strand that extends in the $5' \rightarrow 3'$ direction past the replication fork? This question was answered in 1968 by Reiji Okazaki through the following experiments. If a growing *E. coli* culture is pulse-labeled for 30 s with [³H]thymidine, much of the radioactive and hence newly synthesized DNA has a sedimentation coefficient in alkali of 7S to 11S. These so-called **Okazaki fragments** evidently consist of only 1000 to 2000 nucleotides (**nt;** 100–200 nt in eukaryotes). If, however, following the 30 s [³H]thymidine pulse, the *E. coli* are transferred to an unlabeled medium (a **pulse–chase** experiment), the resulting radioactively labeled DNA sediments at a rate that increases with the

(a)

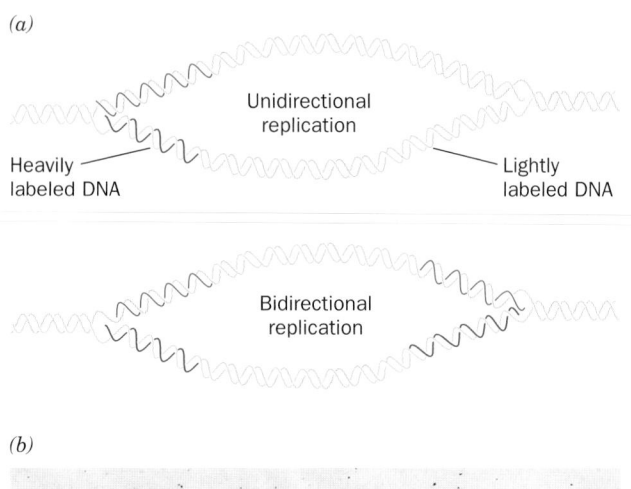

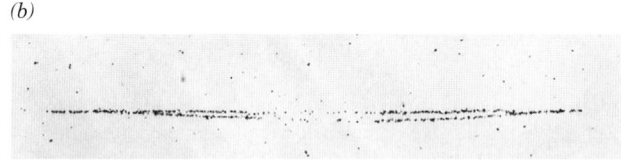

(b)

FIGURE 30-4   Autoradiographic differentiation of unidirectional and bidirectional θ replication of DNA. (*a*) An organism is grown for several generations in a medium that is lightly labeled with [³H]thymidine so that all of its DNA will be visible in an autoradiogram. A large amount of [³H]thymidine is then added to the medium for a few seconds before the DNA is isolated **(pulse labeling)** in order to label only those bases near the replication fork(s). Unidirectional DNA replication will exhibit only one heavily labeled branch point (*above*), whereas bidirectional DNA replication will exhibit two such branch points (*below*). (*b*) An autoradiogram of *E. coli* DNA so treated, demonstrating that it is bidirectionally replicated. [Courtesy of David M. Prescott, University of Colorado.]

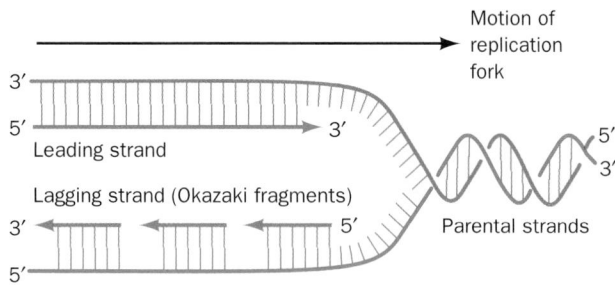

FIGURE 30-5   Semidiscontinuous DNA replication. In DNA replication, both daughter strands (*leading strand red, lagging strand blue*) are synthesized in their $5' \rightarrow 3'$ directions. The leading strand is synthesized continuously, whereas the lagging strand is synthesized discontinuously.

time that the cells had grown in the unlabeled medium. The Okazaki fragments must therefore become covalently incorporated into larger DNA molecules.

Okazaki interpreted his experimental results in terms of the **semidiscontinuous replication** model (Fig. 30-5). The

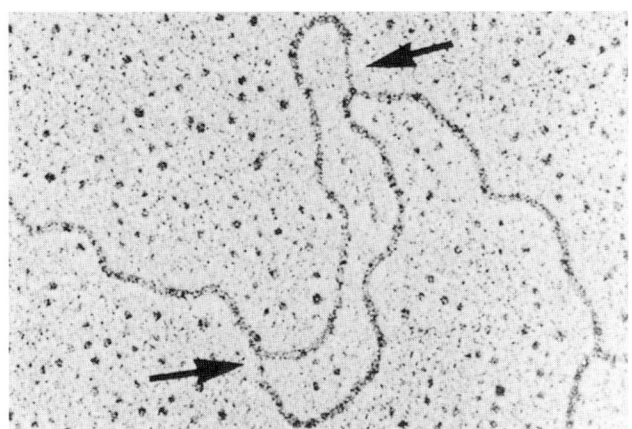

FIGURE 30-6 Electron micrograph of a replication eye in *Drosophila melanogaster* **DNA.** Note that the single-stranded regions (*arrows*) near the replication forks have the trans configuration consistent with the semidiscontinuous model of DNA replication. [From Kreigstein, H.J. and Hogness, D.S., *Proc. Natl. Acad. Sci.* **71**, 173 (1974).]

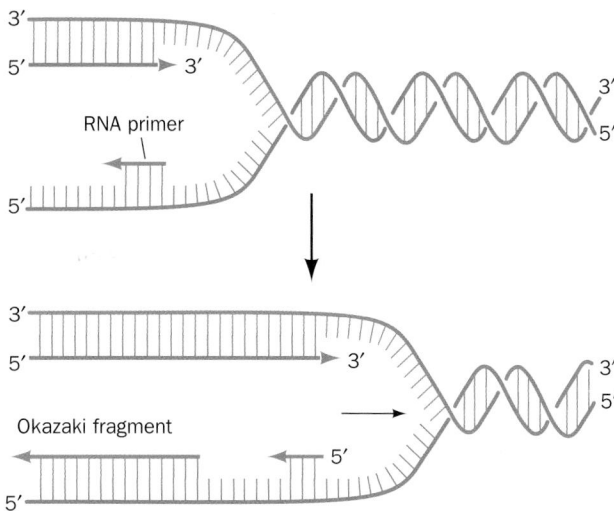

FIGURE 30-7 Priming of DNA synthesis by short RNA segments.

two parent strands are replicated in different ways. *The newly synthesized DNA strand that extends 5′ → 3′ in the direction of replication fork movement, the so-called* **leading strand,** *is essentially continuously synthesized in its 5′ → 3′ direction as the replication fork advances. The other newly synthesized strand, the* **lagging strand,** *is also synthesized in its 5′ → 3′ direction but discontinuously as Okazaki fragments. The Okazaki fragments are only covalently joined together sometime after their synthesis in a reaction catalyzed by the enzyme* **DNA ligase** *(Section 30-2D).*

The semidiscontinuous model of DNA replication is corroborated by electron micrographs of replicating DNA showing single-stranded regions on one side of the replication fork (Fig. 30-6). In bidirectionally replicating DNA, moreover, the two single-stranded regions occur, as expected, on diagonally opposite sides of the replication bubble.

D. *RNA Primers*

DNA polymerases' all but universal requirement for a free 3′-OH group to extend a DNA chain poses a question that was emphasized by the establishment of the semidiscontinuous model of DNA replication: How is DNA synthesis initiated? Careful analysis of Okazaki fragments revealed that *their 5′ ends consist of RNA segments of 1 to 60 nt (a length that is species dependent) that are complementary to the template DNA chain* (Fig. 30-7). E. coli has two enzymes that can catalyze the formation of these **RNA primers: RNA polymerase,** the ~459-kD multisubunit enzyme that mediates transcription (Section 31-2), and the much smaller **primase** (60 kD), the monomeric product of the *dnaG* gene.

Primase is insensitive to the RNA polymerase inhibitor **rifampicin** (Section 31-2C). The observation that rifampicin inhibits only leading strand synthesis therefore indicates that *primase initiates the Okazaki fragment primers.* The initiation of leading strand synthesis in *E. coli,* a much rarer event than that of Okazaki fragments, can be mediated *in vitro* by either RNA polymerase or primase alone but is greatly stimulated when both enzymes are present. It is therefore thought that these enzymes act synergistically *in vivo* to prime leading strand synthesis.

Mature DNA does not contain RNA. The RNA primers are eventually removed and the resulting single-strand gaps are filled in with DNA by a mechanism described in Section 30-2A.

2 ■ ENZYMES OF REPLICATION

DNA replication is a complex process involving a great variety of enzymes. It requires, to list only its major actors in their order of appearance: (1) DNA topoisomerases, (2) enzymes known as helicases that separate the DNA strands at the replication fork, (3) proteins that prevent them from reannealing before they are replicated, (4) enzymes that synthesize RNA primers, (5) a DNA polymerase, (6) an enzyme to remove the RNA primers, and (7) an enzyme to covalently link successive Okazaki fragments. In this section, we describe the properties and functions of many of these proteins.

A. *DNA Polymerase I*

In 1957, Arthur Kornberg reported that he had discovered an enzyme that catalyzes the synthesis of DNA in extracts of *E. coli* through its ability to incorporate the radioactive label from [¹⁴C]thymidine triphosphate into DNA. This enzyme, which has since become known as **DNA polymerase I** or **Pol I,** consists of a monomeric 928-residue polypeptide.

Pol I couples deoxynucleoside triphosphates on DNA templates (Fig. 30-1) in a reaction that occurs through the nucleophilic attack of the growing DNA chain's 3'-OH group on the α-phosphoryl of an incoming nucleoside triphosphate. The reaction is driven by the resulting elimination of PPᵢ and its subsequent hydrolysis by inorganic pyrophosphatase. The overall reaction resembles that catalyzed by RNA polymerase (Fig. 5-23) but differs from it by the strict requirement that the incoming nucleoside be linked to a free 3'-OH group of a polynucleoside that is base paired to the template (RNA polymerase initiates transcription by linking together two ribonucleoside triphosphates on a DNA template; Section 31-2C). The complementarity between the product DNA and the template was at first inferred through base composition and hybridization studies but was eventually directly established by base sequence determinations. The error rate of Pol I in copying the template is extremely low, as was first demonstrated by its *in vitro* replication of the 5386-nt DNA from bacteriophage **φX174** to yield fully infective phage DNA. In fact, its measured error rate is around one wrong base per 10 million.

Pol I is said to be **processive** in that it catalyzes a series of successive polymerization steps, typically 20 or more, without releasing the template. Pol I can, of course, work in reverse by degrading DNA through pyrophosphorolysis. This reverse reaction, however, probably has no physiological significance because of the low *in vivo* concentration of PPᵢ resulting from the action of inorganic pyrophosphatase.

a. Pol I Recognizes the Incoming dNTP According to the Shape of the Base Pair It Forms with the Template DNA

The specificity of Pol I for an incoming base arises from the requirement that it form a Watson–Crick base pair with the template rather than direct recognition of the incoming base (recall that the four base pairs, A · T, T · A, G · C, and C · G, have nearly identical shapes; Fig. 5-12). Thus, as Eric Kool demonstrated, when the "base" **2,4-difluorotoluene (F),**

2,4-Difluorotoluene base (F)   **Thymine (T)**

which is isosteric with (has the same shape as) thymine but does not accept hydrogen bonds, is synthetically inserted

into a template DNA, Pol I incorporates A opposite the F with a similar rate of mismatches as it incorporates A opposite T. Likewise, dFTP is incorporated opposite template A with a similar fidelity as is dTTP. Yet the incorporation of F opposite an A in DNA destabilizes the double helix by 15 kJ/mol relative to T opposite this A. Evidently, *Pol I selects an incoming dNTP largely according to its ability to form a Watson–Crick-shaped pair with the template base but with little regard for its hydrogen bonding properties.* Indeed, the NMR structure of a 12-bp DNA containing a centrally located F opposite an A reveals that it assumes a B-DNA conformation in which the F–A pair closely resembles a T · A base pair in the same position of an otherwise identical DNA.

b. Pol I Can Edit Its Mistakes

In addition to its polymerase activity, Pol I has two independent hydrolytic activities:

1. It can act as a 3' → 5' exonuclease.
2. It can act as a 5' → 3' exonuclease.

The 3' → 5' exonuclease reaction differs chemically from the pyrophosphorolysis reaction (the reverse of the polymerase reaction) only in that H₂O rather than PPᵢ is the nucleotide acceptor. Kinetic and crystallographic studies, however, indicate that these two catalytic activities occupy separate active sites (see below). The 3' → 5' exonuclease function is activated by an unpaired 3'-terminal nucleotide with a free OH group. If Pol I erroneously incorporates a wrong (unpaired) nucleotide at the end of a growing DNA chain, the polymerase activity is inhibited and the 3' → 5' exonuclease excises the offending nucleotide (Fig. 5-36). The polymerase activity then resumes DNA replication. *Pol I therefore has the ability to **proofread** or **edit** a DNA chain as it is synthesized so as to correct its mistakes.* This explains the great fidelity of DNA replication by Pol I: The overall fraction of bases that the enzyme misincorporates, ~10⁻⁷, is the product of the fraction of bases that its polymerase activity misincorporates and the fraction of misincorporated bases that its 3' → 5' exonuclease activity fails to excise. The price of this high fidelity is that ~3% of correctly incorporated nucleotides are also excised.

The Pol I 5' → 3' exonuclease binds to dsDNA at single-strand nicks with little regard to the character of the 5' nucleotide (5'-OH or phosphate group; base paired or not). It cleaves the DNA in a base paired region beyond the nick such that the DNA is excised as either mononucleotides or oligonucleotides of up to 10 residues (Fig. 5-33). In contrast, the 3' → 5' exonuclease removes only unpaired mononucleotides with 3'-OH groups.

c. Pol I's Polymerase and Two Exonuclease Functions Each Occupy Separate Active Sites

The 5' → 3' exonuclease activity of Pol I is independent of both its 3' → 5' exonuclease and its polymerase activi-

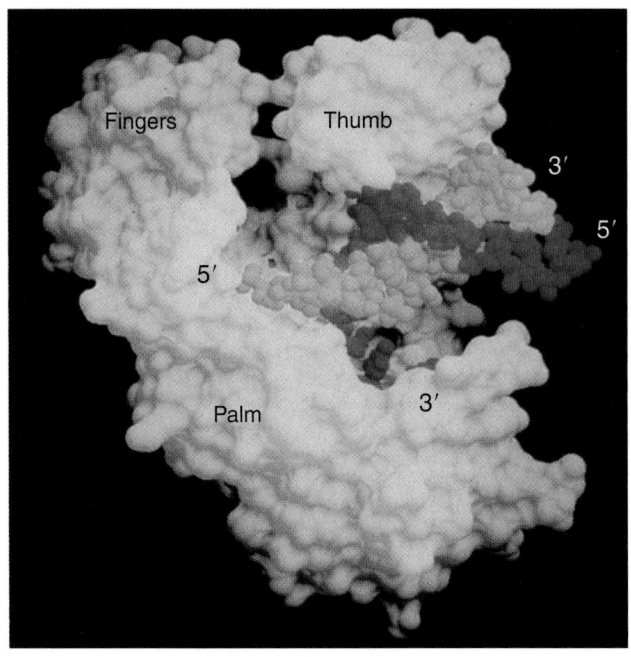

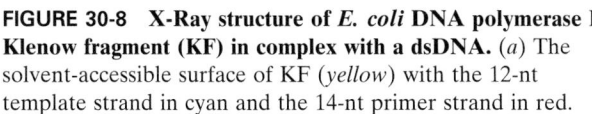

(a)

FIGURE 30-8   X-Ray structure of *E. coli* DNA polymerase I Klenow fragment (KF) in complex with a dsDNA. (*a*) The solvent-accessible surface of KF (*yellow*) with the 12-nt template strand in cyan and the 14-nt primer strand in red.

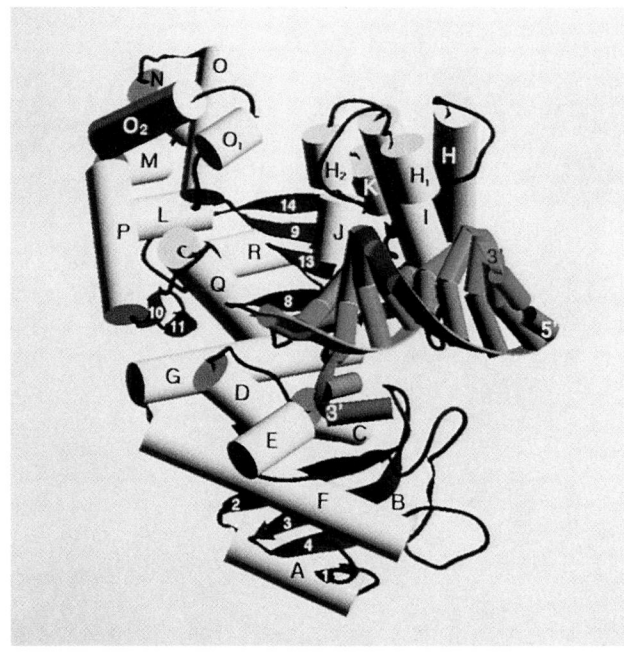

(b)

(*b*) A tube-and-arrow representation of the complex in the same orientation as Part *a* in which the template strand is blue and the primer strand is purple. [Courtesy of Thomas Steitz, Yale University. PDBid 1KLN.] *See the Interactive Exercises.*

ties. In fact, as we saw in Section 7-2A, proteases such as subtilisin or trypsin cleave Pol I into two fragments: a larger C-terminal or **Klenow fragment (KF;** residues 324–928), which contains both the polymerase and the 3′ → 5′ exonuclease activities; and a smaller N-terminal fragment (residues 1–323), which contains the 5′ → 3′ exonuclease activity. Thus Pol I contains three active sites on a single polypeptide chain.

d. The X-Ray Structure of Klenow Fragment Indicates How It Binds DNA

The X-ray structure of KF, determined by Thomas Steitz, reveals that this protein consists of two domains (Fig. 30-8). The smaller domain (residues 324–517) contains the 3′ → 5′ exonuclease site, as was demonstrated by the absence of this function but not polymerase activity in a genetically engineered Klenow fragment mutant that lacks the divalent metal ion–binding sites known to be essential for 3′ → 5′ exonuclease activity but which otherwise has a normal structure. The larger domain (residues 521–928; helix G and beyond in Fig. 30-8*b*) contains the polymerase active site at the bottom of a prominent cleft, a surprisingly large distance (~25 Å) from the 3′ → 5′ exonuclease site. The cleft, which is lined with positively

charged residues, has the appropriate size (~22 Å wide by ~30 Å deep) and shape to bind a B-DNA molecule in a manner resembling a right hand grasping a rod (in which the "thumb" consists of helices H–I, the "fingers" consist of helices L–P, and the remainder of the larger domain, the "palm," includes a 6-stranded antiparallel β sheet that forms the floor of the cleft and contains the polymerase function's active site residues). Indeed, the active sites of all DNA and RNA polymerases of known structure are located at the bottoms of similarly shaped clefts (Sections 30-4B, 30-4C, and 31-2A).

e. DNA Polymerase Distinguishes Watson–Crick Base Pairs via Sequence-Independent Interactions That Induce Domain Movements

The C-terminal domain of the thermostable *Thermus aquaticus (Taq)* DNA polymerase I **(Klentaq1)** is 50% identical in sequence and closely similar in structure to the large domain of Klenow fragment, although Klentaq1 lacks a functional 3′ → 5′ exonuclease site. Gabriel Waksman crystallized Klentaq1 in complex with an 11-bp DNA that has a GGAAA-5′ overhang at the 5′ end of its template strand, and the crystals were incubated with 2′,3′-dideoxy-CTP (ddCTP; which lacks a 3′-OH group). The X-ray

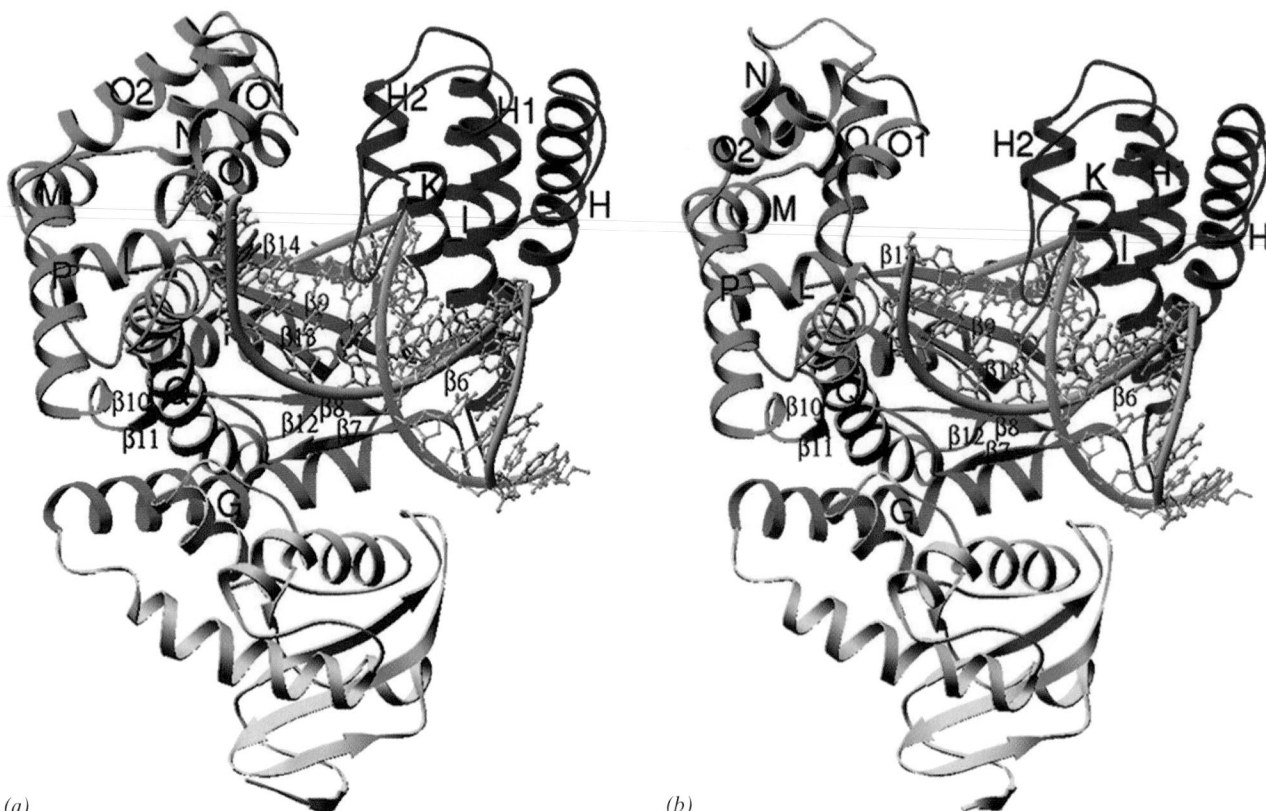

(a)

(b)

FIGURE 30-9   X-Ray structure of Klentaq1 in complex with DNA and ddCTP. (*a*) The closed conformation. (*b*) The open conformation. The protein, which is viewed similarly to that in Fig. 30-8, is represented in ribbon form with its N-terminal, palm, fingers, and thumb domains colored yellow, magenta, green, and dark blue, respectively, and with the O helix in the fingers domain red. The DNA is shown in stick form and its sugar–phosphate backbone is also represented in tube form with the template strand cyan and the primer strand silver. In Part *a*, the bound ddCTP is shown in stick form in black, and its two bound metal ions are represented by orange spheres. [Courtesy of Gabriel Waksman, Washington University School of Medicine. PDBids 3KTQ and 2KTQ.]

structure of these crystals (Fig. 30-9*a*) reveals that a ddC residue had been covalently linked to the 3′ end of the primer and formed a Watson–Crick pair with the template overhang's 3′ G. Moreover, a ddCTP molecule (to which the primer's new 3′-terminal ddC residue is incapable of forming a covalent bond) occupies the enzyme's active site where it forms a Watson–Crick pair with the templates next G. Clearly, Klentaq1 retains its catalytic activity in this crystal.

A DNA polymerase must distinguish correctly paired bases from mismatches and yet do so via sequence-independent interactions with the incoming dNTP. The foregoing X-ray structure reveals that this occurs through an active site pocket that is complementary in shape to Watson–Crick base pairs. This pocket is formed by the stacking of a conserved Tyr side chain on the template base, as well as by van der Waals interactions with the protein and with the preceding base pair. In addition, although the dsDNA is mainly in the B conformation, the 3 base pairs nearest the active site assume the A conformation, as has also been observed in the X-ray structures of sev-

eral other DNA polymerases in their complexes with DNA. The resulting wider and shallower minor groove (Section 29-1B) permits protein side chains to form hydrogen bonds with the otherwise inaccessible N3 atoms of the purine bases and O2 atoms of the pyrimidine bases. The positions of these hydrogen bond acceptors are sequence-independent as can be seen from an inspection of Fig. 5-12 [in contrast, the positions of the hydrogen bonding acceptors in the major groove vary with both the identity (A · T vs G · C) and the orientation (e.g., A · T vs T · A) of the base pair]. However, with a non-Watson–Crick pairing, these hydrogen bonds would be greatly distorted if not completely disrupted. The protein also makes extensive sequence-independent hydrogen bonding and van der Waals interactions with the DNA's sugar–phosphate backbone.

The above Klentaq1 · DNA · ddCTP crystals were partially depleted of ddCTP by soaking them in a stabilizing solution that lacks ddCTP. The X-ray structure of the ddCTP-depleted crystals (Fig. 30-9*b*) revealed that Klentaq1's fingers domain assumed a so-called open conformation, which differs significantly from that in the so-called closed conformation described above. In particular, the O, O_1, and O_2 helices in the closed conformation have moved via a hingelike motion in the direction of the active site relative to their positions in the open complex (Fig. 30-9*a*) so as to bury the bound ddCTP, thereby assembling the productive ternary complex. These observations are consistent with kinetic measurements on Pol I, indicating that the binding of the correct dNTP to the enzyme induces a rate-limiting conformational change that yields a tight ternary complex. It therefore appears that the enzyme rapidly samples the available dNTPs in its open conformation but only when it binds the correct dNTP in a Watson–Crick pairing with the template base does it form the catalytically competent closed conformation. The subsequent reaction steps then rapidly yield the product complex which, following a second conformational change, releases the product PP_i. Finally, the DNA is translocated in the active site, probably via a linear diffusion mechanism, so as to position it for the next reaction cycle.

The comparison of the above X-ray structures with that of Klentaq1 alone indicates that on binding DNA, the thumb domain moves to wrap around the DNA. It is likely that this conformational change is largely responsible for Pol I's processivity. In both Klentaq1 · DNA structures, neither the dsDNA nor the single-stranded DNA (**ssDNA**) passes through the cleft between the thumb and fingers domain as the shape and position of the cleft suggest. Rather, the template strand makes a sharp bend at the first unpaired base, thereby unstacking this base and positioning this ssDNA on the same side of the cleft as the dsDNA. Similar arrangements have been observed in X-ray structures of other DNA polymerases in their complexes with DNA.

f. The DNA Polymerase Catalytic Mechanism Involves Two Metal Ions

The X-ray structures of a variety of DNA polymerases suggest that they share a common catalytic mechanism for nucleotidyl transfer (Fig. 30-10). Their active sites all contain two metal ions, usually Mg^{2+}, that are liganded by two invariant Asp side chains in the palm domain. Metal ion B in Fig. 30-10 is liganded by all three phosphate groups of the bound dNTP, whereas metal ion A bridges the α-phosphate group of this dNTP and the primer's 3'-OH group. Metal ion A presumably activates the primer's 3'-OH group for an in-line nucleophilic attack on the α-phosphate group (Fig. 16-6*b*), whereas metal ion B functions to orient its bound triphosphate group and to electrostatically shield their negative charges as well as the additional negative charge on the transition state leading to the release of the PP_i ion (Section 16-2B).

g. Editing Complexes Contain the Primer Strand in the 3' → 5' Exonuclease Site

Steitz cocrystallized KF with a 12-nt DNA "template" strand (5'-TGCCTCGCGGCC-3'), a 7-nt "primer" strand

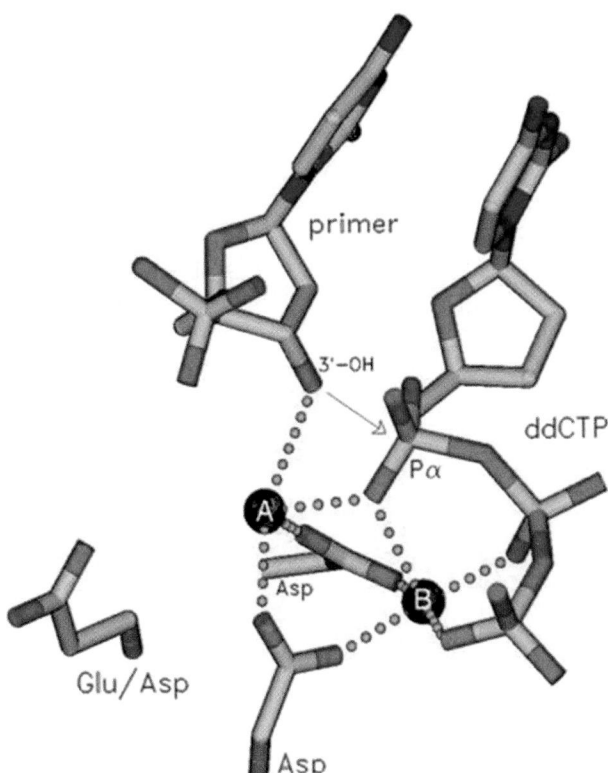

FIGURE 30-10 Schematic diagram for the nucleotidyl transferase mechanism of DNA polymerases. A and B represent enzyme-bound metal ions that usually are Mg^{2+}. Atoms are colored according to atom type (C gray, N blue, O red, and P yellow) and metal ion coordination is represented by green dotted lines. Metal ion A activates the primer's 3'-OH group for in-line nucleophilic attack on the incoming dNTP's α-phosphate group (*arrow*), whereas metal ion B acts to orient and electrostatically stabilize the negatively charged triphosphate group. [Courtesy of Tom Ellenberger, Harvard Medical School.]

(3'-GCGCCGG-5') that is complementary to the 3' end of the template strand, and **2',3'-epoxy-ATP**,

$$^-O—\overset{\overset{\textstyle O}{\|}}{\underset{\underset{\textstyle O^-}{|}}{P}}—O—\overset{\overset{\textstyle O}{\|}}{\underset{\underset{\textstyle O^-}{|}}{P}}—O—\overset{\overset{\textstyle O}{\|}}{\underset{\underset{\textstyle O^-}{|}}{P}}—O—CH_2$$

2',3'-Epoxy-ATP

which promotes the tight binding of DNA to the polymerase site. The X-ray structure of the resulting complex (Fig. 30-8) shows that the primer strand base pairs, as expected, to the 3' end of the template strand to form a distorted segment of B-DNA, and that the polymerase has apparently appended an epoxy-A residue to the primer's 3' end (where it base pairs to a T on the template strand;

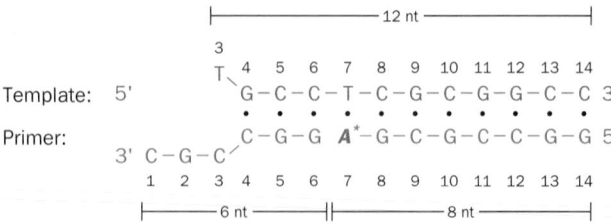

FIGURE 30-11   Probable sequence of the double-stranded DNA seen in the X-ray structure of KF. *A** represents a 2′,3′-epoxy-A residue that KF has appended to the 3′ end of the 7-nt DNA (*red*) with which the KF crystal (Fig. 30-8) was incubated (together with the 12-nt DNA; *blue*). The two-part primer strand's 3′ segment appears to be a second copy of the 7-nt DNA with its 3′ nucleotide removed.

Fig. 30-11). In addition, a second primer strand, whose 3′ G residue has apparently been removed, continues the 3′ end of the primer (after a break in the sugar–phosphate chain) by base pairing, via its 5′-terminal three nucleotides, to the template strand. Thus, this complex contains an 11-bp dsDNA, which has a single nucleotide overhang at the 5′ end of its template strand and a 3-nt overhang at the 3′ end of its primer strand. The 3′-terminal nucleotide of the primer strand (the last one that an active polymerase would have added) is bound at the 3′ → 5′ exonuclease active site, whereas the 5′-terminal nucleotide of the template strand is bound at the entrance of the polymerase active site. Evidently, the KF has bound the DNA in an "editing" complex rather than in the polymerase cleft.

In *E. coli* Pol I, how does the 3′ end of the primer strand transfer between the polymerase active site and the 3′ → 5′ exonuclease active site? This appears to occur through the competition of these sites for the 3′ end of the primer strand, which base pairs to form dsDNA in the polymerase site and binds as a single strand to the exonuclease site. Thus, the formation of a Watson–Crick base pair would tend to bind the primer strand in the polymerase site preparatory for the next round of chain extension, whereas a mismatched base pair would promote the binding of the primer strand as a single strand to the exonuclease site for the subsequent excision of the offending nucleotide. Comparison of the editing complex with those of the Klentaq1 · DNA complexes suggests that the transfer of the primer strand from the polymerase to the editing sites of KF requires that the dsDNA translocate backward (toward the 3′ end of the template strand) by several angstroms along the helix axis.

h. Pol I Functions Physiologically to Repair DNA

For some 13 years after Pol I's discovery, it was generally assumed that this enzyme was *E. coli*'s DNA replicase because no other DNA polymerase activity had been detected in *E. coli*. This assumption was made untenable by Cairns and Paula DeLucia's isolation, in 1969, of a mutant *E. coli* whose extracts exhibit <1% of the normal Pol I activity (although it has nearly normal levels of the 5′ → 3′ exonuclease activity) but which nevertheless reproduce at

the normal rate. This mutant strain, however, is highly susceptible to the damaging effects of UV radiation and **chemical mutagens** (substances that chemically induce mutations; Section 32-1A). *Pol I evidently plays a central role in the repair of damaged (chemically altered) DNA.*

Damaged DNA, as we discuss in Section 30-5, is detected by a variety of DNA repair systems. Many of them endonucleolytically cleave the damaged DNA on the 5′ side of the lesion, thereby activating Pol I's 5′ → 3′ exonuclease. While excising this damaged DNA, Pol I simultaneously fills in the resulting single-strand gap through its polymerase activity. In fact, its 5′ → 3′ exonuclease activity increases 10-fold when the polymerase function is active. Perhaps the simultaneous excision and polymerization activities of Pol I protect DNA from the action of cellular nucleases that would further damage the otherwise gapped DNA.

i. Pol I Catalyzes Nick Translation

Pol I's combined 5′ → 3′ exonuclease and polymerase activities can replace the nucleotides on the 5′ side of a single-strand nick on otherwise undamaged DNA. These reactions, in effect, translate (move) the nick toward the DNA strand's 3′ end without otherwise changing the molecule (Fig. 30-12). This **nick translation** process, in the presence of labeled deoxynucleoside triphosphates, is synthetically employed to prepare highly radioactive DNA (the required nicks may be generated by treating the DNA with a small amount of pancreatic **DNase I**).

j. Pol I's 5′ → 3′ Exonuclease Functions Physiologically to Excise RNA Primers

Pol I's 5′ → 3′ exonuclease also removes the RNA primers at the 5′ ends of newly synthesized DNA while its DNA polymerase activity fills in the resulting gaps (Fig. 5-34). The importance of this function was demonstrated by the isolation of temperature-sensitive *E. coli* mutants that neither are viable nor exhibit any 5′ → 3′ exonuclease activity at the restrictive temperature of ~43°C (the low level of polymerase activity in the Pol I mutant isolated by Cairns and DeLucia is apparently sufficient to carry out this essential gap-filling process during chromosome replication). Thus Pol I has an indispensable role in *E. coli* DNA replication although a different one than was first supposed.

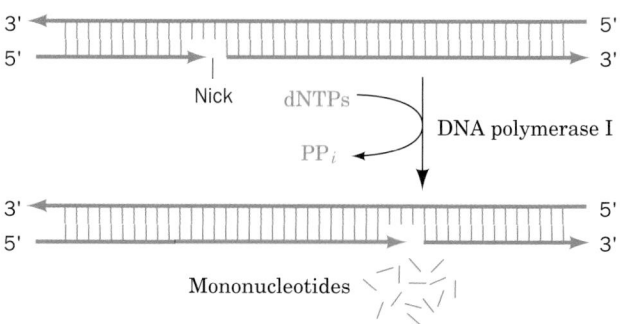

FIGURE 30-12   Nick translation as catalyzed by Pol I.

B. *DNA Polymerase III*

The discovery of normally growing *E. coli* mutants that have very little Pol I activity stimulated the search for an additional DNA polymerizing activity. This effort was rewarded by the discovery of two more enzymes, designated, in the order they were discovered, **DNA polymerase II (Pol II)** and **DNA polymerase III (Pol III).** The properties of these enzymes are compared with that of Pol I in Table 30-1. Pol II and Pol III had not previously been detected because their combined activities in the assays used are normally <5% that of Pol I.

A mutant *E. coli* lacking measurable Pol II activity grows normally. However, Pol II has been implicated as a participant in repairing DNA damage via the **SOS response** (Section 30-5D), as have two additional *E. coli* enzymes that were recently discovered: **DNA polymerase IV (Pol IV)** and **DNA polymerase V (Pol V)** (Section 30-5D).

a. Pol III Is *E. coli's* DNA Replicase

The cessation of DNA replication in temperature-sensitive *polC* mutants above the restrictive (high) temperature demonstrates that *Pol III is E. coli's DNA replicase.* Its **Pol III core** has the subunit composition $\alpha\varepsilon\theta$ where α, the *polC* gene product (Table 30-2), contains the polymerase function. The catalytic properties of Pol III core resemble those of Pol I (Table 30-1) except for Pol III core's inability to replicate primed ssDNA or nicked dsDNA. Rather, Pol III core acts *in vitro* at single-strand gaps of <100 nucleotides, a situation that probably resembles the state of DNA at the replication fork. The Pol III $3' \rightarrow 5'$ exonuclease function, which resides on the enzyme's ε subunit, is DNA's primary editor during replication; it enhances the enzyme's replication fidelity by up to 200-fold. However, the Pol III $5' \rightarrow 3'$ exonuclease acts only on single-stranded DNA, so it cannot catalyze nick translation. θ is an accessory protein that stimulates the editing function of ε.

Pol III core functions in vivo as part of a complicated and labile multisubunit enzyme, the Pol III holoenzyme, which consists of at least 10 types of subunits (Table 30-2). The latter 7 subunits in Table 30-2 act to modulate Pol III core's activity. For example, Pol III core has a processivity of 10 to 15 residues; it can only fill in short single-stranded regions of DNA. However, Pol III core is rendered processive by association with the β **subunit** in the presence of the 7-subunit γ **complex** ($\gamma\tau_2\delta\delta'\chi\psi$). Assembly of the processive enzyme is a two-stage process in which the γ complex transfers the β subunit to the primed template in an ATP-dependent reaction followed by the assembly of Pol III core with the β subunit on the DNA (Section 30-3C). The β subunit confers essentially unlimited processivity (>5000 residues) on the core enzyme even if the γ complex is subsequently removed. In fact, the β subunit is very strongly bound to the DNA, although it can freely slide along it.

b. The β Subunit Forms a Ringlike Sliding Clamp

The observation that a β subunit clamped to a cut circular DNA slides to the break and falls off suggests that the β subunit forms a closed ring around the DNA, thereby preventing its escape. The X-ray structure of the β subunit (Fig. 30-13*a*), determined by John Kuriyan, reveals that it forms a dimer of C-shaped, 366-residue monomer units which associate to form an ~80-Å-diameter doughnut-shaped structure that is equivalently known as the **sliding clamp** and the β **clamp.** The sliding clamp's central hole is ~35 Å in diameter, which is larger than the 20- and 26-Å diameters of B- and A-DNAs (recall that the hybrid helices which RNA primers make with DNA have A-DNA-like conformations; Section 29-2B). Each β subunit consists of six $\beta\alpha\beta\beta\beta$ motifs of identical topology, which associate in pairs to form three pseudo-2-fold symmetric domains of very similar structures (although with <20% sequence identity). The dimeric ring therefore has the shape of a

---

TABLE 30-1 Properties of *E. coli* DNA Polymerases

	Pol I	Pol II	Pol III
Mass (kD)	103	90	130
Molecules/cell	400	?	10–20
Turnover number[a]	600	30	9000
Structural gene	*polA*	*polB*	*polC*
Conditionally lethal mutant	+	−	+
Polymerization: $5' \rightarrow 3'$	+	+	+
Exonuclease: $3' \rightarrow 5'$	+	+	+
Exonuclease: $5' \rightarrow 3'$	+	−	−

[a]Nucleotides polymerized min^{-1} · molecule^{-1} at 37°C.

Source: Kornberg, A. and Baker, T.A., *DNA Replication* (2nd ed.), p. 167, Freeman (1992).

TABLE 30-2 Components of *E. coli* DNA Polymerase III Holoenzyme

Subunit	Mass (kD)	Structural Gene
α^a	130	*polC (dnaE)*
ε^a	27.5	*dnaQ*
θ^a	10	*holE*
τ^b	71	*dnaX*^{c}
γ^b	45.5	*dnaX*^{c}
δ^b	35	*holA*
δ'^b	33	*holB*
χ^b	15	*holC*
ψ^b	12	*holD*
β	40.6	*dnaN*

[a]Components of the Pol III core.
[b]Components of the γ complex.
[c]The γ and τ subunits are encoded by the same gene sequence; the γ subunit comprises the N-terminal end of the τ subunit.

Sources: Kornberg, A. and Baker, T.A., *DNA Replication* (2nd ed.), p. 169, Freeman (1992); *and* Baker, T.A. and Wickner, S.H., *Annu. Rev. Genet.* **26**, 450 (1992).

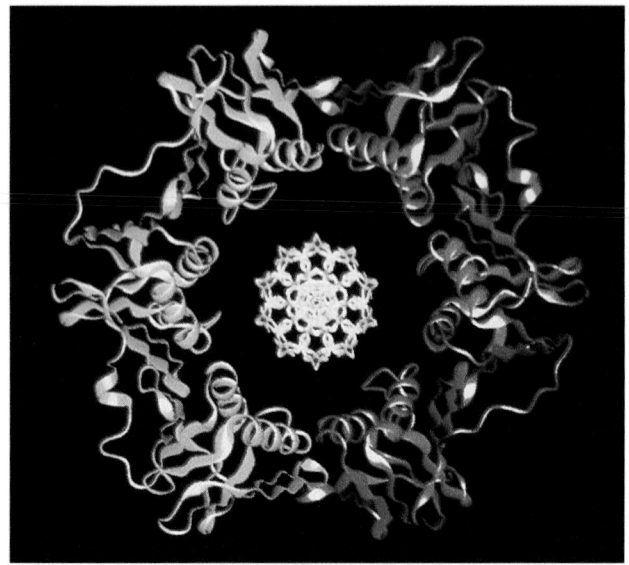

(a)

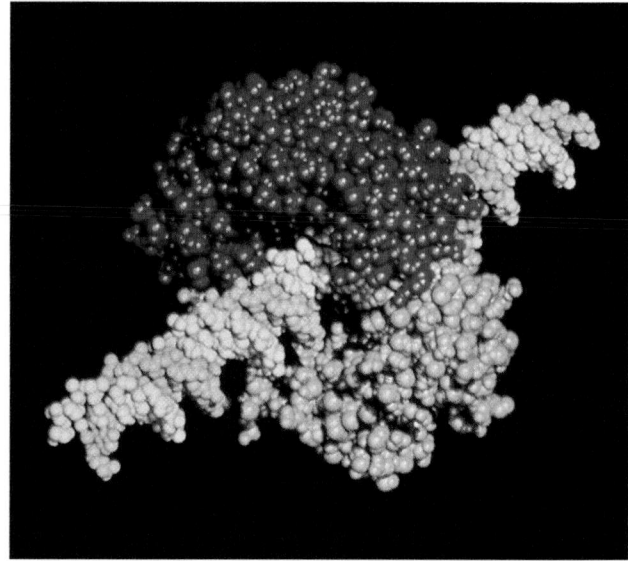

(b)

FIGURE 30-13 X-Ray structure of the β subunit of *E. coli* Pol III holoenzyme. (*a*) A ribbon drawing showing the two monomeric units of the dimeric sliding clamp in yellow and red as viewed along the homodimer's 2-fold axis. A stick model of B-DNA is placed with its helix axis coincident with the sliding clamp's twofold axis. (*b*) A space-filling model of the sliding clamp in the hypothetical complex with the B-DNA shown in Part *a*. The protein is colored as in Part *a* and the DNA is cyan. [Courtesy of John Kuriyan, The Rockefeller University. PDBid 2POL.]

6-pointed star in which the 12 helices line the central hole and the β strands associate in six β sheets that form the protein's outer surface. Electrostatic calculations indicate that the interior surface of the ring is positively charged, whereas its outer surface is negatively charged.

Model building studies in which a B-DNA helix is threaded through the sliding clamp's central hole (Fig. 30-13) indicate that the helices are all oriented such that they are perpendicular to their radially adjacent segments of sugar–phosphate backbone. These helices therefore span the major and minor grooves of the DNA rather than entering into them as do many helices that make sequence-specific interactions with dsDNA (Section 31-3C). Since A- and B-DNAs have 11 and 10 bp per turn, whereas the sliding clamp has a pseudo-12-fold symmetry, it appears that the sliding clamp is designed to minimize its associations with its threaded DNA. This presumably permits the sliding clamp to freely slide along the DNA helix. Indeed, the radius of the sliding clamp's central hole is at least 3.5 Å larger than that of DNA, so any interactions between them are likely to be attenuated by a sheath of intervening water molecules.

C. *Unwinding DNA: Helicases and Single-Strand Binding Protein*

Pol III holoenzyme, unlike Pol I, cannot unwind dsDNA. Rather, *three proteins, DnaB protein (the product of the dnaB gene; proteins may be assigned the name of the gene specifying them but in roman letters with the first letter capitalized), Rep helicase, and single-strand binding protein (SSB) (Table 30-3), work in concert to unwind the DNA be-*

fore an advancing replication fork (Fig. 30-14) in a process that is driven by ATP hydrolysis.

a. Hexameric Helicases Mechanically Separate the Strands of dsDNA by Climbing Up One Strand

Access to the genetic information encoded in a double-helical nucleic acid requires that the double helix be unwound. The proteins that do so, which are known as **helicases,** form a diverse group of enzymes that facilitate a variety of functions including DNA replication, recombination, and repair, as well as transcription termination (Section 31-2A), RNA splicing, and RNA editing (Section 31-4A). Indeed, all forms of life contain helicases, 12 varieties of which occur in *E. coli.* Helicases function by translocating along one strand of a double-helical nucleic acid so as to unwind the double helix in their path. This, of course, requires free energy, and hence helicases are driven by the hydrolysis of NTPs. Helicases have been classified in several ways: whether they translocate along their bound single strand in the $5' \rightarrow 3'$ direction or the $3' \rightarrow 5'$

TABLE 30-3 Unwinding and Binding Proteins of *E. coli* DNA Replication

Protein	Subunit Structure	Subunit Mass (kD)
DnaB protein	hexamer	50
SSB	tetramer	19
Rep protein	monomer	68
PriA protein	monomer	76

Source: Kornberg, A. and Baker, T.A., *DNA Replication* (2nd ed.), p. 366, Freeman (1992).

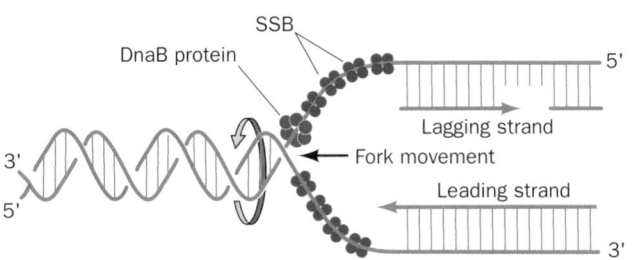

FIGURE 30-14 Unwinding of DNA by the combined action of DnaB and SSB proteins. The hexameric DnaB protein moves along the lagging strand template in the $5' \rightarrow 3'$ direction. The resulting separated DNA strands are prevented from reannealing by SSB binding.

direction, whether they function as hexameric rings or as dimers, and whether their sequences contain certain signature motifs.

E. coli **DnaB** protein, a hexameric helicase of identical 471-residue subunits, separates the strands of dsDNA by translocating along the lagging strand template in the $5' \rightarrow 3'$ direction while hydrolyzing ATP (it can also use GTP and CTP but not UTP). Electron microscopy studies reveal that DnaB forms a hexameric ring that, depending on conditions, exhibits C_3 or C_6 symmetry and which encloses an ~30-Å-diameter central channel. Similarly, the bacteriophage **T7 gene 4 helicase/primase** (bacteriophage T7 infects *E. coli*) forms a two-tiered hexagonal ring (Fig. 30-15) whose smaller N-terminal domains (residues 1–271)

contain its primase activity and whose larger C-terminal domains (residues 272–566) carry out its helicase function. T7 gene 4 helicase/primase (also called **T7 gp4;** gp for *gene product*) preferentially hydrolyzes dTTP but also hydrolyzes dATP and ADP.

The X-ray structure of DnaB has not been determined. However, Dale Wigley has determined that of the mainly C-terminal domain (residues 241–566) of T7 gene 4 helicase/primase in complex with ADPNP. This helicase, as expected, forms a hexagonal ring (Fig. 30-16), which appears to be largely held together by each subunit's N-terminal arm binding to an adjacent subunit. Two loops from each subunit that extend into the hexamer's central channel and which contain several conserved basic residues presumably form the hexamer's DNA-binding surface.

The hexagonal ring exhibits only 2-fold (C_2) rotational symmetry. If the adjacent subunits of the asymmetric half of this ring are labeled A, B, and C, subunit B is related to subunit A by a 15° rotation about an axis lying in the plane of the ring (after a 60° rotation about the 6-fold axis) and subunit C is similarly related to subunit B (a 30° rotation relative to subunit A). The DNA-binding loops thereby form a helical ramp that is approximately complementary in shape to the sugar–phosphate backbone of single-stranded DNA **(ssDNA)** when it is in the B-DNA conformation (although keep in mind that the helical ramp is discontinuous at the interface between adjacent C and A subunits). ADPNP is bound to the A and B subunits but not to the C subunits.

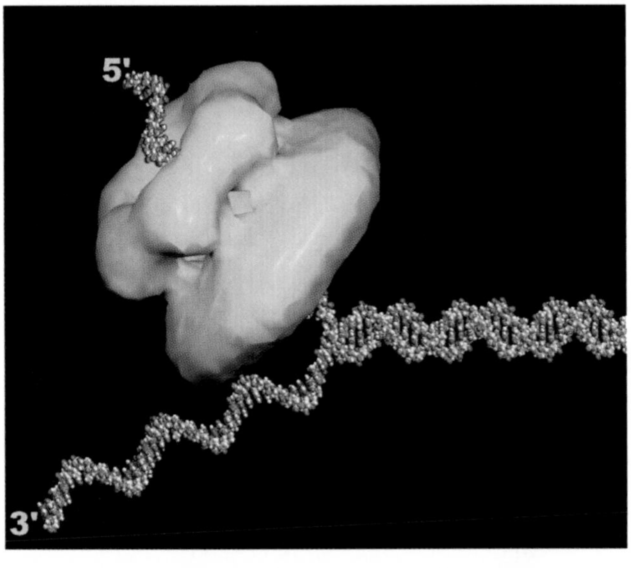

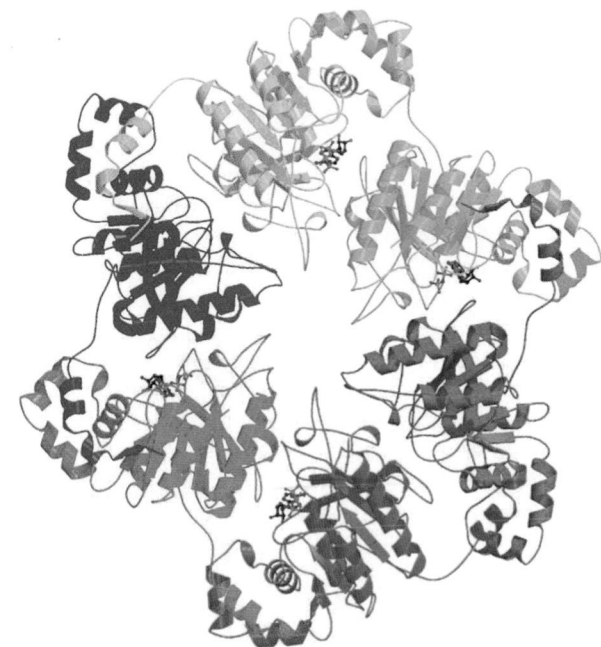

FIGURE 30-15 Electron microscopy–based image reconstruction of T7 gene 4 helicase/primase. In this two-tiered hexameric ring (*yellow*), the smaller lobe of each subunit forms the N-terminal primase domain and the larger lobe forms the C-terminal helicase domain. The protein is postulated to interact with DNA as is depicted by this model of a DNA fork consisting of a 30-bp duplex segment and two 25-nt single-stranded segments with the 5′ tail threaded through the hexameric ring. The way in which the 3′ tail interacts with the protein, if at all, is unknown. [Courtesy of S.S. Patel and K.M. Picha, University of Medicine and Dentistry of New Jersey.]

FIGURE 30-16 X-Ray structure of the helicase domain of T7 gene 4 helicase/primase. Each subunit of this cyclic hexamer is drawn in a different color. The four bound ADPNP molecules are represented in ball-and-stick form. Note that the conformations of adjacent subunits are not identical. [Courtesy of Dale Wigley, Cancer Research U.K. London Research Institute. PDBid 1E0J.]

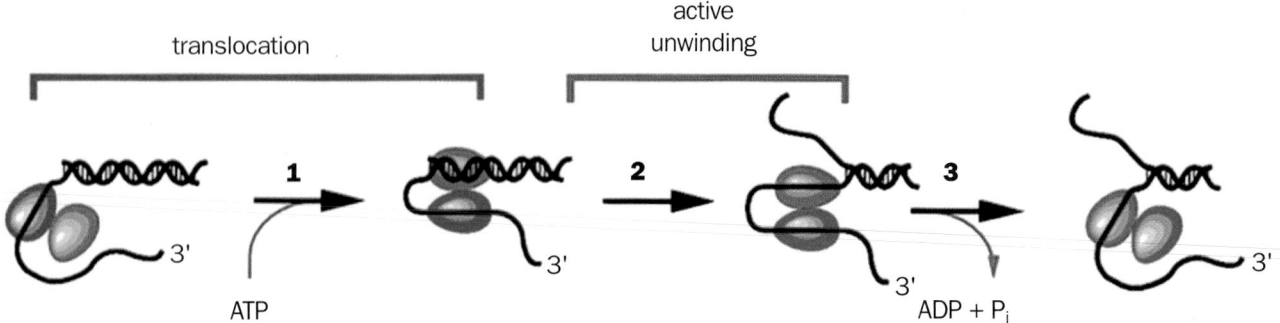

FIGURE 30-17   Active, rolling mechanism for DNA unwinding by Rep helicase. (1) The subunit of dimeric Rep helicase that is not bound to ssDNA binds to dsDNA accompanied by ATP binding. **(2)** The subunit bound to dsDNA unwinds the double strand and remains bound to the 3′-ending strand. **(3)** In a process that is accompanied by the release of the ATP hydrolysis products, the subunit closer to the 3′ end of the bound ssDNA releases it preparatory for a new cycle of dsDNA unwinding. [Courtesy of Gabriel Waksman, Washington University School of Medicine.]

The different conformations and ADPNP-binding properties of the hexamer's chemically identical subunits are reminiscent of the binding change mechanism of ATP synthesis by the F_1F_0-ATPase (Section 22-3C). Wigley has therefore suggested that when a helicase binds and hydrolyzes NTP, it undergoes a conformational change that pulls the ssDNA through the center of the hexameric ring by the lever action of its DNA-binding loops. The conformational states, A, B, and C, are presumed to arise through the sequential binding of an NTP, its hydrolysis, and the release of the product NDP and P_i as similarly occurs in the binding change mechanism (Fig. 22-42). Hence the translocation process is postulated to be motivated by a wave of coupled subunit rotations that ripple around the hexameric ring. The helicase would thereby mechanically separate the strands of dsDNA by essentially pulling itself along the groove of one strand in the $5′ \rightarrow 3′$ direction but without turning relative to the DNA.

b.   Rep Helicase Dimers Separate the Strands of dsDNA via an "Active, Rolling" Mechanism

Two other helicases, **Rep helicase** and **PriA protein,** have been implicated in the replication of various *E. coli* phage DNAs (Section 30-3B) and also participate in certain aspects of *E. coli* DNA replication (Section 30-3C). Both proteins translocate along DNA in the $3′ \rightarrow 5′$ direction (and hence along the opposite strand from DnaB) while hydrolyzing ATP. Rep helicase is not essential for *E. coli* DNA replication but the rate at which *E. coli* replication forks propagate is reduced ~2-fold in *rep⁻* mutants.

Rep helicase is a 673-residue monomer in solution but dimerizes on binding to DNA. Both subunits of the Rep dimer bind to ssDNA or dsDNA such that DNA binding to one subunit strongly inhibits DNA binding to the other (negative cooperativity). This observation led Timothy Lohman to propose the "active, rolling" mechanism for Rep-mediated DNA unwinding in which the two subunits of the dimer alternate in binding dsDNA and the 3′ end of the ssDNA at the ssDNA/dsDNA junction (Fig. 30-17). The two subunits then "walk" up the DNA while unwinding it in an ATP-dependent manner via a subunit switching mechanism in which the helicase subunit that is bound to the dsDNA displaces its 5′-starting strand while remaining bound to its 3′-starting strand. Release of the other subunit from the 3′-starting ssDNA then permits this subunit to bind to and unwind the new end of the dsDNA, thereby continuing the cycle.

The X-ray structure of *E. coli* Rep helicase in complex with the short ssDNA $dT(pT)_{15}$ and ADP (Fig. 30-18), determined by Lohman and Waksman, reveals that the rel-

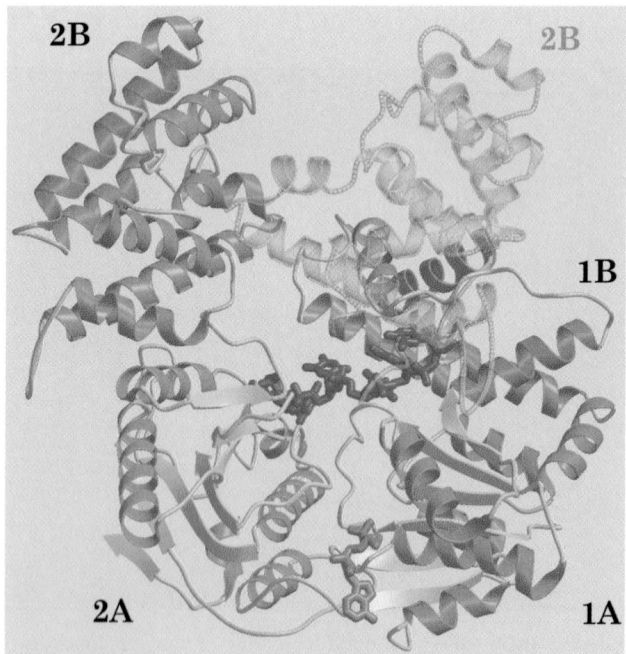

FIGURE 30-18   X-Ray structure of Rep helicase in complex with $dT(pT)_{15}$ and ADP. The monomer in the open conformation is drawn in ribbon form colored according to secondary structure (helices magenta, β sheets yellow, and coil cyan) with its bound ssDNA segment and ADP drawn in stick form in blue and in red. In the closed conformation, subdomain 2B (transparent green ribbon) has rotated via a 130° hinge motion so as to close over the ssDNA. [Courtesy of Gabriel Waksman, Washington University School of Medicine. PDBid 1UAA.]

atively straight ssDNA molecule binds two contacting Rep monomers. A Rep monomer consists of two domains, 1 and 2, each of which is formed by two subdomains, A and B, with the two N-terminal subdomains (1A and 2A) homologous to each other. In the two Rep monomers that are bound to the same ssDNA, subdomain 2B exhibits strikingly different orientations with respect to the other three subdomains (Fig. 30-18). The Rep monomer that is bound to the 5′ end of the ssDNA (which it contacts between bases 1 and 8) assumes the "open" conformation in which the four subdomains form an assembly that is reminiscent of a crab claw with one pincer (subdomain 2B) larger than the other (subdomain 1B). The DNA is bound at the bottom of the resulting cleft, whose floor is formed by subdomains 1A and 2A. In the Rep monomer that binds to the 3′ end of the ssDNA (which it contacts between bases 9 and 16), subdomain 2B has reoriented relative to the other subdomains via a 130° rotation about a hinge region between subdomains 2A and 2B, thereby closing the cleft about the DNA to form the "closed" conformation. This conformation change is consistent with the active, rolling mechanism even though the way in which two Rep monomers form the dimer observed in solution remains unknown. The ADP binds to Rep between its subdomains 1A and 2A in close proximity to the DNA, suggesting that conformation changes at the ATP-binding site arising from ATP hydrolysis are transmitted to the DNA-binding site via the secondary structural elements that contact both sites. The way in which Rep separates the two strands of dsDNA is, as yet, unknown.

c. Single-Strand Binding Protein Prevents ssDNA from Reannealing

If left to their own devices, the separated DNA strands behind an advancing helicase would rapidly reanneal to reform dsDNA. What prevents them from doing so is the binding of **single-strand binding protein (SSB).** It also prevents ssDNA from forming fortuitous intramolecular secondary structures (helical stems) and protects it from nucleases. Numerous copies of SSB cooperatively coat ssDNA, thereby maintaining it in an unpaired state. Note, however, that ssDNA must be stripped of SSB before it can be replicated by Pol III holoenzyme.

E. coli SSB is a homotetramer of 177-residue subunits. SSB binds ssDNA in several distinct modes referred to as $(SSB)_n$, which differ by the number of nucleotides (n) bound to each tetramer. The two major modes are $(SSB)_{35}$, in which only two of the tetramer's subunits strongly interact with the ssDNA, and $(SSB)_{65}$, in which all four subunits interact with the ssDNA. The $(SSB)_{35}$ mode displays unlimited cooperativity in that it forms extended strings of contacting tetramers along the length of a bound ssDNA, whereas the $(SSB)_{65}$ mode has limited cooperativity in that it forms beaded clusters on ssDNA that consist of only a few contacting tetramers.

Proteolysis studies have shown that SSB's ssDNA-binding site is contained within its 115 N-terminal residues. The X-ray structure of *E. coli* SSB's chymotryptic fragment (residues 1–135) in complex with $dC(pC)_{34}$, determined by

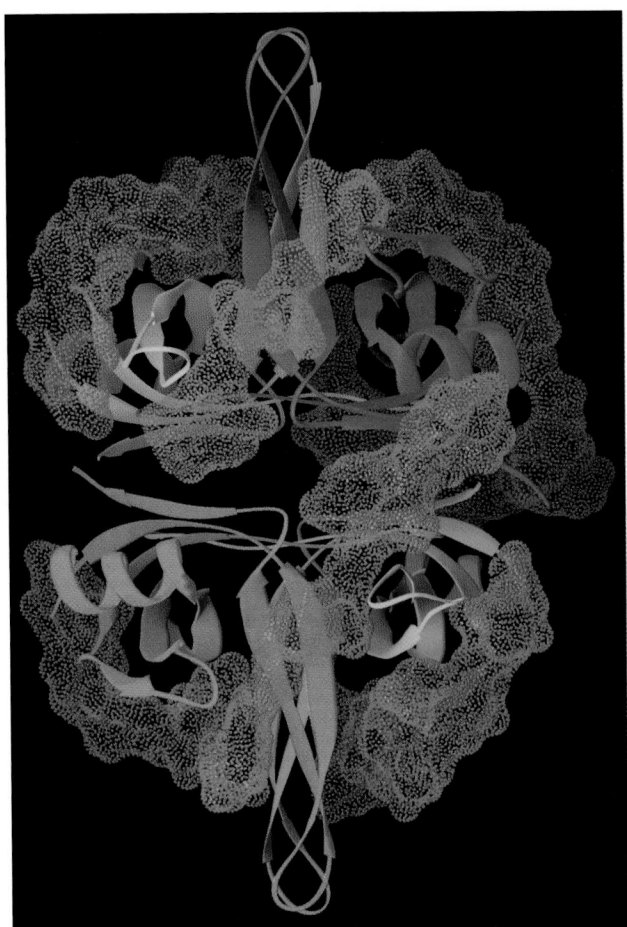

FIGURE 30-19 X-Ray structure of the N-terminal 135 residues of *E. coli* SSB in complex with $dC(pC)_{34}$. The homotetrameric protein is viewed along one of its 2-fold axes with its other 2-fold axes horizontal and vertical. Each of its subunits is differently colored. The two bound ssDNA molecules are represented by dot surfaces with the observed 28-residue segment of one of the ssDNAs green and the observed 14- and 9-residue segments of the other ssDNA red. [Based on an X-ray structure by Timothy Lohman and Gabriel Waksman, Washington University School of Medicine. PDBid 1EYG.]

Lohman and Waksman, reveals that the tetrameric protein has D_2 symmetry and binds two molecules of $dC(pC)_{34}$ (Fig. 30-19). For one of these 35-mers, 28 nucleotides (residues 3–30) were visible and these assumed the shape of an elongated horseshoe that wrapped around two SSB subunits with approximate 2-fold symmetry and with its apex contacting a third subunit. The other bound ssDNA was partially disordered such that only two segments were visible, one with 14 nt (residues 3–16) and the other with 9 nt (residues 19–27). The paths of the ssDNA segments along the surface of the SSB suggested models that rationalize the different properties of $(SSB)_{35}$ and $(SSB)_{65}$. In the $(SSB)_{65}$ model, the two ends of a 65-nt segment emerge from the same side of the tetramer, which would limit the number of SSB tetramers that can bind to contiguous 65-nt segments of ssDNA. However, in the $(SSB)_{35}$ model, the two ends of a 35-nt segment emerge from opposite ends of the tetramer, thereby permitting an unlimited series of SSB tetramers to interact end-to-end along the length of an ssDNA.

D. *DNA Ligase*

Pol I, as we saw in Section 30-2A, replaces the Okazaki fragments' RNA primers with DNA through nick translation. *The resulting single-strand nicks between adjacent Okazaki fragments, as well as the nick on circular DNA after leading strand synthesis, are sealed in a reaction catalyzed by* **DNA ligase.** The free energy required by this reaction is obtained, in a species-dependent manner, through the coupled hydrolysis of either NAD^+ to $NMN^+ + AMP$ or ATP to $PP_i + AMP$. The *E. coli* enzyme, a 671-residue monomer that utilizes NAD^+, catalyzes a three-step reaction (Fig. 30-20):

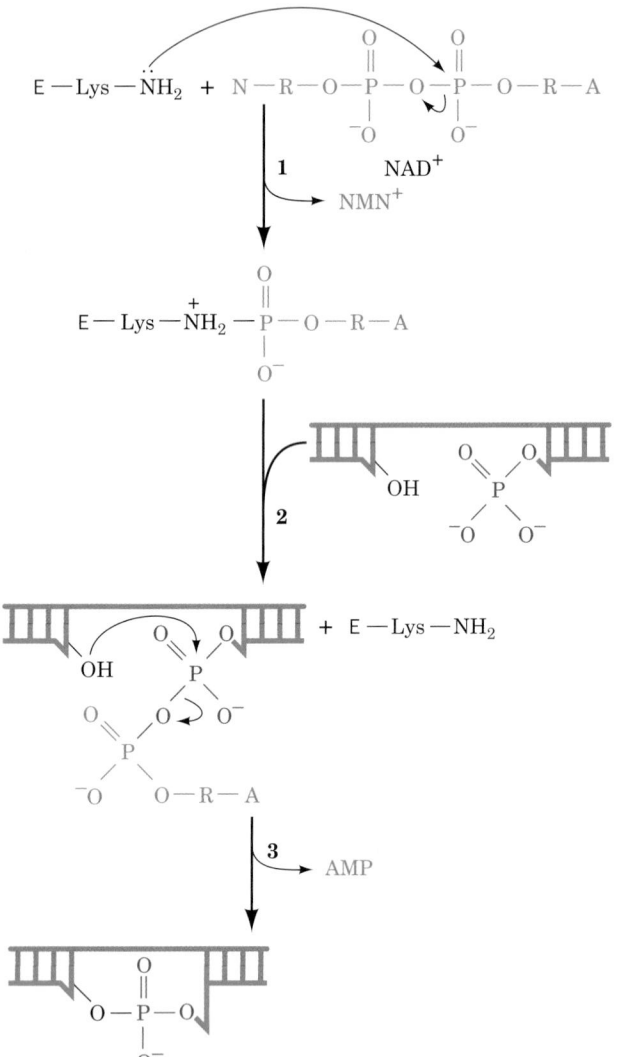

FIGURE 30-20   The reactions catalyzed by *E. coli* DNA ligase. In eukaryotic and T4 DNA ligases, NAD^+ is replaced by ATP so that PP_i rather than NMN^+ is eliminated in the first reaction step. The numbered steps are described in the text.

1. The adenylyl group of NAD^+ is transferred to the ε-amino group of an enzyme Lys residue to form an unusual phosphoamide adduct that is, nevertheless, readily isolated.

2. The adenylyl group of this activated enzyme is transferred to the 5'-phosphoryl terminus of the nick to form an adenylylated DNA. Here, AMP is linked to the 5'-nucleotide via a pyrophosphate rather than the usual phosphodiester bond.

3. DNA ligase catalyzes the formation of a phosphodiester bond by attack of the 3'-OH on the 5'-phosphoryl group, thereby sealing the nick and releasing AMP.

ATP-requiring DNA ligases, such as those of all eukaryotes and bacteriophage T4, release PP_i in the first step of the reaction rather than NMN^+. T4 ligase is also noteworthy in that, at high DNA concentrations, it can link together two duplex DNAs **(blunt end ligation)** in a reaction that is a boon to genetic engineering (Section 5-5C).

The X-ray structure of the 667-residue NAD^+-dependent DNA ligase from *Thermus filiformis*, determined by Se Won Suh, reveals a deeply clefted monomeric protein that consists of four domains (Fig. 30-21): Domain 1 (residues 1–317), which contains the adenylylated Lys residue (Lys 116); Domain 2 (residues 318–403); Domain 3 (residues 404–581), which contains a Zn^{2+} ion that is tetrahedrally liganded by four Cys residues and appears to have a structural but not a catalytic role; and Domain 4 (residues 582–660), which is poorly ordered (its side chains are not visible) and hence appears to be highly mobile. The shape and predominantly positive surface charge of the cleft strongly suggest that it forms the enzyme's DNA-binding site. Nevertheless, this putative DNA-binding cleft contains a highly negative patch in the vicinity of Lys 116 that is formed by the highly conserved side chains of Asp 118, Glu 281, and Asp 283. This suggests that these side chains help form the active site for the ligation reaction, which is chemically similar to the polymerization reaction catalyzed by DNA polymerases and hence is likely to have a similar divalent metal ion mechanism involving acidic side chains (Fig. 30-10). Indeed, several DNA ligases have been shown to require divalent metal ion for activity. The apparent high mobility of Domain 4 suggests that it folds out via a hinge-like mechanism to allow the enzyme's nicked dsDNA substrate to bind to the active site and then folds back over to immobilize the DNA as is drawn in Fig. 30-21.

E. *Primase*

The primases from bacteria and several bacteriophages track the moving replication fork in close association with its DNA helicase. Thus as we have seen (Section 30-2C), the N-terminal domain of T7 gene 4 helicase/primase forms its primase function, whereas *E. coli* primase (DnaG) forms a noncovalent complex with DnaB. Since this DNA helicase translocates along the lagging strand template DNA in its 5' → 3' direction (Fig. 30-14), the primase must reverse its direction of travel in order to synthesize an

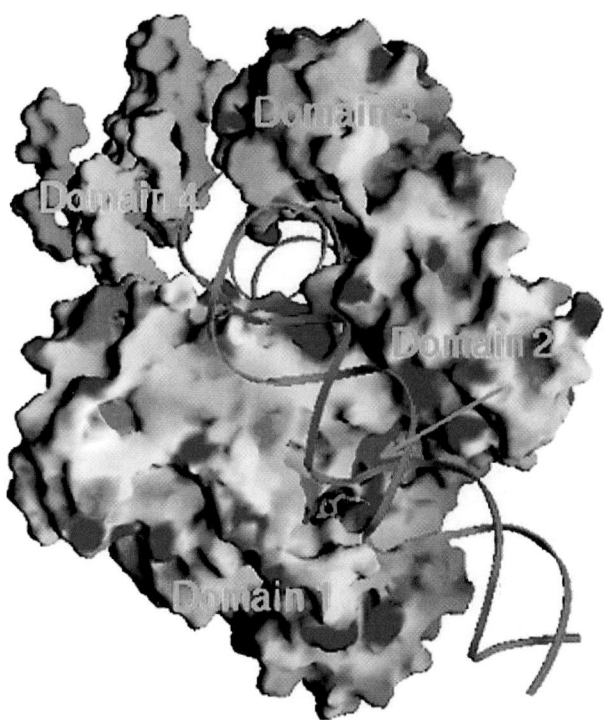

FIGURE 30-21 X-Ray structure of DNA ligase from *Thermus filiformis*. The monomeric protein is represented by its surface diagram, which is colored according to its electrostatic potential with blue, white, and red indicating positive, near neutral, and negative potentials, respectively (Domain 4, which is partially behind Domain 1 in this view, is gray because its side chains were not observed). The exposed ribose–phosphate moiety of the AMP that is covalently bound to the side chain of Lys 116 is drawn in ball-and-stick form with C green, O red, and P yellow. dsDNA has been modeled into the putative DNA-binding cleft (*blue and red ribbons*). The orange arrow points to a highly negative surface patch in the vicinity of the AMP residue that consists of the side chains of the conserved Asp 118, Glu 281, and Asp 283 and that is postulated to form the active site for the divalent metal–catalyzed ligase reaction. [Courtesy of Se Won Suh, Seoul National University, Korea. PDBid 1DGT.]

RNA primer in its $5' \rightarrow 3'$ direction. *E. coli* primase can synthesize up to 60-nt primers *in vitro*, although *in vivo*, primers have the length of 11 ± 1 nt. Since the replication fork in *E. coli* moves at ~1000 nt per second and Okazaki fragments are ~1000 nt in length, primase must synthesize about one RNA primer per second.

E. coli primase is a 581-residue monomeric protein. Proteolysis studies have shown that it consists of three domains: an N-terminal Zn^{2+}-binding domain (residues 1–110), which tetrahedrally ligands a Zn^{2+} ion via three Cys residues and a His residue and is implicated in recognizing ssDNA; a central catalytic domain (residues 111–433) that catalyzes RNA synthesis; and a C-terminal domain (residues 434–581) that interacts with DnaB. The X-ray structure of the catalytic domain (Fig. 30-22), which was independently determined by James Berger and by Kuriyan, reveals a cashew-shaped protein whose fold is un-

related to those of any other DNA or RNA polymerases. However, it contains an ~100-residue segment that is similar in both sequence and structure to segments in types IA and II topoisomerases (Section 29-3C) and has therefore been named the **Toprim fold** (for topoisomerase and primase). The Toprim fold consists of a 4-stranded parallel β sheet flanked by three helices that resembles the nucleotide-binding (Rossmann) fold (Section 8-3B).

The primase catalytic domain contains a groove at the center of its concave surface that is surrounded by residues that are highly conserved in DnaG-type primases. Among them are a Glu and two Asp residues, which are invariant in all known Toprim folds and which in the X-ray structure of a type II topoisomerase coordinate an Mg^{2+} ion. This suggests that these three acidic residues are located at the primase active site, which is known to require Mg^{2+} for activity. Model building by Kuriyan suggests that an RNA–DNA hybrid helix (which has an A-DNA-like conformation; Section 29-1B) binds in the groove with one of its phosphate groups adjacent to the putative Mg^{2+}-binding site (Fig. 30-22). Extension of this double helix into the upper portion of the protein is prevented by a narrowing of the groove. Consequently, it appears that the primase can accommodate only an ~10-bp segment of an RNA–DNA helix, thereby accounting for this enzyme's limited processivity.

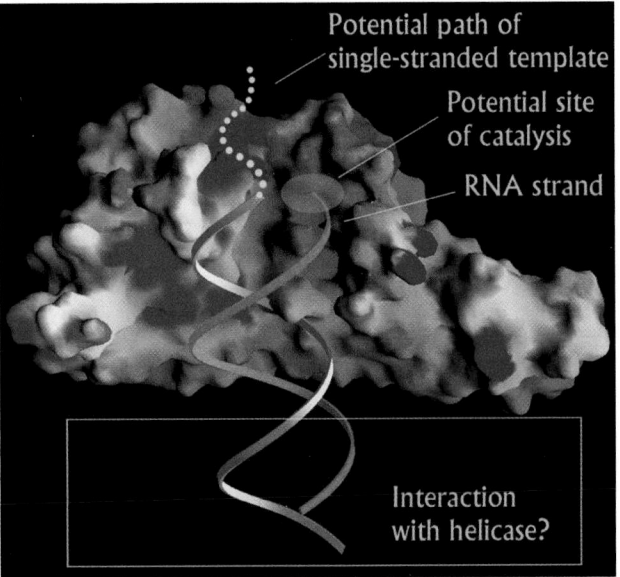

FIGURE 30-22 X-Ray structure of *E. coli* primase. The protein is represented by its molecular surface colored according to its electrostatic potential with red negative, white nearly neutral, and blue positive. The yellow patch marks the enzyme's putative Mg^{2+}-binding site. A segment of an RNA–DNA hybrid double helix has been modeled into the protein so as to place the end of the RNA strand (*red*) near the Mg^{2+}-binding site. Note that the template DNA strand (*blue*) runs along a region of mainly positive charge. [Courtesy of John Kuriyan, The Rockefeller University. PDBid 1EQN.]

3 ■ PROKARYOTIC REPLICATION

Bacteriophages are among the simplest biological entities and their DNA replication mechanisms reflect this fact. Much of what we know about how DNA is replicated therefore stems from the study of this process in various phages. In this section we examine DNA replication in the **coliphages** (bacteriophages that infect *E. coli*) **M13** and φX174 and then consider DNA replication in *E. coli* itself. Eukaryotic DNA replication is discussed in Section 30-4.

A. *Bacteriophage M13*

Bacteriophage M13 carries a 6408-nt single-stranded circular DNA known as its **viral** or (+) strand. On infecting an *E. coli* cell, this strand directs the synthesis of its complementary or (−) strand to form the circular duplex **replicative form (RF),** which may be either nicked **(RF II)** or supercoiled **(RF I).** This replication process (Fig. 30-23)

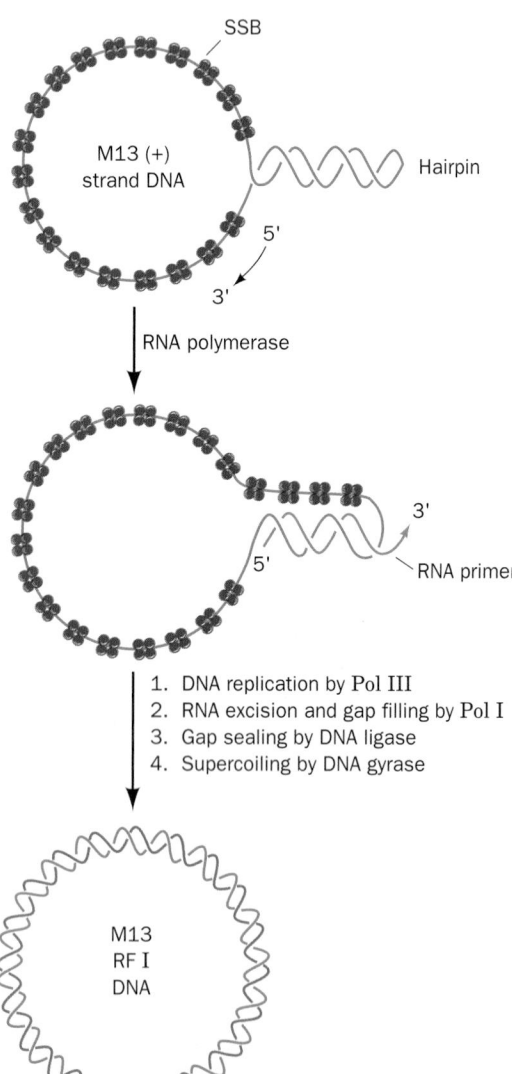

FIGURE 30-23   The synthesis of the M13 (−) strand DNA on a (+) strand template to form M13 RF I DNA.

may be taken as a paradigm for leading strand synthesis in duplex DNA.

As the M13 (+) strand enters the *E. coli* cell, it becomes coated with SSB except at a palindromic 57-nt segment that forms a hairpin. RNA polymerase commences primer synthesis 6 nt before the start of the hairpin and extends the RNA 20 to 30 residues to form a segment of RNA–DNA hybrid duplex. The DNA that is displaced from the hairpin becomes coated with SSB so that when RNA polymerase reaches it, primer synthesis stops. Pol III holoenzyme then extends the RNA primer around the circle to form the (−) strand. The primer is removed by Pol I-catalyzed nick translation, thereby forming RF II, which is converted to RF I by the sequential actions of DNA ligase and DNA gyrase.

B. *Bacteriophage φX174*

Bacteriophage φX174, as does M13, carries a small (5386 nt) single-stranded circular DNA. Curiously, the *in vivo* conversion of the φX174 viral DNA to its replicative form is a much more complex process than that for M13 DNA in that φX174 replication requires the participation of a nearly 600-kD protein assembly known as a **primosome** (Table 30-4).

a. φX174 (−) Strand Replication Is a Paradigm for Lagging Strand Synthesis

φX174 (−) strand synthesis occurs in a six-step process (Fig. 30-24):

1. The reaction sequence begins in the same way as that for M13: The (+) strand is coated with SSB except for a 44-nt hairpin. A 70-nt sequence containing this hairpin, known as *pas* (for *p*rimosome *a*ssembly *s*ite), is then recognized and bound by the PriA, **PriB**, and **PriC** proteins.

2. DnaB and **DnaC** proteins in the form of a DnaB$_6$ · DnaC$_6$ complex add to the DNA with the help of **DnaT protein** in an ATP-requiring process. DnaC protein is then released yielding the **preprimosome.** The preprimosome, in turn, binds primase yielding the primosome.

TABLE 30-4   Proteins of the Primosome[a]

Protein	Subunit Structure	Subunit Mass (kD)
PriA	monomer	76
PriB	dimer	11.5
PriC	monomer	23
DnaT	trimer	22
DnaB	hexamer	50
DnaC[b]	monomer	29
Primase (DnaG)	monomer	60

[a]The complex of all primosome proteins but primase is known as the preprimosome.

[b]Not part of the preprimosome or the primosome.

Source: Kornberg, A. and Baker, T.A., *DNA Replication* (2nd ed.), pp. 286–288, Freeman (1992).

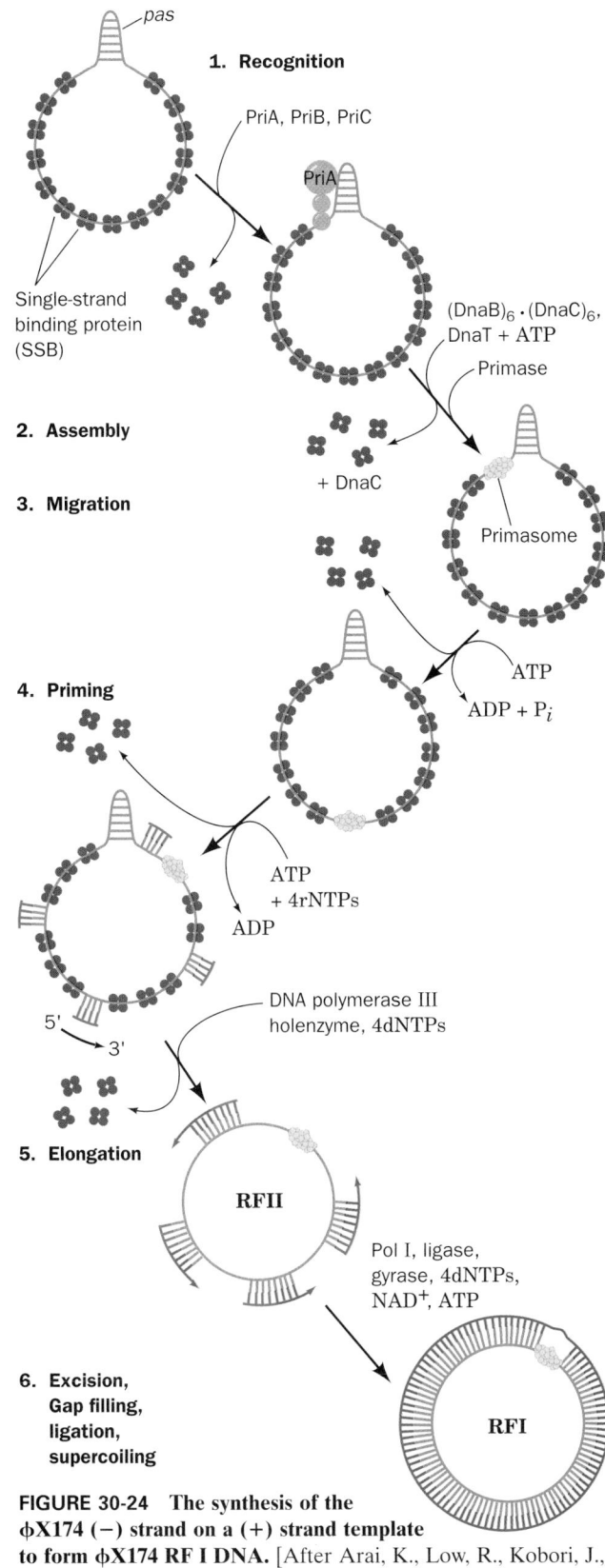

FIGURE 30-24 The synthesis of the
φX174 (−) strand on a (+) strand template
to form φX174 RF I DNA. [After Arai, K., Low, R., Kobori, J.,
Schlomai, J., and Kornberg, A., *J. Biol. Chem.* **256,** 5280 (1981).]

3. The primosome is propelled in the 5′ → 3′ direction
along the (+) strand by the PriA and DnaB helicases at

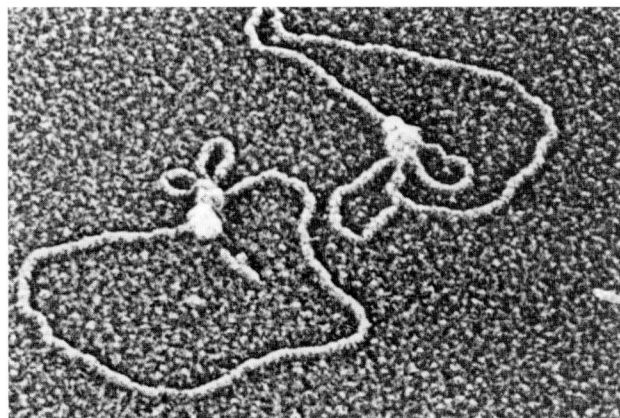

FIGURE 30-25 Electron micrograph of a primosome bound to
a φX174 RF I DNA. Such complexes always contain a single
primosome with one or two associated small DNA loops.
[Courtesy of Jack Griffith, Lineberger Cancer Research Center,
University of North Carolina.]

the expense of ATP hydrolysis. This motion, which displaces the SSB in its path, is opposite in direction to that
of template reading during DNA chain propagation.

4. At randomly selected sites, the primosome reverses
its migration while primase synthesizes an RNA primer.
The initiation of primer synthesis requires the participation of DnaB protein which, through concomitant ATP hydrolysis, is thought to alter template DNA conformation
in a manner required by primase.

5. Pol III holoenzyme extends the primers to form
Okazaki fragments.

6. Pol I excises the primers and replaces them by DNA.
The fragments are then joined by DNA ligase and supercoiled by DNA gyrase to form the φX174 RF I.

The primosome remains complexed with the DNA (Fig.
30-25) where it participates in (+) strand synthesis (see
below).

b. φX174 (+) Strand Replication Serves as a Model for
Leading Strand Synthesis

One strand of a circular duplex DNA may be synthe-
sized via the **rolling circle** or **σ-replication** mode (so called
because of the resemblance of the replicating structure to
the Greek letter sigma; Fig. 30-26). *The φX174 (+) strand*
is synthesized on an RF I template by a variation on this
*process, the **looped rolling circle mode** (Fig. 30-27):*

1. (+) strand synthesis begins with the primosome-
aided binding of the phage-encoded 513-residue enzyme
gene A protein to its ~30-bp recognition site. There, gene
A protein cleaves a specific phosphodiester bond on the
(+) strand nucleotide (near the beginning of gene A) by
forming a covalent bond between a Tyr residue and the
DNA's 5′-phosphoryl group, thereby conserving the
cleaved bond's energy.

2. Rep helicase (Section 30-2C) subsequently attaches
to the (−) strand at the gene A protein and, with the aid

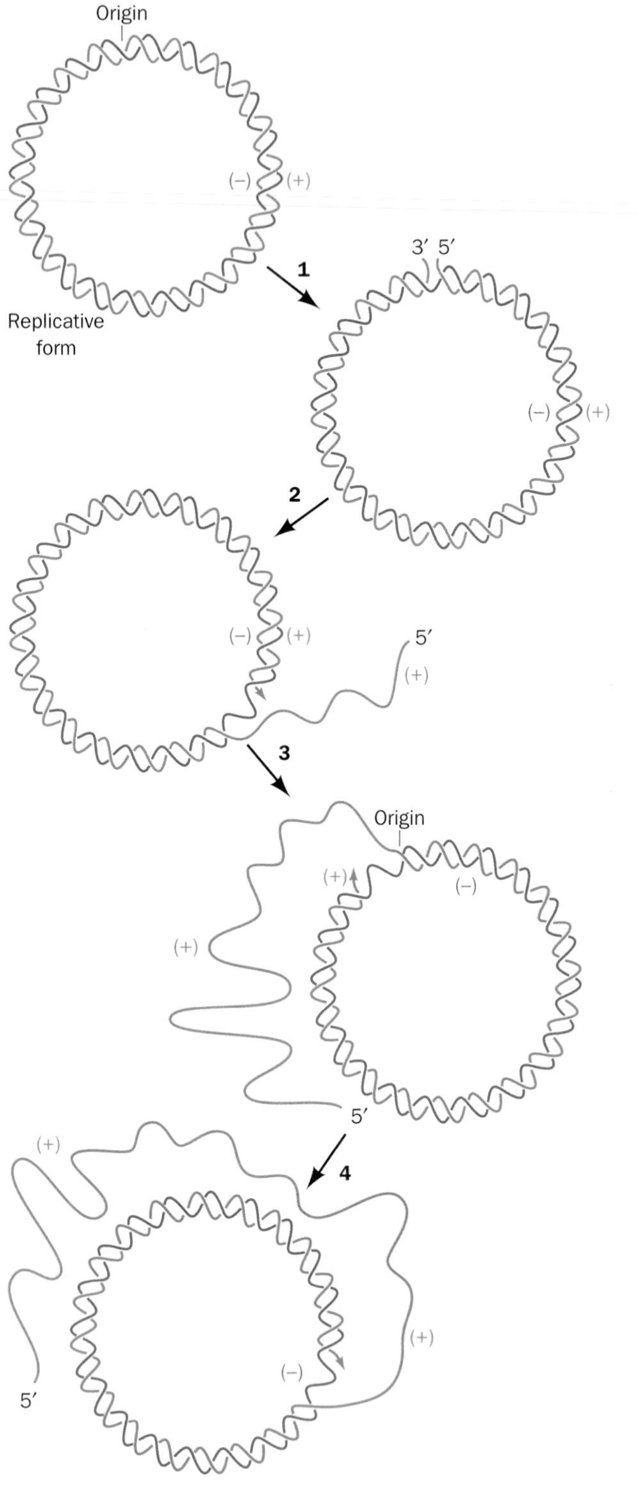

FIGURE 30-26 The rolling circle mode of DNA replication.
The (+) strand being synthesized is extended from a specific
cut made at the replication origin **(1)** so as to strip away the old
(+) strand **(2 and 3).** The continuous synthesis of the (+)
strand on a circular (−) strand template produces a series of
tandemly linked (+) strands **(4),** which may later be separated
by a specific endonuclease.

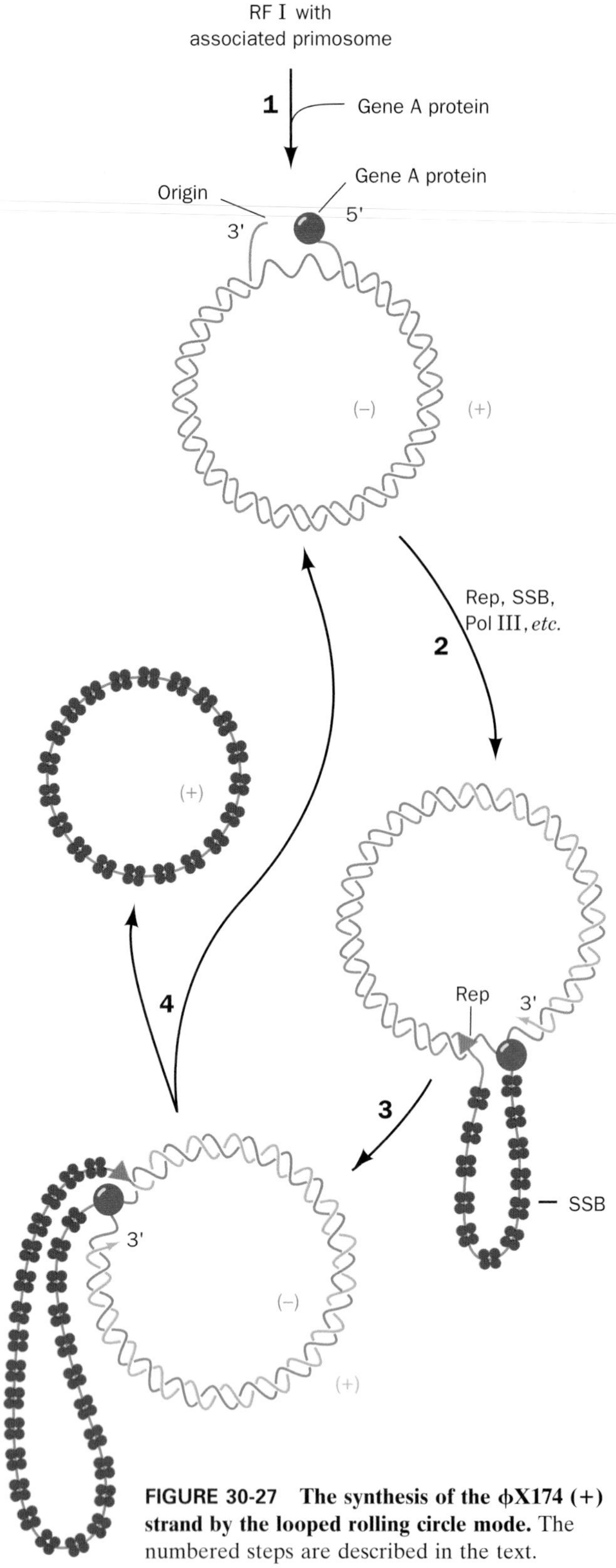

**FIGURE 30-27   The synthesis of the φX174 (+)
strand by the looped rolling circle mode.** The
numbered steps are described in the text.

of the primosome still associated with the (+) strand, commences unwinding the duplex DNA from the (+) strand's

5′ end. The displaced (+) strand is coated with SSB, which
prevents it from reannealing to the (−) strand. Rep helicase is essential for the replication of φX174 DNA, but not
for the *E. coli* chromosome, as is demonstrated by the in-

ability of ɸX174 to multiply in *rep⁻ E. coli*. Pol III holoenzyme extends the (+) strand from its free 3′-OH group.

3. The extension process generates a **looped rolling circle** structure in which the 5′ end of the old (+) strand remains linked to the gene A protein at the replication fork. It is thought that as the old (+) strand is peeled off the RF, the primosome synthesizes the primers required for the later generation of a new (−) strand.

4. When it has come full circle around the (−) strand, the gene A protein again makes a specific cut at the replication origin so as to form a covalent linkage with the new (+) strand's 5′ end. Simultaneously, the newly formed 3′-terminal OH group of the old, looped-out (+) strand nucleophilically attacks its 5′-phosphoryl attachment to the

gene A protein, thereby liberating a covalently closed (+) strand. This is possible because the gene A protein has two closely spaced Tyr residues that alternate in their attachment to the 5′ ends of successively synthesized (+) strands. The replication fork continues its progress about the duplex circle, producing new (+) strands in a manner reminiscent of linked sausages being pulled off a reel.

In the intermediate stages of a ɸX174 infection, each newly synthesized (+) strand directs the synthesis of the (−) strand to form RF I as described above. In the later stages of infection, however, the newly formed (+) strands are packaged into phage particles.

C. *E. coli*

𝄞 **See Guided Exploration 25. The Replication of DNA in *E. coli*.** The *E. coli* chromosome replicates by the bidirectional θ mode from a single replication origin (*Section 30-1A*). The most plausible model for events at the *E. coli* replication fork (Fig. 30-28) is largely derived from studies on the

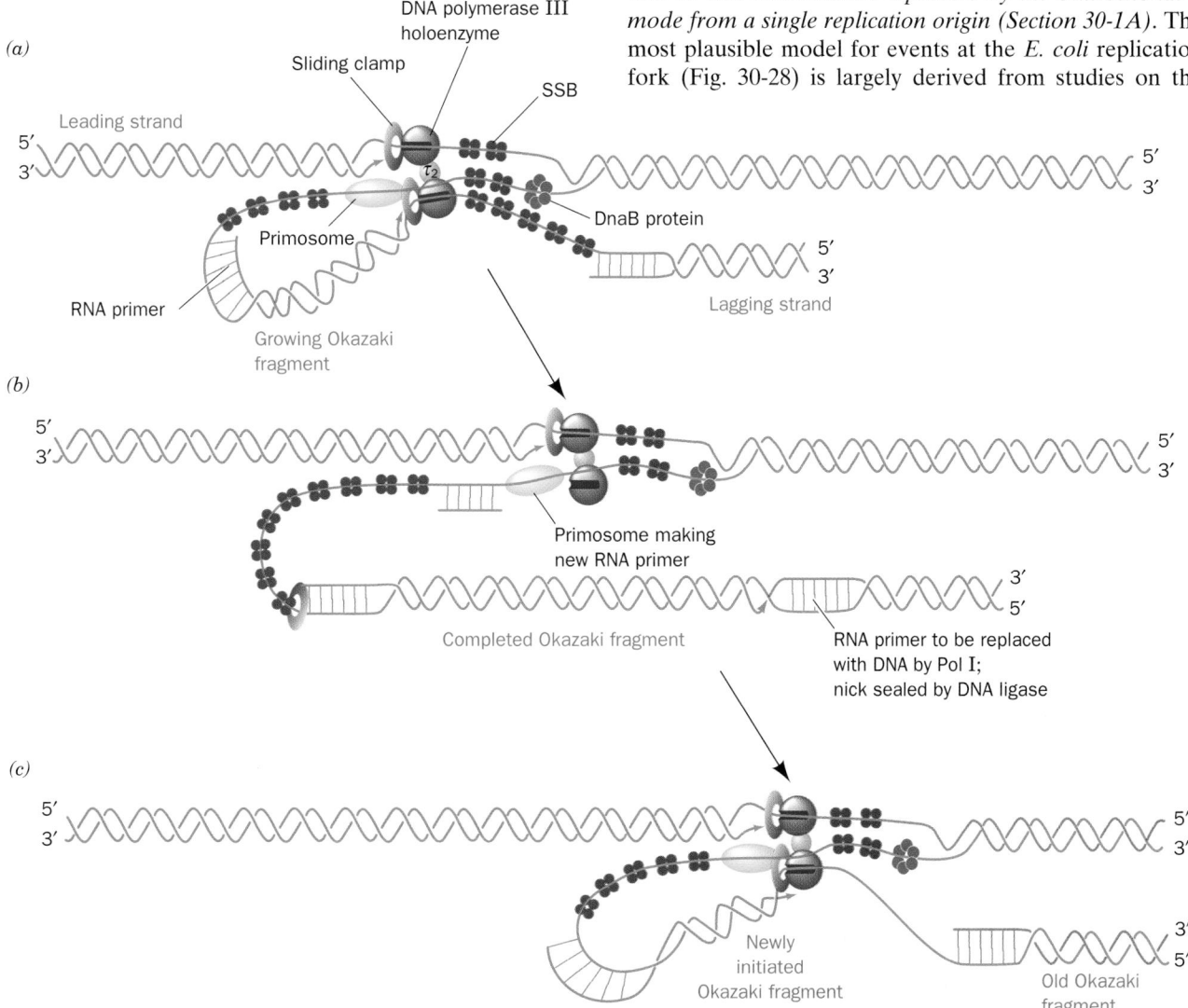

FIGURE 30-28 The replication of *E. coli* DNA. (*a*) The *E. coli* DNA replisome, which contains two DNA polymerase III holoenzyme complexes, synthesizes both the leading and the lagging strands. The lagging strand template must loop around to permit the holoenzyme to extend the primosome-primed lagging strand. (*b*) The holoenzyme releases the lagging strand template when it encounters the previously synthesized

Okazaki fragment. This possibly signals the primosome to initiate the synthesis of lagging strand RNA primer. (*c*) The holoenzyme rebinds the lagging strand template and extends the RNA primer to form a new Okazaki fragment. Note that in this model, leading strand synthesis is always ahead of lagging strand synthesis.

simpler and more experimentally accessible DNA replication mechanisms of coliphages such as M13 and φX174. Duplex DNA is unwound by DnaB helicase on the lagging strand template, where it is joined by the primosome. The separated single strands are immediately coated by SSB. Leading strand synthesis is catalyzed by Pol III holoenzyme, as is that of the lagging strand after priming by primosome-associated primase. Both leading and lagging strand syntheses occur on a single ~900-kD multisubunit particle, the **replisome,** which contains two Pol III cores (αεθ) that are joined together by a dimer of τ subunits that bridges the α subunits. Hence, the lagging strand template must be looped around (Fig. 30-28). The τ₂ dimer also binds the DnaB helicase (not shown in Fig. 30-28), thereby stimulating its helicase action while holding it to the replication fork. After completing the synthesis of an Okazaki fragment, the lagging strand holoenzyme relocates to a new primer near the replication fork, the primer heading the previously synthesized Okazaki fragment is excised by Pol I-catalyzed nick translation, and the nick is sealed by DNA ligase.

a. *E. coli* DNA Replication Is Initiated at *oriC* in a Process Mediated by DnaA Protein

The replication origin of the *E. coli* chromosome consists of a unique 245-bp segment known as the **oriC** locus. This sequence, segments of which are highly conserved among gram-negative bacteria, supports the bidirectional replication of the various plasmids into which it has been inserted. Experiments with such plasmids pioneered by Kornberg indicate that replication initiation in *E. coli* occurs via the following multistep process (Fig. 30-29):

1. DnaA protein (467 residues) recognizes and binds *oriC*'s five so-called **DnaA boxes** (which each contain a highly conserved 9-bp segment of consensus sequence 5'-TTATCCACA-3') to form a complex of negatively supercoiled *oriC* DNA wrapped around a central core of five DnaA protein monomers. This process is facilitated by two related DNA-binding proteins named **HU** and **integration host factor (IHF)** that induce DNA bending (IHF and HU are discussed in Section 33-3C).

2. The DnaA protein subunits then successively melt three tandemly repeated, 13-bp, AT-rich segments (consensus sequence 5'-GATCTNTTNTTTT-3' where N marks nonspecific positions) located near *oriC*'s "left"

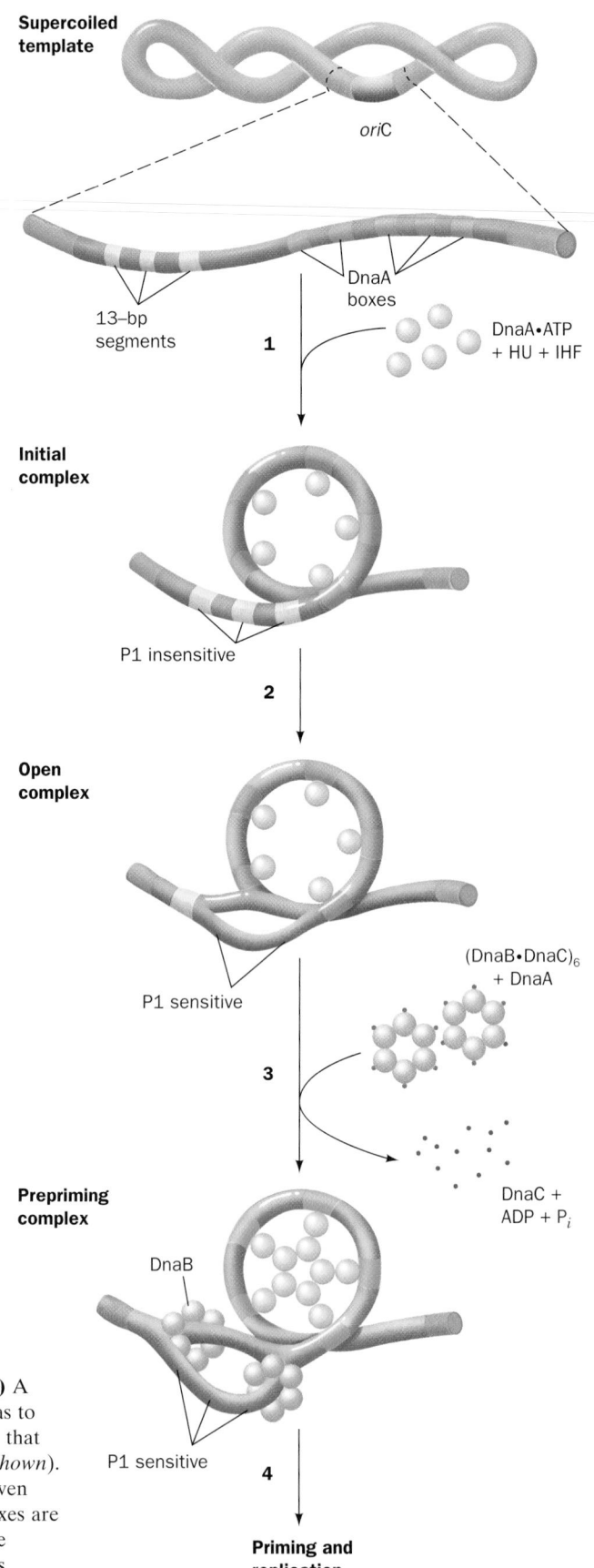

FIGURE 30-29 A model for DNA replication initiation at *oriC*. (1) A DnaA protein subunit binds to each of *oriC*'s five DnaA boxes so as to wrap the suitably supercoiled *oriC* around the proteins in a process that is aided by the binding of HU and IHF proteins to the DNA (*not shown*). **(2)** The three AT-rich 13-bp repeats are then melted in an ATP-driven reaction to form an open complex. **(3)** Two DnaB₆ · DnaC₆ complexes are recruited to opposite ends of the open complex accompanied by the binding of five additional DnaA subunits to form five DnaA dimers. **(4)** The open complex is further unwound through the helicase action of DnaB protein, thereby preparing the complex for priming and bidirectional replication.

boundary. The existence of the resulting ~45-bp open complex was established through its sensitivity to **P1 nuclease,** an endonuclease produced by *Penicillium citrinum* that is specific for single strands. The formation of the open complex requires the presence of DnaA protein and ATP (which DnaA protein tightly binds and hydrolyzes in a DNA-dependent manner). The AT-rich nature of the 13-bp repeats, no doubt, facilitates the melting process.

3. The DnaA complex then recruits two $DnaB_6 \cdot DnaC_6$ complexes to opposite ends of the melted region to form the **prepriming complex** in a process that is accompanied by the binding of five additional DnaA subunits to form five DnaA dimers bound at the DnaA boxes. DnaC, an ATPase that facilitates DnaB loading, is subsequently released.

4. In the presence of SSB and gyrase, DnaB helicase further unwinds the DNA in the prepriming complex in both directions so as to permit the entry of primase and RNA polymerase. The participation of both these enzymes in leading strand primer synthesis (Section 30-1D), together with limitation of this process to the *oriC* site, suggests that the RNA polymerase activates primase to synthesize the primer. This perhaps explains the similarity of *oriC*'s AT-rich 13-mers to RNA polymerase's transcriptional promoters (Section 31-2B).

The stage is thereby set for bidirectional DNA replication by Pol III holoenzyme as described above.

b. The Initiation of *E. coli* DNA Replication Is Strictly Regulated

Chromosome replication in E. coli occurs only once per cell division, so this process must be tightly controlled. The doubling (cell generation) time of *E. coli* at 37°C varies with growth conditions from <20 min to ~10 h. Yet the constant ~1000 nt/s rate of movement of each replication fork fixes the 4.6×10^6-bp *E. coli* chromosome's replication time, C, at ~40 min. Moreover, the segregation of cellular components and the formation of a septum between them, which must precede cell division, requires a constant time, $D = 20$ min, after the completion of the corresponding round of chromosome replication. *Cells with doubling times $<C + D = 60$ min must consequently initiate chromosome replication before the end of the preceding cell division cycle.* This results in the formation of **multiforked chromosomes** as is diagrammed in Fig. 30-30 for a cell division time of 35 min.

Even in cells that contain multiple *oriC* sites, DNA replication is initiated at each such site once and only once per cell generation. However, after initiation has occurred, chain elongation proceeds at a uniform, largely uncontrolled rate. This suggests that a post-initiation *oriC* site is somehow sequestered from (prevented from interacting with) the replication initiation machinery, a phenomenon called **sequestration.** There is extensive morphological evidence, such as shown in Fig. 30-31, that the *E. coli* chromosome is associated with the cell membrane. This attachment would help explain how replicated chromosomes are segregated into different cells during cell division. But what is the mechanism of sequestration?

The sequence most commonly methylated in *E. coli* is the palindrome GATC, which is methylated at N6 of both its A bases by **Dam methyltransferase** (Section 30-7). GATC occurs 11 times in *oriC,* including at the beginning of all four of its 13-bp repeats (see above). Newly replicated GATC segments are hemimethylated, that is, the GATC sequences on the newly synthesized strand are unmethylated. Although Dam methyltransferase begins methylating most hemimethylated GATC segments immediately after their synthesis (within <1.5 min), those on *oriC* remain hemimethylated for around one-third of a cell

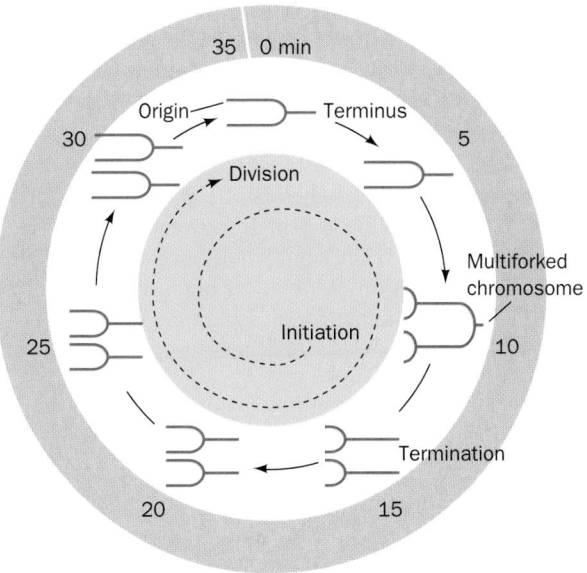

FIGURE 30-30 Multiforked chromosomes in *E. coli.* In cells that are dividing every 35 min, the fixed 60-min interval between the initiation of replication and cell division results in the production of multiforked chromosomes. [After Lewin, B., *Genes VII,* p. 370, Oxford Univ. Press (2000).]

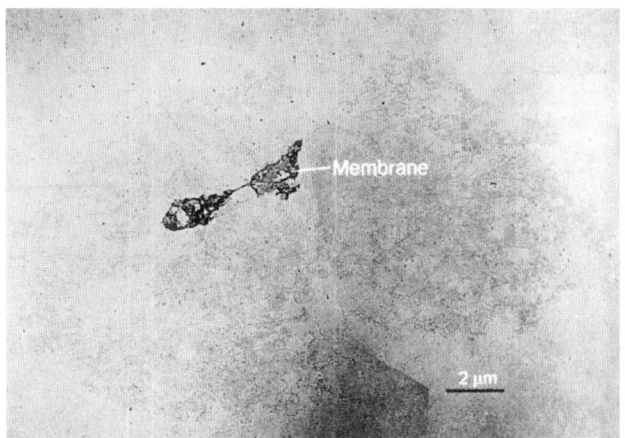

FIGURE 30-31 Electron micrograph of an intact and supercoiled *E. coli* chromosome attached to two fragments of the cell membrane. [From Delius, H. and Worcel, A., *J. Mol. Biol.* **82,** 108 (1974).]

generation. Consequently, the observation that membranes bind hemimethylated *oriC*, but not unmethylated or fully methylated *oriC,* suggests that hemimethylated *oriC* is bound to the membrane in a way that makes it inaccessible to both the initiation machinery and Dam methyltransferase.

The association of hemimethylated *oriC* with membrane requires the presence of the 181-residue **SeqA** protein, the product of *seqA* gene. Thus in *seqA⁻* cells: (1) the time to fully methylate hemimethylated GATC sites in *oriC* is reduced to 5 min, whereas the time to do so for other GATC sites is unaffected; (2) the synchrony of initiation of multiple *oriC* sites is lost; and (3) in the absence of functional Dam methyltransferase, fully methylated *oriC*-containing plasmids are replicated numerous times per cell generation, whereas in the presence of SeqA they are replicated only once. Evidently, sequestration occurs via the SeqA-mediated binding of hemimethylated *oriC* to the membrane. The hemimethylated promoter of the *dnaA* gene is similarly sequestered so as to repress its transcription, thereby providing an additional mechanism for preventing promiscuous initiation of DNA replication.

c. The Clamp Loader Loads the Sliding Clamp onto the DNA

The sliding clamp, which is responsible for Pol III's high processivity, is a ring-shaped dimer of β subunits through which the DNA strand being replicated is threaded (Section 30-2B). The two tightly associated β subunits ($K_D < 50$ nM) that form the sliding clamp dissociate with a half-life of ~100 min at 37°C. Yet, since each replisome synthesizes around one Okazaki fragment per second, a sliding clamp must be loaded onto the lagging strand template at this frequency. This loading function is carried out in an ATP-dependent process by the γ complex (γτ$_2$δδ'χψ). The τ and γ subunits are both encoded by the *dnaX* gene with τ (643 residues) the full-length product and γ (431 residues) its C-terminally truncated form. The γ complex bridges the replisome's two Pol III cores via the C-terminal segments of its two τ subunits, which also bind the DnaB helicase. However, since the χ and ψ subunits are not essential participants in the clamp loading process (their roles are poorly understood), we shall refer to the γτ$_2$δδ' complex as the **clamp loader.** How does the clamp loader do its job?

Of the clamp loader's five subunits, only δ is capable of binding to and opening up the sliding clamp on its own. Kuriyan and Mike O'Donnell determined the X-ray structure of the δ subunit in 1:1 complex with a β subunit that had two residues in its dimerization interface mutated so as to prevent its dimerization. The structure reveals (Fig. 30-32) that δ, which consists of three domains, inserts its β interaction element, a hydrophobic plug that forms the tip of its N-terminal domain, into a hydrophobic pocket on the surface of β. Comparison of δ in this structure with that in the γ$_3$δδ' complex (see below) reveals that the β interaction element undergoes a dramatic conformational change on binding to β in which its α4 helix rotates by 45° and translates by 5.5 Å. Moreover, in forming the β–δ com-

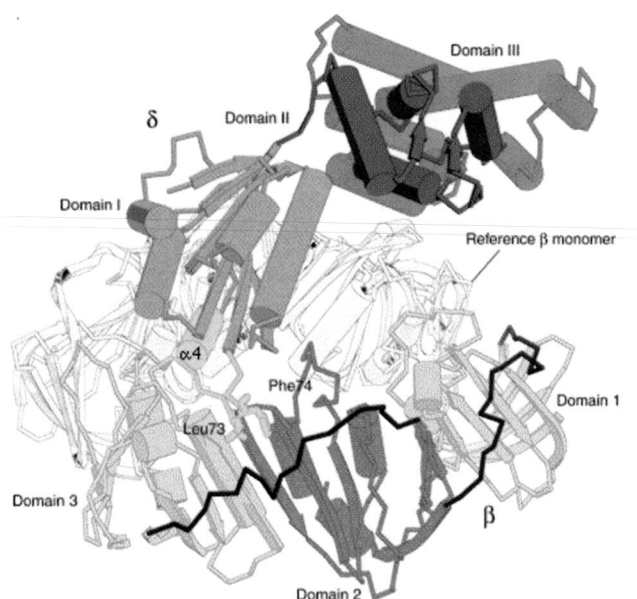

FIGURE 30-32 X-Ray structure of the β–δ complex. A second β subunit taken from the X-ray structure of the sliding clamp (Fig. 30-13), the "Reference β monomer," is drawn in gray. The view is along the edge of the β ring. The δ subunit's β interaction element (*yellow*) consists largely of the α4 helix and two hydrophobic residues, Leu 73 and Phe 74, whose side chains are drawn in stick form. [Courtesy of John Kuriyan, The Rockefeller University. PDB 1JQJ.]

plex, the β subunit increases its radius of curvature relative to that in the β dimer (Fig. 30-13) such that the β–δ interaction would induce the opening of one of the sliding clamp's β–β interfaces by ~15 Å. Such a gap is large enough to permit the passage of ssDNA but not dsDNA. Apparently, the clamp loader functions by trapping one β subunit of the sliding clamp in a conformation that prevents ring closure rather than actively pulling apart the two halves of the ring. This is corroborated by molecular dynamics simulations (Section 9-4) suggesting that a β$_2$ dimer has a stable conformation but that an isolated β subunit with the conformation it has in the β$_2$ dimer rapidly (in ~1.5 ns) converts to a conformation resembling that in the β–δ complex. Thus, the conformational change of the δ subunit's β interaction element on binding to a β subunit is reminiscent of the action of a plumber's wrench in unlatching the nearby β–β interface so as to allow the sliding clamp to spring open.

The X-ray structure of the γ$_3$δδ' complex (the clamp loader with its two τ subunits lacking their C-terminal 212 residues; γ and τ are interchangeable in terms of their clamp loading functions), also determined by Kuriyan and O'Donnell, suggests how the clamp loader functions. The γ, δ, and δ' subunits all have similar folds; they are all members of the widely distributed **AAA⁺** family (for *A*TPases *a*ssociated with a variety of cellular *a*ctivities; DnaA and DnaC proteins are members of this family) even though only the γ (and τ) subunits bind and hydrolyze ATP. The conserved regions of AAA⁺ proteins consist of two do-

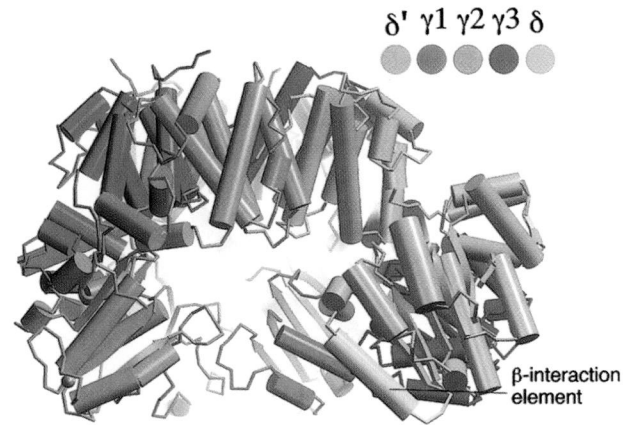

δ' γ1 γ2 γ3 δ

β-interaction
element

FIGURE 30-33 X-Ray structure of the γ₃δδ′ clamp loading complex. The subunits are colored as indicated with the β interaction element yellow. [Courtesy of John Kuriyan, The Rockefeller University. PDB 1JR3.]

mains, an N-terminal ATP-binding domain and a smaller domain composed of a 3-helix bundle, whose relative orientations vary with ATP binding. The γ₃δδ′ complex's C-terminal domains form a ring-shaped collar in which the subunits are arranged in clockwise order δ′-γ1-γ2-γ3-δ (Fig. 30-33). The relative orientations of the three domains differ in each of the five subunits, thereby forming a highly asymmetric structure, particularly in its more N-terminal regions. Even though the X-ray structure of the γ₃δδ′ complex is devoid of bound nucleotides, its γ subunits' ATP-

binding sites have been identified by analogy with the known structures of the nucleotide complexes of other AAA$^+$ proteins, such as NSF protein (Section 12-4D; a supposition that was later confirmed by the X-ray structure of the first two domains of the γ subunit in complex with ATPγS). The γ subunits' ATP-binding sites, which are formed by their N-terminal domains, are all located on the inner surface of the γ₃δδ′ complex and hence the clamp loader.

The clamp loader must tightly bind the sliding clamp prior to its loading on/unloading off the template DNA but must subsequently release the clamp to avoid interfering with its binding to the Pol III core (αεθ). The structures of the clamp loader and the β–δ complex, together with a variety of biochemical evidence, suggest a model of how this might occur (Fig. 30-34): The binding of ATP to γ1 (the γ subunit that contacts δ′) results in a conformational change that exposes the otherwise occluded ATP-binding site of γ2; ATP binding to γ2 likewise exposes γ3; and ATP binding to γ3 exposes the δ subunit's β interaction element, thereby permitting it to bind to a β subunit so as to spring open the sliding clamp. The eventual β- and DNA-stimulated hydrolysis of these bound ATPs reverses this process.

Once the clamp loader has loaded the sliding clamp onto the template DNA, it must dissociate from the clamp to allow the binding of the Pol III core. However, when the synthesis of an Okazaki fragment has been completed, the Pol III core must dissociate from the sliding clamp so that it can initiate the synthesis of the next Okazaki fragment. How does this occur?

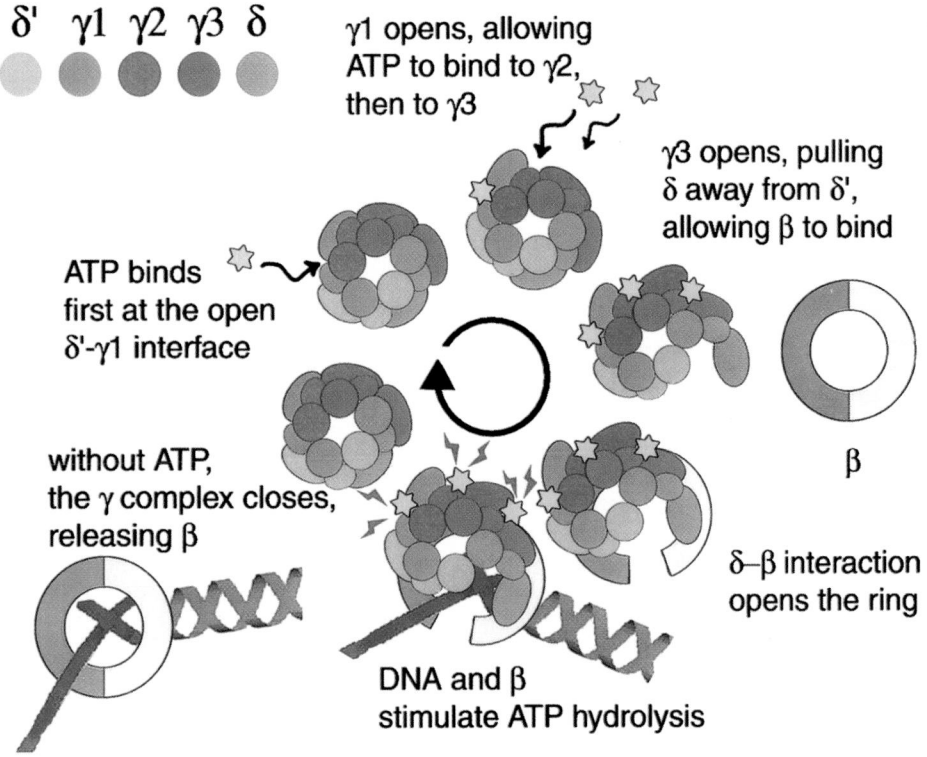

δ' γ1 γ2 γ3 δ

γ1 opens, allowing ATP to bind to γ2, then to γ3

γ3 opens, pulling δ away from δ', allowing β to bind

ATP binds first at the open δ'-γ1 interface

β

without ATP, the γ complex closes, releasing β

δ–β interaction opens the ring

DNA and β stimulate ATP hydrolysis

FIGURE 30-34 Schematic diagram of the clamp loading cycle. This speculative model is based on a combination of structural and biochemical information. [Courtesy of John Kuriyan, The Rockefeller University.]

The α and δ subunits bind to overlapping sites on one face of the β_2 sliding clamp. This was shown by the observations that the phosphorylation of a kinase recognition sequence that had been engineered into the C-terminal segment of β is inhibited by both α and δ. The β subunit has an ~30-fold greater affinity for the γ complex in the presence of ATP than it has for the Pol III core. However, when primed ssDNA is also present, this order of affinity is reversed with β preferring to bind to the Pol III core (possibly due to the additional contacts between the core and the DNA). Thus once the sliding clamp has been loaded onto the primed lagging strand template, the clamp loader is displaced by the Pol III core, which then blocks the clamp loader from unloading the clamp. Instead, the clamp loader loads a new clamp onto the lagging strand template in association with the primer that the primosome had synthesized in preparation for the next round of Okazaki fragment synthesis (Fig. 30-28b). When the Pol III core has completed its synthesis of the Okazaki fragment, that is, when the gap between the two successively synthesized Okazaki fragments has been reduced to a nick, it releases the DNA and the clamp. The Pol III core then binds to the newly primed template and its associated clamp (displacing the clamp loader), where it commences the synthesis of the next Okazaki fragment. Throughout this process, the Pol III holoenzyme is held at the replication fork by the leading strand Pol III core, which remains tethered to the DNA by its associated sliding clamp.

The sliding clamp that remains around the completed Okazaki fragment probably functions to recruit Pol I and DNA ligase so as to replace the RNA primer on the previously synthesized Okazaki fragment with DNA and seal the remaining nick. However, the sliding clamp must eventually be recycled. It was initially assumed that this was the job of the clamp loader. However, it is now clear that the release of the sliding clamp from its associated DNA is carried out by free δ subunit (the "wrench" in the clamp loader that cracks apart the β subunits forming the sliding clamp), which is synthesized in 5-fold excess over that required to populate the cell's few clamp loaders.

d. Replication Termination

The *E. coli* replication terminus is a large (350 kb) region flanked by seven nearly identical nonpalindromic ~23-bp terminator sites, ***TerE, TerD,*** and ***TerA*** on one side and ***TerG, TerF, TerB,*** and ***TerC*** on the other (Fig. 30-35; note that *oriC* is directly opposite the terminus region on the *E. coli* chromosome). A replication fork traveling counterclockwise as drawn in Fig. 30-35 passes through *TerG, TerF, TerB,* and *TerC* but stops on encountering either *TerA, TerD,* or *TerE* (*TerD* and *TerE* are presumably backup sites for *TerA*). Similarly, a clockwise-traveling replication fork transits *TerE, TerD,* and *TerA* but halts at *TerC* or, failing that, *TerB* or *TerF* or *TerG.* Thus, these termination sites act as one-way valves that allow replication forks to enter the termination region but not to leave it. This arrangement guarantees that the two replication forks generated by bidirectional initiation at *oriC* will meet in

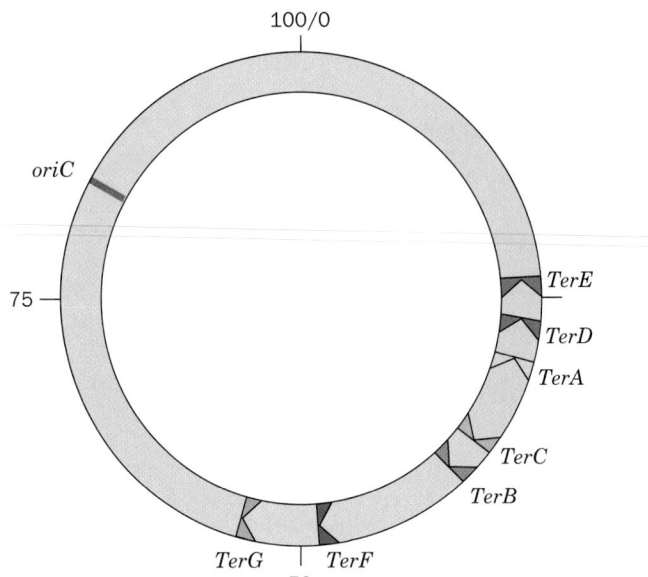

FIGURE 30-35   Map of the *E. coli* chromosome showing the positions of the *Ter* sites and the *oriC* site. The *TerG, TerB, TerF,* and *TerC* sites, in combination with Tus protein, allow a counterclockwise-moving replisome to pass but not a clockwise-moving replisome. The opposite is true of the *TerE, TerD,* and *TerA* sites. Consequently, two replication forks that initiate bidirectional DNA replication at *oriC* will meet between the oppositely facing *Ter* sites.

the replication terminus even if one of them arrives there well ahead of its counterpart.

The arrest of replication fork motion at *Ter* sites requires the action of **Tus** protein, a 309-residue monomer that is the product of the ***tus*** gene (for *t*erminator *u*tilization *s*ubstance). Tus specifically binds to a *Ter* site, where it prevents strand displacement by DnaB helicase, thereby arresting replication fork motion. The X-ray structure of Tus in complex with a 15-bp *Ter* sequence-containing DNA with a single T overhang at each 5′ end, determined by Kosuke Morikawa, reveals that Tus consists of two domains that form a deep positively charged cleft that largely envelops the bound DNA (Fig. 30-36). A 5-bp segment of the DNA near the side of Tus that permits the passage of the replication fork (the lower side of Fig. 30-36) is deformed and underwound relative to canonical (ideal) B-DNA such that its major groove becomes deeper and its minor groove is significantly expanded. The protein makes polar contacts with more than two-thirds of the phosphate groups in a 13-bp region and its interdomain β sheet penetrates the deepened major groove to make sequence-specific contacts with the exposed bases. The importance of this interdomain region for Tus function is demonstrated by the observation that most single residue mutations that reduce the ability of Tus to arrest replication occur in this interdomain region.

When Tus is fused to another DNA-binding protein, replication is inhibited at the other protein's binding site. This suggests that Tus does not act as a simple DNA-

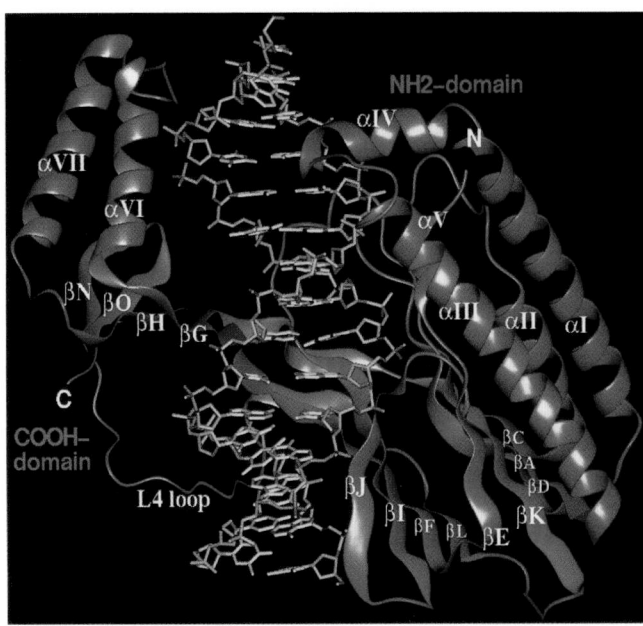

FIGURE 30-36 X-Ray structure of *E. coli* Tus protein in complex with a 15-bp *Ter*-containing DNA. The protein's N- and C-terminal domains are green and blue and its bound DNA is shown in stick form with its bases yellow and its sugar–phosphate backbone gold. [From Kamada, K., Horiuchi, T., Ohsumi, K., Simamato, M., and Morikawa, K., *Nature* **383,** 599 (1996). Used with permission. PDBid 1ECR.] *⨠* **See the Interactive Exercises**

binding clamp, but interacts with DnaB to inhibit its helicase action. Apparently, Tus interferes with the progress of DnaB in unwinding DNA from one side of Tus but not the other, although how Tus and DnaB interact is unknown. Curiously, however, this termination system is not essential for termination. When the replication terminus is deleted, replication simply stops, apparently through the collision of opposing replication forks. Nevertheless, this termination system is highly conserved in gram-negative bacteria.

D. *Fidelity of Replication*

Since a single polypeptide as small as the Pol I Klenow fragment can replicate DNA by itself, why does *E. coli* maintain a battery of >20 intricately coordinated proteins to replicate its chromosome? The answer apparently is *to ensure the nearly perfect fidelity of DNA replication required to preserve the genetic message's integrity from generation to generation.*

The rates of reversion of mutant *E. coli* or T4 phage to the wild type indicates that only one mispairing occurs per 10^8 to 10^{10} base pairs replicated. This corresponds to ~1 error per 1000 bacteria per generation. Such high replication accuracy arises from four sources:

1. Cells maintain balanced levels of dNTPs through the mechanism discussed in Section 28-3A. This is an important aspect of replication fidelity because a dNTP present at aberrantly high levels is more likely to be misincorporated and, conversely, one present at low levels is more likely to be replaced by the dNTPs present at higher levels.

2. The polymerase reaction itself has extraordinary fidelity. This is because, as we have seen (Section 30-2A), the polymerase reaction occurs in two stages: (1) a binding step in which the incoming dNTP base pairs with the template while the enzyme is in an open conformation that cannot catalyze the polymerase reaction; and (2) a catalysis step in which the polymerase forms a closed conformation about the newly formed base pair, which properly positions its catalytic residues (induced fit). Since the formation of the closed conformation requires that the incoming dNTP form a Watson–Crick-shaped base pair with the template, the conformation change constitutes a double check for correct base pairing.

3. The $3' \rightarrow 5'$ exonuclease functions of Pol I and Pol III detect and eliminate the occasional errors made by their polymerase functions. In fact, mutations that increase a DNA polymerase's proofreading exonuclease activity decrease the rates of mutation of other genes.

4. A remarkable battery of enzyme systems, contained in all cells, function to repair residual errors in the newly synthesized DNA as well as any damage that it may incur after its synthesis through chemical and/or physical insults. We discuss these DNA repair systems in Section 30-5.

In addition, *the inability of a DNA polymerase to initiate chain elongation without a primer is a feature that increases DNA replication fidelity.* The first few nucleotides of a chain to be coupled together are those most likely to be mispaired because of the cooperative nature of base pairing interactions (Section 29-2). The editing of a short duplex oligonucleotide is similarly an error-prone process. The use of RNA primers eliminates this source of error since the RNA is eventually replaced by DNA under conditions that permit accurate base pairing to be achieved.

One might wonder why cells have evolved the complex system of discontinuous lagging strand synthesis rather than a DNA polymerase that could simply extend DNA chains in their $3' \rightarrow 5'$ direction. Consideration of the chemistry of DNA chain extension also leads to the con-

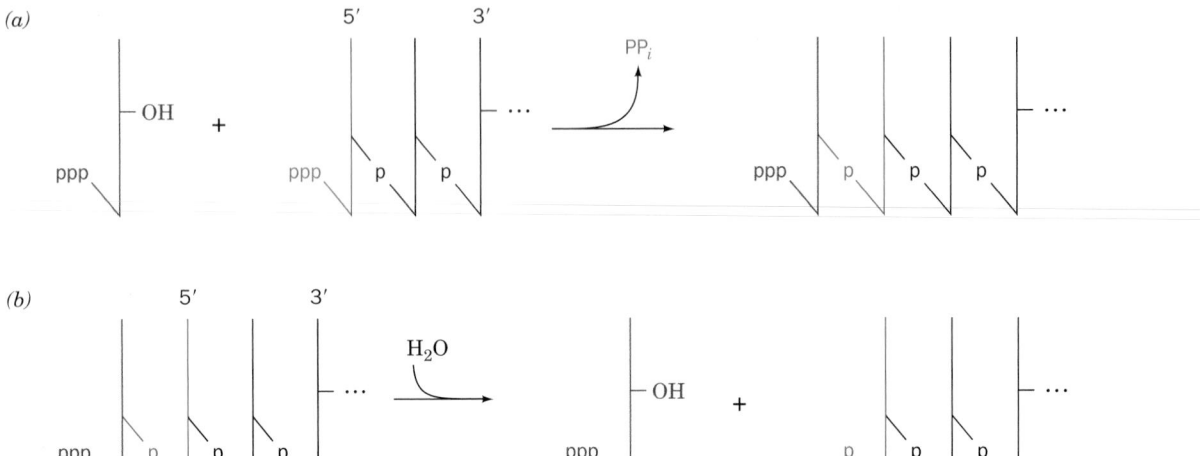

FIGURE 30-37 Chemical consequences if a DNA polymerase could synthesize DNA in its 3′ → 5′ direction. (*a*) The coupling of each nucleoside triphosphate to the growing chain would be driven by the hydrolysis of the previously appended nucleoside triphosphate. (*b*) The editorial removal of an incorrect 5′-terminal nucleoside triphosphate would render the DNA chain incapable of further extension.

clusion that this system promotes high-fidelity replication. The linking of 5′-deoxynucleotide triphosphates in the 3′ → 5′ direction would require the retention of the growing chain's 5′-terminal triphosphate group to drive the next coupling step (Fig. 30-37*a*). On editing a mispaired 5′-terminal nucleotide (Fig. 30-37*b*), this putative polymerase would—in analogy with Pol I, for example—excise the offending nucleotide, leaving either a 5′-OH or a 5′-phosphate group. Neither of these terminal groups is capable of energizing further chain extension. A proofreading 3′ → 5′ DNA polymerase would therefore have to be capable of reactivating its edited product. The inherent complexity of such a system has presumably selected against its evolution.

4 ■ EUKARYOTIC REPLICATION

There is a remarkable degree of similarity between eukaryotic and prokaryotic DNA replication mechanisms. Nevertheless, there are important differences between these two replication systems as a consequence of the vastly greater complexity of eukaryotes in comparison to prokaryotes. For example, eukaryotic chromosomes are structurally complicated and dynamic complexes of DNA and protein (Section 34-1) with which the replication machinery must interact in carrying out its function. Consequently, as is true of most aspects of biochemistry, our knowledge of how DNA is replicated in eukaryotes has lagged well behind that for prokaryotes, although in recent years there has been significant progress in our understanding of this essential process. In this section, we outline what is known about DNA replication in eukaryotes. We also discuss two DNA polymerases that are peculiar to eukaryotic systems: reverse transcriptase and telomerase.

A. *The Cell Cycle*

The **cell cycle,** the general sequence of events that occur during the lifetime of a eukaryotic cell, is divided into four distinct phases (Fig. 30-38):

1. Mitosis and cell division occur during the relatively brief **M phase** (for *m*itosis).

2. This is followed by the **G_1 phase** (for *g*ap), which covers the longest part of the cell cycle. This is the main period of cell growth.

3. G_1 gives way to the **S phase** (for *s*ynthesis) which, in contrast to events in prokaryotes, *is the only period in the cell cycle when DNA is synthesized.*

4. During the relatively short **G_2 phase,** the now tetraploid cell prepares for mitosis. It then enters M phase once again and thereby commences a new round of the cell cycle.

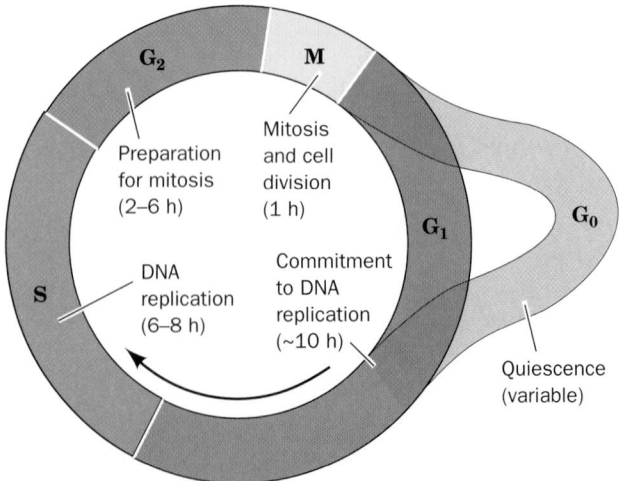

FIGURE 30-38 The eukaryotic cell cycle. Cells in G_1 may enter a quiescent phase (G_0) rather than continuing about the cycle.

The cell cycle for cells in culture typically occupies a 16- to 24-h period. In contrast, cell cycle times for the different types of cells of a multicellular organism may vary from as little as 8 h to >100 days. Most of this variation occurs in the G_1 phase. Moreover, many terminally differentiated cells, such as neurons or muscle cells, never divide; they assume a quiescent state known as the **G_0 phase.**

A cell's irreversible "decision" to proliferate is made during G_1. Quiescence is maintained if, for example, nutrients are in short supply or the cell is in contact with other cells **(contact inhibition).** Conversely, DNA synthesis may be induced by various agents such as carcinogens or tumor viruses, which trigger uncontrolled cell proliferation (cancer; Sections 19-3B and 34-4C); by the surgical removal of a tissue, which results in its rapid regeneration; or by proteins known as **mitogens,** which bind to cell surface receptors and induce cell division (Section 34-4D).

a. The Cell Cycle Is Controlled by Cyclins and Cyclin-Dependent Protein Kinases

The progression of a cell through the cell cycle is regulated by proteins known as **cyclins** and **cyclin-dependent protein kinases (Cdks).** Cyclins are so named because they are synthesized during one phase of the cell cycle and completely degraded during a succeeding phase (protein degradation is discussed in Section 32-6). A particular cyclin specifically binds to and thereby activates its corresponding Cdk(s) to phosphorylate its target proteins, thus activating these proteins to carry out the processes comprising that phase of the cell cycle. In order to enter a new phase in the cell cycle, a cell must satisfy a corresponding **checkpoint,** which monitors whether the cell has satisfactorily completed the preceding phase [e.g., the attachment of all chromosomes to the mitotic spindle must precede mitosis (Section 1-4A); if this were not the case for even one chromosome, one daughter cell would lack this chromosome and the other would have two, both deleterious if not lethal conditions]. If the cell has not met the criteria of the checkpoint, the cell cycle is slowed or even arrested until it does so. We further discuss cell cycle control in Section 34-4C.

B. *Eukaryotic Replication Mechanisms*

Much of what we know about eukaryotic DNA replication has been learned from studies on budding yeast (*Saccharomyces cerevisiae*) and fission yeast (*Schizosaccharomyces pombe*), the simplest eukaryotes, and on simian virus 40 (SV40), which has a 5243-bp circular DNA chromosome (Fig. 5-40) that has only one replication origin. However, studies of DNA replication in the cells of **metazoa** (multicellular animals), particularly *Drosophila, Xenopus laevis* (an African clawed toad, whose eggs are easily studied), and humans, have also led to important advances in our knowledge.

a. Eukaryotic Cells Contain Numerous DNA Polymerases

The many known DNA polymerases can be classified into six families based on phylogenetic relationships: family A (e.g., *E. coli* Pol I), family B (e.g., *E. coli* Pol II), family C (e.g., *E. coli* Pol III), and families D, X, and Y. Animal cells express at least four distinct types of DNA polymerases that are implicated in DNA replication. They are designated, in the order of their discovery, DNA polymerases (pols) α, γ, δ, and ε (alternatively, POLA, POLG, POLD1, and POLE). Their functions were, in part, elucidated by their different responses to various inhibitors (Table 30-5).

Pol α, a B-family enzyme, occurs only in the cell nucleus and participates in the replication of chromosomal DNA. This function was largely established through the use of its specific inhibitor **aphidicolin**

Aphidicolin

TABLE 30-5 Properties of Some Animal DNA Polymerases

	α	β	γ	δ
Location	nucleus	nucleus	mitochondrion	nucleus
Subunit masses (kD)[a]	167, ~83, 58, 48	68	143	125, 55, 40, 22
	(165, 67, 58, 48)	(39)	(125, 43)	(125, 66, 50)
Family	B	X	A	B
Inhibitors:				
Aphidicolin	strong	none	none	strong
Dideoxy NTPs	none	strong	strong	weak
N-Ethylmaleimide (NEM)[b]	strong	none	strong	strong

[a]Yeast *S. cerevisiae* (mammalian cells).
[b]A cysteine-alkylating agent (Section 12-4D).

Source: Kornberg, A. and Baker, T.A., *DNA Replication* (2nd ed.), p. 199, Freeman (1992); *and* Hübscher, U., Nasheuer, H.-P., and Syväoja, J.E., *Trends Biochem. Sci.* **25,** 143 (2000).

and by the observation that pol α activity varies with the rate of cellular proliferation. Pol α, as do all DNA polymerases, replicates DNA by extending a primer in the 5′ → 3′ direction under the direction of a single-stranded DNA template. This enzyme lacks exonuclease activity but tightly associates with a primase (which consists of a 48-kD subunit that contains the primase catalytic site and a 58-kD subunit that is required for full primase activity) and an ~83-kD subunit that is implicated in the regulation of initiation, yielding a protein named **pol α/primase.**

The X-ray structure of the B-family DNA polymerase encoded by bacteriophage RB69 **(RB69 pol),** determined by Steitz, reveals that this enzyme consists of five domains arranged around a central hole that contains its polymerase active site (Fig. 30-39). RB69 pol has the right-hand-like architecture first seen in A-family DNA polymerases (Figs. 30-8 and 30-9), and its palm domain has a structurally similar core that contains the two invariant Asp residues implicated in the nucleotidyl transfer mechanism (Fig. 30-10). However, in comparison with A-family enzymes, the fingers domain of RB69 pol exhibits a 60° rotation and its editing domain lies on the opposite side of its palm domain.

Pol δ, a B-family enzyme, is a nuclear enzyme with inhibitor sensitivities similar to those of pol α (Table 30-5). It lacks an associated primase but exhibits a proofreading 3′ → 5′ exonuclease activity. Moreover, whereas pol α exhibits only moderate processivity (~100 nucleotides), that of pol δ is essentially unlimited (replicates the entire length of a template), but only when it is in complex with a protein named **proliferating cell nuclear antigen (PCNA;** so named because it occurs only in the nuclei of proliferating cells and reacts with antibodies produced by a subset of patients with the autoimmune disease systemic lupus erythematosus). The X-ray structure of PCNA (Fig. 30-40), determined by Kuriyan, reveals that it forms a trimeric ring with almost identical structure (and presumably function) as the *E. coli* β_2 sliding clamp (Fig. 30-13). Thus, each PCNA subunit consists of four rather than six of the structurally similar βαβββ motifs from which the *E. coli* β subunit is constructed. Intriguingly, PCNA and the β subunit exhibit no significant sequence identity, even when their structurally similar portions are aligned.

Pol δ in complex with PCNA is required for both leading and lagging strand synthesis. In contrast, pol α/primase functions to synthesize 7- to 10-nt RNA primers, which it extends by an additional ~15 nt of DNA. Then, in a process called **polymerase switching,** the eukaryotic counterpart of the *E. coli* γ complex (the clamp loader), **replication factor C (RFC),** displaces the pol α and loads PCNA on the template DNA near the primer strand, following which pol δ binds to the PCNA and processively extends the DNA strand.

Pol ε, a B-family nuclear enzyme which superficially resembles pol δ, differs from it in that pol ε is highly processive in the absence of PCNA and has a 3′ → 5′ exonuclease activity that degrades single-stranded DNA to 6- or 7-residue oligonucleotides rather than to mononucleotides, as does that of pol δ. Although pol ε is required for the viability of yeast, its essential function can be carried out by only the noncatalytic C-terminal half of its 256-kD catalytic

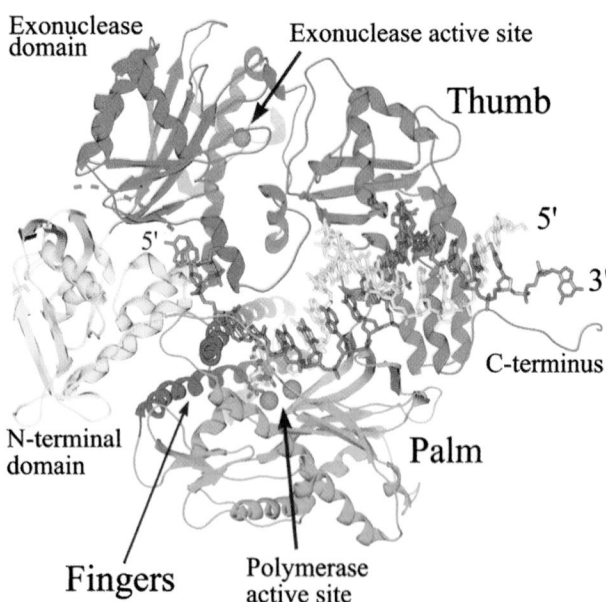

FIGURE 30-39 X-Ray structure of RB69 DNA polymerase (RB69 pol) in complex with primer–template DNA and dTTP. The protein is drawn in ribbon form with its various domains differently colored. The DNA is drawn in stick form with its primer strand gold and its template strand gray. The incoming dTTP, also in stick form, is colored according to atom type (C gold, N blue, O red, and P magenta). The two Ca^{2+} ions at the polymerase site are represented by cyan balls as is the Ca^{2+} ion at the exonuclease site. The probable extension of the path taken by the single-stranded template as it enters the polymerase active site is shown as a dashed gray line. [Courtesy of Thomas Steitz, Yale University. PDBid 1IG9.]

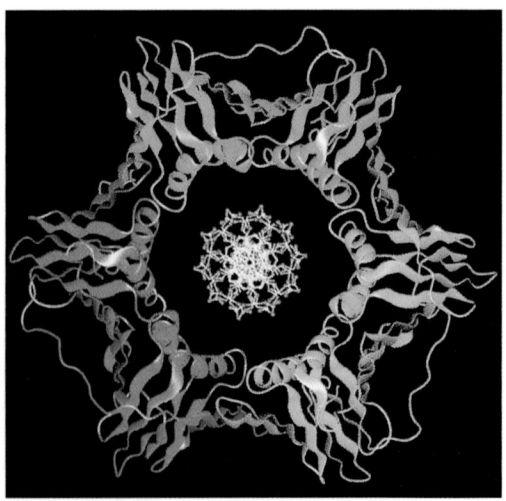

FIGURE 30-40 X-Ray structure of PCNA. Its three subunits (*red, green, and yellow*) form a 3-fold symmetric ring. A model of duplex DNA viewed along its helix axis has been drawn in the center of the PCNA ring. Compare this structure with that of the β subunit dimer of *E. coli* Pol III holoenzyme (Fig. 30-13). [Courtesy of John Kuriyan, The Rockefeller University. PDBid 1PLQ.] 🔖 **See the Interactive Exercises**

subunit, which is unique among B-family DNA polymerases. Moreover, the only DNA polymerases required to replicate SV40 DNA are pol α and pol δ. It therefore appears that, at least in yeast, pol ε has an essential control function but not a catalytic function.

Pol γ, an A-family enzyme, occurs exclusively in the mitochondrion, where it presumably replicates the mitochondrial DNA. Chloroplasts contain a similar enzyme.

Eukaryotic cells contain batteries of DNA polymerases. These include the DNA polymerases that participate in chromosomal DNA replication (pols α, δ, and ε) and several that take part in DNA repair processes (Section 30-5) including **pols β, η, ι, κ,** and ζ (alternatively, POLB, POLH, POLI, POLK, and POLZ). Pol β, an X-family enzyme, is remarkable for its small size (a 335-residue monomer in rat). The X-ray structure of a stable proteolytic fragment (residues 85–335) of rat pol β (Fig. 30-41), independently determined by Zdenek Hostomsky and Joseph Kraut, reveals that this protein has the right-hand-like shape of other polymerases of known structure (e.g., Figs. 30-8, 30-9, and 30-39). However, its folding topology is unique, which suggests that it does not share a common ancestor with these other polymerases.

b. Eukaryotic Chromosomes Consist of Numerous Replicons

Eukaryotic and prokaryotic DNA replication systems differ most obviously in that eukaryotic chromosomes have multiple replication origins in contrast to the single replication origin of prokaryotic chromosomes. Eukaryotic cells replicate DNA at the rate of ~50 nt/s (~20 times slower than does *E. coli*) as was determined by autoradiographically measuring the lengths of pulse-labeled sections of eukaryotic chromosomes. Since a eukaryotic chromosome typically contains 60 times more DNA than those of prokaryotes, its bidirectional replication from a single origin would require ~1 month to complete. Electron micrographs such as Fig. 30-42, however, reveal that eukaryotic chromosomes contain multiple origins, one every 3 to 300 kb depending on both the species and the tissue, so that S phase usually occupies only a few hours.

Cytological observations indicate that the various chromosomal regions are not all replicated simultaneously; rather, clusters of 20 to 80 adjacent **replicons** (replication units; DNA segments that are each served by a replication origin) are activated simultaneously. New replicons are activated throughout S phase until the entire chromosome has been replicated. During this process, replicons that have already been replicated are distinguished from those that have not; that is, *a cell's chromosomal DNA is replicated once and only once per cell cycle.*

c. The Assembly of the Eukaryotic Initiation Complex Occurs in Two Stages

The once-and-only-once replication of eukaryotic DNA per cell cycle is conferred by a type of binary switch. A **pre-replicative complex (pre-RC)** is assembled at each replication origin during the G₁ phase of the cell cycle. This is the only period of the cell cycle during which the pre-

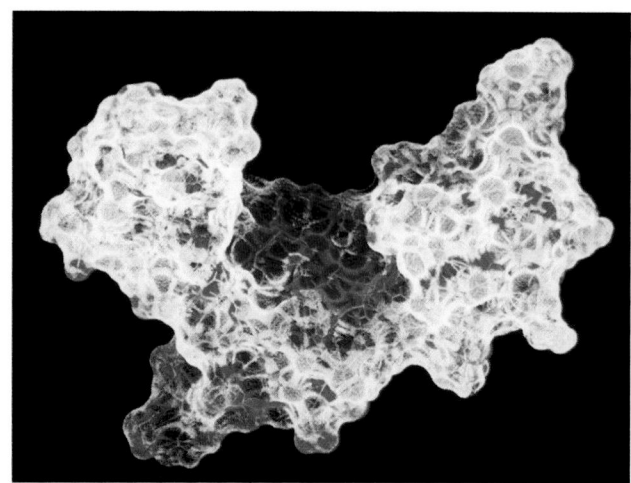

FIGURE 30-41 X-Ray structure of the catalytic domain of rat DNA polymerase β. The protein, which is oriented with its N-terminal fingers subdomain at the top left, is represented by its solvent-accessible surface colored according to charge, with red negative, blue positive, and white neutral. The strong positive charge of the putative DNA binding cleft no doubt facilitates the binding of the polyanionic DNA. [Courtesy of Zdenek Hostomsky, Agouron Pharmaceuticals, San Diego, California. PDBid 1RPL.]

RC can form and hence this process is known as **licensing.** However, a licensed pre-RC cannot initiate DNA replication. Rather, it must be activated to do so, a process that occurs only during S phase. *This temporal separation of pre-RC assembly and origin activation ensures that a new pre-RC cannot assemble on an origin that has already "fired" (commenced replication) so that an origin can only fire once per cell cycle.* How does this occur?

The elucidation of the licensing process and how the pre-RC is activated to form an initiation complex is still in its early stages. Thus, although it appears that most of the proteins forming these complexes have been identified, their structures, interactions, and, in many cases, their functions are largely unknown. Keeping this in mind, let us consider what is known about these processes.

Replication origins are surprisingly variable among species, often within the same organism, and even vary with a given organism's developmental stage. Thus, whereas *S. cerevisiae* origins, which are known as

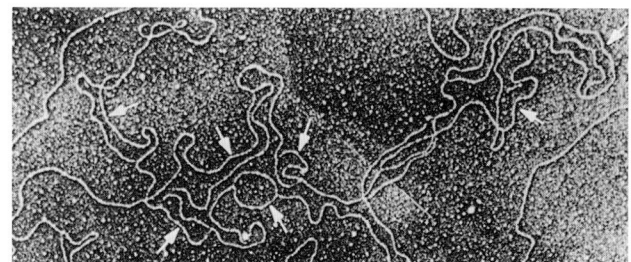

FIGURE 30-42 Electron micrograph of a fragment of replicating *Drosophila* DNA. The arrows indicate its multiple replication eyes. [From Kreigstein, H.J. and Hogness, D.S., *Proc. Natl. Acad. Sci.* **71,** 136 (1974).]

autonomously replicating sequences (ARS), contain a highly conserved 11-bp AT-rich sequence within a less well defined ~125-bp region, some metazoan origins are dispersed over 10 to 50 kb "initiation zones" that contain multiple origins and, in some cases, require no specific DNA sequence at all. Despite this disparity, the proteins that participate in eukaryotic DNA replication are highly conserved from yeast to humans.

The assembly of the pre-RC (Fig. 30-43) begins late in M phase or early in G₁ phase with the binding of the **origin recognition complex (ORC),** a hexamer of related proteins **(Orc1** through **Orc6),** to the origin, where it remains bound during most or all of the cell cycle. ORC, the functional analog of DnaA protein in *E. coli* replication initiation (Section 30-3C), then recruits two proteins, **Cdc6** in *S. cerevisiae* (**Cdc18** in *S. pombe*; Cdc for *c*ell *d*ivision *c*ycle) and **Cdt1.** These proteins then cooperate with the ORC to load the **MCM complex** [named for its *m*inichromosome (plasmid) *m*aintenance functions], a hexamer of related subunits (**Mcm2** through **Mcm7**), onto the DNA to yield the licensed pre-RC. The MCM complex, a ring-

shaped ATP-driven helicase, is the analog of *E. coli* DnaB helicase, whereas Cdc6/Cdc18 together with Cdt1 appears to be an analog of *E. coli* DnaC (which facilitates DnaB loading). With the exception of Cdt1, all of these proteins, Orc1 through Orc6, Cdc6/Cdc18, Mcm2 through Mcm7, as well as *E. coli* DnaA, DnaB, and DnaC, are AAA⁺ ATPases.

The conversion of a licensed pre-RC to an active initiation complex requires the addition of pol α/primase, pol ε, and several accessory proteins, which only occurs at the onset of S phase. This process begins with addition of **Mcm10** protein (which shares no sequence similarity with any of the subunits of the MCM complex) to the pre-RC, which probably displaces Cdt1. This is followed by the addition of at least two protein kinases, a Cdk and **Ddk,** the latter being a heterodimer of the protein kinase **Cdc7** with its activating subunit **Dbf4** (Ddk stands for *D*bf4-*d*ependent *k*inase). Ddk acts to phosphorylate five of the six MCM subunits (all but Mcm2) so as to activate the MCM complex as a helicase. In contrast, the way in which Cdks activate the pre-RC is poorly understood although several ORC and MCM proteins as well as Cdc6/Cdc18 are phosphorylated by Cdks. Ddk together with a Cdk also recruits **Cdc45** to the growing initiation complex. Cdc45, in turn, is required for the assembly of the initiating synthetic machinery at the replication fork, including pol α/primase, pol ε, PCNA, and **replication protein A (RPA),** the heterotrimeric eukaryotic counterpart of SSB, thereby forming an active initiation complex.

d. Re-Replication Is Prevented through the Actions of Cdks and Geminin

Once initiation (priming) has occurred, the initiation complex is joined by RFC and pol δ and, as is described above, is converted to an active replicative complex by polymerase switching. DNA replication then proceeds bidirectionally until each replication fork has collided with an oppositely traveling replication fork, thereby completing the replication of the replicon. An active replication fork will destroy any licensed pre-RCs and unfired initiation complexes in its path, thereby preventing the DNA at such sites from being replicated twice. Eukaryotes appear to lack termination sequences and proteins analogous to the *Ter* sites and Tus protein in *E. coli*.

Several redundant mechanisms ensure that a pre-RC can initiate DNA synthesis only once. Cdks are active from late G₁ phase through late M phase. These elevated Cdk levels, which are required to activate initiation, also prevent reinitiation. The Cdk-mediated phosphorylation of Cdc6/Cdc18, which occurs late in G₁ after the pre-RCs have formed, causes Cdc6/Cdc18 to be proteolytically degraded in yeast and exported from the nucleus in mammalian cells. Evidently, Cdc6/Cdc18 is only required for the assembly of the pre-RC, not its activation. The helicase activity of the MCM complex is inhibited by phosphorylation, at least *in vitro*. Moreover, MCM proteins are exported from the nucleus in G₂ and M phases, a process that is interrupted by Cdk inactivation. However, the function of Cdk-mediated phosphorylation of ORC proteins is unclear.

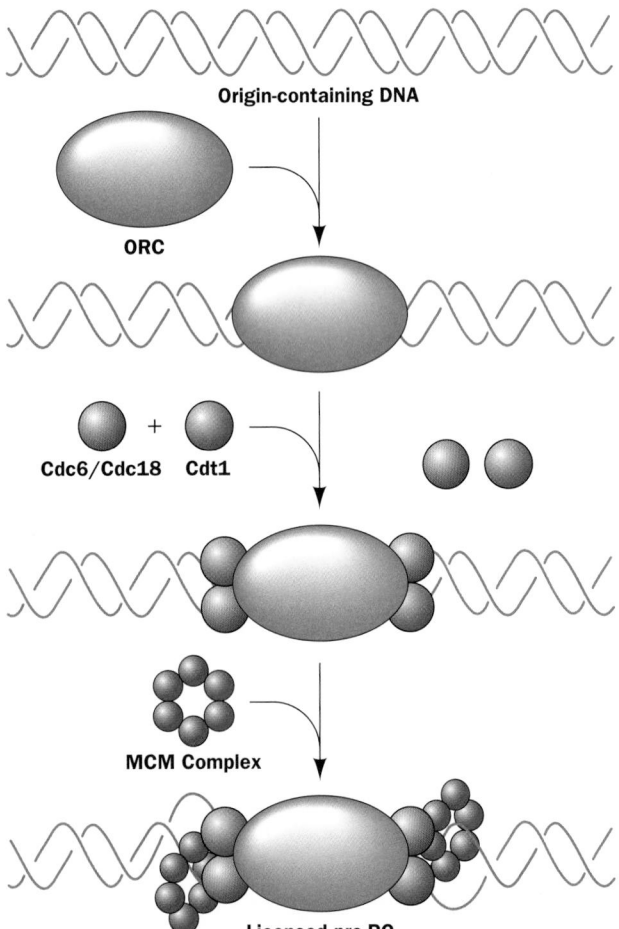

Origin-containing DNA

ORC

Cdc6/Cdc18   **Cdt1**

MCM Complex

Licensed pre-RC

FIGURE 30-43   Schematic diagram for the assembly of the eukaryotic pre-replicative complex (pre-RC). The actual stoichiometries, positions, and interactions of its various components are largely unknown. The pre-RC only forms during the G₁ phase of the cell cycle.

Metazoan cells have yet another mechanism to prevent the assembly of a licensed pre-RC on already replicated DNA. High levels of a protein named **geminin** appear in S phase and continue to accumulate until late M phase, when geminin is degraded. Geminin associates with Cdt1 (which together with Cdc6/Cdc18 loads the MCM complex onto the ORC) so as to inhibit the assembly of the pre-RC. This inhibition can be reversed by the addition of excess Cdt1. It therefore seems likely that the presence of geminin provides protection against DNA re-replication under conditions when Cdks are inhibited by checkpoint activation.

Finally, cells that have shifted to the G_0 (quiescent) phase of the cell cycle (Fig. 30-38)—the majority of cells in the human body—cease making DNA. Such cells are characterized by the absence of Cdk activity. In proliferating cells, this would permit the re-replication of DNA. However, cells in G_0 also lack the proteins of the MCM complex and are therefore incapable of assembling licensed pre-RCs. Since cancerous cells are characterized by being in a state of rapid proliferation (Section 19-3B), the presence of MCM complex proteins in what should be quiescent cells is a promising diagnostic marker for cancer.

e. Primers Are Removed by RNase H1 and Flap Endonuclease-1

The RNA primers of eukaryotic Okazaki fragments are removed through the actions of two enzymes: **RNase H1** removes most of the RNA leaving only a 5′ ribonucleotide adjacent to the DNA, which is then removed through the action of **flap endonuclease-1 (FEN1).** However, as we have seen, pol α/primase extends the RNA primers it has made by ~15 nt of DNA before it is displaced by pol δ. Since pol α lacks proofreading ability, this primer extension is more likely to contain errors than the DNA synthesized by pol δ. However, FEN1 provides what is, in effect, pol α's proofreading function: It is also an endonuclease that excises mismatch-containing oligonucleotides up to 15 nt long from the 5′ end of an annealed DNA strand. Moreover, FEN1 can make several such excisions in succession to remove more distant mismatches. The excised segment is later replaced by pol δ as it synthesizes the succeeding Okazaki fragment.

f. Mitochondrial DNA Is Replicated in D-Loops

Mitochondrial DNA is replicated by a process in which leading strand synthesis precedes lagging strand synthesis (Fig. 30-44). The leading strand therefore displaces the lagging strand template to form a **displacement** or **D-loop.** The 15-kb circular mitochondrial chromosome of mammals normally contains a single 500- to 600-nt D-loop that undergoes frequent cycles of degradation and resynthesis. During replication, the D-loop is extended. When it has reached a point approximately two-thirds of the way around the chromosome, the lagging strand origin is exposed and its synthesis proceeds in the opposite direction around the chromosome. Lagging strand synthesis is therefore only about one-third complete when leading strand synthesis terminates.

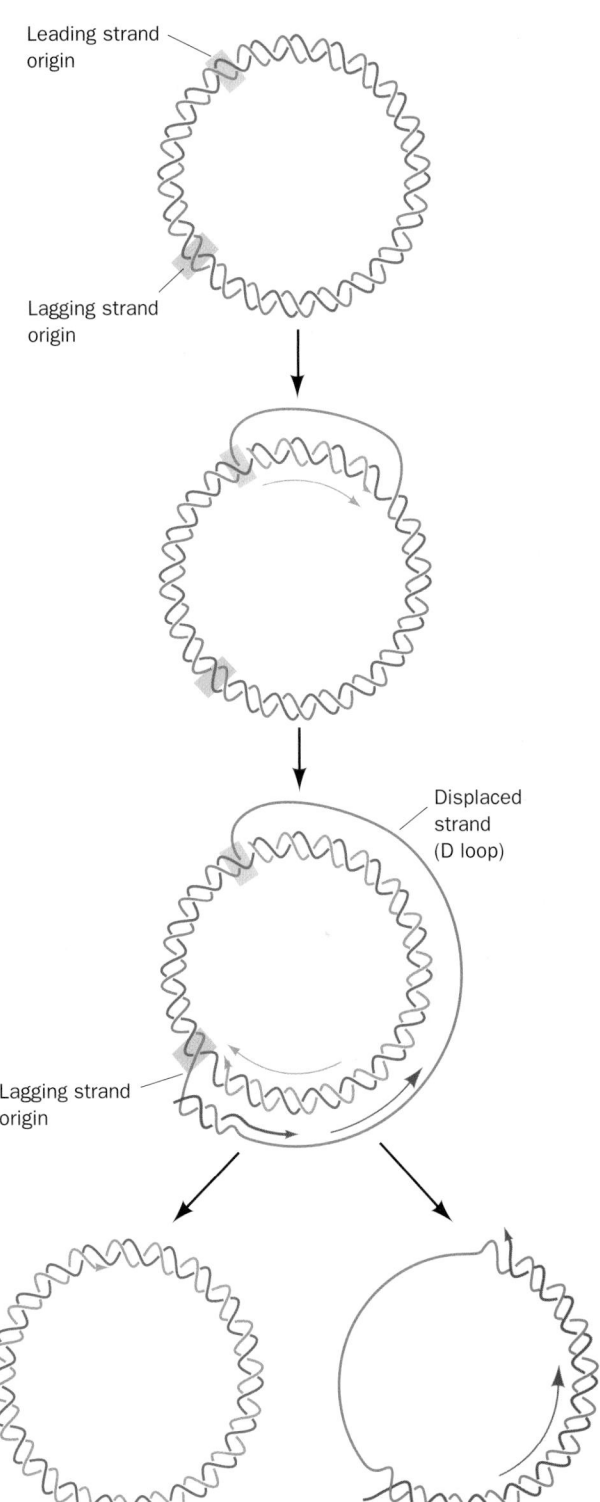

FIGURE 30-44 The D-loop mode of DNA replication.

C. *Reverse Transcriptase*

The **retroviruses,** which are RNA-containing eukaryotic viruses such as certain tumor viruses and human immunodeficiency virus (HIV), contain an **RNA-directed DNA polymerase (reverse transcriptase).** This enzyme, which was independently discovered in 1970 by Howard Temin and David Baltimore, acts much like Pol I in that it synthesizes DNA in the $5' \rightarrow 3'$ direction from primed templates. In the case of reverse transcriptase, however, RNA is the template.

> The discovery of reverse transcriptase caused a mild sensation in the biochemical community because it was perceived by some as being heretical to the central dogma of molecular biology (Section 5-4). There is, however, no thermodynamic prohibition to the reverse transcriptase reaction; in fact, under certain conditions, Pol I can likewise copy RNA templates.

Reverse transcriptase transcribes the retrovirus's single-stranded RNA genome to a double-stranded DNA as follows (Fig. 30-45):

1. The retroviral RNA acts as a template for the synthesis of its complementary DNA (RNA-directed DNA polymerase activity), yielding an RNA–DNA hybrid helix. The DNA synthesis is primed by a host cell tRNA whose 3' end partially unfolds to base pair with a complementary segment of the viral RNA.

2. The RNA strand is then nucleolytically degraded (**RNase H** activity; H for hybrid).

3. Finally, the DNA strand acts as a template for the synthesis of its complementary DNA (DNA-directed DNA polymerase activity), yielding double-stranded DNA.

The DNA is then integrated into a host cell chromosome.

Reverse transcriptase has been a particularly useful tool in genetic engineering because of its ability to transcribe mRNAs to complementary strands of DNA (cDNA). In transcribing eukaryotic mRNAs, which have poly(A) tails (Section 31-4A), the primer can be oligo(dT). cDNAs have been used, for example, as probes in Southern blotting (Section 5-5D) to identify the genes coding for their corresponding mRNAs. An RNA's base sequence can be easily determined by sequencing its cDNA (Section 7-2A).

a. X-Ray Structure of HIV-1 Reverse Transcriptase

HIV-1 reverse transcriptase (RT) is a dimeric protein whose subunits are synthesized as identical 66-kD polypeptides, known as **p66** (p for *protein*), that each contain a polymerase domain and an RNase H domain. However, the RNase H domain of one of the two subunits is proteolytically excised, thereby yielding a 51-kD polypeptide named **p51**. Thus, RT is dimer of p66 and p51.

The first drugs to be clinically approved to treat AIDS, **3′-azido-3′-deoxythymidine (AZT; zidovudine), 2′,3′-dideoxyinosine (ddI; didanosine), 2′,3′-dideoxycytidine (ddC; zalcitabine),** and **2′,3′-didehydro-3′-deoxythymidine (stavudine),**

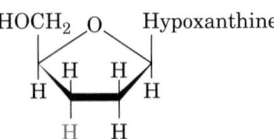

dependently discovered in 1970 by Howard Temin and

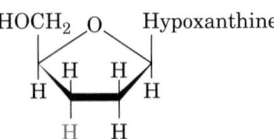

$5' \rule{6cm}{0.5pt} 3'$
Single-stranded RNA

dNTP ⟍ RNA-directed DNA polymerase
1
↓

RNA·DNA hybrid

RNase H
2
NMP

$3' \rule{6cm}{0.5pt} 5'$
Single-stranded DNA

dNTP ⟍ DNA-directed DNA polymerase
3
↓

Double-stranded DNA

FIGURE 30-45   The reactions catalyzed by reverse transcriptase.

are RT inhibitors. Unfortunately, resistant strains of HIV-1 arise quite rapidly because RT lacks a proofreading exonuclease function and hence is highly error prone. Thus, as we have seen (Section 15-4C), effective long-term anti-HIV therapy requires the concurrent administration of at least one RT inhibitor and an HIV protease inhibitor.

The X-ray structure of RT complexed to an 18-bp DNA with a 1-nt overhang at the 5′ end of one strand was determined by Edward Arnold (Fig. 30-46). This complex also contains a monoclonal **Fab fragment** (the antigen-binding segment of an immunoglobulin; Section 35-2B) that specifically binds to RT and presumably facilitated the crystallization of the complex. Steitz independently determined the X-ray structure of RT in the absence of DNA.

The two RT structures are closely similar, although there appear to be shifts in some secondary structural elements, particularly those that contact the DNA and the Fab fragment. The polymerase domains of p66 and p51 each contain four subdomains, which, because of their collective resemblance in p66 to DNA polymerases, are named, from N- to C-terminus, "fingers," "palm," "thumb," and "connection." In p66, the RNase H domain follows the connection.

p51 has undergone a remarkable conformational change relative to p66: The connection has rotated by 155° and translated by 17 Å to bring it from a position in p66 in which it contacts the RNase H domain (Fig. 30-46a) to one in p51 in which it contacts all three other polymerase subdomains (Fig. 30-46b). This permits p66 and p51 to bring different surfaces of their connections into juxtaposition to form, in part, RT's DNA-binding groove. Thus, the chemically identical polymerase domains of p66 and p51 are not related by 2-fold molecular symmetry (a rare but not unprecedented phenomenon), but, rather, associ-

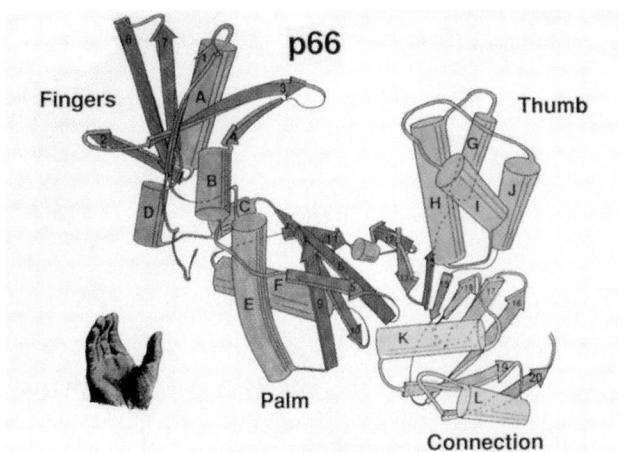

(a)

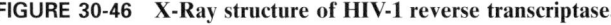

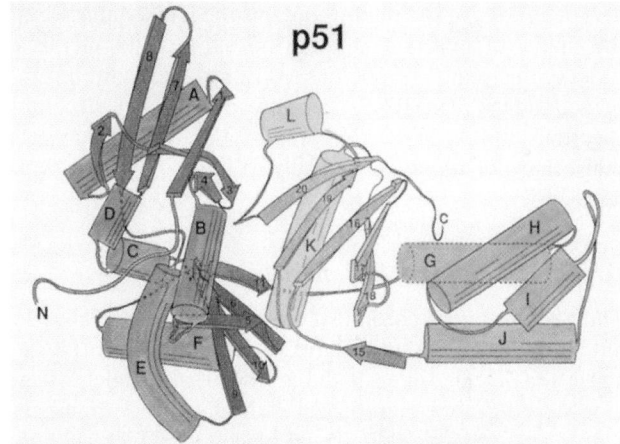

(b)

FIGURE 30-46   X-Ray structure of HIV-1 reverse transcriptase. (*a*) A tube-and-arrow representation of the p66 subunit's polymerase domain in which the N-terminal finger subdomain is blue, the palm is pink, the thumb is green, and the connection is yellow. The RNase H domain (not shown) follows the connection. (*b*) The p51 subunit with its pink palm subunit oriented identically to that in p66. Note the different relative orientations of the four subdomains in the two subunits. The G helix is shown in dashed outline because its electron density is weak and ambiguous. (*c*) A ribbon diagram of the HIV-1 RT p66/p51 heterodimer in complex with DNA. The subdomains of p66 and p51 are colored as in Parts *a* and *b* and the RNase H subdomain of p66 is orange [the labels indicate subunit and (sub)domain; e.g., 51F and 66R denote the p51 finger subdomain and the p66 RNase H domain]. The DNA is shown in ladder representation with the 18-nt primer strand white and the 19-nt template strand blue. The complex is oriented with its p66 polymerase domain toward the top of the figure and viewed from above the protein's template–primer binding cleft (whose floor is largely composed of the connection subdomains of p66 and p51) so as to show the bend in the DNA. [Parts *a* and *b* courtesy of Thomas Steitz, Yale University. PDBid 3HVT. Part *c* courtesy of Edward Arnold, Rutgers University. PDBid 2HMI.] 🖉 **See the Interactive Exercises**

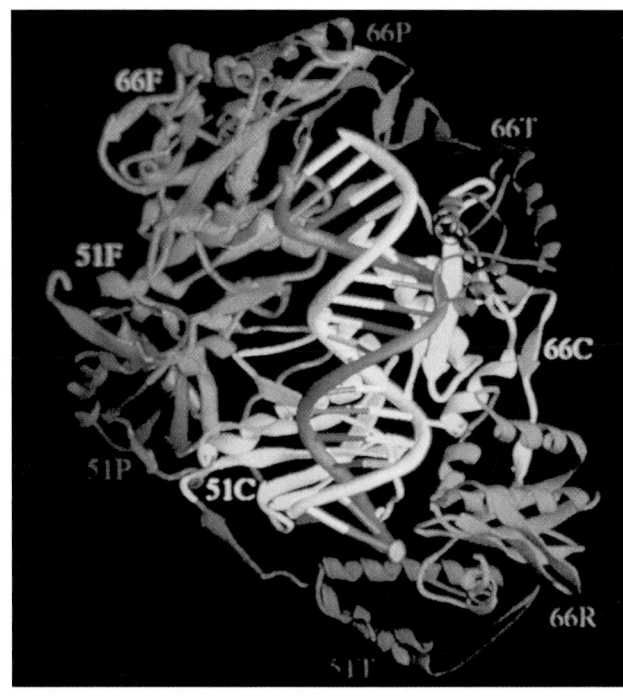

(c)

ate in a sort of head-to-tail arrangement. Consequently, RT has only one polymerase active site and one RNase H active site. This is an example of viral genetic economy: HIV-1, with its limited genome size, has succeeded in using a single polypeptide for what are essentially two different functions.

The DNA assumes a conformation that, near the polymerase active site, resembles A-DNA but, near the RNase H domain, more closely resembles B-DNA (Fig. 30-46c), a phenomenon that also has been observed in several structures of DNA polymerases in their complexes with DNA (Section 30-2A). The 3'-OH group at the end of the 18-nt DNA strand, the so-called primer strand, is near p66's three catalytically essential Asp side chains, where it is properly positioned to nucleophilically attack the α phosphate of an incoming dNTP that had been shown to bind near this site. Most of the protein–DNA interactions involve the DNA's sugar–phosphate backbone and the residues of p66's palm, thumb, and fingers.

The RT active site region contains the few sequence motifs that are conserved among the various polymerases. Indeed, this region of p66 has a striking structural resemblance to DNA and RNA polymerases of known structure (Sections 30-2A, 30-4B, 31-2A, and 31-2F). This suggests that other polymerases are likely to bind DNA in a similar manner.

D. *Telomeres and Telomerase*

The ends of linear chromosomes cannot be replicated by any of the mechanisms we have yet considered. This is because the RNA primer at the 5' end of a completed lagging strand cannot be replaced with DNA; the primer required to do this would have no place to bind. How, then, are the DNA sequences at the ends of eukaryotic chromosomes, the **telomeres** (Greek: *telos,* end), replicated?

Telomeric DNA has an unusual sequence: It consists of up to several thousand tandem repeats of a simple, species-dependent, G-rich sequence concluding the 3'-ending strand of each chromosomal terminus. For example, the ciliated protozoan *Tetrahymena* has the repeating telomeric sequence TTGGGG, whereas in all vertebrates it is TTAGGG. Moreover, this strand ends with an overhang that varies from ~20 nt in yeast to ~200 bp in humans.

Elizabeth Blackburn has shown that telomeric DNA is synthesized by a novel mechanism. The enzyme that synthesizes the G-rich strand of telomeric DNA is named **telomerase.** *Tetrahymena* telomerase, for example, adds tandem repeats of the telomeric sequence TTGGGG to the 3' end of any G-rich telomeric oligonucleotide independently of any exogenously added template. A clue as to how this occurs came from the discovery that telomerases are ribonucleoproteins whose RNA components contain a segment that is complementary to the repeating telomeric sequence. This sequence apparently acts as a template in a kind of reverse transcriptase reaction that synthesizes the telomeric sequence, translocates to the DNA's new 3' end, and repeats the process (Fig. 30-47). This hypothesis is confirmed by the observation that mu-

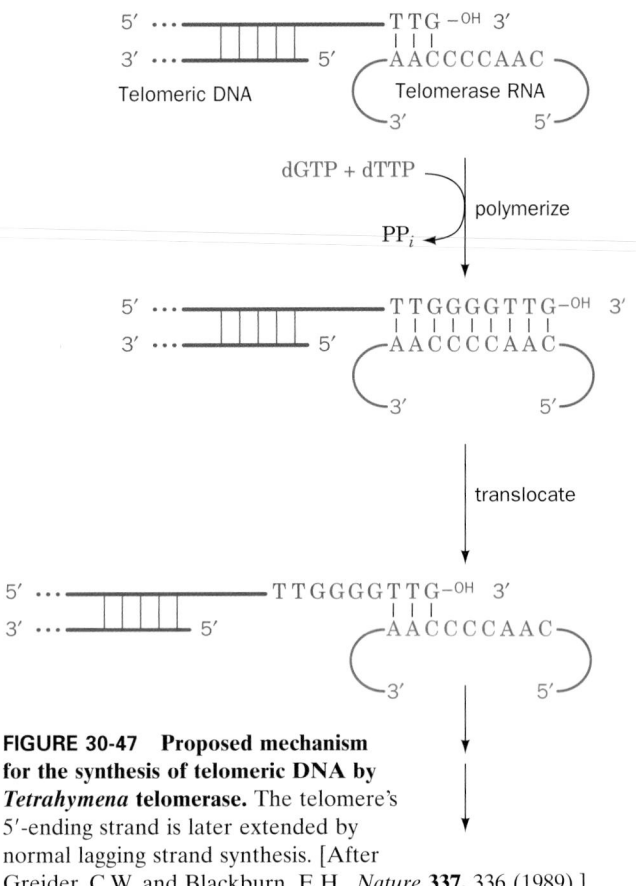

FIGURE 30-47 Proposed mechanism for the synthesis of telomeric DNA by *Tetrahymena* telomerase. The telomere's 5'-ending strand is later extended by normal lagging strand synthesis. [After Greider, C.W. and Blackburn, E.H., *Nature* **337,** 336 (1989).]

tationally altering the telomerase RNA gene segment complementary to telomere DNA results in telomere DNA with the corresponding altered sequence. In fact, telomerase's protein component is homologous to known reverse transcriptases. The DNA strand complementary to the telomere's G-rich strand is apparently synthesized by the normal cellular machinery for lagging strand synthesis, thereby accounting for the 3' overhang of the G-rich strand.

a. Telomeres Must Be Capped

Without the action of telomerase, a chromosome would be shortened at both ends by 50 to 100 nt with every cycle of DNA replication and cell division. It was therefore initially assumed that, in the absence of active telomerase, essential genes located near the ends of chromosomes would eventually be lost, thereby killing the descendents of the originally affected cells. However, it is now evident that telomeres serve a vital chromosomal function that is compromised before this can happen. Free DNA ends, which are subject to nuclease degradation, trigger DNA damage repair systems that normally function to rejoin the ends of broken chromosomes (as well as cell cycle arrest until this has happened). Thus exposed telomeric DNA would result in the end-to-end fusion of chromosomes, a process that leads to chromosomal instability and eventual cell death [fused chromosomes often break in mitosis (their two centromeres may cause them to be pulled in opposite directions), activating DNA damage checkpoints]. However, in a process known as **capping,** telomeric DNA is specifically bound by proteins that sequester the DNA ends. There is mounting evidence that capping is a dynamic process in which the probability of a telomere spontaneously upcapping increases as telomere length decreases. Since most somatic cells in multicellular organisms lack telomerase activity, this explains why such cells in culture can only undergo a limited number of doublings (20–60) before they reach senescence (a stage in which they cease dividing) and eventually die (Section 19-3B). Indeed, otherwise immortal *Tetrahymena* cultures with mutationally impaired telomerases exhibit characteristics reminiscent of senescent mammalian cells before dying off. Apparently, *the loss of telomerase function in somatic cells is a basis for aging in multicellular organisms.*

b. Telomere Length Correlates with Aging

There is strong experimental evidence in support of this theory of aging. The analysis of cultured human fibroblasts from a number of donors between 0 and 93 years old indicates that there is only a weak correlation between the proliferative capacity of a cell culture and the age of its donor. There is, however, a strong correlation, valid over the entire donor age range, between the initial telomere length in a cell culture and its proliferative capacity. Thus, cells that initially have relatively short telomeres undergo significantly fewer doublings than cells with longer telomeres. Moreover, fibroblasts from individuals with **progeria** (a rare disease characterized by rapid and premature aging resulting in childhood death) have short telomeres, an observation that is consistent with their known reduced proliferative capacity in culture. In contrast, sperm (which, being germ cells, are in effect immortal) from donors ranging in age from 19 to 68 years had telomeres that did not vary in length with donor age, which indicates that telomerase is active at some stage of germ cell growth. Likewise, those few cells in a culture that become immortal (capable of unlimited proliferation) exhibit an active telomerase

and a telomere of stable length, as do the cells of unicellular eukaryotes (which are also immortal). It therefore appears that telomere erosion is a significant cause of cellular senescence and hence aging.

c. Cancer Cells Have Active Telomerases

What advantage might multicellular organisms gain by eliminating telomerase activity in their somatic cells? An intriguing possibility is that cellular senescence is a mechanism that protects multicellular organisms from cancer. The two defining characteristics of cancer cells are that they are immortal and grow uncontrollably (Sections 19-3B and 34-4C). If mammalian cells were normally immortal, the incidence of cancer would probably be far greater than it is since immortalization, which requires an active telomerase, is a major step toward **malignant transformation** (cancer formation), which requires several independent genetic changes (Section 19-3B). Indeed, nearly all human cancers exhibit high telomerase activity. Moreover, as Robert Weinberg demonstrated, human fibroblasts in culture can be malignantly transformed by the acquisition of only three genes, those encoding: (1) **TERT,** the protein subunit of telomerase (its 451-nt RNA subunit, **TR,** is normally expressed in somatic cells), (2) an oncogenic variant of H-Ras (an essential participant in intracellular signal transduction pathways; Section 19-3C), and (3) the SV40 **large-T antigen** [SV40 is a tumor virus whose large-T antigen binds and functionally inactivates the tumor suppressor proteins known as **Rb** and **p53** (Section 34-4C; it also functions as a helicase in viral DNA replication)]. This suggests that telomerase inhibitors may be effective antitumor agents.

d. Telomeric DNA Can Dimerize via G-Quartets

It has long been known that guanine forms strong Hoogsteen-type base pairs (Table 29-2) that can further associate to form cyclic tetramers known as **G-quartets** (Fig. 30-48a). Indeed, G-rich polynucleotides are notoriously difficult to work with because of their propensity to ag-

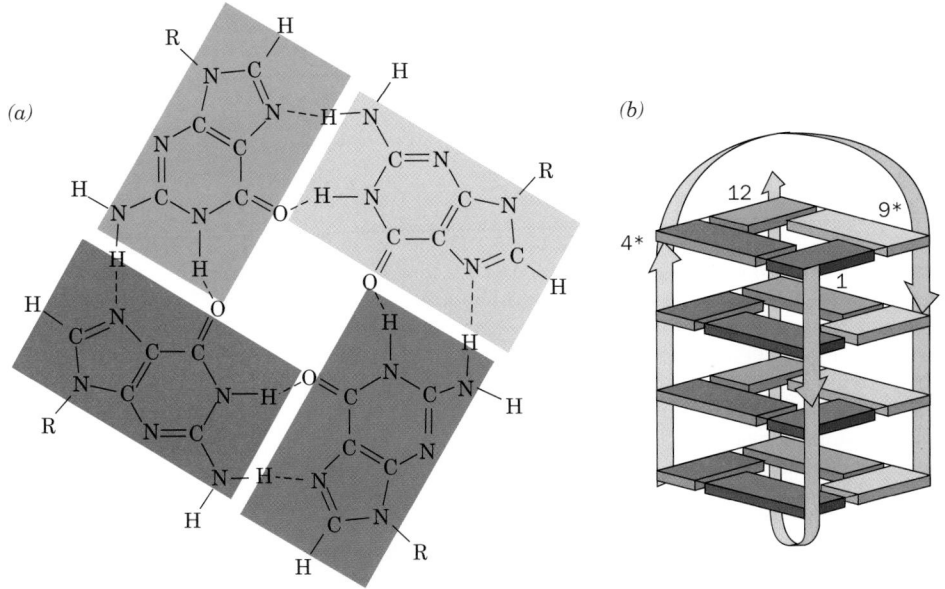

(a)

(b)

FIGURE 30-48 NMR structure of the telomeric oligonucleotide d(GGGGTTTTGGGG). (*a*) The base pairing interactions in the G-quartet at the end of the quadruplex in solution. (*b*) Schematic diagram of the NMR solution structure, in which the strand directions are indicated by arrows. The nucleotides are numbered 1 to 12 in one strand and 1* to 12* in the symmetry-related strand. Guanine residues G1 to G4 are represented by dark blue rectangles, G8 to G12 are light blue, G1* to G4* are red, and G9* to G12* are pink. [After Schultze, P., Smith, F.W., and Feigon, J., *Structure* **2,** 227 (1994). PDBid 156D.]

gregate. The G-rich overhanging strands of telomeres dimerize to form stable complexes in solution, presumably via the formation of G-quartet-containing structures.

The 3'-terminal telomeric overhang of the ciliated protozoan *Oxytricha nova* has the sequence d(T$_4$G$_4$)$_2$, which resembles the repeating telomeric sequences of other organisms. The NMR structure of the dodecamer d(G$_4$T$_4$G$_4$), determined by Juli Feigon (Fig. 30-48*b*), reveals that each oligonucleotide folds back on itself to form a hairpin, two of which associate in an antiparallel fashion to form a structure that contains four stacked G-quartets, with the T$_4$ sequences forming the loops at the ends of each stack.

The **telomere end binding protein (TEBP)** of *O. nova* is a heterodimeric capping protein that binds to and protects the foregoing 3' overhang. The X-ray structure of TEBP in complex with d(G$_4$T$_4$G$_4$), determined by Steve Schultz, reveals that the DNA binds in a deep cleft between the protein's α and β subunits, where it adopts an irregular nonhelical conformation (Fig. 30-49). In addition,

two other d(G$_4$T$_4$G$_4$) molecules form a G-quartet–linked dimer with the same conformation they adopt in solution (Fig. 30-48). The G-quartet assembly fits snugly into a small positively charged cavity formed by the N-terminal domains of three symmetry-related (in the crystal) α subunits at sites distinct from their ssDNA binding sites. The presence of both the ssDNA and the G-quartet assembly in the X-ray structure supports the hypothesis that multiple DNA structures and, in particular, G-quartets, play a role in telomere biology. Nevertheless, no obvious TEBP homologs occur in yeast or vertebrates. However, both humans and fission yeast express a telomere end-binding protein named **Pot1** (for *p*rotection *o*f *t*elomeres), whose deletion causes rapid loss of telomeric DNA and chromosomal end joining.

e. Telomeres Form T-Loops

Mammalian telomeric DNA is also capped by two related proteins, **TRF1** and **TRF2** (TRF for *t*elomere

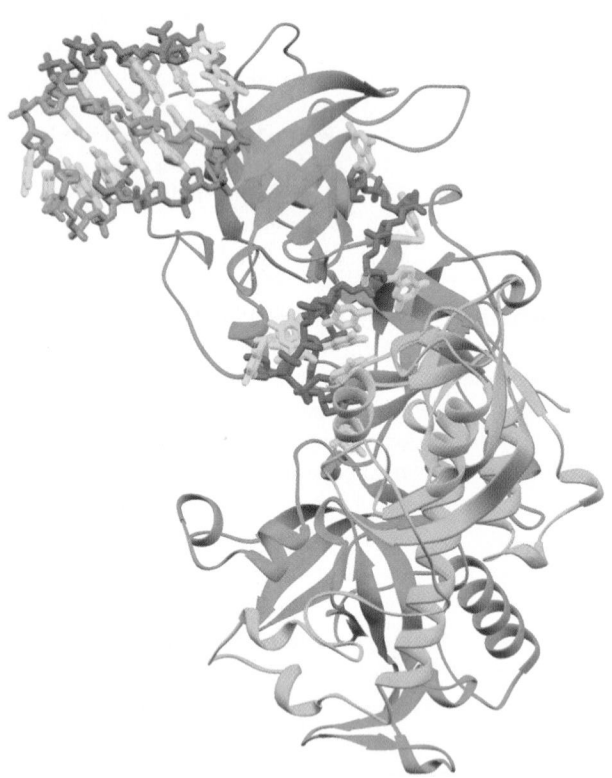

FIGURE 30-49   X-Ray structure of *Oxytricha nova* telomere end binding protein (TEBP) in complex with d(G$_4$T$_4$G$_4$). The TEBP is drawn in ribbon form with its α and β subunits magenta and cyan. The DNA is drawn in stick form with its bases gold, the sugar–phosphate backbone of the single strand that binds in a cleft between the protein's α and β subunits blue, and the backbones of two strands that form a G-quartet–linked dimer red and green. The G-quartet–linked dimer binds in a cavity formed by the N-terminal domains of three symmetry related α chains, although only one of them is shown here. [Based on an X-ray structure by Steve Schultz, University of Colorado. PDBid 1JB7.]

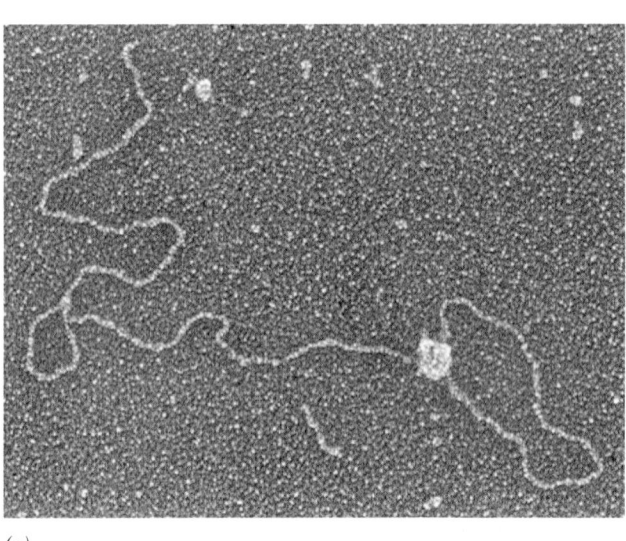

(a)

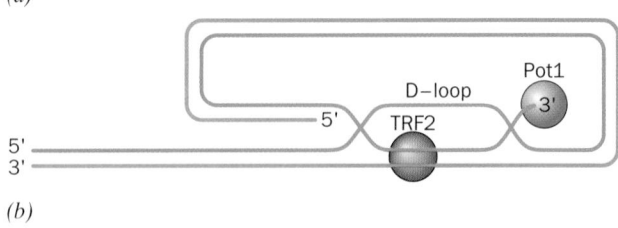

(b)

FIGURE 30-50   The telomeric T-loop. (*a*) An electron micrograph of a dsDNA consisting of a 3-kb unique sequence followed by ~2 kb of the repeating sequence TTAGGG on the strand that ends with a 150- to 200-nt 3' overhang. This model telomeric DNA was then incubated with human TRF2. [Courtesy of Jack Griffith, University of North Carolina at Chapel Hill.] (*b*) The proposed structure of a T-loop. In a process that is mediated by TRF2, the repeating TTAGGG sequence in the DNA's 3' overhang displaces a portion of the same strand (*blue*) in the double-stranded region of the DNA to form a duplex segment with the complementary strand (*red*), thereby generating a D-loop. The telomere end-binding protein Pot1 specifically binds to the end of the 3' overhang.

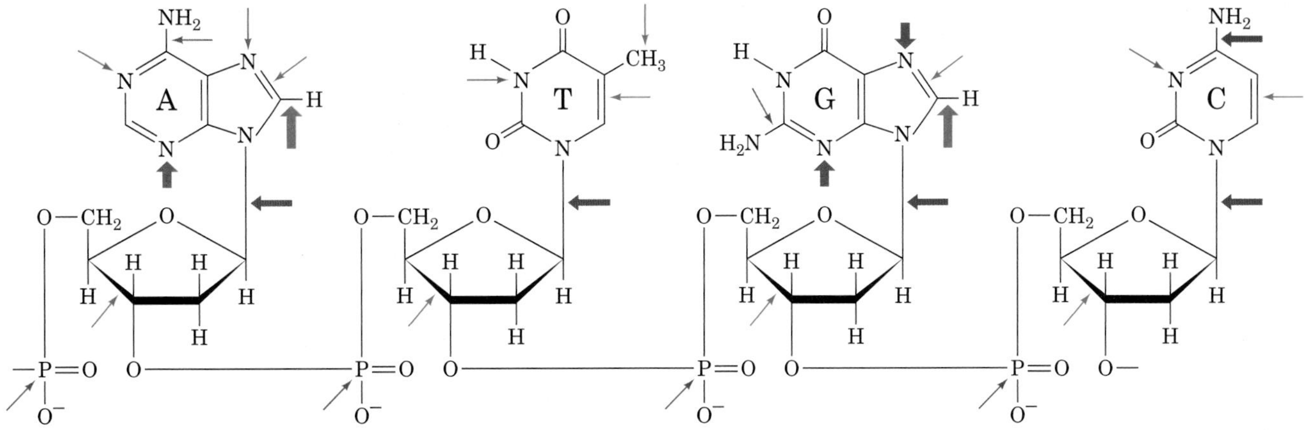

FIGURE 30-51 The types and sites of chemical damage to which DNA is normally susceptible *in vivo*. Red arrows indicate sites subject to oxidative attack, blue arrows indicate sites subject to spontaneous hydrolysis, and green arrows indicate sites subject to nonenzymatic methylation by *S*-adenosylmethionine. The width of an arrow is indicative of the relative frequency of the reaction. [After Lindahl, T., *Nature* **362,** 709 (1993).]

repeat-binding *f*actor). Jack Griffith and Titia de Lange have shown through electron microscopy (EM) studies that, in the presence of TRF2, otherwise linear telomeric DNA forms large duplex end loops named **T-loops** (Fig. 30-50*a*). Moreover, the EM of DNA from mammalian telomeres, whose strands had been chemically cross-linked to preserve their structural relationships on deproteination, likewise revealed the presence of abundant T-loops of varying sizes. These observations suggest that T-loops are formed by the TRF2-induced invasion of the 3′ telomeric overhang into the repeating telomeric dsDNA (Fig. 30-50*b*) to form a D-loop (Section 30-4B). T-loops have also been observed in protozoa, suggesting that T-loops are a conserved feature of eukaryotic telomeres. TRF1 is implicated in controlling telomeric length, presumably by somehow limiting the number of TRF1 molecules that can bind to a telomere.

5 ■ REPAIR OF DNA

DNA is by no means the inert substance that might be supposed from naive consideration of genome stability. Rather, the reactive environment of the cell, the presence of a variety of toxic substances, and exposure to UV or ionizing radiation subjects it to numerous chemical insults that excise or modify bases and alter sugar–phosphate groups (Fig. 30-51). Indeed, some of these reactions occur at surprisingly high rates. For example, under normal physiological conditions, the glycosidic bonds of ~10,000 of the 3.2 billion purine nucleotides in each human cell hydrolyze spontaneously each day.

Any DNA damage must be repaired if the genetic message is to maintain its integrity. Such repair is possible because of duplex DNA's inherent information redundancy. The biological importance of DNA repair is indicated by the identification of at least 130 genes in the human

genome that participate in DNA repair and by the great variety of DNA repair pathways possessed by even relatively simple organisms such as *E. coli*. In fact, *the major DNA repair processes in eukaryotic cells and E. coli are chemically quite similar.* These processes are outlined in this section.

A. Direct Reversal of Damage

a. Pyrimidine Dimers Are Split by Photolyase

UV radiation of 200 to 300 nm promotes the formation of a cyclobutyl ring between adjacent thymine residues on the same DNA strand to form an intrastrand **thymine dimer** (Fig. 30-52). Similar cytosine and thymine–cytosine

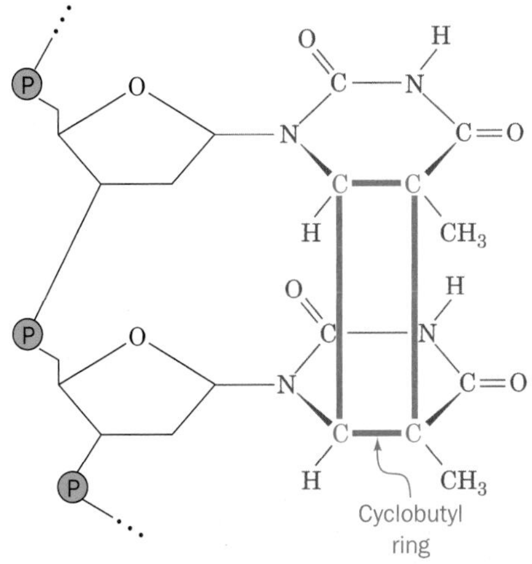

FIGURE 30-52 The cyclobutylthymine dimer that forms on UV irradiation of two adjacent thymine residues on a DNA strand. The ~1.6-Å-long covalent bonds joining the thymine rings (*red*) are much shorter than the normal 3.4-Å spacing between stacked rings in B-DNA, thereby locally distorting the DNA.

dimers are likewise formed but at lesser rates. Such **pyrim-idine dimers** locally distort DNA's base paired structure such that it can be neither transcribed nor replicated. Indeed, a single thymine dimer, if unrepaired, is sufficient to kill an *E. coli.*

Pyrimidine dimers may be restored to their monomeric forms through the action of light-absorbing enzymes named **photoreactivating enzymes** or **DNA photolyases** that are present in many prokaryotes and eukaryotes (including goldfish, rattlesnakes, and marsupials, but not placental mammals). These enzymes are 55- to 65-kD monomers that bind to a pyrimidine dimer in DNA, a process that can occur in the dark. A noncovalently bound chromophore, in some species an N^5,N^{10}-methenyltetrahydrofolate (**MTHF;** Fig. 26-49) and in others a **5-deazaflavin,**

CH$_2$OH

(CHOH)$_3$

CH$_2$

8-Hydroxy-7,8-didemethyl-5-deazariboflavin

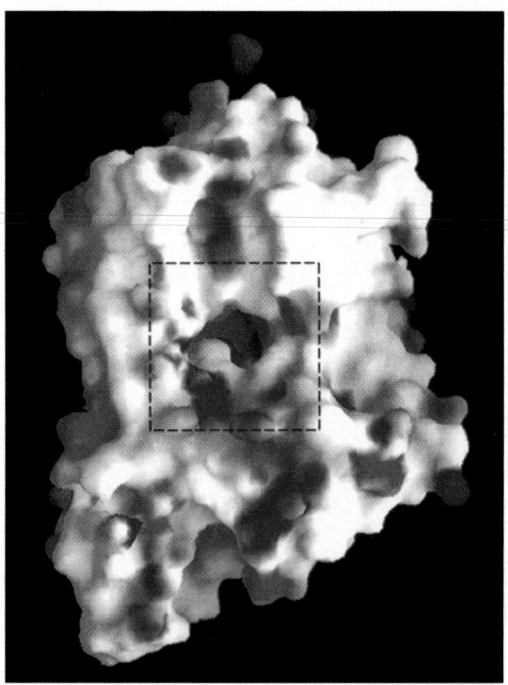

FIGURE 30-53   X-Ray structure of *E. coli* DNA photolyase showing its putative DNA binding surface. The enzyme is represented by its solvent-accessible surface, which is colored according to its electrostatic potential with blue most positive, red most negative, and white nearly neutral. The dashed box encloses the hole in the protein surface that is the presumed pyrimidine dimer binding site. [Courtesy of Johann Diesenhofer, Texas Southwestern Medical Center, Dallas, Texas. PDBid 1DNP.]

then absorbs 300- to 500-nm light and transfers the excitation energy to a noncovalently bound FADH$^-$, which in turn transfers an electron to the pyrimidine dimer, thereby splitting it. Finally, the resulting pyrimidine anion reduces the FADH· and the now unblemished DNA is released, thereby completing the catalytic cycle. DNA photolyases bind either dsDNA or ssDNA with high affinity but without regard to base sequence.

The X-ray structure of the 471-residue *E. coli* DNA photolyase, determined by Johann Deisenhofer, reveals that its MTHF and flavin rings are ~17 Å apart, which permits efficient energy transfer between them. The enzyme's putative DNA binding site (Fig. 30-53) is a positively charged flat surface that is penetrated by a hole that has the size and polarity complementary to that of a pyrimidine dimer–containing dinucleotide. A pyrimidine dimer bound in the hole would be in contact with the flavin ring and hence properly situated for efficient electron transfer from this ring system. This implies that a pyrimidine dimer in a double helix must flip (swing) out of the helix in order to interact with the enzyme. This flip-out is probably facilitated by the relatively weak base pairing interactions of the pyrimidine dimer and the distortions it imposes on the double helix. In the following discussions we shall see that **base flipping** is by no means an unusual process for enzymes that perform chemistry on the bases in dsDNA.

b. Alkyltransferases Dealkylate Alkylated Nucleotides

The exposure of DNA to alkylating agents such as ***N*-methyl-*N*′-nitro-*N*-nitrosoguanidine (MNNG)**

***N*-Methyl-*N*′-nitro-*N*-nitrosoguanidine (MNNG)**

O^6**-Methylguanine residue**

yields, among other products, O^6**-alkylguanine** residues. The formation of these derivatives is highly mutagenic because on replication, they frequently cause the incorporation of thymine instead of cytosine.

O^6**-Methylguanine** and O^6**-ethylguanine** lesions of DNA in all species tested are repaired by O^6**-alkylguanine–DNA alkyltransferase,** which directly transfers the offending alkyl group to one of its own Cys residues. The reaction inactivates this protein, which therefore cannot be strictly

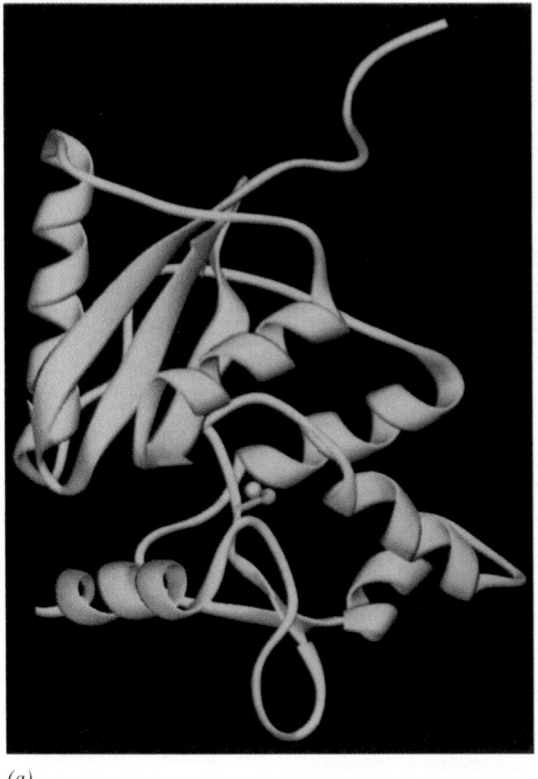

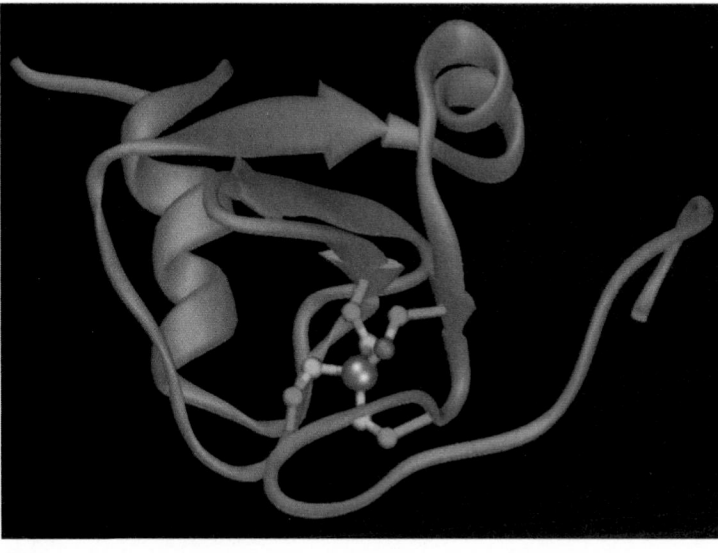

(a)   *(b)*

FIGURE 30-54   The structure of *E. coli* Ada protein. (*a*) The X-ray structure of Ada's 178-residue C-terminal segment, which contains its O^6-alkylguanine–DNA alkyltransferase function. The side chain of Cys 146 (Cys 321 in the intact protein), to which the methyl group is irreversibly transferred, is shown in ball-and-stick form with C green and S yellow. Note that this residue is almost entirely buried within the protein. [Based on an X-ray structure determined by Eleanor Dodson and Peter Moody, University of York, U.K. PDBid 1SFE.] (*b*) The NMR structure of Ada's 92-residue, N-terminal segment, which mediates its methyl phosphotriester repair function. The protein's bound Zn^{2+} ion is represented by a silver sphere and its four tetrahedrally coordinating Cys side chains are shown in ball-and-stick form, with C green and S yellow except for the orange S atom of Cys 69, which becomes irreversibly methylated when the protein encounters a methylated phosphate group on DNA. [Based on an NMR structure determined by Gregory Verdine and Gerhard Wagner, Harvard University. PDBid 1ADN.]

classified as an enzyme. The alkyltransferase reaction has elicited considerable attention because carcinogenesis induced by methylating and ethylating agents is correlated with deficient repair of O^6-alkylguanine lesions.

The *E. coli* O^6-alkylguanine–DNA alkyltransferase activity occurs on the 178-residue C-terminal segment of the 354-residue **Ada protein** (the product of the ***ada*** gene). Its X-ray structure (Fig. 30-54*a*), determined by Eleanor Dodson and Peter Moody, reveals, unexpectedly, that its active site Cys residue, Cys 321, is buried inside the protein. Apparently, the protein must undergo a significant conformation change on DNA binding in order to effect the methyl transfer reaction.

Ada protein's 92-residue N-terminal segment has an independent function: It repairs methyl phosphotriesters in DNA (methylated phosphate groups) by irreversibly transferring the offending methyl group to its Cys 69. The NMR structure of Ada's N-terminal domain (Fig. 30-54*b*), determined by Gregory Verdine and Gerhard Wagner, re-

veals that Cys 69, together with three other Cys residues, tetrahedrally coordinates a Zn^{2+} ion. This presumably stabilizes the thiolate form of Cys 69 over its thiol form, thereby facilitating its nucleophilic attack on the methyl group.

Intact Ada protein that is methylated at its Cys 69 binds to a specific DNA sequence, which is located upstream of the *ada* gene and several other genes encoding DNA repair proteins, thereby inducing their transcription. Evidently, Ada also functions as a chemosensor of methylation damage.

B. *Excision Repair*

Cells employ two types of excision repair mechanisms: (1) **nucleotide excision repair (NER),** which functions to repair relatively bulky DNA lesions; and (2) **base excision repair (BER),** which repairs nonbulky lesions involving a single base.

a. Nucleotide Excision Repair

NER is a DNA repair mechanism found in all cells that eliminates damage to dsDNA by excising an oligonucleotide containing the lesion and filling in the resulting single-strand gap. NER repairs lesions that are characterized by the displacement of bases from their normal positions, such as pyrimidine dimers, or by the addition of a bulky substituent to a base. This system appears to be activated by a helix distortion rather than by the recognition of any particular group. In humans, NER is the major defense against two important carcinogens, sunlight and cigarette smoke. The mechanism of NER in prokaryotes is similar to that in eukaryotes. However, prokaryotic NER employs 3 subunits, whereas eukaryotic NER involves the actions of 16 subunits. The eukaryotic proteins are conserved from yeast to humans but none of them exhibit any sequence similarity to the prokaryotic proteins, suggesting that the two NER systems arose by convergent evolution.

In *E. coli*, NER is carried out in an ATP-dependent process through the actions of the **UvrA, UvrB,** and **UvrC** proteins (the products of the *uvrA, uvrB,* and *uvrC* genes). This system, which is often referred to as the **UvrABC endonuclease** (although, as we shall see, there is no complex that contains all three subunits), cleaves the damaged DNA strand at the seventh and at the third or fourth phosphodiester bonds from the lesion's 5' and 3' sides, respectively (Fig. 30-55). The excised 11- or 12-nt oligonucleotide is displaced by the binding of **UvrD** (also called **helicase II**) and replaced through the actions of Pol I and DNA ligase.

The mechanism of prokaryotic NER was elucidated mainly by Aziz Sancar. It begins with the damage recognition step in which a $(UvrA)_2UvrB$ heterotrimer binds tightly although nonspecifically to dsDNA, which it probes for damage according to its local propensity for bending and unwinding. The presence of a lesion activates the helicase function of UvrB to unwind 5 bp around the lesion in an ATP-driven process. This conformation change induces the dissociation of the UvrA from the complex, which allows the binding of UvrC. UvrB then makes the incision on the 3' side of the lesion following which UvrC makes the incision on its 5' side. UvrD binds to the resulting nicks in the DNA, which displaces UvrC and the lesion-containing oligomer. This makes the 5' incision site accessible to Pol I, which fills in the gap and displaces UvrB. Finally, DNA ligase seals the remaining nick yielding refurbished DNA.

b. Xeroderma Pigmentosum and Cockayne Syndrome Are Caused by Genetically Defective NER

In humans, the rare inherited disease **xeroderma pigmentosum** (**XP;** Greek: *xeros,* dry + *derma,* skin) is mainly characterized by the inability of skin cells to repair UV-induced DNA lesions. Individuals suffering from this autosomal recessive condition are extremely sensitive to sunlight. During infancy they develop marked skin changes such as dryness, excessive freckling, and keratoses (a type of skin tumor; the skin of these children is described as re-

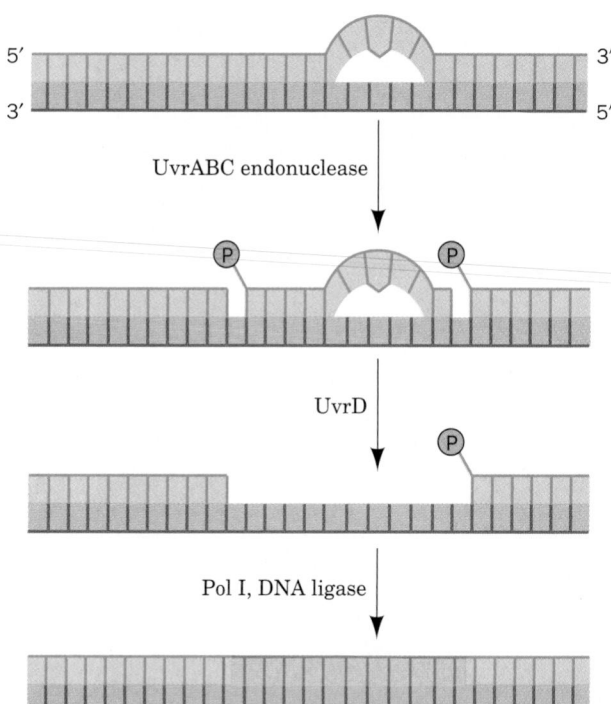

FIGURE 30-55   The mechanism of nucleotide excision repair (NER) of pyrimidine photodimers.

sembling that of farmers with many years of sun exposure), together with eye damage, such as opacification and ulceration of the cornea. Moreover, they develop often fatal skin cancers at a 2000-fold greater rate than normal and internal cancers at a 10 to 20-fold increased rate. Curiously, many individuals with XP also have a bewildering variety of seemingly unrelated symptoms including progressive neurological degeneration and developmental deficits.

Cultured skin fibroblasts from individuals with xeroderma pigmentosum are defective in the NER of pyrimidine dimers. Cell-fusion experiments with cultured cells taken from various patients have demonstrated that this disease results from defects in any of 8 complementation groups (Section 1-4C), indicating that there must be at least 8 gene products, XPA through XPG and XPV, involved in this clearly important UV damage repair pathway. **Cockayne syndrome (CS),** an inherited disease which is also associated with defective NER, arises from defects in XPB, XPD, and XPG as well as in two additional complementation groups, CSA and CSB. Individuals with CS are hypersensitive to UV radiation and exhibit stunted growth as well as neurological dysfunction due to neuron demyelination, but, intriguingly, have a normal incidence of skin cancer. What is the biochemical basis for the diverse group of symptoms associated with impaired NER?

The free radicals produced by oxidative metabolism can damage DNA. Some of these oxidative lesions are repaired via NER. Since neurons have high rates of respiration and are long-lived nondividing cells, it seems likely that they would be particularly susceptible to oxidative damage in the absence of NER. This would explain the progressive neurological deterioration in XP.

The retarded development typical of XPB defects and perhaps the demyelination that occurs in CS appear to be due more to impaired transcription than to defective NER. Moreover, pyrimidine dimers are more efficiently removed from transcribed portions of DNA than from unexpressed sequences. These observations are explained by the discovery that some or all of the subunits of the eukaryotic transcription factor **TFIIH,** a helicase that participates in the initiation of mRNA transcription by RNA polymerase II (Section 34-3B), are required for NER. The coupling of NER and transcription also occurs in *E. coli* in which **Mfd** protein displaces RNA polymerase that has stalled on a damaged template strand (which it cannot transcribe), following which the Mfd recruits the proteins of the UvrABC system to the damage site.

c. Base Excision Repair

DNA bases are modified by reactions that occur under normal physiological conditions as well as through the action of environmental agents. For example, adenine and cytosine residues spontaneously deaminate at finite rates to yield hypoxanthine and uracil residues, respectively. *S*-Adenosylmethionine (SAM), a common metabolic methylating agent (Section 26-3E), occasionally nonenzymatically methylates a base to form derivatives such as 3-methyladenine and 7-methylguanine residues (Fig. 30-51). Ionizing radiation can promote ring opening reactions in bases. Such changes modify or eliminate base pairing properties.

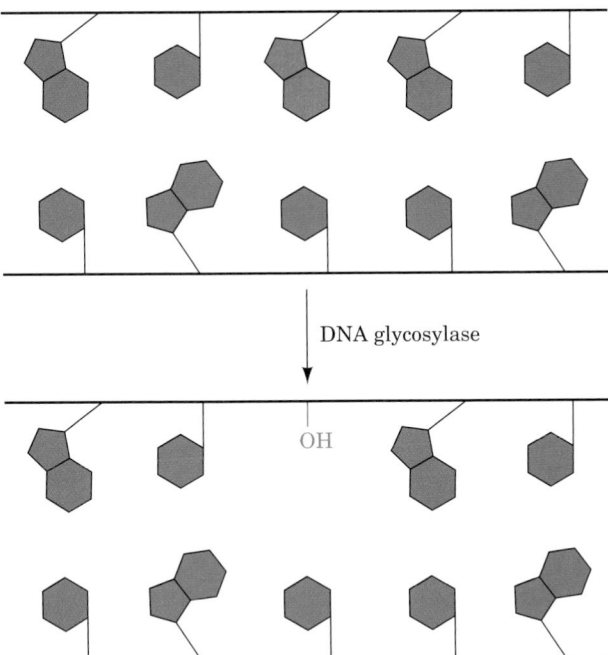

FIGURE 30-56 Action of DNA glycosylases. These enzymes hydrolyze the glycosidic bond of their corresponding altered base (*red*) to yield an AP site.

DNA containing a damaged base may be restored to its native state through base excision repair (BER). Cells contain a variety of **DNA glycosylases** that each cleave the glycosidic bond of a corresponding specific type of altered nucleotide (Fig. 30-56), thereby leaving a deoxyribose residue in the backbone. Such **apurinic** or **apyrimidinic (AP) sites** (also called **abasic sites**) are also generated under normal physiological conditions by the spontaneous hydrolysis of a glycosidic bond. The deoxyribose residue is then cleaved on one side by an **AP endonuclease,** the deoxyribose and several adjacent residues are removed by the action of a cellular exonuclease (possibly associated with a DNA polymerase), and the gap is filled in and sealed by a DNA polymerase and DNA ligase.

d. Uracil in DNA Would Be Highly Mutagenic

For some time after the essential functions of nucleic acids had been elucidated, there seemed no apparent reason for nature to go to the considerable metabolic effort of using thymine in DNA and uracil in RNA when these substances have virtually identical base pairing properties. This enigma was solved by the discovery of cytosine's penchant for conversion to uracil by deamination, either via spontaneous hydrolysis (Fig. 30-51), which is estimated to occur ~120 times per day in each human cell, or by reaction with nitrites (Section 32-1A). If U were the normal DNA base, the deamination of C would be highly mutagenic because there would be no indication of whether the resulting mismatched G · U base pair had initially been G · C or A · U. *Since T is DNA's normal base, however, any U in DNA is almost certainly a deaminated C.* U's that occur in DNA are efficiently excised by **uracil–DNA glycosylase [UDG;** also called **uracil *N*-glycosylase (UNG)]** and then replaced by C through BER.

UDG also has an important function in DNA replication. dUTP, an intermediate in dTTP synthesis, is present in all cells in small amounts (Section 28-3B). DNA polymerases do not discriminate well between dUTP and dTTP (recall that DNA polymerases select a base for incorporation into DNA according to its ability to base pair with the template; Section 30-2A) so that, despite the low dUTP level that cells maintain, newly synthesized DNA contains an occasional U. These U's are rapidly replaced by T through BER. However, since excision occurs more rapidly than repair, all newly synthesized DNA is fragmented. When Okazaki fragments were first discovered (Section 30-1C), it therefore seemed that all DNA was synthesized discontinuously. This ambiguity was resolved with the discovery of *E. coli* defective in UDG. In these *ung⁻* mutants, only about half of the newly synthesized DNA is fragmented, strongly suggesting that DNA's leading strand is synthesized continuously.

e. Uracil–DNA Glycosylase Induces Uridine Nucleotides to Flip Out

The X-ray structure of human UDG in complex with a 10-bp DNA containing a U · G mismatch (which can form a doubly hydrogen bonded base pair whose shape differs from that of Watson–Crick base pairs; Section 32-2D), determined by John Tainer, reveals that the UDG has bound the DNA with its uridine nucleotide flipped out of the dsDNA (Fig. 30-57). Moreover, the enzyme has hydrolyzed uridine's glycosidic bond yielding the free uracil base and an AP site on the DNA, although both remain bound to the enzyme. The cavity in the DNA's base stack that would otherwise be occupied by the flipped out uracil is filled by the side chain of Leu 272, which intercalates into the DNA from its minor groove side. The X-ray structure of a similar complex in which the U · G mismatch was replaced by a U · A base pair contained essentially identical features. However, when the U in the U · A-containing complex was replaced by **pseudouridine** (in which the "glycosidic" bond is made to uracil's C5 atom rather than to N1),

Pseudouridine

the uracil remained covalently linked to the DNA because the UDG could not hydrolyze its now C—C "glycosidic" bond.

How does UDG detect a base paired uracil in the center of DNA and how does it discriminate so acutely between uracil and other bases, particularly the closely similar thymine? The above X-ray structures indicate that the phosphate groups flanking the flipped out nucleotide are 4 Å closer together than they are in B-DNA (8 Å vs 12 Å), which causes the DNA to kink by ~45° in the direction parallel to the view in Fig. 30-57. These distortions arise from the binding of three rigid protein loops to the DNA, which would be unable to simultaneously bind to undistorted B-DNA. This led Tainer to formulate the "pinch–push–pull" mechanism for uracil detection in which he postulated that UDG rapidly scans a DNA for uracil by periodically binding to it so as to compress and thereby slightly bend the DNA's backbone (pinch). The DNA's presumed low resistance to bending at a uracil-containing site (a U · G base pair is smaller than C · G and hence leaves a space in the base stack, whereas a U · A base pair is even weaker than T · A) permits the enzyme to flip out the uracil by intercalating Leu 272 into the minor groove (push), thereby fully bending and kinking the DNA. This process is aided by the tight binding of the flipped out uracil to the enzyme (pull). The exquisite specificity of this binding pocket for uracil prevents the binding and hence hydrolysis of any other base that the enzyme may have induced to flip out. Thus the overall shapes of adenine and guanine exclude them from this pocket, whereas thymine's 5-methyl group is sterically blocked by the rigidly held side chain of Tyr 147. Cytosine, which has approximately the same shape as uracil, is excluded through a set of hydrogen bonds emanating from the protein that mimic those made by adenine in a Watson–Crick A · U base pair.

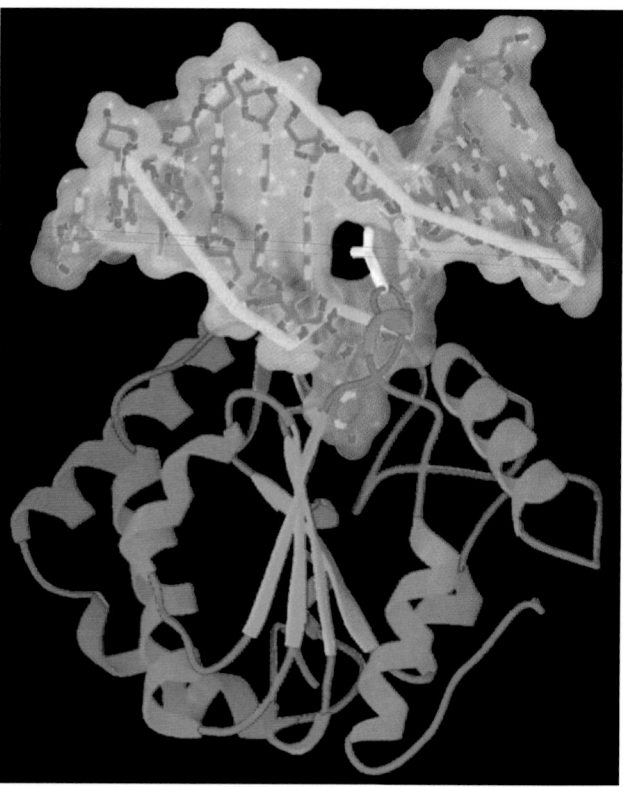

FIGURE 30-57 X-Ray structure of human uracil–DNA glycosylase (UDG) in complex with a 10-bp DNA containing a U · G base pair. The protein (the C-terminal 223 residues of the 304-residue monomer) is colored according to its secondary structure (helices blue, β strands orange, and other segments magenta). The DNA, viewed looking into its minor groove, is drawn in stick form colored according to atom type (C green, N cyan, O red) and with its phosphate backbone traced by yellow tubes (phosphate O atoms have been omitted for clarity). The DNA's transparent solvent-accessible surface is pink and the side chain of Leu 272 is white. The uridine nucleotide can be seen to have flipped out of the double helix (below the DNA) and to have been hydrolyzed to yield an AP nucleotide and uracil, which remains bound in UDG's binding pocket. The side chain of Leu 272 has intercalated into the DNA base stack to fill the space vacated by the flipped out uracil base. [Courtesy of John Tainer, The Scripps Research Institute, La Jolla, California. PDBid 4SKN.]

AP sites in DNA are highly cytotoxic because they irreversibly trap mammalian topoisomerase I in its covalent complex with DNA (Section 29-3C). Moreover, since the ribose at the AP site lacks a glycosidic bond, it can readily convert to its linear form (Section 11-1B), whose reactive aldehyde group can cross-link to other cell components. This rationalizes why AP sites remain tightly bound to UDG in solution as well as in crystals. UDG activity is enhanced by AP endonuclease, the next enzyme in the BER pathway, but the two enzymes do not interact in the absence of DNA. This suggests that UDG remains bound to an AP site it generated until it is displaced by the more tightly binding AP endonuclease, thereby protecting the cell from the AP site's cytotoxic effects. It seems likely that other damage-specific DNA glycosylases function similarly.

C. *Mismatch Repair*

Any replicational mispairing that has eluded the editing functions of the various participating DNA polymerases may still be corrected by a process known as **mismatch repair (MMR).** For example, *E. coli* Pol I and Pol III have error rates of 10^{-6} to 10^{-7} per base pair replicated but the observed mutational rates in *E. coli* are 10^{-9} to 10^{-10} per base pair replicated. In addition, the MMR system can correct insertions or deletions of up to 4 nt (which arise from the slippage of one strand relative to the other in the active site of DNA polymerase). The importance of MMR is indicated by the fact that defects in the human MMR system result in a high incidence of cancer, most notably **hereditary nonpolyposis colorectal cancer (HNPCC;** which affects several organs and may be the most common inherited predisposition to cancer).

If an MMR system is to correct errors in replication rather than perpetuate them, it must distinguish the parental DNA, which has the correct base, from the daughter strand, which has an incorrect although normal base. In *E. coli*, as we have seen (Section 30-3C), this is possible because newly replicated GATC palindromes remain hemimethylated until the Dam methyltransferase has had sufficient time to methylate the daughter strand.

E. coli mismatch repair, which was elucidated in large part by Paul Modrich, requires the participation of three proteins and occurs as follows (Fig. 30-58):

1. MutS (853 residues) binds to a mismatched base pair or unpaired bases as a homodimer.

2. The MutS–DNA complex binds **MutL** (615 residues), also as a homodimer.

3. The MutS–MutL complex translocates along the DNA in both directions, thereby forming a loop in the DNA. The translocation appears to be driven by the ATPase function of MutS.

4. On encountering a hemimethylated GATC palindrome, the MutS–MutL complex recruits **MutH** (228 residues) and activates this single strand endonuclease to make a nick on the 5′ side of the unmethylated GATC. This GATC may be located on either side of the mismatch and over 1000 bp distant from it.

5. MutS–MutL recruits UvrD helicase, which in concert with an exonuclease, separates the strands and degrades the nicked strand from the nick to beyond the mismatch. If the nick is on the 3′ side of the mismatch as shown, the exonuclease is **exonuclease I** (a 3′ → 5′ exonuclease), whereas if the nick is on the 5′ side of the mismatch, the exonuclease can be either **RecJ** or **exonuclease VII** (both 5′ → 3′ exonucleases).

The resulting gap is filled in by Pol III and sealed by DNA ligase, thereby correcting the mismatch. MutL is also an ATPase, which, it is postulated, functions to coordinate the various steps of mismatch repair.

Eukaryotic MMR systems are, not surprisingly, more complicated than those of *E. coli*. Eukaryotes express six homologs of MutS and five homologs of MutL that form

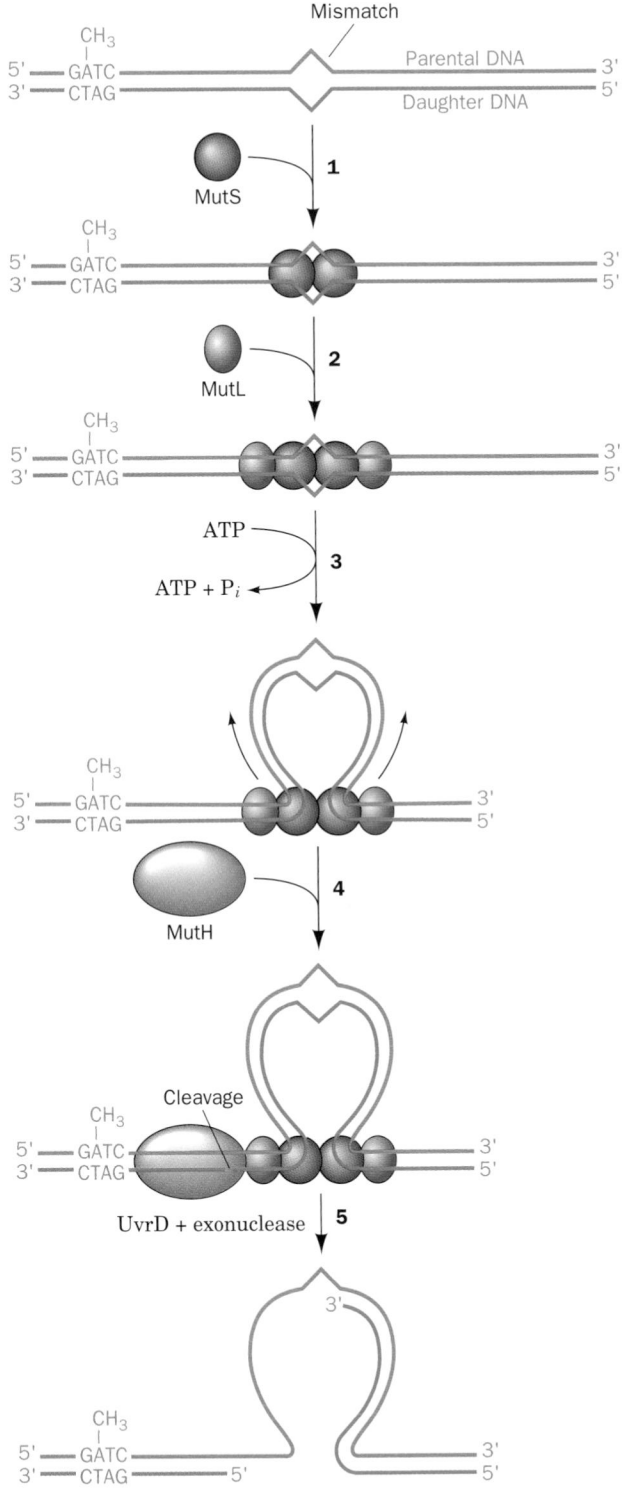

FIGURE 30-58 The mechanism of mismatch repair in *E. coli*.

heterodimers on mismatched DNA. However, homologs of MutH only occur in gram-negative bacteria. Eukaryotes must have some other way of differentiating the parental and daughter DNA strands. Perhaps a newly synthesized daughter strand is identified by its as-yet unsealed nicks.

D. *The SOS Response*

Agents that damage DNA, such as UV radiation, alkylating agents, and cross-linking agents, induce a complex system of cellular changes in *E. coli* known as the **SOS response.** *E. coli so treated cease dividing and increase their capacity to repair damaged DNA.*

a. LexA Protein Represses the SOS Response

Clues as to the nature of the SOS response were provided by the observations that *E. coli* with mutant *recA* or *lexA* genes have their SOS response permanently switched on. **RecA,** a 352-residue protein that coats DNA as a multimeric helical filament, plays a central role, as we shall see, in homologous recombination (Section 30-6A). When *E. coli* are exposed to agents that damage DNA or inhibit DNA replication, their RecA specifically mediates the pro-

teolytic cleavage of **LexA** (202 residues) between its Asp 84 and Gly 85. RecA is activated to do so on binding to ssDNA (it was initially assumed that RecA catalyzes the proteolysis of LexA but subsequent experiments by John Little indicate that activated RecA stimulates LexA to cleave itself). Further investigations indicated that LexA functions as a repressor of 43 genes that participate in DNA repair and the control of cell division, including *recA, lexA, uvrA,* and *uvrB.* DNA sequence analyses of the LexA-repressible genes revealed that they are all preceded by a homologous 20-nt sequence, the so-called **SOS box,** that has the palindromic symmetry characteristic of operators (control sites to which repressors bind so as to interfere with transcriptional initiation by RNA polymerase; Section 5-4A). Indeed, LexA has been shown to specifically bind the SOS boxes of *recA* and *lexA.*

The preceding observations suggest a model for the regulation of the SOS response (Fig. 30-59). During normal growth, LexA largely represses the expression of the SOS

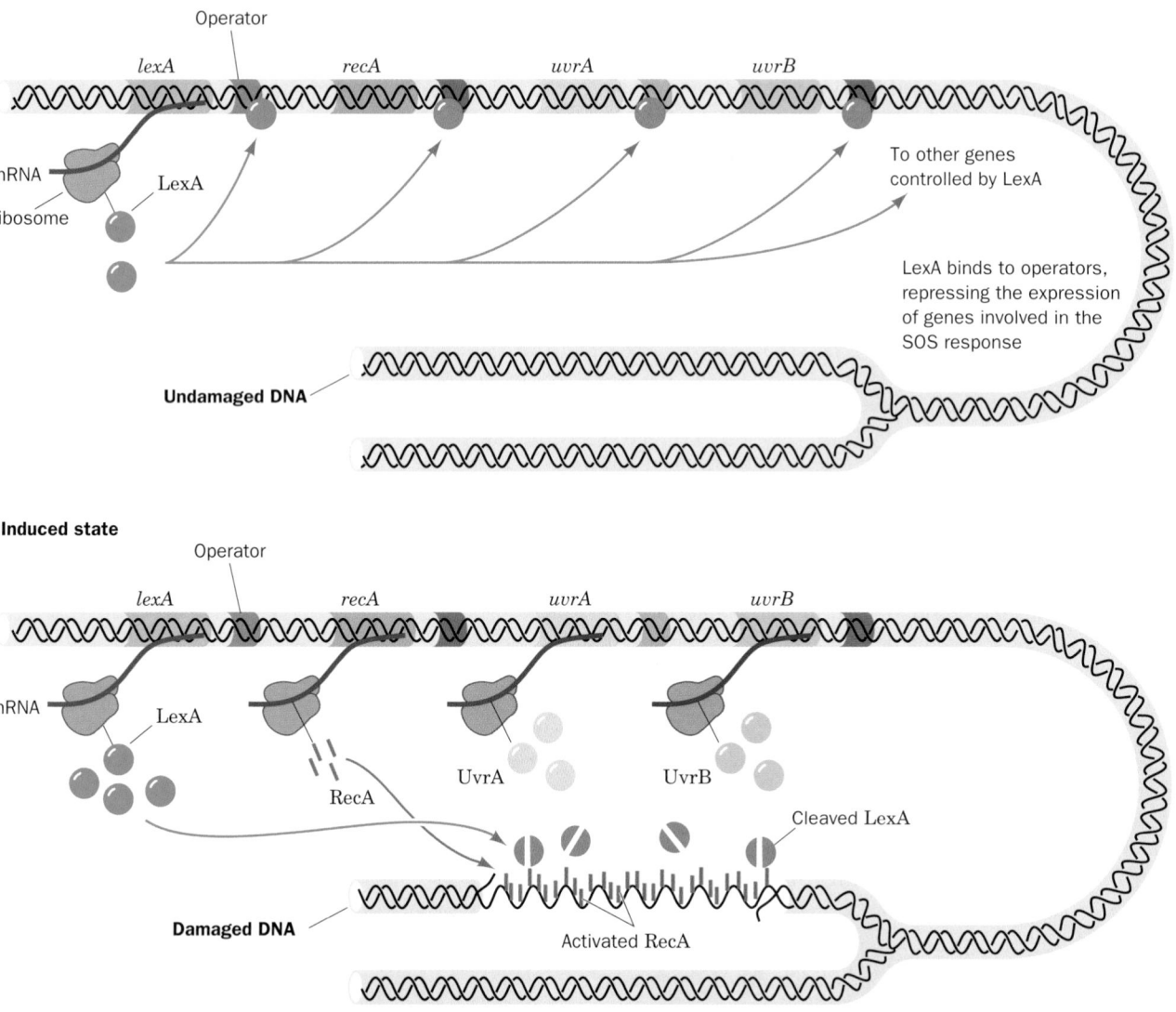

FIGURE 30-59   Regulation of the SOS response in *E. coli.* In a cell with undamaged DNA (*above*), LexA largely represses the synthesis of LexA, RecA, UvrA, UvrB, and other proteins involved in the SOS response. When there has been extensive

DNA damage (*below*), RecA is activated by binding to the resulting single-stranded DNA to stimulate LexA self-cleavage. The consequent synthesis of the SOS proteins results in the repair of the DNA damage.

genes, including the *lexA* gene, by binding to their SOS boxes so as to inhibit RNA polymerase from initiating the transcription of these genes. When DNA damage has been sufficient to produce postreplication gaps, however, this ssDNA binds to RecA so as to stimulate LexA cleavage. The LexA-repressible genes are consequently released from repression and direct the synthesis of SOS proteins including that of LexA (although this repressor continues to be cleaved through the influence of RecA). When the DNA lesions have been eliminated, RecA ceases stimulating LexA's autoproteolysis. The newly synthesized LexA can then function as a repressor, which permits the cell to return to normality.

b. SOS Repair Is Error Prone

The *E. coli* Pol III holoenzyme is unable to replicate through a variety of lesions such as AP sites and thymine dimers. On encountering such lesions, the replisome stalls and disassembles by releasing its Pol III cores, a process that is called replication fork "collapse." Cells have two general modes for restoring collapsed replication forks, **recombination repair** and **SOS repair.** Recombination repair circumvents the damaged template by using a homologous chromosome as its template DNA in a process known as **homologous recombination,** which also functions to generate genetic diversity. Hence we shall postpone our discussion of recombination repair until after our consideration of homologous recombination in Section 30-6A. In the following paragraphs we discuss SOS repair.

In SOS repair, the Pol III core lost from the collapsed replication fork is replaced by one of two so-called **bypass DNA polymerases,** whose synthesis is induced by the SOS response: **DNA polymerase IV (Pol IV,** the 336-residue product of the *dinB* gene) or **DNA polymerase V [Pol V;** the heterotrimeric product of the *umuD* and *umuC* genes, **UmuD'$_2$C** (umu for *UV mu*tagenesis), where UmuD' is produced by the RecA-assisted self-cleavage of the 139-residue **UmuD** to remove its N-terminal 24 residues, and UmuC consists of 422 residues]. Both of these enzymes are Y-family DNA polymerases, all of whose members lack $3' \rightarrow 5'$ proofreading exonuclease activity and replicate undamaged DNA with poor fidelity and low processivity and hence are also known as **error-prone DNA polymerases.**

Translesion synthesis (TLS) by Pol V, which was characterized in large part by O'Donnell and Myron Goodman, requires the simultaneous presence of the β_2 sliding clamp, the γ complex (clamp loader), and SSB, together with a RecA filament in complex with the ssDNA arising from the action of helicase on the dsDNA ahead of the stalled replication fork. This so-called **Pol V mutasome** tends to incorporate G about half as often as A opposite thymine dimers and AP sites, with pyrimidines being installed infrequently. This process is, of course, highly mutagenic. But even in replicating undamaged DNA, Pol V is at least 1000-fold more error prone than is Pol I or Pol III holoenzyme. However, after synthesizing ~7 nt, the Pol V mutasome is replaced by Pol III holoenzyme, which commences normal DNA replication after the now bypassed lesion. Pol II, a TLS participant that accurately replicates DNA, is also in-

duced by the SOS response but it is synthesized well before Pol V appears (see below). The role of Pol II appears to be the mediation of error-free TLS, and only if this process fails, is it replaced by Pol V to carry out error-prone TLS.

There are numerous types of DNA lesions besides AP sites and thymine dimers that interfere with normal DNA replication. Depending on the type of lesion, Pol IV, which is also error prone, may instead be recruited to carry out TLS. With many lesions, TLS may skip over the altered nucleotide, resulting in deletion of one or two bases in the daughter strand opposite the lesion (yielding a **frameshift mutation,** so called because it would change a structural gene's reading frame from that point onward; Section 5-4B). Moreover, Pol IV is prone to generating frameshift mutations even when replicating undamaged DNA.

The Y-family DNA polymerase **Dpo4** from the archaebacterium *Sulfolobus solfataricus* P2, a homolog of *E. coli* Pol IV and Pol V, misincorporates ~1 base per 500 replicated nucleotides. The X-ray structure of a complex of Dpo4 with a primer–template DNA that had been incubated with ddATP (which is complementary to the template base), determined by Wei Yang, reveals the structural basis for this low fidelity (Fig. 30-60). The 352-residue protein contains the fingers, palm, and thumb domains common to all known DNA polymerases (although their orders differ in the sequences of the different families of DNA polymerases) and, in addition, has a C-terminal domain unique to Y-family DNA polymerases that has been

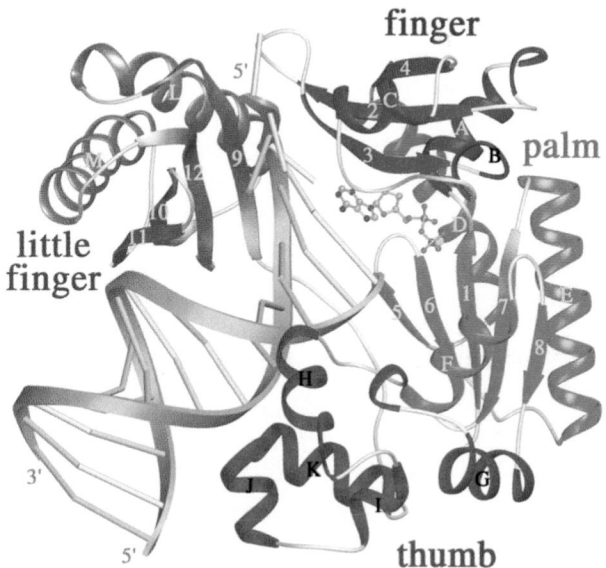

FIGURE 30-60 X-Ray structure of the bypass DNA polymerase Dpo4 from *Sulfolobus solfataricus* P2 in complex with a primer–template DNA and ddADP. The protein is drawn in ribbon form with its fingers, palm, thumb, and little finger domains blue, red, green, and purple, respectively. The DNA is gold with its backbones drawn as ribbons and its bases represented by rods. The ddADP, which is base paired to a template T in the enzyme's active site, is shown in ball-and-stick form colored according to atom type (C pink, N blue, O red, and P magenta). [Courtesy of Wei Yang, NIH, Bethesda, Maryland. PDBid 1JX4.]

dubbed the "little finger" domain. The enzyme, as expected, has incorporated a ddA residue at the 3' end of the primer and, in addition, binds a ddADP in base paired complex to the new template T. The little finger domain binds in the major groove of the DNA. However, the fingers and thumb domains are small and stubby compared to those of replicative DNA polymerases such as Klentaq1 (Fig. 30-9) and RB69 pol (Fig. 30-39), and the residues that contact the base pair in the active site are all Gly and Ala rather than the Phe, Tyr, and Arg that mainly do so in the replicative DNA polymerases. Moreover, the bound DNA is entirely in the B form rather than in the A form at the active site as occurs in replicative DNA polymerases. Since the minor groove is more accessible in A-DNA than in B-DNA (Section 29-1B), this suggests that error-prone DNA polymerases have relatively little facility to monitor the base pairing fidelity of the incoming nucleotide. This accounts for the ability of error-prone DNA polymerases to accommodate distorted template DNA as well as non-Watson–Crick base pairs at their active sites.

SOS repair is an error-prone and hence mutagenic process. It is therefore a process of last resort that is only initiated ~50 min after SOS induction if the DNA has not already been repaired by other means. Yet, DNA damage that normally activates the SOS response is nonmutagenic in the *recA⁻ E. coli* that survive. This is, as we saw, because bypass DNA polymerases will replicate over a DNA lesion even when there is no information as to which bases were originally present. Indeed, *most mutations in E. coli arise from the actions of the SOS repair system*, which is therefore a testimonial to the proposition that survival with a chance of loss of function (and the possible gain of new ones) is advantageous, in the Darwinian sense, over death, although only a small fraction of cells actually survive this process. It has therefore been suggested that, under conditions of environmental stress, the SOS system functions to increase the rate of mutation so as to increase the rate at which the *E. coli* adapt to the new conditions. Finally, it should be noted that the eukaryotic pols η, ι, and κ, all Y-family members, and pol ζ, an X-family member, are implicated in TLS and that pol η, the product of the *XPV* gene, is defective in the XPV form of xeroderma pigmentosum (Section 30-5B).

E. *Double-Strand Break Repair*

Double-strand breaks **(DSBs)** in DNA are produced by ionizing radiation and the free radical by-products of oxidative metabolism. Moreover, DSBs are normal intermediates in certain specialized cellular processes such as meiosis (Section 1-4A) and **V(D)J recombination** in lymphoid cells, which helps generate the vast diversity of antigen-binding sites in antibodies and *T*-cell receptors (Section 35-2C). Unrepaired or misrepaired DSBs can be lethal to cells or cause chromosomal aberrations that may lead to cancer. Hence the efficient repair of DSBs is essential for cell viability and genomic integrity.

Cells have two general modes to repair DSBs, recombination repair and **nonhomologous end-joining (NHEJ).**

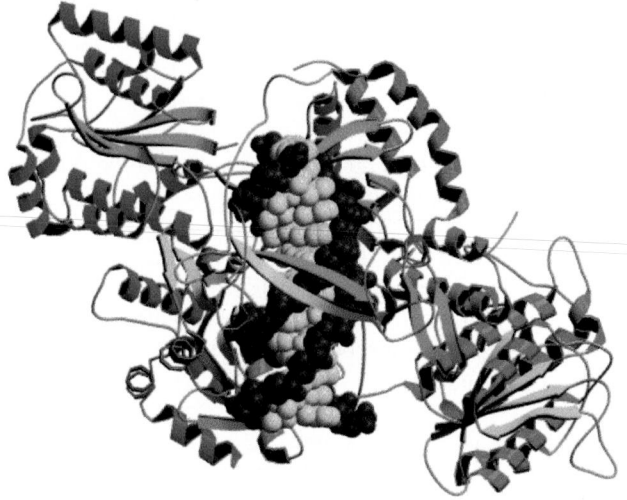

FIGURE 30-61   X-Ray structure of human Ku protein in complex with DNA containing 14 bp. The subunits of Ku70 (*red helices and yellow strands*) and Ku80 (*blue helices and green strands*) are viewed along the pseudo-2-fold axis relating them. The DNA, viewed with its DSB pointing upward, is drawn in space-filling form with its sugar–phosphate backbone dark gray and its base pairs light gray. Note that the DNA is surrounded by a ring of protein. [Courtesy of John Tainer, The Scripps Research Institute, La Jolla, California. Based on an X-ray Structure by Jonathan Goldberg, Memorial Sloan-Kettering Cancer Center, New York, New York. PDBid 1JEY.]

Here we discuss NHEJ, a process which, as its name implies, directly rejoins DSBs. The recombination repair of DSBs is discussed in Section 30-6A.

In NHEJ, the broken ends of the DSB must be aligned, its frayed ends trimmed and/or filled in, and their strands ligated. The core NHEJ machinery in eukaryotes includes the DNA end-binding protein **Ku** (a heterodimer of homologous 70- and 83-kD subunits, **Ku70** and **Ku80**), **DNA ligase IV,** and the accessory protein **Xrcc4.** Ku, an abundant nuclear protein, binds to a DSB, whether blunt or with an overhang, and hence appears to be the cell's primary DSB sensor. The X-ray structure of Ku in complex with a 14-bp DNA, determined by Jonathan Goldberg, reveals that the protein cradles the dsDNA segment along its entire length and encircles its central ~3 bp segment (Fig. 30-61). The protein ring is also present in the closely similar X-ray structure of Ku alone, thereby explaining why Ku that is bound to a dsDNA which is then circularized becomes permanently associated with it. Ku makes no specific contacts with the DNA's bases and few with its sugar–phosphate backbone, but instead fits snugly into the DNA's major and minor grooves so as to precisely orient it.

Ku–DNA complexes have been shown to dimerize so as to align the members of a DSB, both blunt ended and with short (1–4 bp) complementary single strands, for ligation as is diagrammed in Fig. 30-62. The DNA ends are exposed along one face of each Ku–DNA complex, presumably making them accessible to polymerases that fill in gaps and to nucleases that trim excess and inappropriate ends preparatory for ligation by DNA ligase IV in complex with Xrcc4. Nucleotide trimming, which of course generates mutations, appears to be carried out in an ATP-dependent manner by the evolutionarily conserved **Mre11 complex,** which consists of two **Mre11** nuclease subunits and two **Rad50** ATPase subunits. Ku is eventually released from the rejoined DNA, perhaps by proteolytic cleavage.

F. *Identification of Carcinogens*

Many forms of cancer are known to be caused by exposure to certain chemical agents that are therefore known as carcinogens. It has been estimated that as much as 80% of human cancer arises in this fashion. There is considerable evidence that the primary event in carcinogenesis is often damage to DNA (carcinogenesis is discussed in Section 34-4C). Carcinogens are consequently also likely to induce the SOS response in bacteria and thus act as indirect mutagenic agents. In fact, there is a high correlation between carcinogenesis and mutagenesis (recall, e.g., the progress of xeroderma pigmentosum; Section 30-5B).

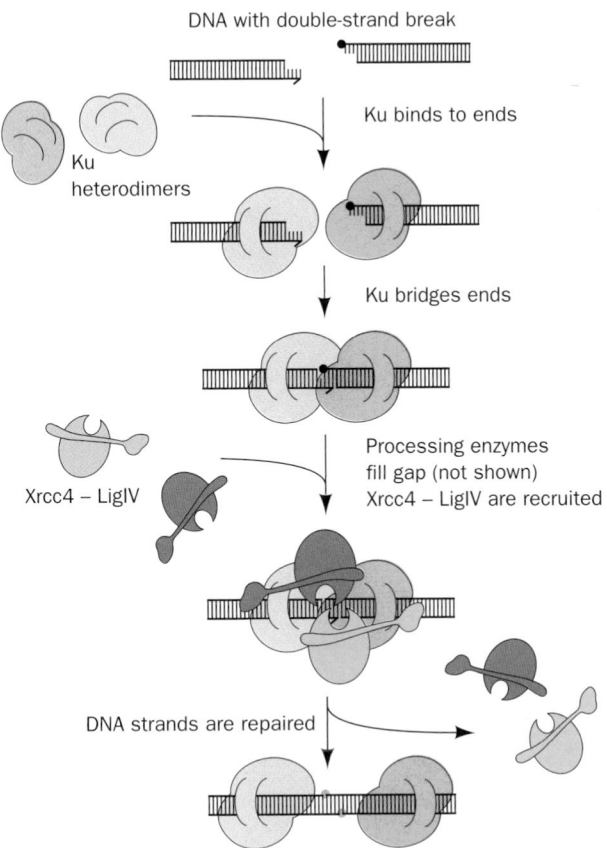

FIGURE 30-62 Schematic diagram of nonhomologous end-joining (NHEJ). The left dsDNA fragment is missing a base and the right fragment is blocked by a nonligatable group (*filled black circle*). The two Ku heterodimers are drawn in two shades of yellow and the Xrcc4–DNA ligase IV complexes are drawn in two shades of blue. The newly repaired links in the DNA are represented by pink circles. [After Jones, J.M., Gellert, M., and Yang, W., *Structure* **9**, 881 (2001).]

There are presently over 60,000 man-made chemicals of commercial importance and ~1000 new ones are introduced each year. The standard animal tests for carcinogenesis, exposing rats or mice to high levels of the suspected carcinogen and checking for cancer, are expensive and require ~3 years to complete. Thus relatively few substances have been tested in this manner.

a. The Ames Test Assays for Probable Carcinogenicity

Bruce Ames devised a rapid and effective bacterial assay for carcinogenicity that is based on the high correlation between carcinogenesis and mutagenesis. He constructed special tester strains of *Salmonella typhimurium* that are *his⁻* (cannot synthesize histidine so that they are unable to grow in its absence), have cell envelopes that lack the lipopolysaccharide coating that renders normal *Salmonella* impermeable to many substances (Section 11-3B), and have inactivated excision repair systems. Mutagenesis in these tester strains is indicated by their reversion to the *his⁺* phenotype.

In the **Ames test,** ~10⁹ tester strain bacteria are spread on a culture plate that lacks histidine. Usually a mixture of several *his⁻* strains is used so that mutations due to both base changes and nucleotide insertions or deletions can be detected. A mutagen placed in the culture medium causes some of these *his⁻* bacteria to revert to the *his⁺* phenotype, which is detected by their growth into visible colonies after 2 days at 37°C (Fig. 30-63). The mutagenicity of a substance is scored as the number of such colonies less the few spontaneously revertant colonies that occur in the absence of the mutagen.

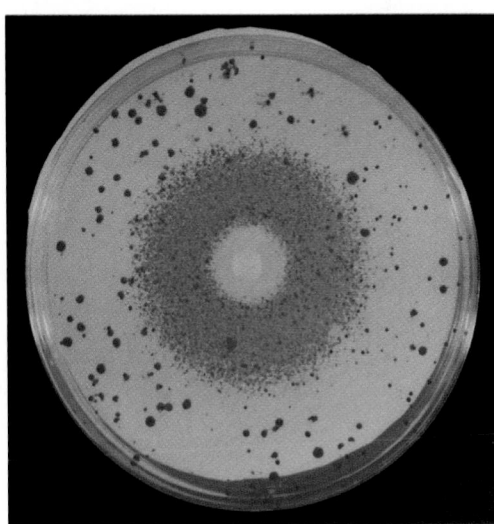

FIGURE 30-63 The Ames test for mutagenesis. A filter paper disk containing a mutagen, in this case the alkylating agent ethyl methanesulfonate, is centered on a culture plate containing *his⁻* tester strains of *Salmonella typhimurium* in a medium that lacks histidine. A dense halo of revertant bacterial colonies appears around the disk from which the mutagen diffused. The larger colonies distributed about the culture plate are spontaneous revertants. The bacteria near the disk have been killed by the toxic mutagen's high concentration. [Courtesy of Raymond Devoret, Institut Curie, Orsay, France.]

Many noncarcinogens are converted to carcinogens in the liver or in other tissues via a variety of detoxification reactions (e.g., those catalyzed by the cytochromes P450; Section 15-4B). A small amount of rat liver homogenate is therefore included in the Ames test medium in an effort to approximate the effects of mammalian metabolism.

b. Both Man-Made and Naturally Occurring Substances Can Be Carcinogenic

There is an ~80% correspondence between the compounds determined to be carcinogenic by animal tests and those found to be mutagenic by the Ames test. Dose–response curves, which are generated by testing a given compound at a number of concentrations, are almost always linear and extrapolate back to zero, indicating that *there is no threshold concentration for mutagenesis.* Several compounds to which humans have been extensively exposed that were found to be mutagenic by the Ames test were later found to be carcinogenic in animal tests. These include tris(2,3-dibromopropyl)phosphate, which was used as a flame retardant on children's sleepwear in the mid-1970s and can be absorbed through the skin; and furylfuramide, which was used in Japan in the 1960s and 1970s as an antibacterial additive in many prepared foods (and had passed two animal tests before it was found to be mutagenic). Carcinogens are not confined to man-made compounds but also occur in nature. For example, carcinogens are contained in many plants that are common in the human diet, including alfalfa sprouts. **Aflatoxin B$_1$,**

Aflatoxin B$_1$

one of the most potent carcinogens known, is produced by fungi that grow on peanuts and corn. Charred or browned food, such as occurs on broiled meats and toasted bread, contains a variety of DNA-damaging agents. Thus, with respect to carcinogenesis, as Ames has written, "Nature is not benign."

6 ■ RECOMBINATION AND MOBILE GENETIC ELEMENTS

The chromosome is not just a simple repository of genetic information. If this were so, the unit of mutation would have to be an entire chromosome rather than a gene because there would be no means of separating a mutated gene from the other genes of the same chromosome. Chromosomes would therefore accumulate deleterious mutations until they became nonviable.

It has been known from some of the earliest genetic studies that pairs of allelic genes may exchange chromo-

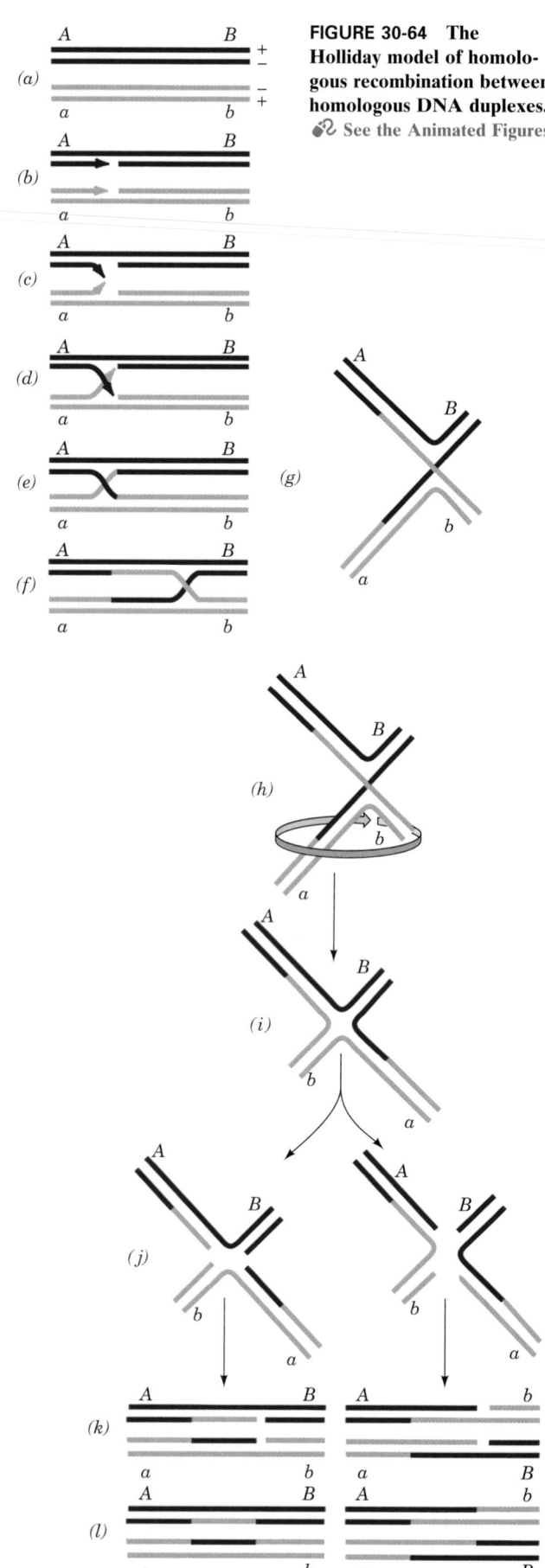

FIGURE 30-64 The Holliday model of homologous recombination between homologous DNA duplexes.

⏩ See the Animated Figures

somal locations by a process known as **genetic recombination** (Section 1-4B). Mutated genes can thereby be individually tested, since their propagation is then not absolutely dependent on the propagation of the genes with which they had been previously associated. In this section, we consider the mechanisms by which genetic elements can move, both between chromosomes and within them.

A. *Homologous Recombination*

Homologous recombination *(also called **general recombination**) is defined as the exchange of homologous segments between two DNA molecules.* Both genetic and cytological studies have long indicated that such a crossing-over process occurs in higher organisms during meiosis (Fig. 1-27). Bacteria, which are normally haploid, likewise have elaborate mechanisms for the interchange of genetic information. They can acquire foreign DNA through transformation (Section 5-2A), through a process called **conjugation** (mating) in which DNA is directly transferred from one cell to another via a cytoplasmic bridge (Section 31-1A), and via **transduction** in which a defective bacteriophage that has erroneously acquired a segment of bacterial DNA rather than the viral chromosome transfers this

DNA to another bacterial cell. In all of these processes, the foreign DNA is installed in the recipient's chromosome or plasmid through homologous recombination (to be propagated, a DNA segment must be part of a replicon; that is, be associated with a replication origin such as occurs in a chromosome, a plasmid, or a virus).

a. Recombination Occurs via a Crossed-Over Intermediate

The prototypical model for homologous recombination (Fig. 30-64) was proposed by Robin Holliday in 1964 on the basis of genetic studies on fungi. The corresponding strands of two aligned homologous DNA duplexes are nicked, and the nicked strands cross over to pair with the nearly complementary strands of the homologous duplex after which the nicks are sealed (Fig. 30-64*a–e*), thereby yielding a four-way junction known as a **Holliday junction** (alternatively, **Holliday intermediate;** Fig. 30-64*e*). A Holliday junction has, in fact, been observed in the X-ray structure of d(CCGGTACCGG), determined Shing Ho (Fig. 30-65), in which, perhaps unexpectedly, all the bases form normal Watson–Crick base pairs without any apparent strain. The crossover point can move in either direction, often thousands of nucleotides, in a process known

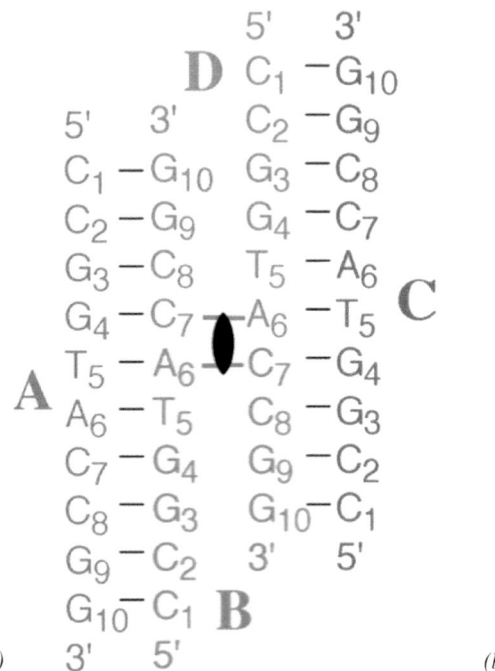
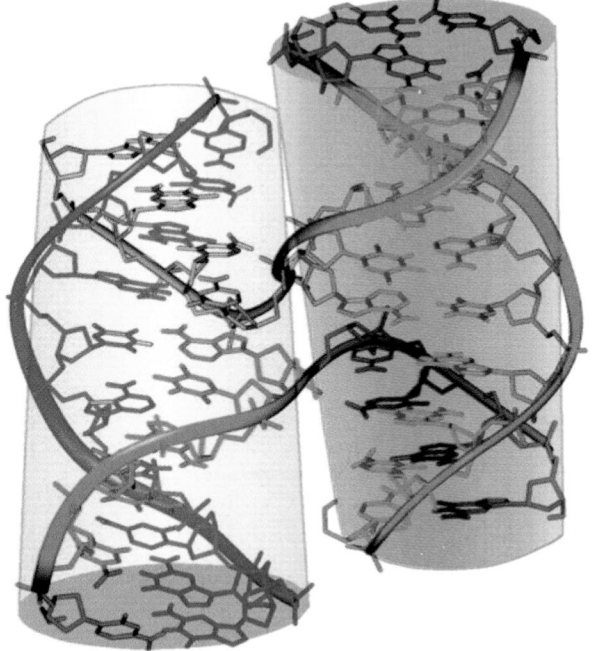

FIGURE 30-65   X-Ray structure of the self-complementary decameric DNA d(CCGGTACCGG). (*a*) The secondary structure of the four-stranded Holliday junction formed by this sequence in which the four strands, A, B, C, and D, are individually colored, their nucleotides are numbered 1 to 10 from their 5′ to 3′ termini, and Watson–Crick base pairing interactions are represented by black dashes. The 2-fold axis relating the two helices of this so-called **stacked-X conformation** is represented by the black lenticular symbol. (*b*) The observed three-dimensional structure of the Holliday junction, as viewed along its 2-fold axis, in which the oligonucleotides are represented in stick form with their backbones traced by ribbons, all

colored as in Part *a*. With the exception of the backbones of strands B and D at the crossovers, the two arms of this structure each form an undistorted B-DNA helix, including the stacking of the base pairs flanking the crossovers. The two helices are inclined to each other by 41°. Note that Fig. 30-64*g* is a schematic representation of the stacked-X conformation as viewed perpendicular to both helices (from the side in this drawing and hence having the projected appearance of the letter X). A Holliday junction can also assume a so-called **open-X conformation,** which is represented by Fig. 30-64*i*. [Courtesy of Shing Ho, Oregon State University. PDBid 1DCW.]

as **branch migration** (Fig. 30-64*e*, *f*) in which the four strands exchange base pairing partners.

A Holliday junction can be resolved into two duplex DNAs in two equally probable ways (Fig. 30-64*g*–*l*):

1. The cleavage of the strands that did not cross over (right branch of Fig. 30-64*j*–*l*) exchanges the ends of the original duplexes to form, after nick sealing, the traditional recombinant DNA (Fig. 1-27*b*).

2. The cleavage of the strands that crossed over (left branch of Fig. 30-64*j*–*l*) exchanges a pair of homologous single-stranded segments.

The recombination of circular duplex DNAs results in the types of structures diagrammed in Fig. 30-66. Electron

microscopic evidence for the existence of the postulated "figure-8" structures is shown in Fig. 30-67*a*. These figure-8 structures were shown not to be just twisted circles by cutting them with a restriction endonuclease to yield **chi structures** (after their resemblance to the Greek letter χ) such as that pictured in Fig. 30-67*b*.

b. Homologous Recombination in *E. coli* Is Catalyzed by RecA

The observation that *recA⁻ E. coli* have a 10^4-fold lower recombination rate than the wild-type indicates that *RecA protein has an important function in recombination*. Indeed, RecA greatly increases the rate at which complementary strands renature *in vitro*. This versatile protein (recall it also stimulates the autoproteolysis of LexA to trigger the SOS response and is an essential participant in the translesion synthesis of DNA; Section 30-5D) polymerizes cooperatively without regard to base sequence on ssDNA or on dsDNA that has a single-stranded gap. The resulting filaments, which may contain up to several thousand RecA monomers, specifically bind the homologous dsDNA, and, in an ATP-dependent reaction, catalyze strand exchange. EM studies by Edward Egelman (Fig. 30-68) reveal that RecA filaments bound to ssDNA or dsDNA form a right-handed helix with ~6.2 RecA monomers per turn and a

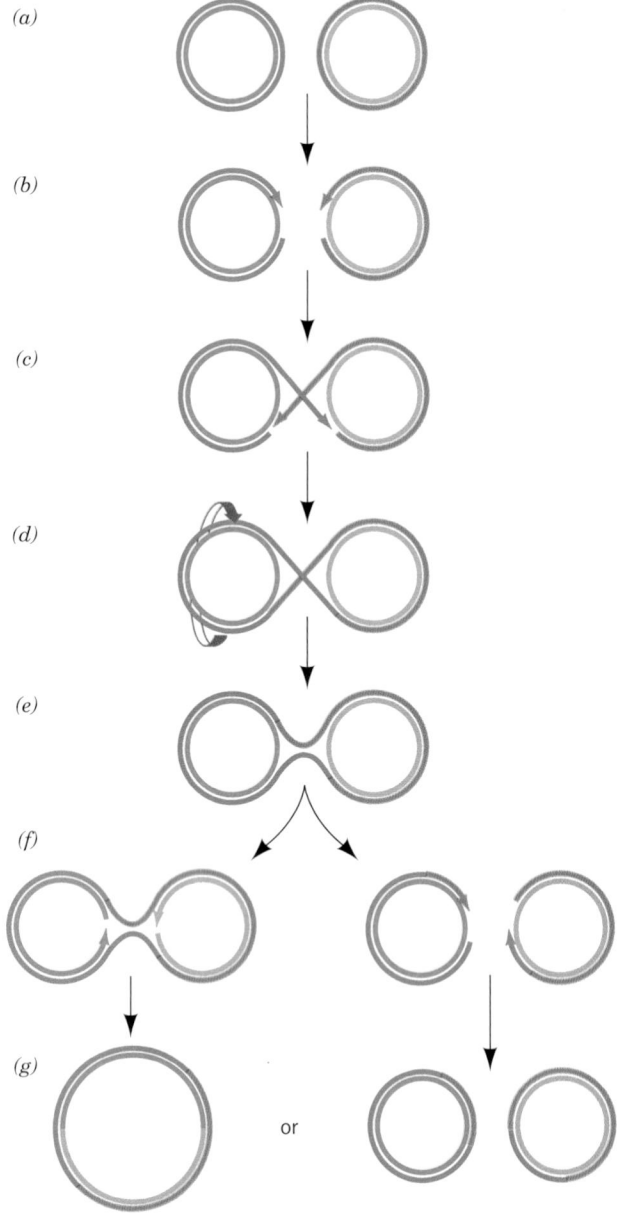

(a)

(b)

(c)

(d)

(e)

(f)

(g)

or

FIGURE 30-66   Homologous recombination between two circular DNA duplexes. This process can result either in two circles of the original sizes or in a single composite circle.

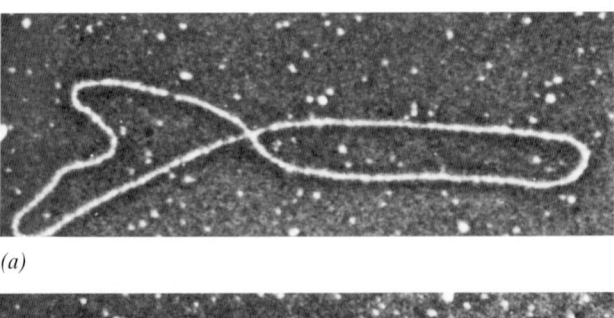

(a)

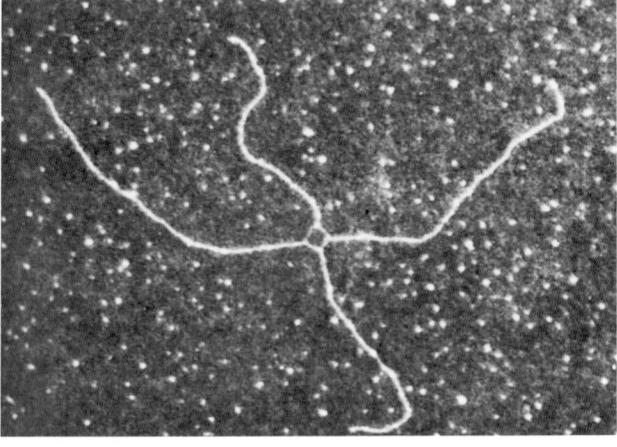

(b)

FIGURE 30-67   Electron micrographs of intermediates in the homologous recombination of two plasmids. (*a*) A figure-8 structure. This corresponds to Fig. 30-66*d*. (*b*) A chi structure that results from the treatment of a figure-8 structure with a restriction endonuclease. Note the thinner single-stranded connections in the crossover region. [Courtesy of Huntington Potter, University of South Florida, and David Dressler, Oxford University, U.K.]

pitch (rise per turn) of 95 Å. The DNA in these filaments, which binds to the protein with 3 nt (or bp) per RecA monomer unit and hence has ~18.6 nt (or bp) per turn, is so extended (having a rise of 5.1 Å/bp vs 3.4 Å/bp in B-DNA) that it must lie near the center of the helical filament as Fig. 30-68 indicates.

The X-ray structure of RecA (Fig. 30-69), determined by Steitz, reveals that the protein consists of a major central domain that is flanked by smaller N- and C-terminal domains. The monomers associate to form an ~120-Å-wide helical filament with six monomer units per turn and a pitch of 82.7 Å. The helical filament is remarkably open,

so much so that there are gaps between the monomer units in successive turns. This arrangement results in a large he-

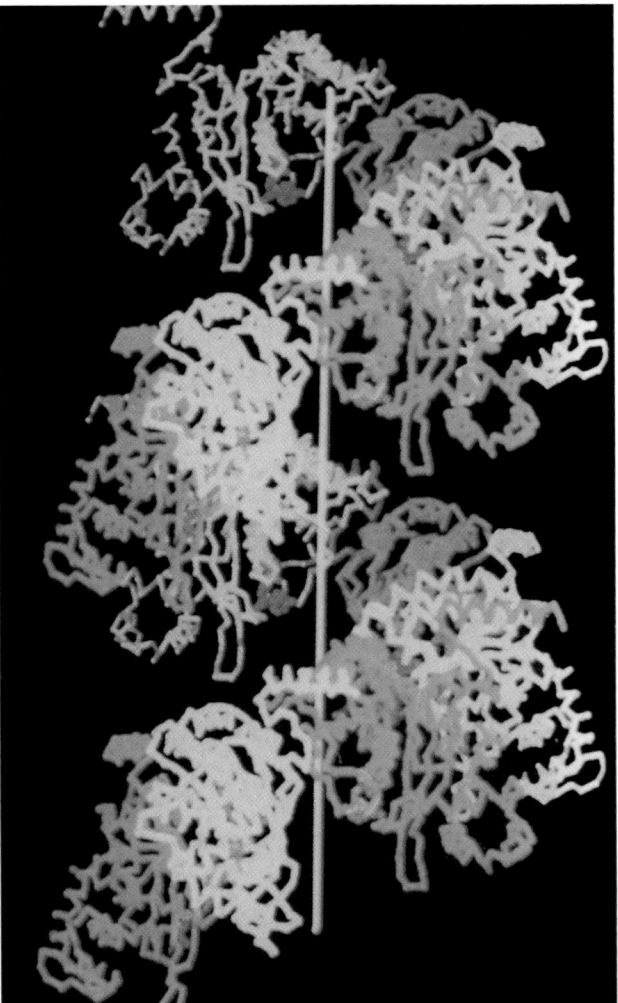

(a)

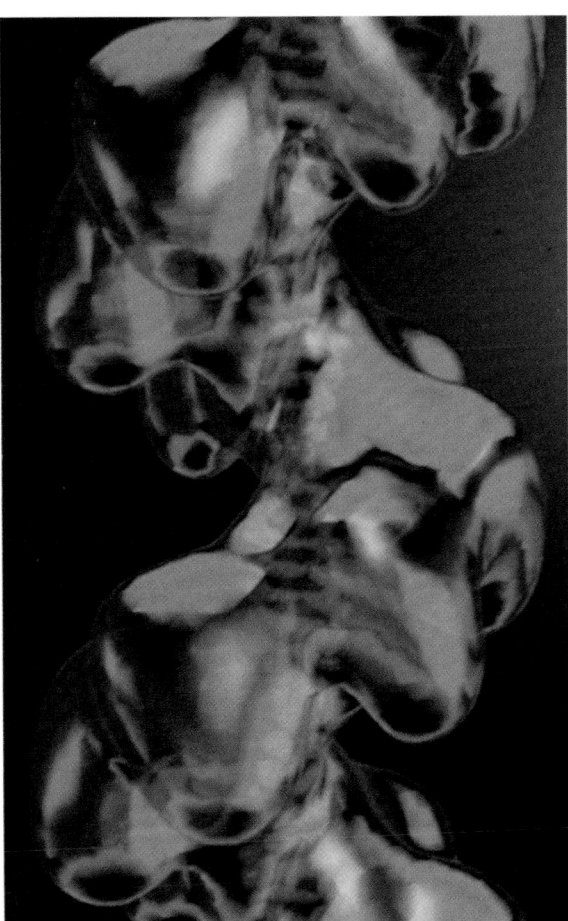

FIGURE 30-68   An electron microscopy–based image (*transparent surface*) of an *E. coli* RecA–dsDNA–ATP filament. The extended and untwisted dsDNA (*red*) has been modeled into this image. [Courtesy of Edward Egelman, University of Minnesota Medical School.]

FIGURE 30-69   X-Ray structure of *E. coli* RecA protein. The RecA monomers are represented by their C_α chains. Alternate RecA monomers are yellow and blue, and their bound ADPs are red. (*a*) View perpendicular to the protein filament's helix axis (*light blue rod*) showing 12 monomers constituting two turns of the helix in the same orientation as in Fig. 30-68. (*b*) View nearly parallel to the helix axis showing one turn of the helix. [Courtesy of Thomas Steitz, Yale University. PDBid 1REA.]

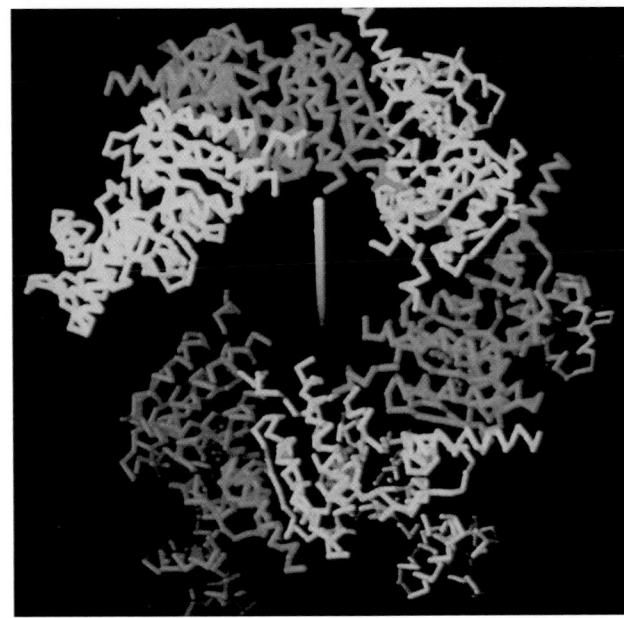

(b)

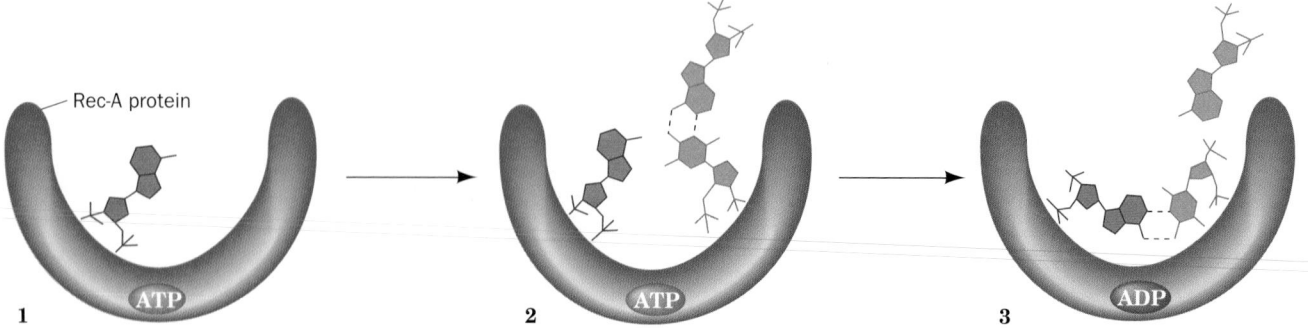

FIGURE 30-70   A model for RecA-mediated pairing and strand exchange between a single-stranded and a duplex DNA.
(1) The ssDNA binds to RecA to form an initiation complex. **(2)** The dsDNA binds to the initiation complex so as to transiently form a three-stranded helix that mediates the correct

pairing of the homologous strands. **(3)** RecA rotates the bases of the aligned homologous strands to effect strand exchange in an ATP-driven process. [After West, S.C., *Annu. Rev. Biochem.* **61,** 618 (1992).]

lical groove running the length of the filament that, when viewed down the helix axis, forms a 25-Å-wide central hole. This helical filament is strikingly similar to that of a RecA–duplex DNA filament as visualized at lower resolution by electron microscopy (Fig. 30-68; EM studies have also shown that LexA protein binds within this helical groove such that it spans the two RecA subunits on successive turns of the helix).

How does RecA mediate DNA strand exchange between single-stranded and duplex DNAs? On encountering a dsDNA with a strand that is complementary to its

bound ssDNA, RecA partially unwinds the duplex and, in a reaction driven by RecA-catalyzed ATP hydrolysis, exchanges the ssDNA with the corresponding strand on the duplex. A model of how RecA might do so is diagrammed in Fig. 30-70. *This process tolerates only a limited degree of mispairing and requires that one of the participating DNA strands have a free end.* The assimilation (exchange) of a single-stranded circle with a strand on a linear duplex (Fig. 30-71) cannot proceed past the 3′ end of a highly mismatched segment in the complementary strand. *The invasion of the single strand must therefore begin with its 5′ end.*

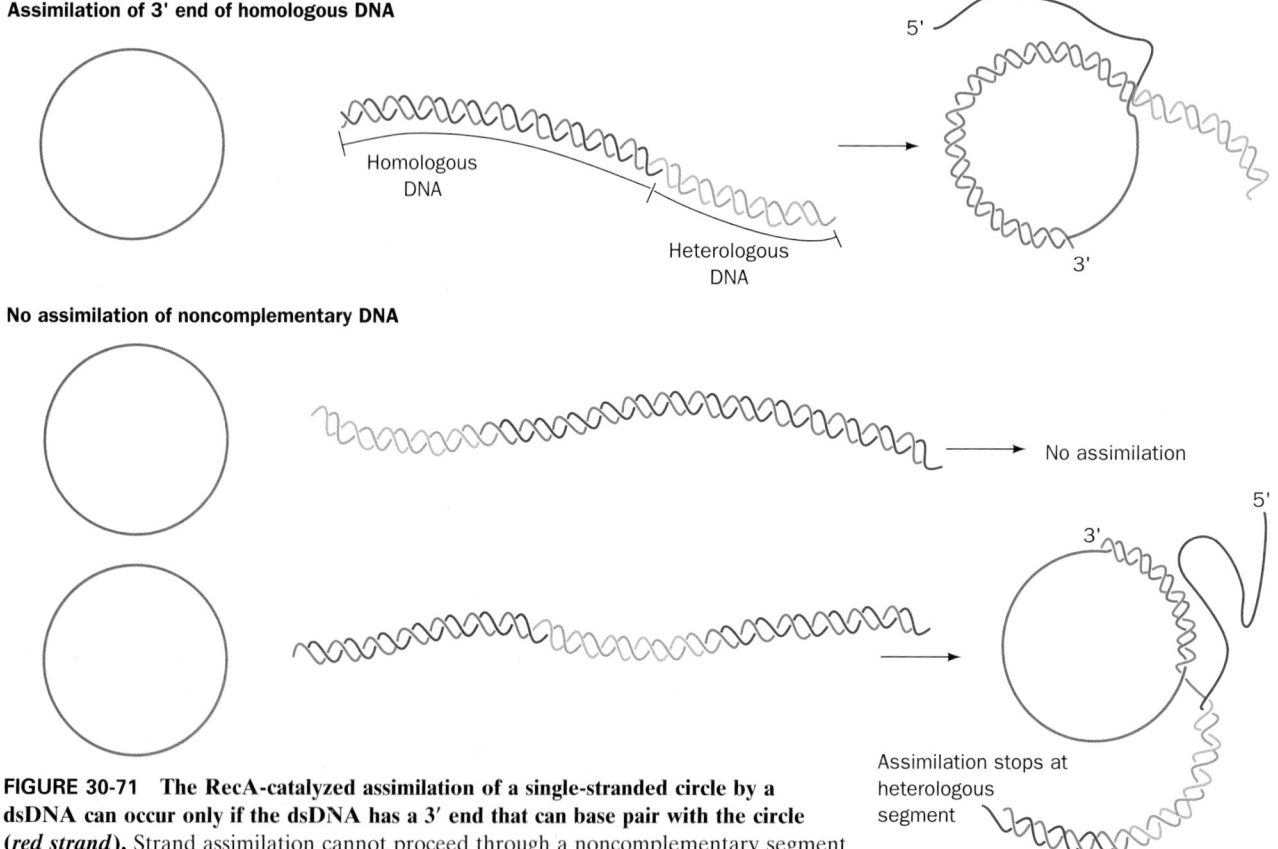

FIGURE 30-71   The RecA-catalyzed assimilation of a single-stranded circle by a dsDNA can occur only if the dsDNA has a 3′ end that can base pair with the circle (*red strand*). Strand assimilation cannot proceed through a noncomplementary segment (*purple and orange strands*).

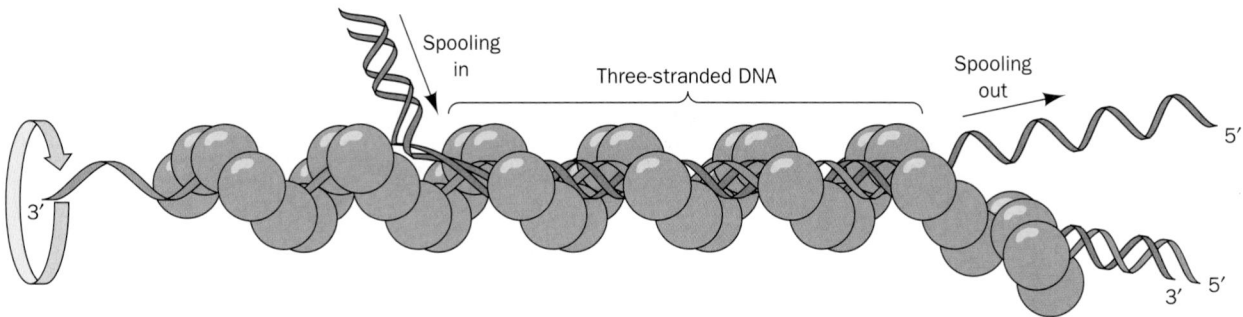

FIGURE 30-72   A hypothetical model for the RecA-mediated strand exchange reaction. Homologous DNA molecules are paired in advance of strand exchange in a three-stranded helix. The ATP-driven rotation of the RecA filament about its helix axis would cause duplex DNA to be "spooled in" to the filament, right to left as drawn. [After West, S.C., *Annu. Rev. Biochem.* **61,** 617 (1992).]

A model for the consequent branch migration process is diagrammed in Fig. 30-72. Of course, two such strand exchange processes must occur simultaneously in a Holliday junction (Figs. 30-64 and 30-66).

c. Eukaryotes Have RecA-Like Proteins

Yeast **RAD51** (339 residues) functions in the ATP-dependent repair and recombination of DNA in much the same way as does the 30% homologous *E. coli* RecA protein. The electron micrograph–based image reconstruction of RAD51 in complex with double-stranded DNA is nearly identical to that of RecA at low resolution: Both complexes form helical filaments in which the DNA has an ~5.1-Å rise per bp and 18.6 bp per turn. Since RAD51 homologs occur in chickens, mice, and humans, it is very likely that such filaments universally mediate DNA repair and recombination.

d. RecBCD Initiates Recombination by Making Single-Strand Nicks

The single-strand nicks to which RecA binds are made by the **RecBCD** protein, the 330-kD heterotrimeric product of the SOS genes *recB, recC,* and *recD,* which has both helicase and nuclease activities (Fig. 30-73). The process begins with RecBCD binding to the end of a dsDNA and then unwinding it via its ATP-driven helicase function. As it does so, it nucleolytically degrades the unwound single strands behind it, with the 3'-ending strand being cleaved more often and hence broken down to smaller fragments than the 5'-ending strand. However, on encountering the sequence GCTGGTGG from its 3' end (the so-called **Chi sequence,** which occurs about every ~5 kb in the *E. coli* genome), it stops cleaving the 3'-ending strand and increases the rate at which it cleaves the 5'-ending strand, thereby yielding the 3'-ending single-strand segment to which RecA binds. This explains the observation that regions containing Chi sequences have elevated rates of recombination.

RecBCD can only commence unwinding DNA at a free duplex end. Such ends are not normally present in *E. coli,* which has a circular genome, but become available during such recombinational processes as bacterial transforma-

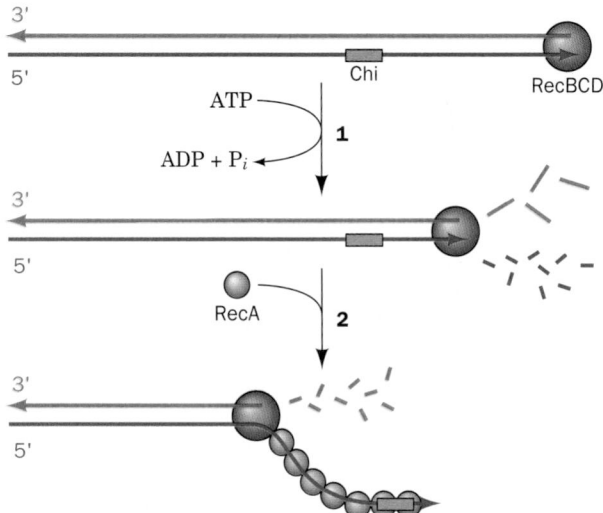

FIGURE 30-73   The generation of a 3'-ending single-strand DNA segment by RecBCD to initiate recombination.
(1) RecBCD binds to a free end of a dsDNA and, in an ATP-driven process, advances along the helix, unwinding the DNA and degrading the resulting single strands behind it, with the 3'-ending strand cleaved more often than the 5'-ending strand.
(2) When RecBCD encounters a properly oriented Chi sequence, it increases the frequency at which it cleaves the 5'-ending strand but stops cleaving the 3'-ending strand, thereby generating the potentially invasive 3'-ending strand segment to which RecA binds.

tion, conjugation, and viral transduction, as well as at collapsed replication forks.

e. RuvABC Mediates the Branch Migration and the Resolution of the Holliday Junction

The branch migration of the RecA-generated Holliday junction (Fig. 30-64*e, f*) requires the breaking and reforming of base pairs as the bases exchange partners in passing from one double helical stem to the other. Since $\Delta G = 0$ for this process, it was initially assumed that it occurs spontaneously. However, such a process moves forward and backward at random and, moreover, is blocked by as little as a single mismatched base pair. In *E. coli,* and most other bacteria, branch migration is an ATP-dependent unidirectional process that is mediated by two proteins whose synthesis is induced by the SOS response (Section 30-5D): **RuvB** (336 residues), an ATP-powered pump that drives branch migration but binds only weakly to DNA; and **RuvA** (203 residues), which binds to both a Holliday junction and to RuvB, thereby targeting RuvB to the DNA.

The X-ray structure of *Mycobacterium leprae* (the cause of leprosy) RuvA in complex with a synthetic and immobile Holliday junction (Fig. 30-74a), determined by Morikawa, reveals that RuvA forms a homotetramer to which the Holliday junction binds in its open-X conformation (Fig. 30-74b). The RuvA tetramer, which has the appearance of a four-petaled flower (it has C_4 symmetry rather than the D_2 symmetry of the vast majority of homotetramers), is relatively flat ($80 \times 80 \times 45$ Å) with one square face concave and the other convex. The concave face (that facing the viewer in Fig. 30-74b), which is highly positively charged and is studded with numerous conserved residues, has four symmetry-related grooves that bind the Holliday junction's four arms. This face's four centrally located projections or "pins" are negatively charged, and hence the repulsive forces between them and the Holliday junction's anionic phosphate groups probably facilitate the separation of the single-stranded DNA segments and guide them from one double helix to another.

RuvB is a member of the AAA^+ family of ATPases (Section 30-3C). The X-ray structure of *Thermus thermophilus* RuvB crystallized in the presence of both ADP and ADPNP, determined by Morikawa, reveals two molecules of RuvB with somewhat different conformations: one binding ADP and the other binding ADPNP. Each RuvB molecule consists of three consecutive domains arranged in a crescentlike configuration with the adenine nucleotides binding at the interface between its N-terminal and middle domains. EM studies indicate that, in the presence of dsDNA, RuvB oligomerizes to form a hexamer (Fig. 30-75a), as do most other AAA^+ family members, including the D2 domain of NSF (Fig. 12-66). A hexameric model of RuvB (Fig. 30-75b), constructed by superimposing the N-terminal domain of the RuvB monomer on the ATPase domains of the NSF-D2 hexamer, agrees well with the EM-based image and contains no serious steric clashes. This 130-Å-diameter hexameric model contains a 30-Å-diameter hole through which a single dsDNA can readily be threaded (see below). Moreover the six β hairpins, one per monomer, that have been implicated in binding to RuvA are located on the top face of the hexamer (that pictured in Fig. 30-75b).

The EM images of the RuvAB–Holliday junction complex indicate that RuvA binds two oppositely located RuvB hexamers. This has led to the model of their interaction depicted in Fig. 30-76 in which RuvA binds the Holliday junction and helps load the RuvB hexameric rings onto two opposing arms of the Holliday junction. The two hexameric rings are postulated to counter-rotate, each in the anticlockwise direction looking toward the center of the junction, so as to screw the horizontal DNA strands

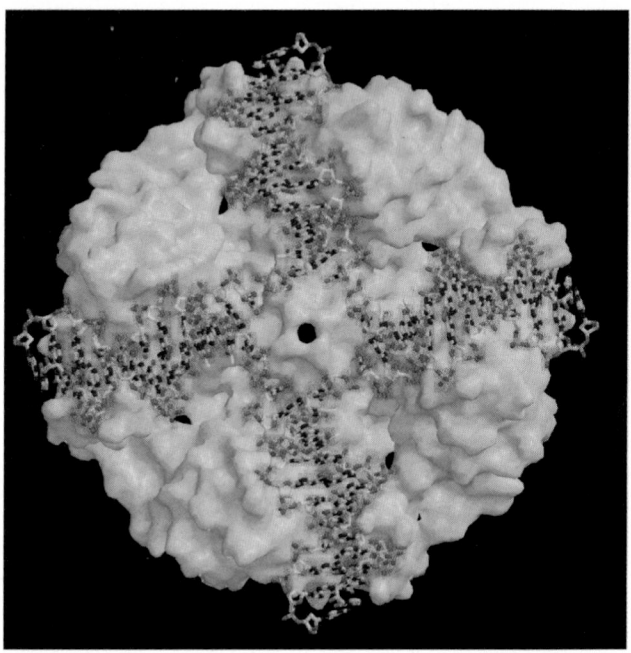

(a)

(b)

FIGURE 30-74 X-Ray structure of a RuvA tetramer in complex with a Holliday junction. (*a*) A schematic drawing of the synthetic and immobile Holliday junction in this structure showing its base sequence. The two A · T base pairs that are disrupted at the crossover (and which, if the Holliday junction consisted of two homologous dsDNAs, as it normally does, would exchange base pairing partners) are magenta. (*b*) The

RuvA–Holliday junction complex as viewed along the protein tetramer's 4-fold axis. The protein is represented as its molecular surface (*gray*) and the DNA is drawn in stick form colored according to atom type (C white, N blue, O red, and P yellow). [Courtesy of Kosuke Morikawa, Biomolecular Engineering Research Institute, Osaka, Japan. PDBid 1C7Y.]

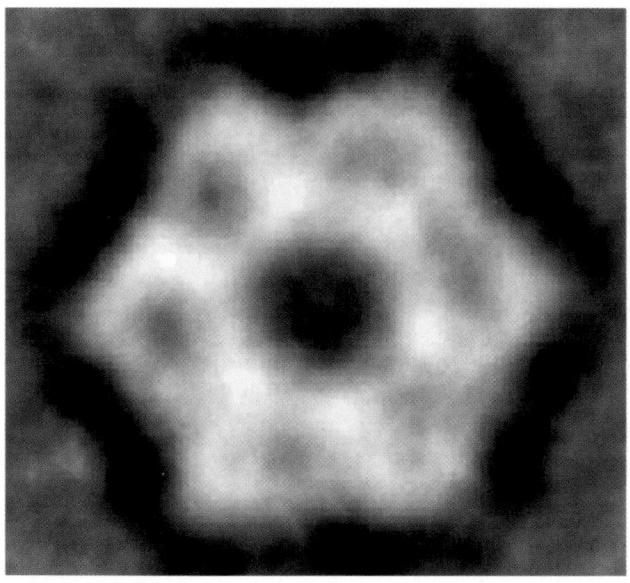

(a)

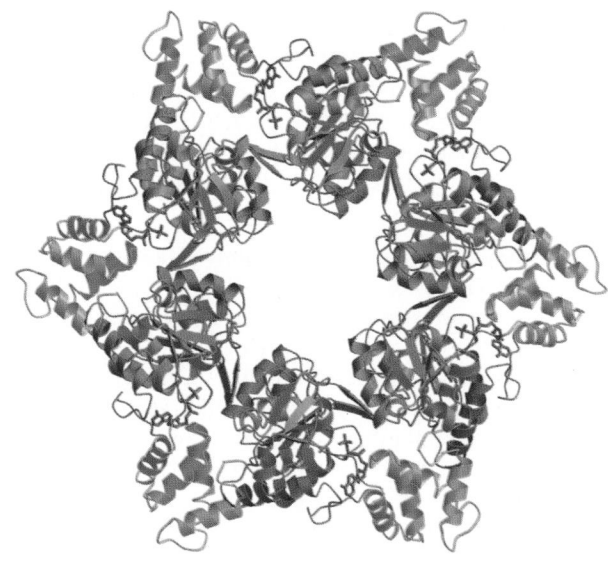

(b)

FIGURE 30-75   Proposed structure of the *T. thermophilus* RuvB hexamer. (*a*) An EM-based image reconstruction of RuvB complexed with a 30-bp DNA (not visible) as viewed along its 6-fold axis. The image resolution is 30 Å. (*b*) A model of the RuvB hexamer that was constructed from the X-ray structure of RuvB monomers by superimposing their

N-terminal domains on the homologous ATPase domains of the NSF-D2 homohexamer (Fig. 12-66). The N-terminal, middle, and C-terminal domains are blue, yellow, and green, respectively, and its bound ADPNP is drawn in stick form in red. [Courtesy of Kosuke Morikawa, Biomolecular Engineering Research Institute, Osaka, Japan. PDBid 1HQC.]

through the center of the junction and into the top and bottom double helices, thereby effecting branch migration (although rather than actually rotating relative to RuvA,

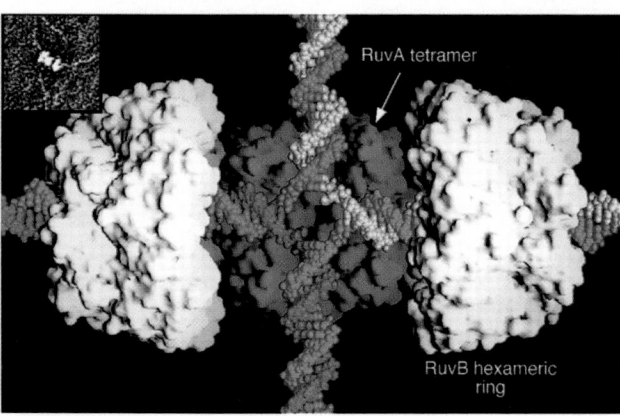

FIGURE 30-76   Model of the RuvAB–Holliday junction complex. The model is based on electron micrographs such as that in the inset. The proteins are represented by their surface diagrams with the RuvA tetramer, as seen in its X-ray structure, green and the two oppositely oriented RuvB hexamers gray. The DNA of the Holliday junction is drawn in space-filling form with its homologous blue and magenta strands complementary to its red and white strands. The complex is postulated to drive branch migration via the ATP-driven counter-rotation of the RuvB hexamers relative to the RuvA tetramer. This pumps (screws) the horizontal dsDNAs through the RuvB hexamers to the center of the Holliday junction, where their strands separate and then base pair with their homologs to form new dsDNAs, which are pumped out vertically. [Courtesy of Peter Artymiuk, University of Sheffield, U.K.]

a RuvB hexamer might pull the dsDNA through its central hole by "walking" up its grooves in a manner resembling that postulated for hexagonal helicases; Section 30-2C). The direction of branch migration depends on which pair of arms the RuvB hexamers are loaded.

The final stage in homologous recombination is the resolution of the Holliday junction into its two homologous dsDNAs. This process is carried out by **RuvC,** a 173-residue homodimeric exonuclease whose active sites are located ~30 Å apart on the same face of the protein. This suggests that RuvC sits down on the open face of the RuvAB–Holliday junction complex, that facing the viewer in Fig. 30-76, to cleave oppositely located strands at the Holliday junction. The resulting single-strand nicks in the now resolved dsDNAs are sealed by DNA ligase.

This so-called RuvABC resolvosome provides a satisfying mechanism for branch migration and Holliday junction resolution. However, there is a fly in this particular ointment. The X-ray structure of an *M. leprae* RuvA–Holliday junction complex crystallized under conditions different from that in Fig. 30-74, determined by Laurence Pearl, resembles the complex in Fig. 30-74*b* but with a second RuvA tetramer in face-to-face contact with the first. Hence the Holliday junction is contained in two intersecting tunnels running through the resulting RuvA octamer. Are both RuvA–Holliday junction structures biologically relevant, or is one an artifact of crystallization? Pearl argues that the extensive complementary contacts between the two RuvA tetramers, which are strongly conserved, are unlikely to be artifactual and that a single RuvA tetramer is unlikely to be able to withstand the torque exerted by the two (in effect) counter-rotating RuvB hexamers. However, if the RuvA octamer is biologically relevant, one of its tetramers would at some point have to dissociate in order to allow RuvC access to the Holliday junction. But modeling studies indicate that the RuvC dimer observed in its X-ray structure cannot contact the DNA strands bound to a RuvB tetramer without a significant conformational change. Further investigations are necessary to resolve these inconsistencies.

f. Recombination Repair Reconstitutes Damaged Replication Forks

Transformation, transduction, and conjugation are such rare events that the vast majority of bacterial cells never participate in these processes. Similarly, the only place in the metazoan life cycle at which gene shuffling through homologous recombination occurs is in meiosis (Section 1-4A). Why then do nearly all cells have elaborate systems for mediating homologous recombination? This is because damaged replication forks occur at a frequency of at least once per bacterial cell generation and perhaps 10 times per eukaryotic cell cycle. The DNA lesions that damage the replication forks can be circumvented via homologous recombination in a process named **recombination repair** [translesion synthesis, which is highly mutagenic, is a process of last resort (Section 30-5D)]. Indeed, the rates of synthesis of RuvA and RuvB are greatly enhanced by the SOS response. Thus, as Michael Cox pointed out, *the primary function of homologous recombination is to repair damaged replication forks.* In what follows, we describe recombination repair as it occurs in *E. coli.*

Recombination repair is called into play when a replication fork encounters an unrepaired single-strand lesion (Fig. 30-77):

1. DNA replication is arrested at the lesion but continues on the opposing undamaged strand for some distance before the replisome fully collapses (Section 30-5D).

2. The replication fork regresses to form a type of Holliday junction dubbed a "chicken foot." This process may occur spontaneously as driven by the positive supercoiling that has built up ahead of the replication fork, it may be mediated by RecA, or it may be promoted by **RecG,** an ATP-driven helicase that catalyzes branch migration at DNA junctions with three or four branches.

3. The single-strand gap at the collapsed replication fork, now an overhang, is filled in by Pol I.

4. Reverse branch migration mediated by RuvAB or RecG yields a reconstituted replication fork, which supports replication restart (see below).

Note that this process does not actually repair the single-strand lesion that has caused the problem but instead reconstructs the replication fork in a way that permits the previously discussed DNA repair systems (Section 30-5) to eventually eliminate the lesion.

A second situation that requires recombination repair is the encounter of a replication fork with an unrepaired single-strand nick (Fig. 30-78):

1. When a single-strand nick is encountered, the replication fork collapses.

2. The repair process begins via the RecBCD plus RecA-mediated invasion of the newly synthesized and undamaged 3′-ending strand into the homologous dsDNA starting at its broken end.

3. Branch migration, as mediated by RuvAB, then yields a Holliday junction, which exchanges the replication fork's 3′-ending strands.

4. RuvC then resolves the Holliday junction yielding a reconstituted replication fork ready for replication restart.

Thus, the 5′-ending strand of the nick has, in effect, become the 5′ end of an Okazaki fragment.

The final step in the recombination repair process is the restart of DNA replication. This process is, of necessity,

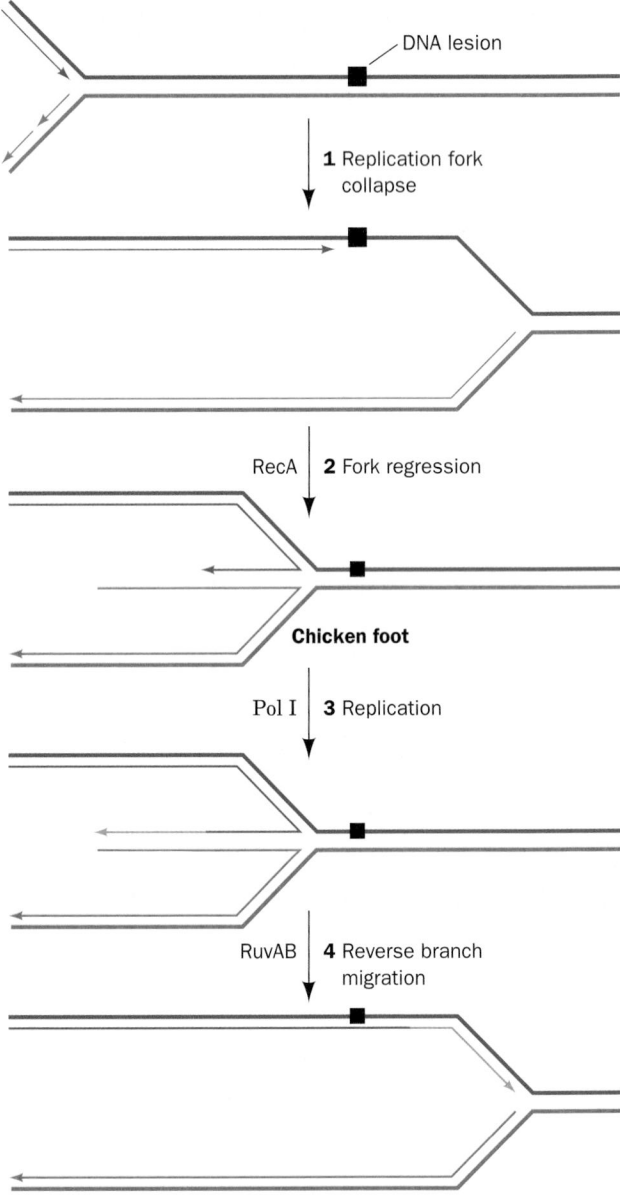

FIGURE 30-77   The recombination repair of a replication fork that has encountered a single-strand lesion. Thick lines indicate parental DNA, thin lines indicate newly synthesized DNA, the cyan lines indicate DNA that was synthesized by Pol I, and the arrows point in the 5′ → 3′ direction. [After Cox, M.M., *Annu. Rev. Genet.* **35,** 53 (2000).]

distinct from the replication initiation that occurs at *oriC* (Section 30-3C). **Origin-independent replication restart** is mediated by the same seven-protein primosome that initiates the minus strand replication of bacteriophage φX174 (Table 30-4), which has therefore been named the **restart primosome.**

g. Recombination Repair Reconstitutes Double-Strand Breaks

We have seen that double-strand breaks (DSBs) in DNA can be rejoined, often mutagenically, by nonhomol-

ogous end-joining (NHEJ; Section 30-5E). DSBs may also be nonmutagenically repaired through a recombination repair process known as **homologous end-joining,** which occurs via two Holliday junctions (Fig. 30-79):

1. The DSB's double-stranded ends are resected to produce single-stranded ends. One of the 3′-ending strands invades the corresponding sequence of a homologous chromosome to form a Holliday junction, a process that, in eukaryotes, is mediated by the RecA homolog RAD51. The other 3′-ending strand pairs with the displaced strand segment on the homologous chromosome to form a second Holliday junction.

2. DNA synthesis and ligation fills in the gaps and seals the joints.

3. Both Holliday junctions are resolved to yield two intact double strands.

Thus, the sequences that may have been expunged in the formation of the DSB are copied from the homologous chromosome. Of course, a limitation of homologous end-joining, particularly in haploid cells, is that a homologous chromosomal segment may not be available.

The importance of recombination repair in humans is demonstrated by the observation that defects in the pro-

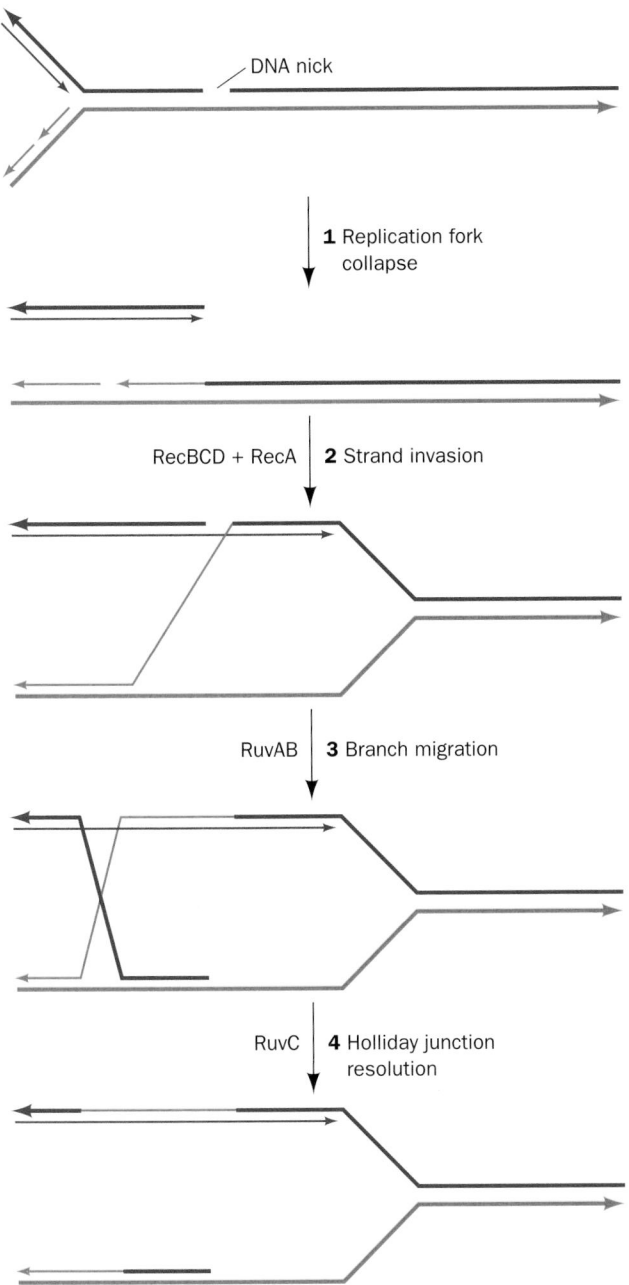

FIGURE 30-78   The recombination repair of a replication fork that has encountered a single-strand nick. Thick lines indicate parental DNA, thin lines indicate newly synthesized DNA, and the arrows point in the 5′ → 3′ direction. [After Cox, M.M., *Annu. Rev. Genet.* **35,** 53 (2000).]

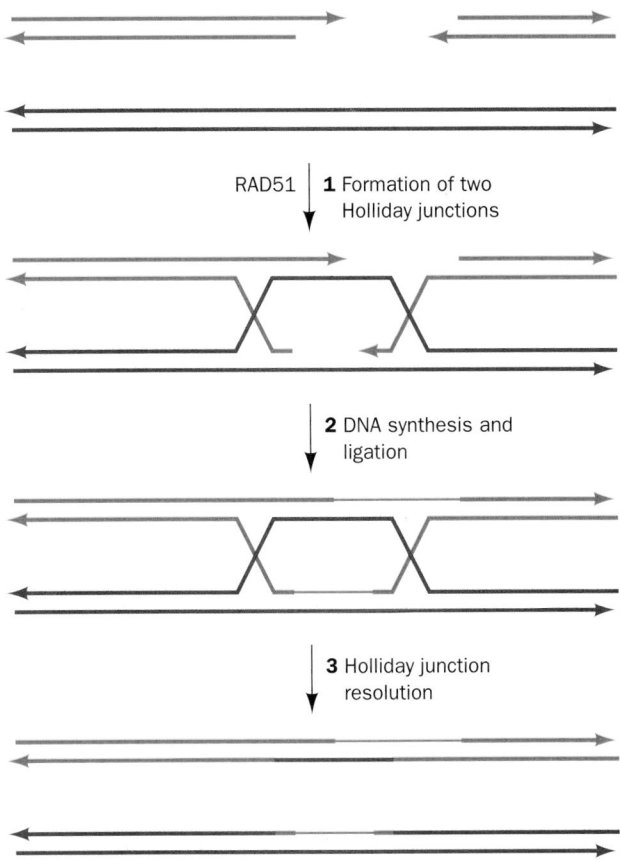

FIGURE 30-79   The repair of a double-strand break in DNA by homologous end-joining. Thick lines indicate parental DNA, thin lines indicate newly synthesized DNA, and the arrows point in the 5′ → 3′ direction. [After Haber, J.E., *Trends Genet.* **16,** 259 (2000).]

teins **BRCA1** (1863 residues) and **BRCA2** (3418 residues), both of which interact with RAD51, are associated with a greatly increased incidence of breast, ovarian, prostate, and pancreatic cancers. Indeed, individuals with mutant *BRCA1* or *BRCA2* genes have up to an 80% lifetime risk of developing cancer.

B. *Transposition and Site-Specific Recombination*

In the early 1950s, on the basis of genetic analysis, Barbara McClintock reported that the variegated pigmentation pattern of maize (Indian corn) kernels results from the action of genetic elements that can move about the maize genome. This proposal was resoundingly ignored because it was contrary to the then held genetic orthodoxy that chromosomes consist of genes linked in fixed order. Another 20 years were to pass before evidence of mobile genetic elements was found in another organism, *E. coli*.

*It is now known that **transposable elements** or **transposons** are common in both prokaryotes and eukaryotes, where they influence the variation of phenotypic expression over the short term and evolutionary development over the long term. Each transposon codes for the enzymes that specifically insert it into the recipient DNA.* This process has been described as **illegitimate recombination** because it requires no homology between donor and recipient DNAs. Since the insertion site is chosen largely at random, transposition is a potentially dangerous process; the insertion of a transposon into an essential gene will kill a cell together with its resident transposons. Hence transposition is tightly regulated; it occurs at a rate of only 10^{-5} to 10^{-7} events per element per generation. The conditions that trigger transposition are, for the most part, unknown.

a. Prokaryotic Transposons

Prokaryotic transposons with three levels of complexity have been characterized:

1. The simplest transposons, and the first to be characterized, are named **insertion sequences** or **IS elements.** They are designated by "IS" followed by an identifying number. IS elements are normal constituents of bacterial chromosomes and plasmids. For example, a common *E. coli* strain has eight copies of **IS1** and five copies of **IS2.** IS elements generally consist of <2000 bp. These comprise a so-called **transposase** gene, and in some cases a regulatory

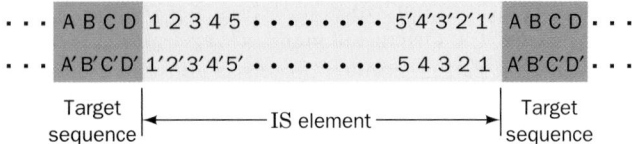

FIGURE 30-80 Structure of IS elements. These and other transposons have inverted terminal repeats (*numerals*) and are flanked by direct repeats of host DNA target sequences (*letters*).

gene, flanked by short inverted (having opposite orientation) terminal repeats (Fig. 30-80 and Table 30-6). The inverted repeats are essential for transposition; their genetic alteration invariably prevents this process. An inserted IS element is flanked by a directly (having the same orientation) repeated segment of host DNA (Fig. 30-80). This suggests that an IS element is inserted in the host DNA at a staggered cut that is later filled in (Fig. 30-81). The length of this target sequence (most commonly 5 to 9 bp), but not its sequence, is characteristic of the IS element.

2. *More complex transposons carry genes not involved in the transposition process, for example, antibiotic resistance genes.* Such transposons are designated "Tn" followed by an identifying number. For example, **Tn3** (Fig. 30-82) consists of 4957 bp and has inverted terminal repeats of 38 bp each. The central region of Tn3 codes for three proteins: (1) a 1015-residue transposase named **TnpA;** (2) a 185-residue protein known as **TnpR,** which mediates the **site-specific recombination** reaction necessary to complete the transposition process (see below) and also functions as a repressor for the expression of both *tnpA* and *tnpR;* and (3) a **β-lactamase** that inactivates ampicillin (Section 11-3B). The site-specific recombination occurs in an AT-rich region known as the **internal resolution site** that is located between *tnpA* and *tnpR.*

3. The so-called **composite transposons** (Fig. 30-83) consist of a gene-containing central region flanked by two identical or nearly identical IS-like modules that have either the same or an inverted relative orientation. It therefore seems that composite transposons arose by the association of two originally independent IS elements. Since the IS-like modules are themselves flanked by inverted repeats, the ends of either type of composite transposon must also be inverted repeats. Experiments demonstrate that composite transposons can transpose any sequence of DNA in their central region.

TABLE 30-6 Properties of Some Insertion Elements

Insertion Element	Length (bp)	Inverted Terminal Repeat (bp)	Direct Repeat at Target (bp)	Number of Copies in *E. coli* Chromosome
IS1	768	23	9	5–8
IS2	1327	41	5	5
IS4	1428	18	11–13	5
IS5	1195	16	4	1–2

Source: Mainly Lewin, B., *Genes VII,* p. 459, Oxford University Press (2000).

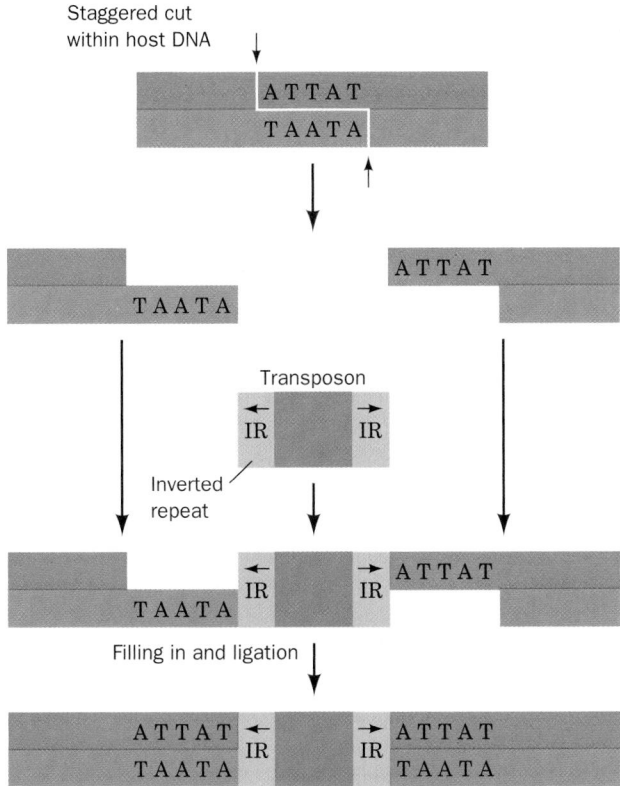

FIGURE 30-81 A model for the generation of direct repeats of the target sequence by transposon insertion.

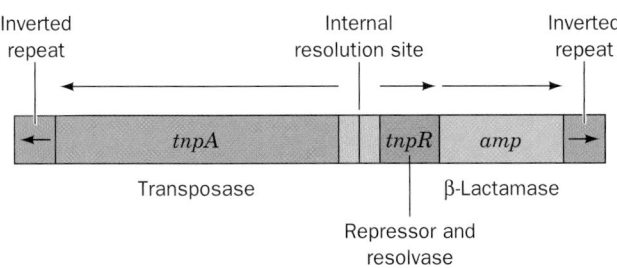

FIGURE 30-82 A map of transposon Tn3.

(a)

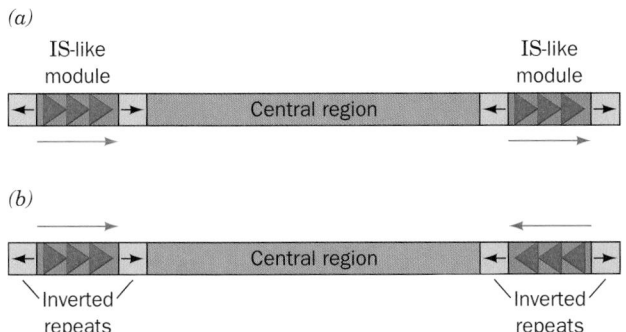

(b)

FIGURE 30-83 A composite transposon. This element consists of two identical or nearly identical IS-like modules (*green*) flanking a central region carrying various genes. The IS-like modules may have either (*a*) direct or (*b*) inverted relative orientations.

There are two modes of transposition: (1) **direct** or **simple transposition,** in which the transposon, as the name implies, physically moves from one DNA site to another; and (2) **replicative transposition,** in which the transposon remains at its original site and a copy of it is inserted at a target site. The two modes, as we shall see, have similar mechanistic features and, indeed, some transposons can move by either mode.

b. Direct Transposition of Tn5 Occurs by a Cut-and-Paste Mechanism

Tn5 is a 5.8-kb composite transposon that contains the gene encoding the 476-residue **Tn5 transposase** together with three antibiotic resistance genes. It is flanked by inverted IS-like modules ending in 19-bp sequences called outside end (OE) sequences. Tn5 undergoes direct transposition via a cut-and-paste mechanism that was elucidated in large part by William Reznikoff (Fig. 30-84):

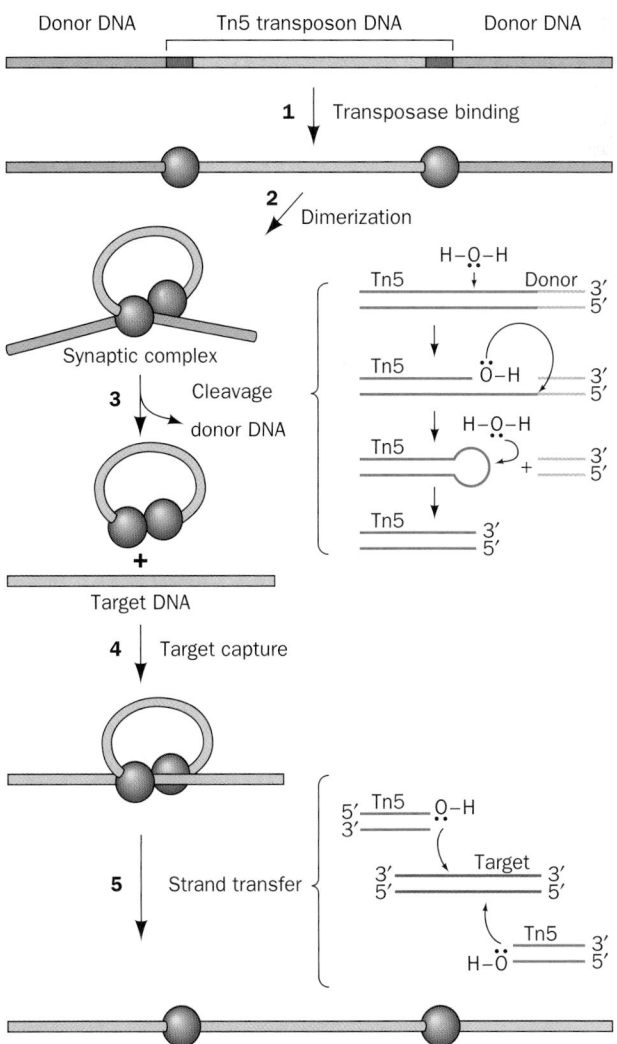

FIGURE 30-84 The cut-and-paste transposition mechanism catalyzed by Tn5 transposase. The reactions comprising Steps 3 and 5 are indicated beside the braces to the right of these steps. [After Davies, D.R., Goryshin, I.Y., Reznikoff, W.S., and Rayment, I., *Science* **289,** 77 (2000).]

1. Each of Tn5's two OE sequences on the donor DNA is bound by a monomer of Tn5 transposase.

2. The transposase dimerizes to form a catalytically active **synaptic complex** in which the transposon is held between the two transposase subunits.

3. Each transposase subunit activates a water molecule to nucleophilically attack the outermost nucleotide of its bound OE sequence, yielding a free 3'-OH group. This 3'-OH group is then activated to attack the opposite strand on the DNA to form a hairpin structure, thereby excising the transposon from the DNA. The hairpin is then hydrolyzed to yield a blunt-ended dsDNA at each end of the transposon, thus completing the "cut" portion of the transposition mechanism.

4. The synaptic complex binds to the target DNA.

5. The transposon's 3'-OH groups nucleophilically attack the target DNA on opposite strands spaced 9 bp apart, thereby installing the transposon at the target site. Remarkably, this reaction and the three preceding lytic reactions are all mediated by the same catalytic site. The repair of the oppositely located single-strand gaps (Fig. 30-81) completes the "paste" portion of the mechanism.

Although, strictly speaking, not part of the transposition process, the double-strand break in the donor DNA left by the excision of the transposon must be repaired if the donor DNA is to be propagated (in bacteria, the donor DNA is often a plasmid so that its loss has little effect on the cell since plasmids are generally present in multiple copies).

The X-ray structure of a Tn5 synaptic complex (Fig. 30-85), determined by Reznikoff and Ivan Rayment, provides a model of the synaptic complex at the stage following its cleavage from the donor DNA (the product of Step 3 in Fig. 30-84). This 2-fold symmetric complex consists of a dimer of Tn5 transposase subunits binding two 20-bp DNA segments containing the Tn5 transposon's 19-bp OE sequence with the outer end of each OE sequence bound to the protein (and whose opposite ends would, *in vivo*, be connected by the looped around transposon; Fig. 30-84). Both transposase subunits extensively participate in binding each DNA segment, thereby explaining why the individual subunits cannot cleave their bound DNA segments before forming the synaptic complex. The protein holds the DNA in a distorted B-DNA conformation with its two end pairs of nucleotides no longer base paired. Indeed, the penultimate base on the nontransferred strand is flipped out of the double helix and binds in a hydrophobic pocket. The transferred strand's free 3'-OH group, which occupies the active site, is bound in the vicinity of a cluster of three catalytically essential acidic residues, the so-called **DDE motif,** which is shared with other transposases. In the X-ray structure the DDE motif binds one Mn^{2+} ion, but physiologically it probably binds two Mg^{2+} ions. This suggests that transposases employ a metal-activated catalytic mechanism similar to that of the DNA polymerases (Section 30-2A). The facing surface of the protein in Fig. 30-85 is positively charged with a prominent groove running from

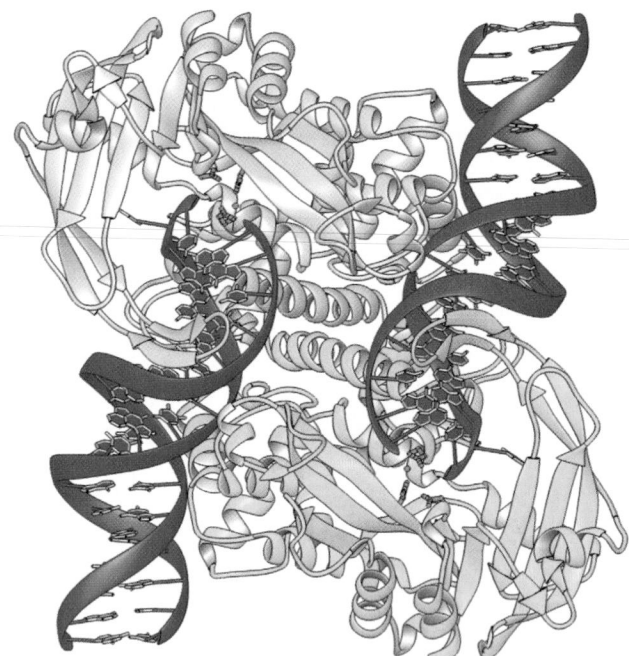

FIGURE 30-85   X-Ray structure of Tn5 transposase in complex with a 20-bp DNA containing the OE sequence. The complex, which represents the product of Step 3 in Fig. 30-84, is viewed along its 2-fold axis with its two identical subunits cyan and yellow. The three acidic residues of each DDE motif are drawn as green ball-and-stick structures with their bound Mn^{2+} ions represented by green balls. The DNAs' sugar–phosphate backbones are represented by purple ribbons and their bases are drawn as purple-filled gray stick figures. The DNAs' reactive 3'-OH groups are located at the ends of the inner strands where they contact the DDE motifs. [Courtesy of Ivan Rayment, University of Wisconsin. PDBid 1F3I.]

upper left to lower right that forms the apparent binding site for the target DNA.

Wild-type Tn5 transposase has such low catalytic activity that it is undetectable *in vitro*. However, that in the X-ray structure is a hyperactive mutant form that contains the mutations E54K and L372P (an unusual circumstance in that it is far more common to mutationally inhibit an enzyme under crystallographic study so as to trap it at some specific stage along its reaction pathway). Lys 54 is hydrogen bonded to O4 of a thymine base on the transferred strand. In the wild-type transposase, Glu 54 would probably have an unfavorable charge–charge repulsion with a nearby phosphate group, thus providing a structural basis for the increased activity of the E54K mutant. The L372P mutation disorders the peptide segment between residues 373 and 391 (it is ordered in the X-ray structure of wild-type Tn5 transposase lacking its N-terminal 55 residues), thereby suggesting that this mutation facilitates a conformation change required for substrate binding.

c. Replicative Transposition

If a plasmid carrying a transposon resembling Tn3 is introduced into a bacterial cell carrying a plasmid that lacks

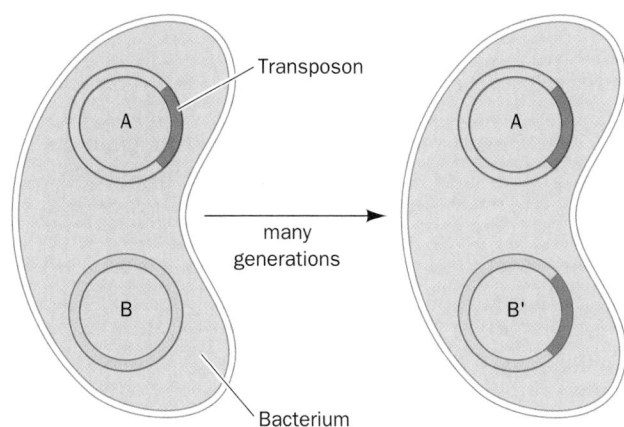

FIGURE 30-86   Replicative transposition. This type of transposition inserts a copy of the transposon at the target site while another copy remains at the donor site.

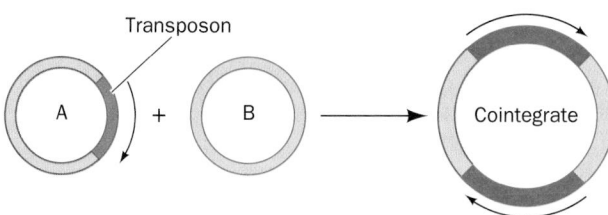

FIGURE 30-87   A cointegrate. This structure forms by the fusion of two plasmids, one carrying a transposon, such that both junctions of the original plasmid are spanned by transposons with the same orientation (*arrows*).

the transposon, in some of the progeny cells both types of plasmid will contain the transposon (Fig. 30-86). Evidently, *such transposition involves the replication of the transposon into the recipient plasmid rather than its transfer from donor to recipient.*

Two plasmids, one containing a replicative transposon, will occasionally fuse to form a so-called **cointegrate** containing like-oriented copies of the transposon at both junctions of the original plasmids (Fig. 30-87). Yet, some of the progeny of a cointegrate-containing cell lack the cointegrate and instead contain both original plasmids, each with one copy of the transposon (Fig. 30-86). The cointegrate must therefore be an intermediate in the transposition process.

Although the mechanism of replicative transposition has not been fully elucidated, a plausible model for this process (and there are several) that accounts for the foregoing observations consists of the following steps (Fig. 30-88):

1. A pair of staggered single-strand cuts, such as is diagrammed in Fig. 30-81, is made by the transposon-encoded transposase at the target sequence of the recipient plasmid so as to liberate 3′-OH ends. Similarly, single-strand cuts are made on opposite strands to either side of the transposon. Note that these reactions resemble those catalyzed by Tn5 transposase (Fig. 30-84).

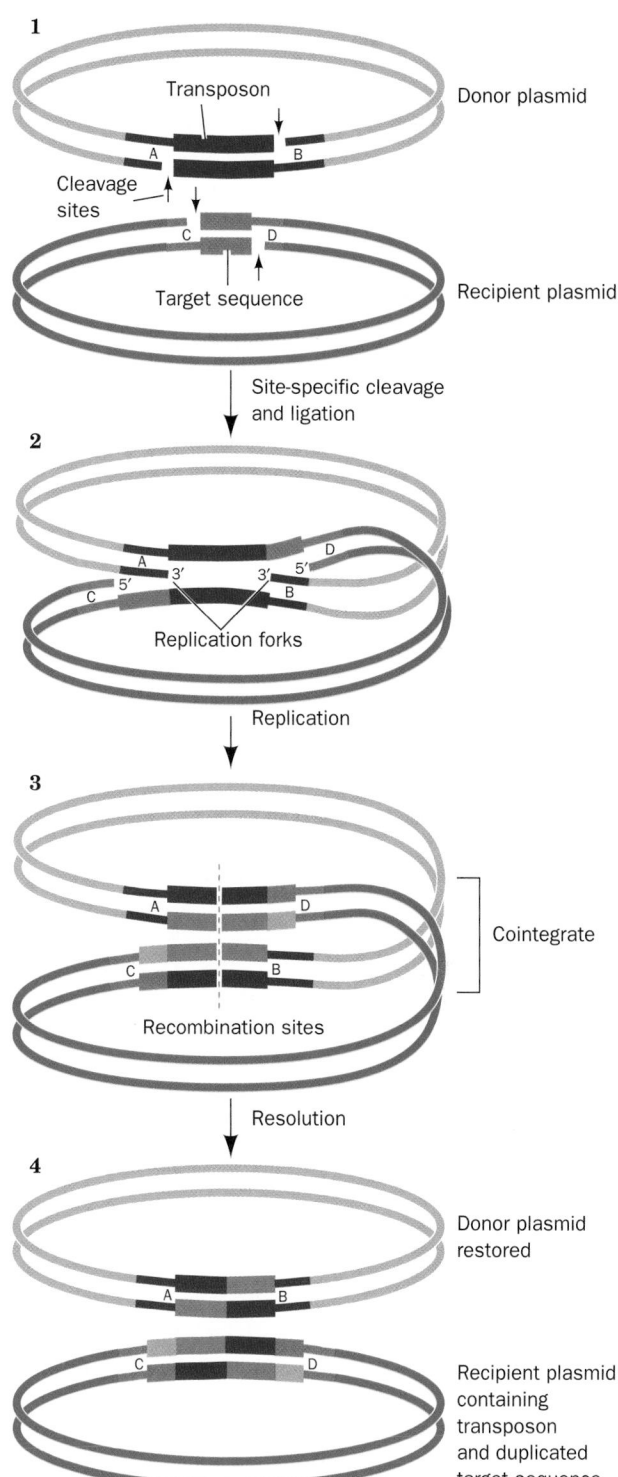

FIGURE 30-88   A model for transposition involving the intermediacy of a cointegrate. Here more lightly shaded bars represent newly synthesized DNA. [After Shapiro, J.A., *Proc. Natl. Acad. Sci.* **76,** 1934 (1979).]

2. Each of the transposon's free ends is ligated to a protruding single strand at the insertion site. This forms a replication fork at each end of the transposon.

3. The transposon is replicated, thereby yielding a cointegrate.

4. Through a site-specific recombination between the internal resolution sites of the two transposons, the cointegrate is resolved into the two original plasmids, each of

which contains a transposon. This crossover process is catalyzed by a transposon-encoded **resolvase** (TnpR in Tn3) rather than RecA; transposition proceeds normally in *recA⁻* cells (although RecA will resolve a cointegrate containing a transposon with a mutant resolvase and/or an altered internal resolution site, albeit at a much reduced rate).

d. γδ Resolvase Catalyzes Site-Specific Recombination

The **γδ resolvase** is a TnpR homolog that is encoded by the **γδ transposon** (a member of the Tn3 family of replicative transposons; Fig. 30-82). It catalyzes a site-specific recombination event in which a cointegrate containing two copies of the γδ transposon is resolved, via double-strand DNA cleavage, strand exchange, and religation (the last step in Fig. 30-88), into two catenated (linked) dsDNA circles that each contain one copy of the γδ transposon (it also serves as its own transcriptional repressor as does TnpR). The γδ transposon contains a 114-bp *res* site that includes three binding sites for γδ resolvase dimers, each of which contains an inverted repeat of the γδ resolvase's 12-bp recognition sequence. The resolution of the cointegrate involves the binding of a γδ resolvase homodimer to all six of these binding sites in the cointegrate (three from each of its two transposons) as is diagrammed in Fig. 30-89. The reaction proceeds via the formation of a transient phosphoSer bond between Ser 10 and the 5'-phosphate at each cleavage site.

The X-ray structure of the γδ resolvase homodimer in complex with a 34-bp palindromic DNA segment containing an inverted repeat of the 12-bp recognition sequence separated by an 8-bp spacer (Fig. 30-90) was determined by Steitz. Each 183-residue resolvase monomer consists of a catalytic domain (residues 1–120) whose structure closely resembles that observed in the absence of DNA, a C-terminal DNA-binding domain (residues 148–183), and an extended arm (residues 121–147) that connects the N- and C-terminal domains.

The centrally located N-terminal domain dimer approaches the DNA from its minor groove side along the

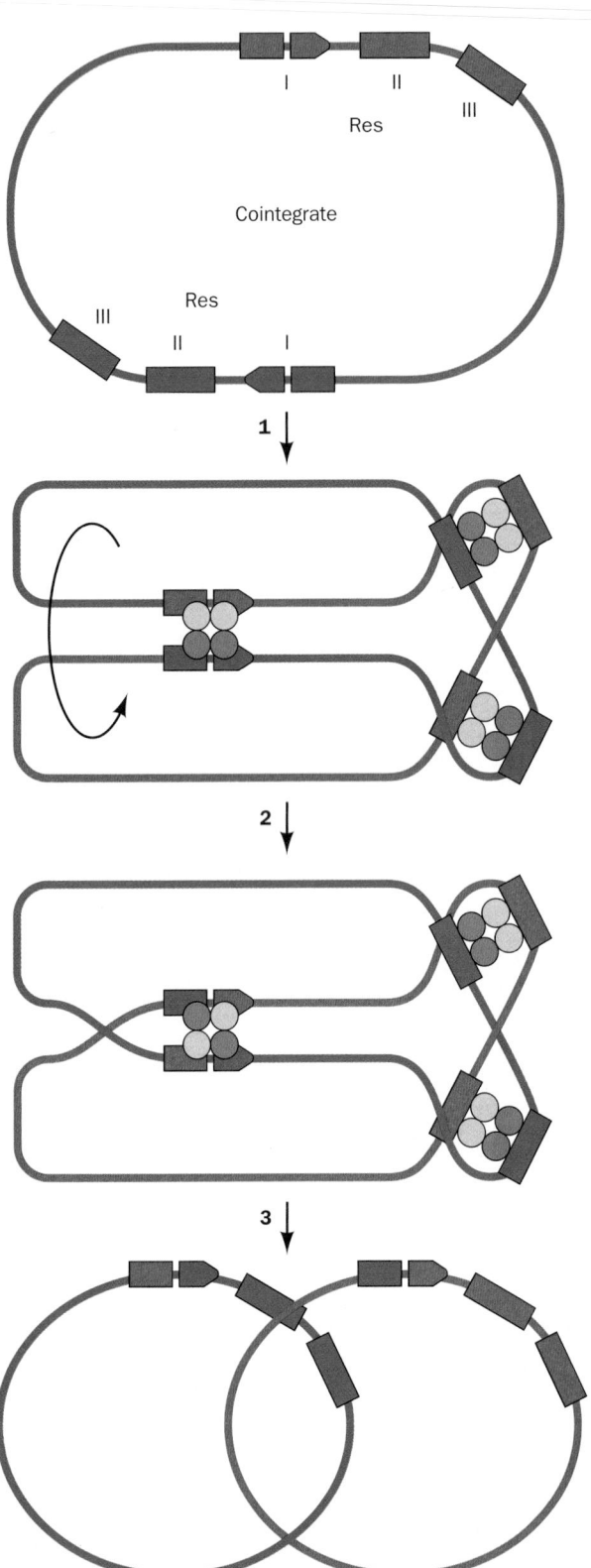

FIGURE 30-89 A model for the resolution of a cointegrate containing two γδ transposons to form two catenated dsDNA circles. (1) The γδ resolvase binds as six homodimers to its binding sites, I, II, and III, in each of the cointegrate's two *res* sites (yellow and green circles represent the γδ resolvase monomers initially bound to the red and blue *res* sites, respectively). Although not shown as such, the pair of dimers bound to sites I associate with the pairs of dimers bound at sites II and III to form, as seen in the electron microscope, a compact globule of unknown structure known as a **synaptosome. (2)** The dsDNA at sites I both undergo staggered (by 2 bp) double strand scissions via the transient formation of phosphoSer bonds between Ser 10 and the 5' phosphates at the cleavage sites. The cleaved strands then exchange places (cross over) in a process that apparently requires the rotation of one of the pairs of resolvase monomers with respect to the other and are then ligated. **(3)** The dissociation of the synaptosome yields the catenated dsDNA circles. [Courtesy of Gregory Mullen, University of Connecticut Health Center.]

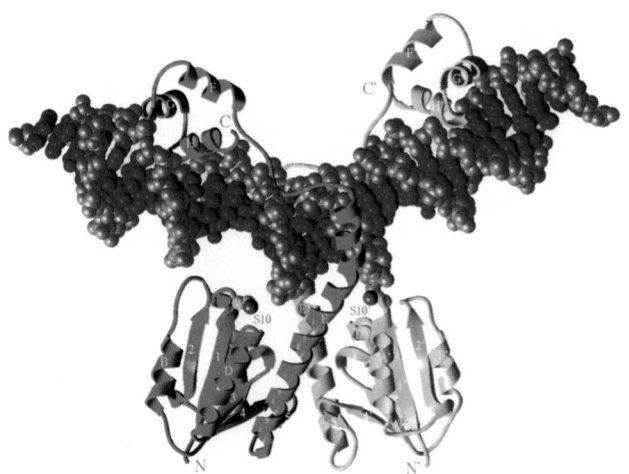

FIGURE 30-90 X-Ray structure of a γδ resolvase homodimer in complex with a 34-bp palindromic DNA containing its binding site. The DNA is drawn in space-filling form with its backbone purple, its bases blue, and its scissile phosphate groups highlighted in magenta. The protein subunits are green and gold and its Ser 10 residues are shown in space-filling form with C yellow and O red. The complex is viewed with its approximate 2-fold axis vertical. [Courtesy of Thomas Steitz, Yale University. PDBid 1GDT.]

structure's local 2-fold axis; the C-terminal domains each bind in the major groove of their target sequence on the opposite side of the DNA from the N-terminal domain dimer such that the two C-terminal domains are separated by two helical turns; and the extended arms which connect the two domains more or less run along the DNA's minor groove. The DNA, which otherwise closely assumes the

B-DNA conformation, is centrally kinked by ~60° such that it bends toward its major groove, away from the N-terminal domain dimer. The C-terminal helix of the N-terminal domain (helix E) binds over the DNA's minor groove such that the dimer's two E helices grip the DNA like a pair of chopsticks (the segment of the E helix that contacts the DNA is disordered in the absence of the DNA). The C-terminal helix (helix H) binds in the major groove and, together with its preceding helix (helix G), forms a **helix–turn–helix (HTH) motif,** a common sequence-specific DNA-binding motif that occurs mainly in prokaryotic transcriptional repressors and activators (Section 31-3D). The structure is asymmetric with the active site Ser 10 in Fig. 30-90's yellow monomer much closer to the DNA than that in the green monomer, although both are quite distant from the scissile bonds in the DNA. This suggests that the two single-strand cleavage reactions catalyzed by the dimer may occur sequentially and, in any case, require significant conformational changes. Of course, a detailed understanding of the mechanism of the γδ resolvase reaction will require the knowledge of how all six γδ resolvase dimers that form the synaptosome participate in the reaction (Fig. 30-89).

e. Replicative Transposons Are Responsible for Much Genetic Remodeling in Prokaryotes

In addition to mediating their own insertion into DNA, *replicative transposons promote inversions, deletions, and rearrangements of the host DNA.* Inversions can occur when the host DNA contains two copies of a transposon in inverted orientation. The recombination of these transposons inverts the region between them (Fig. 30-91a). If, instead, the two transposons have the same orientation, the

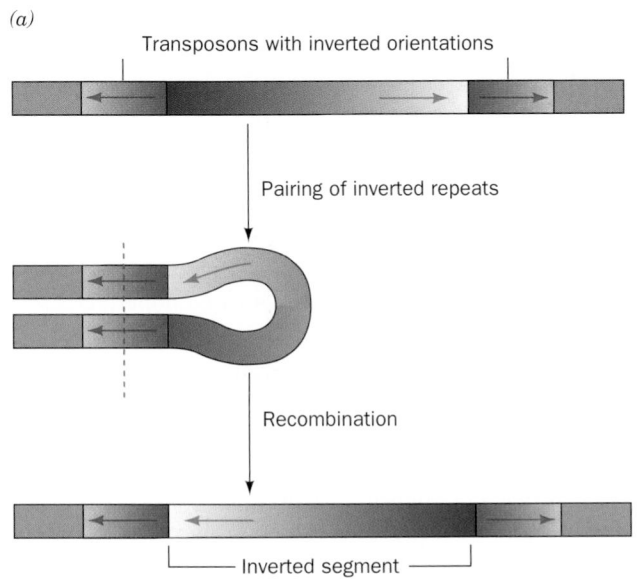
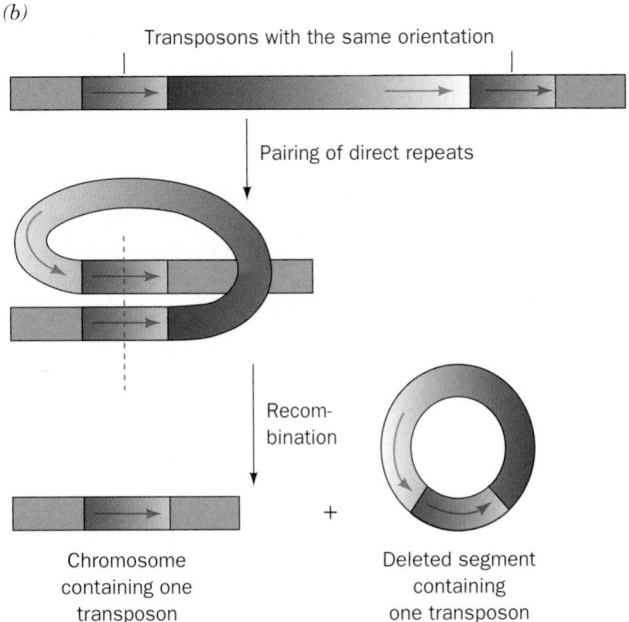

FIGURE 30-91 Chromosomal rearrangement via recombination. (*a*) The inversion of a DNA segment between two identical transposons with inverted orientations. (*b*) The

deletion of a DNA segment between two identical transposons with the same orientation. This process parcels one transposon each to the resulting two DNA segments.

resolution of this cointegrate-like structure deletes the segment between the two transposons (Fig. 30-91*b*; if the deleted segment lacks a replication origin, it will not be propagated). The deletion of a chromosomal segment in this manner, followed by its integration into the chromosome at a different site by a separate recombinational event, results in chromosomal rearrangement.

Transposition appears to be important in chromosomal and plasmid evolution. Indeed, it has been suggested that transposons are nature's genetic engineering "tools." For example, the rapid evolution, since antibiotics came into common use, of plasmids that confer resistance to several antibiotics (Section 5-5B) has resulted from the accumulation of the corresponding antibiotic-resistance transposons in these plasmids. Transposon-mediated rearrangements may well have been responsible for organizing originally distant genes into coordinately regulated operons (Section 5-4A) as well as for forming new proteins by linking two formerly independent gene segments. Moreover, *the occurrence of identical transposons in unrelated bacteria indicates that the transposon-mediated transfer of genetic information between organisms is not limited to related species, in contrast to genetic transfers mediated by homologous recombination.*

f. Phase Variation Is Mediated by Site-Specific Recombination

Phenotypic expression in bacteria can be regulated by site-specific recombination. For example, certain strains of *Salmonella typhimurium* make two antigenically distinct versions of the protein **flagellin** (the major component of the whiplike flagella with which bacteria propel themselves; Section 35-3G) that are designated **H1** and **H2**. Only one of these proteins is expressed by any given cell but about once every 1000 cell divisions, in a process known as **phase variation,** a cell switches the type of flagellin it synthesizes. It is thought that phase variation helps *Salmonella* evade its host's immunological defenses.

What is the mechanism of phase variation? The two flagellin genes reside on different parts of the bacterial chromosome. *H2* is linked to the *rh1* gene that encodes a repressor of H1 expression (Fig. 30-92; *rh1*, *H2*, and *H1* are also known as *fljA*, *fljB*, and *fljC*, respectively). Hence, when the *H2–rh1* transcription unit is expressed, H1 synthesis is repressed; otherwise H1 is synthesized. Melvin Simon has shown that the expression of the *H2–rh1* unit is controlled by the orientation of a 995-bp segment that lies upstream of *H2* (Fig. 30-92) and that contains the following elements:

1. A promoter for *H2–rh1* expression.

2. The *hin* gene, which encodes the 190-residue **Hin DNA invertase.** Hin mediates the inversion of the DNA segment in a manner similar to that diagrammed in Fig. 30-91*a*. In fact, Hin is ~40% identical in sequence with the γδ resolvase, which strongly suggests that these proteins have similar structures.

3. Two closely related 26-bp sites, *hixL* and *hixR*, that form the boundaries of the segment and hence contain its

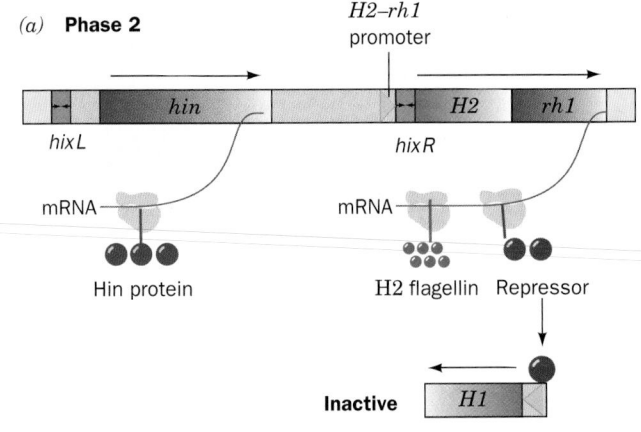

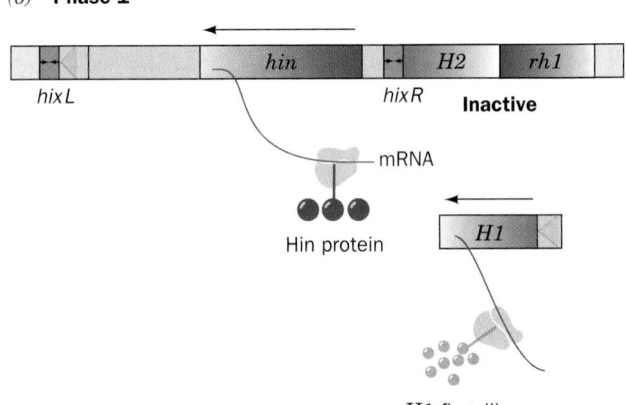

FIGURE 30-92   The mechanism of phase variation in *Salmonella*. (*a*) In Phase 2 bacteria, the *H2–rh1* promoter is oriented so that H2 flagellin and repressor are synthesized. Repressor binds to the *H1* gene, thereby preventing its expression. (*b*) In Phase 1 bacteria, the segment preceding the *H2–rh1* transcription unit has been inverted relative to its orientation in Phase 2 bacteria. Hence this transcription unit cannot be expressed because it lacks a promoter. This releases *H1* from repression and results in the synthesis of H1 flagellin. The inversion of the segment preceding the *H2–rh1* transcription unit is mediated by the Hin protein, which is expressed in either orientation by the *hin* gene.

cleavage sites. They each consist of two imperfect 12-bp inverted repeats separated by 2 nt.

In the Phase 2 orientation (Fig. 30-92*a*), the properly oriented promoter is just upstream of *H2*, so this gene and *rh1* are coordinately expressed, thereby repressing H1 synthesis. In Phase 1 bacteria (Fig. 30-92*b*), however, this segment has the opposite orientation. Consequently, neither *H2* nor *rh1*, which then lacks a promoter, is expressed so that H1 is synthesized.

g. Cre-Mediated Site-Specific Recombination Occurs via 3′-PhosphoTyr Intermediates

Bacteriophages, as we have seen (Fig. 1-31), replicate themselves within their host bacterial cells which, in most cases, they then lyse to release the progeny phage, a lifestyle that is therefore known as the **lytic** mode. However, certain bacteriophages can assume an alternative, nondestructive lifestyle, the **lysogenic** mode, in which they install their DNA, usually in the host chromosome via site-specific recombination, so that the phage DNA is passively replicated with the host DNA. However, if the bacterial host encounters conditions in which it is unlikely to survive, the phage DNA is excised from the bacterial chromosome via a reversal of the site-specific recombination reaction and it reenters the lytic mode so as to escape the doomed host. We discuss the genetic factors that maintain the balance between the lytic and lysogenic lifestyles in **bacteriophage λ** in Section 33-3.

The enzymes that mediate the foregoing site-specific recombination reactions are members of the **λ integrase (λ Int;** alternatively, **tyrosine recombinase)** family, whose >100 known members also occur in prokaryotes and eukaryotes. These include the **XerC** and **XerD** proteins of *E. coli* which, operating in concert, function to decatenate the two linked circular dsDNA products of homologous recombination (Fig. 30-66g, *left*), as well as type 1B topoisomerases (Section 29-3C).

The structurally best characterized member of the λ integrase family is the **Cre recombinase** of *E. coli* **bacteriophage P1.** In its lysogenic state, bacteriophage P1 is a single-copy circular plasmid (rather than being inserted in the host chromosome as is bacteriophage λ), but in the phage head (the lytic mode), P1 DNA is a linear dsDNA that has a 34-bp *loxP* site at each end. The main function of Cre, which is encoded by bacteriophage P1, is to mediate the site-specific recombination between these two *loxP* sites so as to circularize the linear DNA (Fig. 30-93).

The *loxP* site is palindromic except for its central 8-bp crossover region, which confers directionality on the site. In carrying out the recombination reaction, the 343-residue Cre subunits form a homotetramer that binds two *loxP* sites in an antiparallel orientation, with each Cre subunit binding half of a *loxP* site. Then, as is diagrammed in Fig. 30-94, oppositely located Cre subunits catalyze single-strand scissions on the 5′ side of the crossover region on one strand of each of the two dsDNAs. This occurs through the nucleophilic attack of each of these active Cre subunit's conserved Tyr 324 residues on the DNA's scissile phosphoester bond to yield a 3′-phosphoTyr intermediate on one side of the cleaved bond and a free 5′-OH group on the other side (as similarly occurs in the reactions catalyzed by type IB topoisomerases; Section 29-3C). Each of the liberated 5′-OH groups then nucleophilically attacks the 3′-phosphoTyr group on the opposite duplex to form a Holliday junction, thereby releasing the Tyr residues. The Holliday junction is resolved into two recombined dsDNAs when the two Cre subunits that had not yet participated in the reaction mediate the same cleavage and strand exchange reactions on the two heretofore unreacted single strands. This latter process must be preceded by a structural rearrangement (isomerization) of the Cre tetramer that positions the catalytic Tyr residues in the latter pair of subunits to participate in the reaction while those in the

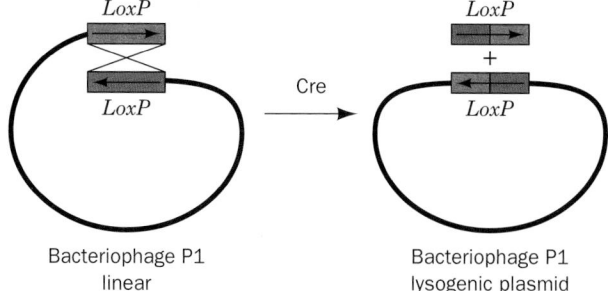

FIGURE 30-93 The circularization of linear bacteriophage P1 DNA. This occurs through the Cre-mediated site-specific recombination between its two terminally located *loxP* sites (*red and green*) to yield its lysogenic plasmid.

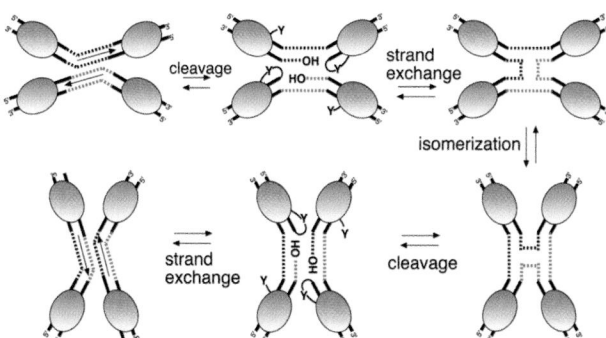

FIGURE 30-94 The mechanism of Cre–*loxP* site-specific recombination. The dashed lines represent the nonpalindromic crossover regions of the *loxP* sites. The green and magenta Cre subunits are active for cleavage in the top and bottom parts of the diagram, respectively, with their roles being switched by the isomerization step. Note that the mechanism does not require branch migration of the Holliday junction intermediate. [Courtesy of Gregory Van Duyne, University of Pennsylvania School of Medicine.]

former pair of subunits are similarly removed from the scene of the action.

The X-ray structures of Cre tetramers in their complexes with several *loxP* model DNAs, determined by Gregory Van Duyne, have helped elucidate the foregoing mechanism. When the DNA had a single-strand nick past the second nucleotide from the 5′ end of the crossover region, Cre-catalyzed strand scission yielded a free nucleotide (a CMP) that diffused away. Since this nucleotide contained the otherwise reactive 5′-OH group, the 3′-phosphoTyr intermediate was irreversibly trapped, that is, Cre could not carry out the strand exchange reaction in Fig. 30-94 (this nicked DNA is a suicide substrate for Cre; Section 28-3B). The X-ray structure of the Cre complex of this nicked DNA confirmed the presence of the 3′-phosphoTyr intermediate and indicated, through model building, that the 5′-OH group on the missing CMP residue would be well positioned to nucleophilically attack the

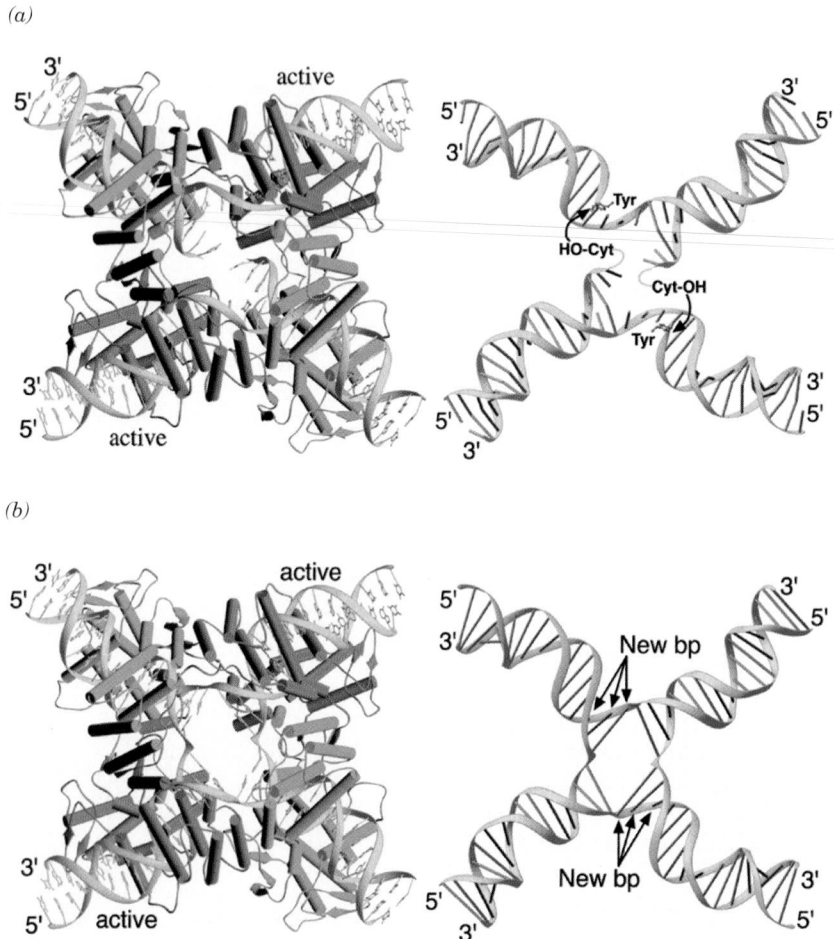

FIGURE 30-95   X-Ray structures of the Cre homotetramer in its complexes with model *loxP* DNAs. (*a*) Two identical dsDNAs that were nicked past the second nucleotide from the 5′ end of their crossover regions; and (*b*) an immobile Holliday junction. The left panels show the Cre–DNA complexes as viewed along their exact 2-fold and pseudo-4-fold axes, with the active and inactive subunits green and magenta, respectively (as in Fig. 30-94), and with the DNA gold. The right panels show only the DNAs in the X-ray structures as viewed from below the left panels. In the right panel of Part *a*, the active site Tyr that is covalently linked to the 3′-OH group of the cleaved DNA strand is shown in stick form (*red*) and the modeled-in position of the cleaved CMP's 5′-OH group is shown positioned to nucleophilically attack the 3′-phosphoTyr group on the opposite dsDNA (*curved arrows*). In the right panel of Part *b*, the three base pairs that form as a consequence of strand exchange are indicated. Note that the vertical strands in the crossovers but not the horizontal strands are distinctly kinked at their centers. [Courtesy of Gregory Van Duyne, University of Pennsylvania School of Medicine. PDBids 2CRX, 3CRX, 4CRX, and 5CRX.]

3′-phosphoTyr bond on the opposite strand (Fig. 30-95*a*). Note that this complex is only 2-fold symmetric although its four Cre subunits and much of the DNA are related by pseudo-4-fold symmetry. When the DNA was, instead, an immobile Holiday junction (Fig. 30-95*b*), the complex was also pseudo-4-fold symmetric with the single strands that had crossed over noticeably kinked at their centers. These structures revealed that the conformational changes necessary to carry out the strand exchange and isomerization reactions (Fig. 30-94) required surprisingly small movements on the part of the Cre subunits and that only the sugar–phosphate backbones of the strand-exchanged nucleotides needed to move in order to form the Holliday junction.

h. Most Transpositions in Eukaryotes Involve RNA Intermediates

Transposons similar to those in prokaryotes also occur in eukaryotes, including yeast, maize, *Drosophila*, and humans. In fact, ~3% of the human genome consists of DNA-based transposons although, in most cases, their sequences have mutated so as to render them inactive, that is, these transposons are evolutionary fossils. However, many eukaryotic transposons exhibit little similarity to those of prokaryotes. Rather, their base sequences resemble those of retroviruses (see below), which suggests that these transposons are degenerate retroviruses. The transposition of these so-called **retrotransposons** occurs via a pathway that resembles the replication of retroviral DNA

(Section 15-4C): (1) their transcription to RNA, (2) the reverse transcriptase–mediated copying of this RNA to cDNA (Section 30-4C), and (3) the largely random insertion of this DNA into the host organism's genome as mediated by enzymes known as **integrases** (which catalyze reactions similar to and structurally resemble cut-and-paste DNA transposases).

The involvement of RNA in retrotransposon-mediated transposition was ingeniously shown by Gerald Fink through his remodeling of **Ty1,** the most common transposable element in budding yeast (which has ~35 copies of this 6.3-kb element comprising ~13% of its 1700 kb genome; Ty stands for *Transposon yeast*), so that it contained a yeast intron (a sequence that is excised from an RNA transcript and hence is absent in the mature RNA; Section 5-4A) and was preceded by a galactose-sensitive yeast promoter. The transposition rate of this remodeled Ty1 element varied with the galactose concentration in the medium and the transposed elements all lacked the intron, thereby demonstrating the participation of an RNA intermediate.

A retroviral genome (Fig. 30-96*a*) is flanked by direct long terminal repeats **(LTRs)** of 250 to 600 bp and typically contains the genes encoding three polyproteins: **gag,** which is cleaved to the proteins comprising the viral core (Fig. 15-34); **pol,** which is cleaved to the above-mentioned reverse transcriptase and integrase, as well as the protease that catalyzes these cleavages; and **env,** which is cleaved to viral outer envelope proteins. Ty1 (Fig. 30-96*b*) is likewise flanked by LTRs (of 330 bp) but expresses only two polyproteins: **TYA** and **TYB,** the counterparts of gag and pol. Moreover, TYA and TYB, together with Ty1 RNA, form viruslike particles in the yeast cytoplasm. However, Ty1 lacks a counterpart of the retroviral *env* gene. Hence Ty1 is an "internal virus" that can only replicate within a genome, albeit at an extremely low rate compared to that of real retroviral infections. *Copia* (Latin for abundance), the most abundant retrotransposon in the *Drosophila* genome (which contains 20–60 copies of copia), resembles Ty1.

The LTRs in retroviruses and retrotransposons such as Ty1 and *copia* are essential elements for their transcription and hence for their transposition. Yet, vertebrate genomes also contain retrotransposons that lack LTRs and hence

cannot be transcribed analogously to retroviruses. A common family of these **nonviral retrotransposons,** the 1- to 7-kb **long interspersed nuclear elements (LINEs),** each contain two open reading frames: *ORF1,* which contains sequences similar to those in *gag;* and *ORF2,* which contains sequences similar to those encoding reverse transcriptase. A proposed mechanism for the transposition of LINEs is diagrammed in Fig. 30-97.

Different types of transposons, DNA-only, retroviral, and nonviral, predominate in different organisms. Thus

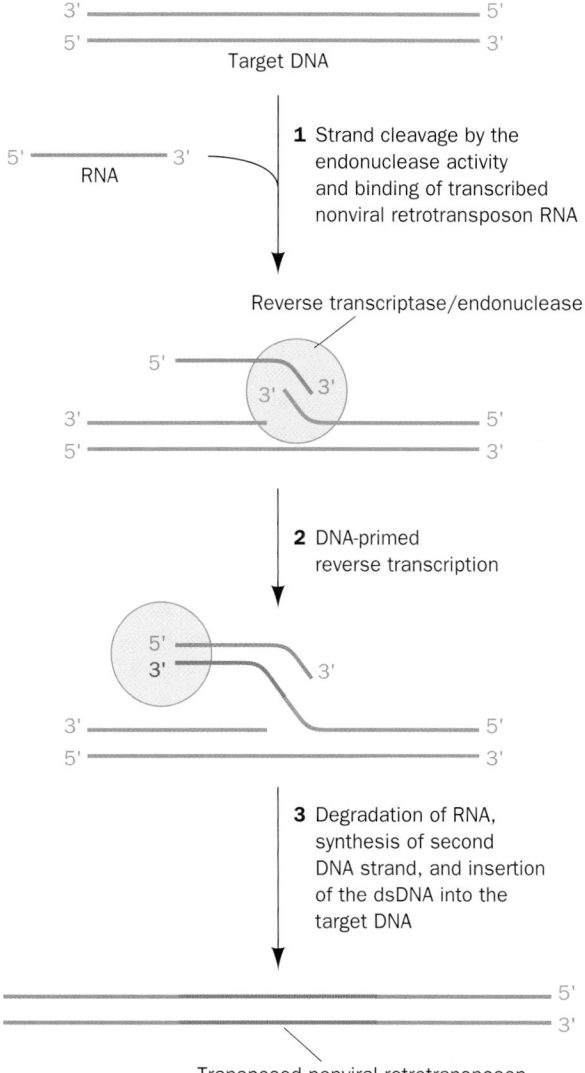

FIGURE 30-97 Proposed mechanism for the transposition of nonviral retrotransposons. (1) The retrotransposon encoded reverse transcriptase/endonuclease nicks one strand of the target DNA and then recruits the RNA transcript of the retrotransposon to this site. **(2)** The DNA-primed reverse transcription of the retrotransposon RNA. **(3)** The RNA is degraded, and the second DNA strand is synthesized using the first strand as its template (normal reverse transcriptase reactions; Section 30-4C), followed by the insertion of the resulting nonviral retrotransposon into the target DNA via a poorly understood process.

(a)

LTR	gag	pol	env	LTR

Retrovirus

(b)

LTR	TYA	TYB	LTR

Ty1

FIGURE 30-96 Gene sequences of (*a*) retroviruses and (*b*) the Ty1 retrotransposon.

bacteria, as we have seen, contain nearly exclusively DNA-only transposons, yeast have mainly retroviral retrotransposons, *Drosophila* have all three types, and in humans LINES predominate. In fact, the human genome contains an estimated 1.4 million LINEs or LINE fragments that comprise ~20% of the 3.2-billion-bp human genome (genomic organization is discussed in Section 34-2). The great majority of these molecular parasites have mutated to the point of inactivity but a few still appear capable of further transposition. Indeed, several hereditary diseases are caused by the insertion of a LINE into a gene. Several other types of retrotransposons also comprise significant fractions of the human genome as we shall see in Section 34-2.

7 ■ DNA METHYLATION AND TRINUCLEOTIDE REPEAT EXPANSIONS

The A and C residues of DNA may be methylated, in a species-specific pattern, to form N^6-methyladenine (m^{6}A), N^4-methylcytosine (m^{4}C), and 5-methylcytosine (m^{5}C) residues, respectively.

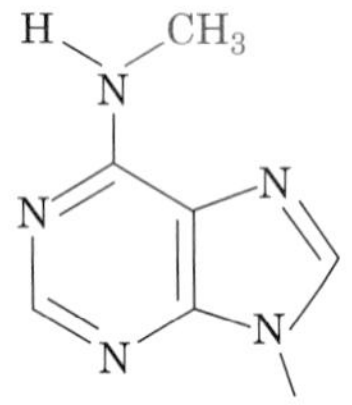

N^6-**Methyladenine (m^{6}A)**
residue

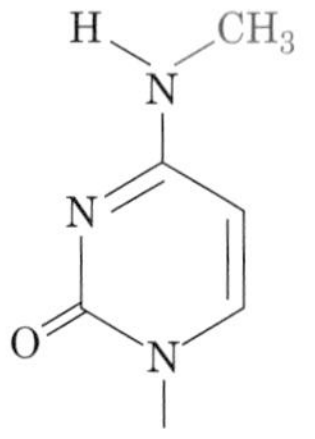

N^4-**Methylcytosine (m^{4}C)**
residue

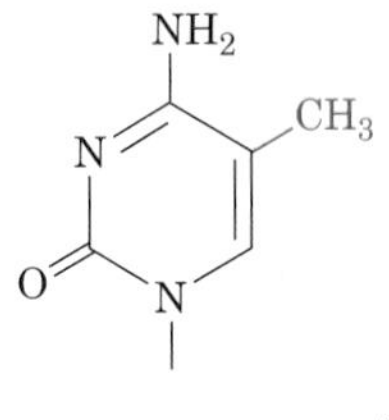

5-Methylcytosine (m^{5}C)
residue

These are the only types of modifications to which DNA is subjected in cellular organisms (although all the C residues of T-even phage DNAs are converted to **5-hydroxymethylcytosine** residues,

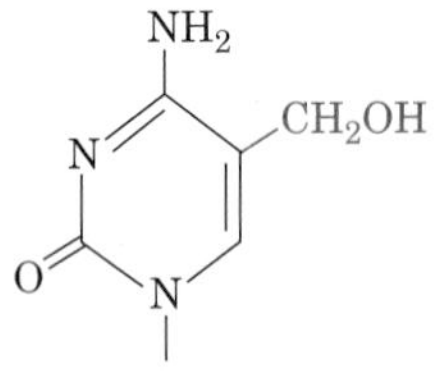

5-Hydroxymethylcytosine residue

which may, in turn, be glycosylated). These methyl groups project into B-DNA's major groove, where they can inter-

act with DNA-binding proteins. In most cells, only a few percent of the susceptible bases are methylated, although this figure rises to >30% of the C residues in some plants.

Bacterial DNAs are methylated at their own particular restriction sites, thereby preventing the corresponding restriction endonucleases from degrading the DNA (Section 5-5A). These restriction–modification systems, however, account for only part of the methylation of bacterial DNAs. In *E. coli*, most DNA methylation is catalyzed by the products of the *dam* and *dcm* genes. The **Dam methyltransferase (Dam MTase)** methylates the A residue in all GATC sequences, whereas the **Dcm MTase** methylates both C residues in CC^A_TGG at their C5 positions. Note that both of these sequences are palindromic. We have seen that *E. coli* uses Dam MTase-mediated methylation to differentiate parental from newly synthesized DNA in mismatch repair (Section 30-5C) and in limiting *oriC*-based DNA replication initiation to once per cell generation via sequestration (Section 30-3C).

a. The MTase Reaction Occurs via a Covalent Intermediate in Which the Target Base Is Flipped Out

The Dam and Dcm MTases, as do all known DNA MTases, use *S*-adenosylmethionine (SAM) as their methyl donor. Indeed, all m^{5}C-MTases share a set of conserved sequence motifs. Daniel Santi has proposed that the catalytic mechanism of these m^{5}C-MTases (Fig. 30-98) is similar to that of thymidylate synthase (Fig. 28-19) in that both types of enzymes transfer methyl groups to pyrimidine C5 atoms via a reaction that is initiated by the nucleophilic attack of a Cys thiolate group on the pyrimidine's C6 position. The pyrimidine's C5 atom is thereby activated as a resonance-stabilized carbanion that nucleophilically attacks the methyl donor's methyl group (which in thymidylate synthase is donated by N^5,N^{10}-methylene-THF rather than SAM) to yield a covalent intermediate. This intermediate subsequently decomposes to products through the enzymatic abstraction of the proton substituent to C5 and elimination of the enzyme. The Cys thiolate nucleophile is a component of a Pro-Cys dipeptide that is invariant in all known m^{5}C-MTases and thymidylate synthases.

This mechanism is supported by the observation that the action of m^{5}C-MTases on a **5-fluorocytosine (f^{5}C)** residue

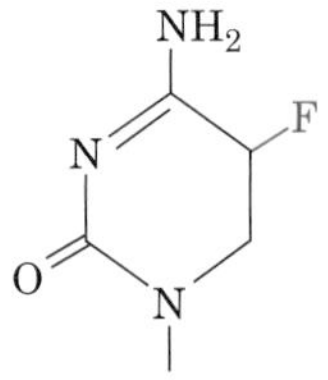

5-Fluorocytosine (f^{5}C) residue

irreversibly traps the covalent intermediate (and hence inactivates the enzyme) because the enzyme cannot abstract fluorine, the most electronegative element, as an F$^+$ ion (5-fluorodeoxyuridylate is likewise a suicide substrate for thymidylate synthase; Section 28-3B). Stereochemical principles dictate that the enzyme's Cys thiolate group can

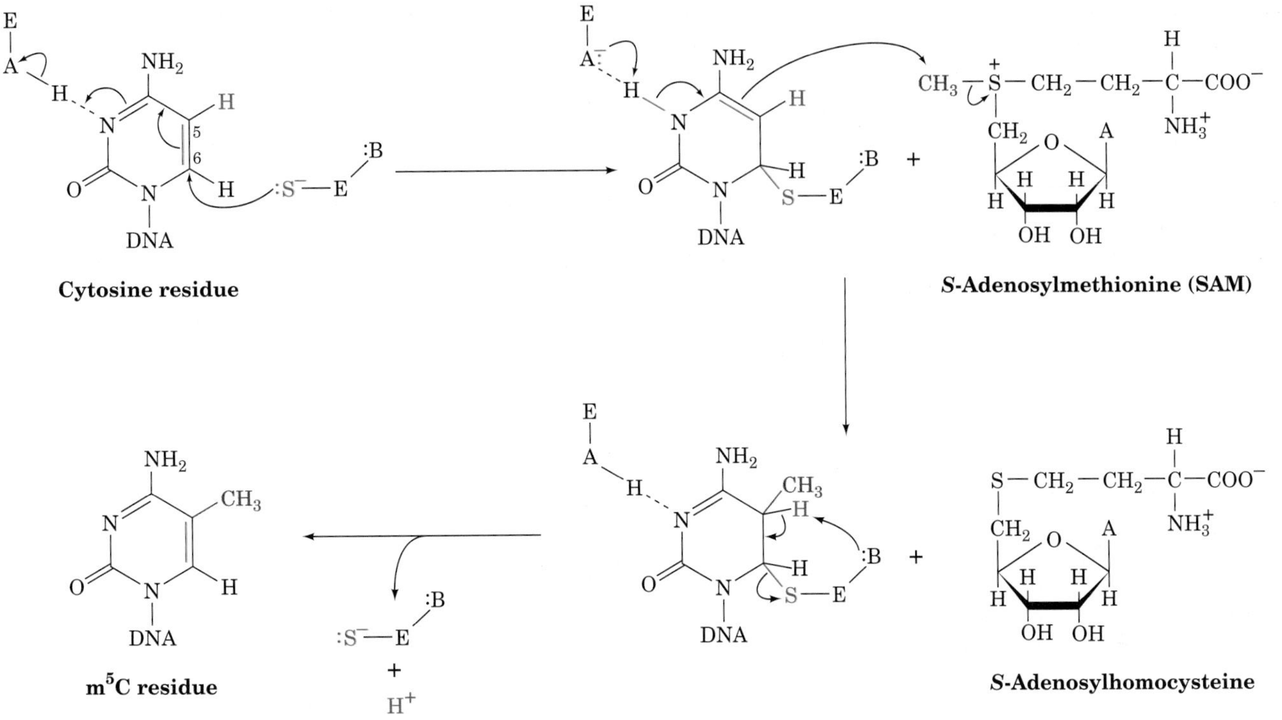

Cytosine residue

S-Adenosylmethionine (SAM)

m⁵C residue

S-Adenosylhomocysteine

FIGURE 30-98   The catalytic mechanism of 5-methylcytosine methyltransferases (m⁵C-MTases). The methyl group is supplied by SAM, which thereby becomes *S*-adenosylhomocysteine. In M.HhaI, the DNA MTase from *Haemophilus haemolyticus,* the active site thiolate group, ⁻S—E, is on Cys 81, the enzyme general acid, E—A, is Glu 119, and the enzyme general base, E—B, has not been identified. [After Verdine, G.L., *Cell* **76,** 198 (1994).]

nucleophilically attack cytosine's C5 position only from above or below the ring. This is possible because, as we shall see below, the enzyme induces its cytosine target to flip out of the DNA double helix.

The DNA MTase from *Haemophilus haemolyticus* **(M.HhaI),** a 327-residue monomer, is a component of this bacterium's restriction–modification system. M.HhaI methylates its recognition sequence, 5'-GCGC-3' in double-stranded DNA, to yield 5'-G-m⁵C-GC-3'. Richard Roberts and Xiaodong Cheng determined the X-ray structure of the inactivated M.HhaI–DNA complex formed by incubating the enzyme with the self-complementary sequence d(TGATA**G-f⁵C-GC**TATC) (in which the enzyme's recognition sequence is in bold) in the presence of SAM. The DNA binds to the enzyme in a large cleft between its two unequally sized domains (Fig. 30-99). The structure's most striking feature is that the f⁵C nucleotide has flipped out of the minor groove in the otherwise largely undistorted B-DNA helix and has inserted into the enzyme's active site. There, the f⁵C has reacted with SAM so as to yield adenosylhomocysteine (SAM without its methyl group) and the methylated intermediate covalently linked to Cys 81. The side chain of Gln 237 fills the cavity in the DNA double helix left by the departure of the f⁵C by hydrogen bonding to the opposing G base. Comparison of this structure with that of M.HhaI in complex only with SAM indicates that on binding the DNA the protein's so-called active site loop (residues 80–99) swings around to

contact the DNA, a movement of up to 25 Å. Nearly all base-specific interactions are made in the major groove by two Gly-rich loops (residues 233–240 and 250–257), the so-called recognition loops. The protein also makes extensive sequence-nonspecific contacts with DNA phosphate groups.

Base flipping was first observed in the above X-ray structure. However, as is now clear from the structures of two other MTases as well as those of variety of DNA repair enzymes (e.g., Sections 30-5A and 30-5B), *base flipping is a common mechanism through which enzymes gain access to the bases in dsDNA on which they perform chemistry.*

b. DNA Methylation in Eukaryotes Functions in Gene Regulation

5-Methylcytosine is the only methylated base in most eukaryotic DNAs, including those of vertebrates. This modification occurs largely in the CG dinucleotide of various palindromic sequences. CG is present in the vertebrate genome at only about one-fifth its randomly expected frequency. The upstream regions of many genes, however, have normal CG frequencies and are therefore known as **CpG islands.**

The degree of eukaryotic DNA methylation and its pattern are conveniently assessed by comparing the Southern blots (Section 5-5D) of DNA cleaved by the restriction endonucleases *Hpa*II (which cleaves CCGG but not C-m⁵C-GG) and *Msp*I (which cleaves both). Such

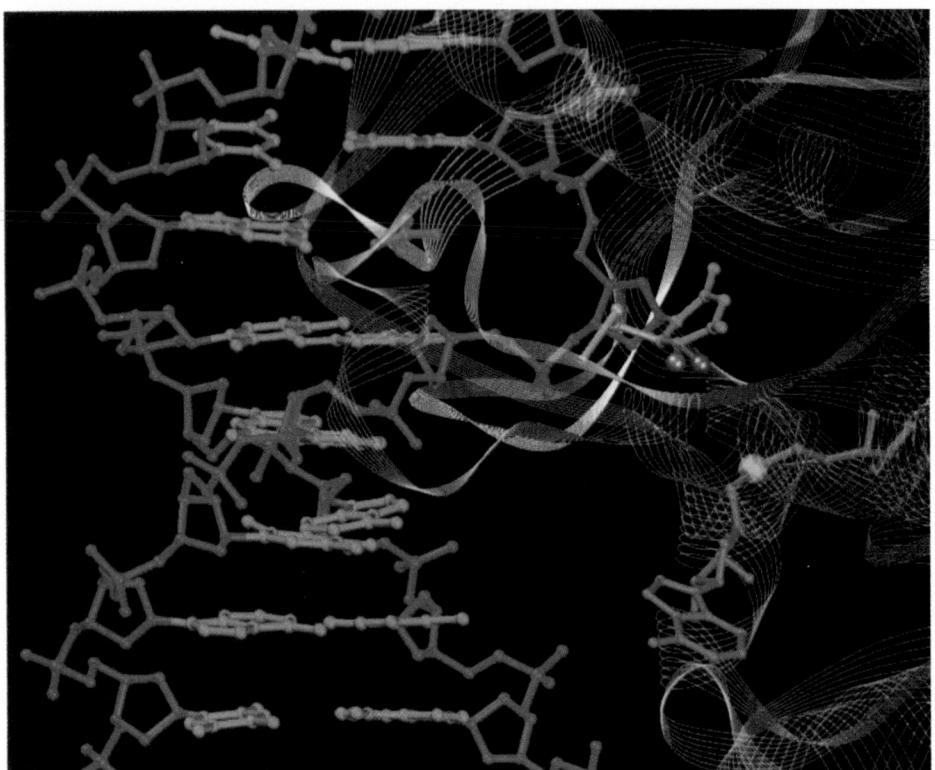

FIGURE 30-99 **X-Ray structure of M.HhaI in complex with *S*-adenosylhomocysteine and a duplex 13-mer DNA containing a methylated f^{5}C residue at the enzyme's target site.** The DNA is shown in ball-and-stick form with its bases green and its sugar–phosphate backbone purple. The protein backbone is represented by a multiline orange ribbon with its active site loop (residues 80–89) cyan and its two recognition loops (residues 233–240 and 250–257) white. The latter interact with the DNA's target sequence in its major groove from the back of the drawing. The methylated f^{5}C residue has swung out of the DNA into the enzyme's active site pocket, where its C6 forms a covalent bond with the S atom of Cys 81 (*yellow*). Both the methyl group and the F atom substituent to C5 of the f^{5}C base are represented by silver spheres because the X-ray structure cannot confidently differentiate them. The flipped out f^{5}C base is replaced in the DNA double helix by the side chain of Gln 237 (*magenta*), which hydrogen bonds to the "orphaned" guanine base. The adenosylhomocysteine (*red*) is shown in ball-and-stick form with its S atom, the methyl donor in the SAM that methylated the f^{5}C, represented by a yellow sphere. [Based on an X-ray structure determined by Richard Roberts, New England Biolabs, Beverly, Massachusetts, and Xiaodong Cheng, Cold Spring Harbor Laboratory, Cold Spring Harbor, New York. PDBid 1MHT.]

studies indicate that eukaryotic DNA methylation varies with the species, the tissue, and the position along a chromosome. The m^{5}C residues in a given DNA segment can be identified through **bisulfite sequencing,** in which the DNA is reacted with **bisulfite ion** (HSO$_3^-$), which selectively deaminates C (but not m^{5}C) residues to U, followed by PCR amplification (Section 5-5F), which copies these U's to T's and the m^{5}C's to C's. Comparison of the sequences of the amplified DNA with that of untreated DNA (as determined by the chain-terminator method; Section 7-2A) reveals which C's in the untreated DNA are methylated.

There is clear evidence that *DNA methylation switches off eukaryotic gene expression, particularly when it occurs in the promoter regions upstream of a gene's transcribed sequences.* For example, globin genes are less methylated in erythroid cells than they are in nonerythroid cells and, in fact, the specific methylation of the control region in a recombinant globin gene inhibits its transcription in transfected cells. In further support of the inhibitory effect of

DNA methylation is the observation that **5-azacytosine (5-azaC),**

**5-Azacytosine
(5-azaC)**

a base analog that cannot be methylated at its N5 position and that inhibits DNA MTases, stimulates the synthesis of several proteins and changes the cellular differentiation patterns of cultured eukaryotic cells. The observation that repetitive intragenic parasites such as LINEs are highly methylated in somatic tissues has led to the hypothesis that CpG methylation in mammals arose to prevent the spurious transcriptional initiation of these retrotransposons.

The way in which DNA methylation prevents gene expression is poorly understood. However, in many cases, DNA methylation is recognized by a family of proteins that contain a conserved **methyl-CpG binding domain (MBD).** Since the methyl groups of m^5C residues extend into dsDNA's major groove, MBDs can bind to them without perturbing DNA's double helical structure. MBD-containing proteins inhibit the transcription of their bound promoter-methylated genes by recruiting protein complexes that induce the alteration of the local chromosome structure in a way that prevents the transcription of the associated gene (eukaryotic chromosome structure is discussed in Section 34-1). Another possibility has been raised by the observation that the methylation of synthetic poly(GC) stabilizes its Z-DNA conformation. Perhaps the formation of Z-DNA, which has been detected *in vivo* (Section 29-1B), acts as a conformational switch to turn off local gene expression.

c. DNA Methylation in Eukaryotes Is Self-Perpetuating

The palindromic nature of DNA methylation sites in eukaryotes permits the methylation pattern on a parental DNA strand to direct the generation of the same pattern in its daughter strand (Fig. 30-100). This **maintenance methylation** would result in the stable "inheritance" of a methylation pattern in a cell line and hence cause these cells to all have the same differentiated phenotype. Such changes to the genome are described as being **epigenetic** (Greek: *epi,* upon or beside) because they provide an additional layer of information that specifies when and where specific portions of the otherwise fixed genome are expressed (an epigenetic change that we have already encountered is the lengthening of telomeres in germ cells; Section 30-4A). Epigenetic characteristics, as we shall see, are not bound by the laws of Mendelian inheritance.

There is considerable experimental evidence favoring the existence of maintenance methylation, including the observation that artificially methylated viral DNA, on transfection into eukaryotic cells, maintains its methylation pattern for at least 30 cell generations. Maintenance methylation in mammals appears to be mediated mainly by the **DNMT1** protein, which has a strong preference for methylating hemimethylated substrate DNAs. In contrast, prokaryotic DNA MTases such as M.HhaI do not differentiate between hemimethylated and fully methylated substrate DNAs. The importance of maintenance methylation is demonstrated by the observation that mice that are homozygous for deletion of the *DNMT1* gene die early in embryonic development.

The pattern of DNA methylation in mammals varies in early embryological development. DNA methylation levels are high in mature gametes (sperm and ova) but are nearly eliminated by the time a fertilized ovum has become a **blastocyst** (a hollow ball of cells, the stage at which the embryo implants into the uterine wall; embryonic development is discussed in Section 34-4A). After this stage, however, the embryo's DNA methylation levels globally rise until, by the time the embryo has reached the developmental stage known as a **gastrula,** its DNA methylation

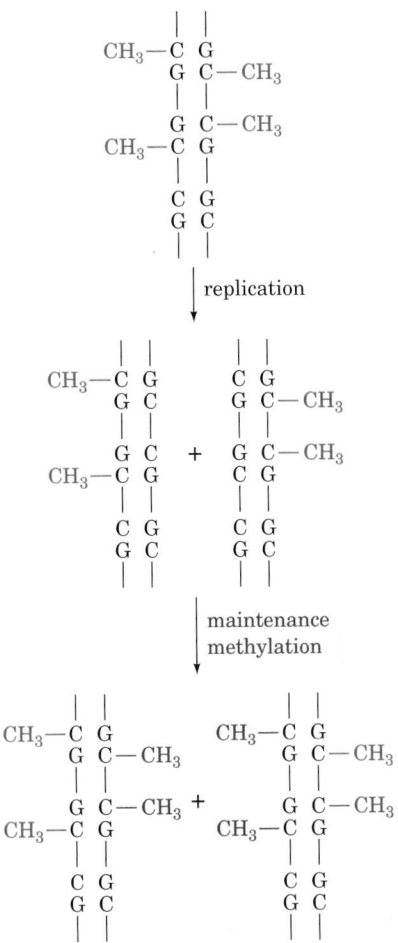

FIGURE 30-100 Maintenance methylation. The pattern of methylation on a parental DNA strand induces the corresponding methylation pattern in the complementary strand. In this way, a stable methylation pattern may be maintained in a cell line.

levels have risen to adult levels, where they remain for the lifetime of the animal. This *de novo* (new) methylation appears to be mediated by two DNA MTases distinct from DNMT1 named **DNMT3a** and **DNMT3b.** An important exception to this remethylation process is that the CpG islands of germline cells (cells that give rise to sperm or ova) remain unmethylated. This ensures the faithful transmission of the CpG islands to the succeeding generation in the face of the strong mutagenic pressure of m^5C deamination (which yields T, a mutation that mismatch repair occasionally fails to correct).

The change in DNA methylation levels (epigenetic reprogramming) during embryonic development suggests that the pattern of genetic expression differs in embryonic and somatic cells. This explains the observed high failure rate in cloning mammals (sheep, mice, cattle, etc.) by transferring the nucleus of an adult cell into an enucleated oocyte (immature ovum). Few of these animals survive to birth, many of those that do so die shortly thereafter, and most of the ~1% that do survive have a variety of abnormalities, most prominently an unusually large size.

However, the survival of any embryos at all is indicative that the oocyte has the remarkable capacity to epigenetically reprogram somatic chromosomes (although it is rarely entirely successful in doing so) and that mammalian embryos are relatively tolerant of epigenetic abnormalities. Presumably, the reproductive cloning of humans from adult nuclei would result in similar abnormalities and for this reason (in addition to social and ethical prohibitions) should not be attempted.

d. Genomic Imprinting Results from Differential DNA Methylation

It has been known for thousands of years that maternal and paternal inheritance can differ. For example, a mule (the offspring of a mare and a male donkey) and a hinny (the offspring of a stallion and a female donkey) have obviously different physical characteristics, a hinny having shorter ears, a thicker mane and tail, and stronger legs than a mule. This is because, in mammals only, certain maternally and paternally supplied genes are differentially expressed, a phenomenon termed **genomic imprinting.** The genes that are subject to genomic imprinting are, as Rudolph Jaenisch has shown, differentially methylated in the two parents during gametogenesis and the resulting different methylation patterns are resistant to the wave of demethylation that occurs during the formation of the blastocyst and to the wave of *de novo* methylation that occurs thereafter.

The importance of genomic imprinting is demonstrated by the observation that an embryo derived from the transplantation of two male or two female pronuclei into an ovum fails to develop (pronuclei are the nuclei of mature sperm and ova before they fuse during fertilization). Inappropriate imprinting is also associated with certain diseases. For example, **Prader-Willi syndrome (PWS),** which is characterized by the failure to thrive in infancy, small hands and feet, marked obesity, and variable mental retardation, is caused by a >5000-kb deletion in a specific region of the paternally inherited chromosome 15. In contrast, **Angelman syndrome (AS),** which is manifested by severe mental retardation, a puppetlike ataxic (uncoordinated) gait, and bouts of inappropriate laughter, is caused by a deletion of the same region from the maternally inherited chromosome 15. These syndromes are also exhibited by those rare individuals who inherit both their chromosomes 15 from their mothers for PWS and from their fathers for AS. Evidently, certain genes on the deleted chromosomal region must be paternally inherited to avoid PWS and others must be maternally inherited to avoid AS. Several other human diseases are also associated with either maternal or paternal inheritance or lack thereof.

e. DNA Methylation Is Associated with Cancer

The mutation of an m^5C residue to T (with its associated G to A mutation on the complementary strand) is, by far, the most prevalent mutational change in human cancers. Such mutations usually convert proto-oncogenes to oncogenes (Section 19-3B) or inactivate tumor suppressors (Fig. 34-4C). In addition, the hypomethylation of proto-oncogenes and the hypermethylation of genes encoding tumor suppressors are associated with cancers, although it is unclear whether these are initiating or consolidating events for malignancies.

f. Several Neurological Diseases Are Associated with Trinucleotide Repeat Expansions

Fragile X syndrome, whose major symptoms include mental retardation and a characteristic long, narrow face with large ears, afflicts 1 in 4500 males and 1 in 9000 females. Fragile X syndrome is so named because, in affected individuals, the tip of the X chromosome's long arm is connected to the rest of the chromosome by a slender thread that is easily broken. The genetics of this condition are bizarre. The maternal grandfathers of individuals having fragile X syndrome may be asymptomatic, both clinically and cytogenetically. Their daughters are likewise asymptomatic, but these daughters' children of either sex may have the syndrome. Evidently, the fragile X defect is activated by passage through a female. Moreover, the probability of a child having fragile X syndrome and the severity of the disease increase with each succeeding generation, a phenomenon termed **genetic anticipation.**

The affected gene in fragile X syndrome, *FMR1* (for *f*ragile X *m*ental *r*etardation *1*), encodes a 632-residue RNA-binding protein named **FMRP** (for *FMR* protein), which apparently functions in the transport of certain mRNAs from the nucleus to the cytoplasm (Section 34-3C), where it probably regulates their translation. FMRP, which is highly conserved in vertebrates, is expressed in most tissues but most heavily in brain neurons, where a variety of evidence indicates that its participation is required for the proper formation and/or function of synapses.

In the general population, the 5' untranslated region of *FMR1* contains a polymorphic $(CGG)_n$ sequence with n ranging from 6 to 60 and often punctuated by one or two AGG interruptions. However, in certain asymptomatic individuals, n has increased from 60 to 200, a so-called premutation that males transmit in unchanged form to their daughters (they transmit a Y rather than an X chromosome to their sons). In the daughters' children, however, ~80% of the individuals inheriting a premutant *FMR1* gene exhibit an astonishing expansion (amplification) of the triplet repeat with n ranging from >200 to several thousand, as well as the symptoms of the disease, a so-called full mutation. These triplet repeats differ in size among siblings and often exhibit heterogeneity within an individual, suggesting that they are somatically generated.

These **dynamic mutations,** which expand more often than they contract, probably arise through slippage of the template DNA during replication. Slippage is thought to occur through the formation of loop-outs (Fig. 30-101) on either the newly synthesized strand (causing expansions) or on the template strand (causing contractions). Since the lagging strand has more single-stranded character than the leading strand, most slippage probably occurs during lagging strand synthesis. As expected, the frequency of slippage increases with the number of repeats.

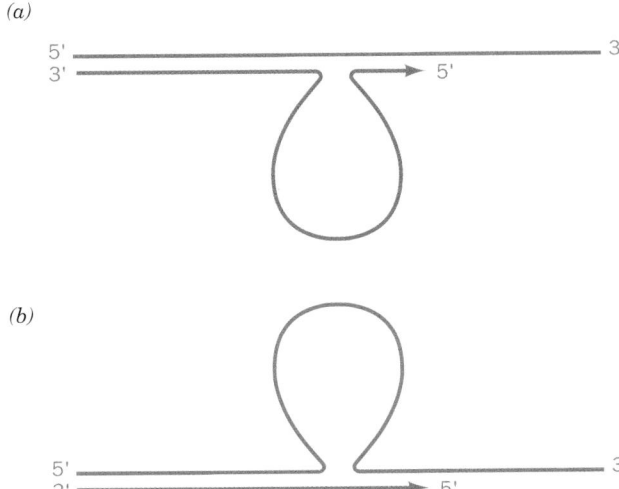

FIGURE 30-101 The loop-out mechanism for the alteration of the number of consecutive triplet repeats in DNA through its replication. Here the template strand is red and its nascent (newly synthesized) daughter strand is blue. With long tracts of repeating sequences, the probability of a loop-out occurring increases because its neighboring sequences will remain base paired. (*a*) If the daughter strand loops out, the number of repeats increases. (*b*) If the template strand loops out, the number of repeats decreases.

The peculiar genetics of fragile X syndrome is a consequence of genomic imprinting through methylation. The *FMR1* gene is unmethylated in normal individuals. However, it is hypermethylated in individuals with a maternally transmitted full mutation. This maintains the *FMR1* gene in a transcriptionally silent (inactive) state, thereby accounting for the symptoms of the disease. The lesser frequency and severity of fragile X syndrome in females are accounted for by the fact that females have two X chromosomes, one of which is unlikely to be mutated.

Thirteen other pathological instances of the expansion of a GC-rich trinucleotide repeat, all also neurological diseases, are known including the following:

1. Myotonic dystrophy (DM), the most common adult form of **muscular dystrophy** (its estimated incidence is 1 in 8000). It is a multisystem autosomal dominant disorder that is mainly characterized by progressive muscle weakness and wasting, the severity of which increases with successive generations while the age of onset decreases (genetic anticipation). Its most severe form, congenital DM, is exclusively maternally transmitted. DM arises from a trinucleotide expansion in the 3' untranslated region of the gene encoding **myotonic dystrophy protein kinase (MDPK),** which is expressed in the neurons affected by DM. The repeating triplet, $(CAG)_n$, is present in between 5 and 30 copies in the MDPK gene of normal individuals but expands from at least 50 repeats in minimally affected individuals to ~2000 repeats in severely affected individuals.

2. Huntington's disease (HD; previously called **Huntington's chorea),** a devastating neurodegenerative disorder characterized by progressively choreic (disordered) movements, cognitive decline, and emotional disturbances over an average 18-year course that is inevitably fatal. This dominant autosomal disorder, which affects ~1 in 10,000 individuals and has an average age of onset of ~40 years, is a consequence of the selective loss of certain groups of neurons in the brain. The *HD* gene, which encodes a widely expressed 3145-residue polypeptide named **huntingtin,** contains a polymorphic trinucleotide repeat, $(CAG)_n$, within its polypeptide coding sequence. The *HD* genes from 150 independent families with HD all contained between 37 and 86 repeat units, whereas those from normal individuals had 11 to 34 repeats. Moreover, the *HD* repeat length is unstable: >80% of meiotic transmissions show either increases or decreases, with the largest increases occurring in paternal transmissions (genomic imprinting). The number of repeats in afflicted individuals is inversely correlated with the age of onset of HD.

CAG is the codon for Gln (Table 5-3) and hence mutant huntingtin contains a long poly(Gln) tract. Synthetic poly(Gln) aggregates as β sheets that are linked by hydrogen bonds involving both their main chain and side chain amide groups. Indeed, the nuclei of HD-affected neurons contain inclusions that presumably consist of aggregates of huntingtin or its proteolytic products. It is these inclusions, as Max Perutz pointed out, that apparently kill the neurons in which they are contained, although the mechanism of how they do so is unknown. The long incubation period before the symptoms of HD become evident is attributed to the lengthy nucleation time for aggregate formation, much like what we have seen occurs in the formation of amyloid fibrils (Section 9-5A).

3. Spinocerebellar ataxia (SCA) type 1, a progressive neurodegenerative disease whose age of onset is typically in the third or fourth decade, although it exhibits genetic anticipation. Like HD, it is caused by selective neuronal loss and is associated with an expansion of a CAG repeat in a coding region, in this case of a neuronal protein named **ataxin-1.** There it expands from ~28 to between 43 and 81 copies, thereby yielding a poly(Gln) tract of increased length (and tendency to aggregate). Four similar diseases, **SCA types 2, 3, 6,** and **7,** are caused by CAG expansions in different neuronal proteins.

CHAPTER SUMMARY

1 and 2 ■ DNA Replication DNA is replicated in the 5' → 3' direction by the assembly of deoxynucleoside triphosphates on complementary DNA templates. Replication is initiated by the generation of short RNA primers, as mediated in *E. coli* by primase and RNA polymerase. The DNA is then extended from the 3' ends of the primers through the action of DNA polymerase (Pol III in *E. coli*). The leading strand at a replication fork is synthesized continuously, whereas the lagging strand is synthesized discontinuously by the formation of Okazaki fragments. RNA primers on newly synthesized DNA are excised and replaced by DNA through Pol I-catalyzed (in *E. coli*) nick translation. The single-strand nicks are then sealed by DNA ligase. Mispairing errors during DNA synthesis are corrected by the $3' \rightarrow 5'$ exonuclease functions of both Pol I and Pol III. The Klenow fragment of Pol I and other DNA polymerases of known structure have a right-hand-like structure with the active site located in the palm domain. Pol I recognizes the incoming nucleotide according to the shape of the base pair it forms with the template base and catalyzes the formation of a phosphodiester bond via a mechanism involving two metal ions. DNA synthesis in *E. coli* requires the participation of many auxiliary proteins including DNA gyrase, DnaB helicase, single-strand binding protein (SSB), primase, the β_2 sliding clamp, and DNA ligase.

3 ■ Prokaryotic Replication DNA synthesis commences from specific sites known as replication origins. In the synthesis of the bacteriophage M13 (−) strand on the (+) strand template, the origin is recognized and primer synthesis is initiated by RNA polymerase. The analogous process in bacteriophage φX174, as well as in *E. coli,* is mediated by a complex primase-containing particle known as the primosome. φX174 (+) strands are synthesized according to the looped rolling circle mode of DNA replication on (−) strand templates of the replicative form in a process that is directed by the virus-specific gene *A* protein.

The *E. coli* chromosome is bidirectionally replicated in the θ mode from a single origin, *oriC*, which is recognized by DnaA protein. Leading strand synthesis is probably primed by RNA polymerase and primase working together, whereas Okazaki fragments are primed by primase in the primosome. The uncontrolled initiation of DNA replication is prevented by the sequestration of newly synthesized and hence hemimethylated *oriC* by membrane-associated SeqA protein, which prevents the *oriC* from becoming fully methylated at its multiple GATC sites. The β_2 sliding clamp, which is responsible for Pol III's processivity, is loaded onto the DNA by the $\gamma_3\delta\delta'$ clamp loader in an ATP-driven process. The δ subunit, when unmasked by ATP binding to the γ subunits, acts as a molecular "wrench" to spring open the sliding clamp, thereby permitting the entry of a single-stranded template DNA. Replication termination is facilitated by Tus protein which, on binding to an appropriately oriented *Ter* site, arrests the motion of a replication fork by binding to DnaB helicase. The great complexity of the DNA replication process functions to ensure the enormous fidelity necessary to maintain genome integrity.

4 ■ Eukaryotic Replication Progression through the eukaryotic cell cycle is mediated by cyclins complexed to their cognate cyclin-dependent protein kinases (Cdks). Chromosomal DNA replication is initiated by pol α/primase, which synthesizes a primer followed by a short length of DNA. Then, via polymerase switching mediated by replication factor C (RFC), the eukaryotic clamp loader, pol δ processively synthesizes both the lagging and leading strands in complex with PCNA, the eukaryotic sliding clamp.

Eukaryotic chromosomal DNA is synthesized in multiple origin-containing segments known as replicons. Nevertheless, chromosomal DNA is synthesized once and only once per cell cycle. The re-replication of DNA is prevented because replication initiation is licensed only in the G_1 phase of the cell cycle by the formation of the pre-replicative complex (pre-RC) but DNA is synthesized only in S phase by the activation of the pre-RC. The pre-RC is assembled in G_1 phase by the binding of the origin recognition complex (ORC) to an origin, which recruits Cdc6/Cdc18 and Cdt1 followed by the MCM complex, the replicative helicase. The activation of the pre-RC begins in S phase with the addition of Mcm10 followed by the phosphorylation of many of the pre-RC's subunits by Cdks and Ddk. Cdc45 then binds followed by pol α/primase, pol ε, PCNA, and the replication protein A (RPA), the SSB counterpart, to yield the active initiation complex. Re-replication is prevented through the actions of Cdks, which cause the elimination of Cdc6/Cdc18 and inhibit the helicase activity of the MCM complex. In metazoan cells, re-replication is also prevented by the binding of geminin to Cdt1.

Mitochondrial DNA is replicated in the D-loop mode by DNA polymerase γ. Retroviruses produce DNA on RNA templates in a reaction sequence catalyzed by reverse transcriptase. Telomeric DNA, a G-rich repeating octamer on the 3'-ending strand, is synthesized by the RNA-containing enzyme telomerase. Telomerase is active in germ cells but not somatic cells, a phenomenon that is at least in part responsible for cellular senescence and aging. The observation that telomerase is active in nearly all cancer cells suggests that telomerase inactivation is a defense against the development of cancer. The free DNA ends of telomeres are capped to prevent them from triggering DNA damage checkpoints. The *O. nova* telomere end binding protein (TEBP) binds both single strands of telomere DNA and a G-quartet-containing dimer, whereas humans and yeast have an unrelated telomere end-binding protein named Pot1. Telomeric DNA forms T-loops which are formed by the TRF2-mediated invasion of the 3' telomeric overhang into repeating telomeric dsDNA to form a D-loop.

5 ■ DNA Repair Cells have a great variety of DNA repair mechanisms. DNA damage may be directly reversed such as in the photoreactivation of UV-induced pyrimidine dimers or in the repair of O^6-alkylguanine lesions by the transfer of the offending alkyl group to a repair protein. Pyrimidine dimers, as well as many other types of DNA lesions, may also be removed by nucleotide excision repair (NER), which in *E. coli* involves the UvrABC system. Xeroderma pigmentosum, an inherited human disease characterized by marked UV-induced skin changes and a greatly increased incidence of cancer, is caused by defects in any of seven complementation groups that participate in NER. In base excision repair (BER), DNA glycosylases specifically remove the corresponding chemically altered bases, including uracil, through mechanisms that involve base flip-outs to form AP sites. The AP sites are cleaved on one side by an AP endonuclease, removed together with adjacent residues by an exonuclease, and replaced through the actions of a DNA polymerase and a DNA ligase. In mismatch repair (MMR), base pairing mismatches arising from replication errors are corrected. In *E. coli* MMR, MutS and MutL bind to the mismatch and then identify the daughter strand, which contains the error, according to which strand of the nearest hemimethylated GATC palindrome is unmethylated. MutH then cleaves this strand, which is excised past the mismatch and replaced.

DNA damage in *E. coli* induces the SOS response, a LexA- and RecA-mediated process in which the error-prone bypass DNA polymerases Pol IV and Pol V replicate a damaged template DNA even if it provides no information as to which base to incorporate. Double-strand break (DSB) repair by nonhomologous end-joining (NHEJ) is facilitated by Ku protein, which holds two dsDNA ends together for ligation by DNA ligase IV in complex with Xrcc4. The high correlation between mutagenesis and carcinogenesis permits the detection of carcinogens by the Ames test.

6 ■ Recombination and Mobile Genetic Elements Genetic information may be exchanged between homologous DNA sequences through homologous recombination, a process that occurs according to the Holliday model. In *E. coli*, strand invasion to form Holliday junctions is mediated by RecA after the RecBCD-mediated generation of the single-strand nicks to which RecA binds. Branch migration is mediated by RuvAB, which consists of a homotetramer (or a homooctamer) of RuvA, which binds both a Holliday junction and two oppositely located RuvB hexamers. In an ATP-driven process, the RuvB hexamers (effectively) counter-rotate to pump the dsDNA stems into the center of the RuvA-bound Holliday junction, where each of its single strands exchange base pairing partners to form new dsDNA stems, which are translocated toward the periphery of the complex. The Holliday junction is eventually resolved to its component dsDNAs by RuvC.

The primary function of homologous recombination is to repair damaged replication forks resulting from the encounters of replisomes with unrepaired single-strand lesions or breaks. DSBs may be rejoined via a recombination repair process called homologous end-joining.

Chromosomes and plasmids may be rearranged through the action of transposons. These DNA segments carry the genes that encode the proteins that mediate the transposition process as well as other genes. Tn5 transposase catalyzes the cut-and-paste transposition of the Tn5 transposon. Replicative transposition proceeds via the intermediacy of cointegrates, which are resolved through the action of enzymes such as the γδ resolvase. Transposition may be important in chromosomal and plasmid evolution and has been implicated in the control of phenotypic expression such as phase alternation in *Salmonella,* a process that is catalyzed by the Hin DNA invertase, a homolog of the γδ resolvase. Members of the λ integrase family of proteins, such as Cre recombinase, insert dsDNA segments into their target sites via a Holliday junction intermediate in which transient covalent bonds are formed between active site Tyr side chains and the 3′-OH groups at the cleavage sites. Retrotransposons undergo transposition through an RNA intermediate. Many retrotransposons, such as yeast Ty1, are "internal" retroviruses that can only replicate within a genome. Nonviral retrotransposons, such as LINEs, the dominant transposons in the human genome, have a different transpositional mechanism.

7 ■ DNA Methylation and Trinucleotide Repeat Expansions Prokaryotic DNA may be methylated at its A or C bases. This prevents the action of restriction endonucleases and permits the correct mismatch repair of newly replicated DNA. In most eukaryotes, DNA methylation, which occurs, mainly at CpG islands, through the formation of m^5C, has been implicated in the control of gene expression and, via maintenance methylation, in genomic imprinting.

Several inherited neurological diseases, including fragile X syndrome, myotonic dystrophy, and Huntington's disease, are characterized by the genetically bizarre expansion of segments of repeating GC-rich triplets. If an expanded triplet repeat occurs in an upstream noncoding region of a gene, its aberrant methylation, perhaps through genomic imprinting, may lead to the gene's transcriptional silencing, and if the expanded repeat is instead manifested as a poly(Gln) tract in a protein, the resulting protein aggregates may kill the neurons in which it occurs.

REFERENCES

GENERAL

Adams, R.L.P., Knowler, J.T., and Leader, D.P., *The Biochemistry of the Nucleic Acids* (11th ed.), Chapters 6 and 7, Chapman & Hall (1992).

Kornberg, A., *For Love of Enzymes: The Odyssey of a Biochemist,* Harvard University Press (1989). [A scientific autobiography.]

Kornberg, A. and Baker, T.A., *DNA Replication* (2nd ed.), Freeman (1992). [A compendium of information about DNA replication whose first author is the founder of the field.]

Lewin, B., *Genes VII,* Chapters 12–17 and 33–36, Oxford University Press (2000).

PROKARYOTIC DNA REPLICATION

Baker, T.A. and Wickner, S.H., Genetics and enzymology of DNA replication in *Escherichia coli, Annu. Rev. Genet.* **26,** 447–477 (1992).

Beese, L.S., Derbyshire, V., and Steitz, T.A., Structure of DNA polymerase I Klenow fragment bound to duplex DNA, *Science* **260,** 352–355 (1993).

Benkovic, J.J., Valentine, A.M., and Salinas, F., Replisome-mediated DNA replication, *Annu. Rev. Biochem.* **70,** 181–208 (2001).

Carr, K.M. and Kaguni, J.M., Stoichiometry of DnaA and DnaB protein in initiation at the *Escherichia coli* chromosomal origin, *J. Biol. Chem.* **276,** 44919–44925 (2001).

Caruthers, J.M. and McKay, D.B., Helicase structure and mechanism, *Curr. Opin. Struct. Biol.* **12,** 123–133 (2002).

Crooke, E., Regulation of chromosomal replication in *E. coli:* sequestration and beyond, *Cell* **82,** 877–880 (1995).

Davey, M.J., Jeruzalmi, D., Kuriyan, J., and O'Donnell, M., Motors and switches: AAA$^+$ machines within the replisome, *Nature Rev. Mol. Cell Biol.* **3,** 1–10 (2002).

Doublié, S., Sawaya, M.R., and Ellenberger, T., An open and closed case for all polymerases, *Structure* **7,** R31–R35 (1999). [Reviews the mechanisms of DNA polymerases.]

Frick, D.N. and Richardson, C.C., DNA primases, *Annu. Rev. Biochem.* **70,** 39–80 (2001).

Guckian, K.M., Krugh, T.R., and Kool, E.T., Solution structure of a DNA duplex containing a replicable difluorotoluene–adenine pair, *Nature Struct. Biol.* **5,** 954–959 (1998); *and* Moran, S., Ren, R.X.-F., and Kool, E.T., A thymine triphosphate shape analog lacking Watson–Crick pairing ability is replicated with high sequence specificity, *Proc. Natl. Acad. Sci.* **94,** 10506–10511 (1997).

Jeruzalmi, D., O'Donnell, M., and Kuriyan, J., Clamp loaders and sliding clamps, *Curr. Opin. Struct. Biol.* **12,** 217–224 (2002); Jeruzalmi, D., Yurieva, O., Zhao, Y., Young, M., Stewart, J., Hingorani, M., O'Donnell, M., and Kuriyan, J., Mechanism of processivity clamp opening by the delta subunit wrench of the clamp loader complex of *E. coli* DNA polymerase III, *Cell* **106,** 417–428 (2001); Jeruzalmi, D., O'Donnell, M., and Kuriyan, J., Crystal structure of the processivity clamp loader gamma (γ) complex of *E. coli* DNA polymerase III, *Cell* **106,** 429–441 (2001); *and* Podobnik, M., Weitze, T.F., O'Donnell, M., and Kuriyan, J., Nucleotide-induced conformational change in an isolated *Escherchia coli* DNA polymerase III clamp loader subunit, *Structure* **11,** 253–263 (2003).

Johnson, K.A., Conformational coupling in DNA polymerase fidelity, *Annu. Rev. Biochem.* **62,** 685–713 (1993).

Kamada, K., Horiuchi, T., Ohsumi, K., Shimamoto, N., and Morikawa, K., Structure of a replication–terminator protein complexed with DNA, *Nature* **383,** 598–603 (1996).

Keck, J.L., Roche, D.D., Lynch, A.S., and Berger, J.M., Structure of the RNA polymerase domain of *E. coli* primase, *Science* **287,** 2482–2486 (2000); *and* Podobnik, M., McInerney, P., O'Donnell, M., and Kuriyan, J., A TOPRIM domain in the crystal structure of the catalytic core of *Escherichia coli* primase confirms a structural link to DNA topoisomerases, *J. Mol. Biol.* **300,** 353–362 (2000).

Kelman, Z. and O'Donnell, M., DNA polymerase III holoenzyme: Structure and function of a chromosomal replicating machine, *Annu. Rev. Biochem.* **64,** 171–200 (1995).

Kiefer, J.R., Mao, C., Braman, J.C., and Beese, L.S., Visualizing DNA replication in a catalytically active *Bacillus* DNA polymerase crystal, *Nature* **391,** 304–307 (1998).

Kong, X.-P., Onrust, R., O'Donnell, M., and Kuriyan, J., Three-dimensional structure of the β subunit of *E. coli* DNA polymerase III holoenzyme: A sliding DNA clamp, *Cell* **69,** 425–437 (1992).

Kool, E.T., Active site tightness and substrate fit in DNA replication, *Annu. Rev. Biochem.* **71,** 191–219 (2002); *and* Hydrogen-bonding, base stacking, and steric effects in DNA replication, *Annu. Rev. Biophys. Biomol. Struct.* **30,** 1–2 (2001).

Korolev, S., Hsieh, J., Gauss, G.H., Lohman, T.M., and Waksman, G., Major domain swiveling revealed by the crystal structures of complexes of *E. coli* Rep helices bound to single-stranded DNA and ADP, *Cell* **90,** 635–647 (1997).

Kunkel, T.A. and Bebenek, K., DNA replication fidelity, *Annu. Rev. Biochem.* **69,** 497–529 (2000).

Lee, J.Y., Chang, C., Song, H.K., Moon, J., Yang, J.K., Kim, H.-K., Kwon, S.-T., and Suh, S.W., Crystal structure of NAD$^+$-dependent DNA ligase: modular architecture and functional implications, *EMBO J.* **19,** 1119–1129 (2000).

Li, Y., Korolev, S., and Waksman, G., Crystal structures of open and closed forms of binary and ternary complexes of the large fragment of *Thermus aquaticus* DNA polymerase I: structural basis for nucleotide incorporation, *EMBO J.* **17,** 7514–7525 (1998).

Naktinis, V., Turner, J., and O'Donnell, M., A molecular switch in the replication machine defined by internal competition for protein rings, *Cell* **84,** 137–145 (1996).

Patel, S.S. and Picha, K.M., Structure and function of hexameric helicases, *Annu. Rev. Biochem.* **69,** 651–697 (2000).

Raghunathan, S., Kozlov, A.G., Lohman, T.M., and Waksman, G., Structure of the DNA binding domain of *E. coli* SSB bound to ssDNA, *Nature Struct. Biol.* **7,** 648–652 (2000).

Singleton, M.R., Sawaya, M.R., Ellenberger, T., and Wigley, D.B., Crystal structure of T7 gene 4 ring helicase indicates a mechanism for sequential hydrolysis of nucleotides, *Cell* **101,** 589–600 (2000).

Soultanas, P. and Wigley, D.B., Unwinding the 'Gordian knot' of helicase action, *Trends Biochem. Sci.* **26,** 47–54 (2001).

Steitz, T.A., DNA polymerases: structural diversity and common mechanisms, *J. Biol. Chem.* **274,** 17395–17398 (1999).

Watson, J.D. and Crick, F.H.C., Genetical implications of the structure of deoxyribonucleic acid, *Nature* **171,** 964–967 (1953). [The paper in which semiconservative DNA replication was first postulated.]

EUKARYOTIC DNA REPLICATION

Allsopp, R.C., Vaziri, H., Patterson, C., Goldstein, S., Younglai, E.V., Futcher, A.B., Greider, C.W., and Harley, C.B., Telomere length predicts replicative capacity of human fibroblasts, *Proc. Natl. Acad. Sci.* **89,** 10114–10118 (1992).

Arezi, B. and Kuchta, R.D., Eukaryotic DNA primase, *Trends Biochem. Sci.* **25,** 572–576 (2000).

Bell, S.P. and Dutta, A., DNA replication in eukaryotic cells, *Annu. Rev. Biochem.* **71,** 333–374 (2002).

Blackburn, E.H., Telomerases, *Annu. Rev. Biochem.* **61,** 113–129 (1992).

Blackburn E.H., Switching and signaling at the telomere, *Cell* **106,** 661–673 (2001); *and* Telomere states and cell fates, *Nature* **408,** 53–56 (2000).

Blow, J.J. and Hodgson, B., Replication licensing—defining the proliferative state? *Trends Cell Biol.* **12,** 72–78 (2002).

Cech, T.R., Life at the end of the chromosome: Telomeres and telomerase, *Angew. Chemie* **39,** 34–43 (2000).

Clayton, D.A., Replication and transcription of vertebrate mitochondrial DNA, *Annu. Rev. Cell Biol.* **7,** 453–478 (1991).

Davies, J.F., II, Almassey, R.J., Hostomska, Z., Ferre, R.A., and Hostomsky, Z., 2.3 Å crystal structure of the catalytic domain of DNA polymerase β, *Cell* **76,** 1123–1133 (1994).

DePamphilis, M.L., Replication origins in metazoan chromosomes: fact or fiction, *BioEssays* **21,** 5–16 (1999).

Diffley, J.F.X., DNA replication: Building the perfect switch, *Curr. Biol.* **11,** R367–R370 (2001).

Ding, J., Das, K., Hsiou, Y., Sarafianos, S.G., Clark, A.D., Jr., Jacobo-Molina, A., Tantillo, C., Hughes, S.H., and Arnold, E., Structure and functional implications of the polymerase active site region in a complex of HIV-1 RT with a double-stranded DNA template-primer and an antibody Fab fragment at 2.8 Å resolution, *J. Mol. Biol.* **284,** 1095–1111 (1998); *and* Jacobo-Molina, A., Ding, J., Nanni, R.G., Clark, A.D., Jr., Lu, X., Tantillo, C., Williams, R.L., Kamer, G., Ferris, A.L., Clark, P., Hizi, A., Hughes, S.H., and Arnold, E., Crystal structure of human immunodeficiency virus type 1 reverse transcriptase complexed with double-stranded DNA at 3.0 Å resolution shows bent DNA, *Proc. Natl. Acad. Sci.* **90,** 6320–6324 (1993).

Franklin, M.C., Wang, J., and Steitz, T.A., Structure of the replicating complex of a pol α family DNA polymerase, *Cell* **105,** 657–667 (2001).

Gilbert, D.M., Making sense out of eukaryotic DNA replication origins, *Science* **294,** 96–100 (2001).

Griffith, J.D., Comeau, L., Rosenfield, S., Stansel, R.M., Bianchi, A., Moss, H., and de Lange, T., Mammalian telomeres end in a large duplex loop, *Cell* **97,** 503–514 (1999).

Hahn, W.C., Counter, C.M., Lundberg, A.S., Beijersbergen, R.L., Brooks, M.W., and Weinberg, R.A., Creation of human tumour cells with defined genetic elements, *Nature* **400,** 464–468 (1999).

Horvath, M.P. and Schultz, S.C., DNA G-quartets in a 1.86 Å resolution structure of an *Oxytricha nova* telomeric protein–DNA complex, *J. Mol. Biol.* **310**, 367–377 (2001).

Hübscher, U., Maga, G., and Spadari, S., Eukaryotic DNA polymerases, *Annu. Rev. Biochem.* **71**, 133–163 (2002); *and* Hübscher, U., Nasheuer, H.-P., and Syväoja, J.E., Eukaryotic DNA polymerases. A growing family, *Trends Biochem. Sci.* **25**, 143–147 (2000).

Jäger, J. and Pata, J.D., Getting a grip: polymerases and their substrate complexes, *Curr. Opin. Struct. Biol.* **9**, 21–28 (1999).

Kelleher, C., Teixeira, M.T., Förstemann, K., and Lingner, J., Telomerase: Biochemical considerations for enzyme and substrate, *Trends Biochem. Sci.* **27**, 572–579 (2002).

Kelly, T.J. and Brown, G.W., Regulation of chromosome replication, *Annu. Rev. Biochem.* **69**, 829–880 (2000).

Kohlstaedt, L.A., Wang, J., Friedman, J.M., Rice, P.A., and Steitz, T.A., Crystal structure at 3.5 Å resolution of HIV-1 reverse transcriptase complexed with an inhibitor, *Science* **256**, 1783–1790 (1992).

McEachern, M.J., Krauskopf, A., and Blackburn, E.H., Telomeres and their control, *Annu. Rev. Genet.* **34**, 331–358 (2000).

Neidle, S. and Parkinson, G., Telomere maintenance as a target for anticancer drug discovery, *Nature Rev. Drug Discov.* **1**, 383–393 (2002); *and* The structure of telomeric DNA, *Curr. Opin. Struct. Biol.* **13**, 275 (2003).

Schultze, P., Smith, F.W., and Feigon, J., Refined solution structure of the dimeric quadruplex formed from the *Oxytricha* telomeric oligonucleotide d(GGGGTTTTGGGG), *Structure* **2**, 221–233 (1994).

Takisawa, H., Mimura, S., and Kubota, Y., Eukaryotic DNA replication: from pre-replication complex to initiation complex, *Curr. Opin. Cell Biol.* **12**, 690–696 (2000).

Tye, B.K. and Sawyer, S., The hexameric eukaryotic MCM helicase: building symmetry from nonidentical parts, *J. Biol. Chem.* **275**, 34833–34836 (2000); *and* Tye, B.K., MCM proteins in DNA replication, *Annu. Rev. Biochem.* **68**, 649–686 (1999).

Urquidi, V., Tarin, D., and Goddison, S., Role of telomerase in cell senescence and oncogenesis, *Annu. Rev. Med.* **51**, 65–79 (2000).

Waga, S. and Stillman, B., The DNA replication fork in eukaryotic cells, *Annu. Rev. Biochem.* **67**, 721–751 (1998).

REPAIR OF DNA

Ames, B.N., Identifying environmental chemicals causing mutations and cancer, *Science* **204**, 587–593 (1979).

Beckman, K.B. and Ames, B.N., Oxidative decay of DNA, *J. Biol. Chem.* **272**, 19633–19636 (1997).

Devoret, R., Bacterial tests for potential carcinogens, *Sci. Am.* **241**(2), 40–49 (1979).

Friedberg, E.C., Wagner, R., and Radman, M., Specialized DNA polymerases, cellular survival, and the genesis of mutations, *Science* **296**, 1627–1630 (2002).

Friedberg, E.C., Walker, G.C., and Siede, W., *DNA Repair and Mutagenesis,* ASM Press (1995).

Goodman, M.F., Error-prone repair DNA polymerases in prokaryotes and eukaryotes, *Annu. Rev. Biochem.* **71**, 17–50 (2002).

Hall, J.G., Genomic imprinting: Nature and clinical relevance, *Annu. Rev. Med.* **48**, 35–44 (1997).

Harfe, B.D. and Jinks-Robertson, S., DNA mismatch repair and genetic instability, *Annu. Rev. Genet.* **34**, 359–399 (2000).

Hopfner, K.-P., Putnam, C.D., and Tainer, J.A., DNA double-strand break repair from head to tail, *Curr. Opin. Struct. Biol.* **12**, 115–122 (2002).

Jaenisch, R., DNA methylation and imprinting: why bother? *Trends Genet.* **13**, 322–329 (1997).

Jiricny, J., Replication errors: cha(lle)nging the genome, *EMBO J.* **17**, 6427–6436 (1998). [A review of mismatch repair.]

Kenyon, C.J., The bacterial response to DNA damage, *Trends Biochem. Sci.* **8**, 84–87 (1983).

Lalande, M., Parental imprinting and human disease, *Annu. Rev. Genet.* **30**, 173–195 (1997).

Lindahl, T., Instability and decay of the primary structure of DNA, *Nature* **363**, 709–715 (1993).

Lindahl, T. and Wood, R.D., Quality control by DNA repair, *Science* **286**, 1897–1905 (1999). [A review.]

Ling, H., Boudsocq, F., Woogate, R., and Yang, W., Crystal structure of a Y-family DNA polymerase in action: A mechanism for error-prone and lesion-bypass replication, *Cell* **107**, 91–102 (2001).

Marra, G. and Schär, P., Recognition of DNA alterations by the mismatch repair system, *Biochem. J.* **338**, 1–13 (1999).

McCullough, A.K., Dodson, M.L., and Lloyd, R.S., Initiation of base excision repair: glycosylase mechanism and structures, *Annu. Rev. Biochem.* **68**, 255–285 (1999).

Mitra, S. and Kaina, B., Regulation of repair of alkylation damage in mammalian genomes, *Prog. Nucleic Acid Res. Mol. Biol.* **44**, 109–142 (1993).

Modrich, P., Mismatch repair in replication fidelity, genetic recombination, and cancer biology, *Annu. Rev. Biochem.* **65**, 101–133 (1996).

Mol, C.D., Parikh, S.S., Putnam, C.D., Lo, T.P., and Tainer, J.A., DNA repair mechanism for the recognition and removal of damaged DNA bases, *Annu. Rev. Biophys. Biomol. Struct.* **28**, 101–128 (1999).

Moore, M.H., Gulbis, J.M., Dodson, E.J., Demple, B., and Moody, P.C.E., Crystal structure of a suicidal DNA repair protein: Ada O^6-methylguanine-DNA methyltransferase from *E. coli*, *EMBO J.* **13**, 1495–1501 (1994).

Myers, L.C., Verdine, G.L., and Wagner, G., Solution structure of the DNA methyl triester repair domain of *Escherichia coli* Ada, *Biochemistry* **32**, 14089–14094 (1993).

Parikh, S.S., Mol, C.D., Slupphaug, G., Bharati, S., Krokan, H.E., and Tainer, J.A., Base excision repair initiation revealed by crystal structures and binding kinetics of human uracil–DNA glycosylase with DNA, *EMBO J.* **17**, 5214–5226 (1998).

Park, H.-W., Kim, S.-T., Sancar, A., and Diesenhofer, J., Crystal structure of DNA photolyase from *Escherichia coli*, *Science* **268**, 1866–1872 (1995).

Pegg, A.E., Dolan, M.E., and Moschel, R.C., Structure, function, and inhibition of O^6-alkylguanine–DNA alkyltransferase, *Prog. Nucleic Acid Res. Mol. Biol.* **51**, 167–223 (1995).

Pham, P., Rangarajan, S., Woodgate, R., and Goodman, M.F., Roles of DNA polymerases V and II in SOS-induced error-prone and error-free repair in *Escherichia coli*, *Proc. Natl. Acad. Sci.* **98**, 8350–8354 (2001); *and* Goodman, M.F., Coping with replication 'train wrecks' in *Escherichia coli* using Pol V, Pol II, and RecA proteins, *Trends Biochem. Sci.* **25**, 189–195 (2000).

Sancar, A., DNA excision repair, *Annu. Rev. Biochem.* **65**, 43–81 (1996).

Scriver, C.R., Beaudet, A.L., Sly, W.S., and Valle, D. (Eds.), *The Metabolic & Molecular Bases of Inherited Disease* (8th ed.), Chaps. 28 and 32, McGraw-Hill (2001). [Discussions of xeroderma pigmentosum, Cockayne syndrome, and hereditary nonpolyposis colorectal cancer.]

Sutton, M.D., Smith, B.T., Godoy, V.G., and Walker, G.C., The SOS response: recent insights into *umuDC*-dependent mutagenesis and DNA damage tolerance, *Annu. Rev. Genet.* **34**, 479–497 (2000).

Tainer, J.A. and Friedberg, E.C. (Eds.), *Biological Implications from Structures of DNA Repair Proteins, Mutation Research* **460**, 139–335 (2000). [A series of authoritative reviews.]

Walker, J.R., Corpina, R.A., and Goldberg, J., Structure of the Ku heterodimer bound to DNA and its implications for double-strand break repair, *Nature* **412,** 607–614 (2001).

Wood, R.D., Nucleotide excision repair in mammalian cells, *J. Biol. Chem.* **272,** 23465–23468 (1997); *and* DNA repair in eukaryotes, *Annu. Rev. Biochem.* **65,** 135–167 (1996).

Yang, W., Damage repair DNA polymerases Y, *Curr. Opin. Struct. Biol.* **13,** 23–30 (2003).

RECOMBINATION AND MOBILE GENETIC ELEMENTS

Ariyoshi, M., Nishino, T., Iwasaki, H., Shinagawa, H., and Morikawa, K., Crystal structure of the Holliday junction DNA in complex with a single RuvA tetramer, *Proc. Natl. Acad. Sci.* **97,** 8257–8262 (2000).

Changela, A., Perry, K., Taneja, B., and Mondragón, A., DNA manipulators: caught in the act, *Curr. Opin. Struct. Biol.* **13,** 15–22 (2003).

Cox, M.M., Recombinational DNA repair of damaged replication forks in *Escherichia coli*: questions, *Annu. Rev. Genet.* **35,** 53–82 (2001); *and* Recombinational DNA repair in bacteria and the RecA protein, *Prog. Nucleic Acid Res. Mol. Biol.* **63,** 311–366 (2000).

Cox, M.M., Goodman, M.F., Kreuzer, K.N., Sherratt, D.J., Sandler, S.J., and Marians, K.J., The importance of repairing stalled replication forks, *Nature* **404,** 37–41 (2000).

Craig, N.L., Target site selection in transposition, *Annu. Rev. Biochem.* **66,** 437–474 (1997).

Craig, N.L., Craigie, R., Gellert, M., and Lambowitz, A.M. (Eds.), *Mobile DNA II,* ASM Press (2002). [A compendium of authoritative articles.]

Davies, D.R., Gorshin, I.Y., Reznikoff, W.S., and Rayment, I., Three-dimensional structure of the Tn5 synaptic complex transposition intermediate, *Science* **289,** 77–85 (2000); *and* Reznikoff, W.S., Bhasin, A., Davies, D.R., Gorshin, I.Y., Mahnke, L.A., Naumann, T., Rayment, I., Steiniger-White, M., and Twining, S.S., Tn5: a molecular window on transposition, *Biochem. Biophys. Res. Commun.* **266,** 729–734 (1999).

Egelman, E.H., What do X-ray crystallographic and electron microscopic structural studies of RecA protein tell us about recombination? *Curr. Opin. Struct. Biol.* **3,** 189–197 (1993).

Eichman, B.F., Vargason, J.M., Mooers, B.H.M., and Ho, P.S., The Holliday junction in an inverted repeat DNA sequence: sequence effects on the structure of four-way junctions, *Proc. Natl. Acad. Sci.* **97,** 3971–3976 (2000).

Feng, J.-A., Dickerson, R.E., and Johnson, R.C., Proteins that promote DNA inversion and deletion, *Curr. Opin. Struct. Biol.* **4,** 60–66 (1994).

Haber, J.E., Partners and pathways. Repairing a double-strand break, *Trends Genet.* **16,** 259–264 (2000); *and* DNA recombination: the replication connection, *Trends Biochem. Sci.* **24,** 271–275 (1999).

Haren, L., Ton-Hoang, B., and Chandler, M., Integrating DNA: transposases and retroviral integrases, *Annu. Rev. Microbiol.* **53,** 245–281 (1999).

Ho, P.S. and Eichman, B.F., The crystal structures of Holliday junctions, *Curr. Opin. Struct. Biol.* **11,** 302–308 (2001).

Kuzminov, A., Recombinational repair of DNA damage in *Escherichia coli* and bacteriophage λ, *Microbiol. Mol. Biol. Rev.* **63,** 751–813 (1999).

Lusetti, S.L. and Cox, M.M., The bacterial RecA protein and the recombinational DNA repair of stalled replication forks, *Annu. Rev. Biochem.* **71,** 71–100 (2002).

Marians, K.J., PriA-directed replication fork restart in *Escherichia coli, Trends Biochem. Sci.* **25,** 185–189 (2000).

Rice, P.A. and Baker, T.A., Comparative architecture of trans-

posase and integrase complexes, *Nature Struct. Biol.* **8,** 302–307 (2001).

Roe, S.M., Barlow, T., Brown, T., Oram, M., Keeley, A., Tsaneva, I.R., and Pearl, L.H., Crystal structure of an octameric RuvA–Holliday junction complex, *Molecular Cell* **2,** 361–372 (1998).

Simon, M., Zieg, J., Silverman, M., Mandel, G., and Doolittle, R., Phase variation: evolution of a controlling element, *Science* **209,** 1370–1374 (1980).

Story, R.M., Weber, I.T., and Steitz, T.A., The structure of the *E. coli recA* protein monomer and polymer, *Nature* **355,** 318–325 (1992); *and* the erratum for this paper, *Nature* **355,** 367 (1992). [These two papers should be read together.]

Van Duyne, G.D., A structural view of Cre–*loxP* site-specific recombination, *Annu. Rev. Biophys. Biomol. Struct.* **30,** 87–104 (2001).

Yamada, K., Kunishima, N., Mayanagi, K., Ohnishi, T., Nishino, T., Iwasaki, H., Shinagawa, H., and Morikawa, K., Crystal structure of the Holliday junction migration motor protein RuvB from *Thermus thermophilus* HB8, *Proc. Natl. Acad. Sci.* **98,** 1442–1447 (2001).

West, S.C., Processing of recombination intermediates by the RuvABC proteins, *Annu. Rev. Genet.* **31,** 213–244 (1997).

Yang, W. and Steitz, T.A., Crystal structure of the site-specific recombinase γδ resolvase complexed with a 34 bp cleavage site, *Cell* **82,** 193–207 (1995).

DNA METHYLATION AND TRINUCLEOTIDE REPEAT EXPANSIONS

Bowater, R.P. and Wells, R.D., The intrinsically unstable life of DNA repeats associated with human hereditary disorders, *Prog. Nucleic Acid Res. Mol. Biol.* **66,** 159–202 (2001).

Cheng, X., Structure and function of DNA methyltransferases, *Annu. Rev. Biophys. Biomol. Struct.* **24,** 293–318 (1995).

Cummings, C.J. and Zoghbi, H.Y., Trinucleotide repeats: mechanisms and pathophysiology, *Annu. Rev. Genomics Hum. Genet.* **1,** 281–328 (2002); *and* Zoghbi, H.Y. and Orr, H.T., Glutamine repeats and neurodegeneration, *Annu. Rev. Neurosci.* **23,** 217–247 (2000).

Goodman, J. and Watson, R.E., Altered DNA methylation: a secondary mechanism involved in carcinogenesis, *Annu. Rev. Pharmacol. Toxicol.* **42,** 501–525 (2002).

Jones, P.A. and Baylin, S.B., The fundamental role of epigenetic events in cancer, *Nature Rev. Genet.* **3,** 415–428 (2002); *and* Jones, P.A. and Takai, D., The role of DNA methylation in mammalian epigenetics, *Science* **293,** 1068–1070 (2001).

Klimasauskas, S., Kumar, S., Roberts, R.J., and Cheng, X., HhaI methyltransferase flips its target base out of the DNA helix, *Cell* **76,** 357–369 (1994).

Marinus, M.G., DNA methylation in *Escherichia coli, Annu. Rev. Genet.* **21,** 113–131 (1987).

O'Donnell, W.T. and Warren, S.T., A decade of molecular studies of fragile X syndrome, *Annu. Rev. Neurosci.* **25,** 315–338 (2002).

Perutz, M.F. and Windle, A.H., Causes of neural death in neurodegenerative diseases attributable to expansion of glutamine repeats, *Nature* **12,** 143–144 (2001); *and* Perutz, M.F., Glutamine repeats and neurodegenerative diseases: molecular aspects, *Trends Biochem. Sci.* **24,** 58–63 (1999).

Reik, W., Dean, W., and Walter, J., Epigenetic reprogramming in mammalian development, *Science* **293,** 1089–1093 (2001).

Rideout, W.M., III, Eggan, K., and Jaenisch, R., Nuclear cloning and epigenetic reprogramming of the genome, *Science* **293,** 1093–1098 (2001).

Roberts, R.J. and Cheng, X., Base flipping, *Annu. Rev. Biochem.* **67,** 181–198 (1998).

Scriver, C.R., Beaudet, A.L., Sly, W.S., and Valle, D. (Eds.), *The Metabolic & Molecular Bases of Inherited Disease* (8th ed.),

Chaps. 64, 223, and 226, McGraw-Hill (2001). [Discussions of fragile X syndrome, Huntington's disease, and the spinocerebellar ataxias.]

Szyf, M. and Detich, N., Regulation of the DNA methylation machinery and its role in cellular transformation, *Prog. Nucleic Acid Res. Mol. Biol.* **69**, 47–79 (2001).

PROBLEMS

1. Explain how certain mutant varieties of Pol I can be nearly devoid of DNA polymerase activity but retain almost normal levels of $5' \rightarrow 3'$ exonuclease activity.

2. Why haven't Pol I mutants been found that completely lack $5' \rightarrow 3'$ activity at all temperatures?

3. Why aren't type I topoisomerases necessary in *E. coli* DNA replication?

*4. The $3' \rightarrow 5'$ exonuclease activity of Pol I excises only unpaired 3'-terminal nucleotides from DNA, whereas this enzyme's pyrophosphorolysis activity removes only properly paired 3'-terminal nucleotides. Discuss the mechanistic significance of this phenomenon in terms of the polymerase reaction.

5. You have isolated *E. coli* with temperature-sensitive mutations in the following genes. What are their phenotypes above their restrictive temperatures? Be specific. (a) *dnaB*, (b) *dnaE*, (c) *dnaG*, (d) *lig*, (e) *polA*, (f) *rep*, (g) *ssb*, and (h) *recA*.

6. About how many Okazaki fragments are synthesized in the replication of an *E. coli* chromosome?

*7. What are the minimum and maximum number of replication forks that occur in a contiguous chromosome of an *E. coli* that is dividing every 25 min; every 80 min?

8. To put the *E. coli* replication system on a human scale, let us imagine that the 20-Å-diameter B-DNA was expanded to 1 m in diameter. If everything were proportionally expanded, then each DNA polymerase III holoenzyme would be about the size of a medium-sized truck. In such an expanded system: (a) How fast would each replisome be moving? (b) How far would each replisome travel during a complete replication cycle? (c) What would be the length of an Okazaki fragment? (d) What would be the average distance a replisome would travel between each error it made? Provide your answers in km/hr and km.

9. Why can't linear duplex DNAs, such as occur in bacteriophage T7, be fully replicated by only *E. coli*-encoded proteins?

*10. What is the half-life of a particular purine base in the human genome assuming that it is subject only to spontaneous depurination? What fraction of the purine bases in a human genome will have depurinated in the course of a single generation (assume 25 years)? The DNAs of ~4000-year-old Egyptian mummies have been sequenced. Assuming that mummification did not slow the rate of DNA depurination, what fraction of the purine bases originally present in the mummy would still be intact today.

11. Why is the methylation of DNA to form O^6-methylguanine mutagenic?

12. A replication fork encountering a single-strand lesion may either dissociate or leave a single-strand gap. The latter process is more likely to occur during lagging strand synthesis than during leading strand synthesis. Explain.

13. The *E. coli* genome contains 1009 Chi sequences. Do these sequences occur at random, and, if not, how much more or less frequently than random do they occur?

14. *Deinococcus radiodurans*, which the *Guiness Book of World Records* has dubbed the world's toughest bacterium, can tolerate doses of ionizing radiation ~3000-fold greater than those

that are lethal to humans (it was first discovered growing in a can of ground meat that had been "sterilized" by radiation). It appears to have several strategies to repair radiation damage to its DNA (which large doses of ionizing radiation fragment to many pieces) including a particularly large number of genes encoding proteins involved in DNA repair and 4 to 10 copies per cell of its genome, which consists of two circular chromosomes and two circular plasmids. Yet, these strategies, alone, do not account for *D. radiodurans'* enormously high radiation resistance. However, in an additional strategy, it organizes its multiple identical dsDNA circles into stacks in which, it is thought, the identical genes in the neighboring circles are aligned side by side. How would this latter strategy help *D. radiodurans* efficiently repair its fragmented DNA?

15. CpG islands occur in eukaryotic genomes at about one-fifth their expected random frequency. Suggest an evolutionary (mutational) process that eliminates CpG islands.

16. Explain why the brief exposure of a cultured eukaryotic cell line to 5-azacytosine results in permanent phenotypic changes to these cells.

17. Explain why chi structures, such as that shown in Fig. 30-67b, have two pairs of equal length arms.

*18. Single-stranded circular DNAs containing a transposon have a characteristic stem-and-double-loop structure such as that shown in Fig. 30-102. What is the physical basis of this structure?

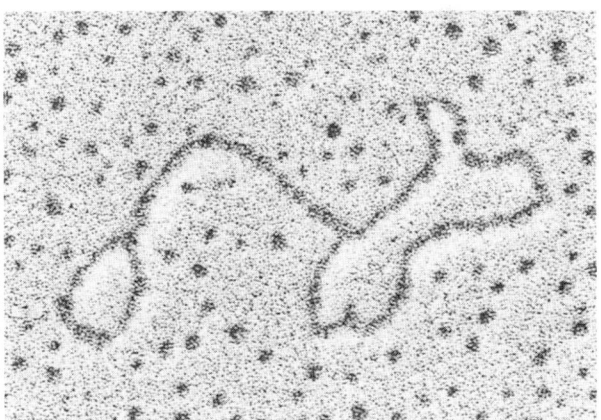

FIGURE 30-102 Electron micrograph of a single-stranded circular DNA containing a transposon. [Courtesy of Stanley N. Cohen, Stanford University School of Medicine.]

19. A composite transposon integrated in a circular plasmid occasionally transposes the DNA comprising the original plasmid rather than the transposon's central region. Explain how this is possible.

*20. Cre recombinase has an additional function to that of circularizing the linear P1 dsDNA (Fig. 30-93). It is also required to resolve the circular dimers of P1 plasmids that result from their recombinational repair during replication, thereby permitting both daughter cells to receive a copy of the P1 plasmid. Using simple line diagrams, outline how these plasmids become dimerized and how Cre resolves them to circular monomers.

Chapter

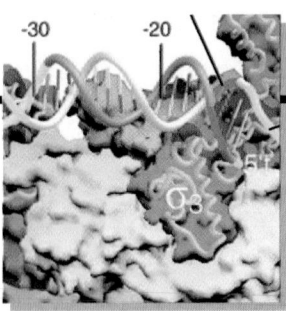

31

Transcription

There are three major classes of RNA, all of which participate in protein synthesis: **ribosomal RNA (rRNA), transfer RNA (tRNA),** and **messenger RNA (mRNA).** All of these RNAs are synthesized under the direction of DNA templates, a process known as **transcription.**

RNA's involvement in protein synthesis became evident in the late 1930s through investigations by Torbjörn Caspersson and Jean Brachet. Caspersson, using microscopic techniques, found that DNA is confined almost exclusively to the eukaryotic cell nucleus, whereas RNA occurs largely in the cytosol. Brachet, who had devised methods for fractionating cellular organelles, came to similar conclusions based on direct chemical analyses. He found, in addition, that the cytosolic RNA-containing particles are also protein rich. Both investigators noted that the concentration of these RNA–protein particles (which were later named ribosomes) is correlated with the rate at which a cell synthesizes protein, implying a relationship between RNA and protein synthesis. Indeed, Brachet even suggested that *the RNA–protein particles are the site of protein synthesis.*

Brachet's suggestion was shown to be valid when radioactively labeled amino acids became available in the 1950s. A short time after injection of a rat with a labeled amino acid, most of the label that had been incorporated in proteins was associated with ribosomes. This experiment also established that *protein synthesis is not immediately directed by DNA because, at least in eukaryotes, DNA and ribosomes are never in contact.*

In 1958, Francis Crick summarized the then dimly perceived relationships among DNA, RNA, and protein by what he called the **central dogma** of molecular biology: *DNA directs its own replication and its transcription to RNA which, in turn, directs its translation to proteins (Fig. 5-21).*

> *The peculiar use of the word "dogma," one definition of which is a religious doctrine that the true believer cannot doubt, stemmed from a misunderstanding. When Crick formulated the central dogma, he was under the impression that dogma meant "an idea for which there was no reasonable evidence."*

We begin this chapter by discussing experiments that led to the elucidation of mRNA's central role in protein synthesis. We then study the mechanism of transcription and its control in prokaryotes. Finally, in the last section, we consider posttranscriptional processing of RNA in both prokaryotes and eukaryotes. Translation is the subject of Chapter 32. Note that these subjects were outlined in Section 5-4. Here we shall delve into much greater detail.

1 ■ THE ROLE OF RNA IN PROTEIN SYNTHESIS

The idea that proteins are specified by mRNA and synthesized on ribosomes arose from the study of **enzyme induction,** a phenomenon in which bacteria vary the synthesis rates of specific enzymes in response to environmental changes. In this section, we discuss the classic experiments that explained the basis of enzyme induction and revealed the existence of mRNA. We shall see that *enzyme induction occurs as a consequence of the regulation* *of mRNA synthesis by proteins that specifically bind to the mRNA's DNA templates.*

A. *Enzyme Induction*

E. coli can synthesize an estimated ~4300 different polypeptides. There is, however, enormous variation in the amounts of these different polypeptides that are produced. For instance, the various ribosomal proteins may each be present in over 10,000 copies per cell, whereas certain regulatory proteins (see below) normally occur in <10 copies per cell. Many enzymes, particularly those involved in basic cellular "housekeeping" functions, are synthesized at a more or less constant rate; they are called **constitutive enzymes.** Other enzymes, termed **adaptive** or **inducible enzymes,** are synthesized at rates that vary with the cell's circumstances.

a. Lactose-Metabolizing Enzymes Are Inducible

Bacteria, as has been recognized since 1900, adapt to their environments by producing enzymes that metabolize certain nutrients, for example, lactose, only when those substances are available. *E. coli* grown in the absence of lactose are initially unable to metabolize this disaccharide. To do so they require the presence of two proteins: **β-galactosidase,** which catalyzes the hydrolysis of lactose to its component monosaccharides,

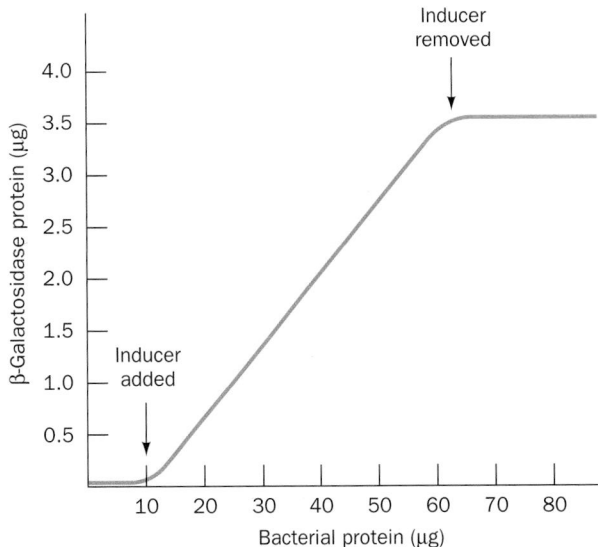

and **galactoside permease** (also known as **lactose permease;** Section 20-4B), which transports lactose into the cell. *E. coli* grown in the absence of lactose contain only a few (<5) molecules of these proteins. Yet, a few minutes after lactose is introduced into their medium, *E. coli* increase the rate at which they synthesize these proteins by ~1000-fold (such that β-galactosidase can account for up to 10% of their soluble protein) and maintain this pace until lactose is no longer available. The synthesis rate then returns to its miniscule **basal level** (Fig. 31-1). *This ability to produce a series of proteins only when the substances they metabolize are present permits bacteria to adapt to their environment without the debilitating need to continuously synthesize large quantities of otherwise unnecessary substances.*

Lactose or one of its metabolic products must somehow trigger the synthesis of the above proteins. Such a substance is known as an **inducer.** The physiological inducer of the lactose system, the lactose isomer **1,6-allolactose,**

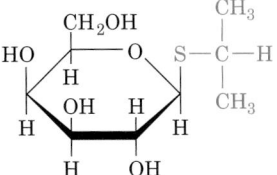

1,6-Allolactose

arises from lactose's occasional transglycosylation by β-galactosidase. Most studies of the lactose system use **isopropylthiogalactoside (IPTG),**

Isopropylthiogalactoside (IPTG)

a potent inducer that structurally resembles allolactose but that is not degraded by β-galactosidase.

Lactose system inducers also stimulate the synthesis of **thiogalactoside transacetylase,** an enzyme that, *in vitro*, transfers an acetyl group from acetyl-CoA to the C6-OH group of a β-thiogalactoside such as IPTG. Since lactose fermentation proceeds normally in the absence of thiogalactoside transacetylase, however, this enzyme's physiological role is unknown.

b. *lac* System Genes Form an Operon

The genes specifying wild-type β-galactosidase, galactoside permease, and thiogalactoside transacetylase are designated Z^+, Y^+, and A^+, respectively. Genetic mapping of the defective mutants Z^-, Y^-, and A^- indicated that

FIGURE 31-1   The induction kinetics of β-galactosidase in *E. coli*. [After Cohn, M., *Bacteriol. Rev.* **21,** 156 (1957).]

these *lac* **structural genes** (genes that specify polypeptides) are contiguously arranged on the *E. coli* chromosome (Fig. 31-2). *These genes, together with the control elements P and O, form a genetic unit called an* **operon,** *specifically the* **lac operon.** The nature of the control elements is discussed below. The role of operons in prokaryotic gene expression is examined in Section 31-3.

c. Bacteria Can Transmit Genes via Conjugation

An important clue as to how *E. coli* synthesizes protein was provided by a mutation that causes the proteins of the *lac* operon to be synthesized in large amounts in the absence of inducer. This so-called **constitutive mutation** occurs in a gene, designated *I*, that is distinct from although closely linked to the genes specifying the *lac* enzymes (Fig. 31-2). What is the nature of the *I* gene product? This riddle was solved in 1959 by Arthur Pardee, Francois Jacob, and Jacques Monod through an ingenious experiment that is known as the **PaJaMo experiment.** To understand this experiment, however, we must first consider **bacterial conjugation.**

Bacterial conjugation is a process, discovered in 1946 by Joshua Lederberg and Edward Tatum, through which some bacteria can transfer genetic information to others. The ability to conjugate ("mate") is conferred on an otherwise indifferent bacterium by a **plasmid** named **F factor** (for *fertility*). Bacteria that possess an F factor (designated F⁺ or male) are covered by hairlike projections known as **F pili.** These bind to cell-surface receptors on bacteria that lack

the F factor (F⁻ or female), which leads to the formation of a cytoplasmic bridge between these cells (Fig. 31-3). The F factor then replicates and, as the newly replicated single strand is formed, it passes through the cytoplasmic bridge to the F⁻ cell where the complementary strand is synthesized (Fig. 31-4). This converts the F⁻ cell to F⁺ so that the F factor is an infectious agent (a bacterial venereal disease?).

On very rare occasions, the F factor spontaneously integrates into the chromosome of the F⁺ cell. In the result-

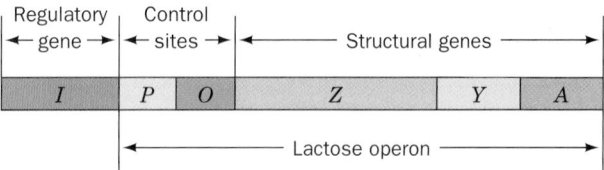

FIGURE 31-2   **Genetic map of the *E. coli lac* operon.** The map shows the genes encoding the proteins mediating lactose metabolism and the genetic sites that control their expression. The *Z*, *Y*, and *A* genes, respectively, specify β-galactosidase, galactoside permease, and thiogalactoside transacetylase.

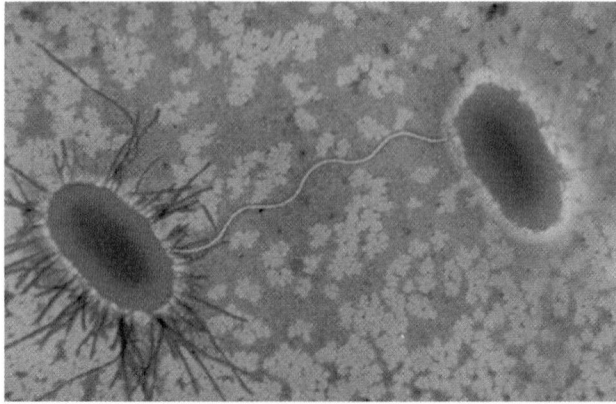

FIGURE 31-3   **Bacterial conjugation.** An electron micrograph shows an F⁺ (*left*) and an F⁻ (*right*) *E. coli* engaged in sexual conjugation. [Dennis Kunkel/Phototake.]

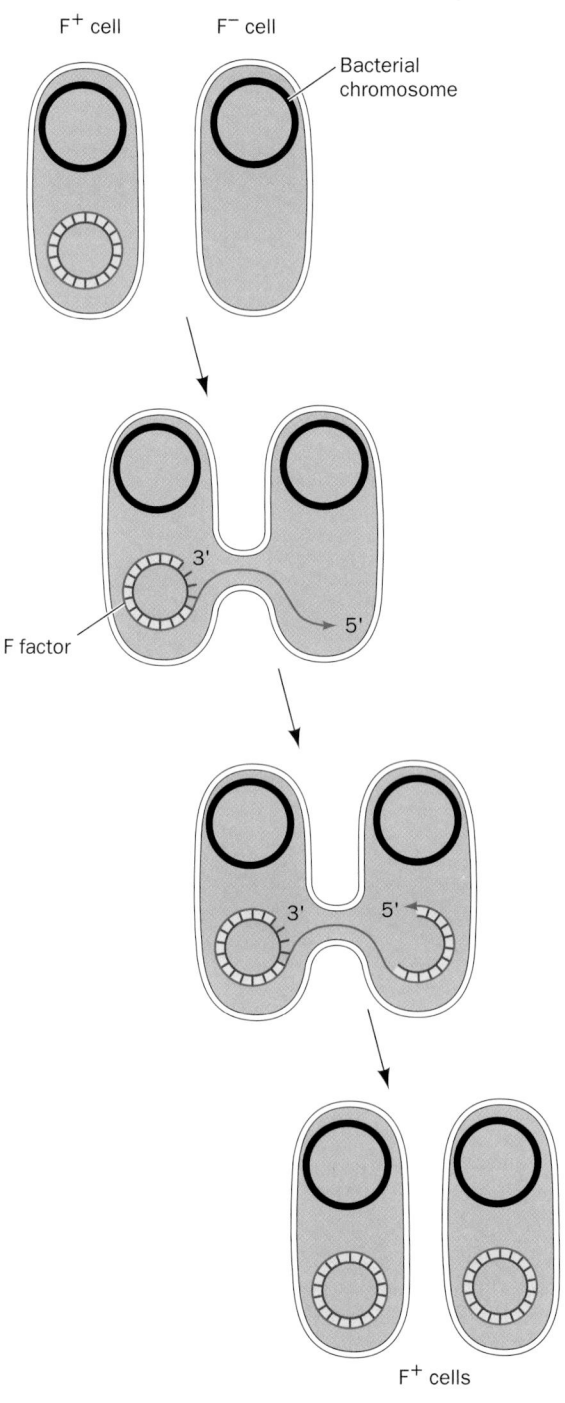

FIGURE 31-4   **Diagram showing how an F⁻ cell acquires an F factor from an F⁺ cell.** A single strand of the F factor is replicated, via the rolling circle mode (Section 30-3B), and is transferred to the F⁻ cell where its complementary strand is synthesized to form a new F factor.

ing **Hfr** (for *h*igh *f*requency of *r*ecombination) cells, the F factor behaves much as it does in the autonomous state. Its replication commences at a specific internal point in the F factor, and the replicated section passes through a cytoplasmic bridge to the F⁻ cell, where its complementary strand is synthesized. In this case, however, the replicated chromosome of the Hfr cell is also transmitted to the F⁻ cell (Fig. 31-5). *Bacterial genes are transferred from the Hfr cell to the F⁻ cell in fixed order.* This is because the F factor in a given Hfr strain is integrated into the bacterial chromosome at a specific site and because only a particular strand of the Hfr chromosomal DNA is replicated and

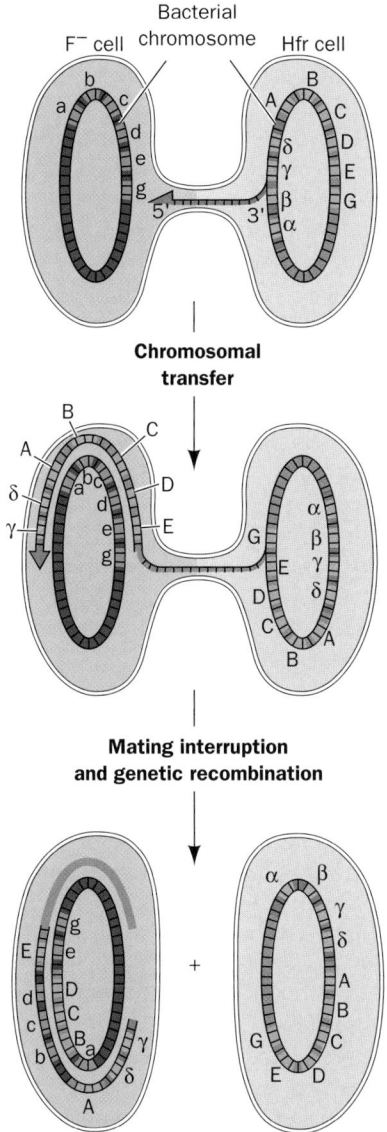

FIGURE 31-5   Transfer of the bacterial chromosome from an Hfr cell to an F⁻ cell and its subsequent recombination with the F⁻ chromosome. Here, Greek letters represent F factor genes, uppercase Roman letters represent bacterial genes from the Hfr cell, and lowercase Roman letters represent the corresponding alleles in the F⁻ cell. Since chromosomal transfer, which begins within the F factor, is rarely complete, the entire F factor is seldom transferred. Hence the recipient cell usually remains F⁻.

transferred to the F⁻ cell. Usually, only part of the Hfr bacterial chromosome is transferred during sexual conjugation because the cytoplasmic bridge almost always breaks off sometime during the ~90 min required to complete the transfer process. In the resulting **merozygote** (a partially diploid bacterium), the chromosomal fragment, which lacks a complete F factor, neither transforms the F⁻ cell to Hfr nor is subsequently replicated. However, the transferred chromosomal fragment recombines with the chromosome of the F⁻ cell (Section 30-6A), thereby permanently endowing the F⁻ cell with some of the traits of the Hfr strain.

The integrated F factor in an Hfr cell occasionally undergoes spontaneous excision to yield an F⁺ cell. In rare instances, the F factor is aberrantly excised such that a portion of the adjacent bacterial chromosome is incorporated in the subsequently autonomously replicating F factor. Bacteria carrying such a so-called **F′ factor** are permanently diploid for its bacterial genes.

d. *lac* Repressor Inhibits the Synthesis of *lac* Operon Proteins

In the PaJaMo experiment, Hfr bacteria of genotype I^+Z^+ were mated to an F⁻ strain of genotype I^-Z^- in the absence of inducer while the β-galactosidase activity of the culture was monitored (Fig. 31-6). At first, as expected,

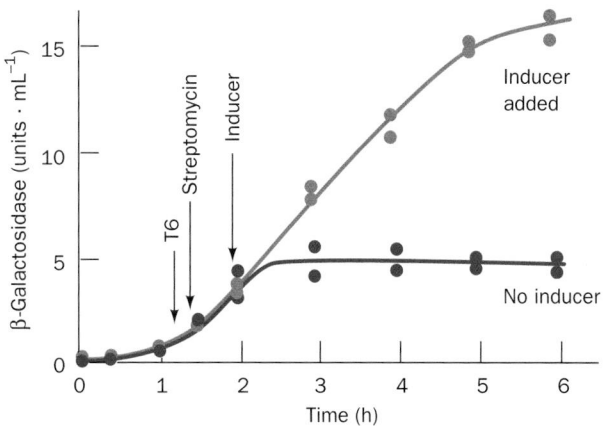

FIGURE 31-6   The PaJaMo experiment. This experiment demonstrated the existence of the *lac* repressor through the appearance of β-galactosidase in the transient merozygotes (partial diploids) formed by mating I^+Z^+ Hfr donors with I^-Z^- F⁻ recipients. The F⁻ strain was also resistant to both **bacteriophage T6** and **streptomycin,** whereas the Hfr strain was sensitive to these agents. Both types of cells were grown and mated in the absence of inducer. After sufficient time had passed for the transfer of the *lac* genes, the Hfr cells were selectively killed by the addition of T6 phage and streptomycin. In the absence of inducer (*lower curve*), β-galactosidase synthesis commenced at around the time at which the *lac* genes had entered the F⁻ cells but stopped after ~1 h. If inducer was added shortly after the Hfr donors had been killed (*upper curve*), enzyme synthesis continued unabated. This demonstrates that the cessation of β-galactosidase synthesis in uninduced cells is not due to the intrinsic loss of the ability to synthesize this enzyme but to the production of a repressor specified by the I^+ gene. [After Pardee, A.B., Jacob, F., and Monod, J., *J. Mol. Biol.* **1,** 173 (1959).]

there was no β-galactosidase activity because the Hfr donors lacked inducer and the F⁻ recipients were unable to produce active enzyme (only DNA passes through the cytoplasmic bridge connecting mating bacteria). About 1 h after conjugation began, however, when the I^+Z^+ genes had just entered the F⁻ cells, β-galactosidase synthesis began and only ceased after about another hour. The explanation for these observations is that the donated Z^+ gene, on entering the cytoplasm of the I^- cell, directs the synthesis of β-galactosidase in a constitutive manner. Only after the donated I^+ gene has had sufficient time to be expressed is it able to repress β-galactosidase synthesis. *The I^+ gene must therefore give rise to a diffusible product, the lac repressor, which inhibits the synthesis of β-galactosidase (and the other lac proteins).* Inducers such as IPTG temporarily inactivate *lac* repressor, whereas I^- cells constitutively synthesize *lac* enzymes because they lack a functional repressor. *Lac* repressor, as we shall see in Section 31-3B, is a protein.

B. *Messenger RNA*

The nature of the *lac* repressor's target molecule was deduced in 1961 through a penetrating genetic analysis by Jacob and Monod. A second type of constitutive mutation in the lactose system, designated O^c (for **operator constitutive**), which complementation analysis (Section 1-4C) has shown to be independent of the I gene, maps between the I and Z genes (Fig. 31-2). In the partially diploid F′ strain O^cZ^-/F O^+Z^+, β-galactosidase activity is inducible by IPTG, whereas the strain O^cZ^+/F O^+Z^- constitutively synthesizes this enzyme. *An O^+ gene can therefore only control the expression of a Z gene on the same chromosome.* The same is true with the Y^+ and A^+ genes.

Jacob and Monod's observations led them to conclude that the proteins are synthesized in a two-stage process:

1. The structural genes on DNA are transcribed onto complementary strands of **messenger RNA (mRNA).**

2. The mRNAs transiently associate with ribosomes, which they direct in polypeptide synthesis.

This hypothesis explains the behavior of the *lac* system that we previously outlined in Section 5-4A (Fig. 5-25; *See* **Guided Exploration 2: Regulation of Gene Expression by the *lac* Repressor System**). *In the absence of inducer, the lac repressor specifically binds to the O gene (the **operator**) so as to prevent the enzymatic transcription of mRNA. On binding inducer, the repressor dissociates from the operator, thereby permitting the transcription and subsequent translation of the lac enzymes.* The operator–repressor–inducer system thereby acts as a molecular switch so that the *lac* operator can only control the expression of *lac* enzymes on the same chromosome. The O^c mutants constitutively synthesize *lac* enzymes because they are unable to bind repressor. The **coordinate** (simultaneous) expression of all three *lac* enzymes under the control of a single operator site arises, as Jacob and Monod theorized, from the transcription of the *lac* operon as a single **polycistronic mRNA** which directs the ribosomal synthesis of each of these proteins (the term **cistron** is a somewhat archaic synonym for gene). This transcriptional control mechanism is further discussed in Section 31-3. [DNA sequences that are on the same DNA molecule are said to be "in cis" (Latin: on this side), whereas those on different DNA molecules are said to be "in trans" (Latin: across). Control sequences such as the O gene, which are only active on the same DNA molecule as the genes they control, are called **cis-acting elements.** Genes such as *lacI*, which specify the synthesis of diffusible products and can therefore be located on a different DNA molecule from the genes they control, are said to direct the synthesis of **trans-acting factors.**]

a. mRNAs Have Their Predicted Properties

The kinetics of enzyme induction, as indicated, for example, in Figs. 31-1 and 31-6, requires that the postulated mRNA be both rapidly synthesized and rapidly degraded. An RNA with such quick turnover had, in fact, been observed in T2-infected *E. coli*. Moreover, the base composition of this RNA fraction resembles that of the viral DNA rather than that of the bacterial RNA (keep in mind that base sequencing techniques would not be formulated for another ~15 years). Ribosomal RNA, which comprises up to 90% of a cell's RNA, turns over much more slowly than mRNA. Ribosomes are therefore not permanently committed to the synthesis of a particular protein (a once popular hypothesis). Rather, *ribosomes are nonspecific protein synthesizers that produce the polypeptide specified by the mRNA with which they are transiently associated.* A bacterium can therefore respond within a few minutes to changes in its environment.

Evidence favoring the Jacob and Monod model rapidly accumulated. Sydney Brenner, Jacob, and Matthew Meselson carried out experiments designed to characterize the RNA that *E. coli* synthesized after T4 phage infection. *E. coli* were grown in a medium containing ¹⁵N and ¹³C so as to label all cell constituents with these heavy isotopes. The cells were then infected with T4 phages and immediately transferred to an unlabeled medium (which contained only the light isotopes ¹⁴N and ¹²C) so that cell components synthesized before and after phage infection could be separated by equilibrium density gradient ultracentrifugation in CsCl solution (Section 6-5B). No "light" ribosomes were observed, which indicates, in agreement with the above-mentioned T2 phage results, that no new ribosomes are synthesized after phage infection.

The growth medium also contained either ³²P or ³⁵S so as to radioactively label the newly synthesized and presumably phage-specific RNA and protein, respectively. Much of the ³²P-labeled RNA was associated, as was postulated for mRNA, with the preexisting "heavy" ribosomes (Fig. 31-7). Likewise, the ³⁵S-labeled proteins were transiently associated with, and therefore synthesized by, these ribosomes.

Sol Spiegelman developed the RNA–DNA hybridization technique (Section 5-3C) in 1961 to characterize the RNA synthesized by T2-infected *E. coli*. He found that this

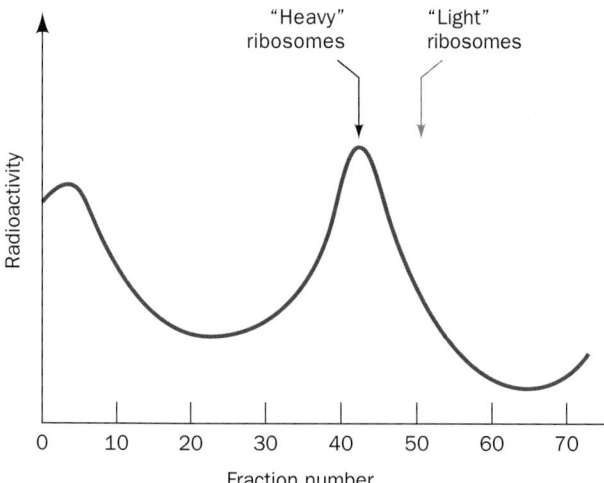

FIGURE 31-7 The distribution, in a CsCl density gradient, of ^{32}**P-labeled RNA that had been synthesized by** *E. coli* **after T4 phage infection.** Free RNA, being relatively dense, bands at the bottom of the centrifugation cell (*left*). Much of the RNA, however, is associated with the ^{15}N- and ^{13}C-labeled "heavy" ribosomes that had been synthesized before the phage infection. The predicted position of unlabeled "light" ribosomes, which are not synthesized by phage-infected cells, is also indicated. [After Brenner, S., Jacob, F., and Meselson, M., *Nature* **190**, 579 (1961).]

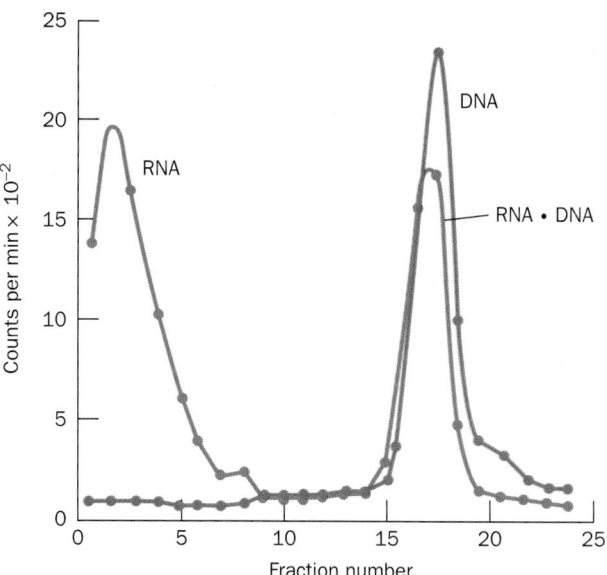

FIGURE 31-8 The hybridization of ^{32}**P-labeled RNA produced by T2-infected** *E. coli* **with** ^{3}**H-labeled T2 DNA.** On radioactive decay, ^{32}P and ^{3}H emit β particles (electrons) with characteristically different energies so that these isotopes can be independently detected. Although free RNA (*left*) in a CsCl density gradient is denser than DNA, much of the RNA bands with the DNA (*right*). This indicates that the two polynucleotides have hybridized and are therefore complementary in sequence. [After Hall, B.D. and Spiegelman, S., *Proc. Natl. Acad. Sci.* **47**, 141 (1961).]

phage-derived RNA hybridizes with T2 DNA (Fig. 31-8) but does not hybridize with DNAs from unrelated phage nor with the DNA from uninfected *E. coli*. This RNA must therefore be complementary to T2 DNA in agreement with Jacob and Monod's prediction; that is, the phage-specific RNA is a messenger RNA. Hybridization studies have likewise shown that mRNAs from uninfected *E. coli* are complementary to portions of *E. coli* DNA. In fact, other RNAs, such as transfer RNA and ribosomal RNA, have corresponding complementary sequences on DNA from the same organism. Thus, *all cellular RNAs are transcribed from DNA templates.*

2 ■ RNA POLYMERASE

RNA polymerase (RNAP), the enzyme responsible for the DNA-directed synthesis of RNA, was discovered independently in 1960 by Samuel Weiss and Jerard Hurwitz. The enzyme couples together the ribonucleoside triphosphates ATP, CTP, GTP, and UTP on DNA templates in a reaction that is driven by the release and subsequent hydrolysis of PP_i:

$$(\text{RNA})_{n \text{ residues}} + \text{NTP} \rightleftharpoons (\text{RNA})_{n+1 \text{ residues}} + \text{PP}_i$$

All cells contain RNAP. In bacteria, one species of this enzyme synthesizes all of the cell's RNA except the RNA primers employed in DNA replication (Section 30-1D). Various bacteriophages encode RNAPs that synthesize

only phage-specific RNAs. Eukaryotic cells contain four or five RNAPs that each synthesize a different class of RNA. In this section we first consider the properties of the bacterial RNAPs and then consider the eukaryotic enzymes.

E. coli RNAP's so-called **holoenzyme** is an ~459-kD protein with subunit composition $\alpha_2\beta\beta'\omega\sigma$ (Table 31-1) in which the β and β′ subunits contain several colinearly arranged homologous segments. Once RNA synthesis has been initiated, however, the σ subunit (also called **σ factor** or σ^{70} since its molecular mass is 70 kD) dissociates from the **core enzyme**, $\alpha_2\beta\beta'\omega$, which carries out the actual polymerization process (see below).

TABLE 31-1. Components of *E. coli* **RNA Polymerase Holoenzyme**

Subunit	Number of Residues	Structural Gene
α	329	*rpoA*
β	1342	*rpoB*
β′	1407	*rpoC*
ω	91	*rpoZ*
σ^{70}	613	*rpsD*

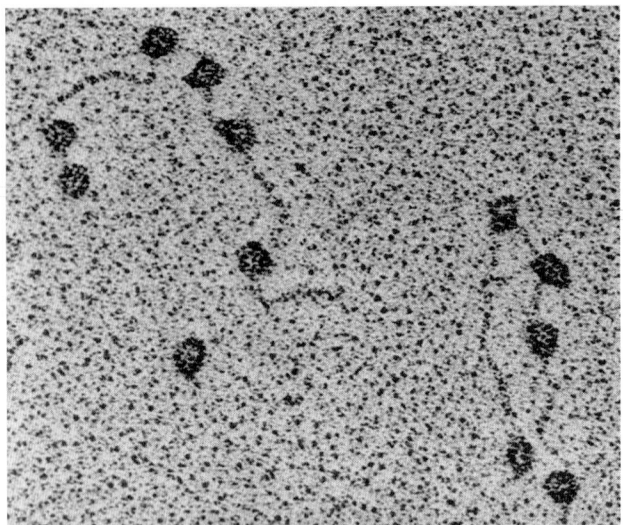

FIGURE 31-9   An electron micrograph of *E. coli* RNA polymerase (RNAP) holoenzyme attached to various promoter sites on bacteriophage T7 DNA. RNAP is one of the largest known soluble enzymes. [From Williams, R.C., *Proc. Natl. Acad. Sci.* **74,** 2313 (1977).]

Electron micrographs (Fig. 31-9) clearly indicate that RNAP, which has a characteristic large size, binds to DNA as a protomer. This large size is presumably a consequence of the holoenzyme's several complex functions including (1) template binding, (2) RNA chain initiation, (3) chain elongation, and (4) chain termination. We discuss these various functions below.

A. *Template Binding*

RNA synthesis is normally initiated only at specific sites on the DNA template. This was first demonstrated through hybridization studies of bacteriophage φX174 DNA with the RNA produced by φX174-infected *E. coli.* Bacteriophage φX174 carries a single strand of DNA known as the (+) strand. On its injection into *E. coli,* the (+) strand directs the synthesis of the complementary (−) strand with which it combines to form a circular duplex DNA known as the replicative form (Section 30-3B). The RNA produced by φX174-infected *E. coli* does not hybridize with DNA from intact phages but does so with the replicative form. Thus only the (−) strand of φX174 DNA, the so-called **antisense strand,** is transcribed, that is, acts as a template; the (+) strand, the **sense strand** (or **coding strand;** so called because it has the same sequence as the transcribed RNA), does not do so. Similar studies indicate that in larger phages, such as T4 and λ, the two viral DNA strands are the antisense (template) strands for different sets of genes. The same is true of cellular organisms.

a. Holoenzyme Specifically Binds to Promoters

*RNA polymerase binds to its initiation sites through base sequences known as **promoters** that are recognized by the corresponding σ factor.* The existence of promoters was first recognized through mutations that enhance or diminish the transcription rates of certain genes, including those of the *lac* operon. *Genetic mapping of such mutations indicated that the promoter consists of an ~40-bp sequence that is located on the 5′ side of the transcription start site.* [By convention, the sequence of template DNA is represented by its sense (nontemplate) strand so that it will have the same directionality as the transcribed RNA. A base pair in a promoter region is assigned a negative or positive number that indicates its position, upstream or downstream in the direction of RNAP travel, from the first nucleotide that is transcribed to RNA; this start site is +1 and there is no 0.] RNA, as we shall see, is synthesized in the 5′ → 3′ direction (Section 31-2C). Consequently, the promoter lies on the "upstream" side of the RNA's starting nucleotide. Sequencing studies indicate that the *lac* promoter (*lacP*) overlaps the *lac* operator (Fig. 31-2).

The holoenzyme forms tight complexes with promoters (dissociation constant $K \approx 10^{-14}M$) and thereby protects the bound DNA segments from digestion by DNase I. The region from about −20 to +20 is protected against exhaustive DNase I degradation. The region extending upstream to about −60 is also protected but to a lesser extent, presumably because it binds holoenzyme less tightly.

Sequence determinations of the protected regions from numerous *E. coli* and phage genes have revealed the "consensus" sequence of *E. coli* promoters (Fig. 31-10). *Their most conserved sequence is a hexamer centered at about the −10 position, the so-called **Pribnow box*** (named after David Pribnow, who pointed out its existence in 1975). It has a consensus sequence of TATAAT in which the leading TA and final T are highly conserved. *Upstream sequences around position −35 also have a region of sequence similarity,* TTGACA, which is most evident in efficient promoters. The sequence of the segment between the −10 and the −35 sites is unimportant but its length is critical; it ranges from 16 and 19 bp in the great majority of promoters. The initiating (+1) nucleotide, which is nearly always A or G, is centered in a poorly conserved CAT or CGT sequence. Most promoter sequences vary considerably from the consensus sequence (Fig. 31-10). Nevertheless, a mutation in one of the partially conserved regions can greatly increase or decrease a promoter's initiation efficiency. In addition, Richard Gourse discovered that certain highly expressed genes contain an A + T-rich segment between positions −40 and −60, the **upstream promoter (UP) element,** which binds to the C-terminal domain of RNAP's α subunits (αCTD; Section 31-3C). The UP element-containing genes include those encoding the ribosomal RNAs, the *rrn* genes (e.g., Fig. 31-10), which collectively account for 60% of the RNA synthesized by *E. coli.* *The rates at which genes are transcribed, which span a range of at least 1000, vary directly with the rate at which their promoters form stable initiation complexes with the holoenzyme.* Promoter mutations that increase or decrease the rate at which the associated gene is transcribed are known as **up mutations** and **down mutations.**

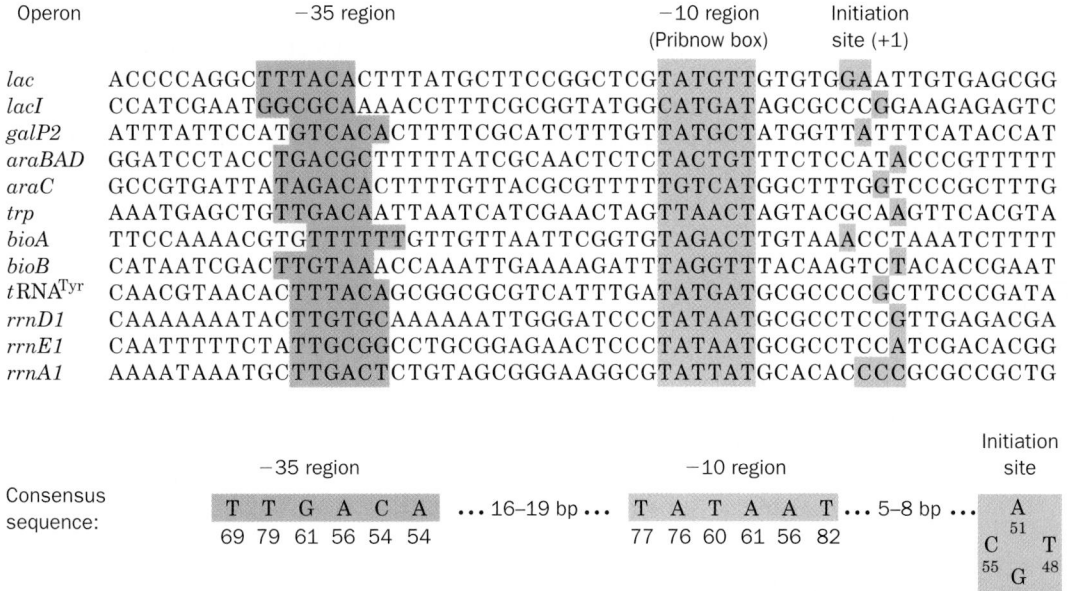

FIGURE 31-10 The sense (nontemplate) strand sequences of selected *E. coli* promoters. A 6-bp region centered around the −10 position (*red shading*) and a 6-bp sequence around the −35 region (*blue shading*) are both conserved. The transcription initiation sites (+1), which in most promoters occurs at a single purine nucleotide, are shaded in green. The bottom row shows the consensus sequence of 298 *E. coli* promoters with the number below each base indicating its percentage occurrence. The downstream portions of the *rrn* genes' UP elements can be seen. [After Rosenberg, M. and Court, D., *Annu. Rev. Genet.* **13**, 321–323 (1979). Consensus sequence from Lisser, S. and Margalit, H., *Nucleic Acids Res.* **21**, 1512 (1993).]

b. Initiation Requires the Formation of an Open Complex

The promoter regions in contact with the holoenzyme were identified by determining where the enzyme alters the susceptibility of the DNA to alkylation by agents such as dimethyl sulfate (DMS), a procedure named **DMS footprinting** (Section 34-3B). These experiments demonstrated that the holoenzyme contacts the promoter mainly around its −10 and −35 regions. These protected sites are both on the same side of the B-DNA double helix as the initiation site, which suggests that holoenzyme binds to only one face of the promoter.

DMS methylates G residues at N7, A residues at N1 and N3, and C residues at N3. Since N1 on A and N3 on C participate in base pairing interactions, however, they can only react with DMS in single-stranded DNA. This differential methylation of single- and double-stranded DNAs provides a sensitive test for DNA strand separation or "melting." Such chemical footprinting studies indicate that the binding of holoenzyme "melts out" the promoter in a region of ~14 bp extending from the middle of the −10 region to just past the initiation site. The need to form this "open complex" explains why promoter efficiency tends to decrease with the number of G · C base pairs in the −10 region; this presumably increases the difficulty in opening the double helix as is required for chain initiation (recall that G · C pairs are more stable than A · T pairs).

Core enzyme, which does not specifically bind promoter (except when it has an UP element), tightly binds duplex DNA (the complex's dissociation constant is $K \approx 5 \times 10^{-12}M$ and its half-life is ~60 min). Holoenzyme, in contrast, binds to nonpromoter DNA comparatively loosely ($K \approx 10^{-7}M$ and a half-life >1 s). Evidently, the σ subunit allows holoenzyme to move rapidly along a DNA strand in search of the σ subunit's corresponding promoter. Once transcription has been initiated and the σ subunit jettisoned, the tight binding of core enzyme to DNA apparently stabilizes the ternary enzyme–DNA–RNA complex.

B. Chain Initiation

The 5′-terminal base of prokaryotic RNAs is almost always a purine with A occurring more often than G. The initiating reaction of transcription is simply the coupling of two nucleoside triphosphates in the reaction

$$pppA + pppN \rightleftharpoons pppApN + PP_i$$

and hence, unlike DNA replication, does not require a primer. Bacterial RNAs therefore have 5′-triphosphate groups as was demonstrated by the incorporation of radioactive label into RNA when it was synthesized with [γ-^{32}P]ATP. Only the 5′ terminus of the RNA can retain the label because the internal phosphodiester groups of RNA are derived from the α-phosphate groups of nucleoside triphosphates.

The difficulty in forming an open complex is reflected in the observation that RNA synthesis is frequently aborted after usually 2 or 3 but up to 12 nucleotides have been joined. However, the holoenzyme does not release the promoter but, rather, reinitiates transcription. Eventually, the open complex forms and processive (continuous) RNA synthesis commences. At this point, σ factor dissociates from the core–DNA–RNA complex and can join with another core to form a new initiation complex. This is demonstrated by a burst of RNA synthesis on addition of core enzyme to a transcribing reaction mixture that initially contained holoenzyme.

a. RNAP Has a Highly Complex Structure

The X-ray structure of *E. coli* RNAP has not been determined. However, Seth Darst has elucidated the X-ray structures of the closely related *Thermus aquaticus* (*Taq*) RNAP core enzyme and holoenzyme. The X-ray structure of *Taq* core enzyme, in agreement with EM studies of *E. coli* RNAP, has the overall shape of a crab claw whose two "pincers" are formed by the β and β′ subunits (Fig. 31-11*a*). The protein is ~150 Å long (parallel to the pincers), ~115 Å high, and ~110 Å deep with the channel (really a cavern) between the two pincers ~27 Å high. A large internal segment of the β′ subunit as well as the C-terminal domains of both α subunits are disordered. The β and β′ subunits extensively interact with one another, particularly at the base of the channel where the active site Mg^{2+} ion is located, which is also where their homologous segments converge. The β′ subunit binds a Zn^{2+} ion via four Cys residues that are invariant in prokaryotes but not in eukaryotes.

The X-ray structure of the *Taq* holoenzyme indicates that its σ subunit ($σ^A$) consists of three flexibly linked domains, $σ_2$, $σ_3$, and $σ_4$, that extend across the top of the holoenzyme (Fig. 31-11*b*). The binding of $σ^A$ causes the core enzyme's pincers to come together so as to narrow the channel between them by ~10 Å. The outer surface of the holoenzyme is almost uniformly negatively charged, whereas those surfaces presumed to interact with nucleic acids, particularly the inner walls of the main channel, are positively charged.

The X-ray structure of *Taq* holoenzyme in complex with a so-called fork-junction promoter DNA fragment (Fig. 31-12*a*) reveals that the DNA lies across one face of the holoenzyme, completely outside of the active site channel (Fig. 31-12*b*). All sequence-specific contacts that the holoenzyme makes with the DNA (with the −10 and −35 regions as well as the so-called extended −10 region just upstream of the −10 region) are mediated by the $σ^A$ subunit via conserved residues. This structure presumably resembles the closed complex.

The foregoing X-ray structures, together with footprinting data, have led Darst to construct models for the closed (RP_c) and open (Rp_o) complexes (Fig. 31-13). RP_c (Fig. 31-13*a*) resembles the holoenzyme–DNA complex (Fig. 31-12*b*) but with an extended length of dsDNA (from −60 to +25) whose upstream end is bent around the enzyme such that the DNA's UP element contacts the 80-residue C-terminal domains of the α subunits (which are disordered in the above X-ray structures). In Rp_o (Fig. 31-13*b*), the template strand of the transcription bubble has slipped into a tunnel formed by the $σ^A$, β, and β′ subunits that is lined with universally conserved basic residues. This tunnel directs the template strand to the active site channel, where it base pairs with the initiating ribonucleotides at the i and $i + 1$ sites near the Mg^{2+} ion.

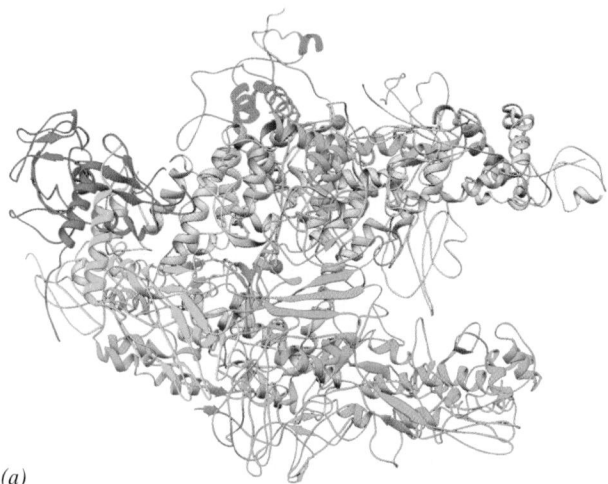

(a)

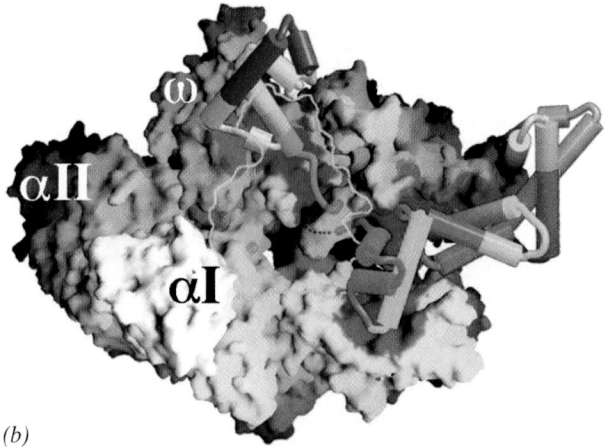

(b)

FIGURE 31-11   X-Ray structure of *Taq* RNAP. (*a*) The core enzyme in which the two α subunits are yellow and green, the β subunit is cyan, the β′ subunit is pink, and the ω subunit is gray. The bound Mg^{2+} and Zn^{2+} ions are represented by red and orange spheres, respectively. Note that residues 156 through 452 of the 1524-residue β′ subunit, which extend from the tip of its "pincer," are disordered and hence not visible. (*b*) The holoenzyme viewed as in Part *a*. Its core enzyme component is represented by its molecular surface with its α and ω subunits gray, its β subunit blue-green, and its β′ subunit pink. The $σ^A$ subunit is represented by its $C_α$ backbone with its helices drawn as cylinders, its various conserved segments in different colors, and its N-terminal end at the right. A portion of the β subunit, which is outlined in yellow, has been deleted to expose the connectivity of the $σ^A$ subunit. Those portions of the β and β′ subunits within 4 Å of the $σ^A$ subunit are colored green and red, respectively. The active site Mg^{2+} ion is represented by a magenta sphere. [Part *a* based on an X-ray structure by and Part *b* courtesy of Seth Darst, The Rockefeller University.]

(a)

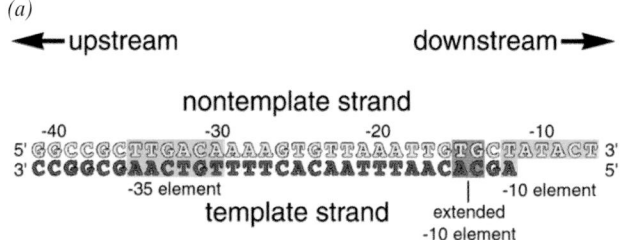

FIGURE 31-12   X-Ray structure of *Taq* holoenzyme in complex with a fork-junction promoter DNA fragment.
(*a*) The sequence of the fork-junction DNA with the numbers indicating the base position relative to the transcription start site, +1. (*b*) The structure of the holoenzyme–fork-junction DNA complex viewed as in Fig. 31-11 but rotated counterclockwise by 45° about an axis perpendicular to the page. The DNA is drawn in ladder form with its template strand green and its nontemplate (sense) strand yellow-green, except that its −35 and −10 elements are yellow and its extended −10 element is red. The holoenzyme is represented by its molecular surface colored according to subunit with αI, αII, and ω gray, β light green, β′ pink, and σ^A orange and partially transparent to reveal its C$_α$ backbone. Those portions of the holoenzyme <4 Å from the DNA (which occurs only on σ^A) are dark green. [Courtesy of Seth Darst, The Rockefeller University. PDBid 1L9Z.]

(b)

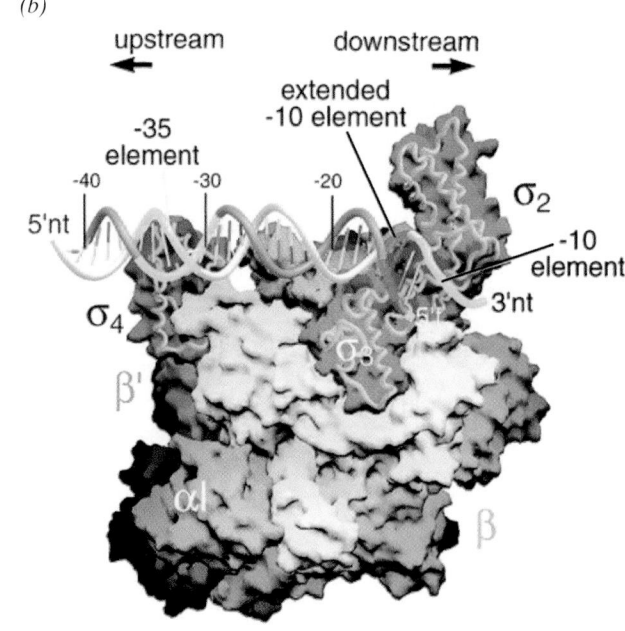

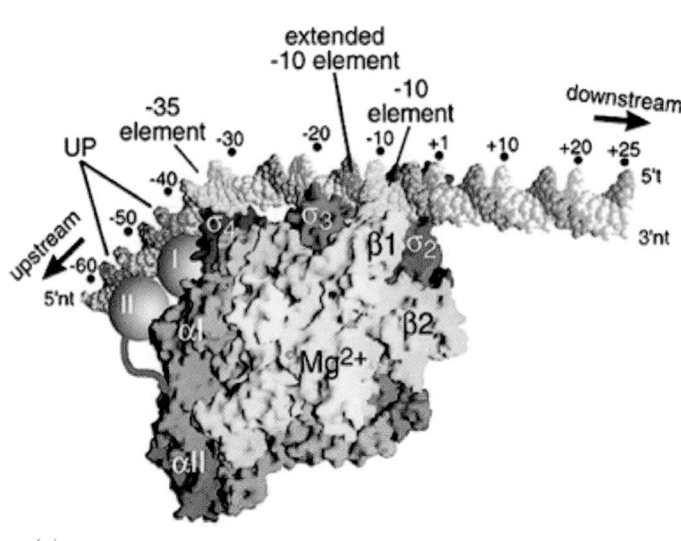

(a)

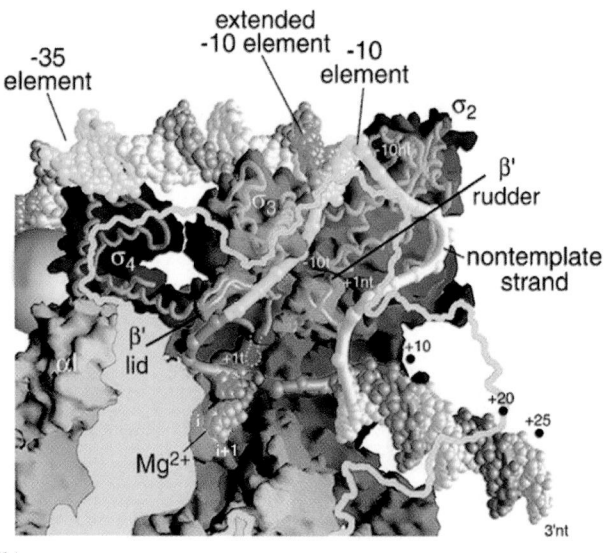

(b)

FIGURE 31-13   Models of the closed (RP$_c$) and open (Rp$_o$) complexes of *Taq* RNAP with promoter-containing DNA extending between positions −60 and +25. (*a*) A model of RP$_c$ in which the protein is represented similarly to that in Fig. 31-12*b* but rotated by 65° about the horizontal axis and the β subunit is partially transparent to show the position of the active site Mg^{2+} ion (*magenta sphere*). The dsDNA is drawn in space-filling form with its template (t) strand green and its nontemplate (nt) strand light green except for the −35 and −10 elements, which are yellow, and the UP element, the extended −10 element, and the transcription start site (+1), which are red. The UP element bends around the enzyme to interact with the C-terminal domains of the α subunits (*gray spheres labeled*

I and II). However, there is no evidence that the downstream segment of the dsDNA interacts with the protein, and hence this DNA is represented as extending away from the protein. (*b*) A model of Rp$_o$ shown in magnified view relative to Part *a* and with obscuring portions of the β subunit removed (its outline is shown as a light green line) to reveal the transcription bubble and its interactions with the active site. Single-stranded portions of the DNA are represented by their linked P atoms. The initiating ribonucleotides at positions *i* and *i* + 1 are drawn in space-filling form (*red and orange*). Note how the downstream segment of the dsDNA emanates from the end of the active site channel. [Courtesy of Seth Darst, The Rockefeller University.]

b. Rifamycins Inhibit Prokaryotic Transcription Initiation

Two related antibiotics, **rifamycin B,** which is produced by *Streptomyces mediterranei,* and its semisynthetic derivative **rifampicin**

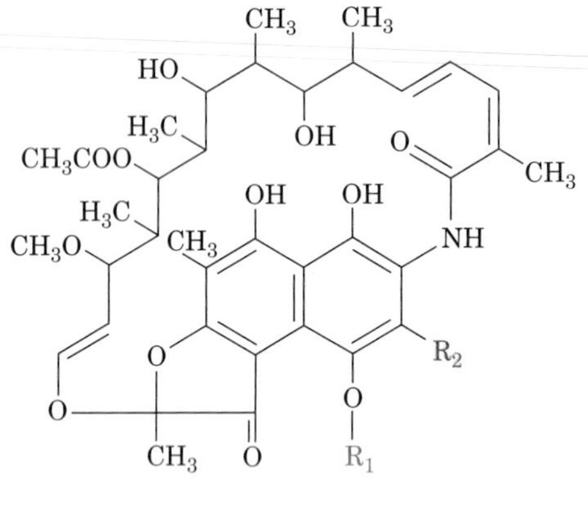

Rifamycin B $R_1 = CH_2COO^-$; $R_2 = H$

Rifampicin $R_1 = H$; $R_2 = CH = N$ N—CH$_3$

specifically inhibit transcription by prokaryotic, but not eukaryotic, RNAPs. This selectivity and their high potency (bacterial RNAP is 50% inhibited by $2 \times 10^{-8}M$ rifampicin) has made them medically useful bacteriocidal agents against gram-positive bacteria and tuberculosis. Indeed, few other antibiotics are effective against tuberculosis, which is reaching epidemic levels in some parts of the world.

The finding that the β subunits of rifamycin-resistant mutants have altered electrophoretic mobilities first demonstrated that this subunit contains the rifamycin-binding site. Rifamycins inhibit neither the binding of RNAP to the promoter nor the formation of the first phosphodiester bond, but they prevent further chain elongation. The inactivated RNAP remains bound to the promoter, thereby blocking its initiation by uninhibited enzymes. Once RNA chain initiation has occurred, however, rifamycins have no effect on the subsequent elongation process. The rifamycins are therefore useful research tools because they permit the transcription process to be dissected into its initiation and its elongation phases.

The X-ray structure of *Taq* core enzyme in complex with rifampicin reveals how this antibiotic inhibits RNAP. Rifampicin binds with close complementary fit but little conformational change in a pocket in the β subunit that is located within the main DNA–RNA channel, ~12 Å distant from the active site Mg^{2+} ion. Model building indicates that the bound rifampicin would sterically interfere with the RNA transcript at positions -2 to -5 in the transcription bubble. Thus, as is observed, rifampicin would not interfere with the initiation of transcription but would mechanically block the extension of the RNA transcript. The residues lining the pocket in which rifampicin binds are highly conserved among prokaryotes but not in eukaryotes, thereby explaining why rifamycins inhibit only prokaryotic RNAPs.

C. *Chain Elongation*

What is the direction of RNA chain elongation; that is, does it occur by the addition of incoming nucleotides to the 3′ end of the nascent (growing) RNA chain (5′ → 3′ growth; Fig. 31-14a) or by their addition to its 5′ terminus

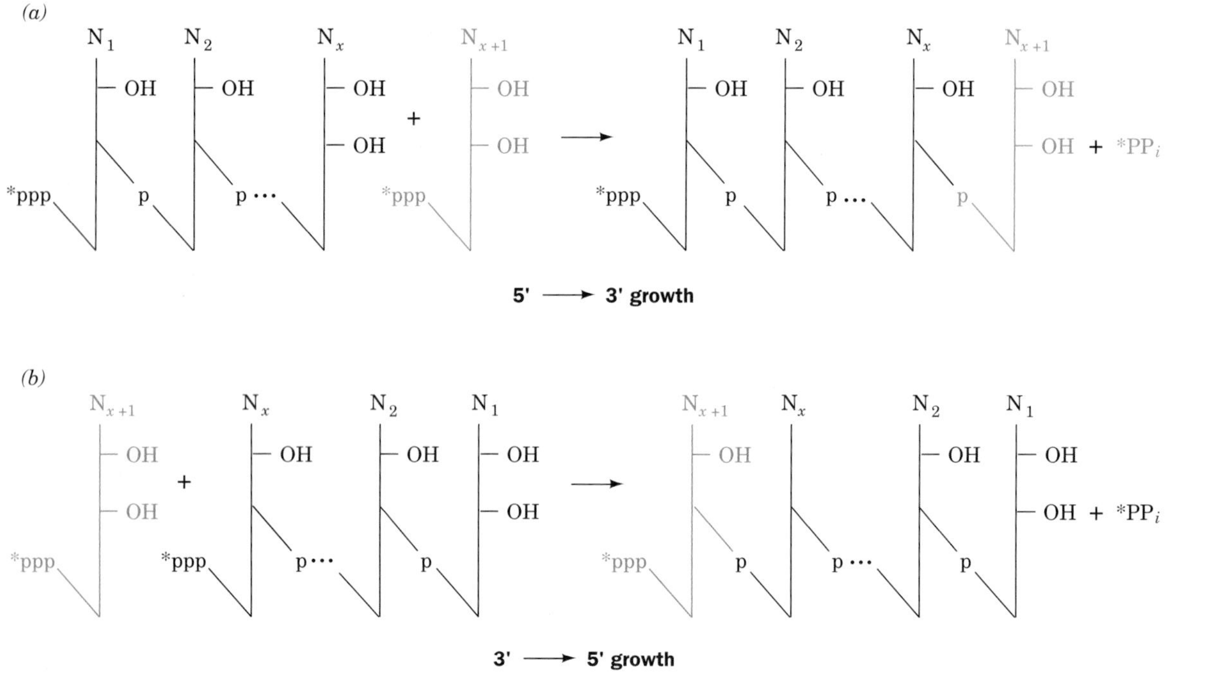

FIGURE 31-14 The two possible modes of RNA chain growth. Growth may occur (*a*) by the addition of nucleotides to the 3′ end and (*b*) by the addition of nucleotides to the 5′ end. RNA polymerase catalyzes the former reaction.

$(3' \rightarrow 5'$ growth; Fig. 31-14*b*)? This question was answered by determining the rate at which the radioactive label from $[\gamma\text{-}^{32}P]GTP$ is incorporated into RNA. For $5' \rightarrow 3'$ elongation, the $5'$ γ-P is permanently labeled and, hence, the chain's level of radioactivity would not change on replacement of the labeled GTP with unlabeled GTP. However, for $3' \rightarrow 5'$ elongation, the $5'$ γ-P is replaced with the addition of every new nucleotide so that, on replacement of labeled with unlabeled GTP, the nascent RNA chains would lose their radioactivity. The former was observed. *Chain growth must therefore occur in the $5' \rightarrow 3'$ direction (Fig. 31-14a), the same direction as DNA is synthesized.* This conclusion is corroborated by the observation that the antibiotic **cordycepin,**

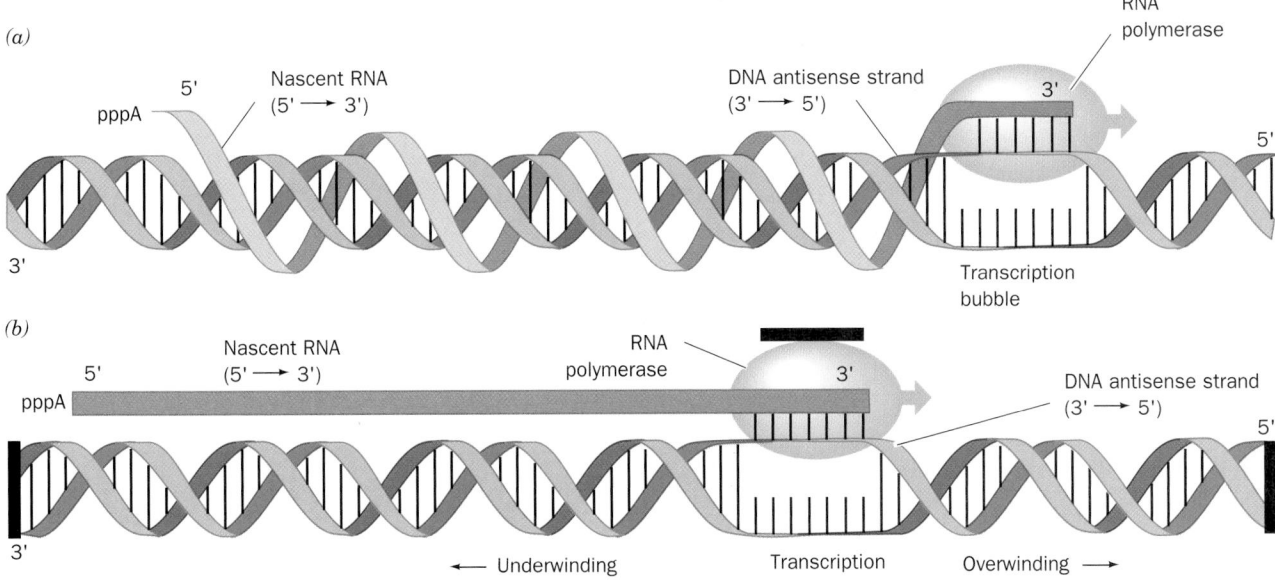

**Cordycepin
(3'-deoxyadenosine)**

an adenosine analog that lacks a $3'$-OH group, inhibits bac-terial RNA synthesis. Its addition to the $3'$ end of RNA, as is expected for $5' \rightarrow 3'$ growth, prevents the RNA chain's further elongation. Cordycepin would not have this effect if chain growth occurred in the opposite direction because it could not be appended to an RNA's $5'$ end.

a. Transcription Supercoils DNA

RNA chain elongation requires that the double-stranded DNA template be opened up at the point of RNA synthesis so that the template strand can be transcribed to its complementary RNA strand. In doing so, the RNA chain only transiently forms a short length of RNA–DNA hybrid duplex, as is indicated by the observation that transcription leaves the template duplex intact and yields single-stranded RNA. The unpaired "bubble" of DNA in the open initiation complex apparently travels along the DNA with the RNAP. There are two ways this might occur (Fig. 31-15):

1. If the RNAP followed the template strand in its helical path around the DNA, the DNA would build up little supercoiling because the DNA duplex would never be unwound by more than about a turn. However, the RNA transcript would wrap around the DNA, once per duplex turn. This model is implausible since it is unlikely that its DNA and RNA could be readily untangled: The RNA would not spontaneously unwind from the long and often circular DNA in any reasonable time, and no known topoisomerase can accelerate this process.

FIGURE 31-15 RNA chain elongation by RNA polymerase. In the region being transcribed, the DNA double helix is unwound by about a turn to permit the DNA's sense strand to form a short segment of DNA–RNA hybrid double helix with the RNA's $3'$ end. As the RNAP advances along the DNA template (here to the right), the DNA unwinds ahead of the RNA's growing $3'$ end and rewinds behind it, thereby stripping the newly synthesized RNA from the template (antisense) strand. (*a*) One way this might occur is by the RNAP following the path of the template strand about the DNA double helix, in which case the transcript would become wrapped about the DNA once per duplex turn. (*b*) A second and more plausible possibility is that the RNA moves in a straight line while the DNA rotates beneath it. In this case the RNA would not wrap around the DNA but the DNA would become overwound ahead of the advancing transcription bubble and unwound behind it (consider the consequences of placing your finger between the twisted DNA strands in this model and pushing toward the right). The model presumes that the ends of the DNA, as well as the RNAP, are prevented from rotating by attachments within the cell (*black bars*). [After Futcher, B., *Trends Genet.* **4,** 271, 272 (1988).]

2. If the RNAP moves in a straight line while the DNA rotates, the RNA and DNA will not become entangled. Rather, the DNA's helical turns are pushed ahead of the advancing transcription bubble so as to more tightly wind the DNA ahead of the bubble (which promotes positive supercoiling), and the DNA behind the bubble becomes equivalently unwound (which promotes negative supercoiling, although note that the linking number of the entire DNA remains unchanged). This model is supported by the observations that the transcription of plasmids in *E. coli* causes their positive supercoiling in gyrase mutants (which cannot relax positive supercoils; Section 29-3C) and their negative supercoiling in topoisomerase I mutants (which cannot relax negative supercoils). In fact, by tethering RNAP to a glass surface and allowing it to transcribe DNA that had been fluorescently labeled at one end, Kazuhiko Kinosita demonstrated, through fluorescence microscopy (using techniques similar to those showing that the F_1F_0-ATPase is a rotary engine; Section 22-3C), that single DNA molecules rotated in the expected direction during transcription.

Inappropriate superhelicity in the DNA being transcribed halts transcription (Section 29-3C). Quite possibly the torsional tension in the DNA generated by negative superhelicity behind the transcription bubble is required to help drive the transcriptional process, whereas too much such tension prevents the opening and maintenance of the transcription bubble.

b. Transcription Occurs Rapidly and Accurately

The *in vivo* rate of transcription is 20 to 50 nucleotides per second at 37°C as indicated by the rate at which *E. coli* incorporate ^{3}H-labeled nucleosides into RNA (cells cannot take up nucleoside triphosphates from the medium). Once an RNAP molecule has initiated transcription and moved away from the promoter, another RNAP can follow suit. The synthesis of RNAs that are needed in large quantities, ribosomal RNAs, for example, is initiated as often as is sterically possible, about once per second (Fig. 31-16).

RNA polymerase, unlike DNA polymerase, cannot rebind a polynucleotide that it has released. Hence, RNA synthesis must be entirely processive, even for the largest eukaryotic genes (which are longer than 2000 kb). Thus, RNAP does not exonucleolytically correct its mistakes as do many DNA polymerases. This accounts for the observations that the error frequency of RNA synthesis, as es-

timated from the analysis of transcripts of simple templates such as poly[d(AT)]·poly[d(AT)], is one wrong base incorporated for every ~10^4 transcribed, whereas, for example, *E. coli* Pol I incorporates one incorrect base in 10^7 (Section 30-2A). The former rate is tolerable because (1) most genes are repeatedly transcribed, (2) the genetic code contains numerous synonyms (Table 5-3), (3) amino acid substitutions in proteins are often functionally innocuous, and (4) large portions of many eukaryotic transcripts are excised in forming the mature mRNAs (intron excision; Section 31-4A).

c. Intercalating Agents Inhibit Both RNA and DNA Polymerases
    Actinomycin D,

Actinomycin D

a useful antineoplastic (anticancer) agent produced by

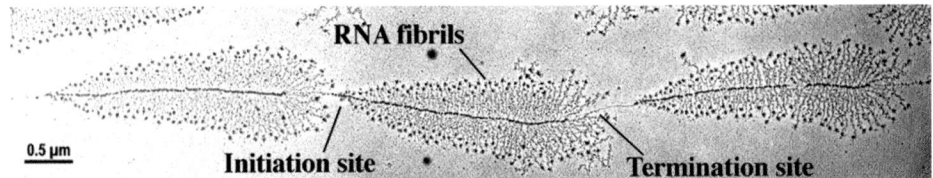

FIGURE 31-16   An electron micrograph of three contiguous ribosomal genes from oocytes of the salamander *Pleurodeles waltl* undergoing transcription. The "arrowhead" structures result from the increasing lengths of the nascent RNA chains as the RNAP molecules synthesizing them move from the initiation site on the DNA to the termination site. [Courtesy of Ulrich Scheer, University of Würzburg, Germany.]

Streptomyces antibioticus, tightly binds to duplex DNA and, in doing so, strongly inhibits both transcription and DNA replication, presumably by interfering with the passage of RNA and DNA polymerases. The X-ray structure of actinomycin D in complex with a duplex DNA composed of two strands of the self-complementary octamer d(GAAGCTTC) reveals that the DNA assumes a B-like conformation in which the actinomycin's **phenoxazone ring system,** as had previously been shown, is intercalated between the DNA's central G · C base pairs (Fig. 31-17). Consequently, the DNA helix is unwound by 23° at the intercalation site and the central G · C base pairs are separated by 7.0 Å. The DNA helix is severely distorted from the normal B-DNA conformation such that its minor groove is wide and shallow in a manner resembling that of A-DNA. Actinomycin D's two chemically identical cyclic **depsipeptides** (having both peptide bonds and ester linkages), which assume different conformations, extend in opposite directions from the intercalation site along the minor groove of the DNA. The complex is stabilized through the formation of base–peptide and phenoxazone–sugar-phosphate backbone hydrogen bonds, as well as by hydrophobic interactions, in a way that explains the preference of actinomycin D to bind to DNA with its phenoxazone ring intercalated between the base pairs of a 5'-GC-3' sequence. Several other intercalation agents, including ethidium and proflavin (Sections 6-6C and 29-3A), also inhibit nucleic acid synthesis, presumably by similar mechanisms.

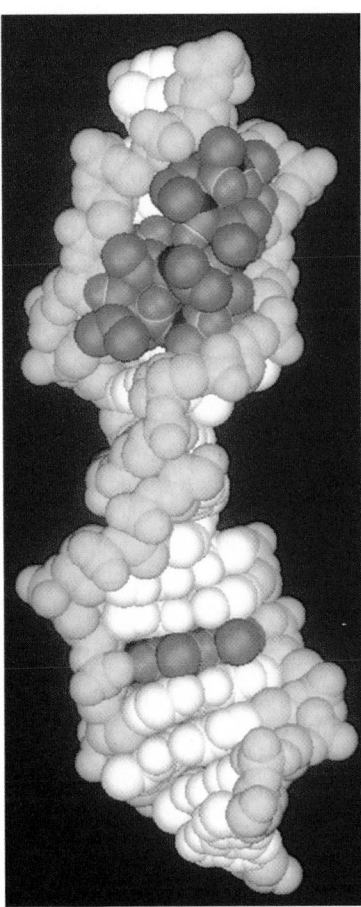

FIGURE 31-17   X-Ray structure of actinomycin D in complex with a duplex DNA of self-complementary sequence d(GAAGCTTC). The complex is shown in space-filling form in which the DNA's sugar–phosphate backbone is yellow, its bases are white, and the actinomycin D is colored according to atom type with C green, N blue, and O red. The two symmetry-related DNA molecules that are shown stack vertically to form a pseudocontinuous helix. The upper DNA is viewed toward its minor groove into which its bound actinomycin D's two cyclic depsipeptides are tightly wedged. The lower DNA (which is turned 180° about its helix axis relative to the upper DNA) is viewed toward its major groove, into which its bound actinomycin D's intercalated phenoxazone ring system projects from the minor groove side. [Based on an X-ray structure by Fusao Takusagawa, University of Kansas. PDBid 172D.]

D. *Chain Termination*

Electron micrographs such as Fig. 31-16 suggest that DNA contains specific sites at which transcription is terminated. The transcriptional termination sequences of many *E. coli* genes share two common features (Fig. 31-18*a*):

1. A series of 4 to 10 consecutive A · T's with the A's on the template strand. The transcribed RNA is terminated in or just past this sequence.

2. A G + C-rich region with a palindromic (2-fold symmetric) sequence that is immediately upstream of the series of A · T's.

(a)

G · C   A · T
rich region rich region

```
5' ··· NNAAGCGCCGNNNNCCGGCGCTTTTTTNNN ··· 3'
3' ··· NNTTCGCGGCNNNNGGCCGCGAAAAAANNN ··· 5'   DNA template

5' ··· NNAAGCGCCGNNNNCCGGCGCUUUUUU—OH 3'      RNA transcript
```

FIGURE 31-18   A hypothetical strong (efficient) *E. coli* terminator. The base sequence was deduced from the sequences of several transcripts. (*a*) The DNA sequence together with its corresponding RNA. The A · T-rich and G · C-rich sequences are shown in blue and red, respectively. The 2-fold symmetry axis (*green lenticular symbol*) relates the flanking shaded segments that form an inverted repeat. (*b*) The RNA hairpin structure and poly(U) tail that trigger transcription termination. [After Pribnow, D., *in* Goldberger, R.F. (Ed.), *Biological Regulation and Development,* Vol. 1, p. 253, Plenum Press (1979).]

(b)

```
        N
      /   \
    N       N
    |       |
    N       C
     \     /
      G · C
      C · G
      C · G
      G · C
      C · G
      G · C
      A · U
      A · U
···NNNN        UUUU—OH 3'
```

The RNA transcript of this region can therefore form a self-complementary "hairpin" structure that is terminated by several U residues (Fig. 31-18*b*).

The stability of a terminator's G + C-rich hairpin and the weak base pairing of its oligo(U) tail to template DNA appear to be important factors in ensuring proper chain termination. In fact, model studies have shown that oligo(dA · rU) forms a particularly unstable hybrid helix although oligo(dA · dT) forms a helix of normal stability. The formation of the G + C-rich hairpin causes RNAP to pause for several seconds at the termination site. This, it has been proposed, induces a conformational change in the RNAP, which permits the nontemplate DNA strand to displace the weakly bound oligo(U) tail from the template strand, thereby terminating transcription. Consistent with this notion is the observation that mutations in the termination site that alter the strengths of these associations reduce the efficiency of chain termination and often eliminate it. Termination is similarly diminished when *in vitro* transcription is carried out with GTP replaced by **inosine triphosphate (ITP):**

Inosine triphosphate (ITP)

I · C pairs are weaker than those of G · C because the hypoxanthine base of I, which lacks the 2-amino group of G, can only make two hydrogen bonds to C, thereby decreasing the hairpin's stability.

Despite the foregoing, experiments by Michael Chamberlin in which segments of highly efficient terminators were swapped via recombinant DNA techniques indicate that the RNA terminator hairpin and U-rich 3′ tail do not function independently of their upstream and downstream flanking regions. Indeed, terminators that lack a U-rich segment can be highly efficient when joined to the appropriate sequence immediately downstream from the termination site. Moreover, mutations in the β subunit of RNAP can both increase and decrease termination efficiency. This has led to an alternative model for termination in which the RNA transcript is stably bound to the RNAP via interactions with two single strand-specific RNA binding sites. Termination then occurs through the formation of a stable RNA hairpin that reduces the RNA's binding to one RNA-binding site. In this latter model, termination can occur on both single-stranded and double-stranded templates.

To distinguish between the above two models, Chamberlin constructed a system in which termination took place on a single-stranded template. Because RNA polymerase can only initiate transcription on double-stranded DNA, this was done by allowing RNAP to initiate transcription on dsDNA containing an initiation and a termination site, "walking" the RNAP past the start site to a specific site upstream of the terminator by adding and removing appropriate subsets of NTPs, and then adding **exonuclease III** (a processive 3′ → 5′ double strand-specific exonuclease), which digested both the template and the nontemplate strands from their 3′ ends to the stalled RNAP. The exonuclease III was then removed and NTPs were added, thereby permitting the RNA polymerase to resume transcription on the now single-stranded template. Since the terminator preceded the 5′ end of the template, transcripts that had normally terminated would be shorter than unterminated transcripts that had simply run off the end of the template. The results of these experiments, using three different terminators, revealed that termination is essentially as efficient on single-stranded templates as it is on double-stranded templates. Evidently, neither the nontemplate strand nor the transcription bubble is required for termination by *E. coli* RNAP.

a. Termination Often Requires the Assistance of Rho Factor

The above termination sequences induce the spontaneous termination of transcription. Around half the termination sites, however, lack any obvious similarities and are unable to form strong hairpins; *they require the participation of a protein known as **Rho factor** to terminate transcription.* The existence of Rho factor was suggested by the observation that *in vivo* transcripts are often shorter than the corresponding *in vitro* transcripts. Rho factor, a hexameric protein of identical 419-residue subunits, enhances the termination efficiency of spontaneously terminating transcripts as well as inducing the termination of nonspontaneously terminating transcripts.

Several key observations have led to a model of Rho-dependent termination:

1. Rho factor is a helicase that unwinds RNA–DNA and RNA–RNA double helices. This process is powered by the hydrolysis of NTPs to NDPs + P_i with little preference for the identity of the base. NTPase activity is required for Rho-dependent termination as is demonstrated by its *in vitro* inhibition when the NTPs are replaced by their β,γ-imido analogs,

β,γ-Imido nucleoside triphosphate

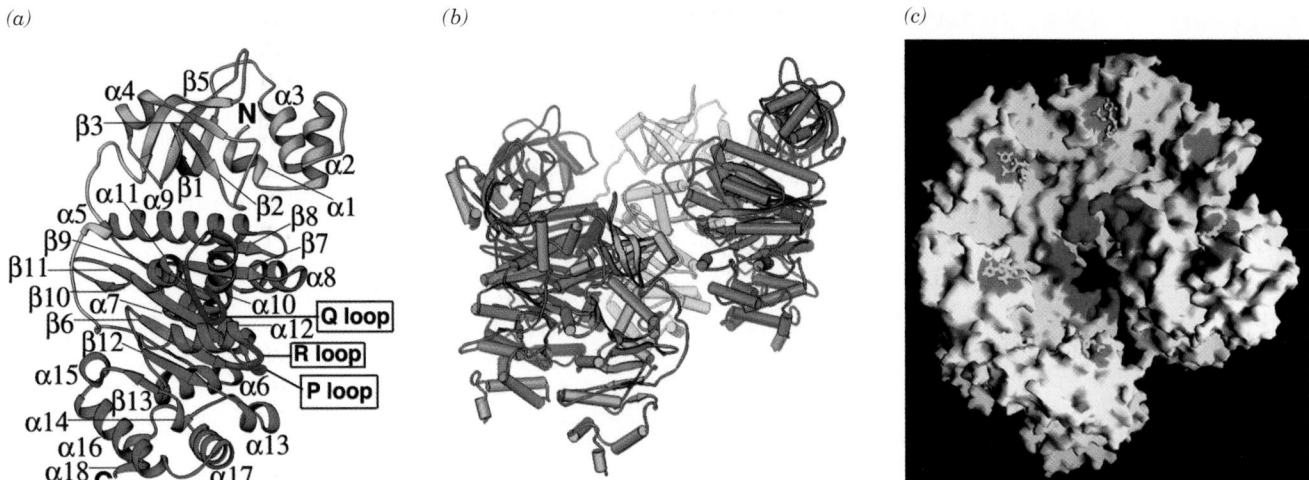

FIGURE 31-19   X-Ray structure of Rho factor in complex with RNA. (*a*) The Rho protomer with its N-terminal domain cyan, its C-terminal domain red, and their connecting linker yellow. The P loop (*blue*), Q loop (*magenta*), and R loop (*green*) have been implicated in mRNA binding and translocation as well as NTP binding. (*b*) The Rho hexamer. Its six subunits, each of which are drawn in a different color, form an open lock washer-shaped hexagonal ring. The yellow subunit is viewed similarly to that in Part *a*. (*c*) The solvent-accessible surface of the Rho hexamer as viewed from the top of Part *b*. The primary RNA binding sites, which occur on the N-terminal domain are cyan and the secondary RNA binding sites, which occur on the C-terminal domain, are magenta. The RNA, which is drawn in yellow stick form, is bound to the primary RNA binding sites. It is only partially visible in the X-ray structure. [Courtesy of James Berger, University of California at Berkeley. PDBid 1PVO.]

substances that are RNAP substrates but cannot be hydrolyzed by Rho factor.

2. Genetic manipulations indicate that Rho-dependent termination requires the presence of a specific recognition sequence on the newly transcribed RNA upstream of the termination site. The recognition sequence must be on the nascent RNA rather than the DNA as is demonstrated by Rho's inability to terminate transcription in the presence of pancreatic RNase A. The essential features of this termination site have not been fully elucidated; the construction of synthetic termination sites indicates that it consists of 80 to 100 nucleotides that lack a stable secondary structure and contain multiple regions that are rich in C and poor in G.

These observations suggest that Rho factor attaches to nascent RNA at its recognition sequence and then migrates along the RNA in the 5′ → 3′ direction until it encounters an RNAP paused at the termination site (without the pause, Rho might not be able to overtake the RNA polymerase). There, Rho unwinds the RNA–DNA duplex forming the transcription bubble, thereby releasing the RNA transcript. Rho-terminated transcripts have 3′ ends that typically vary over a range of ~50 nucleotides. This suggests that Rho somehow pries the RNA away from its template DNA rather than "pushing" an RNA release "button."

Each Rho subunit consists of two domains that can be separated by proteolysis: Its N-terminal domain binds sin-gle-stranded polynucleotides and its C-terminal domain, which is homologous to the α and β subunits of the F_1-ATPase (Section 22-3C), binds an NTP. The X-ray structure of Rho in complex with ADPNP and an 8-nt RNA, $(UC)_4$, determined by James Berger, reveals that the N-terminal domain consists of a 3-helix bundle mounted on a 5-stranded β barrel and that the C-terminal domain consists of seven parallel β strands sandwiched between several helices (Fig. 31-19*a*). Rho forms the expected hexamer but, unlike the ring-shaped F_1-ATPase, forms a lock washer-shaped helix that is 120 Å in diameter with an ~30-Å diameter central hole and whose first and sixth subunits are separated by a 12-Å gap and rise of 45 Å along the helix axis (Fig. 31-19*b*). The RNA, which is only partially visible in the structure, binds to the so-called primary RNA binding sites on the N-terminal domains that face the interior of the helix, as do the so-called secondary RNA binding sites on the C-terminal domain that have been implicated in mRNA translocation and unwinding (Fig. 31-19*c*). Since electron microscopic images of Rho show both closed as well as notched hexameric rings, the X-ray structure probably represents an open state that is poised to bind mRNA that has entered its central cavity through the notch. Presumably, mRNA binding would cause the hexameric ring to close. Ensuing cycles of ATP hydrolysis, with their attendant conformational changes, would then propel Rho along the mRNA in the 5′→3′ direction.

E. *Eukaryotic RNA Polymerases*

Eukaryotic nuclei, as Robert Roeder and William Rutter discovered, contain three distinct types of RNAPs that differ in the RNAs they synthesize:

1. RNA polymerase I (RNAP I; also called **Pol I** and **RNAP A),** which is located in the nucleoli (dense granular bodies in the nuclei that contain the ribosomal genes; Section 31-4B), synthesizes precursors of most ribosomal RNAs (rRNAs).

2. RNA polymerase II (RNAP II; also called **Pol II** and **RNAP B),** which occurs in the nucleoplasm, synthesizes mRNA precursors.

3. RNA polymerase III (RNAP III; also called **Pol III** and **RNAP C),** which also occurs in the nucleoplasm, synthesizes the precursors of 5S ribosomal RNA, the tRNAs, and a variety of other small nuclear and cytosolic RNAs.

Eukaryotic nuclear RNAPs have considerably greater subunit complexity than those of prokaryotes. These enzymes have molecular masses of up to 600 kD and, as is indicated in Table 31-2, each contains two nonidentical "large" (>120 kD) subunits comprising ~65% of its mass that are homologs of the prokaryotic RNAP β' and β subunits, and up to 12 additional "small" (<50 kD) subunits, two of which are homologs of prokaryotic RNAP α, and one of which is a homolog of prokaryotic RNAP ω. Of these small subunits, five are identical in all three eukaryotic RNAPs and two others (the RNAP α homologs) are identical in RNAPs I and III. Two of the RNAP II subunits, Rbp4 and Rbp7, are not essential for activity and, in fact, are present in RNAP II in less than stoichiometric amounts. (Curiously, Rbp7 has a 102-residue segment that is 30% identical to a portion of σ^{70}, the predominant *E. coli* σ factor.) Thus 10 of the 12 RNAP II subunits are either identical or closely similar to subunits of RNAPs I and III (Table 31-2). Moreover, the sequences of these subunits are highly conserved (~50% identical) across species from yeast to humans (and to a lesser extent between eukaryotes and bacteria). In fact, in all ten cases tested, a human RNAP II subunit could replace its counterpart in yeast without loss of cell viability.

Rpb1, the β' homolog in RNAP II, has an extraordinary C-terminal domain **(CTD).** In mammals, it contains 52 highly conserved repeats of the heptad PTSPSYS (26 repeats in yeast with other eukaryotes having intermediate values). Five of the seven residues in these particularly hydrophilic repeats bear hydroxyl groups and at least 50 of them, predominantly those on Ser residues, are subject to reversible phosphorylation by **CTD kinases** and **CTD phosphatases.** RNAP II initiates transcription only when the CTD is unphosphorylated but commences elongation only after the CTD has been phosphorylated, which sug-

TABLE 31-2.   RNA Polymerase Subunits[a]

S. cerevisiae RNAP I (14 subunits)	*S. cerevisiae* RNAP II (12 subunits)	*S. cerevisiae* RNAP III (15 subunits)	*E. coli* RNAP Core (5 subunits)	Class[b]
Rpa1 (A190)	Rbp1 (B220)	Rpc1 (C160)	β'	Core
Rpa2 (A135)	Rbp2 (B150)	Rpc2 (C128)	β	Core
Rpc5 (AC40)	Rpb3 (B44.5)	Rpc5 (AC40)	α	Core
Rpc9 (AC19)	Rpb11 (B13.6)	Rpc9 (AC19)	α	Core
Rbp6 (ABC23)	Rbp6 (ABC23)	Rpb6 (ABC23)	ω	Core/common
Rpb5 (ABC27)	Rpb5 (ABC27)	Rpb5 (ABC27)		Common
Rpb8 (ABC14.4)	Rpb8 (ABC14.4)	Rpb8 (ABC14.4)		Common
Rbp10 (ABC10β)	Rpb10 (ABC10β)	Rpb10 (ABC10β)		Common
Rbp12 (ABC10α)	Rpb12 (ABC10α)	Rpb12 (ABC10α)		Common
Rpa9 (A12.2)	Rpb9 (B12.6)	Rpc12 (C11)		
Rpa8 (A14)[c]	Rpb4 (B32)	—		
Rpa4 (A43)[c]	Rpb7 (B16)	Rpc11 (C25)		
+2 others[d]		+4 others[d]		

[a]Homologous subunits occupy the same row. In the alternative subunit names in parentheses, the letter(s) indicates the RNAPs in which the subunit is a component (A, B, and C for RNAPs I, II, and III) and the numbers indicate its approximate molecular mass in kD.

[b]Core: sequence partially homologous in all RNAPs; common: shared by all eukaryotic RNAPs.

[c]Potential homologs of Rbp4 and Rbp7.

[d]Rpa3 (A49) and Rpa5 (A34.5) in RNAP I and Rpc3 (C74), Rpc4 (C53), Rpc6 (C34), and Rpc8 (C31) in RNAP III.

Source: Mainly Cramer, P., *Curr. Opin. Struct. Biol.* **12,** 89 (2002).

gests that this process triggers the conversion of RNAP II's initiation complex to its elongation complex. Charge–charge repulsions between nearby phosphate groups probably cause a highly phosphorylated CTD to project as far as 500 Å from the globular portion of RNAP II. Indeed, as we shall see, the phosphorylated CTD provides the binding sites for numerous auxiliary factors that have essential roles in the transcription process.

In contrast to the somewhat smaller prokaryotic RNAP holoenzymes, eukaryotic RNAPs do not independently bind their target DNAs. Rather, as we shall see in Section 34-3B, they are recruited to their target promoters through the mediation of complexes of transcription factors and their ancillary proteins that, in the case of RNAP II–transcribed genes, are so large and complicated that they collectively dwarf RNAP II.

In addition to the foregoing nuclear enzymes, eukaryotic cells contain separate mitochondrial and (in plants) chloroplast RNAPs. These small (~100 kD) single-subunit RNAPs, which resemble those encoded by certain bacteriophages, are much simpler than the nuclear RNAPs although they catalyze the same reaction.

a. X-Ray Structures of Yeast RNAP II Reveal a Transcribing Complex

In a crystallographic tour de force, Roger Kornberg determined the X-ray structure of yeast (*S. cerevisiae*) RNAP II that lacks its nonessential Rpb4 and Rpb7 subunits (Fig. 31-20). This enzyme, as expected, resembles *Taq* RNAP (Fig. 31-11) in its overall crab claw-like shape and in the positions and core folds of their homologous subunits although, of course, RNAP II is somewhat larger than and has several subunits that have no counterpart in *Taq* RNAP. RNAP II binds two Mg^{2+} ions at its active site (although one appears to be weakly bound and hence is only faintly visible in the X-ray structure; perhaps it accompanies the incoming NTP) in the vicinity of five conserved acidic residues, which suggests that RNAPs catalyze RNA elongation via a 2-metal ion mechanism similar to that employed by DNA polymerases (Section 30-2A). As is the case with *Taq* RNAP, the surface of RNAP II is almost entirely negatively charged except for the DNA-binding cleft and the region about the active site, which are positively charged.

Although, as mentioned above, RNAP II does not normally initiate transcription by itself, Kornberg found that it will do so on a dsDNA bearing a 3′ single-stranded tail at one end. Consequently, incubating yeast RNAP II with

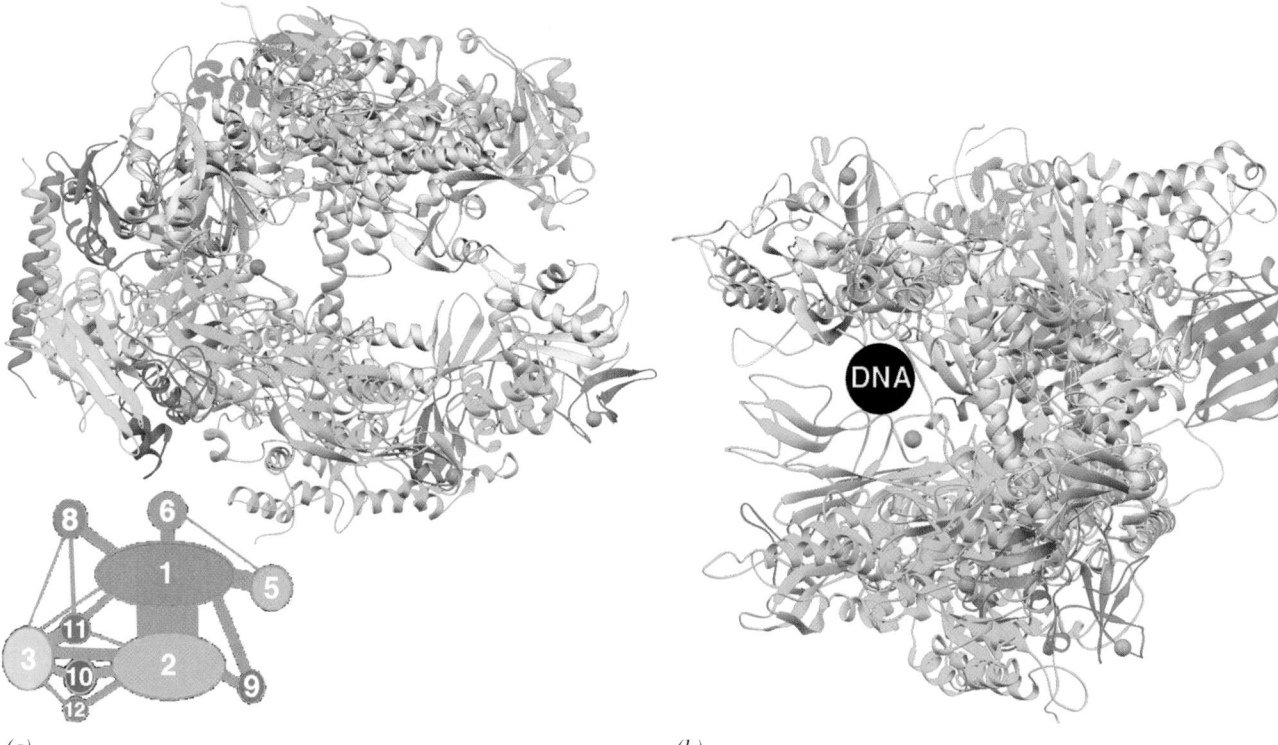

(a)   *(b)*

FIGURE 31-20 The X-Ray structure of yeast RNAP II that lacks its Rpb4 and Rpb7 subunits. (*a*) The enzyme is oriented similarly to *Taq* RNAP in Fig. 31-11 and its subunits are colored as is indicated in the accompanying diagram, with the subunits homologous to those of *Taq* RNAP given the same colors. The strongly bound Mn^{2+} ion (physiologically Mg^{2+}) that marks the active site is shown as a red sphere and the enzyme's 8 bound Zn^{2+} ions are shown as orange spheres. The

Rpb1 C-terminal domain (CTD) is not visible due to disorder. In the accompanying diagram, the area of each numbered ellipsoid is proportional to the corresponding subunit's size and the width of each gray line connecting a pair of subunits is proportional to the surface area of their interface. (*b*) View of the enzyme from the right in Part *a* showing its DNA binding cleft. The black circle has the approximate diameter of B-DNA. [Based on an X-ray structure by Roger Kornberg. PDBid 1I50.]

the DNA shown in Fig. 31-21*a* and all NTPs but UTP yielded the DNA · RNA hybrid helix diagrammed in Fig. 31-21*a* bound to RNAP II. The X-ray structure of this paused transcribing complex revealed, as expected, that the dsDNA had bound in the enzyme's cleft (Fig. 31-21*b, c;* transcription resumed on soaking the crystals in UTP, thereby demonstrating that the crystalline complex was active). In comparison with the X-ray structure of RNAP II alone, a massive (~50 kD) portion of Rpb1 and Rpb2 named the "clamp" has swung down over the DNA to trap it in the cleft, in large part accounting for the enzyme's essentially infinite processivity. The mainly rigid motion of the clamp is mediated by conformational changes at five so-called switch regions at the base of the clamp in which three of these switches, which are disordered in the structure of RNAP II alone, become ordered in the transcribing complex.

The DNA unwinds by three bases before entering the active site (which is contained on Rpb1). Past this point, however, a portion of Rpb2 dubbed the "wall" directs the

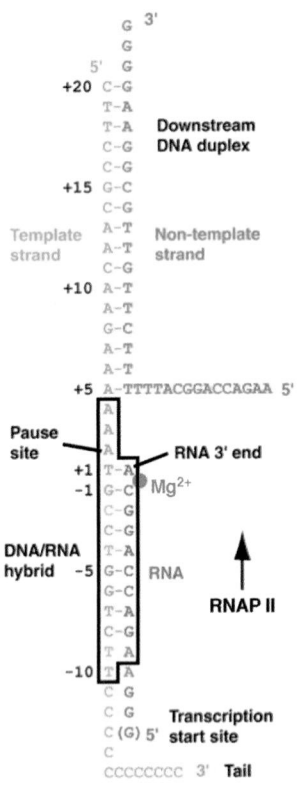

(a)

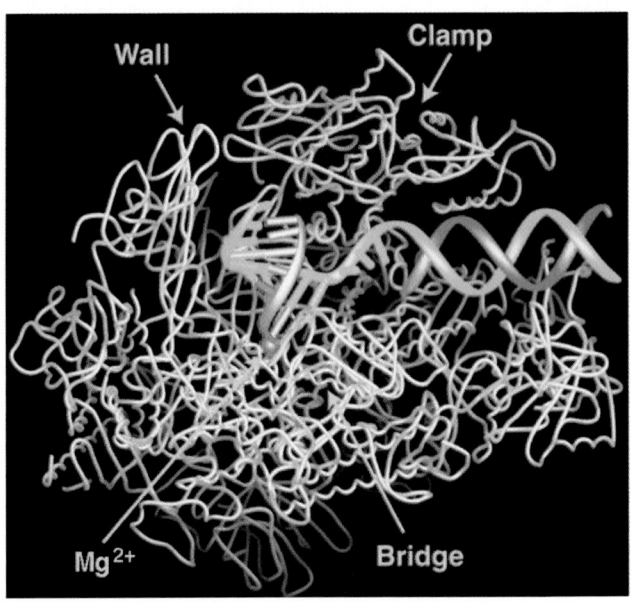

(b)

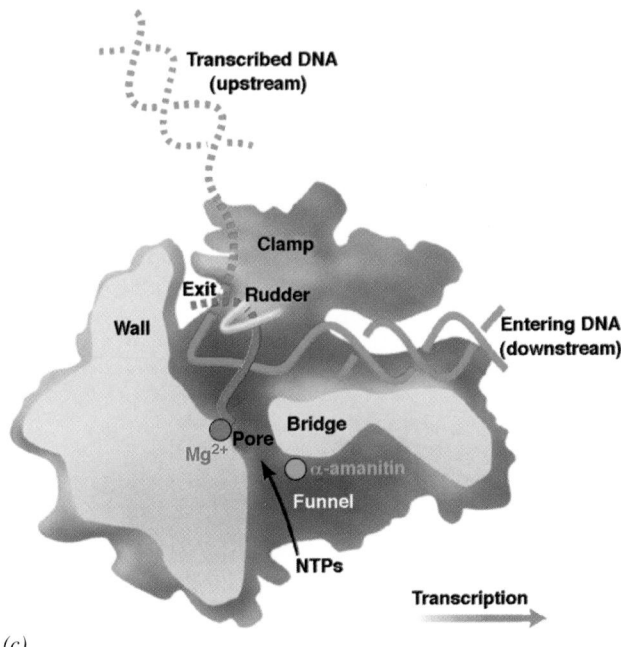

(c)

FIGURE 31-21   X-Ray structure of an RNAP II elongation complex. (*a*) The RNA · DNA complex in the structure with the template DNA cyan, the nontemplate DNA green, and the newly synthesized RNA red. The magenta dot marked Mg^{2+} represents the strongly bound active site metal ion. The black box encloses those portions of the complex that are clearly visible in the structure; the double-stranded portion of the DNA marked "Downstream DNA duplex" is poorly ordered, and the remaining portions of the complex are disordered. (*b*) View of the transcribing complex from the bottom of Fig. 31-20*a* in which portions of Rpb2 that form the near side of the cleft have been removed to expose the bound RNA · DNA complex. The protein is represented by its backbone in which the clamp, which is closed over the downstream DNA duplex, is yellow, the bridge helix is green, and the remaining portions of the protein are gray. The DNA and RNA are colored as in Part *a* with their well-ordered portions drawn in ladder form and their less ordered portions drawn in backbone form. The active site Mg^{2+} ion is represented by a magenta sphere. (*c*) Cutaway schematic diagram of the transcribing complex in Part *b* in which the cut surfaces of the protein are light gray, its remaining surfaces are darker gray, and several of its functionally important structural features are labeled. The DNA, RNA, and active site Mg^{2+} ion are colored as in Part *a* with portions of the DNA and RNA that are not visible in the X-ray structure represented by dashed lines. The α-amanitin binding site is marked by an orange dot. [Modified from diagrams by Roger Kornberg, Stanford University. PDBid 1I6H.]

template strand out of the cleft in an ~90° turn. As a consequence, the template base at the active site (+1) points toward the floor of the cleft where it can be read out by the active site. This base is paired with the ribonucleotide at the 3′ end of the RNA, which is positioned above a "pore" at the end of a "funnel" to the protein exterior through which NTPs presumably gain access to the otherwise sealed off active site. The RNA · DNA hybrid helix adopts a nonstandard conformation intermediate between those of A- and B-DNAs, which is underwound relative to that in the X-ray structure of an RNA · DNA hybrid helix alone (Fig. 29-6). Nearly all contacts that the protein makes with the RNA and DNA are to their sugar–phosphate backbones; none are with the edges of their bases. The specificity of the enzyme for a ribonucleotide rather than a deoxyribonucleotide is attributed to the enzyme's recognition of both the incoming ribose sugar and the RNA · DNA hybrid helix. After about one turn of hybrid helix, a loop extending from the clamp called the "rudder" separates the RNA and template DNA strands, thereby permitting the DNA double helix to reform as it exits the enzyme (although the unpaired 5′ tail of the nontemplate strand and the 3′ tail of the template strand are disordered in the X-ray structure). The models of the RNAP II transcribing complex and the *Taq* RNAP holoenzyme open complex differ mainly in the placement of the downstream segment of dsDNA (compare Figs. 31-13*b* and 31-21*c*). Of course, to become a transcribing complex, the *Taq* open complex must first jettison its σ subunit, to which the downstream dsDNA segment is bound.

How does RNAP translocate its bound RNA–DNA assembly in preparation for a new round of synthesis? The highly conserved helical segment of Rpb1, dubbed the "bridge" because it bridges the two pincers forming the enzyme's cleft (Figs. 31-20 and 31-21), nonspecifically contacts the template DNA base at the +1 position. Although this helix is straight in all X-ray structures of RNAP II yet determined, it is bent in that of *Taq* RNAP (Fig. 31-11*a*). If the bridge helix, in fact, alternates between its straight and bent conformations, it would move by 3 to 4 Å. Kornberg has therefore speculated that translocation occurs through the bending of the bridge helix so as to push the paired nucleotides at position +1 to position −1 (Fig. 31-22). The recovery of the bridge helix to its straight conformation would then yield an empty site at position +1 for entry of the next NTP, thereby preparing the enzyme for a new round of nucleotide addition.

b. Amatoxins Specifically Inhibit RNA Polymerases II and III

The poisonous mushroom ***Amanita phalloides* (death cap),** which is responsible for the majority of fatal mushroom poisonings, contains several types of toxic substances, including a series of unusual bicyclic octapeptides known as **amatoxins. α-Amanitin,**

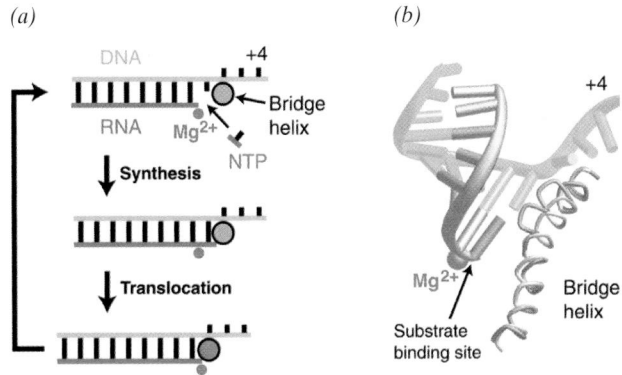

α-Amanitin

which is representative of the amatoxins, forms a tight 1:1 complex with RNAP II ($K = 10^{-8}M$) and a looser one with RNAP III ($K = 10^{-6}M$). Its binding slows an RNAP's rate of RNA synthesisis from several thousand to only a few nucleotides per minute. α-Amanitin is therefore a useful tool for mechanistic studies of these enzymes. RNAP I as well as mitochondrial, chloroplast, and prokaryotic RNAPs are insensitive to α-amanitin.

The X-ray structure of RNAP II in complex with α-amanitin, also determined by Kornberg, reveals that α-amanitin binds in the funnel beneath the protein's bridge helix (Fig. 31-21*c*) such that it interacts almost exclusively with the residues of the bridge helix and the adjacent part of Rpb1. The observation that RNAP II mutations that affect α-amanitin inhibition also map to this site indicates that this binding mode is not just an artifact of crystallization. The α-amanitin binding site is too far away from the enzyme active site to directly interfere with NTP entry or

FIGURE 31-22 The proposed transcription cycle and translocation mechanism of RNAP. (*a*) The nucleotide addition cycle in which the enzyme active site is marked by its strongly bound Mg²⁺ ion (*magenta*). The translocation of the transcribing RNA · DNA complex is proposed to be motivated by a conformational change of the bridge helix from straight (*gray circle*) to bent (*violet circle*). The relaxation of the bridge helix back to its straight form would complete the cycle by yielding an empty NTP binding site at the active site. (*b*) The RNA · DNA complex in RNAP II viewed and colored as in Fig. 31-21*b*. The RNAP II bridge helix is gray and the superimposed (and bent) *Taq* polymerase bridge helix is violet. The side chains extending from the bent helix would sterically clash with the hybrid base pair at position +1. [Courtesy of Roger Kornberg, Stanford University.]

RNA synthesis, consistent with the observation that α-amanitin does not influence the affinity of RNAP II for NTPs. Most probably, α-amanitin binding impedes the conformational change of the bridge helix postulated to motivate the RNAP translocation step (Fig. 31-22), which further supports this mechanism.

Despite the amatoxins' high toxicity (5–6 mg, which occurs in ~40 g of fresh mushrooms, is sufficient to kill a human adult), they act slowly. Death, usually from liver dysfunction, occurs no earlier than several days after mushroom ingestion (and after recovery from the effects of other mushroom toxins). This, in part, reflects the slow turnover of eukaryotic mRNAs and proteins.

c. Mammalian RNA Polymerase I Has a Bipartite Promoter

Since, as we shall see in Section 31-4B, the numerous rRNA genes in a given eukaryotic cell have essentially identical sequences, its RNAP I only recognizes one promoter. Yet, in contrast to the case for RNAPs II and III, RNAP I promoters are species specific, that is, an RNAP I only recognizes its own promoter and those of closely related species. This is because only closely related species exhibit recognizable sequence identities near the transcriptional start sites of their rRNA genes. RNAP I promoters were therefore identified by determining how the transcription rate of an rRNA gene is affected by a series of increasingly longer deletions approaching its start site from either its upstream or its downstream sides. Such studies have indicated, for example, that mammalian RNAPs I require the presence of a so-called **core promoter element,** which spans positions −31 to +6 and hence overlaps the transcribed region. However, efficient transcription additionally requires an **upstream promoter element,** which is located between residues −187 and −107. These elements, which are G + C-rich and ~85% identical, are bound by specific transcription factors which then recruit RNAP I to the transcription start site.

d. RNA Polymerase II Promoters Are Complex and Diverse

The promoters recognized by RNAP II are considerably longer and more diverse than those of prokaryotic genes but have not yet been fully described. The structural genes expressed in all tissues, the so-called housekeeping genes, which are thought to be constituitively transcribed, have one or more copies of the sequence GGGCGG or its complement (the **GC box**) located upstream from their transcription start sites. The analysis of deletion and point mutations in eukaryotic viruses such as SV40 indicates that GC boxes function analogously to prokaryotic promoters. On the other hand, structural genes that are selectively expressed in one or a few types of cells often lack these GC-rich sequences. Rather, *many contain a conserved AT-rich sequence located 25 to 30 bp upstream from their transcription start sites (Fig. 31-23).* Note that this so-called **TATA box** resembles the −10 region of prokaryotic promoters (TATAAT), although they differ in their locations relative to the transcription start site (−27 vs −10). The functions of these two promoter elements are not strictly analogous, however, since the deletion of the TATA box does not necessarily eliminate transcription. Rather, TATA box deletion or mutation generates heterogeneities in the transcriptional start site, thereby indicating that the TATA box participates in selecting this site.

The gene region extending between about −50 and −110 also contains promoter elements. For instance, many eukaryotic structural genes, including those encoding the various globins, have a conserved sequence of consensus CCAAT (the **CCAAT box**) located between about −70 and −90 whose alteration greatly reduces the gene's transcription rate. Globin genes have, in addition, a conserved **CACCC box** upstream from the CCAAT box that has also been implicated in transcriptional initiation. Evidently, the promoter sequences upstream of the TATA box form the initial DNA-binding sites for RNA polymerase II and the other proteins involved in transcriptional initiation (see below).

e. Enhancers Are Transcriptional Activators That Can Have Variable Positions and Orientations

Perhaps the most surprising aspect of eukaryotic transcriptional control elements is that some of them need not have fixed positions and orientations relative to their corresponding transcribed sequences. For example, the SV40 genome, in which such elements were first discovered, contains two repeated sequences of 72 bp each that are located upstream from the promoter for early gene expres-

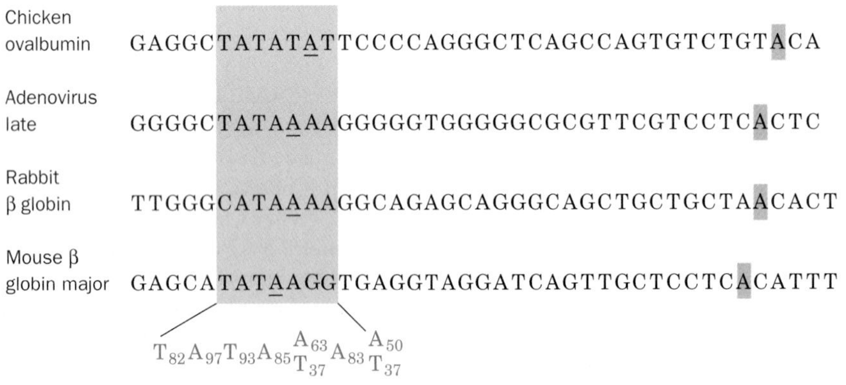

Chicken
ovalbumin   GAGGC**TATATA**TTCCCCAGGGCTCAGCCAGTGTCTGT**A**CA

Adenovirus
late   GGGGC**TATAA**AAGGGGGTGGGGGCGCGTTCGTCCTC**A**CTC

Rabbit
β globin   TTGGGC**ATAAA**AGGCAGAGCAGGGCAGCTGCTGCTA**A**CACT

Mouse β
globin major GAGCA**TATAA**GGTGAGGTAGGATCAGTTGCTCCTC**A**CATTT

$T_{82}A_{97}T_{93}A_{85}\begin{smallmatrix}A_{63}\\T_{37}\end{smallmatrix}A_{83}\begin{smallmatrix}A_{50}\\T_{37}\end{smallmatrix}$

FIGURE 31-23   The promoter sequences of selected eukaryotic structural genes. The homologous segment, the TATA box, is shaded in red with the base at position −27 underlined and the initial nucleotide to be transcribed (+1) shaded in green. The bottom row indicates the consensus sequence of several such promoters with the subscripts indicating the percent occurrence of the corresponding base. [After Gannon, F., et al., *Nature* **278,** 433 (1978).]

sion. Transcription is unaffected if one of these repeats is deleted but is nearly eliminated when both are absent. The analysis of a series of SV40 mutants containing only one of these repeats demonstrated that its ability to stimulate transcription from its corresponding promoter is all but independent of its position and orientation. Indeed, transcription is unimpaired when this segment is several thousand base pairs upstream or downstream from the transcription start site. Gene segments with such properties are named **enhancers** to indicate that they differ from promoters, with which they must be associated in order to trigger site-specific and strand-specific transcription initiation (although the characterization of numerous promoters and enhancers indicates that their functional properties are similar). Enhancers occur in both eukaryotic viruses and cellular genes.

Enhancers are required for the full activities of their cognate promoters. It was originally thought that enhancers somehow acted as entry points on DNA for RNAP II (perhaps by altering DNA's local conformation or through a lack of binding affinity for the histones that normally coat eukaryotic DNA; Section 34-1A). However, it is now clear that *enhancers are recognized by specific transcription factors that*

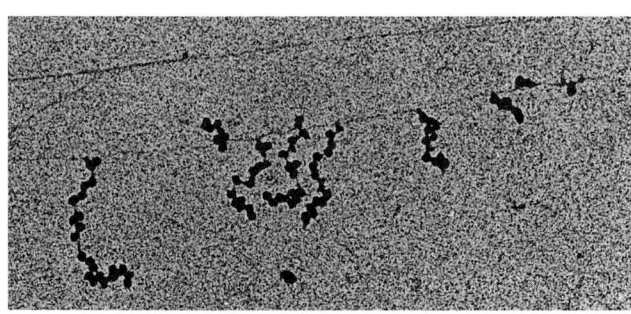

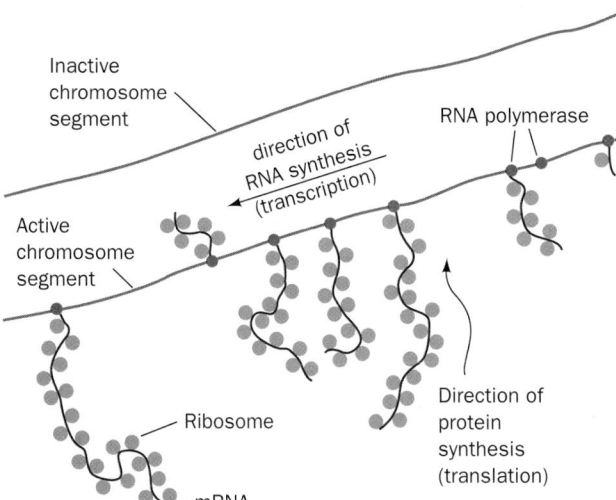

FIGURE 31-24 An electron micrograph and its interpretive drawing showing the simultaneous transcription and translation of an *E. coli* gene. RNA polymerase molecules are transcribing the DNA from right to left while ribosomes are translating the nascent RNAs (mostly from bottom to top). [Courtesy of Oscar L. Miller, Jr. and Barbara Hamkalo, University of Virginia.]

stimulate RNA polymerase II to bind to the corresponding but distant promoter. This requires that the DNA between the enhancer and promoter loop around so that the transcription factor can simultaneously contact the enhancer and the RNAP II and/or its associated proteins at the promoter. Most cellular enhancers are associated with genes that are selectively expressed in specific tissues. It therefore seems, as we discuss in Section 34-3B, that *enhancers mediate much of the selective gene expression in eukaryotes.*

f. RNA Polymerase III Promoters Can Be Located Downstream from Their Transcription Start Sites

The promoters of genes transcribed by RNAP III can be located entirely within the genes' transcribed regions. Donald Brown established this through the construction of a series of deletion mutants of a *Xenopus borealis* 5S RNA gene. Deletions of base sequences that start from outside one or the other end of the transcribed portion of the 5S gene only prevent transcription if they extend into the segment between nucleotides $+40$ and $+80$. Indeed, a fragment of the 5S RNA gene consisting of only nucleotides 41 to 87, when cloned in a bacterial plasmid, is sufficient to direct specific initiation by RNAP III at an upstream site. This is because, as was subsequently demonstrated, the sequence contains the binding site for transcription factors that stimulate the upstream binding of RNAP III. Further studies have shown, however, that the promoters of other RNAP III-transcribed genes lie entirely upstream of their start sites. These upstream sites also bind transcription factors that recruit RNAP III.

3 ■ CONTROL OF TRANSCRIPTION IN PROKARYOTES

Prokaryotes respond to sudden environmental changes, such as the influx of nutrients, by inducing the synthesis of the appropriate proteins. This process takes only minutes because transcription and translation in prokaryotes are closely coupled: *Ribosomes commence translation near the 5' end of a nascent mRNA soon after it is extruded from RNA polymerase (Fig. 31-24).* Moreover, *most prokaryotic mRNAs are enzymatically degraded within 1 to 3 min of their synthesis,* thereby eliminating the wasteful synthesis of unneeded proteins after a change in conditions (protein degradation is discussed in Section 32-6). In fact, the 5' ends of some mRNAs are degraded before their 3' ends have been synthesized.

In contrast, the induction of new proteins in eukaryotic cells frequently takes hours or days, in part because transcription takes place in the nucleus and the resulting mRNAs must be transported to the cytoplasm, where translation occurs. However, eukaryotic cells, particularly those of multicellular organisms, have relatively stable environments; major changes in their transcription patterns usually occur only during cell differentiation.

In this section we examine some of the ways in which prokaryotic gene expression is regulated through transcriptional control. Eukaryotes, being vastly more complex creatures than are prokaryotes, have a correspondingly more complicated transcriptional control system whose general outlines are beginning to come into focus. We therefore defer discussion of eukaryotic transcriptional control until Section 34-3B, where it can be considered in light of what we know about the structure and organization of the eukaryotic chromosome.

A. *Promoters*

In the presence of high concentrations of inducer, the *lac* operon is rapidly transcribed. In contrast, the *lacI* gene is transcribed at such a low rate that a typical *E. coli* cell contains <10 molecules of the *lac* repressor. Yet, the *I* gene has no repressor. Rather, it has such an inefficient promoter (Fig. 31-10) that it is transcribed an average of about once per bacterial generation. *Genes that are transcribed at high rates have efficient promoters.* In general, the more efficient a promoter, the more closely its sequence resembles that of the corresponding consensus sequence.

a. Gene Expression Can Be Controlled by a Succession of σ Factors

The processes of development and differentiation involve the temporally ordered expression of sets of genes according to genetically specified programs. Phage infections are among the simplest examples of developmental processes. Typically, only a subset of the phage genome, often referred to as *early* genes, are expressed in the host immediately after phage infection. As time passes, *middle* genes start to be expressed, and the *early* genes as well as the bacterial genes are turned off. In the final stages of phage infection, the *middle* genes give way to the *late* genes. Of course some phage types express more than three sets of genes and some genes may be expressed in more than one stage of an infection.

One way in which families of genes are sequentially expressed is through "cascades" of σ factors. In the infection of *Bacillus subtilis* by bacteriophage SP01, for example, the *early* gene promoters are recognized by the bacterial RNAP holoenzyme. Among these *early* genes is gene 28, whose gene product is a new σ subunit, designated σ^{gp28}, that displaces the bacterial σ subunit from the core enzyme. This reconstituted holoenzyme recognizes only the phage *middle* gene promoters, which all have similar −35 and −10 regions, but bear little resemblance to the corresponding regions of bacterial and phage *early* genes. The *early* genes therefore become inactive once their corresponding mRNAs have been degraded. The phage *middle* genes include genes 33 and 34, which together specify yet another σ factor, σ$^{gp33/34}$, which, in turn, permits the transcription of only *late* phage genes.

Several bacteria, including *E. coli* and *B. subtilis*, likewise have several different σ factors. These are not necessarily utilized in a sequential manner. Rather, those that differ from the predominant or primary σ factor (σ^{70} in *E. coli*) control the transcription of coordinately expressed groups of special purpose genes, whose promoters are quite different from those recognized by the primary σ factor. For example, sporulation in *B. subtilis*, a process in which the bacterial cell is asymmetrically partitioned into two compartments, the **forespore** (which becomes the **spore**, a germline cell from which subsequent progeny arise) and the **mother cell** (which synthesizes the spore's protective cell wall and is eventually discarded), is governed by five σ factors in addition to that of the **vegetative** (nonsporulating) cell: one that is active before cell partition occurs, two that are sequentially active in the forespore, and two that are sequentially active in the mother cell. Cross-regulation of the compartmentalized σ factors permits the forespore and mother cell to tightly coordinate this differentiation process.

B. *lac Repressor I: Binding*

In 1966, Beno Müller-Hill and Walter Gilbert isolated *lac* repressor on the basis of its ability to bind ^{14}C-labeled IPTG (Section 31-1A) and demonstrated that it is a protein. This was an exceedingly difficult task because *lac* repressor comprises only ~0.002% of the protein in wild-type *E. coli*. Now, however, *lac* repressor is available in quantity via molecular cloning techniques (Section 5-5G).

a. *lac* Repressor Finds Its Operator by Sliding along DNA

The *lac* repressor is a tetramer of identical 360-residue subunits, each of which binds one IPTG molecule with a dissociation constant of $K = 10^{-6}M$. In the absence of inducer, the repressor tetramer nonspecifically binds duplex DNA with a dissociation constant of $K \approx 10^{-4}M$. However, it specifically binds to the *lac* operator with far greater affinity: $K \approx 10^{-13}M$. Limited proteolysis of *lac* repressor with trypsin reveals that each subunit consists of two functional domains: Its 58-residue N-terminal peptide binds DNA but not IPTG, whereas the remaining "core tetramer" binds only IPTG.

The observed rate constant for the binding of *lac* repressor to *lac* operator is $k_f \approx 10^{10}M^{-1} \text{ s}^{-1}$. This "on" rate is much greater than that calculated for the diffusion-controlled process in solution: $k_f = 10^7 M^{-1} \text{ s}^{-1}$ for molecules the size of *lac* repressor. Since it is impossible for a reaction to proceed faster than its diffusion-controlled rate, the *lac* repressor must not encounter operator from solution in a random three-dimensional search. Rather, *it appears that lac repressor finds operator by nonspecifically binding to DNA and diffusing along it in a far more efficient one-dimensional search.*

b. *lac* Operator Has a Nearly Palindromic Sequence

The availability of large quantities of *lac* repressor made it possible to characterize the *lac* operator. *E. coli* DNA that had been sonicated to small fragments was mixed with *lac* repressor and passed through a nitrocellulose filter.

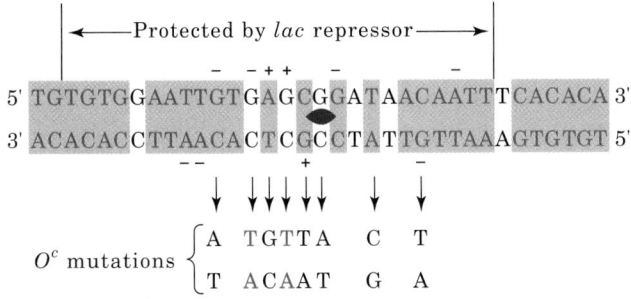

FIGURE 31-25 The base sequence of the *lac* operator. The symmetry related regions (*red*) comprise 28 of its 35 bp. A "+" denotes positions at which repressor binding enhances methylation by dimethyl sulfate (which methylates G at N7 and A at N3) and a "−" indicates where this footprinting reaction is inhibited. The bottom row indicates the positions and identities of different point mutations that prevent *lac* repressor binding (*O^c* mutants). Those in red increase the operator's symmetry. [After Sobell, H.M., in Goldberger, R.F. (Ed.), *Biological Regulation and Development*, Vol. 1, p. 193, Plenum Press (1979).]

Protein, with or without bound DNA, sticks to nitrocellulose, whereas duplex DNA, by itself, does not. The DNA was released from the filter-bound protein by washing it with IPTG solution, recombined with *lac* repressor, and the resulting complex treated with DNase I. The DNA fragment that *lac* repressor protects from nuclease degradation consists of a run of 26 bp that is embedded in a nearly 2-fold symmetric sequence of 35 bp (Fig. 31-25, *top*). *Such palindromic symmetry is a common feature of DNA segments that are specifically bound by proteins* (recall, for example, that restriction endonuclease recognition sites are also palindromic; Section 5-5A).

Palindromic DNA sequences, as we have seen, bind to proteins that have matching 2-fold symmetry. However, methylation protection experiments on the *lac* repressor–operator system do not fully support this model: There is an asymmetric pattern of differences between free and

repressor-bound operator in the susceptibility of its bases to reaction with DMS (Fig. 31-25). Furthermore, point mutations in the operator that render it operator constitutive (*O^c*), and that invariably weaken the binding of repressor to operator, may increase as well as decrease the operator's 2-fold symmetry (Fig. 31-25).

c. *lac* Repressor Prevents RNA Polymerase from Forming a Productive Initiation Complex

Operator occupies positions −7 through +28 of the *lac* operon relative to the transcription start site (Fig. 31-26). Nuclease protection studies, it will be recalled, indicate that, in the initiation complex, RNA polymerase tightly binds to the DNA between positions −20 and +20 (Section 31-2B). Thus, *the lac operator and promoter sites overlap.* It was therefore widely assumed for many years that *lac* repressor simply physically obstructs the binding of RNA polymerase to the *lac* promoter. However, the observation that *lac* repressor and RNA polymerase can simultaneously bind to the *lac* operon indicates that *lac* repressor must act by somehow interfering with the initiation process. Closer investigation of this phenomenon revealed that, in the presence of bound *lac* repressor, RNA polymerase holoenzyme still abortively synthesizes oligonucleotides, although they tend to be shorter than those made in the absence of repressor. Evidently, *lac repressor acts by somehow increasing the already high kinetic barrier for RNA polymerase to generate the open complex and commence processive elongation.*

We discuss the *lac* repressor structure and further aspects of *lac* operator organization in Section 31-3F.

C. *Catabolite Repression: An Example of Gene Activation*

Glucose is E. coli's metabolite of choice; the availability of adequate amounts of glucose prevents the full expression of >100 genes that encode proteins involved in the fermentation of numerous other catabolites, including lactose (Fig.

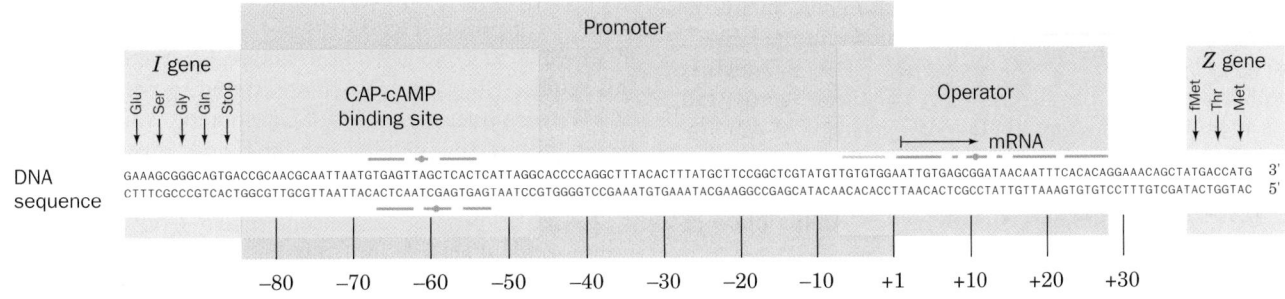

FIGURE 31-26 The nucleotide sequence of the *E. coli lac* promoter–operator region. The region extends from the C-terminal portion of *lacI* (*left*) to the N-terminal portion of *lacZ* (*right*). The palindromic sequences of the operator and the CAP-binding site (Section 31-3C) are overscored or underscored. [After Dickson, R.C., Abelson, J., Barnes, W.M., and Reznikoff, W.A., *Science* **187,** 32 (1975).]

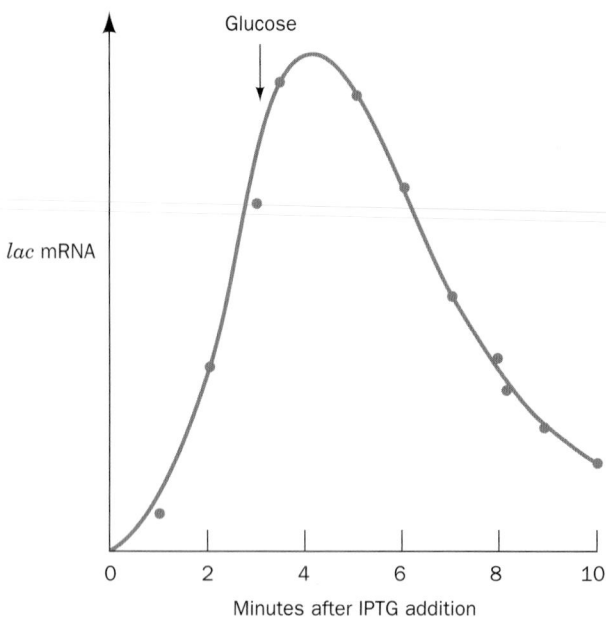

FIGURE 31-27   The kinetics of *lac* operon mRNA synthesis following its induction with IPTG, and of its degradation after glucose addition. *E. coli* were grown on a medium containing glycerol as their only carbon-energy source and ³H-labeled uridine. IPTG was added to the medium at the beginning of the experiment to induce the synthesis of the *lac* enzymes. After 3 min, glucose was added to stop the synthesis. The amount of ³H-labeled *lac* RNA was determined by hybridization with DNA containing the *lacZ* and *lacY* genes. [After Adesnik, M. and Levinthal, C., *Cold Spring Harbor Symp. Quant. Biol.* **35,** 457 (1970).]

31-27), arabinose, and galactose, even when these metabolites are present in high concentrations. This phenomenon, which is known as **catabolite repression,** prevents the wasteful duplication of energy-producing enzyme systems.

a. cAMP Signals the Lack of Glucose

The first indication of the mechanism of catabolite repression was the observation that, in *E. coli,* the level of cAMP, which was known to be a second messenger in animal cells (Section 18-3E), is greatly diminished in the presence of glucose. This observation led to the finding that the addition of cAMP to *E. coli* cultures overcomes catabolite repression by glucose. Recall that, in *E. coli,* adenylate cyclase is activated by the phosphorylated enzyme EIIA*glc* (or possibly inactivated by dephospho-EIIA*glc*), which is dephosphorylated on the transport of glucose across the cell membrane (Section 20-3D). *The presence of glucose, therefore, normally lowers the cAMP level in E. coli.*

b. CAP–cAMP Complex Stimulates the Transcription of Catabolite Repressed Operons

Certain *E. coli* mutants, in which the absence of glucose does not relieve catabolite repression, are missing a cAMP-binding protein that is synonymously named **catabolite gene activator protein (CAP)** and **cAMP receptor protein (CRP).** CAP is a homodimer of 210-residue subunits that

undergoes a large conformational change on binding cAMP. Its function was elucidated by Ira Pastan, who showed that *CAP–cAMP complex, but not CAP itself, binds to the lac operon (among others) and stimulates transcription from its otherwise low-efficiency promoter in the absence of lac repressor.* CAP is therefore a **positive regulator** (turns on transcription), in contrast to *lac* repressor, which is a **negative regulator** (turns off transcription).

The X-ray structure, by Thomas Steitz, of CAP–cAMP in complex with a palindromic 30-bp segment of duplex DNA whose sequence resembles that of the CAP binding sequence (Fig. 31-26) reveals that the DNA is bent by ~90° around the protein (Fig. 31-28*a*). The bend arises from two ~45° kinks in the DNA between the fifth and sixth bases out from the complex's 2-fold axis in both directions. This distortion results in the closing of the major groove and an enormous widening of the minor groove at each kink.

Why is the CAP–cAMP complex necessary to stimulate the transcription of its target operons? And how does it do so? The *lac* operon has a weak (low-efficiency) promoter; its −10 and −35 sequences (TATGTT and TTTACA; Fig. 31-10) differ significantly from the corresponding consensus sequences of strong (high-efficiency) promoters (TATAAT and TTGACA; Fig. 31-10). Such weak promoters evidently require some sort of help for efficient transcriptional initiation.

Richard Ebright has shown that CAP interacts directly with RNAP via the C-terminal domain of its 85-residue α subunit (αCTD) in a way that stimulates RNAP to initiate transcription from a nearby promoter. The αCTD also binds dsDNA nonspecifically but does so with higher affinity at A + T-rich sites such as those of UP elements (Section 31-2A). It is flexibly linked to the rest of the α subunit and hence is not seen in the X-ray structure of *Taq* RNAP (Fig. 31-11) due to disorder.

Three classes of CAP-dependent promoters have been characterized:

1. Class I promoters, such as that of the *lac* operon, require only CAP–cAMP for transcriptional activation. The CAP binding site on the DNA can be located at various distances from the promoter provided that CAP and RNAP bind to the same face of the DNA helix. Thus, CAP–cAMP activates the transcription of the *lac* operon if its DNA binding site is centered near positions −62 (its wild-type position; Fig. 31-26), −72, −83, −93, or −103, all of which are one helical turn apart. For the latter sites, this requires that the DNA loop around to permit CAP–cAMP to contact the αCTD. Such looping is likely to be facilitated by the bending of the DNA around CAP–cAMP.

2. Class II promoters also require only CAP–cAMP for transcriptional activation. However, in class II promoters, the CAP binding site only occupies a fixed position that overlaps the RNAP binding site, apparently by replacing the promoter's −35 promoter region. CAP then interacts with RNAP via interactions with both the αCTD and the α subunit's N-terminal domain.

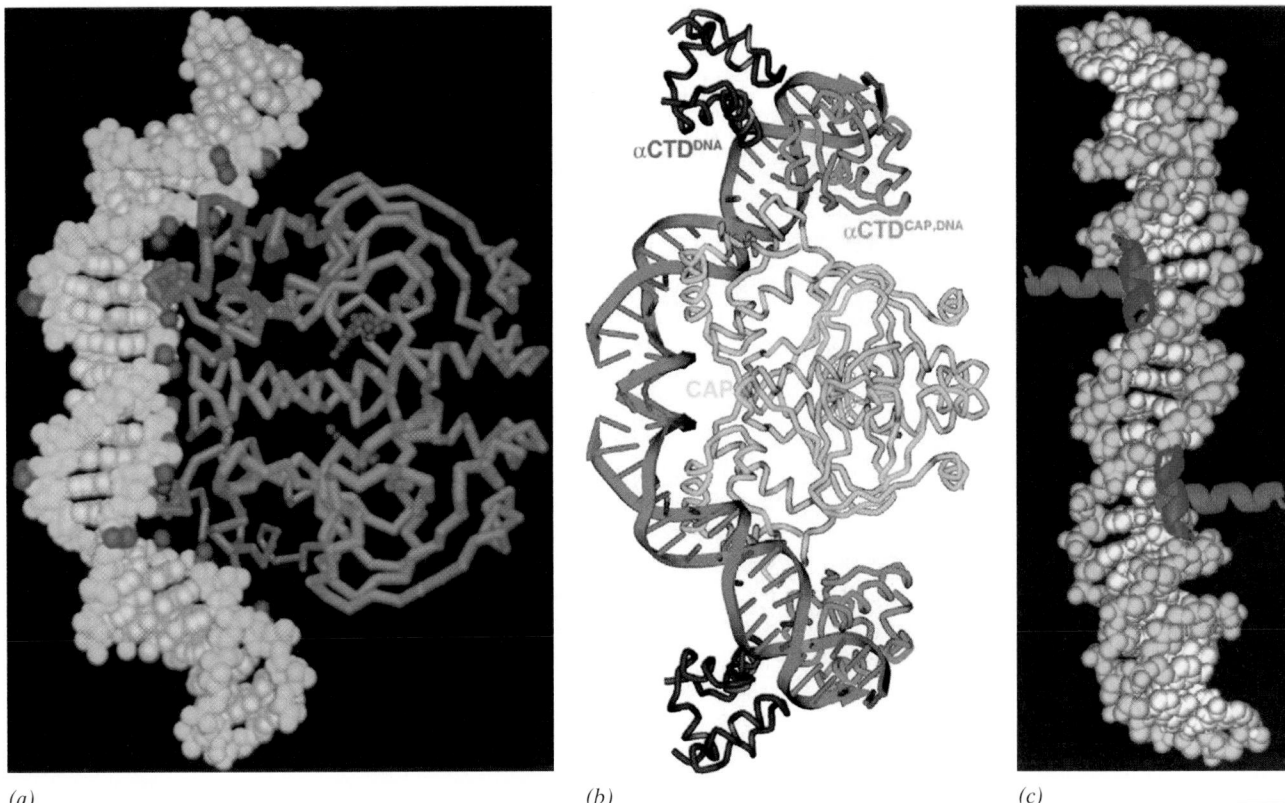

(a)   *(b)*   *(c)*

FIGURE 31-28   X-Ray structures of CAP–cAMP complexes.
(*a*) CAP–cAMP in complex with a palindromic 30-bp duplex DNA. The complex is viewed with its molecular 2-fold axis horizontal. The protein is represented by its C_α backbone with its N-terminal cAMP-binding domain blue and its C-terminal DNA-binding domain purple. The DNA is shown in space-filling form with its sugar–phosphate backbone yellow and its bases white (the atoms are drawn with slightly less than van der Waals radii). The DNA phosphates whose ethylation interferes with CAP binding are red. Those in the complex that are hypersensitive to DNase I are blue (these latter phosphates bridge the CAP-induced kinks and hence occur where the minor groove has been dramatically widened, which apparently increases their susceptibility to DNase I digestion). The bound cAMPs are shown in ball-and-stick form in red. (*b*) CAP–cAMP in complex with a 44-bp palindromic DNA and the

αCTD oriented similarly to Part *a*. The DNA is shown in ladder form in red (it contains two symmetrically related single-phosphate gaps that are separated by 4 bp); the proteins are represented by their backbones with the CAP dimer blue-green, each αCTD^{CAP,DNA} light green, and each αCTD^{CAP} dark green; and cAMP is drawn in wireframe form in red. (*c*) The same structure as in Part *a* showing the binding of the CAP dimer's two helix–turn–helix (HTH) motifs in successive major grooves of the DNA. The HTH motif's N-terminal helix is blue and its C-terminal recognition helix is red. The view is along the molecular 2-fold axis and is related to that in Part *a* by a 90° rotation about its vertical axis. [Part *a* courtesy of and Part *c* based on an X-ray structure by Thomas Steitz, Yale University. PDBid 1CGP. Part *b* courtesy of Helen Berman and Richard Ebright, Rutgers University. PDBid 1LB2.] **See the Interactive Exercises**

3. Class III promoters require multiple activators to maximally stimulate transcription. These may be two or more CAP–cAMP complexes or a CAP–cAMP complex acting in concert with promoter-specific activators as occurs in the *araBAD* operon (Section 31-3E).

The X-ray structure of CAP–cAMP in complex with the *E. coli* αCTD and a 44-bp palindromic DNA containing the 22-bp CAP–cAMP binding site and 5'-AAAAAA-3' at each end, determined by Helen Berman and Ebright, reveals how these components interact (Fig. 31-28*b*). The 2-fold symmetric CAP–cAMP–αCTD complex contains two differently located pairs of αCTDs. Each member of the pair designated αCTD^{CAP,DNA} binds to both CAP and to the DNA. CAP and αCTD^{CAP,DNA} interact over a sur-

prisingly small surface area involving only six residues on each protein that mutagenesis experiments had previously implicated. αCTD^{CAP,DNA} also interacts with the minor groove of a 6-bp segment of the DNA (5'-AAAAAG-3') centered 19 bp from the center of the DNA. Each member of the other pair of αCTDs, designated αCTD^{DNA}, interacts with the minor groove of an UP element-like sequence (5'-GAAAAA-3') that is fortuitously present in the DNA but it makes no contacts with other protein molecules. The common portions of the two CAP complexes pictured in Fig. 31-28*a*, *b* are closely superimposable, thereby indicating that the conformation of CAP and its interaction with DNA are not significantly altered by its association with the αCTD. Evidently, CAP–cAMP transcriptionally activates RNAP via a simple "adhesive" mechanism that facilitates and/or stabilizes its interaction with the promoter DNA. The structures of αCTD^{CAP,DNA} and αCTD^{CAP} and their interactions with DNA are nearly identical, thereby suggesting that they are representative of the interaction of an αCTD with an UP element.

D. *Sequence-Specific Protein–DNA Interactions*

Since genetic expression is controlled by proteins such as CAP and *lac* repressor, an important issue in the study of gene regulation is how these proteins recognize their target base sequences on DNA. Sequence-specific DNA-binding proteins generally do not disrupt the base pairs of the duplex DNA to which they bind. Consequently, these proteins can only discriminate among the four base pairs (A · T, T · A, G · C, and C · G) according to the functional groups of these base pairs that project into DNA's major and minor grooves. An inspection of Fig. 5-12 reveals that the groups exposed in the major groove have a greater variation in their types and arrangements than do those that are exposed in the minor groove. Indeed, the positions of the hydrogen bonding acceptors in the major groove vary with both the identity and orientation of the base pair, whereas in the minor groove they are largely sequence independent. Moreover, the ~5-Å-wide and ~8-Å-deep minor groove of canonical (ideal) B-DNA is too narrow to admit protein structural elements such as an α helix, whereas its ~12-Å-wide and ~8-Å-deep major groove can do so. Thus, in the absence of major conformational changes to B-DNA, it would be expected that proteins could more readily differentiate base sequences from its major groove than from its minor groove. We shall see below that this is, in fact, the case.

a. The Helix–Turn–Helix Motif Is a Common DNA Recognition Element in Prokaryotes

✍ **See Guided Exploration 30. Transcription factor-DNA Interactions** The CAP dimer's two symmetrically disposed F helices protrude from the protein surface in such a way that they fit into successive major grooves of B-DNA (Fig. 31-28). *CAP's E and F helices form a* **helix–turn–helix (HTH) motif** *(supersecondary structure) that conformationally resembles analogous HTH motifs in numerous other prokaryotic repressors of known X-ray and NMR structure,* including the *lac* repressor, the *E. coli* **trp repressor** (Section 31-3F), and the **cI repressors** and **Cro proteins** from **bacteriophages** λ and **434** (Section 33-3D). HTH motifs are ~20-residue polypeptide segments that form two α helices which cross at ~120° (Fig. 31-28c). They occur as components of domains that otherwise have widely varying structures, although all of them bind DNA. Note that HTH motifs are structurally stable only when they are components of larger proteins.

The X-ray and NMR structures of a number of protein–DNA complexes (see below) indicate that *DNA-binding proteins containing an HTH motif associate with their target base pairs mainly via the side chains extending from the second helix of the HTH motif, the so-called* **recognition**

helix (helix F in CAP, E in *trp* repressor, and α3 in the phage proteins). Indeed, replacing the outward-facing residues of the 434 repressor's recognition helix with the corresponding residues of the related **bacteriophage P22** yields a hybrid repressor that binds to P22 operators but not to those of 434. Moreover, the HTH motifs in all these proteins have amino acid sequences that are similar to each other and to polypeptide segments in numerous other prokaryotic DNA-binding proteins, including *lac* repressor. Evidently, *these proteins are evolutionarily related and bind their target DNAs in a similar manner.*

How does the recognition helix recognize its target sequence? Since each base pair presents a different and presumably readily differentiated constellation of hydrogen bonding groups in DNA's major groove, it seemed likely that there would be a simple correspondence, analogous to Watson–Crick base pairing, between the amino acid residues of the recognition helix and the bases they contact in forming sequence-specific associations. The above X-ray structures, however, indicate this idea to be incorrect. Rather, base sequence recognition arises from complex structural interactions. For instance:

1. The X-ray structures of the closely similar N-terminal domain of 434 repressor (residues 1–69) and the entire 71-residue 434 Cro protein in their complexes with the identical 20-bp target DNA (434 phage expression is regulated through the differential binding of these proteins to the same DNA segments; Section 33-3D) were both determined by Stephen Harrison. Both dimeric proteins, as seen for CAP (Fig. 31-28), associate with the DNA in a 2-fold symmetric manner with their recognition helices bound in successive turns of the DNA's major groove (Figs. 31-29 and 31-30). In both complexes, the protein closely conforms to the DNA surface and interacts with its paired bases and sugar–phosphate chains through elaborate networks of hydrogen bonds, salt bridges, and van der Waals contacts. Nevertheless, the detailed geometries of these associations are significantly different. In the repressor–DNA complex (Fig. 31-29), the DNA bends around the protein in an arc of radius ~65 Å which compresses the minor groove by ~2.5 Å near its center (between the two protein monomers) and widens it by ~2.5 Å toward its ends. In contrast, the DNA in complex with Cro (Fig. 31-30), although also bent, is nearly straight at its center and has a less compressed minor groove (compare Figs. 31-29a and 31-30a). This explains why the simultaneous replacement of three residues in the repressor's recognition helix with those occurring in Cro does not cause the resulting hybrid protein to bind DNA with Cro-like affinity: *The different conformations of the DNA in the repressor and Cro complexes prevents any particular side chain from interacting identically with the DNA in the two complexes.*

2. Paul Sigler determined the X-ray structure of *E. coli trp* repressor in complex with a DNA containing an 18-bp palindrome (TGT<u>ACTAGTTA</u><u>ACTAGT</u>AC, where the *trp* repressor's target sequence is underlined) that closely resembles the *trp* operator (Section 31-3F). The dimeric

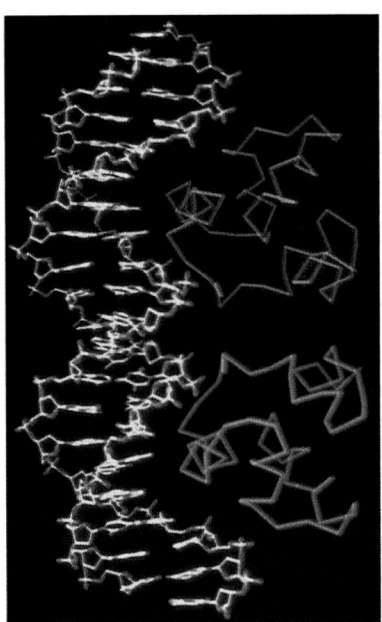

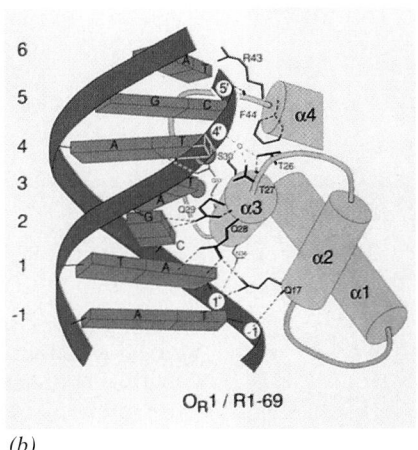

(b)

(a)

(c)

FIGURE 31-29 X-Ray structure of the 69-residue N-terminal domain of 434 phage repressor in complex with a 20-bp fragment of its target sequence. One strand of the DNA (*left*) has the sequence d(TATACAAGAAAGTTTGTACT). The complex is viewed perpendicular to its 2-fold axis of symmetry. (*a*) A skeletal model with the protein's two identical subunits (*blue and red*; C_α backbone only). Only the first 63 residues of the protein are visible. (*b*) A schematic drawing indicating how the helix–turn–helix motif, which encompasses helices α2 and α3, interacts with its target DNA. Short bars emanating from the polypeptide chain represent peptide NH groups, hydrogen bonds are represented by dashed lines, and DNA phosphates are represented by numbered circles. The small circle is a water molecule. (*c*) A space-filling model corresponding to Part *a*. The DNA is colored according to atom type (C gray, N blue, O red, and P green) and the protein's non-H atoms are all yellow. [Courtesy of Aneel Aggarwal, John Anderson, and Stephen Harrison, Harvard University. PDBid 2OR1.] *See the* **Interactive Exercises and Kinemage Exercise 19-1**

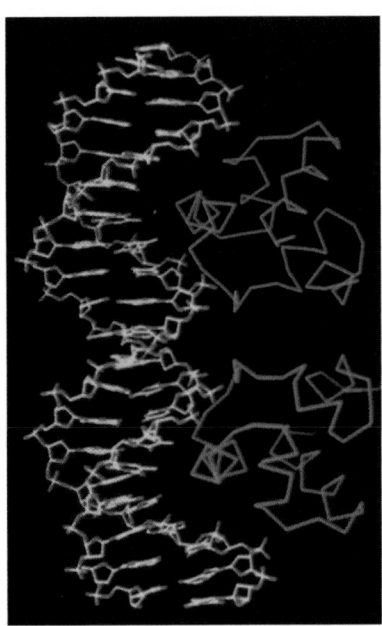

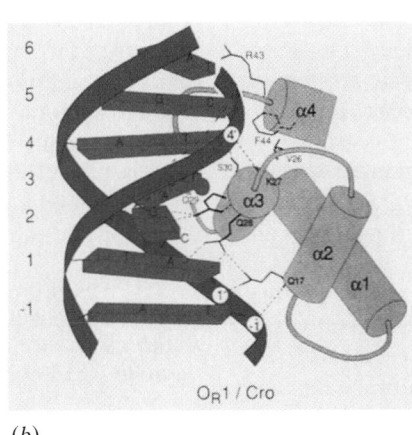

(b)

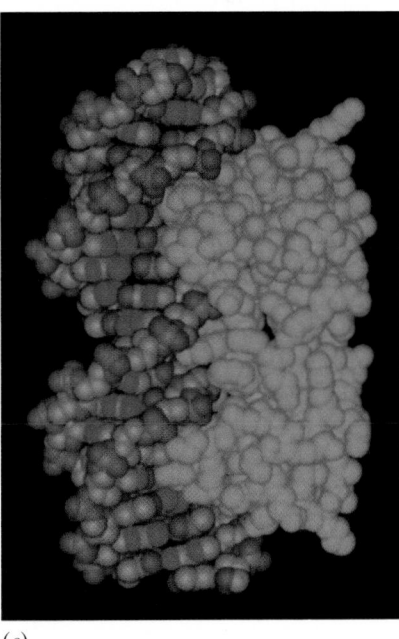

(a)

(c)

FIGURE 31-30 X-Ray structure of the 72-residue 434 Cro protein in complex with the same 20-bp DNA shown in Fig. 31-29. The complex is viewed perpendicular to its 2-fold axis of symmetry. Only the first 64 residues of the protein are visible. Parts *a*, *b*, and *c* correspond to those in Fig. 31-29 with the protein in Part *c* shown in cyan. Note the close but not identical correspondence between the two structures. [Courtesy of Alfonso Mondragón, Cynthia Wolberger, and Stephen Harrison, Harvard University. PDBid 3CRO.]

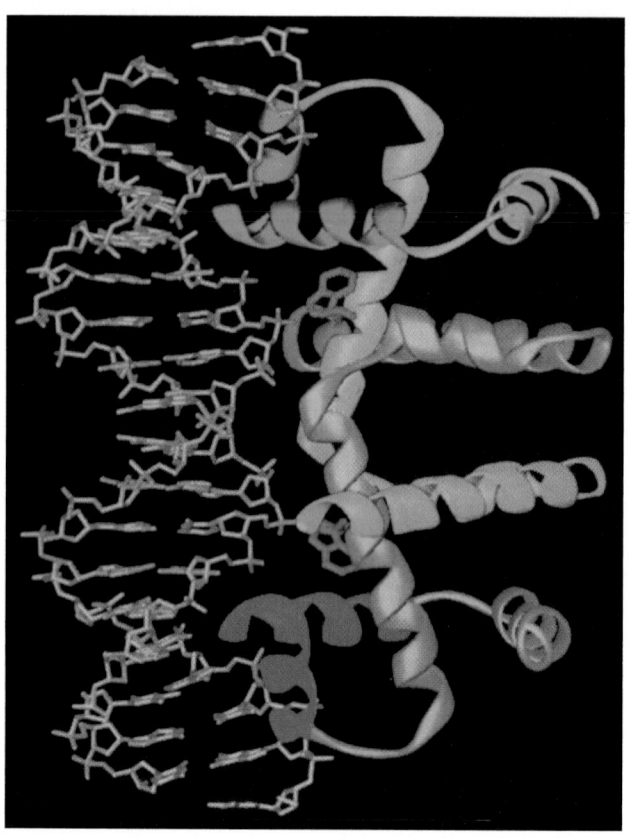

FIGURE 31-31 **X-Ray structure of an *E. coli trp* repressor–operator complex.** The complex is viewed with its molecular 2-fold axis horizontal. The protein's two identical subunits are green and blue with their HTH motifs (helices D and E) more deeply colored. The 18-bp-containing self-complementary DNA is gold. *trp* repressor binds its operator only when L-tryptophan (*red*) is simultaneously bound. Note that the protein's recognition helices (E) bind, as expected, in successive major grooves of the DNA but extend approximately perpendicular to the DNA duplex axis. In contrast, the recognition helices of 434 repressor and Cro proteins are nearly parallel to the major grooves of their bound DNAs (Figs. 31-29 and 31-30), whereas those of CAP assume an intermediate orientation (Fig. 31-28). [Based on an X-ray structure by Paul Sigler, Yale University. PDBid 1TRO.] *See the Interactive Exercises*

protein's recognition helices bind, as expected, in successive major grooves of the DNA, each in contact with an operator half-site (<u>ACTAGT</u>; Fig. 31-31). There are numerous hydrogen bonding contacts between the *trp* repressor and its bound DNA's nonesterified phosphate oxygens. Astoundingly, however, *there are no direct hydrogen bonds or nonpolar contacts that can explain the repressor's specificity for its operator. Rather, all but one of the side chain–base hydrogen bonding interactions are mediated by bridging water molecules* (the one direct interaction involves a base that can be mutated without greatly affecting repressor binding affinity). Such buried water molecules have therefore been described as "honorary" protein side chains. In addition, the operator contains several base pairs that are not in contact with the repressor but whose mutation nevertheless greatly decreases repressor binding affinity. This suggests that the operator assumes a sequence-specific conformation that makes favorable contacts with the repressor. Indeed, comparison of the X-ray structure of an uncomplexed 10-bp self-complementary DNA containing the *trp* operator's half-site (CC<u>ACTAGT</u>GG) with that of the DNA in the *trp* repressor–operator complex reveals that the ACTAGT half-site assumes nearly identical idiosyncratic conformations and patterns of hydration in both structures. However, the B-DNA helix, which is straight in the DNA 10-mer, is bent by 15° toward the major groove in each operator half-site of the repressor–operator complex. Other

DNA sequences could conceivably assume the repressor-bound operator's conformation but at too high an energy cost to form a stable complex with repressor (*trp* repressor's measured 10^4-fold preference for its operator over other DNAs implies an ~23 kJ · mol^{-1} difference in their binding free energies). This phenomenon, in which a protein senses the base sequence of DNA through the DNA's backbone conformation and/or flexibility, is referred to as **indirect readout.** 434 repressor apparently also employs indirect readout: Replacing the central A · T base pair of the operator shown in Fig. 31-29 with G · C reduces repressor binding affinity by 50-fold even though 434 repressor does not contact this region of the DNA.

It therefore appears that *there are no simple rules governing how particular amino acid residues interact with bases. Rather, sequence specificity results from an ensemble of mutually favorable interactions between a protein and its target DNA.*

b. *met* Repressor Contains a Two-Stranded Antiparallel β Sheet That Binds in Its Target DNA's Major Groove

The *E. coli* ***met* repressor (MetJ),** when complexed with *S*-adenosylmethionine (SAM; Fig. 26-18), represses the transcription of its own gene and those encoding enzymes involved in the synthesis of methionine (Fig. 26-60) and SAM. The X-ray structure of the *met* repressor–SAM–operator complex (Fig. 31-32), determined by Simon

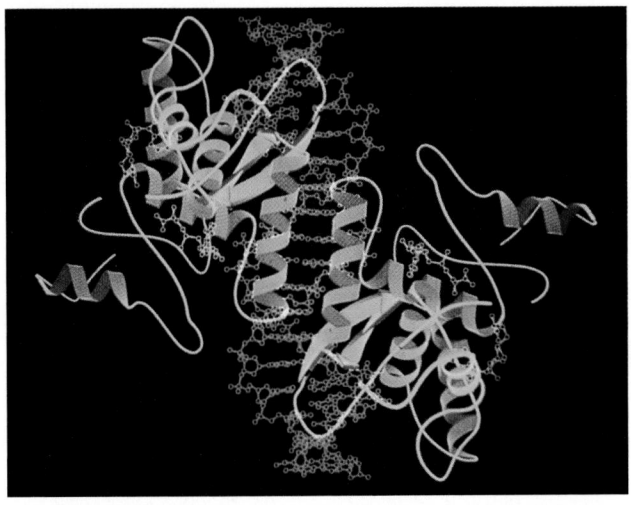

(a)

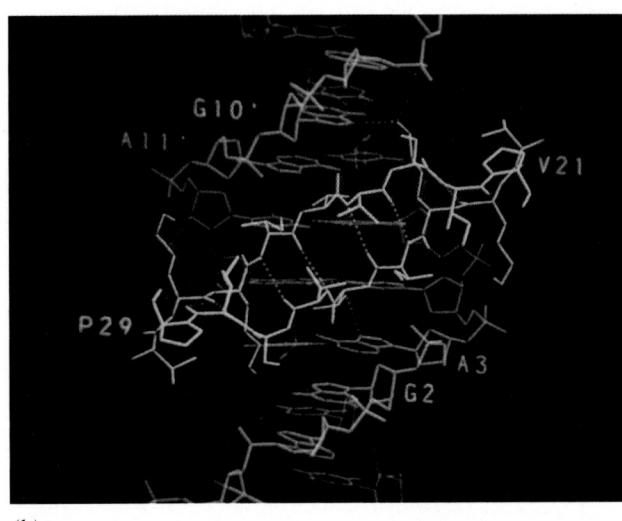

(b)

FIGURE 31-32 X-Ray structure of the *E. coli met* repressor–SAM–operator complex. (*a*) The overall structure of the complex as viewed along its 2-fold axis of symmetry. The 104-residue repressor subunits are shown in gold. The self-complementary 19-bp DNA and SAM, which must be bound to the repressor for it to also bind DNA, are shown in ball-and-stick form with the DNA blue and SAM green. Note that the DNA has four bound repressor subunits: Pairs of subunits form symmetric dimers in which each subunit donates one strand of the 2-stranded antiparallel β ribbon that is inserted in the DNA's major groove (*upper left and lower right*). Two such dimers pair across the complex's 2-fold axis via their antiparallel N-terminal helices which contact one another over the DNA's minor groove. (*b*) Detailed view of the 2-stranded antiparallel β ribbon (*yellow*, residues 21–29) inserted into the DNA's major groove (*blue*). Hydrogen bonds are indicated by dashed lines. [Courtesy of Simon Phillips, University of Leeds, U.K. PDBid 1CMA.]

🔎 **See the Interactive Exercises**

Phillips, reveals a symmetric homodimer of intertwined monomers that lacks an HTH motif. Rather, *met* repressor binds to its palindromic target DNA sequence through a symmetry-related pair of symmetrical two-stranded antiparallel β sheets (called **β ribbons**) that are inserted in successive major grooves of the DNA. Each β ribbon makes sequence-specific contacts with its target DNA sequence via hydrogen bonding and, probably, indirect readout.

Phillips first determined the X-ray structure of *met* repressor in the absence of DNA. Model building studies aimed at elucidating how *met* repressor binds to its palindromic target DNA assumed that the 2-fold rotation axes of both molecules would be coincident, as they are in all prokaryotic protein–DNA complexes of known structure. There were, consequently, two reasonable choices: (1) The protein could dock to the DNA with the above pairs of β ribbons entering successive major grooves; or (2) a symmetry-related pair of protruding α helices on the opposite face of the protein could do so in a manner resembling the way in which the recognition helices of HTH motifs interact with DNA. A variety of structural criteria suggested that the α helices make significantly better contacts with the DNA than do the β ribbons. Thus, the observation that it is, in fact, the β ribbons that bind to the DNA provides an important lesson: *The results of model building studies must be treated with utmost caution.* This is because our imprecise understanding of the energetics of intermolecular interactions (Sections 8-4 and 29-2) prevents us from reliably predicting how associating macromolecules conform to one another. In the case of the Met repressor, unpredicted mutual structural accommodations of the protein and DNA yielded a significantly more extensive interface than had been predicted by simply docking the uncomplexed Met repressor to canonical B-DNA.

The numerous prokaryotic transcriptional regulators of known structure either contain an HTH motif or pairs of β ribbons like the *met* repressor (although numerous prokaryotic DNA-binding proteins, including CAP, contain an elaboration of the HTH motif known as the **winged helix** motif in which two protein loops, one of which contacts the DNA's minor groove, flank the HTH recognition helix like the wings of a butterfly). Moreover, most of these proteins are homodimers that bind to palindromic or pseudopalindromic DNA target sequences. However, eukaryotic transcription factors, as we shall see in Section 34-3B, employ a much wider variety of structural motifs to bind their target DNAs, many of which lack symmetry.

E. *araBAD* Operon: Positive and Negative Control by the Same Protein

Humans neither metabolize nor intestinally absorb the plant sugar L-arabinose. Hence, the *E. coli* that normally inhabit the human gut are periodically presented with a banquet of this pentose. Three of the five *E. coli* enzymes

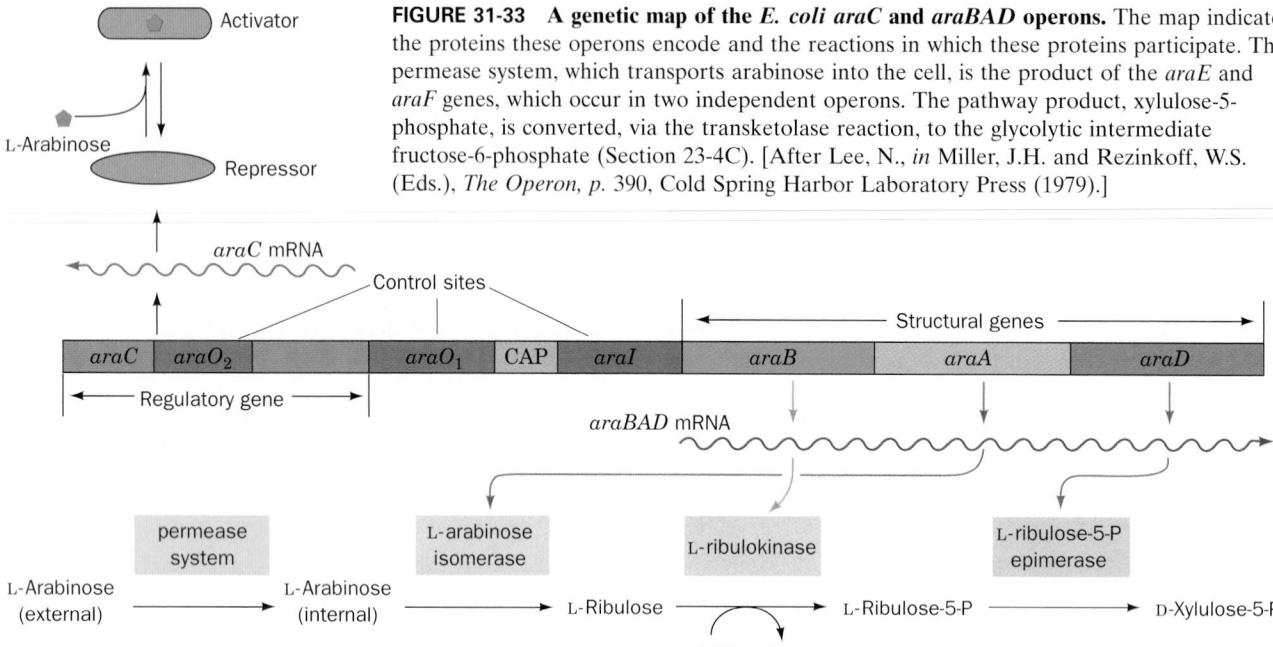

FIGURE 31-33 A genetic map of the *E. coli araC* and *araBAD* operons. The map indicates the proteins these operons encode and the reactions in which these proteins participate. The permease system, which transports arabinose into the cell, is the product of the *araE* and *araF* genes, which occur in two independent operons. The pathway product, xylulose-5-phosphate, is converted, via the transketolase reaction, to the glycolytic intermediate fructose-6-phosphate (Section 23-4C). [After Lee, N., *in* Miller, J.H. and Rezinkoff, W.S. (Eds.), *The Operon*, p. 390, Cold Spring Harbor Laboratory Press (1979).]

that metabolize arabinose are products of the catabolite repressible ***araBAD* operon** (Fig. 31-33).

The *araBAD* operon, as Robert Schleif has shown, contains, moving upstream from its transcriptional start site, the *araI*, *araO₁*, and *araO₂* control sites (Fig. 31-34*a*). The *araI* site (*I* for inducer) consists of two closely similar 17-bp half-sites, *araI₁* and *araI₂*, that are direct repeats separated by 4 bp and are oriented such that *araI₂*, which overlaps the −35 region of the *araBAD* promoter, is downstream of *araI₁*. Likewise, *araO₁* consists of two directly repeating half-sites, O_{1L} and O_{1R}. Intriguingly, however, *araO₂* consists of a single half-site that is located in a noncoding upstream region of the *araC* gene (see below), at position −270 relative to the *araBAD* start site.

The transcription of the *araBAD* operon is regulated by both CAP–cAMP and the arabinose-binding protein **AraC.** Each 292-subunit of the homodimeric AraC consists of an N-terminal, arabinose-binding, dimerization domain (residues 1–170) connected via a flexible linker to a C-terminal DNA-binding domain (residues 178–292). Regulation of the *araBAD* operon occurs as follows (Fig. 31-34):

1. In the absence of AraC, RNA polymerase initiates transcription of the *araC* gene in the direction away from its upstream neighbor, *araBAD*. The *araBAD* operon is expressed at a low basal level.

2. When AraC is present, but neither arabinose nor CAP–cAMP (high glucose), AraC binds to *araO₂* and *araI₁*. The binding of AraC to *araI₁* prevents RNAP from initiating transcription of the *araBAD* operon (negative control). A series of deletion mutations indicate that the presence of *araO₂* is also required for the repression of *araBAD*. The remarkably large 211-bp separation between *araO₂* and *araI₁* therefore strongly suggests that the DNA

between them is looped such that a dimeric molecule of AraC protein simultaneously binds to both *araO₂* and *araI₁*. This is corroborated by the observation that the level of repression is greatly diminished by the insertion of 5 bp (half a turn) of DNA between these two sites, thereby transferring *araO₂* to the opposite face of the DNA helix relative to *araI₁* in the putative loop. Yet, the insertion of 11 bp (one turn) of DNA has no such effect. Moreover, looping does not readily occur unless the DNA is supercoiled, which presumably drives the looping process. The AraC dimer also binds to *araO₁*, the operator of the *araC* gene, so as to block the transcription of *araC* but only at high concentrations. Thus, it is likely that DNA looping itself represses the transcription of *araC*. In either case, the expression of *araC* is autoregulatory.

3. When arabinose is present, it allosterically induces the AraC subunit bound to *araO₂* to instead bind to *araI₂*. This activates RNAP to transcribe the *araBAD* genes (positive control). When the cAMP level is high (low glucose), CAP–cAMP, whose presence is required to achieve the maximum level of transcriptional activation, binds to a site between *araO₁* and *araI₁*, where it functions to help break the loop between *araO₂* and *araI₁* and hence to increase the affinity of AraC for *araI₂*. The orientation of *araO₁* with respect to *araC* is opposite to that of *araI* with respect to *araBAD*, and hence the binding of AraC–arabinose at *araO₁* blocks RNAP binding at the *araC* promoter, that is, it represses the expression of AraC.

If the *araI₂* subsite is mutated so as to increase AraC's affinity for it, arabinose is no longer required for transcriptional activation. This suggests that arabinose does not conformationally transform AraC to an activator but, rather, weakens its binding affinity for *araO₂*. If the *araI* site is turned around or if it is moved upstream so that *araI₂*

(a) **When AraC is absent, *araC* is transcribed and *araBAD* is transcribed at a basal level**

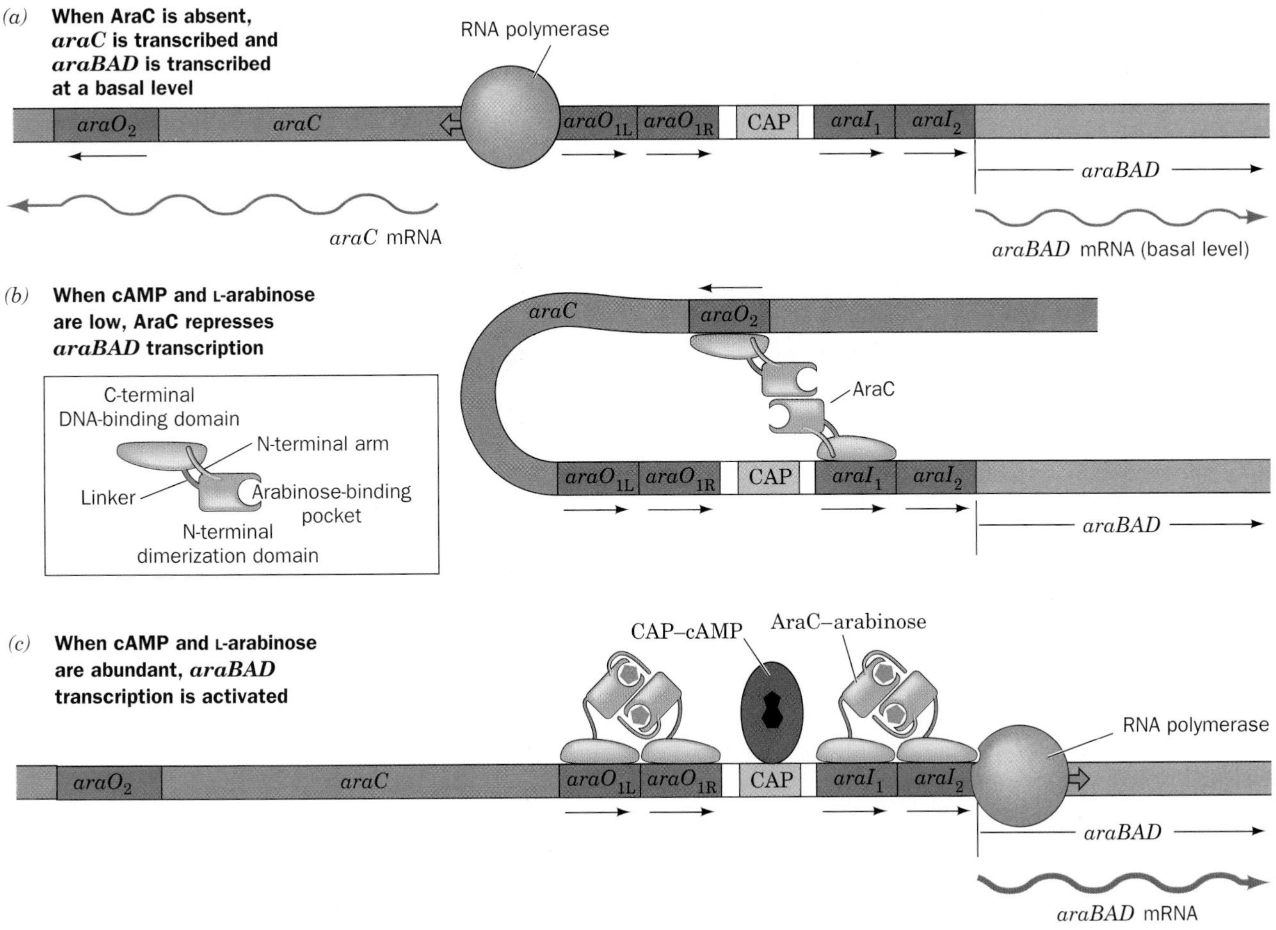

RNA polymerase

| $araO_2$ | $araC$ | | $araO_{1L}$ | $araO_{1R}$ | CAP | $araI_1$ | $araI_2$ | |

araBAD

araC mRNA

araBAD mRNA (basal level)

(b) **When cAMP and L-arabinose are low, AraC represses *araBAD* transcription**

C-terminal
DNA-binding domain

N-terminal arm

Linker

Arabinose-binding pocket

N-terminal
dimerization domain

$araC$

$araO_2$

AraC

| $araO_{1L}$ | $araO_{1R}$ | CAP | $araI_1$ | $araI_2$ |

araBAD

(c) **When cAMP and L-arabinose are abundant, *araBAD* transcription is activated**

CAP–cAMP   AraC–arabinose

RNA polymerase

| $araO_2$ | $araC$ | | $araO_{1L}$ | $araO_{1R}$ | CAP | $araI_1$ | $araI_2$ | |

araBAD

araBAD mRNA

FIGURE 31-34   The mechanism of *araBAD* regulation. *(a)* In the absence of AraC, RNAP initiates the transcription of *araC*. *araBAD* is also expressed but at a low basal level. *(b)* When AraC is present, but not L-arabinose or cAMP, AraC links together *araO_2* and *araI_1* to form a DNA loop, thereby repressing both *araC* and *araBAD*. *(c)* When AraC and L-arabinose are both present and cAMP is abundant, the resulting AraC–arabinose complex releases *araO_2* and instead binds *araI_2*, thereby activating *araBAD* transcription. This process is facilitated by the binding of CAP–cAMP. *araC* is repressed by

does not overlap the *araBAD* promoter, AraC cannot stimulate transcription. Evidently, *AraC activates RNAP through specific and relatively inflexible protein–protein interactions.*

The X-ray structures of the N-terminal domain of AraC (residues 2–178), in both the presence and the absence of arabinose, were determined by Schleif and Cynthia Wolberger. In the presence of arabinose, this domain consists of an 8-stranded β barrel followed by two antiparallel α helices (Fig. 31-35). Two such domains associate via an antiparallel coiled coil between each of their C-terminal helices to form the protein's dimerization interface. An arabinose molecule binds in a pocket of each β barrel via a network of direct and water-mediated hydrogen bonds with side chains that line the pocket. Residues 7 to 18 of the N-terminal arm lie across the mouth of the sugar-binding pocket (residues 2–6 are disordered), thereby fully enclosing the arabinose. The structure of the N-terminal domain in the absence of arabinose is largely superimposable on that in the complex with arabinose, with the

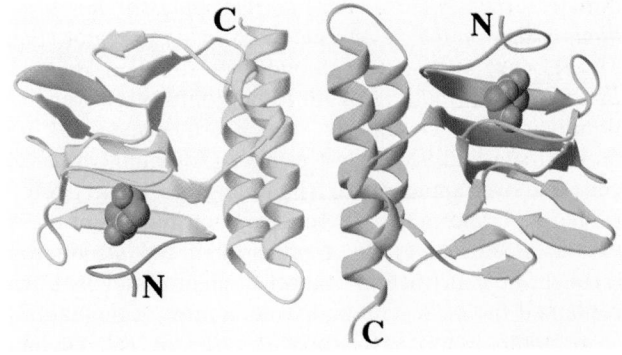

FIGURE 31-35   X-Ray structure of *E. coli* AraC in complex with L-arabinose. The homodimeric protein is viewed along its 2-fold axis with its subunits cyan and gold except for their N-terminal arms, which are orange and magenta. The arabinose is drawn in space-filling form with C green and O red. [Based on an X-ray structure by Robert Schleif and Cynthia Wolberger, Johns Hopkins University. PDBid 2ARC.]

exception that the N-terminal arm is disordered, a not unexpected observation considering that it interacts with bound arabinose via a series of hydrogen bonds.

How does arabinose binding induce the AraC subunit bound at *araO$_2$* to instead bind to *araI$_2$*? Several lines of evidence indicate that, in the absence of arabinose, AraC's N-terminal arm binds to its DNA-binding domain in a way that favors loop formation: (1) the deletion of the N-terminal arm beyond its sixth residue makes AraC act as if arabinose is present; (2) mutations to surface residues on the DNA-binding domain that presumably eliminate its binding of the N-terminal arm also constitutively activate AraC; and (3) mutations in the DNA-binding domain that weaken the binding of arabinose to the protein, presumably by strengthening the binding of the N-terminal arm, can be suppressed by a second mutation in the N-terminal arm or by the deletion of its five N-terminal residues. Evidently, *the binding of the N-terminal arms to the DNA-binding domains in the absence of arabinose rigidifies the AraC dimer such that it cannot simultaneously bind to the directly repeated araI$_1$ and araI$_2$ and hence induce the transcription of araBAD.* This is corroborated by the observations that (1) joining two AraC DNA-binding domains by flexible polypeptide linkers yields proteins that behave like AraC in the presence of arabinose, and (2) a construct consisting of two double-stranded *araI$_1$* half-sites flexibly connected by a 24-nt segment of ssDNA binds wild-type AraC with an affinity that is unaffected by arabinose.

F. *lac Repressor II: Structure*

Here we continue our discussions of the *lac* repressor, but now in terms of the concepts learned in Sections 31-3CDE.

a. Loop Formation Is Important in the Expression of the *lac* Operon

DNA loop formation, which is now known to occur in numerous bacterial and eukaryotic systems, apparently permits several regulatory proteins and/or regulatory sites on one protein to simultaneously influence transcription initiation by RNAP. In fact, *the lac repressor has three binding sites on the lac operon:* the primary operator (Fig. 31-25), now known as O_1, and two so-called pseudo-operators (previously thought to be nonfunctional evolutionary fossils), O_2 and O_3, which are located 401 bp downstream and 92 bp upstream of O_1 (within the *lacZ* gene and overlapping the CAP binding site, respectively). Müller-Hill determined the relative contributions of these various operators to the repression of the *lac* operon through the construction of a set of eight plasmids: Each contained the *lacZ* gene under the control of the natural *lac* promoter as well as the three *lac* operators (O_1, O_2, and O_3), which were either active or mutagenically inactive in all possible combinations. When all three operators are active, *lacZ* expression is repressed 1300-fold relative to when all three operators are inactive. The inactivation of only O_1 results in almost complete loss of repression whereas the inactivation of only O_2 or O_3 causes only a ~2-fold loss in repression. However, when O_2 and O_3 are

both inactive, repression is decreased ~70-fold. These results suggest that efficient repression requires the formation of a DNA loop between O_1 and either O_2 or O_3. Indeed, such loop formation, and/or the cooperativity of repressor binding arising from it, appears to be a greater contributor to repression than repressor binding to O_1 alone, which provides only 19-fold repression.

b. The *lac* Repressor Is a Dimer of Dimers

Ponzy Lu and Mitchell Lewis determined the X-ray structures of the *lac* repressor alone, in its complex with IPTG, and in its complex with a 21-bp duplex DNA segment whose sequence is a palindrome of the left half of O_1 (Fig. 31-25). Each repressor subunit consists of five functional units (Fig. 31-36): (1) an N-terminal DNA-binding domain (residues 1–49) which is known as the "headpiece" because it is readily proteolytically cleaved away from the remaining still tetrameric "core" protein; (2) a hinge helix (residues 50–58) that also binds to the DNA; (3 & 4) a sugar-binding domain (residues 62–333) that is divided into an N-subdomain and a C-subdomain; and (5) a C-terminal tetramerization helix (residues 340–360).

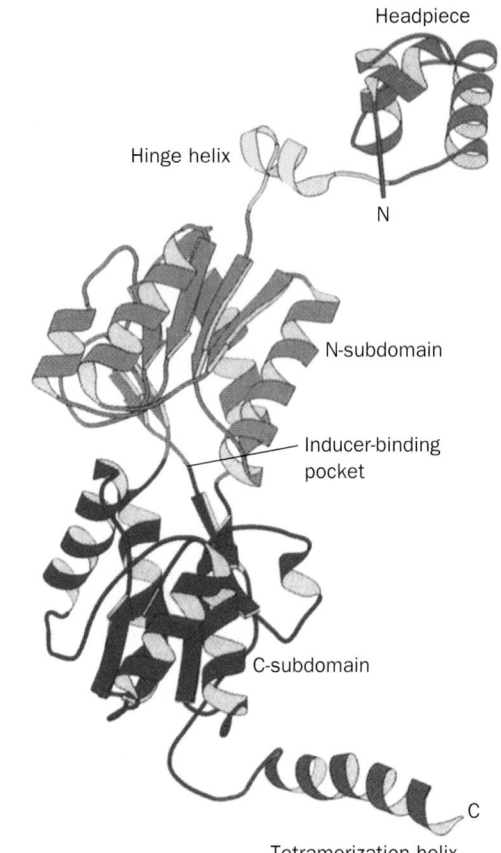

FIGURE 31-36   X-Ray structure of the *lac* repressor subunit. The DNA-binding domain (the headpiece), which contains an HTH motif, is red, the DNA-binding hinge helix is yellow, the N-subdomain of the sugar-binding domain is light blue, its C-subdomain is dark blue, and the tetramerization helix is purple. [Courtesy of Ponzy Lu and Mitchell Lewis, University of Pennsylvania. PDBid 1LBI.]

The *lac* repressor has an unusual quaternary structure (Fig. 31-37a). Whereas nearly all homotetrameric non-membrane proteins of known structure have D_2 symmetry (three mutually perpendicular 2-fold axes; Fig. 8-64b), *lac* repressor is a V-shaped protein that has only 2-fold symmetry. Each leg of the V consists of a locally symmetric dimer of closely associated repressor subunits. Two such

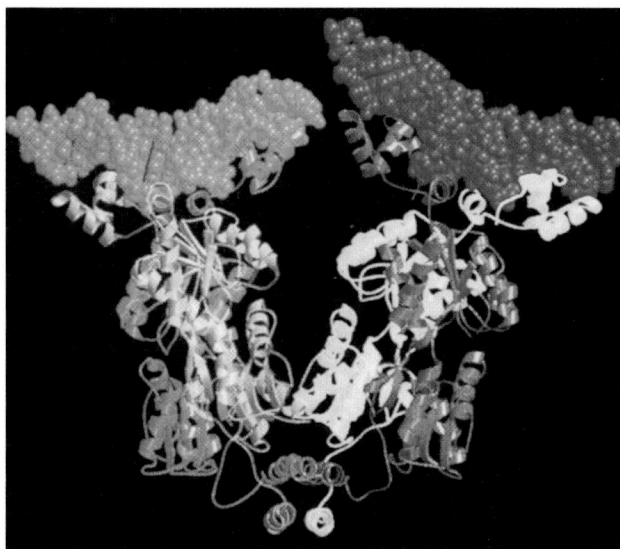

(a)

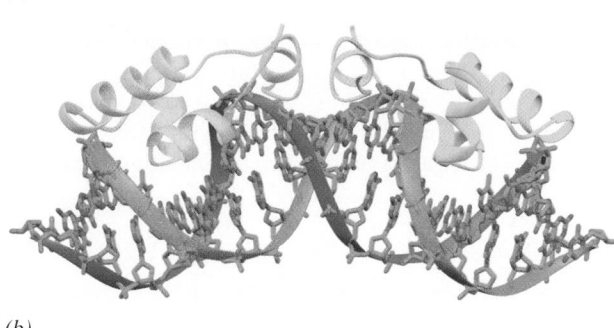

(b)

FIGURE 31-37 The structure of the *lac* repressor in complex with DNA. (*a*) The X-ray structure of the *lac* repressor tetramer bound to two 21-bp segments of symmetric *lac* operator DNA. The protein monomers are green, pink, yellow, and red and the DNA segments, drawn in space-filling form, are cyan and blue. [Courtesy of Ponzy Lu and Mitchell Lewis, University of Pennsylvania. PDBid 1LBG.] (*b*) The NMR structure of a 22-bp symmetric *lac* operator DNA in complex with two segments of the *lac* repressor consisting of its DNA-binding domain and its hinge helix. The two protein subunits are cyan and gold. The DNA is represented in stick form colored according to atom type (C green, N blue, O red, and P magenta) with its sugar–phosphate backbones traced by red and blue ribbons. The complex is viewed with its 2-fold axis vertical. Note that the protein dimer's two HTH motifs are inserted in successive major grooves at the periphery of the complex and that the insertion of the two centrally located hinge helices into the DNA's minor groove greatly widens and flattens the minor groove at this point and kinks the DNA in a downward bend. [Based on an NMR structure by Robert Kaptein, Utrecht University, The Netherlands. PDBid 1CJG.]

dimers associate rather tenuously, but with 2-fold symmetry, at the base (point) of the V to form a dimer of dimers.

In the structures of *lac* repressor alone and that of its IPTG complex, the DNA-binding domain is not visible, apparently because the hinge region that loosely tethers it to the rest of the protein is disordered. However, in the DNA complex, in which one DNA duplex binds to each of the two dimers forming the repressor tetramer, the DNA domain forms a compact globule containing three helices, the first two of which form a helix–turn–helix (HTH) motif. The two DNA-binding domains extending from each repressor dimer (at the top of each leg of the V) bind in successive major grooves of a DNA molecule via their HTH motifs, much as is seen, for example, in the complexes of 434 phage repressor and *trp* repressor with their target DNAs (Figs. 31-29 and 31-31). The binding of the *lac* repressor distorts the operator DNA such that it bends away from the DNA-binding domain with an ~60 Å radius of curvature due to an ~45° kink at the center of the operator that widens the DNA's minor groove to over 11 Å and reduces its depth to less than 1 Å. These distortions permit the now ordered hinge helix to bind in the minor groove so as to contact the identically bound hinge helix from the other subunit of the same dimer. NMR structures by Robert Kaptein reveal that the DNA-binding domain, when cleaved from the repressor, binds to the *lac* operator without distorting the DNA. However, the DNA-binding domain together with the hinge helix forms a complex with the *lac* operator in which the hinge helix binds in the DNA's distorted minor groove (Fig. 31-37b) as in the X-ray structure. Thus, the binding of the two hinge helices to the *lac* operator appears necessary for DNA distortion. The two DNA duplexes that are bound to each repressor tetramer are ~25 Å apart and do not interact.

The sugar-binding domain consists of two topologically similar subdomains that are bridged by three polypeptide segments (Fig. 31-36). The two sugar-binding domains of a dimer make extensive contacts (Fig. 31-37a). IPTG binds to each sugar-binding domain between its subdomains. This does not significantly change the conformations of these subdomains, but it changes the angle between them. Although the hinge helix is not visible in the IPTG complex, model building indicates that, since the dimer's two hinge helices extend from its sugar-binding domains, this conformation change levers apart these hinge helices by 3.5 Å such that they and their attached HTH motifs can no longer simultaneously bind to their operator half-sites. Thus, inducer binding, which is allosteric within the dimer (has a positive homotropic effect; Section 10-4B), greatly loosens the repressor's grip on the operator.

The C-terminal helices from each subunit, which are located on the opposite end of each subunit from the DNA-binding portion (at the point of the V), associate to form a 4-helix bundle that holds together the two repressor dimers, thereby forming the tetramer (Fig. 31-37a). The allosteric effects of inducer binding within each dimer are apparently not transmitted between dimers. Moreover, the *E. coli* **purine repressor (PurR),** which is homologous to the *lac* repressor but lacks its C-terminal helix, crystallizes

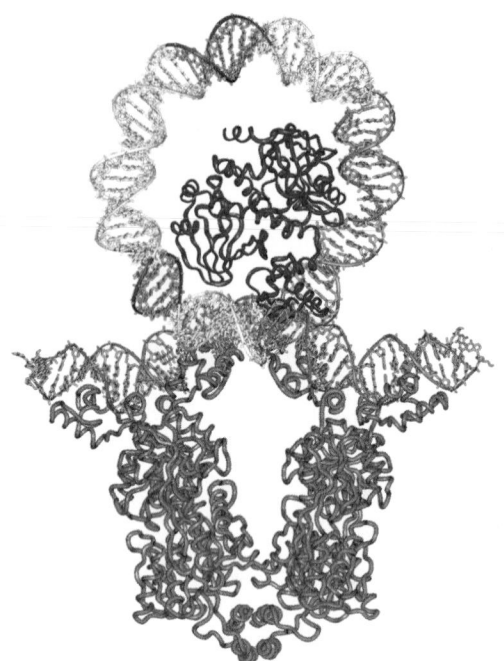

FIGURE 31-38 Model of the 93-bp DNA loop formed when *lac* repressor binds to O_1 and O_3. The proteins are represented by their C_α backbones and the DNA is drawn in stick form with its sugar–phosphate backbones traced by helical ribbons. The model was constructed from the X-ray structure of the *lac* repressor (*magenta*) in complex with two 21-bp operator DNA segments (*red*) and the X-ray structure of CAP–cAMP (*blue*) in complex with its 30-bp target DNA (*cyan*; Fig. 31-28). The remainder of the DNA loop was generated by applying a smooth curvature to canonical B-DNA (*white*) with the −10 and −35 regions of the *lac* promoter highlighted in green. [Courtesy of Ponzy Lu and Mitchell Lewis, University of Pennsylvania.]

it from fully engaging the promoter in this looped complex, thereby maximizing repression.

c. Combining Genetic and Structural Studies of the *lac* Repressor Reveals Its Allosterically Important Residues

The phenotypes of 4042 point mutations of the *lac* repressor, which encompass nearly all of its 360 residues (making the *lac* repressor the most exhaustively mutationally characterized protein known) have been mapped onto its X-ray structure. Mutations with an I$^-$ phenotype (*lac* repressors that fail to bind to the *lac* operator, so that β-galactosidase is constitutively synthesized) are located at the *lac* repressor's DNA-binding interface, at its dimer interface, or at internal residues of its inducer-binding core domain. Residues whose mutations result in the I^S phenotype (S for super-repressed; *lac* repressors that, in the presence of inducer, continue to repress the synthesis of β-galactosidase) appear to be of two types: (1) residues that are in direct contact with the inducer, whose alteration therefore interferes with inducer binding; and (2) residues at the dimer interface that are >8 Å from (not in direct contact with) the inducer-binding site. These latter observations reveal which residues mediate the *lac* repressor's allosteric mechanism rather than directly binding the inducer or the DNA. Most of the allosterically important residues are located at the dimer interface and are members of the N-subdomain of the core domain, which links the inducer-binding sites to the operator DNA-binding sites. This is consistent with the observation that inducer binding causes a relative twist and translation of the N-subdomain, a movement which is propagated to the hinge helix and DNA-binding domain. This study demonstrates the power of combining genetic analysis with structural studies to elucidate structure–function relationships.

G. *trp* Operon: Attenuation

We now discuss a sophisticated transcriptional control mechanism named **attenuation** through which bacteria regulate the expression of certain operons involved in amino acid biosynthesis. This mechanism was discovered through the study of the *E. coli* **trp operon** (Fig. 31-39), which encodes five polypeptides comprising three enzymes that mediate the synthesis of tryptophan from chorismate (Section 26-5B). Charles Yanofsky established that the *trp* operon genes are coordinately expressed under the control of the *trp* repressor, a dimeric protein of identical 107-residue subunits that is the product of the *trpR* gene (which forms an independent operon). *The trp repressor binds L-tryptophan, the pathway's end product, to form a complex that specifically binds to trp operator (trpO, Fig. 31-40) so as to reduce the rate of trp operon transcription 70-fold.* The X-ray structure of the *trp* repressor–operator complex (Section 31-3D) indicates that tryptophan binding allosterically orients *trp* repressor's two symmetry related helix–turn–helix "DNA reading heads" so that they can simultaneously bind to *trpO* (Fig. 31-31). Moreover, the bound tryptophan forms a hydrogen bond to a DNA

as a dimer whose X-ray structure closely resembles that of the *lac* repressor dimer. What then is the function of *lac* repressor tetramerization?

Model building suggests that when the *lac* repressor tetramer simultaneously binds to both the O_1 and O_3 operators, the 93-bp DNA segment containing them forms a loop ~80 Å in diameter (Fig. 31-38). Furthermore, the CAP–cAMP binding site is exposed on the inner surface of the loop. Adding the CAP–cAMP at its proper position to this model reveals that the ~90° curvature which CAP–cAMP binding imposes on DNA (Fig. 31-28) has the correct direction and magnitude to stabilize the DNA loop, thereby stabilizing this putative CAP–cAMP–*lac* repressor–DNA complex. It may seem paradoxical that the binding of CAP–cAMP, a transcriptional activator, stabilizes the repressor–DNA complex. However, when both glucose and lactose are in short supply, it is important that the bacterium lower its basal rate of *lac* operon expression in order to conserve energy. The binding site (promoter) for RNAP is also located on the inner surface of the loop. Thus, the large size of the RNAP molecule would prevent

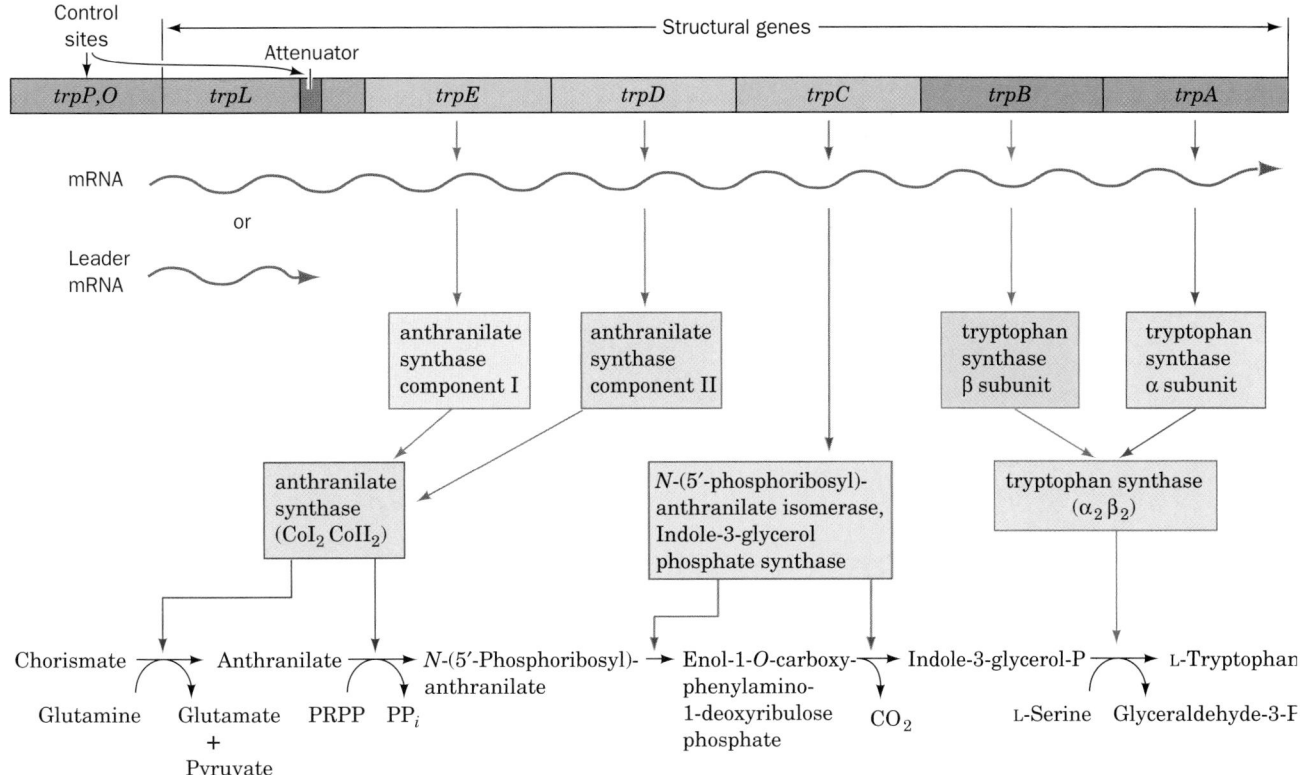

FIGURE 31-39   A genetic map of the *E. coli* *trp* operon indicating the enzymes it specifies and the reactions they catalyze. The gene product of *trpC* catalyzes two sequential reactions in the synthesis of tryptophan. [After Yanofsky, C., *J. Am. Med. Assoc.* **218**, 1027 (1971).]

phosphate group, thereby strengthening the repressor–operator association. Tryptophan therefore acts as a **corepressor;** its presence prevents what is then superfluous tryptophan biosynthesis (SAM similarly functions as a corepressor with the *met* repressor; Fig. 31-32*a*). The *trp* repressor also controls the synthesis of at least two other operons: the **trpR operon** and the **aroH operon** (which encodes one of three isozymes that catalyze the initial reaction of chorismate biosynthesis; Section 26-5B).

a. Tryptophan Biosynthesis Is Also Regulated by Attenuation

The *trp* repressor–operator system was at first thought to fully account for the regulation of tryptophan biosynthesis in *E. coli*. However, the discovery of *trp* deletion

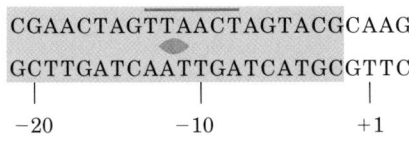

FIGURE 31-40   The base sequence of the *trp* operator. The nearly palindromic sequence is boxed and its −10 region is overscored.

mutants located downstream from *trpO* that increase *trp* operon expression 6-fold indicated the existence of an additional transcriptional control element. Sequence analysis established that *trpE*, the *trp* operon's leading structural gene, is preceded by a 162-nucleotide **leader sequence** (*trpL*). Genetic analysis indicated that the new control element is located in *trpL*, ~30 to 60 nucleotides upstream of *trpE* (Fig. 31-39).

When tryptophan is scarce, the entire 6720-nucleotide polycistronic *trp* mRNA, including the *trpL* sequence, is synthesized. As the tryptophan concentration increases, the rate of *trp* transcription decreases as a result of the *trp* repressor–corepressor complex's consequent greater abundance. Of the *trp* mRNA that is transcribed, however, an increasing proportion consists of only a 140-nucleotide segment corresponding to the 5' end of *trpL*. *The availability of tryptophan therefore results in the premature termination of trp operon transcription.* The control element responsible for this effect is consequently termed an **attenuator.**

b. The *trp* Attenuator's Transcription Terminator Is Masked when Tryptophan Is Scarce

What is the mechanism of attenuation? The attenuator transcript contains four complementary segments that can form one of two sets of mutually exclusive base paired hair-

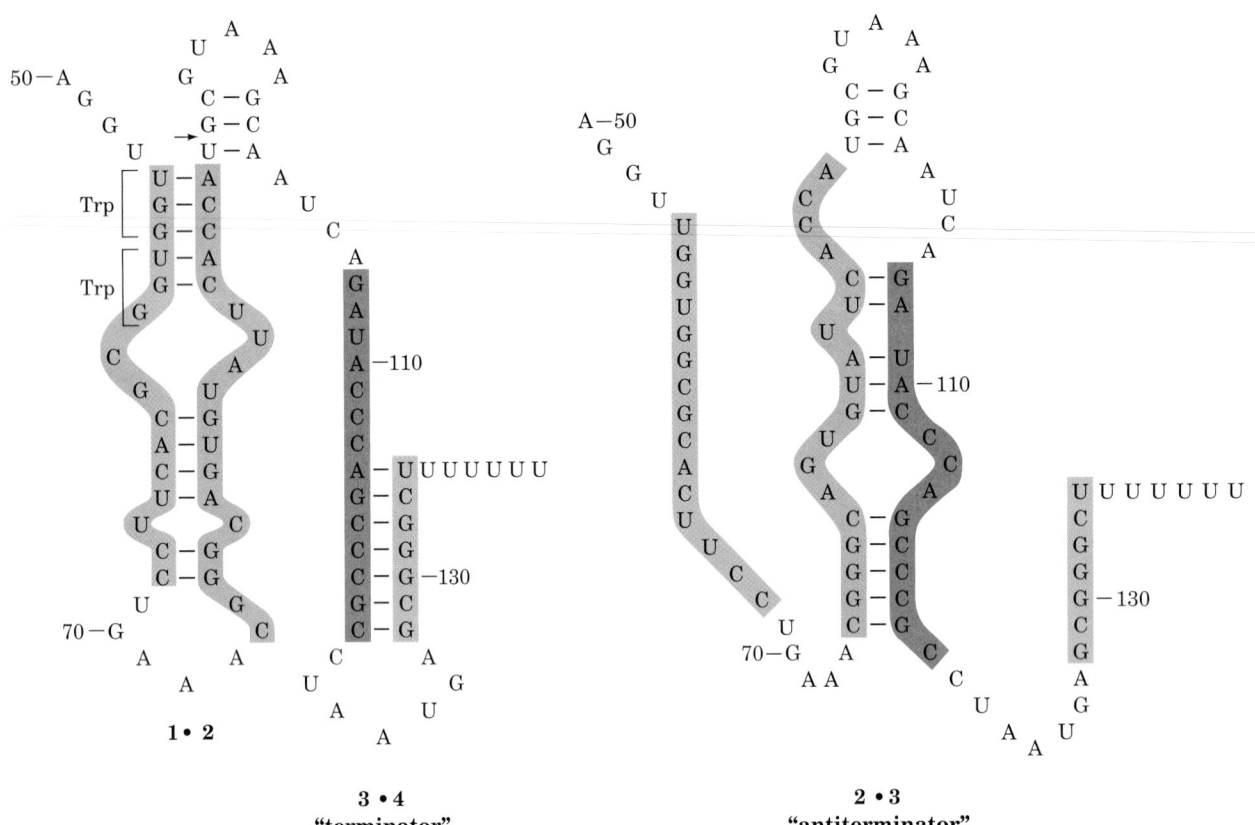

FIGURE 31-41   The alternative secondary structures of *trpL* mRNA. The formation of the base paired 2 · 3 (antiterminator) hairpin (*right*) precludes the formation of the 1 · 2 and 3 · 4 (terminator) hairpins (*left*) and vice versa. Attenuation results in the premature termination of transcription immediately after nucleotide 140 when the 3 · 4 hairpin is present. The arrow indicates the mRNA site past which RNA polymerase pauses until approached by an active ribosome. [After Fisher, R.F. and Yanofsky, C., *J. Biol. Chem.* **258**, 8147 (1983).]

pins (Fig. 31-41). *Segments 3 and 4 together with the succeeding residues comprise a normal rho-independent transcription terminator (Section 31-2D): a G + C-rich sequence that can form a self-complementary hairpin structure followed by several sequential U's (compare with Fig. 31-18). Transcription rarely proceeds beyond this termination site unless tryptophan is in short supply.*

A section of the leader sequence, which includes segment 1 of the attenuator, is translated to form a 14-residue polypeptide that contains two consecutive Trp residues (Fig. 31-41, *left*). The position of this particularly rare dipeptide segment (~1% of the residues in *E. coli* proteins are Trp) provided an important clue to the mechanism of attenuation. An additional essential aspect of this mechanism is that ribosomes commence the translation of a prokaryotic mRNA shortly after its 5' end has been synthesized.

The above considerations led Yanofsky to propose the following model of attenuation (Fig. 31-42). An RNA polymerase that has escaped repression initiates *trp* operon transcription. Soon after the ribosomal initiation site of the *trpL* gene has been transcribed, a ribosome attaches to it and begins translation of the leader peptide. When tryptophan is abundant, so that there is a plentiful supply of **tryptophanyl–tRNA**[Trp] (the transfer RNA specific for Trp

with an attached Trp residue; Section 32-2C), the ribosome follows closely behind the transcribing RNA polymerase so as to sterically block the formation of the 2 · 3 hairpin. Indeed, RNA polymerase pauses past position 92 of the transcript and only continues transcription on the approach of a ribosome, thereby ensuring the proximity of these two entities at this critical position. The prevention of 2 · 3 hairpin formation permits the formation of the 3 · 4 hairpin, the transcription terminator pause site, which results in the termination of transcription (Fig. 31-42*a*). When tryptophan is scarce, however, the ribosome stalls at the tandem UGG codons (which specify Trp; Table 5-3) because of the lack of tryptophanyl–tRNA[Trp]. As transcription continues, the newly synthesized segments 2 and 3 form a hairpin because the stalled ribosome prevents the otherwise competitive formation of the 1 · 2 hairpin (Fig. 31-42*b*). The formation of the transcriptional terminator's 3 · 4 hairpin is thereby preempted for sufficient time for RNA polymerase to transcribe through it and consequently through the remainder of the *trp* operon. The cell is thus provided with a regulatory mechanism that is responsive to the tryptophanyl–tRNA[Trp] level, which, in turn, depends on the protein synthesis rate as well as on the tryptophan supply.

There is considerable evidence supporting this model of attenuation. The *trpL* transcript is resistant to limited

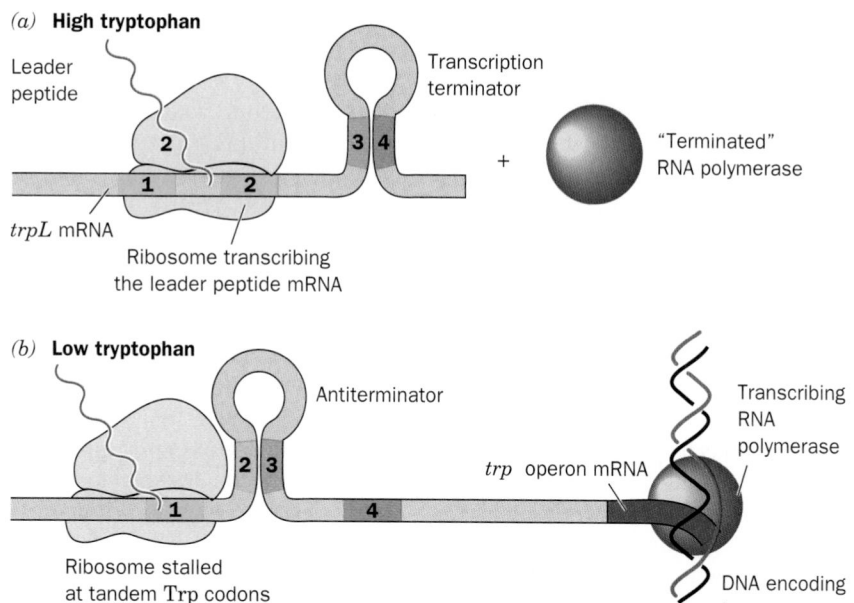

(a) **High tryptophan**

Leader peptide

Transcription terminator

2

3 4

1 2

+

"Terminated" RNA polymerase

trpL mRNA

Ribosome transcribing the leader peptide mRNA

(b) **Low tryptophan**

Antiterminator

Transcribing RNA polymerase

2 3

trp operon mRNA

1

4

Ribosome stalled at tandem Trp codons

DNA encoding *trp* operon

FIGURE 31-42 **Attenuation in the *trp* operon.** (*a*) When tryptophanyl–tRNA^Trp is abundant, the ribosome translates *trpL* mRNA. The presence of the ribosome on segment 2 prevents the formation of the base paired 2 · 3 hairpin. The 3 · 4 hairpin, an essential component of the transcriptional terminator, can thereby form, thus aborting transcription.

(*b*) When tryptophanyl–tRNA^Trp is scarce, the ribosome stalls on the tandem Trp codons of segment 1. This situation permits the formation of the 2 · 3 hairpin which, in turn, precludes the formation of the 3 · 4 hairpin. RNA polymerase therefore transcribes through this unformed terminator and continues *trp* operon transcription.

RNase T1 digestion, indicating that it has extensive secondary structure. The significance of the tandem Trp codons in the *trpL* transcript is corroborated by their presence in *trp* leader regions of several other bacterial species. Moreover, the leader peptides of the five other amino acid–biosynthesizing operons known to be regulated by attenuation (most exclusively so) are all rich in their corresponding amino acid residues (Table 31-3). For example, the *E. coli* **his operon,** which specifies enzymes synthesizing histidine (Fig. 26-65), has seven tandem His residues in its leader peptide whereas the **ilv operon,** which specifies enzymes participating in isoleucine, leucine, and valine biosynthesis (Fig. 26-61), has five Ile's, three Leu's, and

six Val's in its leader peptide. Finally, the leader transcripts of these operons resemble that of the *trp* operon in their capacity to form two alternative secondary structures, one of which contains a trailing termination structure.

H. *Regulation of Ribosomal RNA Synthesis: The Stringent Response*

E. coli cells growing under optimal conditions divide every 20 min. Such cells contain up to 70,000 ribosomes and hence must synthesize ~35,000 ribosomes per cell division cycle. Yet RNAP can initiate the transcription of an rRNA

TABLE 31-3. **Amino Acid Sequences of Some Leader Peptides in Operons Subject to Attenuation**

Operon	Amino Acid Sequence[a]
trp	Met-Lys-Ala-Ile-Phe-Val-Leu-Lys-Gly-TRP-TRP-Arg-Thr-Ser
pheA	Met-Lys-His-Ile-Pro-PHE-PHE-PHE-Ala-PHE-PHE-PHE-Thr-PHE-Pro
his	Met-Thr-Arg-Val-Gln-Phe-Lys-HIS-HIS-HIS-HIS-HIS-HIS-HIS-Pro-Asp
leu	Met-Ser-His-Ile-Val-Arg-Phe-Thr-Gly-LEU-LEU-LEU-LEU-Asn-Ala-Phe-Ile-Val-Arg-Gly-Arg-Pro-Val-Gly-Gly-Ile-Gln-His
thr	Met-Lys-Arg-ILE-Ser-THR-THR-ILE-THR-THR-THR-ILE-THR-ILE-THR-THR-Gln-Asn-Gly-Ala-Gly
ilv	Met-Thr-Ala-LEU-LEU-Arg-VAL-ILE-Ser-LEU-VAL-VAL-ILE-Ser-VAL-VAL-VAL-ILE-ILE-ILE-Pro-Pro-Cys-Gly-Ala-Ala-Leu-Gly-Arg-Gly-Lys-Ala

[a]Residues in uppercase are synthesized in the pathway catalyzed by the operon's gene products.

Source: Yanofsky, C., *Nature* **289,** 753 (1981).

gene no faster than about once per second. If *E. coli* contained only one copy of each of the three types of rRNA genes (those specifying the so-called 23S, 16S, and 5S rRNAs; Section 32-3A), fast-growing cells could synthesize no more than ~1200 ribosomes during their cell division cycle. However, *the E. coli genome contains seven separately located rRNA operons, all of which contain one nearly identical copy of each type of rRNA gene.* Moreover, rapidly growing cells contain multiple copies of their replicating chromosomes (Section 30-3C), thereby accounting for the observed rRNA synthesis rate.

Cells have the remarkable ability to coordinate the rates at which their thousands of components are synthesized. For example, *E. coli* adjust their ribosome content to match the rate at which they can synthesize proteins under the prevailing growth conditions. The rate of rRNA synthesis is therefore proportional to the rate of protein synthesis. One mechanism by which this occurs is known as the **stringent response:** *A shortage of any species of amino acid-charged tRNA (usually a result of "stringent" or poor growth conditions) that limits the rate of protein synthesis triggers a sweeping metabolic readjustment.* A major facet of this change is an abrupt 10- to 20-fold reduction in the rate of rRNA and tRNA synthesis. This **stringent control,** moreover, depresses numerous metabolic processes (including DNA replication and the biosynthesis of carbohydrates, lipids, nucleotides, proteoglycans, and glycolytic intermediates) while stimulating others (such as amino acid biosynthesis). The cell is thereby prepared to withstand nutritional deprivation.

a. (p)ppGpp Mediates the Stringent Response

*The stringent response is correlated with a rapid intracellular accumulation of two unusual nucleotides, **ppGpp** and **pppGpp** [known collectively as **(p)ppGpp**], and their prompt decay when amino acids become available.* The observation that mutants, designated *relA⁻*, which do not exhibit the stringent response (they are said to have **relaxed control**) lack (p)ppGpp suggests that these substances mediate the stringent response. This idea was corroborated by *in vitro* studies demonstrating, for example, that (p)ppGpp inhibits the transcription of rRNA genes but stimulates the transcription of the *trp* and *lac* operons as does the stringent response *in vivo.* Apparently, (p)ppGpp acts by somehow altering RNAP's promoter specificity at stringently controlled operons, a hypothesis that is supported by the isolation of RNAP mutants that exhibit reduced responses to (p)ppGpp. In addition, (p)ppGpp causes an increased frequency of pausing in RNAPs engaged in elongation, thereby reducing the rate of transcription.

The protein encoded by the wild-type *relA* gene, named **stringent factor (RelA),** catalyzes the reaction

$$\text{ATP} + \text{GTP} \rightleftharpoons \text{AMP} + \text{pppGpp}$$

and, to a lesser extent,

$$\text{ATP} + \text{GDP} \rightleftharpoons \text{AMP} + \text{ppGpp}$$

However, several ribosomal proteins convert pppGpp to

ppGpp so that ppGpp is the stringent response's usual effector. Stringent factor is only active in association with a ribosome that is actively engaged in translation. (p)ppGpp synthesis occurs when a ribosome binds its mRNA-specified but uncharged tRNA (lacking an appended amino acid residue). The binding of a specified and charged tRNA greatly reduces the rate of (p)ppGpp synthesis. *The ribosome apparently signals the shortage of an amino acid by stimulating the synthesis of (p)ppGpp which, acting as an intracellular messenger, influences the rates at which a great variety of operons are transcribed.*

(p)ppGpp degradation is catalyzed by the *spoT* gene product. The *spoT⁻* mutants show a normal increase in (p)ppGpp level on amino acid starvation but an abnormally slow decay of (p)ppGpp to basal levels when amino acids again become available. The *spoT⁻* mutants therefore exhibit a sluggish recovery from the stringent response. *The (p)ppGpp level is apparently regulated by the countervailing activities of stringent factor and the spoT gene product.*

4 ■ POSTTRANSCRIPTIONAL PROCESSING

The immediate products of transcription, the **primary transcripts,** are not necessarily functional entities. In order to acquire biological activity, many of them must be specifically altered in several ways: (1) by the exo- and endonucleolytic removal of polynucleotide segments; (2) by appending nucleotide sequences to their 3′ and 5′ ends; and (3) by the modification of specific nucleosides. The three major classes of RNAs, mRNA, rRNA, and tRNA, are altered in different ways in prokaryotes and in eukaryotes. In this section we shall outline these **posttranscriptional modification** processes.

A. *Messenger RNA Processing: Caps, Tails, and Splicing*

In prokaryotes, most primary mRNA transcripts function in translation without further modification. Indeed, as we have seen, ribosomes in prokaryotes usually commence translation on nascent mRNAs. In eukaryotes, however, mRNAs are synthesized in the cell nucleus whereas translation occurs in the cytosol. Eukaryotic mRNA transcripts can therefore undergo extensive posttranscriptional processing while still in the nucleus.

a. Eukaryotic mRNAs Are Capped

*Eukaryotic mRNAs have a peculiar enzymatically appended **cap structure** consisting of a 7-methylguanosine residue joined to the transcript's initial (5′) nucleoside via a 5′–5′ triphosphate bridge* (Fig. 31-43). The cap, which is added to the growing transcript before it is ~30 nucleotides long, defines the eukaryotic translational start site (Section 32-3C). A cap may be $O^{2'}$-methylated at the transcript's leading nucleoside (**cap-1,** the predominant cap in multi-

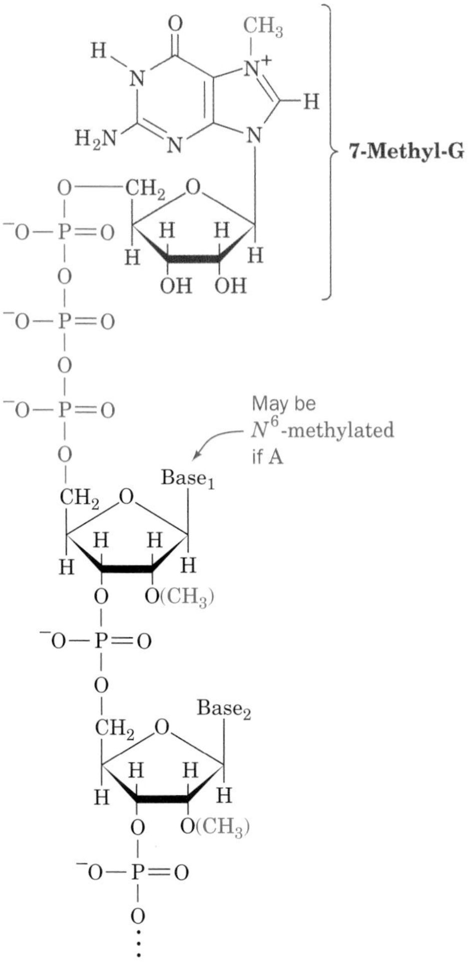

FIGURE 31-43 The structure of the 5′ cap of eukaryotic mRNAs. It is known as cap-0, cap-1, or cap-2, respectively, if it has no further modifications, if the leading nucleoside of the transcript is $O^{2'}$-methylated, or if its first two nucleosides are $O^{2'}$-methylated.

cellular organisms), at its first two nucleosides **(cap-2),** or at neither of these positions **(cap-0,** the predominant cap in unicellular eukaryotes). If the leading nucleoside is adenosine (it is usually a purine), it may also be N^6-methylated.

Capping involves several enzymatic reactions: (1) the removal of the leading phosphate group from the mRNA's 5′ terminal triphosphate group by an **RNA triphosphatase;** (2) the guanylation of the mRNA by **capping enzyme,** which requires GTP and yields the 5′–5′ triphosphate bridge and PP_i; (3) the methylation of guanine by **guanine-7-methyltransferase** in which the methyl group is supplied by *S*-adenosylmethionine (SAM); and possibly (4) the O2′ methylation of the mRNA's first and perhaps its second nucleotide by a SAM-requiring **2′-*O*-methyltransferase.** Both the capping enzyme and the guanine-7-methyltransferase bind to RNAP II's phosphorylated CTD (Section 31-2E). Hence it is likely that capping marks the completion of RNAP II's switch from transcription initiation to elongation.

b. Eukaryotic mRNAs Have Poly(A) Tails

Eukaryotic mRNAs, in contrast to those of prokaryotes, are invariably monocistronic. Yet, the sequences signaling transcriptional termination in eukaryotes have not been identified. This is largely because the termination process is imprecise; that is, the primary transcripts of a given structural gene have heterogeneous 3′ sequences. Nevertheless, mature eukaryotic mRNAs have well-defined 3′ ends; *almost all of them in mammals have 3′-poly(A) tails of ~250 nucleotides (~80 in yeast).* The poly(A) tails are enzymatically appended to the primary transcripts in two reactions that are mediated by a 500- to 1000-kD complex that consists of at least six proteins:

1. A transcript is cleaved to yield a free 3′-OH group at a specific site that is 15 to 25 nucleotides past an AAUAAA sequence and within 50 nucleotides before a U-rich or G + U-rich sequence. The AAUAAA sequence is highly conserved in higher eukaryotes (but not yeast) in which its mutation abolishes cleavage and polyadenylation. The precision of the cleavage reaction has apparently eliminated the need for accurate transcriptional termination. Nevertheless, the identity of the endonuclease that cleaves the RNA is uncertain although **cleavage factors I** and **II (CFI** and **CFII)** are required for this process.

2. The poly(A) tail is subsequently generated from ATP through the stepwise action of **poly(A) polymerase (PAP).** This enzyme, which by itself only weakly binds RNA, is recruited by **cleavage and polyadenylation specificity factor (CPSF)** on this heterotetramer's recognition of the AAUAAA sequence, which it does with almost no tolerance for sequence variation. The downstream G + U-rich element is recognized by the heterotrimer **cleavage stimulation factor (CstF),** which increases the affinity with which CPSF binds the AAUAAA sequence. However, once the poly(A) tail has grown to ~10 residues, the AAUAAA sequence is no longer required for further chain elongation. This suggests that CPSF becomes disengaged from its recognition site in a manner reminiscent of the way σ factor is released from the transcriptional initiation site once the elongation of prokaryotic mRNA is under way (Section 31-2C). The final length of the poly(A) tail is controlled by **poly(A)-binding protein II (PAB II),** multiple copies of which bind to successive segments of poly(A). PAB II also increases the processivity of PAP.

Both CPSF and CstF bind to the phosphorylated RNAP II CTD (Section 31-2E); deleting the CTD inhibits polyadenylation. Evidently, the CTD couples polyadenylation to transcription.

PAP is a template-independent RNA polymerase that elongates an mRNA primer with a free 3′-OH group. The X-ray structures of yeast and bovine PAPs in complex with 3′-dATP (cordycepin), respectively determined by Andrew Bohm and Sylvie Doublié, reveal that these closely similar monomeric proteins each consist of three domains that enclose an ~20 × 25 Å U-shaped cleft (Fig.

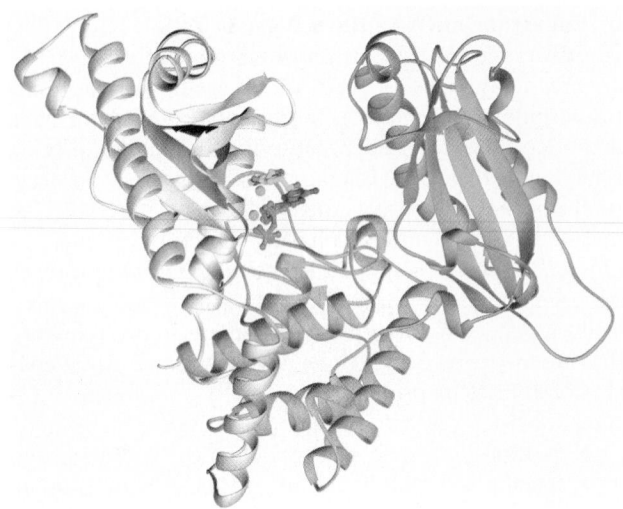

FIGURE 31-44   X-Ray structure of the 568-residue yeast poly(A) polymerase (PAP) in complex with two molecules of 3′-dATP. The N-terminal (palm) domain is lavender, the central (fingers) domain is orange, and the C-terminal domain is light blue. The 3′-dATP molecules are drawn in ball-and-stick form and colored according to atom type (incoming nucleotide C green, mRNA 3′ end C yellow, N blue, O red, and P magenta). The two Mn^{2+} ions that are bound at the active site are represented by cyan balls. [Based on an X-ray structure by Andrew Bohm, Tufts University School of Medicine. PDBid 1FA0.]

31-44). Hence they superficially appear to have the hand-like domain arrangement of the DNA polymerases (Section 30-2A). Indeed, PAP's N-terminal domain, which contains the enzyme's active site, is structurally homologous to the palm domain of DNA polymerase β, although it forms one side of the cleft rather than its base. PAP's central domain, which forms the base of the cleft, is functionally analogous to the polymerase fingers domain in that it interacts with the β and γ phosphates of the incoming nucleotide. However, the C-terminal domain shows no resemblance to a thumb domain. Rather, it is topologically similar to the **RNA-recognition motif (RRM)** that occurs in >200 different RNA-binding proteins (see below). Yeast PAP binds two molecules of 3′-dATP: One occupies the position of the incoming nucleotide and the other, whose triphosphate group is not observed, is presumed to mimic the 3′ end of the mRNA primer on the basis of its similarity to the X-ray structure of DNA polymerase β. The incoming base interacts with the protein so as to differentiate adenine from other bases, whereas in templated polymerases, the incoming base only contacts the template (Section 30-2A).

In vitro studies indicate that a poly(A) tail is not required for mRNA translation. Rather, the observations that an mRNA's poly(A) tail shortens as it ages in the cytosol and that unadenylated mRNAs have abbreviated cytosolic lifetimes suggest that poly(A) tails have a protective role. In fact, the only mature mRNAs that lack poly(A) tails, those of histones (which, with few exceptions, lack the AAUAAA cleavage–polyadenylation signal), have lifetimes of <30 min in the cytosol, whereas most other mRNAs last hours or days. The poly(A) tails are specifically complexed in the cytosol by **poly(A) binding protein (PABP;** not related to PAB II), which organizes poly(A)-bearing mRNAs into ribonucleoprotein particles. PABP is thought to protect mRNA from degradation as is suggested, for example, by the observation that the addition of PABP to a cell-free system containing mRNA and mRNA-degrading nucleases greatly reduces the rate at which the mRNAs are degraded and the rate at which their poly(A) tails are shortened.

All known PABPs contain four tandem and highly conserved RNA-recognition motifs (RRMs) followed by a less conserved Pro-rich C-terminal segment of variable length. A variety of evidence suggests that PABP's first two RRMs support most of the biochemical functions of full length PABP. The X-ray structure of the first two RRMs of human PABP (RRM1/2; the N-terminal 190 residues of this 636-residue protein) in complex with A_{11}, determined by Stephen Burley, reveals that RRM1/2 forms a continuous trough-shaped surface in which the poly(A) binds in an extended conformation via interactions with conserved residues (Fig. 31-45). Each RRM, as also seen in the structures of a variety of other RNA-binding proteins, consists of a compact globule made of a 4-stranded antiparallel sheet that forms the RNA-binding surface backed by two helices.

c. Eukaryotic Genes Consist of Alternating Expressed and Unexpressed Sequences

The most striking difference between eukaryotic and prokaryotic structural genes is that the coding sequences of most eukaryotic genes are interspersed with unexpressed regions. Early investigations of eukaryotic structural gene transcription found, quite surprisingly, that primary transcripts are highly heterogeneous in length (from ~2000 to well over 20,000 nucleotides) and are much larger than is expected from the known sizes of eukaryotic proteins. Rapid labeling experiments demonstrated that little of this so-called **heterogeneous nuclear RNA (hnRNA)** is ever transported to the cytosol; most of it is quickly turned over (degraded) in the nucleus. Yet, the hnRNA's 5′ caps and 3′ tails eventually appear in cytosolic mRNAs. *The straightforward explanation of these observations, that pre-mRNAs are processed by the excision of internal sequences, seemed so bizarre that it came as a great surprise in 1977 when Phillip Sharp and Richard Roberts independently demonstrated that this is actually the case.* In fact, pre-

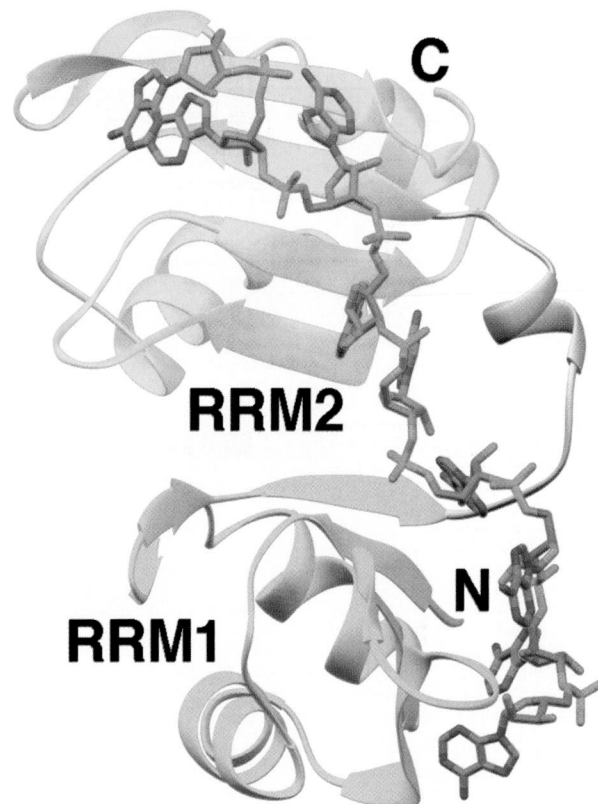

FIGURE 31-45   X-Ray structure of the N-terminal two RNA-recognition motifs (RRMs) of human PABP in complex with A_{11}. RRM1 is cyan, RRM2 is gold, and their linking segment is lavender. The poly(A), only nine of whose nucleotides are observed, is drawn in stick form with C green, N blue, O red, and P magenta. [Based on an X-ray structure by Stephen Burley, The Rockefeller University. PDBid 1CVJ.]

mRNAs typically contain eight noncoding **intervening sequences (introns)** whose aggregate length averages four to ten times that of its flanking **expressed sequences (exons).** This situation is graphically illustrated in Fig. 31-46, which is an electron micrograph of chicken **ovalbumin** mRNA hybridized to the antisense strand of the ovalbumin gene (ovalbumin is the major protein component of egg white).

Exons have lengths that range up to 17,106 nt (in the gene encoding the 29,926-residue muscle protein **titin,** the largest known single-chain protein; Section 35-3A) but with most <300 nt (and averaging 150 nt in humans).

Introns, in contrast, are usually much longer, with lengths averaging ~3500 nt and as high 2.4 million nt (in the gene encoding the muscle protein dystrophin; Section 35-3A) with no obvious periodicity. Moreover, the corresponding introns from genes in two vertebrate species can vary extensively in both length and sequence so as to bear little resemblance to one another. The number of introns in a gene averages 7.8 in the human genome and varies from none to 234 (with the latter number also occurring in the gene encoding titin).

The formation of eukaryotic mRNA begins with the transcription of an entire structural gene, including its in-

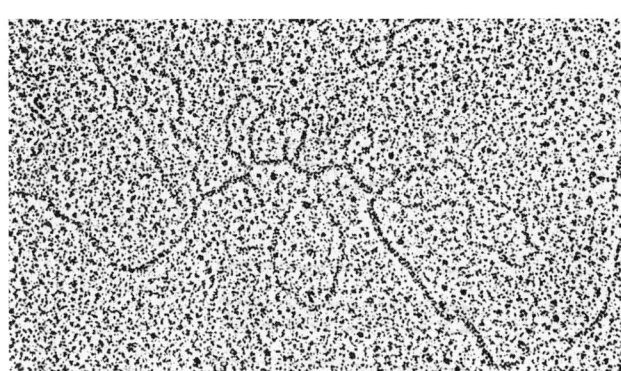

FIGURE 31-46   An electron micrograph and its interpretive drawing of a hybrid between the antisense strand of the chicken ovalbumin gene and its corresponding mRNA. The complementary segments of the DNA (*purple line in drawing*)

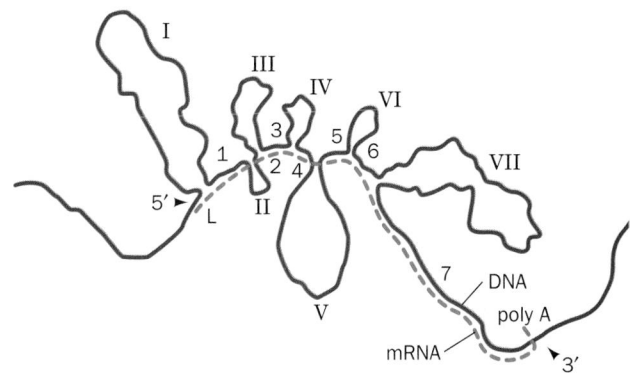

and mRNA (*red dashed line*) have annealed to reveal the exon positions (*L*, 1–7). The looped-out segments (I–VII), which have no complementary sequences in the mRNA, are the introns. [From Chambon, P., *Sci. Am.* **244**(5), 61 (1981).]

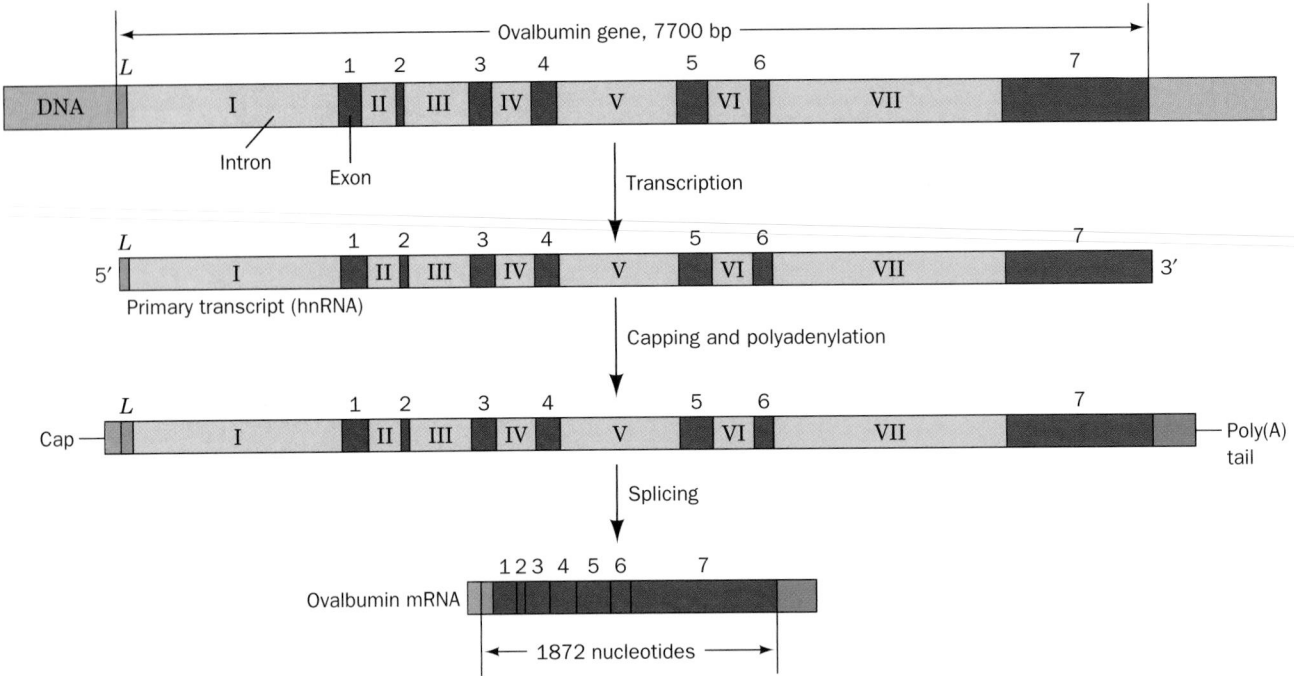

FIGURE 31-47 The sequence of steps in the production of mature eukaryotic mRNA as shown for the chicken ovalbumin gene. Following transcription, the primary transcript is capped and polyadenylated. The introns are then excised and the exons spliced together to form the mature mRNA. However, splicing may also occur cotranscriptionally.

trons, to form pre-mRNA (Fig. 31-47). Then, following capping, the introns are excised and their flanking exons are connected, a process called **gene splicing** or just **splicing,** that often occurs cotranscriptionally. *The most striking aspect of gene splicing is its precision; if one nucleotide too few or too many were excised, the resulting mRNA could not be translated properly (Section 32-1B). Moreover, exons are never shuffled, their order in the mature mRNA is exactly the same as that in the gene from which it is derived.*

d. Exons Are Spliced in a Two-Stage Reaction

Sequence comparisons of exon–intron junctions from a diverse group of eukaryotes indicate that they have a high degree of homology (Fig. 31-48, including, as Richard Breathnach and Pierre Chambon first pointed out, *an invariant GU at the intron's 5′ boundary and an invariant AG at its 3′ boundary. These sequences are necessary and sufficient to define a splice junction:* Mutations that alter the sequences interfere with splicing, whereas mutations that change a nonjunction to a consensus-like sequence can generate a new splice junction.

Investigations of both cell free and *in vivo* splicing systems by Argiris Efstradiadis, Tom Maniatis, Michael Rosbash, and Sharp established that intron excision occurs via two transesterification reactions that are remarkably similar from yeast to humans (Fig. 31-49):

1. The formation of a 2′,5′-phosphodiester bond between an intron adenosine residue and its 5′-terminal phosphate group with the concomitant liberation of the 5′ exon's 3′-OH group. *The intron thereby assumes a novel **lariat** structure.* The adenosine residue at the lariat branch has been identified in yeast as the last A in the highly conserved sequence UACUAAC and in vertebrates as the A in the equivalent but more permissive sequence YNCURAY [where R represents purines (A or G), Y represents pyrimidines (C or U), and N represents any nucleotide]. In yeast and vertebrates, the branch point A occurs ~50 and 18 to 40 residues upstream of the associated 3′ splice site, respectively. In yeast, which have relatively few introns, mutations that change this branch point A residue abolish splicing at that site. However, in higher eukaryotes, the mutation or deletion of a branch site often activates a so-called

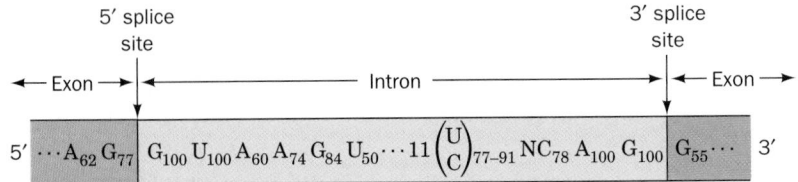

FIGURE 31-48 The consensus sequence at the exon–intron junctions of vertebrate pre-mRNAs. The subscripts indicate the percentage of pre-mRNAs in which the specified base(s) occurs. Note that the 3′ splice site is preceded by a tract of 11 predominantly pyrimidine nucleotides. [Based on data from Padgett, R.A., Grabowski, P.J., Konarska, M.M., Seiler, S.S., and Sharp, P.A., *Annu. Rev. Biochem.* **55,** 1123 (1986).]

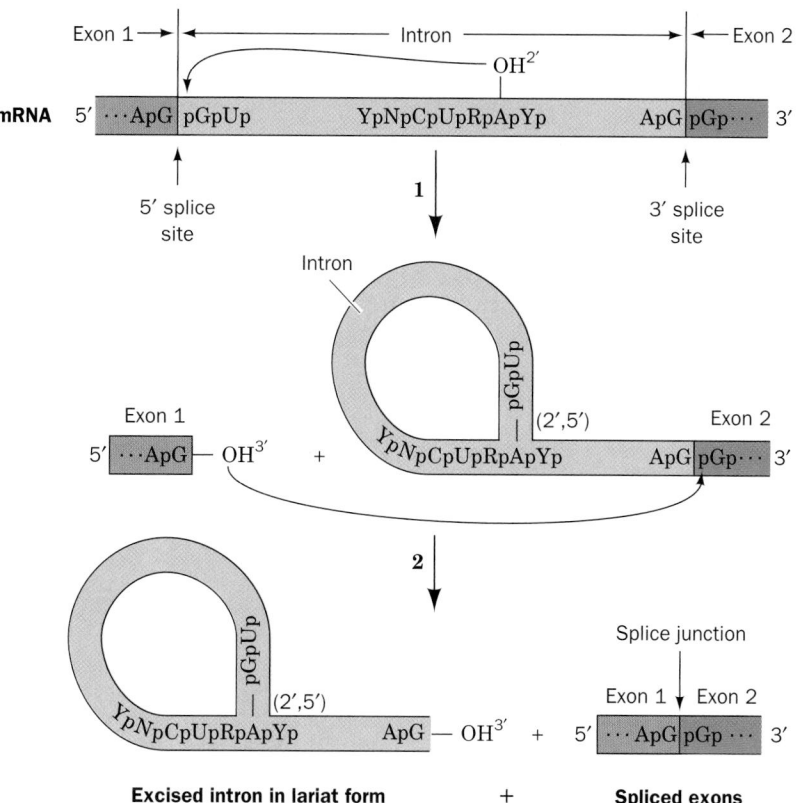

FIGURE 31-49 The sequence of transesterification reactions that splice together the exons of eukaryotic pre-mRNAs. The exons and introns are drawn in blue and orange, and R and Y represent purine and pyrimidine residues. **(1)** The 2'-OH group of a specific intron A residue nucleophilically attacks the 5'-phosphate at the 5' intron boundary to yield an unusual 2',5'-phosphodiester bond and thus form a lariat structure. **(2)** The liberated 3'-OH group forms a 3',5'-phosphodiester bond with the 5' terminal residue of the 3' exon, thereby splicing the two exons together and releasing the intron in lariat form with a free 3'-OH.

Excised intron in lariat form   +   Spliced exons

cryptic branch site that is also near the 3' splice site. Evidently, the branch site functions to identify the nearest 3' splice site as a target for linkage to the 5' splice site.

2. The now free 3'-OH group of the 5' exon forms a phosphodiester bond with the 5'-terminal phosphate of the 3' exon yielding the spliced product and releasing the intron lariat with a free 3'-OH group. The intron lariat is then debranched (linearized) and, *in vivo,* is rapidly degraded. Mutations that alter the conserved AG at the 3' splice site block this second step, although they do not interfere with lariat formation.

Note that the splicing process proceeds without free energy input; its transesterification reactions preserve the free energy of each cleaved phosphodiester bond through the concomitant formation of a new one.

The sequences required for splicing are the short consensus sequences at the 3' and 5' splice sites and at the branch site. Nevertheless, these sequences are poorly conserved. However, other short sequence elements within exons that are known as **exonic sequence enhancers (ESEs)** also play important roles in splice site selection although their characteristics are poorly understood (even highly sophisticated computer programs are only ~50% successful in predicting actual splice sites over apparently equally good candidates that are not). In contrast, large portions of most introns can be deleted without impeding splicing.

e. Some Eukaryotic Genes Are Self-Splicing

It is now recognized that there are eight distinct types of introns, seven of which occur in eukaryotes (Table 31-4). **Group I introns** occur in the nuclei, mitochondria, and

TABLE 31-4. Types of Introns

Intron Type	Where Found
GU–AG introns	Eukaryotic nuclear pre-mRNA
AU–AC introns	Eukaryotic nuclear pre-mRNA
Group I	Eukaryotic nuclear pre-mRNA, organelle RNAs, a few bacterial RNAs
Group II	Organelle RNAs, a few prokaryotic RNAs
Group III	Organelle RNAs
Twintrons (composites of two and/or more group II or III introns)	Organelle RNAs
Pre-tRNA introns	Eukaryotic nuclear pre-tRNAs
Archaeal introns	Various RNAs

Source: Brown, T.A., *Genomes* (2nd ed.), Wiley-Liss, *p.* 287 (2002).

chloroplasts of diverse eukaryotes (but not vertebrates), and even in some bacteria. Thomas Cech's study of how group I introns are spliced in the ciliated protozoan *Tetrahymena thermophila* led to an astonishing discovery: *RNA can act as an enzyme. When the isolated pre-rRNA of this organism is incubated with guanosine or a free guanine nucleotide (GMP, GDP, or GTP), but in the absence of protein, its single 421-nucleotide intron excises itself and splices together its flanking exons; that is, this pre-rRNA is self-splicing.* The three-step reaction sequence of this process (Fig. 31-50) resembles that of mRNA splicing:

1. The 3′-OH group of the guanosine forms a phosphodiester bond with the intron's 5′ end, liberating the 5′ exon.

2. The 3′-terminal OH group of the newly liberated 5′ exon forms a phosphodiester bond with the 5′-terminal phosphate of the 3′ exon, thereby splicing together the two exons and releasing the intron.

3. The 3′-terminal OH group of the intron forms a phosphodiester bond with the phosphate of the nucleotide 15 residues from the intron's 5′ end, yielding the 5′-terminal fragment with the remainder of the intron in cyclic form.

This self-splicing process consists of a series of transesterifications and therefore does not require free energy input. Cech further established the enzymatic properties of the *Tetrahymena* intron, which presumably stem from its three-dimensional structure, by demonstrating that it catalyzes the *in vitro* cleavage of poly(C) with an enhancement factor of 10^{10} over the rate of spontaneous hydrolysis. Indeed, this RNA catalyst even exhibits Michaelis–Menten kinetics ($K_M = 42 \ \mu M$ and $k_{cat} = 0.033 \ s^{-1}$ for C_5). Such RNA enzymes have been named **ribozymes.**

Although the idea that an RNA can have enzymatic properties may seem unorthodox, *there is no fundamental reason why an RNA, or any other macromolecule, cannot have catalytic activity* (recall that it was likewise once generally accepted that nucleic acids lack the complexity to carry hereditary information; Section 5-2). Of course, in order to be an efficient catalyst, a macromolecule must be able to assume a stable structure but, as we shall see below and in Sections 32-2B and 32-3A, RNAs, including tRNAs and rRNAs, can do so. [Synthetic ssDNAs are also

known to have catalytic properties although such "deoxyribozymes" are unknown in biology.]

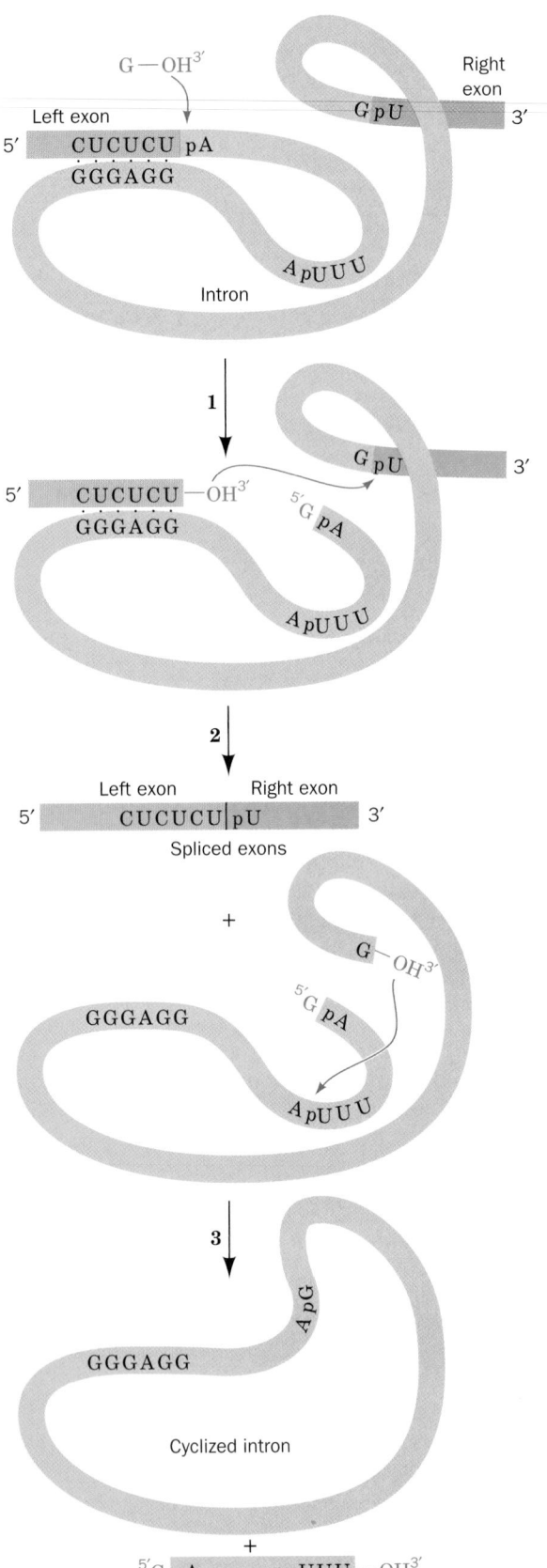

FIGURE 31-50 The sequence of reactions in the self-splicing of *Tetrahymena* group I intron. (1) The 3′-OH group of a guanine nucleotide attacks the intron's 5′-terminal phosphate so as to form a phosphodiester bond and release the 5′ exon. **(2)** The newly generated 3′-OH group of the 5′ exon attacks the 5′-terminal phosphate of the 3′ exon, thereby splicing the two exons and releasing the intron. **(3)** The 3′-OH group of the intron attacks the phosphate of the nucleotide that is 15 residues from the 5′ end so as to cyclize the intron and release its 5′-terminal fragment. Throughout this process, the RNA maintains a folded, internally hydrogen bonded conformation that permits the precise excision of the intron.

The **group II introns,** which occur in the mitochondria of fungi and plants and comprise the majority of the introns in chloroplasts, are also self-splicing. They react via a lariat intermediate and do not utilize an external nucleotide, a process that resembles the splicing of nuclear pre-mRNAs (Fig. 31-49). We shall see below that nuclear pre-mRNA splicing is mediated by complex ribonucleoprotein particles known as **spliceosomes.** The chemical similarities of the pre-mRNA and group II intron splicing reactions therefore suggest that *spliceosomes are ribozymal systems whose RNA components have evolved from primordial self-splicing RNAs and that their protein components serve mainly to fine-tune ribozymal structure and function.* Similarly, the RNA components of ribosomes, which are two-thirds RNA and one-third protein, clearly have a catalytic function in addition to the structural and recognition roles traditionally attributed to them (Section 32-3). Thus, the observations that nucleic acids but not proteins can direct their own synthesis, that cells contain batteries of protein-based enzymes for manipulating DNA but few for processing RNA, and that many coenzymes are ribonucleotides (e.g., ATP, NAD$^+$, and CoA), led to the hypothesis that *RNAs were the original biological catalysts in precellular times (the so-called RNA world) and that the chemically more versatile proteins were relative latecomers in macromolecular evolution (Section 1-5C).*

f. The X-Ray Structures of a Group I Ribozyme

The sequence of the *Tetrahymena* group I intron, together with phylogenetic comparisons, indicates that it contains nine double helical segments that are designated P1 through P9 (Fig. 31-51*a*). Such analysis further indicates that the conserved catalytic core of group I introns consists of sets of coaxially stacked helices interspersed with internal loops that are organized into three domains, P1-P2, P4-P6 and P3-P9. Chemical protection experiments suggest that the isolated P4-P6 domain of the *Tetrahymena*

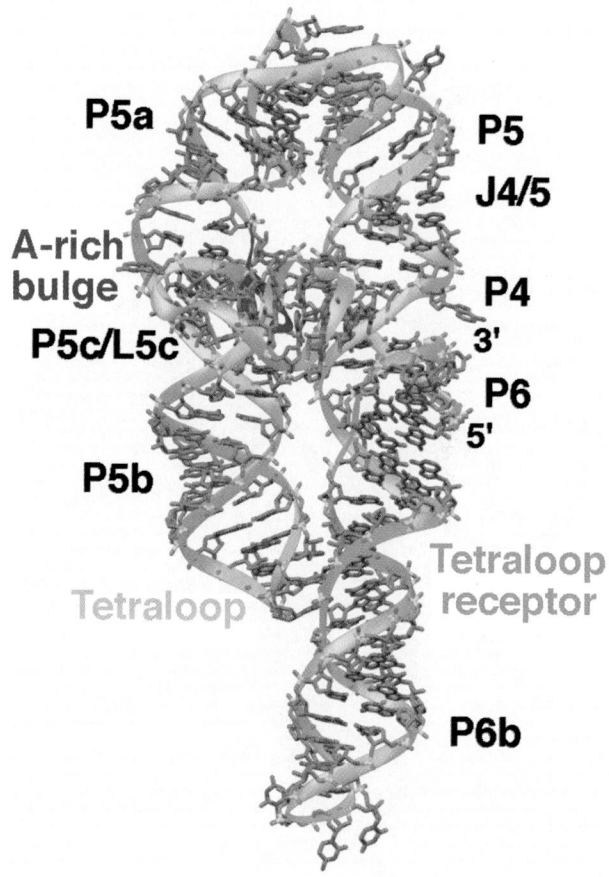

(a)

(b)

FIGURE 31-51 The self-splicing group I intron from *Tetrahymena thermophila.* (*a*) The secondary structure of the entire 413-nt intron is shown on the right with its phylogenetically conserved catalytic core shaded in blue-gray. Helical regions are numbered sequentially along the intron's sequence with P for base *p*aired segment, J for *j*oining region, and L for *l*oop; the arrows indicate the 5′ and 3′ splice sites. The sequence of the independently folded 160-nt P4-P6 domain is enlarged at left. Segments of functional interest are highlighted as follows: the GAAA tetraloop is cyan; the conserved tetraloop receptor is magenta; the A-rich bulge, which is required for the proper folding of P4-P6, is dark blue; segments of the conserved core

are light green and red; and P5c is blue-gray. Watson–Crick and non-Watson–Crick base pairing interactions are represented by short horizontal lines and small circles. (*b*) The X-ray structure of P4-P6 viewed as in Part *a*. The structure is drawn in stick form with C green, N blue, O red, and P yellow. The sugar–phosphate backbone is traced by a ribbon that is colored as in Part *a* for the tetraloop, tetraloop receptor, and A-rich bulge and is gold elsewhere. Note the numerous interactions between the various segments of this RNA molecule. [Part *a* based on a drawing by and Part *b* based on an X-ray structure by Jennifer Doudna, Yale University. PDBid 1GID.] *See the Interactive Exercises*

group I intron folds as an independent unit. Indeed, when the P4-P6 domain is combined with the remaining portion of the *Tetrahymena* intron, it forms a catalytically active complex.

Jennifer Doudna and Cech determined the X-ray structure of the 160-nt P4-P6 domain of the *Tetrahymena* group I intron (Fig. 31-51*b*). When this structure was reported in 1997, it was over twice the size of the largest RNA molecule whose structure had previously been determined, that of a 76-nt tRNA (Section 32-2B). For the most part, the P4-P6 domain consists of two coaxially stacked sets of A-RNA-like helices, one of 29 bp and the other of 23 bp. Its dimensions are around $25 \times 50 \times 110$ Å, the first two dimensions being the widths of one A-RNA helix and two side-by-side A-RNA helices.

P4-P6 was the first RNA of known structure that is large enough to exhibit side-by-side helical packing. Of particular note are its so-called A-rich bulge, a 7-nt sequence about halfway along the short arm of the U-shaped macromolecule, and the 6-nt sequence at the tip of the short arm of the U, whose central GAAA assumes a characteristic

conformation known as a **tetraloop.** In both of these substructures, the bases are splayed outward so as to stack on each other and to associate in the minor groove of specific segments of the long arm of the U via hydrogen bonding interactions involving ribose residues as well as bases. In the interaction involving the A-rich bulge, the close packing of phosphates from adjacent helices is mediated by hydrated Mg^{2+} ions. Throughout this structure, the defining characteristic of RNA, its 2'-OH group, is both a donor and an acceptor of hydrogen bonds to phosphates, bases, and other 2'-OH groups.

Subsequently, Cech designed a 247-nt RNA that encompasses both the P4-P6 and the P3-P9 domains of the *Tetrahymena* group I intron, with the addition of a 3' G, which functions as an internal guanosine nucleophile. This RNA is catalytically active; it binds the P1-P2 domain via tertiary interactions and, with the assistance of its 3' G, cleaves P1 in a manner similar to the intact intron.

The X-ray structure of this ribozyme was determined to a resolution of 5 Å (Fig. 31-52). At this low resolution, the sugar–phosphate backbone can be traced and stacked bases often appear as continuous tubes of electron density, but such atomic-level features as hydrogen bonding interactions are not apparent. Within this structure the P4-P6 domain appears to be essentially unchanged from that of the domain alone (Fig. 31-51*b*). Moreover, the structure is compatible with a large body of biochemical data. The close packing of the two domains forms a shallow cleft that appears capable of binding the short helix that contains the 5' splice site. The intron's guanosine binding site is located on P7, which deviates significantly from A-form geometry in a way that provides a snug binding site for the guanosine substrate. Evidently, this ribozyme is largely preorganized for substrate binding and catalysis, much as are protein enzymes.

g. Hammerhead Ribozymes Catalyze an In-Line Nucleophilic Attack

The simplest and perhaps best characterized ribozymes, which are embedded in the RNAs of certain plant viruses, are named **hammerhead ribozymes** due to the superficial resemblance of their secondary structures, as customarily laid out, to a hammer (Fig. 31-53*a*). Hammerhead ribozymes have three duplex stems and a conserved core of two nonhelical segments.

Hammerhead ribozymes catalyze a transesterification reaction in which the 3',5'-phosphodiester bond between nucleotides C-17 and A-1.1 is cleaved so as to yield a cyclic 2',3'-phosphodiester on C-17 with inversion of configuration about the P atom, together with a free 5'-OH on A-1.1, much like the intermediate product in the RNA hydrolysis reaction catalyzed by RNase A (Section 15-1A). This suggests that the reaction proceeds via an "in-line" mechanism such as that diagrammed in Fig. 16-6*b* in which the transition state forms a trigonal bipyramidal intermediate in which the attacking nucleophile, the 2'-OH group (Y in Fig. 16-6*b*), and the leaving group, which forms the free 5'-OH group (X in Fig. 16-6*b*), occupy the axial positions. Under physiological conditions the reaction requires the

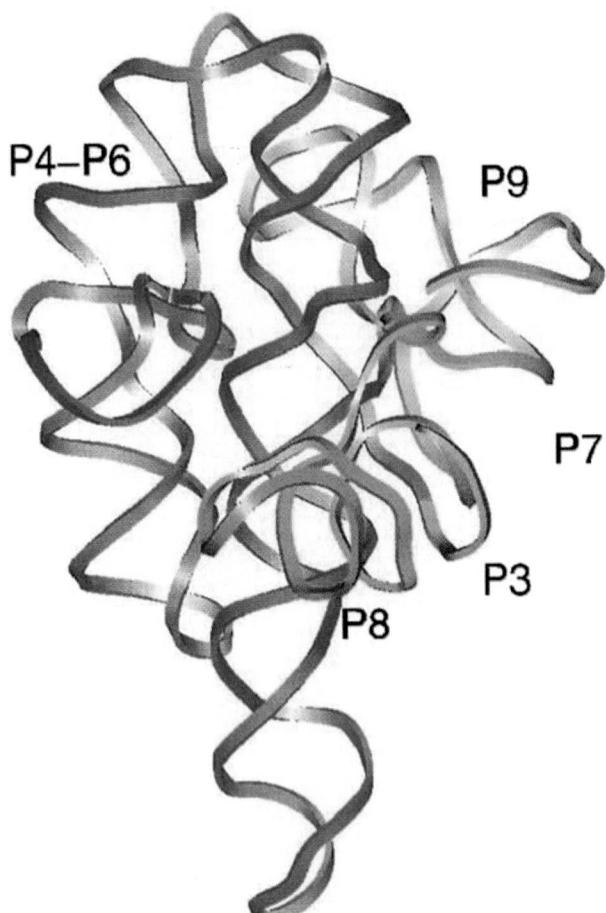

FIGURE 31-52 Low resolution X-ray structure of the *Tetrahymena* **group I intron encompassing its P4-P6 (***violet***) and its P3-P9 (***green***) domains.** The ribozyme, which is represented by a ribbon drawn through its phosphate groups, is oriented such that its P4-P6 domain is viewed similarly to that in Fig. 31-51*b*. [Courtesy of Thomas Cech, University of Colorado. PDBid 1GRZ.]

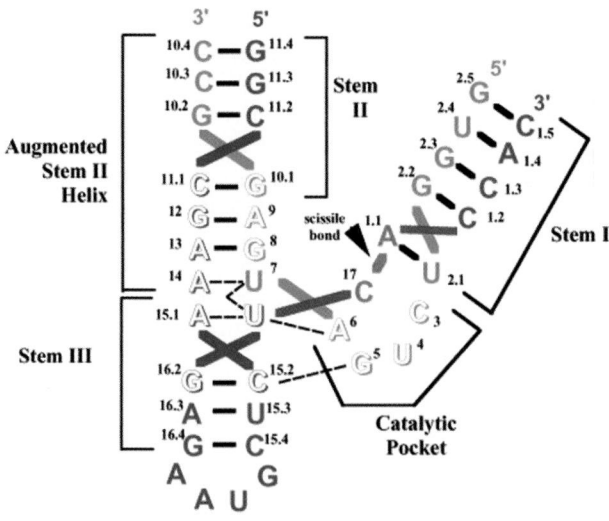

(a)

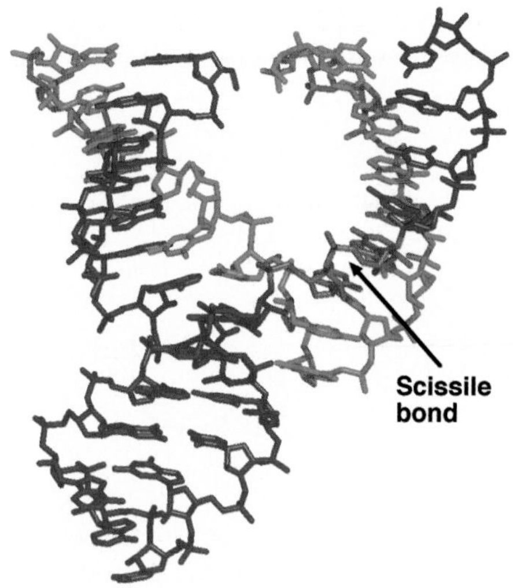

(b)

FIGURE 31-53 X-Ray structure of a hammerhead ribozyme. (a) The sequence and schematic structural representation of the ribozyme drawn with its 16-nt enzyme strand green, with its 25-nt substrate strand blue, but with the nucleotides spanning its cleavage site (C-17 and A-1.1) red. Essential and highly conserved nucleotides are represented by hollow letters, and the universal numbering scheme is provided. Watson–Crick as well as two G · A Hoogsteen base pairing interactions (Section 29-2C) are shown as single black dashes, and single hydrogen bonds between bases or between bases and backbone riboses are indicated by black dashed lines. (b) A stick model of the ribozyme in its ground state colored as in Part a. [Courtesy of William Scott, University of California at Santa Cruz, PDBid 1MME.] 🖱 **See the Interactive Exercises**

presence of a divalent cation, preferentially Mg^{2+} or Mn^{2+}, in roughly millimolar concentrations. However, the ribozyme is active in the absence of divalent cations when the concentration of monovalent cations is very high (e.g., $4M\ Na^+$, Li^+, or NH_4^+), which suggests that metal ions serve a structural rather than a catalytic role in the hammerhead ribozyme.

In order to crystallize a hammerhead ribozyme in its native conformation without it self-destructing, William Scott and Aaron Klug synthesized it with C-17 at its cleavage site replaced by 2′-methoxy-C, thereby blocking the formation of the 2′,3′-cyclic reaction product at this position. The X-ray structure of this catalytically inactive ribozyme (Fig. 31-53b) reveals that it has the expected secondary structure of three A-form helical segments although its overall shape more closely resembles a wishbone than a hammer. The nucleotides in the helical stems form normal Watson–Crick base pairs, whereas nucleotides U-7 through A-9 form non-Watson–Crick base pairs with nucleotides G-12 through A-14 in which ribose oxygens participate as both hydrogen bond donors and acceptors. This explains the observations that most helical positions can be occupied by any Watson–Crick base pair but that few core bases can be changed without reducing ribozymal activity. The absolutely conserved CUGA tetranucleotide loop forms a catalytic pocket into which the cleavage site base, that of C-17, is inserted.

The X-ray structure of this so-called ground state ribozyme (Fig. 31-53b) indicates that its C-17 and A-1.1 have the standard A-RNA conformation, which is incompatible with an in-line attack of the C-17 O2′ atom on the scissile phosphate group to form the cyclic 2′,3′-phosphodiester product (Fig. 31-54a). In order to trap the hammerhead ribozyme in a conformation capable of forming this

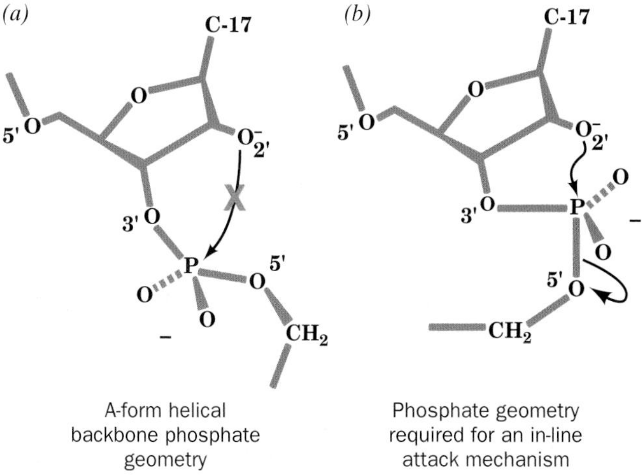

(a)

(b)

FIGURE 31-54 The conformation required for an in-line nucleophilic attack in the hammerhead ribozyme. (a) In the ribozyme's ground state conformation, C-17 has the standard A-RNA conformation in which its O2′ atom is 90° from the proper position for an in-line attack on the scissile phosphate group. (b) In the ribozyme's kinetically trapped active conformation, the scissile phosphate group has moved into the proper position for an in-line attack by O2′.

covalent intermediate, Scott synthesized it with a 5'-*C*-methyl-ribose modification on A-1.1:

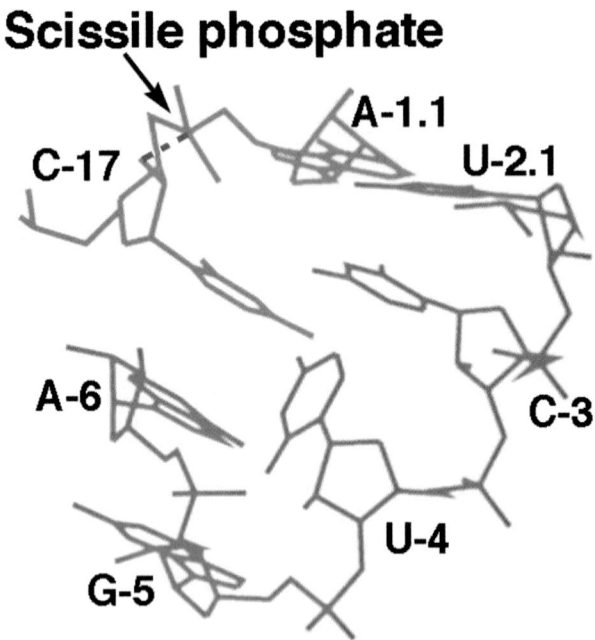

This creates a "kinetic bottleneck" for the reaction by stabilizing the otherwise scissile bond, presumably by altering the electronic properties of the leaving group. A crystal of this modified ribozyme that was soaked in a Co^{2+}-containing buffer at pH 8.5 for 30 min and then flash-frozen to near liquid nitrogen temperatures ($-196°C$; which arrests all molecular motion) had an X-ray structure that was indistinguishable from that of the ground state ribozyme. However, the X-ray structure of a crystal that was flash-frozen after a 2.5 hour soak revealed extensive changes in the region of the active site (Fig. 31-55). In par-

FIGURE 31-55 The X-ray structure of the catalytic pocket in the hammerhead ribozyme's kinetically trapped intermediate. The invariant CUGA tetranucleotide segment of the enzyme strand that forms the catalytic pocket is green and the residues on the substrate strand that span the scissile phosphate group are red. The dashed blue line, which marks the trajectory that atom O2' of C-17 must take to nucleophilically attack the phosphorus atom of the scissile phosphate group, is in line with the bond joining this phosphorus atom to atom O5' of A-1.1, the leaving group. [Adapted from Murray, J.B., Terwey, D.P., Maloney, L., Karpeisky, A., Usman, N., Beigleman, L., and Scott, W.G., *Cell* **92,** 665 (1998). PDBid 379D.]

ticular, the base and ribose of C-17 have rotated by ~60° so as to cause the base to stack on that of A-6 (which remains stacked on the base of G-5), a movement of 8.7 Å by the C-17 base. Moreover, the furanose oxygen of A-1.1 has stacked on the C-17 base. This movement has resulted in a conformational change at the scissile phosphate so as to properly position it for an in-line nucleophilic attack by the C-17 O2' atom (Figs. 31-54*b* and 31-55). In fact, similar conformation changes had been predicted by molecular dynamics simulations of the hammerhead ribozyme, starting from its ground state structure.

h. Splicing of Pre-mRNAs Is Mediated by snRNPs in the Spliceosome

How are the splice junctions of pre-mRNAs recognized and how are the two exons to be joined brought together in the splicing process? Part of the answer to this question was established by Joan Steitz going on the assumption that one nucleic acid is best recognized by another. The eukaryotic nucleus, as has been known since the 1960s, contains numerous copies of several highly conserved 60- to 300-nucleotide RNAs called **small nuclear RNAs (snRNAs),** which form protein complexes termed **small nuclear ribonucleoproteins (snRNPs;** pronounced "snurps"). Steitz recognized that the 5' end of one of these snRNAs, **U1-snRNA** (so called because it is a member of a U-rich subfamily of snRNAs), is partially complementary to the consensus sequence of the 5' splice site. The consequent hypothesis, that *U1-snRNA recognizes the 5' splice site,* was corroborated by the observations that splicing is inhibited by the selective destruction of the U1-snRNA sequences that are complementary to the 5' splice site or by the presence of anti-U1-snRNP antibodies (produced by patients suffering from **systemic lupus erythematosus,** an often fatal autoimmune disease). Three other snRNPs are also implicated in splicing: **U2-snRNP, U4–U6-snRNP** (in which the **U4-** and **U6-snRNAs** associate via base pairing), and **U5-snRNP.**

*Splicing takes place in an as yet poorly characterized ~45S particle dubbed the **spliceosome** (Fig. 31-56).* The spliceosome brings together a pre-mRNA, the foregoing four snRNPs, and a variety of pre-mRNA binding proteins. Note that the spliceosome, which consists of 5 RNAs and ~65 polypeptides, is comparable in size and complexity to the ribosome (which in *E. coli* consists of 3 RNAs and 52 polypeptides; Section 32-3A). Although it had been generally accepted that the spliceosome's component snRNPs assembled anew on each pre-mRNA substrate, John Abelson has recently demonstrated that, at least with yeast, this is an experimental artifact arising from the use of unphysiologically high salt concentrations. Rather, it appears that the preassembled spliceosome associates as a whole with the pre-mRNA. The spliceosome is, nevertheless, a highly dynamic entity whose machinations in carrying out the splicing process are ATP-driven. For example, to carry out the first transesterification reaction yielding the lariat structure (Fig. 31-49), the spliceosome undergoes a complex series of rearrangements that are schematically diagrammed in Fig. 31-57. Similarly extensive rearrange-

FIGURE 31-56 An electron micrograph of spliceosomes in action. A *Drosophila* gene that is ~6 Kb long enters from the upper left of the micrograph and exits at the lower left. Transcription initiates near the point marked by an asterisk. The growing RNA chains appear as fibrils of increasing lengths that emanate from the DNA. The transcripts are undergoing cotranslational splicing as revealed by the progressive formation and loss of intron loops near the 5′ ends of the RNA transcripts (*arrows*). The beads at the base of each intron loop as well as elsewhere on the transcripts are the spliceosomes. The large arrow points to a transcript near the 3′ end of the gene that is no longer attached to the DNA template and hence appears to have recently been terminated and released. The bar is 200nm long. [Courtesy of Ann Beyer and Yvonne Osheim, University of Virginia.]

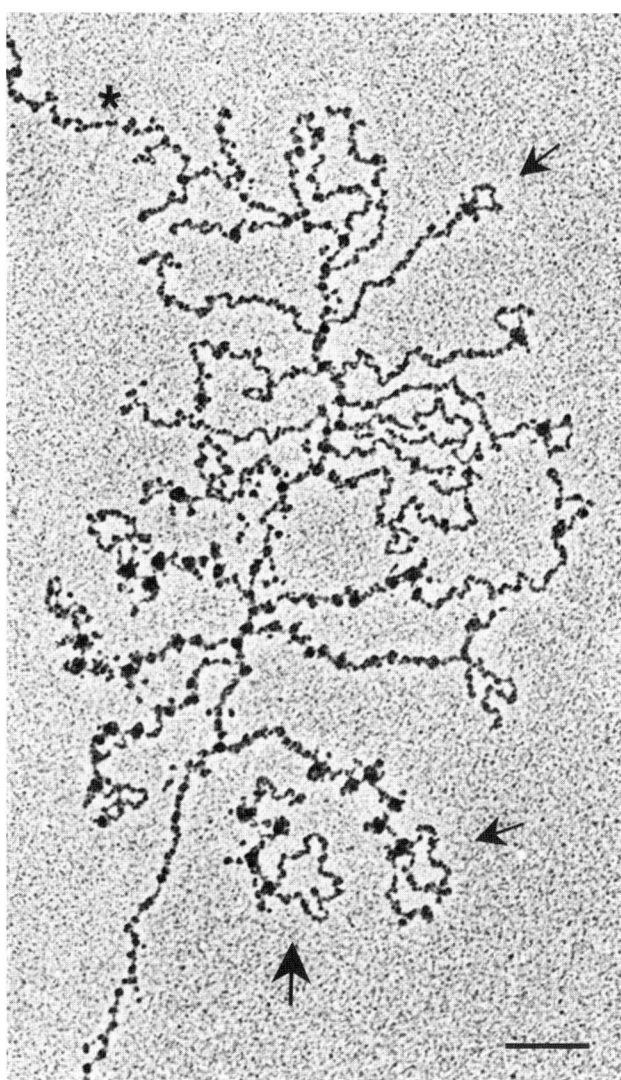

ments are required to carry out the second transesterification reaction and to recycle the spliceosome for subsequent splicing reactions.

Although spliceosomal transesterification reactions were initially assumed to be mediated by protein catalysts, their chemical resemblance to the reactions carried out by the self-splicing group II introns suggests, as is noted above, that it is really the snRNAs that catalyze the splicing of pre-mRNAs (pre-mRNA introns have such varied sequences outside of their splice and branch sites that they are unlikely to play an active role in splicing). In fact, James Manley has shown that, in the absence of protein, segments of human U2- and U6-snRNAs catalyze an Mg^{2+}-dependent reaction in an intron branch site sequence-containing RNA that resembles splicing's first transesterification reaction.

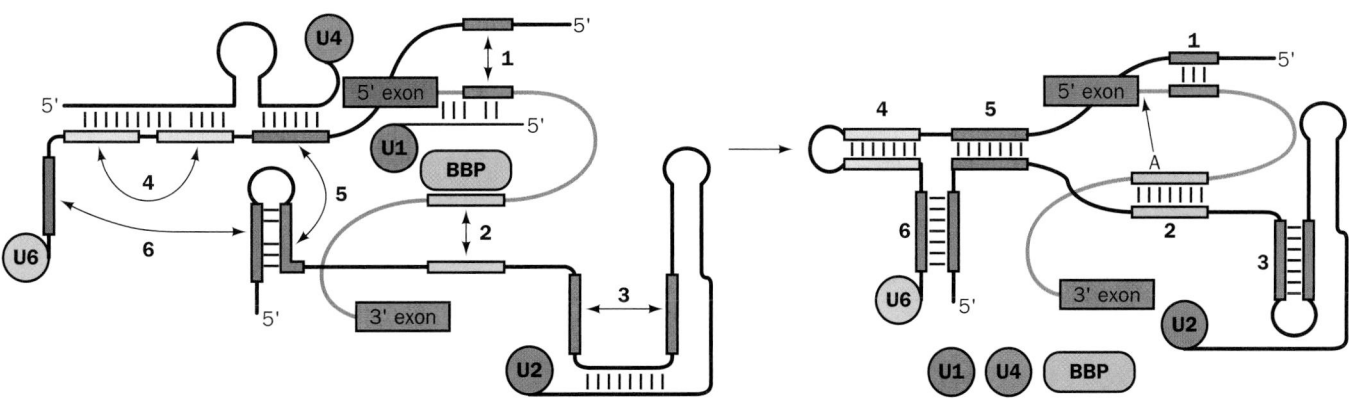

FIGURE 31-57 A schematic diagram of six rearrangements that the spliceosome undergoes in mediating the first transesterification reaction in pre-mRNA splicing. The RNA is color coded to indicate segments that become base paired. The black and green lines represent snRNA and pre-mRNA, and BBP stands for branch point-binding protein. U5, which participates in the second transesterification reaction, has been omitted for clarity. **(1)** Exchange of U1 for U6 in base pairing to the intron's 5′ splice site. **(2)** Exchange of BBP for U2 in binding to the intron's branch site. **(3)** Intramolecular rearrangement in U2. **(4)** Disruption of a base paired stem between U4 and U6 to form a stem–loop in U6. **(5)** Disruption of a second stem between U4 and U6 to form a stem between U2 and U6. **(6)** Disruption of a stem–loop in U2 to form a second stem between U2 and U6. The order of these rearrangements is unclear. The transesterification reaction is represented by the arrow from the A in the yellow segment of the pre-mRNA (*right panel*) to the 3′ end of the 5′ exon. [Adapted from Staley, J.P. and Guthrie, C., *Cell* **92,** 315 (1998).]

i. Splicing Also Requires the Participation of Splicing Factors

A variety of proteins known as **splicing factors** that are extrinsic to spliceosomes also participate in splicing. Among them are **branch point-binding protein [BBP; also known as splicing factor 1 (SF1)]** and **U2-snRNP auxiliary factor (U2AF),** which cooperate to select the intron's branch point. U2AF binds to the polypyrimidine tract upstream of the 3′ splice site (Fig. 31-48), whereas BBP recognizes the nearby branch point sequence (Figs. 31-49 and 31-57). The NMR structure of the 131-residue RNA-binding segment of the 638-residue BBP in complex with an 11-nt RNA containing a branch point sequence, determined by Michael Sattler, reveals that the RNA assumes an extended conformation and is largely buried in a groove that is lined with both aliphatic and basic residues (Fig. 31-58). The branch point adenosine, whose mutation abolishes BBP binding, is deeply buried and binds to BBP via hydrogen bonds that mimic Watson–Crick base pairing with uracil.

Other splicing factors include **SR** proteins and several members of the **heterogeneous nuclear ribonucleoprotein (hnRNP)** family. SR proteins each have one or more RRMs near their N-terminus and a distinctive C-terminal domain that is rich in Ser and Arg (SR) and which participates in protein–protein interactions. SR proteins specifically bind to exonic splicing enhancers (ESEs) and thereby recruit the splicing machinery to the flanking 5′ and 3′ splice sites. hnRNP proteins are highly abundant RNA-binding proteins that lack RS domains and whose functions are discussed below.

A simplistic interpretation of Fig. 31-49 suggests that any 5′ splice site could be joined with any following 3′ splice site, thereby eliminating all the intervening exons together with the introns joining them. However, such **exon skipping** does not normally take place (but see below). Rather, all of a pre-mRNA's introns are individually excised in what appears to be a largely fixed order that more or less proceeds in the 5′ → 3′ direction. This occurs, at least in part, because splicing occurs cotranscriptionally. Thus, as a newly synthesized exon emerges from an RNAP II, it is bound by splicing factors that are also bound to the RNAP II's highly phosphorylated C-terminal domain (CTD; Section 31-2E). This tethers the exon and its associated spliceosome to the CTD so as to ensure that splicing occurs when the next exon emerges from the RNAP II.

j. Spliceosomal Structures

All four snRNPs involved in pre-mRNA splicing contain the same so-called **snRNP core protein,** which consists of seven **Sm proteins** (so called because they react with autoantibodies of the Sm serotype from patients with systemic lupus erythematosis), which are named **B/B′, D₁, D₂, D₃, E, F,** and **G proteins** [B and B′ are alternatively spliced products of a single gene (see below) that differ only in their C-terminal 11 residues]. Each of these Sm proteins contains two conserved segments, Sm1 and Sm2, that are separated by a linker of variable length. The seven Sm proteins collectively bind to a conserved RNA sequence, the **Sm RNA motif,** which occurs in U1-, U2-, U4-, and U5-snRNAs and which has the single-stranded sequence AAUUUGUGG. However, in the absence of a U-snRNA,

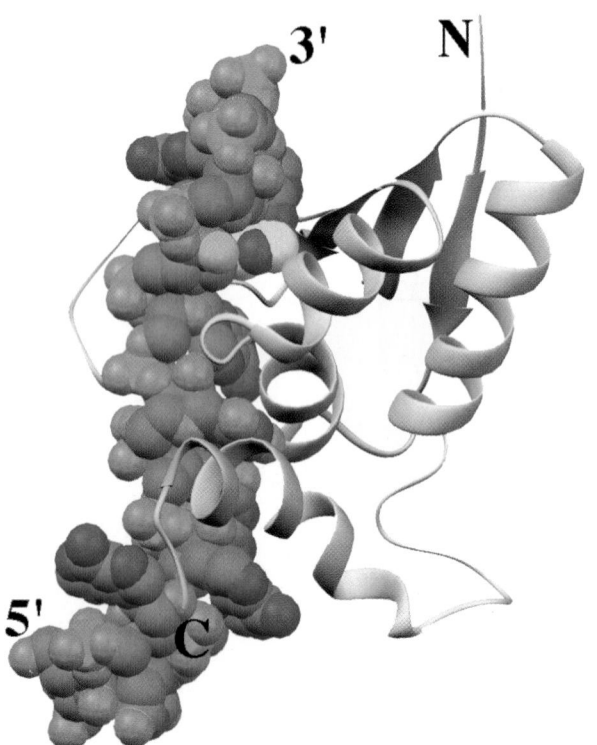

FIGURE 31-58 The NMR structure of the RNA binding portion of human branch point-binding protein (BBP) in complex with its target RNA. The 11-nt RNA has the sequence 5′-UA<u>UACUAAC</u>AA-3′ in which the branch site sequence for both yeast and vertebrates is underlined. The RNA is drawn in space-filling form with C green, N blue, O red, and P magenta except for the C's of the branch point adenine, which are yellow, and the branch point O2′, which is cyan. [Based on an NMR structure by Michael Sattler, European Molecular Biology Laboratory, Heidelberg, Germany. PDBid 1K1G.]

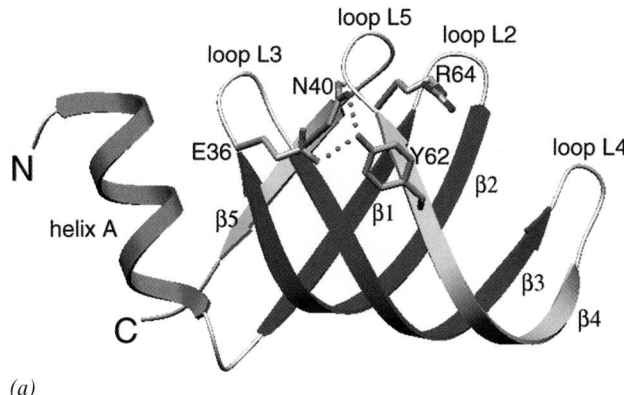

(a)

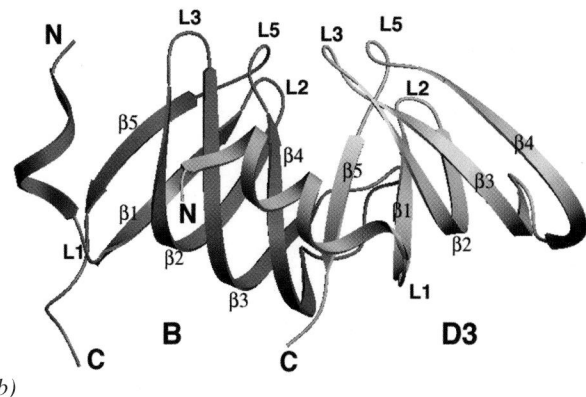

(b)

FIGURE 31-59   X-Ray structures of Sm proteins. (*a*) The structure of D_3 protein. The N-terminal helix and the β strands of its Sm1 domain are red and blue and the β strands of its Sm2 domain are yellow. The B, D_1, and D_2 Sm proteins have similar structures with their L4 loops and N-terminal segments, including helix A, comprising their most variable portions. Several highly conserved residues are shown in stick form (with

C gray, N blue, and O red), and a conserved hydrogen bonding network is represented by green dotted lines. (*b*) The D_3B dimer with D_3 gold and B blue. The β5 strand of D_3 associates with the β4 strand of B to form a continuous antiparallel β sheet. Note that their corresponding loops extend in similar directions. [Courtesy of Kiyoshi Nagai, MRC Laboratory of Molecular Biology, Cambridge, U.K. PDBid 1D3B.]

the Sm proteins form three stable complexes consisting of D_1 and D_2; D_3 and B/B′; and E, F, and G. None of these complexes alone bind U-snRNA. However, the D_1D_2 and

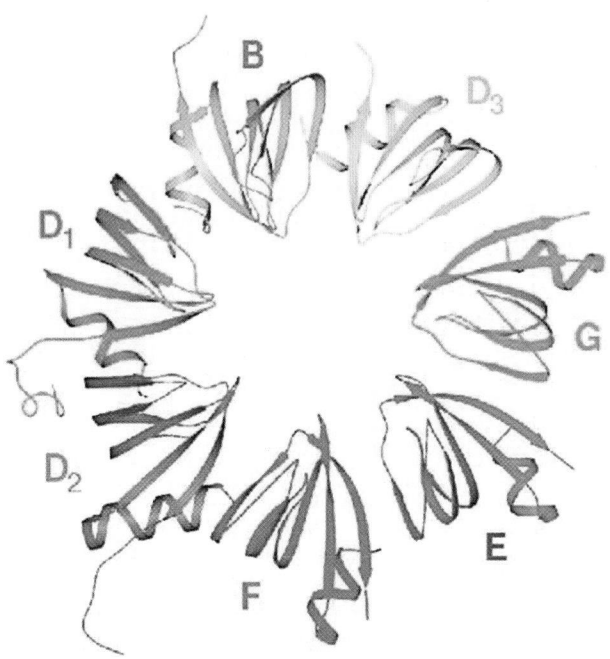

FIGURE 31-60   A model of the snRNP core protein. Its seven Sm proteins, which are each differently colored, are arranged in a 7-membered ring based on the structures of its component D_3B and D_1D_2 complexes with other pairwise interactions deduced from biochemical and mutagenic evidence. This heptameric ring has an outer diameter of 70 Å and a central hole with a diameter of 20 Å when only the main chain atoms are considered. [Courtesy of Kiyoshi Nagai, MRC Laboratory of Molecular Biology, Cambridge, U.K.]

EFG complexes form a stable subcore snRNP with U-snRNA, to which D_3B binds to form the complete **Sm core domain.**

The X-ray structures of the D_3B and D_1D_2 heterodimers, determined by Reinhard Lührmann and Kiyoshi Nagai, reveals that these four proteins share a common fold which consists of an N-terminal helix followed by a 5-stranded antiparallel β sheet that is strongly bent so as to form a hydrophobic core (Fig. 31-59*a*). The subunits of both dimers associate in a similar manner with the β5 strands of D_3 and D_1 binding to the β4 strands of B and D_2, respectively, so as to join their β sheets (Fig. 31-59*b*).

The structures of the D_1D_2 and D_3B complexes suggest that all seven Sm proteins could form a heptameric ring. Biochemical and mutagenic evidence indicates that in the EFG complex, E protein is centrally located and binds to the β5 side of F and β4 side of G. In addition, yeast two-hybrid experiments (Section 19-3C) indicate that D_2 and F interact as do D_3 and G. This has led to the model for the snRNP core protein that is drawn in Fig. 31-60. Its funnel-shaped central hole is positively charged and is large enough to permit the passage of single-stranded but not double-stranded RNA. The loops L2, L3, L4, and L5 from all seven proteins protrude into the central hole where they appear poised to bind ligands. This model is corroborated by the X-ray structure of an Sm-like protein from the hyperthermophilic archeon *Pyrobaculum aerophilum*, determined by David Eisenberg, that forms a homoheptameric ring that is structurally similar to the heteroheptameric model. This structure also supports the hypothesis that the seven eukaryotic Sm proteins arose through a series of duplications of an archaeal Sm-like protein gene.

Mammalian U1-snRNP consists of U1-snRNA and ten proteins, the seven Sm proteins that are common to all U-snRNPs, as well as three that are specific to U1-snRNP:

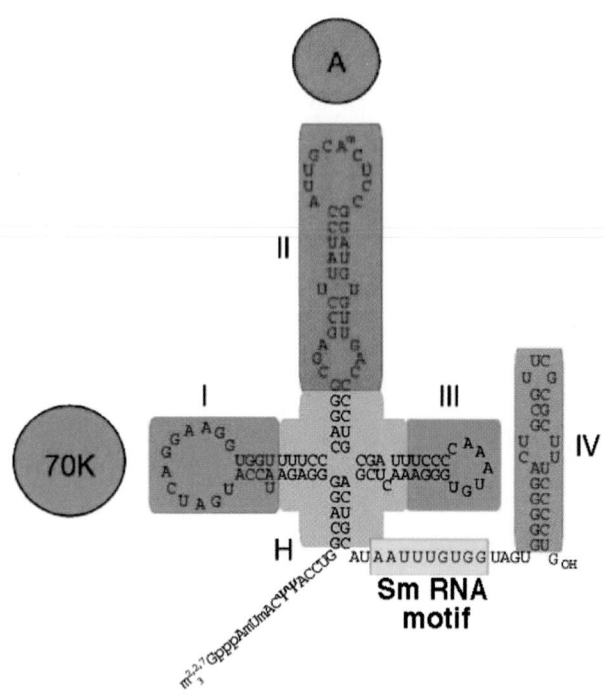

(a)

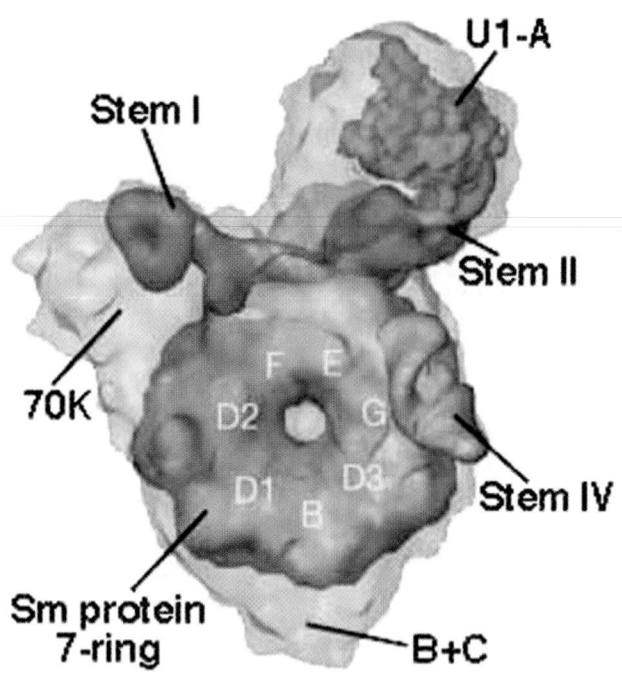

(b)

FIGURE 31-61 The electron microscopy–based structure of U1-snRNP at 10 Å resolution. (*a*) The predicted secondary structure of U1-snRNA with the positions at which the proteins U1-70K and U1-A bind the RNA indicated. (*b*) The molecular outline of U1-snRNP in light blue with its component ring-shaped Sm core protein yellow (and viewed oppositely from Fig. 31-60) and U1-snRNA colored as in Part *a*. (*c*) The U1-snRNA colored as in Part *a*. [Courtesy sof Holgar Stark, Max-Planck-Institut für biophysikalische Chemie, Göttingen, Germany.]

U1-70K, U1-A, and **U1-C.** The predicted secondary structure of the 165-nt U1-snRNA (Fig. 31-61*a*) contains five double helical stems, four of which come together at a 4-way junction. U1-70K and U1-A bind directly to RNA stem–loops I and II, respectively, whereas U1-C is bound by other proteins.

The 10-Å-resolution electron microscopy–based image of U1 snRNP (Fig. 31-61*b*) was elucidated by Holgar Stark and Lührmann. Its most obvious feature is a ring-shaped body that is 70 to 80 Å in diameter with a funnel-shaped central hole that closely matches the model of the Sm core protein in Fig. 31-60. Proteins were assigned to U1-snRNP's various protuberances based on cross-linking and binding studies as well electron micrographs of U1-snRNP lacking U1-A or U1-70K. The positions of the U1-snRNA's various structural elements (Fig. 31-61*c*), which were identified on the basis of known protein–RNA interactions, indicates that the Sm RNA motif, in fact, passes through the central hole in Sm core protein.

k. The Significance of Gene Splicing

The analysis of the large body of known DNA sequences reveals that introns are rare in prokaryotic struc-

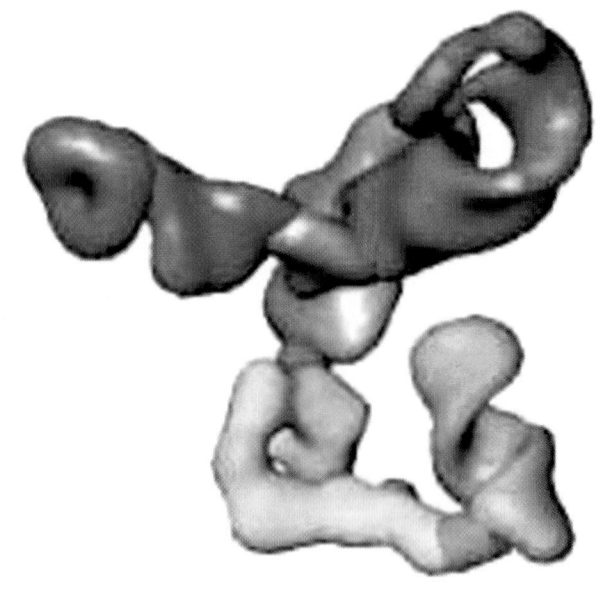

(c)

tural genes, uncommon in lower eukaryotes such as yeast (which has a total of 239 introns in its ~6000 genes and, with two exceptions, only one intron per polypeptide), and abundant in higher eukaryotes (the only known vertebrate structural genes lacking introns are those encoding histones and the antiviral proteins known as interferons). Pre-mRNA introns, as we have seen, can be quite long and many genes contain large numbers of them. Consequently, unexpressed sequences constitute ~80% of a typical vertebrate structural gene and >99% of a few of them.

The argument that introns are only molecular parasites (**junk DNA**) seems untenable since it would then be difficult to rationalize why the evolution of complex splicing machinery offered any selective advantage over the elimination of the split genes. What then is the function of gene splicing? Although, since its discovery, the significance of gene splicing has been often vehemently debated, two important roles for it have emerged: (1) It is an agent for rapid protein evolution; and (2) through **alternative splicing,** it permits a single gene to encode several (sometimes many) proteins that may have significantly different functions. In the following paragraphs, we discuss these aspects of gene splicing.

l. Many Eukaryotic Proteins Consist of Modules That Also Occur in Other Proteins

The 839-residue LDL receptor is a plasma membrane protein that functions to bind low-density lipoprotein (LDL) to coated pits for transport into the cell via endocytosis (Section 12-5B). LDL receptor's 45-kb gene contains 18 exons, most of which encode specific functional domains of the protein. *Moreover, 13 of these exons specify polypeptide segments that are homologous to segments in other proteins:*

1. Five exons encode a 7-fold repeat of a 40-residue sequence that occurs once in **complement C9** (an immune system protein; Section 35-2F).

2. Three exons each encode a 40-residue repeat similar to that occurring four times in **epidermal growth factor (EGF;** Section 19-3C) and once each in three blood clotting system proteins: **factor IX, factor X,** and **protein C** (Section 35-1).

3. Five exons encode a 400-residue sequence that is 33% identical with a polypeptide segment that is shared only with EGF.

Evidently, the LDL receptor gene is modularly constructed from exons that also encode portions of other proteins. Numerous other eukaryotic proteins are similarly constituted including, as we have seen, many of the proteins involved in signal transduction (e.g., those containing SH2 and SH3 domains; Chapter 19). *It therefore appears that the genes encoding these modular proteins arose by the stepwise collection of exons that were assembled by (aberrant) recombination between their neighboring introns.*

m. Alternative Splicing Greatly Increases the Number of Proteins Encoded by Eukaryotic Genomes

The expression of numerous cellular genes is modulated by the selection of alternative splice sites. Thus, certain exons in one type of cell may be introns in another. For example, a single rat gene encodes seven tissue-specific variants of the muscle protein **α-tropomyosin** (Section 35-3B) through the selection of alternative splice sites (Fig. 31-62).

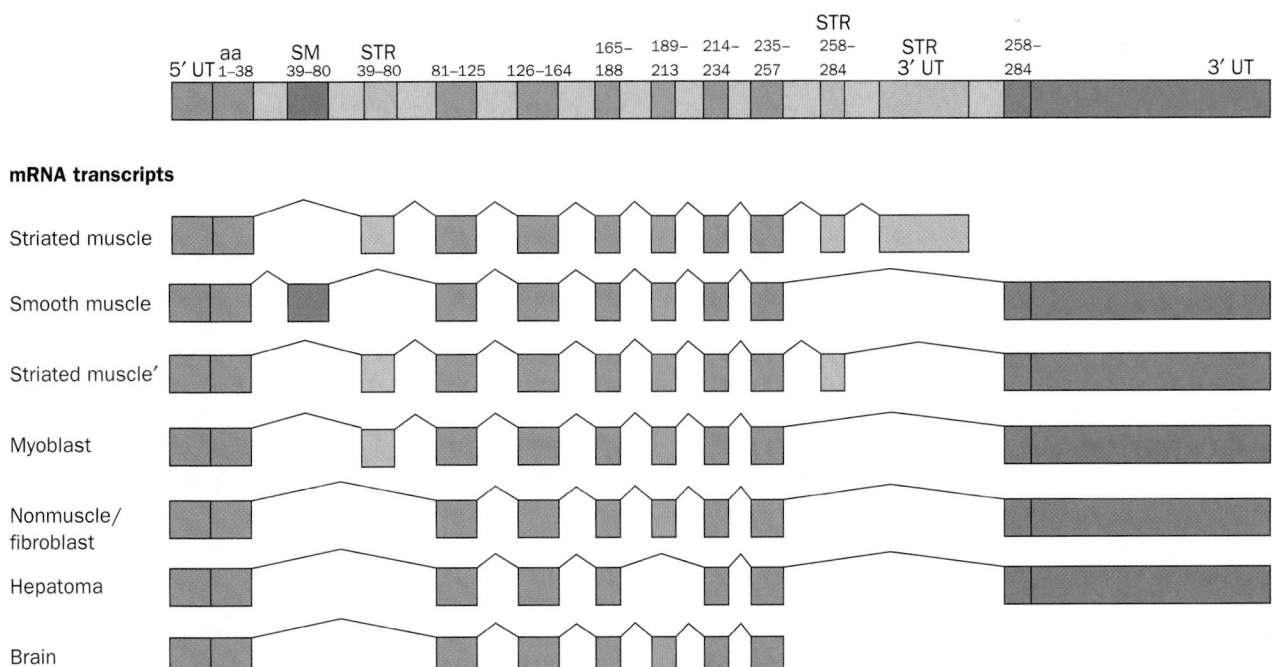

FIGURE 31-62 The organization of the rat α-tropomyosin gene and the seven alternative splicing pathways that give rise to cell-specific α-tropomyosin variants. The thin kinked lines indicate the positions occupied by the introns before they are spliced out to form the mature mRNAs. Tissue-specific exons are indicated together with the amino acid (aa) residues they encode: "constitutive" exons (those expressed in all tissues) are green, those expressed only in smooth muscle (SM) are brown, those expressed only in striated muscle (STR) are purple, and those variably expressed are yellow. Note that the smooth and striated muscle exons encoding amino acid residues 39 to 80 are mutually exclusive; likewise, there are alternative 3'-untranslated (UT) exons. [After Breitbart, R.E., Andreadis, A., and Nadal-Ginard, B., *Annu. Rev. Biochem.* **56,** 481 (1987).]

Alternative splicing occurs in all metazoa and is especially prevalent in vertebrates. In fact, it is estimated that up to 60% of human structural genes are subject to alternative splicing. This perhaps rationalizes the discrepancy between the ~30,000 genes identified in the human genome (Section 7-2B) and earlier estimates that it contains 50,000 to 140,000 structural genes.

The variation in mRNA sequence can take several different forms: Exons can be retained in an mRNA or they can be skipped; introns may be excised or retained; and the positions of 5' and 3' splice sites can be shifted to make exons shorter or longer. Alterations in the transcriptional start site and/or the polyadenylation site can further contribute to the diversity of the mRNAs that are transcribed from a single gene. In a particularly striking example, the *Drosophila* **DSCAM** protein, which functions in neuronal development, is encoded by 24 exons of which there are 12 mutually exclusive variants of exon 4, 48 of exon 6, 33 of exon 9, and 2 of exon 17 (which are therefore known as **cassette exons**) for total of 38,016 possible variants of this protein (compared to ~13,000 identified genes in the *Drosophila* genome). Although it is unknown if all possible DSCAM variants are produced, experimental evidence suggests that the *Dscam* gene expresses many thousands of them. Clearly, the number of genes in an organism's genome does not by itself provide an adequate assessment of its protein diversity. Indeed, it has been estimated that, on average, each human structural gene encodes three different proteins.

The types of changes that alternative splicing confers on expressed proteins spans the entire spectrum of protein properties and functions. Entire functional domains or even single amino acid residues may be inserted into or deleted from a protein, and the insertion of a stop codon may truncate the expressed polypeptide. Splice variations may, for example, control whether a protein is soluble or membrane bound, whether it is phosphorylated by a specific kinase, the subcellular location to which it is targeted, whether an enzyme binds a particular allosteric effector, and the affinity with which a receptor binds a ligand. Changes in an mRNA, particularly in its noncoding regions, may also influence the rate at which it is transcribed and its susceptibility to degradation. Since the selection of alternative splice sites is both tissue- and developmental stage-specific, splice site choice must be tightly regulated in both space and time. In fact, ~15% of human genetic diseases are caused by point mutations that result in pre-mRNA splicing defects. Some of these mutations delete functional splice sites, thereby activating nearby pre-existing **cryptic splice sites.** Others generate new splice sites that are used instead of the normal ones. In addition, tumor progression is correlated with changes in levels of proteins implicated in alternative splice site selection.

How are alternative splice sites selected? The best understood examples of such processes occur in the pathway responsible for sex determination in *Drosophila,* two of which we discuss here:

1. Exon 2 of *transformer (tra)* pre-mRNA contains two alternative 3' splice sites (which suceed the excised intron),

with the proximal (close; to exon 1) site used in males and the distal (far) site used in females (Fig. 31-63*a*). The region between these two sites contains a stop codon (UAG). In males, the splicing factor U2AF binds to the proximal 3' splice site to yield an mRNA containing this premature stop codon, which thereby directs the synthesis of truncated and hence nonfunctional **TRA** protein. In females, however, the proximal 3' splice site is bound by the female-specific **SXL** protein, the product of the *sex-lethal (sxl)* gene (which is only expressed in females), so as to block the binding of U2AF, which then binds to the distal 3' splice site, thereby excising the UAG and inducing the expression of functional TRA protein.

2. In *doublesex (dsx)* pre-mRNA, the first three exons are constitutively spliced in both males and females. However, the branch site immediately upstream of exon 4 has a suboptimal pyrimidine tract to which U2AF does not bind (Fig. 31-63*b*). Hence in males, exon 4 is not included in *dsx* mRNA, leading to the synthesis of male-specific **DSX-M** protein that functions as a repressor of female-specific genes. However, in females, TRA protein promotes the cooperative binding of the SR protein **RBP1** and the SR-like protein **TRA2** [the product of the *transformer 2 (tra-2)* gene] to six copies of an exonic splice enhancer (ESE) within exon 4. This heterotrimeric complex recruits the splicing machinery to the upstream 3' splice site of exon 4, leading to its inclusion in *dsx* mRNA. The resulting female-specific **DSX-F** protein is a repressor of male-specific genes.

Thus, the synthesis of functional TRA protein involves the repression of a splice site, whereas the synthesis of female-specific DSX-F protein involves the activation of a splice site. Similar mechanisms of alternative splice site selection have been identified in vertebrates.

n. AU–AC Introns Are Excised by a Novel Spliceosome

A small fraction of introns (~0.3%) have AU rather than GU at their 5' ends and AC rather than AG at their 3' ends, but are nevertheless excised via a lariat structure to an internal intron A. These so-called **AU–AC introns** (alternatively, **AT–AC introns** after their DNA sequences), which occur in organisms as diverse as *Drosophila,* plants, and humans, are excised by a novel so-called **AU–AC spliceosome** (alternatively, an **AT–AC spliceosome)** that has one snRNP, U5, in common with the major (GU–AG) spliceosome, and three others, **U11, U12,** and **U4atac–U6atac,** which are distinct from but structurally and functionally analogous to U1, U2, and U4–U6. Curiously, all genes known to contain AU–AC introns also contain multiple major class introns. Moreover, AU–AC introns are not conserved in either length or position in their host genes. Thus, the functional and evolutionary significance of the AU–AC spliceosome and introns is obscure.

o. Trans-Splicing

The types of splicing we have so far considered occur within single RNA molecules and hence are known as **cis-**

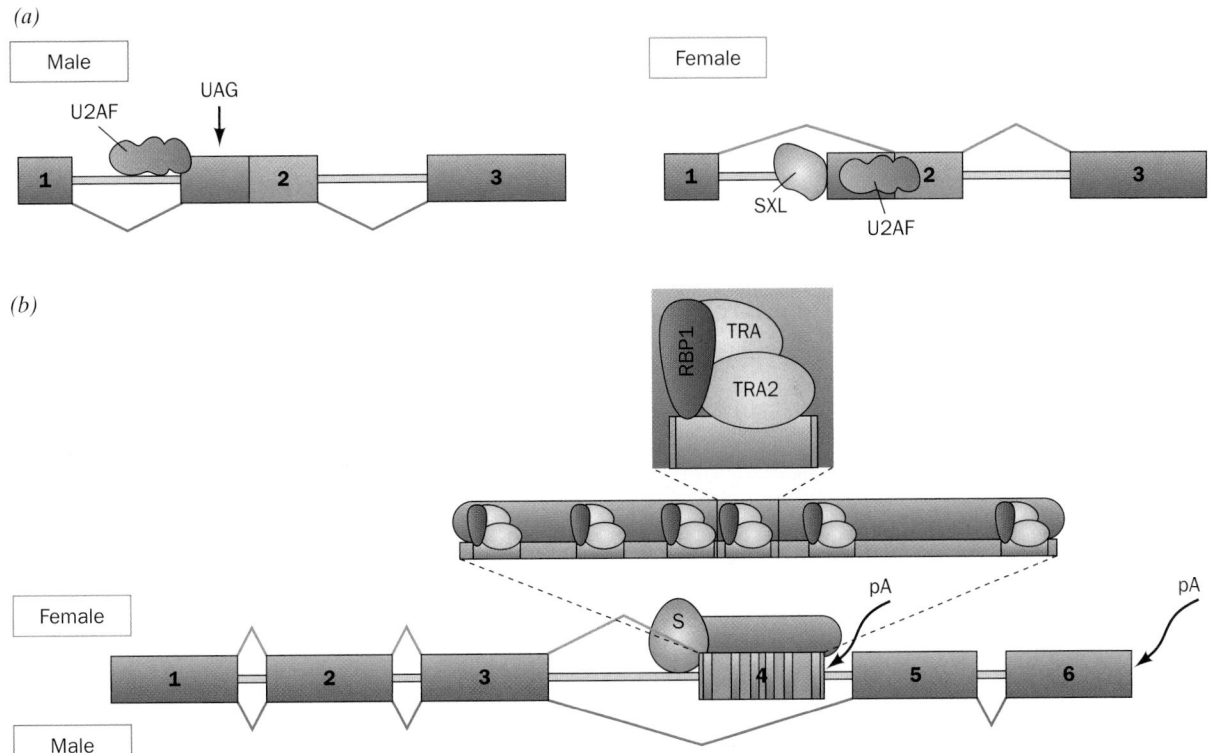

FIGURE 31-63 Mechanisms of alternative splice site selection in the *Drosophila* sex-determination pathway as described in the text. In all panels, exons are represented by colored rectangles and introns are shown as pale gray lines. (*a*) Alternative splicing in *tra* pre-mRNA. UAG is a stop codon. (*b*) Alternative splicing in *dsx* pre-mRNA. The six ESEs (exonic splice enhancers) in exon 4 are indicated by green rectangles and S represents the splicing machinery. In females, polyadenylation (pA) of *dsx* mRNA occurs downstream of exon 4, whereas in males, it occurs downstream of exon 6. [After a drawing by Maniatis, T. and Tasic, B., *Nature* **418**, 236 (2002).]

splicing. The chemistry of the spliceosomal cis-splicing reaction, however, is the same as would occur if the two exons to be joined initially resided on two different RNA molecules, a process called **trans-splicing.** This, in fact, occurs in trypanosomes (kinetoplastid protozoa; the cause of African sleeping sickness). Trypanosomal mRNAs all have the same 35-nt noncoding leader sequence, although this leader sequence is not present in the corresponding genes. Rather, this sequence is part of a so-called **spliced leader (SL) RNA** that is transcribed from an independent gene. The 5′ splice site that succeeds the SL RNA leader sequence, and the branch site and 3′ splice site that precede the exon sequence have the same consensus sequences as occur in the RNAs spliced by the major spliceosome. Consequently, the SL RNA leader and the pre-mRNA are joined in a trans-splicing reaction that resembles the spliceosomal cis-splicing reaction (Fig. 31-49) with the exception that the product of the first transesterification reaction is necessarily Y-shaped rather than lariat-shaped (Fig. 31-64). Trypanosomes, whose pre-mRNAs lack introns, nevertheless have U2- and U4–U6-snRNPs but lack U1- and U5-snRNPs. However, the SL RNA, which is predicted to fold into three stem–loops and a single-stranded Sm RNA-like motif as does U1-snRNA (Fig. 31-61*a*), ap-

parently carries out the functions of U1-snRNA in the trans-splicing reaction.

Trans-splicing has been shown to occur in nematodes (roundworms; e.g., *C. elegans*) and flatworms. These organisms also carry out cis-splicing and, indeed, perform both types of splicing on the same pre-mRNA. There are also several reports that trans-splicing occurs in higher eukaryotes such as *Drosophila* and vertebrates, but if it does occur, it does so in only a few pre-mRNAs and at a very low level.

p. mRNA Is Methylated at Certain Adenylate Residues

During or shortly after the synthesis of vertebrate pre-mRNAs, ~0.1% of their A residues are methylated at their N6 atoms. These m^6A's tend to occur in the sequence RRm^6ACX, where X is rarely G. Although the functional significance of these methylated A's is unknown, it should be noted that a large fraction of them are components of the corresponding mature mRNAs.

q. hnRNP Proteins Coat mRNAs

Throughout their residency in the nucleus, hnRNAs (pre-mRNAs) are coated with a great variety of proteins, thereby forming hnRNPs. Although the functions of these

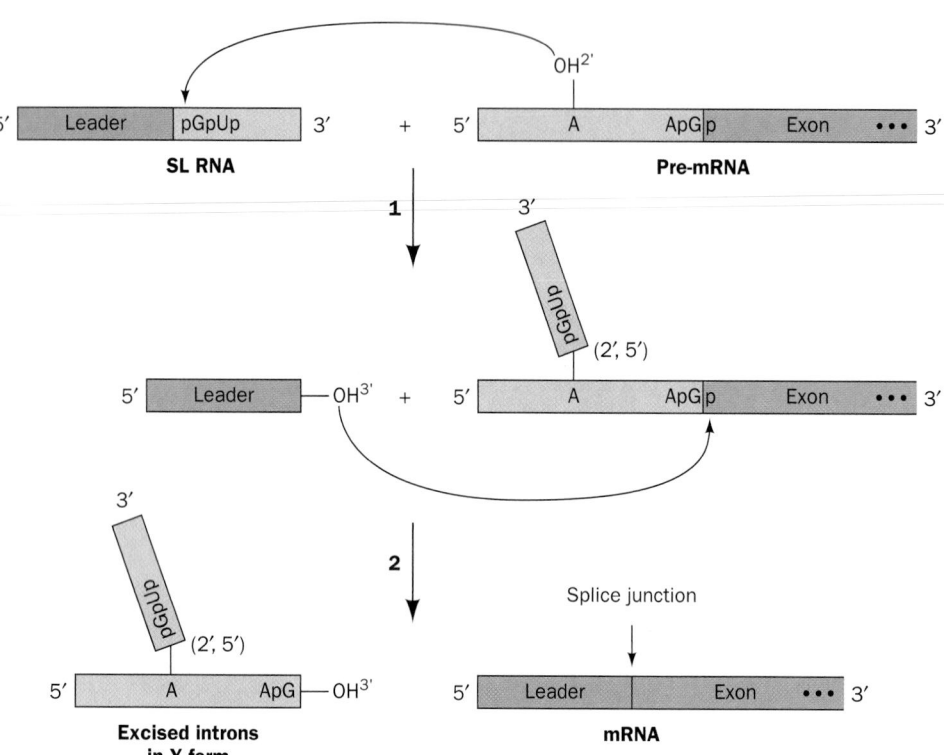

FIGURE 31-64 The sequence of transesterification reactions that occurs in trans-splicing. The chemistry is closely similar to that of pre-mRNA cis-splicing (Fig. 31-49).

hnRNP proteins are only beginning to come into focus, it is clear that they facilitate the processing of the mRNAs, including, as described above, constitutive and alternative splicing, their stability, and their transport to different regions of the nucleus and, ultimately, as Gideon Dreyfuss demonstrated, to the cytoplasm, where they regulate the localization, translation, and turnover of their associated mRNAs.

r. RNA Can Be Edited by the Insertion or Deletion of Specific Nucleotides

Certain mRNAs from a variety of eukaryotic organisms have been found to differ from their corresponding genes in several unexpected ways, including C → U and U → C changes, the insertion or deletion of U residues, and the insertion of multiple G or C residues. The most extreme examples of this phenomenon, which occur in the mitochondria of trypanosomes (whose DNA encodes only 20 genes), involve the addition and removal of up to hundreds of U's to and from 12 otherwise untranslatable mRNAs. The process whereby a transcript is altered in this manner is called **RNA editing** because it originally seemed that the required enzymatic reactions occurred without the direction of a nucleic acid template and hence violated the central dogma of molecular biology (Fig. 5-21). Eventually, however, a new class of trypanosomal mitochondrial transcripts called **guide RNAs (gRNAs)** was identified. gRNAs, which consist of 50 to 70 nucleotides, have 3' oligo(U) tails, an internal segment that is precisely complementary to the edited portion of the pre-edited mRNA (if G · U pairs, which are common in RNAs, are taken to

be complementary), and a 10- to 15-nt so-called anchor sequence near the 5' end that is largely complementary in the Watson–Crick sense to a segment of the mRNA that is not edited.

An unedited transcript presumably associates with the corresponding gRNA via its anchor sequence (Fig. 31-65). Then, in a process mediated by the appropriate enzymatic machinery in an ~20S RNP named the **editosome,** the gRNA's internal segment is used as a template to "correct" the transcript, thereby yielding the edited mRNA. Insertion editing requires at least three enzymatic activities that, somewhat surprisingly, are encoded by nuclear genes (Fig. 31-66a): (1) an endonuclease at a mismatch between the gRNA and the pre-edited mRNA to cleave the pre-edited mRNA on the 5' side of the insertion point; (2) **terminal uridylyltransferase (TUTase)** to insert the new U(s); and (3) an **RNA ligase** to reseal the RNA. Deletion requires similar enzymatic apparatus with the exceptions that the endonuclease cleaves the RNA being edited on the 3' side of the U(s) to be deleted and TUTase is replaced by **3'-U-exonuclease (3'-U-exo),** which excises the U(s) at the deletion site (Fig. 31-66b). A single gRNA mediates the editing of a block of 1 to 10 sites. Thus, the genetic information specifying an edited mRNA is derived from two or more genes. The functional advantage of this complicated process, either presently or more likely in some ancestral organism, is obscure.

s. RNA Can Be Edited by Base Deamination

Humans express two forms of **apolipoprotein B (apoB):** **apoB-48,** which is made only in the small intestine and

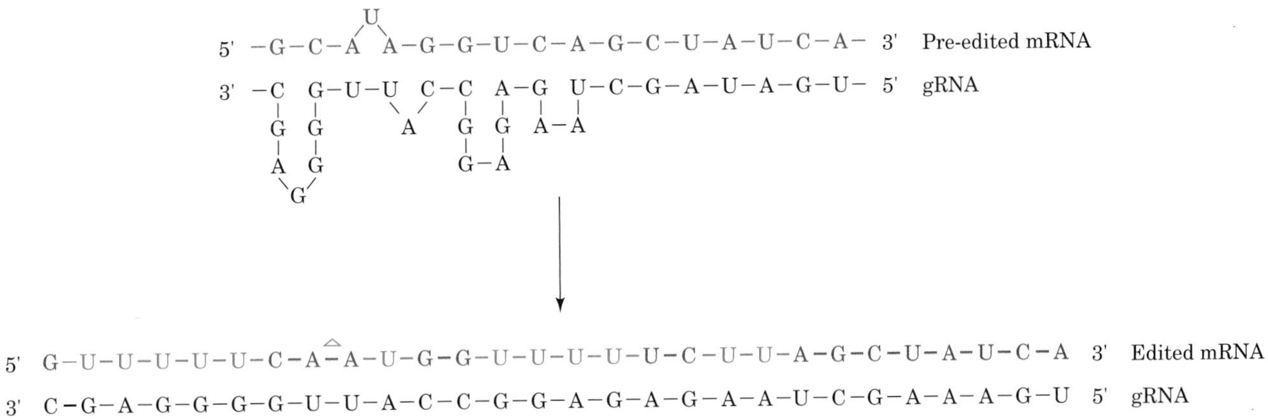

5' –G–C–A $\overset{U}{\diagup}$ A–G–G–U–C–A–G–C–U–A–U–C–A– 3' Pre-edited mRNA

3' –C G–U–U C–C A–G U–C–G–A–U–A–G–U– 5' gRNA

(gRNA stem-loop structures below:)
G G   A G G A–A
A G   G–A
 G

5' G–U–U–U–U–U–C–A–$\overset{\triangle}{\_}$A–U–G–G–U–U–U–U–U–C–U–U–A–G–C–U–A–U–C–A 3' Edited mRNA

3' C–G–A–G–G–G–G–U–U–A–C–C–G–G–A–G–A–G–A–A–U–C–G–A–A–A–G–U 5' gRNA

FIGURE 31-65 A schematic diagram indicating how gRNAs direct the editing of trypanosomal pre-edited mRNAs. The red U's in the edited mRNA are insertions and the triangle (△) marks a deletion. Several gRNAs may be necessary to direct the editing of consecutive segments of a pre-edited mRNA. [After Bass, B.L., *in* Gesteland, R.F. and Atkins, J.F. (Eds.), *The RNA World, p.* 387, Cold Spring Harbor Laboratory Press (1993).]

functions in chylomicrons to transport triacylglycerols from the intestine to the liver and peripheral tissues; and **apoB-100,** which is made only in the liver and functions in

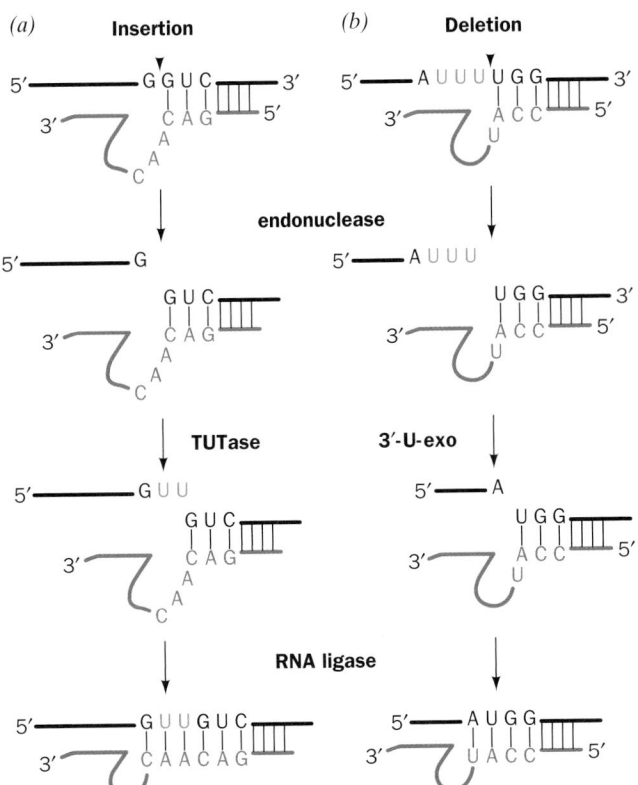

FIGURE 31-66 Trypanosomal RNA editing pathways. The RNAs being edited (*black*) are shown base paired to the gRNAs (*blue*) with the U's that are (*a*) inserted by TUTase or (*b*) deleted by 3'-U-exo drawn in red. The arrowheads indicate the positions that are cleaved by the endonuclease. [After Madison-Antenucci, S., Grams, J., and Hajduk, S.L., *Cell* **108,** 435 (2002).]

VLDL, IDL, and LDL to transport cholesterol from the liver to the peripheral tissues (Sections 12-5A and 12-5B). ApoB-100 is an enormous 4536-residue protein, whereas apoB-48 consists of apoB-100's N-terminal 2152 residues and therefore lacks the C-terminal domain of apoB-100 that mediates LDL receptor binding.

Despite their differences, both apoB-48 and apoB-100 are expressed from the same gene. How does this occur? Comparison of the mRNAs encoding the two proteins indicates that they differ by a single C → U change: The codon for Gln 2153 (CAA) in apoB-100 mRNA is, in apoB-48 mRNA, a UAA stop codon. The activity that catalyzes this conversion is a protein: It is destroyed by proteases and protein-specific reagents but not by nucleases. When apoB mRNA is synthesized with [α-^{32}P]CTP, *in vitro* editing yields a [^{32}P]UMP residue solely at the editing site. Evidently, the editing activity is a site-specific **cytidine deaminase.** This type of RNA editing differs in character from that in trypanosomal mitochondria, which inserts and deletes multiple U's into mRNAs under the direction of gRNAs. ApoB mRNA editing therefore falls into a different class of RNA editing that is called **substitutional editing.**

The several other known examples of pre-mRNA substitutional editing all occur on pre-mRNAs that encode ion channels and G protein-coupled receptors in nerve tissue. Among them is vertebrate brain **glutamate receptor** pre-mRNA, which undergoes an A → I deamination [where I is inosine (guanosine lacking its 2-amino group), which the translational apparatus reads as G] that transforms a Gln codon (CAG) to that of a functionally important Arg (CIG; normally CGG). The enzymes that catalyze such A → I RNA editing, **ADAR1** (1200 residues) and **ADAR2** (729 residues; ADAR for *a*denosine *d*eaminases *a*cting on *R*NA), have the curious requirement that their target A residues must be members of RNA double helices that are formed between the editing site and a complementary sequence that is usually located in a downstream intron (Fig.

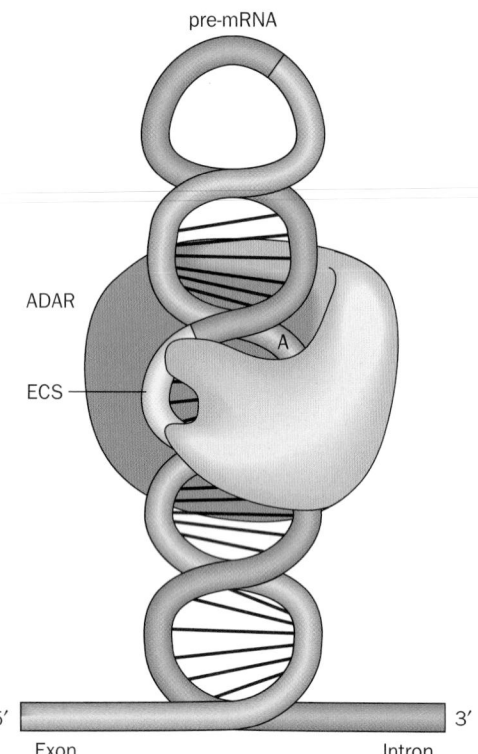

pre-mRNA

ADAR

ECS

A

5′   Exon   Intron   3′

FIGURE 31-67   The recognition of ADAR editing sites. Both ADAR1 and ADAR2 bind to a 9- to 15-bp double-stranded RNA that is formed between the editing site (*orange*) on a pre-mRNA exon and a so-called **editing site complementary sequence (ECS;** *magenta*) that is often located in a downstream intron (*brown*). A represents the adenosine that the ADAR (*green*) converts to inosine. [After Keegan, L.P., Gallo, A., and O'Connell, M.A., *Nature Rev. Genet.* **2,** 869 (2001).]

is the function of Zab? Alexander Rich has proposed that since the negative supercoiling of the DNA immediately behind actively transcribing RNAP stimulates the transient formation of Z-DNA (recall that Z-DNA has a left-handed helix), Zab targets ADAR1 to genes that are undergoing transcription. This would facilitate rapid A → I editing, which must take place before the next splicing reaction occurs.

t. RNA Interference

In recent years it has become increasingly clear that noncoding RNAs can have important roles in controlling gene expression. One of the first indications of this phenomenon occurred in Richard Jorgensen's attempt to genetically engineer more vividly purple petunias by introducing extra copies of the gene that directs the synthesis of the purple pigment. Surprisingly, the resulting transgenic plants had variegated and often entirely white flowers. Apparently, the purple-making genes somehow switched each other off. Similarly, it is well known that **antisense RNA** (RNA that is complementary to at least a portion of an mRNA) prevents the translation of the corresponding mRNA because the ribosome cannot translate double-stranded RNA (Section 32-4E). Yet, injecting **sense RNA** (RNA with the same sequence as an mRNA) into the nematode *C. elegans* also blocks protein production. Since the added RNA somehow interferes with gene expression, this phenomenon is known as **RNA interference [RNAi; posttranscriptional gene silencing (PTGS)** in plants]. RNAi/PTGS is now known to occur in all eukaryotes except perhaps yeast.

The mechanism of RNAi began to come to light in 1998 when Andrew Fire and Craig Mello showed that double-stranded RNA **(dsRNA)** was substantially more effective in causing RNAi in *C. elegans* than were either of its component strands alone. RNAi is induced by only a few molecules of dsRNA per affected cell, suggesting that RNAi is a catalytic rather than a stoichiometric effect. Further investigations, in large part in *Drosophila*, have led to the elucidation of the following pathway mediating RNAi (Fig. 31-68):

1. The trigger dsRNA, as Phillip Zamore discovered, is chopped up into ~21- to 23-nt-long double-stranded fragments known as **small interfering RNAs (siRNAs),** each of whose strands has a 2-nt overhang at its 3′ end and a 5′ phosphate. This reaction is mediated by an ATP-dependent RNase named **Dicer,** a homodimer of 2249-residue subunits that is a member of the **RNase III** family of double-strand–specific RNA endonucleases.

2. An siRNA is transferred to a 250- to 500-kD multisubunit complex known as **RNA-induced silencing complex (RISC),** which contains an endoribonuclease that is distinct from Dicer. The antisense strand of the siRNA guides the RISC complex to an mRNA with the complementary sequence.

3. RISC cleaves the mRNA, probably opposite the bound siRNA. The cleaved mRNA is then further degraded by cellular nucleases, thereby preventing its translation.

31-67). Hence, ADAR-mediated editing must precede splicing.

Substitutional editing may contribute to protein diversity. For example, *Drosophila cacophony* pre-mRNA that encodes a voltage-gated Ca^{2+} channel subunit contains 10 different substitutional editing sites and hence has the potential of generating 1000 different isoforms in the absence of alternative splicing.

Substitutional editing can also generate alternative splice sites. For example, rat ADAR2 edits its own pre-mRNA by converting an intronic AA dinucleotide to AI, which mimics the AG normally found at 3′ splice sites (Fig. 31-49). The consequent new splice site adds 47 nucleotides near the 5′ end of the *ADAR2* mRNA so as to generate a new translational initiation site. The resulting ADAR2 isozyme is catalytically active but is produced in smaller amounts than that from unedited transcripts, perhaps due to a less efficient translational initiation site. Thus, rat ADAR2 appears to regulate its own rate of expression.

ADAR1 contains an N-terminal Z-DNA-binding domain, Zab, that is composed of two subdomains, Zα and Zβ. We have seen that in the X-ray structure of Zα in complex with Z-DNA (Fig. 29-5), Zα binds Z-DNA via sequence-independent complementary surfaces (Section 29-1B). What

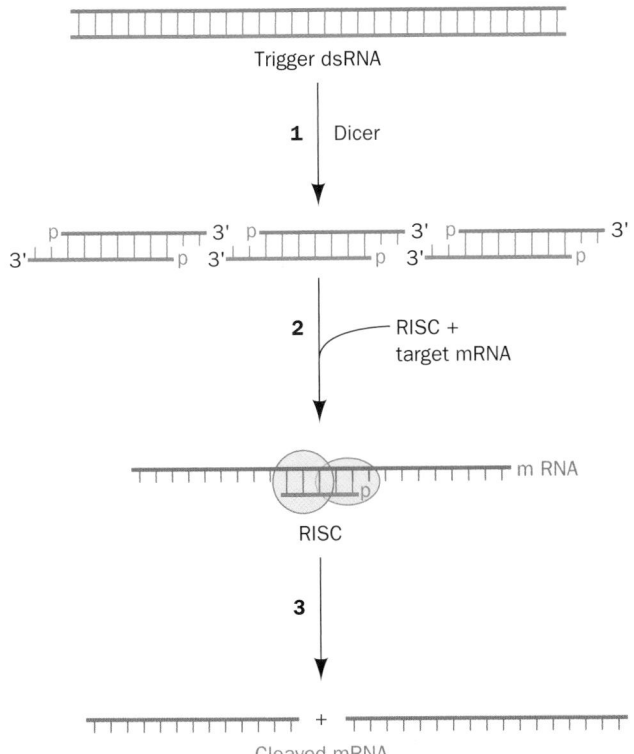

FIGURE 31-68 A model for RNA interference (RNAi). See the text for details.

RNAi requires that the trigger dsRNA be copied so as to permit the siRNAs to reach sufficient concentrations to cleave the target mRNAs. This amplification process is mediated by an **RNA-dependent RNA polymerase (RdRP).** Moreover, an siRNA strand can act as a primer for the RdRP-catalyzed synthesis of secondary trigger dsRNA, which is subsequently "diced" to yield secondary siRNAs (Fig. 31-69). Since the secondary trigger dsRNA may extend beyond the sequence complementary to the original trigger dsRNA, some of the resulting secondary siRNAs could be complementary to segments of other mRNAs that have no complementarity to the original trigger dsRNA. This would cause the silencing of genes with segments resembling portions of the original target mRNA but with no similarity to any portion of the original trigger dsRNA, a phenomenon known as **transitive RNAi.**

FIGURE 31-69 A model for transitive RNAi. The siRNA resulting from the action of Dicer on dsRNA is unwound and binds its target mRNA. There it acts as a primer for RNA-directed RNA polymerase (RdRP), which extends it in the $5' \rightarrow 3'$ direction (as do all known RNA and DNA polymerases). The resulting secondary dsRNA is subsequently cleaved by Dicer to form secondary siRNA. Note that those secondary siRNAs that contain segments of the mRNA downstream of those complementary to the original dsRNA may silence genes containing sequences similar to those of these downstream mRNA segments but yet have no sequence in common with the original dsRNA.

The ease with which RNAi/PTGS may be induced has made it the method of choice for generating null mutant (knockout) phenotypes of specific genes in plants and nonvertebrates, although care must be taken that transitive RNAi does not cause spurious results. It also appears that RNAi can be of similar use in mammalian systems, even though mammals lack the mechanisms that amplify silencing in plants and nonvertebrates so that the effects of RNAi in mammals are transient. But what is the physiological function of RNAi/PTGS? Since most eukaryotic viruses store and replicate their genomes as RNA (Chapter 33), it seems likely that RNAi arose as a defense against viral infections. Indeed, many plant viruses contain genes that suppress various steps of PTGS and which are essential for pathogenesis. RNAi/PTGS may also inhibit the movement of retrotransposons (Section 30-6B). Finally, the exquisite specificity of RNAi may make it possible to prevent viral infections and to silence disease-causing mutant genes such as oncogenes.

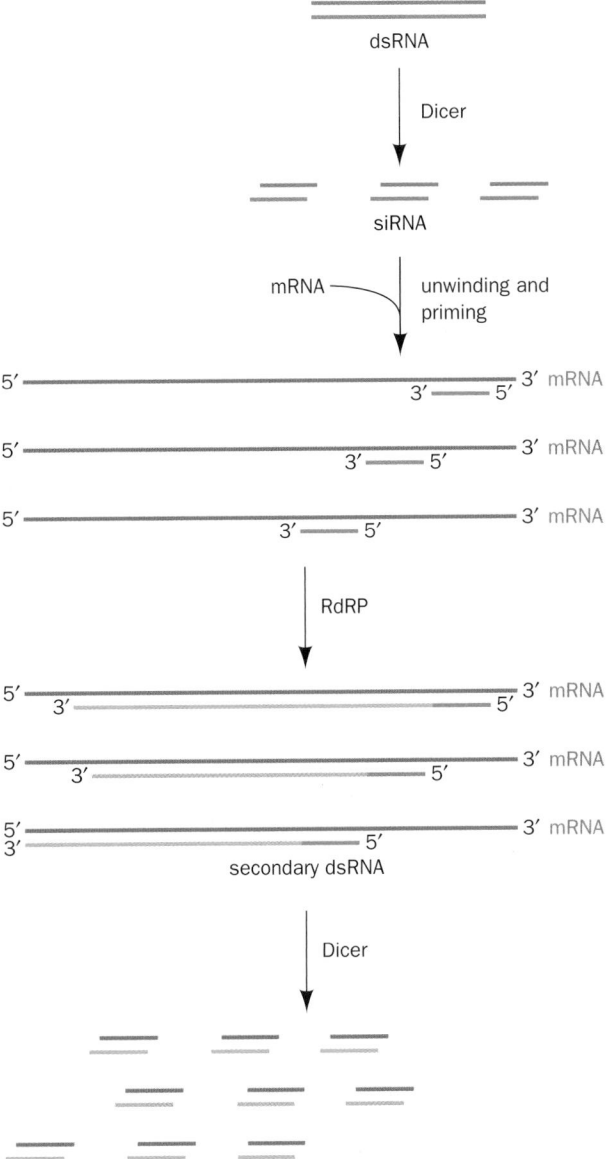

B. *Ribosomal RNA Processing*

The seven *E. coli* rRNA operons all contain one (nearly identical) copy of each of the three types of rRNA genes (Section 32-3F). Their polycistronic primary transcripts, which are >5500 nucleotides in length, contain 16S rRNA at their 5′ ends followed by the transcripts for 1 or 2 tRNAs, 23S rRNA, 5S rRNA, and, in some rRNA operons, 1 or 2 more tRNAs at the 3′ end (Fig. 31-70). The steps in processing these primary transcripts to mature rRNAs were elucidated with the aid of mutants defective in one or more of the processing enzymes.

The initial processing, which yields products known as **pre-rRNAs,** commences while the primary transcript is still being synthesized. It consists of specific endonucleolytic cleavages by **RNase III, RNase P, RNase E,** and **RNase F** at the sites indicated in Fig. 31-70. The base sequence of the primary transcript suggests the existence of several base paired stems. The RNase III cleavages occur in a stem consisting of complementary sequences flanking the 5′ and 3′ ends of the 23S segment (Fig. 31-71) as well as that of the 16S segment. Presumably, certain features of these stems constitute the RNase III recognition site.

The 5′ and 3′ ends of the pre-rRNAs are trimmed away in secondary processing steps (Fig. 31-70) through the action of **RNases D, M16, M23,** and **M5** to produce the mature rRNAs. These final cleavages only occur after the pre-rRNAs become associated with ribosomal proteins.

a. Ribosomal RNAs Are Methylated

During ribosomal assembly, the 16S and 23S rRNAs are methylated at a total of 24 specific nucleosides. The methylation reactions, which employ *S*-adenosylmethionine (Section 26-3E) as a methyl donor, yield N^6,N^6-dimethyladenine and $O^{2'}$-methylribose residues. $O^{2'}$-methyl groups may protect adjacent phosphodiester bonds from degradation by intracellular RNases (the mechanism of RNase

hydrolysis involves utilization of the free 2′-OH group of ribose to eliminate the substituent on the 3′-phosphoryl group via the formation of a 2′,3′-cyclic phosphate intermediate; Figs. 5-3 and 15-3). However, the function of base methylation is unknown.

b. Eukaryotic rRNA Processing Is Guided by snoRNAs

The eukaryotic genome typically has several hundred tandemly repeated copies of rRNA genes that are contained in small, dark-staining nuclear bodies known as **nucleoli** (the site of rRNA transcription and processing and ribosomal subunit assembly; Fig. 1-5; note that nucleoli are not membrane enveloped). The primary rRNA transcript is an ~7500-nucleotide 45S RNA that contains, starting from its 5′ end, the 18S, 5.8S, and 28S rRNAs separated by spacer sequences (Fig. 31-72). In the first stage of its processing, 45S RNA is specifically methylated at numerous sites (106 in humans) that occur mostly in its rRNA sequences. About 80% of these modifications yield $O^{2'}$-methylribose residues and the remainder form methylated bases such as N^6,N^6-dimethyladenine and 2-methylguanine. In addition, many pre-rRNA U's (95 in humans) are converted to pseudouridines (Ψ's) (Section 30-5B), which may contribute to the rRNA's tertiary stability through hydrogen bonding involving its newly acquired ring NH group. The subsequent cleavage and trimming of the 45S RNA superficially resembles that of prokaryotic rRNAs. In fact, enzymes exhibiting RNase III- and RNase P-like activities occur in eukaryotes. The 5S eukaryotic rRNA is separately processed in a manner resembling that of tRNA (Section 31-4C).

The methylation sites in eukaryotic rRNAs occur exclusively within conserved domains that are therefore likely to participate in fundamental ribosomal processes. Indeed, the methylation sites generally occur in invariant sequences among yeast and vertebrates although the methylations themselves are not always conserved. These

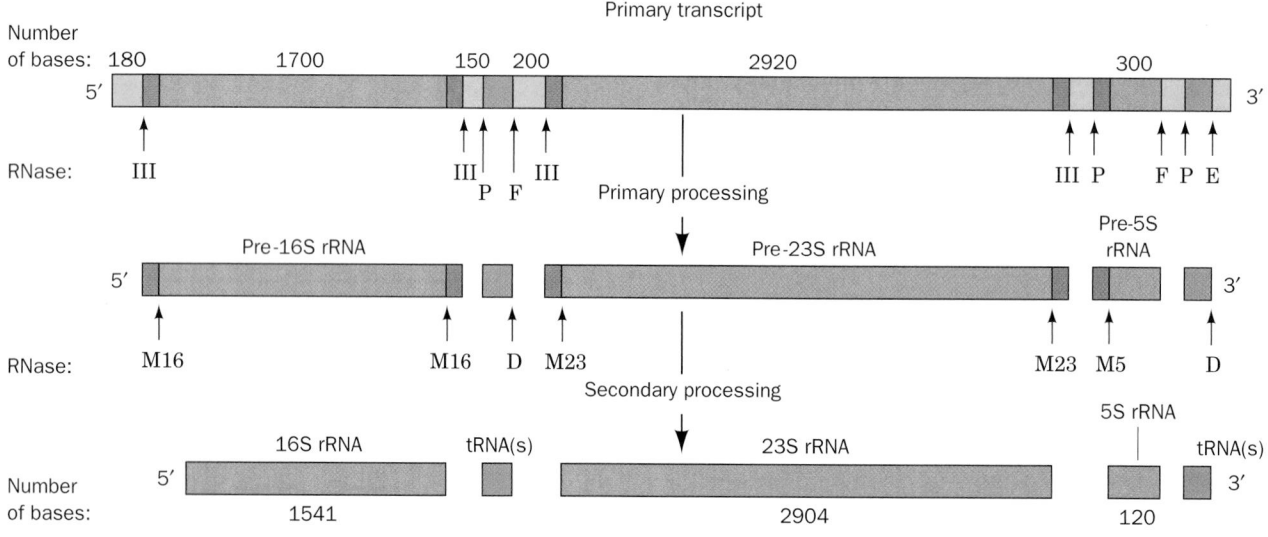

FIGURE 31-70 The posttranscriptional processing of *E. coli* rRNA. The transcriptional map is shown approximately to scale. The labeled arrows indicate the positions of the various nucleolytic cuts and the nucleases that generate them. [After

Apiron, D., Ghora, B.K., Plantz, G., Misra, T.K., and Gegenheimer, P., *in* Söll, D., Abelson, J.N., and Schimmel P.R. (Eds.), *Transfer RNA: Biological Aspects, p.* 148, Cold Spring Harbor Laboratory Press (1980).]

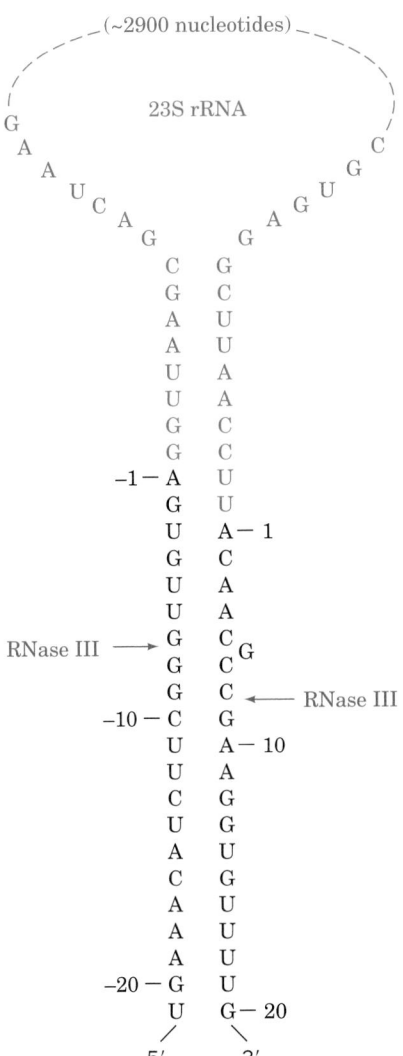

FIGURE 31-71 The proposed stem-and-giant-loop secondary structure in the 23S region of the *E. coli* primary rRNA transcript. The RNase III cleavage sites are indicated. [After Young, R.R., Bram, R.J., and Steitz, J.A., *in* Söll, D., Abelson, J.N., and Schimmel, P.R. (Eds.), *Transfer RNA: Biological Aspects*, p. 102, Cold Spring Harbor Laboratory Press (1980).]

methylation sites do not appear to have a consensus structure that might be recognized by a single methyltransferase. How, then, are these methylation sites targeted?

An important clue as to how the methylation sites on rRNA are selected came from the observation that pre-rRNA interacts with the members of a large family of **small nucleolar RNAs (snoRNAs;** ~100 in yeast and ~200 in mammals). These snoRNAs, whose lengths vary from 70 to 100 nt, contain segments of 10 to 21 nt that are precisely

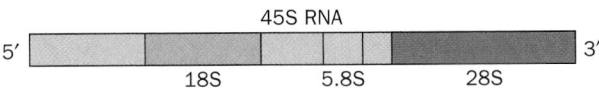

FIGURE 31-72 The organization of the 45S primary transcript of eukaryotic rRNA.

complementary to segments of the mature rRNAs that contain the O2'-methylation sites. These snoRNA sequences are located between the conserved sequence motifs known as box C (RUGAUGA) and box D (CUGA), which are respectively located on the 5' and 3' sides of the complementary segments. In intron-rich organisms such as vertebrates, most snoRNAs are encoded by the introns of structural genes so that not all excised introns are discarded.

The snoRNA nucleotide that pairs with the nucleotide to be O2'-methylated always precedes box D by exactly 5 nt. Evidently, each of these so-called **box C/D snoRNAs** act to guide the methylation of a single site. In fact, in those cases in which two adjacent ribose residues are methylated, two box C/D snoRNAs with overlapping sequences occur. The methylation is mediated by a complex of at least six nucleolar proteins, including **fibrillarin** (~325 residues), the likely methyltransferase, which together with a box C/D snoRNA form **snoRNPs.** The conversion of specific rRNA U's to Ψ's is similarly mediated by a different subgroup of snoRNAs, the **box H/ACA snoRNAs,** so called because they contain the sequence motifs ACANNN at the snoRNA's 3' end and box H (ANANNA) at its 5' end, so as to flank a sequence that partially base pairs to the pre-rRNA segment containing the U to be converted to Ψ. Archaea also modify their rRNAs via RNA-guided methylations and U to Ψ conversions, but interestingly, the analogous reactions in eubacteria are mediated by protein enzymes that lack RNA.

C. *Transfer RNA Processing*

tRNAs, as we discuss in Section 32-2A, consist of ~80 nucleotides that assume a secondary structure with four base paired stems known as the **cloverleaf structure** (Fig. 31-73). All tRNAs have a large fraction of modified bases

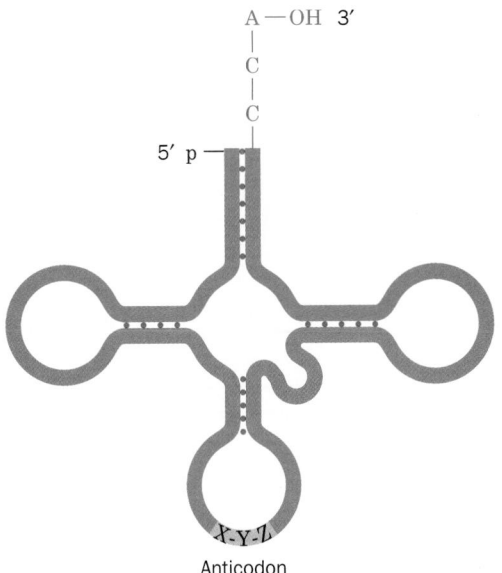

FIGURE 31-73 A schematic diagram of the tRNA cloverleaf secondary structure. Each dot indicates a base pair in the hydrogen bonded stems. The position of the anticodon triplet and the 3'-terminal —CCA are indicated.

(whose structure, function, and synthesis is also discussed in Section 32-2A) and each has the 3′-terminal sequence —CCA to which the corresponding amino acid is appended in the amino acid–charged tRNA. The **anticodon** (which is complementary to the codon specifying the tRNA's corresponding amino acid) occurs in the loop of the cloverleaf structure opposite the stem containing the terminal nucleotides.

The *E. coli* chromosome contains ~60 tRNA genes. Some of them are components of rRNA operons (Section 31-4B); the others are distributed, often in clusters, throughout the chromosome. The primary tRNA transcripts, which contain from one to as many as four or five identical tRNA copies, have extra nucleotides at the 3′ and 5′ ends of each tRNA sequence. The excision and trimming of these tRNA sequences resemble those for *E. coli* rRNAs (Section 31-4B) in that both processes employ some of the same nucleases.

a. RNase P Is a Ribozyme

RNase P, which generates the 5′ ends of tRNAs (Fig. 31-70), is a particularly interesting enzyme because it has, in *E. coli*, a 377-nucleotide RNA component (~125 kD vs 14 kD for its 119-residue protein subunit) that is essential for its enzymatic activity. The enzyme's RNA was, quite understandably, first proposed to function in recognizing the substrate RNA through base pairing and to thereby guide the protein subunit, which was presumed to be the actual nuclease, to the cleavage site. However, Sidney Altman demonstrated that *the RNA component of RNase*

P is, in fact, the enzyme's catalytic subunit by showing that protein-free RNase P RNA catalyzes the cleavage of substrate RNA at high salt concentrations. RNase P protein, which is basic, evidently functions at physiological salt concentrations to electrostatically reduce the repulsions between the polyanionic ribozyme and substrate RNAs. The argument that trace quantities of RNase P protein are really responsible for the RNase P reaction was disposed of by showing that catalytic activity is exhibited by RNase P RNA that has been transcribed in a cell-free system. RNase P activity occurs in eukaryotes (nuclei, mitochondria, and chloroplasts) as well as in prokaryotes although eukaryotic nuclear RNase P's have 9 or 10 protein subunits. Indeed, RNase P mediates one of the two ribozymal activities that occur in all cellular life, the other being associated with ribosomes (Section 32-3D).

The 400-residue RNase P from *B. subtilis*, which differs greatly in sequence from that of *E. coli*, is predicted to form two secondary structural domains (Fig. 31-74*a*): the specificity domain, which consists of nucleotides 86 to 239, and the catalytic domain, which comprises the remainder of the molecule. The X-ray structure of the specificity domain, determined by Alfonso Mondragón, is in excellent agreement with this secondary structural prediction (Fig. 31-74*b*). It consists mainly of two sets of stacked helical stems (P7-P10-P12 and P8-P9) coming together at a junction together with an unusual folded module (J11/12-J12/11) that links stems P11 and P12. Sequence analysis reveals that many of this RNA domain's conserved residues participate in interactions that are necessary for its proper

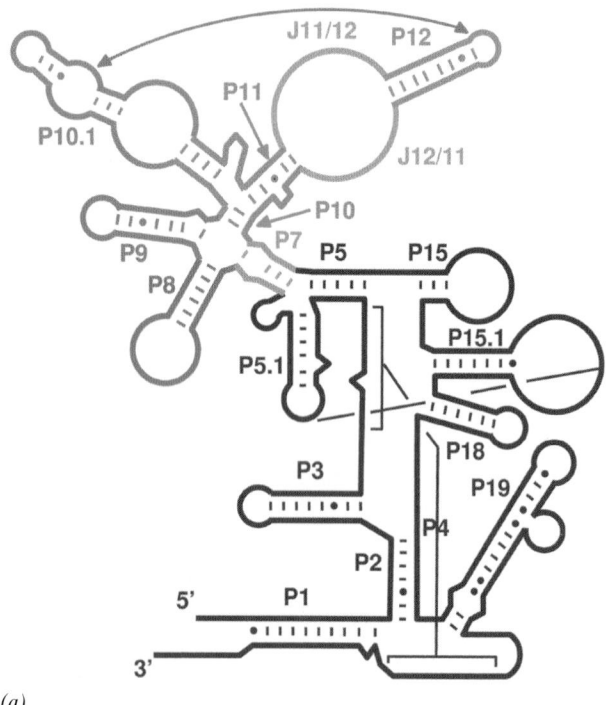

(a)

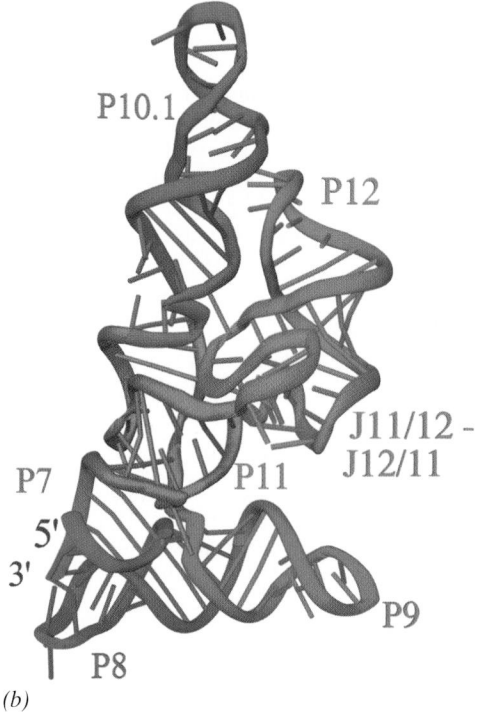

(b)

FIGURE 31-74 The structure of the RNA of B. *subtilis* RNase P. (*a*) Its predicted secondary structure in which the specificity domain is drawn in various colors and the catalytic domain is black. (*b*) The X-ray structure of the specificity domain in which its various segments are colored as in Part *a*. [Courtesy of Alfonso Mondragón, Northwestern University, PDBid 1NBS.]

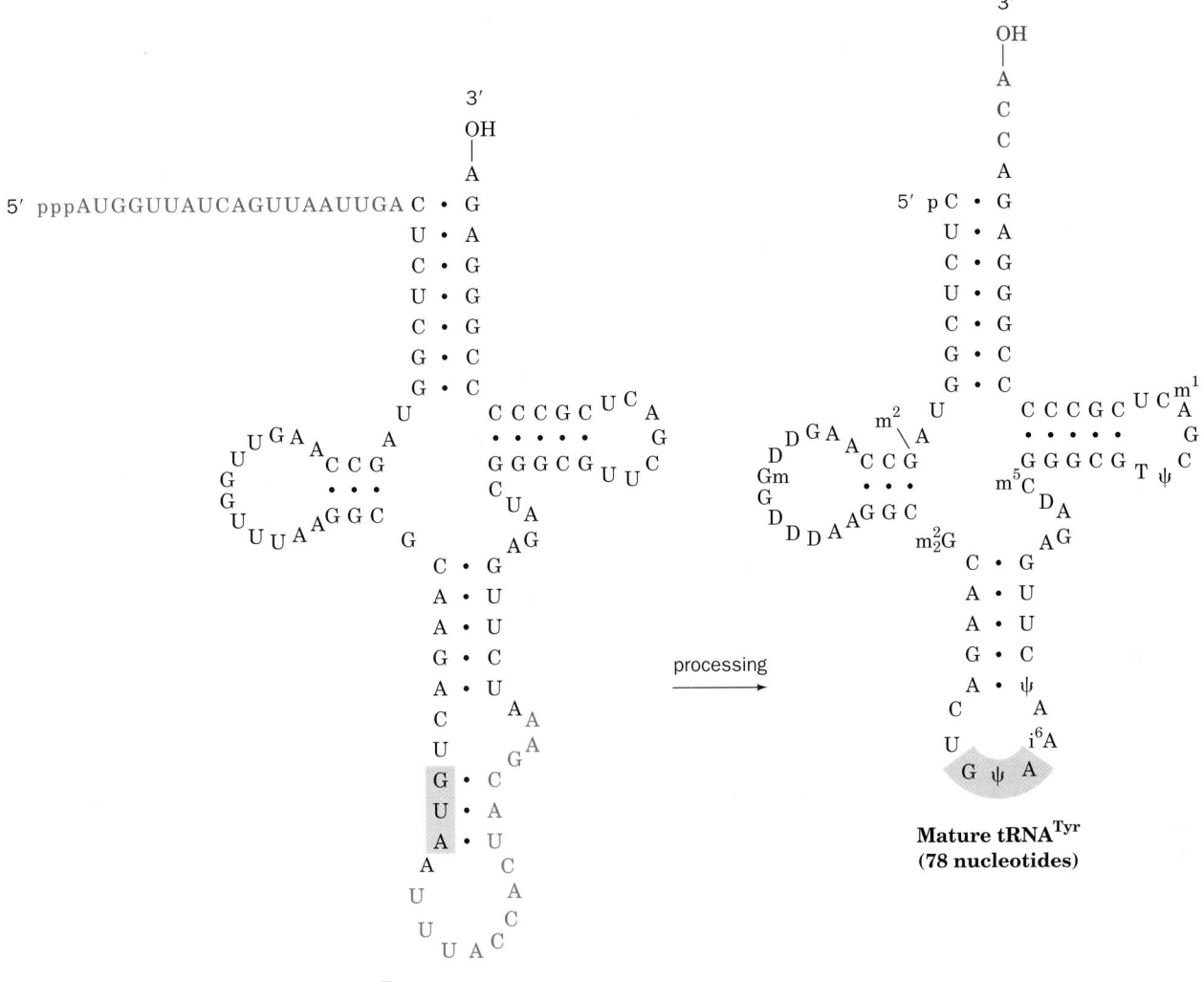

**tRNA^Tyr primary transcript
(108 nucleotides)**

**Mature tRNA^Tyr
(78 nucleotides)**

FIGURE 31-75   The posttranscriptional processing of yeast tRNA^Tyr. A 14-nucleotide intervening sequence and a 19-nucleotide 5′-terminal sequence are excised from the primary transcript, a —CCA is appended to the 3′ end, and several of the bases are modified (their symbols are defined in Fig. 32-13) to form the mature tRNA. The anticodon is shaded. [After DeRobertis, E.M. and Olsen, M.V., *Nature* **278,** 142 (1989).]

folding. Chemical modification studies indicate that the binding of pre-tRNA to RNase P protects the stacked bases of A130 and A230 that respectively bulge out of the P9 and P11 stems, together with the base of G220, which extends from J12/11. This identifies the side of the specificity domain to which its pre-tRNA substrate binds as that facing the viewer in Fig. 31-74*b*.

b. Many Eukaryotic Pre-tRNAs Contain Introns

Eukaryotic genomes contain from several hundred to several thousand tRNA genes. Many eukaryotic primary tRNA transcripts, for example, that of yeast tRNA^Tyr (Fig. 31-75), contain a small intron adjacent to their anticodons as well as extra nucleotides at their 5′ and 3′ ends. Note

that this intron is unlikely to disrupt the tRNA's cloverleaf structure.

c. The —CCA Ends of Eukaryotic tRNAs Are Posttranscriptionally Appended

Eukaryotic tRNA transcripts lack the obligatory —CCA sequence at their 3′ ends. This is appended to the immature tRNAs by the enzyme **tRNA nucleotidyltransferase,** which sequentially adds two C's and an A to tRNA using CTP and ATP as substrates. This enzyme also occurs in prokaryotes, although, at least in *E. coli,* the tRNA genes all encode a —CCA terminus. The *E. coli* tRNA nucleotidyltransferase is therefore likely to function in the repair of degraded tRNAs.

CHAPTER SUMMARY

1 ■ The Role of RNA in Protein Synthesis The central dogma of molecular biology states that "DNA makes RNA makes protein" (although RNA can also "make" DNA). There is, however, enormous variation among the rates at which the various proteins are made. Certain enzymes, such as those of the *lac* operon, are synthesized only when the substances whose metabolism they catalyze are present. The *lac* operon consists of the control sequences *lacP* and *lacO* followed by the tandemly arranged genes for β-galactosidase *(lacZ)*, galactoside permease *(lacY)*, and thiogalactoside transacetylase *(lacA)*. In the absence of inducer, physiologically allolactose, the *lac* repressor, the product of the *lacI* gene, binds to operator *(lacO)* so as to prevent the transcription of the *lac* operon by RNA polymerase. The binding of inducer causes the repressor to release the operator that allows the *lac* structural genes to be transcribed onto a single polycistronic mRNA. The mRNAs transiently associate with ribosomes so as to direct them to synthesize their encoded polypeptides.

2 ■ RNA Polymerase The holoenzyme of *E. coli* RNA polymerase (RNAP) has the subunit structure $\alpha_2\beta\beta'\omega\sigma$. It initiates transcription on the antisense (template) strand of a gene at a position designated by its promoter. The most conserved region of the promoter is centered at about the -10 position and has the consensus sequence TATAAT. The -35 region is also conserved in efficient promoters. DMS footprinting studies indicate that the RNAP holoenzyme forms an "open" initiation complex with the promoter. *Taq* RNAP holoenzyme has the shape of a crab claw with its σ subunit extended along its "top." In the closed complex, the DNA binds across the holoenzyme's "top" with all sequence-specific contacts made by the σ subunit. This suggests a model of the open complex in which the template strand in the transcription bubble passes through a tunnel in the protein to the active site channel where it pairs with incoming ribonucleotides. After the initiation of RNA synthesis, the σ subunit dissociates from the core enzyme, which then autonomously catalyzes chain elongation in the $5' \rightarrow 3'$ direction. RNA synthesis is terminated by a segment of the transcript that forms a G + C-rich hairpin with an oligo(U) tail that spontaneously dissociates from the DNA. Termination sites that lack these sequences require the assistance of Rho factor for proper chain termination.

In the nuclei of eukaryotic cells, RNAPs I, II, and III, respectively, synthesize rRNA precursors, mRNA precursors, and tRNAs + 5S RNA. The structure of yeast RNAP II resembles that of *Taq* RNAP but is somewhat larger and has more subunits. The structure of its transcribing complex reveals a one-turn segment of RNA · DNA hybrid helix at the active site, which is in contact with the solvent via a pore leading into a funnel through which NTPs presumably pass. The minimal RNA polymerase I promoter extends between nucleotides -31 and $+6$. Many RNA polymerase II promoters contain a conserved TATAAAA sequence, the TATA box, located around position -27. Enhancers are transcriptional activators that can have variable positions and orientations relative to the transcription start site. RNA polymerase III promoters are located within the transcribed regions of their gene between positions $+40$ and $+80$.

3 ■ Control of Transcription in Prokaryotes Prokaryotes can respond rapidly to environmental changes, in part because the translation of mRNAs commences during their transcription and because most mRNAs are degraded within 1 to 3 min of their synthesis. The ordered expression of sets of genes in some bacteriophages and bacteria is controlled by a succession of σ factors. The *lac* repressor is a tetrameric protein of identical subunits that, in the absence of inducer, nonspecifically binds to duplex DNA but binds much more tightly to *lac* promoter. The promoter sequence that *lac* repressor protects from nuclease digestion has nearly palindromic symmetry. Yet, methylation protection and mutational studies indicate that repressor is not symmetrically bound to promoter. *lac* repressor prevents RNA polymerase from properly initiating transcription at the *lac* promoter.

The presence of glucose represses the transcription of operons specifying certain catabolic enzymes through the mediation of cAMP. On binding cAMP, which accumulates only in the absence of glucose, catabolite gene activator protein (CAP) binds at or immediately upstream of the promoters of these operons, including the *lac* operon, thereby activating their transcription through the binding to the C-terminal domain of the associated RNAP's α subunit (αCTD). CAP's two symmetry equivalent DNA-binding domains each bind in the major groove of their target DNA via a helix–turn–helix (HTH) motif that also occurs in numerous prokaryotic repressors. The binding between these repressors and their target DNAs is mediated by mutually favorable associations between these macromolecules rather than any specific interactions between particular base pairs and amino acid side chains analogous to Watson–Crick base pairing. Sequence-specific interactions between the *met* repressor and its target DNA occur through a 2-fold symmetric antiparallel β ribbon that this protein inserts into the DNA's major groove. *araBAD* transcription is controlled by the levels of L-arabinose and CAP–cAMP through a remarkable complex of the control protein AraC to two binding sites, *araO₂* and *araI₁*, that forms an inhibitory DNA loop. On binding L-arabinose and when CAP–cAMP is adjacently bound, AraC releases *araO₂* and instead binds *araI₂*, thereby releasing the loop and activating RNA polymerase to transcribe the *araBAD* operon. The expression of the *lac* operon is also in part controlled by DNA loop formation. The *lac* repressor is a dimer of dimers, one of which binds to the operator *lacO₁* and the other to *lacO₂* or *lacO₃* to form a DNA loop that may interfere with RNAP binding to the *lac* promoter. The binding of an inducer such as IPTG to a *lac* repressor dimer core domain alters the angle between its two attached DNA-binding domains such that they cannot simultaneously bind to the *lac* operator, thereby weakening the repressor's grip on the DNA.

The expression of the *E. coli trp* operon is regulated by both attenuation and repression. On binding tryptophan, its corepressor, *trp* repressor binds to the *trp* operator, thereby blocking *trp* operon transcription. When tryptophan is available, much of the *trp* transcript that has escaped repression is prematurely terminated in the *trpL* sequence because its transcript contains a segment that forms a normal terminator structure. When tryptophanyl–tRNA^{Trp} is scarce, ribosomes stall at the transcript's two tandem Trp codons. This permits the newly synthesized RNA to form a base paired stem and loop that prevents the formation of the terminator structure. Several other operons are similarly regulated by attenuation. The stringent response is another mechanism by which *E. coli* match the rate of transcription to charged tRNA availability. When a specified charged tRNA is scarce, stringent factor on active ribosomes synthesizes ppGpp, which inhibits the transcription of rRNA and some mRNAs while stimulating the transcription of other mRNAs.

4 ■ Posttranscriptional Processing Most prokaryotic mRNA transcripts require no additional processing. However, eukaryotic mRNAs have an enzymatically appended 5′ cap and, in most cases, an enzymatically generated poly(A) tail. Moreover, the introns of eukaryotic mRNA primary transcripts (hnRNAs) are precisely excised via lariat intermediates and their flanking exons are spliced together. Group I and Group II introns are self-splicing, that is, their RNAs function as ribozymes (RNA enzymes). Ribozymes, such as the *Tetrahymena* pre-rRNA and hammerhead ribozymes, have complex structures containing several base paired stems. Pre-mRNAs are spliced by large and complex particles named spliceosomes that consist of four different small nuclear ribonucleoproteins (snRNPs) and which are assisted by the participation of a variety of protein splicing factors. Many eukaryotic proteins consist of modules that also occur in other proteins and hence appear to have evolved via the stepwise collection of exons through recombination events. The alternative splicing of pre-mRNAs greatly increases the variety of proteins expressed by eukaryotic genomes. Certain mRNAs are subject to RNA editing, either by the replacement, insertion, or deletion of specific bases in a process that is directed by guide RNAs (gRNAs), or by substitutional editing mediated by cytidine deaminases or adenosine deaminases. In RNA interference (RNAi), dsRNA is cleaved by the endoribonuclease Dicer to small interfering RNAs (siRNAs) that guide the hydrolytic cleavage of the complementary mRNAs by the RNA-induced silencing complex (RISC), thereby preventing the mRNAs' transcription.

The primary transcripts of *E. coli* rRNAs contains all three rRNAs together with some tRNAs. These are excised and trimmed by specific endonucleases and exonucleases. The eukaryotic 18S, 5.8S, and 28S rRNAs are similarly transcribed as a 45S precursor, which is processed in a manner resembling that of *E. coli* rRNAs. Eukaryotic rRNAs are modified by the methylation of specific nucleosides, as are prokaryotic rRNA, and by the conversion of certain U's to pseudouridines (Ψ's). These processes are guided by small nucleolar RNAs (snoRNAs). Prokaryotic tRNAs are excised from their primary transcripts and trimmed in much the same way as are rRNAs. In RNase P, one of the enzymes mediating this process, the catalytic subunit is an RNA. Eukaryotic tRNA transcripts also require the excision of a short intron and the enzymatic addition of a 3′-terminal —CCA to form the mature tRNA.

REFERENCES

GENERAL

Adams, R.L.P., Knowler, J.T., and Leader, D.P., *The Biochemistry of the Nucleic Acids* (11th ed.), Chapters 9–11, Chapman and Hall (1992).

Brown, T.A., *Genomes* (2nd ed.), Chapter 10, Wiley-Liss (2002).

Gesteland, R.F., Cech, T.R., and Atkins, J.F. (Eds.), *The RNA World* (2nd ed.), Cold Spring Harbor Laboratory (1999). [A series of authoritative articles on the nature of the prebiotic "RNA world" as revealed by the RNA "relics" in modern organisms.]

Hodgson, D.A. and Thomas, C.M. (Eds.), *Signals, Switches, Regulons and Cascades: Control of Bacterial Gene Expression*, Cambridge University Press (2002).

Lewin, B., *Genes VII*, Chapters 5, 9, 10, and 20, Oxford (2000).

THE GENETIC ROLE OF RNA

Brachet, J., Reminiscences about nucleic acid cytochemistry and biochemistry, *Trends Biochem. Sci.* **12**, 244–246 (1987).

Brenner, S., Jacob, F., and Meselson, M., An unstable intermediate carrying information from genes to ribosomes for protein synthesis, *Nature* **190**, 576–581 (1960). [The experimental verification of mRNA's existence.]

Crick, F., Central dogma of molecular biology, *Nature* **227**, 561–563 (1970).

Hall, B.D. and Spiegelman, S., Sequence complementarity of T2-DNA and T2-specific RNA, *Proc. Natl. Acad. Sci.* **47**, 137–146 (1964). [The first use of RNA–DNA hybridization.]

Jacob, F. and Monod, J., Genetic regulatory mechanisms in the synthesis of proteins, *J. Mol. Biol.* **3**, 318–356 (1961). [The classic paper postulating the existence of mRNA and operons and explaining how the transcription of operons is regulated.]

Pardee, A.B., Jacob, F., and Monod, J., The genetic control and cytoplasmic expression of "inducibility" in the synthesis of β-galactosidase by *E. coli*, *J. Mol. Biol.* **1**, 165–178 (1959). [The PaJaMo experiment.]

Thieffry, D., Forty years under the central dogma, *Trends Biochem. Sci.* **23**, 312–316 (1998).

RNA POLYMERASE AND mRNA

Campbell, E.A., Korzheva, N., Mustaev, A., Murakami, K., Nair, S., Goldfarb, A., and Darst, S.A., Structural mechanism for rifampicin inhibition of bacterial RNA polymerase, *Cell* **104**, 901–912 (2001).

Cramer, P., Multisubunit RNA polymerases, *Curr. Opin. Struct. Biol.* **12**, 89–97 (2002).

Cramer, P., Bushnell, D.A., and Kornberg, R.D., Structural basis of transcription: RNA polymerase at 2.8 Å resolution, *Science* **292**, 1863–1876 (2001); *and* Gnatt, A.L., Cramer, P., Fu, J., Bushnell, D.A., and Kornberg, R.D., Structural basis of transcription: An RNA polymerase II elongation complex at 3.3 Å resolution, *Science* **292**, 1876–1882 (2001).

Dahmus, M.E., Reversible phosphorylation of the C-terminal domain of RNA polymerase II, *J. Biol. Chem.* **271**, 19009–19012 (1996).

Darst, S.A., Bacterial RNA polymerase, *Curr. Opin. Struct. Biol.* **11**, 155–162 (2001).

Das, A., Control of transcription termination by RNA-binding proteins, *Annu. Rev. Biochem.* **62**, 893–930 (1993).

DeHaseth, P.L., Zupancic, M.L., and Record, M.T., Jr., RNA polymerase-promoter interactions: the comings and goings of RNA polymerase, *J. Bacteriol.* **180**, 3019–3025 (1998).

Erie, D.A., Yager, T.D., and von Hippel, P.H., The single nucleotide addition cycle in transcription, *Annu. Rev. Biophys. Biomol. Struct.* **21**, 379–415 (1992).

Estrem, S.T., Gaal, T., Ross, W., and Gourse, R.L., Identification of an UP element consensus sequence for bacterial promoters, *Proc. Natl. Acad. Sci.* **95**, 9761–9766 (1998).

Futcher, B., Supercoiling and transcription, or vice versa? *Trends Genet.* **4**, 271–272 (1988).

Gannan, F., O'Hare, K., Perrin, F., LePennec, J.P., Benoist, C., Cochet, M., Breathnach, R., Royal, A., Garapin, A., Cami, B., and Chambon, P., Organization and sequences of the 5′ end of a cloned complete ovalbumin gene, *Nature* **278**, 428–434 (1979).

Geiduschek, E.P. and Tocchini-Valentini, G.P., Transcription by RNA polymerase III, *Annu. Rev. Biochem.* **57**, 873–914 (1988).

Harada, Y., Ohara, O., Takatsuki, A., Itoh, H., Shimamoto, N., and Kinosita, K., Jr., Direct observation of DNA rotation during transcription by *Escherichia coli* RNA polymerase, *Nature* **409**, 113–115 (2001).

Huffman, J.L. and Brennan, R.G., Prokaryotic transcriptional regulators: more than just the helix-turn-helix motif, *Curr. Opin. Struct. Biol.* **12**, 98–106 (2002).

Kamitori, S. and Takusagawa, F., Crystal structure of the 2:1 complex between d(GAAGCTTC) and the anticancer drug actinomycin D, *J. Mol. Biol.* **225**, 445–456 (1992).

Khoury, G. and Gruss, P., Enhancer elements, *Cell* **33**, 313–314 (1983).

Murikami, K.S., Masuda, S., and Darst, S., Structural basis of transcription initiation: RNA polymerase at 4 Å resolution, *Science* **296**, 1280–1284 (2002); Murikami, K.S., Masuda, S., Campbell, E.A., Muzzin, O., and Darst, S., Structural basis of transcription initiation: An RNA polymerase holoenzyme-DNA complex, *Science* **296**, 1285–1290 (2002); *and* Marakami, K.S. and Darst, S.A., Bacterial RNA polymerases: the whole story, *Curr. Opin. Struct. Biol.* **13**, 31–39 (2003).

Reynolds, R., Bermúdez-Cruz, R.M., and Chamberlin, M.J., Parameters affecting transcription termination by *Escherichia coli* RNA. I. Analysis of 13 rho-independent terminators, *J. Mol. Biol.* **224**, 31–51 (1992); *and* Reynolds, R. and Chamberlin, M.J., Parameters affecting transcription termination by *Escherichia coli* RNA. II. Construction of hybrid terminators, *J. Mol. Biol.* **224**, 53–63 (1992).

Richardson, J.P., Transcription termination, *Crit. Rev. Biochem. Mol. Biol.* **28**, 1–30 (1993).

Richardson, J.P., Structural organization of transcription termination factor rho, *J. Biol. Chem.* **271**, 1251–1254 (1996).

Shilatifard, A., Conway, R.C., and Conway, J.W., The RNA polymerase II elongation complex, *Annu. Rev. Biochem.* **72**, 693–715 (2003).

Skordalakes, E. and Berger, J.M., The structure of Rho transcription terminator: Mechanism of mRNA recognition and helicase loading, *Cell.* **114**, 135–146 (2003).

Uptain, S.M., Kane, C.M., and Chamberlin, M.J., Basic mechanisms of transcription elongation and its regulation, *Annu. Rev. Biochem.* **66**, 117–172 (1997).

Willis, I.M., RNA polymerase III, *Eur. J. Biochem.* **212**, 1–11 (1993).

Zhang, G., Campbell, E.A., Minakhin, L., Richter, C., Severinov, K., and Darst, S.A., Crystal structure of *Thermus aquaticus* core RNA polymerase at 3.3 Å resolution, *Cell* **98**, 811–824 (1999).

CONTROL OF TRANSCRIPTION

Anderson, J.E., Ptashne, M., and Harrison, S.C., The structure of the repressor–operator complex of bacteriophage 434, *Nature* **326**, 846–852 (1987).

Bell, C.E. and Lewis, M., The Lac repressor: a second generation of structural and functional studies, *Curr. Opin. Struct. Biol.* **11**, 19–25 (2001).

Bennoff, B., Yang, H., Lawson, C.L., Parkinson, G., Liu, J., Blatter, E., Ebright, Y.W., Berman, H.M., and Ebright, R.H., Structural basis of transcription activation: The CAP-αCTD-DNA complex, *Science* **297**, 1562–1566 (2002).

Busby, S. and Ebright, R.H., Transcription activation by catabolite activator protein (CAP), *J. Mol. Biol.* **293**, 199–213 (1999).

Gallant, J.A., Stringent control in *E. coli*, *Annu. Rev. Genet.* **13**, 393–415 (1979).

Gartenberg, M.R. and Crothers, D.M., Synthetic DNA bending sequences increase the rate of *in vitro* transcription initiation at the *Escherichia coli lac* promoter, *J. Mol. Biol.* **219**, 217–230 (1991).

Gilbert, W. and Müller-Hill, B., Isolation of the lac repressor, *Proc. Natl. Acad. Sci.* **56**, 1891–1898 (1966).

Harmor, T., Wu, M., and Schleif, R., The role of rigidity in DNA looping–unlooping by AraC, *Proc. Natl. Acad. Sci.* **98**, 427–431 (2001).

Kolb, A., Busby, S., Buc, H., Garges, S., and Adhya, S., Transcriptional regulation by cAMP and its receptor protein, *Annu. Rev. Biochem.* **62**, 749–795 (1993).

Kolter, R. and Yanofsky, C., Attenuation in amino acid biosynthetic operons, *Annu. Rev. Genet.* **16**, 113–134 (1982).

Lamond, A.I. and Travers, A.A., Stringent control of bacterial transcription, *Cell* **41**, 6–8 (1985).

Lee, J. and Goldfarb, A., *lac* repressor acts by modifying the initial transcribing complex so that it cannot leave the promoter, *Cell* **66**, 793–798 (1991).

Lewis, M., Chang, G., Horton, N.C., Kercher, M.A., Pace, H.C., Schumacher, M.A., Brennan, R.G., and Lu, P., Crystal structure of the lactose operon repressor and its complexes with DNA and inducer, *Science* **271**, 1247–1254 (1996).

Lobel, R.B. and Schleif, R.F., DNA looping and unlooping by AraC protein, *Science* **250**, 528–532 (1990).

Luisi, B.F. and Sigler, P.B., The stereochemistry and biochemistry of the *trp* repressor-operator complex, *Biochim. Biophys. Acta* **1048**, 113–126 (1990).

McKnight, S.L. and Yamamoto, K.R. (Eds.), *Transcriptional Regulation*, Cold Spring Harbor Laboratory Press (1992). [A two-volume compendium that contains authoritative articles on many aspects of prokaryotic transcriptional control.]

Mondragón, A. and Harrison, S.C., The phage 434 Cro/O$_{R}$1 complex at 2.5 Å resolution, *J. Mol. Biol.* **219**, 321–334 (1991); *and* Wolberger, C., Dong, Y., Ptashne, M., and Harrison, S.C., Structure of phage 434 Cro/DNA complex, *Nature* **335**, 789–795 (1988).

Oehler, S., Eismann, E.R., Krämer, H., and Müller-Hill, B., The three operators of the *lac* operon cooperate in repression, *EMBO J.* **9**, 973–979 (1990).

Pace, H.C., Kercher, M.A., Lu, P., Markiewicz, P., Miller, J.H., Chang, G., and Lewis, M., *Lac* repressor genetic map in real space, *Trends Biochem. Sci.* **22**, 334–339 (1997).

Reeder, T. and Schleif, R., AraC protein can activate transcription from only one position and when pointed in only one direction, *J. Mol. Biol.* **231**, 205–218 (1993).

Rogers, D.W. and Harrison, S.C., The complex between phage 434 repressor DNA-binding domain and operator site O$_R$3: structural differences between consensus and non-consensus half-sites, *Structure* **1**, 227–240 (1993).

Schleif, R., DNA looping, *Annu. Rev. Biochem.* **61**, 199–223 (1992).

Schleif, R., Regulation of the L-arabinose operon of *Escherichia coli*, *Trends Genet.* **16**, 559–565 (2000).

Schultz, S.C., Shields, G.C., and Steitz, T.A., Crystal structure of a CAP-DNA complex: The DNA is bent by 90°, *Science* **253**, 1001–1007 (1991).

Shakked, Z., Guzikevich-Guerstein, G., Frolow, F., Rabinovich, D., Joachimiak, A., and Sigler, P.B., Determinants of repressor/operator recognition from the structure of the *trp* operator binding site, *Nature* **368**, 469–473 (1994).

Soisson, S.M., MacDougall-Shackleton, B., Schleif, R., and Wolberger, C., Structural basis for ligand-regulated oligomerization of AraC, *Science* **276**, 421–425 (1997). [The X-ray structure of AraC alone and in complex with arabinose.]

Somers, W.S. and Phillips, S.E.V., Crystal structure of the *met* repressor-operator complex at 2.8 Å resolution reveals DNA recognition by β-strands, *Nature* **359**, 387–393 (1992).

Spronk, C.A.E.M., Bonvin, A.M.J.J., Radha, P.K., Melacini, G., Boelens, R., and Kaptein, R., The solution structure of *Lac* repressor headpiece 62 complexed to a symmetrical *lac* operator, *Structure* **7**, 1483–1492 (1999).

Steitz, T.A., Structural studies of protein–nucleic acid interaction: the sources of sequence-specific binding, *Quart. Rev. Biophys.* **23**, 205–280 (1990). [Also published as a book of the same title by Cambridge University Press (1993).]

Yanofsky, C., Transcription attenuation, *J. Biol. Chem.* **263**, 609–612 (1988); *and* Attenuation in the control of expression of bacterial operons, *Nature* **289**, 751–758 (1981).

POSTTRANSCRIPTIONAL PROCESSING

Apiron, D. and Miczak, A., RNA processing in prokaryotic cells, *BioEssays* **15**, 113–119 (1993).

Bachellerie, J.-P. and Cavaillé, J., Guiding ribose methylation of rRNA, *Trends Biochem. Sci.* **22**, 257–261 (1997).

Bard, J., Zhelkovsky, A.M., Helmling, S., Earnest, T.N., Moore, C.L. and Bohm, A., Structure of yeast poly(A) polymerase alone and in complex with 3′-dATP, *Science* **289**, 1346–1349 (2000); *and* Martin, G., Keller, W., and Doublié, S., Crystal structure of mammalian poly(A) polymerase in complex with an analog of ATP, *EMBO J.* **19**, 4193–4203 (2000).

Bass, B.L., RNA editing by adenosine deaminases that act on RNA, *Annu. Rev. Biochem.* **71**, 817–846 (2002).

Black, D.L., Mechanism of alternative pre-messenger RNA splicing, *Annu. Rev. Biochem.* **72**, 291–336 (2003).

Brantl, S., Antisense regulation and RNA interference, *Biochim. Biophys. Acta* **1575**, 15–25 (2002).

Cate, J.H., Gooding, A.R., Podell, E., Zhou, K., Golden, B.L., Kundrot, C.E., Cech, T.R., and Doudna, J.A., Crystal structure of a group I ribozyme domain: principles of RNA packing, *Science* **273**, 1678–1690 (1996); Cate, J.H., Gooding, A.R., Podell, E., Zhou, K., Golden, B.L., Szewczak, A.A., Kundrot, C.E., Cech, T.R., and Doudna, J.A., RNA tertiary mediation by adenosine platforms, *Science* **273**, 1696–1699 (1996); *and* Cate, J.H. and Doudna, J.A., Metal-binding sites in the major groove of a large ribozyme domain, *Structure* **4**, 1221–1229 (1996).

Cech, T.R., Self-splicing of group I introns, *Annu. Rev. Biochem.* **59**, 543–568 (1990).

Chambon, P., Split genes, *Sci. Am.* **244**(5), 60–71 (1981).

Davis, R.E., Spliced leader RNA *trans*-splicing in metazoa, *Parasitology Today* **12**, 33–40 (1996).

Decatur, W.A. and Fournier, M.J., RNA-guided nucleotide modification of ribosomal and other RNAs, *J. Biol. Chem.* **278**, 695–698 (2003).

Denli, A.M. and Hannon, G.J., RNAi: An evergrowing puzzle, *Trends Biochem. Sci.* **28**, 196–201 (2003).

Doherty, E.A. and Doudna, J.A., Ribozyme structures and mechanisms, *Annu. Rev. Biophys. Biomol. Struct.* **30**, 457–475 (2001); and *Annu. Rev. Biochem.* **69**, 597–615 (2000).

Dreyfuss, G., Kim, V.N., and Kataoka, N., Messenger-RNA-binding proteins and the messages they carry. *Nature Rev. Cell Biol.* **3**, 195–205 (2002) ; *and* Shyu, A.-B. and Wilkinson, M.F., The double lives of shuttling mRNA binding proteins, *Cell* **102**, 135–138 (2000).

Ehretsmann, C.P., Carpousis, A.J., and Krisch, H.M., mRNA degradation in prokaryotes, *FASEB J.* **6**, 3186–3192 (1992).

Frank, D.N. and Pace, N.R., Ribonuclease P: unity and diversity in a tRNA processing ribozyme, *Annu. Rev. Biochem.* **67**, 153–180 (1998).

Gerber, A.P. and Keller, W., RNA editing by base deamination: more enzymes, more targets, new mysteries, *Trends Biochem. Sci.* **26**, 376–384 (2001).

Golden, B.L., Gooding, A.R., Podell, E.R., and Cech, T.R., A preorganized active site in the crystal structure of the *Tetrahymena* ribozyme, *Science* **282**, 259–264 (1998).

Gopalan, V., Vioque, A., and Altman, S., RNase P: variations and uses, *J. Biol. Chem.* **277**, 6759–6762 (2002).

Gott, J.M. and Emeson, R.B., Functions and mechanisms of RNA editing, *Annu. Rev. Genet.* **34**, 499–531 (2000).

Hannon, G.J., RNA interference, *Nature* **418**, 244–251 (2002).

Grosjean, H. and Benne, R. (Eds.), *Modification and Editing of RNA*, ASM Press (1998).

Kambach, C., Walke, S., Young, R., Avis, J.M., de la Fortelle, E., Raker, V.A., Lührmann, R., and Nagai, K., Crystal structures of two Sm protein complexes and their implications for the assembly of the spliceosomal snRNPs, *Cell* **96**, 375–387 (1999).

Keegan, L.P., Gallo, A., and O'Connell, M.A., The many roles of an RNA editor, *Nature Rev. Genet.* **2**, 869–878 (2001).

Krämer, A., The structure and function of proteins involved in mammalian pre-mRNA splicing, *Annu. Rev. Biochem.* **65**, 367–409 (1996).

Krasilnikov, A.S., Yang, X., Pan, T., and Mondragón, A., Crystal structure of the specificity domain of ribonuclease P, *Nature* **421**, 760–764 (2003).

Li, Y. and Breaker, R.R., Deoxyribozymes: new players in an ancient game of biocatalysis, *Curr. Opin. Struct. Biol.* **9**, 315–323 (1999).

Liu, Z., Luyten, I., Bottomley, M.J., Messais, A.C., Houngninou-Molango, S., Sprangers, R., Zanier, K., Krämer, A., and Satler, M., Structural basis for recognition of the intron branch site RNA by splicing factor 1, *Science* **294**, 1098–1102 (2001).

Maas, S., Rich, A., and Nishikura, K., A-to-I RNA editing: Recent news and residual mysteries, *J. Biol. Chem.* **278**, 1391–1394 (2003); *and* Blanc, V. and Davidson, N.O., C-to-U RNA editing: Mechanisms leading to genetic diversity, *J. Biol. Chem.* **278**, 1395–1398 (2003).

Madison-Antenucci, S., Grams, J., and Hajduk, S.L., Editing machines: the complexities of trypanosome editing, *Cell* **108**, 435–438 (2002).

Maniatis, T. and Tasic, B., Alternative pre-mRNA splicing and proteome expansion in metazoans, *Nature* **418**, 236–243 (2002).

McManus, M.T. and Sharp, P.A., Gene silencing in mammals by small interfering RNAs, *Nature Rev. Genet.* **3**, 737–747 (2002).

Mura, C., Cascio, D., Sawaya, M.R., and Eisenberg, D.S., The crystal structure of a heptameric archaeal Sm protein: Implication for the eukaryotic snRNP core, *Proc. Natl. Acad. Sci.* **98**, 5532–5537 (2001).

Nishikura, K., A short primer on RNAi: RNA-directed RNA polymerase acts as a key catalyst, *Cell* **107**, 415–418 (2001).

Proudfoot, N., Connecting transcription to messenger RNA processing, *Trends Biochem. Sci.* **25,** 290–293 (2000); *and* Proudfoot, N.J., Furger, A., and Dye, M.J., Integrating mRNA processing with transcription, *Cell* **108,** 501–512 (2002).

Rio, D.C., RNA processing, *Curr. Opin. Cell Biol.* **4,** 444–452 (1992).

Scott, W.G., Biophysical and biochemical investigations of RNA catalysis in the hammerhead ribozyme, *Quart. Rev. Biophys.* **32,** 241–284 (1999); Murray, J.B., Terwey, D.P., Maloney, L., Karpeisky, A., Usman, N., Beigleman, L., and Scott, W.G., The structural basis of hammerhead ribozyme self-cleavage, *Cell* **92,** 665–673 (1998); *and* Scott, W.G., Murray, J.B., Arnold, J.R.P., Stoddard, B.L., and Klug, A., Capturing the structure of a catalytic RNA intermediate: the hammerhead ribozyme, *Science* **274,** 2065–2069 (1996).

Sharp, P.A., Split genes and RNA splicing, *Cell* **77,** 805–815 (1994).

Smith, C.W.J. and Valcárcel, J., Alternative pre-mRNA splicing: the logic of combinatorial control, *Trends Biochem. Sci.* **25,** 381–388 (2000).

Staley, J.P. and Guthrie, C., Mechanical devices of the spliceosome: motors, clocks, springs, and things, *Cell* **92,** 315–326 (1998).

Stark, H., Dube, P., Lührmann, R., and Kastner, B., Arrangement of RNA and proteins in the spliceosomal U1 small nuclear ribonucleoprotein particle, *Nature* **409,** 539–543 (2001).

Stevens, S.W., Ryan, D.E., Ge, H.Y., Moore, R.E., Young, M.K., Lee, T.D., and Abelson, J., Composition and functional characterization of the yeast spliceosomal penta-snRNP, *Molec. Cell* **9,** 31–44 (2002).

Tanaka Hall, T.M., Poly(A) tail synthesis and regulation: recent structural insights, *Curr. Opin. Struct. Biol.* **12,** 82–88 (2002).

Tarn, W.-Y. and Steitz, J.A., Pre-mRNA splicing: the discovery of a new spliceosome doubles the challenge, *Trends Biochem. Sci.* **22,** 132–137 (1997).

Valadkhan, S. and Manley, J.L. Splicing-related catalysis by protein-free snRNAs, *Nature* **413,** 701–707 (2001).

Wahle, E. and Kühn, U., The mechanism of cleavage and polyadenylation of eukaryotic pre-RNA, *Prog. Nucl. Acid Res. Mol. Biol.* **57,** 41–71 (1997); *and* Wahle, E. and Keller, W., The biochemistry of polyadenylation, *Trends Biochem. Sci.* **21,** 247–250 (1996).

Weinstein, L.B. and Steitz, J.A., Guided tours: from precursor to snoRNA to functional snoRNP, *Curr. Opin. Cell Biol.* **11,** 378–384 (1999).

Xiao, S., Scott, F., Fierke, C.A., and Enelke, D.R., Eukaryotic ribonuclease P: A plurality of ribonucleoprotein enzymes, *Annu. Rev. Biochem.* **71,** 165–189 (2002).

Zamore, P.D., Ancient pathways programmed by small RNAs, *Science* **296,** 1265–1269 (2002); *and* Hutvágner and Zamore, P.D., RNAi: Nature abhors a double strand, *Carr. Opin. Genet. Dev.* **12,** 225–232 (2002).

PROBLEMS

1. Indicate the phenotypes of the following *E. coli lac* partial diploids in terms of inducibility and active enzymes synthesized.

a. $I^-P^+O^+Z^+Y^-/I^+P^-O^+Z^+Y^+$

b. $I^-P^+O^cZ^+Y^-/I^+P^+O^+Z^-Y^+$

c. $I^-P^+O^cZ^+Y^+/I^-P^+O^+Z^+Y^+$

d. $I^+P^-O^cZ^+Y^+/I^-P^-O^cZ^-Y^-$

2. Superrepressed mutants, I^S, encode *lac* repressors that bind operator but do not respond to the presence of inducer. Indicate the phenotypes of the following genotypes in terms of inducibility and enzyme production.

a. $I^SO^+Z^+$ b. $I^SO^cZ^+$ c. $I^+O^+Z^+/I^SO^+Z^+$

3. Why do *lacZ⁻ E. coli* fail to show galactoside permease activity after the addition of lactose in the absence of glucose? Why do *lac Y⁻* mutants lack β-galactosidase activity under the same conditions?

4. What is the experimental advantage of using IPTG instead of 1,6-allolactose as an inducer of the *lac* operon?

5. Indicate the −10 region, the −35 region, and the initiating nucleotide on the sense strand of the *E. coli* tRNA^Tyr promoter shown below.

5′ CAACGTAACACTTTACAGCGGCGCGTCATTTGATATGATGCGCCCCGCTTCCCGATA 3′

3′ GTTGCATTGTGAAATGTCGCCGCGCAGTAAACTATACTACGCGGGGCGAAGGGCTAT 5′

6. Why are *E. coli* that are diploid for rifamycin resistance and rifamycin sensitivity (*rif^R/rif^S*) sensitive to rifamycin?

7. What is the probability that the 4026-nucleotide DNA sequence coding for the β subunit of *E. coli* RNA polymerase will be transcribed with the correct base sequence? Perform the calculations for the probabilities of 0.0001, 0.001, and 0.01 that each base is incorrectly transcribed.

8. If an enhancer is placed on one plasmid and its corresponding promoter is placed on a second plasmid that is catenated (linked) with the first, initiation is almost as efficient as when the enhancer and promoter are on the same plasmid. However, initiation does not occur when the two plasmids are unlinked. Explain.

9. What is the probability that the symmetry of the *lac* operator is merely accidental?

10. Why does the inhibition of DNA gyrase in *E. coli* inhibit the expression of catabolite-sensitive operons?

11. Describe the transcription of the *trp* operon in the absence of active ribosomes and tryptophan.

12. Why can't eukaryotic transcription be regulated by attenuation?

13. Charles Yanofsky and his associates have synthesized a 15-nucleotide RNA that is complementary to segment 1 of *trpL* mRNA (but only partially complementary to segment 3). What is its effect on the *in vitro* transcription of the *trp* operon? What is its effect if the *trpL* gene contains a mutation in segment 2 that destabilizes the 2 · 3 stem and loop?

14. Why are *relA⁻* mutants defective in the *in vivo* transcription of the *his* and *trp* operons?

15. Why aren't primary rRNA transcripts observed in wild-type *E. coli*?

16. Why can't hammerhead ribozymes catalyze the cleavage of ssDNA?

Chapter
32

Translation

In this chapter we consider **translation,** the mRNA-directed biosynthesis of polypeptides. Although peptide bond formation is a relatively simple chemical reaction, the complexity of the translational process, which involves the coordinated participation of over 100 macromolecules, is mandated by the need to link 20 different amino acid residues accurately in the order specified by a particular mRNA. Note that we previewed this process in Section 5-4B.

We begin by discussing the **genetic code,** the correspondence between nucleic acid sequences and polypep-

tide sequences. Next, we examine the structures and properties of **tRNAs,** the amino acid–bearing entities that mediate the translation process. Following this, we consider the structure and functions of **ribosomes,** the complex molecular machines that catalyze peptide bond formation between the mRNA-specified amino acids. Peptide bond formation, however, does not necessarily yield a functional protein; many polypeptides must first be posttranslationally modified as we discuss in the subsequent section. Finally, we study how cells degrade proteins, a process that must balance protein synthesis.

1 ■ THE GENETIC CODE

How does DNA encode genetic information? According to the one gene–one polypeptide hypothesis, the genetic message dictates the amino acid sequences of proteins. Since the base sequence of DNA is the only variable element in this otherwise monotonously repeating polymer, the amino acid sequence of a protein must somehow be specified by the base sequence of the corresponding segment of DNA.

A DNA base sequence might specify an amino acid sequence in many conceivable ways. With only 4 bases to code for 20 amino acids, a group of several bases, termed a **codon,** is necessary to specify a single amino acid. A triplet code, that is, one with 3 bases per codon, is minimally required since there are $4^3 = 64$ different triplets of bases, whereas there can be only $4^2 = 16$ different doublets, which is insufficient to specify all the amino acids. In a triplet code, as many as 44 codons might not code for amino acids. On the other hand, many amino acids could be specified by more than one codon. Such a code, in a term borrowed from mathematics, is said to be **degenerate.**

Another mystery was, how does the polypeptide synthesizing apparatus group DNA's continuous sequence of bases into codons? For example, the code might be overlapping; that is, in the sequence

$$ABCDEFGHIJ\cdots$$

ABC might code for one amino acid, BCD for a second, CDE for a third, and so on. Alternatively, the code might be nonoverlapping, so that ABC specifies one amino acid, DEF a second, GHI a third, and so on. The code might also contain internal "punctuation" such as in the nonoverlapping triplet code

$$ABC,DEF,GHI,\cdots$$

in which the commas represent particular bases or base sequences. A related question is, how does the genetic code specify the beginning and the end of a polypeptide chain?
The genetic code is, in fact, a nonoverlapping, comma-free, degenerate, triplet code. How this was determined and how the genetic code dictionary was elucidated are the subjects of this section.

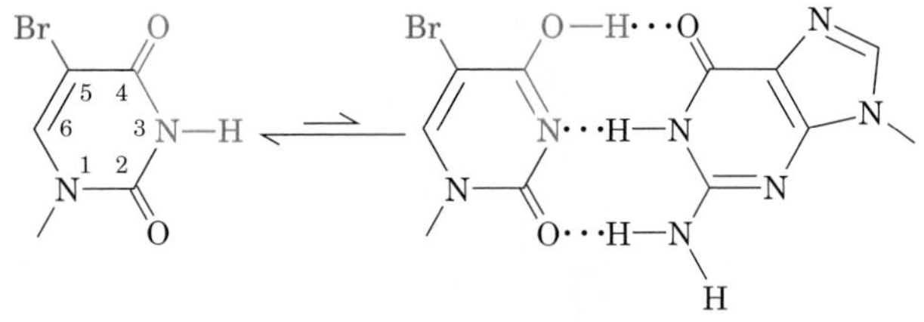

A. Chemical Mutagenesis

The triplet character of the genetic code, as we shall see below, was established through the use of **chemical mutagens,** substances that chemically induce mutations. We therefore precede our study of the genetic code with a discussion of these substances. There are two major classes of mutations:

   1. Point mutations, in which one base pair replaces another. These are subclassified as

      (a) Transitions, in which one purine (or pyrimidine) is replaced by another.
      (b) Transversions, in which a purine is replaced by a pyrimidine or vice versa.

   2. Insertion/deletion mutations, in which one or more nucleotide pairs are inserted in or deleted from DNA.

A mutation in any of these three categories may be reversed by a subsequent mutation of the same but not another category.

a. Point Mutations Are Generated by Altered Bases
Point mutations can result from the treatment of an organism with base analogs or with substances that chemically alter bases. For example, the base analog **5-bromouracil (5BU)** sterically resembles thymine (5-methyluracil) but, through the influence of its electronegative Br atom, frequently assumes a tautomeric form that base pairs with guanine instead of adenine (Fig. 32-1). Consequently, when 5BU is incorporated into DNA in place of thymine, as it usually is, it occasionally induces an A · T → G · C transition in subsequent rounds of DNA replication. Occasionally, 5BU is also incorporated into DNA in place of cytosine, which instead generates a G · C → A · T transition.
   The adenine analog **2-aminopurine (2AP),** normally base pairs with thymine (Fig. 32-2a) but occasionally forms an undistorted but singly hydrogen bonded base pair with cytosine (Fig. 32-2b). Thus 2AP generates A · T → G · C transitions.
   In aqueous solutions, **nitrous acid** (HNO$_2$) oxidatively deaminates aromatic primary amines so that it converts cytosine to uracil (Fig. 32-3a) and adenine to the guanine-like **hypoxanthine** (which forms two of guanine's three hydrogen bonds with cytosine; Fig. 32-3b). Hence, treat-

5-Bromouracil (5BU)   **5BU**
(keto tautomer)   **(enol tautomer)**   **Guanine**

FIGURE 32-1   5-Bromouracil. Its keto form (*left*) is its most common tautomer. However, it frequently assumes the enol form (*right*), which base pairs with guanine.

ment of DNA with nitrous acid, or compounds such as **nitrosamines**

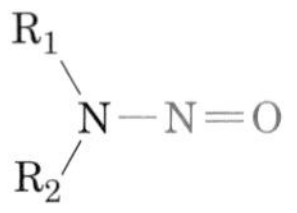

Nitrosamines

that react to form nitrous acid, results in both A · T → G · C and G · C → A · T transitions.

   Nitrite, the conjugate base of nitrous acid, has long been used as a preservative of prepared meats such as frankfurters. However, the observation that many mutagens are also carcinogens (Section 30-5F) suggests that the consumption of nitrite-containing meat is harmful to humans. Proponents of nitrite preservation nevertheless argue that to stop it would

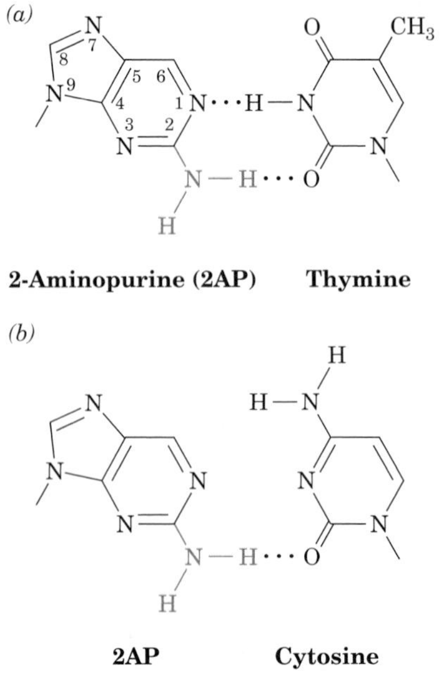

2-Aminopurine (2AP)   **Thymine**

2AP   **Cytosine**

FIGURE 32-2   Base pairing by the adenine analog 2-aminopurine. It normally base pairs with thymine (*a*) but occasionally also does so with cytosine (*b*).

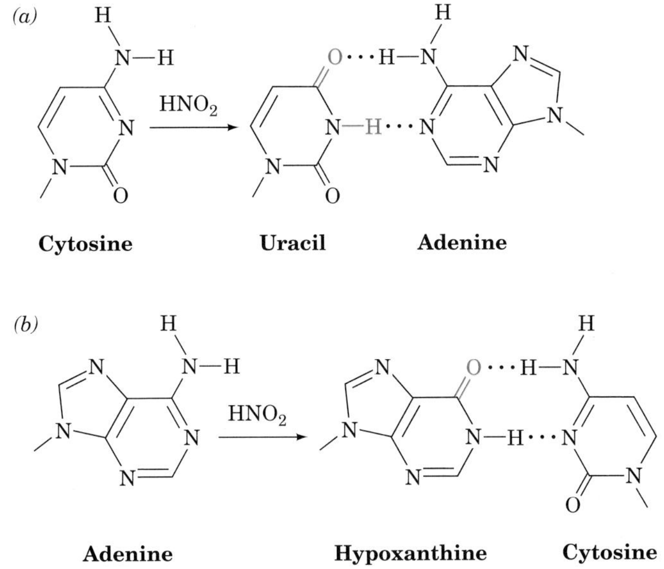

(a)

Cytosine   Uracil   Adenine

(b)

Adenine   Hypoxanthine   Cytosine

FIGURE 32-3   Oxidative deamination by nitrous acid.
(*a*) Cytosine is converted to uracil, which base pairs with
adenine. (*b*) Adenine is converted to hypoxanthine, a guanine
derivative (it lacks guanine's 2-amino group) that base pairs
with cytosine.

result in far more fatalities. This is because lack of such treatment would greatly increase the incidence of **botulism,** an often fatal form of food poisoning caused by the ingestion of
protein neurotoxins secreted by the anaerobic bacterium
Clostridium botulinum (Section 12-4D).

Hydroxylamine (NH_2OH) also induces $G \cdot C \rightarrow A \cdot T$ transitions by specifically reacting with cytosine to convert it
to a compound that base pairs with adenine (Fig. 32-4).

The use of alkylating agents such as dimethyl sulfate,
nitrogen mustard, and **ethylnitrosourea**

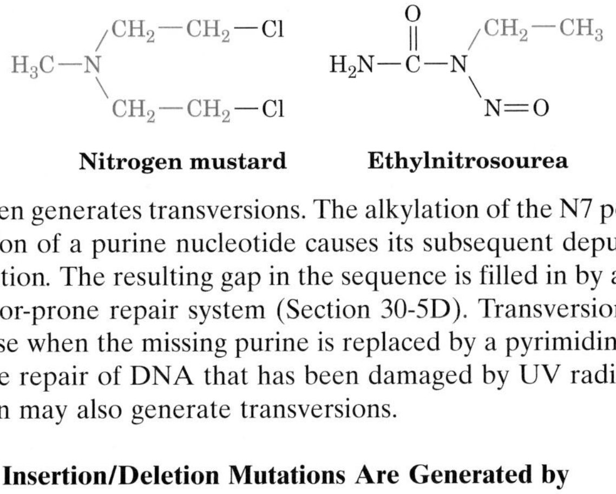

Nitrogen mustard   **Ethylnitrosourea**

often generates transversions. The alkylation of the N7 position of a purine nucleotide causes its subsequent depurination. The resulting gap in the sequence is filled in by an
error-prone repair system (Section 30-5D). Transversions
arise when the missing purine is replaced by a pyrimidine.
The repair of DNA that has been damaged by UV radiation may also generate transversions.

**b. Insertion/Deletion Mutations Are Generated by
Intercalating Agents**

*Insertion/deletion mutations may arise from the treat-
ment of DNA with intercalating agents such as acridine orange or proflavin (Section 6-6C).* The distance between two
consecutive base pairs is doubled by the intercalation of
such a molecule between them. The replication of such a

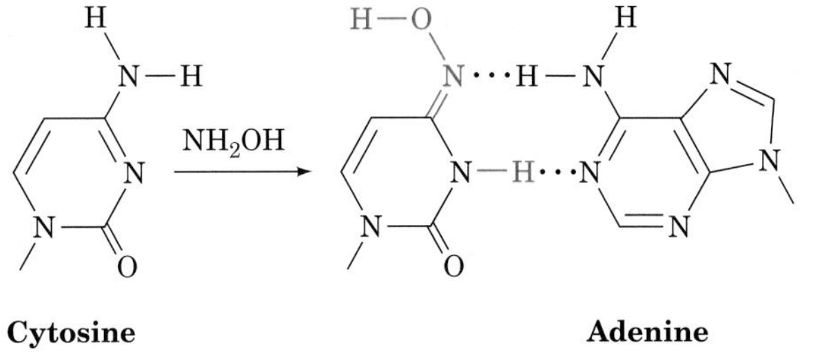

**FIGURE 32-4   Reaction with hydroxylamine converts cytosine
to a derivative that base pairs with adenine.**

distorted DNA occasionally results in the insertion or deletion of one or more nucleotides in the newly synthesized
polynucleotide. (Insertions and deletions of large DNA
segments generally arise from aberrant crossover events;
Section 34-2C.)

B. *Codons Are Triplets*

In 1961, Francis Crick and Sydney Brenner, through genetic investigations into the previously unknown character
of proflavin-induced mutations, determined the triplet
character of the genetic code. In bacteriophage T4, a particular proflavin-induced mutation, designated *FC*0, maps
in the *rIIB* cistron (Section 1-4E). The growth of this mutant phage on a permissive host (*E. coli* B) resulted in the
occasional spontaneous appearance of phenotypically
wild-type phages as was demonstrated by their ability to
grow on a restrictive host [*E. coli* K12(λ); recall that *rIIB*
mutants form characteristically large plaques on *E. coli* B
but cannot lyse *E. coli* K12(λ); Section 1-4E]. Yet, these
doubly mutated phages are not genotypically wild type; the
simultaneous infection of a permissive host by one of them
and true wild-type phage yielded recombinant progeny
that have either the *FC*0 mutation or a new mutation designated *FC*1. Thus the phenotypically wild-type phage is a
double mutant that actually contains both *FC*0 and *FC*1.
These two genes are therefore **suppressors** *of one another;
that is, they cancel each other's mutant properties. Furthermore, since they map together in the rIIB cistron, they are*
mutual **intragenic suppressors** (suppressors in the same
gene).

The treatment of *FC*1 in a manner identical to that described for *FC*0 provided similar results: the appearance
of a new mutant, *FC*2, that is an intragenic suppressor of
*FC*1. By proceeding in this iterative manner, Crick and
Brenner collected a series of different *rIIB* mutants, *FC*3,
*FC*4, *FC*5, etc., in which each mutant *FC*(*n*) is an intragenic
suppressor of its predecessor, *FC*(*n* − 1). Recombination
studies showed, moreover, that odd-numbered mutations
are intragenic suppressors of even-numbered mutations,
but neither pairs of different odd-numbered mutations nor
pairs of different even-numbered mutations suppress each
other. However, recombinants containing three odd-
numbered mutations or three even-numbered mutations
all are phenotypically wild type.

Crick and Brenner accounted for these observations by the following set of hypotheses:

1. The proflavin-induced mutation *FC*0 is either an insertion or a deletion of one nucleotide pair from the *rIIB* cistron. If it is a deletion then *FC*1 is an insertion, *FC*2 is a deletion, and so on, and vice versa.

2. *The code is read in a sequential manner starting from a fixed point in the gene.* The insertion or deletion of a nucleotide shifts the **frame** (grouping) in which succeeding nucleotides are read as codons (insertions or deletions of nucleotides are therefore also known as **frameshift mutations**). Thus the code has no internal punctuation that indicates the reading frame; that is, *the code is comma free.*

3. *The code is a triplet code.*

4. All or nearly all of the 64 triplet codons code for an amino acid; that is, *the code is degenerate.*

These principles are illustrated by the following analogy. Consider a sentence (gene) in which the words (codons) each consist of three letters (bases).

THE   BIG   RED   FOX   ATE   THE   EGG

(Here the spaces separating the words have no physical significance; they are only present to indicate the reading frame.) The deletion of the fourth letter, which shifts the reading frame, changes the sentence to

THE   IGR   EDF   OXA   TET   HEE   GG

so that all words past the point of deletion are unintelligible (specify the wrong amino acids). An insertion of any letter, however, say an X in the ninth position,

THE   IGR   EDX   FOX   ATE   THE   EGG

restores the original reading frame. Consequently, only the words between the two changes (mutations) are altered. As in this example, such a sentence might still be intelligible (the gene could still specify a functional protein), particularly if the changes are close together. Two deletions or two insertions, no matter how close together, would not suppress each other but just shift the reading frame. However, three insertions, say X, Y, and Z in the fifth, eighth, and twelfth positions, respectively, would change the sentence to

THE   BXI   GYR   EDZ   FOX   ATE   THE   EGG

which, after the third insertion, restores the original reading frame. The same would be true of three deletions. As before, if all three changes were close together, the sentence might still retain its meaning.

Crick and Brenner did not unambiguously demonstrate that the genetic code is a triplet code because they had no proof that their insertions and deletions involved only single nucleotides. Strictly speaking, they showed that a codon consists of $3r$ nucleotides where r is the number of nucleotides in an insertion or deletion. Although it was generally assumed at the time that $r = 1$, proof of this as-

sertion had to await the elucidation of the genetic code (Section 32-1C).

C. Deciphering the Genetic Code

The genetic code could, in principle, be determined by simply comparing the base sequence of an mRNA with the amino acid sequence of the polypeptide it specifies. In the 1960s, however, techniques for isolating and sequencing mRNAs had not yet been developed. The elucidation of the genetic code dictionary therefore proved to be a difficult task.

a. UUU Specifies Phe

The major breakthrough in deciphering the genetic code came in 1961 when Marshall Nirenberg and Heinrich Matthaei established that UUU is the codon specifying Phe. They did so by demonstrating that the addition of poly(U) to a cell-free protein synthesizing system stimulates only the synthesis of poly(Phe). The cell-free protein synthesizing system was prepared by gently breaking open *E. coli* cells by grinding them with powdered alumina and centrifuging the resulting cell sap to remove the cell walls and membranes. This extract contained DNA, mRNA, ribosomes, enzymes, and other cell constituents necessary for protein synthesis. When fortified with ATP, GTP, and amino acids, the system synthesized small amounts of proteins. This was demonstrated by the incubation of the system with ^{14}C-labeled amino acids followed by the precipitation of its proteins by the addition of trichloroacetic acid. The precipitate proved to be radioactive.

A cell-free protein synthesizing system, of course, produces proteins specified by the cell's DNA. On addition of DNase, however, protein synthesis stops within a few minutes because the system can no longer synthesize mRNA, whereas the mRNA originally present is rapidly degraded. Nirenberg found that crude mRNA-containing fractions from other organisms were highly active in stimulating protein synthesis in a DNase-treated protein synthesizing system. This system is likewise responsive to synthetic mRNAs.

The synthetic mRNAs that Nirenberg used in subsequent experiments were synthesized by the *Azotobacter vinelandii* enzyme **polynucleotide phosphorylase.** This enzyme, which was discovered by Severo Ochoa and Marianne Grunberg-Manago, links together nucleotides in the reaction

$$(RNA)_n + NDP \rightleftharpoons (RNA)_{n+1} + P_i$$

In contrast to RNA polymerase, however, polynucleotide phosphorylase does not utilize a template. Rather, it randomly links together the available NDPs so that the base composition of the product RNA reflects that of the reactant NDP mixture.

Nirenberg and Matthaei demonstrated that poly(U) stimulates the synthesis of poly(Phe) by incubating poly(U) and a mixture of 1 radioactive and 19 unlabeled amino acids in a DNase-treated protein synthesizing

TABLE 32-1   Amino Acid Incorporation Stimulated by a Random Copolymer of U and G in Mole Ratio 0.76:0.24

Codon	Probability of Occurrence	Relative Incidence[a]	Amino Acid	Relative Amount of Amino Acid Incorporated
UUU	0.44	100	Phe	100
UUG	0.14	32	Leu	36
UGU	0.14	32	Cys	35
GUU	0.14	32	Val	37
UGG	0.04	9	Trp	14
GUG	0.04	9		
GGU	0.04	9	Gly	12
GGG	0.01	2		

[a]Relative incidence is defined here as 100 × probability of occurrence/0.44.

Source: Matthaei, J.H., Jones, O.W., Martin, R.G., and Nirenberg, M., *Proc. Natl. Acad. Sci.* **48,** 666 (1962).

system. Significant radioactivity appeared in the protein precipitate only when phenylalanine was labeled. *UUU must therefore be the codon specifying Phe.* In similar experiments using poly(A) and poly(C), it was found that poly(Lys) and poly(Pro), respectively, were synthesized. Thus *AAA specifies* Lys *and CCC specifies Pro.* [Poly(G) cannot function as a synthetic mRNA because, even under denaturing conditions, it aggregates to form a four-stranded helix (Section 30-4D). An mRNA must be single stranded to direct its translation; Section 32-2D.]

Nirenberg and Ochoa independently employed ribonucleotide copolymers to further elucidate the genetic code. For example, in a poly(UG) composed of 76% U and 24% G, the probability of a given triplet being UUU is 0.76 × 0.76 × 0.76 = 0.44. Likewise, the probability of a particular triplet consisting of 2U's and 1G, that is, UUG, UGU, or GUU, is 0.76 × 0.76 × 0.24 = 0.14. The use of this poly(UG) as an mRNA therefore indicated the base compositions, but not the sequences, of the codons specifying several amino acids (Table 32-1). Through the use of copolymers containing two, three, and four bases, the base compositions of codons specifying each of the 20 amino acids were inferred. Moreover, *these experiments demonstrated that the genetic code is degenerate since, for example, poly(UA), poly(UC), and poly(UG) all direct the incorporation of Leu into a polypeptide.*

b. The Genetic Code Was Elucidated through Triplet Binding Assays and the Use of Polyribonucleotides with Known Sequences

In the absence of GTP, which is necessary for protein synthesis, trinucleotides but not dinucleotides are almost as effective as mRNAs in promoting the ribosomal binding of specific tRNAs. This phenomenon, which Nirenberg and Philip Leder discovered in 1964, permitted the various codons to be identified by a simple binding assay. Ribosomes, together with their bound tRNAs, are retained

by a nitrocellulose filter but free tRNA is not. The bound tRNA was identified by using charged tRNA mixtures in which only one of the pendent amino acid residues was radioactively labeled. For instance, it was found, as expected, that UUU stimulates the ribosomal binding of only Phe tRNA. Likewise, UUG, UGU, and GUU stimulate the binding of Leu, Cys, and Val tRNAs, respectively. Hence UUG, UGU, and GUU must be codons that specify Leu, Cys, and Val, respectively. In this way, the amino acids specified by some 50 codons were identified. For the remaining codons, the binding assay was either negative (no tRNA bound) or ambiguous.

The genetic code dictionary was completed and previous results confirmed through H. Gobind Khorana's synthesis of polyribonucleotides with specified repeating sequences (Section 7-6A). In a cell-free protein synthesizing system, UCUCUCUC···, for example, is read

UCU   CUC   UCU   CUC   UCU   C···

so that it specifies a polypeptide chain of two alternating amino acid residues. In fact, it was observed that this mRNA stimulated the production of

Ser—Leu—Ser—Leu—Ser—Leu—···

which indicates that either UCU or CUC specifies Ser and the other specifies Leu. This information, together with the tRNA-binding data, permitted the conclusion that UCU codes for Ser and CUC codes for Leu. These data also prove that codons consist of an odd number of nucleotides, thereby relieving any residual suspicions that codons consist of six rather than three nucleotides.

Alternating sequences of three nucleotides, such as poly(UAC), specify three different homopolypeptides because ribosomes may initiate polypeptide synthesis on these synthetic mRNAs in any of the three possible reading frames (Fig. 32-5). Analyses of the polypeptides specified by various alternating sequences of two and three nucleotides confirmed the identity of many codons and filled out missing portions of the genetic code.

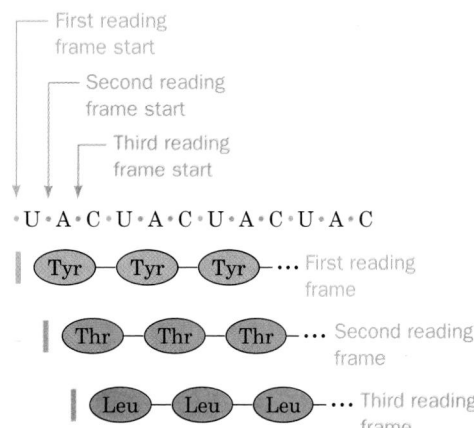

FIGURE 32-5   The three potential reading frames of an mRNA. Each reading frame would yield a different polypeptide.

c. mRNAs Are Read in the 5′ → 3′ Direction

The use of repeating tetranucleotides indicated the reading direction of the code and identified the chain termination codons. Poly(UAUC) specifies, as expected, a polypeptide with a tetrapeptide repeat:

$$\overset{5'}{\text{UAU}} \ \text{CUA} \ \text{UCU} \ \text{AUC} \ \text{UAU} \ \text{CUA} \overset{3'}{\cdots}$$

$$\text{Tyr} - \text{Leu} - \text{Ser} - \text{Ile} - \text{Tyr} - \text{Leu} - \cdots$$

The amino acid sequence of this polypeptide indicates that the mRNA's 5′ end corresponds to the polypeptide's N-terminus; that is, *mRNA is read in the 5′ → 3′ direction.*

d. UAG, UAA, and UGA Are Stop Codons

In contrast to the above results, poly(AUAG) yields only dipeptides and tripeptides. This is because *UAG is a signal to the ribosome to terminate protein synthesis:*

$$\text{AUA} \ \text{GAU} \ \text{AGA} \ \text{UAG} \ \text{AUA} \ \text{GAU} \cdots$$

$$\text{Ile} - \text{Asp} - \text{Arg} \quad \text{Stop} \quad \text{Ile} - \text{Asp} - \cdots$$

Likewise, poly(GUAA) yields dipeptides and tripeptides because UAA is also a chain termination signal:

$$\text{GUA} \ \text{AGU} \ \text{AAG} \ \text{UAA} \ \text{GUA} \ \text{AGU} \cdots$$

$$\text{Val} - \text{Ser} - \text{Lys} \quad \text{Stop} \quad \text{Val} - \text{Ser} - \cdots$$

UGA is a third stop signal. These Stop codons, whose existence was first inferred from genetic experiments, are known, somewhat inappropriately, as **nonsense codons** because they are the only codons that do not specify amino acids. UAG, UAA, and UGA are sometimes referred to as *amber, ochre,* and *opal* codons. [They were so named as the result of a laboratory joke: The German word for amber is Bernstein, the name of an individual who helped discover *amber* mutations (mutations that change some other codon to UAG); *ochre* and *opal* are puns on *amber.*]

e. AUG and GUG Are Chain Initiation Codons

The codons AUG, and less frequently GUG, form part of the chain initiation sequence (Section 32-3C). However, they also specify the amino acid residues Met and Val, respectively, at internal positions of polypeptide chains. (Nirenberg and Matthaei's discovery that UUU specifies Phe was only possible because ribosomes indiscriminately initiate polypeptide synthesis on an mRNA when the Mg^{2+} concentration is unphysiologically high as it was, serendipitously, in their experiments.)

D. *The Nature of the Code*

The genetic code dictionary, as elucidated by the above methods, is presented in Table 32-2 as well as in Table 5-3. Examination of the table indicates that the genetic code has several remarkable features:

1. *The code is highly degenerate.* Three amino acids, Arg, Leu, and Ser, are each specified by six codons, and

TABLE 32-2   The "Standard" Genetic Code[a]

First position (5′ end)	Second position				Third position (3′ end)
	U	**C**	**A**	**G**	
U	UUU Phe / UUC Phe	UCU / UCC / UCA / UCG Ser	UAU Tyr / UAC Tyr	UGU Cys / UGC Cys	U / C
	UUA Leu / UUG Leu		UAA Stop / UAG Stop	UGA Stop / UGG Trp	A / G
C	CUU / CUC / CUA / CUG Leu	CCU / CCC / CCA / CCG Pro	CAU His / CAC His / CAA Gln / CAG Gln	CGU / CGC / CGA / CGG Arg	U / C / A / G
A	AUU / AUC / AUA Ile / AUG Met[b]	ACU / ACC / ACA / ACG Thr	AAU Asn / AAC Asn / AAA Lys / AAG Lys	AGU Ser / AGC Ser / AGA Arg / AGG Arg	U / C / A / G
G	GUU / GUC / GUA / GUG Val	GCU / GCC / GCA / GCG Ala	GAU Asp / GAC Asp / GAA Glu / GAG Glu	GGU / GGC / GGA / GGG Gly	U / C / A / G

[a] Nonpolar amino acid residues are tan, basic residues are blue, acidic residues are red, and nonpolar uncharged residues are purple.
[b] AUG forms part of the initiation signal as well as coding for internal Met residues.

most of the rest are specified by either four, three, or two codons. Only Met and Trp, two of the least common amino acids in proteins (Table 4-1), are represented by a single codon. Codons that specify the same amino acid are termed **synonyms.**

2. *The arrangement of the code table is nonrandom.* Most synonyms occupy the same box in Table 32-2; that is, they differ only in their third nucleotide. The only exceptions are Arg, Leu, and Ser, which, having six codons each, must occupy more than one box. XYU and XYC always specify the same amino acid; XYA and XYG do so in all but two cases. Moreover, changes in the first codon position tend to specify similar (if not the same) amino acids, whereas codons with second position pyrimidines encode mostly hydrophobic amino acids (tan in Table 32-2), and those with second position purines encode mostly polar amino acids (blue, red, and purple in Table 32-2). Apparently *the code evolved so as to minimize the deleterious effects of mutations.*

Many of the mutations causing amino acid substitutions in a protein can be rationalized, according to the genetic code, as single point mutations. *As a consequence of the genetic code's degeneracy, however, many point mutations at a third codon position are phenotypically silent; that is, the mutated codon specifies the same amino acid as the wild type.* Degeneracy may account for as much as 33% of the

25 to 75% range in the G + C content among the DNAs of different organisms (Section 5-1B). The frequent occurrence of Arg, Ala, Gly, and Pro also tends to give a high G + C content, whereas Asn, Ile, Lys, Met, Phe, and Tyr contribute to a low G + C content.

a. Some Phage DNA Segments Contain Overlapping Genes in Different Reading Frames

Since any nucleotide sequence may have three reading frames, it is possible, at least in principle, for a polynucleotide to encode two or even three different polypeptides. This idea was never seriously entertained, however, because it seemed that the constraints on even two overlapping genes in different reading frames would be too great for them to evolve so that both could specify sensible proteins. It therefore came as a great surprise, in 1976, when Frederick Sanger reported that the 5386-nucleotide DNA of bacteriophage φX174 (which, at the time, was the largest DNA to have been sequenced) contains two genes that are completely contained within larger genes of different reading frames (Fig. 32-6). Moreover, the end of the overlapping D and E genes contains the control sequence for the ribosomal initiation of the J gene so that this short DNA segment performs triple duty. Bacteria also exhibit such coding economy; the ribosomal initiation sequence of one gene in a polycistronic mRNA often overlaps the end of the preceding gene. Nevertheless, completely overlapping genes have only been found in small single-stranded DNA phages, which presumably must make maximal use of the little DNA that they can pack inside their capsids.

b. The "Standard" Genetic Code Is Widespread but Not Universal

For many years it was thought that the "standard" genetic code (that given in Table 32-2) is universal. This assumption was, in part, based on the observations that one kind of organism (e.g., *E. coli*) can accurately translate the genes from quite different organisms (e.g., humans). This phenomenon is, in fact, the basis of genetic engineering. Once the "standard" genetic code had been established, presumably during the time of prebiotic evolution (Section 1-5B), any mutation that would alter the way the code is translated would result in numerous, often deleterious, protein sequence changes. Undoubtedly there is strong selection against such mutations. DNA sequencing studies in 1981 nevertheless revealed that *the genetic codes of certain mitochondria (mitochondria contain their own genes and protein synthesizing systems but produce only a few mitochondrial proteins; Section 12-4E) are variants of the "standard" genetic code (Table 32-3)*. For example, in mammalian mitochondria, AUA, as well as the standard AUG, is a Met/initiation codon; UGA specifies Trp rather than "Stop"; and AGA and AGG are "Stop" rather than Arg. Note that all mitochondrial genetic codes except those of plants simplify the "standard" code by increasing its degeneracy. For example, in the mammalian mitochondrial code, each amino acid is specified by at least two codons that differ only in their third nucleotide. Apparently the constraints preventing alterations of the

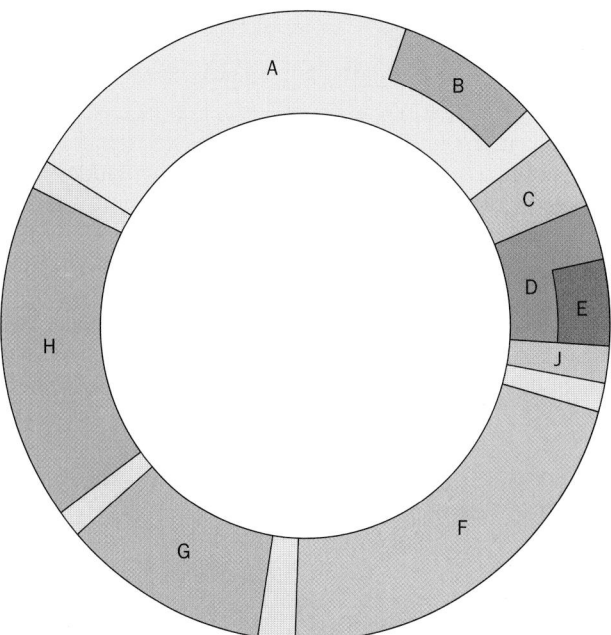

FIGURE 32-6   Genetic map of bacteriophage φX174 as determined by DNA sequence analysis. Genes are labeled A, B, C, etc. Note that gene B is wholly contained within gene A and gene E is wholly contained within gene D. These pairs of genes are read in different reading frames and therefore specify unrelated proteins. The unlabeled regions correspond to untranslated control sequences.

genetic code are eased by the small sizes of mitochondrial genomes. More recent studies, however, have revealed that in ciliated protozoa, the codons UAA and UAG specify Gln rather than "Stop." Perhaps UAA and UAG were sufficiently rare codons in a primordial ciliate (which molecular phylogenetic studies indicate branched off very early in eukaryotic evolution) to permit the code change without unacceptable deleterious effects. At any rate, *the "standard" genetic code, although very widely utilized, is not universal.*

TABLE 32-3   Mitochondrial Deviations from the "Standard" Genetic Code

Mitochondrion	UGA	AUA	CUN[a]	AG$_G^A$	CGG
Mammalian	Trp	Met[b]		Stop	
Baker's yeast	Trp	Met[b]	Thr		
Neurospora crassa	Trp				
Drosophila	Trp	Met[b]		Ser[c]	
Protozoan	Trp				
Plant					Trp
"Standard" code	Stop	Ile	Leu	Arg	Arg

[a]N represents any of the four nucleotides.
[b]Also acts as part of an initiation signal.
[c]AGA only; no AGG codons occur in *Drosophila* mitochondrial DNA.

Source: Mainly Breitenberger, C.A. and RajBhandary, U.L., *Trends Biochem. Sci.* **10,** 481 (1985).

2 ■ TRANSFER RNA AND ITS AMINOACYLATION

The establishment of the genetic function of DNA led to the realization that cells somehow "translate" the language of base sequences into the language of polypeptides. Yet, nucleic acids originally appeared unable to bind specific amino acids [more recently RNA aptamers for specific amino acids have been generated; aptamers are nucleic acids that have been selected for their ability to bind specific ligands (Section 7-6C)]. In 1955, Crick, in what became known as the **adaptor hypothesis,** postulated that translation occurs through the mediation of "adaptor" molecules. Each adaptor was postulated to carry a specific enzymatically appended amino acid and to recognize the corresponding codon (Fig. 32-7). Crick suggested that these adaptors contain RNA because codon recognition could then occur by complementary base pairing. At about this time, Paul Zamecnik and Mahlon Hoagland discovered that in the course of protein synthesis, ^{14}C-labeled amino acids became transiently bound to a low molecular mass fraction of RNA. Further investigations indicated that these RNAs, which at first were called "soluble RNA" or "sRNA" but are now known as **transfer RNA (tRNA),** are, in fact, Crick's putative adaptor molecules.

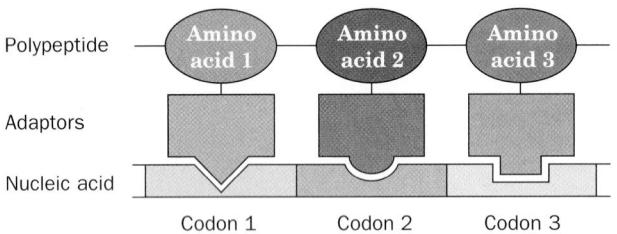

FIGURE 32-7   The adaptor hypothesis. It postulates that the genetic code is read by molecules that recognize a particular codon and carry the corresponding amino acid.

A. *Primary and Secondary Structures of tRNA*

✎ **See Guided Exploration 26: The Structure of tRNA.** In 1965, after a 7-year effort, Robert Holley reported the first known base sequence of a biologically significant nucleic acid, that of yeast **alanine tRNA (tRNA^{Ala};** Fig. 32-8). To do so Holley had to overcome several major obstacles:

   1. All organisms contain many species of tRNAs (usually at least one for each of the 20 amino acids) which, because of their nearly identical properties (see below), are not easily separated. Preparative techniques had to be developed to provide the gram or so of pure yeast tRNA^{Ala} Holley required for its sequence determination.

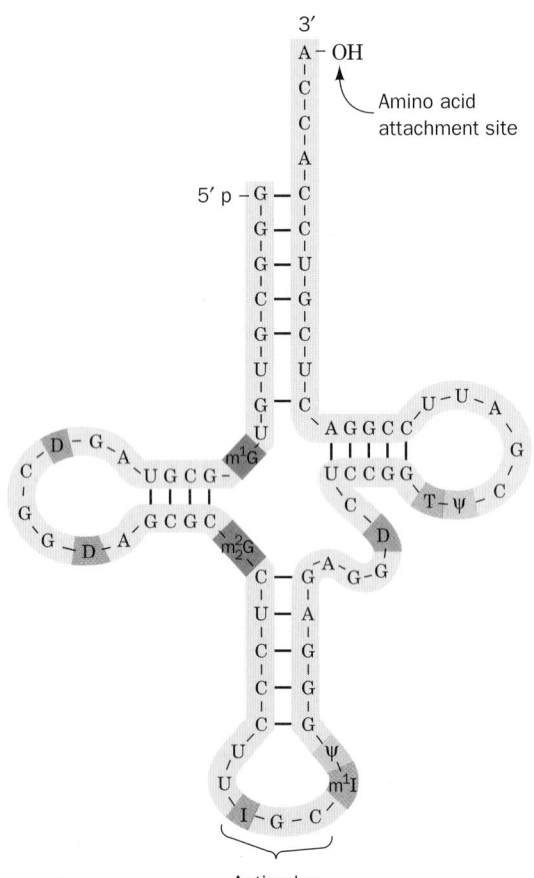

FIGURE 32-8   Base sequence of yeast tRNA^{Ala} drawn in the cloverleaf form. The symbols for the modified nucleosides (*color*) are explained in Fig. 32-10.

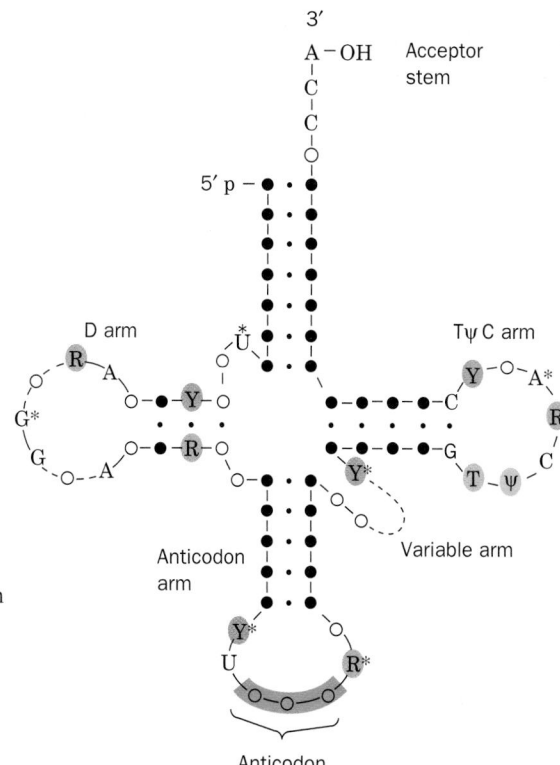

FIGURE 32-9 Cloverleaf secondary structure of tRNA. Filled circles connected by dots represent Watson–Crick base pairs, and open circles in the double-helical regions indicate bases involved in non-Watson–Crick base pairing. Invariant positions are indicated: R and Y represent invariant purines and pyrimidines and ψ signifies pseudouridine. The starred nucleosides are often modified. The dashed regions in the D and variable arms contain different numbers of nucleotides in the various tRNAs.

2. Holley had to invent the methods that were initially used to sequence RNA (Section 7-2).

3. Ten of the 76 bases of yeast tRNA^{Ala} are modified (see below). Their structural formulas had to be elucidated although they were never available in more than milligram quantities.

Since 1965, the techniques for tRNA purification and sequencing have vastly improved. A tRNA may now be sequenced in a few days' time with only ~1 μg of material. Presently, the base sequences of more than 4000 tRNAs from over 200 organisms and organelles are known (most from their corresponding DNA sequences). They vary in length from 54 to 100 nucleotides (18–28 kD) although most have ~76 nucleotides.

Almost all known tRNAs, as Holley first recognized, may be schematically arranged in the so-called cloverleaf secondary structure (Fig. 32-9). Starting from the 5′ end, they have the following common features:

1. A 5′-terminal phosphate group.

2. A 7-bp stem that includes the 5′-terminal nucleotide and that may contain non-Watson–Crick base pairs such as G · U. This assembly is known as the **acceptor** or **amino acid stem** because the amino acid residue carried by the tRNA is appended to its 3′-terminal OH group (Section 32-2C).

3. A 3- or 4-bp stem ending in a loop that frequently contains the modified base **dihydrouridine (D;** see below).

This stem and loop are therefore collectively termed the **D arm.**

4. A 5-bp stem ending in a loop that contains the **anticodon,** the triplet of bases that is complementary to the codon specifying the tRNA. These features are known as the **anticodon arm.**

5. A 5-bp stem ending in a loop that usually contains the sequence TψC (where ψ is the symbol for **pseudouridine;** see below). This assembly is called the **TψC** or **T arm.**

6. All tRNAs terminate in the sequence CCA with a free 3′-OH group. The —CCA may be genetically specified or enzymatically appended to immature tRNA (Section 31-4C).

7. There are 15 invariant positions (always have the same base) and 8 **semi-invariant** positions (only a purine or only a pyrimidine) that occur mostly in the loop regions. These regions also contain **correlated invariants,** that is, pairs of nonstem nucleotides that are base paired in all tRNAs. The purine on the 3′ side of the anticodon is invariably modified. The structural significance of these features is examined in Section 32-2B.

The site of greatest variability among the known tRNAs occurs in the so-called **variable arm.** It has from 3 to 21 nucleotides and may have a stem consisting of up to 7 bp. The D loop also varies in length from 5 to 7 nucleotides.

a. Transfer RNAs Have Numerous Modified Bases

One of the most striking characteristics of tRNAs is their large proportion, up to 25%, of posttranslationally

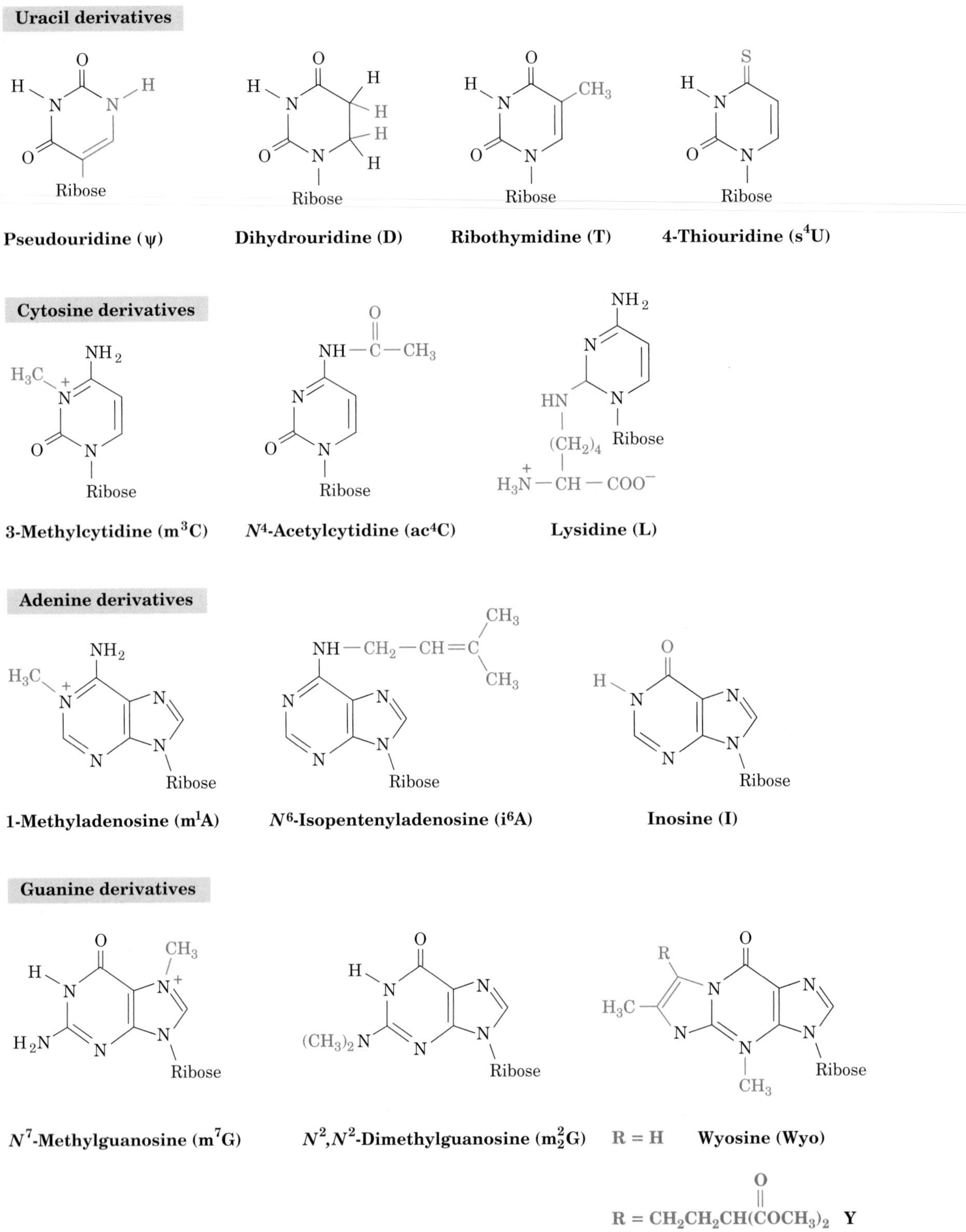

Uracil derivatives

Pseudouridine (ψ)   **Dihydrouridine (D)**   **Ribothymidine (T)**   **4-Thiouridine (s⁴U)**

Cytosine derivatives

3-Methylcytidine (m³C)   **N⁴-Acetylcytidine (ac⁴C)**   **Lysidine (L)**

Adenine derivatives

1-Methyladenosine (m¹A)   **N⁶-Isopentenyladenosine (i⁶A)**   **Inosine (I)**

Guanine derivatives

N⁷-Methylguanosine (m⁷G)   **N²,N²-Dimethylguanosine (m₂²G)**   R = H   **Wyosine (Wyo)**

$$R = CH_2CH_2CH(COCH_3)_2 \quad \mathbf{Y}$$

FIGURE 32-10   A selection of the modified nucleosides that occur in tRNAs together with their standard abbreviations. Note that although inosine chemically resembles guanosine, it is biochemically derived from adenosine. Nucleosides may also be methylated at their ribose 2′ positions to form residues symbolized, for instance, by Cm, Gm, and Um.

modified or hypermodified bases. Nearly 80 such bases, found at >60 different tRNA positions, have been characterized. A few of them, together with their standard abbreviations, are indicated in Fig. 32-10. Hypermodified nucleosides, such as i⁶A, are usually adjacent to the anticodon's 3′ nucleotide when it is A or U. Their low polarities probably strengthen the otherwise relatively weak pairing associations of these bases with the codon, thereby increasing translational fidelity. Conversely, certain methylations block base pairing and hence prevent inappropriate structures from forming. Some of these modifications form important recognition elements for the enzyme that attaches the correct amino acid to a tRNA (Section 32-2C). However, none of them are essential for maintaining a tRNA's structural integrity (see below) or for its proper binding to the ribosome. Nevertheless, mutant bacteria unable to form certain modified bases compete poorly against the corresponding normal bacteria.

B. *Tertiary Structure of tRNA*

🔲 **See Guided Exploration 26: The Structure of tRNA.** The earliest physicochemical investigations of tRNA indicated that it has a well-defined conformation. Yet, despite numerous hydrodynamic, spectroscopic, and chemical cross-linking studies, its three-dimensional structure remained

an enigma until 1974. In that year, the 2.5-Å resolution X-ray crystal structure of yeast **tRNA**[Phe] was separately elucidated by Alexander Rich in collaboration with Sung Hou Kim and, in a different crystal form, by Aaron Klug. *The molecule assumes an L-shaped conformation in which one leg of the L is formed by the acceptor and T stems folded into a continuous A-RNA-like double helix (Section 29-1B) and the other leg is similarly composed of the D and anticodon stems (Fig. 32-11). Each leg of the L is ~60 Å long and the anticodon and amino acid acceptor sites are at opposite ends of the molecule, some 76 Å apart. The narrow 20- to 25-Å width of native tRNA is essential to its biological function: During protein synthesis, two RNA molecules must simultaneously bind in close proximity at adjacent codons on mRNA (Section 32-3D).*

a. tRNA's Complex Tertiary Structure Is Maintained by Hydrogen Bonding and Stacking Interactions

The structural complexity of yeast tRNA[Phe] is reminiscent of that of a protein. Although only 42 of its 76 bases occur in double helical stems, *71 of them participate in*

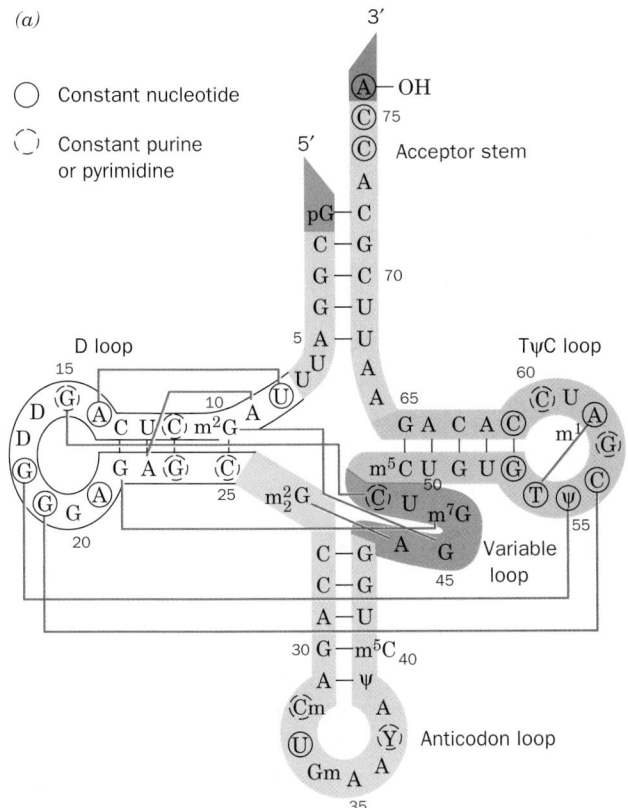

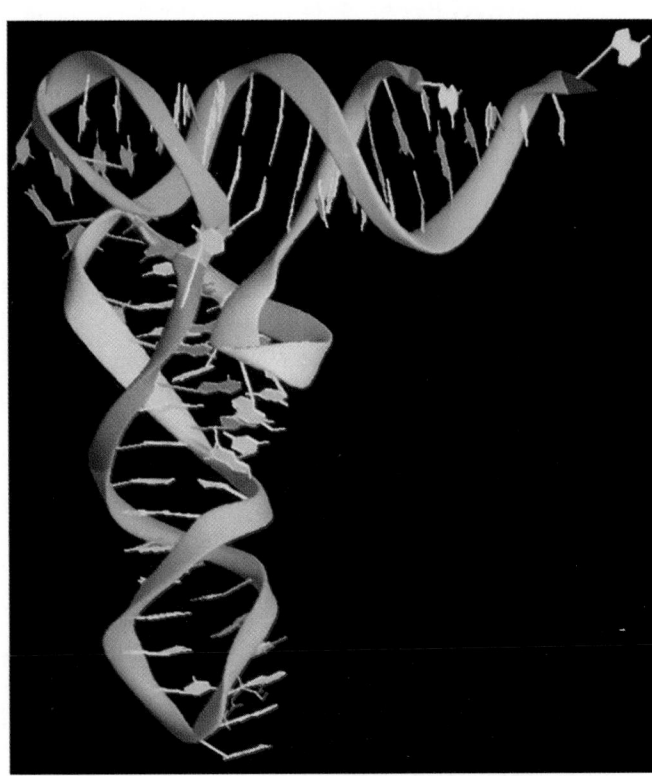

FIGURE 32-11 Structure of yeast tRNA[Phe]**.** (*a*) The base sequence drawn in cloverleaf form. Tertiary base pairing interactions are represented by thin red lines connecting the participating bases. Bases that are conserved or semiconserved in all tRNAs are circled by solid and dashed lines, respectively. The 5′ terminus is colored bright green, the acceptor stem is yellow, the D arm is white, the anticodon arm is light green, the variable arm is orange, the TψC arm is cyan, and the 3′ terminus is red. (*b*) The X-ray structure drawn to show how its base paired stems are arranged form the L-shaped molecule. The sugar–phosphate backbone is represented by a ribbon with the same color scheme as that in Part *a*. [Courtesy of Mike Carson, University of Alabama at Birmingham. PDBid 6TNA.]
🔲 **See the Kinemage Exercise 20-1**

stacking associations (Fig. 32-12). The structure also contains 9 base pairing interactions that cross link its tertiary structure (Figs. 32-11a and 32-12). Remarkably, all but one of these tertiary interactions, which appear to be the mainstays of the molecular structure, are non-Watson–Crick associations. Moreover, most of the bases involved in these interactions are either invariant or semi-invariant, which strongly suggests that all tRNAs have similar conformations (see below). The structure is also stabilized by several unusual hydrogen bonds between bases and either phosphate groups or the 2′-OH groups of ribose residues.

The compact structure of yeast tRNA^{Phe} results from its large number of intramolecular associations, which renders most of its bases inaccessible to solvent. The most notable exceptions to this are the anticodon bases and those of the amino acid–bearing —CCA terminus. Both of these groupings must be accessible in order to carry out their biological functions.

The observation that the molecular structures of yeast tRNA^{Phe} in two different crystal forms are essentially identical lends much credence to the supposition that its crystal structure closely resembles its solution structure. Transfer RNAs other than yeast tRNA^{Phe} have, unfortunately, been notoriously difficult to crystallize. As yet, the X-ray structures of only three other uncomplexed tRNAs have been reported (although the X-ray structures of numerous tRNAs in complex with the enzymes that append their corresponding amino acids and with ribosomes have been elucidated; Sections 32-2C and 32-3D). The major structural differences among them result from an apparent flexibility in the anticodon loop and the —CCA terminus as well as from a hingelike mobility between the two legs of the L that gives, for instance, yeast **tRNA^{Asp}** a boomerang-like shape. Such observations are in accord with the expectation that all tRNAs fit into the same ribosomal cavities.

C. Aminoacyl–tRNA Synthetases

🔹 See Guided Exploration 27: The Structures of aminoacyl-tRNA synthetases and their interaction with tRNAs. *Accurate translation requires two equally important recognition steps: (1) the choice of the correct amino acid for covalent attachment to a*

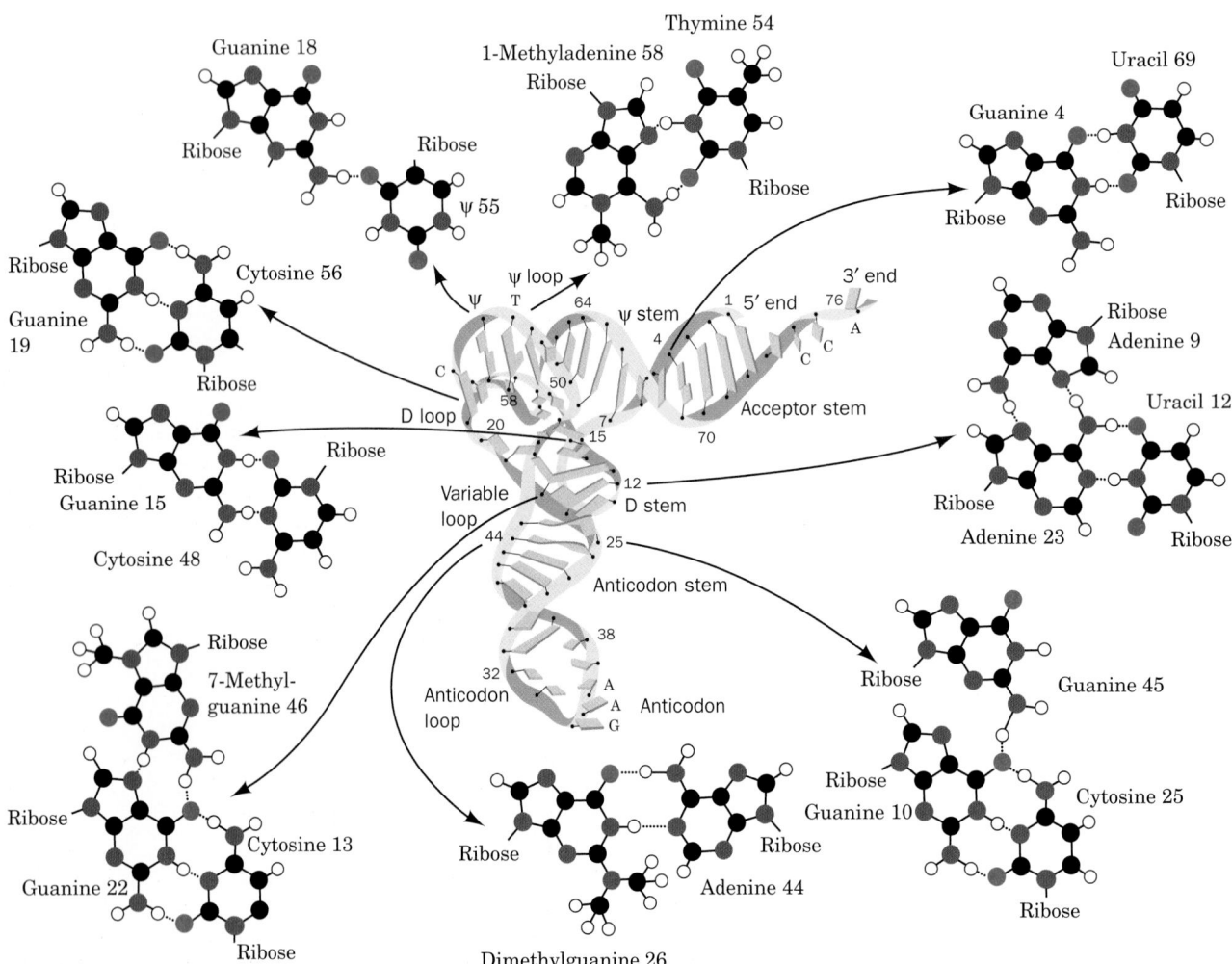

FIGURE 32-12 Tertiary base pairing interactions in yeast tRNA^{Phe}. Note that all but one of these nine interactions involve non-Watson–Crick pairs and that they are all located near the corner of the L. [After Kim, S.H., *in* Schimmel, P.R., Söll, D., and Abelson, J.N. (Eds.), *Transfer RNA: Structure, Properties and Recognition*, p. 87, Cold Spring Harbor Laboratory Press (1979). Drawing of tRNA copyrighted © by Irving Geis.] 🔹 See the Kinemage Exercise 20-3

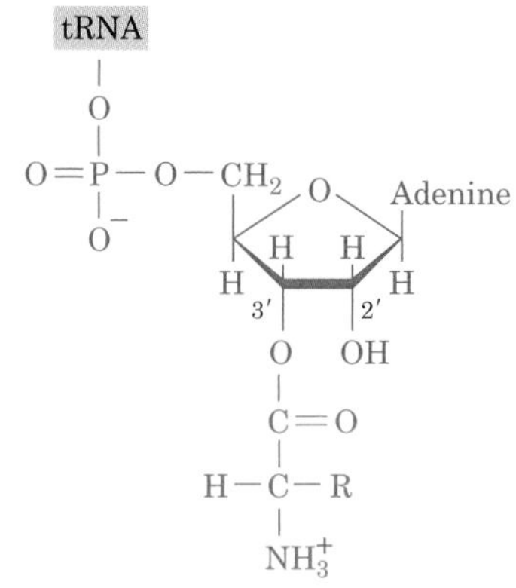

Aminoacyl–tRNA

FIGURE 32-13 An Aminoacyl–tRNA. The amino acid residue is esterified to the tRNA's 3'-terminal nucleoside at either its 3'-OH group, as shown here, or its 2'-OH group.

tRNA; and (2) the selection of the amino acid-charged tRNA specified by mRNA. The first of these steps, which is catalyzed by amino acid–specific enzymes known as **aminoacyl–tRNA synthetases (aaRSs),** appends an amino acid to the 3'-terminal ribose residue of its cognate tRNA to form an **aminoacyl–tRNA** (Fig. 32-13). This otherwise unfavorable process is driven by the hydrolysis of ATP in two sequential reactions that are catalyzed by a single enzyme.

1. The amino acid is first "activated" by its reaction with ATP to form an **aminoacyl–adenylate:**

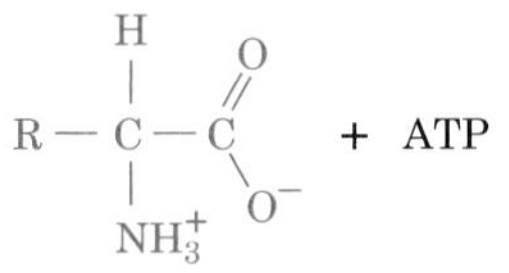

Amino acid

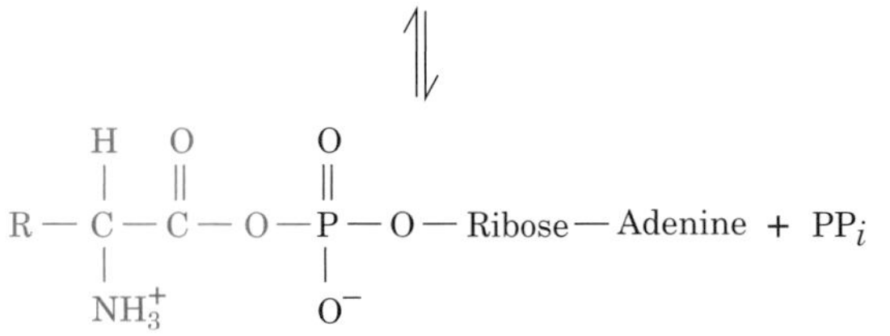

**Aminoacyl–adenylate
(aminoacyl–AMP)**

which, with all but three aaRSs, can occur in the absence of tRNA. Indeed, this intermediate may be isolated although it normally remains tightly bound to the enzyme.

2. This mixed anhydride then reacts with tRNA to form the aminoacyl–tRNA:

Aminoacyl–AMP + tRNA $\rightleftharpoons$ aminoacyl–tRNA + AMP

Some aaRSs exclusively append an amino acid to the terminal 2'-OH group of their cognate tRNAs, and others do so at the 3'-OH group. This selectivity was established with the use of chemically modified tRNAs that lack either the 2'- or 3'-OH group of their 3'-terminal ribose residue. The

use of these derivatives was necessary because, in solution, the aminoacyl group rapidly equilibrates between the 2' and 3' positions.

The overall aminoacylation reaction is

Amino acid + tRNA + ATP $\rightleftharpoons$

aminoacyl–tRNA + AMP + PP$_i$

These reaction steps are readily reversible because the free energies of hydrolysis of the bonds formed in both the aminoacyl–adenylate and the aminoacyl–tRNA are comparable to that of ATP hydrolysis. The overall reaction is driven to completion by the inorganic pyrophosphatase-catalyzed hydrolysis of the PP$_i$ generated in the first reaction step. Amino acid activation therefore chemically resembles fatty acid activation (Section 25-2A); the major difference between these two processes, which were both elucidated by Paul Berg, is that tRNA is the acyl acceptor in amino acid activation, whereas CoA performs this function in fatty acid activation.

a. There Are Two Classes of Aminoacyl–tRNA Synthetases

Most cells have one aaRS for each of the 20 amino acids. The similarity of the reactions catalyzed by these enzymes and the structural resemblance of all tRNAs suggests that all aaRSs evolved from a common ancestor and should therefore be structurally related. This is not the case. In fact, *the aaRSs form a diverse group of enzymes.* The over 1000 such enzymes that have been characterized each have one of four different types of subunit structures, α, α_2 (the predominant forms), α_4, and $\alpha_2\beta_2$, with known subunit sizes ranging from ~300 to ~1200 residues. Moreover, there is little sequence similarity among synthetases specific for different amino acids. Quite possibly, aminoacyl–tRNA synthetases arose very early in evolution, before the development of the modern protein synthesis apparatus other than tRNAs.

Detailed sequence and structural comparisons of aminoacyl–tRNA synthetases by Dino Moras indicate that these enzymes form two unrelated families, termed **Class I** and **Class II aaRSs,** that each have the same 10 members in nearly all organisms (Table 32-4). The Class I enzymes, although of largely dissimilar sequences, share two homologous polypeptide segments, not present in other proteins, that have the consensus sequences His-Ile-Gly-His (HIGH) and Lys-Met-Ser-Lys-Ser (KMSKS). The X-ray structures of Class I enzymes indicate that both of these segments are components of a dinucleotide-binding fold (Rossmann fold, which is also possessed by many NAD$^+$- and ATP-binding proteins; Section 8-3B) in which they participate in ATP binding and are implicated in catalysis. The Class II synthetases lack the foregoing sequences but have three other sequences in common. Their X-ray structures reveal that these sequences occur in a so-called signature motif, a fold found only in Class II enzymes that consists of a 7-stranded antiparallel β sheet with three flanking helices, which forms the core of their catalytic domains.

Many Class I aaRSs require anticodon recognition to aminoacylate their cognate tRNAs. In contrast, several

TABLE 32-4 Characteristics of Bacterial Aminoacyl–tRNA Synthetases

Amino Acid	Quaternary Structure	Number of Residues
Class I		
Arg	α	577
Cys	α	461
Gln	α	553
Glu	α	471
Ile	α	939
Leu	α	860
Met	α, α_2	676
Trp	α_2	325
Tyr	α_2	424
Val	α	951
Class II		
Ala	α, α_4	875
Asn	α_2	467
Asp	α_2	590
Gly	$\alpha_2\beta_2$	303/689
His	α_2	424
Lys	α_2	505
Pro	α_2	572
Phe	$\alpha_2\beta_2, \alpha$	327/795
Ser	α_2	430
Thr	α_2	642

Source: Mainly Carter, C.W., Jr., *Annu. Rev. Biochem.* **62,** 715 (1993).

Class II enzymes, including **AlaRS** and **SerRS,** do not interact with their bound tRNA's anticodon. Indeed, several class II aaRSs accurately aminoacylate "microhelices" derived from only the acceptor stems of their cognate tRNAs. Another difference between Class I and Class II synthetases is that all Class I enzymes aminoacylate their bound tRNA's 3'-terminal 2'-OH group, whereas Class II enzymes, with the exception of **PheRS,** all charge the 3'-OH group. The amino acids for which the Class I synthetases are specific tend to be larger and more hydrophobic than those used by Class II synthetases. Finally, as Table 32-4 indicates, Class I aaRSs are mainly monomers, whereas most Class II aaRSs are homodimers.

LysRS has been classified as a Class II aaRS. However, a search of the genome sequences of *Methanococcus jannaschii* and *Methanobacterium thermoautotrophicum* failed to reveal the presence of such a LysRS. This led to the discovery that the LysRSs expressed by these archaebacteria are Class I rather than Class II enzymes. This raises the interesting question of how Class I LysRS evolved.

Prokaryotic aaRSs occur as individual protein molecules. However, in many higher eukaryotes (e.g., *Drosophila* and mammals), 9 aaRSs, some of each class, associate to form a multienzyme particle in which the glutamyl and prolyl synthetase functions are fused into a single polypeptide named **GluProRS.** The advantages of these systems are unknown.

b. The Structural Features Recognized by Aminoacyl–tRNA Synthetases May Be Quite Simple

As we shall see in Section 32-2D, ribosomes select aminoacyl–tRNAs only via codon–anticodon interactions, not according to the identities of their aminoacyl groups. *Accurate translation therefore requires not only that each tRNA be aminoacylated by its cognate aaRS but that it not be aminoacylated by any of its 19 noncognate aaRSs.* Considerable effort has therefore been expended, notably by LaDonne Schulman, Paul Schimmel, Olke Uhlenbeck, and John Abelson, in elucidating how aaRSs manage this feat, despite the close structural similarities of nearly all tRNAs. The experimental methods employed involved the use of specific tRNA fragments, mutationally altered tRNAs, chemical cross-linking agents, computerized sequence comparisons, and X-ray crystallography. The most common synthetase contact sites on tRNA occur on the inner (concave) face of the L. Other than that, there appears to be little regularity in how the various tRNAs are recognized by their cognate synthetases. Indeed, as we shall see, some aaRSs recognize only their cognate tRNA's acceptor stem, whereas others also interact with its anticodon region. Additional tRNA regions may also be recognized.

Genetic manipulations by Schimmel revealed that the tRNA features recognized by at least one type of aaRS are surprisingly simple. Numerous sequence alterations of *E. coli* tRNA^{Ala} do not appreciably affect its capacity to be aminoacylated with alanine. Yet, most base substitutions in the G3 · U70 base pair located in the tRNA's acceptor stem (Fig. 32-14a) greatly diminish this reaction. Moreover, the introduction of a G · U base pair into the analogous position of **tRNA^{Cys}** and tRNA^{Phe} causes them to be aminoacylated with alanine even though there are few other sequence identities between these mutant tRNAs and tRNA^{Ala} (e.g., Fig. 32-15). In fact, *E. coli* AlaRS even efficiently aminoacylates a 24-nt "microhelix" derived from only the G3 · U70-containing acceptor stem of *E. coli* tRNA^{Ala}. Since the only known *E. coli* tRNAs that normally have a G3 · U70 base pair are the tRNA^{Ala}, and this base pair is also present in the tRNA^{Ala} from many organisms including yeast (Fig. 32-8), the foregoing observations strongly suggest that *the G3 · U70 base pair is a major feature recognized by AlaRSs.* These enzymes presumably recognize the distorted shape of the G · U base pair (Fig. 32-12), a hypothesis corroborated by the observation that base changes at G3 · U70 which least affect the acceptor identity of tRNA^{Ala} yield base pairs that structurally resemble G · U.

The elements of three other tRNAs, which are recognized by their cognate tRNA synthetases, are indicated in Fig. 32-14. As with tRNA^{Ala}, these identity elements appear to comprise only a few bases. Note that the anticodon forms an identity element in two of these tRNAs. In another example of an anticodon identifier, the *E. coli* **tRNA^{Ile}** specific for the codon AUA has the anticodon LAU, where L is **lysidine,** a modified cytosine whose 2-keto group is replaced by the amino acid lysine (Fig. 32-10). The L in this context pairs with A rather than G, a rare instance of base modification altering base pairing specificity. The replacement of this L with unmodified C,

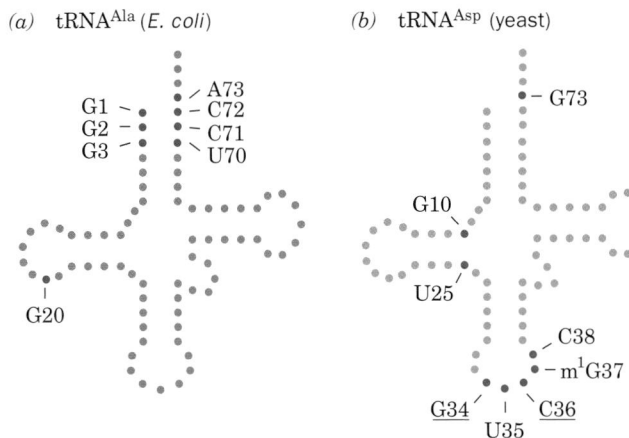

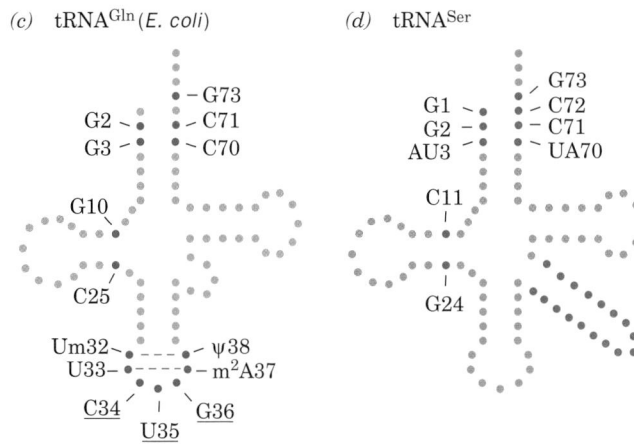

FIGURE 32-14 Major identity elements in four tRNAs. Each base in the tRNA is represented by a filled circle. Red circles indicate positions that have been shown to be identity elements for the recognition of the tRNA by its cognate aminoacyl–tRNA synthetase. The anticodon bases that are identity elements are underlined. In each case, additional identity elements may yet be discovered. The base at position 73, which is an identity element in all four tRNAs shown here, is known as the **discriminator base.**

as expected, yields a tRNA that recognizes the Met codon AUG (codons bind anticodons in an antiparallel fashion). Surprisingly, however, this altered tRNA^{Ile} is also a much better substrate for **MetRS** than it is for **IleRS.** Thus, both the codon and the amino acid specificity of this tRNA are changed by a single posttranscriptional modification. The N^1-methylation of G37 in yeast tRNA^{Asp} (Fig. 32-14b) provides another example of a base modification forming an identity element. In the absence of this N^1-methyl group,

tRNA^{Asp} is recognized by **ArgRS,** largely via its C36 and G37, whereas ArgRS normally recognizes only **tRNA^{Arg},** mainly via its C35 and U36.

The available experimental evidence has largely located the various tRNA identifiers in the acceptor stem and the anticodon loop (Fig. 32-16). The X-ray structures of sev-

FIGURE 32-15 Three-dimensional model of *E. coli* tRNA^{Ala}. This model is based on the X-ray structure of yeast tRNA^{Phe} (Fig. 32-11b) in which the nucleotides that are different in *E. coli* tRNA^{Cys} are highlighted in cyan and the G3 · U70 base pair is highlighted in ivory. [Courtesy of Ya-Ming Hou, MIT.]

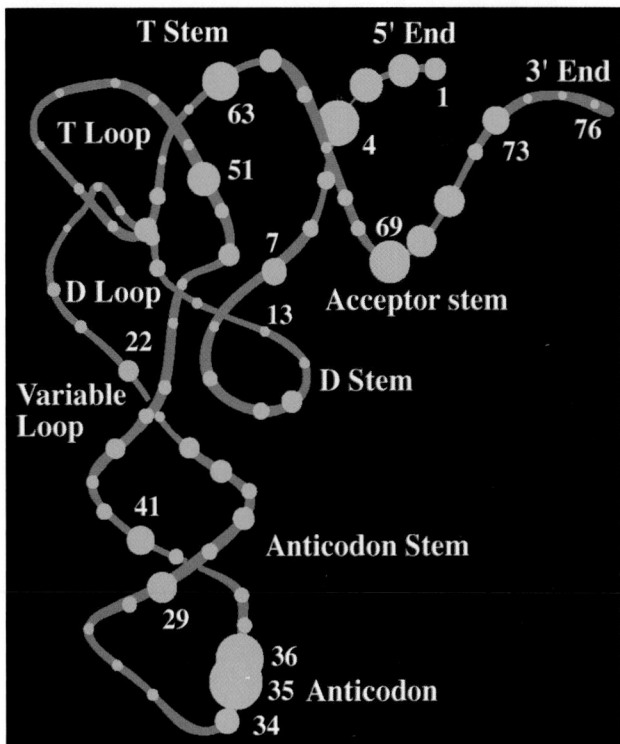

FIGURE 32-16 Experimentally observed identity elements of tRNAs. The tRNA backbone is cyan and each of its nucleotides is represented by a yellow circle whose diameter is proportional to the fraction of the 20 tRNA acceptor types for which the nucleotide is an observed determinant. [Courtesy of William McClain, University of Wisconsin.]

eral aaRS · tRNA complexes, which we consider next, have structurally rationalized some of these observations.

c. The X-Ray Structure of GlnRS · tRNA^Gln, a Class I Complex

The X-ray structures of all but two (Ala and Leu) of the 20 different amino acid–specific aaRSs have been determined, many of which are in complex with ATP, their cognate amino acids, or their analogs. These structures reveal that the active sites of these enzymes bind the ATP and target amino acid in optimal positions for in-line nucleophilic displacement (Section 16-2B) during amino acid activation and that the specificity of an aaRS for its target amino acid is determined by idiosyncratic contacts with the side chain of the amino acid.

The X-ray structures of 12 different aaRSs in their complexes with their cognate tRNAs have so far been reported. The first of them to be elucidated, that of *E. coli* **GlnRS**, a Class I synthetase, in its complex with **tRNA^Gln** and ATP (Fig. 32-17), was determined by Thomas Steitz. The tRNA^Gln assumes an L-shaped conformation that resembles those of tRNAs of known structures (e.g., Fig. 32-11*b*).

GlnRS, a 553-residue monomeric protein that consists of four domains arranged to form an elongated molecule, interacts with the tRNA along the entire inside face of the L such that the anticodon is bound near one end of the protein and the acceptor stem is bound near its other end.

Genetic and biochemical data indicate that the identity elements of tRNA^Gln are largely clustered in its anticodon loop and acceptor stem (Fig. 32-14*c*). The anticodon loop of tRNA^Gln is extended by two novel non-Watson–Crick base pairs (2′-*O*-methyl-U32 · ψ38 and U33 · m²A37), thereby causing the bases of the anticodon to unstack and splay outward in different directions so as to bind in separate recognition pockets of GlnRS. These structural features suggest that GlnRS uses all seven bases of the anticodon loop to discriminate among tRNAs. Indeed, changes to any one of the bases of residues C34 through ψ38 yield tRNAs with decreases in k_{cat}/K_M for aminoacylation by GlnRS by factors ranging from 70 to 28,000.

The GCCA at the 3′ end of the tRNA^Gln makes a hairpin turn toward the inside of the L rather than continuing

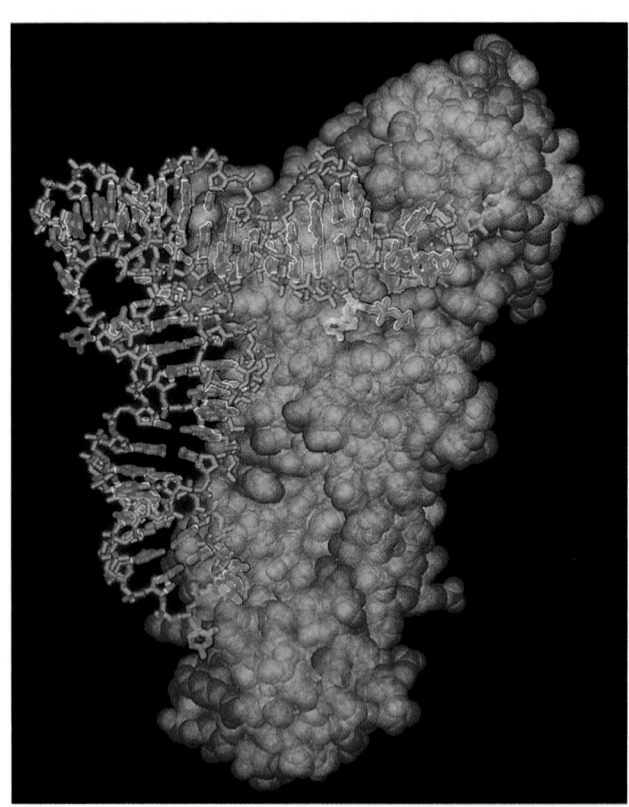

(a)

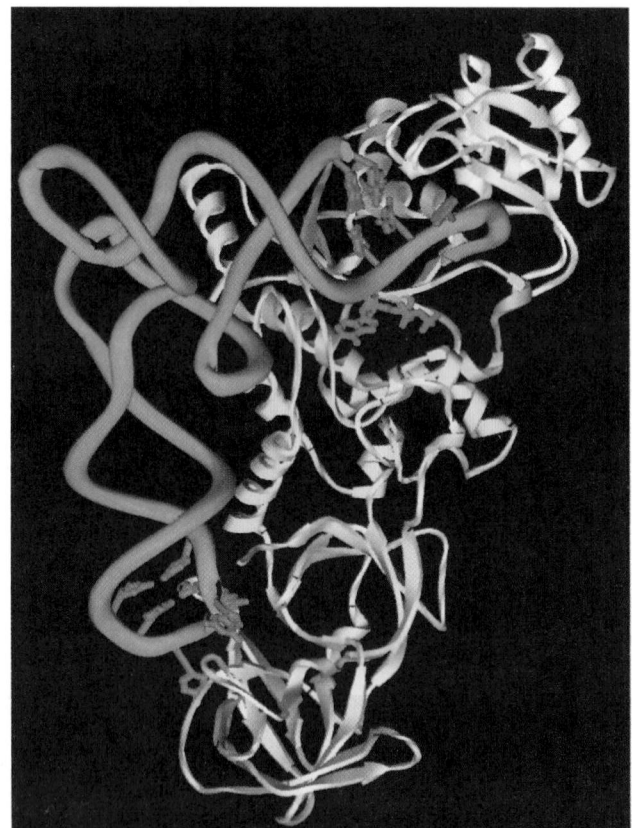

(b)

FIGURE 32-17   X-Ray structure of *E. coli* GlnRS · tRNA^Gln · ATP. (*a*) The tRNA and ATP are shown in skeletal form with the tRNA sugar–phosphate backbone green, its bases magenta, and the ATP red. The protein is represented by a translucent cyan space-filling model that reveals the buried portions of the tRNA and ATP. Note that both the 3′ end of the tRNA (*top right*) and its anticodon bases (*bottom*) are inserted into deep pockets in the protein. (*b*) A ribbon drawing of the complex

viewed as in Part *a*. The tRNA's sugar–phosphate backbone is represented by a green worm and the bases forming its identity elements (Fig. 32-14*c*) are magenta. The protein's four domains are differently colored with the dinucleotide binding fold gold and the remainder of the catalytic domain that contains it yellow. The ATP is shown in skeletal form (*red*). [Based on an X-ray structure by Thomas Steitz, Yale University. PDBid 1GSG.] 🔗 **See the Kinemage Exercise 21**

helically onward (as does the ACCA at the 3′ end in the X-ray structure of tRNA^{Phe}; Fig. 32-11*b*). This conformation change is facilitated by the insinuation of a Leu side chain between the 5′ and 3′ ends of the tRNA so as to disrupt the first base pair of the acceptor stem (U1 · A72). The GlnRS reaction is therefore relatively insensitive to base changes in these latter two positions except when base pairing is strengthened by their conversion to G1 · C72. The GCCA end of the tRNA^{Gln} plunges deeply into a protein pocket that also binds the enzyme's ATP and glutamine substrates. Three protein "fingers" are inserted into the minor groove of the acceptor stem to make sequence-specific interactions with base pairs G2 · C71 and G3 · C70 [recall that double helical RNA has an A-DNA-like structure (Section 29-1B) whose wide minor groove read-ily admits protein but whose major groove is normally too narrow to do so].

The GlnRS domain that binds glutamine, ATP, and the GCCA end of tRNA^{Gln}, the so-called catalytic domain, contains, as we previously discussed, a dinucleotide-binding fold. Much of this domain is nearly superimposable with and thus evolutionarily related to the corresponding domains of other Class I aaRSs.

d. The X-Ray Structure of AspRS · tRNA^{Asp}, a Class II Complex

Yeast **AspRS,** a Class II synthetase, is an α_2 dimer of 557-residue subunits. Its X-ray structure in complex with tRNA^{Asp}, determined by Moras, reveals that the protein symmetrically binds two tRNA molecules (Fig. 32-18).

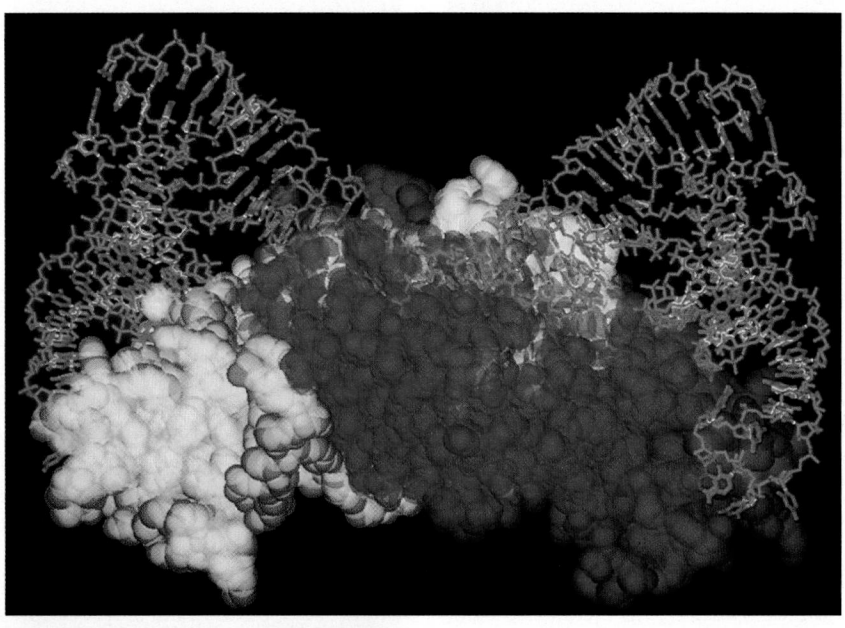

(a)

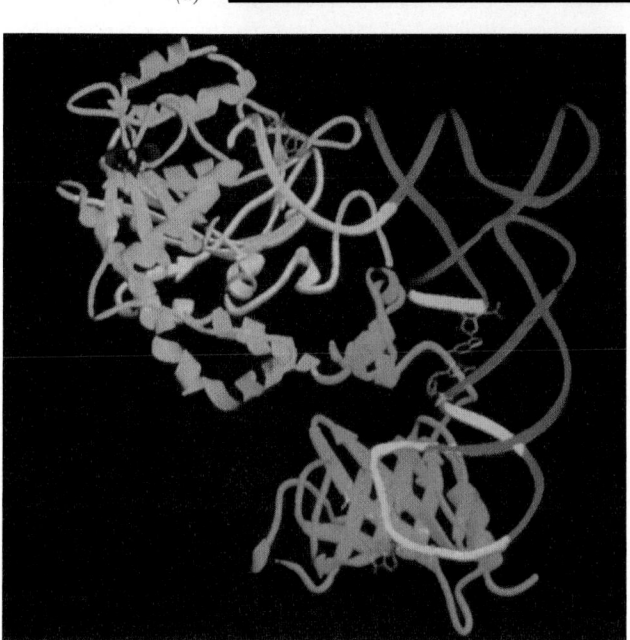

(b)

FIGURE 32-18 X-Ray structure of yeast AspRS · tRNA^{Asp} · ATP. (*a*) The homodimeric enzyme with its two symmetrically bound tRNAs viewed with its 2-fold axis approximately vertical. The tRNAs are shown in skeletal form with their sugar–phosphate backbones green and their bases magenta. The two protein subunits are represented by translucent yellow and blue space-filling models that reveal buried portions of the tRNAs. (*b*) A ribbon diagram of the AspRS · tRNA^{Asp} · ATP monomer. The tRNA's contact regions with the protein are yellow and its identity elements are shown in stick form in red as is the ATP. The protein's N-terminal domain is blue-green, the central domain is cyan, and the C-terminal catalytic domain is orange with its component signature motif (the 7-stranded antiparallel β sheet with 3 flanking helices characteristic of Type II aaRSs) in white. [Part *a* based on an X-ray structure by and Part *b* courtesy of Dino Moras, CNRS/INSERM/ULP, Illkirch Cédex, France. PDBid 1ASY.]

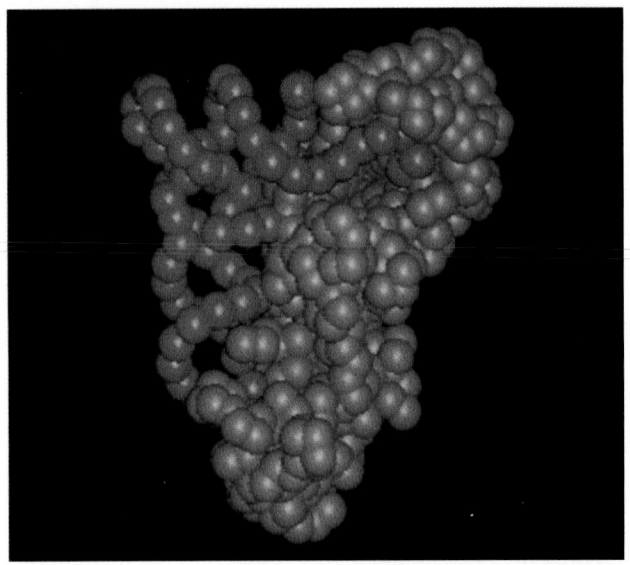

(a)

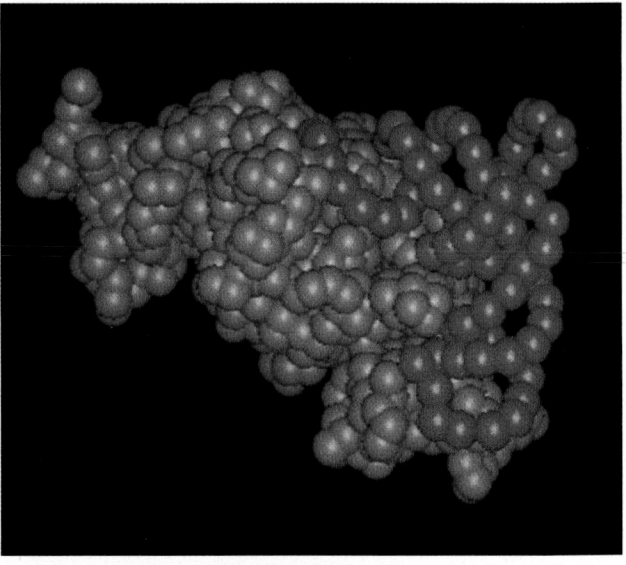

(b)

FIGURE 32-19   Comparison of the modes by which GlnRS and AspRS bind their cognate tRNAs. The proteins and tRNAs are represented by blue and red spheres centered on their C_α and P atom positions. Note how GlnRS (*a*), a Class I synthetase, binds tRNA^{Gln} from the minor groove side of its acceptor stem so as to bend its 3′ end into a hairpin conformation. In contrast, AspRS (*b*), a Class II synthetase, binds tRNA^{Asp} from the major groove side of its acceptor stem so that its 3′ end continues its helical path on entering the active site. [Courtesy of Dino Moras, CNRS/INSERM/ULP, Illkirch Cédex, France.]

Like GlnRS, AspRS principally contacts its bound tRNA both at the end of its acceptor stem and in its anticodon region. The contacts in these two enzymes are, nevertheless, quite different in character (Fig. 32-19): Although both tRNAs approach their cognate synthetases along the inside of their L shapes, tRNA^{Gln} does so toward the direction of the minor groove of its acceptor stem, whereas tRNA^{Asp} does so toward the direction of its major groove. The GCCA at the 3′ end of tRNA^{Asp} thereby continues its helical track as it plunges into AspRS's catalytic site, whereas, as we saw, the GCCA end of tRNA^{Gln} bends backward into a hairpin turn that opens up the first base pair (U1 · A72) of its acceptor stem. Although the deep major groove of an A-RNA helix is normally too narrow to admit groups larger than water molecules (Section 29-1B), the major groove at the end of the acceptor stem in AspRS · tRNA^{Asp} is sufficiently widened for its base pairs to interact with a protein loop.

The anticodon arm of tRNA^{Asp} is bent by as much as 20 Å toward the inside of the L relative to that in the X-ray structure of uncomplexed tRNA^{Asp} and its anticodon bases are unstacked. The hinge point for this bend is a G30 · U40 base pair in the anticodon stem which, in nearly all other species of tRNA, is a Watson–Crick base pair. The anticodon bases of tRNA^{Gln} are also unstacked in contacting GlnRS but with a backbone conformation that differs from that in tRNA^{Asp}. Evidently, the conformation of a tRNA in complex with its cognate synthetase appears to be dictated more by its interactions with the protein (induced fit) than by its sequence.

Structural analyses of complexes of AspRS · tRNA^{Asp} with ATP and aspartic acid, and of GlnRS · tRNA^{Gln} with ATP, have permitted models of the aminoacyl–AMP complexes of these enzymes to be independently formulated. Comparison of these models reveals that the 3′-terminal A residues of tRNA^{Gln} and tRNA^{Asp} (to which the aminoacyl groups are appended; Fig. 32-13) are positioned on opposite sides of the enzyme-bound aminoacyl–AMP intermediate (Fig. 32-20). The 3′-terminal ribose residues are puckered C2′-*endo* for tRNA^{Asp} and C3′-*endo* for tRNA^{Gln}; see Fig. 29-10) such that the 2′-hydroxyl group of tRNA^{Gln} (Class I) is stereochemically positioned to attack the aminoacyl–AMP's carboxyl group, whereas for tRNA^{Asp} (Class II), only the 3′ hydroxyl group is situated to do so. This clearly explains the different aminoacylation specificities of the Class I and Class II aaRSs.

e. Proofreading Enhances the Fidelity of Amino Acid Attachment to tRNA

The charging of a tRNA with its cognate amino acid is a remarkably accurate process: aaRSs display an overall error rate of about 1 in 10,000. We have seen that aaRSs bind only their cognate tRNAs through an intricate series of specific contacts. But how do they discriminate among the various amino acids, some of which are quite similar?

Experimental measurements indicate, for example, that IleRS transfers as many as 40,000 isoleucines to **tRNA^{Ile}** for every valine it so transfers. Yet, as Linus Pauling first pointed out, *there are insufficient structural differences between Val and Ile to permit such a high degree of discrimination in the direct generation of aminoacyl–tRNAs.* The X-ray structure of *Thermus thermophilus* IleRS, a monomeric Class I aaRS, in complex with isoleucine, determined by Shigeyuki Yokoyama and Schimmel, indicates that isoleucine fits snugly into its binding site in the enzyme's Rossmann fold domain and hence that this binding site would sterically exclude leucine as well as larger amino acids. However, valine, which differs from isoleucine by only the lack of a single methylene group, fits into this isoleucine-binding site. The binding free energy of a methylene group is estimated to be ~12 kJ · mol^{-1}. Equation [3.16] indicates that the ratio f of the equilibrium constants, K_1 and K_2, with which two substances bind to a given binding site is given by

$$f = \frac{K_1}{K_2} = \frac{e^{-\Delta G_1^{\circ\prime}/RT}}{e^{-\Delta G_2^{\circ\prime}/RT}} = e^{-\Delta\Delta G^{\circ\prime}/RT} \qquad [32.1]$$

where $\Delta\Delta G^{\circ\prime} = \Delta G_1^{\circ\prime} - \Delta G_2^{\circ\prime}$ is the difference between the free energies of binding of the two substances. It is therefore estimated that isoleucyl–tRNA synthetase could discriminate between isoleucine and valine by no more than a factor of ~100.

Berg resolved this apparent paradox by demonstrating that, in the presence of tRNA^{Ile}, IleRS catalyzes the nearly quantitative hydrolysis of valyl-aminoacyl–adenylate to valine + AMP rather than forming Val-tRNA^{Ile}. Moreover, the few Val-tRNA^{Ile} molecules that do form are hydrolyzed to valine + tRNA^{Ile}. Thus, *IleRS subjects both aminoacyl–adenylate and aminoacyl–tRNA^{Ile} to a* **proofreading** *or*

editing step that occurs at a separate catalytic site. This site binds Val residues but excludes the larger Ile residues. The enzymes's overall selectivity is therefore the product of the selectivities of its synthesis and proofreading steps, thereby accounting for the high fidelity of aminoacylation. Note that in this so-called **double-sieve** *mechanism, editing occurs at the expense of ATP hydrolysis, the thermodynamic price of high fidelity (increased order).*

The X-ray structure of *Staphylococcus aureus* IleRS in complex with tRNA^{Ile} and the clinically useful antibiotic **mupirocin**

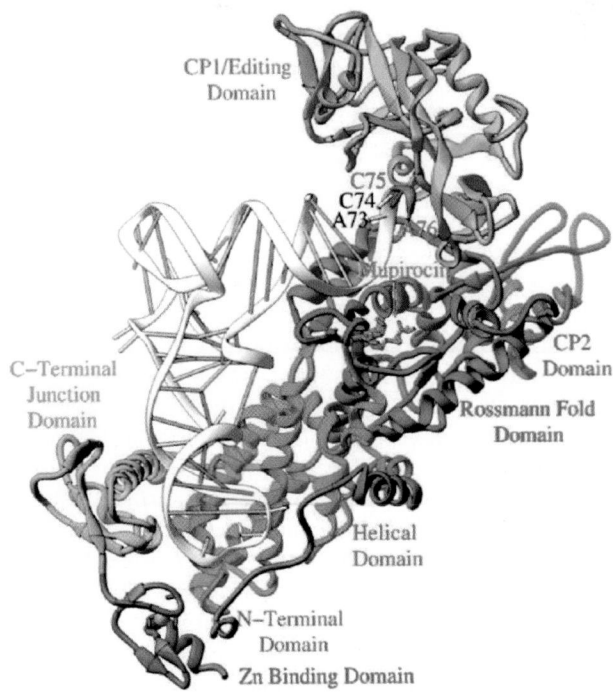

Mupirocin

(a product of *Pseudomonas fluorescens* that acts by specifically binding to bacterial IleRS so as to inhibit bacterial protein synthesis), determined by Steitz, suggests how IleRS carries out its editing process. The X-ray structure (Fig. 32-21) reveals that this complex resembles the GlnRS · tRNA^{Gln} · ATP complex (Fig. 32-17) but with IleRS having an additional editing domain (also called CP1 for *connective peptide 1*) inserted in its Rossmann fold domain. The two 3' terminal residues of the tRNA^{Ile}, C75 and A76, are disordered but, when modeled so as to continue

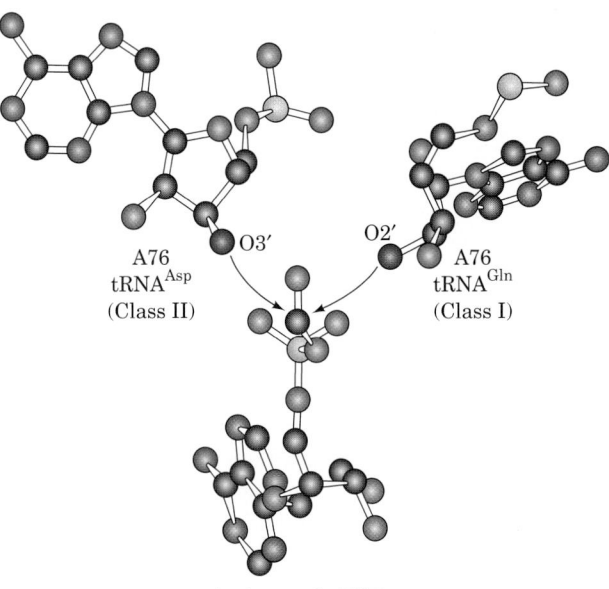

Aminoacyl–AMP

FIGURE 32-20 Comparison of the stereochemistries of aminoacylation by Class I and Class II aaRSs. The positions of the 3' terminal adenosine residues (A76) of AspRS (Class II, *left*) and GlnRS (Class I, *right*) are drawn relative to that of the enzyme-bound aminoacyl–AMP (*below*; only the carbonyl group of its aminoacyl residue is shown). Note how only O3' of tRNA^{Gln} and O2' of tRNA^{Asp} are suitably positioned to attack the aminoacyl residue's carbonyl group and thereby transfer the aminoacyl residue to the tRNA. [After Cavarelli, J., Eriani, G., Rees, B., Ruff, M., Boeglin, M., Mitschler, A., Martin, F., Gangloff, J., Thierry, J.-C., and Moras, D., *EMBO J.* **13**, 335 (1994).]

FIGURE 32-21 X-Ray structure of *T. thermophilus* isoleucyl–tRNA synthetase in complex with tRNA^{Ile} and mupirocin. The tRNA is white, the protein is colored by domain, and the mupirocin is shown in stick form in pink. [Courtesy of Thomas Steitz, Yale University. PDBid 1QU2.]

(a)

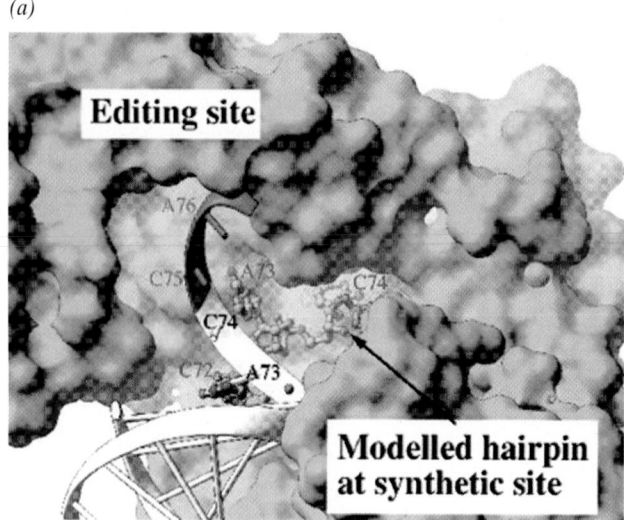

FIGURE 32-22 Comparison of the putative aminoacylation and editing modes of IleRS · tRNA^Ile. (*a*) The superposition of tRNA^Ile in these two binding modes on the solvent-accessible surface of IleRS (*cyan*). The acceptor strand of tRNA^Ile in the editing mode observed in the X-ray structure of IleRS · tRNA^Ile · mupirocin (Fig. 32–21) is drawn in ribbon form in white with the modeled positions of C75 and A76 in red. This places the tRNA's 3′ end in the editing site. In contrast, the three 3′ terminal residues of tRNA^Ile, as positioned through homology modeling based on the X-ray structure of GlnRS · tRNA^Gln · ATP (Fig. 32-17) and drawn in ball-and-stick form with C yellow, N blue, O red, and P magenta, places the tRNA's 3′ end in the synthetic (aminoacylation) site, 34 Å distant from its position in the editing site. Note that there is a cleft running between the editing and synthetic sites and that the 3′ end of the tRNA continues its A-form helical path in the editing mode but assumes a hairpin conformation in the synthetic mode. (*b*) A cartoon comparing the positions of the 3′ end of tRNA^Ile in its complex with IleRS in its synthetic mode (*left*) and in its editing mode (*right*). [Part *a* courtesy of and Part *b* based on a drawing by Thomas Steitz, Yale University.]

(b)

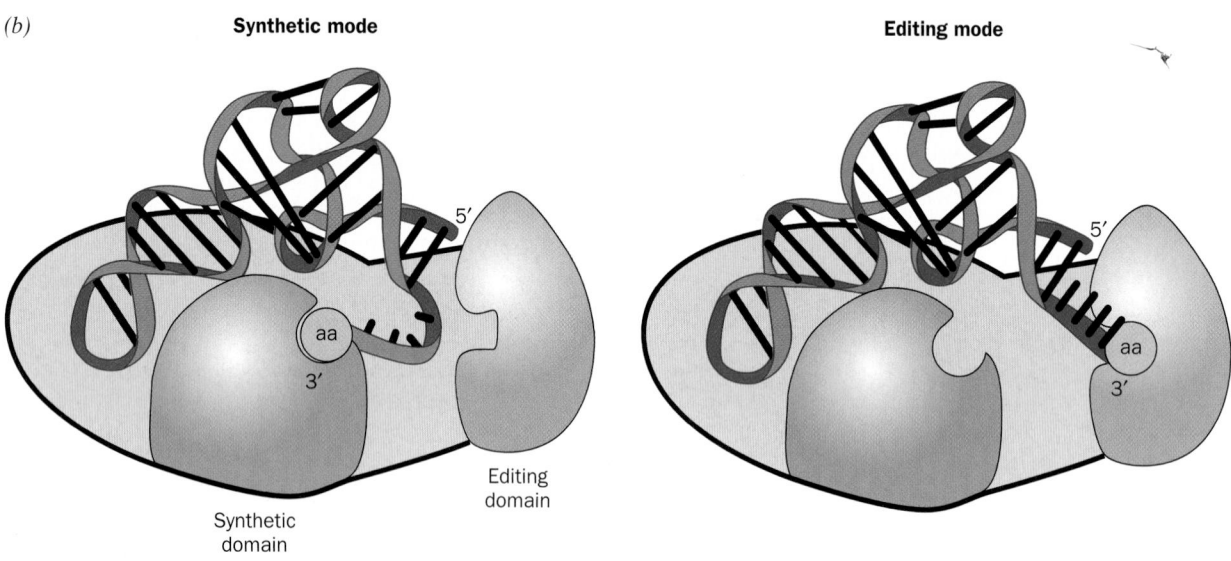

the acceptor stem's stacked A-form helix, extend into a cleft in the editing domain that has been implicated as its hydrolytic site (Fig. 32-22a, *left*). Thus, this IleRS complex appears to resemble an "editing complex" instead of a "transfer complex" as seen in the GlnRS structure. However, a transfer complex would form if the 3′ ending segment of the tRNA^Ile assumes a hairpin conformation (Fig. 32-22a, *right*) similar to that in the GlnRS structure (Fig. 32-17b; recall that IleRS and GlnRS are both Class I aaRSs). Steitz has therefore postulated that the aminoacyl group is shuttled between the IleRS's aminoacylation site and its editing site by such a conformational change (Fig. 32-22b). This process functionally resembles the way in which DNA polymerase I edits its newly synthesized strand (Section 30-2A), which Steitz also elucidated.

ValRS is a monomeric Class I aaRS that resembles IleRS. The X-ray structure of the complex of *T.*

thermophilus ValRS, **tRNA^Val,** and the nonhydrolyzable **valyl-aminoacyl–adenylate** analog **5′-O′-[N-(L-valyl)sul-famoyl]adenosine (Val-AMS),**

5′-O-[N-(L-valyl)sulfamoyl]adenosine (Val-AMS)

Valyl-aminoacyl–adenylate

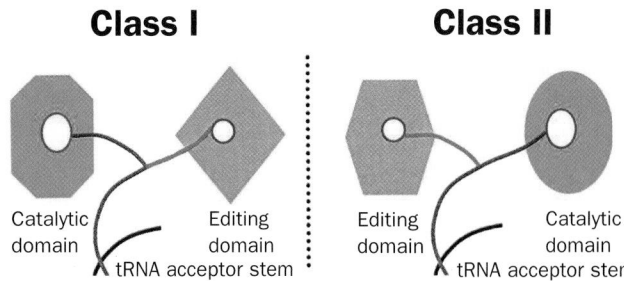

FIGURE 32-23 Schematic diagram of the aminoacylation and editing mechanisms of Class I and Class II aaRSs emphasizing the "mirror symmetry" of their overall mechanisms. With Class I aaRSs (*left;* e.g., IleRS), the 3′ end of the bound tRNA's acceptor stem assumes a hairpin conformation in the synthetic mode and a helical conformation in the editing mode, whereas the converse occurs with Class II aaRSs (*right;* e.g., ThrRS). [Courtesy of Dino Moras, CNRS/INSERM/ULP, Illkirch Cedex, France.]

determined by Yokoyama, reveals that the Val-AMS is bound in the aminoacylation pocket in the Rossmann fold domain, which accommodates the isosteric Val and Thr moieties but sterically excludes Ile. Modeling studies based on the IleRS · tRNA^{Ile} · mupirocin structure indicate that the Thr side chain would fit into the ValRS editing pocket with its side chain hydroxyl group hydrogen bonded to the side chain of Asp 279 of ValRS, which protrudes into the pocket in contrast to the corresponding Asp 328 of IleRS, which does not. Consequently, a Val side chain would be excluded from the ValRS editing pocket because it cannot form such a hydrogen bond, thereby explaining why this editing pocket hydrolyzes **threonyl-aminoacyl–adenylate** and Thr-tRNA^{Val} but not the corresponding Val derivatives. The ValRS · tRNA^{Val} structure also indicates that ValRS and tRNA^{Val} together form a tunnel connecting the ValRS's aminoacylation pocket with its editing pocket. Improperly formed threonyl-aminoacyl–adenylate is proposed to be channeled through this tunnel for hydrolysis in the editing pocket, thereby explaining why tRNA^{Val} must be bound to ValRS for this pretransfer editing reaction to occur. Valyl-aminoacyl–adenylate is presumably channeled through the similar IleRS · tRNA^{Ile} complex for its hydrolysis.

ThrRS, a Class II homodimer, has the opposite problem of ValRS: It must synthesize **Thr–tRNA^{Thr}** but not Val-tRNA^{Thr}. The X-ray structure of *E. coli* ThrRS that lacks its N-terminal domain but remains catalytically active in a complex with either threonine or the threonyl-aminoacyl–adenylate analog **Thr-AMS,** determined by Moras, reveals that ThrRS's aminoacylation pocket contains a Zn^{2+} ion that is coordinated by the side chain hydroxyl and amino groups of the threonyl group as well as by three protein side chains. The isosteric valine could not coordinate the Zn^{2+} ion in this way and hence does not undergo adenylylation by ThrRS. However, what prevents ThrRS from synthesizing Ser–tRNA^{Thr}? In fact, the truncated ThrRS synthesizes Ser–tRNA^{Thr} at more than half the rate it synthesizes Thr–tRNA^{Thr}, thereby indicating that the N-terminal domain of wild-type ThrRS contains the enzyme's editing site. Mutational analysis of ThrRS has localized this editing site to a cleft in the N-terminal domain of wild-type ThrRS, whose X-ray structure in complex with tRNA^{Thr} was also determined by Moras. In this latter structure, the tRNA's 3′ end follows a regular helical path similar to that seen in the X-ray structure of AspRS · tRNA^{Asp} · ATP (Fig. 32-18) so as to enter the

aminoacylation site. However, if the 3′ end of the bound tRNA^{Thr} assumed a hairpin conformation similar to that seen in X-ray structure of tRNA^{Gln} in complex with the Class I enzyme GlnRS and ATP (Fig. 32-17), its covalently linked aminoacyl group would enter the editing site. This indicates an intriguing "mirror symmetry" (Fig. 32-23): In Class I aaRSs that mediate a double-sieve editing mechanism, the 3′ end of the bound cognate tRNA assumes a hairpin conformation when it enters the aminoacylation site and a helical conformation when it enters the editing site, whereas the converse holds for Class II aaRSs. Finally, ThrRS does not appear to mediate pretransfer editing (does not hydrolyze **seryl-aminoacyl–adenylate**), and, in fact, the ThrRS · tRNA^{Thr} complex lacks a channel connecting its aminoacylation and editing sites such as is seen in the ValRS · tRNA^{Val} complex.

Synthetases that have adequate selectivity for their corresponding amino acid lack editing functions. Thus, for example, the TyrRS aminoadenylylation site discriminates between tyrosine and phenylalanine through hydrogen bonding with the tyrosine —OH group. The cell's other amino acids, standard as well as nonstandard, have even less resemblance to tyrosine which rationalizes why TyrRS lacks an editing site.

f. Gln–tRNA^{Gln} May Be Formed via an Alternative Pathway

Although it was long believed that each of the 20 standard amino acids is covalently linked to a tRNA by its corresponding aaRS, it is now clear that gram-positive bacteria, archaebacteria, cyanobacteria, mitochondria, and chloroplasts all lack GlnRS. Rather glutamate is linked to tRNA^{Gln} by the same GluRS that synthesizes **Glu–tRNA^{Glu}.** The resulting **Glu–tRNA^{Gln}** is then transamidated to Gln–tRNA^{Gln} by the enzyme **Glu-tRNA^{Gln} amidotransferase (Glu-AdT)** in an ATP-requiring reaction in which glutamine is the amide donor. Some microorganisms use a similar transamidation pathway for the synthesis of Asn–tRNA^{Asn} from **Asp-tRNA^{Asn}.**

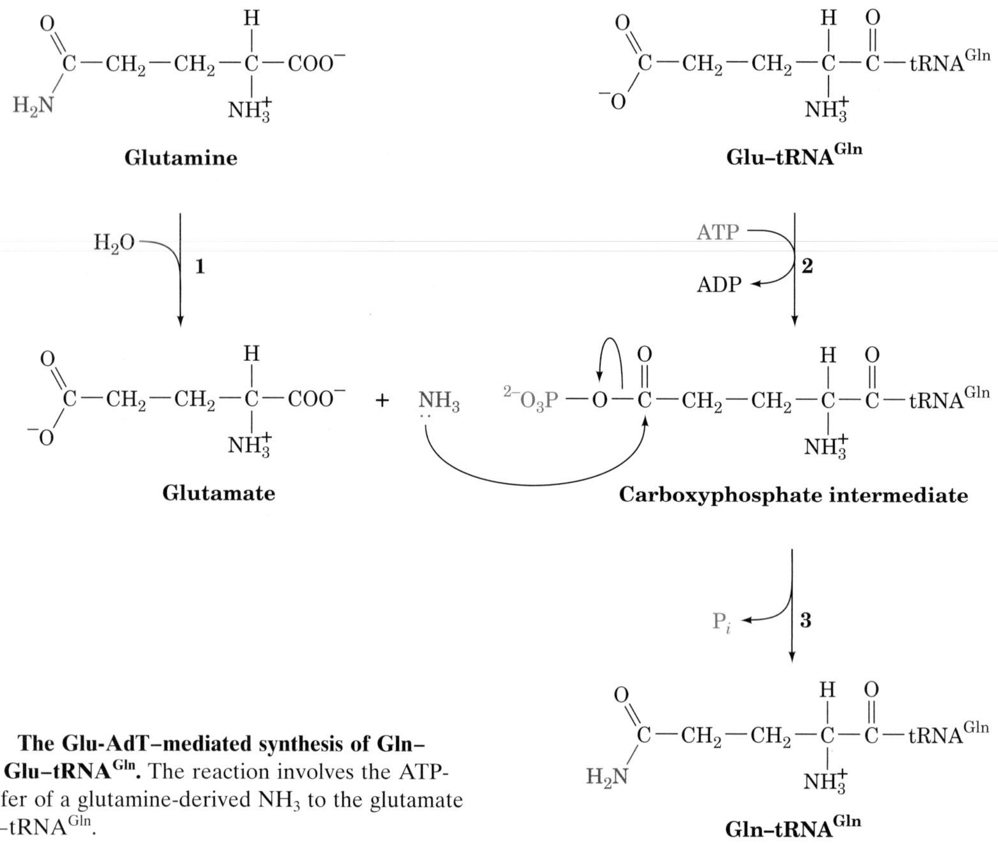

FIGURE 32-24   The Glu-AdT–mediated synthesis of Gln–tRNA^{Gln} from Glu–tRNA^{Gln}. The reaction involves the ATP-activated transfer of a glutamine-derived NH_3 to the glutamate moiety of Glu–tRNA^{Gln}.

The overall reaction catalyzed by Glu-AdT occurs in three stages (Fig. 32-24): (1) Glutamine is hydrolyzed to glutamate and the resulting NH_3 sequestered; (2) ATP reacts with the Glu side chain of Glu–tRNA^{Gln} to yield an activated acylphosphate intermediate and ADP; and (3) the acylphosphate intermediate reacts with the NH_3 to yield Gln–tRNA^{Gln} + P_i. Glu-AdT from *Bacillus subtilis*, which was characterized by Söll, is a heterotrimeric protein, none of whose subunits exhibit significant sequence similarity to GlnRS. The genes encoding these subunits, *gatA, gatB,* and *gatC,* form a single operon whose disruption is lethal, thereby demonstrating that *B. subtilis* has no alternative pathway for Gln–tRNA^{Gln} production. The **GatA** subunit of Glu-AdT appears to catalyze the activation of the side chain carboxyl of glutamic acid via a reaction resembling that catalyzed by carbamoyl phosphate synthetase (Section 26-2A). Nevertheless, GatA exhibits no sequence similarity with other known glutamine amidotransferases (members of the triad or Ntn families; Section 26-5A). The **GatB** subunit may be used to select the correct tRNA substrate. The role of the **GatC** subunit is unclear, although the observation that its presence is necessary for the expression of GatA in *E. coli* suggests that it participates in the modification, folding, and/or stabilization of GatA.

Since Glu is not misincorporated into *B. subtilis* proteins in place of Gln, the Glu–tRNA^{Gln} product of the above aminoacylation reaction must not be transported to the ribosome. It is likely that this does not occur because, as has been shown in chloroplasts, **EF-Tu,** the elongation factor that binds and transports most aminoacyl–tRNAs to the ribosome in a GTP-dependent process (Section 32-3D), does not bind Glu–tRNA^{Gln}. It is unclear why two independent routes have evolved for the synthesis of Gln–tRNA^{Gln}.

g. Some Archaebacteria Lack a Separate CysRS

The genomes of certain archaebacteria such as *M. jannaschii* lack an identifiable gene for CysRS. This is because the enzyme responsible for synthesizing **Pro-tRNA^{Pro}** in these organisms also synthesizes **Cys-tRNA^{Cys}.** Interestingly, this enzyme, which is named **ProCysRS,** does not synthesize Pro-tRNA^{Cys} or Cys–tRNA^{Pro}. Although ProCysRS synthesizes **cysteinyl-aminoacyl–adenylate** only in the presence of tRNA^{Cys}, it synthesizes **prolyl-aminoacyl–adenylate** in the absence of tRNA^{Pro}. The binding of tRNA^{Cys} to ProCysRS blocks the activation of proline so that only cysteine can be activated. Conversely, the activation of proline facilitates the binding of tRNA^{Pro} while preventing the binding of tRNA^{Cys}. However, the mechanism through which ProCysRS carries out these mutually exclusive syntheses is unknown. In any case, it appears that some organisms can get by with as few as 17 different aaRSs; they may lack GlnRS, AspRS, and a separate CysRS.

D. Codon–Anticodon Interactions

In protein synthesis, the proper tRNA is selected only through codon–anticodon interactions; the aminoacyl group does not participate in this process. This phenomenon was demonstrated as follows. Cys–tRNA^{Cys}, in which the Cys residue was ^{14}C labeled, was reductively desulfurized with Raney nickel so as to convert the Cys residue to Ala:

$$HS-CH_2-\underset{\underset{NH_3^+}{|}}{\overset{\overset{H}{|}}{C}}-\overset{\overset{O}{||}}{C}-O-tRNA^{Cys} \quad + \quad Ni(H)_x$$

Cys–tRNA^{Cys}   **Raney nickel**

$$H-CH_2-\underset{\underset{NH_3^+}{|}}{\overset{\overset{H}{|}}{C}}-\overset{\overset{O}{||}}{C}-O-tRNA^{Cys} \quad + \quad H_2S \quad + \quad Ni$$

Ala–tRNA^{Cys}

The resulting ^{14}C-labeled hybrid, Ala–tRNA^{Cys}, was added to a cell-free protein synthesizing system extracted from rabbit reticulocytes. The product hemoglobin α chain's only radioactive tryptic peptide was the one that normally contains the subunit's only Cys. No radioactivity was found in the peptides that normally contain Ala but no Cys. Evidently, *only the anticodons of aminoacyl–tRNAs participate in codon recognition.*

a. Genetic Code Degeneracy Is Largely Due to Variable Third Position Codon–Anticodon Interactions

One might naively guess that each of the 61 codons specifying an amino acid would be read by a different tRNA. Yet, even though most cells contain several groups of **isoaccepting tRNAs** (different tRNAs that are specific for the same amino acid), *many tRNAs bind to two or three of the codons specifying their cognate amino acids.* For example, yeast tRNA^{Phe}, which has the anticodon GmAA, recognizes the codons UUC and UUU (remember that the anticodon pairs with the codon in an antiparallel fashion),

	3′		5′	3′		5′
Anticodon:	—A	—A	—Gm—	—A	—A	—Gm—
	:	:	:	:	:	:
Codon:	5′		3′	5′		3′
	—U	—U	—C—	—U	—U	—U—

and yeast tRNA^{Ala}, which has the anticodon IGC, recognizes the codons GCU, GCC, and GCA.

	3′		5′	3′		5′
Anticodon:	—C	—G	—I —	—C	—G	—I —
	:	:	:	:	:	:
Codon:	5′		3′	5′		3′
	—G	—C	—U—	—G	—C	—C—

	3′		5′
Anticodon:	—C	—G	—I —
	:	:	:
Codon:	5′		3′
	—G	—C	—A—

It therefore seems that non-Watson–Crick base pairing can occur at the third codon–anticodon position (the anticodon's first position is defined as its 3′ nucleotide), the site of most codon degeneracy (Table 32-2). Note also that

the third (5′) anticodon position commonly contains a modified base such as Gm or I.

b. The Wobble Hypothesis Structurally Accounts for Codon Degeneracy

By combining structural insight with logical deduction, Crick proposed, in what he named the **wobble hypothesis,** how a tRNA can recognize several degenerate codons. He assumed that the first two codon–anticodon pairings have normal Watson–Crick geometry. The structural constraints that this places on the third codon–anticodon pairing ensure that its conformation does not drastically differ from that of a Watson–Crick pair. Crick then proposed that there could be a small amount of play or "wobble" in the third codon position which allows limited conformational adjustments in its pairing geometry. This permits the formation of several non-Watson–Crick pairs such as U · G and I · A (Fig. 32-25a). The allowed "wobble" pairings are indicated in Fig. 32-25b. Then, by analyzing the known pat-

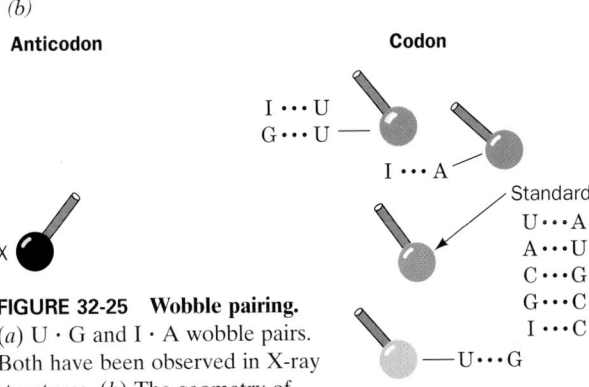

FIGURE 32-25 Wobble pairing.
(*a*) U · G and I · A wobble pairs. Both have been observed in X-ray structures. (*b*) The geometry of wobble pairing. The spheres and their attached bonds represent the positions of ribose C1′ atoms with their accompanying glycosidic bonds. X (*left*) designates the nucleoside at the 5′ end of the anticodon (tRNA). The positions on the right are those of the 3′ nucleoside of the codon (mRNA) in the indicated wobble pairings. [After Crick, F.H.C., *J. Mol. Biol.* **19**, 552 (1966).]

TABLE 32-5   Allowed Wobble Pairing Combinations in the Third Codon–Anticodon Position

5′-Anticodon Base	3′-Codon Base
C	G
A	U
U	A or G
G	U or C
I	U, C, or A

tern of codon–anticodon pairing, Crick deduced the most plausible sets of pairing combinations in the third codon–anticodon position (Table 32-5). Thus, an anticodon with C or A in its third position can only pair with its Watson–Crick complementary codon. If U, G, or I occupies the third anticodon position, two, or three codons are recognized, respectively.

No prokaryotic or eukaryotic cytoplasmic tRNA is known to participate in a nonwobble pairing combination. There is, however, no known instance of such a tRNA with an A in its third anticodon position, which suggests that the consequent A · U pair is not permitted. The structural basis of wobble pairing is poorly understood, although it is clear that it is influenced by base modifications.

A consideration of the various wobble pairings indicates that at least 31 tRNAs are required to translate all 61 coding triplets of the genetic code (there are 32 tRNAs in the minimal set because translational initiation requires a separate tRNA; Section 32-3C). Most cells have >32 tRNAs, some of which have identical anticodons. In fact, mammalian cells have >150 tRNAs. Nevertheless, *all isoaccepting tRNAs in a cell are recognized by a single aminoacyl–tRNA synthetase.*

c. Some Mitochondrial tRNAs Have More Permissive Wobble Pairings than Other tRNAs

The codon recognition properties of mitochondrial tRNAs must reflect the fact that mitochondrial genetic codes are variants of the "standard" genetic code (Table 32-3). For instance, the human mitochondrial genome, which consists of only 16,569 bp, encodes 22 tRNAs (together with 2 ribosomal RNAs and 13 proteins). Fourteen of these tRNAs each read one of the synonymous pairs of codons indicated in Tables 32-2 and 32-3 (MNX, where X is either C or U or else A or G) according to normal G · U wobble rules: The tRNAs have either a G or a modified U in their third anticodon position that, respectively, permits them to pair with codons having X = C or U or else X = A or G. The remaining 8 tRNAs, which, contrary to wobble rules, each recognize one of the groups of four synonymous codons (MNY, where Y = A, C, G, or U), all have anticodons with a U in their third position. Either this U can somehow pair with any of the four bases or these tRNAs read only the first two codon positions and ignore the third. Thus, not surprisingly, many mitochondrial tRNAs have unusual structures in which, for example, the GTΨCRA sequence (Fig. 32-9) is missing, or, in the most bizarre case, a tRNA^Ser lacks the entire D arm.

d. Frequently Used Codons Are Complementary to the Most Abundant tRNA Species

The analysis of the base sequences of several highly expressed structural genes of *S. cerevisiae* has revealed a remarkable bias in their codon usage. Only 25 of the 61 coding triplets are commonly used. *The preferred codons are those that are most nearly complementary, in the Watson–Crick sense, to the anticodons in the most abundant species in each set of isoaccepting tRNAs.* Furthermore, codons that bind anticodons with two consecutive G · C pairs or three A · U pairs are avoided so that the preferred codon–anticodon complexes all have approximately the same binding free energies. A similar phenomenon occurs in *E. coli*, although several of its 22 preferred codons differ from those in yeast. The degree with which the preferred codons occur in a given gene is strongly correlated, in both organisms, with the gene's level of expression (the measured rates of aminoacyl–tRNA selection in *E. coli* span a 25-fold range). This, it has been proposed, permits the mRNAs of proteins that are required in high abundance to be rapidly and smoothly translated.

e. Selenocysteine Is Carried by a Specific tRNA

Although it is widely stated, even in this text, that proteins are synthesized from the 20 "standard" amino acids, that is, those specified by the "standard" genetic code, some organisms, as Theresa Stadtman discovered, use a twenty-first amino acid, **selenocysteine (Sec;** alternatively **SeCys),** in synthesizing a few of their proteins:

$$\begin{array}{c} | \\ NH \\ | \\ CH-CH_2-Se-H \\ | \\ C{=}O \\ | \end{array}$$

The selenocysteine (Sec) residue

Selenium, a biologically essential trace element, is a component of several enzymes in both prokaryotes and eukaryotes. These include thioredoxin reductase (Section 28-3A) and the **thyroid hormone deiodinases** (which participate in thyroid hormone synthesis; Section 19-1D) in mammals and three forms of **formate dehydrogenases** in *E. coli,* all of which contain Sec residues. The Sec residues are ribosomally incorporated into these proteins by a unique tRNA, **tRNA^Sec,** bearing a UCA anticodon that is specified by a particular (in the mRNA) UGA codon (normally the *opal* Stop codon). The Sec-tRNA^Sec is synthesized by the aminoacylation of tRNA^Sec with L-serine by the same SerRS that charges tRNA^Ser, followed by the enzymatic selenylation of the resulting Ser residue.

How does the ribosomal system differentiate a Sec-specifying UGA codon from a normal opal Stop codon? As we saw to be the case with Glu-tRNA^Gln (Section 32-2C), EF-Tu, the elongation factor that conducts most aminoacyl–tRNAs to the ribosome in a GTP-dependent process, does not bind Sec-tRNA^Sec. Instead it is bound by a specific elongation factor named **SELB,** which, in the presence of GTP, recognizes a ribosomally bound mRNA hairpin structure on the 3′ side of the UGA codon specifying Sec.

E. *Nonsense Suppression*

Nonsense mutations are usually lethal when they prematurely terminate the synthesis of an essential protein. An organism with such a mutation may nevertheless be "rescued" by a second mutation on another part of the genome. For many years after their discovery, the existence of such **intergenic suppressors** was quite puzzling. It is now known, however, that they usually arise from mutations in a tRNA gene that cause the tRNA to recognize a nonsense codon. Such a **nonsense suppressor** tRNA appends its amino acid (which is the same as that carried by the corresponding wild-type tRNA) to a growing polypeptide in response to the recognized Stop codon, thereby preventing chain termination. For example, the *E. coli amber* suppressor known as *su*3 is a tRNA^{Tyr} whose anticodon has mutated from the wild-type GUA (which reads the Tyr codons UAU and UAC) to CUA (which recognizes the *amber* Stop codon UAG). An *su*3$^+$ *E. coli* with an otherwise lethal *amber* mutation in a gene coding for an essential protein would be viable if the replacement of the wild-type amino acid residue by Tyr does not inactivate the protein.

There are several well-characterized examples of *amber* (UAG), *ochre* (UAA), and *opal* (UGA) suppressors in *E. coli* (Table 32-6). Most of them, as expected, have mutated anticodons. UGA-1 tRNA, however, differs from the wildtype only by a G → A mutation in its D stem, which changes a G · U pair to a stronger A · U pair. This mutation apparently alters the conformation of the tRNA's CCA anticodon so that it can form an unusual wobble pairing with UGA as well as with its normal codon, UGG. Nonsense suppressors also occur in yeast.

a. Suppressor tRNAs Are Mutants of Minor tRNAs

How do cells tolerate a mutation that both eliminates a normal tRNA and prevents the termination of polypeptide synthesis? They survive because the mutated tRNA is usually a minor member of a set of isoaccepting tRNAs and because nonsense suppressor tRNAs must compete for Stop codons with the protein factors that mediate the termination of polypeptide synthesis (Section 32-3E). Consequently, the rate of suppressor-mediated synthesis of active proteins with either UAG or UGA nonsense mutations rarely exceeds 50% of the wild-type rate, whereas mutants with UAA, the most common termination codon, have suppression efficiencies of <5%. Many mRNAs, moreover, have two tandem Stop codons so that even if their first Stop codon were suppressed, termination could occur at the second. Nevertheless, many suppressor-rescued mutants grow relatively slowly because they cannot make an otherwise prematurely terminated protein as efficiently as do wild-type cells.

Other types of suppressor tRNAs are also known. **Missense suppressors** act similarly to nonsense suppressors but substitute one amino acid in place of another.

TABLE 32-6 **Some *E. coli* Nonsense Suppressors**

Name	Codon Suppressed	Amino Acid Inserted
*su*1	UAG	Ser
*su*2	UAG	Gln
*su*3	UAG	Tyr
*su*4	UAA, UAG	Tyr
*su*5	UAA, UAG	Lys
*su*6	UAA	Leu
*su*7	UAA	Gln
UGA-1	UGA	Trp
UGA-2	UGA	Trp

Source: Körner, A.M., Feinstein, S.I., and Altman, S., *in* Altman, S. (Ed.), *Transfer RNA*, p. 109, MIT Press (1978).

Frameshift suppressors have eight nucleotides in their anticodon loops rather than the normal seven. They read a four base codon beyond a base insertion thereby restoring the wild-type reading frame.

3 ■ RIBOSOMES AND POLYPEPTIDE SYNTHESIS

Ribosomes were first seen in cellular homogenates by dark-field microscopy in the late 1930s by Albert Claude who referred to them as "microsomes." It was not until the mid-1950s, however, that George Palade observed them in cells by electron microscopy, thereby disposing of the contention that they were merely artifacts of cell disruption. The name ribosome derives from the fact that these particles in *E. coli* consist of approximately two-thirds RNA and one-third protein. (**Microsomes** are now defined as the artifactual vesicles formed by the endoplasmic reticulum on cell disruption. They are easily isolated by differential centrifugation and are rich in ribosomes.) The correlation between the amount of RNA in a cell and the rate at which it synthesizes protein led to the suspicion that ribosomes are the site of protein synthesis. This hypothesis was confirmed in 1955 by Paul Zamecnik, who demonstrated that ^{14}C-labeled amino acids are transiently associated with ribosomes before they appear in free proteins. Further research showed that ribosomal polypeptide synthesis has three distinct phases: (1) chain initiation, (2) chain elongation, and (3) chain termination.

In this section we examine the structure of the ribosome and then outline the ribosomal mechanism of polypeptide synthesis. In doing so we shall compare the properties of ribosomes from prokaryotes with those of eukaryotes.

A. *Ribosome Structure*

The *E. coli* ribosome, which has a particle mass of ~2.5 × 10^6 D and a sedimentation coefficient of 70S, is a spheroidal particle that is ~250 Å across in its largest dimension. It may be dissociated, as James Watson discovered,

TABLE 32-7   Components of *E. coli* Ribosomes

	Ribosome	Small Subunit	Large Subunit
Sedimentation coefficient	70S	30S	50S
Mass (kD)	2520	930	1590
RNA			
Major		16S, 1542 nucleotides	23S, 2904 nucleotides
Minor			5S, 120 nucleotides
RNA mass (kD)	1664	560	1104
Proportion of mass	66%	60%	70%
Proteins		21 polypeptides	31 polypeptides
Protein mass (kD)	857	370	487
Proportion of mass	34%	40%	30%

into two unequal subunits (Table 32-7). The small (30S) subunit consists of a 16S rRNA molecule and 21 different polypeptides, whereas the large (50S) subunit contains a 5S and a 23S rRNA together with 31 different polypeptides. The up to 20,000 ribosomes in an *E. coli* cell account for ~80% of its RNA content and ~10% of its protein.

Structural studies of the ribosome began soon after its discovery through electron microscopy. Three-dimensional (3D) structures of the ribosome and its subunits at low (~50 Å) resolution first became available in the 1970s through image reconstruction techniques, pioneered by Klug, in which electron micrographs of a single particle or ordered sheets of particles taken from several directions are combined to yield its 3D image. The small subunit is a roughly mitten-shaped particle, whereas the large subunit is spheroidal with three protuberances on one side (Fig. 32-26).

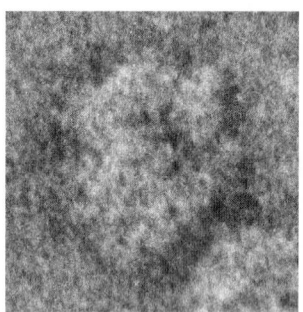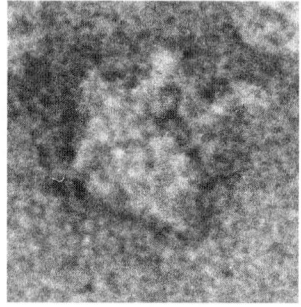

(a)

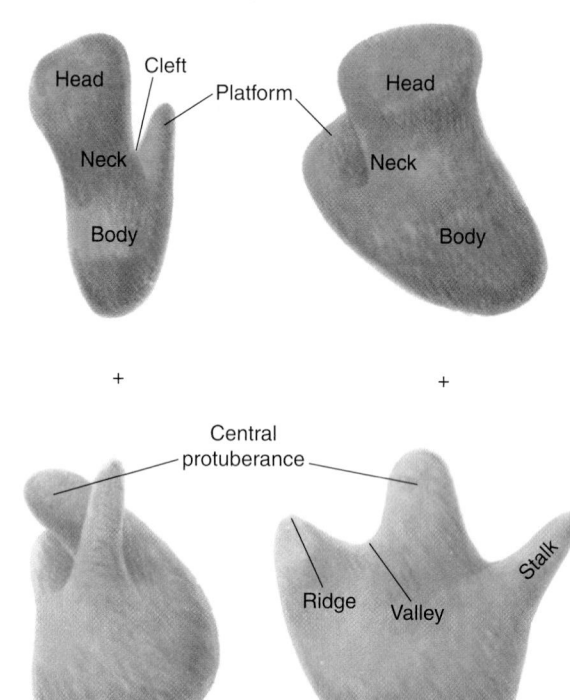

(b)

FIGURE 32-26   *E. coli* ribosome at low resolution.
(*a*) Transmission electron micrographs of negatively stained ribosomes. In **negative staining,** the particle being imaged is embedded in electron-absorbing heavy metal salts, thereby providing contrast between the relatively electron-transparent particle and the background. [Courtesy of James Lake, UCLA.] (*b*) A three-dimensional model of the *E. coli* ribosome that was mathematically deduced from series of two-dimensional electron micrographs such as those in Part *a* taken from different directions, a process known as **image reconstruction.** The small subunit (*top*) combines with the large subunit (*middle*) to form the complete ribosome (*bottom*). The two views of the complete ribosome match those seen in Part *a*.

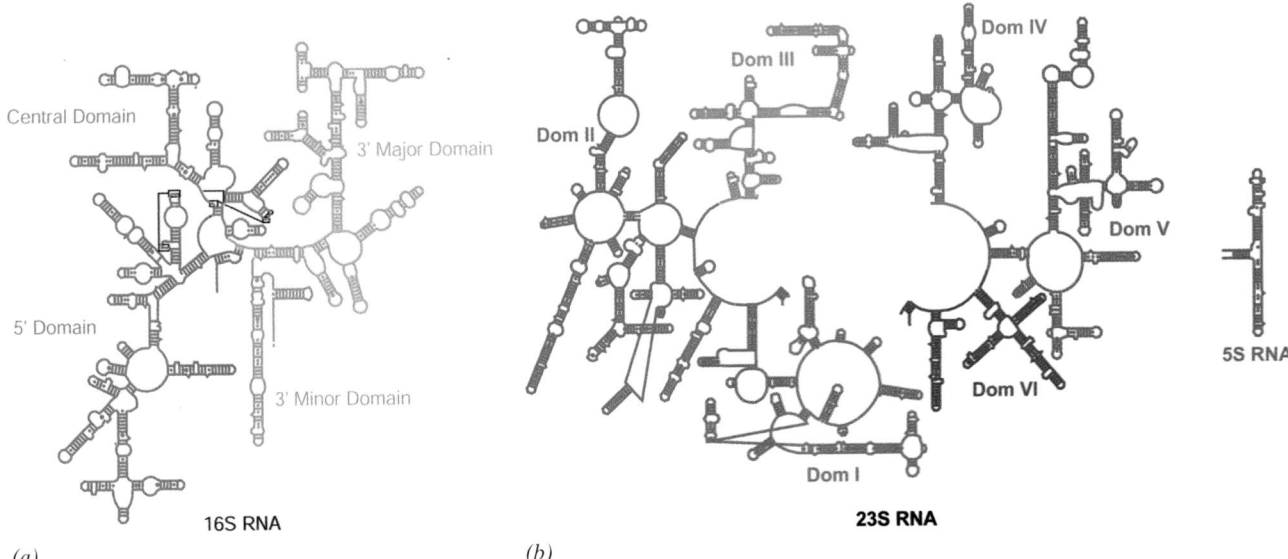

FIGURE 32-27 Secondary structures of the *E. coli* ribosomal RNAs. (*a*) 16S RNA and (*b*) 23S and 5S RNAs. The rRNAs are colored by domain with short lines spanning a stem representing Watson–Crick base pairs, small dots representing G · U base pairs, and large dots representing other non-Watson–Crick base pairs. Note the flowerlike series of stems and loops forming each domain. [Courtesy of V. Ramakrishnan, MRC Laboratory of Molecular Biology, Cambridge, U.K., and Peter Moore, Yale University. Adapted from diagrams in http://www.rna.icmb.utexas.edu.]

a. Ribosomal RNAs Have Complicated Secondary Structures

The *E. coli* 16S rRNA, which was sequenced by Harry Noller, consists of 1542 nucleotides. A computerized search of this sequence for stable double helical segments yielded many plausible but often mutually exclusive secondary structures. However, the comparison of the sequences of 16S rRNAs from several prokaryotes, under the assumption that their structures have been evolutionarily conserved, led to the flowerlike secondary structure for 16S rRNA seen in Fig. 32-27*a*. This four-domain structure, which is 54% base paired, is reasonably consistent with the results of nuclease digestion and chemical modification studies. Its double helical stems tend to be short (<8 bp) and many of them are imperfect. Intriguingly, electron micrographs of the 16S rRNA resemble those of the complete 30S subunit, thereby suggesting that the 30S subunit's overall shape is largely determined by the 16S rRNA. The large ribosomal subunit's 5S and 23S rRNAs, which consist of 120 and 2904 nucleotides, respectively, have also been sequenced. As with the 16S rRNA, they have extensive secondary structures (Fig. 32-27*b*).

b. Ribosomal Proteins Have Been Partially Characterized

Ribosomal proteins are difficult to separate because most of them are insoluble in ordinary buffers. By convention, ribosomal proteins from the small and large subunits are designated with the prefixes S and L, respectively, followed by a number indicating their position, from upper left to lower right, on a two-dimensional gel electrophoretogram (roughly in order of decreasing molecular mass; Fig. 32-28). Only protein S20/L26 appears to be common to both subunits. One of the large subunit proteins is partially acetylated at its N-terminus so that it gives rise to two electrophoretic spots (L7/L12). Four copies of this protein, a dimer of dimers, are present in the large subunit. Moreover, these four copies of L7/L12 aggregate with L10 to form a stable complex that was initially thought to be a unique protein, "L8." All the other ribosomal proteins occur in only one copy per subunit.

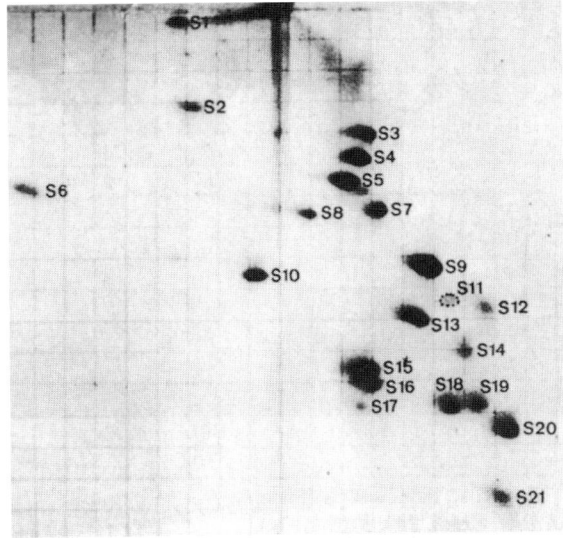

FIGURE 32-28 Two-dimensional gel electrophoretogram of *E. coli* small ribosomal subunit proteins. First dimension (*vertical*): 8% acrylamide, pH 8.6; second dimension (*horizontal*): 18% acrylamide, pH 4.6. [From Kaltschmidt, E. and Wittmann, H.G., *Proc. Natl. Acad. Sci.* **67**, 1277 (1970).]

The amino acid sequences of all 52 *E. coli* ribosomal proteins were elucidated, mainly by Heinz-Günter Wittmann and Brigitte Wittmann-Liebold. They range in size from 46 residues for L34 to 557 residues for S1. Most of these proteins, which exhibit little sequence similarity with one another, are rich in the basic amino acids Lys and Arg and contain few aromatic residues as is expected for proteins that are closely associated with polyanionic RNA molecules.

The X-ray and NMR structures of around half of the ribosomal proteins or their fragments have been independently determined. These proteins form a wide variety of structural motifs although most of their folds occur in other proteins of known structure. Around one-third of these ribosomal proteins contain the **RNA-recognition motif (RRM;** Fig. 32-29), which occurs in >200 RNA-binding proteins including rho protein (the transcriptional termination factor, which contains four such motifs; Section 31-2D), poly(A) polymerase, poly(A)-binding protein (PABP), several proteins involved in gene splicing (Section 31-4A), and the translational initiation factor **eIF4B** (Section 32-3C). All of these proteins presumably evolved from an ancient RNA-binding protein.

c. Ribosomal Subunits Are Self-Assembling

Ribosomal subunits form, under proper conditions, from mixtures of their numerous macromolecular components. *Ribosomal subunits are therefore self-assembling entities.* Masayasu Nomura determined how this occurs through partial reconstitution experiments. If one macromolecular component is left out of an otherwise self-assembling mixture of proteins and RNA, the other

components that fail to bind to the resulting partially assembled subunit must somehow interact with the omitted component. Through the analysis of a series of such partial reconstitution experiments, Nomura constructed an assembly map of the small (30S) subunit (Fig. 32-30). This map indicates that the initial steps in small subunit assembly are the independent binding to naked 16S rRNA of six so-called primary binding proteins (S4, S7, S8, S15, S17, and S20). The resulting assembly intermediates provide the molecular scaffolding for binding secondary binding proteins, which in turn generate the attachment sites for tertiary binding proteins. At one stage in the assembly process, an intermediate particle must undergo a marked conformational change before assembly can continue. The large subunit self-assembles in a similar manner. The observation that similar assembly intermediates occur *in vivo* and *in vitro* suggests that *in vivo* and *in vitro* assembly processes are much alike.

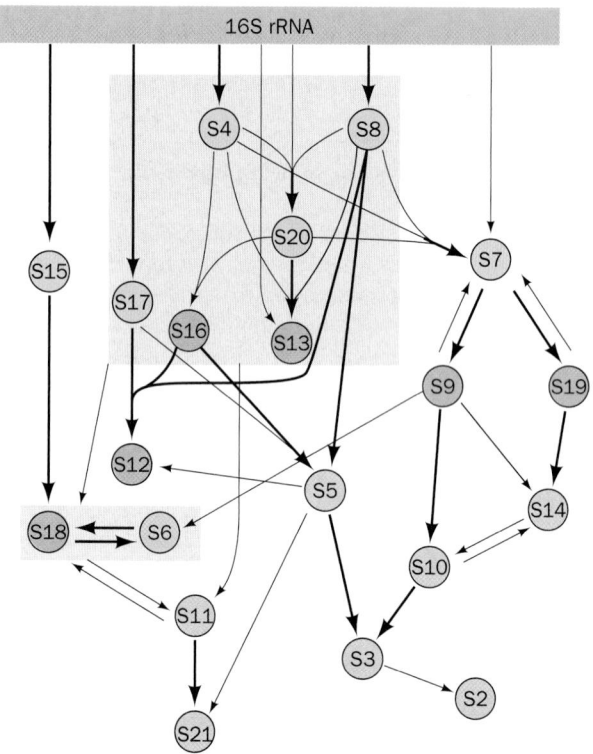

FIGURE 32-30   Assembly map of the *E. coli* small subunit. Primary, secondary, and tertiary binding proteins are represented by green, blue, and red circles, respectively, and thick and thin arrows between components indicate strong and weak facilitation of binding. For example, the thick arrow from the 16S rRNA to S15 indicates that S15 binds directly to the 16S rRNA in the absence of other proteins and is therefore a primary binding protein, the thick arrow from S15 to S18 indicates that S18 is a secondary binding protein, and the arrows between S18 and S6 and S11 indicate that S6 and S11 are tertiary binding proteins. The arrows from the shaded boxes to S11 indicate that the proteins in the boxes collectively bind S11. [After Held, W.A., Ballou, B., Mizushima, S., and Nomura, M., *J. Biol. Chem.* **249,** 3109 (1974).]

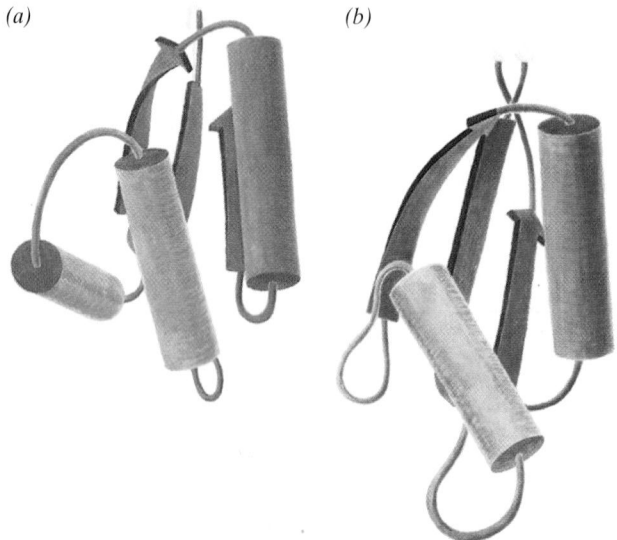

FIGURE 32-29   X-Ray structures of two ribosomal proteins. (*a*) The 74-residue C-terminal fragment of *E. coli* L7/L12. (*b*) *Bacillus stearothermophilus* L30 (61 residues). The two protein molecules are oriented so as to show their closely similar RNA-recognition motifs (RRMs, *darker shading*). [After Leijonmarck, M., Appelt, K., Badger, J., Liljas, A., Wilson, K.S., and White, S.W., *Proteins* **3,** 244 (1988).]

d. The Atomic Structure of the Prokaryotic Ribosome Has Been Long in Coming

The elucidation of the ribosome's atomic structure has been a tortuous affair extending over four decades in which slow incremental improvements were occasionally punctuated by significant technical gains. The process began in the 1960s with shadowy transmission electron micrographs such as Fig. 32-26*a* that provided only rough 2D shapes. This was followed in the 1970s by image reconstruction techniques that generated 3D models although still at low resolution (Fig. 32-26*b*). Later in the 1970s, the sites of many of the ribosome's proteins were determined by James Lake and Georg Stöffler through **immune electron microscopy,** a technique in which antibodies raised against a particular ribosomal protein are used to mark its position in electron micrographs of the antibody complexed to a ribosomal subunit. These results were improved and extended in the 1980s by neutron scattering experiments conducted by Donald Engleman and Peter Moore on the 30S subunit, which indicated the distances between the centers of mass of its component proteins and hence their three-dimensional distribution. These structural studies were supplemented by a variety of chemical cross-linking and fluorescence transfer studies that demonstrated the proximity of various ribosomal components.

The molecular structure of the prokaryotic ribosome began to come into focus in the mid-1990s through the development of **cryoelectron microscopy (cryo-EM).** In this technique, the sample is cooled to near liquid N_2 temperatures ($-196°C$) so rapidly (in a few milliseconds) that the water in the sample does not have time to crystallize but, rather, assumes a vitreous (glasslike) state. Consequently, the sample remains hydrated and hence retains its native shape to a greater extent than in conventional electron microscopy (in which the sample is vacuum dried). Studies, carried out in large part by Joachim Frank, revealed the positions where tRNAs and mRNA as well as various soluble protein factors bind to the ribosome (Fig. 32-31). The highest resolution achieved by cryo-EM of ribosomes has gradually improved over the years to $\sim$10 Å.

Ribosomal subunits were first crystallized by Ada Yonath in 1980 although they diffracted X-rays poorly. Over the course of several years, however, the quality of these crystals were incrementally improved until, in 1991, Yonath reported crystals of the 50S subunit that diffracted X-rays to 3-Å resolution. It was not until later in the 1990s, however, that technology was up to the task of determining the X-ray structures of these gargantuan molecular complexes. In 2000, the *annus mirabilis* (miracle year) of ribosomology, Moore and Steitz reported the X-ray structure of the 50S ribosomal subunit of the halophilic (salt-loving) bacterium *Haloarcula marismortui* at atomic (2.4-Å) resolution and V. Ramakrishnan and Yonath independently reported the X-ray structure of the 30S subunit of *T. thermophilus* at $\sim$3-Å resolution. In 2001, Noller reported the 5.5-Å resolution structure of the entire *T. thermophilus* ribosome. In the following paragraphs we discuss the properties of these ground-breaking structures. We consider their functional implications starting in Section 32-3C.

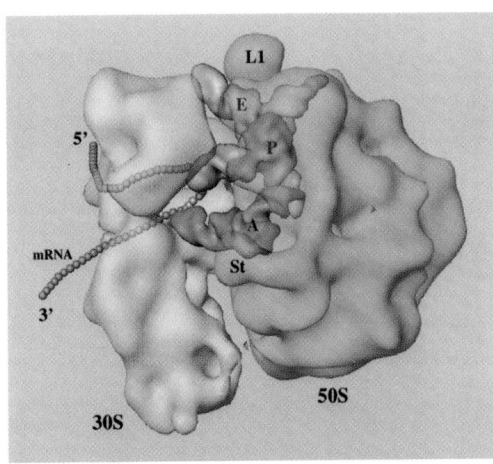

FIGURE 32-31 Cryoelectron microscopy–based image of the *E. coli* ribosome at $\sim$25 Å resolution. In this semitransparent 3D model, the 30S subunit (*yellow*) is on the left and the 50S subunit (*blue*) is on the right. The tRNAs that occupy the A, P, and E sites (Section 32-3B) are colored magenta, green, and gold. The inferred path of the mRNA is represented by a chain of orange beads with the six nucleotides contacting the A and P sites colored blue and purple, respectively. A portion of the tunnel in the 50S subunit through which the growing polypeptide chain is extruded is visible (*center right*). The symbols L1 and St indicate the positions of the L1 protein and the stalk in the 50S subunit (Fig. 32-26*b*). [Courtesy of Joachim Frank, State University of New York at Albany.]

e. Ribosomal Architecture

Several generalizations can be made about ribosomal architecture based on the structures of the 30S and 50S subunits:

1. Both the 16S and 23S rRNAs are assemblies of helical elements connected by loops, most of which are irregular extensions of helices (Fig. 32-32). These structures, which are in close accord with previous secondary structure predictions (Fig. 32-27), are stabilized by interactions between helices such as minor groove to minor groove packing, which has also been seen in the structure of the group I intron (Section 31-4A; recall that A-form RNA has a very shallow minor groove); the insertion of a phosphate ridge into a minor groove; and adenines that are distant in sequence but often highly conserved that are inserted into minor grooves. Although the determination of the structures of the 30S and 50S ribosomal subunits increased the amount of RNA structure that is known at atomic resolution by $\sim$10-fold, nearly all of the secondary structural motifs seen in the ribosome also occur in these smaller RNA structures. This suggests that the repertoire of RNA secondary structural motifs is limited.

2. Each of the 16S RNA's four domains, which extend out from a central junction (Fig. 32-27*a*), forms a morphologically distinct portion of the 30S subunit (Fig. 32-32*a*): The 5′ domain forms most of the body (Fig. 32-26*b*), the central domain forms the platform, the 3′ major domain forms the entire head, and the 3′ minor domain, which consists of just two helices, is located at the

(a)

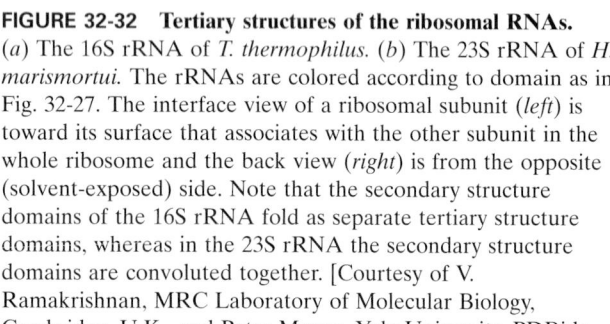

(b)

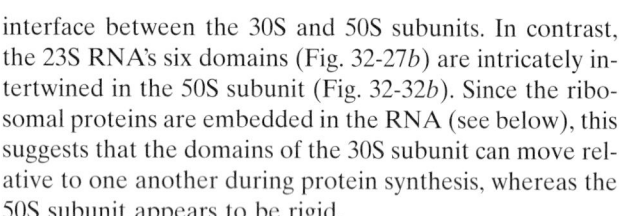

FIGURE 32-32   Tertiary structures of the ribosomal RNAs.
(*a*) The 16S rRNA of *T. thermophilus*. (*b*) The 23S rRNA of *H. marismortui*. The rRNAs are colored according to domain as in Fig. 32-27. The interface view of a ribosomal subunit (*left*) is toward its surface that associates with the other subunit in the whole ribosome and the back view (*right*) is from the opposite (solvent-exposed) side. Note that the secondary structure domains of the 16S rRNA fold as separate tertiary structure domains, whereas in the 23S rRNA the secondary structure domains are convoluted together. [Courtesy of V. Ramakrishnan, MRC Laboratory of Molecular Biology, Cambridge, U.K., and Peter Moore, Yale University. PDBids

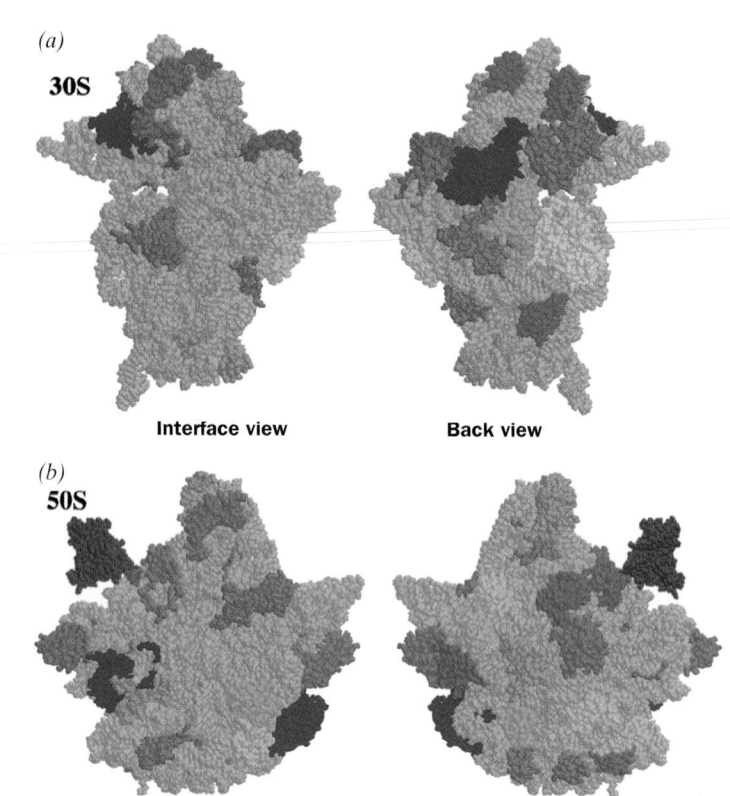

FIGURE 32-33   Distribution of protein and RNA in the ribosomal subunits. (*a*) The 30S subunit of *T. thermophilus*. (*b*) The 50S subunit of *H. marismortui*. The subunits are drawn in space-filling form with their RNAs gray and their proteins in various colors. Note that the interface side of each subunit is largely free of protein, particularly in its regions that interact with mRNA and tRNAs. [Part *a* based on an X-ray structure by V. Ramakrishnan, MRC Laboratory of Molecular Biology, Cambridge, U.K. Part *b* based on an X-ray structure by Peter Moore and Thomas Steitz, Yale University. PDBids 1J5E and 1JJ2.]

interface between the 30S and 50S subunits. In contrast, the 23S RNA's six domains (Fig. 32-27*b*) are intricately intertwined in the 50S subunit (Fig. 32-32*b*). Since the ribosomal proteins are embedded in the RNA (see below), this suggests that the domains of the 30S subunit can move relative to one another during protein synthesis, whereas the 50S subunit appears to be rigid.

3. The distribution of the proteins in the two ribosomal subunits is not uniform (Fig. 32-33). The vast majority of the ribosomal proteins are located on the back and sides of their subunits. In contrast, the face of each subunit that forms the interface between the two subunits, particularly those regions that bind the tRNAs and mRNA (see below), is largely devoid of proteins.

4. Most ribosomal proteins consist of a globular domain, which is, for the most part, located on a subunit surface (Fig. 32-33), and a long segment that is largely devoid of secondary structure and unusually rich in basic residues that infiltrates between the RNA helices into the subunit interior (Fig. 32-

34). Indeed, a few ribosomal proteins lack a globular domain altogether (e.g., L39e in Fig. 32-34*b*). Ribosomal proteins make far fewer base-specific interactions than do other known RNA-binding proteins. They tend to interact with the RNA through salt bridges between their positively charged side chains and the RNAs' negatively charged phosphate oxygen atoms, thereby neutralizing the repulsive charge–charge interactions between nearby RNA segments. This is consistent with the hypothesis that the primordial ribosome consisted entirely of RNA (the RNA world) and that the proteins that were eventually acquired stabilized its structure and fine-tuned its function.

The X-ray structure of the entire *T. thermophilus* ribosome in complex with three tRNAs and a 36-nt mRNA fragment was determined by Noller at 5.5 Å resolution. At this low resolution, the RNA backbones can be confidently traced and proteins of known structure can be properly positioned. Stereo diagrams of this enormous molecular machine are presented in Fig. 32-35. The structures of the

(a)

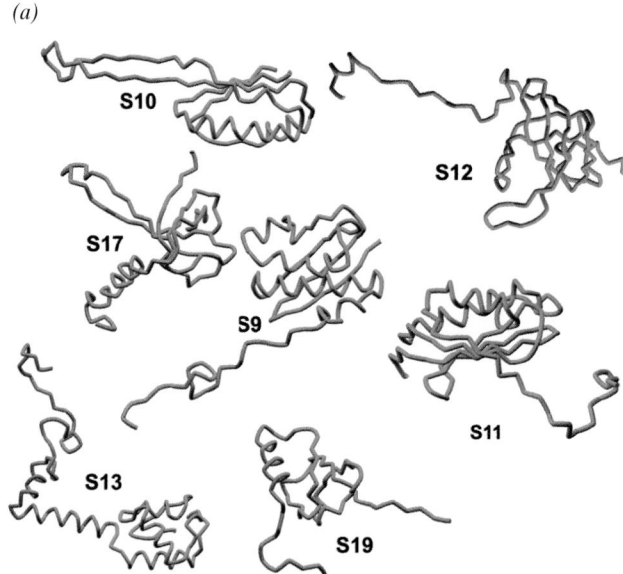

(b)

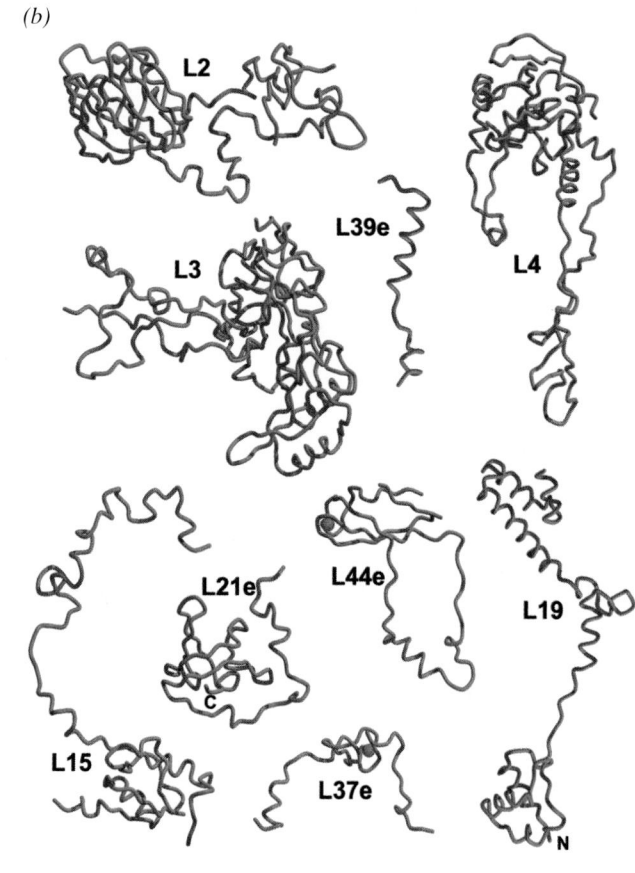

FIGURE 32-34   Gallery of ribosomal protein structures.
Proteins from (a) the 30S subunit and (b) the 50S subunit. The proteins are represented by their backbones with their globular portions green and their highly extended segments red. The globular portions are exposed on the surface of their associated subunit (Fig. 32-33), whereas the extended segments are largely buried in the RNA. The Zn²⁺ ions bound by L37e and L44e are represented by magenta spheres. [Courtesy of V. Ramakrishnan, MRC Laboratory of Molecular Biology, Cambridge, U.K., and Peter Moore, Yale University.]

(a)

(b)

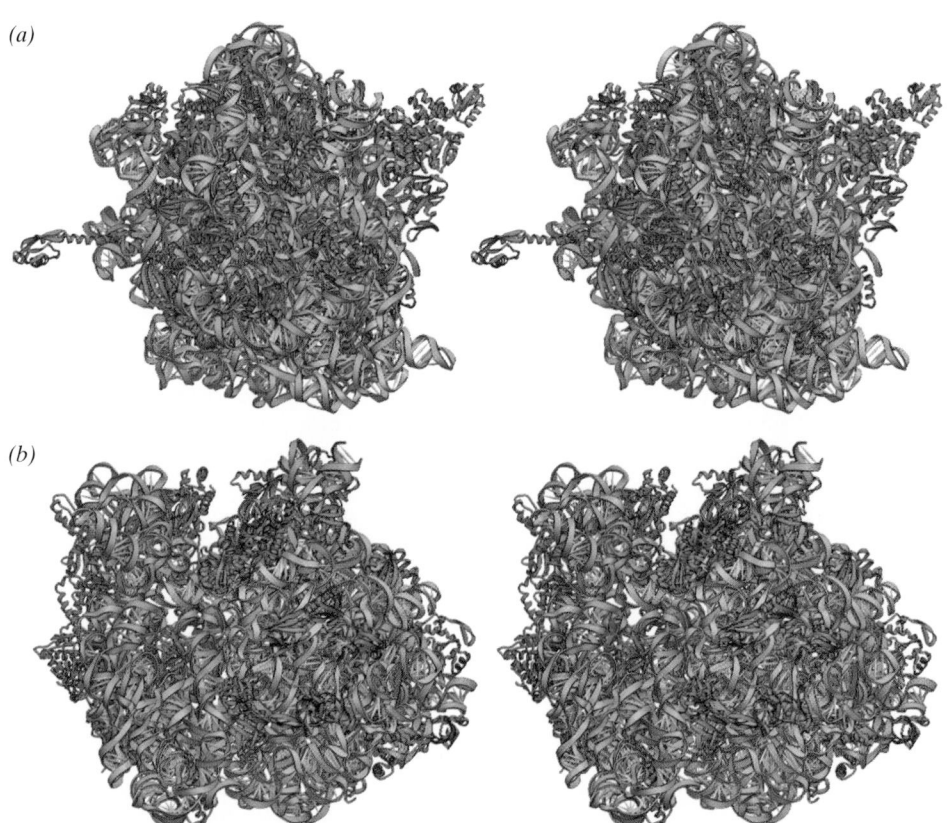

FIGURE 32-35   X-Ray structure of the *T. thermophilus* 70S ribosome in complex with three tRNAs and an mRNA fragment.
In these stereo diagrams (whose viewing is described in the Appendix to Chapter 8), the 16S RNA is cyan, the 23S RNA is gray, the 5S RNA is light blue, the small subunit proteins are dark blue, the large subunit proteins are violet, and the tRNAs bound to the A, P, and E sites (which are largely occluded) are gold, orange, and red, respectively. (a) View similar to that on the lower right of Fig. 32-26b in which the small subunit is in front of the large subunit. (b) A view rotated 90° around the vertical axis relative to Part a which resembles that in Fig. 32-31. Here the A-site tRNA is more clearly visible at the bottom of a funnel in which elongation factors bind (Section 32-3D). [Courtesy of Harry Noller, University of California at Santa Cruz. PDBids 1GIX and 1GIY.]

👁 See the Interactive Exercises

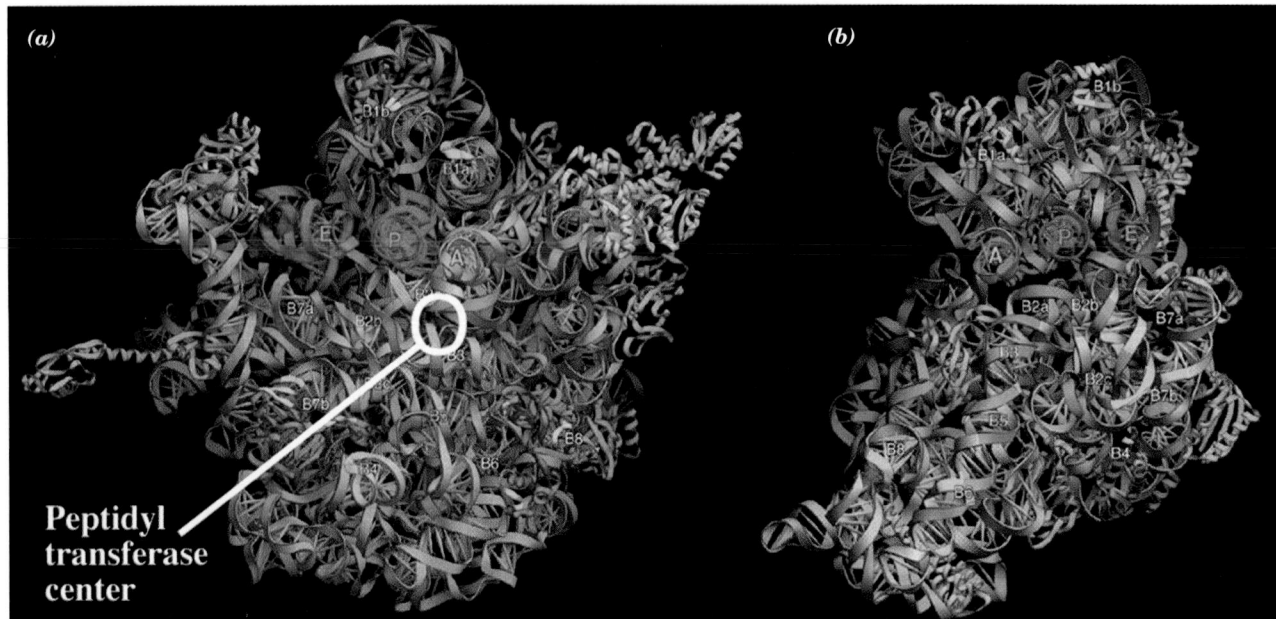

FIGURE 32-36   **Ribosomal subunits in the X-ray structure of the *T. thermophilus* 70S ribosome in complex with three tRNAs and an mRNA.** (*a*) Interface view of the large subunit (similar to Fig. 32-33*b, left*). (*b*) Interface view of the small subunit (similar to Fig. 32-33*a, left*). Here the RNA is gray with its segments that participate in intersubunit contacts magenta and the protein is blue with its segments that participate in intersubunit contacts yellow. The tRNAs bound in the A, P, and E sites are gold, orange, and red, respectively. [Courtesy of Harry Noller, University of California at Santa Cruz. PDBids 1GIX and 1GIY.]

associated 30S and 50S subunits closely resemble those of the isolated subunits although there are several regions at the subunit interface that exhibit significant conformational shifts (between 3.5 and 10 Å), which suggests that these changes occur as a consequence of subunit association. In addition, several disordered portions of the isolated *H. marismortui* 50S subunit are ordered in the intact *T. thermophilus* ribosome, although this may be a consequence of the latter's greater thermal stability.

The two subunits in the intact ribosome contact each other at 12 positions via RNA–RNA, protein–protein, and RNA–protein bridges (Fig. 32-36). These intersubunit bridges have a distinct distribution: The RNA–RNA bridges are centrally located adjacent to the three bound tRNAs, whereas the protein–protein and RNA–protein bridges are peripherally located away from the ribosome's functional sites. The RNA–RNA contacts consist mainly of minor groove–minor groove interactions although major groove, loop, and backbone contacts also occur. In the RNA–protein bridges, the proteins contact nearly all types of RNA features including major groove, minor groove, backbone, and loop elements.

Ribosomes, as we shall see in Section 32-3B, have three functionally distinct tRNA-binding sites known as the A, P, and E sites. The ribosome binds all three tRNAs in a similar manner with their anticodon stem–loops bound to the 30S subunit and their remaining portions, the D-stem, elbow, and acceptor stem, bound to the 50S subunit. These interactions, which mainly consist of RNA–RNA contacts,

are made to the tRNAs' universally conserved segments, thereby permitting the ribosome to bind different species of tRNAs in the same way. Nevertheless, the three tRNAs have somewhat different conformations: The A-site tRNA closely resembles crystallized tRNA^Phe (Fig. 32-11*b*), the P-site tRNA is slightly kinked around the junction of the D and anticodon stems so as to angle the anticodon loop toward the A site, and the E-site tRNA is further distorted with its elbow angle more open and its anticodon loop making an unusually sharp turn.

We discuss the path of the mRNA and how it interacts with the tRNAs in Section 32-3D. There we shall see that *the large subunit is mainly involved in mediating biochemical tasks such as catalyzing the reactions of polypeptide elongation, whereas the small subunit is the major actor in ribosomal recognition processes such as mRNA and tRNA binding* (although, as we have seen, the large subunit also participates in tRNA binding). We shall also see that *rRNA has the major functional role in ribosomal processes* (recall that RNA has demonstrated catalytic properties; Sections 31-4A and 31-4C).

f. Eukaryotic Ribosomes Are Larger and More Complex than Prokaryotic Ribosomes

Although eukaryotic and prokaryotic ribosomes resemble one another in both structure and function, they differ in nearly all details. Eukaryotic ribosomes have particle masses in the range 3.9 to 4.5 × 10^6 D and have a nominal sedimentation coefficient of 80S. They dissociate

TABLE 32-8   Components of Rat Liver Cytoplasmic Ribosomes

	Ribosome	Small Subunit	Large Subunit
Sedimentation coefficient	80S	40S	60S
Mass (kD)	4220	1400	2820
RNA			
Major		18S, 1874 nucleotides	28S, 4718 nucleotides
Minor			5.8S, 160 nucleotides
			5S, 120 nucleotides
RNA mass (kD)	2520	700	1820
Proportion of mass	60%	50%	65%
Proteins		33 polypeptides	49 polypeptides
Protein mass (kD)	1700	700	1000
Proportion of mass	40%	50%	35%

into two unequal subunits that have compositions that are distinctly different from those of prokaryotes (Table 32-8; compare with Table 32-7). The small **(40S)** subunit of the rat liver cytoplasmic ribosome, the most well-characterized eukaryotic ribosome, consists of 33 unique polypeptides and an **18S rRNA.** Its large **(60S)** subunit contains 49 different polypeptides and three rRNAs of **28S, 5.8S,** and **5S.** The additional complexity of the eukaryotic ribosome relative to its prokaryotic counterpart is presumably due to the eukaryotic ribosome's additional functions: Its mechanism of translational initiation is more complex (Section 32-3C); it must be transported from the nucleus, where it is formed, to the cytoplasm, where translation occurs; and

the machinery with which it participates in the secretory pathway is more complicated (Section 12-4B).

Sequence comparisons of the corresponding rRNAs from various species indicates that evolution has conserved their secondary structures rather than their base sequences (Figs. 32-27a and 32-37). For example, a G · C in a base paired stem of *E. coli* 16S rRNA has been replaced by an A · U in the analogous stem of yeast 18S rRNA. The **5.8S rRNA,** which occurs in the large eukaryotic subunit in base paired complex with the **28S rRNA,** is homologous in sequence to the 5′ end of prokaryotic 23S rRNA. Apparently 5.8S RNA arose through mutations that altered rRNA's posttranscriptional processing producing a fourth rRNA.

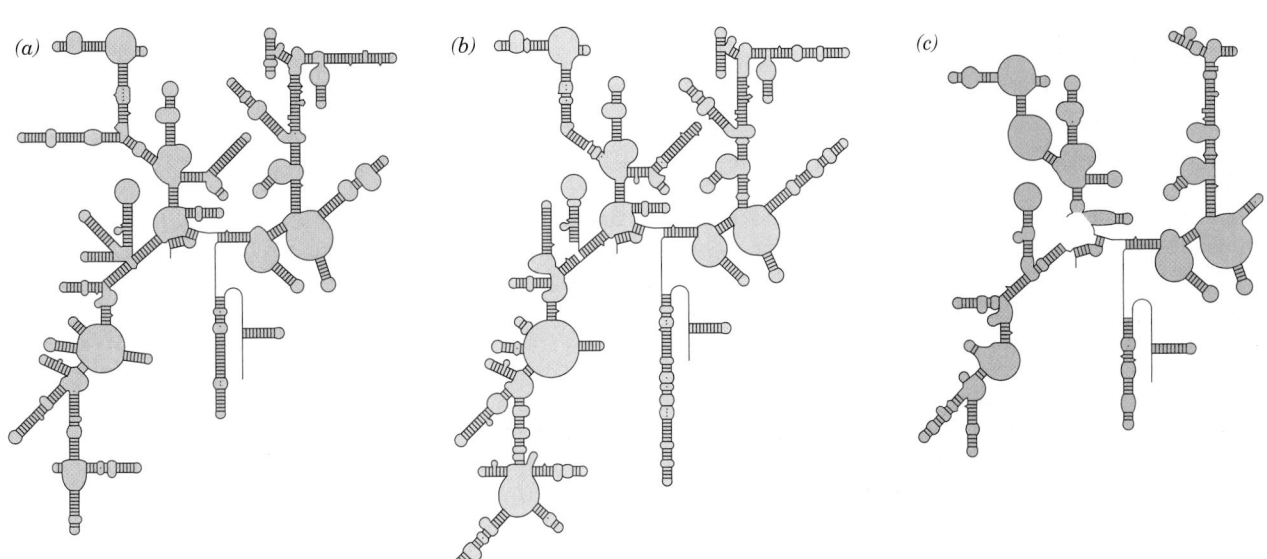

FIGURE 32-37   Predicted secondary structures of evolutionarily distant 16S-like rRNAs. (*a*) Archaebacteria (*Halobacterium volcanii*), (*b*) eukaryotes (*S. cerevisiae*), and (*c*) mammalian mitochondria (bovine). Compare them with Fig. 32-27a, the secondary structure of 16S RNA from eubacteria (*E. coli*). Note the close similarities of these assemblies; they differ mostly by insertions and deletions of stem-and-loop structures. The 23S-like rRNAs from a variety of species likewise have similar secondary structures. [After Gutell, R.R., Weiser, B., Woese, C.R., and Noller, H.F., *Prog. Nucleic Acid Res. Mol. Biol.* **32,** 183 (1985).]

(a)

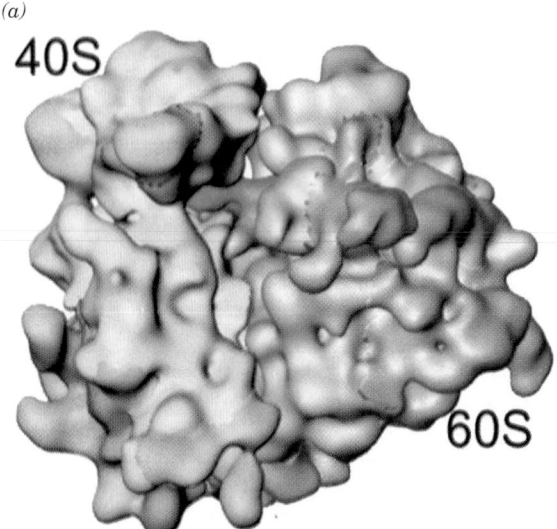

(b)

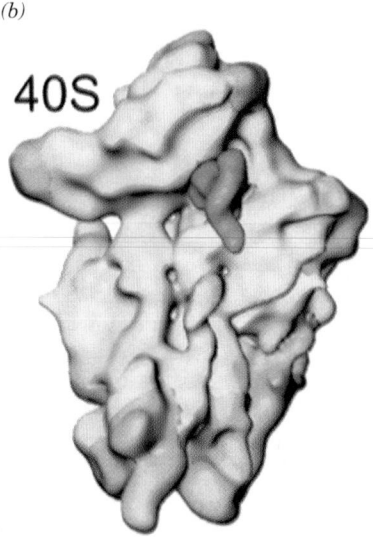

(c)

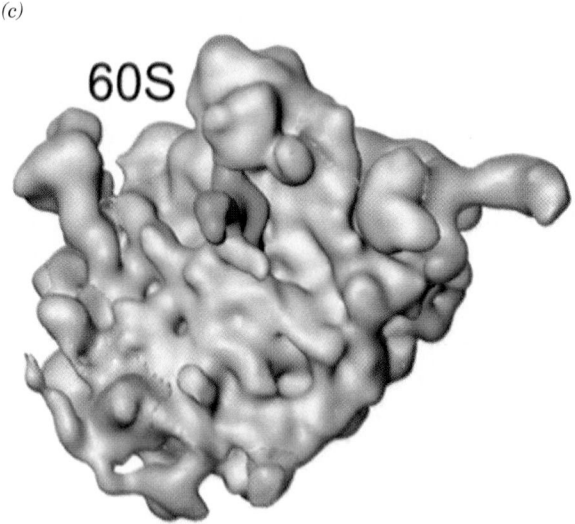

FIGURE 32-38   Cryo-EM–based image of the yeast 80S ribosome at 15 Å resolution. (*a*) The ribosome shown in side view analogous to Fig. 32-31 of the *E. coli* ribosome. The small (40S) subunit is yellow, the large (60S) subunit is blue, and the tRNA that is bound in the ribosomal P site is green. Portions of this ribosome that are not homologous to the RNA or proteins of the *E. coli* ribosome are shown in gold for the small subunit and magenta for the large subunit. (*b*) The computationally isolated small subunit shown in interface view analogous to the left panel of Fig. 32-33*a*. (*c*) The computationally isolated large subunit shown in interface view analogous to the left panel of Fig. 32-33*b*. [Courtesy of Joachim Frank, State University of New York at Albany.]

The cryo-EM–based image of the yeast 80S ribosome (Fig. 32-38), determined at 15 Å resolution by Andrej Sali, Günter Blobel, and Frank, reveals that there is a high degree of structural conservation between eukaryotic and prokaryotic ribosomes. Although the yeast 40S subunit (which consists of a 1798-nt 18S rRNA and 32 proteins) contains an additional 256 nt of RNA and 11 proteins relative to the *E. coli* 30S subunit (Table 32-8; 15 of the *E. coli* proteins are homologous to those of yeast), both exhibit a similar division into head, neck, body, and platform (Fig. 32-38*b* vs Figs. 32-26*b* and 32-33*a*). Many of the differences between these two small ribosomal subunits are accounted for by the 40S subunit's additional RNA and proteins, although their homologous portions exhibit several distinct conformational differences. Similarly, the yeast 60S subunit (Fig. 32-38*c;* which consists of an aggregate of 3671 nt and 45 proteins) structurally resembles the considerably smaller (Table 32-7) prokaryotic 50S subunit (Figs. 32-26*b* and 32-33*b*). The yeast ribosome exhibits 16 intersubunit bridges, 12 of which match the 12 that were observed in the X-ray structure of the *T. thermophilus* ribosome (Fig. 32-36), a remarkable evolutionary conservation that indicates the importance of these bridges. Moreover, the tRNA that occupies the P site of the yeast ribosome has a conformation that more closely resembles that of the P-site tRNA in the *T. thermophilus* ribosome than that of crystallized tRNA[Phe].

B. *Polypeptide Synthesis: An Overview*

Before we commence our detailed discussion of polypeptide synthesis, it will be helpful to outline some of its major features.

a. Polypeptide Synthesis Proceeds from N-Terminus to C-Terminus

The direction of ribosomal polypeptide synthesis was established, in 1961 by Howard Dintzis, through radioactive labeling experiments. He exposed reticulocytes that were actively synthesizing hemoglobin to ^{3}H-labeled leucine for times less than that required to make an entire polypeptide. The extent to which the tryptic peptides from the soluble (completed) hemoglobin molecules were labeled increased with their proximity to the C-terminus (Fig. 32-39). Incoming amino acids must therefore be appended to a growing polypeptide's C-terminus; that is, *polypeptide synthesis proceeds from N-terminus to C-terminus.*

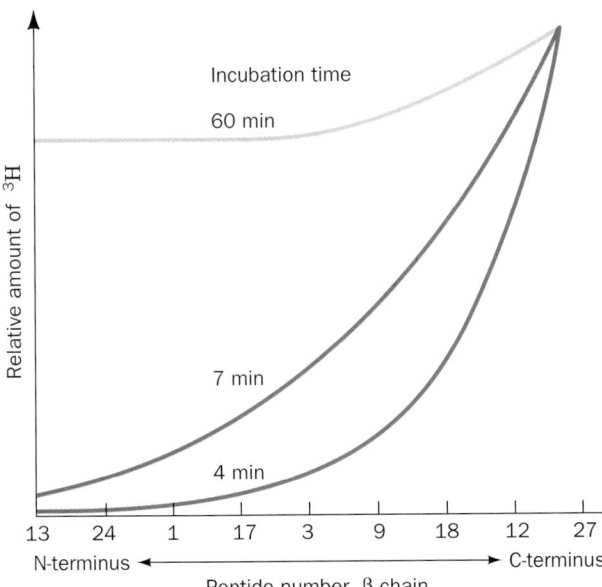

FIGURE 32-39   Demonstration that polypeptide synthesis proceeds from the N-terminus to the C-terminus. Rabbit reticulocytes were incubated with [³H]leucine. The curves show the distribution of [³H]Leu among the tryptic peptides from the β subunit of soluble rabbit hemoglobin after the indicated incubation times. The numbers on the horizontal axis are peptide identifiers arranged from N-terminus to C-terminus. [After Dintzis, H.M., *Proc. Natl. Acad. Sci.* **47**, 255 (1961).]

b. Ribosomes Read mRNA in the 5′ → 3′ direction

The direction in which the ribosome reads mRNAs was determined through the use of a cell-free protein synthesizing system in which the mRNA was poly(A) with a 3′-terminal C.

$$5'\ \ A—A—A—\cdots—A—A—A—C\ \ 3'$$

Such a system synthesizes a poly(Lys) that has a C-terminal Asn.

$$\overset{+}{H_3N}—Lys—Lys—Lys—\cdots—Lys—Lys—Asn—COO^-$$

This, together with the knowledge that AAA and AAC code for Lys and Asn and the polarity of polypeptide synthesis, indicates that *the ribosome reads mRNA in the 5′ → 3′ direction.* Since mRNA is synthesized in the 5′ → 3′ direction, this accounts for the observation that, in prokaryotes, ribosomes initiate translation on nascent mRNAs (Section 31-3).

c. Active Translation Occurs on Polyribosomes

Electron micrographs, as Rich discovered, reveal that ribosomes engaged in protein synthesis are tandemly arranged on mRNAs like beads on a string (Figs. 32-40 and 31-24). The individual ribosomes in these **polyribosomes (polysomes)** are separated by gaps of 50 to 150 Å so they have a maximum density on mRNA of ~1 ribosome per 80 nucleotides. Polysomes arise because once an active ribosome has cleared its initiation site, a second ribosome can initiate at that site.

d. Chain Elongation Occurs by the Linkage of the Growing Polypeptide to the Incoming tRNA's Amino Acid Residue

During polypeptide synthesis, amino acid residues are sequentially added to the C-terminus of the nascent, ribosomally bound polypeptide chain. If the growing polypeptide is released from the ribosome by treatment with high

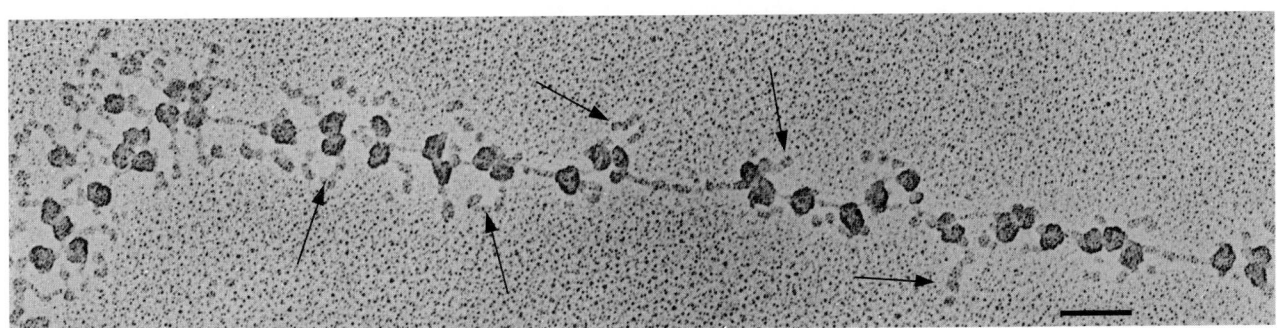

FIGURE 32-40   Electron micrographs of polysomes from silk gland cells of the silkworm *Bombyx mori.* The 3′ end of the mRNA is on the left. Arrows point to the silk fibroin polypeptides. The bar represents 0.1 μm. [Courtesy of Oscar L. Miller, Jr. and Steven L. McKnight, University of Virginia.]

salt concentrations, its C-terminal residue is invariably esterified to a tRNA molecule as a **peptidyl–tRNA:**

$$
\begin{array}{c}
\text{tRNA} \\
|\\
\text{O} \\
|\\
\text{O}=\text{P}-\text{O}-\text{CH}_2 \quad \text{Adenine} \\
|\\
\text{O}^- \\
\quad \text{H} \quad \text{OH} \\
\text{O} \\
|\\
\text{C}=\text{O} \\
|\\
\text{CH}-\text{R}_n \\
|\\
\text{NH} \\
|\\
\text{C}=\text{O} \\
|\\
\text{CH}-\text{R}_{n-1} \\
|\\
\text{NH} \\
\vdots \\
|\\
\text{C}=\text{O} \\
|\\
\text{CH}-\text{R}_1 \\
|\\
\text{NH}_3^+
\end{array}
$$

Peptidyl–tRNA

The nascent polypeptide must therefore grow by being transferred from the peptidyl–tRNA to the incoming aminoacyl–tRNA to form a peptidyl–tRNA with one more residue (Fig. 32-41). Apparently, the ribosome has at least two tRNA-binding sites: the so-called **peptidyl** or **P site,** which binds the peptidyl–tRNA, and the **aminoacyl** or **A site,** which binds the incoming aminoacyl–tRNA (Fig.

32-41). Consequently, after the formation of a peptide bond, the newly deacylated P-site tRNA must be released and replaced by the newly formed peptidyl–tRNA from the A site, thereby permitting a new round of peptide bond formation. The finding by Knud Nierhaus that each ribosome can bind up to three deacylated tRNAs but only two aminoacyl–tRNAs indicates, however, that the ribosome has a third tRNA-binding site, the **exit** or **E site,** which transiently binds the outgoing deacylated tRNA. All three sites, as we shall see, extend over both ribosomal subunits.

The details of the chain elongation process are discussed in Section 32-3D. Chain initiation and chain termination, which are special processes, are examined in Sections 32-3C and 32-3E, respectively. In all of these sections we shall first consider the process of interest in *E. coli* and then compare it with the analogous eukaryotic activity.

C. *Chain Initiation*

a. fMet Is the N-Terminal Residue of Prokaryotic Polypeptides

The first indication that the initiation of translation requires a special codon, since identified as AUG (and, in prokaryotes, occasionally GUG), was the observation that almost half of the *E. coli* proteins begin with the otherwise uncommon amino acid Met. This was followed by the discovery of a peculiar form of Met–tRNA^{Met} in which the Met residue is *N*-formylated:

$$
\begin{array}{c}
\text{S}-\text{CH}_3 \\
|\\
\text{CH}_2 \\
|\\
\text{O} \qquad \text{CH}_2 \quad \text{O} \\
\|\qquad\qquad |\qquad \| \\
\text{HC}-\text{NH}-\text{CH}-\text{C}-\text{O}-\text{tRNA}_f^{Met}
\end{array}
$$

***N*-Formylmethionine–tRNA$_f^{Met}$
(fMet–tRNA$_f^{Met}$)**

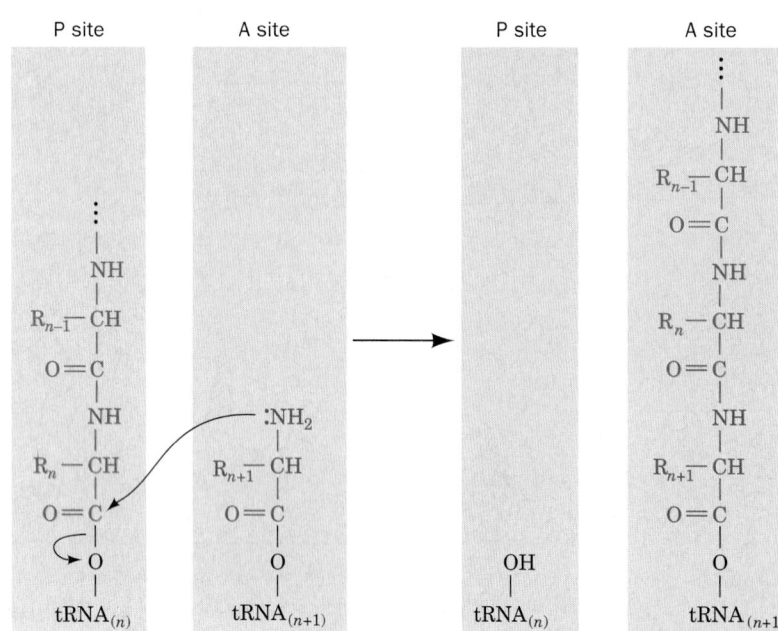

FIGURE 32-41 Ribosomal peptidyl transferase reaction forming a peptide bond. The ribosome catalyzes the nucleophilic attack of the amino group of the aminoacyl–tRNA in the A site on the peptidyl–tRNA ester in the P site, thereby forming a new peptide bond and transferring the nascent polypeptide to the A-site tRNA, while displacing the P-site tRNA.

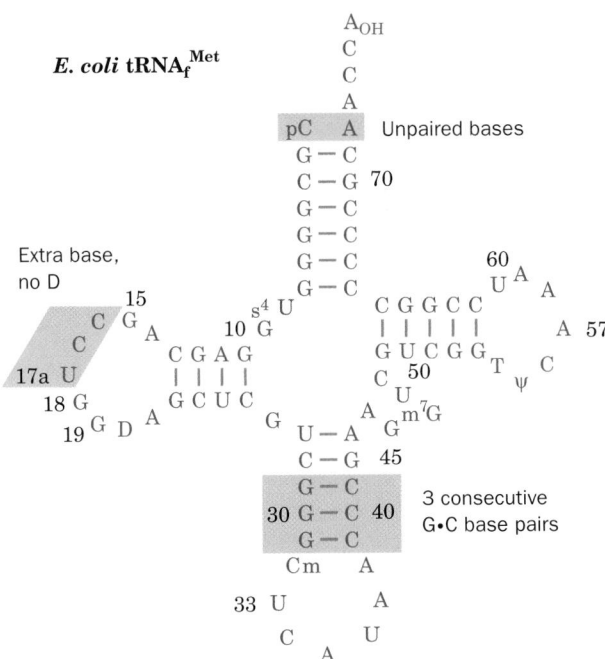

FIGURE 32-42   Nucleotide sequence of *E. coli* tRNAf^Met shown in cloverleaf form. The shaded boxes indicate the significant differences between this initiator tRNA and noninitiator tRNAs such as yeast tRNA^Ala (Fig. 32-8). [After Woo, N.M., Roe, B.A., and Rich, A., *Nature* **286**, 346 (1980).]

The ***N*-formylmethionine** residue **(fMet)** already has an amide bond and can therefore only be the N-terminal residue of a polypeptide. In fact, polypeptides synthesized in an *E. coli*–derived cell-free protein synthesizing system always have a leading fMet residue. *fMet must therefore be E. coli's initiating residue.*

The tRNA that recognizes the initiation codon, **tRNAf^Met** (Fig. 32-42), differs from the tRNA that carries internal Met residues, **tRNAm^Met**, although they both

recognize the same codon. In *E. coli*, uncharged (deacylated) tRNAf^Met is first aminoacylated with methionine by the same MetRS that charges tRNAm^Met. The resulting **Met–tRNAf^Met** is specifically *N*-formylated to yield **fMet–tRNAf^Met** in an enzymatic reaction that employs N^{10}-formyltetrahydrofolate (Section 26-4D) as its formyl donor. The formylation enzyme does not recognize Met–tRNAm^Met. The X-ray structures of *E. coli* tRNAf^Met and yeast tRNA^Phe (Fig. 32-11b) are largely similar but differ conformationally in their acceptor stems and anticodon loops. Perhaps these structural differences permit tRNAf^Met to be distinguished from tRNAm^Met in the reactions of chain initiation and elongation (see Section 32-3D).

E. coli proteins are posttranslationally modified by deformylation of their fMet residue and, in many proteins, by the subsequent removal of the resulting N-terminal Met. This processing usually occurs on the nascent polypeptide, which accounts for the observation that mature *E. coli* proteins all lack fMet.

b. Base Pairing between mRNA and the 16S rRNA Helps Select the Translational Initiation Site

AUG codes for internal Met residues as well as the initiating Met residue of a polypeptide. Moreover, mRNAs usually contain many AUGs (and GUGs) in different reading frames. Clearly, *a translational initiation site must be specified by more than just an initiation codon.* This occurs in two ways: (1) the masking of AUGs that are not initiation codons by mRNA secondary structure; and (2) interactions between the mRNA and the 16S rRNA that select the initiating AUG as we now discuss.

The 16S rRNA contains a pyrimidine-rich sequence at its 3′ end. This sequence, as John Shine and Lynn Dalgarno pointed out in 1974, is partially complementary to a purine-rich tract of 3 to 10 nucleotides, the **Shine–Dalgarno sequence,** that is centered ~10 nucleotides upstream from the start codon of nearly all known prokaryotic mRNAs (Fig. 32-43). *Base pairing interactions between an mRNA's Shine–Dalgarno sequence and the 16S rRNA apparently*

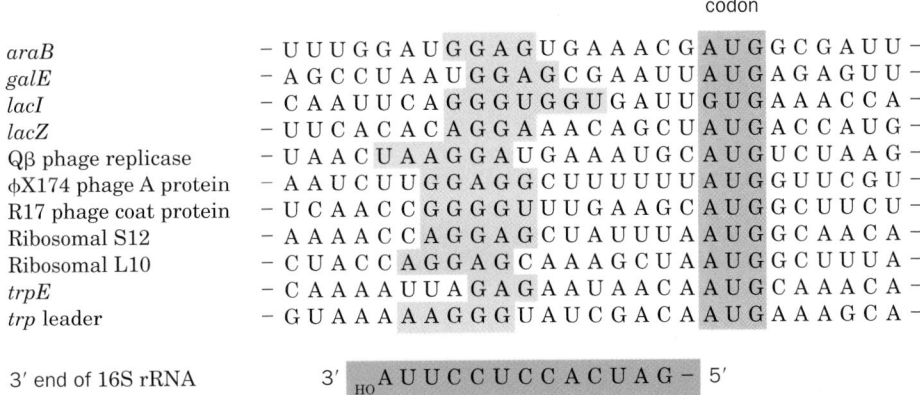

FIGURE 32-43   Some translational initiation sequences recognized by *E. coli* ribosomes. The mRNAs are aligned according to their initiation codons (*blue shading*). Their Shine–Dalgarno sequences (*red shading*) are complementary, counting G · U pairs, to a portion of the 16S rRNA's 3′ end

(*below*). [After Steitz, J.A., *in* Chambliss, G., Craven, G.R., Davies, J., Davis, K., Kahan, L., and Nomura, M. (Eds.), *Ribosomes. Structure, Function and Genetics*, pp. 481–482, University Park Press (1979).]

permit the ribosome to select the proper initiation codon. Thus ribosomes with mutationally altered anti-Shine–Dalgarno sequences often have greatly reduced ability to recognize natural mRNAs, although they efficiently translate mRNAs whose Shine–Dalgarno sequences have been made complementary to the altered anti-Shine–Dalgarno sequences. Moreover, treatment of ribosomes with the bactericidal protein **colicin E3** (produced by *E. coli* strains carrying the E3 plasmid), which specifically cleaves a 49-nucleotide fragment from the 3′ terminus of 16S rRNA, yields ribosomes that cannot initiate new polypeptide synthesis but can complete the synthesis of a previously initiated chain.

The X-ray structure of the 70S ribosome reveals, in agreement with Fig. 32-31, that an ~30-nt segment of the mRNA is wrapped in a groove that encircles the neck of the 30S subunit (Fig. 32-44). The mRNA codons in the A and P sites are exposed on the interface side of the 30S subunit, whereas its 5′ and 3′ ends are bound in tunnels

composed of RNA and protein. The mRNA's Shine–Dalgarno sequence, which is located near its 5′ end, is base paired, as expected, with the 16S RNA's anti-Shine–Dalgarno sequence, which is situated close to the E site. The resulting double helical segment is accommodated in a cleft formed by both RNA and protein elements of the 16S subunit's head, neck, and platform.

c. Prokaryotic Initiation Is a Three-Stage Process That Requires the Participation of Soluble Protein Initiation Factors

See Guided Exploration 28: Translational Initiation. Intact ribosomes do not directly bind mRNA so as to initiate polypeptide synthesis. Rather, *initiation is a complex process in which the two ribosomal subunits and fMet–tRNA$_f^{Met}$ assemble on a properly aligned mRNA to form a complex that is competent to commence chain elongation. This assembly process also requires the participation of protein* **initiation factors** *that are not permanently associated with the ribosome.* Initiation in *E. coli* involves three initiation factors designated **IF-1, IF-2,** and **IF-3** (Table 32-9). Their existence was discovered when it was found that

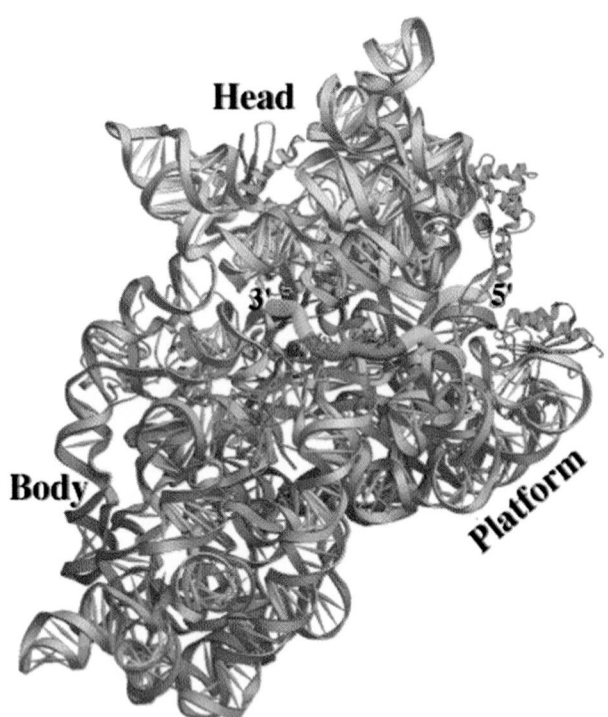

FIGURE 32-44 Path of mRNA through the ribosomal 30S subunit as viewed from its interface side. The 16S RNA is cyan, the mRNA is represented in worm form with its A- and P-site codons orange and red, the Shine–Dalgarno helix (which includes a segment of 16S RNA) magenta, and its remaining segments yellow. The S3, S4, and S5 proteins are green, the S7, S11, and S12 proteins are purple, and the remaining ribosomal proteins have been omitted for clarity. The S3, S4, and S5 proteins, which in part form the tunnel through which the mRNA enters the ribosome, may function as a helicase to remove secondary structure from the mRNA that would otherwise interfere with tRNA binding. [Courtesy of Gloria Culver, Iowa State University. Based on an X-ray structure by Harry Noller, University of California at Santa Cruz. PDBid 1JGO.]

TABLE 32-9 The Soluble Protein Factors of *E. coli* Protein Synthesis

Factor	Number of Residues[a]	Function
Initiation Factors		
IF-1	71	Assists IF-3 binding
IF-2	890	Binds initiator tRNA and GTP
IF-3	180	Releases mRNA and tRNA from recycled 30S subunit and aids new mRNA binding
Elongation Factors		
EF-Tu	393	Binds aminoacyl–tRNA and GTP
EF-Ts	282	Displaces GDP from EF-Tu
EF-G	703	Promotes translocation through GTP binding and hydrolysis
Release Factors		
RF-1	360	Recognizes UAA and UAG Stop codons
RF-2	365	Recognizes UAA and UGA Stop codons
RF-3	528	Stimulates RF-1/RF-2 release via GTP hydrolysis
RRF	185	Together with EF-G, induces ribosomal dissociation to small and large subunits

[a]All *E. coli* translational factors are monomeric proteins.

washing small ribosomal subunits with 1*M* ammonium chloride solution, which removes the initiation factors but not the "permanent" ribosomal proteins, prevents initiation.

The initiation sequence in *E. coli* ribosomes has three stages (Fig. 32-45):

1. On completing a cycle of polypeptide synthesis, the 30S and 50S subunits remain associated as an inactive 70S ribosome (Section 32-3E). IF-3 binds to the 30S subunit of

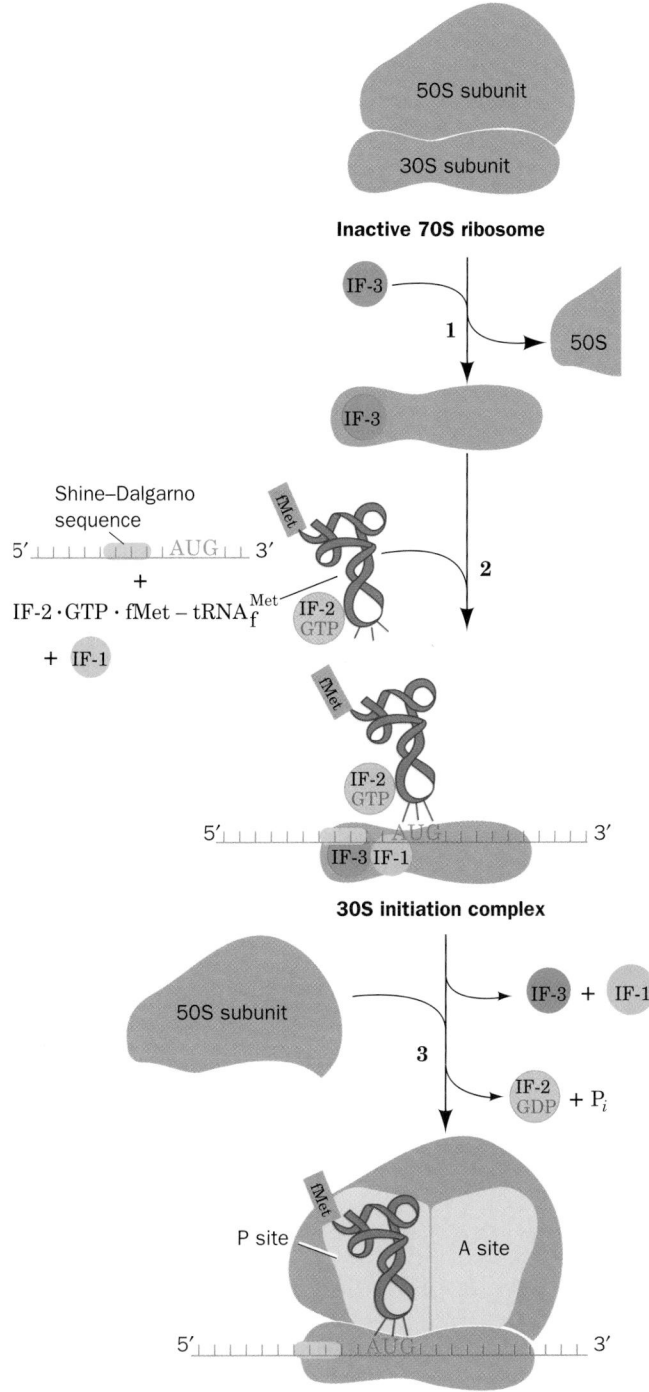

Inactive 70S ribosome

30S initiation complex

70S initiation complex

this complex so as to promote its dissociation. The X-ray structure of the 30S subunit in complex with the C-terminal domain of IF-3 (which by itself prevents the association of the 30S and 50S subunits), determined by Yonath and François Franceschi, indicates that IF-3 binds to the upper end of the platform (Fig. 32-26) on its solvent (back) side. Hence IF-3 does not function by physically blocking the binding of the 50S subunit.

2. mRNA and IF-2 in a ternary complex with GTP and fMet–tRNA$_\mathrm{f}^\mathrm{Met}$ that is accompanied by IF-1 subsequently bind to the 30S subunit in either order. Hence, fMet–tRNA$_\mathrm{f}^\mathrm{Met}$ recognition must not be mediated by a codon–anticodon interaction; it is the only tRNA–ribosome association that does not require one. This interaction, nevertheless, helps bind fMet–tRNA$_\mathrm{f}^\mathrm{Met}$ to the ribosome. IF-1 binds in the A site where it may function to prevent the inappropriate or premature binding of a tRNA. IF-3 also functions in this stage of the initiation process: it destabilizes the binding of tRNAs that lack the three G · C pairs in the anticodon stem of RNA$_\mathrm{f}^\mathrm{Met}$ (Fig. 32-42) and helps discriminate between matched and mismatched codon–anticodon interactions.

3. Last, in a process that is preceded by IF-1 and IF-3 release, the 50S subunit joins the 30S initiation complex in a manner that stimulates IF-2 to hydrolyze its bound GTP to GDP + P$_i$. This irreversible reaction conformationally rearranges the 30S subunit and releases IF-2 for participation in further initiation reactions.

IF-2 is a member of the superfamily of regulatory GTPases such as Ras and hence is a **G-protein** (Section 19-2A). The 30S initiation complex therefore functions as its **GAP** (GTPase-activating protein; Section 19-3C).

Initiation results in the formation of an fMet–tRNA$_\mathrm{f}^\mathrm{Met}$ · mRNA · ribosome complex in which the fMet–tRNA$_\mathrm{f}^\mathrm{Met}$ occupies the ribosome's P site while its A site is poised to accept an incoming aminoacyl–tRNA (an arrangement analogous to that at the conclusion of a round of elongation; Section 32-3D). In fact, tRNA$_\mathrm{f}^\mathrm{Met}$ is the only tRNA that directly enters the P site. All other tRNAs must do so via the A site during chain elongation (Section 32-3D). This arrangement was established through the use of the antibiotic **puromycin** as is discussed in Section 32-3D.

d. Eukaryotic Initiation Is Far More Complicated than That of Prokaryotes

Although translational initiation in eukaryotes superficially resembles that in prokaryotes, it is, in fact, a far more complicated process. Whereas prokaryotic initiation only requires the assistance of three monomeric initiation factors, that in eukaryotes involves the participation of at least 11 initiation factors (designated eIF*n*; "e" for eukaryotic)

FIGURE 32-45   Translational initiation pathway in *E. coli*.

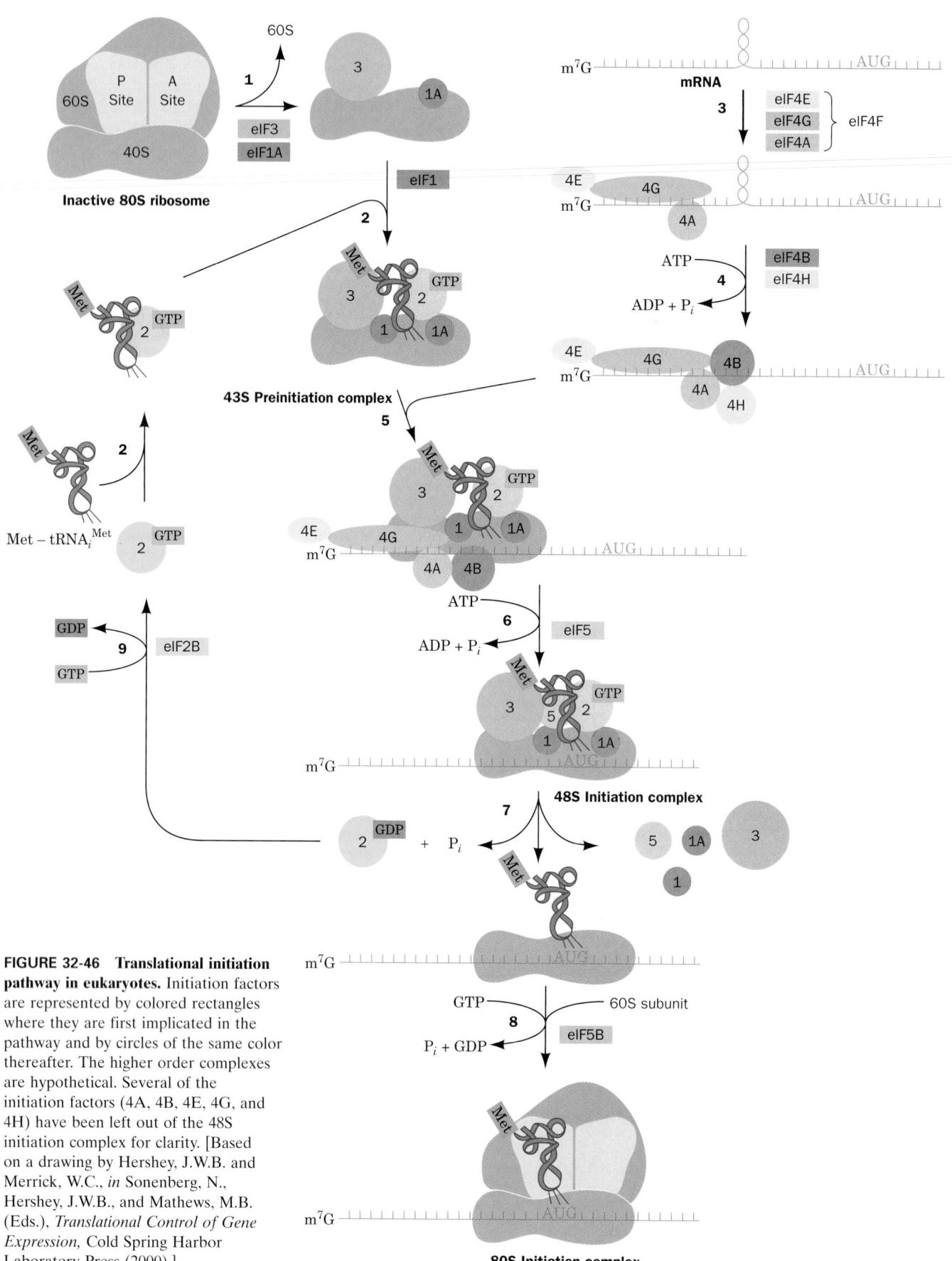

FIGURE 32-46 Translational initiation pathway in eukaryotes. Initiation factors are represented by colored rectangles where they are first implicated in the pathway and by circles of the same color thereafter. The higher order complexes are hypothetical. Several of the initiation factors (4A, 4B, 4E, 4G, and 4H) have been left out of the 48S initiation complex for clarity. [Based on a drawing by Hershey, J.W.B. and Merrick, W.C., *in* Sonenberg, N., Hershey, J.W.B., and Mathews, M.B. (Eds.), *Translational Control of Gene Expression,* Cold Spring Harbor Laboratory Press (2000).]

that consist of at least 26 polypeptide chains. Eukaryotic initiation occurs as follows (Fig. 32-46):

1. The process begins with the binding of **eIF3** (which consists of 11 different subunits) and **eIF1A** (a monomer and homolog of bacterial IF-1) to the 40S subunit in the inactive 80S ribosome (which had terminated elongation in its previous elongation cycle) so that it releases the 50S subunit.

2. The ternary complex of **eIF2** (a heterotrimer), GTP, and **Met–tRNA$_i^{Met}$** binds to the 40S ribosomal subunit accompanied by **eIF1** (a monomer) to form the so-called **43S preinitiation complex.** Here the subscript "i" on tRNA$_i^{Met}$ distinguishes this eukaryotic initiator tRNA, whose appended Met residue is never *N*-formylated, from that of prokaryotes; both species are, nevertheless, readily interchangeable *in vitro*.

3. Eukaryotic mRNAs lack the complementary sequences to bind to the 18S rRNA in the Shine–Dalgarno manner. Rather, they have an entirely different mechanism for recognizing the mRNA's initiating AUG codon. *Eukaryotic mRNAs, nearly all of which have an m^{7}G cap and a poly(A) tail (Section 31-4A), are invariably monocistronic and almost always initiate translation at their leading AUG.* This AUG, which occurs at the end of a 5'-untranslated region of 50 to 70 nt, is embedded in the consensus sequence GCCRCCAUGG, with changes in the purine (R) 3 nt before the AUG and the G immediately following it reducing translational efficiency by ~10-fold each and with other changes having much smaller effects. In addition, secondary structure (stem–loops) in the mRNA upstream of the initiation site may affect initiation efficiency. The recognition of the initiation site begins by the binding of **eIF4F** to the m^{7}G cap. eIF4F is a heterotrimeric complex of **eIF4E, eIF4G,** and **eIF4A** (all monomers), in which eIF4E **(cap-binding protein)** recognizes the mRNA's m^{7}G cap and eIF4G serves as a scaffold to join eIF4E with eIF4A. Both the X-ray and NMR structures of eIF4E in complex with **m^{7}GDP,** determined by Nahun Sonenberg and Stephen Burley and by Sonenberg and Gerhard Wagner, reveal that the protein binds the m^{7}G base by intercalating it between two highly conserved Trp residues (Fig. 32-47*a*) in a region that is adjacent to a positively charged cleft that forms the putative mRNA-binding site (Fig. 32-47*b*). The m^{7}G base is specifically recognized by hydrogen bonding to protein side chains in a manner reminiscent of G · C base pairing. eIF4G also binds poly(A)-binding protein (PABP; Section 31-4A) bound to the mRNA's poly(A) tail, thereby circularizing the mRNA (not shown in Fig. 32-46). Although this explains the synergism between an mRNA's m^{7}A cap and its poly(A) tail in stimulating translational initiation, the func-

(a)

(b)

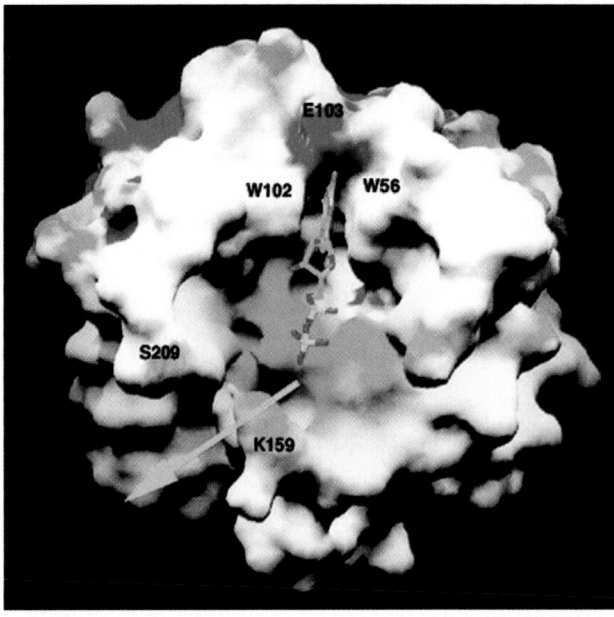

FIGURE 32-47 X-Ray structure of murine eIF4E in complex with the m^{7}G cap analog m^{7}GDP. (*a*) The m^{7}GDP-binding site with the m^{7}GDP and the side chains that bind it drawn in ball-and-stick form with the atoms of the m^{7}GDP colored according to type (C green, N dark yellow, O red, and P bright yellow) and the protein side chains with which the m^{7}GDP interacts drawn in various colors. Hydrogen bonds, salt bridges, and van der Waals interactions are represented by dashed lines and bridging water molecules are drawn as black spheres. The m^{7}G base is intercalated between the indole rings of Trp 56 and Trp 102, where it specifically interacts with protein side chains through hydrogen bonds and van der Waals interactions. The GDP's phosphate groups interact directly and indirectly with three basic side chains. (*b*) The solvent-accessible surface of eIF4E colored according to its electrostatic potential (red negative, blue positive, and white neutral) and viewed approximately as in Part *a.* The m^{7}GDP is drawn in ball-and-stick form colored as in Part *a.* The mRNA presumably binds in the positively charged cleft (*yellow arrow*) that is adjacent to the m^{7}G binding site and which passes between Lys 159 and Ser 209. [Courtesy of Nahum Sonenberg, McGill University, Montréal, Québec, Canada. PDBid 1EJ1.]

tion of this circle is unclear. However, an attractive hypothesis is that it enables a ribosome that has finished translating the mRNA to reinitiate translation without having to disassemble and then reassemble. Another possibility is that it prevents the translation of incomplete (broken) mRNAs.

4. eIF4B (an RRM-containing homodimer) and **eIF4H** (a monomer) join the eIF4F–mRNA complex where they stimulate the RNA helicase activity of eIF4A to unwind the mRNA's helical segments in an ATP-dependent process. This presumably also strips away the proteins that are bound to the mRNA (Section 32-4A). eIF4A is the prototype of the so-called **DEAD-box family** of proteins, which is named after one of the sequence motifs shared by the diverse members of this family, all of which have NTPase activity.

5. The eIF4F–mRNA–eIF4B–eIF4H complex joins the 43S preinitiation complex through a protein–protein interaction between eIF4G and the 40S subunit-bound eIF3. This differs substantially from the corresponding prokaryotic process (Fig. 32-45) in which the mRNA is bound to the 30S ribosomal subunit via associations between RNA molecules (that involving the Shine–Dalgarno sequence and the codon–anticodon interaction).

6. eIF5 (a monomer) joins the growing assembly. The 43S preinitiation complex then translocates along the mRNA, an ATP-dependent process called **scanning,** until it encounters the mRNA's AUG initiation codon, thereby yielding the **48S initiation complex.** The recognition of the AUG occurs mainly through base pairing with the CUA anticodon on the bound Met–tRNA$_i^{Met}$, as was demonstrated by the observation that mutating this anticodon results in the recognition of the new cognate codon instead of AUG.

7. The formation of the 48S initiation complex induces eIF2 to hydrolyze its bound GTP to GDP + P$_i$, which results in the release of all the initiation factors, thereby leaving the Met–tRNA$_i^{Met}$ in the small subunit's P site. The hydrolysis reaction is stimulated by eIF5, acting as a GAP (Section 19-3C).

8. The 60S subunit then joins the mRNA-bound Met–tRNA$_i^{Met}$–40S subunit complex in a GTPase reaction mediated by **eIF5B** (a monomer and homolog of bacterial IF-2), thereby yielding the 80S ribosomal initiation complex. Thus eukaryotic translation initiation consumes two GTPs versus one for prokaryotic initiation (Fig. 32-45).

9. What remains is to recycle the eIF2 · GDP complex by exchanging its GDP for GTP. This reaction is mediated by **eIF2B** (a heteropentamer), which therefore functions as eIF2's **GEF** (guanine nucleotide exchange factor; Section 19-3C).

Many eukaryotic initiation factors are subject to phosphorylation/dephosphorylation and are therefore likely to participate in the control of eukaryotic translation, a subject we discuss in Section 32-4.

Although the initiation sites on most eukaryotic mRNAs are identified by the above-described scanning mechanism, a few mRNAs have an **internal ribosome entry site (IRES)** to which the 40S subunit can directly bind in a process reminiscent of prokaryotic initiation. However, little is yet known about the mechanism of IRES-based initiation. Indeed, IRESs lack clearly identifiable consensus sequences.

D. *Chain Elongation*

🔎 **See Guided Exploration 29: Translational Elongation.** *Ribosomes elongate polypeptide chains in a three-stage reaction cycle that adds amino acid residues to a growing polypeptide's C-terminus (Fig. 32-48):*

1. Decoding, in which the ribosome selects and binds an aminoacyl–tRNA, whose anticodon is complementary to the mRNA codon in the A site.

2. Transpeptidation, in which the peptidyl group on the P-site tRNA is transferred to the aminoacyl group in the A site through the formation of a peptide bond (Fig. 32-41).

3. Translocation, in which A-site and P-site tRNAs are respectively transferred to the P site and E site accompanied by their bound mRNA; that is, the mRNA, together with its base paired tRNAs, is ratcheted through the ribosome by one codon.

This process, which occurs at a rate of 10 to 20 residues/s, involves the participation of several nonribosomal proteins known as **elongation factors** (Table 32-9). We describe these processes in the following paragraphs.

a. Decoding

In the decoding stage of the *E. coli* elongation cycle, a binary complex of GTP with the elongation factor **EF-Tu** (also called **EF1A**) combines with an aminoacyl–tRNA. The resulting ternary complex binds to the ribosome, and, in a reaction that hydrolyzes the GTP to GDP + P$_i$, the aminoacyl–tRNA is bound in a codon–anticodon complex to the ribosomal A site and EF-Tu · GDP + P$_i$ is released. In the remainder of this stage, the bound GDP is replaced by GTP in a reaction mediated the elongation factor **EF-Ts** (also called **EF1B**). EF-Tu, as are several other GTP-binding ribosomal factors, is a G-protein, and hence the ribosome functions as its GAP and EF-Ts is its GEF.

Aminoacyl–tRNAs can bind to the ribosomal A site without the mediation of EF-Tu but at a rate too slow to support cell growth. The importance of EF-Tu is indicated by the fact that it is the most abundant *E. coli* protein; it is present in ~100,000 copies per cell (>5% of the cell's protein), which is approximately the number of tRNA molecules in the cell. Consequently, *the cell's entire complement of aminoacyl–tRNAs is essentially sequestered by EF-Tu.*

b. EF-Tu Is Sterically Prevented from Binding Initiator tRNA

The X-ray structure of the Phe–tRNA^{Phe} · EF-Tu · GDPNP ternary complex (GDPNP is a nonhydrolyzable GTP analog; Section 19-3C), determined by Brian Clark and Jens Nyborg, reveals that these two macromolecules associate to form a corkscrew-shaped complex in which the

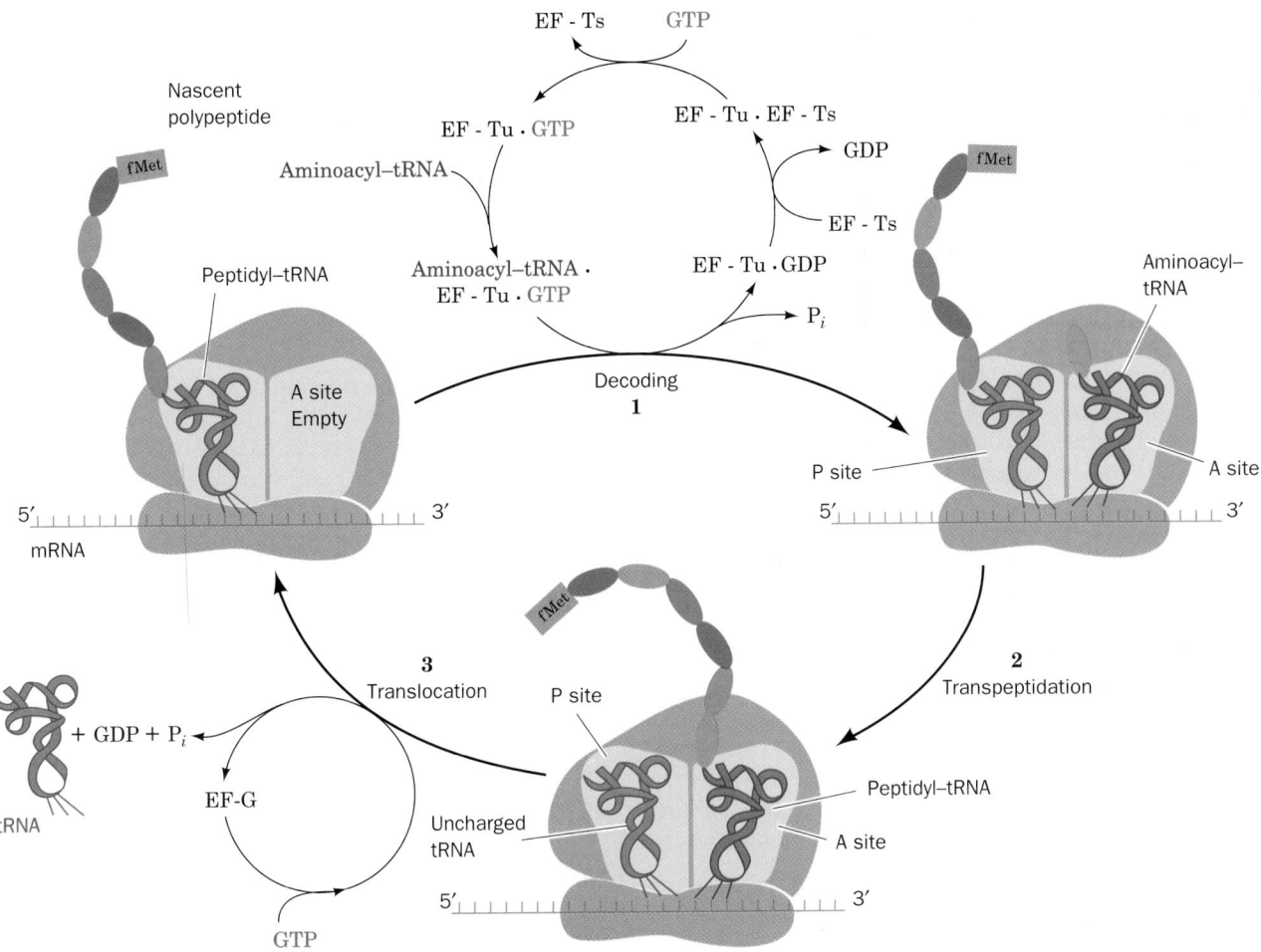

FIGURE 32-48 Elongation cycle in *E. coli* ribosomes. The E site, to which discharged tRNAs are transferred before being released to solution, is not shown. Eukaryotic elongation follows a similar cycle but EF-Tu and EF-Ts are replaced by a single multisubunit protein, eEF-1, and EF-G is replaced by eEF2.

EF-Tu and the tRNA's acceptor stem form a knoblike handle and the tRNA's anticodon helix forms the screw (Fig. 32-49). The conformation of the tRNA^{Phe} closely resembles that of the uncomplexed molecule (Fig. 32-11*b*). The EF-Tu folds into three distinct domains that are connected by flexible peptides, rather like beads on a string. The N-terminal domain 1, which binds guanine nucleotides and catalyzes GTP hydrolysis, structurally resembles other known G-proteins.

The two macromolecules associate rather tenuously via three major regions: (1) the CCA—Phe segment at the 3′ end of the Phe–tRNA^{Phe} binds in a cleft between domains 1 and 2 of the EF-Tu (red and green in Fig. 32-49) that ends in a pocket large enough to accommodate all amino acid residues; (2) the 5′-phosphate of the tRNA binds in a depression at the junction of EF-Tu's three domains; and (3) one side of the tRNA's TψC stem contacts exposed main chain and side chains of EF-Tu domain 3 (blue in Fig. 32-49). The tight association of the aminoacyl group with EF-Tu appears to greatly increase the affinity of EF-Tu for the otherwise loosely bound tRNA, which explains why EF-Tu does not bind uncharged elongator tRNAs.

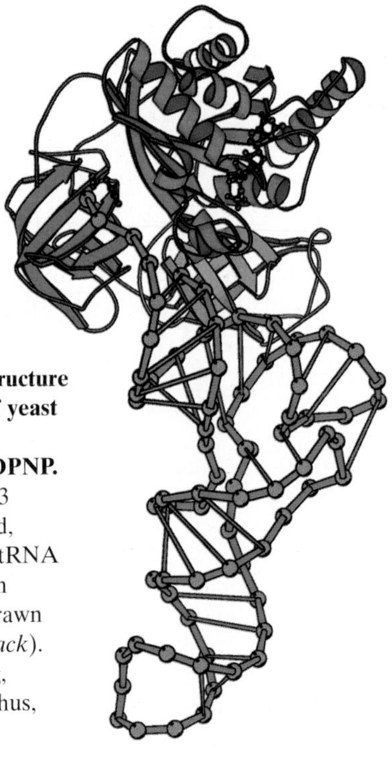

FIGURE 32-49 X-Ray structure of the ternary complex of yeast Phe–tRNA^{Phe}, *Thermus aquaticus* EF-Tu, and GDPNP. EF-Tu domains 1, 2, and 3 (N- to C-terminal) are red, green, and cyan, and the tRNA is shown in ladder form in purple. The GDPNP is drawn in ball-and-stick form (*black*). [Courtesy of Jens Nyborg, University of Aarhus, Århus, Denmark. PDBid 1TTT.]

EF-Tu binds neither formylated aminoacyl–tRNAs nor unformylated Met–tRNA$_f^{Met}$, which is why the initiator tRNA never reads internal AUG or GUG codons. The first base pair of tRNA$_f^{Met}$ is mismatched (Fig. 32-42) and hence this initiator tRNA has a 3′ overhang of 5 nt vs 4 nt for an elongator tRNA. It seems likely that this mismatch, together with the formyl group, prevents fMet–tRNA$_f^{Met}$ from binding to EF-Tu. Indeed, EF-Tu binds to *E. coli* tRNA$_f^{Met}$ whose 5′-terminal C residue has been deaminated by bisulfite treatment (Section 30-7), which reestablishes the "missing" base pair as U · A. Similarly, Sec–tRNA^{Sec}, which is also not bound by EF-Tu (but rather by SELB; Section 32-2D), has 8 bp in its acceptor stem vs 7 bp in those of other elongator tRNAs. However, initiator tRNAs from several sources have fully base paired acceptor stems, and the U1 · A72 base pair of tRNA^{Gln} is opened up on binding to GlnRS (Section 32-2C).

c. EF-Tu Undergoes a Major Conformational Change on Hydrolyzing GTP

Morten Kjeldgaard and Nyborg determined the X-ray structures of *T. aquaticus* EF-Tu (405 residues) in complex with GDPNP and the 70% identical *E. coli* EF-Tu (393 residues) in complex with GDP (Fig. 32-50). The conformation of EF-Tu in its complex with only GDPNP closely resembles that in its ternary complex with Phe–tRNA^{Phe} and GDPNP (Fig. 32-49). However, comparison of the GDPNP and GDP complexes indicates that, on hydrolyz-

ing its bound GTP, EF-Tu undergoes a major structural re-organization. Its greatest local conformation changes occur in the Switch I and Switch II regions of domain 1, which in all G-proteins signal the state of the bound nucleotide to interacting partners (Section 19-2B; here domains 2 and 3): Switch I converts from a β hairpin to a short α helix and the α helix of Switch II shifts toward the C-terminus by 4 residues. As a consequence, this latter helix reorients by 42°, which results in domain 1 rigidly changing its orientation with respect to domains 2 and 3 by a dramatic 91° rotation. The tRNA binding site is thereby eliminated.

d. EF-Ts Disrupts the Binding of GDP to EF-Tu

EF-Tu has a 100-fold higher affinity for GDP than GTP. Hence, replacement of the EF-Tu–bound GDP by GTP must be facilitated by the interaction of EF-Tu with EF-Ts (Fig. 32-48, *top*). The X-ray structure of the EF-Tu · EF-Ts complex, determined by Stephen Cusack and Reuben Leberman, reveals that the EF-Tu has a conformation resembling that of its GDP complex (Fig. 32-51) but with its domains 2 and 3 swung away from domain 1 by ~18°. EF-Ts is an elongated molecule that binds along the right side of EF-Tu as shown in Fig. 32-51, where it contacts EF-Tu's domains 1 and 3. The intrusive interactions of EF-Ts side chains with the GDP binding pocket on EF-Tu disrupts the Mg^{2+} ion binding site. This reduces the affinity of EF-Tu for GDP, thereby facilitating its exchange for GTP (after EF-Ts has dissociated), which has a 10-fold higher concentration in the cell than does GDP (the GEF-containing segment of Sos similarly interferes with Mg^{2+} binding and hence guanine nucleotide binding by Ras; Section 19-3C). EF-Tu's subsequent binding of a charged elongator tRNA increases its affinity for GTP.

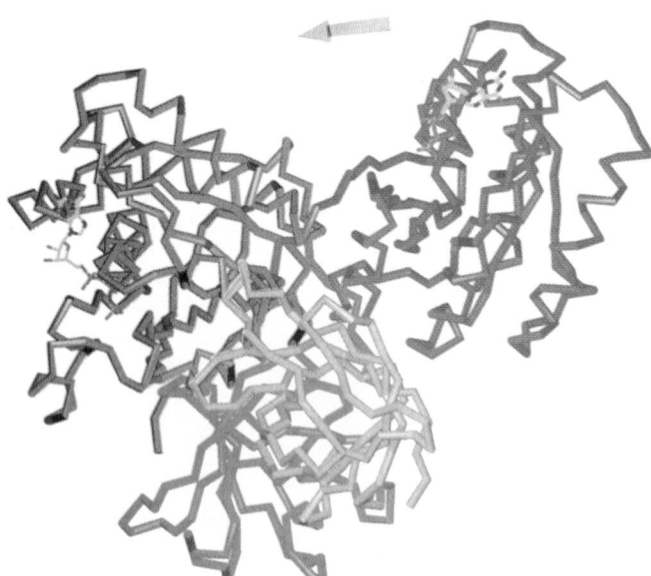

FIGURE 32-50 Comparison of the X-ray structures of EF-Tu in its complexes with GDP and GDPNP. The protein is represented by its C$_\alpha$ backbone with domain 1, its GTP-binding domain, purple in the GDP complex and red in the GDPNP complex. Domain 2 and domain 3, which have the same orientation in both complexes, are green and cyan. The bound GDP and GDPNP are shown in stick form with C yellow, N blue, O red, and P green. [Courtesy of Morten Kjeldgaard and Jens Nyborg, University of Aarhus, Århus, Denmark. PDBid 1EFT.] **◈ See the Interactive Exercises**

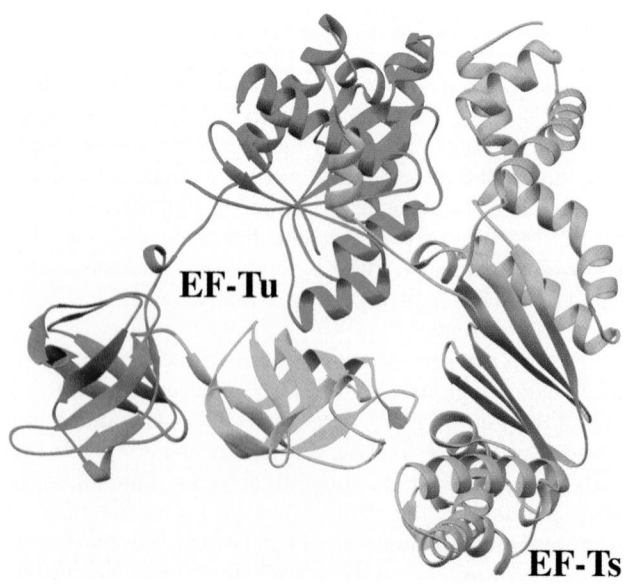

FIGURE 32-51 X-Ray structure of the *E. coli* EF-Tu · EF-Ts complex. Domains 1, 2, and 3 of EF-Tu are magenta, green, and cyan, respectively, and EF-Ts is orange. [Based on an X-ray structure by Stephen Cusack and Reuben Leberman, EMBL, Grenoble Cedex, France. PDBid 1EFU.]

e. Transpeptidation

In the transpeptidation stage of the elongation cycle (Fig. 32-48), the peptide bond is formed through the nucleophilic displacement of the P-site tRNA by the amino group of the 3′-linked aminoacyl–tRNA in the A site (Fig 32-41). The nascent polypeptide chain is thereby lengthened at its C terminus by one residue and transferred to the A-site tRNA. The reaction occurs without the need of activating cofactors such as ATP because the ester linkage between the nascent polypeptide and the P-site tRNA is a "high-energy" bond. The **peptidyl transferase** center that catalyzes peptide bond formation is located entirely on the large subunit as is demonstrated by the observation that in high concentrations of organic solvents such as ethanol, the large subunit alone catalyzes peptide bond formation. The organic solvent apparently distorts the large subunit in a way that mimics the effect of small subunit binding.

f. Puromycin Is an Aminoacyl–tRNA Analog

The ribosomal elongation cycle was originally characterized through the use of the antibiotic **puromycin** (Fig. 32-52). This product of *Streptomyces alboniger*, which resembles the 3′ end of Tyr–tRNA, causes the premature termination of polypeptide chain synthesis. Puromycin, in competition with the mRNA-specified aminoacyl–tRNA but without the need of elongation factors, binds to the ribosomal A site which, in turn, catalyzes a normal transpeptidation reaction to form peptidyl–puromycin. Yet, the ribosome cannot catalyze the transpeptidation reaction in the next elongation cycle because puromycin's "amino acid residue" is linked to its "tRNA" via an amide rather than an ester bond. Polypeptide synthesis is therefore aborted and the peptidyl–puromycin is released.

In the absence of the elongation factor EF-G (see below), an active ribosome cannot bind puromycin because its A site is at least partially occupied by a peptidyl–tRNA. A newly initiated ribosome, however, violates this rule; it catalyzes fMet–puromycin formation. *These observations* demonstrated the functional existence of the ribosomal P and A sites and established that fMet–tRNA$_f^{Met}$ binds directly to the P site, whereas other aminoacyl–tRNAs must first enter the A site.

g. The Ribosome Is a Ribozyme

What is the nature of the peptidyl transferase center, that is, does it consist of RNA, protein, or both? Since all proteins, including those associated with ribosomes, are ribosomally synthesized, the primordial ribosome must have preceded the primordial proteins and hence consisted entirely of RNA. Despite this (in hindsight) obvious evolutionary argument, the idea that rRNA functions catalytically was not seriously entertained until after it had been discovered that RNA can, in fact, act as a catalyst (Section 31-4A). Several other observations further indicate that the ribosome is a ribozyme:

1. The absence of any of all but three ribosomal proteins (L2, L3, and L4) from the 50S subunit does not abolish peptidyl transferase function.

2. rRNAs are more highly conserved throughout evolution than are ribosomal proteins.

3. Most mutations that confer resistance to antibiotics that inhibit protein synthesis occur in genes encoding rRNAs rather than ribosomal proteins.

Nevertheless, the unambiguous demonstration that rRNA functions catalytically in polypeptide synthesis proved to be surprisingly elusive. Noller succeeded in showing that the *T. thermophilus* large ribosomal subunit from which ~95% of the protein had been removed by treatment with SDS and **proteinase K** followed by phenol extraction (which denatures proteins; Section 6-6A) maintained >80% of its peptidyl transferase activity in a model reaction. Moreover, this activity was abolished by RNase treatment. However, since the remaining protein was due to several intact ribosomal proteins (which are presumably

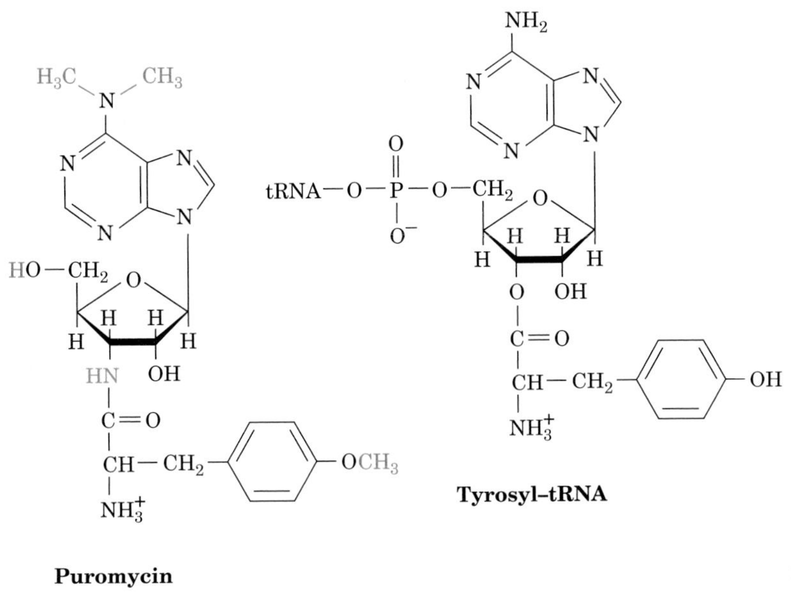

FIGURE 32-52 Puromycin. This antibiotic (*left*) resembles the 3′-terminus of tyrosyl–tRNA (*right*).

Puromycin

Tyrosyl–tRNA

sequestered within the 23S RNA), it could be argued that these proteins are essential for ribosomal catalytic function, a reasonable expectation in light of the >3.5 billion years that ribosomal proteins and RNAs have coevolved.

The nature of the peptidyl transferase center was unequivocally revealed through its identification in the X-ray structure of the 50S subunit. Peptide bond formation presumably resembles the reverse of peptide bond hydrolysis such as that catalyzed by serine proteases (Section 15-3C). The ribosomal reaction's tetrahedral intermediate (Fig. 32-53a) is mimicked by a compound synthesized by Michael Yarus that consists of the trinucleotide CCdA linked to puromycin via a phosphoramidite group (Fig. 32-53b). This compound, which is named **CCdA-p-Puro,** binds tightly to the ribosome so as to inhibit its peptidyl transferse activity. The X-ray structure of the 50S subunit in complex with CCdA-p-Puro reveals that the inhibitor binds at the bottom of a deep cleft (Fig. 32-36a) at the entrance to the 100-Å-long polypeptide exit tunnel that runs through to the back of the subunit (Fig. 12-50b). There, *the inhibitor is completely enveloped in RNA with no protein side chain approaching closer than ~18 Å to the inhibitor's phosphoramidite group.* Moreover, all the nucleotides that contact the CCdA-p-Puro are >95% conserved among all three kingdoms of life. Clearly, *the ribosomal transpeptidase reaction is catalyzed by RNA.*

How does the ribosome catalyze the transpeptidase reaction? Certainly the ribosome's greatest catalytic influence, as is true of all enzymes, is that it correctly positions its substrates for reaction. In addition, inspection of the peptidyl transferase active site reveals that the only acid–base group within 5 Å of the bound CCdA-p-Puro's phosphoramidite group is atom N3 of the invariant rRNA base A2486 (A2451 in *E. coli*). It is ~3 Å from and hence hydrogen bonded to the phosphoramidite oxygen that corresponds to the oxyanion of the tetrahedral intermediate. Moore and Steitz therefore postulated that A2486-N3 acts as a general base in the peptidyl transferase reaction (Fig. 32-54). This would require A2486-N3 to have a pK of at least 7 (recall that proton transfers between hydrogen bonded groups only occur at physiologically significant rates when the pK of the proton acceptor is not less than 2 or 3 pH units below that of the proton donor; Section 15-3D). Yet, the pK of N3 in AMP is <3.5. However, since the pH of the 50S subunit crystals is 5.8, A2486-N3 could only be hydrogen bonded to the phosphoramidite oxygen if its pK is >6. Moreover, kinetic investigations of 70S ribosomes have identified a titratable group with a pK of 7.5 that affects catalysis and which disappears when A2486 is mutated to U, a mutation that reduces the rate of peptide bond formation by 130-fold. Apparently, the pK of A2486-N3 is greatly increased by its ribosomal environment. It is postulated that this occurs via a charge relay system in which the phosphate group of A2485, one of the most solvent-inaccessible phosphate groups in the 50S subunit, electrostatically increases the pK of A2486-N3 through a hydrogen bonded network involving the highly conserved G2482 (Fig. 32-55).

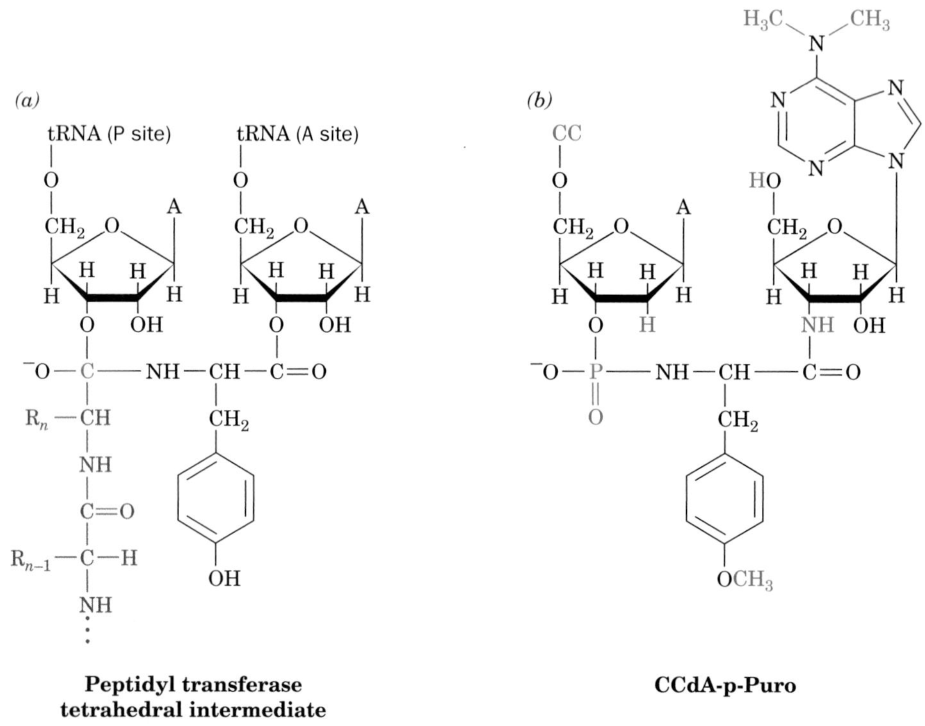

(a)

**Peptidyl transferase
tetrahedral intermediate**

(b)

CCdA-p-Puro

FIGURE 32-53   Ribosomal tetrahedral intermediate and its analog. (*a*) The chemical structure of the tetrahedral intermediate (*red C*) in ribosomally mediated peptide bond formation in which the A-site aminoacyl residue is Tyr.

(*b*) CCdA-p-Puro, the transition state analog of the tetrahedral intermediate in Part *a* produced by linking the 3′-OH group of CCdA to the amino group of puromycin's *O*-methyltyrosine residue via a phosphoryl group.

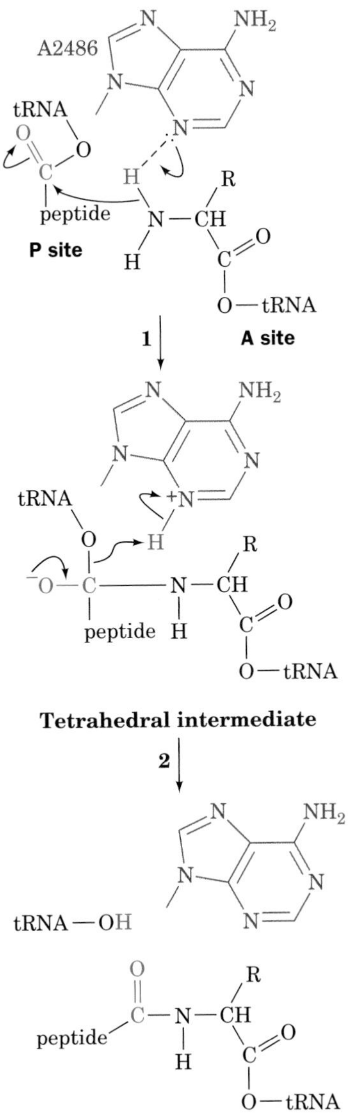

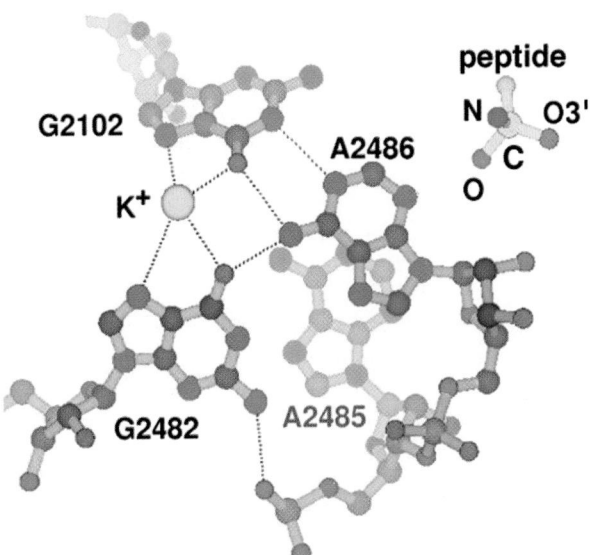

FIGURE 32-55 Catalytic apparatus of the peptidyl transferase active site. Atoms are colored according to type with C gray, N blue, O red, and P magenta. A K⁺ ion is drawn as a yellow sphere and hydrogen bonds are represented by dotted lines. Note that the buried phosphate group of A2485 is hydrogen bonded to the base of G2482, which in turn is hydrogen bonded to the base of A2486. It is these interactions that are proposed to relay the electronic charge from the anionic phosphate group to A2486-N3, thereby greatly increasing the latter atom's pK and enabling it to function as a general base. [Courtesy of Peter Moore and Thomas Steitz, Yale University.]

P site

A site

1

Tetrahedral intermediate

2

FIGURE 32-54 Proposed mechanism of ribosomal peptide synthesis. The reaction is catalyzed, in part, through the abstraction of a proton from the attacking amino group by A2486-N3 (general base catalysis) in Step 1 and its donation of this proton to the tetrahedral intermediate (general acid catalysis) in Step 2.

The X-ray structures of the 50S subunit in its complexes with several substrate, reaction intermediate, and product analogs have led to a reaction model that begins with the reactant complex shown in Fig. 32-56. Here, the 2′ O atom of the P-site tRNA's A76 acts to properly orient the attacking amino group via a hydrogen bond. Indeed, a P-site substrate with a 2′-deoxyribose residue at A76 is unreactive (which is why the CCdA-p-Puro inhibitor was synthesized with dA rather than A). However, this model also indicates that the reaction intermediate's oxyanion points away from and hence is not stabilized by A2486-N3 (Fig. 32-54; the hydrogen bond between A2486-N3 and the phosphoramidite oxyanion of CCdA-p-Puro occurs because this inhibitor assumes a conformation in the 50S subunit that would be sterically prevented if its dA residue

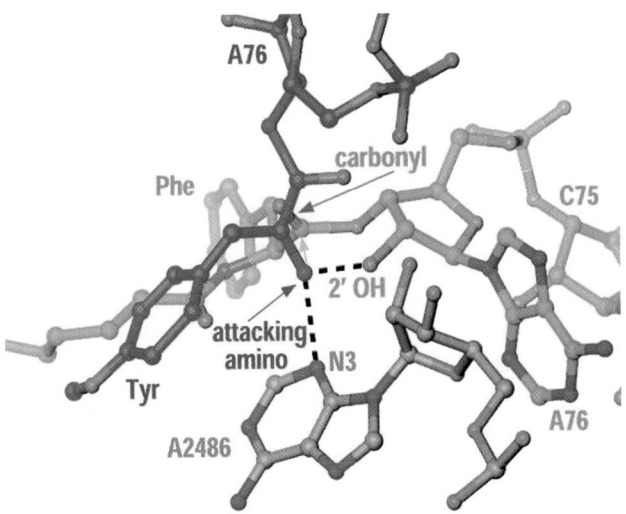

FIGURE 32-56 Model of the substrate complex of the 50S ribosomal subunit. Atoms are colored according to type with the A-site substrate C and P purple, P-site substrate C and P green, 23S rRNA C and P orange, N blue, and O red. The attacking amino group of the A-site aminoacyl residue is held in position for nucleophilic attack (*cyan arrow*) on the carbonyl C of the P-site aminoacyl ester through hydrogen bonds (*dashed black lines*) to A2486-N3 and the 2′-O of the P-site A76. [Courtesy of Peter Moore and Thomas Steitz, Yale University.]

were replaced by ribo-A, as occurs in all tRNAs). Thus, considering the ribosome's apparent conformational flexibility and the many pitfalls in identifying an enzymatic group responsible for an observed pK, the foregoing catalytic model should be treated with caution.

h. Translocation

In the translocation stage of the elongation cycle, the now uncharged P-site tRNA (at first tRNA$_f^{Met}$ but subsequently an elongator tRNA) is transferred to the E site (not shown in Fig. 32-48), its former occupant having been previously expelled (see below). Simultaneously, the peptidyl–tRNA in the A site, together with its bound mRNA, is moved to the P site. This prepares the ribosome for the next elongation cycle. The maintenance of the peptidyl–tRNA's codon–anticodon association is no longer necessary for amino acid specification. Rather, it probably acts as a place-keeper that permits the ribosome to precisely step off the three nucleotides along the mRNA required to preserve the reading frame. Indeed, the observation that frameshift suppressor tRNAs induce a four-nucleotide translocation (Section 32-2E) indicates that mRNA movement is directly coupled to tRNA movement.

i. EF-G Structurally Mimics the EF-Ts · tRNA Complex

The translocation process requires the participation of elongation factor **EF-G** (also called **EF2**), which binds to the ribosome together with GTP and is only released on hydrolysis of the GTP to GDP + P$_i$ (Fig. 32-48). EF-G release is a prerequisite for beginning the next elongation cycle because EF-G and EF-Tu bind to the same site of the ribosome and hence their binding is mutually exclusive.

The X-ray structure of *Thermus thermophilus* EF-G · GDP, determined by Steitz and Moore, reveals a tadpole-shaped monomeric protein that consists of five domains (Fig. 32-57). The first two domains closely resemble those in EF-Tu · GDPNP rather than those in EF-Tu · GDP (Fig. 32-50). This, it is argued, is because the two elongation factors have reciprocal functions with EF-Tu · GTP facilitating the conversion of the ribosome from its post- to its pretranslocational state and EF-G · GTP promoting the reverse transition. This idea is supported by the intriguing observation that the Phe–tRNA^{Phe} · EF-Tu · GDPNP and EF-G · GDP complexes are almost identical in appearance: EF-G's three C-terminal domains (magenta in Fig. 32-57), which have no counterparts in EF-Tu, closely resemble the EF-Tu–bound tRNA in shape, a remarkable case of **macromolecular mimicry.**

EF-G is unusual among G-proteins in that it has no corresponding GEF. However, its N-terminal guanine nucleotide-binding domain contains a unique α helical insert (dark blue in Fig. 32-57) that contacts the domain's conserved core at sites analogous to those in EF-Tu that interact with EF-Ts. This suggests that this subdomain acts as an internal GEF.

j. Translocation Occurs via Intermediate States

Chemical footprinting studies (Section 31-2A) by Noller revealed that certain bases in the 16S rRNA are protected

FIGURE 32-57 X-Ray structure of EF-G in complex with GDP. Domain 1 is red with its α helical insert dark blue, domain 2 is green, and domains 3, 4, and 5 are magenta. The GDP is drawn in ball-and-stick form (*black*). A 25-residue segment of EF-G's domain 1 is not visible in the X-ray structure and domain 3 is poorly defined, thereby accounting for the several peptide segments seen in this model. Note the remarkable resemblance in shape between this structure and that of Phe–tRNA^{Phe} · EF-Tu · GDPNP (Fig. 32-49). [Courtesy of Jens Nyborg, University of Aarhus, Århus, Denmark. Based on an X-ray structure by Thomas Steitz and Peter Moore, Yale University. PDBid 2EFG.]

by tRNAs bound in the ribosomal A and P-sites and that certain bases in the 23S rRNA are protected by tRNAs in the A, P, and E sites. Almost all of these protected bases are absolutely conserved in evolution and many of them have been implicated in ribosomal function through biochemical or genetic studies.

Variations in chemical footprinting patterns during the elongation cycle together with the more recently determined X-ray and cryo-EM studies indicate that the translocation of tRNA occurs in several discrete steps (Fig. 32-58):

1. Let us start with the ribosome in its **posttranslocational state**: a deacylated tRNA bound to the E subsites of both the 30S and 50S subunits (the E/E binding state), a peptidyl–tRNA bound in the P subsites of both subunits (the P/P state), and the A site empty. An aminoacyl–tRNA (aa–tRNA) in ternary complex with EF-Tu and GTP binds to the A site accompanied by the release of the E-site tRNA (but see below). This yields a complex in which the incoming aa–tRNA is bound in the 30S subunit's A subsite via a codon–anticodon interaction (recall that the mRNA is bound to the 30S subunit) but with the EF-Tu

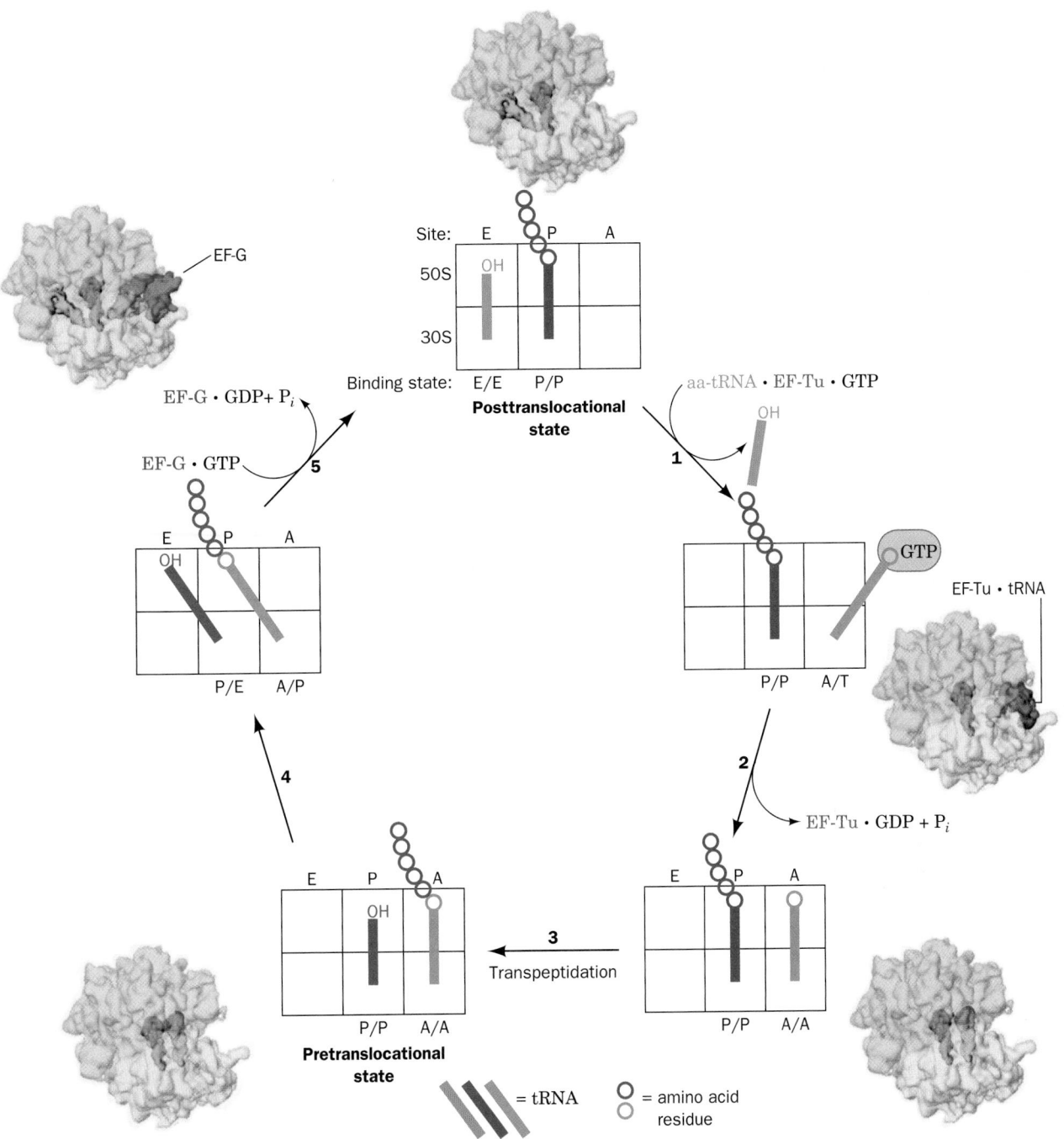

FIGURE 32-58 Ribosomal binding states in the elongation cycle. Note how this scheme elaborates the classical elongation cycle diagrammed in Fig. 32-48. The drawings are accompanied by 17-Å-resolution cryo-EM–based images of the *E. coli* 70S ribosome in the corresponding binding states in which the 30S subunit is transparent yellow, the 50S subunit is transparent blue, and the tRNAs and elongation factors are colored as in the drawing they accompany. [Cryo-EM images courtesy of Knud Nierhaus, Max-Planck-Institut für Molekulare Genetik, Berlin, Germany, and Joachim Frank, Wadsworth Center, State University of New York at Albany.]

preventing the entry of the tRNA's aminoacyl end into the 50S subunit's A subsite, an arrangement termed the A/T state (T for EF-*T*u).

2. EF-Tu hydrolyzes its bound GTP to GDP + P_i and is released from the ribosome, permitting the aa–tRNA to fully bind to the A site (the A/A state), a process called **accommodation.**

3. The transpeptidation reaction occurs, yielding the **pretranslocational state.**

4. The acceptor end of the new peptidyl–tRNA shifts from the A subsite of the 50S subunit to its P-subsite, while the tRNA's anticodon end remains associated with the A subsite of the 30S subunit (yielding the A/P hybrid binding state). The acceptor end of the newly deacylated tRNA simultaneously moves from the P subsite to the E subsite of the 50S subunit while its anticodon end remains associated with the P subsite of the 30S subunit (the P/E state).

5. The ribosomal binding of the EF-G · GTP complex and the subsequent GTP hydrolysis impel the anticodon ends of these tRNAs, together with their bound mRNA, to move relative to the small ribosomal subunit such that the peptidyl–tRNA assumes the P/P state and the deacylated tRNA assumes the E/E state (the posttranslocational state), thereby completing the elongation cycle

The binding of tRNA to the A and E sites, as Nierhaus has shown, exhibits negative allosteric cooperativity. In the pretranslocational state, the E site binds the newly deacylated tRNA with high affinity, whereas the empty A site has low affinity for aminoacyl–tRNA. However, in the posttranslocational state, the ribosome has undergone a conformational change that converts the A site to a high-affinity state and the E site to a low-affinity state, which consequently releases the deacylated tRNA when aa–tRNA · EF-Tu · GTP binds to the A site. Thus, the E site is not simply a passive holding site for spent tRNAs but performs an essential function in the translation process. The GTP hydrolysis by the elongation factors EF-Tu and EF-G as well as the transpeptidation reaction apparently function to reduce the activation barriers between these conformational states. The unidirectional A → P → E flow of tRNAs through the ribosome is thereby facilitated.

Certain aspects of the foregoing mechanism are not fully resolved. Thus, Noller has proposed that the E-site tRNA is not released from the ribosome until Step 4 of Fig. 32-58, when it is displaced by the preceding tRNA. This latter model is largely based on the X-ray structure of the 70S ribosome in complex with three tRNAs (Figs. 32-35 and 32-36). However, Nierhaus and Frank argue that this complex was crystallized in the presence of an unphysiologically high tRNA concentration and, since the tRNA sites are only partially occupied, the X-ray structure is likely to be a superposition of different tRNA binding states (e.g., E/E + P/P and P/P + A/A). Whatever the case, it is clear that the changes in binding states result in large scale tRNA movements, in some instances >50 Å. Moreover, cryo-EM studies indicate that on binding EF-G · GDP(CH₂)P (like GDPNP but with a CH₂ group rather than an NH group bridging its β and γ phosphates), the 30S subunit rotates with respect to the 50S subunit by 6° clockwise when viewed from the 30S subunit's solvent side, which results in a maximum displacement of ~19 Å at the periphery of the ribosome. This rotation is accompanied by many smaller conformational changes in both subunits, particularly in the regions about the entrance and exit to the mRNA channel. Clearly, we are only beginning to understand how the ribosome works at the molecular level.

k. The Eukaryotic Elongation Cycle Resembles That of Prokaryotes

The eukaryotic elongation cycle closely resembles that of prokaryotes. In eukaryotes, the functions of EF-Tu and EF-Ts are respectively assumed by the eukaryotic elongation

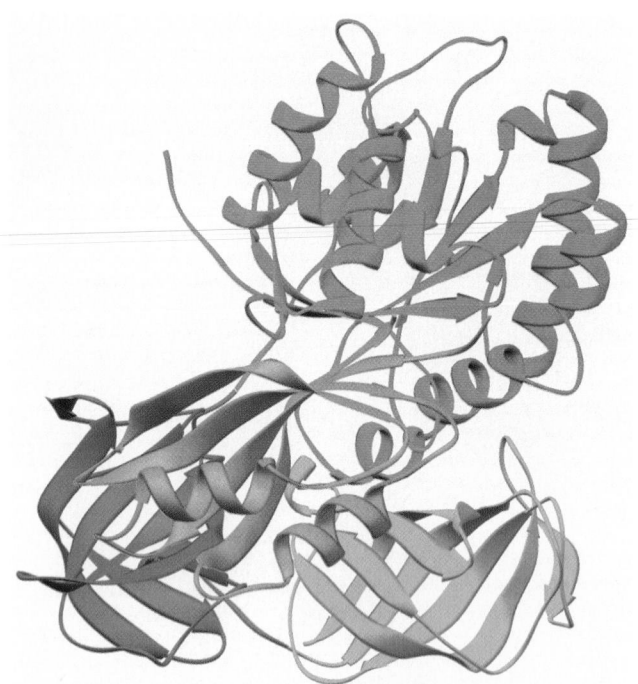

FIGURE 32-59   X-Ray structure of yeast eEF1A · eEF1Bα Domains 1, 2, and 3 of eEF1A are magenta, green, and cyan, respectively, and eEF1Bα is orange. The complex is oriented so as to emphasize the structural resemblance between eEF1A and the similarly colored EF-Tu in its complex with EF-Ts (Fig. 32-51). Note the lack of resemblance between eEF1Bα and EF-Ts. [Based on an X-ray structure by Morten Kjeldgaard and Jens Nyborg, University of Aarhus, Århus, Denmark. PDBid 1F60.]

factors **eEF1A** and **eEF1B,** with yeast eEF1B consisting of two subunits: **eEF1Bα,** which catalyzes nucleotide exchange, and **eEF1Bγ,** which has unknown function (in higher eukaryotes, eEF1B contains a third subunit, **eEF1Bβ,** that possesses a nucleotide exchange activity similar to that of eEF1Bα). Likewise, **eEF2** functions in a manner analogous to EF-G. However, the corresponding eukaryotic and prokaryotic elongation factors are not interchangeable.

The X-ray structure of yeast eEF1A · eEF1Bα (Fig. 32-59), determined by Kjeldgaard and Nyborg, reveals that eEF1A structurally resembles the homologous EF-Tu (Fig. 32-51), whereas eEF1Bα exhibits no resemblance to EF-Ts, either in sequence or in structure. Nevertheless, eEF1Bα functionally interacts with eEF1A much as EF-Ts interacts with EF-Tu: Both GEFs associate with their corresponding G-protein so as to disrupt the Mg^{2+} binding site associated with its bound guanine nucleotide.

E. *Chain Termination*

Polypeptide synthesis under the direction of synthetic mRNAs such as poly(U) terminates with a peptidyl–tRNA in association with the ribosome. However, *the translation*

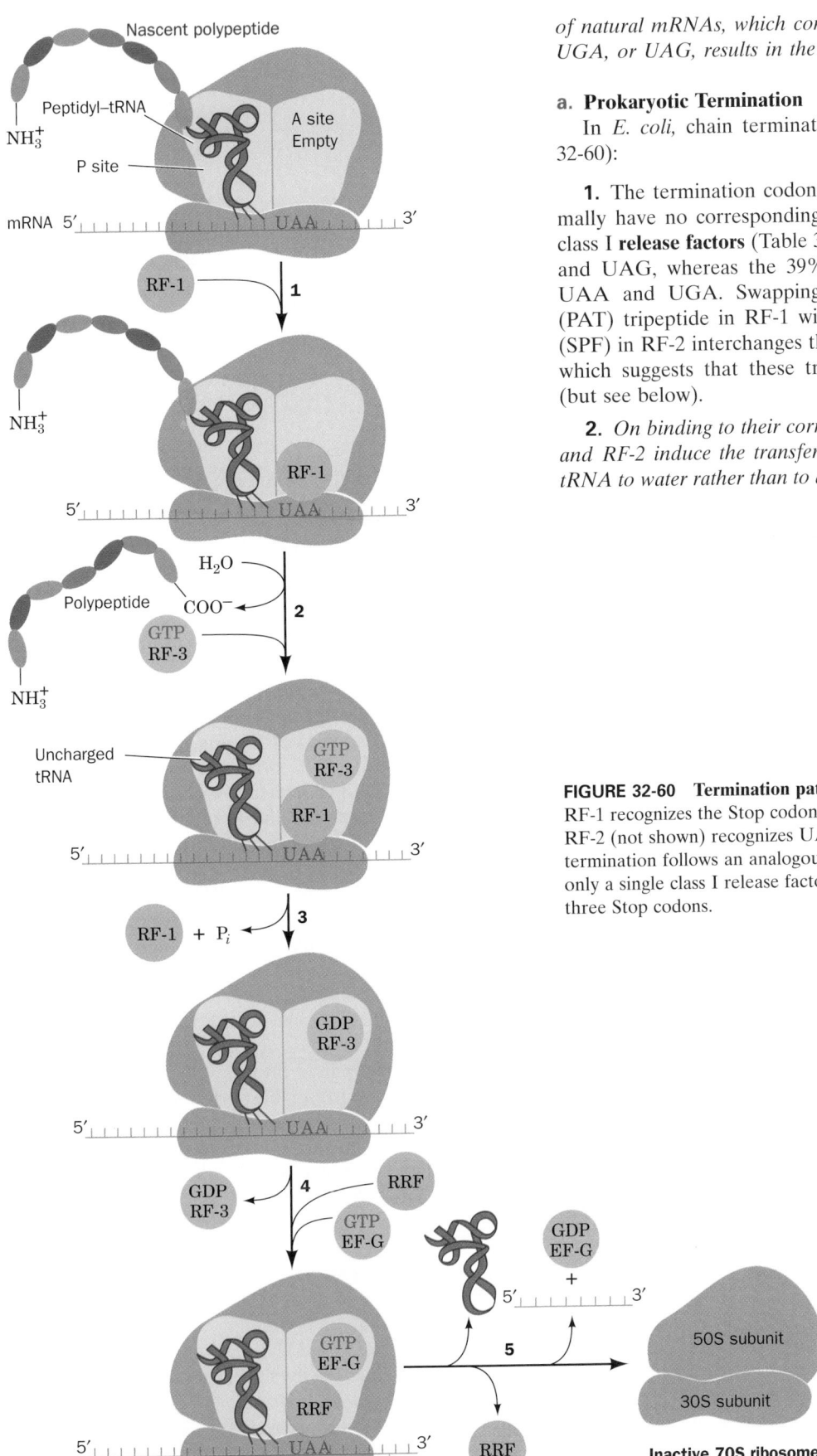

of natural mRNAs, which contain the Stop codons UAA, UGA, or UAG, results in the release of free polypeptides.

a. Prokaryotic Termination

In *E. coli*, chain termination has several stages (Fig. 32-60):

1. The termination codons, the only codons that normally have no corresponding tRNAs, are recognized by class I **release factors** (Table 32-9): **RF-1** recognizes UAA and UAG, whereas the 39% identical **RF-2** recognizes UAA and UGA. Swapping a conserved Pro-Ala-Thr (PAT) tripeptide in RF-1 with a conserved Ser-Pro-Phe (SPF) in RF-2 interchanges their Stop codon specificities, which suggests that these tripeptides mimic anticodons (but see below).

2. *On binding to their corresponding Stop codon, RF-1 and RF-2 induce the transfer of the peptidyl group from tRNA to water rather than to an aminoacyl–tRNA, thereby*

FIGURE 32-60 Termination pathway in *E. coli* ribosomes. RF-1 recognizes the Stop codons UAA and UAG, whereas RF-2 (not shown) recognizes UAA and UGA. Eukaryotic termination follows an analogous pathway but requires only a single class I release factor, eRF1, that recognizes all three Stop codons.

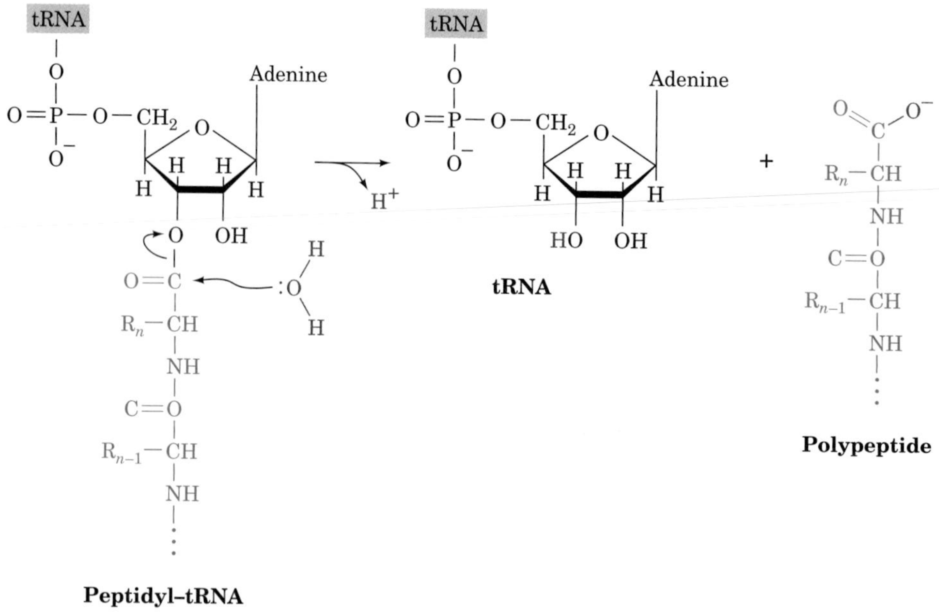

FIGURE 32-61   Ribosome-catalyzed hydrolysis of peptidyl–tRNA to form a polypeptide and free tRNA.

releasing the completed polypeptide (Fig 32-61). The class I release factors act at the ribosomal A site as is indicated by the observations that they compete with suppressor tRNAs for termination codons and that they cannot bind to the ribosome simultaneously with EF-G.

3. Once the newly synthesized polypeptide has been released from the ribosome, the class II release factor, **RF-3,** a G-protein, binds to the ribosome in complex with GTP and, on hydrolyzing its bound GTP to GDP + P_i, induces the ribosome to release its bound class I release factor. RF-3 is not required for cell viability although it is necessary for maximum growth rate.

4. Ribosomal recycling factor (RRF) binds in the ribosomal A site followed by EF-G · GTP. RRF, which was characterized largely by Akira Kaji, is essential for cell viability.

5. EF-G hydrolyzes its bound GTP, which causes RRF to be translocated to the P site and the tRNAs previously in the P and E sites (the latter not shown in Fig. 32-60) to be released. Finally, the RRF and then the EF-G · GDP and mRNA are released, yielding an inactive 70S ribosome ready for reinitiation (Fig. 32-45).

b. Eukaryotic Termination

Chain termination in eukaryotes resembles that in prokaryotes, but has only one class I release factor, **eRF1,** that recognizes all three Stop codons. It is unrelated in sequence to RF-1 and RF-2. However, the eukaryotic class II release factor, **eRF3,** resembles RF-3 in both sequence and function. Nevertheless, eRF3 is essential for eukaryotic cell viability.

c. RF-2, eRF1, and RRF Are Structural but Not Functional Mimics of tRNA

The class I release factors' functional mimicry of tRNAs suggests that they structurally mimic tRNAs. Indeed the X-ray structure of human eRF1 (Fig. 32-62*a*), determined by David Barford, and *E. coli* RF-2 (Fig. 32-62*b*), determined by Richard Buckingham, Nyborg, and Kjeldgaard, indicate that these two unrelated proteins both structurally resemble tRNA. However, the SPF tripeptide of RF-2 that was postulated to be an anticodon mimic (see above) is located in a different domain of RF-2 than the putative anticodon loop mimic (Fig. 32-62*b*). Moreover, cryo-EM studies of the *E. coli* ribosome in complex with RF2, independently carried out by Frank and Marin van Heel, indicate that RF2 undergoes a large conformational change on binding to the ribosome so that it no longer mimics tRNA. The X-ray structure of *Thermatoga maritima* RRF (Fig. 32-62*c*), determined by Kaji and Anders Liljas, reveals that it too resembles tRNA in appearance. Nevertheless, footprinting studies of RRF bound to the 70S ribosome indicate that RRF binds to the A site in an orientation that differs markedly from any previously observed for a tRNA. Evidently, things are not necessarily what they appear to be.

d. GTP Hydrolysis Speeds Up Ribosomal Processes

What is the role of the GTP hydrolysis reactions mediated by the various GTP-binding factors (IF-2, EF-Tu, EF-G, and RF-3 in *E. coli*)? Translation occurs in the absence of GTP, albeit slowly, so that the free energy of the transpeptidation reaction is sufficient to drive the entire translational process. Moreover, none of the GTP hydrolysis reactions yields a "high-energy" covalent intermediate as does, say, ATP hydrolysis in numerous biosynthetic reactions. It therefore appears that the ribosomal binding of

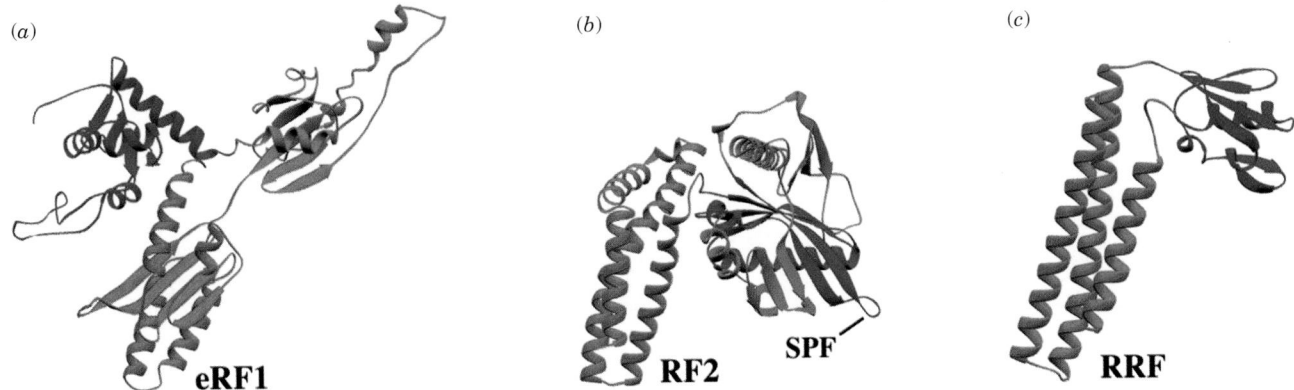

FIGURE 32-62   X-Ray structures of putative tRNA mimics that participate in translational termination. (*a*) Human eRF1, (*b*) *E. coli* RF-2, and (*c*) *T. maritima* RRF. The various domains in these proteins are differently colored with the domains that appear to mimic the tRNA anticodon stem drawn in red. The position of the RF-2 SPF tripeptide is indicated. Compare these structures to Figs. 32-49 and 32-57. [Courtesy of V. Ramakrishnan, MRC Laboratory of Molecular Biology, Cambridge, U.K. Part *a* is based on an X-ray structure by David Barford, Institute of Cancer Research, London, U.K.; Part *b* is based on an X-ray structure by Richard Buckingham, CNRS, Paris, France, and Jens Nyborg and Morten Kjeldgaard, University of Aarhus, Århus, Denmark; and Part *c* is based on an X-ray structure by Akira Kaji, University of Pennsylvania, and Anders Liljas, Lund University, Lund, Sweden. PDBids (*a*) 1DT9, (*b*) 1GQE, and (*c*) 1DD5.]

a GTP-binding factor (G-protein) in complex with GTP allosterically facilitates a change in ribosomal conformation so as to permit a particular process such as tRNA binding to occur (e.g., Fig. 32-50). This conformation change also catalyzes GTP hydrolysis, which, in turn, permits the ribosome to relax to its initial conformation with the concomitant release of products, including GDP + P_i (with the aid of EF-Ts for EF-Tu). *The high rate and irreversibility of the GTP hydrolysis reaction therefore ensures that the various complex ribosomal processes to which it is coupled, initiation, elongation, and termination, will themselves be fast and irreversible.* In essence, G-protein · GTP complexes act as Maxwell's demons to trap the ribosome in functionally productive conformations. Hence, the ribosome utilizes the free energy of GTP hydrolysis to gain a more ordered (lower entropy) state rather than a higher energy state as often occurs in ATP-dependent processes. In the same way, GTP hydrolysis facilitates translational accuracy (see below).

F. Translational Accuracy

The genetic code is normally expressed with remarkable fidelity. We have already seen that transcription and tRNA aminoacylation both proceed with high accuracy (Sections 31-2C and 32-2C). The accuracy of ribosomal mRNA decoding was estimated from the rate of misincorporation of [^{35}S]Cys into highly purified **flagellin,** an *E. coli* protein (Section 35-3G) that normally lacks Cys. These measurements indicated that the mistranslation rate is ~10^{-4} errors per codon. This rate is greatly increased in the presence of **streptomycin,** an antibiotic that increases the rate of ribosomal misreading (Section 32-3G). From the types of reading errors that streptomycin is known to induce, it was concluded that the mistranslation arose almost entirely from the confusion of the Arg codons CGU and CGC for the Cys codons UGU and UGC. The above error rate is therefore largely caused by mistakes in ribosomal decoding.

An aminoacyl–tRNA is selected by the ribosome only according to its anticodon. Yet the binding energy loss arising from a single base mismatch in a codon–anticodon interaction is estimated to be ~12 kJ · mol^{-1}, which, according to Eq. [32.1], cannot account for a ribosomal decoding accuracy of less than ~10^{-2} errors per codon. Moreover, the base pairing interaction between the UUU codon for Phe and the GAA anticodon of tRNA^{Phe} would be naively expected to be less stable than the incorrect pairing between the UGC codon for Ser and the GCG anticodon of tRNA^{Arg}. This is because both interactions have one G · U base pair and the former correct interaction's remaining two A · U base pairs are weaker than the latter incorrect interaction's remaining two G · C base pairs. Evidently, the ribosome has some sort of proofreading mechanism that increases its overall decoding accuracy.

a. The Ribosome Monitors the Formation of a Correct Codon–Anticodon Complex

As we have seen (Fig. 32-58), the aminoacyl–tRNA · EF-Tu · GTP ternary complex initially binds to the ribosome with the tRNA in the A/T binding state. The tRNA only assumes the fully bound A/A state (accommodation) after the GTP has been hydrolyzed and the EF-G · GDP complex has been released from the ribosome. These two states presumably permit the ribosome to double-check (proofread) the codon–anticodon complex that the mRNA makes with the incoming tRNA.

The X-ray structure of the *T. thermophilus* 30S subunit in complex with a U_6 hexanucleotide mRNA and a 17-nt RNA consisting of the tRNA^{Phe} anticodon stem–loop (Fig. 32-11, although its nucleotides are unmodified), deter-

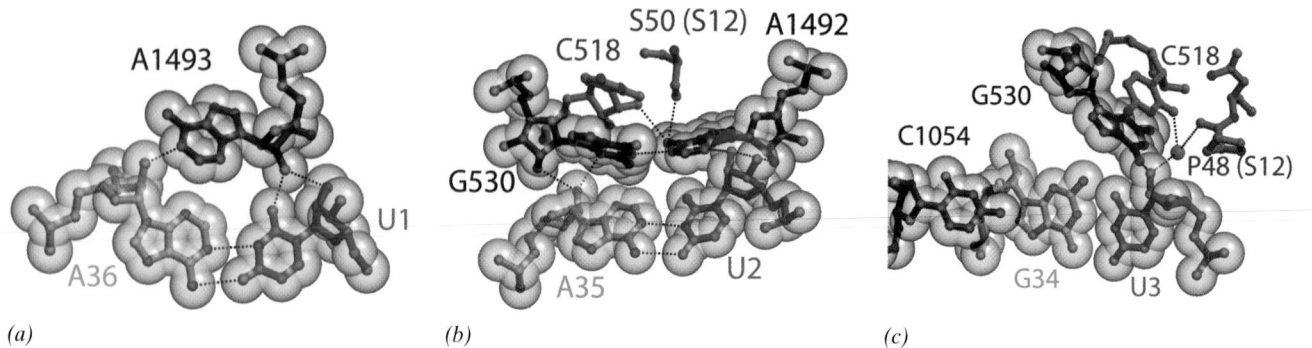

(a)   (b)   (c)

FIGURE 32-63 Codon–anticodon interactions in the ribosome. The (a) first, (b) second, and (c) third codon–anticodon base pairs as seen in the X-ray structure of the *T. thermophilus* 30S subunit in complex with U$_6$ (a model mRNA) and the 17-nt anticodon stem–loop of tRNA^{Phe} (whose anticodon is GAA). The structures are drawn in ball-and-stick form embedded in their semitransparent van der Waals surfaces. Codons are purple, anticodons are yellow, and rRNA is brown or gray with non-C atoms colored according to type (N blue, O red, and P green). Protein segments are gray and Mg^{2+} ions are represented by magenta spheres. [Courtesy of V. Ramakrishnan, MRC Laboratory of Molecular Biology, Cambridge, U.K. PDBid 1IBM.]

mined by Ramakrishnan, reveals how an mRNA-specified tRNA initially binds to the ribosome. The codon–anticodon association is stabilized by its interactions with three universally conserved ribosomal bases, A1492, A1493, and G530 (Fig. 32-63):

1. The first codon–anticodon base pair, that between mRNA U1 and tRNA A36, is stabilized by the binding of the rRNA A1493 base in the base pair's minor groove (Fig. 32-63a).

2. The second codon–anticodon base pair, that between U2 and A35, is bolstered by A1492 and G530, which both bind in this base pair's minor groove (Fig. 32-63b).

3. The third codon–anticodon base pair (the wobble pair; Section 32-2D), that between U3 and G34, is reinforced through minor groove binding by G530 (Fig. 32-63c). This latter interaction appears to be less stringent than those in the first and second codon–anticodon positions, which is consistent with the need for the third codon–anticodon pairing to tolerate non-Watson–Crick base pairs (Section 32-2D).

Comparison of this structure with that of the 30S subunit alone reveals that the foregoing rRNA nucleotides undergo conformational changes on the formation of a codon–anticodon complex (Fig. 32-64). In the absence of tRNA, the bases of A1492 and A1493 stack in the interior of an RNA loop but flip out of this loop to form the codon–anticodon complex, whereas the G530 base switches from the syn to the anti conformation (Section 29-2A). These interactions enable the ribosome to monitor whether an incoming tRNA is cognate to the codon in the A site; a non-Watson–Crick base pair could not bind these ribosomal bases in the same way. Indeed, any mutation of A1492 or A1493 is lethal because pyrimidines in these positions could not reach far enough to interact with the codon–anticodon complex or G530 and because a G in either position would be unable to form the required hydrogen bonds and its N2

would be subjected to steric collisions. An incorrect codon–anticodon provides insufficient free energy to bind the tRNA to the ribosome and it therefore dissociates from it, still in its ternary complex with EF-Tu and GTP.

b. GTP Hydrolysis by EF-Tu Is a Thermodynamic Prerequisite to Ribosomal Proofreading

A proofreading step must be entirely independent of the initial selection step. Only then can the overall probability of error be equal to the product of the probabilities of error of the individual selection steps. We have seen that DNA polymerases and aminoacyl–tRNA synthetases maintain the independence of their two selection steps by carrying them out at separate active sites (Sections 30-2A and 32-2C). Yet the ribosome only recognizes the incoming aminoacyl–tRNA according to its anticodon's complementarity to the codon in the A site. Consequently, the ribosome must somehow examine this codon–anticodon interaction in two separate ways.

The formation of a correct codon–anticodon complex triggers EF-Tu to hydrolyze its bound GTP, although how this occurs is unclear (note that EF-Tu's GTPase domain is bound in the 50S subunit which, together with the observation that GTP hydrolysis requires an intact tRNA, suggests that the hydrolysis signal is at least in part transmitted through the tRNA). The resulting conformational change in EF-Tu (Fig. 32-50) presumably swings its bound tRNA into the A/A state, a process in which the codon–anticodon interaction is subjected to a second screening that only permits a cognate aminoacyl–tRNA to enter the peptidyl transferase center. The irreversible GTPase reaction must precede this proofreading step because otherwise the dissociation of a noncognate tRNA (the release of its anticodon from the codon) would simply be the reverse of the initial binding step, that is, it would be part of the initial selection step rather than proofreading. *GTP hydrolysis therefore provides the second context necessary for proofreading; it is the entropic price the system must pay for*

(a)

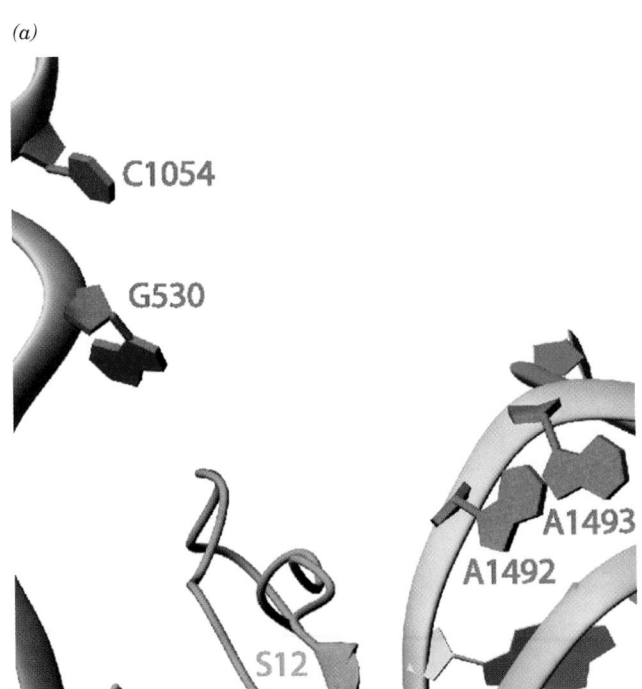

(b)

FIGURE 32-64 Ribosomal decoding site. The X-ray structures of *T. thermophilus* 30S subunit (*a*) alone and (*b*) in its complex with U$_6$ and the 17-nt anticodon stem–loop of tRNA^{Phe}. The RNAs are drawn as ribbons with their nucleotides in paddle form with tRNA gold, A-site mRNA purple, P-site mRNA green, rRNA gray, and nucleotides that undergo conformational changes red. Protein S12 is tan and Mg^{2+} ions are represented by red spheres. Compare Part *b* with Figure 32-63. [Courtesy of V. Ramakrishnan, MRC Laboratory of Molecular Biology, Cambridge, U.K. PDBids (*a*) 1FJF and (*b*) 1IBM.]

accurate tRNA selection. Further high resolution structural studies will be necessary to elucidate how this occurs.

G. Protein Synthesis Inhibitors: Antibiotics

Antibiotics are bacterially or fungally produced substances that inhibit the growth of other organisms. Antibiotics are known to inhibit a variety of essential biological processes, including DNA replication (e.g., ciprofloxacin, Section 29-3C), transcription (e.g., rifamycin B; Section 31-2B),

and bacterial cell wall synthesis (e.g., penicillin; Section 11-3B). However, *the majority of known antibiotics, including a great variety of medically useful substances, block translation.* This situation is presumably a consequence of the translational machinery's enormous complexity, which makes it vulnerable to disruption in many ways. Antibiotics have also been useful in analyzing ribosomal mechanisms because, as we have seen for puromycin (Section 32-3D), the blockade of a specific function often permits its biochemical dissection into its component steps. Table 32-10

TABLE 32-10 Some Ribosomal Inhibitors

Inhibitor	Action
Chloramphenicol	Inhibits peptidyl transferase on the prokaryotic large subunit
Cycloheximide	Inhibits peptidyl transferase on the eukaryotic large subunit
Erythromycin	Inhibits translocation by the prokaryotic large subunit
Fusidic acid	Inhibits elongation in prokaryotes by binding to EF-G · GDP in a way that prevents its dissociation from the large subunit
Paromomycin	Increases the ribosomal error rate
Puromycin	An aminoacyl–tRNA analog that causes premature chain termination in prokaryotes and eukaryotes
Streptomycin	Causes mRNA misreading and inhibits chain initiation in prokaryotes
Tetracycline	Inhibits the binding of aminoacyl–tRNAs to the prokaryotic small subunit
Diphtheria toxin	Catalytically inactivates eEF2 by ADP-ribosylation
Ricin/abrin/α-sarcin	**Ricin** and **abrin** are poisonous plant glycosidases that catalytically inactivate the eukaryotic large subunit by hydrolytically depurinating a specific highly conserved A residue of the 28S RNA, which is located on the so-called **sarcin–ricin loop** that forms a critical part of the ribosomal factor-binding center; **α-sarcin** is a fungal protein that cleaves a specific phosphodiester bond in the sarcin–ricin loop

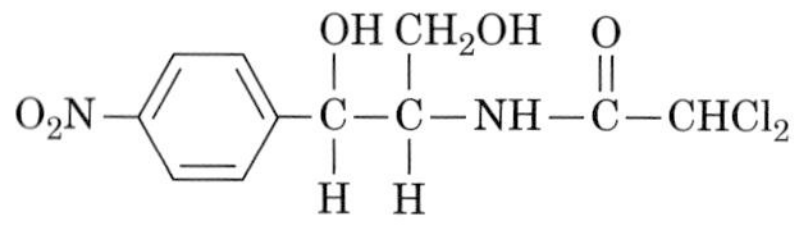

Chloramphenicol

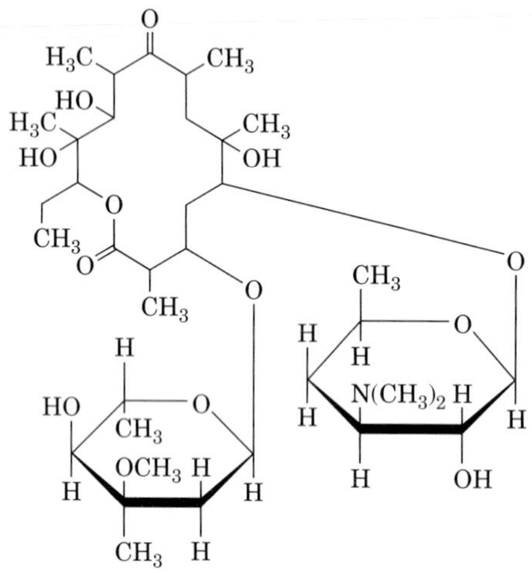

Erythromycin

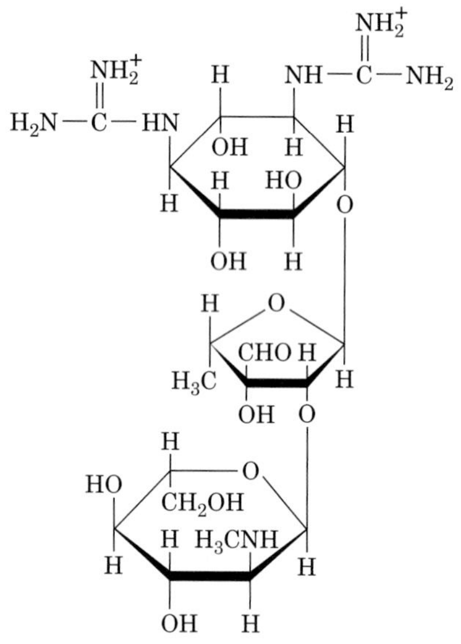

Streptomycin

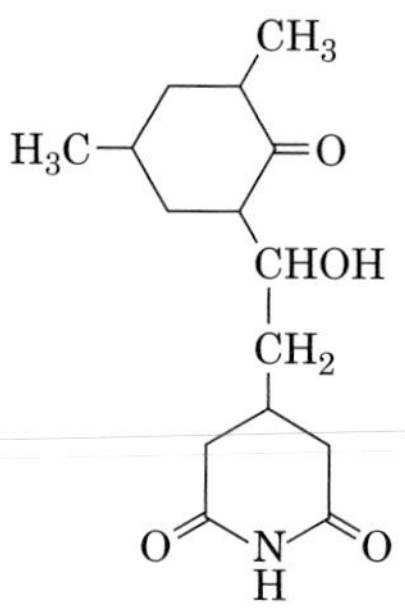

Cycloheximide

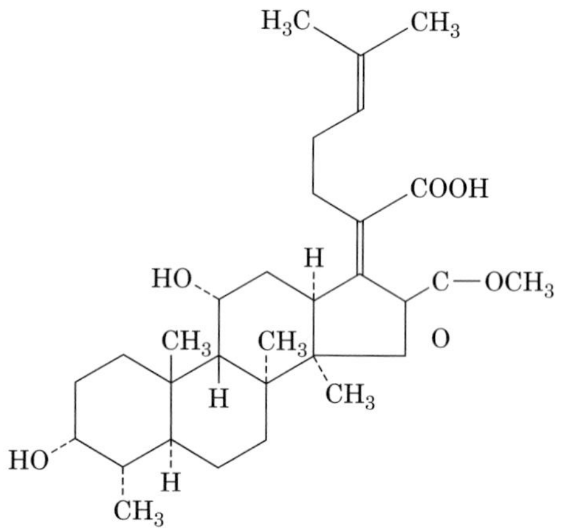

Fusidic acid

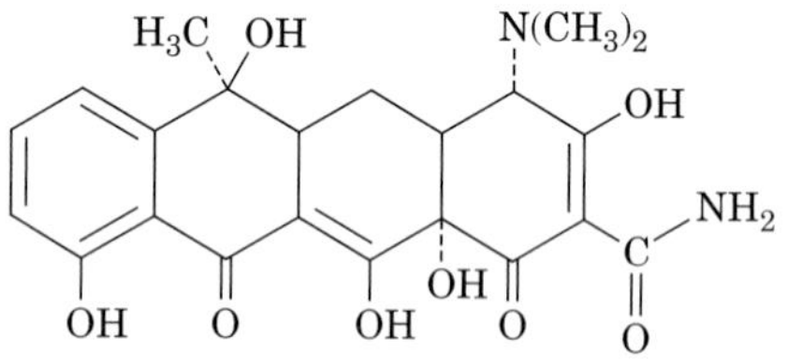

Tetracycline

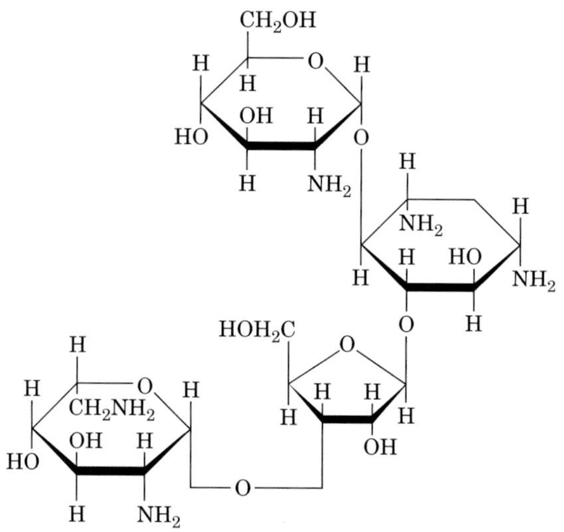

Paromomycin

FIGURE 32-65 Selection of antibiotics that act as translational inhibitors.

and Fig. 32-65 present several medically significant and/or biochemically useful translational inhibitors. We study the mechanisms of a few of the best characterized of them below.

a. Streptomycin

Streptomycin, which was discovered in 1944 by Selman Waksman, is a medically important member of a family of antibiotics known as **aminoglycosides** that inhibit prokaryotic ribosomes in a variety of ways. At low concentrations, streptomycin induces the ribosome to characteristically misread mRNA: One pyrimidine may be mistaken for the other in the first and second codon positions and either pyrimidine may be mistaken for adenine in the first position. This inhibits the growth of susceptible cells but does not kill them. At higher concentrations, however, streptomycin prevents proper chain initiation and thereby causes cell death.

Certain streptomycin-resistant mutants (str^R) have ribosomes with an altered protein S12 compared with streptomycin-sensitive bacteria (str^S). Intriguingly, a change in base C912 of 16S rRNA (which lies in its central loop; Fig. 32-27a) also confers streptomycin resistance. (Some mutant bacteria are not only resistant to streptomycin but dependent on it; they require it for growth.) In partial diploid bacteria that are heterozygous for streptomycin resistance (str^R/str^S), streptomycin sensitivity is dominant. This puzzling observation is explained by the finding that, in the presence of streptomycin, str^S ribosomes remain bound to initiation sites, thereby excluding str^R ribosomes from these sites. Moreover, the mRNAs in these blocked complexes are degraded after a few minutes, which allows the str^S ribosomes to bind to newly synthesized mRNAs as well.

b. Chloramphenicol

Chloramphenicol, the first of the "broad-spectrum" antibiotics, inhibits the peptidyl transferase activity on the large subunit of prokaryotic ribosomes. However, its clinical uses are limited to only severe infections because of its toxic side effects, which are caused, at least in part, by the chloramphenicol sensitivity of mitochondrial ribosomes. The 23S RNA is implicated in chloramphenicol binding by the observation that some of its mutants are chloramphenicol resistant. Indeed, X-ray studies indicate that chloramphenicol binds in the large subunit's polypeptide exit tunnel in the vicinity of the A site. This explains why chloramphenicol competes for binding with the 3′ end of aminoacyl–tRNAs as well as with puromycin (whose ribosomal binding site overlaps that of chloramphenicol) but not with peptidyl–tRNAs. These observations suggest that chloramphenicol inhibits peptidyl transfer by interfering with the interactions of ribosomes with A site–bound aminoacyl–tRNAs.

c. Paromomycin

Paromomycin, a clinically useful aminoglycoside antibiotic, increases the ribosomal error rate. The X-ray structure of the 30S subunit in complex with paromomycin

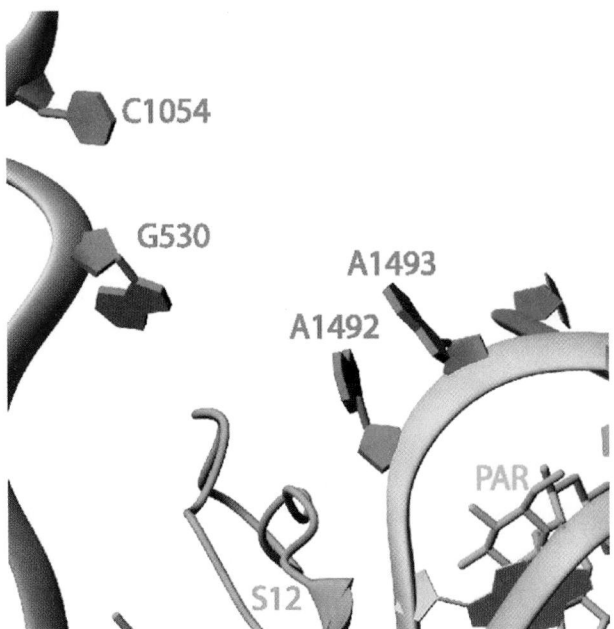

FIGURE 32-66   X-Ray structure of the 30S ribosome in complex with the antibiotic paromomycin. The view and coloring are the same as those in Fig. 32-64 with the paromomycin (PAR) drawn in stick form in yellow-green. [Courtesy of V. Ramakrishnan, MRC Laboratory of Molecular Biology, Cambridge, U.K. PDBid 1IBK.]

(Fig. 32-66) reveals that it binds to the interior of the RNA loop in which the bases of A1492 and A1493 are normally stacked (Fig. 32-64a). This causes these bases to flip out of the loop and assume a conformation resembling that in the codon–anticodon–30S subunit complex (Fig. 32-64b). Indeed, this codon–anticodon–30S subunit complex is not significantly disturbed by the binding of paromomycin. As we have seen in Section 32-3F, the 30S subunit employs A1492 and A1493 to ascertain whether the first two codon–anticodon base pairs are Watson–Crick base pairs, that is, whether the incoming tRNA is cognate to the codon in the A site. Noncognate tRNAs normally have insufficient codon–anticodon binding energy to flip A1492 and A1493 out of the loop and consequently are rejected by the ribosome. However, the binding of paromomycin to the 30S subunit pays the energetic price of these base flips. This facilitates the ribosomal acceptance (stabilizes the binding) of near-cognate aminoacyl–tRNAs and hence the erroneous incorporation of their amino acid residues into the polypeptide being synthesized.

d. Tetracycline

Tetracycline and its derivatives are broad-spectrum antibiotics that bind to the small subunit of prokaryotic ribosomes, where they inhibit aminoacyl–tRNA binding. An X-ray structure of tetracycline in complex with the 30S subunit reveals that tetracycline mainly binds in a crevice comprised of only the 3′ major domain of 16S RNA (Fig. 32-27a) and which is located in the neck of the 30S subunit just above its A site. This permits the initial screening of the aminoacyl–tRNA to proceed but physically blocks its accommodation into the peptidyl transferase (A/A) site after EF-Tu–catalyzed GTP hydrolysis has occurred, resulting in the release of the tRNA. Hence, in addition to preventing protein synthesis, tetracycline binding causes the unproductive hydrolysis of GTP which, since this occurs every time a cognate aminoacyl–tRNA binds to the ribosome, poses an enormous energetic drain on the cell. The nucleotides forming the tetracycline binding site are poorly conserved in eukaryotic ribosomes, thereby accounting for tetracycline's bacterial specificity.

Tetracycline also blocks the stringent response (Section 31-3H) by inhibiting (p)ppGpp synthesis. This indicates that deacylated tRNA must bind to the A site in order to activate stringent factor.

Tetracycline-resistant bacterial strains have become quite common, thereby precipitating a serious clinical problem. Resistance is often conferred by a decrease in bacterial cell membrane permeability to tetracycline rather than any alteration of ribosomal components.

e. Diphtheria Toxin

Diphtheria is a disease resulting from bacterial infection by *Corynebacterium diphtheriae* that harbor the bacteriophage **corynephage β.** Diphtheria was a leading cause of childhood death until the late 1920s when immunization became prevalent. Although the bacterial infection is usually confined to the upper respiratory tract, the bacteria secrete a phage-encoded protein, known as **diphtheria toxin (DT),** which is responsible for the disease's lethal effects. *Diphtheria toxin specifically inactivates the eukaryotic elongation factor eEF2, thereby inhibiting eukaryotic protein synthesis.*

The pathogenic effects of diphtheria are prevented, as was discovered in the 1880s, by immunization with **toxoid** (formaldehyde-inactivated toxin). Individuals who have contracted diphtheria are treated with antitoxin from horse serum, which binds to and thereby inactivates DT, as well as with antibiotics to combat the bacterial infection.

DT is a member of the family of bacterial toxins that includes cholera toxin (CT) and pertussis toxin (PT; Section 19-2C). It is a monomeric 535-residue protein that is readily cleaved past its Arg residues 190, 192, and 193 by trypsin and trypsinlike enzymes. This hydrolysis occurs around the time diphtheria toxin encounters its target cell, yielding two fragments, A and B, which, nevertheless, remain linked by a disulfide bond. The B fragment's C-terminal domain binds to a specific receptor on the plasma membrane of susceptible cells, thereby inducing DT's uptake into the endosome (Fig. 12-79) via receptor-mediated endocytosis (Section 12-5B; free fragment A is devoid of toxic activity). The endosome's low pH of 5 triggers a conformational change in the B fragment's N-terminal domain, which then inserts into the endosomal membrane so as to facilitate the entry of the A fragment into the cytoplasm. The disulfide bond linking the A and B subunits is then cleaved by the cytoplasm's reducing environment.

Within the cytosol, the A fragment catalyzes the **ADP-ribosylation** of eEF2 by NAD$^+$,

$$\text{eEF2} \; + \; \text{NAD}^+$$
$$(\textit{active})$$

$$\downarrow \text{diphtheria toxin}$$

$$\text{ADP-ribosyl-eEF2} \; + \; \text{Nicotinamide} \; + \; \text{H}^+$$
$$(\textit{inactive})$$

thereby inactivating this elongation factor. Since the A fragment acts catalytically, *one molecule is sufficient to ADP-ribosylate all of a cell's eEF2s, which halts protein synthesis and kills the cell.* Only a few micrograms of diphtheria toxin are therefore sufficient to kill an unimmunized individual.

Diphtheria toxin specifically ADP-ribosylates a modified His residue on eEF2 known as **diphthamide:**

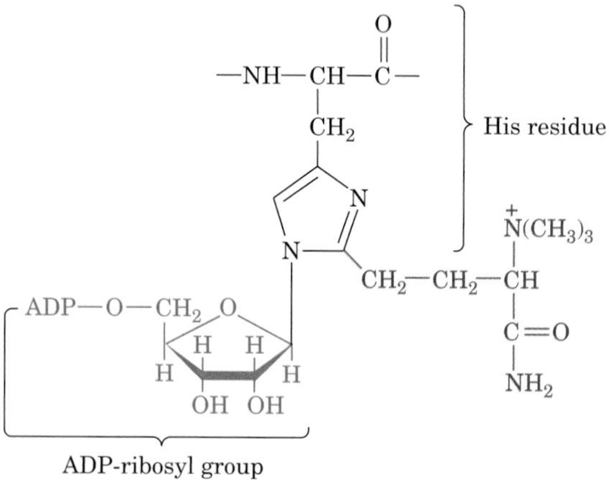

ADP-Ribosylated diphthamide

Diphthamide occurs only in eEF2 (not even in its bacterial counterpart, EF-G), which accounts for the specificity of diphtheria toxin in exclusively modifying eEF2 (recall that CT ADP-ribosylates a specific Arg residue on G$_{s\alpha}$ and PT ADP-ribosylates a specific Cys residue on G$_{i\alpha}$; Section 19-2C). Since diphthamide occurs in all eukaryotic eEF2s, it probably is essential to eEF2 activity. Yet, certain mutant cultured animal cells, which have unimpaired capacity to synthesize proteins, lack the enzymes that posttranslationally modify His to diphthamide (although mutating the diphthamide His to Asp, Lys, or Arg inactivates translation). Perhaps the diphthamide residue has a control function.

4 ■ CONTROL OF EUKARYOTIC TRANSLATION

The rates of ribosomal initiation on prokaryotic mRNAs vary by factors of up to 100. For example, the proteins specified by the *E. coli lac* operon, β-galactosidase, galactose permease, and thiogalactoside transacetylase, are synthesized in molar ratios of 10:5:2. This variation is probably a consequence of their different Shine–Dalgarno sequences. Alternatively, ribosomes may attach to *lac* mRNA only at its β-galactosidase gene and occasionally detach in response to a chain termination signal, thereby accounting for the decreasing translational rates along the operon. In any case, there is no evidence that prokaryotic translation rates are responsive to environmental changes. *Genetic expression in prokaryotes is therefore almost entirely transcriptionally controlled (Section 31-3).* Of course, since their mRNAs have lifetimes of only a few minutes, it would seem that prokaryotes have little need of translational controls.

It is clear, however, that eukaryotic cells can respond to at least some of their needs through translational control. This is feasible because the lifetimes of eukaryotic mRNAs are generally hours or days. In this section, we examine how translation is regulated via the phosphorylation/dephosphorylation of eIF2 and eIF4E. We then consider translational control by mRNA masking and cytoplasmic polyadenylation and end by discussing the uses of antisense oligonucleotides.

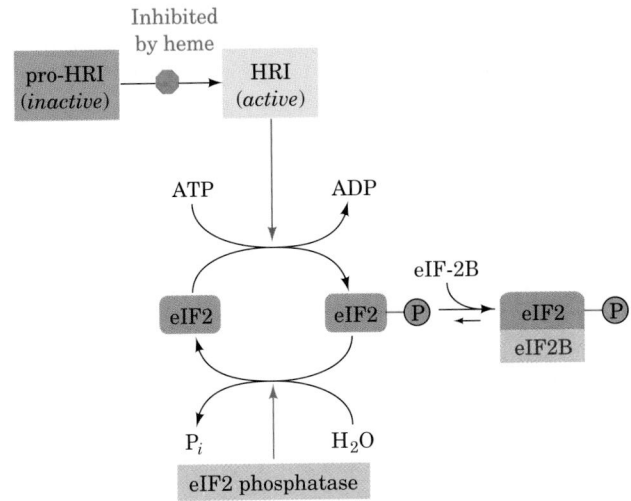

FIGURE 32-67   Model for heme-controlled protein synthesis in reticulocytes.

A. *Regulation of eIF2*

Four important pathways for the regulation of translation in eukaryotes involve the phosphorylation of the conserved Ser 51 on the α subunit of eIF2 (**eIF2α;** recall that eIF2 is an αβγ trimer that conducts Met–tRNA$_i^{Met}$ to the 40S ribosomal subunit, and the resulting complex scans the bound mRNA for the initiating AUG codon to form the 48S initiation complex; Section 32-3C). The so-called **eIF2α kinases** that do so share a conserved kinase domain but have unique regulatory domains

a. Heme Availability Controls Globin Translation

Reticulocytes synthesize protein, almost exclusively hemoglobin, at an exceedingly high rate and are therefore a favorite subject for the study of eukaryotic translation. Hemoglobin synthesis in fresh reticulocyte lysates proceeds normally for several minutes but then abruptly stops because of the inhibition of translational initiation and the consequent polysome disaggregation. This process is prevented by the addition of heme [a mitochondrial product (Section 26-4A) that this *in vitro* system cannot synthesize], thereby indicating that *globin synthesis is regulated by heme availability.* The inhibition of globin translational initiation is also reversed by the addition of the eukaryotic initiation factor eIF2 and by high levels of GTP.

In the absence of heme, reticulocyte lysates accumulate an eIF2α kinase named **heme-regulated inhibitor [HRI; also called heme-controlled repressor (HCR)].** HRI is a homodimer whose 629-residue subunits each contain two heme-binding sites. When heme is plentiful, both of these sites are occupied and the protein, which is autophosphorylated at several Ser and Thr residues, is inactive. However, when heme is scarce, one of these sites loses its bound heme, thereby activating HRI to autophosphorylate itself at several additional sites and to phosphorylate Ser 51 of eIF2α.

Phosphorylated eIF2 can participate in the ribosomal initiation process in much the same way as unphosphoryl-ated eIF2. This puzzling observation was clarified by the discovery that GDP does not dissociate from phosphorylated eIF2 at the completion of the initiation process as it normally does through a process facilitated by eIF2B acting as a GEF (Fig. 32-46). This is because phosphorylated eIF2 forms a much tighter complex with eIF2B than does unphosphorylated eIF2. This sequesters eIF2B (Fig. 32-67), which is present in lesser amounts than eIF2, thereby preventing the regeneration of the eIF2 · GTP required for translational initiation. The presence of heme reverses this process by inhibiting HRI, whereupon the phosphorylated eIF2 molecules are reactivated through the action of **eIF2 phosphatase,** which is unaffected by heme. The reticulocyte thereby coordinates its synthesis of globin and heme.

b. Interferons Protect against Viral Infection

Interferons are cytokines that are secreted by virus-infected vertebrate cells. On binding to surface receptors of other cells, interferons convert them to an antiviral state, which inhibits the replication of a wide variety of RNA and DNA viruses. Indeed, the discovery of interferons in the 1950s arose from the observation that virus-infected individuals are resistant to infection by a second type of virus.

There are three families of interferons: **type α** or **leukocyte interferon** (165 residues; leukocytes are white blood cells), the related **type β** or **fibroblast interferon** (166 residues; fibroblasts are connective tissue cells), and **type γ** or **lymphocyte interferon** (146 residues; lymphocytes are immune system cells). *Interferon synthesis is induced by the double-stranded RNA (dsRNA) that is generated during infection by both DNA and RNA viruses, as well as by the synthetic dsRNA poly(I) · poly(C).* Interferons are effective antiviral agents in concentrations as low as $3 \times 10^{-14} M$, which makes them among the most potent biological substances known. Moreover, they have far wider specificities than antibodies raised against a particular virus. They have therefore elicited great medical interest, particularly since some cancers are virally induced (Section 19-3B). Indeed, they are in clinical use against certain tumors and viral infections. These treatments are made possible by the production of large quantities of these otherwise quite scarce proteins through recombinant DNA techniques (Section 5-5D).

Interferons prevent viral proliferation largely by inhibiting protein synthesis in infected cells (lymphocyte interferon also modulates the immune response). They do so in two independent ways (Fig. 32-68):

1. Interferons induce the production of an eIF2α kinase, **double-stranded RNA-activated protein kinase [PKR;** also known as **double-stranded RNA-activated inhibitor (DAI);** 551 residues], which on binding dsRNA,

dimerizes and autophosphorylates itself. This activates PKR to phosphorylate eIF2α at its Ser 51, thereby inhibiting ribosomal initiation and hence the proliferation of viruses in virus-infected cells. The importance of PKR to cellular antiviral defense is indicated by the observation that many viruses express inhibitors of PKR.

2. Interferons also induce the synthesis of **(2′,5′)-oligoadenylate synthetase (2,5A synthetase).** In the presence of dsRNA, this enzyme catalyzes the synthesis from ATP of the unusual oligonucleotide **pppA(2′p5′A)$_n$** where $n = 1$ to 10. *This compound, **2,5-A,** activates a preexisting endonuclease, **RNase L,** to degrade mRNA, thereby inhibiting protein synthesis.* 2,5-A is itself rapidly degraded by an enzyme named **(2′,5′)-phosphodiesterase** so that it must be continually synthesized to maintain its effect.

The independence of the 2,5-A and PKR systems is demonstrated by the observation that the effect of 2,5-A on protein synthesis is reversed by added mRNA but not by added eIF2. [Recall that RNA interference (RNAi; Section 31-4A) constitutes an alternative dsRNA-based antiviral defense.]

c. PERK Prevents the Buildup of Unfolded Proteins in the ER

PKR-like endoplasmic reticulum kinase (PERK), a 1087-residue transmembrane protein, resides in the endoplasmic reticulum (ER) membrane of all multicellular eukaryotes. It is repressed by its binding to the ER-resident chaperone BiP (Section 12-4B). When the ER contains an excessive amount of unfolded proteins (caused by various forms of stress such as high temperatures), BiP dissociates from PERK, thereby activating PERK to phosphorylate eIFα at its Ser 51 and hence inhibit translation. Thus PERK functions to protect the cell from the irreversible damage caused by the accumulation of unfolded proteins in the ER.

Wolcott–Rallison syndrome is a genetic disease characterized mainly by insulin-dependent (type I) diabetes that develops in early infancy (type I diabetes usually first appears in childhood; Section 27-3B). It is caused by mutations in the catalytic domain of PERK. This results in the death of pancreatic β cells, in which PERK is particularly abundant. Multiple systemic disorders subsequently occur including **osteoporosis** (reduction in the quantity of bone) and growth retardation.

d. GCN2 Regulates Amino Acid Biosynthesis

GCN2 (1590 residues), the sole eIF2α kinase in yeast, is a transcriptional activator of the gene encoding **GCN4,** a transcriptional activator of numerous yeast genes, many of which encode enzymes that participate in amino acid biosynthetic pathways. The C-terminal domain of GCN2, which resembles histidyl–tRNA synthetase (HisRS), preferentially binds uncharged tRNAs (whose presence is indicative of an insufficient supply of amino acids). The binding of an uncharged tRNA to this HisRS-like domain activates the adjacent eIF2α kinase domain and thereby inhibits translational initiation, although at only a modest level.

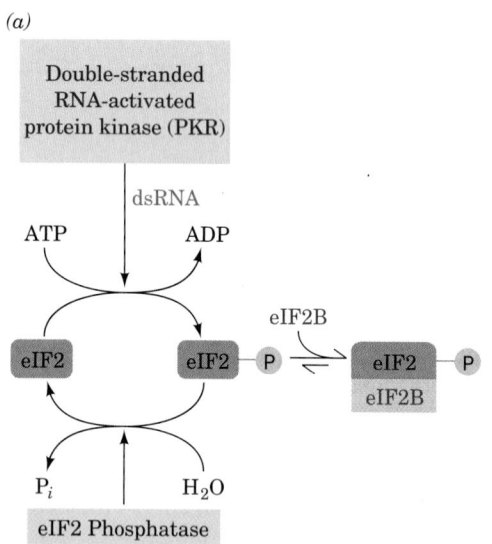

Inhibition of Translation

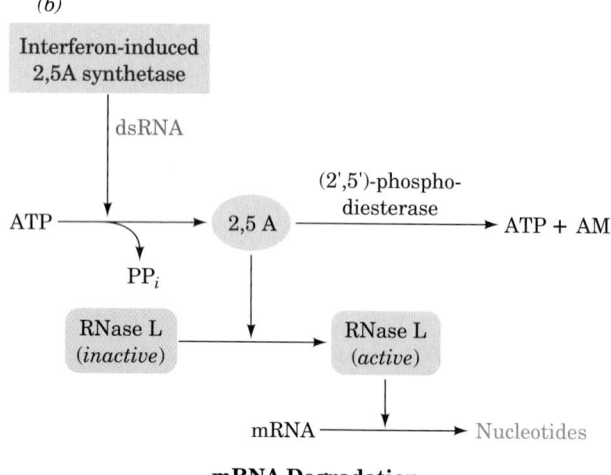

mRNA Degradation

FIGURE 32-68   The action of interferon. In interferon-treated cells, the presence of dsRNA, which normally results from a viral infection, causes (*a*) the inhibition of translational initiation and (*b*) the degradation of mRNA, thereby blocking translation and preventing virus replication.

Despite this inhibition of yeast protein synthesis, activated GCN2 induces the expression of GCN4. This seemingly paradoxical property of GCN2, as Alan Hinnebusch explained, arises from the fact that GCN4 mRNA contains four short so-called **upstream open reading frames (uORFs),** uORF1 to uORF4, in its 5′ leader that precedes the sequence encoding GCN4. Under the normal nutrient conditions in which GCN2 is inactive, the ribosome binds to the mRNA near its 5′ cap and scans for the nearest AUG initiation codon (which is in uORF1), where it forms the 48S initiation complex (Fig. 32-46) and commences the translation of uORF1 (Section 32-3C). On terminating translation at uORF1's Stop codon, the presence of the surrounding A + U-rich sequences causes the ribosome to resume scanning for the next AUG codon, where it initiates the translation of uORF2. This process repeats until the ribosome terminates at the end of uORF4, where its Stop codon's surrounding G + C-rich sequences induce the ribosome to disengage from the mRNA. Hence GNC4 is only expressed at a low basal level. However, under the low nutrient conditions in which GCN2 phosphorylates eIF2α at its Ser 51, the resulting reduced level of the eIF2 · Met–tRNA$_i^{Met}$ · GTP ternary complex causes the 40S subunit to scan longer distances before it can form the 48S initiation complex. Consequently, ~50% of the ribosomes scan past uORF2, uORF3, and uORF4 and only initiate translation at the *GCN4* AUG codon, which is therefore translated at a high level (uORF2 and uORF3 can be mutationally eliminated without significantly affecting translational control).

Mammalian homologs of GCN2 are activated under conditions of amino acid starvation. This suggests that the foregoing process has been conserved throughout eukaryotic evolution.

B. *Regulation of eIF4E*

eIF4E (cap-binding protein) binds to the m^{7}G cap of eukaryotic mRNAs and thereby participates in translational initiation by helping to identify the initiating AUG codon (Section 32-3C). When mammalian cells are treated with hormones, cytokines, **mitogens** (substances that induce mitosis), and/or growth factors, Ser 209 of human eIF4E is phosphorylated via a Ras-activated MAP kinase cascade (Sections 19-3C and 19-3D), thereby increasing eIF4E's affinity for capped mRNA and hence stimulating translational initiation. Ser 209 occupies a surface position on eIF4E adjacent to the binding site for the β phosphate group of the m^{7}GDP and flanking the putative binding cleft for mRNA (Fig. 32-47b). The structure of eIF4E suggests that the phosphoryl group of phosphorylated Ser 209 forms a salt bridge with Lys 159, which occupies the other side of the putative mRNA-binding cleft, so as to form a clamp that would help stabilize the bound mRNA. The importance of regulating eIF4E activity is indicated by the observations that the overexpression of eIF4E causes the malignant transformation of rodent cell lines and that eIF4E expression is elevated in several human cancers.

The homologous ~120-residue proteins known as **4E-BP1, 4E-BP2,** and **4E-BP3** (BP for *binding protein*; the first two are also known as **PHAS-I** and **PHAS-II**) inhibit cap-dependent translation. They do so by binding on the opposite side of eIF4E from its mRNA-binding site, presumably to a patch of seven highly conserved surface residues, and hence do not prevent eIF4E from binding the m^{7}G cap. Rather, they block eIF4E from binding to eIF4G and thereby interfere with the formation of the eIF4F complex that positions the 40S ribosomal subunit-bound Met–tRNA$_i^{Met}$ on the mRNA's initiating AUG codon (Section 32-3C). In fact, the 4E-BPs and eIF4G all possess the sequence motif YXXXXLφ (where φ is an aliphatic residue, most often L but also M or F) through which they bind to eIF4E.

The treatment of responsive cells with insulin or any of several protein growth factors causes the 4E-BPs to dissociate from eIF4E. This is because the presence of these hormones induces the phosphorylation of the 4E-BPs at six Ser/Thr residues via the signal transduction pathway involving PI3K, PKB, and mTOR (Fig. 19-64). Evidently, the phosphorylation of eIF4E and the 4E-BPs have similar if not synergistic effects in the hormonal regulation of translation in eukaryotes.

C. *mRNA Masking and Cytoplasmic Polyadenylation*

It has been known since the nineteenth century that early embryonic development in animals such as sea urchins, insects, and frogs is governed almost entirely by information present in the oocyte (egg) before fertilization. Indeed, sea urchin embryos exposed to sufficient actinomycin D (Section 31-2C) to inhibit RNA synthesis without blocking DNA replication develop normally through their early stages without a change in their protein synthesis program. This is in part because an unfertilized egg contains large quantities of mRNA that is "masked" by associated proteins to form ribonucleoprotein particles, thereby preventing the mRNAs' association with the ribosomes that are also present. On fertilization, this mRNA is "unmasked" in a controlled fashion, quite possibly by the dephosphorylation of the associated proteins, and commences directing protein synthesis. Development of the embryo can therefore start immediately on fertilization rather than wait for the synthesis of paternally specified mRNAs. Thus, gene expression in the early stages of development is entirely translationally controlled; transcriptional control only becomes important when transcription is initiated.

a. Cytoplasmic Polyadenylation

Another mechanism of translational control in oocytes and early embryos involves the polyadenylation of mRNAs in the cytoplasm (polyadenylation usually occurs in the nucleus, following which the mRNA is exported to the cytoplasm; Section 31-4A). A substantial number of maternally supplied mRNAs in oocytes have relatively short poly(A) tails (20–40 nt versus a usual length of ~250 nt). The 3′ untranslated region of these mRNAs contains both the AAUAAA polyadenylation signal (which is required for polyadenylation in the nucleus; Section 31-4A) together with a so-called **cytoplasmic polyadenylation element (CPE),** which has the consensus sequence UUUU-UAU. The CPE is recognized by **CPE-binding protein (CPEB),** which contains two RNA recognition motifs (RRMs) as well as a **zinc finger** (Fig. 9-27a) that contribute to its binding to the mRNA. Joel Richter discovered that CPEB recruits a 931-residue protein named **maskin** which, in turn, binds the eIF4E (cap-binding protein) that is bound to the mRNA's 5' cap (Fig. 32-69a). Maskin contains the same YXXXXLφ motif through which the 4E-BPs and eIF4G bind to eIF4E (Section 32-4B), thereby blocking the binding of eIF4G to eIF4E and hence preventing the formation of the 48S initiation complex (Fig. 32-46).

In the maturation of *Xenopus laevis* oocytes, a process that precedes fertilization and is stimulated by the steroid hormone progesterone (Section 19-1I), a variety of mRNAs, including those encoding several cyclins (which participate in cell cycle control; Section 30-4A) are translationally activated. Soon after exposure to progesterone, a protein kinase named **aurora** phosphorylates the mRNA-bound CPEB at its Ser 174. This increases CPEB's affinity for cleavage and polyadenylation specificity factor (CPSF; Section 31-4A), which then binds to the mRNA's AAUAAA sequence, where it recruits poly(A) polymerase (PAP) to lengthen the mRNA's poly(A) tail (Fig. 32-69b).

Translational initiation and cytoplasmic polyadenylation occur simultaneously, which suggests that these processes are linked. Indeed, Richter has shown that this occurs through the binding to poly(A) of poly(A)-binding protein (PABP; Section 31-4A), which as we saw (Section 32-3C), also binds to eIF4G to circularize the mRNA. The eIF4G in this complex displaces maskin from eIF4E, thereby permitting the formation of the 48S initiation complex and hence the mRNA's translation (Fig. 32-69b).

Mammalian cells also exhibit cell cycle–dependent cytoplasmic polyadenylation of mRNAs. This suggests that

translational control by polyadenylation is a general feature in animal cells.

D. *Antisense Oligonucleotides*

Since ribosomes cannot translate double-stranded RNA or DNA–RNA hybrid helices, the translation of a given mRNA can be inhibited by a segment of its complementary RNA or DNA, that is, an **antisense RNA** or an **antisense oligodeoxynucleotide,** which are collectively known as **antisense oligonucleotides.** Moreover, endogenous RNase H's (enzymes that cleave the RNA strand of an RNA–DNA duplex; Section 31-4C) cleave an mRNA–oligodeoxynucleotide duplex on its mRNA strand, leaving the antisense oligodeoxynucleotide intact for binding to another mRNA. Alternatively, an antisense RNA could be incorporated into a ribozyme that would destroy the target mRNA (Section 31-4A).

Since the human genome consists of ~3.2 billion bp, an ~15-nt oligonucleotide (which is easily synthesized; Section 7-6A) should ideally be able to target any segment of the human genome. This exquisite specificity provides the delivery of an antisense oligonucleotide to, or its expression in a selected tissue or organism with enormous biomedical and biotechnological potential. However, care must be taken that an antisense oligonucleotide does not also eliminate nontarget mRNAs.

Methods for the delivery of a therapeutically useful antisense oligonucleotide to a target tissue are as yet in their infancy. This is in large part because oligonucleotides are readily degraded by the many nucleases present in an organism and because they do not readily pass through cell membranes. Moreover, a target mRNA is likely to be associated with cellular proteins and hence not available for binding to other molecules. The nuclease resistance of oligonucleotides can be increased by derivitizing them, for example, by replacing a nonbridging oxygen at each phosphate group with a methyl group or an S atom so as to

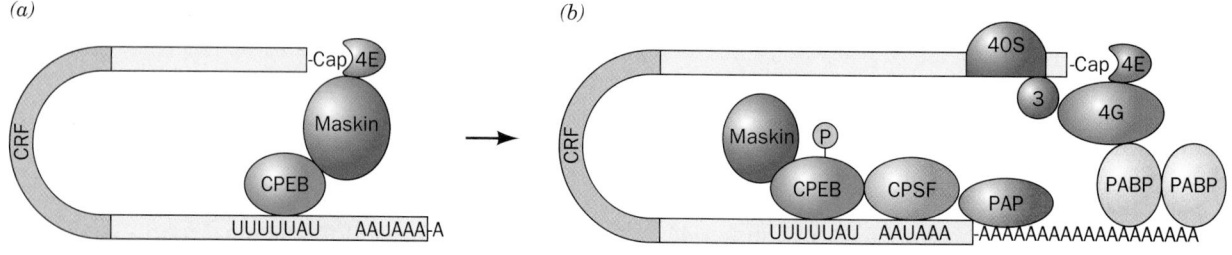

FIGURE 32-69 CPEB-mediated translational control. (*a*) In immature *Xenopus* oocytes, an mRNA containing the CPE (UUUUUAU) is bound by CPEB, which binds maskin, which then binds eIF4E so as to prevent it from binding eIF4G, thereby maintaining the mRNA in a translationally dormant (masked) state. (*b*) In the maturation process, CPEB is phosphorylated by an aurora protein kinase. The

phosphorylated CPEB binds CPSF, which recruits PAP to extend the mRNA's heretofore short poly(A) tail. PABP binds to the newly lengthened poly(A) tail and simultaneously binds to eIF4G so as to displace maskin. This permits the 48S initiation complex to assemble and hence the translation of the mRNA to proceed. [Based on a drawing by Mendez, R. and Richer, J.D., *Nature Rev. Mol. Cell Biol.* **2,** 521 (2001).]

yield **methylphosphonate** or **phosphorothioate oligonucleotides,** although this reduces their antisense activity. The expression of antisense oligonucleotides in the specified tissues would, of course, circumvent the delivery problem but has all the difficulties associated with gene therapy (Section 5-5H).

Despite the foregoing, antisense technology is beginning to show success. **Fomivirsen** (trade name **Vitravene**), a 21-nt phosphorothioate oligonucleotide that is complementary to an mRNA expressed by **cytomegalovirus (CMV),** is effective in the treatment of retinitis (inflammation of the retina) caused by CMV infection in individuals with AIDS (CMV is an opportunistic pathogen that rarely infects individuals with normal immune systems). It was approved for human use in 1998 by the FDA, the first antisense drug so approved. In addition, a number of antisense oligonucleotides that are mainly targeted against genes that are overexpressed in specific cancers and autoimmune diseases as well as other viral infections are in Phase II or Phase III clinical trials (Section 15-4B).

Antisense technology has also had some success in the arena of biotechnology. For example, in tomatoes and other fruits, the enzyme **polygalacturonase (PG),** which is expressed during ripening, depolymerizes the pectin (mainly polygalacturonic acid) in the cell wall. This results in a softening of tomatoes to the point that vine-ripened (and hence better tasting) tomatoes are unable to withstand the rigors of shipping and hence must be picked before they are ripe. The introduction into a tomato, via genetic engineering techniques, of a gene expressing antisense PG RNA yielded the so-called Flavr Savr tomato that had substantially reduced PG expression and hence remained firm after vine ripening.

5 ■ POSTTRANSLATIONAL MODIFICATION

To become mature proteins, polypeptides must fold to their native conformations, their disulfide bonds, if any, must form, and, in the case of multisubunit proteins, the subunits must properly combine. Moreover, as we have seen throughout this text, many proteins are modified in enzymatic reactions that proteolytically cleave certain peptide bonds and/or derivatize specific residues. In this section we shall review some of these **posttranslational modifications.**

A. *Proteolytic Cleavage*

Proteolytic cleavage is the most common type of posttranslational modification. Probably all mature proteins have been so modified, if by nothing else than the proteolytic removal of their leading Met (or fMet) residue shortly after it emerges from the ribosome. Many proteins, which are involved in a wide variety of biological processes, are synthesized as inactive precursors that are activated under proper conditions by limited proteolysis. Some examples of this phenomenon that we have encountered are

the conversion of trypsinogen and chymotrypsinogen to their active forms by tryptic cleavages of specific peptide bonds (Section 15-3E), and the formation of active insulin from the 84-residue proinsulin by the excision of its internal 33-residue C chain (Section 9-1A). Inactive proteins that are activated by removal of polypeptides are called **proproteins,** whereas the excised polypeptides are termed **propeptides.**

a. Propeptides Direct Collagen Assembly

Collagen biosynthesis is illustrative of many facets of posttranslational modification. Recall that collagen, a major extracellular component of connective tissue, is a fibrous triple-helical protein whose polypeptides each contain the amino acid sequence $(\text{Gly-X-Y})_n$ where X is often Pro, Y is often 4-hydroxyproline (Hyp), and $n \approx 340$ (Section 8-2B). The polypeptides of **procollagen** (Fig. 32-70) differ from those of the mature protein by the presence of both N-terminal and C-terminal propeptides of ~100 residues whose sequences, for the most part, are unlike those of mature collagen. The procollagen polypeptides rapidly assemble, *in vitro* as well as *in vivo*, to form a collagen triple helix. In contrast, polypeptides extracted from mature collagen will reassemble only over a period of days, if at all. *The collagen propeptides are apparently necessary for proper procollagen folding.*

The N- and C-terminal propeptides of procollagen are respectively removed by **amino-** and **carboxyprocollagen**

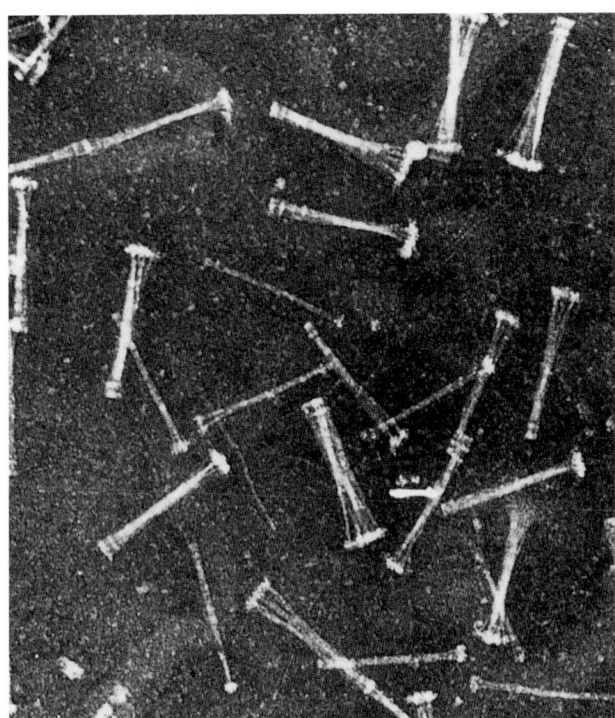

FIGURE 32-70 Electron micrograph of procollagen aggregates that have been secreted by fibroblasts into the extracellular medium. [Courtesy of Jerome Gross, Massachusetts General Hospital, Harvard Medical School.]

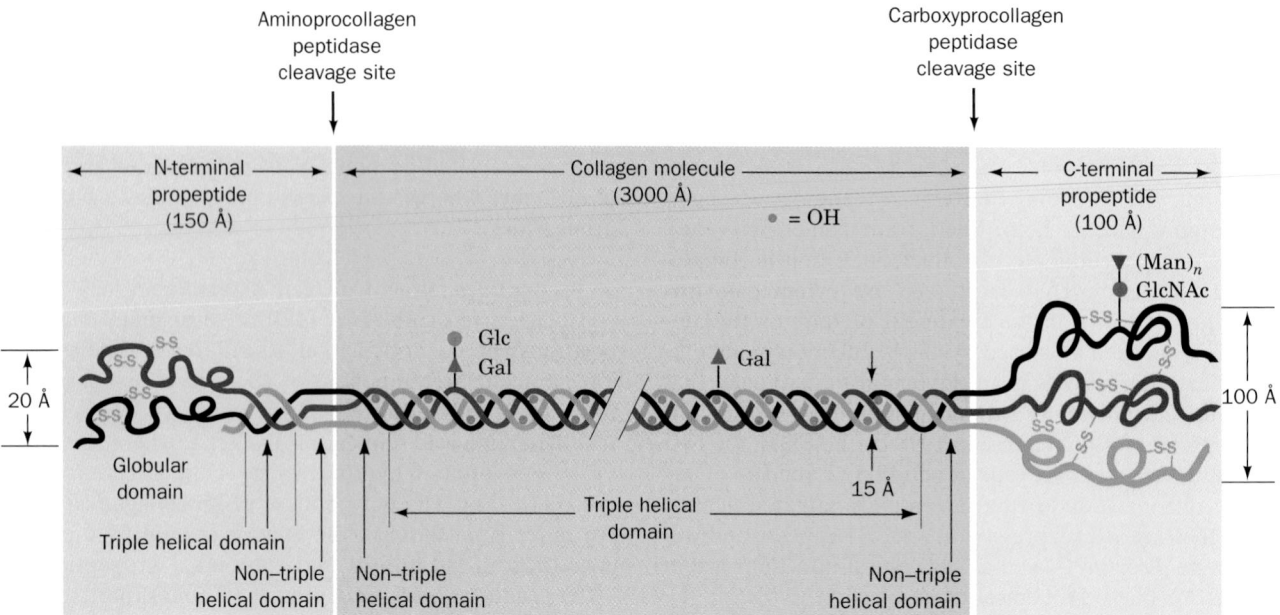

FIGURE 32-71 Schematic representation of the procollagen molecule. Gal, Glc, GlcNAc, and Man, respectively, denote galactose, glucose, *N*-acetylglucosamine, and mannose residues. Note that the N-terminal propeptide has intrachain disulfide bonds while the C-terminal propeptide has both intrachain and interchain disulfide bonds. [After Prockop, D.J., Kivirikko, K.I., Tuderman, L., and Guzman, N.A., *New Engl. J. Med.* **301**, 16 (1979).]

peptidases (Fig. 32-71), which may also be specific for the different collagen types. An inherited defect of aminoprocollagen peptidase in cattle and sheep results in a bizarre condition, **dermatosparaxis,** that is characterized by extremely fragile skin. An analogous disease in humans, **Ehlers–Danlos syndrome VII,** is caused by a mutation in one of the procollagen polypeptides that inhibits the enzymatic removal of its aminopropeptide. Collagen molecules normally spontaneously aggregate to form collagen fibrils (Figs. 8-31 and 8-32). However, electron micrographs of dermatosparaxic skin show sparse and disorganized collagen fibrils. *The retention of collagen's aminopropeptides apparently interferes with proper fibril formation.* (The dermatosparaxis gene was bred into some cattle herds because heterozygotes produce tender meat.)

b. Signal Peptides Are Removed from Nascent Proteins by a Signal Peptidase

Many transmembrane proteins or proteins that are destined to be secreted are synthesized with an N-terminal **signal peptide** of 13 to 36 predominantly hydrophobic residues. As described by the **signal hypothesis** (Section 12-4B), a signal peptide is recognized by a **signal recognition particle (SRP).** The SRP binds a ribosome synthesizing a signal peptide to a protein pore known as the **translocon** that is embedded in the membrane [the rough endoplasmic reticulum (RER) in eukaryotes and the plasma membrane in bacteria] and conducts the signal peptide and its following nascent polypeptide through the translocon.

Proteins bearing a signal peptide are known as **preproteins** or, if they also contain propeptides, as **preproproteins.**

Once the signal peptide has passed through the membrane, it is specifically cleaved from the nascent polypeptide by a membrane-bound **signal peptidase.** Both insulin and collagen are secreted proteins and are therefore synthesized with leading signal peptides in the form of **preproinsulin** and **preprocollagen.** These and many other proteins are therefore subject to three sets of sequential proteolytic cleavages: (1) the deletion of their initiating Met residue, (2) the removal of their signal peptides, and (3) the excision of their propeptides.

c. Polyproteins

Some proteins are synthesized as segments of **polyproteins,** polypeptides that contain the sequences of two or more proteins. Examples include many polypeptide hormones (Section 34-3C); the proteins synthesized by many viruses, including those causing polio (Section 33-2C) and AIDS (Section 15-4C); and **ubiquitin,** a highly conserved eukaryotic protein involved in protein degradation (Section 32-6B). Specific proteases posttranslationally cleave polyproteins to their component proteins, presumably through the recognition of the cleavage site sequences. Some of these proteases are conserved over remarkable evolutionary distances. For instance, ubiquitin is synthesized as several tandem repeats **(polyubiquitin)** that *E. coli* properly cleave even though prokaryotes lack ubiquitin. Other proteases have more idiosyncratic cleavage sequences. This has allowed medicinal chemists to design inhibitors of **HIV protease** (which catalyzes an essential step in the viral life cycle) that have been highly effective in attenuating if not preventing the progression of AIDS (Section 32-6B).

B. *Covalent Modification*

Proteins are subject to specific chemical derivatizations, both at the functional groups of their side chains and at their terminal amino and carboxyl groups. Over 150 different types of side chain modifications, involving all side chains but those of Ala, Gly, Ile, Leu, Met, and Val, are known (Section 4-3A). These include acetylations, glycosylations, hydroxylations, methylations, nucleotidylations, phosphorylations, and ADP-ribosylations as well as numerous "miscellaneous" modifications.

Some protein modifications, such as the phosphorylation of glycogen phosphorylase (Section 18-1A) and the ADP-ribosylation of eEF-2 (Section 32-3G), modulate protein activity. Several side chain modifications covalently bond cofactors to enzymes, presumably to increase their catalytic efficiency. Examples of linked cofactors that we have encountered are N^{ε}-lipoyllysine in dihydrolipoyl transacetylase (Section 21-2A) and 8α-histidylflavin in succinate dehydrogenase (Section 21-3F). The attachment of complex carbohydrates, which occur in almost infinite variety, alter the structural properties of proteins and form recognition markers in various types of targeting and cell–cell interactions (Sections 11-3C, 12-3E, and 23-3B). Modifications that cross-link proteins, such as occur in collagen (Section 8-2B), stabilize supramolecular aggregates. The functions of most side chain modifications, however, remain enigmatic.

a. Collagen Assembly Requires Chemical Modification

Collagen biosynthesis (Fig. 32-72) is illustrative of protein maturation through chemical modification. As the nascent procollagen polypeptides pass into the RER of the fibroblasts that synthesized them, the Pro and Lys residues are hydroxylated to Hyp, 3-hydroxy-Pro, and 5-hydroxy-Lys (Hyl). The enzymes that do so are sequence specific: **Prolyl 4-hydroxylase** and **lysyl hydroxylase** act only on the Y residues of the Gly-X-Y sequences, whereas **prolyl 3-hydroxylase** acts on the X residues but only if Y is Hyp. Glycosylation, which also occurs in the RER, subsequently attaches sugar residues to Hyl residues (Section 8-2B). The folding of three polypeptides into the collagen triple helix must follow hydroxylation and glycosylation because the hydroxylases and glycosyl transferases do not act on helical substrates. Moreover, the collagen triple helix denatures below physiological temperatures unless stabilized by hydrogen bonding interactions involving Hyp residues (Section 8-2B). Folding is also preceded by the formation of specific interchain disulfide bonds between the carboxyl-propeptides. This observation bolsters the previously discussed conclusion that collagen propeptides help select and align the three collagen polypeptides for proper folding.

The procollagen molecules pass into the Golgi apparatus where they are packaged into **secretory vesicles** (Sections 12-4C, 12-4D, and 23-3B) and secreted into the extracellular spaces of connective tissue. The amino-propeptides are excised just after procollagen leaves the cell and the carboxypropeptides are removed sometime later. The collagen molecules then spontaneously assemble into fibrils, which suggests that an important propeptide function is to prevent intracellular fibril formation. Finally, after the action of the extracellular enzyme lysyl oxidase, the collagen molecules in the fibrils spontaneously cross-link (Fig. 8-33).

C. *Protein Splicing: Inteins and Exteins*

Protein splicing is a posttranslational modification process in which an *int*ernal pro*tein* segment (an **intein**) excises itself from a surrounding *ext*ernal pro*tein*, which it ligates to form the mature **extein**. The portions of the unspliced extein on the N- and C-terminal sides of the intein are called the **N-extein** and the **C-extein**. Over 130 putative inteins, ranging in length from 134 to 600 residues, have so far been identified in archaebacteria, eubacteria, and single-celled

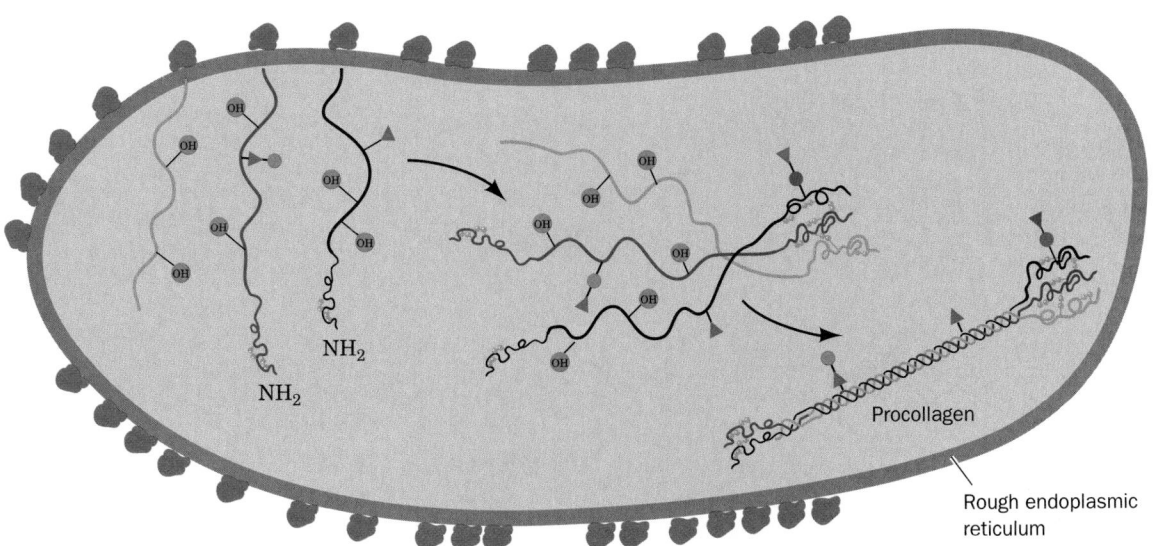

FIGURE 32-72   Schematic representation of procollagen biosynthesis. The diagram does not indicate the removal of signal peptides. [After Prockop, D.J., Kivirikko, K.I., Tuderman, L., and Guzman, N.A., *New Engl. J. Med.* **301,** 18 (1979).]

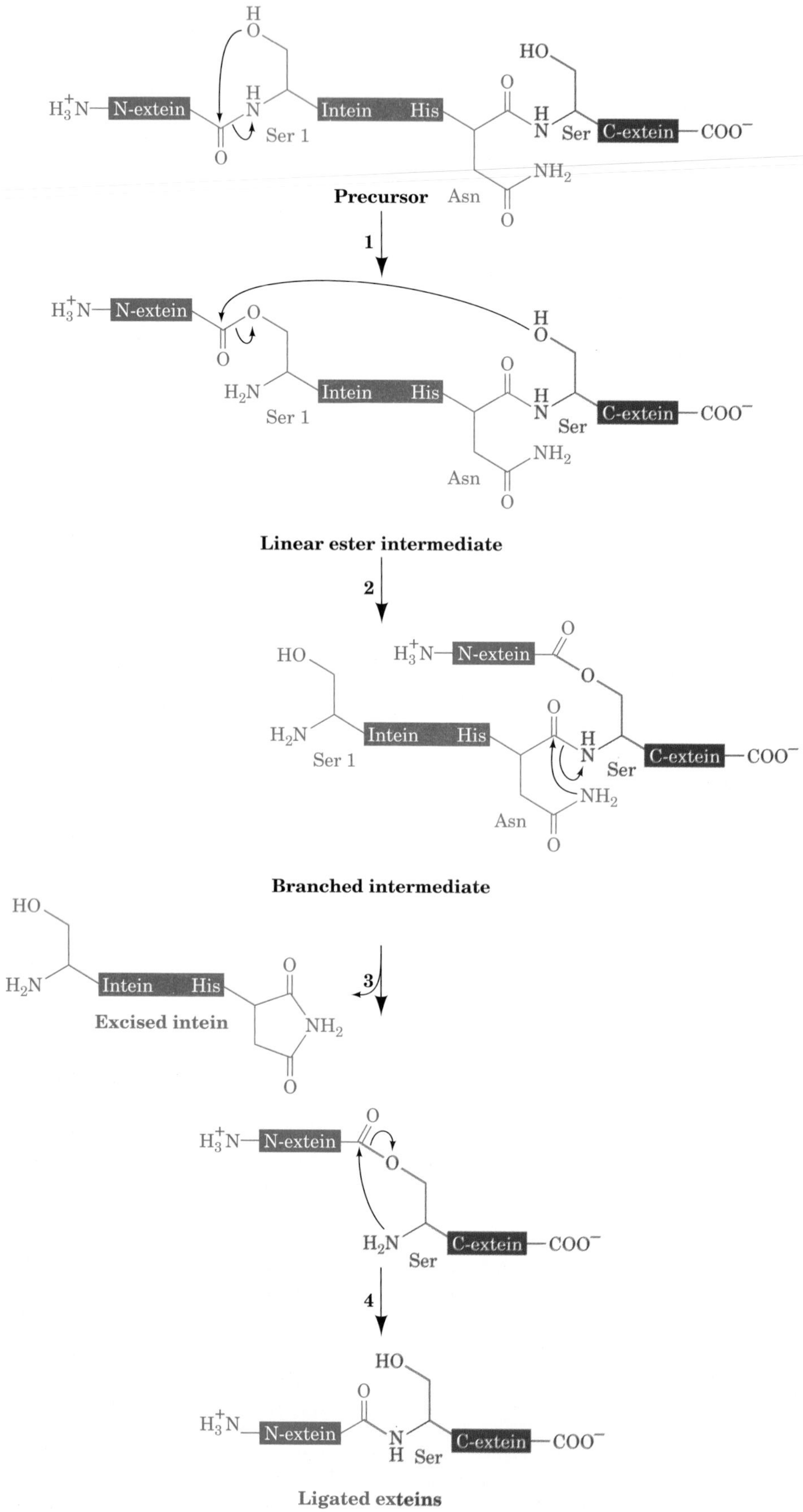

FIGURE 32-73 Series of reactions catalyzed by inteins to splice themselves out of a polypeptide chain.

eukaryotes (and are registered in the Intein Database at http://www.neb.com/neb/inteins.html/). The various exteins in which these inteins are embedded exhibit no significant sequence similarity and, in fact, can be replaced by other polypeptides, thereby indicating that exteins do not contain the catalytic elements that mediate protein splicing. In contrast, the ~150-residue splicing elements of inteins exhibit significant sequence similarity. All of them have four conserved splice-junction residues: (1) a Ser/Thr/Cys at the intein's N-terminus; and (2 and 3) a His–Asn/Gln dipeptide at the intein's C-terminus; which is immediately followed by (4) a Ser/Thr/Cys at the N-terminus of the C-extein.

Protein splicing occurs via a reaction sequence that involves four successive nucleophilic displacements, the first three of which are catalyzed by the intein (Fig. 32-73):

1. Attack by the N-terminal intein residue (Ser, Thr, or Cys; shown in Fig. 32-73 as Ser) on its preceding carbonyl group, yielding a linear (thio)ester intermediate.

2. A transesterification reaction in which the -OH or -SH group on the C-extein's N-terminal residue (shown in Fig. 32-73 as Ser) attacks the above (thio)ester linkage, thereby yielding a branched intermediate in which the N-extein has been transferred to the C-extein.

3. Cleavage of the amide linkage connecting the intein to the C-extein by cyclization of the intein's C-terminal Asn or Gln (shown in Fig. 32-73 as Asn).

4. Spontaneous rearrangement of the (thio)ester linkage between the ligated exteins to yield the more stable peptide bond.

The X-ray structure of the 198-residue **GyrA intein** from *Mycobacterium xenopi*, determined by James Sacchetini, indicates how this intein catalyzes the foregoing splicing reactions. This intein's N-terminal residue, Cys 1, was replaced by an Ala–Ser dipeptide with the expectation that the mutant protein would resemble the intein's presplicing state (the new N-terminal residue, Ala 0, presumably represents the C-terminal residue of the N-extein). The X-ray structure reveals that this monomeric protein consists primarily of β strands, two of which curve about the periphery of the entire protein to give it the shape of a flattened horseshoe (Fig. 32-74). The intein's catalytic site is located at the bottom of a broad and shallow cleft near the center of this so-called **β-horseshoe,** where the intein's N-terminal and C-terminal residues are in close proximity. The Ala 0—Ser 1 peptide bond, the bond cleaved in Reaction 1 of the protein splicing process (Fig. 32-73) assumes the cis conformation (Fig. 8-2), a rare high-energy conformation (except when the peptide bond is followed by Pro) that destabilizes this bond. Its amide nitrogen atom is hydrogen bonded to the side chain of the highly conserved His 75. Hence His 75 is well positioned to donate a proton that would promote the breakdown of the tetrahedral intermediate in Reaction 1. The side chains of Thr 72 and Asn 74 appear well positioned to stabilize this tetrahedral intermediate in a manner resembling that of the

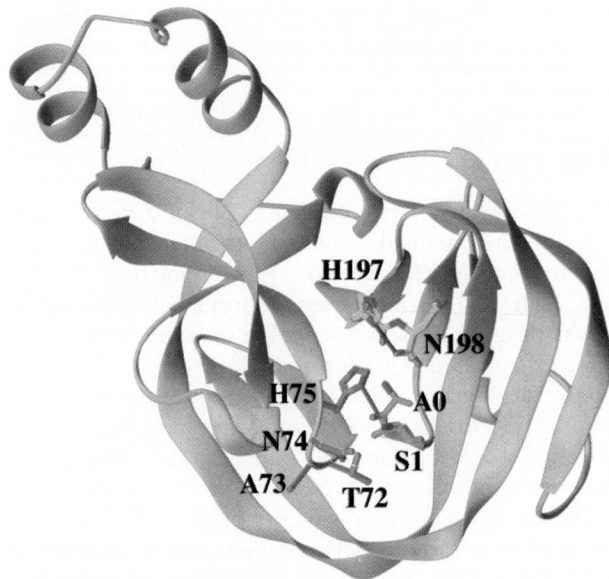

FIGURE 32-74   X-Ray structure of the *M. xenopi* Gyr A intein in which Cys 1 was replaced by an Ala 0–Ser 1 dipeptide. The protein is drawn in ribbon form with its N-terminal Ala 0–Ser 1 dipeptide and its C-terminal His 197–Asn 198 dipeptide as well as the side chains of residues 72 through 75 drawn in stick form colored according to atom type (C of residues 0–1 magenta, C of residues 72–75 green, C of residues 197–198 cyan, N blue, and O red). Hydrogen bonds are represented by thin gray bonds. [Based on an X-Ray structure by James Sacchetini, Texas A&M University. PDBid 1AM2.]

oxyanion hole in serine proteases (Section 15-3D). The position of Ser 1 and a modeled Thr at the intein's C-terminus is consistent with Reaction 2 of the splicing process. The side chain of the invariant His 197 is hydrogen bonded to the carboxylate of the C-terminal Asn 198 and hence is positioned to protonate the peptide bond cleaved in Reaction 3.

a. Most Inteins Encode a Homing Endonuclease

What is the biological function of inteins? Nearly all inteins contain polypeptide inserts forming so-called **homing endonucleases.** These are site-specific endonucleases that make a double-strand break in genes that are homologous to their corresponding extein but which lack inteins. The break initiates the double-strand break repair of the DNA via recombination (Section 30-6A). Since the intein-containing gene is likely to be the only other gene in the cell containing extein-like sequences, the intein gene is copied into the break. Thus, most inteins mediate a highly specific transposition or "homing" of the genes that insert them in similar sites. The intein's protease and endonuclease activities appear to have a symbiotic relationship: The protease activity excises the intein from the host protein, thereby preventing deleterious effects on the host, whereas the endonuclease activity assures the mobility of the intein gene. Thus intein genes appear to be molecular parasites (junk DNA) that only function to propagate themselves. Indeed, homing endonucleases are also encoded by certain types of introns.

6 ■ PROTEIN DEGRADATION

The pioneering work of Henry Borsook and Rudolf Schoenheimer around 1940 demonstrated that the components of living cells are constantly turning over. Proteins have lifetimes that range from as short as a few minutes to weeks or more. In any case, *cells continuously synthesize proteins from and degrade them to their component amino acids.* The function of this seemingly wasteful process is twofold: (1) to eliminate abnormal proteins whose accumulation would be harmful to the cell, and (2) to permit the regulation of cellular metabolism by eliminating superfluous enzymes and regulatory proteins. Indeed, since the level of an enzyme depends on its rate of degradation as well as its rate of synthesis, *controlling a protein's rate of degradation is as important to the cellular economy as is controlling its rate of synthesis.* In this section we consider the processes of intracellular protein degradation and their consequences.

A. *Degradation Specificity*

Cells selectively degrade abnormal proteins. For example, hemoglobin that has been synthesized with the valine analog **α-amino-β-chlorobutyrate**

α-Amino-β-Chlorobutyrate   **Valine**

has a half-life in reticulocytes of ~10 min, whereas normal hemoglobin lasts the 120-day lifetime of the red cell (which makes it perhaps the longest lived cytoplasmic protein). Likewise, unstable mutant hemoglobins are degraded soon after their synthesis, which, for reasons explained in Section 10-3A, results in the hemolytic anemia characteristic of these molecular disease agents. Bacteria also selectively degrade abnormal proteins. For instance, *amber* and *ochre* mutants of β-galactosidase have half-lives in *E. coli* of only a few minutes, whereas the wild-type enzyme is almost indefinitely stable. Most abnormal proteins, however, probably arise from the chemical modification and/or spontaneous denaturation of these fragile molecules in the cell's reactive environment rather than by mutations or the rare errors in transcription or translation. *The ability to eliminate damaged proteins selectively is therefore an essential recycling mechanism that prevents the buildup of substances that would otherwise interfere with cellular processes.*

Normal intracellular proteins are eliminated at rates that depend on their identities. A given protein is elimi-

TABLE 32-11   Half-lives of Some Rat Liver Enzymes

Enzyme	Half-life (h)
Short-Lived Enzymes	
Ornithine decarboxylase	0.2
RNA polymerase I	1.3
Tyrosine aminotransferase	2.0
Serine dehydratase	4.0
PEP carboxylase	5.0
Long-Lived Enzymes	
Aldolase	118
GAPDH	130
Cytochrome *b*	130
LDH	130
Cytochrome *c*	150

Source: Dice, J.F. and Goldberg, A.L., *Arch. Biochem. Biophys.* **170,** 214 (1975).

nated with first-order kinetics, indicating that the molecules being degraded are chosen at random rather than according to their age. The half-lives of different enzymes in a given tissue vary substantially as is indicated for rat liver in Table 32-11. Remarkably, *the most rapidly degraded enzymes all occupy important metabolic control points, whereas the relatively stable enzymes have nearly constant catalytic activities under all physiological conditions. The susceptibilities of enzymes to degradation have evidently evolved together with their catalytic and allosteric properties so that cells can efficiently respond to environmental changes and metabolic requirements.* The criteria through which native proteins are selected for degradation are considered in Section 32-6B.

The rate of protein degradation in a cell also varies with its nutritional and hormonal state. Under conditions of nutritional deprivation, cells increase their rate of protein degradation so as to provide the necessary nutrients for indispensable metabolic processes. The mechanism that increases degradative rates in *E. coli* is the stringent response (Section 31-3H). A similar mechanism may be operative in eukaryotes since, as happens in *E. coli*, increased rates of degradation are prevented by antibiotics that block protein synthesis.

B. *Degradation Mechanisms*

Eukaryotic cells have dual systems for protein degradation: lysosomal mechanisms and ATP-dependent cytosolically based mechanisms. We consider both mechanisms below.

a. Lysosomes Mostly Degrade Proteins Nonselectively

Lysosomes are membrane-encapsulated organelles (Section 1-2A) that contain ~50 hydrolytic enzymes, including a variety of proteases known as **cathepsins.** The lysosome maintains an internal pH of ~5 and its enzymes have acidic pH optima. This situation presumably protects the cell against accidental lysosomal leakage since lysosomal enzymes are largely inactive at cytosolic pH's.

Lysosomes recycle intracellular constituents by fusing with membrane-enclosed bits of cytoplasm known as **autophagic vacuoles** and subsequently breaking down their contents. They similarly degrade substances that the cell takes up via endocytosis (Section 12-5B). The existence of these processes has been demonstrated through the use of lysosomal inhibitors. For example, the antimalarial drug **chloroquine**

$$Cl \longrightarrow \overset{N}{\underset{\quad}{\bigcirc\bigcirc}}$$

$$NH-\underset{\underset{CH_3}{|}}{CH}-CH_2-CH_2-CH_2-N(C_2H_5)_2$$

Chloroquine

is a weak base that, in uncharged form, freely penetrates the lysosome where it accumulates in charged form, thereby increasing the intralysosomal pH and inhibiting lysosomal function. The treatment of cells with chloroquine reduces their rate of protein degradation. Similar effects arise from treatment of cells with cathepsin inhibitors such as the polypeptide antibiotic **antipain.**

$$^-OOCCHNH-\overset{O}{\underset{\underset{Phe}{|}}{C}}-NH\overset{O}{\underset{\underset{Arg}{|}}{CHC}}-NH\overset{O}{\underset{\underset{Val}{|}}{CHC}}-NH\overset{O}{\underset{\underset{Arg}{|}}{CHC}}-H$$

Antipain

Lysosomal protein degradation in well nourished cells appears to be nonselective. Lysosomal inhibitors do not affect the rapid degradation of abnormal proteins or short-lived enzymes. Rather, they prevent the acceleration of nonselective protein breakdown on starvation. However, the continued nonselective degradation of proteins in starving cells would rapidly lead to an intolerable depletion of essential enzymes and regulatory proteins. Lysosomes therefore also have a selective pathway, which is activated only after a prolonged fast, that takes up and degrades proteins containing the pentapeptide Lys-Phe-Glu-Arg-Gln (KFERQ) or a closely related sequence. Such KFERQ proteins are selectively lost in fasting animals from tissues that atrophy in response to fasting (e.g., liver and kidney) but not from tissues that do not do so (e.g., brain and testes). KFERQ proteins are specifically bound in the cytosol and delivered to the lysosome by a 73-kD **peptide recognition protein (prp73),** a member of the 70-kD heat shock protein (Hsp70) family (Section 9-2C).

Both normal and pathological processes are associated with increased lysosomal activity. **Diabetes mellitus** (Section 27-3B) stimulates the lysosomal breakdown of proteins. Similarly, muscle wastage caused by disuse, denervation,

or traumatic injury arises from increased lysosomal activity. The regression of the uterus after childbirth, in which this muscular organ reduces its mass from 2 kg to 50 g in 9 days, is a striking example of this process. Many chronic inflammatory diseases, such as **rheumatoid arthritis,** involve the extracellular release of lysosomal enzymes that break down the surrounding tissues.

b. Ubiquitin Marks Proteins Selected for Degradation

It was initially assumed that protein degradation in eukaryotic cells is primarily a lysosomal process. Yet, reticulocytes, which lack lysosomes, selectively degrade abnormal proteins. The observation that protein breakdown is inhibited under anaerobic conditions led to the discovery of a cytosolically based ATP-dependent proteolytic system that is independent of the lysosomal system. This phenomenon was thermodynamically unexpected since peptide hydrolysis is an exergonic process.

The analysis of a cell-free rabbit reticulocyte system demonstrated that **ubiquitin** (Fig. 32-75) is required for ATP-dependent protein degradation. *This 76-residue monomeric protein, so named because it is ubiquitous as well as abundant in eukaryotes, is the most highly conserved protein known:* It is identical in such diverse organisms as humans, toad, trout, and *Drosophila* and differs in only three residues between humans and yeast. Evidently, ubiquitin is all but uniquely suited to an essential cellular function.

Proteins that are selected for degradation are so marked by covalently linking them to ubiquitin. This process, which is reminiscent of amino acid activation (Section 32-2C), oc-

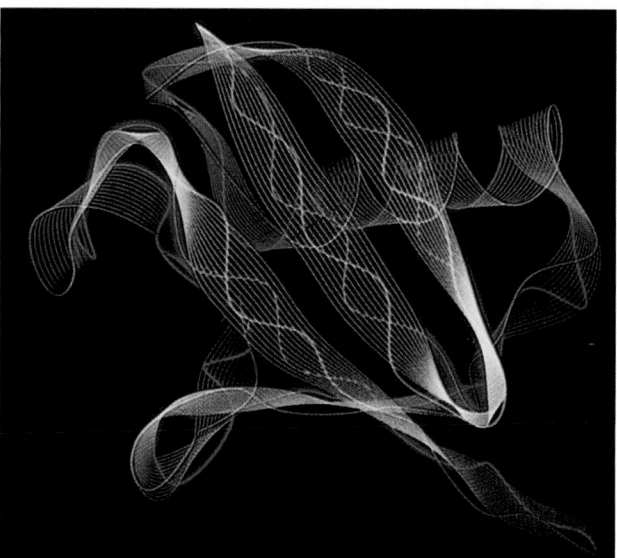

FIGURE 32-75 X-Ray structure of ubiquitin. The multistranded white ribbon represents the polypeptide backbone, and the red and blue curves, respectively, indicate the directions of the carbonyl and amide groups. [Courtesy of Mike Carson, University of Alabama at Birmingham. X-Ray structure determined by Charles Bugg, University of Alabama at Birmingham. PDBid 1UBQ.] 🔎 See the Interactive Exercises

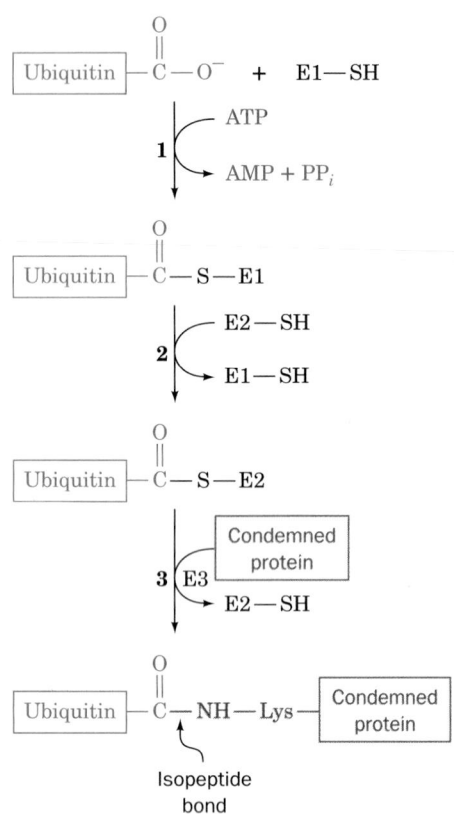

FIGURE 32-76   Reactions involved in the attachment of ubiquitin to a protein. In the first part of the process, ubiquitin's terminal carboxyl group is joined, via a thioester linkage, to E1 in a reaction driven by ATP hydrolysis. The activated ubiquitin is subsequently transferred to a sulfhydryl group of an E2 and then, in a reaction catalyzed by an E3, to a Lys ε-amino group on a condemned protein, thereby flagging the protein for proteolytic degradation by the 26S proteasome.

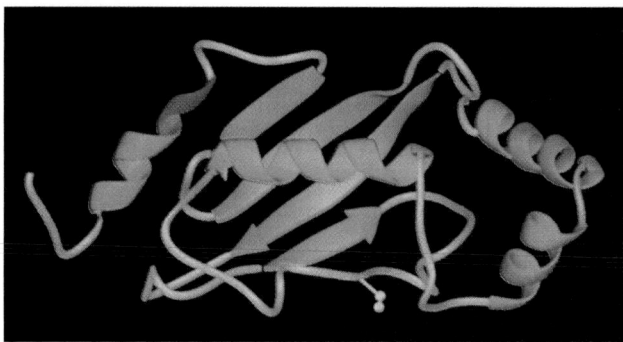

FIGURE 32-77   X-Ray structure of an E2 protein from *Arabidopsis thaliana.* α Helices are blue, the 3_{10} helical segment is purple, β strands are green, and the remainder of the molecule is light blue. The side chain of Cys 88, to which ubiquitin is covalently linked, is shown in ball-and-stick form in yellow. [Courtesy of William Cook, University of Alabama at Birmingham. PDBid 2AAK.]

curs in a three-step pathway that was elucidated notably by Avram Hershko and Aaron Ciechanover (Fig. 32-76):

1. In an ATP-requiring reaction, ubiquitin's terminal carboxyl group is conjugated, via a thioester bond, to **ubiquitin-activating enzyme (E1),** a homodimer of ~1050-residue subunits. Most organisms, including yeast and humans, have only one type of E1.

2. The ubiquitin is then transferred to a specific Cys sulfhydryl group on one of numerous proteins named **ubiquitin-conjugating enzymes (E2s;** 11 in yeast and over 20 in mammals). The various E2's are characterized by ~150-residue catalytic cores containing the active site Cys that exhibit at least 25% sequence identities and which mainly vary by the presence or absence of N- and/or C-terminal extensions that exhibit little sequence identity to each other. The X-ray and NMR structures of several species of E2 reveal that their catalytic cores all assume closely similar α/β structures (e.g., Fig. 32-77) in which most of the identical residues are clustered on one surface near the ubiquitin-accepting Cys residue, where they presumably interact with ubiquitin and E1.

3. *Ubiquitin–protein ligase (E3) transfers the activated ubiquitin from E2 to a Lys ε-amino group of its target protein, thereby forming an isopeptide bond.* Cells contain many species of E3s, each of which mediates the ubiquitination (alternatively, ubiquitylation) of a specific set of proteins and thereby marks them for degradation. Each E3 is served by one or a few specific E2s. The known E3s are members of two unrelated families, those containing a **HECT** domain (HECT for *h*omologous to *E*6AP *C-t*erminus) and those containing a so-called **RING finger** (RING for *r*eally *i*nteresting *n*ew *g*ene), although some E2s react well with members of both families. HECT domain E3s are modularly constructed with a unique N-terminal domain that interacts with its target proteins via their so-called **ubiquitination signals** (usually short polypeptide segments; see below) and an ~350-residue HECT domain that mediates E2 binding and catalyzes the ubiquitination reaction. The RING finger, which is implicated in recognizing the substrate protein's ubiquitination signal, is a 40- to 60-residue motif that binds two structurally but not catalytically implicated Zn^{2+} ions via a total of 8 Cys and His residues in a characteristic consensus sequence (much like the zinc finger motifs in certain DNA-binding proteins, Section 34-3B). RING finger–containing E3s may consist of a single subunit or may be multisubunit proteins in which the RING finger is contained in one subunit. RING finger E3s mediate the direct transfer of ubiquitin from E2 to a substrate protein's Lys residue, whereas HECT E3s do so via the intermediate transfer of ubiquitin from E2 to a catalytically essential Cys residue that is located ~35 residues from the N-terminus of the HECT domain.

In order for a target protein to be efficiently degraded, it must be linked to a chain of at least four tandemly linked ubiquitin molecules in which Lys 48 of each ubiquitin forms an isopeptide bond with the C-terminal carboxyl group of the succeeding ubiquitin (Fig. 32-78). These **polyubiquitin (polyUb)** chains, which can reach lengths of 50 or more ubiquitin molecules, are generated by the E3s, although how they

(a)

(b)

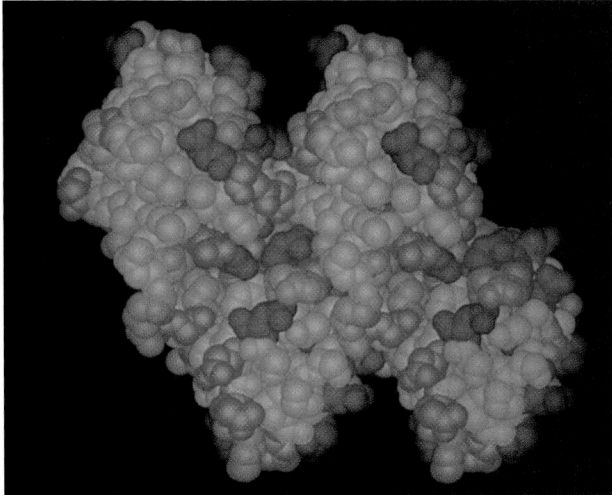

FIGURE 32-78   X-Ray structure of tetraubiquitin. (*a*) A ribbon drawing in which the isopeptide bonds connecting successive ubiquitin molecules, together with the Lys side chains making them, are orange. However, since the isopeptide bond connecting ubiquitins 2 and 3 is not visible in the X-ray structure, it is represented by a stick bond (this isopeptide bond nevertheless exists, as was demonstrated by SDS–PAGE of dissolved crystals). It seems likely that the monomer units in a multiubiquitin chain of any length would be arranged with the repeating symmetry of the tetraubiquitin structure. (*b*) A space-filling model, viewed as in Part *a*, in which basic residues (Arg, Lys, His) are blue, acidic residues (Asp, Glu) are red, uncharged polar residues (Gly, Ser, Thr, Asn, Gln) are purple, and hydrophobic residues (Ile, Leu, Val, Ala, Met, Phe, Tyr, Pro) are green. Note the unusually large solvent-exposed surface occupied by hydrophobic residues. [Courtesy of William Cook, University of Alabama at Birmingham. PDBid 1TBE.]

switch from transferring a ubiquitin to the target protein to processively synthesizing a polyubiquitin chain is unknown.

c. Ubiquitinated Proteins Are Hydrolyzed in the Proteasome

A ubiquitinated protein is proteolytically degraded to short peptides in an ATP-dependent process mediated by a

*large (2000 kD, 26S) multisubunit protein complex named the **26S proteasome** (sometimes spelled "proteosome") that electron micrographic studies reveal has the shape of a bi-capped hollow barrel (Fig. 32-79).* Proteolysis occurs inside the barrel, which permits this process to be extensive and processive, while preventing nonspecific proteolytic damage to other cellular components. PolyUb chains are the signals that target a protein to the proteasome; the identity of the target protein has little effect on the efficiency with which it is degraded by the proteasome. Nevertheless, the proteasome does not degrade ubiquitin molecules; they are returned to the cell. The size and functional complexity of this entire proteolytic system, which occurs in the nucleus as well as the cytosol, rivals that of the ribosome (Section 32-3) and the spliceosome (Section 31-4A) and hence is indicative of the importance of properly managing protein degradation. Indeed, ~5% of the proteins expressed by yeast participate in protein degradation. We discuss the structure and function of the 26S proteasome below.

d. Many E3s Have Elaborate Modular Structures

The proto-oncogene product **c-Cbl** (906 residues) is a single-subunit, RING finger–containing E3 that functions to ubiquitinate certain activated receptor tyrosine kinases (RTKs; Section 19-3A), thereby terminating their signaling. Nikola Pavletich determined the X-ray structure of the N-terminal half of c-Cbl (residues 47–447) in its ternary complex with the E2 protein **UbcH7** (which consists of little more than the ~150-residue E2 catalytic core) and an 11-residue peptide containing the ubiquitination signal from a nonreceptor tyrosine kinase (NRTK) named

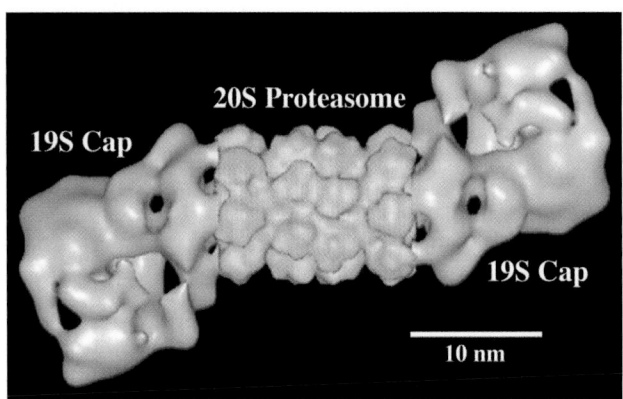

FIGURE 32-79   Electron microscopy–based image of the *Drosophila melanogaster* 26S proteasome. The complex is around 450 × 190 Å. The central portion of this 2-fold symmetric multiprotein complex (*yellow*), the **20S proteasome,** consists of four stacked 7-membered rings of subunits that form a hollow barrel in which the proteolysis of ubiquitin-linked proteins occurs. The **19S caps** (*blue*), which may attach to one or both ends of the 20S proteasome, control the access of condemned proteins to the 20S proteasome (see below). [Courtesy of Wolfgang Baumeister, Max-Planck-Institut für Biochemie, Martinsried, Germany.]

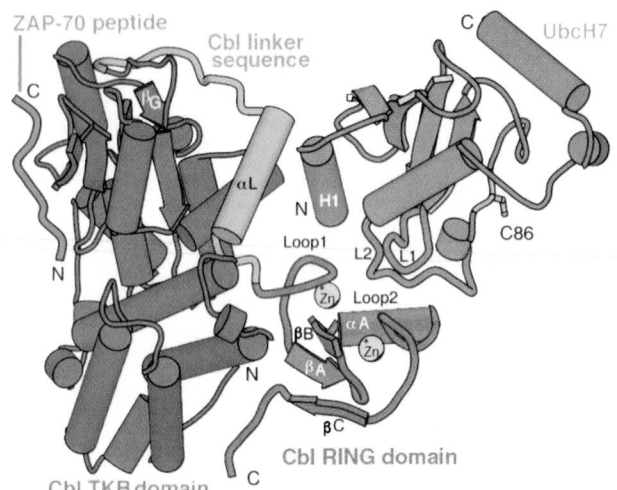

FIGURE 32-80   X-Ray structure of the human c-Cbl–UbcH7–ZAP-70 peptide ternary complex. UbcH7, an E2 that consists almost entirely of the E2 catalytic core, is colored blue with its active site Cys 86 shown in stick form (*yellow*). The 11-residue ubiquitination site of the RTK ZAP-70 is magenta. c-Cbl (residues 47–447 of the 903-residue protein), a RING finger, single-subunit E3, is colored according to domain with its RING finger domain red, its TKB domain green, and the linker joining them yellow. The RING finger's two bound Zn^{2+} ions are represented by gray spheres. [Courtesy of Nikola Pavletich, Memorial Sloan-Kettering Cancer Center, New York, New York. PDBid 1FBV.]

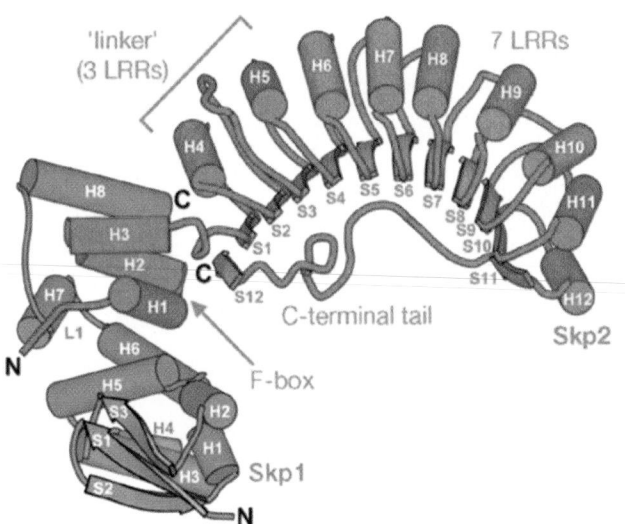

FIGURE 32-81   X-Ray structure of the human Skp1–Skp2 complex. Skp1 is blue and Skp2 is red. Skp2 consists of an N-terminal F-box that forms three helices, followed by 3 noncanonical so-called linker leucine-rich repeats (LRRs) that are contiguous with 7 LRRs that were predicted from their amino acid sequences for a total of 10 LRRs. After the tenth LRR, Skp2's ~30-residue C-terminal tail extends back past the first LRR by packing under the concave surface of the LRR domain. [Courtesy of Nikola Pavletich, Memorial Sloan-Kettering Cancer Center, New York, New York. PDBid 1FQB.]

 ZAP-70 (Fig. 19-42). The structure (Fig. 32-80) reveals that UbcH7, and c-Cbl's RING finger and SH2-containing tyrosine kinase-binding (TKB) domains interact with one another across multiple interfaces to form a compact and apparently rigid structure. The RING finger domain consists of a 3-stranded β sheet, an α helix, and two large loops that are held together by the two tetrahedrally coordinated Zn^{2+} ions. UbcH7 adopts the characteristic α/β fold of other E2s of known structure (Fig. 32-77). The ZAP-70 peptide is bound on the opposite side of the TKB domain from the UbcH7 active site Cys residue (Cys 86) and is ~60 Å distant from it.

SCF complexes are multisubunit RING finger E3s that consist of **Cul1** (a member of the **cullin** family; 776 residues), **Rbx1** (which contains the complex's RING finger domain; 108 residues), **Skp1** (163 residues), and a member of the **F-box protein** family (~430 to >1000 residues; SCF for *Skp1–cullin–F-box protein*). Rbx1 and Cul1 form the complex's catalytic core that binds E2; F-box proteins consist of an ~40-residue **F-box** that binds Skp1 followed by protein–protein interaction modules such as **leucine-rich repeats (LRRs)** or WD40 repeats (Section 19-2C) that bind substrate protein; and Skp1 functions as an adapter that links the F-box to Cul1. Cells contain numerous different F-box proteins (at least 38 in humans) that presumably permit the specific ubiquitination of a diverse variety of protein substrates (see below).

Pavletich has also determined the X-ray structures of two segments of the **SCF^{Skp2}** complex (where the super-

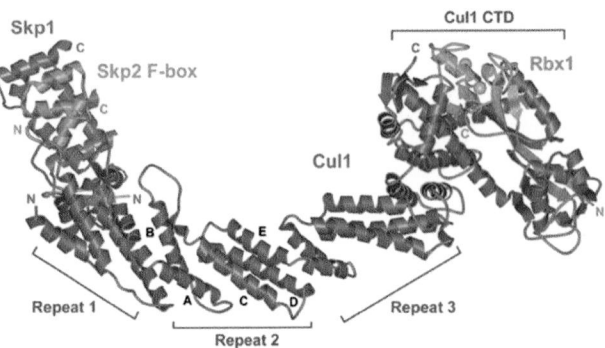

FIGURE 32-82   X-Ray structure of the human Cul1–Rbx1–Skp1–F-box^{Skp2} quaternary complex. Cul1, Rbx1, Skp1, and the Skp2 F-box are respectively colored green, red, blue, and magenta. The three cullin repeats of Cul1 are indicated with the five helices of the second repeat labeled A through E. The Zn^{2+} ions bound to Rbx1 are represented by yellow spheres. [Courtesy of Nikola Pavletich, Memorial Sloan-Kettering Cancer Center, New York, New York. PDBid 1LDK.]

script identifies the complex's F-box protein, here **Skp2**; 436 residues). The structure of the Skp1–Skp2 complex (Fig. 32-81) reveals that it has the shape of a sickle with the Skp1 and the 3-helix F-box of Skp2 forming the handle and its 10 LRRs (~26 residues each) forming the curved blade. The structure of the Cul1–Rbx1–Skp1–F-box^{Skp2} quaternary complex (Fig. 32-82) shows that Cul1 is an elongated protein that consists of a long stalk formed by

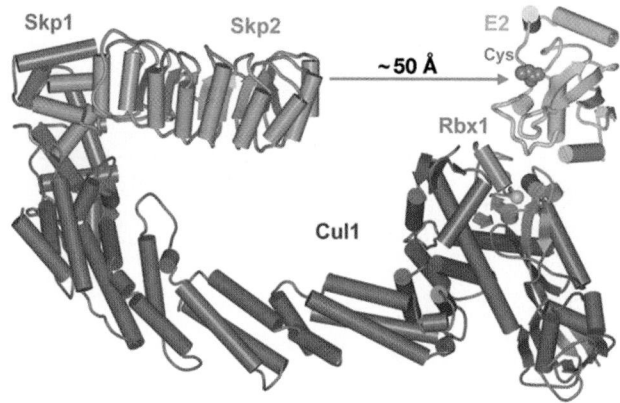

FIGURE 32-83 Model of the SCF^{Skp2}–E2 complex. The model, which is based on the X-ray structures in Figs. 32-80, 32-81, and 32-82, is colored and viewed as in Fig. 32-82. E2 is yellow with its active site Cys residue, to which ubiquitin would be covalently linked, drawn in space-filling form in cyan. The Zn^{2+} ions associated with the Rbx1 RING finger are represented by yellow spheres. The gray arrow indicates the 50-Å gap between the tip of the Skp2 LRR domain and the E2 active site. [Courtesy of Nikola Pavletich, Memorial Sloan-Kettering Cancer Center, New York, New York.]

TABLE 32-12 Half-lives of Cytoplasmic Enzymes as a Function of Their N-Terminal Residues

N-Terminal Residue	Half-life
Stabilizing	
Met	>20 h
Ser	
Ala	
Thr	
Val	
Gly	
Destabilizing	
Ile	~30 min
Glu	
Tyr	~10 min
Gln	
Highly destabilizing	
Phe	~3 min
Leu	
Asp	
Lys	
Arg	~2 min

Source: Bachmair, A., Finley, D., and Varshavsky, A., *Science* **234**, 180 (1986).

three repeats of a novel five-helix motif known as a cullin repeat followed by a globular domain that binds Rbx1. Apparently Cul1 acts like a rigid scaffold that organizes the Skp1–F-box^{Skp2} complex and Rbx1 so as to hold them over 100 Å apart. The Rbx1 RING finger contains a 20-residue insert that forms the binding site for a third tetrahedrally liganded Zn^{2+} ion.

The apparent rigidity of the foregoing three structures has enabled Pavletich to construct a model of the intact SCF^{Skp2}–E2 complex by superimposing Skp1–Skp2 on Cul1–Rbx1–Skp1–F-box^{Skp2} and docking the E2 UbcH7 onto the Rbx1 RING finger based on the c-Cbl–UbcH7 structure (Fig. 32-83). The model indicates that E2 and the LRR-containing domain of Skp2 are on the same side of the SCF complex but separated by a distance of ~50 Å. This suggests that Cul1's long stalk functions to separate the complex's substrate-binding and catalytic sites so that substrates with different sizes and various distances between their ubiquitinated Lys residues and their ubiquitination signals can be accommodated.

e. The Ubiquitin System Has Both Housekeeping and Regulatory Functions

Until the mid-1990s, it appeared that the ubiquitin system functioned mainly in a "housekeeping" capacity to maintain the proper balance among metabolic proteins and to eliminate damaged proteins. Indeed, as Alexander Varshavsky discovered, *the half-lives of many cytoplasmic proteins vary with the identities of their N-terminal residues (Table 32-12).* Thus, in a selection of 208 cytoplasmic proteins known to be long lived, all have a "stabilizing" residue, Met, Ser, Ala, Thr, Val, or Gly, at their N-termini. This so-called **N-end rule** is true for both eukaryotes and

prokaryotes, which suggests the system that selects proteins for degradation is conserved in eukaryotes and prokaryotes, even though prokaryotes lack ubiquitin.

The N-end rule results from the actions of the single-subunit, RING finger E3 named **E3α** (~1950 residues; also known as **Ubr1**) whose ubiquitination signals are the destabilizing N-terminal residues in Table 32-12. However, it is now clear that the ubiquitin system is far more sophisticated than a simple garbage disposal system. Thus, the growing list of known E3s have a variety of ubiquitination signals that often occur on a quite limited range of target proteins, many of which have regulatory functions. For example, the transcription factor **NF-κB** (NF for *n*uclear *f*actor), which plays a central role in immune and inflammatory responses, is maintained in an inactive state in the cytoplasm through its binding to the inhibitor **IκBα** (Fig. 12-38) in a way that occludes the short internal basic sequence that directs NF-κB's import into the nucleus (its **nuclear localization signal; NLS**). However, the stimulation of cell-surface receptors by certain cytokines initiates a signal transduction phosphorylation cascade (Section 19-3D) that phosphorylates IκBα bound to NF-κB at both Ser residues in the sequence DSGLDS. This phosphorylated sequence is the ubiquitination signal for the SCF complex containing the F-box protein **β-TrCP** (605 residues), which mediates the ubiquitination of the phosphorylated IκBα. The consequent destruction of IκBα exposes the NLS of NF-κB, which is then translocated to the nucleus where it activates the transcription of its target genes (Section 34-3B).

It has long been known that proteins with segments rich in Pro (P), Glu (E), Ser (S), and Thr (T), the so-called

PEST proteins, are rapidly degraded. It is now realized that this is because these PEST elements often contain phosphorylation sites that target their proteins for ubiquitination.

The ubiquitination system also has an essential function in cell cycle progression. The cell cycle, as we have seen in Section 30-4A and will further discuss in Section 34-4D, is regulated by a series of proteins known as cyclins. A given cyclin, which is expressed immediately preceding and/or during a specific phase of the cell cycle, binds to a corresponding cyclin-dependent protein kinase (Cdk), which then phosphorylates its target proteins so as to activate them to carry out the processes of that phase of the cell cycle. In addition, many cyclins also inhibit the transition to the subsequent phase of the cell cycle (e.g., DNA replication or mitosis). Thus, in order for a cell to progress from one phase of the cell cycle to the next, the cyclin(s) governing that phase must be eliminated. This occurs via the specific ubiquitination of the cyclin, thereby condemning it to be destroyed by the proteasome. The E3s responsible for this process are the SCF complexes containing F-box proteins targeted to a corresponding cyclin and a multisubunit complex known as the **anaphase-promoting complex (APC;** alternatively the **cyclosome).** APC, an ~1500-kD RING finger–containing particle that in yeast consists of 11 subunits, specifically ubiquitinates proteins that contain the 9-residue consensus sequence RTALGDIGN, the so-called **destruction box,** near their N-termini.

Some viruses usurp the ubiquitination system. Oncogenic forms of **human papillomavirus (HPV),** the cause of nearly all cervical cancers (a leading cause of death of women in developing countries), encode the ~150-residue **E6 protein,** which combines with the 875-residue cellular protein named **E6-associated protein (E6AP;** the first E3 known to contain a HECT domain) to ubiquitinate **p53,** thereby marking it for destruction. This latter protein is a transcription factor that monitors genome integrity and hence is important in preventing malignant transformation and the proliferation of cancer cells (Section 34-4C), that is, it is a **tumor suppressor** (a protein whose loss of function is a cause of cancer). Consequently, HPV provokes the uncontrolled growth of the cells it infects and hence its own proliferation. E6AP normally functions to ubiquitinate certain members of the Src family of protein tyrosine kinases (Section 19-3B), including Src itself. The deletion of the segment of chromosome 15 that contains the E6AP gene causes Angelman syndrome, which as we have seen (Section 30-7) is characterized by severe mental retardation and is exclusively maternally inherited due to genomic imprinting.

BRCA1 is a tumor suppressor, many of whose mutations greatly increase the predisposition for a number of cancers, notably breast and ovarian cancers (Section 30-6A; BRCA for *br*east *ca*ncer). Around 20% of BRCA1's clinically relevant mutations occur in this 1863-residue protein's N-terminal 100 residues, which contains a RING finger domain. BRCA1 forms a heterodimer with the RING finger domain-containing protein **BARD1** (for *B*RCA1-*a*ssociated *R*ing *d*omain protein 1; 777 residues) that collectively functions as a ubiquitin-protein ligase (E3), whose

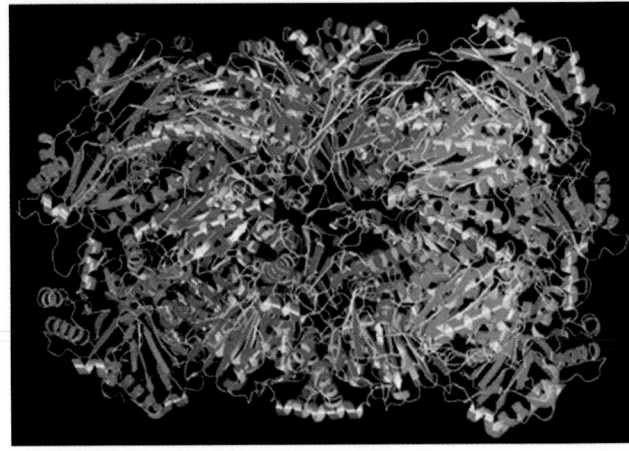

(a)

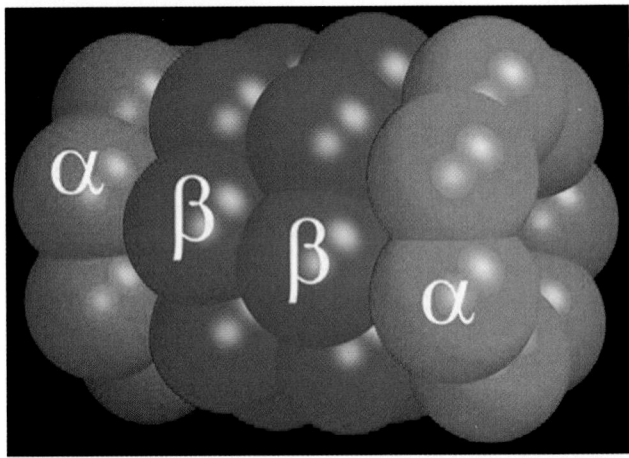

(b)

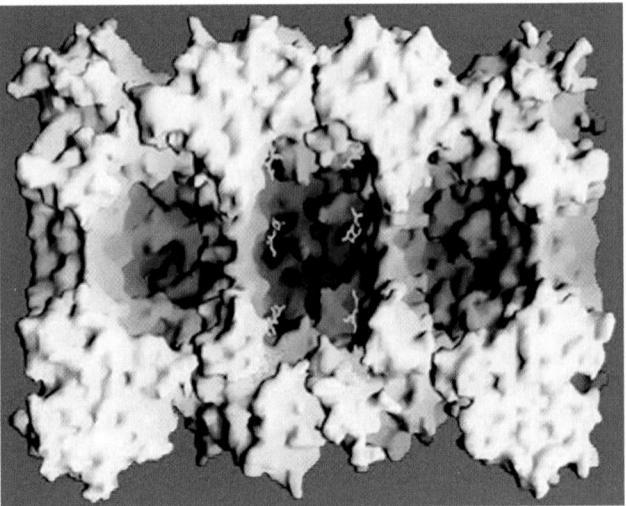

(c)

FIGURE 32-84   X-Ray structure of the *T. acidophilum* 20S proteasome. (*a*) A ribbon diagram, in which α subunits are red and β subunits are blue, viewed with a 2-fold axis tilted to the left and its 7-fold axis horizontal. Only subunits with a forward exposure are shown for clarity. (*b*) A diagram in which subunits are represented by equal-sized spheres that is viewed and colored as in Part *a*. (*c*) A cutaway diagram of the solvent-accessible surface showing its three internal cavities. The active sites on the β subunits are marked by the bound inhibitor, LLnL, which is drawn in stick form in yellow. [Courtesy of Robert Huber, Max-Planck-Institut für Biochemie, Martinsried, Germany. PDBid 1PMA.]

cognate ubiquitin-conjugating enzyme (E2), **UbcH5c,** binds only to the BRCA1 RING finger domain. All known cancer-associated mutations of the BRCA1 RING finger domain abrogate the heterodimer's E3 activity *in vitro*.

f. The 20S Proteasome Catalyzes Proteolysis Inside a Hollow Barrel

The 26S proteasome (Fig. 32-79) is an ~2100-kD multisubunit protein that catalyzes the ATP-dependent hydrolysis of ubiquitin-linked proteins. This yields oligopeptides averaging from 7 to 9 residues in length that are subsequently degraded to their component amino acids by cytosolic exopeptidases. The 26S proteasome consists of a **20S proteasome** (~670 kD), the barrel-shaped catalytic core of the 26S proteasome, and its **19S caps** (~700 kD; also known as **PA700** and the **19S regulator**), which associate with the ends of the 20S proteasome and stimulate its activity (PA for *proteasome activator*). The 20S proteasome only hydrolyzes unfolded proteins in an ATP-independent manner; the 19S caps function to identify and unfold the ubiquitinated protein substrates.

The 20S proteasome occurs in the nuclei and cytosol of all eukaryotic cells and in all Archaebacteria yet examined. However, the only eubacteria in which it occurs are those of the class Actinomyces (Fig. 1-4), which suggests that they obtained it via horizontal gene transfer from some other organism.

The 20S proteasome of *Thermoplasma acidophilum* (an archaebacterium) consists of 14 copies each of α and β subunits (233 and 203 residues) that electron microscopy studies reveal form a 150-Å long and 110-Å in-diameter barrel in which the subunits are arranged in four stacked rings (as is evident in the central portion of the 26S proteasome seen in Fig. 32-79). The α and β subunits are 26% identical in sequence except for an ~35-residue N-terminal segment of the α subunit, which the β subunit lacks. Eukaryotic 20S proteasomes are more complex in that they consist of 7 different α-like and 7 different β-like subunits versus only one of each type for the *T. acidophilum* 20S proteasome.

The X-ray structure of the *T. acidophilum* 20S proteasome, determined by Wolfgang Baumeister and Robert Huber, reveals that its two inner rings each consist of 7 β subunits and its two outer rings each consist of 7 α subunits arranged with D_7 symmetry (Fig. 32-84). Thus the overall structure of the 20S proteasome superficially resembles that of the unrelated molecular chaperone GroEL (Section 9-2C). The structures of the α and β subunits are remarkably similar (Fig. 32-85) except, of course, for the α subunit's N-terminal segment, which contacts the N-terminal segment in an adjacent α subunit. This accounts for the observation that α subunits alone spontaneously assemble into 7-membered rings (a capacity that is abolished by the deletion of their N-terminal 35 residues), whereas β subunits alone remain monomeric.

The central channel of the *T. acidophilum* 20S proteasome, which has a maximum diameter of 53 Å, consists of three large chambers (Fig. 32-84c: Two are located at the interfaces between adjoining rings of α and β subunits, with

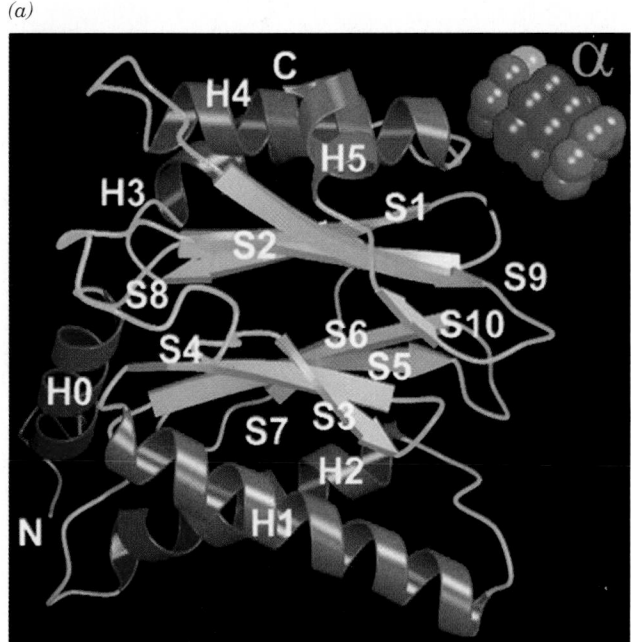

(a)

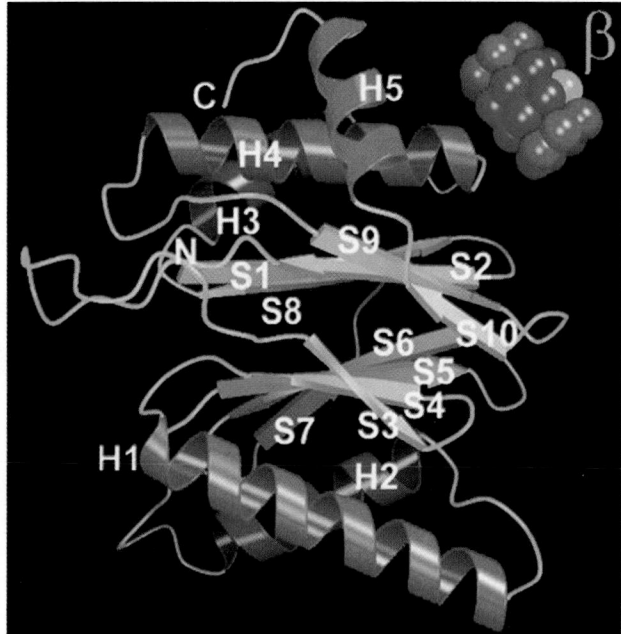

(b)

FIGURE 32-85 X-Ray structures of the subunits of the *T. acidophilum* 20S proteasome. (*a*) The α subunit which is colored according to secondary structure with helices red, β strands blue, and other segments yellow. The N-terminal helix, H0, occupies a position at the end of the proteasome.

The identically oriented subunit is represented by the yellow sphere in the inset diagram of the entire 20S subunit (*upper right*). (*b*) The similarly colored and oriented β subunit. [Courtesy of Robert Huber, Max-Planck-Institut für Biochemie, Martinsried, Germany. PDBid 1PMA.]

the third, larger chamber centrally located between the two rings of β subunits. Unfolded polypeptide substrates appear to enter the central chamber of the barrel (where the proteasome's active sites are located; see below) through ~13-Å in diameter axially located apertures in the α rings that are lined with hydrophobic residues. This allows only unfolded proteins to enter the central chamber, thereby protecting properly folded proteins from indiscriminant degradation by this omnivorous protein-dismantling machine.

The X-ray structure of the yeast 20S proteasome, also determined by Huber, demonstrates that its outer and inner rings respectively consist of seven different α-type subunits and seven different β-type subunits, all of which are uniquely arranged (Fig. 32-86). The α-like subunits have folds that are similar to one another as well as to that of the *T. acidophilum* 20S proteasome and likewise for the β-like subunits. Consequently, this 28-subunit, 6182-residue protein complex has exact 2-fold rotational symmetry relating the two pairs of rings but only pseudo-7-fold rotational symmetry relating the subunits within each ring. The narrow axial apertures in the α rings through which unfolded polypeptides almost certainly enter the hydrolytic chamber in the *T. acidophilum* 20S proteasome (Fig. 32-84c) are closed in the yeast 20S proteasome (Fig. 32-86c) by a plug formed by the interdigitation of its α subunits' N-terminal tails. This suggests that the 19S caps of the 26S proteasome, which have been shown to activate the 20S proteasome, control the access to it by inducing conformational changes in its α rings. The X-ray structure of the bovine 20S proteasome, determined by Tomitake Tsukihara, reveals that its arrangement of seven α-type and seven β-type subunits is similar to that in yeast.

g. The Proteasome Catalyzes Peptide Hydrolysis via a Novel Mechanism

The X-ray structure of the *T. acidophilum* 20S proteasome in complex with the aldehyde inhibitor **acetyl-Leu-Leu-norleucinal (LLnL)**

Acetyl-Leu-Leu-norleucinal (LLnL)

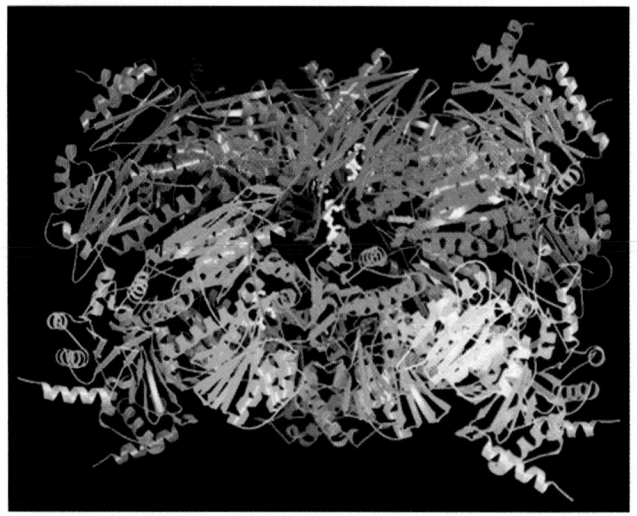

(a)

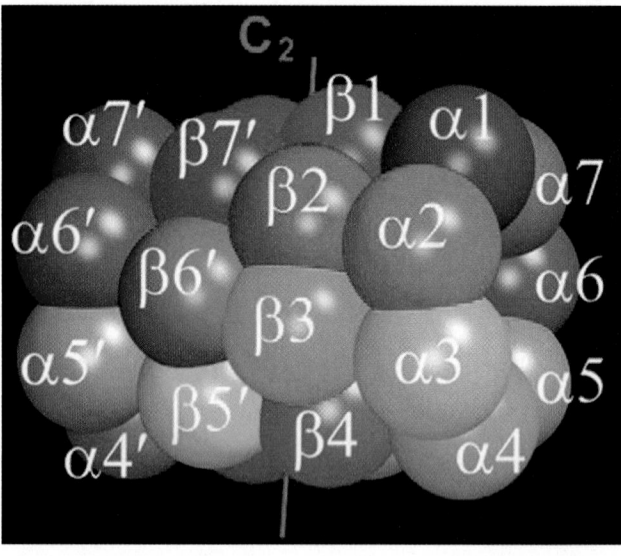

(b)

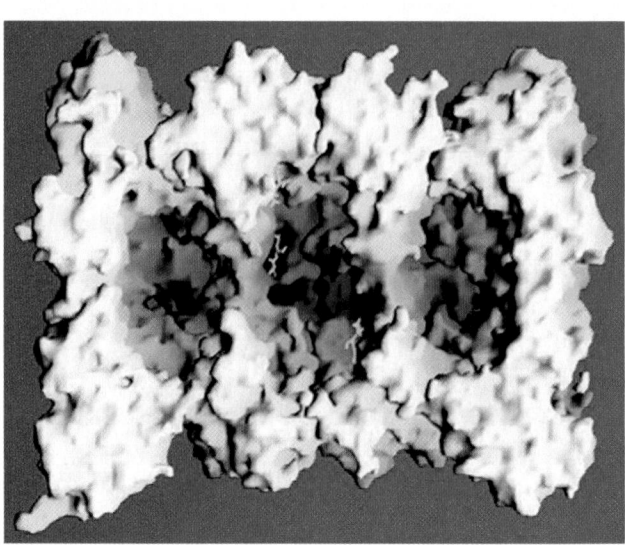

(c)

FIGURE 32-86 X-Ray structure of the yeast 20S proteasome. (a) A ribbon diagram in which the various subunits are differently colored and viewed with a 2-fold axis tilted to the left and its pseudo-7-fold axis horizontal. Only subunits with a forward exposure are shown for clarity. (b) A diagram in which subunits are represented by equal-sized spheres that are viewed and colored as in Part a. (c) A cutaway diagram of the solvent-accessible surface showing its three internal cavities. The active sites on the β1, β2, and β5 subunits are marked by the bound inhibitor, LLnL, which is drawn in stick form in yellow. Compare these diagrams with those in Fig. 32-84. [Courtesy of Robert Huber, Max-Planck-Institut für Biochemie, Martinsried, Germany. PDBid 1RYP.]

reveals that its active sites are on the inner surfaces of its rings of β subunits, with the aldehyde function of the LLnL close to the side chain of the highly conserved Thr 1β. Deletion of this Thr or its mutation to Ala yields properly assembled 20S proteasomes that are completely inactive. Evidently, 20S proteasomes catalyze peptide hydrolysis by a novel mechanism in which the hydroxyl group of its Thr 1β is the attacking nucleophile. This as yet poorly understood mechanism in which the amino group at the N-terminus presumably acts to increase the nucleophilicity of the hydroxyl sidechain, is now known to be employed by other hydrolases (e.g., glutamate synthase; Section 26-5A), which are collectively known as the **N-terminal nucleophile (Ntn) family** of hydrolases. The *T. acidophilum* β subunits preferably cleave polypeptides after hydrophobic residues. However, in the yeast and the bovine 20S proteasomes, only subunits β1, β2, and β5 are catalytically active. Their respective preferences for cleavage after acidic, basic (trypsin-like), and hydrophobic (chymotrypsin-like) residues are explained by the basic, acidic, and nonpolar characters of their pockets that bind the side chain of the residue preceding the scissile peptide bond. The functions of the four different catalytically inactive β subunits is unknown, although mutagenically modifying an inactive β subunit can abolish the catalytic activity of an active β subunit.

h. The 19S Caps Control the Access of Ubiquitinated Proteins to the 20S Proteasome

The 20S proteasome probably does not exist alone *in vivo;* it is most often in complex with two 19S caps that function to recognize ubiquitinated proteins, unfold them, and feed them into the 20S proteasome in an ATP-dependent manner (it may also associate with other regulatory complexes; see below). The 19S cap, which consists of ~18 different subunits, is poorly characterized due in large part to its low intrinsic stability. Its so-called base complex consists of 9 different subunits, 6 of which are ATPases that form a ring that abuts the α ring of the 20S proteasome (Fig. 32-79). Each of these ATPases contains an ~230-residue **AAA** module (AAA for *ATPases associated* with a variety of *activities*, a highly diverse family of proteins; also called **AAA+**). Cecile Pickart has demonstrated via cross-linking experiments that one of these ATPases, named **S6′** (alternatively **Rpt5**), contacts the polyUb signal that targets a condemned protein to the 26S proteasome. This suggests that the recognition of the polyUb chain as well as substrate protein unfolding are ATP-driven processes. Moreover, the ring of ATPases must function to open (gate) the otherwise closed axial aperture of the 20S proteasome so as to permit the entry of the unfolded substrate protein.

Eight additional subunits form the so-called lid complex, the portion of the 19S cap that is more distal to (distant from) the 20S proteasome. The functions of the lid subunits are largely unknown, although a truncated 26S proteasome that lacks the lid subunits is unable to degrade polyubiquitinated substrates. Several other subunits may be transiently associated with the 19S cap and/or with the 20S proteasome.

i. Deubiquitinating Enzymes Have Several Functions

The enzymes that hydrolytically cleave the isopeptide bonds linking successive ubiquitin units in polyUb are known as **deubiquitinating enzymes (DUBs).** Cells contain a surprisingly large number of DUBs (at least 17 in yeast and nearly 100 in humans). Nearly all known DUBs are **cysteine proteases,** enzymes whose catalytic mechanism resembles that of serine proteases (Section 15-3C) but whose attacking nucleophile is Cys—S$^-$ rather than Ser—OH.

DUBs may release entire polyUb chains from a condemned protein or sequentially release ubiquitin units from the chain terminus. It has been proposed that this latter process functions as a clock to time the protein degradation process. If a polyUb chain is trimmed to less than four ubiquitin units before degradation begins, then its attached protein is likely to escape destruction. This would spare proteins that had been inappropriately tagged with only short polyUb chains.

The mammalian 19S lid subunit known as **POH1** (**Rpn11** for the 65% identical yeast subunit) appears to be responsible for the deubiquitination of target proteins prior to their degradation; its inactivation prevents target protein degradation. Curiously, this DUB is a Zn^{2+}-dependent protease (as is carboxypeptidase A; Fig. 15-42) rather than a cysteine protease.

Certain DUBs function to dismember polyUb chains that have been released from substrate proteins by sequentially removing ubiquitin units from the end of the chain that is nearest to the substrate protein (that with a free C-terminus). Consequently, these DUBs cannot remove ubiquitin units from polyUb chains that are still attached to substrate proteins, thereby preventing their premature removal.

Cells express ubiquitin as polyproteins containing several ubiquitin units or with ubiquitin fused to certain ribosomal subunits (there is no gene that encodes a single ubiquitin unit). These polyproteins are rapidly processed by as yet unidentified DUBs to yield free ubiquitin.

j. The 11S Regulator Forms a Heptameric Barrel That Opens the 20S Proteasome

Higher eukaryotes contain an **11S regulator** that functions to open the channel into the 20S proteasome in an ATP-independent manner so as to permit the entrance of polypeptides (but not folded proteins). The mammalian 11S regulator, which functions in the generation of peptides for presentation to the immune system (Section 35-2E), is named **REG** (alternatively **PA28**). It is a heteroheptameric complex of two ~245-residue subunits, **REGα** and **REGβ,** that exhibit ~50% sequence identity except for a highly variable internal 18-residue segment that is thought to confer subunit-specific properties. Indeed REGα alone forms a heptamer whose biochemical properties are similar to that of REG (although both subunits must be present *in vivo*).

The trypanosome *Trypanosoma brucei*, which lacks 19S caps, expresses a homoheptameric 11S regulator named **PA26** that is only 14% identical to human REGα. Nevertheless, the various 11S regulators activate 20S proteasomes from widely divergent species. Thus, rat 20S proteasome is activated by PA26 and the yeast 20S proteasome is activated by human REGα despite the fact that yeast lacks 11S regulators.

The X-ray structure of PA26 in complex with the yeast 20S proteasome, determined by Christopher Hill, reveals that each PA26 monomer consists of an up–down–up–down 4-helix bundle. These monomers form a 7-fold symmetric heptameric barrel that is 90 Å in diameter, 70 Å long, and has a 33-Å in diameter central pore (Fig. 32-87*a*) and which closely resembles the previously determined X-ray structure of human REGα. Two PA26 barrels associate coaxially with the 20S proteasome, one at each end (Fig. 32-87*b*). The conformation of the 20S proteasome in this complex, for the most part, is closely similar to that of the 20S proteasome alone (Fig. 32-86). However, the C-terminal tails of the PA26 subunits insert into pockets on the 20S proteasome's α subunits in a way that induces conformational changes in its N-terminal tails that clear the 20S proteasome's otherwise blocked central aperture, thus permitting unfolded polypeptides to enter the proteasome's central chamber.

k. Bacteria Contain a Variety of Self-Compartmentalized Proteases

Nearly all eubacteria lack 20S proteasomes. Nevertheless they have ATP-dependent proteolytic assemblies

that share the same barrel-shaped architecture and carry out similar functions. For example, in *E. coli*, two proteins known as **Lon** and **Clp** mediate up to 80% of the bacterium's protein degradation, with additional contributions from at least three other proteins including **heat shock locus UV (HslUV)**. Thus, *all cells appear to contain proteases whose active sites are only available from the inner cavity of a hollow particle to which access is controlled.* These so-called **self-compartmentalized proteases** appear to have arisen early in the history of cellular life, before the ad-

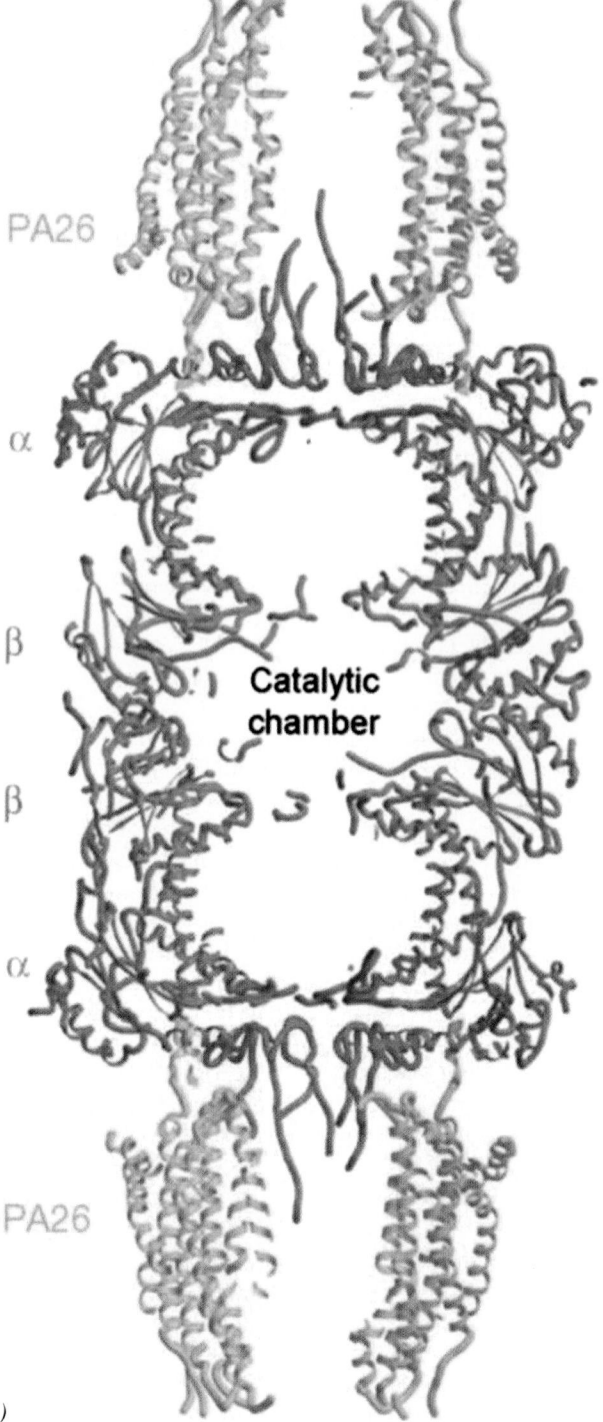

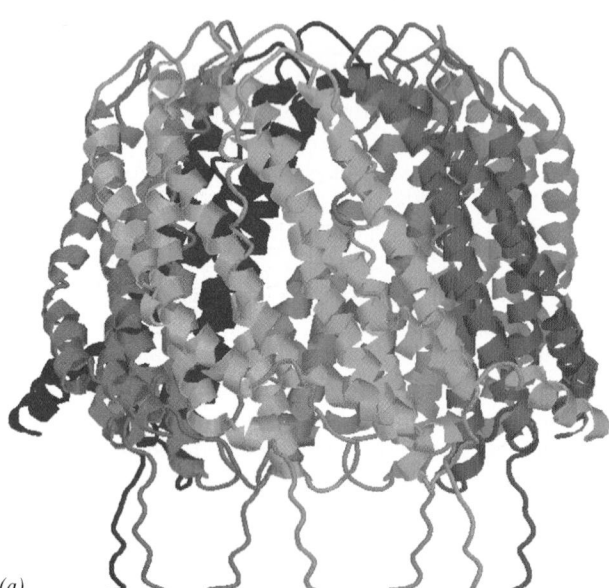

(a)

FIGURE 32-87 X-Ray structure of *T. brucei* PA26 in complex with the yeast 20S proteasome. (*a*) The PA26 heptamer viewed with its 7-fold axis vertical. Each of its subunits are differently colored. (*b*) Cutaway diagram of the entire complex viewed with its 7-fold axis vertical. The PA26 is yellow, the α and β subunits of the 20S proteasome are magenta and blue, its α-annulus is green, and its N-terminal segments that are ordered and partially disordered are red and pink. [Part *a* based on an X-ray structure by and Part *b* courtesy of Christopher Hill, University of Utah. PDBid 1FNT.]

(b)

vent of eukaryotic membrane-bound organelles such as the lysosome, which similarly carry out degradative processes in a way that protects the cell contents from indiscriminant destruction.

Clp protease consists of two components, the proteolytically active **ClpP** and one of several ATPases, which in *E. coli* are **ClpA** and **ClpX.** The X-ray structure of ClpP, determined by John Flanagan, reveals that it oligomerizes to

form a ~90-Å long and wide hollow barrel that consists of two back-to-back 7-fold symmetric rings of 193-residue subunits and thereby has the same D_7 symmetry as does the 20S proteasome (Fig. 32-88). Nevertheless, the ClpP subunit has a novel fold that is entirely different from that of the 20S proteasome's homologous α and β subunits. The ClpP active site, which is only exposed on the inside of the barrel, contains a catalytic triad composed of Ser 97, His 122, and Asp 171, and hence is a serine protease (Section 15-3B).

HslUV protease appears to be a hybrid of Clp and the 26S proteasome. Its **HslV** subunits in *E. coli* (145 residues) are 18% identical to the β subunits of the *T. acidophilum* 20S proteasome, whereas its regulatory **HslU** caps (443 residues) have ATPase activity and are homologous to *E. coli* ClpX. The X-ray structure of HslUV, determined by Huber, indicates that HslV forms a dimer of hexameric rather than heptameric rings (Fig. 32-89). A hexameric ring

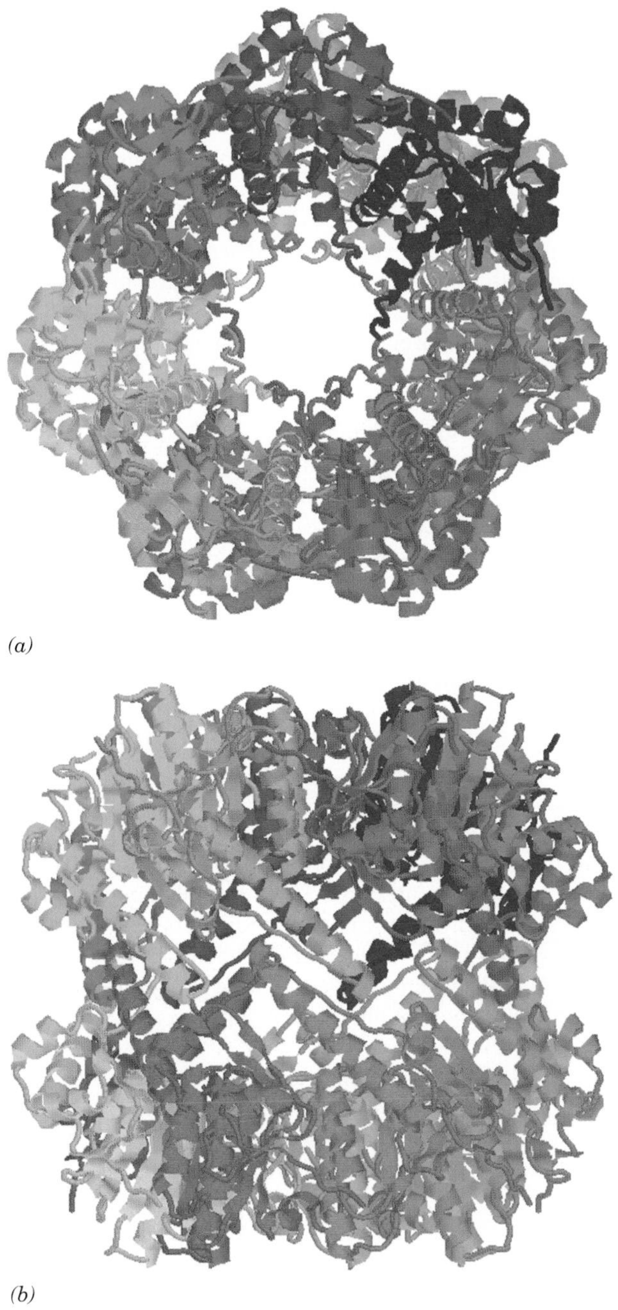

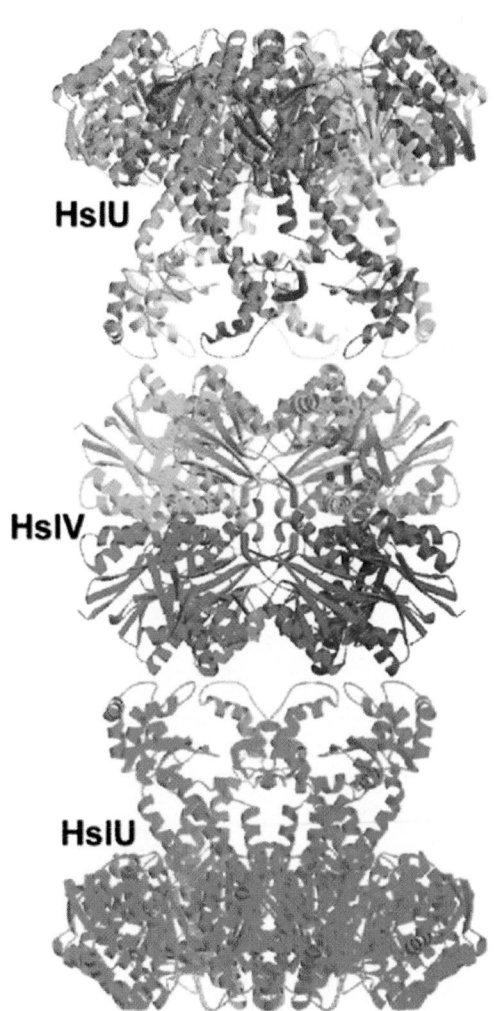

FIGURE 32-89   X-Ray structure of *E. coli* HslVU. The 820-kD complex is viewed along a 2-fold axis with its 6-fold axis vertical. The D_6 symmetric dodecamer of HslV subunits is coaxially bound at both ends by C_6 symmetric HslU hexamers to yield a complex with overall D_6 symmetry. [Courtesy of Robert Huber, Max-Planck-Institut für Biochemie, Martinsried, Germany. PDBid 1E94.]

(a)

(b)

FIGURE 32-88   X-Ray structure of *E. coli* ClpP. (*a*) View of the heptameric complex along its 7-fold axis with each of its subunits differently colored. (*b*) View along the complex's 2-fold axis (rotated 90° about a horizontal axis with respect to Part *a*). [Based on an X-ray structure by John Flannagan, Brookhaven National Laboratory, Upton, New York. PDBid 1TYF.]

(a)

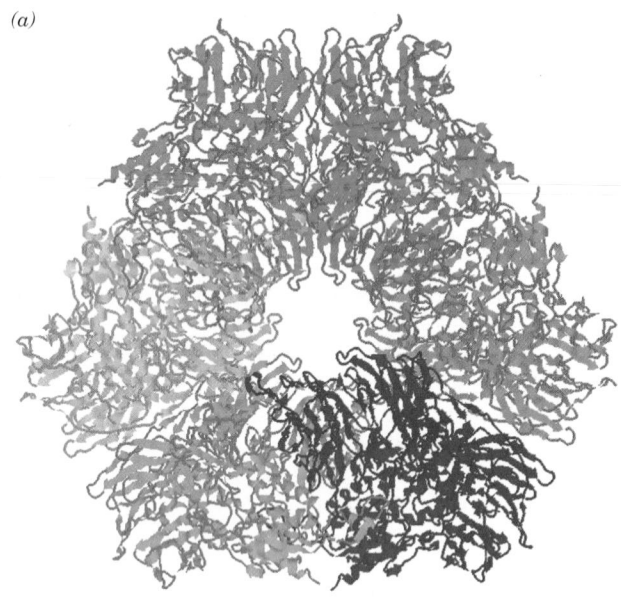

(b)

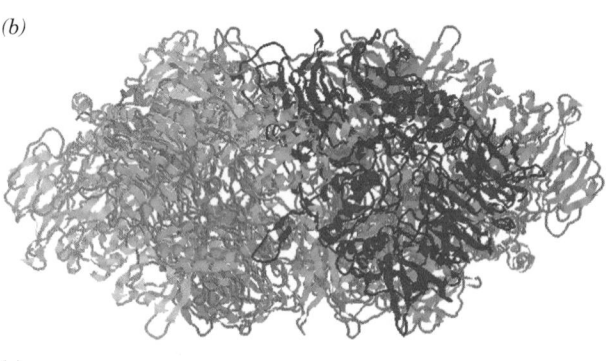

(c)

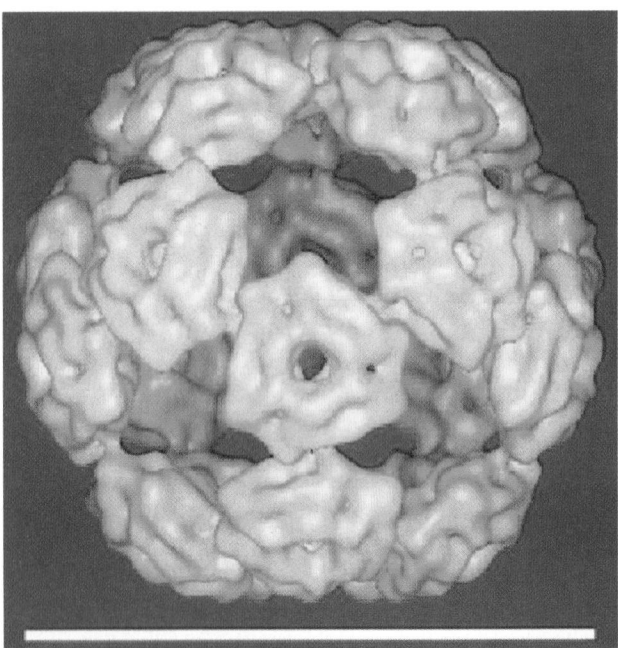

FIGURE 32-90   Structure of the tricorn protease. (*a*) The X-ray structure of the hexameric complex as viewed along its 3-fold axis with its subunits differently colored. (*b*) View along the hexameric complex's 2-fold axis (rotated 90° about a horizontal axis with respect to Part *a*). (*c*) A cryo-EM–based image of the icosahedral complex. Each of its component "plates" represents a hexameric complex such as those drawn in Parts *a* and *b*. Note how neighboring plates are related by both 3-fold axes (which are coincident with those of the hexameric complexes) and 5-fold axes. The white scale bar is 50 nm long. [Parts *a* and *b* are based on an X-ray structure by Robert Huber, Max-Planck-Institut für Biochemie, Martinsried, Germany. PDBid 1K32. Part *c* is courtesy of Wolfgang Baumeister, Max-Planck-Institut für Biochemie, Martinsried, Germany.]

Nevertheless, both the fold and the intersubunit contacts of the HslV subunits are closely similar to those of the 20S proteasome β subunits. In addition, both have N-terminal Thr residues. Thus, HslV can be regarded as the eubacterial homolog of archaebacterial and eukaryotic 20S proteasomes.

T. acidophilum contains another large proteolytic complex, which is unrelated to the proteasome. The X-ray structure of this protease (Fig. 32-90*a,b*), determined by Huber, indicates that it forms a 730-kD toroidal hexameric ring with D_3 symmetry that has a peculiar triangular shape reminiscent of a tricorn (a hat whose brim is turned up on three sides) and hence was named **tricorn protease.** Cryo-EM studies indicate that 20 of these tricorn hexamers associate to form a 14,600-kD hollow icosahedron (Fig. 32-90*c*; an icosahedron is shown in Fig. 8-64*c*), making it by far the largest homooligomeric enzyme complex known (it is even larger than some virus particles, many of which also have icosahedral symmetry; Section 33-2A).

I. Ubiquitinlike Modifiers Participate in a Variety of Regulatory Processes

Eukaryotic cells express several proteins that are related in sequence to ubiquitin and are similarly conjugated to other proteins. These **ubiquitinlike modifiers (Ubls),** which participate in a variety of fundamental cellular processes, each have a corresponding activating enzyme (E1), at least one conjugating enzyme (E2), and in many cases, one or more ligases (E3s), that function to link the Ubl to its target protein(s) in a manner closely resembling that of ubiquitin.

Two of the most extensively studied Ubls are **SUMO** (*s*mall *u*biquitin-related *mo*difier; 18% identical to ubiquitin) and **RUB1** (*r*elated-to-*ub*iquitin 1; called **NEDD8** in vertebrates; 50% identical to ubiquitin), proteins that are highly conserved from yeast to humans. One of SUMO's target proteins is IκBα, which binds to the transcription factor NF-κB so as to occlude its nuclear localization signal (see above). IκBα is sumoylated at the same residue (Lys 21) at which it is ubiquitinated, thereby blocking its ubiquitination and subsequent degradation and hence preventing the translocation of NF-κB to the nucleus. Evidently, there is a complex regulatory interplay between the ubiquitination and sumoylation of IκBα. SUMO also modifies two mammalian glucose transporters, GLUT1 and

of HslU subunits binds to both ends of the HslV dodecamer to form a 24-subunit assembly with D_6 symmetry, rather than the D_7 symmetry of the 20S proteasome.

GLUT4 (Section 20-2E), and in doing so, increases the availability of GLUT4 but decreases that of GLUT1.

All known RUB1 targets are cullins, all of which are subunits of SCF complexes, the multisubunit RING finger E3s (see above). In fact, β-TrCP, the E3 that directs the ubiquitination of IκBα, must be conjugated to RUB1 before it can do so, thereby adding further complexity to the control of NF-κB.

■ CHAPTER SUMMARY

1 ■ The Genetic Code Point mutations are caused by either base analogs that mispair during DNA replication or by substances that react with bases to form products that mispair. Insertion/deletion (frameshift) mutations arise from the association of DNA with intercalating agents that distort the DNA structure. The analysis of a series of frameshift mutations that suppress one another has established that the genetic code is an unpunctuated triplet code. In a cell-free protein synthesizing system, poly(U) directs the synthesis of poly(Phe), thereby demonstrating that UUU is the codon specifying Phe. The genetic code was elucidated through the use of polynucleotides of known composition but random sequence, by the ability of defined triplets to promote the ribosomal binding of tRNAs bearing specific amino acids, and through the use of synthetic mRNAs of known alternating sequences. The latter investigations have also demonstrated that the 5′ end of mRNA corresponds to the N-terminus of the polypeptide it specifies and have established the sequences of the Stop codons. Degenerate codons differ mostly in the identities of their third base. Small single-stranded DNA phages such as φX174 contain overlapping genes in different reading frames. The genetic code used by mitochondria differs in several codons from the "standard" genetic code.

2 ■ Transfer RNA and Its Aminoacylation Transfer RNAs consist of 54 to 100 nucleotides that can be arranged in the cloverleaf secondary structure. As many as 10% of a tRNA's bases may be modified. Yeast tRNA^{Phe} forms a narrow, L-shaped, three-dimensional structure that resembles that of other tRNAs. Most of the bases are involved in stacking and base pairing associations including nine tertiary interactions that appear to be essential for maintaining the molecule's native conformation. Amino acids are appended to their cognate tRNAs in a two-stage reaction catalyzed by the corresponding aminoacyl–tRNA synthetase (aaRS). There are two classes of aaRSs, each containing 10 members. Class I aaRSs have two conserved sequence motifs that occur in the Rossmann fold common to the catalytic domain of these enzymes. Class II aaRSs have three conserved sequence motifs that occur in the 7-stranded antiparallel β sheet-containing fold that forms the core of their catalytic domains. In binding only their cognate tRNAs, aaRSs recognize only an idiosyncratic but limited number of bases (identity elements) that are, most often, located at the anticodon and in the acceptor stem. The great accuracy of tRNA charging arises from the proofreading of the bound amino acid by certain aminoacyl–tRNA synthetases via a double-sieve mechanism and at the expense of ATP hydrolysis.

Many organisms and organelles lack a GlnRS and instead synthesize Gln–tRNA^{Gln} by the GluRS-catalyzed charging of tRNA^{Gln} with glutamate followed by its transamidation using glutamine as the amido group source in a reaction mediated by Glu–tRNA^{Gln} amidotransferase (Glu-AdT). Ribosomes select tRNAs solely on the basis of their anticodons. Sets of degenerate codons are read by a single tRNA through wobble pairing. The UGA codon, which is normally the *opal* Stop codon may, depending on its context in mRNA, specify a selenoCys (Sec) residue, which is carried by a specific tRNA (tRNA^{Sec}), thereby forming a selenoprotein. Nonsense mutations may be suppressed by tRNAs whose anticodons have mutated to recognize a Stop codon.

3 ■ Ribosomes and Polypeptide Synthesis The ribosome consists of a small and a large subunit whose complex shapes have been revealed by cryoelectron microscopy and X-ray crystallography. The three RNAs and 52 proteins comprising the *E. coli* ribosome self-assemble under proper conditions. Both ribosomal subunits consist of an RNA core in which the proteins are embedded, mainly as globular domains on the back and sides of the particle, with long basic polypeptide segments that infiltrate between the RNA helices so as to neutralize their anionic charges. Eukaryotic ribosomes are larger and more complex than those of prokaryotes.

Ribosomal polypeptide synthesis proceeds by the addition of amino acid residues to the C-terminal end of the nascent polypeptide. The mRNAs are read in the 5′ → 3′ direction. mRNAs are usually simultaneously translated by several ribosomes in the form of polysomes. The ribosome has three tRNA-binding sites: the A site, which binds the incoming aminoacyl–tRNA; the P-site, which binds the peptidyl–tRNA; and the E site, which transiently binds the outgoing deacylated tRNA. During polypeptide synthesis, the nascent polypeptide is transferred to the aminoacyl–tRNA, thereby lengthening the nascent polypeptide by one residue. The newly deacylated tRNA is translocated to the E site and the new peptidyl–tRNA, with its associated codon, is translocated to the P site. In prokaryotes, the initiation sites on mRNA are recognized through their Shine–Dalgarno sequences and by their initiating codon. Prokaryotic initiating codons specify fMet–tRNA$_f^{Met}$. Initiation involves the participation of three initiation factors that induce the assembly of the ribosomal subunits with fMet–tRNA$_f^{Met}$ in the P site and mRNA. Eukaryotic initiation is a far more complicated process that requires the participation of at least 11 initiation factors. The system binds the mRNA's 5′ cap and scans along the mRNA until it finds its AUG initiation codon, usually the mRNA's first AUG, through codon–anticodon interactions with the initiating tRNA, Met–tRNA$_i^{Met}$.

Polypeptides are elongated in a three-part cycle, consisting of aminoacyl–tRNA decoding, transpeptidation, and translocation, that requires the participation of elongation factors and is vectorially driven by GTP hydrolysis. EF-Tu, which functions to escort aminoacyl–tRNA into the ribosomal A site, undergoes a major conformational change on hydrolyzing its bound GTP. The X-ray structure of the 50S subunit clearly shows that the ribosomal peptidyl transferase center is distant from any protein and hence that the ribosome is a ribozyme. A2486 appears to be properly positioned and activated to function as a general base in the peptidyl transferase reaction. Translocation is motivated through the EF-G–catalyzed hydrolysis of GTP. EF-G · GDP, which binds to the same ribosomal site as aminoacyl–tRNA · EF-Tu · GTP, is a macromolecular mimic of this complex. Translocation occurs via intermediate states, the A/P and P/E states, in which the newly formed peptidyl–tRNA and the newly deacylated tRNA are respectively bound to the A and P subsites of the 30S subunit and to the P and E subsites of the 50S subunit, following which EF-G hydrolyzes its bound GTP and shifts these tRNAs to the P/P and E/E states. Termination codons bind release factors that induce the peptidyl transferase to hydrolyze the peptidyl–tRNA bond. Eukaryotic elongation and termination resemble that of prokaryotes.

The ribosome initially selects an aminoacyl–tRNA whose anticodon is cognate to its A-site–bound codon through interactions involving three universally conserved 30S subunit bases while the tRNA is in the A/T binding state. The codon–anticodon interaction is then proofread in an independent process that follows the hydrolysis of the EF-Tu–bound GTP and which occurs when the tRNA has shifted to the A/A binding state. Ribosomal inhibitors, many of which are antibiotics, are medically important and biochemically useful in elucidating ribosomal function. Streptomycin causes mRNA misreading and inhibits prokaryotic chain initiation, chloramphenicol inhibits prokaryotic peptidyl transferase, paromomycin causes codon misreading, tetracycline inhibits aminoacyl–tRNA binding to the prokaryotic 30S subunit, and diphtheria toxin ADP-ribosylates eEF2.

4 ■ Control of Eukaryotic Translation Several mechanisms of translational control have been elucidated in eukaryotes. eIF2α kinases catalyze the phosphorylation of eIF2α, which then tightly binds eIF2B so as to prevent it from exchanging eIF2-bound GDP for GTP and hence inhibits translational initiation. These eIF2α kinases include heme-regulated inhibitor (HRI), which functions to coordinate globin synthesis with heme availability; double-stranded RNA-activated protein kinase (PKR), an interferon-induced protein that functions to inhibit viral proliferation; and PKR-like endoplasmic reticulum kinase (PERK), which functions to protect the cell from the irreversible damage caused by the accumulation of unfolded proteins in the ER. GCN2, in contrast, is an eIF2α kinase that, when amino acids are scarce, stimulates the translation of the transcriptional activator GCN4 by causing the 40S ribosomal subunit to scan across four upstream open reading frames (uORFs) in the *GCN4* mRNA, thereby permitting the ribosome to initiate translation at the GCN4 coding sequence. The phosphorylation of eIF4E (cap-binding protein) by a MAP kinase cascade increases eIF4E's affinity for capped mRNA and thereby stimulates translational initiation. The binding of 4E-BPs to eIF4E blocks its binding of eIF4G and hence prevents initiation. However, the insulin-induced phosphorylation of the 4E-BPs causes them to dissociate from eIF4E. mRNAs in certain animal oocytes are masked by the binding of proteins, which prevents their translation.

Many oocyte mRNAs have short poly(A) tails that are preceded by a cytoplasmic polyadenylation element (CPE) that is bound by CPE-binding protein (CPEB). CPEB binds maskin which binds eIF4E, thereby inhibiting translational initiation. However, when CPEB is phosphorylated, it recruits poly(A) polymerase (PAP), which extends the mRNA's poly(A) tail such that it is bound by poly(A)-binding protein (PABP). PABP then binds to eIF4G, which in turn displaces maskin from eIF4E, thereby permitting the translation of the mRNA. Antisense oligonucleotides can be used to inhibit the translation of their complementary mRNAs. Although the delivery of antisense oligonucleotides to their sites of action has proved to be a difficult problem, their use is starting to show some medical and biotechnological successes.

5 ■ Posttranslational Modification Proteins may be posttranslationally modified in a variety of ways. Proteolytic cleavages, usually by specific peptidases, activate proproteins. The signal peptides of preproteins are removed by signal peptidases. Covalent modifications alter many types of side chains in a variety of ways that modulate the catalytic activities of enzymes, provide recognition markers, and stabilize protein structures. Protein splicing occurs via the intein-catalyzed self-excision between an N-extein and a C-extein accompanied by the ligation of the N- and C-exteins via a peptide bond. Most inteins contain a homing endonuclease that makes a double-strand nick in a gene similar to that encoding the corresponding extein, thereby triggering a recombinational double-strand DNA repair process that copies the gene encoding the intein into the break. Inteins therefore appear to be molecular parasites.

6 ■ Protein Degradation Proteins in living cells are continually turning over. This controls the level of regulatory enzymes and disposes of abnormal proteins that would otherwise interfere with cellular processes. Proteins are degraded by lysosomes via a nonspecific process as well via a process specific for KFERQ proteins that is stimulated during starvation. A cytosolically based ATP-dependent system degrades normal as well as abnormal proteins in a process that flags these proteins by the covalent attachment of polyubiquitin chains to their Lys residues. This process is mediated by three consecutively acting enzymes: ubiquitin-activating enzyme (E1), ubiquitin-conjugating enzyme (E2), and ubiquitin–protein ligase (E3). Most cells have one species of E1, several species of E2, and many species of E3, each of which is served by one or a few E2s. The polyubiquitinated protein is proteolytically degraded in the 26S proteasome.

E3s can have complicated modular structures, each having different specificities for target proteins. SCF complexes, one of whose several subunits contains a RING finger, are particularly elaborate. The RING finger E3 known as E3α functions to ubiquitinate proteins that satisfy the N-end rule. The transcription factor NF-κB is activated through the ubiquitination and subsequent destruction of its otherwise bound inhibitor IκBα by the SCF β-TrCP, which is activated through phosphorylation via a signal transduction cascade. Cyclins, which mediate the cell cycle, are destroyed in a programmed manner through ubiquitination by their cognate E3s, one of which is anaphase-promoting complex (APC).

The 26S proteasome consists of a hollow protein barrel formed by two rings of seven α subunits flanking two rings of seven β subunits known as the 20S proteasome, which is bound at each end by 19S caps that each consist of ~18 subunits. The active sites of the β subunits, which are members of the N-terminal nucleophile (Ntn) family of hydrolases, are inside the barrel. Ubiquitinated proteins are selected by the 19S caps, which unfold them in an ATP-dependent matter and then feed them into the 20S proteasome via its axial channel.

The polyubiquitin (polyUb) chains are excised from the condemned protein by proteasome-associated deubiquitinating enzymes (DUBs), while other DUBs dismember the polyUb chains to their component ubiquitin units, thereby recycling them. The 11S regulator is a heptameric complex that, on binding to one end of a 20S proteasome, opens its axial channel in an ATP-independent manner, thereby permiting polypeptides, but not folded proteins, to enter the 20S proteasome.

Eubacteria, nearly all of which lack proteasomes, nevertheless express a variety of self-compartmentalized proteases, including ClpP, heat shock locus UV (HslUV), and tricorn protease, that function to proteolytically dispose of their cellular proteins. Ubiquitinlike modifiers (Ubls), such as SUMO and RUB1, participate in numerous regulatory processes.

REFERENCES

GENERAL

Adams, R.L.P., Knowler, J.T., and Leader, D.P., *The Biochemistry of the Nucleic Acids* (11th ed.), Chapter 12, Chapman & Hall (1992).

Lewin, B., *Genes VII,* Chapters 5–7, Wiley (2000).

THE GENETIC CODE

Attardi, G., Animal mitochondrial DNA: an extreme example of genetic economy, *Int. Rev. Cytol.* **93,** 93–145 (1985).

Benzer, S., The fine structure of the gene, *Sci. Am.* **206**(1), 70–84 (1962). *The Genetic Code, Cold Spring Harbor Symp. Quant. Biol.* **31** (1966). [A collection of papers describing the establishment of the genetic code. See especially the articles by Crick, Nirenberg, and Khorana.]

Crick, F.H.C., The genetic code, *Sci. Am.* **207**(4), 66–74 (1962) [The structure of the code as determined by phage genetics]; *and* The genetic code: III, *Sci. Am.* **215**(4), 55–62 (1966). [A description of the nature of the code after its elucidation was almost complete.]

Crick, F.H.C., Burnett, L., Brenner, S., and Watts-Tobin, R.J., General nature of the genetic code for proteins, *Nature* **192,** 1227–1232 (1961).

Fox, T.D., Natural variation in the genetic code, *Annu. Rev. Genet.* **21,** 67–91 (1987).

Judson, J.F., *The Eighth Day of Creation,* Expanded Edition, Part II, Cold Spring Harbor Laboratory Press (1996). [A fascinating historical narrative on the elucidation of the genetic code.]

Khorana, H.G., Nucleic acid synthesis in the study of the genetic code, *Nobel Lectures in Molecular Biology, 1933–1975,* pp. 303–331, Elsevier (1977).

Knight, R.D., Freeland, S.J., and Landweber, L.F., Selection, history and chemistry: the three faces of the genetic code, *Trends Biochem. Sci.* **24,** 241–247 (1999).

Nirenberg, M., The genetic code, *Nobel Lectures in Molecular Biology, 1933–1975,* pp. 335–360, Elsevier (1977).

Nirenberg, M.W., The genetic code: II, *Sci. Am.* **208,** 80–94 (1963). [Discusses the use of synthetic mRNAs to analyze the genetic code.]

Nirenberg, M. and Leder, P., RNA code words and protein synthesis, *Science* **145,** 1399–1407 (1964). [The determination of the genetic code by the ribosomal binding of tRNAs using specific trinucleotides.]

Nirenberg, M.W. and Matthaei, J.H., The dependence of cell-free protein synthesis in *E. coli* upon naturally occurring or synthetic polyribonucleotides, *Proc. Natl. Acad. Sci.* **47,** 1588–1602 (1961). [The landmark paper reporting the finding that poly(U) stimulates the synthesis of poly(Phe).]

Singer, B. and Kusmierek, J.T., Chemical mutagenesis, *Annu. Rev. Biochem.* **51,** 655–693 (1982).

TRANSFER RNA AND ITS AMINOACYLATION

Alexander, R.W. and Schimmel, P., Domain–domain communication in aminoacyl–tRNA synthetases, *Prog. Nucleic Acid Res. Mol. Biol.* **69,** 317–349 (2001).

Björk, G.R., Ericson, J.U., Gustafsson, C.E.D., Hagervall, T.G., Jösson, Y.H., and Wikström, P.M., Transfer RNA modification, *Annu. Rev. Biochem.* **56,** 263–287 (1987).

Böck, A., Forschhammer, K., Heider, J., and Baron, C., Selenoprotein synthesis: an expansion of the genetic code, *Trends Biochem. Sci.* **16,** 463–467 (1991).

Carter, C.W., Jr., Cognition, mechanism, and evolutionary relationships in aminoacyl–tRNA synthetases, *Annu. Rev. Biochem.* **62,** 715–748 (1993).

Crick, F.H.C., Codon–anticodon pairing: the wobble hypothesis, *J. Mol. Biol.* **19,** 548–555 (1966).

Cusack, S., Aminoacyl–tRNA synthetases, *Curr. Opin. Struct. Biol.* **7,** 881–889 (1997).

Fukai, S., Nureki, O., Sekine, S., Shimada, A., Tao, J., Vassylyev, D.G., and Yokoyama, S., Structural basis for double-sieve discrimination of L-valine from L-isoleucine and L-threonine by the complex of tRNA^Val and valyl–tRNA synthetase, *Cell* **103,** 793–803 (2000).

Geigé, R., Puglisi, J.D., and Florentz, C., tRNA structure and aminoacylation efficiency, *Prog. Nucleic Acid Res. Mol. Biol.* **45,** 129–206 (1993). [A detailed review.]

Gesteland, R.F., Weiss, R.B., and Atkins, J.F., Recoding: Reprogrammed genetic coding, *Science* **257,** 1640–1641 (1992). [Discusses contextual signals in mRNA that alter the way the ribosome reads certain codons.]

Hatfield, D.L., Lee, B.J., and Pirtle, R.M. (Eds.), *Transfer RNA in Protein Synthesis,* CRC Press (1992). [Contains articles on such subjects as the role of modified nucleosides in tRNAs, variations in reading the genetic code, patterns of codon usage, and tRNA identity elements.]

Hou, Y.-M., Discriminating among the discriminator bases of tRNAs, *Chem. Biol.* **4,** 93–96 (1997).

Ibba, M. and Söll, D., Aminoacyl–tRNA synthesis, *Annu. Rev. Biochem.* **69,** 617–650 (2000).

Ibba, M., Becker, H.D., Stathopoulos, C., Tumbula, D.L., and Söll, D., The adaptor hypothesis revisited, *Trends Biochem. Sci.* **25,** 311–316 (2000).

Ibba, M., Morgan, S., Curnow, A.W., Pridmore, D.R., Vothknecht, U.C., Gardner, W., Lin, W., Woese, C.R., and Söll, D., A euryarchaeal lysyl–tRNA synthetase: Resemblance to class I synthetases, *Science* **278,** 1119–1122 (1997).

Jacquin-Becker, C., Ahel, I., Ambrogelly, A., Ruan, B., Söll, D., and Stathopoulos, C., Cysteinyl–tRNA formation and prolyl-tRNA synthetase, *FEBS Lett.* **514,** 34–36 (2002).

Kim, S.H., Suddath, F.L., Quigley, G.J., McPherson, A., Sussman, J.L., Wang, A.M.J., Seeman, N.C., and Rich, A., Three-dimensional tertiary structure of yeast phenylalanine transfer RNA, *Science* **185,** 435–440 (1974); *and* Robertus, J.D., Ladner, J.E., Finch, J.T., Rhodes, D., Brown, R.S., Clark, B.F.C., and Klug, A., Structure of yeast phenylalanine tRNA at 3 Å resolution, *Nature* **250,** 546–551 (1974). [The landmark papers describing the high-resolution structure of a tRNA.]

McClain, W.H., Rules that govern tRNA identity in protein synthesis, *J. Mol. Biol.* **234**, 257–280 (1993).

Moras, D., Structural and functional relationships between aminoacyl–tRNA synthetases, *Trends Biochem. Sci.* **17**, 159–169 (1992). [Discusses Class I and Class II enzymes.]

Nureki, O., Vassylyev, D.G., Tateno, M., Shimada, A., Nakama, T., Fukai, S., Konno, M., Hendrickson, T.L., Schimmel, P., and Yokoyama, S., Enzyme structure with two catalytic sites for double-sieve selection of substrate, *Science* **280**, 578–582 (1998). [The X-ray structures of IleRS in complexes with isoleucine and valine.]

Rould, M.A., Perona, J.J., and Steitz, T.A., Structural basis of anticodon loop recognition by glutaminyl–tRNA synthetase, *Nature* **352**, 213–218 (1991).

Ruff, M., Krishnaswamy, S., Boeglin, M., Poterszman, A., Mitschler, A., Podjarny, A., Rees, B., Thierry, J.C., and Moras, D., Class II aminoacyl transfer RNA synthetases: Crystal structure of yeast aspartyl–tRNA synthetase complexed with tRNA^Asp, *Science* **252**, 1682–1689 (1991).

Saks, M.E., Sampson, J.R., and Abelson, J.N., The transfer identity problem: A search for rules, *Science* **263**, 191–197 (1994).

Sankaranarayanan, R., Dock-Bregeon, A.-C., Rees, B., Bovee, M., Caillet, J., Romby, P., Francklyn, C.S., and Moras, D., Zinc ion-mediated amino acid discrimination by threonyl–tRNA synthetase, *Nature Struct. Biol.* **7**, 461–465 (2000); *and* Dock-Bregeon, A.-C., Sankaranarayanan, R., Romby, P., Caillet, J., Springer, P., Rees, B., Francklyn, C.S., Ehresmann, C., and Moras, D., Transfer RNA-mediated editing in threonyl–tRNA synthetase: the Class II solution to the double discrimination problem, *Cell* **103**, 877–884 (2000).

Schimmel, P., Giegé, R., Moras, D., and Yokoyama, S., An operational RNA code for amino acids and possible relationship to genetic code, *Proc. Natl. Acad. Sci.* **90**, 8763–8768 (1993).

Schulman, L.H., Recognition of tRNAs by aminoacyl–tRNA synthetases, *Prog. Nucleic Acid Res. Mol. Biol.* **41**, 23–87 (1991).

Silvian, L.F., Wang, J., and Steitz, T.A., Insights into editing from an Ile-tRNA synthetase structure with tRNA^Ile and mupirocin, *Science* **285**, 1074–1077 (1999).

Söll, D. and RajBhandary, U.L. (Eds.), *tRNA: Structure, Biosynthesis, and Function,* ASM Press (1995).

Stadtman, T.C., Selenocysteine, *Annu. Rev. Biochem.* **65**, 83–100 (1996).

Steege, D.A. and Söll, D.G., Suppression, *in* Goldberger, R.F. (Ed.), *Biological Regulation and Development,* Vol. 1, pp. 433–485, Plenum Press (1979).

RIBOSOMES AND POLYPEPTIDE SYNTHESIS

Agrawal, R.K., Spahn, C.MT., Penczek, P., Grassuci, R.A., Nierhaus, K.H., and Frank, J., Visualization of tRNA movements in the *Escherichia coli* 70S ribosome during the elongation cycle, *J. Cell Biol.* **150**, 447–459 (2000).

Ban, N., Nissen, P., Hansen, J., Moore, P.B., and Steitz, T., The complete atomic structure of the large ribosomal subunit at 2.4 Å resolution, *Science* **289**, 905–920 (2000).

Bell, C.E. and Eisenberg, D.E., Crystal structure of diphtheria toxin bound to nicotinamide adenine dinucleotide, *Biochemistry* **35**, 1137–1149 (1996).

Brodersen, D.E., Clemons, W.M., Jr., Carter, A.P., Morgan-Warren, R.J., Wimberly, B.T., and Ramakrishnan, V., The structural basis for the action of the antibiotics tetracycline, pactamycin, and hygromycin B on the 30S ribosomal subunit, *Cell* **103**, 1143–1154 (2000).

Czworkowski, J., Wang, J., Steitz, J.A., and Moore, P.B., The crystal structures of elongation factor G complexed with GDP, at 2.7 Å resolution; *and* Ævarsson, A., Brazhnikov, E., Garber, M., Zheltonosova, J., Chirgadze, Yu., Al-Karadaghi, S., Svensson, L.A., and Liljas, A., Three-dimensional structure of the ribosomal translocase: elongation factor G from *Thermus thermophilus, EMBO J.* **13**, 3661–3668 *and* 3669–3677 (1994).

Dintzis, H.M., Assembly of the peptide chains of hemoglobin, *Proc. Natl. Acad Sci.* **47**, 247–261 (1961). [The determination of the direction of polypeptide biosynthesis.]

Fersht, A., *Structure and Mechanism in Protein Science,* Chapter 13, Freeman (1999). [A discussion of enzymatic specificity and editing mechanisms.]

Frank, J., Single-particle imaging of macromolecules by cryo-electron microscopy, *Annu. Rev. Biophys. Biomol. Struct.* **31**, 303–319 (2002).

Frank, J. and Agrawal, R.K., A ratchet-like inter-subunit reorganization of the ribosome during translocation, *Nature* **406**, 318–322 (2000).

Gingras, A.-C., Raught, B., and Sonnberg, N., eIF4 initiation factors: effectors of mRNA recruitment to ribosomes and regulators of translation, *Annu. Rev. Biochem.* **68**, 913–963 (1999).

Green, R. and Lorsch, J.R., The path to perdition is paved with protons, *Cell* **110**, 665–668 (2002). [Discusses the difficulties in characterizing the catalytic mechanism of the ribosomal peptidyl transferase.]

Hansen, J., Schmeing, T.M., Moore, P.B., and Steitz, T., Structural insights into peptide bond formation, *Proc. Natl. Acad. Sci.* **99**, 11670–11675 (2002); *and* Nissen, P., Hansen, J., Ban, N., Moore, P.B., and Steitz, T., The structural basis of ribosome activity in peptide bond synthesis, *Science* **289**, 920–930 (2000).

Held, W.A., Ballou, B., Mizushima, S., and Nomura, M., Assembly mapping of 30S ribosomal proteins from *Escherichia coli, J. Biol. Chem.* **249**, 3103–3111 (1974).

Jenni, S. and Ban, N., The chemistry of protein synthesis and voyage through the ribosomal tunnel, *Curr. Opin. Struct. Biol.* **13**, 212–219 (2003).

Kawashima, T., Berthet-Colominas, C., Wulff, M., Cusack, S., and Leberman, R., The structure of the *Escherichia coli* EF-Tu · EF-Ts complex at 2.5 Å resolution, *Nature* **379**, 511–518 (1996).

Kisselev, L.L. and Buckingham, R.H., Translation termination comes of age, *Trends Biochem. Sci.* **25**, 561–566 (2000).

Kjeldgaard, M. and Nyborg, J., Refined structure of elongation factor EF-Tu from *Escherichia coli, J. Mol. Biol.* **223**, 721–742 (1992); *and* Kjeldgaard, M., Nissen, P., Thirup, S., and Nyborg, J., The crystal structure of elongation factor EF-Tu from *Thermus aquaticus* in the GTP conformation, *Structure* **1**, 35–50 (1993).

Lake, J.A., Evolving ribosome structure: domains in archaebacteria, eubacteria, eocytes and eukaryotes, *Annu. Rev. Biochem.* **54**, 507–530 (1985).

Lancaster, L., Kiel, M.C., Kaji, A., and Noller, H.F., Orientation of ribosome recycling factor in the ribosome from directed hydroxyl radical probing, *Cell* **111**, 129–140 (2002).

Marcotrigiano, J., Gingras, A.-C., Sonenberg, N., and Burley, S.K., Cocrystal structure of the messenger RNA 5′ cap-binding protein (eIF4E) bound to 7-methyl-GDP, *Cell* **89**, 951–961 (1997).

Moazed, D. and Noller, H.F., Intermediate states in the movement of transfer RNA in the ribosome, *Nature* **342**, 142–148 (1989).

Moore, P.B. and Steitz, T.A., The involvement of RNA in ribosome function, *Nature* **418**, 229–235 (2002); After the ribosome: How does peptidyl transferase work, *RNA* **9**, 155–159 (2003); *and* The structural basis of large ribosomal subunit function, *Annu. Rev. Biochem.* **72**, 813–850 (2003).

Nakamura, Y., Ito, K., and Ehrenberg, M., Mimicry grasps reality in translation terminator, *Cell* **101**, 349–352 (2000).

Nissen, P., Kjeldgaard, M., Thirup, S., Polekhina, G., Reshetnikova, L., Clark, B.F.C., and Nyborg, J., Crystal structure of the ternary complex of Phe–tRNA^{Phe}, EF-Tu, and a GTP analog, *Science* **270**, 1464–1472 (1995).

Noller, H.F., Hoffarth, V., and Zimniak, L., Unusual resistance of peptidyl transferase to protein extraction procedures, *Science* **256**, 1416–1419 (1992); *and* Noller, H.F., Peptidyl transferase: protein, ribonucleoprotein, or RNA? *J. Bacteriol.* **175**, 5297–5300 (1993).

Nollar, H.F., Yusupov, M.M., Yusupova, G.Z., Baucom, A., and Cate, J.H.D., Translocation of tRNA during protein synthesis, *FEBS Lett.* **514**, 11–16 (2002).

Ogle, J.M., Brodersen, D.E., Clemons, W.M., Jr., Tarry, M.J., Carter, A.P., and Ramakrishnan, V., Recognition of cognate transfer RNA by the 30S ribosomal subunit, *Science* **292**, 897–902 (2001); *and* Ogle, J.M., Carter, A.P., and Ramakrishnan, V., Insights into the decoding mechanism from recent ribosome structures, *Trends Biochem. Sci.* **28**, 259–266 (2003).

Pioletti, M., et al., Crystal structure of complexes of the small ribosomal subunit with tetracycline, edeine and IF3, *EMBO J.* **20**, 1829–1839 (2001).

Poole, E. and Tate, W., Release factors and their role as decoding proteins: specificity and fidelity for termination of protein synthesis, *Biochim. Biophys. Acta* **1493**, 1–11 (2000).

Ramakrishnan, V., Ribosome structure and the mechanism of translocation, *Cell* **108**, 557–572 (2002). [A detailed and incisive review.]

Ramakrishnan, V. and Moore, P.B., Atomic structure at last: the ribosome in 2000, *Curr. Opin. Struct. Biol.* **11**, 144–154 (2001).

Rané, H.A., Klootwijk, J., and Musters, W., Evolutionary conservation of structure and function of high molecular weight ribosomal RNA, *Prog. Biophys. Mol. Biol.* **51**, 77–129 (1988).

Rawat, U.B.S., Zavialov, A.V., Sengupta, J., Valle, M., Grassucci, R.A., Linde, J., Vestergaard, B., Ehrenberg, M., and Frank, J., A cryo-electro microscopic study of ribosome-bound termination factor RF2; *and* Klaholz, B.P., Pape, T., Zavialov, A.V., Myasnikov, A.G., Orlova, E.V., Vestergaard, B., Ehrenberg, M., and van Heel, M., Structure of Eschericia coli ribosomal termination complex with release factor 2, *Nature* **421**, 87–90 and 90–94 (2003).

Rodnina, M.V. and Wintermeyer, W., Ribosome fidelity: tRNA discrimination, proofreading and induced fit, *Trends Biochem. Sci.* **26**, 124–130 (2001); Fidelity of aminoacyl–tRNA selection on the ribosome: kinetic and structural mechanisms, *Annu. Rev. Biochem.* **70**, 415–435 (2001); *and* Peptide bond formation on the ribosome: Structure and mechanism, *Curr. Opin. Struct. Biol.* **13**, 334–340 (2003).

Schluenzen, F., Tocilj, A., Zarivach, R., Harms, J., Gluehmann, M., Janell, D., Bashan, A., Bartels, H., Agmon, I., Franceschi, F., and Yonath, A., Structure of functionally activated small ribosomal subunit at 3.3 Å resolution, *Cell* **102**, 615–623 (2000).

Selmer, M., Al-Karadaghi, S., Hirokawa, G., Kaji, A., and Liljas, A., Crystal structure of *Thermatoga maritima* ribosome recycling factor: a tRNA mimic, *Science* **286**, 2349–2352 (1999).

Sonenberg, N. and Dever, T.E., Eukaryotic translation initiation factors and regulators, *Curr. Opin. Struct. Biol.* **13**, 56–63 (2003).

Song, H., Mugnier, P., Das, A.K., Webb, H.M., Evans, D.R., Tuite, M.F., Hemmings, B.A., and Barford, D., The crystal structure of human eukaryotic release factor eRF1—Mechanism of stop codon recognition and peptidyl–tRNA hydrolysis, *Cell* **100**, 311–321 (2000).

Spahn, C.M.T., Beckmann, R., Eswar, N., Penczek, P.A., Sali, A., Blobel, G., and Frank, J., Structure of the 80S ribosome from *Saccharomyces cerevisiae*—tRNA-ribosome and subunit-subunit interactions, *Cell* **107**, 373–386 (2001).

Spirin, A.S., Ribosome as a molecular machine, *FEBS Lett.* **514**, 2–10 (2002). [Discusses the role in GTP hydrolysis in ribosomal processes.]

Steitz, J.A. and Jakes, K., How ribosomes select initiator regions in mRNA: base pair formation between the 3′ terminus of 16S RNA and the mRNA during initiation of protein synthesis in *Escherichia coli*, *Proc. Natl. Acad. Sci.* **72**, 4734–4738 (1975).

The Ribosome, Cold Spring Harbor Symposium on Quantitative Biology, Volume LXVI, Cold Spring Harbor Laboratory Press (2001). [The latest "bible" of ribosomology.]

Vestergaard, B., Van, L.B., Andersen, G.R., Nyborg, J., Buckingham, R.H., and Kjeldgaard, M., Bacterial polypeptide release factor RF2 is structurally distinct from eukaryotic eRF1, *Mol. Cell* **8**, 1375–1382 (2001).

Wimberly, B.T., Broderson, D.E., Clemons, W.M., Jr., Morgan-Warren, R., von Rhein, C., Hartsch, T., and Ramakrishnan, V., Structure of the 30S ribosomal subunit, *Nature* **407**, 327–339 (2000); *and* Broderson, D.E., Clemons, W.M., Jr., Carter, A.P., Wimberly, B.T., and Ramakrishnan, V., Crystal structure of the 30S ribosomal subunit from *Thermus thermophilus:* Structure of the proteins and their interactions with 16S RNA, *J. Mol. Biol.* **316**, 725–768 (2002).

Yonath, A., The search and its outcome: High resolution structures of ribosomal particles from mesophilic, thermophilic, and halophilic bacteria at various functional states, *Annu. Rev. Biophys. Biomol. Struct.* **31**, 257–273 (2002).

Yusupova, G.Z., Yusupov, M.M., Cate, J.D.H., and Noller, H.F., The path of messenger RNA through the ribosome, *Cell* **106**, 233–241 (2001).

CONTROL OF TRANSLATION

Branch, A.D., A good antisense molecule is hard to find, *Trends Biochem. Sci.* **23**, 45–50 (1998).

Calkhoven, C.F., Müller, C., and Leutz, A., Translational control of gene expression and disease, *Trends Molec. Med.* **8**, 577–583 (2002).

Chen, J.-J., and London, I.M., Regulation of protein synthesis by heme-regulated eIF-2α kinase, *Trends Biochem. Sci.* **20**, 105–108 (1995).

Clemens, M.J., PKR—A protein kinase regulated by double-stranded RNA, *Int. J. Biochem. Cell Biol.* **29**, 945–949 (1997).

Dever, T.E., Gene-specific regulation by general translation factors, *Cell* **108**, 545–556 (2002).

Gray, N.K. and Wickens, M., Control of translation initiation in animals, *Annu. Rev. Cell Dev. Biol.* **14**, 399–458 (1998).

Hershey, J.W.B., Translational control in mammals, *Annu. Rev. Biochem.* **60**, 717–755 (1991).

Kozak, M., Regulation of translation in eukaryotic systems, *Annu. Rev. Cell Biol.* **8**, 197–225 (1992).

Lawrence, J.C., Jr. and Abraham, R.T., PHAS/4E-BPs as regulators of mRNA translation and cell proliferation, *Trends Biochem. Sci.* **22**, 345–349 (1997).

Lebedeva, I. and Stein, C.A., Antisense oligonucleotides: promise and reality, *Annu. Rev. Pharmacol. Toxicol.* **41**, 403–419 (2001).

Mendez, R. and Richter, J.D., Translational control by CPEB: a means to the end, *Nature Rev. Mol. Cell Biol.* **2**, 521–529 (2001).

Phillips, M.I. (Ed.), *Antisense technology: Part A. General Methods, Methods of Delivery, and RNA Studies;* and *Part B. Applications, Meth. Enzymol.* **313** and **314** (2000).

Sen, G.C. and Lengyel, P., The interferon system, *J. Biol. Chem.* **267**, 5017–5020 (1992).

Sheehy, R.E., Kramer, M., and Hiatt, W.R., Reduction of poly-galacturonase activity in tomato fruit by antisense RNA, *Proc. Natl. Acad. Sci.* **85,** 8805–8809 (1988).

Sonenberg, N., Hershey, J.W.B., and Mathews, M.B., *Translational Control of Gene Expression,* Cold Spring Harbor Laboratory Press (2000). [A compendium of authoritative articles.]

Tafuri, S.R. and Wolffe, A.P., Dual roles for transcription and translation factors in the RNA storage particles of *Xenopus* oocytes, *Trends Cell. Biol.* **3,** 94–98 (1993).

Tamm, I., Dörken, B., and Hartmann, G., Antisense therapy in oncology: new hope for an old idea? *Lancet* **358,** 489–497 (2001).

Weiss, B., Davidkova, G., and Zhou, L.-W., Antisense RNA therapy for studying and modulating biological processes, *Cell. Mol. Life Sci.* **55,** 334–358 (1999).

POSTTRANSLATIONAL MODIFICATION

Fessler, J.H. and Fessler, L.I., Biosynthesis of procollagen, *Annu. Rev. Biochem.* **47,** 129–162 (1978).

Harding, J.J., and Crabbe, M.J.C. (Eds.), *Post-Translational Modifications of Proteins,* CRC Press (1992).

Klabunde, T., Sharma, S., Telenti, A., Jacobs, W.R., Jr., and Sacchetini, J.C., Crystal structure of Gyr A protein from *Mycobacterium xenopi* reveals structural basis of splicing, *Nature Struct. Biol.* **5,** 31–36 (1998).

Liu, X.-Q., Protein-splicing intein: genetic mobility, origin, and evolution, *Annu. Rev. Genet.* **34,** 61–76 (2000).

Noren, C.J., Wang, J., and Perler, F.B., Dissecting the chemistry of protein splicing and its applications, *Angew. Chem. Int. Ed.* **39,** 450–466 (2000).

Wold, F., In vivo chemical modification of proteins, *Annu. Rev. Biochem.* **50,** 783–814 (1981).

Wold, F. and Moldave, K. (Eds.), Posttranslational Modifications, Parts A and B, *Methods Enzymol.* **106** and **107** (1984). [Contains extensive descriptions of the amino acid "zoo."]

PROTEIN DEGRADATION

Bochtler, M., Ditzel, L., Groll, M., Hartmann, C., and Huber, R., The proteasome, *Annu. Rev. Biophys. Biomol. Struct.* **28,** 295–317 (1999).

Brandstetter, H., Kim, J.-S., Groll, M., and Huber, R., Crystal structure of the tricorn protease reveals a protein disassembly line, *Nature* **414,** 466–470 (2001); Walz, J., Tamura, T., Tamura, N., Grimm, R., Baumeister, W., and Koster, A.J., Tricorn protease exists as an icosahedral supermolecule *in vivo, Mol. Cell* **1,** 59–65 (1997); *and* Walz, J., Koster, A.J., Tamura, T., and Baumeister, W., Capsids of tricorn protease studied by cryo-microscopy, *J. Struct. Biol.* **128,** 65–68 (1999).

Cook, W.J., Jeffrey, L.C., Kasperek, E., and Pickart, C.M., Structure of tetraubiquitin shows how multiubiquitin chains can be formed, *J. Mol. Biol.* **236,** 601–609 (1994).

Cook, W.J., Jeffrey, L.C., Sullivan, M.L., and Vierstra, R.D., Three-dimensional structure of a ubiquitin-conjugating enzyme (E2), *J. Biol. Chem.* **267,** 15116–15121.

Deshaies, R.J., SCF and cullin/RING H2-based ubiquitin ligases, *Annu. Rev. Cell Dev. Biol.* **15,** 435–467 (1999).

Dice, F., Peptide sequences that target cytosolic proteins for lysosomal proteolysis, *Trends Biochem. Sci.* **15,** 305–309 (1990).

Ferrell, K., Wilkinson, C.R.M., Dubiel, W., and Gordon, C., Regulatory subunit interactions of the 26S proteasome, a complex problem, *Trends Biochem. Sci.* **25,** 83–88 (2000).

Glickman, M.H. and Ciechanover, A., The ubiquitin-proteasome proteolytic pathway: destruction for the sake of construction, *Physiol. Rev.* **82,** 373–428 (2002).

Goldberg, A.L., and Rock, K.L., Proteolysis, proteasomes and antigen presentation, *Nature* **357,** 375–379 (1992).

Hartmann-Petersen, R., Seeger, M., and Gordon, C., Transferring substrates to the 26S proteasome, *Trends Biochem. Sci.* **28,** 26–31 (2003).

Hershko, A. and Ciechanover, A., The ubiquitin system, *Annu. Rev. Biochem.* **67,** 425–479 (1998).

Jentsch, S. and Pyrowalakis, G. Ubiquitin and its kin: how close are the family ties, *Trends Cell Biol.* **10,** 335–342 (2003). [Discusses Ubls.]

Lam, Y.A., Lawson, T.G., Velayutham, M., Zweier, J.L., and Pickart, C.M., A proteasomal ATPase subunit recognizes the polyubiquitin degradation signal, *Nature* **416,** 763–767 (2002).

Laney, J.D. and Hochstrasser, M., Substrate targeting in the ubiquitin system, *Cell* **97,** 427–430 (1999).

Löwe, J., Stock, D., Jap, B., Zwicki, P., Baumeister, W., and Huber, R., Crystal structure of the 20S proteasome from the archeon *T. acidophilum* at 3.4 Å resolution, *Science* **268,** 533–539 (1995); *and* Groll, M., Ditzel, L., Löwe, J., Stock, D., Bochtler, M., Bartunik, H.D., and Huber, R., Structure of 20S proteasome from yeast at 2.4 Å resolution, *Nature* **386,** 463–471 (1997).

Page, A.M. and Hieter, P., The anaphase-promoting complex: new subunits and regulators, *Annu. Rev. Biochem.* **68,** 583–609 (1999).

Pickart, C.M., Mechanisms underlying ubiquitination, *Annu. Rev. Biochem.* **70,** 503–533 (2001); *and* Ubiquitin in chains, *Trends Biochem. Sci.* **25,** 544 (2000).

Schwartz, A.L. and Ciechanover, A., The ubiquitin-proteasome pathway and the pathogenesis of human disease, *Annu. Rev. Med.* **50,** 57–74 (1999).

Senahdi, V.-J., Bugg, C.E., Wilkinson, K.D., and Cook, W.J., Three-dimensional structure of ubiquitin at 2.8 Å resolution, *Proc. Natl. Acad. Sci.* **82,** 3582–3585 (1985).

Song, H.K., Hartmann, C., Ramachandran, R., Bochtler, M., Behrendt, R., Moroder, L., and Huber, R., Mutational studies on HslU and its docking mode with HsIV, *Proc. Natl. Acad. Sci.* **97,** 14103–14108 (2000).

Unno, M., Mizushima, T., Morimoto, Y., Tomisugi, Y., Tanaka, K., Yasuoka, N., and Tsukihara, T., The structure of the mammalian proteasome at 2.75 Å resolution, *Structure* **10,** 609–618 (2002).

VanDemark, A.P. and Hill, C.P., Structural basis of ubiquitylation, *Curr. Opin. Struct. Biol.* **12,** 822–830 (2002).

Varshavsky, A., Turner, G., Du, F., and Xie, Y., The ubiquitin system and the N-end rule, *Biol. Chem.* **381,** 779–789 (2000); *and* Varshavsky, A., The N-end rule, *Cell* **69,** 725–735 (1992).

Voges, D., Zwickl, P., and Baumeister, W., The 26S proteasome: a molecular machine designed for controlled proteolysis, *Annu. Rev. Biochem.* **68,** 1015–1068 (1999).

Wang, J., Hartling, J.A., and Flanagan, J.M., The structure of ClpP at 2.3 Å resolution suggests a model for ATP-dependent proteolysis, *Cell* **91,** 447–456 (1997).

Whitby, F.G., Masters, E.I, Kramer, L., Knowlton, J.R., Yao, Y., Wang, C.C., and Hill, C.P., Structural basis for the activation of 20S proteasomes by 11S regulators, *Nature* **408,** 115–120 (2000).

Yao, T. and Cohen, R.E., A cryptic protease couples deubiqitination and degradation by the proteasome, *Nature* **419,** 403–407 (2002); *and* Verma, R., Aravind, L., Oania, R., McDonald, W.H., Yates, J.R., III, Koonin, E.V., and Deshaies, R.J., Role of Rpn11 metalloprotease in deubiquitination and degradation by the 26S proteasome, *Science* **298,** 611–615 (2002).

Zheng, N., et al., Structure of the Cul1–Rbx1–Skp1–F-box^Skp2 SCF ubiquitin ligase complex, *Nature* **41,** 703–709 (2002);

Schulman, B.A., et al., Insights into SCF ubiquitin ligases from the structure of the Skp1–Skp2 complex, *Nature* **408**, 381–386 (2000); *and* Zheng, N., Wang, P., Jeffrey, P.D., and Pavletich, N.P., Structure of a c-Cbl–UbcH7 complex: RING domain function in ubiquitin-protein ligases, *Cell* **102**, 533–539 (2000).

Zwickl, P., Seemüller, E., Kapelari, B., and Baumeister, W., The proteasome: A supramolecular assembly designed for controlled proteolysis, *Adv. Prot. Chem.* **59**, 187–222 (2002).

PROBLEMS

1. What is the product of the reaction of guanine with nitrous acid? Is the reaction mutagenic? Explain.

2. What is the polypeptide specified by the following DNA antisense strand? Assume translation starts after the first initiation codon.

5'-TCTGACTATTGAGCTCTCTGGCACATAGCA-3'

*3. The fingerprint of a protein from a phenotypically revertant mutant of bacteriophage T4 indicates the presence of an altered tryptic peptide with respect to the wild type. The wild-type and mutant peptides have the following sequences:

Wild type   Cys-Glu-Asp-His-Val-Pro-Gln-Tyr-Arg
Mutant   Cys-Glu-Thr-Met-Ser-His-Ser-Tyr-Arg

Indicate how the mutant could have arisen and give the base sequences, as far as possible, of the mRNAs specifying the two peptides. Comment on the function of the peptide in the protein.

4. Explain why the various classes of mutations can reverse a mutation of the same class but not a different class.

5. Which amino acids are specified by codons that can be changed to an *amber* codon by a single point mutation?

6. The mRNA specifying the α chain of human hemoglobin contains the base sequence

···UCCAAAUACCGUUAAGCUGGA···

The C-terminal tetrapeptide of the normal α chain, which is specified by part of this sequence, is

-Ser-Lys-Tyr-Arg

In hemoglobin Constant Spring, the corresponding region of the α chain has the sequence

-Ser-Lys-Tyr-Arg-Gln-Ala-Gly-···

Specify the mutation that causes hemoglobin Constant Spring.

7. Explain why a minimum of 32 tRNAs are required to translate the "standard" genetic code.

8. Draw the wobble pairings not in Fig. 32-25a.

9. A colleague of yours claims that by exposing *E. coli* to HNO₂ she has mutated a tRNA^Gly to an *amber* suppressor. Do you believe this claim? Explain.

*10. Deduce the anticodon sequences of all suppressors listed in Table 32-6 except UGA-1 and indicate the mutations that caused them.

11. How many different types of macromolecules must be minimally contained in a cell-free protein synthesizing system from *E. coli*? Count each type of ribosomal component as a different macromolecule.

12. Why do oligonucleotides containing Shine–Dalgarno sequences inhibit translation in prokaryotes? Why don't they do so in eukaryotes?

13. Why does m⁷GTP inhibit translation in eukaryotes? Why doesn't it do so in prokaryotes?

14. What would be the distribution of radioactivity in the completed hemoglobin chains on exposing reticulocytes to ³H-labeled leucine for a short time followed by a chase with unlabeled leucine?

15. Design an mRNA with the necessary prokaryotic control sites that codes for the octapeptide Lys-Pro-Ala-Gly-Thr-Glu-Asn-Ser.

16. Indicate the translational control sites in and the amino acid sequence specified by the following prokaryotic mRNA.

5'-CUGAUAAGGAUUUAAAUUAUGUGUCAAUCACGA-AUGCUAAUCGAGGCUCCAUAAUAACACUU CGAC-3'

17. What is the energetic cost, in ATP equivalents, for the *E. coli* synthesis of a polypeptide chain of 100 residues starting from amino acids and mRNA? Assume that no losses are incurred as a result of proofreading.

*18. It has been suggested that Gly-tRNA synthetase does not require an editing mechanism. Why?

19. An antibiotic named fixmycin, which you have isolated from a fungus growing on ripe passion fruit, is effective in curing many types of venereal disease. In characterizing fixmycin's mode of action, you have found that it is a bacterial translational inhibitor that binds exclusively to the large subunit of *E. coli* ribosomes. The initiation of protein synthesis in the presence of fixmycin results in the generation of dipeptides that remain associated with the ribosome. Suggest a mechanism of fixmycin action.

20. Heme inhibits protein degradation in reticulocytes by allosterically regulating ubiquitin-activating enzyme (E1). What physiological function might this serve?

21. Genbux Inc., a biotechnology firm, has cloned the gene encoding an industrially valuable enzyme into *E. coli* such that the enzyme is produced in large quantities. However, since the firm wishes to produce the enzyme in ton quantities, the expense of isolating it would be greatly reduced if the bacterium could be made to secrete it. As a high-priced consultant, what general advice would you offer to solve this problem?

Chapter 33

Viruses: Paradigms For Cellular Functions

Viruses are parasitic entities, consisting of nucleic acid molecules with protective coats, that are replicated by the enzymatic machinery of suitable host cells. Since they lack metabolic apparatus, viruses are not considered to be alive (although this is a semantic rather than a scientific distinction). They range in complexity from **satellite tobacco necrosis virus** (**STNV;** Section 33-2B), whose genome encodes only one protein, to **Paramecium bursaria Chlorella virus** (**PBCV;** Section 33-2G), which encodes 377 proteins [to put this latter number into perspective, the genome of the smallest free-living organism, *Mycoplasma genitalium* (Table 7-3), encodes ~470 proteins].

Viruses were originally characterized at the end of the nineteenth century as infectious agents that could pass through filters that held back bacteria. Yet viral diseases, varying in severity from smallpox and rabies to the common cold, have no doubt plagued mankind since before the dawn of history. It is now known that viruses can infect plants and bacteria as well as animals. Each viral species has a very limited **host range;** that is, it can reproduce in only a small group of closely related species.

An intact virus particle, which is referred to as a **virion,** consists of a nucleic acid molecule encased by a protein **capsid.** In some of the more complex virions, the capsid is surrounded by a lipid bilayer and glycoprotein-containing **envelope,** which is derived from a host cell membrane. Since the small size of a viral nucleic acid severely limits the number of proteins that can be encoded by its genome, its capsid, as Francis Crick and James Watson pointed out in 1957, must be built up of one or a few kinds of protein subunits that are arranged in a symmetrical or nearly symmetrical fashion. There are two ways that this can occur:

1. In the **helical viruses** (Section 33-1), the coat protein subunits associate to form helical tubes.

2. In the **icosahedral viruses** (also known as **spherical viruses;** Section 33-2), coat proteins aggregate as closed polyhedral shells.

In both cases, the viral nucleic acid occupies the capsid's central region. In many viruses, the coat protein subunits may be "decorated" by other proteins so that the capsid exhibits spikes and, in larger bacteriophages, a complex tail. These assemblies are involved in recognizing the host cell and delivering the viral nucleic acid into its interior. Figure 33-1 is a "rogues' gallery" of viruses of varying sizes and morphologies.

The great simplicity of viruses in comparison to cells makes them invaluable tools in the elucidation of gene structure and function, as well as our best characterized models for the assembly of biological structures. Although all viruses use ribosomes and other host factors for the RNA-instructed synthesis of proteins, their modes of genome replication are far more varied than that of cellular life. In contrast to cells, in which the hereditary molecules are invariably double-stranded DNA, viruses contain either single- or double-stranded DNA or RNA. In RNA viruses, the viral RNA may be directly replicated or act as a template in the synthesis of DNA. The RNA of single-stranded RNA viruses may be the positive strand (the mRNA) or the negative strand (complementary to the mRNA). Viral

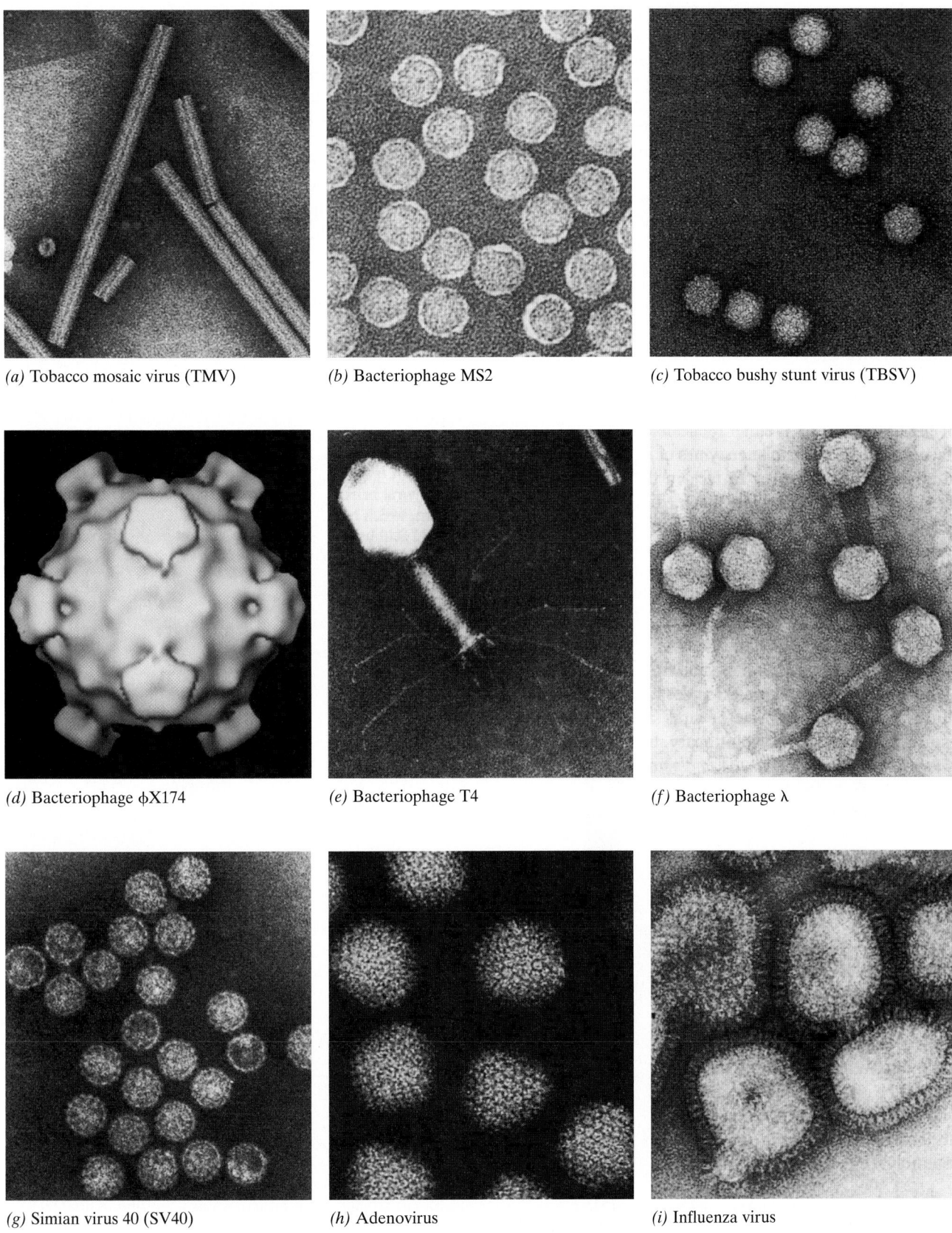

(a) Tobacco mosaic virus (TMV)

(b) Bacteriophage MS2

(c) Tobacco bushy stunt virus (TBSV)

(d) Bacteriophage φX174

(e) Bacteriophage T4

(f) Bacteriophage λ

(g) Simian virus 40 (SV40)

(h) Adenovirus

(i) Influenza virus

FIGURE 33-1 Electron micrographs of a selection of viruses. TMV, MS2, TBSV, and influenza virus are single-stranded RNA viruses; φX174 is a single-stranded DNA virus; and λ, T4, SV40, and adenovirus are double-stranded DNA viruses. Bacteriophage M13, a filamentous, single-stranded DNA coliphage, is shown in Fig. 5-45. [Parts *a–c* and *f–i* courtesy of Robley Williams, University of California at Berkeley, and Harold Fisher, University of Rhode Island; Part *d* courtesy of Michael Rossmann, Purdue University; and Part *e* courtesy of John Finch, Cambridge University.]

DNA may replicate autonomously or be inserted in the host chromosome for replication with the host DNA. The DNA of eukaryotic viruses is either replicated and transcribed in the cell nucleus by cellular enzymes or in the cytoplasm by virally specified enzymes. In fact, in the case of negative strand RNA viruses, enzymes that mediate viral RNA transcription must be carried by the virion because most cells lack the ability to transcribe RNA.

This chapter is a discussion of the structures and biology of a variety of viruses. In it, we examine mainly **tobacco mosaic virus (TMV)**, a helical RNA virus; several **icosahedral viruses**; **bacteriophage λ**, a tailed DNA bacteriophage; and **influenza virus**, an enveloped RNA virus. These examples have been chosen to illustrate important aspects of viral structure, assembly, molecular genetics, and evolutionary strategy. *Much of this information is relevant to the understanding of the corresponding cellular phenomena.*

1 ■ TOBACCO MOSAIC VIRUS

Tobacco mosaic virus causes leaf mottling and discoloration in tobacco and many other plants. It was the first virus to be discovered (by Dmitri Iwanowsky in 1892), the first virus to be isolated (by Wendell Stanley in 1935), and even now is among the most extensively investigated and well-understood viruses from the standpoint of structure and assembly. In this section, we discuss these aspects of TMV.

A. *Structure*

TMV is a rod-shaped particle (Fig. 33-1*a*) that is ~3000 Å long, 180 Å in diameter, and has a particle mass of 40 million D. Its ~2130 identical copies of coat protein subunits (158 amino acid residues; 17.5 kD) are arranged in a hollow right-handed helix that has 161/3 subunits/turn, a pitch (rise per turn) of 23 Å, and a 40-Å-diameter central cavity (Fig. 33-2). TMV's single RNA strand (~6400 nt; 2 million D) is coaxially wound within the turns of the coat protein helix such that 3 nt are bound to each protein subunit (Fig. 33-2).

a. TMV Coat Protein Aggregates to Form Viruslike Helical Rods

The aggregation state of TMV coat protein is both pH and ionic strength dependent (Fig. 33-3). At slightly alkaline pH's and low ionic strengths, the coat protein forms complexes of only a few subunits. At higher ionic strengths, however, the subunits associate to form a double-layered disk of 17 subunits/layer, a number that is nearly equal to the number of subunits per turn in the intact virion. At neutral pH and low ionic strengths, the subunits form short helices of slightly more than two turns (39 ± 2 subunits) termed "protohelices" (also known as "lockwashers"). If

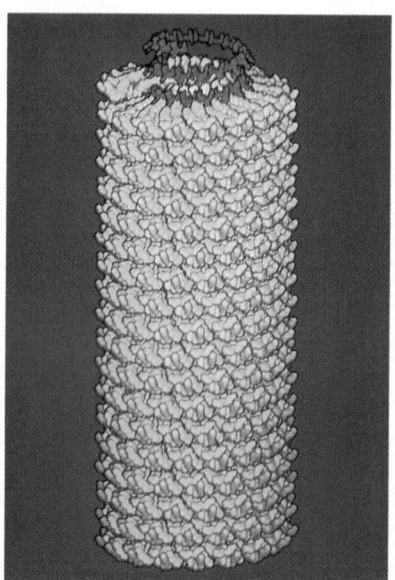

FIGURE 33-2   Model of tobacco mosaic virus (TMV) illustrating the helical arrangement of its coat protein subunits and RNA molecule. The RNA is represented by the red chain exposed at the top of the viral helix. Only 18 turns (415 Å) of the TMV helix are shown, which represent ~14% of the TMV rod. [Courtesy of Gerald Stubbs and Keiichi Namba, Vanderbilt University; and Donald Caspar, Brandeis University.]

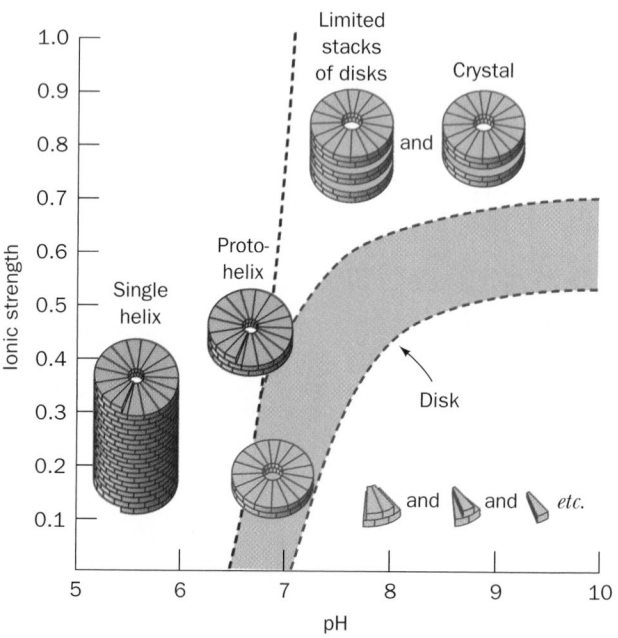

FIGURE 33-3   Aggregation state of TMV coat protein as a function of pH and ionic strength. Under basic conditions, the subunits aggregate into small clusters. Around neutrality and at high ionic strengths, the protein forms a 34-subunit double-layered disk. Under acidic conditions and at low ionic strengths, the subunits form protohelices that stack to form long helices. At neutral pH and low ionic strength, which resembles physiological conditions, the protein forms helices only in the presence of TMV RNA. [After Durham, A.C.H., Finch, J.T., and Klug, A., *Nature New Biol.* **229**, 38 (1971).]

the pH of these protohelices is shifted to ~5, they stack in imperfect register and eventually anneal to form indefinitely long helical rods that, although they lack RNA, resemble intact virions (Fig. 33-4). These observations, as we shall see below, lead to the explanation of how TMV assembles.

b. TMV Coat Protein Interacts Flexibly with Viral RNA

X-Ray studies of TMV have been pursued on two fronts. The virus itself does not crystallize but forms a highly oriented gel of parallel viral rods. The X-ray analysis of this gel by Kenneth Holmes and Gerald Stubbs yielded a structure of sufficient resolution (2.9 Å) to re-

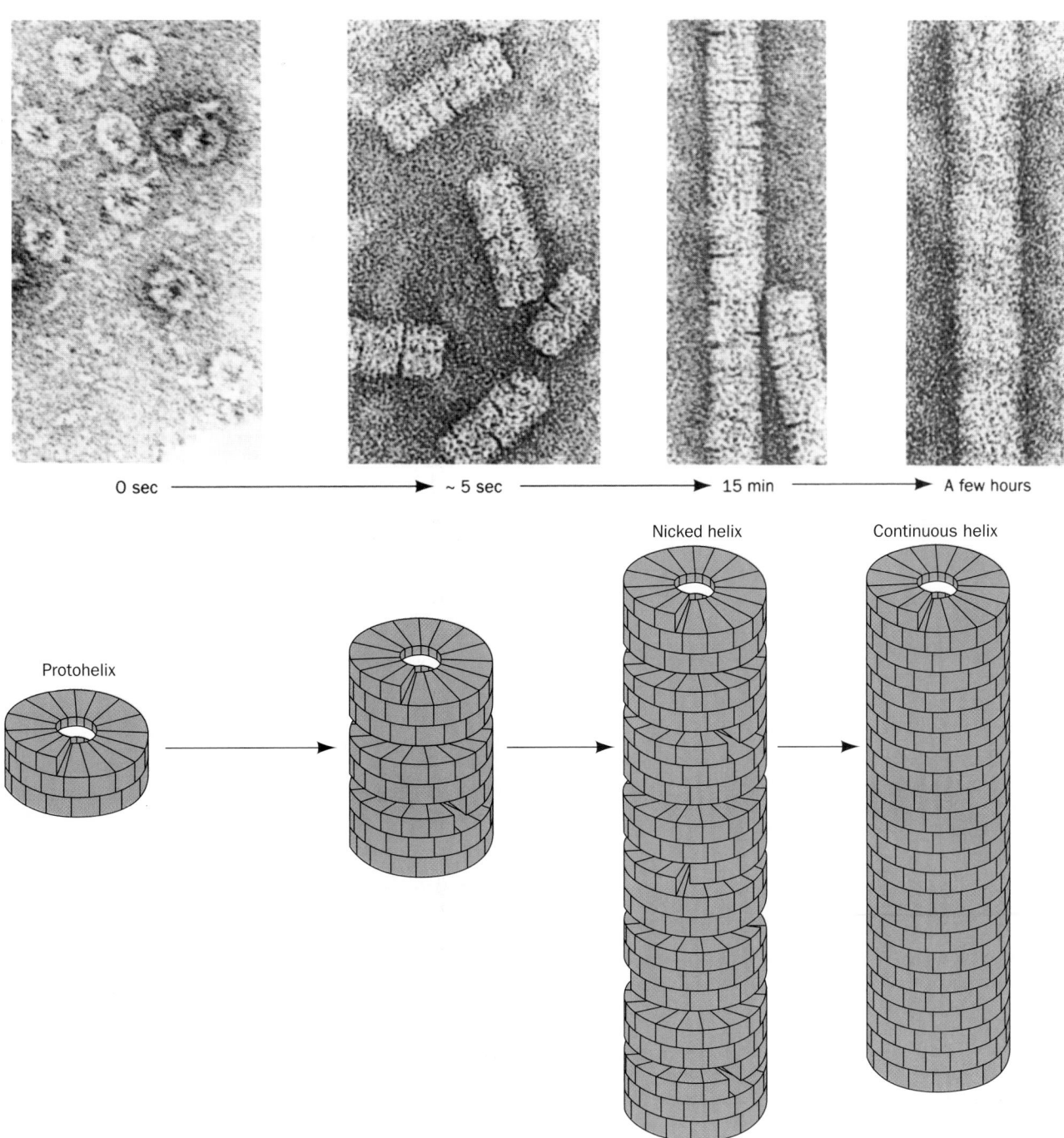

FIGURE 33-4 Growth of TMV coat protein rods. Electron micrographs (*above*) and their interpretive diagrams (*below*) show TMV coat protein aggregates following a rapid change in pH from 7 to 5 at low ionic strength. This pH shift causes the protohelices to form "nicked" (imperfectly stacked) helices that, within a few hours, anneal to yield continuous helical protein rods. [Courtesy of Aaron Klug, MRC Laboratory of Molecular Biology, Cambridge, U.K.]

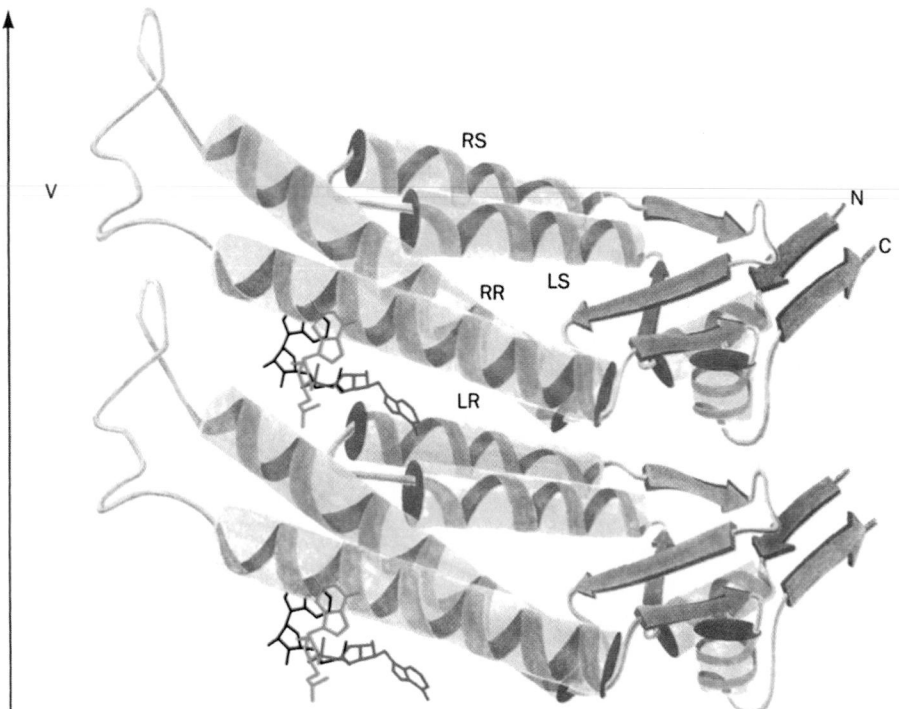

FIGURE 33-5 X-Ray structure of two vertically stacked TMV subunits. The structure is viewed perpendicular to the virus helix axis (*vertical arrow on the left*). Each subunit has four approximately radially extending helices (LR, RR, LS, and RS), as well as a short vertical segment (V), which comprises part of the flexible loop in the disk structure (dashed lines in Fig. 33-7). Portions of two successive turns of RNA are shown passing through their binding sites. Each subunit binds three nucleotides, here represented by GAA with each of its nucleotides differently colored, such that their three bases lie flat against the LR helix so as to grasp it in a clawlike manner. [After Namba, K., Pattanayek, R., and Stubbs, G., *J. Mol. Biol.* **208,** 314 (1989). PDBid 2TMV.]

veal the folding of the protein and the RNA (Figs. 33-5 and 33-6). This study is complemented by Aaron Klug's X-ray crystal structure determination, at 2.8-Å resolution, of the 34-subunit coat protein disk (Fig. 33-7).

A major portion of each subunit consists of a bundle of four alternately parallel and antiparallel α helices that project more or less radially from the virus axis (Figs. 33-5 to 33-7). In the disk, one of the inner connections between these α helices, a 24-residue loop (residues 90–113; dashed line in Fig. 33-7), is not visible, apparently because it is highly mobile. This disordered loop is also present in the protohelix as shown by NMR studies. In the virus, however, the loop adopts a definite conformation containing

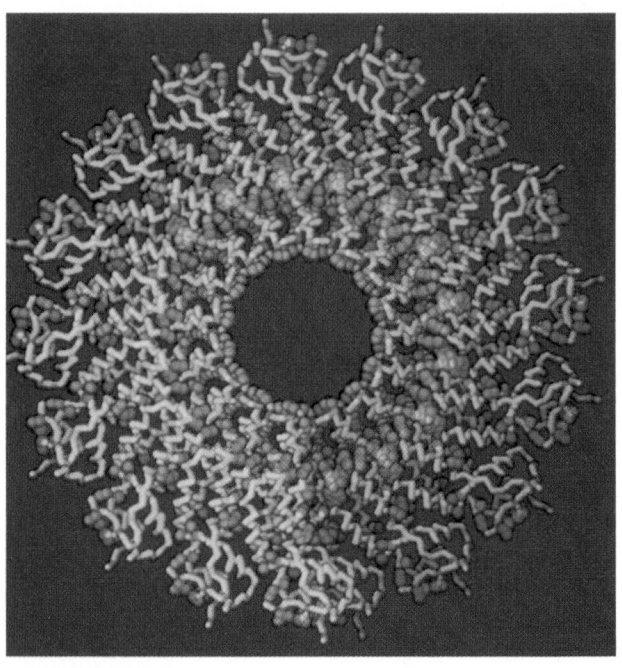

FIGURE 33-6 Top view of 17 TMV coat protein subunits comprising slightly more than one helical turn in complex with a 33-nucleotide RNA segment. The protein is represented by its Cα atoms, shown as connected yellow rods, together with its acidic side chains (Asp and Glu) as red balls and its basic side chains (Arg and Lys) as blue balls. The RNA's phosphate atoms are green and its bases are magenta. Note that the acidic side chains form a 25-Å-radius helix that lines the virion's inner cavity and the basic side chains form a 40-Å-radius helix that interacts with the RNA's anionic sugar–phosphate chain. [Courtesy of Gerald Stubbs and Keiichi Namba, Vanderbilt University; and Donald Caspar, Brandeis University. PDBid 2TMV.]

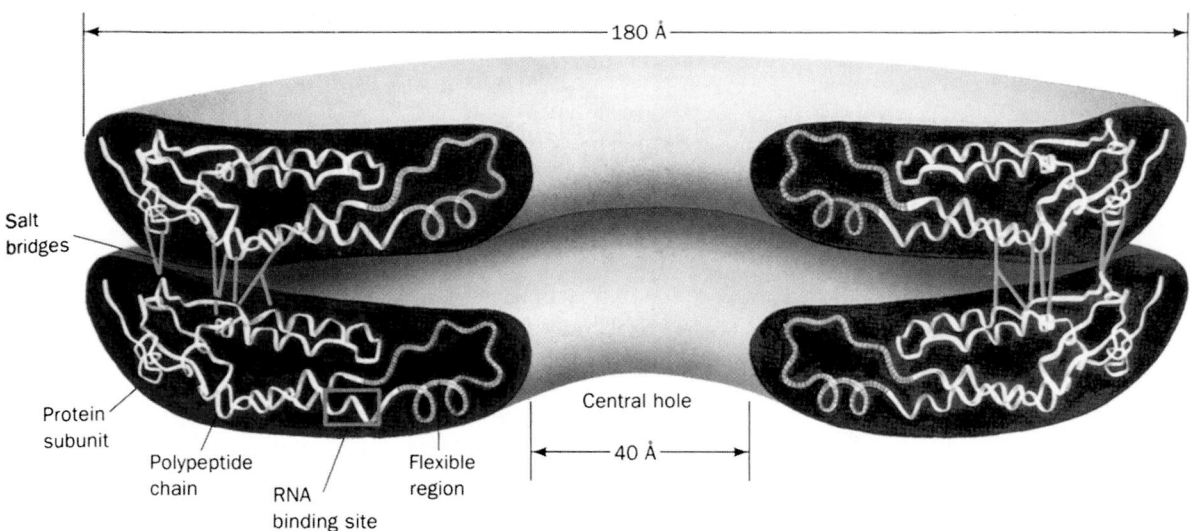

FIGURE 33-7   Structure of the TMV protein disk in cross section showing its polypeptide chains as ribbon diagrams. The dashed lines represent disordered loops of polypeptide chain that are therefore not visible in the disk X-ray structure. The stacked protein rings interact along their outer rims through a system of salt bridges (*red lines*). [After Butler, P.J.G. and Klug, A., *Sci. Am.* **239**(5): 67 (1978). Copyright © 1978 by Scientific American, Inc.]

a series of reverse turns arranged such that the overall direction of this polypeptide segment is approximately parallel to the virus axis (V in Fig. 33-5). This conformational change, as we shall see, is an important aspect of virus assembly.

In the virus, the RNA is helically wrapped between the coat protein subunits at a radius of ~40 Å. The triplet of bases binding to each subunit forms a clawlike structure around one of the radial helices (LR in Fig. 33-5) with each base occupying a hydrophobic pocket in which it lies flat against LR. Arg residues 90 and 92, which are invariant in the several known TMV strains and which are part of the disk and protohelix's disordered loop, as well as Arg 41, form salt bridges to the RNA phosphate groups.

B. *Assembly*

How is the TMV virion assembled from its component RNA and coat protein subunits? *The assembly of any large molecular aggregate, such as a crystal or a virus, generally occurs in two stages: (1)* **nucleation,** *the largely random aggregation of subunits to form a quasi-stable nucleation complex, which is almost always the rate-determining step of the assembly process; followed by (2)* **growth,** *the cooperative addition of subunits to the nucleation complex in an orderly arrangement that usually proceeds relatively rapidly.* For TMV, it might reasonably be expected that the nucleation complex minimally consists of the viral RNA in association with the 17 or 18 subunits necessary to form a stable helical turn, which could then grow by the accumulation of subunits at one or both ends of the helix. The low probability for the formation of such a complicated nucleation complex from disaggregated subunits accounts for the observed 6-h time necessary to complete this *in vitro* assembly process. Yet, the *in vivo* assembly of TMV probably occurs much faster. A clue as to the nature of this *in vivo* process was provided by the observation that if protohelices rather than disaggregated subunits are mixed with TMV RNA, complete virus particles are formed in 10 min. Other RNAs do not have this effect. Evidently, *the in vivo nucleation complex in TMV assembly is the association of a protohelix with a specific segment of TMV RNA.* (Although it was originally assumed that the double-layered disk rather than the protohelix formed the nucleating complex, experimental evidence indicates that the disk does not form under physiological conditions and that its rate of conversion to the protohelix under these conditions is too slow to account for the rate of TMV assembly. Other experiments, however, suggest that it is the disk that predominates at pH 7.0, the pH at which TMV most rapidly assembles from its component protein and RNA. Thus, keep in mind that the question as to whether TMV assembles from protohelices, as we state here, or from double-layered disks has not been fully resolved.)

a. TMV Assembly Proceeds by the Sequential Addition of Protohelices

The specific region of the TMV RNA responsible for initiating the virus particle's growth was isolated using the now classic nuclease protection technique. The RNA is mixed with a small amount of coat protein so as to form a nucleation complex that cannot grow because of the lack of coat protein. The RNA that is not protected by coat protein is then digested away by RNase, leaving intact only the initiation sequence. This RNA fragment forms a hairpin loop whose 18-nucleotide apical sequence, AGAA-GAAGUUGUUGAUGA, has a G at every third residue

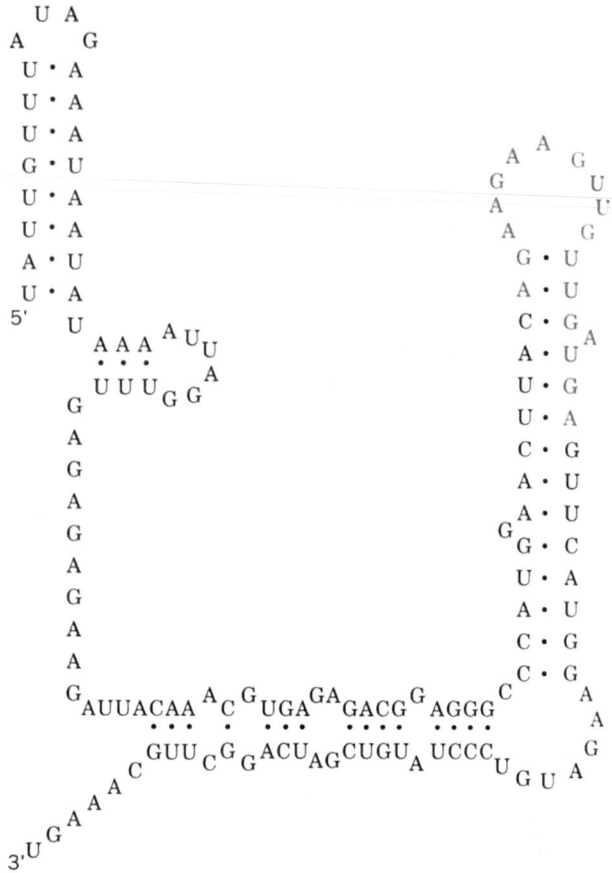

FIGURE 33-8   Initiation segment of TMV RNA. It probably forms a weakly base paired hairpin, as drawn, that is thought to begin TMV assembly by specifically binding to a coat protein protohelix. Note that this RNA's loop region has an 18-nt segment (*red*) with a G every third residue (each coat protein subunit binds three nucleotides) but no C's.

(recall that each coat protein subunit binds three nucleotides) but no C's (Fig. 33-8). Site-directed mutagenesis studies have confirmed that this initiation sequence is sufficient to direct TMV assembly and that the regularly spaced G's and lack of C's are important for its function. TMV's high binding affinity for this initiation sequence is explained, in part, by the observations that coat protein subunits bind every third nucleotide in the unusual syn conformation and that G assumes this conformation more easily than any other nucleotide (Section 29-2A). The lack of C's perhaps prevents the involvement of these G's in base pairing associations.

The above initiation complex is located some 1000 nucleotides from the 3' end of the TMV RNA. Hence, the simple model of viral assembly in which the RNA is sequentially coated by protein from one end to the other cannot be correct. Rather, the RNA initiation hairpin must insert itself between the protohelix's protein layers from its central cavity (Fig. 33-9a). The RNA binding, for reasons explained below, induces the ordering of the disordered loop, thereby trapping the RNA (Fig. 33-9b). Growth then proceeds by a repetition of this process at the "top" of the complex, thereby incrementally pulling the RNA's 5' end up through the central cavity of the growing viral helix (Fig. 33-9c).

The above assembly model has been corroborated by several experimental observations:

1. Electron micrographs reveal that partially assembled rods (Fig. 33-10) have two RNA "tails" projecting from one end.

2. The length of the longer tail, presumably the 5' end, decreases linearly with the length of the rod, whereas the shorter tail maintains a more or less constant length.

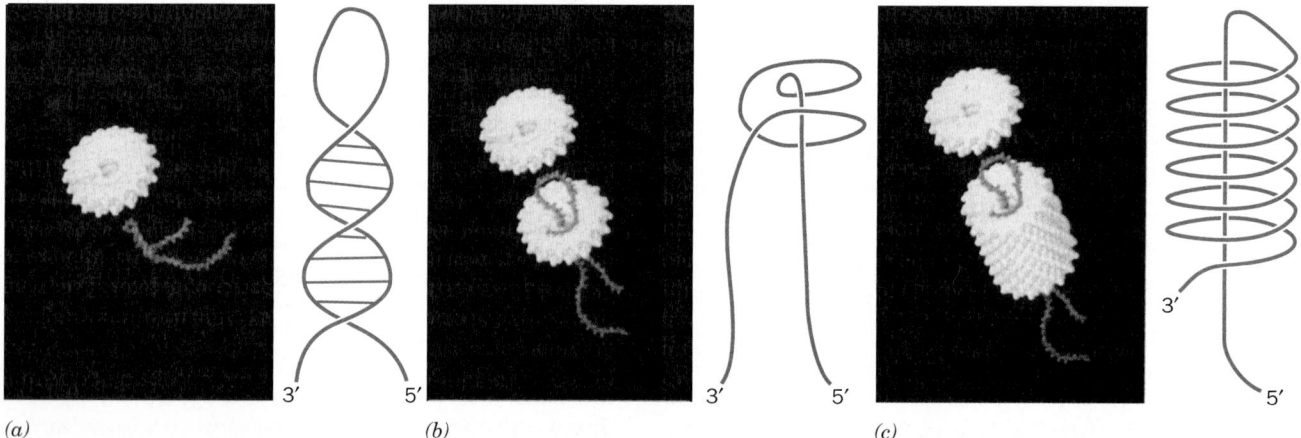

(a)   *(b)*   *(c)*

FIGURE 33-9   Assembly of TMV. (*a*) The process begins by the insertion of the hairpin loop formed by the initiation sequence of the viral RNA into the protohelix's central cavity. (*b*) The RNA then intercalates between the layers of the protohelix, thereby ordering the disordered loop and trapping the RNA. (*c*) Elongation proceeds by the stepwise addition of protohelices to the "top" of the viral rod. The consequent binding of the RNA to each protohelix, which converts it to the helical form, pulls the RNA's 5' end up through the virus' 40-Å-diameter central cavity to form a traveling loop at the viral rod's growing end. [Viral images courtesy of Hong Wang and Gerald Stubbs, Vanderbilt University.]

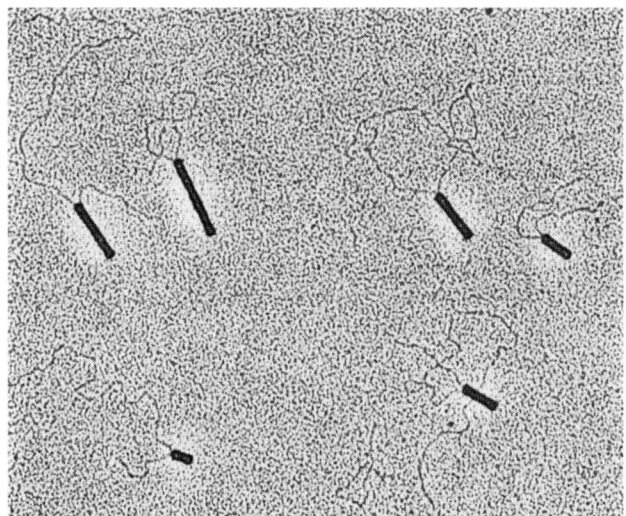

FIGURE 33-10 Electron micrograph of partially reconstituted TMV particles showing that their two RNA tails emerge from the same end of the growing viral rod. An analysis of these particles indicates that the length of one of the tails, probably the 3′ end, is constant (720 ± 80 nucleotides), whereas that of the other tail is inversely proportional to the length of its incomplete rod. [Courtesy of K.E. Richards, CNRS, France.]

3. Nuclease digestion experiments on partially assembled rods indicate that the RNA is protected in increments of ~100 nucleotides, as is expected for elongation steps consisting of the addition of a protohelix to the growing rod.

The coating of the 3′ end of the RNA is a much slower process than the coating of its 5′ end and hence probably occurs by the successive addition of single subunits. The RNA, which acts as the viral mRNA, carries the gene specifying the coat protein near its 3′ end. Perhaps this assembly mechanism allows coat protein synthesis during all but the final stages of assembly, thereby permitting the completion of this process.

b. Electrostatic Repulsions and Steric Interactions Prevent Helix Formation in the Absence of RNA

What is the mechanism that prevents the formation of TMV coat protein helices in the absence of viral RNA but triggers virus assembly in its presence (and, conversely, how does intact TMV disassemble to initiate an infection)? Structural considerations suggest that the coat protein subunit's disordered loop sterically prevents the protohelix from growing longer. Moreover, as we have seen (Fig. 33-3), the state of coat protein aggregation varies with pH. Titration studies show that each subunit has two ionizations with pK's near 7, which must each be attributed to anomalously basic carboxyl groups because coat protein has no His residues. The most plausible candidates for these anomalously basic carboxyls are two intersubunit pairs of carboxyl groups: Glu 95–Glu 106, disordered loop members which interact across a side-to-side subunit interface; and Glu 50–Asp 77, which interact across a top-to-bottom subunit interface. Moreover, Asp 116 is close to an RNA phosphate group. The electrostatic repulsions between these closely spaced negative charges promotes the formation of the disordered loop and therefore favors the protohelix conformation. The binding of the RNA initiation sequence to the protein apparently provides sufficient free energy to overcome these repulsions, thereby triggering helix formation (a process that partially protonates the anomalously basic carboxyl groups; recall the similar conformationally induced pK changes in the Bohr effect of hemoglobin; Section 10-2E). Indeed, site-directed mutagenesis of Glu 50 → Gln or Asp 77 → Asn both increases virion stability and decreases its infectivity (presumably by inhibiting viral disassembly). Further growth of the viral rod can occur on RNA segments that lack this sequence as a consequence of the additional binding interactions between adjacent protohelices. *The carboxyl groups evidently act as a negative switch to prevent the formation of a protein helix in the absence of RNA under physiological conditions.*

2 ■ ICOSAHEDRAL VIRUSES

The simpler **icosahedral viruses,** being uniform molecular assemblies, crystallize in much the same way as proteins. The techniques of X-ray crystallography can therefore be brought to bear on determining virus structures. In this section we consider the results of such studies.

A. *Virus Architecture*

The very limited genomic resources of the simpler viruses in many cases limit them to having but one type of protein in their capsid. Since these coat protein subunits are chemically identical, they must all assume the same or nearly the same conformations and have similar interactions with their neighbors. What geometrical constraints does this limitation impose on viral architecture?

We have already seen that TMV solves this problem by assuming a helical geometry (Fig. 33-2). The coat protein subunits in such a long but finite helix, although geometrically distinguishable, have, with the exception of the subunits at the helix ends, virtually identical environments. Such subunits are said to be **quasi-equivalent** to indicate that they are not completely indistinguishable as they would be in an object whose elements are all related by exact symmetry.

a. Icosahedral Viruses Have Icosahedral Capsids

A second arrangement of equivalent subunits that can encapsulate a nucleic acid is that of a polyhedral shell. There are only three polyhedral symmetries in which all the elements are indistinguishable: those of a tetrahedron, a cube, and an icosahedron (Fig. 8-64c). Capsids with these symmetries would have 12, 24, or 60 subunits identically arranged on the surface of a sphere. For example, an

(a)

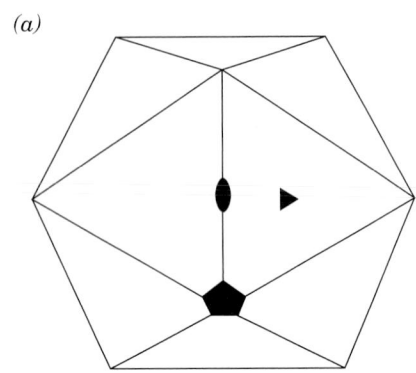

(a)

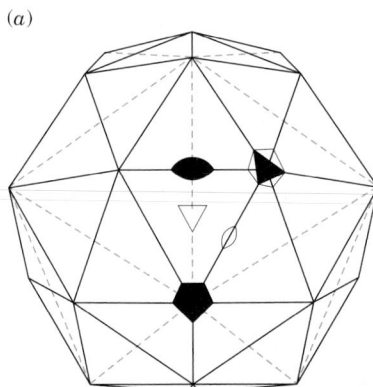

(b)

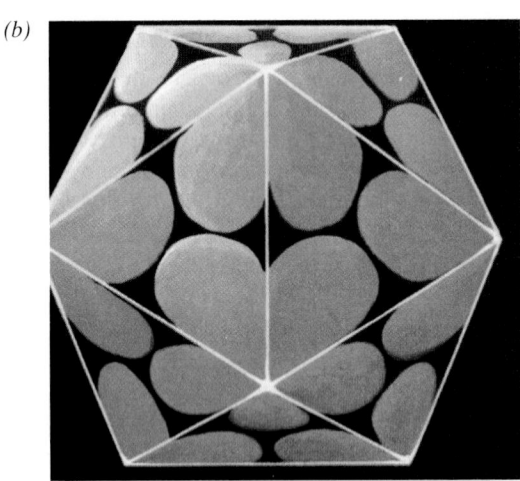

(b)

FIGURE 33-11 Icosahedron. (*a*) This regular polyhedron has 12 vertices, 20 equilateral triangular faces of identical size, and 30 edges. It has a 5-fold axis of symmetry through each vertex, a 3-fold axis through the center of each face, and a 2-fold axis through the center of each edge (also see Fig. 8-64*c*). (*b*) A drawing of 60 identical subunits (*lobes*) arranged with icosahedral symmetry. [Illustration, Irving Geis/Geis Archives Trust, Copyright Howard Hughes Medical Institute. Reproduced with permission.]

FIGURE 33-12 *T* = 3 icosadeltahedron. (*a*) This polyhedron has the exact rotational symmetry of an icosahedron (*solid symbols*) together with local 6-fold, 3-fold, and 2-fold rotational axes (*hollow symbols*). Note that the edges of the underlying icosahedron (*dashed red lines*), are not edges of this polyhedron and that its local 6-fold axes are coincident with its exact 3-fold axes. (*b*) A drawing of a *T* = 3 icosadeltahedron showing its arrangement of 3 quasi-equivalent sets of 60 icosahedrally related subunits (*lobes*). The A lobes (*orange*) pack about the icosadeltahedron's exact 5-fold axes, whereas the B and C lobes (*blue and green*) alternate about its local 6-fold axes. TBSV's chemically identical coat protein subunits are arranged in this manner. [Illustration, Irving Geis/Geis Archives Trust, Copyright Howard Hughes Medical Institute. Reproduced with permission.]

icosahedron (Fig. 33-11*a*) has 20 triangular faces, each with 3-fold symmetry, for a total of 20 × 3 = 60 equivalent positions (each represented by a lobe in Fig. 33-11*b*). Of these polyhedra, the icosahedron encloses the greatest volume per subunit. Indeed, electron microscopy of the so-called icosahedral viruses (such as Fig. 33-1*b–h*) first demonstrated that *they have icosahedral symmetry.*

b. Viral Capsids Resemble Geodesic Domes

A viral nucleic acid, if it is to be protected effectively against a hostile environment, must be completely covered by coat protein. Yet, many viral nucleic acids occupy so large a volume that their coat protein subunits would have to be prohibitively large if their capsids were limited to the 60 subunits required by exact icosahedral symmetry. In fact, nearly all viral capsids have considerably more than 60 chemically identical subunits. How is this possible?

Donald Caspar and Klug pointed out the solution to this dilemma. *The triangular faces of an icosahedron can be subdivided into integral numbers of equal sized equilateral*

triangles (e.g, Fig. 33-12*a*). The resulting polyhedron, an **icosadeltahedron,** has "local" symmetry elements relating its subunits (lobes in Fig. 33-12*b*) in addition to its exact icosahedral symmetry. By local symmetry, we mean that the symmetry is only approximate so that, in contrast to the case for exact symmetry, it breaks down over larger distances. For instance, the subunits (lobes) in Fig. 33-12*b* that are distributed about each exact triangular vertex form clusters whose members are related by a local 6-fold axis of symmetry. *Adjacent subunits in these clusters are not exactly equivalent; they are quasi-equivalent.* In contrast, the subunits clustered about the twelve 5-fold axes of icosahedral symmetry are exactly equivalent. The interac-

FIGURE 33-13 Geodesic dome built on the plan of a *T* = 36 icosadeltahedron. Two of its pentagonal vertices are visible in this photograph. [Stanley Schoenberger/Grant Heilman.]

tions between the subunits clustered about the local 6-fold axes are therefore essentially distorted versions of those about the exact 5-fold axes. Consequently, *the coat protein subunits of any viral capsid with icosadeltahedral symmetry must make alternative sets of intersubunit associations and/or have sufficient conformational flexibility to accommodate these distortions.*

Icosadeltahedra are familiar figures. The faceted surface of a soccer ball is an icosadeltahedron. Likewise, **geodesic domes** (Fig. 33-13), which were originally designed by Buckminster Fuller, are portions of icosadeltahedra. It was, in fact, Fuller's designs that inspired Caspar and Klug. *Geodesic domes are inherently rigid shell-like structures that are constructed from a few standard parts, make particularly efficient use of structural materials, and can be rapidly and easily assembled. Presumably the evolution of icosahedral virus capsids was guided by these very principles.*

The number of subunits in an icosadeltahedron is 60*T*, where *T* is called the **triangulation number.** The permissible values of *T* are given by $T = h^2 + hk + k^2$, where *h* and *k* are positive integers. An icosahedron, the simplest icosadeltahedron, has *T* = 1 (*h* = 1, *k* = 0) and therefore 60 subunits. The icosadeltahedron with the next level of complexity has a triangulation number of *T* = 3 (*h* = 1, *k* = 1) and hence 180 subunits (Fig. 33-12). A capsid with this geometry has three different sets of icosahedrally related subunits that are quasi-equivalent to each other (lobes A, B, and C in Fig. 33-12*b*). The X-ray structures of viruses with capsids consisting of *T* = 1, 3, 4, 7, and 13 icosadeltahedra have been determined. Some of the larger icosahedral viruses form icosadeltahedra with even greater triangulation numbers (see below). However, some of them are based on somewhat different assembly principles (Section 33-2D). The *T* value for any particular capsid, presumably, depends on its subunit's innate curvature.

B. *Tomato Bushy Stunt Virus*

Tomato Bushy Stunt Virus (TBSV); Fig. 33-1*c* is a *T* = 3 icosahedral virus that is ~175 Å in radius. It consists of 180 identical coat protein subunits, each of 386 residues (43 kD), encapsulating a single-stranded RNA molecule of ~4800 nt (1500 kD; the positive or message strand) and a single copy of an ~85-kD protein. The X-ray crystal structure of TBSV, the first of a virus to be determined at high resolution, was reported in 1978 by Stephen Harrison. TBSV's coat protein subunits have three domains (Fig. 33-14): P, the C-terminal domain, which projects outward from the virus; S, which forms the viral shell; and R, the protein's inwardly extending N-terminal domain, which is attached to the S domain via a connecting arm. The S domain is almost entirely composed of an 8-stranded antiparallel β barrel, which we shall see occurs in the coat proteins of the majority of icosahedral viruses with known structures.

a. TBSV's Identical Subunits Associate through Nonidentical Contacts

The chemically identical TBSV coat protein subunits occupy three symmetrically distinct environments denoted

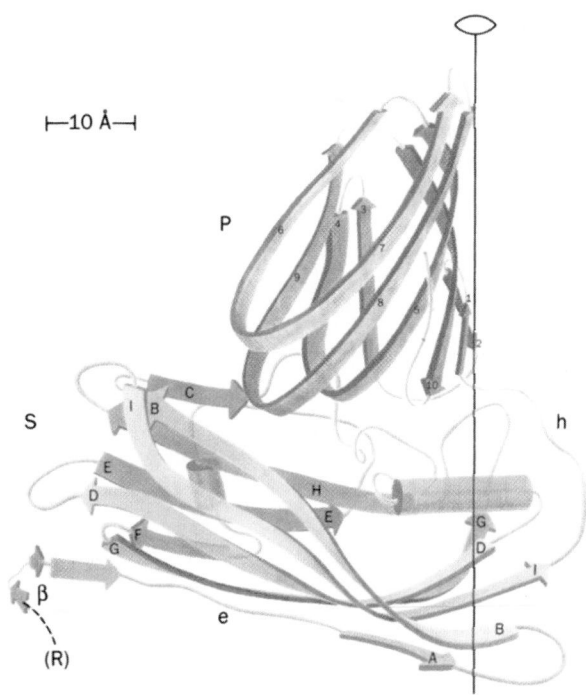

FIGURE 33-14 X-Ray structure of the tomato bushy stunt virus (TBSV) coat protein subunit. It consists of three domains: P, which projects from the virion's surface (*purple*); S, which forms the capsid (*green*); and R, which extends below the capsid surface where it participates in binding the viral RNA. The S domain is largely comprised of an 8-stranded antiparallel β barrel that has the jelly roll or Swiss roll topology (Section 8-3B). The P domain is also composed largely of an antiparallel β sheet, whereas the R domain is not visible in the viral X-ray structure so its tertiary structure is unknown. [After Olsen, A.J., Bricogne, G., and Harrison, S.C., *J. Mol. Biol.* **171,** 78 (1983). PDBid 2TBV.]

A, B, and C (Fig. 33-15). How does the protein accommodate the different contacts required by its several sets of analogous but nonidentical associations? TBSV's structure reveals that *analogous intersubunit contacts vary both*

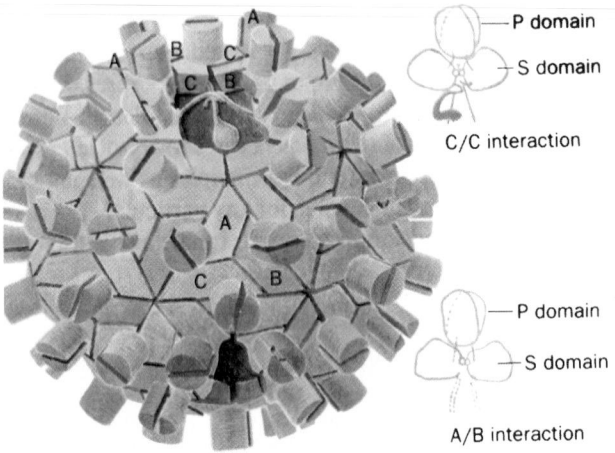

FIGURE 33-15 *T = 3* **icosadeltahedral arrangement of TBSV's coat protein subunits.** The subunits occur in three quasi-equivalent packing environments, A, B, and C. The A subunits (*orange*) pack around exact 5-fold axes, whereas the B subunits (*blue*) alternate with the C subunits (*green*) about the exact 3-fold axes (local 6-fold axes). The C subunits are also disposed about the strict 2-fold axes, whereas the A and B subunits are related by local 2-fold axes. The subunits respond to the different conformational requirements of their three quasi-equivalent positions through flexion at the hinge region between their S and P domains (*right and in cutouts*). Compare this drawing to Fig. 33-12. [After Harrison, S.C., *Trends Biochem. Sci.* **9**, 348, 349 (1984).]

through alternative sets of interactions and by conformational distortions of the same interactions. Perhaps the most remarkable alternative interaction is the interdigitation of the arms connecting the R and S domains of the C subunits. These arms extend toward each icosahedral 3-fold axis (quasi-6-fold axis) in the clefts between the adjacent C and B subunits and then spiral downward about this 3-fold axis to form a β-sheetlike arrangement that resembles the overlapping flaps of a cardboard carton: chain 1 over chain 2 over chain 3 over chain 1 (Fig. 33-16*a*). This interaction, together with a strong association between neighboring C subunits across the icosahedral 2-fold axis (Fig. 33-15), organizes the 60 C subunits into a coherent network (Fig. 33-16*b*) that determines the triangulation number of the TBSV capsid: *The C subunits can be thought of as forming a T = 1 icosahedral shell whose gaps are filled in by the A and B subunits.* In response, the three sets of quasi-equivalent subunits assume somewhat different conformations: The three- or four-residue "hinge" connecting the S and P domains (h in Fig. 33-14) has an ~30° greater dihedral angle in the A and B subunits than in the C subunits (Fig. 33-15, *right*). This, in turn, permits the interactions between P domains to be identical in the AB and CC dimers (projecting dimeric knobs in Fig. 33-15). Evidently, interdomain associations between subunits are stronger in TBSV than those within subunits.

b. TBSV's RNA-Containing Core Is Disordered

The entire connecting arm between the R and S domains in the A and B subunits, as well as their first few residues in the C subunits, are not visible in TBSV's X-ray structure, thereby indicating that these polypeptide segments have no fixed conformations. The R domains are therefore flexibly tethered to the S domains so that they are also absent from the X-ray structure, even though these domains probably have a fixed conformation. Neutron scattering studies, nevertheless, suggest that protein, con-

(a)   *(b)*

FIGURE 33-16 **Architecture of the TBSV capsid.** (*a*) The C subunit arms of TBSV protein pack about the capsid's exact 3-fold axes (*triangle*) and associate as β sheets. The view is from outside the capsid. (*b*) A stereo cutaway drawing showing the capsid's internal scaffolding of C subunit arms. The chemically identical A (*blue*), B (*cyan*), and C subunits (*red*) are represented by large spheres, whereas the residues comprising the C subunit arms are represented by small yellow spheres.

The C subunit arms associate to form an icosahedral (*T* = 1) framework that apparently plays a major role in holding together the viral capsid. Directions for viewing stereo drawings are given in the Appendix to Chapter 8. [Part *a* after a drawing by Jane Richardson, Duke University, and Part *b* courtesy of Arthur Olson, The Scripps Research Institute, La Jolla, California. Based on an X-ray structure by Stephen Harrison, Harvard University. PDBid 2TBV.]

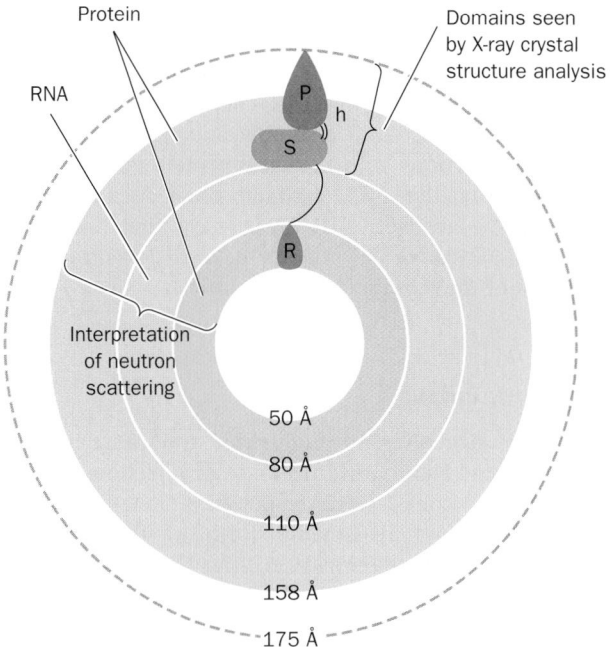

FIGURE 33-17 Radial organization of TBSV indicating the distribution of its protein and RNA components. The R domain positions are inferred from their known chain length. Only about half of the R domains are contained in the inner protein shell. [After Harrison, S.C., *Biophys. J.* **32,** 140 (1980).]

stituting perhaps half of the R domains, forms a 50- to 80-Å radius inner shell. The remaining R domains are thought to project into the space between the inner and outer shells.

The viral RNA is absent from the X-ray structure, which indicates that it too is disordered. The above neutron scattering studies reveal that this RNA is sandwiched between the virus' inner and outer protein shells (Fig. 33-17). The volume constraints imposed by this arrangement require that the RNA be tightly packed. This packing is made possible because most of the negative charges of the RNA phosphate groups are neutralized by the numerous positively charged Arg and Lys residues of the R domains, the inner faces of the S domains, and their connecting arms.

c. Many Other RNA Viruses Are Remarkably Similar to TBSV

The structures of numerous other RNA plant viruses have been elucidated, including those of **southern bean mosaic virus (SBMV)** by Michael Rossmann and **satellite tobacco mosaic virus (STMV)** by Alexander McPherson. SBMV is a *T* = 3 virus that closely resembles TBSV in its quaternary structure. Moreover, SBMV's 260-residue coat protein subunit, although it entirely lacks a P domain, has an S domain whose polypeptide backbone is nearly superimposable on that of TBSV (Fig. 33-18*a*). The RNA in SBMV, as is that in TBSV, is disordered.

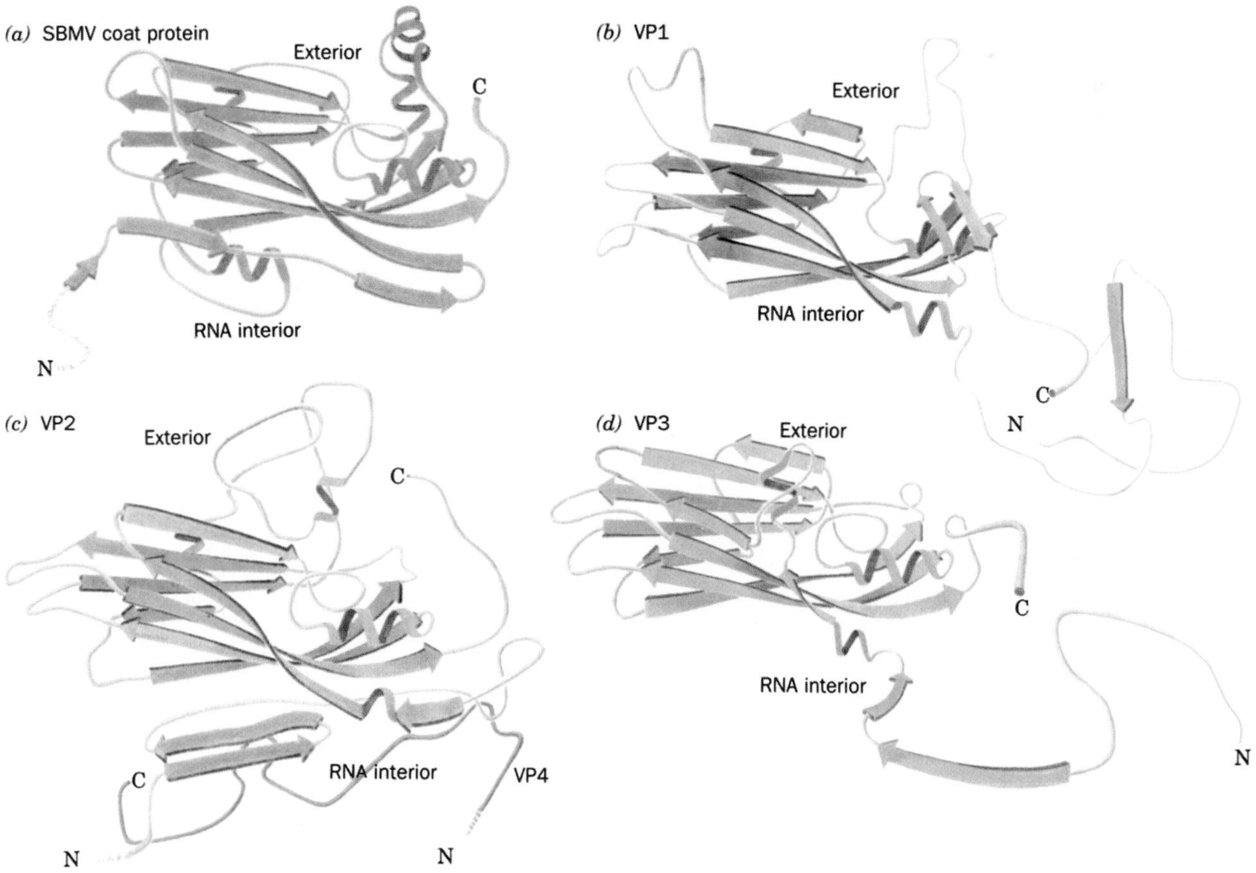

(a) SBMV coat protein

(b) VP1

(c) VP2

(d) VP3

FIGURE 33-18 Comparison of the X-ray structures of southern bean mosaic virus (SBMV) and human rhinovirus coat proteins. (*a*) SBMV coat protein, and (*b*) VP1, (*c*) VP2 (together with VP4), and (*d*) VP3 proteins of human rhinovirus. Note the close structural similarities of their 8-stranded β-barrel cores and that of TBSV's S domain (Fig. 33-14). The VP1, VP2, and VP3 proteins of poliovirus also have this fold. [After Rossmann, M.G., et al., *Nature* **317,** 148 (1985). PDBids 4BSV and 4RHV.]

STMV's quaternary structure differs from those of TBSV or SBMV: It is a $T = 1$ RNA virus. This 172-Å-diameter particle, which is among the smallest of known virions, encloses a 1058-nt RNA that encodes only one protein, its 196-residue viral coat protein (STMV can only multiply in cells that are coinfected with the more complex TMV, the only known example of such a parasitic relationship between an icosahedral virus and a rod-shaped virus). Nevertheless, STMV's coat protein, which also lacks a P domain, has an S domain that structurally resembles those of SBMV and TBSV. Evidently, these biochemically dissimilar viruses arose from a common ancestor.

d. Most of STMV's RNA Is Visible

The most striking aspect of the STMV structure is that nearly 80% of its RNA is visible (Fig. 33-19). The RNA takes the form of 30 largely double helical segments that lie on the icosahedron's 2-fold axes and which are linked by mainly disordered single-stranded regions. Computerized searches indicate up to 68% of STMV RNA could simultaneously form base pairs. However, it seems unlikely that a unique structure with extensive and nonrepetitive pairing between bases that are distant in sequence could be made to fold in a manner consistent with STMV's icosahedral symmetry. Rather, it appears that the double helical segments seen in Fig. 33-19 represent a series of somewhat

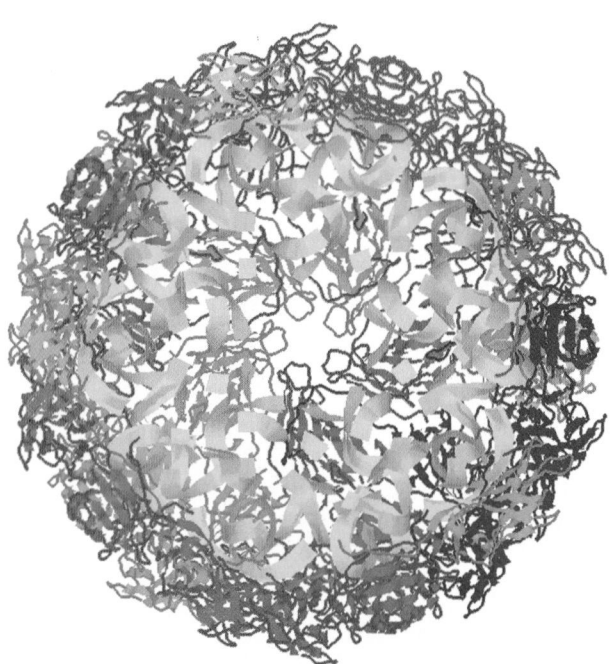

FIGURE 33-19   X-Ray structure of satellite tobacco mosaic virus (STMV). The virion is shown in cutaway view along its icosahedral ($T = 1$) 5-fold axis. Its coat protein subunits, which form a shell between the radii of 57 to 86 Å, are drawn in different colors, whereas its 30 mainly double helical RNA segments that are located on the icosahedral 2-fold axes are yellow. [Based on an X-ray structure by Alexander McPherson, University of California at Irvine. PDBid 1A34.]

different local stem–loop structures. Thus, inside the viral capsid, STMV RNA assumes a structure that is probably not its lowest free energy state. More likely, it assumes one of numerous relatively low free energy states that transiently form during viral assembly and become trapped through interactions with the viral protein coat. Indeed, the observation that STMV coat protein does not form capsids in the absence of RNA suggests that the RNA, which lacks icosahedral symmetry, nevertheless directs the formation of the icosahedral viral particle.

C. *Picornaviruses*

The X-ray structures of **poliovirus,** the cause of **poliomyelitis,** and **rhinovirus,** the cause of **infectious rhinitis** (the common cold), were respectively determined by James Hogle and Rossmann. Both of these human pathogens are **picornaviruses,** a large family of animal viruses that also includes the agents causing human **hepatitis A** and **foot-and-mouth disease.** Picornaviruses (*pico,* small + *rna*) are among the smallest RNA-containing animal viruses: They have a particle mass of $\sim 8.5 \times 10^6$ D of which $\sim 30\%$ is a single-stranded RNA of ~ 7500 nucleotides. Their icosahedral protein shell, which is ~ 300 Å in diameter, contains 60 protomers, each consisting of four structural proteins, **VP1, VP2, VP3,** and **VP4.** These four proteins are synthesized by an infected cell as a single polyprotein, which is cleaved to the individual subunits during virion assembly. Picornaviruses can be highly specific as to the cells they infect; for example, poliovirus binds to receptors that occur only on certain types of primate cells.

The structures of poliovirus, rhinovirus, and **foot-and-mouth disease virus (FMDV;** determined by David Stuart) are remarkably alike, both to each other and to TBSV and SBMV. Although VP1, VP2, and VP3 of picornaviruses have no apparent sequence similarities with each other or with the coat proteins of TBSV and SBMV, these proteins all exhibit striking structural similarities (Figs. 33-14 and 33-18; VP4, which is much smaller than the other subunits, forms, in effect, an N-terminal extension of VP2). Indeed, the picornaviruses' chemically distinct VP1, VP2, and VP3 subunits are pseudosymmetrically related by pseudo-3-fold axes passing through the center of each triangular face of the icosahedral ($T = 1$) virion, which therefore has pseudo-$T = 3$ symmetry (Fig. 33-20). The chemically identical but conformationally distinct A, B, and C subunits of the $T = 3$ plant viruses are likewise quasi-symmetrically related by analogously located local 3-fold axes (Fig. 33-15). These structural similarities strongly suggest that the picornaviruses and the icosahedral plant viruses all diverged from a common ancestor.

The protein capsids of poliovirus, rhinovirus, and FMDV form a hollow shell enclosing a disordered core composed of the viral RNA and some protein, much as in the icosahedral plant viruses. This arrangement is vividly illustrated in Fig. 33-21, which shows both the inner and outer views of the poliovirus capsid. Note that VP4 largely

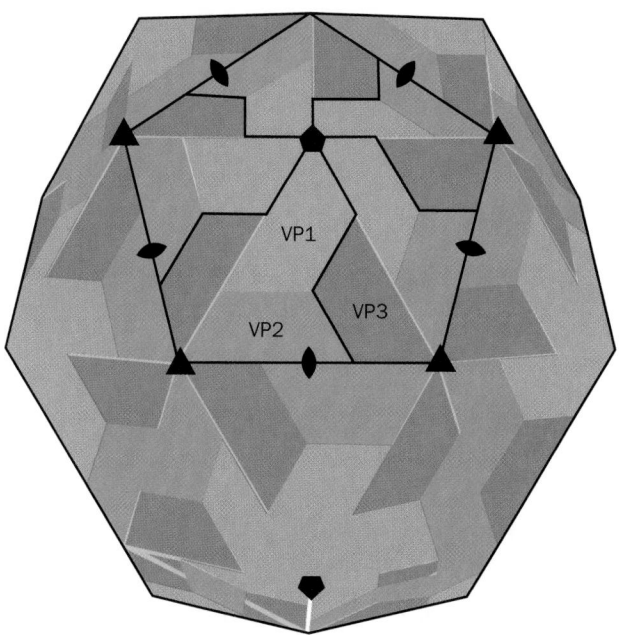

FIGURE 33-20 Arrangement of the 60 trimers (*triangles*) of pseudo-equivalent VP1, VP2, and VP3 subunits in the icosahedral capsid of human rhinovirus. This arrangement resembles that of TBSV in which 180 chemically identical subunits are quasi-symmetrically related to form a $T = 3$ icosadeltahedron (Figs. 33-12 and 33-15). The positions of the icosahedron's exact 5-, 3-, and 2-fold axes are marked. [After Rossmann, M.G., et al., *Nature* **317**, 147 (1985).]

lies inside of the capsid. Also note the rugged topography of the capsid's outer surface. Some of its crevices form the receptor-binding site through which the virus is targeted to specific cells.

D. *Simian Virus 40 (SV40)*

Simian virus 40 (SV40) is a **polyomavirus,** the simplest class of viruses containing double-stranded DNA. This ~500-Å external diameter icosahedral virus (Fig. 33-1*g*) functions

to transfer a 5243-bp circular "minichromosome" (DNA in complex with histone-containing particles known as **nucleosomes;** Section 34-1B) from the nucleus of one cell to that of another. The viral capsid consists of 360 copies of a 361-residue protein, **VP1,** that are arranged with icosahedral symmetry. However, this number of particles cannot be arranged with the icosadeltahedral symmetry characteristic of TBSV, for example, because $T = 360/60 = 6$ is a forbidden value for icosadeltahedra (for which $T = h^2 + hk + k^2$). Rather, as Caspar demonstrated through low-resolution X-ray studies of polyomaviruses, VP1 exclu-

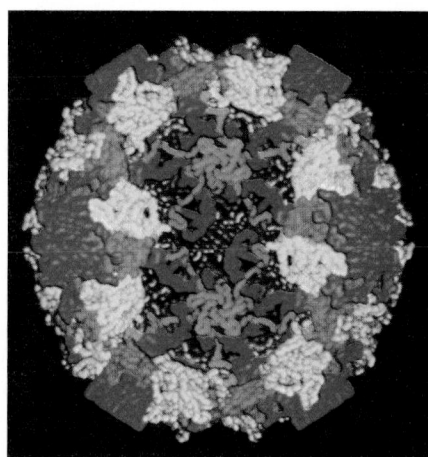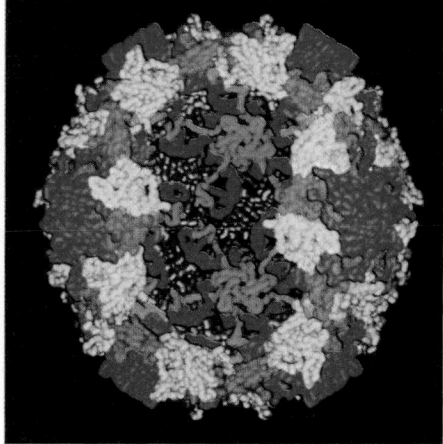

FIGURE 33-21 Stereo diagram of the poliovirus capsid in which the inner surface is revealed by the removal of two pentagonal faces. Here, the polypeptide chain is represented by a folded tube that approximates the volume of the protein and which is blue in VP1, yellow in VP2, red in VP3, and green in VP4. The VP4 subunits, which line the capsid's inner surface,

associate about its 5-fold axes of symmetry to form a framework similar to although geometrically distinct from that formed by the C subunit arms in TBSV (Fig. 33-16). [Courtesy of Arthur Olson, The Scripps Research Institute, La Jolla, California. Based on an X-ray structure by James Hogle, Harvard Medical School. PDBid 2PLV.]

(a)   *(b)*   *(c)*

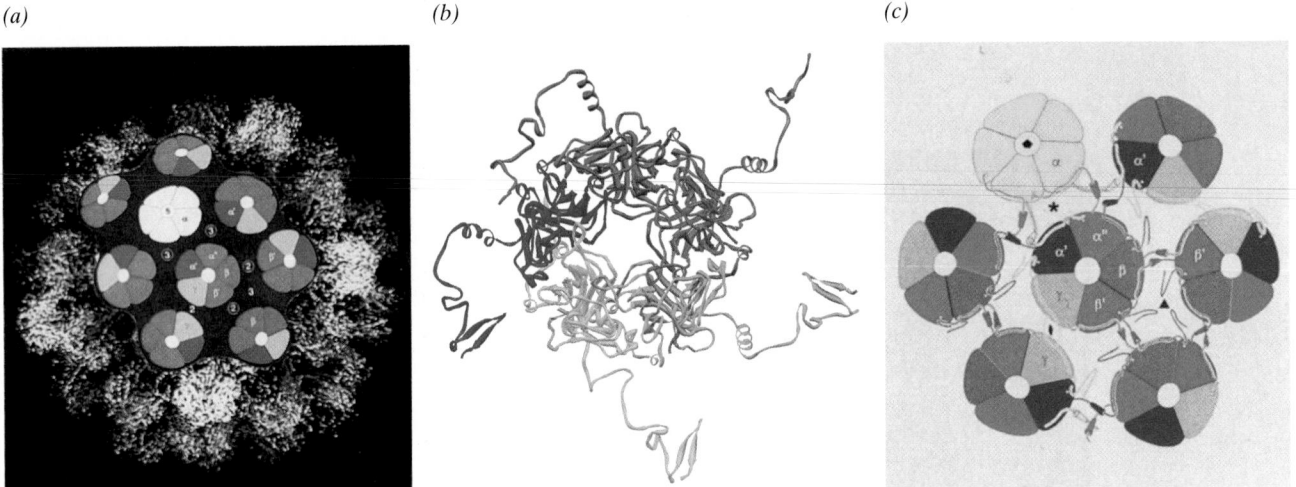

FIGURE 33-22   X-Ray structure of simian virus 40 (SV40).
(*a*) The SV40 virion consists of 360 copies of VP1 that are organized into 72 pentamers of which 12 (*white*) are 5-coordinated and 60 (*colored*) are 6-coordinated. Three types of interpentamer clustering are indicated on the schematic part of the drawing: The white (α), purple (α′), and green (α″) subunits form a 3-fold interaction ③; the red (β) and cyan (β′) subunits form one type of 2-fold interaction ②; and the yellow subunits (γ) form a second type of 2-fold interaction (2). The icosahedral axes of symmetry are indicated by the numerals 5, 3, and 2. (*b*) A 6-coordinated pentamer as viewed from outside the virion. The VP1 subunits are colored as in Part *a*. Note the C-terminal arms extending out from each subunit. (*c*) Schematic diagram showing how the C-terminal arms tie the pentamers together. The C-terminal arms are represented by lines and small cylinders (helices). The icosahedral particle's exact 5-, 3-, and 2-fold axes are represented by the conventional symbols, whereas the asterisk indicates a local 2-fold axis relating 5- and 6-coordinated pentamers. [Parts *a* and *c* courtesy of and Part *b* based on an X-ray structure by Stephen Harrison, Harvard University. PDBid 1SVA.]

sively forms pentamers that take up two nonequivalent positions (Fig. 33-22*a*). In fact, the SV40 capsid consists of 72 VP1 pentamers that are centered on the vertices of a $T = 7$ icosadeltahedron. Twelve of these pentamers lie on the icosahedron's twelve 5-fold rotation axes, each surrounded by 5 pentamers of a different geometric class. This latter class of 60 pentamers (which in a true $T = 7$ icosadeltahedron would have to be hexamers) are each surrounded by 6 pentamers, 5 of its own class and one of the former class. As a consequence, each capsid contains six symmetry-inequivalent classes of the chemically identical VP1 subunits. What conformational adjustments must the subunits make to form such a structure and, in particular, how does a pentameric structure coordinate with 6 other such pentamers?

The X-ray structure of SV40, determined by Harrison, indicates that VP1 consists of three modules: (1) an N-terminal arm that extends across the inside of the pentamer beneath the clockwise neighboring subunit (looking from the outside in) and whose first 15 residues are not visible in the structure (they probably extend inward to interact with the minichromosome which is likewise not visible); (2) an antiparallel β barrel with the same topology as that in RNA plant viruses and picornaviruses (Figs. 33-14 and 33-18), although oriented more or less radially with respect to the capsid rather than tangentially; and (3) a 45- to 50-residue C-terminal arm, the site of the only major conformational variation among the six symmetry-inequivalent sets of VP1 subunits. The C-terminal arms form the principal interpentamer contacts by extending from their pentamer of origin so as to invade a neighboring pentagon (Fig. 33-22*b,c*). Each pentamer thereby receives five invading arms from adjacent pentamers as well as donating five such arms. *It is the differing patterns of C-terminal arm exchange among the various pentamers that determines how they associate in forming the capsid.* Since these C-terminal arms are probably flexible and unstructured on a free pentamer, the capsid's pentameric building blocks probably behave, so to speak, as if they are tied together with ropes rather than being cemented together across extended complementary surfaces. Indeed, deletion of the C-terminal arms from recombinant VP1 subunits does not prevent their associating into pentamers but precludes these pentamers from assembling into the viruslike shells that they would otherwise form.

There are many other viruses of known structure whose capsid proteins contain the foregoing 8-stranded antiparallel β barrel. These include the $T = 1$ bacteriophage φX174 (a single-stranded DNA virus; Section 30-3B; Fig. 33-1*d*), the $T = 3$ **Norwalk virus** (which is responsible for >96% of nonbacterial gastroenteritis in the United States) and **black beetle virus** (both single-stranded RNA viruses), the $T = 4$ **Nudaurelia ω Capensis virus** (a single-stranded RNA insect virus), and the $T = 13$ **bluetongue virus** (Section 33-2F). Information concerning all icosahedral viruses of known X-ray structure is available over the Web from the Virus Particle ExploreR (VIPER) at http://chagall.scripps.edu/viper/.

E. *Bacteriophage MS2*

The RNA bacteriophage MS2 infects only F^+ (male) *E. coli* (Section 31-1A) because infection is initiated by viral attachment to bacterial F pili. The 275-Å-diameter MS2 virion (Fig. 33-1*b*) consists of 180 identical 129-residue coat protein subunits arranged with $T = 3$ icosadeltahedral symmetry encapsidating a 3569-nt single-stranded RNA molecule. The virion also contains a single copy of the 44-kD A-protein, which is thought to be responsible for viral attachment to the F pili and must therefore be exposed on the phage surface.

The X-ray structure of MS2, determined by Karin Valegård and Lars Liljas, reveals that it has a $T = 3$ protein shell formed by 60 icosahedrally related triangular protomers, each of which consists of three chemically identical subunits with slightly different conformations, much as in TBSV (Fig. 33-15). However, *the MS2 coat protein does not contain the 8-stranded antiparallel β barrel present in all the previously discussed icosahedral viruses.* Rather, each subunit consists of a 5-stranded antiparallel β sheet facing the interior of the particle overlaid with a short β hairpin and two α helices facing the viral exterior (Fig. 33-23). This protein fold resembles those of several other bacteriophage coat proteins.

F. *Bluetongue Virus*

Bluetongue virus is a member of the **orbivirus** genus within the *Reoviridae* family, one of the largest families of viruses. Members of the *Reoviridae* are responsible for significant levels of child mortality in developing countries (by the **rotaviruses,** which cause diarrhea) as well as a variety of economically important diseases of both plants and animals. The orbiviruses are icosahedral viruses whose capsids are made of two shells: an outer shell consisting of the viral proteins **VP2** and **VP5,** which is lost on cell entry; and a transcriptionally active core that is released into the cytoplasm. The core consists of two layers, an inner $T = 2$ shell constructed from 120 copies of the 100-kD protein **VP3(T2),** and an outer $T = 13$ shell consisting of 780 copies of the 38-kD protein **VP7(T13).** The capsid contains an aggregate of ~20 kb of double-stranded RNA (dsRNA) in usually 10 different segments, nearly all of which encode only one protein each. During infection, the dsRNA is maintained within the core because its release into the cell would trigger the interferon-mediated shutdown of translation (Section 32-4A), which would prevent viral proliferation. Consequently, the core also contains multiple copies of virally encoded **dsRNA-dependent RNA polymerase [VP1(Pol)],** helicase **[VP6(Hel)],** and capping enzyme **[VP4(Cap)],** which are associated with each dsRNA segment so as to form active transcription complexes. The

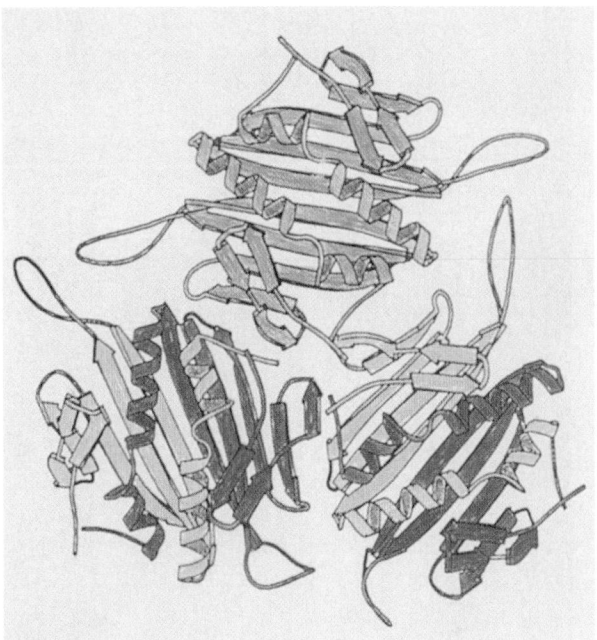

FIGURE 33-23 **X-Ray structure of bacteriophage MS2 showing three dimers related by a quasi-3-fold axis of the $T = 3$ icosadeltahedral particle.** The A, B, and C subunits, as defined in Fig. 33-12*b*, are, respectively, yellow, red, and orange (those in Fig. 33-12*b* are differently colored). The two C monomers shown are related by the particle's exact 2-fold axis, whereas closely associated A and B monomers are related by quasi-2-fold axes. In all cases, each monomer's five-stranded antiparallel β sheet is extended across the 2-fold axis and its helices interlock with those of its dimeric mate. Note the lack of structural resemblance between the MS2 subunits and the eight-stranded antiparallel β barrels that form the coat proteins of nearly all other icosahedral viruses with known structures (e.g., Figs. 33-14 and 33-18). [Courtesy of Karin Valegård, Uppsala University, Sweden. PDBid 2MS2.]

resulting mRNAs are extruded into the host cell's cytoplasm, where they direct the ribosomal synthesis of viral proteins. In addition, the mRNAs are encapsidated within growing cores, where they act as templates for the synthesis of negative strand RNA segments, thus forming progeny dsRNAs. Nevertheless, how each core is packaged with precisely one copy of each dsRNA segment is unknown.

Bluetongue virus **(BTV)** infects ungulates (hoofed mammals; e.g., sheep) and is transmitted by certain blood-feeding insects. It is named for the cyanotic tongues (due to swelling) that many BTV-infected animals have. The X-ray structure of the BTV core, determined by Stuart, reveals both layers of the 700-Å-diameter, ~55,000-kD particle and hence BTV is the largest particle of known X-ray structure (although the ribosome constitutes the largest asymmetric assembly of known X-ray structure; Section 32-3A). The asymmetric portion of the outer shell consists of 13 independent copies of VP7(T13) arranged as five geo-

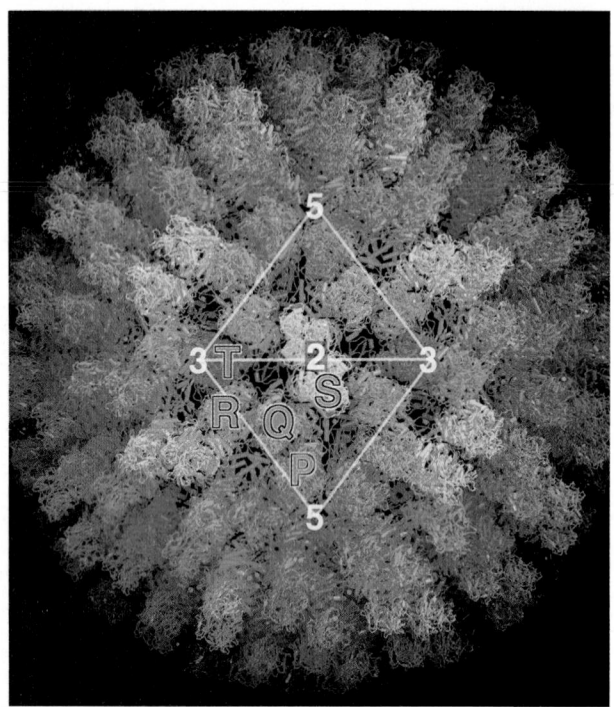

(a)

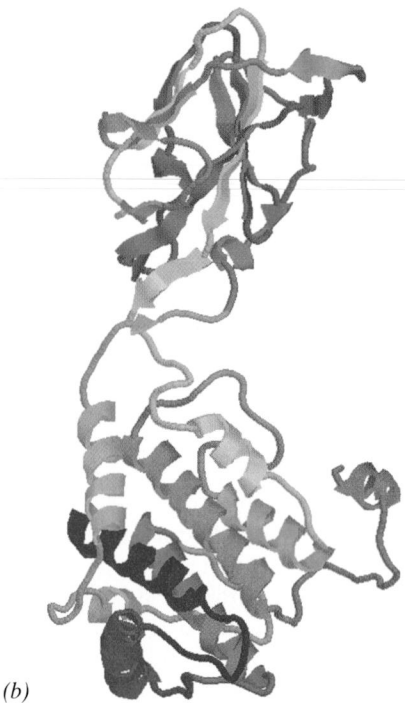

(b)

FIGURE 33-24   X-Ray structure of bluetongue virus core.
(*a*) Its *T* = 13 outer shell. The triangular icosahedral
asymmetric unit, whose edges (*white lines*) link the
icosahedron's symmetry axes, contains 13 copies of VP7
arranged as five trimers, P, Q, R, S, and T, which are colored
red, orange, green, yellow, and blue, respectively. Trimer T sits
on an icosahedral 3-fold axis and hence contributes a monomer
to the asymmetric unit. (*b*) The structure of VP7, which is
color-ramped in rainbow order from its N-terminus (*blue*) to its
C-terminus (*red*). The 8-stranded antiparallel β barrel domain
(*above*) forms the viral core's outer projections and the helical
domain (*below*) forms its outer shell. [Part *a* courtesy of and
Part *b* based on an X-ray structure by David Stuart, Oxford
University, U.K. PDBid 2BTV.]

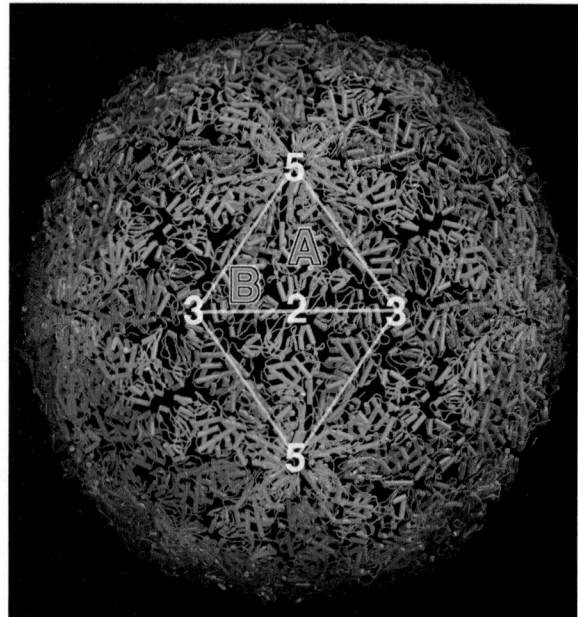

(a)

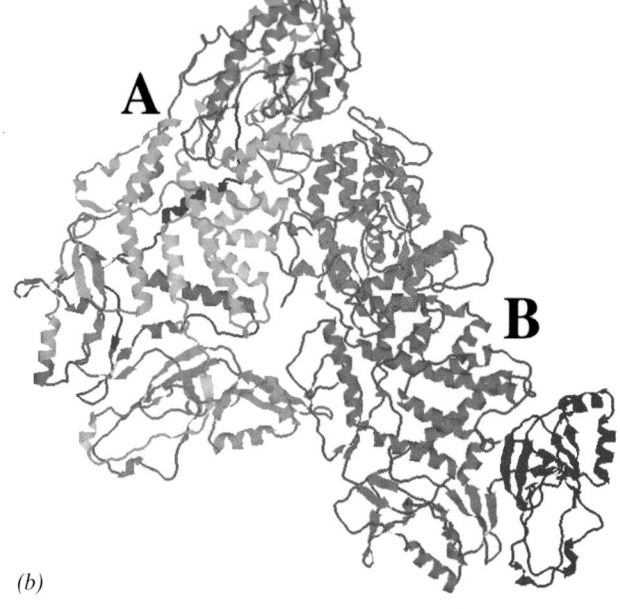

(b)

FIGURE 33-25   X-Ray structure of bluetongue virus core.
(*a*) Its *T* = 2 inner shell. It is constructed from homodimers
of VP3 subunits arranged with icosahedral (*T* = 1) symmetry
so that the two subunits forming the homodimer, A (*green*)
and B (*red*), are symmetically inequivalent. The triangular
icosahedral asymmetric unit is indicated (*white lines*). Compare
this structure with that of the *T* = 13 outer shell (Fig. 33-24*a*).
(*b*) The structure of the VP3 asymmetric dimer. Its A subunit is
color-ramped in rainbow order from its N-terminus (*blue*) to its
C-terminus (*red*) and its B subunit is colored according to domain
with its apical domain red, its carapace domain green, and its
dimerization (across the 2-fold axis) domain blue. Note the
somewhat different conformations and the very different
structural environments of these two chemically identical
subunits. [Part *a* courtesy of and Part *b* based on an X-ray
structure by David Stuart, Oxford University, U.K. PDBid 2BTV.]

metrically distinct trimers, P, Q, R, S, and T (Fig. 33-24*a*). The 349-residue VP7(T13) consists of two domains (Fig. 33-24*b*): an 8-stranded antiparallel β barrel common to many icosahedral viruses that forms the core's outer projections (and is responsible for its hedgehoglike appearance) and presumably contacts the virion's outer layer; and a helical domain that forms the core's outer shell. Note that in most icosahedral viruses that we have encountered, the β barrel domain forms the viral shell (Section 33-2B). The various geometrically distinct copies of VP7(T13) have nearly identical conformations, with maximum deviations between pairs of equivalent C_α positions of only 0.3 Å, even though there are significant differences in the way neighboring subunits contact one another.

BTV's inner shell consists of 60 asymmetric homodimers of VP3(T2) subunits, A and B, arranged with icosahedral ($T = 1$) symmetry (Fig. 33-25*a*). The 901-residue VP3(T2) consists of three domains (Fig. 33-25*b*): an apical domain, which contains 11 helices and 10 β strands in A but 10 helices and 11 β strands in B; a carapace domain, which contains 20 helices in A and 21 in B; and a dimerization domain, which contains 5 helices and 13 β strands in A and 4 helices and 14 β strands in B. The inner shell is relatively smooth and has few charged residues. It has 9-Å-wide pores at its icosahedral 5-fold vertices. These pores, which are lined with conserved Arg residues, are too narrow to permit the exit of mRNA, but in the presence of Mg^{2+} they open sufficiently to do so (see below).

The BTV core's outer and inner shells interact through relatively flat, predominantly hydrophobic surfaces. The symmetry mismatch between these two shells requires that they have 13 different sets of contacts, which makes the closely similar conformations of the 13 geometrically distinct VP7(T13) subunits all the more remarkable. VP3(T2) self-assembles to form a subcore but VP7(T13), although it forms trimers in solution, does not self-assemble to form an icosahedral shell. It is therefore likely that the inner shell forms a permanent scaffolding on which 260 VP7(T13) trimers crystallize in two dimensions to form the outer shell.

The X-ray structure of BTV also reveals the paths of ~80% of its 19,219-bp dsRNA (Fig. 33-26*a*). The dsRNA

appears to be partially ordered about the icosahedral 5-fold axes through its interactions with the inside of the VP3(T2) shell (although note that the dsRNA, whose 10 segments range in size from 822 to 3954 bp, cannot be

(a)

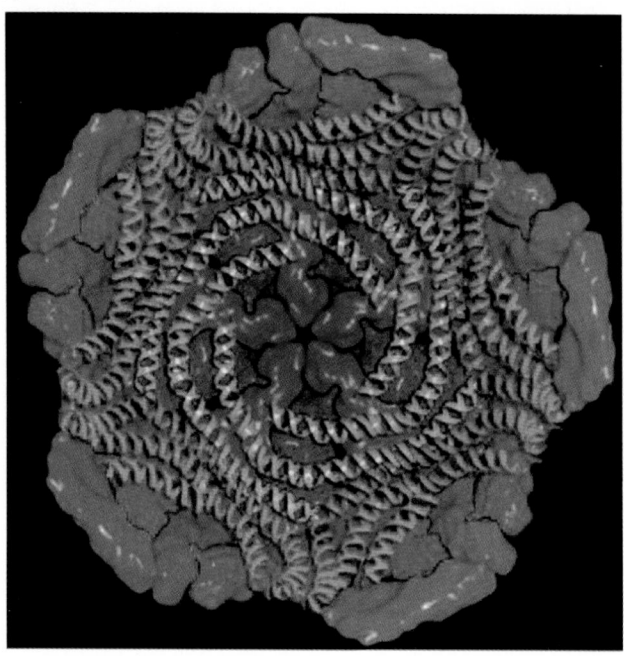

(b)

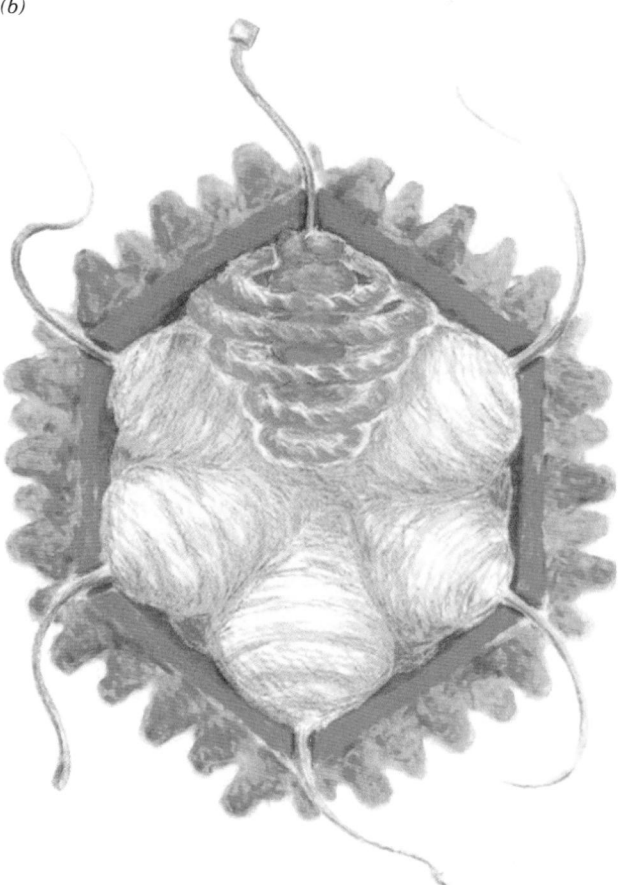

FIGURE 33-26 Arrangement of RNA in BTV. (*a*) The X-ray structure–based packing of dsRNA in the BTV core's inner shell as viewed along an icosahedral 5-fold axis and showing only the core's lower hemisphere. The relatively poorly resolved electron density has been modeled as A-form RNA with the RNA that is packed about the centrally located 5-fold axis blue and that packed about other 5-fold axes orange. The A and B subunits of the VP3(T2) forming the core's inner shell are green and red. (*b*) A cartoon model for the arrangement of RNA (*blue*) in the BTV core. The dsRNA is drawn as a coil that is wrapped about its associated transcription complex (*green*). Newly synthesized mRNA tails are shown exiting the pores at the core's 5-fold vertices. Note that each transcription complex is associated with one 5-fold vertex. [Courtesy of David Stuart, Oxford University, U.K.]

(a)

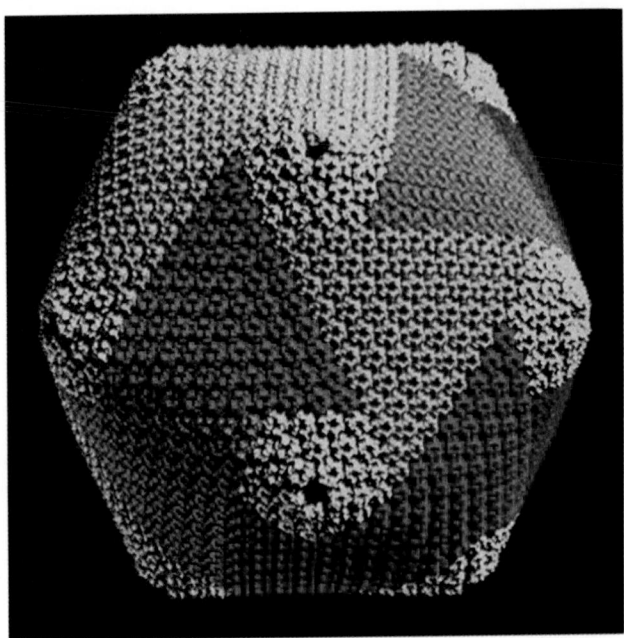

(b)

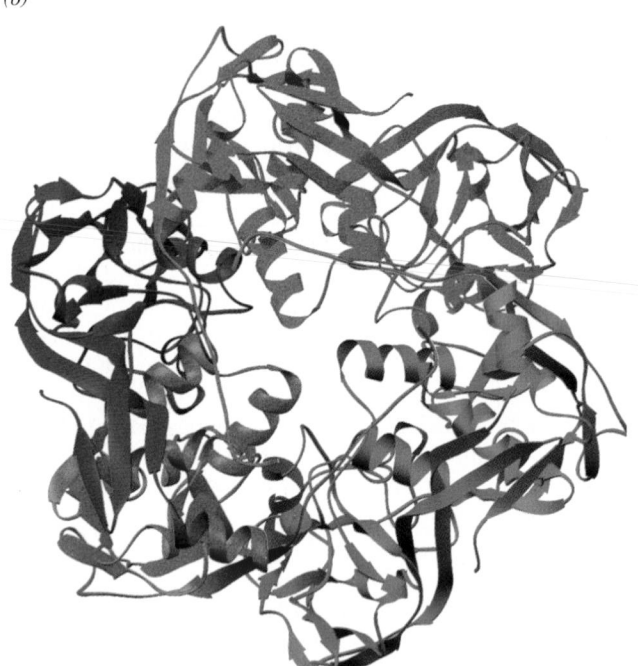

FIGURE 33-27 Structure of the PBCV-1 capsid. (*a*) A quasi-atomic model based on fitting the X-ray structure of Vp54 to the cryo-EM–based image of the capsid. The pentasymmetrons are yellow and the trisymmetrons are variously colored. (*b*) The X-ray structure of the Vp54 homotrimer as viewed along its 3-fold axis. The N-terminal β barrel domain of the leftmost subunit is red, its C-terminal β barrel is blue, and the remaining monomers are green and magenta. [Part *a* courtesy of and Part *b* based on an X-ray structure by Michael Rossmann, Purdue University. PDBid 1M4X.]

arranged with true icosahedral symmetry). This suggests that each transcription complex is organized about a 5-fold axis near the inner surface of the core with its associated dsRNA segment spiraling toward the center of the core. Indeed, X-ray structures of BTV crystals that had been soaked in solutions containing 20-nt oligonucleotides reveal electron density emanating from the viral core along its 5-fold axes that presumably mimics newly synthesized mRNAs (Fig. 33-26*b*).

G. *Paramecium bursaria Chlorella Virus*

Viruses of the **chlorovirus** genus are among the largest and most complex known icosahedral viruses. These viruses have a layered structure comprising dsDNA surrounded by a protein core, a lipid membrane, and finally an icosahedral protein shell. **Paramecium bursaria Chlorella virus type 1 (PBCV-1)** infects certain *Chlorella*-like algae. It attaches to its host cell and, through the mediation of viral enzymes to digest the host cell wall around the point of attachment, injects its dsDNA into the cell, leaving the empty capsid on the cell surface. PBCV-1, which has a molecular mass of $\sim 10^9$ D, has a 331-kb genome that encodes 377 proteins and 10 tRNAs. Its major capsid protein, **Vp54** (a 437-residue glycoprotein), accounts for $\sim 40\%$ of the virion's protein mass.

The cryo-EM–based image of PBCV-1 (Fig. 33-27*a*), determined by Timothy Baker and Rossmann, reveals that its outer shell forms a 1900-Å-diameter, $T = 169$ icosadelta-

hedron ($h = 7$ and $k = 8$). This enormous capsid is constructed from 20 triangular units named **trisymmetrons** and 12 pentagonal caps named **pentasymmetrons,** which respectively consist of pseudohexagonal arrays of 66 and 30 trimers of Vp54 for a total of 1680 trimers and hence 5040 monomers of Vp54 in the capsid (see below). The trisymmetrons do not correspond to the capsid's icosahedral faces. Instead, they bend around the edges of the icosahedron, leaving openings at its 5-fold vertices that are filled by the pentasymmetrons. Each pentasymmetron also contains a pentamer of a different protein at its 5-fold vertex.

The X-ray structure of Vp54 (Fig. 33-27*b*) reveals that it forms cyclic trimers in which each monomer consists of two consecutive antiparallel β barrels similar to those in other icosahedral viruses. The two β barrels in the Vp54 monomer are related by a 53° rotation about the trimer's 3-fold axis and hence the trimer has pseudohexagonal symmetry. These trimers have been fitted to the cryo-EM–based image of the PBCV-1 capsid, thereby yielding its quasi-atomic model (Fig. 33-27*a*).

3 ■ BACTERIOPHAGE λ

Bacteriophage λ (Figs. 33-1*f* and 33-28), a midsize (58 million D) **coliphage** (bacteriophage that infects *E. coli*), has a 55-nm-diameter icosahedral head and a flexible 15- to 135-nm-long tail that bears a single thin fiber at its end. The virion contains a 48,502-bp linear double-stranded

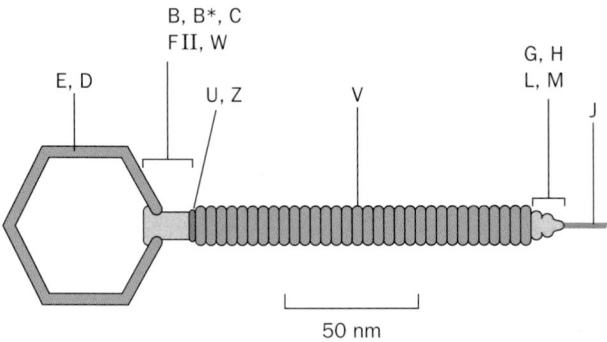

FIGURE 33-28 A sketch of bacteriophage λ indicating the locations of its protein components. The letters refer to specific proteins (gene products; see text). The bar represents 50 nm. [After Eiserling, F.A., *in* Frankel-Conrat, H. and Wagner, R.R. (Eds.), *Comparative Virology,* Vol. 13, *p.* 550, Plenum (1979).]

B-DNA molecule of known sequence. Phage λ is among the most extensively characterized complex viruses with respect to its molecular biology. Indeed, as we shall see in this section, *its genetic regulatory mechanisms form one of our best paradigms for the control of development in higher organisms and its assembly is among our most well-characterized examples of the morphogenesis of biological structures.*

Bacteriophage λ adsorbs to *E. coli* through a specific interaction between the viral tail fiber and **maltoporin** (Section 20-2D; the product of the *E. coli lamB* gene), which is a component of the bacterium's outer membrane. This interaction initiates a complex and poorly understood process in which the phage DNA is injected through the viral tail into the host cell. Soon after entering the host, the λ DNA, which has complementary single-stranded ends of 12 nucleotides (cohesive ends), circularizes and is covalently closed and supertwisted by the host DNA ligase and DNA gyrase (Fig. 33-29, Stages 1 and 2).

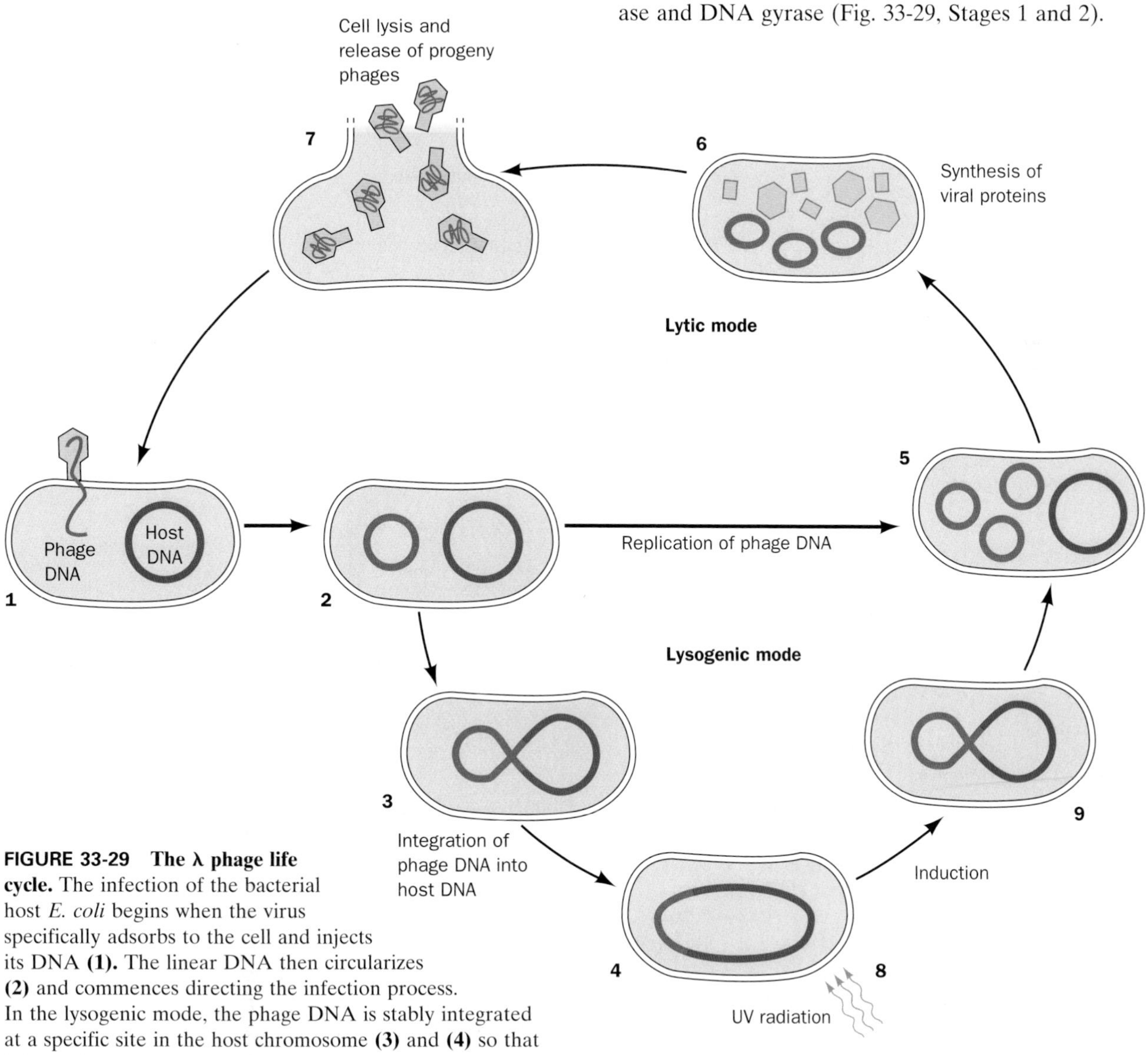

Cell lysis and
release of progeny
phages

7

6

Synthesis of
viral proteins

Lytic mode

5

Replication of phage DNA

Phage
DNA

Host
DNA

1

2

Lysogenic mode

3

9

Integration of
phage DNA into
host DNA

Induction

4

8

UV radiation

FIGURE 33-29 The λ phage life cycle. The infection of the bacterial host *E. coli* begins when the virus specifically adsorbs to the cell and injects its DNA **(1).** The linear DNA then circularizes **(2)** and commences directing the infection process. In the lysogenic mode, the phage DNA is stably integrated at a specific site in the host chromosome **(3)** and **(4)** so that it is passively replicated with the bacterial cell. Alternatively, the phage may take up the lytic mode in which the DNA directs its own replication **(5),** as well as the synthesis of viral proteins **(6)** so as to result in the lysis of the host cell with the release of ~100 progeny phages **(7).** DNA damage, as is caused, for example, by UV radiation **(8),** induces the excision of the prophage DNA from the lysogenic bacterial chromosome **(9)** and causes the phage to take up the lytic mode.

At this stage the virus has a "choice" of two alternative life styles (Fig. 33-29):

1. It can follow the familiar **lytic** mode in which the phage is replicated by the host such that, after 45 min at 37°C, the host lyses to release ~100 progeny phages.

2. *The phage may take up the so-called* **lysogenic** *life cycle, in which its DNA is inserted at a specific site in the host chromosome such that the phage DNA passively replicates with the host DNA. Nevertheless, even after many bacterial generations, if conditions warrant, the phage DNA will be excised from the host DNA to initiate a lytic cycle in a process known as* **induction.**

How the phage chooses between the lytic and lysogenic modes is the subject of Section 33-3D.

Phage DNA that is following a lysogenic life cycle is described as a **prophage,** whereas its host is called a **lysogen.** An intriguing property of lysogens is that they cannot be reinfected by phages of the type with which they are lysogenized: *They are* **immune** *to superinfection.* A bacterio-

phage that can follow either a lytic or a lysogenic life style is known as a **temperate phage,** whereas those that have only a lytic mode are said to be **virulent.** Bacteriophages that are reproducing lytically are said to be engaged in **vegetative growth.**

Over 90% of the thousands of known types of phages are temperate and, conversely, most bacteria in nature are lysogens. Yet, the presence of prophages has frequently gone unnoticed because they have little apparent affect on their hosts. For example, the K12 strain of *E. coli* had been the subject of intensive investigations for >20 years before 1951 when Ester Lederberg found it to be lysogenic for bacteriophage λ (which marks the discovery of this phage as well as the phenomenon of lysogeny).

The advantage of lysogeny is clear. A parasite that can form a stable association with its host has a better chance of long-term survival than one that invariably destroys its host. A virulent phage, on encountering a colony of its host bacteria, will multiply prodigiously. After the colony has been wiped out, however, it may be some time, if at all, before any of the progeny encounter another suitable host

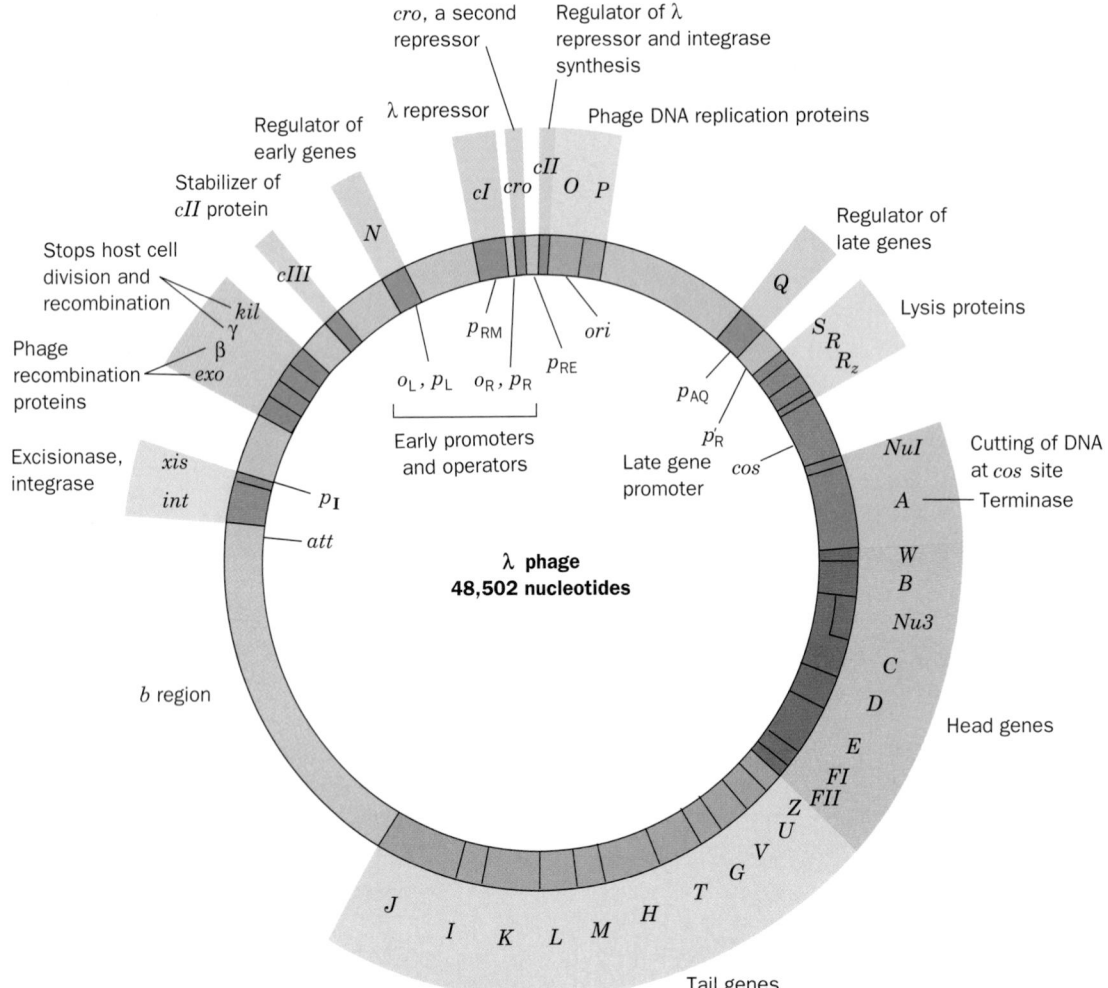

FIGURE 33-30   A genetic map of bacteriophage λ. Most of its structural genes (indicated outside the circle) and control sites (indicated inside the circle) are shown, with the genes encoding regulatory proteins shaded in red. On packaging into the virion, the circular chromosome is cut at the *cos* site yielding a linear DNA.

in a generally hostile world. In contrast, a prophage will multiply with its host indefinitely so long as the host remains viable.

But what if the host is fatally injured? Does the parasite die with the host? In the case of bacteriophage λ, it is precisely such traumatic conditions, exposure to agents that damage the host DNA or disrupt its replication, that induce the lytic phase. This has been described as the "lifeboat" response: The prophage escapes a doomed host through the formation of infectious viral particles that have at least some chance of further replication. Conversely, lysogeny is triggered by poor nutritional conditions for the host (phages can lytically replicate only in an actively growing host) or a large number of phages infecting each host cell (which signals that the phages are on the verge of eliminating the host).

This section describes the genetic system that controls the orderly formation of phage particles in the lytic mode, the mechanism through which these phage particles are assembled, and the regulatory mechanism through which bacteriophage λ selects and maintains its life cycle. *Analogous systems are believed to underlie many cellular processes.*

A. *The Lytic Pathway*

The bacteriophage λ genome, as its genetic map indicates (Fig. 33-30), encodes ~50 gene products and contains numerous control sites. Note the λ chromosome's organization. Its genes are clustered according to function. For example, the genes concerned with the synthesis of phage tail proteins are tandemly arranged on the bottom of Fig. 33-30. This organization, as we shall see, enables these genes to be transcribed together, that is, as an operon. The functions of many of the λ genes and control sites, together with those of the host that are important in phage function, are tabulated in Table 33-1.

In the lytic replication of phage λ, as in love and war, proper timing is essential. This is because the DNA must be replicated in sufficient quantity before it is made unavailable by packaging into phage particles and because packaging must be completed before the host cell is enzymatically lysed. The transcription of the λ genome, which is carried out by host RNA polymerase, is controlled in both the lytic and the lysogenic programs by the regulatory genes that are shaded in red in Fig. 33-30.

a. The Lytic Mode Has Early, Delayed Early, and Late Phases

The lytic transcriptional program has three phases (Fig. 33-31):

1. Early transcription *Soon after phage infection or induction, E. coli RNA polymerase commences "leftward" transcription of the phage DNA starting at the promoter p_L and "rightward" transcription (and thus from the opposite DNA strand) from the promoters p_R and p'_R (Fig. 33-31a):*

(i) The "leftward" transcript, L1, which terminates at termination site t_{L1}, encodes the **N** gene.

TABLE 33-1   Important Genes and Genetic Sites for Bacteriophage λ

Gene or Site	Function
Phage genes	
cI	λ repressor; establishment and maintenance of lysogeny
cII, cIII	Establishment of lysogeny
cro	Repressor of *cI* and early genes
N, Q	Antiterminators for early and delayed early genes
O, P	Origin recognition in DNA replication
γ	Inhibits host RecBCD
int	Prophage integration and excision
xis	Prophage excision
B, C, D, E, W, Nu3, FI, FII	Head assembly
G, H, I, J, K, L, M, U, V, Z	Tail assembly
A, Nu1	DNA packaging
R, R_z, S	Host lysis
b	Accessory gene region
Phage sites	
*att*P	Attachment site for prophage integration
*att*L, *att*R	Prophage excision sites
cos	Cohesive end sites in linear duplex DNA
o_L, o_R	Operators
p_I, p_L, p_R, p_{RM}, p_{RE}, p'_R	Promoters
t_{L1}, t_{R1}, t_{R2}, t_{R3}, t'_R	Transcriptional termination sites
*nut*L, *nut*R	N utilization sites
qut	Q utilization site
ori	DNA replication origin
Host genes[a]	
lamB	Host recognition protein
dnaA, dnaB	DNA replication initiation
lig	DNA ligase
gyrA, gyrB	DNA gyrase
rpoA, rpoB, rpoC	RNA polymerase core enzyme
rho	Transcription termination factor
nusA, nusB, nusE	Necessary for gpN function
groEL, groES	Head assembly
himA, himD	Integration host factor
hflA, hflB	Degrades gpcII
cap, cya	Catabolite repressor system
*att*B	Prophage integration site
recA	Induction of lytic growth

[a]The genes encoding DNA polymerase I and the subunits of DNA polymerase III (Table 30-2) and the primosome (Table 30-4) are also required.

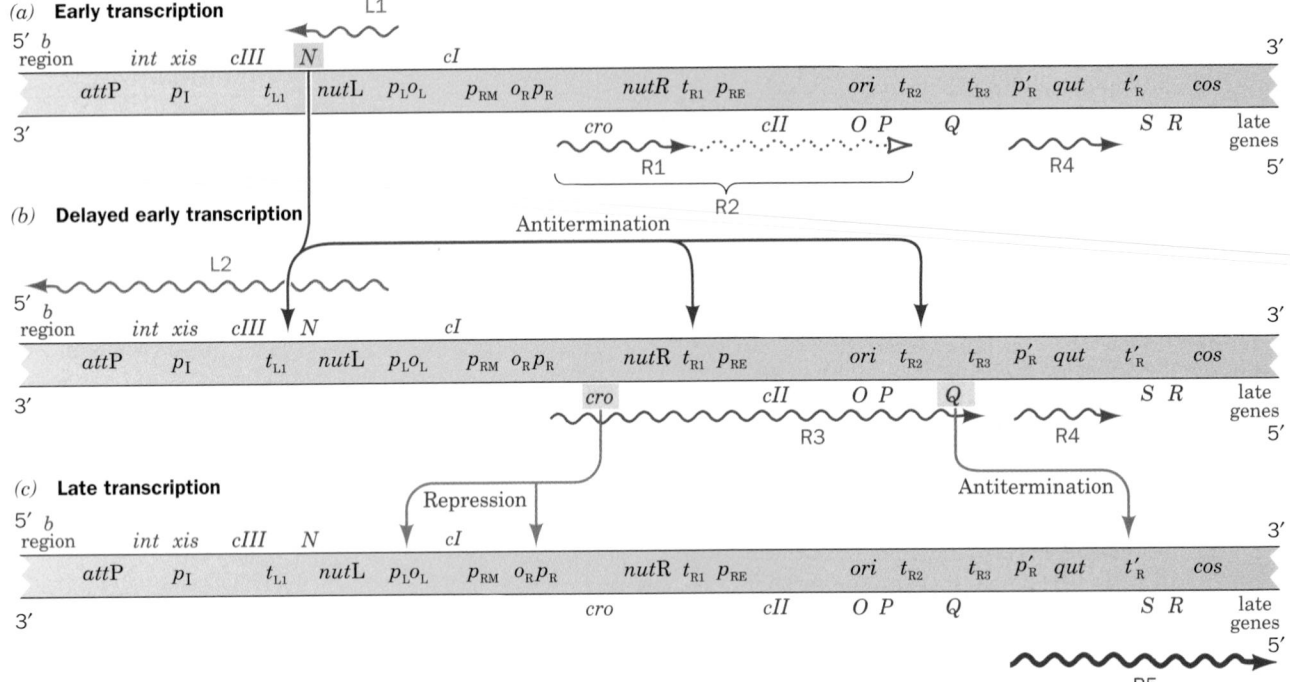

FIGURE 33-31 Gene expression in the lytic pathway of phage λ. Genes specifying proteins that are transcribed to the "left" and "right" are shown above and below the phage chromosome. Control sites are indicated between the DNA strands. The genetic map is not drawn to scale and not all of the genes or control sites are indicated. Transcripts are represented by wiggly arrows pointing in the direction of mRNA elongation; the actions of regulatory proteins are denoted by arrows pointing from each regulatory protein to the site(s) it controls. The lytic pathway has three transcriptional phases: (*a*) early transcription, (*b*) delayed early transcription, and (*c*) late transcription. Gene expression in each of the latter two phases is regulated by proteins synthesized in the preceding phase as is explained in the text. [After Arber, W., *in* Hendrix, R.W., Roberts, J.W., Stahl, F.W., and Weisberg, R.A. (Eds.), *Lambda II, p.* 389, Cold Spring Harbor Laboratory (1983).]

(ii) "Rightward" transcription from p_R terminates with ~50% efficiency at t_{R1}, to yield transcript R1, and otherwise at t_{R2} to yield transcript R2. R1 contains only the *cro* gene transcript, whereas R2 also contains the *cII, O,* and *P* gene transcripts.

(iii) "Rightward" transcription from p'_R terminating at t'_R yields a short transcript, R4, that specifies no protein.

L1, R1, and R2 are translated by host ribosomes to yield proteins whose functions are described below.

2. Delayed early transcription The second transcriptional phase commences as soon as a significant quantity of the protein **gpN** (gp for gene product) accumulates. *This protein, through a mechanism considered below, acts as a transcriptional antiterminator at termination sites t_{L1}, t_{R1}, and t_{R2} (Fig. 33-31b):*

(i) Leftward transcript L1 is extended to form L2, which additionally contains the transcripts of the *cIII, xis,* and *int* genes (which encode proteins involved in switching between the lytic and lysogenic modes; Sections 33-3C and 33-3D) together with the *b* **region** gene transcripts (which specify the so-called **accessory proteins** that, although not essential for lytic growth, increase its efficiency).

(ii) Transcript R2 is extended to form transcript R3, which additionally encodes a second antiterminator, **gpQ,** whose function is discussed below. The continuing translation of R2 and later R3 to yield **gpO** and **gpP,** proteins that are both required for λ DNA replication, stimulates viral DNA production. Similarly, the translation of R1 and later R3 yields **Cro protein (gpcro),** a repressor of both the "rightward" and "leftward" genes (see below; *cro* stands for *c*ontrol of *r*epressor and *o*ther things).

At this stage, ~15 min postinfection, Cro protein has accumulated in sufficient quantity to bind to operators o_L and o_R, thereby shutting off transcription from p_L and p_R. This is more than just efficient use of resources; the overexpression of the early genes, as occurs in λcro⁻ phage, poisons the lytic cycle's late phase.

3. Late transcription In the final transcriptional phase (Fig. 33-31c), *the antiterminator gpQ acts to extend transcript R4 through t'_R to form transcript R5.* The "gene dosage" effect of the ~30 copies of phage DNA that have accumulated by the beginning of this stage results in the rapid synthesis of the capsid-forming proteins (which are all encoded by late genes; their assembly to form mature phage particles is described in Section 33-3B), as well as

gpR, gpR$_z$, and **gpS,** which catalyze host cell lysis [gpR is a transglycosidase that cleaves the bond between NAG and NAM in the host cell wall peptidoglycan (Section 11-3B); gpR$_z$ is an endopeptidase that hydrolyzes a peptidoglycan peptide bond; and gpS forms pores in the cell membrane, thereby providing gpR and gpR$_z$ with access to their peptidoglycan substrate]. The first phage particle is completed ~22 min postinfection.

b. Antitermination Requires the Action of Several Proteins

Transcriptional control in the λ *lytic phase is exerted by* gpN*- and* gpQ*-mediated antitermination rather than by repressor binding at an operator site* through which, for example, *lac* operon expression (Section 31-1B) is regulated. gpN (107 residues) acts at both rho-dependent and rho-independent termination sites [t_{L1} and t_{R1} are rho dependent (and are, in fact, the terminators with which rho was originally identified), whereas t_{R2} is rho independent; transcriptional termination is discussed in Section 31-2D]. Yet, gpN does not act at just any transcriptional termination site. Rather, genetic analysis of mutant phage defective for antitermination has established the existence of two so-called *nut* (for *N utilization*) sites that are required for antitermination: **nutL,** which is located between p_L and *N*, and **nutR,** which occurs between *cro* and t_{R1} (Fig. 33-31). These sites have closely similar sequences consisting of two elements, *boxB*, whose transcripts can form hydrogen-bonded hairpin loops, and *boxA* (Fig. 33-32*a*).

What is the mechanism of gpN-mediated antitermination? The observation that some *E. coli* defective in antitermination have mutations that map in the *rpoB* gene (which encodes the RNA polymerase β subunit) suggests that gpN acts at *nut* sites to render core RNA polymerase

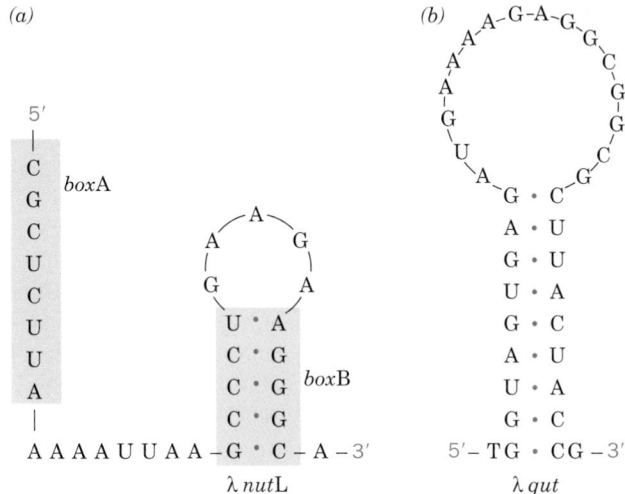

FIGURE 33-32   The RNA sequences of the phage λ control sites: (*a*) *nut*L, which closely resembles *nut*R, and (*b*) *qut.* Each of these control sites is thought to form a base-paired hairpin.

(lacking a σ subunit) resistant to termination. Indeed, gpN-modulated RNA polymerase will pass over many different terminators that it encounters either naturally or by experimental design. A variety of evidence, including the observation that covering *nut* RNA with ribosomes prevents antitermination, indicates that gpN recognizes this site on RNA, not DNA.

Genetic analyses have revealed that antitermination requires several other host factors termed **Nus** (for *N utilization substance*) **proteins** (Fig. 33-33): **NusA,** which specifically binds to both gpN and RNA polymerase; **NusE**

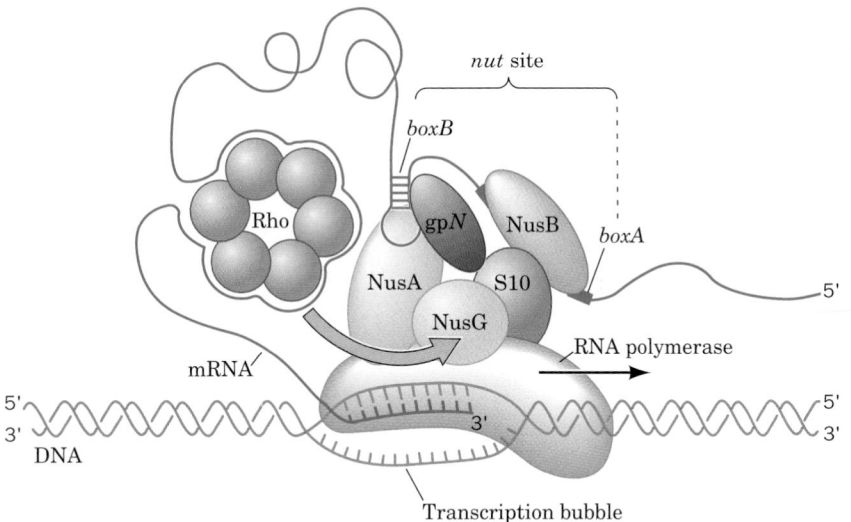

FIGURE 33-33   Schematic model of the antitermination complex between transcribing RNA polymerase, gpN, and Nus proteins. gpN and the Nus proteins form a complex on a *nut* site of the nascent RNA that binds to the transcribing RNA polymerase further along the looped-out RNA. This complex inhibits RNA polymerase from pausing at a transcriptional termination site, which may prevent rho factor from overtaking the RNA polymerase so as to release the transcript. Another possibility is that transcript release may be inhibited by a gpN-modulated direct interaction between NusG and rho factor (*curved arrow*). [After Greenblatt, J., Nodwell, J.R., and Mason, S.W., *Nature* **364,** 402 (1993).]

(which, interestingly, is ribosomal protein S10) and **NusG,** which both bind to RNA polymerase; and **NusB,** which binds to S10. On encountering a *nut* site, gp*N* forms a complex with the Nus proteins and RNA polymerase that travels with this enzyme during elongation and inhibits it from pausing at termination sites. At rho-independent terminators, this deters the release of the transcript at the terminator's weakly bound poly(U) segment, whereas at rho-dependent terminators, it may prevent rho factor from overtaking RNA polymerase, thereby stopping it from unwinding and thus releasing the transcript at the transcription bubble. Alternatively, since it has been shown that NusG binds directly to rho, this interaction, as modulated by gp*N*, may inhibit rho from releasing the nascent transcript.

The *boxB* RNA is recognized by gp*N* via the latter's ~18-residue, Arg-rich, N-terminal segment. The NMR structure of the 15-nt *boxB* hairpin from the lambdoid (λ-like) **bacteriophage P22** (which grows on *Salmonella typhimurium*), in complex with the 20-residue N-terminal segment of its gp*N*, was determined by Dinshaw Patel. It reveals that the peptide forms a helix that binds against the major groove face of the *boxB* hairpin via electrostatic and hydrophobic interactions (Fig. 33-34). This presumably orients the opposite face of the RNA for interactions with host factors.

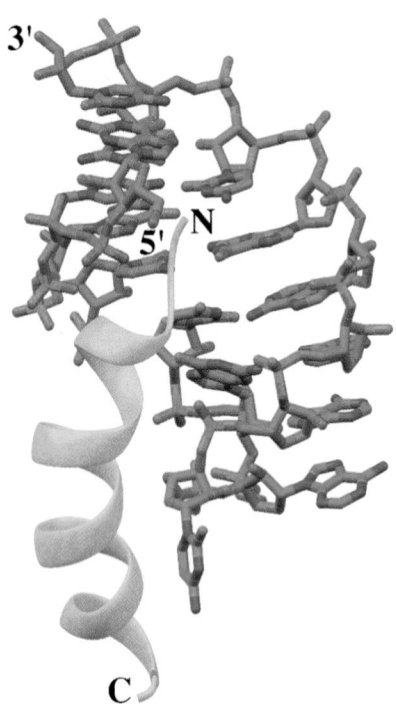

FIGURE 33-34   NMR structure of the bacteriophage P22 *boxB* **RNA in complex with the 20-residue, Arg rich, N-terminal segment of its gp*N*.** The peptide is represented by a gold ribbon and the RNA is drawn in stick form colored according to atom type (C green, N blue, O red, and P magenta). [Based on an NMR structure by Dinshaw Patel, Memorial Sloan-Kettering Cancer Center, New York, New York. PDBid 1A4T.]

Transcriptional antitermination is not limited only to certain bacteriophage. Indeed, the 7 ribosomal RNA (*rrn*) operons of λ's host organism, *E. coli* (which encode its 5S, 16S, and 23S RNAs; Section 31-4B), each contain a *box*A-like element which, together with the Nus proteins, mediates antitermination at *rrn* (which probably explains the function of S10 as a Nus protein). This suggests that λ *box*A is a defective form of *rrn box*A that requires the presence of gp*N* bound to *box*B in addition to the Nus proteins to inhibit termination.

gp*Q* (207 residues), which overrides t'_R to permit late transcription, acts at a ***qut*** site (analogous to the *nut* sites) that is located some 20 bp downstream from p'_R and that can form an RNA hairpin similar to those of the *nut* sites (Fig. 33-32*b*). Curiously, however, gp*Q*-mediated antitermination occurs via a mechanism that is quite different from that mediated by gp*N*. In fact, gp*Q* binds specifically to *qut* DNA, not to RNA, where together with NusA it binds to RNA polymerase holoenzyme that is paused at p'_R during the initiation phase, thereby accelerating it out of this promoter site and somehow inducing it not to terminate transcription at t'_R.

c.   gp*O* and gp*P* Participate in λ DNA Replication

The course of DNA replication in phage λ is diagrammed in Fig. 33-35. Electron microscopy indicates that in the early stages of lytic infection, λ DNA replication occurs via the bidirectional θ mode (Section 30-1A) from a single replication origin ***(ori).*** However, by the late stage of the lytic program, when ~50 λ DNA circles have been synthesized, θ mode DNA replication ceases, probably due to exhaustion of one or more of the required host proteins. At this point, around 3 of the ~50 DNA circles commence replication via the rolling circle (σ) mode (Section 30-3B), with the accompanying synthesis of the complementary strand, although the mechanism of the switchover between the two modes of DNA replication is unclear. The host RecBCD protein (Section 30-6A), a nuclease that would rapidly fragment the resulting concatemeric (consisting of tandemly linked identical units) linear duplex DNA, is inactivated by the phage **γ protein.**

In the process of phage assembly (Section 33-3B), the concatemeric DNA is specifically cleaved in its ***cos*** (for *co*hesive-end *s*ite) site to yield the linear duplex DNA with complementary 12-nt single-stranded ends that are contained in mature phage particles. The staggered double-stranded scission is made by the so-called **terminase,** which is a complex of the phage proteins **gp*A*** (641 residues) and **gp*Nu1*** (181 residues).

Phage λ DNA is replicated by the host DNA replication machinery (Sections 30-1, to 30-3) with the participation of only two phage proteins, gp*O* (333 residues) and gp*P* (233 residues). gp*O*, presumably as dimers, specifically binds to four repeated 18-bp palindromic segments within the phage DNA *ori* region, whereas gp*P* interacts with both gp*O* and the DnaB protein of the host primosome. gp*O* and gp*P*, it is thought, act analogously to host DnaA and DnaC proteins, which as we saw, are required for the initiation of replication of *E. coli* DNA (Section 30-3C).

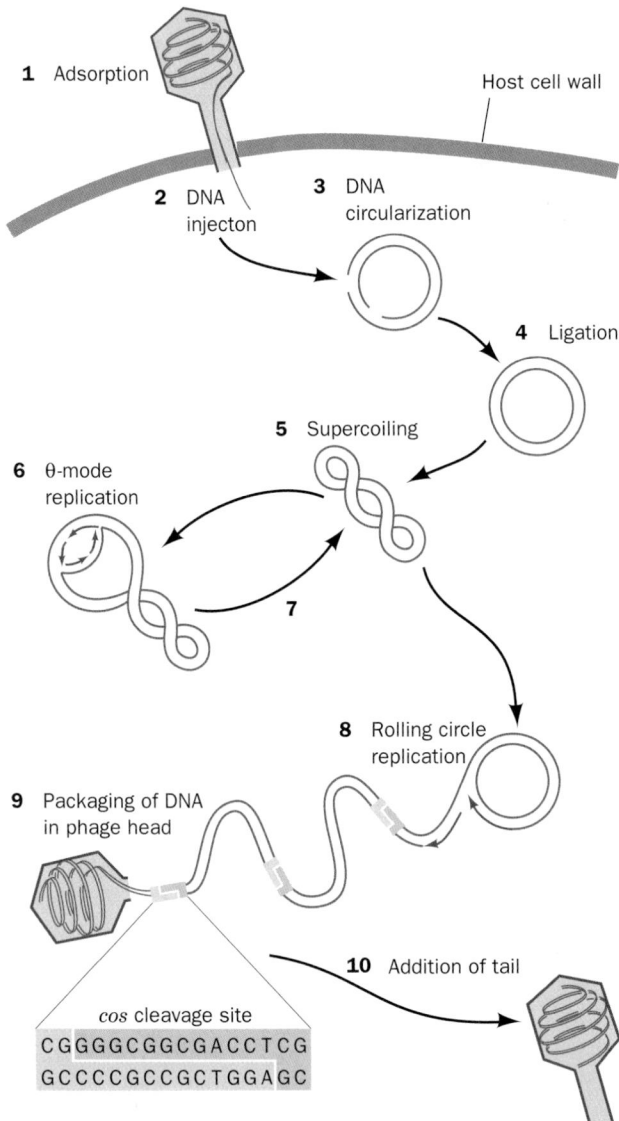

FIGURE 33-35   DNA replication in the lytic mode of bacteriophage λ. The phage particle adsorbs to the host cell **(1)** and injects its linear duplex DNA chromosome **(2)**. The DNA circularizes by base pairing at its complementary single-stranded ends **(3)**, and the resulting nicked circle is covalently closed **(4)** and supercoiled **(5)** by the sequential actions of host DNA ligase and host DNA gyrase. DNA replication commences according to both the bidirectional θ mode **(6** and **7)** and the rolling circle mode **(8)** but in the later stages of infection occurs exclusively by the rolling circle mode. Here curved blue arrows indicate the most recently synthesized DNA at the replication forks and the arrowheads represent the 3′ ends of the growing DNA chains. The concatemeric DNA produced by the rolling circle mode is specifically cleaved at its *cos* sites *(shaded boxes)* and is packaged into phage heads **(9)**. The addition of tails **(10)** completes the assembly of the mature phage particles, which are each capable of initiating a new round of infection. [After Furth, M.E. and Wickner, S.H., *in* Hendrix, R.W., Roberts, J.W., Stahl, F.W., and Weisberg, R.A. (Eds.), *Lambda II, p.* 146, Cold Spring Harbor Laboratory (1983).]

Nevertheless, DnaA is required for λ DNA replication. The gp*O* and gp*P* proteins apparently function to recognize the λ *ori* site, which, curiously, lies within the *O* gene.

B. Virus Assembly

The mature λ phage head contains two major proteins: **gp*E*** (341 residues), which forms its polyhedral shell, and **gp*D*** (110 residues), which "decorates" its surface. Electron microscopy indicates that these proteins, which are present in equal numbers, are arranged on the surface of a $T = 7$ icosadeltahedron. However, the λ head also contains four major proteins, **gp*B*, gp*C*, gp*FII*,** and **gp*W*,** which form a cylindrical structure that attaches the tail to the head. This **head–tail connector** occurs at one of the head's 5-fold vertices and thereby breaks its icosahedral symmetry. Hence, gp*E* and gp*D* are present in somewhat fewer than the 420 copies/phage in a perfect $T = 7$ icosadeltahedron.

The tail is a tubular entity that consists of 32 stacked hexagonal rings of **gp*V*** (246 residues) for a total of 192 subunits. The tail begins with a complex adsorption organelle composed of five different proteins, **gp*G*, gp*H*, gp*L*, gp*M*,** and **gp*J*,** and ends with an assembly of **gp*U*** and **gp*Z*** (Fig. 33-28).

The study of complex virus assembly has been motivated by the conviction that it will provide a foundation for understanding the assembly of cellular organelles. Phage assembly is studied through a procedure developed by Robert Edgar and William Wood that combines genetics, biochemistry, and electron microscopy. Conditionally lethal mutations (either temperature-sensitive mutants, which appear normal at low temperatures but exhibit a mutant phenotype at higher temperatures; or suppressor-sensitive *amber* mutants, Section 32-2E) are generated that, under nonpermissive conditions, block phage assembly at various stages. This process results in the accumulation of intermediate assemblies or side products that can be isolated and structurally characterized through electron microscopy. The mutant protein can be identified, through a process known as ***in vitro* complementation** (in analogy with *in vivo* genetic complementation; Section 1-4C), by mixing cell-free extracts containing these structural intermediates with the corresponding normal protein to yield infectious phage particles.

The assembly of bacteriophage λ occurs through a branched pathway in which the phage heads and tails are formed separately and then joined to yield mature virions.

a. Phage Head Assembly

λ Phage head assembly occurs in five stages (Fig. 33-36, *right*):

1. Two phage proteins, gp*B* (533 residues) and **gp*Nu3*,** together with two host-supplied chaperonin proteins, GroEL and GroES, interact to form an "initiator" that consists of 12 copies of gp*B* arranged in a ring with a central orifice. This precursor of the mature phage head–tail connector (Fig. 33-28) apparently organizes the phage

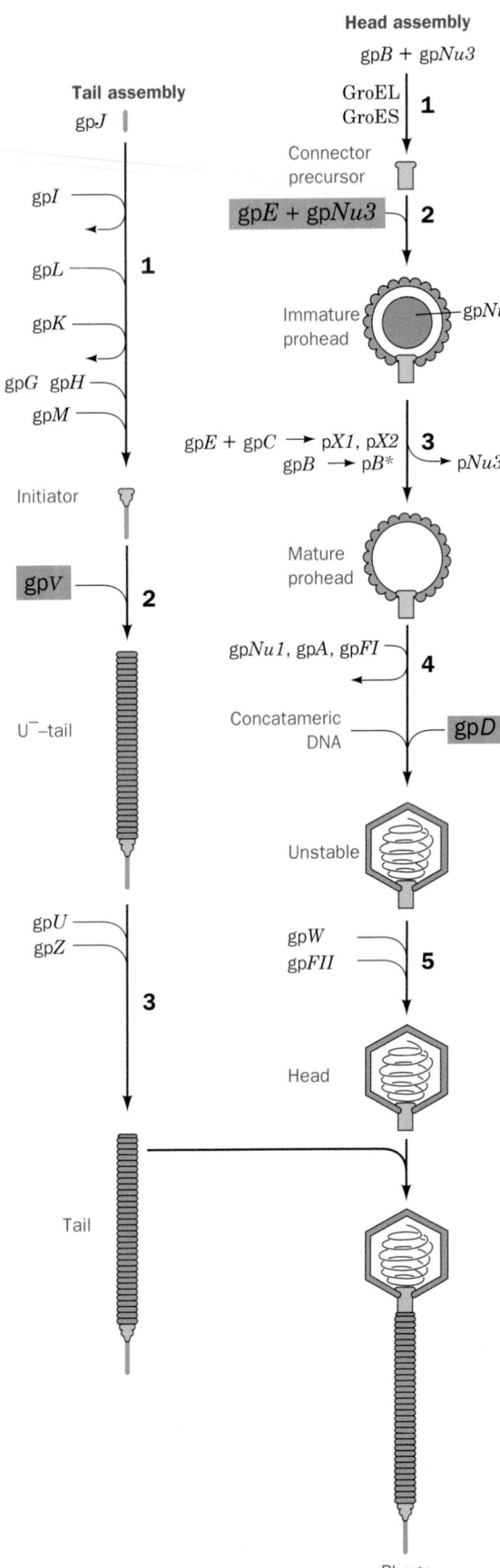

FIGURE 33-36   The assembly of bacteriophage λ. The heads and tails are assembled in separate pathways before joining to form the mature phage particle. Within each pathway the order of the various reactions is obligatory for proper assembly to occur. gpE, gpNu3, gpD, and gpV are highlighted in red boxes to indicate that relatively large numbers of these proteins are required for phage assembly. The numbered steps are described in the text.

head's subsequent formation. GroEL and GroES, it will be recalled, provide a protected environment that facilitates the proper folding and assembly of proteins and protein complexes such as the connector precursor (Section 9-2C). In fact, these chaperonins were discovered through their role in λ assembly. gpNu3, as we shall see, also functions as a molecular chaperone in that it has but a transient role in phage head assembly.

2. gpE and additional gpNu3 associate to form a structure called an immature **prohead.** If gpB, GroEL, or GroES is defective or absent, some gpE assembles into spiral or tubular structures, which indicates that the missing proteins guide the formation of a proper shell. The absence of gpNu3 results in the formation of but a few shells that contain only gpE. gpNu3 evidently facilitates proper shell construction and promotes the association of gpE with gpB.

3. In the formation of the mature prohead, the N-terminal 22-residue segment of ~75% of the gpB is excised to form **gpB***; the gpNu3 is degraded and lost from the structure; and 10 copies of gpC (439 residues) participate in a fusion–cleavage reaction with 10 additional copies of gpE to yield the hybrid proteins **pX1** and **pX2** (p for protein), which form the collar that apparently holds the connector in place. This maturation process, which involves only phage gene products that are part of the immature prohead, requires that all of the prohead components be present and functional, that is, that the immature prohead be correctly assembled to start with. The enzyme(s) that catalyzes this process has, nevertheless, not been identified.

4. The concatemeric viral DNA is packaged in the phage head and cleaved by mechanisms discussed below. During this process, the capsid proteins undergo a conformational change that results in an expansion of the phage head to twice its original volume (a process that occurs in 4M urea in the absence of DNA). gpD then attaches to newly exposed binding sites on gpE, thereby partially stabilizing the capsid's expanded structure.

5. In the final stage of phage head assembly, gpW (68 residues) and gpFII (117 residues) add in that order to stabilize the head and form the tail-binding site.

These stages of phage head assembly, as well as some of their component reactions, must proceed in an obligatory order for proper assembly to occur. Of particular interest is that *the components of the mature phage head are not entirely self-assembling as are, for example, TMV (Section 33-1B) and ribosomes (Section 32-3A).* Rather, the

E. coli proteins GroEL and GroES, as we saw, facilitate head–tail connector assembly. Moreover, gp*Nu3*, which occurs in ~200 copies inside the immature prohead but is absent from the mature prohead, evidently acts as a "scaffolding" protein that organizes gp*E* to form a properly assembled phage head. Finally, *since phage assembly involves several proteolytic reactions, it must also be considered to occur via enzyme-directed processes.*

b. DNA Is Tightly Packed in the Phage Head

An intriguing question of tailed phage assembly is, how does a relatively small phage head (55 nm in diameter in λ) package a far longer (16,500 nm in λ), stiff dsDNA molecule? Cryo-EM studies of DNA-filled phage heads from **bacteriophage T7** (a tailed coliphage) appear to answer this question. The images of these phage heads seen in axial view (along the line through the capsid vertex containing the head–tail connector to the center of the particle) reveal a striking pattern of at least 10 concentric rings, with only the outer ring, which is slightly thicker than the others, representing the protein shell (Fig. 33-37*a*). In contrast, the side views of these phage heads show only punctate (marked with dots or spots) patterns that in some places form linear features. Computer modeling indicates that these patterns can be accounted for by the spooling of the DNA in concentric shells around the axis through the connector (Fig. 33-37*b*). Six such shells are required to accommodate the entire 40-kb T7 DNA into its 55-nm-diameter T7 phage head. This does not contradict the observation of at least nine concentric rings of DNA because, in a given shell, the DNA is coiled more tightly toward the poles of the phage head and therefore appears in projection as multiple rings of lower radii. Since the DNA linearly enters the phage prohead through the head–tail connector (see below), it has been proposed that its stiffness causes it to first coil against the inner wall of the rigid protein shell and then to wind concentrically inward, much like a spool of twine. Nevertheless, the DNA's detailed winding path varies randomly from particle to particle as is indicated by the observation that packaged DNA can be cross-linked to the capsid along its entire length.

c. DNA Is "Pumped" into the Phage Head by an ATP-Driven Process

The packaging of λ DNA begins when terminase (gp*A* + gp*Nu1*) binds to its recognition sequences on a randomly selected ~200-bp *cos* site. The resulting complex then binds to the prohead so as to introduce the DNA into it through the orifice in its head–tail connector. The "left" end of the DNA chromosome enters the prohead first as is indicated by the observation that only this end of the chromosome is packaged by an *in vitro* system when λ DNA restriction fragments are used. Whether the cutting of the initial *cos* site precedes or follows the initiation of packaging is unknown. However, at least *in vitro*, this process requires the binding to *cos* of the *E. coli* histone-like protein known as **integration host factor (IHF).** IHF binds specific sequences of duplex DNA, which it wraps around its surface, thereby inducing a sharp bend in the DNA (see below).

(a)

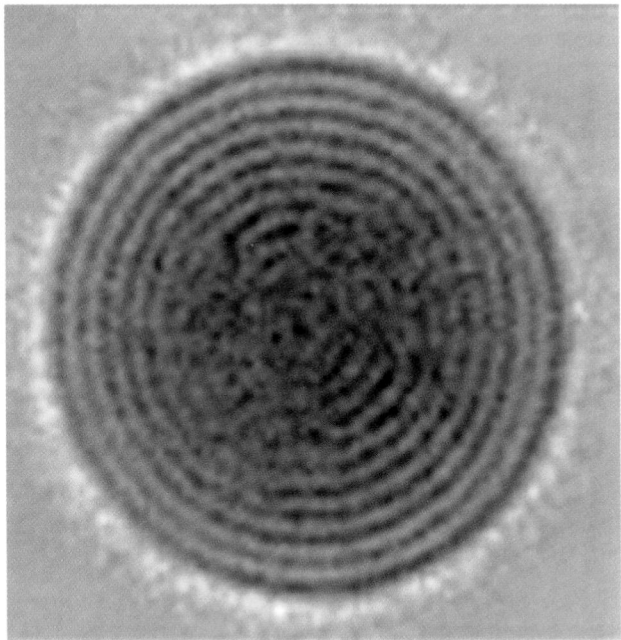

(b)

FIGURE 33-37   The packing of dsDNA inside a T7 phage head. (*a*) A cryo-EM–based image of a dsDNA-filled T7 phage head as viewed along a line from the head–tail connector to the center of the particle. The outer somewhat thicker ring represents the phage's protein capsid and the nine inner rings, whose spacing is 2.5 nm, represent coiled dsDNA. [Courtesy of Alasdair Steven, NIH, Bethesda, Maryland.] (*b*) A drawing of the concentric shell model in which the DNA is wound inward like a spool of twine about the phage's long axis. [After Harrison, S.C., *J. Mol. Biol.* **171,** 579 (1983).]

The packing of double-stranded DNA (dsDNA) inside a phage head must be an enthalpically as well as entropically unfavorable process because of dsDNA's stiffness and its intramolecular charge repulsions. The observation that DNA packaging requires the presence of ATP therefore strongly suggests that dsDNA is actively "pumped" into the phage head by an ATP-driven process. The injection of λ DNA into a host bacterium by a mature phage is presumably a spontaneous process that, once it has been triggered, is driven by the free energy stored in the compacted DNA.

The structure of the head–tail connector of bacteriophage λ is unknown. However, Baker and Rossmann have determined the X-ray structure of the head–tail connector of **bacteriophage φ29** (a tailed dsDNA-containing phage that infects *Bacillus subtilis*), which as in bacteriophage λ is the portal through which dsDNA enters and exits the phage head. The structure reveals that the φ29 head–tail connector consists of a funnel-shaped cyclic dodecamer of identical 309-residue subunits that is 75 Å high, has a maximum width of 69 Å, and encloses a central channel whose diameter is 36 Å at its narrow end and 60 Å at its wide end (Fig. 33-38). Two Arg-rich peptide segments of each subunit that project into the central channel are disordered, and are therefore presumed to flexibly interact with the anionic dsDNA as it is translocated through the channel.

(a)

(b)

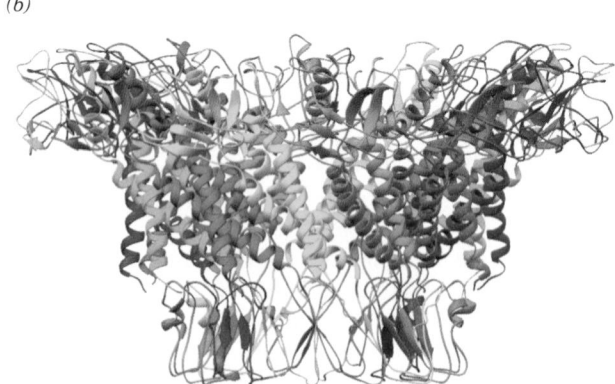

FIGURE 33-38 X-Ray structure of the bacteriophage φ29 head–tail connector. The dodecameric protein is shown in ribbon form with its identical subunits in different colors. (*a*) View along the 12-fold axis. (*b*) View perpendicular to Part *a*. [Based on an X-ray structure by Timothy Baker and Michael Rossmann, Purdue University. PDBid 1FOU.]

Cryo-EM studies of φ29 indicate that the connector is mounted at a pentagonal vertex of the phage prohead with its narrow end protruding into the exterior. The prohead, in addition, contains a virally encoded 174-nt RNA known as **pRNA** (p for *p*rohead), whose presence is essential for DNA packaging but which is absent in the mature virion. Cryo-EM studies indicate that the pRNA forms a cyclic homopentamer that surrounds the narrow end of the dodecameric head–tail connector (Fig. 33-39) and, moreover, suggest that the viral ATPase that drives DNA packaging is associated with the prohead's pentagonal vertex. This led Baker and Rossmann to propose that this entire assembly together with the dsDNA being pumped into the prohead acts as a rotary engine with the prohead–pRNA–ATPase complex as its stator (the stationary part of a rotary machine), the head–tail connector as its ball-race (a groove along which a ball bearing slides), and the dsDNA as its spindle. In this model, the five ATPases around the stator successively fire so as to motivate a traveling wave of conformational changes about the connector that, in essence, causes it to walk along the dsDNA's helical groove(s) so as to translocate it through the connector into the prohead (a mechanism that is reminiscent of that postulated for hexagonal helicases such as T7 gene 4 helicase/primase; Section 30-2C). This would translocate the DNA by one-fifth of its helical pitch for every ATP hydrolyzed, which for canonical B-DNA (which has 10 bp/turn) is the height of two base pairs, a quantity that is consistent with the observed ATP consumption of φ29 during packaging. The

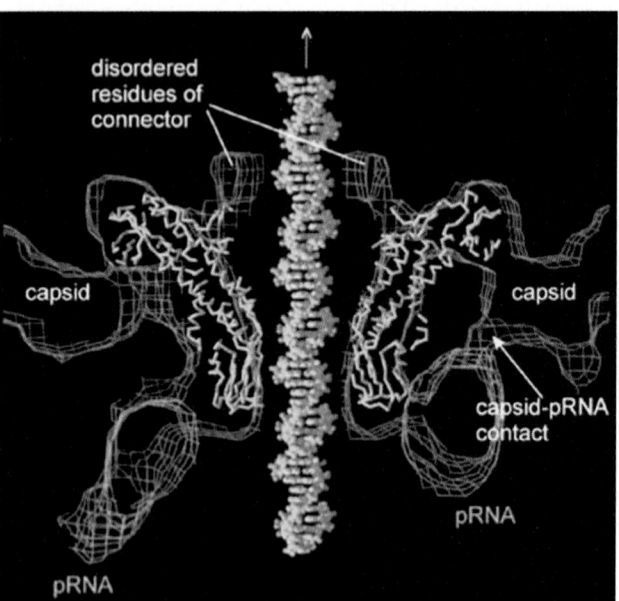

FIGURE 33-39 Cross section of the cryo-EM–based electron density of the bacteriophage φ29 prohead. The capsid (*red meshwork*) is fitted with the C_α backbone of the head–tail connector (*yellow*) and the pRNA (*green meshwork*). Canonical B-DNA is placed in the connector's central channel. The positions of the partially disordered and highly basic peptide segments (residues 229–246 and 287–309) that presumably contact the DNA are indicated. [Courtesy of Timothy Baker and Michael Rossmann, Purdue University.]

DNA packaging engine of bacteriophage λ most likely has a similar mechanism, although φ29 and its close relatives are the only phages known to have a pRNA.

The final step in the bacteriophage λ DNA packaging process is the recognition and cleavage of the next *cos* site (Fig. 33-35) on the concatemeric DNA by terminase, possibly with the participation of **gp*FI***. Phage λ therefore contains a unique segment of DNA (in contrast to some phages in which the amount of DNA packaged is limited by a "headful" mechanism that results in their containing somewhat more DNA than an entire chromosome). Indeed, the λ packaging system will efficiently package a DNA that is 75 to 105% the length of the wild-type λ DNA as long as it is flanked by *cos* sites (the central third of the phage DNA, which encodes the dispensable accessory genes, can be replaced by other sequences, thereby making phage λ a useful cloning vector; Section 5-5B).

d. Tail Assembly

Tail assembly, which occurs independently of head assembly, proceeds, as a comparison of Figs. 33-28 and 33-36 indicates, from the 200-Å-long tail fiber toward the head-binding end. This strictly ordered series of reactions can be considered to have three stages (Fig. 33-36, *left*):

1. The formation of the "initiator," which ultimately becomes the adsorption organelle, requires the sequential actions on gp*J* (the tail fiber protein) of the products of phage genes ***I, L, K, G, H,*** and ***M.*** Of these, only **gp*I*** and **gp*K*** are not components of the mature tail.

2. The initiator forms the nucleus for the polymerization of gp*V*, the major tail protein, to form a stack of 32 hexameric rings. The length of this stack is thought to be regulated by gp*H* (853 residues), which, the available evidence suggests, becomes extended along the length of the growing tail and somehow limits its growth. λ tail length is apparently specified in much the same way that the helical length of TMV is governed (Section 33-1B), although in TMV the regulating template is an RNA molecule rather than a protein.

3. In the termination and maturation stage of tail assembly, **gp*U*** attaches to the growing tail, thereby preventing its further elongation. The resultant immature tail has the same shape as the mature tail and can attach to the head. In order to form an infectious phage particle, however, the immature tail must be activated by the action of **gp*Z*** before joining the head.

The completed tail then spontaneously attaches to a mature phage head to form an infectious λ phage particle (Fig. 33-36, *bottom*).

e. The Assembly of Other Double-Stranded DNA Phages Resembles That of λ

The assembly of several other double-stranded DNA bacteriophages has been studied in detail, notably that of coliphages **T4** (Fig. 33-1*e*) and T7 and phage P22. All of them are formed in assembly processes that closely resemble that of phage λ. For example, their head assembly processes proceed in obligatory reaction sequences through

an initiation stage; the scaffolded assembly of a prohead; an ATP-driven DNA packaging process, in which the DNA assumes a tightly packed conformation and the prohead undergoes an expansion; and a final stabilization. The mature phages then form by the attachment of separately assembled tails to the completed and DNA-filled heads.

C. *The Lysogenic Mode*

Lysogeny is established by the integration of viral DNA into the host chromosome accompanied by the shutdown of all lytic gene expression. With phage λ, integration takes place through a site-specific recombination process that differs from homologous recombination (Section 30-6A) in that it occurs only between the chromosomal sites designated *att*P on the phage and *att*B on the bacterial host (Fig. 33-40). These two *att*achment sites have a 15-bp identity

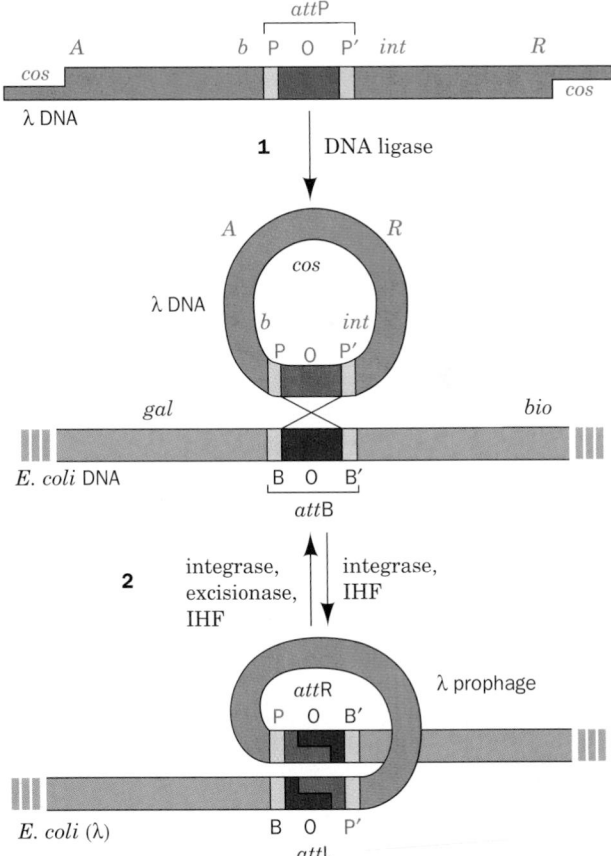

FIGURE 33-40   Site-specific recombination in bacteriophage λ. This schematic diagram shows **(1)** the circularization of the linear phage λ DNA through base pairing between its complementary ends to form the *cos* site; and **(2)** the integration/excision of this DNA into/from the *E. coli* chromosome through site-specific recombination between the phage *att*P and host *att*B sites. The darker colored regions in the *att* sites represent the identical 15-bp crossover sequences (O), whereas the lighter colored regions symbolize the unique sequences of bacterial (B and B′) and phage (P and P′) origin. [After Landy, A. and Weisberg, R.A., *in* Hendrix, R.W., Roberts, J.W., Stahl, F.W., and Weisberg, R.A. (Eds.), *Lambda II, p.* 212, Cold Spring Harbor Laboratory (1983).]

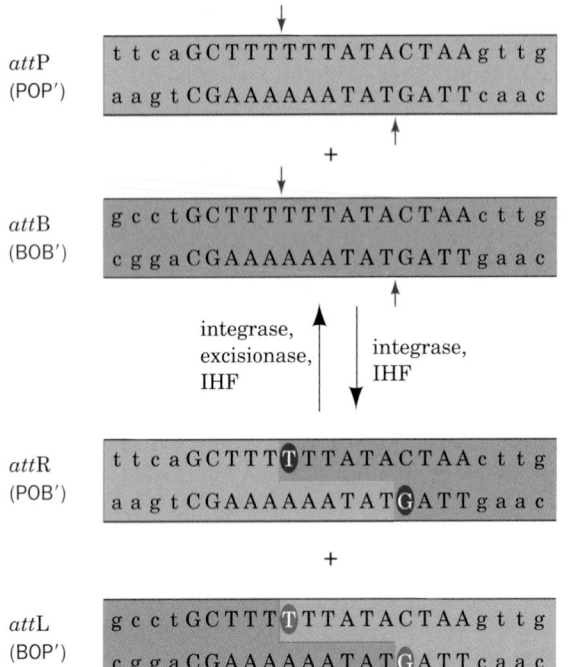

*att*P
(POP′)

```
ttcaGCTTTTTTATACTAAgttg
aagtCGAAAAAATATGATTcaac
```

+

*att*B
(BOB′)

```
gcctGCTTTTTTATACTAActtg
cggaCGAAAAAATATGATTgaac
```

integrase,
excisionase,
IHF   integrase,
             IHF

*att*R
(POB′)

```
ttcaGCTTTTTTATACTAActtg
aagtCGAAAAAATATGATTgaac
```

+

*att*L
(BOP′)

```
gcctGCTTTTTTATACTAAgttg
cggaCGAAAAAATATGATTcaac
```

FIGURE 33-41   The site-specific recombination process that inserts/excises phage λ DNA into/from the chromosome of its *E. coli* host. Exchange occurs between the phage *att*P site (*red*) and the bacterial *att*B site (*blue*), and the prophage *att*L and *att*R sites. The strand breaks occur at the approximate positions indicated by the short blue arrows. The sources of the more darkly shaded bases in *att*R and *att*L are uncertain. The uppercase letters represent bases in the O region common to the phage and bacterial DNAs, whereas lowercase letters symbolize bases in the flanking B, B′, P, and P′ sites.

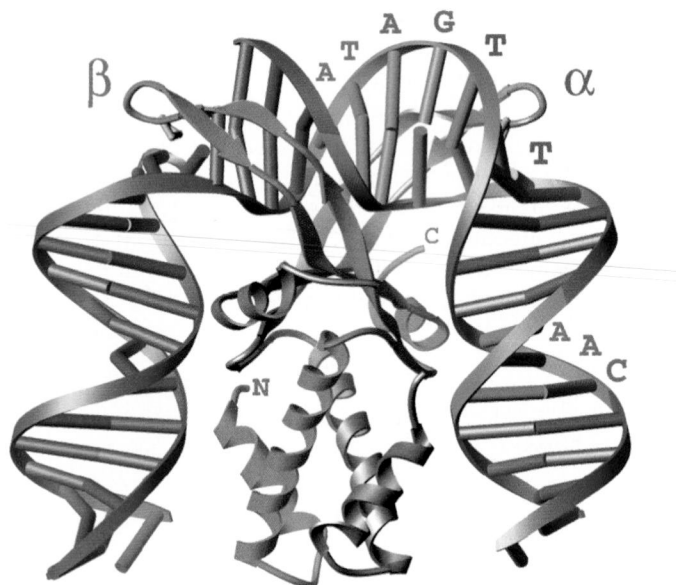

FIGURE 33-42   X-Ray structure of integration host factor (IHF) in complex with a 35-bp target DNA. The structure is viewed with its pseudo-2-fold axis vertical. The α subunit of IHF is gray and its β subunit is magenta. The "top" strand of the 35-bp *att* site DNA, which is shown in ladder form, was synthesized in two segments. The consensus sequence to which IHF binds is highlighted in green and interacts mainly with the β ribbon arm of the α subunit and the body of the β subunit. The Pro side chains near the tip of each arm that intercalate between base pairs so as to kink the DNA are drawn in yellow. [Courtesy of Phoebe Rice, University of Chicago. PDBid 1IHF.]

(Fig. 33-41), so they can be represented as having the sequences POP′ for *att*P and BOB′ for *att*B, where O denotes their common sequence. Phage integration occurs through a process that yields the inserted phage chromosome flanked by the sequence BOP′ on the "left" (the *att*L site) and POB′ on the "right" (the *att*R site; Fig. 33-40). The nature of the crossover site was determined through the use of ³²P-labeled bacterial DNA and unlabeled phage DNA. The crossover site occurs at a unique position on each strand that is displaced with respect to its complementary strand so as to form a staggered recombination joint (Fig. 33-41).

a. Integrase Mediates λ DNA Integration, whereas Excisionase Is Additionally Required for λ DNA Excision

Phage integration is mediated by **λ integrase** (356 residues; the **λ *int*** gene product) acting in concert with IHF. The λ integrase is homologous to and mediates a similar site-specific recombination reaction as does the Cre recombinase of bacteriophage P1, whose structure and mechanism are discussed in Section 30-6B. This, together with a variety of biochemical and genetic evidence, indicates that λ integrase functions via a similar mechanism.

IHF, a heterodimer of 30% identical 99- and 94-residue subunits, has no demonstrable endonuclease or topoisomerase activity but specifically binds to a dsDNA bearing an *att* sequence. The X-ray structure of IHF in complex with a 35-bp target DNA, determined by Phoebe Rice and Howard Nash, reveals that the pseudo-2-fold symmetric protein wraps the DNA around it in a >160° bend (Fig. 33-42). Most of this bend arises from two large kinks, separated by 9 bp, that are each formed by the intercalation of a highly conserved Pro side chain between two consecutive base pairs. IHF presumably facilitates the action of λ integrase by bending the DNA in a U-turn so as to bring the two DNA segments of an *att* site which λ integrase binds into close proximity. [The bacterial protein **HU** (Section 30-3C) is closely related to IHF, both in sequence and structure, but functionally differs from it in that HU binds to dsDNA nonspecifically.]

Since viral integration is not an energy-consuming process, why is phage integration not readily reversible? The answer is that the prophage excision requires the participation of **excisionase** (72 residues; the **λ *xis*** gene product) in concert with integrase, IHF, and **Fis** (a DNA-binding host protein that also stimulates Hin-mediated gene inversion; Section 30-6B). Apparently the λ recombination system has an inherent asymmetry that ensures the kinetic stability of the lysogenic integration product.

The mechanism by which excisionase reverses the integration process in unknown, although it has been shown that this protein specifically binds to POB′, where it induces a sharp bend in this DNA.

b. The Relative Levels of Cro Protein and cI Repressor Determine the λ Phage Life Cycle

The establishment of lysogeny in phage λ *is triggered by high concentrations of* **gpcII** *(see below).* This early gene product stimulates "leftward" transcription from two promoters, p_I (I for integrase) and p_{RE} (RE for *repressor establishment*; Fig. 33-43*a*):

1. Transcription initiated from p_I, which is located within the *xis* gene, results in the production of integrase but not excisionase. λ DNA is consequently integrated into the host chromosome to form the prophage.

2. The transcript initiated from p_{RE} encodes the ***cI*** gene whose product is called the **λ** or **cI repressor.** The λ repressor, as does Cro protein (Section 33-3A), binds to the o_L and o_R operators, thereby blocking transcription from p_L and p_R, respectively (Fig. 33-43; note that these operators are upstream from their corresponding promoters rather than downstream as in the *lac* operon; Fig. 31-2). *Both repressors therefore act to shut down the synthesis of early gene products, including Cro protein and gpcII.*

gpcII is metabolically unstable with a half-life of ~1 min (see below) so that *cI* transcription from p_{RE} soon ceases. λ repressor bound at o_R, but not Cro protein, however, stimulates "leftward" transcription of *cI* from p_{RM} (RM for repressor *maintenance*; Fig. 33-43*b*). In other words, *Cro protein represses all mRNA synthesis, whereas* λ *repressor stimulates transcription of its own gene while repressing all other mRNA synthesis. This conceptually simple difference between the actions of* λ *repressor and Cro protein forms the basis of a genetic switch that stably maintains phage* λ *in either the lytic or the lysogenic state.* The molecular mech-

anism of this switch is described in Section 33-3D. In the following paragraphs we discuss how this switch is "thrown" from one state to another. You should recognize, however, that, *once the switch is thrown in favor of the lytic cycle, that is, when Cro protein occupies* o_L *and* o_R, *the phage is irrevocably committed to at least one generation of lytic growth.*

c. gpcII Is Activated when Phage Multiplicity Is High or Nutritional Conditions Are Poor

The reason why a high gpcII concentration is required to establish lysogeny is that this early gene product can stimulate transcription from p_I and p_{RE} only when it is in oligomeric form. This phenomenon accounts for the observation that lysogeny is induced when the **multiplicity of infection** (ratio of infecting phages to bacteria) is large (≥ 10) since this gene dosage effect results in gpcII being synthesized at a high rate.

gpcII is metabolically unstable because it is preferentially proteolyzed by host proteins, notably **gp*hflA*** and **gp*hflB*.** However, gp*cIII* somehow protects gpcII from the action of gp*hflA*, which is why its presence enhances lysogenation (Fig. 33-43*a*). The activity of gp*hflA* is dependent on the host cAMP-activated catabolite repression system (Section 31-3C) as is indicated by the observation that *E. coli* mutants defective in this system lysogenize with less than normal frequency. Yet, if these mutant strains are also *hflA⁻*, they lysogenize with greater than normal frequency. Apparently the *E. coli* catabolite repression system, which is known to regulate the transcription of many bacterial genes, controls *hflA* activity, perhaps by directly repressing this protein's synthesis at high cAMP concentrations. *This explains why poor host nutrition, which results in elevated cAMP concentrations, stimulates lysogenation.*

Once a prophage has been integrated in the host chromosome, lysogeny is stably maintained from generation to generation by λ *repressor.* This is because λ repressor stimulates its own synthesis at a rate sufficient to maintain

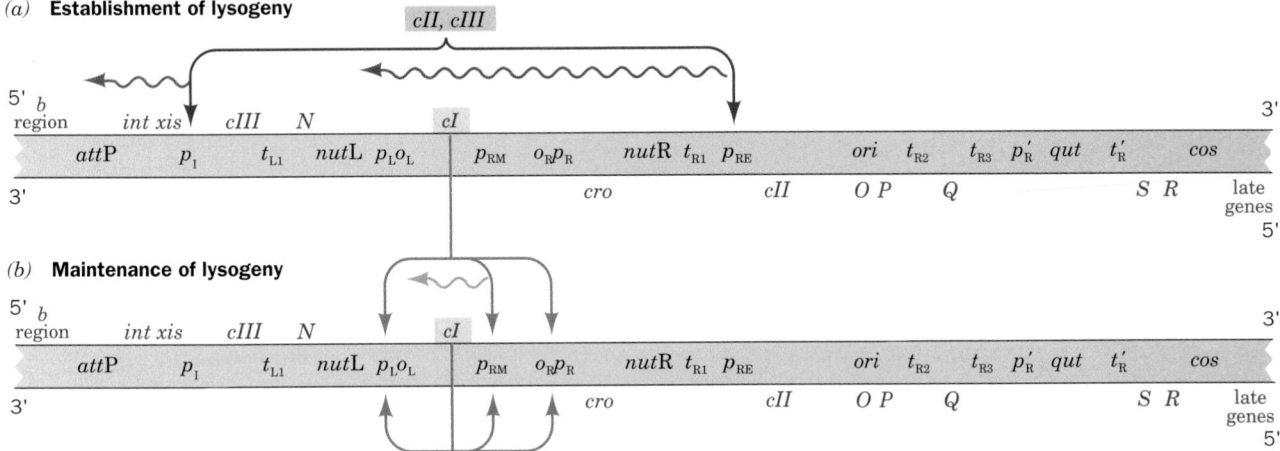

FIGURE 33-43   Control of gene expression in bacteriophage λ.
(*a*) The establishment lysogeny. (*b*) The maintenance of lysogeny. The symbols used are described in the legend of Fig.

33-31. [After Arber, W., *in* Hendrix, R.W., Roberts, J.W., Stahl, F.W., and Weisberg, R.A. (Eds.), *Lambda II,* p. 389, Cold Spring Harbor Laboratory (1983).]

lysogeny in the bacterial progeny while repressing the transcription of all other phage genes. In fact, λ *repressor is synthesized in sufficient excess to also repress transcription from superinfecting* λ *phage, thereby accounting for the phenomenon of immunity.* We shall see below how induction occurs.

D. *Mechanism of the λ Switch*

The lysogenic cycle is a highly stable mode of phage λ replication; under normal conditions lysogens spontaneously induce only about once per 10^5 cell divisions. Yet, transient exposure to inducing conditions triggers lytic growth in almost every cell of a lysogenic bacterial culture. In this section, we consider how this genetic switch, whose mechanism was largely elucidated by Mark Ptashne, can so tightly repress lytic growth and yet remain poised to turn it on efficiently.

a. o_R Consists of Three Homologous Palindromic Subsites

Both of the operators to which λ *repressor and Cro protein bind,* o_L *and* o_R, *consist of three subsites (Fig. 33-44).* These are designated o_{L1}, o_{L2}, and o_{L3} for o_L, and o_{R1}, o_{R2}, and o_{R3} for o_R. *Each of these subsites consists of a similar 17-bp segment that has approximate palindromic symmetry.* However, as we shall see, o_L plays only a minor role in the λ switch relative to that of o_R.

b. λ Repressor and Cro Protein Structurally Resemble Other Repressors

λ repressor binds to DNA as a dimer so that its 2-fold symmetry matches those of the operator subsites to which it binds. The monomer's 236-residue polypeptide chain is folded into two roughly equal sized domains connected by an ~30-residue segment that is readily cleaved by proteolytic enzymes. The isolated N-terminal domains retain their ability to bind specifically to operators (although with only half of the binding energy of the intact repressor) but do not dimerize in solution. The C-terminal domains can still dimerize but lack the capacity to bind DNA. Evidently,

repressor's N-terminal domain binds operator, whereas its C-terminal domain provides the contacts for dimer formation.

Although the λ repressor has not been crystallized, its N-terminal domain comprising residues 1 to 92, as excised by treatment with the papaya protease **papain,** does crystallize. The X-ray structure of this protein, both alone and in complex with a 20-bp DNA containing the o_{L1} sequence, was determined by Carl Pabo. The N-terminal domain crystallizes as a symmetric dimer with each subunit containing an N-terminal arm and five α helices (Fig. 33-45*a*). Two of these helices, α2 and α3, form a helix–turn–helix (HTH) motif, much like those in other prokaryotic repressors of known structure (Section 31-3D). The α3 helix, the recognition helix, protrudes from the protein surface such that the two α3 helices of the dimeric protein fit into successive major grooves of the operator DNA. Similar associations are observed in the X-ray structures of the closely related **bacteriophage 434 repressor** N-terminal fragment in complex with a 20-bp DNA containing its operator sequence (Fig. 31-29). The X-ray structure of the C-terminal domain (residues 132–236) of the λ repressor, determined by Mitchell Lewis, reveals how this domain dimerizes (Fig. 33-45*b*). The intact λ repressor presumably dimerizes on its target DNA as is drawn in Fig. 33-45*c*.

Cro protein also forms dimers. In contrast to λ or 434 repressor, however, this 66-residue polypeptide forms but one domain that contains both its operator recognition site and its dimerization contacts. The X-ray structure of Cro in complex with a 17-bp tight-binding operator DNA, determined by Brian Matthews, reveals that this dimer likewise contains a pair of HTH units (Fig. 33-46), but which bind to the DNA such that they induce it to bend about the protein by 40°. The sequence-specific binding predicted by this structure is supported by Robert Sauer's genetic studies indicating that mutant varieties of Cro, in which the proposed DNA-contacting residues have been changed, are defective in operator binding. Moreover, this structure closely resembles that of the related **phage 434 Cro protein** in complex with a 20-bp DNA containing its operator sequence (Fig. 31-30).

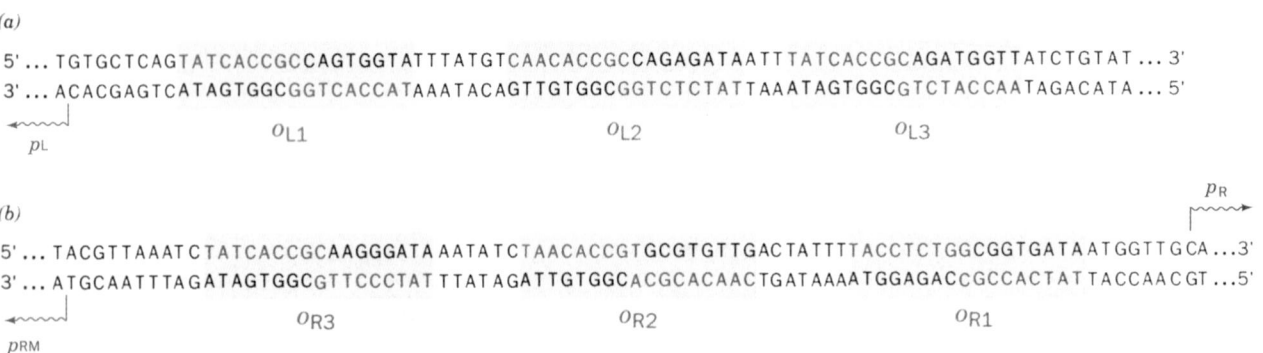

FIGURE 33-44   Base sequences of the operator regions of the phage λ chromosome. (*a*) o_L and (*b*) o_R. Each of these operators consists of three homologous 17-bp subsites separated by short AT-rich spacers. Each subsite has approximate palindromic (2-fold) symmetry as is demonstrated by the comparison of the two sets of red letters in each subsite. The wiggly arrows mark the transcriptional start sites and directions at the indicated promoters.

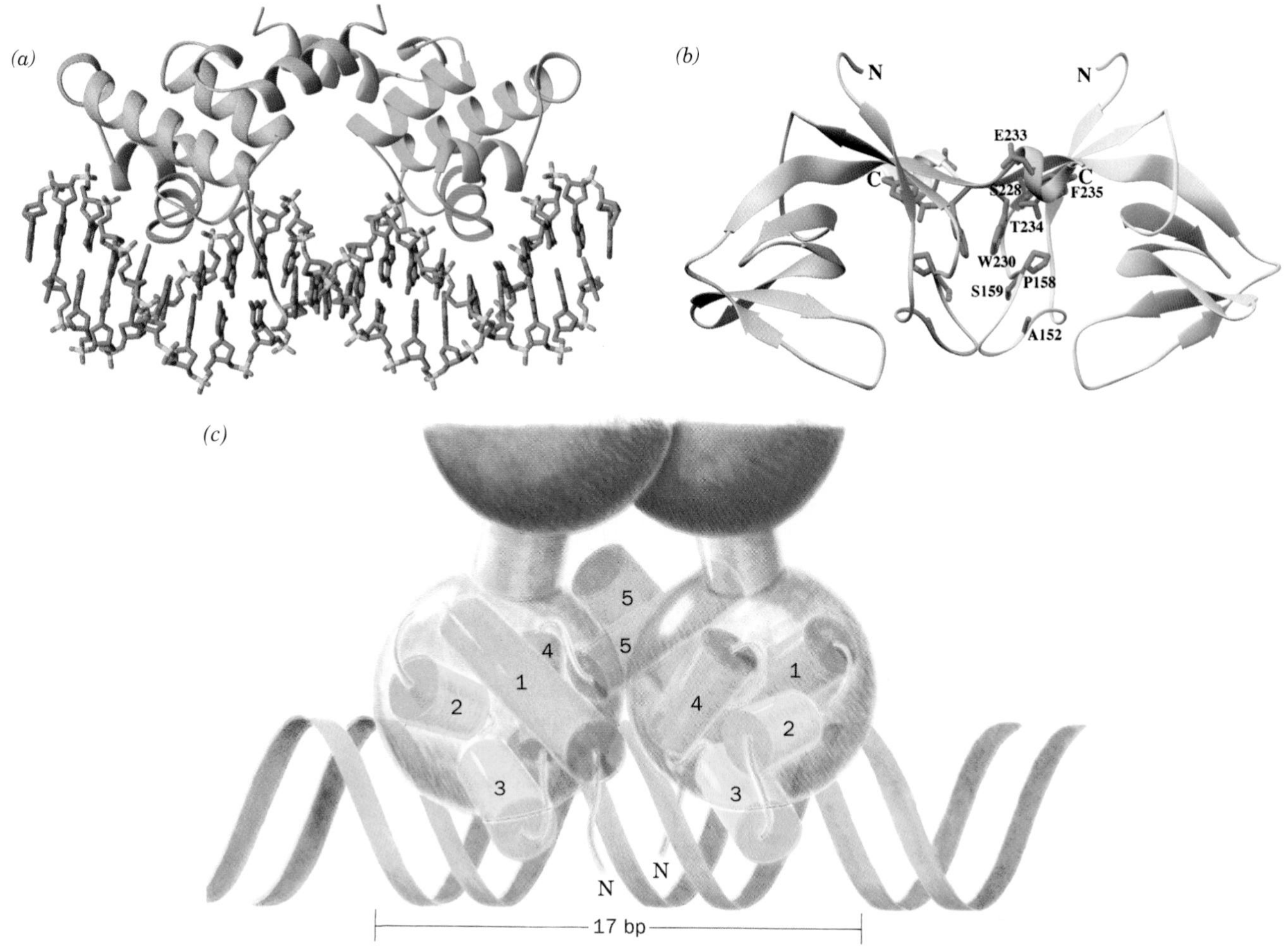

(a)

(b)

(c)

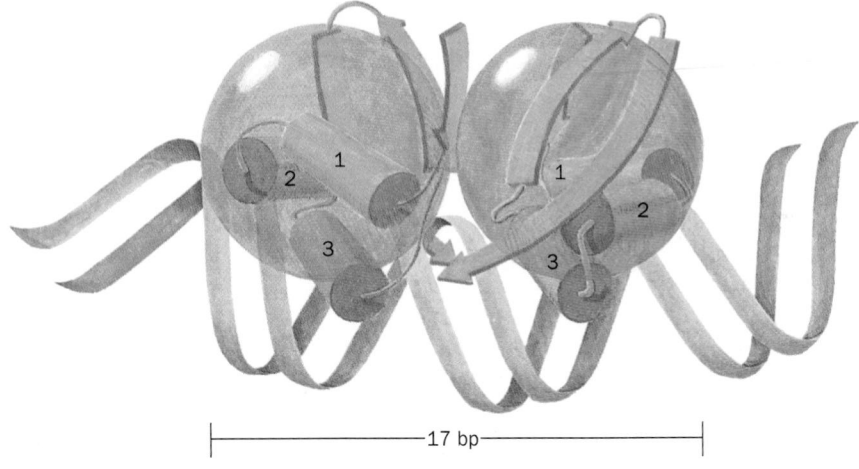

FIGURE 33-45   Structure of the λ repressor. (*a*) The X-ray structure of the N-terminal domain homodimer in complex with B-DNA. The DNA is drawn in stick form colored according to atom type (C green, N blue, O red, and P yellow). The two protein subunits are drawn as orange and cyan ribbons with their recognition helices magenta. Note that the protein's N-terminal arms wrap around the DNA. This accounts for the observation that the G residues in the major groove on the repressor–operator complex's "back side" are protected from methylation only when these N-terminal arms are intact. [Based on an X-Ray structure by Carl Pabo, The Johns Hopkins University. PDBid 1LMB.] (*b*) The X-ray structure of the C-terminal domain dimer. Mutations of the residues that are drawn in ball-and-stick form (and which are labeled for the green subunit) interfere with dimerization. [Based on an X-ray structure by Mitchell Lewis, University of Pennsylvania. PDBid 1F39.] (*c*) An interpretive drawing indicating how contacts between the repressor's C-terminal domains (*upper lobes*) maintain the intact protein's dimeric character. The λ repressor binds to the 17-bp operator subsites of o_L and o_R as symmetric dimers with the N-terminal domain of each subunit specifically binding to a half-subsite. Note how the α3 recognition helices of the symmetry related α2–α3 HTH units (*light yellow*) fit into successive turns of the DNA's major groove. [After Ptashne, M., *A Genetic Switch* (2nd ed.), *p.* 38, Cell Press & Blackwell Scientific Publications (1992).]   🐭 See the Interactive Exercises

FIGURE 33-46   X-Ray structure of the Cro protein dimer in its complex with B-DNA. Note that the λ repressor (Fig. 33-45), although otherwise dissimilar, contains HTH units that also bind in successive turns of the DNA's major groove. [After Ptashne, M., *A Genetic Switch* (2nd ed.), *p.* 40, Cell Press & Blackwell Scientific Publications (1992).]
🐭 See the Interactive Exercises

c. Repressor Stimulates Its Own Synthesis While Repressing All Other λ Genes

Chemical and nuclease protection experiments have indicated that λ repressor has the following order of intrinsic affinities for the subsites of o_R (Fig. 33-47):

$$o_{R1} > o_{R2} > o_{R3}$$

Despite this order, o_{R1} and o_{R2} are filled nearly together. This is because λ *repressor bound at o_{R1} cooperatively binds*

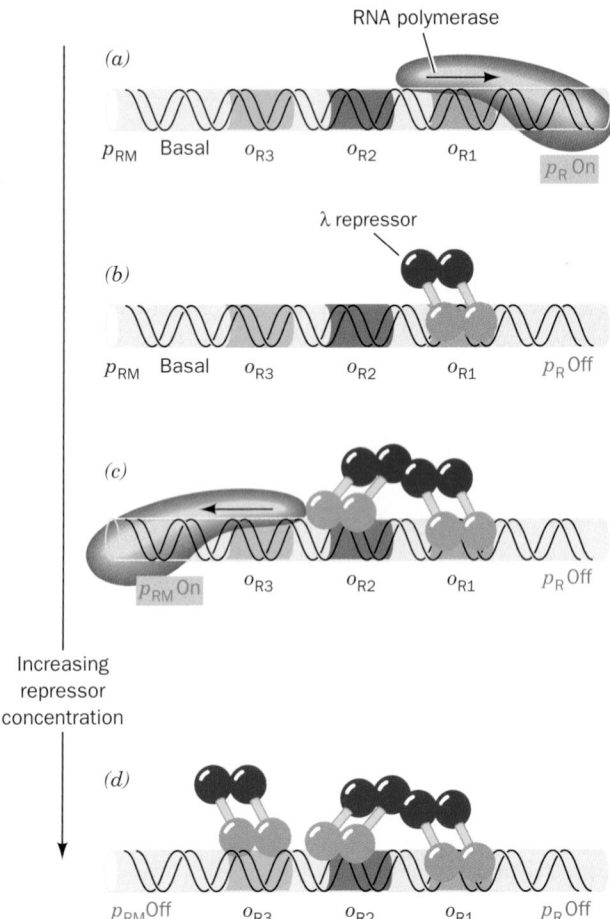

FIGURE 33-47   The binding of λ repressor to the three subsites of o_R. (*a*) In the absence of repressor, RNA polymerase initiates transcription at a high level from p_R (*right*) and at a basal level from p_{RM}. (*b*) Repressor has ~10 times higher affinity for o_{R1} than it does for o_{R2} or o_{R3}. Repressor dimer therefore first binds to o_{R1} so as to block transcription from p_R. (*c*) A second repressor dimer binds to o_{R2} at only slightly higher repressor concentrations due to specific binding between the C-terminal domains of neighboring repressors. In doing so, it stimulates RNA polymerase to initiate transcription from p_{RM} at a high level (*left*). (*d*) At high repressor concentrations, repressor binds to o_{R3} so as to block transcription from p_{RM}. Note that although Parts *c* and *d* are drawn with interdimer contacts between the C-terminal domains of only two repressor monomers, this interaction may involve contacts between the C-terminal domains of all four repressor monomers. [After Ptashne, M., *A Genetic Switch* (2nd ed.), p. 23, Cell Press & Blackwell Scientific Publications (1992).]

*repressor at o_{R2} through associations between their C-terminal domains (Fig. 33-47c). o_{R1} and o_{R2} are therefore both occupied at low λ repressor concentrations, whereas o_{R3} becomes occupied only at higher repressor concentrations.

The binding of λ repressor to o_R, as we previously mentioned, abolishes transcription from p_R and stimulates it from p_{RM} (Fig. 33-47c). At high concentrations of λ repressor, however, transcription from p_{RM} is also repressed (Fig. 33-47d). These phenomena have been clearly demonstrated through the construction of a series of hybrid operons that permit the effect of λ repressor on a promoter to be studied in a controlled manner. The system has two elements (Fig. 33-48):

1. A plasmid bearing the *lacI* gene (which encodes *lac* repressor; Section 31-1A) and the *lac* operator–promoter sequence fused to the *cI* gene. This construct permits the amount of λ repressor produced to be directly controlled by varying the concentration of the *lac* inducer IPTG (Section 31-1A).

2. A prophage containing o_R and either p_{RM}, as Fig. 33-48 indicates, or p_R fused to the *lacZ* gene. The amount of the *lacZ* gene product, β-galactosidase, produced, which can be readily assayed, reflects the activity of p_{RM} (or p_R).

The manipulation of these systems has demonstrated that at intermediate λ repressor concentrations (when o_{R1} and

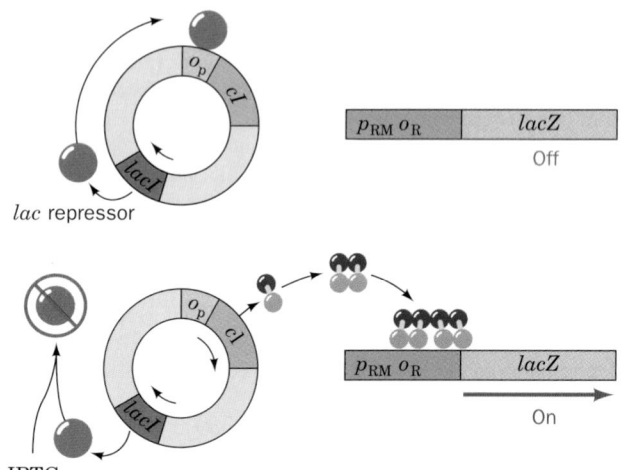

FIGURE 33-48   The genetic system used to study the effect of λ repressor on p_{RM}. The bacterium contains two hybrid operons. The first (*left*) is a plasmid bearing the *lac* operator–promoter (*Op*) fused to the λ *cI* gene so as to provide a source of repressor. The *lacI* gene, which encodes *lac* repressor, is also incorporated in the plasmid so that the level of λ repressor in the bacterium may be controlled by the concentration of the *lac* inducer IPTG. The second operon (*right*) is carried on a prophage that contains the promoter p_{RM} fused to the *lacZ* gene. The level of β-galactosidase (gp*lacZ*) in these cells therefore reflects the activity of p_{RM}. In similar experiments, the *cro* gene was substituted for λ *cI* and/or p_{RM} was replaced by p_R. [After Ptashne, M., *A Genetic Switch* (2nd ed.), p. 89, Cell Press & Blackwell Scientific Publications (1992).]

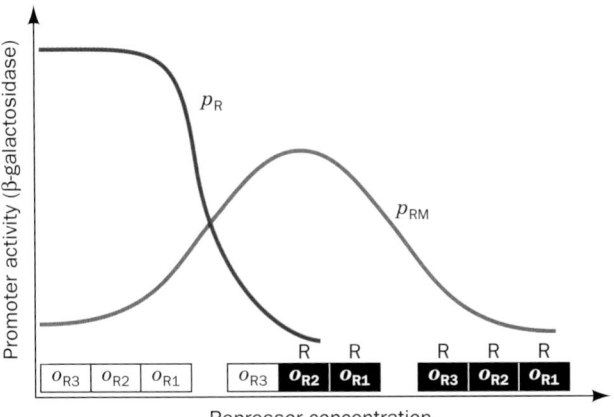

FIGURE 33-49   The response of p_{RM} and p_R to the λ repressor level. The p_{RM} curve was derived using the system diagrammed in Fig. 33-48, whereas the p_R curve was obtained using a similar system but with p_R rather than p_{RM} fused to *lacZ*. The amount of λ repressor that maximally stimulates p_{RM} is approximately that which occurs in a λ lysogen. At least 5-fold more repressor is required to half-maximally repress p_{RM}. The boxes indicate the states of each o_R subsite at the various repressor concentrations; black represents repressor occupancy. [After Ptashne, M., *A Genetic Switch* (2nd ed.), p. 90, Cell Press & Blackwell Scientific Publications (1992).]

o_{R2} are occupied), transcription from p_R is indeed repressed, whereas that from p_{RM} is stimulated (Fig. 33-49). Transcription from p_{RM} only becomes repressed at high levels of λ repressor (when o_{R3} is also occupied). The stimulation of transcription from p_{RM} is abolished by mutations in o_{R2} that prevent repressor binding, whereas its repression at high repressor concentrations is relieved by mutations in o_{R3}. Thus, *occupancy of o_{R2} by λ repressor stimulates transcription from p_{RM}, whereas occupancy of o_{R3} prevents it by excluding RNA polymerase from p_{RM} (Fig. 33-47c,d). By the same token, occupancy of o_{R1} and/or o_{R2} prevents transcription from p_R.* In this way, λ repressor prevents the synthesis of all phage gene products but itself. Yet, at high repressor concentrations, its synthesis is also repressed, thereby maintaining the repressor concentration within reasonable limits.

What is the basis of λ repressor's remarkable property of inhibiting transcription from one promoter while stimulating it from another? Knowledge of the sizes and shapes of repressor and RNA polymerase, as well as their positions on the DNA as demonstrated by chemical protection experiments, indicate that repressor at o_{R2} and RNA polymerase at p_{RM} are in contact (Fig. 33-50). Evidently, *repressor stimulates RNA polymerase activity through their cooperative binding to DNA.* This model was corroborated by the analysis of repressor mutants that bind normally (or nearly so) to operators but fail to stimulate the binding of RNA polymerase: All of the mutated residues occur either in helix α2 or in the link connecting it to helix α3 and lie on the surface of the protein that is thought to face the RNA polymerase-binding site (Fig. 33-50).

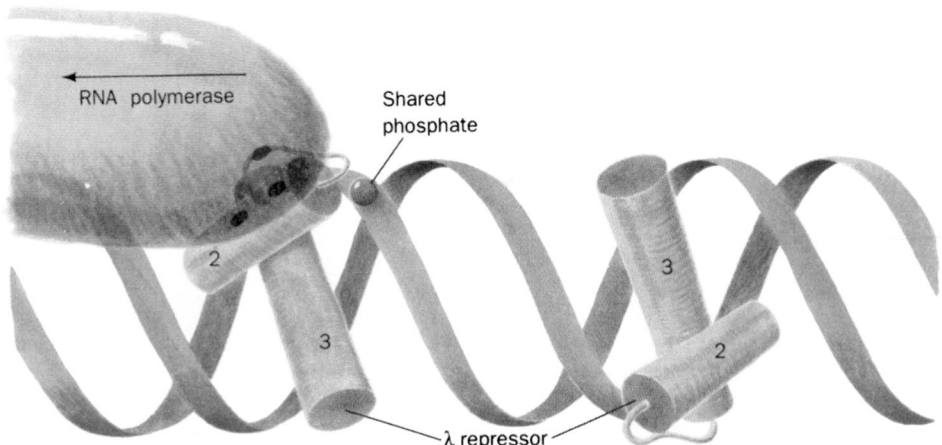

FIGURE 33-50   Interactions between the λ repressor and RNA polymerase. Repressor bound at o_{R2} is proposed to stimulate transcription at p_{RM} through a specific association with RNA polymerase that helps the polymerase bind to the promoter. This model is supported by the locations of the altered residues (*blue dots*) in three mutant repressors that bind normally to o_{R2} but fail to stimulate transcription at p_{RM}. The relative positions of repressor and RNA polymerase are established by the location of a phosphate group (*orange sphere*) whose ethylation interferes with the binding of both proteins to the DNA. For the sake of clarity, only the α_2–α_3 helix–turn–helix units of the repressor dimer are shown.

d. Cro Protein Binding to o_R Represses All λ Genes

Cro protein binds to the subsites of o_R in an order opposite to that of λ repressor (Fig. 33-51):

$$o_{R3} > o_{R2} \approx o_{R1}$$

This binding is noncooperative. Through experiments similar to that diagrammed in Fig. 33-48, but with *cro* in place of *cI*, the binding of Cro protein to o_{R3} was shown to abolish transcription from p_{RM}. Additional Cro binding to o_{R2} and/or o_{R1} turns off transcription from p_R.

e. The SOS Response Induces the RecA-Mediated Cleavage of λ Repressor

A final piece of information allows us to understand the workings of the λ switch. *The lytic phase is induced by agents that damage host DNA or inhibit its replication.* These are just the conditions that induce *E. coli*'s SOS response: The resulting fragments of single-stranded DNA activate RecA protein to stimulate the self-cleavage of LexA protein, the SOS gene repressor, at an Ala—Gly bond (Section 30-5D). *Activated RecA protein likewise stimulates the autocatalytic cleavage of λ repressor monomer's Ala 111—Gly 112 bond, which occurs in the polypeptide segment linking the λ repressor's two domains.* The ability of λ repressor to cooperatively bind to o_{R2} is thereby abolished (Fig. 33-52a,b); the C-terminal domains can still dimerize but they no longer link the DNA-binding N-terminal domains. The consequent reduction in concentration of intact free monomers shifts the monomer–dimer equilibrium such that the operator-bound dimers dissociate to form monomers, which are then cleaved through the influence of activated RecA before they can rebind to their target DNA.

In the absence of repressor at o_R, the λ early genes, including *cro*, are transcribed (Fig. 33-52c). As Cro accumulates, it first binds to o_{R3} so as to block even basal levels of λ repressor synthesis (Fig. 33-52d). Thus, *there being no*

mechanism for selectively inactivating Cro, the phage irreversibly enters the lytic mode: The λ switch, once thrown, cannot be reset. The prophage is subsequently excised from the host chromosome by the integrase and excisionase that are produced in the delayed early phase (Fig. 33-31).

f. The λ Switch's Responsiveness to Conditions Arises from Cooperative Interactions among Its Components

The complexity of the above switch mechanism endows it with a sensitivity that is not possible in simpler systems. The degree of repression at p_R is a steep function of repressor concentration (Fig. 33-53, *right*): The repression of p_R in a lysogen is normally 99.7% complete but drops to half this level on inactivation of 90% of the repressor. This steep sigmoid binding curve arises from the much greater operator affinity of repressor dimers compared to monomers. This situation, in turn, results from the cooperative linking of the monomer–dimer equilibrium, the binding of dimer to operator, and the association of dimers bound at o_{R1} and o_{R2} to form a tetramer. In fact, this cooperative effect is further enhanced by the similar binding of repressor tetramer to o_{L1} and o_{L2}, which through DNA looping, forms an octamer with the repressor tetramer bound at o_{R1} and o_{R2} (a phenomenon that likewise increases the repression of p_L). In contrast, a 99.7% repressed promoter controlled by a stably oligomeric repressor binding to a single operator site, such as occurs in the *lac* system, requires 99% repressor inactivation for 50% expression (Fig. 33-53, *left*). *The cooperativity of λ repressor oligomerization and multiple operator site binding are therefore responsible for the remarkable responsiveness of the λ switch to the health of its host.*

4 ■ INFLUENZA VIRUS

Influenza is one of the few common infectious diseases that is poorly controlled by modern medicine. Its annual epidemics, one of which was recorded by Hippocrates in

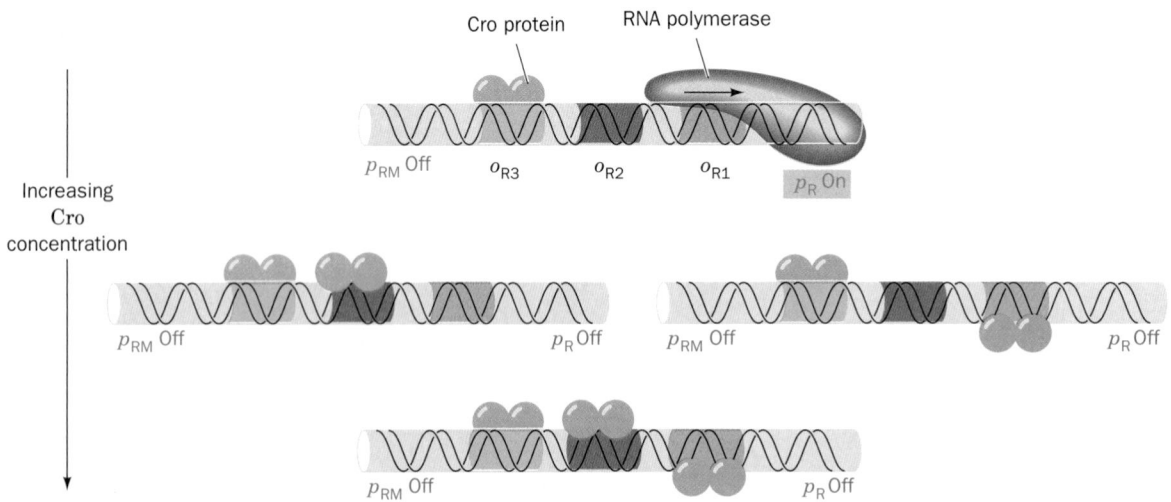

FIGURE 33-51   The binding of Cro protein to the three o_R subsites. o_{R3} binds Cro ~10 times more tightly than does o_{R1} or o_{R2}. Cro dimer therefore first binds to o_{R3}. A second dimer then binds to either o_{R1} or o_{R2} and in each case blocks transcription from p_R. At high Cro concentrations, all three

operator subsites are occupied. Compare this binding sequence with that of λ repressor (Fig. 33-47). [After Ptashne, M., *A Genetic Switch* (2nd ed.), p. 27, Cell Press & Blackwell Scientific Publications (1992).]

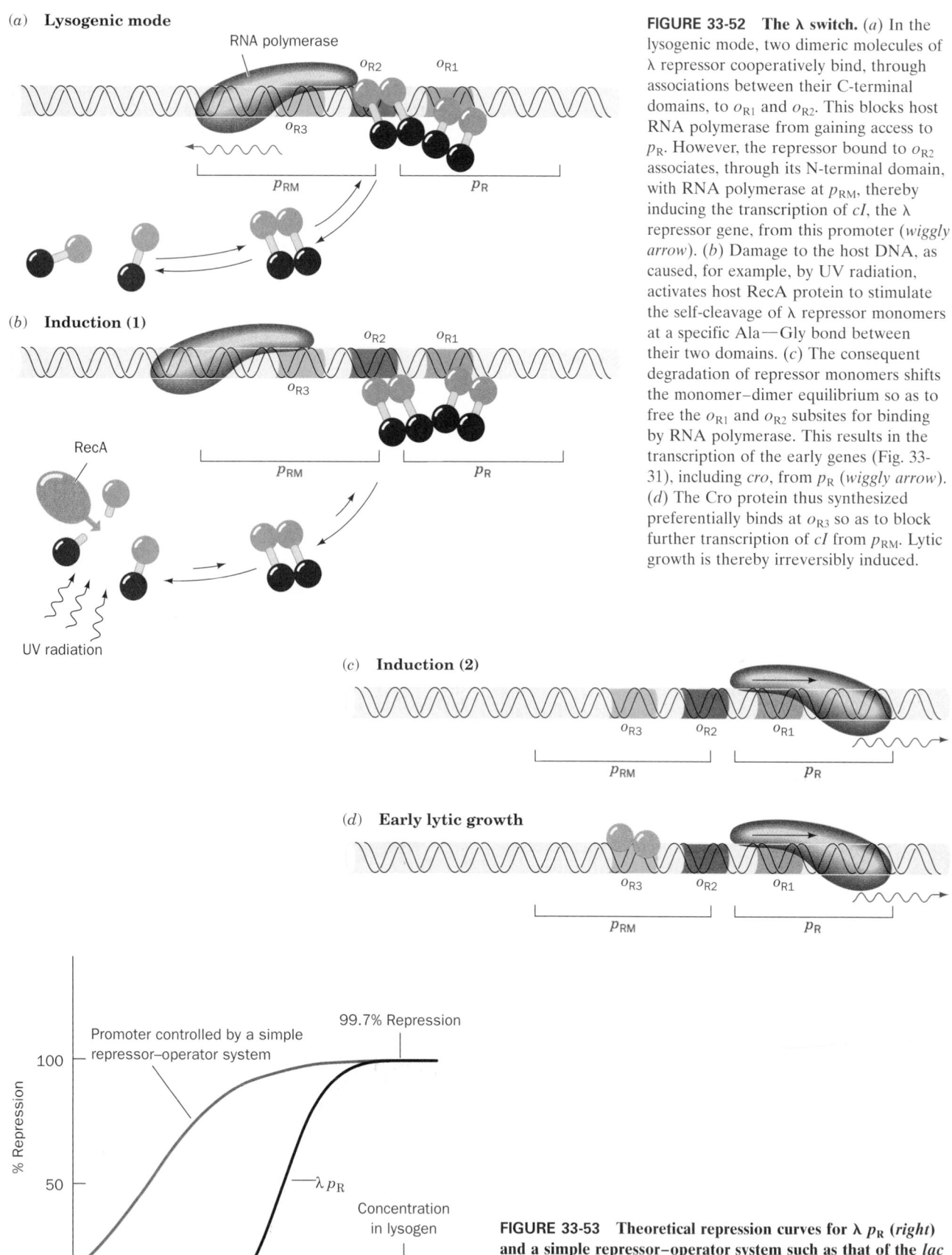

(a) **Lysogenic mode**

RNA polymerase

o_{R2} o_{R1}

o_{R3}

p_{RM} p_R

(b) **Induction (1)**

o_{R2} o_{R1}

o_{R3}

RecA

p_{RM} p_R

UV radiation

(c) **Induction (2)**

o_{R3} o_{R2} o_{R1}

p_{RM} p_R

(d) **Early lytic growth**

o_{R3} o_{R2} o_{R1}

p_{RM} p_R

FIGURE 33-52 The λ switch. (*a*) In the lysogenic mode, two dimeric molecules of λ repressor cooperatively bind, through associations between their C-terminal domains, to o_{R1} and o_{R2}. This blocks host RNA polymerase from gaining access to p_R. However, the repressor bound to o_{R2} associates, through its N-terminal domain, with RNA polymerase at p_{RM}, thereby inducing the transcription of *cI*, the λ repressor gene, from this promoter (*wiggly arrow*). (*b*) Damage to the host DNA, as caused, for example, by UV radiation, activates host RecA protein to stimulate the self-cleavage of λ repressor monomers at a specific Ala—Gly bond between their two domains. (*c*) The consequent degradation of repressor monomers shifts the monomer–dimer equilibrium so as to free the o_{R1} and o_{R2} subsites for binding by RNA polymerase. This results in the transcription of the early genes (Fig. 33-31), including *cro*, from p_R (*wiggly arrow*). (*d*) The Cro protein thus synthesized preferentially binds at o_{R3} so as to block further transcription of *cI* from p_{RM}. Lytic growth is thereby irreversibly induced.

% Repression

Promoter controlled by a simple repressor–operator system

99.7% Repression

100

50

λ p_R

Concentration in lysogen

Log repressor concentration

FIGURE 33-53 Theoretical repression curves for λ p_R (*right*) and a simple repressor–operator system such as that of the *lac* operon (*left*). Note the greater sensitivity of the λ system to a decrease in repressor concentration. [After Johnson, A.D., Poteete, A.R., Lauer, G., Sauer, R.T., Ackers, G.K., and Ptashne, M., *Nature* **294**, 221 (1981).]

412 B.C., are occasionally punctuated by devastating pandemics that infect 20 to 40% of the world's population. For example, the influenza pandemic of 1918, the so-called Spanish flu, which killed 40 to 50 million people worldwide (often previously healthy young adults; around 2% of the world's population at the time) was among the most lethal plagues ever recorded (it lowered the average life expectancy in the United States by 10 years). Since that time there have been three other pandemics of lesser severity, the so-called Asian flu of 1957, the Hong Kong flu of 1968, and the Russian flu of 1977 (the historical record suggests that there have been 12 pandemics in the past 400 years). All of these pandemics were characterized by the appearance of a new strain of influenza virus to which the human population had little resistance and against which previously existing influenza virus vaccines were ineffective. Moreover, between pandemics, influenza virus undergoes a gradual antigenic variation that degrades the level of immunological resistance against renewed infection. Even in nonpandemic years, influenza is responsible for the deaths of one-half to one million mainly elderly people; it is among the ten leading causes of death in the United States. What characteristics of the influenza virus permit it to evade human immunological defenses? In this section we shall discuss this question and, in doing so, examine the structure and life cycle of the influenza virus.

A. *Virus Structure and Life Cycle*

Electron micrographs of influenza virus (Fig. 33-1*i*) reveal a collection of nonuniform spheroidal particles that are ~100 nm in diameter and whose surfaces are densely studded with radially projecting "spikes." The influenza virion, which grows by budding from the plasma membrane of an infected cell (Fig. 33-54), is an example of an **enveloped virus**. *Its outer envelope consists of a lipid bilayer of cellular origin that is pierced by virally specified integral membrane glycoproteins, the "spikes."* There are two types of these surface spikes (Fig. 33-55):

1. A rod-shaped spike composed of **hemagglutinin (HA),** so named because it causes erythrocytes to agglutinate (clump together). HA mediates influenza target cell recognition by specifically binding to cell-surface receptors (glycophorin A molecules in erythrocytes; Section 12-3A) bearing terminal *N*-acetylneuraminic acid (sialic acid; Fig. 11-11) residues. Each virion contains ~500 copies of HA.

2. A mushroom-shaped spike known as **neuraminidase (NA),** which catalyzes the hydrolysis of the linkage joining a terminal sialic acid residue to a D-galactose or a D-galactosamine residue. NA probably facilitates the transport of the virus to and from the infection site by permitting its passage through mucin (mucus) and preventing viral self-aggregation. Each virion incorporates ~100 copies of NA.

In addition, the membranous outer envelope contains small amounts of **matrix protein 2 (M2).**

Just beneath the viral membrane is a 6-nm-thick protein shell composed of ~3000 copies of **matrix protein 1**

(M1), the virion's most abundant protein. M1 interacts with **nuclear export protein [NEP;** formerly known as **nonstructural protein 2 (NS2)].**

The influenza virus genome is unusual in that it consists of eight different sized segments of single-stranded RNA. These RNA molecules are negative strands; that is, they are complementary to the viral mRNAs. In the viral core,

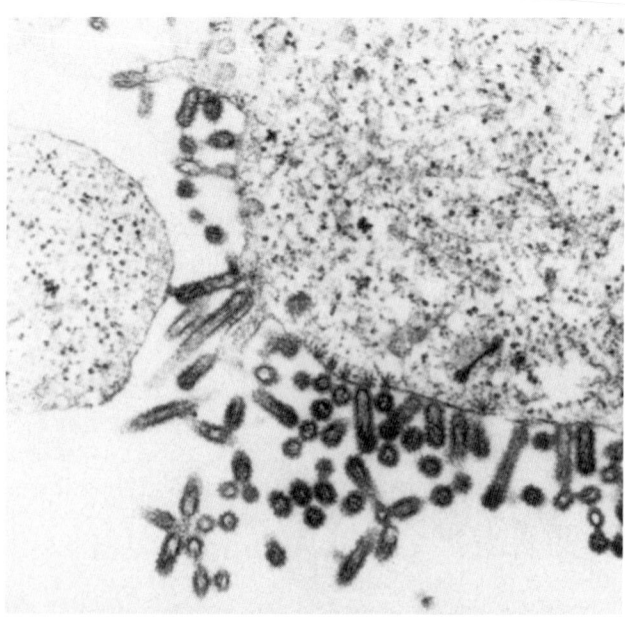

FIGURE 33-54   Electron micrograph of influenza viruses budding from infected chick embryo cells. [From Sanders, F.K., *The Growth of Viruses, p.* 15, Oxford University Press (1975).]

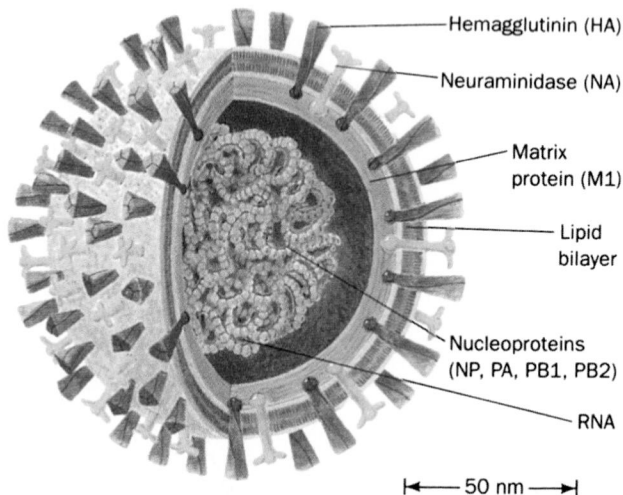

FIGURE 33-55   Cutaway diagram of the influenza virion. The HA and NA spikes are embedded in a lipid bilayer that forms the virion's outer envelope. M1 coats the underside of this membrane. The virion core contains the eight single-stranded RNA segments that comprise its genome in complex with the proteins NP, PA, PB1, and PB2 to form helical structures named nucleocapsids. [After Kaplan, M.M. and Webster, R.G., *Sci. Am.* **237**(6): 91 (1977). Copyright © 1977 by Scientific American, Inc.]

these RNAs occur in complex with four different proteins: **nucleocapsid protein (NP),** which occurs in ~1000 copies, and **polymerase acidic protein (PA), polymerase basic protein 1 (PB1),** and **polymerase basic protein 2 (PB2),** which are present in 30 to 60 copies each. The resulting **nucleocapsids** have the appearance of flexible rods.

The eight viral RNAs, which vary in length from 890 to 2341 nucleotides, have all been sequenced. They encode the virus' nine structural proteins (HA, NA, M1, M2, NEP, NP, PA, PB1, and PB2) and **nonstructural protein 1 (NS1),** which occurs only in infected cells. The sizes of the RNAs and the proteins they encode are listed in Table 33-2. About 10% of the viral mRNAs encoding M1 and NS1 are processed by the host cell splicing machinery to yield smaller mRNAs that respectively encode M2 and NEP but in mainly different reading frames from M1 and NS1.

a. Virus Life Cycle

The influenza infection of a susceptible cell begins with the HA-mediated adsorption of the virus to specific cell-surface receptors. The virus is then taken into the cell via endocytosis (Section 12-5B), whereupon the endocytotic vesicle fuses with the endosome (Fig. 12-79). In the acidic (pH ~5) medium of the endosome, the viral M2 protein, a proton channel, admits protons into the virion, which induces the separation of the nucleocapsids from M1. The viral and endosome membranes then fuse through a mechanism discussed in Section 33-4C, thereby introducing the nucleocapsids into the cytosol. By ~20 min postinfection, in a process mediated by NP, the still intact nucleocapsids have been transported to the cell nucleus, where they commence transcription of the viral RNAs **(vRNAs).** Cellular enzyme systems are incapable of mediating such RNA-directed RNA synthesis. Rather, it is carried out by a viral RNA transcriptase system that consists of the nucleocapsid proteins.

The transcription of the influenza virus genome is terminated if infected cells are treated with inhibitors of RNA polymerase II (which synthesizes cellular mRNA precur-

TABLE 33-2   The Influenza Virus Genome

RNA Segment	Length (nt)	Polypeptide(s) Encoded
1	2341	PB2
2	2341	PB1
3	2233	PA
4	1778	HA
5	1565	NP
6	1413	NA
7	1027	M1, M2
8	890	NS1, NEP

Source: Lamb, R.A. and Choppin, P.W., *Annu. Rev. Biochem.* **52,** 473 (1983).

sors; Section 31-2E) such as actinomycin D or α-amanitin. Yet, none of these agents affects the viral transcriptase's *in vitro* activity. The resolution of this seeming paradox is that *in vivo* viral mRNA synthesis is primed by newly synthesized cellular mRNA fragments consisting of a 7-methyl-G cap (Section 31-4A) followed by a 9- to 17-nt chain ending in A or G (Fig. 33-56, *top*). Viral mRNAs, as do most mature cellular mRNAs, have poly(A) tails appended to their 3' ends by the cellular polyadenylation machinery (Section 31-4A).

The synthesis of the viral mRNAs is terminated 16- to 17-nt from the 5' ends of their vRNA templates by the presence of a sequence of 5 to 7 U's on the vRNAs. Consequently, the viral mRNAs cannot act as templates in vRNA replication. Rather, in an alternative transcription process that begins some 30 min postinfection, complete vRNA complements are synthesized. These so-called **cRNAs,** whose synthesis does not require a primer, begin with pppA at their 5' ends and lack poly(A) tails (Fig. 33-56, *bottom*). Hence cRNAs, unlike viral mRNAs, do not associate with ribosomes in infected cells. The synthesis of cRNAs, in contrast to that of viral mRNAs, requires the

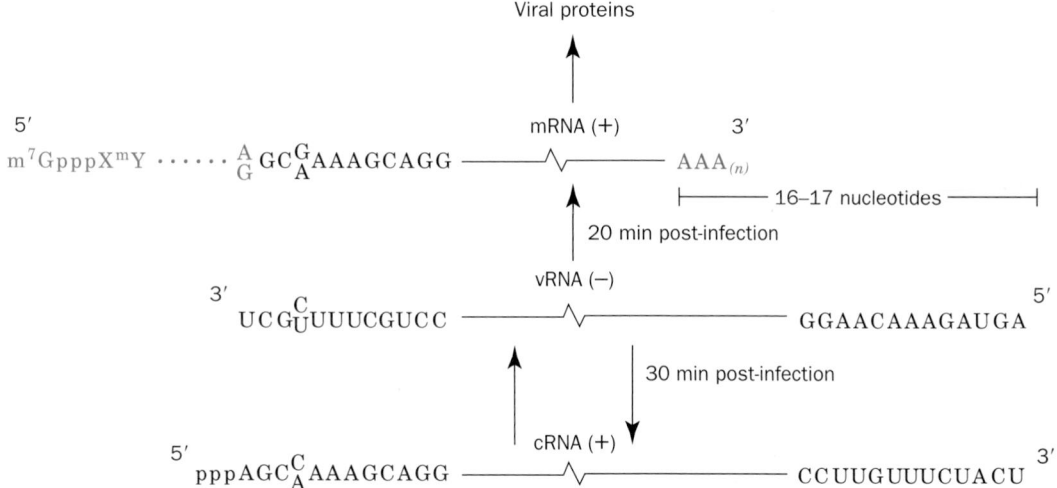

FIGURE 33-56   The biosynthesis of influenza vRNA, mRNA, and cRNA. The conserved nucleotides at the ends of the RNA segments are indicated. The viral mRNA's host-derived capped 5' head and 3' poly(A) tail are shown in color. [After Lamb, R.A. and Choppin, P.W., *Annu. Rev. Biochem.* **52,** 490 (1983).]

presence of NS1, which inhibits the processing of cellular pre-mRNAs and interferes with the synthesis of poly(A) tails. The cRNAs are the templates for vRNA synthesis. The resulting dsRNA would normally induce an interferon-mediated antiviral state in the infected cell (Section 32-4A). However, NS1 also functions as an interferon antagonist, thereby permitting viral proliferation.

The influenza virus transcription complex is a trimer consisting of PB1, PB2, and PA, which, together with NP, bind the RNA template. PB1 is the polymerase that catalyzes both the initiation and the elongation of the RNA transcript. PB2 binds to the 5' caps of cellular pre-mRNAs,

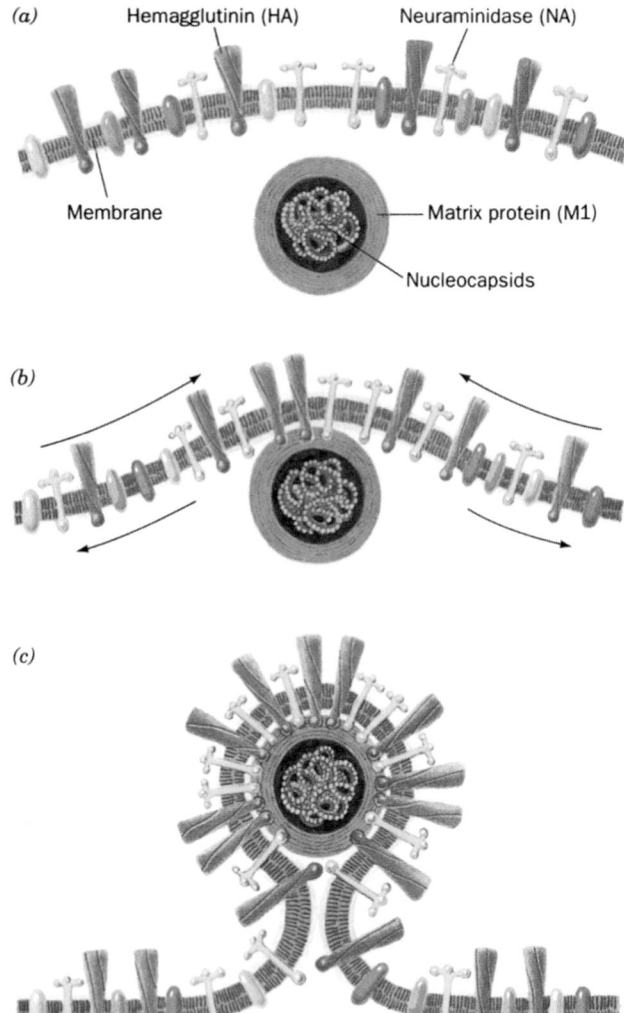

FIGURE 33-57 The budding of influenza virus from the host cell membrane. (*a*) The viral glycoproteins, HA and NA, are inserted into the plasma membrane of the host cell and the matrix protein, M1, forms the nucleocapsid-containing shell. (*b*) The binding of the matrix protein to the cytoplasmic domains of HA and NA results in the aggregation of these glycoproteins so as to exclude host cell membrane proteins (*arrows*). (*c*) This binding process induces the membrane to envelop the matrix protein shell such that the mature virion buds from the host cell surface. [After Wiley, D.C., Wilson, I.A., and Skehel, J.J., *in* Jurnak, F.A. and McPherson, A. (Eds.), *Biological Macromolecules and Assemblies,* Vol. 1: *Virus Structures,* Wiley (1984).]

although the endonuclease function that cleaves the capped primers from them appears to reside on PB1. Mutational experiments indicate that PA is required for vRNA but not mRNA synthesis, although its role in vRNA synthesis is poorly understood. The abundance of NP suggests that it has a structural role in the nucleocapsid, although it has also been implicated in the antitermination required to synthesize cRNAs rather than vRNAs.

The mechanism of influenza virus assembly is not well characterized. The viral spike glycoproteins, HA and NA, are ribosomally synthesized on the rough endoplasmic reticulum, further processed in the Golgi apparatus (Section 12-4B), and then transported, presumably in clathrin-coated vesicles, to areas of the plasma membrane containing lipid rafts (Section 12-4D). There, they aggregate in sufficient numbers to exclude host proteins (Fig. 33-57*a,b*). In the nucleus, the vRNAs combine with PA, PB1, and PB2 to form the nucleocapsids, which then interact with M1 protein. Nuclear export protein (NEP), as its name implies, mediates the export of nucleocapsids from the nucleus, which it does in partnership with both M1 and a cellular export factor. M1 then forms a nucleocapsid-enclosing shell that binds to HA and NA on the inside of the plasma membrane (Fig. 33-57*b*). This binding process causes the entire assembly to bud from the cell surface, thereby forming the mature virion (Fig. 33-57*c*). The complete infection cycle occupies ~8 to 12 h.

One of the mysteries of influenza virus assembly is how each virion acquires a complete set of the eight vRNAs. There is no evidence that the newly formed nucleocapsids are physically linked. On the contrary, in mixed infections with various influenza strains, the reassortment of their genomic segments occurs with high frequency. It has therefore been suggested that the nucleocapsids are randomly selected but that each virion contains sufficient numbers of vRNAs to ensure a reasonable probability that a given particle be infectious. This proposal is in agreement with the observation that aggregates of influenza virus have enhanced infectivity, a process that presumably occurs through the complementation of their vRNAs. Alternatively, the eight vRNAs may be selected by an ordered process, a hypothesis that is supported by the observation that mature viruses, but not infected cells, contain roughly equimolar amounts of the vRNAs.

B. *Mechanism of Antigenic Variation*

Influenza viruses are classified into three immunological types, A, B, and C, depending on the antigenic properties of their differing nucleoproteins and matrix proteins. The A virus has caused all of the major pandemics in humans and has therefore been more extensively investigated than the B and C viruses. The B and C viruses infect mainly humans. However, the A virus infects a wide variety of mammalian and avian species in addition to humans. Indeed, it is thought that migratory birds (and, more recently, jet planes) are the major vectors that transport influenza A viruses around the world. The species specificity of a particular viral strain presumably arises from the binding specificity of its HA for cell-surface glycolipids.

a. HA Residue Changes Are Responsible for Most of the Antigenic Variation in Influenza Viruses

HA, being the influenza virus' major surface protein, is largely responsible for stimulating the production of the antibodies that neutralize the virus. Consequently, the different influenza virus subtypes arise mainly through the variation of HA. Antigenic variation in NA, the virus' other major surface protein, also occurs but this has lesser immunological consequences.

Two distinct mechanisms of antigenic variation have been observed in influenza A viruses:

1. Antigenic shift, in which the gene encoding one HA species is replaced by an entirely new one. This change may or may not be accompanied by a replacement of NA. It is thought that these new viral strains arise from the reassortment of genes among animal and human flu viruses. *Antigenic shift is responsible for influenza pandemics because the human population's immunity against previously existing viral strains is ineffective against the newly generated strain.* Evidently, these viruses had retained the (largely unknown) genetic traits responsible for their virulence in humans.

2. Antigenic drift, which occurs through a succession of point mutations in the HA gene, resulting in an accumulation of amino acid residue changes that attenuate the host's immunity. This process occurs in response to the selective pressure brought about by the buildup in the human population of immunity to the extant viral strains. HA varies in this manner by an average of 3.5 accepted amino acid changes per year.

Influenza A viruses are classified into subtypes according to the similarities of their HA and NA. There are 15 known subtypes of HA (H1 through H15) and 9 of NA (N1 through N9) that occur in mammals and birds. Avian virus subtypes occur in nearly all combinations, whereas only a few combinations have been found in humans. For example, human influenza A viruses circulating before 1957 were designated H1N1, those of the 1957 pandemic were H2N2, those of the 1968 pandemic were H3N2, and those of the 1977 pandemic were again H1N1 (and hence affected mainly young people who not been exposed to pre-1918 viruses). Since 1977, H3N2 and H1N1 viruses have been cocirculating.

Humans are rarely infected by avian flu viruses and such viruses do not appear to be transmitted between humans. However, phylogenetic studies indicate that pigs (swine) and birds can exchange influenza viruses as can pigs and humans. This suggests that pigs serve as "mixing vessels" for the creation of new pandemic flu viruses, which in turn, explains why Southeast Asia, where humans, pigs, and birds (ducks and chickens) often live in close proximity, is where most flu pandemics appear to have originated.

In early 1976, at Fort Dix, New Jersey, there was an outbreak of an H1N1 influenza strain whose HA subtype occurs in swine flu virus. This viral strain is thought to have caused the great pandemic of 1918 (although influenza virus was not isolated until 1933, individuals who had contracted influenza during the 1918 pandemic have antibodies against swine flu virus in their serum; the sequencing, in 1999, by Ann Reid and Jeffery Taubenberger, of the HA gene obtained from preserved tissues of individuals who had died of the 1918 flu confirmed that its HA resembles that of swine flu). If this new strain had been virulent, no one under the age of 50 at the time would have been immune to it. There was, consequently, grave concern that a deadly influenza pandemic would ensue. This situation led to a crash program in which well over one million people deemed to be at high risk (such as pregnant women and the elderly) were vaccinated against swine flu. Fortunately, the 1976 swine flu was not virulent; it did not spread beyond Fort Dix.

b. HA Is an Elongated Trimeric Transmembrane Glycoprotein

HA plays a central role in both the viral infection process and in the immunological measures and countermeasures taken in the continuing biological contest between host and parasite. This has motivated considerable efforts to elucidate the structural basis of its properties. HA is a homotrimer of 550-residue subunits that is 19% carbohydrate by weight. The protein has three domains (Fig. 33-58):

1. A large hydrophilic, carbohydrate-bearing N-terminal domain that occupies the viral membrane's external surface and that contains its sialic acid–binding site.

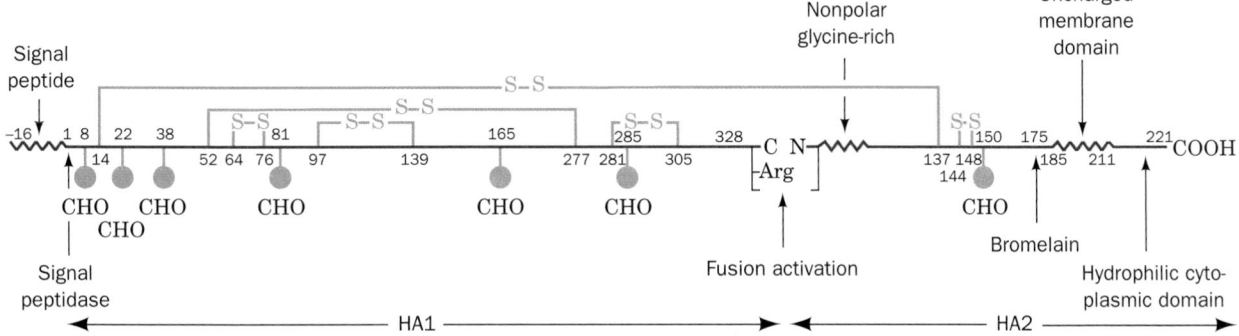

FIGURE 33-58 The primary structure of the 1968 Hong Kong influenza virus hemagglutinin. Its external domain (all of HA1 and HA2 through 185), its membrane anchoring domain (185–211 of HA2), and its cytoplasmic domain (212–221 of HA2) are indicated as are the positions of the signal peptide directing the protein's insertion into the membrane, the S—S bridges, the carbohydrate (CHO) attachment sites, the fusion activation site, and the bromelain cleavage site. [After Wilson, I.A., Skehel, J.J., and Wiley, D.C., *Nature* **289**, 367 (1981).]

2. A hydrophobic 24- to 28-residue membrane-spanning domain that is located near the polypeptide's C-terminus.

3. A hydrophilic C-terminal domain that occurs on the membrane's inner side and that consists of the protein's 10 C-terminal residues.

HA, which is synthesized as a single polypeptide designated HA0, is posttranslationally cleaved by host-secreted proteases by the excision of Arg 329, thereby yielding two chains, HA1 and HA2, that are linked by a disulfide bond. This cleavage, which does not affect HA's receptor-binding affinity, is required for the fusion of the virus with the host cell and therefore activates viral infectivity (see below). Indeed, the cleavability of HA is one of the major factors determining the virulence of influenza viruses.

HA can be removed from the virion by treatment with detergent but the resulting solubilized protein has not been made to crystallize. However, treatment of HA from a Hong Kong–type (H3) virus with the pineapple protease **bromelain,** which cleaves the polypeptide 9 residues before the membrane-spanning segment, yields a water-soluble protein named **BHA** that has been crystallized. X-Ray analysis of these crystals by John Skehel and Don Wiley revealed an unusual structure (Fig. 33-59). The monomer consists of a long fibrous stalk extending from the membrane surface on which is perched a globular region. The fibrous stalk consists of segments from HA1 and HA2 and includes a remarkable 76-Å-long (53 residues in 14 turns) α helix. The globular region, which is comprised of only HA1 residues, contains an 8-stranded antiparallel β-sheet

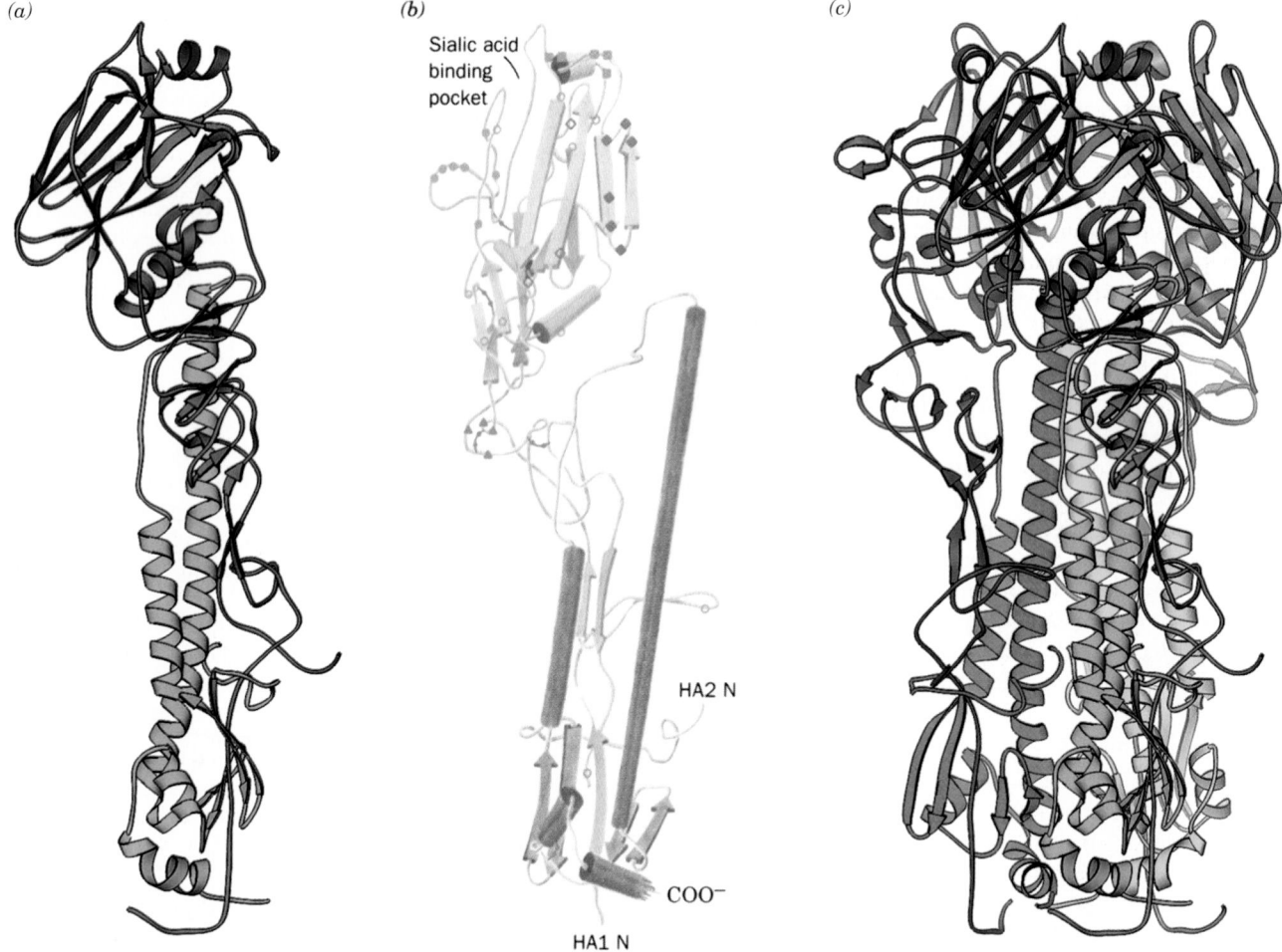

(a)   *(b)*   *(c)*

Sialic acid binding pocket

HA2 N

COO⁻

HA1 N

FIGURE 33-59   X-Ray structure of influenza hemagglutinin.
(*a*) The polypeptide backbone of the monomer drawn as a ribbon. HA1 is green and HA2 is cyan. (*b*) A cartoon diagram of the monomer from a somewhat different point of view than Part *a* but similarly colored. The pairs of linked, small, filled circles represent disulfide groups. The positions of the mutant residues at the four antigenic sites are indicated by filled circles, squares, triangles, and diamonds. Open symbols represent antigenically neutral residues. Note the position of the sialic acid–binding pocket. (*c*) A ribbon diagram of the HA trimer. Each HA1 and HA2 chain is drawn in a different color. The orientation of the green HA1 and cyan HA2 are the same as in Part *a*. [Based on an X-ray structure by John Skehel, National Institute for Medical Research, London, U.K., and Don Wiley, Harvard University. Parts *a* and *c* courtesy of Michael Carson, University of Alabama at Birmingham; Part *b* after a drawing by Hidde Ploegh, *in* Wilson, I.A., Skehel, J.J., and Wiley, D.C., *Nature* **289,** 366 (1981). PDBid 4HMG.]

structure (a distorted jelly roll barrel; Section 8-3B) that forms the sialic acid–binding pocket.

The dominant interaction stabilizing BHA's trimeric structure is a triple-stranded coiled coil consisting of the 76-Å α helices from each of its subunits (Fig. 33-59c). The BHA trimer is therefore an elongated molecule, some 135 Å in length, with a triangular cross-section that varies in radius from 15 to 40 Å. The carbohydrate chains, which are attached to the protein via *N*-glycosidic linkages at each of its subunit's seven Asn-X-Thr/Ser sequences (Section 11-3C), are located almost entirely along the trimer's lateral surfaces. The role of the carbohydrates is unclear despite the fact that they cover some 20% of the protein's surface. However, the observation that the mutational generation of a new oligosaccharide attachment site blocks antibody binding to HA suggests that carbohydrates modulate HA's antigenicity.

c. Antigenic Variation Results from Surface Residue Changes

HA's antigenic sites have been identified by mapping HA sequence changes on the protein's three-dimensional structure. The HA residues that mutated in an antigenically significant manner in Hong Kong–type viruses during the period 1968 to 1977 are indicated in Fig. 33-59b. *These residues all occur on the protein's surface, often in polypeptide loops, where their mutational variation affects the protein's surface character but apparently not its overall structure or stability. The variable residues are clustered in four sites surrounding HA's receptor-binding pocket, which is formed from amino acid residues that are largely conserved in numerous influenza virus strains.* The strains responsible for the major flu epidemics between 1968 and 1975 had at least one mutation in each of these four antigenic sites. This degree of antigenic variation appears necessary to reinfect individuals previously infected with the same viral type. Evidently, *antibodies directed against even conserved residues in HA's receptor-binding pocket, which would otherwise prevent HA from binding to its receptor, are dislodged by the antigenic variation that so readily occurs about the rim of this binding pocket (we study antibody–antigen interactions in Section 35-2B).*

d. NA Is a Tetrameric Transmembrane Glycoprotein

Influenza virus neuraminidase (NA) is a homotetrameric glycoprotein of 469-residue subunits. It has a box-shaped globular head attached to a slender stalk that is anchored in the viral membrane. Pronase digestion cleaves NA before residues 74 or 77, after the membrane attachment site, to yield an enzymatically active and crystallizable protein. The X-ray structure of this protein (Fig. 33-60), determined by Peter Colman and Graeme Laver, shows it to have 4-fold symmetry. Each subunit is composed of six topologically identical 4-stranded antiparallel β sheets arranged like the blades of a propeller. This so-called β propeller structurally resembles the 7-bladed β propellers of the clathrin heavy chain (Fig. 12-56b) and the heterotrimeric G protein G_β subunit (Fig. 19-18b).

Sugar residues are linked to NA at four of its five potential Asn-X-Ser/Thr *N*-glycosylation sites.

NA's sialic acid–binding site is located in a large pocket on the top of each monomer (star on the upper right subunit in Fig. 33-60). Its bound sialic acid residue interacts, through an extensive hydrogen bonding network, with 16 polar residues that are conserved in all known NA sequences (HA, in contrast, has only 2 polar residues in its sialic acid–binding site). Sequence changes in antigenic variants of NA occur in 7 chain segments that form a nearly continuous surface that encircles the catalytic site (squares on the lower right subunit in Fig. 33-60) in a manner similar to that of HA's receptor-binding site. Between 1968 and 1975, NA exhibited the same number of residue changes in its putative antigenic determinants as did HA. Antibodies against NA, nevertheless, do not neutralize infectivity. Rather, they restrict multiple cycles of viral replication and thus probably attenuate illness.

The realization that the structure of the NA catalytic site is strain-invariant together with the X-ray structure of NA in complex with sialic acid has led to the design of clinically effective inhibitors of NA (drug design is discussed in Section 15-4). For example, **zanamivir** (trade name, **Relenza**),

Zanamivir (Relenza)

***N*-Acetylneuraminic acid (sialic acid)**

a sialic acid mimic and a potent inhibitor of NA ($K_I = 0.1$ nM), is an effective antiviral agent, both in tissue culture and when delivered as an orally inhaled powder. Although the use of zanamivir does not prevent influenza infection, if it is administered within 2 days of the onset of flulike symptoms, it significantly diminishes the length and severity of these symptoms and reduces the incidence of secondary bacterial infections.

C. Mechanism of Membrane Fusion

An influenza virus infection begins with the binding of HA to its cell-surface receptor. Then, as we discussed in Section 12-4B, the bound virus is taken into the cell via receptor-

FIGURE 33-60 X-Ray structure of the influenza neuraminidase tetramer. The view is along the 4-fold axis of the homotetrameric protein, looking toward the viral membrane. In each monomer unit, each of the six topologically equivalent 4-stranded antiparallel β sheets are differently shaded. The positions of the disulfide bonds are indicated in the upper left subunit. In the lower left subunit, the four carbohydrate attachment sites are indicated by filled purple circles and the Asp residues that ligand Ca²⁺ ions are represented by red arrows. In the upper right monomer, the filled red circles and blue triangles, respectively, represent the conserved acidic and basic residues surrounding the enzyme's sialic acid–binding site, which is represented by a red star. In the lower right monomer, the positions of the mutated residues in NA's antigenic variants are flagged by filled brown squares. [After Varghese, J.N., Laver, W.G., and Colman, P.M., *Nature* **303,** 35 (1983). PDBid 1NN2.]

mediated endocytosis, a process that is accompanied by the fusion of the virus' enveloping membrane with that of the cell, thereby injecting the viral nucleocapsids into the cytosol. What is the mechanism of this membrane fusion?

Membrane fusion is mediated by HA but only after it has been exposed to the ~5.0 pH of the endosomal vesicle (Fig. 12-79). A variety of studies have implicated an ~25-residue conserved hydrophobic segment at the N-terminus of HA2, the so-called fusion peptide, with mediating mem-

brane fusion by inserting into the cellular membrane. Yet, BHA's X-ray structure indicates that the fusion peptide is buried in the protein's hydrophobic interior, ~100 Å from the receptor binding site at the "top" of protein, the region in closest proximity to the cellular membrane. Thus, HA must undergo an extensive conformational change before it can initiate membrane fusion.

At pH 5.0, BHA indeed undergoes a conformational change but one that causes it to aggregate in a manner un-

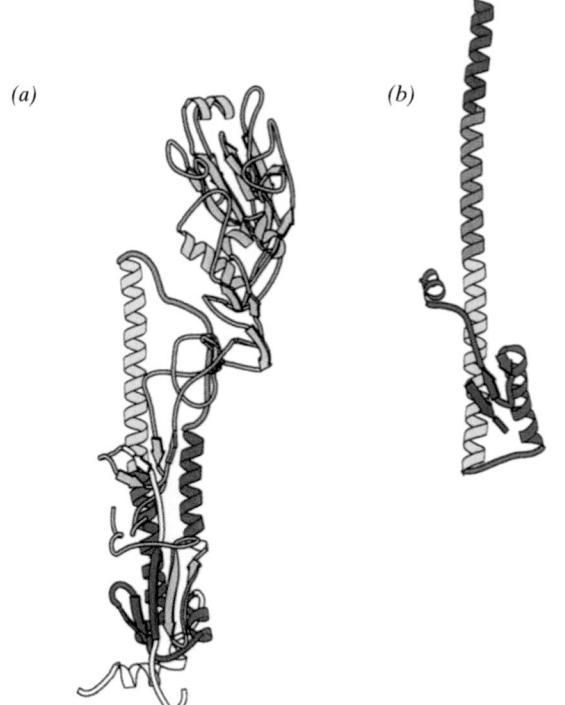

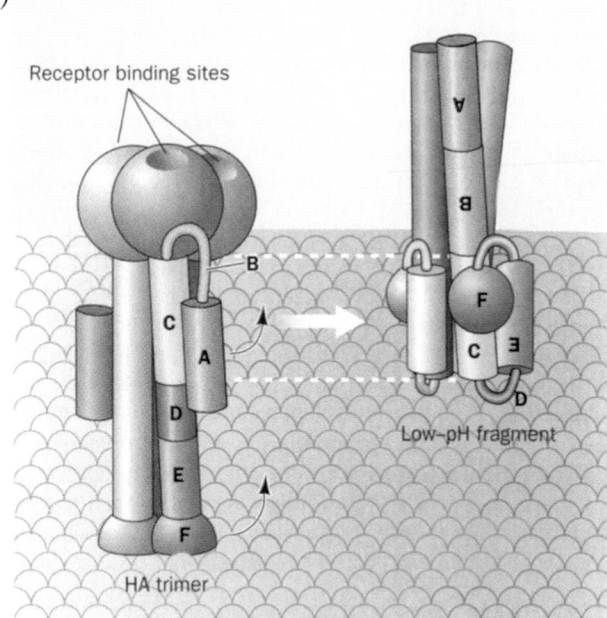

FIGURE 33-61 Comparison of the X-ray structures of BHA and TBHA₂. (*a*) Ribbon diagram of BHA in which the structural elements of the HA2 chain in TBHA₂ are colored in rainbow order (*red to violet*) from N- to C-terminus and the HA1 segment of TBHA₂, which is disulfide-linked to HA2, is blue. Regions of BHA that are proteolytically excised to form TBHA₂ are gray and those that are apparently disordered in TBHA₂ are white. (*b*) Ribbon diagram of TBHA₂ colored as in Part *a*. The heights of the various structural element relative to the yellow helix segment, which is common to both BHA and TBHA₂, are indicated. (*c*) Schematic diagram showing the positions and heights above the viral membrane surface of TBHA₂'s various structural elements in the HA trimer and in the low-pH fragment. The structural elements are colored as in Parts *a* and *b*. In the low-pH fragment, the fusion peptide (*not shown*) would protrude well above the receptor-binding heads where it would presumably insert itself into the cellular membrane. [Parts *a* and *b* courtesy of Don Wiley, Harvard University. PDBids 4HMG and 1HTM.

suitable for crystallographic studies. However, the successive proteolytic digestion of BHA at pH 5.0 by trypsin and thermolysin (Table 7-2) yields a crystallizable protein fragment named **TBHA₂,** which consists of residues 1 to 27 of HA1 and residues 38 to 175 of HA2 that are linked by a disulfide bond.

The X-ray structure of TBHA₂, determined by Skehel and Wiley, reveals that this protein has dramatically re-folded relative to BHA in a way that involves extensive changes in both its secondary and tertiary structural elements (Fig. 33-61). Thus, segments A and B at the N-terminus of TBHA₂ (Fig. 33-61*c;* the red and orange segments in Fig. 33-61) undergo a jackknife-like movement of ~100 Å in a way that extends the top of the long helix by ~10 helical turns toward the cell membrane [although the long helix is shortened from the bottom by a similar but not so extensive shift of segment D (green in Fig. 33-61) so as to partially replace the flipped out helix A]. This con-

formational change is irreversible (HA does not revert to its original form when the pH is raised) and hence has been described as occurring via a "spring-loaded" mechanism. The rearrangement of segments A and B had been predicted by Peter Kim who noted that segments A, B, and C have the heptad repeat characteristic of coiled coils (Section 8-2A). Such conformational shifts in intact HA translocate the fusion peptide (which would extend beyond the N-terminus of TBHA₂ at the top of segment A) by at least 100 Å from its position in BHA. There it could bridge the viral and cellular membranes so as to facilitate their fusion in a manner similar to that postulated for SNARE complexes (Section 12-4D). The conformational change is probably triggered, at least in part, by the protonation at pH 5 of the six Asp side chains in the 19-residue segment B, thereby reducing the charge–charge repulsions that apparently prevent this segment from forming a coiled coil at higher pH's.

CHAPTER SUMMARY

1 ■ Tobacco Mosaic Virus Viruses are complex molecular aggregates that exhibit many attributes of living systems. Their structural and genetic properties have therefore served as valuable paradigms for the analogous cellular functions. The tobacco mosaic virus (TMV) virion consists of a helix of identical and therefore largely quasi-equivalent coat protein subunits enclosing a coaxially wound single strand of RNA. X-Ray studies of TMV gels reveal that this RNA is bound, with three nucleotides per subunit, between the subunits of the protein helix. In the absence of TMV RNA, the subunits aggregate at high ionic strengths to form double-layered disks, and at low ionic strengths to form protohelices, which stack to form helical rods under acidic conditions. The virus' innermost polypeptide loop is disordered in both the disk and the protohelix. Virus assembly is initiated when a protohelix (or possibly a double-layered disk) binds to the initiation sequence of TMV RNA, which is located ~1000 nucleotides from the RNA's 3′ end. Interactions between the RNA and protohelix trigger the ordering of the disordered loop, thereby converting the protohelix to the helical form. Elongation of the virus particle then proceeds by the sequential addition of protohelices (disks) to the "top" of the assembly so as to pull the 5′ end of the RNA up through the center of the growing viral helix.

2 ■ Icosahedral Viruses Viral capsids are formed from one or a few types of coat protein subunits. These must be either helically arranged, as in TMV, or quasi-equivalently arranged in a polyhedral shell so as to enclose the viral nucleic acid. The coat proteins of many icosahedral viruses are arranged in icosadeltahedra consisting of $60T$ subunits, where T is the triangulation number. The coat protein of tomato bushy stunt virus (TBSV) is arranged in a $T = 3$ icosadeltahedron so that TBSV subunits occupy three symmetrically distinct positions. The subunits must therefore associate through several sets of nonidentical intersubunit contacts. Some of the R domains form a structurally disordered inner protein shell. The viral RNA together with the remaining R domains are tightly packed in the space between the inner and outer protein shells. Other spherical plant viruses, southern bean mosaic virus (SBMV) and satellite tobacco mosaic virus (STMV), have tertiary and quaternary structures that are clearly related to those of TBSV. STMV's single-stranded RNA appears to form a series of stem and loop structures with the viral capsid. The structurally similar VP1, VP2, and VP3 coat proteins of poliovirus, rhinovirus, and foot-and-mouth disease virus (FMDV) are likewise icosahedrally arranged. However, the simian virus 40 (SV40) capsid consists of 72 pentagons of identical subunits in two different environments that are linked together by differering arrangements of their C-terminal arms in a nonicosadeltahedral arrangement. Although the coat proteins of most spherical viruses consist mainly of structurally similar 8-stranded antiparallel β barrels, that of bacteriophage MS2, a $T = 3$ virion, has an unrelated fold. Among the large icosahedral viruses of known structure are the bluetongue virus (BTV) core, which has a $T = 13$ outer shell and a $T = 2$ inner shell that envelops its largely visible dsRNA genome; and Paramecium bursaria Chlorella virus type 1 (PBCV-1), which has a $T = 169$ outer shell consisting of 1680 trimers of Vp54.

3 ■ Bacteriophage λ Lytic growth of bacteriophage λ in *E. coli* is controlled by the sequential syntheses of antiterminators, which inhibit both rho-independent and rho-dependent transcriptional terminators. Thus gp*N*, which is synthesized in the early stage of growth, permits the synthesis of gp*Q* in the delayed early stage which, in turn, permits the synthesis of the capsid proteins in the late stage. Early gene transcription is repressed in the delayed early stage by Cro protein. DNA replication, which commences in the early stage, is mediated by the host DNA replication machinery with the aid of the phage proteins gp*O* and gp*P*. DNA synthesis initially occurs by both the θ and rolling circle (σ) modes but eventually switches entirely to the rolling circle mode.

The λ virion heads and tails are separately assembled. Head assembly is a complex process involving the participation of many phage gene products, not all of which are part of the mature virion. Phage heads are not self-assembling in that their formation is guided by host chaperonins and a viral scaffolding protein and requires several enzymatically catalyzed protein modification reactions. The mature phage head is a $T = 7$ icosadeltahedron of gp*E*, which is decorated by an equal number of gp*D* subunits. Just before the final stage of its assembly, the phage head is filled with a linear double strand of DNA in a process that is driven by ATP hydrolysis. The packaged DNA appears to be wound in a spool that winds from outside to inside. Tail assembly occurs in a stepwise process from the tail fiber to the head-binding end. The body of the tail consists of a stack of hexameric rings of gp*V*. The completed heads and tails spontaneously join to form the mature virion.

Lysogeny is established by site-specific recombination between the phage *att*P site and the bacterial *att*B sites in a process mediated by phage integrase (gp*int*) and integration host factor (IHF). Induction, in which this process is reversed, requires the additional action of phage excisionase (gp*xis*). Lysogeny is established by a high level of gp*cII*, which stimulates the transcription of *int* and the λ repressor gene, *cI*. Repressor, as does Cro, binds to the o_L and o_R operators to shut down early gene transcription, including that of *cro* and *cII*. Each of these dimeric proteins, like other repressors of known structure, contains two symmetrically related helix–turn–helix (HTH) units that bind in successive turns of B-DNA's major groove. However, repressor, but not Cro, induces its own synthesis from the promoter p_{RM} by binding to o_{R2} so as to interact with RNA polymerase. The induction of repressor synthesis therefore throws the genetic switch that stably maintains the phage in the lysogenic state from generation to generation. Damage to host DNA, nevertheless, stimulates host RecA protein to mediate λ repressor cleavage so as to release repressor from o_L and o_R. This initiates the synthesis of early gene products, including gp*int* and gp*xis*, from p_L and p_R and thus triggers induction. If sufficient Cro protein is then synthesized to repress the synthesis of repressor, the phage becomes irrevocably committed to at least one generation of lytic growth. The tripartite character of o_R, the site of the λ switch, together with the cooperative nature of repressor binding to o_R, confers the λ switch with a remarkable sensitivity to the health of its host.

4 ■ Influenza Virus The influenza virion's enveloping membrane is studded with protein spikes consisting of hemagglutinin (HA), which mediates host recognition, and neuraminidase (NA), which facilitates the passage of the virus to and from the infection site. Inside the membrane is a shell of matrix protein that contains the virus' genome of eight single-stranded RNAs, each in a separate protein complex known as a nucleocapsid. These vRNAs are templates for the transcription of mRNAs as catalyzed by the nucleocapsid proteins. This process is primed by host-derived 7-methyl-G-capped mRNA fragments. The viral mRNAs, which have poly(A) tails, lack the sequences complementary to the vRNA's 5′ ends. The vRNAs, however, also act as templates for the transcription of the corresponding cRNAs which, in turn, are the templates for vRNA synthesis. The virus is assembled in and near the plasma membrane and forms by budding from the cell surface.

Influenza viruses infect a variety of mammals besides humans as well as many birds. Variation in the antigenic character of HA has been mainly responsible for the different influenza subtypes. Antigenic variation in HA occurs by either antigenic shift, in which the HA gene from an animal virus replaces that from a human virus, or antigenic drift, which occurs by a succession of point mutations in the HA gene. NA may vary in a similar fashion. HA is an elongated trimeric glycoprotein. Its surface has four antigenic sites that surround its sialic acid–binding pocket and that, in the viruses which caused the major epidemics between 1968 and 1975, all exhibit at least one mutational change. NA is a mushroom-shaped tetrameric glycoprotein. Its antigenic variations occur on a surface that also encircles its active site. HA mediates the fusion of the viral and host endosome membranes through a dramatic conformational change that translocates its fusion peptides to the vicinity of the endosome membrane into which they then insert.

REFERENCES

GENERAL

Cann, A.J., *Principles of Modern Virology,* Academic Press (1993).

Chiu, W., Burnett, R.M., and Garcea, R.L. (Eds.), *Structural Biology of Viruses,* Oxford University Press (1997).

Dimmock, N.J., Easton, A.J., and Leppard, K.N., *Introduction to Modern Virology* (5th ed.), Blackwell Science (2001).

Levine, A.J., *Viruses,* Scientific American Library (1992).

Radetsky, P., *The Invisible Invaders. The Story of the Emerging Age of Viruses,* Little, Brown and Co. (1991).

Voyles, B.A., *The Biology of Viruses,* Mosby (1993).

TOBACCO MOSAIC VIRUS

Bloomer, A.C., Champness, J.N., Bricogne, G., Staden, R., and Klug, A., Protein disk of tobacco mosaic virus at 2.8 Å showing the interactions within and between subunits, *Nature* **276,** 362–368 (1978).

Butler, P.J.G., Self-assembly of tobacco mosaic virus: The role of an intermediate aggregate in generating both specificity and speed, *Phil. Trans. R. Soc. Lond.* **B354,** 537–550 (1999).

Butler, P.J.G., Bloomer, A.C., and Finch, J.T., Direct visualization of the structure of the "20 S" aggregate of coat protein of tobacco mosaic virus, *J. Mol. Biol.* **224,** 381–394 (1992). [Evidence indicating that the TMV coat protein double-layered disk predominates over the protohelix at pH 7.0.]

Butler, P.J. and Klug, A., The assembly of a virus, *Sci. Am.* **239**(5): 62–69 (1978).

Klug, A., The tobacco mosaic virus particle: Structure and assembly, *Phil. Trans. R. Soc. Lond.* **B354,** 531–535 (1999).

Lomonosoff, G.P. and Wilson, T.M.A., Structure and in vitro assembly of tobacco mosaic virus, *in* Davis, J.W. (Ed.), *Molecular Plant Virology,* Vol. I, pp. 43–83, CRC Press (1985).

Namba, K., Pattanayek, R., and Stubbs, G., Visualization of protein–nucleic acid interactions in a virus. Refined structure of intact tobacco mosic virus at 2.9 Å by X-ray fiber diffraction, *J. Mol. Biol.* **208,** 307–325 (1989).

Raghavendra, K., Kelly, J.A., Khairallah, L., and Schuster, T.M., Structure and function of disk aggregates of the coat protein of tobacco mosaic virus, *Biochemistry* **27,** 7583–7588 (1988). [Evidence indicating that the TMV coat protein disks do not convert to the protohelices that nucleate TMV assembly.]

Stubbs, G., Molecular structures of viruses from the tobacco mosaic group, *Sem. Virol.* **1,** 405–412 (1990).

Stubbs, G., Tobacco mosaic virus particle structure and the initiation of disassembly, *Phil. Trans. R. Soc. Lond.* **B354,** 551–557 (1999).

ICOSAHEDRAL VIRUSES

Abad-Zapetero, C., Abdel-Meguid, S.S., Johnson, J.E., Leslie, A.G.W., Rayment, I., Rossmann, M.G., Suck, D., and Tsukihara, T., Structure of southern bean mosaic virus at 2.8 Å resolution, *Nature* **286,** 33–39 (1980).

Acharya, R., Fry, E., Stuart, D., Fox, G., and Brown, F., The three-dimensional structure of foot-and-mouth disease virus at 2.9 Å resolution, *Nature* **337,** 709–716 (1989).

Arnold, E. and Rossmann, M.G., Analysis of the structure of a common cold virus, human rhinovirus 14, refined at a resolution of 3.0 Å, *J. Mol. Biol.* **211,** 763–801 (1990); *and* Rossmann, M.G., et al., Structure of a human common cold virus and relationship to other picornaviruses, *Nature* **317,** 145–153 (1985).

Caspar, D.L.D. and Klug, A., Physical principles in the construction of regular viruses, *Cold Spring Harbor Symp. Quant. Biol.* **27,** 1–24 (1962). [The classic paper formulating the geometric principles governing the construction of icosahedral viruses.]

Dokland, T., Freedom and restraint: Themes in virus capsid assembly, *Structure* **8,** R157–R162 (2000).

Grimes, J.M., Burroughs, J.N., Gouet, P., Diprose, J.M., Malby, R., Ziéntara, S., Mertens, P.P.C., and Stuart, D.I., The atomic structure of the bluetongue virus core, *Nature* **395,** 470–478 (1998); Gouet, P., Diprose, J.M., Grimes, J.M., Malby, R., Burroughs, J.N., Ziéntara, S., Stuart, D.I., and Mertens, P.P.C., The highly ordered double-stranded RNA genome of bluetongue virus revealed by crystallography, *Cell* **97,** 481–490 (1999); *and* Diprose, J.M., et al., Translocation portals for the substrates and products of a viral transcription complex: the bluetongue virus core, *EMBO J.* **20,** 7229–7239 (2001).

Harrison, S.C., Common features in the structures of some icosahedral viruses: a partly historical view, *Sem. Virol.* **1,** 387–403 (1990).

Harrison, S.C., The familiar and unexpected in structures of icosahedral viruses, *Curr. Opin. Struct. Biol.* **11,** 195–199 (2001).

Harrison, S.C., Olson, A.J., Schutt, C.E., Winkler, F.K., and

Bricogne, G., Tomato bushy stunt virus at 2.9 Å resolution, *Nature* **276**, 368–373 (1978). [The first report of a high-resolution virus structure.]

Hogle, J.M., Chow, M., and Filman, D.J., The structure of poliovirus, *Sci. Am.* **256**(3): 42–49 (1987); *and* Three-dimensional structure of poliovirus at 2.9 Å resolution, *Science* **229**, 1358–1365 (1985).

Hurst, C.J., Benton, W.H., and Enneking, J.M., Three dimensional model of human rhinovirus type 14, *Trends Biochem. Sci.* **12**, 460 (1987). [A "paper doll"-type cutout with accompanying assembly directions for constructing an icosahedral model of human rhinovirus. This useful learning device may also be taken as a *T* = 3 icosadeltahedron.]

Larson, S.B. and McPherson, A., Satellite tobacco mosaic virus RNA: structure and implications for assembly, *Curr. Opin. Struct. Biol.* **11**, 59–65 (2001); *and* Larson, S.B., Day, J., Greenwood, A., and McPherson, A., Refined structure of satellite tobacco mosaic virus at 1.8 Å resolution, *J. Mol. Biol.* **277**, 37–59 (1998).

Munshi, S., Liljas, L., Cavarelli, J., Bomu, W., McKinney, B., Reddy, V, and Johnson, J.E., The 2.8 Å structure of a *T* = 4 animal virus and its implications for membrane translocation of RNA, *J. Mol. Biol.* **261**, 1–10 (1996). [The X-ray structure of Nudaurelia ω Capensis virus.]

Nandhagopal, N., Simpson, A.A., Gurnon, J.R., Yan, X., Baker, T.S., Graves, M.V., Van Etten, J.L., and Rossmann, M.G., The structure and evolution of the major capsid protein of a large, lipid-containing DNA virus, *Proc. Natl. Acad. Sci.* **99**, 14758–14763 (2002); *and* Yan, X., Olson, N.H., Van Etten, J.L., Bergoin, M., Rossmann, M.G., and Baker, T.S., Structure and assembly of large lipid-containing dsDNA viruses, *Nature Struct. Biol.* **7**, 101–103 (2000). [The structure of PBCV-1.]

Rossmann, M.G. and Johnson, J.E., Icosahedral RNA virus structure, *Annu. Rev. Biochem.* **58**, 533–573 (1989).

Stehle, T., Gamblin, S.J., Yan, Y., and Harrison, S.C., The structure of simian virus 40 refined at 3.1 Å, *Structure* **4**, 165–182 (1996); *and* Liddington, R.C., Yan, Y., Moulai, J., Sahli, R., Benjamin, T.L., and Harrison, S.C., Structure of simian virus 40 at 3.8-Å resolution, *Nature* **354**, 278–284 (1991).

Valegård, K., Liljas, L., Fridborg, K., and Unge, T., The three-dimensional structure of the bacterial virus MS2, *Nature* **345**, 36–41 (1990); *and* Golmohammadi, R., Valegård, K., Fridborg, K., and Liljas, L., The refined structure of bacteriophage MS2 at 2.8 Å resolution, *J. Mol. Biol.* **234**, 620–639 (1993).

BACTERIOPHAGE λ

Albright, R.A. and Matthews, B.W., Crystal structure of λ-Cro bound to a consensus operator at 3.0 Å resolution, *J. Mol. Biol.* **280**, 137–151 (1998); *and* Brennan, R.G., Roderick, S.L., Takeda, Y., and Matthews, B.W., Protein–DNA conformational changes in the crystal structure of λ Cro–operator complex, *Proc. Natl. Acad. Sci.* **87**, 8165–8169 (1990).

Azaro, M.A. and Landy, A., λ Integrase and λ Int family, *in* Craig, N.L., Craigie, R., Gellert, M. and Lambowitz, A.M. (Eds.), *Mobile DNA II*, 118–148, ASM Press (2002).

Beamer, L.J. and Pabo, C.O., Refined 1.8 Å crystal structure of the λ repressor–operator complex, *J. Mol. Biol.* **227**, 177–196 (1992); *and* Jordan, S.R. and Pabo, C.O., Structure of the lambda complex at 2.5 Å resolution: details of the repressor–operator interactions, *Science* **242**, 893–899 (1988).

Bell, C.E., Frescura, P., Hochschild, A., and Lewis, M., Crystal structure of the λ repressor C-terminal domain provides a model for cooperative operator binding, *Cell* **101**, 801–811 (2000).

Brüssow, H. and Hendrix, R.W., Phage genomics: Small is beautiful, *Cell* **108**, 13–16 (2002).

Cai, Z., Gorin, A., Frederick, R., Ye, X., Hu, W., Majumdar, A., Kettani, A., and Patel, D.J., Solution structure of P22 transcriptional antitermination N peptide–box B RNA complex, *Nature Struct. Biol.* **5**, 203–212 (1998); *and* Legault, P., Li, J., Mogridge, J., Kay, L.E., and Greenblatt, J., NMR structure of the bacteriophage λ N peptide/*boxB* RNA complex: Recognition of a GNRA fold by an arginine-rich motif, *Cell* **93**, 289–299 (1998).

Cerritelli, M.E., Cheng, N., Rosenberg, A.H., McPherson, C.E., Booy, F.P., and Steven, A.C., Encapsidated conformation of bacteriophage T7 DNA, *Cell* **91**, 271–280 (1997).

Echols, H., Bacteriophage λ development: temporal switches and the choice of lysis or lysogeny, *Trends Genet.* **2**, 26–30 (1986).

Greenblatt, J., Nodwell, J.R., and Mason, S.W., Transcriptional antitermination, *Nature* **364**, 401–406 (1993).

Hendrix, R.W., and Garcea, R.L., Capsid assembly of dsDNA viruses, *Sem. Virol.* **5**, 15–26 (1994).

Hendrix, R.W., Roberts, J.W., Stahl, F.W., and Weisberg, R.A. (Eds.), *Lambda II*, Cold Spring Harbor Laboratory (1982). [A compendium of review articles on many aspects of bacteriophage λ.]

Murialdo, H., Bacteriophage lambda DNA maturation and packaging, *Annu. Rev. Biochem.* **60**, 125–153 (1991).

Oppenheim, A.B., Kornitzer, D., and Altuvia, S., Posttranscriptional control of the lysogenic pathway in bacteriophage lambda, *Prog. Nucleic Acid Res. Mol. Biol.* **46**, 37–49 (1993).

Pabo, C.O. and Lewis, M., The operator-binding domain of λ repressor: structure and DNA recognition, *Nature* **298**, 443–447 (1982).

Ptashne, M., *A Genetic Switch* (2nd ed.), Cell Press & Blackwell Scientific Publications (1992). [An authoritative review of the λ switch.]

Rice, P.A., Yang, S., Mizuuchi, K., and Nash, H., Crystal structure of an IHF-DNA complex: A protein-induced DNA turn, *Cell* **87**, 1295–1306 (1996); *and* Rice, P.A., Making DNA do a U-turn: IHF and related proteins, *Curr. Opin. Struct. Biol.* **7**, 86–93 (1997).

Roberts, J.W., RNA and protein elements of *E. coli* and λ transcription antitermination complexes, *Cell* **72**, 653–655 (1993).

Simpson, A.A., et al., Structure of the bacteriophage φ29 DNA packaging motor, *Nature* **409**, 745–750 (2000).

Taylor, K. and Wegrzyn, G., Replication of coliphage lambda DNA, *FEMS Microbiol. Rev.* **17**, 109–119 (1995).

INFLUENZA VIRUS

Air, G.M. and Laver, W.G., The molecular basis of antigenic variation in influenza virus, *Adv. Virus Res.* **31**, 53–102 (1986).

Bullough, P.A., Hughson, F.M., Skehel, J.J., and Wiley, D.C., Structure of influenza haemagglutinin at the pH of membrane fusion, *Nature* **371**, 37–43 (1994).

Carr, C.M., and Kim, P.S., A spring-loaded mechanism for the conformational change of influenza hemagglutinin, *Cell* **73**, 823–832 (1994).

Colman, P., Influenza virus neuraminidase: Structure, antibodies, and inhibitors, *Protein Sci.* **3**, 1687–1696 (1994).

Colman, P.M., Neuraminidase inhibitors as antivirals, *Vaccine* **20**, S55–S58 (2002).

Cox, N.J. and Subbaro, K., Global epidemiology of influenza: Past and present, *Annu. Rev. Med.* **51**, 407–421 (2000).

Eckert, D.M. and Kim, P.S., Mechanisms of viral membrane fusion and its inhibition, *Annu. Rev. Biochem.* **70**, 777–810 (2001).

Kolata, G.B., *Flu: The Story of the Great Influenza Pandemic of 1918 and the Search for the Virus that Caused It,* Farrar, Straus and Giroux (1999).

Nicholson, K.G., Webster, R.G., and Hay, A.J., *Textbook of Influenza,* Blackwell Science (1998).

Potter, C.W. (Ed.), *Influenza,* Elsevier (2002).

Skehel, J.J. and Wiley, D.C., Receptor binding and membrane fusion in virus entry: The influenza hemagglutinin, *Annu. Rev. Biochem.* **69,** 531–569 (2000).

Skehel, J.J., Stevens, D.J., Daniels, R.S., Douglas, A.R., Knossow, M., Wilson, I.A., and Wiley, D.C., A carbohydrate side chain on hemagglutinins of Hong Kong influenza viruses inhibits recognition by a monoclonal antibody, *Proc. Natl. Acad Sci.* **81,** 1779–1783 (1984).

Varghese, J.N., Laver, W.G., and Colman, P.M., Structure of the influenza virus glycoprotein antigen neuramimidase at 2.9 Å resolution, *Nature* **303,** 35–40 (1983); *and* Colman, P.M., Varghese, J.N., and Laver, W.G., Structure of the catalytic and antigenic sites in influenza virus neuraminidase, *Nature* **303,** 41–44 (1983).

Varghese, J.N., McKimm-Breschkin, J.L., Caldwell, J.B., Kortt, A.A., and Colman, P.M., The structure of the complex between influenza virus neuraminidase and sialic acid, the viral receptor, *Proteins* **14,** 327–332 (1992).

Webster, R.G., Laver, W.G., Air, G.M., and Schild, G.C., Molecular mechanisms of variation in influenza viruses, *Nature* **296,** 115–121 (1982).

Weis, W., Brown, J.H., Cusack, S., Paulson, J.C., Skehel, J.J., and Wiley, D.C., Structure of the influenza virus haemagglutinin complexed with its receptor, sialic acid, *Nature* **333,** 426–431 (1988).

Wilson, I.A., Skehel, J.J., and Wiley, D.C., Structure of the haemagglutinin membrane glycoprotein of influenza virus at 3 Å resolution, *Nature* **289,** 366–373 (1981); *and* Wiley, D.C., Wilson, I.A., and Skehel, J.J., Structural identification of the antibody-binding sites of Hong Kong influenza haemaglutinin and involvement in antigenic variation, *Nature* **289,** 373–378 (1981).

PROBLEMS

1. Why does a pH shift from 7 to 5 at low ionic strengths cause TMV double-layered disks to aggregate as helical rods?

2. Can a nucleic acid encode a monomeric protein large enough to enclose it? Explain.

*3. Explain why the number of vertices in an icosadeltahedron always ends in the numeral "2" (e.g., 12 for $T = 1$).

4. Sketch a $T = 9$ icosadeltahedron.

5. The coat protein pentagons of SV40 are arranged at the vertices of a $T = 7$ icosadeltahedron. Yet the SV40 virion cannot have icosadeltahedral symmetry. Explain.

6. Why is it necessary to use conditionally lethal mutations in studying phage assembly rather than just lethal mutations?

7. Compare the volume contained by a λ phage head to that of λ DNA.

8. Virulent phages form clear plaques on a bacterial lawn, whereas bacteriophage λ forms turbid (cloudy) plaques. Explain.

9. What is the mutual consensus sequence of the o_L and o_R half-subsites?

10. λ repressor binds cooperatively to o_{R1} and o_{R2} but independently to o_{R3}. However, if o_{R1} is mutationally altered so that it does not bind repressor, then repressor binds cooperatively to o_{R2} and o_{R3}. Explain.

11. Bacteriophage 434 is a lambdoid phage that has both a repressor and a Cro protein. You have constructed a hybrid repressor that consists of the 434 repressor with its α3 helix replaced by that from 434 Cro protein. Compare the pattern of contacts this hybrid protein makes with its operator, as indicated by chemical protection experiments, with those of the native 434 repressor and Cro proteins.

12. What is the probability that an influenza virion will have its proper complement of eight different RNAs if it has room for only eight nucleocapsids and binds them at random?

Chapter 34

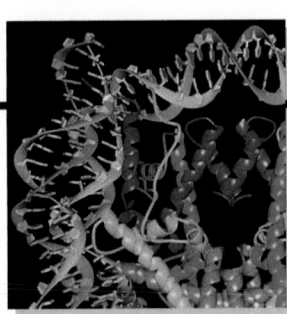

Eukaryotic Gene Expression

How does a fertilized ovum give rise to a highly differentiated multicellular organism? This question, of course, is just a sophisticated version of one that every child has asked: Where did I come from? Biologists began rational attempts to answer this question in the late nineteenth century and since that time have assembled an impressive body of knowledge concerning the general patterns of cellular differentiation and organismal development. Yet we have had the technical ability to study embryogenesis on the molecular level only in the last 30 years or so.

In order to understand cellular differentiation we must first understand the workings of the eukaryotic cell.

Eukaryotic cells are, for the most part, much larger and far more complex than prokaryotic cells (Section 1-2). However, *the basic difference between these two types of cells is that eukaryotes have a nuclear membrane that separates their chromosomes from their cytoplasm, thereby physically divorcing the eukaryotic transcriptional process from that of translation.* In contrast, the prokaryotic chromosome is embedded in the cytosol so that the initiation of protein synthesis often occurs on mRNAs that are still being transcribed. The transcriptional and translational control processes in eukaryotes are consequently fundamentally different from those of prokaryotes. This situation is reflected in both the packaging and the genetic organization of eukaryotic DNA in comparison with that of prokaryotes. We therefore begin this chapter with a physical description of the eukaryotic chromosome. We then consider how the eukaryotic genome is organized and how it is expressed. Finally, we discuss cell differentiation, its aberration, cancer, how the cell cycle is controlled, and programmed cell death. In all these subjects, as we shall see, our knowledge is quite fragmentary. Eukaryotic molecular biology is under such intense scrutiny, however, that significant advances in its understanding are made almost daily. Thus, perhaps more so than for other subject matter considered in this text, it is important that the reader supplement the material in this chapter with that in the recent biochemical literature.

1 ■ CHROMOSOME STRUCTURE

Eukaryotic chromosomes, which consist of a complex of DNA, RNA, and protein called **chromatin,** are dynamic entities whose appearance varies dramatically with the stage of the cell cycle. The individual chromosomes assume their familiar condensed forms (Figs. 1-18 and 34-1) only during cell division (M phase of the cell cycle; Section 30-4A). During interphase, the remainder of the cell cycle, when the chromosomal DNA is transcribed and replicated, the chromosomes of most cells become so highly dispersed that they cannot be individually distinguished (Fig. 34-2). Cytologists have long recognized that there are two types of this dispersed chromatin: a less densely packed variety named **euchromatin** and a more densely packed variety termed **hetero-**

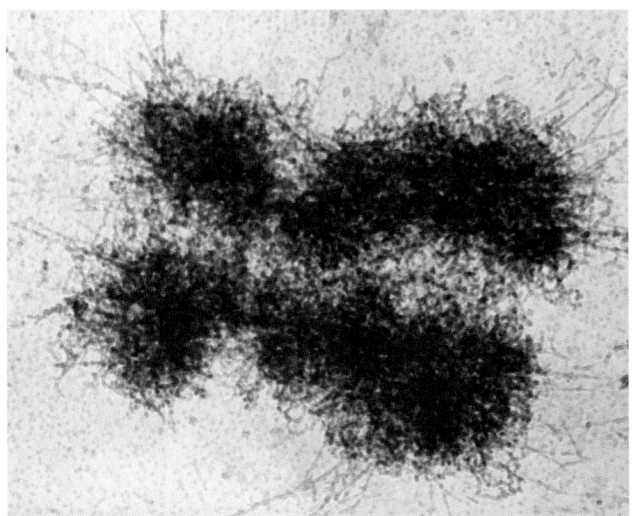

FIGURE 34-1 Electron micrograph of a human metaphase chromosome. It consists of two sister (identical) chromatids joined at their centromeres (the constricted portion near the left end of the chromosome). [Courtesy of Gunther Bahr, Armed Forces Institute of Pathology.]

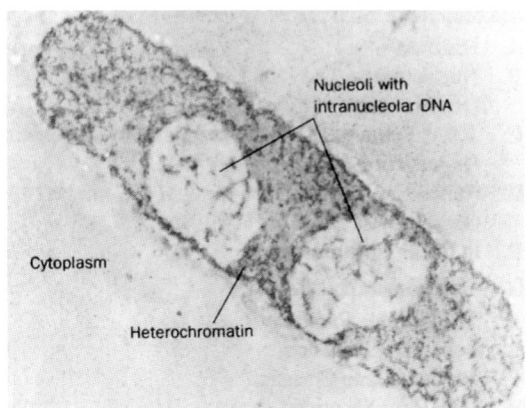

FIGURE 34-2 Thin section through a cell nucleus treated with Feulgen reagent (which reacts with DNA to form an intense red stain). Heterochromatin appears as dark-staining regions near the nucleolus and the nuclear membrane. The less darkly staining material is euchromatin. [Courtesy Edmund Puvion, CNRS, France.]

chromatin (Fig. 34-2). These two types of chromatin differ, as we shall see, in that euchromatin is genetically expressed, whereas heterochromatin is not expressed.

The 46 chromosomes in a human cell each contain between 44 and 246 million bp, so their DNAs, which are continuous (Section 5-3D), have contour lengths between 1.5 and 8.4 cm (3.4 Å/bp). Yet in metaphase, their most condensed state (Fig. 34-1), these chromosomes range in length from 1.3 to 10 μm. *Chromosomal DNA therefore has a **packing ratio** (ratio of its contour length to the length of its container) of >8000.* How does the DNA in chromatin attain such a high degree of condensation? Structural studies have revealed that this results from three levels of folding. We discuss these levels below, starting with the lowest level. We begin, however, by studying the proteins responsible for much of this folding.

A. Histones

*The protein components of chromatin, which comprise somewhat more than half its mass, consist mostly of **histones,*** which were discovered in 1884 by Albrecht Kossel and for many years were believed to be the genetic material itself. There are five major classes of these proteins, **histones H1, H2A, H2B, H3,** and **H4,** all of which have a large proportion of positively charged residues (Arg and Lys; Table 34-1). These proteins therefore ionically bind DNA's negatively charged phosphate groups. Indeed, histones may be extracted from chromatin by 0.5*M* NaCl, a salt solution of sufficient concentration to interfere with these electrostatic interactions.

a. Histones Are Evolutionarily Conserved

The amino acid sequences of histones H2A, H2B, H3, and H4 have remarkably high evolutionary stability (Table 34-1). For example, histones H4 from cows and peas, species that diverged 1.2 billion years ago, differ by only two conservative residue changes (Fig. 34-3), which makes histone H4, the most invariant histone, among the most evolutionarily conserved proteins known (Section 7-3B). *Such rigid evolutionary stability implies that the above four histones have critical functions to which their structures are so well tuned that they are all but intolerant to change.* The fifth histone, histone H1, is more variable than the other histones; we shall see below that its role differs from that of the other histones.

TABLE 34-1 Calf Thymus Histones

Histone	Number of Residues	Mass (kD)	% Arg	% Lys	UEP[a] (× 10⁻⁶ year)
H1	215	23.0	1	29	8
H2A	129	14.0	9	11	60
H2B	125	13.8	6	16	60
H3	135	15.3	13	10	330
H4	102	11.3	14	11	600

[a]*U*nit *e*volutionary *p*eriod: The time for a protein's amino acid sequence to change by 1% after two species have diverged (Section 7-3B).

```
Ac—Ser—Gly—Arg—Gly—Lys—Gly—Gly—Lys—Gly—Leu—10
     Gly—Lys—Gly—Gly—Ala—Lys—Arg—His—Arg—Lys—20
     Val—Leu—Arg—Asp—Asn—Ile—Gln—Gly—Ile—Thr—30
     Lys—Pro—Ala—Ile—Arg—Arg—Leu—Ala—Arg—Arg—40
     Gly—Gly—Val—Lys—Arg—Ile—Ser—Gly—Leu—Ile—50
     Tyr—Glu—Glu—Thr—Arg—Gly—Val—Leu—Lys—Val—60
     Phe—Leu—Glu—Asn—Val—Ile—Arg—Asp—Ala—Val—70
     Thr—Tyr—Thr—Glu—His—Ala—Lys—Arg—Lys—Thr—80
     Val—Thr—Ala—Met—Asp—Val—Val—Tyr—Ala—Leu—90
     Lys—Arg—Gln—Gly—Arg—Thr—Leu—Tyr—Gly—Phe—100
     Gly—Gly                                      102
```

FIGURE 34-3 The amino acid sequence of calf thymus histone H4. This 102-residue protein's 25 Arg and Lys residues are indicated in red. Pea seedling H4 differs from that of calf thymus by conservative changes at the two shaded residues: Val 60 → Ile and Lys 77 → Arg. The underlined residues are subject to posttranslational modification: Ser 1 is invariably *N*-acetylated and may also be *O*-phosphorylated; Lys residues 5, 8, 12, and 16 may be *N*-acetylated; and Lys 20 may be mono- or di-*N*-methylated. [After DeLange, R.J., Fambrough, D.M., Smith, E.L., and Bonner, J., *J. Biol. Chem.* **244,** 5678 (1969).]

b. Histones May Be Modified

Histones are subject to posttranslational modifications that include methylations, acetylations, and phosphorylations of specific Arg, His, Lys, and Ser side chains. These modifications, many of which are reversible, all decrease the histones' positive charges, thereby significantly altering histone–DNA interactions. Yet, despite the histones' great evolutionary stability, their degree of modification varies enormously with the species, tissue, and the stage of the cell cycle. A particularly intriguing modification is that 10% of the H2As have an isopeptide bond between the ε-amino group of their Lys 119 and the terminal carboxyl group of the protein ubiquitin. Although ubiquitination, really polyubiquitination, marks cytosolic proteins for degradation by cellular proteases (Section 32-6B), this is not the case for H2A. Rather, as we shall see in Section 34-3B, mono-ubiquitination as well as the other histone modifications serve to modulate eukaryotic gene expression.

Many, if not all, eukaryotes have genetically distinct subtypes of histones H1, H2A, H2B, and H3, many of whose syntheses are switched on or off during specific stages of embryogenesis and in the development of certain cell types. The sequence variations of these subtypes are limited to only a few residues in H2A, H2B, and H3 but are much more extensive in H1. Indeed, the erythroid cells of chick embryos contain an H1 variant that differs so greatly from other H1s that it is named **histone H5** (avian erythrocytes, unlike those of mammals, have nuclei). Histone switching seems to be related to cell differentiation, but the nature of this relationship is unknown.

B. *Nucleosomes: The First Level of Chromatin Organization*

The first level of chromatin organization was pointed out by Roger Kornberg in 1974 through the synthesis of several lines of evidence:

1. Chromatin contains roughly equal numbers of molecules of histones H2A, H2B, H3, and H4, and no more than half that number of histone H1 molecules.

2. X-Ray diffraction studies indicate that chromatin fibers have a regular structure that repeats about every 10 nm along the fiber direction. This same X-ray pattern is observed when purified DNA is mixed with equimolar amounts of all the histones except histone H1.

3. Electron micrographs of chromatin (Fig. 34-4) reveal that it consists of ~10-nm-diameter particles connected by thin strands of apparently naked DNA, rather like beads on a string. These particles are presumably responsible for the foregoing X-ray pattern.

4. Brief digestion of chromatin by **micrococcal nuclease** (which cleaves double-stranded DNA) cleaves the DNA between some of the above particles (Fig. 34-5*a*); apparently the particles protect the DNA closely associated with them from nuclease digestion. Gel electrophoresis indicates that each particle *n*-mer contains ~200*n* bp of DNA (Fig. 34-5*b*).

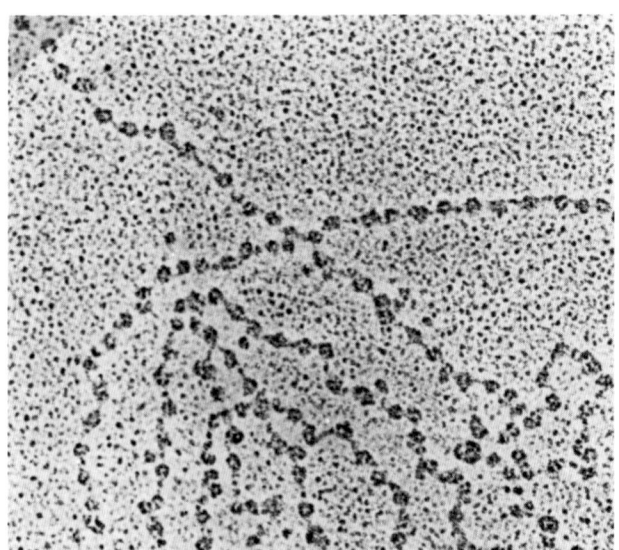

FIGURE 34-4 Electron micrograph of *D. melanogaster* chromatin showing that its 10-nm fibers are strings of closely spaced nucleosomes. [Courtesy of Oscar L. Miller, Jr., University of Virginia.]

(a)

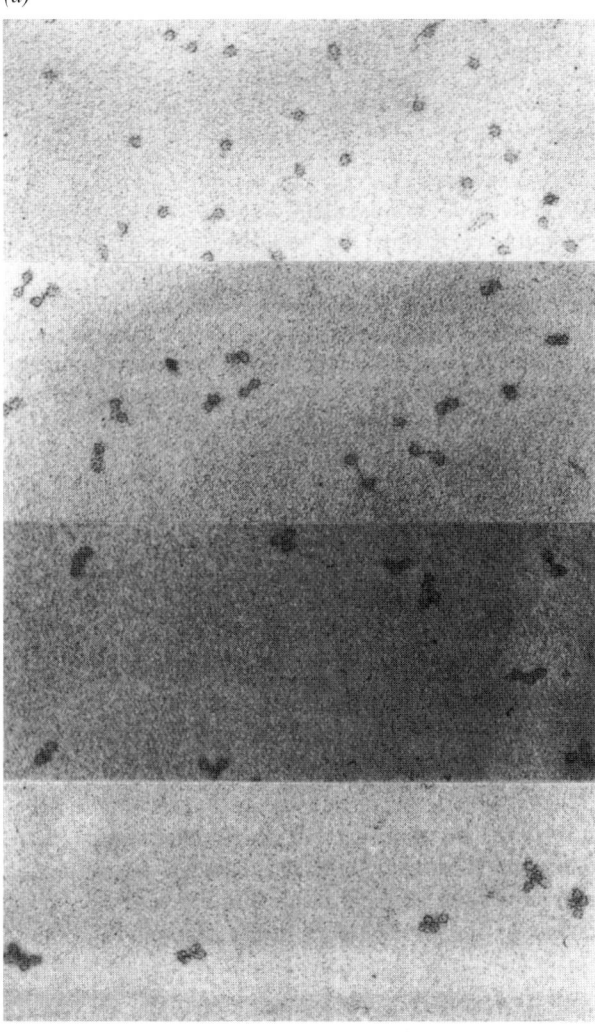

(b)

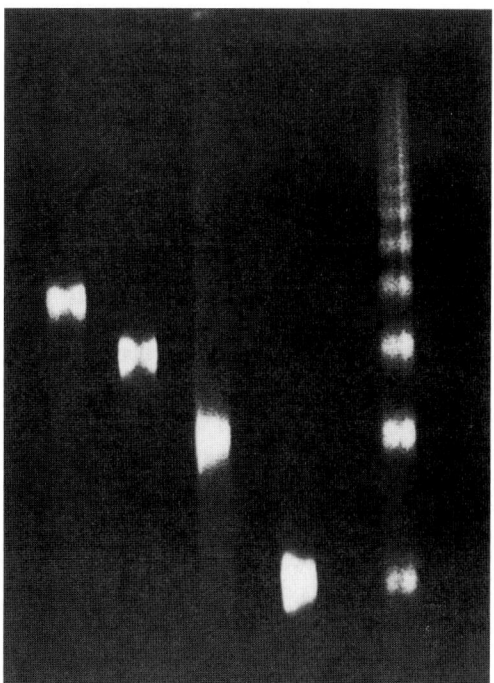

FIGURE 34-5   Defined lengths of calf thymus chromatin obtained by sucrose density gradient ultracentrifugation of chromatin that had been partially digested by micrococcal nuclease. (*a*) Electron micrographs of sucrose density gradient fractions containing, from top to bottom, nucleosome monomers, dimers, trimers, and tetramers. (*b*) Gel electrophoresis of DNA extracted from the nucleosome multimers indicates that they are the corresponding multiples of ~200 bp. The rightmost lane contains DNA from the unfractionated nuclease digest. [Courtesy of Roger Kornberg, Stanford University School of Medicine.]

5. Chemical cross-linking experiments, such as are described in Section 8-5C, indicate that histones H3 and H4 associate to form the tetramer (H3)₂(H4)₂ (Fig. 34-6).

These observations led Kornberg to propose that *the chromatin particles, which are called **nucleosomes**, consist of the octamer (H2A)₂(H2B)₂(H3)₂(H4)₂ in association with ~200 bp of DNA.* The fifth histone, H1, was postulated to be associated in some manner with the outside of the nucleosome (see below).

a. DNA Coils around a Histone Octamer to Form the Nucleosome Core Particle

Micrococcal nuclease, as described above, initially degrades chromatin to particles known as **chromatosomes** that each consist of 166 bp of DNA in complex with a histone octamer and one molecule of histone H1. On further digestion, some of the chromatosome's DNA is trimmed away in a process that releases histone H1. This yields the 205-kD **nucleosome core particle,** which consists of a 145- to 147-bp strand of DNA in association with the above histone octamer. The DNA cumulatively removed by this digestion, which had previously joined neighboring nucle-

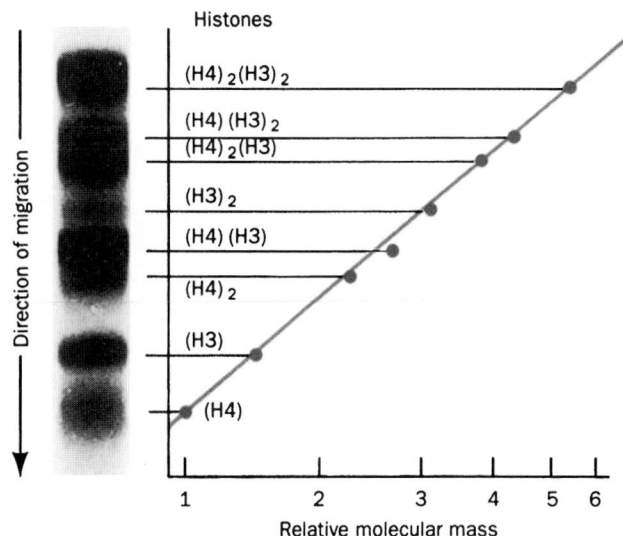

FIGURE 34-6   SDS–gel electrophoresis of a mixture of calf thymus histones H3 and H4 that had been cross-linked by dimethylsuberimidate. The electrophoretogram contains all the bands expected from an (H3)₂(H4)₂ tetramer. [Courtesy of Roger Kornberg, Stanford University School of Medicine.]

osome core particles, is named **linker DNA.** Its length has been found to vary between 8 and 114 bp from organism to organism and tissue to tissue although it is usually ~55 bp.

The X-ray structures of nucleosome core particles containing 146- or 147-bp palindromic DNAs of defined sequence and histones from *X. laevis,* chicken, and *S. cerevisiae* were respectively determined by Timothy Richmond, Gerard Bunick, and Karolin Luger. These structures reveal the nucleosome core particle to be a nearly 2-fold symmetric wedge-shaped disk that has a diameter of ~110 Å and a maximum thickness of ~60 Å. The DNA, which assumes the B form, is wrapped around the outside of the histone octamer in 1.65 turns of a left-handed superhelix (Fig. 34-7). This is the origin of supercoiling in eukaryotic DNA.

Despite having only weak sequence similarity, all four types of histones in the histone octamer share a similar ~70-residue **histone fold** near their C-termini in which a long central helix is flanked on each side by a loop and a shorter helix (Fig. 34-8). Pairs of histone folds interdigitate in head-to-tail arrangements to form the crescent-shaped heterodimers H2A–H2B and H3–H4, each of which binds 2.5 turns (27–28 bp) of duplex DNA that curves around it in a 140° arc. Successive arcs are joined by 3- or 4-bp segments. The H3–H4 pairs interact, via a bundle of four helices from the two H3 histones, to form an (H3–H4)$_2$ tetramer with which each H2A–H2B pair interacts, via a

similar four-helix bundle between H2B and H4, to form the histone octamer (Fig. 34-7*b*).

The histones bind exclusively to the inner face of the DNA, primarily via its sugar–phosphate backbones, through hydrogen bonds, salt bridges, and helix dipoles (their positive N-terminal ends), all interacting with phosphate oxygens, as well as through hydrophobic interactions with the deoxyribose rings. There are few contacts between the histones and the bases, in accord with the nucleosome's lack of sequence specificity. However, an Arg side chain is inserted into the DNA's minor groove at each of the 14 positions it faces the histone octamer. The DNA superhelix has an average radius of 42 Å and a pitch (rise per turn) of 26 Å. However, the DNA does not follow a uniform superhelical path but, rather, is bent fairly sharply at several locations due to outward bulges of the histone core. Moreover, the DNA double helix exhibits considerable conformational variation along its length such that its twist, for example, varies from 7.5 to 15.2 bp/turn with an average value of 10.4 bp/turn (vs 10.4 bp/turn for B-DNA in solution). ~75% of the DNA surface is accessible to solvent and hence appears to be available for interactions with DNA-binding proteins.

The histones of the nucleosome core contain N-terminal tails that emanate from their central histone folds (Fig. 34-8) and vary in length from 23 to 43 residues (they comprise ~25% of the mass of these histones). These highly positively charged polypeptide segments, in agree-

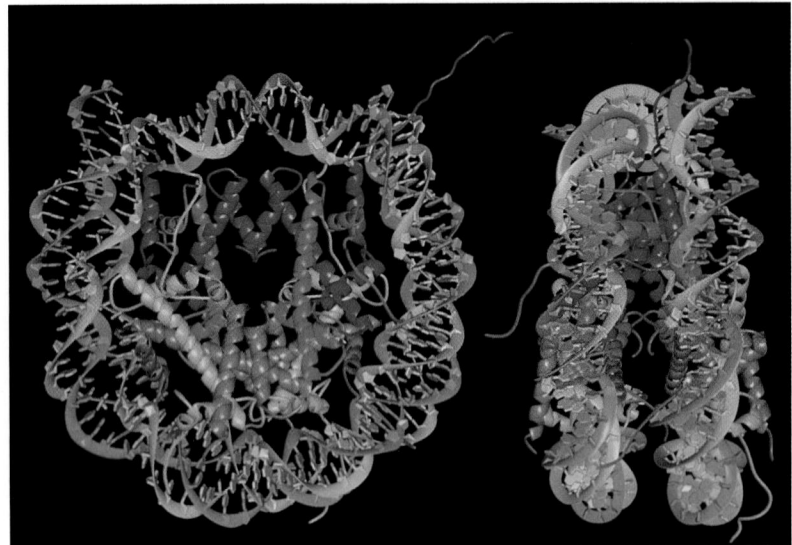

(a)

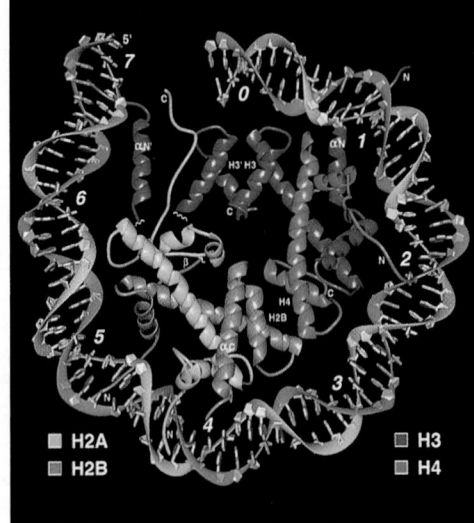

(b)

FIGURE 34-7 X-Ray structure of the nucleosome core particle.
(*a*) The entire core particle as viewed (*left*) along its superhelical axis and (*right*) rotated 90° about the vertical axis. The proteins of the histone octamer are drawn in ribbon form with H2A yellow, H2B red, H3 blue, and H4 green. The sugar–phosphate backbones of the 146-bp DNA are drawn as tan and cyan ribbons whose attached bases are represented by polygons of the same color. In both views, the pseudo-twofold axis is vertical and passes through the DNA center at the top. (*b*) The top half of the nucleosome core particle as viewed in

Part *a, left,* and identically colored. The numbers 0 through 7 arranged about the inside of the 73-bp DNA superhelix mark the positions of sequential double helical turns. Those histones that are drawn in their entirety are primarily associated with this DNA segment, whereas only fragments of H3 and H2B from the other half of the particle are shown. The two four-helix bundles shown are labeled H3′ H3 and H2B H4. [Courtesy of Timothy Richmond, Eidgenössische Technische Hochschule, Zürich, Switzerland. PDBid 1AOI.]

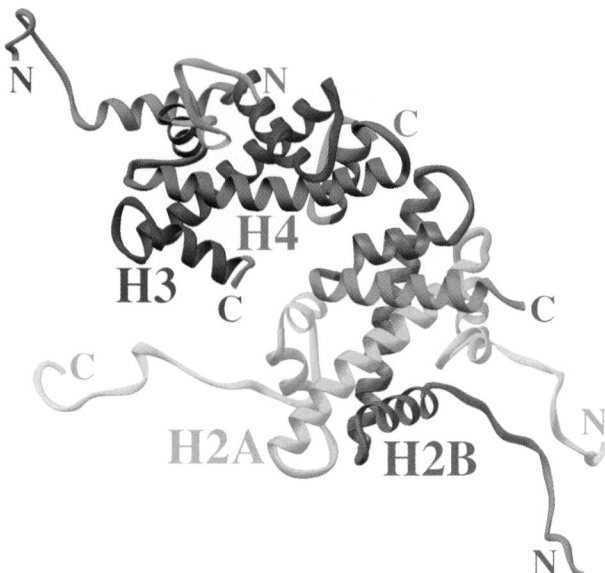

FIGURE 34-8   X-Ray structure of a histone octamer within the nucleosome core particle. Those portions of H2A, H2B, H3, and H4 that form the histone folds are yellow, red, blue, and green, respectively, with their N- and C-terminal tails colored in lighter shades. [Based on an X-ray structure by Gerard Bunick, University of Tennessee and Oak Ridge National Laboratory, Oak Ridge, Tennessee. PDBid 1EQZ.]

ment with previous biochemical studies, extend beyond the DNA; they exit the nucleosome between the gyres of the DNA superhelix, with those from H2B and H3 doing so in channels formed by two vertically aligned minor grooves. Those portions of the N-terminal tails that extend past the DNA are largely unstructured, that is, they are devoid of secondary structure, and substantial portions of their N-terminal segments are disordered. Nevertheless, one of the H4 N-terminal tails makes multiple hydrogen bonds and salt bridges with a highly negatively charged region on an H2A–H2B dimer of an adjacent nucleosome in the crystal structure. In addition, an H2A N-terminal tail of one nucleosome interacts with both the DNA and the H2A N-terminal tail of a neighboring nucleosome. Solution studies indicate, moreover, that the N-terminal tails interact with linker DNA. We shall see in Section 34-3B that the modulation of these interactions by the extensive and varied posttranslational modifications of the N-terminal tails listed above are implicated in facilitating chromatin unfolding to make its component DNA available to participate in such essential processes as transcription, DNA replication, recombination, and DNA repair. Among the foregoing histones, only H2A has an extensive C-terminal tail (39 residues), although it is entirely contained within the body of the nucleosome core and hence is unlikely to participate in internucleosomal interactions.

The archaeon *Methanothermus fervidus* expresses two closely related proteins that form a spheroidal complex with DNA, which presumably functions to prevent the thermal denaturation of this hyperthermophile's DNA.

These proteins are ~30% identical in sequence to the histone fold domains of the histone octamer but lack their N- and C-terminal tails. Evidently, these tails have been added to the histone fold during the course of evolution.

b.   Histone H1 "Seals Off" the Nucleosome

In the micrococcal nuclease digestion of nucleosomes, the ~200-bp DNA is first degraded to 166 bp. Then there is a pause before histone H1 is released from the chromatosomes and the DNA is further shortened to 146 bp. The 2-fold symmetry of the core particle suggests that the reduction in length of the 166-bp DNA comes about by the removal of 10 bp from each of its two ends. Since the 146-bp DNA of the core particle makes 1.8 superhelical turns, the 166-bp intermediate should be able to make two full superhelical turns, which would bring its two ends as close together as possible. Aaron Klug therefore proposed that histone H1 binds to nucleosomal DNA in a cavity formed by the central segment of its DNA and the segments that enter and leave the core particle (Fig. 34-9). This model is supported by the observation that in chromatin filaments containing H1, the DNA enters and leaves the nucleosome on the same side (Fig. 34-10*a*), whereas in H1-depleted chromatin, the entry and exit points are more randomly distributed and tend to occur on opposite sides of the nucleosome (Fig. 34-10*b*). The model also suggests that the length of the linker DNA is controlled by the subspecies of histone H1, which are collectively known as **linker histones,** bound to it.

Histone H5 is a variant of histone H1 that has several Lys → Arg substitutions and binds chromatin more tightly. The observations that the expression of histone H5 in rat sarcoma cells inhibits DNA replication, thereby arresting cells in the G1 phase of the cell cycle, and that histone H5

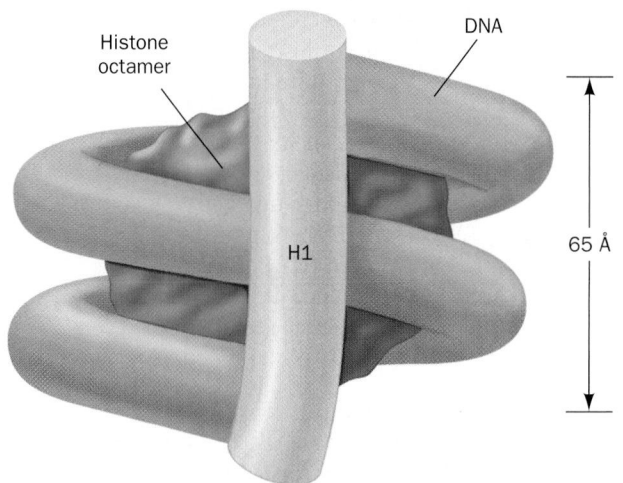

FIGURE 34-9   Model of the interaction of histone H1 with the DNA of the 166-bp chromatosome. The DNA's two complete superhelical turns enable H1 to bind to the DNA's two ends and its middle. Here the histone octamer is represented by the central spheroid (*green*) and the H1 molecule is represented by the cylinder (*yellow*).

(a)   (b)

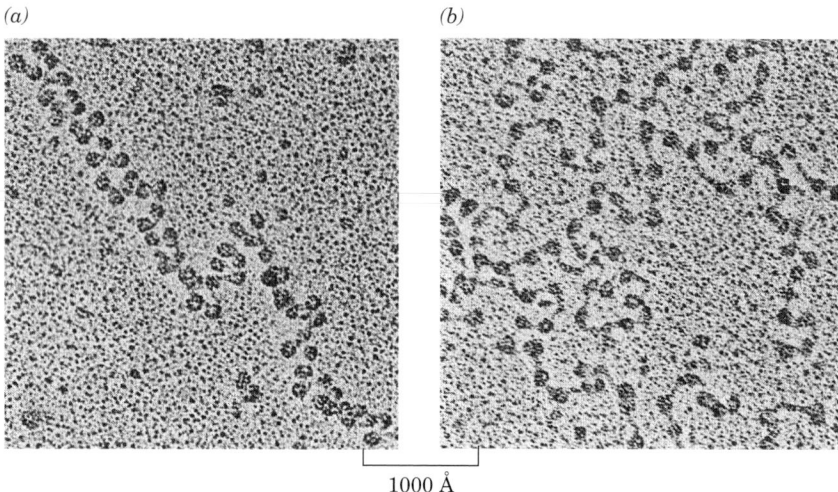

1000 Å

FIGURE 34-10   Electron micrographs of chromatin. (*a*) H1-containing chromatin and (*b*) H1-depleted chromatin, both in 5 to 15 m*M* salt. [Courtesy of Fritz Thoma, Eidgenössische Technische Hochschule, Zürich, Switzerland.]

more closely resembles **histone H1°** (a histone H1 variant that occurs in terminally differentiated cells) than does histone H1 itself, suggest that histone H5 is associated with replicationally and transcriptionally inactive chromatin.

Linker histones consist of a highly conserved globular, trypsin-resistant domain that is flanked by extended N- and C-terminal arms that are rich in basic residues. These basic arms, which comprise more than half of the intact protein, are therefore thought to interact with the linker DNA connecting adjacent nucleosomes even though it is the globular domain that is required for the binding of histone H1 to the nucleosome.

c. The Globular Domain of Histone H5 Structurally Resembles CAP Protein

V. Ramakrishnan has determined the X-ray structure of **GH5,** an 89-residue polypeptide that contains the 81-residue globular domain of histone H5 (although its five N-terminal and eleven C-terminal residues are disordered). The polypeptide chain folds into a 3-helix bundle with a 2-stranded β sheet at its C-terminus (Fig. 34-11). This structure and, in particular, its 3-helix bundle, is strikingly similar in conformation to that of the helix–turn–helix (HTH) motif-containing DNA-binding domain of *E. coli* catabolite activator protein (CAP; Fig. 31-28). Thus, even though there is little sequence identity between GH5 and CAP, their similar structures suggest that GH5 binds DNA in a manner analogous to CAP. Indeed, a model of the GH5–DNA complex based on the known X-ray structure of the CAP–DNA complex (Section 31-3C) positions GH5's highly conserved Lys 69, Arg 73, and Lys 85 side chains to interact with the DNA (Fig. 34-11). These residues, which all have counterparts in CAP, are protected against chemical modification in chromatin. Moreover, GH5 contains a cluster of four conserved basic residues on the opposite face of the protein from its "recognition helix," which could interact with a second segment of duplex

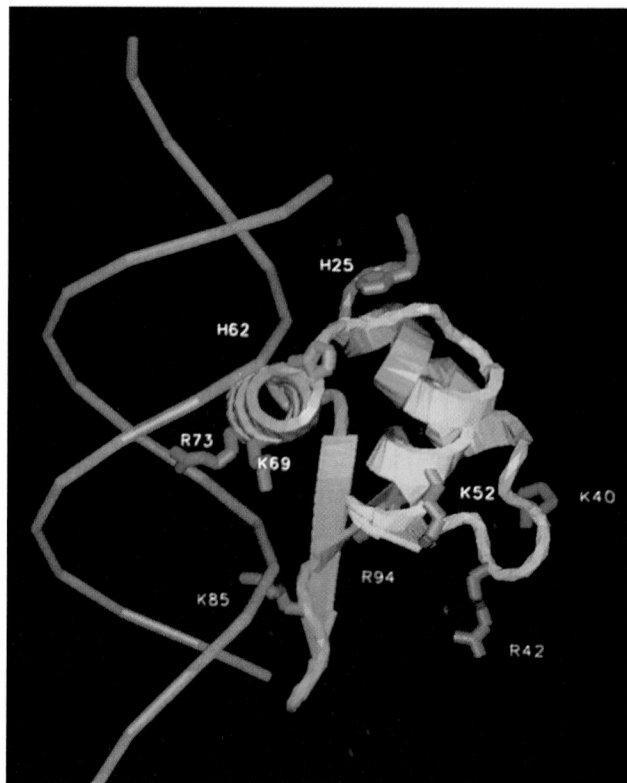

FIGURE 34-11   X-Ray structure of GH5 shown in hypothetical complex with DNA. This model was constructed by superimposing the structure of GH5 on that of CAP in the CAP–DNA structure (Fig. 31-28*a*). However, to avoid any presumptions about the nature of the DNA, that in the CAP structure, which is bent, was replaced by ideal B-DNA, which is represented here by its phosphate backbone (*red*). GH5 is shown in ribbon form and is color-ramped from red to blue going from its N- to its C-terminus. Conserved basic residues, as well as two His residues that have been cross-linked to DNA, are shown in stick form (*blue*). [Courtesy V. Ramakrishnan, MRC Laboratory of Molecular Biology, Cambridge, U.K. PDBid 1HST.]

DNA in agreement with the experimental evidence that GH5 simultaneously binds two DNA duplexes.

d. Parental Nucleosomes Are Transferred to Daughter Duplexes on DNA Replication

The *in vivo* replication of eukaryotic DNA is accompanied by its packaging into chromatin; that is, it is the chromatin that actually is replicated. What, then, is the fate of the histone octamers originally associated with the parental DNA? There are several possibilities: The "parental" octamers may remain associated with either the leading strand or the lagging strand, or they may be partitioned between the two daughter DNA duplexes, either at random or in some systematic way. Attempts to resolve this issue have yielded contradictory results. However, the weight of the evidence now indicates that parental octamers are distributed at random between the daughter duplexes. Moreover, the parental octamers remain associated with DNA during the replication process instead of dissociating from the parental DNA and later rebinding the daughter duplexes. Thus, nucleosomes either open up to permit the passage of a replication fork or parental histone octamers immediately in front of an advancing replication fork are somehow transferred to the daughter duplexes immediately behind the replication fork.

e. Nucleosome Assembly Is Facilitated by Molecular Chaperones

How are nucleosomes formed *in vivo*? *In vitro,* at high salt concentrations, nucleosomes self-assemble from the proper mixture of DNA and histones. In fact, when only H3, H4, and DNA are present, the mixture forms nucleosome-like particles by the deposition of $(H3)_2(H4)_2$ tetramers onto the DNA. Nucleosome cores are then formed by the recruitment of H2A–H2B dimers to these particles.

At physiological salt concentrations, *in vitro* nucleosome assembly occurs much more slowly than at high salt concentrations and, unless the histone concentrations are carefully controlled, is accompanied by considerable histone precipitation. However, in the presence of **nucleoplasmin,** an acidic protein that has been isolated from *X. laevis* oocyte nuclei, and DNA topoisomerase I (Section 29-3C), nucleosome assembly proceeds rapidly without histone precipitation. Nucleoplasmin binds to histones but neither to DNA nor to nucleosomes. Evidently, *nucleoplasmin functions as a molecular chaperone (Section 9-2C) to bring histones and DNA together in a controlled fashion, thereby preventing their nonspecific aggregation through their otherwise strong electrostatic interactions.* The topoisomerase I, no doubt, acts to provide the nucleosome with its preferred level of supercoiling.

C. *30-nm Filaments: The Second Level of Chromatin Organization*

The 166-bp nucleosomal DNA has a packing ratio of ~7 (its 560-Å contour length is wound into an ~80-Å-high su-

percoil). Clearly, the filament of nucleosomes, which only occurs at low ionic strengths (and hence is unlikely to have an independent existence *in vivo*), represents only the first level of chromosomal DNA compaction. Only at physiological ionic strengths does the next level of chromosomal organization become apparent.

As the salt concentration is raised, the H1-containing nucleosome filament initially folds to a zigzag conformation (Fig. 34-10a), whose appearance suggests that nucleosomes interact through contacts between their H1 molecules. Then, as the salt concentration approaches the physiological range, chromatin forms a 30-nm-thick filament in which the nucleosomes are visible (Fig. 34-12). Klug proposed that the 30-nm filament is constructed by winding the 10-nm nucleosome filament into a solenoid with ~6 nucleosomes per turn and a pitch of 110 Å (the diameter of a nucleosome; Fig. 34-13). The solenoid is stabilized by H1 molecules whose relatively variable, extended N-terminal and C-terminal arms (which are absent in GH5; Fig. 34-11) are thought to contact adjacent nucleosomes, at least in part by interacting with neighboring H1s in a head-to-tail fashion. This model, which is consistent with the X-ray diffraction pattern of the 30-nm filaments, has a packing ratio of ~40 (6 nucleosomes, each with ~200 bp DNA, rising a total of 110 Å). Note, however, that several other plausible models for the 30-nm chromatin filament have also been formulated. Indeed, Kensal van Holde has argued that the 30-nm filament does not have a regular structure but rather, because of the varying lengths of the presumably straight and stiff linker DNAs connecting the nucleosome cores, has an irregular helix-like structure that simulations indicate forms a filament with an average diameter of 30 nm. This would account for

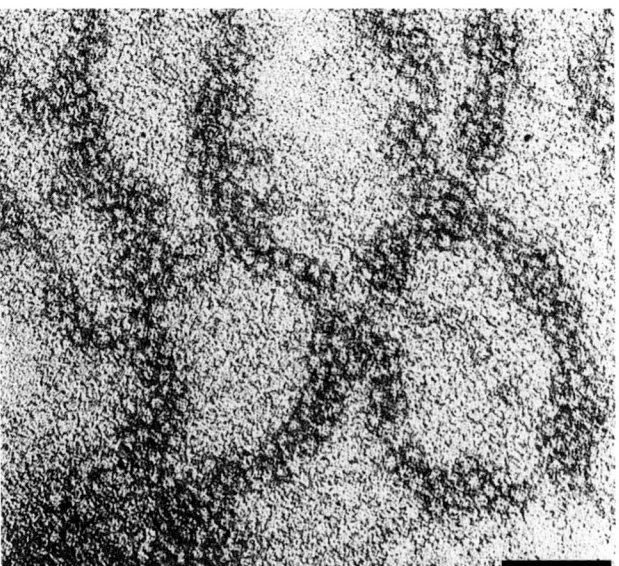

FIGURE 34-12 Electron micrograph of the 30-nm chromatin filaments. Note that the filaments are two to three nucleosomes across. The bar represents 1000 Å. [Courtesy of Jerome B. Rattner, University of Calgary, Canada.]

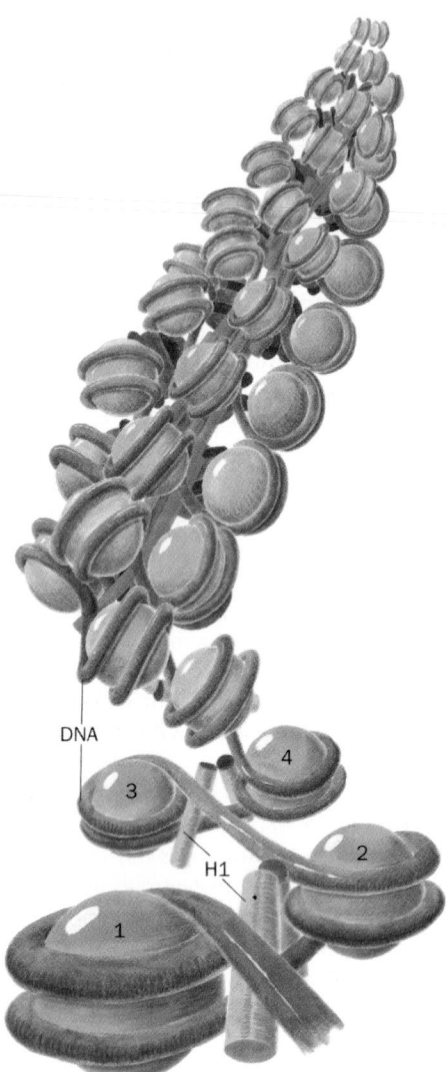

FIGURE 34-13   Model of the 30-nm chromatin filament. The
filament is represented (*bottom to top*) as it might form with
increasing salt concentration. The zigzag pattern of nucleosomes
(*1, 2, 3, 4*) closes up to form a solenoid with ~6 nucleosomes
per turn. The H1 molecules (*yellow cylinders*), which stabilize
the structure, are thought to form a helical polymer running
along the center of the solenoid.

the difficulty in experimentally determining the structure
of the 30-nm filament despite numerous attempts to do so
over nearly three decades.

D. *Radial Loops: The Third Level of Chromatin Organization*

Histone-depleted metaphase chromosomes exhibit a central fibrous protein matrix or scaffold surrounded by an
extensive halo of DNA (Fig. 34-14a). The strands of DNA
that can be followed are observed to form loops that enter
and exit the scaffold at nearly the same point (Fig. 34-14b).
Most of these loops have lengths in the range 15 to 30 μm
(which corresponds to 45–90 kb), so that when condensed
as 30-nm filaments they would be ~0.6 μm long. Electron

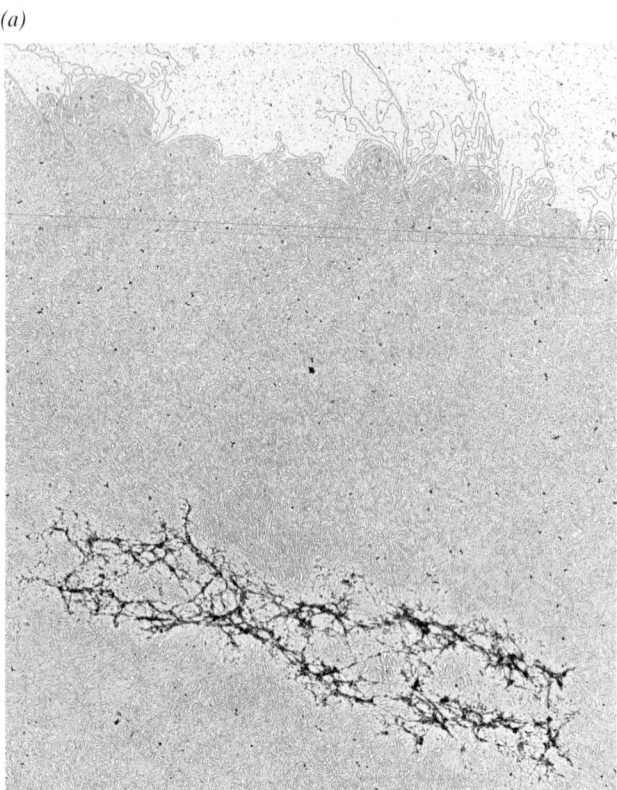

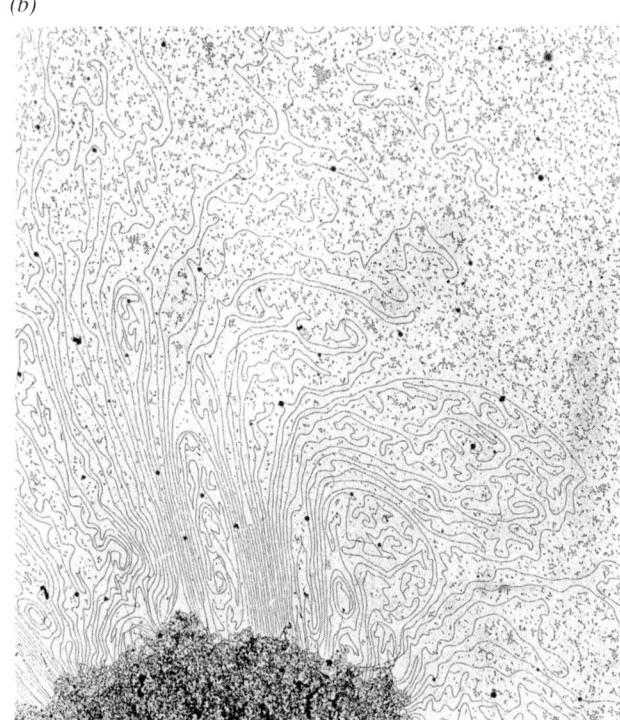

**FIGURE 34-14   Electron micrographs of a histone-depleted
metaphase human chromosome.** (*a*) The central protein matrix
(scaffold) serves to anchor the surrounding DNA. (*b*) At higher
magnification it can be seen that the DNA is attached to the
scaffold in loops. [Courtesy of Ulrich Laemmli, University of
Geneva, Switzerland.]

micrographs of chromosomes in cross section, such as Fig. 34-15a, strongly suggest that the chromatin fibers of metaphase chromosomes are radially arranged. If the observed loops correspond to these radial fibers, they would each contribute 0.3 μm to the diameter of the chromosome (a fiber must double back on itself to form a loop). Taking into account the 0.4-μm width of the scaffold, this model predicts the diameter of the metaphase chromosome to be 1.0 μm, in agreement with observation (Fig. 34-15b). A typical human chromosome, which contains ~140 million bp, would therefore have ~2000 of these ~70-kb radial loops. The 0.4-μm-diameter scaffold of such a chromosome

(a)

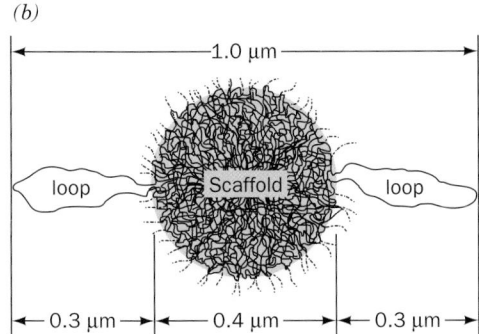

(b)

FIGURE 34-15   **Organization of DNA in a metaphase chromosome.** (*a*) Electron micrograph of a human metaphase chromosome in cross section. Note the mass of chromatin fibers radially projecting from the central scaffold. [Courtesy of Ulrich Laemmli, University of Geneva, Switzerland.] (*b*) Interpretive diagram indicating how the 0.3-μm-long radial loops are thought to combine with the 0.4-μm-wide scaffold to form the 1.0-μm-diameter metaphase chromosome.

has sufficient surface area along its 6-μm length to bind this number of radial loops. The radial loop model therefore accounts for DNA's observed packing ratio in metaphase chromosomes.

The radial DNA loops are attached to the matrix via AT-rich **matrix-associated regions [MARs;** alternatively, **scaffold attachment regions (SARs)].** The radial loops are therefore also known as **structural domains.** Nevertheless, little is known about how the matrix is composed, how the radial loops are organized, and how metaphase chromosomes and the far more dispersed interphase chromosomes interconvert. Certainly, **nonhistone proteins,** whose thousands of varieties constitute ~10% of chromosomal proteins, must participate in these processes. Moreover, since the protein machinery controlling gene expression in a given structural domain is unlikely to directly affect expression in a neighboring structural domain, the structural domains probably comprise the chromosomal transcriptional units.

E. *Polytene Chromosomes*

The diffuse structure of most interphase chromosomes (Fig. 34-2) makes it all but impossible to characterize them at the level of individual genes. Nature, however, has greatly ameliorated this predicament through the production of "giant" banded chromosomes in certain terminally differentiated (nondividing) secretory cells of dipteran (two-winged) flies (Fig. 34-16). These chromosomes, of which those from the salivary glands of *D. melanogaster* larvae are the most extensively studied, are produced by multiple replications of a synapsed (joined in parallel) diploid pair in which the replicas remain attached to one another and in register (see below). Each diploid pair may replicate in this manner as many as nine times so that the final **polytene chromosome** contains up to $2 \times 2^9 = 1024$ DNA strands. The function(s) of polytene chromosomes is unknown although perhaps this permits a greatly increased rate of transcription of certain genes.

D. melanogaster's four giant chromosomes have an aggregate length of ~2 mm so that its haploid genome of 1.37×10^8 bp has an average packing ratio in these chromosomes of almost 25. About 95% of this DNA is concentrated in chromosomal bands (Fig. 34-17). These bands (more properly, **chromomeres**), as microscopically visualized through staining, form a pattern that is characteristic of each *D. melanogaster* strain. Indeed, chromosomal rearrangements such as duplications, deletions, and inversions result in a corresponding change in the banding pattern. *A polytene chromosome's banding pattern therefore forms a cytological map that parallels its genetic map.*

The characteristic banding pattern of each polytene chromosome suggests that its component DNA molecules are precisely aligned. This hypothesis was corroborated by the application of ***in situ*** (on site) **hybridization.** In this technique, developed by Mary Lou Pardue and Joseph Gall, an immobilized chromosome preparation is treated with NaOH to denature its DNA; it is then hybridized with a purified species of radioactively labeled mRNA (or its

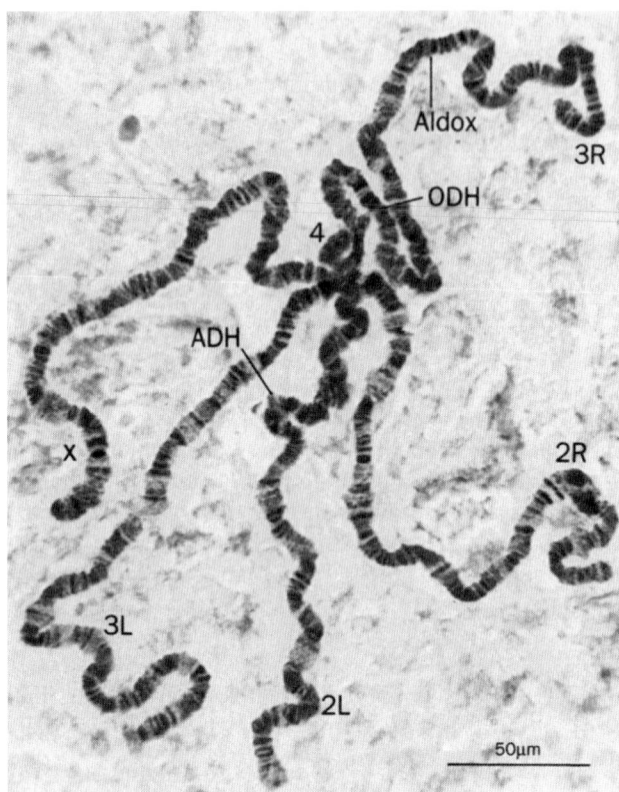

FIGURE 34-16 Photomicrograph of the stained polytene chromosomes from the *D. melanogaster* salivary gland. Such chromosomes consist of darkly staining bands interspersed with light-staining interband regions. All four chromosomes in a single cell are held together by their centromeres. The chromosomal positions for the genes specifying alcohol dehydrogenase (ADH), aldehyde oxidase (Aldox), and octanol dehydrogenase (ODH) are indicated. [Courtesy of B.P. Kaufmann, University of Michigan.]

corresponding cDNA), and the chromosomal binding site of the radioactive probe is determined by autoradiography. A given mRNA hybridizes with one, or no more than a few, chromosomal bands (Fig. 34-18).

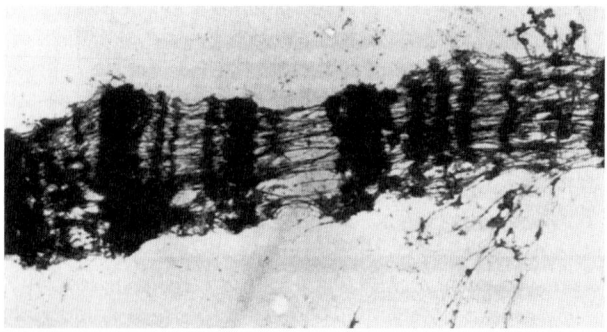

FIGURE 34-17 Electron micrograph of a segment of polytene chromosome from *D. melanogaster*. Note that its interband regions consist of chromatin fibers that are more or less parallel to the long axis of the chromosome, whereas its bands, which contain ~95% of the chromosome's DNA, are much more highly condensed. [Courtesy of Gary Burkholder, University of Saskatechewan, Canada.]

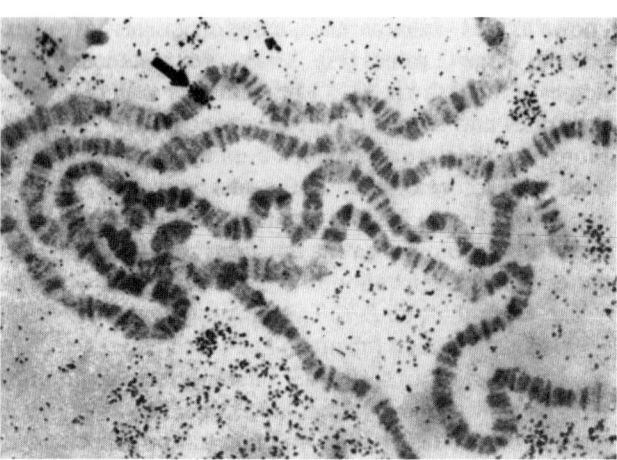

FIGURE 34-18 Autoradiograph of a *D. melanogaster* polytene chromosome that has been *in situ* hybridized with yolk protein cDNA. The dark grains (*arrow*) identify the chromosomal location of the yolk protein gene. [From Barnett, T., Pachl, C., Gergen, J.P., and Wensink, P.C., *Cell* **21**, 735 (1980). Copyright © 1980 by Cell Press.]

D. melanogaster's four polytene chromosomes exhibit an aggregate of ~5000 bands. It originally appeared that the number of *D. melanogaster* genes was roughly equal to this number of bands and hence it was thought that each band corresponds to a single gene. However, the recently determined genome sequence of *D. melanogaster* indicates that it has ~13,000 genes, nearly three times its number of bands. In fact, genes have been shown to be located in both band and interband regions, with some bands containing several genes and others containing none. Thus, it is likely that the banding pattern of polytene chromosomes is a consequence of different levels of gene expression due to variations in chromatin structure (see Section 34-3B), with the genes in the relatively open interband regions presumably more highly expressed than those in the more condensed and hence less accessible bands.

2 ■ GENOMIC ORGANIZATION

Higher organisms contain a great variety of cells that differ not only in their appearances (e.g., Fig. 1-10) but in the proteins they synthesize. Pancreatic acinar cells, for example, synthesize copious quantities of digestive enzymes, including trypsin and chymotrypsin, but no insulin, whereas the neighboring pancreatic β cells produce large amounts of insulin but no digestive enzymes. Clearly, each of these different types of cells expresses different genes. Yet most of a multicellular organism's somatic cells contain the same genetic information as the fertilized ovum from which they are descended (a phenomenon described as **totipotency**) as is demonstrated, for example, by the ability to raise a mammal such as a sheep, cow, or mouse from an enucleated oocyte into which the nucleus from an adult cell had been inserted. Similarly, a single cell from a plant can give rise to the normal plant. Evidently, cells have enormous expressional flexibility. Nevertheless, only a small fraction of the DNA in higher eukaryotic genomes is expressed. What is the nature of the remaining unexpressed sequences and do they have any function? In this section we describe the genetic organization of the eukaryotic chromosome. How eukaryotic gene expression is controlled is the subject of Section 34-3.

A. *The C-Value Paradox*

One might reasonably expect the morphological complexity of an organism to be roughly correlated with the amount of DNA in its haploid genome, its **C value.** After all, the morphological complexity of an organism must reflect an underlying genetic complexity. Nevertheless, in what is known as the **C-value paradox,** many organisms have unexpectedly large C values (Fig. 34-19). For instance, the genomes of lungfish are 10 to 15 times larger than of those of mammals and those of some salamanders are yet larger. Moreover, the C-value paradox even applies to closely related species; for example, the C values for several species of *Drosophila* have a 2.5-fold spread. Does the "extra" DNA in the larger genomes have a function, and if not, why is it preserved from generation to generation?

The 4.6 million-bp *E. coli* genome encodes ~4300 proteins. In contrast, the 3.2 billion-bp haploid human genome, which is ~700 times larger than that of *E. coli*, is estimated to encode ~30,000 proteins; that is, humans have only ~7 times as many structural genes as do *E. coli*. Certainly the control of genetic expression in eukaryotes is a far more elaborate process than it is in prokaryotes. Yet does much of the unexpressed DNA in the human genome, at least 98% of the total, function in the control of genetic expression? The recent and ongoing eukaryotic genome sequence determinations are beginning to provide answers to the foregoing questions.

a. C_0t Curve Analysis Indicates DNA Complexity

The rate at which DNA renatures is indicative of the lengths of its unique sequences. If DNA is sheared into uniform fragments of 300 to 10,000 bp (Section 5-3D), denatured, and kept at a low concentration so that the effects of mechanical entanglement are small, the rate-determining step in renaturation is the collision of complementary sequences. Once the complementary sequences have found each other through random diffusion, they rapidly zip up to form duplex molecules. The rate of re-

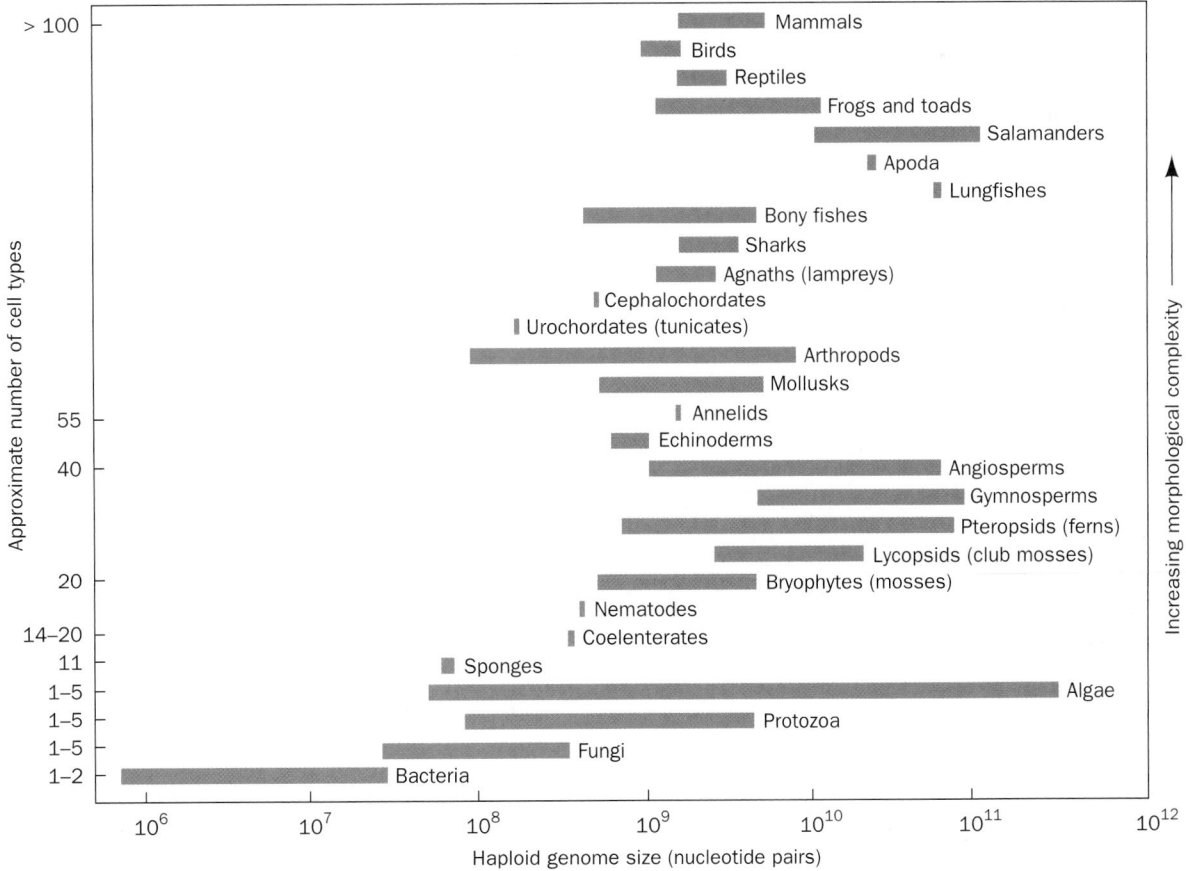

FIGURE 34-19 The range of haploid genome DNA contents in various categories of organisms indicating the C-value paradox. The morphological complexity of the organisms, as estimated according to their number of cell types, increases from bottom to top. [After Raff, R.A. and Kaufman, T.C., *Embryos, Genes, and Evolution*, p. 314, Macmillan (1983).]

naturation of denatured DNA is therefore expressed

$$\frac{d[A]}{dt} = -k[A][B] \qquad [34.1]$$

where A and B represent complementary single-stranded sequences and k is a second-order rate constant (Section 14-1B). Since $[A] = [B]$ for duplex DNA, Eq. [34.1] integrates to

$$\frac{1}{[A]} = \frac{1}{[A]_0} + kt \qquad [34.2]$$

where $[A]_0$ is the initial concentration of A.

It is convenient to measure the fraction f of unpaired strands:

$$f = \frac{[A]}{[A]_0} \qquad [34.3]$$

Combining Eqs. [34.2] and [34.3] yields

$$f = \frac{1}{1 + [A]_0 kt} \qquad [34.4]$$

The concentration terms in these equations refer to unique sequences since the collision of noncomplementary sequences does not lead to renaturation. Hence, if C_0 is the initial concentration of base pairs in solution, then

$$[A]_0 = \frac{C_0}{x} \qquad [34.5]$$

where x is the number of base pairs in each unique sequence and is known as the DNA's **complexity.** For example, the repeating sequence $(AGCT)_n$ has a complexity of 4, whereas an *E. coli* chromosome, which consists of ~4.6 million bp of unrepeated sequence, has a complexity of ~4.6 million. Combining Eqs. [34.4] and [34.5] yields

$$f = \frac{1}{1 + C_0 kt/x} \qquad [34.6]$$

When half of the molecules in the sample have renatured, $f = 0.5$, so that

$$C_0 t_{1/2} = \frac{x}{k} \qquad [34.7]$$

where $t_{1/2}$ is the time for this to occur. The rate constant k is characteristic of the rate at which single strands collide in solution under the conditions employed, so it is independent of the complexity of the DNA and, for reasonably short DNA fragments, the length of a strand. Consequently, *for a given set of conditions, the value of $C_0 t_{1/2}$ depends only on the complexity x of the DNA.* This situation is indicated in Fig. 34-20, which is a series of plots of f versus $C_0 t$ for various DNAs. Such plots are referred to as $C_0 t$ (pronounced "cot") curves. The complexities of the DNAs in Fig. 34-20 vary from 1 for the synthetic duplex poly(A) · poly(U) to ~3×10^9 for some fractions of mammalian DNAs. Their corresponding values of $C_0 t_{1/2}$ vary accordingly.

The speed and sensitivity of $C_0 t$ curve analysis is greatly enhanced through the hydroxyapatite fractionation of the renaturing DNA. Hydroxyapatite, it will be recalled (Section 6-6B), binds double-stranded DNA at a higher phosphate concentration than it binds single-stranded DNA. The single- and double-stranded DNAs in a solution of renaturing DNA may therefore be separated by hydroxyapatite chromatography and the amounts of each measured. The single-stranded DNA can then be further renatured and the process repeated. If the renaturing DNA is radioactively labeled, much smaller quantities of it can be detected than is possible by spectroscopic means. Thus,

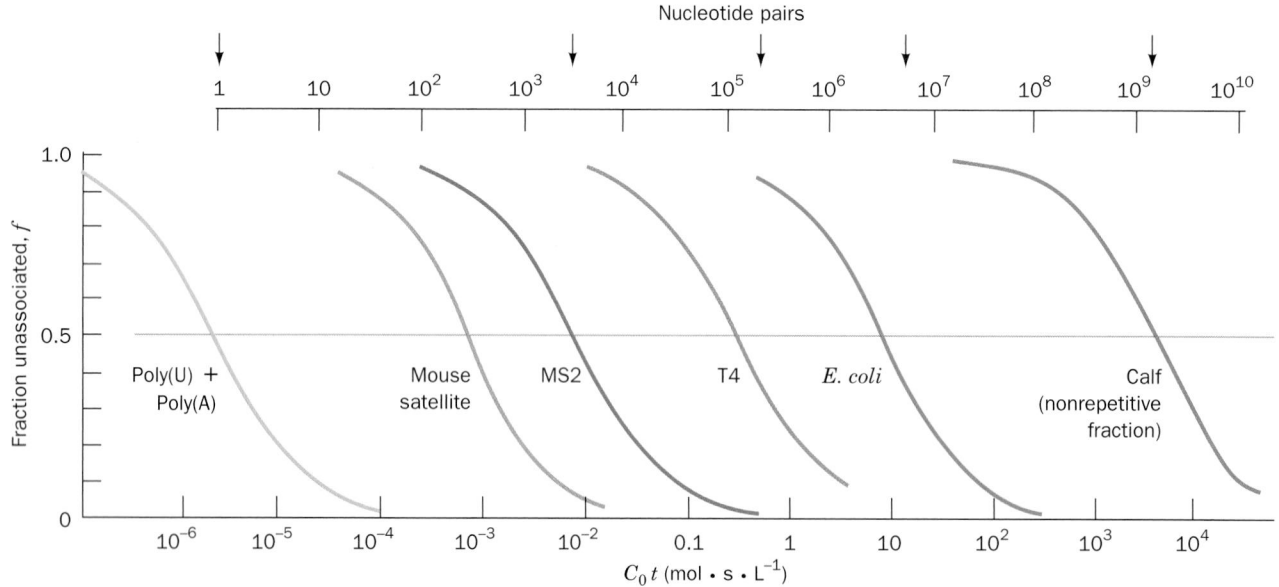

FIGURE 34-20 The reassociation ($C_0 t$) curves of duplex DNAs from the indicated sources. The DNA was dissolved in a solution containing $0.18M$ Na$^+$ and sheared to an average length of 400 bp. The upper scale indicates the genome sizes of some of the DNAs (**MS2** and **T4** are bacteriophages). [After Britten, R.J. and Kohne, D.E., *Science* **161**, 530 (1968).]

through the hydroxyapatite chromatography of radioactively labeled DNA, the C_0t curve analysis of a DNA of such a high complexity that its $t_{1/2}$ is days or weeks can be conveniently measured in a small fraction of that time.

B. *Repetitive Sequences*

Consider a sample of DNA that consists of sequences with varying degrees of complexity. Its C_0t curve, Fig. 34-21 for example, is the sum of the individual C_0t curves for each complexity class of DNA. *C_0t curve analysis (and more recently, genome sequencing) has demonstrated that viral and prokaryotic DNAs have few, if any, repeated sequences (e.g., Fig. 34-20 for MS2, T4, and E. coli). In contrast, eukaryotic DNAs exhibit complicated C_0t curves (e.g., Fig. 34-22) that* must arise from the presence of DNA segments of several different complexities.

Eukaryotic C_0t curves may be attributed to the presence of five somewhat arbitrarily defined classes of DNAs: (1) **inverted repeats,** (2) **highly repetitive sequences** ($>10^6$ copies per haploid genome), (3) **moderately repetitive sequences** ($<10^6$ copies per haploid genome), (4) **segmental duplications** (blocks of 1–200 kb that have been copied to one or more regions of the genome that may be within the same chromosome or on different chromosomes; they constitute ~5% of the human genome), and (5) **unique sequences** (~1 copy per haploid genome). The sequences and chromosomal distributions of these DNA segments vary with the species, so a unifying description of their arrangements cannot be made. Nevertheless, several broad generalizations are possible, as we shall see below.

a. Inverted Repeats Form Foldback Structures

The most rapidly reassociating eukaryotic DNA, which represents as much as 10% of some genomes, renatures with first-order kinetics. This DNA contains inverted (self-complementary) sequences in close proximity, which can

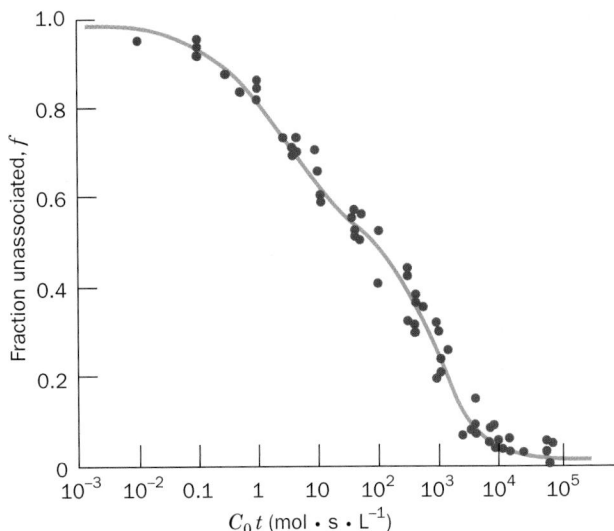

FIGURE 34-22 The C_0t curve of *Strongylocentrotus purpuratus* (a sea urchin) DNA. [After Galau, G.A., Britten, R.J., and Davidson, E.H., *Cell* **2,** 11 (1974).]

fold back on themselves to form hairpinlike **foldback structures** (Fig. 34-23*a*). Inverted sequences may be isolated by adsorbing the duplex DNA formed at very low C_0t values to hydroxyapatite and subsequently degrading its single-stranded loop and tails with **S1 nuclease** (an endonuclease from *Aspergillus oryzae* that preferentially cleaves single strands). The resulting inverted repeats range in length from 100 to 1000 bp, sizes much too large to have evolved at random. *In situ* hybridization studies on metaphase chromosomes using these inverted repeats as probes indicate that they are distributed at many chromosomal sites.

The function of inverted repeats, some 2 million of which occur in the human genome, is unknown. However, since the cruciform structures formed by paired foldback structures (Fig. 34-23*b*) are only slightly less stable than the corresponding normal duplex DNA, it has been suggested that the inverted repeats function in chromatin as some sort of molecular switch.

b. Highly Repetitive DNA Is Clustered at Telomeres and Centromeres

Highly repetitive DNA consists of short sequences that are tandemly repeated, either perfectly or slightly imperfectly, often thousands of times. Such **simple sequence repeats [(SSRs);** alternatively, **short tandem repeats (STRs)]** can often be separated from the bulk of the chromosomal DNA by shear degradation followed by density gradient ultracentrifugation in CsCl since their distinctive base compositions cause them to form "satellites" to the main DNA band (Fig. 34-24; recall that the buoyant density of DNA in CsCl increases with its G + C content; Section 6-6D). The sequences of these SSRs, which are also known as **satellite DNAs,** are species specific (SSRs with a short repeat unit of n = 1–13 nt are often called **microsatellites,** whereas those with n = 14–500 are often called **minisatellites**). For example, the crab *Cancer borealis* has an SSR comprising 30% of its genome in which the repeating unit is the dinucleotide

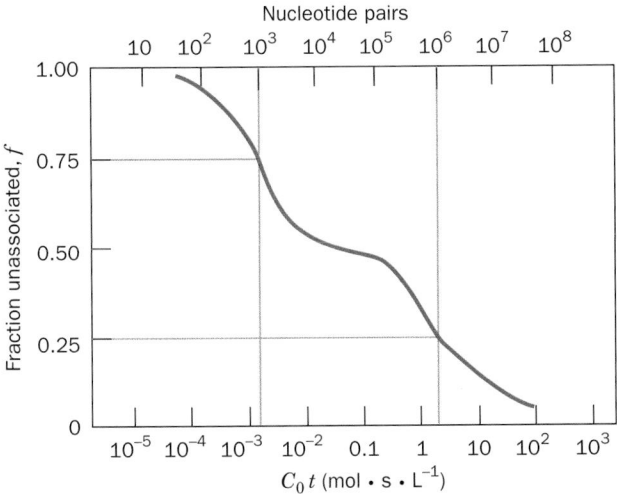

FIGURE 34-21 *C_0t curve of a hypothetical DNA molecule.* Before fragmentation, this DNA was 2 million bp in length and consisted of a unique sequence of 1 million bp and 1000 copies of a 1000-bp sequence. Note the curve's biphasic nature.

(a)

(b)

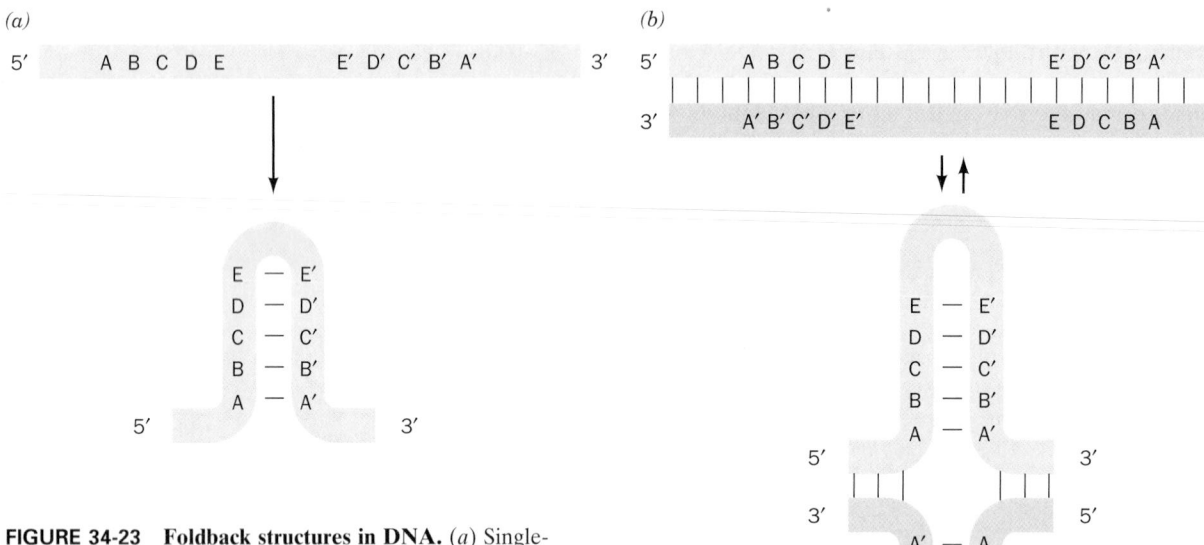

FIGURE 34-23   Foldback structures in DNA. (*a*) Single-stranded DNA containing an inverted repeat will, under renaturing conditions, form a base paired loop known as a foldback structure. Here A is complementary to A′, B is complementary to B′, etc. (*b*) An inverted repeat in duplex DNA could assume a cruciform conformation consisting of two opposing foldback structures. The stability of this structure would be less than that of the corresponding duplex but only by the loss of the base pairing energy in the unpaired loops.

AT. The DNA of *Drosophila virilis* exhibits three satellite bands (Fig. 34-24), which each consist of a different although closely related repeating heptanucleotide sequence:

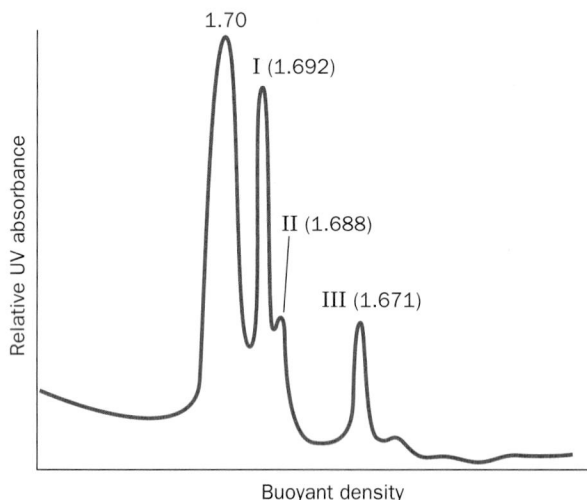

FIGURE 34-24   The buoyant density pattern of *Drosophila virilis* DNA centrifuged to equilibrium in neutral CsCl. Three prominent bands of satellite DNA (ρ = 1.692, 1.688, and 1.671 g/cm^3) are present, in addition to the main DNA band (ρ = 1.70 g/cm^3). [After Gall, J.G., Cohen, E.H., and Atherton, D.D., *Cold Spring Harbor Symp. Quant. Biol.* **38**, 417 (1973).]

$$5' - \text{ACAAACT} - 3'$$
$$3' - \text{TGTTTGA} - 5'$$
Satellite I

$$5' - \text{ATAAACT} - 3'$$
$$3' - \text{TATTTGA} - 5'$$
Satellite II

$$5' - \text{ACAAATT} - 3'$$
$$3' - \text{TGTTTAA} - 5'$$
Satellite III

These comprise 25, 8, and 8% of the 3.1×10^8-bp *D. virilis* genome, so that these sequences are repeated 11, 3.6, and 3.6 million times, respectively.

Telomeres, as we have seen (Section 30-4D), consist of G-rich SSRs. In addition, the *in situ* hybridization of mouse chromosomes with ^{3}H-labeled RNA synthesized on mouse simple sequence DNA templates established that SSR DNA is concentrated in the heterochromatic region associated with the chromosomal centromere [Fig. 34-25; the centromere is the constricted segment of the chromosome at which sister chromatids are joined (Fig. 34-1) and at which the chromosome attaches to the mitotic spindle (Fig. 1-19)]. This observation suggests that centromeric SSR DNA, which is not transcribed *in vivo*, functions to align homologous chromosomes during meiosis (Fig. 1-20) and/or to facilitate their recombination. This hypothesis is supported by the observation that satellite DNAs are largely or entirely eliminated from the somatic cells of a variety of eukaryotes (which are consequently no longer totipotent) but not from their germ cells.

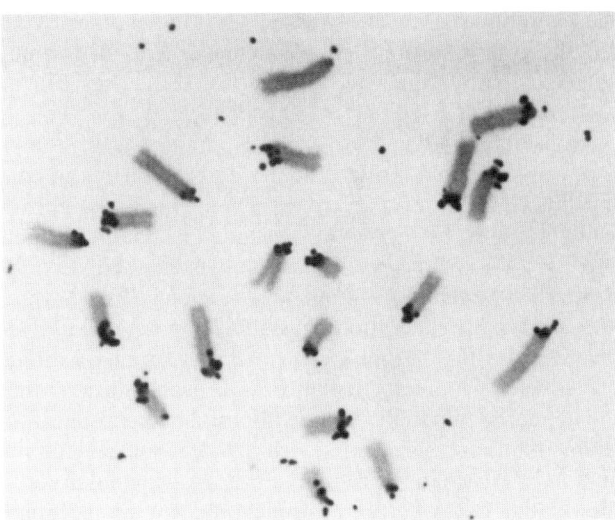

FIGURE 34-25 Autoradiograph of mouse chromosomes showing the centromeric location of their SSR DNA through *in situ* **hybridization.** Note that the centromeres in mouse chromosomes are all located at one end of the chromosome (no genes lie beyond the mouse centromeres). In human and yeast chromosomes, however, centromeres occupy more internal positions (e.g., Fig. 34-1), whereas *D. melanogaster* has chromosomes of both types. [Courtesy of Joseph Gall, Carnegie Institution of Washington.]

SSRs comprise ~3% of the human genome, with the greatest contribution (0.5%) provided by dinucleotide repeats, most frequently $(CA)_n$ and $(TA)_n$. SSRs appear to have arisen by template slippage during DNA replication.

This occurs more frequently with short repeats, which therefore have a high degree of length polymorphism in the human population. Consequently, genetic markers based on the lengths of SSRs, particularly $(CA)_n$ repeats, have been a mainstay of human genetic studies (Section 5-5F).

c. Moderately Repetitive DNAs Are Arranged in Dispersed Repeats

Moderately repetitive DNAs occur in segments of 100 to several thousand base pairs that are interspersed with larger blocks of unique DNA. Some of this repetitive DNA consists of tandemly repeated groups of genes that specify products that cells require in large quantities, such as rRNAs, tRNAs, and histones. The organization of these repeated genes is discussed in Section 34-2D.

Around 42% of the human genome consists of retrotransposons (transposable elements that propagate through the intermediate synthesis of RNA; Section 30-6B). Three major types of retrotransposons inhabit the human genome (Table 34-2):

1. Long interspersed nuclear elements (LINEs), which comprise 20.4% of the human genome, are 6- to 8-kb-long segments that encode the proteins that mediate their transposition (Section 30-6B), although the vast majority (>99%) of LINEs have accumulated mutations that render them transpositionally inactive. LINEs, which are derived from RNA polymerase II–generated transcripts, are dispersed throughout the genomes of all mammals, which suggests that the ancestral LINE became associated with the mammalian genome very early in its evolution. The

TABLE 34-2   Moderately Repetitive Sequences in the Human Genome[a]

Type of Repeat	Number of Copies (× 1000)	Total Number of Nucleotides (Mb)	Fraction of the Genome Sequence (%)
LINEs	868	559	20.4
L1	516	462	16.9
L2	315	88	3.2
L3	37	8	0.3
SINEs	1558	360	13.1
Alu	1090	290	10.6
MIR	393	60	2.2
MIR3	75	9	0.3
LTR Retrotransposons	443	227	8.3
ERV-class I	112	79	2.9
ERV(K)-class II	8	8	0.3
ERV(L)-class III	83	40	1.4
MaLR	240	100	3.6
DNA Transposons	294	78	2.8
HAT group	195	42	1.6
Tc-1 group	75	32	1.2
Unclassified	22	3.2	0.1
Total		1227	44.8

[a]These numbers are approximate and are likely to be underestimates.

Source: International Human Genome Sequencing Consortium, *Nature* **409,** 880 (2001).

most common LINE in the human genome, **LINE-1 (L1),** consists of ~6.1 kb containing a 5′ untranslated region (UTR), two open reading frames (ORFs), the second of which contains a reverse transcriptase gene, and a 3′ UTR ending in a poly(A) tail. However, the average L1 has a length of ~900 bp because most L1s consist of truncated fragments of the full length retrotransposon. Two other LINEs occur in the human genome, **L2** and **L3,** which are distantly related to L1. However, L1 is the only LINE in the human genome that is still transpositionally active.

2. Short interspersed nuclear elements (SINEs), which comprise 13.1% of the human genome, consist of 100- to 400-bp elements that are derived from RNA polymerase III–generated transcripts. SINEs each contain an RNA polymerase III promoter but, in contrast to LINEs, do not encode proteins; they are apparently propagated by LINE-encoded enzymes. The most common SINEs in the human genome are members of the *Alu* **family,** which are so named because most of their ~300-bp segments contain a cleavage site for the restriction endonuclease *Alu*I (AGCT; Table 5-4). *Alu* elements consist of two imperfect tandem repeats that are ~90% identical in sequence to portions of the 7S RNA of the signal recognition particle (SRP; Section 12-4B) but that both end in poly(A) segments that are not present in SRP 7S RNA. *Alu* elements occur only in primates, which indicates that they are of relatively recent origin. However, *Alu*-like elements occur in such distantly related organisms as slime molds, echinoderms, amphibians, and birds. All other types of SINEs are derived from tRNA sequences.

3. LTR retrotransposons, which contain long terminal repeats (LTRs) flanking *gag* and *pol* genes, are propagated via cytoplasmic retrovirus-like particles (Section 30-6B). They comprise 8.3% of the human genome. Only the vertebrate-specific **endogenous retroviruses (ERVs)** appear to have been active in the mammalian genome.

In addition, the human genome contains **DNA transposons** (Table 34-2) that resemble bacterial transposons (Section 30-6B). They comprise 2.8% of the human genome. Hence, *a total of ~45% of the human genome consists of widely dispersed and almost entirely inactive transposable elements.*

d. Moderately Repetitive DNAs Are Probably Selfish DNA

It would seem likely, considering their ranges of segment lengths and copy numbers, that nonexpressed, moderately repetitive DNAs have several different functions. There is, however, little experimental evidence in support of any of the various proposals that have been put forward in this regard. The proposal that is usually given the most credence is that moderately repetitive DNAs function as control sequences that participate in coordinately activating nearby genes. Another possibility, which is based on the observation that *Alu* elements contain a segment that is homologous to the **papovavirus** replication origin, is that certain families of moderately repetitive DNAs act as DNA replication origins. A third class of proposed functions for moderately repetitive DNAs is that they increase

the evolutionary versatility of eukaryotic genomes by facilitating chromosomal rearrangements and/or forming reservoirs from which new functional sequences can be recruited.

Considering both the enormous amount of repetitive DNA in most eukaryotic genomes and the dearth of confirmatory evidence for any of the above proposals, a possibility that must be seriously entertained is that most repetitive DNA serves no useful purpose whatever for its host. Rather, it is **selfish** or **junk DNA,** molecular parasites that, over many generations, have disseminated themselves throughout the genome via various transpositional processes. The theory of natural selection indicates that the increased metabolic burden imposed by the replication of an otherwise harmless selfish DNA would eventually lead to its elimination. Yet, for slowly growing eukaryotes, the relative disadvantage of replicating, say, an additional 1000 bp of selfish DNA in an ~1 billion-bp genome would be so slight that its rate of elimination would be balanced by its rate of propagation. The C-value paradox may therefore simply indicate that a significant fraction, if not the great majority, of each eukaryotic genome is selfish DNA.

C. *Distribution of Genes*

The major goal of the human genome project is to provide a catalog of all human genes and their encoded proteins. Even with the finished sequence now in hand, this is by no means a simple task. In organisms with small genomes, such as bacteria and yeast, gene identification is quite straightforward because these genomes contain relatively little unexpressed DNA. However, *only 1.1 to 1.4% of the human genome consists of expressed sequences,* with ~24% of the genome consisting of introns and ~75% consisting of **intragenic sequences** (untranscribed sequences between genes). Consequently, our incomplete knowledge of the features through which cells recognize genes combined with the fact that human genes consist of relatively short exons (averaging ~150 nt) interspersed by much longer introns (averaging ~3500 nt and often much longer) greatly increases the difficulty (decreases the signal-to-noise ratio) of identifying genes. Hence, computer programs for sequence-based gene identification have had but limited success. Gene prediction algorithms therefore rely on sequence alignments with **expressed sequence tags (ESTs;** cDNAs that have been reverse transcribed from mRNAs; Section 7-2B) together with alignments with known genes from other organisms (which is often successful for highly conserved genes but is less so for genes that are rapidly evolving).

An important clue as to the occurrence of a gene is provided by the presence of a **CpG island.** 5-Methylcytosine (m^5C), as we have seen, occurs largely in the CG dinucleotides of various eukaryotic palindromic sequences, where it is implicated in switching off gene expression (Section 30-7). Since the spontaneous deamination of m^5C yields a normal T and thereby often results in a CG → TA mutation, CG dinucleotides occur in the human genome at about one-fifth of their randomly expected frequency

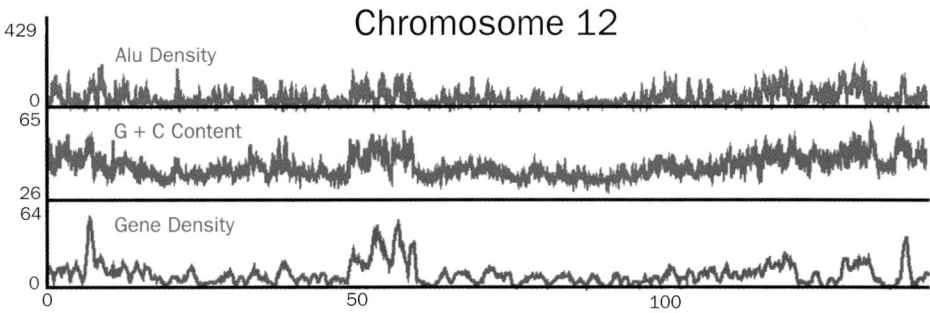

FIGURE 34-26   Density of structural features along the length of human chromosome 12. Gene density (*orange*) in this 133-megabase pair **(Mb)** chromosome is calculated per 1-Mb window, the percent G + C content (*green*) is calculated per 100-kb window, and the density of *Alu* elements (*magenta*) is calculated per 100-kb window. [Courtesy of Craig Venter, Celera Genomics, Rockville, Maryland.]

(which, since human DNA is 42% G + C, is 0.21 × 0.21 × 100 = 4% of dinucleotides). Nevertheless, the human genome contains ~29,000 ~1-kb regions known as CpG islands in which unmethylated CG dinucleotides occur at close to their expected frequency. About 56% of human genes are associated with CpG islands, which overlap the promoter regions of these genes and extend up to ~1 kb into their coding regions. Hence the presence of a CpG island in a vertebrate chromosome is strongly indicative of the presence of an associated gene.

Around 30,000 putative genes have been identified in the human genome. The discrepancy between this number and previous estimates of 50,000 to 140,000 genes is largely attributed to a much greater prevalence of alternative splicing than had previously been surmised (Section 31-4A). The gene density along the lengths of the various chromosomes is highly variable. Thus, although the average gene frequency in the human genome is ~1 gene per 100 kb

of DNA, this value varies from 0 to 64 genes per 100 kb (e.g., Fig. 34-26).

Genes may be classified as those that encode proteins (structural genes) and those that are transcribed to RNAs that are not translated. These latter so-called **noncoding RNAs (ncRNAs)** consist of tRNAs, rRNAs, small nuclear RNAs (snRNAs, which are components of spliceosomes; Section 31-4A), small nucleolar RNAs (snoRNAs, which participate in nucleolar RNA processing and base modification; Section 31-4B), as well as a variety of miscellaneous RNAs including the RNA components of the signal recognition particle (Section 12-4B), RNase P (Section 31-4B), and telomerase (Section 30-4D). The distribution of the rRNA and tRNA genes is discussed in Section 34-2D.

A total of 26,383 predicted structural genes in the human genome have been classified according to molecular function through sequence comparisons at both the level of protein families and of domains (Fig. 34-27). Note that

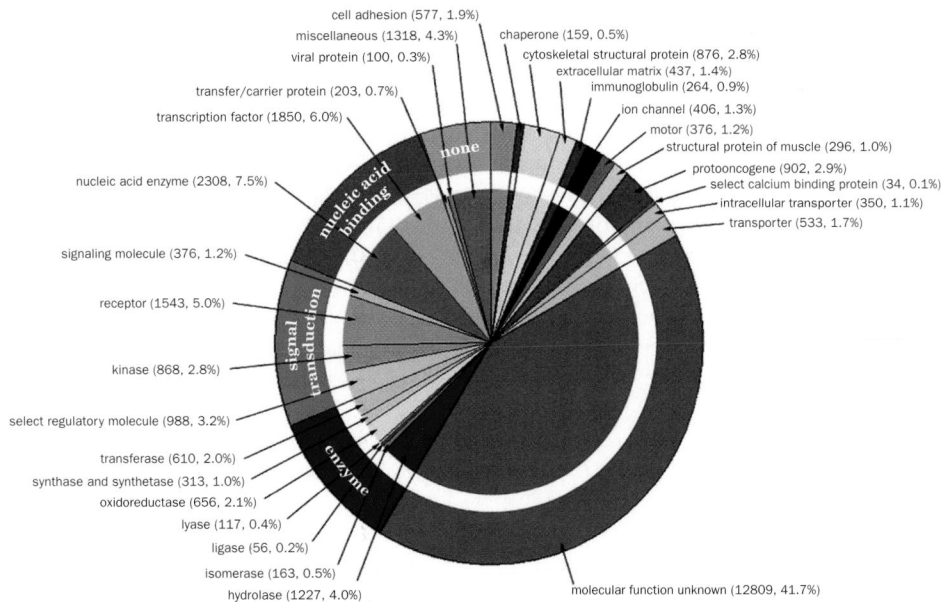

FIGURE 34-27   Distribution of molecular functions of 26,383 putative structural genes in the human genome. Each wedge of this pie chart lists, in parentheses, the number and percentage of the genes assigned to the indicated category of molecular function. The outer circle indicates the general functional categories whereas the inner circle provides a more detailed breakdown of these categories. [Courtesy Craig Venter, Celera Genomics, Rockville, Maryland.]

nearly 42% of them are classified as having unknown functions, as is likewise the case with most other genomes of known sequence, including those of prokaryotes. It can be seen from Fig. 34-27 that the most common molecular functions are those of transcriptions factors, proteins that mediate nucleic acid metabolism (nucleic acid enzymes), and receptors. Other common functions are those of kinases, hydrolases (most of which are proteases), proto-oncogenes (Section 19-3B), and select regulatory proteins (proteins that participate in signal transduction). Comparison of the structural genes in the human genome with those in the genomes of *D. melanogaster* and the nemotode worm *Caenorhabditis elegans* reveals that the greatest expansions of gene families occurred in those encoding proteins involved in developmental regulation (Section 34-4B), neuronal structure and function, hemostasis (blood clotting and related processes; Section 35-1), the acquired immune response (Section 35-2), and cytoskeletal complexity.

D. Tandem Gene Clusters

Most genes occur but once in an organism's haploid genome. This is sufficient, even for genes specifying proteins required in large amounts, through the accumulation of their corresponding mRNAs. However, the great cellular demand for rRNAs (which comprise ~80% of a cell's RNA) and tRNAs, which are all ncRNAs, can only be satisfied through the expression of multiple copies of the genes specifying them. In this subsection we discuss the organization of the genes coding for rRNAs and tRNAs. We shall also consider the organization of histone genes, the only protein-encoding genes that occur in multiple identical copies.

a. rRNA Genes Are Organized into Repeating Sets

We have seen in Sections 31-4B and 31-4C that even the *E. coli* genome, which otherwise consists of unique sequences, contains multiple copies of rRNA and tRNA genes. In eukaryotes, the genes specifying the 18S, 5.8S, and 28S rRNAs are invariably arranged in this order, reading $5' \rightarrow 3'$ on the RNA strand, and separated by short transcribed spacers to form a single transcription unit of ~7500 bp (Fig. 34-28). (Recall that the primary transcript of this gene cluster is a 45S RNA from which the mature rRNAs are derived by posttranscriptional cleavage;

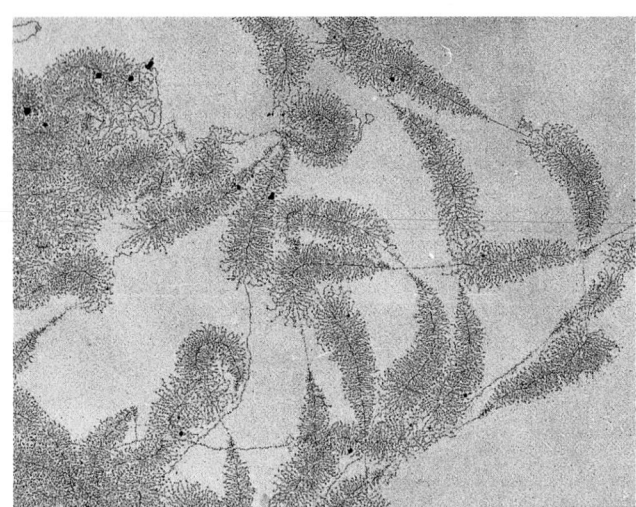

FIGURE 34-29 Electron micrograph of tandem arrays of actively transcribing 18S, 5.8S, and 28S rRNA genes from the nucleoli of the newt *Notophthalmus viridescens*. The axial fibers are DNA. The fibrillar "Christmas tree" matrices, which consist of newly synthesized RNA strands in complex with proteins, outline each transcriptional unit. Note that the longest ribonucleoprotein branches of each "Christmas tree" are only ~10% the length of their corresponding DNA stem. Apparently, the RNA strands are compacted through secondary structure interactions and/or protein associations. The matrix-free segments of DNA are the untranscribed spacers. [Courtesy of Oscar L. Miller, Jr., and Barbara R. Beatty, University of Virginia.]

Section 31-4B.) *Indeed, this rRNA gene arrangement is universal since the 5' end of prokaryotic 23S rRNA is homologous to eukaryotic 5.8S rRNA (Section 32-3A).*

Electron micrographs, such as Fig. 34-29, indicate that *the blocks of transcribed eukaryotic rRNA genes are arranged in tandem repeats that are separated by untranscribed spacers (Fig. 34-28).* These tandem repeats are typically ~12,000 bp in length, although the untranscribed spacer varies in length between species and, to a lesser extent, from gene to gene. Quantitative measurements of the amounts of radioactively labeled rRNAs that can hybridize with the corresponding nuclear DNA (**rDNA**) and more recently genomic sequencing indicate that these rRNA genes, which may be distributed among several chromosomes, vary in haploid number from less than 50 to over 10,000, depending on the species. Humans, for example, have 150 to 200 blocks of rDNA spread over 5 chromosomes.

b. The Nucleolus Is the Site of rRNA Synthesis and Ribosome Assembly

In a typical interphase cell nucleus, the rDNA condenses to form a single nucleolus (Fig. 1-5). There, as Fig. 34-29 suggests, these genes are rapidly and continuously transcribed by RNA polymerase I (Section 31-2E). The nucleolus, as demonstrated by radioactive labeling experiments, is also the site where these rRNAs are posttranscriptionally processed and assembled with cytoplasmically synthesized ribosomal proteins into immature ribosomal

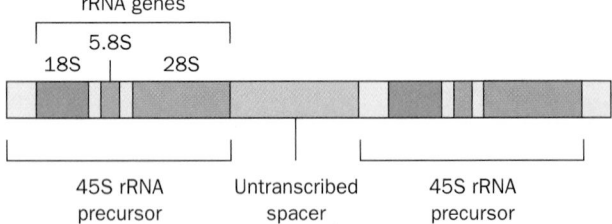

FIGURE 34-28 The 18S, 5.8S, and 28S rRNA genes are organized in tandem repeats in which sequences encoding the 45S rRNA precursor are interspersed by untranscribed spacers.

subunits. Final assembly of the ribosomal subunits only occurs as they are being transferred to the cytoplasm, which presumably prevents the premature translation of partially processed mRNAs (hnRNAs) in the nucleus.

c. 5S rRNA and tRNA Genes Occur in Multiple Clusters

The genes encoding the 120-nucleotide 5S rRNAs, much like the other rRNA genes, are arranged in clusters that contain a total of several hundred to several hundred thousand tandem repeats distributed among one or more chromosomes. In *X. laevis*, for example, the repeating unit consists of the 5S rRNA gene, a nearby **pseudogene** (a 101-bp segment of the 5S rRNA gene that, curiously, is not transcribed), and an untranscribed spacer of variable length but averaging ~400 bp (Fig. 34-30). The 5S rRNA genes are transcribed outside of the nucleolus by RNA polymerase III (Section 31-2E). 5S rRNA must therefore be transported into the nucleolus for incorporation into the large ribosomal subunit.

The 497 tRNA genes that have been identified in the human genome are likewise transcribed by RNA polymerase III. They are also multiply reiterated and clustered, with >25% of them occurring in a 4-Mb region on chromosome 6 and most of the remainder clustered on numerous but not all chromosomes.

d. Histone Genes Are Reiterated

Histone mRNAs have relatively short cytoplasmic lifetimes because of their lack of the poly(A) tails that are appended to other eukaryotic mRNAs (Section 31-4A). Yet histones must be synthesized in large amounts during S phase of the cell cycle (when DNA is synthesized). *This process is made possible through the multiple reiteration of histone genes, which in most organisms are the only identically repeated genes that code for proteins.* This organization, it is thought, permits the sensitive control of histone synthesis through the coordinate transcription of sets of histone genes. Histone genes also differ from nearly all other eukaryotic genes in that almost all histone sequences lack introns. The significance of this observation is unknown.

There is little relationship between a genome's size and its total number of histone genes. For example, birds and mammals have 10 to 20 copies of each of the five histone genes, *D. melanogaster* has ~100, and sea urchins have several hundred. This suggests that the efficiency

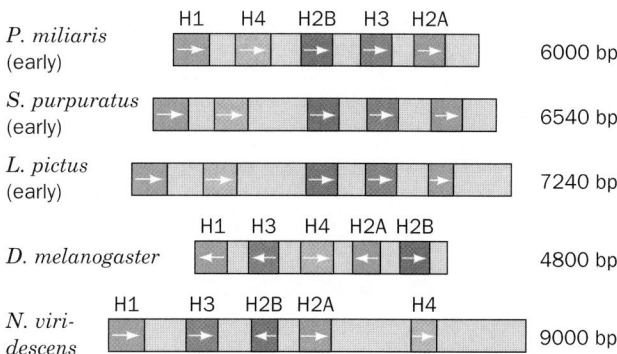

FIGURE 34-31 **The organization and lengths of the histone gene cluster repeating units in a variety of organisms.** Coding regions are indicated in color and spacers are gray. The arrows denote the directions of transcription (the top three organisms are distantly related sea urchins).

of histone gene expression varies with species. In many organisms the histone genes are organized into tandemly repeated quintets consisting of a gene coding for each of the five different histones interspersed by untranscribed spacers (Fig. 34-31). The gene order and the direction of transcription in these quintets are preserved over large evolutionary distances. Corresponding spacer sequences vary widely among species and, to a limited extent, among the repeating quintets within a genome. In birds and mammals, this repetitious organization has broken down; their histone genes occur in clusters but in no particular order.

e. Reiterated Sequences May Be Generated and Maintained by Unequal Crossovers and/or Gene Conversion

How do reiterated genes maintain their identity? The usual mechanism of Darwinian selection would seem ineffective in accomplishing this since deleterious mutations in a few members of a multiply repeated set of identical genes would have little phenotypic effect. Indeed, many mutations do not affect the function of a gene product and are therefore selectively neutral. Reiterated gene sets must therefore maintain their homogeneity through some additional mechanism. Two such mechanisms seem plausible:

1. In the **unequal crossover** mechanism (Fig. 34-32*a*), recombination occurs between homologous segments of misaligned chromosomes, thereby excising a segment from one of the chromosomes and adding it to the other. Computer simulations indicate that such repeated expansions and contractions of a chromosome will, by random processes, generate a cluster of reiterated sequences that have been derived from a much smaller ancestral cluster. Unequal crossing-over is also thought to be the mechanism that generated segmental duplications (Section 34-2B).

2. In the **gene conversion** mechanism (Fig. 34-32*b*), one member of a reiterated gene set "corrects" a nearby variant through a process resembling recombination repair (Section 30-6A).

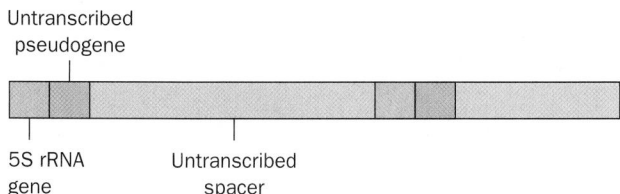

FIGURE 34-30 **The organization of the 5S RNA genes in** *Xenopus laevis.* Each of the ~750-nt tandemly repeated units consists of a 5S rRNA gene trailed by an untranscribed spacer in which a pseudogene closely follows the 5S gene.

(a) Crossing over

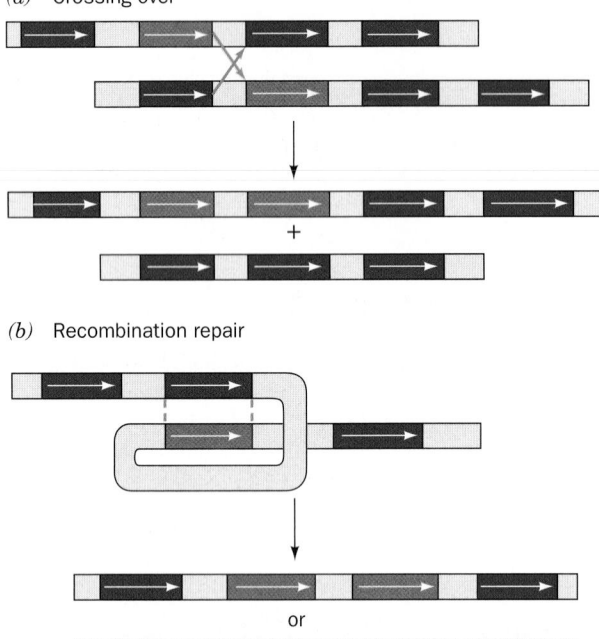

(b) Recombination repair

or

FIGURE 34-32 Two possible mechanisms for maintaining the homogeneity of a tandem multigene family. (*a*) Unequal crossing-over between mispaired but similar genes results in an unpaired DNA segment being deleted from one chromosome and added to the other. (*b*) Gene conversion "corrects" one member of a tandem array with respect to the other via a recombination repair mechanism. Repeated cycles of either process may either eliminate a variant gene or spread it throughout the entire tandem array.

Since point mutations are rare events compared to crossovers, either mechanism would eventually result in a newly arisen variant copy of a repeated sequence either being eliminated or taking over the entire cluster. If a mutation that has been so concentrated is deleterious, it will be eliminated by Darwinian selection. In contrast, variant spacers, which are not as subject to selective pressure, would be eliminated at a slower rate. The existence of reiterated sets of identical genes separated by somewhat heterogeneous spacers may therefore be reasonably attributed to either homogenization model.

E. *Gene Amplification*

The selective replication of a particular set of genes, a process known as **gene amplification,** normally occurs only at specific stages of the life cycle of certain organisms. In the following subsections, we outline what is known about this phenomenon.

a. rRNA Genes Are Amplified during Oogenesis

The rate of protein synthesis during the early stages of embryonic growth is so great that in some species the normal genomic complement of rRNA genes cannot satisfy the demand for rRNA. In these species, notably certain insects, fish, and amphibians, the rDNA is differentially replicated in developing oocytes (immature egg cells). In one of the most spectacular examples of this process, the rDNA in *X. laevis* oocytes is amplified by ~1500 times its amount in somatic cells to yield some 2 million sets of rRNA genes comprising nearly 75% of the total cellular DNA. The amplified rDNA occurs as extrachromosomal circles, each containing one or two transcription units, that are organized into hundreds of nucleoli (Fig. 34-33). Mature *Xenopus* oocytes therefore contain ~10^{12} ribosomes, 200,000 times the number in most larval cells. This is so many that mutant zygotes (fertilized ova) that lack nucleoli (and thus cannot synthesize new ribosomes; the oocyte's extra nucleoli are destroyed during its first meiotic division) survive to the swimming tadpole stage with only their maternally supplied ribosomes.

What is the mechanism of rDNA amplification? An important clue is that the untranscribed spacers from a given extrachromosomal nucleolus all have the same length, whereas we have seen that the corresponding chromosomal spacers exhibit marked length heterogeneities. This observation suggests that the rDNA circles in a single nucleolus are all descended from a single chromosomal gene. Gene amplification has been shown to occur in two stages: A low level of amplification in the first stage followed by massive amplification in the second stage. It therefore seems likely that, in the first stage, no more than a few chromosomal rRNA genes are replicated by an unknown mechanism and the daughter strands are released as extrachromosomal circles. Then, in the second stage, these circles are multiply replicated by the rolling circle mechanism (Section 30-3B). In support of this hypothesis are electron micrographs of amplified genes showing the

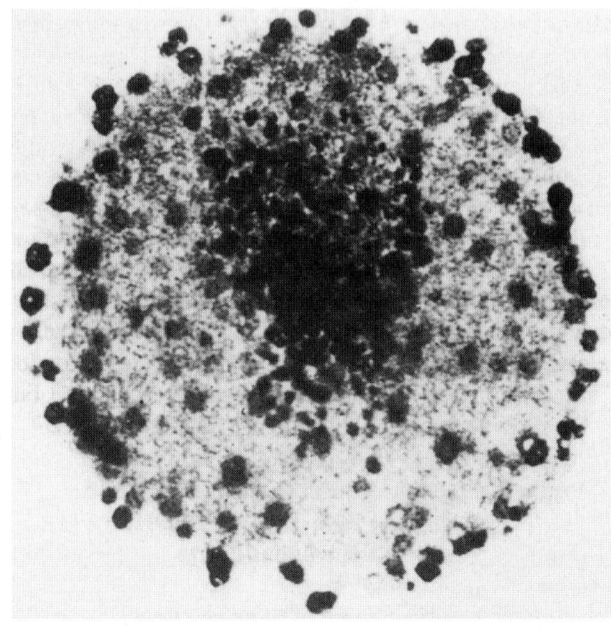

FIGURE 34-33 Photomicrograph of an isolated oocyte nucleus from *X. laevis*. Its several hundred nucleoli, which contain amplified rRNA genes, appear as darkly staining spots. [Courtesy of Donald Brown, Carnegie Institution of Washington.]

"lariat" structures postulated to be rolling circle intermediates (Fig. 30-25).

b. Chorion Genes Are Amplified

The only other known example of programmed gene amplification is that of the *D. melanogaster* ovarian follicle cell genes that code for **chorion** (egg shell) **proteins** (ovarian follicle cells surround and nourish the maturing egg). Prior to chorion synthesis, the entire haploid genome of each ovarian follicle cell is replicated 16-fold. This process is followed by an ~10-fold selective replication of only the chorion genes to form a multiply branched (partially polytene) structure in which the amplified chorion genes remain part of the chromosome (Fig. 34-34). Interestingly, chorion gene amplification does not occur in silk moth oocytes. Rather, this organism's genome has multiple copies of chorion genes.

c. Drug Resistance Can Result from Gene Amplification

In cancer chemotherapy, a common observation is that the continued administration of a cytotoxic drug causes an initially sensitive tumor to become increasingly drug resistant to the point that the drug loses its therapeutic efficacy. One mechanism by which a cell line can acquire such drug resistance is through the overproduction of the drug's target enzyme. Such a process can be observed, for example, by exposing cultured animal cells to the dihydrofolate analog methotrexate. This substance, it will be recalled, all but irreversibly binds to dihydrofolate reductase (DHFR), thereby inhibiting DNA synthesis (Section 28-3B). Slowly increasing the methotrexate dose yields surviving cells that ultimately contain up to 1000 copies of the DHFR gene and are thereby capable of tremendous overproduction of this enzyme, a clear laboratory demonstration of Darwinian selection. Members of some of these cell lines contain extrachromosomal elements known as **double minute chromosomes** that each bear one or more copies of the DHFR gene, whereas in other cell lines the additional DHFR genes are chromosomally integrated. The mechanism of gene amplification in either cell type is not well understood, although it is worth noting that this phenomenon is only known to occur in cancer cells. Both types of amplified genes are genetically unstable; further cell growth in the absence of methotrexate results in the gradual loss of the extra DHFR genes.

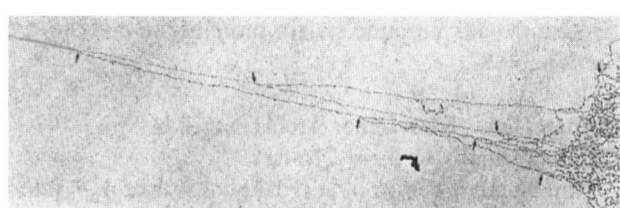

FIGURE 34-34   An electron micrograph of a chorion gene-containing chromatin strand from an oocyte follicle cell of *D. melanogaster.* The strand has undergone several rounds of partial replication (*arrows at replication forks*) to yield a multiforked structure containing several parallel copies of chorion genes. [Courtesy of Oscar L. Miller, Jr., University of Virginia.]

F. *Clustered Gene Families: Hemoglobin Gene Organization*

Few proteins in a given organism are really unique. Rather, like the digestive enzymes trypsin, chymotrypsin, and elastase (Section 15-3), or the various collagens (Section 8-2B), they are usually members of families of structurally and functionally related proteins. In many cases, the family of genes specifying such proteins are clustered together in a single chromosomal region. In the following subsections, we consider the organization of two of the best characterized clustered gene families, those encoding the two types of human hemoglobin subunits. The clustered gene families that encode immune system proteins are discussed in Section 35-2C.

a. Human Hemoglobin Genes Are Arranged in Two Developmentally Ordered Clusters

Human adult hemoglobin (HbA) consists of $\alpha_2\beta_2$ tetramers in which the α and β subunits are structurally related. The first hemoglobin made by the human embryo, however, is a $\zeta_2\varepsilon_2$ tetramer **(Hb Gower 1)** in which ζ and ε are α- and β-like subunits, respectively (Fig. 34-35). By around 8 weeks postconception, the embryonic subunits have been supplanted (in newly formed erythrocytes) by the α subunit and the β-like γ subunit to form fetal hemoglobin (HbF), $\alpha_2\gamma_2$ (the hemoglobins present during the changeover period, $\alpha_2\varepsilon_2$ and $\zeta_2\gamma_2$, are named **Hb Gower 2** and **Hb Portland,** respectively). The γ subunit is gradually superseded by β starting a few weeks before birth. Adult blood normally contains ~97% HbA, 2% **HbA$_2$** ($\alpha_2\delta_2$ in which δ is a β variant), and 1% HbF.

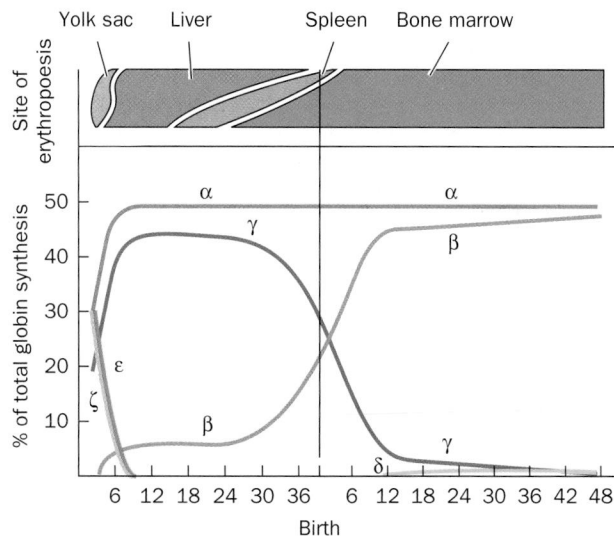

FIGURE 34-35   The progression of human globin chain synthesis with embryonic and fetal development. Note that any red blood cell contains only one type each of α- and β-like subunits. The progression in the sites of **erythropoiesis** (red cell formation), which is indicated in the upper panel, corresponds roughly to the major switches in hemoglobin types. [After Weatherall, D.J. and Clegg, J.B., *The Thalassaemia Syndromes* (3rd ed.), *p.* 64, Blackwell Scientific Publications (1981).]

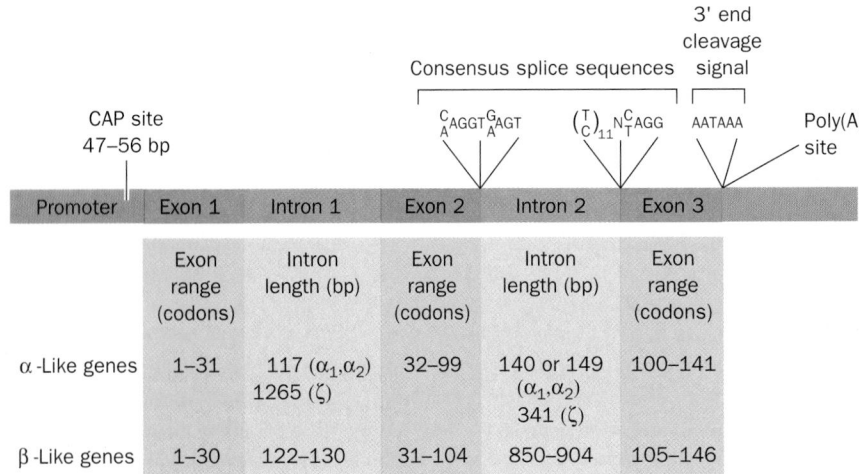

α-Like genes
Active gene   Pseudogene (ψ)   *Alu* element
Chromosome 16
ζ   ψζ ψα2   ψα1   α2   α1   ψθ

β-Like genes
L1
Chromosome 11
ε   Gγ   Aγ   ψβ   δ   β

FIGURE 34-36   The organization of human globin genes on their respective sense strands. Red boxes represent active genes; green boxes represent pseudogenes; yellow boxes represent L1 sequences, with the arrows indicating their relative orientations; and triangles represent *Alu* elements in their relative orientations. [After Karlsson, S. and Nienhuis, A.W., *Annu. Rev. Biochem.* **54,** 1074 (1985).]

In mammals, the genes specifying the α- and β-like hemoglobin subunits form two different gene clusters that occur on separate chromosomes. In humans and many other mammals, the genes in each globin cluster are arranged, 5′ → 3′ on the coding strands, in the order of their developmental expression (Fig. 34-36). This ordering is common in mammals but not universal; in the mouse β gene cluster, for instance, the adult genes precede the embryonic genes.

The β-globin gene cluster (Fig. 34-36), which spans ~100 kb, contains five functional genes: the embryonic ε gene, two fetal genes, Gγ and Aγ (duplicated genes that encode polypeptides differing only by having either Gly or Ala at their positions 136), and the two adult genes, δ and β. The β-globin cluster also contains one **pseudogene**, ψβ (an untranscribed relic of an ancient gene duplication that is ~75% identical to the β gene), eight *Alu* elements, and two L1 elements (Section 34-2B).

The α-globin gene cluster (Fig. 34-36), which spans ~28 kb, contains three functional genes: the embryonic ζ gene and two slightly different α genes, α1 and α2, which encode identical polypeptides. The α cluster also contains four pseudogenes, ψζ, ψα2, ψα1, and ψθ, and three *Alu* elements.

b. Hemoglobin Genes All Have the Same Exon–Intron Structure

Protein-coding sequences represent <5% of either globin gene cluster. This situation is largely a consequence of the heterogeneous collection of untranscribed spacers separating the genes in each cluster. In addition, *all known vertebrate globin genes, including that of myoglobin and most hemoglobin pseudogenes, consist of three nearly identically placed exons separated by two somewhat variable introns (Fig. 34-37).* This gene structure apparently arose quite early in vertebrate history, well over 500 million years ago. Indeed, much of this structure even predates the divergence of plants and animals. The structure of the gene encoding leghemoglobin (a plant globin that functions in legumes to protect nitrogenase from O_2 poisoning; Section 26-6) differs from that of vertebrates only in that the central exon of vertebrate globins is split by a third intron in the leghemoglobin gene. Quite possibly the central exon in vertebrate globins arose through the fusion of the two interior exons in a leghemoglobin-like ancestral gene.

c. DNA Polymorphisms Can Establish Genealogies

Unexpressed sequences, which are subject to little selective pressure, evolve so much faster than expressed

3′ end
cleavage
Consensus splice sequences   signal
CAP site
47–56 bp ^C_AAGGT^G_AAGT $(^T_C)_{11}$N^C_TAGG   AATAAA   Poly(A)
site

Promoter	Exon 1	Intron 1	Exon 2	Intron 2	Exon 3

	Exon range (codons)	Intron length (bp)	Exon range (codons)	Intron length (bp)	Exon range (codons)
α-Like genes	1–31	117 (α₁,α₂) 1265 (ζ)	32–99	140 or 149 (α₁,α₂) 341 (ζ)	100–141
β-Like genes	1–30	122–130	31–104	850–904	105–146

FIGURE 34-37   Structure of the prototypical hemoglobin gene. The conserved sequences at the exon–intron boundaries (splice sequences) and at the 3′ end of the gene (polyadenylation site) are indicated. The length range of each exon (in codons) and each intron (in base pairs) is given. [After Karlsson, S. and Nienhuis, A.W., *Annu. Rev. Biochem.* **54,** 1079 (1985).]

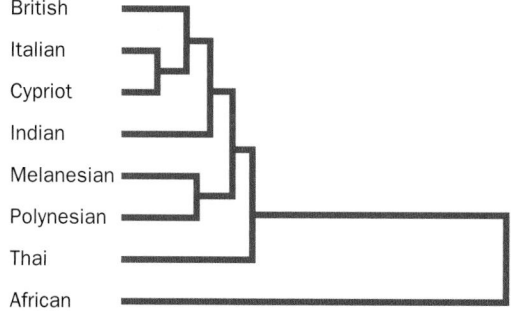

FIGURE 34-38 A family tree showing the lines of descent among eight human population groups. These were determined from the distribution of five restriction-fragment length polymorphisms (RFLPs) in their β-globin gene clusters. The horizontal axis is indicative of the genetic distances between related populations and therefore of the times between their divergence. [After Wainscoat, J.S., Hill, A.V.S., Boyce, A.L., Flint, J., Hernandez, M., Thein, S.L., Old, J.M., Lynch, J.R., Falusi, A.G., Weatherall, D.J., and Clegg, J.B., *Nature* **319**, 493 (1986).]

sequences that they even accumulate significant numbers of sequence **polymorphisms** (variations) within a single species. Consequently, the evolutionary relationships among populations within a species can be established by determining how a series of polymorphic DNA sequences are distributed among them. For example, the genealogy of several diverse human populations has been inferred from the presence or absence of certain restriction sites [restriction-fragment length polymorphisms (RFLPs); Section 5-5A] in five segments of their β-globin gene clusters. This study has led to the construction of a "family tree" (Fig. 34-38), which indicates that non-African (Eurasian) populations are much more closely related to each other than they are to African populations. Fossil evidence indicates that anatomically modern man arose in Africa about 100,000 years ago and rapidly spread throughout that continent. This family tree therefore suggests that all Eurasian populations are descended from a surprisingly small "founder population" (perhaps only a few hundred individuals) that left Africa ~50,000 years ago. A similar analysis indicates that the sickle-cell variant of the β gene arose on at least three separate occasions in geographically distinct regions of Africa.

G. *The Thalassemias: Genetic Disorders of Hemoglobin Synthesis*

The study of mutant hemoglobins (Section 10-3) has provided invaluable insights into structure–function relationships in proteins. Likewise, the study of defects in hemoglobin expression has greatly facilitated our understanding of eukaryotic gene expression.

The most common class of inherited human disease results from the impaired synthesis of hemoglobin subunits. These anemias are named **thalassemias** (Greek: *thalassa*, sea) because they commonly occur in the region surrounding the Mediterranean Sea (although they are also prevalent in Central Africa, India, and the Far East). The

observation that malaria is or was endemic in these same areas (Fig. 7-20) led to the realization that heterozygotes for thalassemic genes (who appear normal or are only mildly anemic; a condition known as **thalassemia minor**) are resistant to malaria. Thus, as we have seen in our study of sickle-cell anemia (Section 10-3B), mutations that are seriously debilitating or even lethal in homozygotes (who are said to suffer from **thalassemia major**) may offer sufficient selective advantage to heterozygotes to ensure the propagation of the mutant gene.

Thalassemia can arise from many different mutations, each of which causes a disease state of characteristic severity. In α^0- and β^0-thalassemias, the indicated globin chain is absent, whereas in α^+- and β^+-thalassemias, the normal globin subunit is synthesized in reduced amounts. In what follows, we shall consider thalassemias that are illustrative of several different types of genetic lesions.

a. α-Thalassemias

Most α-thalassemias are caused by the deletion of one or both of the α-globin genes in an α gene cluster (Fig. 34-36). A variety of such mutations have been cataloged. In the absence of equivalent numbers of α chains, the fetal γ chains and the adult β chains form homotetramers: **Hb Bart's** (γ_4) and **HbH** (β_4). Neither of these tetramers exhibits any cooperativity or Bohr effect (Sections 10-1C and 10-1D), which makes their oxygen affinities so high that they cannot release oxygen under physiological conditions. Consequently, α^0-thalassemia occurs with four degrees of severity depending on whether an individual has 1, 2, 3, or 4 missing α-globin genes:

1. Silent-carrier state: The loss of one α gene is an asymptomatic condition. The rate of expression of the remaining α genes largely compensates for the less than normal α gene dosage so that, at birth, the blood contains only ~1 to 2% Hb Bart's.

2. α-Thalassemia trait: With two missing α genes (either one each deleted from both α gene clusters or both deleted from one cluster), only minor anemic symptoms occur. The blood contains ~5% Hb Bart's at birth.

3. Hemoglobin H disease: Three missing α genes results in a mild to moderate anemia. Affected individuals can usually lead normal or nearly normal lives.

4. Hydrops fetalis: The lack of all four α genes is lethal. Unfortunately, the synthesis of the embryonic ζ chain continues well past the 8 weeks postconception when it normally ceases (Fig. 34-35), so the fetus usually survives until around birth.

α-Thalassemias caused by nondeletion mutations are relatively uncommon. One of the best characterized such lesions changes the UAA stop codon of the α2-globin gene to CAA (a Gln codon), so that protein synthesis continues for the 31 codons beyond this site to the next UAA. The resultant **Hb Constant Spring** is produced in only small amounts because, for unknown reasons, its mRNA is rapidly degraded in the cytosol. Another point mutation in the α2 gene changes Leu H8(125)α to Pro, which no doubt disrupts the H helix. The consequent α^+-thalassemia results from the rapid degradation of this abnormal **Hb Quong Sze.**

b. β-Thalassemias

Heterozygotes of β-thalassemias are usually asymptomatic. Homozygotes become so severely anemic, however, that once their HbF production has diminished, many require frequent blood transfusions to sustain life and all require them to prevent the severe skeletal deformities caused by bone marrow expansion. The anemia results not only from the lack of β chains but also from the surplus of α chains. The latter form insoluble membrane-damaging precipitates that cause premature red cell destruction (Section 10-3A). The coinheritance of α-thalassemia therefore tends to lessen the severity of β-thalassemia major.

In β-thalassemia, there may be an increased production of the δ and γ chains, so that the consequent extra HbA_2 and HbF can compensate for some of the missing HbA. In δβ-**thalassemia,** the neighboring δ and β genes have both been deleted, so that only increased production of the γ chain is possible. Yet many adult δβ-thalassemics, for reasons that are not understood, produce so much HbF that they are asymptomatic. Such individuals are said to have **hereditary persistence of fetal hemoglobin (HPFH).** This condition is therefore of medical interest because it could also alleviate the symptoms of β-thalassemia and sickle-cell anemia.

The so-called Greek form of HPFH is associated with a G → A mutation at position −117 of the γ-globin gene (its promoter region). In an effort to establish whether this mutation does, in fact, cause HPFH, the mutated γ-globin gene was introduced into mice. The resulting fetal and adult transgenic animals synthesized γ-globin at a high level, with a concomitant decrease in the synthesis of the β-globin gene. These changes in gene expression correlate with the loss of binding of the transcription factor **GATA-1** to the γ-globin promoter, thereby suggesting that this protein is a negative regulator of the γ-globin gene expression in normal human adults (transcription factors are discussed in Section 34-3B).

β^0-Thalassemias caused by deletions are rare compared to those causing α^0-thalassemias. This is probably because the long repeated sequences in which the α-globin genes are embedded make them more prone to unequal crossing-over than the β-globin gene. Nevertheless, a β-thalassemic lesion causing the production of **Hb Lepore** is a particularly clear instance of this deletion mechanism. This lesion, the consequence of a deletion extending from within the δ gene to the corresponding position of its neighboring β gene, yields a δ/β hybrid subunit. Such deletions almost certainly arose through unequal crossovers between the β gene on one chromosome and the δ gene on another (Fig. 34-39; the two genes are 93% identical in sequence). The second product of such crossovers, a chromosome containing a β/δ hybrid flanked by normal δ and β genes (Fig. 34-39), is known as **Hb anti-Lepore.** Homozygotes for Hb Lepore have symptoms similar to those of β-thalassemia major, whereas homozygotes for Hb anti-Lepore, which have the full complement of normal globin genes, are symptom free and have only been detected through blood tests.

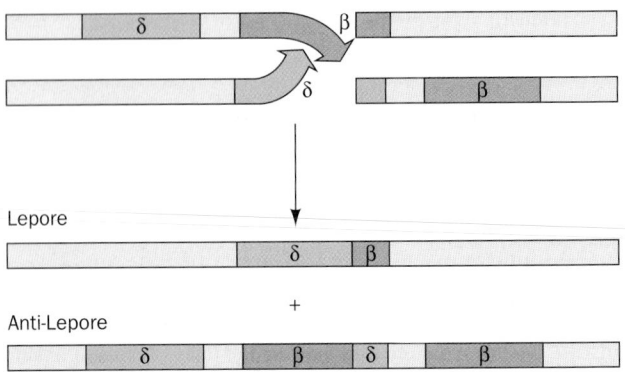

FIGURE 34-39 The formation of Hb Lepore and Hb anti-Lepore. This occurs by unequal crossing-over between the β-globin gene on one chromosome and the δ-globin gene on its homolog.

Most β-thalassemias are caused by a wide variety of point mutations that affect the production of β chains. These include:

1. Nonsense mutations that convert normal codons to the stop codon UAG.

2. Frameshift mutations that insert/delete one or more base pairs into/from an exon.

3. Point mutations in the β gene's promoter region, either in its TATA box or in its CACCC box (Section 31-2E). These attenuate transcriptional initiation.

4. Point mutations that alter the sequence at an exon–intron junction (Section 31-4A). These diminish/abolish splicing and/or activate a **cryptic splice site** (an exon–intron junction-like sequence that normally is not spliced) to pair with the altered intron's unaltered end.

5. A point mutation that alters an intron's lariat-branch site (Section 31-4A). This activates a cryptic 3′ splice upstream of the original site, leading to the excision of a shorter than normal intron.

6. Point mutations that create new splice sites. These either compete with the neighboring normal splice site or pair with a nearby cryptic splice site.

7. A point mutation that alters the AAUAAA cleavage signal at the mRNA's 3′ end (Section 31-4A).

Consideration of the effects of these mutations, particularly those involving gene splicing, has confirmed and extended our understanding of how eukaryotic genes are constructed and expressed.

3 ■ CONTROL OF EXPRESSION

The elucidation of the mechanisms controlling gene expression in eukaryotes had lagged at least 20 years behind that of prokaryotes. In addition to the far greater complexity of eukaryotic systems, this is largely because the types of genetic analyses that have been so useful in characterizing prokaryotic systems (which require the detection of very rare events) are precluded in metazoa (multicellular animals) by their much slower reproductive rates.

Compounding this problem are the difficulties in selecting for mutations in essential genes; the missing product of a defective enzyme in a metazoan usually cannot be replaced by simply adding that product to the diet as is often possible with, say, *E. coli.* This latter difficulty can be partially overcome by the rather laborious task of growing cells from metazoa in tissue culture. Since somatic cells do not normally undergo genetic recombination, however, genetic manipulations cannot be carried out in tissue culture the way they can in a bacterial culture.

What has made genetic manipulations of metazoa feasible is the development, in the 1970s, of molecular cloning techniques (Section 5-5). The gene encoding a particular eukaryotic protein can be identified in genomic or cDNA libraries through Southern blotting (Section 5-5D) or PCR (Section 5-5F) using an oligonucleotide probe or primer whose sequence encodes a segment of the protein (a process termed reverse genetics; Section 7-2C). Alternatively, if the organism's genome has been sequenced, the gene may be identified *in silico* (computationally). The gene may then be modified, for example, through site-directed mutagenesis (Section 5-5G), and the effects of the modification analyzed in an expression vector such as *E. coli* or yeast, or alternatively, *in vitro.*

The expression of foreign genes in metazoans (a gain of function) has been made possible through the development of a process in which DNA is microinjected into the nucleus of a fertilized ovum (Fig. 5-59). Such DNA often integrates into the chromosome of the resulting zygote, that then undergoes normal development to form a **transgenic** individual whose cells each contain the foreign genes (in *Xenopus,* this merely involves allowing the transfected egg to hatch, whereas in mice the fertilized ovum must be implanted in the uterus of a properly prepared foster mother; see Fig. 5-5 for a striking example of a transgenic mouse). Alternatively, a normal gene may be selectively inactivated (knocked out; a loss of function) through the use of DNA specifying the defective gene, which then recombines with the normal gene. The genome of an already multicellular organism may be altered, in a technique that holds great promise for gene therapy, through the use of defective (unable to reproduce) retroviruses that contain the genes to be transferred (Section 5-5H). Thus, the genomes of metazoans can now be manipulated, albeit with considerable clumsiness. We are, however, rapidly becoming more adept at these procedures as we gain further understanding of how eukaryotic chromosomes are organized and expressed.

Single-celled eukaryotes, particularly yeasts, are exceptions to the foregoing discussion because they can be grown and manipulated in much the same way as bacteria. Indeed, much of our knowledge of eukaryotic molecular biology has been obtained through molecular genetic analyses of budding (baker's) yeast (*Saccharomyces cerevisiae*). In this chapter, unless otherwise indicated, the term "yeast" refers to *S. cerevisiae.*

In this section we consider the molecular basis of the enormous expressional variation that eukaryotic cells exhibit. In doing so, we shall first study the nature of transcriptionally active chromatin, then discuss how genetic expression in eukaryotes is mainly regulated through the control of transcriptional initiation, and finally consider the other means by which eukaryotes control genetic expression. Eukaryotic gene regulation, as we shall see, is an astoundingly complex process that requires the participation of well over 100 polypeptides that form assemblies with molecular masses of several million daltons. In the following section, we take up the molecular basis of normal cell differentiation, its aberration, cancer, and programmed cell death.

A. *Chromosomal Activation and Deactivation*

Interphase chromatin, as is mentioned in Section 34-1, may be classified in two categories: the highly condensed and transcriptionally inactive heterochromatin, and the diffuse and transcriptionally active or activatable euchromatin (Fig. 34-2). Two types of heterochromatin have been distinguished:

1. Constitutive heterochromatin, which is permanently condensed in all cells and consists mostly of the highly repetitive sequences clustered near the chromosomal centromeres (Section 34-2B) and telomeres (Section 30-4D). Constitutive heterochromatin is transcriptionally inert.

2. Facultative heterochromatin, which varies in a tissue-specific manner. Presumably the condensation of facultative heterochromatin functions to transcriptionally inactivate large chromosomal blocks.

a. Most Mammalian Cells Have Only One Active X Chromosome

Female mammalian cells contain two X chromosomes, whereas male cells have one X and one Y chromosome. *Female somatic cells, however, maintain only one of their X chromosomes in a transcriptionally active state.* Consequently, males and females make approximately equal amounts of X chromosome–encoded gene products, a phenomenon known as **dosage compensation.** The inactive X chromosome is visible during interphase as a heterochromatin structure known as a **Barr body** (Fig. 34-40). In marsupials (pouched mammals), the Barr body is always the paternally inherited X chromosome, an epigenetic phenomenon (Section 30-7). In placental mammals, however, one randomly selected X chromosome in every somatic cell is inactivated when the embryo consists of only a few cells. The progeny of each of these cells epigenetically maintain the same inactive X chromosomes. *Female placental mammals are therefore mosaics composed of clonal groups of cells in which the active X chromosome is either paternally or maternally inherited.* This situation is particularly evident in human females who are heterozygotes for the X-linked congenital sweat gland deficiency **anhidrotic ectodermal dysplasia.** The skin of these women consists of patches lacking sweat glands, in which only the X chromosome containing the mutant gene is active, alternating with normal patches in which only the other X chromosome is active. Similarly, calico cats, whose coats consist of patches of black fur and yellow fur, are almost always females whose two X chromosomes are allelic for black and yellow furs.

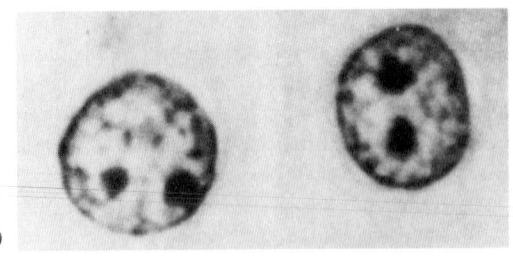

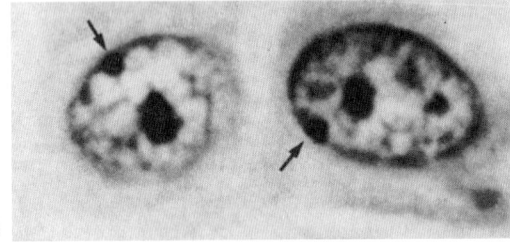

FIGURE 34-40 Photomicrographs of stained nuclei from human oral epithelial cells. (*a*) From a normal XY male showing no Barr body. (*b*) From a normal XX female showing a single Barr body (*arrow*). The presence of Barr bodies permits the rapid determination of an individual's chromosomal sex. [From Moore, K.L. and Barr, M.L., *Lancet* **2,** 57 (1955).]

The mechanism of X chromosome inactivation is only beginning to come to light. Inactivation appears to be triggered by the transcription of the ***Xist*** gene in the inactive chromosome only. The consequent *Xist* RNA "paints" the inactive X chromosome over its entire length but does not bind to the active X chromosome. This localized *Xist* RNA, through a cascade of poorly understood changes, recruits DNA-binding proteins that repress transcription (Section 34-3B) as well as variant histones, particularly the H2A variant **macroH2A1,** which has a large C-terminal globular domain that H2A lacks. X inactivation also results in modification of its bound core histones, particularly the methylation of H3 Lys 4, the demethylation of H3 Lys 9, and the hypoacetylation of histones H3 and H4 (histone methylation and acetylation are discussed in Section 34-3B). In addition, the DNA of the *Xist* gene on the active X chromosome becomes methylated, whereas the active *Xist* gene on the otherwise inactive X chromosome remains unmethylated (DNA methylation is discussed in Section 30-7). Conversely, the CpG islands within many promoters on the inactive X chromosome become methylated, whereas those on the active X chromosome remain unmethylated. Some or all of these changes are essential for X chromosome inactivation and, moreover, almost certainly provide the epigenetic imprint that is responsible for maintaining the inactive X chromosome's state of inactivity in subsequent cell generations.

b. Chromosome Puffs and Lampbrush Chromosomes Are Transcriptionally Active

The condensed state of facultative heterochromatin presumably renders it transcriptionally inactive by making its DNA inaccessible to the proteins mediating transcription. Conversely, *transcriptionally active chromatin must*

have a relatively open structure. Such decondensed chromatin occurs in the **chromosome puffs** that emanate from single bands of giant polytene chromosomes (Fig. 34-41). These puffs reproducibly form and regress as part of the normal larval development program and in response to such physiological stimuli as hormones and heat. Autoradiographic studies with ³H-labeled uridine and immunofluorescence studies using antibodies against RNA polymerase II clearly demonstrate that *puffs are the major sites of RNA synthesis in polytene chromosomes.*

In amphibian oocytes, the analogous decondensation of nonpolytene chromosomes occurs most conspicuously in the so-called **lampbrush chromosomes** (Fig. 34-42). During their prolonged meiotic prophase I (Fig. 1-20), these previously condensed chromosomes loop out segments of transcriptionally active DNA that electron micrographs such as Fig. 34-43 indicate are often single transcription units.

B. *Regulation of Transcriptional Initiation*

The foregoing observations suggest that selective transcription is mainly responsible for the differential protein synthesis among the various types of cells in the same organism. It was not until 1981, however, that James Darnell actually demonstrated this to be the case, as follows. Experimentally useful amounts of mouse liver genes were obtained by inserting the cDNAs of mouse liver mRNAs (some 95% of which are cytosolic) into plasmids and replicating them in *E. coli* (Section 5-5B). By hybridizing the resulting cloned cDNAs with radioactively labeled mRNAs from various mouse cell types, the *E. coli* colonies

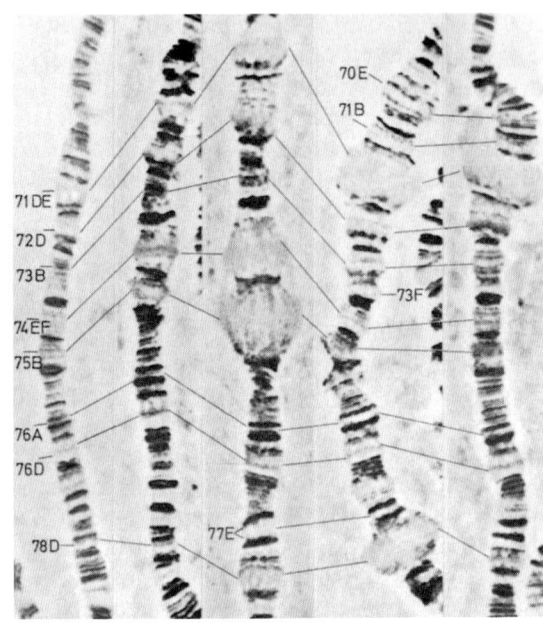

FIGURE 34-41 Formation and regression of chromosome puffs (*lines*) in a *D. melanogaster* polytene chromosome over a 22-h period of larval development. The very large puffs in this series of photomicrographs are also known as **Balbiani rings.** [Courtesy of Michael Ashburner, Cambridge University.]

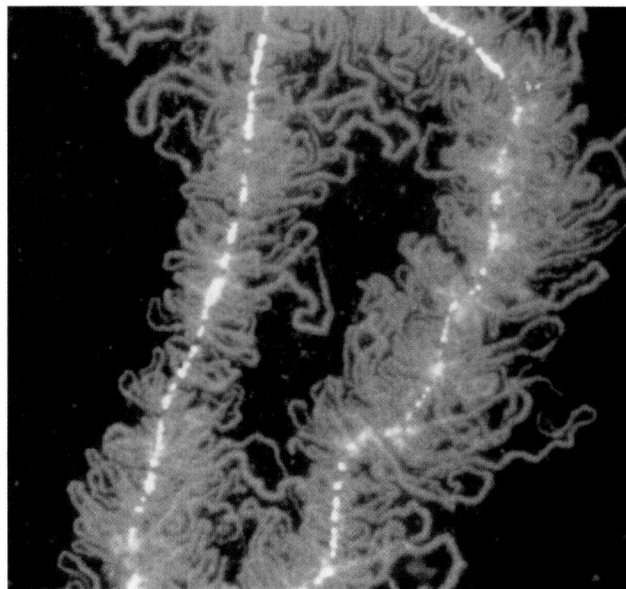

FIGURE 34-42   Immunofluorescence micrograph of a lampbrush chromosome from an oocyte nucleus of the newt *Notophthalmus viridescens*. The chromosome's numerous transcriptionally active loops give rise to the name "lampbrush" (an obsolete implement for cleaning kerosene lamps). [From Roth, M.B. and Gall, J.G., *J. Cell Biol.* **105,** 1049 (1987). Copyright © 1987 by Rockefeller University Press.]

containing liver-specific genes were distinguished from colonies containing genes common to most mouse cells. In this way, 12 liver-specific cDNA clones and three common cDNA clones were obtained. The question was then asked, does a eukaryotic cell transcribe only the genes encoding the proteins it synthesizes, or does it transcribe all of its genes but only process properly the transcripts it translates? This question was answered by hybridizing the cloned mouse genes with freshly synthesized and therefore unprocessed RNAs (hnRNAs) obtained from the nuclei of mouse liver, kidney, and brain cells (Fig. 34-44). Only the

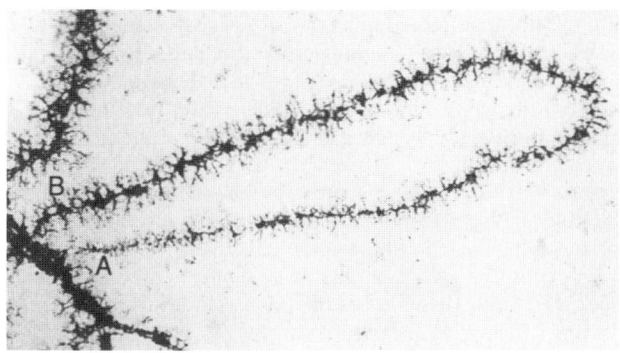

FIGURE 34-43   Electron micrograph of a single loop of a lampbrush chromosome. The ribonucleoprotein matrix coating the loop increases in thickness from one end of the loop (A) to the other (B), which indicates that the loop comprises a single transcriptional unit. [Courtesy of Oscar L. Miller, Jr., University of Virginia.]

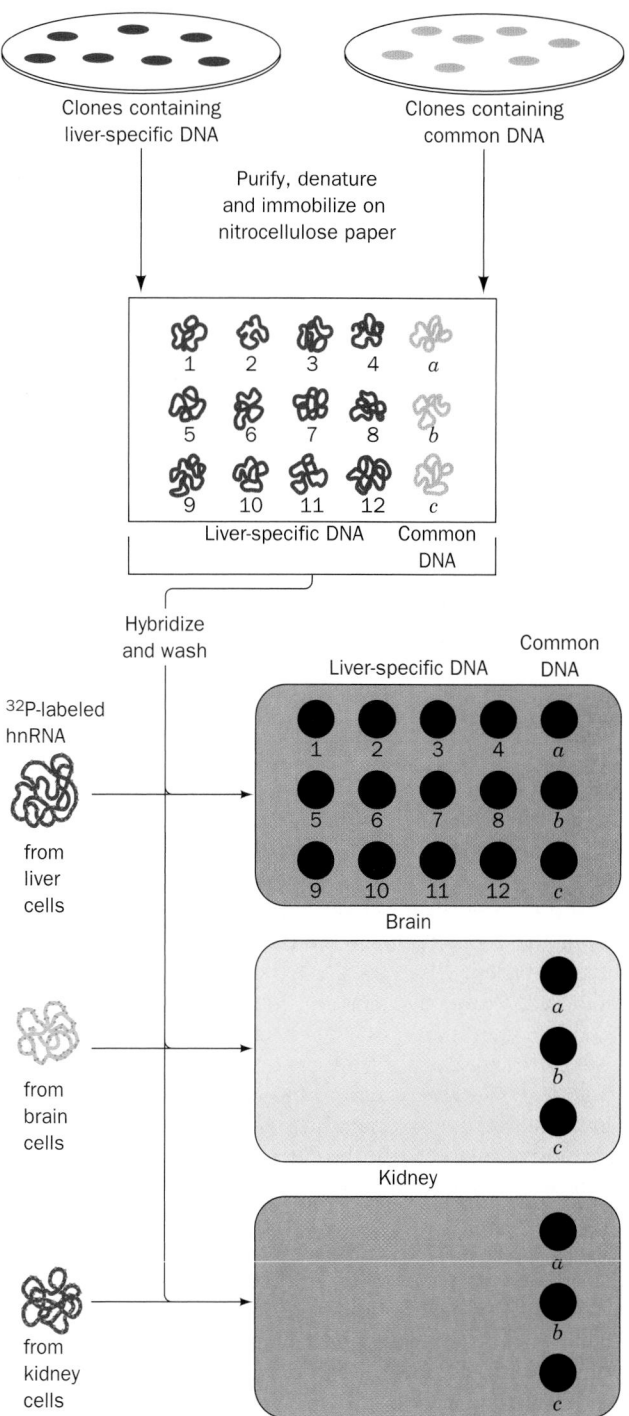

FIGURE 34-44   Determination of the primary role of selective transcription in the control of eukaryotic gene expression. This was established as follows. Cloned cDNAs encoding 12 different mouse liver-specific proteins (1–12) and 3 different proteins common to most mouse cells (*a–c*) were purified, denatured, and spotted onto filter paper (*top*). The DNAs were hybridized with newly formed and therefore unprocessed radioactively labeled RNAs produced by either mouse liver, kidney, or brain nuclei (*lower left*). Autoradiography showed that the liver RNAs hybridized with all 12 liver-specific cDNAs and all 3 common cDNAs but that the kidney and brain RNAs only hybridized with the common cDNAs (*right*).

RNAs extracted from liver nuclei hybridized with the 12 liver-specific cDNAs that were probed. The RNAs from all three cell types, however, hybridized with the DNA from the three clones containing the common mouse genes. Evidently, *liver-specific genes are not transcribed by brain or kidney cells. This strongly suggests that the control of genetic expression in eukaryotes is primarily exerted at the level of transcription.*

In more recent times, the use of DNA microarray technology (DNA chips; Section 7-6B) has enormously increased the number of genes whose levels of transcription may be simultaneously monitored as well as greatly reduced the effort required to do so. For example, **hepatocellular carcinoma (HCC),** the most common liver malignancy, is among the five leading causes of cancer deaths in the world and is closely associated with chronic infections by hepatitis B or C viruses although the nature of this association is unclear. David Botstein and Patrick Brown characterized the gene expression patterns in HCC tumors and normal liver tissues by isolating their mRNAs and reverse transcribing them to cDNAs, which were then coupled to a fluorescent dye. These labeled cDNAs were then hybridized to DNA microarrays containing ~17,400 human genes. The level of transcription of each of these genes in a given tissue sample was determined from the fluorescence intensity at the corresponding position on the microarray relative to that of a reference cDNA that had a different fluorescent label. The results of this exhaustive analysis are presented in Fig. 34-45, which clearly indicates that HCC tumors have transcriptional patterns that differ from those of nontumor liver tissues. Surprisingly, however, different HCC nodules from the same patient exhibited gene expression patterns whose similarities were no greater than those of tumors from different patients. Nevertheless, certain genes are consistently expressed at high levels in HCC tumors, and hence those whose products are secreted or membrane associated may serve as serological markers for the early detection of liver cancers and/or as potential therapeutic targets.

a. The Transcriptional Initiation of Structural Genes Involves Three Classes of Transcription Factors

Transcriptional initiation in eukaryotes has been most widely studied in protein-encoding genes, that is, genes that are transcribed by RNA polymerase II **(RNAP II).** In the following paragraphs, we concentrate on the major findings of these studies.

Differentiated eukaryotic cells possess a remarkable capacity for the selective expression of specific genes. The synthesis rates of a particular protein in two cells of the same organism may differ by as much as a factor of 10^9; that is, unexpressed eukaryotic genes are completely turned off. In contrast, simply repressible prokaryotic systems such as the *E. coli lac* operon (Section 31-3B) exhibit no more than a thousand-fold range in their transcriptional rates; they have significant basal levels of expression. Nevertheless, as we shall see below, *the basic mechanism of expressional control in eukaryotes resembles that in prokaryotes: the selective binding of proteins to specific*

0.25   0.5   1   2   4

FIGURE 34-45 Relative transcriptional activities of the genes in hepatocellular carcinoma (HCC) tumors as determined using DNA microarrays. The data are presented in matrix form, with each column representing one of 156 tissue samples (82 HCC tumors and 74 nontumor liver tissues) and each row representing one of 3180 genes (those of the ~17,400 genes on the DNA microarray with the greatest variation in transcriptional activity among the various tissue samples). The data are arranged so as to group the genes as well as the tissue samples on the basis of similarities of their expression patterns. The color of each cell indicates the expression level of the corresponding gene in the corresponding tissue relative to its mean expression level in all the tissue samples with bright red, black, and bright green indicating expression levels of 4, 1, and 1/4 times that of the mean for that gene (as indicated on the scale below). The dendrogram at the top of the matrix indicates the similarities in expression patterns among the various tissue samples. [Courtesy of David Botstein and Patrick Brown, Stanford University School of Medicine.]

genetic control sequences so as to modulate the rate of transcriptional initiation.

RNAP II, unlike prokaryotic RNA polymerase holoenzyme (Section 31-2), has little if any inherent ability to bind to its promoters. Rather, three different classes of so-called **transcription factors** have been implicated in regulating transcriptional initiation by RNAP II:

1. General transcription factors (GTFs), which are required for the synthesis of all mRNAs, select the transcriptional initiation site and deliver RNAP II to it, thereby forming a complex that initiates transcription at a basal rate.

2. Upstream transcription factors are proteins that bind to specific DNA sequences upstream of the initiation site so as to stimulate or repress transcriptional initiation by GTF-complexed RNAP II. The binding of upstream factors to DNA is unregulated; that is, they bind to any available DNA containing their target sequence. Those that are present in a cell vary with its developmental state and its needs; their synthesis is also regulated.

3. Inducible transcription factors function similarly to upstream transcription factors but must be activated (or inhibited), either by phosphorylation or by specific ligands, in order bind to their target DNA sites and influence transcriptional initiation. They are synthesized and/or activated in specific tissues at particular times and therefore mediate gene expression in a positionally and temporally specific manner.

We discuss these transcription factors below.

b. The Preinitiation Complex Is a Large and Complex Assembly

Extensive research in numerous laboratories has revealed that *the accurate transcriptional initiation of most* *structural genes requires the presence of six GTFs, most of which are multiprotein complexes, named* **TFIIA, TFIIB, TFIID, TFIIE, TFIIF,** *and* **TFIIH** (Table 34-3; TF for transcription factor and II for **class II genes,** those that are transcribed by RNAP II). These GTFs combine, in an ordered pathway, with RNAP II and promoter-containing DNA near the transcriptional start site to form a so-called **preinitiation complex (PIC)** that supports a basal level of transcription. The so-called **core promoters** in structural genes are largely upstream of the transcriptional start site and often contain a **TATA box,** a segment of the sense strand (the DNA strand with the same sequence as its corresponding mRNA; Section 31-2A) that is centered at around position -27 and whose consensus sequence is $TATA_T^A A_T^A$ (Fig. 31-23). The sequence motifs in a typical core promoter are indicated in Fig. 34-46. The remaining portions of the promoter, to which various transcription factors are idiosyncratically targeted, are known as **upstream activation sequences (UASs).** An entire eukaryotic promoter typically extends over ~100 bp.

The assembly of the preinitiation complex, which is diagrammed in Fig. 34-47, begins with the binding of **TATA box–binding protein (TBP)** *to the TATA box, thereby identifying the transcriptional start site* (recall that eliminating the TATA box does not necessarily eliminate transcription but does result in heterogeneities in the transcriptional start site; Section 31-2E). TBP is then joined by a series of ~10 **TBP-associated factors (TAFs;** previously called $TAF_{II}s$ to indicate they are associated with class II genes) to form the ~700-kD multisubunit complex TFIID. TFIIA then binds to the TFIID–DNA complex so as to stabilize it, followed by TFIIB. At this point, TFIIF recruits RNAP II to the promoter in a manner reminiscent of the way that σ factors interact with core RNA polymerase in bacteria (Sections 31-2A and 31-2B). Indeed, the smaller of human TFIIF's two

TABLE 34-3 **Properties of the General Transcription Factors**

Factor	Number of Unique Subunits in Yeast	Mass in Yeast (kD)	Number of Unique Subunits in Humans	Mass in Humans (kD)	Functions
TFIIA	2	46	3	69	Stabilizes TBP and TAF binding
TFIIB	1	38	1	35	Stabilizes TBP binding; recruits RNAP II; influences start site selection
TFIID					Recognizes TATA box; recruits TFIIA and TFIIB; has positive and negative regulatory functions
TBP	1	27	1	38	
TAFs	14	~1050	≥12[a]	≥960	
TFIIE	2	184	2	165	An $\alpha_2\beta_2$ heterotetramer; recruits TFIIH and stimulates its helicase activity; enhances promoter melting
TFIIF	3	156	2	87	Facilitates promoter targeting; stimulates elongation
TFIIH	9	518	9	470	Contains an ATP-dependent helicase that functions in promoter melting and clearance

[a]Although only 12 human TAFs have been identified versus 14 yeast TAFs, the close correspondence between each known human TAF and a yeast TAF suggests that two more human TAFs are yet to be identified.

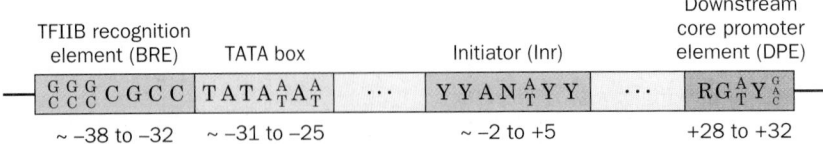

	TFIIB recognition element (BRE)	TATA box		Initiator (Inr)		Downstream core promoter element (DPE)

FIGURE 34-46 The sequence elements in a typical class II core promoter. The approximate positions of its various elements relative to the transcriptional initiation site (+1) are indicated below. Note that any or all of these elements may be absent in a given class II promoter.

subunits exhibits substantial sequence homology with σ^{70} (the predominant bacterial σ factor) and, moreover, can specifically interact with bacterial RNAPs (although it does not participate in promoter recognition). Finally, TFIIE and TFIIH join, in that order, to form the PIC. Once this complex has been assembled, an ATP-dependent activation

step, probably mediated by TFIIH's helicase function to melt the promoter, is required to initiate transcription at a basal rate. You should note that the human PIC, exclusive of the ~12-subunit, ~600-kD RNAP II, contains at least 25 subunits with an aggregate mass of ~1600 kD. Indeed, many of the proteins in the PIC are the targets of transcriptional regulators.

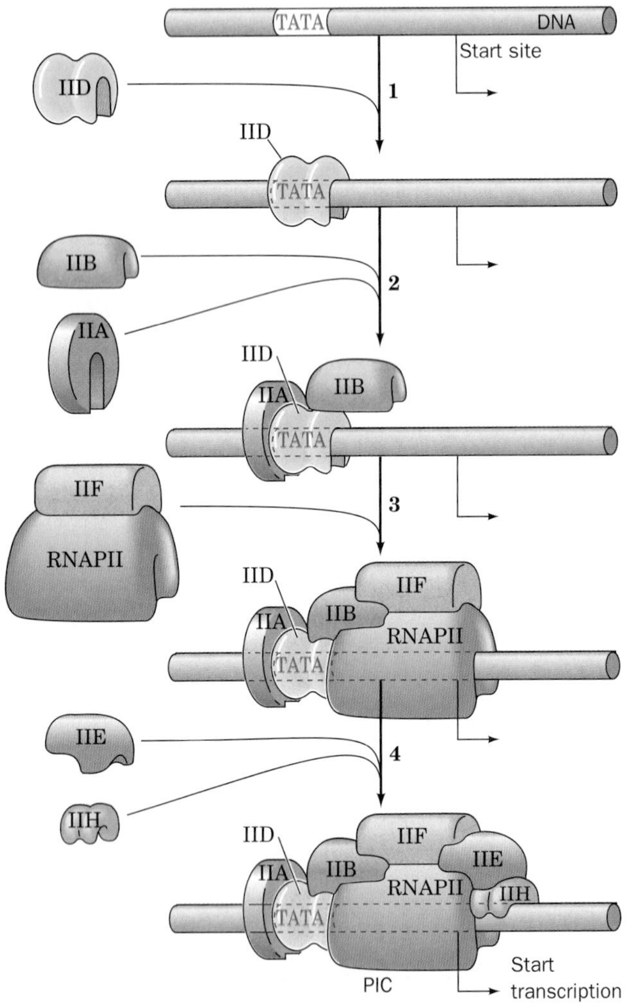

c. TBP Greatly Distorts Its Bound TATA Box DNA

TBP has a highly conserved (81% identical between yeast and humans) C-terminal domain of 180 residues that contains two ~40% identical direct repeats of 66 or 67 residues separated by a highly basic segment. In contrast, TBP's N-terminal domain is widely divergent, both in length and sequence, and, in fact, is unnecessary for TBP function *in vitro*. Curiously, the human N-terminal domain contains an uninterrupted run of 38 Gln residues, whereas that of *D. melanogaster* contains two blocks of 6 and 8 Gln residues separated by 32 residues, and that of yeast entirely lacks such sequences. Perhaps the N-terminal domains of TBPs have evolved to satisfy species-specific functions.

The X-ray structures of TBP from yeast (only its C-terminal domain) and from the flowering plant *Arabidopsis thaliana* (whose N-terminal domain consists of only 18 residues), by Kornberg and by Stephen Burley, reveal a saddle-shaped molecule (Fig. 34-48a) that consists of two structurally similar and topologically identical domains, each composed of one of the direct repeats. These are arranged with pseudo-twofold symmetry such that the protein consists of a 10-stranded antiparallel β sheet, 5 strands from each domain, flanked at each end by two α helices and a loop that is reminiscent of a stirrup hanging from the protein saddle. The curvature of the β pleated sheet saddle is such that it appears that TBP, in agreement with biochemical and genetic evidence, could fit snugly astride the DNA. However, the X-ray structures of the DNA complexes tell quite a different story.

Two closely similar X-ray structures of TBP–DNA complexes have been determined: one by Paul Sigler of yeast TBP in complex with a 27-nt DNA that forms an 11-bp TATA box–containing stem whose ends are joined by a 5-nt loop; and one by Burley of *Arabidopsis* TBP in complex with a 14-bp TATA box–containing duplex DNA. The DNA indeed binds to the concave surface of TBP but with its duplex axis nearly perpendicular rather than parallel to the saddle's "cylindrical" axis (Fig. 34-48b). The DNA is kinked by ~45° between the first two and the last two base pairs of its 8-bp TATA element. Between these kinks, the DNA is severely, although smoothly, bent with a radius of

FIGURE 34-47 Assembly of the preinitiation complex (PIC) on a TATA box–containing promoter. (1) TFIID assembles on the TATA box beginning with the binding of TATA box–binding protein (TBP) to the TATA box. **(2)** TFIIA and TFIIB then bind to the growing complex. **(3)** TFIIF then binds to RNAP II and escorts it to the complex. **(4)** Finally, TFIIE and TFIIH are sequentially recruited to the complex, thereby completing the PIC. [After Zawel, L. and Reinberg, D., *Curr. Opin. Cell Biol.* **4,** 490 (1992).]

(a)

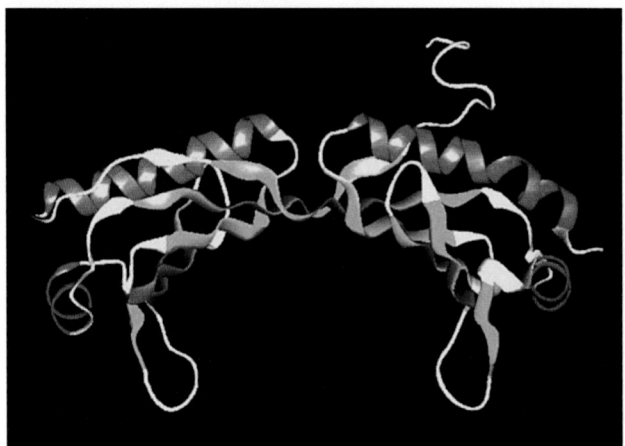

(b)

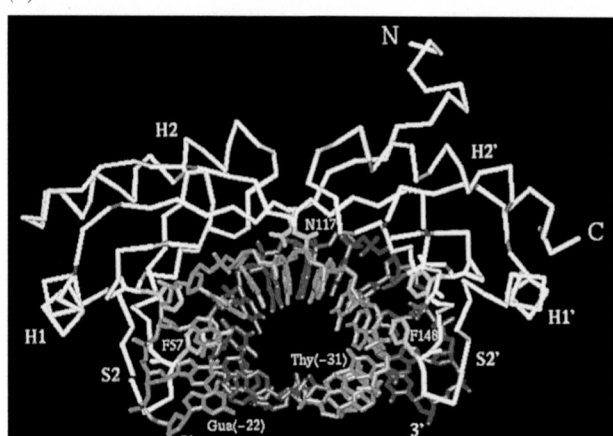

FIGURE 34-48 X-Ray structure of *Arabidopsis thaliana* TATA box–binding protein (TBP). (*a*) A ribbon diagram of the protein in the absence of DNA in which α helices are red, β strands are blue, and the remainder of the polypeptide backbone is white. The protein's pseudo-twofold axis of symmetry is vertical. Note that the protein seems to be precisely the proper size and shape to sit astride a 20-Å-diameter cylinder of B-DNA. This, however, is not what happens. (*b*) TBP in complex with a 14-bp TATA box–containing segment of the adenovirus major late promoter (single-strand sequence GC**TATAAAAG**GGCA, with its TATA box in bold) viewed as in Part *a*. The protein is represented as its C$_\alpha$ backbone (*white*); together with the side chains of Phe residues 57, 74, 148, and 165 (*yellow*), which

induce sharp kinks in the DNA; Asn residues 27 and 117 (*also yellow*), which make hydrogen bonds in the minor groove; and Ile 152 and Leu 163 (*blue*), which are implicated in specific DNA recognition. The DNA is drawn in stick form with the sense and antisense strands in green and red, respectively. B-form DNA enters its binding site with the 5′ end of the sense strand below the saddle on the right and exits on the left with its helix axis nearly perpendicular to the page (the last two base pairs have been removed for clarity). Between the kinks, which are located at each end of the TATA box, the DNA is partially unwound with the protein's central 8 strands of its 10-stranded antiparallel β sheet inserted into the DNA's greatly widened minor groove. [Courtesy of Stephen Burley, The Rockefeller University.]

curvature of ~25 Å and unwound by ~1/3 of a turn. This permits the protein's antiparallel β sheet to bind in the DNA's greatly widened and more shallow minor groove through hydrogen bonding and van der Waals interactions (the protein does not contact the DNA's major groove). A noteworthy aspect of this remarkable structure is that each kink in the DNA is stabilized by a wedge of two Phe side chains extending from the adjacent stirrup that pries apart the base pairs flanking the kink from their minor groove side and severely buckles the interior base pair. As a result of these unprecedented distortions to the DNA (the protein undergoes only slight conformational adjustments on binding DNA), there is an ~100° angle and a lateral 18-Å displacement between the helix axes of the B-form DNA entering and leaving TBP's binding site, thereby giving the DNA a cranklike shape. The DNA, nevertheless, maintains normal Watson–Crick pairing throughout the distorted region.

d. TFIIA and TFIIB Both Bind to DNA and TBP

The X-ray structures of ternary complexes of yeast TFIIA, TBP, and a TATA box–containing promoter DNA were independently determined by Richmond and Sigler and those of ternary complexes of human TFIIB, human or *Arabadopsis* TBP, and a TATA box–containing promoter DNA were independently determined by Sigler and by Robert Roeder and Burley. The TBP–DNA complexes in all the foregoing binary and ternary com-

plexes are closely similar. A plausible model of the TFIIA–TFIIB–TBP–DNA quaternary complex was therefore constructed by superimposing the TBP–DNA complexes in TFIIA- and TFIIB-containing ternary complexes (Fig. 34-49). TFIIA, a heterodimer in yeast, consists of a 6-stranded β barrel and a 4-helix bundle that, together, have a bootlike shape. The β barrel domain of TFIIA binds to TBP's N-terminal stirrup so as to extend TBP's β sheet to form a continuous 16-stranded β sheet. In addition, the TFIIA β barrel binds to the DNA over its major groove through salt bridges between four of its Lys and Arg side chains and the DNA's phosphate groups. TFIIB, a monomer, consists of two similar α helical domains that are rotated by 90° with respect to one another so as to form a cleft that clamps the TBP's C-terminal stirrup. TFIIB binds to the DNA via both its domains through several salt bridges with the DNA's phosphate groups as well as base-specific contacts in both the major and minor grooves to the consensus sequence $^{GGG}_{CCC}$CGCC, which occurs just upstream of the TATA box in many core promoters (Fig. 34-46). The formation of these interactions requires the distortions that TBP binding imposes on the DNA structure and hence TFIIB binding is synergistic with TBP binding. Since the pseudosymmetric TBP has been shown to bind to the TATA box in either orientation, it appears that the base-specific interactions between TFIIB and the promoter at its so-called **TFIIB recognition element (BRE;** Fig. 34-46) function to position TFIIB to properly orient

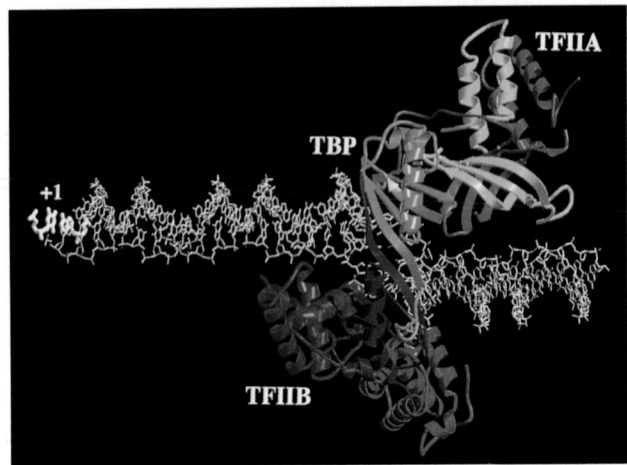

FIGURE 34-49 Model of the TFIIA–TFIIB–TBP–TATA box–containing DNA quaternary complex. The arrangement of the proteins (*ribbons*) and DNA (*white stick model*) is based on the independently determined X-ray structures of the TFIIA–TBP–DNA and TFIIB–TBP–DNA ternary complexes. In the model, the DNA has been extended in both directions beyond the TATA box, with its transcription start site (+1) on the left. The TBP's pseudosymmetrically related N- and C-terminal domains are cyan and purple, the TFIIA's two subunits are yellow and green, and the TFIIB's similar N- and C-terminal domains are red and magenta. The TBP binds to both TFIIA and TFIIB and all three proteins bind to the DNA at independent sites. [Courtesy of Stephen Burley, The Rockefeller University. Based on PDBids 1YTF and 1VOL.]

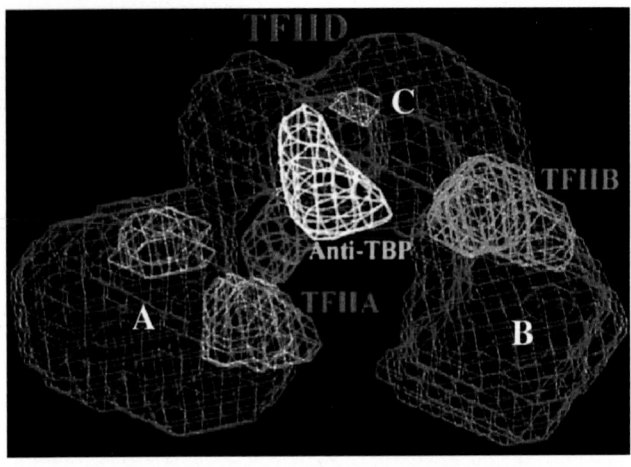

FIGURE 34-50 EM-based image of the human TFIID–TFIIA–TFIIB complex at 35-Å resolution. The blue mesh outlines the entire ternary complex, which consists of three domains, A, B, and C, arranged in a horseshoe shape and roughly 200 Å wide, 135 Å high, and 110 Å thick. The red and green meshes indicate the positions of TFIIA and TFIIB as determined by comparison with the EM-based images of the TFIID–TFIIA complex and TFIID alone. The yellow mesh indicates the binding position of an anti-TBP antibody. The different shapes of TFIIA here and in Fig. 34-49 are probably due to the fact that the TFIIA in Fig. 34-49 consists only of residues 56 to 209 of the 286-residue protein. [Courtesy of Eva Nogales, University of California at Berkeley.]

the TBP on the promoter. The model of the quaternary complex (Fig. 34-49) indicates that its three proteins all bind to the DNA upstream of the transcriptional start site, leaving ample room for the additional protein–DNA and protein–protein interactions that regulate the frequency with which RNAP II is recruited to the promoter.

e. TFIID Is a Horseshoe-Shaped Complex That Probably Contains a Histonelike Octamer

The electron microscopy–based structure of the human TFIID–TFIIA–TFIIB complex was determined at 35-Å resolution by Robert Tjian and Eva Nogales. TFIID is a horseshoe-shaped trilobal complex to which TFIIA and TFIIB are bound on opposite lobes that flank the central cavity (Fig. 34-50). This, together with the foregoing model of the TFIIA–TFIIB–TBP–DNA quaternary complex (Fig. 34-49), strongly suggests that TBP is located at the top of the cavity where it can contact both TFIIA and TFIIB and that the core promoter DNA passes through the cavity, where it is bound by TBP, TFIIA, and TFIIB. Indeed, the EM-based image of an anti-TBP antibody in complex with TFIID reveals that the antibody binds to TFIID in the expected position (Fig. 34-50).

The various TAFs are highly conserved from yeast to humans. Moreover, portions of 9 of the 14 known species of TAFs are homologous to nonlinker histones. For example, segments consisting of residues 17 to 86 of the 268-residue **dTAF9** (d for *Drosophila;* previously called

dTAF$_{\text{II}}$42, where the number indicates its nominal molecular mass in kD) and residues 1 to 70 of the 592-residue **dTAF6** (previously called **dTAF$_{\text{II}}$60**) are, respectively, homologous to histones H3 and H4. The X-ray structure of the dTAF9(17-86)–dTAF6(1-70) complex, determined by Roeder and Burley, reveals that both of these polypeptide segments assume the histone fold (Fig. 34-51*a*): two short helices flanking a long central helix (Fig. 34-8). In fact, TAF9(17-86) and TAF6(1-70) associate quite similarly to H3 and H4 in the nucleosome (Fig. 34-7) to form an $\alpha_2\beta_2$ heterotetramer (Fig. 34-51*a*). In addition, hTAF12(57-128) (h for human; **hTAF12,** which has 161 residues, was previously called **hTAF$_{\text{II}}$20**), which is homologous to histone H2B, forms a complex with hTAF4(870-943) **(hTAF4,** which has 1083 residues, was previously called **hTAF$_{\text{II}}$135),** which is homologous to H2A. The X-ray structure of this complex, determined by Dino Moras, reveals that it forms a histonelike heterodimer (Fig. 34-51*b*) but not a histonelike tetramer. **TAF11** and **TAF13** also form a histonelike heterodimer.

Gel filtration chromatography and sedimentation measurements (Sections 6-3B and 6-5A) by Stephen Buratowski and Song Tan indicate that the heterotetramer of **yTAF6** (y for yeast) and **yTAF9** associates with two heterodimers of **yTAF12** and **yTAF4** to form a heterooctamer. The mutation to Ala or Tyr of the highly conserved Leu 464 of yTAF12 (the homolog of H2B residue Leu 77, which is located near the C-terminus of this histone's long

(a)

(b)

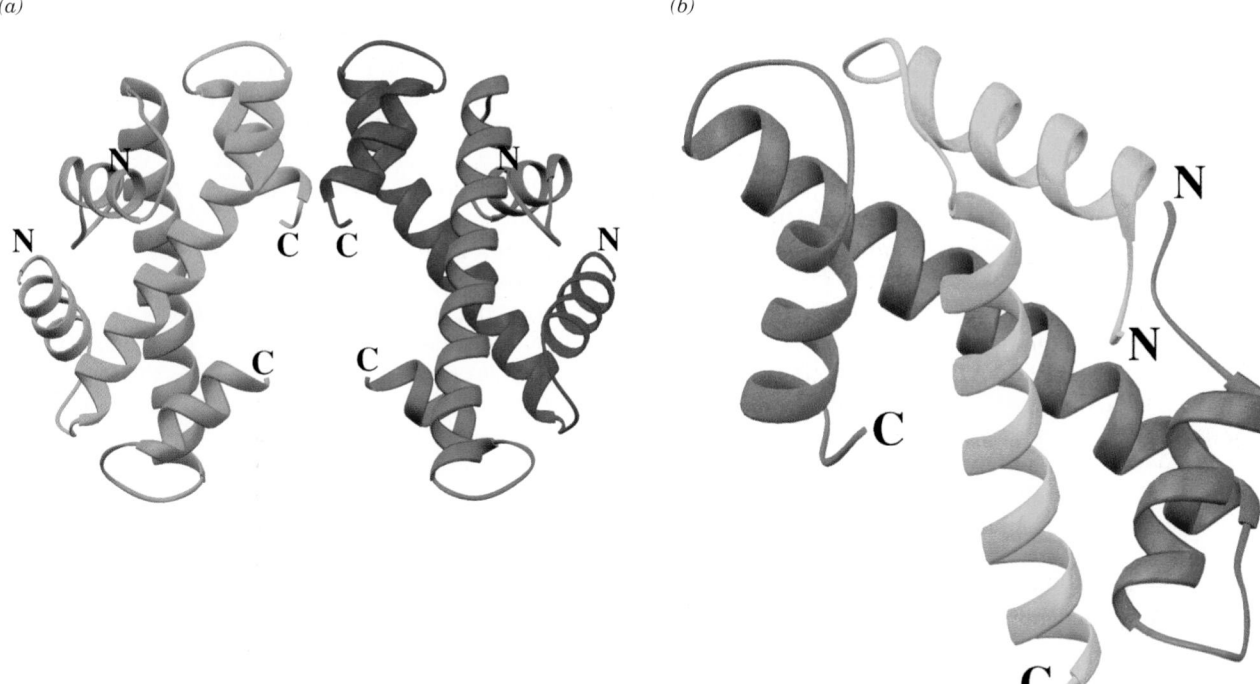

FIGURE 34-51   X-Ray structures of TAFs that form histonelike complexes. *(a)* The dTAF9(17-86)–dTAF6(1-70) $\alpha_2\beta_2$ heterotetramer as viewed with its twofold axis vertical. Note how the H3-like TAF9 segments (*blue and cyan*) and the H4-like TAF6 segments (*green and olive*) all assume the histone fold, how TAF9–TAF6 pairs interdigitate in head-to-tail arrangements to form heterodimers, and how the two TAF9 segments interact via a four-helix bundle to form the heterotetramer, much as do histones H3 and H4 in nucleosome cores (Figs. 34-7 and 34-8). [Based on an X-ray structure by

Robert Roeder and Stephen Burley, The Rockefeller University. PDBid 1TAF.] *(b)* The hTAF12(57-128)–hTAF4(870-943) heterodimer. Note how the H2B-like TAF12 segment (*red*) forms a regular histone fold but that the H2A-like TAF4 segment (*gold*) lacks the histone fold's C-terminal loop and helix. This is because TAF4 residues 918 to 943 are disordered. Nevertheless, the two subunits interdigitate to form a heterodimer, much as do histones H2B and H2A in nucleosome cores (Fig. 34-8). [After an X-ray structure by Dino Moras, CNRS/INSERM/ULP, Illkirch Cédex, France. PDBid 1H3O.]

central helix and hence occupies the hydrophobic core of the H4–H2B four-helix bundle; Fig. 34-8) prevents the formation of this octamer. This suggests that the octamer is held together by 4-helix bundles between yTAF6 and yTAF12 similar to those between H4 and H2B in nucleosomes (Figs. 34-7b and 34-8). Indeed, a model of this interface constructed from the above two X-ray structures suggests that its putative 4-helix bundle is remarkably similar to that of the H4–H2B interface. Nevertheless, it seems unlikely that this putative TAF octamer is wrapped with DNA in the PIC as is the histone octamer in the nucleosome. This is because most of the histone residues that make critical contacts with DNA in the nucleosome have not been conserved in the foregoing TAFs and, in fact, many of them have been replaced in these TAFs by highly conserved (in the TAFs) acidic residues, which would repel the anionic DNA.

f. Many Class II Core Promoters Lack a TATA Box

The core promoters of 65% of class II genes lack TATA boxes. They are mostly "housekeeping" genes; that is, genes that are constitutively expressed in all cells at relatively low rates. How can RNAP II properly initiate transcription at these TATA-less promoters? Investigations have shown

that TATA-less promoters often contain a so-called **initiator (Inr)** element that extends from positions −6 to +11 and that contains the loose consensus sequence YYAN${}_{\mathrm{T}}^{\mathrm{A}}$YY, where Y is a pyrimidine (C or T), N is any nucleotide, and A is the initiating (+1) nucleotide (Fig. 34-46). The presence of the Inr element is sufficient to direct RNAP II to the correct start site. These systems require the participation of many of the same GTFs that initiate transcription from TATA box–containing promoters. Surprisingly, they also require TBP. This suggests that with TATA-less promoters, Inr recruits TFIID such that its component TBP binds to the −30 region in a sequence-nonspecific manner. Indeed, in Inr-containing promoters that also contain a TATA box, the two elements act synergistically to promote transcriptional initiation. Nevertheless, a mutant TBP that is defective in TATA box binding will support efficient transcription from some TATA-less promoters although not from others. This suggests that the former promoters do not require a stable interaction with TBP. Some TATA-less promoters have a so-called **downstream core promoter element (DPE)**, which has the consensus sequence RG${}_{\mathrm{T}}^{\mathrm{A}}Y{}_{\mathrm{C}}^{\mathrm{G}}$, where R is a purine (A or G) and is located precisely from +28 to +32 (Fig. 34-46).

The foregoing suggests that there are variants of at least some of the GTFs and TAFs. In fact, the human genome contains multiple sequences related to TFIIA and TFIID subunits as well as alternative genes for several TAFs. Some of these variant genes are only expressed in certain cell types and/or at specific developmental stages. The resulting variant transcription factors probably recognize alternative core promoter elements and/or mediate selective interactions with upstream transcription factors.

g. Class I and Class III Genes Also Require TBP for Transcriptional Initiation

RNA polymerase I (**RNAP I,** which synthesizes most rRNAs) and RNA polymerase III (**RNAP III,** which synthesizes 5S rRNA and tRNAs) require different sets of GTFs from each other and from RNAP II to initiate transcription at their respective promoters. This is not unexpected considering the very different organizations of these three classes of promoters (Section 31-2E). Indeed, the promoters recognized by RNAP I (class I promoters) and nearly all those recognized by RNAP III (class III promoters) lack TATA boxes. Thus, it came as a surprise when it was demonstrated that *TBP is required for initiation by both RNAP I and RNAP III.* It participates by combining with different sets of TAFs to form the GTFs **SLI** (with class I promoters) and **TFIIIB** (with class III promoters). As with certain class II TATA-less promoters, a TBP mutant that is defective for TATA-box binding can still support *in vitro* transcriptional initiation by both RNAP I and RNAP III. Clearly, TBP, the only known universal transcription factor, is an unusually versatile protein.

h. Transcriptional Initiation of Class II Genes Is Mediated by Cell-Specific Upstream Transcription Factors Bound to Promoter and Enhancer Elements

The use of molecular cloning procedures has permitted the demonstration that *eukaryotic promoter and enhancer elements mediate the expression of cell-specific genes* (recall that an enhancer is a gene sequence that is required for the full activity of its associated promoter but that may have a variable position and orientation with respect to that promoter; Section 31-2E). For example, William Rutter linked the 5′-flanking sequences of either the insulin or the chymotrypsin gene to the sequence encoding **chloramphenicol acetyltransferase (CAT),** an easily assayed enzyme not normally present in eukaryotic cells. A plasmid containing the insulin gene recombinant elicits expression of the CAT gene only when introduced into cultured cells that normally produce insulin. Likewise, the chymotrypsin recombinant is only active in chymotrypsin-producing cells. Dissection of the insulin control sequence indicates that the segment between its positions −103 and −333 contains an enhancer: In insulin-producing cells only, it stimulates the transcription of the CAT gene with little regard to the enhancer's position and orientation relative to its promoter.

The foregoing indicates that cells contain specific transcription factors, the upstream transcription factors, that recognize the promoters and enhancers in the genes they transcribe. For instance, Tjian isolated a protein, **Sp1** (for

specificity protein-1), from cultured human cells that stimulates, by factors of 10 to 50, the transcription of cellular and viral genes containing at least one properly positioned GC box [GGGCGG (Section 31-2E); Fig. 34-52]. This protein binds, for example, to the 5′-flanking region of the SV40 virus early genes so as to protect its GC boxes from DNase I digestion (Fig. 34-53a; **DNase I footprinting**) and from methylation by dimethyl sulfate (Fig. 34-53b; **DMS footprinting**). Likewise, Sp1 specifically interacts with the four GC boxes in the upstream region of the mouse dihydrofolate reductase gene and with the single GC boxes in the human **metallothionein I$_A$** and **II$_A$** promoters (metallothioneins are metal ion–binding proteins that participate in heavy metal ion detoxification processes and whose synthesis is triggered by heavy metal ions).

Upstream transcription factors are essential participants in controlling the differential expression of the various globin genes in the human embryo, fetus, and adult (Section 34-2F). A typical β-globin gene promoter, in addition to its TATA box, has two positive-acting promoter elements: a CCAAT box near the −70 to −90 region and a CACCC motif at variable sites but often near positions −95 to −120 (Section 31-2E). Their importance is demonstrated by the observations that individuals with point mutations in their TATA or CACCC elements have reduced β-globin levels. These promoter elements are specifically bound by upstream transcription factors. Thus, the CCAAT box is bound by the ubiquitous transcription factor **CP1** and the CACCC element is bound by Sp1, which also binds to other globin promoter sequences that resemble Sp1's consensus binding sequence. Four erythroid-specific upstream transcription factors have also been implicated in globin gene expression: **GATA-1** (so named because it binds to sequences that contain the conserved core GATA), **NF-E2** (NF-E for *nuclear factor-erythroid*), **NF-E3,** and **NF-E4** (GATA-1 was previously named NF-E1).

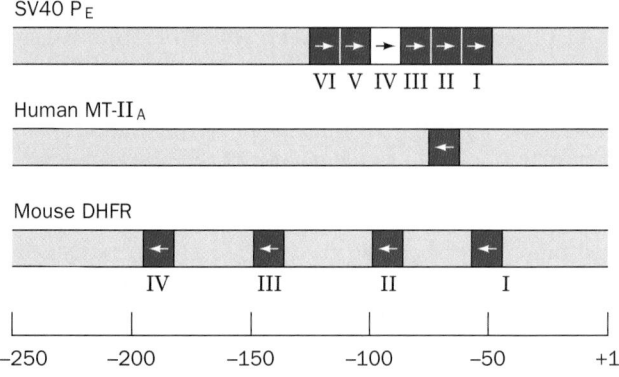

FIGURE 34-52   Arrangement and relative orientations of the GC boxes in the indicated promoters. Each arrow indicates the relative orientation of a GC box, which has the sequence NGGGCGGNNN. The blue boxes represent Sp1-binding sites, whereas SV40 GC box IV is shown as a white box because Sp1 bound at GC box V prevents this transcription factor from efficiently binding to GC box IV. The transcription start site is designated by +1. DHFR = dihydrofolate reductase; MT = metallothionein. [After Kadonaga, J.T., Jones, K.A., and Tjian, R., *Trends Biochem. Sci.* **11,** 21 (1986).]

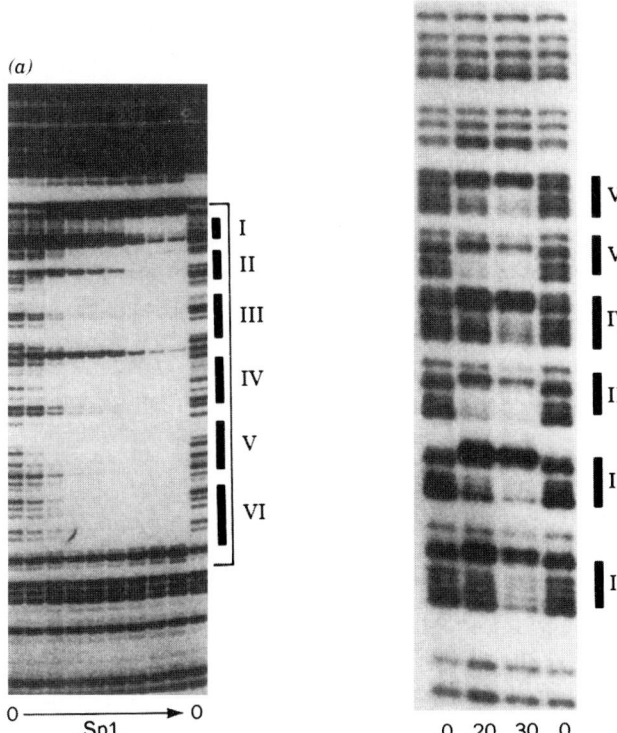

FIGURE 34-53 Identification of the Sp1-binding sites on the SV40 early promoter (Fig. 34-52, *top*). (*a*) Pancreatic DNase I is a relatively nonspecific endonuclease. In a DNase I footprinting assay, a DNA segment that is ^{32}P end-labeled on one strand is incubated with a binding protein and then lightly digested with DNase I such that, on average, each labeled DNA strand is cleaved only once. The DNA is then denatured, the resulting labeled fragments separated according to size by electrophoresis on a sequencing gel (Section 7-2A), and detected by autoradiography. Unprotected DNA is cleaved more or less at random and therefore appears as a "ladder" of bands, each representing an additional nucleotide (as in a sequencing ladder; Figs. 7-14 and 7-15). In contrast, the DNA sequences that the protein protects from DNase I cleavage have no corresponding bands. In the above footprint, the lanes labeled "0" are the DNase I digestion pattern in the absence of Sp1 and in the other lanes the amount of Sp1 increases from left to right. The footprint boundary is delineated by the bracket and the positions of SV40 GC boxes I to VI are indicated. [From Kadonaga, J.T., Jones, K.A., and Tjian, R., *Trends Biochem. Sci.* **11,** 21 (1986). Copyright © 1986 by Elsevier Biomedical Press.] (*b*) Dimethyl sulfate (DMS) methylates DNA's G residues at their N7 positions, which on treatment with weak base, excises the methylated G nucleosides from the DNA, thereby cleaving its sugar–phosphate backbone. In DMS footprinting, a protein-complexed ^{32}P end-labeled DNA segment is lightly treated with DMS such that each labeled DNA strand is, on average, cleaved only once. The resulting fragments are electrophoretically separated on a sequencing gel and detected by autoradiography. The DNA regions that the protein protects from methylation are not cleaved by this procedure and therefore are not represented in the resulting G residue "ladder." In the above autoradiogram, the number below each lane indicates the amount, in μL, of an Sp1 fraction added to a fixed quantity of SV40 early promoter DNA. The positions of its GC boxes are indicated. [From Gidoni, D., Katonaga, J.T., Barrera-Saldana, H., Takahashi, K., Chambon, P., and Tjian, R., *Science* **230,** 516 (1985). Copyright © 1985 by the American Society for the Advancement of Science.]

Analysis of hereditary persistence of fetal hemoglobin (HPFH), a syndrome characterized by the inappropriate expression of γ genes in human adults (Section 34-2G), has provided valuable insights into the basis of stage-specific globin expression. There are several HPFH variants that differ from normal only by a point mutation in the γ gene promoter. Such mutations might result in either tighter binding of a positive transcription factor or looser binding of a negative regulator. Thus, an HPFH mutation at position −117, which is located in the more upstream of the γ gene's two CCAAT boxes, increases the resemblance of this site to CP1's consensus binding sequence and results in a twofold tighter binding of CP1 to the mutant site. Similarly, HPFH mutations in a GC-rich region close to position −200 result in tighter Sp1 binding.

i. Upstream Transcription Factors Interact Cooperatively with Each Other and the PIC

How do upstream transcription factors stimulate (or inhibit) transcription? *Evidently, when these proteins bind to their target DNA sites in the vicinity of a PIC (in some cases, many thousands of base pairs distant), they somehow activate (or repress) its component RNAP II to initiate transcription.* Transcription factors may bind cooperatively to each other and/or the PIC in a manner resembling the binding of two λ repressor dimers and RNA polymerase to the o_R operator of bacteriophage λ (Section 33-3D), thereby synergistically stimulating (or repressing) transcriptional initiation. Indeed, molecular cloning experiments indicate that many enhancers and **silencers** (the analogs of enhancers that function in the transcriptional repression of their associated gene) consist of segments (modules) whose individual deletion reduces but does not eliminate enhancer/silencer activity. *Such complex arrangements presumably permit transcriptional control systems to respond to a variety of stimuli in a graded manner.* In some cases, however, several transcription factors together with so-called **architectural proteins** cooperatively assemble on an ~100-bp enhancer to form a multisubunit complex, known as an **enhanceosome,** in which the absence of a single subunit all but eliminates its ability to stimulate transcriptional initiation at the associated promoter. Thus, enhanceosomes function more like on/off switches rather than providing a graded response. Architectural proteins function to bend and/or otherwise deform enhancers so as to promote the assembly of the other enhanceosome proteins. Enhanceosomes may also contain **coactivators** and/or **corepressors,** proteins that do not bind to DNA but, rather, interact with proteins that do so to activate or repress transcription.

The functional properties of many upstream transcription factors are surprisingly simple. They appear to have (at least) two domains:

1. A DNA-binding domain that binds to the protein's target DNA sequence (and whose structural properties are discussed below).

2. A domain containing the transcription factor's activation function. Sequence analysis indicates that many of these **activation domains** (also called **transactivation domains** because they act in trans with the genes they control) have conspicuously acidic surface regions whose negative charges, if mutationally increased or decreased, respectively raise or lower the transcription factor's activity. This suggests that the associations between these transcription factors and a PIC are mediated by relatively nonspecific electrostatic interactions rather than by conformationally more demanding hydrogen bonds. Other types of activation domains have also been characterized, including those with Gln-rich regions, such as Sp1, and those with Pro-rich regions.

The DNA-binding and activation functions of eukaryotic transcription factors can be physically separated (which is why they are thought to occur on different domains). Thus, a genetically engineered hybrid protein, containing the DNA-binding domain of one transcription factor and the activation domain of a second, activates the same genes as the first transcription factor. Indeed, it makes little functional difference as to whether the activation domain is placed on the N-terminal side of the DNA-binding domain or on its C-terminal side. This geometric permissiveness in the binding between the activation domain and its target protein is also indicated by the observation that transcription factors are largely insensitive to the orientations and positions of their corresponding enhancers relative to the transcriptional start site [Section 31-2E; it is also the basis of the two-hybrid system for identifying proteins that interact *in vivo* (Section 19-3C)]. Of course, *the DNA between an enhancer and its distant transcriptional start site must be looped around for an enhancer-bound transcription factor to interact with the promoter-bound PIC (Section 31-2E).*

The synergy (cooperativity) of multiple transcription factors in initiating transcription may be understood in terms of a simple recruitment model. Suppose an enhancer-bound transcription factor increases the affinity with which a PIC binds to the enhancer's associated promoter so as to increase the rate at which the PIC initiates transcription there by a factor of 10. Then, if another transcription factor binding to a different enhancer subsite likewise increases the initiation rate by a factor of 20, both transcription factors acting together will increase the initiation rate by a factor of 200. *In this way, a limited number of transcription factors can support a much larger number of transcription patterns.* Transcriptional activation, according to this model, is essentially a mass action effect: The binding of a transcription factor to an enhancer increases the transcription factor's effective concentration at the associated promoter (the DNA holds the transcription factor in the vicinity of the promoter), which consequently increases the rate at which the PIC binds to the promoter. This explains why a transcription factor that is not bound to DNA (or even lacks a DNA-binding domain) inhibits transcriptional initiation. Such unbound transcription factors compete with DNA-bound transcription factors for their target sites and thereby reduce the rate at which the PIC is recruited to the associated promoter. This phenomenon, which is known as **squelching,** is apparently why transcription factors in the nucleus are almost always bound to inhibitors unless they are actively engaged in transcriptional initiation.

j. Steroid Receptors Are Examples of Inducible Transcription Factors

Eukaryotic cells express many cell-specific proteins in response to the presence of various hormones. Many of these hormones are **steroids** (Section 25-6C), cholesterol derivatives that mediate a wide variety of physiological and developmental responses (Section 19-1G). For example, the administration of **estrogens** (female sex hormones) such as **β-estradiol** causes chicken oviducts to increase their ovalbumin mRNA level from ~10 to ~50,000 molecules per cell, and the amount of ovalbumin they produce rises from unde-

β-Estradiol

Ecdysone

tectable levels to a majority of their newly synthesized protein. Similarly, the insect steroid hormone **ecdysone** mediates several aspects of larval development (the temporal sequence of chromosome puffing shown in Fig. 34-41 can be induced by ecdysone administration).

Steroids, which are nonpolar molecules, spontaneously pass through the membranes of their target cells to bind to their corresponding steroid receptors. In the absence of their cognate steroid, these receptors are bound in large multiprotein complexes that contain chaperone proteins such as Hsp90 and Hsp70 as well as immunophilins (Sections 9-2B and 9-2C), which presumably function to maintain the receptor in its native conformation, ready to bind its cognate steroid. Depending on the identity of the receptor, these complexes mainly inhabit the nucleus or the cytosol. Steroid binding releases the receptors from these complexes, whereupon they dimerize. In the case of cytosolically located receptors, steroid binding is thought to also unmask their previously sequestered nuclear localization signals (NLS; Section 32-6B), thereby causing the steroid–receptor complexes to be transported to the nucleus. [Most NLSs consist of a 48-residue segment of mainly basic residues or two such segments separated by an 8- to 12-residue linker that is mutation-resistant; the precise location of an NLS within a polypeptide is unimportant, unlike the case for other types of signal peptides (Section 12-4).]

In the nucleus, steroid–receptor complexes bind to specific segments of chromosomal enhancers known as **hormone response elements (HREs)** *so as to induce, or in some cases repress, the transcription of their associated genes.* For example, receptors for **glucocorticoids** (a class of steroids that affect carbohydrate metabolism; Section 19-1G) bind to specific 15-bp **glucocorticoid response elements (GREs)** in the upstream regions of many genes, including those of metallothioneins. Thus, eukaryotic steroid receptors are inducible transcription factors: Their actions resemble those of prokaryotic transcriptional regulators such as the *E. coli* CAP–cAMP complex (Section 31-3C). However, eukaryotic systems are much more complex. For instance, different cell types may have the same receptor for a given steroid hormone and yet synthesize different proteins in response to the hormone. Apparently, only some of the genes inducible by a given steroid are made available for activation in each type of cell responsive to that steroid. Consequently, a given eukaryotic sequence-specific regulator may function as an activator or a repressor depending on the identities of the proteins with which it is interacting. The structures of steroid receptors are discussed below.

k. Eukaryotic Transcription Factors Have a Great Variety of DNA-Binding Motifs

How do DNA-binding transcription factors recognize their target DNA sequences? In prokaryotes, as we have seen (Section 31-3D), most repressors and activators do so via helix–turn–helix (HTH) motifs and, in a few cases, via β ribbon motifs. Eukaryotes, as we shall see, employ a far greater variety of DNA-binding motifs in their transcription factors. In the following paragraphs we discuss the

structures of several of the more common of these motifs and how they bind their target DNAs.

l. Zinc Finger DNA-Binding Motifs

The first of the predominantly eukaryotic DNA-binding motifs was discovered by Aaron Klug in *Xenopus* **transcription factor IIIA (TFIIIA),** a protein that binds to the internal control sequence of the 5S rRNA gene (Section 31-2E). This complex then sequentially binds TFIIIB (which contains TBP), **TFIIIC,** and RNA polymerase III, which, in turn, initiates transcription of the 5S rRNA gene. The 344-residue TFIIIA contains nine similar, tandemly repeated, ~30-residue modules, each of which contains two invariant Cys residues, two invariant His residues, and several conserved hydrophobic residues (Fig. 34-54*a*). Each of these units binds a Zn^{2+} ion, which X-ray absorption measurements indicate is tetrahedrally liganded by the invariant Cys and His residues. Sequence analyses have since revealed that these so-called **zinc fingers** occur from 2 to over 60 times in a variety of eukaryotic transcription factors, including Sp1, several *D. melanogaster* developmental regulators (Section 34-4B), and certain proto-oncogene proteins (proteins whose mutant forms promote cancerous growth; Section 19-3B), as well as the *E. coli* UvrA protein (Section 30-5B). In fact, it is estimated that ~1% of mammalian proteins contain zinc fingers. In some zinc fingers, the two Zn^{2+}-liganding His residues are replaced by two additional Cys residues, whereas others have six Cys residues ligating two Zn^{2+} ions. Indeed, as we shall see, structural diversity is a hallmark of zinc finger proteins. In all cases, however, the Zn^{2+} ions appear to knit together relatively small globular domains, thereby elim-

(*a*)

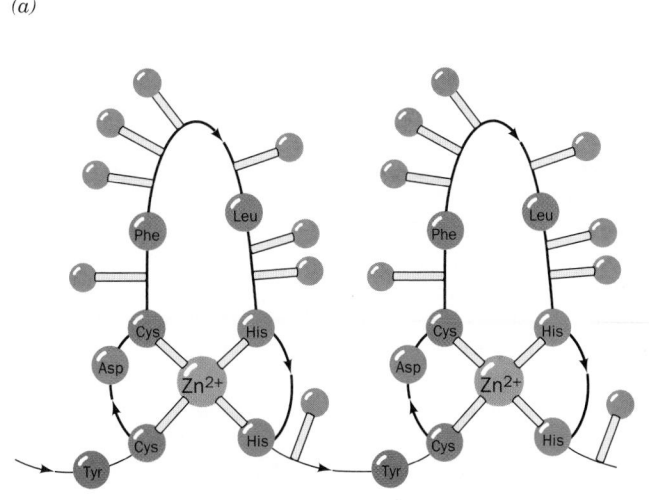

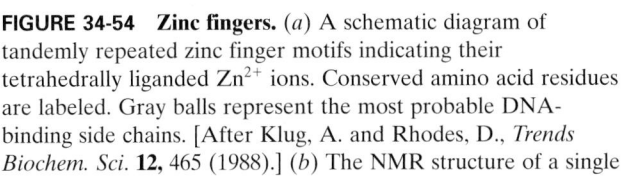

(*b*)

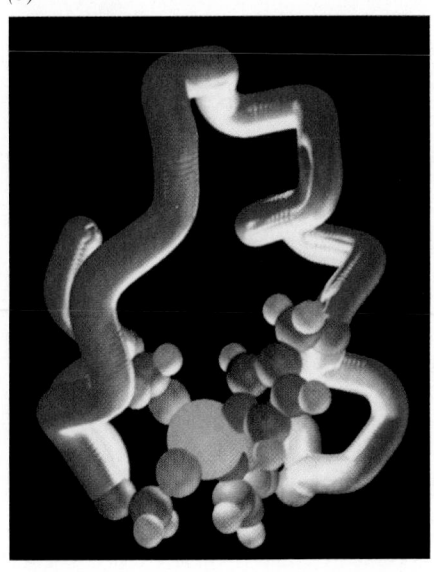

FIGURE 34-54   Zinc fingers. (*a*) A schematic diagram of tandemly repeated zinc finger motifs indicating their tetrahedrally liganded Zn^{2+} ions. Conserved amino acid residues are labeled. Gray balls represent the most probable DNA-binding side chains. [After Klug, A. and Rhodes, D., *Trends Biochem. Sci.* **12,** 465 (1988).] (*b*) The NMR structure of a single zinc finger from the *Xenopus* protein Xfin. The Zn^{2+} ion together with the atoms of its His and Cys ligands are represented as spheres with Zn^{2+} cyan, C gray, N blue, S yellow, and H white. [Courtesy of Michael Pique, The Scripps Research Institute, La Jolla, California. Based on an NMR structure by Peter E. Wright, The Scripps Research Institute. PDBid 1ZNF.]

inating the need for much larger hydrophobic cores (although see Section 9-3C).

m. Cys₂–His₂ Zinc Fingers: Xfin and Zif268 Proteins

The first reported zinc finger structure, an NMR structure by Peter Wright of the 31st of the 37 tandemly repeated zinc fingers in the *Xenopus* **Xfin protein,** revealed that this 25-residue peptide forms a compact globule containing a 2-stranded antiparallel β sheet and one α helix (a ββα unit) that are held together by the tetrahedrally liganded Zn^{2+} ion (Fig. 34-54b). This was followed by Carl Pabo's X-ray structure of a 72-residue segment of the mouse protein **Zif268** that incorporates the protein's three zinc fingers in complex with a DNA segment containing the protein's 9-bp consensus binding sequence. The structures of the three Zif268 zinc finger motifs (Fig. 34-55a) are closely superimposable and are nearly identical to that of the Xfin zinc finger (Fig. 34-54b). The three Zif268 zinc fingers are arranged as separate domains in a C-shaped structure that fits snugly into the DNA's major groove (Fig. 34-55b). Each zinc finger interacts in a conformationally identical manner with successive 3-bp segments of the DNA, predominantly through hydrogen bonding interactions between the zinc finger's α helix and one strand of the DNA (here, a G-rich strand). Each zinc finger makes specific hydrogen bonding contacts with two bases in the major groove. Interestingly, five of these six associations involve interactions between Arg and G residues. In addition to these sequence-specific interactions, each zinc finger hydrogen bonds with the DNA's phosphate groups via conserved Arg and His residues.

The Cys₂–His₂ zinc finger broadly resembles the prokaryotic HTH motif as well as most other DNA-binding motifs we shall encounter (including other types of zinc finger modules) in that *all of these DNA-binding motifs provide a platform for inserting an α helix into the major groove of B-DNA.* However, Cys₂–His₂ zinc finger proteins, unlike those containing other DNA-binding motifs, possess repeated protein modules that each contact successive DNA segments. Such a modular system can recognize extended asymmetric base sequences.

n. Cys₂–Cys₂ Zinc Fingers: The Glucocorticoid Receptor and Estrogen Receptor DNA-Binding Domains

The **nuclear receptor superfamily,** which occurs in animals ranging from worms to humans, is composed of >150 proteins that bind a variety of hormones such as steroids (glucocorticoids, mineralocorticoids, progesterone, estrogens, and androgens; Section 19-1G), thyroid hormones (Section 19-1D), vitamin D (Section 19-1E), and **retinoids** (Section 34-4B). However, the ligands, if any, that many superfamily members bind are, as yet, unknown and hence these proteins are known as **orphan receptors.** The nuclear receptors, many of which activate distinct but overlapping sets of genes, share a conserved modular organization that includes, from N- to C-terminus, a poorly conserved transactivation domain, a highly conserved DNA-binding domain, a connecting hinge region, and a ligand-binding domain. The DNA-binding domains contain 8 Cys residues that, in groups of four, tetrahedrally coordinate two Zn^{2+} ions. Many members of the nuclear receptor superfamily recognize hormone re-

(a)

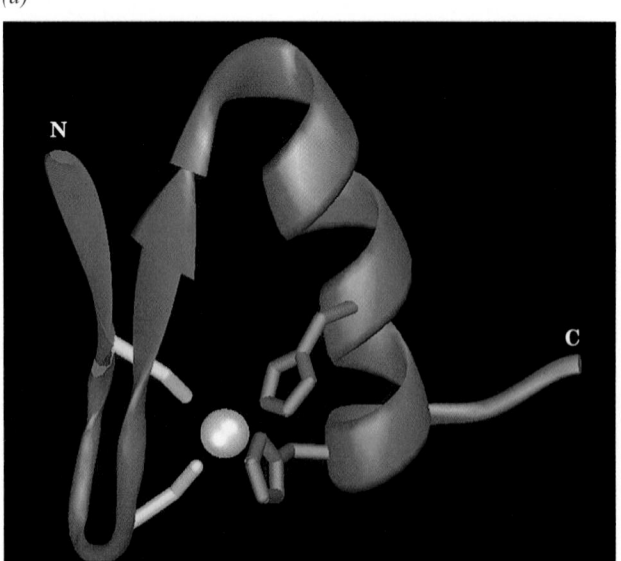

(b)

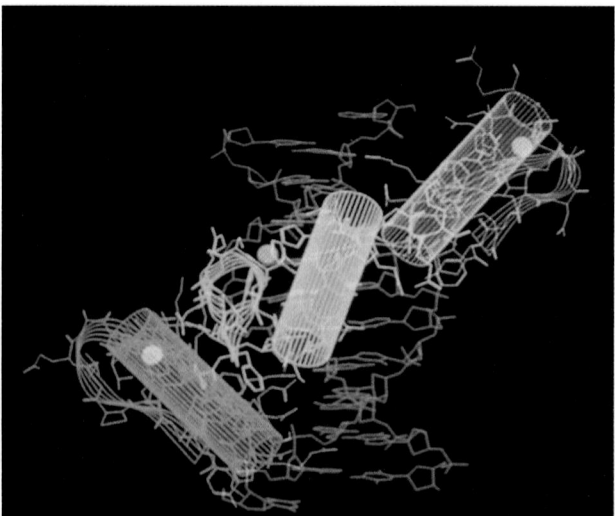

FIGURE 34-55   X-Ray structure of a three-zinc finger segment of Zif268 in complex with a 10-bp DNA. (*a*) A ribbon diagram of a single zinc finger motif (finger 1) with its Zn^{2+} ion's tetrahedrally liganding His (*cyan*) and Cys (*yellow*) side chains shown in stick form and its Zn^{2+} ion represented by a silver sphere. (*b*) The complex of the entire protein segment with DNA. The protein and DNA are shown in stick form, with

superimposed cylinders and ribbons marking the protein's α helices and β sheets. Finger 1 is orange, finger 2 is yellow, finger 3 is pink, the DNA is blue, and the Zn^{2+} ions are represented by white spheres. Note how the N-terminal end of each zinc finger's helix extends into the DNA's major groove to contact three base pairs. [Part *a* based on an X-ray structure by and Part *b* courtesy of Carl Pabo, MIT. PDBid 1ZAA.]

(a)

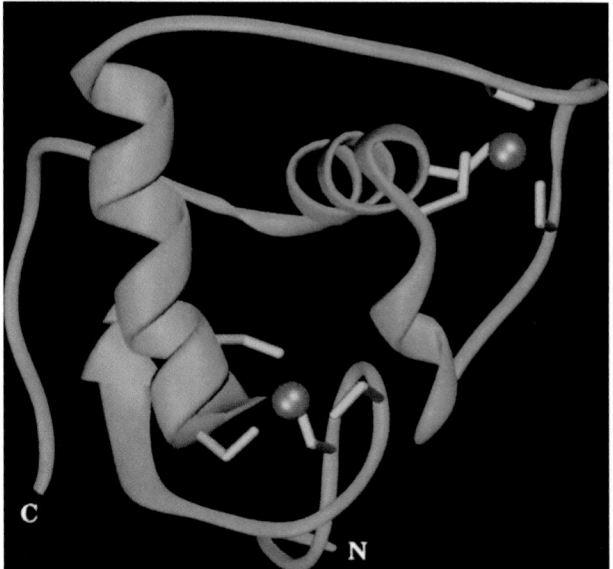

(b)

FIGURE 34-56   X-Ray structure of the dimeric glucocorticoid receptor (GR) DNA-binding domain in complex with an 18-bp DNA. The DNA contains two inverted repeats of the 6-bp glucocorticoid response element (GRE) half-sites (5′-AGAACA-3′) separated by a 4-bp spacer (GRE$_{4S}$). (*a*) A ribbon diagram of a single subunit of the GR with its two Zn^{2+} ions represented by silver spheres and their tetrahedrally liganding Cys side chains shown in stick form (*yellow*). Compare this structure with Fig. 34-55*a*. (*b*) The complex of the dimeric protein with GRE$_{4S}$ DNA as viewed with its approximate 2-fold molecular axis horizontal. The protein is shown in ribbon form with its two subunits differently colored and its bound Zn^{2+} ions represented by silver spheres. The DNA is drawn in stick form with its two 6-bp GRE half-sites magenta and the remainder cyan. Note how the GR's two N-terminal helices are inserted into adjacent major grooves of the DNA. However, only the upper (*green*) subunit binds to the DNA in a sequence-specific manner; the lower (*gold*) subunit binds to the palindromic DNA one base pair closer to the center of the DNA molecule than does the upper subunit and hence does not make sequence-specific contacts with the DNA. [Based on an X-ray structure by Paul Sigler, Yale University. PDBid 1GLU.]

sponse elements that have the half-site consensus sequences 5′-AGAACA-3′ for steroid receptors and 5′-AGGTCA-3′ for other nuclear receptors. These sequences are arranged in direct repeats ($\rightarrow n \rightarrow$), inverted repeats ($\rightarrow n \leftarrow$), and everted repeats ($\leftarrow n \rightarrow$), where n represents a 0- to 8-bp spacer (usually 1–5 bp) to whose length a specific receptor is targeted. Steroid receptors bind to their hormone response elements as homodimers, whereas other nuclear receptors do so as homodimers, heterodimers, and in a few cases, as monomers.

The X-ray structures of two related DNA segments complexed with the 86-residue DNA-binding domain of rat **glucocorticoid receptor (GR)** were determined by Sigler and Keith Yamamoto. One segment, designated GRE$_{4S}$, contains two ideal 6-bp glucocorticoid response element (GRE) half-sites arranged in inverted repeats about a 4-bp (nonnative) spacer ($n = 4$), whereas the other DNA, GRE$_{3S}$, differs from GRE$_{4S}$ in that its spacer has the nat-

urally occurring length of $n = 3$ bp. In both complexes, the protein forms a symmetric dimer involving protein–protein contacts even though it exhibits no tendency to dimerize in the absence of DNA (NMR measurements indicate that the contact region is flexible in solution).

The X-ray structure of the DNA-binding domain of the GR subunit complexed to DNA resembles that of its NMR structure in the absence of DNA: It consists of two structurally distinct modules, each nucleated by a Zn^{2+} coordination center, that closely associate to form a compact globular fold (Fig. 34-56*a*). The C-terminal module provides the entire dimerization interface as well as making several contacts with the phosphate groups of the DNA backbone. The N-terminal module, which is also anchored to the phosphate backbone, makes all of the GR's sequence-specific interactions with the GRE via three side chains extending from the N-terminal α helix, its recognition helix, which is inserted into the GRE's major groove.

In the GRE$_{3S}$ complex, a subunit of the GR DNA-binding domain binds to each GRE half-site in a structurally identical manner, making sequence-specific contacts even though the odd number of base pairs in its spacer, which the protein does not contact, renders the DNA sequence nonpalindromic. However, in the GRE$_{4S}$ complex (Fig. 34-56b), the protein dimer maintains a structure that is essentially identical to that in the GRE$_{3S}$ complex so that only one of its subunits can bind to a GRE half-site in a manner resembling that in the GRE$_{3S}$ complex. The other subunit is shifted out of register with the GRE sequence by 1 bp and hence only makes nonspecific contacts with the DNA. The dimer interactions are apparently stronger than the protein–DNA interactions, a surprising finding in light of the protein's failure to dimerize in the absence of DNA. Thus, the two subunits and the DNA associate in a cooperative fashion that favors the binding of the glucocorticoid receptor to targets with properly spaced half-sites.

The **estrogen response element (ERE),** the DNA segment to which the **estrogen receptor (ER)** specifically binds, differs from the GRE only by changes in the central two base pairs in their otherwise identical 6-bp half-sites. The X-ray structure of the ER DNA-binding domain in complex with an ERE-containing DNA segment, determined by Daniela Rhodes, closely resembles that of the GR–GRE complex. However, the side chains that make base-specific contacts with each ERE half-site are quite differently arranged from those contacting the GRE$_{4S}$ half-sites. Evidently, the discrimination of a half-site sequence is not simply a matter of substituting one or more different amino acid residues into a common framework but, rather, involves considerable side chain rearrangement.

Members of the nuclear receptor superfamily often recognize hormone response elements with similar or even identical half-site sequences as well as different spacings. The foregoing observations provide a structural basis for the graded affinities of these receptors toward their various target genes.

o. Binuclear Cys$_6$ Zinc Fingers: The GAL4 DNA-Binding Domain

The yeast protein **GAL4** is a transcriptional activator of several genes that encode galactose-metabolizing proteins. This 881-residue protein binds to a 17-bp DNA segment as a homodimer. Residues 1 to 65, which contain six Cys residues that collectively ligand two Zn^{2+} ions (Fig. 34-57), have been implicated in DNA binding; residues 65 to 94 participate in dimerization (although, as we shall see, residues 50–64 also have a weak dimerization function); and residues 94 to 106, 148 to 196, and 768 to 881 function as acidic transcriptional activating regions. The X-ray crystal structure of the 65-residue N-terminal fragment of GAL4 in complex with a symmetrical 19-bp DNA containing GAL4's palindromic 17-bp consensus sequence has been determined by Mark Ptashne, Ronen Mamorstein, and Stephen Harrison.

The protein binds to the DNA as a symmetric dimer (Fig. 34-57a), although in the absence of DNA it is only monomeric. Each subunit folds into three distinct modules: a compact Zn^{2+}-liganding domain that binds specific sequences of DNA (residues 8–40), an extended linker (residues 41–49), and a short α helical dimerization element (residues 50–64). In the Zn^{2+}-liganding module (Fig. 34-57b and top and bottom of Fig. 34-57a), the two Zn^{2+} ions are each tetrahedrally coordinated by four of the six Cys residues, with two of these residues ligating both metal ions so as to form a binuclear cluster. This module's polypeptide chain forms two short α helices connected by a loop such that the module, together with its bound Zn^{2+} ions, has pseudo-2-fold symmetry. The N-terminal helix is inserted into the DNA's major groove, thereby making sequence-specific contacts with a highly conserved CCG sequence at each end of the consensus sequence. The DNA's conformation deviates little from that of ideal B-DNA.

The dimerization helices (center of Fig. 34-57a) associate to form a short segment of parallel coiled coil in which the contact region between the coiled coil's component helices is hydrophobically stabilized by three pairs of Leu residues and a pair of Val residues (an arrangement similar to that in the so-called **leucine zipper** described below). The coiled coil is positioned over the minor groove of the DNA such that its superhelix axis coincides with the DNA's 2-fold axis. The linkers connecting the coiled coil to the DNA-binding modules wrap around the DNA, largely following its minor groove while making several nonspecific contacts with DNA phosphate groups until, on reaching the DNA-binding module, they shift over into the DNA's major groove. The two symmetrically related DNA-binding modules thereby approach the major groove from opposite sides of the DNA, ~1.5 helical turns apart, rather than from the same side of the DNA, ~1 helical turn apart, as do, for example, HTH motifs and the glucocorticoid receptor. The resulting relatively open structure could permit other proteins to bind simultaneously to the DNA.

p. Leucine Zippers Mediate Transcription Factor Dimerization

Transcriptional activation requires, as we have seen, the cooperative association of several proteins that bind to specific sequences on DNA. Steven McKnight discovered one way in which such associations occur. We have seen (Section 8-2A) that α helices with the 7-residue pseudorepeating sequence (a-b-c-d-e-f-g)$_n$, in which the a and d residues are hydrophobic, have a hydrophobic strip along one side, which induces them to dimerize so as to form a coiled coil. McKnight noticed that the rat liver transcription factor named **C/EBP** (for *CCAAT/enhancer binding protein*), which specifically binds to the CCAAT box (Section 31-2E), has a Leu at every seventh position of a 28-residue segment in its DNA-binding domain. Similar heptad repeats occur in a number of known dimeric DNA-binding proteins, including the yeast transcriptional activator **GCN4** and several DNA-binding proteins encoded by proto-oncogenes (Section 34-4C). McKnight suggested that these proteins form coiled coils in which the Leu side chains are interdigitated, much like the teeth of a zipper. He therefore named this motif the **leucine zipper.** The leucine zipper, as we shall see, mediates both the homodimerization and the heterodimerization of DNA-binding proteins (but note that it is not, in itself, a DNA-binding motif).

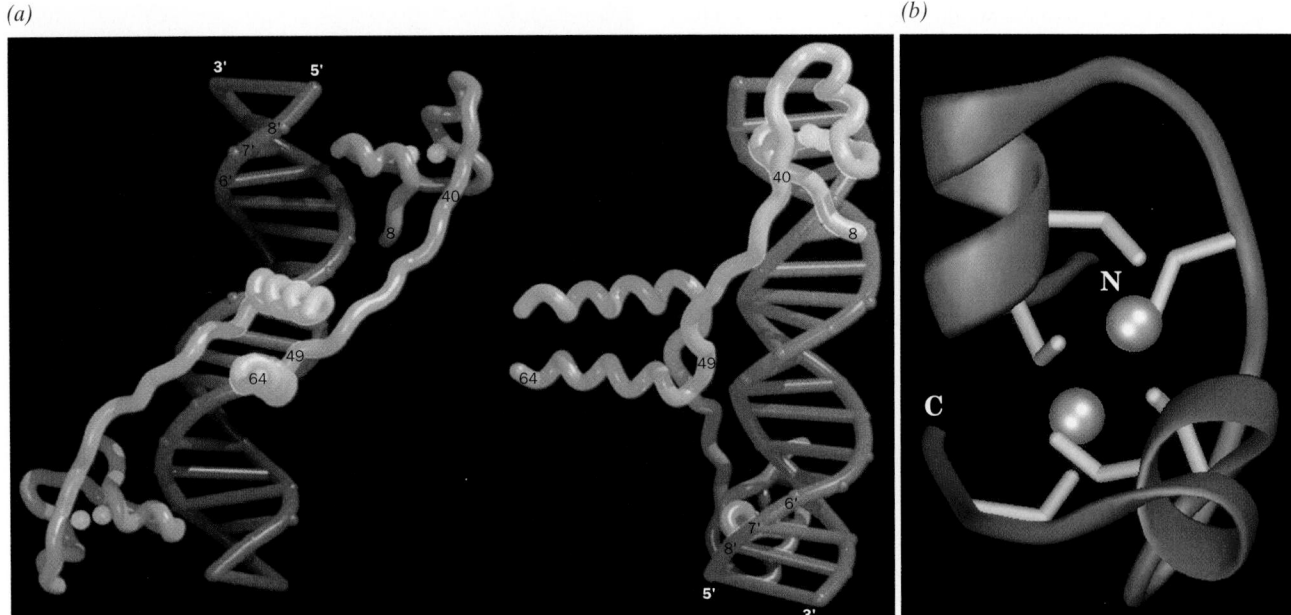

(a)   (b)

FIGURE 34-57   X-Ray structure of the yeast GAL4 DNA-binding domain in complex with a palindromic 19-bp DNA (except for the central base pair) containing the protein's consensus binding sequence. (*a*) The complex of the dimeric protein with the DNA as shown in tube form and with the DNA red, the protein backbone cyan, and the Zn^{2+} represented by yellow spheres. The views are along the complex's 2-fold axis (*left*) and turned 90° with the 2-fold axis horizontal (*right*). Note

how the C-terminal end of each subunit's N-terminal helix extends into the DNA's major groove. (*b*) A ribbon diagram of the protein's zinc finger domain (residues 8–40) with the Cys side chains of its $Zn_2^{2+}Cys_6$ complex shown in stick form (*yellow*) and its Zn^{2+} ions shown as silver spheres. Compare this structure with Figs. 34-55*a* and 34-56*a*. [Part *a* courtesy of and Part *b* based on an X-ray structure by Stephen Harrison and Ronen Mamorstein, Harvard University. PDBid 1D66.]

The X-ray structure of the 33-residue polypeptide corresponding to the leucine zipper of the 281-residue GCN4 was determined by Peter Kim and Thomas Alber. Its first 30 residues, which contain ~3.6 heptad repeats (Fig. 34-58*a*),

coil into an ~8-turn α helix that dimerizes as McKnight predicted to form ~1/4 turn of a parallel left-handed coiled coil (Fig. 34-58*b*). The dimer can be envisioned as a twisted ladder whose sides consist of the helix backbones and whose

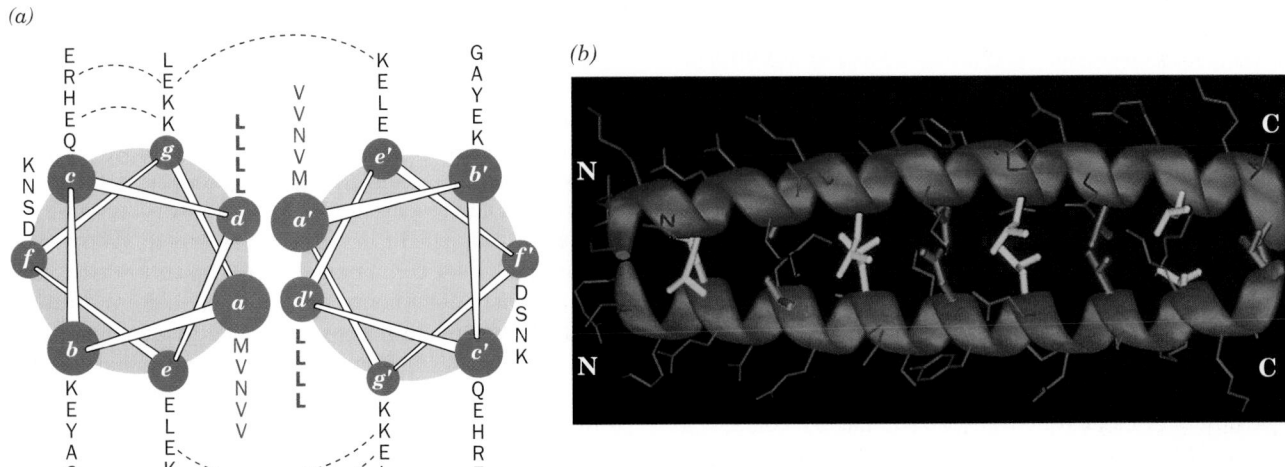

(a)   (b)

FIGURE 34-58   The GCN4 leucine zipper motif. (*a*) A helical wheel representation of the motif's two helices as viewed from their N-termini. The sequences of residues at each position are indicated by the adjacent column of one-letter codes. Residues that form ion pairs in the crystal structure are connected by dashed lines. Note that all residues at positions *d* and *d'* are Leu (L), those at positions *a* and *a'* are mostly Val (V), and those at other positions are mostly polar. [After O'Shea, E.K., Klemm,

J.D., Kim, P.S., and Alber, T., *Science* **254,** 540 (1991).] (*b*) The X-ray structure, in side view, in which the helices are shown in ribbon form. Side chains are shown in stick form with the contacting Leu residues at positions *d* and *d'* yellow and residues at positions *a* and *a'* green. [Based on an X-ray structure by Peter Kim, MIT, and Tom Alber, University of Utah School of Medicine. PDBid 2ZTA.]

rungs are formed by the interacting hydrophobic side chains. The conserved Leu residues at heptad position *d*, which comprise every second rung, are not interdigitated as McKnight originally suggested but, instead, make side-to-side contacts. The alternate rungs are likewise formed by the *a* residues of the heptad repeat (which are mostly Val) in side-to-side contact. Each Leu side chain at position *d*, in addition to packing against the symmetry-related Leu side chain, *d'*, from the other polypeptide, packs against the side chain of the succeeding residue, *e'*. Similarly, each side chain at position *a* packs between its symmetry mate, *a'*, and the preceding residue, *g'*. These two sets of alternating layers thereby form an extensive hydrophobic interface between the coiled coil's component helices.

q. bZIP Motifs: The GCN4 DNA-Binding Domain

In many but not all leucine zipper proteins, a DNA-binding region, which is rich in basic residues, is immediately N-terminal to the leucine zipper. Sequence comparisons among 11 of these so-called **basic region leucine zipper (bZIP) proteins** revealed that the 16-residue basic sequence invariably ends 7 residues before the leucine zipper's N-terminal Leu residue. Moreover, all of these basic regions, as well as the 6-residue segment linking them to the leucine zipper, are devoid of the two strongest helix-destabilizing residues, Pro and Gly (Section 9-3A), thereby suggesting that each bZIP polypeptide is entirely α helical.

The C-terminal 56 residues of GCN4 constitute its bZIP element. Harrison and Kevin Struhl determined the X-ray structure of this polypeptide segment in complex with a 19-bp-containing duplex DNA whose central 9 bp consist of GCN4's symmetrized target sequence (Fig. 34-59). The bZIP element forms a symmetric dimer in which each subunit consists, almost entirely, of a continuous α helix. The C-terminal 25 residues of two such helices associate via a leucine zipper whose geometry closely resembles that of the 33-residue GCN4 leucine zipper element alone (Fig. 34-58*b*). Past this point, the two α helices smoothly diverge to bind in the DNA's major groove on opposite sides of the helix, thereby clasping the DNA in a sort of scissors grip. The DNA, whose helix axis is nearly perpendicular to that of the coiled coil, maintains what is essentially a straight and undistorted B-form conformation. The basic region residues that are conserved in bZIP proteins thereby make numerous contacts with both the bases and with phosphate oxygens of the DNA target sequence.

r. bHLH Motifs: The Max DNA-Binding Domain

The **basic helix–loop–helix (bHLH) motif,** which occurs in a variety of eukaryotic transcription factors, contains a conserved DNA-binding basic region. This is immediately followed by two amphipathic helices connected by a loop that mediates the protein's dimerization. The bHLH motif in many proteins is followed by a conserved leucine zipper (Z) motif that presumably augments protein dimerization. The transcription factor **Max** is such a **bHLH/Z** protein, which, *in vivo*, forms a heterodimer with the proto-oncogene protein **Myc** and is required for both its normal and cancer-inducing activities. Max, by itself, readily ho-

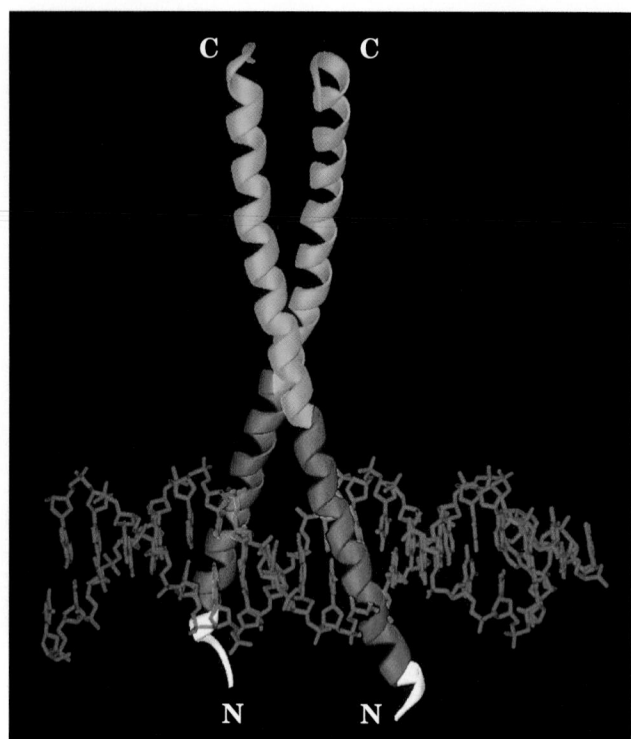

FIGURE 34-59   X-Ray structure of the GCN4 bZIP region in complex with its target DNA. The DNA (*red*) is represented in stick form and is viewed with its molecular 2-fold axis vertical. It consists of a 19-bp segment with a single nucleotide overhang at each end and contains the protein's palindromic (except for the central base pair) 7-bp target sequence. The two identical subunits, shown in ribbon form, each contain a continuous 52-residue α helix. At their C-terminal ends (*yellow*), the two subunits associate in a parallel coiled coil (a leucine zipper), and at their basic regions (*green*), they smoothly diverge to each engage the DNA in its major groove at the target sequence. The N-terminal ends are white. [Based on an X-ray structure by Stephen Harrison, Harvard University. PDBid 1YSA.]

modimerizes and binds DNA with high affinity but Myc does not do so.

The X-ray structure of a truncated version of the 160-residue Max, Max(22-113), which contains the parent protein's bHLH and leucine zipper elements, was determined, by Edward Ziff and Burley, in complex with a 22-bp quasi-palindromic DNA containing Max's 6-bp central recognition element. Each subunit of this homodimeric protein consists of two long α helices connected by a loop to form a novel protein fold (Fig. 34-60). The N-terminal α helix (b/H1) contains residues from the protein's basic region (b) followed, without interruption, by those of the HLH motif's leading helix (H1). The C-terminal α helix (H2/Z), which is composed of the second HLH helix (H2) and the leucine zipper (Z), mediates the protein's homodimerization through the formation of a parallel left-handed coiled coil similar to that in GCN4 (Fig. 34-59). Each of the dimer's two b/H1 helices projects from the resulting parallel 4-helix bundle to engage the DNA in a manner reminiscent of a pair of forceps by binding in its major groove on opposite sides of the helix (much like the way GCN4

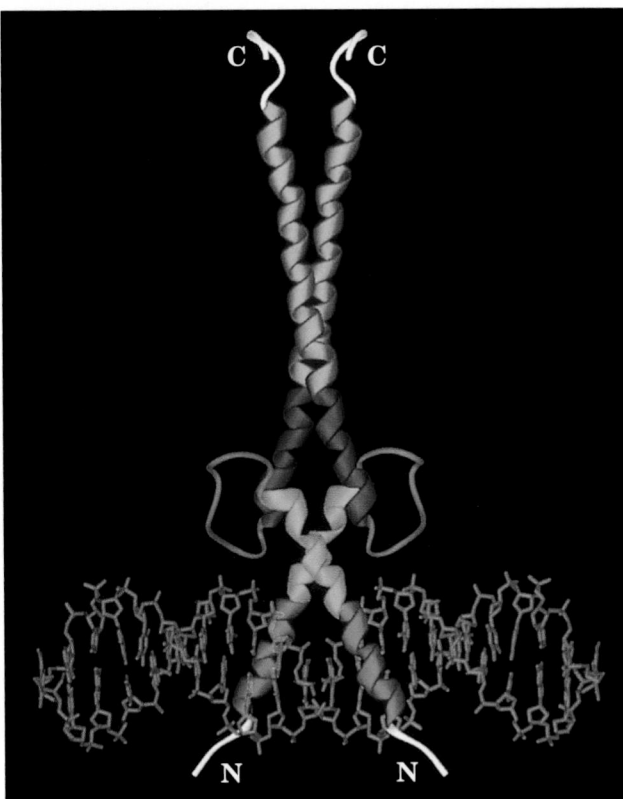

FIGURE 34-60 **X-Ray structure of the Max(22-113) dimer in complex with a 22-bp DNA containing the protein's palindromic 6-bp target sequence.** The DNA (*red*) is shown in stick form and the homodimeric protein is shown in ribbon form. The protein's N-terminal basic region (*green*) forms an α helix that engages its target sequence in the DNA's major groove and then merges smoothly with the H1 helix (*yellow*) of the helix–loop–helix (HLH) motif. Following the loop (*magenta*), the protein's two H2 helices (*purple*) of the HLH motif form a parallel left-handed four-helix bundle with the two H1 helices. Each H2 helix then merges smoothly with the leucine zipper (Z) motif (*cyan*) to form a parallel coiled coil. The protein's N- and C-terminal ends are white. [Based on an X-ray structure by Edward Ziff and Stephen Burley, The Rockefeller University. PDBid 1AN2.]

grips its target DNA, although GCN4's bZIP element consists of only two α helices rather than the four of Max). The DNA helix is essentially straight with only small deviations from the ideal B-DNA structure. Each basic region makes several sequence-specific interactions with the bases of the DNA's 6-bp recognition element as well as numerous contacts with its phosphate groups. Side chains of both the loop and the N-terminal end of the H2 helix also contact DNA phosphate groups.

s. NF-κB Binds DNA Differently from Other Transcription Factors

Nuclear factor κB (NF-κB), a transcription factor that was originally identified as an inducible nuclear activity that binds to the κB sequence in the immunoglobulin κ light chain gene enhancer (immunoglobulin genes are discussed in Section 35-2), is present in nearly all animal cells, although its role is particularly prominent in the immune system. It is present *in vivo* mainly as a DNA-binding heterodimer of the **p50** and **p65** (alternatively **RelA**) proteins (p for *protein* with the number indicating its nominal molecular mass in kD), both of which contain an ~300-residue segment known as the **Rel homology region (RHR)** because it also occurs in the product of the *rel* **oncogene.** However, p50 and p65 can also form DNA-binding homodimers. RHRs, which mediate protein dimerization and DNA binding, and which contain nuclear localization signals (NLSs), are present in a variety of proteins that serve as regulators of cellular defense mechanisms against stress, injury, and external pathogens, as well as of differentiation. Moreover, certain viruses, including HIV, have subverted RHRs to activate the expression of their genes. There are two classes of **Rel proteins:** those such as p65, **c-Rel,** and the *Drosophila* morphogen proteins (proteins that mediate development; Section 34-4B) **Dorsal** and **Dif,** whose N-terminal domains contain an RHR and whose highly variable C-terminal domains are strong transcriptional activators; and those such as p50 and the closely related **p52,** which are generated by the proteolytic processing of larger precursors and lack transactivation domains so that their homodimers function primarily as repressors.

The activity of NF-κB is largely regulated by proteins known as **inhibitor-κBs (IκBs),** which by binding to an NF-κB mask its NLS so that IκB–NF-κB complexes reside in the cytoplasm. The IκBs contain multiple ankyrin repeats (Section 12-3D) through which they bind the NF-κBs. The extracellular presence of a remarkable variety of external stimuli, including certain bacterial and viral products, several cytokines (Section 19-1L), phorbol esters (Section 19-4C), and oxidative and physical stress (e.g, free radicals and UV radiation), results, via signaling cascades, in IκBs being phosphorylated by **IκB kinase (IKK).** This, in turn, induces ubiquitination of the IκBs and their subsequent degradation by the proteosome (Section 32-6B). The liberated NF-κB is thereupon translocated to the nucleus, where it mediates transcriptional initiation by binding to 10-bp κB DNA segments that have the consensus sequence GGGRNNYYCC. Additional specificity may be achieved through the synergistic interaction of the NF-κB with other DNA-bound transcription factors such as Sp1. This activation process is self-limiting: The transcription of the gene encoding the most common IκB protein, **IκBα** (whose X-ray structure is shown in Fig. 12-38), is induced by the binding of NF-κB to the κB sites in this gene's promoter. The resulting newly synthesized IκBα enters the nucleus, where it releases the NF-κB from its complex with DNA and directs its export to the cytoplasm.

In a related mode of NF-κB activation, p50 is synthesized as the N-terminal domain of **p105,** a protein whose C-terminal domain is an IκB. The IκB domain of p105 prevents both the nuclear localization and the DNA binding of p105 as well as other RHR-containing proteins. The above external stimuli also accelerate the proteolytic processing of p105 to yield a free NF-κB and the IκB-containing C-terminal domain of p105, which as discussed above, is phosphorylated and proteolytically degraded.

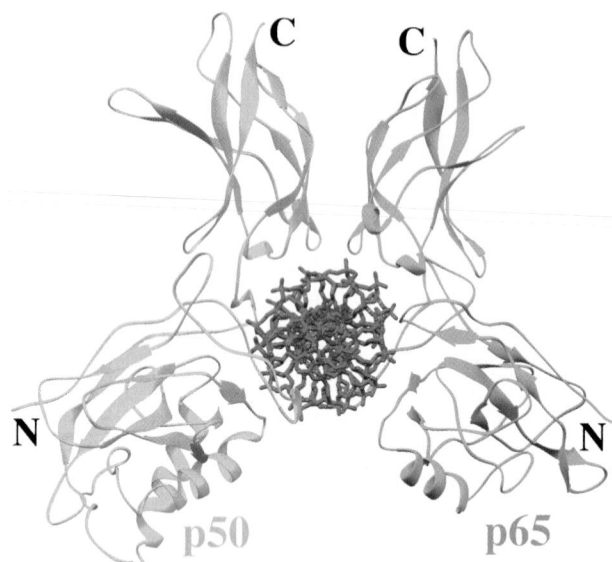

FIGURE 34-61   X-Ray structure of the mouse NF-κB p50–p65 heterodimer bound to κB DNA from the interferon β enhancer. The structure is viewed along the helix axis of the DNA, whose two strands have the sequences 5′-TGGGAAATTCCT-3′ and 5′-AAGGAATTTCCC-3′ (the duplex DNA consists of 11 bp with a 1-nt overhang at each end) and are drawn in stick form colored according to atom type (C green, N blue, O red, and P magenta). The protein is represented by ribbons with p50 (residues 39–364 of 435 residues) gold and p65 (residues 19–291 of 549 residues) cyan. [Based on an X-ray structure by Gourisankar Ghosh, University of California at San Diego. PDBid 1LE5.]

The X-ray structure of the heterodimer of mouse p50 and p65 in complex with the κB segment of the β-interferon enhancer, determined by Gourisankar Ghosh, bears a striking resemblance to a butterfly with its homologous protein subunits forming its outspread wings and the DNA its torso (Fig. 34-61). The two protein subunits have similar structures that each consists of two domains, with the C-terminal domains forming the dimerization interface and both domains interacting with the DNA. Both their N- and C-terminal domains have immunoglobulin-like folds (a sandwich of a 3- and a 4-stranded antiparallel β sheet; Section 35-2B) and interact with the DNA exclusively through 10 loops, 5 from each subunit, that link their β strands and fill the DNA's major groove. The p50 and p65 bind to 5-bp and 4-bp subsites at the 5′ and 3′ ends of the consensus sequence, respectively, with the two subsites separated by a single base pair. The protein's DNA-binding surface is much more extensive than those of other transcription factors, which accounts for the unusually high affinity of NF-κBs for their target sequences. This also explains the inability of deletion mutagenesis to localize NF-κB's DNA-binding region, since changes anywhere in its structure are likely to affect the disposition of its DNA-binding loops.

Comparisons of the X-ray structures of the p50–p65 heterodimer bound to several κB DNA segments with different sequences, all determined by Ghosh, reveal small but significant structural differences among these various complexes. These arise mainly from the different degrees of bending of the various κB DNAs as well as the different interactions of the proteins with the different sequences of bases, all of which result in small conformational differences among the chemically identical proteins in these complexes. This is probably why the substitution in an enhancer of one κB segment for another does not produce the same level of transcription, even though the NF-κB binds to the isolated κB segments with equal affinity. Evidently, the way in which NF-κB interacts with other proteins that are bound to the enhancer (such as the glucocorticoid receptor, which interacts with p65) affects its activational potency and hence fine-tunes the expressional levels of its target genes.

t. Mediator Provides the Interface between Transcriptional Activators and RNAP II

Eukaryotic genomes encode as many as several thousand transcriptional regulators for class II genes (e.g., Fig. 34-27). How does the binding of these various regulators to their cognate enhancers/silencers influence the rate at which RNAP II initiates transcription? Genetic studies have implicated TFIIB, TFIID, and TFIIH in this process *in vivo*. Nevertheless, activators fail to stimulate transcription by a reconstituted PIC *in vitro*. Evidently, an additional factor is required to do so. *Indeed, genetic studies in yeast by Kornberg led him to discover an ~20-subunit, ~1000-kD complex named **Mediator**, whose presence is required for transcription from nearly all class II gene promoters in yeast.* Mediator, which is therefore considered to be a coactivator, binds to the C-terminal domain (CTD) of RNAP II's β′ subunit (Section 31-2E) to form the so-called **RNAP II holoenzyme.** Further investigations revealed that metazoans contain several multisubunit complexes that function similarly to yeast Mediator. These include complexes known as **CRSP, NAT, ARC/DRIP, TRAP/SMCC, mMED,** and **PC2,** which share many of their numerous subunits. Moreover, many of their subunits are related, albeit distantly, to those of yeast Mediator. *Mediators apparently function as adaptors that bridge DNA-bound transcriptional regulators and RNAP II so as to influence (induce or inhibit) the formation of a stable PIC at the associated promoter. They thereby function to integrate the various signals implied by the binding of these transcriptional regulators to their target DNAs.* The different metazoan mediators presumably relay signals from different sets of transcriptional regulators.

Kornberg and Francisco Asturias have determined the EM-based low resolution (30–35 Å) structure of yeast Mediator and human TRAP (Fig. 34-62). The two particles are similarly shaped with nearly perpendicular "head" and "middle-tail" domains. The EM-based image of yeast RNAP II holoenzyme (Fig. 34-63) reveals that Mediator has assumed a more extended conformation in which the "middle" and "tail" domains are clearly separated. The head domain interacts closely with the RNAP II, although >75% of RNAP II's surface remains accessible for interaction with other components of the PIC. However, the tail domain appears not to contact RNAP II at all.

Altogether, then, the transcriptional machinery for class II genes comprises nearly 60 polypeptides with an aggregate

(a)   *(b)*

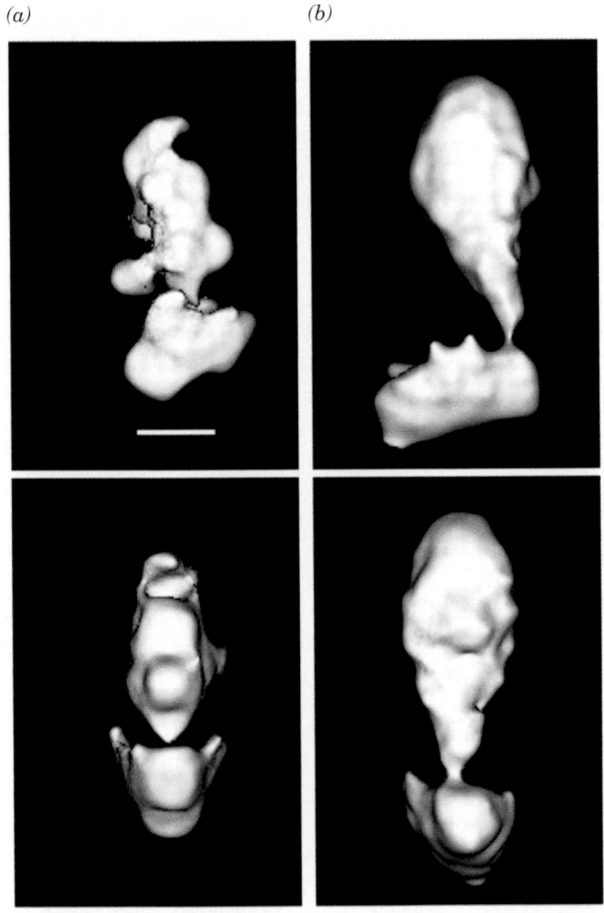

FIGURE 34-62 **EM-based structures of (*a*) yeast Mediator and (*b*) human TRAP complex.** The orientations of the complexes in the upper and lower rows differ by 90° rotations about the vertical direction. The bottom part of each image forms the "head" domain of the complex, which is nearly perpendicular to its top portion, the "middle-tail" domain. The bar is 100 Å in length. [Courtesy of Francisco Asturias, The Scripps Research Institute, La Jolla, California.]

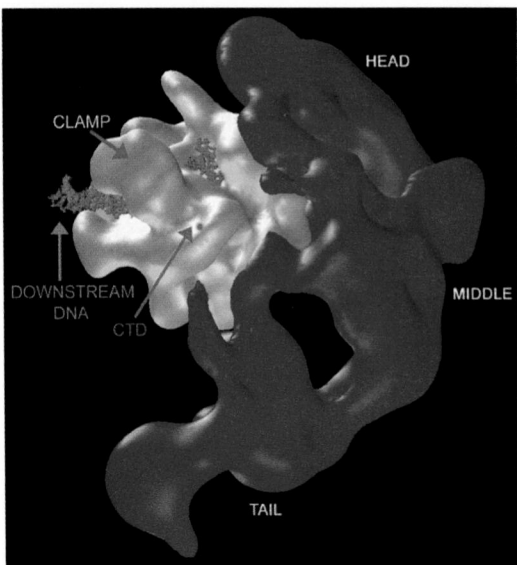

FIGURE 34-63 **EM-based projection of the yeast RNAP II holoenzyme at 35-Å resolution.** Mediator (*blue*) assumes a more extended conformation than that in Fig. 34-62*a*, such that its head, middle, and tail domains are clearly distinguishable. The independently determined EM-based image of yeast RNAP II (*white*) is oriented to best match the RNAP II density in the lower resolution holoenzyme image. Promoter DNA (*orange*) was modeled in, based on the structure of the yeast RNAP II elongation complex (Fig. 31-21*b*). Note that the RNAP II's DNA binding cleft remains fully accessible in the holoenzyme complex. [Courtesy of Francisco Asturias, The Scripps Research Institute, La Jolla, California, and Roger Kornberg, Stanford University School of Medicine.]

molecular mass of ~3 million D. Nevertheless, as we see below, this ribosome-sized assembly (the eukaryotic ribosome has a molecular mass of ~4.2 million; Table 32-8) requires considerable assistance from yet other large macromolecular assemblies to gain access to the DNA in chromatin.

u. Transcriptionally Active Chromatin Is Sensitive to Nuclease Digestion

Early research on the mechanism of eukaryotic transcription largely ignored the influence of chromatin. Yet, as we have seen (Section 34-2A), euchromatin but not heterochromatin is transcriptionally active. *Indeed, investigations over the past decade have revealed that eukaryotic cells contain elaborate systems that participate in controlling transcriptional initiation by altering chromatin structure.* In the remainder of this subsection we discuss the nature of these systems.

The open structure of transcriptionally active chromatin presumably gives the transcriptional machinery access to the active genes. This hypothesis was corroborated by Harold Weintraub's demonstration that *transcriptionally active chromatin is about an order of magnitude more susceptible to cleavage by DNase I than is transcriptionally inactive chromatin.* For example, globin genes from chicken erythrocytes (avian red cells have nuclei) are more sensitive to DNase I digestion than are those from chicken oviduct (where eggs are made), as was indicated by the loss of the abilities of these genes to hybridize with a complementary DNA probe after DNase I treatment. Conversely, the gene encoding **ovalbumin** (the major egg white protein) from oviduct is more sensitive to DNase I than is that from erythrocytes. Thus, nuclease sensitivity appears to delineate chromatin's **functional domains,** although their relationship to chromatin's structural domains (Section 34-3A) is unclear. Nevertheless, nuclease sensitivity reflects a gene's potential for transcription rather than transcription itself: The DNase I sensitivity of the oviduct ovalbumin gene is independent of whether or not the oviduct has been hormonally stimulated to produce ovalbumin.

The variation of a given gene's transcriptional activity with the cell in which it is located indicates that chromosomal proteins participate in the gene activation process. Yet histones' chromosomal abundance and lack of variety make it highly unlikely that they have the specificity required for this role. Among the most common nonhistone proteins are the members of the **high mobility group (HMG),** so named because of their high electrophoretic mobilities in polyacrylamide gels (and possibly because they were discovered by H.M. Goodwin). These highly conserved, low molecular mass (<30 kD) proteins, which have the unusual amino acid composition of ~25% basic side chains and 30% acidic side chains, are relatively abundant, with ~1 HMG molecule per 10 to 15 nucleosomes. The HMG proteins can be eluted from chick erythrocyte chromatin by 0.35*M* NaCl without gross structural changes to the nucleosomes. This treatment eliminates the preferential nuclease sensitivity of the erythrocyte globin genes.

v. HMG Proteins Are Architectural Proteins That Participate in Regulating Gene Expression

The HMG proteins consist of three superfamilies, **HMGB, HMGA,** and **HMGN,** which have the following properties:

1. The mammalian HMGB proteins, **HMGB1** and **HMGB2** (~210 residues; previously known as **HMG1** and **HMG2**), which bind DNA without regard to sequence, each consist of two tandem ~80-residue **HMG boxes,** A and B, followed by an acidic tail consisting of ~30 (HMGB1) or ~20 (HMGB2) consecutive Asp or Glu residues. However, *Drosophila* **HMG-D** and yeast **NHP6A** proteins each contain only one HMG box, which is closely similar to the B domain of HMGB1. The NMR structure of NHP6A in complex with a 15-bp DNA, determined by Juli Feigon, reveals, in agreement with the structures of several other HMG box–containing proteins, that the HMG box consists of three helices arranged in an L-shape with the inside of the L inserted into the minor groove of the DNA (Fig. 34-64). This induces the DNA to bend by as much as 130° toward its major groove. Apparently, nuclear HMGB proteins function as architectural proteins that induce the binding of other proteins, including various steroid receptors, to DNA and hence facilitate the as-

sembly of nucleoprotein complexes. Indeed, NHP6A and HMG-D can functionally replace the bacterial DNA-bending protein HU even though HMGB and HU proteins have no structural or sequence similarity (HU is discussed in Section 33-3C). HMG boxes also occur in several sequence-specific transcription factors, including the mammalian male sex determining factor SRY (Section 19-1G).

2. The HMGA superfamily consists of four proteins: the 107-, 96-, and 179-residue splice variants **HMGA1a, HMGA1b,** and **HMGA1c** (previously named **HMG-I, HMG-Y,** and **HMG-I/R**) and the homologous 109-residue **HMGA2** (previously named **HMG-C**). Each of these proteins contains three similar so-called **AT hooks** that have the invariant core sequence Arg-Gly-Arg-Pro flanked by positively charged residues and that bind to AT-rich DNA sequences. The NMR spectrum of a truncated form of HMGA1a that contains only its second and third AT hooks (residues 51–90) is indicative of a random coil. However, the NMR structure of this truncated HMGA1a in complex with a 12-bp DNA containing an AT-rich segment of the β-interferon enhancer, determined by Angela Gronenborn and Marius Clore, reveals that each of its AT hooks binds in an extended conformation in the minor groove of a separate DNA molecule (Fig. 34-65). Despite the relatively undistorted DNA in this structure, it has been shown that full-length HMGA proteins can bend, straighten, unwind, and induce loop formation in dsDNA. HMGA proteins have been implicated in regulating the transcription of numerous genes. For example, HMGA1 proteins recruit the transcription factors NF-κB and **c-Jun** (Section 34-4C) to the enhanceosome at the β-interferon enhancer by a combination of DNA bending and protein–protein interactions.

3. The HMGN proteins, **HMGN1** and **HMGN2** (98 and 89 residues; previously known as **HMG14** and **HMG17**), occur in mammals but not in *Drosophila* or yeast. They are 60% identical in sequence and consist of three functional motifs: a bipartite nuclear localization signal (NLS; see above), a conserved ~30-residue, positively charged **nucleosome-binding domain (NBD),** and a **chromatin-unfolding domain (CHUD).** The ~30-residue, positively charged NBD, as its name implies, targets HMGN proteins to bind to nucleosome core particles as homodimers of HMGN1 or HMGN2 (but not as heterodimers) without preference for the underlying DNA sequence. This stabilizes the nucleosome core particle by bridging its two adjacent dsDNA strands. Nevertheless, HMGN proteins increase the rate of transcription and DNA replication, presumably because they loosen the structure of chromatin fibers. This apparently occurs because the CHUD domain interacts with the N-terminal tail of histone H3 (see below) and because nucleosome-bound HMGN proteins compete with histone H1 for its nucleosomal binding site. HMGN-containing nucleosomes occur as clusters averaging six adjacent nucleosomes, thereby confirming that they alter internucleosomal structure. The presumably decondensed chromatin in these clusters could provide gateways through which regulatory proteins gain access to their target DNAs.

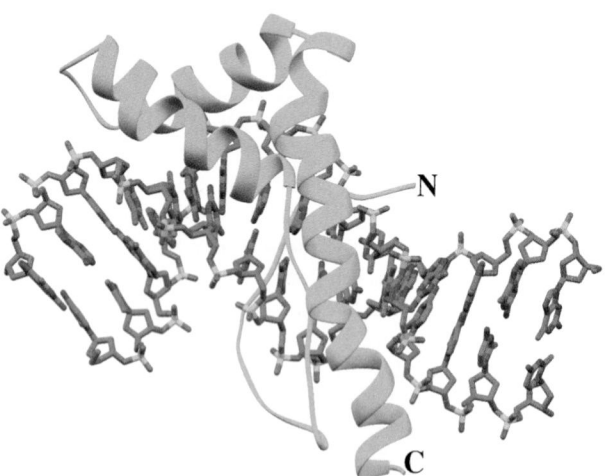

FIGURE 34-64 NMR structure of yeast NHP6A protein in complex with a 15-bp DNA. The protein is drawn in ribbon form (*cyan*) and the DNA is drawn in stick form colored according to atom type (C green, N blue, O red, and P yellow). The NHP6A's L-shaped HMG box binds in the DNA's minor groove so as to bend the DNA by ~70° toward its major groove. [Based on an NMR structure by Juli Feigon, University of California at Los Angeles. PDBid 1J5N.]

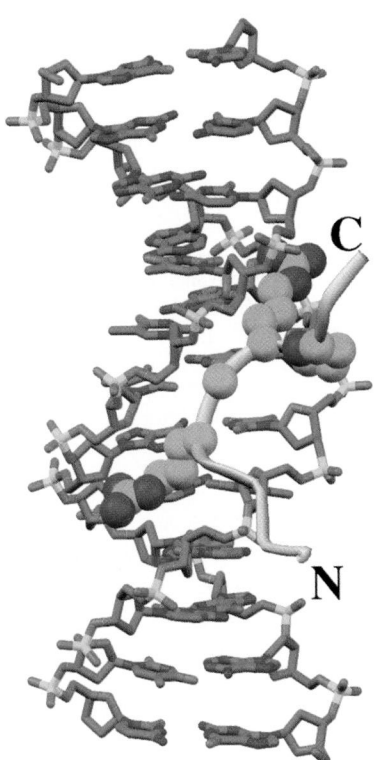

FIGURE 34-65 NMR structure of a truncated HMGA1a consisting of only its second and third AT hooks in complex with a 12-bp AT-rich DNA. The protein is drawn in ribbon form (*lavender*) with the side chains and C$_\alpha$ atoms of the invariant core sequence of its AT hook, Arg-Gly-Arg-Pro, drawn in space-filling form with C cyan and N blue. The DNA is drawn in stick form colored according to atom type (C green, N blue, O red, and P yellow). The protein's two 10-residue AT hooks bind to separate DNA dodecamers. Nevertheless, only one set of DNA resonances was observed, which indicates that the two AT hook–DNA structures are closely similar. The peptide segment that links the two AT hooks is not observed and hence must be highly mobile. Consequently, only the structure shown was observed. Note that the AT hook binds in the DNA's minor groove but does not cause it to bend. [Based on an NMR structure by Angela Gronenborn and Marius Clore, NIH, Bethesda, Maryland. PDBid 2EZF.]

w. Nucleosome Cores Are Transferred Out of the Path of an Advancing RNA Polymerase

Since nucleosomes bind their component DNA tightly and quite stably, how does an actively transcribing RNA polymerase, which is roughly the size of a nucleosome and must separate the strands of duplex DNA to transcribe it, get access to the DNA? Two classes of models have been proposed: The advancing RNA polymerase either (1) induces a conformational change in the nucleosome that permits its DNA to be transcribed while still associated with the nucleosome or (2) displaces the nucleosome from the DNA. These models were differentiated by Gary Felsenfeld as follows: A single nucleosome core was assembled onto a short DNA segment of defined sequence. Then, under conditions in which nucleosome cores are stable (don't decompose or move) in the absence of tran-

scription, the resulting assembly was ligated into a plasmid between a promoter and terminators for the RNA polymerase from **bacteriophage SP6** and the DNA between these two sites was transcribed by this enzyme. This treatment caused the nucleosome to move to a different site on the same plasmid, with a small preference for the untranscribed region preceding the promoter. However, the use of a very short (227-bp) DNA template containing the SP6 promoter and a bound nucleosome revealed that nucleosome transfer occurred only to the same template molecule, 40 to 95 bp upstream of its original site, even in the presence of a large excess of competitor DNA. Evidently, the histone octamer somehow steps around a transcribing RNA polymerase so as to transfer to a nearby segment of the same DNA. Felsenfeld has proposed that this occurs via a DNA looping mechanism in which the histone octamer incrementally spools onto its new position behind the advancing RNA polymerase as the polymerase peels the octamer away from its original position (Fig. 34-66).

How does RNA polymerase displace nucleosomes from DNA? SP6 RNA polymerase, being a phage enzyme, cannot have evolved to interact with histones but, nevertheless, appears to do so. Other prokaryotic RNA polymerases can likewise transcribe through nucleosomes. A plausible mechanism for this phenomenon is that it is promoted by the transcriptionally induced supercoiling of DNA. A moving transcription bubble, it will be recalled (Section 31-2C), generates positive supercoils in the DNA ahead of it and negative supercoils behind it. However, nucleosomal DNA is wound around its histone core in a left-handed toroidal coil and is therefore negatively supercoiled (Section 29-3A). Consequently, an advancing RNA polymerase molecule should destabilize the nucleosomes ahead of it while facilitating nucleosome assembly in its wake, precisely what is observed (although, as we shall see below, the cell employs several methods for loosening the grip of nucleosomes on DNA). Subsequent investigations revealed that yeast RNA polymerase III, which is considerably larger than SP6 RNA polymerase, interacts with nucleosomes in a similar manner.

x. Locus Control Regions Are Nuclease Hypersensitive

The very light digestion of transcriptionally active chromatin with DNase I and other nucleases has revealed the presence of **DNase I hypersensitive sites** that are about an order of magnitude more susceptible to cleavage by DNase I than are DNase I sensitive sites. These specific 100- to 200-bp DNA segments are mostly located in the 5′-flanking regions of transcriptionally active or activatable genes as well as in sequences involved in replication and recombination. Nuclease hypersensitive sites, as we shall see, are apparently the "open windows" that allow the transcriptional machinery access to DNA control sequences. This is because *DNase I hypersensitive gene segments are free of nucleosomes*. For example, in SV40-infected cells, none of the ~24 nucleosomes that are complexed to the virus' 5.2-kb circular DNA (Fig. 34-67) incorporate the ~250-bp viral transcription initiation site, thereby rendering that site nuclease hypersensitive.

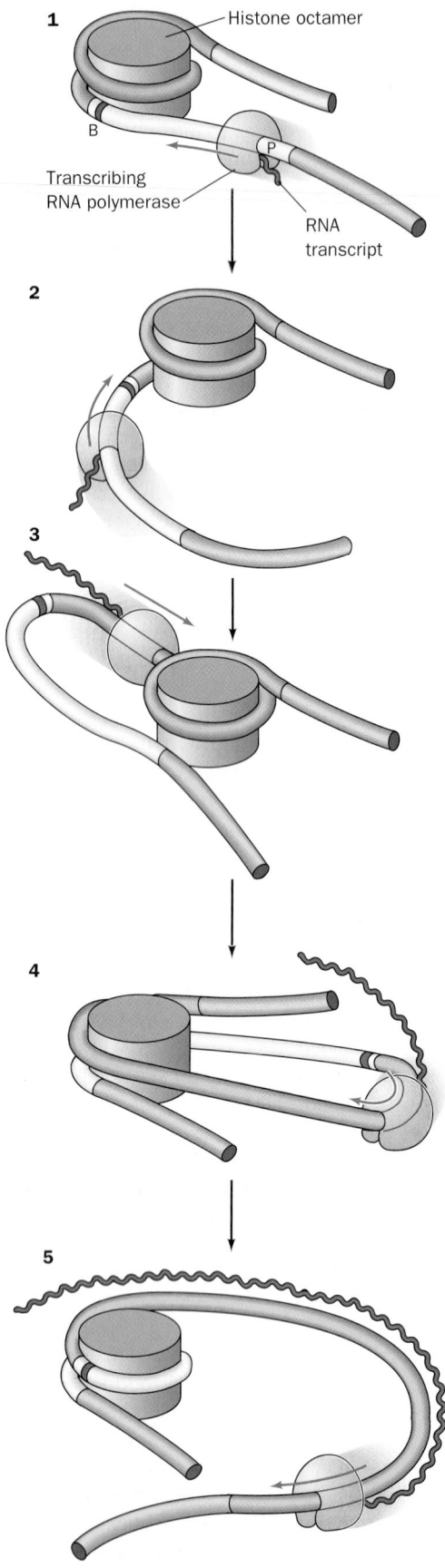

FIGURE 34-66 Spooling model for transcription through a nucleosome. (1) RNA polymerase commences transcription at a promoter, P; the border of the nucleosome is indicated by B. **(2)** As the RNA polymerase approaches the nucleosome, it induces the dissociation of the proximal (nearest) DNA, thereby exposing part of the histone octamer surface. **(3)** The exposed histone surface binds to the DNA behind the RNA polymerase, thus forming a loop. Note that this loop is topologically isolated from the rest of the DNA and, consequently, is subject to the superhelical stress that the advancing RNA polymerase generates (see the text). **(4)** As the RNA polymerase continues to advance, the DNA ahead of it peels off the histone octamer while the trailing DNA spools onto it. **(5)** The nucleosome is thereby re-formed behind the RNA polymerase, thus permitting the transcript to be completed. [After Studitsky, V.M., Clark, D.J., and Felsenfeld, G., *Cell* **76,** 379 (1994).]

However, since naked DNA is not DNase I hypersensitive, the special properties of nuclease hypersensitive chromatin must arise from the sequence-specific binding of proteins so as to exclude nucleosomes.

The human β-globin cluster (Section 34-2F) has five nuclease hypersensitive sites in a region 6 to 22 kb on the 5′ side of the ε gene as well as one hypersensitive site 20 kb on the 3′ side of the β gene (Fig. 34-68). These hypersensitive sites appear to demarcate the boundaries of a large segment of transcriptionally active chromatin. Individuals with an extensive upstream deletion that eliminates the 5′ hypersensitive sites, the so-called Hispanic deletion, but with normal β-like genes, have **(γδβ)⁰-thalassemia** (severely reduced synthesis of γ, δ, and β globins). Similarly, mice that are transgenic for the human β-globin gene together with its local regulatory sites either fail to express or express very low levels of human β-globin. This is because a DNA segment that is randomly inserted into a genome will most often occupy a position in transcriptionally inactive heterochromatin, a phenomenon known as a **position effect.** However, mice transgenic for the entire region of the human β-globin cluster between its hypersensitive sites express high levels of human β-globin in

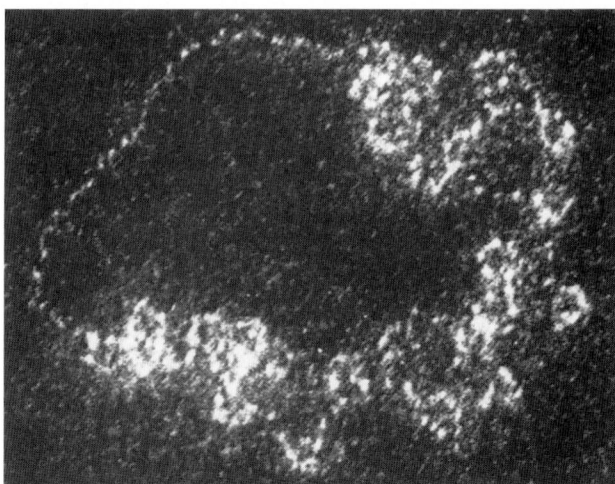

FIGURE 34-67 Electron micrograph of an SV40 minichromosome that has a nucleosome-free DNA segment. [Courtesy of Moshe Yaniv, Institut Pasteur, Paris, France.]

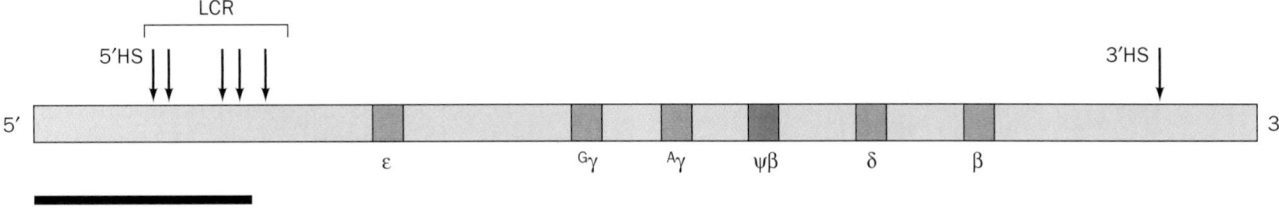

Hispanic deletion

FIGURE 34-68 The β-globin cluster showing the positions (*arrows*) of its genes and DNase I hypersensitive sites (HSs). The hypersensitive sites on the 5′ side of the ε-globin gene (5′HSs) form the locus control region (LCR), whose presence is required for the expression of the β-like genes. The deletion of the LCR, as occurs in the Hispanic deletion, all but eliminates the expression of the β-like genes. The products of the β-globin cluster are discussed in Section 34-2F.

erythroid tissues. Thus the β-globin cluster's 5′ nuclease hypersensitive sites, which are collectively known as a **locus control region (LCR),** function to suppress position effects over large distances (e.g., the ~100 kb length of the β-globin cluster). LCRs have enhancerlike properties but, unlike enhancers, are orientation- and position-specific.

LCRs are apparently activated by proteins that are expressed only in specific cell lineages (e.g., only in erythroid cells for genes controlled by the β-globin LCR) so as to render the gene(s) under an LCR's control susceptible to activation by transcription factors. In support of this contention, it has been shown that nonglobin genes that have been put under the control of the β-globin cluster LCR are expressed in erythroid cells but not in nonerythroid cells. LCRs occur in a growing list of mammalian genes. However, the way that they permit the expression of the genes under their control remains largely conjectural.

y. Insulators Isolate Genes from Distant Regulatory Elements

We have seen that enhancers function independently of their position and orientation. But then, what prevents an enhancer from affecting the transcription of all the genes in its chromosome? Conversely, the formation of heterochromatin appears to be self-nucleating. What prevents heterochromatin from spreading into neighboring segments of euchromatin so as to prevent the transcription of their component genes? In many cases, this appears to be the job of short (<2 kb) DNA sequences known as **insulators** that thereby define the boundaries of functional domains.

Among the best characterized insulators are the *Drosophila* sequences **scs** and **scs′** (for *specialized chromatin structure*), which normally flank two consecutive *hsp70* heat shock genes. The transformation into *Drosophila* of the *white* gene (which confers white eye color; Section 1-4C) together with a minimal promoter yielded lines of flies that varied in eye color, a manifestation of the position effect. However, when the construct was flanked by scs and scs′, it yielded only flies with white eyes. Evidently, these insulators overcome the position effect. In addition, if scs or scs′ is inserted between a gene and its upstream regulatory sequences, then the expression of this gene is no longer influenced by these sequences. Several other insulators have been characterized, both in *Drosophila* and in vertebrates.

The foregoing indicates that insulators resemble LCRs in that they suppress position effects. However, unlike LCRs, insulators have no enhancerlike properties; that is, they do not have positive or negative effects on the expression of the genes they control. Rather, *insulators only function to prevent regulatory elements outside the region they control from influencing the expression of the genes inside the region.* LCRs lack this property; they do not protect their associated genes from the influence of control sequences that are upstream of the LCR. In fact, the chicken β-globin cluster has an upstream insulator named **HS4** (for *hypersensitive site 4*) that prevents regulatory elements that are further upstream from influencing the expression of its genes.

The way that insulators work remains enigmatic. Presumably, it is not the insulators themselves but the proteins that bind to them that form the active insulator elements. For example, in *Drosophila,* the insertion of the transposable element *gypsy* between the promoter of the gene *yellow* (which gives flies a pale yellow body rather than the wild-type yellow-brown) and its upstream enhancers prevents these enhancers from activating *yellow* but does not affect downstream enhancers. The 12-zinc finger protein named **Su(Hw)** (for *suppressor of hairy wing*) specifically binds to the *gypsy* insulator and is required for its enhancer-blocking properties. Su(Hw) also binds to the protein **Mod(mdg4)** (for *modifier of mdg4*), and together they bind to the **nuclear matrix** (the nuclear equivalent of the cytoskeleton). In fact, immunostaining studies by Victor Corces indicate that Su(Hw) and Mod(mdg4) colocalize to several hundred sites in *Drosophila* polytene chromosomes (Fig. 34-69) and are distributed in a punctate pattern around the nuclear matrix. Similar distributions were seen with the protein **BEAF-32** (for *boundary element-associated factor of 32 kD*), which specifically binds to scs′ but not to scs. These observations suggest that these proteins each bind to numerous insulator sites on *Drosophila* chromosomes, which, in turn, suggests that insulators function as matrix-associated regions (MARs; Section 34-1D) to form structural domains. These proteins may prevent enhancers that are outside such a domain from influencing the expression of the genes that are inside the domain and, furthermore, may inhibit heterochromatin from encroaching on and thereby transcriptionally inactivating the domain. Similarly, enhancer blocking in the chicken β-globin cluster is associated with the binding of the 11-zinc finger

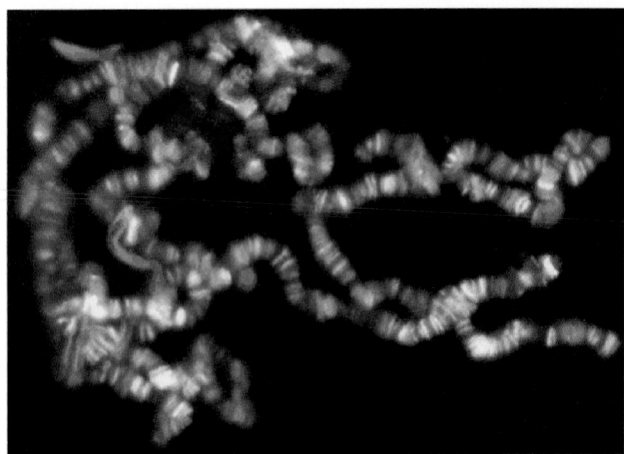

FIGURE 34-69   Colocalization of the Su(Hw) and Mod(mdg4) proteins on the polytene chromosomes of *Drosophila* larvae. The DNA has been stained blue, whereas the proteins have been immunostained such that sites containing Su(Hw) are green, those containing Mod(mdg4) are red, and sites where both proteins colocalize are yellow. [Courtesy of Victor Corces, The Johns Hopkins University.]

protein **CTCF** to HS4 and, moreover, CTCF-binding sites occur throughout the genome.

z. Chromatin Immunoprecipitation Reveals the DNA Binding Sites of Proteins

Throughout this section, we discuss the DNA sequences in chromatin to which specific proteins bind. In many cases, these DNA sequences have been identified using a procedure known as **chromatin immunoprecipitation (ChIP).** ChIP involves the following steps:

1. Living cells are treated with formaldehyde, which rapidly cross-links the amino and imino groups of Arg, His, and Lys residues to nearby (within ~2 Å) amino groups of bases, primarily those of adenine and cytosine, while preserving the chromatin structure:

$$R_1—NH_2 \quad + \quad \underset{\substack{| \\ H}}{\overset{\substack{O \\ || \\ C}}{}}\!\!{\overset{}{\underset{H}{}}} \quad + \quad H_2N—R_2$$

Formaldehyde

$$R_1—NH—CH_2—HN—R_2$$

2. The cells are lysed and the cross-linked chromatin is sheared into manageable (~500-nt) fragments by sonication.

3. The chromatin fragments are treated with antibodies (Section 35-2B) raised against the protein of interest (which may be a histone with a specific modification such as an acetylation or a methylation at a particular site; see below). The mixture is then absorbed to agarose gel beads to which *Staphylococcus aureus* **Protein A** has been cross-linked (Section 6-3C). Protein A binds antibodies, but only when they are bound to their target antigens, thereby permitting the isolation of only those chromatin fragments in which the DNA is cross-linked to the antibody-bound protein.

4. The DNA is released from the protein to which it is cross-linked by acidification, which reverses the formaldehyde cross-linking reaction, and the DNA is isolated.

5. The DNA is identified by PCR using specific primers (Section 5-5F), by Southern blotting (Section 5-5D), or by hybridization to a DNA microarray (Section 7-6B), thereby revealing the DNA segments to which the protein of interest binds.

aa. Histone Modification and Remodeling Play an Essential Role in Transcriptional Activation

The DNA packaged by chromatin must be accessible by the transcriptional machinery in order for it to be expressed. This, as we shall see, occurs via two types of processes acting in synergy: (1) the posttranslational modifications of core histones, mainly their N-terminal tails, and (2) the remodeling of chromatin through the ATP-driven alteration of the position and/or properties of its nucleosomes.

The posttranslational modifications to which core histones are subject include the acetylation/deacetylation of specific Lys side chains, the methylation of specific Lys and Arg side chains, the phosphorylation/dephosphorylation of specific Ser and possibly Thr side chains, and the ubiquitination of specific Lys side chains (Fig. 34-70). Moreover, Lys side chains can be mono-, di-, and trimethylated and Arg

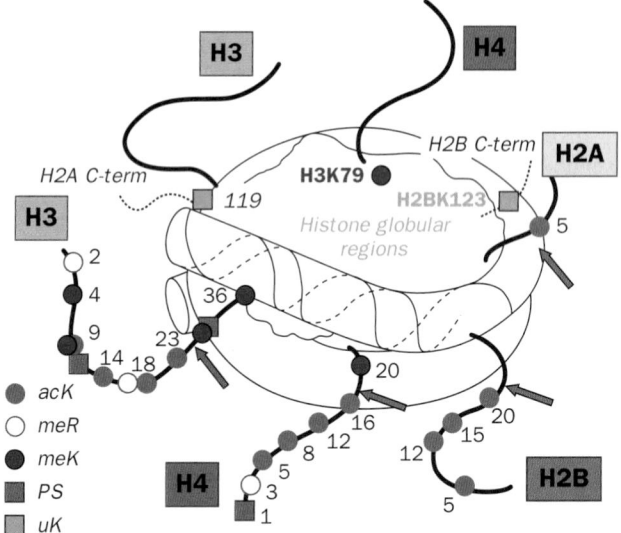

FIGURE 34-70   Histone modifications on the nucleosome core particle. Posttranslational modification sites are indicated by the residue numbers and the colored symbols, which are defined in the key at the lower left (*acK* = acetyl-Lys, *meR* = methyl-Arg, *meK* = methyl-Lys, *PS* = phospho-Ser, and *uK* = ubiquitinated Lys). Note that H3 Lys 9 can be either methylated or acetylated. The N-terminal tail modifications are shown on only one of the two copies of H3 and H4 and only one molecule each of H2A and H2B are shown. The C-terminal tails of one H2A and one H2B are represented by dashed lines. The green arrows indicate the sites in intact nucleosomes that are susceptible to trypsin cleavage. This cartoon summarizes data from several organisms, some of which may lack particular modifications. [Courtesy of Bryan Turner, University of Birmingham School of Medicine, U.K.]

side chains can be mono- and both symmetrically and asymmetrically dimethylated. The core histones' N-terminal tails, as we have seen (Section 34-1B), are implicated in stabilizing the structures of both core nucleosomes and higher order chromatin. All of these modifications but methylations reduce (make more negative) the electronic charge of the side chains to which they are appended and hence are likely to weaken histone–DNA interactions so as to promote chromatin decondensation, although as we shall see, this is not always the case. Methyl groups, in contrast, increase the basicity and hydrophobicity of the side chains to which they are linked and hence tend to stabilize chromatin structure. Modified histone tails also interact with specific chromatin-associated nonhistone proteins in a way that changes the transcriptional accessibility of their associated genes.

The characterization of a variety of histone tail modifications led David Allis to hypothesize that *there is a "histone code" in which specific modifications evoke certain chromatin-based functions and that these modifications act sequentially or in combination to generate unique biological outcomes.* For example, uncondensed and hence transcriptionally active chromatin is associated with the acetylation of H3 Lys 9 and 14 and H4 Lys 5 and the methylation of H3 Lys 4 and H4 Arg 3; condensed and hence transcriptionally inactive chromatin is associated with the acetylation of H4 Lys 12 and the methylation of H3 Lys 9; and nucleosome deposition is associated with the phosphorylation of H3 Ser 10 and 28. It can be seen from Fig. 34-70 that there are a vast number of possible combinations of histone modifications.

The growth of a multicellular organism requires the proliferation of the cells in its various tissues without changing their identities (the process whereby cells progressively and irreversibly change their identities, which is known as differentiation, is discussed in Section 34-4). A particular cell type is largely defined by its characteristic pattern of gene expression. Since most cells in a multicellular organism have the same complement of DNA, how do cells maintain their identities (patterns of gene expression) from one cell generation to the next? Evidently, histone modifications are largely preserved between cell generations, that is, they are epigenetic markings in much the same way as are the methylation patterns of DNA (Section 30-7). The way in which a cell confers its histone epigenetic markings on its progeny is unknown, although it almost certainly involves the recruitment of histone-modifying enzymes to newly assembled nucleosomes on recently replicated DNA.

bb. Histone Acetyltransferases (HATs) Are Components of Multisubunit Transcriptional Coactivators

Although Vincent Allfrey discovered, in the late 1960s, that histone acetylation and deacetylation are respectively correlated with transcriptional activation and repression, it was not until the mid-1990s that the proteins that mediate histone acetylation and deacetylation were identified and characterized. Histone Lys side chains are acetylated in a sequence-specific manner by enzymes known as **histone acetyltransferases (HATs),** all of which employ acetyl-CoA (Fig. 21-2) as their acetyl group donors:

$$CoA-S-\overset{\overset{\displaystyle O}{\|}}{C}-CH_3 \;+\; Lys-(CH_2)_4-\overset{+}{N}H_3$$
$$\textbf{Acetyl-CoA} \qquad\qquad \textbf{Histone Lys}$$

$$\big\downarrow \text{HAT}$$

$$CoA-SH \;+\; Lys-(CH_2)_4-NH-\overset{\overset{\displaystyle O}{\|}}{C}-CH_3$$
$$\textbf{Histone Acetyl-Lys}$$

The large number of known HATs are members of five families: (1) the **GNAT family** (for *G*cn5-relatated *N-a*cetyl*transferase; **Gcn5,** first found in yeast, is one of the best characterized HATs), whose members include Gcn5, its homologs **Gcn5L** (for *G*cn5-*l*ike protein) and **PCAF** [for *p*300/*CBP-a*ssociated *f*actor; **p300** and **CBP** (for *c*AMP *r*esponse *b*inding *e*lement *p*rotein) are homologous transcriptional coactivators], and **Hat1** (which acetylates histones in the cytoplasm before they are imported to the nucleus); (2) the **MYST family** (named for its founding members, **MOZ, Ybf2/Sas3, Sas2,** and **Tip60**); (3) the **p300/CBP family;** (4) the **TAF1 family** (TAF1, the largest subunit of TFIID, was formerly named **TAF$_{II}$250**); and (5) the **SRC family** (for *s*teroid *r*eceptor *c*oactivator). Most HATs besides Hat1 function as transcriptional coactivators or silencers but some are implicated in regulating cell cycle progression, DNA regulation, and transcriptional elongation.

Most if not all HATs function *in vivo* as members of often large (10–20 subunits) multisubunit complexes, many of which were initially characterized as transcriptional regulators. These include **SAGA** (for *S*pt/*A*da/*G*cn5/*a*cetyltransferase), the closely similar **PCAF complex** (which contains PCAF), **STAGA** (*S*pt3/*TAF*/*G*cn5L *a*cetyltransferase; its HAT is Gcn5L), **ADA** (transcriptional *ada*ptor), TFIID (which contains TAF1), **TFTC** (*T*BP-*f*ree *T*AF-*c*ontaining complex), **NuA3,** and **NuA4** (*nu*cleosomal *a*cetyltransferases of H*3* and H*4*).

Many **HAT complexes** share subunits. For example, three of ADA's four subunits, Gcn5, **Ada2,** and **Ada3,** are common to the 14-subunit SAGA. Likewise, SAGA and NuA4 both contain **Tra1,** a homolog of the phosphoinositide 3-kinases (Section 19-4D) that interacts with specific transcriptional activators, including **Myc** (Section 34-4C). Intriguingly, several HAT complexes besides TFIID and TFTC contain TAFs. For example, SAGA contains **TAF5,** TAF6, TAF9, **TAF10,** and TAF12, as does the PCAF complex with the exception that TAF5 and TAF6 are replaced in the PCAF complex by their close homologs **PAF65β** and **PAF65α** (PAF for *PCAF a*ssociated *f*actor). TAF6, TAF9, and TAF12, as we discussed above, are structural homologs of histones H3, H4, and H2B, respectively. Consequently, these TAFs probably associate to form an architectural element that is common to TFIID, SAGA, and the PCAF complex and hence these complexes are likely to interact with TBP in a similar manner.

The various HAT complexes presumably target their component HATs to the promoters of active genes. Moreover, they alter the specificities of these HATs. For example, a general property of HATs is that although they can acetylate at least one type of free histone, they can only acetylate histones in nucleosomes as members of HAT complexes. Thus, free Gcn5 acetylates H3 Lys 14 and, to a lesser extent, H4 Lys 8 and 16. However, Gcn5 in SAGA expands its H3 sites to Lys 9, 14, and 18 and also acetylates H2B, whereas Gcn5 in ADA acetylates H3 at Lys 14 and 18, as well as H2B. Neither complex acetylates H4.

The X-ray structures of several HATs have been determined. That of the HAT domain of *Tetrahymena thermophila* GCN5 (residues 48–210 of the 418-residue protein) in complex with a bisubstrate inhibitor was determined by Ronen Marmorstein. The bisubstrate inhibitor (Fig. 34-71a) consists of CoA covalently linked from its S atom via an isopropionyl group (which mimics an acetyl group) to the side chain of Lys 14 of the 20-residue N-terminal segment of histone H3. The structure (Fig. 34-71b) reveals the enzyme to be deeply clefted and to contain a core region common to all HATs of known structure (magenta in Fig. 34-71b) that consists of a 3-stranded antiparallel β sheet connected via an α helix to a fourth β strand that forms a parallel interaction with the β sheet. Only 6 residues of the histone tail, Gly 12 through Arg 17, are visible in the X-ray structure. The CoA moiety binds in the enzyme's cleft such that it is mainly contacted by core residues. The comparison of this structure

with other Gcn5-containing structures indicates that the cleft has closed down about the CoA moiety.

cc. Bromodomains Recruit Coactivators to Acetylated Lys Residues in Histone Tails

The different patterns of histone acetylation required for different functions (the histone code) suggest that the function of histone acetylation is more complex than merely attenuating the charge–charge interactions between the cationic histone N-terminal tails and anionic DNA. In fact, there is growing evidence that specific acetylation patterns are recognized by protein modules of transcriptional coactivators in much the same way that specific phosphorylated sequences are recognized by protein modules such as the SH2 and PTB domains that mediate signal transduction via protein kinase cascades (Section 19-3). Thus, nearly all HAT-associated transcriptional coactivators contain ~110-residue modules known as **bromodomains** that specifically bind acetylated Lys residues on histones. For example, Gcn5 essentially consists of its HAT domain followed by a bromodomain, whereas TAF1 consists mainly of an N-terminal kinase domain followed by a HAT domain and two tandem bromodomains.

The X-ray structure of human TAF1's double bromodomain (residues 1359–1638 of the 1872-residue protein), determined by Tjian, reveals that it consists of two nearly

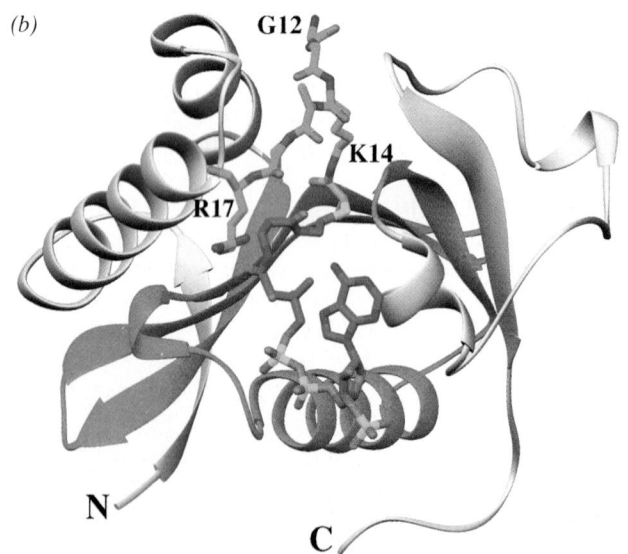

(a)

Histone H3 N-terminal peptide

$$\text{Ac—A}_1\text{—R—T—K—Q}_5\text{—T—A—R—K—S}_{10}\text{—T—G—G—K—A}_{15}\text{—P—R—K—Q—L}_{20}\text{—COO}^-$$

$$\begin{array}{c} | \\ CH_2 \\ | \\ CH_2 \\ | \\ CH_2 \\ | \\ CH_3 \quad O \quad CH_2 \\ | \quad\quad || \quad\quad | \\ CoA—S—CH—C—NH \end{array}$$

Isopropionyl group

(b)

FIGURE 34-71 X-Ray structure of *Tetrahymena* GCN5 in complex with a bisubstrate inhibitor. (*a*) The bisubstrate inhibitor, a peptide–CoA conjugate, consists of CoA covalently linked from its S atom via an isopropyl group to the side chain of Lys 14 of the 20-residue N-terminal segment of histone H3. (*b*) The protein is drawn in ribbon form in lavender with its core region magenta. The peptide–CoA conjugate is drawn in stick form colored according to atom type (histone C blue, isopropionyl group C orange, CoA C green, N blue, O red, S yellow, and P gold). [Part *b* based on an X-ray structure by Ronen Marmorstein, The Wistar Institute, Philadelphia, Pennsylvania. PDBid 1M1D.]

identical antiparallel 4-helix bundles (Fig. 34-72). A variety of evidence, including NMR structures of single bromodomains in complex with their target acetyl-Lys–containing peptides, indicates that the acetyl-Lys binding site of each bromodomain occurs in a deep hydrophobic pocket that is located at the end of its 4-helix bundle opposite its N- and C-termini. The double bromodomain's two binding pockets are separated by ~25 Å, which makes them ideally positioned to bind two acetyl-Lys residues that are separated by 7 or 8 residues. In fact, the N-terminal tail of histone H4 contains Lys residues at its positions 5, 8, 12, and 16 (Fig. 34-70), whose acetylation is correlated with increased transcriptional activity. Moreover, the 36-residue N-terminal peptide of histone H4, when fully acetylated, binds to the TAF1 double bromodomain in 1:1 ratio with 70-fold higher affinity than do single bromodomains but fails to bind when it is unacetylated.

The foregoing structure suggests that the TAF1 bromodomains serve to target TFIID to promoters that are within or near nucleosomes (in contrast to the widely held notion that TFIID targets PICs to nucleosome-free regions). Tjian has therefore postulated that the transcriptional initiation process may begin with the recruitment of a HAT-containing coactivator complex by an upstream DNA-binding protein (Fig. 34-73). The HAT could then acetylate the N-terminal histone tails of nearby nucleosomes, which would recruit TFIID to an appropriately located promoter via the binding of its TAF1 bromodomains to the acetyl-Lys residues. Moreover, the TAF1 HAT activity could acetylate other nearby nucleosomes, thereby initiating a cascade of acetylation events that would render the DNA template competent for transcriptional initiation.

dd. Histone Deacetylases (HDACs)

Histone acetylation is a reversible process. The enzymes that remove the acetyl groups from histones, the **histone deacetylases (HDACs),** promote transcriptional repression and gene silencing. Eukaryotic cells from yeast to

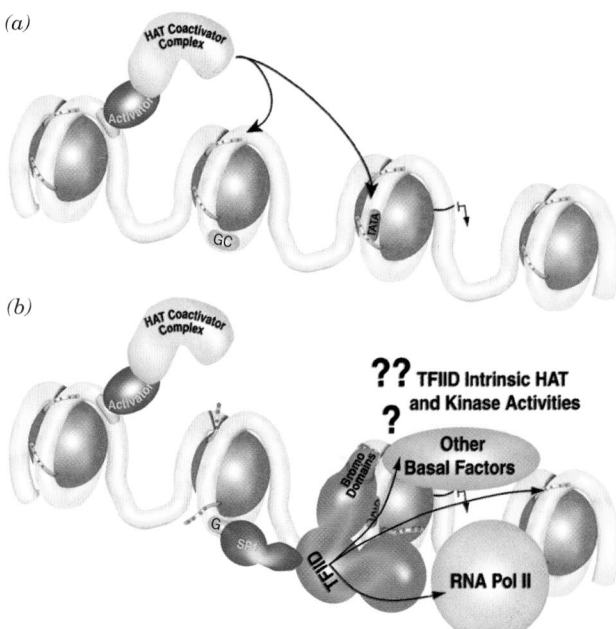

FIGURE 34-73 Simplified model for the assembly of a transcriptional initiation complex on chromatin-bound templates. Here the DNA is represented by a yellow worm, the histone octamers around which the DNA wraps to form nucleosomes are shown as red spheres, and their N-terminal histone tails are drawn as short cyan rods with the red and green dots representing unacetylated and acetylated Lys residues. The transcription initiation site is represented by the black ring about the DNA from which the squared-off arrow points downstream. (*a*) The process begins by the recruitment of a HAT-containing transcriptional coactivator complex (*yellow*) through its interactions with a DNA-binding activator protein (*purple*) that is bound to an upstream enhancer (*light blue*). The coactivator HAT is thereby positioned to acetylate the N-terminal tail on nearby nucleosomes (*curved arrows*). (*b*) The binding of TAF1's bromodomains to the acetylated histone tails could then help recruit TFIID (*magenta*) to a nearby TATA box (*orange patch*). Further acetylation of nearby histone tails by TAF1's HAT domain could help recruit other basal factors (*cyan*) and RNAP II (*orange*) to the promoter, thus stimulating PIC formation. Note that this model does not preclude other activation pathways such as the binding of enhancer-bound SP1 (*purple*) to TFIID. [Courtesy of Robert Tjian, University of California at Berkeley.]

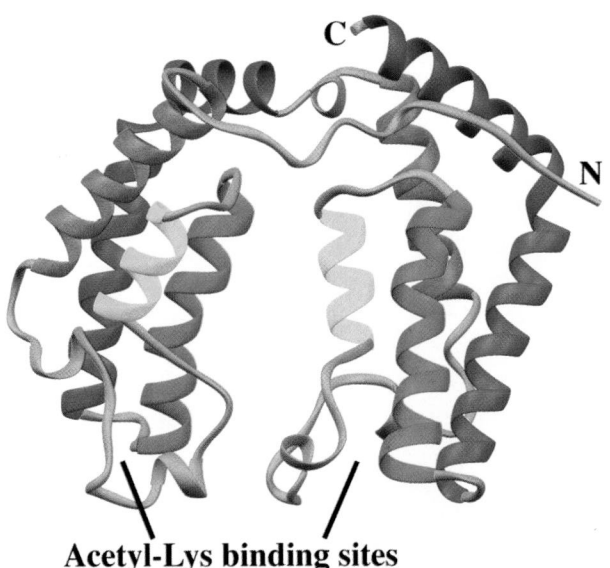

Acetyl-Lys binding sites

FIGURE 34-72 X-Ray structure of the human TAF1 double bromodomain. Each bromodomain consists of an antiparallel four-helix bundle whose helices are colored, from N- to C-termini, red, yellow, green, and blue, with the remaining portions of the protein orange. The two four-helix bundles are related by an ~108° rotation about an axis that is approximately parallel to the principal axes of the four-helix bundles (the vertical direction in this drawing). The acetyl-Lys binding sites occupy deep hydrophobic pockets at the end of each four-helix bundle opposite its N- and C-termini. [Based on an X-ray structure by Robert Tjian, University of California at Berkeley. PDBid 1EQF.]

humans typically contain numerous different HDACs; 10 HDACs have been identified in yeast and 17 in humans. The HDACs consist of three protein families: Class I, which in humans contains **HDAC1, 2, 3,** and **8;** Class II, which in humans contains **HDAC4–7, 9,** and **10;** and Class III, which in humans contains the so-called **sirtuins, SIRT1–7** (SIR for *silent information regulator*). Most, if not all, of the Class I HDACs are members of several multisubunit complexes. Thus, HDAC1 and HDAC2 form the catalytic cores of three complexes, **Sin3, NuRD** (*nu*cleosome *r*emodeling histone *d*eacetylase), and **CoREST** (*co*repressor to *RE*1 *s*ilencing *t*ranscription factor), whereas HDAC3 is the catalytic core of **N-CoR** (*n*uclear hormone receptor *co*repressor) and **SMRT** (*s*ilencing *m*ediator of *r*etinoid and *t*hyroid hormone receptor). These complexes serve as **transcriptional corepressors** for numerous transcriptional repressors as well as cooperating with each other. For example, the repressor **REST** (neuron-*rest*rictive repressor), on binding to its target DNA site, recruits CoREST and Sin3, which together repress transcription from nearby nucleosomes. Many if not all of the Class II HDACs function as transcriptional corepressors although few of them appear to be members of multisubunit complexes.

The Class III HDACs, the sirtuins, are unusual in that they contain an essential NAD^+ cofactor. Rather than simply hydrolyzing the amide bond linking the acetyl group to their target Lys side chain, they transfer it to the ADP–ribosyl group of NAD^+, thereby yielding **O-acetyl-ADP-ribose,** nicotinamide, and the deacetylated Lys residue:

The X-ray structure of only one eukaryotic HDAC has yet been elucidated, that of human SIRT2, a component of transcriptionally silent chromatin that is required for gene silencing. This structure, determined by Nikola Pavletich, reveals that the 323-residue catalytic core of this 389-residue monomer consists of an *N*-terminal, NAD^+-binding, Rossmann fold–like domain and a smaller

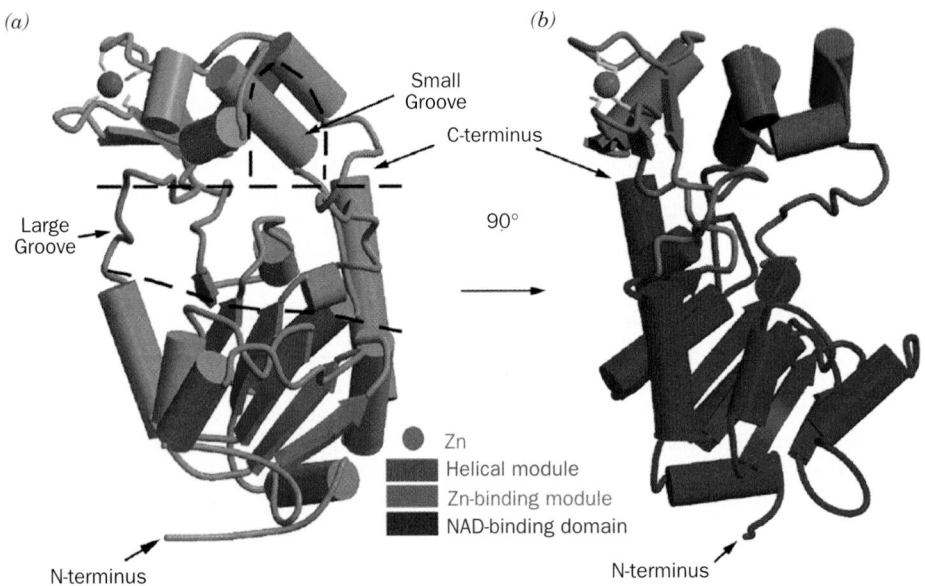

FIGURE 34-74   X-Ray structure of the human sirtuin SIRT2.
(*a*) View of the protein in which the β-strands are green, the helices and most loops are cyan, the loops forming the major structural elements of the large groove are tan, the bound Zn^{2+} ion is represented by a magenta sphere, and its four liganding Cys side chains are drawn in stick form in yellow. The position of the large groove, which forms the enzyme's catalytic site, is indicated as is that of a smaller groove. (*b*) A view of the protein rotated by 90° about the vertical axis relative to that of Part *a,* in which the NAD^+-binding domain is blue, the helical module is red, and the Zn^{2+}-binding module is gray. [Courtesy of Nikola Pavletich, Memorial Sloan-Kettering Cancer Center, New York, New York. PDBid 1J8F.]

C-terminal domain that consists of a helical module and a Zn^{2+}-binding module (Fig. 34-74). Mutagenic studies indicate that a large groove between the two domains, which is lined with conserved hydrophobic residues, is the enzyme's likely catalytic site.

ee. Histone Methylation

Histone methylation at both the Lys and Arg side chains of histone H3 and H4 N-terminal tails (Fig. 34-70) tends to silence the associated genes by inducing the formation of heterochromatin. The enzymes mediating these methylations, the **histone methyltransferases (HMTs),** all utilize S-adenosylmethionine (SAM; Section 26-3E) as their methyl donor. Thus, the lysine HMTs, the most extensively characterized HMTs, catalyze the reaction

Histone Lys   ***S*-Adenosylmethionine**

lysine methyltransferase

***S*-Adenosylhomocysteine**

These enzymes all have a so-called **SET domain** [*Su*(var)3-9, *E*(Z), *Trithorax*], which contains their catalytic sites.

The human lysine HMT named **SET7/9** monomethylates Lys 4 of histone H3. The X-ray structure of the SET domain of SET7/9 (residues 108–366 of the 366-residue protein) in complex with SAM and the N-terminal decapeptide of histone H3 in which Lys 4 is monomethylated, was determined by Steven Gamblin. Interestingly, SAM and the peptide substrate bind to opposite sides of the protein (Fig. 34-75). However, there is a narrow tunnel through the protein into which the Lys 4 side chain is inserted such that its amine group is properly positioned for methylation by SAM. The arrangement of the hydrogen bonding acceptors for the Lys amine group stabilizes the methyl-Lys side chain in its observed orientation about the C_ε—N_ζ bond, thus sterically precluding the methyl-Lys group from assuming a conformation in which it could be further methylated by SAM.

No enzymes that demethylate histones have been identified despite considerable effort to do so. This suggests that histone methylation is irreversible. However, there are several known instances in which demethylation appears to occur, albeit at a low level. For example, in yeast, H3 Lys 4 is trimethylated at active promoters but dimethylated at repressed promoters, which is indicative that histones can reversibly change their methylation states. This demethylation may be mediated by an as yet uncharacterized demethylase, by the replacement of a methylated histone by one that is unmodified, or possibly by the proteolytic excision of a methylated histone tail, although this may be the first step in histone replacement.

Methylated histones are recognized by so-called **chromodomains.** For example, methylated H3 Lys 9 is bound by the chromodomain-containing **heterochromatin protein 1 (HP1),** which thereupon recruits proteins that control chromatin structure and gene expression. The NMR structure of the mouse HP1 chromodomain (residues 8–80 of the 185-residue protein) in complex with the N-terminal 18-residue tail of H3 in which Lys 4 and 9 are dimethylated (Fig. 34-76), determined by Natalia Murzina and Earnest Laue, reveals that HP1 binds the H3 tail in an extended β-strandlike conformation in a groove on its surface (Fig. 34-76). The chromodomain buries the side chain of H3 Lys 9 (but not that of H3 Lys 4) such that its two methyl groups are contained in a hydrophobic box formed by three conserved aromatic residues. In contrast, the unmethylated H3 tail does not bind to HP1.

As previously mentioned, heterochromatin has a tendency to spread, thus silencing the newly heterochromatized genes. One way in which this appears to occur is via the binding of HP1 to nucleosomes whose H3 Lys 9 residues have been methylated (which is associated with transcriptionally inactive chromatin). The bound HP1 recruits the HMT **Suv39h,** which methylates nearby nucleosomes at their H3 Lys 9 residues, which thereupon recruit additional HP1, etc. Such heterochromatin spreading, as we discussed above, is prevented by the presence of an insulator. The HS4 insulator in the chicken β-globin cluster recruits HATs that acetylate H3 Lys 9 on nearby nucleosomes (which is associated with transcriptional activity), thereby blocking their methylation. Note that this activity is distinct from the enhancer-blocking function of HS4, a process that is mediated by the binding of CTCF to a different subsite of HS4 than that to which HATs bind.

ff. Histone Ubiquitination Functions to Regulate Transcription

Although ubiquitination mainly functions to mark proteins for destruction by the proteasome (Section 32-6B), it is also implicated in the control of transcription. In yeast, for example, the monoubiquitination of H2B Lys 123 (in contrast to polyubiquitination, which marks proteins for destruction), which is mediated by the ubiquitin-conjugating enzyme (E2) **Rad6** and the RING-finger–containing ubiquitin-protein ligase (E3) **Bre1,** is a prerequisite for the methylation of H3 Lys 4 and Lys 79. Together, these modifications are implicated in the silencing of genes located near telomeres. It has therefore been suggested that H2B ubiquitination functions as a master switch that controls

(a)

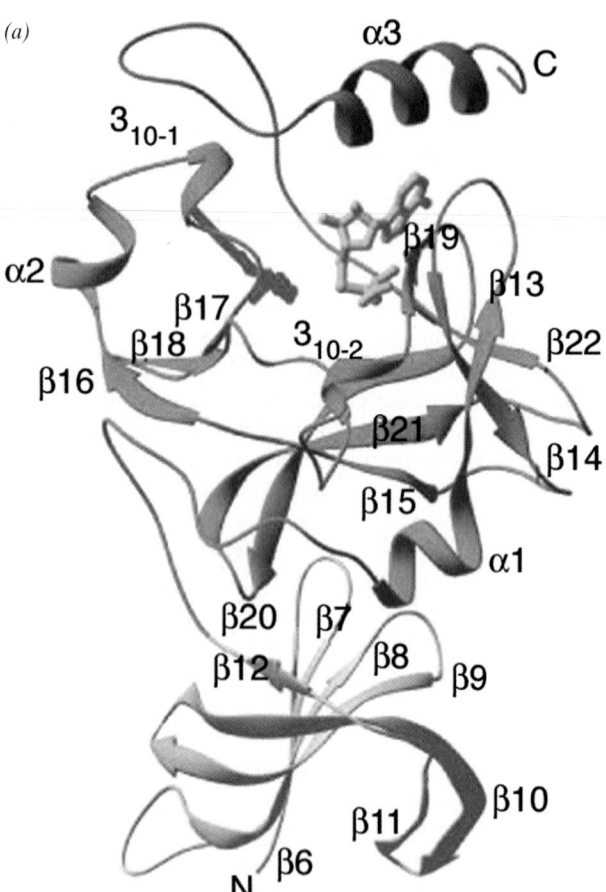

(b)

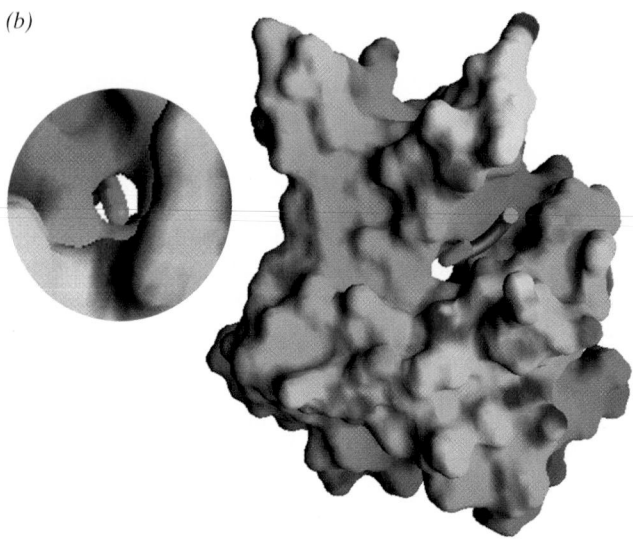

domain is pink, the SET domain is blue, and the C-terminal segment is gray. The H3 N-terminal decapeptide with its methylated Lys 4 is green. The SAM cofactor is drawn in stick form in yellow. (*b*) Surface representation of the protein as seen from the back side of Part *a*. The protein surface is colored according to charge (blue most positive, red most negative, and gray neutral) and the H3 decapeptide is represented by a green ribbon. Note the narrow tunnel through the protein in which the methyl-Lys side chain is inserted. The inset at the left shows a close-up of this Lys access channel containing the methyl-Lys side chain (*green*) as viewed from the SAM-binding site. Note that the dimethylated side chain of Lys 9 is bound to the protein whereas that of Lys 4 is not. [Courtesy of Steven Gamblin, National Institute for Medical Research, London, U.K.]

FIGURE 34-75   X-Ray structure of the human histone methyltransferase SET7/9 in complex with SAM and the histone H3 N-terminal decapeptide with its Lys 4 monomethylated. (*a*) Ribbon diagram in which the N-terminal

the site-selective histone methylation patterns responsible for telomeric gene silencing. Similarly, the TAF1 subunit of TFIID, functioning as both a ubiquitin-activating enzyme (E1) and an E2, monoubiquitinates H1, a post-translational modification that is required for the expression of genes in the correct order during *Drosophila* development. Conversely, a histone deubiquitinating enzyme (DUB) is associated with the SAGA chromatin modifying complex. Thus, although the role of histone ubiquitination is only beginning to come to light, it is clear that it is an essential transcriptional regulator.

gg. Chromatin-Remodeling Complexes

Sequence-specific DNA-binding proteins must gain access to their target DNAs before they can bind to them. Yet nearly all DNA in eukaryotes is sequestered by nucleosomes if not by higher order chromatin. How then do the proteins that bind to DNA segments gain access to their target DNAs? The answer, which has only become apparent since the mid-1990s, is that *chromatin contains ATP-driven complexes that remodel nucleosomes,* that is, they somehow disrupt the interactions between histones and DNA in nucleosomes to make the DNA more accessible. This may cause the histone octamer to slide along the DNA strand to a new location (a cis transfer) or relo-

cate to a different DNA (a trans transfer). Thus, *these chromatin-remodeling complexes impose a "fluid" state on chromatin that maintains its DNA's overall packaging but transiently exposes individual sequences to interacting factors.*

Chromatin-remodeling complexes consist of multiple subunits. The first of them to be characterized was the yeast **SWI/SNF** complex, so-called because it was discovered through genetic screens as being essential for the expression of the *HO* gene, which is required for mating type switching (SWI for *swi*tching defective), and for the expression of the *SUC2* gene, which is required for growth on sucrose (SNF for sucrose *n*onfermenter). SWI/SNF, an 1150-kD complex of 11 different types of subunits, is only essential for the expression of ~3% of yeast genes and is not required for cell viability. However, a related complex named **RSC** (for *r*emodels the *s*tructure of *c*hromatin) is ~100 times more abundant in yeast and is required for cell viability. RSC shares two subunits with SWI/SNF and many of their remaining subunits are homologs, including their ATPase subunits, which are named **Swi2/Snf2** in SWI/SNF and **Sth1** in RSC. The Swi2/Snf2 ATPase, as well as two of RSC's subunits, contain bromodomains that are likely to facilitate the binding of their complexes to acetylated histones.

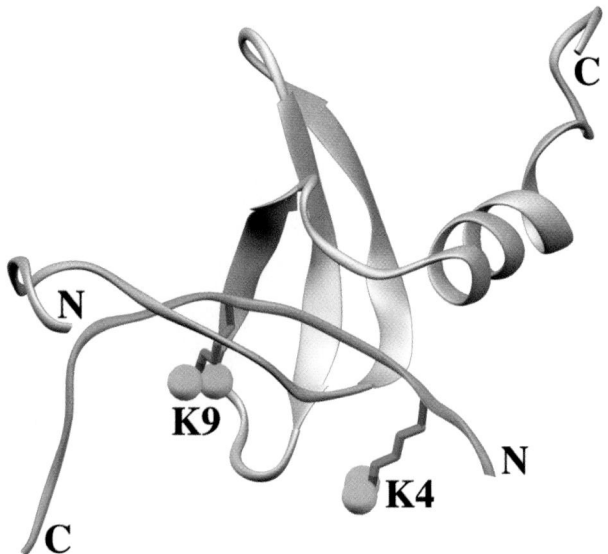

FIGURE 34-76   X-Ray structure of the mouse HP1 chromodomain in complex with the 18-residue N-terminal tail of histone H3 in which Lys 4 and Lys 9 are dimethylated. The 80-residue chromodomain is lavender, the H3 N-terminal tail is orange, and its two dimethyl-Lys side chains are drawn in stick form with C green, N blue, and with their methyl groups represented by cyan spheres. The side chain of H3 Lys 9, but not that of H3 Lys 4, is buried by the chromodomain. [Based on an X-ray structure by Natalia Murzina and Earnest Laue, University of Cambridge, U.K. PDBid 1GUW.]

All eukaryotes contain multiple chromatin-remodeling complexes. They have been classified into three main groups on the basis of the similarities of their component ATPase subunits: (1) the SWI/SNF complexes, whose ATPases are homologous to yeast Swi2/Snf2 and include yeast RSC, *Drosophila* **Brahma,** and human **BRM** (*Brahma* protein homolog) and **BRG1** (*Brahma*-related *gene 1*); (2) the **ISWI** (for *imitation switch*) complexes, whose ATPases are homologs of yeast **ISW1** and include the yeast **ISW1** and **ISW2** complexes, *Drosophila* **ACF** (*ATP-utilizing chromatin assembly and remodeling factor*), **CHRAC** (*chromatin accessibility complex*), **NURF** (*nucleosome-remodeling factor*), and human **RSF** (*remodeling and spacing factor*); and (3) the **Mi-2** complexes, whose ATPases are homologs of the *Xenopus* **Mi-2** complex and include human NuRD (which, as is discussed above, also contains HDAC1 and HDAC2). Many of these complexes contain bromodomains, chromodomains, and/or AT hooks that presumably recruit the complexes to their target genes. Moreover, some complexes are bound by specific transcriptional activators.

An electron microscopy–based image of yeast RSC, determined by Asturias and Kornberg at ~28 Å resolution, reveals that it consists of four modules surrounding a central cavity (Fig. 34-77*a*). Biochemical studies indicate that RSC binds tightly to nucleosomes in a 1:1 complex. Indeed, the size and shape of RSC's central cavity appear to be appropriate for binding a single nucleosome core particle, as Fig. 34-77*b* indicates. This would explain how, in the presence of ATP, RSC could loosen the DNA in a nucleosome without the loss of its associated histones.

The simultaneous release of all of the many interactions holding DNA to a histone octamer would require an enormous free energy input and hence is unlikely to occur. Then, how do chromatin-remodeling complexes function? Their various ATPase subunits share a region of homology with helicases (Section 30-2C), although they lack helicase activity. Nevertheless, it seems plausible that, like helicases, chromatin-remodeling complexes "walk" up DNA strands as driven by ATP hydrolysis. If such a complex were directly or indirectly tethered to a histone, this would put torsional strain on the DNA in the nucleosome, thereby decreasing its local twist (DNA supercoiling is discussed in Section 29-3A). The region of decreased twist could diffuse along the DNA wrapped around the nucleosome, thereby transiently loosening the histone octamer's grip on a segment of DNA. The torsional strain might also

(a)

(b)

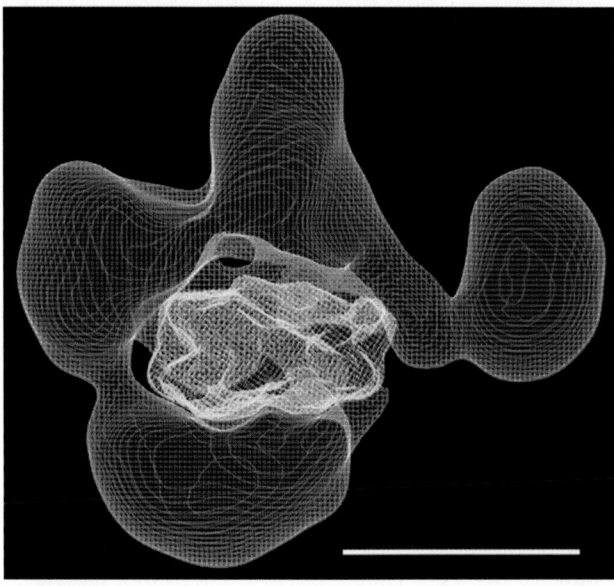

FIGURE 34-77   EM-based image of yeast RSC. (*a*) Two views of the structure (*front and back*) at ~28 Å resolution revealing that it consists of four modules surrounding a central cavity. (*b*) A model made by manually fitting the X-ray structure of the core nucleosome (Fig. 34-7) reduced to 25 Å resolution into the central cavity of the RSC structure in Part *a*. The complex is shown in mesh outline with RSC red and the nucleosome, which is viewed edgewise, yellow. The nucleosome fits snugly into the cavity with no steric clash. The scale bar in both parts represents 100 Å. [Courtesy of Francisco Asturias, The Scripps Research Institute, La Jolla, California.]

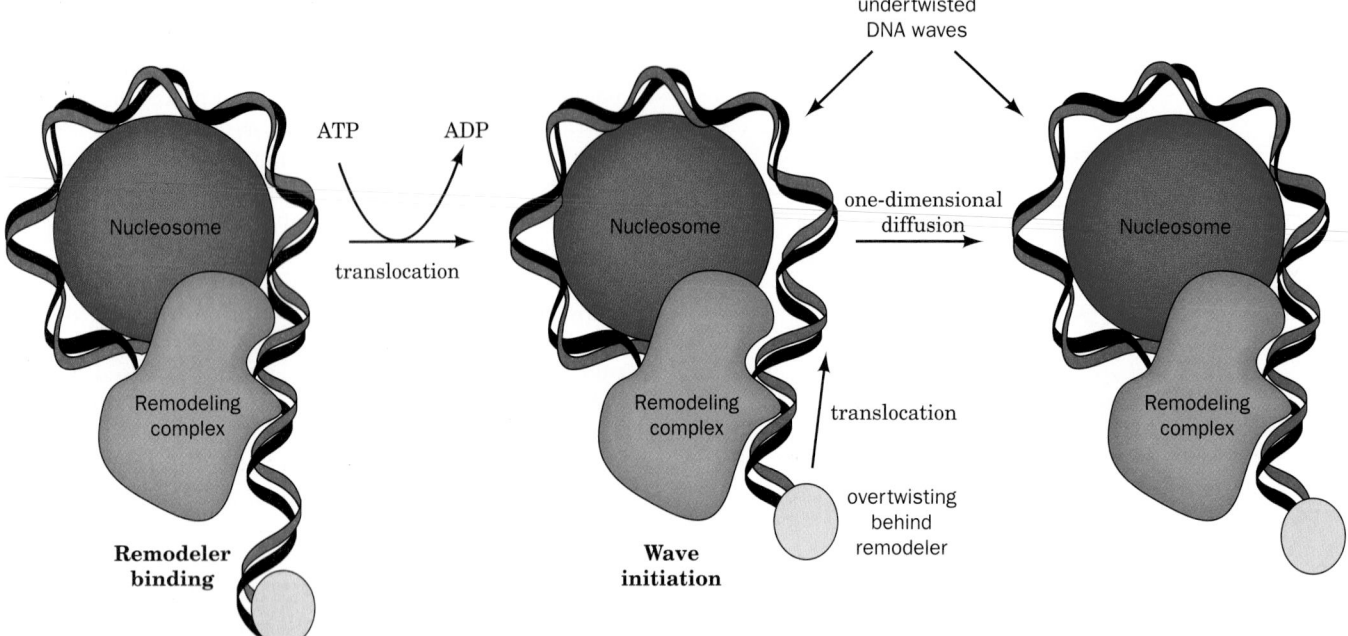

FIGURE 34-78 Model for nucleosome remodeling by chromatin-remodeling complexes. The chromatin-remodeling complex (*green*) couples the free energy of ATP hydrolysis to the translocation and concomitant twisting of the DNA in the nucleosome (*blue,* only half of which is shown for clarity) as depicted by the movement of a fixed point on the DNA (*yellow ellipsoid*). This locally breaks the contacts between the histones and the DNA. The position of the undertwisted and/or bulged DNA propagates around the nucleosome in a one-dimensional wave that transiently releases the DNA from the histone as it passes, thereby providing DNA-binding factors access to the DNA. [After a drawing by Saha, A., Wittmeyer, J., and Cairns, B.R., *Genes Dev.* **16,** 2120 (2002).]

be partially accommodated as a writhe, which would lift a segment of DNA off the nucleosome's surface. In either case, the resulting DNA distortion could diffuse around the surface of the nucleosome in a wave that would locally and transiently release the DNA from the histone octamer as it passed (Fig. 34-78) and hence permit the DNA to bind to its cognate DNA-binding factors. This latter mechanism resembles that proposed for the passage of RNAPs through nucleosomes (Fig. 34-66). Note that multiple cycles of ATP hydrolysis would send multiple DNA-loosening waves around the nucleosome, thereby sliding the nucleosome along the DNA.

hh. Afterword

As we have seen, eukaryotic transcriptional initiation is an astoundingly complex process that involves the synergistic participation of numerous multisubunit complexes comprising several hundred often loosely or sequentially interacting polypeptides (i.e., histones of various types and subtypes; the PIC; Mediator-like complexes; a variety of transcription factors, architectural factors, coactivators, and corepressors that in some cases form enhanceosomes; several types of histone modification complexes; and chromatin-remodeling complexes), as well as large segments of DNA. Intensive investigations in many laboratories over the past two decades have, as we have discussed, identified many of these complexes, characterized their component polypepides, and in many cases, elucidated their general functions. However, we are far from having more

than a rudimentary understanding of how these various components interact *in vivo* to transcribe only those genes required by their cell under its particular circumstances in the appropriate amounts and with the proper timing. It is likely to require several additional decades of research to gain a detailed understanding of how this remarkable molecular machinery functions.

C. Other Expressional Control Mechanisms

The expression of many eukaryotic genes is regulated only by the control of transcriptional initiation. However, many cellular and viral genes additionally respond to other types of control processes. The various mechanisms employed by these secondary systems are outlined below.

1. Selection of Alternative Initiation Sites: *The expression of several eukaryotic genes is controlled, in part, through the selection of alternative transcriptional initiation sites.* For example, identical molecules of α-amylase are produced by mouse liver and salivary gland but the corresponding mRNAs synthesized by these two organs differ at their 5' ends. Comparison of the sequences of these mRNAs with that of their corresponding genomic DNA indicates that the different mRNAs arise from separate initiation sites that are ~2.8 kb apart (Fig. 34-79). Thus, after being spliced, the liver and salivary gland α-amylase mRNAs have different untranslated 5' leaders but the same coding sequences. The two initiation sites, it is

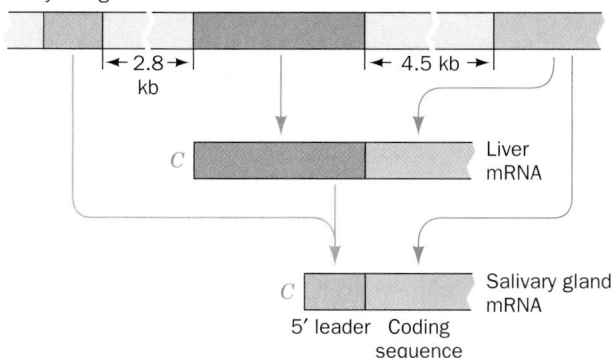

FIGURE 34-79   The transcription start site of the mouse α-amylase gene is subject to tissue-specific selection so as to yield mRNAs with different cap (C) and leader segments but the same coding sequences. [After Young, R.A., Hagenbüchle, O., and Schibler, U., *Cell* **23,** 454 (1981).]

thought, support different rates of initiation. This hypothesis accounts for the observation that α-amylase mRNA comprises 2% of the polyadenylated mRNA in salivary gland but only 0.02% of that in liver.

2. Selection of Alternative Splice Sites: Numerous cellular genes, as we have seen (Section 31-4A), are subject to alternative splicing. Thus, certain exons in one type of cell may be introns in another (e.g., Fig. 31-62).

3. Translocational Control: The observation that only ~5% of nuclear RNA ever makes its way to the cytosol, probably less than can be accounted for by gene splicing, suggests that differential mRNA translocation to the cytosol may be an important expressional control mechanism in eukaryotes. Evidence is accumulating that this is, in fact, the case. Cellular RNA is never "naked" but rather is always in complex with a variety of conserved proteins (Section 31-4A). Intriguingly, nuclear and cytosolic mRNAs are associated with different sets of proteins, indicating that there is protein exchange on translocating mRNA out of the nucleus.

4. Control of mRNA Degradation: The rates at which eukaryotic mRNAs are degraded in the cytosol vary widely. Whereas most have half-lives of hours or days, some are degraded within 30 min of entering the cytosol. A given mRNA may also be subject to differential degradation. For example, the major egg yolk protein **vitellogenin** is synthesized in chicken liver in response to estrogens (in roosters as well as in hens) and transported via the bloodstream to the oviduct. Radioactive-labeling experiments established that estrogen stimulation increases the rate of vitellogenin mRNA transcription by several hundredfold and that this mRNA has a cytosolic half-life of 480 h. When estrogen is withdrawn, the synthesis of vitellogenin mRNA returns to its basal rate and its cytosolic half-life falls to 16 h.

The poly(A) tails appended to nearly all eukaryotic mRNAs apparently help protect them from degradation (Section 31-4A). For example, histone mRNAs, which lack poly(A) tails, have much shorter half-lives than most other

mRNAs. Histones, in contrast to most other cellular proteins, are largely synthesized during the relatively short S phase of the cell cycle, when they are required in massive amounts for chromatin replication (the small amounts of histones synthesized during the rest of the cell cycle are thought to be used for repair purposes). The short half-lives of histone mRNAs ensure that the rate of histone synthesis closely parallels the rate of histone gene transcription.

A structural feature that increases the rate at which mRNAs are degraded is the presence of certain AU-rich sequences in the untranslated 3′ segments. These sequences, when grafted to mRNAs that lack them, decrease the mRNAs' cytosolic lifetimes. By and large, however, the nature of the signals through which mRNAs are selected for degradation are poorly understood, in part, no doubt, because the nucleases that do so have not been identified.

5. Control of Translational Initiation Rates: The rates of translational initiation of eukaryotic mRNAs, as we have seen (Section 32-4), are responsive to the presence of certain substances, including heme (in reticulocytes) and interferon, as well as to mRNA masking.

6. Selection of Alternative Posttranslational Processing Pathways: Polypeptides synthesized in both prokaryotes and eukaryotes are subject to proteolytic cleavage and covalent modification (Section 32-5). These posttranslational processing steps are important regulators of enzyme activity (e.g., see Section 15-3E) and, in the case of glycosylations, are major determinants of a protein's final cellular destination (Sections 12-4C and 23-3B). The selective degradation of proteins (Section 30-6) is also a significant factor in eukaryotic gene expression.

In addition to the foregoing, most eukaryotic polypeptide hormones (whose functions are discussed in Section 19-1) are synthesized as segments of large precursor polypeptides known as **polyproteins.** These are posttranslationally cleaved to yield several, not necessarily different, polypeptide hormones. *The cleavage pattern of a particular polyprotein may vary among different tissues so that the same gene product can yield different sets of polypeptide hormones.* For example, the polyprotein **pro-opiomelanocortin (POMC),** which, in the rat, is synthesized in both the anterior and intermediate lobes of the pituitary, contains seven different polypeptide hormones (Fig. 34-80). In both of these lobes, which are functionally separate glands, posttranslational processing of POMC yields an N-terminal fragment, **ACTH** and **β-LPH.** Processing in the anterior lobe ceases at this point. In the intermediate lobe, however, the N-terminal fragment is further cleaved to yield **γ-MSH,** ACTH is converted to **α-MSH** and **CLIP,** and β-LPH is split to **γ-LPH** and **β-END** (Fig. 34-80). These various hormones have different activities, so that the products of the anterior and intermediate lobes of the pituitary are physiologically distinct.

Most of the cleavage sites in POMC and other polyproteins consist of pairs of basic amino acid residues, Lys–Arg,

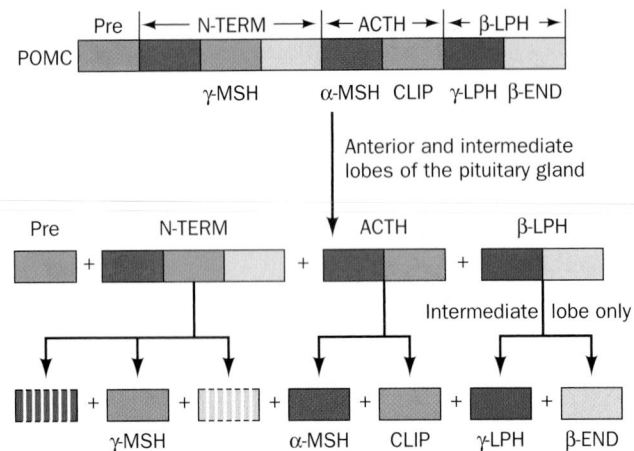

FIGURE 34-80   The tissue-specific posttranscriptional processing of POMC yields two different sets of polypeptide hormones. In both the anterior and intermediate lobes of the pituitary gland, POMC is proteolytically cleaved to yield its N-terminal fragment (N-TERM), **adrenocorticotropic hormone (ACTH;** Section 19-H) and **β-lipotropin (β-LPH).** In the intermediate lobe only, these polypeptide hormones are further cleaved to yield **γ-melanocyte stimulating hormone (γ-MSH), α-MSH, corticotropin-like intermediate lobe peptide (CLIP), γ-LPH,** and **β-endorphin (β-END;** Section 19-K). [After Douglass, J., Civelli, O., and Herbert, E., *Annu. Rev. Biochem.* **53,** 698 (1984).]

for example, which suggests that cleavage is mediated by enzymes with trypsin-like activity. Indeed, the enzymes that process POMC also activate other prohormones such as proinsulin. Moreover, the observation that a yeast protease that normally functions to activate a yeast prohormone also properly processes POMC suggests that prohormone processing enzymes are evolutionarily conserved.

4 ■ CELL DIFFERENTIATION AND GROWTH

Perhaps the most awe inspiring event in biology is the growth and development of a fertilized ovum to form an extensively differentiated multicellular organism. No outside instruction is required to do so; *fertilized ova contain all the information necessary to form complex multicellular organisms such as human beings.* Since, contrary to the beliefs of the earliest microscopists, zygotes do not contain miniature adult structures, these structures must somehow be generated through genetic specification. We begin this section by outlining how embryos develop, followed by a discussion of the best understood example of embryological development, that of the basic body plan in *Drosophila.* We then consider the genetic basis of cancer, a group of diseases caused by the proliferation of cells that have lost some of their developmental constraints. We end by discussing how the cell cycle is controlled and how unneeded or irreparably damaged cells commit suicide through programmed cell death.

A. *Embryological Development*

The formation of multicellular animals can be considered as occurring in four somewhat overlapping stages (Fig. 34-81):

1. Cleavage, in which the zygote undergoes a series of rapid mitotic divisions to yield many smaller cells arranged in a hollow ball known as a **blastula.**

2. Gastrulation, whereby the blastula, through a structural reorganization that includes the blastula's invagination, forms a triple-layered bilaterally symmetric structure called a **gastrula.** Cleavage and gastrulation together take from a few hours to several days depending on the organism.

3. Organogenesis, in which the body structures are formed in a process requiring various groups of proliferating cells to migrate from one part of the embryo to another in a complicated but reproducible choreography. Organogenesis occupies hours to weeks.

4. Maturation and growth, whereby the embryonic structures achieve their final sizes and functional capacities. This stage stretches into and sometimes throughout adulthood.

a. Cell Differentiation Is Mediated by Developmental Signals

As an embryo develops, its cells become progressively and irreversibly committed to specific lines of development. What this means is that these cells undergo sequences of self-perpetuating internal changes that distinguish them and their progeny from other cells. A cell and its descendents therefore "remember" their developmental changes even when placed in a new environment. For example, the dorsal (upper) ectoderm (outer layer) of an amphibian embryo (Fig. 34-82) is normally fated to give rise to brain tissue, whereas its ventral (lower) ectoderm becomes epidermis. If a block of an early gastrula's dorsal ectoderm is cut out and exchanged with a block of its ventral ectoderm, both blocks develop according to their new locations to yield a normal adult. If, however, this experiment is performed on the late gastrula, the transplanted tissues will differentiate as they had originally been fated, that is, as misplaced brain and epidermal tissues. Evidently, the dorsal and ventral ectoderms become committed to form brain and epidermal tissues sometime between the early and late gastrula stages.

How are developmental changes triggered; that is, what are the signals that induce two cells with identical genomes to follow different developmental pathways? To begin with, the zygote is not spherically symmetric. Rather, its yolk, as well as other substances, is concentrated toward one end. Consequently, the various cells in the early cleavage stages inherit different cytoplasmic determinants that apparently govern their further development. Even as early as an embryo's eight-cell stage, some of its cells are demonstrably different in their developmental potential from others. However, as the above transplantation ex-

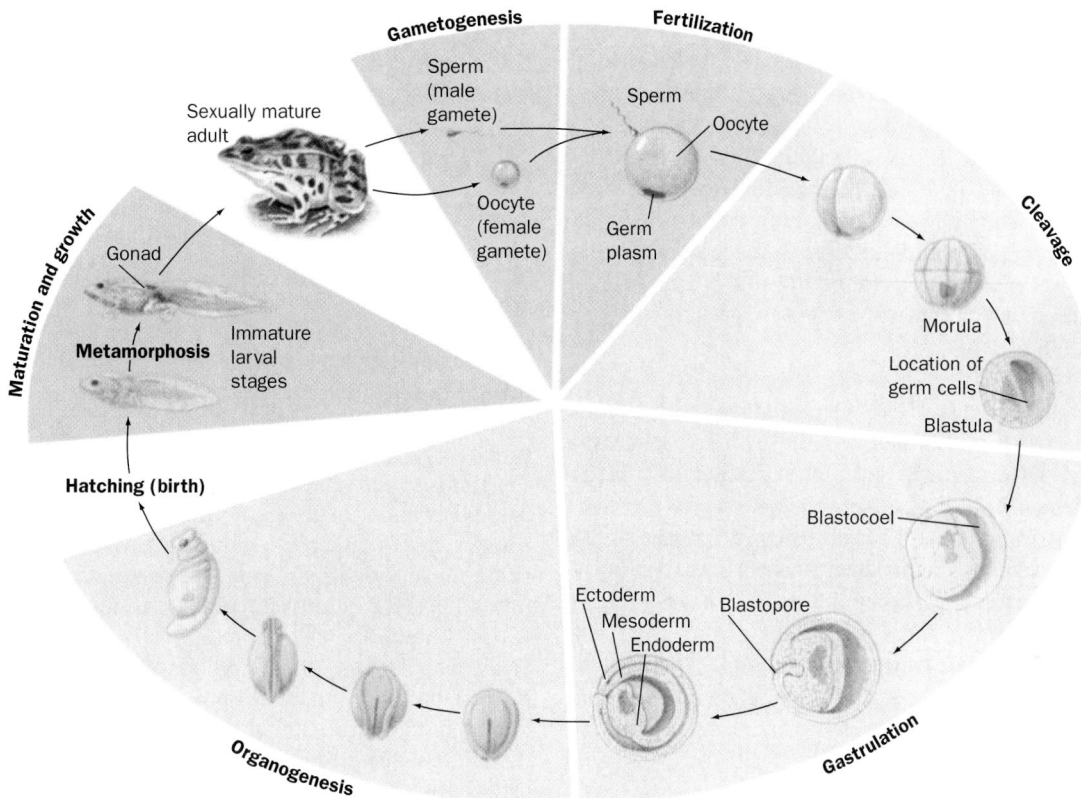

FIGURE 34-81   Embryogenesis in a representative animal, the frog.

periments indicate, cells in later stages of development also obtain developmental cues from their embryonic positions.

Cells may obtain spatial information in two ways:

1. Through direct intercellular interactions.

2. From the gradients of diffusible substances called **morphogens** released by other cells.

For most developmental programs, the interacting tissues must be in direct contact, but this is not always the case. For example, mouse ectoderm fated to become eye lens will only do so in the presence of mesenchyme (em-

bryonic tissue that gives rise to the muscle, skeleton, and connective tissue) but this process still occurs if the interacting tissues are separated by a porous filter. Lens development must therefore be mediated by diffusible substances.

Developmental signals may be recognized over great evolutionary distances. For instance, the epidermis from the back of a chick embryo, through interactions with the underlying dermis, forms feather buds that are arrayed in a characteristic hexagonal pattern. If embryonic chick epidermis is instead combined with dermis from the whiskered region of mouse embryo snout, the chick epidermis still forms feather buds but arranged in the pattern of mouse whiskers.

Even though mammals and birds diverged ~300 million years ago, mouse inducers can still activate the appropriate chicken genes, although, of course, they cannot alter the products these genes specify. In an intriguing example of this phenomenon, combining epithelium from the jaw-forming region of a chick embryo with molar mesenchyme from mouse embryo induces the chick tissue to grow teeth that are unlike those of mammals (Fig. 34-83). Apparently chickens, whose ancestors have been toothless for ~60 million years (a primordial bird, *Archaeopterix,* had teeth), retain the genetic potential to grow teeth even though they lack the developmental capacity to activate these genes. This observation corroborates the hypothesis that organismal evolution proceeds largely via mutations that alter developmental programs rather than the structural genes whose expression they control (Section 7-3B).

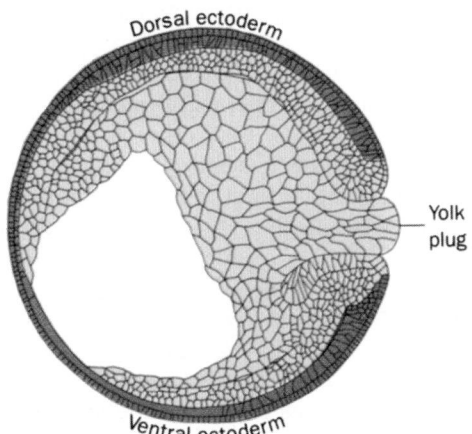

FIGURE 34-82   The dorsal and ventral ectoderm of an amphibian embryo.

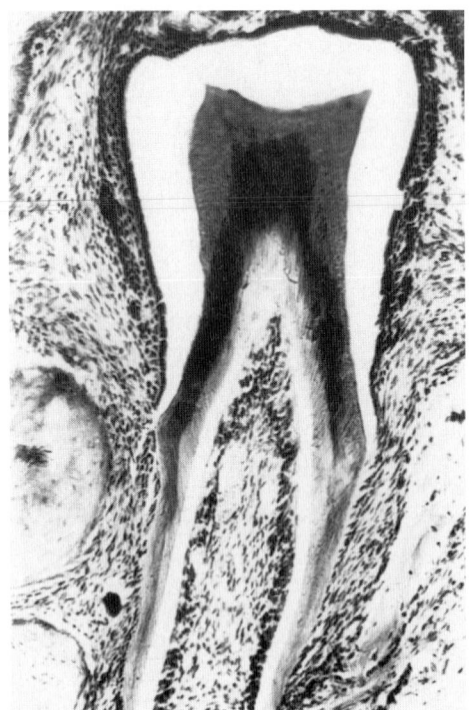

FIGURE 34-83   The proverbial "hen's tooth" forms in chick embryo jaw-forming epithelium under the influence of mouse embryo molar mesenchyme tissue. [Courtesy of Edward Kollar, University of Connecticut Health Center.]

b.   Developmental Signals Act in Combination

An additional developmental stimulus to a previously determined cell will modulate, but not reverse, its developmental state. Consider, for example, what happens in a chicken embryo if undifferentiated tissue from the base of a leg bud, which normally gives rise to part of the thigh, is transplanted beneath the end of a wing bud, which normally develops into the handlike wing tip. The transplant does not become a wing tip or even misplaced thigh tissue; instead it forms a foot (Fig. 34-84). Apparently the same stimulus that causes the end of a wing bud to form a wing tip causes tissue that is already committed to be part of a leg to form a leg's morphological equivalent to a wing tip, a foot. Evidently, the many different tissues of a higher organism do not each form in response to a tissue-specific developmental stimulus. Rather, *a given tissue results from the effects of a particular combination of relatively nonspecific developmental stimuli.* This situation, of course,

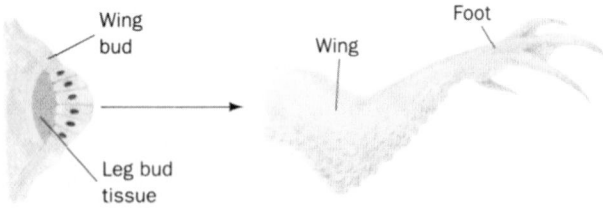

FIGURE 34-84   Presumptive thigh tissue from a chicken leg bud develops into a misplaced foot when implanted beneath the tip of a chicken wing bud.

greatly reduces the number of different developmental stimuli necessary to form a complex organism and therefore simplifies the regulation of the developmental process.

B.   *The Molecular Basis of Development*

The study of the molecular basis of cell differentiation has only become possible in recent decades with the advent of modern methods of molecular genetics. Much of what we know about this subject is based on studies of the fruit fly *D. melanogaster.* We therefore begin this section with a synopsis of embryogenesis in this genetically best characterized multicellular organism.

a.   *Drosophila* Development

Almost immediately after the *Drosophila* egg (Fig. 34-85*a*) is laid (which, rather than the earlier fertilization, triggers development), it commences a series of synchronized nuclear divisions, one every 6 to 10 min. The DNA must therefore be replicated at a furious rate, among the fastest known for eukaryotes. Most probably each of its replicons (Section 30–4B) are simultaneously active. The nuclear division process is unusual in that it is not accompanied by the formation of new cell membranes; the nuclei continue sharing their common cytoplasm to form a so-called **syncytium** (Fig. 34-85*b*), which facilitates the rapid pace of nuclear division because there is no need to increase cell mass. After the 8th round of nuclear division, the ~256 nuclei begin to migrate toward the cortex (outer layer) of the egg where, by around the 11th nuclear division, they have formed a single layer surrounding a yolk-rich core known as a **syncytial blastoderm** (Fig. 34-85*c*). At this stage, the mitotic cycle time begins to lengthen while the nuclear genes, which have heretofore been fully engaged in DNA replication, become transcriptionally active (a freshly laid egg contains an enormous store of mRNA that has been contributed, via cytoplasmic bridges, by the developing oocyte's 15 surrounding "nurse" cells). In the 14th nuclear division cycle, which lasts ~60 min, the egg's plasma membrane invaginates around each of the ~6000 nuclei to yield a cellular monolayer surrounding a yolk-rich core called a **cellular blastoderm** (Fig. 34-85*d*). At this point, after ~2.5 h of development, genomic transcriptional activity reaches its maximum in the embryo, mitotic synchrony is lost, the cells become motile, and gastrulation begins.

Until the cellular blastoderm is formed, most of the embryo's nuclei maintain the ability to colonize any portion of the cortical cytoplasm and hence to form any part of the larva or adult except its germ cells [the germ cell progenitors, the five **pole cells** (Fig. 34-85*c*), are set aside after the 9th nuclear division]. *It is therefore a nucleus' location within the syncytium that determines the types of cells its descendents will be become. Once the cellular blastoderm has formed, however, its cells become progressively committed to ever narrower lines of development.* This has been demonstrated, for example, by tracing the developmental fates of small clumps of cells by excising them or ablating (destroying) them with a laser microbeam and characterizing the resultant deformity.

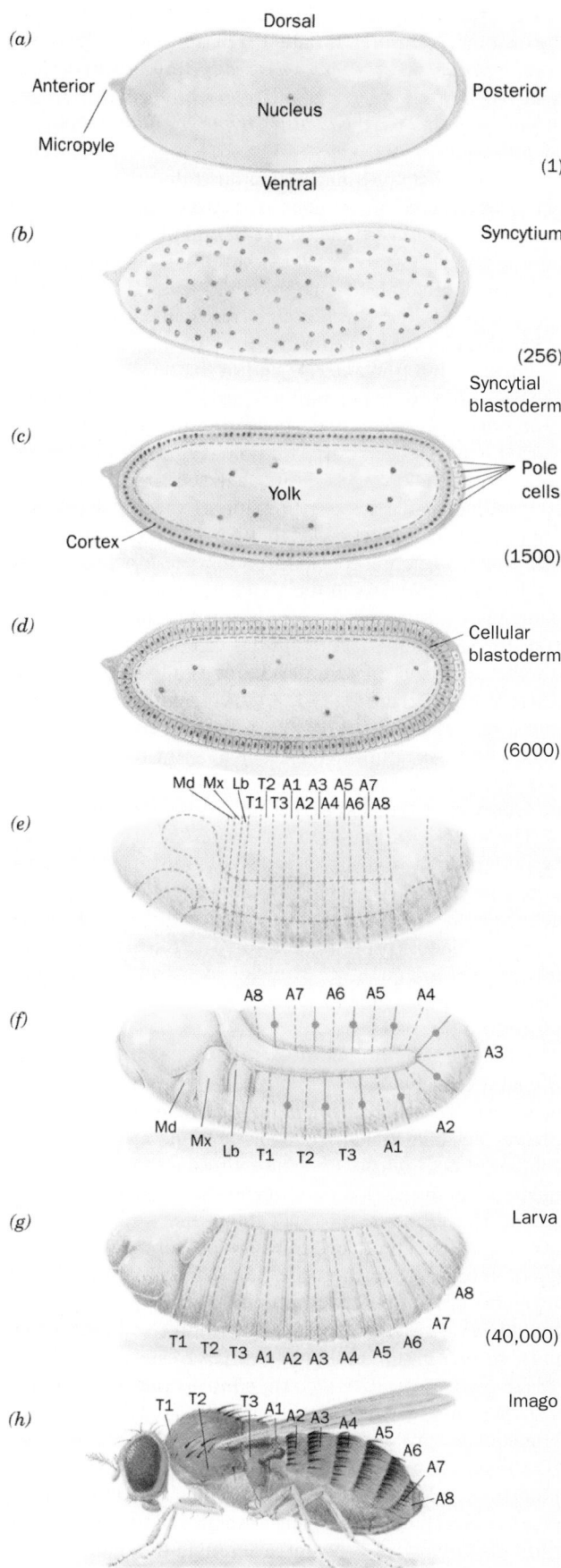

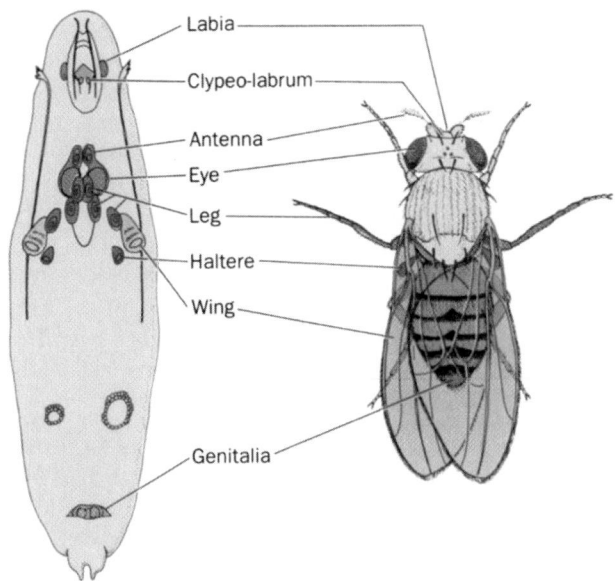

FIGURE 34-85   Development in *Drosophila*. The various stages are explained in the text. Note that the embryos and newly hatched larva are all the same size, ~0.5 mm long. The adult is, of course, much larger. The approximate number of cells in the early stages of development are given in parentheses.

During the embryo's next few hours, it undergoes gastrulation and organogenesis. A striking aspect of this remarkable process, in *Drosophila* as well as in higher animals, is the division of the embryo into a series of segments corresponding to the adult organism's organization (Fig. 34-85*e*). The *Drosophila* embryo has three segments that eventually merge to form its head (Md, Mx, and Lb for mandibulary, maxillary, and labial), three thoracic segments (T1–T3), and eight abdominal segments (A1–A8). As development continues, the embryo elongates and several of its abdominal segments fold over its thoracic segments (which permits it to fit inside the eggshell; Fig. 35-85*f*). At this stage, the segments become subdivided into anterior (forward) and posterior (rear) compartments. The embryo then shortens and straightens to form a larva that consists of ~40,000 cells that hatches ~20 hours after beginning development (Fig. 34-85*g*). Over the next 5 days, the larva feeds, grows, molts twice, pupates, and commences metamorphosis to form an adult (**imago**; Fig. 34-85*h*). In this latter process, the larval epidermis is almost entirely replaced by the outgrowth of apparently undifferentiated patches of larval epithelium known as **imaginal disks** that are committed to their developmental fates as early as the cellular blastoderm stage. These structures, which maintain the larva's segmental boundaries, form the adult's legs, wings, antennae, eyes, etc. (Fig. 34-86). About

FIGURE 34-86   Locations and developmental fates, in *Drosophila*, of the imaginal disks (*left*), pouches of larval tissue that form the adult's outer structures. [After Fristrom, J.W., Raikow, R., Petri, W., and Stewart, D., *in* Hanly, E.W. (Ed.), *Problems in Biology: RNA in Development*, p. 382, University of Utah Press (1970).]

10 days after commencing development, the adult emerges and, within a few hours, initiates a new reproductive cycle.

b. Developmental Patterns Are Genetically Mediated

What is the mechanism of embryonic pattern formation? In what follows, we discuss only the anteroposterior (head to tail) differentiation system. Keep in mind, however, that *Drosophila* also have a system that imposes dorsoventral (back to belly) differentiation.

Much of what we know about anteroposterior pattern formation stems from genetic analyses of a series of bizarre mutations in three classes of *Drosophila* genes that normally specify progressively finer regions of cellular specialization in the developing embryo:

1. *Maternal-effect genes, which define the embryo's polarity,* that is, its anteroposterior axis. Mutations of these genes globally alter the embryonic body pattern regardless of the paternal genotype. For instance, females homozygous for the *dicephalic* (two-headed) **mutation** lay eggs that develop into nonviable two-headed monsters. These are embryos with two anterior ends pointing in opposite directions and completely lacking posterior structures. Similarly, the *bicaudal* (two-tailed) and *snake* **mutations** give rise to mirror-symmetric embryos with two abdomens (Fig. 34-87*a*).

2. *Segmentation genes, which specify the correct number and polarity of embryonic body segments.* Investigations by Christiane Nüsslein-Volhard and Eric Wieschaus led to their subclassification as follows:

  a. **Gap genes,** the first of a developing embryo's to be transcribed, divide the embryo into several broad regions. Gap genes are so named because their mutations result in gaps in the embryo's segmentation pattern. Embryos with defective *hunchback (hb)* genes, for example, lack mouthparts and thorax structures.

  b. **Pair-rule genes** specify the division of the embryo's broad gap regions into segments. These genes are so named because their mutations usually delete portions of every second segment. This occurs, for example, in embryos that are homozygous for mutations in the *fushi tarazu* (*ftz;* Japanese for segment deficient) gene (Fig. 34-87*b*).

  c. **Segment polarity genes** specify the polarities of the developing segments. Thus, homozygous *engrailed* (*en;* indented with curved notches) mutants lack the posterior compartment of each segment.

3. *Homeotic selector genes, which specify segmental identity.* Homeotic mutations transform one body part into another. For instance, *Antennapedia* (*antp,* antenna-foot) mutants have legs in place of antennae (Fig. 34-87*c,d*), whereas the mutations *bithorax (bx), anteriorbithorax (abx),* and *postbithorax (pbx)* each transform sections of halteres (vestigial wings that function as balancers; Fig. 34-86), which normally occur only on segment T3, to the corresponding sections of wings, which normally occur only on segment T2 (Fig. 34-87*e*).

c. Maternal-Effect Gene Products Specify the Egg's Directionality through Gradient Formation

The properties of maternal-effect gene mutants suggest that maternal-effect genes specify morphogens whose distributions in the egg cytoplasm define the future embryo's spatial coordinate system. Indeed, immunofluorescence studies by Nüsslein-Volhard demonstrated that the product of the *bicoid (bcd)* gene is distributed in a gradient that decreases toward the posterior end of the normal embryo (Figs. 34-88 and 34-89*a*), whereas embryos with *bcd*-deficient mothers lack this gradient. The gradient, which is facilitated by the syncytium's lack of cellular boundaries, arises through the secretion, by the ovarian nurse cells, of *bcd* mRNA into the anterior end of the oocyte during oogenesis and its translation in the early embryo. The *nanos* gene mRNA is similarly desposited in the egg but it is localized near the egg's posterior pole (Fig. 34-89*a*). The *bcd* and *nanos* gene products, as we shall see, are transcription factors that regulate the expression of specific gap genes. Other maternal-effect genes that participate in anteroposterior axis formation specify proteins that function to trap the localized mRNAs in their area of deposition. This explains why early embryos produced by females homozygous for maternal-effect mutations can often be "rescued" by the injection of cytoplasm, or sometimes just the mRNA, from early wild-type embryos. With some of these mutations, the polarity of the rescued embryo is determined by the site of the injection.

d. Gap Genes Are Expressed in Specific Regions

The mRNA of the gap gene *hunchback (hb)* is maternally deposited in a uniform distribution throughout the unfertilized egg (Fig. 34-89*a*). However, **Bicoid protein** activates the transcription of the embryonic *hb* gene, whereas **Nanos protein** inhibits the translation of *hb* mRNA. Consequently, **Hunchback protein** becomes distributed in a gradient that decreases from anterior to posterior (Fig. 34-89*b*).

DNase I footprinting studies have demonstrated that Bicoid protein binds to five homologous sites (consensus sequence TCTAATCCC) in the *hb* gene's upstream promoter region. Nüsslein-Volhard demonstrated the ability of Bicoid protein to activate the *hb* gene by fusing the *hb* promoter upstream of the CAT reporter gene (Section 34-3B) and injecting the resulting construct into early *Drosophila* embryos. CAT was produced in wild-type but not in *bcd*-deficient embryos. Moreover, the use of progressively shorter segments of the *hb*-derived promoter region demonstrated that at least three of the five Bicoid protein-binding sites must be present to obtain full CAT expression.

Hunchback protein, in turn, controls the expression of several other gap genes (Fig. 34-89*c,d*): High levels of Hunchback protein induce *giant* expression, *Krüppel* (German: cripple) is expressed where the level of Hunchback protein begins to decline, *knirps* (German: pygmy) is expressed at even lower levels of Hunchback protein, and *giant* is again activated in regions where Hunchback protein is undetectable.

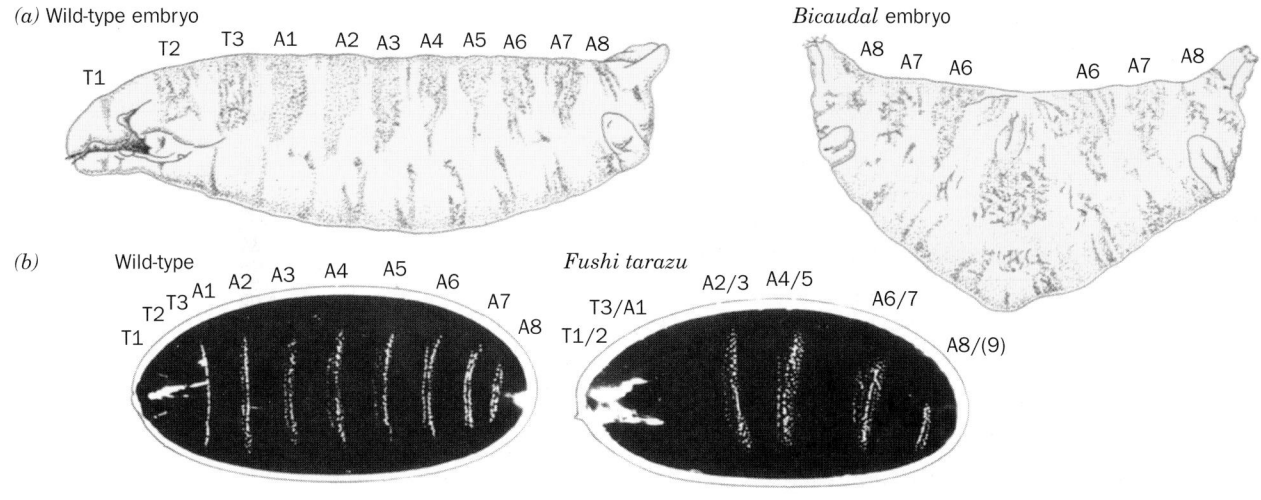

(a) Wild-type embryo

Bicaudal embryo

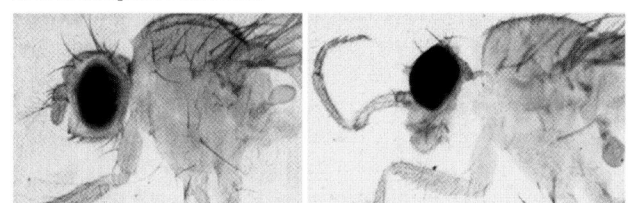

(b) Wild-type

Fushi tarazu

(c) Antennapedia mutation

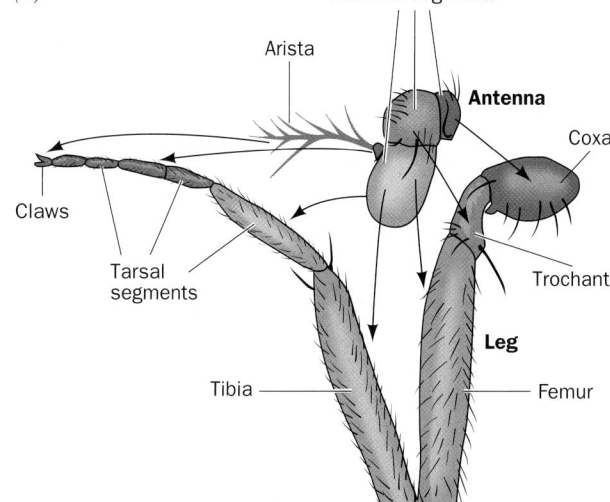

(d)

(e)

FIGURE 34-87   Developmental mutants of *Drosophila*. (*a*) The cuticle patterns of wild-type embryos (*left*) exhibit 11 body segments, T1 to T3 and A1 to A8 (the head segments have retracted into the body and hence are not visible here). In contrast, the nonviable "monsters" produced by homozygous *bicaudal* mutant females (*right*) develop only abdominal segments arranged with mirror symmetry. [After Gergen, P.J., Coulter, D., and Weischaus, E., *in* Gall, J.G., *Gametogenesis and the Early Embryo*, p. 200, Liss (1986).] (*b*) In the wild-type embryo (*left*), the anterior edge of each of the 11 abdominal and thoracic segments has a belt of tiny projections known as denticles (which help larvae crawl) that appear in these photomicrographs as white

stripes. *Fushi tarazu* mutants (*right*) lack portions of alternate segments and the remaining segments are fused together (e.g., A2/3), yielding a nonviable embryo with only half of the normal number of denticle belts. [Courtesy of Walter Gehring, University of Basel, Switzerland.] (*c*) Head and thorax of a wild-type adult fly (*left*) and one that is heterozygous for a mutant form of the homeotic *Antennapedia* (*antp*) gene (*right*). The mutant gene is inappropriately expressed in the imaginal disks that normally form antennae (where the wild-type *antp* gene is not expressed) so that they develop as the legs that normally occur only on segment T2. [Courtesy of Ginés Morata, Universidad Autónoma de Madrid, Spain.] (*d*) The correspondence (*arrows*) between antennae and the legs to which the *Antp* mutation transforms them. [After Postlethwait, J.H. and Schneiderman, H.A., *Devel. Biol.* **25,** 622 (1971).] (*e*) A four-winged *Drosophila* (it normally has two wings; Fig. 34–86) that results from the presence of three mutations in the bithorax complex, *abx, bx,* and *pbx*. These mutations cause the normally haltere-bearing segment T3 to develop as if it were the wing-bearing segment T2. This striking architectural change may reflect evolutionary history: *Drosophila* evolved from more primitive insects that had four wings. [Courtesy of Edward B. Lewis, Caltech.]

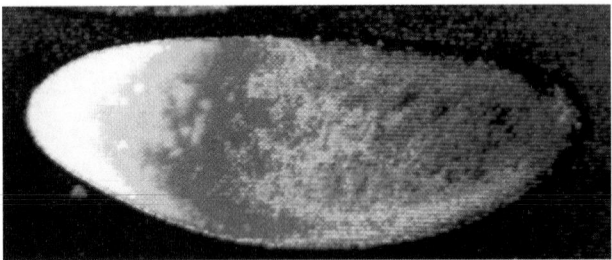

FIGURE 34-88 The distribution of Bicoid protein in a *Drosophila* syncytial blastoderm as revealed by immunofluorescence. High concentrations of the protein are yellow, lower concentrations are red, and its absence is black. [Courtesy of Christiane Nüsslein-Volhard, Max-Planck-Institut für Entwicklungsbiologie, Germany.]

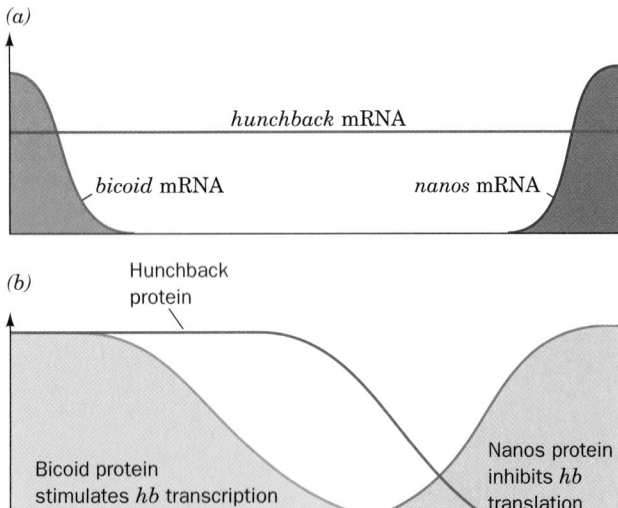

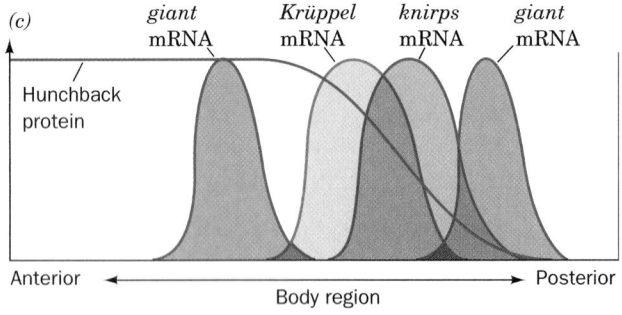

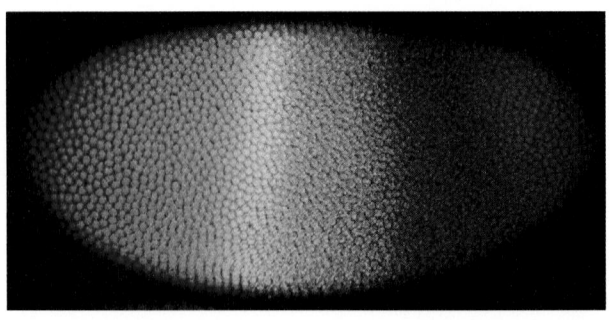

Although the original positions of the proteins encoded by these latter gap genes are elicited by the appropriate concentrations of Hunchback protein, these positions are stabilized and maintained through their mutual interactions. Thus **Krüppel protein** binds to the promoters of the *hb* gene, which it activates, and the *knirps* gene, which it represses. Conversely, **Knirps protein** represses the *Krüppel* gene. This mutual repression is thought to be responsible for the sharp boundaries between the various gap domains.

e. Pair-Rule Genes Are Expressed in "Zebra Stripes"

Pair-rule genes are expressed in sets of 7 stripes, each just a few nuclei wide, along the embryo's anterior–posterior axis. The embryo (which, at this stage, is just beginning to cellularize) is thereby divided into 15 domains (Fig. 34-90). These "zebra stripe" expression patterns for the various pair-rule genes are offset relative to one another.

The gap gene products are transcription factors for three **primary pair-rule genes: *hairy*, *even-skipped* (*eve*),** and ***runt*.** The striped pattern of expression arises because the control regions of most primary pair-rule genes comprise a series of enhancer modules, each of which induce their gene's expression in a particular stripe (Fig. 34-91*a*). For example, the transformation of an embryo by the *lacZ* gene preceded by a specific enhancer module in the *eve* gene resulted in *lacZ* transcription in only stripe 1 (Fig. 34-91*b*), whereas a different module did so in only stripe 5 (Fig. 34-91*c*), and both of these modules together induced the production of the *lacZ* transcript in both stripes 1 and 5 (Fig. 34-91*d*). Each of these modules contains a particular arrangement of activating and inhibitory binding sites for the various gap gene proteins so as to enable the expression of the associated pair-rule gene under the particular combination of gap gene proteins present in

FIGURE 34-89 Formation and effects of the Hunchback protein gradient in *Drosophila* embryos. (*a*) The unfertilized egg contains maternally supplied *bicoid* and *nanos* mRNAs placed at its anterior and posterior poles, together with a uniform distribution of *hunchback* mRNA. (*b*) On fertilization, the three mRNAs are translated. Bicoid and Nanos proteins are not bound to the cytoskeleton as are their mRNAs and hence their gradients are broader than those of the mRNAs. Bicoid protein stimulates the transcription of the *hunchback* gene, whereas Nanos protein inhibits its translation, resulting in a gradient of Hunchback protein that decreases nonlinearly from anterior to posterior. (*c*) Specific concentrations of Hunchback protein induce the transcription of the *giant*, *Krüppel*, and *knirps* genes. The gradient of Hunchback protein thereby specifies the positions at which these latter mRNAs are synthesized. (*d*) A photomicrograph of a *Drosophila* embryo (*anterior end left*) that has been immunofluorescently stained for both Hunchback (*green*) and Krüppel proteins (*red*). The region where these proteins overlap is yellow. [Parts *a, b,* and *c* after Gilbert, S.F., *Developmental Biology* (4th ed.), p. 543, Sinauer Associates (1994); Part *d* courtesy of Jim Langeland, Steve Paddock, and Sean Carroll, Howard Hughes Medical Institute, University of Wisconsin–Madison.]

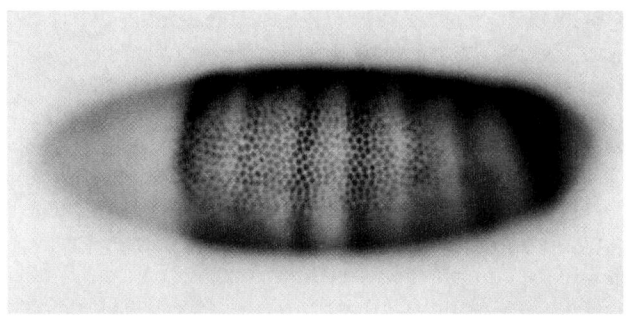 

FIGURE 34-90 *Drosophila* **embryos stained for Ftz (*brown*) and Eve (*gray*) proteins.** These proteins are each expressed in seven stripes which, at first, are relatively blurred (*left*) but within a short time become sharply defined (*right*). [Courtesy of Peter Lawrence, MRC Laboratory of Molecular Biology, U.K.]

the corresponding stripe. Thus, in *giant*-deficient embryos, the posterior border of stripe 5 is missing (Fig. 34-91*e*). As with the gap genes, the patterns of expression of the primary pair-rule genes become stabilized through interactions among themselves.

The primary pair-rule gene products similarly induce or inhibit the expression of five **secondary pair-rule genes** including *ftz*. Thus, as Walter Gehring demonstrated, *ftz* transcripts first appear in the nuclei lining the cortical cytoplasm during the embryo's 10th nuclear division cycle.

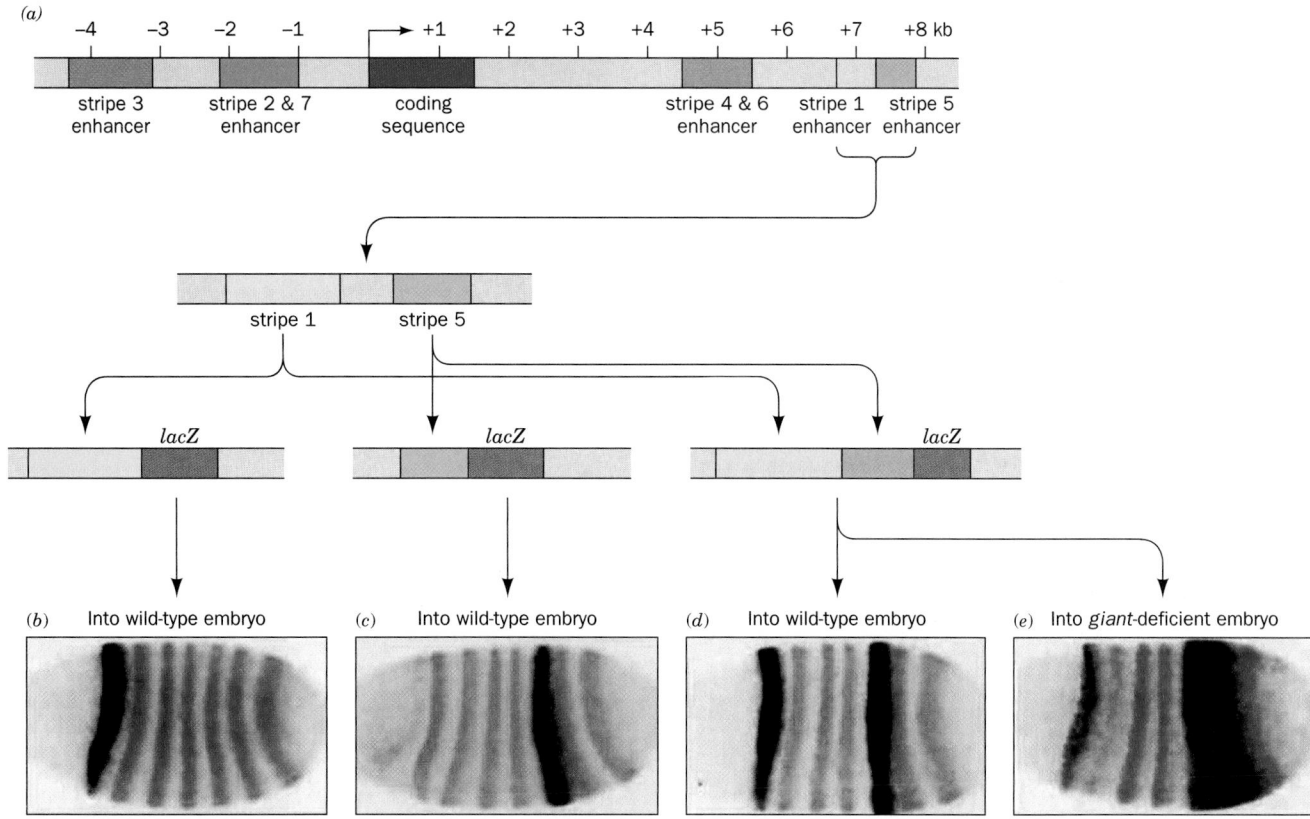

FIGURE 34-91 **Expression of the *even-skipped* (*eve*) gene in a pattern of seven stripes in the *Drosophila* embryo.** (*a*) Diagram of the *eve* gene, which contains a series of enhancer modules, some upstream of the coding region (*blue*) and others downstream, that on binding the particular combination of gap genes present in their corresponding stripe, induce the expression of *eve* in that stripe. The positions of various elements in the gene, in kb, relative to the transcriptional start site (*squared off arrow*), are indicated. The *lacZ* reporter gene (*magenta*) under control of (*b*) the stripe 1 enhancer (*yellow*), (*c*) the stripe 5 enhancer (*cyan*), or (*d*) both the stripe 1 and 5 enhancers was injected into wild-type *Drosophila* oocytes. The resulting embryos were hybridized *in situ* with dye-labeled *lacZ* antisense RNA to yield a blue band where *lacZ* had been transcribed and then stained with anti-Eve antibodies (*orange*), thereby demonstrating that *lacZ* is expressed only in the corresponding stripe(s). (*e*) When an oocyte deficient in the gap gene *giant* was injected with *lacZ* under the control of the stripe 1 and 5 enhancers, stripe 1 was normal but stripe 5 lacked its posterior border, which indicates that Giant protein normally functions to inhibit *eve* expression past the end of stripe 5. [Part *a* based on a drawing by Scott Gilbert, Swarthmore College. Parts *b, c, d,* and *e* courtesy of James Jaynes, Thomas Jefferson University.]

The rate of *ftz* expression then increases as the embryo develops until the 14th division cycle, when the cellular blastoderm forms. At this stage, as immunochemical staining dramatically shows, *ftz* is expressed in a pattern of 7 belts around the cellular blastoderm, each 3 or 4 cells wide (Fig. 34-90), which correspond precisely to the missing regions in homozygous *ftz⁻* embryos. Then, as the embryonic segments form, *ftz* expression subsides to undetectable levels (although it is later reactivated during the differentiation of specific nerve cells in which it is required to specify their correct "wiring" pattern). Evidently, the *ftz* gene must be expressed in alternate sections of the embryo for normal segmentation to occur.

f. Segment Polarity Genes Define Parasegment Boundaries

The expression of eight known segment polarity genes is initiated by pair-rule gene products. For example, by the 13th nuclear division cycle, as Thomas Kornberg demonstrated, *engrailed (en)* transcripts become detectable but are more or less evenly distributed throughout the embryonic cortex. However, since *en* is expressed in nuclei containing high concentrations of either **Eve** or **Ftz** proteins (Fig. 34-90), by the 14th cycle they form a striking pattern of 14 stripes around the cellular blastoderm (half the spacing of *ftz* expression). Continuing development reveals that these stripes are localized in the primordial posterior compartment of every segment (Fig. 34-92), just those compartments that are missing in homozygous *en⁻* embryos. Thus, much like we saw for *ftz*, the *en* gene product induces the posterior half of each segment to develop in a different fashion from its anterior half.

Another segment polarity gene, ***wingless (wg)***, is expressed simultaneously with *en* but in narrow bands on the anterior side of most *en* bands (Fig. 34-93). Cells expressing *en* and *wg* genes thereby define the boundaries of the so-called **parasegments**, embryonic regions that consist of the posterior portion of one segment and the anterior portion of the segment behind it. Parasegments do not become morphological units in the larva or adult but, nevertheless, are thought to be the embryo's actual developmental units.

g. Homeotic Selector Genes Direct the Development of the Individual Body Segments

The structural components of developmentally analogous body parts, say *Drosophila* antennae and legs, are nearly identical; only their organizations differ (Fig. 34-87d). *Consequently, developmental genes must function to control the pattern of structural gene expression rather than simply turning these genes on or off.* Thus, as we saw for the segmentation genes, the expression of the structural genes characteristic of any given tissue must be controlled by a complex network of regulatory genes. The homeotic selector genes, as we shall see, are the "master" genes in the control networks governing segmental differentiation.

Most homeotic mutations in *Drosophila* (which were first described in 1894 by William Bateson, who coined the

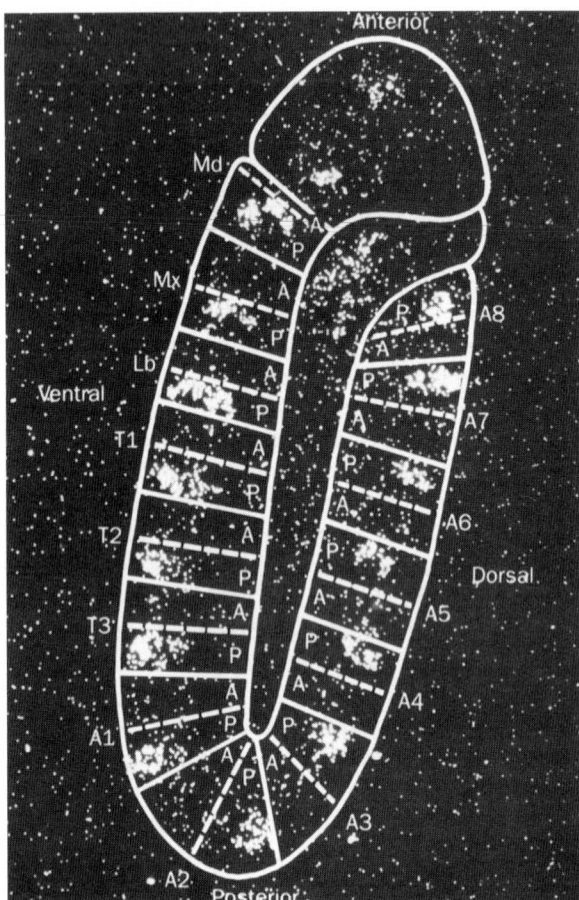

FIGURE 34-92 *In situ* **hybridization demonstrates that the *Drosophila* engrailed gene is expressed in the posterior compartment of every embryonic segment.** [Courtesy of Walter Gehring, University of Basel, Switzerland.]

name "homeosis" to indicate something that has been changed into the likeness of something else) map into eight related genes that are distributed in two clusters that function as a single cluster: the **bithorax complex (BX-C)**, which controls the development of the thoracic and abdominal segments, and the **antennapedia complex (ANT-C)**, which primarily affects head and thoracic segments. *Recessive mutations in BX-C, when homozygous, cause one or more segments to develop as if they were more anterior segments.* Thus, the combined ***bx, abx***, and ***pbx*** mutations cause segment T3 to develop as if it were segment T2 (Fig. 34-87e). Similarly, the entire deletion of *BX-C* causes all segments posterior to T2 to resemble T2; apparently T2 is the developmental "ground state" of these 10 segments. The evolution of such gene families, it is thought, permitted arthropods (the phylum containing insects) to arise from the more primitive annelids (segmented worms) in which all segments are nearly alike.

Detailed genetic analysis of *BX-C* led Edward B. Lewis to formulate a model for segmental differentiation (Fig. 34-94). *BX-C*, Lewis proposed, contains at least one gene for each segment from T3 to A8, which for simplicity are numbered 0 to 8 in Fig. 34-94. These genes, for reasons that are not understood, are arranged in the same order,

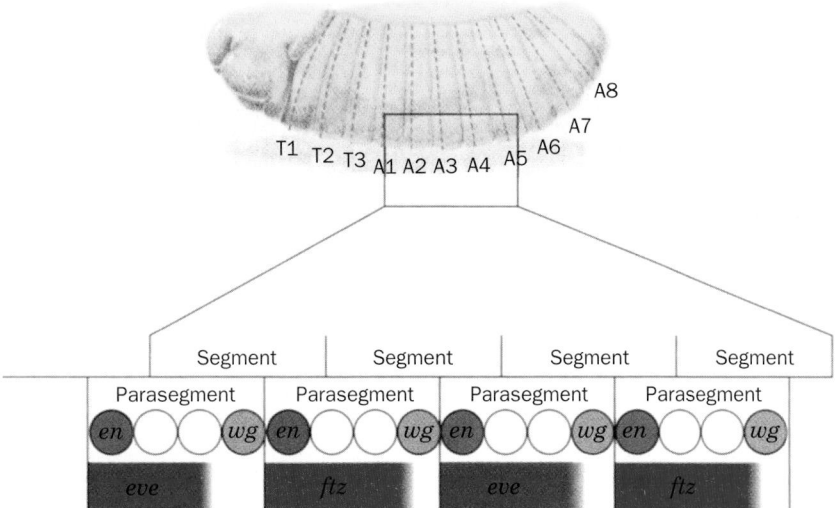

FIGURE 34-93   The pair-rule proteins Eve and Ftz regulate the expression of the segment polarity genes *engrailed* **(***en***) and** *wingless* **(***wg***).** When either Eve or Ftz is present, *en* is expressed, whereas when both proteins are absent, *wg* is expressed. The parasegment boundaries are thereby defined. Other pair-rule proteins are thought to inhibit *en* and *wg* expression in nuclei not at the parasegment boundaries.

from "left" to "right," as the segments whose development they influence. Starting with segment T3, progressively more posterior segments express successively more *BX-C* genes until, in segment A8, all of these genes are expressed. The developmental fate of a segment is thereby determined by its position in the embryo.

Sequence analysis of the *BX-C* region led to a difficulty with Lewis' model: The *BX-C* contains only three protein-encoding genes, ***Ultrabithorax (Ubx), Abdominal-A (Abd-A),*** and ***Abdominal-B (Abd-B).*** However, further analysis indicated, for example, that mutations such as *bx, abx,* and *pdx,* which were previously assumed to occur on separate genes, are actually mutations of enhancer elements that enable the position-specific expression of the *Ubx* gene. Thus, the nine "genes" in Lewis' model have turned out to be enhancer elements on the three *BX-C* genes.

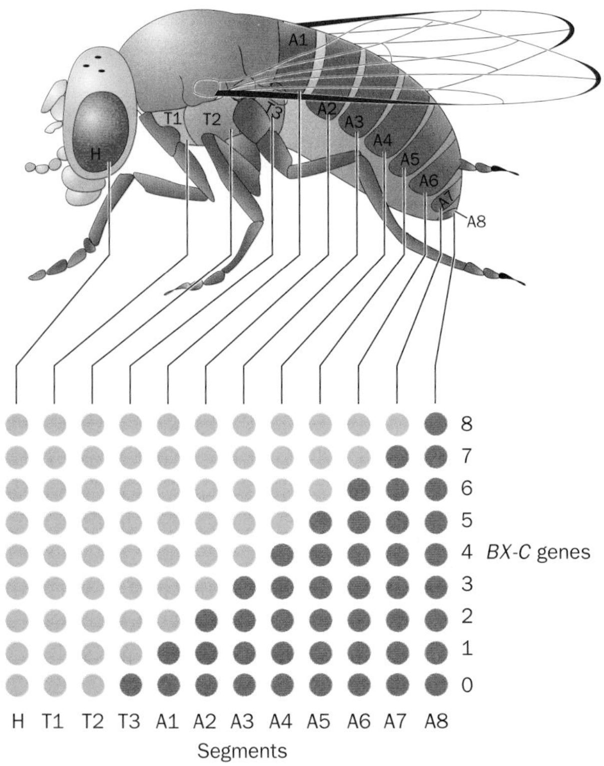

h. Developmental Genes Have Common Sequences

In characterizing the ***Antennapedia (Antp)*** gene, Gehring and Matthew Scott independently discovered that *Antp* cDNA hybridizes to both the *Antp* and the *ftz* gene and that, therefore, *these genes share a common base sequence.* This startling observation rapidly led to the discovery that the *Drosophila genome contains numerous such sequences, many of which occur in the homeotic gene complexes ANT-C and BX-C.* DNA sequencing studies of

FIGURE 34-94   Model for the differentiation of embryonic segments in *Drosophila* **as directed by the genes of the bithorax complex (***BX-C***).** Segments T2, T3, and A1–8 in the embryo, as the lower drawing indicates, are each characterized by a unique combination of active (*purple circles*) and inactive (*yellow circles*) *BX-C* "genes." These "genes" (which sequencing studies later demonstrated are really enhancer elements), here numbered 0 to 8, are thought to be sequentially activated from anterior to posterior in the embryo so that segment T2, the developmentally most primitive segment, has no active *BX-C* genes, while in segment A8, all of them are active. Such a pattern of gene expression may result from a gradient in the concentration of a *BX-C* repressor that decreases from the anterior to the posterior of the embryo. [After Ingham, P., *Trends Genet.* **1,** 113 (1985).]

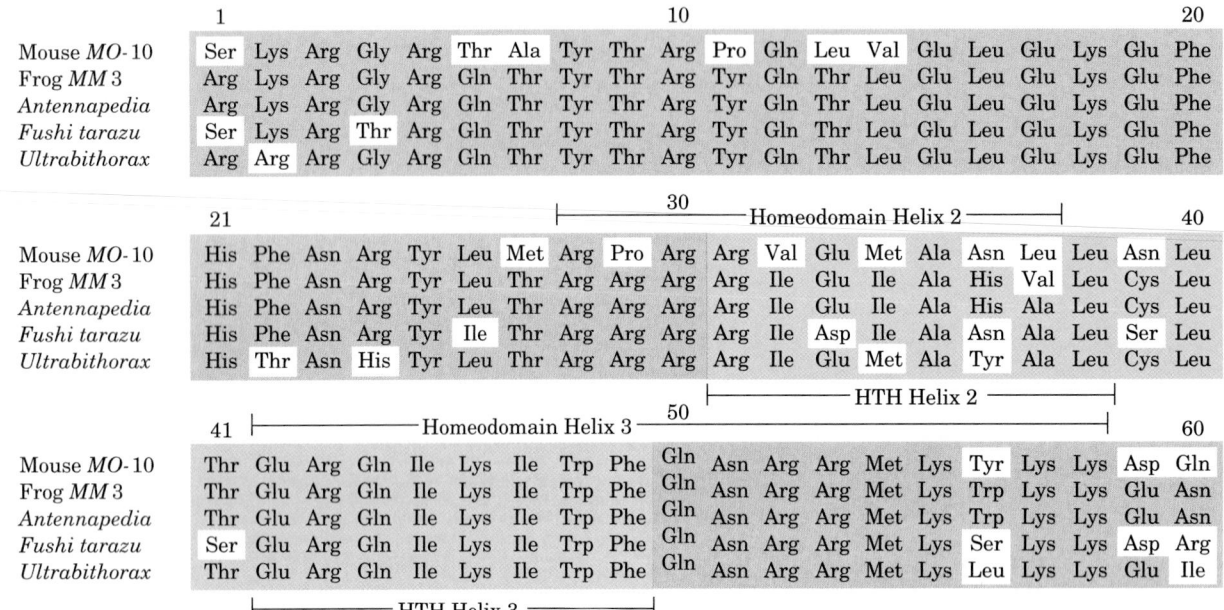

FIGURE 34-95 Amino acid sequences of the polypeptides encoded by the homeodomains of five genes from mouse, *Xenopus*, and *Drosophila* (*Ultrabithorax* is a *BX-C* gene). Discrepancies between the polypeptide specified by the *Antp* homeobox and those of the other genes lack shading. Each polypeptide has a 19-residue segment (*red shading*), which is homologous to the DNA-binding HTH fold of prokaryotic repressors. The positions of the helices observed in the X-ray and NMR structures of homeodomains, together with the corresponding positions of HTH helices in prokaryotic proteins, are indicated.

these genes revealed that each contains a 180-bp sequence, the so-called **homeodomain** or **homeobox,** which are 70 to 90% identical to one another and which encode even more identical 60-residue polypeptide segments (Fig. 34-95).

Further hybridization studies using homeodomain probes led to the truly astonishing finding that *multiple copies of the homeodomain are also present in the genomes of segmented animals ranging from annelids to vertebrates such as Xenopus, mice, and humans.* In some of these sequences the degree of homology is remarkably high; for example, the homeodomains of the *Drosophila Antp* gene and the *Xenopus MM3* **gene** encode polypeptides that have 59 of their 60 amino acids in common (Fig. 34-95). The individuality of these homeodomain-containing proteins is presumably imparted by their other segments.

i. The Homeodomain's DNA-Binding Motif Resembles a Helix–Turn–Helix Motif

Since vertebrates and invertebrates diverged over 600 million years ago, this strongly suggest that the gene product of the homeodomain has an essential function. What might this function be? The ~30% Arg + Lys content of homeodomain polypeptides suggests that they bind DNA. Sequence comparisons and NMR studies further suggest that these polypeptide segments form helix–turn–helix (HTH) motifs resembling those of prokaryotic gene regulators such as the *E. coli trp* repressor (Section 31-3D) and the λ Cro protein (Section 33-3D). Indeed, the polypeptide encoded by the homeodomain of the *Drosophila engrailed* gene specifically binds to the DNA sequences just

upstream from the transcription start sites of both the *en* and the *ftz* genes. Moreover, fusing the *ftz* gene's upstream sequence to other genes imposes *ftz*'s pattern of stripes (Fig. 34-90) on the expression of these genes in *Drosophila* embryos. *These observations suggest, in agreement with the idea that the products of developmental genes act to regulate the expression of other genes, that homeodomain-containing genes encode transcription factors.* In fact, not all homeodomain-encoded proteins are involved in regulating development. The homeodomain is apparently a widespread genetic motif that specifies the DNA-binding segments of a variety of proteins.

Thomas Kornberg and Pabo have determined the X-ray structure of the 61-residue homeodomain from the *Drosophila* Engrailed protein in complex with a 21-bp DNA (Fig. 34-96). Two copies of the protein bind to the DNA, one near the center of the DNA and the other near one end, where it also contacts a second DNA molecule that, in the crystal, forms a pseudocontinuous helix with the first. The conformations of the two protein molecules, and the contacts they make with the DNA, are nearly identical. The two homeodomains are not in contact so, in contrast to other DNA-binding motifs of known structure, *they bind to their target DNAs as monomers.* The X-ray structure is largely consistent with the NMR structure of the *Antennapedia* homeodomain in complex with a 14-bp DNA determined by Gehring and Kurt Wüthrich.

The homeodomain consists largely of three α helices, the last two of which, as sequence comparisons had previously suggested, form an HTH motif that is closely super-

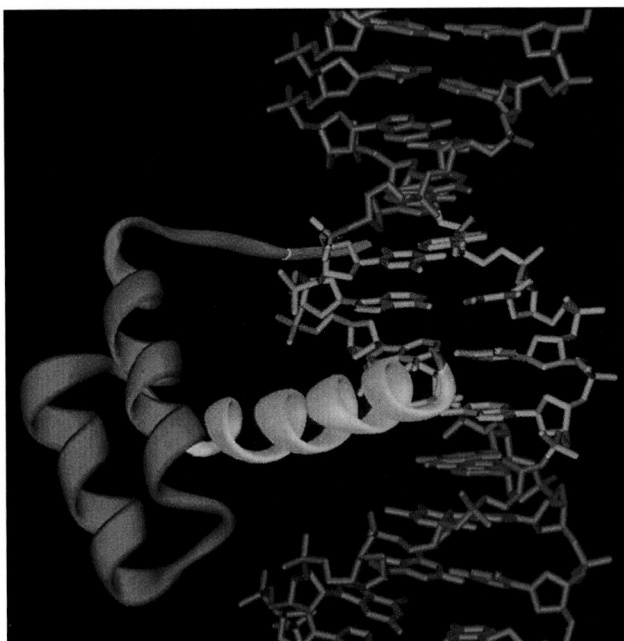

FIGURE 34-96   X-Ray structure of the Engrailed protein homeodomain in complex with a 21-bp DNA containing its target sequence. The 60-residue protein is shown in ribbon form (*green*) with its recognition helix (helix 3, residues 42–58), which is bound in the DNA's major groove, highlighted in gold. The DNA is shown in stick form (*light blue*) with the base pairs comprising its TAAT subsite highlighted in magenta. A second homeodomain that binds to the lower end of the DNA in a nearly identical manner but does not contact the homeodomain shown has been omitted for clarity. Note how the N-terminal segment (*red,* residues 3–5; residues 1 and 2 are disordered) binds in the minor groove of the DNA. [Based on an X-ray structure by Carl Pabo, MIT. PDBid 1HDD.]

imposable with the HTH motifs of prokaryotic repressors such as that of the λ repressor (Fig. 33-45*a*). However, although helix 3, the HTH motif's recognition helix, fits into the major groove of its corresponding DNA, it does so quite differently in the two complexes. In the λ repressor complex, for example, the N-terminal end of the recognition helix is inserted into the DNA's major groove, whereas in the homeodomain complex the DNA is shifted toward the C-terminal end of the helix, which is longer than that of the λ repressor (it extends from residues 42 to 58 in Fig. 34-95). As a consequence, the way in which the first helix of the HTH motif (helix 2; residues 28 to 37 in Fig. 34-95) contacts the DNA also differs between the two complexes.

Most homeodomain binding sites have the subsequence TAAT. The recognition helix in the X-ray structure makes base-specific hydrogen bonding contacts with this subsequence in the major groove through residues that are highly conserved in higher eukaryotic homeodomains. It therefore appears that these interactions function to align the homeodomain with the other bases that it contacts. In addition, two conserved Arg residues located in the N-terminal tail of the homeodomain make base-specific hydrogen bonding contacts with the TAAT subsequence in the minor groove of the DNA. The protein thereby grips

the TAAT subsequence from two sides. Note that few other sequence-specific DNA-binding proteins contact bases in the minor groove. Finally, the homeodomain makes extensive contacts with the DNA backbone that, it is presumed, also play an important part in binding and recognition.

j. Homeodomain Genes Function Analogously in Vertebrates and *Drosophila*

Homeodomain-encoding genes have collectively become known as **Hox genes.** In vertebrates, they are organized in four clusters of 9 to 11 genes, each located on a separate chromosome and spanning more than 100 kb. In contrast, *Drosophila,* as we saw, has a split *Hox* cluster, whereas in nematodes (roundworms), which are evolutionarily more primitive than insects, the single *Hox* cluster remains unsplit. The genes in the primordial *Hox* cluster presumably arose in some more primitive ancestral organism through a series of gene duplications, as did the four vertebrate *Hox* clusters. The genes in each vertebrate *Hox* cluster, as in *Drosophila,* are activated in the same order, left to right, as they are expressed from the anterior end of the embryo to its posterior end. Perhaps this arrangement is necessary for the homeodomain genes to be activated in the proper order, although, at least in *Drosophila,* gap and pair-rule proteins can still act on *Hox* control regions that have been transplanted to other parts of the genome. Whatever the case, the various *Hox* clusters, as well as their component genes, almost certainly arose through a series of gene duplications and diversifications starting with a single *Hox* gene in a primitive ancestral organism.

Vertebrate *Hox* genes, like those of *Drosophila,* are expressed in specific patterns and at particular stages during embryogenesis. Most *Hox* genes are expressed at a gestational time when organogenesis prevails. That the *Hox* genes directly specify the identities and fates of embryonic cells, that is, are homeotic in character, was shown, for example by the following experiment. Mouse embryos were made transgenic for the *Hox-1.1* gene that had been placed under the control of a promoter that is active throughout the body even though *Hox-1.1* is normally expressed only below the neck. The resulting mice had severe craniofacial abnormalities such as a cleft palate and an extra vertebra and an intervertebral disk at the base of the skull. Some also had an extra pair of ribs in the neck region. Thus, this *Hox* gene's "gain of function" resulted in a homeotic mutation, that is, a change in the development pattern, analogous to those observed in *Drosophila.*

Homozygotic mice resulting from the replacement of their *Hox-3.1* gene coding sequence in embryonic stem cells with that of *lacZ* are born alive but usually die within a few days. They exhibit skeletal deformities in their trunk regions in which several skeletal segments are transformed into the likenesses of more anterior segments. The pattern of β-galactosidase activity (Fig. 34-97), as colorimetrically detected through the use of X-Gal (Section 5-5C), in both homozygotes and heterozygotes, indicates that *Hox-3.1* deletion modifies the properties but not the positions of the embryonic cells that normally express *Hox-3.1.*

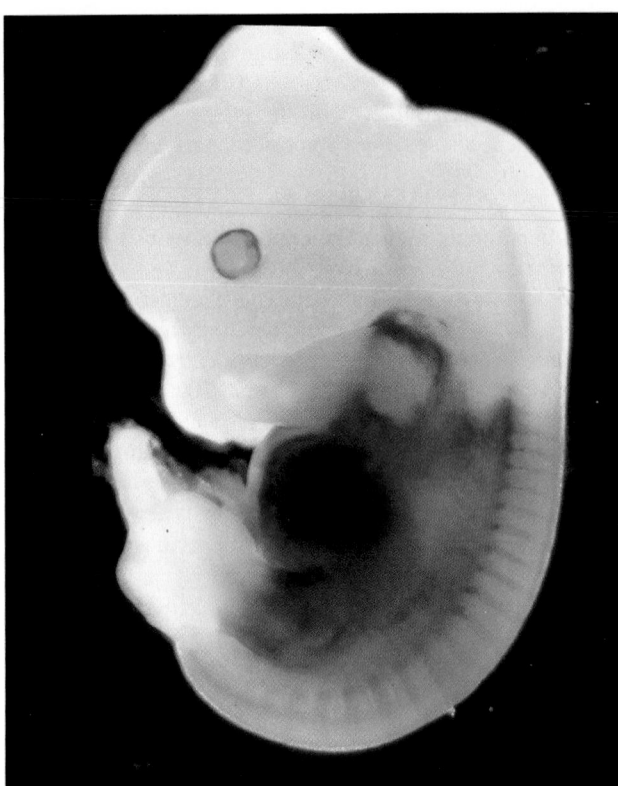

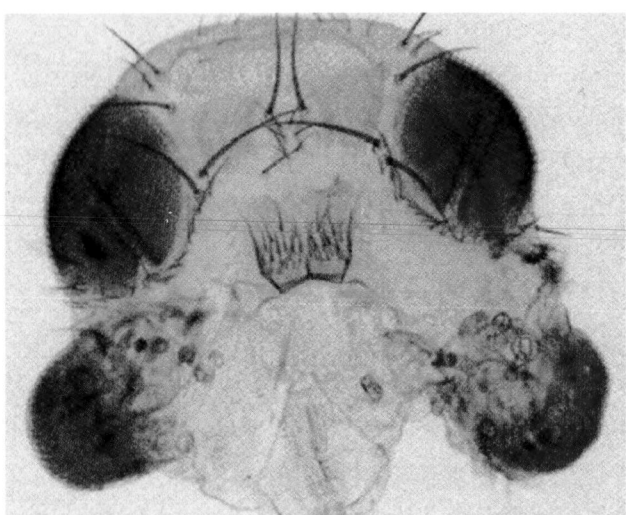

FIGURE 34-98   Ectopic eyes result from the targeted expression of the *Drosophila ey* gene in its imaginal disk primordia. Shown here is the cuticle of an adult *Drosophila* head in which both antennae have formed eye structures that exhibit the morphology and red pigmentation of normal eyes. Such eye structures have been similarly expressed on wings and legs. [Courtesy of Walter Gehring, University of Basel, Switzerland.]

FIGURE 34-97   Pattern of expression of the *Hox-3.1* gene in a 12.5-day postconception mouse embryo. The protein-encoding portion of the embryo's *Hox-3.1* gene was replaced by the *lacZ* gene. The regions of this transgenic embryo in which *Hox-3.1* is expressed are revealed by the blue color that develops on soaking the embryo in X-Gal–containing buffer. [Courtesy of Phillipe Brûlet, Collège de France and the Pasteur Institute, Paris, France.]

k. Expression of the *Drosophila eyeless* Gene Induces the Ectopic Formation of Eyes

Mutations in the *Drosophila eyeless* (*ey*) gene, first described in 1915, result in flies whose compound eyes are reduced in size or completely absent. The expression of *ey*, which contains a homeodomain, is first detected in the embryonic nervous system and later in the embryonic primordia of the eye. In subsequent larval stages, it is expressed in the developing eye imaginal disks. Mutant forms of four other *Drosophila* genes that have similar phenotypes do not affect the expression of the *ey* gene, which indicates that *ey* acts before these other genes. These observations led to the suggestion that the *ey* gene is the master control gene for eye development.

Genetic engineering studies by Gehring have confirmed this hypothesis. Through the targeted expression of *ey* cDNA in various imaginal disk primordia of *Drosophila*, ectopic (inappropriately positioned) compound eyes were induced to form on the wings, legs, and antennae (Fig. 34-98) of various flies. Moreover, in many cases, these eyes appeared morphologically normal in that they consisted of fully differentiated ommatidia (the simple eye elements that form a compound eye) with a complete set of photoreceptor cells that appear to be electrically active when illuminated (although it is unknown if the flies could see

with these ectopic eyes, that is, whether these eyes made appropriate neural connections to the brain).

The mouse *Small eye* (*Sey* or *Pax-6*) gene and the human *Aniridia* gene are closely similar in sequence to the *Drosophila ey* gene and are similarly expressed during morphogenesis. Mice with mutations in one of their two *Sey* genes have underdeveloped eyes, whereas those with mutations in both *Sey* genes are eyeless. Similarly, humans that are heterozygotes for a defective *Aniridia* gene have defects in their iris, lens, cornea, and retina. Evidently, the *ey, Sey,* and *Aniridia* genes all function as master control genes for eye formation in their respective organisms, a surprising result considering the enormous morphological differences between insect and mammalian eyes. Thus, despite the 500 million years since the divergence of insects and mammals, their developmental control mechanisms appear to be closely related.

l. Retinoic Acid Is a Vertebrate Morphogen

Retinoic acid (RA), a derivative of **vitamin A (retinol),**

X = COOH: **Retinoic acid (RA)**

X = CH₂OH: **Retinol (vitamin A)**

has been found to have a graded distribution in developing chick limbs and is therefore thought to be a morphogen. The systematic administration of RA during mouse embryogenesis results in severe malformations, notably skeletal deformities that appear to arise from anterior or posterior shifts of their normal characteristics. A variety of evidence suggests that the expression of *Hox* genes mediates the positional information that RA disrupts. The *Hox* genes are differentially activated by RA according to their positions in their Hox clusters: Those toward the 3′ end of a cluster are maximally induced by as little as $10^{-8}M$ RA, those toward the 5′ end of the cluster require $10^{-5}M$ RA to do so, and those at the 5′ ends are insensitive to RA. Moreover, $10^{-5}M$ RA sequentially activates the *Hox* genes from the 3′ to the 5′ end of a cluster, the same order as their expression patterns in developing axial systems such as the skeleton and the central nervous system.

The foregoing explains why the RA analog **13-cis-retinoic acid,** which, taken orally, has been invaluable in the treatment of severe **cystic acne,** induces birth defects if used by pregnant women. The characteristic pattern of cranial deformities in the resulting infants, whose analog is induced in mouse embryos that had been exposed to low concentrations ($2 \times 10^{-6}M$) of this drug, indicates that its presence alters the expression of *Hox* genes early in gestation (~1 month postfertilization in humans, ~9 days in mice).

C. *The Molecular Basis of Cancer*

Cancer, being one of the major human health problems, has received enormous biomedical attention over the past several decades. Around 100 different types of human cancers are recognized, methods of cancer detection and treatment are highly developed, and cancer epidemiology has been extensively characterized. Nevertheless, we are far from fully understanding the biochemical basis of this collection of diseases. In Section 19-3B we discussed the general nature of cancer, its causes, and how tumor viruses cause cancer. In this section we outline how genetic alterations cause cancer.

a. Malignancies May Result from Specific Genetic Alterations

Although much of what we know concerning oncogenes stems from the study of retroviral oncogenes (Section 19-3B), few human cancers are caused by retroviruses. Nevertheless, *it seems likely that all cancers are caused by genetic alterations.* Robert Weinberg demonstrated this to be the case for mouse fibroblasts that had been transformed by a known carcinogen: Normal mouse fibroblasts in culture are transformed on transfection with DNA from the transformed cells. Moreover, these newly transformed cells, when inoculated into mice, form tumors. Similar investigations indicate that DNAs from a wide variety of malignant tumors likewise have transforming activity.

What sorts of genetic changes can give rise to cancer? Several types of changes have been observed:

1. Altered Proteins: *An oncogene, as we have seen, may give rise to a protein product with an anomalous activity relative to that of the corresponding proto-oncogene.* This may even result from a simple point mutation. For example, Weinberg, Michael Wigler, and Mariano Barbacid showed

that the *ras* oncogene isolated from a human bladder **carcinoma** (a malignant tumor arising from epithelial tissue) differs from its corresponding proto-oncogene by the mutation of the Gly 12 codon (GGC) to a Val codon (GTC). The resulting amino acid change attenuates the GTPase activity of Ras protein (Section 19-3C), evidently without affecting its ability to stimulate protein phosphorylation, thereby prolonging the time this G-protein remains in the "on" state. Indeed, comparison of the X-ray structures of normal human Ras and its oncogenic counterpart (Gly 12→Val), both in complex with GDP, indicate that the mutation mainly alters the normal protein structure in the vicinity of its GTPase function. Most other *ras* oncogene–activating mutations also change residues close to this site. Ras, which plays a central role in MAP kinase cascades (Fig. 19-38), as might be expected, is one of the most commonly implicated proto-oncogenes in human cancers.

2. Altered Regulatory Sequences: *Malignant transformation can result from the inappropriately high expression of a normal cellular protein.* For example, the proto-oncogene **c-fos,** which encodes the transcription factor **Fos** (which is activated by MAP kinase cascades; Fig. 19-38), differs from the retroviral oncogene **v-fos** mainly in regulatory sequences: v-fos has an efficient enhancer, whereas c-fos has a 67-nucleotide AT-rich segment in its unexpressed 3′-terminal end that, when transcribed, promotes rapid mRNA degradation (Section 34-3C). Thus, c-fos can be converted to an oncogene by deleting its 3′ end and adding the v-fos enhancer.

3. Loss of Degradation Signals: *An oncogene protein that is degraded more slowly than the corresponding normal cellular protein may cause malignant transformation through its consequent inappropriately high concentration in the cell.* For example, the transcription factor **c-Jun** (which is also activated by MAP kinase cascades; Fig. 19-38), but not **v-Jun,** is efficiently multiubiquinated and hence proteolytically degraded by the cell (Section 32-6B). This is because v-Jun lacks a 27-residue segment present in c-Jun that mutagenesis experiments indicate is essential for the efficient ubiquination of c-Jun even though this segment does not contain the protein's principal ubiquitin attachment sites.

4. Chromosomal Rearrangements: *An oncogene may be inappropriately transcribed when brought under the control of a foreign regulatory sequence through chromosomal rearrangement* (a position effect). For example, Carlo Croce found that the human cancer **Burkitt's lymphoma** (a lymphoma is an immune system cell malignancy) is characterized by an exchange of chromosomal segments in which the proto-oncogene **c-myc** is translocated from its normal position at one end of chromosome 8 to the end of chromosome 14 adjacent to certain immunoglobulin genes. The misplaced c-myc gene is thereby brought under the transcriptional control of the highly active (in immune system cells) immunoglobulin regulatory sequences. The consequent overproduction of the normal c-myc gene product **Myc** (a transcription factor that is also activated by MAP kinase cascades and whose transient increase is normally correlated with the onset of cell division), or alternatively, its production at the wrong time in the cell cycle, is apparently a major factor in cell transformation.

5. Gene Amplification: *Oncogene overexpression can also occur when the oncogene is replicated multiple times, either as sequentially repeated chromosomal copies or as extrachromosomal particles.* The amplification of the c-*myc* gene, for example, has been observed in several types of human cancers. Gene amplification is usually an unstable genetic condition that can only be maintained under strong selective pressure such as that conferred by cytotoxic drugs (Section 34-2D). It is not known how oncogene amplification is stably maintained.

6. Viral Insertion into a Chromosome: *Inappropriate oncogene expression may result from the insertion of a viral genome into a cellular chromosome such that the proto-oncogene is brought under the transcriptional control of a viral regulatory sequence.* For instance, **avian leukosis virus,** a retrovirus that lacks an oncogene but that nevertheless induces lymphomas in chickens, has a chromosomal insertion site near c-*myc.* Some DNA tumor viruses also transform cells in this manner.

7. Inappropriate Inactivation or Activation of Chromatin Modification Enzymes: Heterozygotes for a defective *CPB* gene, whose gene product activates the PCAF HAT complex (Section 34-3B), have **Rubinstein–Tabi syndrome,** a condition that predisposes to cancer. In a related example, the **retinoic acid receptor (RAR),** which is important for myeloid (blood-forming) tissue differentiation, helps recruit HDAC complexes such as NCoR and SMRT (Section 34-3B) to **retinoic acid response elements (RAREs),** but on binding ligand, releases them. However, in **promyelocytic leukemia,** a chromosomal translocation yields a defective RAR that binds to RAREs and recruits HDACs but is unresponsive to the presence of retinoids.

8. Loss or Inactivation of Tumor Suppressor Genes: The high incidence of particular cancers in certain families suggests that there are genetic predispositions toward these diseases. A particularly clear-cut example of this phenomenon occurs in **retinoblastoma,** a cancer of the developing retina that therefore afflicts only infants and young children. The offspring of surviving retinoblastoma victims also have a high incidence of this disease, as well as several other types of malignancies. In fact, retinoblastoma is associated with the inheritance of a copy of chromosome 13 from which a particular segment has been deleted. Retinoblastoma develops, as Alfred Knudson first explained, through a somatic mutation in a **retinoblast** (a retinal precursor cell) that alters the same segment of the second, heretofore normal copy of chromosome 13. This is because *the affected chromosomal segment contains a gene, the **Rb gene,** which specifies a factor that restrains uninhibited cell proliferation; that is, the Rb gene product, **pRb,** is a **tumor suppressor** (alternatively, an **anti-oncogene** protein).* Indeed, the *Rb* gene is frequently mutated in diverse types of human cancers. The structure and function of pRb is further discussed below.

Several other tumor suppressors have been characterized including **p53,** which is encoded by the most com-

monly altered gene in human cancers (~50% of cancers contain a mutation in p53 and many other oncogenic mutations occur in genes encoding proteins that directly or indirectly interact with p53; the structure and function of p53 is further discussed in Section 34-4D); **neurofibromatosis type 1 (NF1) protein,** whose defect causes benign tumors of the peripheral nerves, such as those of the famous "Elephant Man" of Victorian England, that occasionally become malignant; **BRCA1,** which forms a portion of a ubiquitin-protein ligase (E3) and whose defect predisposes to breast and ovarian cancers (Section 32-6B); **BRCA2,** a DNA-binding protein that participates in the repair of double-strand breaks and whose defect also predisposes to breast and ovarian cancers; and **PTEN,** an inositol polyphosphate 3-phosphatase, whose structure and function are discussed in Section 19-4E.

Mutations altering normal gene products, causing chromosomal rearrangements and deletions, and perhaps gene amplification can all result from the actions of carcinogens on cellular DNA. Thus, normal cells bear the seeds of their own cancers. To date, over 100 viral and cellular oncogenes and tumor suppressors have been identified.

b. pRb Functions by Binding to Certain Transcription Factors

pRb is a 928-residue DNA-binding protein that is localized in the nucleus of normal retinal cells but is absent in retinoblastoma cells. It is a phosphoprotein that is phosphorylated in a cell cycle–dependent manner, as is discussed in Section 34-4D. Hypophosphorylated forms of pRb form complexes with certain transcription factors, including **E2F,** which regulates the expression of several cellular and viral genes. E2F was first identified as a cellular factor involved in the regulation of the **adenovirus** early *E2* gene by the adenovirus oncogene product **E1A,** although further investigation revealed that the adenovirus *E4* gene product also participates in this process. E1A protein, which does not bind DNA, promotes the dissociation of pRb from E2F by complexing pRb. It thereby frees E2F protein to combine with **E4 protein** on the adenovirus *E2* promoter so as to stimulate the transcription of the *E2* gene. These observations suggest that *the interaction of pRb with E2F and other transcription factors to which it binds plays an important role in the suppression of cellular proliferation* and that the dissociation of this complex is, at least in part, the means by which E1A inactivates pRb function. Thus, *an additional way that oncogenes can cause cancer is by inactivating the products of normal cellular tumor suppressor genes.* We continue our discussion of pRb below.

D. The Regulation of the Cell Cycle

The cell cycle, as we discussed in Section 30-4A, is the sequence of major events that occur during the life of a eukaryotic cell. It is divided into four phases (Fig. 30-38): M phase, during which mitosis and cell division occur; G_1 phase, the main period of cell growth; S phase, during which DNA is synthesized; and G_2 phase, the interval in which the cell prepares for the next M phase. The progression through the cell cycle is regulated by external as well as internal signals. Thus, yeast have a regulatory point known as **START** that occurs late in G_1 beyond which they are committed to enter S phase, that is, replicate their DNA. However, if there are insufficient nutrients available or if the cell has not reached some minimum size, the cell cycle is arrested at START and the cell assumes a resting state until these criteria have been met. Animal cells have a similar decision point in G_1 named the **restriction point,** but it responds mainly to the extracellular presence of the appropriate **mitogens,** protein growth factors that signal the cell to proliferate. The cell cycle also has a series of **checkpoints** that monitor its progress and/or the health of the cell and arrest the cell cycle if certain conditions have not been satisfied. Thus, G_2 has a checkpoint that prevents the initiation of M until all of the cell's DNA has been replicated, thereby ensuring that both daughter cells will receive a full complement of DNA. Similarly, a checkpoint in M prevents mitosis until all chromosomes have properly attached to the mitotic spindle. Checkpoints in G_1 and S, as well as that in G_2, also arrest the cell cycle in response to damaged DNA so as to give the cell time to repair the damage (Section 30-5). In the cells of multicellular organisms, if after a time the checkpoint conditions have not been satisfied, the cell may be directed to commit suicide, a process named **apoptosis** (Section 34-4E), thereby preventing the proliferation of a genetically irreparably damaged and hence dangerous (e.g., cancerous) cell. However, single-celled eukaryotes such as yeast lack such a mechanism, presumably because, in their case, the survival of a genetically damaged cell is preferable in a Darwinian sense to its death.

a. The Activation of Cdk1 Triggers Mitosis

What are the molecular events that drive and coordinate the cell cycle? The first clues to this process came from studies of sea urchin embryos by Tim Hunt, which revealed that a class of proteins named **cyclins** accumulate steadily throughout the cell cycle and then abruptly disappear just before the anaphase portion of mitosis (Fig. 1-19). Homologs of these proteins have since been discovered in all eukaryotic cells examined. Indeed, mammals encode at least 20 different cyclins, many but not all of which participate in cell cycle control or appear and disappear cyclically with the cell cycle. The cyclins form a diverse protein family that are 30 to 50% similar over an ~100-residue segment known as the **cyclin box.**

Further indications as to the way the cell cycle is controlled came from experiments in which human cells at different stages of the cell cycle were fused to yield a single cell with two nuclei. When a cell in G_1 phase was fused with one in S phase, the G_1 nucleus immediately entered S phase, whereas the S nucleus continued replicating its DNA. However, when S-phase and G_2-phase cells were fused, the S nucleus continued replicating its DNA and the G_2 nucleus remained in G_2. Similarly, when G_1 and G_2 cells were fused, the G_1 nucleus entered S phase according to its own schedule and the G_2 nucleus remained in G_2. Evidently, S-phase cells contain a diffusible activator of

DNA replication, only G_1 cells can initiate DNA replication, and cells that have transited S phase are unable to re-replicate their DNA until they have passed through M phase.

Many of the proteins that participate in cell cycle regulation were identified in the 1970s by Lee Hartwell through his study of temperature-sensitive mutants in *S. cerevisiae* (budding yeast) that were defective in cell cycle progression and by similar studies in *Schizosaccharomyces pombe* (fission yeast) by Paul Nurse. In what is perhaps the best characterized portion of the cell cycle, that inducing M phase, **cyclin B** combines with **Cdc2** (Cdc for *c*ell *d*ivision *c*ycle), whose sequence clearly indicates that it is a member of the Ser/Thr protein kinase family (Section 19-3C) and which is highly conserved from yeasts to humans. *It is Cdc2 that is the central cell cycle regulator in species ranging from yeasts to humans. It does so by phosphorylating a variety of nuclear proteins, among them histone H1, several oncogene proteins (see below), and proteins involved in nuclear disassembly, cytoskeletal rearrangement, spindle assembly, chromosome condensation, and Golgi fragmentation. This initiates a cascade of cellular events that culminates in mitosis.*

The binding of cyclin B to Cdc2 forms an activated complex that is alternatively called **cyclin-dependent protein kinase 1 (Cdk1)** and **maturation promoting factor (MPF).** This, however, is by no means the entire activation story. Cdc2 is a phosphoprotein that can be phosphorylated on Tyr 15, Thr 161, and in higher eukaryotes, Thr 14. Cdk1 is active only when both Thr 14 and Tyr 15, which occupy the region of its ATP-binding site, are dephosphorylated and when Thr 161 is phosphorylated. Moreover, the phosphorylation of Tyr 15 requires that cyclin B be present. Thus, mitosis is triggered through the following series of events (Fig. 34-99):

1. The cell enters G_1 with cyclin B absent and with Cdc2, which is present at a constant level throughout the cell cycle, dephosphorylated. **Cdk-activating kinase (CAK)** then phosphorylates Cdc2's Thr 161. Curiously, CAK, which consists of the heterodimer **Cdk7−cyclin H,** is also a component of the general transcription factor TFIIH (Section 34-3B), where its function is unknown.

2. In S phase, newly synthesized cyclin B binds to Cdc2, whereupon its Tyr 15 is phosphorylated by **Wee1** (so named because, in fission yeast, its inactivation causes cells to enter mitosis prematurely and hence at an unusually small, that is, wee, size) and Thr 14 is phosphorylated by the Wee1 homolog **Myt1,** which can also phosphorylate Tyr 15. The resulting triply phosphorylated cyclin B−Cdc2 complex is enzymatically inactive because Thr 14 and Tyr 15 prevent Cdc2 from binding ATP. Thus, the entire system appears designed to maintain Cdc2 in an inactive state while cyclin B gradually accumulates during S phase.

3. At the cell cycle's G_2/M boundary, Cdc2's Thr 14 and Tyr 15 are rapidly and specifically dephosphorylated by **Cdc25C,** a dual-specificity protein tyrosine phosphatase (PTP; Section 19-3F) that thereby activates Cdk1, which in turn triggers mitosis (M phase). This process is initiated by the activating phosphorylation of Cdc25C by **plk1** (for

FIGURE 34-99   Regulation of Cdk1 in the animal cell cycle. Details are described in the text. [After Norbury, C. and Nurse, P., *Annu. Rev. Biochem.* **61,** 451 (1992).]

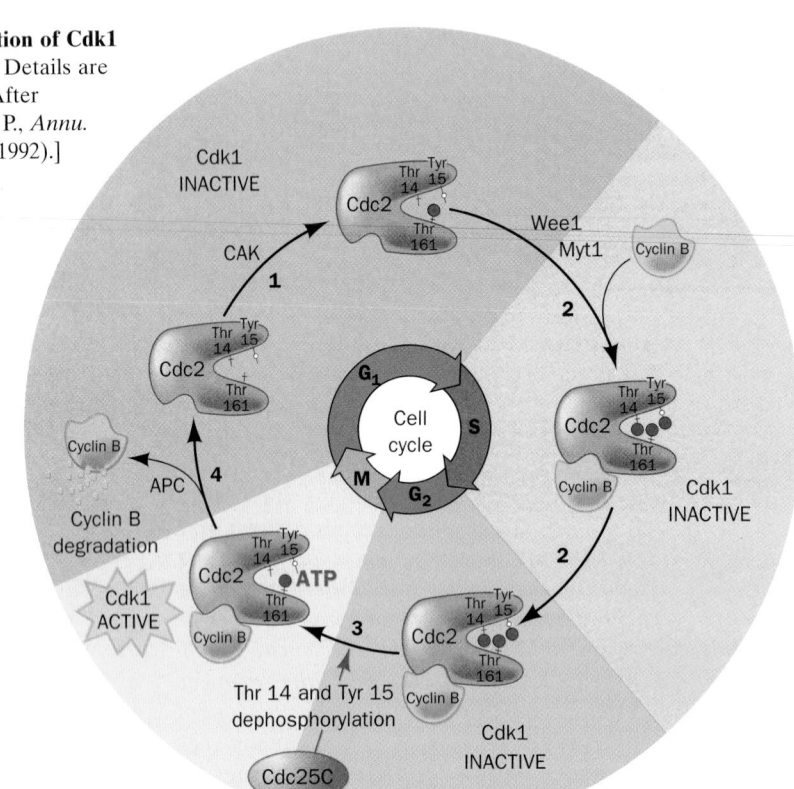

polo-like kinase 1; so named because it is a homolog of *Drosophila* **polo kinase**). The resulting activated Cdk1 also activates Cdc25C so that, through the intermediacy of Cdc25C, Cdk1 activates itself in a rapid burst. Moreover, Cdk1 inhibits its own inactivation by phosphorylating and thereby inactivating Wee1 and Myt1.

4. Cyclin B is quickly proteolyzed by the proteosome in a ubiquitin-mediated pathway whose E3 is a multisubunit complex known as the **anaphase-promoting complex (APC;** Section 32-6B), followed by the rapid dephosphorylation of Cdc2 Thr 161. This inactivates Cdk1, thereby returning the now divided cell to G_1. APC, which is inactive during S and G_2 phases, is activated, at least in part, by Cdk1, which thereby brings about the destruction of its own cyclin B component.

Other combinations of Cdks and cyclins that similarly mediate specific portions of the cell cycle include the following (Fig. 34-100): **Cdk4** and its close isoform **Cdk6,** which in complex with D-type cyclins **(cyclins D1, D2,** and **D3)** drive events in G_1; **Cdk2** and **cyclin E,** which are required for progression through the G_1/S boundary and for the initiation of DNA synthesis; and Cdk2 and **cyclin A,** which control passage through S phase. Thus, *Cdk–cyclin complexes form the engines that drive the various processes of the cell cycle as well as the clocks that time them.*

b. The X-Ray Structure of Cdk2 Resembles That of PKA

Cdk2, which is closely similar to Cdc2, is activated by the binding of a cyclin A, which is closely similar to cyclin B,

followed by the CAK-catalyzed phosphorylation of Thr 160. Cdk2 is also negatively regulated by the phosphorylation of Tyr 15 and, to a lesser extent, the adjacent Thr 14.

The X-ray structure of Cdk2 in complex with ATP (Fig. 34-101*a*), determined by Sung-Hou Kim, indicates that this

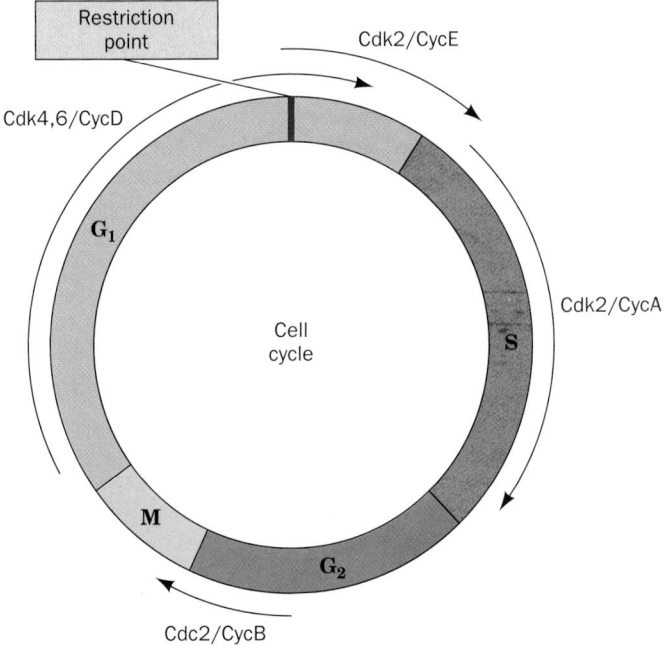

FIGURE 34-100   Complexes of cyclin-dependent kinases (Cdks) and cyclins (Cycs) that mediate passage through specific segments of the cell cycle.

(a)

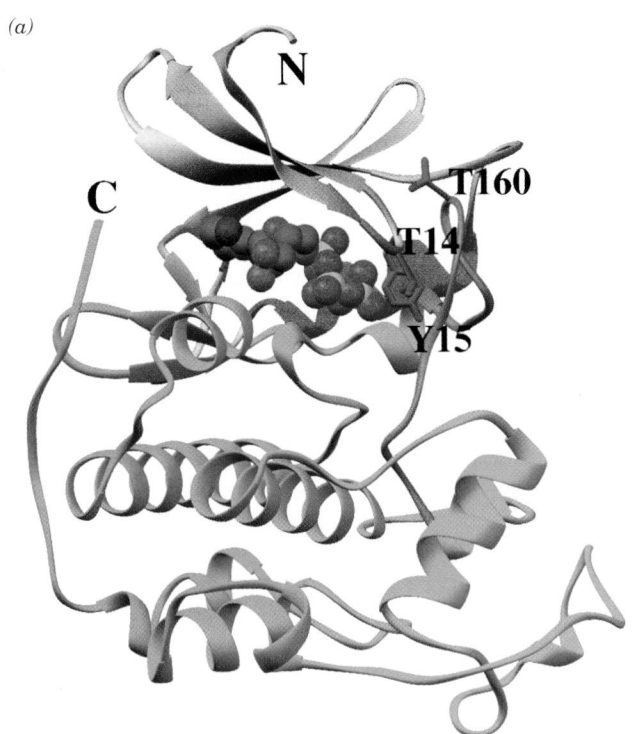

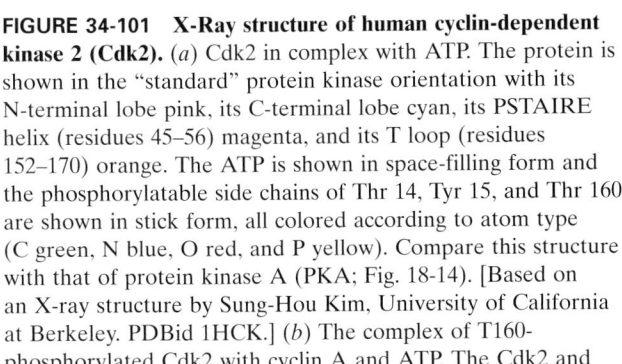

(b)

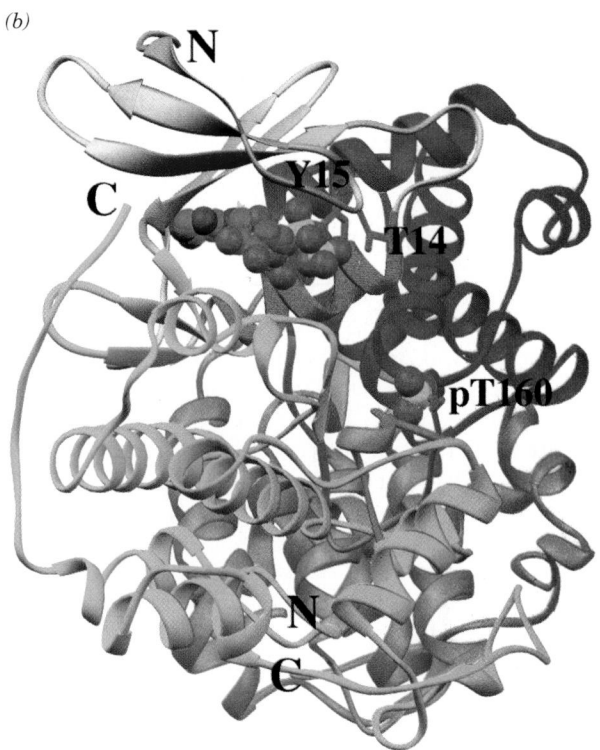

FIGURE 34-101   X-Ray structure of human cyclin-dependent kinase 2 (Cdk2). (*a*) Cdk2 in complex with ATP. The protein is shown in the "standard" protein kinase orientation with its N-terminal lobe pink, its C-terminal lobe cyan, its PSTAIRE helix (residues 45–56) magenta, and its T loop (residues 152–170) orange. The ATP is shown in space-filling form and the phosphorylatable side chains of Thr 14, Tyr 15, and Thr 160 are shown in stick form, all colored according to atom type (C green, N blue, O red, and P yellow). Compare this structure with that of protein kinase A (PKA; Fig. 18-14). [Based on an X-ray structure by Sung-Hou Kim, University of California at Berkeley. PDBid 1HCK.] (*b*) The complex of T160-phosphorylated Cdk2 with cyclin A and ATP. The Cdk2 and

ATP are represented as in Part *a* and viewed similarly. The cyclin A is colored yellow-green with its cyclin box (residues 206–306) dark green. The Cdk2 phosphoThr 160 phosphoryl group is drawn in space-filling form. Note how the binding of cyclin A together with the phosphorylation of Cdc2 Thr 160 has caused a major structural reorganization of the T loop together with significant conformational adjustments of the Cdk2 N-terminal lobe, including its PSTAIRE helix. Also note the different conformations of the ATP triphosphate group in the two structures. [Based on an X-ray structure by Nikola Pavletich, Memorial Sloan-Kettering Cancer Center, New York, New York. PDBid 1JST.]

298-residue monomeric Ser/Thr protein kinase closely resembles the catalytic subunit of protein kinase A (PKA; Section 18-3C), a protein whose sequence is 24% identical to that of Cdk2. However, there are functionally significant structural differences between these two kinases:

1. The relative arrangement of ATP's β- and γ-phosphate groups in Cdk2 is likely to greatly reduce the reactivity of the γ-phosphate relative to that in the PKA–ATP complex (stereoelectronic control), thereby rationalizing, in part, why Cdk2 alone is catalytically inactive, whereas the catalytic subunit of PKA alone is catalytically active.

2. Access to the γ-phosphate of Cdk2's bound ATP by its protein substrates appears to be blocked by a 19-residue protein loop (residues 152–170) that has been named the "T loop" because it contains Thr 160.

The X-ray structure also explains why the phosphorylation of Thr 14 inactivates Cdk2: The hydroxyl group of this side chain is close to the ATP's γ-phosphate so that phospho-

rylation of Thr 14 is likely to disrupt the conformation of the ATP's phosphate groups. It is unclear, however, how the phosphorylation of Tyr 15 affects Cdk2 activity.

c. Cyclin A Binding and Thr 160 Phosphorylation Conformationally Reorganize Cdk2

The X-ray structure of human Cdk2 that is phosphorylated at its Thr 160 in complex with ATP and the C-terminal portion of human cyclin A (its residues 173–432), determined by Pavletich, indicates that cyclin A binds to Cdk2's "back" side (Fig. 34-101*b*). There it interacts with both lobes of the Cdk2 to form an extensive and continuous protein–protein interface. Cyclin A consists mainly of a bundle of 12 α helices (and no β sheets) in which its cyclin box forms helices 2 to 6. Interestingly, helices 7 to 11 form a bundle that is nearly superimposable on that of the cyclin box even though these two motifs exhibit little sequence similarity. Comparison of the X-ray structures of free cyclin A with that in its complex with Cdk2 indicates that cyclin A does not undergo significant conformational change on binding Cdk2. In contrast, cyclin A binding causes Cdk2 to undergo significant conformational shifts in the region around its catalytic cleft. In particular, the N-terminal α helix of Cdk2, which contains the PSTAIRE sequence motif characteristic of the Cdk family, rotates about its axis by 90° and moves several angstroms into the catalytic cleft relative to its position in free Cdk2, where it contacts the cyclin box segment of cyclin A. This movement brings Glu 51 (the E in PSTAIRE) from its solvent-exposed position outside the catalytic cleft of free Cdk2 to a position inside the catalytic cleft, where it forms a salt bridge with Lys 33, which in free Cdk2 instead forms a salt bridge to Asp 145. These three side chains (Lys 33, Glu 51, and Asp 145), which are conserved in all eukaryotic protein kinases, participate in ATP phosphate coordination and Mg^{2+} ion coordination. Their conformational reorientation on cyclin A binding apparently places them in a catalytically active arrangement.

The binding of cyclin A also induces Cdk2's T loop to undergo extensive conformational reorganization involving positional shifts of up to 21 Å, such that the T loop, which now also contacts the cyclin box, adopts a backbone conformation which closely resembles that of the analogous region of the catalytically active PKA. These movements greatly increase the access of a protein substrate to the ATP bound in the catalytic cleft, which has assumed a more reactive conformation than that in free Cdk2. The phosphate group on Thr 160 fits snugly into a positively charged pocket composed of three Arg residues that forms, in part, on cyclin A binding. Indeed, comparison of this structure with that in which Thr 160 is unphosphorylated indicates that Thr 160 phosphorylation induces activating conformational changes in the catalytic cleft of the cyclin A–Cdk2 complex as well as contributing to the reorganization of the T loop.

d. Cdk Inhibitors Function to Arrest the Cell Cycle

In addition to their control by phosphorylation/ dephosphorylation and by the binding of the appropriate cyclin, *Cdk activities are regulated by **cyclin-dependent kinase inhibitors (CKIs)**, which induce cell cycle arrest in response to such antiproliferative signals as contact with other cells, DNA damage, terminal differentiation, and senescence (in which cell cycle arrest is permanent).* The known CKIs have been grouped, according to their sequence and functional similarities, into two families: (1) the **Kip/Cip family** (*k*inase *i*nteracting *p*rotein/*c*ytokine-*i*nducible *p*rotein), whose members inhibit most Cdk–cyclin complexes (but not Cdk4/6–cyclin D) and can bind to isolated Cdks or cyclins, although with lower affinity than for Cdk–cyclin complexes; and (2) the **INK4 family** (*i*nhibitor of Cd*k4* and Cdk6), whose members specifically inhibit Cdk4 and Cdk6 (which, together with cyclin D, mediate the cell's progression through G_1; Fig. 34-100) and can bind to either the isolated Cdk or its complex with cyclin D. The importance of CKIs is indicated by their frequent alterations in cancer. For example, p16[INK4a] is mutated in about one-third of all human cancers, p21[Cip1] arrests the cell cycle on behalf of p53 (see below), and **p27[Kip1]**

(also known as **p27[Cip2]**) may be degraded in several types of cancers and its low levels are correlated with a poor clinical prognosis. Moreover, certain herpes viruses, including **Kaposi's sarcoma-associated herpes virus,** express an oncogenic D-type cyclin known as **K-cyclin** that binds to and thereby activates Cdk4/6, thus contributing to the deregulation of the cell cycle.

The members of the Kip/Cip family have homologous N-terminal ~65-residue segments that are necessary and sufficient to bind to and inhibit Cdk–cyclin complexes, whereas their C-terminal segments are divergent in length, sequence, and function. p27[Kip1] is an inhibitor of Cdk2 that occurs in cells that have been treated with the antimitotic protein named **transforming growth factor β (TGFβ).** The X-ray structure of the N-terminal inhibitory domain (residues 25–93) of the 198-residue human p27[Kip1] bound to the human cyclin A–Cdk2 complex in which Cdk2 is phosphorylated at Thr 160 was determined by Pavletich (Fig. 34-102). The p27 inhibitory domain is draped across both Cdk2 and cyclin A, where it assumes an extended conformation that does not have a hydrophobic core of its own and whose secondary structural elements do not interact with each other. p27's N-terminal end interacts with cyclin A, whose conformation is essentially unaffected by this interaction. In contrast, the binding of the p27 inhibitory domain's C-terminal segment to Cdk2 causes extensive conformational changes in Cdk2 that appear likely to destabilize its binding of ATP. More importantly, this C-terminal segment extends into Cdk2's active site cleft, where its conserved Tyr 88 side chain mimics the binding

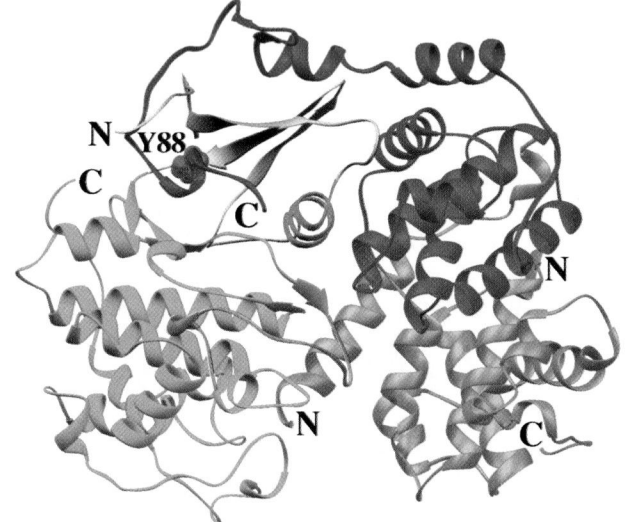

FIGURE 34-102 X-Ray structure of the Cdk2–Cyclin A–p27[Kip1] ternary complex. The Cdk2 and cyclin A are colored as in Fig. 34-101 and p27[Kip1] is blue. The complex is viewed along the Cdk2 PSTAIRE helix (approximately from the right of Fig. 34-101). The p27[Kip1] Tyr 88 side chain, which is drawn in space-filling form with C green and O red, occupies the binding site for the adenine moiety of ATP. [Based on an X-ray structure by Nikola Pavletich, Memorial Sloan-Kettering Cancer Center, New York, New York. PDBid 1JSU.]

of ATP's adenine moiety in both its position and the contacts it makes to the active site groups, thereby eliminating any possibility of ATP binding. Last, p27's N-terminus occupies the peptide-binding groove on cyclin A's conserved cyclin box that is probably a docking site for a number of the cyclin A–Cdk2 complex's tight-binding substrates, thereby reducing the ability of substrate binding to reverse the effects of p27-induced conformational changes.

INK4 proteins can bind to monomeric Cdk4/6 so as to prevent its association with cyclin D or bind to a Cdk4/6–cyclin D binary complex to form a catalytically inactive ternary complex. *In vivo,* INK4–Cdk4/6 binary complexes are more abundant than INK4–Cdk4/6–cyclin D ternary complexes, which suggests that INK4 binding increases the rate of cyclin dissociation from the ternary complex. The X-ray structure of the ternary complex of human p18INK4c, human Cdk6, and K-cyclin (Fig. 34-103), determined by Pavletich, reveals that p18INK4c binds to the Cdk6 –K-cyclin complex in an entirely different manner from the way p27Kip1 binds to the Cdk2–cyclin A complex (Fig. 34-102). The 160-residue p18INK4c, which consists of five ankyrin repeats (which participate in protein–protein interactions; Section 12-3E), binds to the 301-residue Cdk6 in the region of its ATP-binding site, where it interacts with Cdk6's N- and C-terminal lobes via the second and third ankyrin repeats of p18INK4c. This rotates the N-terminal lobe by 13° with respect to the C-terminal lobe relative to their orientations in the Cdk2–cyclin A complex (Fig. 34-101*b*) and thereby distorts the Cdk6 ATP-binding site and misaligns its catalytic residues. Moreover, p18INK4c binding distorts Cdk6's cyclin binding site such that their interface is reduced in area by ~30%: The cyclin box of

the 253-residue K-cyclin binds to the N-terminal lobe of Cdk6 centered on its PSTAIRE sequence motif (which has the sequence PLSTIRE) but, unlike in the structure of the Cdk2–cyclin A complex (Figs. 34-101*b* and 34-102), there are no significant contacts between the Cdk6 C-terminal lobe and K-cyclin. Apparently, INK4 binding reduces the stability of the Cdk–cyclin interface.

e. Cell Cycle Arrest at the G₂ Checkpoint Is Mediated by a Phosphorylation/Dephosphorylation Cascade

How does failure to satisfy a checkpoint cause cell cycle arrest? For the G_2 checkpoint, the process is initiated, as diagrammed in Fig. 34-104, by at least six poorly characterized **sensor proteins** that have been identified through mutations in the genes *rad1, rad3, rad9, rad17, rad26,* and *hus1* that are defective in repair and replication checkpoints. The sensor proteins bind to damaged or unreplicated DNA, which causes them to activate two related large (~3000 residue) protein kinases known as **ATM** and **ATR** [ATM for *ataxia telangiectasia mutated* (**ataxia telangiectasia** is a rare genetic disease characterized by a progressive loss of motor control, growth retardation, immune system deficiencies, premature aging, and a greatly increased risk of cancer); ATR for *ATM and Rad3-related* (**Rad3** is the ATR homolog in *S. pombe*)]. Activated ATM and ATR respectively phosphorylate and thereby activate **Chk2** and **Chk1** (Chk for *checkpoint kinase*). These latter protein kinases phosphorylate Cdc25C which, it will be recalled, functions to activate Cdk1 by dephosphorylating Cdc2's Thr 14 and Tyr 15. This phosphorylation, which occurs on Ser 216 of Cdc25C, does not inactivate this protein phosphatase. Rather, it provides a binding site for members of the **14-3-3 family** of adaptor proteins (which bind certain phosphorylated motifs in a wide variety of phosphorylated proteins; the name "14-3-3" is based on these proteins' column fractionation and electrophoretic mobility properties), and the resulting

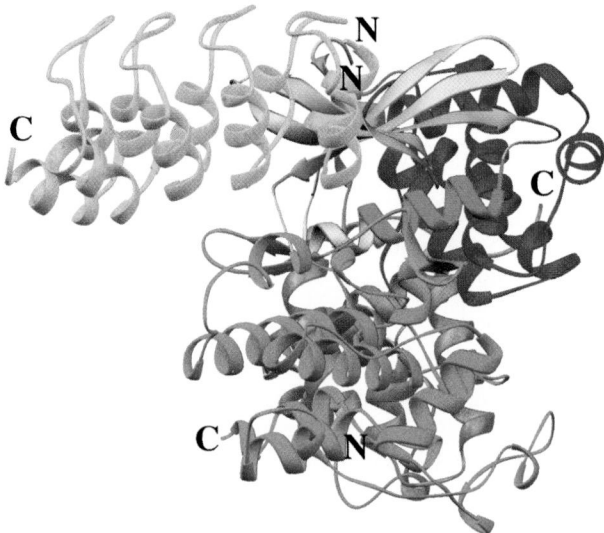

FIGURE 34-103   X-Ray structure of the Cdk6–K-cyclin–p18INK4c ternary complex. The Cdk6 and K-cyclin are colored as in the homologous Cdk2 and cyclin A in Figs. 34-101 and 34-102 and the p18INK4c is gold. The structure is viewed with the Cdk6 in the "standard" protein kinase orientation as in Fig. 34-101. [Based on an X-ray structure by Nikola Pavletich, Memorial Sloan-Kettering Cancer Center, New York, New York. PDBid 1G3N.]

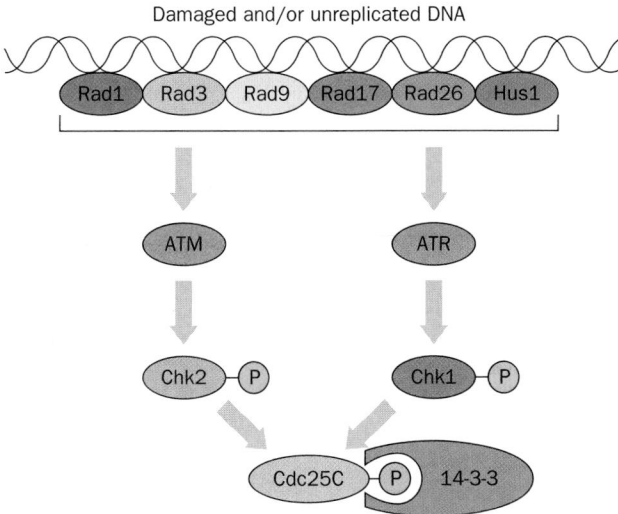

FIGURE 34-104   The G₂ checkpoint phosphorylation/dephosphorylation cascade that results in cell cycle arrest. Details are given in the text.

complex is sequestered in the cytoplasm out of contact with the Cdk1 in the nucleus. Since Cdk1 is the protein kinase that activates mitosis, the cell remains in G_2 until its DNA is repaired and/or fully replicated.

f. p53 Is a Transcriptional Activator That Arrests the Cell Cycle in G_2

The idea that p53 is a tumor suppressor first arose from the discovery that germ line mutations in the *p53* gene often occur in individuals with the rare inherited condition known as **Li–Fraumeni syndrome** that renders them highly susceptible to a variety of malignant tumors, particularly breast cancer, which they often develop before their 30th birthdays. That p53 is indeed a tumor suppressor has been clearly demonstrated in mice in which the *p53* gene has been inactivated. These knockout mice (Section 5-5H) appear to be developmentally normal but spontaneously develop a variety of cancers by the age of 6 months. Indeed, p53 functions as a "molecular policeman" in monitoring genome integrity: On the detection of DNA damage, p53 arrests the cell cycle until the damage is repaired, or failing that, induces apoptosis.

Despite the central role of p53 in preventing tumor formation, the way it does so has only gradually come to light. This tumor suppressor is specifically bound in humans by the homolog of the mouse **Mdm2 protein.** The *mdm2* gene is the dominant transforming oncogene present on *mouse* *double* *minute* chromosomes (amplified extrachromosomal segments of DNA; Section 34-2E). Mdm2 protein is a ubiquitin-protein ligase (E3) that specifically ubiquitinates p53, thereby marking it for proteolytic degradation by the proteasome. Consequently, the amplification of the *mdm2* locus, which occurs in >35% of human **sarcomas** (none of which have a mutated *p53* gene; sarcomas are malignancies of connective tissues such as muscle, tendon, and bone), results in an increased rate of degradation of p53, thereby predisposing cells to malignant transformation. Similarly, as we have seen (Section 32-6B), E6 protein from human papilloma virus, which causes the great majority of cervical cancers, functions to ubiquitinate p53. Certain DNA tumor virus oncoproteins, such as SV40 **large T antigen** and adenovirus **E1B protein,** inactivate p53 by specifically binding to it. Thus, *an additional way that oncogenes can cause cancer is by inactivating normal tumor suppressors.*

p53 protein is an efficient transcriptional activator. Indeed, all point mutated forms of p53 that are implicated in cancer have lost their sequence-specific DNA-binding properties. But then, how does p53 function as a tumor suppressor? A clue to this riddle came from the observation that the treatment of cells with DNA-damaging ionizing radiation induces the accumulation of normal p53. This led to the discovery that both ATM and Chk2 phosphorylate p53, which prevents its binding by Mdm2 and hence increases the otherwise low level of this protein in the nucleus. *Although p53 does not initiate cell cycle arrest in G_2, its presence is required to prolong this process. It does so by activating the transcription of the gene encoding the CKI p21^{Cip1}, which binds to several Cdk–cyclin complexes so as to inhibit both the G_1/S and G_2/M transitions.* p21^{Cip1}

also binds to PCNA, the homotrimeric sliding clamp in DNA replication (Section 30-4B), so as to prevent its participation in DNA replication but not in DNA repair. Thus, p21^{Cip1} has a dual role in cell cycle arrest in that it both blocks cell cycle progression and inhibits DNA replication in S-phase cells.

p53 also induces the transcription of the gene encoding the 14-3-3 family member **14-3-3σ,** which binds to Cdk1, thereby confining it to the cytoplasm. Moreover, the 14-3-3σ–Cdk1 complex binds the protein kinase Wee1 (which, as discussed above, inactivates Cdk1 by phosphorylating its Cdc2 component at its Tyr 15), thereby ensuring that Cdk1 remains in its inactive state. Thus, the disruption of the gene encoding 14-3-3σ is fatal for cells that sustain DNA damage. Chk2 also phosphorylates Wee1, which inhibits its proteasomal degradation. Consequently, germ line mutations in the *chk2* gene are also associated with Li–Fraumeni syndrome. Excessive levels of p53 are toxic, which explains why loss of the *mdm2* gene in mice is lethal unless the *p53* gene is also knocked out. In the absence of p53 activation, cells control the level of p53 through a feedback loop in which p53 stimulates the transcription of the *mdm2* gene.

Cells that are irreparably damaged are induced by p53 to commit suicide via apoptosis (Section 34-4E), thereby preventing the proliferation of potentially cancerous cells. p53 does so by transactivating the expression of several of the proteins that participate in apoptosis (Section 34-4E) and repressing the expression of others that inhibit this process.

g. The X-Ray Structure of p53 Explains Its Oncogenic Mutations

p53 is a tetramer of identical 393-residue subunits. Each subunit consists of four domains: an N-terminal transactivation domain (residues 1–99), a sequence-specific, DNA-binding core domain (residues 100–300, which binds two half-site decamers, each with the consensus palindromic sequence $RRRC_T^A{}_A^TGYYY$, that are separated by 0–13 nt and with a p53 dimer binding to each such decamer), a tetramerization domain (residues 301–356), and a nonspecific DNA-binding domain (residues 357–393, which binds a wide variety of DNAs including short single strands, irradiated DNA, Holliday junctions, and insertions/deletions). Although the entire protein has so far resisted crystallization, Pavletich has determined the X-ray structure of the DNA-binding core (residues 102–313) in complex with a 21-bp DNA segment containing its 5-bp target sequence (AGACT). *The vast majority of the >1000 p53 mutations that have been found in human tumors occur in this core.*

The structure of the p53 DNA-binding core domain (Fig. 34-105) contains a sandwich of two antiparallel β pleated sheets, one with four strands and the other with five, and a loop–sheet–helix motif that packs against one edge of the β sandwich. This edge of the β sandwich also contains two large loops running between the two β sheets that are held together through their tetrahedral coordination of a Zn^{2+} ion via one His and three Cys side chains.

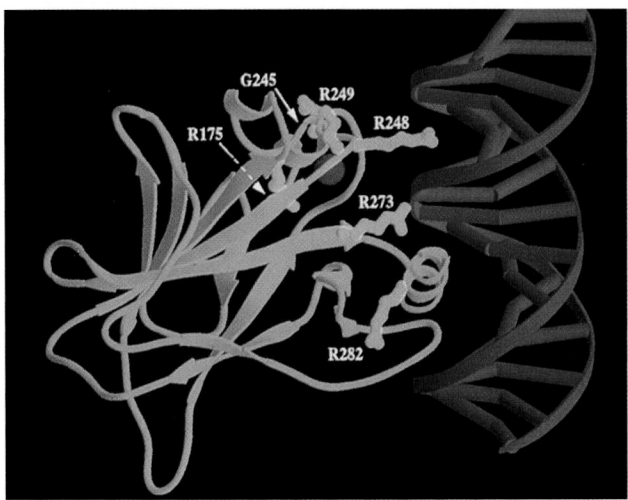

FIGURE 34-105   X-Ray structure of the DNA-binding domain of human p53 in complex with its target DNA. The protein is shown in ribbon form (*cyan*), the DNA in ladder form with its bases represented by cylinders (*blue*), the tetrahedrally liganded Zn^{2+} ion is shown as a red sphere, and the side chains of the six most frequently mutated residues in human tumors are shown in stick form (*yellow*) and identified with their one-letter codes. [Courtesy of Nikola Pavletich, Memorial Sloan-Kettering Cancer Center, New York. PDBid 1TSR.]

The p53 DNA binding motif does not resemble any other that has previously been characterized. The helix and loop from the loop—sheet—helix motif are inserted in the DNA's major groove, where they make sequence-specific contacts with the bases (lower right of Fig. 34-105). One of the large loops provides a side chain (Arg 248) that fits in the minor groove (upper right of Fig. 34-105). The protein also contacts the DNA backbone between the major and minor grooves in this region (notably with Arg 273).

The structure's most striking feature is that *its DNA-binding motif consists of conserved regions comprising the most frequently mutated residues in the p53 variants found in tumors.* Among them are one Gly and five Arg residues (highlighted in yellow in Fig. 34-105) whose mutations collectively account for over 40% of the *p53* variants in tumors. The two most frequently mutated residues, Arg 248 and Arg 273, as we saw, directly contact the DNA. The other four "mutational hotspot" residues appear to play a critical role in structurally stabilizing p53's DNA-binding surface. The relatively sparse secondary structure in the polypeptide segments forming this surface (one helix and three loops) accounts for this high mutational sensitivity: Its structural integrity mostly relies on specific side chain—side chain and side chain—backbone interactions.

h. p53 Is a Sensor That Integrates Information from Several Pathways

p53 may be activated by several other pathways. For example, aberrant growth signals, such as those generated by oncogenic variants of MAP kinase cascade components such as Ras, stimulate the expression of a variety of transcription factors (Fig. 19-38), many of which are proto-oncogene products. One of them, **Myc**, activates the transcription of the gene encoding **p19ARF** (in mice; **p14ARF** in humans), which also encodes **p16INK4a**. This is because these two proteins, which have no sequence similarity, are expressed through alternative splicing of their first exons and share second and third exons that are translated in different reading frames for the two proteins (ARF for *a*lternative *r*eading *f*rame protein), an unprecedented economy of genomic resources in higher eukaryotes, although a common phenomenon in bacteriophage (Section 32-1D). p19ARF binds to Mdm2 and inhibits its activity, thereby preventing the degradation of p53 and hence triggering the p53-dependent transcriptional programs leading to cell cycle arrest as well as apoptosis (Section 34-4E). Evidently, p19ARF acts as part of a p53-dependent fail-safe system to counteract hyperproliferative signals.

A third activation pathway for p53 is induced by a wide variety of DNA-damaging chemotherapeutic agents, protein kinase inhibitors, and UV radiation. These activate ATR to phosphorylate p53 so as to reduce its affinity for Mdm2 in much the same way as do ATM and Chk2.

p53 is also subject to a rich variety of reversible post-translational modifications that markedly influence the expression of its target genes. These include acetylation at several Lys residues, glycosylation, ribosylation, and sumoylation (Section 32-6B), as well as phosphorylation at multiple Ser/Thr residues and ubiquitination. p53 does not bind to short DNA fragments that contain its target sites unless its C-terminal domain has either been deleted or modified by phosphorylation and/or acetylation. Yet NMR evidence indicates that the C-terminal domain does not interact with other p53 domains. This suggests that p53's C-terminal and core domains compete for DNA binding unless the C-terminal domain has been modified.

p53, as we have only glimpsed, is the recipient of a vast number of intracellular signals and, in turn, controls the activities of a large number of downstream regulators. One way to understand the operation of this highly complex and interconnected network is in analogy with the Internet. In the Internet (cell), a small number of highly connected servers or hubs ("master" proteins) transmit information to/from a large number of computers or nodes (other proteins) that directly interact with only a few other nodes (proteins). In such a network, overall performance is largely unperturbed by the inactivation of one of the nodes (other proteins). However, the inactivation of a hub ("master" protein) will greatly impact system performance. p53 is a "master" protein, that is, it is analogous to a hub. Inactivation of one of the many proteins that influences its performance or one of the many proteins whose activity it influences usually has little effect on cellular events due to the system's redundant and highly interconnected components. However, the inactivation of p53 or several of its most closely associated proteins (e.g., Mdm2) disrupts the cell's responses to DNA damage and tumor-predisposing stresses. Nevertheless, a quantitative understanding of the functions and malfunctions of the p53-based network will require a complete description of all the proteins with which p53 directly or indirectly interacts and how they do so under the conditions present in the cell, something that we are far from having. Thus, for the foreseeable future, we will be limited to qualitative descriptions of the functioning of the p53 network. Our understanding of other cellular signal transduction systems is similarly vague (Section 19-4F).

i. pRb Regulates the Cell Cycle's G₁/S Transition

The tumor suppressor pRb **(retinoblastoma-associated protein),** a 928-residue monomer that is localized in the nucleus of normal animal cells but is defective or absent in retinoblastoma cells (Section 34-4C), is an important regulator of the cell cycle's G_1/S transition, that is, the cell's passage through the restriction point. The effects of pRb are largely manifested through its interactions with the members of the **E2F** family of transcription factors, which, in the absence of pRb, activate their target promoters in complex with a member of the **DP** (for E2F *dimerization partner*) family. The mammalian E2F family consists of six ~440-residue members, of which **E2F-1** through **E2F-4** interact with pRb via a conserved 18-residue polypeptide segment contained in their ~70-residue, C-terminal, transactivation domains. The mammalian DP family consists of two ~430-residue members, **DP-1** and **DP-2.** The E2F–DP heterodimers induce the transcription of a variety of genes that encode proteins required for S-phase entry (e.g., Cdc2, Cdk2, and cyclins A and E) and for DNA synthesis [e.g., DNA polymerase α (pol α; Section 30-4B), Orc1 and several Mcm proteins (which participate in initiating DNA replication; Section 30-4B), ribonucleotide reductase (Section 28-3A), thymidylate synthase, and dihydrofolate reductase (Section 28-3B)].

How does pRb mediate cell cycle progression? pRb, which is synthesized throughout the cell cycle, is a phosphoprotein that is phosphorylated at as many as 16 of its Ser/Thr residues by Cdk4/6–cyclin D in mid to late G_1, by Cdk2–cyclin E in late G_1, by Cdk2–cyclin A in S, and by Cdk1 (Cdc2–cyclin B) in M, with different Cdk–cyclin complexes phosphorylating different sets of sites on pRb. *Hypophosphorylated but not hyperphosphorylated pRb binds to the transactivation domain of E2F so as to prevent it from activating transcription at the promoter to which it is bound.* In nonproliferating cells (those in early G_1), pRb remains hypophosphorylated because, unless such cells receive mitogenic signals, the highly unstable D-type cyclins (they have half-lives of ~10 min) do not accumulate to levels sufficient to generate significant amounts of Cdk4/6–cyclin D (mitogens trigger MAP kinase cascades that stimulate the expression of D-type cyclins). Moreover, since hypophosphorylated pRb prevents E2F–DP from activating the expression of Cdk2 and cyclins E and A, Cdk2–cyclin E and Cdk2–cyclin A do not accumulate to sufficient levels to hyperphosphorylate pRb. In fact, the small amounts of Cdk2–cyclin A and Cdk2–cyclin E that are present are inhibited by p27^Kip1, which occurs in high levels in pre-restriction point G_1 cells.

Mitogens also activate the expression of p27^Kip1 and p21^Cip1 which, contrary to what might be expected, do not inhibit Cdk4/6–cyclin D complexes but instead stimulate

their activities by enhancing their assembly and promoting their nuclear import. Thus, mitogenic signals break the pRb-imposed blockade to cell cycle progression by inducing the formation of Cdk4/6–cyclin D–p27^Kip1/p21^Cip1 complexes that begin the phosphorylation of pRb. This releases a small amount of E2F, which thereupon induces the expression of Cdk2 and cyclins E and A. The Cdk4/6–cyclin D complexes also sequester p27^Kip1 and p21^Cip1, which permits the resulting Cdk2–cyclin E complex to catalyze a second wave of pRb phosphorylation and import [although when large amounts of p21^Cip1 are produced through the influence of activated p53 (see above), it inhibits Cdk2–cyclin E so as to arrest the G_1/S transition]. This frees large amounts of E2F, resulting in a surge in the transcription of the genes that promote cell cycle progression. As the cell cycle continues, pRb is increasingly phosphorylated, first by Cdk2–cyclin A and then by Cdk1, until the exit from M phase, whereupon pRb is abruptly dephosphorylated, probably by the protein Ser/Thr phosphatase PP1 (Section 19-3F), permitting pRb to again arrest cell cycle progression by inhibiting E2F.

A variety of proteins that contain the LXCXE sequence motif bind to pRb. These comprise several cellular proteins, including the D-type cyclins, which may thereby be directed to pRb, and certain viral oncoproteins, whose binding to pRb prevents it from binding E2F (which lacks an LXCXE motif but, as we saw above, binds pRb via an 18-residue sequence). Indeed, E2F was first identified (and named) as a cellular factor involved in the regulation of the adenovirus early *E2* gene by the adenovirus oncogene product **E1A,** although further investigations revealed that the adenovirus *E4* gene product also participates in this process. E1A, which does not bind DNA, binds, via its LXCXE motif, to pRb. This causes the pRb to release its bound E2F, which permits the E2F, in combination with **E4,** to activate the transcription of the *E2* gene from the adenovirus *E2* promoter. The freed E2F also drives the infected cell into S phase, which facilitates adenovirus DNA replication. The SV40 large T antigen and the human papilloma virus **E7** protein, which also contain LXCXE motifs, similarly activate E2F. Over 100 proteins have been reported to bind to pRb, although in most cases by other means than via LXCXE motifs. The functions of these interactions are largely unknown.

j. E2F and the LXCXE Motif Bind to Separate Sites on the pRb Pocket Domain

pRb's so-called **pocket domain** forms the binding site for both E2F and the LXCXE motif and is the major site of genetic alterations in tumors. The pocket domain consists of its conserved A- and B-boxes (residues 379–572 and 646–772) linked by a poorly conserved spacer. However, when the spacer is excised, the A- and B-boxes nevertheless associate noncovalently.

The X-ray structure of the pRb pocket domain lacking its spacer in complex with the 18-residue pRb-binding peptide of E2F was independently determined by Marmorstein and Gamblin and by Yunje Cho. It reveals that the 18-residue E2F peptide binds in a boomerang-shaped con-

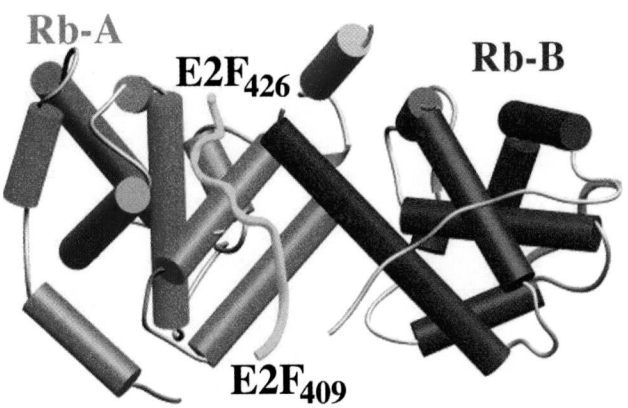

FIGURE 34-106 X-Ray structure of the pRb pocket domain in complex with the 18-residue pRb-binding peptide of E2F. The helices of the A- and B-boxes are respectively drawn as red and blue cylinders and the main chain of the E2F peptide is shown as a gold worm. The superimposed structure of an LXCXE-containing nonapeptide segment of the human papilloma E7 protein in its complex with the pRb pocket domain is represented by a green worm. [Courtesy of Steven Gamblin, National Institute for Medical Research, London, U.K. PDBids 1O9K and 1GUX.]

formation at the highly conserved interface between the A- and B-boxes, both of which contain the five-helix cyclin fold (Fig. 34-106). However, the X-ray structure of the pRb pocket domain in complex with the 9-residue LXCXE-containing peptide from human papilloma virus E7 protein, determined by Pavletich, indicates that the E7 LXCXE peptide binds, in an extended conformation, in a shallow groove on the B-box that is ~30 Å distant from the E2F-binding site (Fig. 34-106). This latter binding site, which is formed by highly conserved residues, closely resembles the primary Cdk2 binding site of cyclin A (Fig. 34-101b) and the TBP binding site of the 20% identical TFIIB (Fig. 34-49). The corresponding portion of the A-box participates in forming the A–B interface.

k. pRb Also Represses Transcription by Recruiting HDACs and SWI/SNF Homologs

Binding experiments reveal that pRb associates with the histone deacetylases (HDACs; Section 34-3B) HDAC1 and HDAC2, both of which contain an LXCXE sequence motif (actually IXCXE). Consequently, the presence of human papilloma virus E7 protein or mutations that disrupt the pRb pocket domain abolish the binding of these HDACs to pRb. These observations suggest that pRb also functions to recruit HDAC1 and HDAC2 to DNA-bound E2F so as to facilitate the histone deacetylation and hence the transcriptional inactivation of the chromatin containing E2F's target genes. This explains the observation that pRb can repress transcription from promoters to which it is artificially linked by a DNA-binding domain that differs from that of E2F. HDAC3 also associates with pRb, although it lacks an LXCXE motif.

The human SWI/SNF homologs BRM and BRG1, which both have LXCXE motifs, also bind to pRb. BRM

and BRG1, it will be recalled (Section 34-3B), are DNA-dependent ATPases that are components of chromatin-remodeling complexes. Thus, the observation that pRb can simultaneously bind BRG1 and an HDAC (despite both having LXCXE motifs) suggests that pRb recruits chromatin remodeling complexes to facilitate the action of HDACs at E2F promoters.

E. *Apoptosis: Programmed Cell Death*

The maxim that death is part of life is even more appropriate on the cellular level than it is on the organismal level. **Programmed cell death** or **apoptosis** (Greek: falling off, as leaves from a tree), which was first described by John Kerr in the late 1960s, is a normal part of development as well as the maintenance and defense of the adult animal body. For example, in the nematode worm *Caenorhabditis elegans,* a transparent organism whose cell lineages have been microscopically elucidated, precisely 131 of its 1090 somatic cells undergo apoptosis in forming the normal adult body. In many vertebrates, the digits of the developing hands and feet are initially connected by webbing that is eliminated by programmed cell death (Fig. 34-107), as are the tails of tadpoles and the larval tissues of insects during their metamorphoses into adults (Figs. 34-81 and 34-86). Apoptosis is particularly prevalent in the developing mammalian nervous system in which an approximately threefold excess of neurons is produced. However, only those neurons that make adequate synaptic connections are retained; the remainder are eliminated via apoptosis (Fig. 34-108).

In the adult human body, which consists of nearly 10^{14} cells, an estimated 10^{11} cells are eliminated each day through programmed cell death (which closely matches the number of new cells produced by mitosis). Indeed, the mass of the cells that we annually lose in this manner approaches that of our entire body. A particularly obvious manifestation of this phenomenon is the monthly sloughing off of the uterine lining in menstruation (Section 19-1I). Similarly, the immune system cells known as *T* lymphocytes (*T* cells) undergo apoptosis in the thymus if the *T* cell receptors they produce recognize antigens that are normally present in the body or are improperly formed (Sections 35-2A and 35-2D); ~95% of immature *T* cells are eliminated in this way. Autoimmune diseases such as rheumatoid arthritis and insulin-dependent diabetes (Section 27-3B) arise when this process goes awry. Apoptosis is also an essential part of the body's defense systems. The immune system eliminates virus-infected cells, in part, by inducing them to undergo apoptosis, thereby preventing viral replication. Cells with irreparably damaged DNA and hence at risk for malignant transformation undergo apoptosis, thereby protecting the entire organism from cancer. Indeed, *one of the defining characteristics of malignant cells is their ability to evade apoptosis.* Cells that become detached from their normal positions in the body likewise commit suicide. Indeed, as Martin Raff pointed out, *apoptosis appears to be the default option for metazoan cells: Unless they continually receive external*

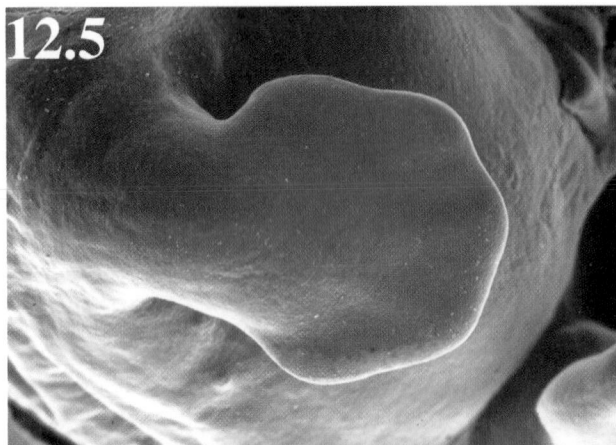

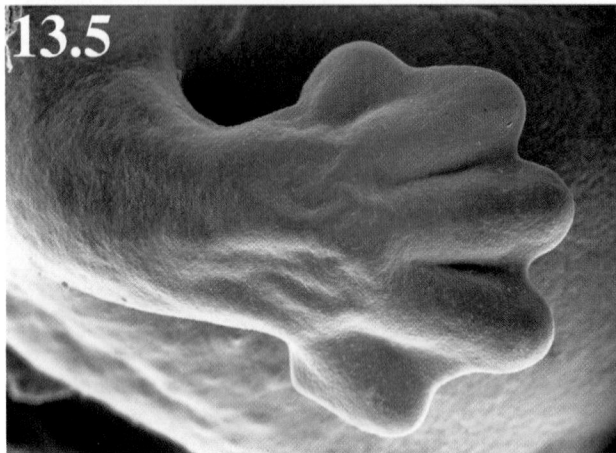

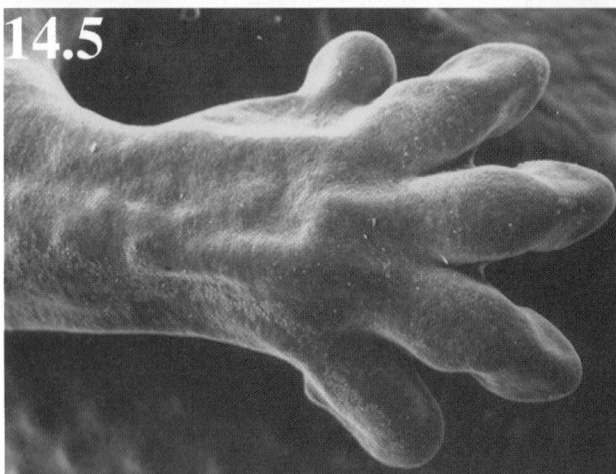

FIGURE 34-107   Programmed cell death in the embryonic mouse paw. At day 12.5 of development, its digits are fully connected by webbing. At day 13.5, the webbing has begun to die. By day 14.5, this apoptotic process is complete. [Courtesy of Paul Martin, University College of London, U.K.]

(a)   *(b)*

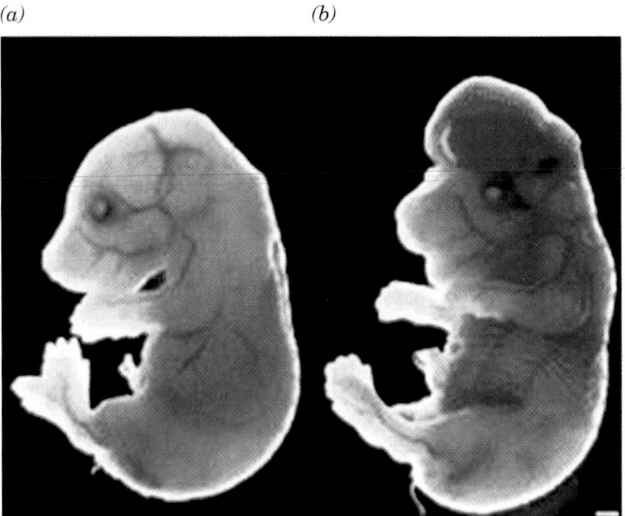

FIGURE 34-108   Brain development in 16.5-day-old mouse embryos. (*a*) A wild-type embryo. (*b*) An embryo in which **caspase-9,** an enzyme that mediates apoptosis (see below) has been knocked out. Note the protruding and morphologically abnormal brain in the knockout embryo due to the overproliferation of brain neurons. [Courtesy of Richard Flavell, Yale University Medical School.]

Alzheimer's disease (Section 9-5B), Parkinson's disease (Section 26-4B), and Huntington's disease (Section 30-7), as well as much of the damage caused by stroke and heart attacks. Consequently, the signaling systems that mediate apoptosis have become targets for therapeutic intervention. In fact, many of the chemotherapeutic agents in present use do not kill their target cancer cells outright but, rather, damage them so as to induce their apoptosis.

Apoptosis is qualitatively different from **necrosis,** the type of cell death caused by trauma (e.g., lack of oxygen, extremes of temperature, and mechanical injury). Cells undergoing necrosis essentially explode: They and their membrane-enclosed organelles swell as water rushes in through their compromised membranes, releasing lytic enzymes that digest the cell contents until the cell lyses, spilling its contents into the surrounding region (Section 22-4B). The cytokines that the cell releases often induce an inflammatory response (which can damage surrounding cells) that attracts **phagocytotic cells** (cells, such as the white blood cells known as **macrophages,** that ingest foreign particles and waste matter) to "mop up" the resulting cell debris. In contrast, apoptosis begins with the loss of intercellular contacts by an apparently healthy cell followed by its shrinkage, the condensation of its chromatin at the nuclear periphery, the collapse of its cytoskeleton, the dissolution of its nuclear envelope, the fragmentation of its DNA, and violent blebbing (blistering) of its plasma membrane. Eventually, the cell disintegrates into numerous membrane-enclosed **apoptotic bodies** that are phagocytosed by neighboring cells as well as by roving macrophages with-

hormonal and/or neuronal signals not to commit suicide, they will do so. Thus, adult organs maintain their constant size by balancing cell proliferation with apoptosis. Not surprisingly, therefore, inappropriate apoptosis has been implicated in several neurodegenerative diseases including

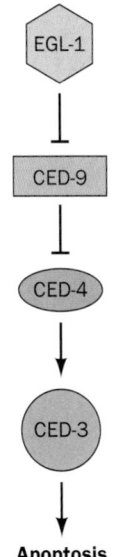

FIGURE 34-109 The pathway initiating apoptosis in
C. elegans. Arrows indicate activation and blunted lines
indicate inhibition.

out spilling the cell contents and hence not inducing an inflammatory response.

a. Apoptosis Is Induced by Signaling Cascades

The pathway for apoptosis, which was first elucidated
in *C. elegans* through genetic studies by John Sulston and
Robert Horvitz, involves three so-called *ced* (for *cell death
abnormal*) gene products (Fig. 34-109): **CED-4** protein, a
protease, activates the protease **CED-3,** which then initiates the destruction of the cell; **CED-9** functions to inactivate CED-4. In fact, mutations that inactivate CED-9 result in numerous embryonic cells that would normally
survive in the adult organism to inappropriately activate
its CED-4 and CED-3 and hence die, thereby killing the
embryo. Conversely, if CED-9 is expressed at abnormally
high levels or CED-3 or CED-4 is inactivated, cells that
normally die will survive (which, curiously, has little ap-

parent effect on the health of the adult organism). Later
investigations revealed that a fourth protein, **EGL-1,** functions to inhibit CED-9 and hence its overexpression
induces apoptosis.

Apoptotic pathways in mammals are considerably more
complex than that in *C. elegans.* Nevertheless, the above
CED proteins and EGL-1 all have counterparts in mammalian pathways:

1. CED-3 is the prototype of a family of proteases
known as **caspases** (for *c*ysteinyl *asp*artate-specific
prote*ases*) because they are **cysteine proteases** [whose
mechanism resembles that of serine proteases (Section 15-
3C) but with Cys replacing the active site Ser] that cleave
after an Asp residue. Their target cleavage sites are specified mainly by this Asp and its three preceding residues.

2. CED-4 is a scaffolding protein that plays an essential
role in caspase activation. Its mammalian counterpart is
called **Apaf-1** (for *a*poptotic *p*rotease-*a*ctivating *f*actor-*1*).

3. CED-9 is a member of the **Bcl-2** family (so named
because its founding member, Bcl-2, was initially characterized as a gene involved in *B* cell *l*ymphoma). Some of
the numerous members of this family, including CED-9,
protect cells from death and hence are said to be **anti-
apoptotic.** Others promote cell death and are therefore
said to be **pro-apoptotic.**

4. EGL-1 is a pro-apoptotic member of the Bcl-2
family.

b. Caspases Have Closely Similar Structures

Caspases are $\alpha_2\beta_2$ heterotetramers that consist of two
large α subunits (~300 residues) and two small β subunits
(~100 residues). They are expressed as zymogens (**procaspases**) that have three domains (Fig. 34-110): an N-
terminal prodomain that is proteolytically excised on
activation, followed by sequences comprising the active enzyme's α and β subunits that are proteolytically separated
on activation. The activating cleavage sites all follow Asp
residues and are, in fact, targets for caspases (the only
other eukaryotic protease known to cleave after an Asp

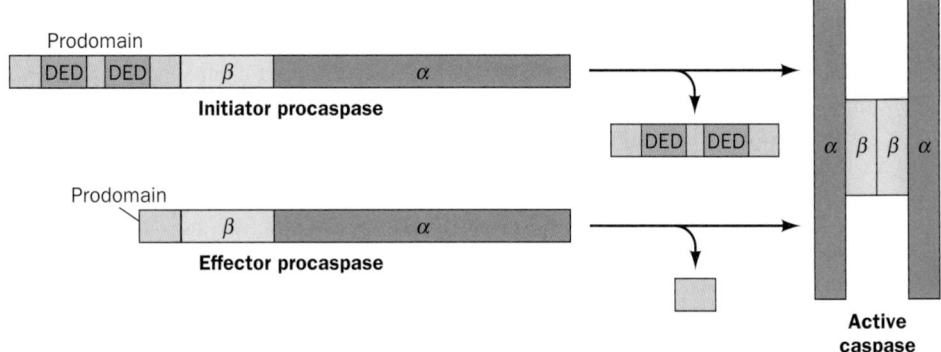

FIGURE 34-110 Caspase domain structure and activation.
The zymogens of initiator caspases have long N-terminal
prodomains, which in several cases contain two death
effector domains (DEDs), whereas the zymogens of effector
caspases have only short prodomains. Procaspases are
activated by proteolytic cleavages that excise their
prodomains and separate their α and β subunits to yield the
active $\alpha_2\beta_2$ caspases.

residue is **granzyme B,** a chymotrypsin-like serine protease expressed by cytotoxic *T* cells that functions to induce apoptosis in tumor and virally infected cells; Section 35-2A). Thus, caspase activation, as we shall see, may either be autocatalytic or be catalyzed by another caspase.

Humans express eleven caspases, six of which participate exclusively in apoptosis [with the remainder being involved mainly in the cytokine activation and hence the control of inflammation; the founding member, **caspase-1,** is also known as **interleukin-1β-converting enzyme (ICE)** because it proteolytically activates the cytokine **interleukin-1β** (Section 35-2A)]. There are two classes of apoptotic caspases (Fig. 34-110):

1. Initiator caspases (caspases-8, -9, and **-10)** are characterized by long prodomains (129–219 residues) that target their zymogens to scaffolding proteins that promote their autoactivation. The prodomains of caspases-8 and -10 each contain two ~80-residue **death effector domains (DEDs),** through which they bind to DEDs on their target adaptor proteins (see below). The prodomain of caspase-9 instead contains the structurally similar ~90-residue **caspase recruitment domain (CARD)** that promotes the interaction of this caspase with certain scaffolding and regulatory proteins.

2. Effector caspases (caspases -3, -6, and **-7)** have short (~25 residue) prodomains and are activated by initiator caspases. The activated effector caspases, which have been described as the cell's executioners, cleave a wide variety of cellular proteins (see below), thereby bringing about apoptosis.

The X-ray structure of caspase-7 (Fig. 34-111), determined by Keith Wilson and Paul Charifson, closely resembles those of the several other caspases of known X-ray structures. Each αβ heterodimer of this twofold symmetric $\alpha_2\beta_2$ heterotetramer contains a six-stranded β sheet, five of whose strands are parallel, and which is flanked by five α helices, two on one side and three on the other, that are approximately parallel to the β strands. The β sheet is continued across the protein's 2-fold axis to form a twisted 12-stranded β sheet. Each αβ heterodimer contains an active site that is located at the C-terminal ends of its parallel β strands and which recognizes a tetrapeptide on the N-terminal side of its Asp–X cleavage site. The structures of the various caspases differ mainly in the conformations of the four loops forming their active sites. Comparison of the X-ray structure of caspase-7 with that of **procaspase-7** (with its active site Cys 186 mutated to Ser to prevent its autoactivation), independently determined by Weigon Shi and Wolfram Bode, reveals that, although the two proteins are otherwise closely superimposable, the four active site loops in procaspase-7 have undergone large conformational changes relative to those in caspase-7 that essentially obliterates the active site. In particular, the loop containing both the catalytically essential Cys residue and the activating cleavage site between the α and β subunits changes its orientation by 90° after this cleavage so as to expose

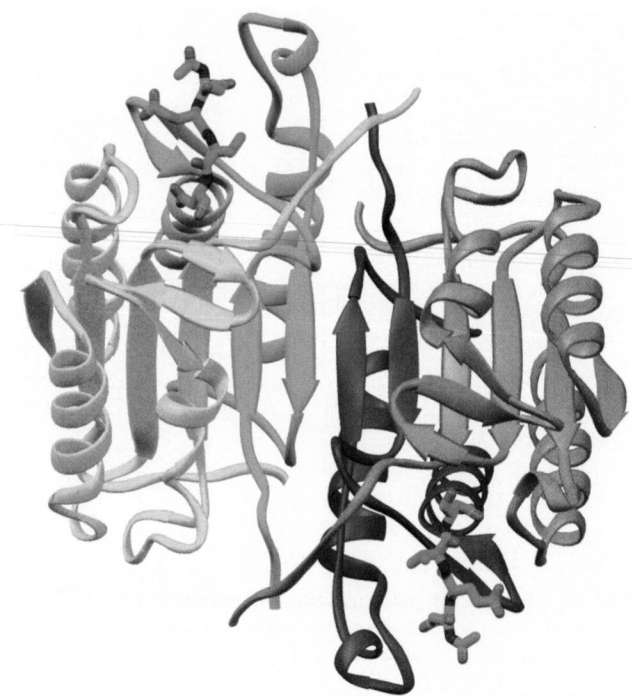

FIGURE 34-111   X-Ray structure of caspase-7 in complex with the tetrapeptide aldehyde inhibitor acetyl-Asp-Glu-Val-Asp-CHO (Ac-DEVD-CHO). The $\alpha_2\beta_2$ heterotetrameric enzyme is viewed along its 2-fold axis with its large (α) subunits orange and gold and its small (β) subunits cyan and light blue. The Ac-DEVD-CHO inhibitor is drawn in stick form with C green, N blue, and O red. [Based on an X-ray structure by Keith Wilson and Paul Charifson, Vertex Pharmaceuticals, Cambridge, Massachusetts. PDBid 1F1J.]

and properly position the previously buried catalytic Cys residue.

c. Caspases Cleave a Wide Variety of Proteins and Activate the Degradation of Chromosomal DNA

Over 60 cellular proteins have been identified as caspase substrates. These include cytoskeletal proteins [e.g., actins (Section 35-3E) and **lamins** (intermediate filaments that form the meshwork lining the inner nuclear envelope)], proteins involved in cell cycle regulation (e.g., cyclin A, Wee1, p21^{Cip1}, ATM, and pRb; Section 34-4D), proteins that participate in DNA replication [e.g., topoisomerase I (Section 29-3C) and Mcm3 (Section 30-4B)], transcription factors (e.g., Sp-1 and NF-κB; Section 34-3B), and proteins that participate in signal transduction [e.g., RasGAP (Section 19-3C) and protein kinase C (Section 19-4C)]. Nevertheless, how the cleavage of these numerous proteins causes the morphological changes that cells undergo during apoptosis is unclear.

The induction of apoptosis also causes the rapid degradation of chromosomal DNA. Chromosomal DNA is attached to the chromosomal protein matrix at intervals of ~70 kb via AT-rich matrix-associated regions (MARs; Section 34-1D). During apoptosis, **caspase-activated**

DNase **(CAD)** cleaves the chromosomal DNA at these sites, which is often followed by its cleavage between nucleosomes to yield a series of DNA fragments that differ in their lengths by increments of ~200 bp. CAD is ubiquitously expressed in all tissues in complex with its inhibitor **ICAD** (*i*nhibitor of *CAD*), which on induction of apoptosis is cleaved by caspases-3 and -7, thereby releasing active CAD. ICAD also functions as a chaperone that must be present when CAD is being ribosomally synthesized in order for CAD to fold to its native conformation. This ensures that native CAD can only form in complex with ICAD and hence prevents inappropriate DNA cleavage. Although the cleavage of a cell's chromosomal DNA would certainly cause its death, cells containing mutant ICAD undergo apoptosis even though their chromosomal DNA remains intact. This suggests that DNA cleavage during apoptosis functions to prevent the cells that have phagocytosed apoptotic bodies from being transformed by the intact viral or damaged chromosomal DNA that the apoptotic bodies might otherwise contain.

d. The Death-Inducing Signal Complex Activates Apoptosis

Apoptosis in a given cell may be induced either by externally supplied signals in the so-called **extrinsic pathway** *(death by commission) or by the absence of external signals that inhibit apoptosis in the so-called* **intrinsic pathway** *(death by omission).* The extrinsic pathway is initiated by the association of a cell destined to undergo apoptosis with a cell that has selected it to so. In what is perhaps the best characterized such pathway (Fig. 34-112), a 281-residue, single-pass, transmembrane protein named **Fas ligand (FasL)** that projects from the plasma membrane of the inducing cell, a so-called **death ligand,** binds to a 335-residue, single-pass, transmembrane protein known as **Fas** (alternatively, **CD95** and **Apo1**) that projects from the plasma membrane of the apoptotic cell, a so-called **death receptor.** FasL is a cytokine that is predominantly expressed by certain immune system cells, including activated *T* cells (although the association between the apoptotic cell and the immune system cell is mainly mediated by antigen-containing complexes; Section 35-2E); it is a member of the **tumor necrosis factor (TNF)** family (so named because its founding member, **TNFα,** was originally characterized as a cytokine that kills tumor cells—but by inducing their apoptosis, not their necrosis).

FasL is a homotrimeric protein, whose extracellular C-terminal domains associate with the extracellular N-terminal domains of three Fas molecules to form a 3-fold symmetric complex, thereby causing the Fas cytoplasmic domains to trimerize. This is the triggering event of the extrinsic pathway; it can also be induced by cross-linking Fas molecules using antibodies. Fas, which is abundantly expressed in a variety of tissues, is a member of the **TNF receptor (TNFR)** family. Consequently, the arrangement of the FasL–Fas complex almost certainly resembles that observed in the X-ray structure of the **TNFβ** trimer in its complex with the extracellular domains from three **TNF**

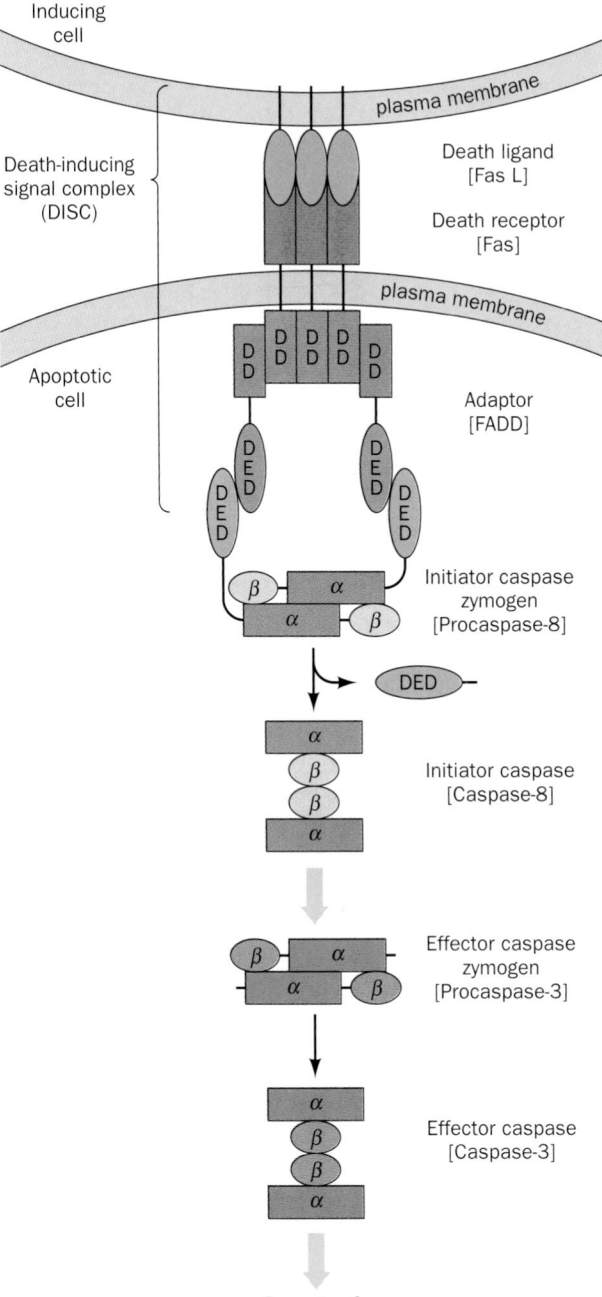

FIGURE 34-112 The extrinsic pathway of apoptosis. Flat arrows indicate activation. The binding of a trimeric death ligand (e.g., FasL) on the inducing cell to the death receptor (e.g., Fas) on the apoptotic cell causes the death receptor's cytoplasmic death domains (DDs) to trimerize. This recruits adaptors (e.g., FADD), which bind via their DDs to the DDs of the death receptor. The adaptors, in turn, recruit initiator procaspases (e.g., procaspase-8) via the interactions between the death effector domains (DEDs) on the adaptors and the initiator procaspases, which induces the autoactivation of the initiator procaspases to form the corresponding heterotetrameric initiator caspases (e.g., caspase-8). The initiator caspases then proteolytically activate effector procaspases (e.g., procaspase-3) to yield the heterotetrameric effector caspases (e.g., caspase-8), which catalyze the proteolytic cleavages, resulting in apoptosis.

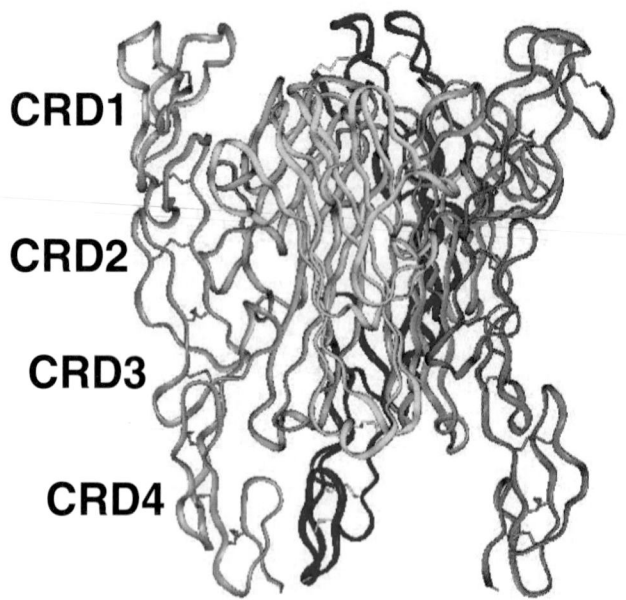

FIGURE 34-113 X-Ray structure of the TNFβ homotrimer in complex with the extracellular domains of three TNFR1 molecules. The centrally located TNFβ subunits are orange, yellow, and green, and the peripherally located TNFR1 domains are red, blue, and magenta. The TNFR1 domains each consist of four ~40-residue pseudorepeats known as **cysteine-rich domains (CRDs)** that each contain three disulfide bonds formed by six Cys residues that are drawn here in stick form (*gray*). The elongated TNFR1 domains each bind at an interface between two TNFβ subunits. This complex, whose formation also induces apoptosis, presumably resembles that between the extracellular domains of the homologous proteins FasL and Fas (whose extracellular domain contains only three CRDs). [Courtesy of Stephen Fesik, Abbott Laboratories, Abbott Park, Illinois. Based on an X-ray structure by David Banner, F. Hoffmann-La Roche Ltd., Basel, Switzerland. PDBid 1TNR.]

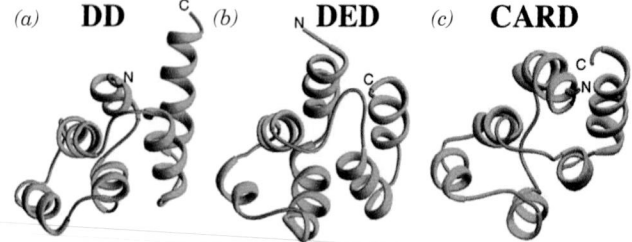

FIGURE 34-114 NMR structures of modules that transduce the death signal. (*a*) The death domain (DD) from Fas. (*b*) The death effector domain (DED) from FADD. (*c*) The caspase recruitment domain (CARD) from **RAIDD**, an adapter protein that is similar to FADD. Each of these domains consists of a bundle of six antiparallel α helices that associates with a domain of the same type but not with one of a different type. Nevertheless, the similarities of their structures and functions suggests that these domains are distantly related. [Courtesy of Stephen Fesik, Abbott Laboratories, Abbott Park, Illinois. Part *c* based on an NMR structure by Gerhard Wagner, Harvard University. PDBids 1DDF, 1A1Z, and 3CRD.]

receptor 1 **(TNFR1)** molecules that was determined by David Banner (Fig. 34-113). The cytoplasmic C-terminal domain of Fas consists mostly of an ~80-residue **death domain (DD)** that occurs in all of the six known mammalian death receptors (one of which is TNFR1), each of which are TNFR family members. The DD consists of six antiparallel, amphipathic α helices that have an unusual arrangement and whose structure resembles those of the death effector domain (DED) and the caspase recruitment domain (CARD), as Fig. 34-114 indicates.

Trimerized Fas recruits three molecules of the 208-residue adaptor protein known as **FADD** (for *Fas-associating death domain-containing protein*; alternatively **MORT1** for *mediator of receptor-induced toxicity 1*) via interactions between FADD's C-terminal DD and that on Fas. The remaining portion of FADD consists almost entirely of a DED that, in turn, recruits procaspases-8 and -10 via the DEDs in their prodomains (Fig. 34-110) to form the **death-inducing signal complex (DISC).** The consequent clustering of the procaspases-8 and -10 molecules results in their proteolytic autoactivation, yielding cas-

pases-8 and -10. These initiator caspases, in turn, activate the effector (executioner) caspase, caspase-3, whose actions cause the cell to undergo apoptosis.

Cells also express a protein named **c-FLIP** [for *cellular FLICE inhibitory protein*; **FLICE** (for *FADD-like ICE*) is an alternative name for procaspase-8] that resembles caspase-8 but is catalytically inactive. It associates with FADD via its two DEDs and thereby inhibits the autoactivation of caspases-8 and -10. FLIP apparently functions to dampen the cell's response to Fas so as to prevent inappropriate apoptosis. Certain herpes viruses and poxviruses encode **v-FLIPs** that function similarly to c-FLIPs to prevent apoptosis, thereby permitting the virus to propagate in the infected cell.

e. The Intrinsic Pathway Is Controlled by Bcl-2 Family Proteins

Most metazoan cells are continuously bathed in an extracellular soup, generated in part by neighboring cells, that contains a wide variety of cytokines that regulate the cell's growth, differentiation, activity, and survival. The withdrawal of this chemical support for its survival or the loss of direct cell–cell interactions induces a cell to undergo apoptosis via the intrinsic pathway. The initial step of this pathway (Fig. 34-115) appears to be the activation of one or more of the cell's several pro-apoptotic Bcl-2 family members.

The 15 known members of the ~180-residue mammalian Bcl-2 family have been classified into three groups (Fig. 34-116):

1. Group I members, which include Bcl-2 and **Bcl-x$_L$,** all have four short regions of homology, **BH1** to **BH4** (BH for *Bcl-2 homology region*), and a C-terminal hydrophobic segment that inserts into the outer mitochondrial membrane, or less frequently the endoplasmic reticulum, such that the bulk of these proteins face the cytosol. All Group I Bcl-2 family members are anti-apoptotic.

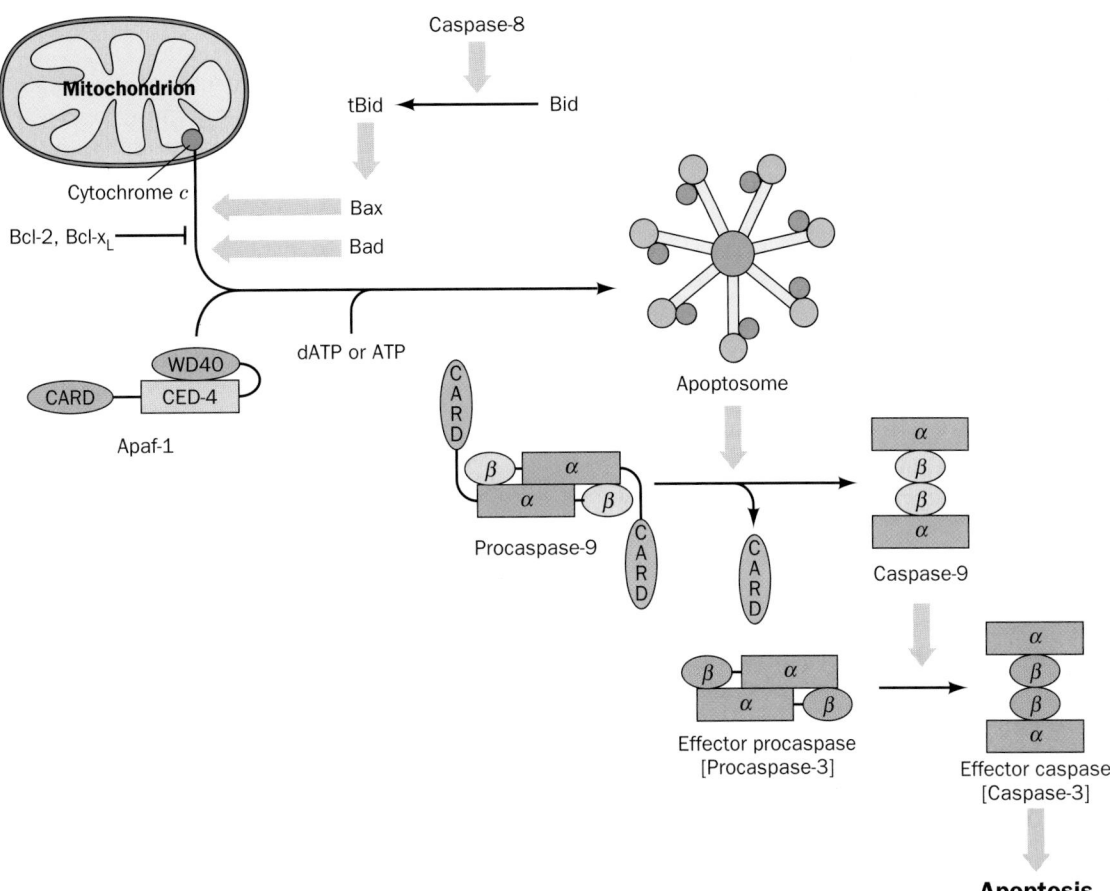

Figure 34-115 The intrinsic pathway of apoptosis. Flat arrows indicate activation and a blunted line indicates inhibition. A variety of stimuli or lack of them causes the mitochondrion to release cytochrome *c* from its intermembrane space. This process is induced by activated pro-apoptotic Bcl-2 family members such as Bid after it has been proteolytically cleaved by caspase-8 to yield tBid or dephosphorylated Bad; the process is inhibited by anti-apoptotic Bcl-2 family members such as Bcl-2 and Bcl-x_L. The liberated cytochrome c binds to Apaf-1, which on additionally binding dATP or ATP, forms the wheel-shaped heptameric apoptosome. The apoptosome binds procaspase-9, which it activates to cleave initiator procaspases (e.g., procaspase-3) to yield the corresponding effector caspases (e.g., caspase-3), which catalyze the proteolytic cleavages, resulting in apoptosis.

2. Group II members, which include **Bax** and **Bak,** resemble Group I proteins but lack a BH4 region. All Group II members are pro-apoptotic.

3. Group III members, which include **Bad, Bid, Bik, Bim,** and **Blk** (and the *C. elegans* protein EGL-1), all possess only one BH region, the ~15-residue BH3, and have

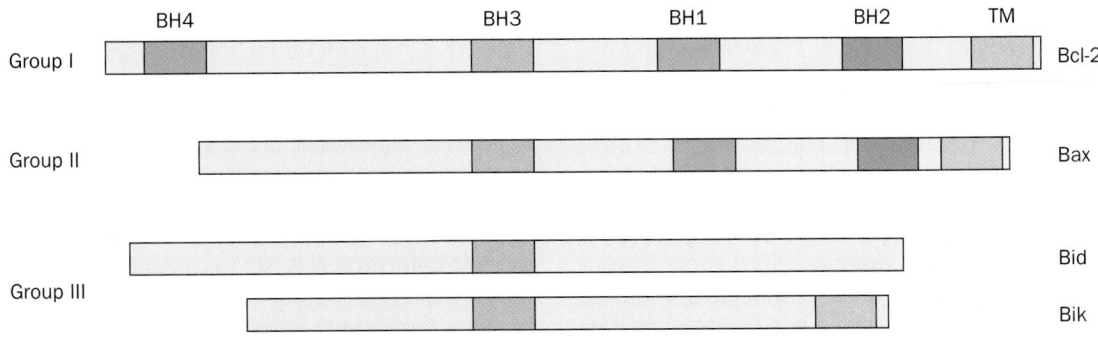

FIGURE 34-116 Sequence comparisons of members of the Bcl-2 family of proteins. The BH1 through BH4 homology regions are blue, purple, red, and green, respectively, and the hydrophobic transmembrane (TM) region is yellow. Group I proteins are anti-apoptotic, whereas Group II and Group III proteins are pro-apoptotic. [After a drawing by Michael Hengartner, *Nature* **407,** 770 (2000).]

no other sequence resemblance to Bcl-2. These so-called **BH3-only proteins** are all pro-apoptotic.

The activities of the various BH3-only proteins are controlled by specific posttranslational modifications. For example, Bad is phosphorylated at two Ser residues by protein kinase A (PKA; Section 18-3C), mitogen-activated protein kinase (MAPK; Section 19-3D), and Akt (Section 19-4D) (which themselves are activated by complex signal transduction pathways), thereby generating a binding site for 14-3-3 proteins that then sequester Bad in the cytosol. On appropriate stimulation, calcineurin and PP1 (Section 19-3F) dephosphorylate Bad, which permits it to interact with the mitochondrion, where it initiates apoptosis (see below). In contrast, Bid is activated by caspase-8–catalyzed proteolytic cleavage to **tBid** (*truncated Bid*), thereby providing a link between the extrinsic and intrinsic pathways of apoptosis.

f. Cytochrome *c* Is an Essential Participant in the Intrinsic Pathway

The association of members of the pro-apoptotic Bcl-2 family with the mitochondrion causes it to release cytochrome c from its intermembrane space into the cytosol. There, as Xiaodong Wang unexpectedly discovered, this well-characterized component of the mitochondrial electron transport chain (Section 22-2C) functions to induce apoptosis. It does so by combining with Apaf-1 and dATP or ATP to form an ~1100-kD complex named the **apoptosome** (Fig. 34-115). The apoptosome binds several molecules of procaspase-9 in a manner that induces their autoactivation to yield caspase-9, which remains bound to the apoptosome. This caspase-9 then activates procaspase-3 to instigate cell death.

g. The Apoptasome Has a Wheel-like Structure

Apaf-1, the apoptosome's major component, is a 1248-residue scaffolding protein that consists of an N-terminal caspase recruitment domain (CARD), a central nucleotide-binding domain that is homolgous to CED-4, and a C-terminal domain that consists of seven WD40 repeats (Section 25-6B), a short linker, and six additional WD40 repeats. Procaspase-9 binds to Apaf-1's CARD in the apoptosome via its CARD, thereby placing several procaspase-9 molecules in close proximity such that they proteolytically activate one another. More importantly, however, caspase-9's association with the apoptosome increases its catalytic activity by three orders of magnitude, presumably via an allosteric mechanism. Indeed, a procaspase-9 mutant (D315A) that is noncleavable between its α and β domains, in complex with the apoptosome, nevertheless efficiently activates procaspase-3. Apaf-1's WD40 repeats function to bind cytochrome *c* (WD40 repeats usually participate in protein–protein interactions); their excision from Apaf-1 permits it to bind and activate procaspase-9 in the absence of cytochrome *c*. This suggests that Apaf-1's WD40 repeats bind to its CARD so as to prevent it from binding procaspase-9 but preferentially bind cytochrome *c* and thereby release the CARD. Evolution presumably selected cytochrome *c* for this function because it is normally absent from the cytosol.

The 27-Å resolution structure of the apoptosome, determined by Chistopher Akey through cryoelectron microscopy, reveals a wheel-like assembly with seven spokes ending in two lobes each that radiate from a central hub (Fig. 34-117*a*). Modeling studies suggest that the larger and smaller of these lobes respectively consist of a 7- and a 6-bladed β propeller (WD40 repeats form β propellers of various sizes; e.g., Fig. 19-18*b*) that are bridged by a cytochrome *c* molecule and that the Apaf-1 CARD occupies the apoptosome's hub region so that the nucleotide-binding domain must form at least the arm of the spoke. The cryoelectron microscopy–based image of the apoptosome in complex with the noncleavable D315A mutant of procaspase-9 revealed a previously unobserved domelike feature above the central hub that presumably represents the bound procaspase-9 (Fig. 34-117*b*). However, this dome is too small to accommodate seven procaspase-9 monomers, which suggests that the procaspase-9 in this complex is partially disordered, perhaps due to a flexible link between its CARD and its α domain.

h. Several Mechanisms Have Been Proposed for Mitochondrial Cytochrome *c* Release

The way in which pro-apoptotic Bcl-2 family members cause the mitochondrion to release its cytochrome *c* is unclear. However, based on largely circumstantial evidence, three models for this process, which are not mutually exclusive, have been proposed. The first model is based on the structural resemblance of the anti-apoptotic Bcl-2 family member Bcl-x_L to membrane-inserting bacterial toxins such as diphtheria toxin (Fig. 34-118; the mechanism of diphtheria toxin is discussed in Section 32-3G). This suggests that one or more molecules of Bcl-x_L and/or its homologs can insert into the outer mitochondrial membrane to form a pore. Indeed, Bcl-x_L, Bcl-2, and Bax have been shown to form pores in synthetic lipid bilayers. Moreover, tBid is targeted to mitochondria, where it triggers Bax oligomerization and its insertion into the outer mitochondrial membrane. However, it is unclear if the resulting pores are large enough to permit the passage of cytochrome *c*; in addition, it is not known how pores formed by anti-apoptotic proteins such as Bcl-x_L would inhibit apoptosis.

In the second model, Bcl-2 family members induce preexisting mitochondrial outer membrane proteins to form channels through which cytochrome *c* is released. An attractive candidate for such a protein is the **voltage-dependent anion channel (VDAC; alternatively mitochondrial porin)** because several Bcl-2 family members can bind to it and alter its channel activity. However, the known size of the VDAC channel is too small to allow the passage of cytochrome *c*, so that this model requires that VDAC undergo significant conformational change on binding Bcl-2 family members.

In the third model, Bcl-2 family members perturb or stabilize the preexisting pores through which ATP and ADP are exchanged between the mitochondrial matrix and the cytosol. This exchange process is mediated by the ATP–

(a)

(b)

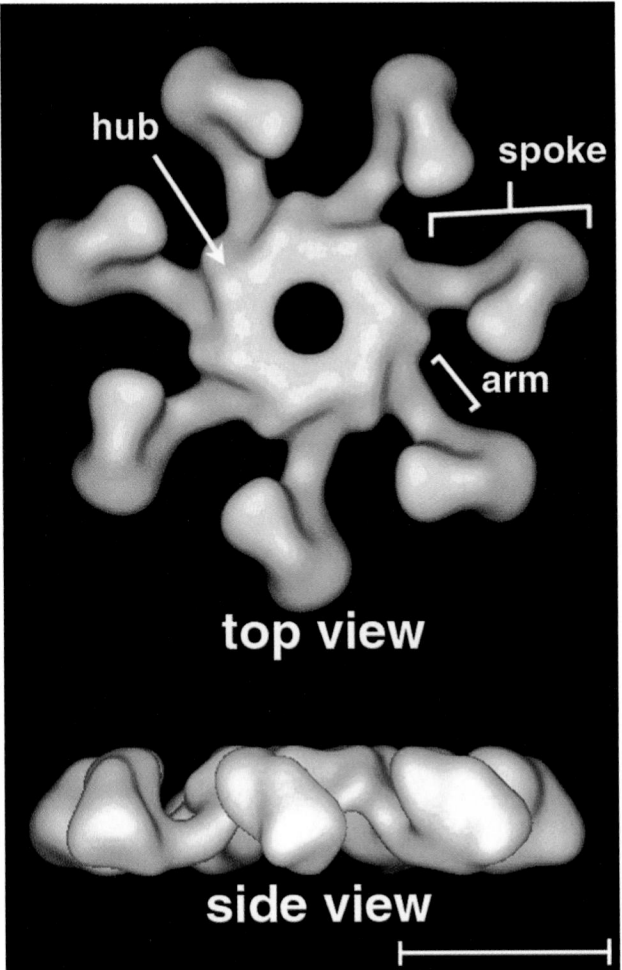

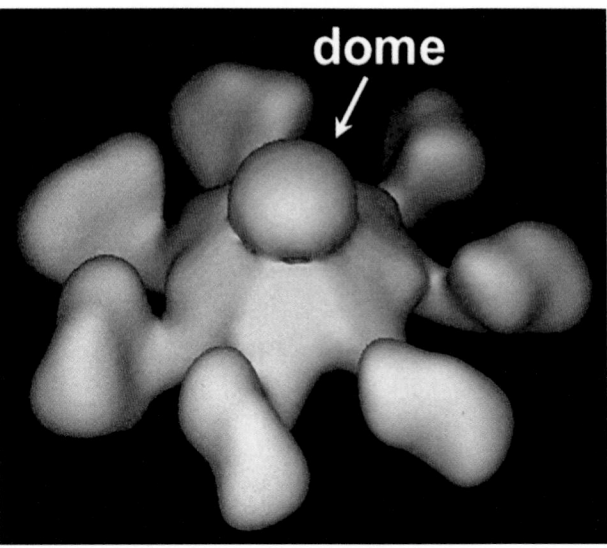

FIGURE 34-117 Cryoelectron microscopy–based images of the apoptosome at 27-Å resolution. (*a*) The free apoptosome. In its top view, the particle is viewed along its 7-fold axis of symmetry. The side view reveals the flattened nature of this wheel-like particle. The scale bar represents 100 Å. (*b*) The apoptosome in complex with a noncleavable mutant of procaspase-9 in oblique top view. Note the domelike feature above the central hub, which presumably represents the bound procaspase-9. [Courtesy of Christopher Akey, Boston University School of Medicine.]

ADP translocator in the inner mitochondrial membrane (Section 20-4C) and by VDAC in the outer mitochondrial membrane. The opening of these pores, it is postulated, would result in chemical equilibration between the cytosol and the matrix, which would cause the highly concentrated matrix to osmotically swell until it ruptured the outer mitochondrial membrane (which has considerably less surface area than the inner mitochondrial membrane due to the latter's cristae; Section 22-1A), thereby releasing cytochrome *c* into the cytosol. Consistent with this model, atractyloside and bongkregic acid (Section 20-4C), inhibitors of the ATP–ADP translocator that respectively open and close its pore, respectively induce and inhibit apoptosis. However, the rupture of the outer mitochondrial membrane has rarely been observed in cells undergoing apoptosis.

The way in which anti-apoptotic Bcl-2 family members antagonize the functions of pro-apoptotic family members

is better understood. The members of these opposing factions readily form homodimers in which the BH3 region of the pro-apoptotic protein, which forms an amphipathic α helix, binds in a hydrophobic groove on the anti-apoptotic protein. Such an arrangement occurs in the NMR structure, determined by Fesik, of Bcl-x_L in its complex with the 16-residue segment from the BH3 region of the pro-apoptotic Bcl-2 protein Bak (Fig. 34-119). Since the BH3 regions of pro-apoptotic Bcl-2 proteins are necessary and probably even sufficient for their killing activity, the sequestering of

FIGURE 34-118 Comparison of the structures of (*a*) Bcl-x_L and (*b*) the pore-forming domain of diphtheria toxin. In both proteins, the two central hydrophobic helices (*red*) are surrounded by amphipathic α helices. [Courtesy of Stephen Fesik, Abbott Laboratories, Abbott Park, Illinois. Part *b* based on an X-ray structure by David Eisenberg, UCLA. PDBids 1LXL and 1DDT.]

(a)   (b)

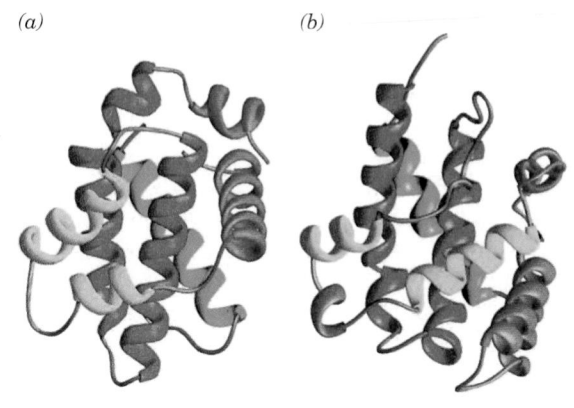

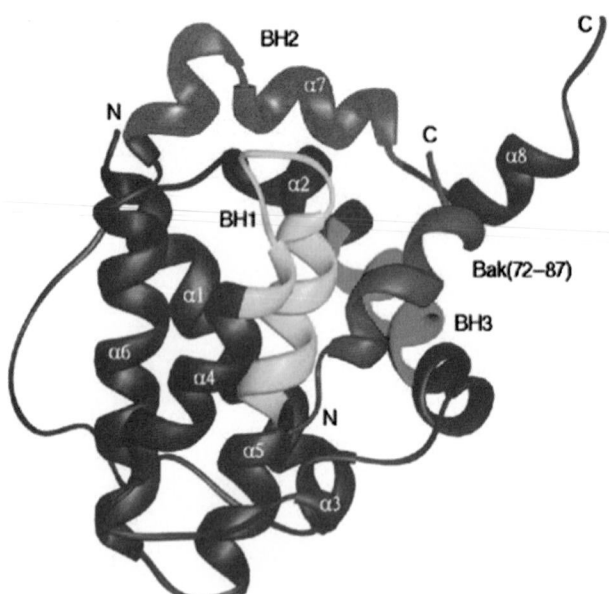

FIGURE 34-119 NMR structure of Bcl-x_L in complex with the 16-residue BH3 region of Bak. The BH1, BH2, and BH3 regions of Bcl-x_L are yellow, red, and green, respectively, and the Bak peptide is violet. [Courtesy of Stephen Fesik, Abbott Laboratories, Abbott Park, Illinois. PDBid 1BXL.]

these BH3 regions by anti-apoptotic Bcl-2 proteins at least partially explains their anti-apoptotic properties.

i. IAPs Regulate Apoptosis by Inhibiting Caspases

As might be expected, cells have elaborate systems that prevent their inadvertent apoptosis. We have previously discussed how the anti-apoptotic Bcl-2 family proteins keep their pro-apoptotic cousins in check, and how c-FLIP inhibits the death-inducing signal complex (DISC)-mediated activation of caspases-8 and -10. In addition, the members of the **IAP** (for *i*nhibitors of *ap*optosis) family of proteins, which are conserved from *Drosophila* to humans, regulate apoptosis by directly inhibiting caspases. Humans express eight IAPs, which vary in length from 236 to 4829 residues.

All IAPs have one to three ~70-residue **BIR domains** (for *b*aculovirus *I*AP *r*epeat; so named because they were discovered in the baculovirus protein **p35,** which functions to inhibit apoptosis in host cells during viral infection). BIRs contain a characteristic signature sequence, $CX_2CX_{16}HX_6C$, which forms a novel Zn^{2+}-binding motif. In addition, many IAPs (five in humans) have a C-terminal RING finger domain [a ubiquitin–protein ligase (E3); Section 32-6B]. **BIR2** [the second BIR domain of the human protein **XIAP** (for *X*-linked *IAP*)] and its surrounding regions specifically bind and inhibit the effector caspases-3 and -7, whereas **BIR3** and its surrounding regions do so for the initiator caspase-9; the function of **BIR1** is unknown. An IAP that has a RING finger domain may also ubiquitinate its bound caspase, thereby condemning it to destruction in the proteasome:

The X-ray structure of caspase-3 in complex with the XIAP BIR2 domain and its 38-residue N-terminal extension, determined by Fesik, Robert Liddington, and Guy

Salvesen, reveals, unexpectedly, that the BIR2 domain makes only limited contacts with caspase-3 (Fig. 34-120). Rather, most of the contacts to caspase-3 are made by the N-terminal extension, which spans the enzyme's active site so as to sterically block substrate binding. Curiously, the N-terminal extension extends across that active site in the reverse direction relative to that taken by polypeptide inhibitors of caspases such as Ac-DEVD-CHO (Fig. 34-111). The structures of both caspase-3 and the BIR2 domain in the complex are largely unperturbed relative to their uncomplexed structures.

The induction of apoptosis requires that the inhibitory effects of IAPs on caspases be relieved. This is the task of a homodimeric protein that is alternatively named **Smac** (for *s*econd *m*itochondria-derived *a*ctivator of *c*aspases) and **DIABLO** (for *di*rect *IAP*-*b*inding protein with *lo*w pI). Smac/DIABLO, which binds to the BIR domains of IAPs so as to prevent them from binding to caspases, is released from the mitochondrion together with cytochrome *c*, thereby ensuring that the intrinsic pathway of apoptosis will generate active caspases.

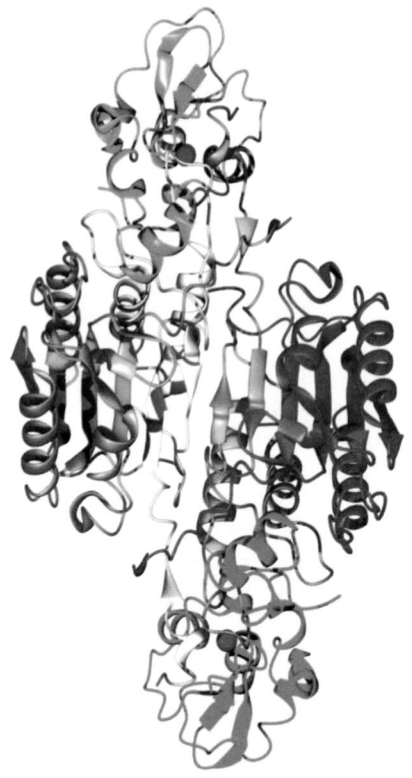

FIGURE 34-120 X-Ray structure of caspase-3 in complex with the XIAP BIR2 domain and its N-terminal extension. The complex is viewed along its 2-fold axis (as is caspase-7 in Fig. 34-111). The two αβ protomers of the heterotetrameric caspase-3 are blue and purple, with its β subunits shaded more lightly than its α subunits. The globular BIR2 domains are green and their bound Zn^{2+} ions represented by magenta spheres. Each of their N-terminal extensions, which begin with an α helix, are bound across the active site face of a caspase protomer, thereby blocking the binding of substrate proteins. [Courtesy of Guy Salvesen, The Burnham Institute, La Jolla, California. PDBid 1I3O.]

CHAPTER SUMMARY

1 ■ Chromosome Structure There are two types of eukaryotic chromatin: euchromatin, which can be transcriptionally active, and the more densely packed heterochromatin, which is transcriptionally inactive. Chromatin consists of DNA and proteins, the majority of which are the highly conserved histones. Chromatin is structurally organized in a hierarchical manner. In the first level of chromatin organization, ~200 bp of DNA are doubly wrapped around a histone octamer, $(H2A)_2(H2B)_2(H3)_2(H4)_2$, to form a nucleosome. Each nucleosome is associated with one molecule of histone H1. The passage of transcribing RNA polymerase causes nucleosomes to dissociate from the DNA and then to rebind it in a process that appears to be driven by supercoiling. DNA replication causes the parental nucleosomes to be randomly distributed between the daughter duplexes. The assembly of nucleosomes from their components is mediated by the molecular chaperone nucleoplasmin. In the second level of chromatin organization, the nucleosome filaments coil into 30-nm-thick filaments that probably contain six nucleosomes per turn. Then, in the third and final level of chromatin organization, the 30-nm-thick filaments form 15- to 30-μm-long radial loops that project from the axis of the metaphase chromosome. This accounts for DNA's packing ratio of >8000 in the metaphase chromosome. The larvae of certain dipteran flies, including *Drosophila*, contain banded polytene chromosomes, which consist of up to 1024 identical DNA strands in parallel register.

2 ■ Genomic Organization The complexity of a DNA sample can be determined from its renaturation rate through C_0t curve analysis. Eukaryotic DNAs have complex C_0t curves that arise from the presence of unique, moderately repetitive and highly repetitive sequences, as well as from inverted repeats. The function of inverted repeats, which form foldback structures, is unknown. Highly repetitive sequences, which occur in the heterochromatic regions near the chromosomal centromeres, probably function to align homologous chromosomes during meiosis and/or to facilitate their recombination. Moderately repetitive DNAs, which consist largely of inactive retrotransposons, mainly LINEs and SINEs, make up ~42% of the human genome. For the most part, they have unknown functions; they may simply be selfish DNA. Expressed sequences comprise only 1.1 to 1.4% of the human genome, which has made the identification of genes an uncertain process. However, CpG islands are often associated with the 5′ ends of genes.

Around 30,000 putative genes have been identified in the human genome. Some of them are transcribed to noncoding RNAs. The structural genes in the human genome have been classified according to function through sequence comparisons. Around 42% of them have unknown functions. The genes specifying rRNAs and tRNAs are organized into tandemly repeated clusters. The rDNA condenses to form nucleoli, the sites of rRNA transcription by RNA polymerase I and of partial ribosomal assembly. The 5S RNA and tRNAs are transcribed outside the nucleoli by RNA polymerase III. The genes specifying histones, which are required in large quantities only during S phase of the cell cycle, are the only repeated structural genes. The identity of a series of repeated genes is probably maintained through unequal crossing-over and/or gene conversion. Certain genes are amplified such as the *Xenopus* genes for rRNAs during oogenesis, chorion genes in *Drosophila*, and genes targeted by cancer chemotherapy.

Many families of genes specifying related proteins are clustered into gene families. In mammals, the gene clusters encoding the α- and β-like hemoglobin subunits occur on separate chromosomes. Nevertheless, all vertebrate globin genes have the same exon–intron structure: three exons separated by two introns. The thalassemias are inherited diseases caused by the genetic impairment of hemoglobin synthesis. Many α-thalassemias are caused by the deletion of one or more of the α-globin genes, whereas many β-thalassemias arise from point mutations that affect the transcription or the posttranscriptional processing of the β-globin mRNAs.

3 ■ Control of Expression Heterochromatin may be subclassified as constitutive heterochromatin, which is never transcriptionally active, and facultative heterochromatin, whose activity varies in a tissue-specific manner. The Barr bodies in the cells of female mammals constitute a common form of facultative heterochromatin: One of each cell's two X chromosomes is permanently condensed via the binding of *Xist* RNA and epigenetically confers its state of inactivity on its progeny through histone modification and DNA methylation. Active chromatin has a relatively open structure that makes it available to the transcriptional machinery. Two well-characterized examples of transcriptionally active chromatin are the chromosome puffs that emanate from single bands in polytene chromosomes and the lampbrush chromosomes of amphibian oocytes.

The differential protein synthesis characteristic of the cells in a multicellular organism largely stems from the selective transcription of the expressed genes. The first step in the transcriptional initiation of RNAP II-transcribed genes is often the binding of TATA-box binding protein (TBP) to the promoter's TATA box, which is located around position −27. This is followed by the addition of TBP-associated factors (TAFs) to form transcription factor IID (TFIID), together with general transcription factors (GTFs) and RNAP II to form the preinitiation complex (PIC), which is capable of a basal rate of transcription. Several TAFs assume the histone fold and associate with one another in TFIID, as do hisones in the histone octamer. TBP, together with other GTFs, is also required for the transcriptional initiation of class I and III genes. Promoters of class II genes that lack a TATA box often contain an initiator (Inr) sequence that spans the transcription start site and may also have a downstream promoter element (DPE).

The cell-specific expression of a gene is mediated by the gene's promoter and enhancer elements. Consequently, cells contain specific upstream transcription factors that recognize these genetic elements. For example, Sp1 binds to the GC box that precedes many genes. Likewise, steroid hormones bind to their cognate receptors, which in turn bind to specific enhancers so as to modulate the transcriptional activitiy of the associated gene. The cooperative binding of several transcriptional factors to their target promoter and enhancer sites stimulates the associated PIC to increase the rate at which it initiates the transcription of the associated gene. The binding of transcription factors to a silencer represses the transcription of the associated gene. Several transcription factors may bind an enhancer and associate with architectural factors and coactivators to form an enhanceosome. Many transcription factors have two domains, a DNA-binding domain targeted to a specific sequence, and an activation domain, which interacts with the PIC in a largely nonspecific manner, often via a negatively charged surface region. Eukaryotic transcription factors have a great variety of DNA-binding motifs, including several types of zinc fingers, the bZIP motif, and the bHLH/Z motif. Many transcription factors, including those with the latter two types of motif, dimerize through the formation of a leucine zipper. Nuclear factor κB (NF-κB) is activated in the cytoplasm by the destruction of its bound inhibitor IκB, whereupon NF-κB is translocated to the nucleus, where it binds to a κB DNA segment so as to activate the associated gene. Mediator is an ~20-subunit yeast complex that binds to the RNAP II β′ subunit's C-terminal domain (CTD), where it influences RNAP II's activity through its binding of DNA-bound transcriptional regulators. Metazoa contain several mediator-like complexes that presumably relay signals from different sets of transcriptional activators.

Transcriptionally poised or active genes contain nuclease-hypersensitive sites that occur in nucleosome-free regions of DNA. Nuclease hypersensitivity is conferred on DNA by the binding of specific proteins that presumably make the genes accessible to the proteins mediating transcriptional initiation. This is largely due to the presence of the nonhistone, DNA-binding, high mobility group (HMG) proteins, which are architectural proteins that function to activate gene expression by decondensing chromatin and recruiting transcription factors. RNA polymerase transcribes through nucleosomes by inducing the histone octamers it encounters to step around it. Locus control regions (LCRs), which are DNase I hypersensitive sites that function to suppress position effects, that is, the encroachment of heterochromatin, are activated by proteins that are expressed only in specific cell lineages such as erythroid cells. Insulators are DNA segments that, through the binding of specific proteins, inhibit heterochromatin from spreading into neighboring segments of euchromatin and prevent regulatory elements outside the region controlled by the insulator from influencing the expression of the genes inside the region.

The transcriptional machinery gains access to the DNA packaged by chromatin through the posttranscriptional modification of the core histones' N-terminal tails and the ATP-driven remodeling of chromatin. The modifications to which histone N-terminal tails are subject include the acetylation/deacetylation of specific Lys side chains, the methylation of specific Lys and Arg side chains, the phosphorylation/dephosphorylation of specific Ser side chains, and the ubiquitination of specific Lys side chains. There appears to be a histone code in which specific modifications, acting sequentially or in combination, evoke certain chromatin-based functions that result in unique biological outcomes such as transcriptional activation or silencing. In addition, some of these histone modifications may act as epigenetic markers through which cells confer their identities on their progeny. Histone acetylation is catalyzed by histone acetyltransferases (HATs) that are components of multisubunit transcriptional activators such as SAGA, PCAF, and TFIID, which all contain histonelike TAFs. Nearly all HAT-associated coactivators contain bromodomains that specifically bind acetylated histone Lys residues and hence are likely to recruit HATs to

acetylate the N-terminal tails of nearby nucleosomes. Histone deacetylases (HDACs), many of which are also members of multisubunit complexes, serve as transcriptional corepressors. Histone methylation, which is largely if not entirely irreversible, is catalyzed by histone methyltransferases (HMTs), which often function as transcriptional corepressors. Methylated histones are recognized by chromodomains such as that in heterochromatin protein 1 (HP1). The spreading of heterochromatin appears to be mediated, at least in part, by HP1's recruitment of the HMT Suv39, which methylates nearby nucleosomes such that additional HP1 can bind to them, etc. Histone monoubiquitination functions as an essential transcriptional regulator. Chromatin-remodeling complexes, such as yeast SWI/SNF and RSC, contain helicase-like ATPases that, it appears, "walk" up the DNA in a nucleosome so as to decrease its helical twist. The resulting DNA distortion, it is postulated, diffuses around the nucleosome in a wave that locally and transiently releases the DNA from the histone octamer, thereby permitting the nucleosome to slide along the DNA and providing transcriptional activators access to their target sequences that would otherwise be sequestered by the nucleosome.

Other forms of selective gene expression in eukaryotes include the use of alternative initiation sites in a single gene, the selection of alternative splice sites, the possible regulation of mRNA translocation across the nuclear membrane, the control of mRNA degradation, the control of translational initiation rates, and the selection of alternative posttranslational processing pathways.

4 ■ Cell Differentiation and Growth Embryogenesis occurs in four stages: cleavage, gastrulation, organogenesis, and maturation and growth. One of the most striking characteristics of embryological development is that cells become progressively and irreversibly committed to specific lines of development. The signals that trigger developmental changes, which are recognized over great evolutionary distances, may be transmitted through direct intercellular contacts or from the gradients of substances, known as morphogens, released by other embryonic cells. Developmental signals act combinatorially; that is, the developmental fate of a specific tissue is determined by several not necessarily unique developmental stimuli. In *Drosophila*, early embryonic development is governed by maternal-effect genes whose distribution imposes the embryo's spatial coordinate system. These encode transcription factors that regulate the expression of gap genes, which in turn regulate the expression of pair-rule genes, which in turn regulate the expression of segment polarity genes. Sequentially finer domains of the embryonic body are thereby defined in a way that specifies the number and polarity of the larval and adult body segments. Homeotic selector or *Hox* genes, whose mutations transform one body part into another, then regulate the differentiation of the individual segments. These regulatory genes, which occur in two gene clusters, are, as the preceding genes, selectively expressed in the embryonic tissues whose development they control. They have closely related base sequences that encode ~60-residue polypeptide segments known as homeodomains, which bind their target DNA sequences in a manner similar to but distinct from that of the homologous HTH module. In vertebrates, *Hox* genes occur in four clusters and likewise control development.

Cancers result from specific genetic alterations to cells. The types of genetic changes that give rise to malignancies include the generation of altered proteins such as a Ras variant that lacks GTPase activity; altered regulatory sequences that, for example, result in the overexpression of key transcription factors; the loss of degradation signals that cause an oncogene protein such as Jun to be degraded abnormally slowly; chromosomal rearrangements that place proto-oncogenes such as c-*myc* under the control of inappropriately active regulatory sequences; gene amplification that results in the overexpression of a proto-oncogene; the insertion of a virus into a chromosome such that a proto-oncogene is brought under the control of viral regulatory sequences; the inappropriate activation or inactivation of chromatin modification enzymes such as HAT and HDAC; and the loss or inactivation of tumor suppressor genes such as those encoding p53 and pRb. The mutations causing these gene alterations often arise from the actions of carcinogens on cellular DNA.

The progression of a cell through the cell cycle is regulated mainly by the presence of the appropriate mitogens together with a series of checkpoints that monitor the cell's health as well as its progress through the cell cycle. Checkpoints arrest the cell cycle until the proper conditions for its progression are met, for example, that the replication of DNA has been successfully completed and it is undamaged. The cell cycle is characterized by the accumulation of cyclins, which abruptly disappear at the end of mitosis. For example, M phase is induced when cyclin B combines with Cdc2 to form cyclin-dependent kinase 1 (Cdk1), which is preceded by Cdc2's phosphorylation at Thr 161 by Cdk-activating kinase (CAK) and succeeded by Cdc2's inactivating phosphorylation at Thr 14 and Tyr 15 by Wee1 and Myt1. At the G_2/M boundary, Thr 14 and Tyr 15 are rapidly dephosphorylated by Cdc25C, yielding active Cdk1, which phosphorylates a variety of nuclear proteins. The structures of Cdks resemble those of other protein kinases. However, cyclin binding to a Cdk and its phosphorylation at Thr 160 conformationally reorganize its active site. Members of the Kip/Cip family, such as p21^{Cip1}, inhibit most Cdk–cyclin complexes except Cdk4/6–cyclin D, whereas members of the INK4 family, such as p16^{INK4a}, inhibit Cdk4/6–cyclin D. Cell cycle arrest at the G_2 checkpoint is initiated by sensor proteins that bind to damaged and unreplicated DNA. These activate ATM and ATR to respectively phosphorylate Chk2 and Chk1, which phosphorylate Cdc25C, thereby providing a binding site for 14-3-3 proteins such that Cdc25C is sequestered in the cytoplasm, where it cannot dephosphorylate and hence activate Cdc2.

p53, a tumor suppressor that is implicated in 50% of human cancers, is bound by Mdm2, which ubiquitinates it so as to mark it for destruction in the proteasome. Hence the cell normally has a low level of p53. However, when p53 is phosphorylated by ATM or Chk2, it no longer binds to Mdm2 and thereupon transactivates the expression of p21^{Cip1}, which binds to several Cdk–cyclin complexes and to PCNA, thereby inhibiting both the G_1/S and G_2/M transitions and DNA replication. If the cell is irreparably damaged, p53 induces it to undergo apoptosis, thereby preventing the proliferation of potentially cancerous cells. The X-ray structure of p53's DNA-binding core in complex with its target DNA reveals that many of its residues that participate in DNA binding are frequently mutated in tumors. p53 is also activated by a variety of pathways such as MAP kinase cascades, the activation of ATR by DNA-damaging agents, and a variety of posttranslational modifications. Thus, p53 is the recipient of numerous

intracellular signals and activates a variety of downstream regulators.

The tumor suppressor pRb, a regulator of the cell cycle's G_1/S transition, functions by inhibiting E2F, a transcription factor for many proteins required for S-phase entry. pRb, a phosphoprotein that is phosphorylated at numerous Ser/Thr sites by various Cdk–cyclin complexes, must be in its hypophosphorylated form to bind E2F. A variety of proteins that have an LXCXE sequence motif bind to pRb at a separate site from E2F on pRb's pocket domain, a major site of genetic alteration in tumors. Viral proteins with the LXCXE motif, including adenovirus E1A and papillomavirus E7, cause pRb to release its bound E2F, thereby driving the infected cell into S phase, which facilitates viral DNA replication. The histone deacetylases HDAC1 and HDAC2 each contain an LXCXE motif, which suggests that pRb functions to recruit these proteins to E2F's target promoters, thereby deactivating them. The SWI/SNF homologs BRM and BRG1, which both have LXCXE motifs, can bind to pRb simultaneously with HDACs, which suggests that these chromatin-remodeling complexes are recruited to E2F promoters, where they facilitate the action of HDACs.

Apoptosis (programmed cell death) occurs normally during embryogenesis and in many adult processes. In fact, it is the default option for metazoan cells. Insufficient apoptosis can cause autoimmune diseases and cancer, whereas inappropriate apoptosis is responsible for several neurodegenerative diseases and much of the damage caused by stroke and heart attacks. In apoptosis, the cell dismantles itself in an orderly program to yield membrane-enclosed apoptotic bodies that are phagocytosed by surrounding cells without inducing an inflammatory response. The executioners in apoptosis are cysteine proteases known as caspases that specifically cleave polypeptides after Asp residues. Caspases are synthesized as zymogens called procaspases that are proteolytically activated via apoptotic pathways, ending in the activation of initiator caspases that activate effector caspases, which cleave a wide variety of cellular proteins. Among the latter is ICAD, which is an inhibitor of caspase-activated DNase (CAD) that, in the absence of ICAD, functions to fragment the cell's DNA.

In the extrinsic pathway of apoptosis (death by commission), a trimeric transmembrane cytokine of the tumor necrosis factor (TNF) family, such as Fas ligand (FasL), which is on the inducing cell, binds to a transmembrane so-called death receptor of the TNFR family, such as Fas, which is on the apoptotic cell. The binding of trimeric ligand to a death receptor causes its cytoplasmic death domain (DD) to form a trimer to which three molecules of the adaptor protein FADD then bind via their DDs. FADD, in turn, recruits procaspases-8 and -10 via interactions between the two proteins' death effector domains (DEDs) to form the death-inducing signal complex (DISC). This results in the proteolytic autoactivation of the bound procaspases-8 and -10, which then activate procaspase-3, an effector caspase. In the intrinsic pathway (death by omission), pro-apoptotic members of the Bcl-2 family are activated in various ways, including the withdrawal of cytokines and contact with other cells, to induce the release of cytochrome c from the mitochondrion. The cytochrome c binds to the scaffolding protein Apaf-1 to form a wheel-shaped heptameric complex called the apoptosome. The apoptosome binds several molecules of procaspase-9 through interactions between the two proteins' CARD domains, which activates procaspase-9 to activate procaspase-3. Pro-apoptotic Bcl-2 family members are kept in check through their heterodimerization with anti-apoptotic Bcl-2 family members. In addition, the members of the IAP family inhibit apoptosis by directly binding to caspases so as to block their active sites and, in some cases, also ubiquitinating them so as to mark them for destruction in the proteasome. Smac/DIABLO, which is released from the mitochondrion together with cytochrome *c*, reverses this inhibition by binding to IAPs, thereby permitting apoptosis to commence.

■ REFERENCES

GENERAL

Alberts, B., Johnson, A., Lewis, J., Raff, M., Roberts, K., and Walter, P., *Molecular Biology of the Cell* (4th ed.), Chapters 4, 7, 17, 21, and 23, Garland Publishing (2002).

Brown, T.A., *Genomes 2,* Chapters 1, 2, 8, 9, and 12, Wiley-Liss (2002).

Elgin, S.C.R. and Workman, J.L. (Eds.), *Chromatin Structure and Gene Expression* (2nd ed.), Oxford University Press (2000).

Lewin, B., *Genes VII,* Chapters 18–21, Oxford (2000).

Lodish, H., Berk, A., Zipursky, S.L., Matsudaira, P., Baltimore, D., and Darnell, J., *Molecular Cell Biology* (4th ed.), Chapters 9, 10, 13, and 14, Freeman (2000).

Sumner, A.T., *Chromosomes, Organization and Function,* Blackwell Science (2003).

Wolffe, A., *Chromatin, Structure and Function,* Academic Press (1998).

CHROMOSOME STRUCTURE

Bustin, M., Chromatin unfolding and activation by HMGN chromosomal proteins, *Trends Biochem. Sci.* **26,** 431–437 (2001).

Carey, M. and Smale, S.T., *Transcriptional Regulation in Eukaryotes. Concepts, Strategies, and Techniques,* Cold Spring Harbor Laboratory Press (2000). [A comprehensive guide to the methods used in analyzing transcriptional regulatory mechanisms.]

Cohen, D.E. and Lee, J.T., X-Chromosome inactivation and the search for chromosome-wide silencers, *Curr. Opin. Genet. Dev.* **12,** 219–224 (2002).

Earnshaw, W.C., Large scale chromosome structure and organization, *Curr. Opin. Struct. Biol.* **1,** 237–244 (1991).

Felsenfeld, G. and McGhee, J.D., Structure of the 30 nm chromatin fiber, *Cell* **44,** 375–377 (1986).

Hansen, J.C., Conformation dynamics of the chromatin fiber in solution: Determinants, mechanisms, and functions, *Annu. Rev. Biophys. Biomol. Struct.* **31,** 361–392 (2002).

Harp, J.M., Hanson, B.L., Timm, D.E., and Bunick, G.J., Asymmetries in the nucleosome core particle at 2.5 Å resolution, *Acta Cryst.* D**56,** 1513–1534 (2000).

Kornberg, R.D., Chromatin structure: a repeating unit of histones and DNA, *Science* **184,** 868–871 (1974). [The classic paper first indicating the constitution of nucleosomes.]

Kornberg, R.D. and Lorch, Y., Twenty-five years of the nucleosome, fundamental particle of the eukaryotic chromosome, *Cell* **98,** 285–295 (1999).

Locker, J. (Ed.), *Transcription Factors,* Academic Press (2001).

Luger, K., Mäder, A.W., Richmond, R.K., Sargent, D.F., and Richmond, T.J., Crystal structure of the nucleosome particle at 2.8 Å resolution, *Nature* **389,** 251–260 (1997); Davey, C.A., Sargent, D.F., Luger, K., Maeder, A.W., and Richmond, T.J., Solvent mediated interactions in the structure of the nucleosome core particle at 1.9 Å resolution, *J. Mol. Biol.* **319,** 1097–1113 (2002); *and* Davey, C.A. and Richmond, T.J., The structure of DNA in the nucleosome core, *Nature* **423,** 145–150 (2003).

Luger, K., Structure and dynamic behaviour of nucleosomes, *Curr. Opin. Genet. Dev.* **13,** 127–135 (2003); *and* Akey, C.W. and Luger, K., Histone chaperones and nucleosome assembly, *Curr. Opin. Struct. Biol.* **13,** 6–14 (2003).

Masse, J.E., Wong, B., Yen, Y.-M., Allain, F.H.-T., Johnson, R.C., and Feigon, J., The *S. cerevisiae* architectural HMGB protein NHP6A complexed with DNA: DNA and protein conformational changes upon binding, *J. Mol. Biol.* **323,** 263–284 (2002).

Merika, M. and Thanos, D., Enhanceosomes, *Curr. Opin. Genet. Dev.* **11,** 205–208 (2001).

Ramakrishnan, V., Histone structure, *Curr. Opin. Struct. Biol.* **4,** 44–50 (1994).

Ramakrishnan, V., Finch, J.T., Graziano, V., Lee, P.L., and Sweet, R.M., Crystal structure of globular domain of histone H5 and its implications for nucleosome binding, *Nature* **362,** 219–223 (1993); *and* Ramakrishnan, V., Histone H1 and chromatin higher-order structure, *Crit. Rev. Euk. Gene Exp.* **7,** 215–230 (1997).

Reeves, R. and Beckerbauer, L., HMGI/Y proteins: flexible regulators of transcription and chromatin structure, *Biochim. Biophys. Acta* **1519,** 13–29 (2001).

Thomas, J.O. and Travers, A.A., HMG1 and 2, and related 'architectural' DNA-binding proteins, *Trends Biochem. Sci.* **26,** 167–174 (2001).

van Holde, K. and Zlatanova, J., Chromatin higher order structure: Chasing a mirage? *J. Biol. Chem.* **270,** 8373–8376 (1995). [Presents the arguments that the 30-nm chromatin filament has an irregular structure.]

Weatherall, D.J., Clegg, J.B., Higgs, D.R., and Wood, W.G., The hemoglobinopathies, *in* Scriver, C.R., Beaudet, A.L., Sly, W.S., and Valle, D. (Eds.), *The Metabolic & Molecular Bases of Inherited Disease* (8th ed.), *pp.* 4571–4636, McGraw-Hill (2001). [Contains a discussion of the thalassemias.]

White, C.L., Suto, R.K., and Luger, K., Structure of the yeast nucleosome core particle reveals fundamental changes in internucleosome interactions, *EMBO J.* **20,** 5207–5218 (2001).

Widom, J., Structure, dynamics, and function of chromatin in vitro, *Annu. Rev. Biophys. Biomol. Struct.* **27,** 285–327 (1998).

GENOMIC ORGANIZATION

Berry, M., Grosveld, F., and Dillon, N., A single point mutation is the cause of the Greek form of hereditary persistence of fetal hemoglobin, *Nature* **358,** 499–502 (1992).

Craig, N.L., Craigie, R., Gellert, M., and Lambowitz, A.M., *Mobile DNA II,* Chapters 35, 47, 48, and 49, ASM Press (2002). [Discussions of transposable elements in eukaryotic genomes.]

Deininger, P.L., Batzer, M.A., Hutchinson, C.A., III, and Edgell, M.H., Master genes in mammalian repetitive DNA amplification, *Trends Genet.* **8,** 307–311 (1992).

Hamlin, J.L., Leu, T.-H., Vaughn, J.P., Ma, C., and Dijkwel, P.A., Amplification of DNA sequences in mammalian cells, *Prog. Nucleic Acid Res. Mol. Biol.* **41,** 203–239 (1991).

International Human Genome Sequencing Consortium, Initial sequencing and analysis of the human genome, *Nature* **409,**

860–921 (2001); *and* Venter, J.C., et al., The sequence of the human genome, *Science* **291,** 1304–1351 (2001). [The landmark papers describing the base sequence of the human genome. They contain descriptions of repeating elements and gene distributions in the human genome.]

Kafatos, F.C., Orr, W., and Delidakis, C., Developmentally regulated gene amplification, *Trends Genet.* **1,** 301–306 (1985).

Li, W.-H., Gu, Z., Wang, H., and Nekrutenko, A., Evolutionary analysis of the human genome, *Nature* **409,** 847–849 (2001). [A survey of repetitive elements in the human genome.]

Mandal, R.K., The organization and transcription of eukaryotic ribosomal RNA genes, *Prog. Nucleic Acid Res. Mol. Biol.* **31,** 115–160 (1984).

Maxson, R., Cohn, R., and Kedes, L., Expression and organization of histone genes. *Annu. Rev. Genet.* **17,** 239–277 (1983).

Orgel, L.E. and Crick, F.H.C., Selfish DNA: the ultimate parasite, *Nature* **284,** 604–607 (1980).

Orr-Weaver, T.L., *Drosophila* chorion genes: Cracking the eggshell's secrets, *BioEssays* **13,** 97–105 (1991).

Saccone, C. and Pesole, G., *Handbook of Comparative Genomics. Principles and Methods,* Wiley-Liss (2003).

Schimke, R.T., Gene amplification in cultured cells, *J. Biol. Chem.* **263,** 5989–5992 (1988).

Stamatoyannopoulos, G., Majerus, P.W., Permutter, R.M., and Varmus, H., (Eds.), *The Molecular Basis of Blood Diseases* (3rd ed.), Chapters 2–5, Elsevier (2001). [Discusses hemoglobin genes and their normal and thalassemic expression.]

Südhof, T.C., Goldstein, J.L., Brown, M.S., and Russell, D.W., The LDL receptor gene: a mosaic of exons shared with different proteins, *Science* **228,** 815–828 (1985).

Wainscoat, J.S., Hill, A.V.S., Boyce, A.L., Flint, J., Hernandez, M., Thein, S.L., Old, J.M., Lynch, J.R., Falusi, A.G., Weatherall, D.J., and Clegg, J.B., Evolutionary relationships of human populations from an analysis of nuclear DNA polymorphisms, *Nature* **319,** 491–493 (1986).

CONTROL OF EXPRESSION

Adams, C.C. and Workman, J.L., Nucleosome displacement in transcription, *Cell* **72,** 305–308 (1993).

Andel, F., III, Ladurner, A.G., Inouye, C., Tjian, R., and Nogales, E., Three-dimensional structure of the human TFIID–IIA–IIB complex, *Science* **286,** 2153–2156 (1999).

Andres, A.J. and Thummel, C.S., Hormones, puffs and flies: the molecular control of metamorphosis by ecdysone, *Trends Genet.* **8,** 132–138 (1992).

Ashburner, M., Puffs, genes, and hormones revisited, *Cell* **61,** 1–3 (1990).

Asturias, F.J., Chung, W-H., Kornberg, R.D., and Lorch, Y., Structural analysis of the RSC chromatin-remodeling complex, *Proc. Natl. Acad. Sci.* **99,** 13477–13480 (2002).

Bach, I. and Ostendorff, H.P., Orchestrating nuclear functions: Ubiquitin sets the rhythm, *Trends Biochem. Sci.* **28,** 189–195 (2003); *and* Muratani, M. and Tansey, W.P., How the ubiquitin–proteasome system controls transcription, *Nature Rev. Mol. Cell Biol.* **4,** 192–201 (2003).

Becker, P.B. and Hörz, W., ATP-dependent nucleosome remodeling, *Annu. Rev. Biochem.* **71,** 247–273 (2002); Flaus, A. and Owen-Hughes, T., Mechanisms for ATP-dependent chromatin remodeling, *Curr. Opin. Genet. Dev.* **11,** 148–154 (2001); *and* Fry, C.J. and Peterson, C.L., Chromatin remodeling enzymes: Who's on first? *Curr. Biol.* **11,** R185–R197 (2001).

Bell, A.C., West, A.G., and Felsenfeld, G., Insulators and boundaries: Versatile regulatory elements in the eukaryotic genome, *Science* **291,** 447–450 (2001).

Berger, S., Histone modifications in transcriptional regulation, *Curr. Opin. Genet. Dev.* **12,** 142–148 (2002).

Buratowski, S., The basics of basal transcription by RNA polymerase II, *Cell* **77,** 1–3 (1994).

Burgess-Beusse, B., Farrell, C., Gaszner, M., Litt, M., Mutskov, V., Recillas-Targa, F., Simpson, M., West, A., and Felsenfeld, G., The insulation of genes from external enhancers and silencing chromatin, *Proc. Natl. Acad. Sci.* **99,** 16433–16437 (2002).

Branden, C. and Tooze, J., *Introduction to Protein Structure* (2nd ed.), Chapters 9 and 10, Garland Publishing (1999).

Carey, M. and Smale, S.T., *Transcriptional Regulation in Eukaryotes. Concepts, Strategies, and Techniques,* Cold Spring Harbor Laboratory Press (2000).

Chasman, D.I., Flaherty, K.M., Sharp, P.A., and Kornberg, R.D., Crystal structure of yeast TATA-binding protein and model for interaction with DNA, *Proc. Natl. Acad. Sci.* **90,** 8174–8178 (1993); *and* Nikolov, D.B., Hu, S.-H., Lin, J., Gasch, A., Hoffmann, A., Horikoshi, M., Chua, N.-H., Roeder, R.G., and Burley, S.K., Crystal structure of TFIID TATA-box binding protein, *Nature* **360,** 40–46 (1992).

Chen, F.E. and Ghosh, G., Regulation of DNA binding by Rel/NF-κB transcription factors: structural view, *Oncogene* **18,** 6845–6852 (1999); *and* Berkowitz, B., Huang, D.-B., Chen-Park, F.E., Sigler, P.B., and Ghosh, G., The X-ray crystal structure of the NF-κB p50·p65 heterodimer bound to the interferon β-κB site, *J. Biol. Chem.* **277,** 24694–24700 (2002).

Chen, X., et al., Gene expression patterns in human liver cancers, *Mol. Biol. Cell* **13,** 1929–1939 (2002).

Conway, R.C. and Conway, J.W., General initiation factors for RNA polymerase II, *Annu. Rev. Biochem.* **62,** 161–190 (1993).

Dotson, M.R., Yuan, C.X., Roeder, R.G., Myers, L.C., Gustafsson, C.M., Jiang, Y.W., Li, Y., Kornberg, R.D., and Asturias, F.J., Structural organization of yeast and mammalian mediator complexes, *Proc. Natl. Acad. Sci.* **97,** 14307–14310 (2000); *and* Davis, J.A., Takagi, Y., Kornberg, R.D., and Asturias, F.J., Structure of the yeast RNA polymerse II holoenzyme: Mediator conformation and polymerase interaction, *Mol. Cell* **10,** 409–415 (2002).

Elgin, S.C.R., The formation and function of DNase I hypersensitive sites in the process of gene activation. *J. Biol. Chem.* **263,** 19259–19262 (1988).

Ellenberger, T.E., Getting a grip on DNA recognition: structures of the basic region leucine zipper, and the basic region helix-loop-helix DNA-binding domains, *Curr. Opin. Struct. Biol.* **4,** 12–21 (1994).

Ellenberger, T.E., Brandl, C.J., Struhl, K., and Harrison, S.C., The GCN4 basic region leucine zipper binds DNA as a dimer of uninterrupted α helices: Crystal structure of the protein–DNA complex, *Cell* **71,** 1223–1237 (1992).

Evans, R.M., The steroid and thyroid hormone receptor superfamily, *Science* **240,** 889–895 (1988).

Ferré-d'Amaré, A.R., Prendergast, G.C., Ziff, E.B., and Burley, S.K., Recognition by Max of its cognate DNA through a dimeric b/HLH/Z domain, *Nature* **363,** 38–45 (1993).

Finnin, M.S., Donigian, J.R., and Pavletich, N.P., Structure of the histone deacetylase SIRT2, *Nature Struct. Biol.* **8,** 621–625 (2001).

Funder, J.W., Glucocorticoid and mineralocorticoid receptors: Biology and clinical relevance, *Annu. Rev. Med.* **48,** 231–240 (1997).

Gangloff, Y.-G., Romier, C., Thuault, S., Werten, S., and Davidson, I., The histone fold is a key structural motif of transcription factor TFIID, *Trends Biochem. Sci.* **26,** 250–257 (2001).

Garvie, C.W. and Wolberger, C., Recognition of specific DNA complexes, *Mol. Cell* **8,** 937–946 (2001).

Geiger, J.H., Hahn, S., Lee, S., and Sigler, P.B., Crystal structure of the yeast TFIIA/TBP/DNA complex, *Science* **272,** 830–836 (1996); *and* Tan, S., Hunziker, Y., Sargent, D.F., and Richmond, T.J., Crystal structure of a yeast TFIIA/TBP/DNA complex, *Nature* **381,** 127–134 (1996).

Grewal, S.I.S. and Moazed, D., Heterochromatin and epigenetic control of gene expression, *Science* **301,** 798–802 (2003).

Gross, D.S. and Garrard, W.T., Nuclease hypersensitive sites in chromatin, *Annu. Rev. Biochem.* **57,** 159–197 (1988).

Gustafsson, C.M. and Samuelsson, T., Mediator—a universal complex in transcription regulation, *Mol. Microbiol.* **41,** 1–8 (2001).

Huth, J.R., Bewley, C.A., Nissen, M.S., Evans, J.N.S., Reeves, R., Gronenborn, A.M., and Clore, G.M., The solution structure of an HMG-I(Y)–DNA complex defines a new architectural minor groove binding motif, *Nature Struct. Biol.* **4,** 657–665 (1997).

Iizuka, M. and Smith, M.M., Functional consequences of histone modifications, *Curr. Opin. Genet. Dev.* **13,** 154–160 (2003).

Jacobson, R.H., Ladurner, A.G., King, D.S., and Tjian, R., Structure and function of the human TAF$_{II}$250 double bromodomain module, *Science* **288,** 1422–1425 (2000).

Khorasanizadeh, S. and Rastinejad, F., Nuclear-receptor interactions on DNA-response elements, *Trends Biochem. Sci.* **26,** 384–390 (2001).

Kim, Y., Geiger, J.H., Hahn, S., and Sigler, P.B., Crystal structure of a yeast TBP/TATA-box complex; Kim, J.L., Nikolov, D.B., and Burley, S.K., Co-crystal structure of TBP recognizing the minor groove of a TATA element, *Nature* **365,** 512–520 *and* 520–527 (1993); *and* Nikolov, D.B. and Burley, S.K., 2.1 Å resolution refined structure of TATA box-binding protein (TBP), *Nature Struct. Biol.* **1,** 621–637 (1994).

Klug, A. and Rhodes, D., 'Zinc fingers': a novel protein motif for nucleic acid recognition, *Trends Biochem. Sci.* **12,** 464–469 (1987).

Kouzarides, T., Histone methylation in transcriptional control, *Curr. Opin. Genet. Dev.* **12,** 198–209 (2002); *and* Bannister, A.J., Schneider, R., and Kouzarides, T., Histone methylation: Dynamic or static, *Cell* **109,** 801–806 (2002).

Lee, T.I. and Young, R.A., Transcription of eukaryotic protein-coding genes, *Annu. Rev. Genet.* **34,** 77–137 (2000).

Lemon, B. and Tjian, R., Orchestrated response: A symphony of transcription factors for gene control, *Genes Devel.* **14,** 2551–2569 (2000); *and* Näär, A.M., Lemon, B.D., and Tjian, R., Transcriptional coactivator complexes, *Annu. Rev. Biochem.* **70,** 475–501 (2001).

Li, Q., Peterson, K.R., Fang, X., and Stamatoyannopoulos, G., Locus control regions, *Blood* **100,** 3077–3086 (2002).

Locker, J. (Ed.), *Transcription Factors,* Academic Press (2001).

Luisi, B.F., Xu, W.X., Otwinowski, Z., Freedamn, L.P., Yamamoto, K.R., and Sigler, P.B., Crystallographic analysis of the interaction of the glucocorticoid receptor with DNA, *Nature* **352,** 497–505 (1991).

Malik, S. and Roeder, R.G., Transcriptional regulation through mediator-like coactivators in yeast and metazoan cells, *Trends Biochem. Sci.* **25,** 277–283 (2000).

Marmorstein, R., Structure and function of histone acetyltransferases, *Cell. Mol. Life Sci.* **58,** 693–703 (2001); Marmorstein, R., Protein modules that manipulate histone tails for chromatin regulation, *Nature Rev. Mol. Cell. Biol.* **2,** 422–432 (2001); *and* Marmorstein, R. and Roth, S.Y., Histone acetyltransferases: function, structure, and catalysis, *Curr. Opin. Genet. Dev.* **11,** 155–161 (2001).

Marmorstein, R., Structure of histone deacetylases: Insights into substrate recognition and catalysis, *Structure* **9,** 1127–1133 (2001).

Marmorstein, R., Carey, M., Ptashne, M., and Harrison, S.C., DNA recognition by GAL4: structure of a protein–DNA complex, *Nature* **356,** 408–414 (1992).

Marmorstein, R. and Fitzgerald, M.X., Modulation of DNA-binding domains for sequence-specific DNA recognition, *Gene* **304,** 1–12 (2003).

Martin, G.M., X-Chromosome inactivation in mammals, *Cell* **29,** 721–724 (1982).

McKnight, S.L. and Yamamoto, K.R. (Eds.), *Transcriptional Regulation,* Cold Spring Harbor Laboratory Press (1992). [A two-volume compendium.]

Myers, L.C. and Kornberg, R.D., Mediator of transcriptional regulation, *Annu. Rev. Biochem.* **69,** 729–749 (2000).

Nielsen, P.R., Nietlspach, D., Mott, H.R., Callaghan, J., Bannister, A., Kouzarides, T., Murzin, A.G., Murzin, N.V., and Laue, E.D., Structure of the HP1 chromodomain bound to histone H3 methylated at lysine 9, *Nature* **416,** 103–107 (2002).

Nikolov, D.B., Chen, H., Halay, E.D., Usheva, A.A., Hisatake, K., Lee, D.K., Roeder, R.G., and Burley, S.K., Crystal structure of a TFIIB–TBP–TATA-element ternary complex, *Nature* **377,** 119–128 (1995); *and* Tsai, F.T.F. and Sigler, P.B., Structural basis of preinitiation complex assembly on human Pol II promoters, *EMBO J.* **19,** 25–36 (2000).

Orlando, V., Mapping chromosomal proteins *in vivo* by formaldehyde-crosslinked-chromatin immunoprecipitation, *Trends Biochem. Sci.* **25,** 99–104 (2000).

O'Shea, E.K., Klemm, J.D., Kim, P.S., and Alber, T., X-Ray structure of the GCN4 leucine zipper, a two-stranded, parallel coiled coil, *Science* **254,** 539–544 (1991).

Patikoglou, G. and Burley, S.K., Eukaryotic transcription factor-DNA complexes, *Annu. Rev. Biophys. Biomol. Struct.* **26,** 289–325 (1997); *and* Burley, S.K. and Kamada, K., Transcription factor complexes, *Curr. Opin. Struct. Biol.* **12,** 225–230 (2002).

Pavletich, N.P. and Pabo, C.O., Zinc finger–DNA recognition: Crystal structure of a Zif268-DNA complex at 2.1 Å, *Science* **252,** 809–817 (1991).

Pelz, S.W., Brewer, G., Bernstein, P., Hart, P.A., and Ross, J., Regulation of mRNA turnover in eukaryotic cells, *Crit. Rev. Euk. Gene Express.* **1,** 99–126 (1991); *and* Atwater, J.A., Wisdom, R., and Verma, I.M., Regulated mRNA stability, *Annu. Rev. Genet.* **24,** 519–541 (1990).

Poux, A.N., Cebrat, M., Kim, C.M., Cole, P.A., and Marmorstein, R., Structure of the GCN5 histone acetyltransferase bound to a bisubstrate inhibitor, *Proc. Natl. Acad. Sci.* **99,** 14065–14070 (2002).

Ptashne, M. and Gann, A., *Genes & Signals,* Cold Spring Harbor Laboratory Press (2002). [Discusses mechanisms of genetic regulation.]

Pugh, B.F., Control of gene expression through the regulation of the TATA-binding protein, *Gene* **255,** 1–14 (2000).

Raghow, R., Regulation of messenger RNA turnover in eukaryotes, *Trends Biochem. Sci.* **12,** 358–360 (1987).

Riggs, A.D. and Pfeifer, G.P., X-chromosome inactivation and cell memory, *Trends Genet.* **8,** 169–174 (1992).

Roth, S.Y., Denu, J.M., and Allis, C.D., Histone acetyltransferases, *Annu. Rev. Biochem.* **70,** 81–120 (2001).

Schmiedeskamp, M. and Klevit, R.E., Zinc finger diversity, *Curr. Opin. Struct. Biol.* **4,** 28–35 (1994).

Schwabe, J.W.R., Chapman, L., Finch, J.T., and Rhodes, D., The crystal structure of the estrogen receptor DNA-binding domain bound to DNA: How receptors discriminate between their response elements, *Cell* **75,** 567–578 (1993).

Schwabe, J.W.R. and Klug, A., Zinc mining for protein domains, *Nature Struct. Biol.* **1**, 345–349 (1994). [Discusses the varieties of zinc finger proteins.]

Smale, S.T. and Kadonaga, J.T., The RNA polymerase II core promoter, *Annu. Rev. Biochem.* **72**, 449–479 (2003).

Stamatoyannopoulos, G. and Nienhuis, A.W. (Eds.), *The Regulation of Hemoglobin Switching*, The Johns Hopkins University Press (1991).

Strahl, B.D. and Allis, C.D., The language of covalent histone modifications, *Nature* **403**, 41–45 (2000); *and* Rice, J.C. and Allis, C.D., Histone methylation versus histone acetylation: New insights into epigenetic regulation, *Curr. Opin. Cell Biol.* **13**, 263–273 (2001).

Struhl, K., Duality of TBP, the universal transcription factor, *Science* **263**, 1103–1104 (1994); Rigby, P.W.J., Three in one and one in three: It all depends on TBP, *Cell* **72**, 7–10 (1993); *and* White, R.J. and Jackson, S.P., The TATA-binding protein: a central role in transcription by RNA polymerases I, II, and III, *Trends Genet.* **8**, 284–288 (1992).

Studitsky, V.M., Clark, D.J., and Felsenfeld, G., A histone octamer can step around a transcribing polymerase without leaving the template, *Cell* **76**, 371–382 (1994); Studitsky, V.M., Kassavetis, G.A., Geiduschek, E.P., and Felsenfeld, G., Mechanism of transcription through the nucleosome by eukaryotic RNA polymerase, *Science* **278**, 1960–1965 (1997); *and* Felsenfeld, G., Clark, D., and Studitsky, V., Transcription through nucleosomes, *Biophys. Chem.* **86**, 231–237 (2000).

Tora, L., A unified nomenclature for TATA box binding protein (TBP)-associated factors (TAFs) involved in RNA polymerase II transcription, *Genes Dev.* **16**, 673–675 (2002).

Tsai, M.J. and O'Malley, B., Molecular mechanisms of action of steroid/thyroid receptor superfamily members, *Annu. Rev. Biochem.* **63**, 451–486 (1994).

Turner, B.M., Cellular memory and the histone code, *Cell* **111**, 285–291 (2002).

Veenstra, G.J.C. and Wolffe, A.P., Gene-selective developmental roles of general transcription factors, *Trends Biochem. Sci.* **26**, 665–671 (2001).

Wolffe, S.A., Nekludova, L., and Pabo, C.O., DNA recognition by Cys$_2$His$_2$ zinc finger proteins, *Annu Rev. Biophys. Biomol. Struct.* **3**, 183–212 (1999).

Xiao, B., et al., Structure and catalytic mechanism of the human histone methyltransferase SET7/9, *Nature* **421**, 652–656 (2003).

Xie, X., Kokubo, K., Cohen, S.L., Mirza, U.A., Hoffmann, A., Chait, B.T., Roeder, R.G., Nakatani, Y., and Burley, S.K., Structural similarities between TAFs and the heterotetrameric core of the histone octamer, *Nature* **380**, 316–322 (1996); Selleck, W., Howley, R., Fang, Q., Podolny, V., Fried, M.G., Buratowski, S., and Tan, S., A histone fold TAF octamer within the yeast TFIID transcriptional coactivator, *Nature Struct. Biol.* **8**, 695–700 (2001); *and* Werten, S., Mitschler, A., Romier, C., Gangloff, Y.-G., Thuault, S., Davidson, I., and Moras, D., Crystal structure of a subcomplex of human transcription factor TFIID formed by TATA binding protein-associated factors hTAF4 (hTAF$_{II}$135) and hTAF12 (hTAF$_{II}$20), *J. Biol. Chem.* **277**, 45502–45509 (2002).

Yang, X.-J. and Seto, E., Collaborative spirit of histone deacetylases in regulating chromatin structure and gene expression, *Curr. Opin. Genet. Dev.* **13**, 143–153 (2003).

DEVELOPMENT

Bate, M. and Arias, A.M. (Eds.), *The Development of Drosophila melanogaster*, Cold Spring Harbor Laboratory Press (1993).

Blau, H.M., Differentiation requires continuous active control, *Annu. Rev. Biochem.* **61**, 1213–1230 (1992).

Fujioka, M., Emi-Sarker, Y., Yusibova, G.L., Goto, T., and Jaynes, J.B., Analysis of an *even-skipped* rescue transgene reveals both composite and discrete neuronal and early blastoderm enhancers and multi-stripe positioning by gap gene repressor gradients, *Development* **126**, 2527–2538 (1999).

Gehring, W.J., Affolter, M., and Bürglin, T., Homeodomain proteins, *Annu. Rev. Biochem.* **63**, 487–526 (1994); *and* Gehring, W.J., Qian, Y.Q., Billeter, M., Furukobu-Tokunaga, K., Schier, A.F., Resendez-Perez, D., Affolter, M., Otting, G., and Wüthrich, K., Homeodomain–DNA recognition, *Cell* **78**, 211–223 (1994).

Gilbert, S.F., *Developmental Biology* (7th ed.), Sinauer Associates (2003).

Gossler, A. and Balling, R., The molecular and genetic analysis of mouse development, *Eur. J. Biochem.* **204**, 5–11 (1992).

Gurdon, J.B., The generation of diversity and pattern in animal development, *Cell* **68**, 185–199 (1992).

Halder, G., Callaerts, P., and Gehring, W.J., Induction of ectopic eyes by targeted expression of the *eyeless* gene in *Drosophila*, *Science* **267**, 1788–1792 (1995).

Kenyon, C., If birds can fly, why can't we? Homeotic genes and evolution, *Cell* **78**, 175–180 (1994).

Kissinger, C.R., Liu, B., Martin-Blanco, E., Kornberg, T.B., and Pabo, C.O., Crystal structure of an engrailed homeo-domain–DNA complex at 2.8 Å resolution: A framework for understanding homeodomain–DNA interactions, *Cell* **63**, 579–590 (1990).

Krumlauf, R., *Hox* genes in development, *Cell* **78**, 191–201 (1994).

Lawrence, P.A., *The Making of a Fly*, Blackwell Scientific Publications (1992).

Lawrence, P.A. and Morata, G., Homeobox genes: Their function in Drosophila segmentation and pattern formation, *Cell* **78**, 181–191 (1994).

Le Mouellic, H., Lallemand, Y., and Brûlet, P., Homeosis in the mouse induced by a null mutation in the *Hox-3.1* gene, *Cell* **69**, 251–264 (1992).

Mann, R.S. and Morata, G., The developmental and molecular biology of genes that subdivide the body of *Drosophila*, *Annu. Rev. Cell Dev. Biol.* **16**, 243–271 (2000).

Mavilio, F., Regulation of vertebrate homeobox-containing genes by morphogens, *Eur. J. Biochem.* **212**, 273–288 (1993).

Nüsslein-Volhard, C., Axis determination in the *Drosophila* embryo, *Harvey Lect.* **86**, 129–148 (1992); *and* St. Johnston, D. and Nüsslein-Volhard, C., The origin of pattern and polarity in the *Drosophila* embryo, *Cell* **68**, 201–219 (1992). [Detailed reviews.]

Nüsslein-Volhard, C., The identification of genes controlling development in flies and fishes (Nobel lecture); *and* Wieschaus, E., From molecular patterns to morphogenesis—The lessons from studies on the fruit fly *Drosophila* (Noble lecture), *Angew. Chem. Int. Ed. Engl.* **35**, 2177–2187 and 2189–2194 (1996).

Scott, M.P., Development: The natural history of genes, *Cell* **100**, 27–40 (2000).

CANCER AND THE REGULATION OF THE CELL CYCLE

Adams, P.D., Regulation of the retinoblastoma tumor suppressor protein by cyclin/cdks, *Biochim. Biophys. Acta* **1471**, M123–M133 (2001).

Cho, Y., Gorina, S., Jeffrey, P.D., and Pavletich, N.P., Crystal structure of a p53 tumor suppressor–DNA complex: Understanding tumorigenic mutations, *Science* **265**, 346–355 (1994).

Cooper, G.M. and Hausman, R.E., *The Cell. A Molecular Approach* (3rd ed.), Chapter 14, ASM Press (2004).

De Bondt, H.L., Rosenblatt, J., Jancarik, J., Jones, H.D., Morgan, D.O., and Kim, S.-H., Crystal structure of cyclin-dependent kinase 2, *Nature* **363**, 595–602 (1993); Jeffrey, P.D., Russo, A.A., Polyak, K., Gibbs, E., Hurwitz, J., Massagu, J., and Pavletich, N.P., Mechanism of CDK activation revealed by the structure of a cyclin A–CDK2 complex, *Nature* **376**, 313–320 (1995); Russo, A.A., Jeffrey, P.D., Patten, A.K., Massagué, J., and Pavletich, N.P., Crystal structure of the p27^{Kip1} cyclin-dependent-kinase inhibitor bound to the cyclin A–Cdk2 complex, *Nature* **382**, 325–331 (1996); *and* Russo, A.A., Jeffrey, P.D., and Pavletich, N.P., Structural basis of cyclin-dependent kinase activation by phosphorylation, *Nature Struct. Biol.* **3**, 696–700 (1996).

Donehower, L.A., Harvey, M., Slagle, B.L., McArthur, M.J., Montgomery, C.A., Jr., Butel, J.S., and Bradley, A., Mice deficient for p53 are developmentally normal but susceptible to spontaneous tumours, *Nature* **356**, 215–221 (1992).

Haluska, F.G., Tsujimoto, Y., and Croce, C.M., Oncogene activation by chromosome translocation in human malignancy, *Annu. Rev. Genet.* **21**, 321–345 (1987).

Harbour, J.W. and Dean, D.C., The Rb/E2F pathway: expanding roles and emerging paradigms, *Genes Dev.* **14**, 2393–2409 (2000).

Harper, J.W. and Adams, P.D., Cyclin-dependent kinases, *Chem. Rev.* **101**, 2511–2526 (2001).

Hickman, E.S., Moroni, M.C., and Helin, K., The role of p53 and pRB in apoptosis and cancer, *Curr. Opin. Genet. Dev.* **12**, 60–66 (2002).

Hunter, T. and Pines, J., Cyclins and cancer, *Cell* **66**, 1071–1074 (1991).

Jeffrey, P.D., Tong, L., and Pavletich, N.P., Structural basis of inhibition of CDK-cyclin complexes by INK4 inhibitors, *Genes Dev.* **14**, 3115–3125 (2000).

Johnson, D.G. and Walker, C.L., Cyclins and cell cycle checkpoints, *Annu. Rev. Pharmacol Toxicol.* **39**, 295–312 (1999).

Johnstone, R.W., Histone-deacetylase inhibitors: Novel drugs for the treatment of cancer, *Nature Rev. Drug Disc.* **1**, 287–299 (2002).

Lee, E.Y.-H., Chang, C.-Y., Hu, N., Wang, Y.-C.J., Lai, C.-C., Herrup, K., Lee, W.-H., and Bradley, A., Mice deficient for Rb are nonviable and show defects in neurogenesis and haematopoiesis, *Nature* **359**, 288–394 (1992).

Morgan, D.O., Cyclin-dependent kinases: Engines, clocks, and microprocessors, *Annu. Rev. Cell Dev. Biol.* **13**, 261–291 (1997).

Morris, E.J. and Dyson, N.J., Retinoblastoma protein partners, *Adv. Cancer Res.* **82**, 1–54 (2001).

Nigg, E.A., Mitotic kinases as regulators of cell division and its checkpoints, *Nature Rev. Mol. Biol.* **2**, 21–32 (2001).

Norbury, C. and Nurse, P., Animal cell cycles and their control, *Annu. Rev. Biochem.* **61**, 441–470 (1992); *and* Forsburg, S.L. and Nurse, P., Cell cycle regulation in the yeasts *Saccharomyces cerevisiae* and *Schizosaccharomyces pombe*, *Annu. Rev. Cell Biol.* **7**, 227–256 (1991).

Pavletich, N.P., Mechanisms of cyclin-dependent kinase regulation: Structures of Cdks, their cyclin activators, and Cip and INK4 inhibitors, *J. Mol. Biol.* **287**, 821–828 (1999).

Pollard, T.D. and Earnshaw, W.C., *Cell Biology,* Chapters 43–47 and 49, Saunders (2002).

Russell, P., Checkpoints on the road to mitosis, *Trends Biochem. Sci.* **23**, 399–402 (1998).

Russo, A.A., Jeffrey, P.D., Patten, A.K., Massagué, J., and Pavletich, N.P., Crystal structure of the p27^{Kip1} cyclin-dependent-kinase inhibitor bound to the cyclin A–Cdk2 complex, *Nature* **382**, 325–331 (1996).

Sherr, C.J. and Weber, J.D., The ARF/p53 pathway, *Curr. Opin. Genet. Dev.* **10**, 94–99 (2000).

Vogelstein, B. and Kinzler, K.W., The multistep nature of cancer, *Trends Genet.* **9**, 138–140 (1993).

Volgelstein, B., Lane, D., and Levine, A.J., Surfing the p53 network, *Nature* **408**, 307–310 (2000). [Discusses how p53 integrates the various signals that control cell life and death.]

Xiao, B., Spencer, J., Clements, A., Ali-Khan, N., Mittnacht, S., Broceño, C., Burghammer, M., Parrakis, A., Marmorstein, M., and Gamblin, S.J., Crystal structure of the retinoblastoma tumor suppressor protein bound to E2F and the molecular basis of its regulation, *Proc. Natl. Acad. Sci.* **100**, 2363–2368 (2003); *and* Lee, C., Chang, J.H., Lee, H.S., and Cho, Y., Structural basis for the recognition of the E2F transactivation domain by the retinoblastoma tumor suppressor, *Genes Dev.* **16**, 3199–3212 (2002).

APOPTOSIS

Acehan, D., Jiang, X., Morgan, D.G., Heuser, J.E., Wang, X., and Akey, C.W., Three-dimensional structure of the apoptosome: Implications for assembly, procaspase-9 binding, and activation, *Mol. Cell* **9**, 423–432 (2002).

Chai, J., Wu, Q., Shiozaki, E., Srinivasula, S.M., Alnemri, E.S., and Shi, Y., Crystal structure of a procaspase-7 zymogen: Mechanisms of activation and substrate binding, *Cell* **107**, 399–407 (2001); *and* Riedl, S.J., Fuentes-Prior, P., Renatus, M., Kairies, N., Krapp, S., Huber, R., Salvesen, G.S., and Bode, W., Structural basis for the activation of human procaspase-7, *Proc. Natl. Acad. Sci.* **98**, 14790–14795 (2001).

Desagher, S. and Martinou, J.C., Mitochondria as the central control point of apoptosis, *Trends Cell Biol.* **10**, 369–377 (2000).

Earnshaw, W.C., Martins, L.M., and Kaufmann, S.H., Mammalian caspases: Structure, activation, substrates, and functions during apoptosis, *Annu. Rev. Biochem.* **68**, 383–424 (1999); *and* Grütter, M.G., Caspases: key players in programmed cell death, *Curr. Opin. Struct. Biol.* **10**, 649–655 (2000).

Fesik, S.W., Insights into programmed cell death through structural biology, *Cell* **103**, 272–282 (2000).

Hengartner, M.O., The biochemistry of apoptosis, *Nature* **407**, 770–776 (2000).

Jacobson, M.D. and McCarthy, N. (Eds.), *Apoptosis,* Oxford (2002).

Nagata, S., Fas ligand-induced apoptosis, *Annu. Rev. Genet.* **33**, 29–55 (1999).

Riedl, S.J, Renatus, M., Schwarzenbacher, R., Zhou, Q., Sun, C., Fesik, S.W., Liddington, R.C., and Salvesen, G.S., Structural basis for the inhibition of caspase-3 by XIAP, *Cell* **104**, 791–800 (2001).

Strasser, A., O'Connor, L., and Dixit, V.M., Apoptosis signaling, *Annu. Rev. Biochem.* **69**, 217–245 (2000).

Wei, Y., Fox, T., Chambers, S.P., Sinchak, J., Coll, J.T., Golec, J.M.C., Swenson, L., Wilson, K.P., and Charifson, P., The structures of capsases-1, -3, -7, and -8 reveal the basis for substrate and inhibitor selectivity, *Chem. Biol.* **7**, 423–432 (2000).

Yin, X.M. and Dong, Z. (Eds.), *Essentials of Apoptosis,* Humana Press (2003).

PROBLEMS

1. What is the maximum possible packing ratio of a 10^6-bp segment of DNA; of a 10^9-bp segment of DNA? Assume the DNA is a 20-Å-diameter cylinder with a contour length of 3.4 Å/bp.

2. When an SV40 minichromosome (a closed circular duplex DNA in complex with nucleosomes) is relaxed so that it forms an untwisted circle and is then deproteinized, the consequent closed circular DNA has about –1 superhelical turn for each of the nucleosomes that it originally had. Explain the discrepancy between this observation and the fact that the DNA in each nucleosome is wrapped nearly twice about its histone octamer in a left-handed superhelix.

3. Explain why acidic polypeptides such as polyglutamate facilitate *in vitro* nucleosome assembly.

*4. Consider a 1 million-bp DNA molecule that has 1500 tandem repeats of a 400-bp sequence with the remainder of the DNA consisting of unique sequences. Sketch the C_0t curve of this DNA when it is sheared into pieces averaging 1000 bp long; when they are 100 bp long.

5. Why do isolated foldback structures, when treated by an endonuclease that cleaves only single-stranded DNA and then denatured, yield complicated C_0t curves?

6. During its 2-month period of maturation, the *Xenopus* oocyte synthesizes ~10^{12} ribosomes. The consequent tremendous rate of rRNA synthesis is only possible because the normal genomic complement of rDNA has been amplified 1500-fold. (a) Why is it unnecessary to likewise amplify the genes encoding the ribosomal proteins? (b) Assuming that rRNA gene amplification occurs in a short time at the beginning of the maturation period, how long would oogenesis require if the rDNA were not amplified?

7. Hb Kenya is a β-thalassemia in which the β-globin cluster is deleted between a point in the $^A\gamma$-globin gene and the corresponding position in the β-globin gene. Describe the most probable mechanism for the generation of this mutation.

8. Red–green color blindness is caused by an X-linked recessive genetic defect. Hence females rarely exhibit the red–green color-blind phenotype but may be carriers of the defective gene. When a narrow beam of red or green light is projected onto some areas of the retina of such a female carrier, she can readily differentiate the two colors but on other areas she has difficulty in doing so. Explain.

9. Figure 34-53a contains a single band just above the bracketed region that increases in density as the Sp1 concentration increases. What is origin of this band?

10. Why do the rare instances of male calico cats all have the abnormal XXY genotype?

11. In *Drosophila*, an *esc⁻* homozygote develops normally unless its mother is also an *esc⁻* homozygote. Explain.

12. The fusion of cancer cells with normal cells often suppresses the expression of the tumorigenic phenotype. Explain.

NOTES

NOTES

Index

Page references in **bold face** refer to a major discussion of the entry. Positional and configurational designations in chemical names (e.g. 3-, α, *N*-, *p*-, *trans*, D-, *sn*-) are ignored in alphabetizing. Numbers and Greek letters are otherwise alphabetical as if they were spelled out.